Comprehensive Organic Functional Group Transformations

Comprehensive Organic Functional Group Transformations

Editors-in-Chief

Alan R. Katritzky, FRS
University of Florida, Gainesville, FL, USA

Otto Meth-Cohn
University of Sunderland, UK

Charles W. Rees, FRS
Imperial College of Science, Technology and Medicine, London, UK

Volume 7
INDEXES

PERGAMON

UK	Elsevier Science Ltd., The Boulevard, Langford Lane, Kidlington, Oxford OX5 1GB, UK
USA	Elsevier Science Inc., 660 White Plains Road, Tarrytown, New York 10591-5153, USA
JAPAN	Elsevier Science Japan, Tsunashima Building Annex, 3-20-12 Yushima, Bunkyo-ku, Tokyo 113, Japan

First edition 1995

Library of Congress Cataloging in Publication Data
Comprehensive organic functional group transformations / editors-in-chief, Alan R. Katritzky, Otto Meth-Cohn, Charles W. Rees. — 1st ed.
p. cm.
Includes indexes.
Contents: v. 1. Synthesis: carbon with no attached heteroatoms / edited by Stanley M. Roberts — v. 2. Synthesis: carbon with one heteroatom attached by a single bond / edited by Steven V. Ley — v. 3. Synthesis: carbon with one heteroatom attached by a multiple bond / edited by Gerald Pattenden — v. 4. Synthesis: carbon with two heteroatoms, each attached by a single bond / edited by Gordon W. Kirby — v. 5. Synthesis: carbon with two attached heteroatoms with at least one carbon-to-heteroatom multiple link / edited by Christopher J. Moody — v. 6. Synthesis: carbon with three or four attached heteroatoms / edited by Thomas L. Gilchrist — v. 7. Indexes.
1. Organic compounds—Synthesis. I. Katritzky, Alan R. II. Meth-Cohn, Otto. III. Rees, Charles W.
QD262.C534 1995
547'.2—dc20 95–31088

British Library Cataloguing in Publication Data
A catalogue record for this book is available from the British Library.

ISBN 0–08–040604–1 (set : alk. paper)
ISBN 0–08–042705–7 (Volume 7)

∞™ The paper used in this publication meets the minimum requirements of the American National Standard for Information Sciences—Permanence of Paper for Printed Library Materials, ANSI Z39.48–1984.

Printed and bound in Great Britain by Cambridge University Press, Cambridge, UK.

Contents

Contents of All Volumes

Volume 1 Synthesis: Carbon with No Attached Heteroatoms

(Part I Tetracoordinated Carbon with No Attached Heteroatoms)

1.01 One or More CH Bond(s) Formed by Substitution: Reduction of C–Halogen and C–Chalcogen Bonds
1.02 One or More CH Bond(s) Formed by Substitution: Reduction of Carbon–Nitrogen, –Phosphorus, –Arsenic, –Antimony, –Bismuth, –Carbon, –Boron, and –Metal Bonds
1.03 Two or More CH Bond(s) Formed by Addition to CC Multiple Bonds
1.04 One or More CC Bond(s) Formed by Substitution: Substitution of Halogen
1.05 One or More CC Bond(s) Formed by Substitution: Substitution of Chalcogen
1.06 One or More CC Bond(s) Formed by Substitution: Substitution of Carbon–Nitrogen, –Phosphorus, –Arsenic, –Antimony, –Boron, –Silicon, –Germanium, and –Metal Functions
1.07 One or More CC Bond(s) Formed by Addition: Addition of Carbon Electrophiles and Nucleophiles to CC Multiple Bonds
1.08 One or More CC Bond(s) Formed by Addition: Addition of Carbon Radicals and Electrocyclic Additions to CC Multiple Bonds
1.09 One or More CH and/or CC Bond(s) Formed by Rearrangement

(Part II Tricoordinated Carbon with No Attached Heteroatoms)

1.10 One or More =CH Bond(s) Formed by Substitution or Addition
1.11 One or More =CC Bond(s) Formed by Substitution or Addition
1.12 One or More C=C Bond(s) Formed by Addition
1.13 One or More C=C Bond(s) by Elimination of Hydrogen, Carbon, Halogen, or Oxygen Functions
1.14 One or More C=C Bond(s) by Elimination of S, Se, Te, N, P, As, Sb, Bi, Si, Ge, B, or Metal Functions
1.15 One or More C=C Bond(s) Formed by Condensation: Condensation of Nonheteroatom Linked Functions, Halides, Chalcogen, or Nitrogen Functions
1.16 One or More C=C Bond(s) Formed by Condensation: Condensation of P, As, Sb, Bi, Si, Ge, B, or Metal Functions
1.17 One or More C=C Bond(s) by Pericyclic Processes
1.18 One or More =CH, =CC and/or C=C Bond(s) Formed by Rearrangement
1.19 Tricoordinate Anions, Cations, and Radicals

(Part III Dicoordinate and Monocoordinate Carbon with No Attached Heteroatoms)

1.20 Allenes and Cumulenes
1.21 Alkynes
1.22 Ions, Radicals, Carbenes and Other Monocoordinated Systems

Volume 2 Synthesis: Carbon with One Heteroatom Attached by a Single Bond

(Part I Functions Linked by a Single Bond to an sp^3 Carbon Atom)

2.01 Alkyl Halides
2.02 Alkyl Chalcogenides: Oxygen-based Functional Groups
2.03 Alkyl Chalcogenides: Sulfur-based Functional Groups
2.04 Alkyl Chalcogenides: Selenium- and Tellurium-based Functional Groups
2.05 Alkylnitrogen Compounds: Amines and Their Salts
2.06 Alkylnitrogen Compounds: Compounds with N—Halogen, N—O, N—S, N—Se, and N—Te Functional Groups
2.07 Alkylnitrogen Compounds: Compounds with N—N, N—P, N—As, N—Sb, N—Bi, N—Si, N—Ge, N—B, and N—Metal Functional Groups

2.08 Alkylphosphorus Compounds
2.09 Alkylarsenic, -antimony, and -bismuth Compounds
2.10 Alkylboron and -silicon Compounds
2.11 Alkyl Metals

(Part II Functions Linked by a Single Bond to an sp^2 Carbon Atom)

2.12 Vinyl and Aryl Halides
2.13 Alkenyl and Aryl Chalcogenides: Oxygen-based Functional Groups
2.14 Vinyl and Aryl Chalcogenides: Sulfur-, Selenium-, and Tellurium-based Functional Groups
2.15 Vinyl- and Arylnitrogen Compounds
2.16 Vinyl- and Arylphosphorus Derivatives
2.17 Vinyl- and Arylarsenic, -antimony, and -bismuth Compounds
2.18 Vinyl- and Arylsilicon, -germanium, and boron Compounds
2.19 Vinyl- and Arylmetals
2.20 Stabilized Substituted Ions and Radicals Bearing One Heteroatom ($R^1R^2C^-X$, $R^1R^2C^+X$, $R^1R^2C^{\cdot}X$)

(Part III Functions Linked by a Single Bond to an *sp* Carbon Atom)

2.21 Alkynyl Halides and Chalcogenides
2.22 Alkynylnitrogen and -phosphorus Compounds
2.23 Alkynylarsenic, -antimony, -bismuth, -boron, -silicon, -germanium, and -metal Compounds

Volume 3 Synthesis: Carbon with One Heteroatom Attached by a Multiple Bond

(Part I Tricoordinated Carbon Functions, $R_2C{=}Y$)

3.01 Aldehydes: Alkyl Aldehydes
3.02 Aldehydes: α,β-Unsaturated Aldehydes
3.03 Aldehydes: Aryl and Heteroaryl Aldehydes
3.04 Ketones: Dialkyl Ketones
3.05 Ketones: α,β-Unsaturated Ketones
3.06 Ketones Bearing an α,β-Aryl or -Hetaryl Substituent
3.07 Aldehyde and Ketone Functions Further Substituted on Oxygen
3.08 Thioaldehydes and Thioketones
3.09 Seleno- and Telluroaldehydes and -ketones
3.10 Imines and Their *N*-Substituted Derivatives: NH, NR, and N-Haloimines
3.11 Imines and Their *N*-Substituted Derivatives: Oximes and Their *O*-R Substituted Analogues
3.12 Imines and Their *N*-Substituted Derivatives: Hydrazones and Other =NN Derivatives Including Diazo Compounds
3.13 Synthesis of P, As, Sb and Bi Ylides ($R_3P{=}CR_2$, etc.)
3.14 Doubly Bonded Metalloid Functions (Si, Ge, B)
3.15 Doubly Bonded Metal Functions

(Part II Dicoordinated Carbon Functions, $R_2C{=}C{=}Y$)

3.16 Ketenes, Their Cumulene Analogues and Their S, Se, and Te Analogues
3.17 Ketenimines and Their P, As, Sb and Bi Analogues

(Part III Dicoordinated Carbon Functions, $R{-}C{\equiv}Z$)

3.18 Nitriles: General Methods and Aliphatic Nitriles
3.19 α,β-Unsaturated and Aryl Nitriles
3.20 *N*-Substituted Nitriles and Other Heteroanalogues of Nitriles of the Type RCZ

(Part IV Monocoordinated Carbon Functions)

3.21 Isocyanides and Their Heteroanalogues (RZC)

Volume 4 Synthesis: Carbon with Two Heteroatoms, Each Attached by a Single Bond

(Part I Tetracoordinated Carbon Functions Bearing Two Heteroatoms, R_2CXX')

4.01 Dihalo Alkanes, $R_2C(Hal)_2$
4.02 Functions Incorporating a Halogen and a Chalcogen
4.03 Functions Incorporating a Halogen and Another Heteroatom Group Other Than a Chalcogen
4.04 Functions Bearing Two Oxygens, $R^1{}_2C(OR^2)_2$

4.05 Functions Incorporating Oxygen and Another Chalcogen
4.06 Functions Incorporating Two Chalcogens Other Than Oxygen
4.07 Functions Incorporating a Chalcogen and a Group 15 Element
4.08 Functions Incorporating a Chalcogen and a Silicon, Germanium, Boron, or Metal
4.09 Functions Bearing Two Nitrogens
4.10 Functions Incorporating a Nitrogen and Another Group 15 Element
4.11 Functions Incorporating a Nitrogen and a Silicon, Germanium, Boron, or a Metal
4.12 Functions Containing One Phosphorus and Either Another Phosphorus or As, Sb, Bi, Si, Ge, B, or a Metal
4.13 Functions Containing at Least One As, Sb, or Bi with or without a Metalloid (Si or Ge) or a Metal
4.14 Functions Containing at Least One Metalloid (Si, Ge, or B) Together with Another Metalloid or Metal
4.15 Functions Containing Two Atoms of the Same Metallic Element
4.16 Functions Containing Two Atoms of Different Metallic Elements

(Part II Tricoordinated Carbon Functions Bearing Two Heteroatoms, $R_2C{=}CXX'$)

4.17 Functions Incorporating Two Halogens or a Halogen and a Chalcogen
4.18 Functions Incorporating a Halogen or Another Group Other Than a Halogen or a Chalcogen
4.19 Functions Bearing Two Chalcogens
4.20 Functions Containing a Chalcogen and Any Group Other Than a Halogen or a Chalcogen
4.21 Functions Containing at Least One Nitrogen and No Halogen or Chalcogen
4.22 Functions Containing at Least One Phosphorus, Arsenic, Antimony or Bismuth and No Halogen, Chalcogen or Nitrogen
4.23 Functions Containing at Least One Metalloid (Si, Ge, or B) and No Halogen, Chalcogen or Group 15 Element; Also Functions Containing Two Metals

(Part III)

4.24 Tri- and Dicoordinated Ions, Radicals and Carbenes Bearing Two Heteroatoms ($RC^+X^1X^2$, $RC^-X^1X^2$, $RC^\bullet X^1X^2$, $:CX^1X^2$)

Volume 5 Synthesis: Carbon with Two Attached Heteroatoms with at Least One Carbon-to-Heteroatom Multiple Link

(Part I Tricoordinate Carbon Functions with One Doubly Bonded and One Singly Bonded Heteroatom, $RC{=}YX$)

5.01 Acyl Halides
5.02 Carboxylic Acids
5.03 Carboxylic Esters and Lactones
5.04 Other Acyloxy Compounds
5.05 Acylsulfur, -selenium, or -tellurium Functions
5.06 Amides
5.07 *N*-Heterosubstituted Amides
5.08 Acylphosphorus, -arsenic, -antimony, or -bismuth Functions
5.09 Acylsilicon, -germanium, or -boron Functions
5.10 Acyl Metal Functions
5.11 Thio-, Seleno-, and Telluroacyl Halides
5.12 Thio, Seleno, and Telluro Acyloxy Functions, $R^1C(S)OR^2$, $R^1C(Se)\ OR^2$, $R^1C(Te)OR^2$, etc
5.13 Functions with Two Chalcogens Other Than Oxygen
5.14 Thionoamides and Their Se and Te Analogues
5.15 *N*–Substituted Thionoamides and Their Se and Te Analogues
5.16 Thioacyl Functions linked to a Metalloid (Si, Ge, or B) or Metal; and Their Seleno and Telluro Analogues
5.17 Iminoacyl Halides and Oxy Functions
5.18 Iminoacyl Functions Linked to Chalcogens Other Than Oxygen
5.19 Amidines and *N*-Substituted Amidines
5.20 Iminoacyl Functions Linked to Any Heteroatom Other Than Halogen, Chalcogen or Nitrogen
5.21 *N*-Heterosubstituted Iminoacyl Functions
5.22 Diazo Functions with an α-Heteroatom ($RC(X)N_2$)

5.23 Phosphoacyl Functions and Their As, Sb, and Bi Analogues
5.24 Doubly Bonded Metalloid Functions, $R^1C(X){=}SiR^2_2$, $R^1C(X){=}BR^2$, $R^1C(X){=}GeR^2_2$
5.25 Functions Doubly Bonded to a Metal

(Part II Dicoordinate Carbon Functions with Two Doubly Bonded Heteroatoms, Y=C=Y′)

5.26 Functions with at Least One Oxygen, Y=C=O
5.27 Functions with at Least One Chalcogen Other Than Oxygen
5.28 Functions with at Least One Nitrogen and No Chalcogens
5.29 Functions with Heteroatoms Other Than Chalcogen or Nitrogen (Y=C=Y′)

(Part III Dicoordinate Carbon Functions with One Singly Bonded and One Triply Bonded Heteroatom, X—C≡Z)

5.30 Nitriles with a Heteroatom Attached to the Cyanocarbon
5.31 Triply Bonded Heteroatom Derivatives Other Than Nitriles with Another Heteroatom Attached to the *sp*-Carbon Atom

Volume 6 Synthesis: Carbon with Three or Four Attached Heteroatoms

(Part I Tetracoordinated Carbon with Three Attached Heteroatoms, RCXX′X″)

6.01 Trihalides
6.02 Functions Containing Halogens and Any Other Elements
6.03 Functions Containing Three Chalcogens (and No Halogens)
6.04 Functions Containing a Chalcogen and Any Other Heteroatoms Other Than a Halogen
6.05 Functions Containing at Least One Group 15 Element (and No Halogen or Chalcogen)
6.06 Functions Containing at Least One Metalloid (Si, Ge, or B) and No Halogen, Chalcogen, or Group 15 Element; Also Functions Containing Three Metals

(Part II Tetracoordinated Carbon with Four Attached Heteroatoms, CXX′X″X‴)

6.07 Functions Containing Four Halogens or Three Halogens and One Other Heteroatom Substituent
6.08 Functions Containing Two Halogens and Two Other Heteroatom Substituents
6.09 Functions Containing One Halogen and Three Other Heteroatom Substituents
6.10 Functions Containing Four or Three Chalcogens (and No Halogens)
6.11 Functions Containing Two or One Chalcogens (and No Halogens)
6.12 Functions Containing at Least One Group 15 Element (and No Halogen or Chalcogen)
6.13 Functions Containing at Least One Metalloid (Si, Ge, or B) and No Halogen, Chalcogen, or Group 15 Element; Also Functions Containing Four Metals

(Part III Tricoordinated Carbon with Three Attached Heteroatoms, Y=CXX′)

6.14 Functions Containing a Carbonyl Group and at Least One Halogen
6.15 Functions Containing a Carbonyl Group and at Least One Chalcogen (but No Halogen)
6.16 Functions Containing a Carbonyl Group and Two Heteroatoms Other Than a Halogen or Chalcogen
6.17 Functions Containing a Thiocarbonyl Group and at Least One Halogen; Also at Least One Chalcogen and No Halogen
6.18 Functions Containing a Thiocarbonyl Group Bearing Two Heteroatoms Other Than a Halogen or Chalcogen
6.19 Functions Containing a Selenocarbonyl or Tellurocarbonyl Group—Se(X)X′ and TeC(X)X′
6.20 Functions Containing an Iminocarbonyl Group and at Least One Halogen; Also One Chalcogen and No Halogen
6.21 Functions Containing an Iminocarbonyl Group and Any Elements Other Than a Halogen or Chalcogen
6.22 Functions Containing Doubly Bonded P, As, Sb, Bi, Si, Ge, B, or a Metal

(Part IV)

6.23 Tricoordinated Stabilized Cations and Radicals, ^+CXYZ and $^{\bullet}CXYZ$

Volume 7 Indexes

Author Index
Cumulative Subject Index

Explanation of the Reference System

Throughout this work, references are designated by a number–lettering coding of which the first two numbers denote tens and units of the year of publication, the next one to three letters denote the journal, and the final numbers denote the page. This code appears in the text each time a reference is quoted. The system has been used previously in "Comprehensive Heterocyclic Chemistry," eds A. R. Katritzky and C. W. Rees, Pergamon, Oxford, 1984 and is based on that used in the following two monographs: (a) A. R. Katritzky and J. M. Lagowski, "Chemistry of the Heterocyclic N-Oxides," Academic Press, New York, 1971; (b) J. Elguero, C. Marzin, A. R. Katritzky and P. Linda, "The Tautomerism of Heterocycles," in "Advances in Heterocyclic Chemistry," Supplement 1, Academic Press, New York, 1976.

The following additional notes apply:

1. A list of journal codes in alphabetical order, together with the journals to which they refer, is given immediately following these notes. Journal names are abbreviated throughout using the CASSI (Chemical Abstracts Service Source Index) system.
2. Each volume contains all the references cited *in that volume*; no separate lists are given for individual chapters.
3. The list of references is arranged in order of (a) year, (b) journal in alphabetical order of journal code, (c) part letter or number if relevant, (d) volume number if relevant, (e) page number.
4. In the reference list the code is followed by (a) the complete literature citation in the conventional manner and (b) the number(s) of the page(s) on which the reference appears, whether in the text or in tables, schemes, etc.
5. For nontwentieth-century references the year is given in full in the code.
6. For journals which are published in separate parts, the part letter or number is given (when necessary) in parentheses immediately after the journal code letters.
7. Journal volume numbers are *not* included in the code numbers unless more than one volume was published in the year in question, in which case the volume number is included in parentheses immediately after the journal code letters.
8. Patents are assigned appropriate three-letter codes.
9. Frequently cited books are assigned codes.
10. Less common journals and books are given the code "MI" for miscellaneous with the whole code for books prefixed by the letter "B-".
11. Where journals have changed names, the same code is used throughout, e.g. CB refers to both *Chem. Ber.* and to *Ber. Dtsch. Chem. Ges.*

Journal Codes

AAC	Antimicrob. Agents Chemother.
ABC	Agric. Biol. Chem.
AC	Appl. Catal.
AC(P)	Ann. Chim. (Paris)
AC(R)	Ann. Chim. (Rome)

ACH	Acta Chim. Acad. Sci. Hung.
ACR	Acc. Chem. Res.
ACS	Acta Chem. Scand.
ACS(A)	Acta Chem. Scand., Ser. A
ACS(B)	Acta Chem. Scand., Ser. B
AF	Arzneim.-Forsch.
AFC	Adv. Fluorine Chem.
AG	Angew. Chem.
AG(E)	Angew. Chem., Int. Ed. Engl.
AHC	Adv. Heterocycl. Chem.
AHCS	Adv. Heterocycl. Chem. Supplement
AI	Anal. Instrum.
AJC	Aust. J. Chem.
AK	Ark. Kemi
AKZ	Arm. Khim. Zh.
AM	Adv. Mater. (Weinheim, Ger.)
AMLS	Adv. Mol. Spectrosc.
AMS	Adv. Mass. Spectrom.
ANC	Anal. Chem.
ANL	Acad. Naz. Lncei
ANY	Ann. N. Y. Acad. Sci.
AOC	Adv. Organomet. Chem.
AP	Arch. Pharm. (Weinheim, Ger.)
APO	Adv. Phys. Org. Chem.
AQ	An. Quim.
AR	Annu. Rep. Prog. Chem.
AR(A)	Annu. Rep. Prog. Chem., Sect. A
AR(B)	Annu. Rep. Prog. Chem., Sect. B
ARP	Annu. Rev. Phys. Chem.
ASI	Acta Chim. Sin. Engl. Ed.
ASIN	Acta Chim. Sin.
AX	Acta Crystallogr.
AX(A)	Acta Crystallogr., Part A
AX(B)	Acta Crystallogr., Part B
B	Biochemistry
BAP	Bull. Acad. Pol. Sci., Ser. Sci. Chim.
BAU	Bull. Acad. Sci. USSR, Div. Chim. Sci.
BBA	Biochim. Biophys. Acta
BBR	Biochim. Biophys. Res. Commun.
BCJ	Bull. Chem. Soc. Jpn.
BEP	Belg. Pat.
BJ	Biochem. J.
BJP	Br. J. Pharmacol.
BMC	Bioorg. Med. Chem. Lett.
BP	Biochem. Biopharmacol.
BPJ	Br. Polym. J.
BRP	Br. Pat.
BSB	Bull. Soc. Chim. Belg.
BSF	Bull. Soc. Chim. Fr.
BSF(2)	Bull. Soc. Chim. Fr., Part 2
C	Chimia
CA	Chem. Abstr.
CAN	Cancer
CAR	Carbohydr. Res.
CAT	Chim. Acta Turc.

CB	Chem. Ber.
CBR	Chem. Br.
CC	J. Chem. Soc., Chem. Commun.
CCA	Croat. Chem. Acta
CCC	Collect. Czech. Chem. Commun.
CCR	Coord. Chem. Rev.
CE	Chem. Express
CEN	Chem. Eng. News
CHE	Chem. Heterocycl. Compd. (Engl. Transl.)
CHEC	Comp. Heterocycl. Chem.
CI(L)	Chem. Ind. (London)
CI(M)	Chem. Ind. (Milan)
CJC	Can. J. Chem.
CJS	Can. J. Spectrosc.
CL	Chem. Lett.
CLY	Chem. Listy
CM	Chem. Mater.
CMC	Comp. Med. Chem.
COC	Comp. Org. Chem.
COMC-I	Comp. Organomet. Chem., 1st edn.
COS	Comp. Org. Synth.
CP	Can. Pat.
CPB	Chem. Pharm. Bull.
CPH	Chem. Phys.
CPL	Chem. Phys. Lett.
CR	C. R. Hebd. Seances Acad. Sci.
CR(A)	C. R. Hebd. Seances Acad. Sci., Ser. A
CR(B)	C. R. Hebd. Seances Acad. Sci., Ser. B
CR(C)	C. R. Hebd. Seances Acad. Sci., Ser. C
CRAC	Crit. Rev. Anal. Chem.
CRV	Chem. Rev.
CS	Chem. Scr.
CSC	Cryst. Struct. Commun.
CSR	Chem. Soc. Rev.
CT	Chem. Tech.
CZ	Chem.-Ztg.
CZP	Czech. Pat.
DIS	Diss. Abstr.
DIS(B)	Diss. Abstr. Int. B.
DOK	Dokl. Akad. Nauk SSSR
DP	Dyes Pigm.
E	Experientia
EC	Educ. Chem.
EF	Energy Fuels
EGP	Ger. (East) Pat.
EJM	Eur. J. Med. Chem.
EUP	Eur. Pat.
FCF	Forschr. Chem. Forsch.
FCR	Fluorine Chem. Rev.
FES	Farmaco Ed. Sci.
FOR	Forschr. Chem. Org. Naturst.
FRP	Fr. Pat.
G	Gazz. Chim. Ital.
GAK	Gummi Asbest Kunstst.
GEP	Ger. Pat.

GEP(O)	Ger. Pat. Offen.
GSM	Gen. Synth. Methods
H	Heterocycles
HAC	Heteroatom Chem.
HC	Chem. Heterocycl. Compd.
HCA	Helv. Chim. Acta
HOU	Methoden Org. Chem. (Houben-Weyl)
HP	Hydrocarbon Process
IC	Inorg. Chem.
ICA	Inorg. Chim. Acta
IEC	Ind. Eng. Chem. Res.
IJ	Isr. J. Chem.
IJC	Indian J. Chem.
IJC(A)	Indian J. Chem., Sect. A
IJC(B)	Indian J. Chem., Sect. B
IJM	Int. J. Mass Spectrom. Ion Phys.
IJQ	Int. J. Quantum Chem.
IJS	Int. J. Sulfur Chem.
IJS(A)	Int. J. Sulfur Chem., Part A
IJS(B)	Int. J. Sulfur Chem., Part B
IS	Inorg. Synth
IZV	Izv. Akad. Nauk SSSR Ser. Khim.
JA	J. Am. Chem. Soc.
JAN	J. Antibiot.
JAP	Jpn. Pat.
JAP(K)	Jpn. Kokai
JBC	J. Biol. Chem.
JC	J. Chromatogr.
JCC	J. Coord. Chem.
JCE	J. Chem. Ed.
JCED	J. Chem. Eng. Data
JCI	J. Chem. Inf. Comput. Sci.
JCP	J. Chem. Phys.
JCPB	J. Chim. Phys. Physico-Chim. Biol.
JCR(M)	J. Chem. Res. (M)
JCR(S)	J. Chem. Res. (S)
JCS	J. Chem. Soc.
JCS(A)	J. Chem. Soc. (A)
JCS(B)	J. Chem. Soc. (B)
JCS(C)	J. Chem. Soc. (C)
JCS(D)	J. Chem. Soc., Dalton Trans.
JCS(F1)	J. Chem. Soc., Faraday Trans. 1
JCS(F2)	J. Chem. Soc., Faraday Trans. 2
JCS(P1)	J. Chem. Soc., Perkin Trans. 1
JCS(P2)	J. Chem. Soc., Perkin Trans. 2
JCS(S2)	J. Chem. Soc. (Suppl. 2)
JEC	J. Electroanal. Chem. Interfacial Electrochem.
JEM	J. Energy Mater.
JES	J. Electron. Spectrosc.
JFA	J. Sci. Food. Agri.
JFC	J. Fluorine Chem.
JGU	J. Gen. Chem. USSR (Engl. Transl.)
JHC	J. Heterocycl. Chem.
JIC	J. Indian Chem. Soc.
JINC	J. Inorg. Nucl. Chem.

JLC	J. Liq. Chromatogr.
JMAS	J. Mat. Sci.
JMC	J. Med. Chem.
JMOC	J. Mol. Catal.
JMR	J. Magn. Reson.
JMS	J. Mol. Sci.
JOC	J. Org. Chem.
JOM	J. Organomet. Chem.
JOU	J. Org. Chem. USSR (Engl. Transl.)
JPC	J. Phys. Chem.
JPJ	J. Pharm. Soc. Jpn.
JPO	J. Phys. Org. Chem.
JPP	J. Pharm. Pharmacol.
JPR	J. Prakt. Chem.
JPS	J. Pharm. Sci.
JPS(A)	J. Polym. Sci., Polym. Chem., Part A
JPU	J. Phys. Chem. USSR (Engl. Transl.)
JSC	J. Serbochem. Soc.
JSP	J. Mol. Spectrosc.
JST	J. Mol. Struct.
K	Kristallografiya
KFZ	Khim. Farm. Zh.
KGS	Khim. Geterotsikl. Soedin.
KO	Kirk-Othmer Encyc.
KPS	Khim. Prir. Soedin.
L	Langmuir
LA	Liebigs Ann. Chem.
LC	Liq. Cryst.
LS	Life Sci.
M	Monatsh. Chem.
MAC	Macromol. Chem.
MC	Mendeleev Chem. J. (Engl. Transl.)
MCLC	Mol. Cryst. Liq. Cryst.
MI	Miscellaneous [book/journal]
MIP	Miscellaneous Pat.
MM	Macromolecules
MP	Mol. Phys.
MRC	Magn. Reson. Chem.
N	Naturwissenschaften
NAT	Nat.
NEP	Neth. Pat.
NJC	Nouv. J. Chim.
NKK	Nippon Kagaku Kaishi (J. Chem. Soc. Jpn.)
NKZ	Nippon Kagaku Zasshi
NZJ	N. Z. J. Sci. Technol.
OCS	Organomet. Synth.
OM	Organometallics
OMR	Org. Magn. Reson.
OMS	Org. Mass Spectrom.
OPP	Org. Prep. Proced. Int.
OR	Org. React.
OS	Org. Synth.
OSC	Org. Synth., Coll. Vol.
P	Phytochemistry
PA	Polym. Age

PAC	Pure Appl. Chem.
PAS	Pol. Acad. Sci.
PB	Polym. Bull.
PC	Personal Communication
PCS	Proc. Chem. Soc.
PHA	Pharmazi
PHC	Prog. Heterocycl. Chem.
PIA	Proc. Indian Acad. Sci.
PIA(A)	Proc. Indian Acad. Sci., Sect. A
PJC	Pol. J. Chem.
PJS	Pak. J. Sci. Ind. Res.
PMH	Phys. Methods Heterocycl. Chem.
PNA	Proc. Natl. Acad. Sci. USA
POL	Polyhedron
PP	Polym. Prepr.
PRS	Proceed. Roy. Soc.
PS	Phosphorus Sulfur
QR	Q. Rev., Chem. Soc.
QRS	Quart. Rep. Sulfur. Chem.
QSAR	Quant. Struct. Act. Relat. Pharmacol. Chem. Biol.
RC	Rubber Chem. Technol.
RCM	Rapid Commun. Mass Spectrom.
RCP	Rec. Chem. Prog.
RCR	Russ. Chem. Rev. (Engl. Transl.)
RHA	Rev. Heteroatom Chem.
RJ	Rubber J.
RP	Rev. Polarogr.
RRC	Rev. Roum. Chim.
RS	Ric. Sci.
RTC	Recl. Trav. Chim. Pays-Bas
RZC	Rocz. Chem.
S	Synthesis
SA	Spectrochim. Acta
SA(A)	Spectrochim. Acta, Part A
SAP	S. Afr. Pat.
SC	Synth. Commun.
SCI	Science
SL	Synlett
SM	Synth. Met.
SR	Sulfur Reports
SRI	Synth. React. Inorg. Metal-Org. Chem.
SS	Sch. Sci. Rev.
SST	Org. Compd. Sulphur, Selenium, Tellurium [R. Soc. Chem. series]
SUL	Sulfur Letters
SZP	Swiss Pat.
T	Tetrahedron
T(S)	Tetrahedron, Suppl.
TA	Tetrahedron Asymmetry
TAL	Talanta
TCA	Theor. Chim. Acta
TCC	Top. Curr. Chem.
TCM	Tetrahedron, Comp. Method
TFS	Trans. Faraday Soc.
TH	Thesis
TL	Tetrahedron Lett.

TS	Top. Stereochem.
UK	Usp. Khim.
UKZ	Ukr. Khim. Zh. (Russ. Ed.)
UP	Unpublished Results
URP	USSR Pat.
USP	US Pat.
WCH	Wiadom. Chem.
YGK	Yuki Gosei Kagaku Kyokaishi
YZ	Yakugaku Zasshi (J. Pharm. Soc. Jpn.)
ZAAC	Z. Anorg. Allg. Chem.
ZAK	Zh. Anal. Khim.
ZC	Z. Chem.
ZN	Z. Naturforsch.
ZN(A)	Z. Naturforsch., Teil A
ZN(B)	Z. Naturforsch., Teil B
ZOB	Zh. Obshch. Khim.
ZOR	Zh. Org. Khim.
ZPC	Hoppe–Seyler's Z. Physiol. Chem.
ZPK	Zh. Prikl. Khim.

Author Index

This Author Index comprises an alphabetical listing of the names of over 50,000 authors cited in the references which appear in Volumes 1–6.

Each entry consists of the author's name, volume numbers in bold, page numbers and reference codes. For example

Abbott, D. E., **2**, 174 ⟨60JOC736⟩, 234 ⟨73JA2738⟩; **6**, 752 ⟨74CPB2201⟩

The reference codes employed throughout the text volumes are included in this index so that the reader can proceed to the point on the text page where an author's work is cited, to the reference itself in the list at the end of each volume, or directly to the original publication.

The accuracy of the spelling of authors' names has been affected by the use, by some authors, of different initials or a different spelling of their name in different papers or review articles (sometimes this may arise from a transliteration process) and by those journals which give only one initial to each author.

A

Aalbersberg, W. G. L., **2**, 944 ⟨75JA5600⟩
Aalten, H. L., **2**, 684 ⟨89T5565⟩
Aamlid, K. H., **6**, 550 ⟨93CAR(242)281⟩
Aantadosi, C., **2**, 252 ⟨86JMC2114⟩
Aarnts, M. P., **4**, 528 ⟨90RTC305⟩
Aaron, A. J., **3**, 359 ⟨78JOC3374⟩
Aaron, D., **1**, 895 ⟨53JA3955⟩
Aaron, H. S., **1**, 19 ⟨67TL2329⟩
Aasen, S. M., **1**, 802 ⟨77JA2342, 78JOC1493⟩
Abad, A., **1**, 49 ⟨86T2429⟩, 1082 ⟨92JOC50⟩
Abakumov, G. A., **2**, 569 ⟨88MI 211-02⟩
Abarca, B., **2**, 1003 ⟨91T5277⟩; **4**, 993 ⟨92H(34)1005⟩
Abatjoglou, A. G., **1**, 52 ⟨84OM(3)923⟩, 859 ⟨77JA8262⟩
Abbas, I. M., **5**, 532 ⟨89MI 512-01⟩
Abbaspour, A., **3**, 220 ⟨78JOC4650⟩, 244 ⟨81TL3565⟩
Abbenhuis, H. C. L., **3**, 512 ⟨90RTC446⟩; **4**, 723 ⟨93OM2227⟩
Abbink, J., **4**, 94 ⟨88GEP3639877A1⟩
Abblard, J., **4**, 103 ⟨71BSF942⟩
Abbott, D. E., **2**, 27 ⟨84JOC2795⟩, 78 ⟨84TL3927⟩
Abbott, P. J., **1**, 1114 ⟨77JA3769⟩
Abbott, T. W., **2**, 30 ⟨43OSC(2)270⟩; **5**, 103 ⟨43OSC(2)10, 43OSC(2)515⟩
Abboud, K. A., **5**, 938 ⟨94OM971⟩
Abbou-Hadeed, K., **5**, 677 ⟨91PS(60)119⟩
Abdallah, M. A., **6**, 654 ⟨85JHC813⟩
Abdallah, S. O., **3**, 250 ⟨73IJC128⟩
Abd Allah, S. O., **5**, 532 ⟨78ZN(B)75, 84JHC253⟩
Abdallah, Y. M., **4**, 355 ⟨92JOC6552⟩
Abd Elall, E. H. M., **2**, 24 ⟨87CC1577⟩; **5**, 1108 ⟨87JCS(P1)2729⟩
Abdel-Baky, S., **1**, 578 ⟨88SC1795⟩; **5**, 196 ⟨86JOC3390⟩
Abdel-Galil, F. M., **3**, 669 ⟨86H(24)2023⟩; **5**, 676 ⟨82GSM100⟩
Abdel Ghani, N. T., **5**, 532 ⟨89MI 512-01⟩
Abdelhamid, A. O., **5**, 676 ⟨82H(19)319⟩, 676, 677 ⟨82JHC73, 83JHC719⟩; **6**, 654 ⟨85JHC813⟩
Abdelhamid, H., **5**, 676 ⟨84JHC797⟩
Abdelkader, M., **4**, 827 ⟨82TL603⟩
Abdelkafi, M. M., **4**, 190 ⟨90SC2527⟩
Abdel-Lateef, G., **6**, 127, 128 ⟨90H(31)47⟩
Abdel-Latif, F. F., **5**, 366 ⟨93PJC711⟩, 616 ⟨90PS(53)145⟩
Abdel-Mageed, M. F., **5**, 478, 494, 582 ⟨87SUL153⟩
Abdel-Magid, A. F., **2**, 300, 310 ⟨90TL5595, 90SL537⟩; **3**, 439 ⟨87JOC702⟩
Abdel-Megeed, M. F., **5**, 483, 590 ⟨86T739⟩; **6**, 274 ⟨86T739⟩
Abdel-Meguid, S. S., **4**, 692 ⟨76JA8289⟩
Abdel-Rahman, M. O., **2**, 377 ⟨81LA943⟩
Abd El-Rahman, N. M., **4**, 464 ⟨89JPR906⟩
Abderhalden, R., **5**, 92 ⟨B-47MI 502-03⟩
Abdesaken, F., **5**, 418, 924 ⟨82OM994⟩, 923, 924 ⟨81CC191, 85AX(C)1632, 82JA5667⟩
Abdou, S. E., **4**, 420 ⟨83H(20)1615⟩; **5**, 532 ⟨84JHC253⟩, 615 ⟨87SUL105⟩
Abdou, W. M., **4**, 464 ⟨89JPR906⟩
Abduganiev, E. G., **3**, 588 ⟨75JCS(P1)2513⟩; **4**, 894 ⟨72IZV1347, 74IZV358⟩
Abduganiev, Yo. G., **4**, 731 ⟨75JFC(6)227⟩
Abdulla, F., **3**, 264 ⟨79T1675⟩
Abdulla, R. F., **1**, 486 ⟨78JOC4248⟩; **3**, 261 ⟨76TL521⟩; **5**, 746 ⟨79T1675⟩; **6**, 104 ⟨79T1675⟩
Abdullaev, N. P., **5**, 977 ⟨89ZOR885⟩
Abdullah, A. H., **1**, 455 ⟨84TL4867⟩
Abdul-Majid, Q.,**3**, 430 ⟨86JCS(P1)1691⟩
Abdulzanieva, K. A., **6**, 54 ⟨67ZOB152⟩
Abe, A., **4**, 419 ⟨90BCJ1160⟩
Abe, H., **1**, 226 ⟨81JOC5379⟩; **2**, 70 ⟨90JA6118⟩, 731 ⟨81CL1117⟩, 784 ⟨92S1119⟩; **3**, 221 ⟨91TA178⟩, 629 ⟨92CL2443⟩; **4**, 390 ⟨90JA6118⟩, 651 ⟨90TL2925⟩
Abe, K., **3**, 286 ⟨77CL645⟩; **6**, 465 ⟨93EUP559212, 94EUP581240⟩
Abe, M., **1**, 946 ⟨82JOC1775⟩; **2**, 18 ⟨87JFC(34)337⟩, 399 ⟨73BCJ2156⟩
Abe, N., **3**, 252 ⟨76JCS(P1)1146⟩, 625 ⟨66TL1509, 67CPB1811⟩, 625 ⟨71JOC2948⟩; **4**, 850 ⟨80JCS(P1)2693⟩; **5**, 652 ⟨88CE215⟩, 1025 ⟨89NKK822⟩
Abe, O., **2**, 212, 214 ⟨74JOC253⟩
Abe, R., **2**, 90 ⟨88TL4139⟩
Abe, S., **1**, 489 ⟨92BCJ2863⟩
Abe, T., **2**, 17 ⟨83CL1135⟩, 265 ⟨78BCJ2183⟩, 598 ⟨70JOM(25)353⟩; **4**, 3 ⟨66BCJ219⟩, 90 ⟨91COS(7)762⟩, 153 ⟨72JOM(43)117⟩, 563 ⟨88JAP63185993⟩, 794 ⟨88CL1887, 89CL905⟩; **5**, 223 ⟨85JCS(P1)2361⟩, 412, 422 ⟨76CL993⟩, 413 ⟨78TL259⟩, 414 ⟨76TL1591⟩, 414, 425 ⟨78JOM(145)C1⟩, 420 ⟨78BCJ2391⟩, 423 ⟨84BCJ2167⟩; **6**, 36 ⟨76BCJ1888⟩, 37 ⟨79JFC(13)519, 79JFC(15)353⟩, 39 ⟨78JFC(12)1, 78JFC(12)359⟩, 39 ⟨83JFC(23)123⟩, 51 ⟨77JFC(9)279⟩, 116 ⟨86JAP61158986, 86JOC4737, 87NKK1447⟩, 125 ⟨76TL1591, 78BCJ2391⟩, 219 ⟨90CL813⟩, 230 ⟨90JFC(50)173⟩, 253 ⟨73JFC(3)17⟩, 257 ⟨90JFC(48)257⟩, 409 ⟨70JAP7026611⟩, 410 ⟨76JAP7630034⟩, 440 ⟨89JFC(45)293⟩
Abe, Y., **1**, 481, 1028 ⟨83CPB4346⟩, 933 ⟨90JOM(381)155⟩, 1010 ⟨78H(9)1397, 80BCJ3273⟩, 1134 ⟨89S898⟩; **2**, 211 ⟨70BCJ1890⟩, 688 ⟨72CL441⟩; **5**, 226 ⟨81TL1257⟩, 247 ⟨70BCJ1890⟩, 513 ⟨90JA3040⟩, 540 ⟨81BCJ790⟩, 1111 ⟨83MI 530-01⟩
Abecassis, J., **1**, 790 ⟨82TL3277, 83TL1247⟩
Abed, D. E., **1**, 978, 1061 ⟨86JOC1199⟩
Abed, N. M., **5**, 703 ⟨88H(27)695⟩
Abed, O. H., **1**, 51 ⟨86TL1209⟩
Abegaz, B., **3**, 344 ⟨71JOC3442⟩; **5**, 565, 614 ⟨87ZC8⟩
Abegg, V. P., **5**, 354, 359 ⟨73TL4091⟩
Abel, E., **4**, 591 ⟨82COMC-I(1)⟩
Abel, E. W., **1**, 447 ⟨58JCS1451⟩; **2**, 111, 112 ⟨58JCS1451⟩, 231 ⟨64JCS5975⟩, 263 ⟨61JCS4933, 65JCS7098⟩, 264 ⟨60JCS4406⟩, 278 ⟨67JCS(A)2014⟩, 399 ⟨64JCS3122⟩, 418 ⟨58JCS3746⟩, 508, 509, 894 ⟨68JCS(A)2623⟩, 544 ⟨70MI 210-01⟩, 851, 852 ⟨64JCS5585⟩, 872 ⟨82COMC-I(1)43⟩, 890 ⟨66JOM(5)326⟩; **3**, 460 ⟨80MI 312-01⟩, 570 ⟨69CC967⟩, 570, 571 ⟨72JCS(D)787⟩; **4**, 283 ⟨87JCS(D)757⟩, 344 ⟨69JOM(17)161, 69TFS1697⟩, 533, 534 ⟨75JOM(97)159⟩, 573 ⟨68JCS(A)1105⟩, 588 ⟨76JOM(107)73⟩, 1034, 1037, 1042 ⟨79JCS(D)1552, 73CC258⟩, 1042 ⟨79JCS(D)814⟩; **5**, 937 ⟨82COMC-I⟩; **6**, 582 ⟨68JCS(A)1105⟩
Abeles, R. H., **1**, 1135 ⟨76ACR313⟩
Abell, A. D., **2**, 1004 ⟨91JCS(P2)1077⟩; **4**, 802 ⟨93JOC1531⟩
Abell, P. I., **1**, 948, 950 ⟨B-73MI 119-17⟩; **2**, 4, 29 ⟨62CRV599⟩
Abell, R. D., **3**, 452 ⟨53JCS2834⟩
Abelman, M. M., **1**, 494 ⟨87JOC4130⟩, 831 ⟨82JA4030, 86JOC3247⟩; **3**, 536 ⟨88TL1493⟩, 540 ⟨89SL36⟩; **5**, 334, 336 ⟨92TL153⟩
Abeln, R., **4**, 673 ⟨86AG(E)909⟩
Abelt, C. J., **1**, 597 ⟨87JOC1218⟩
Aben, R. W. M., **2**, 29 ⟨80SC821⟩; **4**, 414 ⟨87JOC365⟩,

826 ⟨90RTC399⟩, 830 ⟨74TL1019, 77JOC3128⟩; **6**, 75 ⟨77JOC3128⟩
Abenhaim, D., **5**, 63 ⟨82TL1259⟩
Abeywickregma, A. N., **1**, 331 ⟨85JA4594⟩
Abeywickrema, R. S., **2**, 13 ⟨80JOC4226⟩
Abeywickreyma, A. N., **1**, 936 ⟨85JA4595⟩
Abgott, R. A., **5**, 772, 850, 855 ⟨73JOC1344⟩
Abicht, H. P., **4**, 454 ⟨74SRI191⟩
Abicht, H.-P., **4**, 454, 481 ⟨76ZAAC(424)97⟩
Abicht, H. P., **4**, 551, 552, 564 ⟨70JPR456⟩
Abicht, H.-P., **4**, 566, 575 ⟨76ZAAC(422)237⟩, 570, 575, 576 ⟨78ZAAC(447)53, 81ZC341⟩, 575 ⟨82ZAAC(494)55⟩
Abicht, H. P., **6**, 151, 701 ⟨70JPR456⟩
Abidi, S. L., **1**, 1063 ⟨86TL267⟩
Abiko, A., **5**, 74 ⟨86TL4537⟩
Abiko, S., **1**, 279, 1033 ⟨73JA3080⟩, 1033, 1034 ⟨86JOC4507⟩; **2**, 1086 ⟨86JOC4507⟩
Abiko, T., **1**, 822 ⟨89TL1575, 90CC418, 90T4887⟩
Abinu, T., **1**, 1010 ⟨92JMC241⟩
Abiru, T., **1**, 1010 ⟨85CPB1761⟩
Ableman, M. M., **1**, 831 ⟨86T2831⟩
Ablenas, F. J., **1**, 106 ⟨86TL4853⟩
Abley, P., **1**, 73, 74 ⟨71CC840⟩, 507 ⟨69CC1503⟩
Abood, N. A., **1**, 210 ⟨93SC811⟩
Abood, N., **3**, 221 ⟨89JOC6008⟩
Abou-Gharbia, M. A., **3**, 432 ⟨79OPP95⟩, 669 ⟨79H(12)819⟩; **6**, 51 ⟨79H(12)637⟩
Abou-Gharbia, M., **5**, 334 ⟨89JMC1024⟩, 350 ⟨85JOC2224⟩
Aboujaoude, E. E., **3**, 49 ⟨83S634, 84JOM(264)9⟩, 50 ⟨85T427⟩; **4**, 557 ⟨85TL4435, 86JOM(312)283⟩, 557, 576 ⟨84JOM(264)9⟩, 557, 576, 801 ⟨86JOM(304)283⟩, 570 ⟨86S934⟩; **6**, 288 ⟨86JOM(304)283, 87SC1559⟩
Aboulezz, A. F., **5**, 215 ⟨86IJC(B)373⟩
About-Jaudet, E., **3**, 50 ⟨92S1221⟩
Abraham, A. C., **1**, 327, 328 ⟨93T4821⟩, 328, 927 ⟨91JOC4335⟩
Abraham, D. J., **4**, 770, 779 ⟨65T2899⟩, 979 ⟨67JOC990⟩
Abraham, M. H., **2**, 112 ⟨59JCS429, 60JCS4130⟩
Abraham, N. A., **1**, 99 ⟨67JA1044⟩
Abraham, W. D., **1**, 302, 303 ⟨87JA7923⟩
Abrahams, C. B., **1**, 1023 ⟨58JA4593⟩
Abrahart, E. N., **6**, 501 ⟨36JCS1273⟩
Abraitys, V. Y., **3**, 538 ⟨72JA663⟩
Abramivitch, A., **2**, 901, 904 ⟨91T343⟩
Abramo, J., **4**, 757, 759, 820 ⟨58JOC1661⟩
Abramova, N. A., **4**, 706, 1036 ⟨79JOM(168)C33⟩
Abramova, Z. A., **2**, 1076 ⟨74RCR657⟩
Abramov, M. A., **6**, 116 ⟨90ZOR1129⟩
Abramov, V. S., **2**, 463 ⟨69ZOB1543⟩; **5**, 403 ⟨69JGU597⟩, 404, 405 ⟨67JGU826⟩
Abramovich, R. A., **2**, 249 ⟨65CJC3407⟩
Abramovich, V. B., **5**, 46 ⟨70MI 502-01⟩
Abramovitch, A., **1**, 488, 983, 1033, 1034 ⟨91T343⟩, 1102 ⟨77TL411⟩; **3**, 447 ⟨64CRV149⟩; **4**, 815 ⟨91T343⟩
Abramovitch, R. A., **2**, 365 ⟨75JA676⟩, 765 ⟨67JA1537, 69JHC989⟩, 765 ⟨72JOC1690, 72JOC3584⟩, 765 ⟨B-74MI 215-01⟩; **3**, 310 ⟨72JOC3584⟩; **4**, 905, 911 ⟨85CC845⟩
Abrams, C., **4**, 205 ⟨75JHC749⟩
Abrams, S. R., **1**, 1085 ⟨84CJC1333⟩; **2**, 423 ⟨87OS(66)127⟩
Abramtsev, G. I., **2**, 411 ⟨62ZOB3070⟩
Abrazhanova, E. A., **6**, 619 ⟨75ZOR2243⟩, 651 ⟨74ZOR42⟩
Abrecht, S., **1**, 546, 946 ⟨87PAC363⟩
Abrosimova, A. T., **4**, 507 ⟨87JOM(326)159⟩
Absalon, M. J., **1**, 808 ⟨88JOC196⟩
Abshire, C. J., **2**, 782 ⟨60JA1801⟩
Abskharoun, G. A., **4**, 25 ⟨73TL1367⟩
Abt, D. C., **5**, 344 ⟨92SC2729⟩
Abu-El-Halawa, R., **2**, 1009 ⟨84CB1900⟩; **3**, 687 ⟨84CB1900⟩; **5**, 1146 ⟨83CB1834⟩, 1146, 1149 ⟨84CB1161⟩; **6**, 643 ⟨84CB1161⟩
Abuin, E., **1**, 657 ⟨68JCS(B)1044⟩
Abul-Hajj, Y. J., **2**, 28 ⟨86JOC3380⟩
Abul'khanov, A. G., **4**, 915 ⟨88CA(109)211147t⟩
Abu-Orabi, S. T., **2**, 903 ⟨89T1119⟩
Aburatani, M., **2**, 305 ⟨89TL367⟩, 318 ⟨89TL739⟩
Abu Salah, O. M., **5**, 452 ⟨74JCS(D)2302⟩
Abushanab, E., **1**, 212 ⟨93TL3539⟩
Abu-Taha, A., **3**, 194 ⟨76CB154⟩; **5**, 1046 ⟨76CB154, 78CB3750⟩
Acar, M., **4**, 8 ⟨79TL1913⟩
Acemoglu, M., **1**, 193, 288 ⟨89JA7487⟩, 193, 288 ⟨89TL1495⟩
Acharya, S. P., **1**, 74 ⟨84JCS(D)1477⟩; **3**, 4 ⟨90S127⟩
Acharyulu, P. V. R., **3**, 142 ⟨93TL8157⟩
Achatz, D., **4**, 1017 ⟨82ZN(B)1044⟩
Acher, R., **5**, 85 ⟨74AG(E)186⟩
Acheson, M., **2**, 1034 ⟨87JCS(P2)1275⟩
Acheson, R. M., **1**, 102 ⟨76JCS(P1)45⟩, 518 ⟨73JCS(P1)2818⟩; **2**, 764 ⟨78JCS(P1)1117⟩; **4**, 804, 840 ⟨87JCS(P1)1275⟩
Achilles, W., **4**, 473 ⟨65ZC109⟩
Achiwa, K., **1**, 17 ⟨69JOC3667⟩, 367 ⟨85TL3011, 86TL4039⟩, 368 ⟨88H(27)981⟩, 1134 ⟨69JA4318, 70TL2245⟩; **2**, 54 ⟨90JA5876⟩, 54, 305 ⟨89TL363⟩, 331 ⟨87CPB3845⟩; **3**, 367 ⟨87CL1851⟩, 713 ⟨85CC1584⟩; **4**, 369 ⟨85TL3011, 86H(24)1571⟩, 369 ⟨87CPB1734⟩, 509 ⟨85CPB4596⟩, 937 ⟨86CL2089⟩; **5**, 329 ⟨84CL369, 89CPB1179⟩, 354 ⟨75TL2701⟩; **6**, 129 ⟨86H(24)1571⟩, 129, 282 ⟨87CPB1734⟩
Achmatowicz, B., **1**, 210 ⟨93TA399⟩, 697, 698, 700, 703 ⟨85TL5597⟩
Acken, B. J., **3**, 434 ⟨89JOC1736⟩
Acker, D. S., **2**, 224 ⟨57JA6483⟩; **4**, 893, 979, 985 ⟨62JA3387⟩
Acker, R. D., **1**, 204 ⟨78TL2399⟩
Acker, R.-D., **1**, 204, 206, 478 ⟨77TL3407⟩
Acker, S., **3**, 252 ⟨63JOC2533⟩
Ackerfeldt, F., **5**, 856, 860 ⟨15JPR1⟩
Ackermann, E., **4**, 990 ⟨90CB115⟩, 991, 993, 996 ⟨85BSB475⟩
Ackermann, H., **4**, 129, 473 ⟨49HCA1175⟩
Ackermann, K., **3**, 553 ⟨81AG(E)593, 85CB873⟩; **5**, 800 ⟨85CB1517⟩
Ackermann, M. N., **2**, 402 ⟨68JA7173⟩; **4**, 681 ⟨73JOM(60)C57⟩
Ackermann, P., **1**, 200, 238 ⟨88JOC1475⟩
Ackrell, J., **3**, 179, 194 ⟨86JOC3374⟩
Ackroyd, J., **3**, 223, 273 ⟨85HCA338⟩
Acksteiner, B., **2**, 404 ⟨58LA(618)173⟩; **3**, 444 ⟨58CB502⟩
Acott, B., **5**, 999 ⟨65TL4039⟩
Acquadro, M. A., **1**, 785 ⟨75CC206⟩
Acton, E. M., **1**, 625, 628 ⟨58JA355⟩; **5**, 476, 477 ⟨68JOC1783, 68JOC3727⟩; **6**, 502 ⟨66JMC444⟩, 617 ⟨84JOC528⟩
Acton, N., **1**, 811 ⟨70JA6643⟩
Acuña A., C., **4**, 660 ⟨93JOM(448)69⟩
Acyutha Rao, S., **2**, 969 ⟨87CC495⟩
Adachi, I., **1**, 1077 ⟨69TL4447⟩
Adachi, K., **5**, 138 ⟨84T145⟩
Adachi, M., **2**, 289 ⟨79TL2251⟩; **3**, 282 ⟨78JA4842⟩, 664 ⟨90SC71⟩, 669 ⟨80JCS(P1)1132⟩; **5**, 595 ⟨79S887⟩, 1041, 1107 ⟨77TL4417⟩, 1055, 1114 ⟨79TL2251⟩
Adachi, S., **1**, 143 ⟨87JOC1414⟩; **4**, 391 ⟨87JOC1414⟩
Adaevskaya, L. V., **4**, 38 ⟨90IZV1802⟩
Adair, N. K., **4**, 224 ⟨88JOC845⟩; **5**, 540 ⟨88JOC845⟩
Adam, F., **5**, 148 ⟨80AG(E)1029⟩
Adam, G., **3**, 707 ⟨88CB507⟩; **5**, 179 ⟨89LA1233⟩, 658 ⟨88CB507⟩; **6**, 622 ⟨88CB507⟩
Adam, M. A., **1**, 221 ⟨83TL1377⟩; **2**, 530 ⟨86JA3422⟩
Adam, M. J., **2**, 627 ⟨82CC625⟩
Adam, W., **1**, 520 ⟨87CB531⟩, 566 ⟨93CB1481⟩, 566, 681,

682 ⟨81JOC3359⟩, 681 ⟨78JOC4574⟩, 681, 682, 785 ⟨72JA2000, 79S388⟩, 785 ⟨91JOC5782⟩; **3**, 11 ⟨79S388⟩, 156 ⟨89TL6497⟩, 424 ⟨92CB2361⟩, 562, 563 ⟨79AG(E)788⟩; **4**, 188 ⟨89TL6497⟩, 209 ⟨89TL4223⟩, 832 ⟨77JOC38⟩; **5**, 161 ⟨72JA2000⟩, 227 ⟨90TL1245, 90TL2259⟩; **6**, 230 ⟨91JA7654⟩, 426 ⟨86S330⟩
Adamczyk, M., **3**, 29 ⟨84JOC1378⟩, 280 ⟨84JOC4226⟩, 632 ⟨83JOC366⟩
Adamek, F., **6**, 7 ⟨92CCC1291⟩
Adamek, J. P., **2**, 366 ⟨74JOC3525⟩
Adamov, A. V., **5**, 369 ⟨88ZOB2628⟩
Adamovich, S. N., **1**, 61 ⟨83ZOB806⟩; **2**, 1091 ⟨86JGU849⟩; **4**, 617 ⟨87IZV1424⟩
Adamowicz, H., **5**, 344 ⟨85T2441⟩
Adams, C. E., **2**, 177 ⟨48IEC1129⟩
Adams, C. M., **3**, 424 ⟨90SC2359⟩; **5**, 361, 362 ⟨93TL6669⟩
Adams, H., **1**, 94 ⟨84JCS(D)1747⟩; **2**, 202 ⟨93CC1195⟩; **4**, 580 ⟨91JCS(D)1117⟩; **5**, 387, 388 ⟨86JCS(D)813⟩, 943 ⟨90OM2621⟩, 955 ⟨95JOM(494)C12⟩; **6**, 247 ⟨91CC165⟩
Adams, H.-N., **6**, 341 ⟨80AG411, 81AG94⟩
Adams, J., **1**, 514 ⟨84TL2179⟩, 1002, 1134 ⟨83TL5185⟩; **2**, 7 ⟨86JOC789⟩, 241 ⟨82JOC2785⟩, 411 ⟨78TL3989⟩; **3**, 65, 232 ⟨83TL5185, 84TL35⟩; **5**, 338 ⟨82JOC2785⟩
Adams, J. A., **3**, 129 ⟨91T1765⟩
Adams, J. B., **2**, 275 ⟨66JOC2179⟩
Adams, J. L., **1**, 68 ⟨80JA337⟩
Adams, J. S., **1**, 924 ⟨86JA7935⟩
Adams, L. L., **3**, 4, 115 ⟨89JOC5387⟩
Adams, M. L., **5**, 944 ⟨95OM2102⟩
Adams, R., **2**, 52 ⟨23JA1071, 24JA1675⟩, 192 ⟨41OSC(1)84⟩, 273 ⟨73OSC(5)107⟩, 498 ⟨22JA805⟩, 608 ⟨50JA2454⟩, 779 ⟨47OSC(1)214⟩; **3**, 612 ⟨41OSC(1)107⟩; **4**, 29 ⟨32OSC(1)358⟩, 53 ⟨21JA651, 21JA660⟩; **5**, 10 ⟨20JA599⟩, 34 ⟨32OSC(1)436⟩, 728 ⟨23JA1744⟩
Adams, R.,**5**, 767 ⟨56JA3825⟩
Adams, R. D., **1**, 818 ⟨88TL4913⟩; **2**, 327 ⟨88OM2250⟩, 579 ⟨82COMC-I(4)967⟩; **4**, 677 ⟨91OM1278, 92OM1473⟩, 677 ⟨92OM1480⟩, 678 ⟨91OM2541⟩, 955 ⟨91JA9406, 93OM3431⟩; **5**, 798, 799, 800 ⟨79JOM(172)251⟩; **6**, 716 ⟨89CRV1703⟩
Adams, R. M., **2**, 110 ⟨61USP2992266⟩
Adams, R. R., **3**, 638 ⟨44JA725⟩
Adams, W., **1**, 800 ⟨B-83MI 118-01⟩
Adamson, C. L., **5**, 375, 376 ⟨90OM1557⟩
Adamson, D. W., **3**, 465 ⟨37JCS1551⟩, 465, 490 ⟨35JCS286⟩
Adamsone, B. J., **6**, 555 ⟨92KGS1117⟩
Adamzyk, M., **5**, 666, 671 ⟨83JOC366⟩
Adank, K., **2**, 227 ⟨54HCA375⟩
Adcock, J. I., **6**, 251 ⟨92MI 608-01⟩
Adcock, J. L., **1**, 664 ⟨79JOC3431⟩; **4**, 52 ⟨74JA7588⟩; **6**, 36 ⟨73JOC3617⟩, 37 ⟨75JOC3271⟩, 230 ⟨92JOC4749⟩, 301 ⟨87JFC(37)327⟩
Adcock, W., **1**, 895 ⟨79JOC2652⟩; **2**, 16 ⟨75JOM(91)C20⟩, 438 ⟨88JOC3016⟩
Addison, C. E., **2**, 150 ⟨56JCS2705⟩
Addy, W. D., **5**, 276 ⟨66JOC3829⟩
Adelaere, B., **3**, 339 ⟨87BSF517⟩
Adelman, A. M., **2**, 249 ⟨70S545⟩
Adelstein, S. J., **2**, 632 ⟨80JOM(186)C33⟩
Adembri, G., **2**, 316 ⟨92JHC321⟩; **5**, 532 ⟨67JHC54⟩
Ademek, J. P., **1**, 28 ⟨74JOC3525⟩
Adenis, J. C., **6**, 231 ⟨80JFC(15)201⟩, 441 ⟨80JFC(15)37⟩
Adhikary, P., **1**, 486 ⟨69CB2966⟩
Adickes, F., **3**, 616 ⟨43JPR271⟩
Adickes, H. W., **3**, 61 ⟨73JOC36⟩, 707 ⟨69TL5151⟩; **5**, 1014 ⟨69TL5151⟩
Adinolfi, M., **2**, 45 ⟨75G377, 76TL3661⟩
Adiwidjaja, G., **1**, 653 ⟨89T3163⟩; **2**, 938 ⟨91AG(E)1687⟩; **3**, 353 ⟨91CB1397⟩, 369 ⟨90CB2065⟩, 546 ⟨79CB2698⟩, 548 ⟨84AG(E)439⟩, 550 ⟨91CB1485⟩; **4**, 843, 846, 854 ⟨88CB1165⟩; **5**, 344 ⟨88CB689⟩, 344, 639 ⟨80TL4247⟩, 485 ⟨80AG(E)563⟩, 581, 589, 604 ⟨84CB2205⟩; **6**, 337 ⟨81LA264⟩, 637 ⟨82CB2516⟩
Adkins, H., **1**, 90, 92 ⟨41JA1563⟩, 881 ⟨43OSC(2)606⟩; **2**, 29 ⟨27JA2517⟩, 30 ⟨55OSC(3)785⟩, 45 ⟨55OSC(3)217⟩, 52 ⟨48JA695, 54OR(8)1⟩, 595 ⟨51JA2854⟩; **3**, 446 ⟨49JA3836⟩, 526 ⟨49JA2242⟩; **5**, 195 ⟨49JA2242⟩; **6**, 524 ⟨48JA383⟩
Adkins, J. D., **3**, 82 ⟨70JA5774⟩
Adkins, R. L., **2**, 95 ⟨83JOC3126⟩; **3**, 131 ⟨82JOC3006⟩
Adkins, S., **5**, 709 ⟨83JOC888⟩
Adlef, W., **6**, 71 ⟨57MI 603-01⟩
Adler, E., **3**, 231 ⟨71TL3803⟩
Adler, J., **3**, 618 ⟨18M240⟩
Adler, M., **4**, 235 ⟨78CB2859⟩
Adler, M. E., **3**, 8 ⟨89S940⟩
Adley, T. J., **6**, 309 ⟨67JCS(C)807⟩
Adlington, R. M., **1**, 154 ⟨83CC1040⟩, 154 ⟨86CC176⟩, 264 ⟨89CC1852⟩, 340 ⟨92T3413⟩, 630, 632, 1094 ⟨83ACR55⟩, 633 ⟨81JCS(P1)2848⟩, 634 ⟨81CC65, 81T3935⟩, 931 ⟨92T3385⟩; **3**, 143 ⟨88CC1404⟩, 293 ⟨83CC1040⟩, 714, 718 ⟨85CC816⟩; **4**, 536 ⟨83CC1040, 86T4223⟩; **5**, 329 ⟨90CC720⟩, 376 ⟨90T4733⟩
Adlkofer, J., **4**, 585 ⟨73AG(E)415, 74CB3697⟩
Adolfsson, L., **6**, 85, 89 ⟨76CS122⟩
Adolph, H. G., **4**, 55, 57 ⟨68CB3604⟩, 435 ⟨82JOC2474⟩; **6**, 56, 261, 279, 287 ⟨68JOC3073⟩, 57 ⟨69JOC45, 70JOC3188⟩, 145 ⟨89MI 605-01, 90SC3303⟩
Adrian, F. J., **1**, 322 ⟨64JCP(40)213⟩
Adrian, G., **2**, 28 ⟨81SC669⟩
Adrian, M. A. P., **5**, 1028 ⟨89CAR(188)35⟩
Advani, B. G., **3**, 323 ⟨67TL2689⟩, 361 ⟨82T1673⟩; **4**, 94 ⟨82T1673⟩, 973 ⟨77IJC(B)297⟩; **5**, 760 ⟨79JCS(P1)1762⟩, 1073 ⟨65JOC3369⟩
Advena, J., **4**, 134 ⟨74TL303⟩
Aeberli, P., **2**, 382, 384 ⟨69JOC2720⟩; **4**, 415 ⟨69JOC2720⟩, 416 ⟨75JMC177, 75JMC182⟩
Aebischer, B., **2**, 397 ⟨83HCA789⟩
Aehyutha Rao, S., **2**, 594 ⟨93JOC2694⟩
Aeppli, L., **1**, 432 ⟨84JMC1575⟩
Aerssens, M. H. P. J., **1**, 457, 508 ⟨84CC735⟩
Afanas'ev, M. M., **5**, 644 ⟨79ZOB2323, 86ZOB2359⟩, 738 ⟨79JGU2323⟩
Afanas'ev, P. S., **5**, 1044 ⟨91ZOR340⟩
Afanas'ev, Yu., **4**, 911 ⟨68ZOB1523⟩
Afanasova, O. B., **4**, 723 ⟨72IZV2594⟩
Afanassova, O. B., **4**, 696 ⟨74JOM(65)131⟩
Afarinkia, K., **2**, 665 ⟨92T9111⟩; **3**, 50 ⟨90T7175⟩, 196 ⟨92CC285⟩; **4**, 461 ⟨90SL415⟩; **5**, 394, 396 ⟨92H(34)369⟩
Affleck, J. G., **2**, 227 ⟨50JOC865⟩
Afghahi, F., **1**, 399 ⟨79JOC2348⟩
Afonin, A. V., **4**, 777 ⟨90ZOR2056⟩
Afonso, A., **1**, 431 ⟨60JOC2217⟩
Afonso, C. A. M., **1**, 692 ⟨92TL3899⟩
Afonso, C. M., **3**, 8, 121 ⟨87JCS(P1)1221⟩
Afrashteh, A., **4**, 775 ⟨88AP(321)903⟩; **5**, 555 ⟨88AP(321)903⟩
Afridi, A. S., **2**, 710 ⟨91HCA1931⟩
Afshar, S., **6**, 25 ⟨83M813⟩
Afshari, G. M., **3**, 4 ⟨86S249⟩
Afza, N., **4**, 413 ⟨82ZN(B)512⟩
Afzal, D., **5**, 801 ⟨83IC3954⟩
Agabekyan, R. S., **5**, 669 ⟨89KFZ1317⟩
Agadzhanyan, Z. E., **6**, 104 ⟨65T3537⟩
Agafonov, S. V., **4**, 119 ⟨86JGU2394⟩, 916 ⟨80ZOB1225⟩
Agakhanova, T. B., **2**, 915 ⟨77IZV2389⟩
Agami, C., **4**, 202 ⟨93SL349⟩
Agarrabeitia, A. R., **1**, 800 ⟨90CC934⟩
Agarwal, A., **4**, 707, 720 ⟨77JOM(141)85⟩
Agarwala, U., **5**, 1026 ⟨70IJC759⟩
Agarwal, K. P., **4**, 56 ⟨82JOC2582⟩

Agarwal, S., **3**, 4, 115 ⟨90T4417⟩
Agarwal, V. K., **6**, 443 ⟨87IJC(B)748⟩
Agati, V., **1**, 573 ⟨93CC1038⟩, 955 ⟨93CC270, 93JOC5702⟩
Agawa, T., **1**, 608 ⟨79CB355⟩, 1038 ⟨86TL933⟩, 1137 ⟨61JA449⟩; **2**, 14 ⟨83BCJ1881⟩, 244 ⟨69T4283, 72JOC3615⟩, 251 ⟨73S667⟩, 303 ⟨71S89⟩, 399 ⟨73BCJ2156⟩, 456 ⟨85BCJ3104, 85BCJ3104⟩, 843 ⟨79CB355⟩, 850 ⟨82BCJ909⟩, 932 ⟨81TL3079⟩; **3**, 62 ⟨82CL1997⟩, 127 ⟨85JA7179⟩, 130, 285, 295 ⟨86TL929⟩, 216 ⟨78CC12⟩, 273 ⟨86TL933⟩, 527 ⟨79JOC2067⟩, 557 ⟨79CC900, 80JOC5377⟩, 574 ⟨81CL1197⟩, 593 ⟨70JOC2076⟩; **4**, 13 ⟨83BCJ1772⟩, 66 ⟨82CL587⟩, 326 ⟨74JOC3412⟩, 328 ⟨85S276⟩, 348 ⟨79CB355⟩, 517, 538, 1015 ⟨82BCJ1163⟩, 538 ⟨81AG(E)126⟩, 979 ⟨74JOC3198⟩; **5**, 274, 286 ⟨80S933⟩, 364 ⟨85H(23)677⟩, 652 ⟨75BCJ3259⟩, 989 ⟨85S276⟩, 1032 ⟨74JOC922⟩, 1033 ⟨68TL2651⟩; **6**, 123, 697 ⟨79CB355⟩, 162 ⟨82BCJ1163⟩
Ageeva, A. B., **4**, 478 ⟨69IZV1757⟩, 561 ⟨64ZOB3938⟩; **5**, 221 ⟨86IZV2535⟩
Agenäs, L.-B., **5**, 1113 ⟨70ACS3751⟩, 1114 ⟨67ACS837⟩
Ager, D. J., **1**, 160 ⟨90OR(38)1⟩, 160, 648, 758, 982 ⟨84S384⟩, 248 ⟨83JCS(P1)1131⟩, 653 ⟨78CC177⟩; **2**, 539 ⟨77JCR(S)6⟩, 908 ⟨81TL587⟩, 908, 1000, 1005 ⟨83JCS(P1)1131⟩, 936, 1005 ⟨84S384⟩, 997, 1005 ⟨90OR(38)1⟩; **3**, 20 ⟨80TL1677⟩, 20, 136, 137 ⟨82CSR493⟩, 28, 31, 43 ⟨B-87MI 301-01, 84CC486⟩, 157, 160, 167 ⟨B-87MI 304-01⟩, 166 ⟨81TL2803⟩, 182 ⟨B-87MI 304-07⟩, 271 ⟨82TL1945⟩; **4**, 212 ⟨89CC178⟩, 250 ⟨80TL4763⟩, 352 ⟨89CC1256⟩, 365 ⟨80TL1677⟩, 365, 366, 392 ⟨82CSR493⟩, 366, 392, 394, 652 ⟨81TL2923⟩, 366, 392, 938 ⟨81TL587⟩, 366, 936 ⟨83TL95⟩, 392, 938, 957 ⟨83JCS(P1)1131⟩, 618, 652, 945, 1048, 1049 ⟨84JOC168⟩, 646 ⟨84S384⟩, 662, 962 ⟨86OM1906⟩, 773 ⟨82TL1945⟩; **6**, 127, 135 ⟨86JCS(P1)195, 86JOC3983⟩, 325, 327, 391 ⟨84JOC168⟩
Ager, E., **5**, 1111 ⟨70JCS(C)1530⟩
Ager, I. R., **4**, 804 ⟨74JCS(P1)1644⟩
Ager, J. H., **1**, 2, 8 ⟨58JOC1938⟩
Aggarwal, A., **4**, 900 ⟨80S748⟩
Aggarwal, A. K., **2**, 891 ⟨89POL581⟩; **5**, 318, 369 ⟨89POL581⟩
Aggarwal, M., **6**, 390 ⟨91MI 613-01⟩
Aggarwal, P., **1**, 282 ⟨91JOC3908⟩
Aggarwal, R. C., **5**, 385 ⟨86JMC199⟩
Aggarwal, V. K., **3**, 35 ⟨91JCS(P1)451⟩; **4**, 268 ⟨91JCS(P1)662⟩, 269 ⟨90TL135, 91TL7743, 92JOC6390⟩, 863 ⟨92SL730⟩
Agha, B. J., **6**, 462 ⟨86MI 615-01⟩
Agmon, I., **1**, 988 ⟨87JA782⟩
Agnel, G., **2**, 980 ⟨90JA8590⟩, 988 ⟨91JA7424⟩; **4**, 140 ⟨89S687⟩
Agnello, E. J., **3**, 226 ⟨60JA4293⟩
Agniant, P., **3**, 72 ⟨75CR(281)35⟩
Agopovich, J. W., **6**, 410 ⟨80JA7572⟩
Agosta, W. C., **1**, 19 ⟨85JA5732⟩, 596 ⟨86JA3385⟩; **3**, 241 ⟨85TL703⟩, 413 ⟨83JOC1925⟩, 538 ⟨69TL4517⟩; **5**, 204 ⟨65TL2681⟩, 277 ⟨85JA5490⟩
Agostinho, R., **3**, 620 ⟨45JA2152⟩
Agoston, G. E., **2**, 73 ⟨91TL3001⟩
Agranat, I., **1**, 248 ⟨87CC381⟩, 909 ⟨76JOC2379⟩
Agrawal, S., **5**, 385 ⟨86JMC199⟩
Agrawal, S. P., **4**, 416 ⟨63JIC777⟩
Agre, C. L., **4**, 38 ⟨52JA3895⟩; **6**, 80 ⟨48JA2268⟩
Agrebi, A., **4**, 607, 621, 622, 628, 629 ⟨89OM1585⟩, 628, 629 ⟨87OM659⟩
Aguero, A., **2**, 681 ⟨86CC531⟩; **3**, 515 ⟨85CC793⟩
Aguiar, A. M., **1**, 999 ⟨66JOM(5)205⟩; **2**, 435, 1060, 1064 ⟨69JOC2684⟩, 452 ⟨68JOC579⟩, 739 ⟨73JOC2908⟩, 739, 1072 ⟨71JOC2719⟩, 827, 828 ⟨64JA2299⟩, 841 ⟨65JOC2826⟩, 874 ⟨67JOC2627, 68JOC1681⟩, 875 ⟨67TL4447⟩, 1060 ⟨73JOC1611, 76MI 222-01⟩, 1070 ⟨69JOC3349⟩, 1072 ⟨71JOC2720⟩; **3**, 493 ⟨66JA4090⟩; **4**, 455, 477 ⟨67JOC2383⟩, 563 ⟨66TL3187⟩
Aguiar, H., **3**, 493 ⟨66JA4090⟩
Aguiar, H. J., **1**, 999 ⟨66JOM(5)205⟩; **2**, 874 ⟨68JOC1681⟩; **4**, 563 ⟨66TL3187⟩
Aguilá, S., **5**, 570, 618 ⟨81JCR(S)58⟩
Aguilar, A. M., **2**, 438 ⟨91JOC5919⟩
Aguilar, E., **2**, 322 ⟨89CC1132⟩
Aguilar, M. A., **1**, 24 ⟨81S55⟩; **6**, 120 ⟨91JOC5919⟩
Aguirre, G., **6**, 416 ⟨91SA(A)1619⟩
Agulló, C., **1**, 1082 ⟨92JOC50⟩
Agviar, A. M., **1**, 445 ⟨67JOC2627⟩
Agwaramgbo, E. L. O., **1**, 648 ⟨B-88MI 114-03⟩; **4**, 619, 620 ⟨89JOC5613⟩; **6**, 125 ⟨89JOC5613⟩
Aharon-Shalom, E., **1**, 909 ⟨76JOC2379⟩; **4**, 876 ⟨82JA1154⟩
Ahbala, M., **3**, 457 ⟨87JOM(331)11⟩
Ahern, E. P., **4**, 990 ⟨80JOC4302⟩
Ahern, T. P., **4**, 267 ⟨80CJC878⟩; **6**, 254, 255 ⟨80CJC1996, 83CJC610⟩
Ahlberg, P., **1**, 918 ⟨72JA7063, 83JCS(P2)291⟩, 918 ⟨86JA3819⟩
Ahlbrecht, H., **1**, 148 ⟨84TL1353⟩, 560 ⟨79S127⟩; **2**, 302 ⟨88S775⟩, 739 ⟨80S320⟩, 741 ⟨79S127⟩, 748 ⟨75S512, 77S495⟩, 748 ⟨78S897⟩, 751 ⟨74S672, 82S717⟩, 751 ⟨89S117, 92S1018⟩, 789 ⟨73S109⟩, 812 ⟨76S746, 80S630⟩, 812 ⟨82S273⟩, 1001 ⟨84TL1353, 90CB1853⟩, 1001, 1002 ⟨93CB1887⟩; **3**, 227 ⟨88S210⟩, 493 ⟨67AG(E)249⟩, 638, 657 ⟨79S127, 77S495⟩, 657 ⟨78S897⟩, 658 ⟨73S109⟩; **4**, 530 ⟨74S672, 83S61⟩, 530 ⟨88S775, 91S637⟩, 913, 918, 925 ⟨84CB2622⟩, 1003, 1004 ⟨84CB1, 77S336⟩; **5**, 900 ⟨67AG(E)249⟩; **6**, 561 ⟨94S983⟩, 561, 563 ⟨91S637⟩
Ahlbrect, H., **6**, 563 ⟨88S775⟩
Ahlers, H., **1**, 647, 667, 717 ⟨78TL4399⟩; **2**, 280, 908 ⟨83CB1001⟩; **4**, 401 ⟨77AG(E)710⟩, 565 ⟨80TL2803⟩, 947, 965 ⟨83CB1001⟩
Ahlers, K.-D., **5**, 1054 ⟨67CB1367⟩; **6**, 596 ⟨63ZC348, 67CB1459⟩, 596, 598 ⟨67CB1373⟩
Ahlfaenger, B., **1**, 534 ⟨75S632⟩
Ahlgrén, M., **5**, 1028 ⟨91JHC1003⟩
Ahlgren, M., **6**, 263 ⟨92JCS(P2)835, 92PS(70)183⟩
Ahlrichs, R., **1**, 1109 ⟨73CPL309⟩
Ahmad, A., **1**, 999 ⟨87IJC(B)77⟩
Ahmad, M. A., **6**, 558 ⟨91PJS422⟩
Ahmad, M. M., **4**, 786 ⟨89SA(A)1011⟩
Ahmad, S., **1**, 834 ⟨85JA6731⟩; **2**, 172 ⟨63JCS2636⟩, 172 ⟨63JCS2636⟩; **3**, 214 ⟨88CC692⟩, 229 ⟨89TL1033⟩, 369 ⟨85JA6731⟩; **5**, 177 ⟨87JOC3937⟩, 264, 275 ⟨87CC114⟩
Ahmad, S. S., **5**, 1125 ⟨88IJC583⟩
Ahmad, Y., **5**, 360, 361 ⟨67CJC1539⟩
Ahmed, A., **3**, 221 ⟨84CC179⟩
Ahmed, F. U., **6**, 491 ⟨84IC4322, 86S833⟩, 491 ⟨89IS79⟩
Ahmed, G., **5**, 606 ⟨93TL5397⟩
Ahmed, J., **2**, 442 ⟨77JA7876⟩
Ahmed, K. J., **2**, 1009 ⟨89JA1299⟩; **4**, 675 ⟨85OM1312, 86OM2171⟩; **6**, 517 ⟨93OM2401⟩
Ahmed, M., **3**, 361 ⟨69CJC2039⟩; **4**, 54 ⟨75JCS(P1)2048⟩; **5**, 634 ⟨70CJC1991⟩
Ahmed, R., **3**, 363 ⟨69TL3611⟩
Ahmed, S. A., **5**, 199 ⟨88TL1595⟩
Ahmed, S., **4**, 154 ⟨74JOM(78)395⟩
Ahmed, S. S., **3**, 300 ⟨76TL3217⟩
Ahmed, S. Z., **6**, 647 ⟨89TL7313⟩
Ahmed, Z., **1**, 218 ⟨88SC151⟩
Ahn, J. H., **2**, 60 ⟨93JOC1941⟩; **3**, 86 ⟨93JOC1941⟩
Ahn, K. H., **1**, 39 ⟨85JOC1927⟩; **3**, 85 ⟨84JOC1717⟩
Ahn, S. H., **1**, 183 ⟨86TL943⟩
Ahn, Y., **3**, 131 ⟨94TL203⟩
Ahnfelt-Rønne, I., **6**, 645 ⟨80JMC13⟩
Ahnoff, M., **6**, 426 ⟨90JC593⟩
Ahond, A., **2**, 579 ⟨73JOM(55)375⟩
Ahra, M., **4**, 605 ⟨85JOM(280)313⟩

Ahrens, F. B., **1**, 103 ⟨1895MI 103-01⟩
Ahrens, G., **4**, 860 ⟨82LA1022⟩, 903 ⟨82LA585⟩
Ahrens, H., **1**, 138 ⟨92TL5323⟩; **2**, 742 ⟨66CB1580⟩; **4**, 164 ⟨67JHC625⟩
Ahrens, K. H., **5**, 685 ⟨78ZN(B)636⟩
Ahrens, U., **2**, 852 ⟨81ZAAC(480)90⟩
A. Hudson, **2**, 489 ⟨80JCS(D)2428⟩
Ahuja, A. S., **6**, 559 ⟨82JCS(P1)653⟩
Ahuja, H. S., **2**, 384 ⟨61JOC1819⟩
Ahuja, J. R., **1**, 801, 802 ⟨93SL875, 91T8155⟩, 802 ⟨91TL1107⟩
Ahuja, V. K., **1**, 503 ⟨73JOC2226⟩
Aiba, S., **5**, 332 ⟨80JCS(P1)2122⟩
Aichenegg, P. C., **4**, 777 ⟨71USP3592896⟩
Aicher, T. D., **3**, 244 ⟨87TL3463⟩
Aichholz, R., **3**, 631 ⟨91LA47⟩; **5**, 135 ⟨91LA47⟩
Aida, T., **2**, 32 ⟨73CC212⟩, 212, 215, 231 ⟨86TL441⟩, 231, 723 ⟨83TL5173⟩, 244 ⟨84S1037⟩, 312 ⟨89BCJ3845⟩, 709 ⟨85TL1795⟩; **4**, 161 ⟨77JCS(P1)372⟩, 831 ⟨79TL4029⟩; **5**, 242, 337 ⟨87S1122, 84S1037⟩
Aidhen, I. S., **3**, 285 ⟨92TL5431⟩
Aiga, M., **5**, 173 ⟨78JOM(159)137⟩
Aig, E. R., **1**, 1134 ⟨83JA3656⟩
Aigami, K., **5**, 282 ⟨77S632⟩
Aigbirhio, F., **6**, 394 ⟨90CC1471⟩
Aigbirhio, F. I., **6**, 174, 176, 381 ⟨92JCS(D)1015⟩, 176, 394 ⟨91JOM(405)149⟩
Aihara, M., **2**, 630 ⟨88S236⟩; **4**, 750 ⟨88S236⟩
Aihara, S., **6**, 70 ⟨76JAP(K)76108012⟩
Aikawa, H., **6**, 126 ⟨87TL2147⟩
Aikawa, Y., **1**, 308 ⟨76BCJ779⟩; **3**, 119 ⟨88JCS(P1)2423⟩; **5**, 291 ⟨76CL57⟩
Ailman, D. E., **2**, 405 ⟨38JA1930⟩
Aime, S., **4**, 682 ⟨89JOM(365)163⟩, 718 ⟨83POL77⟩
Aimetti, J. A., **1**, 4, 9 ⟨79TL4631⟩
Aimi, N., **5**, 324 ⟨92GSM79⟩, 1127 ⟨90TL1287⟩
Aimino, D., **2**, 708 ⟨89JOC4458⟩
Aimo, G., **6**, 97 ⟨79S223⟩
Ainley, D. A., **1**, 447 ⟨30JCS2171⟩
Ainslie, R. D., **2**, 561 ⟨83JOM(250)1⟩
Ainsworth, C., **1**, 257 ⟨50JA2134⟩; **4**, 832 ⟨72JOM(46)73⟩, 1047 ⟨79TL87⟩
Aiqiao, M., **3**, 415 ⟨89SC1423⟩
Airoldi, G., **2**, 190 ⟨74JCS(P1)771⟩
Aitchison, A. A., **6**, 339, 341 ⟨87OM819⟩, 339, 346, 347 ⟨86OM1103⟩
Ait-Haddou-Mouloud, H., **6**, 24 ⟨88TL1699⟩
Aitken, D. J., **1**, 224 ⟨93SL599⟩; **3**, 638 ⟨93SL599⟩
Aitken, R. A., **1**, 790 ⟨93PHC1⟩, 811 ⟨82CC1164⟩, 1079 ⟨85CC1140, 90SL211⟩, 1079 ⟨90SL212, 93TL5621⟩; **5**, 900 ⟨94CC805⟩
Aiube, Z. H., **6**, 174, 390 ⟨84JOM(269)217⟩
Aiura, A., **2**, 325 ⟨87H(26)929⟩
Aiura, H., **5**, 287 ⟨87TL3377⟩
Aizaka, T., **5**, 1078 ⟨76CL811⟩
Aizawa, M., **4**, 286 ⟨81T2547⟩
Aizawa, S., **4**, 902 ⟨67BCJ2641⟩; **6**, 105, 106 ⟨67BCJ2641⟩
Aizawa, T., **5**, 1024 ⟨74JOC1970, 75BCJ2981⟩, 1035 ⟨77BCJ425⟩
Aizenshtadt, I. N., **6**, 145 ⟨71ZOR1304⟩, 413 ⟨68ZPK1380⟩
Aizpurua, J. M., **2**, 10 ⟨86JOC4941⟩, 14 ⟨84TL1103, 85T4657⟩, 14 ⟨86CJC2342⟩, 34 ⟨85T2903⟩, 36 ⟨84TL3123⟩; **3**, 4, 115 ⟨83TL4367, 85T2903⟩, 88 ⟨86CJC225⟩, 162 ⟨87TL5361⟩, 528 ⟨92TL4779⟩; **6**, 162, 163 ⟨91JOM(419)C1⟩, 163 ⟨92TL3903⟩
Aizpurua, J. M. C., **2**, 262 ⟨85TL475⟩
Ajimura, N., **5**, 189, 195, 282 ⟨86JOC4150⟩
Ajjou, A. N., **3**, 294 ⟨92SC1967⟩
Akabori, S., **1**, 76 ⟨56MI 103-01⟩; **3**, 88 ⟨68JOC3277⟩
Akaboshi, S., **5**, 567 ⟨69CPB550⟩
Akada, W., **5**, 254, 537, 539, 540 ⟨73BCJ244⟩, 537 ⟨72IJS279⟩
Akadegawa, N., **6**, 135 ⟨92CC870⟩
Akahane, A., **1**, 150 ⟨91TL5501⟩, 150 ⟨91TL5509⟩; **3**, 350 ⟨90EUP379979⟩
Akai, S., **1**, 1061, 1062 ⟨90JOC4853⟩, 1140 ⟨85TL1549, 86TL195, 87CPB1405, 87TL4583, 88CPB3897⟩; **2**, 941 ⟨90JOC1108⟩; **4**, 661 ⟨91SL911, 92JCS(P1)2813⟩; **5**, 189, 195, 282 ⟨86JOC4150⟩, 189, 199 ⟨84TL6027⟩, 215 ⟨89S334⟩, 223 ⟨90CPB1836⟩
Akaishi, R., **2**, 279 ⟨89CC1789⟩
Akaji, K., **1**, 37, 441 ⟨83JA4017⟩; **2**, 227 ⟨91CC167⟩, 478 ⟨90CPB673⟩
Akama, K., **5**, 995 ⟨88CL1385⟩
Akamanchi, K. G., **3**, 4, 116 ⟨91SC419⟩
Akam, T. M., **1**, 457, 508 ⟨80TL1069⟩
Akamatsu, H., **1**, 907 ⟨88JA633⟩
Akamone, M., **3**, 405, 426 ⟨90CL635⟩
Akamsin, V. D., **2**, 460 ⟨66DOK(168)807⟩, 462 ⟨66IZV493⟩, 463 ⟨67IZV1976⟩
Akano, M., **5**, 547 ⟨69JOC730⟩, 580 ⟨70JOC2438⟩
Akasaka, K., **5**, 155 ⟨79JA259⟩
Akasaka, T., **4**, 161 ⟨77JCS(P1)372⟩, 258 ⟨90T5093⟩, 624 ⟨93OM1514⟩
Akashi, C., **2**, 7, 835 ⟨68CPB1009⟩, 20 ⟨68CPB1784⟩
Akashi, K., **3**, 5, 117 ⟨76TL2503⟩
Akatsuka, R., **3**, 177 ⟨80BCJ817⟩
Akazawa, T., **3**, 549 ⟨71T1781⟩
Akbar Ali, M., **6**, 563 ⟨91POL823⟩
Akcamur, Y., **5**, 217 ⟨87OPP52⟩
Akchurua, R. A., **2**, 359 ⟨81ZOB609⟩
Akelah, A., **3**, 557 ⟨80MI 317-03⟩; **5**, 122 ⟨81S413⟩
Åkerfeldt, S., **2**, 120 ⟨60ACS1980⟩
Åkermark, B., **2**, 308 ⟨88JOC3845⟩; **3**, 65 ⟨88ACS569⟩; **5**, 375 ⟨89JOC4649⟩
Akers, J. A., **2**, 518 ⟨89TL2187⟩
Akgün, E., **4**, 193 ⟨85LA2472⟩, 193 ⟨84M587⟩, 193 ⟨84S227⟩
Akgun, H., **3**, 162 ⟨90SC2279⟩
Akhmedov, A. A., **5**, 63 ⟨84MI 502-01⟩
Akhmedov, I. M., **2**, 136 ⟨90KGS1477⟩
Akhmedov, M. A., **2**, 136 ⟨90KGS1477⟩
Akhmedov, Sh. T., **4**, 913 ⟨76ZOB1652⟩; **5**, 484 ⟨74IZV55⟩
Akhmetova, F. V., **2**, 885 ⟨72JGU1960⟩
Akhnoukh, T., **4**, 690 ⟨93JOM(459)325⟩
Akhobadze, D. Sh., **5**, 720 ⟨70IZV1157, 72IZV1150⟩
Akhrem, A. A., **1**, 248 ⟨83RCR61⟩, 391, 884 ⟨70RCR732⟩, 794 ⟨81RCR666⟩; **2**, 350 ⟨76MI 206-01⟩; **3**, 239 ⟨85S784⟩
Akhrem, I. S., **1**, 281 ⟨80CC748⟩, 493 ⟨74JOM(72)163⟩
Akhtar, M., **1**, 935 ⟨68CJC633⟩; **3**, 40 ⟨79CC875⟩; **4**, 153, 811 ⟨68CJC2165⟩; **5**, 207 ⟨87T5341⟩
Akhtar, M. H., **3**, 584 ⟨73TL3727⟩; **4**, 912 ⟨82MI 420-01⟩
Akhtar, N., **2**, 621 ⟨86IJC328⟩
Akhverdiev, I. R., **6**, 412 ⟨72URP345094⟩
Aki, L. Y., **1**, 808 ⟨88JOC196⟩
Aki, O., **2**, 228 ⟨70CPB235⟩; **3**, 76 ⟨72CPB1325⟩
Akiba, E., **4**, 17 ⟨83JA2771⟩
Akiba, K., **1**, 1041 ⟨84TL1083⟩; **2**, 325 ⟨87H(26)929⟩, 840 ⟨69BCJ2975⟩, 853 ⟨72BCJ506⟩; **3**, 439 ⟨77H(7)1131⟩; **5**, 287 ⟨87TL3377⟩; **6**, 120, 698 ⟨76TL3695⟩, 581 ⟨69BCJ2975⟩, 655 ⟨74BCJ935⟩
Akiba, K.-Y., **2**, 695 ⟨88TL4719⟩; **3**, 186 ⟨84TL5405⟩; **6**, 496 ⟨73CC13⟩
Akiba, M., **4**, 872 ⟨84SC1119⟩
Akiba, T., **6**, 447 ⟨92TL3483⟩
Akihiko, F., **3**, 357 ⟨91EUP409164⟩
Akimoto, H., **1**, 55 ⟨69TL3105⟩
Akimoto, K., **5**, 1023 ⟨92OPP346⟩; **6**, 550 ⟨79JA6116⟩
Akimoto, N., **1**, 188 ⟨89JOC4055⟩
Akimova, L. I., **2**, 433 ⟨87URP1294811⟩
Akinaga, K., **3**, 496 ⟨89JOC974⟩
Akita, H., **1**, 212 ⟨93CPB55⟩
Akita, M., **2**, 630 ⟨87JOC1100⟩, 923, 924

⟨82JOM(226)C9⟩; **4**, 674 ⟨93OM2925⟩, 679 ⟨92CC1031⟩, 1052 ⟨83JOM(254)13⟩; **5**, 223 ⟨88OM366⟩
Akita, T., **5**, 226 ⟨80CC1029⟩
Akita, Y., **1**, 318 ⟨82H(19)329⟩, 1010 ⟨86CPB1447⟩; **5**, 327 ⟨81JGU2125⟩
Akiyama, A., **4**, 67 ⟨86JHC1163⟩
Akiyama, F., **1**, 692 ⟨78JOC2477⟩; **3**, 10 ⟨79TL3929⟩
Akiyama, I., **3**, 352, 366 ⟨93TL115⟩; **5**, 548 ⟨92CC1522⟩, 554, 563, 605 ⟨93BCJ623, 93TL115⟩, 627 ⟨94CL77⟩
Akiyama, M., **1**, 5, 7 ⟨84S949⟩; **5**, 332 ⟨80JCS(P1)2122⟩
Akiyama, S., **1**, 1063 ⟨91CC948⟩; **3**, 54, 62, 252 ⟨79TL2809⟩; **6**, 215 ⟨81MI 607-01⟩
Akiyama, T., **1**, 769 ⟨80JOC2005⟩; **3**, 168 ⟨84CL1531⟩
Akiyoshi, K., **1**, 491 ⟨87JOM(334)181⟩, 1006, 1035 ⟨87CC477⟩; **3**, 228 ⟨87CL1007⟩; **4**, 655, 658, 659 ⟨88JA646⟩, 658 ⟨89JA3089⟩
Akiyoshi, S., **4**, 212 ⟨54JA693⟩
Akkerman, O. S., **3**, 508 ⟨83TL3935⟩; **4**, 605, 626, 636, 637 ⟨88JOM(338)159⟩, 608, 626, 669, 696, 700 ⟨85JOM(288)13⟩, 618 ⟨89TL6195⟩, 657, 663, 701 ⟨84TL5191⟩, 669, 670 ⟨86JOM(308)1, 87TL6493⟩; **6**, 190, 192, 198 ⟨86TL6123⟩, 399 ⟨89TL6195, 92JA7302⟩, 406 ⟨91AOC(32)147⟩
Aklalay, Y., **2**, 471 ⟨83MI 208-01⟩
Akmedov, Sh. T., **4**, 884, 913 ⟨77CA121332u⟩
Akopyan, N. E., **5**, 690 ⟨77KFZ38⟩
Akopyan, R. A., **5**, 690 ⟨77KFZ38, 90CHE1170⟩, 690 ⟨90KGS1403⟩
Akram, M., **6**, 556 ⟨85PS(21)291⟩
Aksela, R., **2**, 903 ⟨91T1163⟩
Aksenes, W., **2**, 889 ⟨73ACS1337, 73ACS2135⟩
Aksinenko, A. Yu., **4**, 494 ⟨87DOK(297)1132, 87IZV1680⟩, 494 ⟨88IZV936⟩; **5**, 1076 ⟨87IZV1177⟩
Aksnes, G., **1**, 51 ⟨78CJC1933⟩, 751 ⟨83PS(16)339⟩; **2**, 444 ⟨61ACS438⟩, 453 ⟨66ACS2508⟩; **3**, 498 ⟨61ACS692⟩
Aksoy, H., **2**, 385 ⟨92SC2607⟩
Aktaev, N. P., **2**, 183, 249 ⟨75IZV2530⟩; **5**, 386 ⟨74ZOR1428⟩
Aktogu, N., **5**, 453 ⟨82CC1303, 84JOM(262)49⟩
Akutagawa, S., **1**, 77 ⟨87JA1596, 87JA4129⟩; **2**, 53 ⟨87JA5856, 93JA3318⟩, 841 ⟨88OS(67)20⟩; **3**, 127 ⟨92TL635⟩
Akutagawa, S., **3**, 158 ⟨93JOC2929⟩
Akuzawa, K., **1**, 194 ⟨87CL1237⟩
Alabaster, R. J., **2**, 906 ⟨88S950⟩; **5**, 8 ⟨89S598⟩
Aladzheva, I. M., **4**, 551 ⟨80TL2931, 92ZOB263⟩, 557 ⟨84ZOB36, 85ZOB1234⟩, 557 ⟨86ZOB1220, 88ZOB1987⟩, 557 ⟨88ZOB1998, 88ZOB2012⟩, 561 ⟨63ZOB1045⟩
Alajarin, M., **1**, 835 ⟨91TL4041⟩; **2**, 121 ⟨85TL469⟩; **3**, 585 ⟨90CC829, 91JOC4008⟩, 623 ⟨93T5091⟩; **5**, 622, 1067 ⟨89TL2847⟩, 1066 ⟨89T4263, 90CC7, 94TL3817⟩, 1068 ⟨89JA355⟩, 1079 ⟨82SC573⟩
Alajarin, R., **2**, 249 ⟨92TL3677⟩; **3**, 171 ⟨92TL3677⟩
Alami, M., **1**, 469 ⟨93TL6403⟩, 1009 ⟨93SL217⟩, 1012 ⟨91TL6109⟩, 1066 ⟨92TL2681⟩; **2**, 587 ⟨90TL7423⟩
Alanine, A. I. D., **5**, 685, 726 ⟨89TL4443⟩
Alarcón, P., **5**, 1066 ⟨88TL379⟩
Alías, A., **5**, 1065 ⟨93JOC5264⟩
Al-Aseer, M., **1**, 148 ⟨83JA2080⟩
Al Ashmawy, M. I., **5**, 1108 ⟨87JCS(P1)2729⟩
Alauddin, M. M., **1**, 418 ⟨86CJC793⟩
Al-Awadi, N., **1**, 777 ⟨82JCS(P2)1775⟩, 86JCS(P2)1255>, 777 ⟨88JCS(P2)177⟩, 782 ⟨89JCS(P2)579, 90JCS(P2)2187⟩
Al-Azawi, F. W., **3**, 275 ⟨74JHC1101⟩
Albanbauer, J., **4**, 340 ⟨77LA624⟩, 493 ⟨76LA36⟩; **5**, 634 ⟨77CB2114⟩
Albanese, D., **2**, 25, 31, 35 ⟨92JCS(P1)2309⟩; **3**, 427, 430 ⟨90S333⟩, 643 ⟨90SC965⟩; **5**, 1030 ⟨91S1001⟩
Albanese, J. A., **4**, 584 ⟨89JOM(375)265⟩
Albano, C., **1**, 905 ⟨80JCS(P2)1447⟩
Albano, E. L., **2**, 35 ⟨69JOC3519⟩
Albano, V. G., **5**, 1088 ⟨91AG(E)847⟩; **6**, 332 ⟨86G101⟩
Albanov, A. I., **4**, 140 ⟨89JOM(361)147⟩, 507 ⟨87JOM(326)159⟩
Al-Bashir, R. F., **1**, 782 ⟨89JCS(P2)579⟩
Al-Bayati, R., **1**, 204 ⟨82TL5229⟩; **3**, 247 ⟨82TL353⟩; **4**, 206 ⟨82TL353⟩
Albeck, M.,**4**, 877 ⟨81TL4199⟩
Alberghina, G., **3**, 549 ⟨84JCR(S)370⟩
Albericio, F., **5**, 259 ⟨93T11065⟩
Alberola, A., **2**, 306 ⟨90SL763⟩, 312 ⟨83JOC2255⟩, 320 ⟨90S153, 90SC1149⟩; **3**, 252 ⟨90SC2537⟩, 653 ⟨86TL2027⟩
Albers, H., **5**, 397 ⟨52CB239⟩
Albers, M. O., **1**, 456 ⟨85JMOC213⟩
Albert, A., **3**, 615 ⟨16CB1383⟩; **5**, 579 ⟨15CB470⟩, 690 ⟨76JCS(P1)291⟩
Albert, A. H., **6**, 149 ⟨81JHC1477⟩
Albert, C., **2**, 108 ⟨05CB53⟩
Albert, D. J., **2**, 379 ⟨81JINC3389⟩
Albert, M. S., **5**, 690 ⟨84JOC1212⟩
Albert, P., **2**, 19 ⟨89JOC5380⟩, 611 ⟨86BSF910⟩
Albert, R., **1**, 785 ⟨91JOC5782⟩; **4**, 65 ⟨73JOC2160⟩
Albert, W., **2**, 511, 512 ⟨80CB3318⟩; **5**, 911 ⟨80CB3318⟩
Alberti, A., **2**, 238 ⟨92JOM(434)151⟩; **6**, 734 ⟨93JA8444⟩
Alberts, A. H., **1**, 970 ⟨88CC748⟩; **5**, 51 ⟨72SC79⟩
Alberts, V., **1**, 67, 69 ⟨81JOC563⟩
Albertson, N. F., **5**, 278 ⟨62OR(12)157⟩
Albery, W. J., **1**, 844 ⟨67MI 119-07⟩
Albinak, M. J.,**5**, 998 ⟨60JOC1837⟩
Albinati, A., **4**, 566 ⟨91HCA983⟩
Albini, A., **1**, 782 ⟨93S263⟩; **2**, 346, 347, 350, 355 ⟨93S263⟩; **3**, 317 ⟨78CJC2985⟩
Albini, E., **2**, 793 ⟨83T689⟩
Al'bitskaya, V. M., **4**, 343 ⟨87ZOB2300⟩
Albizati, K. F., **1**, 1091 ⟨77JA7365⟩; **2**, 676 ⟨77JA7365⟩; **3**, 60 ⟨77JA7365⟩, 119 ⟨82TL4647⟩, 160 ⟨90S1059⟩; **4**, 171, 176 ⟨89CRV1617⟩; **6**, 79 ⟨88TL2023⟩
Albizzati, E., **4**, 725 ⟨71JOM(26)357⟩
Albonico, S. M., **2**, 675 ⟨72CI(L)425⟩
Albrecht, B., **4**, 19 ⟨66ACS253⟩
Albrecht, E., **5**, 1040 ⟨90JHC1059⟩
Albrecht, F. X., **6**, 573 ⟨68JOC851⟩
Albrecht, H., **4**, 528 ⟨84TL1353⟩
Albrecht, I., **2**, 564 ⟨86OM1296⟩; **4**, 578 ⟨84AG(E)522⟩
Albrecht, J., **2**, 421 ⟨67AG(E)633⟩
Albrecht, M., **4**, 725 ⟨91OM3791, 92CB1953⟩, 725 ⟨92JA8531, 92JOM(427)C21⟩, 725 ⟨92OM3517, 93OM4979⟩
Albrecht, P., **1**, 180 ⟨93TL1197⟩
Albrecht, R., **2**, 100 ⟨62AG(E)97⟩, 793 ⟨62AG(E)89⟩; **4**, 304 ⟨64CB483⟩, 304, 311 ⟨64CB490⟩; **5**, 94 ⟨78LA617⟩, 652 ⟨64CB483⟩
Albrecht, T., **5**, 934 ⟨94TL561⟩
Albright, C. F., **2**, 906 ⟨63OSC(4)68⟩
Albright, J. A., **3**, 666 ⟨66JOC4100⟩
Albright, J. D., **1**, 165 ⟨83T3207⟩; **3**, 137, 292, 627 ⟨83T3207⟩, 672 ⟨73JHC899⟩
Albright, J. O., **3**, 62 ⟨80JA1436⟩
Albright, T. A., **3**, 494 ⟨76JOC1168⟩
Albritton, D. L., **1**, 1143 ⟨75JCP(63)3238, 76JCP(64)4544⟩
Albuquerque, P., **4**, 580 ⟨88ZN(B)1293⟩, 581 ⟨89JOM(371)329⟩
Alcaide, B., **1**, 217 ⟨90H(31)1997⟩; **3**, 414 ⟨85TL4403⟩, 417 ⟨84JCR(S)144⟩; **4**, 304 ⟨84JCR(M)144⟩; **5**, 696 ⟨91OM11, 93JOC3886⟩
Alcantara, J., **3**, 623 ⟨93T5153⟩
Alcaraz, C., **4**, 858, 938, 961 ⟨91TL1385⟩
Alcaraz, G., **4**, 589 ⟨93PS(76)49⟩; **5**, 888 ⟨93CC1354⟩; **6**, 691, 704 ⟨93CC1354⟩
Alcaraz, J.-M., **4**, 341 ⟨82TL1565⟩
Alcazar, V., **4**, 232 ⟨90T1783⟩, 266 ⟨89AQ22⟩; **5**, 554 ⟨90T1057⟩

Alcock, N. W., **1**, 803 ⟨79JCS(P2)962⟩; **2**, 998 ⟨90AX(C)1300⟩; **4**, 151, 585 ⟨90JCS(D)1553⟩, 867 ⟨91PS(60)131⟩, 1081 ⟨88JCS(P1)2839⟩; **5**, 445, 447 ⟨80JA5093⟩
Alcudia, F., **2**, 152 ⟨92JOC6789⟩
Aldenhoven, H., **2**, 474 ⟨90CB1837⟩
Alder, K.,**1**, 524 ⟨62LA(651)141⟩
Alder, R. W., **1**, 847 ⟨B-71MI 119-01⟩, 848 ⟨B-71MI 119-02⟩, 851 ⟨B-71MI 119-03, B-71MI 119-04⟩; **2**, 383 ⟨84JCS(P2)411⟩, 383, 384 ⟨82JCS(P1)603⟩
Alderdice, M., **1**, 482 ⟨84OS(62)14⟩; **5**, 103 ⟨84OS(62)14⟩
Alderman, R. M., **5**, 835 ⟨92JOC4648⟩
Aldinger, F., **5**, 1077 ⟨93CB2569⟩
Aldokhina, T. F., **4**, 992 ⟨79ZOB1087⟩
Aldous, D. J., **1**, 1042 ⟨90JOC4657⟩; **3**, 712, 718, 719 ⟨89SL9⟩; **4**, 102 ⟨90JOC4657⟩
Aldous, G., **1**, 895 ⟨79JOC2652⟩
Aldous, S. C., **1**, 1042 ⟨90JOC4657⟩; **4**, 102 ⟨90JOC4657⟩
Aldrich, P. E., **4**, 780 ⟨91JMC2525⟩; **5**, 285 ⟨71JMC535⟩; **6**, 229, 230, 424 ⟨64JOC11⟩
Aldridge, C. L., **6**, 70 ⟨53JA3987⟩
Aldridge, M., **2**, 769 ⟨11JCS404⟩
Al-Dulaymmi, M. F. M., **4**, 547 ⟨88JOM(338)C31⟩
Al Dulayymi, J. R., **4**, 775 ⟨93JCS(P1)1945⟩
Aleinikov, N. N., **6**, 261 ⟨88IZV2639, 88IZV466⟩
Alekperov, R. K., **4**, 323 ⟨87IZV2399⟩; **6**, 558 ⟨83ZOR92, 84ZOR2158⟩, 558 ⟨85MI 617-01, 90MI 617-02⟩
Aleksandrova, I. A., **6**, 563 ⟨81ZOB530⟩, 564 ⟨81ZOB28⟩
Aleksandrov, A. M., **4**, 777 ⟨73ZOR69⟩, 861 ⟨68ZOB1503⟩; **6**, 19 ⟨82ZOR228⟩
Aleksandrov, G. G., **4**, 461 ⟨80IZV1125⟩
Aleksandrowicz, P., **2**, 395 ⟨82T1321⟩
Aleksankin, M. M., **5**, 988 ⟨65ZOB923⟩
Alekseev, A. A., **4**, 833 ⟨71KGS1645⟩
Alekseeva, L. A., **6**, 19 ⟨B-89MI 601-02⟩, 37 ⟨83JFC(22)105⟩, 38, 39 ⟨70ZOR144, 70ZOR2498⟩, 39 ⟨75ZOR1672, 76ZOR1798⟩, 39, 43 ⟨76ZOR1287⟩, 61 ⟨68ZOB2813⟩
Alekseev, V. V., **6**, 581 ⟨75ZOB1484⟩
Alekseiko, L. N., **6**, 671, 672 ⟨92MI 621-01⟩
Aleksiev, D., **4**, 797 ⟨77UKZ1000⟩
Alemagna, A., **2**, 692 ⟨87S192⟩; **3**, 430 ⟨85CC417⟩; **5**, 476, 477 ⟨57MI 511-01, 57MI 511-02⟩, 676, 677, 906 ⟨85T2503⟩, 831 ⟨58MI 521-01⟩, 905 ⟨81G285, 84T2165, 83CC337⟩, 906 ⟨85T3321⟩
Aleshkova, M. M., **2**, 108 ⟨91JGU999⟩
Alewood, P. E., **6**, 445 ⟨78JCS(P1)1066⟩
Alewood, P. F., **5**, 328 ⟨81S121⟩
Alexakis, A., **1**, 230, 476 ⟨80TL935⟩, 312 ⟨84PAC91⟩, 313 ⟨86TL1047⟩, 465, 1017, 1018 ⟨79S826⟩, 468 ⟨81TL959, 83BSF332, 84T2741⟩, 471, 1097, 1107 ⟨81TL3851⟩, 478, 479 ⟨80T1961, 78S528, 78TL2027, 84TL3075⟩, 479 ⟨86T5607⟩, 496 ⟨84IJ113, 92BSF171⟩, 533, 1107 ⟨80TL3763⟩, 538, 539, 540, 541, 542, 1093, 1106, 1139 ⟨81S841⟩, 540 ⟨75JOM(96)471⟩, 541 ⟨82TL5151, 84OS(62)1⟩, 651 ⟨89T381⟩, 670, 967 ⟨90JA8042, 85TL4197⟩, 967 ⟨91T1677⟩, 1024 ⟨89T6197⟩, 1092 ⟨78TL3711⟩; **2**, 86 ⟨89TL2387, 89TL2391⟩, 315 ⟨91SL625⟩, 381 ⟨91TL1171, 92JOC4563⟩, 629, 977 ⟨84OS(62)1⟩, 676 ⟨77BSF693⟩, 903, 909, 910, 917, 919, 924, 925, 976, 977, 978 ⟨81S841⟩, 919 ⟨79S826⟩, 977 ⟨82TL5151⟩; **3**, 42 ⟨91TL1171, 92JOC4563⟩, 130 ⟨79SC639, 84T683⟩, 212 ⟨83TL5081⟩, 212, 228 ⟨86T1369⟩, 220 ⟨78TL4205⟩, 222 ⟨80T1961⟩; **4**, 176 ⟨88PAC49⟩, 176 ⟨90TA477⟩, 192 ⟨84TL3075⟩, 405, 408 ⟨91SL111⟩, 963 ⟨77BSF693⟩; **5**, 152 ⟨88TL4411⟩
Alexander, B., **6**, 415 ⟨89JOC3231⟩
Alexander, C. A., **2**, 439 ⟨85IC2344⟩
Alexander, D. L., **4**, 171 ⟨82JOC615⟩
Alexander, E. R., **6**, 78 ⟨52JA554⟩
Alexander, E. S., **2**, 339 ⟨68JCS(C)796⟩; **4**, 979 ⟨68JCS(C)796⟩; **6**, 239 ⟨68JCS(C)796⟩
Alexander, J. J., **5**, 799 ⟨89OM1523, 89OM2589⟩
Alexander, J. R., **4**, 77 ⟨31JCS1913⟩
Alexander, M. D., **2**, 310 ⟨90JOC2563⟩
Alexander, S. M., **2**, 378, 379 ⟨71JMC1052⟩; **4**, 539 ⟨72JOC215⟩
Alexandre, C., **5**, 199 ⟨91T5481⟩
Alexandrou, N. E., **5**, 356 ⟨82S824⟩, 366 ⟨90JHC1741⟩, 765 ⟨83CCC596⟩
Alexanian, V., **5**, 124 ⟨78TL4475⟩
Al-Farhan, E., **6**, 416, 417 ⟨91JOC891⟩
Al-Fekri, D. M., **4**, 5 ⟨87JOC1872⟩
Alferyev, I. S., **4**, 560 ⟨85BBA(818)96⟩, 560 ⟨83IZV2802, 84DOK(277)371⟩, 560 ⟨84IZV1122, 87IZV860⟩, 561 ⟨92ZN(B)1213⟩
Al'fonsov, V. A., **2**, 254 ⟨93PS(79)179⟩, 464 ⟨88JGU1131⟩; **4**, 345 ⟨86ZOB1905, 86ZOB216⟩, 345 ⟨87DOK(292)360⟩, 921 ⟨84ZOB1324⟩; **5**, 369 ⟨91IZV2103⟩, 405 ⟨77JGU505, 89JGU2388⟩, 406 ⟨85JGU1956, 88JGU1548⟩, 535 ⟨89JGU1548⟩, 737 ⟨85ZOR559⟩
Alford, J. A., **4**, 162 ⟨68JOC2368⟩
Algharib, M. S., **3**, 681 ⟨87JHC577⟩; **5**, 676, 703 ⟨88OPP521, 87JHC577⟩
Alhadeff, E. S., **2**, 403 ⟨70JOC3401⟩
Al-Hashimi, S., **6**, 186, 195 ⟨78JOM(153)253⟩
Al-Hassan, M., **2**, 922, 931 ⟨83TL2055⟩
Al-Hassan, M. I., **1**, 471 ⟨87S816⟩; **2**, 925, 926 ⟨84JOC725⟩; **4**, 748, 757, 759 ⟨86MI 417-01⟩, 786 ⟨89SA(A)1011⟩, 807 ⟨89JOM(372)183⟩
Al Holly, M. M., **3**, 539 ⟨68CC764⟩
Ali, A., **1**, 347 ⟨92TL2851⟩
Ali, J., **4**, 281 ⟨81SA(A)819⟩
Ali, M. A., **6**, 561 ⟨91CAR(216)271⟩, 563 ⟨72ICA11, 78JINC451⟩
Ali, M., **5**, 94 ⟨79SC363⟩
Ali, M. B., **1**, 357 ⟨82JA1742⟩
Ali, N. M., **1**, 489 ⟨92T8117⟩
Ali, R., **2**, 476 ⟨88JCS(D)2077⟩; **5**, 652 ⟨90JCS(D)2593⟩, 1129 ⟨88JCS(D)2077⟩
Ali, R. Mohd., **5**, 1041 ⟨78SC497⟩
Ali, S. A., **5**, 173 ⟨81JA1568⟩, 201 ⟨89T5979⟩
Ali, Sk. A., **2**, 350 ⟨79JA7114⟩
Ali, S. M., **2**, 391 ⟨88S805⟩; **3**, 428 ⟨88S805⟩, 456 ⟨89JCS(P1)279⟩, 458 ⟨88JCS(P1)945⟩; **6**, 550 ⟨89CC684, 89JOC5603⟩
Alicot, M. J. C., **2**, 360 ⟨80EUP10477⟩
Alieva, S. V., **6**, 547 ⟨85KPS844⟩
Aliev, F. Y., **5**, 369 ⟨90ZOB1183⟩
Aliev, R. Z., **4**, 556 ⟨76ZOB266⟩
Aliev, Z. G., **2**, 1003 ⟨86IZV134⟩
Alimov, M. P., **6**, 656 ⟨69IZV119⟩
Alimov, P. I., **5**, 404 ⟨66BAU1432⟩, 404, 405 ⟨71BAU723⟩; **6**, 656 ⟨69IZV119⟩
Aliocco, A., **1**, 255 ⟨90CC1532⟩
Alioto, R. L., **6**, 647 ⟨78AF1435⟩
Alipou, E., **6**, 25 ⟨83M813⟩
Al-Isa, F. S., **5**, 774 ⟨89JPR149⟩
Al-Jabar, N. A. A., **2**, 872 ⟨85JOM(287)57⟩
Aljadeff, G., **4**, 205 ⟨89JOC4866⟩
Al-Jallo, H. N., **3**, 275 ⟨74JHC1101⟩
Al-Jibori, S., **4**, 1023, 1028, 1030 ⟨82CC286⟩, 1028, 1030 ⟨82CC287⟩; **6**, 151 ⟨82CC286⟩
Al-Juaid, S. S., **5**, 886 ⟨93CC267⟩; **6**, 165, 289, 372, 373 ⟨89JOM(366)39⟩, 173, 394 ⟨93JOM(462)45⟩, 176, 394 ⟨91JOM(405)149⟩, 387 ⟨89JCS(D)447⟩, 392, 394 ⟨92JOM(437)41⟩, 393 ⟨89CC273⟩, 393 ⟨94AG(E)1268⟩, 393 ⟨94JOM(469)129⟩, 393 ⟨94JOM(480)199, 88CC1389⟩, 394 ⟨95UP 613-01⟩, 394, 395 ⟨86CC908⟩, 396 ⟨87JOM(325)117⟩, 515 ⟨91JOM(403)293⟩
Alkabets, R., **1**, 80 ⟨76JOC3682⟩
Al-Kazimi, H. R., **6**, 540 ⟨55JA2479⟩
Alker, D., **2**, 15 ⟨80NJC239⟩; **5**, 326 ⟨90JCS(P1)1623⟩
Al-Khaidar, Kh. Ya., **6**, 518 ⟨80MI 616-01⟩
Alkinson, R. S., **2**, 775 ⟨79COC(2)245⟩

Alks, V., **4**, 99 ⟨89SC1479⟩
Alla, M., **2**, 407 ⟨80ZOB2621⟩
Allain, R., **3**, 526 ⟨70HCA120⟩, 527 ⟨70HCA2159⟩
Allais, A., **4**, 174 ⟨80T1763⟩
Allala, R., **3**, 620 ⟨83SC35⟩
Allan, F. J., **3**, 444 ⟨58JOC639⟩
Allan, G. G., **3**, 444 ⟨58JOC639⟩; **5**, 199 ⟨71CI(L)545⟩
Allan, R. D., **2**, 299 ⟨83JCS(P1)2983⟩
Allard, M., **2**, 799 ⟨69CC1515⟩; **3**, 601 ⟨77JOC4261⟩
Allavena, C., **4**, 146 ⟨88CC333⟩; **5**, 152 ⟨88TL5369⟩
Allaway, J. R., **1**, 28 ⟨75JA6873⟩
Allcock, H. R., **4**, 577 ⟨63JPS(A)3627⟩, 706 ⟨84JA2337⟩; **5**, 368 ⟨B-72MI 507-01⟩; **6**, 558 ⟨91IC1776⟩
Allcock, S. J., **5**, 351, 363 ⟨91T10053⟩; **6**, 515 ⟨91T10053⟩
Alleman, S., **5**, 584 ⟨93SL801⟩
Allen, A. A., **1**, 78 ⟨78T1261⟩
Allen, A. D., **2**, 928, 941 ⟨93JA10097⟩, 1022 ⟨87CJC1719⟩
Allen, C. F., **2**, 29 ⟨63OSC(4)398, 63OSC(4)608⟩; **5**, 98 ⟨63OSC(4)608⟩, 99 ⟨63OSC(4)616⟩
Allen, C. F. H., **2**, 23 ⟨43OSC(2)336⟩, 117 ⟨66CJC2315⟩; **3**, 67 ⟨55OSC(3)731⟩, 67, 79 ⟨45OS(25)92⟩, 612 ⟨41OSC(1)156⟩, 616 ⟨55OSC(3)275⟩, 633 ⟨44OSC(2)498⟩; **4**, 47 ⟨49JOC754⟩; **5**, 34 ⟨63OSC(4)804⟩, 184 ⟨55OSC(3)28⟩, 302 ⟨55OSC(3)765⟩
Allen, C. W., **2**, 864 ⟨91CRV119, B-90MI 216-01⟩
Allen, D. G., **2**, 442 ⟨82IC1007⟩
Allen, D. S., **3**, 2, 113 ⟨56JOC478⟩
Allen, D. W., **1**, 739 ⟨76JOM(110)315⟩; **2**, 810, 892 ⟨72JCS(P1)2793⟩, 824 ⟨76JCS(P1)2529⟩, 844, 846 ⟨77JCS(P2)1705⟩, 847, 848 ⟨77JCS(P2)789⟩, 889 ⟨70JCS(C)810⟩
Allen, G., **1**, 870 ⟨66JCS(B)161⟩; **5**, 24 ⟨53QR255⟩
Allen, G. R., **5**, 25, 64 ⟨73JOC4044⟩
Allen, J. C., **1**, 925 ⟨62JCS4468⟩
Allen, J. F., **2**, 412 ⟨55JA2871⟩
Allen, Jr., G. R., **1**, 785 ⟨70JA1763⟩
Allen, L. C., **1**, 891 ⟨73TS253⟩, 892 ⟨69JA1037⟩, 1109 ⟨69JA5350, 70JA2141⟩
Allen, M. J., **1**, 548 ⟨60JCS1045⟩
Allen, P., **2**, 186, 193 ⟨42JOC23⟩, 193 ⟨51JOC767⟩
Allen, R. E., **1**, 637 ⟨58JA4588⟩; **2**, 412 ⟨61JOC1533⟩; **4**, 184, 189 ⟨66JMC127⟩
Allen, R. P., **1**, 331, 936 ⟨89CC1387⟩; **2**, 536 ⟨89CC1387⟩
Allen, S. T., **3**, 159 ⟨92JOC426⟩
Allen, T., **1**, 327, 328 ⟨91JA7002⟩
Allen, T. C., **5**, 965 ⟨56USP2733254⟩
Allen, T. G., **2**, 8 ⟨79JOC359⟩
Allenstein, E., **4**, 792 ⟨63CB1035⟩, 792, 971 ⟨63CB3230⟩, 894 ⟨67CB2604⟩, 972 ⟨68CB1232⟩; **6**, 51 ⟨81ZAAC(474)7⟩, 52 ⟨78CB921⟩, 238, 613, 727 ⟨64CB1286⟩, 452, 728 ⟨64CB3162⟩, 624, 730 ⟨66ZAAC(344)113⟩, 632 ⟨67CB2604⟩
Allentoff, A. J., **1**, 810 ⟨91JOC321⟩; **3**, 528 ⟨91JOC321⟩
Aller, E., **5**, 680 ⟨88H(27)1935, 89JCR(S)262⟩, 680 ⟨89S843⟩, 1066 ⟨94TL3817⟩; **6**, 504 ⟨92S297⟩
Aller, R. T. von, **2**, 179 ⟨66JOC2357⟩
Allies, P. G., **4**, 697 ⟨73JOM(54)35⟩
Alliger, G., **5**, 589, 595, 620 ⟨49JOC962⟩
Alligood, D. B., **1**, 811 ⟨87T4265⟩
Allin, S. M.,**3**, 181 ⟨94TL2427⟩
Allinger, G., **2**, 358 ⟨49JOC921⟩
Allinger, J., **1**, 883 ⟨59JA5740⟩
Allinger, N. L., **1**, 503, 505 ⟨56JA2518⟩, 869 ⟨67JOC1994⟩, 927 ⟨86JA2381⟩
Allison, C. G., **4**, 45 ⟨71FCR77⟩
Allison, J. A. C., **6**, 217, 225 ⟨59JA1089⟩
Allison, N. T., **6**, 403 ⟨83JA1384⟩
Allison, W. S., **2**, 184 ⟨76ACR293⟩
Allman, R., **6**, 340 ⟨75JOM(102)109⟩
Allman, T., **5**, 1133 ⟨79JOM(166)365⟩
Allmang, G., **3**, 226 ⟨91TL1275⟩
Allmann, R., **3**, 417 ⟨85CB3089⟩; **4**, 229 ⟨84AG(E)890⟩; **5**, 694 ⟨85CB3089⟩; **6**, 625 ⟨86CB3236, 87CB1271⟩
Allner, K., **3**, 245 ⟨63HCA1259⟩
Allocco, A., **2**, 302 ⟨90CC1532⟩
Alloum, A. B., **3**, 193 ⟨89SC2567⟩, 251 ⟨91S301⟩
Allpress, C. F., **2**, 881 ⟨53JCS16⟩; **6**, 467 ⟨24JCS(125)2259⟩
Allred, A. L., **5**, 432 ⟨65IC1747⟩
Allred, E. L., **1**, 660 ⟨66JOC3498⟩
Allspach, T., **4**, 479, 480 ⟨86JOM(306)39⟩, 484 ⟨86S31⟩; **5**, 839, 878, 879 ⟨86S31⟩, 878, 879, 1155 ⟨86JOM(306)39⟩, 1148 ⟨86MI 531-01, 87PS(30)479⟩; **6**, 668 ⟨85JOM(290)33⟩
Allwohn, J., **4**, 633 ⟨89AG(E)784⟩, 634 ⟨88AG(E)1370⟩; **5**, 928 ⟨88AG(E)961, 90AG(E)398⟩, 928, 930 ⟨92ZN(B)983⟩, 929 ⟨90AG(E)399, 92AG(E)1384⟩; **6**, 184, 389, 398, 714 ⟨89AG(E)781⟩, 188 ⟨92AG(E)1384⟩, 201, 397, 715, 716 ⟨90AG(E)1030⟩, 201, 400 ⟨88AG(E)1370⟩, 389, 398, 401, 713 ⟨89AG(E)784⟩, 397, 398 ⟨90AG(E)1032⟩, 397, 714 ⟨88AG(E)961, 90AG(E)399⟩, 713, 714 ⟨90AG(E)398, 90ZN(B)290⟩
Allwörden, U. von, **4**, 94 ⟨88CZ69⟩
Almagro, V., **5**, 1044 ⟨93TL6717⟩
Alm, J., **6**, 207 ⟨74JPR557⟩
Alm, J. C., **4**, 554 ⟨65T5⟩
Alm, R. M., **2**, 126 ⟨50JA731⟩
Almansa, C., **1**, 1085 ⟨88S707⟩; **3**, 218 ⟨88T2657⟩; **4**, 196 ⟨88S707⟩
Almasi, L., **2**, 361 ⟨69CB1489⟩
Al-Masoudi, N. A. L., **4**, 227 ⟨86CAR(148)25⟩
Almásy, A., **5**, 199 ⟨86JCS(P1)927⟩
Almdal, K., **4**, 221 ⟨84SUL1⟩
Almeida, M. L. S., **5**, 977 ⟨88S992⟩
Almeida Saavedra, M., **3**, 357 ⟨91EGP294941⟩
Almena, J., **2**, 420, 553 ⟨93TL1649, 94T5775⟩
Almetkina, L. A., **2**, 889 ⟨89JGU2254⟩; **5**, 369, 373 ⟨88ZOB829⟩; **6**, 564 ⟨89ZOB2520⟩
Almog, J., **5**, 735 ⟨74JCS(P1)853⟩
Almond, Jr., H. R., **1**, 722 ⟨89TL1361⟩
Almond, M. R., **4**, 422 ⟨84JOC4272⟩; **5**, 999 ⟨84JOC4272⟩
Almond, S. W., **1**, 401, 800 ⟨79TL2779⟩
Almond, V., **3**, 337 ⟨77CC483⟩
Almousawi, S., **5**, 358 ⟨78BCJ512⟩
Almquist, R. G., **3**, 180 ⟨82TL2533⟩
Almy, J., **1**, 797 ⟨69JA4459⟩
Al-Najjar, I. M., **4**, 786 ⟨89SA(A)1011⟩
Al Neirabeyeh, M.,**1**, 829 ⟨91TL5525⟩; **3**, 118 ⟨76S811⟩
Alnima, H. H., **5**, 676, 677 ⟨82JCED103⟩
Al-Omran, F., **2**, 11 ⟨80JCS(P1)1890⟩
Alonso, C., **1**, 434 ⟨75TL4005⟩
Alonso, D. E., **1**, 498 ⟨92JOC1146⟩; **5**, 765, 766 ⟨93JA7584⟩
Alonso, J. H., **1**, 801 ⟨72JA2544⟩
Alonso, J. L. T., **2**, 761 ⟨78JOC2639⟩
Alonso, J. M., **2**, 9 ⟨91S353⟩
Alonso, M. E., **2**, 220 ⟨78OS(58)147⟩, 220, 270 ⟨88OSC(4)235⟩
Alonso, M. S., **2**, 30 ⟨91TL3217, 94T6433⟩
Alonso, R. A., **2**, 874 ⟨82JOC77, 84JOM(270)177⟩
Alonso, T., **4**, 654 ⟨87OM2110⟩
Alonso-Cires, L., **5**, 132 ⟨83S649⟩
Aloup, J. C., **4**, 229 ⟨87JMC24⟩; **5**, 558 ⟨87JMC24⟩
Alpan, R. S., **6**, 561 ⟨92ACR90⟩
Alparova, M. V., **6**, 659 ⟨87ZOB2467⟩
Alpatova, N. M., **1**, 248 ⟨87TCC149⟩
Alpegiani, M., **2**, 6 ⟨85G393⟩
Alper, H., **1**, 13 ⟨90TL4101⟩, 22, 437 ⟨80CC169⟩, 24 ⟨75JOC2694⟩, 80 ⟨83CC365⟩, 94 ⟨83OM1055⟩, 202 ⟨88JOC3306⟩, 225 ⟨88TL1763⟩, 239 ⟨85JOC147, 85TL2609⟩, 242 ⟨87OM1625, 89JOC20⟩, 297 ⟨83CC1270⟩, 298 ⟨90JA2803⟩, 427 ⟨91OM1620⟩, 437 ⟨74JOM(73)359, 83TL5329⟩, 1002 ⟨92JOC2188⟩; **2**, 125, 221, 228 ⟨69AG(E)677⟩, 171 ⟨81JOC289⟩, 753 ⟨80TL2603⟩; **3**, 122 ⟨80JOC2269⟩, 129 ⟨79JOC2568⟩,

233 ⟨78CC311, 79TL2665⟩, 666 ⟨73JOC4365⟩; **4**, 519, 520 ⟨75JOC437⟩, 519, 520 ⟨75JOM(99)385⟩; **5**, 38 ⟨86JOC4354⟩, 142 ⟨84MIP84/1376A1⟩, 237 ⟨85TL2609⟩, 238 ⟨86OM596⟩, 356 ⟨90OM1245⟩, 568, 613 ⟨78AG(E)689⟩, 655 ⟨79JA4245⟩, 801 ⟨79JOM(168)97⟩; **6**, 351 ⟨74IC225⟩, 445 ⟨72S81⟩
Al'pert, M. L., **6**, 565 ⟨86ZOR2145⟩
Alpoim, C. M., **5**, 1004 ⟨80NJC127⟩
Al-Sader, B. H., **2**, 376, 403 ⟨70CJC2745⟩
Al-Saleh, B., **1**, 595 ⟨89JCR(S)30⟩
Al-Salim, N., **6**, 563 ⟨88JCS(D)2363⟩
Alsdorf, H., **1**, 104 ⟨67AG(E)801⟩
Al-Shaar, A. H., **4**, 866 ⟨86EUP174832A2⟩
Al Shaar, A. H. M., **5**, 690 ⟨92JCS(P1)2789⟩
Al-Shoura, A. M., **4**, 775 ⟨90MI 417-01⟩
Alsop, D. J., **1**, 426 ⟨62JCS1801⟩
Alsop, W. G., **6**, 217 ⟨37JOC76⟩
Alston, W. B., **5**, 24 ⟨72MI 502-02⟩
Altabev, N., **4**, 28, 39, 40 ⟨73CI(L)331⟩
Alt, G. H., **3**, 266 ⟨64JOC794⟩, 408 ⟨B-88MI 310-01⟩
Alt, H., **6**, 173 ⟨67AG(E)942⟩
Alt, H. G., **2**, 963 ⟨⟨86JOM(310)27⟩; **3**, 386 ⟨85JOM(289)C21, 87CB1905⟩; **4**, 581 ⟨89JOM(362)117⟩; **5**, 944 ⟨87OM1424⟩; **6**, 720 ⟨88JOM(355)437⟩
Alt, K. O., **6**, 612, 616 ⟨69HCA812⟩
Al-Talib, M., **4**, 109 ⟨92S710⟩, 434 ⟨87S745⟩; **5**, 350 ⟨90SC1811⟩
Altamura, M., **2**, 9 ⟨93JOC272⟩
Alt, R., **4**, 679 ⟨81AG(E)1049⟩
Al-Taweel, S. M., **2**, 896 ⟨91OM2068⟩; **4**, 1038 ⟨92JA372⟩, 1041 ⟨91OM2068⟩, 1038, 1041 ⟨92OM1491, 92OM2743⟩
Altenbach, H., **4**, 173 ⟨85TL6329⟩
Altenbach, H.-J., **3**, 200 ⟨82AG(E)371⟩, 220 ⟨79AG1005⟩
Altenbach, H. J., **3**, 61 ⟨83AG(E)410⟩, 171 ⟨82LA1096⟩
Altenburger, J. M., **5**, 328 ⟨92TL5055⟩
Altenkirk, B., **5**, 316 ⟨62JOC4532⟩
Altepeter, B., **4**, 135, 592, 596, 599, 600 ⟨85CB2353⟩; **6**, 170, 210 ⟨79TL501⟩
Althoff, U., **4**, 112 ⟨89ZN(B)175⟩; **6**, 242, 685 ⟨91HAC385⟩
Althoff, W., **4**, 282 ⟨76ZN(B)153⟩, 282, 563 ⟨73ZN(B)98⟩, 563, 564 ⟨81CB1082, 75CC468⟩
Althousen, D., **2**, 30 ⟨43OSC(2)270⟩
Altland, H. W., **2**, 652 ⟨70JA3520⟩
Altman, J., **2**, 304 ⟨89S825⟩
Altnau, G., **2**, 918 ⟨80TL4069⟩, 918, 923 ⟨82CZ441⟩; **4**, 355 ⟨81AG(E)581, 83TL45⟩; **6**, 125 ⟨81AG(E)581⟩
Altpeter, B., **4**, 401, 599, 708, 724 ⟨82CB1810⟩
Altukhov, K. V., **4**, 428 ⟨76RCR1052⟩; **6**, 144 ⟨78MI 605-01⟩, 147 ⟨69ZOR2246⟩, 149 ⟨66ZOR1902⟩, 149 ⟨69ZOR1313⟩, 149 ⟨69ZOR220⟩, 149 ⟨73ZOR269⟩, 149 ⟨86ZOR2618⟩, 149 ⟨87ZOR2624⟩, 149 ⟨88ZOR644⟩
Alul, H. R., **2**, 764 ⟨64JA1139⟩
Aluotto, P. F., **6**, 27 ⟨68CJC2233⟩
Alvanipour, A., **1**, 455 ⟨80JOM(201)233⟩
Alvarez, B., **6**, 331 ⟨91OM3005, 91OM384⟩
Alvarez, C., **5**, 798 ⟨87JOM(327)339, 89OM2253⟩, 798, 952 ⟨87JOM(333)245, 91COS(6)485⟩; **6**, 716 ⟨91CRV35⟩
Alvarez, E., **1**, 195 ⟨88JOM(339)199⟩, 485 ⟨88T111⟩
Alvarez, F. S., **5**, 132 ⟨68JOC2143⟩
Alvarez, M., **3**, 361 ⟨92JCS(P1)351⟩
Alvarez, M. A., **4**, 674 ⟨93JA3786⟩
Alvarez, M. P., **2**, 44 ⟨92TL4637⟩
Alvarez, R. M., **1**, 433 ⟨84S384⟩; **2**, 36 ⟨87TL6441⟩, 630 ⟨86S222⟩; **3**, 92, 101 ⟨90CC1571⟩; **4**, 214 ⟨87S49⟩
Alvarez-Builla, J., **2**, 249 ⟨92TL3677⟩, 301 ⟨91SC535⟩; **3**, 171 ⟨92TL3677⟩, 306 ⟨93JOC5862⟩; **4**, 843 ⟨89H(29)1877⟩
Alvarez-Insv́a, A. S., **3**, 305 ⟨81JHC1345⟩
Alvarez-Larena, A., **4**, 689 ⟨95JOM(489)C65⟩, 763 ⟨94JA2153⟩
Alvernhe, G., **1**, 780 ⟨92JOC714⟩; **2**, 18 ⟨86TL4449, 87S562⟩, 280 ⟨90TL2127, 90TL663⟩; **4**, 787 ⟨90TL2127⟩
Alves, A. J., **4**, 231 ⟨92JOC2217⟩
Alves, M. J., **5**, 690 ⟨90JCS(P1)1705⟩
Al-Wassil, A. I., **6**, 381 ⟨86JCS(P2)1363⟩
Aly, A. A. M., **2**, 444 ⟨91ZN(B)775⟩
Aly, M. F., **5**, 603, 621 ⟨88T7271⟩
Alyea, E. C., **2**, 421 ⟨68CC495⟩; **4**, 566 ⟨88SRI163⟩, 566, 573 ⟨85MI 412-01⟩
Am, T., **3**, 361, 558 ⟨77ZC93⟩
Amamchyan, A. R., **5**, 778 ⟨87IZV165, 89IZV2353⟩
Amamoto, H., **1**, 974, 1056 ⟨82JOC2225⟩
Amamria, A., **2**, 910, 911, 916, 917, 932, 940, 946 ⟨87JOC4868⟩; **4**, 813, 1066 ⟨83JOM(256)37⟩, 1062 ⟨81JOM(210)C17⟩, 1065, 1066, 1068, 1069 ⟨83JOM(252)47⟩; **6**, 209 ⟨83JOM(252)47⟩
Aman, H., **3**, 475 ⟨64LA(675)47⟩
Amann, P., **2**, 473 ⟨86ZAAC(535)106⟩
Amano, A., **1**, 411 ⟨65JPC1278⟩
Amano, S., **3**, 20 ⟨89TL6371⟩
Amano, T., **1**, 469 ⟨86BCJ1656⟩; **5**, 30, 35 ⟨86SC499⟩
Amarasekara, A. S., **4**, 313, 314 ⟨88JOC22⟩; **5**, 697 ⟨88JOC22⟩
Amaratunga, S., **2**, 753 ⟨80TL2603⟩
Amaratunga, W., **3**, 14 ⟨83TL1143⟩
Amaravathi, M., **3**, 293 ⟨83IJC(B)1246⟩
Amarskii, E. G., **2**, 841 ⟨76ZOB1701⟩
Amatatsu, Y., **3**, 575 ⟨87JSP(123)476⟩
Amato, J. A., **4**, 48 ⟨79S970⟩
Amato, J. S., **3**, 131 ⟨90SC2683⟩; **4**, 247 ⟨89JOC3718⟩, 302 ⟨84TL4337⟩
Amatore, C., **1**, 474 ⟨88CC203⟩
Ambasht, S., **4**, 352 ⟨80S318⟩
Ambekar, S. Y., **5**, 690 ⟨90JIC260⟩
Ambelang, J. C., **5**, 1048 ⟨39JA632⟩
Amberg, W., **2**, 48 ⟨92JOC2768⟩, 49 ⟨93JOC844⟩
Amberger, E., **1**, 446 ⟨61CB1447⟩; **2**, 481, 490 ⟨61CB1447⟩; **6**, 173 ⟨64CB1999⟩
Amberger, H.-D., **5**, 782 ⟨92CB2171⟩
Ambrose, M. G., **6**, 22 ⟨83JOC1776⟩
Ambrosius, H. P. P. M., **2**, 209 ⟨77JCS(P1)1469⟩
Ambrosius, K., **5**, 389 ⟨78IJ132⟩
Ambrush, I., **3**, 239 ⟨58JGU3024⟩
Ambrush, T. I., **3**, 239 ⟨58JGU3024⟩
Amburn, H. W., **1**, 869, 871 ⟨69JA2797, 69JA530⟩
Amedio, J. C., **1**, 829 ⟨89JOC736⟩
Amel, R. T., **1**, 737 ⟨68CJC461⟩; **2**, 1000 ⟨69JOC1240⟩
Amemiya, M., **5**, 221 ⟨88BCJ4467⟩
Amemiya, S., **3**, 35 ⟨85T4449⟩
Amemiya, T., **3**, 293 ⟨77CL587⟩
Amer, A. M., **3**, 311 ⟨92TL3003⟩
Amer, I., **1**, 94 ⟨83TL4139⟩, 95 ⟨87JOC2804⟩
Amer, M. I., **3**, 282 ⟨83JCS(P1)1075⟩
Amer, S., **4**, 984 ⟨77G555⟩
Ames, A., **2**, 278 ⟨78JOC2735⟩; **3**, 282, 295, 302, 305, 307, 309 ⟨80JA860⟩
Ames, A., **3**, 295, 302, 305, 307, 309 ⟨85T4821⟩
Ames, A., **4**, 538 ⟨80JOC2548⟩; **5**, 239 ⟨78JOC2735, 85T4821⟩
Ames, D. E., **1**, 1001 ⟨63JCS5889, 63JCS775⟩, 1001 ⟨65JCS894⟩, 1010, 1012 ⟨81S364⟩; **2**, 31 ⟨50JCS406⟩
Ames, G. R., **5**, 46 ⟨58CRV895⟩
Ames, M. M., **4**, 419, 427 ⟨93JMC305⟩, 528 ⟨90USP4963655⟩
Amey, R. L., **3**, 6, 119 ⟨79JA5294⟩
Amezaga, J. M., **3**, 226 ⟨89JOC3984⟩
Amiard, G., **1**, 403 ⟨49BSF501⟩, 419 ⟨55BSF1341⟩
Amica, P., **3**, 67, 235 ⟨76S196⟩
Amice, P., **2**, 28 ⟨76S194⟩; **3**, 154 ⟨76S194⟩, 224 ⟨81S289, 81S291⟩
Amick, D. R., **3**, 97 ⟨74TL3463⟩; **4**, 775 ⟨91USP5023275⟩

Amiedo, J. C., **1**, 582 ⟨92JOC2270⟩
Amiet, G., **4**, 724 ⟨70CC161⟩
Amiet, R. G., **4**, 681 ⟨74JA7108, 68JA1057⟩
Amin, J. H., **2**, 18 ⟨73OSC(5)136⟩; **5**, 59 ⟨73OSC(5)580⟩
Amin, M., **4**, 1031 ⟨85JHC513⟩
Amin, N. V., **3**, 684, 686 ⟨84JOC4595⟩; **4**, 333 ⟨81JOC765⟩
Amin, S. R., **5**, 955 ⟨95OM547⟩
Amine, M. El., **4**, 141 ⟨88SRI317⟩
Amino, Y., **5**, 607 ⟨81TL3413, 82JA4018⟩
Amirabadi, H. M., **2**, 687 ⟨78T1231⟩
Amishiro, N., **2**, 939 ⟨91JA7778⟩
Amitani, H., **2**, 455, 846 ⟨79S110⟩, 459 ⟨78TL3741⟩
Amma, E. L., **2**, 954 ⟨83OM101⟩
Ammadi, F., **3**, 627 ⟨93SC2389⟩
Ammanamanchi, R., **2**, 752 ⟨83TL4733⟩
Ammann, A. A., **3**, 237 ⟨87HCA321⟩
, M. Marraud, **5**, 22 ⟨91TL2765⟩
Ammerlahn, G., **6**, 213 ⟨B-59MI 607-01⟩
Ammi, L., **3**, 457 ⟨88TL1287⟩
Ammon, H. L., **1**, 835 ⟨84JOC2708, 90JA7682⟩; **2**, 1004 ⟨73JA6158⟩; **3**, 92, 97 ⟨87TL2473⟩; **4**, 428 ⟨87JOC4781⟩; **6**, 370 ⟨85IC2889⟩
Amoo, V. E., **1**, 755 ⟨88TL2401⟩
Amor, A. B. H., **5**, 700, 701 ⟨92SC1985⟩; **6**, 558 ⟨86MI 617-02⟩
Amorose, D. M., **4**, 656 ⟨91OM2191⟩
Amoroso, R., **4**, 99 ⟨92JOC1082⟩
Amos, R. A., **1**, 968 ⟨78JOC555⟩
Amosova, S. V., **2**, 225 ⟨92ZOR927⟩, 708 ⟨83IZV230⟩, 1036 ⟨90ZOR2635⟩; **4**, 775 ⟨92ZOR1780⟩, 777 ⟨90ZOR2056⟩; **6**, 565 ⟨86ZOR2145, 90ZOR1364⟩
Amougay, A., **3**, 294 ⟨91TL3993⟩
Amouroux, R., **1**, 1089 ⟨84H(22)1489⟩; **3**, 40 ⟨88BSF95⟩, 286 ⟨75JOM(99)41⟩
Amri, H., **1**, 191 ⟨87TL5521⟩; **3**, 627 ⟨93SC2389⟩
Amrollah-Madjdabadi, A., **2**, 25, 26, 31, 32 ⟨89S614⟩, 398 ⟨86S826⟩; **4**, 108 ⟨86S826, 86S828⟩
Amschler, H., **1**, 866 ⟨64AG(E)312⟩; **4**, 979 ⟨63CB2100⟩
Amstutz, E. D., **3**, 179 ⟨64RTC1047⟩; **6**, 506 ⟨43OSC(2)462⟩
Amstutz, R., **1**, 873 ⟨81HCA2617, 85JA5403⟩; **2**, 999 ⟨84HCA224⟩
Amsutz, E. D., **1**, 591 ⟨44JOC310⟩
Amupitan, J., **3**, 218 ⟨78CC852⟩
Amupitan, J. O., **5**, 600, 621 ⟨83S730⟩
An, D. K., **2**, 60 ⟨93JOC1941⟩; **3**, 84 ⟨91MI 303-01⟩, 86 ⟨93JOC1941⟩
An, T. D., **2**, 615 ⟨79HCA1194, 79HCA635⟩
An, Y.-Z., **3**, 198 ⟨92JOC317⟩
Anacona, J. R., **4**, 30 ⟨86CI(L)490⟩
Anagnostopoulos, C. E., **4**, 12 ⟨58JA6233⟩
Ananchenko, S. N., **1**, 79 ⟨56ZOB1175⟩
Anand, N., **6**, 443 ⟨87IJC(B)748⟩, 577 ⟨75JMC913⟩
Anand, S. M., **3**, 324 ⟨87TL1011⟩
Anand, S. P., **2**, 613 ⟨75JOC807⟩
Anani, A., **3**, 3, 114 ⟨79JCS(P1)2493⟩
Ananoff, J., **2**, 451 ⟨1874CB1688⟩
Ananthan, S., **4**, 988 ⟨84TL1291⟩
Anantharamaiah, G. M., **1**, 74 ⟨79JOC3442⟩
Anastasis, P., **2**, 617 ⟨85JCS(P1)1127⟩
Anastassiou, A. G., **1**, 865 ⟨75CC201⟩; **2**, 153 ⟨71CC979⟩; **4**, 303 ⟨73CC601⟩; **5**, 1119 ⟨65JA2296⟩
Anazawa, A., **4**, 902 ⟨87JCS(P1)1501⟩
Anciaux, A., **4**, 238, 239, 283, 375 ⟨75TL1613⟩
Anciaux, A. J., **1**, 549 ⟨78TL1239⟩
Ancillotti, M., **6**, 325, 335, 520 ⟨86TL5985⟩
Ancos, B. de, **3**, 658 ⟨88S136⟩, 689 ⟨88AG(E)1534⟩
Andell, O. S., **3**, 614 ⟨81CC1098⟩
Andemichael, Y. W., **1**, 497, 1057 ⟨93JOC1651⟩; **2**, 904, 939 ⟨92JOC794⟩
Andermann, G., **2**, 753 ⟨88TL635⟩
Anders, B., **2**, 726 ⟨65AG(E)430⟩; **3**, 699 ⟨69AG(E)20⟩; **5**, 989 ⟨B-71MI 526-03⟩, 1071 ⟨65AG(E)430⟩; **6**, 602 ⟨67AG(E)649⟩, 602, 604, 605, 606 ⟨B-71MI 620-01⟩
Anders, E., **2**, 446 ⟨92S577⟩; **3**, 310 ⟨86CB279⟩, 494 ⟨86CB1350⟩, 531 ⟨73AG(E)25⟩; **4**, 604, 647 ⟨91S1221⟩; **5**, 226 ⟨87S931⟩; **6**, 463 ⟨80S485⟩
Anders, K., **1**, 925 ⟨89JOC3571⟩
Anders, M. W., **3**, 550 ⟨91JA5120⟩
Anders, R. J., **2**, 491, 496, 498 ⟨47JA927⟩
Anders, R. T., **6**, 51 ⟨78JOC1346⟩
Andersch, J., **5**, 600, 621 ⟨89S202⟩; **6**, 619 ⟨89LA931⟩
Andersen, C. C., **1**, 21 ⟨28CB1750⟩
Andersen, H. M., **2**, 125 ⟨54JA1082⟩, 127 ⟨53JA6019⟩
Andersen, J. A., **5**, 943 ⟨95AOC(37)169⟩
Andersen, J. A. M., **2**, 573 ⟨92JOM(439)C25⟩
Andersen, K. K., **2**, 151 ⟨62TL93⟩, 186, 189, 196, 721 ⟨79COC(3)317⟩, 248, 720, 726 ⟨79COC(3)331⟩, 366, 797 ⟨79COC(3)363⟩; **4**, 776 ⟨62TL93⟩; **5**, 521 ⟨81JA7055⟩, 521, 523 ⟨82JCS(P2)473⟩
Andersen, N. H., **4**, 245 ⟨82JOC1145⟩, 252 ⟨75TL4547⟩; **6**, 134 ⟨82JOC1145⟩
Andersen, R. A., **2**, 574 ⟨76JCS(D)2204⟩, 575 ⟨75JOM(99)C19⟩, 603 ⟨81IC2991⟩; **4**, 656, 677 ⟨76JCS(D)2204⟩, 672 ⟨78JCS(D)446⟩, 750 ⟨79IS262⟩, 966 ⟨76AG(E)609⟩
Andersen, T. P., **5**, 573 ⟨83T3419⟩
Anderskewitz, R., **1**, 463 ⟨87AG(E)683⟩; **6**, 604 ⟨86JFC(32)415⟩
Anderson, A. B., **4**, 692 ⟨78JA2090⟩
Anderson, A. G., **2**, 1020, 1028 ⟨87JA228⟩
Anderson, A. L., **1**, 431 ⟨71JFC(1)121⟩
Anderson, B. A., **1**, 551 ⟨93JA10671⟩; **5**, 952, 958 ⟨93JA4602⟩, 956 ⟨89TL4061, 94ICA215⟩
Anderson, B. C., **6**, 2 ⟨68JOC1016⟩
Anderson, B. F., **3**, 543 ⟨73CC247⟩
Anderson, C. B., **2**, 914, 939 ⟨88TL6719⟩; **4**, 49, 51 ⟨67JOC607⟩
Anderson, C. E., **4**, 178, 184 ⟨58JA6613⟩
Anderson, C. H., **3**, 468 ⟨61JA399⟩
Anderson, C. L., **2**, 58, 526 ⟨90TL4677⟩, 936 ⟨88TL2425, 88TL2777, 92TL6915⟩; **4**, 353 ⟨90TL4677⟩; **5**, 897 ⟨92TL6915⟩
Anderson, D., **5**, 727 ⟨83JOC1544⟩
Anderson, D. A., **1**, 834 ⟨91JCS(P1)3199⟩; **2**, 394 ⟨90JOC511⟩, 622 ⟨89CB2213⟩; **3**, 369 ⟨91JCS(P1)3199⟩; **4**, 182 ⟨87JOC188⟩
Anderson, D. G., **1**, 759 ⟨70CJC561⟩; **2**, 417 ⟨89JCS(D)779⟩; **4**, 138, 141, 646, 651, 652, 654, 663 ⟨70CJC561⟩, 222, 370, 371, 570 ⟨68CJC2115⟩, 647, 652, 663 ⟨70JA7567⟩; **6**, 192 ⟨70CJC561⟩
Anderson, D. J., **3**, 269 ⟨74JOC3076⟩
Anderson, D. K., **1**, 209 ⟨83JOC1455⟩; **4**, 646 ⟨81CC179⟩
Anderson, D. W., **1**, 881 ⟨70CC1325⟩; **4**, 472 ⟨90TL1759⟩
Anderson, E., **4**, 189 ⟨71JA1701, 71JOC2357⟩, 218 ⟨70JA5464⟩
Anderson, E. L., **4**, 420, 421, 426 ⟨55JA745⟩
Anderson, G. H., **4**, 985 ⟨84JA6015⟩
Anderson, G. J., **3**, 88 ⟨59JA4113⟩; **4**, 161 ⟨57JA6562⟩
Anderson, G. K., **2**, 437 ⟨88JOM(342)263⟩; **5**, 464 ⟨84ACR67⟩, 465 ⟨79JCS(D)1246, 80JCS(D)712⟩, 1139 ⟨82OM64⟩
Anderson, G. L., **4**, 893 ⟨77JOC4159⟩
Anderson, G. W., **5**, 84 ⟨60ANY(88)676⟩, 277 ⟨64JA1839⟩, 279 ⟨67JA7151⟩; **6**, 487 ⟨57JA6180⟩
Anderson, H. H., **2**, 417, 418 ⟨52JA1421⟩; **3**, 711 ⟨51JA5439⟩; **5**, 1015, 1016, 1017 ⟨40JA761⟩, 1016 ⟨47JA1241⟩
Anderson, H. H., **5**, 1016 ⟨47JA3048, 48JA1043, 48JA1222⟩, 1017 ⟨45JA2176, 50JA193⟩, 1017 ⟨50JA2761, 54JOC1300⟩, 1136 ⟨51JA5439, 51JA5440⟩, 1136 ⟨56JA1692, 64IC910⟩
Anderson, H. J., **1**, 19 ⟨83CJC2415⟩; **3**, 307 ⟨80CJC2527, 81TL4899⟩, 307 ⟨85S353⟩, 669, 670 ⟨81CJC2673⟩; **6**, 648 ⟨71CJC2315⟩

Anderson, H. V., **4**, 209 ⟨54JA743⟩
Anderson, J. C., **1**, 780 ⟨91T6813⟩; **2**, 44 ⟨90SL107⟩; **3**, 68 ⟨64T2091⟩, 156 ⟨90SL107⟩
Anderson, J. E., **4**, 9 ⟨69JA1386⟩
Anderson, J. L., **4**, 5 ⟨61JA382⟩
Anderson, J. M., **2**, 122 ⟨66JA5245⟩; **3**, 612 ⟨44JA218⟩
Anderson, Jr., A. G., **2**, 25 ⟨76S398⟩, 32 ⟨64JA5037, 72JOC626⟩; **5**, 17 ⟨73TL5121⟩
Anderson, J. W., **2**, 278, 294 ⟨73JCS(D)1716⟩, 293 ⟨75JIN233⟩, 857 ⟨75INC233⟩
Anderson, K. R., **2**, 28 ⟨88TL6835⟩
Anderson, K. W., **2**, 632 ⟨83JOC1542⟩
Anderson, L., **4**, 176 ⟨82CA127400p⟩, 176 ⟨82CC1274⟩
Anderson, L. B., **1**, 776 ⟨90JA228⟩
Anderson, L. C., **3**, 490 ⟨45JA955⟩
Anderson, L. G., **1**, 362, 555, 558 ⟨92JA7375⟩
Anderson, L. L., **4**, 683 ⟨92OM3736⟩, 1015 ⟨89OM1371⟩
Anderson, L. R., **2**, 94 ⟨69JA1310⟩, 98 ⟨70CC216⟩, 99 ⟨67JA4313⟩; **4**, 582 ⟨77IC3187⟩; **6**, 42, 43 ⟨70JOC3730⟩, 231 ⟨71CC784⟩, 240 ⟨70CC395⟩, 428 ⟨69TL723⟩
Anderson, L. W., **2**, 410 ⟨93JOC4606⟩
Anderson, M., **2**, 792 ⟨94JCS(P1)1935⟩
Anderson, M. B., **1**, 652 ⟨87SC621⟩; **4**, 365, 393, 394 ⟨91SC(21)1675⟩
Anderson, M. W., **4**, 414 ⟨82CC282⟩
Anderson, O. P., **1**, 630 ⟨85JOC5460⟩; **3**, 29 ⟨84JOC1378⟩; **4**, 679 ⟨82JA7325⟩
Anderson, P. C., **1**, 243 ⟨82JOC1641⟩, 688, 690 ⟨88JA6914⟩; **2**, 525 ⟨88PAC1705⟩; **4**, 199 ⟨87TL2225⟩
Anderson, P. S., **4**, 97 ⟨79JOC3117⟩
Anderson, R., **1**, 648 ⟨85S717⟩, 879 ⟨83JOC2957⟩; **2**, 537 ⟨85S717⟩; **4**, 506, 646 ⟨85S717⟩; **6**, 174 ⟨85S717⟩
Anderson, R. A., **5**, 797 ⟨79IC2928⟩
Anderson, R. C., **1**, 218, 831 ⟨83JA1988⟩
Anderson, R. H., **2**, 260 ⟨70CC425⟩
Anderson, R. J., **1**, 216 ⟨72JA8225⟩, 226, 479 ⟨70JA4978⟩, 727 ⟨75JA4327⟩; **3**, 494 ⟨75JA4327⟩
Anderson, R. L., **1**, 165 ⟨74JA1230⟩; **4**, 955 ⟨74JA1230⟩; **5**, 943 ⟨71JA3554⟩, 946 ⟨74JOM(73)C28⟩, 955 ⟨72JA8947, 74JA1230⟩
Anderson, R. M., **4**, 122 ⟨67USP3306937⟩
Anderson, R. S., **2**, 341, 764 ⟨65JOC2880⟩
Anderson, T. J., **1**, 66, 69, 454 ⟨82JOC3707⟩; **4**, 206 ⟨82JOC3707⟩
Anderson, V. E., **5**, 551 ⟨88BBA269⟩
Anderson, W. K., **4**, 167 ⟨77JOC1045⟩, 517 ⟨86JMC2241⟩
Andersson, B., **2**, 652 ⟨88JMC1080⟩, 804 ⟨69ACS2983⟩
Andersson, C. M., **2**, 237, 238 ⟨92JA9737⟩
Andersson, C.-M., **2**, 682 ⟨89JOC1502⟩; **3**, 245 ⟨87TL4215, 88JOC4257⟩
Andersson, F. O., **6**, 546 ⟨92MI 617-04⟩
Andersson, K., **2**, 646 ⟨87JOC5205⟩; **4**, 28 ⟨72ACS1735⟩
Andersson, P. H., **1**, 192 ⟨93JA6609⟩
Andersson, R., **6**, 85, 89 ⟨76CS122⟩
Andes Hess, Jr., B., **3**, 703 ⟨88AG(E)938⟩; **5**, 1020 ⟨90JA4183⟩
Andisik, D., **4**, 314 ⟨84JOC4237⟩
Ando, A., **6**, 9 ⟨88JOC3637⟩, 233, 234 ⟨86BCJ447⟩
Ando, D. J., **4**, 993 ⟨92H(34)1005⟩
Ando, H., **5**, 516, 519, 522, 523, 626 ⟨90BCJ825⟩
Ando, K., **1**, 236 ⟨85CL1149⟩; **2**, 75 ⟨90JA6348⟩, 75, 529, 557 ⟨90JA6339⟩; **6**, 135 ⟨92CC870⟩
Ando, M., **1**, 831 ⟨90CC1001⟩, 1082 ⟨90TL6205⟩; **2**, 769 ⟨89BCJ128⟩; **5**, 731 ⟨83JA2831⟩; **6**, 14 ⟨81CL1719⟩
Ando, N., **4**, 328 ⟨85S276⟩; **5**, 989 ⟨85S276⟩
Ando, R., **1**, 485, 1083 ⟨84JOC1230⟩, 620 ⟨81JOC5246⟩; **3**, 147 ⟨81JOC5246⟩
Ando, T., **1**, 434 ⟨87JOC300⟩, 1079 ⟨84TL1377⟩, 1124 ⟨73TL4767⟩; **2**, 18, 19 ⟨88CL1877, 86CC793⟩, 177, 179 ⟨92CL891⟩; **3**, 116 ⟨83CL379⟩, 294 ⟨88CL1387⟩, 613 ⟨84CL725⟩; **4**, 426 ⟨82TL2665⟩, 735 ⟨83TL507, 83TL5657⟩, 750 ⟨67BCJ1275⟩, 752 ⟨67TL1123, 68BCJ756⟩; **5**, 285 ⟨81CL1143⟩, 1039 ⟨90CC1213⟩, 1105 ⟨87JOC681⟩, 1107 ⟨87CC1301⟩; **6**, 423, 438 ⟨86CC793⟩
Ando, W., **1**, 63 ⟨85OM584⟩, 233, 400 ⟨72JOC1721⟩, 242 ⟨84TL4011, 89JOC4660⟩, 400, 821, 823 ⟨77ACR179⟩, 408, 409, 410, 820, 821, 822 ⟨72JA3870⟩, 409, 821, 822 ⟨72JOC3791⟩, 656 ⟨85TL4523⟩, 820 ⟨71JOC1732⟩, 821 ⟨B-78MI 118-01⟩, 983 ⟨86T1989, 87TL1803⟩, 983 ⟨89TL4271⟩; **2**, 146 ⟨86BCJ1275, 86TL1591⟩, 232 ⟨83TL3631⟩, 266 ⟨83TL4033⟩, 936 ⟨79JOC413⟩, 941 ⟨93T11389⟩, 947 ⟨85TL4523, 86TL3251⟩, 1089 ⟨93MI 202-02, 93T6359⟩; **3**, 337 ⟨91JA7782⟩, 351 ⟨83JA6151, 87TL4833, 89TL2955⟩, 351, 551 ⟨90TL3571⟩, 353 ⟨89JOC4660, 89TL4271⟩, 353 ⟨89TL4807⟩, 366 ⟨90TL7641⟩, 373, 378 ⟨72IJS(B)189, 72JOC1721, 77ACR179⟩, 417 ⟨82TL3073⟩, 483 ⟨81JA5573⟩, 531, 532 ⟨B-78MI 316-01⟩, 551 ⟨89TL1249⟩; **4**, 25 ⟨84JA8174⟩, 226 ⟨85T2133⟩, 258 ⟨90T5093⟩, 318 ⟨85CL939, 86S139⟩, 398 ⟨79CC1121⟩, 437 ⟨84S259⟩, 437, 438 ⟨84S256⟩, 612, 1049 ⟨87OM1857⟩, 624 ⟨93OM1514⟩, 628, 629 ⟨90OM2061, 88OM1882⟩, 629 ⟨89OM2286⟩, 836 ⟨84TL4011⟩, 837 ⟨89TL1249⟩, 1082 ⟨B-95MI 424-01⟩; **5**, 234, 242 ⟨82CL885⟩, 323 ⟨83SC1053⟩, 413, 430 ⟨79TL871⟩, 606 ⟨85S92⟩, 873 ⟨77JA6995, 83CL871, 83TL4245⟩, 925 ⟨81JA5573, 85TL2337, 86CL2025, 87OM2337⟩, 1020 ⟨87OM2270, 88OM558⟩; **6**, 129 ⟨87CL2177⟩, 187 ⟨94OM434⟩, 383 ⟨91JA7790⟩, 672 ⟨87OM1857⟩
Andrade, J. G., **5**, 1135 ⟨78JOC2280⟩
Andre, C., **1**, 615, 698, 703 ⟨86JA2776⟩
André, D., **1**, 995 ⟨79CC859⟩
Andreades, S., **2**, 194 ⟨63USP3081350⟩, 688 ⟨64JOC898⟩; **3**, 638, 665 ⟨69JA4181⟩; **4**, 43, 164 ⟨61JA4670⟩; **6**, 233 ⟨64JOC898⟩, 540 ⟨69JOC3011⟩
Andreae, S., **2**, 374 ⟨88S529⟩, 379 ⟨85JPR445⟩
Andreasch, R., **5**, 1023 ⟨06M1221⟩
Andree, R., **2**, 340, 342, 350, 353, 355, 356 ⟨90HOU(E16)1⟩; **6**, 232 ⟨90HOU(E16a)214, 90HOU(E16a)271⟩
Andree, T., **5**, 334 ⟨89JMC1024⟩
Andreetta, A., **1**, 79 ⟨74JOM(70)89⟩
Andreetti, G. D., **2**, 247 ⟨81CC350⟩; **3**, 311 ⟨84JOC590⟩
Andreev, A. A., **4**, 541 ⟨80ZOB1427⟩
Andreev, L. N., **5**, 565, 579 ⟨69RCR21⟩
Andreev, N. A., **2**, 838 ⟨80ZOB803⟩, 859 ⟨79ZOB2230⟩
Andreev, S. A., **2**, 386 ⟨78ZOR240⟩; **5**, 361 ⟨77ZOR1144⟩
Andreeva, L. M., **6**, 149 ⟨69ZOR1313⟩, 149 ⟨69ZOR220⟩
Andreeva, P., **5**, 376 ⟨90AF1030⟩
Andreichikov, Y. S., **3**, 475 ⟨78ZOR2234⟩
Andreini, B. P., **1**, 1007 ⟨88TL2239⟩
Andrejevik, V., **3**, 620 ⟨65TL461⟩
Andre-Louisfert, J., **1**, 42 ⟨76T829⟩
Andreoli, P., **2**, 316 ⟨90JOC4199⟩, 544 ⟨90PAC605⟩; **3**, 456, 457, 458 ⟨90PAC605⟩, 456, 458 ⟨86TL1695, 90JOC4199⟩; **5**, 287 ⟨86TL1695⟩
Andrés, C., **2**, 312 ⟨83JOC2255⟩, 320 ⟨90S153, 90SC1149⟩
Andres, H., **5**, 1121 ⟨93JMC2004⟩
Andrés, J. A., **5**, 371 ⟨82BSF(2)369⟩
Andres, R. J., **5**, 221 ⟨47JA925⟩
Andretti, G. D., **5**, 497, 499 ⟨74JCS(P2)1483⟩
Andrew, S. S., **4**, 525 ⟨86JOC1610⟩
Andrewes, A. G., **1**, 747, 748 ⟨73ACS1401⟩
Andrews, B. K., **1**, 924 ⟨86JA7935⟩
Andrews, D., **4**, 166 ⟨72HCA249⟩
Andrews, D. M., **2**, 145 ⟨94TL9629⟩, 145 ⟨95SL773⟩
Andrews, D. R., **1**, 812 ⟨85JOC201⟩
Andrews, G., **1**, 822 ⟨72TL5121⟩
Andrews, G. C., **1**, 136 ⟨74JA5560⟩, 140, 143, 609, 788 ⟨74ACR147, 77JA5453⟩; **2**, 154, 196 ⟨74ACR147⟩; **3**, 36 ⟨74JA5560⟩; **5**, 553 ⟨77JA5453⟩
Andrews, G. D., **1**, 801 ⟨76JA6705⟩
Andrews, L., **1**, 921 ⟨71ARP109⟩, 923 ⟨67JCP(47)3637⟩; **4**, 42 ⟨85IC3285⟩, 697, 698 ⟨88JA3231⟩; **5**, 1020 ⟨92JPC10195⟩, 1096 ⟨92JA9239⟩; **6**, 244 ⟨75JA187⟩

Andrews, L. J., **4**, 189 ⟨52JA6189⟩; **5**, 204 ⟨59JA2374⟩
Andrews, P. R., **1**, 826 ⟨73B3492, 79AJC1921⟩
Andrews, P. S., **4**, 689, 690 ⟨73IC2396⟩
Andrews, R. C., **5**, 207 ⟨86TL5197⟩
Andrews, S. A., **3**, 264 ⟨82SC35⟩
Andrews, S. B., **2**, 998 ⟨71JOM(30)151⟩; **4**, 154, 155, 156 ⟨71JOM(30)151⟩, 623 ⟨71JA3709⟩
Andrews, S. W., **1**, 216 ⟨89JOC3114⟩
Andriamiadanarivo, R., **2**, 668 ⟨93TL7923⟩
Andriamizaka, J. D., **2**, 473, 474 ⟨82PS(12)265, 82PS(12)279⟩
Andrianov, A. A., **5**, 530 ⟨B-82MI 507-01⟩
Andrianov, K. A., **5**, 720 ⟨70IZV1157, 72IZV1150⟩
Andrianov, V., **5**, 668 ⟨92SC453⟩
Andrianov, V. G., **4**, 696 ⟨77JOM(137)217⟩, 696, 723 ⟨78JOM(153)115⟩; **6**, 57 ⟨80IZV2356⟩, 625 ⟨77MI 620-02⟩
Andrianova, G. M., **2**, 748 ⟨68T4677⟩
Andries, K., **6**, 598 ⟨91JMC3187⟩
Andrieu, C., **3**, 363 ⟨69BSF827⟩
Andrieu, C. G., **3**, 351 ⟨77BSF688⟩, 352 ⟨85RTC9⟩
Andrievskii, V. N., **1**, 1006 ⟨70IZV1144⟩
Andringa, H., **1**, 95, 889 ⟨90SC2165⟩, 464 ⟨90SC2349⟩; **4**, 138 ⟨87JOM(336)C41⟩
Andrist, A. H., **1**, 511 ⟨77JOC3981⟩
Androes, G., **2**, 212 ⟨61JA4357⟩
Androsenko, S. I., **4**, 660 ⟨89MI 414-01⟩, 662 ⟨91MI 414-01⟩
Andrus, A., **4**, 59 ⟨84JA1809⟩
Andrus, D. W., **2**, 32 ⟨55OSC(3)692⟩
Andrus, W. A., **3**, 9, 123 ⟨77TL4175⟩
Andrushchenko, V. V., **4**, 326 ⟨82ZOR841⟩
Andruski, S. W., **5**, 991 ⟨86JOC5123⟩
Andruszkiewicz, R., **2**, 315 ⟨89S953⟩
Andrzejewski, D., **1**, 624, 629 ⟨71JA7118⟩, 624, 629 ⟨72JA8608⟩, 624, 782 ⟨73JOC1742, 74CC819⟩, 629 ⟨74CC820⟩
Aneja, R., **2**, 7 ⟨73CC110⟩
Anelli, P. L., **3**, 7 ⟨87JOC2559⟩, 31 ⟨89JOC2970⟩
Anello, L. G., **4**, 160 ⟨68JOC2692⟩; **6**, 16 ⟨64FRP1369784⟩
Aneser, E., **5**, 832 ⟨61AG656⟩
Anet, F. A. L., **1**, 906 ⟨82JA7105⟩
Anet, R., **3**, 66 ⟨61TL720⟩
Anfilogova, S. N., **5**, 433 ⟨80MI 509-01⟩
Anfinogenov, V. A., **4**, 310 ⟨78ZOR1723⟩
Ang, H. G., **2**, 504 ⟨68JINC1715, 93JCS(D)663⟩; **4**, 116 ⟨68JINC1715⟩; **6**, 242 ⟨93JOM(448)219⟩, 243 ⟨92JST(268)389, 93JCS(D)663⟩
Angara, G. J., **4**, 760 ⟨92TL2285⟩
Angel, H. S., **5**, 194 ⟨50JA3874⟩
Angel, R. L., **5**, 385 ⟨90IC20⟩
Angelastro, M., **1**, 983 ⟨80JOC3522⟩
Angelastro, M. R., **3**, 240 ⟨89JOC3913⟩; **5**, 327 ⟨89JOC3913⟩
Angelastro, R., **1**, 487 ⟨80S945⟩
Angeletakis, C. N., **2**, 724 ⟨84JOC1931⟩
Angeletti, E., **2**, 25, 31 ⟨87JCS(P1)2157⟩; **3**, 653 ⟨89JCS(P1)105⟩
Angeli, C., **5**, 1078 ⟨85S509⟩
Angeli, P., **4**, 232 ⟨87JMC1934⟩
Angelici, R. A., **6**, 719, 732 ⟨75IC1513⟩
Angelici, R. J., **1**, 774 ⟨86JOC113⟩; **3**, 689 ⟨93OM3864⟩
Angell, E. C., **5**, 325 ⟨88SC1331⟩
Angell, R., **1**, 115 ⟨93SL189⟩; **2**, 931 ⟨93SL189⟩; **4**, 807 ⟨93SL189⟩
Angelos, G. H., **2**, 311 ⟨82TL4651⟩
Angelova, O., **2**, 1004 ⟨88AX(C)626⟩; **3**, 494 ⟨89JCS(P2)1405⟩
Angelov, C., **5**, 344 ⟨89SC3113⟩
Angelucci, F., **3**, 284 ⟨84CC530⟩
Anger, V., **5**, 864 ⟨34MI 521-01⟩
Angerer, J., **3**, 35 ⟨74LA2085⟩
Angermann, A., **3**, 40 ⟨86JOC5294⟩
Angermund, K., **6**, 154 ⟨87AG(E)908⟩
Angibaud, P., **2**, 22 ⟨91SL807⟩; **5**, 315, 316, 318 ⟨93SC2423⟩
Angier, R. B., **4**, 312 ⟨69CC369⟩
Angiolini, L., **1**, 635 ⟨90T1791⟩; **4**, 96 ⟨90T1791⟩
Angle, S. R., **1**, 287, 789 ⟨86TL1445⟩, 830, 831 ⟨89TL515, 93TL3985⟩; **2**, 930 ⟨91TL2327⟩; **5**, 227 ⟨89TL515⟩
Anglin, J. R., **6**, 520 ⟨76JA4678⟩
Angoh, A. G., **1**, 244 ⟨87JOC4943⟩, 505, 651 ⟨84CC534⟩, 547 ⟨85CC980⟩, 942 ⟨91JCS(P1)1433⟩; **3**, 141 ⟨82CC828⟩, 144 ⟨84CC534⟩; **4**, 286, 779 ⟨87JOC4943⟩
Angoletta, M., **4**, 688 ⟨84JOM(267)199⟩
Angres, I., **1**, 108 ⟨76JA2580⟩; **2**, 392 ⟨77JOC563⟩
Angst, C., **4**, 174 ⟨87JOC622⟩
Anguiano, J., **1**, 130 ⟨87JCPB765⟩
Angus, Jr., R. O., **1**, 522 ⟨83JOC273⟩, 990 ⟨84JOC2880⟩
Angus, P. M., **5**, 385 ⟨93IC5285⟩, 385, 386 ⟨91IC1564⟩, 387 ⟨91IC4806⟩
Angyal, S. J., **3**, 88 ⟨54OR(8)197⟩, 104 ⟨53JCS1742⟩, 108 ⟨53JCS1740⟩, 428 ⟨59JA5251⟩; **4**, 222 ⟨78AJC1031⟩
Anh, N. T., **2**, 693 ⟨69CC112⟩
Anikina, E. V., **1**, 462 ⟨87BAU327⟩
Anisimov, A. V., **4**, 775 ⟨88ZOR443, 90MI 417-01⟩
Anisimov, K. N., **2**, 464 ⟨55IZV823⟩, 478 ⟨56IZV927⟩; **4**, 672 ⟨68CC1365, 69IZV100⟩, 706 ⟨68IZV686⟩, 724 ⟨62IZV1683⟩, 726 ⟨73IZV2815⟩; **6**, 522 ⟨40DOK(26)57, 59MI 616-02⟩
Anisuzzaman, A. K. M., **2**, 7 ⟨78CAR(61)511⟩; **6**, 309 ⟨67JCS(C)807⟩
Anjaneyulu, A. S. R., **1**, 827 ⟨91JCS(P1)2089⟩
Anjaneyulu, B., **3**, 527 ⟨67T4769⟩
Anjaneyulu, P. S. R., **5**, 332 ⟨87MI 507-02⟩
Anjanmurthy, C., **3**, 684 ⟨92OPP91⟩
Anker, D., **1**, 780 ⟨92JOC714⟩; **2**, 280 ⟨90TL2127, 90TL663⟩; **4**, 787 ⟨90TL2127⟩
Ankianiec, B. C., **4**, 658 ⟨89POL57⟩
Anklekar, T. V., **4**, 361, 379 ⟨92JOC5078⟩
Ankner, K., **3**, 303 ⟨87JA3981⟩
Annaka, M., **1**, 1010 ⟨88CPB1890⟩
Annan, T. A., **4**, 153, 588 ⟨91OM2159⟩
Annapurna, G. S., **1**, 6 ⟨87JOC4784⟩; **3**, 709 ⟨83SC331⟩; **4**, 428 ⟨87JOC4781⟩
Annarelli, D. C., **1**, 656 ⟨75JA7162⟩; **4**, 1051, 1052 ⟨84JOM(272)123⟩
Annen, K., **1**, 5, 9 ⟨83LA705⟩; **2**, 1018 ⟨84AG(E)727⟩; **3**, 227 ⟨82S34⟩
Annen, U., **5**, 1155 ⟨87PS(30)479⟩
Anner, G., **1**, 1082 ⟨67HCA2101⟩; **2**, 95 ⟨61E475⟩; **3**, 226 ⟨67HCA269⟩
Anneser, E., **6**, 309 ⟨61AG656⟩
Annis, G. D., **1**, 226 ⟨81JCS(P1)270⟩, 653 ⟨84JOC1317⟩; **3**, 234 ⟨83JA7358⟩
Annoura, H., **1**, 827, 828 ⟨89CPB2047⟩
Annunziata, R., **1**, 85 ⟨74JOC3195⟩; **3**, 37 ⟨82S805⟩, 431 ⟨89G253⟩, 437 ⟨82JCS(P1)339⟩, 634 ⟨81JCS(P1)614⟩; **4**, 262 ⟨86JOC891⟩, 272 ⟨79S535, 82S767⟩; **5**, 237 ⟨92JOC6339⟩
Anoda, T., **3**, 234 ⟨76CL495⟩
Anon., **5**, 15 ⟨82MI 501-01⟩
Anoshina, N. P., **4**, 343, 561 ⟨78ZOB1001⟩; **5**, 369 ⟨78ZOB2468⟩, **5**, 373 ⟨73ZOB1451⟩
Ansari, F., **1**, 1144 ⟨67CB2011⟩
Ansari, H. R., **5**, 142 ⟨73T1559⟩
Ansari, S. A., **4**, 268 ⟨91JIC368⟩
Ansbacher, S., **2**, 753 ⟨55OSC(3)307⟩
Anschütz, L., **2**, 847 ⟨56CB688⟩, 881 ⟨56N59⟩
Anschutz, R., **5**, 1023 ⟨09LA(371)201⟩
Anschütz, W., **3**, 480 ⟨71LA(748)207⟩, 481 ⟨69LA(730)194⟩; **4**, 489 ⟨69CB2216⟩; **5**, 869 ⟨68CB3734, 69CB2216⟩, 871, 874 ⟨71LA(748)101⟩; **6**, 666 ⟨68CB3734, 68TL3171⟩, 670 ⟨71LA(748)207⟩
Ansell, G. B., **4**, 579 ⟨83OM154⟩; **6**, 707 ⟨83OM154⟩

Ansell, J. M., **1**, 1076 ⟨76JOC3650⟩
Ansell, M. F., **1**, 815 ⟨67TL4537⟩; **3**, 61 ⟨82TL2811⟩; **5**, 46 ⟨B-56MI 502-03⟩, 130 ⟨B-72MI 503-02⟩
Ansell, P. J., **2**, 1034 ⟨87JCS(P2)1275⟩; **4**, 804, 840 ⟨87JCS(P1)1275⟩
Anselmann, R., **2**, 406 ⟨89IS8⟩
Anselme, G., **3**, 503 ⟨91JOM(403)93⟩; **6**, 724 ⟨92OM2748⟩
Anselme, J.-P., **2**, 379 ⟨83JHC121⟩, 385 ⟨84CC451, 84TL2619⟩, 386 ⟨79JOC1881⟩, 403 ⟨88BSB461⟩, 420 ⟨70JOC960⟩; **3**, 464 ⟨67JA5312, 75CC114⟩, 464 ⟨75TL1565⟩, 666 ⟨70TL3851⟩, 721, 722 ⟨70TL3309⟩; **4**, 161 ⟨77JOC754⟩; **5**, 991 ⟨71JOC2387⟩
Anslow, W. K., **5**, 85 ⟨32OSC(1)298⟩
Anslyn, E. V., **3**, 508 ⟨89OM583⟩
Ansorge, A., **5**, 382 ⟨93JOM(446)25⟩; **6**, 243 ⟨93AG429, 93JOM(446)25⟩
Anstock, M., **6**, 517 ⟨87IC526⟩
Anstoff, P. A., **3**, 220 ⟨85JA7967⟩
Antberg, M., **2**, 431 ⟨84IC4170⟩, 439 ⟨86PS(26)143⟩
Antczak, K., **4**, 460 ⟨78PJC1315⟩
Antebi, S., **1**, 239 ⟨85JOC147, 85TL2609⟩; **5**, 237 ⟨85TL2609⟩, 238 ⟨86OM596⟩
Antel, J., **6**, 726 ⟨85CB5007⟩
Antes, I., **2**, 73 ⟨92JOC6384⟩
Anteunis, M., **4**, 185 ⟨74S23⟩, 231 ⟨75T327⟩
Anteunis, M. J., **1**, 647, 738, 740, 742 ⟨88BSB271⟩
Anteunis, M. J. O., **3**, 638 ⟨81TL141⟩; **5**, 1133 ⟨87BSB115⟩
Antezak, K., **4**, 460 ⟨79ZC253⟩
Anthes, E., **3**, 534, 541 ⟨13CB3539⟩; **5**, 19 ⟨13CB1423⟩; **6**, 412, 417, 418 ⟨13CB1426⟩
Anthes, H. I., **6**, 69, 74, 80 ⟨42JA2525⟩
Anthoine, G., **5**, 337 ⟨90TL873⟩
Anthoni, U., **6**, 570, 571 ⟨69ACS324⟩, 574 ⟨68ACS1898⟩, 592 ⟨70ACS2055⟩, 594 ⟨66ACS2742⟩, 594, 595 ⟨70ACS959⟩
Anthonsen, J. W., **3**, 344 ⟨77ACS(B)40⟩
Anthony, J., **1**, 999 ⟨92AG(E)1240⟩; **2**, 1091 ⟨91JA6943⟩
Anthony, N. J., **1**, 697, 700, 703 ⟨87TL5763⟩, 700, 703 ⟨87TL5755⟩; **4**, 286 ⟨89T7161⟩
Anthony, T.,**4**, 11 ⟨78KO686⟩
Anthony, W. C., **5**, 15 ⟨54JA6208⟩
Antia, M. B., **6**, 611 ⟨61MI 620-01⟩
Antipin, M. Y., **2**, 444 ⟨91ZOB1075⟩; **6**, 116 ⟨88JOC3609⟩
Antipin, M. Yu., **2**, 878 ⟨87JOU1417⟩; **4**, 557 ⟨88ZOB1987, 88ZOB1998⟩; **5**, 703 ⟨87JOU75, 87ZOR86⟩, 1075 ⟨87ZOR2390⟩; **6**, 138 ⟨89T3487⟩
Antipova, V. V., **5**, 1128 ⟨91IZV2847⟩, 1131 ⟨92IZV187⟩
Antokhina, L. A., **5**, 369 ⟨78IZV2412⟩, 404 ⟨66BAU1432⟩, 404, 405 ⟨71BAU723⟩, 679 ⟨91JGU2372, 91ZOB2257⟩
Antonakis, K., **3**, 4, 115 ⟨82JCS(P1)1967⟩, 231 ⟨91TL1791⟩
Anton, M., **5**, 1137 ⟨92ICA173⟩
Antonini, I., **4**, 206 ⟨81JMC1181⟩
Antonioletti, R., **3**, 67 ⟨81TL1041⟩, 222, 303 ⟨93TL7089⟩, 245 ⟨87TL2297⟩
Antonisse, A. J. J., **4**, 264 ⟨92RTC163⟩
Antonov, D. M., **3**, 341 ⟨90KGS1285⟩
Antonov, D. S., **6**, 502 ⟨73GEP2241471⟩
Antonov, V. K., **6**, 104 ⟨62TL701, 63TL439⟩, 104 ⟨64TL47, 65T3537⟩
Antonova, V. V., **5**, 105 ⟨81MI 502-04⟩
Antonovich, V. A., **4**, 706 ⟨81JOM(209)233⟩; **6**, 208, 210 ⟨83IZV636⟩
Antonsen, O., **4**, 232 ⟨88ACS(B)515⟩
Anvarova, G. Y., **5**, 438 ⟨66JGU1945⟩
Anwar, S., **2**, 62 ⟨86CC831, 88T3761⟩; **3**, 183 ⟨84TL2977⟩
Anwer, M. K., **1**, 427 ⟨85TL1381⟩
Anzai, K., **5**, 1122 ⟨70BCJ2539⟩
Anzalone, L., **1**, 19, 24 ⟨85JOC2607⟩
Anzel, J., **2**, 322 ⟨91OPP721⟩; **6**, 632 ⟨91OPP721⟩
Ao, M. S., **3**, 211, 576 ⟨73JOC2129⟩
Aoai, T., **1**, 603 ⟨81JOC4727⟩; **2**, 282 ⟨80JOC1953, 81JOC3021⟩, 284 ⟨81JOC4727⟩, 286 ⟨81CC546⟩; **5**, 282 ⟨81CC546⟩, 282, 300 ⟨81JOC4727⟩
Aoe, K., **5**, 284 ⟨87CC1627⟩
Aoki, A., **4**, 856 ⟨78CPB685⟩
Aoki, B., **5**, 223 ⟨87CPB1903⟩; **6**, 447 ⟨86JAN230⟩
Aoki, D., **2**, 1017 ⟨61CI(L)1619⟩
Aoki, H., **3**, 356 ⟨74CL1101⟩
Aoki, K., **1**, 108 ⟨69JOC3974, 69JOC3978⟩; **6**, 465 ⟨93JAP05255201⟩
Aoki, M., **1**, 934 ⟨91CL1691, 91TL6559⟩, 934 ⟨93JCS(P1)2417⟩; **5**, 1036 ⟨91TL3503⟩
Aoki, O., **2**, 244 ⟨69TL447⟩
Aoki, S., **1**, 122 ⟨87JA8056, 89JA3091⟩, 483 ⟨88JA3296⟩, 615, 698 ⟨81TL2675⟩; **2**, 69 ⟨87JA8056⟩, 87 ⟨91JA8981⟩; **4**, 388 ⟨92JA9715⟩; **5**, 953, 956 ⟨92JA2985⟩, 958 ⟨92JA9715, 93JA9015⟩; **6**, 491 ⟨89TL7141⟩
Aoki, T., **2**, 281 ⟨80CC412⟩; **3**, 309 ⟨90CPB3261⟩; **4**, 673 ⟨69BCJ545⟩; **6**, 107 ⟨82H(18)201⟩
Aoki, Y., **6**, 60, 62 ⟨93T8487⟩
Aono, M., **1**, 367 ⟨86TL4039⟩, 368 ⟨88H(27)981⟩; **3**, 367 ⟨87CL1851⟩; **4**, 369 ⟨86H(24)1571, 87CPB1734⟩, 937 ⟨86CL2089⟩; **5**, 125, 126 ⟨82S833⟩; **6**, 129 ⟨86H(24)1571⟩, 129, 282 ⟨87CPB1734⟩
Aonuma, S., **1**, 867 ⟨90JOC996⟩
Aotsuka, T., **4**, 979 ⟨91JHC485⟩
Aoyagi, E. I., **4**, 776 ⟨85CA(102)78589t⟩
Aoyagi, K., **2**, 927 ⟨93CC1042⟩
Aoyagi, S., **1**, 1067 ⟨92JA10653⟩; **5**, 223 ⟨85JCS(P1)2361⟩, 330 ⟨91TL4325, 94JOC1358⟩
Aoyagi, T., **1**, 188 ⟨85CC1452⟩, 666 ⟨73JOC1130⟩
Aoyama, A., **1**, 1044 ⟨72BCJ1846⟩
Aoyama, H., **3**, 225 ⟨79JA494⟩, 343 ⟨73BCJ2253⟩, 366 ⟨84JOC1837⟩; **5**, 630 ⟨88JCS(P1)2203⟩; **6**, 4 ⟨91EUP421322, 92EUP473105⟩, 114 ⟨86TL1335⟩
Aoyama, M., **2**, 902 ⟨90TL247⟩
Aoyama, T., **2**, 683 ⟨84CPB3759⟩, 753 ⟨84TL429⟩, 944 ⟨91S1163⟩, 945 ⟨76JOC3559⟩; **3**, 125 ⟨82CPB119, 82CPB3380⟩, 202 ⟨88S228⟩, 475, 483 ⟨80TL4461⟩, 531, 532 ⟨82CPB526⟩, 557,558 ⟨91S1163⟩; **4**, 509 ⟨74JOM(82)21, 75CC640, 76JOM(118)1, 78JOM(153)193⟩; **5**, 132, 873 ⟨81CPB1475⟩, 252 ⟨87NKK1457⟩, 413, 873 ⟨86TL2005⟩, 867 ⟨82CPB899, 80TL4461⟩, 873 ⟨80TL4619, 81CPB3249, 81H(15)975, 82CPB119, 82CPB3380, 90O(68)S1⟩, 874 ⟨91S1163⟩; **6**, 675 ⟨85H(23)2371⟩
Aoyama, Y., **1**, 925, 946 ⟨93TL1317⟩; **2**, 213 ⟨76CL857⟩; **3**, 286 ⟨91TL6731⟩, 459 ⟨92JOC4777⟩
Aoyana, T., **2**, 89 ⟨90TL5507⟩
Aparicio, D. M., **4**, 230 ⟨85TL1927⟩
Apasov, E. T., **2**, 394 ⟨86BAU2442⟩; **5**, 362 ⟨91IZV1108⟩, 362 ⟨91MI 507-06⟩, 380 ⟨83IZV1413⟩; **6**, 147 ⟨85IZV2635⟩
Apel, G., **2**, 391, 392, 394 ⟨71HOU(10/1)12⟩
Apel, H. G., **2**, 784 ⟨71HOU(10/1)9⟩
Apeloig, Y., **1**, 905 ⟨88JOC661⟩, 1110 ⟨79JA5095⟩, 1110, 1112 ⟨75JA821, 75JA836⟩, 1112, 1113 ⟨77JOC3004⟩, 1122 ⟨81JA2770⟩; **3**, 502 ⟨92OM2326⟩; **5**, 924 ⟨84CC768⟩; **6**, 386, 401, 404 ⟨76JA5419⟩
Apitz, G., **2**, 140 ⟨91TL3163⟩; **4**, 100, 311 ⟨91TL3163⟩
Apjok, A., **6**, 73 ⟨67T4181⟩
Apodaca, L., **1**, 145 ⟨90JOC2234⟩
Apotecher, B., **1**, 545 ⟨84JOM(270)C15⟩
Appapillai, Y., **5**, 301 ⟨85JA4577⟩
Appa Rao, J., **4**, 168 ⟨89JOC2751⟩
Apparao, S., **3**, 267 ⟨87S896⟩; **4**, 704 ⟨88T1667⟩; **5**, 555 ⟨85JCS(P1)641⟩, 731 ⟨88JCS(P1)169, 88T1667⟩
Apparu, M., **1**, 497 ⟨84JOC2125⟩; **2**, 85 ⟨83OR(29)345⟩
Appel, H., **4**, 738 ⟨64LA(679)51⟩
Appel, R., **1**, 837 ⟨87CB131⟩; **2**, 7 ⟨75AG(E)801, 76CB3446⟩, 165, 402 ⟨71CB2023⟩, 259 ⟨81ZN(B)447⟩, 384 ⟨66CB3118, 70CB3733⟩, 399 ⟨70CB2152,

71CB2648, 75CB2329⟩, 399, 400 ⟨76CB2442, 71ZN(B)73⟩, 402 ⟨71CB3875⟩, 453 ⟨76CB805⟩, 483 ⟨69CB3955⟩, 809 ⟨60AG209⟩, 835, 837 ⟨77CB2382⟩, 835 ⟨B-79MI 216-01⟩, 864 ⟨68ZAAC(363)176⟩, 936 ⟨86CB535⟩; **3**, 354 ⟨71AG(E)132⟩, 495 ⟨77S679⟩, 495, 687 ⟨75AG(E)801, B-79MI 313-03⟩, 605, 607 ⟨86CB2466, 84AG(E)619⟩, 609 ⟨86JOM(307)93⟩, 618 ⟨71CB2025⟩, 618, 619 ⟨71CB1030⟩, 688 ⟨89AG(E)53⟩; **4**, 111 ⟨B-89MI 403-01⟩, 112 ⟨77CB2382, 79ZAAC(459)7, 83CB114⟩, 146 ⟨86CB1977⟩, 547, 564 ⟨79CB648⟩, 552 ⟨71CB2250⟩, 559 ⟨83CB2275⟩, 563 ⟨74ZAAC(406)131, 75CB919⟩, 566 ⟨81ZAAC(475)18, 83TL3591, 84CB2693, 85CB3419, 85CB814⟩, 566, 568 ⟨86IS110⟩, 567 ⟨85CB4068, 86CB535⟩, 738 ⟨75AG(E)801⟩, 1026 ⟨86CB2466⟩; **5**, 275 ⟨83CB2037⟩, 397, 880 ⟨87AG(E)779⟩, 675 ⟨85CB1632⟩, 876 ⟨89MI 523-01⟩, 876, 1154 ⟨81AG(E)731⟩, 876, 877, 878 ⟨87ZAAC(553)7, B-90MI 523-02⟩, 877, 878 ⟨85TL3551, 84TL4109⟩, 880 ⟨83CB938⟩, 887 ⟨80AG(E)556⟩, 888 ⟨81ZAAC(475)18⟩, 889 ⟨85CB1352⟩, 894 ⟨77AG(E)180⟩, 910 ⟨79CB1093⟩, 994 ⟨58AG271⟩, 994, 1018 ⟨58CB1200, 64CB849⟩, 1019 ⟨83AG(E)785, 83TL2639⟩, 1058 ⟨84AG(E)970⟩, 1063 ⟨71CB1335⟩, 1086 ⟨87ZAAC(555)23⟩, 1086, 1092 ⟨80TL2497, 84AG(E)619⟩, 1092, 1093, 1095 ⟨83CB2275⟩, 1093 ⟨74CB2169, 76AG(E)702, 78TL2689, 78CB2054⟩, 1154 ⟨81AG(E)197, 81TL2159⟩; **6**, 164 ⟨81ZAAC(475)18⟩, 164, 373 ⟨84CB2063⟩, 164, 703 ⟨79CB648⟩, 166 ⟨85CB814⟩, 166, 689 ⟨84CB2693⟩, 166, 693 ⟨86CB1977⟩, 265 ⟨77S699⟩, 288 ⟨83CB114, 86IS117⟩, 288, 693 ⟨82TL2017⟩, 293, 693 ⟨82AG(E)80⟩, 371 ⟨86CB2609⟩, 454 ⟨83AG(E)785⟩, 454, 686 ⟨83CB109, 83TL2639⟩, 511 ⟨89AG(E)53⟩, 512 ⟨77CB2368⟩, 519 ⟨83CB2371⟩, 618 ⟨76CB810⟩, 620 ⟨73CB2093, 74CB698⟩, 665, 686 ⟨82AG(E)448⟩, 678, 689, 692 ⟨81AG(E)731⟩, 680, 682, 683 ⟨87ZAAC(553)7⟩, 680, 684 ⟨84TL4447⟩, 680, 687 ⟨84AG(E)619⟩, 680, 687 ⟨84AG(E)970⟩, 680, 689, 692, 694 ⟨84TL4109⟩, 681, 682, 689, 690 ⟨85TL3551⟩, 683 ⟨84AG(E)895⟩, 686 ⟨79AG(E)469, 79AG(E)873⟩, 686 ⟨82CB1617, 83CB1873⟩, 687 ⟨82CB2371, 86CB2748, 86TL1661, 88ZAAC(556)7⟩, 688 ⟨81TL2159⟩, 688, 689 ⟨81TL4957, 82AG(E)219⟩, 692 ⟨B-89MI 622-01⟩, 693 ⟨84ZC384, 85AG(E)419, 86CB535, 88PS(36)147⟩, 695 ⟨77TL399⟩, 703 ⟨79CB1068⟩
Appelman, E., **2**, 94 ⟨91JA2648⟩
Appelman, E. H., **2**, 94 ⟨93JA1379⟩, 613 ⟨81JA1964⟩
Appelt, A., **2**, 474, 475 ⟨85AG(E)402⟩, 1003 ⟨85OM231⟩; **4**, 566, 588 ⟨83ZN(B)1399⟩, 587 ⟨85OM231, 86OM1664⟩; **6**, 164, 371, 373 ⟨88ZN(B)1416⟩, 169 ⟨86JOM(312)C1⟩
Apperson, L. D., **2**, 489 ⟨39JOC162⟩
Appl, M., **3**, 407 ⟨58CB12⟩
Applegate, H. E., **5**, 346 ⟨82JOC179⟩
Applequist, D. E., **1**, 846 ⟨63JA743⟩, 928 ⟨73JA856⟩
Applequist, J., **6**, 79 ⟨88JA4144⟩
Appleton, D., **1**, 791 ⟨94TL1071⟩
Appleton, D. C., **2**, 161 ⟨80JCS(P2)385⟩
Appleton, T. G., **5**, 385 ⟨89IC815⟩; **6**, 247 ⟨88JOM(342)399, 93JOM(453)299, 93JOM(453)307⟩
Applewhite, T. H., **5**, 189 ⟨63JOC1905⟩, 189 ⟨63MI 504-01⟩
ApSimon, J. W., **2**, 337 ⟨72CJC1167⟩; **3**, 404 ⟨86T5157⟩
Aquila, W., **1**, 296 ⟨74CA(81)120222t⟩
Aqullo, C., **1**, 49 ⟨86T2429⟩
Arad, D., **1**, 135, 159 ⟨84JA6467⟩, 905 ⟨88JOC661⟩; **2**, 999, 1006 ⟨84JA6467⟩; **4**, 378, 391, 816 ⟨84JA6467⟩
Aradi, A. A., **4**, 726 ⟨91OM2726⟩; **6**, 338 ⟨88OM812⟩
Arad-Yellin, R., **2**, 304 ⟨89S825⟩
Arafa, I. M., **2**, 578 ⟨88JA5228⟩
Aragona, M., **2**, 220 ⟨78OS(58)147⟩
Aragonès, M., **5**, 361 ⟨91JOC7038⟩
Aragono, H., **2**, 220, 270 ⟨88OSC(4)235⟩
Arai, H., **1**, 1010 ⟨78H(9)1397⟩; **2**, 50 ⟨88JA3712⟩; **3**, 663 ⟨88T1827⟩; **5**, 325 ⟨85JA5310⟩
Arai, I., **1**, 1055 ⟨86JA483⟩; **2**, 905 ⟨86JA483⟩
Arai, K., **1**, 1029 ⟨91TL3405⟩; **4**, 66 ⟨76BCJ553⟩, 988 ⟨82JCS(P1)1059⟩; **5**, 547 ⟨80S566⟩; **6**, 48, 82 ⟨76BCJ553⟩
Arai, M., **1**, 122 ⟨89JA3091⟩, 124 ⟨92T5709⟩, 881 ⟨76TL1817⟩; **2**, 587 ⟨92T5709⟩; **4**, 932 ⟨77JA4181⟩; **5**, 413 ⟨77JA5827⟩, 414 ⟨77JA4181⟩
Arai, N., **4**, 252 ⟨92CL1229⟩
Arai, T., **6**, 465 ⟨93JAP05255201, 93JAP05310644⟩
Arai, Y., **1**, 767, 769 ⟨88BCJ467⟩; **4**, 270, 854, 859, 860 ⟨86SC233⟩, 285 ⟨85TL6205⟩; **6**, 504 ⟨82CPB534⟩
Arakawa, Y., **2**, 227 ⟨88CPB791⟩
Arakelyan, A. S., **1**, 629 ⟨87ZOR1660⟩
Araki, H., **5**, 718 ⟨68JPS(B)389⟩
Araki, K., **1**, 613, 700, 1073 ⟨84JA3670⟩, 830 ⟨93TL2251⟩; **2**, 35 ⟨81TL1017⟩
Araki, M., **2**, 35 ⟨87CL715⟩; **3**, 286 ⟨73JA4763, 74BCJ1777⟩
Araki, N., **1**, 299, 354 ⟨77TL109⟩
Araki, S., **2**, 26 ⟨84S841⟩, 73 ⟨88JOC1831, 88JOM(347)5⟩, 73, 597 ⟨89JOM(369)291, 91JOC2538⟩; **3**, 622 ⟨81BCJ629⟩; **4**, 166 ⟨92JOC742⟩
Araki, T., **1**, 666 ⟨73JOC1130⟩, 1010 ⟨88CPB1890⟩; **2**, 210 ⟨86TL215⟩; **5**, 383 ⟨66MI 507-01⟩, 607 ⟨76JA7084⟩
Araki, Y., **2**, 35 ⟨87CL715⟩; **3**, 708 ⟨85MI 321-01⟩; **6**, 495 ⟨71BCJ1393⟩
Aranchalam, T., **3**, 234 ⟨83JOC1007⟩
Arancibia, L., **5**, 513 ⟨88SC323⟩
Arase, A., **1**, 273 ⟨67JA5708⟩, 663 ⟨80BCJ1652⟩, 1018 ⟨83BCJ2855, 84CL1029⟩; **2**, 20 ⟨79CL961⟩, 43 ⟨91CC205⟩, 273 ⟨85CC715⟩, 289 ⟨76CL1115, 76CL785⟩, 623 ⟨78CL665⟩, 748 ⟨74CC346⟩, 922 ⟨86BCJ659, 92BCJ685⟩; **3**, 416, 645 ⟨91CC748⟩, 622 ⟨84CC398⟩, 622, 645 ⟨89CC266, 74BCJ2511⟩; **4**, 815 ⟨88BSJ3764⟩, 1056, 1061 ⟨86BCJ659⟩; **5**, 1114 ⟨76CL1115, 76CL785⟩
Arase, H., **3**, 78, 271 ⟨85TL2677⟩
Aratani, M., **1**, 263 ⟨75JOC2009⟩; **3**, 696, 708 ⟨84TL5075⟩
Aratani, T., **1**, 353 ⟨85PAC1839⟩
Arbasino, M., **5**, 1004 ⟨65TL4645⟩
Arbenz, U., **2**, 4744 ⟨70JOM(21)351⟩
Arbor, A., **3**, 224 ⟨74TL909⟩
Arbuckle, B. W., **2**, 441, 444 ⟨91PS(56)27⟩
Arbusov, A. E., **4**, 133 ⟨36JGU283⟩
Arbuzov, A. E., **2**, 450 ⟨48ZOB2008⟩, 451 ⟨45MI 208-01⟩, 461 ⟨29MI 208-01⟩; **3**, 446 ⟨58CA(52)9980⟩; **4**, 576 ⟨27CB291⟩
Arbuzov, B. A., **2**, 439 ⟨59DOK(127)1217, 52MI 208-01⟩, 452, 452 ⟨31JPR357⟩, 454 ⟨62MI 208-01⟩, 462 ⟨57MI 208-01⟩, 693 ⟨64PAC307⟩; **3**, 200 ⟨90S515⟩, 689 ⟨91HAC577⟩; **4**, 336 ⟨94AHC(61)59⟩, 339 ⟨76IZV1416⟩, 341 ⟨79DOK(244)610, 79IZV866, 80IZV1626⟩, 342 ⟨83IZV1374, 85IZV1102, 86IZV2502, 88IZV159⟩, 342 ⟨89IZV946⟩, 343 ⟨79IZV1863, 86IZV640, 87IZV418⟩, 346 ⟨57ZOB2360⟩, 453, 456 ⟨80IZV2417⟩, 457 ⟨80IZV952, 80IZV1438⟩, 478 ⟨83IZV1379⟩, 502 ⟨79DOK(246)1130⟩, 549 ⟨82IZV1196, 82IZV2718, 84IZV2023, 84ZOB1511, 85ZOB1471⟩, 553 ⟨57ZOB2356⟩, 558 ⟨81IZV1113⟩, 569 ⟨90PS(53)1⟩, 1006 ⟨73IZV1112⟩; **5**, 395 ⟨90BAU2424⟩, 395, 396 ⟨91PS(55)133⟩; **6**, 78 ⟨50DOK(70)231⟩, 160 ⟨91PS(61)361⟩, 297 ⟨49DOK(68)515⟩, 691 ⟨92PS(66)257⟩
Arcadi, A., **1**, 546 ⟨85T5121⟩, 1029 ⟨89TL2581⟩; **2**, 983 ⟨92JOC976⟩; **3**, 146 ⟨83JOC4236⟩; **4**, 169 ⟨91T1525⟩
Arcamone, F., **3**, 284 ⟨84CC530⟩
Arcand, H. R., **3**, 623 ⟨93TL1453⟩
Arcas, A., **5**, 1138 ⟨78MI 530-02⟩
Arce, A. J., **2**, 582 ⟨84JOM(276)C45⟩; **4**, 682 ⟨82JCS(D)1155⟩, 683 ⟨92JCS(D)2423⟩
Arcelli, A., **3**, 86 ⟨89JOC949⟩

Arceneaux, R. L., **2**, 413 ⟨59JOC1419⟩
Archavlis, A., **1**, 499 ⟨90TL4859⟩
Archelas, A., **2**, 51 ⟨89JOC4686⟩
Archer, C. M., **3**, 158 ⟨92JOC5620⟩
Archer, D. A., **1**, 627 ⟨71JCS(C)1327⟩; **3**, 225 ⟨90JCS(P1)353⟩
Archer, S., **4**, 950 ⟨91T6539⟩; **5**, 303 ⟨65OR(14)1⟩
Archer, W. L., **3**, 91, 99 ⟨63OSC(4)331⟩
Archibald, T. G., **1**, 445 ⟨67JOC2627⟩, 999 ⟨66JOM(5)205⟩; **2**, 317 ⟨89S450⟩, 874 ⟨67JOC2627, 68JOC1681⟩, 875 ⟨67TL4447⟩; **4**, 108 ⟨89JOC2869⟩, 109 ⟨88JOC4645⟩, 428 ⟨90JOC2920⟩, 563 ⟨66TL3187⟩, 731 ⟨90JOC3562⟩, 795 ⟨91JOC537⟩, 970, 993 ⟨92JOC235⟩; **6**, 55 ⟨90MI 602-01⟩
Arco, M. J., **1**, 410, 821 ⟨78JOC4831⟩, 410, 823 ⟨78TL523⟩
Arcoleo, J. P., **4**, 357 ⟨77TL591⟩
Arcus, C. L., **2**, 30 ⟨52JCS2669⟩; **5**, 994 ⟨54JCS4018, 57JCS1091⟩
Ardakani, A. A., **5**, 336 ⟨71JCS(C)3550⟩
Ardecky, R. J., **1**, 569 ⟨82JOC409⟩
Ardisson, J., **1**, 634 ⟨87TL1997⟩
Ardoin, N., **1**, 474 ⟨90BSF401⟩
Ardra, P. C., **2**, 229 ⟨75CJC455⟩
Arduengo, III, A. J., **3**, 314 ⟨83JA5929⟩, 679 ⟨84TL405⟩; **4**, 478 ⟨82JA2497⟩; **6**, 722, 733 ⟨93OM3405⟩
Areces, P., **1**, 217 ⟨90H(31)1997⟩
Aregullin, M., **4**, 779 ⟨74TL4157⟩
Arenal, I., **5**, 87 ⟨85S773⟩
Arendt, V., **2**, 783 ⟨66JCE144⟩
Arens, G., **6**, 254 ⟨86CB3631⟩
Arens, J. F., **1**, 785 ⟨59RTC551⟩, 954 ⟨70QRS1⟩, 955 ⟨68RTC916⟩, 955, 1125 ⟨65QR274⟩, 971 ⟨68RTC1179⟩, 986 ⟨67RTC129⟩, 999 ⟨62RTC993, 67RTC1138, B-69MI 121-06⟩, 999, 1000 ⟨63RTC183⟩, 1023 ⟨68RTC1249⟩, 1030 ⟨68RTC69⟩, 1031 ⟨B-67MI 121-02⟩, 1037 ⟨65RTC1200⟩; **2**, 434 ⟨63RTC302⟩, 676 ⟨48RTC973⟩, 939 ⟨68RTC916⟩, 1021, 1025, 1027 ⟨B-60MI 221-01⟩, 1030 ⟨56RTC1459, 56RTC481⟩, 1030 ⟨59RTC252, 61RTC237⟩, 1030, 1031 ⟨62RTC510, 60RTC866⟩, 1032 ⟨61RTC244, 62RTC533, 62RTC583, 66RTC580, 66RTC952, 68RTC97⟩, 1033 ⟨69RTC179⟩, 1059, 1079, 1081 ⟨64RTC1301⟩, 1061 ⟨62RTC993⟩, 1066 ⟨69RTC71, 73RTC762⟩; **3**, 62 ⟨48RTC973⟩, 247 ⟨58RTC761⟩, 540 ⟨61RTC810⟩, 548 ⟨60RTC866, 67RTC907⟩, 548, 552 ⟨66RTC889⟩; **4**, 765 ⟨60MI 417-01, 69MI 417-02⟩, 836 ⟨68RTC1179⟩, 843 ⟨74RTC184⟩, 844, 846 ⟨59RTC354⟩, 884 ⟨54MI 420-01, 56RTC1377⟩; **5**, 14 ⟨57RTC969⟩, 423 ⟨68MI 509-01⟩, 550, 563 ⟨66NEP6606456⟩; **6**, 41 ⟨57RTC969⟩, 74 ⟨53MI 603-01, 60RTC888⟩, 83, 84 ⟨59RTC354⟩, 84, 86, 303 ⟨62RTC1009⟩, 88 ⟨67M1043⟩, 110 ⟨57RTC949⟩
Arens, J. H., **1**, 956 ⟨63RTC1040⟩
Arenz, S., **5**, 871 ⟨89CB565⟩, 924 ⟨88CB1363⟩
Arenz, T., **2**, 624 ⟨86TL1995⟩; **4**, 574 ⟨84AG(E)381, 86AG(E)559⟩; **5**, 896 ⟨86TL1995⟩, 916 ⟨84AG(E)381, 86AG(E)559⟩; **6**, 704 ⟨84AG(E)381, 86AG(E)559⟩
Aresta, M., **6**, 428 ⟨93JOM(451)243⟩
Aretakis, A. J., **1**, 920 ⟨89JOC308⟩
Argay, G., **6**, 557 ⟨83T1829⟩
Arguëlles, M., **1**, 521 ⟨66AG(E)1039⟩, 522 ⟨66AG1057⟩
Arguelles, R., **4**, 357 ⟨80JOM(198)15⟩
Argyle, J. C., **2**, 660 ⟨74JA2121⟩
Argyropoulos, J. N., **1**, 884 ⟨82JA6788, 86JOC472⟩
Arham, W. E., **3**, 73 ⟨53JA1647, 54JA1068, 60JOC53⟩
Arhart, R. J., **1**, 576 ⟨72JA5003, 77OS(57)22⟩
Arias, G., **2**, 329 ⟨88SC2055⟩
Arias, L. A., **5**, 709 ⟨83JOC888⟩
Arias Encarnacion, L. A., **1**, 785 ⟨79S388⟩; **3**, 11 ⟨79S388⟩
Aries, R., **2**, 194 ⟨67FRP1493679⟩
Arif, A. M., **1**, 988 ⟨88JOC3122, 91T4539⟩; **2**, 511 ⟨86CC1543⟩, 512 ⟨86JA1703, 87IC4065⟩, 1020 ⟨90AG(E)287⟩, 1021 ⟨91JA8997, 92MI 221-01⟩, 1036 ⟨93JOC5235⟩; **4**, 678 ⟨92OM3617⟩, 681 ⟨94OM2423⟩; **5**, 444 ⟨89JA149⟩, 884 ⟨88CC171⟩, 944 ⟨93OM3510⟩, 1097 ⟨92JA5890, 93JA3824⟩; **6**, 338 ⟨89JA149⟩, 689 ⟨85CC1687, 86OM593⟩
Ariga, M., **3**, 428 ⟨88BCJ461⟩
Ariga, T., **6**, 298, 300 ⟨92MM3829⟩
Arigoni, D., **2**, 292 ⟨73JA7917⟩
Ari-izumi, A., **3**, 57 ⟨77TL2425⟩
Arikawa, Y., **4**, 259 ⟨90JCS(P1)2035⟩
Arima, M., **1**, 907 ⟨82BCJ3257⟩
Arimoto, M., **1**, 490 ⟨81TL4491⟩, 1080 ⟨85CC697⟩; **2**, 1020, 1063 ⟨85TL4501⟩; **3**, 143, 225 ⟨81TL1115⟩
Arison, B. H., **6**, 496 ⟨68JA3254⟩
Arista, L., **3**, 222, 303 ⟨93TL7089⟩
Aristoff, P. A., **3**, 220 ⟨85JOC1765⟩
Aristova, N. V., **4**, 132 ⟨91JGU1459⟩
Arita, M., **5**, 138 ⟨84T145⟩
Ariura, S., **4**, 708 ⟨93AG(E)105⟩
Ariyan, Z. S., **2**, 216 ⟨66QRS285⟩
Ariyasu, J., **3**, 307 ⟨88H(27)1855⟩
Ariyoshi, K., **3**, 89 ⟨84CL891⟩
Arkell, A., **3**, 463 ⟨59JOC385⟩, 526, 527, 546 ⟨60JA2498⟩; **5**, 47, 275 ⟨60JA2498⟩
Arkhipova, O. G., **4**, 451, 469 ⟨68PAC235⟩
Arlt, D., **5**, 28 ⟨81AG(E)703⟩, 809 ⟨69AG572⟩, 1004 ⟨70S20⟩, 1012 ⟨71BRP1252099⟩; **6**, 606, 607 ⟨70S20, 69AG(E)606⟩
Armacost, L. M., **5**, 329 ⟨94JOC5292⟩
Armand, J., **2**, 782 ⟨65CR(260)2512⟩; **5**, 637 ⟨66BSF882⟩
Armanetti, L., **1**, 926 ⟨93TL1701⟩
Armani, E., **2**, 28 ⟨84T2035⟩; **3**, 27, 154 ⟨84T2035⟩
Armantrout, J., **3**, 545 ⟨84JA5178⟩
Armarego, W. L. F., **2**, 395 ⟨69JCS(C)986, 71JCS(C)1812⟩; **4**, 855 ⟨82AJC785⟩
Armas, P. de, **1**, 927 ⟨88JCS(P1)3255⟩
Armbrecht, B. H., **2**, 28 ⟨53JA3513⟩
Armbrecht, Jr., F. M., **6**, 267, 270, 293, 399 ⟨70JOM(23)361⟩, 270 ⟨67JA2790, 69JA1954⟩
Armbruster, C. W., **3**, 538 ⟨62TL1297⟩
Armer, B., **2**, 590 ⟨70AG(E)101⟩
Armer, R., **1**, 144 ⟨93JCS(P1)3099⟩
Armesto, D., **1**, 800 ⟨90CC934⟩; **3**, 413 ⟨86JCR(S)46⟩, 414 ⟨86JCS(P1)2021⟩, 418 ⟨87OPP181⟩; **4**, 1079 ⟨B-92MI 424-02⟩; **5**, 215 ⟨91JCS(P1)223⟩
Armistead, D. M., **1**, 243 ⟨82JOC3348⟩, 583 ⟨94JOC332⟩, 780 ⟨87TL4959⟩, 831 ⟨84JOC4320, 85TL1163⟩
Armitage, D. A., **2**, 231 ⟨64JCS5975⟩, 263 ⟨65JCS7098⟩, 358 ⟨71JCS(C)2840⟩, 358, 361 ⟨71JCS(C)3867⟩, 360, 361 ⟨71CC1413⟩, 399 ⟨64JCS3122, 74JINC993⟩, 416, 907 ⟨82COMC-I(2)1⟩, 720 ⟨84S1042⟩, 725 ⟨70JOM(24)629⟩, 851, 852 ⟨64JCS5585⟩, 890 ⟨66JOM(5)326⟩; **5**, 374 ⟨B-91MI 507-03⟩; **6**, 172 ⟨82COMC-I(2)1⟩
Armitage, J. B., **1**, 1035 ⟨52JCS2014, 54JCS147⟩, 1069 ⟨52JCS2010⟩
Armour, M. A., **2**, 431 ⟨91BCJ2398⟩
Arms, P. de, **1**, 931 ⟨92AG(E)772⟩
Armsen, R., **5**, 894 ⟨70S590⟩
Armstrong, A., **2**, 90 ⟨88TL2483⟩; **4**, 286 ⟨89T7161⟩
Armstrong, D. E., **4**, 18 ⟨33JCS496⟩
Armstrong, D. R., **3**, 459 ⟨87JCS(D)1071⟩
Armstrong, G. P., **2**, 655 ⟨68MI 213-01⟩
Armstrong, III, J. D., **1**, 417, 830, 876 ⟨91JOC650⟩; **4**, 831 ⟨91JOC650⟩
Armstrong, J. C., **1**, 1009 ⟨86AG(E)268⟩
Armstrong, L. J., **2**, 643 ⟨45JA386⟩, 645 ⟨44JA1274⟩, 645, 646 ⟨44JA1464⟩
Armstrong, P., **1**, 369, 524 ⟨85T3547⟩
Armstrong, R. J., **1**, 216 ⟨83CJC214, 86CJC584⟩
Armstrong, R. W., **1**, 489 ⟨87JA4756⟩; **3**, 717 ⟨91TL3687⟩
Armstrong, V. W., **2**, 1004 ⟨75JCS(P1)2513⟩
Arnaiz, D. O., **1**, 830, 831 ⟨89TL515⟩; **5**, 227 ⟨89TL515⟩
Arnason, B., **3**, 494 ⟨62CB1513⟩

Arnason, I., **6**, 267 ⟨76ZAAC(419)213⟩
Arnason, J. T., **4**, 175 ⟨91JOC1788⟩
Arndt, C., **1**, 1142 ⟨66CB1642⟩
Arndt, F., **1**, 394 ⟨35CB200⟩; **3**, 466 ⟨43OSC(2)461⟩, 479 ⟨33CB1012⟩, 565 ⟨38CB1627⟩, 566 ⟨36LA(521)95⟩; **6**, 302 ⟨13LA(396)1⟩, 306 ⟨11LA(384)322⟩
Arndt, G., **3**, 355 ⟨76S445⟩, 356 ⟨76S692⟩
Arndt, H. C., **1**, 698 ⟨76TL3477⟩
Arndt, S., **1**, 970 ⟨93CB251, 93CB261⟩
Arne, K., **4**, 372, 373, 525 ⟨78JA1325⟩; **6**, 124, 135 ⟨78JA1325⟩
Arne, K. H., **3**, 167 ⟨82OM280⟩
Arneri, R., **3**, 78 ⟨77IC3124⟩
Arnett, E. M., **1**, 859 ⟨78JA3514⟩, 883 ⟨78JOC815⟩, 891 ⟨87MI 119-10⟩, 905 ⟨79JA522, 85MI 119-06⟩
Arnhold, M., **6**, 227 ⟨1887LA(240)192⟩
Arnó, M., **1**, 49 ⟨86T2429⟩, 1082 ⟨92JOC50⟩
Arnold, B., **3**, 476 ⟨81AG118, 81AG(E)113⟩, 476, 477 ⟨81BSB615⟩, 476, 490, 579 ⟨80TL909, 79AG(E)320⟩
Arnold, B. J., **1**, 420 ⟨72CC1034⟩
Arnold, C., **3**, 613 ⟨60JOC257⟩
Arnold, D. P., **3**, 568 ⟨80JOM(199)119⟩; **5**, 386 ⟨80JOM(202)107⟩
Arnold, D. R., **1**, 921 ⟨84CJC1850, 84CJC1860⟩; **2**, 92 ⟨68MI 202-03⟩; **3**, 314 ⟨70JA1404, 71JHC1097⟩, 314, 318 ⟨70PAC523⟩, 317 ⟨78CJC2985⟩
Arnold, E. V., **1**, 971 ⟨80JA2134⟩; **4**, 603 ⟨81TL7⟩
Arnold, G., **6**, 111 ⟨69T5995⟩
Arnold, H.-S., **2**, 492, 494 ⟨64CB449⟩
Arnold, J., **4**, 355 ⟨87JA3318⟩; **5**, 444 ⟨89JA149⟩; **6**, 338 ⟨89JA149⟩, 349, 355, 357, 358 ⟨93IC5126⟩, 355, 357, 358 ⟨93JA6777⟩, 357, 358 ⟨93JA545⟩, 521 ⟨85JA6409, 86JA5355⟩
Arnold, L. D., **1**, 230 ⟨87JA4649⟩; **5**, 85 ⟨88JA2237, 91OS(70)1⟩, 1078 ⟨89JA3973⟩
Arnold, R. C., **2**, 126 ⟨50JA731⟩
Arnold, R. D., **1**, 988 ⟨84CC152⟩
Arnold, R. G., **5**, 962 ⟨57CRV47⟩
Arnold, R. T., **1**, 417, 829 ⟨49JA1150⟩, 417, 829 ⟨49JA2439⟩, 417, 829 ⟨72SC27⟩, 795, 1023 ⟨60JA4918⟩; **2**, 22, 29, 34 ⟨78JOC3687⟩, 558 ⟨77SC223⟩; **3**, 616 ⟨42JOC164⟩; **5**, 103 ⟨43OSC(2)10⟩, 201 ⟨57JA419⟩
Arnold, S., **2**, 30 ⟨78JOC2793⟩
Arnold, T., **1**, 955 ⟨92AG(E)1033⟩
Arnold, W., **3**, 362 ⟨01JPR182⟩
Arnold, Z., **2**, 745 ⟨90S39⟩; **3**, 48, 473 ⟨70CCC3618⟩, 68, 69 ⟨59CCC2378, 59CCC2385⟩, 71 ⟨88TL2861⟩, 74 ⟨73CCC1168⟩, 75 ⟨88CCC1519, 66TL1109⟩, 473 ⟨67CC299⟩; **6**, 51 ⟨59CCC4048⟩, 54 ⟨61CCC3059, 62CCC2886⟩, 54 ⟨63CCC2047⟩, 57 ⟨58CCC452⟩, 104 ⟨63CCC2040⟩, 104, 105 ⟨64CCC645⟩, 452 ⟨59CCC760⟩, 512 ⟨69RZC1443⟩, 635 ⟨92CCC1947⟩
Arnoldi, A., **2**, 632 ⟨81SC639⟩
Arnone, A., **3**, 32 ⟨93CC984⟩
Arnost, M. J., **2**, 213 ⟨79JOC3230⟩; **3**, 334 ⟨80JOC2601⟩; **5**, 554 ⟨80JOC2601, 82JOC4384⟩
Arnould, D., **1**, 612 ⟨73BSF746⟩
Arnswald, M., **2**, 726 ⟨91CB1997⟩; **5**, 344, 593 ⟨93JOC7022⟩, 1006 ⟨91CB1997⟩
Aroca, R., **4**, 281 ⟨81SA(A)819⟩
Aron, A. J., **2**, 232 ⟨78JOC3374⟩
Aronson, J. R., **6**, 212 ⟨59JA806⟩
Arora, B., **6**, 561 ⟨82FES205⟩
Arora, R. K., **4**, 982 ⟨66JIC650⟩
Arora, S. K., **3**, 85 ⟨70JA7161⟩; **4**, 184 ⟨86OS(64)138⟩, 632 ⟨69T1557⟩, 632, 634 ⟨67TL723, 69T1571⟩; **5**, 82 ⟨83JOC4076⟩; **6**, 448 ⟨85EUP136147⟩
Arotsky, J., **2**, 631 ⟨70JCS(C)1480⟩
Arouzmanian, H., **2**, 40 ⟨71S527⟩
Arpe, H.-J., **6**, 215, 220, 224 ⟨B-93MI 607-02⟩
Arques, A., **5**, 711 ⟨85SC643⟩, 975, 1031 ⟨82S596⟩, 1065 ⟨93JOC5264⟩, 1066 ⟨88JOC4654, 89CB307⟩, 1079 ⟨82SC573⟩
Arques, J. S., **3**, 106, 308 ⟨82SC231⟩
Arranaghi, M., **3**, 90, 101 ⟨81AG(E)878⟩
Arrhenius, T. S., **3**, 209, 241 ⟨83JA2030⟩
Arrick, B. A., **3**, 221, 227, 275 ⟨84JA5585⟩
Arrieta, A., **2**, 28, 30 ⟨84SC939⟩; **3**, 154 ⟨84SC939⟩, 618 ⟨84TL3365⟩; **5**, 329 ⟨87JCS(P1)845⟩, 367 ⟨83SC289, 84TL3365⟩
Arrigoni-Martelli, E., **6**, 645 ⟨80JMC13⟩, 648 ⟨78JMC773⟩
Arrington, C. A., **3**, 502 ⟨83JA6176, 84JA525⟩; **5**, 1019 ⟨88JA6240⟩
Arrojo, P., **4**, 683 ⟨92JCS(D)2423⟩
Arrowsmith, J., **3**, 229 ⟨80JCS(P1)2851⟩; **4**, 540 ⟨82JCS(P1)143⟩
Arsenault, G. J., **4**, 150 ⟨89JCS(D)761⟩, 151 ⟨88JCS(D)1773⟩
Arsenijevic, L., **3**, 15, 90, 97, 101 ⟨84TL1843⟩
Arsenijevic, V., **3**, 15, 90, 97, 101 ⟨84TL1843⟩
Arseniyadis, S., **1**, 68, 454 ⟨83TL3997⟩, 155 ⟨84OR(31)1⟩, 495 ⟨78JCR(S)416⟩; **2**, 325 ⟨88T2457⟩; **3**, 292 ⟨84OR47⟩, 422 ⟨88T2457⟩, 616 ⟨87TL547⟩, 620 ⟨84OR1⟩; **4**, 249 ⟨93TL1141⟩, 938 ⟨83JOC383⟩
Arshady, R., **3**, 696 ⟨82AG(E)374⟩
Arshinova, R. P., **4**, 341 ⟨80IZV1626⟩, 342 ⟨83IZV1374⟩, 343 ⟨79IZV1863⟩
Artamkina, G. A., **1**, 883 ⟨87RCR983⟩; **4**, 642 ⟨80IZV2429⟩; **5**, 113 ⟨84IZV1567⟩, 116 ⟨84TL4989⟩
Artamkina, O. B., **4**, 104 ⟨89JGU1625⟩; **5**, 374 ⟨88ZOB1093⟩
Arth, G. E., **1**, 881 ⟨60JA1684, 60JA5704⟩
Arthur, C. D., **2**, 613 ⟨81JA1964⟩
Arthur, M.-P., **3**, 485 ⟨91JA5856⟩, 682 ⟨91OM3205, 92S43⟩; **5**, 872 ⟨91JA5856⟩, 872, 1148 ⟨91OM3205⟩; **6**, 670, 675 ⟨91OM3205⟩
Arthur, P., **5**, 122 ⟨68CI(L)1568⟩
Arthur, S. D., **1**, 710 ⟨79JA5060⟩; **4**, 71 ⟨80JOC752⟩
Arthur, Jr., P., **3**, 614 ⟨54JA5364⟩
Artico, M., **4**, 856 ⟨82JHC1493⟩
Artschwager-Perl, U., **3**, 394 ⟨90AG(E)609⟩
Artyushin, O. I., **4**, 559, 766 ⟨90ZOB1539⟩
Arukwe, J., **3**, 303 ⟨91ACS914⟩
Arunachalam, T., **2**, 658 ⟨68CJC3625⟩
Arvanaghi, M., **1**, 5, 8, 9 ⟨80JOC3531⟩, 904 ⟨80JA6640, 82JA1628⟩, 906 ⟨82JA7105⟩, 1046 ⟨81AG925⟩; **2**, 22 ⟨84JOC2032⟩, 620 ⟨86S868⟩; **3**, 14, 90, 101 ⟨84S228⟩, 18, 134 ⟨80S220⟩, 93, 94 ⟨87CRV671⟩, 151, 153 ⟨79TL3653⟩, 152 ⟨84JOC2032⟩; **4**, 42 ⟨81JOC571⟩, 1072 ⟨83CB3926⟩; **5**, 367, 992 ⟨83JOC3358⟩
Arwin, S., **5**, 706 ⟨85JHC1435⟩
Arya, F., **3**, 424 ⟨90JOC412⟩
Arya, P., **2**, 141, 265 ⟨90JOC6248⟩, 141, 265 ⟨91TL2853⟩
Arya, V. P., **5**, 690 ⟨82IJC(B)989⟩
Arykova, L. A., **5**, 389 ⟨76ZOB75⟩
Arzoumanian, H., **1**, 242 ⟨87OM1625⟩, 278, 487, 661, 1103 ⟨67JA3652⟩, 487, 1100, 1103 ⟨67JA5086⟩, 509 ⟨67JA291⟩, 976 ⟨92OM493, 93OM1871⟩; **2**, 33 ⟨83JOM(243)175⟩, 41, 900 ⟨67JA291, 67JA3652⟩, 901 ⟨67JA5086⟩, 1005 ⟨66TL2535⟩; **4**, 580 ⟨86CC1208⟩, 631 ⟨67JA291⟩, 664, 665 ⟨66TL2535⟩, 814 ⟨67JA5086⟩; **5**, 52 ⟨67JA291⟩, 918 ⟨85CC1151⟩
Arzpura, J. M., **3**, 410 ⟨91CC524⟩
Asaad, M. M., **3**, 165 ⟨88JMC2199⟩
Asaba, H., **5**, 547 ⟨69JOC730⟩
Asada, S., **3**, 628 ⟨85MAC1755, 86BCJ893⟩
Asada, T., **2**, 743 ⟨92BCJ1379⟩
Asadorian, A. A., **6**, 215 ⟨51USP2553518⟩
Asahina, K., **4**, 863 ⟨92MI 419-01⟩
Asahina, M., **4**, 52, 101, 109 ⟨88JCS(P1)1149⟩, 101 ⟨87JCS(P1)2203⟩; **6**, 43, 47, 54, 57 ⟨86JOC955⟩
Asai, H., **5**, 70, 73 ⟨83CC1487⟩
Asai, M., **3**, 363 ⟨92JOM(423)329⟩
Asai, N., **3**, 701, 704 ⟨92TL6823⟩
Asai, T., **6**, 14 ⟨80JCS(P1)2755⟩

Asai, Y., **6**, 322, 337 ⟨70MI 611-01⟩, 576 ⟨70MI 618-02⟩
Asakura, T., **3**, 605 ⟨90TL2311⟩
Asami, M., **2**, 85 ⟨89TL2125⟩; **4**, 408 ⟨78CL1253, 80CL17⟩; **5**, 282 ⟨83CL683⟩
Asano, F., **3**, 118 ⟨89JOC1531⟩
Asano, K., **2**, 320 ⟨89CL463⟩; **5**, 318 ⟨89JCS(P1)1702⟩
Asano, M., **5**, 340 ⟨88CPB877⟩; **6**, 546, 561 ⟨87S349⟩
Asano, O., **2**, 657 ⟨87H(26)65⟩
Asano, S.,**4**, 833 ⟨62BCJ71⟩
Asanuma, M., **2**, 543 ⟨89TL6047⟩
Asao, N., **2**, 73 ⟨93CRV2207⟩
Asao, T., **1**, 248 ⟨88BCJ1281⟩, 899, 909 ⟨90PAC507⟩; **3**, 64 ⟨78CL41⟩
Asaoka, M., **1**, 133 ⟨79BCJ1241⟩, 558 ⟨94T655⟩; **3**, 166 ⟨78BCJ3008⟩, 224 ⟨88CL921⟩; **4**, 206 ⟨80BCJ1061⟩; **5**, 179 ⟨80TL4611⟩
Asaro, M. F., **3**, 516 ⟨87OM433⟩
Asato, A. E., **1**, 505 ⟨83JA2923⟩, 613 ⟨86TL3337⟩; **3**, 231 ⟨85TL2873⟩
Asberom, T., **3**, 308 ⟨87JA7553⟩; **4**, 952 ⟨87JA7553⟩
Asbury, R. L., **4**, 706 ⟨65JOC1600⟩
Ascherl, B., **5**, 670, 671 ⟨87BSB51⟩
Ase, K., **2**, 858 ⟨73ACS1939⟩
Asensio, G., **1**, 68 ⟨80JCS(P1)1420, 83T2609⟩, 914 ⟨78JA6299⟩; **2**, 18, 24, 31, 34 ⟨85AG(E)319⟩, 31, 35 ⟨87CC1491⟩, 631 ⟨84JCS(P1)2591⟩, 1019 ⟨87S661⟩; **4**, 152, 153 ⟨79S893⟩, 300 ⟨88JCS(P1)1631⟩, 412 ⟨88CB1813⟩, 453, 477 ⟨90SL261⟩; **4**, 749, 819, 1064 ⟨88JA5567⟩; **5**, 132 ⟨83S649⟩, 152 ⟨83T2863⟩, 1044 ⟨87CC1491⟩
Asequell, G., **2**, 1092 ⟨90TL2739⟩
Asgharian, R., **4**, 870 ⟨80JHC117⟩; **6**, 630 ⟨80JHC117⟩
Ash, A. B., **3**, 108 ⟨70JOC841⟩; **5**, 1078 ⟨67JOC2895⟩
Ash, D. K., **2**, 199 ⟨71JOC322⟩, 217 ⟨70TL3551⟩, 233 ⟨80JOC5155⟩, 725 ⟨79JOC4135⟩; **5**, 338 ⟨82JOC2785⟩
Ash, M. L., **1**, 861 ⟨81JHC939⟩
Ashare, R., **6**, 573 ⟨91CRV1⟩
Ashbrook, C. W., **1**, 410, 823 ⟨71JA2342⟩
Ashburn, S. P., **5**, 329 ⟨85TL4855⟩
Ashby, B. A., **4**, 520 ⟨67USP3346588⟩
Ashby, C. E., **3**, 618 ⟨45JA161⟩
Ashby, E. C., **1**, 4, 7, 8, 10, 17 ⟨78JOC1263⟩, 75, 457, 514 ⟨78JOC2567⟩, 458, 515 ⟨78JOC757, 77TL4481, 78CC30⟩, 636 ⟨78JOC4750⟩, 884 ⟨82JA6788, 83TL1667, 86JOC472⟩, 1039 ⟨75CRV521⟩; **2**, 25, 26, 31, 32 ⟨89S614⟩, 67 ⟨75CRV521, 80PAC545⟩, 68 ⟨79JOC4371⟩, 418, 419 ⟨71IC900⟩, 421 ⟨71IC1888, 71IC893, 71IC906, 78IC1862⟩, 521 ⟨57JA5190⟩, 558 ⟨77CC169⟩, 561 ⟨78CC30, 83JOM(250)1⟩, 969 ⟨78JOC757⟩, 1084 ⟨64JOC3225⟩; **3**, 127 ⟨76JOC1939, 76TL3865⟩; **4**, 698 ⟨82JOM(225)71⟩
Ashcroft, M. R., **2**, 29 ⟨88OSC(6)512⟩
Ashcroft, P. L., **3**, 685 ⟨83JCS(P1)293⟩
Ashe, A. J., **2**, 499 ⟨86JOM(308)289⟩, 505 ⟨82OM1408⟩, 508 ⟨86JOM(303)197⟩, 971 ⟨71JA1804⟩; **3**, 269 ⟨78TL2537⟩
Ashe, III, A. J., **1**, 894 ⟨68JA63⟩; **2**, 491, 505, 893 ⟨83OM1859⟩, 505, 893 ⟨90AOC(30)77⟩, 872 ⟨82TCC125⟩, 879 ⟨71JA3293, 71JA6690, 72JA7596⟩, 880 ⟨79CC880, 80JOM(202)C95, 84OM495, 87AG(E)229⟩, 880, 893 ⟨83OM1005, 83OM1573⟩, 893 ⟨87OM1185⟩, 896 ⟨91OM2068⟩, 915 ⟨93JOM(447)197⟩; **4**, 1038 ⟨92JA372⟩, 1038, 1041 ⟨92OM1491⟩, 1039 ⟨85OM1478⟩, 1039 ⟨87AG(E)229, 86AG(E)571⟩, 1041 ⟨91OM2068, 92OM2743⟩, 1042 ⟨80JOM(202)C95, 82TCC125⟩
Asher, B. M., **6**, 485 ⟨92JMC3731⟩
Asher, V., **3**, 638 ⟨81TL141⟩
Ashida, T., **3**, 616 ⟨87CL687⟩
Ashida, Y., **5**, 278 ⟨82JCS(P1)2909⟩
Ashimori, A., **4**, 410 ⟨93JOC6949⟩
Ashiya, H., **3**, 119 ⟨87BCJ4143⟩
Ashkinadze, L. D., **4**, 587 ⟨68ZOR1685⟩
Ashley-Smith, J., **5**, 1139 ⟨70JCS(A)1847⟩
Ashmawy, M. I., **2**, 24 ⟨87CC1577⟩
Ashton, C. P., **5**, 280 ⟨84TL4825⟩
Ashton, D. P., **2**, 615 ⟨85JFC(27)263⟩; **6**, 409, 419 ⟨88EUP253527⟩
Ashton, D. S., **4**, 30 ⟨71JCS(B)1723⟩
Ashton, L., **5**, 652 ⟨68JMC534⟩
Ashton, P. R., **1**, 571 ⟨92JA6330⟩
A-Shura, A. M., **4**, 775 ⟨88ZOR443⟩
Ashwell, M. A., **2**, 393 ⟨88S229⟩; **3**, 190 ⟨88S229⟩; **4**, 173, 237 ⟨91JCS(P1)897⟩; **5**, 238 ⟨88CC282, 90T7429⟩, 423 ⟨91JCS(P1)897⟩; **6**, 126 ⟨88CC645, 91JCS(P1)897⟩
Ashwood, M. S., **2**, 783 ⟨88S379⟩
Ashworth, D. R., **5**, 842 ⟨33JCS1389⟩
Ashworth, P. A., **1**, 481 ⟨92JCS(P1)3431⟩
Asinger, F., **1**, 46 ⟨69M724⟩; **2**, 167 ⟨60LA(637)146⟩, 175 ⟨42CB34⟩, 175 ⟨58CB2130⟩, 178 ⟨42CB344⟩, 233 ⟨59LA(627)195⟩; **4**, 317 ⟨58LA(619)169⟩, 318 ⟨59LA(622)83, 59LA(622)94⟩, 328 ⟨63AG1050, 67AG(E)907⟩; **5**, 579 ⟨63AG(E)397, 65M1474⟩, 579, 602 ⟨65M1278⟩, 599 ⟨64AG(E)19, 67AG(E)907⟩
Asirvatham, E., **1**, 23 ⟨85JOC2589⟩
Askani, R., **2**, 340, 341, 790, 1003 ⟨91COS(6)103⟩
Askani, U., **6**, 651 ⟨77AP820⟩
Asker, W., **6**, 114 ⟨64T25⟩
Askew, B., **5**, 151 ⟨87JA2426⟩
Askew, W. B., **6**, 340 ⟨68IC2265⟩
Askham, F. R., **2**, 439 ⟨85JA7423⟩; **4**, 550 ⟨85JA7423⟩, 581 ⟨90JA3713⟩, 716 ⟨90JOM(387)C31⟩
Askin, D., **3**, 243 ⟨85JA1246⟩; **4**, 174 ⟨87JOC622⟩
Asknes, G., **1**, 737 ⟨64ACS655⟩
Aslam, F. M., **3**, 101 ⟨72JCS(P1)892⟩
Aslam, M., **1**, 141 ⟨85JA6729⟩, 619, 707 ⟨83JA6164⟩, 619, 707 ⟨87OS(65)90⟩, 619, 707, 949 ⟨86JA4568, 83JA6165⟩; **2**, 9 ⟨79TL3615⟩, 177 ⟨80S285⟩, 177 ⟨82TL4203⟩, 204 ⟨81TL3573⟩, 205 ⟨78JA1637⟩, 262 ⟨88T281⟩; **4**, 82, 83, 85 ⟨86JA4568⟩, 364 ⟨88T281⟩, 365, 392 ⟨85JA6729⟩, 368 ⟨82TL4203⟩; **5**, 1106 ⟨78JA1637⟩; **6**, 349 ⟨85CC1674, 85JA6729⟩, 349 ⟨87IC1488⟩, 349, 585 ⟨85TL2259, 88T281⟩
Aslanidis, P., **4**, 593 ⟨91ZAAC(605)151⟩; **5**, 222 ⟨88CZ106, 90CZ71⟩
Aslanidis, R., **2**, 496 ⟨91ZAAC(605)151⟩
Asmus, K. D., **1**, 5 ⟨B-83MI 101-01⟩, 935 ⟨B-83MI 119-01⟩, 937 ⟨B-83MI 119-02⟩; **2**, 132, 202 ⟨90MI 201-01⟩
Aso, Y., **1**, 232 ⟨89BCJ164⟩, 594, 611, 621 ⟨89JOC4391⟩, 611 ⟨89CL607, 89JCS(P1)1775⟩; **2**, 753 ⟨84CL853⟩; **3**, 39, 176 ⟨89CL607⟩, 89 ⟨84CL891⟩, 114 ⟨86TL6099⟩; **4**, 238 ⟨91CL415⟩, 850 ⟨88CC822⟩; **5**, 210 ⟨87TL1281, 89JOC4391⟩, 242 ⟨86CL977⟩, 1078 ⟨86TL6099⟩
Asokan, C. V., **5**, 555 ⟨90T5423⟩
Asomaning, W. A., **1**, 452 ⟨73JCS(P1)137⟩
Aspinall, G. O., **5**, 62 ⟨54MI 502-02⟩
Asprou, C. M., **3**, 305 ⟨80JHC87⟩
Asquith, R. S., **3**, 362 ⟨48JCS1181⟩
Assadi-Far, H., **1**, 108 ⟨70BSF991⟩; **3**, 236 ⟨70BSF991⟩
Assaf, Y., **2**, 683 ⟨75CI(L)281⟩; **5**, 374 ⟨85JOC1712⟩
Asscher, M., **2**, 157 ⟨64JCS4962⟩, 157 ⟨72JCS(P1)1543⟩; **4**, 82 ⟨64JCS4962⟩; **6**, 7 ⟨63JCS1887⟩
Asselin, A. A., **2**, 627 ⟨90JHC2151⟩
Assercq, J.-M., **1**, 434 ⟨85JOC2668⟩; **3**, 220 ⟨87JOC1440⟩
Assil, H. I., **5**, 1078 ⟨89JA3973⟩
Assony, S. J., **5**, 1022 ⟨B-61MI 527-01⟩
Astafeva, L. I., **2**, 1090 ⟨63HOU(6/3)1⟩
Ast, W., **1**, 674 ⟨76MAC1349⟩
Asthana, A., **5**, 402 ⟨89JOM(366)281⟩
Astill, B. D., **5**, 1126 ⟨55JA4079⟩
Astin, S., **2**, 101 ⟨35JCS901⟩
Astles, D. P., **3**, 297 ⟨92T7581⟩
Astles, P. C., **1**, 595, 601, 624 ⟨91COS(6)1011⟩; **2**, 184, 196 ⟨91COS(6)1011⟩
Astoin, J., **3**, 304 ⟨77JHC861⟩
Aston, J. G., **1**, 391 ⟨40JA2590⟩, 392 ⟨51JA3900⟩; **2**, 405 ⟨38JA1930⟩; **5**, 66 ⟨55OSC(3)538⟩

Astorga, C., **5**, 353 ⟨91S350, 92IJC(B)906⟩, 359 ⟨93S287⟩
Astratyan, G. V., **4**, 730 ⟨75ZOR961⟩
Astruc, D., **1**, 164 ⟨89CC614⟩, 164 ⟨92SL57⟩, 165 ⟨82JA7549⟩, 474 ⟨90BSF401⟩; **2**, 575 ⟨B-87MI 211-04⟩; **4**, 581 ⟨90OM1036⟩
Asveld, E. W. H., **1**, 706 ⟨74TL2805⟩
Ataka, K., **5**, 801, 802 ⟨76JCS(D)327⟩
Ataka, Y., **4**, 193 ⟨90BCJ166⟩
Atanassova, I. A., **5**, 345 ⟨90SC2083⟩
Atarashi, Y., **4**, 239, 334, 660 ⟨90OM1355⟩
Atavin, A. S., **2**, 674 ⟨68RCR907⟩
Ates, M., **2**, 489, 490 ⟨89JOM(364)67⟩, 893 ⟨88POL2601⟩, 893, 894 ⟨86ZN(B)321, 89PS(44)129⟩; **4**, ⟨85TL2769, 85TL2445⟩; **6**, 169 ⟨92OM145⟩
Athar, M., **5**, 341 ⟨86IJC(A)57⟩, 734 ⟨86IJC57⟩
Atherthon, F. R., **5**, 712 ⟨55CI(L)1183⟩
Atherton, F. R., **2**, 105 ⟨45JCS382⟩, 412 ⟨45JCS660, 47JCS674⟩; **5**, 975 ⟨45JCS660⟩; **6**, 221 ⟨47JCS674⟩
Atherton, J. I., **1**, 1079 ⟨85CC1140⟩
Atkins, A. R., **1**, 67, 69 ⟨81JOC563⟩
Atkins, Jr., G. M., **2**, 401, 793 ⟨72JA6135⟩; **6**, 106 ⟨67JA2502, 72JA6135⟩
Atkins, K. E., **6**, 25, 31 ⟨67JOC2166⟩
Atkins, M. P., **6**, 79 ⟨80CC207⟩
Atkins, P. J., **6**, 25, 31 ⟨83CC283⟩, 25, 31 ⟨84JCS(P2)1247⟩, 30 ⟨83CC140, 84JCS(P2)1239⟩
Atkins, P. R., **6**, 629 ⟨73JCS(P1)2644⟩
Atkins, R. L., **4**, 295, 407, 412 ⟨73JOC3288, 74JOC1349⟩, 434, 435, 440 ⟨84JOC5147⟩, 444 ⟨76JOC3221⟩
Atkins, Jr., G. M., **2**, 367 ⟨67JA2502⟩
Atkinson, B., **4**, 5 ⟨53JCS2083⟩
Atkinson, J., **4**, 175 ⟨91JOC1788⟩
Atkinson, J. G., **1**, 850 ⟨69CC1067⟩; **2**, 142 ⟨84JOC4538⟩
Atkinson, M. R., **5**, 849 ⟨54JCS3319⟩
Atkinson, R. E., **1**, 1006 ⟨67JCS(C)2011, 67JCS(C)578⟩, 1006 ⟨69JCS(C)2173⟩; **2**, 872, 880 ⟨84CHEC(1)539⟩; **3**, 78 ⟨67CC718, 69JCS2173⟩
Atkinson, R. F., **1**, 777 ⟨81JOC2804⟩, 778 ⟨89JOC389⟩; **6**, 608 ⟨73JA2982⟩
Atkinson, R. H., **6**, 411, 412 ⟨20JCS1410⟩
Atkinson, R. S., **2**, 373, 387, 402, 404 ⟨79COC(2)219⟩, 642 ⟨92CC1256⟩, 769 ⟨79COC(2)224⟩, 772 ⟨79COC(2)233⟩, 773 ⟨79COC(2)236⟩, 776, 777 ⟨79COC(2)267, 79COC(2)268⟩; **4**, 515 ⟨89TL2703⟩
Atoh, A., **3**, 290, 409 ⟨86BCJ677⟩
Atoh, M., **2**, 437 ⟨85BCJ2793, 90BCJ2051⟩
Atovmyan, L. O., **2**, 1003 ⟨86IZV134⟩; **5**, 685 ⟨90JOU1408, 90ZOR1629⟩, 1070 ⟨87IZV1829⟩
Atrees, S. S., **2**, 856 ⟨93T1271⟩
Atsumi, K., **1**, 980 ⟨79JA2208⟩; **2**, 921 ⟨77TL2001⟩, 1091 ⟨77CC76⟩; **3**, 230 ⟨79JA2208⟩; **5**, 236 ⟨74CL187⟩
Atta, A. H., **5**, 690 ⟨90S704⟩
Atta, F. M., **4**, 927 ⟨86CB472⟩
Atta, S. M. S., **5**, 690 ⟨85LA1910⟩
Attaby, F. A., **5**, 615 ⟨87SUL105⟩
Attah-Poku, S. K., **3**, 124 ⟨79TL4121⟩
Attanasi, O. A., **2**, 770 ⟨86OPP299⟩; **5**, 354 ⟨87T4249⟩
Attanasi, O., **3**, 618 ⟨83S740⟩; **5**, 367 ⟨85OPP385⟩
Attar-Bashi, M. T., **4**, 136 ⟨82LA1946⟩
Attardo, G., **3**, 89, 96 ⟨93TL3519⟩
Attaway, J. A., **5**, 808 ⟨59JA3599⟩; **6**, 257 ⟨59JA3599⟩
Attenberger, P., **1**, 1012 ⟨88TL3651⟩
Attenburrow, J., **3**, 229 ⟨52JCS1094⟩
Attina, M., **2**, 614 ⟨83CC108⟩
Attisso, M., **5**, 772 ⟨83EJM413⟩
Attridge, C. J., **4**, 136 ⟨76JCS(D)694⟩, 605, 636 ⟨70JOM(21)103⟩
Attwood, S. V., **1**, 700, 703 ⟨86JOC4840⟩; **3**, 244 ⟨81CC556⟩; **4**, 172 ⟨81CC556⟩; **5**, 177 ⟨86JOC4840⟩
Atwa, A., **6**, 127, 128 ⟨90H(31)47⟩
Atwal, K. S., **3**, 21 ⟨82TL2073⟩; **4**, 905 ⟨90JMC1510⟩; **6**, 647 ⟨89TL7313⟩
Atwater, N. W., **1**, 881 ⟨60JA2847⟩
Atwell, L., **1**, 432 ⟨84JMC1575⟩
Atwell, W. H., **4**, 608 ⟨64JOM(2)277⟩; **5**, 419 ⟨73USP3772347⟩
Atwood, D. A., **2**, 895 ⟨92JOM(439)C33⟩
Atwood, J. D., **5**, 461 ⟨87JA5145⟩
Atwood, J. L., **1**, 666 ⟨90OM3086⟩; **2**, 473 ⟨92CC726⟩, 475 ⟨90AG(E)1150⟩, 512 ⟨87IC4065⟩, 564 ⟨81CC292⟩, 565 ⟨79JCS(D)54⟩, 566 ⟨77JA6645⟩, 866, 893 ⟨84IC2582⟩, 968 ⟨83OM1555⟩, 969 ⟨84JA1291⟩; **3**, 358 ⟨87JOC1874⟩, 519 ⟨84JOM(264)327⟩, 545 ⟨85JA3728⟩; **4**, 656 ⟨78CC1081, 78CC140⟩, 656, 663, 709, 724 ⟨81JCS(D)1593⟩, 702 ⟨79JCS(D)54⟩, 709 ⟨69JA2538⟩, 725, 726 ⟨79JCS(D)45⟩; **5**, 439 ⟨86JA1722⟩, 440 ⟨81CC706⟩, 443 ⟨84JCS(D)399⟩, 445 ⟨84JA3214⟩; **6**, 168 ⟨84IC2582⟩, 187 ⟨81JOM(212)C4⟩, 195 ⟨77JA6645, 78CC140, 81JAP(K)56125326⟩, 395 ⟨87JCS(D)747⟩, 522 ⟨79JA5079⟩, 603 ⟨88IC570⟩, 689 ⟨84JA7015⟩
Au, A., **4**, 69 ⟨75JA5957⟩
Au, A. T., **3**, 292 ⟨84SC749⟩, 292, 303, 311 ⟨84SC743⟩
Au, K. G., **1**, 520 ⟨84JA2443⟩
Au, M.-K., **1**, 709 ⟨78TL4269⟩
Aubé, J., **2**, 342, 354 ⟨91TL4431⟩
Aube, J., **3**, 410 ⟨92JA5466⟩; **5**, 129 ⟨94JA9009⟩
Aube, P., **4**, 406 ⟨88BSF1009⟩
Aubert, C., **1**, 362 ⟨93JOC4298⟩; **2**, 944 ⟨93JOC4298⟩
Aubert, F., **2**, 753 ⟨86TL6193⟩
Aubert, T., **3**, 76 ⟨89JCS(P1)2095⟩, 77 ⟨88SC1475⟩
Aubrey, D. W., **2**, 418, 419 ⟨62JCS1786⟩, 419, 533 ⟨59JCS2927, 62JCS4088⟩
Aubry, A., **5**, 22 ⟨91TL2765⟩, 327, 351 ⟨93JA8898⟩; **6**, 488 ⟨93JA8898⟩
Auburn, P. R., **6**, 472, 473 ⟨86IJ250⟩
Auchus, R. J., **1**, 803 ⟨81JA2443⟩
Audia, J. E., **1**, 374 ⟨84JOC5277⟩, 831 ⟨88TL1371, 89JA8231⟩
Audouin, M., **5**, 952 ⟨91CSR503⟩; **6**, 716 ⟨91CSR503⟩
Audrain, H., **1**, 818 ⟨92TL4563⟩
Audrieth, J. F., **2**, 366 ⟨44JOC89⟩
Audrieth, L. F., **2**, 126 ⟨30JA4500⟩, 374 ⟨54JA4869⟩, 374, 379 ⟨55JA3131, 55JA790⟩, 388 ⟨58JA2516⟩, 412, 413 ⟨42JA1337, 42JA1553⟩, 691 ⟨57IS122⟩, 861 ⟨57JOC265⟩; **5**, 275 ⟨55OSC(3)516⟩, 298 ⟨38JA579⟩
Aue, D. H., **1**, 6 ⟨68JA7271⟩, 811 ⟨86JA5901⟩, 1109 ⟨76JA6700⟩; **4**, 829 ⟨75JOC2552⟩; **6**, 107 ⟨73TL1807⟩
Auerbach, J., **5**, 273, 296 ⟨76JOC725⟩
Auerbach, M., **4**, 332 ⟨70JOC2551⟩
Auerbach, R. A., **2**, 675 ⟨73JOC514⟩
Auf Dem Keller, H. H., **3**, 280 ⟨71AP(304)543⟩
Aufderhaar, E., **6**, 107 ⟨66AG(E)875⟩
Aufranc, P., **2**, 309 ⟨93TL4193⟩
Aufrand, M., **3**, 226 ⟨91TL1433⟩
Augé, J., **1**, 826 ⟨90TL4147⟩; **3**, 192 ⟨92TL8073⟩; **4**, 603, 609 ⟨88SC85⟩
Augenstein, L. L., **6**, 51 ⟨71JOC3627⟩
Auger, J., **4**, 1065 ⟨77JOM(140)237⟩
Auger, V., **2**, 495 ⟨03CR(137)925⟩, 506 ⟨06CR(142)1151⟩, 507 ⟨04CR(138)1705⟩
Augstein, J., **6**, 642 ⟨65JMC446⟩
August, J., **3**, 579 ⟨89JCS(P2)1841⟩
Augustin, M., **3**, 251 ⟨77S472⟩, 265 ⟨77JPR545⟩; **4**, 837 ⟨84T381⟩, 899 ⟨79JPR699⟩, 900 ⟨77ZC289, 78ZC345⟩, 969 ⟨79T551⟩, 970 ⟨77JPR545⟩, 983 ⟨78JPR585⟩; **6**, 564 ⟨89M871⟩, 565 ⟨84JPR633⟩
Augustine, R. L., **1**, 71, 87 ⟨B-65MI 103-01⟩, 73 ⟨70CC495, 70CC497⟩, 73 ⟨70CC571⟩, 502 ⟨B-65MI 112-01⟩, 884 ⟨67JOC2257⟩
Augustín, J., **5**, 1022, 1038 ⟨B-77MI 527-01⟩
Augy-Dorey, S., **1**, 828 ⟨93CC960⟩
Aulmich, G., **1**, 922 ⟨78TL4871⟩
Aumann, R., **1**, 500 ⟨87AG(E)357, 91CB2335⟩; **3**, 233 ⟨87CB23⟩, 556, 596, 597 ⟨68CB954, 88AG(E)1456⟩, 597 ⟨78CB1223, 85CB952⟩, 597 ⟨86CB3141, 87CB1939⟩,

597 ⟨88CB1085, 89CB1139⟩, 597 ⟨90CB2053, 93CB1867⟩
Aumann, R., **5**, 931 ⟨88AG(E)1456⟩, 946, 957 ⟨87CB537⟩, 946, 957 ⟨94OM3510⟩, 950, 951 ⟨90CB2053⟩, 953 ⟨86CB3801⟩, 953 ⟨95OM2447⟩, 954 ⟨87CB1297⟩, 955 ⟨92CB1141, 89CB365⟩, 957 ⟨90CB611, 90CB2047⟩, 957 ⟨92CB2765⟩, 958 ⟨94CB725, 94CB2475⟩, 958 ⟨94OM3502⟩; **6**, 719 ⟨69CB1495⟩
Aumüller, A., **6**, 304 ⟨88EUP293749⟩
Auner, N., **2**, 547 ⟨B-94MI 210-01⟩; **3**, 501 ⟨80JOM(188)151, 80JOM(197)13⟩, 502 ⟨91AG(E)444⟩; **4**, 567, 622, 623 ⟨83ZAAC(500)132⟩, 606 ⟨92OM1137⟩, 608, 610, 621 ⟨80JOM(188)151⟩, 611 ⟨88ZAAC(558)55, 89JOM(363)7, 90JOM(393)33, 92ZN(B)805, 93CB575⟩, 651 ⟨87JOM(336)59, 88ZAAC(558)87, 92JOM(426)1, 93OM4123, 93OM4135⟩; **5**, 925 ⟨92JA4910⟩, 926 ⟨92OM2494⟩; **6**, 174, 192 ⟨92ZN(B)805⟩, 659 ⟨92CC1274⟩
Auräth, B., **3**, 483 ⟨68TL4973⟩; **5**, 873 ⟨68TL4973⟩
Auret, B. J., **2**, 151 ⟨68JCS(C)2374⟩
Aurich, H. G., **2**, 343, 346, 355 ⟨B-82MI 206-02⟩; **3**, 431 ⟨86CB2756⟩, 432 ⟨85CB1105⟩, 577 ⟨74CB13⟩, 578 ⟨88JOC4997⟩; **6**, 445 ⟨80T3543⟩
Aurrecoechea, J. M., **1**, 1039 ⟨89JOC4072⟩; **4**, 392 ⟨87JOC844⟩
Aurrekoetxea, N., **3**, 632 ⟨90TL2205⟩
Austad, T., **2**, 858 ⟨73ACS1939⟩
Austenat, L., **2**, 887 ⟨70JPR10⟩
Austin, A. T., **2**, 88 ⟨60NAT1086⟩
Austin, D. J., **1**, 470 ⟨92JOC1161⟩; **2**, 629 ⟨89JCR(S)313⟩
Austin, G. T., **6**, 214, 215 ⟨74MI 607-01⟩
Austin, M., **4**, 700 ⟨87OM834⟩
Austin, T. E., **6**, 53 ⟨66IC488⟩, 284 ⟨68IC1647⟩, 284, 363, 616 ⟨72IC418⟩
Austin, W. B., **1**, 1137 ⟨81JOC2280⟩
Autenrieth-Ansorge, L., **5**, 131 ⟨87JOC3784⟩
Autenrieth, W., **5**, 184 ⟨01CB168⟩
Auterhoff, G., **4**, 406 ⟨71CB3354⟩
Auteri, S. C., **2**, 478 ⟨77AJC2479⟩
Autrey, R. L., **3**, 663 ⟨68JA4924⟩
Autze, V., **5**, 174 ⟨87AG(E)362⟩
Auvrey, P., **2**, 159 ⟨85TL4455⟩
Auwers, K., **3**, 452 ⟨25CB1974⟩; **5**, 592 ⟨1894CB1733⟩
Auyang, K., **6**, 647 ⟨78AF1435⟩
Au-Yeung, B. W., **2**, 436 ⟨90POL2901⟩
Auzou, G., **2**, 749 ⟨86S419⟩
Auzzi, G., **5**, 583 ⟨66G125⟩
Avadhani, C. V., **5**, 990 ⟨92MI 526-01⟩
Avakian, S., **5**, 276 ⟨40JOC606⟩
Åvall, A.-K. C., **4**, 553 ⟨90SC1865⟩
Avanesyan, K. V., **5**, 340 ⟨89ZOB611⟩, 1078 ⟨88ZOR1581⟩
Avasthi, K., **4**, 949 ⟨80CJC130⟩; **5**, 322 ⟨81JOC4188⟩
Avdonin, V. V., **6**, 250 ⟨93IZV2513⟩
Avendano, C., **5**, 584, 587 ⟨85H(23)649⟩, 773 ⟨82TL3357⟩
Avent, A. G., **6**, 381 ⟨92JOM(427)9⟩, 390 ⟨93JCS(D)3259⟩
Averbakh, K. O., **2**, 383 ⟨71ZOR258⟩
Averko-Antonovich, I. G., **2**, 25, 31 ⟨69JOU971⟩
Avery, M. A., **4**, 182 ⟨94TL969⟩, 359 ⟨89JOC1789⟩
Avetisyan, A. A., **6**, 25 ⟨91RCR1318⟩
Avetisyan, E. A., **3**, 560, 561, 588 ⟨73BAU1740⟩; **4**, 894 ⟨74IZV358⟩
Avignon-Tropis, M., **1**, 1009 ⟨89TL4951⟩, 1064 ⟨92JOC651⟩
Avila, L., **6**, 362 ⟨84OPP299⟩
Avni, E., **5**, 28 ⟨79MI 502-05⟩
Avram, M., **1**, 518 ⟨73TCC47⟩
Avramovitch, B., **4**, 759, 764 ⟨82JOC1397, 87JA6687⟩
Avrutov, I. M., **2**, 695 ⟨74JOM(78)115⟩
Avvakumova, L. V., **2**, 891 ⟨74JGU1506⟩
Awaad, M. A. A., **2**, 83, 553 ⟨90JOC1528⟩
Awad, S. B., **2**, 150 ⟨82JA5412⟩, 718 ⟨83JOC3071⟩, 959 ⟨77JOM(127)127⟩
Awaji, H., **6**, 124, 698 ⟨84TL4227, 83TL3469⟩
Awang, D. V. C., **5**, 320 ⟨67JA5287⟩, 813 ⟨71CJC1384⟩
Awang, M. R., **4**, 716 ⟨85JCS(D)2009⟩
Awano, Y., **6**, 533 ⟨87JAP01075462, 87JAP01102058, 90JAP04066566, 90JAP04128263, 90JAP05032616⟩
Awasthi, A. K., **1**, 900 ⟨90JA3228⟩; **3**, 163 ⟨89TL667⟩
Awata, N., **1**, 961 ⟨75CC362⟩
Awen, B. Z. E., **3**, 192 ⟨89CC841⟩, 410 ⟨90CL239⟩, 435 ⟨90T7471⟩
Axelrad, G., **2**, 619 ⟨81SC405⟩
Axelrod, E., **3**, 493 ⟨70JA2139⟩
Axelrod, E. H., **1**, 49, 51, 157 ⟨70JA2139⟩
Axelrod, J., **2**, 650 ⟨54JBC(208)731⟩
Axelrod, L. R., **3**, 219 ⟨65JCS1356⟩
Axen, R., **2**, 219 ⟨72BJ723⟩
Axen, U., **1**, 729 ⟨69CC303, 69JA5364⟩; **3**, 494 ⟨69CC303⟩
Axenrod, T., **4**, 429, 430, 431 ⟨93TL6677⟩
Axiotis, G. P., **3**, 40 ⟨80S889⟩
Axmacher, W., **2**, 504 ⟨80CB2928⟩
Ayabe, A., **3**, 145 ⟨92TL861⟩
Ayada, A., **5**, 1106 ⟨72CPB427⟩
Ayaguchi, Y., **1**, 505, 507 ⟨83BCJ353⟩, 515 ⟨85T2387⟩
Ayala, A. D., **1**, 950 ⟨88JOM(340)317⟩
Ayalon-Chass, D., **4**, 648 ⟨77CC772⟩
Ay, M., **2**, 305 ⟨89CR1947⟩; **3**, 181 ⟨89CRV1947⟩
Aycock, D. F., **5**, 244, 546 ⟨79JOC569⟩
Aydin, O., **6**, 488 ⟨92GEP4219418⟩
Ayed, A., **2**, 1022, 1023 ⟨88JPC37⟩
Ayer, W. A., **3**, 331, 341 ⟨60JA3800, 90T7599⟩
Ayers, T. A., **1**, 495 ⟨92TL5311⟩, 497 ⟨91TL3659⟩
Ayestarán, J., **6**, 563 ⟨91S147⟩
Aygen, S., **3**, 382 ⟨84CB187⟩
Ayi, A. I., **2**, 10 ⟨78TL4507, 81JFC(17)127⟩; **6**, 52, 54 ⟨81JFC(18)93⟩
Aylett, B. J., **2**, 416 ⟨67JCS(A)1918, 67JCS(A)652, 70CI(L)499⟩, 481 ⟨B-79MI 209-01⟩, 872 ⟨B-79MI 217-01⟩
Aylward, F., **1**, 74 ⟨62CI(L)484⟩
Aylward, J. B., **2**, 388 ⟨72JCS(P1)1692⟩; **3**, 681 ⟨69JCS(C)2587⟩; **5**, 356 ⟨69JCS(C)2587⟩, 356, 703, 845 ⟨65TL841⟩, 366 ⟨72JCS(D)869, 70JCS(C)968⟩, 999 ⟨71QR411⟩
Ayman, H. I., **6**, 630 ⟨89EGP270530⟩
Aymonino, P. J., **6**, 230 ⟨67MI 607-02, 68MI 607-01⟩, 424 ⟨65CC241⟩
Aynsley, E. E., **2**, 278 ⟨66CI(L)379⟩
Ayoko, G. A., **6**, 290, 381, 392 ⟨87JCS(P2)1047⟩, 380 ⟨86JCS(P2)1289, 87JCS(P2)381⟩, 394, 395 ⟨87JOM(320)137⟩
Ayotte, C., **3**, 37 ⟨90TL63⟩
Ayoubi, S. E., **1**, 160 ⟨93JOM(451)C1⟩
Äyräs, P., **4**, 213, 213 ⟨70TL4095⟩
Ayrebi, H., **4**, 581 ⟨90IC1674⟩
Ayres, A. M. J., **4**, 197 ⟨62JOC281⟩
Ayres, G. B., **5**, 89 ⟨32OSC(1)23⟩
Ayres, R. L., **2**, 35 ⟨71JA1389⟩
Ayrey, G., **2**, 287, 291 ⟨62JCS2089⟩
Ayrey, P. M., **1**, 643 ⟨92JCS(P1)3407⟩, 757 ⟨89TL4581⟩
Ayyangar, N. R., **1**, 898 ⟨89JA7659⟩; **2**, 42, 57, 519 ⟨64JA397⟩, 516, 517 ⟨63JA2072⟩, 716 ⟨92S825⟩; **3**, 283, 295, 296, 305 ⟨91S322⟩; **5**, 384 ⟨91SC1889⟩; **6**, 557 ⟨90OPP128⟩
Ayyar, K. S., **3**, 226 ⟨73CC161⟩
Azadani, M. N., **3**, 3, 114 ⟨86SC43⟩
Azam, K. A., **4**, 585, 691 ⟨82CC614⟩
Azarov, A. G., **4**, 861 ⟨92ZOR1711⟩
Azegami, I., **2**, 1091 ⟨77CC76⟩
Azerad, R., **1**, 8 ⟨92TL7337⟩
Azerbaev, I. N., **2**, 1076 ⟨74RCR657⟩
Azhayev, A., **4**, 222 ⟨91TL7593⟩
Aziz, M. A., **3**, 339, 361 ⟨85T1885⟩
Aznar, F., **1**, 68 ⟨88S144⟩, 833 ⟨84CC1427⟩; **2**, 676 ⟨88S144⟩, 740 ⟨86S960⟩, 744 ⟨84S304⟩, 1097

⟨83JCS(P1)1087, 88CC121⟩; **3**, 75 ⟨86CC1180⟩, 413 ⟨85CC1375⟩, 417 ⟨83JCS(P1)1093⟩
Aznur, F., **4**, 1067 ⟨83JCS(P1)1087⟩
Azoulay, M., **2**, 127 ⟨93JA2139⟩
Azran, J., **3**, 89, 95 ⟨90BSB345⟩
Azuhata, T., **2**, 34 ⟨83S461⟩, 457 ⟨83S916⟩
Azuma, K., **1**, 405, 816, 817 ⟨84T2303⟩, 405, 816, 817 ⟨84TL565⟩, 816 ⟨83CL1379⟩
Azuma, T., **1**, 442 ⟨87JCS(P1)1951⟩
Azuma, Y., **3**, 350 ⟨83CPB1929⟩
Azzaro, M., **2**, 744 ⟨81S880⟩; **3**, 263 ⟨81S880⟩
Azzem, M. A., **4**, 206 ⟨90H(31)1959⟩
Azzena, U., **1**, 182, 475 ⟨91JOC183⟩, 398 ⟨90JOC5532⟩, 432 ⟨90JOC5386⟩; **2**, 9 ⟨89TL4555⟩, 553 ⟨94SC591⟩; **4**, 179 ⟨90S313⟩
Azzenza, U., **1**, 432 ⟨89S28⟩

B

Baache, M., **2**, 428, 431 ⟨80CB1356⟩
Baake, H., **1**, 708 ⟨77CB2978⟩
Baar, B. von, **2**, 1022 ⟨86AG(E)282⟩
Baardman, F., **1**, 498 ⟨75AG(E)709⟩; **5**, 991 ⟨73S413⟩
Baas, J. L., **1**, 884 ⟨66T285⟩
Baasner, B., **6**, 238 ⟨82GEP3044216⟩, 437 ⟨83HOU(E4)1078⟩, 438, 439 ⟨80EUP52842, 82JFC(19)553⟩
Baba, A. B., **5**, 802, 803 ⟨89CL1603⟩
Baba, H., **4**, 3 ⟨66BCJ219⟩; **5**, 614 ⟨75JA1608⟩; **6**, 36 ⟨76BCJ1888⟩, 37 ⟨79JFC(15)353, 78JFC(12)1⟩, 39 ⟨83JFC(23)123⟩, 51 ⟨77JFC(9)279⟩, 230 ⟨90JFC(50)173⟩, 257 ⟨90JFC(48)257⟩, 409 ⟨70JAP7026611⟩, 410 ⟨76JAP7630034⟩
Baba, S., **1**, 468 ⟨88OS(66)60⟩, 468, 533, 1106 ⟨76JA6729⟩, 468, 533, 536 ⟨87JA2393⟩, 471, 533 ⟨76CC596⟩, 479, 534, 1106 ⟨76CC17⟩, 534 ⟨76TL1927⟩, 1003, 1021 ⟨75JA7385⟩; **2**, 992 ⟨87JA2393⟩
Baba, T., **3**, 135 ⟨83SC367⟩; **6**, 634 ⟨83SC367⟩
Baba, Y., **1**, 2, 3, 8 ⟨78JA1786⟩, 458, 516 ⟨91TL5559⟩; **4**, 820 ⟨86CC1794⟩
Babad, H., **5**, 965 ⟨73CRV75⟩; **6**, 460 ⟨73CRV75⟩
Babaeva, T. A., **4**, 884, 913 ⟨77CA135599v, 76ZOB1652⟩
Babasheva, K. K., **2**, 1033 ⟨83IZV1662, 84DOK(274)822⟩
Babayan, A. T., **1**, 1040 ⟨86AKZ306⟩
Babayan, K. N., **1**, 986 ⟨84AKZ116⟩
Babb, D. P., **3**, 456 ⟨71MI 312-02⟩, 456, 458, 459 ⟨72IC242⟩; **6**, 253 ⟨79JA5949⟩
Babb, R. M., **2**, 219 ⟨61JOC1152⟩
Babers, F. H., **2**, 393 ⟨43OSC(2)512⟩
Babiak, K. A., **1**, 120 ⟨88JA2641⟩; **2**, 610 ⟨77JOC3173⟩, 992 ⟨88JA2641⟩; **3**, 140 ⟨90JA7441⟩; **5**, 557 ⟨83JA5015⟩
Babič, K., **5**, 764 ⟨76JHC487⟩
Babich, E. D., **4**, 622 ⟨69DOK(189)334⟩, 658 ⟨75IZV1600⟩
Babichev, F. S., **6**, 104 ⟨66UKZ204⟩
Babin, F., **2**, 767 ⟨85TL4443⟩
Babin, J. E., **2**, 327 ⟨88OM2250⟩
Babin, P., **5**, 373 ⟨87BSF997⟩
Babine, R. E., **1**, 206 ⟨86TL5791⟩
Babirad, S. A., **1**, 978 ⟨87JOC1370⟩, 1099 ⟨87JOC4825⟩
Babkina, E. I., **2**, 448 ⟨72ZOB1285⟩; **4**, 561 ⟨64ZOB2897⟩
Babler, J. H., **1**, 600, 788 ⟨82JOC4801⟩, 660 ⟨70TL3861⟩; **2**, 31 ⟨84SC1313⟩, 554 ⟨84TL4323⟩, 753 ⟨81SC925⟩; **3**, 13, 83 ⟨82SC839, 81TL11⟩, 642 ⟨72TL669⟩
Babot, O., **2**, 699 ⟨87T2075⟩; **4**, 624 ⟨88JOM(341)125⟩
Baboulene, M., **1**, 531, 662 ⟨82T2355⟩; **2**, 457 ⟨79PS(7)101⟩; **3**, 44 ⟨90SC2091⟩
Babston, R. E., **2**, 699 ⟨84TL699⟩; **4**, 831 ⟨84TL699⟩
Babudri, F., **1**, 1061 ⟨86S638⟩; **2**, 740 ⟨91S225⟩; **3**, 230 ⟨91CC237⟩; **4**, 321 ⟨86S638⟩, 476 ⟨91S225⟩
Baburina, V. A., **3**, 336 ⟨79ZOB1084⟩; **5**, 538, 539 ⟨79JGU940⟩
Bacaloglu, I., **5**, 1126 ⟨76JCS(P2)524⟩
Bacaloglu, R., **5**, 1126 ⟨76JCS(P2)524⟩
Bac, N. V., **1**, 626 ⟨90TL1865⟩, 626, 630, 655, 789 ⟨82JA7666⟩
Baccar, B., 22 ⟨92PS(69)163⟩, **5**, 690 ⟨92BSB445⟩, 700, 701 ⟨92SC1985⟩, 705 ⟨78CR(C)69⟩; **6**, 558 ⟨86MI 617-02⟩
Baccaro, H. R., **5**, 640 ⟨61ACS1109⟩
Bacchetti, T., **5**, 476, 477 ⟨57MI 511-01, 57MI 511-02⟩, 831 ⟨58MI 521-01⟩
Baccolini, G., **3**, 480 ⟨75S44⟩, 618 ⟨73JOC1060⟩; **6**, 121 ⟨90JCS(P1)19⟩
Baceiredo, A., **2**, 1067 ⟨88JA6463⟩; **3**, 485 ⟨91AG1174, 91AG(E)871⟩
Bacha, J. D., **1**, 561 ⟨68T2215⟩
Bach, H., **3**, 468 ⟨32CB265⟩
Bach, J., **1**, 2, 6 ⟨85JOC3408⟩; **3**, 621 ⟨85JOC3408⟩
Bachand, C., **5**, 236 ⟨88CJC1537⟩, 316, 318 ⟨74JOC3136⟩
Bachand, P. C., **1**, 990 ⟨91CC566⟩; **4**, 746, 748, 760, 817, 821 ⟨93JA5430⟩
Bach, R. D., **1**, 66, 69, 454 ⟨82JOC3707⟩, 387 ⟨78JA1605⟩, 388 ⟨85TL985⟩, 624, 629 ⟨71JA7118⟩, 624, 629 ⟨72JA8608⟩, 624, 782 ⟨73JOC1742⟩, 628 ⟨73OSC(5)315, 74CC819⟩, 629 ⟨74CC820, 79TL3815⟩, 629 ⟨91JA4712⟩; **4**, 206 ⟨82JOC3707⟩; **5**, 282 ⟨80JOC165⟩, 709 ⟨83JOC888⟩
Bach, R. O., **6**, 172, 204 ⟨B-85MI 606-02⟩
Bach, S. R., **5**, 166 ⟨58JA3079⟩
Bacheiredo, A., **3**, 689 ⟨88JA6463⟩
Bachhawat, J. M., **2**, 45 ⟨71TL691⟩
Bachi, M. D., **1**, 547 ⟨82TL2505⟩, 548, 1131 ⟨86TL641⟩, 938 ⟨88JCS(P1)1517, 89JOC1234⟩, 938 ⟨92JOC6803⟩, 1016 ⟨91JOC3912, 87TL2637⟩, 1131 ⟨87TL2887⟩; **2**, 166 ⟨80JCS(P1)2228⟩, 230 ⟨72JOC3550⟩, 1020 ⟨91JOC3912⟩; **5**, 513 ⟨89JOC1234, 92JOC6803⟩, 593 ⟨90JOC3442⟩, 731 ⟨72JCS(P1)2326⟩, 739 ⟨89JA1886⟩; **6**, 543, 550, 557 ⟨92JOC6803⟩, 548 ⟨88JCS(P1)1517⟩
Bachinskii, T. P., **2**, 256 ⟨80ZOB2459⟩
Bachman, G. B., **1**, 156 ⟨54JA3972⟩; **4**, 18 ⟨34JA2730⟩, 23, 38 ⟨35JA1088⟩; **5**, 1125 ⟨47JA365⟩
Bachman, G. L., **1**, 76 ⟨77JA5946⟩
Bachmann, B., **4**, 672, 673 ⟨90ZN(B)221⟩
Bachmann, G. B., **2**, 397 ⟨68JOC313, 69JOC4121⟩, 397 ⟨70JOC4229⟩, 398 ⟨72JOC2810⟩, 614 ⟨47JA2022⟩, 800 ⟨58JOC1800⟩
Bachmann, P., **2**, 507 ⟨82ZAAC(485)129⟩, 508 ⟨81AG(E)123⟩
Bachmann, W. E., **1**, 393 ⟨42OR(1)38⟩; **2**, 125 ⟨33JA2932⟩; **3**, 218 ⟨53JA3268⟩, 474, 475 ⟨42OR(1)38⟩; **4**, 434, 435 ⟨51JA2773⟩, 434, 435, 443 ⟨51JA2777⟩, 435, 443 ⟨51JA2769⟩; **5**, 276 ⟨42OR(1)38⟩
Bachovchin, W. W., **4**, 527, 528 ⟨90JBC(265)3738⟩
Baciocchi, E., **1**, 290 ⟨92CC59⟩, 926 ⟨93JOC7610⟩; **2**, 14 ⟨94TL2763⟩, 21, 28 ⟨88T6525⟩; **4**, 138 ⟨90AG(E)658⟩
Back, T. G., **1**, 25, 440 ⟨84CC1417⟩, 248 ⟨87JOC4265⟩, 436 ⟨90CC819⟩, 439, 545 ⟨86TL2187, 88JOC3815⟩, 441 ⟨92PS(67)203⟩, 486, 545 ⟨87JOC4258⟩, 601, 603, 607 ⟨B-87MI 114-02⟩, 601 ⟨B-87MI 114-01⟩, 605 ⟨81TL5111, 83JOC4776, 87TL1737, 89JOC4146⟩, 609 ⟨92TL4725⟩, 685, 686 ⟨75CC539, 76JCS(P1)2079⟩, 712, 713, 715 ⟨B-87MI 115-01⟩; **2**, 217 ⟨70TL3551, 71JOC3828⟩, 238 ⟨80TL2213⟩, 241, 358 ⟨71TL4953⟩, 280 ⟨B-87MI 204-01⟩, 285 ⟨80TL2215, 81JOC3249, 81TL5111, 83JOC4776⟩, 285, 1034, 1035 ⟨83JOC3077⟩, 338 ⟨91CJC1482⟩, 363 ⟨72TL5313⟩, 363, 688 ⟨73JOC4328⟩, 388 ⟨82CJC2711⟩, 743 ⟨85CJC2313⟩, 858 ⟨75CC539, 76JCS(P1)2079⟩; **3**, 38 ⟨B-87MI 301-03⟩, 88 ⟨78CC952⟩, 171 ⟨83JOC3077⟩, 173 ⟨B-87MI 304-04⟩, 174 ⟨85T4759⟩, 391 ⟨75CC539⟩, 391, 527 ⟨76JCS(P1)2079⟩, 637 ⟨81JOC3249⟩; **4**, 290, 335 ⟨85JOM(286)171⟩, 874 ⟨85CJC2313⟩, 947 ⟨83JOC3077⟩, 953 ⟨87JOC4258⟩; **5**, 315 ⟨94H(38)159⟩, 316 ⟨91H(32)481⟩, 335 ⟨87CC1233, 89JOC1904⟩, 335, 338 ⟨86CJC308⟩, 342 ⟨72TL5313, 73JOC4328⟩, 1053 ⟨76JCS(P1)2079⟩, 1056, 1108 ⟨85JOM(286)171⟩
Backeberg, O. G., **2**, 94 ⟨23JCS2999⟩
Backenstrass, F., **5**, 330 ⟨90TL2139⟩
Backer, H. J., **2**, 126, 127 ⟨50RTC638⟩, 226 ⟨53RTC612⟩,

251 ⟨52RTC1228⟩, 364 ⟨28RTC942, 51RTC254⟩, 723 ⟨54RTC129⟩, 724 ⟨53BSB3⟩, 1034 ⟨54RTC565⟩; **3**, 467 ⟨54RTC229, 63OSC(4)250⟩, 566 ⟨54RTC575⟩; **4**, 38 ⟨28JCS2125⟩, 83 ⟨48RTC451⟩, 770, 776 ⟨53RTC813⟩; **5**, 749 ⟨51RTC638⟩; **6**, 81, 91 ⟨33RTC437⟩, 90 ⟨48RTC884, 30RTC1107⟩, 91, 275 ⟨46RTC53⟩, 114 ⟨50RTC1223, 51RTC733, 51RTC892⟩, 301 ⟨52RTC1071⟩, 306 ⟨33RTC923⟩, 306, 307 ⟨33RTC1039, 48RTC894⟩, 307 ⟨52RTC409⟩
Backhaus, P., **4**, 305, 311 ⟨75LA1790⟩
Bäcklund, B., **1**, 617 ⟨40MI 114-01⟩, 704, 706 ⟨40AK(13A)1⟩
Backlund, S. J., **1**, 488, 1103 ⟨78JOM(156)159⟩, 977, 1056 ⟨78JA5561⟩; **2**, 901 ⟨77JA5192⟩; **4**, 814 ⟨77JA5192⟩, 1056 ⟨77JA3184⟩; **5**, 417 ⟨77JA3184⟩
Backmann, M., **5**, 800 ⟨87JOM(332)371⟩
Bäckvall, J.-E., **1**, 187 ⟨90JA6615⟩, 192 ⟨93JA6609⟩, 193 ⟨89TL617⟩, 237, 238 ⟨87JA6396⟩, 595 ⟨81TL1919, 83JOC3918⟩; **2**, 289 ⟨81TL1919⟩, 301 ⟨83TL2745⟩, 312 ⟨92CC980⟩, 392 ⟨90TL4199⟩, 395 ⟨92TL131⟩, 113 ⟨88TL2885⟩; **3**, 119, 294 ⟨92CC285⟩, 122 ⟨93TL5459⟩, 192 ⟨92TL5641⟩, 614 ⟨81CC1098⟩
Bacon, C. C., **2**, 164, 401, 402 ⟨79JOC2055⟩
Bacon, N., **5**, 725 ⟨65JCS5230⟩
Bacon, R. G., **2**, 361 ⟨60JCS5079⟩
Bacon, R. G. R., **2**, 753 ⟨69CC578⟩; **5**, 1105 ⟨61JCS2447⟩, 1105, 1110 ⟨61JCS2428⟩, 1105, 1110 ⟨61JCS2436⟩; **6**, 607, 609 ⟨58JCS764⟩
Bacon, S. W., **6**, 462 ⟨84JOC1122⟩
Bacos, D., **3**, 411 ⟨87TL2353⟩
Bacquet, C., **1**, 1067 ⟨74BSF(2)1731⟩; **2**, 420 ⟨75JOM(97)325⟩; **3**, 210 ⟨81CJC945⟩; **4**, 22 ⟨75BSF1797⟩
Bacso, I., **1**, 961 ⟨74S344⟩
Badaoui, E., **1**, 713 ⟨89T1579, 89T2023⟩; **2**, 281 ⟨89T2023⟩, 478 ⟨89T1859⟩; **4**, 284 ⟨89T2023⟩, 400 ⟨89T1859⟩
Badar, Y., **1**, 725 ⟨73JCS(P1)1416⟩
Baddar, F. G., **1**, 181 ⟨63JCS4522⟩
Baddeley, G., **1**, 422 ⟨53JCS124⟩; **3**, 295 ⟨56JCS4943⟩
Baddiley, J., **2**, 127 ⟨52JCS800⟩
Baddley, W. H., **3**, 568 ⟨72JOM(39)217⟩, 570 ⟨70JA3489, 72JOM(35)403⟩; **5**, 465 ⟨72CC672⟩
Badea, F., **4**, 152 ⟨92RRC393⟩
Badejo, I., **4**, 353 ⟨91JOC5213⟩
Bader, A., **1**, 810 ⟨68FRP1519701⟩; **2**, 438 ⟨92JCS(D)1751⟩
Bäder, E., **4**, 234 ⟨54CB129⟩, 235 ⟨54CB784⟩
Bader, E., **4**, 331 ⟨55CB41⟩
Bader, F. E., **5**, 164 ⟨58T1⟩
Bader, J., **2**, 440, 442, 446 ⟨62LA(659)49⟩; **4**, 112 ⟨62LA(659)49⟩; **5**, 901 ⟨61TL724⟩
Bader, R. F. W., **1**, 776 ⟨61CJC348⟩, 883 ⟨61JA1951⟩
Badet, B., **1**, 24 ⟨76BSF525⟩, 623 ⟨88T2913⟩; **3**, 56 ⟨75BSF1363⟩
Badger, G. M., **3**, 293 ⟨41JCS535⟩
Badia, C., **2**, 1010 ⟨71JA2215⟩
Badia, M. C., **2**, 456 ⟨86TL1751⟩
Badia, M. D., **2**, 89 ⟨88T6677⟩
Badiger, V. V., **3**, 179 ⟨82JHC69⟩
Badika, N., **3**, 119 ⟨92JOC1600⟩
Badley, E. M., **6**, 721 ⟨69CC1322, 71JCS(A)21⟩
Badone, D., **2**, 9 ⟨87S920⟩
Badoni, E., **2**, 554 ⟨92TL3381⟩
Badoud, R., **1**, 218, 831 ⟨83JA1988⟩
Badovskaya, L. A., **5**, 350 ⟨86CC607, 92ZOR568⟩
Badrudin, S. P., **4**, 485 ⟨91TL2125⟩
Badrutdinov, Sh. Kh., **2**, 885, 890 ⟨87JGU2496⟩, 889 ⟨89JGU2254⟩, 891 ⟨91JGU405⟩; **6**, 564 ⟨89ZOB2520⟩
Bae, S. K., **3**, 88 ⟨84S866⟩
Baeckström, P., **1**, 828 ⟨90SC423⟩
Baehler, B., **2**, 456 ⟨61HCA1812⟩
Baens, N. P., **2**, 23 ⟨93T3193⟩
Baenziger, N. C., **1**, 898 ⟨67JA3405⟩, 990 ⟨91CC566⟩
Baer, D. R., **1**, 389 ⟨60OR(11)157⟩
Baer, E., **2**, 693 ⟨59JBC(234)1⟩
Baer, H. H., **2**, 398 ⟨65CJC3074, 65CJC840⟩, 784, 785, 787 ⟨B-70MI 215-02⟩; **4**, 176 ⟨88CJC187⟩, 540 ⟨70MI 411-02⟩
Baerends, E. J., **2**, 1003 ⟨84T765⟩
Baerhausen, D., **5**, 386 ⟨91JOM(403)401⟩
Baeyer, A., **1**, 897 ⟨02CB3031⟩; **2**, 492 ⟨58LA(617)257⟩, 779 ⟨1874CB809⟩
Baeza, J., **1**, 681, 682, 785 ⟨72JA2000⟩; **5**, 161 ⟨72JA2000⟩
Bag, A. K., **5**, 107 ⟨86IJC433⟩
Bagal, L. I., **5**, 270, 283 ⟨71QR411⟩; **6**, 145 ⟨71ZOR1304⟩, 145 ⟨72KGS713⟩
Baganz, H., **2**, 166, 167 ⟨60CB782⟩; **4**, 198 ⟨66AG(E)420⟩, 979 ⟨62CB2095⟩; **6**, 39 ⟨59CB3170⟩, 41, 46 ⟨58CB806⟩, 79 ⟨58CB650⟩
Bagby, M. O., **1**, 955 ⟨65JOC2983⟩
Baggaley, K. H., **4**, 308 ⟨75JCS(P1)1670⟩; **5**, 334 ⟨85JMC1661⟩
Baggett, N., **2**, 478 ⟨91COC(1)819⟩
Baggio, R., **1**, 926 ⟨93TL1701⟩
Baggiolini, E. G., **1**, 576, 1021 ⟨86JOC3098⟩, 697, 700, 703 ⟨88JOC4282⟩, 749, 828 ⟨87JOC1372⟩
Baggott, J., **4**, 1079 ⟨B-91MI 424-01⟩
Baghdanov, V. M., **2**, 393 ⟨88JOC2872⟩
Bagheri, V., **1**, 400, 408, 409, 822, 823, 825 ⟨81JOC5094⟩, 464 ⟨88JA5383⟩, 544 ⟨84JA6105⟩, 820 ⟨88TL5119⟩, 1045, 1047 ⟨83TL5181⟩, 1070, 1094 ⟨87TL5793⟩; **2**, 30 ⟨87TL5793⟩, 539 ⟨90JOC6082⟩, 926, 931, 973, 976 ⟨84JA6105⟩; **3**, 130, 288 ⟨83TL5181⟩, 203 ⟨88JOC6158⟩; **4**, 809 ⟨88JA5383⟩, 1048, 1062 ⟨84JA6105⟩
Bagli, J. F., **4**, 323 ⟨79CJC1153⟩
Bagnall, R. D., **4**, 3 ⟨73JFC(3)329⟩, 4 ⟨79JFC(13)325⟩
Bagnell, Jr., J. J., **4**, 846 ⟨61JOC1987⟩; **6**, 509 ⟨63JOC991⟩
Bagno, A., **4**, 1075 ⟨87MI 424-01⟩; **5**, 322 ⟨94JA916⟩
Bagnoli, N., **3**, 119 ⟨92JOC1600⟩
Bagrov, F. V., **4**, 588 ⟨78ZOB1422, 79ZOB1772⟩
Bahadir, M., **5**, 1037 ⟨79ZN(B)768⟩
Bahar, E., **3**, 431 ⟨89IJ187⟩
Bahari, Z. B., **4**, 540 ⟨82JCS(P1)143⟩
Bahir-Hashemi, A., **4**, 94 ⟨84JOC1084⟩
Bahn, C. A., **1**, 1110, 1117 ⟨71JA1513⟩
Bahn, H., **6**, 312 ⟨73ZC465⟩
Bahner, C. T., **6**, 46 ⟨50JA3642⟩
Bähr, G., **2**, 563 ⟨73HOU(13/2a)4, 73HOU(13/2a)533⟩, 597 ⟨70HOU(13/4)315, 70HOU(13/4)345, 70HOU(13/4)363⟩, 599 ⟨78HOU(13/6)181⟩, 601 ⟨75HOU(13/7)1⟩, 1098 ⟨70HOU(13/4)154⟩, 1101 ⟨75HOU(13/7)73⟩; **6**, 172 ⟨75HOU(13/7)1⟩, 172, 204 ⟨78HOU(13/6)181⟩
Bahrmann, H., **1**, 202 ⟨B-80MI 105-02⟩
Bahry, M., **5**, 560 ⟨77TL4607⟩
Bahsoun, A. A., **6**, 155 ⟨82OM1114⟩
Bai, L. I., **5**, 1018 ⟨64ZOB3609⟩
Baibulatova, N. Z., **3**, 336 ⟨91URP1680699, 92IZV160⟩, 336 ⟨92URP1710560⟩; **5**, 603 ⟨89IZV1319⟩
Baier, H., **4**, 1026, 1030 ⟨75TL3171⟩; **5**, 148 ⟨80AG(E)1029⟩
Baig, M. I., **5**, 360, 361 ⟨67CJC1539⟩
Baigrie, B. D., **2**, 777 ⟨75JCS(P1)1065⟩
Bailer, G., **3**, 635 ⟨91S1205⟩
Bäiler, T., **1**, 1114 ⟨74CB1676⟩
Bailey, A., **4**, 876 ⟨87SM647⟩
Bailey, A. B., **4**, 876 ⟨88SMB425⟩
Bailey, A. R., **1**, 429 ⟨89JFC(44)1637⟩
Bailey, D., **3**, 280 ⟨71CI(L)682⟩
Bailey, D. L., **1**, 450 ⟨54JA1613⟩, 652 ⟨48JA2869⟩; **5**, 421 ⟨54JA1613⟩
Bailey, D. S., **2**, 161 ⟨72JA208⟩
Bailey, G. C., **1**, 674 ⟨64IEC170⟩
Bailey, J. H., **2**, 212, 232 ⟨47JA1710⟩
Bailey, J. H. E., **2**, 239 ⟨93CJC42⟩

Bailey, J. R., **2**, 388, 403 ⟨22JA2556⟩
Bailey, Jr., W. C., **5**, 168 ⟨68JOC2819⟩
Bailey, Jr., W. I., **4**, 673 ⟨78JA802⟩, 674 ⟨78JA5764⟩
Bailey, K., **5**, 83 ⟨51MI 502-02⟩, 85 ⟨B-52MI 502-09⟩
Bailey, N. A., **2**, 202 ⟨93CC1195⟩; **4**, 580 ⟨91JCS(D)1117⟩, 687 ⟨64PCS401, 68JCS(A)1293⟩; **5**, 387, 388 ⟨86JCS(D)813⟩, 943 ⟨82CC215, 90OM2621⟩, 944 ⟨83JCS(D)2397⟩, 955 ⟨95JOM(494)C12⟩; **6**, 247 ⟨91CC165⟩
Bailey, P. D., **2**, 326 ⟨88TL2231⟩; **5**, 263 ⟨93JCS(P1)451⟩
Bailey, P. L., **1**, 622 ⟨91TL3119⟩; **4**, 862 ⟨93JCS(P1)577⟩, 938 ⟨93JCS(P1)343⟩
Bailey, P. M., **2**, 585 ⟨81CC808⟩; **4**, 687 ⟨83JCS(D)1441⟩, 693 ⟨78JCS(D)1825⟩
Bailey, P. S., **1**, 45 ⟨46JA1813⟩; **2**, 348 ⟨58CRV925, 78JOC2662⟩, 782 ⟨60JA1801⟩; **3**, 326 ⟨B-78MI 307-01, B-82MI 307-01⟩; **5**, 28 ⟨B-78MI 502-01, B-82MI 502-01⟩, 115 ⟨73OSC(5)493⟩
Bailey, R. A., **3**, 712, 714, 715 ⟨68JA3761⟩
Bailey, R. B., **2**, 163 ⟨71JOC2207⟩, 793 ⟨76JOC2207⟩
Bailey, R. J., **2**, 1003 ⟨83JA6096⟩
Bailey, S. I., **4**, 647, 656 ⟨86JCS(D)603⟩
Bailey, T. R., **1**, 149 ⟨84JA3270⟩, 150 ⟨85JA7974⟩, 422 ⟨79TL1999, 80JOC3017⟩, 451 ⟨79JA4420⟩, 1136 ⟨85JOC1019⟩; **2**, 911 ⟨78JOC1620, 80JOC3017⟩, 1002 ⟨85JOC1019⟩; **3**, 126 ⟨80JOC3028⟩, 182 ⟨85JOC1019⟩, 303 ⟨91S243⟩; **4**, 357 ⟨78JOC1620, 80JOC3028⟩, 530 ⟨85JOC1019⟩, 532 ⟨86JOC872⟩
Bailey, W. F., **1**, 316 ⟨87JA2442, 92JA8053, 94T5957⟩; **2**, 68 ⟨89TL3901⟩, 112 ⟨94JA6577⟩, 554 ⟨90JOC5404⟩, 556 ⟨89JA765, 91JA5720, 92JA8053, 94T5957⟩, 972 ⟨90TL627⟩; **4**, 212 ⟨84JOC4958⟩; **6**, 466 ⟨82JA1769⟩
Bailey, W. J., **1**, 51 ⟨57JA3567⟩, 521 ⟨55JA73⟩, 637 ⟨58JOC996⟩, 637, 773, 782 ⟨59JA651⟩, 772 ⟨78JOC1421⟩, 773 ⟨77JOC3895⟩, 774 ⟨59JA2126⟩, 777 ⟨77JOC3899, 91JOC846⟩, 958 ⟨55JOC1337, 55JOC95⟩; **2**, 442 ⟨57JA3567⟩, 442, 444 ⟨60JOC1996⟩, 1099 ⟨55JOC1337⟩; **4**, 826 ⟨91TL1539⟩
Baillargeon, M., **4**, 360, 382, 388 ⟨78JOC4255⟩
Baillargeon, M. M., **2**, 4, 23, 30 ⟨93JA3071⟩
Baillargeon, V. P., **3**, 23, 89, 94, 95, 97, 103 ⟨86JA452⟩, 58, 89 ⟨83JA7175⟩
Baina, N. F., **6**, 58 ⟨59JGU1115⟩
Bain, J. D., **2**, 409, 410 ⟨91JOC4608⟩
Bain, S., **4**, 607, 621 ⟨88JA2611⟩
Baine, N. H., **1**, 320, 322, 1127, 1129, 1130 ⟨82JA2321⟩, 1129 ⟨85TL5927⟩; **2**, 689 ⟨94JOC6683⟩
Baines, K. M., **1**, 924 ⟨89T951⟩; **3**, 92 ⟨89T951⟩, 501, 502 ⟨86AOC(25)1⟩; **5**, 418, 923 ⟨86AOC(25)1⟩, 418, 924, 925 ⟨85JA4338, 89OM693, 89OM709⟩
Bains, S., **3**, 27 ⟨90TL179⟩
Bainton, H. P., **1**, 1091 ⟨81TL1437⟩; **3**, 58, 273 ⟨81TL1437⟩
Baiocchi, L., **5**, 681, 771 ⟨83JHC225⟩
Bairakova, R. S., **2**, 359 ⟨81ZOB609⟩
Bair, K. W., **1**, 624, 629 ⟨72JA8608, 74CC819⟩, 629 ⟨74CC820⟩; **3**, 618 ⟨76TL603⟩; **6**, 235 ⟨77JOC3875⟩
Bairamov, A. A., **2**, 136 ⟨90KGS1477⟩
Baird, G. J., **5**, 453 ⟨83CC1202, 83JOM(248)C1⟩, 453 ⟨84JOM(262)49⟩
Baird, Jr., W. C., **2**, 23, 24, 30 ⟨71JOC3324⟩, 24 ⟨71JOC2088⟩, 631 ⟨70JOC3436⟩; **4**, 14 ⟨71JOC3324⟩
Baird, Jr., W. K., **1**, 72 ⟨72JOC1182⟩
Baird, M., **1**, 985 ⟨92SL635⟩
Baird, M. C., **1**, 669 ⟨78TL3307⟩; **2**, 478 ⟨67JCS(A)1347⟩; **5**, 464 ⟨67JCS(A)865⟩
Baird, M. S., **1**, 549 ⟨88TCC137⟩, 811 ⟨70CC1519⟩, 990 ⟨84TL2887⟩; **2**, 449 ⟨91SL433⟩; **3**, 271 ⟨86TL5143⟩; **4**, 29 ⟨76JCS(P1)54⟩, 299, 406 ⟨70TL1729⟩, 742 ⟨82TL3795, 85JCR(S)182, 90JCS(P1)1881⟩, 742, 747 ⟨88JCR(S)292⟩, 748 ⟨91JCS(P1)2575⟩, 775 ⟨93JCS(P1)1945⟩; **5**, 873 ⟨87T215⟩; **6**, 124 ⟨86TL5143⟩
Baitz-Gács, E., **2**, 374 ⟨88S529⟩
Baiwir, M., **5**, 241 ⟨80SA(A)769⟩
Baize, M. W., **5**, 1147 ⟨91IC1046⟩; **6**, 515 ⟨91IC1046, 92IC4911⟩
Baizer, M. M., **1**, 427 ⟨B-83MI 110-01⟩; **2**, 395 ⟨73TL3597⟩; **3**, 498, 651 ⟨65JOC1357⟩
Bajgur, C. S., **4**, 657 ⟨89JOM(364)105⟩
Bajpai, B., **5**, 1143 ⟨76ICA-L64⟩
Bajpai, K. K., **5**, 1143 ⟨76ICA-L64⟩, 1143, 1144 ⟨72JIC1143⟩
Bajt, O., **5**, 836 ⟨84H(22)1545⟩
Bajwa, B. S., **3**, 433 ⟨83SC649⟩
Bajwa, J. S., **1**, 74 ⟨92TL2299⟩; **2**, 516 ⟨85TL6297⟩; **4**, 952 ⟨84JOC4091⟩
Bak, B., **3**, 553 ⟨79CC99⟩
Bak, C., **3**, 365 ⟨79CB2744, 80JHC1655, 80ZN(B)372⟩; **4**, 232 ⟨80JHC1655⟩
Bakale, R. P., **1**, 668 ⟨90JA6729⟩
Bakasse, M., **3**, 210 ⟨88S564⟩; **5**, 633 ⟨88T139⟩
Baker, B. E., **1**, 494 ⟨88TL905⟩
Baker, B. R., **2**, 129 ⟨58JA1680⟩, 861 ⟨60JOC666⟩; **5**, 748, 757 ⟨67JMC1123⟩
Baker, C. J., **1**, 564 ⟨94JCS(P1)793⟩
Baker, D. C., **3**, 32, 33 ⟨83JOC5180⟩, 190 ⟨78S478⟩; **6**, 364 ⟨90S151⟩
Baker, D. R., **2**, 1093 ⟨67JCP(46)658⟩; **5**, 738 ⟨87EUP243970⟩; **6**, 254 ⟨68USP3376314⟩
Baker, E. B., **1**, 850 ⟨66JA1272⟩, 893, 899 ⟨64JA1360, 63JA1328⟩; **3**, 282 ⟨63JA1328⟩
Baker, E. C., **4**, 703 ⟨74JA7586⟩
Baker, III, T. N., **2**, 391 ⟨64JA4074⟩
Baker, Jr., J. D., **1**, 19, 41 ⟨75JOC1834, 77S124⟩, 41, 635 ⟨76JOC574⟩
Baker, J. T., **5**, 24 ⟨72MI 502-01⟩
Baker, J. W., **2**, 103 ⟨52JCS1193⟩; **4**, 920 ⟨60CA(54)18356h⟩
Baker, K. M., **1**, 870 ⟨68T1651⟩
Baker, K. V., **2**, 558, 560 ⟨91JOC698⟩
Baker, L. A., **3**, 363 ⟨64JOC1355⟩
Baker, M. T., **4**, 45 ⟨93MI 402-01⟩
Baker, P. K., **5**, 952 ⟨80JA7811⟩
Baker, R., **1**, 240 ⟨66CC757⟩, 674 ⟨77TL441⟩, 677 ⟨89JCS(P1)190⟩, 697, 700, 703 ⟨87JCS(P1)1623, 85CC1326⟩, 745 ⟨81JCS(P1)3087⟩, 847 ⟨B-71MI 119-01⟩, 848 ⟨B-71MI 119-02⟩, 851 ⟨B-71MI 119-03, B-71MI 119-04⟩; **2**, 87 ⟨69JCS(C)1605⟩, 628 ⟨90JCS(P1)47⟩; **3**, 121, 294 ⟨83S572⟩, 273 ⟨89TL985⟩; **6**, 472, 473 ⟨85CC1326⟩
Baker, R. H., **1**, 15 ⟨48JA1490⟩
Baker, R. T., **2**, 43 ⟨93CJC930⟩, 527 ⟨92JA9350⟩
Baker, S. R., **1**, 1004 ⟨90JOC5833⟩; **3**, 653 ⟨94TL445⟩
Baker, W. A., **5**, 802, 803 ⟨89CL1603⟩
Baker, W., **1**, 21 ⟨51JCS556⟩; **2**, 651 ⟨48JCS2303⟩; **3**, 328 ⟨48JCS2138⟩, 612 ⟨41OSC(1)181⟩; **5**, 35 ⟨32OSC(1)451⟩, 95 ⟨32OSC(1)181⟩, 639 ⟨50JCS3389⟩
Baker, W. R., **2**, 118, 124 ⟨93JMC460⟩; **4**, 279 ⟨92TL1573⟩
Bakhitov, M. I., **6**, 366 ⟨81ZOB1035⟩, 512 ⟨73JGU1019⟩
Bakhmutou, Yu. A., **3**, 385 ⟨91ZOR408⟩
Bakhmutov, V. I., **5**, 385 ⟨88T5507⟩, 652 ⟨88JCS(P1)305⟩
Bakhtiyarova, F. A., **6**, 563 ⟨82MI 617-02⟩
Bakhtiyarova, I. V., **4**, 339 ⟨86ZOB711⟩
Bakke, J., **2**, 690 ⟨78MI 213-02⟩; **3**, 462 ⟨68ACS1833⟩
Bakke, J. M., **3**, 462 ⟨81ACS(B)179, 82ACS(B)127⟩
Bakker, B. H., **2**, 208 ⟨87TL1703⟩; **3**, 234 ⟨82CJC2268⟩; **4**, 98 ⟨75TL4065⟩, 468 ⟨84TL4259⟩
Bakker, C. G., **4**, 760, 765 ⟨93HOU(E15/2)1599⟩, 825 ⟨79TL2925⟩; **6**, 75 ⟨79TL2925, 81RTC13⟩
Bakker, C. N. M., **2**, 613 ⟨86JOC1886⟩
Bakker, P. H., **2**, 346, 349 ⟨79TL4391⟩
Bakker, S., **2**, 167 ⟨74JA3146⟩
Baklein, S., **3**, 370 ⟨75JCS(P1)2099⟩
Baklouti, A., **2**, 18 ⟨89TL3167⟩; **3**, 627 ⟨93SC2389⟩; **4**, 190 ⟨90SC2527⟩

Bakó, E. M., **5**, 219, 833 ⟨88CB723⟩
Bakola-Christianopoulou, M., **3**, 500 ⟨87JCR(S)374⟩; **5**, 1089 ⟨87JOM(331)193⟩
Bakos, J., **2**, 330 ⟨91CC1684⟩, 438 ⟨83JOM(253)249⟩, 439 ⟨79JOC3095⟩
Bakshi, R. K., **1**, 831 ⟨89JA3441⟩, 1034 ⟨88JA1529⟩; **2**, 519, 531 ⟨87JOC310⟩, 528 ⟨87JA5551, 87JA7925⟩, 530 ⟨85JOC5586⟩, 532, 698 ⟨89JA3441⟩; **3**, 135 ⟨87JA5420, 88JA1529⟩
Baksova, R. A., **2**, 458 ⟨65ZOB732⟩
Bakthan, S., **3**, 228 ⟨84JOC5024⟩
Bakulev, V. A., **5**, 567, 571 ⟨87KGS845⟩
Bakuzis, M. L. F., **3**, 494 ⟨77JOC2362⟩
Bakuzis, P., **3**, 494 ⟨77JOC2362⟩
Bal, B. S., **4**, 198 ⟨79JOC3727⟩, 203 ⟨83JOC1765⟩
Bal, S. Y., **4**, 260 ⟨89TL4007, 90JCS(P1)3217⟩
Balaban, A. T., **1**, 1132 ⟨83T3943⟩; **2**, 1008 ⟨82AHC(2)1, 90AHC(50)157⟩; **3**, 232 ⟨61JCS3566, 62T257⟩, 238 ⟨B-64MI 305-01⟩
Balabane, M., **2**, 301 ⟨83TL2745⟩
Balabanova, L. V., **5**, 687 ⟨91CHE299, 91KGS369⟩
Balabanov, E. Y., **6**, 619 ⟨93JPO319⟩
Balabanov, G. P., **2**, 478 ⟨65ZOR1256⟩
Balabukha, V. S., **2**, 464 ⟨60IZV651⟩
Balachandran, K. S., **5**, 28 ⟨75CRV491⟩
Balahura, R. J., **5**, 385 ⟨74CJC1762⟩, 386 ⟨79IC937⟩
Balakrishnan, P., **3**, 174, 207 ⟨81TL3043⟩
Balakumar, A., **1**, 810 ⟨93JOC5482⟩; **3**, 144 ⟨93JOC5482⟩
Balandin, A. A., **1**, 56 ⟨65MI 102-01⟩
Balas, L., **1**, 474 ⟨90BSF401⟩; **3**, 303 ⟨91OM366⟩; **6**, 449 ⟨91OM366⟩
Baláspiri, L., **3**, 475 ⟨70HCA1057⟩
Balasubrahmanyam, S. N., **5**, 1131 ⟨84IJC(B)254⟩
Balasubramanian, K. K., **1**, 827 ⟨88TL6797⟩, 961 ⟨86TL3777⟩
Balasubramanian, M., **4**, 310 ⟨94H(37)1951⟩
Balasubramanian, N., **3**, 431 ⟨85OPP23⟩, 433 ⟨85TL4331⟩; **5**, 990 ⟨89JOC224⟩, 993 ⟨84JOC4569⟩
Balasubramaniyan, V., **5**, 131 ⟨83T1475⟩
Balatoni, J. A., **2**, 4 ⟨83HCA1018⟩
Balavoine, E., **2**, 212, 225 ⟨92T1883⟩
Balavoine, G., **1**, 597 ⟨87TL5509⟩; **2**, 64 ⟨89TL5141⟩, 915 ⟨90JOC1857⟩; **4**, 813 ⟨90JOC1857⟩
Balazy, M., **6**, 304 ⟨82BCJ1106⟩
Balba, M. H., **4**, 310 ⟨68MI 407-01⟩
Balbach, B., **3**, 401 ⟨84CC686⟩
Balbi, A., **4**, 979 ⟨82FES387⟩
Balcerzak, P., **6**, 488 ⟨92JA9386⟩
Balch, A. L., **4**, 453 ⟨90JA6139⟩, 565 ⟨85JA5272⟩, 691 ⟨81JA3764⟩; **5**, 696 ⟨74JOM(71)453⟩; **6**, 721 ⟨72JA417, 73JA4769⟩, 721 ⟨74IC921⟩, 732 ⟨72IC2069⟩
Balchunis, R. J., **2**, 139, 140 ⟨77JOC3236⟩, 761 ⟨77TL2235⟩; **5**, 338 ⟨77TL2235⟩
Balci, M., **1**, 248 ⟨87AG(E)928⟩
Balczewski, P., **3**, 35 ⟨87S659⟩, 197 ⟨84S691⟩; **4**, 346 ⟨87PS(30)225⟩, 347 ⟨B-94MI 407-02⟩, 925 ⟨89S101⟩; **6**, 120, 121, 123 ⟨80S127⟩, 121 ⟨92T8697⟩, 129, 326, 350, 351 ⟨89S101⟩
Bald, E., **2**, 656 ⟨75CL1041⟩
Baldassarre, A., **1**, 1094 ⟨76S801⟩; **2**, 911 ⟨76S801⟩
Baldinger, H., **5**, 360 ⟨82S592⟩
Baldo, B., **3**, 256 ⟨88JCS(P1)1029⟩
Baldoli, C., **2**, 691 ⟨88S344⟩, 692 ⟨87S192⟩; **3**, 430 ⟨85CC417⟩; **5**, 953, 958 ⟨95JOM(486)279⟩, 956 ⟨94TL6377⟩; **6**, 723 ⟨93OM2994⟩
Baldrauf, L., **4**, 552 ⟨60MI 412-01⟩
Balducci, R., **2**, 1036, 1037 ⟨93SL211⟩
Baldwin, J. C., **5**, 917 ⟨80ZN(B)1289⟩; **6**, 706 ⟨80ZN(B)1289⟩
Baldwin, J. E., **1**, 130, 304 ⟨77CC233⟩, 154 ⟨83CC1040⟩, 154 ⟨86CC176⟩, 213, 304, 336 ⟨76CC734⟩, 264 ⟨82T2939, 89CC1852⟩, 340 ⟨92T3413⟩, 345 ⟨87CC166⟩, 354, 355 ⟨91COS(5)63⟩, 399 ⟨70CC576⟩, 403, 420 ⟨87JA8051⟩, 410, 823 ⟨71JA2342⟩, 498 ⟨69CC1225⟩, 574 ⟨89TL3335⟩, 601 ⟨81T263⟩, 634 ⟨81CC1121⟩, 801 ⟨76JA6705⟩, 806 ⟨70CC165⟩, 812 ⟨72JA5910⟩, 816, 818 ⟨70TL353, 71JA3556⟩, 822 ⟨72CC354⟩, 831 ⟨73CC117⟩, 931 ⟨92T3385⟩, 946 ⟨92T9373⟩, 988 ⟨84CC152, 87JOC443⟩, 1089 ⟨74JA7125⟩; **2**, 223 ⟨71JA2870⟩, 230 ⟨91T5603⟩, 346 ⟨84T4363⟩, 389, 765 ⟨69JCS(C)1073⟩, 681, 908, 999 ⟨74JA7125⟩; **3**, 124 ⟨77CC233⟩, 137 ⟨81CC1121⟩, 143 ⟨88CC1404⟩, 157 ⟨74JA7125⟩, 171 ⟨93CC1434⟩, 244 ⟨77JOC3846⟩, 293 ⟨83CC1040⟩, 334 ⟨82CC1029⟩, 334, 335 ⟨83T1487⟩, 429 ⟨85CC126, 85T699⟩, 696, 704 ⟨90TL2047⟩, 696, 713 ⟨90SL603⟩, 696, 719 ⟨94CC85⟩, 698 ⟨83T2989⟩, 698, 721 ⟨82CC942⟩, 703, 714 ⟨89TL3335⟩, 712 ⟨85T1931⟩, 712, 718, 719 ⟨89SL9, 91SL551⟩, 713 ⟨90TL2051⟩, 714, 718 ⟨84CC133, 85CC816⟩; **4**, 69 ⟨75JA5957⟩, 536 ⟨83CC1040, 86T4223⟩, 914, 920 ⟨70JOC3583⟩, 948 ⟨76JOC2312, 76JOC2874⟩, 949 ⟨74JA7125⟩, 953 ⟨75CC519⟩; **5**, 329 ⟨90CC720⟩, 341 ⟨81T2181⟩, 346 ⟨87TL4319⟩, 376 ⟨90T4733⟩, 553 ⟨72CC354⟩, 675 ⟨81CC1121⟩, 1106 ⟨90TA881⟩; **6**, 515 ⟨83T2989⟩, 608 ⟨68CC968⟩, 627 ⟨72CC354⟩
Baldwin, M. A., **1**, 627 ⟨67JCS(B)509⟩
Baldwin, R. A., **2**, 453, 837, 845 ⟨67JOC2172⟩, 832 ⟨61JOC3547⟩, 844 ⟨67JOC1572⟩, 863 ⟨61JA4466⟩, 863, 864 ⟨65JOC3866⟩
Baldwin, S. W., **2**, 342, 354 ⟨91TL4431⟩; **4**, 212 ⟨80JA1198⟩
Baldwin, W. E., **4**, 423 ⟨44JA222⟩, 609 ⟨47JA2247⟩
Baldwin, W. H., **1**, 778 ⟨61JOC846⟩
Baldy, A., **4**, 580 ⟨86CC1208⟩; **5**, 918 ⟨85CC1151⟩
Balenkova, E. S., **5**, 433 ⟨80MI 509-01⟩
Balenovic, K., **2**, 12 ⟨53JOC868⟩, 147 ⟨60TL20⟩, 478 ⟨61CI(L)469⟩; **4**, 259 ⟨73RTC117⟩
Bales, S. E., **2**, 558, 560, 959 ⟨74JA1775⟩
Bales, S. F., **2**, 958 ⟨73CC879⟩
Balezina, G. G., **2**, 1064 ⟨75JGU1832⟩
Balf, R. J., **1**, 1124 ⟨71CJC3135⟩
Balfe, M. P., **2**, 118 ⟨50JCS3309⟩, 288, 291 ⟨38JCS341, 41JCS70⟩
Balhar, A., **6**, 5 ⟨67CC1089⟩
Baliah, V., **4**, 272, 277 ⟨81S995⟩
Balicki, R., **5**, 690 ⟨83JOC3⟩
Balina, G., **4**, 203 ⟨86JOC3811⟩
Bálint, J., **3**, 368 ⟨77LA1529⟩
Balitskii, Yu. V., **2**, 467 ⟨79JGU34⟩; **5**, 713 ⟨90JGU849, 90ZOB963⟩
Balko, T. W., **1**, 777 ⟨81JOC2804⟩; **6**, 555 ⟨73JOC2106⟩
Balkovec, J. M., **1**, 287, 789 ⟨86TL1445⟩
Ball, Jr., A. L., **1**, 110 ⟨69JOC641⟩
Ball, M., **2**, 686 ⟨75S42⟩, 687 ⟨76LA2331⟩
Ball, R. E., **5**, 275 ⟨71JOC3048⟩
Ball, R. G., **1**, 829 ⟨93TL1851⟩; **4**, 547 ⟨88IC4038⟩
Ball, W. J., **1**, 990 ⟨67JCS(C)194⟩
Ballabio, M., **5**, 676, 677, 906 ⟨85T2503⟩
Ballam, J., **1**, 777 ⟨82JCS(P2)1175⟩
Ballantine, J. A., **3**, 308 ⟨66T(S)241⟩
Ballard, D., **2**, 941 ⟨B-69MI 218-02⟩
Ballard, D. A., **5**, 74 ⟨32OSC(1)89⟩
Ballard, D. G. H., **2**, 565 ⟨79JCS(D)54⟩; **4**, 702 ⟨79JCS(D)54⟩, 725 ⟨75CC621⟩, 725, 726 ⟨79JCS(D)45⟩; **6**, 297 ⟨72MI 610-01⟩
Ballard, D. H., **2**, 98 ⟨71JCS(C)945⟩
Ballesta, J. P. G., **4**, 264 ⟨92RTC163⟩
Ballester, M., **2**, 1010 ⟨71JA2215, 85ACR380⟩; **4**, 764 ⟨86JOC1413⟩; **5**, 46 ⟨55CRV283⟩
Ballester, P., **1**, 326 ⟨91JA5918⟩, 326, 327, 328 ⟨92JA7007⟩
Ballesteros, A., **2**, 920 ⟨87CC1195⟩; **4**, 421 ⟨89TL4573⟩; **6**, 571 ⟨89S228⟩
Ballesteros, P., **5**, 104 ⟨85OS(64)63⟩, 327 ⟨87T2557⟩
Ballesteros, R., **2**, 1003 ⟨91T5277⟩; **4**, 993 ⟨92H(34)1005⟩
Ballestri, M., **1**, 5, 7, 9, 10, 17, 936, 943, 945 ⟨91JOC678⟩, 933 ⟨92TL1787⟩; **5**, 1135 ⟨91JOC678⟩

Balli, H., **3**, 473 ⟨61LA(647)11⟩, 477, 479 ⟨78HCA97, 66TL5821⟩; **5**, 869 ⟨61LA(647)11, 61LA(647)1⟩
Ballini, B., **2**, 391, 783 ⟨88S833⟩
Ballini, M., **2**, 393, 394 ⟨86S849⟩
Ballini, R., **1**, 37 ⟨83S137, 85S269⟩, 39 ⟨90JOC5159⟩, 578 ⟨92JOC2160⟩; **2**, 318 ⟨83S1014, 87S713⟩, 395 ⟨86S237, 86S1024, 87S711⟩, 396 ⟨86SC1781, 90T7531, 92S355⟩, 397 ⟨92TL4835⟩; **3**, 18, 133 ⟨88JCS(P1)2563⟩, 46 ⟨84T3809⟩, 46, 191 ⟨90OPP707⟩, 47 ⟨86S1024⟩, 47, 192 ⟨86S237⟩, 190 ⟨92SL64⟩, 191 ⟨84S607⟩, 191 ⟨88S833⟩
Ballivet-Tkatchenko, D., **5**, 886 ⟨93PS(76)41⟩
Bally, T., **1**, 517 ⟨80T343⟩
Balme, G., **1**, 225 ⟨79S508⟩, 808 ⟨78TL1803⟩, 975 ⟨93TL3129⟩, 1020 ⟨91TL6541, 92TL2811⟩; **2**, 11 ⟨86TL1907⟩
Balocchi, L., **6**, 341 ⟨81AG215⟩
Balogh, D. W., **3**, 528 ⟨91TL1039⟩
Balogh, M., **4**, 429, 431 ⟨90JOC6198⟩
Balon, W. J., **2**, 435 ⟨58USP2853518⟩
Bal'on, Y. G., **4**, 794 ⟨78JOU135⟩; **6**, 504, 508 ⟨79JOU1077⟩
Balsamini, C., **3**, 273 ⟨74S357⟩
Balssa, F., **5**, 955 ⟨94JOM(476)C25⟩
Balszuweit, A., **4**, 461 ⟨78ZAAC(444)249⟩, 463 ⟨80ZAAC(469)109, 83PS(14)171⟩; **6**, 490 ⟨87ZAAC(546)142⟩, 491 ⟨85ZAAC(530)16⟩, 512 ⟨87EGP242812⟩
Balthazor, T. M., **5**, 321 ⟨79JOC1447⟩
Balturk, I. M., **2**, 84 ⟨90TL735⟩
Baltzly, R., **1**, 2, 6, 8, 12, 427 ⟨46JA261⟩; **5**, 140 ⟨47JA200⟩, 756 ⟨48JA1904⟩
Balucani, D., **3**, 105 ⟨67JCS(C)779⟩
Balueva, A. S., **4**, 342 ⟨86IZV2502, 92IZV196⟩
Balyaer, N. N., **4**, 941 ⟨79ZOB812⟩
Balyeat, J. R., **3**, 134 ⟨77JOC3205⟩
Balykova, I. A., **4**, 462 ⟨76ZOB1012⟩
Balz, E. H., **4**, 297 ⟨21JA341⟩
Balzarini, J., **6**, 264 ⟨91BMC357⟩
Balzer, B. L., **5**, 936 ⟨92JA8735, 92OM1759⟩
Balzer, W. D., **2**, 441 ⟨65TL1157⟩
Balzer, W. P., **4**, 575 ⟨66TL3315⟩
Balzuweit, A., **1**, 446 ⟨75ZAAC(418)158⟩
Bamberger, E., **2**, 764 ⟨1894CB1548⟩, 782 ⟨01LA(316)292, 03CB3803⟩, 805 ⟨1894CB914, 1897CB366⟩, 806 ⟨04CB629⟩; **3**, 620 ⟨1895CB1682⟩; **5**, 838 ⟨02CB3884⟩, 852 ⟨1893CB2385⟩, 855 ⟨02CB1084⟩, 857 ⟨01JPR199, 1893CB2978⟩, 858 ⟨01CB2001, 01JPR222⟩, 858 ⟨1892CB3547⟩; **6**, 653 ⟨26LA(446)260⟩
Bambury, R. E., **3**, 102 ⟨68JHC95⟩
Bamezai, S., **5**, 355 ⟨85TL5739⟩
Bamfield, P., **1**, 537 ⟨65JCS4355⟩; **2**, 662 ⟨84CSR441⟩
Bamford, C. H., **5**, 972 ⟨B-56MI 526-01⟩
Bamford, W. R., **1**, 39, 630, 631 ⟨52JCS4735⟩; **3**, 446, 464 ⟨52JCS4735⟩; **4**, 610 ⟨66JCS(C)1137⟩; **6**, 267 ⟨67JCS(C)1470⟩
Bamkole, T. O., **1**, 774 ⟨69JCS(B)187⟩
Bán, L., **5**, 769, 848 ⟨66CB2600⟩, 773 ⟨81CB3684⟩
Ban, T., **5**, 694 ⟨87NKK1304⟩
Ban, Y., **1**, 102 ⟨80CRV99⟩, 439, 450 ⟨82JOC4713⟩, 494 ⟨85TL1519⟩, 813 ⟨86TL219⟩; **2**, 402 ⟨78TL3435⟩; **3**, 264 ⟨65TL2261⟩, 310 ⟨81JA6990, 87CPB4700⟩; **4**, 414 ⟨76S201⟩, 512 ⟨76JOM(113)115⟩, 881 ⟨68TL497⟩; **5**, 157 ⟨77TL1037⟩, 157 ⟨79JCS(P1)771⟩, 157 ⟨79TL1133⟩, 296 ⟨83JOC4058⟩, 301 ⟨78JOC1684, 82CC741⟩, 301 ⟨86CC841⟩; **6**, 106 ⟨68TL497⟩
Banait, N. S., **1**, 623 ⟨90JA6950⟩; **2**, 1022 ⟨87CJC441⟩
Banaszek, A., **4**, 176 ⟨85CAR(144)342⟩
Banba, F., **5**, 332 ⟨80JCS(P1)2122⟩
Banba, Y., **1**, 834 ⟨86H(24)3347⟩
Banciu, A., **3**, 326 ⟨75CB1642⟩
Bancroft, G. M., **6**, 247 ⟨90IC2496⟩
Bandai, T., **5**, 411 ⟨87JAP6261987⟩; **6**, 126 ⟨87JAP6261987⟩
Bandal, H., **2**, 578, 579 ⟨90JCS(P1)2691, 90JCS(P1)2709⟩
Bandara, B. M. R., **5**, 96 ⟨82JCS(P1)1763⟩
Bandiera, T., **2**, 793 ⟨83T689⟩
Bandini, E., **2**, 544 ⟨90MI 201-01⟩; **3**, 456, 457, 458 ⟨90PAC605⟩, 457 ⟨91SL243⟩; **5**, 377 ⟨91SL243⟩
Bandler, M., **2**, 478 ⟨85ZN(B)1090⟩
Bando, K., **3**, 3 ⟨78BCJ3016⟩
Bando, T., **1**, 192 ⟨87TL3497⟩; **2**, 206 ⟨90JOC1823⟩; **3**, 412 ⟨79TL3453⟩, 418 ⟨86CC980⟩; **6**, 674 ⟨86CC980⟩
Bandodakar, B. S., **1**, 574 ⟨90S843⟩; **2**, 622 ⟨90S843⟩
Bandoli, G., **4**, 677 ⟨81AG(E)887⟩
Bandurco, V. T., **2**, 404 ⟨69TL4643⟩, 773 ⟨74JA5158⟩
Bandy, J. A., **5**, 453 ⟨83CC1202⟩
Bandyopadhyay, S., **5**, 601, 621 ⟨89IJC(B)673⟩
Banerjee, S. K., **2**, 647 ⟨62CJC2175⟩
Banerji, A., **3**, 431 ⟨86MI 311-01⟩; **6**, 143 ⟨77JCS(P1)1162⟩
Banert, K., **1**, 519 ⟨89CB123⟩, 993 ⟨92AG(E)90⟩; **2**, 770 ⟨92TL7331⟩; **5**, 1041 ⟨92AG(E)90⟩, 1101 ⟨92AG(E)866⟩; **6**, 436, 533, 534 ⟨92AG(E)866⟩
Banfi, L., **2**, 75 ⟨88JA3979⟩; **3**, 37 ⟨82S829⟩
Banfi, S., **3**, 31 ⟨89JOC2970⟩; **4**, 262 ⟨86JOC891⟩
Bánhidai, B., **3**, 474, 484, 485, 486 ⟨74LA1767⟩, 484 ⟨72LA(761)137⟩; **5**, 872 ⟨74LA1767⟩, 873, 874 ⟨72LA(761)137⟩
Banholzer, R., **6**, 444 ⟨75LA2227⟩
Bani-Akoto, E., **6**, 111 ⟨91JHC481⟩
Banitt, E. H., **1**, 681, 785 ⟨66JOC4043⟩; **3**, 497 ⟨64TL245⟩; **6**, 697 ⟨64TL245⟩
Bankaitis, D. M., **3**, 244 ⟨83TL5303⟩
Bank, S., **1**, 797 ⟨68JOC221⟩, 881 ⟨68JA4505⟩, 882 ⟨65JA3244⟩
Bankert, R. A., **5**, 1106 ⟨52JA1323⟩
Banko, K., **4**, 82, 83, 237 ⟨75JOC266⟩
Bankow, Y. I., **5**, 162 ⟨78JGU1249⟩
Banks, A. A., **6**, 212, 214, 217, 218, 219, 220, 222, 225, 226, 246 ⟨48JCS2188⟩
Banks, B. J., **1**, 596 ⟨91SL873⟩; **4**, 906 ⟨84CC670⟩; **5**, 238 ⟨84CC670⟩
Banks, C. H., **2**, 261 ⟨80JCS(P1)2287⟩, 289 ⟨79JMC572⟩
Banks, C. K., **2**, 260 ⟨48JA1762⟩, 491, 496, 498 ⟨47JA927⟩, 736 ⟨40JA1859⟩, 884, 888 ⟨40JA3142⟩; **5**, 1125 ⟨46JA2102⟩
Banks, C. M., **1**, 2 ⟨79T885⟩
Banks, C. V., **4**, 554, 563 ⟨61JA1722⟩, 557, 575 ⟨63JOC123⟩
Banks, H. D., **6**, 528 ⟨93JFC(60)85⟩
Banks, M. R., **2**, 357 ⟨86JCS(P2)151⟩; **3**, 305 ⟨86JCS(P1)507⟩; **5**, 347 ⟨86JCS(P2)1211⟩
Banks, R., **6**, 445 ⟨75JCS(P1)1836⟩
Banks, R. B., **1**, 491 ⟨76JA3732⟩
Banks, R. E., **2**, 15 ⟨86JFC(33)227, 92CC595⟩, 15 ⟨94CC343⟩, 16 ⟨88JFC(41)297⟩, 16, 336 ⟨88JCS(P1)2805⟩, 19 ⟨81JFC(17)93⟩, 183, 194 ⟨66JCS(C)1171⟩, 334 ⟨91JFC(52)389, 92USP5083178⟩, 1013 ⟨68TL3909⟩; **4**, 108 ⟨86JFC(32)461⟩, 736 ⟨70TL5215⟩, 799 ⟨66JCS(C)2304⟩; **5**, 22 ⟨91JFC(52)389⟩, 313 ⟨66JCS(C)1514⟩; **6**, 54 ⟨65JCS6149, 67JCS(C)2263⟩, 241 ⟨73JCS(P1)1092, 75JCS(P1)2033, 80JFC(16)391, 81JCS(P1)455, 81JFC(17)331, 81JFC(17)85, 82JCS(P1)685⟩, 257 ⟨65JCS6077⟩, 604 ⟨64JCS4066⟩
Banks, R. L., **1**, 674 ⟨64IEC170, 67HP232⟩, 674 ⟨79CT494⟩
Bankston, D., **5**, 54 ⟨87OS(66)180⟩
Banlish, B. K., **3**, 286 ⟨69JA5886⟩
Bannai, K., **3**, 192 ⟨87T813⟩
Bannard, A. B., **2**, 456 ⟨53CJC976⟩
Bannard, R. A. B., **6**, 641, 642, 643 ⟨58CJC1541⟩, 646 ⟨61CJC1017⟩
Banner, B. L., **1**, 980 ⟨80JOC2080⟩; **4**, 168, 170 ⟨83JA3661⟩; **5**, 159 ⟨80JOC2080⟩
Bannet, D. M., **1**, 751 ⟨78T997⟩
Bannikova, O. B., **4**, 140 ⟨89JOM(361)147⟩, 772

⟨90ZOB227, 91ZOB1084⟩, 804 ⟨91JGU983⟩; **5**, 718 ⟨84JGU426, 84ZOB477⟩, 718 ⟨88BAU378, 88IZV460⟩, 719 ⟨83JGU1851, 83JGU200, 83ZOB229⟩, 719, 720 ⟨83BAU1373, 83IZV1515, 83ZOB2051⟩
Bannister, R. M., **6**, 255 ⟨90JCS(P1)509⟩
Bannister, W., **2**, 1034 ⟨60JA3799⟩
Banno, H., **1**, 415, 826 ⟨88JA7922⟩, 415, 826 ⟨90JA316⟩, 825, 826 ⟨90JA7791, 89TL1265⟩, 826 ⟨90TL377⟩; **5**, 421 ⟨90JA7791⟩, 421, 431 ⟨91TA647⟩
Bannwart, E., **2**, 1007 ⟨93JA1731⟩
Bannwarth, W., **1**, 363 ⟨74AG(E)468⟩
Bannworth, W.,**5**, 279, 281 ⟨89TL1927⟩
Banoub, J., **6**, 71 ⟨80MI 603-01⟩
Bansal, H. S., **3**, 712 ⟨85T1931⟩
Bansal, R. C., **1**, 522 ⟨69CJC2391⟩
Bansal, R. K., **3**, 609 ⟨81TL3839⟩; **4**, 584, 587 ⟨85IJC(A)398⟩, 588 ⟨86JOM(303)351⟩
Bänsch, S., **2**, 876 ⟨71JPR254⟩
Banta, W. E., **2**, 554 ⟨84TL4323⟩
Bantel, H., **4**, 711 ⟨90CB661⟩
Banthorpe, D. V., **1**, 627 ⟨67JCS(B)509⟩; **5**, 989 ⟨B-71MI 526-01, B-71MI 526-02⟩
Banville, J., **1**, 804 ⟨79CC1138⟩; **2**, 667 ⟨76JCS(P1)1852⟩; **3**, 10 ⟨79TL3929⟩
Banwell, M. G., **1**, 549 ⟨B-87MI 112-01⟩; **3**, 113 ⟨93CC954⟩
Banyon, P. J., **2**, 254 ⟨62JCS2953⟩
Bao, J., **1**, 551 ⟨93JA10671⟩; **2**, 671 ⟨91JA9873⟩
Bao, L. Q., **2**, 3 ⟨75JOC3420⟩; **5**, 8 ⟨88OSC(6)190⟩
Bao, R., **5**, 370 ⟨89S618⟩
Baog, N. M., **6**, 517 ⟨82IC1704⟩
Baozhen Yang, **1**, 742 ⟨89JOM(375)45⟩
Bapat, J. B., **5**, 1002 ⟨76TL2691⟩
Bapuji, S. A., **1**, 527 ⟨89TL7107⟩
Barabanov, V. I., **2**, 463 ⟨69ZOB1543⟩; **5**, 370 ⟨70ZOB2464⟩
Barabara, C., **6**, 106 ⟨70CR(C)468⟩
Bar, E., **5**, 1088, 1089 ⟨88JOM(347)101⟩
Barager, H. J., **5**, 867 ⟨75JA3238⟩
Barager, III, H. J., **4**, 73 ⟨74OPP77⟩, 74, 75, 76, 78 ⟨73JOC17, 73CC319⟩
Barajas, L., **1**, 1073, 1085 ⟨90SC2733⟩
Barak, G., **3**, 117 ⟨88JOC3553⟩
Barakat, M. Z., **3**, 464 ⟨55JA1670⟩
Baraldi, P. G., **2**, 744 ⟨83S902⟩; **3**, 220 ⟨81S199⟩, 263 ⟨83S902⟩, 648 ⟨85S969⟩; **4**, 475 ⟨82S653⟩
Baram, G. O., **6**, 242 ⟨93ZOB1240⟩, 242, 682 ⟨92ZOB948⟩
Baramee, A., **1**, 787 ⟨94CC889⟩
Baranaev, M. K., **2**, 440 ⟨63ZOB1919⟩
Baran, Jr., J. R., **4**, 668 ⟨90JA9415⟩; **6**, 204 ⟨92MI 606-02⟩
Baranetskaya, N. K., **6**, 456 ⟨84MI 607-01⟩
Baraniak, J., **5**, 694 ⟨91TL4193⟩
Barani, M., **3**, 181 ⟨94T3829⟩
Baranin, S. V., **1**, 58 ⟨84JOM(270)9⟩
Baran Marszak, M.,**1**, 988 ⟨74TL1905⟩
Baranov, G. M., **4**, 451, 490, 493 ⟨92RCR1220⟩, 490 ⟨81ZOB757⟩
Baranov, N. N., **5**, 698 ⟨82IZV937⟩
Baranov, S. M., **5**, 848 ⟨72UKZ171⟩
Baranov, V. A., **2**, 256 ⟨80ZOB2459⟩
Baranov, Yu. I., **2**, 449 ⟨68DOK(182)337⟩; **4**, 556 ⟨69ZOB839⟩
Baranowska, E., **1**, 697, 698, 700, 703 ⟨85TL5597⟩
Baranski, A., **6**, 149, 369 ⟨73RZC1243⟩
Barany, G., **2**, 219, 230 ⟨89JOC244⟩, 233 ⟨84S657⟩; **6**, 275, 535 ⟨84SUL241⟩, 276, 435 ⟨83TL5683⟩, 434, 435, 474, 534, 550, 551, 567 ⟨83JOC4750, 84JCS(P1)2615⟩, 435 ⟨84JOC1043, 84MI 614-02⟩, 435, 480, 497, 548, 560 ⟨86JOC1866⟩, 496, 546 ⟨78JOC2930⟩, 548 ⟨90JOC1475⟩
Barash, L., **2**, 15, 613 ⟨70JA7494⟩
Barashenkov, G. G., **2**, 369 ⟨74ZOR2619⟩, 370 ⟨78ZOB1110⟩; **5**, 341 ⟨86ZOR1189⟩
Baratti, J., **5**, 226 ⟨85TL1857⟩
Barauah, J. N., **3**, 417 ⟨85SC335⟩
Barbachyn, M. R., **1**, 711 ⟨85PS(24)531⟩
Barbalat-Rey, F., **5**, 211 ⟨87MI 504-02⟩, 212, 213 ⟨85CAR375⟩; **6**, 695 ⟨76CAR(46)9⟩
Barbara, C., **1**, 416, 828 ⟨66TL6425⟩, 518 ⟨75TL3151⟩; **2**, 1047 ⟨64BSF871, 65BSF2787⟩, 1048 ⟨67CR(C)1496⟩, 1050 ⟨74BSF1535⟩
Barbaro, G., **3**, 369 ⟨86CC964, 90JOC3744⟩, 560 ⟨84JOC2200⟩; **5**, 647, 648 ⟨86JCS(P1)381, 90JOC3744⟩, 1073 ⟨77JOC3372⟩; **6**, 127 ⟨87JCS(P1)2647⟩, 127, 128 ⟨90H(31)47⟩
Barbaro, S., **5**, 734 ⟨82SC865⟩
Barbas, III, C. F., **3**, 48 ⟨89JA3924⟩, 159 ⟨91S499⟩
Barbe, J., **3**, 280 ⟨91TL6709⟩
Barbeaux, E., **2**, 556 ⟨94T7177⟩
Barbeaux, P., **1**, 245 ⟨86CC457⟩, 246 ⟨87CC1214, 91TL417⟩; **2**, 141 ⟨91TL2509⟩
Barbe, W., **1**, 925 ⟨83CB1042⟩
Barbee, R. B., **2**, 213, 215 ⟨69JOC36⟩, 216, 218 ⟨69JOC1792⟩
Barber, C., **4**, 655 ⟨90PAC1933⟩
Barber, E. J., **4**, 3 ⟨51JA4241⟩
Barber, G. N., **4**, 60 ⟨76TL3381⟩
Barber, G. W., **3**, 59 ⟨49JA3310⟩
Barber, H. J., **5**, 703, 767, 813 ⟨43JCS101⟩
Barber, J., **2**, 11 ⟨82TL1549⟩
Barber, J. J., **2**, 596 ⟨79JOC3603⟩
Barber, M., **4**, 698 ⟨82OM1307⟩
Barbero, A., **2**, 929 ⟨91JCS(P1)2811⟩, 930 ⟨90CC1030, 92JCS(P1)327⟩
Barbero, M., **2**, 180 ⟨89S957⟩; **4**, 846, 854 ⟨87G227⟩; **6**, 302, 554, 729 ⟨88S22⟩
Barbier, G., **5**, 22 ⟨91JHC1511⟩, 355 ⟨92JHC1081⟩, 705 ⟨84JHC1689⟩; **6**, 444 ⟨89JHC231⟩
Barbier, M., **3**, 294 ⟨84HCA866⟩
Barbieri, G., **3**, 62 ⟨88JA4735⟩
Barboni, L., **2**, 346 ⟨90JCS(P1)2133⟩
Barbosa, M. C. N., **1**, 934 ⟨91JOC2866⟩; **3**, 311 ⟨91JOC2866⟩
Barbot, F., **1**, 311 ⟨83TL5089⟩, 318 ⟨77JOM(132)445⟩, 961 ⟨85S768⟩; **4**, 192 ⟨81JCR(M)4016⟩, 321 ⟨87JOM(335)283⟩
Barbour, A. K., **5**, 34, 63, 100 ⟨61JCS808⟩; **6**, 16, 38 ⟨63AFC(3)181⟩, 225 ⟨B-59MI 607-02⟩
Barbry, D., **2**, 742 ⟨87S727⟩
Barbulescu, N., **6**, 71 ⟨57MI 603-01⟩
Barcelo, G., **4**, 58, 59 ⟨86S627, 85JOC3953⟩, 60 ⟨87S1027⟩; **6**, 643, 645 ⟨90T1839⟩
Barchielli, G., **3**, 284 ⟨84CC530⟩
Barcina, J. O., **2**, 36 ⟨87TL6441⟩; **3**, 92, 101 ⟨90CC1571⟩
Barclay, L. R. C., **1**, 923 ⟨84JA1793⟩
Barco, A., **2**, 181 ⟨74S877⟩, 301 ⟨91S479⟩, 356 ⟨91TL2517⟩, 395 ⟨72S44⟩; **3**, 220 ⟨81S199⟩, 648 ⟨85S969⟩
Barcza, S., **4**, 520 ⟨72USP3692798⟩
Bard, A. J., **1**, 934 ⟨78JA4903⟩; **6**, 375 ⟨87JCS(D)249⟩
Bard, R. R., **2**, 850 ⟨79JOC4918⟩
Bardenhagen, J., **3**, 719 ⟨85LA1825⟩
Bardi, R., **5**, 142 ⟨79ICA345⟩, 573 ⟨85T5595⟩
Bardisso, M., **2**, 354 ⟨90JOC3427⟩
Bardone-Gaudemar, F., **1**, 1060 ⟨56CR(243)1895⟩
Bardos, T. J., **2**, 260 ⟨66JOC221⟩, 413 ⟨88JOC1761⟩
Bardshiri, E., **5**, 199 ⟨88TL1595⟩
Barea, F., **1**, 777 ⟨81JOC2804⟩
Bare, T. M., **1**, 871, 881 ⟨68JOC943⟩
Barelski, P. M., **2**, 420, 422 ⟨73JOC489⟩
Bares, J. E., **1**, 142, 155 ⟨75JA7006⟩
Barette, B., **4**, 423, 427 ⟨55BSF669⟩
Barfield, M., **4**, 24 ⟨83JA3411⟩
Barfknecht, G. W., **1**, 1110 ⟨70JA207⟩
Bargagna, A., **2**, 204 ⟨83JHC839⟩; **3**, 80 ⟨77JHC249⟩; **5**, 501 ⟨83JHC1549, 86JHC455⟩, 501 ⟨87JHC633⟩, 597 ⟨83FES101⟩

Bargamova, M. D., **4**, 731 ⟨80CA71025t⟩, 799 ⟨91BAU1705⟩, 966 ⟨74IZV2545⟩; **5**, 22 ⟨91BAU1705⟩, 679 ⟨91IZV1918⟩
Bargar, T. M., **1**, 967 ⟨87TL2207⟩; **3**, 618, 619 ⟨80SC479⟩
Barger, G., **5**, 90 ⟨43OSC(2)384⟩
Barger, P. T., **3**, 545 ⟨84JA5178⟩
Bargurie, M., **3**, 469 ⟨80JCS(P1)2670⟩
Barhdadi, R., **3**, 620 ⟨93T5091⟩
Bari, S. S., **5**, 274, 286 ⟨92S439⟩
Barieux, J.-J., **1**, 660 ⟨71BSF3978⟩; **2**, 43 ⟨71BSF1649⟩; **6**, 430 ⟨87TL6443⟩
Barillier, D., **3**, 351 ⟨77BSF688⟩; **4**, 233 ⟨86JOC2276⟩; **5**, 552 ⟨79PS(7)69⟩
Barillier, Z., **1**, 833, 834 ⟨79PS(7)69⟩
Baringhaus, K. H., **4**, 206 ⟨92AG(E)1335⟩
Barinov, I. V., **4**, 674 ⟨85JOM(297)C25, 86ZOR2457⟩, 674 ⟨91JOM(418)C24, 93JOM(455)C9⟩
Barion, D., **5**, 885 ⟨89TL459⟩
Barishnikova, A. N., **2**, 188 ⟨64DOK(157)139⟩
Bark, D. S., **5**, 13 ⟨79MI 501-01⟩
Barkallah, S.,**5**, 22 ⟨92PS(69)163⟩
Barkash, V. A., **3**, 536 ⟨72ZOR654⟩
Barkemeyer, H., **5**, 972 ⟨66JA3163⟩; **6**, 496 ⟨68JA3254⟩
Barkenbus, C., **2**, 100 ⟨34JA1204⟩
Barker, A. J., **6**, 121 ⟨90TL4359⟩
Barker, B. J., **6**, 507 ⟨79AG(E)503⟩
Barker, G. K., **2**, 278, 294 ⟨73JCS(D)1716⟩, 566 ⟨77JA6645⟩; **5**, 952 ⟨80JA7811⟩; **6**, 195 ⟨74JOM(76)C45⟩, 195 ⟨77JA6645⟩, 195 ⟨78JCS(D)734⟩
Barker, H. C., **5**, 201 ⟨56JOC1473⟩
Barker, J., **5**, 781 ⟨86JCS(D)1359, 87JCS(D)2687⟩
Barker, J. M., **5**, 8 ⟨90S1143⟩
Barker, M. W., **3**, 556 ⟨B-80MI 317-02⟩, 558 ⟨74JHC633⟩, 560 ⟨74JHC241⟩; **5**, 730 ⟨73JOC3951⟩
Barker, P. L., **5**, 742, 758 ⟨81JOC2455⟩
Barker, R. L., **2**, 752 ⟨39JCS1299⟩
Barker, S. J., **3**, 579 ⟨90JCS(P1)485⟩
Barker, W. D., **1**, 986 ⟨70S543⟩
Barkley, L. B., **1**, 860 ⟨51JA4301⟩
Barkóczy, J., **6**, 564 ⟨90JHC1249⟩
Barkovich, A. J., **1**, 1064 ⟨76JA2667⟩; **2**, 944 ⟨75JA5600⟩
Barleunga, J., **3**, 414 ⟨85CB3652⟩
Barlin, G. B., **3**, 108 ⟨53JCS1740⟩
Barlos, K., **5**, 223 ⟨89LA387⟩
Barlow, J. L., **2**, 118, 124 ⟨93JMC460⟩
Barlow, L., **3**, 494 ⟨76JCS(P1)1466⟩
Barlow, M. G., **2**, 859 ⟨66JCS(C)1592⟩, 1013 ⟨68TL3909⟩; **4**, 46 ⟨80CC158⟩, 136 ⟨76JCS(D)694⟩; **5**, 681 ⟨83JFC(23)293⟩; **6**, 54 ⟨65JCS6149⟩, 238 ⟨80JCS(P1)2254⟩, 241 ⟨80JCS(P1)1960⟩
Barltrop, J. A., **1**, 517 ⟨67JCS(C)1625⟩
Barluenga, J., **1**, 68 ⟨80JCS(P1)1420, 83T2863⟩, 68 ⟨88S144⟩, 332, 343, 948 ⟨88CRV487⟩, 466 ⟨93JOC5976⟩, 681 ⟨86CC1665⟩, 833 ⟨84CC1427, 84JCS(P1)721⟩, 990 ⟨88JA5567⟩, 1040 ⟨89S33⟩, 1090 ⟨88CC1135, 88TL4859, 89JCR(S)200, 90CC1521, 90CC587, 91CC1001, 92TL831⟩; **2**, 18, 24, 31, 34 ⟨85AG(E)319⟩, 18, 610 ⟨91JOC2234⟩, 24, 31 ⟨86JCR(S)274, 85CC1422⟩, 31, 35 ⟨87CC1491, 90JCS(P1)2807⟩, 322 ⟨89CC1132⟩, 553 ⟨88TL2859, 90PAC595, 93JOC5976⟩, 629 ⟨90JOC3104⟩, 631 ⟨84JCS(P1)2623⟩, 676 ⟨88S144⟩, 740 ⟨86S960⟩, 744 ⟨84S304⟩, 920 ⟨87CC1195⟩, 936 ⟨88S234⟩, 1019 ⟨87S661⟩, 1097 ⟨83JCS(P1)1087, 88CC121⟩; **3**, 27, 155 ⟨86S678⟩, 58 ⟨85JCS(P1)447⟩, 75 ⟨86CC1180⟩, 152 ⟨91JCS(P1)2890⟩, 152, 154 ⟨90S1003⟩, 188 ⟨83JOC2255⟩, 413 ⟨85CC1375⟩, 414 ⟨86CC1179, 88TL4863⟩, 417 ⟨83JCS(P1)1093⟩, 638 ⟨81S375⟩; **4**, 152, 153 ⟨79S893⟩, 300 ⟨88JCS(P1)1631⟩, 408 ⟨88CC410⟩, 412 ⟨88CB1813⟩, 421 ⟨89TL4573, 91CC1704⟩, 453, 477 ⟨90SL261⟩, 741, 747, 756, 758 ⟨89JCS(P1)691⟩, 749 ⟨93AG(E)893⟩, 749, 819, 1064 ⟨88JA5567⟩, 760 ⟨90TL2751⟩, 772 ⟨90TL7375⟩, 808 ⟨87JCS(P1)1017⟩, 1067 ⟨83JCS(P1)1087⟩; **5**, 71 ⟨83JCS(P1)3019⟩, 132 ⟨83S649⟩, 152 ⟨83T2863⟩, 284 ⟨81CC1178, 81CC670⟩, 371 ⟨90JCS(P1)2193, 90TL3497⟩, 946, 958 ⟨93CC1068, 95TL3937⟩, 1044 ⟨87CC1491⟩, 1066 ⟨88TL4863⟩, 1108 ⟨90TL7375⟩; **6**, 121 ⟨88S562⟩, 571 ⟨89S228⟩, 575 ⟨79CC675⟩
Barna, P. M., **4**, 732 ⟨66CI(L)2054⟩
Barnard, C. F. J., **5**, 455 ⟨76CC1032⟩
Barnard, D., **2**, 144, 173 ⟨61JCS5339⟩, 192, 287 ⟨57JCS4547⟩, 287, 291 ⟨62JCS2089⟩; **6**, 594 ⟨61JCS2922⟩
Barnáth, G., **6**, 104 ⟨68TL2713⟩
Barner, B. A., **2**, 976 ⟨82JA4708⟩; **4**, 382 ⟨89TL5413⟩; **5**, 173 ⟨82JA4708⟩; **6**, 472, 473 ⟨82JA4708⟩
Bar-Ner, N., **1**, 1016 ⟨91JOC3912⟩; **2**, 1020 ⟨91JOC3912⟩
Barner, R., **1**, 827 ⟨63HCA2444, 68HCA1603⟩; **2**, 656 ⟨63HCA2444, 68HCA1603⟩, 656 ⟨73HCA14⟩
Barnes, C. E., **4**, 709 ⟨87JOM(321)257⟩
Barnes, C. L., **4**, 214 ⟨91JA8168⟩
Barnes, D., **2**, 314 ⟨90TL5555⟩
Barnes, G. H., **2**, 536 ⟨57JA974⟩
Barnes, H. M., **4**, 887, 889 ⟨40JA1281⟩
Barnes, J. C., **3**, 363 ⟨90JOM(385)369⟩
Barnes, J. J., **3**, 685 ⟨83JCS(P1)293⟩
Barnes, J. L., **4**, 279 ⟨90TL6819⟩
Barnes, J. M., **2**, 1097 ⟨68MI 223-01⟩
Barnes, N. J., **1**, 215 ⟨81TL1751⟩
Barnes, R. K., **2**, 650 ⟨69JOC3441⟩; **4**, 298 ⟨70JOC3140, 70T2555⟩
Barnes, R. M., **5**, 864 ⟨80ANC967⟩
Barnes, R. N., **6**, 262 ⟨85JCS(P1)1191⟩
Barnes, R. P., **5**, 8, 112 ⟨55OSC(3)555⟩
Barnett, B. L., **4**, 681 ⟨74JA7108⟩, 711 ⟨73CSC347⟩
Barnett, C., **1**, 871 ⟨68JA4629⟩
Barnett, K. W., **5**, 445 ⟨72JOM(44)1⟩, 453 ⟨66IC1177⟩
Barnett, W. E., **5**, 161 ⟨72CC472⟩
Barnette, W. E., **1**, 603 ⟨81JA3486⟩, 604 ⟨79JA3704⟩; **2**, 16 ⟨85JOC3698⟩, 16, 365, 611, 613 ⟨84JA452⟩, 281 ⟨79JA3704⟩; **5**, 203 ⟨85JOC3698⟩
Barney, C. L., **1**, 967 ⟨87TL2207⟩; **2**, 611 ⟨90TL973⟩; **4**, 52 ⟨90TL973⟩, 1080 ⟨87CC469⟩
Barnick, J. W. F. K., **4**, 522 ⟨68RTC188⟩
Barnikow, G., **4**, 442 ⟨83PHA449⟩, 903 ⟨66ZC417⟩, 979 ⟨67CB1661⟩; **5**, 476, 477, 486, 1050 ⟨68ZC142⟩, 516 ⟨68ZC335⟩, 583 ⟨87ZC90⟩, 588, 632 ⟨85ZC219⟩, 595 ⟨67JPR302, 74JPR534, 84ZC438⟩, 674, 1053 ⟨80ZC97⟩, 1051 ⟨71ZC420⟩, 1052 ⟨74ZC356⟩; **6**, 559 ⟨81ZC438, 85ZC219⟩, 560 ⟨90ZC90⟩
Barnish, I. T., **4**, 901 ⟨89TL4449, 91TL405⟩; **5**, 356 ⟨65JCS2999⟩, 566, 613 ⟨89T6771, 89TL4449⟩, 613 ⟨91TL405⟩
Barnum, C., **2**, 282 ⟨80TL3643⟩; **3**, 38, 174, 207 ⟨81JOC2920⟩, 174 ⟨80TL3643⟩, 174 ⟨85T4881⟩, 174, 207 ⟨81TL3043⟩
Barnum, C. S., **4**, 401 ⟨80JOC2736, 81JOC4301⟩
Barnum, H. W., **2**, 220 ⟨70JA7629⟩; **5**, 247 ⟨70JA7629⟩
Baron, L., **4**, 987 ⟨81CB2001⟩
Baron, P., **6**, 489 ⟨93MI 615-03⟩
Baron, W. J., **1**, 549 ⟨73JA6286⟩
Barone, A. D., **3**, 48 ⟨78JOC2066⟩
Barone, G., **2**, 45 ⟨75G377, 76TL3661⟩
Barot, B. C., **4**, 185 ⟨81JOC2981⟩; **6**, 566 ⟨87JHC749⟩
Barrabis, J., **2**, 1015 ⟨68BSF193⟩
Barrack, S. A., **1**, 1029 ⟨88JOC1790⟩
Barraclough, C. G., **3**, 397 ⟨69AJC891⟩
Barraclough, P., **5**, 342 ⟨89JCS(P1)1815⟩, 1126 ⟨76JCS(P1)716⟩
Barr, D. A., **2**, 158 ⟨89JCS(P1)1550⟩; **5**, 988 ⟨56JCS3428⟩; **6**, 218 ⟨57JCS30⟩, 238 ⟨55JCS2532⟩, 604 ⟨55JCS1881, 56JCS3416⟩, 604 ⟨64JCS1351⟩
Barr, D., **3**, 459 ⟨87JCS(D)1071, 87JCS(D)2141⟩
Barrage, A.-K., **2**, 19 ⟨81JFC(17)93⟩
Barr, J. T., **6**, 46 ⟨50JA3642⟩

Barr, K. J., **2**, 927 ⟨91TL5465⟩
Barr, P. J., **1**, 1010 ⟨81TL421, 83JOC1854⟩
Barr, R. D., **4**, 716 ⟨85JCS(D)2009⟩
Barral, I., **2**, 243 ⟨73S795⟩
Barrans, J., **5**, 708, 861 ⟨70BSF1363⟩, 775 ⟨77JCS(P1)1806, 68CR(C)1241⟩
Barras, J.-P., **5**, 342 ⟨87HCA1666⟩
Barratt, D., **5**, 927 ⟨87JOM(325)185⟩
Barrau, J., **1**, 656 ⟨83JOM(243)281, 90CRV283⟩; **3**, 501 ⟨90CRV283⟩; **4**, 141 ⟨88SRI317⟩, 141, 372 ⟨84SRI21⟩, 362 ⟨88SRI317⟩, 607, 621, 622, 628, 629 ⟨89OM1585⟩, 627 ⟨88JOM(345)39⟩, 628, 629 ⟨87OM659, 85JOM(282)315⟩, 629 ⟨90ICA141, 90ICA159⟩, 629 ⟨90JOM(395)27, 91JOM(406)123⟩, 629 ⟨91NJC657⟩, 640, 646 ⟨90SRI1373⟩, 641 ⟨90JOM(387)65⟩; **5**, 927 ⟨90CRV283⟩; **6**, 187, 709, 724 ⟨90CRV283⟩
Barraus, M. T., **4**, 478 ⟨89MI 410-01⟩
Barre, R., **6**, 77 ⟨49MI 603-1⟩
Barre, V., **3**, 172 ⟨89TL7389⟩
Barreau, J., **4**, 581 ⟨90IC1674⟩
Barreau, M., **1**, 156 ⟨91TL4125⟩; **3**, 357 ⟨81TL4507⟩, 496 ⟨85TL5451⟩; **5**, 688 ⟨87S262⟩
Barreiro, E., **1**, 960 ⟨76CC183⟩, 966 ⟨75TL4615⟩
Barren, E., **4**, 821 ⟨89CC1159⟩
Barrero, A. F., **2**, 617 ⟨91JHC939⟩
Barret, R., **2**, 148 ⟨85SUL127⟩
Barrett, A., **2**, 783 ⟨91CSR95⟩
Barrett, A. G., **4**, 906 ⟨90OS(68)8⟩
Barrett, A. G. M., **1**, 16 ⟨81JCS(P1)1501⟩, 156 ⟨88TL5733⟩, 630, 632, 1094 ⟨83ACR55⟩, 633 ⟨81JCS(P1)2848⟩, 634 ⟨81CC65, 81T3935⟩, 648 ⟨91SL764⟩, 700, 703 ⟨86JOC4840⟩, 825 ⟨87OM2249⟩; **2**, 75 ⟨93CC339⟩, 75, 521 ⟨91JOC5243⟩, 89 ⟨91COS(8)235⟩, 393, 396, 398, 783 ⟨86CRV751⟩, 394 ⟨89JOC1233⟩, 398 ⟨89OS(68)8, 93T7857⟩, 943 ⟨88JOC1815⟩; **3**, 192 ⟨86CRV751⟩, 203 ⟨92JOC386⟩, 244 ⟨81CC556⟩, 309 ⟨84JOC4409⟩, 713 ⟨93CC1760⟩; **4**, 138, 355 ⟨92JOC386⟩, 139, 652, 655 ⟨91JOC638⟩, 172 ⟨81CC556⟩, 327 ⟨93OSC(8)550⟩, 660 ⟨91TL3285⟩, 795 ⟨91CSR95⟩, 897 ⟨89TL2349⟩, 898 ⟨89JOC992, 90JOC5110⟩, 906 ⟨84CC670, 85JOC2603, 86JOC1012, 87JOC4693, 89JOC4723⟩, 1019 ⟨86TL3811, 88OM2553⟩; **5**, 177 ⟨86JOC4840⟩, 238 ⟨84CC670, 89JOC4723⟩, 344, 1005 ⟨85JOC169⟩, 526 ⟨79CC645, 80JCS(P1)2191⟩, 526, 685, 694 ⟨80JCS(P1)2184⟩, 652 ⟨88T5615⟩, 735 ⟨80JCS(P1)665⟩, 937 ⟨87OM2249⟩, 955 ⟨88OM2553⟩, 1004 ⟨80NJC127⟩
Barrett, G. C., **2**, 175 ⟨B-75MI 203-01⟩; **4**, 3, 37 ⟨71JCS(C)279⟩, 390 ⟨79COC(3)33⟩; **5**, 586 ⟨69JCS(C)1117⟩, 1022 ⟨70SST83, 73SST73⟩, 1022 ⟨75SST65, 77SST59⟩, 1022 ⟨79SST52, 81SST66⟩
Barrett, J. H., **1**, 96 ⟨73OSC(5)467⟩
Barrette, E. P., **3**, 116 ⟨85TL5855⟩
Barrick, P. L., **4**, 5 ⟨49JA490⟩; **6**, 2 ⟨49JOC747⟩
Barrientos-Penna, C. F., **6**, 517 ⟨85OM478⟩
Barrier, C., **3**, 429 ⟨89TL5763⟩
Barrière, F., **1**, 223 ⟨85TL3121⟩
Barrière, J.-C., **1**, 223 ⟨85TL3121⟩; **3**, 59 ⟨81HCA1140⟩
Barringer, C. M., **6**, 220 ⟨56USP2739989⟩
Barrio, J. R., **2**, 25, 31, 35 ⟨90TL4973⟩, 613 ⟨88TL1501, 90JOC3373⟩, 632 ⟨81JOC5239, 83JOC4394⟩
Barrish, J. C., **1**, 697, 700, 703 ⟨88JOC4282⟩, 749, 828 ⟨87JOC1372⟩
Barriuo, B., **4**, 309 ⟨74CR(C)221⟩
Barron, A. R., **1**, 666 ⟨90OM3086⟩; **2**, 511 ⟨88POL77⟩; **3**, 127 ⟨90TL265⟩
Barron, C. A., **3**, 528 ⟨87CC1728⟩
Barron, L. B., **6**, 242 ⟨55JA2869⟩
Barros, M. T., **3**, 8, 121 ⟨87JCS(P1)1221⟩
Barros, S. M., **1**, 245, 441 ⟨89TL7353⟩, 999, 1035 ⟨90SC2181⟩; **2**, 730 ⟨89OM1661⟩, 1036, 1038 ⟨90SC2181⟩
Barrow, M. J., **3**, 685 ⟨83JCS(P1)293⟩; **5**, 539 ⟨82JCS(D)1131⟩
Barry, C. N., **2**, 830 ⟨85JA5210⟩
Barry, J., **5**, 132 ⟨85S40⟩
Barsa, E. A., **6**, 445 ⟨86JOC4483⟩
Barsantai, P., **1**, 187 ⟨78CC1085⟩
Barsegyen, S. K., **2**, 446 ⟨86AKZ463⟩
Barstow, J. F., **1**, 831 ⟨82JOC3933⟩
Barstow, L. E., **5**, 300 ⟨71JOC1305⟩
Barsukov, L. I., **1**, 722 ⟨63IZV1053⟩, 727, 733 ⟨67T2709⟩; **3**, 494 ⟨67T2709⟩; **4**, 752 ⟨64TL2669⟩
Barta, M., **2**, 317 ⟨87TL5941⟩
Barta, N. S., **1**, 835 ⟨92JOC7188⟩
Barta, T. E., **2**, 938 ⟨85JOC3421⟩; **3**, 198 ⟨91TL2581⟩
Bart, J. C. J., **2**, 1004 ⟨69JCS(B)350⟩
Bartal, A. von, **6**, 417, 420 ⟨06LA(345)334, 07MI 614-01⟩
Bartch, R., **2**, 475 ⟨82JFC(20)85⟩
Bartczak, T. J., **3**, 543 ⟨73CC247⟩
Bartel, J., **3**, 618, 662 ⟨74CB1221⟩
Bartel, K., **3**, 519, 707 ⟨74AG(E)599⟩, 709 ⟨77AG(E)707⟩; **4**, 1017 ⟨75JOM(97)61, 82ZN(B)1044⟩; **6**, 431 ⟨71JCS(C)3766⟩, 732 ⟨74AG(E)599⟩
Bartel, S., **1**, 218, 221 ⟨89TL685⟩
Bartels, A. P., **1**, 97 ⟨71JOC2588⟩
Bartels, G., **6**, 32 ⟨92GEP4025227⟩
Bartels, H. M., **1**, 325, 920 ⟨87JOC205⟩
Bartels, K., **1**, 498 ⟨89JA3717⟩
Bartetzko, R., **3**, 371 ⟨78JOC3893⟩, 372 ⟨79CB260⟩
Barth, J., **5**, 225 ⟨81SC763⟩
Barth, L., **3**, 194 ⟨92TL2709⟩
Barth, R. C., **2**, 827 ⟨75JOM(94)327⟩; **4**, 112 ⟨75JOM(94)327⟩
Barth, S., **4**, 549 ⟨93PS(76)1⟩; **5**, 1152 ⟨92AG(E)758⟩; **6**, 153 ⟨92AG(E)584⟩
Barth, V., **5**, 880 ⟨83CB938⟩; **6**, 686 ⟨79AG(E)469, 79AG(E)873⟩, 686 ⟨82CB1617⟩
Barth, W., **1**, 208 ⟨88S943⟩; **4**, 1081 ⟨83JCS(P2)1687⟩
Bartho, B., **3**, 361 ⟨75TL2683⟩; **4**, 983, 986 ⟨77JPR149⟩
Barthold, K. P., **4**, 45 ⟨71FCR77⟩
Bartholomew, D., **1**, 769 ⟨84TL2035⟩; **2**, 401 ⟨77JCR(S)237⟩
Bartholomew, D. G., **5**, 350 ⟨78JHC439⟩
Bartl, A., **2**, 361, 792 ⟨78TL4003⟩
Bartl, V., **6**, 5 ⟨67CCC3888⟩
Bartlett, E. H., **3**, 664 ⟨72JOM(46)267⟩
Bartlett, P. A., **1**, 67, 68 ⟨84T2317, 80JA337⟩, 175 ⟨88JA1633⟩, 295 ⟨B-84MI 107-01⟩, 414, 806, 825, 836 ⟨80T3⟩, 826 ⟨85JA7792, 88JOC3195⟩, 827, 828 ⟨79JOC882, 81JOC3896⟩, 831 ⟨82JOC3933, 85JOC5179⟩, 1015 ⟨86JOC2230⟩, 1073 ⟨78JA4852⟩; **2**, 351 ⟨88JA1633⟩, 469 ⟨83JA1613⟩, 702 ⟨87OM2069⟩; **3**, 168 ⟨90JOC5966⟩, 482 ⟨77JA1267, 82JOC1284⟩; **5**, 39 ⟨70TL4459⟩, 168, 172 ⟨B-84MI 503-02⟩, 174 ⟨78JA4858⟩; **6**, 118, 119, 636, 666 ⟨82JOC1284⟩, 468 ⟨82JOC4013⟩, 469 ⟨84JA5304, 77JA1267⟩, 666 ⟨82CC536⟩
Bartlett, P. D., **1**, 785 ⟨58JA3585⟩, 853 ⟨34JA967, 40JA2933⟩, 853, 898 ⟨50JA1005⟩, 892 ⟨55JA2801⟩, 900 ⟨39JA3184⟩, 905 ⟨59LA(623)217⟩; **2**, 98 ⟨70JA3223, 72IJ101⟩, 180 ⟨73OSC(5)196⟩, 403 ⟨66JA137⟩; **3**, 449 ⟨32JA2881⟩
Bartlett, R. A., **6**, 194, 195 ⟨88CC1007⟩, 349 ⟨85CC1674⟩
Bartlett, R. J., **1**, 1143 ⟨78JA6930⟩; **3**, 582 ⟨90JA3779⟩
Bartlett, R. K., **1**, 630 ⟨67JCS(C)1964⟩
Bartlett, R. S., **2**, 341, 764 ⟨65JOC2880⟩, 763 ⟨65JOC2877⟩
Bartlett, W. R., **1**, 414, 416, 827 ⟨70JA741⟩; **2**, 1006 ⟨93HOU(E/19d)713⟩; **5**, 159 ⟨70JA741⟩
Bartley, J. P., **1**, 870 ⟨68T1651⟩; **2**, 25 ⟨85AJC1879⟩
Bartling, D., **1**, 1023 ⟨56CB853⟩
Bartmann, E. A., **2**, 232, 724, 725 ⟨93S490⟩
Bartmann, E., **1**, 14 ⟨86AG(E)653⟩; **2**, 553 ⟨86AG(E)653⟩, 558 ⟨90CB1517⟩; **6**, 421, 422 ⟨92GEP4023107⟩
Bartmess, J. E., **1**, 142, 155 ⟨75JA7006⟩, 142, 165 ⟨78JOC3101⟩, 155 ⟨78JOC3095⟩, 865 ⟨86TL5931⟩

Bartocha, B., **2**, 421 ⟨62CJC342⟩, 904 ⟨59ZN(B)809⟩; **4**, 816 ⟨61JA2202⟩
Bartók, M., **1**, 502, 505, 506 ⟨B-85MI 112-02⟩
Bartoletti, I., **5**, 143 ⟨74JOC3318⟩
Bartoli, D., **4**, 92 ⟨91JOC4529⟩
Bartoli, G., **2**, 318 ⟨88TL2251⟩, 346 ⟨93CC1373⟩; **3**, 190 ⟨92SL64⟩, 262, 263 ⟨90S895⟩, 434 ⟨90JOC4456⟩; **4**, 646, 653 ⟨88CC807⟩
Bartoli, J. F., **3**, 699, 700 ⟨78TL2955⟩; **5**, 1097 ⟨78AG(E)781, 81IC3119⟩
Bartolini, O., **2**, 147 ⟨86TL6257⟩
Bartolotti, L. J., **1**, 152 ⟨89JOC175⟩
Barton, A. D., **3**, 459 ⟨48JA2012⟩
Barton, A. E., **3**, 62 ⟨80JA1436⟩
Barton, D., **3**, 492 ⟨B-79MI 313-02⟩; **5**, 11 ⟨91S18⟩
Barton, D. H., **1**, 828 ⟨93CC960⟩
Barton, D. H. R., **1**, 11, 16, 935, 938 ⟨75JCS(P1)1574⟩, 13 ⟨54JCS3045⟩, 16 ⟨81JCS(P1)1501⟩, 16, 938 ⟨81S743⟩, 16, 940 ⟨93TL2733⟩, 17, 938 ⟨91TL7187⟩, 17, 938, 946 ⟨90TL4681⟩, 38, 945 ⟨79TL2291, 80S68, 88TL2571⟩, 172, 940 ⟨85TL757⟩, 223 ⟨85TL3121⟩, 231 ⟨86JCS(P1)1613⟩, 240 ⟨81JOC5027⟩, 247 ⟨88TL3533⟩, 332, 935 ⟨85TL5943, 92T7091⟩, 332, 935, 1130 ⟨86JCS(P1)39⟩, 333 ⟨92TL5017, 93TL6505⟩, 333, 935 ⟨92T7083⟩, 344 ⟨87TL1413⟩, 348 ⟨88CC285⟩, 438 ⟨73JCS(P1)654⟩, 443 ⟨93TL8051⟩, 587 ⟨91TL2569⟩, 595 ⟨70CC1683⟩, 602 ⟨85T4347⟩, 615, 700, 939 ⟨91TL2703⟩, 684, 685, 686 ⟨72JCS(P1)305, 74JCS(P1)1794⟩, 685, 686 ⟨75CC539, 76JCS(P1)2079⟩, 700, 939 ⟨92T7109⟩, 773 ⟨51JCS1048⟩, 775 ⟨49JCS2174⟩, 777 ⟨94T639⟩, 871, 881 ⟨69CC1497⟩, 919 ⟨B-92MI 119-16⟩, 927 ⟨68PAC1⟩, 929, 936 ⟨81JCS(P1)2363⟩, 934 ⟨80CC732⟩, 935 ⟨83CC939, 83TL5889, 84CC1298, 84TL1055, 85T3901, 87T2733, 88T5479, 92T712⟩, 935, 938 ⟨81PAC15, 86PAC675⟩, 935, 939, 946 ⟨92T2529⟩, 937, 938 ⟨92TL5709⟩, 937, 938, 945 ⟨93JOC6838, 91SL435⟩, 938, 946 ⟨90TL3991, 91T8969⟩, 939 ⟨89TL2619, 92T7435, 92TL6629⟩, 940 ⟨82TL2019, 84CC774, 86JCS(P1)1603⟩, 940, 941 ⟨77JCS(P1)1718⟩, 944 ⟨88TL6581, 90JA891, 92JA5904, 92T9261, 93JA8050⟩, 945 ⟨80JCS(P1)2657, 80JCS(P1)2665, 82CC447⟩, 1075 ⟨88T147⟩; **2**, 2 ⟨92ACR504, 92T9195, 92TL3413, 93TL1871, 93TL5689, 94T31⟩, 3 ⟨94TL1431⟩, 8 ⟨76JCS(P1)2112⟩, 13 ⟨65JCS2438, 83TL4979⟩, 15 ⟨70PAC285, 76JA3034, 76JA3036, 80NJC239⟩, 16 ⟨72CC122⟩, 88, 104 ⟨77JCS(P1)1114⟩, 108 ⟨91TL4671⟩, 131 ⟨62JCS1967⟩, 139 ⟨91JOC6697⟩, 195, 688 ⟨70CC1466⟩, 226 ⟨88T1153⟩, 240, 360 ⟨91JOC6702⟩, 284 ⟨78JCS(P1)1682⟩, 309 ⟨74JCS(P1)2101⟩, 335 ⟨74JCS(P1)732, 75CC97⟩, 357 ⟨75CC855⟩, 364 ⟨86TL1153⟩, 390 ⟨80JCS(P1)1212⟩, 391 ⟨60JA2640⟩, 536 ⟨91SL435, 92TL6629⟩, 539 ⟨92TL5041⟩, 608, 612 ⟨68CC806⟩, 627 ⟨87T4321⟩, 630 ⟨62JCS470, 83TL1605, 88T147⟩, 675 ⟨86TL3619⟩, 684 ⟨86TL3615⟩, 690 ⟨81JCS(P1)1473⟩, 700 ⟨88TL1153⟩, 721 ⟨91T9167⟩, 747 ⟨75JCS(P1)1237⟩, 765 ⟨75JCS(P1)1764⟩, 785 ⟨82CC551, 83BSF(2)61⟩, 858 ⟨75CC539, 76JCS(P1)2079⟩; **3**, 8, 121 ⟨81CC1232⟩, 18, 134 ⟨80JCS(P1)1212⟩, 88 ⟨78CC952⟩, 113 ⟨86T5627, B-91MI 304-01⟩, 118, 464 ⟨79CC705⟩, 132 ⟨80JCS(P1)1654, 82TL957⟩, 145 ⟨85JCS(P1)2667⟩, 215 ⟨63JCS3675⟩, 343, 348, 464 ⟨74JCS(P1)1794⟩, 352 ⟨75CC557, 79JCS(P1)1166⟩, 358 ⟨75CC756, 77JCS(P1)515⟩, 366 ⟨72CC632, 73JCS(P1)1580⟩, 368 ⟨89T3741⟩, 391 ⟨75CC539⟩, 391, 527 ⟨76JCS(P1)2079⟩, 405 ⟨84CC337, 84TL3707⟩, 408 ⟨85TL4603, 86JCS(P1)2243⟩, 427, 430 ⟨87T551⟩, 464 ⟨62JCS470⟩, 540 ⟨60JCS1, 77CB3582⟩, 627 ⟨84TL1055⟩, 635 ⟨87T4297, 91TL4713, 92TL5017⟩, 652, 654 ⟨73CC781⟩, 713 ⟨88TL3343⟩; **4**, 190 ⟨75CC432⟩, 206 ⟨71JCS2166⟩, 240 ⟨90JA891⟩, 266 ⟨91T7091⟩, 327 ⟨85T5507⟩, 841 ⟨77JCS(P1)1723⟩, 1077 ⟨83CC939, 83TL4979⟩, 1079 ⟨72JCS(P1)1103⟩; **5**, 247 ⟨91T6127⟩, 278 ⟨71JCS(C)3540⟩, 289 ⟨75JCS(P1)1237⟩, 313, 314, 316 ⟨74JCS(P1)732⟩, 366 ⟨80JCS(P1)1212⟩, 516, 521, 522 ⟨77JCS(P1)1723⟩, 516, 522, 526 ⟨76JCS(P1)2112⟩, 516, 522, 526, 627 ⟨75JCS(P1)1574⟩, 516, 523 ⟨92T7109⟩, 516, 526 ⟨77JPC915, 91TL2703, 78CC393⟩, 523 ⟨79CC755, 80JCS(P1)1650, 81T(S)213⟩, 526 ⟨79CC645, 80JCS(P1)2191⟩, 526, 685, 694 ⟨80JCS(P1)2184⟩, 560 ⟨73JCS(P1)1571⟩, 727 ⟨83TL5889, 86JCS(P1)39⟩, 728 ⟨88T3501⟩, 735 ⟨74JCS(P1)853, 80JCS(P1)665, 91JOC6697⟩, 846 ⟨76JCS(P1)38⟩, 1004 ⟨80NJC127⟩, 1053 ⟨76JCS(P1)2079⟩, 1054 ⟨80JCS(P1)2657⟩, 1110 ⟨92T2613⟩, 1115 ⟨86T4983⟩; **6**, 231 ⟨74JCS(P1)732⟩, 309 ⟨76JCS(P1)2112⟩, 541, 550 ⟨91TL2703⟩, 546, 567 ⟨85TL5943, 86T2329⟩, 550 ⟨93TL2733⟩, 550, 557 ⟨81JCS(P1)2363, 75JCS(P1)1574⟩, 560 ⟨84JCS(P1)1005⟩, 641, 643 ⟨82JCS(P1)2085⟩
Barton, D. L., **2**, 11 ⟨75CC303, 80JOC2566⟩
Barton, J. C., **1**, 120 ⟨86T2873⟩; **2**, 588 ⟨86T2873⟩
Barton, J. W., **3**, 227 ⟨86JCS(P1)961⟩, 468 ⟨71JCS(C)1384⟩
Barton, R. J., **2**, 897 ⟨82IC3731⟩
Barton, T. J., **1**, 64, 522 ⟨73JA3078, 72JA6229, 72JOM(42)C21⟩; **2**, 939 ⟨93JA3794⟩, 1023 ⟨87JA7568⟩, 1091 ⟨83JA6316⟩; **3**, 483, 483, 503 ⟨80JA1584⟩, 502 ⟨76JA7844⟩, 502 ⟨82JA6140⟩, 502 ⟨82OM210, 73JA3078⟩; **4**, 602, 612, 619, 626, 627, 628, 637, 640, 652, 663 ⟨80JA1584⟩, 603 ⟨81TL7⟩, 610 ⟨72JOM(42)C21⟩, 613 ⟨79JOM(168)23⟩, 614 ⟨76JA7746, 78JA6236⟩, 614, 616 ⟨81JA6788⟩, 617 ⟨80JA7979⟩, 627 ⟨73JA3078⟩, 647 ⟨79JOM(172)11⟩, 1048 ⟨85JA3935⟩, 1051, 1052 ⟨82COMC-I(1)221⟩, 1055 ⟨80PAC615⟩; **5**, 873 ⟨77JA6995⟩, 926, 927 ⟨80JA1584⟩; **6**, 179, 190, 202, 203, 672 ⟨80JA1584⟩
Bartosevich, J. F., **5**, 1056 ⟨81EJM317⟩
Bartra, M., **2**, 341 ⟨90T587⟩
Bartroli, J., **1**, 878 ⟨81JA2127, 81PAC1109⟩, 878 ⟨82TL807⟩; **5**, 148 ⟨81PAC1109⟩, 329 ⟨86TL4957⟩
Bartroli, R. M., **4**, 967 ⟨88MI 421-01⟩
Bartsch, B., **4**, 298 ⟨85AP(318)473⟩
Bartsch, H.-J., **6**, 443 ⟨85CB2294⟩
Bartsch, R. A., **1**, 28 ⟨75JA6873⟩; **2**, 26 ⟨83JOC3353⟩, 622 ⟨83JOC876⟩, 632 ⟨79TL2503⟩
Bartsch, R., **2**, 10 ⟨82S326⟩, 440 ⟨83IC3624, 84IC3304⟩; **4**, 548 ⟨90CC1307, 93PS(77)274⟩; **6**, 154 ⟨89CC1046⟩, 154 ⟨92JOM(433)C11⟩, 154 ⟨93CC311⟩
Bartsch, W., **2**, 936 ⟨75CB1732⟩
Bartulin, J., **6**, 627 ⟨66CB3268⟩
Bartz, W., **6**, 666 ⟨68TL3171⟩
Barua, N. C., **4**, 936, 950, 951 ⟨87CB213⟩
Barua, N. G., **3**, 12, 127 ⟨91T7713
Baruah, J. B., **2**, 308 ⟨91T9449⟩
Baruah, P. D., **2**, 744 ⟨89S631⟩
Baru, V., **1**, 926 ⟨84JOC1603⟩
Barve, M. V., **3**, 90 ⟨79S906⟩
Baryshnikov, A. T., **6**, 283, 349, 355 ⟨89IZV2106⟩
Barz, M., **2**, 1031 ⟨85ZN(B)645⟩
Barzen, R., **4**, 970 ⟨82AP(315)680⟩
Barzoukas, M., **1**, 474 ⟨88CC203⟩
Basak, A., **1**, 450, 527, 1061 ⟨83T485⟩; **2**, 930, 940, 943 ⟨83T925⟩
Basak, S., **1**, 988 ⟨81JA4638⟩
Basalgina, T. A., **4**, 637, 641 ⟨75JOM(96)213⟩
Basato, M., **3**, 268 ⟨84CC1593, 84JOC4696⟩, 268 ⟨88JCR(S)246⟩; **4**, 687 ⟨74JCS(D)1856, 74JCS(D)607⟩
Basavaiah, D., **1**, 487, 488 ⟨88JOC239⟩, 509 ⟨82JOM(225)63⟩, 531 ⟨82JOM(225)C1⟩, 531, 662 ⟨82JOC171, 82JOC3806, 82JOC5407⟩, 531, 663 ⟨82JOC1792, 82JOC754⟩; **2**, 900 ⟨86S674⟩, 901 ⟨82JOC3808, 82JOC754, 84S920, 88JOC239⟩; **3**, 135 ⟨82JOC3808, 82JOC754⟩, 275 ⟨83S885⟩; **4**, 815 ⟨82JOC3808⟩; **5**, 229 ⟨84S920⟩
Basch, H., **5**, 685 ⟨B-70MI 517-01⟩

Basche, R. W., **3**, 208 ⟨65JOC2942⟩
Basel, A., **2**, 500 ⟨79ZAAC(448)89⟩
Basenko, S. V., **4**, 140 ⟨91IZV1039⟩
(BASF AG)826, **4**, ⟨82IC645⟩ ⟨82GEP3240287⟩
Basha, A., **1**, 1042 ⟨88TL3455⟩; **2**, 343 ⟨91TL3783⟩, 343, 351 ⟨88TL3455⟩; **3**, 300 ⟨76TL3217⟩, 306 ⟨91TL2003⟩; **5**, 263 ⟨80OS(59)49⟩, 277 ⟨77TL4171⟩
Basha, F. Z., **3**, 311 ⟨82JA536⟩; **5**, 687 ⟨87JHC789⟩
Bashall, A. P., **4**, 190 ⟨75TL3489⟩
Bash, S. F., **4**, 32 ⟨84JFC(26)29⟩
Bashe, R. W., **1**, 118, 119, 469, 473 ⟨69JA4871⟩; **3**, 287 ⟨69JA4871⟩
Bashford, L. A., **2**, 1014, 1017 ⟨38JCS1358⟩
Bashiardes, G., **1**, 1075 ⟨88T147⟩; **2**, 630 ⟨83TL1605, 88T147⟩; **5**, 328, 329 ⟨93JCS(P1)459⟩, 329, 330 ⟨88TL6509⟩
Bashilov, V. V., **4**, 152 ⟨89MI 403-03⟩, 152 ⟨91MI 403-01⟩, 1067 ⟨78JOM(162)271⟩
Bashirova, L. A., **2**, 459 ⟨74ZOB2621⟩
Bashkirov, A. N., **5**, 46 ⟨70MI 502-01⟩
Basi, J. S., **2**, 421 ⟨68CC495, 71JCS(A)1433⟩
Basil, J. D., **2**, 434 ⟨90IS305⟩
Baskakov, Yu. A., **3**, 434 ⟨84ZOR988⟩
Baskar, A. J., **5**, 801 ⟨81JA1467⟩
Baskaran, S., **3**, 113 ⟨89JOC5182⟩
Basolo, F., **5**, 447 ⟨64JA3994⟩
Basquet, C., **4**, 224 ⟨80CAR(80)C17⟩
Bass, A. G., **5**, 680 ⟨90BAU294, 90IZV352⟩
Bass, J. D., **1**, 498 ⟨64JA5570⟩
Bass, K. C., **1**, 101 ⟨73OSC(5)398⟩, 889 ⟨62OS(42)48⟩
Bass, L., **1**, 971 ⟨80JA2134⟩
Bass, L. W., **3**, 455 ⟨24CR(178)811⟩
Bass, R. J., **4**, 211 ⟨85SC225⟩
Bass, S. W., **2**, 215 ⟨80JOC710⟩
Basse, W., **3**, 211 ⟨77CB1007⟩
Bassedas, M., **3**, 182 ⟨88S314⟩
Basselier, J.-J., **2**, 626 ⟨89CJC2061⟩; **6**, 231 ⟨80JFC(15)201⟩, 441 ⟨80JFC(15)37⟩
Basset, J., **1**, 674 ⟨77RTC(M)91⟩
Basset, J.-M., **1**, 674 ⟨77CC502, 85CC1816⟩; **4**, 722 ⟨91OM4005⟩
Bassett, H. L., **5**, 277 ⟨54JCS1188⟩
Bassett, K. E., **2**, 619 ⟨87JOC1619⟩
Bassignani, L., **4**, 205, 210 ⟨79CB148⟩
Bassindale, A. R., **1**, 447, 450 ⟨B-89MI 110-01⟩; **2**, 132 ⟨70JOM(25)389⟩, 700 ⟨75JOM(90)C6⟩, 813 ⟨78JOM(152)C25⟩, 943 ⟨B-89MI 218-03⟩; **4**, 137 ⟨75CJC332⟩, 653 ⟨84TL2705⟩; **5**, 421 ⟨75JOM(99)379⟩, 720 ⟨79JOM(175)273⟩; **6**, 734 ⟨70CC559⟩
Bässler, T., **1**, 1110 ⟨70JA4985⟩, 1114 ⟨69JA2117, 71TL2171⟩, 1114, 1115, 1124 ⟨74JA6686⟩
Bassner, S. L., **5**, 449 ⟨89PAC1723⟩
Basso, N., **4**, 637 ⟨93JA6025⟩
Basso-Bert, M., **1**, 456 ⟨86CC1317⟩
Bassova, G. I., **1**, 76 ⟨67T2235⟩
Bast, K., **5**, 360 ⟨73OSC(5)650⟩
Bast, M. J., **3**, 494 ⟨82SC469⟩
Bast, P., **1**, 876 ⟨87AG(E)1212⟩
Bastiaansen, L. A. M., **3**, 311 ⟨78S675⟩
Bastian, H., **3**, 59, 75 ⟨87S1⟩, 71 ⟨77S1⟩
Bastian, V., **2**, 399 ⟨75CB2329⟩; **4**, 564 ⟨77AG(E)718⟩
Bastide, J., **1**, 516, 549 ⟨B-78MI 112-03⟩
Bastien, I. J., **1**, 893, 899 ⟨64JA1360, 63JA1328⟩; **3**, 282 ⟨63JA1328⟩
Bastos, C., **1**, 76 ⟨89T6901⟩
Bastos, C. M., **5**, 446 ⟨92CB1583⟩
Batail, P., **3**, 699 ⟨78JA4312⟩
Batcheller, S. A., **2**, 947, 949 ⟨88TL3383, 89JA1888⟩; **4**, 628 ⟨88TL3383⟩
Batchelor, R., **2**, 820 ⟨82JA674⟩
Batchelor, R. J., **4**, 291 ⟨87OM2164⟩, 552 ⟨90IC808⟩
Batcho, A. D., **1**, 576, 1021 ⟨86JOC3098⟩
Bateman, L., **2**, 144, 173 ⟨61JCS5339⟩, 146 ⟨54PRS(A)389⟩, 180 ⟨58JCS2834⟩
Bateman, L. R., **4**, 689 ⟨69JA7292⟩
Bateman, L. W., **6**, 343 ⟨82CC773, 83JCS(D)2599⟩
Bates, A. C., **5**, 772, 850, 852, 855 ⟨71JOC1155⟩; **6**, 55 ⟨71JOC1155⟩, 573 ⟨68JOC851⟩
Bates, A. D., **5**, 374, 375 ⟨86TL5285⟩
Bates, E. B., **1**, 513 ⟨54JCS1854⟩
Bates, G. S., **1**, 178, 265 ⟨80CJC716⟩, 722 ⟨⟨75JA3512⟩; **2**, 436 ⟨88JOC4425⟩, 992 ⟨86TL1537⟩; **3**, 636 ⟨83CJC2006⟩; **5**, 234, 239 ⟨76TL4423⟩, 235 ⟨75CJC3693⟩
Bates, H. A., **2**, 8 ⟨86JOC2637⟩; **4**, 414 ⟨86JOC2228⟩
Bates, R. B., **1**, 178 ⟨93JOC4469⟩, 844 ⟨B-83MI 119-03⟩, 890 ⟨B-80MI 119-04⟩, 955 ⟨65JOC2983⟩; **3**, 124 ⟨93JOC4469⟩; **6**, 206 ⟨74JA5640⟩
Bates, R. W., **5**, 164 ⟨91T9929⟩
Bates, T. F., **4**, 602, 650 ⟨89JOM(359)285⟩, 611 ⟨86JA3122⟩, 620 ⟨89JOC1784⟩, 1048 ⟨87JA913⟩; **5**, 926 ⟨87JA913⟩; **6**, 190 ⟨89JOC1784⟩
Bateson, J. H., **4**, 208 ⟨94SL152⟩
Batey, R. A., **1**, 928 ⟨92T8031⟩; **6**, 557 ⟨92T8031⟩
Bathe, A., **1**, 576, 1010 ⟨88LA543, 88LA69⟩
Bathelt, W., **5**, 952 ⟨68JOM(14)P9⟩
Bather, P. A., **3**, 419 ⟨71JCS(C)3060⟩
Bathgate, A., **5**, 330 ⟨87TL5937⟩
Batista de Carvalho, L. A. E., **5**, 506, 507 ⟨89JCS(F2)1945⟩, 506, 508 ⟨90JST67⟩
Batka, H., **5**, 1147 ⟨70M627⟩; **6**, 622 ⟨69M1823⟩
Batlle de Pabon, H. N., **1**, 731 ⟨65JOC1473⟩; **3**, 494 ⟨65JOC1473⟩
Baton, T. J., **4**, 607, 621 ⟨88JA2611⟩
Batra, M. S., **3**, 165 ⟨87IJC(B)1111⟩
Batres, E., **3**, 234 ⟨56JOC1432⟩
Bats, J. W., **1**, 558 ⟨91TL3357⟩; **5**, 174 ⟨87AG(E)362⟩; **6**, 616 ⟨85CB1371⟩
Batsanov, A. S., **6**, 698 ⟨93PS(74)279⟩
Batt, D. G., **1**, 434, 635 ⟨84JA3353⟩
Battaglia, A., **2**, 247 ⟨81CC350⟩; **3**, 368 ⟨71JCS(B)1547⟩, 369 ⟨86CC964, 90JOC3744⟩, 558 ⟨80JOC3766, 82JOC3998⟩, 560 ⟨84JOC2200⟩; **5**, 647, 648 ⟨86JCS(P1)381, 90JOC3744⟩, 665 ⟨72JCS(P2)1911⟩, 1038 ⟨71S378⟩, 1073 ⟨77JOC3372⟩; **6**, 127 ⟨87JCS(P1)2647⟩
Battais, A., **6**, 7 ⟨89JFC(45)215⟩
Battegay, M., **5**, 592, 593 ⟨33HCA999⟩
Batten, R. J., **2**, 232 ⟨80S910⟩
Battersby, A. R., **3**, 106 ⟨81CC797⟩, 419 ⟨55CI(L)1039⟩
Bättig, K., **1**, 599, 779 ⟨77HCA2388⟩
Battin, D. E., **6**, 25, 31 ⟨67JOC2166⟩
Battioni, J. P., **2**, 587 ⟨78BSF(2)575⟩; **5**, 59 ⟨66CR(C)761, 69BSF911⟩
Battioni, P., **2**, 977 ⟨75JA7171⟩
Battiste, M., **1**, 226 ⟨83S284⟩
Battiste, M. A., **1**, 763 ⟨84TL5603⟩, 907 ⟨61CI(L)550, 61JA4101⟩; **4**, 411 ⟨90JOC4777⟩
Battistel, E., **2**, 151 ⟨92TA921⟩
Battisti, A. J., **6**, 573 ⟨68JOC851⟩
Battoni, P., **3**, 230 ⟨80T1331⟩
Batty, D., **3**, 136 ⟨90JCS(P1)2875⟩; **5**, 351 ⟨93TL457⟩
Battye, P. J., **1**, 403, 800 ⟨84CC990⟩, 841 ⟨86CC1807⟩
Batyeva, É. S., **2**, 254 ⟨93PS(79)179⟩, 820 ⟨84ZOB1917⟩; **4**, 345 ⟨86ZOB1905, 86ZOB216⟩, 345 ⟨87DOK(292)360⟩, 476 ⟨69ZOB1235⟩, 921 ⟨84ZOB1324⟩; **5**, 369 ⟨91IZV2103⟩, 373 ⟨73ZOB1451⟩, 405 ⟨77JGU505, 89JGU2388⟩, 406 ⟨85JGU1956, 88JGU1548⟩, 535 ⟨89JGU1548⟩, 737 ⟨85ZOR559⟩; **6**, 581 ⟨88ZOB1468⟩, 582 ⟨76IZV455⟩, 583 ⟨77IZV1177⟩
Batyrbaev, N. A., **4**, 1080 ⟨83ZOB416⟩; **6**, 478 ⟨81ZOR1539⟩
Batz, M., **4**, 553 ⟨92ZN(B)725⟩
Bätzel, V., **6**, 340 ⟨72JOM(46)149, 75JOM(102)109⟩, 340 ⟨76JOM(113)67, 76ZN(B)342⟩, 340 ⟨78CB1603⟩, 341 ⟨76CB3339⟩
Bau, R., **4**, 696 ⟨82JA2072⟩, 697, 698 ⟨88JA3231⟩

Bau, Y., **2**, 365 ⟨70CC777⟩
Bauchat, P., **2**, 179 ⟨91CL1391⟩
Baudat, R., **2**, 282 ⟨79HCA1406⟩; **3**, 38, 174 ⟨79HCA1406⟩
Baude, E. A., **3**, 451 ⟨51JCS1762⟩
Bauder, A., **2**, 636 ⟨83JST(97)47⟩
Bauder, M., **5**, 386 ⟨70CB583⟩
Baudet, P., **4**, 203 ⟨70HCA1330⟩
Baudin, J.-B., **1**, 621 ⟨88TL3255⟩, 622 ⟨93SL839⟩, 963 ⟨90TL213⟩; **2**, 201 ⟨88TL3255⟩, 247, 361, 363 ⟨88TL3251⟩, 359 ⟨87BSF181⟩, 359 ⟨87T881, 91T6655⟩; **3**, 16 ⟨85TL2333⟩, 369 ⟨92SL909⟩
Baudis, U., **2**, 496 ⟨78ZAAC(439)139⟩, 496, 501 ⟨76ZAAC(423)40⟩
Baudisch, H., **1**, 657 ⟨73TL401⟩
Baudisch, O., **2**, 779 ⟨40SCI336⟩
Baudler, M., **2**, 471 ⟨76ZN(B)1305, 77ZN(B)1244, 83MI 208-01, 83ZAAC(498)57, 85CB3268⟩, 472 ⟨78CB3838, 83MI 208-02, 85AG(E)504, 85ZN(B)1274, 91CB1153, 91CB69, 92ZAAC(611)47⟩, 473 ⟨83AG(E)775⟩, 474 ⟨86ZAAC(533)146⟩, 507 ⟨82ZAAC(485)129⟩, 508 ⟨81AG(E)123⟩; **4**, 545 ⟨71AG(E)940, 73ZN(B)224, 77ZN(B)1490, 78ZN(B)1208, 82ZAAC(486)39, 87ZAAC(544)87⟩, 546 ⟨78ZN(B)691⟩, 547 ⟨78GEP2705994⟩, 1025 ⟨84CB1542⟩; **6**, 366 ⟨83AG(E)632⟩, 605 ⟨87CB421⟩, 680, 682, 685 ⟨88CB281⟩
Baudoux, D., **3**, 566 ⟨88JOC18⟩
Baudouy, R., **1**, 197 ⟨81JCR(S)278⟩, 464 ⟨79TL937⟩, 515, 1025 ⟨74S573⟩, 958 ⟨79NJC321⟩, 960 ⟨74TL1593⟩, 962 ⟨75BSF2159, 75BSF2166, 73TL3361⟩
Baudry, D., **2**, 675, 676 ⟨78CC694⟩
Bauduin, G., **4**, 191 ⟨73T4225⟩, 191 ⟨77T3105⟩
Baudy, M., **6**, 629 ⟨78JOC3732⟩
Baudy-Floc'h, M., **5**, 652 ⟨87S306⟩
Bauer, B., **6**, 553 ⟨88PS(35)93⟩
Bauer, C., **4**, 679 ⟨81JOM(204)C21⟩, 686 ⟨82CB878, 82JOM(226)C59⟩, 686 ⟨82JOM(236)C18, 83JOM(258)81⟩, 689 ⟨83JOM(256)147⟩
Bauer, D. P., **2**, 28 ⟨75JOC1990⟩
Bauer, H., **2**, 1094 ⟨B-87MI 223-02, B-87MI 223-03, B-87MI 223-04⟩, 1094, 1095 ⟨B-86MI 223-01, B-87MI 223-01⟩; **4**, 425 ⟨61JOC1649⟩; **5**, 838 ⟨06CB1480⟩; **6**, 268 ⟨83AG(E)730⟩
Bauer, I., **5**, 874 ⟨86AG(E)1093⟩
Bauer, K., **4**, 120 ⟨86ACH145⟩
Bauer, L., **4**, 492 ⟨72CZ691⟩; **5**, 90 ⟨56JOC1182⟩, 332 ⟨74AG(E)419⟩, 1000 ⟨59JOC1293, 74AG(E)376⟩
Bauer, L. N., **4**, 47, 48, 49, 51 ⟨35JA2364⟩
Bauer, M., **5**, 1103 ⟨79GEP142333⟩
Bauer, O. H., **5**, 623, 624 ⟨76LA1584⟩, 624, 697 ⟨75LA305⟩
Bauer, P., **6**, 653 ⟨00CB1058⟩
Bauer, P. E., **1**, 630 ⟨85JOC5460⟩; **4**, 938 ⟨83JOC34⟩
Bauer, R., **1**, 665 ⟨52CB593⟩; **2**, 403 ⟨78AG(E)370⟩, 403, 420 ⟨78AG(E)202⟩
Bauer, R. F., **1**, 1008 ⟨87JMC193⟩
Bauer, S., **6**, 692 ⟨90OM793⟩
Bauer, V. J., **5**, 164 ⟨61JA606⟩; **6**, 650 ⟨68JA6846⟩, 655 ⟨67JMC273⟩
Bauer, W., **1**, 873 ⟨85CB764⟩, 1088 ⟨88JA6033⟩; **2**, 999, 1000 ⟨89OM1688⟩, 1002 ⟨86AG(E)1103⟩; **3**, 607 ⟨93TL2915⟩; **4**, 604, 647 ⟨91S1221⟩; **5**, 232, 233 ⟨85HOU(E5)832⟩, 629 ⟨85HOU(E5)1266⟩, 635, 636 ⟨85HOU(E5)1279⟩, 639 ⟨85HOU(E5)1291⟩, 699 ⟨85HOU(E5/2)1218⟩, 846 ⟨85HOU(E5)1300⟩; **6**, 469 ⟨79HCA398⟩
Bauer, W. V., **6**, 412 ⟨68FRP1526958⟩
Bäuerler, P., **2**, 1006, 1009 ⟨87JA882⟩
Bauermeister, S., **1**, 828 ⟨91JCS(P1)561⟩; **5**, 371, 373, 707 ⟨93HAC11⟩, 714 ⟨92PS(72)201⟩
Bauker, Y. I., **4**, 827 ⟨84ZOB1925⟩
Bauknight, Jr., C. W., **5**, 809 ⟨84JA4266, 88JOC4443⟩; **6**, 256, 261 ⟨84JOC3590⟩, 259, 261, 615 ⟨90JA728⟩, 261, 603, 615 ⟨88JOC4443⟩, 603 ⟨84JA4266⟩
Baukova, G. G., **6**, 626 ⟨91ZOB1202⟩
Baukova, T. V., **4**, 696 ⟨74JOM(65)131⟩, 723 ⟨69IZV2032⟩
Baukov, I. Y., **4**, 661 ⟨70ZOB2607⟩
Baukov, Y. I., **2**, 172 ⟨80ZOB1059⟩, 209 ⟨84ZOB1842⟩; **3**, 484 ⟨75ZOB819⟩, 534 ⟨69USP3462489, 70ZOB707⟩, 540 ⟨73ZOB2088⟩; **4**, 418 ⟨84JGU635⟩, 605 ⟨84ZOB1842⟩, 637 ⟨76MI 414-01⟩; **5**, 374 ⟨88ZOB1093⟩, 376 ⟨69JOM(17)241, 73ZOB1753⟩, 376, 389 ⟨74ZOB1753, 73ZOB603⟩, 389, 717 ⟨78ZOB1146⟩, 390 ⟨70ZOB707⟩, 716, 1006 ⟨72ZOB1868⟩, 1016 ⟨66ZOB952⟩; **6**, 132 ⟨84ZOB1842⟩
Baukov, Yu. I., **1**, 62 ⟨70MI 102-02⟩; **2**, 700 ⟨68JOM(14)229⟩, 701 ⟨70JOM(24)359⟩; **4**, 98, 99 ⟨91JGU1875⟩, 100 ⟨89JGU200⟩, 104 ⟨89JGU1625⟩, 140 ⟨89JOM(361)147⟩, 886 ⟨69JOM(17)241⟩, 915 ⟨75ZOB556⟩; **5**, 702 ⟨91JGU1268, 91ZOB1389⟩, 716 ⟨72JGU1857⟩, 717 ⟨78JGU1046⟩, 723 ⟨89JGU2055, 89ZOB2293⟩
Bauld, N. L., **1**, 248 ⟨87ACR371⟩, 805 ⟨84JA2730⟩
Bauldler, M., **2**, 865 ⟨77ZAAC(432)67⟩
Baum, G., **2**, 1000 ⟨86AG(E)947⟩; **4**, 672, 673 ⟨90ZN(B)221⟩, 1082 ⟨87AG(E)546⟩; **5**, 228 ⟨70JOM(22)269⟩, 685 ⟨86TL2743, 89CB657⟩, 780 ⟨91USP4997967⟩, 781 ⟨90ZN(B)1273⟩, 928 ⟨87AG(E)546, 87AG(E)798, 88AG(E)961, 88ZN(B)801⟩; **6**, 188, 711 ⟨87AG(E)798⟩, 202, 724 ⟨87AG(E)546⟩, 397, 714 ⟨88AG(E)961⟩, 398 ⟨86AG(E)1111, 87PAC1011⟩, 714 ⟨88ZN(B)801⟩
Baum, J., **3**, 357 ⟨73CC833, 76JOC818⟩; **6**, 115 ⟨76JOC813⟩, 644 ⟨75C514⟩
Baum, J. C., **6**, 255 ⟨91CJC2127⟩
Baum, J. S., **1**, 907 ⟨74CRV189⟩; **3**, 193 ⟨87SC1709⟩
Baum, K., **1**, 1130 ⟨82JOC2251⟩; **2**, 95 ⟨75JOC2536⟩, 317 ⟨89S450⟩, 335 ⟨69JA1679, 70JOC1545⟩, 339 ⟨74JA3237⟩; **4**, 108 ⟨89JOC2869⟩, 109 ⟨88JOC4645⟩, 196, 810 ⟨82JOC2251⟩, 428 ⟨85JOC1699, 90JOC2920⟩, 439 ⟨68JA7083, 69JOC2046⟩, 439 ⟨69JOC2049⟩, 731 ⟨90JOC3562⟩, 795 ⟨91JOC537⟩, 970, 993 ⟨92JOC235⟩, 994 ⟨85JOC2736⟩; **5**, 313, 314, 989 ⟨70JOC1545⟩, 967 ⟨80JOC347, 82JOC2251⟩, 967 ⟨91JOC3043⟩; **6**, 55 ⟨70JOC846, 90MI 602-01⟩, 56, 282 ⟨68JOC3080⟩, 57 ⟨79S311⟩, 150 ⟨92JOC3026⟩, 440 ⟨70JA2096⟩, 649 ⟨69JOC2840⟩
Baum, M. W., **2**, 528 ⟨91JOC751⟩; **6**, 446 ⟨91JOC751⟩
Bauman, B. A., **2**, 675 ⟨78JOC2073⟩; **6**, 429 ⟨78JOC752⟩
Baumann, B. C., **1**, 826 ⟨71HCA1589⟩
Baumann, E., **4**, 244 ⟨1885CB883⟩
Baumann, H., **5**, 138 ⟨71MI 503-01⟩
Baumann, J. B., **2**, 692 ⟨75S782⟩
Baumann, M., **1**, 422 ⟨76TL3585⟩; **6**, 213 ⟨B-59MI 607-01⟩, 573 ⟨72TL2939⟩
Baumann, N., **2**, 496, 695 ⟨75ZAAC(418)167⟩, 501 ⟨74ZAAC(408)261, 75ZAAC(418)279⟩
Baumann, T., **2**, 454, 457 ⟨83CB2936⟩
Baumann, W. J., **2**, 272 ⟨64JOC3055⟩
Baumanns, J., **6**, 220 ⟨76JFC(8)177⟩
Baumbach, E.-A., **3**, 453 ⟨69LA(726)81⟩
Baumberger, F.,**1**, 321, 946 ⟨83HCA2210⟩
Baumegger, A., **5**, 923, 924 ⟨92OM3088⟩
Baumeister, U., **2**, 258 ⟨93PS(79)131⟩; **4**, 559 ⟨83CB2275⟩, 575 ⟨82ZAAC(494)55⟩; **5**, 1092, 1093, 1095 ⟨83CB2275⟩
Baumeister, W., **4**, 699 ⟨72JOM(36)267, 72JOM(36)277⟩
Baumgardner, C. L., **3**, 224 ⟨82TL2379⟩
Baumgarten, H. E., **1**, 637 ⟨58JA4588⟩; **2**, 374 ⟨76JOC3805⟩, 412 ⟨61JOC1533⟩; **3**, 468 ⟨61JA399⟩; **4**, 440 ⟨54JA3489⟩; **5**, 315 ⟨61JA4469⟩, 698, 813 ⟨66JOC3708⟩, 999 ⟨65JA1141, 75JOC3554⟩
Baumgarten, K. G., **4**, 397 ⟨80JOM(194)9⟩
Baumgarten, R. J., **1**, 28 ⟨74JOC3525⟩, 28, 35 ⟨B-82MI 102-01⟩, 32, 35 ⟨66JCE398⟩, 636 ⟨81TL199⟩; **2**, 366 ⟨74JOC3525⟩; **3**, 471 ⟨64JOC2070⟩

Baumgarten, R. L., **2**, 764 ⟨64JA1139⟩
Baumgartner, G., **6**, 167, 371 ⟨92CB1333⟩
Baumgärtner, J., **4**, 660, 663 ⟨78JOM(148)247⟩
Baumgartner, M. T., **1**, 474 ⟨87TL4653⟩
Bäuml, E., **4**, 192 ⟨89S128⟩; **6**, 252 ⟨88S961⟩
Baumstark, A. L., **1**, 688, 690 ⟨76TL3265⟩, 688, 690 ⟨78JOC3609⟩
Baune, J., **4**, 673 ⟨93CB89⟩
Baunitzer, G., **5**, 85 ⟨57AG189⟩
Baur, K., **4**, 186, 187 ⟨35USP2000252⟩
Baur, R., **3**, 292 ⟨88HCA531⟩; **6**, 143 ⟨82CB1721⟩
Baures, P. W., **4**, 277 ⟨90SL359⟩
Bauroth, B., **6**, 520 ⟨92AG(E)1064⟩
Baus, U., **1**, 153 ⟨84T1345⟩; **4**, 890 ⟨93JPO59⟩
Bausch, M. J., **1**, 922 ⟨90JOC58⟩
Bausher, L. P., **2**, 807 ⟨73JA8066⟩
Bauta, W. E., **2**, 607 ⟨86JOC277⟩; **5**, 934 ⟨89JOC3249⟩
Bautista, J., **1**, 145 ⟨90JOC2234⟩
Bavaro, L. M., **6**, 343, 344 ⟨85JOM(287)357⟩
Bavert, K., **2**, 468 ⟨89AG(E)1675⟩
Bavin, E. M., **4**, 206 ⟨50JPP764⟩
Bavry, R. H., **2**, 157 ⟨70JOC4217⟩
Bawa, A., **2**, 785 ⟨85SC1253⟩; **3**, 155 ⟨89CL1241⟩
Bax, H.-J., **1**, 725 ⟨74CB1773⟩
Baxi, A. J., **6**, 561 ⟨85MI 617-02, 86MI 617-01⟩
Baxter, A. D., **1**, 826 ⟨83CC932⟩
Baxter, A. G. W., **4**, 748 ⟨91JCS(P1)2575⟩
Baxter, A. J. G., **5**, 235 ⟨80CC429, 80TL5071⟩
Baxter, E. W., **1**, 835 ⟨89JOC2893, 90JA7682⟩
Baxter, G. J., **3**, 541 ⟨77AJC459⟩
Baxter, J. N., **5**, 1035 ⟨54CI(L)785, 56JCS659⟩
Baxter, R. L., **2**, 127 ⟨92JCS(P1)255⟩
Baxter, S. G., **2**, 446 ⟨87TL603⟩
Baxter, S. M., **2**, 969 ⟨87JA203⟩
Bay, E., **1**, 18, 564 ⟨84JA6702⟩; **2**, 7, 621 ⟨90JOC3415, 88JOC2858⟩; **6**, 410 ⟨89EUP310255⟩, 432 ⟨88USP4754072⟩, 533 ⟨87USP4754072⟩
Bayandina, E. V., **2**, 857 ⟨78ZOB2673⟩
Bay, P. G., **5**, 748 ⟨55MI 519-01⟩
Bay, W. E., **2**, 440 ⟨86EUP194716⟩
Baydar, A. E., **5**, 367 ⟨76CC718⟩
Baye, L. J., **4**, 697 ⟨64JINC1531⟩
Bayer, A. G., **2**, 241 ⟨85GEP3332269⟩; **6**, 657 ⟨77GEP2547512⟩
Bayer, C., **3**, 38, 174, 207 ⟨81JOC2920⟩
Bayer, E., **6**, 213 ⟨B-72MI 607-02⟩
Bayer, M., **3**, 147 ⟨80JA2508⟩
Bayer, O., **4**, 920 ⟨59CA(53)10040f⟩; **5**, 738 ⟨59USP3053876⟩; **6**, 55 ⟨61CB2594⟩, 73 ⟨60GEP1084733⟩, 316 ⟨63BEP627486⟩
Bayer, P., **2**, 281 ⟨74AG(E)804⟩, 288, 1001 ⟨74AG(E)274⟩
Bayer, R., **5**, 689 ⟨76S46⟩
Bayer, R. P., **6**, 88 ⟨62JA1751⟩
Bayha, H., **4**, 574 ⟨76ZAAC(421)1⟩
Bayler, A., **4**, 313 ⟨95T755⟩
Bayles, R., **5**, 292 ⟨77S31, 77S33⟩
Bayless, J., **1**, 630 ⟨67JA5736⟩, 631 ⟨65JA659⟩
Bayless, J. H., **1**, 631 ⟨68JA531⟩
Bayless, P. L., **1**, 399 ⟨55JOC1119⟩, 881 ⟨54JOC1490⟩
Baylis, E. K., **2**, 451 ⟨84JCS(P1)2845⟩; **4**, 464 ⟨84JCS(P1)2845⟩
Baylouny, R. A., **1**, 774 ⟨59JA2126⟩
Bayod, M., **1**, 68 ⟨88S144⟩, 833 ⟨84CC1417⟩; **2**, 676 ⟨88S144⟩, 1097 ⟨88CC121⟩
Bayomi, S. M., **3**, 635 ⟨78CPB784⟩
Bayón, A. M., **4**, 300 ⟨88JCS(P1)1631⟩, 412 ⟨88CB1813⟩
Bayreuther, N., **5**, 1018 ⟨71CB2588⟩
Baysdon, S. L., **5**, 457 ⟨84JA6451⟩
Bayyuk, S., **4**, 317, 416 ⟨86S804⟩; **6**, 561 ⟨85H(23)3099⟩
Bayyuk, S. I., **4**, 301 ⟨91S279⟩
Bazan, G. C., **3**, 513 ⟨90JA3875⟩, 514 ⟨90JA8378⟩
Bazanova, I. M., **4**, 864 ⟨90UKZ1310⟩
Bazant, V., **2**, 478 ⟨68AG(E)112⟩, 535 ⟨B-65MI 210-01, B-73MI 210-02⟩
Bazavova, I. M., **4**, 272 ⟨91ZOR216⟩, 275, 277 ⟨89UKZ1216⟩, 865 ⟨89ZOR1102⟩; **5**, 593 ⟨74ZOR1992, 76ZOR681⟩, 593 ⟨76ZOR69⟩; **6**, 254 ⟨88ZOR1327⟩
Bazazova, I. M., **6**, 564 ⟨81ZOR191⟩
Bazlen, M., **4**, 234 ⟨05CB1057, 09CB4634⟩
Beaber, N. J., **2**, 135 ⟨25JA1449⟩
Beach, J. W., **4**, 231 ⟨92JOC2217⟩
Beach, R. G., **2**, 421 ⟨71IC1888, 71IC906⟩
Beacham, L. M., **2**, 219 ⟨71JOC488⟩
Beachell, H. C., **3**, 711 ⟨52JA5247⟩; **5**, 1133 ⟨52JA5247⟩
Beachem, M. T., **2**, 271, 273 ⟨59JA5430⟩
Beachley, Jr., O. T., **2**, 511 ⟨65JCS3241⟩; **4**, 642, 659 ⟨80IC1021⟩, 652 ⟨86IS95⟩, 659 ⟨88OM1516⟩
Beadle, E. C., **2**, 530 ⟨87TL4499⟩
Beagley, B., **2**, 197 ⟨92JCS(P1)2371⟩, 881 ⟨87AG(E)264⟩; **4**, 274 ⟨90JCS(P1)773⟩, 281 ⟨92JCS(P1)2371⟩
Beak, P., **1**, 137 ⟨81JOC2363⟩, 148 ⟨83JA2080, 84CRV471, 88JA8145, 89JOC458⟩, 151 ⟨89TL1197, 90JOC2578, 93JOC1109, 93JOC823⟩, 240 ⟨87JOC218⟩, 363 ⟨86JOC4627⟩, 613 ⟨89JOC1647⟩, 889 ⟨67JOC2778⟩; **2**, 302 ⟨84CR471⟩, 305 ⟨82JOC2822⟩, 328 ⟨89JOC5574⟩, 965 ⟨82ACR306⟩, 999 ⟨81JOC2363⟩, 1002 ⟨84CRV471⟩; **3**, 90 ⟨82ACR306⟩, 182 ⟨81JOC4316⟩, 182 ⟨89TL1197⟩, 183 ⟨84CRV471⟩, 322 ⟨83TL2945⟩; **4**, 138 ⟨86JOC3325⟩, 255, 394 ⟨72JA597⟩, 360, 382, 388 ⟨78JOC4255⟩, 382 ⟨77JA5213, 81JOC2363⟩, 382, 392 ⟨78CRV275, 78JA5428⟩, 392 ⟨87JA5403⟩, 394 ⟨75JOC3052⟩, 511, 530 ⟨89TL1197⟩, 529, 535, 537, 1014, 1015, 1016 ⟨84CRV471, 84JA1010⟩; **5**, 237 ⟨77TL1839⟩; **6**, 555 ⟨75JOC3052⟩
Beal, D. A., **2**, 13 ⟨74S560, 76JOC1627⟩, 619 ⟨70JOC719⟩
Beal, G. D., **5**, 30 ⟨32OSC(1)379⟩
Beal, III, P. F., **3**, 474 ⟨49JA1506⟩; **5**, 229 ⟨50JA5163⟩
Beal, R. B., **1**, 815, 832 ⟨88JOC4508⟩; **3**, 528 ⟨88JOC4508⟩
Beam, C. F., **4**, 449 ⟨82JCED475⟩
Beames, D. J., **2**, 587 ⟨72JA7210⟩; **4**, 845 ⟨73JA5829⟩; **5**, 171 ⟨74TL2439⟩, 235 ⟨73JA5829⟩
Bean, G. P., **1**, 114 ⟨62JA4660⟩; **3**, 308 ⟨65JHC473⟩
Bean, H. J., **3**, 615, 616 ⟨44OSC(2)29⟩; **5**, 85 ⟨43OSC(2)29⟩
Bean, M., **6**, 479 ⟨83JOC5033⟩
Beanan, L. R., **6**, 343 ⟨83OM1179⟩
Beard, C. C., **1**, 967 ⟨91JOC1083⟩; **6**, 314 ⟨77USP4031228⟩
Beard, C. D., **1**, 1118 ⟨69JA5677, 70JA7564⟩, 1144 ⟨70JA4309⟩; **2**, 95 ⟨75JOC2536⟩, 339 ⟨74JA3237⟩, 630, 801 ⟨70JA4309, 70JA7564⟩, 801 ⟨70JOC2412⟩
Beard, R. D., **1**, 144 ⟨73JOM(59)53⟩; **4**, 393 ⟨73JOM(59)53⟩
Beard, R. L., **1**, 149 ⟨91JOC2091⟩
Beard, R. M., **1**, 709 ⟨76TL1243⟩
Bearman, H., **2**, 449, 466 ⟨74S119⟩
Bearn, L., **4**, 302 ⟨78BSF(2)83⟩
Bearse, A. E., **2**, 89 ⟨51IEC1596⟩
Beaton, J. M., **2**, 391 ⟨60JA2640⟩; **3**, 331, 341 ⟨60JA3800, 90T7599⟩
Beatson, R. P., **4**, 78 ⟨77CJC2323⟩
Beattie, M. S., **1**, 587 ⟨90JA6408⟩
Beattie, T. R., **4**, 923 ⟨71CA(75)88759m⟩
Beau, J., **1**, 1038 ⟨91TL3171⟩
Beau, J.-M., **1**, 489 ⟨87JA4756⟩; **4**, 950, 952 ⟨86TL6201⟩
Beaucage, S. L., **2**, 107 ⟨77JA1277⟩, 409, 410 ⟨81TL1859, 85JOC2019⟩
Beauchamp, P. D., **1**, 811 ⟨86JA5901⟩
Beaucourt, J.-P., **1**, 536 ⟨82TL63⟩; **2**, 47 ⟨89AG(E)755⟩, 753 ⟨86TL6193⟩, 968 ⟨82TL63⟩; **3**, 212 ⟨73S537⟩; **6**, 480 ⟨90MI 614-02⟩
Beaudet, I., **1**, 1020 ⟨92TL3647⟩
Beaudet, R. A., **6**, 183 ⟨70JA4158⟩
Beaudoin, S., **1**, 645 ⟨84JA5754⟩; **2**, 50 ⟨93JOC1991⟩
Beaujon, J. H. R., **5**, 90 ⟨73OSC(5)376⟩
Beaulieu, F., **2**, 707 ⟨92S112⟩, 942 ⟨92SL112⟩; **6**, 556 ⟨92S112⟩

Beaulieu, N., **1**, 624, 626 ⟨86JA1039⟩; **5**, 278 ⟨86JA1039⟩

Beaulieu, P. L., **1**, 174, 547, 1131 ⟨84JOC1313⟩; **3**, 621 ⟨84JOC1313⟩

Beaupere, D., **2**, 7 ⟨91TL3977⟩, 25 ⟨92TL3129, 92TL4901⟩

Beaven, G. H., **5**, 84 ⟨52MI 502-03⟩

Beaven, W., **2**, 381 ⟨84JHC1709⟩

Beaver, D. J., **5**, 1027 ⟨57JA1236⟩

Beavers, W. A., **6**, 206 ⟨74JA5640⟩

Bebenburg, W. von, **3**, 467 ⟨53CB278⟩

Bebernitz, G. E., **4**, 961 ⟨85TL169⟩

Beccalli, E. M., **1**, 1076 ⟨85JOC2372⟩; **2**, 1034, 1035 ⟨85JOC2372⟩

Becerra, R., **2**, 18 ⟨60JA4001, 60JA4007⟩

Bechara, E. J. H., **1**, 688, 690 ⟨76TL3265⟩

Becher, H. J., **2**, 110 ⟨55ZAAC(282)86⟩, 828, 866 ⟨73CB177⟩; **5**, 400 ⟨76CB359⟩; **6**, 518 ⟨73AG(E)842⟩

Becher, J., **3**, 66 ⟨81T2375⟩, 342 ⟨91S609⟩; **4**, 436 ⟨72ACS1659, 74ACS539⟩; **5**, 593 ⟨76TL3347, 77T341⟩, 1066 ⟨88JOC4654⟩

Becher, R., **2**, 478 ⟨85ZN(B)1090⟩

Becherer, J., **6**, 538, 539 ⟨89EUP355578⟩

Bechgaard, K., **3**, 39 ⟨84CC89⟩; **6**, 561 ⟨91JOC5684⟩, 590 ⟨83CC295⟩, 591 ⟨75JOC746, 84CC89⟩, 592 ⟨89CC1520⟩

Bechtold, R. A., **4**, 1054 ⟨78JOC950⟩; **6**, 182, 199 ⟨78JOC950⟩

Bechtolsheimer, H. H., **2**, 853 ⟨77ZN(B)584⟩

Bechtolt, S. G., **4**, 413 ⟨59JA3789⟩

Beck, A., **2**, 134 ⟨28JPR198⟩

Beck, A. K., **1**, 441 ⟨75CB314⟩, 716 ⟨77LA846⟩; **2**, 279 ⟨75CB314⟩, 281 ⟨74AG(E)806, 77LA846⟩, 394 ⟨81HCA2264, 82HCA1101⟩, 420 ⟨93OS(72)62⟩; **3**, 192 ⟨85HCA162⟩; **4**, 250 ⟨71OS(51)39⟩, 290, 402 ⟨75CB314⟩, 394 ⟨74AG(E)806⟩, 979 ⟨91CB2897⟩; **6**, 82, 86, 88, 302, 303, 317 ⟨72CB3280⟩, 88, 317 ⟨72CB3892⟩, 302 ⟨72CB3905, 78CB3644⟩

Beck, B. M., **5**, 482 ⟨64LA(673)124⟩; **6**, 535 ⟨64LA(674)124⟩, 578 ⟨61LA(646)96⟩

Beck, F., **5**, 654, 655, 657 ⟨62CB126⟩

Beck, G., **3**, 467 ⟨50CB137⟩, 680 ⟨87CB2031⟩, 702 ⟨88CB461⟩, 716 ⟨89JOM(379)97⟩; **5**, 658 ⟨89ZN(B)1414⟩, 1087 ⟨87CB461, 88AG(E)1344, 89CB1907⟩; **6**, 606 ⟨79GEP2753204, 79GEP2754604⟩, 606, 619, 721 ⟨88AG(E)1344, 89CB1907⟩, 617 ⟨76GEP2451635⟩

Beck, H., **2**, 724 ⟨60CB2736⟩

Beck, H. P., **5**, 396 ⟨77ZAAC(430)77, 77ZAAC(430)91⟩

Beck, I., **2**, 788 ⟨69MIP156590⟩

Beck, J. A., **4**, 674 ⟨80JOM(202)C49⟩

Beck, J. R., **1**, 486 ⟨78T2057⟩; **5**, 291 ⟨68JA4706⟩, 1111 ⟨78JOC1604⟩

Beck, P., **2**, 427, 444 ⟨B-72MI 208-03⟩, 428, 430, 431, 441, 822, 827 ⟨58CB1583⟩, 454 ⟨61CB1317⟩, 462, 821, 827 ⟨59CB2088⟩, 822, 823 ⟨68CB2899⟩, 834, 835 ⟨61CB2122⟩; **4**, 576 ⟨60CB1371, 61CB1323⟩

Beck, T. M., **2**, 833, 837 ⟨55JA929⟩

Beck, W., **1**, 517 ⟨84CB3169⟩; **2**, 252 ⟨83ZN(B)365⟩; **3**, 568 ⟨66ZAAC(344)285, 67JOM(8)547⟩, 568 ⟨67ZAAC(354)69, 73CB2144⟩; **4**, 710 ⟨86AG(E)734, 89CB1901⟩, 710 ⟨93AG(E)923⟩, 1019 ⟨76JOM(121)211⟩; **5**, 386 ⟨70CB583⟩, 446, 448 ⟨73AG(E)77⟩, 936 ⟨92JOM(426)343⟩, 944 ⟨93JOM(460)C25⟩, 955 ⟨91CB2431⟩, 983 ⟨67AG(E)169⟩, 1088, 1089 ⟨88JOM(347)101⟩, 1097 ⟨90AG(E)318⟩

Becke, F., **5**, 601 ⟨69CZ474⟩

Becke-Goehring, M., **2**, 361 ⟨59CB1149⟩; **4**, 592, 593 ⟨69ZAAC(370)31⟩

Becker, A., **6**, 187, 387 ⟨91OM1647, 91OM3838⟩

Becker, B., **4**, 94 ⟨88GEP3639877A1⟩, 136, 368 ⟨90JOM(388)57⟩, 141 ⟨92JOM(438)45⟩, 142, 362 ⟨88JOM(354)147⟩, 352, 353 ⟨92CB591⟩

Becker, C., **6**, 174 ⟨93IC2308⟩

Becker, D., **1**, 498 ⟨82JOC3297⟩, 622 ⟨92T2515⟩; **2**, 422 ⟨84JOC2494⟩

Becker, D. A., **5**, 555 ⟨82SC829⟩

Becker, D. P., **2**, 926 ⟨87JA7495⟩; **3**, 114 ⟨93TL2087⟩

Becker, E. I., **1**, 428 ⟨60JOC2203⟩; **2**, 753 ⟨73OSC(5)552⟩; **3**, 452 ⟨55JA984⟩, 459 ⟨66JOC3375⟩, 698, 721 ⟨63JOC1707⟩

Becker, G., **2**, 188 ⟨75S55⟩, 428 ⟨78ZAAC(443)42⟩, 473, 474 ⟨70ZAAC(372)171⟩, 478 ⟨90IS249⟩, 507 ⟨82ZAAC(486)70⟩, 508, 509 ⟨80ZAAC(462)113⟩, 509, 511 ⟨89ZAAC(569)62⟩; **3**, 688 ⟨81ZN(B)16⟩; **4**, 479, 480 ⟨84CC1634⟩, 484 ⟨86S31⟩, 549 ⟨81ZAAC(479)41⟩, 568 ⟨87PS(30)349⟩; **5**, 395, 879 ⟨81ZAAC(473)7⟩, 396 ⟨77ZAAC(430)77, 77ZAAC(430)91, 92ZAAC(608)33⟩, 396, 878, 879, 880 ⟨77ZAAC(430)66⟩, 396, 879, 891 ⟨91ZAAC(605)7⟩, 399 ⟨77AG(E)463⟩, 400 ⟨80ZAAC(468)55⟩, 839, 878, 879 ⟨86S31⟩, 878, 879 ⟨76ZAAC(423)242, 81ZAAC(480)21⟩, 879, 880 ⟨85MI 523-01, 92ZAAC(612)72⟩, 882 ⟨81ZAAC(479)41⟩, 891 ⟨80ZAAC(470)131, 80ZAAC(470)144⟩, 892 ⟨81ZN(B)1080⟩, 1086 ⟨83PS(18)7⟩, 1086, 1087, 1092 ⟨83JA7194⟩, 1154 ⟨75CB2484, 81ZN(B)16⟩, 1155 ⟨85MI 531-01⟩; **6**, 117 ⟨84CC1634⟩, 160 ⟨80ZAAC(462)130, 82ZAAC(485)23⟩, 195 ⟨92CB1547⟩, 324, 684 ⟨84ZAAC(517)75⟩, 582, 684 ⟨85ZAAC(520)120, 85ZAAC(520)139⟩, 584, 680, 684 ⟨84ZAAC(518)21⟩, 678 ⟨83PS(14)267, 80ZAAC(463)144⟩, 684 ⟨84ZAAC(517)89⟩, 684, 687 ⟨79ZAAC(459)87⟩

Becker, G. von, **4**, 594 ⟨80ZAAC(470)144, 80ZAAC(470)157⟩

Becker, H., **2**, 570 ⟨90ZN(B)191⟩; **4**, 172 ⟨64JOC3070⟩

Becker, H.-D., **2**, 646 ⟨87JOC5205⟩; **4**, 221 ⟨64JOC1358⟩; **6**, 507 ⟨70JOC2099⟩

Becker, H. G. O., **3**, 262 ⟨61JPR294⟩; **5**, 773 ⟨70JPR601⟩, 856 ⟨70JPR610⟩; **6**, 624 ⟨73JPR640⟩

Becker, J., **3**, 474, 475 ⟨16CB1978⟩

Becker, J. Y., **1**, 540 ⟨74JOM(67)467⟩; **4**, 668 ⟨78JOM(157)1⟩, 873 ⟨88TL6177, 92CC1048⟩; **5**, 1074 ⟨88JOC1744⟩; **6**, 206 ⟨73JCS(P2)599⟩

Becker, K. B., **1**, 618, 707, 708 ⟨83HCA1090⟩; **2**, 13 ⟨73S493⟩, 23 ⟨73S789⟩

Becker, M., **1**, 979 ⟨85TL1831⟩

Becker, P., **4**, 568, 573 ⟨92CB771⟩

Becker, R., **2**, 314 ⟨86OM739⟩, 324 ⟨89JA7500⟩

Becker, R. F., **5**, 592, 609 ⟨71LA(753)187⟩, 608 ⟨70LA(733)195⟩, 608 ⟨71LA(755)145, 69LA(725)234⟩; **6**, 238, 240, 540 ⟨72LA(755)145⟩, 537, 539, 540 ⟨70LA(733)195⟩

Becker, R. H., **1**, 399 ⟨65JOC3888⟩

Becker, S., **1**, 437 ⟨88TL2963⟩, 437, 438, 439 ⟨90JOC6194⟩

Becker, V. G., **5**, 395 ⟨82ZAAC(495)73⟩, 396, 400 ⟨92ZAAC(607)101⟩

Becker, W., **2**, 474, 509 ⟨72CB1962⟩; **3**, 628 ⟨65AG(E)1079⟩; **4**, 484 ⟨86S31⟩, 568 ⟨87PS(30)349⟩; **5**, 396, 879, 891 ⟨91ZAAC(605)7⟩, 839, 878, 879 ⟨86S31⟩, 879, 880 ⟨85MI 523-01⟩, 1086 ⟨83PS(18)7⟩, 1155 ⟨85MI 531-01⟩; **6**, 584, 680, 684 ⟨84ZAAC(518)21⟩, 678 ⟨83PS(14)267⟩

Becker, Y., **5**, 290 ⟨80JOC2139⟩

Beckers, H., **6**, 243 ⟨86JOM(316)41, 90JOM(385)207⟩

Beckert, R., **2**, 386, 399 ⟨85HOU(E11)558, 82ZC237⟩; **3**, 353 ⟨92S751⟩, 695, 721 ⟨77S275⟩; **6**, 500 ⟨82JPR227⟩

Beckett, A. H., **2**, 341, 342, 350 ⟨75T2595⟩, 390 ⟨76T1267⟩; **4**, 422 ⟨81T4337⟩

Beck-Goehring, M., **2**, 414 ⟨57CB2075⟩

Beckh, H. J., **4**, 549 ⟨87AG(E)236⟩

Beckhaus, H., **3**, 474, 484, 485, 486 ⟨74LA1767⟩; **5**, 872 ⟨74LA1767⟩

Beckhaus, H.-D., **1**, 321 ⟨86TCC1⟩, 637, 638 ⟨88TL2805, 88C106⟩, 922 ⟨78CB72⟩, 922 ⟨87AG(E)770⟩, 923 ⟨78AG(E)594⟩, 925 ⟨83CB1042, 83TL185⟩; **2**, 404 ⟨74TL265⟩

Becking, L., **1**, 934 ⟨88TL2797, 88TL2801⟩
Beckley, R. S., **1**, 69 ⟨72JOC4341⟩
Beckman, J. C., **5**, 597 ⟨76SC387⟩
Beckman, L., **6**, 326 ⟨80LA1751⟩
Beckmann, E., **6**, 570, 571 ⟨1897JPR71⟩
Beckmann, L., **4**, 392 ⟨80LA1765⟩, 938, 957 ⟨80AG(E)303⟩
Beckmann, W., **1**, 1000 ⟨75S423⟩
Beckwith, A. L. J., **1**, 245, 1127 ⟨88JA2565⟩, 331 ⟨85JA4594⟩, 336, 337, 338, 547 ⟨85T3925⟩, 337 ⟨83AJC545⟩, 339 ⟨74CC472⟩, 426 ⟨83CC907⟩, 547 ⟨79JCS(P2)1535⟩, 848 ⟨61JCS73⟩, 919 ⟨81T3073⟩, 919, 927 ⟨B-80MI 119-05⟩, 931 ⟨88JA4415, 89JA230⟩, 936 ⟨85JA4595, 86JA5893⟩, 942 ⟨86AJC77⟩, 949 ⟨91TL49⟩, 1127, 1129 ⟨86TL4525⟩; **2**, 348 ⟨83AJC719⟩; **3**, 104 ⟨81CC595⟩; **5**, 259 ⟨B-70MI 506-01⟩, 317, 698, 722 ⟨87AJC701⟩, 341 ⟨81T2181⟩, 999 ⟨65TL4039, 68AJC2756⟩; **6**, 543, 550, 551 ⟨91T121⟩
Becu, C., **3**, 638 ⟨81TL141⟩; **4**, 185 ⟨74S23⟩; **5**, 1133 ⟨87BSB115⟩
Beddoes, R. L., **3**, 419 ⟨86JCS(P1)619⟩
Bedell, L., **2**, 538, 539 ⟨80JOC4797⟩; **3**, 684 ⟨83JOC2780⟩; **5**, 765 ⟨85JOM(281)135⟩
Bedell, L. J., **4**, 108 ⟨91USP5043489⟩
Bedenbaugh, A. O., **3**, 82 ⟨70JA5774⟩
Bedenbaugh, J. H., **3**, 82 ⟨70JA5774⟩
Bederke, K., **2**, 100 ⟨62AG(E)97⟩, 793 ⟨62AG(E)89⟩
Bedeschi, A., **1**, 494 ⟨92JOC3558⟩; **2**, 6 ⟨85G393⟩; **3**, 293 ⟨90JOC3654⟩
Bedford, C. D., **1**, 1130 ⟨82JOC2251⟩; **2**, 386 ⟨87JOC2292⟩, 1053 ⟨87JOC2294, 90JOC1916⟩; **4**, 196, 810 ⟨82JOC2251⟩, 199 ⟨86JOC4711⟩, 430, 431 ⟨78JOC2460⟩, 807 ⟨86S132⟩, 995 ⟨79JOC633⟩; **5**, 664 ⟨84JMC1201⟩, 668, 830 ⟨86JMC1689⟩, 690 ⟨79JOC633⟩, 967 ⟨80JOC347, 82JOC2251⟩
Bedford, M. A., **5**, 842 ⟨66JHC155⟩
Bednarski, M. D., **4**, 176 ⟨91TL7369⟩, 202 ⟨91COS(2)661⟩
Bednarski, P. J., **1**, 635 ⟨85JOC2798⟩
Bednyagina, N. P., **2**, 361 ⟨60ZOB1431⟩
Bédos, P., **5**, 593 ⟨67CR(C)99⟩
Bedoukian, P. Z., **2**, 28 ⟨55OSC(3)127⟩
Bedoya-Zurita, M., **1**, 195 ⟨88TL4709⟩, 195 ⟨88TL6433⟩, 195 ⟨89TL5883⟩
Bedworth, P. V., **1**, 680 ⟨89TL3927⟩
Bee, J. A., **6**, 650 ⟨66JCS(C)2031⟩
Beears, W. L., **5**, 1106 ⟨52JA1323⟩
Beebe, T. R., **2**, 95 ⟨83JOC3126⟩; **3**, 131 ⟨82JOC3006⟩; **5**, 203, 320 ⟨70JOC2056⟩
Beeby, P. J., **1**, 104 ⟨72JA2128⟩; **5**, 13 ⟨77TL3379⟩
Beech, W. F.,**3**, 428 ⟨54JCS1297⟩
Beecken, H., **2**, 245 ⟨62T1527, 67CB2159⟩
Beedle, E. C., **1**, 787 ⟨87JOC5034⟩; **4**, 526 ⟨87TL4499⟩, 813, 934 ⟨90HAC135⟩
Beekes, M. L., **3**, 430 ⟨86RTC161⟩, 430, 433 ⟨86RTC22⟩, 432 ⟨80RTC246, 80RTC278⟩
Beeley, N. R. A., **1**, 108 ⟨77CC321⟩
Beelitz, K., **3**, 349 ⟨78ZN(B)417⟩
Beer, D., **3**, 428 ⟨85HCA2254⟩
Beer, G., **5**, 381, 382 ⟨77M1279⟩
Beer, H., **4**, 563 ⟨72CB3261⟩
Beer, P. D.,**1**, 1137 ⟨93JCS(D)2629⟩
Beer, R. J. S., **3**, 378 ⟨72TL1835⟩; **4**, 863 ⟨80ACS(A)577⟩
Beereboom, J. J., **1**, 381 ⟨65JOC4230⟩
Beermann, C., **1**, 999 ⟨56ZAAC(287)261⟩; **2**, 1058, 1063, 1066, 1069 ⟨56ZAAC(287)261⟩
Beers, O. C. P., **4**, 716 ⟨93OM315, 93OM3187⟩
Beesley, T. E., **5**, 972 ⟨66JA3163⟩
Beets, M. G. J., **3**, 231 ⟨50RTC307⟩
Beetz, T., **4**, 179 ⟨73RTC1047⟩
Befoure, E. L., **2**, 357 ⟨83S34⟩
Befurt, R., **6**, 173, 383 ⟨93ZAAC(619)1494⟩
Beger, C., **5**, 592 ⟨1894CB1733⟩
Beger, J., **2**, 778 ⟨65JPR123⟩; **5**, 658 ⟨81M(112)959⟩; **6**, 606, 607, 618 ⟨69JPR15⟩
Begland, R. W., **4**, 438 ⟨72JOC4136⟩; **5**, 818 ⟨72JOC4136⟩
Begley, M. J., **1**, 144 ⟨93JCS(P1)3099⟩, 347 ⟨94TL2417⟩, 690 ⟨88JCS(P1)1085⟩, 831 ⟨84CC827⟩; **5**, 222 ⟨92JOM(426)49⟩, 227 ⟨84CC827⟩
Begoon, A., **1**, 1065 ⟨54JA4558⟩
Bégué, J. P., **2**, 751 ⟨89S309, 92JOC3807⟩; **3**, 148 ⟨91T3207⟩, 150 ⟨92JOC3807⟩; **6**, 9 ⟨91T3207⟩
Begum, S., **5**, 1125 ⟨82ZN(B)1481⟩
Beh, R. A., **6**, 37 ⟨75JOC3271⟩
Behagel, O., **2**, 733 ⟨32CB812, 35CB1540⟩
Behar, V., **1**, 826 ⟨90TL3241⟩
Behforouz, M., **2**, 795 ⟨79TL4493⟩; **3**, 696 ⟨79JA1608⟩; **5**, 336 ⟨69JOC51, 71JCS(C)3550⟩
Behl, H., **2**, 1088 ⟨89AG(E)1219⟩
Behling, J. R., **1**, 120 ⟨88JA2641⟩, 207 ⟨88TL893, 90JA4063⟩; **2**, 992 ⟨88JA2641⟩; **3**, 140 ⟨90JA7441⟩
Behling, Jr., J. R., **2**, 589 ⟨90JA4063⟩
Behlug, J. R., **2**, 555 ⟨92S191⟩
Behm, J., **5**, 926 ⟨92OM2494⟩
Behnam, B. A., **5**, 418, 924, 925 ⟨89OM693⟩
Behnam-Dahkordy, M., **6**, 622 ⟨77JCS(D)2015, 79JOM(181)69⟩
Behnke, C., **5**, 1086 ⟨87ZAAC(555)23⟩
Behnke, M., **1**, 314 ⟨83TL1909⟩; **3**, 624 ⟨86JOC1745⟩
Beholz, L. G., **1**, 835 ⟨93JOC5095⟩; **5**, 572 ⟨94TL1669⟩
Behr, H., **3**, 351, 548 ⟨83CB66⟩
Behr, J. P., **5**, 114 ⟨76JA1743⟩, 360 ⟨85TL215⟩
Behr, L. D., **5**, 86 ⟨43OSC(2)19⟩, 305 ⟨43OSC(2)562⟩
Behr, O. M., **1**, 1064 ⟨60JCS3614, 64JCS1151⟩
Behrend, E., **5**, 1005 ⟨73GEP(O)2153795⟩
Behrend, R., **2**, 103 ⟨1890CB203⟩; **5**, 597, 599 ⟨03LA(329)341⟩, 652 ⟨00LA(314)200⟩
Behrendt, W., **6**, 560 ⟨B-77MI 617-01⟩
Behrens, C. H., **1**, 14 ⟨85JOC5678⟩; **2**, 82 ⟨83PAC589⟩
Behrens, G., **4**, 1078 ⟨80JCS(P2)883, 81JCS(P2)143⟩, 1078 ⟨83AG(E)500⟩
Behrens, H., **6**, 517 ⟨80AOC(18)1⟩
Behrens, M., **2**, 906 ⟨1894CB244⟩
Behrens, O. K., **5**, 584, 586 ⟨48JA2837⟩
Behrens, U., **3**, 351 ⟨77AG(E)722⟩, 552 ⟨76JOM(118)C41⟩, 553 ⟨82CB1332, 84JOM(270)93, 84JOM(272)C40, 85JOM(288)C47⟩; **4**, 679 ⟨94JOM(476)101⟩, 719 ⟨88ZN(B)1461⟩; **5**, 605 ⟨78AG(E)455⟩, 782 ⟨90JOM(338)21⟩
Behringer, H., **1**, 1078 ⟨66TL1663⟩; **3**, 255 ⟨60AG415⟩, 366, 371, 378 ⟨65LA(682)188⟩, 373 ⟨65TL1757⟩; **5**, 354 ⟨75LA1264⟩, 581 ⟨59CB910⟩, 605, 616, 634 ⟨68CB1428⟩, 1049 ⟨67TL1013⟩; **6**, 511 ⟨57LA(607)67⟩
Behrle, E., **6**, 213 ⟨B-59MI 607-01⟩
Behrman, E. J., **2**, 651 ⟨88OR(35)421⟩; **5**, 275 ⟨71JOC3048⟩
Behrooz, M., **1**, 146 ⟨85JOC3674⟩, 485, 1032 ⟨84TL4851⟩; **3**, 224 ⟨85JOC3674⟩; **4**, 393 ⟨85JOM(285)121⟩
Behun, J. D., **6**, 75 ⟨54JA5736⟩
Bei, M. Z., **3**, 699, 721 ⟨92CL1143⟩
Beier, B. F., **4**, 692 ⟨78JA2090⟩
Beijer, B., **3**, 696, 721, 722 ⟨72AG(E)929⟩
Beilan, H. S., **2**, 160 ⟨74JOC2148⟩
Beil, V., **6**, 317 ⟨91TL6329⟩
Beilin, S. I., **2**, 566 ⟨77JOM(142)145⟩
Beillmann, J. F., **6**, 5 ⟨85T4503⟩
Beim, H. J., **4**, 408 ⟨70JHC355⟩
Beimling, P., **3**, 105 ⟨86CB3198⟩
Beinhofer, E., **2**, 492, 494 ⟨64CB449⟩
Beinoravichute, Z. A., **4**, 706 ⟨57DOK(112)439, 58DOK(120)1263⟩
Beirich, C., **2**, 569, 570, 571 ⟨93CB2093, 82AG(E)144⟩
Beisiegel, E., **3**, 668 ⟨74S22⟩
Beissner, C., **3**, 272 ⟨71LA(754)64⟩
Beiter, A. H., **2**, 409, 410 ⟨89S497⟩
Bejenke, V., **4**, 583, 585, 586, 587 ⟨77CB3536⟩
Beji, M., **2**, 367 ⟨86PS(27)293⟩
Bekarek, V., **5**, 630 ⟨87CCC2087⟩

Beke, S., **5**, 4 ⟨56CB862⟩
Beker, D. A., **2**, 930, 944 ⟨89JA389⟩
Bekhazi, M., **3**, 323 ⟨83CJC619⟩
Bekker, A., **3**, 270 ⟨81JGU1105⟩
Bekker, R. A., **2**, 168 ⟨82IZV1123⟩, 640 ⟨74DOK(217)1320, 75ZOR1370⟩, 641 ⟨75DOK(220)146, 76DOK(229)870, 76DOK(231)864, 77DOK(233)591, 78IZV1193, 78IZV2775, 78IZV493⟩; **4**, 730 ⟨75ZOR961⟩, 761, 763 ⟨67ZOR1006⟩; **5**, 59 ⟨78DOK(239)330⟩, 475 ⟨90BAU1725⟩, 583 ⟨90IZV1898⟩, 811 ⟨67IZV1060⟩; **6**, 65 ⟨75DOK(221)1331, 76ZOR1377⟩
Bekova, N. S., **5**, 691 ⟨92ZPK1430⟩
Belai, I., **4**, 120 ⟨86ACH145⟩
Belakhov, V. V., **4**, 802 ⟨87JGU844⟩
Belanger, A., **3**, 63 ⟨72CC863⟩; **4**, 827 ⟨72CC863⟩; **6**, 75 ⟨72CC863⟩
Belanger, P., **1**, 850 ⟨69CC1067⟩; **3**, 307 ⟨79TL2505⟩
Bélanger, P. C., **1**, 13, 18, 20 ⟨86JOC3038⟩; **2**, 612 ⟨88CJC1479⟩
Belatskaya, I. P., **1**, 486 ⟨90IZV2665⟩
Belaud, C., **5**, 296 ⟨85SC1233⟩, 296, 297 ⟨87TL59⟩
Belavenster, M. A., **2**, 208 ⟨72IZV2510⟩
Belavin, I. Y., **4**, 1063 ⟨78ZOB2147⟩; **5**, 374 ⟨88ZOB1093⟩
Belavin, I. Yu., **2**, 700 ⟨68JOM(14)229⟩, 701 ⟨70JOM(24)359⟩
Belaya, V. P., **6**, 438 ⟨68ZOR720⟩
Belcher, A. I., **3**, 131 ⟨82JOC3006⟩
Belčič, B., **5**, 836 ⟨84H(22)1545⟩
Belder, A. N. de, **1**, 18 ⟨64CI(L)1689⟩
Belen'kii, D. N., **5**, 666 ⟨90BAU1538⟩
Belen'kii, G. G., **2**, 876, 881 ⟨72IZV213⟩; **4**, 134 ⟨72IZV215⟩, 730 ⟨81IZV1920⟩
Belen'kii, H., **2**, 143 ⟨B-90MI 203-02⟩
Belen'kii, L. I., **2**, 13 ⟨77RCR891⟩; **5**, 363 ⟨91T447, 90IZV1692⟩; **6**, 29 ⟨89CS81⟩
Belen'kii, L. T.,**3**, 429 ⟨91T447⟩
Beletokaya, I. P., **5**, 113 ⟨84IZV1567⟩
Beletskaya, A. A., **5**, 116 ⟨84TL4989⟩
Beletskaya, I. P., **1**, 318 ⟨78RCR425⟩, 492 ⟨84IZV2580⟩, 844 ⟨B-78MI 119-04⟩, 845 ⟨B-68MI 119-01⟩, 859 ⟨77MI 119-01, 81JOU381⟩, 883 ⟨87RCR983⟩, 1004, 1006 ⟨81IZV2836, 81JOU18⟩, 1006 ⟨83DOK(272)1384, 83JOM(250)551⟩, 1010 ⟨84IZV1561, 85DOK(105)630⟩, 1012 ⟨84S728⟩; **2**, 471, 472 ⟨91JGU460⟩; **3**, 290 ⟨82IZV221⟩, 290, 305 ⟨85TL4819⟩, 430 ⟨B-83MI 311-01⟩, 647 ⟨89JOM(371)397⟩; **4**, 627 ⟨80IZV2668⟩, 642 ⟨80IZV2429⟩, 766 ⟨91ZOB1014⟩; **5**, 467 ⟨85JOM(282)421⟩; **6**, 187, 710 ⟨83IZV959⟩, 368 ⟨74ZOR1793⟩, 387, 395 ⟨79BAU2222⟩, 406 ⟨68BAU414⟩, 562 ⟨88ZOR2019⟩
Beletskeya, I. P., **1**, 1015 ⟨81CS97⟩
Belew, J. S., **3**, 2 ⟨B-69MI 301-01⟩, 113 ⟨B-69MI 304-01⟩
Belf, L. J., **6**, 16, 38 ⟨63AFC(3)181⟩
Belfield, K., **2**, 4 ⟨93S735⟩
Belforte, A., **2**, 893 ⟨84IC1504⟩; **5**, 975, 976 ⟨88CB1891⟩
Belica, P. S., **2**, 708 ⟨90S539⟩; **5**, 345, 346 ⟨88JOC5507⟩
Belik, M. Y., **1**, 492 ⟨86UKZ1090⟩
Belikov, V. M., **5**, 385 ⟨85CC171, 88T5507⟩, 652 ⟨88JCS(P1)305⟩
Belin, B., **2**, 308, 325 ⟨79S161⟩
Belinski, C., **2**, 387 ⟨64CR(259)3737⟩
Belkhouya, N., **2**, 7 ⟨91TL3977⟩
Belkina, O. B., **6**, 65 ⟨76IZV2547⟩
Bell, A. P., **2**, 968 ⟨71JOM(27)373⟩
Bell, A. J., **4**, 979 ⟨64JOC2932⟩; **5**, 599 ⟨64JOC2932⟩
Bell, A. N., **4**, 737 ⟨80JCS(P1)487⟩
Bell, D., **5**, 681 ⟨83JFC(23)293⟩
Bell, E. V., **2**, 147 ⟨27JCS1798⟩
Bell, E. W., **2**, 94 ⟨63OSC(4)125⟩
Bell, H. C., **1**, 490 ⟨74TL3391⟩
Bell, H. M., **1**, 4, 7 ⟨66JA1473⟩, 4, 7, 9, 10, 17, 429 ⟨69JOC3923⟩
Bell, I., **5**, 74, 102, 103 ⟨58JCS1313⟩, 102 ⟨56CI(L)548⟩
Bell, J. A., **4**, 440 ⟨66JCS(C)867, 66JCS(C)870⟩, 440, 442 ⟨66JCS(C)862⟩, 442, 443 ⟨67JCS(C)562⟩
Bell, K. H., **1**, 795 ⟨67CC397⟩
Bell, L. A., **2**, 783 ⟨88S379⟩
Bell, M. R., **5**, 342 ⟨91BMC733⟩
Bell, N. A., **1**, 452, 456 ⟨82COMC-I(1)121⟩; **2**, 563 ⟨82COMC-I(2)121⟩
Bell, R. A., **4**, 822 ⟨76CC200, 76JA6046⟩; **5**, 940 ⟨77IC677⟩
Bell, R., **2**, 184 ⟨81JCS(P1)2106⟩
Bell, R. E., **6**, 63 ⟨57JA3351⟩
Bell, R. H., **2**, 171 ⟨69CAR(9)201⟩
Bell, R. P., **1**, 853 ⟨46JCS636⟩, 869 ⟨66JCS(B)241⟩, 869, 871 ⟨67JCS(B)827⟩, 869, 871 ⟨70JCS(B)194⟩, 870 ⟨B-59MI 119-03⟩
Bell, S. I., **2**, 245 ⟨88TL4233⟩
Bell, T. N., **2**, 486 ⟨63AJC636⟩; **6**, 49 ⟨63AJC722⟩
Bell, W., **3**, 363 ⟨90JCS(D)3697, 90JOM(385)369⟩, 363 ⟨91JCR(S)44⟩; **4**, 4 ⟨79JFC(13)325⟩
Bell, W. J., **3**, 494 ⟨75JOC3456⟩
Bellachioma, G., **5**, 799 ⟨91G101⟩
Belladen, L., **6**, 412 ⟨28G443⟩
Bellama, J. M., **2**, 507 ⟨75JOM(102)445⟩; **4**, 134, 135 ⟨75JOM(102)437, 75JOM(102)445⟩, 135 ⟨75SRI199⟩, 140 ⟨75IC1614⟩, 141, 142 ⟨71MI 403-02⟩, 361 ⟨75IC1618⟩; **6**, 173 ⟨75JOM(92)163⟩
Bellamy, A. J., **3**, 448 ⟨65JCS2788, 65JCS3528⟩
Bellamy, F., **3**, 308 ⟨91S1079⟩
Bellamy, F. D., **1**, 738 ⟨90AG(E)1454⟩; **2**, 753 ⟨84TL839⟩; **6**, 558 ⟨91S265⟩
Bellan, J., **4**, 547 ⟨86TL1145, 87T1793⟩; **6**, 264 ⟨87T1793⟩, 373, 688 ⟨89CC593⟩
Bellanger, N., **1**, 826 ⟨90TL4147⟩
Bellasio, E., **2**, 182, 248 ⟨62HCA717⟩
Bellassoued, M., **1**, 544 ⟨79JOM(166)1⟩; **3**, 461 ⟨74JOM(81)139⟩; **4**, 442 ⟨83S951⟩, 604 ⟨93JOC2517⟩, 832 ⟨84JOM(263)C21⟩
Bellato, R., **6**, 126 ⟨92T2501⟩
Belle, A. J., **2**, 399 ⟨65TL1487⟩
Belleau, B., **2**, 749 ⟨88TL2295⟩; **5**, 219, 574 ⟨85CJC3089, 83TL3815⟩, 574 ⟨86JA182⟩, 588 ⟨93BMC619⟩, 652 ⟨68JA1651⟩
Bellec, C., **3**, 269 ⟨79JHC1657⟩
Bellegarde, B., **2**, 701 ⟨68JOM(11)97⟩; **3**, 640 ⟨67PNA(265)939⟩
Bellemin, A. R., **4**, 25 ⟨89S188⟩
Bellenbaum, M., **4**, 654 ⟨84CB383⟩
Beller, M., **2**, 412 ⟨90AG(E)782⟩; **4**, 182 ⟨88LA559⟩
Beller, N. R., **3**, 98 ⟨77JOC3514⟩; **4**, 853 ⟨81JOC196⟩; **5**, 546 ⟨81JOC196⟩
Bellesia, F., **2**, 22 ⟨86JCR(S)426, 90JCR(S)188⟩, 22 ⟨92T4579⟩, 24 ⟨89JCR(S)108, 89JCR(S)360⟩, 24 ⟨91G559, 91SC489⟩, 28 ⟨86JCR(S)428⟩, 150 ⟨93SC1759⟩; **3**, 27 ⟨86JCR(S)426⟩; **4**, 249 ⟨93T199⟩
Bellet, E., **3**, 614 ⟨12MI 318-01, 14MI 318-01⟩
Belletire, J. L., **1**, 568 ⟨83TL1475⟩; **2**, 319 ⟨88SC29⟩; **3**, 494 ⟨82SC469⟩; **4**, 846 ⟨71TL871, 84TL5729⟩
Bellett, E., **3**, 614 ⟨15AC(R)28, 20AC(R)364⟩
Bellew, D. R., **6**, 45 ⟨89S463⟩
Belley, M., **2**, 137 ⟨89JOC1230⟩
Belli, A., **4**, 297 ⟨75S789⟩
Bellin, S. A., **1**, 726 ⟨68JOC3566⟩
Bellinger, G. C. A., **4**, 148 ⟨89JOM(366)175⟩
Bellinger, O., **6**, 540 ⟨80CB750⟩
Bellon, P. L., **4**, 688 ⟨84JOM(267)199⟩
Bellucci, G., **2**, 30 ⟨86JOC4224⟩
Belluco, U., **3**, 314 ⟨72JOM(46)C29⟩
Bellur, N. S., **1**, 801, 802 ⟨93SL875, 91T8259⟩
Bellus, D., **1**, 571 ⟨88OSC(6)427, 78HCA3096⟩, 834 ⟨89TL3517, 91AG(E)1465⟩, 837 ⟨93S729⟩, 7, 833 ⟨83JOC860⟩; **2**, 14 ⟨85PAC1827⟩, 749 ⟨77HCA2379⟩; **3**, 297 ⟨67CRV599⟩, 527 ⟨79JOC1208⟩; **4**, 12

⟨85T4057⟩, 14 ⟨80HCA1947⟩, 830 ⟨78HCA1784⟩; **6**, 7 ⟨85PAC1827⟩, 64 ⟨88JOM(339)17⟩
Bellut, H., **2**, 418 ⟨68LA(720)1⟩
Bellville, D. J., **1**, 248 ⟨87ACR371⟩, 805 ⟨84JA2730⟩; **2**, 880 ⟨79CC880⟩
Belmonte, F. G., **1**, 101 ⟨83JOC2796, 83SC1103⟩, 457, 508 ⟨84JOC1662⟩
Belmore, K. A., **2**, 954 ⟨83OM101⟩, 978 ⟨84OM134⟩
Beloeil, J., **1**, 1039 ⟨81CC363⟩
Belokon', Y. N., **5**, 385 ⟨85CC171, 85JA4252, 88T5507, 93S117⟩ 652 ⟨88JCS(P1)305⟩
Belosludtev, Y. Y., **3**, 64 ⟨91S320⟩
Belotti, D., **1**, 195, 548 ⟨87TL4547⟩, 287, 362 ⟨90T1859⟩
Belous, V. M., **6**, 39 ⟨75ZOR1672, 76ZOR1798⟩, 39, 43 ⟨76ZOR1287⟩
Belousova, L. I., **4**, 140 ⟨89JOM(361)147⟩; **5**, 391, 718 ⟨87ZOB1199, 87JGU1073⟩
Belov, B. I., **1**, 36 ⟨63RCR59⟩
Belov, Yu. P., **4**, 461 ⟨73IZV955, 80IZV1125⟩
Belova, N. V., **5**, 212 ⟨85ZOR196⟩
Below, H., **5**, 624, 625 ⟨84ZC99⟩
Bel Rhlid, R., **3**, 16, 132 ⟨90TL2599⟩
Belser, R. B., **3**, 170 ⟨88JOC906⟩
Bel'skii, S. M., **2**, 350 ⟨76MI 206-01⟩
Bel'skii, V. E., **4**, 556 ⟨67DOK(177)340⟩
Bel'skii, V. K., **4**, 94 ⟨86ZOR1923⟩, 704 ⟨91ZOR2362⟩; **6**, 187, 710 ⟨83IZV959⟩
Belsky, I., **4**, 899, 902, 903 ⟨69T4649⟩
Belsky, Y., **2**, 626 ⟨82CC778⟩
Belt, H.-J., **4**, 567 ⟨89JOM(368)167⟩
Belter, R. K., **1**, 191 ⟨92TL2617⟩; **3**, 200 ⟨87JOC4810⟩
Beltrame, P., **1**, 1110 ⟨68G367⟩, 1144 ⟨61G889, 63G757⟩
Beltrame, P. L., **1**, 1110 ⟨68G367⟩
Beltrami, R. T., **2**, 381, 382, 383 ⟨56JA2467⟩
Beltzer, M., **6**, 558 ⟨93OPP83⟩
Belyaev, N. N., **1**, 748 ⟨80ZOB76⟩
Belyaev, P. G., **6**, 145 ⟨76KFZ35⟩
Belyalow, R., **2**, 254 ⟨87ZOB1429⟩
Belyk, K., **1**, 974, 1056 ⟨92JOC4070⟩; **2**, 69, 72, 591 ⟨93TL3115⟩
Belz, J., **2**, 409, 503 ⟨90CB975⟩
Belzecki, C., **4**, 174 ⟨84TL4797⟩; **5**, 344 ⟨85T2441⟩; **6**, 650, 653 ⟨70CC806⟩, 651 ⟨70BAP569⟩
Belzner, J., **4**, 1052 ⟨93TL6541⟩; **6**, 125 ⟨91OM3466⟩
Bembech, P., **2**, 918 ⟨93TL3311⟩
BeMiller, J. N., **2**, 1003 ⟨87CAR(168)7⟩
Benac, B. L., **2**, 511 ⟨86CC1543⟩
Benack, H., **2**, 360 ⟨76S308⟩; **3**, 374 ⟨76S310⟩; **5**, 339 ⟨76S308, 79S97⟩
Benaim, J., **2**, 1076 ⟨68CR(C)937⟩, 1076, 1079, 1080 ⟨65CR(261)1996⟩
Ben Akacha, A.,**5**, 22 ⟨92PS(69)163⟩
Ben Alloum, A., **4**, 257 ⟨92SC1359⟩, 281 ⟨90SC925⟩; **6**, 121, 325 ⟨92SC1359⟩
Benani, Y., **2**, 50 ⟨93JOC1991⟩
Benary, E., **1**, 486 ⟨31CB2543⟩
Benassi, R., **2**, 825 ⟨74JCS(P2)1338⟩
Benati, L., **3**, 166 ⟨88TL2381⟩; **5**, 561 ⟨76JOC2639⟩
Benazza, M., **2**, 7 ⟨91TL3977, 92TL3129⟩, 25 ⟨92TL4901⟩
Ben Cheikh, A., **3**, 541 ⟨88TL5919⟩
Bencheqroun, M., **3**, 405 ⟨93SC1683⟩
Bencini, E., **5**, 143 ⟨85JOM(285)293⟩
Bencivengo, D. J., **1**, 65, 66 ⟨84JA3703⟩
Bencze, W. L., **4**, 922 ⟨80HCA402⟩
Benda, A. F., **6**, 42 ⟨86ZOR1842⟩, 42, 46 ⟨86ZOR1834⟩
Bendas, H., **5**, 833 ⟨53JOC64⟩
Ben-David, Y., **1**, 861 ⟨81JOC2059⟩; **3**, 89 ⟨84S504⟩, 90 ⟨89CC1816⟩
Bende, Z., **6**, 625 ⟨77ACH77⟩
Bender, A., **2**, 249 ⟨85S66⟩; **3**, 171 ⟨85S66⟩
Bender, D., **5**, 605, 616, 634 ⟨68CB1428⟩
Bender, D. D., **1**, 1096 ⟨81JOC1061⟩
Bender, D. R., **2**, 342, 346 ⟨88JOC1768⟩; **3**, 258 ⟨68JOC2504⟩; **6**, 85 ⟨68JOC2504⟩
Bender, J., **3**, 547 ⟨81LA1073⟩
Bender, M. L., **1**, 6 ⟨49JA1767⟩; **3**, 88 ⟨49JA1767, 63OSC(4)932⟩; **6**, 104 ⟨60CRV54⟩
Bender, R. H. W., **6**, 504 ⟨90H(31)1393⟩
Bender, S. L., **2**, 554, 593 ⟨88JA2506⟩; **4**, 93, 290 ⟨82TL1531, 83TL237⟩, 877 ⟨81TL1495⟩
Benderley, A., **2**, 533, 697 ⟨87JOC1347⟩; **4**, 949 ⟨80CJC130⟩; **5**, 350 ⟨88TL739⟩
Benders, P. H., **1**, 518 ⟨86JOC2004⟩; **5**, 1124 ⟨78S775⟩, 1125 ⟨72RTC343⟩; **6**, 642 ⟨73TL3653⟩
Bendich, A., **3**, 283 ⟨63JOC1379⟩, 450 ⟨59JA2515⟩
Bendly, M., **6**, 477 ⟨83MIP8303603⟩
Bénéchie, M., **1**, 1024 ⟨90TL7145, 92SL266⟩
Benecke, B., **4**, 536 ⟨88AG(E)1194⟩
Benedetti, F., **4**, 273, 274 ⟨91JOC3530⟩; **5**, 354 ⟨85G663⟩, 355 ⟨91G401⟩
Benedict, G. E., **1**, 513 ⟨51JA5444⟩
Benedikt, G., **2**, 1083, 1084, 1087, 1088, 1097 ⟨68LA(717)21⟩
Benedusi, A., **5**, 762 ⟨87T4785⟩
Ben-Efraim, D. A., **1**, 998, 1068, 1069 ⟨B-78MI 121-03⟩; **2**, 1021, 1025, 1027 ⟨B-78MI 221-01⟩
Benend, H., **1**, 560 ⟨92M939⟩
Benetollo, F., **6**, 247 ⟨90OM1449⟩, 663 ⟨90JCS(D)1197⟩
Benetti, S., **2**, 181 ⟨74S877⟩, 301 ⟨91S479⟩; **3**, 220 ⟨81S199⟩, 648 ⟨85S969⟩
Benezra, C., **2**, 10 ⟨87TL1697⟩; **4**, 489 ⟨72CJC1078⟩
Benfield, F. W. S., **2**, 572 ⟨74JCS(D)1324⟩
Bengelsdorf, I. S., **2**, 904 ⟨66JOC2766⟩; **6**, 242 ⟨55JA2869⟩
Bengtsson, M., **1**, 1000 ⟨88S250⟩
Bengtsson, S., **4**, 189 ⟨89JOC4549⟩
Benham, J., **3**, 538 ⟨69TL4517⟩
Ben Hamida, N., **4**, 581 ⟨90IC1674⟩, 607, 621, 622, 628, 629 ⟨89OM1585⟩, 628, 629 ⟨87OM659, 85JOM(282)315, 90ICA141, 90ICA159, 90JOM(395)27, 91JOM(406)123, 91NJC657⟩, 640, 646 ⟨90SRI1373⟩, 641 ⟨90JOM(387)65⟩
Benhamou, M. C., **1**, 68, 454 ⟨79S891⟩
Benhamza, R., **2**, 341, 344 ⟨90TL531⟩; **5**, 214 ⟨85HCA1893⟩
Benhaoua, H., **1**, 637 ⟨89BSF409⟩
Benhaqua, H., **4**, 993 ⟨87BSF325⟩
Ben Hassine-Coniac, N., **3**, 75 ⟨71BSF3985⟩
Benigni, D. A., **1**, 1004 ⟨90JOC5833⟩
Ben-Ishai, D., **4**, 98 ⟨73JOC2251⟩, 100, 309, 327 ⟨75T863⟩, 304 ⟨67T2869⟩
Benitez, A., **2**, 129 ⟨58JA1680⟩
Benjamin, A., **2**, 458 ⟨88SC285⟩
Benjamin, B. M., **1**, 1124 ⟨77JA1669⟩
Benkeser, A., **2**, 913, 914 ⟨61JA4385⟩
Benkeser, R. A., **1**, 21 ⟨70JA3232, 73JOC3660⟩, 21 ⟨79JOC2185⟩, 51, 52, 101 ⟨55JA3230, 79JOC3737⟩, 101 ⟨83JOC2796, 83SC1103, 84TL2089, 88OSC(6)852⟩, 115 ⟨68JOC2203⟩, 401, 800 ⟨69JA5162⟩, 401, 800, 883 ⟨78JA2134⟩, 457, 508 ⟨55JA3378, 84JOC1662⟩, 881 ⟨45JA2106⟩; **2**, 910, 913, 923 ⟨B-69MI 218-01⟩, 913 ⟨B-66MI 218-01⟩, 914 ⟨74JOM(69)193⟩; **5**, 27 ⟨57CRV867⟩
Ben Koud, M. L., **5**, 690 ⟨92BSB445⟩
Benkovic, P. A., **2**, 807 ⟨67JA4714⟩; **4**, 412 ⟨69JA1860⟩
Benkovic, S. J., **2**, 807 ⟨67JA4714, 71JA4009⟩, 807 ⟨B-66MI 215-06⟩; **4**, 412 ⟨69JA1860⟩
Benmaarouf-Khallaayoun, Z., **3**, 44 ⟨90SC2091⟩
Benn, H., **4**, 676 ⟨92ICA741⟩
Benn, M., **5**, 637 ⟨77CC597⟩
Benn, M. H., **5**, 828, 832 ⟨68CJC1873, 64CJC2393⟩
Bennani, Y. L., **2**, 48 ⟨92JOC2768⟩, 49 ⟨93TL2079, 93TL2083⟩, 86 ⟨93JOC844⟩, 459 ⟨90TL6461⟩; **4**, 130 ⟨90TL6461⟩, 473 ⟨93SL35⟩, 492 ⟨90TL6465⟩
Benn, R., **2**, 984 ⟨89AG(E)610⟩; **3**, 509 ⟨89AG(E)610⟩; **4**, 659 ⟨85CB4248⟩, 698 ⟨87JOM(333)155⟩; **5**, 908 ⟨84T3273⟩
Benneche, T., **3**, 271 ⟨87ACS(B)536⟩; **4**, 54, 57 ⟨89ACS74⟩, 202 ⟨89ACS706⟩, 232 ⟨88ACS(B)515⟩; **5**, 324 ⟨93ACS1141⟩

Benner, C. W., **1**, 412 ⟨87JOC5198⟩
Benner, L. S., **4**, 566, 568 ⟨81IC3200⟩
Benner, S. A., **5**, 581, 611, 643 ⟨81TL1851⟩
Bennetau, B., **1**, 447, 447 ⟨93SL171⟩, 448 ⟨90TL6179⟩, 963 ⟨81CC1094⟩, 963, 978 ⟨81TL3401⟩; **2**, 940 ⟨81TL3401⟩, 943 ⟨93SL171⟩, 945 ⟨90TL6179⟩; **3**, 230 ⟨81TL3401⟩
Bennett, A. J., **2**, 383, 384 ⟨82JCS(P1)603⟩; **3**, 150 ⟨92SL483⟩; **4**, 949 ⟨92SL483⟩
Bennett, C. F., **2**, 180 ⟨63JOC2485⟩
Bennett, C. R., **4**, 657 ⟨74CC29⟩
Bennett, D. A., **4**, 66, 68 ⟨75TL4433⟩
Bennett, D. J., **3**, 657 ⟨67CC218⟩
Bennett, F. S., **5**, 183 ⟨85S681⟩
Bennett, F. W., **6**, 242 ⟨54JCS3598⟩
Bennett, G. B., **3**, 21, 141 ⟨77S589⟩
Bennett, G. M., **2**, 147 ⟨27JCS1798⟩
Bennett, Jr., F. C., **6**, 596 ⟨53JOC292, 63OSC(4)359⟩
Bennett, M. A., **2**, 579 ⟨82COMC-I(4)691⟩, 874 ⟨68ICA379⟩, 963 ⟨69JA6983⟩; **3**, 568 ⟨80JOM(199)119⟩; **4**, 150 ⟨86AJC1363⟩, 682 ⟨89AG(E)1296⟩, 688 ⟨76IC90, 76IC97⟩, 689 ⟨76IC107, 78IC1995⟩; **5**, 386 ⟨80JOM(202)107⟩
Bennett, R., **4**, 799 ⟨27JCS2850⟩
Bennett, R. H., **4**, 66 ⟨69JHC115⟩
Bennett, R. L., **5**, 447 ⟨75JOM(94)62⟩
Bennett, R. P., **2**, 752 ⟨61JA221⟩; **5**, 984 ⟨67TL961, 68JA3295⟩
Bennett, S. M., **1**, 1131 ⟨86CC878⟩
Bennett, W., **2**, 625 ⟨81TL3745⟩
Bennett, W. D., **4**, 445 ⟨85TL695⟩
Bennett, W. E., **4**, 802 ⟨90JFC(49)75⟩; **6**, 266 ⟨90JFC(49)75⟩
Bennetts, W. W., **5**, 652 ⟨68JMC534⟩
Benneville, P. L. de, **5**, 580 ⟨56JOC772⟩
Benning, A. F., **6**, 214 ⟨44USP2351390⟩
Benning, W. F., **3**, 426 ⟨59JA2151⟩
Bennua, B., **3**, 220 ⟨86S41⟩
Bennyarto, F., **4**, 558 ⟨81LA1044⟩
Benoiton, N. L., **5**, 185 ⟨78S928⟩
Benoy, G. J., **4**, 11 ⟨64T1567⟩
Benrashid, R., **2**, 795 ⟨79TL4493⟩
Bens, E. M., **2**, 388 ⟨59JA5546⟩
Benschop, H. P., **2**, 855 ⟨71JA3792⟩
Ben-Shoshan, R., **4**, 681 ⟨71CC1241⟩
Bensoam, J., **2**, 19 ⟨79TL353⟩; **3**, 149 ⟨77TL2797⟩; **4**, 10 ⟨71T3965⟩, 58, 59 ⟨86S627⟩; **6**, 229, 230 ⟨73TL2253⟩
Benson, G. A., **2**, 366 ⟨80CRV151⟩, 366, 367, 368 ⟨B-91MI 206-01⟩
Benson, R., **2**, 30 ⟨55OSC(3)105⟩
Benson, R. E., **1**, 863 ⟨65JA1941⟩; **3**, 317 ⟨65JA3657⟩; **4**, 325 ⟨48JA2115⟩, 893, 979, 985 ⟨62JA3387⟩
Benson, S. C., **2**, 152 ⟨91TL5885⟩
Benson, S. W., **1**, 1109 ⟨69CRV279⟩
Benson, W. R., **1**, 469 ⟨66CRV161⟩; **3**, 238, 266 ⟨64JOC385⟩, 238, 658 ⟨66CRV161⟩; **4**, 970 ⟨66CRV161⟩; **5**, 320 ⟨73OSC(5)663⟩
Bentley, K. W., **1**, 627 ⟨B-73MI 114-01⟩; **5**, 1126 ⟨67JA3281⟩
Bentley, M. D., **2**, 232 ⟨72JOC333⟩, 358 ⟨77JOC875⟩
Bentley, P., **4**, 205 ⟨76TL3739⟩
Bentley, P. H., **1**, 604 ⟨78CC518⟩; **3**, 708 ⟨74CC278⟩; **4**, 824 ⟨79TL391⟩
Bentley, T. W., **1**, 1114, 1116 ⟨72JA992, 76JA7658, 76JA7667⟩; **2**, 904 ⟨76JCS(P1)2419⟩
Bentolila, A., **5**, 403 ⟨92PS(70)109⟩
Benton, F. L., **2**, 33 ⟨42JA1128⟩
Bentov, M., **2**, 838 ⟨64JCS4750⟩
Bentrude, W. G., **2**, 256 ⟨87JA1227⟩, 413 ⟨85JOC1271⟩, 440, 441, 453 ⟨83JA4727⟩; **4**, 914 ⟨68JA5924, 72JA3058⟩
Bentz, H., **1**, 1111, 1113 ⟨78LA1894⟩; **2**, 689 ⟨74S293⟩
Ben-Yakov, H., **2**, 19 ⟨79TL1823⟩
Benz, W., **5**, 973 ⟨61AG657, 61LA(648)72⟩; **6**, 417 ⟨61AG66⟩
Benzing, E., **2**, 404 ⟨60LA(631)1⟩; **3**, 261 ⟨57CB2833⟩; **5**, 597, 599, 614 ⟨62CB926⟩
Benzinger, H., **2**, 1034 ⟨12LA(394)325⟩; **4**, 772 ⟨12LA(394)325⟩
Ben-Zvi, Z., **5**, 112 ⟨69CC343⟩
Beppu, F., **2**, 312 ⟨90CL203⟩
Beppu, K., **3**, 308 ⟨86T6465⟩
Beranek, I., **1**, 325 ⟨B-89MI 108-01⟩, 921 ⟨B-89MI 119-08⟩, 927 ⟨B-89MI 119-07⟩
Berbalk, H., **1**, 1004, 1005 ⟨83JCR(M)167⟩
Berbing, U., **3**, 386 ⟨88JOM(358)229, 92CC563⟩
Bercaw, J. E., **2**, 969 ⟨87JA203, 90SL74⟩; **3**, 511 ⟨86JA5347⟩, 545 ⟨84JA5178⟩; **4**, 679 ⟨88POL759⟩; **6**, 519 ⟨85JA4670⟩
Berchadsky, Y., **4**, 485 ⟨91TL2125⟩
Berchet, G. J., **1**, 965 ⟨32JA4066⟩, 1043 ⟨33JA1094⟩
Berchtold, G. A., **1**, 520 ⟨84JA2443⟩; **4**, 245 ⟨59JA3148⟩
Bercin, E., **6**, 564 ⟨85AP(318)848⟩
Berclaz, T., **6**, 335 ⟨92C130⟩
Bercovici, S., **5**, 701 ⟨92LA997⟩
Bercson, G., **2**, 238 ⟨64ACS815⟩
Berdinskii, I. S., **5**, 354 ⟨81ZOR2153⟩
Berdnikov, E. A., **4**, 489 ⟨82ZOB2444⟩, 861 ⟨78ZOR478⟩, 926 ⟨84ZOB1758⟩
Berdt, A., **4**, 633 ⟨89AG(E)784⟩
Berdyeva, É. A., **6**, 478 ⟨87JGU365⟩
Berenbaum, M. B., **2**, 378 ⟨63OSC(4)274⟩, 403 ⟨63OSC(4)66⟩
Berenblit, V. V., **6**, 36 ⟨74ZOR2031⟩, 251 ⟨76ZOR767⟩
Berenbom, M., **3**, 465, 474 ⟨49JA1629⟩
Berencsi, P., **5**, 547 ⟨72JOC1727⟩
Berenjian, N., **1**, 356 ⟨82CJC425⟩
Berens, M. E., **2**, 252 ⟨86JMC2114⟩
Beres, J., **2**, 413 ⟨85JOC1271⟩
Béress, L., **1**, 1072 ⟨65AG(E)1080⟩
Berezina, L. I., **6**, 55 ⟨68IZV2307⟩
Bereznak, J. F., **6**, 611 ⟨90JOC3045⟩
Bereznitskii, G. K., **1**, 492 ⟨72ZOR1484⟩
Berg, A., **3**, 620 ⟨1895AC(R)289⟩
Berg, C., **3**, 553 ⟨79CC99⟩
Berg, D., **4**, 141 ⟨92JOM(438)45⟩
Berg, F. J., **4**, 659 ⟨89OM2461, 93OM3890⟩
Berg, G., **4**, 310 ⟨66CB2127⟩; **5**, 289 ⟨66CB2127⟩
Berg, H. A. J., **1**, 809 ⟨76JOC2048⟩
Berg, H., **4**, 535 ⟨70AG(E)163, 77CB2659⟩
Berg, J., **6**, 580 ⟨66TL3695⟩
Bergamini, P., **4**, 151 ⟨92OM3879⟩, 151, 585 ⟨90JCS(D)1553⟩
Berg, U., **3**, 344 ⟨83JCS(P2)1321⟩; **5**, 631, 632 ⟨66ACS689⟩
Bergbreiter, D. E., **1**, 5, 9 ⟨87JOC473⟩; **2**, 7 ⟨85CC337⟩ 35 ⟨85JOC5828⟩, 395 ⟨87JOC1601⟩, 527 ⟨76JOC3031⟩, 589 ⟨74JA2806⟩, 770 ⟨80JOC4666⟩; **3**, 10 ⟨B-83MI 301-01⟩, 124 ⟨B-83MI 304-01⟩, 404 ⟨B-83MI 310-01⟩; **5**, 274, 294 ⟨80JOC3413⟩
Bergdahl, M., **1**, 311 ⟨93JOC7238⟩; **2**, 331 ⟨88TL6493⟩; **5**, 329 ⟨93TL7371⟩
Berge, J. M., **1**, 936 ⟨79S471⟩
Bergelson, L. D., **1**, 639 ⟨64AG(E)250⟩, 721 ⟨64PAC271⟩, 722 ⟨63IZV1053⟩, 725 ⟨64IZV2003⟩, 725, 733 ⟨63T149⟩, 727, 733 ⟨67T2709⟩; **3**, 494 ⟨67T2709⟩; **4**, 752 ⟨63T149, 64TL2669⟩; **5**, 100 ⟨64AG(E)250⟩
Bergemann, K., **2**, 258 ⟨93PS(79)131⟩
Bergens, S. H., **1**, 133 ⟨91JA958⟩; **2**, 640 ⟨91JA958⟩
Bergenthal, M. D., **1**, 1081 ⟨69AG(E)429⟩
Berger, A., **4**, 74 ⟨70JPR683⟩; **5**, 90 ⟨73JOC457⟩; **6**, 624 ⟨73JPR289⟩
Berger, B., **2**, 658 ⟨82S683⟩
Berger, C. R. A., **1**, 20 ⟨58JA2896⟩
Berger, D., **2**, 421 ⟨67AG(E)633⟩, 936 ⟨93JA9305⟩; **4**, 536 ⟨77CB3034⟩
Berger, G., **5**, 33, 108 ⟨27RTC600⟩
Berger, H., **2**, 192 ⟨63RTC773⟩

Berger, H. O., **2**, 1048 ⟨78JOM(145)17⟩
Berger, J., **1**, 177 ⟨87H(26)1291⟩
Berger, S., **4**, 1082 ⟨87AG(E)546⟩; **5**, 928 ⟨87AG(E)546⟩; **6**, 202, 724 ⟨87AG(E)546⟩, 414 ⟨90MI 614-01⟩, 711 ⟨87PAC1011⟩
Bergerhoff, G., **4**, 594 ⟨76AG(E)56, 78JCR(S)252⟩
Bergeron, R., **2**, 300, 303 ⟨73TL3839⟩
Bergeron, R. J., **3**, 417 ⟨80JOC161⟩; **5**, 328 ⟨92JOC7140⟩
Bergesen, K., **2**, 453 ⟨66ACS2508⟩
Bergfeld, M., **2**, 494 ⟨68ZAAC(363)84⟩
Bergin, W. A., **3**, 82 ⟨70JA5774⟩
Berglund, C., **4**, 207 ⟨63AK(20)225⟩
Bergman, E., **6**, 64, 225, 226, 245 ⟨57JA4159⟩
Bergman, E. D., **5**, 46 ⟨59OR(10)179⟩
Bergman, J., **1**, 484 ⟨68ACS1883⟩, 595 ⟨83JOC3918⟩, 1072 ⟨79TL1509⟩; **2**, 289 ⟨81JA2715⟩; **3**, 309 ⟨87TL3741⟩; **4**, 426 ⟨80T2505⟩, 877 ⟨81TL4199⟩; **5**, 199 ⟨87TL4441⟩, 242 ⟨80ZN(B)217⟩, 244 ⟨78OPP289⟩
Bergman, R., **4**, 299 ⟨77JCS(P1)684⟩
Bergman, R. G., **1**, 453, 545 ⟨81JA3002⟩, 806, 807 ⟨77JA1499⟩, 866 ⟨70JA2163⟩, 1114 ⟨71JA1941, 73CC589⟩, 1126 ⟨73CC679⟩; **2**, 962 ⟨82JA352, 83JA3929⟩, 978 ⟨80JA7003⟩; **4**, 687 ⟨81JA2489, 82OM219⟩, 710 ⟨92ICA377, 93JA2743, 93OM4269, 93OM65, 94OM1899, 94OM2668⟩; **5**, 386 ⟨89JA3258⟩, 450, 935 ⟨89JA1285⟩
Bergmann, A., **3**, 75, 265 ⟨89CB83⟩
Bergmann, E., **2**, 847, 1073 ⟨33CB278⟩; **4**, 178 ⟨48JA2827⟩; **5**, 66 ⟨48JA1153⟩
Bergmann, E. D., **1**, 108 ⟨61JCS3452⟩, 888 ⟨59OR(10)179⟩; **2**, 614 ⟨56JA6037⟩, 838 ⟨64JCS4750⟩; **3**, 294 ⟨67TL3665⟩; **4**, 33 ⟨61JCS3452⟩, 540 ⟨59OR(10)179⟩, 743 ⟨62JOC1426⟩, 751 ⟨57JA4174⟩; **5**, 66 ⟨50JA5012⟩, 833 ⟨53JOC64⟩; **6**, 25 ⟨50JA5012⟩
Bergmann, F., **1**, 108 ⟨54JA4137⟩; **5**, 98 ⟨52JCS4521⟩
Bergmann, H.-J., **5**, 374 ⟨86AP(319)203⟩, 384 ⟨86AP(319)216⟩
Bergmann, M., **2**, 387 ⟨23CB679⟩; **6**, 487 ⟨32CB1192⟩
Bergmann, P., **6**, 687 ⟨85ZAAC(529)216⟩
Bergmann, R. G., **1**, 1114 ⟨69JA2115, 71JA1925⟩
Bergmark, W. R., **2**, 378, 381 ⟨69JOC2254⟩, 403 ⟨69JOC2263⟩; **6**, 450, 509 ⟨69JOC2254, 69JOC2263⟩
Bergmeier, S. C., **3**, 411 ⟨90TL5541⟩; **4**, 22 ⟨90JOC5719⟩
Bergquist, K.-E., **3**, 355 ⟨91JOC4919⟩
Bergson, G., **2**, 237 ⟨61AK(18)1143⟩, 292 ⟨57ACS571⟩
Bergstorm, R. G., **4**, 176 ⟨B-80MI 404-02⟩
Bergstra, R. J., **1**, 244 ⟨90JOC1786⟩
Bergsträier, U., **1**, 985 ⟨92SL635⟩; **6**, 512, 582 ⟨91S1099⟩
Bergsträsser, U., **2**, 449 ⟨91AG(E)207⟩; **4**, 483 ⟨91S1099⟩, 549 ⟨92AG(E)1055, 93CC569, 93PS(76)1⟩; **5**, 395 ⟨91S1099⟩, 885 ⟨92AG(E)1055⟩; **6**, 153 ⟨91AG(E)207⟩, 153 ⟨92AG(E)879⟩, 154 ⟨92AG(E)1055⟩
Bergstrom, F. W., **1**, 881 ⟨45JA2110, 55OSC(3)778⟩; **2**, 806 ⟨42JA19⟩; **3**, 616 ⟨34JA2095⟩, 620 ⟨45JA2152⟩; **4**, 23 ⟨46JOC55⟩
Bergveld, P., **6**, 562 ⟨92JA10573⟩
Beringer, F. M., **1**, 108 ⟨55JA5533⟩; **2**, 193 ⟨49JA2221⟩, 1019 ⟨65JOC1930⟩; **6**, 73 ⟨67JOC2630⟩
Beringhelli, T., **4**, 693 ⟨95JOM(488)39⟩
Berjeaud, J. M., **1**, 1064 ⟨92JOC651⟩
Berk, S. C., **2**, 82 ⟨91JA5093, 92JOC3751⟩; **3**, 287 ⟨88JOC2390⟩
Berke, H., **5**, 449 ⟨74ZN(B)275⟩
Berkessel, A., **5**, 867 ⟨89CB1147⟩
Berkhimer, E. R., **5**, 57, 64 ⟨29JA260⟩
Berkman, Y. P., **5**, 1108 ⟨58MI 530-01⟩
Berkova, G. A., **4**, 490 ⟨81ZOB757⟩; **6**, 149 ⟨88ZOR644⟩
Berkovic, S., **2**, 614 ⟨56JA6037⟩
Berkowitz, D. B., **1**, 3, 6 ⟨90S649⟩, 178 ⟨93JOC6174⟩
Berkowitz, L. M., **5**, 307 ⟨58JA6682⟩
Berkowitz, W. F., **1**, 84 ⟨79TL1641⟩; **4**, 175 ⟨82JOC824⟩
Berks, A. H., **5**, 437 ⟨88JA957⟩
Berlage, U., **1**, 690 ⟨87TL3091⟩
Berl, H., **2**, 212 ⟨55JA704⟩
Berlan, J., **1**, 313 ⟨86TL1047⟩; **2**, 587 ⟨78BSF(2)575⟩; **4**, 802 ⟨71CR(C)1107, 75BSF2259⟩; **6**, 264 ⟨91TL5965⟩
Berlin, A., **4**, 1076 ⟨86CC1191⟩
Berlin, A. A., **5**, 24 ⟨73RCR939⟩
Berlin, K. D., **2**, 445 ⟨75JOC2801⟩, 462 ⟨60CRV243⟩; **3**, 639 ⟨88OSC(6)932⟩; **4**, 472 ⟨68JA4495, 68JOC3090⟩, 473 ⟨72JOC3304⟩; **5**, 403, 404 ⟨68JOC3090⟩, 709, 861 ⟨66T975⟩
Berlin, T., **2**, 656 ⟨39JCS398⟩
Berliner, E., **1**, 318 ⟨77JA290⟩; **5**, 98 ⟨49OR(5)229⟩
Berliner, E. M., **4**, 621 ⟨75ZOB2672⟩, 639 ⟨70ZOB2601⟩
Berlinguet, L., **5**, 34, 275, 305 ⟨63OSC(4)496⟩
Berman, D. A., **5**, 992 ⟨72JOC1738⟩
Berman, E. M., **1**, 603 ⟨85JA3891⟩
Berman, H. M., **4**, 493 ⟨75JA38⟩
Berman, S. T., **2**, 446 ⟨75IZV2845⟩, 877 ⟨84DOK(274)606, 84IZV700⟩, 877 ⟨86IZV181⟩; **4**, 587 ⟨68ZOR1685⟩
Bermann, M., **2**, 468 ⟨69JINC271⟩
Bérmont, B., **5**, 145 ⟨59MI 503-01⟩
Bernabé, M., **1**, 187 ⟨90JOC2294⟩; **5**, 87 ⟨85S773⟩
Bernad, Jr., P. L., **5**, 958 ⟨95TL3937⟩
Bernadi, F., **1**, 879 ⟨75CJC3365⟩
Bernadinelli, G., **2**, 341, 344 ⟨90TL531⟩
Bernady, K. F., **1**, 534 ⟨72TL4083⟩, 556 ⟨80JOC4702⟩; **2**, 28 ⟨80JOC4702⟩; **3**, 140, 156 ⟨79JOC1438⟩
Bernal, I., **4**, 677 ⟨82IC645⟩, 686 ⟨77JOM(140)73, 79JOM(165)C17⟩; **5**, 1097 ⟨85CB340⟩; **6**, 396 ⟨93JCS(D)2891⟩
Bernard, C., **2**, 175 ⟨76BSF483⟩; **4**, 990 ⟨80CC940⟩
Bernard, D., **1**, 491 ⟨76JOM(113)99⟩, 1090 ⟨76S245⟩; **2**, 186 ⟨69JOM(121)123⟩; **3**, 230 ⟨85TL4923, 87T2721⟩; **4**, 478 ⟨74MI 410-01⟩
Bernard, G., **1**, 713 ⟨89T1579⟩; **2**, 478 ⟨89T1859⟩; **4**, 400 ⟨89T1859⟩
Bernard, H., **4**, 420 ⟨92PHA754⟩
Bernard, J. M., **2**, 304 ⟨93SL680⟩
Bernard, M. K., **6**, 558 ⟨93OPP83⟩
Bernardes, V., **4**, 763 ⟨94JA2153⟩
Bernard-Henriet, C., **4**, 791 ⟨81H(15)1179⟩, 897 ⟨86RTC456⟩
Bernardi, A., **2**, 46 ⟨88T491⟩, 698 ⟨92JOC5173⟩
Bernardi, F., **1**, 1109 ⟨77JCS(P2)542⟩; **4**, 610 ⟨93JA3322⟩; **6**, 691 ⟨86JA5395⟩
Bernardi, G., **6**, 522 ⟨59G809⟩
Bernardi, G. C., **2**, 175 ⟨70TL1313⟩
Bernardi, L., **4**, 327 ⟨75CC320⟩; **5**, 300 ⟨75CC320⟩
Bernardi, R., **1**, 934 ⟨71T3575⟩
Bernardin, J. A., **6**, 241 ⟨81JFC(17)331⟩
Bernardinelli, G., **1**, 195 ⟨90TL6995⟩, 305 ⟨87HCA2201⟩; **2**, 202 ⟨84TL5889⟩; **3**, 701 ⟨92TL1593⟩; **5**, 347 ⟨90TL4117⟩; **6**, 335 ⟨92C130⟩
Bernardini, A., **5**, 615 ⟨70MI 513-01⟩
Bernardou, F., **1**, 543 ⟨76TL3083⟩
Bernasconi, C., **4**, 1079 ⟨78NJC79⟩; **5**, 955 ⟨89OM2282⟩
Bernasconi, C. F., **1**, 888 ⟨89T4017⟩; **5**, 953 ⟨93JA1341, 93JA12526⟩
Bernasconi, S., **3**, 169 ⟨87S1126⟩; **5**, 295 ⟨80S385⟩
Bernát, J., **5**, 1041, 1049 ⟨90SC509⟩, 1043 ⟨88ZC141⟩; **5**, 612 ⟨85ZC324⟩
Bernath, G., **4**, 273 ⟨89MRC760⟩, 319 ⟨87H(26)2381⟩, 416, 417 ⟨85S1148⟩; **6**, 111 ⟨91JHC481⟩, 557 ⟨83T1829⟩, 558 ⟨82H(19)1191, 83JHC1181⟩, 558 ⟨84JHC1373, 85S1149⟩, 558 ⟨85T1353, 90MRC1045⟩
Bernath, T., **2**, 652 ⟨73JA7114⟩; **4**, 304 ⟨67T2869⟩
Bernatowicz, M. S., **3**, 373 ⟨83JOC1999⟩; **6**, 641 ⟨92JOC2497, 93SC657⟩, 641 ⟨93TL3389⟩
Bernaus, C., **3**, 315 ⟨91T7713⟩
Bernd, A., **5**, 858 ⟨21JPR1⟩
Berndt, A., **1**, 922 ⟨79TL929⟩, 981 ⟨82AG(E)924⟩; **2**, 778 ⟨75AG(E)70⟩, 903 ⟨84ZN(B)1042⟩; **3**, 501 ⟨93AG(E)985⟩; **4**, 634 ⟨85AG(E)788, 88AG(E)1370⟩,

1048, 1054, 1057 ⟨84ZN(B)19042⟩, 1054, 1063 ⟨84AG(E)313⟩, 1082 ⟨84AG(E)826, 87AG(E)546⟩; **5**, 927, 928 ⟨83AG(E)877, 85AG(E)788, 84AG(E)369, 84AG(E)372, 84AG(E)826, 87AG(E)546, 87AG(E)798, 88AG(E)961, 88ZN(B)801, 90AG(E)398⟩, 928, 929 ⟨92AG(E)1238⟩, 928, 930 ⟨88AG(E)960, 92ZN(B)983, 90AG(E)399⟩, 929 ⟨92AG(E)1384⟩; **6**, 184, 389, 398, 714 ⟨89AG(E)781⟩, 187, 188, 201, 712, 714, 715, 716 ⟨93AG(E)985⟩, 188 ⟨92AG(E)1384⟩, 188, 711 ⟨87AG(E)798⟩, 201 ⟨85AG(E)788⟩, 201, 397, 715, 716 ⟨90AG(E)1030⟩, 201, 400 ⟨88AG(E)1370⟩, 201, 715 ⟨92AG(E)1238⟩, 202, 724 ⟨87AG(E)546⟩, 389, 398, 401, 713 ⟨89AG(E)784⟩, 389, 714 ⟨91AG(E)594⟩, 397, 398 ⟨90AG(E)1032⟩, 397, 714 ⟨88AG(E)961, 86AG(E)1111⟩, 400 ⟨90AG(E)399⟩, 711 ⟨87PAC1011⟩, 712, 713 ⟨90AG(E)401, 82JOM(234)C17⟩, 713 ⟨84ZN(B)1042⟩, 713, 714 ⟨90AG(E)398, 88ZN(B)75⟩, 714 ⟨90ZN(B)290⟩
Berndt, D. C., **5**, 322 ⟨84JOC106⟩
Berndt, J., **2**, 947 ⟨83TL4083⟩
Berndt, M., **4**, 621 ⟨72ZAAC(391)219⟩; **6**, 267 ⟨71AG(E)510, 71ZN(B)480⟩, 267 ⟨72ZAAC(391)219⟩
Bernert, C.-R., **2**, 97 ⟨70T3279⟩
Bernet, B., **4**, 226 ⟨93HCA995⟩, 229 ⟨93HCA1779⟩; **6**, 549 ⟨93HCA995⟩
Berney, D., **1**, 512 ⟨86JMC112⟩; **3**, 219 ⟨89JOC4673⟩, 282 ⟨75HCA74⟩
Bernhard, K., **5**, 28 ⟨B-47MI 502-04⟩
Bernhard, S. A., **3**, 541 ⟨48JA3426⟩; **4**, 204 ⟨48JA3426⟩
Bernhard, W., **1**, 654 ⟨84JOM(271)281⟩; **3**, 52, 203 ⟨84CC28⟩
Bernhardt, J. C., **1**, 128 ⟨78JOM(156)45⟩, 492, 1095 ⟨77JOC1680⟩; **3**, 212 ⟨76TL3097⟩
Bernheim, M., **2**, 753 ⟨80AG(E)1010⟩, 1041 ⟨83AG(E)53⟩
Berninger, C. J., **1**, 734 ⟨66JOC2907⟩
Bernocchi, E., **4**, 169 ⟨91T1525⟩
Bernotas, R. C., **2**, 303 ⟨90SC1209⟩
Bernstein, D., **6**, 615, 728 ⟨87JFC(37)259⟩
Bernstein, J., **4**, 873 ⟨88TL6177, 92CC1048⟩; **5**, 354 ⟨53JA1933⟩
Bernstein, M. A., **4**, 199 ⟨87TL2225⟩
Bernstein, M. D., **2**, 763 ⟨71JA2897⟩
Bernstein, P. A., **6**, 231, 232 ⟨71JA3882⟩
Bernstein, P. R., **5**, 342 ⟨90JMC2621⟩
Bernstein, S., **2**, 368 ⟨87OPP466⟩, 753 ⟨68MI 215-01⟩; **4**, 179 ⟨64JA2183⟩
Bernthsen, A., **5**, 609, 619 ⟨1877CB1238⟩
Beroza, M., **3**, 328 ⟨70SCI87⟩
Berrada, S., **4**, 255 ⟨85BSF881⟩; **5**, 556 ⟨85BSF881, 86BSF817⟩, 556 ⟨87TL409, 88T3575⟩
Berre, A. Le, **2**, 176 ⟨73BSF210⟩, 381 ⟨78BSF(2)602⟩, 381, 383, 384 ⟨75BSF1627⟩
Berre, N. Le, **2**, 749 ⟨88TL2299⟩
Berre-Cosquer, N. Le, **4**, 674 ⟨92OM721⟩
Berrée, F., **6**, 617 ⟨92TL6155⟩
Berrer, R., **3**, 247 ⟨76CB2928⟩
Berridge, M., **6**, 414 ⟨83MI 614-01⟩
Berriel, J., **1**, 245, 441 ⟨89TL7353, 90SC1681⟩
Berrier, A. L., **1**, 899 ⟨81PNA(78)1998⟩, 906 ⟨82JOC3903⟩
Berrier, C.,**3**, 429 ⟨91BSF730⟩
Berrino, L., **5**, 597 ⟨83FES101, 85FES120⟩
Berris, B. C., **1**, 448 ⟨83CC502⟩; **2**, 944 ⟨85JA5670⟩
Berry, A. D., **2**, 483 ⟨92OM2163⟩
Berry, D. H., **4**, 679 ⟨88POL759⟩
Berry, F. J., **2**, 690 ⟨80JOM(188)255⟩
Berry, J. P., **2**, 457 ⟨72JOC4396⟩; **4**, 476 ⟨72JOC4397⟩
Berry, M., **4**, 673 ⟨80JCS(D)29⟩
Berry, M. B., **5**, 223 ⟨92SL41⟩
Berry, P. D., **4**, 419 ⟨91TL2683⟩
Berry, R. D., **6**, 247 ⟨88JOM(342)399⟩
Bersch, H. W., **1**, 628 ⟨50AP(283)36⟩
Berseneva, L. H., **2**, 449 ⟨84ZOB2703⟩
Berseneva, L. S., **2**, 252 ⟨84ZOB2545⟩
Bershas, J. P., **1**, 470, 736 ⟨75TL1359, 73JOC3625⟩; **3**, 494 ⟨75TL1359⟩
Bersier, P. M., **5**, 314 ⟨92T1595⟩
Berson, J. A., **1**, 104 ⟨71JA1544⟩, 358, 359 ⟨78ACR446⟩, 400, 799, 805 ⟨67JA5503⟩, 400, 799, 805, 838 ⟨68ACR152⟩, 412 ⟨64JA5017⟩, 412, 809, 815 ⟨64JA5019⟩, 799 ⟨70JA1096⟩, 799, 805 ⟨72ACR406, 72JA8917⟩, 810, 815 ⟨68JA4729⟩, 811 ⟨76JA5937, 74JA3468⟩; **4**, 746 ⟨82JA2223⟩
Berstermann, H.-M., **3**, 578 ⟨80AG(E)564⟩, 680 ⟨86AG(E)85⟩
Bert, A., **2**, 100 ⟨24CR(178)1827⟩
Bert, G., **3**, 547 ⟨77JA1663⟩
Bertani, R., **4**, 1018 ⟨84JOM(269)C15⟩; **6**, 247 ⟨90OM1449⟩, 663 ⟨90JCS(D)1197⟩, 721 ⟨88IC2809⟩
Bertelli, D. J., **1**, 854, 906 ⟨73MI 119-20⟩
Bertelo, C. A., **3**, 23 ⟨76JA262⟩; **5**, 441 ⟨75JA228⟩, 442 ⟨76JA262⟩
Bertelson, R. C., **4**, 307 ⟨B-71MI 407-01⟩
Bertenshaw, S., **3**, 498 ⟨89TL2731⟩; **5**, 366 ⟨89TL2317⟩
Berthauer, H., **6**, 488 ⟨92GEP4215648⟩
Berthe, M. C., **5**, 1105 ⟨92SC617⟩
Berthelot, J., **2**, 29, 30 ⟨84BSB157, 91TL4135⟩, 209 ⟨69BSF3609⟩, 626 ⟨89CJC2061⟩
Berther, C., **3**, 427 ⟨60CB132⟩
Berthiaume, G., **1**, 538 ⟨86TL5455⟩; **3**, 222 ⟨86TL5455⟩
Berthiaume, S., **2**, 525 ⟨88PAC1627⟩; **4**, 217 ⟨90CJC897⟩
Berthmann, A., **2**, 104 ⟨63HOU(6/2)325⟩
Bertho, A., **6**, 105 ⟨49LA(562)229⟩
Bertho, C., **5**, 199 ⟨91T5481⟩
Berthold, H., **3**, 347 ⟨63CB3096⟩
Berthold, H. J., **2**, 566 ⟨63ZAAC(319)230⟩
Berti, F., **4**, 273, 274 ⟨91JOC3530⟩
Berti, G., **2**, 90 ⟨73TS93⟩
Bertin, D., **3**, 269 ⟨79JHC1657⟩
Bertin, D. M., **4**, 873 ⟨77JCS(P2)775⟩
Bertin, J., **2**, 25 ⟨74TL763⟩; **5**, 122 ⟨74JA8113⟩
Bertinato, P., **1**, 106 ⟨93CC617⟩, 558, 828 ⟨93CC619⟩
Bertini, F., **1**, 679 ⟨70T1281⟩, 680 ⟨67TL1581⟩, 934 ⟨71T3575⟩; **2**, 1006 ⟨70T1281⟩; **4**, 146 ⟨71T6109⟩, 669 ⟨70T1281⟩
Bertini, V., **3**, 620 ⟨93T8423⟩
Bertino, R. J., **4**, 697 ⟨74JOM(67)C61⟩
Bertleff, W., **6**, 335, 336 ⟨85ZN(B)1351⟩
Bertocchio, R., **6**, 7 ⟨89JFC(45)215⟩
Bertolasi, V., **3**, 560 ⟨84JOC2200⟩
Bertolini, G., **2**, 375 ⟨86JA6394⟩; **5**, 88 ⟨86JA6394⟩
Bertoni, G., **2**, 104 ⟨1888G434⟩
Bertorello, H. E., **5**, 589 ⟨78JCS(P1)325⟩
Bertran, J., **1**, 107 ⟨92JA9576⟩, 130 ⟨87JCPB765⟩
Bertrand, G., **1**, 656 ⟨84OM1340⟩; **2**, 470 ⟨86IC2695⟩, 1067 ⟨88JA6463⟩, 1068 ⟨92BSF367⟩; **3**, 485 ⟨91AG1174, 91AG(E)1154⟩, 485 ⟨91JA5856⟩, 486 ⟨87OM1822⟩, 502 ⟨81T2875⟩, 596 ⟨89AG(E)621⟩, 680, 682 ⟨94AG(E)527⟩, 681 ⟨88JA2663, 91OM3205⟩, 682 ⟨92JA6059, 92S43⟩, 689 ⟨88JA6463, 90PS(49)301⟩; **4**, 548 ⟨90PS(47)327, 93AG(E)1167⟩, 552 ⟨86IC712⟩, 567 ⟨88JA6463⟩, 567, 1082 ⟨89AG(E)621⟩, 567, 1082 ⟨90AG(E)1429⟩, 589 ⟨93PS(76)49⟩; **5**, 871, 872, 1148 ⟨92JA6059⟩, 871, 872, 873 ⟨88JA2663, 88JA6463⟩, 871, 872, 873, 1071, 1148 ⟨91JOC1801⟩, 871, 872, 873, 1148 ⟨90JA6277⟩, 872 ⟨91AG(E)1154, 91JA5856, 92BSF367⟩, 872, 1148 ⟨91OM3205⟩, 888 ⟨93CC1354⟩, 1076, 1077, 1148 ⟨89AG(E)1250⟩, 1131 ⟨88ACS(B)605⟩; **6**, 125 ⟨89AG617⟩, 125, 668, 669 ⟨91NJC393⟩, 187 ⟨94OM436⟩, 659 ⟨92CC1274⟩, 659, 667 ⟨91CB1739⟩, 665 ⟨89JOM(372)201⟩, 665, 667, 668, 669 ⟨88JA2663⟩, 665, 667, 668, 669, 670 ⟨90JA6277⟩, 665, 670, 675 ⟨92JA6059⟩, 665, 680 ⟨86JA7868⟩, 667 ⟨88AG(E)1350⟩, 668 ⟨91JOC1801⟩, 668, 669, 671 ⟨92BSF367⟩, 669 ⟨89JOC4272, 91AG(E)1154⟩, 670, 675 ⟨91OM3205⟩, 673 ⟨88CC1598, 87OM1822⟩, 680, 681, 689 ⟨92IC3493⟩, 691, 704 ⟨93CC1354⟩, 691, 734 ⟨89JA6853, 88JA6463⟩, 704 ⟨94AG(E)578⟩

Bertrand, J. A., **1**, 886 ⟨76JA7835⟩
Bertrand, J. L., **1**, 713, 714, 716 ⟨84IJ125⟩; **4**, 396, 401 ⟨84IJ125⟩
Bertrand, M., **1**, 498 ⟨72TL1189, 80JCR(S)98⟩, 498 ⟨85T2759⟩, 805 ⟨77TL1779⟩, 957 ⟨57CR(C)1790⟩, 959 ⟨74CR(C)149, 75BSF1407⟩, 959 ⟨76S755, 79TL1845⟩, 1114 ⟨74T243⟩, 1125 ⟨64CR(259)2251⟩, 1126 ⟨68CC718, 69TL3699⟩, 1126 ⟨73BSF1800, 74T227⟩, 1126 ⟨74T235⟩; **3**, 216 ⟨75T849⟩
Bertrand, M. E., **1**, 805 ⟨80JCR1551⟩
Bertrand, M. P., **1**, 926, 935, 942 ⟨88PAC1659⟩, 949 ⟨92JOC6118⟩
Bertrand, M.-T., **1**, 543 ⟨75CR(C)999⟩
Bertrand, R. D., **4**, 342 ⟨73PS(3)1⟩
Bertrand, W. S., **2**, 13 ⟨86CJC138⟩
Bertranne, M., **1**, 1039 ⟨81CC363⟩
Bertsch, H., **2**, 176, 181 ⟨55CB201⟩
Bertsch, R. J., **1**, 4, 7 ⟨71JOC1568⟩
Bertz, R. T., **4**, 203 ⟨63OSC(4)489⟩
Bertz, S. H., **1**, 118, 119 ⟨91JA631⟩, 205 ⟨86AG(E)760⟩; **2**, 586 ⟨89T425⟩, 588 ⟨82JA5284, 90JA4031⟩, 662 ⟨80S708⟩; **3**, 287 ⟨82JA5824, 84JOC1119⟩; **4**, 195 ⟨82JOC2216⟩; **5**, 556 ⟨83CC388⟩
Berube, G., **1**, 810 ⟨91CJC77⟩, 990 ⟨84CJC1558⟩
Beruben, D., **2**, 594 ⟨93TL2303, 93TL7575⟩
Beruda, H., **6**, 702 ⟨92CB1053⟩
Berven, H. M., **1**, 949 ⟨93JOC6851, 93SL827⟩
Besace, Y., **1**, 313 ⟨86TL1047⟩; **4**, 534 ⟨76JOM(113)C13⟩, 674 ⟨94OM2244⟩
Bésán, J., **5**, 745 ⟨80S883⟩
Besbes, R., **3**, 363 ⟨91S539⟩
Besendorf, H., **5**, 276 ⟨61HCA1546⟩
Besenyei, G., **2**, 726 ⟨93TL6105⟩; **5**, 987 ⟨90AG(E)1147, 93TL2839⟩
Besidsky, E., **4**, 222 ⟨91TL7593⟩
Besl, G., **3**, 518 ⟨76AG(E)543⟩
Beslin, P., **1**, 834 ⟨89CC414, 91T6275, 93T3131⟩; **2**, 229 ⟨83T3839⟩, 313 ⟨92TL5339⟩, 375 ⟨92TL935⟩; **3**, 338 ⟨83JHC1753⟩, 343 ⟨80JOC2517⟩, 343, 351 ⟨79TL2677⟩; **4**, 257 ⟨80TL4657⟩; **5**, 550 ⟨87S835⟩, 552 ⟨83TL3617, 85T2691⟩, 557 ⟨89CC414, 89T4445, 91T6275, 92T4135, 93T3131⟩
Besolova, E. A., **4**, 765 ⟨77ZOR1675⟩
Bespalko, G. K., **3**, 494 ⟨90JGU613⟩
Bespalova, A. M., **5**, 105 ⟨81MI 502-04⟩
Bespalova, N. B., **4**, 610 ⟨85IZV1917⟩, 658 ⟨85ZOB1629⟩
Besprozvannaya, M. M., **5**, 727 ⟨70ZOR805⟩
Bessard, Y., **3**, 495 ⟨90T5213⟩
Besser, S., **6**, 252 ⟨92CC1017⟩
Bessey, E., **5**, 236 ⟨80CB2100⟩
Besseyre, J., **1**, 518 ⟨75TL3151⟩; **6**, 106 ⟨70CR(C)468⟩
Bessho, K., **1**, 189 ⟨93T9479⟩
Bessiére, Y., **3**, 67 ⟨77HCA1739⟩
Bessinger, W. E., **2**, 100 ⟨47JA2158⟩
Besso, H., **3**, 357, 358 ⟨78BCJ179⟩
Besson, A., **6**, 222 ⟨1892CB188⟩, 420 ⟨1895BSF444⟩
Besson, T., **2**, 6 ⟨90SC3039⟩; **3**, 90, 97 ⟨93T1421⟩
Bestian, H., **2**, 129, 413 ⟨50LA(566)210⟩; **3**, 534, 541 ⟨63AG(E)608⟩; **5**, 84 ⟨68AG(E)278⟩, 1006 ⟨69GEP1289526⟩, 1010 ⟨50LA(566)210⟩; **6**, 483 ⟨50LA(566)229⟩
Bestman, H. J., **2**, 444 ⟨89S204⟩; **3**, 245 ⟨76CB403⟩; **5**, 43 ⟨74TL1275⟩
Bestman, J. H., **1**, 52 ⟨B-79MI 102-01⟩
Bestmann, H. J., **1**, 157 ⟨62ZN(B)787⟩, 229 ⟨76CB3375⟩, 248 ⟨83CB2378⟩, 268 ⟨62AG(E)116⟩, 393 ⟨60AG535⟩, 395 ⟨59CB528⟩, 446 ⟨79CB28⟩, 723, 733 ⟨62CB1894⟩, 727 ⟨65AG(E)583, 74S798, 74TL207, 76CB1694⟩, 988 ⟨74TL1275, 75TL4025⟩, 1016 ⟨80LA2061⟩, 1041 ⟨84AG(E)53, 87S848⟩, 1079 ⟨77LA282⟩, 1134 ⟨83AG(E)780⟩; **2**, 1004 ⟨77AG(E)349, 77CB3168⟩, ⟨91COS(6)171, 93AG(E)1746⟩, 1051, 1063 ⟨83CB1320⟩, 1064 ⟨78TL3343⟩, 1088 ⟨89AG(E)1219⟩; **2**, 447, 1064 ⟨77AG(E)45⟩, 624 ⟨86TL1995⟩, 831 ⟨79AG945⟩, 997 ⟨83TCC85⟩, 1004 ⟨74CB1949⟩; **3**, 35 ⟨74LA2085⟩, 48, 195 ⟨91COS(6)171⟩, 85 ⟨59CB530⟩, 220 ⟨82TL3543, 86TL1999⟩, 220 ⟨88S49⟩, 245, 481, 602 ⟨77AG(E)349⟩, 347, 363 ⟨92TL5955⟩, 391, 586 ⟨61CB2477⟩, 475, 489 ⟨79AG1012, 79AG(E)947⟩, 481 ⟨59CB1345, 77AG361⟩, 493 ⟨61AG27⟩, 494 ⟨62CB1513, 76CB1694, 78CB248, 80HCA438⟩, 497 ⟨86CB1331⟩, 531, 532 ⟨59CB528, 63CB1948⟩, 547, 581 ⟨75AG(E)53⟩, 547, 603, 609 ⟨75AG(E)634⟩, 557 ⟨91TL4279⟩, 571 ⟨68LA(718)24, 74AG(E)473⟩, 592 ⟨83CB2708⟩, 600 ⟨69CB1816⟩; **3**, 602 ⟨82HOU(E1)759⟩, 603 ⟨69AG(E)216, 73CB2601⟩, 603 ⟨77AG(E)877, 81TL1681⟩, 603 ⟨82CB161, 85CB1720⟩, 603, 604 ⟨80CB274, 74AG(E)273⟩, 604, 605 ⟨75TL4025⟩, 604, 605 ⟨80CB3369, 93CB2151⟩, 609 ⟨81TL3839⟩; **4**, 69 ⟨55ZN296, 60CB2340⟩, 245 ⟨93S149⟩, 313 ⟨84AG(E)53⟩, 559 ⟨91TL4279⟩, 571 ⟨82AG(E)545, 85TL2769⟩, 571 ⟨92CB2081, 92S787⟩, 574 ⟨81AG(E)1038, 84AG(E)381⟩, 574 ⟨86AG(E)559, 88CB1509⟩, 574 ⟨91CB199⟩, 587, 588 ⟨93PAC617⟩, 802, 825 ⟨69AG(E)216⟩, 825, 834 ⟨77AG(E)349⟩, 843 ⟨79CB28⟩, 966 ⟨80LA2061⟩; **5**, 17 ⟨66LA(693)132⟩, 223 ⟨86AG(E)994⟩, 271, 292 ⟨59CB528⟩, 343, 730 ⟨80CB912⟩, 413 ⟨85AG(E)1068⟩, 419 ⟨85JOC130⟩, 481 ⟨74AG(E)273⟩, 691 ⟨91TL4279⟩, 894 ⟨70S590, 82HOU(E1)616⟩, 896 ⟨86TL1995, 89TL5261⟩, 916 ⟨84AG(E)381, 86AG(E)559⟩, 1078 ⟨68LA(718)24⟩, 1092 ⟨77AG(E)349⟩; **6**, 75 ⟨86S397, 88T5095⟩, 520 ⟨92AG(E)1064⟩, 520, 702 ⟨85AG(E)229⟩, 694, 695, 697, 698 ⟨82HOU(E1)616⟩, 694, 697, 700, 701, 702 ⟨91COS(6)171⟩, 697, 700 ⟨86CB1331⟩, 701 ⟨87AG(E)79⟩, 701 ⟨87TL2111⟩, 701 ⟨B-72MI 622-01, 82AG(E)545⟩, 702 ⟨92S787⟩, 704 ⟨84AG(E)381⟩, 704 ⟨86AG(E)559⟩
Betancor, C., **1**, 931 ⟨93TL4865⟩; **5**, 1119 ⟨87JCS(P1)937⟩
Bethea, III, T. W., **5**, 229 ⟨68JOC75⟩
Bethel, D., **1**, 891 ⟨B-67MI 119-04⟩, 896 ⟨B-67MI 119-05⟩
Bethell, D., **1**, 891 ⟨78MI 119-02, 79COC(2)411, 81MI 119-02⟩, 1109 ⟨B-67MI 122-01⟩; **2**, 89 ⟨69JCS(B)749⟩, 145 ⟨94TL9629, 95SL773⟩, 145, 156 ⟨93JCS(P2)2161, 93SC1507⟩; **4**, 1072 ⟨B-65MI 424-01⟩
Betkouski, M. F., **5**, 845 ⟨73CC932⟩
Betschart, C., **1**, 687 ⟨89C39⟩; **3**, 186 ⟨84HCA1593⟩
Bett, W., **5**, 537 ⟨80M(111)193⟩
Betterton, K., **2**, 403 ⟨72S632⟩
Betterton-Cruz, K., **4**, 846 ⟨84TL995⟩
Bettinetti, G. F., **4**, 416 ⟨76S748⟩
Bettler, C. R., **4**, 622, 624 ⟨70IC2372⟩, 636 ⟨70IC1060⟩
Bettolo, R. M., **1**, 382 ⟨82HCA371⟩; **6**, 550 ⟨82HCA371⟩
Betts, M. J., **5**, 344, 1005 ⟨85JOC169⟩
Betyeva, E. S., **5**, 398 ⟨88JGU1965⟩
Betz, H., **6**, 355, 357 ⟨85CC1800⟩
Betz, K., **6**, 439, 452 ⟨40CB177⟩
Betz, P., **5**, 395 ⟨91CB2843⟩, 890 ⟨90CB1617⟩
Betz, R., **4**, 927 ⟨86CB472⟩, 1016 ⟨82S748⟩
Betz, W., **3**, 696 ⟨61CB2814⟩, 721 ⟨64CA(60)5638⟩
Betzing, H., **2**, 235 ⟨86EUP165534⟩
Beugelmann, R., **2**, 398 ⟨86S826⟩
Beugelmans, R., **1**, 474 ⟨88TL1289⟩, 871 ⟨64JA2832⟩; **4**, 108 ⟨86S826, 86S828⟩; **5**, 302 ⟨81S729⟩
Beugnies, D., **3**, 579 ⟨88JCS(P2)1071⟩
Beumel, Jr., O. F., **1**, 1000, 1001 ⟨74S441⟩; **2**, 420 ⟨63JOC2775⟩
Beumel, O. F., **1**, 1037 ⟨64JOC1872⟩
Beutel, R. H., **1**, 795 ⟨55JA5166⟩
Beuttler, T., **4**, 465 ⟨82AG(E)776⟩
Beuze, A. Le, **1**, 165 ⟨82JA7549⟩
Bevan, W. I., **4**, 136 ⟨76JCS(D)694⟩
Beveridge, A. D., **2**, 835 ⟨66JCS(A)520⟩
Beverly, G. M., **2**, 478 ⟨67JOC3645⟩
Beverung, W., **3**, 596 ⟨69CC1377⟩
Beverung, W. N., **1**, 89 ⟨71OS(51)103⟩

Bevinakatti, H. S., **2**, 392 ⟨86JHC825⟩; **3**, 179 ⟨82JHC69⟩
Bevins, C. L., **1**, 797 ⟨B-89MI 118-01⟩
Bewersdorf, M., **4**, 146 ⟨91CB1259⟩, 288 ⟨92LA643⟩
Bewick, A., **2**, 284 ⟨80TL3827⟩
Bey, A. E., **4**, 604 ⟨66JOC2036⟩
Bey, P., **1**, 440 ⟨91JA7439⟩; **3**, 240 ⟨89JOC3913⟩; **4**, 73 ⟨89JA1127⟩, 812 ⟨91JA7439⟩; **5**, 327 ⟨89JOC3913⟩; **6**, 487 ⟨91JHC1013⟩
Beyer, E., **2**, 96 ⟨58CB1212⟩; **5**, 1023 ⟨65CB2425⟩; **6**, 39 ⟨67OS(47)47⟩, 79 ⟨58CB1942⟩
Beyer, G., **5**, 652 ⟨78PHA375⟩
Beyer, H., **2**, 109 ⟨60CB1078⟩; **4**, 622 ⟨62CB3030⟩; **5**, 318 ⟨61CB480⟩; **6**, 383 ⟨59CB1018⟩, 504 ⟨66CB2937⟩
Beyer, L., **5**, 633 ⟨81ZAAC(480)21⟩, 655 ⟨88TL3475⟩, 657 ⟨84T405, 89ZAAC(576)203⟩, 657, 1053 ⟨88JPR241⟩; **6**, 565 ⟨81ZAAC(476)7⟩
Beyer, R. D., **5**, 43 ⟨73JOC1491⟩
Beyerle, G., **2**, 826, 866 ⟨77IS186⟩
Beyerlin, H. P., **4**, 881, 882, 972 ⟨64CB3081⟩
Beyerstedt, F., **6**, 74 ⟨36JA529, 37JA2266⟩, 80 ⟨37JA1273⟩
Beyl, V., **2**, 689 ⟨73LA20⟩
Beyler, R. E., **2**, 1016 ⟨61JA4663⟩
Bezemer, L., **1**, 420 ⟨87RTC613⟩
Beziat, Y., **2**, 445 ⟨87S648⟩; **3**, 196 ⟨87S648⟩
Bezmenova, T. E., **6**, 562 ⟨81KGS907⟩
Beznosko, B. K., **6**, 613 ⟨88KFZ143⟩
Bezuidenhoudt, B. C. B., **3**, 24, 146 ⟨92TL2383⟩
Bezwada, R. S., **6**, 504 ⟨80JCED292⟩
Bezzi, S., **2**, 213 ⟨50G180⟩; **3**, 255 ⟨58G1226⟩, 397, 399 ⟨62G859⟩
Bhacca, N. S., **2**, 1070 ⟨69JOC3349⟩
Bhadti, V. S., **1**, 585 ⟨89JOC2217⟩; **4**, 988 ⟨84TL1291⟩; **5**, 351 ⟨90S1095⟩
Bhaduri, S., **2**, 156 ⟨81JCS(D)447⟩, 156 ⟨83JCS(D)415⟩
Bhagwat, M. M., **1**, 459 ⟨72TL1391⟩
Bhagwat, S. S., **1**, 1010 ⟨92BMC1619⟩
Bhakuni, D. S., **4**, 202 ⟨90SC687⟩, 306 ⟨84IJC(B)1286⟩
Bhala, A. R., **5**, 697 ⟨84H(22)2487⟩
Bhalerao, U. T., **2**, 303 ⟨90TL1199⟩; **3**, 58, 494 ⟨71JA5311⟩, 305 ⟨84TL5439, 70JA3429⟩
Bhan, A., **3**, 673 ⟨85JOC5892⟩
Bhandal, H., **1**, 335 ⟨88CC1110⟩, 335, 346, 495, 946 ⟨90JCS(P1)2709⟩, 345, 946 ⟨90JCS(P1)2691⟩
Bhandari, A. M., **5**, 226 ⟨80IJC(A)69⟩
Bhandari, K., **2**, 320 ⟨90CI547⟩
Bhandari, N. S., **2**, 690 ⟨85JOM(291)185⟩
Bhanu, M. N., **1**, 1134 ⟨93TL707⟩
Bhanu, S., **1**, 1053 ⟨75CC817⟩
Bharadwaj, M. M., **4**, 258 ⟨78CJC1183⟩; **5**, 348 ⟨78CJC1183⟩
Bharadwaj, P. K., **2**, 441, 444 ⟨91PS(56)27⟩
Bhar, S., **3**, 168 ⟨92JOC7349⟩; **4**, 258 ⟨92JOC7349⟩
Bharathi, S. N., **1**, 694 ⟨90JOC2128⟩; **5**, 720 ⟨90JOC3682⟩
Bhargava, P. N., **6**, 643 ⟨69JMC558⟩
Bhargava, S. K., **4**, 283 ⟨87JCS(D)757⟩
Bhasin, K. K., **5**, 210 ⟨89IJC(A)1528⟩
Bhaskar, K. V., **1**, 342, 1132 ⟨89TL225⟩
Bhaskar Kanth, J. V., **2**, 300 ⟨93JOC3156⟩
Bhaskar Reddy, D., **4**, 272, 275 ⟨89SUL79⟩
Bhat, H. B., **5**, 49 ⟨69JOC112⟩
Bhat, K. S., **2**, 75 ⟨89JOC1570⟩, 75, 521 ⟨88JA1535⟩, 75, 521 ⟨90JA2389⟩, 520 ⟨85JA2564⟩, 520 ⟨86JOC445⟩, 520 ⟨91JCS(P1)2633, 86JOC432⟩
Bhat, N. E., **4**, 815 ⟨91TA277⟩
Bhat, N. G., **1**, 449 ⟨89JOC6064⟩, 487 ⟨91TA277⟩, 487, 488 ⟨88JOC239, 88JOC6009, 86JOC3398⟩, 660, 661 ⟨88TL21⟩, 999 ⟨88TL2631⟩, 1034 ⟨86JOC4521, 91JOC1500⟩; **2**, 530 ⟨93TL7845⟩, 623 ⟨86S480⟩, 623, 628 ⟨89JOC6068⟩, 900 ⟨86JOC3398, 86S674⟩, 900, 901 ⟨89JOC6075, 88JOC239⟩, 901, 904 ⟨88JOC6009⟩, 901, 906 ⟨86JOC5277, 88TL21⟩, 1086 ⟨86JOC4521, 88TL2631⟩; **3**, 275 ⟨83S885⟩
Bhat, S. V., **3**, 433 ⟨83SC649⟩
Bhatia, B., **1**, 320, 336, 345, 346, 925, 926, 937, 942, 946 ⟨94CRV519⟩; **3**, 139 ⟨93TL6301⟩
Bhatia, V. G., **5**, 131 ⟨83T1475⟩
Bhatki, K. S., **3**, 432 ⟨79OPP249⟩
Bhatnagar, I., **2**, 388 ⟨68JOC2407⟩
Bhattacharjee, G., **3**, 533 ⟨86TL3913⟩
Bhattacharjee, M. N., **3**, 4, 88, 115 ⟨82S588⟩
Bhattacharjee, P., **6**, 563 ⟨91POL823⟩
Bhattacharjee, S. S., **5**, 555 ⟨82S693, 85JCS(P1)641⟩
Bhattacharjya, A., **2**, 398 ⟨85S886⟩
Bhattacharya, A., **1**, 558 ⟨88JA3318⟩; **3**, 131 ⟨90SC2683⟩
Bhattacharya, A. K., **2**, 232 ⟨78JOC3374⟩, 469 ⟨81CRV415⟩; **3**, 196 ⟨81CRV415⟩, 359 ⟨78JOC3374⟩; **4**, 167 ⟨81JOC2260⟩, 340 ⟨81CRV415⟩
Bhattacharya, P. K., **5**, 601, 621 ⟨89IJC(B)673⟩
Bhattacharya, S., **1**, 486 ⟨93JCS(P1)1943⟩
Bhattacharya, S. K., **3**, 527 ⟨67T4769⟩
Bhattacharya, S. N., **2**, 176 ⟨68JCS(C)1265⟩; **5**, 390 ⟨84IJC(A)952⟩, 1132 ⟨78IJC(A)778⟩, 1143, 1144 ⟨72JIC1143⟩
Bhattacharyya, N. K., **5**, 407 ⟨91IC1174⟩
Bhattacharyya, P., **1**, 239 ⟨90IJC(B)390⟩
Bhattacharyya, S., **3**, 127 ⟨89SC673⟩
Bhattacharyya, S. C., **2**, 660 ⟨62JOC1430⟩
Bhatt, M. V., **1**, 657, 658 ⟨67JA567, 66JA1440⟩; **2**, 10, 658 ⟨83S249⟩, 33 ⟨78JOM(156)221⟩, 659 ⟨74CC584⟩, 747 ⟨76CC103⟩; **4**, 54 ⟨71JCS(C)1772⟩, 58 ⟨71JCS(C)3344⟩
Bhatt, R. K., **4**, 175 ⟨94TL1825⟩
Bhawal, B. M., **3**, 91 ⟨77IJC(B)1056⟩; **4**, 901, 904 ⟨90JCS(P1)1217⟩, 901, 904 ⟨93T2101⟩
Bhide, S. R., **5**, 110 ⟨76S123⟩
Bhikkappa, S. A., **4**, 528 ⟨88EUP293881⟩
Bhongle, N. N., **6**, 489 ⟨90AX2374⟩
Bhuma, V., **3**, 294 ⟨92JOC5841⟩
Bhuniya, D., **3**, 151 ⟨92TL6021⟩
Bhupathy, M., **1**, 236 ⟨89ACR152⟩, 805 ⟨83JA520, 83JA6798⟩; **2**, 83, 553 ⟨90JOC1528, 89ACR152⟩; **4**, 222 ⟨87TL4793⟩, 247 ⟨89JOC3718⟩
Bhusan, V., **3**, 88 ⟨84CL2131⟩
Bhushan, V., **3**, 32, 156 ⟨88TL2437⟩; **6**, 426 ⟨86S330⟩
Biagini, M. C., **4**, 677 ⟨81AG(E)887⟩
Bialecka, E., **3**, 634 ⟨72TL2391⟩
Biali, S. E., **2**, 636, 643 ⟨88ACR442⟩, 645 ⟨84JA477⟩, 646 ⟨85JA1007⟩; **4**, 11 ⟨87JOC3713⟩
Bianchetti, G., **3**, 467 ⟨63CB802, 65CB2715⟩
Bianchi, D., **2**, 151 ⟨92TA921⟩
Bianchi, G., **3**, 678, 681 ⟨79AG(E)721⟩
Bianchi, G. E., **4**, 205, 210 ⟨79CB148⟩
Bianchi, R. G., **1**, 1008 ⟨87JMC193⟩
Bianchi, T. A., **2**, 183, 717 ⟨77JOC2031⟩
Bianchini, C., **1**, 507 ⟨88IC4429⟩; **4**, 851 ⟨83CC1309⟩; **6**, 325 ⟨80AG1055, 81JOM(218)81⟩, 327 ⟨82OM778⟩, 332 ⟨83JOM(246)C13, 84JOM(270)251⟩
Bianchini, R., **2**, 30 ⟨86JOC4224⟩
Biancini, B., **4**, 205, 210 ⟨79CB148⟩
Bianco, A., **3**, 92 ⟨90SC2537⟩
Bianco, E. J., **3**, 618 ⟨55JA1701⟩
Bianco, R., **6**, 596 ⟨59G693⟩
Bianco, V. D., **1**, 52 ⟨76MI 102-01⟩; **2**, 435 ⟨78IS169⟩, 437 ⟨76IS155⟩, 822, 823 ⟨76IS161⟩, 824 ⟨78IS170, 78IS172⟩, 824, 825 ⟨78IS171⟩, 827 ⟨76IS157, 76IS158⟩, 827 ⟨76IS159, 76IS160⟩
Bibler, J. P., **5**, 452 ⟨66IC889⟩
Bicke, F. F., **4**, 557 ⟨64JOC2036⟩
Bickel, A. F., **1**, 901 ⟨67CC635, 69RTC371⟩; **2**, 127 ⟨60T76⟩; **4**, 16 ⟨62RTC925⟩
Bickel, C. L., **3**, 244 ⟨47JA73⟩
Bickel, H., **5**, 164 ⟨58T1⟩
Bickelhaupt, F., **1**, 194 ⟨88TL3579⟩, 194 ⟨93TL3309⟩, 415, 826 ⟨86TL6267⟩, 445 ⟨83T4225⟩, 795 ⟨68CC35⟩; **2**, 324 ⟨88TL3579⟩, 441 ⟨83T4225⟩, 521 ⟨93TL4655⟩, 523 ⟨73JOM(47)33⟩, 925 ⟨87TL2889, 92T6087, 92T6105⟩, 1003 ⟨84T765⟩; **3**, 508 ⟨83TL3935⟩, 608 ⟨93BSF185⟩,

689, 726 ⟨91CB2609⟩; **5**, 890 ⟨92OM3844⟩, 890 ⟨93PS(77)252⟩; **6**, 190, 192, 198 ⟨86TL6123⟩, 325, 379, 388, 398 ⟨88TL5237⟩, 399 ⟨89TL6195, 92JA7302⟩, 406 ⟨91AOC(32)147⟩, 610 ⟨93JOC6930⟩, 682, 690 ⟨91CB2677⟩
Bickelhaupt, F. B., **5**, 1089 ⟨90JOM(388)47⟩
Bicking, J. B., **1**, 34 ⟨53JA1128⟩
Bickley, H. T., **4**, 220 ⟨72JOC2885⟩
Bida, G. T., **2**, 632 ⟨90JOC4560⟩
Bidaine, J., **5**, 476 ⟨61BSB519⟩
Bidd, I., **2**, 25, 31 ⟨84TL5949⟩
Biddle, J. A., **6**, 491 ⟨89TL7141⟩
Biddlecom, W. G., **1**, 821 ⟨73JOC3438⟩
Bidell, W., **1**, 500 ⟨90CC858⟩; **4**, 1018 ⟨91OM389⟩
Bieber, J. B., **1**, 396, 397 ⟨66JA78⟩
Bieber, L., **5**, 1073 ⟨75AG(E)428⟩
Bieber, T. I., **5**, 995 ⟨52JA4700⟩
Bieber, W., **2**, 132 ⟨78CB2445⟩
Bieberbach, A., **1**, 886 ⟨78CB2480⟩
Bied, C., **2**, 565 ⟨92T3877⟩
Biedenbach, B., **5**, 890 ⟨88TH17, 90CB1617⟩, 890 ⟨90PS(49)337⟩
Biedermann, A. G., **3**, 366 ⟨84AG(E)800⟩
Biedrzycka, Z., **4**, 144 ⟨92JOM(431)255⟩
Biedrzycki, M., **4**, 144 ⟨92JOM(431)255⟩
Biehl, E. H., **3**, 626 ⟨83JOC4397⟩
Biehl, E. R., **2**, 753 ⟨83JOC4397, 85JOC1334⟩, 753 ⟨87JOC2619⟩; **3**, 626 ⟨85JOC1334, 87JOC1333⟩, 626 ⟨87JOC2619⟩
Biehl, F. R., **2**, 753 ⟨86JOC5157⟩
Biekert, E., **5**, 24 ⟨75M1019⟩
Biel, J. H., **2**, 373, 374, 379, 381 ⟨59JA2805⟩
Bielak, L., **6**, 637 ⟨89MI 620-01⟩
Bielefeldt, D., **6**, 433 ⟨88JFC(40)365⟩, 440 ⟨83CB1257⟩
Biellmann, J.-F., **1**, 107 ⟨82OR(27)1⟩, 590 ⟨79JA3283⟩, 630 ⟨67BSF675⟩; **2**, 143 ⟨71T5861⟩, 143 ⟨68TL5629⟩, 143 ⟨69TL3707⟩, 157 ⟨88SC1061⟩, 997 ⟨82OR(27)1⟩; **3**, 375 ⟨80JA2460⟩; **4**, 381 ⟨82OR(27)1⟩
Biemann, K., **5**, 84 ⟨60C293⟩
Bien, S., **3**, 320 ⟨82T1477⟩; **4**, 267 ⟨90JCS(P2)1987⟩
Biener, H., **4**, 310 ⟨65LA(686)102⟩
Bieniek, A., **4**, 317 ⟨89CS33⟩
Bienz, S., **1**, 985 ⟨94TL1161⟩, 1001, 1134 ⟨90JOC791⟩; **4**, 140, 352 ⟨91HCA1477⟩; **5**, 272, 284 ⟨85HCA484⟩, 416 ⟨91HCA1477⟩
Bieräugel, H., **3**, 90, 95 ⟨79CC117⟩
Biere, H., **2**, 740 ⟨79TL1361⟩
Bierer, D. E., **5**, 223 ⟨89OM655⟩
Bieri, J. H., **3**, 680 ⟨88HCA1177⟩; **4**, 683 ⟨86OM2413⟩; **5**, 622 ⟨84HCA534⟩
Bierl, B. A., **3**, 328 ⟨70SCI87⟩
Bierling, V. B., **2**, 659 ⟨72JPR170⟩
Biermacher, U., **6**, 641 ⟨63JMC275⟩
Biermann, T. F., **2**, 397 ⟨70JOC4229⟩
Biermann, U., **5**, 1007, 1018 ⟨67CB1082⟩; **6**, 260 ⟨67MI 608-01⟩, 409, 410 ⟨67CB1184⟩, 409, 419 ⟨67CB2484⟩, 442 ⟨67CB1082⟩
Biernacki, W., **2**, 11 ⟨79S37⟩
Biernbaum, M. S., **4**, 1054 ⟨78JOC950⟩; **6**, 182, 199 ⟨78JOC950⟩
Bieron, J. F., **2**, 613 ⟨85JOC4576⟩; **5**, 313, 360, 361 ⟨B-70MI 507-01⟩
Biersack, H., **4**, 676 ⟨75JOM(97)245⟩
Bierschenk, T. R., **6**, 246 ⟨86JA4103, 86JOM(307)C58⟩
Biersmith, E. L., **1**, 842 ⟨69CC1049⟩
Biesalski, E., **6**, 412 ⟨24MI 614-01⟩
Biessels, H. W. A., **1**, 784 ⟨84JCS(P1)1701⟩
Biezais, A., **2**, 237 ⟨61AK(18)1143⟩, 238 ⟨64ACS815⟩
Biffar, W., **6**, 405 ⟨81JMR(44)54⟩
Biffi, C., **3**, 7 ⟨87JOC2559⟩
Bigelli, C., **1**, 1035 ⟨85TL523⟩
Bigelow, H. E., **2**, 773 ⟨31CRV117⟩
Bigelow, L. A., **2**, 609, 627 ⟨32OSC(1)136⟩; **4**, 3 ⟨40JA1171⟩; **5**, 807 ⟨52JA710⟩, 808 ⟨59JA3599⟩; **6**, 257 ⟨59JA3599⟩, 262 ⟨62JA2751⟩, 284 ⟨68IC1647⟩, 284, 363, 616 ⟨72IC418⟩
Bigelow, S. S., **2**, 529 ⟨82JOC2498⟩; **4**, 970, 993 ⟨92JOC235⟩
Bigg, D., **5**, 871, 872, 873, 1071, 1148 ⟨91JOC1801⟩; **6**, 668 ⟨91JOC1801⟩
Biggadike, K., **2**, 19 ⟨88JCS(P1)549⟩
Bigge, C. F., **6**, 58 ⟨89TL7013⟩
Biggs, B. S., **3**, 619 ⟨55OSC(3)768⟩
Bigham, E., **4**, 194 ⟨73JA3635⟩
Bigler, P., **4**, 53 ⟨78S593⟩, 53, 54, 56 ⟨78HCA2047⟩, 852 ⟨89HCA1506⟩
Bigley, D. B., **1**, 75 ⟨71JINC1157⟩, 772 ⟨64JCS3897, 65JCS6202⟩, 772 ⟨67JCS(B)557⟩, 777 ⟨72JCS(P2)1744, 72JCS(P2)2359⟩, 777 ⟨72JCS(P2)926⟩, 777 ⟨73JCS(P2)1293, 75JCS(P2)317⟩; **3**, 85 ⟨70JA7161⟩
Biglino, G., **4**, 259 ⟨89CZ349⟩
Bignardi, G., **2**, 204 ⟨67M1518, 83JHC839⟩; **5**, 344 ⟨86FES539⟩, 501 ⟨83JHC1549⟩
Bigorra, J., **1**, 579 ⟨93T6717⟩
Bigot, B., **3**, 313 ⟨79JA1095, 79JA1101⟩
Bigotto, A., **5**, 354 ⟨79TL4761⟩
Biguard, D., **3**, 615 ⟨32MI 318-01⟩
Bihari, M., **2**, 34 ⟨93T5133⟩
Bihátsi-karsai, E., **5**, 199 ⟨86JCS(P1)927⟩
Bihlmayer, C., **2**, 903 ⟨81ZN(B)1265⟩; **4**, 665 ⟨83CB3182⟩
Bijpost, E. A., **2**, 521 ⟨93TL4655⟩
Bilbo, A. J., **4**, 816 ⟨61JA2202⟩
Bil'dinov, K. N., **6**, 501 ⟨82JOU19582220⟩
Bildstein, B., **2**, 465 ⟨90PS(47)341⟩; **6**, 355, 356, 357, 358 ⟨89CB2279⟩, 355, 357 ⟨85CC1800⟩, 357, 358 ⟨85JOM(295)C1⟩, 357, 358 ⟨89CB1255⟩
Bilevich, K. A., **2**, 451 ⟨62ZOB1974⟩
Bilevitch, K. A., **2**, 598 ⟨64JOM(2)281⟩
Bilinski, S., **6**, 637 ⟨89MI 620-01⟩
Bill, I., **5**, 168 ⟨59JCS1313⟩
Bill, J. C., **4**, 20 ⟨54OSC(4)807⟩
Billeb, G., **2**, 947 ⟨88TL5245, 90SL113⟩
Billedeau, R., **1**, 1042 ⟨88T5451⟩
Billedeau, R. J., **2**, 942 ⟨83TL4515⟩; **5**, 199 ⟨92T4863⟩
Billen, S., **1**, 232 ⟨89CB1113⟩
Biller, S. A., **1**, 342 ⟨83JA3741⟩; **4**, 557 ⟨90T6645⟩
Billet, D., **5**, 652 ⟨49BSF297⟩
Billeter, O. C., **2**, 249 ⟨04CB690⟩; **5**, 994 ⟨03CB3213, 05CB2013⟩
Billhardt, U. M., **5**, 174 ⟨87AG(E)362⟩
Billi, L., **3**, 458 ⟨90JOC4199⟩
Billiau, F., **6**, 607 ⟨78CB1619⟩
Billica, H. R., **2**, 52 ⟨48JA695⟩
Billig, E., **1**, 52 ⟨84OM(3)923⟩
Billig, F. A., **2**, 906 ⟨63OSC(4)68⟩
Billig, F., **1**, 1072 ⟨69ZC229⟩
Billingham, N. C., **1**, 933 ⟨79JCS(P1)1137⟩
Billington, D. C.,**1**, 529 ⟨88JOM(354)233⟩
Billman, J. H., **3**, 406 ⟨58JOC535⟩
Billmann, E., **2**, 288 ⟨36BSF2310⟩
Billmers, J. M., **2**, 251 ⟨83TL3191⟩, 287 ⟨83TL1213⟩
Billmers, R., **2**, 184 ⟨86JOC1033⟩
Billmers, R. L., **2**, 185 ⟨81JA7016⟩, 718 ⟨83JOC3071⟩
Billups, W. E., **1**, 801 ⟨73JA8096⟩, 907 ⟨B-87MI 119-08⟩; **2**, 20 ⟨71JA438⟩
Biloski, A. J., **2**, 146 ⟨80TL689⟩, 340, 344, 347 ⟨83S537⟩; **5**, 213 ⟨83S537⟩, 285 ⟨82JA3233⟩
Bilow, N., **1**, 1137 ⟨81JOC2280⟩; **3**, 466 ⟨64JA1815⟩
Biltueva, I. S., **5**, 924 ⟨88IZV1947, 89JOM(368)163⟩; **6**, 131 ⟨87IZV2872, 88MI 604-01⟩
Biltz, H., **2**, 134 ⟨28JPR198⟩, 1019 ⟨04CB4415, 1897CB290⟩; **4**, 797 ⟨08CB2190⟩; **5**, 654 ⟨1892CB2533⟩
Bilyard, K. G., **1**, 184 ⟨80S389⟩
Binaghi, M., **3**, 534 ⟨68MI 316-01, 68MI 316-03⟩
Binbahari, Z., **3**, 229 ⟨80JCS(P1)2851⟩
Binder, D., **5**, 354 ⟨77S487⟩

Binder, H., **5**, 179 ⟨89LA1233⟩, 179 ⟨89LA1233⟩
Binder, J., **4**, 916 ⟨86TL5829⟩
Binder, W. H., **2**, 597 ⟨94T749⟩
Bindra, ,R., **2**, 977 ⟨75JA7171⟩
Bindra, R., **6**, 577 ⟨75JMC913⟩
Binger, P., **1**, 363 ⟨87CB1933, 89LA739⟩, 527 ⟨85AG(E)316, 87TCC77⟩; **2**, 449 ⟨91AG(E)207⟩, 520 ⟨74IS134⟩, 523 ⟨74IS141⟩, 984 ⟨89AG(E)610, 91CB2165⟩, 1077 ⟨68LA(717)1⟩, 1083, 1084, 1087, 1088, 1097 ⟨68LA(717)21⟩, 1087, 1088 ⟨65TL1901, 75CB395⟩, 1099 ⟨63AG918⟩; **3**, 273 ⟨67AG(E)84⟩, 509 ⟨89AG(E)610⟩, 688 ⟨88AG(E)1484⟩; **4**, 547, 549 ⟨B-90MI 412-01⟩, 549 ⟨93PS(76)1⟩, 631 ⟨62AG(E)508⟩, 632 ⟨66TL2675, 68AG(E)286⟩, 649 ⟨85TL301⟩, 725 ⟨92OM3517⟩; **5**, 343 ⟨89LA739⟩, 883, 1151, 1152, 1154, 1159 ⟨88AG(E)1484⟩, 885 ⟨91AG(E)310⟩, 890 ⟨90CB1617, 90PS(49)337⟩, 1152 ⟨92AG(E)758⟩; **6**, 152 ⟨88AG(E)1484⟩, 153 ⟨91AG(E)207, 92AG(E)584⟩, 154 ⟨87AG(E)908⟩, 182, 183, 386 ⟨75LA1339⟩, 208 ⟨63AG(E)686⟩
Bingham, E. M., **2**, 16 ⟨80JA4845⟩, 19 ⟨88OSC(6)835⟩, 362 ⟨88OSC(6)440⟩; **3**, 149 ⟨80JA4845⟩
Bingham, R., **1**, 881 ⟨67OS(47)20⟩
Bingham, R. C., **1**, 923 ⟨76JA230⟩, 1114 ⟨70JA2538, 70JA2540⟩
Binh, S. K., **3**, 497 ⟨72BSF3549, 77JOC1242⟩
Binkley, R. W., **3**, 8, 121 ⟨76JOC3030⟩, 8, 121 ⟨76SC281⟩, 8, 122 ⟨77JOC1216⟩; **6**, 22 ⟨83JOC1776⟩
Binkley, W. W., **2**, 102 ⟨38JA2810⟩, 366 ⟨39JA3250⟩
Binks, R., **3**, 419 ⟨55CI(L)1039⟩
Bin Manas, A. R., **3**, 98 ⟨84S166⟩
Binnewies, M., **6**, 264 ⟨85PS(21)349⟩
Binon, F., **2**, 29 ⟨69S39⟩
Binsch, G., **4**, 31, 34 ⟨73JA182⟩, 33, 34, 38 ⟨79T2661⟩
Binshtok, E. V., **1**, 741 ⟨72BAU2056, 72IZV2113⟩
Binz, A., **6**, 258 ⟨35CB1513⟩
Biondi, L. V., **3**, 521 ⟨83OM1505⟩
Biran, C., **2**, 938, 941 ⟨86T669⟩; **5**, 417 ⟨78JOM(161)C23⟩; **6**, 125, 162 ⟨79JOM(177)137⟩, 379 ⟨88JOM(358)31⟩
Birbaum, J.-L., **2**, 710 ⟨87CL1647⟩
Birberg, W., **6**, 546 ⟨91TL7453⟩
Birch, A. J., **1**, 73, 79, 456 ⟨76OR(24)1⟩, 82 ⟨66JCS(C)1894, 71AJC513⟩, 85 ⟨67TL3457⟩, 95, 98 ⟨44JCS430⟩, 96, 890 ⟨80JA3370, 80JA4074, 80JA6430, 81JA284, 51JCS1945⟩, 98 ⟨88OSC(6)936⟩, 99 ⟨47JCS1270, 67TL857, 74TL2395, 75CC480⟩, 99, 890 ⟨59T148⟩, 100 ⟨69AJC1321⟩, 434 ⟨70AJC1641⟩, 797 ⟨68TL3797⟩, 889 ⟨50QR69, 58QR17⟩; **3**, 231 ⟨70CC52⟩; **4**, 413 ⟨55AJC512⟩; **5**, 96 ⟨82JCS(P1)1763⟩, 1126 ⟨85TL501⟩
Birch, A. M., **1**, 357 ⟨80CC1195⟩
Birchall, G. R., **3**, 239 ⟨74CJC610⟩
Birchall, J. M., **4**, 8 ⟨60PCS81⟩; **6**, 32 ⟨65BRP1014252⟩, 222 ⟨59JCS13⟩, 241 ⟨75JCS(P1)2033, 80JFC(16)391⟩, 241 ⟨81JCS(P1)455⟩
Birchall, T., **1**, 897 ⟨64CJC502⟩; **2**, 820 ⟨82JA674⟩
Birch, S. F., **1**, 100 ⟨51JCS2556⟩; **5**, 199 ⟨52JCS1363⟩
Birckenbach, L., **5**, 984 ⟨30CB2528, 30CB2544⟩; **6**, 222, 227 ⟨32CB546⟩, 612 ⟨31LA(489)7⟩
Bird, C. N., **1**, 637 ⟨58JOC996⟩, 773 ⟨77JOC3895⟩
Bird, C. W., **1**, 181 ⟨88T1567⟩, 797 ⟨67MI 118-01⟩; **4**, 212 ⟨91TL2413⟩, 681 ⟨67JCS(C)1862⟩; **5**, 27 ⟨62CRV283, B-67MI 502-01⟩
Bird, G. J., **5**, 937 ⟨84OM1764⟩
Bird, M. L., **2**, 286, 288, 291 ⟨42JCS570⟩
Bird, T. G. C., **5**, 601 ⟨85T693⟩
Birdwhistell, K. R., **5**, 940 ⟨85JA4474⟩, 1079 ⟨93OM1023⟩
Biresaw, G., **1**, 352 ⟨82JOC1615⟩
Birgele, I. S., **4**, 807, 811, 813 ⟨91JGU1295⟩
Birk, U., **4**, 677, 707, 723 ⟨93JOM(458)181⟩
Birkel, M., **4**, 110 ⟨91CB1207⟩, 549 ⟨93PS(76)1⟩; **6**, 153 ⟨92AG(E)879⟩, 153 ⟨93JOC4105⟩
Birkenstock, U., **6**, 465 ⟨93GEP4134688⟩
Birkett, P. R., **5**, 690 ⟨93T11029⟩
Birkhauser, A., **3**, 247 ⟨70HCA1489⟩
Birkhofer, H., **1**, 925 ⟨83TL185⟩
Birkhofer, L., **5**, 374 ⟨60CB2804⟩
Birkinshaw, S., **4**, 934, 935 ⟨91TL6961⟩
Birkinshaw, T. N., **1**, 195 ⟨88TL4705⟩, 195 ⟨90TL6995⟩; **5**, 1125 ⟨84JCS(P1)147⟩
Birkner, H., **2**, 440, 442, 446 ⟨62LA(659)49⟩; **4**, 112 ⟨62LA(659)49⟩
Birkofer, L., **1**, 43 ⟨58CB2383⟩, 1045, 1138 ⟨63CB3280⟩; **2**, 262 ⟨61CB1263⟩, 541 ⟨63CB3289⟩, 907 ⟨B-89MI 218-01⟩, 943 ⟨67CB2681, 72CB1759⟩, 943 ⟨72CB470, 74CB3717⟩, 943 ⟨79CB2829, 81CB2293⟩, 944 ⟨81CB3154, 83CB2564⟩; **3**, 220 ⟨62CB1495⟩, 273 ⟨63CB3280⟩; **4**, 313 ⟨67T2781, 75LA266⟩, 513 ⟨84JOM(264)87⟩; **5**, 84 ⟨44AG135⟩, 376 ⟨61CB1263⟩, 377 ⟨71CB3831, 75LA266⟩, 378 ⟨65TL4007⟩, 379 ⟨63AG93⟩, 718 ⟨63AG96, 63CB1473⟩, 722, 814 ⟨68CB2585⟩; **6**, 507 ⟨60CB2810⟩
Birktimirova, C. G., **4**, 920 ⟨57ZOB2161⟩
Birnbaum, S. M., **5**, 84 ⟨57ANY(69)94⟩
Birner, P., **5**, 641 ⟨88JCS(P2)821⟩
Birot, M., **4**, 624 ⟨88JOM(341)125⟩
Birse, E. F., **4**, 918 ⟨88JCS(P1)1039, 90JCS(P1)2811⟩
Birss, V. I., **1**, 440 ⟨88JOC3815⟩
Birum, G. H., **3**, 547, 549, 604 ⟨68JA3842⟩, 549, 604 ⟨66TL5707⟩, 602 ⟨69ACR373, 66CC736⟩, 604 ⟨67CC137, 67JOC3554⟩, 604 ⟨68CI(L)653⟩; **4**, 64, 65, 80 ⟨52JA3594⟩, 122 ⟨67USP3306937⟩, 466 ⟨74JOC209⟩; **6**, 16, 44, 45, 234 ⟨52JA3594⟩, 661, 733 ⟨72JOC2730⟩
Bisagni, E., **1**, 42 ⟨76T829⟩; **3**, 309 ⟨86T2311⟩, 442 ⟨83T1777⟩; **5**, 199 ⟨92H(34)369⟩
Bisagni, M., **3**, 304 ⟨55JCS3693⟩
Bisaha, J. J., **2**, 152, 363 ⟨92JA5977⟩; **4**, 707 ⟨79JA768⟩
Bis, K. G., **2**, 401, 402 ⟨83JOC1⟩
Bisaria, C. S., **6**, 270 ⟨91SRI401⟩
Bischoff, C., **2**, 392 ⟨72JPR891⟩; **5**, 206 ⟨70JPR(312)2⟩
Bischoff, J. J., **6**, 641 ⟨86S777⟩
Bishop, D., **6**, 233, 234 ⟨90JFC(48)249⟩
Bishop, D. C., **5**, 849 ⟨72JCS(P1)2927⟩
Bishop, P. M., **2**, 560 ⟨82TL3901⟩
Bishop, R., **1**, 685, 686 ⟨93JOC5900⟩; **2**, 1006 ⟨91COS(6)261⟩
Bishop, S. W., **1**, 442 ⟨77JOC3494⟩
Bishop, W. S., **3**, 619 ⟨55OSC(3)768⟩
Bishop, III, K. C., **1**, 841 ⟨73TL2727⟩
Biskup, M., **1**, 540 ⟨69JA1853⟩
Bisnette, M. B., **5**, 448 ⟨64JA1267⟩, 449 ⟨66JOM(5)341⟩, 452 ⟨64JOM(2)15⟩
Biss, J. W., **3**, 86 ⟨59JOC627⟩
Bissell, E. R., **2**, 381, 382, 383 ⟨56JA2467⟩; **4**, 31 ⟨64JOC1591⟩; **6**, 2 ⟨62JOC1482, 64JOC252⟩
Bissey, J. E., **4**, 116 ⟨70OMR(2)81⟩
Bissier, J. R., **5**, 81 ⟨55MI 502-02⟩
Bissing, D. E., **1**, 723 ⟨63JA3878, 65JOC1296⟩; **4**, 738 ⟨65OS(45)33⟩
Bissinger, H. J., **1**, 1077 ⟨86T1711, 88LA221⟩
Bissinger, P., **2**, 437 ⟨92ZN(B)640⟩; **6**, 371 ⟨93OM2757⟩
Bister, K., **5**, 848 ⟨71ZN(B)296⟩
Bistline, R. G., **2**, 190 ⟨63OSC(4)8623⟩
Biswas, A., **5**, 322 ⟨83T2571⟩
Biswas, G. K., **1**, 239 ⟨90IJC(B)390⟩
Biswas, S., **3**, 528 ⟨88CC1421⟩
Bisway, G., **2**, 156 ⟨88JCR308⟩
Biswell, C. B., **2**, 883 ⟨35JA913⟩
Bit, R. A., **5**, 306 ⟨90TL5201⟩
Bitha, P., **2**, 307 ⟨88JHC1035⟩
Bither, T. A., **3**, 709, 711 ⟨58JA4151⟩; **5**, 1133 ⟨58JA4151⟩
Bitit, N., **4**, 621 ⟨89JCR(S)146⟩
Bitler, S. P., **2**, 478 ⟨84JOC4733⟩
Bitonti, A. J., **6**, 487 ⟨91JMC569⟩
Bitter, I., **6**, 503 ⟨81ACH57⟩, 625 ⟨77ACH77⟩
Bitterer, F., **2**, 472 ⟨90JOM(399)C4⟩

Bitterwolf, T. E., **6**, 7 ⟨90JOM(397)51⟩
Bittler, K., **5**, 27 ⟨68AG(E)329⟩, 142 ⟨68AG352⟩
Bittmann, R., **2**, 84 ⟨89JA3077⟩
Bittner, G.,**3**, 680 ⟨68LA(713)1⟩
Bittner, S., **2**, 683 ⟨75CI(L)281⟩; **4**, 843 ⟨87TL3805⟩, 873 ⟨88TL6177⟩; **5**, 219 ⟨76JCS(P1)1708⟩, 374 ⟨85JOC1712⟩, 1000 ⟨74TL1965⟩; **6**, 627 ⟨89S132⟩
Biurrun, C., **1**, 217 ⟨90H(31)1997⟩
Bixon, R., **1**, 1111 ⟨69CC1044⟩
Bizhan, L. N., **1**, 1006 ⟨71IZV1534, 72IZV472⟩
Bizyukova, N. M., **5**, 721, 722 ⟨78JGU1313⟩, 721, 722 ⟨78ZOB1429⟩, 1082 ⟨78ZOB1313⟩
Bizzozero, N., **5**, 211 ⟨87MI 504-02⟩
Bjåmer, K., **3**, 450 ⟨63ACS559⟩
Bjellqvist, B., **2**, 191 ⟨73ACS3180⟩
Bjerre, C., **3**, 553 ⟨79CC99⟩
Björk, P., **1**, 480 ⟨93JOM(460)127⟩
Bjorkling, F., **2**, 1011 ⟨85SC463⟩
Bjørnstad, H. J., **6**, 170 ⟨72ACS1874⟩
B Kamm, **1**, 556 ⟨90T2371⟩
Blache, Y., **5**, 360 ⟨90CPB2352⟩
Black, A. P., **2**, 393 ⟨43OSC(2)512⟩
Black, C. J., **2**, 3 ⟨75JOC3420⟩
Black, D. G., **4**, 656 ⟨93JA8493⟩
Black, D. K., **1**, 808 ⟨65JCS6784⟩; **2**, 677 ⟨65JCS5225⟩
Black, D. S., **2**, 354 ⟨75S205⟩
Black, D. St. C., **2**, 346 ⟨78AJC2317⟩; **4**, 232 ⟨73AJC2491⟩, 409 ⟨89JOC4771⟩; **5**, 605 ⟨72TL4191⟩; **6**, 107 ⟨71TL4519⟩
Black, K. W., **1**, 886 ⟨89JOC1671⟩
Black, L. L., **3**, 686 ⟨78JOC3736⟩
Black, M., **4**, 309 ⟨92JOC547⟩
Black, T. H., **5**, 170 ⟨88JOC2371⟩, 344 ⟨92SC2729⟩
Blackburn, B. K., **2**, 48 ⟨90TL3817⟩
Blackburn, E. V., **1**, 945, 946 ⟨81JA1557⟩
Blackburn, G. M., **1**, 807, 821, 822 ⟨69CC99, 68CC186⟩; **2**, 847 ⟨78CC870, 80JCS(P1)1150⟩; **4**, 119, 129 ⟨86JCS(P1)913, 85JCR(S)92⟩, 131 ⟨83CC886, 87JCS(P1)181⟩, 563 ⟨92TL3927⟩, 801 ⟨82CC1270, 86JCS(P1)1417⟩, 802 ⟨86JCS(P1)1425⟩; **5**, 404 ⟨78CC870, 80JCS(P1)1150⟩; **6**, 121 ⟨87JCS(P1)181⟩, 242 ⟨93TL149⟩, 263 ⟨84JCS(P1)1119⟩, 264 ⟨81CC930, 91TL6425⟩, 282 ⟨85JCS(P1)1935⟩, 289 ⟨83CC886, 86JCS(P1)1425⟩
Blackburn, T. F., **1**, 1105 ⟨75JA679⟩; **2**, 616, 623, 629, 968 ⟨75JA679⟩
Blackett, B. N., **1**, 387 ⟨69T1479⟩; **4**, 182 ⟨70T1311⟩
Blackham, A. U., **2**, 403 ⟨62JA2922⟩
Blacklock, T. J., **1**, 571 ⟨81OS(60)53⟩, 802 ⟨77JA2344⟩, 808 ⟨80JA2797⟩; **2**, 528 ⟨91JOC751, 91JOC763⟩; **6**, 446 ⟨91JOC751⟩
Blackstock, S. C., **2**, 383 ⟨84JA3366, 86T1769⟩
Blackwell, C. M., **1**, 362, 555, 558 ⟨92JA7375⟩
Blackwell, D. S. L., **5**, 634 ⟨74TL91⟩
Blackwood, R. K., **2**, 393 ⟨63OSC(4)454⟩
Blade, R. J., **1**, 536 ⟨87TL4879⟩; **2**, 625 ⟨87TL3857⟩; **5**, 1002 ⟨76TL2691⟩; **6**, 15 ⟨88CC638⟩, 233, 234 ⟨90JFC(48)249⟩
Bladé-Font, A., **1**, 268 ⟨62JA677⟩; **2**, 436, 442, 443 ⟨64JA2378⟩; **5**, 276 ⟨80TL2443⟩, 570, 618 ⟨81JCR(S)58⟩
Bladon, C. M., **3**, 333, 335 ⟨83CC423⟩, 333, 335 ⟨85JCS(P1)1541⟩
Bladon, P., **1**, 21 ⟨50JCS3000⟩, 529 ⟨77JCR(M)153⟩, 530 ⟨88JOM(355)449⟩
Blagbrough, I. S., **2**, 660 ⟨82TL4843⟩; **5**, 572 ⟨94TL2057⟩
Blagg, J., **2**, 532 ⟨90JA2767⟩, 927 ⟨92SL143⟩
Blagoev, B., **4**, 147 ⟨80MI 403-01⟩; **5**, 27 ⟨70S615⟩, 98 ⟨76MI 502-01⟩
Blair, I. A., **5**, 610, 623, 624 ⟨86PS(26)169⟩
Blair, J. A., **2**, 377 ⟨70JCS(C)1714⟩
Blair, L. K., **3**, 119 ⟨92JOC1600⟩
Blair, P. A., **1**, 154 ⟨76TL1439⟩
Blais, C., **1**, 22, 437 ⟨80CC169⟩
Blais, J., **2**, 521 ⟨70CR(C)589⟩
Blaise, E. E., **3**, 461 ⟨01CR(132)478⟩
Blake, A. J., **6**, 457 ⟨88CC529⟩
Blake, D. M., **5**, 460 ⟨71JA1368, 74IC1595⟩
Blake, J. F., **1**, 149 ⟨90JA697⟩
Blakebrough, N., **5**, 117 ⟨B-84MI 502-01⟩
Blakeley, R. L., **3**, 577 ⟨86JA5543⟩; **4**, 892, 979 ⟨86JA5543⟩
Blanc, A., **3**, 40 ⟨88BSF67⟩
Blanc, G., **2**, 54 ⟨03CR(136)1676, 03CR(137)60⟩, 54 ⟨04BSF748, 04CR(138)148⟩
Blanchard, J. M., **1**, 503, 505 ⟨80TL1739⟩
Blanchard, M., **2**, 650 ⟨82JMOC333⟩
Blanchet, D., **6**, 5 ⟨85T4503⟩
Blanchet, G. B., **4**, 291 ⟨85JA675⟩
Blanchette, J. T., **5**, 1063 ⟨89JOC1048⟩
Blanchette, M. A., **1**, 749 ⟨84TL2183⟩; **5**, 155 ⟨84TL2183⟩
Blanchini, J. P., **1**, 1125 ⟨68BSF2120⟩
Blanchot, V., **3**, 240, 245 ⟨90S755⟩
Blanck, K., **2**, 870 ⟨76ZN(B)1058⟩
Blanck, U., **4**, 298 ⟨85AP(318)473⟩
Blanco, L., **2**, 28 ⟨76S194⟩, 395 ⟨91SL907⟩; **3**, 67, 235 ⟨76S196⟩, 154 ⟨76S194⟩, 213 ⟨85TL4195⟩, 224 ⟨81S289, 81S291⟩, 232 ⟨81TL645⟩; **4**, 352 ⟨91TL6325⟩
Blancou, H., **5**, 58 ⟨76CC885⟩
Bland, S. W., **5**, 134 ⟨78GEP2822472⟩
Bland, W. J., **6**, 7 ⟨84JOM(260)C75, 85JOM(280)397⟩
Blaney, W. M., **1**, 596, 780 ⟨87TL221⟩
Blangey, L., **2**, 780 ⟨38HCA1579⟩
Blangley, L., **2**, 803 ⟨B-52MI 215-01⟩
Blank, H. U., **6**, 465 ⟨93GEP4134688⟩
Blanke, E., **2**, 100 ⟨31LA(485)258⟩; **3**, 538 ⟨62TL221, 64CB1799⟩
Blankenship, C., **1**, 284 ⟨84JA6442⟩
Blankenship, R. M., **2**, 14 ⟨74JOC2300⟩
Blankespoor, R. L., **5**, 972 ⟨76JOC2071⟩
Blankley, C. J., **3**, 471 ⟨68JOC53⟩
Blanko, F. F., **2**, 812 ⟨78JOU1101⟩
Blanpain, P., **4**, 1074, 1075 ⟨84JA3785⟩; **6**, 92 ⟨84JA3784⟩
Blanpin, O., **5**, 751, 753 ⟨56MI 519-02⟩
Blanton, C. D., **3**, 308 ⟨81H(16)399⟩
Blanton, J. R., **1**, 5, 9 ⟨87JOC473⟩; **2**, 7 ⟨85CC337⟩, 35 ⟨85JOC5828⟩
Blanzat, J. F., **3**, 375 ⟨80JA2460⟩
Blart, E., **1**, 469, 494, 1013 ⟨92SL715⟩
Blaschek, U., **5**, 166 ⟨84T2211⟩
Blaschette, A., **2**, 100 ⟨91ZN(B)5⟩; **5**, 348 ⟨88ZN(B)1495⟩
Blaschke, G., **2**, 447 ⟨80CB1612⟩, 452 ⟨78ZN(B)1556⟩; **4**, 574 ⟨80CB1480⟩, 582 ⟨78JOM(160)41, 80ZN(B)584⟩, 585 ⟨79JOM(182)251⟩
Blaser, B., **2**, 864 ⟨68ZAAC(363)176⟩
Bläser, E., **1**, 370, 371 ⟨94AG(E)683⟩
Blaser, H.-U., **1**, 76 ⟨91TA843⟩, 494 ⟨82JOM(233)267⟩
Blaser, R., **5**, 700 ⟨69HCA569⟩
Blashek, U., **2**, 422 ⟨84T2211⟩
Blasioli, C., **3**, 3 ⟨76BSF1941⟩, 114 ⟨77BSF1941⟩
Blass, B., **3**, 696, 719 ⟨93TL579⟩
Blaszczak, L., **1**, 479 ⟨76TL4405⟩; **3**, 210 ⟨74JOC2217⟩
Blatcher, P., **3**, 166 ⟨79JCS(P1)1074⟩; **4**, 250 ⟨78TL2345⟩
Blattenberger, J. W., **5**, 595 ⟨41JA314⟩
Blatter, H. M., **5**, 1050 ⟨64TL1087, 64TL855⟩, 1052 ⟨66JOC722⟩
Blatter, K., **5**, 1155 ⟨87PS(30)479⟩
Blattner, R., **1**, 769 ⟨87CC1008⟩; **3**, 621 ⟨87CC1007⟩; **4**, 43, 55 ⟨81CB1951, 81CB1958⟩, 206 ⟨81CB1938⟩
Blau, H., **2**, 512 ⟨80CB1180⟩; **4**, 580, 582 ⟨83ZN(B)711⟩, 581 ⟨83CB690⟩; **5**, 448 ⟨82CB1488⟩
Blau, W., **5**, 1006 ⟨69GEP1289526⟩
Blazejewicz, L., **5**, 48 ⟨60T67⟩
Blazejewski, J. C., **1**, 522 ⟨75TL363⟩; **3**, 145 ⟨85JCS(P1)2667⟩
Blazer, T. A., **4**, 564 ⟨69ZN(B)1081⟩

Blazevic, K., **2**, 421 ⟨68JCS(C)1704⟩
Blazievic, N., **2**, 308, 325 ⟨79S161⟩
Bleasdale, J. L., **4**, 440 ⟨50JA3843⟩
Blechert, S., **1**, 339, 341 ⟨93TL5251⟩, 414, 825, 836 ⟨89S71⟩; **5**, 214 ⟨85LA2073⟩, 711 ⟨82AG375, 85TL5281⟩
Blechschmitt, K., **3**, 387, 390 ⟨83AG(E)314⟩
Bleck, R. D., **2**, 440 ⟨65ZAAC(336)234⟩
Bleck, R. J., **4**, 455 ⟨65ZAAC(336)234⟩
Bledsoe, J. O., **2**, 219 ⟨61JOC850⟩
Bleeke, J. R., **1**, 87, 93 ⟨79ACR324, 78JA2405⟩
Blegen, J. R., **5**, 251 ⟨55OSC(3)116⟩
Bleikolm, A., **4**, 415 ⟨80LA1573⟩
Bleisch, S., **2**, 361, 792 ⟨78TL4003⟩, 398 ⟨78ZC323⟩, 399 ⟨76ZC437, 85HOU(E11)579⟩, 399 ⟨85HOU(E11)584⟩, 401 ⟨85HOU(E11)1031⟩; **3**, 343 ⟨67CB93⟩, 349 ⟨66CB1771⟩
Blenderman, W. G., **1**, 100 ⟨79TL4985⟩
Blenkiron, P., **5**, 400 ⟨92JOM(426)C28⟩
Bleshinskii, S. V., **6**, 205 ⟨82MI 606-01⟩
Bleshinskii, V. S., **6**, 205 ⟨82MI 606-01⟩
Blic, A. de, **2**, 247, 250 ⟨82S281⟩
Blick, K. E., **4**, 107 ⟨70IC975⟩
Blicke, F. F., **2**, 889 ⟨32JA2993⟩, 890 ⟨33JA1161⟩; **5**, 76 ⟨52JA1730⟩
Blicke, F. T., **2**, 882, 883 ⟨41JA575⟩
Blickensderfer, J. R., **4**, 721 ⟨79IC3453⟩
Blinn, D. A., **2**, 439 ⟨90JOM(393)143⟩
Blinne, G., **3**, 577 ⟨74CB13⟩
Bliss, A. D., **2**, 176 ⟨63JOC3537⟩, 259 ⟨66JHC14⟩
Bliznyuk, N. K., **2**, 448 ⟨66CA-P2297, 68CA-P106877⟩, 883 ⟨65JGU1253⟩; **4**, 554 ⟨60ZOB1602⟩; **6**, 263 ⟨60ZOB1602⟩
Blizzard, T. A., **2**, 976 ⟨84JOC1772, 84JOC4332⟩; **5**, 179 ⟨83JOC758⟩
Bloch, A., **4**, 876 ⟨82CC336⟩
Bloch, A. N.,**4**, 869 ⟨77CC505⟩
Bloch, A. N., **4**, 874 ⟨80CC867⟩; **6**, 552 ⟨77CC660⟩, 591 ⟨75JOC746⟩, 591, 594 ⟨80CC866B⟩
Bloch, C., **5**, 834 ⟨1899CB1975⟩
Bloch, F., **5**, 476, 477 ⟨37CR(204)1342⟩
Bloch, H. S., **5**, 106 ⟨55OSC(3)637⟩
Bloch, L. P., **6**, 212 ⟨39JA2962⟩
Bloch, R., **1**, 786 ⟨86TL3511⟩, 790 ⟨82TL3277, 83TL1247⟩, 790 ⟨86T4975⟩, 983 ⟨83TL4691⟩; **3**, 27, 154 ⟨78S140⟩, 539 ⟨73CC795⟩, 632 ⟨89T3731⟩
Block, A. N., **6**, 591 ⟨77CC505⟩
Block, B. P., **2**, 843 ⟨66IS71⟩; **4**, 465 ⟨63JOC2898⟩, 555 ⟨68USP3403176⟩, 556 ⟨65IC198, 67USP3332986⟩
Block, E., **1**, 141 ⟨85JA6729⟩, 232 ⟨69JOC1233⟩, 585 ⟨84OR(30)457⟩, 617 ⟨B-78MI 114-01⟩, 619, 707 ⟨83JA6164⟩, 619, 707 ⟨87OS(65)90⟩, 619, 707, 949 ⟨86JA4568⟩, 623 ⟨B-78MI 114-02, B-78MI 114-03, 83JA6165⟩, 707 ⟨84TL5469⟩, 707, 993 ⟨90JA4072⟩, 834 ⟨85JA6731⟩; **2**, 5 ⟨91COS(4)329⟩, 173 ⟨74JA3921⟩, 177 ⟨82TL4203⟩, 184 ⟨78JA3622⟩, 186 ⟨72JA642⟩, 186 ⟨74JA3929⟩, 195, 196 ⟨85JA1783⟩, 196 ⟨B-78MI 203-01⟩, 262 ⟨88T281⟩; **3**, 334 ⟨72JA642, 74JA3929⟩, 335, 369 ⟨82JA3119⟩, 337 ⟨77CC287⟩, 337 ⟨78JA7436⟩, 337 ⟨90TL5003, 76JA1264⟩, 369 ⟨85JA6731⟩; **4**, 80, 83, 84 ⟨69JOC1233⟩, 82, 83, 85 ⟨86JA4568⟩, 138, 393 ⟨87JOC809⟩, 223 ⟨86JOC3428, 89JA658⟩, 258 ⟨82JA3119⟩, 364 ⟨88T281⟩, 365, 392 ⟨85JA6729⟩, 368 ⟨82TL4203, 85TL1425⟩, 390, 391 ⟨B-78MI 408-01⟩; **6**, 349 ⟨85CC1674, 85JA6729⟩, 349 ⟨87IC1488⟩, 349, 585 ⟨85TL2259, 88T281⟩
Block, H. D., **2**, 294 ⟨70JOM(25)17⟩
Block, H. J., **6**, 360, 361 ⟨78JPR335⟩
Block, Jr., P., **2**, 608 ⟨43JA1430⟩; **3**, 426 ⟨65JOC1307, 73OSC(5)1031⟩; **5**, 76 ⟨73OSC(5)381⟩
Block, L. P., **4**, 2 ⟨37JA1407⟩
Block, M. H., **1**, 831 ⟨88JOC4923⟩; **5**, 235 ⟨88JOC4923⟩
Block, P., Jr., **5**, 93 ⟨73OSC(5)1031⟩
Block, R. J., **5**, 83 ⟨46CRV501⟩
Block, T. F., **5**, 932 ⟨76JA441⟩
Blodgett, J. K., **4**, 422 ⟨84JOC4272⟩; **5**, 999 ⟨84JOC4272⟩
Blodworth, A. J., **2**, 291 ⟨83JCS(P1)471⟩
Bloem, P., **3**, 421 ⟨90JOC1086⟩
Bloink, G. J., **3**, 451 ⟨50JCS1328, 52JCS661⟩
Blokhina, A. N., **1**, 657 ⟨60ZOB3615⟩, 658 ⟨62IZV1373⟩
Blomberg, C., **2**, 73 ⟨77S18⟩
Blomquist, A. T., **1**, 635 ⟨53JA2153⟩, 723 ⟨64JA5041⟩, 773, 776 ⟨55JA1001⟩; **3**, 493 ⟨67JA4996⟩
Blonda, D. S., **6**, 508 ⟨65CJC1798⟩
Bloodworth, A. J., **1**, 68 ⟨82JCS(P1)1797⟩, 1129 ⟨87CC786⟩; **2**, 98 ⟨71JCS(C)945, 82JCS(P1)1807⟩; **4**, 214 ⟨91CC947⟩; **5**, 716 ⟨67JCS(C)1309⟩
Bloom, J. D., **1**, 754, 755 ⟨86JA2662⟩; **5**, 173 ⟨82JA4015⟩
Bloom, R. K., **2**, 619 ⟨69JOC3434⟩
Bloom, S. H., **1**, 316 ⟨88JA4788⟩, 630, 633 ⟨90OR(39)1⟩, 1094 ⟨84TL4901⟩; **2**, 556 ⟨88JA4788⟩
Bloomberg, R. N., **3**, 407 ⟨55USP2700681⟩
Bloomfield, J. J., **1**, 516 ⟨71JA782⟩, 881 ⟨61JOC4112, 68JOC3425⟩, 881, 884 ⟨67OR(15)1⟩; **2**, 749 ⟨67OR(15)1⟩; **4**, 168 ⟨68JCS(C)7⟩; **5**, 46 ⟨67OR(15)1⟩
Bloomfield, P. R., **2**, 437 ⟨59CI(L)541⟩
Bloomquist, A. E., **5**, 997 ⟨67CJC2625⟩
Bloshchitsa, F. A., **2**, 100 ⟨88JOU1473⟩
Bloss, D. E., **3**, 289 ⟨73JA3040⟩
Blossey, E. C., **3**, 215 ⟨73CC56⟩
Blotny, G., **1**, 877 ⟨90JOC1019⟩; **5**, 1026 ⟨82LA1927⟩; **6**, 575 ⟨83S391⟩
Blough, B. E., **1**, 226, 227, 751 ⟨88JOC4274⟩; **2**, 86 ⟨88JOC4274⟩, 327 ⟨93TL7239⟩
Blount, B. K., **2**, 452 ⟨1891JCS1931⟩
Blount, J., **3**, 234 ⟨70JOC4141⟩
Blount, J. F., **1**, 214 ⟨81CI(L)548⟩; **2**, 440, 441, 443 ⟨80JOC3691⟩, 949 ⟨86JA6664⟩; **3**, 294 ⟨71JA4321⟩; **5**, 284 ⟨85JOC2220⟩
Blout, E. R., **3**, 444 ⟨46JA1983⟩
Blower, P. J., **2**, 133 ⟨91CC475⟩
Bloxham, J., **5**, 615 ⟨94JCS(P1)989⟩
Blue, C. D., **1**, 449 ⟨89JOC6064⟩
Blues, E. T., **1**, 3, 6 ⟨63PCS219⟩; **3**, 487 ⟨74CC466⟩, 537 ⟨70JCS(D)699, 71GP2047373⟩, 537 ⟨73CC921, 74CC513⟩; **5**, 874 ⟨74CC466⟩
Blum, D. M., **1**, 710 ⟨79JA5060⟩; **2**, 623 ⟨74JOC3307⟩, 931 ⟨78OS(58)152⟩; **4**, 807, 1060 ⟨88OSC(6)1033⟩
Blum, H., **4**, 553 ⟨86GEP3444678⟩, 561 ⟨79ZAAC(457)203⟩; **6**, 158 ⟨79ZAAC(457)214, 91PS(55)185⟩, 159 ⟨79ZAAC(457)209, 88GEP3611522, 88ZN(B)75⟩
Blum, J., **1**, 73 ⟨66TL1605⟩, 94 ⟨83TL4139⟩, 95 ⟨87JOC2804⟩; **2**, 622 ⟨66TL1605, 66TL3041⟩; **3**, 294 ⟨67TL3665⟩, 619, 666 ⟨70TL1963⟩
Blum, P. M., **1**, 945 ⟨78JCS(P2)1313⟩
Blum, R. B., **1**, 132, 135 ⟨77TL3791⟩
Blumbach, J., **5**, 226 ⟨85AG(E)180⟩
Blumberg, K., **4**, 228 ⟨79CAR(70)217⟩, 238, 287 ⟨77CAR(59)351⟩
Blume, E., **3**, 696 ⟨76JPR127⟩, 706, 714 ⟨75LA533⟩, 708 ⟨73TL629⟩, 709, 713 ⟨72LA(766)130⟩; **4**, 331, 908 ⟨72LA(766)130⟩
Blume, G., **4**, 35 ⟨71TL3869⟩, 36 ⟨73TL611⟩
Blume, R. C., **5**, 161 ⟨69TL1047⟩; **6**, 79 ⟨69TL1047⟩, 298 ⟨70USP3503993⟩
Blumel, J., **4**, 528, 533, 1020 ⟨87CB2081⟩
Blumenkopf, T. A., **1**, 450 ⟨86JA3516⟩
Blumer, R. E., **4**, 1034 ⟨92OM613⟩
Blundell, P., **1**, 935 ⟨92T7121⟩, 938, 946 ⟨91T8969⟩
Bluthe, N., **1**, 413, 810 ⟨83TL1157⟩, 809 ⟨84T3277, 86T1333⟩; **3**, 141 ⟨82TL4263⟩, 144, 217 ⟨86T1333⟩
Bly, R. S., **1**, 1126 ⟨69JA3292, 69JA3299⟩; **5**, 942 ⟨93JOM(445)143⟩
Blyumenfel'd, A. L., **6**, 65 ⟨76IZV2547⟩
Blyznyuk, N. K., **2**, 454 ⟨60ZOB2995⟩

Boag, N. M., **2**, 579 〈82COMC-I(4)161〉; **4**, 692 〈80CC1281, 80JCS(D)2182〉
Boal, J. H., **6**, 489 〈93MI 615-02〉
Boar, R. B., **1**, 16 〈81JCS(P1)1501〉, 438 〈73JCS(P1)654〉; **2**, 747 〈75JCS(P1)1237〉; **3**, 358 〈75CC756, 77JCS(P1)515〉; **4**, 950 〈85JCS(P1)1201〉; **5**, 289 〈75JCS(P1)1237〉
Boardman, L. D., **1**, 464 〈88JA5383〉, 544 〈84JA6105〉, 1094 〈83JA6344〉; **2**, 926, 931, 973, 976 〈84JA6105〉, 926, 931, 976 〈83JA6344〉, 991, 992 〈82TL3327〉; **4**, 809 〈88JA5383〉, 1048, 1062 〈84JA6105〉
Boatman, R. J., **4**, 421 〈76JOC3050〉
Boatman, S., **1**, 881 〈65JA82, 68OS(48)40〉; **3**, 263 〈67JOC3817〉
Boatright, L. G., **5**, 59, 89 〈50JA5137〉
Boaventura, M.-A., **1**, 66, 454, 526 〈85JA1726, 87BSF1006〉, 526 〈87BSF1015〉, 526, 999 〈87TL3923〉, 544 〈83S801〉
Boaventura, M., **3**, 138 〈85JA1726〉
Boaz, N. W., **1**, 983 〈84TL3055〉, 1015, 1016 〈84TL3059〉; **5**, 560 〈93JOC3042〉
Bobbio, F. O., **2**, 130 〈65CB998〉
Bobbio, P. A., **2**, 130 〈65CB998〉
Bobbitt, J. M., **5**, 49 〈56JA2489〉, 298 〈60JOC560〉
Bobbitt, K. L., **3**, 424 〈87OM 1406〉
Boberg, F., **3**, 552 〈75LA1513〉; **4**, 905 〈73LA247〉; **6**, 44 〈58LA(616)1〉, 44, 47 〈59LA(621)8〉, 234 〈56CB1160〉
Bobinski, J., **1**, 168 〈63IC1111〉
Bobkov, S. Yu., **4**, 561 〈92ZN(B)1213〉
Bobošík, V., **3**, 194 〈83CCC3421〉; **5**, 1028 〈83CCC3421〉
Bobrova, N. I., **3**, 209 〈83JOU259〉
Bobrov, M. B., **2**, 227 〈88ZOR1088〉
Boca, J.-P., **2**, 764 〈65CR(260)2851, 67BSF1296〉
Bocelli, G., **5**, 497, 499 〈74JCS(P2)1483〉
Boch, G., **2**, 999 〈93CB1873〉
Boche, E., **5**, 631 〈88ZC182〉
Boche, G., **1**, 140, 146, 876 〈89AG(E)277〉, 421 〈68TL5215〉, 858 〈89AG(E)73〉, 886 〈78CB2480〉; **2**, 340, 357 〈86T2703〉, 610 〈81CB4005〉, 753 〈80AG(E)1010〉, 997, 998 〈93AG(E)1032〉, 997, 999 〈92CB2265, 91AG(E)321〉, 999, 1005 〈91CB543〉, 1000 〈85AG(E)573, 86AG(E)947, 88AG(E)846〉, 1000, 1001 〈89AG(E)277, 90CB1853〉, 1001, 1002 〈93CB1887, 92AG(E)79〉, 1002, 1003 〈87ACR18, 91CB861〉, 1041 〈83AG(E)53〉; **3**, 566 〈86AG(E)373, 89AG(E)1392〉, 704 〈92ACR433〉; **4**, 390, 391, 540 〈89AG(E)277〉, 646 〈91CB543〉, 816 〈93AG(E)1032〉; **5**, 212 〈89AG(E)337, 89AG(E)468〉, 302 〈82TL3255〉
Bocher, S., **1**, 1110, 1117 〈71JA1513〉, 1123 〈70LA(733)5〉; **2**, 689 〈70AG(E)273〉
Bochet, C. G., **2**, 346 〈94JA3139〉
Bochkareva, M. N., **4**, 975 〈89IZV1806〉; **5**, 778 〈87IZV165〉
Bochkarev, M. N., **2**, 492, 509 〈78JOM(162)289〉
Bochkarev, V. N., **4**, 140 〈83JFC(23)339〉, 662 〈91MI 414-01〉
Bochmann, G., **2**, 287 〈68ZAAC(360)293〉
Bochmann, M., **4**, 632 〈83JA2582〉
Bochu, C., **3**, 269 〈86CC839〉
Bochwic, B., **2**, 846 〈51NAT1035〉
Bock, C. R., **4**, 668 〈78JOM(157)41〉, 674 〈77ICA85〉
Bock, H., **2**, 388 〈64ZAAC(332)238〉, 752 〈66CB1347〉, 863 〈68CB1414〉, 1038 〈81CB673〉; **3**, 335, 369 〈82JA3119〉, 337 〈76JA6054, 77CC287, 82JA312〉, 337, 351 〈82CB492〉, 342 〈92MC67〉, 358 〈83CB273〉, 371 〈78JOC3893〉, 380 〈79AG(E)944〉, 382 〈84CB187〉, 527 〈77AG(E)105〉, 547 〈77CPH453, 77JA1663〉; **4**, 258 〈82JA3119〉; **6**, 173 〈67AG(E)942〉, 173 〈78CB3573〉, 174 〈93CB2227, 93JOM(462)31〉, 289 〈68CB2815〉, 349 〈90JOM(398)65〉, 379 〈92JOM(429)1〉, 489 〈66CB2039〉
Bock, K., **1**, 2, 4, 8, 9 〈84CAR(130)125〉
Bock, M., **3**, 452 〈41LA(547)1〉
Bock, M. G., **3**, 60 〈76TL7〉; **4**, 219 〈75TL3269〉, 417 〈86JOC3718, 87JOC3232〉, 649 〈76TL7, 78JA2916〉
Bock, P. A., **1**, 690 〈82TL1777〉
Bock, S., **2**, 885 〈82ZN(B)815〉
Bockerman, G. N., **4**, 115 〈76JINC55〉, 341 〈76JFC(7)1〉, 549 〈76MI 412-01〉
Bockrath, E., **1**, 847 〈74JA5708〉
Bocz, A. K., **4**, 413 〈66CB724〉
Boda, G., **6**, 477 〈83MIP8303603〉
Bodalski, R., **3**, 158 〈84S1025〉, 197, 231 〈80TL2287〉; **4**, 585 〈80PS(9)121〉; **5**, 12 〈89S506〉
Bodansky, A., **5**, 262 〈〈B-94MI 506-03〉
Bodansky, M., **5**, 262 〈B-94MI 506-03〉
Bodanszky, M., **5**, 278 〈74S549〉
Bode, H., **1**, 527, 944 〈85JOC2400〉
Bode, K.-D., **5**, 565, 579 〈66AG(E)447〉, 579, 581, 592, 601 〈66LA(698)131〉, 591, 601, 623, 624 〈62LA(660)74〉, 592, 623 〈65LA(681)64, 65AG(E)1095〉
Bode, K. D., **6**, 533 〈65LA(681)64, 67AG(E)281〉
Bodebenner, K., **6**, 73 〈59LA(623)183〉
Bödeker, J., **3**, 587 〈80ZC211〉; **5**, 643 〈74ZC56〉, 1031 〈75ZC56〉, 1068 〈79ZC59〉
Boden, E. P., **2**, 78 〈84TL1879, 84TL3927〉
Boden, H., **5**, 110, 115 〈61JOC1759〉
Boden, R. M., **1**, 727, 733 〈75S784〉
Bodenbenner, K., **6**, 73 〈60GEP1084733, 60LA(632)38〉
Bodenmann, B., **1**, 1038 〈93TL1467〉
Bodesheim, F., **1**, 38 〈61CB1157〉; **3**, 696, 700 〈73OSC(5)300〉, 721 〈61OS(41)13〉
Bodewes, H. H., **3**, 712 〈95TL2109〉
Bodine, J. J., **5**, 576 〈82SC787〉
Bodnar, T., **5**, 937 〈81JOM(213)C31〉
Bodnarchuk, N. D., **4**, 971 〈79ZOR1381〉, 980, 987 〈77ZOR954〉, 990 〈78ZOB803〉
Bodor, N., **4**, 58, 211 〈83JOC5280〉
Bodurow, C., **5**, 170 〈84JA1496〉
Boduszek, B., **2**, 187 〈63JOC870〉, 847, 848, 850 〈79S452〉
Bodwell, G. J., **5**, 328, 329 〈93JCS(P1)459〉
Bodzay, S. J., **2**, 212, 215, 231 〈86TL441〉
Bøe, B., **1**, 70 〈76JOM(107)139〉
Boeckman, Jr., R. K., **1**, 67 〈83TL4923〉, 307 〈74JA6179〉, 443, 649, 759 〈73TL3497〉, 548 〈79JA7107〉, 618, 707, 709 〈91JA9682〉, 710 〈79JA5060〉, 826 〈85JA4359〉, 1089 〈77TL4187〉; **2**, 22 〈87OS(66)194〉, 70 〈89TL3271〉, 623 〈74JOC3307〉, 651, 652 〈83JA4112〉, 929, 931 〈79JOC4781, 73JA6867〉, 931 〈78OS(58)152, 74TL3365, 83T925〉, 938 〈85JOC3421〉; **3**, 49, 198 〈89TL4787〉, 137 〈81T3997〉, 198 〈91TL2581〉, 308 〈87JA7553〉; **4**, 71 〈80JOC752〉, 188 〈83TL4923〉, 807, 1060 〈88OSC(6)1033〉, 950 〈77TL4187〉, 952 〈87JA7553〉, 953 〈75CC519〉; **5**, 66 〈92OS(71)48〉
Boedeker, J., **3**, 411 〈80JPC336〉
Boeder, C. W., **1**, 966 〈78JOC1950, 78JOC2091〉
Boehm, H., **6**, 219 〈77BEP856233〉
Boehm, J., **1**, 74 〈73JA4346〉
Boehm, P., **3**, 13, 83, 93 〈80JCS(P1)27〉
Boehme, H., **3**, 412 〈79CB1297〉
Boehnel, B., **1**, 181 〈71S203〉
Boehshar, M., **2**, 1028 〈86JA7832, 89JA2225〉, 1029 〈88JA3272〉
Boekelheide, V., **1**, 233 〈75TL219〉, 257 〈50JA2134〉, 537 〈69JOC4134〉, 866 〈50JA1245〉; **2**, 213 〈70TL1203〉; **3**, 89 〈57JOC1135〉; **5**, 1126 〈55JA4079〉
Boekenoogen, H. A., **5**, 566 〈33HCA1323〉
Boelens, H., **1**, 633, 1094 〈76TL2287〉; **2**, 722 〈69RTC519〉; **3**, 64 〈76RTC308, 77TL2129〉, 69 〈77T2027〉, 246 〈73RTC731〉
Boelhouwer, C., **1**, 464 〈69T5771〉, 674 〈74CC793, 77CC198〉
Boer, F. P., **2**, 1008 〈66JA1572〉
Boer, J. L. de, **6**, 195 〈86OM1726〉
Boer, J. Th. de, **1**, 929 〈68RTC1301, 73TL
Boer, R. D. de, **2**, 44 〈92TL4637〉

Boer, Th. J. de, **3**, 430 ⟨86RTC161⟩, 430, 433 ⟨86RTC22⟩, 434 ⟨84RTC328⟩; **4**, 299 ⟨73SC189⟩
Boer, T. J. de, **2**, 389 ⟨70RTC164⟩, 390 ⟨71RTC373⟩, 391 ⟨70RTC151, 70RTC159⟩, 765 ⟨71RTC548⟩; **4**, 299 ⟨74RTC294⟩
Boeré, R. T., **1**, 157 ⟨85IC3680⟩; **5**, 777, 778 ⟨87JOM(331)161⟩
Boerema, J. S., **2**, 226 ⟨53RTC612⟩
Boerma, G. J. M., **4**, 331, 538 ⟨72TL2367⟩
Boerma, J. A., **2**, 400 ⟨76CB1069⟩; **3**, 250 ⟨69RTC994⟩; **6**, 274, 275, 309 ⟨76CB1069⟩
Boersma, B. A. M., **1**, 403, 799 ⟨70CC1168⟩
Boersma, J., **1**, 1018 ⟨88T6699⟩; **2**, 590 ⟨82COMC-I(2)823⟩, 999, 1000 ⟨89OM1688⟩, 1096 ⟨79JOM(174)129⟩; **4**, 656 ⟨93OM3624⟩, 704 ⟨91NJC301⟩, 724 ⟨77JOM(127)391⟩
Boes, M., **1**, 149 ⟨87T5095⟩, 149 ⟨89OS(67)60⟩; **4**, 302 ⟨83JA5390⟩; **5**, 747 ⟨87T5095⟩
Boes, O., **3**, 448, 449 ⟨53LA(579)28⟩; **5**, 857, 858 ⟨53LA(579)28⟩
Boes, S., **2**, 144 ⟨53LA(579)23⟩
Boese, R., **1**, 448 ⟨87AG(E)1035⟩, 1009 ⟨86AG(E)268⟩; **2**, 1088 ⟨91AG(E)1370⟩; **4**, 148 ⟨87OM861⟩, 551 ⟨84ZN(B)1518⟩, 665 ⟨93CB1361⟩; **5**, 381 ⟨93CB565⟩, 400 ⟨86ZN(B)1096⟩, 1092 ⟨90CB733⟩; **6**, 152 ⟨84ZN(B)1518⟩, 186, 387, 712 ⟨89CB1057⟩, 188, 388, 391, 712 ⟨87CB1069⟩, 256 ⟨91CB51, 93JCS(D)2547⟩, 340 ⟨78CB1239, 78CB1603⟩, 388 ⟨86CB1117⟩, 518 ⟨86AG(E)755, 87CB1421⟩, 588 ⟨91CC1378⟩, 729 ⟨94CB597⟩
Boesten, W. H. J., **3**, 40 ⟨94TL1581⟩
Boeters, H. D., **6**, 173 ⟨64CB1999⟩
Boettcher, B., **5**, 81 ⟨58MI 502-03⟩
Boettger, S. D., **3**, 528 ⟨82JA747⟩
Boev, V. I., **2**, 445 ⟨84ZOB1192⟩; **4**, 555 ⟨83ZOB594⟩; **5**, 371 ⟨78ZOB1594⟩; **6**, 208 ⟨81ZOB2241⟩
Boeykens, M., **3**, 12 ⟨92JOC5761⟩; **4**, 189 ⟨94S427⟩; **5**, 348 ⟨94JOC5189⟩
Boga, C., **2**, 79 ⟨88JOM(353)177⟩
Bogacheva, A. V., **6**, 223, 226, 227, 228 ⟨91MI 607-01⟩
Bogan, R. T., **1**, 431 ⟨71JFC(1)121⟩
Boganova, N. N., **2**, 451 ⟨89JGU2130⟩
Boganova, N. V., **4**, 766 ⟨89ZOB955⟩; **6**, 313 ⟨88ZOB2168⟩, 314 ⟨88ZOB1447, 88ZOB1927⟩
Bogardus, C. C., **2**, 95 ⟨83JOC3126⟩
Bogatova, N. G., **1**, 61 ⟨81ZOB420⟩; **4**, 141 ⟨81DOK(261)474⟩
Bogatsky, A. V., **6**, 575 ⟨80TL313⟩
Bogavac, M., **3**, 15, 90, 97, 101 ⟨84TL1843⟩
Bogdal, D., **4**, 761, 764 ⟨88PJC483⟩, 793 ⟨88LA595, 89BAP123⟩
Bogdan, M., **6**, 558 ⟨85JPR1007⟩
Bogdan, S., **4**, 833 ⟨83T2085⟩, 842 ⟨83TL2563⟩
Bogdanova, A., **5**, 348 ⟨92ACH357⟩
Bogdanova, A. V., **6**, 83 ⟨60IZV1901⟩
Bogdanova, G. S., **4**, 616 ⟨79ZOB2785⟩
Bogdanova, I. V., **2**, 436 ⟨67ZOB2080⟩
Bogdanov, B., **5**, 735 ⟨81MI 518-01⟩
Bogdanov, N. N., **4**, 478, 493 ⟨76ZOB1246⟩, 490 ⟨76ZOB1250⟩, 490 ⟨76ZOB1495, 76MI 410-03⟩
Bogdanov, V. S., **4**, 770, 779 ⟨90G235⟩, 866 ⟨84IZV611⟩, 975 ⟨89IZV1806⟩, 979 ⟨90IZV401⟩; **5**, 381, 382 ⟨70DOK(194)595⟩, 778 ⟨89IZV2353⟩
Bogdanović, B., **2**, 559 ⟨88ACR261⟩, 559, 560, 959 ⟨90CB1507⟩, 559, 959 ⟨83AG(E)728⟩
Bogdanovic, B., **2**, 561 ⟨82AG(E)199, 90CB1517⟩; **3**, 458 ⟨67AG818, 67AG(E)803⟩
Bogdanowicz, M. J., **1**, 390 ⟨73JA3068⟩
Bogdanowicz-Szwed, K., **5**, 597, 607 ⟨93M721⟩
Bogdarin, Yu. A., **5**, 1081, 1082 ⟨82ZOB885⟩
Bogentoft, C., **1**, 961 ⟨74ACS(B)765⟩
Boger, D. L., **1**, 375 ⟨91COS(5)451⟩, 523 ⟨87JA2717⟩, 790 ⟨83T2869⟩, 933 ⟨89JOC1777⟩, 1130 ⟨89JOC1238⟩; **2**, 329 ⟨89JA1517⟩, 354, 369 ⟨B-87MI 206-01⟩, 655 ⟨86JOC5436, 87JA2717, 87JOC5283, 87TL1027⟩, 663 ⟨80JOC5002⟩, 665 ⟨82TL4551, 87OS(65)98⟩; **3**, 61 ⟨78TL5⟩, 88 ⟨89TL2037⟩, 136 ⟨89JOC177, 90JA4008⟩, 212 ⟨78TL2461⟩, 220 ⟨78TL4597⟩, 292 ⟨88JOC3377, 89JOC1777⟩, 292 ⟨92JOC1429⟩, 413, 414 ⟨83T2869⟩, 437 ⟨92JOC4777⟩; **4**, 305 ⟨90JOC2999⟩, 321 ⟨78TL5⟩; **5**, 93 ⟨91OS(70)79⟩, 201 ⟨83T2869⟩, 330 ⟨86TL683⟩; **6**, 450 ⟨87JOC2319⟩
Boger, J., **2**, 860 ⟨78HCA2190⟩
Bogert, M. T., **2**, 129 ⟨40JA1173⟩, 136 ⟨41JA2361⟩, 279, 292 ⟨27JA2011⟩, 888 ⟨44JA191⟩; **3**, 531, 532 ⟨41JOC669⟩; **4**, 181 ⟨33JA3741⟩
Boggs, J. E., **3**, 380 ⟨79AG(E)944⟩; **5**, 395 ⟨86JST(139)255⟩
Boggs, R. A., **5**, 955 ⟨72JA8947⟩; **6**, 719 ⟨73SRI249⟩
Bognár, R., **3**, 259 ⟨79LA174⟩, 368 ⟨77LA1529⟩, 634 ⟨78OPP211⟩; **6**, 71 ⟨90LA499⟩
Bognitskii, M. I., **5**, 694 ⟨76JOU1789, 76ZOR1826⟩
Bogolyubov, G. M., **2**, 292 ⟨69ZOB1804⟩, 820, 821 ⟨63ZOB3774⟩
Bogolyubskii, A. B., **3**, 171 ⟨91JOU1231⟩
Bogolyubskii, V. A., **6**, 104 ⟨66UKZ204⟩
Bogomolova, G. S., **3**, 361 ⟨77ZOR2012, 80ZOR443⟩, 361 ⟨80ZOR883⟩, 361, 376 ⟨78ZOR2459⟩
Bogonostseva, N. P., **4**, 339 ⟨76IZV1416⟩, 346 ⟨57ZOB2360⟩, 553 ⟨57ZOB2356⟩
Bogoradovskii, E. T., **2**, 1090 ⟨87JGU2482⟩; **4**, 807, 811, 813 ⟨91JGU1295⟩, 810 ⟨91JGU1306⟩
Bogorodovskii, E. T., **2**, 1090 ⟨92JGU787⟩
Bogucka-Ledóchowska, M., **2**, 1008 ⟨85JA2067⟩
Bogush, E., **2**, 503 ⟨69JOM(19)367⟩
Bogusiak, J., **6**, 562 ⟨88S224⟩
Boguslauskaya, L. S., **2**, 5, 18 ⟨72RCR740⟩, 18 ⟨84RCR1178, 87JOU230⟩
Boguth, W., **1**, 77 ⟨76HCA1832⟩
Bohan, C., **4**, 140 ⟨85SRI321⟩; **5**, 224 ⟨85SRI321⟩
Bohe, L., **1**, 46 ⟨88TL2571⟩
Bohle, D. S., **5**, 400 ⟨88JOM(353)355⟩
Böhler, D., **6**, 184 ⟨88JOM(347)11⟩
Böhlke, H., **4**, 437 ⟨66CB868⟩
Bohlman, E. G., **6**, 221, 226 ⟨42JA1342⟩
Bohlmann, F., **1**, 218, 221 ⟨89TL685⟩, 725 ⟨74CB1773⟩, 736 ⟨73CB3779⟩, 769 ⟨89TL1241⟩, 1069 ⟨51CB785, 53CB657⟩, 1134 ⟨84T2537⟩, 1142 ⟨66CB1642, 67CB1193, 67CB1200⟩; **3**, 64 ⟨82LA1216⟩, 66, 233 ⟨57CB2265⟩, 78 ⟨73CB2755⟩, 247 ⟨74CB2115⟩, 366 ⟨65CB3081⟩
Bohlmann, R., **2**, 2, 6 ⟨91COS(6)203⟩; **3**, 253 ⟨67CB107⟩
Bohm, B. A., **2**, 4, 29 ⟨62CRV599⟩; **5**, 62 ⟨65CRV435⟩
Böhm, I., **5**, 150 ⟨82JA1735⟩
Bohm, M. C., **1**, 451 ⟨79JA4420⟩
Böhm, M. C., **4**, 632 ⟨83JA2582⟩
Bohm, R., **4**, 979 ⟨91PHA741⟩
Böhm, S., **2**, 744 ⟨74CB2186⟩
Böhm, W., **5**, 898 ⟨71S431⟩
Bohme, D. K., **1**, 107 ⟨74JA4027, 76CJC1643⟩
Bohme, H., **2**, 157 ⟨71LA(744)15⟩, 169 ⟨65CB1455⟩, 211 ⟨54LA(585)142, 55HOU(9)4, 57LA(606)75, 58LA(617)62⟩, 212 ⟨54CB1661⟩, 339 ⟨51CB170⟩, 1008 ⟨60CB1305⟩; **3**, 419 ⟨B-76MI 310-01⟩, 601 ⟨B-76MI 317-01⟩; **4**, 49, 310 ⟨71S150⟩, 63, 69, 80, 220 ⟨49LA(563)54⟩, 64 ⟨53LA(581)133⟩, 69 ⟨36CB1610⟩, 69 ⟨64CB179⟩, 96 ⟨63CB600, 70CB104⟩, 99 ⟨63CB595, 81CB3421⟩, 215 ⟨59LA(620)1⟩, 271 ⟨79TL1737⟩, 297 ⟨67AP(300)241⟩, 305 ⟨63CB604, 69CB2651⟩, 305, 311 ⟨75LA1790⟩, 309, 310 ⟨63LA(664)130, 64GEP1183513⟩, 310 ⟨66CB2127, 70CB3918⟩, 321 ⟨67AP(300)647⟩, 327 ⟨70CB2775⟩, 405 ⟨56CB2873⟩, 406 ⟨71CB3354⟩, 412 ⟨67CB2131⟩, 438, 445 ⟨78LA1928⟩, 444 ⟨58CB660⟩, 478, 487 ⟨72CB2233⟩, 796 ⟨77AP(310)30⟩, 860 ⟨82LA1022⟩, 903 ⟨82LA585⟩, 966 ⟨B-76MI 403-01⟩, 979 ⟨61CB3109,

69AP(302)81⟩, 979 ⟨71LA(748)59⟩; **5**, 244 ⟨69LA(730)121⟩, 244, 247, 254 ⟨52LA(576)61⟩, 289 ⟨66CB2127⟩, 558, 587, 593 ⟨73S297⟩, 566, 610, 726 ⟨77LA1291⟩, 599 ⟨71LA(748)59⟩, 606 ⟨67AP241⟩, 655, 660 ⟨78TL1429⟩, 685 ⟨78ZN(B)636⟩, 965 ⟨69AP335, 69AP617⟩; **6**, 44, 91 ⟨53LA(581)133⟩, 44, 95 ⟨62CB1859⟩, 48, 82, 89 ⟨61LA(648)21⟩, 90 ⟨41CB1667⟩, 111 ⟨61LA(641)21⟩, 138 ⟨61CB3109⟩, 275, 278, 279, 307 ⟨72LA(758)132⟩, 435, 546 ⟨81LA1244⟩
Böhme, P., **3**, 194, 704 ⟨71AG(E)491⟩
Böhme, R., **2**, 831 ⟨79AG945⟩; **3**, 59, 75 ⟨87S1⟩, 71 ⟨77S1⟩
Böhmel, T., **6**, 492 ⟨85ZN(C)344⟩
Böhmer, W., **6**, 618 ⟨78LA1704⟩
Böhn, B., **3**, 494 ⟨73AG(E)237⟩
Böhn, D., **4**, 549 ⟨93PS(77)5⟩
Böhn, H., **4**, 309, 310 ⟨63LA(664)130⟩
Böhn, K. H., **2**, 1002, 1003 ⟨87AG(E)78⟩
Bohner, G. E., **5**, 361 ⟨71JOC350⟩
Böhnisch, V., **5**, 856 ⟨71LA(750)12⟩
Bohnstedt, A. C., **1**, 816, 817 ⟨93TL5217⟩
Bohonek, J., **6**, 80 ⟨81JOC886⟩
Böhshar, M., **5**, 871 ⟨85T825, 86T1815⟩, 871 ⟨89CB565⟩
Boie, I., **2**, 256 ⟨73M503⟩, 506, 852 ⟨68CB1381⟩; **5**, 612 ⟨65CB3819⟩
Boiki, V. I., **5**, 669 ⟨91UKZ202⟩
Boiko, V. N., **6**, 49 ⟨77ZOR2008⟩, 233 ⟨79ZOR1245, 79ZOR396⟩, 233 ⟨82JFC(21)365⟩
Boiko, Yu. A., **2**, 1032 ⟨67ZOR2102⟩
Boireau, G., **5**, 63 ⟨82TL1259⟩
Bois, C., **4**, 674 ⟨93JA3786⟩
Bois-Choussy, M., **1**, 474 ⟨88TL1289⟩; **5**, 302 ⟨81S729⟩
Boisseau, M., **4**, 420 ⟨92PHA754⟩
Boisselle, A. P., **1**, 993 ⟨62JOC1828⟩; **2**, 1073 ⟨62JOC1828⟩
Boisvert, L., **1**, 327 ⟨92JA4912⟩
Boit, H. G., **6**, 213 ⟨B-72MI 607-02⟩
Boitsov, E. N., **5**, 716 ⟨76JGU714, 76ZOB714⟩
Boivin, J., **1**, 931 ⟨91JA1055⟩, 935 ⟨91TL199⟩, 1076 ⟨91TL5321⟩; **3**, 150 ⟨92TL1285⟩; **4**, 266 ⟨91T7091⟩, 269 ⟨92TL4913⟩; **6**, 551 ⟨92JA7909⟩, 560 ⟨84JCS(P1)1005⟩
Boivin, J. J., **3**, 453 ⟨53CJC1025⟩
Boivin, T. L. B., **1**, 244 ⟨87JOC4943⟩, 942 ⟨93JOC6857⟩; **4**, 286, 779 ⟨87JOC4943⟩
Bøje, L., **5**, 233, 240, 243, 506, 508 ⟨72ACS1465⟩
Bojilova, A., **5**, 355 ⟨91JHC593⟩
Bojinov, V., **5**, 348 ⟨92ACH357⟩
Bojsova, E., **5**, 984 ⟨92CCC1505⟩
Bokadia, M. M., **3**, 604 ⟨82ACH35⟩; **5**, 821 ⟨58JIC893⟩; **6**, 264 ⟨80IJC(B)610⟩
Bokanov, A. I., **2**, 825 ⟨69ZOB1544, 74ZOB760⟩
Bokelmann, C., **1**, 936 ⟨92JCS(P1)3165⟩
Bokens, H., **3**, 179, 184 ⟨85LA785⟩
Bokerman, G. N., **5**, 419 ⟨73USP3772347⟩
Bokovoi, A. P., **2**, 693 ⟨59CJC1498⟩
Bol, K. M., **1**, 406 ⟨92T6425⟩, 817 ⟨91TL5401⟩
Boland, W., **1**, 5, 9 ⟨77CB1823⟩, 677 ⟨93SL837⟩
Boland-Lussier, B. E., **5**, 950 ⟨82OM628, 82OM635⟩
Bolas, T., **6**, 215 ⟨1870LA(156)60⟩
Bold, G., **3**, 44 ⟨90HCA405⟩
Boldeshul, I. E., **5**, 703 ⟨87JOU75⟩
Boldeshul, I. E., **5**, 703 ⟨87ZOR86⟩
Boldeskul, I. E., **1**, 304 ⟨87T2609⟩; **2**, 1000 ⟨87T2609⟩; **3**, 374 ⟨78ZOR1659⟩, 378 ⟨87T2609⟩; **4**, 116, 133 ⟨89TL2445⟩, 326 ⟨78ZOR1659⟩; **6**, 683 ⟨85ZOB2797⟩
Boldi, A. M., **1**, 999 ⟨92AG(E)1240⟩; **2**, 1091 ⟨91JA6943, 92AG(E)1241⟩
Boldirev, B. G., **4**, 864 ⟨69MI 419-01⟩
Boldrini, G. P., **1**, 688 ⟨85JOM(280)307⟩; **2**, 533, 698 ⟨91JOC5820, 90CC1680⟩; **3**, 621 ⟨76S596⟩
Boldrini, P., **2**, 590 ⟨83JOC4108⟩
Boldt, G., **5**, 340 ⟨85SUL11⟩
Boldt, H. G., **3**, 310 ⟨86CB279⟩
Boldt, P., **1**, 325, 920 ⟨87JOC205⟩; **5**, 692 ⟨84CB1935⟩
Boldyrev, A. I., **6**, 399 ⟨91CC1536⟩
Boldyreva, O. G., **2**, 811 ⟨76BAU1374⟩
Boldyrev, B. G., **2**, 358 ⟨69ZOR1088⟩; **6**, 324, 326 ⟨68MI 611-01⟩
Boldysh, E. E., **6**, 54, 261 ⟨89ZOR2490⟩
Boleslawski, M., **4**, 698 ⟨80JOM(201)97⟩
Boleslawski, M. P., **1**, 13, 20 ⟨92JOC2143⟩, 680 ⟨88JOC2829⟩; **4**, 698 ⟨88JOC2829⟩
Bolestova, G. I., **1**, 281 ⟨80CC748⟩; **3**, 58, 211 ⟨84S991⟩
Bolitt, V., **1**, 1026 ⟨89TL6027⟩; **4**, 186 ⟨88TL4583⟩
Bolko, V. I., **5**, 669 ⟨89UKZ1071⟩
Bollaert, W., **4**, 461 ⟨94S23⟩
Boll, P. M., **3**, 246 ⟨75JOC1927⟩
Böll, W., **2**, 680 ⟨62CB2514⟩; **3**, 489 ⟨63JA3796⟩, 704 ⟨70AG(E)458⟩; **4**, 340 ⟨62CB2514⟩; **6**, 697 ⟨62CB2514⟩
Bollens, E., **5**, 1022 ⟨91JFC(53)1⟩
Bolli, M., **1**, 797 ⟨93HCA418⟩
Bollinger, F. G., **4**, 102 ⟨87JHC945⟩; **6**, 445 ⟨87JHC945⟩, 545 ⟨88JHC1601⟩, 546 ⟨89JHC1245⟩
Bollinger, F. W., **3**, 193 ⟨81SC947⟩
Bollinger, J., **1**, 901, 913 ⟨63JA2998⟩
Bollinger, J. M., **1**, 894 ⟨67JA5687⟩, 909 ⟨69JA3667⟩; **2**, 1007 ⟨67JA2993⟩
Bollinger, L., **2**, 690 ⟨48BSF156⟩
Bollmacher, H., **5**, 221 ⟨83CZ121⟩
Bollman, H. T., **1**, 808 ⟨34JA447⟩
Bollmann, G., **6**, 233 ⟨87JFC(34)475⟩, 252, 274, 275 ⟨92ZN(B)369⟩
Bollyky, L., **1**, 595, 779 ⟨64JOC2699⟩
Bolm, C., **2**, 72 ⟨90AG(E)205⟩; **4**, 285 ⟨90JA5609⟩; **6**, 135 ⟨90JA5609⟩
Bolotova, O. P., **2**, 879 ⟨88JGU629⟩, 887 ⟨86JGU287⟩
Bolourchian, M., **3**, 5 ⟨90T6869⟩
Bolourtchian, M., **4**, 933 ⟨72JOM(43)157⟩, 1014 ⟨72JOM(43)139, 90MI 421-01⟩; **5**, 416 ⟨72JOM(43)157⟩
Bolster, J., **3**, 250 ⟨80JOC4804⟩
Bolt, R. O., **4**, 826 ⟨47IEC412⟩; **6**, 220 ⟨47JA947⟩
Bolton, E. K., **4**, 28 ⟨14JA1899⟩
Bolton, I. J., **1**, 827 ⟨71JCS(C)2950⟩
Bolton, M. A., **1**, 643 ⟨92JCS(P1)3407⟩
Bolton, M., **3**, 366 ⟨72CC632, 73JCS(P1)1580⟩
Bolton, R. E., **3**, 106 ⟨87JCS(P1)931⟩
Bolze, R., **1**, 981 ⟨82AG(E)924⟩
Bombala, M. U., **2**, 239 ⟨79JCS(P1)3013⟩
Bombardieri, C. C., **4**, 442 ⟨55CJC923⟩
Bombieri, G., **5**, 801 ⟨86OM1369⟩; **6**, 247 ⟨90OM1449⟩, 663 ⟨90JCS(D)1197⟩
Bomer, J. G., **6**, 562 ⟨92JA10573⟩
Bomhard, A., **4**, 571 ⟨82AG(E)545, 92S787⟩; **6**, 702 ⟨82AG(E)545, 92S787⟩
Bonadies, F., **3**, 222, 303 ⟨93TL7089⟩, 245 ⟨87TL2297⟩; **5**, 79, 105 ⟨83G421⟩
Bonadies, L., **2**, 28 ⟨88TL6835⟩
Bonadyk, S. V., **5**, 657 ⟨78JOU1457, 78ZOR1559⟩
Bonafede, J. D., **2**, 749 ⟨68T2481⟩
Bonakdar,A., **2**, 973 ⟨79JOM(168)227, 79JOM(168)233⟩
Bonar-Law, R. P., **2**, 543 ⟨90TL6721, 90TL6725⟩
Bonasia, P. J., **6**, 349, 355, 357, 358 ⟨93IC5126⟩, 355, 357, 358 ⟨93JA6777⟩
Bonati, F., **4**, 697 ⟨70JOM(22)5⟩; **6**, 721 ⟨70JOM(25)255, 73G373⟩
Bonazzola, L., **1**, 923 ⟨77JA8348⟩
Boncella, J. M., **3**, 515 ⟨93OM2814⟩
Bond, F. T., **1**, 630, 632, 633, 1091, 1094 ⟨78JOC147⟩, 633 ⟨75TL1815, 81JOC1315⟩, 633, 1094 ⟨76TL2947⟩, 799 ⟨69CC1069, 79S44⟩; **2**, 911 ⟨78JOC147⟩; **3**, 58 ⟨79S44, 81JOC1315⟩
Bond, G. C., **1**, 80 ⟨64MI 103-01⟩, 505 ⟨64MI 112-01⟩
Bond, M. R., **6**, 704 ⟨94AG(E)578⟩
Bondar, V. A., **5**, 683, 707 ⟨87UKZ397⟩, 967 ⟨69ZOR119⟩, 967, 987 ⟨67AG(E)864⟩
Bondar, W. A., **5**, 1032 ⟨73AG(E)842⟩

Bondarenko, N. A., **2**, 430, 437 ⟨86S198⟩, 699 ⟨84URP1035028⟩; **4**, 546, 553 ⟨86S198⟩, 552 ⟨91S125⟩
Bond, R. L., **6**, 212, 217 ⟨40JA3477⟩
Bondavalli, F., **5**, 597 ⟨83FES101, 85FES120⟩, 597 ⟨91FES1033⟩
Bondi, A., **2**, 847, 1073 ⟨33CB278⟩
Bondi, E., **3**, 269 ⟨79HCA497⟩
Bondinell, W. E., **5**, 92 ⟨83EUP66456⟩
Bondon, A., **3**, 703 ⟨92ICA73⟩
Bonds, P. L., **1**, 797 ⟨B-89MI 118-01⟩
Bonete, P., **4**, 393 ⟨92TL4065⟩
Bonfiglio, J. N., **3**, 98 ⟨86JOC2833⟩, 179 ⟨80JOC2307⟩
Bonfrèr, J. M. G., **1**, 420 ⟨75TL3741⟩
Bongers, S. L., **1**, 836 ⟨80JA7587⟩
Bongini, A., **2**, 21, 23 ⟨80CC1278⟩, 28, 30 ⟨80S143⟩; **4**, 698, 708 ⟨74JOM(72)C4⟩; **6**, 469 ⟨82JOC4626⟩
Bonhomme, M., **3**, 102 ⟨70BSF1445⟩, 103 ⟨70BSF1838⟩, 105 ⟨67BSF2495⟩
Boni, M., **4**, 249 ⟨93T199⟩
Bonifacic, M., **1**, 5 ⟨B-83MI 101-01⟩, 935 ⟨B-83MI 119-01⟩, 937 ⟨B-83MI 119-02⟩
Bonini, B. F., **2**, 209 ⟨77JCS(P1)1469⟩, 245 ⟨81JCS(P1)2322⟩, 938 ⟨91TL2971⟩; **3**, 368 ⟨76CC431, 79JCS(P1)1720⟩, 369 ⟨86CC964, 90JOC3744⟩; **4**, 232 ⟨85TL2131, 91TL2971⟩, 371, 944 ⟨89JCS(P1)2083⟩, 859 ⟨92TL6383⟩; **5**, 562 ⟨92TL6383⟩, 647 ⟨81CC822⟩, 647, 648 ⟨86JCS(P1)381, 88CC365⟩, 648 ⟨88JCS(P1)1499, 90JOC3744⟩, 648, 650 ⟨89JCS(P1)2083⟩; **6**, 127 ⟨81CC822, 91TL2971⟩, 127 ⟨91TL6801⟩, 127, 128 ⟨90H(31)47⟩, 127, 309 ⟨85TL2131⟩
Bonini, C., **1**, 10 ⟨91JOC6206, 92TL7429⟩, 215 ⟨88TL815⟩; **2**, 10 ⟨94S225⟩
Bonitz, G. H., **2**, 1008 ⟨80JOC524⟩
Bonjoch, J., **3**, 310 ⟨91CC1687⟩; **5**, 146 ⟨81JOC1538⟩
Bonjouklian, R., **4**, 433 ⟨77JOC4095⟩
Bonnafous, J.-C., **1**, 726 ⟨71BSF(2)4551⟩
Bonnard, H., **2**, 472 ⟨83JA6871⟩
Bonnat, M., **2**, 320 ⟨91SC1579⟩
Bonne, F., **3**, 405 ⟨93SC1683⟩
Bonneau, R., **3**, 324 ⟨90JA744⟩
Bonnema, J., **2**, 1030 ⟨56RTC1459⟩
Bönnemann, H., **2**, 559, 959 ⟨83AG(E)728⟩; **3**, 113 ⟨83HCA177⟩
Bonner, M. P., **3**, 620 ⟨89TL4767⟩
Bonner, W. A., **4**, 235 ⟨48JA2435⟩, 239 ⟨50JA354⟩; **5**, 243 ⟨50JA4270⟩
Bonnert, R. V., **1**, 593, 713 ⟨87CC1540⟩
Bonnet, B., **4**, 749 ⟨93S1071⟩
Bonnet, J. J., **6**, 155 ⟨82OM1114⟩, 665 ⟨89JOM(372)201⟩
Bonnet, P.-A., **5**, 772 ⟨83EJM413⟩
Bonnet, P.-H., **1**, 1142 ⟨67CB1200⟩
Bonnet-Delphon, D., **3**, 148 ⟨91T2683⟩; **6**, 9 ⟨91T3207⟩
Bonnett, R., **2**, 764 ⟨59JCS2094, 62CJC177⟩; **3**, 406 ⟨65JOC4508⟩; **5**, 1044 ⟨69CI(L)1702⟩
Bonrath, W., **2**, 144 ⟨93CB2093⟩
Bonser, S. M., **1**, 709 ⟨83JA667⟩; **4**, 67 ⟨83JA667⟩
Bonsignore, L., **1**, 988 ⟨87G625⟩; **6**, 95 ⟨75S660⟩
Bontemps, J., **3**, 632 ⟨92TL1431⟩
Bonthorne, W., **1**, 910 ⟨59JCS2773⟩; **4**, 190 ⟨69JCS(C)1202⟩
Bonvicino, G. E., **1**, 34 ⟨58JA1654⟩
Bonzougou, Y., **5**, 32 ⟨79T2077⟩
Booher, R. N., **5**, 291 ⟨68JA4706⟩
Bookbinder, D. C., **1**, 420 ⟨79JCS(P2)810⟩
Booker-Milburn, K. I., **1**, 816 ⟨92CC350⟩, 929 ⟨92SL809⟩; **2**, 942 ⟨92SL327⟩; **3**, 224 ⟨93TL7291⟩
Bookham, J. L., **4**, 548 ⟨86CC1041, 88POL129⟩, 548 ⟨88POL239, 92JCS(D)2353⟩, 704 ⟨88JCS(D)503⟩
Boon, W. R., **5**, 579 ⟨45JCS601⟩
Boons, G.-J., **4**, 236 ⟨93TL5649⟩
Boons, G. J. P. H., **1**, 648 ⟨89TL229⟩; **4**, 176, 199, 352 ⟨90TL2197⟩; **6**, 549 ⟨92MI 617-05⟩
Boonstra, H. J., **2**, 1030 ⟨59RTC252⟩, 1031 ⟨60RTC866⟩; **3**, 548 ⟨60RTC866⟩
Boor, Jr., J., **1**, 665 ⟨B-79MI 114-02⟩
Boord, C. E., **2**, 128, 132 ⟨30JA651⟩, 606, 615 ⟨51JA3329⟩; **4**, 51 ⟨30JA651⟩
Booth, B. L., **3**, 282 ⟨83JCS(P1)1075⟩, 687 ⟨83JCS(P1)1067⟩; **5**, 690 ⟨90JCS(P1)1705⟩, 728 ⟨83JCS(P1)1067⟩; **6**, 312, 436 ⟨84JFC(24)485⟩, 403 ⟨79JOM(178)371⟩
Booth, D. J., **4**, 706 ⟨68JCS(C)656⟩
Booth, D. L., **3**, 310 ⟨66JA1049⟩
Booth, G., **5**, 463, 464 ⟨66JCS(A)634⟩
Booth, S. E., **1**, 338 ⟨92TL3543⟩; **2**, 351 ⟨91CC1248⟩
Boothe, J. H., **1**, 34 ⟨58JA1654⟩
Boothe, R., **2**, 34, 629, 631 ⟨88JOC4477⟩
Boothe, T. E., **1**, 321, 591 ⟨76JA951⟩, 600 ⟨78JA3874⟩
Boots, H. A., **4**, 264 ⟨81JOC3273⟩
Bootsma, H., **6**, 74 ⟨60RTC888⟩
Boozer, C. E., **1**, 892 ⟨54JA791⟩; **2**, 25 ⟨52JA308⟩
Bor, G., **6**, 352 ⟨68JOM(11)207, 82IC3781⟩, 352 ⟨83CC1613⟩
Borangazieva, A. K., **2**, 108 ⟨91JGU999⟩
Borchardt, J. K., **1**, 459 ⟨73TL2517⟩
Borchardt, R. T., **1**, 11, 15 ⟨82SC983⟩, 34 ⟨83SC677⟩; **2**, 398 ⟨83TL227⟩
Borchardt, R. T., **4**, 218 ⟨94JOC544⟩
Borch, R. F., **3**, 620 ⟨68TL61⟩
Borch, R. P., **2**, 763 ⟨71JA2897⟩
Borchert, A. E., **1**, 805 ⟨60JA1007⟩; **4**, 698 ⟨65JPC3418⟩
Borcic, S., **4**, 25 ⟨71JA4472⟩
Bordás, B., **5**, 547 ⟨72JOC1727⟩
Bordeau, M., **2**, 537 ⟨86BSF413⟩, 938, 941 ⟨92JOC4705⟩; **4**, 140 ⟨86BSF413⟩; **6**, 379 ⟨88JOM(358)31⟩
Bordeleau, L., **1**, 942 ⟨91JCS(P1)1433⟩
Borden, M. R., **1**, 2, 3, 6 ⟨63JA1609⟩
Borden, W. T., **1**, 415 ⟨88JA2314⟩, 513 ⟨70JA4898⟩, 681 ⟨86T1581⟩, 785 ⟨86JA7121⟩, 799 ⟨80JA4841⟩, 865 ⟨86TL5931⟩, 900 ⟨90JA3227⟩; **2**, 389, 397 ⟨79JOC659⟩
Borders, C. L., **2**, 183 ⟨72JOC3549⟩
Bordi, F., **5**, 347 ⟨84FES788⟩
Bordien, C., **2**, 466 ⟨90T6715⟩
Bordoloi, M., **2**, 55 ⟨93TL1681⟩
Bordoni, S., **5**, 1088 ⟨91AG(E)847⟩
Bordwell, F. G., **1**, 142, 155 ⟨75JA7006⟩, 142, 165 ⟨78JOC3101⟩, 155 ⟨78JOC3095⟩, 592, 646 ⟨59JA578⟩, 617 ⟨70ACR281, B-67MI 114-01⟩, 618 ⟨51JA5187⟩, 620 ⟨68JA429, 68JA435⟩, 704 ⟨B-67MI 115-01⟩, 707 ⟨70JA2581⟩, 861 ⟨81JOC632⟩, 869 ⟨67JA3905⟩, 869, 871 ⟨70JA5926, 70JA5933, 70JA5939, 68JA6749⟩, 919, 922 ⟨89JA7558⟩, 921 ⟨93ACR510⟩, 921 ⟨93JOC6067, 90CJC1714⟩, 922 ⟨90JOC58⟩; **2**, 117 ⟨57JA3493⟩, 118 ⟨58JOC636, 60JA5354⟩, 125 ⟨54JA1082⟩, 127 ⟨53JA6019⟩, 143 ⟨51JA2251⟩, 150 ⟨57JA717⟩, 176, 180 ⟨57JOC980⟩, 190 ⟨54JA3945⟩, 191 ⟨54JA3957⟩, 206 ⟨59JA2002⟩; **4**, 63, 64 ⟨57JA376, 51JA5187⟩, 64, 65, 66, 69 ⟨55JA572⟩, 68, 71, 81, 83, 84 ⟨74JOC2516⟩, 74, 75, 77 ⟨64JA4645⟩, 80 ⟨51JA5184, 74JOC2521⟩, 83 ⟨74JOC2526⟩, 83, 84 ⟨68JA435⟩, 391 ⟨76JOC1883⟩
Borea, P. A., **5**, 22 ⟨93MI 517-02⟩
Borer, A., **2**, 583 ⟨91OM2046⟩
Borer, B. C., **3**, 528 ⟨91TL1039⟩
Borgardt, F. G., **4**, 795 ⟨70JOC4236⟩; **6**, 144, 150 ⟨64CRV19⟩
Borgdorff, J., **6**, 402 ⟨92JOM(437)323⟩
Borgese, J., **5**, 346 ⟨85JOC3462⟩
Borgia, A. L., **5**, 237 ⟨92JOC6339⟩
Borgna, P., **5**, 588 ⟨93JHC1079⟩
Borgulya, J., **1**, 827 ⟨63HCA2444⟩; **2**, 656 ⟨63HCA2444, 73HCA14⟩
Borgwardt, S., **1**, 881 ⟨⟨60JA2895⟩
Borho, B., **4**, 901, 904 ⟨89AP(322)593⟩, 901, 904 ⟨90AP(323)619⟩
Borisenki, A. A., **4**, 115 ⟨84JGU1354⟩, 548 ⟨87ZOB2637⟩, 550 ⟨85ZOB331⟩; **5**, 395, 396, 399 ⟨89JGU1817⟩, 395, 396, 399, 400 ⟨88JGU2372⟩, 396, 399 ⟨87DOK114⟩, 398

⟨91JGU114⟩, 398, 400 ⟨90JGU1810⟩, 399 ⟨91JGU108, 90JGU1817⟩, 400 ⟨90JGU1822⟩, 876, 877 ⟨84ZOB1520⟩, 884 ⟨81DOK(256)1401, 81ZOB482⟩, 1092 ⟨87JGU2350⟩; **6**, 679 ⟨81DOK(256)1401⟩, 680, 681 ⟨84ZOB1520⟩
Borison, A. E., **2**, 971 ⟨67DOK(174)96⟩
Borisova, A. I., **2**, 1077, 1081, 1097 ⟨68IZV2287⟩; **4**, 765 ⟨71IZV162⟩, 766 ⟨83IZV2153⟩
Borisov, A. E., **2**, 876 ⟨67DOK(172)1329, 75IZV897⟩, 881 ⟨45BAU251, 64IZV1197, 64IZV1202, 71IZV2103⟩, 982 ⟨54IZV1008⟩, 1077, 1081, 1097 ⟨68IZV2287⟩
Borisova, I. I., **6**, 187, 710 ⟨83IZV959⟩
Borisova, I. V., **4**, 627 ⟨80IZV2668⟩; **6**, 387, 395 ⟨79BAU2222⟩
Borisova, L., **4**, 704 ⟨91JOM(418)C21⟩
Borisov, A. P., **2**, 442 ⟨83URP1046248, 84ZOB2550⟩
Borisov, G., **4**, 113, 120 ⟨70JGU255⟩, 339 ⟨84PS(21)59⟩, 555 ⟨70ZOB285⟩
Borissova, L. M., **6**, 522 ⟨40DOK(26)54⟩
Borkovec, A. B., **6**, 508 ⟨73JHC639⟩, 655 ⟨67JMC118⟩
Borloo, M., **4**, 461 ⟨94S23⟩
Born, D., **5**, 874 ⟨86AG(E)1093⟩
Born, F., **6**, 545 ⟨60CB3056⟩
Born, L., **4**, 583 ⟨92OM2701⟩
Bornancini, E. R., **2**, 874 ⟨84JOM(270)177⟩
Bornatsch, W., **5**, 848 ⟨71ZN(B)296⟩
Bornemeier, D. A., **6**, 558, 564 ⟨93JMC1802⟩
Bornengo, M., **3**, 536 ⟨63CI(M)1216⟩
Borner, G., **4**, 1075 ⟨91JOC6038⟩
Borner, P., **4**, 881 ⟨56CB2060⟩; **6**, 73 ⟨60LA(632)38⟩, 73, 97, 300 ⟨56CB2060⟩
Bornheim, L. M., **4**, 262 ⟨90JA3191⟩
Bornmann, W. G., **3**, 338, 345 ⟨79JOC486⟩
Bornowski, H., **6**, 298, 300, 312, 321, 646 ⟨64LA(675)142⟩
Bornstein, J., **1**, 2, 3, 6 ⟨63JA1609⟩
Borodin, A. V., **2**, 239 ⟨90ZOR2083⟩
Borodinsky, L., **3**, 715 ⟨85T3235⟩
Boroni, A., **5**, 22 ⟨93MI 517-02⟩
Boros, E. E., **5**, 370 ⟨90PS(50)139⟩, 370, 590, 614, 643 ⟨90JOC6091⟩
Boros, E. J., **4**, 342 ⟨66JA1140⟩
Borovas, D., **2**, 274 ⟨65JA4922⟩, 358 ⟨64JA3660⟩
Borovik, E. I., **5**, 775 ⟨84ZOR1784⟩
Borovikova, G. S., **5**, 337 ⟨86ZOR100, 84ZOR1784⟩
Borovinskaya, I. P., **6**, 55 ⟨68IZV431⟩, 261 ⟨68IZV429⟩
Borowiecka, J., **1**, 261 ⟨90S1173⟩; **3**, 42 ⟨90S1173⟩
Borowitz, I. J., **1**, 392 ⟨60JA4307⟩; **4**, 174 ⟨66JOC3032⟩
Borredon, E., **1**, 217 ⟨90H(31)1997⟩
Borredon, M. E., **3**, 373 ⟨87T3945⟩
Borrelly, S., **2**, 921, 938 ⟨93JOC2714⟩
Borresen, S., **3**, 233 ⟨76JOC678⟩
Borrmann, D., **3**, 526 ⟨68HOU(7/4)65⟩, 546 ⟨68HOU(7/4)312⟩, 556 ⟨68HOU(7/4)323⟩; **4**, 898 ⟨68HOU(7/4)434⟩, 969 ⟨68HOU(7)341⟩
Borror, A. L., **4**, 855 ⟨81USP4348529A⟩
Borrul, F., **5**, 864 ⟨86POL1277⟩
Borsche, W., **1**, 474 ⟨16CB2222⟩; **2**, 782 ⟨26CB818⟩; **3**, 309 ⟨41LA(549)238⟩; **5**, 858 ⟨34LA(512)97⟩
Borselli, G., **4**, 930 ⟨87TL4093⟩; **5**, 416, 423 ⟨87TL4093⟩
Borshberg, H. J., **1**, 831 ⟨83HCA2608⟩
Borsus, J. M., **6**, 629 ⟨65T1537⟩
Borthakur, N., **5**, 370 ⟨91CI(L)353⟩
Borthwick, A. D., **2**, 19 ⟨88JCS(P1)549⟩
Bortolini, M., **2**, 533, 698 ⟨91JOC5820⟩
Bortolini, O., **3**, 5, 117 ⟨87JOC5467⟩; **4**, 261 ⟨86TL6257⟩; **6**, 637 ⟨90H(31)1213⟩
Bortsch, R. A., **3**, 411 ⟨79TL2109⟩
Boruah, R. C., **2**, 34 ⟨91CI(L)175⟩
Borull, F., **5**, 226 ⟨82AQ(B)327⟩
Borum, O. H., **5**, 264 ⟨63OSC(4)5⟩
Bory, S., **1**, 143 ⟨78JA1558⟩
Bos, H., **2**, 682 ⟨75TL3429⟩; **6**, 114 ⟨50RTC1223⟩
Bos, H. J. T., **1**, 981 ⟨81RTC34, 81TL2827⟩, 999 ⟨B-69MI 121-06⟩; **2**, 1021, 1025, 1027, 1030, 1031, 1032, 1033, 1034, 1036 ⟨B-69MI 221-02⟩, 1033 ⟨69RTC179⟩, 1066 ⟨69RTC71, 73RTC762⟩; **3**, 250 ⟨74TL4179⟩, 340 ⟨84TL873⟩; **4**, 765 ⟨69MI 417-02⟩, 843 ⟨69RTC597, 73RTC1067⟩
Bos, H. J. T.,**3**, 548 ⟨74RTC26⟩; **4**, 843 ⟨74RTC184⟩; **5**, 427 ⟨81TL2827⟩, 510 ⟨74RTC51⟩, 554 ⟨69RTC597⟩, 559 ⟨78TL4839, 83RTC103, 83RTC83, 83RTC91⟩, 559, 605, 614 ⟨76TL209⟩, 605, 614 ⟨81RTC414⟩; **6**, 88 ⟨67M1043⟩
Bosbury, P. W. L., **4**, 736 ⟨78JCS(P1)422⟩
Bosch, E., **1**, 334 ⟨89JOC1826⟩, 548, 1131 ⟨86TL641⟩, 938 ⟨88JCS(P1)1517, 89JOC1234, 92JOC6803⟩, 975, 1025 ⟨86JOC4006⟩; **5**, 513 ⟨89JOC1234, 92JOC6803⟩; **6**, 267 ⟨74ZAAC(404)103⟩, 543, 550, 557 ⟨92JOC6803⟩, 548 ⟨88JCS(P1)1517⟩
Bosch, G. K., **1**, 690 ⟨85TL2167, 87JOC4885⟩; **3**, 530, 531 ⟨87JOC4885⟩
Bosch, J., **3**, 310 ⟨91CC1687⟩; **5**, 146 ⟨81JOC1538⟩
Boschan, R., **2**, 103 ⟨55CRV485⟩; **6**, 74 ⟨58JA1247⟩
Bosch, P., **2**, 19 ⟨87TL4733⟩; **3**, 418 ⟨87OPP181⟩
Bosch, R. J., **2**, 11 ⟨78JOC4120⟩; **4**, 748 ⟨84JA3584⟩
Boschelli, D., **1**, 147 ⟨91Txi⟩
Boschelli, D. H., **6**, 558, 564 ⟨93JMC1802⟩
Boschi, T., **3**, 599 ⟨71ICA477⟩; **5**, 801, 802 ⟨81JOM(206)C11⟩; **6**, 721 ⟨70JOM(25)255⟩
Bosco, M., **2**, 318 ⟨88TL2251⟩; **3**, 262, 263 ⟨90S895⟩; **4**, 646, 653 ⟨88CC807⟩
Bose, A. K., **1**, 725 ⟨69JCS(C)2728⟩, 747 ⟨65JOC505⟩; **2**, 8 ⟨73TL3937⟩, 683 ⟨75JCS(P1)461⟩; **3**, 527 ⟨67T4769⟩; **5**, 168 ⟨68JOC2819⟩, 273, 285 ⟨81JOC1229⟩, 274, 286 ⟨73JOC3437⟩, 294 ⟨74TL2633⟩, 305 ⟨73OSC(5)973⟩, 748 ⟨59JOC1309⟩, 1118 ⟨74TL3135⟩; **6**, 113 ⟨73JHC791⟩
Bose, D. S., **2**, 304 ⟨90TL6903⟩
Bose, H. S., **5**, 367 ⟨90JIC172⟩
Böse, R., **4**, 580 ⟨85CB3570⟩, 580, 581 ⟨90CB739⟩
Böshagen, H., **3**, 359 ⟨82LA14⟩
Boshagen, H., **5**, 279 ⟨59CB2813⟩
Boshart, G. L., **5**, 279, 1062 ⟨61JOC2525⟩
Bosin, T. R., **6**, 651 ⟨73JOC1591⟩
Boske, J., **2**, 470 ⟨86IC2695⟩
Böske, J., **4**, 488 ⟨85CB3227⟩
Boskin, M. J., **1**, 268 ⟨59JA6330, 62JA3944⟩; **2**, 853 ⟨59JA6330⟩
Bosl, G., **5**, 388 ⟨92ZN(B)1165⟩
Bosma, R. H. A., **1**, 674 ⟨81CC1081⟩
Bosmans, J.-P., **6**, 14 ⟨92MI 601-02⟩
Bosnich, B., **1**, 133 ⟨91JA958⟩; **2**, 640 ⟨91JA958⟩; **6**, 472, 473 ⟨86IJ250⟩
Bošnjak, J., **1**, 91 ⟨67T215⟩
Bosold, F., **5**, 212 ⟨89AG(E)337⟩, 302 ⟨82TL3255⟩
Bossaller, W., **4**, 797 ⟨43CB88⟩
Bosse, R., **2**, 866 ⟨91CB725⟩
Bossert, F., **4**, 880 ⟨76LA1762⟩, 972 ⟨77LA1895⟩
Bosshard, H. H., **2**, 182 ⟨59HCA1653⟩; **3**, 281 ⟨59HCA1659⟩; **5**, 9 ⟨59HCA1653⟩; **6**, 52 ⟨59HCA1653⟩, 57 ⟨59HCA1659⟩
Bosshard, H. R., **5**, 90 ⟨73JOC457⟩
Bossio, R., **3**, 702 ⟨90H(31)1855⟩, 706 ⟨89S641, 90LA935⟩, 706 ⟨91LA843⟩
Bost, R. W., **2**, 132 ⟨43OSC(2)547⟩, 177 ⟨41JA3530⟩, 191 ⟨40JOC24⟩; **5**, 549 ⟨30JA332, 51JA25⟩, 1027 ⟨46JA2506⟩
Bostrom, L. L., **5**, 334, 336 ⟨92TL153⟩
Boswell, C., **1**, 1010 ⟨92BMC1619⟩
Boswell, C. J., **1**, 1114 ⟨72JA5061⟩
Boswell, G. A., **4**, 9 ⟨71JOC575⟩
Boswell, J. C., **1**, 1114 ⟨71JA6989⟩
Boswell, Jr., G. A., **2**, 19 ⟨74OR(21)1⟩, 610 ⟨71JOC818⟩; **4**, 8 ⟨74OR(21)1⟩, 8, 9 ⟨71JOC818⟩
Bosyakov, Yu. G., **2**, 1076 ⟨74RCR657⟩
Boszczyk, W., **6**, 609 ⟨89ZN(B)1589⟩

Boteju, L. W., **2**, 29 ⟨94T2391⟩
Botha, C., **4**, 582 ⟨83JCS(D)1495⟩
Bothe, E., **4**, 1078 ⟨80JCS(P2)883, 81JCS(P2)143⟩
Bothner-By, A. A., **1**, 95, 96, 889 ⟨59JA3658⟩, 871 ⟨67JOC492⟩, 1144 ⟨55JA3293, 60JPS1271⟩
Bothwell, T. C., **5**, 1047 ⟨81SC781⟩
Boto, A., **1**, 931 ⟨93TL4865⟩
Botta, A., **5**, 1012 ⟨70GEP(O)1913273, 73GEP(O)2145514⟩; **6**, 437 ⟨83HOU(E4)1313⟩
Bottaccio, G., **5**, 68, 95 ⟨77G499⟩
Bott, K., **1**, 186 ⟨65AG967⟩; **2**, 799 ⟨64AG(E)804, 66T1251⟩, 800 ⟨65AG(E)148, 70AG(E)954, 71TL2227, 75CB402, 79AG(E)259⟩; **5**, 24 ⟨80AG169⟩, 26, 38 ⟨66AG(E)870⟩, 28 ⟨80AG(E)171⟩; **6**, 628 ⟨64CB735⟩
Bottaro, J. C., **1**, 154 ⟨83CC1040⟩, 634 ⟨81CC1121⟩; **2**, 386 ⟨87JOC2292⟩, 1053 ⟨87JOC2294, 90JOC1916⟩; **3**, 137 ⟨81CC1121⟩, 293 ⟨83CC1040⟩, 698, 721 ⟨82CC942⟩; **4**, 199 ⟨86JOC4711⟩, 536 ⟨83CC1040, 86T4223⟩; **5**, 668, 830 ⟨86JMC1689⟩, 675 ⟨81CC1121⟩
Bott, R. W., **1**, 448 ⟨64JCS4804⟩, 1000 ⟨64JOM(1)420⟩, 1122 ⟨65JCS384⟩; **2**, 536, 538 ⟨B-68MI 210-01⟩, 841 ⟨65JOM(4)291⟩
Bott, S., **2**, 48 ⟨89JOC2015⟩
Bott, S. G., **1**, 666 ⟨90OM3086⟩; **2**, 473 ⟨92CC726⟩, 475 ⟨90AG(E)1150⟩; **3**, 358 ⟨87JOC1874⟩; **4**, 675 ⟨89CC418⟩; **6**, 381 ⟨92JOM(427)9⟩, 395 ⟨87JCS(D)747⟩, 603 ⟨88IC570⟩
Böttcher, D., **5**, 934 ⟨95CB645⟩
Botteghi, C., **1**, 297 ⟨77JA1058⟩, 297, 414 ⟨93MI 107-01⟩; **3**, 11 ⟨77CI(M)839⟩, 617, 619 ⟨82SC25⟩
Böttger, J., **6**, 630, 731 ⟨84S667⟩
Botthegi, C., **6**, 417 ⟨82SC25⟩
Bottinger, P.,**6**, 268 ⟨73ZAAC(395)159⟩
Bottini, A. T., **1**, 1065 ⟨62JOC3688⟩, 1144 ⟨70JOC1086⟩; **2**, 607, 624 ⟨72T4883⟩, 647 ⟨57JA1458⟩
Bottino, F. A., **1**, 689 ⟨82JCS(P2)77⟩; **5**, 734 ⟨84OMR724⟩
Bottin-Strzalko, T., **1**, 750 ⟨73T4199⟩; **3**, 494 ⟨71CR(272)728⟩
Bottler, R., **5**, 996 ⟨67CB698⟩; **6**, 617 ⟨75AP(308)379⟩
Botto, R. E., **1**, 906 ⟨78JA3847⟩
Bottomley, A. R. H., **2**, 435 ⟨83JOM(250)617⟩
Bottomley, L. A., **5**, 1097 ⟨82CC607⟩
Bottomley, W. E., **4**, 881 ⟨80CC790⟩
Bottoni, A., **4**, 610 ⟨93JA3322⟩; **6**, 691 ⟨86JA5395⟩
Bottorff, E. M., **5**, 652 ⟨73OSC(5)687⟩
Bottrill, M., **2**, 565 ⟨82COMC-I(3)281, 82COMC-I(3)433⟩, 565, 575 ⟨82COMC-I(3)331⟩
Botyeva, E. S., **2**, 461 ⟨88JGU2161⟩
Bou, A., **2**, 1026 ⟨81T1441, 87OS(65)68⟩
Bouachir, F., **2**, 1003 ⟨82ZN(B)1146⟩; **4**, 583 ⟨82ZN(B)1146⟩
Bouas-Laurent, H., **2**, 619 ⟨73BSF720⟩; **4**, 621 ⟨89JCR(S)146⟩
Boubia, B., **1**, 738 ⟨90AG(E)1454⟩
Bouchandon, J., **5**, 558 ⟨87JMC24⟩
Bouchard, F., **6**, 477 ⟨85CJC153⟩
Bouchaudon, J., **4**, 229 ⟨87JMC24⟩
Bouchaut, M., **1**, 656 ⟨83JOM(243)281⟩
Boucher, R., **5**, 1096 ⟨90JA3332⟩
Boucher, T., **5**, 1079 ⟨93OM1023⟩
Boucherot, D., **3**, 194 ⟨80TL931⟩; **5**, 1083 ⟨89TL2379⟩; **6**, 622 ⟨89TL2379⟩
Bouchoule, C., **1**, 127 ⟨68CR(C)1614⟩; **2**, 650 ⟨82JMOC333⟩
Boudakian, M. M., **2**, 1030, 1034 ⟨56JA2760⟩
Boudjouk, P., **1**, 1004, 1005, 1006 ⟨77JOC2626⟩; **2**, 131 ⟨72JA4784⟩; **4**, 368 ⟨72JA4784⟩
Boudreaux, G. J., **1**, 618, 704, 705, 709 ⟨86JA2358⟩, 704, 705 ⟨84TL4617⟩; **2**, 157 ⟨86JA2358⟩; **4**, 342 ⟨75JOC2056⟩, 482 ⟨74JHC1085⟩
Bouffard, F. A., **6**, 128 ⟨80JA6161⟩
Bougauchi, M., **3**, 121 ⟨89JOC3001⟩
Bougault, J., **5**, 763 ⟨20CR(171)38, 21CR(172)452⟩
Bougeard, P., **2**, 176 ⟨82JCS(P1)849⟩; **4**, 713 ⟨85OM1123⟩
Bougeois, J.-L., **2**, 337 ⟨81TL61⟩
Bouguslavskaya, L. S., **2**, 15 ⟨89JOU1835⟩
Bouillon, G., **4**, 43, 55 ⟨81CB1951⟩
Boukduvalas, J., **2**, 52 ⟨89CC1916⟩
Boukouvalas, J., **1**, 133 ⟨92TL1855⟩; **3**, 124 ⟨92TL1855⟩, 307 ⟨90TL995⟩
Boula, J. M., **4**, 655 ⟨81TL2985⟩
Boulet, C. A., **1**, 200, 255 ⟨90SC3347⟩
Boulette, B., **2**, 149 ⟨67JOC3111⟩
Boullanger, P., **3**, 695, 708 ⟨76TL3427⟩, 695, 708 ⟨79TL163⟩; **5**, 747 ⟨80TL1459⟩
Boulton, A. J., **5**, 725 ⟨65JCS5230⟩, 1002 ⟨76TL2691⟩
Boulton, I. J., **5**, 168 ⟨71JCS(C)2944⟩
Bouma, R. J., **3**, 709 ⟨75TL3487⟩; **4**, 331 ⟨75TL3487⟩
Bouma, W. J., **2**, 1022 ⟨82JOC1869⟩
Boumaiza, L., **3**, 231 ⟨91TL1787⟩
Bouman, H., **2**, 569 ⟨76JOM(110)327⟩
Bouman, M. M., **4**, 716 ⟨93OM315⟩
Bouman, T. D., **1**, 634 ⟨81JA5314⟩
Bourdon, F., **2**, 140 ⟨90JOC2596⟩; **3**, 335 ⟨90JOC2596, 93JA324⟩, 351 ⟨90TL6183⟩, 369 ⟨91NJC533⟩
Bourelle-Wargnier, F., **1**, 826 ⟨79CC584⟩
Bourgain, M., **1**, 540 ⟨71TL2583, 74BSF1656⟩, 1045 ⟨70TL2659⟩; **2**, 629, 977 ⟨71TL2583⟩, 679 ⟨75TL3833⟩, 1095 ⟨73BSF2137⟩; **3**, 273 ⟨70TL2659, 73BSF2137⟩
Bourgain-Commerçon, M., **1**, 477 ⟨77JCR(M)2101⟩, 491 ⟨86T1399⟩; **4**, 655 ⟨86T1389⟩
Bourgeois, D., **4**, 933 ⟨74JOM(80)C25⟩
Bourgeois, M.-J., **2**, 98 ⟨93T2477⟩
Bourgeois, P., **1**, 963 ⟨72JOM(39)C44⟩; **2**, 916, 923, 940 ⟨75JOM(87)169⟩, 1090 ⟨69CR(C)72, 72CR(C)1857⟩; **4**, 1014 ⟨72JOM(43)139⟩; **5**, 297 ⟨74BSF(2)1949⟩, 416 ⟨74JOM(80)C25⟩, 419, 793 ⟨74JOM(76)C1⟩; **6**, 125 ⟨72JOM(39)C49⟩, 162 ⟨71JOM(32)79⟩
Bourgeois, R. C., **3**, 104 ⟨63OSC(4)918⟩
Bourguignon, J., **2**, 398 ⟨87BCJ4492⟩
Bourne, E. J., **5**, 193, 199 ⟨54JCS2006⟩
Bourne, G. T., **4**, 404, 418, 419, 424, 427 ⟨91JCS(P1)1693⟩
Bourns, A. N., **1**, 776 ⟨61CJC348⟩; **2**, 320 ⟨63OSC(4)795⟩
Bourrigaud, J., **5**, 760 ⟨80JHC767⟩
Bouska, J. B., **5**, 214 ⟨88JMC1960⟩
Bousquet, C., **3**, 616 ⟨93BSF513⟩
Bousquet, E. N., **5**, 48, 49 ⟨32OSC(1)526⟩
Boussard, G., **5**, 327, 351 ⟨93JA8898⟩; **6**, 488 ⟨93JA8898⟩
Boussoufi, A., **4**, 101 ⟨91TA157⟩
Boustany, K., **2**, 241, 358 ⟨70TL4983⟩
Boustany, K. S., **2**, 217 ⟨70TL3547⟩
Boutagy, J., **1**, 638, 639, 744, 745 ⟨74CRV87⟩
Boutevin, B., **5**, 122 ⟨76MI 503-01⟩; **6**, 7 ⟨77TL4315, 87JPS(A)3025⟩, 7 ⟨89JFC(45)215, 90JFC(47)95⟩
Boutiba, S., **2**, 1007 ⟨89AX(C)1520⟩
Boutin, N. E., **2**, 10 ⟨74T3793⟩
Boutin, R. H., **1**, 1047 ⟨85JOC3972⟩; **2**, 1011 ⟨86JOC5320⟩; **3**, 180, 273 ⟨85JOC3972⟩; **4**, 422 ⟨84JOC4272⟩; **5**, 999 ⟨84JOC4272, 84JOC4277⟩
Boutom, P. J., **2**, 150 ⟨57JA717⟩
Boutonnet, F., **1**, 340, 1127 ⟨93TL2091⟩; **4**, 547, 579 ⟨93AG(E)1735⟩
Boutry, O., **5**, 948 ⟨92JA7288⟩
Boutte, D., **6**, 297 ⟨75FRP2235188⟩
Bouveault, L., **2**, 54 ⟨03CR(136)1676, 03CR(137)60⟩, 54 ⟨04BSF748, 04CR(138)148⟩
Bouvy, D., **1**, 995 ⟨93TL1779⟩
Bouwman, C. T., **1**, 584 ⟨94JOC374⟩
Bouyssi, D., **1**, 975 ⟨93TL3129⟩, 1020 ⟨91TL6541, 92TL2811⟩
Bouyssou, P., **3**, 274 ⟨90S631⟩
Bova, L. M., **5**, 24 ⟨79MI 502-01⟩
Boveleth, W., **4**, 695 ⟨86JOM(308)153⟩
Bovenschulte, E., **2**, 207, 721 ⟨89AG(E)202⟩
Bovete, M., **1**, 155 ⟨74TL1377⟩
Bovey, F. A., **5**, 685 ⟨B-70MI 517-01⟩

Bovicelli, P., **4**, 25 ⟨90S657⟩
Bovin, A. N., **4**, 124, 126, 128 ⟨89JGU1778⟩, 124, 128 ⟨92JGU39⟩, 126 ⟨89IZV850⟩, 134 ⟨90TL5361⟩
Bovonsombat, P., **4**, 760 ⟨92TL2285, 92TL7705⟩
Bowden, J. N., **2**, 3 ⟨49JA3107⟩
Bowden, K., **1**, 519 ⟨46JCS52⟩, 869 ⟨70JCS(B)173, 70JCS(B)179⟩; **3**, 252 ⟨46JCS945⟩, 267 ⟨46JCS45⟩, 275 ⟨46JCS39⟩; **6**, 621, 728 ⟨90JMC434⟩
Bowen, D. H., **6**, 719 ⟨74JCS(D)1189⟩
Bowen, D. M., **5**, 112 ⟨55OSC(3)553⟩
Bowen, J. G., **6**, 308 ⟨90MI 610-01⟩
Bowen, M. W., **3**, 80 ⟨77CC53⟩
Bowen, P., **1**, 927 ⟨86JA2381⟩
Bower, B. K., **2**, 575 ⟨72JA2512⟩
Bower, F. A., **1**, 637 ⟨58JA4588⟩
Bower, J. D., **3**, 372 ⟨69JA6891⟩
Bowers, A., **2**, 18 ⟨60JA4001, 60JA4007⟩
Bowers, C. W., **3**, 613 ⟨74JOC3416⟩
Bowers, G. W., **2**, 884 ⟨36JA1573⟩
Bowers, M. T., **1**, 1109 ⟨76JA6700⟩
Bowers, V. A., **1**, 322 ⟨64JCP(40)213⟩
Bowes, C. M., **1**, 954 ⟨73TL3181⟩
Bowie, R. A., **6**, 61 ⟨69JOM(16)33⟩
Böwing, W. G., **2**, 862 ⟨69ZN(B)1250⟩
Bowler, B. E., **2**, 1009 ⟨89JA1299⟩
Bowler, E. A., **1**, 881 ⟨50JA3704⟩
Bowles, A. J., **6**, 734 ⟨70CC559⟩
Bowles, S. A., **4**, 200 ⟨93SL111⟩
Bowles, T., **2**, 1000 ⟨88JCS(P1)1023⟩; **3**, 713 ⟨88TL3343⟩
Bowlin, T. L., **4**, 73 ⟨89JA1127⟩
Bowlus, S. B., **3**, 426 ⟨75TL3591⟩
Bowmaker, G. A., **2**, 444 ⟨83CB3567⟩, 447 ⟨84ZN(B)618⟩; **5**, 1141 ⟨84JCS(D)2859⟩
Bowman, D. H., **1**, 661 ⟨67JA4531⟩; **2**, 623, 628 ⟨89JOC6068⟩
Bowman, D. W., **2**, 623 ⟨67JA4531⟩
Bowman, M., **4**, 921 ⟨89FP2610926⟩
Bowman, M. P., **4**, 58 ⟨90JOC2240⟩, 920, 921 ⟨90JOC5982⟩; **6**, 429 ⟨90JOC2240, 90JOC5982⟩
Bowman, R. E., **2**, 31 ⟨50JCS406⟩; **3**, 229, 275 ⟨52JCS3945⟩; **5**, 76 ⟨50JCS325⟩
Bowman, S. A., **2**, 657 ⟨73JOC162⟩
Bowman, W. R., **1**, 929 ⟨89TL3343, 92T6883⟩, 931 ⟨92T4027⟩, 950 ⟨92TL4993⟩; **2**, 339, 361 ⟨94UP⟩, 358 ⟨94T1275, 94T1295⟩, 391 ⟨88CSR283⟩; **4**, 327 ⟨77TL4519, 80JCS(P1)1407, 81TL1551, 84JCS(P1)2327⟩; **6**, 544, 557 ⟨92T6883⟩
Bownaker, G. A., **4**, 586 ⟨84JCS(D)2859⟩
Bowry, V. W., **1**, 340 ⟨92JOC4284⟩; **6**, 543, 550, 551 ⟨91T121⟩
Bowser, J. R., **5**, 376 ⟨83JOC4111⟩
Boxer, M., **1**, 693 ⟨78JA2140⟩
Boxer, N. A., **5**, 1076, 1077 ⟨87JOM(327)303⟩
Boxler, D., **1**, 409, 821 ⟨73JOC2572⟩, 601, 604 ⟨75JA1597⟩
Boyadjiev, B., **5**, 1147 ⟨89ZN(B)419⟩
Boyar, E., **2**, 582 ⟨84JOM(276)C45⟩
Boyarinov, G. A., **5**, 1081, 1082 ⟨82ZOB885⟩
Boyce, C. B. C., **4**, 804 ⟨74JCS(P1)1644⟩
Boyd, B. A., **4**, 566 ⟨88OM572, 89IC3219⟩, 573 ⟨87TL6121⟩; **5**, 887 ⟨87PS(30)463⟩; **6**, 129 ⟨87TL6121⟩
Boyd, D. R., **2**, 151 ⟨68JCS(C)2374⟩; **3**, 17, 404 ⟨79TL3201, 85JCS(P1)2123⟩, 405 ⟨86JCS(P1)867⟩, 439 ⟨85CC582⟩; **4**, 262 ⟨92JA8772, 92JCS(P1)1105⟩
Boyd, E. A., **3**, 200 ⟨92TL813⟩
Boyd, G. V., **1**, 888 ⟨B-89MI 119-04⟩; **4**, 302 ⟨84CHEC(6)177⟩, 881 ⟨80CC790⟩; **5**, 367 ⟨76CC718⟩, 642 ⟨68CC549⟩, 716 ⟨78JCS(P1)1338⟩, 742, 755 ⟨B-91MI 519-01⟩, 854 ⟨71JCS(C)409⟩; **6**, 110 ⟨71JCS(C)409⟩, 621 ⟨84CC1105⟩, 621, 728 ⟨87CC99⟩
Boyd, J., **1**, 778 ⟨72JCS(P1)2533⟩
Boyd, L. M., **6**, 247 ⟨90JOM(397)209⟩
Boyd, M., **5**, 613, 615 ⟨93JCS(P1)1835⟩
Boyd, R. H., **1**, 895 ⟨57JA3724⟩, 896 ⟨60JA4729⟩
Boyd, S. A., **5**, 236 ⟨91TL6507⟩
Boyd, S. D., **1**, 637, 638 ⟨71JA4316⟩; **4**, 328 ⟨74JA2580⟩, 332 ⟨73JA3356⟩
Boyd, T., **2**, 31 ⟨41OSC(1)36⟩
Boyer, G., **3**, 280 ⟨91TL6709⟩
Boyer, H.,**5**, 963 ⟨87S907⟩
Boyer, J. H., **2**, 385 ⟨89JCR(S)12, 91CL569⟩, 385, 386 ⟨85S677⟩, 389 ⟨B-69MI 207-01⟩, 753 ⟨72JOC3571⟩, 777, 783 ⟨B-69MI 215-01⟩, 781 ⟨B-61MI 215-01⟩, 1055 ⟨58JOC1051, 69JA6122⟩; **3**, 48 ⟨51JA5248⟩, 566 ⟨70CC889⟩, 596 ⟨69CC1377⟩, 699 ⟨73CC699, 75JCS(P1)1743⟩; **4**, 434 ⟨84H(22)2351⟩, 447 ⟨88JCS(P1)2137⟩, 995 ⟨89AX(C)1751⟩; **5**, 106 ⟨63OSC(4)532⟩, 656 ⟨77JHC765⟩, 658 ⟨76JA1099⟩, 963 ⟨86H(24)2813, 87JCS(P1)2163, 88JCS(P1)2137⟩
Boyer, J. H.,**5**, 964, 1119 ⟨89JCR(S)331⟩, 989 ⟨72JOC724⟩, 1036 ⟨72IJS139⟩, 1036, 1037 ⟨72JOC1360⟩; **6**, 653 ⟨91CL569⟩
Boyer, R. F., **1**, 403, 799 ⟨71JA650, 69JA1578⟩
Boyer, S. K., **1**, 2, 6 ⟨85JOC3408⟩; **3**, 621 ⟨85JOC3408⟩; **5**, 657 ⟨82JOC3658⟩
Boyes, A. L., **3**, 273 ⟨89TL985⟩
Boykin, D. W., **1**, 1090 ⟨77JOC260⟩; **3**, 621 ⟨88SC681⟩
Boyko, W. J., **6**, 632 ⟨91S86⟩
Boyland, E., **2**, 651 ⟨53JCS3623⟩; **6**, 505 ⟨66JCS(C)350⟩
Boyle, B. A., **1**, 927 ⟨86JA2381⟩
Boyle, F. T., **3**, 66 ⟨74JCS(P1)1541⟩; **5**, 1035 ⟨74JCS(P1)1541⟩
Boyle, J., **4**, 704 ⟨93OM3998⟩
Boyle, Jr., W. J., **1**, 869, 871 ⟨70JA5926⟩
Boyle, T. J., **4**, 726 ⟨93JOM(462)141⟩
Boynton, W. A., **1**, 1114 ⟨72JA5061⟩
Boys, M. L., **2**, 612 ⟨88TL3365⟩; **4**, 61, 73, 86 ⟨88TL3365⟩
Boysel, R. M., **4**, 291 ⟨85JA675⟩
Boyton, H. G., **1**, 912, 914 ⟨58T178⟩
Bozec, H. Le, **5**, 940 ⟨91OM2768⟩, 941 ⟨89AOC(29)163, 91CC881⟩, 941 ⟨92CC1220, 92OM809⟩, 941 ⟨95OM1938⟩
Bozell, J. J., **1**, 494 ⟨88JA2655⟩
Bozimo, H. T., **4**, 194 ⟨76JA3037⟩
Bozopoulos, A. P., **2**, 998 ⟨90AX(C)1300⟩
Bozzini, S., **5**, 355 ⟨91G401⟩
B. Page, P. C., **2**, 902 ⟨90T2573⟩
Brabander, H. J., **6**, 508 ⟨66JMC852⟩
Brabec, L., **2**, 636 ⟨88JA7984⟩
Braca, G., **6**, 524 ⟨64CI(M)206⟩
Brace, N. O., **1**, 10 ⟨82JFC(20)313⟩
Brachel, H. V., **4**, 310 ⟨62AG872⟩, 426 ⟨62AG(E)592⟩
Brachel, H. von, **6**, 105 ⟨62AG(E)592, 66GEP1156780⟩
Brachen, S., **2**, 837 ⟨83CC857⟩
Brachmann, H., **3**, 616 ⟨85LA566⟩
Bracho, R. D., **3**, 652, 654 ⟨73CC781⟩
Brachthäuser, B., **6**, 386, 406 ⟨91AG(E)1488⟩
Brachwitz, H., **5**, 830 ⟨66ZC313⟩
Brack, A., **2**, 433 ⟨54ZAAC(277)258⟩
Bracken, G. J., **3**, 411 ⟨79TL2109⟩
Brackenridge, I., **2**, 90 ⟨88TL2483⟩
Bracker-Novak, J., **4**, 682 ⟨89JOM(365)163⟩
Brackman, W., **2**, 621, 627 ⟨66RTC857⟩; **3**, 617 ⟨63RTC757⟩
Bractel, G., **2**, 388 ⟨78HCA1622⟩
Bradaczek, H., **5**, 716 ⟨87LA483⟩
Bradamante, S., **4**, 1076 ⟨86CC1191⟩; **5**, 499 ⟨74JCS(P2)1676⟩
Bradbury, R., **3**, 685 ⟨83JCS(P1)293⟩
Bradbury, R. H., **1**, 811 ⟨81JCS(P1)3225⟩; **4**, 860 ⟨88TL6475, 92T7551⟩
Braden, M. L., **1**, 629 ⟨91JA4712⟩
Braden, R., **6**, 239 ⟨70GEP1926659⟩
Brader, J. J., **5**, 551 ⟨55JA5997⟩
Bradley, C. W., **1**, 663 ⟨38JA2333⟩
Bradley, D. C., **2**, 108 ⟨63JCS2101⟩, 112 ⟨B-78MI 202-

01⟩, 409, 420, 421 ⟨62CJC449⟩, 409, 421 ⟨62CJC1355, 69JCS(A)980, 68CC495⟩, 421 ⟨71JCS(A)1433, 71JCS(C)2741⟩; **4**, 657 ⟨74CC29⟩; **6**, 400 ⟨88POL2049⟩
Bradley, D. D. C., **1**, 623 ⟨92JCS(P1)3225⟩
Bradley, J. S., **4**, 676 ⟨78JOM(150)C1⟩
Bradley, W., **3**, 474 ⟨28JCS1310⟩
Bradley, W. E., **4**, 300 ⟨32JA4172⟩
Bradshaw, J. S., **1**, 35, 39 ⟨63JA1108⟩; **2**, 133 ⟨76JOC134⟩; **5**, 291 ⟨70JOC1219⟩, 510, 511, 512 ⟨84CRV17⟩, 511 ⟨83JOC2635⟩
Bradshaw, R. W., **2**, 299 ⟨68AG986, 68AG(E)919⟩; **3**, 179 ⟨68AG(E)919⟩
Bradsher, C. K., **1**, 213, 478 ⟨78JOC3800⟩, 1094 ⟨81JOC4600, 81JOC4608⟩; **3**, 284 ⟨57JA393, 82ACR300⟩, 284 ⟨82JOC1585⟩, 310 ⟨59JA1938⟩, 626 ⟨81JOC4600⟩; **6**, 314 ⟨67JOC2074⟩
Brady, D. G., **2**, 204 ⟨69JOC3097⟩, 742 ⟨66JOC3543⟩, 1034 ⟨66JOC3545⟩
Brady, F., **4**, 659 ⟨80JOM(193)21⟩
Brady, K., **5**, 716 ⟨80JCS(P2)121⟩; **6**, 620 ⟨80JCS(P2)867⟩
Brady, K. T., **4**, 224 ⟨88JOC845⟩; **5**, 540 ⟨88JOC845⟩
Brady, L. A., **4**, 681 ⟨92CC310⟩
Brady, O. L., **3**, 91 ⟨50JCS767⟩, 446 ⟨22JCS1648⟩
Brady, W. T., **2**, 930 ⟨90JOC4215⟩, 941 ⟨76TL1553⟩; **3**, 151 ⟨81T2949⟩, 246 ⟨81JOC4047⟩, 526 ⟨68TL1977, 73JA7447⟩, 526 ⟨74JOC3790⟩, 527 ⟨66JOC2676, 69JA5679, 70TL2963, 71JOC1486, 74JOC763, 76TL1553⟩, 529 ⟨86JOC2145⟩; **4**, 768 ⟨68TL1977, 70JOC1515⟩, 807 ⟨76TL1553⟩; **6**, 622 ⟨77JOC3220⟩
Braekman, J. C., **3**, 694 ⟨89PAC509⟩; **4**, 414 ⟨88TL1691⟩
Braeunig, C., **2**, 852 ⟨76JPR1019⟩
Braga, A. L., **1**, 1035 ⟨86CC1067⟩; **2**, 1036, 1037 ⟨88SC1979⟩, 1036, 1038 ⟨84S240, 86CC1067⟩; **4**, 284 ⟨82SC595⟩
Braga, D., **2**, 959 ⟨78CC927⟩; **5**, 1088 ⟨91AG(E)847⟩; **6**, 332 ⟨86G101⟩
Bragante, L., **6**, 250 ⟨92EUP499158⟩, 604 ⟨91JFC(53)181⟩
Bragina, I. O., **4**, 899, 902 ⟨81CA96556w⟩; **5**, 223 ⟨90ZOB1928⟩, 727 ⟨82IZV2121⟩
Brahim, S. I., **1**, 883 ⟨89CC161⟩
Brähler, G., **3**, 371 ⟨78JOC3893⟩
Brahmi, R., **3**, 429 ⟨89TL5763⟩
Brahmi, R.,**3**, 429 ⟨91BCJ2948⟩
Braid, M., **2**, 166 ⟨57JA6248⟩
Braid, M.,**4**, 39 ⟨61JA2383⟩
Braid, M., **6**, 2 ⟨58JOC322, 61JA2495⟩, 3 ⟨61JA2383⟩, 5 ⟨58JA851⟩
Brain, E. G., **5**, 9 ⟨63JCS491⟩
Braish, T. F., **4**, 948 ⟨88JOC3647⟩; **5**, 322 ⟨92SC3067⟩
Braker, A. H., **3**, 712 ⟨95TL2109⟩
Brakmann, S., **4**, 353 ⟨91JOM(403)29⟩
Bram, G., **1**, 178 ⟨90SC125⟩; **2**, 19 ⟨88SC1661⟩, 25, 31 ⟨86CC1250, 89BSF247⟩; **3**, 613 ⟨80SC279⟩, 634 ⟨87S56⟩; **5**, 132 ⟨85S40⟩
Brambilla, R., **1**, 91 ⟨81T3615⟩
Bramley, R. K., **1**, 809 ⟨73JCS(P1)1913⟩
Bramm, E., **6**, 645 ⟨80JMC13⟩
Brammer, L., **4**, 676 ⟨86JCS(D)657⟩
Brammer-Petersen, J. V., **5**, 1120 ⟨81USP4304935⟩
Branca, J., **1**, 618 ⟨73TL1843⟩
Branca, J. C., **1**, 618 ⟨74TL1105⟩
Branca, S. J., **3**, 206 ⟨75TL4225, 81JA1996⟩, 533 ⟨76JA7456, 77JOC3165⟩, 533 ⟨84JA3995, 84JA4001⟩
Brancaccio, G., **5**, 746 ⟨64JOC3697⟩
Branch, C. L., **4**, 447 ⟨86JCS(P1)1077⟩; **5**, 345 ⟨89TL3219⟩
Branchaud, B. P., **1**, 335 ⟨88TL167⟩, 495 ⟨91TL6273⟩, 946 ⟨89JOC1320⟩; **2**, 239 ⟨83JOC3538⟩, 239, 359 ⟨83JOC3531⟩, 579 ⟨88TL167⟩; **3**, 436 ⟨83JOC3531⟩
Brancone, L. M., **5**, 1125 ⟨48JOC924⟩
Brancourt, M., **5**, 1102 ⟨70BSF2637⟩
Brand, C., **5**, 138 ⟨86T5961⟩
Brand, E., **5**, 83 ⟨47MI 502-02⟩
Brandäge, S., **1**, 4, 6 ⟨83ACS(B)141⟩
Brand, K., **2**, 764 ⟨22CB875⟩
Brand, M., **2**, 16 ⟨85S665⟩, 17 ⟨86JOC3607⟩, 18 ⟨80TL4543, 85JOC3342⟩, 626 ⟨88JOC5545⟩; **3**, 234 ⟨85JOC3342⟩; **5**, 203 ⟨85S665⟩
Brand, W. W., **1**, 379 ⟨70JOC1828, 70OR(18)99⟩; **2**, 719 ⟨70OR(18)99⟩
Brandange, S., **2**, 325 ⟨88S347⟩
Brandau, D., **2**, 597 ⟨75IC2809⟩
Brandes, B. D., **2**, 92 ⟨94JOC4378⟩
Brandes, E., **1**, 826 ⟨89JOC515, 89JOC5849⟩
Brandes, W., **4**, 141 ⟨92JOM(438)45⟩; **6**, 439 ⟨82EUP46557⟩
Brandi, A., **2**, 320 ⟨90TL3351⟩, 329 ⟨89JOC1748⟩; **3**, 269 ⟨88JOC2426⟩, 431 ⟨91G285⟩, 433 ⟨80JCR(S)122⟩, 686 ⟨82S719⟩
Brandi, G., **2**, 602 ⟨73ICA319⟩
Brandl, A., **5**, 382 ⟨90CB63⟩
Brändle, K. A., **1**, 455 ⟨61JOC2934⟩; **4**, 551, 565 ⟨61JA2055⟩, 811, 812 ⟨60AG77⟩, 812 ⟨61JOC2934⟩
Brändli, U., **1**, 156 ⟨86CB575⟩
Brandoma, L., **4**, 765 ⟨69MI 417-02⟩
Brand-Roth, B., **5**, 1085 ⟨90AG(E)90, 91POL1147⟩
Brandrup, G., **2**, 129 ⟨65JOC3689⟩; **3**, 638 ⟨65JOC3689⟩
Brandsma, L., **1**, 9 ⟨87JOC4299⟩, 95, 889 ⟨90SC2165⟩, 141 ⟨78CC596⟩, 441 ⟨92JOM(441)1⟩, 457, 508 ⟨84CC735⟩, 464 ⟨90SC2349⟩, 880 ⟨86AG(E)465⟩, 954 ⟨70QRS1⟩, 954, 957 ⟨B-84MI 120-01⟩, 955 ⟨68RTC916⟩, 955, 1125 ⟨65RTC31⟩, 956 ⟨63RTC1040⟩, 971 ⟨68RTC1179⟩, 981 ⟨81RTC34, 81TL2827, 82JOM(233)C25⟩, 986 ⟨67RTC129⟩, 998, 1000, 1001 ⟨B-71MI 121-01⟩, 998, 1000, 1034, 1040, 1046, 1047, 1069, 1072, 1084 ⟨B-88MI 121-01⟩, 999 ⟨64RTC307, B-69MI 121-06⟩, 1022 ⟨B-88MI 121-02⟩, 1030 ⟨68RTC69⟩, 1031 ⟨B-67MI 121-02⟩, 1037 ⟨65RTC1200, 81S459⟩, 1044 ⟨B-88MI 121-03⟩, 1045, 1046 ⟨88JOM(338)289⟩, 1048 ⟨73RTC667⟩, 1072 ⟨64RTC456, 71RTC265, 76RTC264⟩, 1133 ⟨B-88MI 122-01⟩; **2**, 220 ⟨69RTC513⟩, 271 ⟨78S359⟩, 278, 279 ⟨63RTC68⟩, 422 ⟨77RTC160⟩, 551 ⟨B-87MI 211-02, B-90MI 211-01⟩, 722 ⟨69RTC513⟩, 907 ⟨78RTC69, 86CC260⟩, 907 ⟨86RTC66⟩, 907, 908, 909, 939, 942, 946 ⟨B-87MI 218-01⟩, 939 ⟨68RTC916, 82JOM(232)C1⟩, 940 ⟨82JOM(233)C25⟩, 941 ⟨79RTC55⟩, 942 ⟨84CC257⟩, 999, 1000 ⟨89OM1688⟩, 1011, 1014, 1016, 1017, 1018, 1025, 1031, 1032, 1034, 1035, 1036 ⟨B-88MI 221-01⟩, 1017 ⟨90S984⟩, 1021, 1025, 1027, 1030, 1031, 1032, 1033, 1034, 1036 ⟨B-69MI 221-02⟩, 1030 ⟨59RTC252, 61RTC237⟩, 1030, 1031 ⟨62RTC510⟩, 1030, 1031 ⟨63RTC1119⟩, 1032 ⟨62RTC583, 64RTC208, 66RTC580, 66RTC952, 68RTC97, 71RTC1098, 77RTC179⟩, 1033 ⟨69RTC179⟩, 1043 ⟨64RTC1211⟩, 1049 ⟨71RTC265⟩, 1076 ⟨B-90MI 223-01⟩; **3**, 58 ⟨86RTC66⟩, 61 ⟨65RTC31⟩, 105 ⟨88S316⟩, 273 ⟨88JOM(338)289⟩, 337 ⟨71IJS(B)85⟩, 337, 352 ⟨70RTC593⟩, 548 ⟨67RTC907, 68RTC1236, 72RTC578, 74RTC26, 77RTC179, 82RTC310⟩, 548, 552 ⟨66RTC889, 68RTC38⟩, 548, 553 ⟨79RTC55, 77S888⟩, 674 ⟨B-71MI 319-01⟩, 675 ⟨73RTC667⟩; **4**, 843 ⟨74RTC184⟩, 843 ⟨85S958⟩, 956, 957 ⟨78RTC69⟩, 1017 ⟨87MI 421-01⟩; **4**, 93 ⟨81PS(11)139⟩, 138 ⟨87JOM(336)C41⟩, 647 ⟨86JOM(316)C4⟩, 792 ⟨87S76⟩, 834 ⟨68RTC929⟩, 836 ⟨68RTC1179⟩, 836, 949 ⟨87JOM(332)99⟩, 842 ⟨79S455⟩, 843 ⟨69RTC597, 73RTC1067⟩; **5**, 423 ⟨68MI 509-01⟩, 427 ⟨81TL2827⟩, 510 ⟨74MI 530-03⟩, 547, 552 ⟨78S303⟩, 550 ⟨73RTC601, 75RTC83⟩, 550, 563 ⟨66RTC889, 68RTC929⟩, 554 ⟨69RTC597⟩, 556 ⟨74RTC258⟩, 589 ⟨82RTC1⟩, 589, 590, 603 ⟨68RTC38⟩, 589, 590, 603, 611, 626 ⟨77S888⟩, 589, 614 ⟨68RTC1236⟩, 589, 614, 618 ⟨72RTC578, 79RTC55⟩, 1105 ⟨78S577⟩; **6**, 88 ⟨67M1043⟩
Brandstadter, S. M., **5**, 274, 286 ⟨87JA1798⟩
Brandstetter, H. H., **6**, 550 ⟨88M127⟩

Brandstetter, T., **1**, 554 ⟨91TL3679⟩, 1038 ⟨92TL7511⟩
Brandstrom, A., **3**, 613 ⟨56ACS1197⟩
Brändström, A., **5**, 991 ⟨74ACS(B)699⟩
Brandt, A., **4**, 205, 210 ⟨79CB148⟩
Brandt, C. A., **1**, 243, 437 ⟨89TL1209⟩; **2**, 279 ⟨82JCR(S)212⟩; **4**, 88 ⟨82JCR(S)212⟩, 291 ⟨83MI 406-01⟩
Brandt, G. A. R., **6**, 233 ⟨52JCS2198⟩
Brandt, G. R. A., **2**, 486, 505 ⟨52JCS2552⟩
Brandt, S., **1**, 905 ⟨84CB2510⟩; **3**, 519 ⟨79JA6473⟩
Brandt, T., **5**, 652 ⟨69MAC144⟩
Brandt, W. E., **3**, 400 ⟨27CB824⟩
Brandvold, T. A., **1**, 551 ⟨87JOM(334)9, 93JA10671⟩
Brandwood, M., **6**, 43 ⟨75JFC(6)37⟩
Brannegan, D. P., **3**, 234 ⟨80JOC1109⟩
Brannen, C. G., **2**, 942 ⟨50JA4280⟩; **5**, 114, 115 ⟨56JA1689⟩
Brannen, Jr., W. T., **4**, 74, 75, 77 ⟨64JA4645⟩
Brannock, K. C., **1**, 355 ⟨64JOC940⟩, 417, 829 ⟨60JOC1815⟩, 825 ⟨59JA3379⟩; **2**, 259 ⟨51JA4953⟩, 751 ⟨63JCS1462⟩; **3**, 546, 547 ⟨68JOC2738⟩, 623, 624 ⟨65JOC2564⟩; **4**, 881 ⟨66JOC2966⟩, 887 ⟨65TL3589, 71JOC2211⟩; **5**, 511 ⟨68JOC2565⟩
Brannon, M. J., **1**, 578 ⟨86TL1449⟩
Bransma, L., **2**, 621 ⟨81TL2495⟩
Branson, P. R., **6**, 664 ⟨72JCS(D)1303⟩
Branton, G. R., **1**, 801 ⟨66JCS(A)1342⟩
Bras, P., **1**, 248 ⟨83JOM(256)C1⟩
Brasca, M. G., **4**, 286 ⟨89T7161⟩
Brasedow, O. H., **2**, 193 ⟨58AC(P)108⟩
Braselton, J. P., **2**, 381 ⟨73JMC736⟩
Brasen, W. R., **1**, 399 ⟨55JOC1119, 63OSC(4)585⟩; **6**, 46 ⟨62JOC3995⟩
Brasili, L., **4**, 232 ⟨87JMC1934⟩
Brasington, R. D., **4**, 156 ⟨72JOM(40)115⟩
Braslau, R., **1**, 457 ⟨89TL4657⟩
Braslavsky, S., **1**, 785 ⟨77CRV473⟩
Brass, H. J., **2**, 845 ⟨73JOC2703⟩
Brassard, P., **2**, 667 ⟨76JCS(P1)1852, 78JOC1435⟩, 667 ⟨81S381⟩; **3**, 63 ⟨72CC863⟩; **4**, 827 ⟨72CC863⟩; **6**, 75 ⟨72CC863⟩
Brassel, B., **1**, 799 ⟨80JA2469⟩
Brassen, W. R., **1**, 560 ⟨56JA82⟩
Bratby, D. M., **1**, 815 ⟨71JCS(C)970, 77T1527⟩
Bratenko, M. K., **5**, 1035 ⟨91UKZ431⟩
Bratkov, A. A., **4**, 775 ⟨90MI 417-01⟩
Bratolyubova, A. G., **4**, 777 ⟨73ZOR69⟩
Brattesani, D. N., **1**, 607, 608 ⟨75JOC2165⟩, 608, 781 ⟨74TL2279⟩, 1000 ⟨73SC245⟩; **3**, 643, 644 ⟨74TL2279⟩
Brauch, G., **4**, 609 ⟨83ZAAC(497)134⟩; **6**, 385 ⟨83ZAAC(497)134⟩
Brauchli, R., **1**, 635 ⟨90HCA640⟩
Braude, E. A., **1**, 104 ⟨46JCS396⟩, 422 ⟨52JCS1430⟩; **3**, 252 ⟨46JCS945⟩, 267 ⟨46JCS45⟩, 275 ⟨56JCS3070⟩
Braude, V., **3**, 502 ⟨92OM84⟩
Brauer, D. J., **2**, 435 ⟨85JOM(296)411⟩, 466 ⟨86CB2767⟩; **4**, 550 ⟨85JOM(296)411⟩, 582, 583 ⟨74CB3706⟩, 693 ⟨77CB1763⟩; **5**, 382 ⟨93JOM(446)25⟩, 605 ⟨87CB751⟩; **6**, 243 ⟨89JOM(378)125, 93AG429, 93JOM(446)25⟩
Brauer, H., **2**, 947 ⟨91SL431⟩
Brault, A., **4**, 317, 322 ⟨69CR(C)1718, 70CR(C)436⟩, 322 ⟨73BSF3499⟩, 405, 411 ⟨75BSF196⟩
Brault, D., **5**, 1097 ⟨77CC648⟩; **6**, 717 ⟨77CC648⟩
Brault, J., **2**, 533 ⟨64CJC2903⟩
Brault, M. A., **4**, 703 ⟨78IC625⟩
Braum, K., **4**, 994 ⟨92TL2949⟩
Brauman, J. I., **1**, 476 ⟨67JA5421⟩, 869 ⟨65JA384⟩, 883 ⟨88JA2714⟩, 917 ⟨66JA2047⟩; **3**, 621 ⟨78JA1119⟩
Braun, A., **3**, 619 ⟨07CB2709⟩
Braun, F., **3**, 547 ⟨81LA1361⟩
Braun, G., **1**, 884 ⟨55OSC(3)101⟩; **2**, 31 ⟨43OSC(2)308⟩; **5**, 449 ⟨60CB901⟩
Braunagel, N., **6**, 268 ⟨73ZAAC(399)280⟩
Braun, H., **2**, 824 ⟨71LA(751)17⟩; **3**, 374 ⟨73TL4033⟩
Braun, J., **3**, 619 ⟨34CB1762⟩; **4**, 145 ⟨72BSF811⟩
Braun, J. D., **4**, 735 ⟨67JOC2749⟩
Braun, J. von, **2**, 33 ⟨41OSC(1)428⟩, 172 ⟨23CB553, 30CB2856⟩; **3**, 661 ⟨31LA(488)111⟩; **4**, 37 ⟨27LA(453)113⟩, 790 ⟨29CB409⟩; **5**,, 654, 751 ⟨34AG611⟩, 674, 767 ⟨34CB1762, 27LA(451)453⟩, 1025 ⟨12CB2188⟩; **6**, 444 ⟨20CB601⟩
Braun, M., **2**, 420 ⟨93OS(72)38⟩; **4**, 172 ⟨81LA2247⟩, 395 ⟨75CB2368⟩, 397 ⟨84CB413⟩, 758 ⟨87TL5145⟩, 758, 818 ⟨91CB1379⟩, 956, 957 ⟨84AG(E)723⟩; **5**, 66 ⟨86CB2191⟩, 138 ⟨84TL5031⟩, 397 ⟨73JOM(49)149⟩; **6**, 72 ⟨91SL160⟩
Braun, R. A., **3**, 445 ⟨58JA3048⟩
Braun, R., **1**, 782 ⟨93LA1133⟩; **2**, 157 ⟨71LA(744)15⟩, 503 ⟨69JOM(19)367⟩; **4**, 564 ⟨88SRI727⟩, 575 ⟨88ZAAC(567)23⟩
Braun, R. O., **6**, 508 ⟨51JOC1829⟩
Braun, S., **1**, 146 ⟨89AG(E)277⟩; **2**, 1000 ⟨89AG(E)1025⟩
Bräuner, H.-J., **5**, 744 ⟨83S35⟩; **6**, 115 ⟨88LA605⟩, 140 ⟨83S905⟩, 643 ⟨84LA108⟩
Braunholtz, W., **3**, 586 ⟨21HCA897⟩
Brauniger, H., **4**, 970 ⟨84MI 421-01⟩
Braunstein, D. M., **2**, 314 ⟨69JOC1817⟩
Braunstein, P., **5**, 962 ⟨89CRV1927⟩
Bräutigam, I., **3**, 373 ⟨79CB3007⟩
Bravdo, T., **5**, 652 ⟨89JOC4263⟩
Braveman, S., **2**, 194, 195 ⟨73TL3563⟩
Braverman, S., **1**, 704 ⟨B-88MI 115-03⟩, 839 ⟨B-88MI 118-01⟩, 955 ⟨76TL3181⟩; **2**, 154 ⟨67CC270⟩, 159, 196 ⟨B-88MI 203-04⟩, 362, 366 ⟨B-90MI 206-02⟩; **3**, 369 ⟨94TL953⟩
Bravetti, D., **1**, 382 ⟨82HCA371⟩; **6**, 550 ⟨82HCA371⟩
Bravo, A. A., **2**, 743 ⟨91S970⟩; **5**, 720 ⟨90JOC3682⟩
Bravo, H. R., **4**, 175 ⟨91JOC1788⟩; **5**, 325 ⟨91H(32)1687⟩
Bravo, P., **1**, 143 ⟨85TL2913⟩, 738 ⟨75G109⟩; **2**, 15 ⟨90TA661⟩, 350 ⟨92JOC2726⟩; **3**, 150 ⟨88CC218⟩; **4**, 170 ⟨89JOC5171⟩, 584 ⟨76JOM(111)361⟩; **5**, 832 ⟨67G1664⟩
Bravo-Borja, S., **1**, 30 ⟨84JCS(P1)1671⟩
Bravo-Zhivotoskii, D. A., **5**, 432 ⟨83JOM(248)51⟩
Bravo-Zhivotovskii, D. A., **4**, 629, 638 ⟨85ZOB2396, 80IZV1451⟩; **5**, 226 ⟨84IZV1897, 82IZV949⟩, 432 ⟨83ZOR1838, 84MI 509-01⟩, 924 ⟨88IZV1947, 89JOM(368)163⟩; **6**, 131 ⟨83JOM(248)51, 87IZV2872⟩, 131 ⟨88MI 604-01⟩, 516 ⟨84IZV1897⟩
Bravo-Zhivotovskii, D., **3**, 502 ⟨92OM2326⟩
Brawner Floyd, M., **4**, 161 ⟨85JOC5022⟩
Bray, B. L., **2**, 1025 ⟨87JOC2332⟩; **3**, 638 ⟨90CJC1305⟩
Bray, E. H., **4**, 1034 ⟨71BSB651⟩
Bray, K., **5**, 1121 ⟨93JMC2004⟩
Bray, L. E., **1**, 110 ⟨69JOC641⟩
Bray, L. S., **4**, 681 ⟨71CC608⟩
Bray, T. L., **1**, 525, 826 ⟨86TL4945⟩
Braye, E. H., **1**, 53 ⟨71T5523⟩; **2**, 465, 989 ⟨61JA4406⟩; **4**, 112 ⟨71T5523, 59JINC250⟩
Braz, G. I., **4**, 135 ⟨41JGU41⟩
Brazier, J. F., **4**, 478 ⟨81PS(11)87⟩
Bream, J. B., **6**, 646, 647 ⟨75AF1477⟩
Breant, P., **3**, 310 ⟨81JOM(216)139⟩
Breau, L., **1**, 418 ⟨91COS(5)675⟩
Brebant, N., **2**, 147 ⟨60TL20⟩
Brechbiel, M., **1**, 624, 783 ⟨81JA4650⟩
Brechbühler, H., **2**, 749 ⟨65HCA1746⟩; **5**, 131 ⟨63AG(E)212⟩; **6**, 104 ⟨63AG296, 65HCA1746⟩
Brechot, P., **2**, 583 ⟨91OM1183⟩
Brecht, A., **1**, 987 ⟨86CB1208⟩, 988 ⟨86AG(E)93⟩
Brecht, E. A., **4**, 178, 184 ⟨58JA6613⟩
Brecht, H., **2**, 293 ⟨69ZN(B)179⟩
Brede, O., **2**, 686 ⟨70JPR43⟩
Bredenkamp, M. W., **4**, 228 ⟨90TL2759⟩; **5**, 573 ⟨90SC2235⟩
Breder, C. V., **2**, 626, 685 ⟨67JOC2358⟩

Bredereck, H., **2**, 29 ⟨62CB803⟩, 724 ⟨60CB2736⟩, 745 ⟨74CB1545⟩, 788 ⟨71CB3475⟩; **3**, 627, 637 ⟨71CB924, 65AG(E)951⟩, 709 ⟨62AG(E)334⟩; **4**, 234 ⟨54CB129⟩, 235 ⟨54CB784⟩, 881 ⟨61AG493, 63CB1350, 71CB3475⟩, 881, 882, 972 ⟨64CB3081⟩, 882 ⟨65GEP1212069⟩, 894, 973, 974 ⟨79LA2096⟩, 972, 979 ⟨80LA372, 61CB2278⟩, 979 ⟨71CB792⟩; **5**, 586, 614, 616 ⟨71CB2709⟩, 744, 745, 751 ⟨59CB837⟩, 747 ⟨80LA344⟩, 753 ⟨57CB1837⟩; **6**, 55 ⟨65CB1078⟩, 104 ⟨68CB41⟩, 106 ⟨62AG(E)331⟩, 111, 115 ⟨73CB3725⟩, 138 ⟨66AG(E)132, 68CB1885, 68CB3058, 69MI 605-01⟩, 138, 140 ⟨68CB51⟩, 139, 140 ⟨59CB329⟩, 140 ⟨60CB1398, 63CB1505⟩, 321, 337 ⟨79LA2089⟩, 634 ⟨70CB256⟩, 645 ⟨61CB2278⟩, 647 ⟨79LA2096⟩
Bredereck, K., **1**, 201 ⟨75BCJ2939⟩; **4**, 974 ⟨61CB2278⟩; **6**, 620 ⟨66CB2454, 61CB2278⟩
Brederick, H., **3**, 709 ⟨65LA(686)92⟩
Bredikhin, A. A., **4**, 129 ⟨78JGU1168⟩
Bredikhina, Z. A., **6**, 660 ⟨93S59⟩
Bredt, J., **1**, 853 ⟨24LA(437)1⟩
Breederveld, H., **2**, 416, 417 ⟨52MI 207-01⟩, 416, 417 ⟨53MI 207-01⟩, 416, 417 ⟨54MI 207-01, 53MI 207-03⟩, 417 ⟨53MI 207-04⟩, 417, 421 ⟨53MI 207-02⟩; **5**, 288 ⟨60RTC401⟩
Bregadze, V. I., **1**, 168 ⟨92CRV209⟩; **6**, 169 ⟨92CRV209⟩, 170 ⟨71IZV1505⟩, 406 ⟨68BAU414⟩
Bregar, Z., **6**, 10 ⟨90TL3357⟩
Bregovic, I., **2**, 478 ⟨61CI(L)469⟩
Brehm, E., **5**, 1147, 1148 ⟨70LA(737)80⟩
Brehm, L., **1**, 445 ⟨86JCS(D)997⟩; **2**, 490 ⟨86JCS(D)997⟩, 491 ⟨86ZAAC(539)50⟩
Brehme, R., **5**, 285 ⟨76S113⟩
Breil, H., **1**, 665 ⟨55AG541⟩; **4**, 671 ⟨66AG(E)898⟩
Breimair, J., **5**, 936 ⟨92JOM(426)343⟩, 955 ⟨91CB2431⟩
Breindel, A., **2**, 485 ⟨64AG(E)66⟩
Breit, B., **4**, 483 ⟨91S1099⟩, 549 ⟨92AG(E)1055, 93PS(76)1⟩, 567 ⟨93S285⟩; **5**, 395 ⟨91CB2843, 91S1099⟩, 885 ⟨92AG(E)1055, 93CB1945⟩; **6**, 154 ⟨92AG(E)1055⟩, 512, 582 ⟨91S1099⟩
Breitbeil, F. W., **5**, 166 ⟨64JOC2810⟩
Breitenbach, J., **3**, 694 ⟨92S41⟩
Breitenback, R., **3**, 234 ⟨80JOC1109⟩
Breitenbucher, J. G., **1**, 831 ⟨93TL3985⟩
Breitenstein, M., **3**, 358 ⟨82JOC1979⟩
Breiter, J. J., **2**, 127 ⟨62JA1621⟩
Breitfeld, B., **5**, 1062 ⟨69TL3191⟩
Breitinger, D., **2**, 817 ⟨64CB1155⟩; **5**, 384, 387 ⟨70CB2388⟩; **6**, 405 ⟨77ZN(B)1022, 79ZN(B)390⟩
Breitinger, D. K., **4**, 697 ⟨80JOM(191)7, 81ZN(B)188, 83JOM(243)245, 86JOM(301)1⟩; **5**, 1141 ⟨83JOM(243)245⟩; **6**, 206, 207 ⟨80JOM(191)7, 83JOM(243)245, 83JOM(256)217⟩
Breitinger, D. K.,**6**, 207 ⟨86JOM(301)1⟩
Breitmaier, E., **3**, 59 ⟨86CB1737⟩, 59, 75 ⟨87S1⟩, 71 ⟨71CB665, 77S1⟩, 131, 284, 303 ⟨92AG(E)1035⟩; **5**, 199 ⟨90S697⟩
Brel, V. K., **5**, 669 ⟨87JGU1478, 87ZOB1658⟩; **6**, 261 ⟨88IZV2639, 88IZV466⟩, 613 ⟨88KFZ143, 89ZOB1455⟩
Bremen, J., **6**, 140 ⟨73CB2523⟩
Bremer, C., **2**, 309 ⟨93TL4193⟩
Bremer, M., **1**, 910 ⟨89JA1147⟩; **2**, 1088 ⟨89AG(E)1219⟩; **4**, 574 ⟨91CB199⟩
Bremner, C., **2**, 111 ⟨40USP2223948⟩
Bremner, D. H., **3**, 479 ⟨76CC538, 77JCS(P1)2298⟩; **5**, 867 ⟨77JCS(P1)2298⟩
Bremner, J., **1**, 439, 700 ⟨82TL3265⟩
Bremner, J. B., **3**, 87 ⟨91AJC135⟩; **5**, 1127 ⟨79CI(L)319, 84AJC1043⟩, 1127 ⟨85JA3910, 91AJC135⟩
Brems, P., **5**, 1040 ⟨90JHC1059⟩
Bremser, W., **4**, 25 ⟨71JA4472⟩
Brendel, G., **6**, 523 ⟨57ZAAC(289)324⟩
Brender, K., **2**, 1014 ⟨61LA(640)5⟩
Brendle, Th., **6**, 138 ⟨66AG(E)132, 68CB1885⟩
Brenemann, C. M., **2**, 586, 588 ⟨85JA3197⟩
Brener, L., **2**, 517 ⟨77JA3427⟩
Brenna, E., **5**, 1045 ⟨90TL7229⟩
Brennan, D. E., **4**, 632 ⟨83JA2582⟩
Brennan, J., **3**, 48, 195 ⟨88CSR1⟩; **4**, 100 ⟨86TL3199⟩
Brennan, T., **1**, 876 ⟨78JOC2705⟩; **2**, 941 ⟨B-69MI 218-02⟩
Brennan, T. M., **3**, 234 ⟨80JOC1109⟩
Brennecke, L., **4**, 1003 ⟨76JPR116⟩; **5**, 788 ⟨76JPR272⟩; **6**, 159 ⟨72JPR969⟩, 159 ⟨74PS(4)241, 76JPR272⟩, 286 ⟨76JPR116⟩, 367 ⟨72JPR969⟩
Brennenstuhl, W., **1**, 1091 ⟨84TL3437⟩
Brenner, G. S., **4**, 48 ⟨75TL3979⟩
Brenner, M., **6**, 430 ⟨57HCA604⟩
Brenner, S., **1**, 155 ⟨74TL1377⟩
Brenner, W., **1**, 809 ⟨67AG(E)800⟩, 812 ⟨69LA(727)194⟩
Brenninger, W., **2**, 739 ⟨73OSC(5)808⟩
Brent, D. A., **3**, 543 ⟨70JOC4204⟩
Breque, A., **4**, 341 ⟨82TL1565⟩
Bresadola, S., **5**, 1038, 1105 ⟨65JA4791⟩
Breshears, S. R., **4**, 413 ⟨59JA3789⟩
Bresinsky, E., **3**, 253 ⟨67CB107⟩
Breslin, H. J., **6**, 598 ⟨91JMC3187⟩
Breslow, D. S., **2**, 208 ⟨57JA5000⟩; **5**, 456 ⟨61JA4023, 62JA2499⟩
Breslow, R., **1**, 5, 936 ⟨90TL2957⟩, 851, 865 ⟨68JA2698, 67JA4383⟩, 865 ⟨82TL795⟩, 907 ⟨60JA2644, 61JA2367, 61JA2375, 67JA5048a, 70JA984⟩, 907, 908, 909 ⟨62JA3168⟩, 907, 909 ⟨57JA5318, 58JA5991⟩, 927 ⟨88MI 119-04, 90JA6842, 91JA8977⟩, 936 ⟨93OS(72)199⟩; **2**, 20 ⟨77JA905, 91COS(7)43⟩, 20 ⟨93TL1107⟩, 1050 ⟨80TL3241⟩; **3**, 473 ⟨72JA4787⟩, 477 ⟨58JA5991⟩; **4**, 271 ⟨62JA684⟩; **6**, 80, 84 ⟨80JOC740⟩
Bressan, M., **4**, 701 ⟨86G471⟩
Bressel, U., **6**, 106 ⟨71CB873⟩
Bresson, C. R., **2**, 130 ⟨67JOC282⟩
Brestensky, D. M., **1**, 86 ⟨88JA291⟩; **3**, 12 ⟨89TL5677⟩, 12, 128 ⟨88JA291⟩, 12, 128 ⟨88TL3749⟩
Brestkin, A. P., **2**, 1033 ⟨91RCR885⟩
Bretinger, H., **4**, 698 ⟨85ZN(B)182⟩
Breton, G. C. Le, **2**, 374 ⟨81JOC5413⟩
Breton, G. W., **2**, 4, 23, 30 ⟨93JA3071⟩
Bretschneider, T., **4**, 99 ⟨88T5403⟩
Bretski, M. M., **5**, 1068, 1069 ⟨89ZOR759⟩
Brett, D., **1**, 185, 200 ⟨67JOC855⟩; **3**, 613 ⟨67JOC855⟩
Brettle, R., **1**, 503 ⟨83JCS(P1)387⟩; **5**, 288 ⟨85JCS(P1)831, 88JCS(P1)2185⟩; **6**, 469 ⟨89JCR(S)15⟩
Brettschneider, H., **1**, 922 ⟨70CB1101⟩
Breu, R., **4**, 687 ⟨57CB1259⟩
Breuer, E., **1**, 751 ⟨78T997⟩; **2**, 305 ⟨64JA3565⟩, 342 ⟨B-82MI 206-01⟩; **5**, 22 ⟨91PS(60)239⟩, 371, 403 ⟨90JCS(P1)3263, 92TL2067, 91JOC4791⟩, 403 ⟨92PS(70)109⟩, 403, 404 ⟨88JCS(P2)2029, 89JCS(P1)765⟩
Breuer, F., **5**, 595 ⟨33JA1262⟩
Breuer, S. W., **2**, 653 ⟨71CC1475, 75CC36⟩; **4**, 304 ⟨67T2869⟩
Breuilles, P., **1**, 231 ⟨90TL357⟩
Breunig, H., **2**, 893 ⟨83JOM(253)C21, 88POL2601⟩
Breunig, H. G., **2**, 501 ⟨88ZN(B)998⟩
Breunig, H. J., **2**, 483 ⟨75JOM(87)83, 86OCS638⟩, 489 ⟨82PS(12)149⟩, 489, 490 ⟨89JOM(364)67⟩, 491 ⟨83JOM(243)417⟩, 493 ⟨82ZN(B)1104, 86ZN(B)1129⟩, 493, 495, 499, 500 ⟨81ZN(B)1105⟩, 499 ⟨88OCS587⟩, 499, 501 ⟨88PS(38)97⟩, 505 ⟨82AG(E)439, 83ZN(B)125, 86OCS625, 86OCS636⟩, 505, 506 ⟨77CZ399⟩, 507 ⟨78CZ263, 83ZAAC(502)175⟩, 507 ⟨86OCS621⟩, 509 ⟨82ZN(B)395⟩, 890, 891 ⟨84JOM(277)257⟩, 893, 894 ⟨86ZN(B)321, 89PS(44)129⟩; **6**, 168, 192 ⟨83POL291⟩, 169 ⟨92OM145⟩
Breunig-Lyriti, V., **2**, 505, 506 ⟨77CZ399⟩
Brevnova, T. N., **5**, 1076, 1077 ⟨87JOM(327)303, 85IZV2824⟩, 1081 ⟨87ZOB405⟩
Brew, S. A., **3**, 690 ⟨93AOC(35)135⟩

Brewbaker, J., **3**, 467 ⟨69JA706⟩
Brewer, D., **3**, 712 ⟨85T1931⟩
Brewer, G. A., **5**, 325 ⟨81IC1823⟩
Brewer, J. D., **1**, 731 ⟨72AJC545⟩; **3**, 102 ⟨71AJC1883⟩
Brewer, P. D., **1**, 1094 ⟨77TL4145⟩
Brewer, R., **1**, 1114 ⟨73JOC3054⟩
Brewer, T. L., **4**, 698 ⟨70JOM(25)C6⟩
Brewis, S., **2**, 445 ⟨65JCS1539⟩
Brewster, A. G., **2**, 690 ⟨81JCS(P1)1473⟩; **3**, 88 ⟨78CC952⟩
Brewster, D., **5**, 142 ⟨73JCS(P1)2796⟩
Brewster, J. H., **1**, 398 ⟨52JA5179⟩; **2**, 89 ⟨91COS(8)211⟩, 302 ⟨53OR(7)99⟩; **5**, 123, 126, 302 ⟨55JA6214⟩
Brewster, R. Q., **1**, 442 ⟨39JA2418⟩; **5**, 1023 ⟨22MI 527-01, 32OSC(1)447⟩
Brey, Jr., W. S., **5**, 144 ⟨63JOC3512⟩
Breyer, R. A., **1**, 326, 327 ⟨91JA7788⟩, 326, 327 ⟨92JA7664⟩
Breyer, R., **1**, 327, 328 ⟨91JA7002⟩
Brezinsky, K., **1**, 1127 ⟨92JPC2151⟩
Briand, S., **1**, 883 ⟨90JOC1052⟩
Brianese, N., **3**, 442 ⟨87ICA275⟩; **6**, 519 ⟨87JCS(D)2039⟩
Bricas, E., **5**, 84 ⟨61BSF2001⟩
Brice, T. J., **4**, 737 ⟨53JA2698⟩; **6**, 218 ⟨46JA968, 49JA2499⟩, 241 ⟨62JOC1064⟩
Brice, W. F., **2**, 274 ⟨58JOC1449⟩
Bricker, D. L., **6**, 375 ⟨85JA8211⟩
Brickmann, K., **1**, 406 ⟨93CB1227⟩; **4**, 231 ⟨93CB1227⟩
Brickner, W., **1**, 380 ⟨1899CB2307⟩
Briddle, H. C., **5**, 825 ⟨00LA(310)11⟩
Bridge, A. W., **1**, 837 ⟨93T5177⟩
Bridger, H. J., **2**, 289 ⟨50JCS1371⟩; **3**, 382 ⟨50JCS1371⟩
Bridges, A. J., **1**, 142, 599 ⟨75JOC2014⟩, 598 ⟨76JA5017⟩, 640, 641, 753 ⟨74CC142⟩; **3**, 90, 96 ⟨92TL7499⟩, 165 ⟨80JA3548⟩, 233 ⟨83TL447⟩, 242 ⟨76JA5017⟩, 250 ⟨83CC485⟩, 310 ⟨90JHC1527⟩
Bridgewater, R. J., **2**, 25 ⟨53JCS1709⟩
Bridon, D., **1**, 935 ⟨87T2733⟩; **2**, 364 ⟨86TL1153⟩, 700 ⟨88TL1153⟩; **5**, 1115 ⟨86T4983⟩
Bridon, D. P., **1**, 925, 935 ⟨89JOC6101⟩; **2**, 13 ⟨89JOC6101, 89TL2461⟩
Bridson, J. N., **1**, 486 ⟨73JOC2574⟩
Brieden, W., **4**, 206 ⟨92AG(E)1335⟩
Brieger, G., **1**, 85, 457 ⟨74CRV567⟩, 881 ⟨70CC1325⟩
Briehl, D. L. H., **3**, 703 ⟨81JOC1046⟩
Briehl, H., **3**, 582 ⟨88JA1337⟩; **5**, 1086, 1087, 1092 ⟨83JA7194⟩
Briel, D., **5**, 622 ⟨85ZC364⟩
Briel, H., **3**, 686 ⟨79AG(E)467⟩
Brielmann, H. L., **5**, 329 ⟨94JOC5292⟩
Brieskorn, C. H., **5**, 1048 ⟨70AP(303)905⟩
Brieva, R., **5**, 263 ⟨93T4007⟩
Brigaud, T., **4**, 62 ⟨90TL2287⟩
Briggs, A. G., **3**, 356, 377 ⟨79JCS(P1)2340⟩
Briggs, D. A., **4**, 586 ⟨87OM1992⟩
Briggs, E. M., **4**, 681 ⟨67JCS(C)1862⟩
Briggs, L. H., **5**, 764 ⟨76AJC357⟩
Briggs, T. N., **5**, 387, 388 ⟨86JCS(D)813⟩
Bright, D. A., **1**, 110 ⟨82JOC3521⟩; **6**, 425 ⟨82JOC3521, 86JOC1180⟩
Bright, D., **3**, 460 ⟨67CC245⟩
Bright, S. T., **1**, 185 ⟨90JOC1338⟩
Brigot, D., **2**, 458 ⟨79T1345⟩
Briks, Yu. L., **3**, 355 ⟨91KGS1432⟩
Brilkina, T. G., **2**, 886 ⟨85JGU63⟩
Brill, G., **3**, 75, 265 ⟨89CB83⟩; **4**, 950 ⟨87LA311⟩
Brill, W. F., **2**, 96 ⟨65JA3286⟩; **5**, 24 ⟨72MI 502-01⟩
Brill, W. K. D., **2**, 203 ⟨91SL163⟩, 253 ⟨89JA2321⟩
Brillon, D., **2**, 750 ⟨90JOC2246⟩; **3**, 361 ⟨90SC3085⟩; **5**, 565 ⟨92SR297⟩, 568, 569 ⟨90SC3085⟩, 582, 611 ⟨92SC1397⟩
Brimble, M. A., **1**, 1045 ⟨92JOC5818⟩; **2**, 27 ⟨90TL7509⟩
Brimelow, H. C., **5**, 114, 115 ⟨51JCS1208⟩
Brimhall, R. S., **2**, 254, 255 ⟨84EUP118109⟩
Brimm, E. O., **6**, 523 ⟨54JA3831⟩
Brinckman, F. E., **2**, 904 ⟨60JA6218⟩
Brindle, J. R., **3**, 620 ⟨71CJC2990⟩
Brindley, J. C., **5**, 655 ⟨87JCS(P1)1153⟩; **6**, 572 ⟨87JCS(P1)1153⟩
Brindley, P. B., **2**, 111 ⟨56JCS824⟩
Bringewski, F., **2**, 431 ⟨90CZ211⟩
Bringhen, A. O., **4**, 97 ⟨89HCA93⟩
Bringmann, G., **1**, 38, 945 ⟨79TL2291, 80S68⟩, 472 ⟨90AG(E)977, 80JCS(P1)2657⟩, 945 ⟨80JCS(P1)2665⟩; **2**, 7 ⟨83S139⟩, 310 ⟨90LA795⟩, 331 ⟨89S608⟩; **3**, 41 ⟨89S608⟩, 422 ⟨91COS(3)744⟩; **5**, 1054 ⟨80JCS(P1)2657⟩
Brink, A. J., **3**, 713, 717 ⟨74MI 321-01⟩
Brink, M., **1**, 232 ⟨75S807⟩; **3**, 57 ⟨75S253⟩
Brinker, U. H., **1**, 117, 1092 ⟨83CB882⟩, 990 ⟨85CC1812⟩; **4**, 737 ⟨92TL4537⟩
Brinkhaus, K-H. G., **6**, 77, 78 ⟨86T553⟩
Brinkman, G. A., **2**, 613 ⟨84CC655, 86JOC1886⟩; **6**, 218 ⟨79MI 607-01⟩, 414 ⟨78MI 614-02⟩
Brinkmann, A., **1**, 363 ⟨89LA739⟩; **5**, 343 ⟨89LA739⟩
Brinkmann, M., **6**, 435, 546 ⟨81LA1244⟩
Brinkmann, R., **2**, 559, 959 ⟨83AG(E)728⟩
Brinkmeyer, R. S., **1**, 197 ⟨78CC876, 80JOC4740⟩; **2**, 940 ⟨80JOC4740⟩; **3**, 90, 97 ⟨74OS(54)42⟩, 264 ⟨79T1675⟩; **4**, 306 ⟨88OSC(6)64⟩; **5**, 746 ⟨79T1675⟩; **6**, 104 ⟨79T1675⟩
Brinnard, M. E., **2**, 484, 492, 497, 888 ⟨32JCS1815⟩
Brinton, R. K., **1**, 1143 ⟨68JPC321⟩
Brintzinger, H., **2**, 167 ⟨50CB87, 53CB557, 54CB325⟩; **4**, 71 ⟨50CB87, 54CB300⟩; **5**, 821 ⟨52CB344⟩; **6**, 233 ⟨53CB557⟩
Brintzinger, H. H., **4**, 689, 690 ⟨77JOM(127)93⟩
Brion, F., **1**, 70 ⟨78AG(E)690⟩, 954 ⟨79AG(E)688⟩
Brion, J.-D., **5**, 615 ⟨93H(36)1979⟩, 743 ⟨76BSF301⟩, 827 ⟨83S220⟩
Bris, M.-T. Le, **6**, 653 ⟨62MI 621-01⟩
Brisbois, R. G., **2**, 668 ⟨90JA3093⟩; **3**, 193, 473 ⟨90JOC1959⟩
Briscoe, H. V. A., **2**, 171 ⟨29JCS1048⟩, 1014, 1017 ⟨38JCS1358⟩
Briscoe, M. W., **1**, 570 ⟨94JCS(P1)3115⟩
Briscoe, P. A., **2**, 366 ⟨56JCS1755⟩
Brisdon, B. J., **1**, 493 ⟨81T173⟩; **2**, 421 ⟨62JCS1330⟩
Brister, A., **2**, 4 ⟨93S973⟩
Bristow, G. S., **5**, 443 ⟨84JCS(D)399⟩
Britcher, S. F., **4**, 97 ⟨79JOC3117, 88EUP264183⟩
Britelli, D. R., **3**, 495 ⟨83JOC3531⟩
Brittain, J., **2**, 541 ⟨93TL3363⟩
Brittain, J. M., **3**, 106, 308 ⟨82SC231⟩
Brittain, W. J., **6**, 51 ⟨78JOC1346⟩
Britten, J. F., **5**, 336 ⟨94JOC806⟩, 388 ⟨92MI 507-01⟩, 389 ⟨92JOM(423)C5⟩
Britten-Kelly, M. R., **1**, 240 ⟨81JOC5027⟩, 685, 686 ⟨75CC539, 76JCS(P1)2079⟩; **2**, 284 ⟨78JCS(P1)1682⟩, 858 ⟨75CC539, 76JCS(P1)2079⟩; **3**, 391 ⟨75CC539⟩, 391, 527 ⟨76JCS(P1)2079⟩; **5**, 1053 ⟨76JCS(P1)2079⟩
Britton, D., **2**, 1002 ⟨74AX(B)147⟩; **4**, 694 ⟨87OM1583⟩
Britton, T. C., **2**, 375 ⟨86JA6395a⟩
Brix, H., **4**, 1019 ⟨76JOM(121)211⟩
Brizzolara, A., **1**, 307 ⟨63JA207⟩; **2**, 738, 758 ⟨63JA207⟩
Brlik, J., **5**, 567 ⟨93JOC6076⟩
Broach, V., **1**, 635 ⟨B-80MI 119-06⟩
Broadbent, H. S., **1**, 21 ⟨63JOC2347⟩, 1037 ⟨75JOC1131⟩
Broaddus, C. D., **1**, 795, 882 ⟨68ACR231⟩; **2**, 554 ⟨70JOC10⟩
Broadhurst, M. D., **4**, 824 ⟨85JOC1117⟩
Broadhurst, M. J., **4**, 303 ⟨82JCS(P1)2227⟩
Brocard, J., **1**, 816, 818 ⟨89TL2549⟩
Brocas, J. M., **2**, 816 ⟨76JOM(120)217⟩
Brock, C. P., **6**, 338 ⟨85JA8136⟩, 456 ⟨87JA6015⟩
Brockelbank, E. L., **2**, 179 ⟨66JOC2357⟩
Brockman, F. J., **4**, 434, 440 ⟨49MI 409-01⟩

Brockman, J. A., **2**, 214 ⟨54JA1828⟩
Brockmann, M., **2**, 813 ⟨86JOM(314)75⟩
Brockmann, W., **4**, 408 ⟨51CB834⟩
Brocksom, T. J., **1**, 187 ⟨88TL5661⟩, 414, 416, 827 ⟨70JA741⟩; **5**, 159 ⟨70JA741⟩, 199 ⟨82JOC3450⟩
Brockson, T. J., **5**, 903 ⟨77JOM(124)1⟩
Brockway, L. O., **3**, 694 ⟨36JA2516⟩
Brocquet, C., **1**, 181 ⟨62BSF1882⟩
Broda, W., **5**, 1078 ⟨85IJ219⟩; **6**, 32 ⟨82CB3894⟩, 554, 555 ⟨83LA1839⟩
Broddus, C. D., **2**, 206 ⟨74JOC2459⟩
Brodelius, P., **5**, 117 ⟨B-84MI 502-01⟩, 118 ⟨81MI 502-07, 82MI 502-01⟩
Brodersen, K., **2**, 817 ⟨64CB1155⟩; **5**, 384, 387 ⟨70CB2388⟩
Brodie, B. B., **2**, 650 ⟨54JBC(208)731⟩
Brodskaya, E. I., **3**, 483 ⟨77JOM(142)155⟩; **4**, 628 ⟨76IZV1887⟩; **5**, 391, 718 ⟨87ZOB1199, 87JGU1073⟩; **6**, 185 ⟨77JOM(142)155⟩
Brodsky, J. L., **5**, 690 ⟨88JOC3852⟩
Brodsky, L., **1**, 736 ⟨74JOC1318⟩
Broekema, R. J., **6**, 74 ⟨75RTC209⟩
Broekhof, N. L. J. M., **2**, 739 ⟨79TL2433, 80TL2671, 84RTC305⟩; **3**, 11 ⟨79TL2433, 80TL2671⟩, 182 ⟨81TL2799⟩; **4**, 64 ⟨77TL885⟩, 210 ⟨76TL1725⟩; **6**, 48, 120 ⟨77TL885⟩
Broekhuysen, M. P., **3**, 54, 253 ⟨82TL4831⟩
Broekma, R. J., **4**, 827 ⟨75RTC209⟩
Broenneke, A., **4**, 382 ⟨82S1045, 85CB2822⟩
Broggini, G., **5**, 677 ⟨92OPP352⟩
Broicher, V., **4**, 137 ⟨90JOM(381)315⟩, 160 ⟨89TL5243⟩
Brois, S. J., **2**, 220 ⟨70JA7629⟩; **5**, 247 ⟨70JA7629⟩
Broka, C. A., **1**, 216 ⟨83JOC2499⟩, 405 ⟨89JA2981⟩, 816 ⟨87TL4993⟩; **2**, 337 ⟨86JOC5045, 88JOC2144⟩; **4**, 380 ⟨89JA2981⟩
Broka, G. A., **2**, 556 ⟨89JA2981⟩
Brokatzky, J., **3**, 315 ⟨80AG(E)47, 80TL4909⟩
Brokatzky-Geiger, J., **3**, 315 ⟨82TL4665, 83CB2383⟩, 315 ⟨84CB2157, 84TL1137⟩
Brokaw, F. C., **1**, 710 ⟨79JOC2880⟩
Brokhovetskii, D. B., **5**, 363 ⟨91T447⟩, 666 ⟨90IZV1692⟩; **6**, 29 ⟨89CS81⟩
Broline, B. M., **1**, 563 ⟨81JOC1874⟩
Brombach, H., **4**, 568, 573 ⟨92CB771⟩
Bromidge, S. M., **3**, 225 ⟨90JCS(P1)353⟩, 324 ⟨86JCS(P1)1725⟩
Bromley, D., **2**, 626 ⟨69TL1623, 72JOC88⟩
Bromm, D., **2**, 1085 ⟨87CB1437⟩; **6**, 183 ⟨91CB2645⟩, 184 ⟨88JOM(347)11⟩, 184 ⟨90CC741⟩, 184 ⟨92CB1807⟩
Broms, M., **1**, 469 ⟨83ACS(B)173⟩
Brondani, D. J., **1**, 160 ⟨93JOM(451)C1⟩, 160 ⟨93TL2111⟩; **4**, 606 ⟨93JOM(451)C1⟩, 607, 608, 654 ⟨93TL2111⟩
Brondum, K., **5**, 1066 ⟨88JOC4654⟩
Bronk, B. S., **2**, 915 ⟨90TL6621⟩; **4**, 700 ⟨91OM936⟩
Bronold, N. M., **5**, 478, 491, 492, 562 ⟨77RTC139⟩
Bronson, J. J., **1**, 400 ⟨91COS(5)999⟩, 803 ⟨85JA4579, 91COC(5)999⟩
Brook, A. G., **1**, 491 ⟨76JOM(113)C11⟩, 652 ⟨76JOM(121)293⟩, 759 ⟨70CJC561⟩; **2**, 700 ⟨75JOM(90)C6⟩, 813 ⟨78JOM(152)C25⟩, 914, 921, 922 ⟨76JOM(121)293⟩, 936 ⟨B-66MI 218-03⟩; **3**, 271 ⟨73CJC2024⟩, 483 ⟨69CJC4353⟩, 501, 502 ⟨86AOC(25)1⟩; **4**, 137 ⟨75CJC332⟩, 137, 138, 141, 811 ⟨76JOM(122)31⟩, 138, 141, 646, 651, 652, 654, 663 ⟨70CJC561⟩, 222, 370, 371, 570 ⟨68CJC2115⟩, 506 ⟨77CJC2589⟩, 609 ⟨68CJC2119⟩, 613 ⟨91OM3292⟩, 647, 652, 663 ⟨70JA7567⟩, 930 ⟨79JA83⟩, 1014 ⟨78CJC2286⟩; **5**, 409, 412, 421, 431 ⟨68AOC(7)95⟩, 410 ⟨63CJC2351, 65CJC1175⟩, 410, 411 ⟨75JOM(87)263⟩, 410, 417 ⟨57JA4373⟩, 410, 431 ⟨67JA431⟩, 412 ⟨65JOC2566⟩, 413 ⟨74ACR77⟩, 418, 923 ⟨86AOC(25)1⟩, 418, 924 ⟨82OM994⟩, 418, 924, 925 ⟨89OM693⟩, 421 ⟨75JOM(99)379⟩, 431 ⟨60JA5102, 66JOM(5)106⟩, 792, 796 ⟨78CJC2286⟩, 923, 924 ⟨81CC191, 85AX(C)1632, 92OM3088, 76JA3381⟩, 924 ⟨79JA6750, 79JA83, 82JA5667, 87OM1246, 87OM2128, 88JOM(341)C9, 88OM2245, 89OM850, 91OM2752, 91OM2758⟩, 925 ⟨85JA4338, 89OM709⟩; **6**, 124 ⟨69CJC4347⟩, 125, 131, 132, 519 ⟨68CJC2119⟩, 127, 132, 325, 326, 520 ⟨67JA431⟩, 192 ⟨70CJC561⟩, 515 ⟨91JOM(403)293⟩
Brook, M. A., **2**, 313 ⟨88SC893⟩; **4**, 138 ⟨89CC957⟩
Brook, M., **2**, 356 ⟨85HCA319⟩
Brooke, G. M., **1**, 426 ⟨62JCS3253⟩
Brooker, L. G. S., **6**, 70 ⟨35JA2480⟩
Brookes, C. J., **6**, 20 ⟨74CC323, 78JCS(P1)202⟩
Brookes, M. H., **2**, 225 ⟨83CC1051, 88JCS(P1)9⟩
Brookes, R. F., **6**, 445 ⟨75JCS(P1)1836⟩
Brookhart, M. A.,**5**, 937 ⟨91JA927⟩
Brookhart, M., **2**, 912 ⟨93JA2151⟩; **3**, 519 ⟨83JA258, 87CR411, B-89MI 315-02⟩, 520 ⟨B-89MI 315-01⟩; **5**, 448 ⟨82OM1730, 83JA258⟩, 937 ⟨85OM943⟩, 942 ⟨87CRV411⟩
Brooks, B. T., **2**, 40 ⟨18JA822⟩
Brooks, D. W., **1**, 1042 ⟨88TL3455⟩; **2**, 7, 10 ⟨87TL767⟩, 343 ⟨91TL3783⟩, 343, 351 ⟨88TL3455⟩, 532, 697 ⟨79TL1665⟩
Brooks, E. H., **2**, 867 ⟨65JCS4283⟩
Brooks, G., **4**, 205 ⟨76TL3739⟩, 824 ⟨79TL391⟩
Brooks, H. G., **1**, 310 ⟨52JA4582⟩; **4**, 651 ⟨54JOC1278⟩; **6**, 192 ⟨52JA4582⟩, 192 ⟨54JOC1278⟩
Brooks, L. A., **2**, 30 ⟨43OSC(2)171⟩; **4**, 407 ⟨42OS(22)65⟩; **5**, 31 ⟨43OSC(2)531⟩
Brooks, P., **2**, 432 ⟨87JOM(323)C1⟩
Brooks, P. R., **1**, 685, 686 ⟨93JOC5900⟩
Brooks, R. J., **2**, 845 ⟨75JOC2059⟩
Brooks, S. G., **4**, 308 ⟨75JCS(P1)1670⟩
Brooks, T. W., **1**, 726 ⟨63JOC372⟩; **3**, 494 ⟨63JOC372⟩
Broom, A. D., **4**, 893 ⟨77JOC4159⟩
Broom, N. J. P., **2**, 664 ⟨78CC162⟩
Broos, R., **1**, 647, 738, 740, 742 ⟨88BSB271⟩
Broquet, C., **3**, 603 ⟨75T1331⟩
Brosa, C., **3**, 226 ⟨89JOC3984⟩
Broschard, R. B., **1**, 34 ⟨58JA1654⟩
Broser, E., **3**, 539 ⟨88LA943⟩
Brosig, H., **3**, 702 ⟨91TL5421⟩
Brossi, A., **3**, 294 ⟨71JA4321⟩, 353 ⟨90JMC567, 90SC637⟩
Brost, R. D., **4**, 685 ⟨89CC498⟩
Brothers, D., **3**, 207 ⟨82JOC1598⟩
Brothers, P. J., **2**, 578 ⟨86ACR209⟩; **6**, 247 ⟨90JOM(394)615⟩, 716, 717, 718, 722, 727 ⟨88CRV1293⟩
Brotherton, R. J., **2**, 418, 419 ⟨60JA6242⟩
Brotherton, T. K., **5**, 847 ⟨59CRV841⟩, 997 ⟨65JOC1260⟩
Brougham, P., **2**, 147, 156 ⟨87S1015⟩
Broughton, B. J., **4**, 866 ⟨86EUP174832A2⟩
Brouillard-Poichet, A., **5**, 330 ⟨89TL7061⟩, 330, 332 ⟨92HCA109⟩
Broussier, R., **4**, 720 ⟨89JOM(362)C27, 90IC1817⟩, 720 ⟨92JOM(427)231⟩
Brouwer, A. C., **3**, 696, 711 ⟨86SC865⟩; **4**, 517 ⟨86SC865⟩; **5**, 559 ⟨78TL4839, 83RTC103, 83RTC83, 83RTC91⟩, 559, 605, 614 ⟨76TL209⟩
Brouwer, A. M., **1**, 420 ⟨87RTC613⟩
Brouwer, D. M., **1**, 893 ⟨64PCS147⟩, 901 ⟨72MI 119-13⟩, 914 ⟨B-70MI 119-05⟩
Brouyere, C., **2**, 624 ⟨71C59⟩
Brovarets, V. S., **4**, 99 ⟨88JGU2192⟩, 771 ⟨91ZOB874⟩, 998 ⟨92ZOB1084, 92ZOB1423⟩, 999 ⟨83ZOB2015, 84ZOB288⟩, 999, 1000 ⟨86ZOB321⟩, 999, 1001, 1002, 1007, 1011 ⟨84ZOB2186⟩, 1005 ⟨92ZOB707⟩; **5**, 655 ⟨86JGU2483, 86ZOB2802, 88JGU2192, 88ZOB2465⟩, 655, 656 ⟨84JGU255, 84ZOB288⟩
Brovko, V. S., **4**, 569 ⟨83ZOB1831⟩
Brower, K. R., **2**, 15 ⟨87JOC798⟩

Brown, A. B., **1**, 1142 ⟨89JA3640⟩
Brown, A. C., **1**, 617 ⟨85JOC1749⟩; **4**, 419 ⟨91TL2683⟩; **5**, 291 ⟨68JA4706⟩
Brown, A. K., **6**, 241 ⟨75JCS(P1)2033, 81JFC(17)85⟩
Brown, B. A., **1**, 889 ⟨65JOC1291⟩
Brown, B. B., **1**, 489 ⟨90TL6509⟩, 788 ⟨93JA2268⟩; **2**, 900, 904 ⟨93JA2268⟩
Brown, B. R., **1**, 772 ⟨51QR131⟩; **4**, 13, 23 ⟨51JCS1145⟩; **5**, 259 ⟨B-94MI 506-01⟩; **6**, 28 ⟨51JCS1145⟩
Brown, C., **1**, 732 ⟨74JCS(P1)2435⟩; **2**, 835 ⟨70JCS(C)878⟩; **5**, 405 ⟨73PS287⟩, 407 ⟨70JCS(C)878⟩
Brown, C. A., **1**, 308 ⟨79CC100, 82H(18)187⟩, 503 ⟨73JOC2226⟩, 506 ⟨69JA5901⟩, 509, 1101 ⟨79JOC2328⟩, 590, 1072, 1085 ⟨78JOC3083⟩, 795 ⟨87OS(65)224, 75JA891⟩, 1085 ⟨76CC959, 77CC318⟩, 1085 ⟨81TL4171⟩; **2**, 423 ⟨73JA982, 87OS(65)224⟩, 423, 696 ⟨74JOC3913⟩, 534 ⟨85JOC549, 86JOC238⟩; **4**, 846 ⟨84TL995⟩; **6**, 128 ⟨79CC100⟩, 317 ⟨84TL991⟩
Brown, C. H., **1**, 1049 ⟨75TL4217⟩, 1050 ⟨76TL235⟩
Brown, C. K., **5**, 461 ⟨71JCS(A)850⟩
Brown, C. N., **6**, 560 ⟨65JMC174⟩
Brown, D., **3**, 229 ⟨87TL1069⟩; **4**, 129 ⟨85JCR(S)92⟩, 131 ⟨87JCS(P1)181⟩; **5**, 130 ⟨53JCS3352⟩; **6**, 121 ⟨87JCS(P1)181⟩
Brown, D. A., **5**, 322 ⟨85SC1159⟩
Brown, D. H., **2**, 465 ⟨77JCS(D)159⟩; **4**, 552 ⟨80JCS(D)871⟩
Brown, D. J., **1**, 90 ⟨62CHE430⟩
Brown, D. M., **2**, 12 ⟨67JCS(C)252⟩, 105 ⟨B-63MI 202-01⟩
Brown, D. S., **1**, 929 ⟨92T6883⟩, 1031 ⟨89T4293, 90SL48⟩; **2**, 959 ⟨77JOM(127)127⟩; **3**, 186 ⟨90SL619⟩; **4**, 200 ⟨91T1329⟩; **6**, 544, 557 ⟨92T6883⟩
Brown, D. W., **3**, 669, 670 ⟨83T3207⟩; **5**, 573 ⟨83T1075, 87TL2171⟩; **6**, 237 ⟨92OM2947, 93CM1321⟩
Brown, E. A., **6**, 107 ⟨66JOC4054⟩
Brown, E., **1**, 412 ⟨66TL6115⟩, 473 ⟨77TL2015⟩; **2**, 64 ⟨92TA841⟩; **5**, 183 ⟨83TL4195⟩
Brown, E. D., **2**, 128 ⟨66JCS(C)415⟩
Brown, E. G., **1**, 835 ⟨88S362⟩; **6**, 468 ⟨82JOC4013⟩
Brown, E. I. G., **3**, 254, 371, 376 ⟨70JCS(C)1202⟩
Brown, E. V., **2**, 776 ⟨67JOC1124⟩; **5**, 276, 599 ⟨75S358⟩
Brown, F. J., **3**, 508 ⟨80MI 315-01⟩; **5**, 342 ⟨90JMC2621⟩, 934 ⟨80MI 525-01⟩; **6**, 716, 719, 721, 722 ⟨80MI 622-01⟩
Brown, F. M., **2**, 304 ⟨88TL6393⟩
Brown, F. W., **3**, 612 ⟨33JA4299⟩
Brown, G. B., **2**, 126, 214 ⟨57MI 203-01⟩
Brown, G. M., **4**, 123 ⟨88IC1787⟩; **6**, 641, 642, 643 ⟨58CJC1541⟩
Brown, G. R., **1**, 47 ⟨82S1036⟩, 571 ⟨92JA6330⟩
Brown, G. W., **2**, 6 ⟨B-71MI 201-01⟩
Brown, H., **2**, 520, 900, 901 ⟨83TL1433⟩; **3**, 3, 114 ⟨83TL1433⟩
Brown, H. C., **1**, 3 ⟨80JOC849⟩, 4, 7 ⟨66JA1473⟩, 4, 7, 8, 10 ⟨82JOC276⟩, 4, 7, 9, 10, 17 ⟨83JOC3085⟩, 14 ⟨68CC1549, 68JA2927, 80JOC1, 83JOC3091⟩, 58 ⟨61JOC631, 83JOM(255)135⟩, 58, 59 ⟨86T5497⟩, 60 ⟨87JOC4014⟩, 67 ⟨81JOC531⟩, 75, 85 ⟨59JA4108⟩, 75 ⟨B-88MI 103-01⟩, 82 ⟨68IJ691⟩, 125 ⟨69JA2147, 69JA6854⟩, 125, 280 ⟨68JA818⟩, 126 ⟨69JA2150, 70JA5790⟩, 273 ⟨67JA5708, 67JA5709, 67JA5710⟩, 274 ⟨68JA4166, 71JA1506, 71JA1508⟩, 274, 546 ⟨70JA3503⟩, 274, 946 ⟨68JA4165⟩, 277 ⟨71JA2792⟩, 277, 977 ⟨73S305⟩, 279, 1033 ⟨73JA3080⟩, 280 ⟨68JA1911⟩, 333, 946 ⟨71JA3777, 72AG(E)692⟩, 428 ⟨69JOC3918⟩, 443 ⟨89JA384, 91JOC1543⟩, 449 ⟨75JOC2845, 89JOC6064⟩, 449, 509 ⟨59JA1512⟩, 449, 510 ⟨86JOC4512⟩, 449 ⟨B-75MI 110-01, B-88MI 110-01⟩, 464 ⟨69JA6852⟩, 485 ⟨83OM779⟩, 486 ⟨73JOC2574, 91SL684⟩, 487 ⟨91TA277⟩, 487, 488 ⟨88JOC239, 88JOC6009, 76JOC1785⟩, 488 ⟨86JOC3398⟩, 488, 1094 ⟨80JOC549⟩, 488, 510, 1102, 1103 ⟨80JOC389⟩, 488, 510, 531, 1103 ⟨73JOC1617⟩, 509 ⟨82JOM(225)63, B-88MI 112-02⟩, 509, 1100, 1102 ⟨61JA3834⟩, 509, 510 ⟨83TL1433, B-88MI 112-01⟩, 510 ⟨83JOC5080⟩, 510 ⟨83JOC644⟩, 510, 1101 ⟨72JA4370⟩, 510, 1101, 1103 ⟨79JA96⟩, 531 ⟨82JOM(225)C1⟩, 531, 1103 ⟨77JOC579⟩, 531, 662 ⟨82JOC171, 82JOC3806, 82JOC5407⟩, 531, 663 ⟨82JOC1792⟩, 532, 1094 ⟨80JOC550⟩, 657 ⟨72JA6549, 81JOC647⟩, 657, 658 ⟨67JA567, 64JA1801⟩, 658 ⟨66JA1440⟩, 658, 659 ⟨68JA2906, 68JA4439⟩, 660, 1103 ⟨73JA6456⟩, 660, 661 ⟨88TL21, 67JA4531⟩, 661, 1103 ⟨73JA5786⟩, 662 ⟨82JOC754⟩, 831 ⟨89JA3441⟩, 838 ⟨77JOM(132)9, 90TL7115⟩, 860, 861 ⟨51JA3308⟩, 895 ⟨73JA2400, 77JA1712⟩, 905 ⟨78JOC3667, 79JOC133, 79JOC3536, 80JOC2113, 83ACR432, 85JOC1165, 85JOC826⟩, 906 ⟨82JOC2089⟩, 946 ⟨70JA712, 70JA714, 71JA1024, 86T5523⟩, 999 ⟨88TL2631⟩, 1018 ⟨81JOC645⟩, 1033, 1034 ⟨86JOC4507, 86JOC4521⟩, 1034 ⟨88JA1529, 91JOC1500⟩, 1035, 1103 ⟨76JOC1078⟩, 1037 ⟨85JOC1577⟩, 1043 ⟨79S104⟩, 1045 ⟨84TL2411⟩, 1050 ⟨77JA954⟩, 1055 ⟨93TL15⟩, 1094, 1100, 1102 ⟨81MI 122-01⟩, 1094, 1102 ⟨76JA1785⟩, 1095 ⟨72JA4371, 72JOM(36)1⟩, 1100 ⟨63JA2063⟩, 1101 ⟨75JA5249, 85TL2187, 56JA2173⟩, 1102 ⟨76JA1798⟩, 1103 ⟨72S555, 77JA7097⟩, ⟨79TCC19⟩ ⟨79TCC1⟩; **2**, 23 ⟨39JA3432, 88T2785⟩, 30 ⟨70JA6660, 88T2763, 88T2773⟩, 34 ⟨68JA5038, 71JA1508, 76JA1290, 76S114, 88T2751⟩, 40 ⟨56JA5694, 57JOC1136, 59JA4108, 60JA3222, 63OR(13)1, 67JA1522, B-62MI 202-01⟩, 41 ⟨73JOC182, 73JOC4092⟩, 42 ⟨88BCJ93, 88JOC2911, 94TL4715⟩, 42, 57 ⟨64JA393⟩, 42, 57, 519 ⟨64JA397, B-83MI 202-01⟩, 57 ⟨88JOC1231, 90JOC6328⟩, 58 ⟨88JA1549, 89JOC1577, 89JOC4504, 94TA1075⟩, 59 ⟨53JA6263⟩, 60 ⟨66JA1464⟩, 72 ⟨89TL5551⟩, 74 ⟨90JOC1868⟩, 75 ⟨89JOC1570, 90T4463⟩, 75, 521 ⟨88JA1535, 90JA2389, 90T4457⟩, 83 ⟨68JA2686, 73JA8486⟩, 83 ⟨76JA3383⟩, 89 ⟨69JA5646⟩, 97 ⟨87T4059⟩, 109 ⟨53JA192, 53JA199, 53JA205, 60JA3866⟩, 109, 111 ⟨39JA673⟩, 109, 532 ⟨56JA3613⟩, 126 ⟨56JA2582, 66JA1458⟩, 126, 708 ⟨84S498⟩, 138, 712 ⟨71JA3291⟩, 268 ⟨84JA8001⟩, 305 ⟨64JA3565, 87T4079⟩, 306, 318 ⟨91JOC1170⟩, 320 ⟨81S441⟩, 428 ⟨52USP2584112⟩, 514 ⟨60JA4708, 85TL2187⟩, 514, 515, 519 ⟨82JOC5065⟩, 514, 516, 520 ⟨B-75MI 210-01⟩, 514, 518, 519, 520, 523, 528, 529, 531, 533 ⟨B-88MI 210-01⟩, 514, 524 ⟨B-72MI 210-01⟩, 514, 900 ⟨82COMC-I(7)111⟩, 514 ⟨B-62MI 210-01, B-83MI 210-02, B-86MI 210-01⟩, 515 ⟨75JA2791, 77JOM(135)C10, 78S146, 79JOC465, 84JOC945⟩, 515, 516, 518, 900 ⟨74S77⟩, 516 ⟨68JA5280, 71JOM(28)C1, 73JA6757, 77JA5514, 79TL4777, 80JOC3543, 80S151, 81JOC5047, 87TL3771⟩, 516, 517 ⟨63JA2072, 81JOC4599⟩, 516, 522 ⟨72JA3561⟩, 516, 523 ⟨78JOC4395, 76JA5297⟩, 517 ⟨77JA3427, 78JOC1058⟩, 517 ⟨79JA2966, 81JOC3978, 85JOC1384, 85JOC1582, 85JOC5583⟩, 518 ⟨61JA1241, 61JA3417, 62JA190, 63JA2063, 63JA2066, 67JA5478, 69JA1224, 69JA1226, 77T2331⟩, 519 ⟨81T3547, 86JOC4296⟩, 519, 531 ⟨86JA2049, 87JOC310, 89JA1754⟩, 520 ⟨70JA2467, 81JOC2988, 85JA2564, 86JOC427, 86JOC445, 91JCS(P1)2633, 92ACR16⟩, 520, 1086 ⟨76JOC1078⟩, 521 ⟨74JOM(73)1, 86JOC432, 90TL455, 91JOC401, 94JOC365⟩, 522 ⟨71JA4062, 78S676, 80JOC4067⟩, 522, 526 ⟨80JOC4542, 82JOC863, 82JOC872⟩, 523 ⟨67JA561, 81JOC4313, 82JOC5069⟩, 524 ⟨72CC71, 72S699, 74JA311, 75JOM(90)C1, 76JA1964⟩, 525 ⟨76JA4917, 77JOC2533, 77S695, 80JOC384, 86JOC895⟩, 525, 900 ⟨76JA1785, 76JA1798, 77IC2938⟩, 525, 900, 901, 904 ⟨80JOC389⟩, 526 ⟨72JOM(44)233, 86JOC5264, 88JA1539, 88JOC2916, 91TL6691, 92JOC2379⟩, 526, 531 ⟨85JOC5446, 86JOC3394⟩, 527 ⟨88JA6246⟩, 527, 528 ⟨71JA1816⟩, 527, 900 ⟨72JA4370⟩, 527, 900, 901 ⟨75JA5249⟩, 529 ⟨91JOC3286, 93SL927⟩, 530 ⟨73JOC2422, 73JOC3968, 73JOM(54)61, 85JOC5586, 86JOC3150, 93TL7845⟩, 531 ⟨70JA6983, 71JA2802, 85JA4980, 85JOC1385,

86JOC4526, 92JOC6608, 94TL2141⟩, 532 ⟨92JOC2716, 92JOC499, 92TL3421⟩, 532, 698 ⟨89JA3441, 93JOC147⟩, 534 ⟨81JOC2712, 85JOC549, 86JOC3278, 86JOC3396⟩, 535 ⟨85JA460, 87JOC4020, 87JOC728⟩, 606, 628 ⟨73JA5786⟩, 623 ⟨67JA4531, 73JA6456, 84S919, 86S480⟩, 623, 628 ⟨89JOC6068⟩, 629 ⟨84JOC4822⟩, 698 ⟨70JA712, 70JA714, 92JOC3767⟩, 740 ⟨89JA384⟩, 900 ⟨61JA3834, 73JOC1617, 76JA7832, 77JOC579, 80JOC550, 86JOC3398, 86S674⟩, 900, 901 ⟨89JOC6075⟩, 900, 901, 906 ⟨79JA96⟩, 900, 903, 904, 906 ⟨B-88MI 218-01, B-72MI 218-01⟩, 900 ⟨B-75MI 218-01, B-81MI 218-01⟩, 901 ⟨80JOC549, 82JOC3808, 82JOC754, 84OM1392, 84S920, 88JOC239⟩, 901, 904 ⟨88JOC6009⟩, 901, 906 ⟨86JOC5277, 88TL21⟩, 906 ⟨91TL3655⟩, 964 ⟨60JA1939, 62JA1650⟩, 966 ⟨B-75MI 219-01⟩, 1084 ⟨77JOM(131)163⟩, 1085 ⟨87OM629, 88JOC1391⟩, 1086 ⟨83OM1316, 86JOC4507, 86JOC4521, 88TL2631⟩, 5266 ⟨80JOC4540⟩; **3**, 13, 82 ⟨79S704, 87JOC5400⟩, 19 ⟨79S701, 72JOC2942⟩, 82 ⟨84JA8001, 86JOC5264⟩, 83 ⟨56JA252, 58JA5377⟩, 83, 93, 94, 98, 100 ⟨93JOC4727⟩
Brown, J., **1**, 865 ⟨67JA4383⟩
Brown, J. B., **2**, 30 ⟨63OSC(4)851⟩
Brown, J. D., **1**, 310 ⟨87TL5279⟩, 389 ⟨92TA1551⟩; **2**, 420 ⟨84JOC1078⟩
Brown, J. H., **2**, 98 ⟨66CRV529⟩
Brown, J. J., **4**, 179 ⟨64JA2183⟩
Brown, J. M., **1**, 77 ⟨80JA3040⟩, 411, 811 ⟨73CC319⟩, 803 ⟨79JCS(P2)962, 65CC226⟩, 847 ⟨B-71MI 119-01⟩, 848 ⟨B-71MI 119-02⟩, 851 ⟨B-71MI 119-03, B-71MI 119-04⟩, 865 ⟨65CC376, 67CC638, 68JCS(B)411, 70JA3821⟩; **2**, 43, 528 ⟨93CC1673, 94JA866⟩, 558, 560 ⟨91JOC698⟩, 862 ⟨91JCS(P2)2081⟩
Brown, J. N., **4**, 458 ⟨72JHC1295⟩
Brown, Jr., A. D., **2**, 450 ⟨68JCS839⟩
Brown, Jr., F., **2**, 377, 382 ⟨70JOC1468⟩
Brown, Jr., P. S., **1**, 835 ⟨91TL179⟩; **3**, 623 ⟨91TL179⟩
Brown, J. T., **4**, 962, 963 ⟨91TL5047⟩
Brown, K. L., **4**, 150 ⟨79JA503⟩
Brown, L. A., **1**, 525 ⟨77TL2831⟩
Brown, L., **1**, 832 ⟨83JOC2122⟩; **5**, 209 ⟨83CC1113, 84JOC3875⟩
Brown, L. D., **2**, 442 ⟨77JA7876⟩
Brown, L. E., **5**, 374 ⟨84JOC2015⟩
Brown, L. R., **2**, 320 ⟨74S290⟩
Brown, M., **1**, 85 ⟨67JOC2013⟩
Brown, M.,**5**, 100 ⟨73OSC(5)1043⟩
Brown, M., **6**, 105 ⟨63USP3092637, 65USP3214412⟩
Brown, M. D., **1**, 591 ⟨88JCS(P1)813, 88JCS(P1)817⟩; **5**, 1125 ⟨82CC444⟩; **6**, 559 ⟨88JCS(P1)813⟩
Brown, M. F., **5**, 935 ⟨90TL2529, 91TL5223⟩, 935, 936 ⟨90JA7806⟩; **6**, 719 ⟨90TL2529⟩
Brown, M. J., **2**, 133 ⟨91JA3176⟩; **5**, 287 ⟨91JOC1933⟩, 652 ⟨89H(29)2225⟩
Brown, M. P., **2**, 134 ⟨88TL4477, 92T5933⟩, 834, 835 ⟨61CI(L)24⟩; **4**, 249 ⟨92T5933⟩, 691 ⟨79IC2808⟩, 1029 ⟨91JCS(D)351⟩; **6**, 183 ⟨73CC532⟩, 183 ⟨75JCS(D)148⟩
Brown, N. C., **2**, 651 ⟨48JCS2303⟩; **4**, 129 ⟨89B8270⟩
Brown, N. M. D., **4**, 1057 ⟨81JOM(209)1⟩
Brown, P. A., **1**, 633 ⟨82TL3733⟩; **3**, 202 ⟨91JCS(P1)1893⟩
Brown, P., **6**, 395 ⟨90JCS(D)2643⟩
Brown, P. E., **2**, 617 ⟨85JCS(P1)1127⟩; **3**, 239 ⟨88JCR(S)208⟩; **4**, 742 ⟨90JCS(P1)139⟩
Brown, P. J., **3**, 169, 172 ⟨83TL405⟩
Brown, P. R., **5**, 511 ⟨83JOC2635⟩
Brown, P. S., **1**, 643 ⟨91JCS(P1)1485⟩
Brown, R., **2**, 117 ⟨51JCS2123⟩; **6**, 27 ⟨80CJC485⟩
Brown, R. C. F., **6**, 107 ⟨71TL4519⟩
Brown, R. D., **3**, 342 ⟨82AJC1747⟩, 550, 552 ⟨87CC573⟩, 582 ⟨84CPL447⟩
Brown, R. F. C., **1**, 772, 784 ⟨B-80MI 117-01⟩, 774 ⟨93AJC1321⟩, 924 ⟨B-80MI 119-06⟩, 988 ⟨87AJC1675⟩; **2**, 764 ⟨59JCS2094⟩; **3**, 445 ⟨67AJC339⟩, 541 ⟨74AJC2373, 74AJC2385, 76JA7421, 77AJC179, 77AJC459⟩, 527, 541 ⟨B-80MI 316-03⟩
Brown, R. K., **2**, 128 ⟨63CJC2671⟩
Brown, R. L., **1**, 45 ⟨43JA1516⟩
Brown, R. S., **3**, 344 ⟨82JA5922⟩; **6**, 104 ⟨80JOC4038⟩
Brown, R. T., **4**, 168 ⟨84TL3127⟩; **6**, 508 ⟨73JHC639⟩
Brown, R. W., **2**, 476 ⟨73CC813⟩; **3**, 120 ⟨78TL2771⟩
Brown, S., **1**, 546 ⟨93TL157⟩
Brown, S. B., **3**, 3, 114 ⟨79JCS(P1)2493⟩, 89 ⟨78S619⟩
Brown, S. J., **2**, 19, 614 ⟨85CC672⟩
Brown, S. L., **5**, 453 ⟨86TL623⟩
Brown, S. P., **5**, 718 ⟨87SC657⟩
Brown, T. J., **5**, 587, 593 ⟨93JMC1604⟩
Brown, T. L., **4**, 677 ⟨82JA4955⟩, 773 ⟨89USP4877899⟩
Brown, U. C., **2**, 525 ⟨83JOC1389⟩
Brownawell, M. L., **1**, 65, 66 ⟨84JA3703⟩
Brown, W. G., **1**, 3, 4, 7, 8, 10 ⟨49JA1675⟩, 871 ⟨40JA113⟩; **5**, 47 ⟨48JA3484⟩
Brown, W. L., **4**, 330 ⟨84CJC586⟩
Brownbridge, P., **1**, 308 ⟨83S1⟩; **2**, 198, 721 ⟨88S252⟩, 662 ⟨79CC578, 80JA3534⟩, 668 ⟨80TL3423⟩, 699 ⟨83S1⟩; **3**, 35 ⟨77JCS(P1)1131⟩; **4**, 831 ⟨83S1⟩
Browne, E. J., **5**, 1127 ⟨84AJC1043, 84AJC367⟩
Browne, L. J., **1**, 811 ⟨76TL3245⟩
Browne, N. R., **1**, 988 ⟨87AJC1675⟩
Browne, R. F., **6**, 312, 436 ⟨84JFC(24)485⟩
Browne, T. E. W., **4**, 799 ⟨31JCS1088⟩
Brownell, K. H., **5**, 738 ⟨87EUP243970⟩
Brown Jr., F., **5**, 350 ⟨68IJ587⟩
Brownlee, P. J. E., **2**, 223 ⟨64JCS3832⟩; **4**, 218 ⟨64JCS3832⟩
Brownlee, R. T. C., **5**, 725 ⟨65JCS5230⟩
Brownlie, G., **1**, 383 ⟨56JCS2419⟩
Brownstein, A. M., **5**, 998 ⟨70USP3483242⟩, 1001 ⟨69USP3465024⟩
Brownstein, S. K., **4**, 1061 ⟨85JA7219⟩
Brown-Wensley, K. A., **1**, 769 ⟨83PAC1733⟩; **3**, 508 ⟨83PAC1733⟩
Broxterman, Q. B., **3**, 40 ⟨94TL1777⟩
Broxterman, W. E., **1**, 401, 800 ⟨69JA5162⟩
Broxton, T. J., **1**, 178 ⟨66AJC521⟩
Bruce, J. M., **1**, 517 ⟨B-74MI 112-01⟩
Bruce, M. I., **1**, 862 ⟨72AOC(10)273⟩, 1143, 1145 ⟨91CRV197⟩; **2**, 579 ⟨82COMC-I(4)691⟩; **4**, 678 ⟨94JOM(464)191⟩, 711 ⟨91JOM(407)391⟩, 713 ⟨82JOM(240)299⟩, 821 ⟨68AG(E)747⟩; **5**, 447 ⟨75JOM(94)62⟩, 452 ⟨74JCS(D)2302⟩, 941 ⟨83AOC(22)59, 91CRV197⟩, 944 ⟨80AJC2605⟩
Bruce, M. R., **4**, 42 ⟨81JOC571⟩, 697, 698 ⟨88JA3231⟩
Bruce, R., **4**, 715 ⟨67CJC2011⟩
Bruce, W. F., **3**, 407 ⟨55USP2700681⟩; **5**, 98 ⟨43OSC(2)12⟩
Bruch, M., **1**, 900 ⟨86JOC2969⟩
Bruché, L., **5**, 371 ⟨86S772⟩, 676, 677 ⟨84JCS(P1)2535⟩, 677 ⟨81JCS(P1)2245, 89T7427⟩, 677 ⟨92OPP352, 93JHC559⟩, 681 ⟨86JCR(S)210⟩, 862 ⟨86JCS(P1)2177⟩
Bruchner, B., **2**, 832, 833 ⟨63OSC(4)784⟩
Brück, B., **2**, 788 ⟨65CB3847⟩
Bruck, D., **1**, 914 ⟨79JA4770⟩
Bruck, M. A., **4**, 589 ⟨88OM1465, 89OM1118⟩
Bruck, P., **1**, 6 ⟨60CI(L)405⟩; **2**, 386, 405 ⟨55JCS3997⟩
Brückmann, R., **2**, 1024, 1025 ⟨85JOC2801⟩; **3**, 483 ⟨87CB635⟩
Bruckmann, R., **3**, 532 ⟨85JOC2801⟩
Brückner, A., **6**, 243 ⟨90ZAAC(588)26⟩
Brückner, C., **1**, 978 ⟨88JOC2450⟩; **5**, 166 ⟨88JOC2440⟩, 553 ⟨88LA465⟩, 729 ⟨88LA471⟩
Brückner, K., **1**, 736 ⟨55CB1415⟩
Bruckner, K., **3**, 234 ⟨61CB1225⟩
Brückner, R., **1**, 1001 ⟨89CB193⟩, 1029 ⟨92CB271, 93TL4177⟩; **1**, 236 ⟨89CB703⟩, 398 ⟨92CB1957, 93TL297⟩, 404, 407 ⟨91COS(6)873⟩, 405, 816

⟨92AG(E)647⟩, 406 ⟨93CB1227⟩, 406, 820 ⟨92TL7323⟩, 552 ⟨90CB555⟩, 816 ⟨90TL4425⟩, 816, 817 ⟨90CB917⟩, 816, 818 ⟨94T3687⟩
Brückner, R., **2**, 154, 196 ⟨91COS(6)873⟩; **4**, 217 ⟨91TL467⟩, 231 ⟨93CB1227⟩, 235 ⟨88TL5233⟩
Bruder, W. A., **5**, 746 ⟨84TL3651⟩
Brudermüller, M., **1**, 504, 623 ⟨88CB2239⟩, 510 ⟨88AG(E)298⟩
Brueckner, R., **4**, 359 ⟨92CB2731⟩, 381 ⟨89CB2023⟩
Bruenner, H. H., **4**, 454 ⟨71MI 410-01⟩
Bruggemann, K., **5**, 580, 592 ⟨24CB981⟩
Brugger, M., **5**, 849 ⟨72LA(755)101⟩
Bruhin, J., **1**, 104 ⟨73C277⟩
Bruhn, J., **1**, 814 ⟨79HCA2630⟩
Bruhn, M., **1**, 1050 ⟨76TL235⟩
Bruhn, M. S., **1**, 1049 ⟨75TL4217⟩
Bruhn, P. R., **2**, 684 ⟨86JOC2137⟩
Bruhnke, J. D., **1**, 326, 327 ⟨91JA7788⟩, 326, 327 ⟨92JA7664⟩
Bruice, T. C., **1**, 749 ⟨86JOC959⟩; **2**, 807 ⟨B-66MI 215-06⟩
Bruin, J. W., **3**, 508 ⟨83TL3935⟩; **4**, 608, 626, 669, 696, 700 ⟨85JOM(288)13⟩
Bruker, A. B., **2**, 430 ⟨66ZOB484⟩, 440 ⟨61DOK(139)1359, 63ZOB1919⟩
Bruker, R., **3**, 496 ⟨91CB1747⟩
Brumby, S., **4**, 1078 ⟨83JCS(P2)1071⟩
Brun, P., **1**, 797 ⟨89BSF799⟩, 929 ⟨B-79MI 119-03⟩
Brunaus, M., **1**, 254 ⟨92SC351⟩
Bruncks, N., **2**, 565 ⟨84OM69⟩
Brune, F., **2**, 566 ⟨71JOM(32)335⟩
Brune, G., **3**, 616 ⟨80LA212⟩
Bruneau, C., **1**, 962 ⟨94SL457⟩; **2**, 676 ⟨91SL755, 93CC850⟩
Brunel, S., **4**, 127 ⟨88BSF699⟩
Brunelet, T., **3**, 4 ⟨80TL4653⟩, 4, 115 ⟨86JOC4016⟩
Brunelle, D. J., **1**, 482, 485 ⟨73CC907⟩, 484 ⟨73JOC2747⟩, 1134 ⟨78JA4618, 78JA4620⟩; **2**, 629 ⟨78JA4618⟩, 711 ⟨84JOC1309⟩; **4**, 863 ⟨73JOC2747⟩; **5**, 173 ⟨78JA4618⟩, 173, 204 ⟨78JA4620⟩
Brunelli, M., **2**, 602 ⟨73ICA319⟩
Bruner, L. B., **2**, 753 ⟨69JOC3430⟩
Brunet, E., **6**, 567 ⟨85H(23)1181⟩
Brunet, J. J., **1**, 254 ⟨83JOC1919⟩
Brunet, J.-J., **1**, 430, 431 ⟨81JOC1270⟩, 458, 515 ⟨77TL1069⟩
Brunet, J. J., **2**, 36 ⟨85TL5445⟩; **3**, 83 ⟨90CRV1041⟩
Brunet, N., **4**, 656 ⟨93OM4572⟩
Brunet, P., **3**, 152 ⟨90S595⟩, 416 ⟨88CC825⟩
Brunetiere, A. P., **4**, 238, 241 ⟨88TL2179⟩
Brungardt, B., **1**, 905 ⟨84TL3403⟩
Brungs, P., **6**, 117 ⟨89T1691⟩
Bruni, E., **5**, 1125 ⟨89TL1679⟩
Bruni, G., **5**, 548 ⟨24G389⟩
Bruni, P., **2**, 803 ⟨72G527⟩
Bruni, R. J., **1**, 3, 6 ⟨64JOC2870⟩; **4**, 424 ⟨55JA2559⟩
Bruni, S., **5**, 946, 959 ⟨94ICA233⟩
Brunn, E., **1**, 524 ⟨69AG(E)556⟩
Brunnelle, W. H., **2**, 89 ⟨90JOC768⟩
Brunner, E., **2**, 324 ⟨89JA7500⟩
Brunner, F., **6**, 370 ⟨89JHC829⟩
Brunner, H., **1**, 530 ⟨88JOM(355)449⟩, 922 ⟨70CB1101⟩; **2**, 64 ⟨88JOM(346)413⟩, 314 ⟨86OM739⟩, 435 ⟨87JOM(334)369⟩, 829 ⟨79AG655⟩; **4**, 548 ⟨91JOM(413)55⟩; **5**, 448 ⟨75JOM(87)223⟩, 453 ⟨72JOM(36)C18⟩
Brunner, K., **5**, 335 ⟨87CC1233, 89JOC1904⟩
Brunner, R. K., **1**, 831 ⟨83HCA2608⟩
Brunnett, J. F., **2**, 850 ⟨74JOC3612, 78JOC1867⟩, 850 ⟨79JOC4918⟩
Bruno, G., **1**, 69 ⟨B-72MI 102-03⟩
Bruno, J. J., **1**, 453, 512 ⟨63JOC1254⟩
Bruno, J. W., **2**, 602 ⟨82JA1860, 83JA6824⟩, 602 ⟨86JA40⟩; **3**, 544 ⟨89JA8738⟩; **4**, 656, 658 ⟨86JA40, 82JA7357⟩
Bruno, M., **1**, 1031 ⟨89T4293⟩
Bruno, O., **5**, 597 ⟨91FES1033⟩
Bruns, A., **6**, 584 ⟨85CB227⟩
Brunskill, J. S. A., **3**, 305 ⟨80JHC87⟩; **5**, 580, 615 ⟨78JCS(P1)629⟩
Brunsvold, W. R., **5**, 951, 958 ⟨77IC3059⟩, 955 ⟨75JOM(102)175⟩, 955 ⟨76JOM(118)309, 74JOM(77)345⟩, 958 ⟨77IC391⟩
Bruntrup, G., **1**, 681, 683 ⟨79CC52⟩
Brüntrup, G., **2**, 422 ⟨84T2211⟩, 677 ⟨77TL4651, 79CC52⟩; **5**, 161 ⟨79AG(E)622⟩, 166 ⟨84T2211⟩, 170 ⟨79AG(E)793⟩
Brüsehaber, L., **2**, 439 ⟨65ZN(B)181⟩
Brüser, W., **2**, 566 ⟨71JOM(32)335⟩
Bruser, W., **2**, 572 ⟨74ZAAC(412)11⟩
Brusilovec, A. I., **5**, 1084 ⟨88ZAAC(564)85⟩
Bruson, H. A., **3**, 621 ⟨45JA601⟩; **5**, 46 ⟨49OR(5)79⟩, 1043 ⟨45JA1178⟩
Brusov, R. V., **5**, 395, 396, 399, 400 ⟨88JGU201⟩
Brussee, J., **2**, 311 ⟨90TA163⟩, 316 ⟨91RTC25⟩, 330 ⟨90T1653⟩; **3**, 628 ⟨88TL4485, 90T979, 91SC1387⟩, 631 ⟨93T965⟩
Brutus, M., **6**, 637, 732 ⟨86NJC51⟩
Bruvo, M., **4**, 268 ⟨80TL3089⟩; **6**, 207 ⟨82JOM(238)327⟩
Bruylants, A., **5**, 558 ⟨75BSB61, 72S622⟩
Bruynes, C. A., **5**, 224, 536 ⟨82JOC3966⟩
Bruza, K. J., **1**, 1089 ⟨77TL4187⟩; **2**, 929, 931 ⟨79JOC4781, 74TL3365⟩; **3**, 137 ⟨81T3997⟩; **4**, 950 ⟨77TL4187⟩, 953 ⟨75CC519⟩
Bryan, E. G., **4**, 682 ⟨77JCS(D)1328⟩
Bryan, J. C., **3**, 517 ⟨87JA7213⟩
Bryan, R. F., **1**, 901 ⟨64JA733⟩
Bryan, W. M., **6**, 234 ⟨77JOC2024⟩
Bryans, J. S., **2**, 326 ⟨88TL2231⟩
Bryant, D. R., **1**, 52 ⟨84OM(3)923⟩
Bryant, G. L., **2**, 711 ⟨91JOC862⟩
Bryant, I. R., **5**, 214 ⟨89AJC2275⟩
Bryant, J. A., **1**, 473 ⟨87CL19⟩
Bryant, Jr., G. L., **6**, 443 ⟨92JOC5136⟩
Bryant, K. E., **1**, 828 ⟨81JOC1485⟩
Bryant, S. R., **6**, 570, 571 ⟨70JMC377⟩
Bry, T. S., **2**, 1016 ⟨61JA4663⟩
Bryce, D. K., **5**, 377 ⟨94BMC1865⟩
Bryce, M. R., **2**, 241, 792 ⟨88CC950⟩, 359 ⟨92AHC(55)1⟩, 361 ⟨84JCS(P1)2591⟩, 613 ⟨86BSF930, 86CC1623, 89TL3835⟩, 792 ⟨94JCS(P1)1935⟩; **3**, 355 ⟨91AG(E)871⟩, 648, 668 ⟨84NAT119⟩; **4**, 847, 848 ⟨91S26⟩, 870 ⟨90CC470, 90PAC473⟩, 873 ⟨90CC816⟩; **6**, 85, 729 ⟨92JOC1696⟩, 120, 123 ⟨90CC470, 91JCS(P1)157⟩, 698 ⟨91S26, 93PS(74)279⟩
Bryce-Smith, D., **1**, 3, 6 ⟨63PCS219⟩, 115 ⟨63JCS577⟩, 889 ⟨66JCS(C)154⟩; **3**, 487 ⟨74CC466⟩, 537 ⟨70JCS(D)699, 71GP2047373⟩, 537 ⟨73CC921, 74CC513⟩; **5**, 874 ⟨74CC466⟩
Brycki, B. E., **4**, 317 ⟨89CS33⟩
Bryndza, H. E., **6**, 516 ⟨85OM939⟩
Bryson, T. A., **1**, 141 ⟨77TL1961⟩; **2**, 1008 ⟨80JOC524⟩; **3**, 169 ⟨77TL1961⟩, 265 ⟨78TL743⟩; **4**, 210 ⟨72SC361⟩, 365, 391, 392 ⟨77TL1961⟩
Brzezowski, C. M., **4**, 280 ⟨92TL717⟩
Bu, X.-R., **4**, 438 ⟨88SC2289, 89S307⟩
Bu, Y.-F., **3**, 260 ⟨91SC1307⟩
Buback, M., **3**, 246 ⟨89CB1179⟩
Bube, T., **5**, 173 ⟨86AG(E)1028⟩
Bubenheim, O., **3**, 431 ⟨86CB2756⟩, 578 ⟨88JOC4997⟩
Bubner, M., **5**, 24 ⟨75MI 502-01⟩; **6**, 490 ⟨87ZAAC(546)142⟩
Bubner, T. P., **3**, 271 ⟨90SC1095⟩
Bubnov, Y. N., **1**, 1030 ⟨80S904⟩; **2**, 269 ⟨60IZV386⟩, 447 ⟨91MI 208-01⟩, 521 ⟨85TL2797⟩; **5**, 1136 ⟨59DOK(127)571⟩
Bubnov, Yu. N., **2**, 514 ⟨B-84MI 210-01⟩
Buc, S. R., **2**, 156 ⟨63JOC1140⟩; **3**, 638 ⟨55OSC(3)93⟩; **4**, 47, 51 ⟨63OSC(4)101⟩, 728 ⟨47JA254⟩

Bucciarelli, M., **2**, 337 ⟨83JOC2640⟩
Bucha, H. C., **6**, 497, 513 ⟨74USP3849102, 71GEP2040367⟩
Büch, H. M., **1**, 527 ⟨87TCC77⟩
Buchanan, D. N., **3**, 538 ⟨73JA6294⟩
Buchanan, G. L., **3**, 181 ⟨88CSR91⟩
Buchanan, J. B., **5**, 829 ⟨69FRP1573102⟩, 830 ⟨68USP3374260⟩
Buchanan, J. G., **1**, 43 ⟨87JCS(P1)2371⟩; **2**, 38, 52, 86, 89 ⟨B-65MI 202-01⟩
Buchanan, R. A., **1**, 821 ⟨87JA5878⟩, 834 ⟨84JOC1840⟩
Buchannan, C., **5**, 24 ⟨B-52MI 502-01⟩, 62 ⟨B-52MI 502-08⟩
Buchardt, O., **5**, 323 ⟨90S1008⟩, 635, 636 ⟨67ACS1936⟩, 640 ⟨61ACS1109⟩
Buchbauer, G., **1**, 564 ⟨86LA1578⟩
Buchdahl, M. R., **5**, 305 ⟨63OSC(4)106⟩
Buchecker, C., **3**, 65, 543 ⟨73TL2875⟩, 473 ⟨69TL2659⟩
Büchel, K. H., **3**, 311 ⟨77LA145⟩; **4**, 413 ⟨66CB724⟩; **5**, 337 ⟨67CB1248⟩
Büchel, T., **5**, 622 ⟨84HCA534⟩
Bucher, A., **1**, 558 ⟨91TL3357⟩
Bücher, D., **4**, 282 ⟨66AG(E)594, 66TL5263⟩; **5**, 495 ⟨66TL5263⟩
Bucherer, H. T., **3**, 616 ⟨34JPR275⟩
Buchert, P., **4**, 540 ⟨87JOM(330)357⟩
Buchholz, B., **2**, 319 ⟨87T565⟩; **5**, 345 ⟨86CB2678⟩
Buchholz, S., **1**, 858 ⟨89AG(E)73⟩
Büchi, G., **1**, 248 ⟨85TL5923⟩, 450 ⟨77TL4305⟩, 554 ⟨81JOC1⟩, 561 ⟨61JA927⟩, 651, 736 ⟨71HCA1767⟩, 706, 707 ⟨74JA3332⟩, 823 ⟨74JA7573⟩, 826 ⟨89HCA996⟩, 832 ⟨85JOC4664⟩, 884 ⟨68JOC460⟩, 983 ⟨79JOC4116⟩, 1082 ⟨79HCA2655⟩, 1131 ⟨79JOC546⟩; **2**, 159 ⟨76JA3332⟩; **3**, 147 ⟨79HCA2655⟩, 187 ⟨71JA2492⟩, 189 ⟨72JA9128⟩, 233 ⟨60JA4333⟩, 260 ⟨71JOC609⟩, 266 ⟨71JA3299⟩, 323 ⟨53JOC1030⟩, 493 ⟨71HCA1767⟩, 496 ⟨75JOC100⟩, 60 ⟨69JOC1122⟩, 141 ⟨85JOC4664⟩; **4**, 209 ⟨71JA746⟩; **5**, 105 ⟨87OS(66)29⟩
Büchi, H., **2**, 749 ⟨65HCA1746⟩; **5**, 131 ⟨63AG(E)212⟩; **6**, 104 ⟨63AG296⟩, 104 ⟨65HCA1746⟩
Büchler, R., **5**, 595, 597, 599 ⟨77CB285⟩
Buchman, O., **1**, 8 ⟨86JOC734⟩; **3**, 89, 95 ⟨90BSB345⟩, 90, 95 ⟨84JOC4009⟩
Buchmeier, W., **4**, 264 ⟨91CB2489⟩
Buchner, E., **3**, 319 ⟨1885CB2371⟩, 488 ⟨1895CB215⟩
Büchner, W., **1**, 38 ⟨66HCA1145⟩; **2**, 1022 ⟨66HCA907⟩, 1022, 1023 ⟨63HCA1121⟩, 1022, 1023 ⟨65HCA1229⟩
Buchner, W., **2**, 488 ⟨73CB1226⟩; **4**, 585 ⟨73AG(E)415⟩; **5**, 458 ⟨79CB834⟩
Bucholz, Y., **5**, 652 ⟨00LA(314)200⟩
Buchshriber, J. M., **3**, 339 ⟨71CJC3299⟩, 361 ⟨69CJC2039⟩; **4**, 78 ⟨77CJC2323⟩; **5**, 634 ⟨70CJC1991⟩
Buchta, E., **1**, 42 ⟨59CB916⟩
Buchwald, S. L., **1**, 397 ⟨89OM1593⟩, 536 ⟨87TL295⟩, 769 ⟨83PAC1733⟩, 980 ⟨82JOC389⟩, 991 ⟨83JA5490⟩; **2**, 82 ⟨91JA5093⟩, 261, 888 ⟨91JA165⟩, 306, 927 ⟨89JA4486⟩, 307 ⟨91JA2321⟩, 330 ⟨92JA7562⟩, 740 ⟨76JA7104⟩, 873 ⟨90AG(E)771, 89JA2870⟩, 927 ⟨91TL5465, 93JA10394⟩, 927, 945 ⟨89JOC2793⟩, 969 ⟨87TL295, 87TL3895, 88TL3445⟩, 979 ⟨83JA5490⟩, 986, 987 ⟨86JA7411⟩, 986, 987 ⟨86JA7441, 87JA2544, 88CRV1047, 89JA776⟩; **3**, 370 ⟨88JA3171⟩, 508 ⟨83PAC1733⟩, 614 ⟨87TL295⟩; **4**, 1061 ⟨89JA2870⟩
Buck, A., **2**, 13 ⟨55OSC(3)195⟩; **4**, 743 ⟨55OSC(3)270⟩; **5**, 38 ⟨55OSC(3)270⟩
Buck, C. E., **1**, 450 ⟨54JA1613⟩; **5**, 421 ⟨54JA1372⟩
Buck, J., **1**, 450 ⟨85JCS(P1)2399⟩
Buck, J. S., **5**, 51 ⟨43OSC(2)333⟩, 275 ⟨43OSC(2)44⟩
Buck, P., **1**, 1068, 1069, 1081 ⟨B-69MI 121-02⟩
Buckel, W., **1**, 458 ⟨B-69MI 122-01⟩
Buckelew, D. Q., **1**, 879 ⟨89JOC773⟩
Buckingham, D. A., **5**, 385 ⟨73JA5649⟩
Buckingham, J., **3**, 446 ⟨69QR37⟩
Buckle, D. R., **2**, 139 ⟨91COS(7)119⟩; **3**, 164 ⟨91COS(7)119⟩
Buckle, J., **1**, 1079 ⟨74JOM(77)C22⟩; **3**, 603 ⟨74JOM(77)C22⟩
Buckle, M. J. C., **1**, 1057 ⟨93TL2383⟩
Buckler, S. A., **1**, 51 ⟨57JA3567⟩; **2**, 96 ⟨53JA4372⟩, 433, 444 ⟨60JA5794⟩, 440 ⟨61USP2969390, 62T1231⟩, 440 ⟨66JOC588⟩, 440, 446 ⟨61JA168⟩, 440, 450 ⟨62T1221⟩, 442 ⟨57JA3567⟩, 442, 444 ⟨60JOC1996⟩, 448 ⟨61JOC4770, 63USP3116334⟩
Buckles, R. E., **1**, 898 ⟨67JA3405⟩; **2**, 23 ⟨60JOC20⟩, 24 ⟨51JA998, 60JOC24⟩; **5**, 24 ⟨57CRV641⟩, 98 ⟨55CRV659⟩; **6**, 71 ⟨43JA613⟩
Buckley, D. J., **3**, 174, 234, 256 ⟨85JCS(P1)2193⟩; **4**, 90 ⟨80CC506, 85JCS(P1)2193⟩
Buckley, G. D., **2**, 783 ⟨47JCS1494, 47JCS1497⟩, 784 ⟨47JCS1471⟩; **3**, 638 ⟨47JCS1500⟩; **6**, 439 ⟨45JCS864⟩
Buckley, L. A., **5**, 772 ⟨77JOC1786⟩
Buckley, T. F., **3**, 17, 132 ⟨82JA4446⟩, 284 ⟨81JA6157⟩
Buckman, B. O., **3**, 170 ⟨91JOC328⟩
Buckman, J. D., **5**, 247 ⟨67JOC3467⟩
Buckschewski, H., **3**, 481 ⟨59CB1345⟩
Buckwalter, B., **1**, 136 ⟨74JA5560⟩; **3**, 36 ⟨74JA5560⟩
Buckwalter, B. L., **3**, 224 ⟨78JA1267⟩
Buda, A. B., **4**, 11 ⟨87JOC3466⟩
Budanov, V. V., **6**, 580 ⟨88MI 618-01⟩
Budding, H. A., **2**, 676, 970, 971 ⟨67JOM(9)285⟩, 970, 971 ⟨67JOM(9)295⟩; **4**, 154 ⟨79JOM(166)339⟩, 156 ⟨78JOM(153)305⟩, 701 ⟨76JOM(111)C33⟩
Buddrus, J., **1**, 722 ⟨68AG(E)536, 72AG(E)1041, 74CB2050⟩; **2**, 45 ⟨73AG(E)163⟩; **4**, 181 ⟨67LA85⟩; **6**, 38 ⟨76CB2351⟩
Bude, D., **4**, 472 ⟨68JA4495⟩
Budenz, R., **2**, 166 ⟨67AG(E)706⟩, 166 ⟨76JFC(7)115⟩
Budesinsky, M., **1**, 4, 10 ⟨84CAR(130)103⟩
Budewitz, E. P., **4**, 47, 50 ⟨63OSC(4)748⟩
Budiansky, S. P., **1**, 516 ⟨79T929⟩
Budilova, I. Yu., **4**, 927 ⟨85ZOB838⟩
Budnick, E. G., **4**, 553 ⟨69USP3463835⟩; **6**, 151, 155 ⟨77USP4020091⟩, 155 ⟨69USP3471406⟩, 155, 366 ⟨75USP3892676⟩, 155, 366 ⟨84USP4440646⟩
Budnik, L. V., **2**, 400 ⟨78ZOR1846⟩; **4**, 1005 ⟨92ZOB707⟩; **5**, 317, 349 ⟨78ZOR1846⟩, 703 ⟨85JOU1166, 85ZOR1281⟩; **6**, 625 ⟨85ZOR1281⟩
Budnik, R. A., **5**, 652 ⟨71CC1514⟩
Budnowski, M., **2**, 168 ⟨61CB1682⟩
Budwig, J., **5**, 753 ⟨42CB1585⟩
Budzelaar, P. H. M., **1**, 910 ⟨86JA561⟩; **4**, 545 ⟨90OM1222⟩; **5**, 927 ⟨84AG(E)825⟩, 928 ⟨85JA2773⟩
Budzikiewicz, H., **1**, 871 ⟨64JA2832⟩; **5**, 244, 246 ⟨86T5969, 85TL4349⟩, 246 ⟨85ZN(B)1563⟩
Buegler, S., **4**, 266 ⟨82SUL63⟩
Buehl, M., **5**, 433 ⟨92MI 509-01⟩
Buehler, C. A., **2**, 38 ⟨B-77MI 202-01⟩, 650 ⟨64JOC2397⟩, 696 ⟨74CRV45⟩; **3**, 279, 281, 295, 296, 297, 298, 301, 305, 309 ⟨72S533⟩, 430 ⟨67JOC261⟩, 526 ⟨B-70MI 316-01⟩; **4**, 48 ⟨B-70MI 402-01⟩, 176 ⟨B-70MI 404-01⟩; **5**, 1 ⟨B-77MI 501-01⟩, 24 ⟨B-70MI 502-01, B-77MI 502-01⟩
Buehler, J. S., **2**, 806 ⟨42JA19⟩
Buell, G. R., **1**, 759 ⟨70JA7424⟩; **4**, 654 ⟨70JA7424⟩
Buenger, G. S., **6**, 15 ⟨89JA8502⟩
Bueno, C., **6**, 343 ⟨83OM1044⟩
Buerger, H., **4**, 607, 621 ⟨76ZAAC(419)157⟩; **6**, 262 ⟨84JFC(24)457⟩
Buerstinghaus, R., **4**, 254, 618, 853 ⟨77LA830⟩, 849 ⟨75S461⟩; **6**, 128, 135, 326, 327 ⟨77CB841⟩, 135, 327 ⟨75AG37⟩
Buescher, H.-H., **6**, 469 ⟨79EUP7498⟩
Buess, C. M., **6**, 233 ⟨50JA3529⟩
Buevich, V. A., **4**, 826 ⟨81ZOR1550⟩; **6**, 74 ⟨78ZOR2229⟩
Bugaeva, S. V., **5**, 380 ⟨83IZV1413⟩
Bugden, G., **4**, 1077 ⟨86TL5033⟩

Bugerenko, E. F., **4**, 569 ⟨62DOK(143)840, 63DOK(148)875⟩, 569, 570, 573 ⟨65IZV286⟩, 570 ⟨62DOK(147)117, 62MI 412-01⟩, 570 ⟨70ZOB609, 71ZOB2186⟩, 588 ⟨77ZOB1664⟩
Bugner, D. E., **2**, 445 ⟨89JOC2580⟩
Bühl, H., **1**, 1077 ⟨78S536, 79CB3728⟩
Buhl, H., **3**, 552 ⟨77T449⟩
Bühl, H., **5**, 511 ⟨77T449, 78JHC697⟩
Bühl, M., **5**, 928, 929 ⟨92AG(E)1238⟩; **6**, 201, 715 ⟨92AG(E)1238⟩
Bühler, U., **5**, 209 ⟨83CB1183⟩
Bühlmeyer, B., **3**, 692 ⟨86JOM(317)187⟩
Buhmann, M., **3**, 459 ⟨93CB957⟩
Buhro, W. E., **1**, 299, 353 ⟨84OM44, 82TL2261⟩; **5**, 451, 937 ⟨83OM1852⟩
Buijink, J. K., **5**, 781 ⟨91ZN(B)1328⟩
Buijink, J.-K. F., **4**, 671 ⟨93OM2268⟩
Buijle, R., **2**, 1046, 1049 ⟨66AG(E)584⟩
Buijs, W., **2**, 36 ⟨83SC387⟩
Buillard, M., **2**, 104 ⟨91BSF222⟩
Buina, N. A., **2**, 410 ⟨67IZV1606⟩, 466 ⟨76MI 208-01⟩, 469 ⟨77IZV1437⟩; **5**, 371 ⟨90IZV1139⟩
Buisson, D., **1**, 8 ⟨92TL7337⟩
Buitenhuis, E. G., **5**, 15 ⟨89RTC51⟩
Bujacz, G., **2**, 184 ⟨89AG(E)97⟩; **6**, 116 ⟨88JOC3609⟩, 124 ⟨91TL4189⟩
Bukac, Z., **5**, 997 ⟨73CZP148911⟩
Bukarev, A. I., **5**, 685 ⟨90JOU1408, 90ZOR1629⟩
Buk, S. C., **2**, 588, 590 ⟨88JOC2390⟩
Bukhmutskaya, S. S., **1**, 422 ⟨47IZV205⟩
Bukowska, M., **3**, 294 ⟨90JOC6161⟩
Bukowski, P., **5**, 803 ⟨75LA1124⟩
Bulatov, P. O., **6**, 100 ⟨86ZOR107⟩
Bulatov, V. P., **5**, 978 ⟨73ZOB2089⟩
Bulka, E., **5**, 624, 625 ⟨84URP570267⟩, 1051 ⟨B-77MI 527-03⟩, 1054 ⟨67CB1367⟩, 1113, 1116 ⟨B-77MI 530-03⟩; **6**, 596 ⟨63ZC348, 67CB1459, 91EGP291752⟩, 596, 598 ⟨67CB1373⟩
Bull, B. A., **5**, 38 ⟨34CRV275⟩
Bull, D., **1**, 1010, 1012 ⟨81S364⟩
Bull, D. C., **2**, 161 ⟨80JCS(P2)385⟩
Bull, H. B., **5**, 83 ⟨52MI 502-04⟩
Bull, J. R., **1**, 25 ⟨86CC451⟩; **4**, 429, 430, 431, 439 ⟨65JCS2601⟩
Bull, W. E., **2**, 388 ⟨58JA2516⟩; **4**, 66 ⟨76BCJ502⟩
Bullee, R. J., **1**, 967 ⟨82JOM(240)329⟩, 1017, 1018 ⟨81JOM(217)267⟩
Bullesbach, P., **2**, 31 ⟨87JOC5560⟩
Bulliard, M., **1**, 327 ⟨91SL423, 91SL425⟩
Bullitt, O. H., **2**, 166 ⟨46JOC469⟩
Bullman Page, P. C., **2**, 936 ⟨90CSR147⟩
Bulloch, G., **4**, 550, 563 ⟨76JCS(D)1113⟩
Bullock, E., **3**, 106 ⟨66CJC1007⟩
Bullock, M. W., **2**, 214 ⟨54JA1828⟩
Bullock, R. M., **3**, 544 ⟨88JA7868⟩; **4**, 679 ⟨88JA7868⟩
Bullock, W. H., **1**, 1089, 1090 ⟨87TL4127⟩; **4**, 938 ⟨91JOC3556⟩
Bullot, J., **4**, 743 ⟨59BSF1828⟩
Bulls, A. R., **2**, 969 ⟨87JA203⟩
Bulman Page, P. C., **1**, 146 ⟨90JOC3840⟩; **4**, 244, 250 ⟨89T7643⟩, 249 ⟨92T5933, 92T7265⟩, 261 ⟨90SL457, 91SL80⟩, 262 ⟨93TA2139⟩, 263 ⟨89JCS(P1)2441, 91SL84⟩, 357 ⟨90T2573⟩, 849 ⟨88JCS(P1)269⟩
Bulmer, G., **2**, 123 ⟨45JCS666⟩
Bülow, B.-G. von, **1**, 630 ⟨66LA(691)41⟩
Bülow, C., **3**, 448 ⟨02CB915⟩; **5**, 841 ⟨12CB3732, 13CB2032, 13CB2370, 17CB1478⟩, 843 ⟨24LA(439)211⟩
Bülow, G., **1**, 485 ⟨92TL461⟩; **4**, 656 ⟨93JA7215⟩
Bulpin, A., **4**, 847 ⟨89TL3415⟩; **6**, 121, 316 ⟨89TL3415⟩, 121, 316 ⟨92JOC4507⟩, 123 ⟨90TL1151⟩
Bulten, E. J., **1**, 72, 88 ⟨90RTC87⟩; **4**, 154 ⟨79JOM(166)339⟩, 156 ⟨78JOM(153)305⟩, 626 ⟨67TL1443, 71JOM(29)409⟩, 699 ⟨76JOM(117)329⟩, 701 ⟨76JOM(111)C33⟩; **6**, 185 ⟨67TL1443⟩
Bulychev, A. G., **5**, 385 ⟨85JA4252⟩
Bulychev, Yu. N., **5**, 690 ⟨82CHE1302, 82KGS1682⟩
Bumagina, I. G., **1**, 1015 ⟨81DOK(261)1141⟩
Bumagin, N. A., **1**, 486 ⟨90IZV2665⟩, 492 ⟨84IZV2580⟩, 1004, 1006 ⟨81IZV2836, 81JOU18⟩, 1006 ⟨83DOK(272)1384⟩, 1010 ⟨84IZV1561, 85DOK(105)630⟩, 1012 ⟨84S728⟩, 1015 ⟨81DOK(261)1141⟩; **3**, 290 ⟨82IZV221⟩, 290, 305 ⟨85TL4819⟩, 647 ⟨89JOM(371)397⟩; **5**, 467 ⟨85JOM(282)421⟩
Bumagins, I. G., **1**, 1006 ⟨81JOU18, 83DOK(272)1384⟩
Bumgardner, C. L., **1**, 210 ⟨82TL2379⟩, 399 ⟨79JOC2348⟩, 624, 625, 788 ⟨57JA960⟩, 625, 783 ⟨57JA4729⟩; **2**, 16 ⟨86TL2715, 87JOC4307⟩, 691 ⟨62CI(L)559⟩; **3**, 26, 148 ⟨86TL2715⟩; **4**, 30 ⟨78JFC(11)527⟩
Bunce, N. J., **2**, 12, 627 ⟨72JOC664, 72CJC3109⟩
Bunce, R. A., **2**, 323 ⟨92JOC1727⟩, 395 ⟨87OPP471⟩, 444 ⟨84JOC4293⟩; **5**, 346 ⟨82JOC5042⟩
Buncel, E., **1**, 844, 918 ⟨B-87MI 119-01⟩, 844 ⟨B-80MI 119-01, B-84MI 119-01⟩, 847 ⟨B-80MI 119-02⟩, 859 ⟨79JOC317⟩, 886 ⟨B-75MI 119-01⟩; **4**, 391 ⟨65JA5498⟩
Bund, J., **1**, 483 ⟨91JA1442⟩
Bundel, Y. G., **2**, 7 ⟨89JOU587, 89JOU642⟩
Bundgaard, H., **5**, 703 ⟨88JMC2066⟩
Bundle, D. R., **2**, 90 ⟨81CC1240⟩
Bunds, J., **2**, 12, 627 ⟨65JOC415⟩
Bundy, J. M., **2**, 30 ⟨78JOC2793⟩
Bunel, E. E., **2**, 969 ⟨90SL74⟩
Bungardt, D., **5**, 400 ⟨86JOM(311)269, 86ZN(B)1096⟩; **6**, 518 ⟨87CB1421⟩
Bunge, K., **3**, 679 ⟨72CB1258⟩; **5**, 657 ⟨72CB1258⟩
Bunina, N. A., **4**, 834 ⟨80ZOR13⟩
Bunke, D., **4**, 551, 553, 564 ⟨88ZAAC(566)90⟩
Bunnage, M. E., **2**, 315 ⟨93JCS(P1)1375, 93SL731⟩
Bunnell, C. A., **3**, 517 ⟨77JA2127⟩; **5**, 447 ⟨76JA1166⟩, 448 ⟨76JA437⟩
Bunnelle, W. H., **1**, 652 ⟨87TL6261⟩; **3**, 210 ⟨87JOC1603⟩; **4**, 214 ⟨91JA8168⟩, 655, 656 ⟨87TL6261⟩
Bunnenberg, R., **6**, 274, 275, 559, 565 ⟨81CB1132⟩, 452, 453 ⟨81CB1746⟩, 559, 565, 605 ⟨81CB2064⟩, 560, 565 ⟨81CB2075⟩, 605, 619 ⟨82CB3587⟩
Bunnet, J. F., **1**, 474, 483 ⟨51CRV273⟩
Bunnett, J. F., **1**, 445 ⟨67JOC2701⟩, 474 ⟨73JOC3020, 75CPB2620⟩, 483 ⟨74JA112⟩, 881 ⟨62JOC3836⟩, 885 ⟨62JOC4152⟩; **2**, 845 ⟨73JOC2703⟩; **3**, 144 ⟨76JOC1702⟩
Bunnistro, K. S., **2**, 756 ⟨88JOU2222⟩
Buntain, G. A., **2**, 357 ⟨85JA4731⟩; **3**, 431 ⟨87S831⟩; **5**, 219 ⟨87S831⟩
Bunt, R. C., **1**, 192 ⟨93TL7513⟩
Bunte, H., **2**, 122 ⟨1874CB646⟩
Buntin, G. A., **5**, 1043 ⟨69JOC616⟩
Bunting, J. W., **1**, 629 ⟨89JA2258, 91JA6950⟩; **4**, 43 ⟨74CJC951⟩
Bunton, C. A., **1**, 384 ⟨58JCS403⟩; **2**, 845 ⟨75JOC2059⟩; **3**, 8 ⟨B-65MI 301-01⟩, 121 ⟨B-65MI 304-01⟩; **5**, 26 ⟨63JCS2918⟩, 26, 37 ⟨60JCS3070⟩, 49 ⟨60JCS1593⟩, 652 ⟨54CI(L)1154⟩
Bunyan, P. J., **6**, 154 ⟨63JCS1527⟩
Buond, G., **3**, 239 ⟨81S872⟩
Buono, G., **1**, 499 ⟨90TL4859⟩, 983 ⟨72TL3257⟩; **2**, 64 ⟨88TL3235⟩; **3**, 558 ⟨73TL5167⟩; **5**, 226 ⟨85TL1857⟩
Buracas, P., **3**, 709 ⟨77AG(E)707⟩
Bur, D., **2**, 72 ⟨90AG(E)205⟩; **5**, 330 ⟨91TA1209⟩
Burakevich, J. V., **5**, 580 ⟨70JOC2102⟩, 832 ⟨70CI(L)238⟩
Burangulova, R. N., **2**, 1074 ⟨89TL577, 89ZOB1979⟩
Burawoy, A., **5**, 1108 ⟨53JCS959⟩, 1111 ⟨56JCS96⟩
Burba, P., **2**, 597 ⟨70HOU(13/4)315, 70HOU(13/4)345⟩, 597 ⟨70HOU(13/4)363⟩, 1098 ⟨70HOU(13/4)154⟩
Burbaum, B. W., **1**, 499 ⟨83JOC4370⟩, 805 ⟨88JOC1848⟩
Bur-Bur, F., **2**, 361 ⟨75CB1365⟩
Burch, G. M., **6**, 58 ⟨63JCS1083⟩
Burch, M. T., **1**, 828 ⟨81JOC1485⟩

Burchardt, A., **5**, 323 ⟨88AP(321)311, 90AP(323)181⟩
Burchardt, B., **3**, 318 ⟨79TL4049⟩
Burch, R. R., **4**, 152 ⟨86JA5359⟩
Burckett-St. Laurent, J. C. T. R., **5**, 1153 ⟨82JST(79)215⟩, 1154 ⟨82JSP(92)158⟩
Burckhalter, J. H., **1**, 256 ⟨58JOC1281⟩, 889 ⟨65JOC1291⟩; **4**, 728 ⟨52JA3868⟩
Burdeska, K., **1**, 19 ⟨82S940⟩
Burdett, J. E., **3**, 246 ⟨69JOC2767⟩
Burdett, K. A., **2**, 14 ⟨74JOC2300⟩; **5**, 8 ⟨91S441⟩
Burdge, D. N., **2**, 127, 188 ⟨60JA2872⟩
Burdon, J., **1**, 426 ⟨62JCS1801, 62JCS3253⟩, 1005 ⟨67CC1259, 72JCS(P1)639⟩; **2**, 182, 364 ⟨57JCS2574⟩; **4**, 439 ⟨73T4195⟩; **6**, 37 ⟨71T4533⟩
Burdsall, D. C., **6**, 123 ⟨90H(30)855⟩
Bureau, R., **3**, 9, 122 ⟨91SL479⟩
Burenko, P. S., **1**, 870 ⟨67TL4525⟩
Bures, E. J., **2**, 942 ⟨87TL5965⟩
Burford, C. S., **4**, 355 ⟨77JA4536⟩
Burford, S. C., **1**, 541 ⟨85TL3285⟩; **4**, 649 ⟨80HCA555⟩
Burg, A. B., **2**, 109 ⟨53JA228⟩, 109, 111 ⟨33JA4020, 39JA673⟩, 409 ⟨58JA1107⟩, 418, 419 ⟨49JA3451⟩, 427, 429 ⟨53JA3869⟩, 431 ⟨60JA2145, 60JA2148⟩, 452 ⟨51JA4590, 62JA3442⟩, 461 ⟨65JA2113⟩, 474 ⟨53JA3872⟩, 481, 482, 509 ⟨54JA386⟩, 481, 491, 510 ⟨59JA1⟩, 492 ⟨65JA1213⟩, 507 ⟨66JA3178, 67JA1040⟩, 1008 ⟨89IC1295⟩; **4**, 111 ⟨81IC2739⟩, 550 ⟨81IC3734⟩; **6**, 246 ⟨60JA5759⟩
Burg, A. B.,**6**, 264 ⟨72CC673⟩
Burg, A. B., **6**, 264, 679 ⟨83IC2573⟩, 265 ⟨85IC148, 81IC3734⟩
Burg, D. A., **1**, 613 ⟨89JOC1647⟩
Burgada, R., **2**, 410, 411 ⟨71BSF136⟩, 414 ⟨70BSF192⟩, 476 ⟨68CR(C)270⟩; **4**, 478 ⟨74MI 410-01, 76T2089, 82PS(13)85⟩; **5**, 708, 861 ⟨70BSF1363, 68CR(C)1241⟩
Burger, A., **2**, 412 ⟨57JA1970⟩; **4**, 297 ⟨50JA5409⟩; **5**, 276 ⟨40JOC606⟩
Burger, B. J., **2**, 969 ⟨87JA203⟩; **3**, 511 ⟨86JA5347⟩
Burger, E., **4**, 493 ⟨73CB3421⟩
Bürger, H., **2**, 512 ⟨70MI 209-02⟩, 545 ⟨69JOM(18)299⟩, 566 ⟨69JOM(20)129, 71JOM(32)313⟩; **4**, 136 ⟨92JOM(427)293⟩, 140 ⟨93OM4930⟩, 719 ⟨73JOM(56)269, 76ZAAC(423)112⟩; **5**, 382 ⟨93JOM(446)25⟩; **6**, 51 ⟨80JFC(17)65⟩, 243 ⟨86JOM(316)41, 89JOM(378)125, 90JOM(385)207, 93AG429, 93JOM(446)25, 93JOM(456)19, 93OM4930, 93ZN(B)935⟩, 602 ⟨88CC105⟩, 615 ⟨83JFC(22)175⟩, 615, 632 ⟨84JFC(26)321⟩
Burger, J. J., **1**, 318 ⟨77TL4528⟩, 619, 707 ⟨81T417⟩, 708 ⟨80H(14)1739⟩
Burger, K., **1**, 1042 ⟨90S115⟩; **2**, 751 ⟨74S816⟩; **3**, 302 ⟨91LA1365⟩, 680 ⟨73AG(E)155, 80CB2699⟩, 680 ⟨86ACR342, 87S924⟩; **4**, 297 ⟨74CB1488⟩, 311, 493 ⟨73AG(E)502⟩, 312, 325 ⟨76JFC(7)471⟩, 340 ⟨77LA624, 71CB1826⟩, 493 ⟨74S816, 76LA36, 78S44, 78S526⟩; **5**, 22 ⟨92S1145⟩, 79 ⟨79LA1547⟩, 631 ⟨88CZ349⟩, 634 ⟨77CB2114⟩
Burger, L. L., **4**, 3 ⟨51JA4241⟩
Burger, P., **5**, 605, 614, 627 ⟨88CB2095⟩
Burger, U., **4**, 8 ⟨79TL1913⟩, 97 ⟨89HCA93⟩, 410, 539 ⟨80HCA1190⟩; **6**, 170 ⟨80HCA1190⟩
Burger, V. K., **5**, 789 ⟨76CZ340⟩
Bürger, W., **2**, 1008 ⟨69AP(302)354⟩; **4**, 47, 48 ⟨73OSC(5)221⟩; **6**, 43 ⟨67LA(707)35⟩
Burgers, P. C., **1**, 1141 ⟨84JA521⟩
Burgert, B. E., **5**, 986 ⟨68TL3341⟩
Burgess, C., **2**, 1016 ⟨62JCS4995⟩, 1016, 1018 ⟨66T2829⟩
Burgess, E. M., **1**, 575 ⟨73JOC26⟩; **2**, 367 ⟨67JA2502⟩, 401, 793 ⟨72JA6135⟩; **3**, 233 ⟨60JA4333⟩, 370 ⟨73JA279⟩, 370, 374 ⟨74JOC2885⟩, 579 ⟨68JA1923⟩; **5**, 335, 336 ⟨73JA279⟩; **6**, 106 ⟨67JA2502, 72JA6135⟩
Burgess, H., **2**, 288 ⟨28JCS321⟩; **5**, 652 ⟨25JCS279⟩
Burgess, K., **1**, 185 ⟨93JOC3767⟩; **2**, 42 ⟨88JOC5178⟩, 43 ⟨89TL395⟩, 43, 527 ⟨89TL5857, 89TL5861⟩, 50 ⟨94AG(E)1182, 91JA6139⟩, 527 ⟨91JOC1020, 92JA9350⟩, 527, 970 ⟨91CRV1179⟩, 528 ⟨91TA613⟩; **5**, 135 ⟨90JA7434⟩
Burgess, V. A., **2**, 394 ⟨88AJC1063⟩
Burget, G., **4**, 112 ⟨84CB1695⟩
Burghard, H., **1**, 689 ⟨80JCR(S)396⟩
Burghardt, R., **2**, 440 ⟨87PS(29)373⟩; **6**, 581 ⟨86ZN(B)1142, 87ZN(B)860⟩
Burgh Norfolk, S. de, **1**, 774 ⟨76JCS(P2)280⟩
Burghstahler, A. W., **1**, 104 ⟨61JA198⟩
Burgis, E., **2**, 751 ⟨74S816⟩; **4**, 297 ⟨74CB1488⟩, 493 ⟨74S816, 76LA36⟩
Burglass, A. J., **2**, 478 ⟨B-90MI 203-10⟩
Burgos-Lepley, C. E., **6**, 118 ⟨93JOC4159⟩
Burgot, J.-L., **3**, 343 ⟨76TL4297⟩, 376 ⟨74CR(C)793⟩; **4**, 255 ⟨76TL4775⟩
Burgoyne, W., **1**, 4, 7, 9, 17, 28 ⟨78JOC2259⟩, 14 ⟨81JOC5214⟩
Burgstahler, A. W., **1**, 415, 825 ⟨60JA4681⟩, 421 ⟨82JA6874⟩, 503 ⟨73JOC3652⟩; **3**, 82 ⟨63JOC2918⟩, 83 ⟨76S767⟩, 494 ⟨75JOC3456⟩; **5**, 76 ⟨73OSC(5)251⟩
Burguete, M. I., **1**, 1081 ⟨90JOC3808⟩
Burham, R. L., **2**, 24 ⟨60JOC24⟩
Burighel, A., **1**, 1112 ⟨72JCS(P2)2026⟩
Burilov, A. R., **2**, 254 ⟨87ZOB1429⟩, 266 ⟨81ZOB2140, 83URP956484⟩; **4**, 345 ⟨81ZOB2140, 83ZOB46⟩, 485 ⟨80ZOB751⟩; **5**, 369 ⟨89ZOB2392⟩
Burini, A., **2**, 983 ⟨92JOC976⟩
Burka, L. T., **1**, 108 ⟨80JA7760⟩
Burk, E. H., **5**, 1013 ⟨70JHC177⟩
Burk, M. J., **1**, 554 ⟨87JA8025⟩; **2**, 62 ⟨91JA8518, 92TL2099⟩, 438 ⟨90OM2653⟩; **5**, 354 ⟨92JA6266⟩
Burk, P., **3**, 210 ⟨73JOC3637⟩
Burk, R., **3**, 422 ⟨83JOC2685⟩
Burkard, U., **5**, 92 ⟨86LA334⟩, 1106 ⟨86CB1594⟩
Burk, R. M., **6**, 467 ⟨93TL395⟩
Burke, Jr., T. R., **5**, 404 ⟨92TL4137⟩
Burke, K. E., **4**, 554, 563 ⟨61JA1722⟩
Burke, M. C., **3**, 202 ⟨80JA5411⟩
Burke, P. L., **2**, 516, 522 ⟨72JA3561⟩
Burke, P. O., **2**, 367 ⟨85S935⟩, 368 ⟨86S1021⟩; **5**, 342 ⟨86S1021⟩
Burke, S., **1**, 602, 607 ⟨75JOC542⟩; **3**, 173 ⟨78JOC4552⟩
Burke, S. D., **1**, 209, 651 ⟨75CC537⟩, 243 ⟨82JOC3348⟩, 504 ⟨83JOC5221⟩, 583 ⟨94JOC332⟩, 604 ⟨76JA1612⟩, 831 ⟨84JOC4320, 85TL1163, 90TL5285⟩; **2**, 932 ⟨91TL855⟩; **3**, 59 ⟨84JA4558⟩, 211 ⟨81JOC2400⟩; **4**, 355 ⟨91TL855⟩
Burke, T. R., **3**, 242 ⟨84JOC1051⟩
Burke, W. J., **4**, 301 ⟨64JOC407⟩, 416 ⟨47JA2136⟩
Burkert, U., **3**, 556, 557, 560, 586 ⟨84T893⟩
Burkett, A. R., **1**, 681 ⟨71JOC1149⟩
Burkey, T. J., **1**, 925 ⟨83JA4701⟩
Burkhard, J., **2**, 617 ⟨88CCC110⟩; **6**, 635 ⟨92CCC1947⟩
Burkhardt, E. R., **2**, 591 ⟨85JOC416⟩
Burkhardt, F. J., **5**, 993 ⟨86JOC1374⟩
Burkhardt, H., **3**, 55 ⟨59CB1564⟩
Burkhardt, J., **6**, 629 ⟨66CB1912⟩
Burkhardt, T. J., **2**, 681 ⟨72JA6543⟩; **3**, 517 ⟨77JA2127⟩
Burkhart, G., **1**, 545 ⟨84JOM(266)203⟩
Burkhart, J. P., **3**, 150 ⟨85TL2861⟩, 472 ⟨80TL15⟩
Burkholder, C. B., **4**, 1082 ⟨88TL6749⟩
Burkholder, C. R., **4**, 32, 1082 ⟨90JOC589⟩, 1083 ⟨90JOC5420⟩
Burkholder, E. G., **1**, 95 ⟨78JOC4283⟩, 97 ⟨80SC627⟩
Burkholder, T. P., **5**, 330 ⟨88JA2341, 90JA9601⟩
Burkholder, T. R., **5**, 1020 ⟨92JPC10195⟩
Burks, Jr., R. E., **2**, 30 ⟨55OSC(3)785⟩
Burlachenko, G. S., **3**, 534 ⟨69ZOB467, 70ZOB707⟩; **4**, 661 ⟨70ZOB2607⟩; **5**, 390 ⟨70ZOB707⟩
Burlant, W., **2**, 729 ⟨56JA4028⟩
Burlant, W. J., **3**, 279 ⟨53JA745⟩

Burlett, D. J., **1**, 253 ⟨81JOC3293⟩
Burling, E. D., **6**, 257 ⟨65JCS6077⟩
Burlitch, J. M., **3**, 498 ⟨63JOC2463⟩; **4**, 689 ⟨89JOM(361)231⟩
Burmakov, A., **2**, 100 ⟨88JOU1473⟩
Burmakov, A. I., **6**, 19 ⟨B-89MI 601-02⟩, 37 ⟨83JFC(22)105⟩, 38, 39 ⟨70ZOR144, 70ZOR2498⟩
Burmeister, J. L., **4**, 584 ⟨89JOM(375)265⟩; **5**, 1140 ⟨71JCS(A)1663⟩
Burmester, A., **1**, 201 ⟨81S125⟩
Burmistrov, S. I., **5**, 638 ⟨66URP181085⟩, 652 ⟨66ZOR1841⟩, 674 ⟨66JOU1809⟩, 703 ⟨69JOU252, 69ZOR263⟩
Burn, A. J., **6**, 154 ⟨63JCS1527⟩
Burn, D., **2**, 1016 ⟨62JCS4995⟩, 1016, 1018 ⟨66T2829⟩; **3**, 71 ⟨64TL733⟩
Burnaeva, L. A., **2**, 1074 ⟨89TL577, 89ZOB1979, 91ZN(B)1578⟩; **4**, 493 ⟨88JGU1148, 88JGU876⟩; **5**, 708 ⟨86JGU1096, 86ZOB1245⟩, 709 ⟨84JGU1969, 84ZOB2203⟩, 786 ⟨75JGU2512⟩
Burn, P. L., **1**, 623 ⟨92JCS(P1)3225⟩
Burnard, R. J., **4**, 152 ⟨68JA7367⟩
Burnell, D. J., **1**, 690 ⟨88TL4369⟩
Burnell-Curty, C., **2**, 814 ⟨92JOC5063⟩; **5**, 329 ⟨93SL850⟩
Burness, D. M., **2**, 138 ⟨46JOC475⟩, 166 ⟨46JOC499⟩; **5**, 62 ⟨63OSC(4)628⟩
Burnett, D. A., **1**, 348 ⟨87JOC2958⟩; **5**, 335 ⟨86JOC1929⟩
Burnett, E., **3**, 184, 185 ⟨79T613⟩
Burnett, F. N., **5**, 690 ⟨84JOC1212, 88JOC382⟩
Burnett, J. N., **1**, 103 ⟨65JOC1154⟩
Burnett, R. M., **4**, 770, 779 ⟨65T2899⟩
Burnett, S. M., **1**, 1143 ⟨83CPL124⟩
Burneva, L. A., **2**, 476 ⟨91MI 208-03⟩
Burney, W., **2**, 89 ⟨50JA2735⟩
Burnham, J. W., **1**, 43 ⟨71JOC737⟩
Burns, B., **1**, 1015 ⟨88TL5565⟩
Burns, B. J., **6**, 647 ⟨78AF1435⟩
Burns, C. A., **1**, 929 ⟨92T6883⟩; **6**, 544, 557 ⟨92T6883⟩
Burns, D., **4**, 1029 ⟨91JCS(D)351⟩
Burns, G. T., **3**, 502 ⟨82JA6140, 82OM210⟩; **4**, 603 ⟨81TL7⟩, 1048 ⟨85JA3935⟩
Burns, H. D., **2**, 632 ⟨80S572⟩
Burns, J. H., **4**, 123 ⟨88IC1787⟩
Burns, S. A., **3**, 502 ⟨82JA6140, 82OM210⟩
Burns, S. J., **1**, 579 ⟨88JOC5576⟩; **4**, 186 ⟨88JOC5574⟩
Burns, T. P., **2**, 558, 559 ⟨87JOC3674⟩, 590 ⟨81JOC4323⟩
Burns, W., **1**, 664 ⟨76JCS(P1)234⟩
Burnside, C. H., **2**, 167 ⟨53JA243⟩
Burotina, L. S., **2**, 255 ⟨87ZOB786⟩
Burpitt, R. D., **1**, 355 ⟨64JOC940⟩; **2**, 717 ⟨79JOC3847⟩, 751 ⟨63JCS1462⟩; **3**, 623, 624 ⟨65JOC2564⟩; **4**, 836, 887 ⟨71JOC2205, 71JOC2211⟩
Burr, L., **6**, 598 ⟨91TL4823⟩
Burr, M., **1**, 907 ⟨60JA2651⟩
Burrell, A. K., **4**, 582 ⟨91JCS(D)609⟩; **6**, 247 ⟨90JOM(388)391, 90JOM(394)615⟩
Burridge, C., **5**, 73 ⟨68CC206⟩
Burrous, M. L., **2**, 913, 914 ⟨61JA4385⟩
Burrows, C. J., **5**, 360 ⟨85TL215⟩, 385 ⟨89JOC1584⟩
Bursch, P., **6**, 243 ⟨86JOM(316)41⟩
Burschka, C., **4**, 232 ⟨87LA451⟩; **6**, 353, 354 ⟨88JOM(394)183⟩, 547 ⟨85ZAAC(525)127⟩
Burski, J., **5**, 1107, 1112 ⟨83T4175⟩
Burstein, C., **5**, 640 ⟨93JMC1157⟩
Bürstinghaus, R., **5**, 411 ⟨77CB841⟩
Burt, D. W., **1**, 999 ⟨69JCS(C)2273⟩; **2**, 1071 ⟨69JCS(C)2273⟩, 1072 ⟨71JCS(C)2872⟩
Burtlitch, J. M., **4**, 16 ⟨65JA4259⟩
Burtner, R. R., **5**, 43, 62, 114 ⟨43JA262⟩
Burton, A., **3**, 16, 132 ⟨79S877⟩; **4**, 288 ⟨79S877⟩
Burton, C. A., **3**, 460 ⟨80MI 312-01⟩; **5**, 248 ⟨76JA6545⟩; **6**, 432 ⟨76JA6545⟩
Burton, C. D., **6**, 284 ⟨67USP3354217⟩
Burton, D. A., **2**, 94 ⟨56JCS3933⟩
Burton, D. J., **1**, 426 ⟨68FCR315⟩, 429 ⟨64JA5361, 89JFC(44)1637⟩, 431 ⟨71JFC(1)121, 82JFC(20)157⟩, 446 ⟨83JA650, 85JA2811⟩, 967 ⟨90TL3699⟩, 990 ⟨91CC566⟩; **2**, 444 ⟨83JFC(23)339, 85JFC(27)85, 88JOC366⟩, 611 ⟨73JFC(3)447, 85JA2811⟩, 612 ⟨79CL983, 83JA650⟩, 849 ⟨79S615⟩; **3**, 197 ⟨88JOC1523⟩, 237 ⟨86BSF876⟩, 495 ⟨73JA8467, 75JOC2796, 83JOC3616⟩, 497 ⟨68TL71⟩; **4**, 7 ⟨73JA8467⟩, 7, 32, 35 ⟨77FCR119⟩, 36 ⟨75JOC2796⟩, 117 ⟨85JFC(27)85⟩, 117, 118 ⟨88JOC366, 90JOC2311⟩, 139 ⟨89JFC(45)435⟩, 151, 152 ⟨92JFC(56)341⟩, 728 ⟨71JFC(1)123⟩, 732 ⟨65TL1883, 67JOC1311, 68JOC1854⟩, 732, 751, 752 ⟨68TL71⟩, 733 ⟨83JFC(23)339⟩, 733, 734 ⟨73SC197, 79CL983⟩, 734 ⟨81JFC(18)293⟩, 735 ⟨76TL895, 83JOC917⟩, 737 ⟨78JOC2643⟩, 746, 748, 760, 817, 821 ⟨93JA5430⟩, 751, 752, 753 ⟨70JOC2125, 77JFC(10)131⟩, 753 ⟨79JFC(13)407, 79JOM(169)123⟩, 754 ⟨71JFC(1)381⟩, 755 ⟨75TL773⟩, 762 ⟨71JOC2351⟩, 801 ⟨83JA650, 85JA2811⟩, 802 ⟨79JA3689, 80JFC(15)543, 90JFC(49)75⟩, 816, 820 ⟨94T2993⟩, 821 ⟨86JA4229⟩; **5**, 14 ⟨75JOC3026⟩, 894 ⟨75JOC2796, 90JOC2311⟩; **6**, 9, 14, 22, 235, 266 ⟨92T189⟩, 15, 246 ⟨85JA5014, 86JA832⟩, 58 ⟨91TL1019⟩, 58, 264 ⟨94JOC2393⟩, 58, 266 ⟨81JFC(18)197, 89JOC613, 88JFC(39)425⟩, 213, 222, 223, 224, 226, 227, 228 ⟨82JFC(20)89⟩, 224 ⟨78JOC2643, 83JFC(23)339⟩, 258 ⟨89JA1773⟩, 264 ⟨80JFC(15)263, 82JFC(20)121, 82JFC(20)617⟩, 265 ⟨88JOC366⟩, 266 ⟨90JFC(49)75⟩, 694 ⟨67JOC1311, 68JOC1854, 71JFC(1)347, 72JA820⟩, 695 ⟨75JOC2796, 75TL3789, 77JFC(10)131, 79CL983, 79JOM(169)123⟩
Burton, G., **3**, 495 ⟨88TL3003⟩, 559 ⟨77GEP(O)2728884⟩, 559, 581 ⟨77TL3831⟩; **4**, 208 ⟨94SL152⟩, 738 ⟨88TL3003⟩
Burton, H., **2**, 203 ⟨45JCS14⟩, 711 ⟨48JCS528⟩, 887 ⟨24JCS2275⟩
Burton, K. A., **1**, 924 ⟨86JA7935⟩
Burton, R., **4**, 706, 709 ⟨62IC227⟩
Burton, S. G., **5**, 197 ⟨89SC3331⟩
Burtsev, Yu. N., **6**, 284 ⟨92MC52⟩
Burwell, Jr., R. L., **1**, 459 ⟨63JA2881⟩
Burwell, R. L., **1**, 89 ⟨57CRV895⟩
Bury, A., **4**, 850 ⟨87S837⟩
Bury, F. W., **2**, 875 ⟨23JCS2489⟩
Burzin, K., **6**, 106 ⟨73AG(E)996⟩
Burzlaff, H., **2**, 831 ⟨79AG945⟩, 1004 ⟨74CB1949, 77CB3168⟩; **3**, 603 ⟨85CB1720⟩
Busacca, C. A., **2**, 321 ⟨93JOC3299⟩
Busato, S., **1**, 946 ⟨90T3155⟩; **3**, 457 ⟨91SL243⟩; **5**, 377 ⟨91SL243⟩
Busby, R., **6**, 522 ⟨77IC822⟩
Busch, A., **3**, 404 ⟨73AG584⟩
Busch, H. M., **6**, 78 ⟨52JA554⟩
Busch, K.-H., **3**, 481 ⟨80LA590⟩
Busch, M., **5**, 856 ⟨26CB1162⟩, 856, 860 ⟨15JPR1⟩, 858 ⟨05JPR366⟩; **6**, 653 ⟨00CB1058⟩
Busch, R. D., **3**, 198 ⟨89JMC1886⟩; **5**, 370 ⟨89JMC1886⟩
Busch, T., **5**, 395 ⟨91CB2843⟩
Buschauer, A., **6**, 632 ⟨87AF1003, 87AF1008⟩
Buschbeck, K.-C., **6**, 181 ⟨B-77MI 606-01⟩
Büschges, U., **4**, 695 ⟨86JOM(312)13⟩, 695, 717 ⟨89JOM(372)53⟩
Buschhoff, M., **3**, 47 ⟨67CB1491⟩
Buschmann, J., **3**, 380 ⟨87AG(E)799, 91JA233⟩, 689 ⟨88AG(E)1534⟩; **4**, 868 ⟨88AG(E)1534⟩; **5**, 138 ⟨91JA910⟩
Buse, C. T., **1**, 874 ⟨80JOC1066⟩, 875 ⟨77JA247⟩
Buser, H.-P., **4**, 913 ⟨92JOC178⟩; **6**, 118, 666 ⟨92JOC178⟩
Buser, K. R., **4**, 391, 393 ⟨54JA3577⟩
Busetto, L., **5**, 975 ⟨69JA3197⟩, 1088 ⟨91AG(E)847⟩; **6**, 332 ⟨86G101⟩
Busev, A. I., **5**, 864 ⟨69MI 521-01⟩

Busev, S. A., **2**, 876 ⟨75IZV897⟩
Bush, J. D., **2**, 110 ⟨61USP3014060⟩
Bush, R. D., **3**, 501 ⟨74JA7105⟩; **4**, 610, 611 ⟨75JA7371, 74JA7105⟩
Bush, R. P., **2**, 263 ⟨65JCS7098⟩, 851, 852 ⟨64JCS5585⟩
Bush, T. L., **4**, 650, 656 ⟨93OM338⟩
Bushby, R. J., **1**, 104 ⟨71JA1544⟩; **2**, 230 ⟨75JCS(P1)2513⟩; **3**, 345 ⟨75JCS(P1)2513⟩
Bushey, D. F., **3**, 623 ⟨79JOC300⟩
Bushkov, A. Y., **2**, 731, 732 ⟨79RCR343⟩
Buslaev, Y. A., **5**, 781 ⟨87MI 519-02⟩
Buslovskaya, E. E., **4**, 970 ⟨82ZOR1835⟩
Buss, A. D., **1**, 641 ⟨81CC100⟩, 641, 642 ⟨83TL5293⟩, 641, 642, 757 ⟨83TL111⟩, 641, 753, 756 ⟨85JCS(P1)2307⟩, 641, 755 ⟨84JCS(P1)243⟩, 643 ⟨84TL357, 92JCS(P1)3407⟩, 753 ⟨83TL3931, 83TL593⟩, 757 ⟨87JCS(P1)2569⟩; **3**, 197 ⟨87JCS(P1)2569⟩
Buss, D., **1**, 660, 765 ⟨87CC297⟩, 764 ⟨85TL5093⟩; **2**, 900, 901 ⟨91TL6239⟩
Buss, H., **3**, 683 ⟨1894CB2193⟩; **5**, 821, 834 ⟨1894CB2193⟩
Buss, V., **1**, 891 ⟨73TS253⟩, 894 ⟨71JA3927⟩, 900 ⟨85JA1421⟩, 1109 ⟨70JA2141⟩; **6**, 417 ⟨80LA122⟩
Bussas, R., **2**, 245 ⟨80LA629, 80LA843⟩, 247 ⟨80AG(E)737⟩, 793 ⟨83SR215⟩
Bussche-Hünnfeld, J. L. Von dem, **2**, 568 ⟨92T5719⟩
Busse, P. J., **1**, 438 ⟨75JOM(88)175⟩; **2**, 279 ⟨75JOM(88)175⟩
Busse, U., **4**, 536 ⟨86S737⟩
Bussman, A., **1**, 1001 ⟨75LA2130⟩
Bussmann, A., **2**, 1087 ⟨75LA2130⟩
Bussolari, J., **1**, 1078 ⟨92TL3715⟩
Busson, R., **2**, 167 ⟨76JOC2561⟩
Buswell, R. L., **1**, 630, 632 ⟨75TL1811⟩
Busygin, I. G., **3**, 356 ⟨89ZOR1865⟩
Buszek, K. R., **3**, 267 ⟨88JA3965⟩; **5**, 173 ⟨94JA5511⟩
Butcher, S., **4**, 216 ⟨91JA7767⟩
Butenandt, A., **1**, 795 ⟨36CB882⟩
Butenschön, H., **2**, 301 ⟨91JCS(P1)483⟩
Butera, J., **1**, 140 ⟨85JOC3676⟩; **3**, 350 ⟨85JOC3676⟩
Buter, J., **1**, 685 ⟨72JOC4045⟩, 706 ⟨74TL2805⟩; **3**, 368 ⟨72JOC4045⟩
Büthe, H., **1**, 76 ⟨68AG(E)942⟩
Buthova, O. L., **2**, 436 ⟨82IZV2390⟩
Butin, K. P., **1**, 844 ⟨B-78MI 119-04⟩; **6**, 406 ⟨68BAU414⟩
Butko, Y. D., **2**, 594 ⟨80JOM(201)39⟩
Butkowskyj-Walkiw, T., **4**, 650 ⟨86T1845⟩
Butler, A. R., **2**, 335, 337, 339, 340, 344, 348 ⟨B-68MI 206-01⟩, 761 ⟨B-68MI 215-02⟩
Butler, D. E., **2**, 378, 379 ⟨71JMC1052⟩; **4**, 539 ⟨72JOC215⟩; **5**, 306 ⟨87JMC498⟩
Butler, D. W., **3**, 620 ⟨59JCS4040⟩
Butler, G. B., **1**, 726 ⟨63JOC372⟩; **2**, 462 ⟨60CRV243⟩; **3**, 494 ⟨63JOC217⟩; **6**, 76 ⟨77JOC679⟩
Butler, H., **6**, 237 ⟨92ZAAC(608)69⟩
Butler, I. R., **4**, 706 ⟨85OM2196⟩, 715 ⟨93JOM(463)C6⟩, 1034 ⟨83CJC2354, 86OM1320⟩
Butler, J., **2**, 120 ⟨87OS(66)150⟩
Butler, J. C., **6**, 69, 296, 297 ⟨65JOC411⟩
Butler, J. R., **1**, 1143, 1144 ⟨70JA7493⟩
Butler, P. E., **2**, 359 ⟨68JOC2111⟩; **4**, 15 ⟨67JOC2651⟩; **5**, 248 ⟨67JOC2925, 68JOC1533⟩
Butler, R., **2**, 631 ⟨70JCS(C)1480⟩
Butler, R. N., **2**, 632 ⟨87JCS(P1)1039⟩; **3**, 18, 134 ⟨79TL4583⟩, 442 ⟨84CHEC(6)373⟩, 681 ⟨68CI(L)437⟩, 681, 685 ⟨84CHEC(5)791⟩; **4**, 313 ⟨75JCS(P1)61, 77JCS(P1)282⟩; **6**, 204, 207 ⟨B-81MI 606-02⟩
Butler, W. M., **2**, 880, 893 ⟨83OM1005⟩
Butlin, R. J., **1**, 573 ⟨92JA9673⟩; **5**, 340 ⟨91JCS(P1)3383⟩, 340, 347 ⟨93JCS(P1)1581⟩
Butsugan, Y., **2**, 26 ⟨84S841⟩, 73 ⟨88JOC1831, 88JOM(347)5⟩, 73, 597 ⟨89JOM(369)291, 91JOC2538⟩; **3**, 622 ⟨81BCJ629⟩; **4**, 166 ⟨92JOC742⟩
Butt, M. A., **6**, 558 ⟨87MI 617-02, 91PJS422⟩
Butt, Y., **1**, 420 ⟨71JA1532⟩
Buttenberg, W. P., **1**, 1144 ⟨1894LA(279)324⟩
Butters, M., **2**, 620 ⟨85S1157, 88TL1319⟩
Büttner, G., **2**, 374, 385, 404 ⟨71CB1088, 71CB1104⟩; **4**, 313 ⟨71CB1104⟩; **6**, 439 ⟨74GEP2311662⟩
Buttner, H., **3**, 61 ⟨58CB2452⟩
Buttrus, N. H., **6**, 174, 176, 381 ⟨92JCS(D)1015⟩, 174, 176, 381, 393 ⟨86CC1043⟩, 174, 391 ⟨86CC969⟩, 194, 195 ⟨90JOM(394)57⟩, 381, 392 ⟨87JCS(P2)891⟩, 393, 394, 396 ⟨88JCS(D)381⟩, 394 ⟨88CC1389⟩, 394, 395 ⟨86CC908, 87JOM(320)137, 85CC1380⟩
Butts, M. D., **4**, 710 ⟨92ICA377, 93JA2743⟩, 710 ⟨93OM4269, 93OM65⟩, 710 ⟨94OM1899, 94OM2668⟩
Butts, S. B., **5**, 445, 447 ⟨80JA5093⟩
Butula, I., **4**, 425 ⟨69LA(729)73⟩; **6**, 444 ⟨77S704⟩
Buu-Hoï, N. P., **1**, 104 ⟨44BSF1944⟩; **3**, 91, 98 ⟨55BSF1594⟩, 105 ⟨67JCS(C)779⟩, 280 ⟨61JOC2401⟩, 304 ⟨55JCS3693⟩
Buurman, D. J., **3**, 419 ⟨84TL3763⟩
Buvashkina, N. I., **6**, 497 ⟨88ZOB26⟩
Buxeraud, J., **6**, 618, 623 ⟨82MI 620-02, 82MI 620-01⟩
Buxton, M. W., **5**, 34, 63, 100 ⟨61JCS808⟩; **6**, 16, 38 ⟨63AFC(3)181⟩
Buxton, S. R., **1**, 990 ⟨84TL2887⟩
Buyakov, A. A., **4**, 609 ⟨72ZOB1361⟩, 621 ⟨75ZOB2672⟩, 628, 639, 640 ⟨72ZOB1521⟩, 639, 640 ⟨73ZOB625, 76ZOB837⟩, 660, 700 ⟨73ZOB801⟩; **6**, 188 ⟨72ZOB1521⟩
Buyck, L. de, **4**, 12 ⟨79SC575⟩
Buyle, R., **2**, 746 ⟨68T4217⟩; **4**, 904, 905 ⟨69T3447⟩; **5**, 220 ⟨66HCA1425⟩; **6**, 57 ⟨68T4217⟩
Buynak, J. D., **4**, 353 ⟨89CC89⟩, 514 ⟨85TL5001⟩; **6**, 128 ⟨85TL5001⟩
Buysch, H. J., **6**, 467 ⟨93EUP543234⟩
Buza, D., **6**, 111 ⟨80PJC145⟩
Buza, M., **2**, 23, 24, 30 ⟨71JOC3324⟩, 24 ⟨71JOC2088⟩; **4**, 14 ⟨71JOC3324⟩
Buzbee, L. R., **6**, 523 ⟨58JA6167⟩
Buzby, J. H., **2**, 390 ⟨89TL6495⟩
Buzilova, S. R., **1**, 1006 ⟨74JOU1990⟩; **4**, 106 ⟨88JOU2003⟩
Buzykin, B. I., **3**, 681 ⟨72RCR495, 76ZOR1676⟩; **4**, 132 ⟨91JGU1459⟩, 804 ⟨92JGU1222⟩; **5**, 356 ⟨75ZOR173, 75ZOR348⟩, 652 ⟨76ZOR1557⟩, 681 ⟨77IZV393⟩, 860 ⟨74ZOR2200⟩; **6**, 366, 659 ⟨90ZOB1980⟩, 622 ⟨90ZOB223⟩, 623 ⟨89ZOB714⟩, 634 ⟨90ZOB222, 86ZOB1427⟩, 659 ⟨87ZOB2467, 90ZOB1288⟩, 659 ⟨90ZOB1293, 90ZOB1986⟩, 659 ⟨90ZOB2694⟩, 660 ⟨91ZOB776, 93S59⟩
By, A. W., **2**, 392 ⟨85TL1193⟩, 631 ⟨90SC877⟩
Bychkova, T. I., **6**, 562 ⟨84ZOR2114⟩
Bychkov, V. T., **2**, 887 ⟨86JGU287⟩, 1093 ⟨73JGU626⟩
Bye, T. S., **1**, 450 ⟨54JA1613⟩; **5**, 421 ⟨54JA1372⟩
Byers, D. J., **2**, 643 ⟨44JA1873⟩, 645 ⟨44JA1272⟩
Byers, J. H., **1**, 243 ⟨90TL5697⟩, 244 ⟨91CC354⟩
Byers, J. R., **2**, 36 ⟨43OSC(2)322⟩, 121 ⟨43OSC(2)31⟩
Byistro, V. K., **4**, 343 ⟨76ZOB783⟩; **5**, 400 ⟨90JGU1817⟩
Bykhovskaya, E. G., **4**, 799 ⟨60MI 418-01⟩, 1023 ⟨72ZOB346⟩; **5**, 1076 ⟨85IZV700⟩; **6**, 238, 242 ⟨75IZV2279⟩, 239, 440 ⟨76IZV209⟩, 240, 615 ⟨80JFC(15)169⟩, 614 ⟨76IZV2381, 76IZV212⟩, 615 ⟨83MI 620-02, 85IZV700⟩, 616 ⟨77IZV2379, 81IZV2632⟩, 643 ⟨78MI 612-01⟩
Bykhovskaya, O. V., **4**, 557 ⟨84ZOB36, 88ZOB1998⟩
Bylina, G. S., **5**, 208 ⟨59DOK(128)1201⟩
Byrd, L. R., **1**, 497 ⟨72JOC3881⟩, 1125 ⟨71JA5758⟩
Byrd, N. R., **2**, 388 ⟨55JA4100⟩
Byriel, K. A., **2**, 29 ⟨93JOC2501⟩
Byriel, K. Y., **1**, 572 ⟨93JOC2501⟩
Byrne, B., **1**, 1026 ⟨86JOC2607⟩; **3**, 119 ⟨87TL769⟩, 208 ⟨86S870⟩

Byrne, E. K., **4**, 691 ⟨89OM2001⟩
Byrne, K. J., **4**, 178 ⟨70JOC3375⟩, 199 ⟨72JOC521⟩
Byrne, L. T., **2**, 1003, 1005 ⟨89JCS(D)105⟩; **4**, 646 ⟨89JCS(D)105⟩
Byrne, M. J., **4**, 524 ⟨81IC1328⟩
Byrnes, P. G., **6**, 573 ⟨68JOC851⟩
Byrom, N. T., **4**, 193 ⟨84JCS(P1)1643⟩
Byron, D. J., **5**, 108 ⟨66JCS(C)840⟩
Bystrova, V. M., **2**, 211 ⟨81IZV1850⟩
Bystrov, V. F., **6**, 233, 234, 237 ⟨65ZOB1628⟩
Byun, H., **3**, 3, 114 ⟨86SC1343⟩
Bywater, S., **1**, 911 ⟨B-63MI 119-03⟩; **3**, 573 ⟨55JA3710⟩
Bzowej, E. I., **1**, 826 ⟨93TL1721⟩

C

Caanova, J., **3**, 698 ⟨66JOC3473⟩
Cabal, M. P., **2**, 740 ⟨86S960⟩; **3**, 413 ⟨85CC1375⟩
Caballeira, N., **1**, 800 ⟨B-83MI 118-01⟩
Caballero, C., **4**, 581 ⟨89JOM(371)329⟩
Caballero, F., **5**, 1137 ⟨87JOM(137)229⟩
Caballero, R. B., **5**, 1028 ⟨89CAR(188)35⟩
Cabello, J. A., **3**, 671 ⟨84JOC5195⟩
Cabeza, J. A., **2**, 581 ⟨85JCS(D)573⟩
Cabiddu, S., **1**, 988 ⟨87G625⟩; **4**, 231 ⟨76S797⟩, 347 ⟨84PS(19)61⟩, 391 ⟨86TL4625⟩; **6**, 95 ⟨75S660⟩, 321 ⟨72G558⟩
Cabildo, P., **6**, 162 ⟨88RTC641⟩
Cabioch, J.-L., **4**, 111 ⟨89MI 403-02⟩
Cable, C. J., **6**, 247 ⟨91CC165⟩
Cabon, O., **1**, 8 ⟨92TL7337⟩
Cabral, J. A., **2**, 553, 555 ⟨92JOC3680⟩; **4**, 648 ⟨93JA1551⟩
Cabre, F., **4**, 851 ⟨90TL2169⟩
Cabre, J., **5**, 280, 295 ⟨84S413⟩
Cabré-Castellvi, J., **5**, 188, 195, 199 ⟨81S616⟩
Cabri, W., **1**, 433 ⟨90JCS(P1)428⟩, 494 ⟨92JOC3558⟩; **3**, 293 ⟨90JOC3654⟩
Cacace, F., **2**, 614 ⟨83CC108⟩
Cacchi, S., **1**, 433 ⟨86TL5541⟩, 435 ⟨84TL4821, 90OS(68)138⟩, 546 ⟨84TL3137, 85T5121⟩, 1029 ⟨86S320, 89TL2581⟩, 1099 ⟨84TL2271⟩; **2**, 28 ⟨79S64⟩, 928 ⟨93JOC4579⟩, 983 ⟨92JOC976⟩; **3**, 4, 57, 115 ⟨79S356⟩, 146 ⟨83JOC4236⟩, 215 ⟨78S848⟩, 618 ⟨73JOC1060⟩; **4**, 168 ⟨82JOC2995⟩, 169 ⟨91T1525⟩; **5**, 275 ⟨80S243⟩, 296, 331 ⟨85TL1109⟩, 301 ⟨86TL3931⟩
Caccia, G., **3**, 11 ⟨77CI(M)839⟩
Cachia, P., **1**, 381 ⟨76JCS(P1)359⟩
Caciagli, V., **4**, 205, 210 ⟨79CB148⟩
Cacripante, G., **1**, 1131 ⟨87CJC104⟩
Cadamuro, S., **3**, 283, 309 ⟨87S311⟩, 306, 307 ⟨93JCS(P1)273⟩; **4**, 846, 854 ⟨87G227⟩; **6**, 302, 554, 729 ⟨88S22⟩
Cadenbach, G., **2**, 959 ⟨26ZAAC246⟩
Caderazzo, F., **2**, 893 ⟨84IC1504⟩
Cadet de Gassincourt, L. C., **2**, 494, 505, 506 ⟨1760MI 209-01⟩
Cadiot, P., **1**, 953 ⟨61BSF2176⟩, 956 ⟨66BSF3016, 66BSF3024⟩, 986 ⟨66BSF2885⟩, 1000 ⟨65BSF35, 68BSF1085⟩, 1017, 1018, 1019, 1034 ⟨B-69MI 121-04⟩, 1025 ⟨65TL1953⟩, 1037 ⟨68BSF205⟩, 1142 ⟨55CR(241)1055⟩; **2**, 517 ⟨74TL3435⟩, 521 ⟨70CR(C)589⟩, 824, 828, 841, 1059, 1061, 1063, 1066, 1069 ⟨66BSF1002⟩, 1015 ⟨68BSF193⟩, 1030, 1031, 1034 ⟨66BSF3016⟩, 1059 ⟨60CR(250)866⟩, 1083, 1085 ⟨66BSF3846⟩, 1084 ⟨66BSF1981⟩, 1093 ⟨67BSF777⟩; **4**, 145 ⟨72BSF811⟩; **6**, 208 ⟨68BSF216⟩
Cadogan, J. I. G., **1**, 36, 442 ⟨73JCS(P1)541⟩, 638 ⟨B-79MI 114-01⟩, 784 ⟨86T2135⟩, 790 ⟨93PHC1⟩, 811 ⟨82CC1164⟩, 881 ⟨67JCS(C)1860⟩, 925 ⟨62JCS4468⟩, 254 ⟨62JCS2953⟩; **2**, 692 ⟨77SC365⟩, 694 ⟨61JCS3079⟩, 777 ⟨75JCS(P1)1065⟩, 809, 892 ⟨74JCS(P1)460⟩, 837 ⟨83CC857⟩, 892 ⟨74JCS(P1)466⟩; **3**, 50 ⟨90T7175⟩; **4**, 461 ⟨90SL415⟩; **6**, 154 ⟨63JCS1527⟩
Cadus, A., **2**, 375 ⟨54LA(590)1⟩
Cady, G. H., **2**, 98 ⟨57JA5628⟩; **4**, 3 ⟨51JA4241⟩; **6**, 217, 225 ⟨59JA1089⟩, 231 ⟨48JA3986, 57JA5628⟩, 231 ⟨66IS165, 69JA4432⟩, 236 ⟨67JINC2819⟩, 250 ⟨67JA1962⟩, 250, 424 ⟨67JA5161⟩, 410 ⟨52JA5792⟩, 430 ⟨73IC769⟩, 433 ⟨68JA3954⟩
Caesar, J. C., **3**, 498 ⟨89JCS(P1)2425⟩
Caesar, P. D., **2**, 125 ⟨50JA1033⟩
Cafaggi, S., **3**, 80 ⟨77JHC249⟩
Caffyn, A. J. M., **1**, 200 ⟨93JA6438⟩
Caglioti, L., **1**, 18, 19 ⟨64CI(L)153⟩, 18, 39 ⟨63T1127⟩, 630 ⟨62TL1261, 64G537⟩; **2**, 28 ⟨79S64⟩, 770 ⟨86OPP299⟩; **3**, 133 ⟨79S207⟩, 451 ⟨66JA3865⟩
Caglioto, L., **1**, 39 ⟨66T487⟩
Cagniant, P., **1**, 104 ⟨44BSF1944⟩
Cagnon, C., **4**, 58 ⟨81EUP40153⟩
Cahiez, G., **1**, 125, 465 ⟨93SL45⟩, 230, 476 ⟨80TL935⟩, 465, 1017, 1018 ⟨79S826⟩, 478, 479 ⟨80T1961, 78S528⟩, 479 ⟨78TL2027⟩, 491 ⟨76JOM(113)99⟩, 491, 1108 ⟨74JOM(77)269⟩, 540 ⟨74BSF1656⟩, 541 ⟨84OS(62)1⟩, 1038 ⟨77TL3383⟩, 1045 ⟨89TL3545⟩, 1090 ⟨76S245⟩, 1092 ⟨78TL3711⟩, 1108 ⟨74JOM(77)281⟩; **2**, 74, 574 ⟨89TL7373⟩, 186 ⟨69JOM(121)123, 89T4163⟩, 574 ⟨B-83MI 211-02⟩, 587 ⟨89TL3541, 90TL7423⟩, 591 ⟨94TL1177⟩, 629 ⟨74JOM(77)269⟩, 629, 977 ⟨84OS(62)1⟩, 676 ⟨77BSF693⟩, 919 ⟨79S826⟩; **3**, 130 ⟨79SC639, 84T683⟩, 130 ⟨92TL4439⟩, 152 ⟨84S37⟩, 211, 273 ⟨85S50⟩, 222 ⟨80T1961, 89TL3545⟩, 303, 306 ⟨92TL5245⟩; **4**, 963 ⟨77BSF693⟩
Cahours, A., **2**, 461, 853 ⟨1857LA(101)1⟩; **5**, 19 ⟨1857LA(104)111⟩
Cai, D., **2**, 66 ⟨93TL3243⟩
Cai, S., **3**, 521 ⟨88CC1489⟩
Cai, Y. M., **6**, 242 ⟨93JOM(448)219⟩
Cai, Z., **1**, 647 ⟨87JOC3558⟩
Caillaud, G., **3**, 254 ⟨70BSF2018⟩
Caillaux, B., **1**, 571 ⟨85T3331⟩; **2**, 1046 ⟨76C387⟩; **5**, 577, 617 ⟨85T3331⟩
Caille, J.-C., **1**, 49 ⟨88JOC477⟩
Caillebourdin, S., **1**, 826 ⟨90TL4147⟩
Caillot, H. J., **2**, 578 ⟨84NJC759⟩
Cain, E. N., **1**, 400, 799 ⟨69JA4322⟩, 865 ⟨70JA3821⟩
Cain, M. E., **2**, 144, 173 ⟨61JCS5339⟩
Cain, P. A., **1**, 651 ⟨78JA1548⟩; **3**, 496 ⟨78JA1548⟩
Cain, P., **1**, 99 ⟨75JA380, 75JOC3606⟩; **3**, 219, 231 ⟨75JA380⟩
Caine, D., **1**, 598 ⟨92T33⟩, 871 ⟨69JOC3070⟩, 873 ⟨B-79MI 119-02⟩, 889 ⟨76OR(23)1⟩; **2**, 696 ⟨76OR(23)1⟩; **3**, 10 ⟨B-79MI 301-01⟩, 12, 127 ⟨76OR(23)1⟩, 124 ⟨B-79MI 304-01⟩, 143 ⟨85JOC175⟩; **4**, 818 ⟨85JOC2195⟩
Cainelli, G., **1**, 659, 764 ⟨66TL4315⟩, 679 ⟨70T1281⟩, 680 ⟨67TL1581⟩; **2**, 19, 25, 31, 35 ⟨76S472⟩, 21, 23 ⟨80CC1278⟩, 25, 31 ⟨83S306⟩, 28, 30 ⟨80S143⟩, 103 ⟨85TL3369⟩, 316 ⟨90JOC4199⟩, 544 ⟨90PAC605⟩, 1006 ⟨70S561⟩; **3**, 4, 115 ⟨76JA6737⟩, 20 ⟨78JOC1598⟩, 253 ⟨83JHC1753⟩, 440, 456 ⟨87TL5369, 91TL2967⟩, 456, 457, 458 ⟨90PAC605⟩, 456, 458 ⟨86TL1695⟩, 457 ⟨91SL243, 90JOC4199⟩, 458 ⟨90TL3481, 93T3809⟩; **4**, 146 ⟨71T6109⟩, 664 ⟨65TL3429⟩, 669 ⟨70T1281⟩; **5**, 287 ⟨86TL1695, 87TL5369⟩, 377 ⟨91SL243, 94S805⟩; **6**, 198 ⟨65TL3429⟩, 198 ⟨66TL4315⟩
Cainelli, G. F., **3**, 621, 625 ⟨75JCS(P1)1273⟩
Caira, M., **6**, 350, 694 ⟨84CC698⟩
Cairns, H., **3**, 297 ⟨85JMC1832⟩

Cairns, N., **3**, 297 ⟨92T7581⟩
Cairns, P. M., **1**, 811 ⟨82TL1405⟩; **5**, 574 ⟨86CC1756⟩
Cairns, S. M., **1**, 50 ⟨87JOC4829⟩
Cairns, T. L., **1**, 299 ⟨73OR(20)1⟩; **2**, 130 ⟨52JA3982⟩; **3**, 352 ⟨60JA1515⟩; **4**, 325 ⟨48JA2115⟩; **5**, 579, 580 ⟨53JOC748⟩
Cajthanu, C. E., **1**, 131 ⟨93TL5213⟩
Calabrese, J. C., **1**, 1013 ⟨90JA4324⟩; **2**, 43 ⟨93CJC930⟩, 206 ⟨93TL3667⟩, 527 ⟨92JA9350⟩, 926, 988 ⟨84JA6422⟩, 979 ⟨84OM1223⟩; **3**, 517 ⟨77JA2127⟩; **4**, 42 ⟨85JA4565⟩, 152 ⟨86JA5359⟩, 582 ⟨87CC1543⟩, 678 ⟨83JA1679⟩; **5**, 447 ⟨76JA1166⟩; **6**, 229 ⟨85JA4565⟩, 722, 733 ⟨93OM3405⟩
Calabretta, P. J., **2**, 889 ⟨65IC973⟩
Cal, K., **3**, 620 ⟨86JA1311⟩
Calas, B., **5**, 185 ⟨85T5331⟩
Calas, P., **1**, 1130 ⟨82CC433⟩
Calas, R., **1**, 284 ⟨76S737⟩, 963 ⟨81CC1094⟩, 963, 978 ⟨81TL3401, 75JOM(93)43⟩, 1038, 1138 ⟨76TL2449⟩; **2**, 190 ⟨76TL1197⟩, 608 ⟨77AG(E)349⟩, 916, 923, 940 ⟨75JOM(87)169, 81TL3401⟩, 943 ⟨85CR(301)1289⟩, 945 ⟨79JOC4275⟩, 1090 ⟨69CR(C)72, 76MI 223-01⟩; **3**, 140 ⟨79JOC3397⟩, 140 ⟨81TL2883⟩, 213 ⟨76TL1871⟩, 230 ⟨81TL3401⟩; **4**, 355 ⟨75JOM(87)151, 77JOM(142)C35⟩, 608 ⟨74JOM(71)377⟩, 624 ⟨88JOM(341)125⟩, 933 ⟨72JOM(43)157⟩, 1014 ⟨72JOM(43)139, 84OM1660⟩; **5**, 416 ⟨72JOM(43)157⟩, 417 ⟨78JOM(161)C23⟩, 419 ⟨69JOM(19)P33, 79JOC420⟩, 419, 420 ⟨71JOM(26)183, 75JOM(93)51⟩; **6**, 27 ⟨75JOM(93)43⟩, 124 ⟨71JOM(26)183, 79JOC420⟩, 125 ⟨72JOM(39)C49⟩, 125, 162 ⟨79JOM(177)137⟩, 162 ⟨71JOM(32)79⟩
Calcagno, M. A., **1**, 267 ⟨78JOC4207⟩
Calder, A., **2**, 390, 397 ⟨88OSC(6)803⟩, 764 ⟨72OS(52)77⟩
Caldera, P., **3**, 349 ⟨82JOC3319⟩
Calder, G. V., **3**, 539 ⟨73JA6134⟩, 542 ⟨73JA4061⟩
Calder, I. C., **5**, 328 ⟨81S121⟩
Calderari, G., **2**, 783, 785, 786 ⟨85T4861⟩; **3**, 192 ⟨85T4861⟩; **4**, 213 ⟨84T1313⟩, 335 ⟨85T4861⟩
Calderazzo, F., **2**, 881, 883, 886, 890 ⟨84JCS(D)2365⟩, 890, 891, 893 ⟨88IC3730⟩; **5**, 441 ⟨77AG(E)299⟩, 447 ⟨66JOM(6)265⟩, 975, 976 ⟨88CB1891⟩; **6**, 412 ⟨82MI 614-01⟩, 517 ⟨79IC1165⟩, 522 ⟨57JA3611, 59G809⟩, 523 ⟨65IC293⟩
Calderon, O., **3**, 282 ⟨84LA1390⟩
Caldin, E. F., **5**, 24 ⟨53QR255⟩
Caldwell, A. G., **5**, 276 ⟨46JCS599⟩, 342 ⟨89JCS(P1)1815⟩
Caldwell, C. G., **2**, 540 ⟨81TL4999, 83JOC3252⟩; **3**, 209 ⟨83JOC3252⟩
Caldwell, J., **1**, 809 ⟨73JCS(P1)1913⟩
Caldwell, J. M., **5**, 655 ⟨87JCS(P1)1153⟩; **6**, 572 ⟨87JCS(P1)1153⟩
Caldwell, R. A., **1**, 847 ⟨69JA529⟩, 848 ⟨69JA527⟩
Calet, S., **1**, 225 ⟨88TL1763⟩, 242 ⟨87OM1625, 89JOC20⟩
Caley, B. A., **4**, 59 ⟨89TL2033⟩; **6**, 416, 428 ⟨89TL2033⟩
Calfee, J. D., **6**, 217 ⟨56USP2748177⟩
Caliman, V., **4**, 549 ⟨94AG(E)2202⟩
Calin, M., **2**, 1007 ⟨68JA938⟩
Calinaud, P., **4**, 226 ⟨83CJC2103⟩
Calingaert, G., **1**, 114 ⟨44JA1389⟩; **2**, 494 ⟨42JA392⟩, 720 ⟨25CRV43⟩
Calisen, L., **5**, 821 ⟨1893LA(274)95⟩
Calkins, A. E., **5**, 10 ⟨55OSC(3)547⟩
Calkins, T. L., **1**, 1009 ⟨92TL2315⟩
Callaghan, P. D., **5**, 641 ⟨70JCS(C)2106, 74CJC879⟩
Callahan, F. M., **5**, 277 ⟨64JA1839⟩
Callahan, J. F., **1**, 78 ⟨83JOC3162⟩
Call, L., **2**, 405 ⟨73JA1677⟩; **5**, 342 ⟨68M1289⟩
Callant, P., **3**, 206 ⟨81T2085⟩, 532 ⟨84SC163⟩
Callaway, J. O., **2**, 597 ⟨75IC2809⟩
Callen, J. E., **3**, 459 ⟨55OSC(3)26⟩
Callens, R., **3**, 638 ⟨81TL141⟩
Calligaris, M., **2**, 783 ⟨74JCS(P1)660⟩; **4**, 683 ⟨92OM154⟩, 711 ⟨94OM1999⟩
Callot, H. J., **4**, 489 ⟨72CJC1078⟩
Calò, V., **1**, 187 ⟨78CC1085⟩, 238 ⟨83JCS(P1)2953⟩, 592 ⟨76S200⟩, 1025 ⟨79TL3873⟩, 1123 ⟨65TL4399⟩; **2**, 28 ⟨77JCS(P1)501⟩, 148 ⟨71IJS(A)130⟩, 626 ⟨71JCS(C)3652⟩; **3**, 153 ⟨78S139⟩, 154 ⟨77JCS(P1)501⟩
Calogeropoulou, T., **3**, 198 ⟨86TL4265⟩, 198 ⟨87JOC4185⟩, 198 ⟨90JOC2842⟩
Calstyanand, L. Kh., **6**, 25 ⟨91RCR1318⟩
Calverley, M. J., **3**, 39 ⟨90SL155⟩
Calvin, M., **2**, 212 ⟨61JA4357⟩, 379 ⟨74JMC948⟩; **3**, 83, 96 ⟨46JA2502⟩; **6**, 110 ⟨71JA4075⟩
Calzada, J. G., **1**, 486 ⟨73JOC2574⟩, 1053 ⟨85TL271⟩; **2**, 7 ⟨88OSC(6)634⟩
Calzavara, P., **4**, 76 ⟨73JA7431⟩
Cama, L., **3**, 287 ⟨80TL2013⟩; **5**, 233 ⟨80TL2013⟩
Cama, L. D., **1**, 214 ⟨74JA5268⟩; **4**, 98 ⟨87H(25)221⟩; **5**, 334 ⟨87H(25)221⟩
Camaggi, G., **4**, 7 ⟨70JCS(C)178⟩, 737 ⟨71JCS(C)925⟩; **6**, 251 ⟨68CI(M)197⟩
Camara, J., **1**, 944 ⟨92T9261⟩; **4**, 249 ⟨93TL1141⟩; **6**, 551 ⟨92JA7909⟩
Camarasa, M. J., **5**, 1039 ⟨84S509⟩
Cambanis, A., **4**, 192 ⟨89S128⟩; **6**, 252 ⟨88S961⟩
Cambell, S. F., **4**, 953 ⟨87JCS(P1)2189⟩
Cambi, L., **5**, 831 ⟨09MI 521-01, 11G166⟩
Cambie, R. C., **1**, 1044 ⟨81JCS(P1)40⟩; **2**, 12 ⟨81JCS(P1)2608⟩, 24 ⟨77JCS(P1)226⟩, 34 ⟨78JCS(P1)126⟩, 45 ⟨74JCS(P1)1858, 80OS(59)169⟩, 631 ⟨76JCS(P1)1161⟩; **3**, 8 ⟨78JCS(P1)1483⟩, 155 ⟨78JCS(P1)126⟩; **4**, 27 ⟨81JCS(P1)2608⟩; **5**, 168 ⟨74JCS(P1)1864⟩, 384 ⟨91JOM(420)387⟩, 729 ⟨81JCS(P1)40⟩, 764 ⟨76AJC357⟩, 1044 ⟨79JCS(P1)757⟩, 1107 ⟨84S1017⟩, 1107, 1109 ⟨81JCS(P1)58⟩, 1115 ⟨78S152⟩; **6**, 131 ⟨88JOM(348)317⟩, 314, 565 ⟨81JCS(P1)52⟩
Cambon, A., **2**, 675 ⟨93BSF664⟩; **5**, 991 ⟨92JFC(56)85⟩, 1022 ⟨91JFC(53)1⟩; **6**, 608 ⟨92TL2339⟩
Cambou, B., **5**, 118 ⟨84MI 502-04⟩, 135 ⟨84JA2687⟩
Camerini, R., **5**, 377 ⟨94S805⟩
Cameron, A. F., **2**, 1004 ⟨75JCS(P2)1030⟩
Cameron, A. F. B., **3**, 229 ⟨52JCS1094⟩
Cameron, A. G., **1**, 831 ⟨84CC827, 86JCS(P1)161⟩; **3**, 254 ⟨82TL561⟩, 496 ⟨83JCS(P1)2979⟩; **4**, 925 ⟨83JCS(P1)2979⟩; **5**, 227 ⟨84CC827⟩
Cameron, A. M., **4**, 314 ⟨69JOC3230⟩
Cameron, D. G., **4**, 129 ⟨88PS(40)183⟩
Cameron, D. W., **2**, 478 ⟨77AJC2479⟩
Cameron, M. L., **4**, 730 ⟨90JOC4448⟩
Cameron, M. P., **5**, 84 ⟨B-58MI 502-01⟩, 85 ⟨B-53MI 502-01⟩
Cameron, N., **5**, 781 ⟨86JCS(D)1359⟩
Cameron, T. S., **1**, 702 ⟨88CJC2860⟩; **2**, 249 ⟨78CJC1183⟩; **4**, 258 ⟨78CJC1183⟩; **5**, 348 ⟨78CJC1183⟩
Caminade, A.-M., **4**, 552 ⟨86IC712⟩; **6**, 373 ⟨84CC1622⟩, 685 ⟨89TL4813⟩, 689 ⟨89TL6869⟩
Camkerten, N., **3**, 162 ⟨90SC2279⟩
Cammenga, H. K., **4**, 352, 353 ⟨92CB591⟩
Camp, D. J., **2**, 127 ⟨92JCS(P1)255⟩
Campagna, F., **3**, 618 ⟨79S56⟩, 618, 619 ⟨77TL1813⟩
Campagnole, M., **2**, 97 ⟨90SC349⟩
Campaigne, E., **2**, 27 ⟨63OSC(4)921⟩, 123 ⟨57JOC561⟩; **3**, 91, 99 ⟨63OSC(4)331⟩, 104 ⟨63OSC(4)918⟩, 342, 360 ⟨B-66MI 308-01⟩, 343 ⟨62JOC3760⟩, 362 ⟨50JOC600⟩;

4, 245 ⟨59JA3148⟩; 5, 83 ⟨63OSC(4)919⟩, 597 ⟨76SC387⟩, 599 ⟨47JA1211⟩
Camparini, A., **2**, 316 ⟨92JHC321⟩
Camp, R. L., **5**, 354, 359 ⟨73TL4091⟩
Campbell, A., **1**, 398 ⟨47JCS93⟩; **5**, 214 ⟨46JCS25⟩
Campbell, A. C., **4**, 56 ⟨85JCS(P1)1567⟩
Campbell, A. D., **2**, 607, 617 ⟨54JCS3659⟩
Campbell, A. L., **1**, 120 ⟨88JA2641⟩, 207 ⟨88TL893⟩; **2**, 992 ⟨88JA2641⟩; **5**, 288, 290, 297, 302 ⟨87S421⟩, 557 ⟨83JA5015⟩
Campbell, B., **2**, 449 ⟨⟨44JA82⟩
Campbell, B. K., **1**, 502 ⟨42CRV77⟩, 1040 ⟨48JA3868⟩; **5**, 25, 108 ⟨46JA1844⟩
Campbell, C. B., **1**, 836, 838 ⟨78JA4822⟩
Campbell, C. D., **2**, 451 ⟨84JCS(P1)2845⟩; **4**, 464 ⟨84JCS(P1)2845⟩
Campbell, C. L., **3**, 442 ⟨84JOC2512⟩
Campbell, G. A., **2**, 761 ⟨72JA3884⟩, 762 ⟨72T2749⟩
Campbell, I. D., **1**, 1103 ⟨65OS(45)39⟩
Campbell, I. G. M., **2**, 858 ⟨65JCS2184⟩, 884 ⟨52JCS4448⟩; **4**, 728 ⟨71JCS(C)1836⟩
Campbell, J. A., **5**, 342 ⟨92JOC5328⟩
Campbell, J. D., **1**, 595 ⟨67JOC1631⟩; **4**, 1054 ⟨78JOC950⟩; **6**, 182, 199 ⟨78JOC950⟩
Campbell, Jr., J. B., **1**, 488 ⟨86JOC3398⟩, 488, 1094 ⟨80JOC549⟩, 488, 510, 1102, 1103 ⟨80JOC389⟩, 532, 1094 ⟨80JOC550⟩, 1094, 1100, 1102 ⟨81MI 122-01⟩; **2**, 525, 900, 901, 904 ⟨80JOC389⟩, 900 ⟨80JOC550, 86JOC3398⟩, 901 ⟨80JOC549⟩
Campbell, J. R., **1**, 562 ⟨79S117⟩; **2**, 23, 26 ⟨66CJC2339⟩
Campbell, K., **2**, 449 ⟨44JA82⟩
Campbell, K. D., **5**, 453 ⟨84JA2853⟩
Campbell, K. N., **1**, 456, 508 ⟨41JA216⟩, 502 ⟨42CRV77⟩, 1040 ⟨48JA3868⟩; **5**, 25, 108 ⟨46JA1844⟩
Campbell, M. M., **2**, 1064 ⟨84CC200⟩; **3**, 351 ⟨73JCS(P1)2866⟩, 479 ⟨76CC538, 77JCS(P1)2298⟩; **4**, 426 ⟨78CRV65⟩, 472 ⟨90TL1759⟩, 487 ⟨80CC730, 82T2513⟩; **5**, 573 ⟨83T1075, 87TL2171⟩, 574 ⟨89TL1997⟩, 867 ⟨77JCS(P1)2298⟩; **6**, 121 ⟨90TL4359⟩, 578 ⟨80JCS(P1)766⟩
Campbell, N., **5**, 199 ⟨49JCS1555⟩
Campbell, R., **5**, 361 ⟨63JOC2294⟩
Campbell, R. D., **2**, 787 ⟨60JOC1877⟩
Campbell, R. H., **6**, 304, 305 ⟨67JOC2567⟩
Campbell, R. W., **2**, 204 ⟨69JOC3097⟩
Campbell, S., **1**, 654, 713, 716 ⟨87CC106⟩
Campbell, S. F., **1**, 700, 703 ⟨87JCS(P1)2171⟩; **2**, 35 ⟨68JA6225⟩; **5**, 173 ⟨85CC1386⟩, 690 ⟨88JMC1031⟩
Campbell, T. W., **2**, 89 ⟨50JA2735⟩, 729 ⟨45JA1965⟩, 753 ⟨49JA740⟩; **3**, 87 ⟨53JA734⟩; **6**, 643 ⟨62JA3673⟩
Campbell, W. H., **3**, 578 ⟨77H(7)247⟩
Campelo, J. M., **3**, 671 ⟨84JOC5195⟩
Campestrini, S., **3**, 5, 117 ⟨87JOC5467⟩, 275 ⟨90JOC3658⟩
Campi, E., **2**, 928 ⟨87S1032⟩; **3**, 77 ⟨87S1033⟩
Campion, B. K., **3**, 598 ⟨87JA2049⟩; **4**, 656, 657 ⟨93JA5527⟩; **5**, 422 ⟨87JA2049⟩; **6**, 521 ⟨87JA2049⟩
Campora, J., **6**, 332 ⟨88OM2577⟩
Campos, M. D. M., **2**, 856 ⟨65CI(L)1076⟩, 857 ⟨68CB3070⟩
Campos, M. P. A., **2**, 629 ⟨79JOC3931⟩
Campos, P., **2**, 31, 35 ⟨90JCS(P1)2807⟩; **4**, 300 ⟨88JCS(P1)1631⟩
Campos, P. J., **1**, 68 ⟨80JCS(P1)1420, 83T2863⟩, 990 ⟨88JA5567⟩, 1040 ⟨89S33⟩; **2**, 18, 24, 31, 34 ⟨85AG(E)319⟩, 18, 610 ⟨91JOC2234⟩, 31, 35 ⟨87CC1491⟩, 629 ⟨90JOC3104⟩, 631 ⟨84JCS(P1)2623⟩, 1019 ⟨87S661⟩; **4**, 152, 153 ⟨79S893⟩, 412 ⟨88CB1813⟩, 453, 477 ⟨90SL261⟩, 749 ⟨93AG(E)893⟩, 749, 819, 1064 ⟨88JA5567⟩, 760 ⟨90TL2751⟩, 772 ⟨90TL7375⟩; **5**, 132 ⟨83S649⟩, 152 ⟨83T2863⟩, 1044 ⟨87CC1491⟩, 1108 ⟨90TL7375⟩
Campostrini, R., **3**, 712 ⟨85JOM(294)C21⟩
Camps, F., **1**, 179 ⟨88SC445⟩, 527 ⟨85TL6397, 87TL4745⟩, 747 ⟨84T2871⟩; **2**, 8 ⟨87S511⟩, 18 ⟨89JOC4294, 87TL4733⟩, 624 ⟨88TL5811⟩, 684 ⟨82S186⟩, 1092 ⟨90TL2739⟩; **3**, 128 ⟨86T4603⟩, 218 ⟨85TL6397, 87TL4745⟩; **4**, 810 ⟨88TL5811⟩; **6**, 695 ⟨83TL3387⟩
Camus, A., **4**, 696 ⟨79JOM(174)121⟩
Canada, Jr., E. D., **4**, 24 ⟨83JA3411⟩
Canal, G., **1**, 1040 ⟨89S33⟩; **4**, 412 ⟨88CB1813⟩, 453, 477 ⟨90SL261⟩
Canary, J. W., **3**, 219 ⟨90JOC2237⟩
Canas, M., **2**, 84, 318 ⟨91TL6931⟩
Canavan, A. E., **4**, 569 ⟨59JCS3751⟩, 569, 570 ⟨62JCS592⟩
Canceill, J., **6**, 33 ⟨85JA6993⟩
Candiani, I., **1**, 494 ⟨92JOC3558⟩; **3**, 293 ⟨90JOC3654⟩
Candlin, J. P., **1**, 82, 84 ⟨68MI 103-01⟩, 456 ⟨68MI 110-01⟩, 507 ⟨68MI 112-01⟩
Cane, D. E., **1**, 643, 644 ⟨69JOC3053⟩, 831 ⟨88JOC4923⟩; **2**, 1045 ⟨70JOC3405⟩; **5**, 235 ⟨88JOC4923⟩
Canepa, C., **3**, 653 ⟨89JCS(P1)105⟩
Canevotti, R., **4**, 168 ⟨87JCS(P1)2301⟩
Canfield, J. H., **2**, 463 ⟨59JA148⟩
Canghlan, C. N., **2**, 831 ⟨71JA5236⟩
Cannan, R. K., **5**, 83 ⟨50MI 502-01⟩
Cann, R. O., **5**, 329 ⟨94JOC5292⟩
Cannell, L. G., **1**, 357 ⟨72JA6867⟩
Cannic, G., **3**, 532 ⟨68BSF4913⟩
Cannizzo, L., **1**, 769 ⟨83PAC1733⟩; **3**, 508 ⟨83PAC1733⟩
Cannizzo, L. F., **1**, 675 ⟨87MM1169⟩, 767 ⟨85JOC2386⟩; **2**, 681 ⟨85JOC2386⟩, 702 ⟨85JOC2316⟩
Cannon, J. B., **5**, 460 ⟨72JA6226⟩
Cannon, J. G., **5**, 277, 299 ⟨70TL1791⟩
Cannon, J. R., **4**, 856 ⟨80AJC1073⟩
Cannone, P., **3**, 131, 405, 459 ⟨80TL155⟩
Cano, F. H., **4**, 408 ⟨88CC410⟩; **5**, 1066 ⟨89CB307, 89T4263⟩, 1068 ⟨89JA355⟩
Cano, M., **5**, 1142 ⟨82ICA127⟩
Cano, M. C., **5**, 366 ⟨91JCS(P1)273⟩
Canonica, S., **3**, 59 ⟨88S697⟩
Cantacuzène, D., **1**, 522 ⟨75TL363⟩
Cantillo, J. H., **2**, 401, 402 ⟨83JOC1⟩
Cantor, P. A., **1**, 42 ⟨56JOC918⟩
Cantrel, J. L.,**4**, 10 ⟨81JOC3917⟩
Cantrell, W. R., **1**, 812 ⟨91JOC723⟩
Canziani, F., **5**, 461 ⟨75CC267⟩
Cao, D., **4**, 717 ⟨93OM996⟩
Cao, S., **6**, 546 ⟨92MI 617-04⟩
Capdevielle, P., **5**, 750 ⟨93TL6395⟩
Capdevila, J. H., **4**, 175 ⟨94TL1825⟩
Capella, L., **2**, 918 ⟨91S1201, 93TL3311⟩, 918, 939 ⟨89JOC1473⟩
Capillon, J., **2**, 69 ⟨73T3659⟩
Caple, G., **1**, 419, 420 ⟨65TL385⟩, 419, 420, 794 ⟨73T3781, 73T3797⟩
Caple, R., **1**, 530 ⟨86TL1241, 89S472⟩
Caplier, I., **1**, 53 ⟨71T5523⟩; **2**, 465, 989 ⟨61JA4406⟩; **4**, 112 ⟨71T5523⟩
Caplow, M., **2**, 807 ⟨67JA6972⟩
Capman, R. D., **2**, 386 ⟨88S743⟩
Capmau, M. L., **1**, 1055 ⟨69CR(C)342, 73BSF3371⟩; **4**, 802 ⟨71CR(C)1107, 75BSF2259⟩
Capocasale, D., **1**, 1110 ⟨73JCS(P2)2078⟩
Capon, B., **1**, 869 ⟨B-90MI 119-03⟩; **2**, 636, 638 ⟨88ACR135, 81JA1761⟩, 639 ⟨86TL3275, 87PAC1577⟩

Caporusso, A. M., **1**, 198 ⟨88G857⟩, 466 ⟨91JOM(417)289⟩, 1016 ⟨87TL6073⟩
Capozzi, G., **1**, 1003 ⟨82CC959⟩, 1111 ⟨70JCS(C)2625⟩, 1112 ⟨68TL4039, 69CC1520⟩, 1112 ⟨71JCS(C)3018, 72JCS(P1)1136⟩, 1112 ⟨72JCS(P1)216⟩; **2**, 192, 212, 216 ⟨B-74MI 203-04⟩
Capparella, G., **1**, 79 ⟨74JOM(70)89⟩
Cappelletti, R., **4**, 852 ⟨74AC(R)305⟩
Capperucci, A., **2**, 140 ⟨91JOC7323⟩, 935 ⟨92SL883⟩; **3**, 336, 340, 346, 363 ⟨91JOC7323⟩, 346, 353 ⟨93TL873⟩; **4**, 944 ⟨92SL499, 92SL883⟩, 951 ⟨91TL1899⟩; **5**, 417 ⟨88JOC3612⟩, 427 ⟨92SL332, 92TL1507⟩, 427, 428 ⟨92SL329⟩, 467 ⟨89JOC2966⟩, 648, 650 ⟨89JOC19⟩
Capponi, M., **2**, 638 ⟨86HCA1331⟩
Caprathe, B. W., **3**, 433 ⟨85JOC3938⟩; **5**, 306 ⟨87JMC498⟩
Capriel, P., **4**, 33, 34, 38 ⟨79T2661⟩
Caproiu, M. T., **1**, 1132 ⟨83T3943⟩
Capuano, L. A., **1**, 630, 632 ⟨75TL1811⟩
Capuano, L., **3**, 417 ⟨86LA132⟩, 547 ⟨81LA1361⟩, 585 ⟨92LA15⟩, 586 ⟨87LA183⟩, 586, 587 ⟨83CB741⟩, 592 ⟨81CB1976, 82LA80⟩, 592, 593 ⟨85LA2305⟩; **4**, 417 ⟨70CB3459⟩, 980, 981, 982 ⟨81CB1976⟩; **5**, 634 ⟨85LA2305⟩, 831 ⟨70CB82⟩, 1064, 1066 ⟨88CB271⟩
Caputo, O., **4**, 259 ⟨89EJM635⟩
Caputo, R., **1**, 587 ⟨81TL3551⟩, 885 ⟨91COS(1)139⟩; **2**, 302 ⟨92SC2305⟩, 565, 566 ⟨91COS(1)139⟩, 688 ⟨93SC1515⟩, 724 ⟨81S888⟩; **3**, 164 ⟨89S464⟩, 170 ⟨93SC1515⟩
Caram, J. A., **2**, 1008 ⟨87CJC2000⟩
Caramella, P., **2**, 793 ⟨83T689⟩; **3**, 681, 683 ⟨B-84MI 320-03⟩
Caranova, Jr., J., **2**, 757 ⟨73OSC(5)273⟩
Carbon, J. A., **6**, 642 ⟨62JOC2504⟩
Carbonaro, A., **1**, 519 ⟨68JOC3948⟩
Carboni, B., **1**, 327 ⟨92TL2673⟩; **2**, 182 ⟨92SL581⟩, 318 ⟨87T1799⟩, 900 ⟨87TL4169⟩; **5**, 199 ⟨87TL4169⟩
Carboni, R. A., **5**, 839 ⟨56USP2817662, 58JA5793⟩
Carceller, R., **4**, 843 ⟨89H(29)1877⟩
Cardaci, G., **5**, 799 ⟨91G101⟩
Cardani, S., **1**, 208 ⟨86S1032⟩; **2**, 46 ⟨88T491⟩
Card, P. J., **2**, 1003 ⟨83JA6096⟩; **4**, 37 ⟨91JOC1663⟩
Carde, A. M., **2**, 152 ⟨77TL4009⟩
Carde, R. T., **2**, 152 ⟨77TL4009⟩
Cardellicchio, C., **1**, 470 ⟨92JOC1718⟩; **3**, 130, 285, 306 ⟨87TL2053⟩, 285 ⟨85TL3595⟩; **4**, 776 ⟨92TL5121⟩
Cárdenas, C. G., **1**, 797 ⟨80JOC900⟩
Cardi, N., **1**, 933 ⟨92TL1787⟩
Cardillo, G., **2**, 23 ⟨94CC735⟩, 87 ⟨81S793⟩; **3**, 4 ⟨76S394⟩, 4, 115 ⟨76JA6737⟩, 57, 89 ⟨76CC190, 76TL3985⟩, 155 ⟨77JOC4268⟩, 158 ⟨84JOC701⟩; **4**, 99 ⟨92JOC1082⟩; **5**, 1125 ⟨89TL1679⟩; **6**, 469 ⟨82JOC4626⟩
Cardin, D. J., **2**, 565 ⟨82COMC-I(3)559, 82COMC-I(3)635⟩; **5**, 931, 940 ⟨72CRV545⟩; **6**, 716 ⟨73CSR99⟩, 716, 721 ⟨72CRV545, 71CC400⟩, 721 ⟨72CC927⟩
Cardinaux, F., **6**, 469 ⟨79HCA398⟩
Cardwell, H. M. E., **1**, 871 ⟨51JCS2430, 51JCS2442⟩
Carey, F. A., **1**, 13 ⟨69JOC4, 71JOC758⟩, 649 ⟨73JOC2664⟩, 760 ⟨82JOC3548⟩; **2**, 709 ⟨72JOC939⟩, 936 ⟨73JOC2664⟩; **4**, 252 ⟨71JOC2731⟩, 260 ⟨76JOC3975⟩, 263 ⟨76JOC3979⟩, 266 ⟨81PS(10)169⟩, 268 ⟨81PS(10)163⟩, 574 ⟨72JOC939⟩; **6**, 128 ⟨76JOC3975⟩, 244 ⟨75JA187⟩, 555 ⟨65JA934⟩
Carey, J. J., **4**, 950 ⟨91T6539⟩
Carey, J. V., **2**, 862 ⟨91JCS(P2)2081⟩
Carey, P. R., **5**, 558 ⟨82CJC199⟩
Carey, R. B., **5**, 84 ⟨48MI 502-02⟩
Carey, R. I., **2**, 127 ⟨93JOC2216⟩, 229 ⟨89JOC3640⟩
Carfagna, C., **1**, 471, 483 ⟨91JOC261⟩
Cargill, R. L., **1**, 74 ⟨73JA4346⟩, 419 ⟨62JOC1910⟩, 795, 801 ⟨72MI 118-01⟩; **3**, 212 ⟨73JOC3829⟩, 623 ⟨79JOC300⟩
Cariati, F., **5**, 946, 959 ⟨94ICA233⟩
Carignano, I., **3**, 280 ⟨91TL6709⟩
Carilla, J., **6**, 30 ⟨91JOC103⟩
Carini, D. J., **1**, 450, 527, 1061 ⟨83T935⟩, 962, 1055 ⟨80JOC3925⟩, 1056, 1058 ⟨86JOC3870⟩; **2**, 930, 940, 943 ⟨83T935⟩; **4**, 780 ⟨91JMC2525⟩
Cariou, M., **3**, 652, 654 ⟨83T1551⟩; **6**, 648 ⟨91CJC861⟩
Carisi, P., **6**, 127 ⟨87JCS(P1)2647⟩
Carithers, R., **3**, 579 ⟨68JA1923⟩
Carius, L., **2**, 100 ⟨1858LA(106)334, 1859LA(111)93⟩, 135 ⟨1861LA(119)313⟩
Carless, H. A. J., **1**, 784 ⟨72JA7235⟩; **2**, 39 ⟨92CC234⟩, 675 ⟨80CC980⟩
Carli, B., **3**, 221 ⟨83S125⟩
Carlier, P. R., **2**, 84 ⟨88JOC5185⟩; **4**, 779 ⟨77CJC3190⟩
Carlin, R. B., **5**, 293 ⟨47JA2007⟩
Carling, R. W., **1**, 177, 178 ⟨92JCS(P1)95⟩, 1021 ⟨86TL6133⟩; **2**, 328 ⟨88TL6985⟩
Carlon, F. E., **3**, 156 ⟨68JOC3294⟩
Carlón, R. P., **4**, 421 ⟨91CC1704⟩
Carlos, D. D., **5**, 1013 ⟨70JHC177⟩
Carlos, G., **2**, 89 ⟨88PS(37)25⟩
Carlsen, L., **2**, 195 ⟨81CC742⟩, 685 ⟨91JCR(S)300⟩; **3**, 344 ⟨79JCS(P2)1532, 82JA5922⟩, 344 ⟨83JCS(P2)1321⟩, 350, 364 ⟨77S256⟩, 365 ⟨77ACS(B)679⟩; **4**, 1081 ⟨83JCS(P2)1687⟩; **5**, 539 ⟨80JCS(P2)1557⟩; **6**, 556, 567 ⟨87CB987⟩, 637 ⟨82CB2516⟩
Carlsen, P., **5**, 1112 ⟨73TL211, 76T745⟩
Carlsen, P. H., **3**, 679 ⟨B-82MI 320-01⟩
Carlsen, P. H. J., **5**, 52 ⟨81JOC3936⟩
Carlsohn, B., **2**, 865 ⟨77ZAAC(432)67⟩; **4**, 545 ⟨77ZN(B)1490⟩
Carlson, A. A., **1**, 89 ⟨64JA118⟩
Carlson, B. A., **2**, 530 ⟨73JOC2422, 73JOC3968⟩, 530 ⟨73JOM(54)61⟩; **4**, 363 ⟨75JOC813⟩
Carlson, G. R., **3**, 212 ⟨73JOC4071⟩; **5**, 282 ⟨73JOC4071⟩
Carlson, H. D., **4**, 297 ⟨88OSC(6)664⟩
Carlson, J. R., **3**, 307 ⟨81JOC839⟩
Carlson, R., **1**, 82 ⟨85ACS(B)187⟩; **2**, 22, 28 ⟨78ACS(B)(32)646⟩, 759 ⟨78ACS(B)646⟩; **5**, 599 ⟨86ACS(B)534, 86ACS(B)694⟩, 599 ⟨87ACS(B)157, 87ACS(B)164⟩
Carlson, R. D., **2**, 27 ⟨62JA769⟩
Carlson, R. G., **1**, 708 ⟨75TL947⟩, 841 ⟨73CC223⟩, 842 ⟨69CC1049⟩, 882 ⟨65JOC2513⟩, 1023 ⟨69CC674⟩
Carlson, R. H., **2**, 191 ⟨55JA6231⟩
Carlson, R. M., **1**, 205 ⟨83SC237⟩, 636 ⟨69TL4001, 74TL2615⟩; **2**, 40 ⟨71TL3361⟩; **3**, 245 ⟨75TL1741⟩; **4**, 184 ⟨73JOC554⟩, 250 ⟨69TL173⟩; **6**, 134 ⟨69TL173⟩
Carlson, S. C., **3**, 649 ⟨73JOC479⟩; **4**, 739 ⟨73JOC479⟩
Carlsson, A., **2**, 652 ⟨88JMC1080⟩
Carlsson, J., **2**, 219 ⟨72BJ723⟩
Carlsson, S., **2**, 738, 751 ⟨82T405⟩
Carlton, L., **4**, 726 ⟨94JOM(468)235⟩
Carlucci, W. F., **1**, 238 ⟨83JCS(P1)2953⟩
Carmack, M., **2**, 133, 135 ⟨49JA84⟩; **4**, 245 ⟨59JA3148⟩; **5**, 272, 276 ⟨45JA1783⟩, 599 ⟨46JA2029, 46OR(3)83⟩, 599 ⟨89JHC1319⟩
Carman, R. M., **2**, 25 ⟨76AJC133, 85AJC1879⟩
Carmody, M. A., **1**, 777 ⟨81JOC2804⟩, 785 ⟨75CC206⟩
Carmona, E., **4**, 583 ⟨84OM1438⟩; **5**, 445 ⟨84JA3214⟩, 800 ⟨89JOM(377)157, 90OM583⟩, 948 ⟨92JA7288, 94JA791⟩; **6**, 327 ⟨89IC2120, 92JCS(D)2307⟩, 327, 330 ⟨93IC5569⟩, 332 ⟨88OM2577⟩

Carmona-Guzman, E., **2**, 574 ⟨76JCS(D)2204⟩, 575 ⟨75JOM(99)C19⟩; **4**, 656, 677 ⟨76JCS(D)2204⟩
Carney, P. A., **6**, 206 ⟨65JA3788⟩
Carney, R. L., **4**, 773 ⟨89USP4877899⟩
Carniato, D., **4**, 465 ⟨85S62⟩
Caro, H., **2**, 779 ⟨1874CB809⟩
Caron, G., **1**, 785 ⟨73CJC981⟩
Caron, M., **2**, 84 ⟨88JOC5185⟩; **3**, 632 ⟨85JOC1557⟩
Carothers, W. H., **1**, 965 ⟨32JA4066⟩, 1043 ⟨33JA1094⟩; **2**, 1016 ⟨33JA4667⟩
Carotti, A., **3**, 618 ⟨79S56⟩, 618, 619 ⟨77TL1813⟩
Carpanelli, C., **5**, 348 ⟨85G265⟩
Carpenter, B. K., **1**, 401 ⟨81JA7661⟩, 785 ⟨78T1877⟩, 801 ⟨B-87MI 118-03⟩, 826 ⟨87JA1170⟩, 1127 ⟨93JA9806⟩; **3**, 291 ⟨83OM467⟩
Carpenter, F. H., **3**, 620 ⟨62B53⟩; **5**, 184 ⟨48JA2964⟩
Carpenter, G. B., **1**, 873 ⟨85JA3345, 86JA462⟩
Carpenter, J. D., **4**, 658 ⟨86JA6805⟩
Carpenter, J. F., **2**, 245 ⟨85CC1449⟩
Carpenter, J. W., **3**, 475 ⟨67CJC1727⟩
Carpenter, N. E., **1**, 825 ⟨87OM2249⟩; **5**, 937 ⟨87OM2249⟩
Carpenter, W., **6**, 54 ⟨66JOC789⟩
Carpenter, W. R., **1**, 884 ⟨65OS(45)25⟩
Carpentier, M., **4**, 340 ⟨73TL173⟩
Carpino, L. A., **1**, 617 ⟨85JOC1749⟩, 620 ⟨71JA476⟩, 636 ⟨70JA1784⟩, 654 ⟨78CC358⟩, 1076 ⟨66JOC2867⟩; **2**, 29 ⟨88OSC(6)403⟩, 147 ⟨79JA390⟩, 377, 379 ⟨60JA2728⟩, 377, 379 ⟨63JA2144, 60JA3133⟩; **3**, 479 ⟨71JA476⟩; **4**, 71, 78 ⟨79JOC1177⟩, 83 ⟨66JA5682, 88OSC(6)403⟩, 83, 85, 86 ⟨71JA476⟩, 275 ⟨90JOC247⟩; **5**, 5 ⟨90JA9651⟩, 206 ⟨55JA6370⟩, 264 ⟨92JOC6371⟩, 279 ⟨93JA4397⟩, 490 ⟨79JOC1177⟩; **6**, 428 ⟨80JOC4250⟩, 487 ⟨70JA5748⟩
Carpino, P., **2**, 672 ⟨89JA5472⟩; **3**, 208, 212 ⟨90TL3691⟩
Carpino, P. A., **1**, 697, 698, 703 ⟨86TL2095, 87JOC3759⟩
Carr, D., **3**, 707, 714 ⟨79JOC2042⟩
Carra, S., **1**, 1144 ⟨61G889⟩
Carr, C. J., **4**, 223 ⟨43JA1477⟩
Carr, C. S., **3**, 156 ⟨86JOC130⟩, 163 ⟨86TL5811⟩; **4**, 183 ⟨86JOC130⟩
Carr, D., **2**, 163 ⟨69TL477⟩
Carr, D. B., **1**, 1049, 1139 ⟨78JA2244, 80JOC3053⟩; **2**, 969, 991, 992 ⟨77JA638⟩, 969, 991, 992 ⟨79JA3521⟩; **3**, 147 ⟨80JOC3053⟩; **5**, 467 ⟨79JA3521⟩
Carr, E. J., **2**, 358 ⟨49JOC921⟩
Carr, E. L., **5**, 589, 595, 620 ⟨49JOC962⟩
Carr, G. E., **6**, 10 ⟨87JOM(325)13⟩, 14 ⟨88JCS(P1)921⟩
Carr, P., **2**, 100 ⟨33BSF1070⟩
Carr, R. A. E., **1**, 700, 703 ⟨86JOC4840⟩; **5**, 177 ⟨86JOC4840⟩
Carr, R. E., **3**, 181 ⟨94TL2427⟩
Carr, R. L. K., **6**, 57 ⟨68JOC1084⟩
Carr, R. V. C., **1**, 654 ⟨83JOC4976⟩
Carr, S. A., **1**, 229 ⟨85JOC2782⟩
Carr, S. W., **6**, 332, 351 ⟨83IC3700⟩
Carraher, Jr., C. E., **5**, 652 ⟨69MAC144⟩
Carrano, C. J., **6**, 704 ⟨94AG(E)578⟩
Carranza, S., **3**, 8, 122 ⟨86T4133⟩
Carrau, R., **3**, 409 ⟨86S1055⟩; **5**, 1119 ⟨87JCS(P1)937⟩, 1125 ⟨86MI 530-01⟩
Carré, J. C., **3**, 318 ⟨76TL3299, 76TL3303⟩
Carre, M.-C., **1**, 539 ⟨88T127⟩
Carrea, G., **2**, 151 ⟨92T9577⟩, 151 ⟨92TA1063⟩
Carreira, E. M., **1**, 98 ⟨91JOC741⟩; **2**, 62 ⟨88JA3560⟩
Carrell, H. L., **4**, 493 ⟨75JA38⟩
Carreno, M. C., **4**, 270, 863 ⟨91TL3695⟩; **6**, 567 ⟨85H(23)1181⟩
Carretero, J. C., **1**, 210, 225, 613 ⟨88TL2059⟩; **2**, 721 ⟨87T5125⟩; **3**, 636 ⟨86TL5099⟩, 695, 709 ⟨85TL3381⟩; **4**, 271, 272 ⟨90T7197⟩, 858, 938, 961 ⟨91TL1385⟩, 864 ⟨91TA93⟩, 936, 938 ⟨93JOC1596, 91TL5159⟩; **5**, 166 ⟨87TL2135⟩
Carreyre, H., **3**, 429 ⟨89TL5763, 91BSF730⟩
Carrie, R., **1**, 637 ⟨89BSF409⟩; **2**, 318 ⟨87T1799⟩, 324 ⟨87TL1757⟩, 1003 ⟨91AX(C)328⟩; **3**, 62 ⟨86TL4983⟩, 474 ⟨81TL441⟩; **4**, 479, 480 ⟨84CC1634⟩, 489 ⟨70BSF1130⟩, 566 ⟨87PS(30)523⟩, 991, 993, 996 ⟨85BSB475⟩, 993 ⟨87BSF325⟩; **5**, 494 ⟨75JOC2990⟩, 785, 792 ⟨83CJC1105⟩, 1153 ⟨86TL5723⟩; **6**, 60 ⟨90T3859⟩, 117 ⟨84CC1634⟩, 156 ⟨89NJC891⟩, 165 ⟨90TL3859⟩
Carriedo, G. A., **4**, 675 ⟨85JCS(D)905⟩
Carrington, D. E. L., **5**, 1112 ⟨71JCS(C)3262⟩
Carroll, A. R., **1**, 688 ⟨90AJC1439⟩
Carroll, F. I., **1**, 84 ⟨87JMC1505⟩; **2**, 219 ⟨61JOC1152⟩, 274 ⟨65JOC33⟩; **3**, 92 ⟨78OPP201⟩
Carroll, G., **5**, 247 ⟨92SUL199⟩
Carroll, G. L., **1**, 944 ⟨81TL4389⟩; **3**, 615 ⟨73CC55, 74JOC914⟩; **5**, 1134 ⟨74JOC914⟩
Carroll, J. D., **2**, 528 ⟨93JOC2791⟩
Carroll, J. J., **4**, 154 ⟨83SC129⟩
Carroll, M. F., **1**, 416, 831 ⟨40JCS1266⟩, 416, 831 ⟨40JCS704⟩, 416, 831 ⟨41JCS507⟩
Carroll, P., **4**, 877 ⟨81TL4199⟩; **5**, 528 ⟨84JA6084⟩
Carroll, P. J., **1**, 147 ⟨93JOC4890⟩; **2**, 150 ⟨92JA1428⟩; **3**, 351 ⟨90JA3029⟩, 365 ⟨85JOC1550⟩, 480 ⟨88TL6031⟩, 634 ⟨92JA1428⟩; **4**, 108 ⟨91JA8807⟩, 262 ⟨92JA1428⟩; **5**, 513 ⟨90JA3029, 90JA3040⟩; **6**, 611 ⟨90JOC3045⟩
Carroll, R. D., **3**, 559 ⟨74TL1515⟩
Carroll, S., **2**, 12, 627 ⟨65JOC415⟩
Carroll, W. A., **1**, 254 ⟨92TL4401⟩
Carrupt, P.-A., **2**, 616 ⟨89HCA1008⟩
Carruth, R. L., **2**, 133 ⟨76JOC134⟩
Carruthers, M. H., **2**, 253 ⟨89JA2321, 90TL1953⟩
Carruthers, N. I., **2**, 1064 ⟨84CC200⟩; **3**, 482 ⟨82JOC1284⟩; **4**, 487 ⟨80CC730, 82T2513⟩; **6**, 118, 119, 636, 666 ⟨82JOC1284, 82CC536⟩
Carruthers, T. F., **4**, 869 ⟨77CC505⟩; **6**, 591 ⟨77CC505⟩
Carruthers, W., **1**, 359, 372 ⟨B-90MI 108-01⟩, 516, 519 ⟨B-90MI 112-01⟩; **2**, 326 ⟨87CC509⟩
Carruthers, W. H., **2**, 52 ⟨23JA1071, 24JA1675⟩
Carry, J.-C., **1**, 788 ⟨91JOC6981⟩; **4**, 764 ⟨91JOC6987⟩
Carson, K. G., **2**, 300, 310 ⟨90TL5595⟩
Carson, L., **2**, 767 ⟨82TL821⟩
Carstens, A., **1**, 138 ⟨90AG(E)1424⟩; **4**, 360, 382 ⟨90AG(E)1424⟩
Cartagena, I., **5**, 711 ⟨85SC643⟩
Cartaya, M., **4**, 154 ⟨83SC129⟩
Cartaya-Marin, C. P., **3**, 39 ⟨91SC951⟩
Carter, A., **4**, 566 ⟨85IC642⟩
Carter, E. A., **5**, 932 ⟨86JA2180, 86JA4746⟩
Carter, G. W., **1**, 1018 ⟨87JMC254⟩
Carter, J. C., **5**, 317 ⟨73SA(A)613⟩
Carter, J. P., **2**, 13 ⟨90SC1011⟩
Carter, J. S., **3**, 620 ⟨87TL464⟩, 620, 621 ⟨87TL4645⟩
Carter, K. M., **4**, 39 ⟨23JCS576⟩
Carter, K. N., **5**, 303 ⟨53JA5905⟩
Carter, L. G., **1**, 137 ⟨81JOC2363⟩; **2**, 999 ⟨81JOC2363⟩; **3**, 686 ⟨78JOC3736⟩; **4**, 360, 382, 388 ⟨78JOC4255⟩, 382 ⟨81JOC2363⟩
Carter, M. J., **2**, 921 ⟨81JCS(P1)2415⟩; **4**, 649 ⟨81JCS(P1)2415⟩
Carter, P., **1**, 868 ⟨65JA914⟩; **3**, 446 ⟨61JCS1743⟩
Carter, P. A., **1**, 1009 ⟨88JA1626⟩
Carter, P. R., **2**, 250 ⟨48JCS147⟩

Carter, S. P., **3**, 320 ⟨88JA2894⟩
Carter, Jr., T. P., **2**, 348 ⟨78JOC2662⟩
Cartledge, F. K., **2**, 33 ⟨86JOC2206⟩; **4**, 138, 139 ⟨92OM859⟩
Carty, A. J., **2**, 1060, 1061, 1076 ⟨71CJC2706⟩; **4**, 692 ⟨77IC172⟩, 711 ⟨90OM2234⟩, 713 ⟨91OM1907⟩; **6**, 187 ⟨81JOM(212)C4⟩, 195 ⟨80IC3637⟩
Carty, D. T., **6**, 69, 296, 297 ⟨65JOC411⟩
Caruso, A. J., **2**, 711 ⟨91JOC862⟩
Caruso, J. A., **6**, 507 ⟨79AG(E)503⟩
Caruso, M., **1**, 196 ⟨90SL224⟩; **2**, 229 ⟨87JOC3154⟩
Caruso, T., **3**, 490 ⟨80JOC4065⟩
Caruso, T. C., **1**, 873 ⟨85TL1599⟩
Caruthers, M. H., **2**, 409, 410 ⟨81TL1859, 83TL245⟩
Carvajal, J., **5**, 1137 ⟨84OM996⟩
Carvalho, E., **5**, 361 ⟨89JCR(S)260⟩, 362, 657, 711 ⟨92JCS(P2)281, 88CC1249⟩
Carver, D. R., **3**, 496 ⟨81JOC3119⟩
Casadebaig, F., **4**, 420 ⟨92PHA754⟩
Casadei, M. A., **5**, 726 ⟨83SC753⟩
Casadevall, E., **1**, 1023 ⟨67CR(C)839⟩
Casaert, F., **3**, 27 ⟨88BSB525⟩
Casal, H. L., **3**, 314 ⟨85JA7204⟩
Casalnuovo, A. L., **1**, 1013 ⟨90JA4324⟩
Casalotti, S. O., **3**, 725 ⟨91SL654⟩; **5**, 1052 ⟨91SL654⟩
Casamitjana, N., **3**, 310 ⟨91CC1687⟩
Casanova, J., **1**, 633 ⟨75T1035⟩; **3**, 696 ⟨63JCS4280⟩
Casanova, Jr., J., **3**, 565 ⟨B-70MI 317-01⟩
Casanova, R., **5**, 144 ⟨49HCA647⟩
Casara, P., **1**, 1138 ⟨78TL1581⟩; **2**, 299 ⟨85JCS(P1)2201⟩
Casares, A., **3**, 624 ⟨86JOC1745⟩
Casares, A. M., **1**, 314 ⟨83TL1909⟩
Casarini, A., **1**, 1048 ⟨91TL2169⟩; **2**, 918, 940 ⟨92SL981⟩, 1018 ⟨91TL2169⟩; **3**, 675 ⟨91TL2169⟩
Casarrubios, L., **5**, 696 ⟨93JOC3886⟩
Casas, J. M., **4**, 692 ⟨88OM2279, 89JCS(D)169⟩, 692 ⟨93JA4145⟩
Casassas, E., **5**, 226 ⟨82AQ(B)327⟩, 864 ⟨86POL1277⟩
Casati, R., **3**, 88 ⟨83S749⟩
Case, F. H., **5**, 769 ⟨68JHC223⟩, 769, 847 ⟨65JOC931, 66JOC2398⟩, 847 ⟨67JHC483⟩
Case, J. R., **2**, 165 ⟨61JCS2066⟩, 165 ⟨61JCS2070⟩
Case, L. C., **5**, 1009 ⟨59NAT675⟩
Case-Green, S. C., **5**, 161 ⟨91SL781⟩
Casella, L., **2**, 151 ⟨88CC1451⟩, 151 ⟨92TA95⟩
Casella, Jr., J., **1**, 954 ⟨58JA1376⟩
Caserio, F. F., **4**, 179 ⟨58JA5837⟩
Caserio, M., **1**, 1125 ⟨68JA6741⟩
Caserio, M. C., **1**, 497 ⟨72JOC3881, 71JA5758⟩; **2**, 89 ⟨58JA2584, 59T36⟩, 138 ⟨82JA3231, 85JOC4390⟩; **3**, 237 ⟨60JA3102⟩
Casey, C., **1**, 12 ⟨64JOC2325⟩
Casey, C. P., **1**, 165 ⟨74JA1230⟩, 313 ⟨79JA4236⟩, 482 ⟨74TL925⟩, 491, 1105 ⟨71JA1379⟩, 668 ⟨70JA1426⟩, 797 ⟨73JA2248⟩; **2**, 572 ⟨78JA7565⟩, 681 ⟨72JA6543⟩; **3**, 517 ⟨77JA2127⟩, 519 ⟨83JOM(254)333, 85JA2924⟩, 520 ⟨82JA3761⟩, 521 ⟨89JA2352⟩, 545 ⟨85JA3172⟩; **4**, 580 ⟨77JA6097, 79JA7282⟩, 581 ⟨90JA3713⟩, 679 ⟨82JA1134, 94OM2527⟩, 680 ⟨88JA6070, 88OM934⟩, 716 ⟨90JOM(387)C31, 92ICA557⟩, 955 ⟨74JA1230⟩; **5**, 447 ⟨76JA1166⟩, 448 ⟨76JA437⟩, 450 ⟨80JA2723, 90JA4083⟩, 932 ⟨76JA441⟩, 942 ⟨85OM736⟩, 943 ⟨70CC1220, 71JA3554⟩, 946 ⟨74JOM(73)C28⟩, 949 ⟨91JA6673⟩, 951, 958 ⟨77IC3059⟩, 955 ⟨72JA8947, 74JA1230, 75JOM(102)175, 76JOM(118)309, 74JOM(77)345⟩, 958 ⟨77IC391⟩; **6**, 719 ⟨73SRI249⟩
Casey, G. C., **6**, 403 ⟨79JOM(178)371⟩
Casey, M., **5**, 655, 657 ⟨87JCS(P1)1389⟩
Cashman, J. R., **4**, 262 ⟨90JA3191, 90MI 406-01⟩, 262 ⟨92JA8772⟩; **6**, 478 ⟨91JMC2049⟩
Cashman, M. P., **3**, 681 ⟨72JCS(P2)44⟩; **5**, 703, 704 ⟨71CC689⟩
Casida, J. E., **2**, 210 ⟨B-81MI 203-01⟩; **4**, 265 ⟨92MI 406-01⟩, 271 ⟨92MI 406-02⟩, 310 ⟨68MI 407-01⟩, 743 ⟨82TL5003⟩
Casini, G., **3**, 469 ⟨65G983⟩, 618, 619 ⟨77TL1813⟩
Casiraghi, G., **3**, 97 ⟨78JCS(P1)318⟩
Casnati, G., **2**, 24, 31, 35 ⟨85G29⟩, 28 ⟨84T2035⟩; **3**, 27, 154 ⟨84T2035⟩, 60 ⟨75JCS(P1)1527, 78JCS(P1)318⟩; **5**, 821 ⟨67TL327⟩
Cason, J., **1**, 127, 128 ⟨57JOC1326⟩, 881 ⟨50JOC850⟩; **3**, 100 ⟨55OSC(3)627⟩, 289 ⟨60JOC477⟩, 461 ⟨53JOC1594⟩; **5**, 32 ⟨63OSC(4)635⟩, 37 ⟨55OSC(3)169⟩, 108 ⟨50JOC617⟩
Cason, L. F., **1**, 310 ⟨52JA4582⟩; **3**, 452 ⟨51JA1074⟩; **4**, 651 ⟨54JOC1278⟩; **6**, 192 ⟨52JA4582⟩, 192 ⟨54JOC1278⟩
Caspari, I., **3**, 260 ⟨62LA(659)64⟩
Caspi, E., **3**, 54 ⟨66TL4113⟩, 234 ⟨83JOC1007⟩, 244 ⟨75JOC1420⟩; **5**, 49 ⟨69JOC112, 69JOC116⟩, 138 ⟨74TL3135⟩
Cassady, D. R., **6**, 509 ⟨64JOC1851⟩
Cassady, J. M., **1**, 752 ⟨87TL3075⟩
Cassady, T. J., **1**, 787 ⟨81JA6691⟩
Cass, J. C., **6**, 143 ⟨77JCS(P1)1162⟩
Cass, M. W., **3**, 530 ⟨82JOC3871⟩
Cass, W. E., **4**, 207 ⟨47JA500⟩
Cassal, A., **6**, 522 ⟨27BSF1041⟩
Cassar, L., **1**, 526 ⟨67AG(E)124⟩; **3**, 661 ⟨73JOM(54)C57, 79JOM(173)335, B-74MI 319-01⟩; **5**, 27 ⟨73S509⟩, 460 ⟨70JA3515⟩, 580 ⟨78S917⟩
Casselman, A. A., **6**, 641, 642, 643 ⟨58CJC1541⟩
Cassenmann, S., **3**, 71 ⟨71CB665⟩
Casser, C., **4**, 566 ⟨84CB2693, 85CB3419⟩; **5**, 877, 878 ⟨85TL3551, 84TL4109⟩, 878 ⟨85TH86⟩; **6**, 166, 689 ⟨84CB2693⟩, 371 ⟨86CB2609⟩, 680, 689, 692, 694 ⟨84TL4109⟩, 681, 682, 689, 690 ⟨85TL3551⟩, 683 ⟨84AG(E)895⟩
Cassey, H. N., **2**, 172 ⟨65JOC933⟩
Cassias, J. B., **4**, 685, 722 ⟨76IC2671⟩
Cassidei, L., **3**, 120 ⟨91JA2205⟩; **4**, 188 ⟨89TL257⟩
Cassidy, H. G., **6**, 522 ⟨47JA1723, 50IS156⟩
Cassidy, J., **2**, 527 ⟨91JOC1020⟩
Cassidy, J. F., **4**, 235, 236 ⟨86TL4355⟩
Cassidy, J. P., **1**, 849 ⟨60JA3652⟩
Cassidy, P. E., **4**, 730 ⟨90JOC4448⟩
Casson, S., **1**, 455 ⟨93SL1133⟩
Castagnani, R., **3**, 190 ⟨92SL64⟩
Castagnini, R., **1**, 578 ⟨92JOC2160⟩
Castagnoli, N., **5**, 1063 ⟨72JOC3372⟩
Cast, J., **1**, 397 ⟨60JCS3521⟩
Castaldi, G., **1**, 391 ⟨84AG(E)413, 85S505⟩; **2**, 29 ⟨86AG(E)259, 87JOC3018, 87S1039⟩, 555 ⟨85JA724⟩; **5**, 28 ⟨84AG(E)413⟩
Castan, F., **5**, 871, 872, 873, 1071, 1148 ⟨91JOC1801⟩, 1076, 1077, 1148 ⟨89AG(E)1250⟩; **6**, 659 ⟨92CC1274⟩, 659, 667 ⟨91CB1739⟩, 668 ⟨91JOC1801⟩
Castaneda, A., **4**, 187, 199 ⟨89JOC5695⟩
Castaner, J., **2**, 1010 ⟨71JA2215⟩; **4**, 17 ⟨91JOC5445⟩
Castaner, J., **4**, 764 ⟨86JOC1413⟩; **6**, 30 ⟨91JOC103⟩
Castedo, L., **1**, 576 ⟨92JOC5208⟩, 606 ⟨92JOC3173⟩, 626 ⟨88H(27)2367⟩, 816, 817 ⟨85TL4959, 87TL2099⟩, 1029 ⟨86TL1523, 88TL1203⟩; **2**, 103 ⟨92SC677⟩; **3**, 57 ⟨87TL2099⟩, 294 ⟨75H(3)449⟩, 412 ⟨82T1569⟩
Castel, A., **1**, 766 ⟨80JA5413⟩; **2**, 815 ⟨83RTC65⟩; **3**, 457

⟨87JOM(331)11⟩, 503 ⟨80JA5413⟩; **4**, 360 ⟨92JOM(433)49⟩, 361 ⟨82JOM(232)137⟩, 521 ⟨79JOM(168)43⟩; **5**, 432 ⟨90JOM(397)7, 92JOM(433)49⟩
Castelhano, A. L., **1**, 921 ⟨82JA3655⟩, 925 ⟨83JA4701⟩, 958 ⟨87JA3491⟩, 980 ⟨84JA1877⟩, 1042 ⟨88T5451⟩
Castell, J. V., **2**, 229 ⟨79HCA2507⟩
Castellani, C. B., **4**, 1018 ⟨84JOM(269)C15⟩
Castellano, S., **6**, 524 ⟨55CI(M)6⟩
Castellato, U., **6**, 519 ⟨87JCS(D)2039⟩
Castellino, A. J., **2**, 691 ⟨84JOC1348⟩
Castells, J., **3**, 182 ⟨88S314⟩
Castells, J. P., **1**, 217 ⟨90H(31)1997⟩
Castellucci, N. T., **3**, 232 ⟨65CI(L)184⟩
Castenada, A., **1**, 450 ⟨86JA3516⟩
Caster, K. C., **2**, 441 ⟨84JA7021⟩; **4**, 322 ⟨87JCS(P1)781⟩
Castiglione, R. de, **4**, 327 ⟨75CC320⟩; **5**, 300 ⟨75CC320⟩
Castillo, C., **4**, 843 ⟨91S481⟩
Castillo, G. E., **5**, 589 ⟨78JCS(P1)325⟩
Castillo, M., **6**, 627 ⟨66CB3268⟩
Castle, J. E., **5**, 59 ⟨73OSC(5)393⟩
Castle, R. B., **4**, 632, 632, 633 ⟨69JOM(20)19, 68JA2194⟩; **6**, 181, 386 ⟨68JA2194⟩, 181, 386 ⟨69JOM(20)19⟩, 405 ⟨70JA231⟩
Castleman, J. K., **1**, 37 ⟨68JOC1421⟩
Castonguay, L. A., **1**, 150 ⟨92JOC3819⟩
Castro, A., **4**, 274, 277 ⟨92JOC3496⟩
Castro, A. M. M., **3**, 170, 631 ⟨92JOC7009⟩
Castro, B., **3**, 180 ⟨86JCS(P1)1895⟩; **4**, 383 ⟨67BSF1533⟩, 395 ⟨64CR(259)830⟩; **5**, 191 ⟨72TL4747⟩, 267 ⟨75TL1219⟩, 333 ⟨88TL2661, 89JOC617⟩
Castro, B. R., **2**, 6, 7 ⟨83OR(29)1⟩
Castro, C. E., **1**, 457, 515, 1004, 1005 ⟨64JA4358⟩, 472 ⟨72OS(52)128⟩, 516 ⟨66JA4964⟩, 1004 ⟨66JOC4071⟩, 1004, 1142 ⟨63JOC3313, 69JA6464⟩; **2**, 621 ⟨69JA6464⟩; **3**, 273 ⟨69JA6464⟩, 621 ⟨66JA4964⟩
Castro, E. A., **5**, 586 ⟨93JOC4908⟩
Castro, I., **4**, 843 ⟨91S481⟩
Castro, J., **1**, 530 ⟨90JA9388⟩
Castro, J. L., **1**, 677 ⟨89JCS(P1)190⟩
Castro, P. P., **4**, 230 ⟨88JOC5179⟩
Castro Dantas, T. N. de, **2**, 444 ⟨82PS(13)97⟩
Casy, G., **1**, 619 ⟨89T455⟩, 708 ⟨88CC454⟩
Catalán, J., **5**, 1068 ⟨89JA355⟩
Catani, V., **1**, 999 ⟨86SC283⟩, 1035 ⟨86CC1067⟩; **2**, 1036, 1038 ⟨86CC1067⟩, 1036, 1038 ⟨86SC283⟩
Cate, L. A., **2**, 183, 717 ⟨77JOC2031⟩
Catellani, M., **1**, 1015, 1138 ⟨82TL4517⟩; **5**, 192 ⟨83TL813⟩
Cates, Jr., H. L., **6**, 145 ⟨61JOC51⟩
Catherwood, B., **3**, 65 ⟨90JCS(P1)89⟩
Caton, M. P. L., **3**, 61 ⟨82TL2811⟩
Catt, J. D., **5**, 334 ⟨86JMC359⟩
Cattran, L. C., **4**, 332 ⟨73JA3356⟩
Caubère, P., **1**, 53 ⟨91COS(8)858⟩, 65, 68, 69 ⟨91COS(8)850⟩, 254 ⟨83JOC1919⟩, 428 ⟨83AG(E)599⟩, 430, 431 ⟨81JOC1270⟩, 437 ⟨88TL2963⟩, 437, 438, 439 ⟨90JOC6194⟩, 437, 444, 445, 454 ⟨91COS(8)835⟩, 458, 515 ⟨77TL1069⟩, 472 ⟨89JOC4840⟩, 539 ⟨88T127⟩; **2**, 36 ⟨85TL5445⟩; **3**, 118 ⟨76S777⟩; **4**, 211 ⟨87SC1467⟩; **5**, 994 ⟨88T6079⟩, 1105 ⟨88T1619, 92SC617⟩
Cauble, R. L., **6**, 250, 424 ⟨67JA1962, 67JA5161⟩
Cauletti, C., **2**, 1100 ⟨88G1⟩
Cauliez, P., **5**, 227 ⟨89TL3073⟩
Caulton, K. G., **2**, 569 ⟨87OM2354⟩
Cauquis, G., **3**, 113 ⟨79TL3⟩; **6**, 666 ⟨75MI 621-02, 75OMS(10)1021⟩
Causey, D. H., **2**, 322 ⟨88SC205⟩
Cava, M. P., **1**, 418 ⟨59JA4266⟩, 569 ⟨82JOC409⟩, 611, 620 ⟨81CC277, 82CC164⟩; **2**, 287, 288 ⟨81CC277⟩, 290 ⟨77JA1470, 81JOC4194⟩, 652 ⟨77JOC2274⟩; **3**, 254 ⟨79JOC930⟩, 339 ⟨88JOC3529⟩, 346, 355 ⟨85T5061⟩, 349, 358 ⟨87JOC169, 86JOC411⟩, 358 ⟨87JOC1874⟩, 365 ⟨85JOC1550⟩, 371, 372 ⟨69JA3952, 84CHEC(4)1037, 74JA1817⟩, 372 ⟨75ACR139⟩, 446, 470 ⟨58JA2257⟩; **4**, 21 ⟨62JOC757⟩, 168 ⟨89JOC2751⟩, 847 ⟨78JOC369⟩, 847, 848 ⟨79JOC930⟩, 847, 848, 870 ⟨80H(14)271⟩, 850, 869 ⟨78ANY355⟩, 855 ⟨78JOC82⟩, 869 ⟨75CC867⟩, 872 ⟨82JOC3946, 84SC1119⟩, 877 ⟨81TL4199⟩; **5**, 95, 306 ⟨73OSC(5)944⟩, 162 ⟨82TL5513⟩, 242, 1117 ⟨81JOC4194⟩, 264 ⟨87TL5099⟩, 336 ⟨87JOC169⟩, 512, 570 ⟨85T5061⟩, 528 ⟨84JA6084⟩, 528, 529 ⟨89JOC4746⟩, 561 ⟨79SC17⟩, 626, 732 ⟨75CC617⟩, 1113 ⟨80JOC2632⟩, 1117 ⟨77JA1470, 80JOC1044⟩; **6**, 101 ⟨80H(14)271⟩, 121, 698 ⟨79JOC930⟩, 538, 539 ⟨83JOC1449, 78JOC369⟩
Cavagnol, J. C., **1**, 881 ⟨50JA3704⟩
Cavalchi, B., **3**, 238 ⟨69JCS(C)1204⟩
Cavalieri, L., **5**, 272, 276 ⟨45JA1783⟩
Cavallini, D., **2**, 211 ⟨63MI 203-01⟩
Cavallito, C. J., **2**, 212, 232 ⟨47JA1710⟩; **3**, 252 ⟨51JA2544⟩
Cavazza, M., **3**, 374 ⟨75TL2085⟩
Cave, A., **5**, 185 ⟨85T5331⟩
Cave, R. J., **1**, 1134 ⟨79JCS(P1)2954⟩
Cavell, K. J., **6**, 560 ⟨79JINC1277⟩
Cavell, R. G., **2**, 472, 508 ⟨68JCS(A)1406⟩, 476 ⟨77JA7379, 83PS(18)89⟩, 477 ⟨90IC5081⟩, 482 ⟨67JCS(A)1308⟩; **4**, 547 ⟨88IC4038⟩, 552 ⟨89IC413, 90IC808⟩; **6**, 622 ⟨90IC5081⟩
Cavezzan, J., **6**, 582 ⟨79TL3507⟩
Cavicchioli, S., **2**, 29 ⟨86AG(E)259, 87JOC3018, 89T4243⟩; **3**, 294 ⟨88TL5561⟩
Caviccholi, S., **1**, 24 ⟨84JOC1246⟩
Cavill, G. W. K., **1**, 747, 748 ⟨69AJC1737⟩, 747, 748 ⟨69AJC2145⟩
Cawkill, E., **6**, 611 ⟨79JCS(P1)724⟩
Cawley, J. J., **4**, 189 ⟨60CI(L)656⟩
Cawse, J. N., **2**, 573 ⟨79JOM(172)405⟩
Cayen, C., **1**, 881 ⟨67OS(47)20⟩
Cazanoue, M., **5**, 936 ⟨92JA8735, 92OM1759⟩
Cazaux, L., **2**, 247, 250 ⟨82S281⟩
Cazaux, M., **1**, 925 ⟨71BSF461⟩
Cazeau, P., **2**, 699 ⟨87T2075⟩; **5**, 1134 ⟨82S212⟩
Cazes, B., **1**, 193 ⟨88S983⟩, 496 ⟨90PAC1867, 92BSF273⟩, 816 ⟨74TL2077, 77BSF925⟩, 958 ⟨81TL103⟩, 962 ⟨84TL845⟩; **4**, 255 ⟨78TL4065⟩, 535 ⟨88S983, 89TL3963⟩, 853 ⟨78JCR(S)68⟩; **5**, 553 ⟨78TL4065⟩
Ceacareanu, D. M., **6**, 274 ⟨83CB3325⟩
Ceccherelli, P., **1**, 44 ⟨79CC959⟩; **2**, 44 ⟨89TL3175⟩; **3**, 129 ⟨91T7403⟩, 533 ⟨84JOC2052⟩
Cecchetti, S., **1**, 44 ⟨85JCS(P1)493⟩
Cecchini, M., **2**, 279 ⟨56JOC1217⟩
Ceccon, A., **3**, 365 ⟨66CC586, 68CC724⟩, 365 ⟨70JCS(B)325⟩
Cech, D., **2**, 416 ⟨70ZC33⟩
Cecil, R., **5**, 85 ⟨59MI 502-02⟩
Ceclan, R., **3**, 539 ⟨88RRC83⟩
Cederbaum, F. E., **1**, 536 ⟨86JOC4080, 86TL2829⟩, 536 ⟨87TL917⟩; **2**, 568, 986, 988 ⟨86TL2829⟩, 927, 986, 988 ⟨89JA3336⟩, 972 ⟨90TL493⟩; **4**, 809 ⟨89JA3336⟩, 1061 ⟨86TL2829⟩, 1061, 1066 ⟨87TL917⟩
Cederbaum, R. E., **3**, 270 ⟨86JOC4080⟩
Cegla, M. T., **4**, 411 ⟨90JOC4777⟩
Ceinini, S., **5**, 386 ⟨76JCS(D)1777⟩

Čeković, Z., **1**, 91 ⟨67T215⟩, 231 ⟨90TL4203, 90TL6085⟩, 348 ⟨86TL5893⟩, 982 ⟨91CC294⟩; **2**, 93 ⟨70S209⟩
Celebi, S. K., **4**, 267 ⟨90JCS(P2)1987⟩
Celebuski, J. E., **2**, 410 ⟨92JOC5535⟩
Celerier, J. P., **3**, 411 ⟨87TL2353⟩, 496 ⟨90S301⟩
Cella, J. A., **1**, 200 ⟨82JOC2125⟩; **6**, 462 ⟨84JOC1122⟩
Cellerin, C., **4**, 328 ⟨91CR(313)517⟩
Celmer, W. D., **3**, 612 ⟨53JA3430⟩
Cen, W., **3**, 79, 274 ⟨85S159⟩, 269, 274 ⟨84S35⟩; **6**, 258 ⟨89JA1773⟩, 258, 278 ⟨89CJC1795⟩
Cenicola, M. L., **5**, 597 ⟨91FES1033⟩
Cenini, S., **5**, 386 ⟨74JOM(76)355, 75JOM(88)237⟩
Centner, R. L., **6**, 263 ⟨78JCE760⟩
Ceprini, M. Q., **5**, 277 ⟨67JA183⟩
Cerar, D., **2**, 12 ⟨53JOC868⟩
Cerdá, V., **5**, 226 ⟨82AQ(B)327⟩, 864 ⟨86POL1277⟩
Ceré, V., **1**, 821 ⟨78JOC4826, 79JOC4128, 81JOC3315⟩; **2**, 143 ⟨81JOC3315⟩, 156 ⟨88JOC5689⟩
Cerfontain, H., **2**, 208 ⟨87TL1703⟩, 720 ⟨85RTC153⟩; **6**, 428 ⟨91EUP415473⟩
Cerioni, G., **2**, 386 ⟨81JCS(P2)1449⟩
Ceriotti, A., **6**, 525 ⟨79JA6110⟩
Cermola, F., **3**, 327 ⟨94JCS(P1)147⟩
Cernia, E., **2**, 28 ⟨79S64⟩
Cerniani, A., **2**, 145 ⟨60G383⟩
Cernigliaro, G., **1**, 979 ⟨77JOC353⟩
Cerny, I., **1**, 4, 10 ⟨84CAR(130)103⟩
Cerny, K. F., **2**, 386 ⟨79JOC1881⟩
Cerny, M., **1**, 4, 10 ⟨84CAR(130)103⟩, 18, 21 ⟨69TL1739⟩; **3**, 84 ⟨72S217⟩; **4**, 223 ⟨59CCC64⟩, 223, 225 ⟨63M290, 59CCC2566⟩, 226 ⟨61CCC2206⟩
Cerreta, F., **2**, 541 ⟨93TL873⟩; **3**, 346, 353 ⟨93TL873⟩; **5**, 587, 611 ⟨93TL6741⟩
Ceruti, M., **3**, 11 ⟨87S79⟩; **4**, 259 ⟨89EJM635⟩, 846, 854 ⟨87G227⟩
Cervantes, A., **3**, 55 ⟨73CC119⟩, 235 ⟨73JA6655⟩; **4**, 8 ⟨73TL1319⟩
Cervantes, H., **4**, 245 ⟨91SC1369⟩, 246 ⟨90SC153⟩
Cerveau, G., **2**, 68 ⟨89T171⟩, 77 ⟨87JOM(328)C17⟩; **5**, 421 ⟨81JOM(205)31⟩
Cervini, L. A., **1**, 316 ⟨88JA4788⟩; **2**, 556 ⟨88JA4788⟩
Cesa, M. C., **1**, 313 ⟨79JA4236⟩
Cesare, F., **1**, 772 ⟨78JOC1421⟩
Cesark, F. F., **2**, 376 ⟨67JA1417⟩
Cesarotti, E., **2**, 9, 19 ⟨79JCS(P1)2248⟩, 53 ⟨91TL4381⟩
Cescon, L. A., **6**, 143 ⟨71JOC2262⟩
Cesson, P., **4**, 884, 891 ⟨70CR(C)1890⟩
Cetini, G., **4**, 683 ⟨72JOM(39)169⟩, 687 ⟨67JOM(8)149⟩, 692 ⟨74HCA1042⟩
Cetinkaya, B., **2**, 852 ⟨82CC691⟩; **3**, 460, 461 ⟨71CC215⟩; **5**, 931, 940 ⟨72CRV545⟩; **6**, 716 ⟨73CSR99⟩, 716, 721 ⟨72CRV545, 71CC400⟩, 721 ⟨72CC851, 74JCS(D)1591, 74JCS(D)1827⟩
Cevasco, G., **2**, 168 ⟨88TL2865⟩, 173, 717 ⟨87TL5193⟩; **5**, 336 ⟨90TA141⟩
Cezza, A., **1**, 738 ⟨75G109⟩
Chaabouni, M. M., **3**, 627 ⟨93SC2389⟩; **4**, 190 ⟨90SC2527⟩
Cha, D. D., **1**, 518 ⟨90OS(68)32⟩
Cha, J. K., **1**, 797 ⟨87TL5473⟩; **2**, 46 ⟨83TL3947⟩
Cha, J. S., **2**, 41 ⟨94TL2⟩, 126, 708 ⟨84S498⟩, 268 ⟨84JA8001⟩, 526 ⟨86JOC5264⟩, 534 ⟨85JOC549⟩; **3**, 3, 114 ⟨93TL5113⟩, 13, 82 ⟨87JOC5400, 87TL2389⟩, 13, 82, 94, 99 ⟨87JOC5030⟩, 13, 85 ⟨91TL6903, 84JA8001⟩, 82 ⟨86JOC5264, 87MI 303-01, 87TL4575, 88H(27)1595, 88MI 303-01, 89OPP451, 90MI 303-01, 92MI 303-01, 92OPP327, 94MI 303-01⟩, 82, 99 ⟨87TL6231⟩, 83, 93, 94, 98, 100 ⟨93JOC4732⟩, 84 ⟨92OPP335⟩, 86 ⟨87MI 303-02, 92MI 303-02⟩, 86 ⟨92OPP331, 93JOC4727⟩, 93 ⟨89TL3677⟩
Cha, Y., **5**, 728 ⟨85JOC4415⟩
Cha, Y. C., **3**, 679, 680 ⟨86JA6739⟩
Chabaeva, N. A., **2**, 889 ⟨67JGU1578⟩
Chabala, J. C., **1**, 1045 ⟨78TL937⟩; **3**, 273 ⟨78TL937⟩
Chabanenko, K. Yu., **6**, 234 ⟨94S145⟩
Chabardes, P., **2**, 21 ⟨92TL495⟩
Chabaud, B., **2**, 711 ⟨81S892⟩, 727 ⟨86JOC875⟩; **3**, 132 ⟨81SC423⟩; **4**, 875 ⟨86JOC875⟩
Chabert, P., **1**, 580 ⟨89TL179⟩; **3**, 48, 57 ⟨89TL179⟩
Chabra, B. R., **3**, 58 ⟨89CI(L)533⟩
Chabrier, P., **5**, 565, 579 ⟨49BSF272⟩
Chabriev, P., **2**, 255 ⟨70CR(C)2086⟩
Chaco, M. C., **3**, 452 ⟨62JOC2765⟩
Chadaeva, N. A., **2**, 500 ⟨57DOK(115)305, 57IZV69⟩, 501 ⟨73JGU825⟩, 889 ⟨61JGU3315, 64DOK(157)371⟩, 889 ⟨66JGU1986, 71IZV1726⟩, 894 ⟨83IZV1152⟩; **4**, 502 ⟨79DOK(246)1130⟩
Chadda, S. K., **2**, 1007 ⟨86JA1694⟩
Chadha, M. S., **3**, 62, 221 ⟨89IJC(B)3⟩
Chadha, R. K., **2**, 49 ⟨93JOC844⟩; **4**, 377 ⟨86IC611⟩
Chadha, V. K., **1**, 521 ⟨68CC552⟩
Chadhu, N. K., **2**, 519 ⟨73JA7171⟩
Chadwick, D. H., **2**, 643 ⟨45JA386⟩, 645, 646 ⟨46JA389⟩; **5**, 965 ⟨56USP2733254⟩
Chadwick, D. J., **3**, 303 ⟨82JCS(P1)1343⟩, 307 ⟨80JCR(S)42, 83JCS(P1)93⟩
Chadwick, J. C., **1**, 815 ⟨77T1527⟩
Chae, Y. B., **1**, 697, 700, 703 ⟨88TL5169⟩
Chaigne, F., **1**, 194, 288 ⟨89TL1803⟩
Chaika, E. A., **4**, 793 ⟨77JOU1918⟩; **6**, 69 ⟨71ZOR2161⟩
Chaimovich, H., **3**, 476 ⟨68JA4088⟩
Chaintreau, A., **2**, 28 ⟨81SC669⟩
Chakalamannil, S., **1**, 343 ⟨82JOC2231⟩
Chakrabarti, J. K., **3**, 618 ⟨72CC1226⟩
Chakraborti, A. K., **4**, 170 ⟨87T4433⟩
Chakraborti, R., **3**, 168 ⟨92JOC7349⟩; **4**, 258 ⟨92JOC7349⟩
Chakraborty, D. P., **5**, 298 ⟨81S977⟩
Chakraborty, R., **2**, 43 ⟨94T6579⟩; **3**, 161 ⟨90TL7641⟩
Chakraborty, S. K., **4**, 132 ⟨91SC1039⟩
Chakraborty, T. K., **1**, 106 ⟨93CC617⟩, 558, 828 ⟨93CC619⟩, 750 ⟨88JA4685⟩; **3**, 5, 115, 116 ⟨80TL1583⟩; **4**, 516 ⟨90TL1335, 91TL679⟩
Chakrasali, R. T., **4**, 253 ⟨90T2195⟩
Chakravarty, P. K., **2**, 223 ⟨78JOC1270⟩; **3**, 197 ⟨87TL611⟩, 198 ⟨89JMC1886⟩; **5**, 370 ⟨89JMC1886⟩
Chalk, A. J., **1**, 669 ⟨64JA1776⟩; **2**, 554 ⟨68JOM(11)615⟩, 913 ⟨65JA16, 67JA1640⟩
Chalkley, G. R., **2**, 402 ⟨78JCS(P1)1580⟩
Challand, B. D., **1**, 356 ⟨68CC982⟩
Challand, S. R., **2**, 777 ⟨75JCS(P1)26⟩
Challenger, C. A., **1**, 551 ⟨87JOM(334)9, 93JA10671⟩
Challenger, F., **2**, 286, 288, 291 ⟨42JCS570⟩, 366 ⟨56JCS1755⟩, 881 ⟨53JCS16⟩, 882 ⟨22JCS104, 22JCS91⟩
Challenger, F. G., **1**, 447 ⟨30JCS2171⟩
Challis, B. C., **2**, 320 ⟨B-70MI 205-02⟩, 335, 337, 339, 340, 344, 348 ⟨B-68MI 206-01⟩, 385 ⟨76CC877, 90CC1598⟩, 385 ⟨B-82MI 207-01⟩, 761 ⟨B-68MI 215-02⟩; **3**, 439 ⟨B-82MI 311-01⟩; **4**, 416 ⟨B-70MI 409-01⟩; **5**, 305 ⟨B-70MI 506-03⟩, 360 ⟨90JCS(P1)3103, 87JCS(P2)1489⟩, 714 ⟨79JCS(P2)634⟩, 723 ⟨B-70MI 517-02⟩
Challis, J. A., **2**, 320 ⟨B-70MI 205-02, B-82MI 207-01⟩; **3**, 439 ⟨B-82MI 311-01⟩; **4**, 416 ⟨B-70MI 409-01⟩; **5**, 714 ⟨79JCS(P2)634⟩
Challis, J., **5**, 305 ⟨B-70MI 506-03⟩, 723 ⟨B-70MI 517-02⟩

Chalmers, A. A., **6**, 333 ⟨84JOM(262)69⟩
Chaloner, P. A., **1**, 77 ⟨80JA3040⟩; **2**, 595 ⟨B-87MI 211-05⟩; **5**, 943 ⟨83CC880⟩
Chaloner, P. M., **1**, 674 ⟨B-86MI 115-01⟩
Chalova, O. B., **6**, 70 ⟨81MI 603-01⟩
Chaloyard, A., **2**, 566 ⟨80CC214⟩
Chaly, T., **2**, 662 ⟨82TL2935⟩
Chaman, E. S., **6**, 463 ⟨71ZOR273⟩
Chamberlain, A. R., **2**, 556 ⟨88JA4788⟩, 911, 919, 956 ⟨90OR(39)1⟩
Chamberlain, D. L., **2**, 165 ⟨55JA1041⟩, 165 ⟨58JOC381⟩
Chamberlain, G., **1**, 357 ⟨85TL4707⟩
Chamberlain, K., **4**, 327 ⟨74AJC1579⟩
Chamberlain, K. B., **1**, 98 ⟨88OSC(6)936⟩
Chamberlain, L. R., **3**, 512 ⟨86JA1502⟩
Chamberlain, T., **1**, 1038, 1140 ⟨91TL1707⟩
Chamberlin, A. R., **1**, 43 ⟨91TL1691⟩, 316 ⟨88JA4788⟩, 630, 632, 633, 1091, 1094 ⟨78JOC147⟩, 630, 633 ⟨90OR(39)1⟩, 633, 1094 ⟨76TL2947, 79S44⟩, 1094 ⟨84TL4901⟩; **2**, 409, 410 ⟨91JOC4608⟩, 660 ⟨88TL4827⟩, 911 ⟨78JOC147⟩; **3**, 58 ⟨79S44⟩
Chamberlin, E. L., **3**, 54 ⟨57JA456⟩
Chamberlin, E. M., **4**, 213 ⟨57JA456⟩
Chambers, D., **1**, 1044 ⟨81JCS(P1)40⟩; **3**, 8 ⟨78JCS(P1)1483⟩; **5**, 729 ⟨81JCS(P1)40⟩
Chambers, G. R., **1**, 1091 ⟨78TL5193⟩
Chambers, J. L., **2**, 183 ⟨72JOC3549⟩
Chambers, J. O., **6**, 555 ⟨76JOC626⟩
Chambers, J. Q., **4**, 846 ⟨74JA945⟩, 874, 875 ⟨76CC148⟩; **5**, 81 ⟨B-78MI 502-03⟩; **6**, 590 ⟨76CC148⟩
Chambers, J. R., **4**, 467 ⟨64JOC832⟩
Chambers, Jr., R. R., **3**, 108 ⟨84JA630⟩
Chambers, M. S., **5**, 573 ⟨87TL2171⟩; **6**, 473, 550 ⟨87CC1228⟩
Chambers, R., **3**, 543 ⟨78TL929⟩
Chambers, R. D., **1**, 107 ⟨79COC(1)493⟩, 428 ⟨65JCS5045⟩, 570 ⟨94JCS(P1)3115⟩, 852 ⟨79COC(1)525⟩; **2**, 2 ⟨79COC(1)493⟩, 613 ⟨86BSF930, 86CC1623⟩, 650 ⟨59JCS1804⟩; **4**, 7 ⟨60JA5298⟩, 34, 36 ⟨61JCS3779⟩, 45 ⟨B-92MI 402-01⟩, 146 ⟨62JCS1993⟩, 737 ⟨78CC475⟩, 1076 ⟨65AFC(4)50⟩; **6**, 2, 3 ⟨61JCS3779⟩, 9, 10 ⟨81AHC(28)1, 87JOM(325)13⟩, 12 ⟨74JCS(P1)108⟩, 13, 22 ⟨B-92MI 601-01⟩, 14 ⟨88JCS(P1)921, B-92MI 601-03⟩, 16 ⟨B-73MI 601-01⟩, 18 ⟨79CC964, B-73MI 601-02⟩, 20 ⟨B-73MI 601-03⟩, 41, 58, 60, 65 ⟨65AFC(4)50⟩, 212, 244 ⟨B-73MI 607-01⟩, 212 ⟨B-92MI 607-02, 60JA5298⟩, 262 ⟨85JCS(P1)1191⟩
Chambers, R. K., **4**, 866 ⟨86EUP174832A2⟩; **5**, 690 ⟨92JCS(P1)2789⟩
Chambers, W. J., **3**, 419 ⟨62JA2337⟩; **5**, 808 ⟨62USP3023208⟩, 808, 989 ⟨62JA2337⟩; **6**, 242 ⟨62JA2337⟩
Chambers, W. T., **1**, 906 ⟨78JA3847⟩
Chamchaang, W., **2**, 319 ⟨90JOC2531⟩; **4**, 849 ⟨82S579⟩; **5**, 346 ⟨88CC710, 90JOC2943⟩
Chamier, C. v., **6**, 732 ⟨69LA(727)228⟩
Chamlers, M. E., **4**, 469 ⟨53JA5278⟩
Chamness, J. T., **2**, 772 ⟨63JOC1232⟩
Chamorro, E., **2**, 18 ⟨89JOC4294⟩, 19 ⟨87TL4733⟩
Chamot, E., **1**, 704, 708 ⟨78JOC4527⟩
Champagne, J., **2**, 675 ⟨64CJC212⟩
Champagne, P. J., **1**, 879 ⟨76CJC3809, 78JA657⟩; **6**, 20 ⟨75CJC529, 79CJC2617⟩
Champetier, M. G., **1**, 1037 ⟨66CR(C)289⟩
Champion, E., **1**, 532, 662 ⟨86TL539⟩
Champion, N., **1**, 980 ⟨92TL4447⟩
Champion, R., **3**, 342 ⟨82AJC1747⟩
Champion, W. C., **3**, 477 ⟨56JA5452⟩
Champney, B., **2**, 95 ⟨83JOC3126⟩
Champseix, A., **2**, 249 ⟨85BSF463⟩
Chan, A. S. C., **1**, 14 ⟨91CC535⟩; **2**, 83 ⟨91CC535⟩
Chan, A. S. K., **6**, 351 ⟨74IC225⟩
Chan, A. W. K., **3**, 552 ⟨70T1493⟩
Chan, C., **1**, 546, 1036 ⟨87JA3486⟩; **2**, 628 ⟨88CC971⟩; **3**, 712, 718, 719 ⟨89SL9⟩
Chan, C.-C., **3**, 4, 88, 116 ⟨82S1091⟩; **5**, 226 ⟨82SC709⟩
Chan, C.-M., **2**, 68 ⟨90TL1985⟩; **4**, 842 ⟨78JOC3065⟩
Chan, D., **4**, 742 ⟨66CC567⟩
Chan, D. M. T., **1**, 287 ⟨79JA6429⟩, 289 ⟨82JA3733⟩, 289, 362 ⟨91JA7350⟩
Chan, F. S. Y., **6**, 562 ⟨85CC1641⟩, 623, 728 ⟨88JCS(P1)899⟩
Chan, J. A., **5**, 138 ⟨75JA4144⟩
Chan, J. B., **1**, 1019 ⟨91SC977⟩
Chan, J. K., **5**, 217 ⟨67JCED151⟩; **6**, 448 ⟨86JOC3781⟩
Chan, K.-K., **1**, 514 ⟨76JOC3497⟩
Chan, K. P., **3**, 309 ⟨90SC3069⟩
Chan, K. S., **1**, 551 ⟨87JOM(334)9⟩; **2**, 607 ⟨86JOC277⟩, 670 ⟨84JOC2293⟩; **5**, 934 ⟨94T2003⟩
Chan, L., **3**, 65 ⟨81CC10, 90JCS(P1)89⟩
Chan, L. C., **4**, 970 ⟨91EUP425030⟩
Chan, L. H., **2**, 815 ⟨67JOM(9)231⟩; **3**, 440, 456, 457 ⟨67JOM(9)231⟩
Chan, M.-C., **1**, 436 ⟨85CC1610, 88JOC4466⟩
Chan, P. C.-M., **1**, 138 ⟨88JOC5584⟩; **4**, 386 ⟨88JOC5584⟩
Chan, S., **1**, 216 ⟨83JOC2499⟩; **2**, 430, 438 ⟨69T1097⟩
Chan, T. H., **1**, 63, 451 ⟨74TL3479⟩, 111 ⟨74TL3511⟩, 161 ⟨91TL2879⟩, 161, 162 ⟨89JOC317⟩, 162 ⟨89JA8737⟩, 162 ⟨93TL1449⟩, 189, 450, 535 ⟨79S761⟩, 318 ⟨75TL3383, 78JOC2994, 77TL3317⟩, 648 ⟨74JOC3264⟩, 648, 651, 762 ⟨78TL2383⟩, 648, 652, 982 ⟨74TL171⟩, 651 ⟨70TL1137⟩, 651 ⟨82CC969⟩, 651, 760, 981 ⟨78JOC1526⟩, 652 ⟨77JA936, 80T2269⟩, 654 ⟨85S817⟩, 692 ⟨84JOM(268)C1⟩, 1091 ⟨88CC1280⟩, 1094 ⟨76S801⟩; **2**, 22 ⟨86TL2563⟩, 78 ⟨89JOC5768, 89TL3041⟩, 85 ⟨92JOC6107⟩, 120 ⟨76TL319⟩, 151, 688 ⟨76JOC3987⟩, 212, 215, 231 ⟨86TL441⟩, 221 ⟨70JA7224⟩, 231, 723 ⟨83TL5173⟩, 236 ⟨78JA1222⟩, 244 ⟨84S1037⟩, 287 ⟨74TL2091⟩, 539 ⟨85S817⟩, 616 ⟨75TL3383⟩, 624, 630 ⟨77TL3317⟩, 662 ⟨79CC578, 80JA3534, 82TL2935⟩, 668 ⟨80TL3423⟩, 674 ⟨91COS(2)595⟩, 699 ⟨91COS(2)599⟩, 709 ⟨85TL1795⟩, 907, 943 ⟨79S761⟩, 911 ⟨76S801⟩, 931, 932 ⟨76JOM(107)C1⟩, 931, 932, 935 ⟨76TL4439, 78TL2383⟩, 935 ⟨84TL717⟩, 4366 ⟨70TL1687⟩; **3**, 156 ⟨91TL715⟩; 58 ⟨77TL3317⟩; **4**, 137 ⟨89JA8737⟩, 216 ⟨76TL319⟩, 506 ⟨89TL3041⟩, 520 ⟨89JOC317⟩, 646 ⟨70TL1137⟩, 646, 651, 653 ⟨74JOC3264⟩, 648 ⟨92JOC6107⟩, 655 ⟨78TL2383⟩, 831 ⟨79TL4029, 91COS(2)604⟩; **5**, 242, 337 ⟨87S1122⟩, 291 ⟨69JOC2766, 84S1037⟩, 338 ⟨78JA1222⟩; **6**, 192 ⟨70TL1137⟩, 296, 297 ⟨74TL4409⟩
Chan, T.-L., **1**, 709 ⟨78TL4269⟩; **4**, 1083 ⟨84JCS(P1)1561⟩
Chan, W. H., **2**, 25 ⟨78CL923⟩; **5**, 278 ⟨86JA1039⟩, 280 ⟨79SC31⟩; **6**, 550 ⟨89SC547⟩, 554 ⟨88SC1531⟩
Chan, W. K., **5**, 161 ⟨76JA7874⟩
Chan, W.-T., **3**, 269 ⟨78TL2537⟩
Chan, Y. N. C., **2**, 312 ⟨90JA9400⟩
Chance, L. H., **4**, 113, 118, 120, 122 ⟨67JCED282⟩, 555 ⟨68JCED585, 71USP3607944⟩; **5**, 1035 ⟨49JA2240⟩
Chancellor, T., **2**, 156 ⟨84S320⟩
Chancharunee, S., **3**, 210 ⟨84TL4141⟩
Chandaeva, N. A., **2**, 881, 883 ⟨72IZV1612⟩

Chander, M. C., **1**, 1002 ⟨92TL135⟩, 1066 ⟨89TL5455⟩
Chandler, A., **5**, 748 ⟨80CC130, 80TL885⟩
Chandler, J. H., **1**, 41 ⟨77S124, 79SC275⟩, 962 ⟨78CC726⟩
Chandler, R., **1**, 515 ⟨90SC3477⟩
Chandra, G., **5**, 277 ⟨69JCS(C)2565⟩
Chandra, H., **2**, 391 ⟨79CC392⟩
Chandra, S., **6**, 443 ⟨87IJC(B)748⟩
Chandra Rao, M. S., **4**, 449 ⟨90S663⟩
Chandrasakar, R., **2**, 649 ⟨84JOC4295⟩
Chandrasegaran, S., **5**, 313, 314 ⟨84JA1543⟩
Chandrasekaran, M. Y., **6**, 128 ⟨85TL5001⟩
Chandrasekaran, R. Y., **4**, 514 ⟨85TL5001⟩
Chandrasekaran, S., **2**, 228 ⟨89JOC2998, 92JOC1699⟩; **3**, 5, 115, 116 ⟨80TL1583⟩, 88 ⟨84CL2131⟩, 113 ⟨89JOC5182⟩, 134 ⟨86JOC3063⟩, 293 ⟨86SC1493⟩, 294 ⟨86JCR(S)458⟩; **4**, 94 ⟨84H(22)585⟩; **5**, 189, 196 ⟨84JCR(S)100⟩
Chandrasekhar, J., **1**, 846 ⟨86TL4411⟩, 1143 ⟨82CPL493⟩
Chandrasekharan, J.,**1**, 510 ⟨83JOC5080, 83JOC644⟩; **2**, 58 ⟨88JA1549⟩, 526 ⟨88JA1539⟩, 526, 531 ⟨85JOC5446⟩, 526, 531 ⟨86JOC3394⟩
Chandrasekhar, S., **3**, 40 ⟨89TL6769⟩
Chandrasekhar, T., **3**, 530 ⟨91JOC7048⟩
Chandresekaran, S., **5**, 252 ⟨94JCS(P1)767⟩
Chandross, E. A., **1**, 811 ⟨70JA703⟩
Chandy, M. J., **1**, 1124 ⟨77TL4377⟩
Chang, B. C., **2**, 830 ⟨69JA5243⟩
Chang, C., **3**, 9 ⟨92SC1217⟩
Chang, C. C., **2**, 655 ⟨81TL3451⟩; **3**, 502 ⟨76JA7844⟩; 542 ⟨75JA6586⟩; **5**, 469 ⟨86ICA107⟩
Chang, C. J., **3**, 226 ⟨76JA3645⟩
Chang, C.-N., **4**, 231 ⟨92JOC2217⟩
Chang, C.-T., **1**, 334, 337, 344, 936 ⟨89JOC3140⟩; **2**, 14, 34 ⟨87TL2477⟩; **4**, 811 ⟨89JOC3140⟩
Chang, D. C. K., **4**, 74 ⟨68TL5415⟩
Chang, D. T., **2**, 555 ⟨84OS(62)58⟩
Chang, E., **1**, 648 ⟨74JOC3264⟩, 651 ⟨70TL1137⟩; **4**, 646 ⟨70TL1137⟩, 646, 651, 653 ⟨74JOC3264⟩; **6**, 192 ⟨70TL1137⟩
Chang, F. C., **5**, 31 ⟨64TL2969⟩
Chang, H., **5**, 138 ⟨85JOC1751⟩, 280 ⟨85TL1341⟩, 1030 ⟨86MI 527-01⟩
Chang, H. F., **6**, 304 ⟨83MI 610-01⟩
Chang, H. M., **3**, 103 ⟨92T1039⟩
Chang, H. W., **1**, 907 ⟨61JA2367, 61JA2375⟩, 907, 908, 909 ⟨62JA3168⟩; **5**, 346 ⟨82JOC179⟩, 370 ⟨83JA3743⟩
Chang, J., **5**, 386 ⟨89JA3258⟩
Chang, J. C., **5**, 322 ⟨81JOC4188⟩
Chang, J. H., **2**, 91 ⟨83JA7719⟩; **6**, 327 ⟨83JA7719⟩
Chang, K., **3**, 197 ⟨89TL3307⟩; **4**, 572 ⟨89SC1891⟩
Chang, L.-J., **1**, 790 ⟨86JCS(P1)1039⟩, 791 ⟨84CC1323⟩
Chang, L. L., **2**, 469 ⟨77JOC782⟩; **3**, 240 ⟨88H(27)651⟩
Chang, L. W., **5**, 316 ⟨86OPP269⟩
Chang, L. W. K., **2**, 689 ⟨76JOC4099⟩
Chang, M. H., **1**, 1130, 1132 ⟨89TL827⟩; **2**, 81 ⟨90TL4765⟩
Chang, M.-K., **1**, 637 ⟨89JOC5424⟩
Chang, M.-M. Y., **2**, 1001 ⟨74JA4099⟩
Chang, M. S., **5**, 821 ⟨63JOC2260⟩
Chang, N. C., **3**, 424 ⟨89JOC3820⟩
Chang, S., **1**, 905 ⟨84JA810⟩; **3**, 694 ⟨71T3069⟩
Chang, S.-C., **1**, 511 ⟨77JOC3981⟩; **5**, 313 ⟨82POL129, 83JOC771⟩; **6**, 256 ⟨83IC805⟩, 261, 615 ⟨83JOC771⟩, 614, 615 ⟨82POL129⟩
Chang, S.-K., **5**, 11 ⟨90JA9586⟩
Chang, T. C. T., **1**, 481 ⟨88OS(66)95⟩; **3**, 138 ⟨80JA5930, 81JOC4103⟩
Chang, T. H., **1**, 216 ⟨84CC1273⟩
Chang, V. H.-T., **1**, 340 ⟨89JA8309⟩
Chang, V. K., **1**, 675 ⟨93S824⟩
Chang, W.-H., **1**, 8 ⟨59JA3634⟩
Chang, Y.-C., **2**, 626 ⟨86CC1159⟩; **3**, 314 ⟨71JHC1097⟩
Chang, Y.-H., **3**, 62 ⟨90S1033⟩
Chang, Y. K., **3**, 88 ⟨84S866⟩
Chang, Y.-M., **5**, 924 ⟨79JA6750, 82JA5667⟩
Chanley, J. D., **2**, 807 ⟨58JA2686⟩
Channareddy, S., **5**, 381, 534 ⟨93CB1119⟩, 927 ⟨90AG(E)199, 92CB73⟩
Chanon, M., **6**, 550 ⟨92JA4111⟩
Chanson, E., **5**, 467 ⟨85TL6075⟩
Chantegral, B., **3**, 246 ⟨84JHC13⟩
Chantegrel, B., **2**, 668 ⟨93TL7923⟩; **3**, 241 ⟨74BSF1043⟩
Chantrapromma, K., **1**, 839 ⟨83JCS(P1)1029⟩
Chanzov, V. A., **2**, 841 ⟨78ZOB91⟩
Chao, B., **2**, 153 ⟨71CC979⟩
Chao, E., **4**, 526 ⟨76JOM(122)145⟩
Chao, H. G., **5**, 5 ⟨90JA9651⟩
Chao, H. S.-I., **1**, 33 ⟨88SC1207⟩
Chao, K.-H., **3**, 442 ⟨85JOC4650⟩
Chao, L. C., **2**, 597 ⟨74JOM(67)C64, 74SRI373⟩
Chao, P., **2**, 153 ⟨73JA920⟩, 153 ⟨73JA922⟩
Chao, S., **1**, 835 ⟨83JOC4262, 84JOC2708, 89JOC2893⟩
Chao, T., **2**, 720 ⟨65CI(L)81⟩
Chao, Y. L., **2**, 11 ⟨81HCA2528⟩
Chapa, O., **1**, 308 ⟨82H(18)187⟩
Chapat, J.-P., **2**, 697 ⟨68BSF4265⟩; **3**, 218 ⟨79BSF381⟩; **5**, 360 ⟨90CPB2352⟩, 772 ⟨83EJM413⟩
Chapdelaine, M. J., **1**, 304 ⟨90OR(38)225, 91COS(4)237⟩, 1010 ⟨89JOC1218⟩; **3**, 119 ⟨77S555⟩, 165 ⟨77TL3227⟩, 220 ⟨78TL4205⟩
Chapeaurouge, A., **4**, 140, 352 ⟨91HCA1477⟩; **5**, 416 ⟨91HCA1477⟩
Chapelet, G., **1**, 674 ⟨77RTC(M)91⟩
Chapelle, J., **4**, 984 ⟨71BSF283⟩
Chapin, E. C., **3**, 638 ⟨44JA725⟩
Chapleo, C. B., **2**, 831 ⟨76HCA695⟩; **5**, 199 ⟨89SL30⟩, 223 ⟨86TL4995⟩, 690 ⟨93T11029⟩, 1120 ⟨87JMC951⟩
Chapleur, Y., **3**, 495 ⟨91SL583⟩
Chaplin, C. A., **2**, 288, 291 ⟨38JCS341⟩
Chaplin, E. J., **2**, 809, 810, 825 ⟨37JCS527⟩
Chapman, A. C., **2**, 407 ⟨61JCS1825⟩
Chapman, D., **1**, 314 ⟨83TL1909⟩; **3**, 624 ⟨86JOC1745⟩
Chapman, F., **4**, 442, 443 ⟨49JCS1638⟩
Chapman, J. A., **5**, 652 ⟨76CC240⟩
Chapman, J. H., **2**, 128 ⟨50JCS579⟩; **3**, 229 ⟨52JCS1094⟩
Chapman, K., **6**, 487 ⟨93EUP547699⟩
Chapman, K. J., **2**, 761 ⟨76AJC367⟩
Chapman, K. T., **2**, 62 ⟨88JA3560⟩; **3**, 119 ⟨80JOC2030, 82TL4647⟩
Chapman, L. E., **6**, 511 ⟨66CI(L)1266⟩
Chapman, O. L., **1**, 404 ⟨62JOC2291⟩, 597 ⟨87JOC1218⟩, 776 ⟨87JA553⟩, 981 ⟨76JOC3326⟩; **2**, 350 ⟨59JA5800⟩; **3**, 382 ⟨85JOC543⟩, 451 ⟨64JA732⟩, 502 ⟨76JA7844⟩, 538, 539 ⟨68JA2449, 73JA6134⟩, 542 ⟨70JA7001, 73JA4061, 75JA6586⟩, 543 ⟨69JA4309, 71CC770, 73JA244, 73JA247⟩; **6**, 75 ⟨71CC383⟩
Chapman, R. C., **3**, 442 ⟨84JOC2512⟩
Chapman, R. D., **2**, 206 ⟨59JA2002⟩; **4**, 435, 447 ⟨88S743⟩; **5**, 361, 362 ⟨88S743⟩
Chapman, R. F., **5**, 587, 593 ⟨93JMC1604⟩
Chapman, S. L., **1**, 21, 23 ⟨86CC1688⟩; **3**, 421 ⟨90JOC1086⟩
Chapman, T. M., **4**, 297, 304, 325 ⟨75JOC2414⟩; **5**, 360 ⟨79S100⟩
Chapovskaya, N. K., **5**, 810 ⟨75ZOR1981⟩
Chappuis, J. L., **1**, 634 ⟨81JA5314⟩

Chapuis, C., **1**, 789 ⟨92TL2461⟩
Chaquin, P., **2**, 626 ⟨89CJC2061⟩
Charandabi, M. R. M. D., **5**, 688 ⟨90IC3218⟩; **6**, 514 ⟨91IC2433⟩, 515 ⟨90IC3218⟩
Charbonell, Y., **5**, 708, 861 ⟨70BSF1363, 68CR(C)1241⟩
Charbonnier, C., **5**, 233 ⟨91S23⟩
Charbonnier, F., **2**, 676, 1026 ⟨87JOC2919⟩, 682 ⟨87JOC2303⟩; **3**, 236 ⟨87JA4752⟩
Charest, M-P., **2**, 27 ⟨71JMC982⟩
Charette, A. B., **1**, 98 ⟨91JOC741⟩; **2**, 79 ⟨93SL81⟩; **3**, 308 ⟨87JA7553⟩; **4**, 952 ⟨87JA7553⟩
Charles, G., **1**, 457, 508 ⟨80TL1069⟩; **2**, 303 ⟨83TL3233⟩, 310 ⟨77TL3469⟩; **4**, 300 ⟨80TL2949⟩
Charles, K. A., **1**, 563 ⟨81JOC1874⟩
Charles, N. G., **2**, 954 ⟨83OM101⟩
Charles, S. W., **3**, 337 ⟨77CC483⟩
Charley, L. M., **6**, 721 ⟨70JOM(24)205⟩
Charlton, J. L., **1**, 418 ⟨86CJC793⟩; **2**, 200, 272 ⟨84TL5287⟩
Charlton, P. T., **5**, 749, 750 ⟨51JCS485⟩
Charoenying, P., **1**, 787 ⟨94CC889⟩
Charpentier-Morize, M., **3**, 24 ⟨82TL2853⟩
Charpiot, B., **3**, 145 ⟨85JCS(P1)2667⟩
Charreau, P., **2**, 623 ⟨89JOM(379)201⟩, 628 ⟨90BSF275⟩; **4**, 742, 746 ⟨89JOM(379)201⟩
Charrier, C., **2**, 472 ⟨83JA6871⟩, 824, 828, 841, 1059, 1061, 1063, 1066, 1069 ⟨66BSF1002⟩, 1072 ⟨65CR(261)1019, 66BSF1707⟩; **4**, 112 ⟨78TL2407⟩, 545, 565 ⟨93PS(77)254⟩
Charrier, J., **4**, 891 ⟨83JOC481⟩
Charrier, M. C., **2**, 1063, 1065, 1066 ⟨67CR(C)995⟩
Charvière, G., **1**, 55 ⟨63BSF1428⟩
Chasar, D. W., **2**, 478 ⟨67JOC2465⟩
Chase, G. O., **3**, 233 ⟨57JOC1611⟩
Chasey, K. L., **1**, 708 ⟨88JA5479⟩
Chasle, M. F., **3**, 568 ⟨72T4431⟩; **4**, 996 ⟨72T5055⟩
Chasle-Pommeret, M.-F., **3**, 570 ⟨72CR(C)1309⟩
Chassaing, G., **2**, 229 ⟨87JOC3154⟩
Chastrette, F., **2**, 381 ⟨90TL1429⟩; **3**, 40 ⟨88BSF67⟩, 41 ⟨90TL1429⟩
Chastrette, M., **3**, 40 ⟨80S889⟩, 286 ⟨75JOM(99)41⟩; **5**, 29, 34, 43 ⟨67BSF2011⟩
Chatani, N., **1**, 202 ⟨90JOC5923⟩, 212 ⟨93JOC4187⟩; **2**, 920, 946 ⟨88JOC3539⟩, 939 ⟨91JA7778, 90JOC3393⟩; **3**, 20 ⟨83JA1370⟩, 33 ⟨93JOC4182⟩, 660 ⟨85CC838, 86TL1841⟩, 661 ⟨86JOC4714⟩, 671 ⟨86TL4201⟩; **4**, 931 ⟨92JA9710⟩; **5**, 346 ⟨90JA9646⟩, 416 ⟨92JA9710⟩
Chateauneuf, J., **1**, 924, 934 ⟨88JA2877⟩
Chatgilialoglu, C., **1**, 4, 17, 331, 342, 919, 936, 938, 943, 945 ⟨92ACR188⟩, 5, 7, 9, 10, 17, 25 ⟨92JOC2427⟩, 5, 7, 9, 10, 17, 25 ⟨92JOC3405⟩, 5, 7, 9, 10, 17, 936, 943, 945 ⟨91JOC678⟩, 5, 7, 9, 10, 331, 936 ⟨88JOC3641⟩, 5, 9, 10, 936 ⟨89TL2733⟩, 17 ⟨93JOC249⟩, 327, 950 ⟨92JOC3994⟩, 331, 342, 936 ⟨89TL681⟩, 331, 936 ⟨87JA5267⟩, 933 ⟨92TL1787, 89JOC2492⟩, 936, 945 ⟨90SL219, 90TL6013⟩; **2**, 132, 202 ⟨B-90MI 203-01⟩, 536 ⟨90SL257⟩, 536, 545 ⟨92ACR188⟩, 541 ⟨90SL219⟩; **5**, 1135 ⟨91JOC678⟩
Chatrenet, B., **6**, 443 ⟨90S1065⟩
Chatrousse, A.-P., **6**, 97, 100 ⟨75JOC2911⟩
Chatt, J., **1**, 668 ⟨68JCS(A)190⟩; **2**, 437 ⟨61JCS896⟩, 962 ⟨65JCS843⟩; **5**, 463, 464 ⟨66JCS(A)634⟩; **6**, 721 ⟨69CC1322, 71JCS(A)21⟩
Chattaway, F. D., **2**, 94 ⟨23JCS2999⟩, 769 ⟨11JCS404⟩; **3**, 451 ⟨30JCS96⟩; **4**, 799 ⟨27JCS2850, 28JCS2756⟩, 799 ⟨31JCS1088⟩; **5**, 840, 860 ⟨25JCS975⟩, 841 ⟨25JCS1687, 30JCS94⟩, 841 ⟨32JA263⟩, 842 ⟨33JCS1389⟩; **6**, 483 ⟨20JCS708⟩, 653 ⟨36JCS1693⟩
Chatterjea, J. N., **3**, 103 ⟨53JIC103⟩, 247 ⟨53JIC379⟩
Chatterjee, A. K., **2**, 105 ⟨68CAR(6)253⟩
Chatterjee, J. N., **3**, 247 ⟨47JIC473⟩
Chatterjee, R. K., **6**, 443 ⟨87IJC(B)748⟩
Chatterjee, S., **1**, 133 ⟨83TL1341⟩, 476 ⟨81TL3737⟩, 1045, 1047 ⟨83TL5181⟩; **3**, 130, 288 ⟨83TL5181⟩
Chatterjee, S. K., **2**, 320 ⟨90CI547⟩
Chatterton, W. J., **5**, 418, 924, 925 ⟨89OM693⟩, 924 ⟨87OM1246, 87OM2128⟩, 924 ⟨91OM2752⟩
Chattha, M. S., **2**, 739 ⟨73JOC2908⟩, 739, 1072 ⟨71JOC2719⟩, 1060 ⟨73JOC1611, 71JOC2720⟩
Chattopadhaya, J., **6**, 488 ⟨85ACS(B)761⟩
Chattopadhaya, J. B., **6**, 572 ⟨74S289⟩
Chattopadhyaya, J., **1**, 244 ⟨91T2237⟩, 946 ⟨93T5189, 93T9079⟩; **4**, 276 ⟨90T2587, 91T3431⟩
Chattopadhyaya, J. B., **5**, 276, 298 ⟨74T2899⟩
Chaturvedi, S. C., **1**, 738 ⟨78S616⟩, 740, 742 ⟨77TL3843⟩, 740, 742 ⟨79IJC(B)859⟩
Chatziiosifidis, I., **3**, 613 ⟨81AG(E)1017⟩, 613, 627, 698 ⟨83T961⟩
Chatzopoulos-Ouar, F., **6**, 472 ⟨85JOC118⟩
Chau, J. H.-L., **5**, 315 ⟨94H(38)159⟩
Chau, K., **2**, 591 ⟨94TL1177⟩
Chau, L. V., **4**, 32 ⟨71CB1921⟩
Chau, M. M., **5**, 204 ⟨74JA3319⟩
Chau, S. H., **6**, 481, 482 ⟨91JCS(P2)501⟩
Chauchard, P., **5**, 92 ⟨47MI 502-03⟩
Chau-DerLi, **6**, 509 ⟨81JMC1089⟩
Chaudhary, A. G., **6**, 550 ⟨93JOC3798⟩
Chaudhary, S. K., **2**, 539 ⟨79TL99⟩; **5**, 374 ⟨79TL99⟩
Chaudhary, S. S., **2**, 306 ⟨76JA5678⟩, 338 ⟨68OS(48)4⟩
Chaudhry, S. C., **6**, 268 ⟨90CB945⟩, 656 ⟨88JOM(339)241⟩
Chaudhuri, A., **5**, 1111 ⟨56JCS96⟩
Chaudhuri, M. K., **3**, 4, 88, 115 ⟨82S588⟩
Chau Diem Dieu, **6**, 415 ⟨87MI 614-02⟩
Chauhan, G. S., **5**, 665 ⟨84IJC(B)728⟩
Chauhan, K., **4**, 175 ⟨94TL1825⟩
Chauhan, M. S., **2**, 229 ⟨75CJC455⟩
Chaurasia, M. R., **6**, 561 ⟨82FES205⟩
Chauret, D. C., **1**, 217 ⟨93TL3695⟩; **3**, 51 ⟨93TL3695⟩
Chaussard, J., **3**, 146 ⟨86TL4175⟩
Chaussin, R., **1**, 833, 834 ⟨79PS(7)69⟩; **5**, 552 ⟨79PS(7)69⟩
Chauvelier, J., **3**, 267 ⟨54BSF734⟩
Chauvet, C., **4**, 269 ⟨92TL4913⟩
Chauvin, J., **2**, 759 ⟨82JCR(M)621, 82JCR(S)48⟩; **3**, 35 ⟨80TL4171⟩
Chauvin, J.,**3**, 442 ⟨82TL1605⟩
Chauvin, R., **2**, 300 ⟨91SC1425⟩
Chaux, R., **5**, 59 ⟨32OSC(1)166⟩
Chauzov, V. A., **4**, 119 ⟨86JGU2394⟩, 451, 469, 472, 473, 475, 493 ⟨74RCR984⟩, 464 ⟨77ZOB579⟩, 471 ⟨76MI 410-02⟩, 478, 493 ⟨76ZOB1246⟩, 490 ⟨76ZOB1250, 76ZOB1495, 76MI 410-03⟩, 556 ⟨87ZOB54⟩, 561 ⟨83ZOB541, 84URP1067006⟩, 753 ⟨77JGU2501⟩, 916 ⟨80ZOB1225⟩; **5**, 716 ⟨72JGU1857⟩, 716, 1006 ⟨72ZOB1868⟩, 1016 ⟨66ZOB952⟩
Chavan, S. P., **1**, 521 ⟨88TL3407⟩
Chavane, V., **4**, 465 ⟨48BSF774⟩
Chavant, P.-Y., **2**, 321 ⟨93SL519⟩, 592 ⟨94SL410⟩; **3**, 303, 306 ⟨92TL5157⟩
Chavasiri, W., **2**, 2 ⟨93TL1871, 94T31⟩
Chavdarian, C. G., **1**, 1024, 1139 ⟨76TL1769⟩, 1089 ⟨75JA3822⟩, 1093 ⟨75JOC2970⟩; **3**, 240 ⟨88H(27)651⟩; **4**, 953 ⟨75JA3822⟩
Chaves das Neves, H. J., **5**, 209 ⟨78TL2171⟩
Chávez, J. A., **4**, 581 ⟨89JOM(371)329⟩
Chavignon, O., **5**, 360 ⟨90CPB2352⟩
Chavis, C., **5**, 560 ⟨73JCS(P1)1571⟩
Chawla, H. M., **2**, 791 ⟨85S70⟩
Chawla, H. P. S., **3**, 226 ⟨76JA3645⟩; **4**, 425 ⟨78JHC949⟩
Chawla, U., **1**, 282 ⟨91JOC3908⟩
Chaykovsky, M., **1**, 51 ⟨64JA1639⟩, 142, 881 ⟨65JA1345⟩, 304 ⟨65JA1353, 61OS(41)5⟩, 881 ⟨62JA866⟩; **2**, 153 ⟨65JA1345⟩, 1000 ⟨65JA1353⟩; **3**, 129 ⟨64JA1639⟩, 372, 378 ⟨65JA1353⟩; **4**, 391, 393 ⟨65JA1345⟩

Cheburkov, Y. A., **3**, 527 ⟨67DOK(173)1117, 67FCR107⟩, 531 ⟨68T1341⟩; **6**, 54 ⟨74IZV2144⟩
Cheburkov, Yu. A., **4**, 966 ⟨74IZV2545⟩
Chechetkina, I. I., **2**, 879 ⟨88JGU1173⟩
Checkhun, A. L., **4**, 913 ⟨76ZOB565⟩
Chedekel, M. R., **3**, 502 ⟨76JA7846⟩
Chefczynska, A., **2**, 713 ⟨78JOC473⟩; **6**, 119 ⟨79JOC2967⟩
Chehayber, J. M., **4**, 377 ⟨86IC611⟩, 377, 401 ⟨86ICA51⟩
Chehidi, I., **2**, 18 ⟨89TL3167⟩
Chehna, M., **3**, 414 ⟨86PS(27)327⟩; **5**, 633 ⟨86PS(27)327⟩, 760 ⟨84CC157⟩
Cheikh, A. B., **1**, 835 ⟨91JOC970⟩; **4**, 900 ⟨91JOC970⟩
Chekhlov, A. N., **2**, 280 ⟨93TL703⟩; **4**, 134 ⟨90TL5361⟩, 494 ⟨87DOK(297)1132, 87IZV1680⟩, 494 ⟨88IZV936⟩; **5**, 370 ⟨88ZOB2424⟩, 669 ⟨89IZV2849⟩, 669, 672 ⟨89BAU2612, 92BAU236⟩, 672 ⟨92IZV315⟩; **6**, 430 ⟨89IZV2849⟩
Chekhun, A. L., **4**, 552 ⟨76ZOB568⟩
Cheklov, A. N., **4**, 124, 128 ⟨92JGU39⟩
Chekulaeva, I. A., **3**, 408 ⟨65RCR669⟩
Chekulaeva, L. A., **6**, 208, 210 ⟨83IZV636⟩
Chekulaeva, V. N., **6**, 368 ⟨68IZV1638⟩
Chelain, E., **5**, 952 ⟨91CSR503⟩, 953 ⟨92JA8088⟩; **6**, 716 ⟨91CSR503⟩
Chell, P. L., **5**, 943 ⟨82CC215⟩, 944 ⟨83JCS(D)2397⟩
Chellappa, K. L., **4**, 240 ⟨73JOC3172⟩; **5**, 210 ⟨73JOC3172⟩
Chelucci, G., **3**, 617, 619 ⟨82SC25⟩; **6**, 417 ⟨82SC25⟩
Chem, M. J., **2**, 348, 349 ⟨90JOC3233⟩
Chem, R., **6**, 598 ⟨91TL4823⟩
Chemburkar, S., **2**, 321 ⟨89JOC3002⟩
Chemburkar, S. R., **4**, 531 ⟨89JOC3002⟩
Chemerda, J. M., **3**, 54 ⟨57JA456⟩; **4**, 213 ⟨57JA456⟩
Chemin, D., **1**, 1066 ⟨92TL2681⟩
Chemla, F., **2**, 594, 998 ⟨93SL665⟩; **3**, 20, 31 ⟨91SL501⟩; **4**, 200 ⟨91SL501⟩
Chemla, P., **3**, 30 ⟨90SL619⟩
Chen, A., **1**, 846 ⟨66JA460⟩, 871 ⟨62JA2899, 62JA2905⟩
Chen, B., **5**, 366 ⟨88JA1638⟩; **6**, 556 ⟨88MI 617-02⟩
Chen, B. C., **1**, 147 ⟨89TL1613, 91T173⟩; **3**, 156 ⟨92CL2443⟩
Chen, B.-F., **6**, 348 ⟨90JCS(D)3033⟩
Chen, C., **1**, 272, 743 ⟨90CC85⟩, 647, 743 ⟨90JCR(S)388⟩, 648 ⟨86TL2903, 89SC501, 89SC83⟩, 648, 743 ⟨89JOM(366)87⟩, 828 ⟨90JOC6236⟩; **2**, 73 ⟨88JCS(P1)2855, 88TL1395⟩, 487 ⟨90CC85⟩, 487, 488 ⟨89JOM(366)87⟩; **3**, 322 ⟨83TL2945⟩, 644 ⟨89SC83⟩; **5**, 241 ⟨92JA8313⟩, 247 ⟨91T6127⟩
Chen, C. C., **3**, 638 ⟨90JCS(P1)3365⟩
Chen, C.-G., **2**, 998 ⟨92CC1487⟩
Chen, C. H., **6**, 79 ⟨73JA278⟩
Chen, C.-K., **3**, 638 ⟨88JA4829⟩
Chen, C. P., **2**, 528 ⟨87JA7925⟩
Chen, C.-S., **5**, 135 ⟨90JCS(P1)2559⟩
Chen, C.-T., **1**, 620 ⟨89JCR(S)275⟩
Chen, C.-W., **4**, 138 ⟨86JOC3325⟩
Chen, D., **4**, 102 ⟨88JA1547⟩, 343 ⟨92S258⟩
Chen, E.-N., **4**, 230 ⟨90TL257⟩
Chen, F., **1**, 232, 234 ⟨71JA676⟩, 564 ⟨71TL2603⟩
Chen, F. M. F., **5**, 185 ⟨78S928⟩
Chen, G., **4**, 677 ⟨91OM1278, 92OM1473⟩, 677 ⟨92OM1480⟩, 678 ⟨91OM2541⟩, 955 ⟨91JA9406, 93OM3431⟩
Chen, G. C., **2**, 261 ⟨80JCS(P1)2287⟩
Chen, H., **2**, 48 ⟨90TL2990⟩, 436 ⟨90POL2901⟩, 1005 ⟨91OM1282⟩; **4**, 578 ⟨90POL875⟩
Chen, H.-B., **4**, 900 ⟨91SC1213⟩
Chen, H. G., **2**, 998 ⟨89JOC5202⟩; **3**, 17, 132 ⟨88TL6701⟩, 303 ⟨89JOC5202⟩; **4**, 385 ⟨89JOC5202⟩
Chen, H. W., **2**, 147 ⟨79JA390⟩
Chen, H. Y., **1**, 247 ⟨88JOM(348)199⟩
Chen, J., **1**, 946 ⟨90JOC3321⟩; **2**, 398 ⟨88JOC3897⟩; **4**, 716 ⟨93OM1213⟩
Chen, J. C., **2**, 517 ⟨78JOM(156)213, 81JOC3978⟩
Chen, J. S., **2**, 362 ⟨91JOC4⟩; **3**, 437 ⟨91JOC4⟩
Chen, K., **1**, 217 ⟨89TL6963⟩; **3**, 163 ⟨93TL779⟩; **6**, 719 ⟨93OM2806⟩
Chen, K. K., **1**, 412, 810 ⟨67JA3462⟩
Chen, K.-M., **3**, 92 ⟨86OPP109⟩
Chen, K. S., **1**, 321, 923 ⟨74JA2201⟩
Chen, L., **1**, 6 ⟨92JA741⟩; **3**, 32 ⟨90JOC4897⟩, 159 ⟨92JA741⟩, 602 ⟨90TL4467⟩; **4**, 678 ⟨91OM2541⟩, 955 ⟨91JA9406, 93OM3431⟩
Chen, L. C., **3**, 108 ⟨82CPB1257⟩
Chen, L. M., **1**, 1091 ⟨88CC1280⟩; **4**, 365 ⟨81CC377⟩, 392 ⟨87CPB1413⟩
Chen, L. S., **6**, 5 ⟨90JFC(47)261⟩
Chen, M. C., **4**, 396 ⟨86JOM(315)135⟩; **5**, 361 ⟨66JOC3038⟩
Chen, M.-H., **1**, 334, 347 ⟨89JA6265⟩, 548, 1132 ⟨85TL4991⟩
Chen, M. H., **4**, 811 ⟨89JA6265⟩
Chen, M. T., **2**, 292 ⟨68JOM(12)401⟩
Chen, M. Y., **3**, 636 ⟨87TL2853⟩; **6**, 134 ⟨92MI 604-01⟩
Chen, O. X.,**6**, 254 ⟨90CC1459⟩
Chen, P. Y.-N., **2**, 374 ⟨76JOC3805⟩
Chen, Q., **1**, 483, 494, 1029 ⟨86TL1171⟩; **4**, 1083 ⟨84MI 424-03⟩; **6**, 38 ⟨89JFC(44)433⟩, 63 ⟨87JFC(37)171⟩
Chen, Q.-Y., **1**, 433 ⟨86CC1452, 88S896⟩, 483, 1028 ⟨87TL2387⟩; **2**, 311 ⟨90JCR(S)268⟩, 1096 ⟨87TL2387⟩; **4**, 281, 1081 ⟨90CC1459⟩; **6**, 15 ⟨89JCS(P1)2385⟩, 219, 227 ⟨92CC807⟩, 232, 234 ⟨89CC705⟩, 232, 234 ⟨93CC918⟩
Chen, R., **5**, 370 ⟨89S618⟩
Chenard, B. L., **2**, 611, 630, 916 ⟨86JOC3561, 85JOC3666⟩, 916, 932 ⟨86TL2801⟩; **3**, 271 ⟨86JOC3561, 86TL2801⟩; **4**, 170 ⟨86JOC1704⟩, 646 ⟨81CC179⟩
Chen, R. H. K., **1**, 540 ⟨73TL1611⟩; **5**, 68, 97 ⟨73JOC4086⟩
Chen, S., **1**, 751 ⟨88JA3929⟩; **3**, 4, 115 ⟨80S223⟩, 244 ⟨88JA3929⟩, 539 ⟨88LA943⟩; **4**, 470, 488 ⟨91S490, 92S1124⟩, 472 ⟨93S955⟩; **6**, 426 ⟨90JC593⟩
Chen, S.-C., **1**, 991 ⟨93OM3971⟩; **2**, 150 ⟨92JOC7274⟩; **5**, 652 ⟨78JOC4393⟩
Chen, S.-F., **3**, 414 ⟨85TL47⟩, 420 ⟨83JA6160⟩; **4**, 518 ⟨85TL47, 88T7013⟩
Chen, S.-H., **1**, 939 ⟨93JOC5028⟩
Chen, S.-L., **4**, 683 ⟨93JOM(461)207⟩
Chen, S.-M., **6**, 68 ⟨80JOC2096⟩
Chen, S.-M. L., **1**, 535 ⟨78JOC3450⟩
Chen, S.-T., **3**, 631 ⟨89TL1917⟩
Chen, T. B. R. A., **1**, 318 ⟨77TL4528⟩, 619, 707 ⟨81T417⟩, 708 ⟨80H(14)1739⟩
Chen, T. L., **1**, 692 ⟨84JOM(268)C1⟩
Chen, T. P., **5**, 864 ⟨71GEP2044574⟩
Chen, T.-S., **3**, 106 ⟨66CJC1007⟩
Chenault, H. K., **3**, 33, 156 ⟨89JOC4249⟩
Chenault, J., **2**, 255 ⟨70CR(C)2086⟩; **3**, 274 ⟨90S631⟩
Chen, W., **2**, 23 ⟨92JOC5700⟩; **5**, 528, 529 ⟨89JOC4746⟩
Chen, W.-J., **5**, 386 ⟨91IC1862⟩
Chen, W. Y., **2**, 378, 386 ⟨68JMC504⟩
Chen, X., **3**, 163 ⟨93TL579⟩
Chen, Y., **1**, 181 ⟨90JOC3961⟩, 835 ⟨83JOC4262, 83TL1021⟩
Chen, Y.,**3**, 300 ⟨90JOC3961⟩
Chen, Y., **5**, 272, 293 ⟨87TL3815⟩, 659 ⟨92IC4917⟩, 696 ⟨85OM1749⟩; **6**, 260 ⟨92IC4917⟩
Chen, Y.-H., **1**, 928 ⟨90CJC1698⟩
Chen, Y. P., **2**, 683 ⟨93H(36)13⟩
Chen, Y.-S., **3**, 215 ⟨85JCS(P1)267⟩; **5**, 284 ⟨87CC1627⟩
Chen, Y. X., **3**, 710 ⟨89TL6657⟩
Chen, Y.-Y., **2**, 147, 155 ⟨82JOC3773⟩; **3**, 106 ⟨80JOC3350⟩, 621 ⟨83TL3447⟩; **4**, 509 ⟨85T3529⟩, 510 ⟨85JOC4006⟩; **6**, 98 ⟨80JOC2236⟩
Chen, Z. C., **2**, 1020, 1028 ⟨87JA228⟩, 1033 ⟨93JOC1924⟩; **6**, 562 ⟨87JOC4117⟩

Chen, Z. X., **4**, 408 ⟨89H(29)865⟩
Chenchaiah, P. C., **2**, 304 ⟨88TL6393⟩
Chenera, B., **2**, 1006 ⟨93JA497⟩; **3**, 696 ⟨92JOC2018⟩
Cheng, A. H-B., **4**, 610, 611, 621 ⟨85OM581⟩
Cheng, C., **3**, 406 ⟨82OR(28)37⟩
Cheng, C. C., **6**, 504 ⟨79JMC28⟩, 653 ⟨63JMC283⟩
Cheng, C. H., **2**, 632 ⟨87JOC691⟩; **6**, 448 ⟨88MI 614-02⟩
Cheng, C. W., **2**, 759, 965 ⟨86JA304⟩
Cheng, C. Y., **2**, 843, 845 ⟨76JCS(P1)1739⟩
Cheng, C.-Y., **3**, 355 ⟨91JOC4919⟩
Cheng, G.-H., **4**, 751 ⟨88JFC(41)425⟩
Cheng, H.-C., **4**, 946 ⟨90TL673⟩
Cheng, J. P., **1**, 922 ⟨90JOC58⟩
Cheng, K. F., **1**, 469 ⟨82CJC1256⟩; **3**, 217 ⟨82CJC1256⟩, 309 ⟨90SC3069⟩
Cheng, K.-M., **1**, 436 ⟨85CC1610, 88JOC4466⟩
Cheng, K.-P. N., **1**, 928 ⟨90CJC1698⟩
Cheng, L., **1**, 638 ⟨77JOC2944⟩
Cheng, L.-T., **2**, 1083 ⟨93JOM(449)27⟩
Cheng, M., **4**, 213 ⟨87JOC1353⟩
Cheng, M.-C., **2**, 938 ⟨93JOC967⟩; **4**, 897 ⟨89TL2349⟩, 898 ⟨89JOC992⟩
Cheng, M.-H., **4**, 711 ⟨93OM108⟩
Cheng, M. T., **2**, 844 ⟨67JOC1572⟩; **5**, 854 ⟨62JOC3240⟩
Cheng, P., **5**, 967 ⟨91JOC3043⟩
Cheng, S. C., **4**, 312 ⟨84JCR(S)108⟩, 444 ⟨83JCR(S)108⟩
Cheng, T., **3**, 304 ⟨63JOC398⟩
Cheng, T-C., **4**, 144 ⟨68JA6259⟩, 144, 373 ⟨68JOC3055⟩, 525, 632, 635 ⟨66JOM(6)100⟩
Cheng, W., **1**, 473 ⟨87TL5093, 87TL5097⟩
Cheng, X., **1**, 944 ⟨92T9261⟩
Cheng, X.-M., **1**, 111 ⟨B-89MI 104-01⟩
Cheng, Y., **3**, 4, 115 ⟨80S223⟩; **4**, 232 ⟨89JA6881⟩
Cheng, Y.-C., **4**, 231 ⟨92JOC2217⟩
Cheng, Y. S. P., **1**, 955 ⟨78TL691⟩
Cheon, S. H., **2**, 651, 652 ⟨83JA4112⟩; **3**, 238 ⟨86TL4759⟩
Cheong, C. S., **2**, 145 ⟨90TL2893⟩
Chepakova, L. A., **6**, 613 ⟨88KFZ143, 89ZOB1455⟩
Cheradame, H., **6**, 410 ⟨79JFC(13)175⟩
Cherbuliez, E., **2**, 104 ⟨64HCA1647⟩, 105 ⟨B-73MI 202-02⟩, 176 ⟨23HCA249⟩, 182 ⟨23HCA256⟩, 451 ⟨63HCA2464⟩, 456 ⟨61HCA1812⟩
Cherednichenko, P. G., **4**, 817 ⟨87JOU246⟩
Cherega, A. N., **2**, 444 ⟨91ZOB1075⟩
Cherepanov, I. A., **2**, 7 ⟨89JOU587⟩
Cherepinnikova, N. F., **2**, 1093 ⟨73JGU626⟩; **5**, 1081 ⟨87ZOB405⟩
Cherepinskii-Malov, V. D., **6**, 57 ⟨80IZV2356⟩
Chérest, M., **1**, 389 ⟨65JCS2513⟩
Cherest, M., **2**, 561 ⟨66TL875⟩
Cherkasova, T. I., **4**, 995 ⟨91IZV1849⟩; **6**, 469 ⟨82JGU131⟩
Cherkasov, L. N., **4**, 1065 ⟨77ZOR457⟩
Cherkasov, R. A., **2**, 259 ⟨82URP891678⟩; **4**, 923 ⟨83PS(16)345⟩; **5**, 370 ⟨88ZOB2424⟩, 581 ⟨79IZV861⟩, 581, 619 ⟨92ZOB71⟩, 642, 644 ⟨91RCR1128⟩, 643 ⟨78ZOB843⟩, 643, 644 ⟨79IZV805, 90ZOB464⟩, 737 ⟨73DOK(211)113⟩; **6**, 583 ⟨93ZOB633⟩, 657 ⟨90ZOB778⟩
Cherkasov, V. K., **2**, 569 ⟨77JOM(131)43, 88MI 211-02⟩
Cherkasov, V. M., **5**, 657 ⟨81JOU1527, 81ZOR1720⟩
Cherkauskas, J. P., **3**, 31, 161 ⟨92JOC6⟩
Cherkinskaya, M. L., **5**, 1078 ⟨88ZOR1581⟩
Cherkofsky, S. C., **3**, 645 ⟨73JOC475⟩
Chermprapai, A., **2**, 11 ⟨79CC238, 80JCS(P1)2901⟩
Chern, C.-I., **2**, 87 ⟨75JOC1678⟩
Chern, C.-Y., **1**, 333 ⟨92TL5017, 93TL6505⟩; **3**, 635 ⟨92TL5017⟩
Chern, J.-W., **5**, 275 ⟨86JOC1065⟩
Chern, L.-J., **2**, 767 ⟨80JOC1688⟩
Chernega, A. N., **6**, 680, 681, 689 ⟨92IC3493⟩, 693 ⟨91ZN(B)1539⟩
Cherney, R. J., **6**, 472 ⟨92JOC2523⟩
Cherng, C.-D., **2**, 936 ⟨92JOC7010⟩; **4**, 930 ⟨92JOC7010⟩; **5**, 410, 411 ⟨92JOC7010⟩; 411 ⟨91TL3515⟩
Chernii, V. N., **2**, 458 ⟨79JGU1714⟩
Chernobrovkina, M. N., **5**, 13 ⟨79MI 501-01⟩
Chernoglazova, N. I., **5**, 385 ⟨85CC171⟩, 652 ⟨88JCS(P1)305⟩
Chernokal'skii, B. D., **2**, 498 ⟨59OOK(128)299⟩, 695 ⟨72MI 213-01⟩, 809 ⟨67CA76112w⟩, 886 ⟨72MI 217-01, 73JGU1921⟩, 891 ⟨74JGU1506⟩
Chernova, A. V., **4**, 489 ⟨82ZOB2444⟩; **5**, 679 ⟨91ZOB2257⟩
Chernova, H. V., **5**, 679 ⟨91JGU2372⟩
Chernov, A. N., **4**, 549 ⟨84ZOB1511⟩; **5**, 398 ⟨88JGU1965⟩; **6**, 564 ⟨83ZOB785⟩
Chernov, N. F., **4**, 152 ⟨88JOM(341)225⟩; **5**, 1135 ⟨77IZV1839⟩, 1142 ⟨76DOK(230)853, 88JOM(341)225⟩
Chernushevich, L. M., **5**, 708 ⟨61JGU473, 61TL807⟩
Chernyaeva, R. F., **4**, 993 ⟨75ZOR452⟩
Chernykh, T. E., **2**, 471 ⟨83JGU2244⟩
Chernysher, E. A., **4**, 654 ⟨92MI 414-01⟩
Chernysheva, O. N., **4**, 98, 99 ⟨87JGU334⟩
Chernysheva, T. I., **4**, 140 ⟨69DOK(185)311⟩
Chernyshev, E. A., **4**, 569 ⟨62DOK(143)840, 63DOK(148)875⟩, 569, 570, 573 ⟨65IZV286⟩, 570 ⟨62DOK(147)117, 70ZOB609, 71ZOB2186⟩, 6540 ⟨92MIP173178⟩
Cherry, D., **2**, 152, 363 ⟨92JA5977⟩
Cherry, P. C., **5**, 304 ⟨93TL2677⟩
Cherry, W. R., **3**, 433 ⟨85JOC1531⟩
Chertanova, L. F., **4**, 804 ⟨92JGU1222⟩; **5**, 395, 396 ⟨91PS(55)133⟩; **6**, 121 ⟨92HAC115⟩, 659 ⟨87ZOB2467⟩
Chertkov, V. A., **4**, 674 ⟨85JOM(297)C25, 93JOM(455)C9⟩; **5**, 1044 ⟨91ZOR340⟩
Chertnova, L. F., **4**, 132 ⟨91JGU1459⟩
Chervin, I. I., **2**, 356, 357 ⟨79ZV924⟩, 356, 357 ⟨81T4097⟩; **4**, 481 ⟨82IZV1611⟩
Chervin, I. O., **4**, 886 ⟨69DOK(186)835⟩
Chervonyi, V. A., **5**, 655 ⟨89JOU2329⟩, 655, 1053 ⟨89ZOR2597⟩, 656 ⟨91UKZ530⟩
Cheshire, D. R., **1**, 135 ⟨91COS(3)193⟩, 1131 ⟨85CC1205⟩; **4**, 377, 381 ⟨91COS(3)193⟩
Chesney, A., **3**, 44 ⟨90SC3167, 92SL77⟩
Chessari, R. J., **5**, 1121 ⟨81JAP39050⟩
Chetcuti, M. J., **4**, 709 ⟨92OM2128⟩, 709, 712 ⟨90OM1345, 91OM3003⟩, 713 ⟨88OM2450, 89OM2077⟩, 713 ⟨90IC1295⟩, 714 ⟨91IC4710⟩
Chettle, S. J., **3**, 83 ⟨86JA2608⟩
Cheung, A. W.-H., **1**, 928 ⟨92TL181⟩; **5**, 142 ⟨94JA7455⟩
Cheung, C.-C. S., **6**, 183 ⟨70JA4158⟩
Cheung, G. K., **3**, 92 ⟨92SL77⟩
Cheung, H. T. A., **6**, 415 ⟨87MI 614-02⟩
Cheung, J. J., **1**, 811 ⟨70JA6643⟩
Cheung, N. H., **5**, 404 ⟨77TL155⟩
Cheung, Y.-F., **5**, 18 ⟨79TL3809⟩
Chevalier, J. L., **2**, 361 ⟨61CR(252)4007⟩
Chevalier, V., **4**, 127 ⟨88BSF699⟩
Chevallier, M. A., **5**, 478, 494 ⟨85SUL177⟩
Chevolot, L., **2**, 517 ⟨74TL3435⟩
Chevrier, B., **2**, 1006 ⟨74AG(E)1⟩; **5**, 1097 ⟨78AG(E)781⟩
Chevrolet, L., **3**, 419 ⟨76BSF1222⟩
Chew, C.,**5**, 757 ⟨27JCS2318⟩
Chia, H.-A., **1**, 498, 518 ⟨74JCS(P1)1209⟩, 518 ⟨71CC1144⟩
Chia, H. L., **1**, 506 ⟨70HCA485⟩
Chiacchio, U., **1**, 550 ⟨90JOC414, 91JOC2523⟩; **2**, 668 ⟨91TL5923⟩; **3**, 250 ⟨88JOC2238⟩, 322 ⟨90JOC414⟩, 549 ⟨84JCR(S)370⟩; **4**, 509 ⟨85T3529⟩; **5**, 734 ⟨84OMR724⟩
Chi, D. Y., **2**, 9 ⟨87JOC658⟩, 18 ⟨86JFC(31)99⟩
Chianelli, D., **1**, 484 ⟨91G59⟩; **2**, 278, 729 ⟨83JOC4289⟩, 709 ⟨83JOC4795⟩, 711 ⟨83S751⟩; **4**, 92 ⟨91JOC4529⟩; **5**, 238 ⟨86T3029⟩
Chiang, C. K., **4**, 45 ⟨93MI 402-01⟩

Chiang, C. S., **3**, 146 ⟨77TL1827⟩; **4**, 191 ⟨77S467⟩, 194 ⟨76JA3037⟩, 194 ⟨76JA6750⟩, 194 ⟨77TL1827⟩
Chiang, G. C., **6**, 538 ⟨92MI 617-02⟩
Chiang, L.-Y., **4**, 850 ⟨83JOC4713⟩, 874 ⟨80CC867⟩; **6**, 591, 594 ⟨80CC866B⟩
Chiang, M.-C., **1**, 860 ⟨45JOC21⟩
Chiang, Y., **2**, 40 ⟨71JA4907⟩, 637 ⟨84JA460, 87JA4000⟩, 637 ⟨91SCI395⟩, 638 ⟨74JA4494, 86HCA1331⟩, 1022 ⟨89JA2355⟩, 1040 ⟨91AG(E)1356⟩; **6**, 611 ⟨84TL487⟩, 646 ⟨91AG(E)1356⟩
Chiang, Y. H., **2**, 362, 363 ⟨69JOC2397⟩; **5**, 1001 ⟨71JOC2155⟩
Chiao, W.-B., **1**, 779, 783 ⟨78JA2802⟩
Chi, P. B., **2**, 494 ⟨80ZAAC(466)179, 81ZAAC(472)75⟩, 502 ⟨80ZAAC(466)183⟩
Chiara, J. L., **1**, 587 ⟨92TL573⟩
Chiarino, D., **6**, 608 ⟨88SC1171⟩
Chiaroni, A., **1**, 626 ⟨90TL1865⟩
Chiavarini, M., **5**, 347 ⟨84FES788⟩
Chi, Y., **2**, 478 ⟨80JOC406⟩; **4**, 677 ⟨92OM1473, 92OM1480⟩; **6**, 348 ⟨90JCS(D)3033, 91OM2485⟩
Chiba, H., **2**, 233 ⟨86CL349⟩
Chiba, K., **5**, 157 ⟨77TL1037⟩, 296 ⟨83JOC4058⟩, 301 ⟨78JOC1684⟩
Chiba, M., **3**, 207 ⟨81CPB2733⟩
Chiba, N.,**6**, 426 ⟨91TL4371⟩
Chiba, S., **5**, 547 ⟨82S457⟩
Chiba, T., **1**, 1107 ⟨85CC1636⟩; **2**, 53 ⟨93TL2351⟩; **3**, 231 ⟨78CPB3877⟩, 638 ⟨77JOC2973⟩; **5**, 698 ⟨93CL521⟩
Chibata, I., **5**, 84 ⟨78AG(E)176⟩
Chiche, B., **3**, 280 ⟨86JOC2128⟩
Chiche, M., **5**, 135 ⟨90TL4875⟩
Chiche, P. L., **4**, 1039 ⟨80AX(B)1344⟩
Chichester, S. V., **3**, 363 ⟨85S639⟩
Chichkanova, T. V., **4**, 476 ⟨79IZV2783⟩; **5**, 221 ⟨86IZV2535⟩, 369 ⟨85IZV173⟩
Chichkareva, G. G., **1**, 986 ⟨87IZV695⟩
Chickos, J., **2**, 807, 845 ⟨68JA4842⟩
Chicote, M. T., **4**, 584 ⟨89JOM(364)407⟩, 586 ⟨87JOM(331)409, 87JOM(333)129⟩, 586 ⟨88OM997, 91JCS(D)2579⟩
Chida, N., **6**, 488 ⟨92JA8008⟩
Chidambara, N., **2**, 228 ⟨92JOC1699⟩
Chidambaram, R., **5**, 457 ⟨90TL3723⟩
Chiesi-Villa, A., **2**, 703 ⟨85IC654⟩; **4**, 696 ⟨83CC1087, 83CC1156⟩; **6**, 342 ⟨86AG283, 88NJC621⟩
Chihaoui, M., **5**, 705 ⟨78CR(C)69⟩
Chihi, A., **2**, 266 ⟨81JOM(210)163⟩
Chihung, N., **3**, 309 ⟨86T2311⟩
Chi Hung, N., **3**, 442 ⟨83T1777⟩
Chikamatsu, H., **1**, 690 ⟨88JA3578⟩
Chikamatsu, K., **5**, 1117 ⟨82SC131⟩
Chikami, E., **6**, 246 ⟨70CPB784⟩
Chikashita, H., **1**, 10 ⟨86JOC5400⟩; **2**, 398 ⟨85SC527⟩; **3**, 12, 13, 83, 98, 129 ⟨86JOC5400⟩, 29, 157 ⟨88BCJ3637⟩, 92 ⟨91JOC1692⟩, 128 ⟨86BCJ1747⟩; **4**, 319 ⟨89BCJ1215⟩, 321 ⟨85H(23)2509, 86S375⟩
Chikovani, E. I., **2**, 1093 ⟨79MI 223-01⟩
Chil, H. D., **2**, 327 ⟨88JHC1383⟩
Childa, N., **1**, 828 ⟨94CC901⟩
Childs, A. F., **2**, 458 ⟨54BRP810930⟩; **4**, 309 ⟨48JCS2174⟩
Childs, R. F., **1**, 794 ⟨82T567⟩, 894 ⟨90JA8912⟩, 910, 917, 918 ⟨84ACR347⟩, 917 ⟨83JOC1431⟩; **2**, 1007 ⟨86JA1692, 86JA1694⟩, 1007 ⟨90JA8912⟩, 1008 ⟨83JOC3043, 85JOC4553, 86JA3613⟩, 1009 ⟨89JA5424⟩; **3**, 229 ⟨73JA3289⟩
Childs, W. V., **6**, 409 ⟨69USP3461050⟩
Chilikin, V. G., **4**, 104 ⟨86IZV810⟩; **6**, 42 ⟨85IZV659⟩
Chilton, W. S., **3**, 351 ⟨74JOC2509⟩
Chim, F. F., **3**, 495 ⟨71JOC3386⟩
Chimiak, A., **2**, 229 ⟨77MI 203-05, 81JOC1868⟩, 342 ⟨74TL2453⟩; **5**, 214 ⟨87AJC1919, 87TL1817⟩, 322 ⟨85S929⟩, 323 ⟨87S418⟩
Chimichi, S., **5**, 419 ⟨85JOC130⟩
Chin, B., **2**, 923 ⟨90TL5829⟩; **4**, 661, 702 ⟨90TL5829⟩
Chin, C. G., **1**, 884 ⟨66CC880⟩; **3**, 494 ⟨71CJC2143⟩; **5**, 577 ⟨76TL967⟩
Chin, C. S., **2**, 573 ⟨86OM2452⟩, 639 ⟨87CC1213, 88JA8244⟩, 639 ⟨91JCS(D)443⟩
Chin, E., **1**, 650 ⟨80JA6894⟩
Chin, M., **4**, 408 ⟨89H(29)865⟩
Chin, R. M., **4**, 690 ⟨92OM2698⟩
Chinchilla, R., **3**, 192 ⟨92TL5641⟩
Chine, P., **2**, 1098, 1099 ⟨62MI 223-01⟩
Chinery, E., **4**, 420 ⟨61JOC4480⟩
Chin-Hsien, W., **6**, 555 ⟨81S622⟩
Chini, M., **2**, 135 ⟨92SL303⟩, 318 ⟨90TL4661⟩
Chini, P., **6**, 524 ⟨60CI(M)133, 60CI(M)137⟩, 524 ⟨67CC440⟩, 525 ⟨59CI(M)132, 79JA6110⟩, 526 ⟨60G1005⟩
Chinn, Jr., J. W., **4**, 668 ⟨82JA2637, 84IC3717⟩; **6**, 192, 205, 398 ⟨84CC1664⟩, 204, 205, 404 ⟨83JOM(249)1⟩, 205 ⟨82JA7345⟩, 206 ⟨81JA5951, 82JA2637⟩
Chinn, M. S., **1**, 400, 823 ⟨84JOC1917⟩
Chinn, R. L., **3**, 321 ⟨89TL1491⟩, 323 ⟨91JOC2911⟩
Chinnasamy, P., **4**, 94 ⟨84H(22)585⟩
Chinn Jr., J. W., **4**, 618 ⟨84CC1664⟩
Chino, K., **1**, 936 ⟨82JA5564⟩, 1130 ⟨82TL2575⟩; **4**, 187 ⟨86JCS(P1)1351⟩
Chino, N., **4**, 526, 527, 528 ⟨92TL4209⟩
Chinone, M., **4**, 682 ⟨75ICA155⟩
Chintani, M., **5**, 199 ⟨87CPB3453⟩
Chiou, ,B. L., **4**, 372 ⟨75JOC814⟩
Chiou, B. L., **2**, 524 ⟨75JOC814⟩
Chiou, H.-S., **1**, 248 ⟨87ACR65⟩
Chirkova, L. P., **5**, 221 ⟨89IZV418⟩, 406 ⟨91BAU661⟩, 1131 ⟨91IZV237⟩
Chiro, N., **2**, 530 ⟨92TL4209⟩
Chisholm, M. H., **1**, 692 ⟨89JA2324⟩; **2**, 421 ⟨68CC495, 71JCS(A)1433⟩, 421 ⟨71JCS(C)2741, 76JA4477⟩; **4**, 673 ⟨78JA802⟩, 674 ⟨78JA5764⟩, 675 ⟨84JA6794, 85OM1312, 86OM2171, 86OM2457, 89OM1626, 89OM80, 91OM535⟩, 676 ⟨84JA6806, 86OM2384⟩, 676 ⟨86OM602, 88POL919, 89OM67⟩, 822 ⟨76CC200, 76JA6046⟩; **5**, 466, 940 ⟨72JA1532⟩, 939 ⟨73ACR202⟩, 939, 944 ⟨70CC763, 71IC1711⟩, 940 ⟨77IC677⟩; **6**, 342 ⟨89JA7283, 90JOM(394)265⟩
Chistokletov, V. N., **2**, 838, 839 ⟨73ZOB2178⟩; **4**, 488 ⟨70ZOB2618⟩, 493 ⟨88JGU1148⟩
Chistoklletov, V. N., **4**, 488 ⟨68ZOB1248⟩
Chistyakov, V. V., **4**, 987 ⟨87KGS1477⟩
Chiswell, B., **2**, 892 ⟨70JPR751⟩
Chittattu, G., **1**, 21, 25, 437, 438, 943 ⟨80JA4438⟩, 609 ⟨78CC770⟩, 942 ⟨78CC41⟩
Chittatu, G., **2**, 286 ⟨77CC725, 78CC441⟩
Chiu, A. T., **4**, 780 ⟨91JMC2525⟩
Chiu, C. K.-F., **1**, 580, 781 ⟨94JOC311⟩; **2**, 679 ⟨94JOC311⟩
Chiu, C. T., **3**, 424 ⟨89JOC3820⟩
Chiu, J. J., **2**, 369 ⟨92JCS(P1)2247⟩
Chiu, K.-W., **1**, 450, 510 ⟨76JOC3484⟩; **3**, 289 ⟨75JOC1675⟩; **4**, 656 ⟨82POL31⟩; **5**, 797, 801 ⟨81JCS(D)2088⟩; **6**, 338 ⟨83CC1557⟩
Chiu, N. W. K., **1**, 747 ⟨68CJC2225⟩
Chiu, P., **4**, 613 ⟨91OM3292⟩
Chiu, S. K., **1**, 978 ⟨80JA5120⟩, 1021 ⟨80TL4047⟩; **4**, 352 ⟨80S318⟩; **6**, 509 ⟨78JOC61⟩
Chiu, W.-H., **4**, 427 ⟨75JHC595⟩
Chiusoli, G. P., **1**, 496 ⟨86JOM(317)373⟩, 526 ⟨67AG(E)124⟩, 1015, 1138 ⟨82TL4517⟩; **5**, 27 ⟨73ACR422, 73S509⟩, 68, 95 ⟨77G499⟩, 192 ⟨83TL813⟩
Chivers, T., **1**, 846 ⟨66JA460⟩; **6**, 93 ⟨92CC1539⟩, 621, 728 ⟨85IC2453⟩
Chiyoshi, K., **3**, 357 ⟨91EUP409164⟩
Chizhov, O. S., **6**, 662 ⟨82IZV2370⟩
Chlebowski, J. F., **2**, 632 ⟨68JOC1636⟩

Chliwner, I., **1**, 473 ⟨80JA790⟩
Chlopin, W., **2**, 279, 292 ⟨14CB1269⟩
Chloupek, F. J., **1**, 905 ⟨78JOC3667, 85JOC826⟩
Chmielewski, J., **6**, 637 ⟨89MI 620-01⟩
Chmielewski, M., **4**, 174 ⟨84TL4797⟩; **5**, 344 ⟨85T2441, 91CAR283⟩
Chmurny, N., **2**, 387 ⟨88JOC1467⟩
Cho, B. T., **2**, 57 ⟨88JOC1231⟩, 312 ⟨90JCS(P1)3200⟩, 534 ⟨86JOC3278, 86JOC3396⟩, 535 ⟨87JOC4020⟩; **3**, 623 ⟨86JOC398⟩
Cho, C.-G., **2**, 307 ⟨92TL3599⟩; **5**, 1044 ⟨94JMC170⟩
Cho, C. S., **3**, 289 ⟨93JOM(443)253⟩, 290 ⟨92CC453⟩
Cho, D.-G., **1**, 967 ⟨92TA1509⟩
Cho, H., **2**, 539, 699 ⟨81TL3455⟩, 657 ⟨76JCS(P1)836⟩, 945 ⟨75JOC3097⟩; **4**, 420 ⟨85JOC4227⟩; **5**, 375 ⟨81TL3455⟩
Cho, H. S., **1**, 179, 183 ⟨93JOC5892⟩; **3**, 31 ⟨91TL367⟩
Cho, I. H., **3**, 6, 118 ⟨84CC762⟩, 116 ⟨89TL2559⟩
Cho, S. H., **2**, 672 ⟨87JOC2631⟩
Cho, W., **2**, 922, 923 ⟨84JOC1838⟩
Cho, Y., **5**, 999 ⟨90TL1559⟩
Cho, Y. R., **5**, 862 ⟨94S1057⟩
Chockalingam, K. N., **4**, 272, 277 ⟨81S995⟩
Chodiewicz, W., **2**, 436 ⟨85IC533⟩
Chodkiewicz, W., **1**, 953 ⟨61BSF2176⟩, 986 ⟨66BSF2885⟩, 1017, 1018, 1019, 1034 ⟨B-69MI 121-04⟩, 1025 ⟨65TL1953⟩, 1055 ⟨69CR(C)342, 73BSF3371⟩, 1142 ⟨55CR(241)1055, 57AC(P)819⟩; **2**, 297 ⟨78JOM(149)C7⟩, 436 ⟨84JOM(269)107⟩, 437, 443 ⟨79TL1069, 84JOM(273)C55⟩, 824, 828, 841, 1059, 1061, 1063, 1066, 1069 ⟨66BSF1002⟩, 1015 ⟨68BSF193⟩, 1059 ⟨60CR(250)866⟩; **4**, 802 ⟨75BSF2259⟩; **5**, 59 ⟨66CR(C)761, 69BSF911⟩
Chodosh, D. F., **5**, 798, 799, 800 ⟨79JOM(172)251⟩
Chodroff, S., **3**, 279 ⟨48JA1647⟩
Chohan, V., **5**, 1127 ⟨84AJC1043⟩
Choi, E. B., **1**, 24, 439 ⟨93TL4541⟩
Choi, H. C., **4**, 201 ⟨93TL3581⟩
Choi, H. D., **3**, 626 ⟨82CPB3574⟩; **4**, 67 ⟨81S324⟩
Choi, J.-K., **1**, 547, 1131 ⟨85T3959⟩
Choi, K. S., **5**, 554, 563, 605 ⟨93BCJ623, 93TL115⟩
Choi, L. S. L., **3**, 352 ⟨75CC557, 79JCS(P1)1166⟩
Choi, N., **1**, 242 ⟨89JOC4660⟩; **2**, 941 ⟨93T1189⟩; **3**, 353 ⟨89JOC4660⟩; **6**, 129 ⟨87CL2177⟩
Choi, O.-S., **5**, 237, 277 ⟨87JOC4531⟩
Choi, S.-C., **1**, 245 ⟨89TL6129⟩, 931 ⟨91T4847⟩; **2**, 223 ⟨71JA2870⟩; **3**, 126 ⟨89T77, 91TL457⟩
Choi, S. Y., **2**, 398 ⟨92JCS(P1)601⟩
Choi, V. M. F., **3**, 631 ⟨83JOC2294, 84TL591⟩
Choi, W., **2**, 532 ⟨85AG(E)1⟩
Choi, W.-B., **4**, 230 ⟨91JA9377⟩
Choi, Y., **1**, 335 ⟨88TL167⟩; **2**, 579 ⟨88TL167⟩
Choi, Y. K., **5**, 135 ⟨92JOC1605⟩
Choi, Y. M., **1**, 601, 781, 785 ⟨93JOC322⟩; **2**, 320 ⟨81S441⟩
Choi, Y. R., **1**, 238 ⟨90TL2713⟩
Choi, Y. W., **1**, 4, 10 ⟨92TL6673⟩
Choileain, N. N., **5**, 322 ⟨85SC1159⟩
Chojnowski, J., **2**, 108, 452 ⟨85TL4965⟩, 467 ⟨85T2471⟩; **4**, 566, 573 ⟨88JOM(356)285⟩
Chokshi, S. K., **1**, 9 ⟨72S483⟩; **2**, 89 ⟨72S483⟩
Cholkiewicz, W., **4**, 802 ⟨71CR(C)1107⟩
Chollet, A., **1**, 807, 809 ⟨81TL1583⟩, 833 ⟨80JOC4413⟩
Chollet-Gravey, A.-M., **4**, 487 ⟨91SC1847⟩
Chon, T.-S., **2**, 998 ⟨89JOC5202⟩
Chondrogianni, J., **3**, 712 ⟨85T1931⟩, 714, 718 ⟨85CC816⟩
Chong, J. M., **1**, 138 ⟨88JOC5584⟩, 217 ⟨93TL3695⟩, 217, 478 ⟨85TL4683⟩, 1000 ⟨86TL5445⟩; **2**, 68 ⟨90TL1985⟩, 83, 85 ⟨92JOC46⟩, 600 ⟨87CJC78⟩; **3**, 51 ⟨93TL3695⟩; **4**, 155, 534 ⟨93JOC523⟩, 386 ⟨88JOC5584⟩, 389 ⟨91TL5683⟩, 531, 533 ⟨92JOC2220⟩; **5**, 160 ⟨89T363⟩
Chong, P. C. S., **2**, 217 ⟨81JBC(256)5064⟩
Chong, W. K. M., **4**, 359 ⟨89JOC1789⟩; **5**, 152 ⟨89JA2984⟩
Chopard, P. A., **1**, 1079 ⟨65JOC1015⟩; **3**, 274 ⟨65JOC1015⟩, 494 ⟨63JOC2446⟩, 498 ⟨63HCA2178⟩
Chopra, G., **2**, 862 ⟨91JCS(P2)2081⟩
Chorbadjiev, S., **2**, 226 ⟨77JPR1036⟩
Chorev, M., **4**, 427 ⟨84JOC821⟩, 446 ⟨83JMC129, 83MI 409-01⟩; **5**, 22 ⟨91PS(60)239⟩, 371, 403 ⟨90JCS(P1)3263⟩; **6**, 206 ⟨76T1839⟩
Chorn, T. A., **3**, 297 ⟨84JCS(P1)1339⟩
Chorvat, R. J., **2**, 441 ⟨67JOC4066⟩; **4**, 528 ⟨92MI 411-02⟩
Chottard, J. C., **1**, 63 ⟨76JOM(110)139⟩; **3**, 699, 700 ⟨78TL3027⟩; **5**, 1097 ⟨77CC648, 78AG(E)781⟩, 1097 ⟨81IC3119⟩; **6**, 717 ⟨77CC648⟩
Chou, C.-H., **1**, 655 ⟨88JA6579⟩, 787 ⟨86JOC4208⟩
Chou, C. S., **3**, 7 ⟨83JA4492⟩, 88 ⟨84JA3374⟩
Chou, C.-Y., **1**, 434 ⟨85JOC2668⟩
Chou, D., **1**, 651 ⟨75SC359⟩
Chou, L., **3**, 311 ⟨92JA5888⟩
Chou, M. H., **5**, 385 ⟨92IC2318⟩
Chou, S. K., **1**, 966 ⟨78JOC1389⟩, 970 ⟨76JOC3496, 78JOC1385⟩
Chou, S. P., **3**, 227 ⟨88JOC5⟩
Chou, S.-S. P., **1**, 489 ⟨89JOC868⟩; **2**, 630 ⟨89JOC868⟩, 938 ⟨93JOC967⟩, 1071 ⟨81TL1875⟩; **3**, 227 ⟨89SC1593⟩
Chou, T., **1**, 24 ⟨91TL3551⟩, 53 ⟨85JOC4329⟩, 790 ⟨86JCS(P1)1039, 87CC934⟩, 791 ⟨84CC1323, 87JOC244⟩; **3**, 114 ⟨78TL1943⟩
Chou, T.-S., **1**, 24 ⟨85TL4495⟩, 145 ⟨92T8963⟩, 439 ⟨87JOC2224⟩, 444 ⟨85JCR(S)18⟩, 445 ⟨86JOM(312)53⟩; **2**, 267 ⟨74TL725⟩, 432 ⟨85JCR(S)18, 85JOC4329, 86JOM(312)53⟩; **3**, 162 ⟨93JOC588⟩, 303 ⟨89JOC5202⟩; **4**, 53, 54, 58 ⟨93JOC588⟩, 316 ⟨79USP771219⟩, 385 ⟨89JOC5202⟩, 396 ⟨92T2025⟩
Chou, W.-C., **4**, 263 ⟨90JOC5515⟩
Choudary, B. M., **2**, 753 ⟨89TL251⟩; **3**, 5, 118 ⟨85TL6257⟩, 59 ⟨90TL7521⟩, 116, 158, 161 ⟨90TL5785⟩, 294 ⟨92JOC5841⟩; **5**, 1106 ⟨90SC2313⟩
Choudhari, S. P., **1**, 612 ⟨87MI 114-01⟩
Choudhary, A. R., **5**, 384 ⟨91SC1889⟩
Choudhry, S. C., **4**, 175 ⟨82JOC824⟩; **5**, 345, 346 ⟨88JOC5507⟩
Choudhury, D. R., **6**, 241 ⟨73JCS(P1)1092⟩
Choudhury, P., **2**, 507 ⟨77JOM(128)155⟩
Choudry, G. G., **3**, 310 ⟨76RTC220⟩
Choudry, S., **1**, 84 ⟨79TL1641⟩
Choukroun, R., **1**, 456 ⟨86CC1317⟩; **2**, 421 ⟨78ICA163⟩
Chounan, Y., **1**, 188 ⟨92JOC1024⟩, 189 ⟨93JOC1207⟩, 189 ⟨93T9479⟩
Chow, A., **4**, 603, 648, 649 ⟨90JA1382⟩, 648 ⟨90JA2582⟩
Chow, F., **1**, 592, 713, 714, 716 ⟨79JA6638⟩, 592, 713, 715, 716 ⟨75CC790⟩, 593, 713, 714, 715, 716 ⟨79JA6648⟩, 601, 602, 604 ⟨78JOC1697⟩; **2**, 279 ⟨79JA6638, 79JA6648⟩, 281 ⟨75CC790, 78JOC1697⟩; **3**, 80 ⟨79JA6638⟩, 256 ⟨79TL4433⟩; **4**, 92 ⟨78JOC1697⟩, 239, 375 ⟨79JA6638⟩, 399 ⟨79JA6648⟩; **6**, 100 ⟨79JA6638⟩, 134 ⟨78MI 604-01⟩
Chow, F.-L., **4**, 578 ⟨90POL875⟩
Chow, J., **4**, 632 ⟨69T1557⟩, 632, 634 ⟨67TL723, 69T1571⟩
Chow, K.-K., **2**, 874 ⟨76JCS(D)1429⟩
Chow, L. W., **1**, 869 ⟨67JOC1994⟩
Chow, M., **1**, 957, 974, 1060 ⟨86JOC2623⟩
Chow, T., **1**, 831 ⟨91JOC353⟩
Chow, T. J., **5**, 384 ⟨90OM560⟩
Chow, V., **3**, 601 ⟨77JOC4261⟩
Chow, Y. L., **1**, 104 ⟨68CJC2821⟩, 442 ⟨79JCS(P1)2409⟩; **3**, 234 ⟨82CJC2268⟩, 427 ⟨85JA3338⟩
Chow, Y. M., **2**, 1002 ⟨74AX(B)147⟩
Chowdhary, M. S., **4**, 176 ⟨82CAR(102)99⟩
Chowdhury, B. L., **4**, 306 ⟨84IJC(B)1286⟩
Choy, T., **3**, 131 ⟨82JOC3006⟩
Choy, W., **1**, 357 ⟨80TL2477⟩, 749 ⟨84TL2183⟩; **2**, 532 ⟨81JA1566⟩; **5**, 155 ⟨84TL2183⟩
Chremos, G. N., **2**, 497, 502 ⟨70JOM(22)637⟩
Christaens, L., **2**, 731 ⟨82TL3905⟩

Christ, W., **5**, 567, 619, 625 ⟨74JHC397⟩
Christ, W. J., **1**, 677, 1038, 1099 ⟨86JA5644⟩; **2**, 46 ⟨83TL3947⟩; **3**, 238 ⟨86TL4759⟩
Christe, K. O., **6**, 220 ⟨74IC2811⟩
Christen, K., **4**, 53, 54, 56 ⟨78HCA2047⟩
Christen, M.-O., **3**, 356 ⟨92MI 308-03⟩
Christensen, A. T., **2**, 1000 ⟨69AX(B)73⟩
Christensen, B. E., **2**, 103 ⟨57JA1222⟩; **4**, 413 ⟨59JA3789⟩
Christensen, B. G., **3**, 287 ⟨80TL2013⟩; **4**, 59 ⟨84JA1809⟩, 346 ⟨68GEP1924135⟩, 463 ⟨73TL4645⟩, 847 ⟨82TL3535⟩; **5**, 233 ⟨80TL2013⟩, 374 ⟨80TL31⟩; **6**, 128 ⟨80JA6161⟩
Christensen, D., **3**, 432 ⟨89JOC126⟩
Christensen, H., **5**, 110 ⟨74SC1⟩
Christensen, J. J., **2**, 133 ⟨76JOC134⟩; **5**, 511 ⟨83JOC2635⟩
Christensen, L. W., **2**, 157 ⟨70JOC4217⟩, 212 ⟨78S50⟩, 213 ⟨79S8⟩; **5**, 251 ⟨78S50⟩
Christensen, N. H., **2**, 157 ⟨64ACS954⟩
Christensen, P. A., **1**, 800 ⟨82JOC4786⟩
Christensen, P. K., **5**, 168 ⟨57ACS582⟩
Christensen, S. B., **1**, 539 ⟨88JCS(P1)839⟩; **2**, 746 ⟨73ACS2802⟩
Christenson, B., **5**, 151 ⟨90T46⟩
Christenson, B. G., **4**, 923 ⟨71CA(75)88759m⟩
Christenson, J. R., **5**, 665 ⟨80LA791⟩
Christiaens, L., **2**, 288 ⟨68BSB153⟩, 368 ⟨B-87MI 206-02⟩; **4**, 205 ⟨68BSB379⟩, 211 ⟨62BSB379⟩; **5**, 22 ⟨91T9053⟩, 239 ⟨76JCS(P1)2452⟩, 333 ⟨B-87MI 507-01⟩, 334 ⟨89BSB395⟩, 337 ⟨90TL873⟩, 595 ⟨91SC85⟩
Christiaens, L. E., **2**, 632 ⟨85JOC1779⟩; **5**, 334 ⟨92TL3863⟩
Christian, B., **1**, 879 ⟨83JOC2957⟩
Christian, D. F., **5**, 802 ⟨76JOM(112)209⟩; **6**, 719 ⟨74JOM(80)C35, 76JOM(112)227⟩, 719, 721 ⟨74JOM(81)C7⟩
Christian, G., **3**, 726 ⟨82CC184⟩; **5**, 1089 ⟨82CC184⟩
Christian, J. F., **5**, 1125 ⟨47JA365⟩
Christiansen, J. V., **2**, 685 ⟨91JCR(S)300⟩
Christiansen, M. L., **3**, 271 ⟨87ACS(B)536⟩
Christie, B., **4**, 935, 949 ⟨78JOM(156)123⟩
Christie, B. D., **3**, 419 ⟨81JOC4914⟩
Christie, K. O., **6**, 419 ⟨64JOC3007⟩, 419, 424 ⟨65JOC1639⟩, 421 ⟨65JOC4104⟩
Christie, M., **4**, 69 ⟨75JA5957⟩
Christie, R. M., **3**, 376 ⟨76JCS(P1)228⟩, 376, 377 ⟨74JCS(P1)722, 76JCS(P1)880⟩
Christl, B., **5**, 560 ⟨76CB740⟩
Christl, M., **3**, 543 ⟨90CB2031⟩, 685 ⟨73CB3291⟩
Christman, D. R., **1**, 896 ⟨60JA4729⟩
Christmann, K. F., **1**, 721 ⟨65AG(E)689⟩, 721, 727, 728, 729, 730 ⟨67LA(708)1⟩, 727, 728 ⟨66AG(E)667⟩, 729, 730 ⟨66AG(E)126, 70CB2814⟩; **2**, 612 ⟨69S38⟩; **3**, 494 ⟨67LA(708)1, 69S38⟩
Christoffers, C., **5**, 934 ⟨95CB645⟩
Christoffers, J., **2**, 670 ⟨93CC1811⟩; **5**, 934 ⟨95CB645⟩, 935 ⟨94OM4189⟩, 944 ⟨95CB157⟩
Christokletov, V. N., **2**, 1074 ⟨89TL577, 89ZOB1979, 91ZN(B)1578⟩
Christol, H., **1**, 50 ⟨79TL349⟩, 444 ⟨83PS(15)359⟩, 795 ⟨62BSF1325⟩, 905 ⟨78JCR(S)62⟩; **2**, 711 ⟨81S892⟩; **3**, 4, 116 ⟨86TL1775⟩, 196 ⟨87S648⟩, 650 ⟨66BSF1315⟩; **4**, 875 ⟨86JOC875⟩, 921 ⟨74BSF2263⟩
Christopfel, W. C., **5**, 1110 ⟨84JOC5198⟩
Christoph, G. G., **1**, 704, 708 ⟨77JA6946⟩; **4**, 822 ⟨76JA6046⟩
Christopher Howell, S., **4**, 165 ⟨83JCS(P1)1579⟩
Christophersen, C., **4**, 842 ⟨88ACS629⟩; **5**, 635, 636 ⟨67ACS1936⟩, 1013 ⟨70ACS1852⟩, 1013, 1101 ⟨70ACS1512⟩, 1112 ⟨73TL211, 76T745⟩; **6**, 570, 571 ⟨69ACS324⟩
Christot, I., **4**, 406 ⟨88BSF1009⟩
Christou, V., **6**, 355, 357, 358 ⟨93JA6777⟩, 357, 358 ⟨93JA545⟩
Christovalova, N. M., **1**, 493 ⟨74JOM(72)163⟩
Christy, M. E., **4**, 97 ⟨79JOC3117⟩
Christy, M. R., **3**, 29 ⟨84JOC1378⟩
Chromniak, E., **6**, 156 ⟨78MIP87868⟩
Chrz, J., **5**, 843 ⟨27BSF1603⟩
Chtara, R., **4**, 442 ⟨83S951⟩
Chu, C. K., **1**, 585 ⟨89JOC2217⟩; **4**, 231 ⟨92JOC2217⟩
Chu, C.-Y., **1**, 1093 ⟨75JOC1460⟩
Chu, D. T. W., **3**, 16 ⟨83JOC3571⟩; **4**, 67 ⟨83TL139⟩; **6**, 617 ⟨90JHC1191⟩
Chu, J. Y. C., **2**, 735 ⟨76CC552⟩
Chu, K., **1**, 455, 1033 ⟨84JOC5175⟩; **3**, 135 ⟨84JOC5175⟩
Chu, K.-H., **2**, 904 ⟨86JOC767⟩, 930 ⟨86JOC745⟩
Chu, N. Y. C., **4**, 307 ⟨B-90MI 407-02⟩, 436 ⟨81CJC679⟩
Chu, S. C., **4**, 514 ⟨85TL5001⟩; **6**, 128 ⟨85TL5001⟩
Chu, S. H., **2**, 278 ⟨63MI 204-01, 68JMC446⟩; **5**, 516, 520, 522 ⟨72JMC760⟩, 563 ⟨70JMC214⟩, 625 ⟨62JOC3671⟩; **6**, 348 ⟨90JCS(D)3033⟩, 637, 732 ⟨62JOC2899⟩
Chu, S. S. C., **5**, 690 ⟨88JOC5309⟩, 765, 824 ⟨86JHC1861⟩
Chua, S.-O., **5**, 758, 767 ⟨74JCS(P2)546⟩
Chuang, C., **3**, 696 ⟨92JOC2018⟩
Chuang, C.-P., **1**, 949 ⟨91JCR(S)1, 91T5425⟩, 950 ⟨91JCR(S)216⟩
Chuang, S.-H., **6**, 348 ⟨91OM2485⟩
Chuang, Y.-H., **2**, 684 ⟨89CC1363⟩
Chubarova, M. A., **2**, 1018, 1019 ⟨65ZOR415⟩
Chubb, F., **5**, 205 ⟨56JA3819⟩
Chuche, J., **1**, 826 ⟨79CC584⟩, 835 ⟨91JOC970⟩, 1018 ⟨92TL5355⟩; **3**, 65 ⟨73TL1635⟩, 541 ⟨85TL833, 86CC369, 87TL885, 88TL5919⟩; **4**, 900 ⟨91JOC970⟩
Chucholowski, A., **1**, 681, 683 ⟨79CC52⟩; **2**, 422 ⟨84T2211⟩, 677 ⟨79CC52⟩; **5**, 161 ⟨79AG(E)622⟩, 166 ⟨84T2211⟩
Chudakova, T. I., **5**, 644 ⟨89ZOB1917, 90ZOB1389⟩, 660 ⟨87ZOB1514⟩, 660, 788 ⟨87JGU1349⟩, 738 ⟨90JGU1389⟩
Chudd, C. C., **4**, 47, 50 ⟨63OSC(4)748⟩
Chugaev, L., **2**, 279, 292 ⟨14CB1269⟩
Chugunov, Yu. V., **4**, 915 ⟨88CA(109)211147t⟩
Chugunov, Y. V., **6**, 583 ⟨90ZOB563⟩
Chuit, C., **1**, 312 ⟨80T2305, 84PAC91⟩, 429 ⟨78S128⟩, 477 ⟨77JOM(127)371⟩, 490 ⟨93CRV1371⟩, 491, 1108 ⟨74JOM(77)269⟩, 540 ⟨74BSF1656, 74JOM(77)281⟩; **2**, 68 ⟨89T171⟩, 77 ⟨87JOM(328)C17⟩, 547 ⟨93CRV1371⟩, 624 ⟨74S803⟩, 629 ⟨74JOM(77)269⟩; **3**, 67, 237 ⟨78S128⟩, 67, 237 ⟨78S458⟩
Chujo, Y., **3**, 139 ⟨80JA6381⟩
Chukhadzhyan, G. A., **1**, 986 ⟨84AKZ116⟩
Chukhovskaya, E. Ts., **4**, 137 ⟨87IZV1087⟩
Chukovskaya, E. C., **4**, 13 ⟨83S773⟩
Chulkova, L. V., **6**, 223, 226, 227, 228 ⟨72MI 607-01⟩
Chulkova, T. I., **5**, 390 ⟨77ZOB2294⟩
Chumaevskii, M. A., **5**, 781 ⟨87MI 519-02⟩
Chum, P. W., **1**, 514 ⟨76TL15⟩
Chun, J., **2**, 639 ⟨91JCS(D)443⟩
Chun, M. W., **1**, 183 ⟨86TL943⟩; **3**, 635 ⟨78CPB784⟩
Chun, Y.-J., **5**, 378 ⟨94S909⟩
Chung, A. C., **6**, 365 ⟨90MI 612-01⟩
Chung, B. C., **2**, 7, 627 ⟨64JA964⟩
Chung, B. Y., **5**, 862 ⟨94S1057⟩
Chung, C., **6**, 204, 379 ⟨72CC1078⟩
Chung, O. Y., **6**, 559 ⟨82MI 617-03⟩
Chung, S., **3**, 116 ⟨89TL2559⟩
Chung, S. K., **1**, 431 ⟨80JOC3513⟩, 75, 458 ⟨79JOC1014⟩; **2**, 142 ⟨82SC903⟩; **3**, 621 ⟨86JA240⟩; **4**, 234 ⟨81JOC5457⟩
Chung, W.-K., **1**, 183 ⟨86TL943⟩
Chung, Y. J., **5**, 343 ⟨94SL197⟩
Chupp, J. P., **3**, 406, 407 ⟨67JOC3246⟩; **4**, 425 ⟨74JOC3745⟩, 843 ⟨89JHC1771⟩; **5**, 597 ⟨68JOC2357⟩, 632 ⟨75JOC66⟩; **6**, 28 ⟨86S224⟩, 508 ⟨71JHC557⟩
Chuprakova, K. G., **2**, 458 ⟨79JGU1714⟩
Chuprikova, T. Yu., **4**, 768 ⟨92ZOR522⟩, 775 ⟨91ZOR1796⟩

Churakov, A. M., **5**, 680 ⟨91MI 517-01⟩
Churashev, D. D., **2**, 448 ⟨91ZOB2161⟩
Church, G., **6**, 429 ⟨89TL1901⟩
Church, J. P., **1**, 832 ⟨69TL3253⟩
Church, K. M., **1**, 949 ⟨90TL5551⟩
Churchill, M. R., **2**, 979 ⟨82JA6808⟩; **3**, 460 ⟨75IC1675⟩, 521 ⟨83OM1505⟩; **4**, 581 ⟨82IC3913⟩, 683 ⟨88JOM(353)103⟩, 687 ⟨64PCS401⟩; **5**, 461 ⟨87JA5145⟩, 950 ⟨82OM628⟩; **6**, 343 ⟨83OM1179, 91OM2384⟩, 345 ⟨86OM1757⟩
Churchley, P., **1**, 506 ⟨72JOC2989⟩
Churusova, S. G.,**3**, 195 ⟨83JGU610⟩
Chuvashev, Yu. A., **3**, 430 ⟨85ZOR766⟩
Chuvashov, D. D., **2**, 440 ⟨85URP1145022⟩
Chuvatkin, N. N., **2**, 15 ⟨89JOU1835⟩, 18 ⟨87JOU230⟩
Chvalovsky, V., **2**, 19 ⟨74CCC2616⟩, 535 ⟨B-65MI 210-01, B-77MI 210-01⟩, 536 ⟨B-72MI 210-02⟩; **4**, 32 ⟨74CCC2616⟩, 137 ⟨76CCC386⟩, 352 ⟨73CCC1522⟩, 362 ⟨72CCC1392⟩
Chvatal, Z., **6**, 410 ⟨90CZP264358⟩
Chvertkin, B. Ya., **4**, 346 ⟨92ZOB1268⟩
Chwastek, H., **1**, 999 ⟨73T883⟩
Chys, J., **3**, 563 ⟨78JOC2670⟩
Ciabattoni, J., **1**, 909 ⟨68JA4495⟩
Ciaccio, J. A., **2**, 84 ⟨91SL248⟩; **3**, 632 ⟨92TL1431⟩
Ciampolini, M., **2**, 438 ⟨84JCS(D)575⟩
Ciani, G. F., **5**, 940 ⟨86OM1976⟩
Ciannotti, C., **2**, 962 ⟨72CC1114⟩
Cias, J., **6**, 647 ⟨78AF1435⟩
Ciattini, P. G., **1**, 433 ⟨86TL5541⟩, 435 ⟨90SC1293⟩, 489 ⟨92TL4815⟩, 1030 ⟨91TL6449⟩; **5**, 301 ⟨86TL3931⟩
Cicchi, S., **2**, 320 ⟨90TL3351⟩; **3**, 431 ⟨91G285⟩
Ciccotto, L., **2**, 360 ⟨80EUP10477⟩
Cichra, D. A., **4**, 435 ⟨82JOC2474⟩
Cid, M. M., **1**, 835 ⟨91TL7233⟩
Cieplik, J., **5**, 528, 529 ⟨86S142⟩
Ciganek, E., **1**, 453, 523 ⟨84OR(32)1⟩, 624 ⟨63OSC(4)612⟩, 625 ⟨59JA2799, 60JA4663⟩, 626 ⟨90JOC3007⟩; **2**, 313, 316 ⟨92JOC4521⟩, 344, 349 ⟨90JOC3007⟩; **3**, 477 ⟨65JOC4198⟩, 584, 586 ⟨69TL5179⟩, 591, 592, 596 ⟨70JOC862⟩; **5**, 278 ⟨63OSC(4)339⟩
Cignarella, S., **2**, 225 ⟨74G455⟩
Cimarelli, C., **3**, 262, 263 ⟨90S895⟩
Cimarusti, C. M., **1**, 563 ⟨68JA113⟩; **5**, 342 ⟨82JOC176⟩, 346 ⟨82JOC179, 82JOC5160⟩, 370 ⟨83JA3743⟩, 575 ⟨75JA5628⟩
Cimatoribus, L., **3**, 280 ⟨61JOC3650⟩
Ciminale, F., **2**, 148 ⟨71IJS(A)130⟩, 626 ⟨71JCS(C)3652⟩
Cini, R., **5**, 386 ⟨93JA5123⟩
Cinqui, M.,**3**, 431 ⟨89G253⟩
Cinquini, M., **2**, 35 ⟨76S516⟩; **3**, 437 ⟨82JCS(P1)339⟩, 634 ⟨81JCS(P1)614⟩; **4**, 74, 75 ⟨72JCS(P2)296, 72S259, 72JCS(P1)1886⟩, 75, 77, 233 ⟨72JCS(P1)1883⟩, 76 ⟨73JA7431, 70CC1441⟩, 233 ⟨72CC734, 76JCS(P2)996⟩, 272 ⟨79S535, 82S767⟩; **5**, 233 ⟨91T8767⟩, 237 ⟨92JOC6339⟩, 607 ⟨85T4929⟩
Cinquinni, M., **2**, 362 ⟨77CC723⟩
Cintas, P., **1**, 676 ⟨92S248⟩
Cintrat, J. C., **4**, 388 ⟨92JOM(437)C19⟩
Cioslowski, J., **5**, 1020 ⟨89JA8333, 90JA4183⟩
Ciotti, Jr., C. J., **5**, 123, 126, 302 ⟨55JA6214⟩
Ciplijauskas, L. J., **2**, 693 ⟨59JBC(234)1⟩
Cipris, D., **6**, 469 ⟨78MI 615-01⟩
Cipullo, M. J., **4**, 853, 855 ⟨82JOC3027⟩
Cirillo, P. F., **2**, 46 ⟨90JA4873⟩, 79 ⟨90JOC6071⟩, 936 ⟨92OPP553⟩; **3**, 19 ⟨91TL457⟩; **4**, 354 ⟨90JOC6071⟩, 355 ⟨90JA4873⟩; **5**, 409 ⟨92OPP553⟩
Cirillo, R., **4**, 259 ⟨89CZ349⟩
Cistone, F., **1**, 4, 7, 9, 17, 28 ⟨78JOC2259⟩
Citerio, L., **5**, 762 ⟨78TL2175⟩, 762, 816 ⟨78JCS(P1)309⟩
Citron, J. D., **1**, 428 ⟨69JOC638⟩; **3**, 83 ⟨69JOC1977⟩
Citterio, A., **1**, 923 ⟨77JA7960⟩, 926 ⟨82TL5587⟩, 934 ⟨83ACR27⟩; **2**, 632 ⟨81SC639⟩; **3**, 428 ⟨86S463⟩; **5**, 301 ⟨86T885⟩
Citterio, Z., **1**, 927 ⟨88TL1441⟩
Ciuciu, M., **3**, 256 ⟨79JPR320⟩
Ciuffarin, E., **1**, 848 ⟨67JA63⟩; **2**, 123 ⟨76S326⟩, 171, 172, 716 ⟨70JOC2006⟩, 195 ⟨70JA3035⟩, 688 ⟨73JA2918⟩; **5**, 766 ⟨73JCS(P2)534⟩
Ciufolini, M. A., **2**, 672 ⟨94TL1127⟩
Ciufolini, M., **1**, 603 ⟨85JA3891⟩
Ciurdaru, G., **3**, 256 ⟨79JPR320⟩
Claassen, II, R. J., **1**, 146 ⟨91TL6497⟩
Clabattoni, J., **2**, 28 ⟨88OSC(6)991⟩
Clader, J. W., **3**, 212, 213 ⟨80TL1205⟩; **4**, 199 ⟨79SC57⟩; **5**, 284 ⟨85JOC2220⟩
Cladingboel, D. E., **1**, 347 ⟨90CC1543⟩
Claesen, C., **2**, 409 ⟨84TL1307⟩
Claeson, G., **2**, 212 ⟨61JA4357⟩, 530 ⟨92TL4209⟩; **4**, 526, 527, 528 ⟨92TL4209⟩
Claesson, A., **1**, 480 ⟨83TL5137⟩, 539 ⟨74TL2161⟩, 961 ⟨74ACS(B)765, 84JOC4120⟩, 970 ⟨83TL1297⟩; **5**, 132 ⟨87JOC3477⟩; **6**, 491 ⟨89CAR(194)209⟩
Claeys, E. G., **2**, 491 ⟨66JOM(5)446⟩
Claff, C. E., **2**, 557 ⟨55JOC981⟩
Claiborne, A. K., **6**, 617 ⟨90JHC1191⟩
Claisen, L., **1**, 414, 825 ⟨12CB3157⟩, 840 ⟨26LA(449)81⟩; **5**, 823 ⟨1887CB252⟩
Clampitt, R. B., **4**, 122 ⟨67USP3306937⟩
Clapp, C. H., **4**, 557 ⟨72IC2340⟩
Clapp, J. W., **2**, 177 ⟨50JA4890⟩
Clar, G., **3**, 246 ⟨91S483⟩
Claramunt, R. M., **5**, 327 ⟨87T2557⟩, 1068 ⟨89JA355⟩, 1122 ⟨84TL5427⟩
Clardy, J., **1**, 521 ⟨79JA7001⟩, 613 ⟨77JOC1349⟩, 704, 708 ⟨77JA2815⟩, 704, 708 ⟨77JA6946⟩, 769 ⟨82PAC1915⟩, 971 ⟨80JA2134⟩, 1016, 1108 ⟨84JA3551⟩; **2**, 152, 363 ⟨92JA5977⟩, 238 ⟨77JA7743⟩; **3**, 92 ⟨91JOC1692⟩, 493 ⟨90JA7001⟩; **4**, 245 ⟨93TL1085⟩, 603 ⟨81TL7⟩; **5**, 355 ⟨85TL5739⟩; **6**, 604 ⟨78TL1391⟩
Clardy, J. C., **1**, 690 ⟨84JA5018, 86JA2932⟩, 690 ⟨86JA515⟩
Clarembeau, M., **1**, 245 ⟨85TL1093, 86CC457⟩, 245 ⟨86TL1723⟩, 713 ⟨89T2023⟩, 713, 714, 716 ⟨84IJ125⟩; **2**, 278, 294 ⟨85T4793⟩, 279 ⟨84TL3625⟩, 281 ⟨89T2023⟩, 478 ⟨89T2005⟩, 554 ⟨85TL1093, 86TL1719⟩, 554 ⟨86TL1723⟩; **4**, 284 ⟨89T2023⟩, 287, 288 ⟨85T4793, 86TL1723⟩, 396, 401 ⟨84IJ125⟩, 399 ⟨84TL3629⟩, 400 ⟨89T2005⟩
Claremon, D. A., **1**, 591, 594, 716 ⟨79CC83⟩, 604 ⟨79JA3704⟩; **2**, 279 ⟨81JOC1215⟩, 281 ⟨79JA3704⟩, 286 ⟨B-87MI 204-02⟩; **3**, 308 ⟨81JA6969, 81TL4647⟩, 618, 619 ⟨88TL2155⟩; **5**, 239, 281 ⟨81JOC1215⟩, 239, 336, 337 ⟨85T4835⟩
Clark, A., **1**, 345 ⟨92T6875⟩
Clark, A. C., **1**, 316 ⟨72JOC4236⟩
Clark, A. D., **6**, 314 ⟨71JCS(C)103⟩
Clark, A. J., **1**, 346 ⟨89TL5485, 94CC41⟩
Clark, Jr., B. C., **1**, 904, 917 ⟨B-72MI 119-11⟩
Clark, C. T., **2**, 299 ⟨78JCS(P1)1088⟩, 650 ⟨54JBC(208)731⟩
Clark, C. W., **5**, 28 ⟨45CRV235⟩
Clark, D. A., **2**, 135 ⟨80JA1436, 80JA3663⟩, 135, 136 ⟨80TL3143⟩
Clark, D. L., **1**, 666 ⟨90OM3086⟩; **4**, 675 ⟨89CC418⟩
Clark, D. M., **1**, 811 ⟨87T4265⟩
Clark, D. N., **1**, 950 ⟨92TL4993⟩; **3**, 514 ⟨78JA6774⟩
Clark, D. R., **4**, 42 ⟨88JCS(P2)1107⟩; **5**, 452 ⟨72JA1788⟩
Clark, D. S., **5**, 652 ⟨79JOC2945⟩
Clark, D. T., **1**, 906 ⟨88JA4112⟩
Clark, Jr., G., **1**, 503, 506 ⟨62JOC4323⟩
Clark, G. M., **1**, 449, 509, 1102 ⟨71JA3395⟩, 509 ⟨71JA1308⟩; **2**, 900 ⟨71JA3395⟩; **4**, 162 ⟨74FCF73⟩; **6**, 208 ⟨71CC1593⟩
Clark, G. R., **1**, 474 ⟨81T3957⟩; **4**, 150 ⟨79JA503⟩, 582

⟨91JCS(D)609⟩, 583 ⟨84JOM(273)C17⟩; **5**, 400 ⟨88JOM(353)355⟩; **6**, 247 ⟨90JOM(388)391, 90JOM(394)615, 90JOM(397)209⟩, 717, 718 ⟨80JA1206⟩, 717, 718, 722 ⟨82JOM(234)C9, 82JOM(236)C7⟩, 718 ⟨83CC719⟩, 719, 721 ⟨74JOM(81)C7⟩, 722 ⟨88JOM(338)393⟩
Clark, H. A., **4**, 617 ⟨61JA3345⟩
Clark, H. C., **1**, 935 ⟨68CJC633⟩; **2**, 597 ⟨67JOM(8)427⟩; **4**, 153, 811 ⟨68CJC2165⟩, 1018 ⟨78JA1164⟩; **5**, 466, 940 ⟨72JA1532⟩, 802 ⟨76JOM(112)209⟩, 939 ⟨73ACR202⟩, 939, 944 ⟨70CC763, 71IC1711⟩, 940 ⟨72IC503, 73JOM(47)C17⟩, 1139 ⟨82OM64⟩; **6**, 244 ⟨60JA5298, 62JA898⟩, 245 ⟨60JA1888⟩, 247 ⟨70IC2556⟩, 719 ⟨76JOM(112)227⟩
Clark, J., **4**, 978 ⟨76JCS(P1)1004⟩, 979 ⟨74JCS(P1)1611⟩
Clark, J. A., **5**, 781 ⟨84ICA85⟩
Clark, J. D., **1**, 190, 402 ⟨91JA5488⟩, 826 ⟨89JOC5849⟩
Clark, J. H., **2**, 19 ⟨86CC791⟩, 19, 614 ⟨85CC672⟩, 609 ⟨85TL2233⟩, 632 ⟨87CC1409⟩; **3**, 293 ⟨89CC1355⟩; **4**, 190 ⟨76TL3361⟩; **5**, 4 ⟨86CC791⟩, 1105 ⟨87JOC681⟩, 1107 ⟨87CC1301⟩; **6**, 15 ⟨88CC638⟩, 233, 234 ⟨90JFC(48)249⟩
Clark, J. R., **4**, 14 ⟨43OSC(2)181⟩
Clark, J. S., **1**, 177, 178 ⟨92JCS(P1)95⟩, 196, 410 ⟨93TL4385⟩, 410, 820 ⟨92TL6193⟩, 769 ⟨88TL4333⟩, 829 ⟨93JA10400⟩; **2**, 532 ⟨92T2127⟩, 702 ⟨90JA866⟩
Clark, J. W., **4**, 31 ⟨70JOC4201⟩
Clark, K. B., **1**, 5, 7, 9, 10, 17, 25 ⟨92JOC2427⟩, 5, 7, 9, 10, 17, 936, 943, 945 ⟨91JOC678⟩, 799 ⟨90CJC1961⟩; **5**, 1135 ⟨91JOC678⟩
Clark, Jr., L. C., **4**, 44 ⟨88JOC78⟩
Clark, M. C., **1**, 971 ⟨80JA1423⟩; **3**, 56 ⟨82JOC1618⟩; **4**, 947 ⟨82JOC1618⟩
Clark, M. J., **2**, 358 ⟨71JCS(C)2840⟩, 358, 361 ⟨71JCS(C)3867⟩, 725 ⟨70JOM(24)629⟩
Clark, M. T., **1**, 45 ⟨46JA1813⟩
Clark, N., **4**, 1034, 1037, 1042 ⟨79JCS(D)1552⟩
Clark, P. D., **2**, 728, 908 ⟨81JOC2775⟩; **4**, 289, 871, 947, 965 ⟨81JOC2775⟩
Clark, P. W., **4**, 545 ⟨81JOM(217)51⟩
Clark, R., **3**, 164 ⟨86CL1809⟩
Clark, R. D., **1**, 144 ⟨59JA2572⟩, 177 ⟨87H(26)1291⟩, 469 ⟨76JOC636⟩, 517 ⟨79JOC248, 79JOC253⟩, 601, 1093 ⟨76JOC1396⟩, 636 ⟨64JOC1316⟩, 1024, 1139 ⟨76TL1769⟩, 1041 ⟨88CC439, 76JOC3144⟩; **2**, 156 ⟨63OSC(4)674⟩; **3**, 239 ⟨74S47⟩, 239, 264 ⟨76JOC636⟩; **5**, 301 ⟨87JOC5378⟩
Clark, R. J. H., **2**, 566 ⟨75IS120⟩
Clark, R. L., **2**, 491, 496, 498 ⟨47JA927⟩
Clark, R. N., **2**, 942 ⟨46JA1675⟩
Clark, T., **1**, 135, 159 ⟨84JA6467⟩; **2**, 999, 1006 ⟨84JA6467⟩, 1046 ⟨81JA6142⟩; **3**, 310 ⟨86CB279⟩, 494 ⟨86CB1350⟩; **4**, 378, 391, 816 ⟨84JA6467⟩; **5**, 781 ⟨92JCS(D)3199, 93JCS(D)2827⟩, 781 ⟨93POL563⟩, 928 ⟨85JA2773⟩
Clark, T. J., **1**, 812 ⟨91JOC3817⟩
Clark, V. M., **2**, 414 ⟨58JCS3039⟩, 764 ⟨59JCS2094⟩
Clark, W. D., **6**, 297, 301 ⟨89JOC1990⟩
Clarke, C. T., **3**, 188 ⟨78JCS(P1)1088⟩
Clarke, D., **5**, 1110 ⟨81JCR(S)234⟩
Clarke, D. A., **2**, 212 ⟨61JOC3401⟩; **5**, 1140 ⟨79JOM(175)303⟩
Clarke, G. M., **4**, 321 ⟨67JCS(C)1411⟩
Clarke, H. C., **4**, 7 ⟨60JA1888, 60JA5298⟩
Clarke, H. T., **2**, 29 ⟨41OSC(1)115⟩; **3**, 615, 616 ⟨44OSC(2)29⟩, 618 ⟨55OSC(3)690⟩; **5**, 30 ⟨32OSC(1)523⟩, 60 ⟨32OSC(1)115⟩, 83 ⟨B-43MI 502-02⟩, 85 ⟨43OSC(2)29⟩, 86 ⟨43OSC(2)19⟩, 108 ⟨43OSC(2)588⟩, 109 ⟨32OSC(1)14⟩, 115 ⟨32OSC(1)543, 43OSC(2)135⟩, 305 ⟨43OSC(2)562⟩
Clarke, J. A., **3**, 104 ⟨75TL4705⟩
Clarke, J. R., **5**, 59 ⟨43OSC(2)181⟩
Clarke, J. T., **5**, 207 ⟨50JA5426⟩
Clarke, K., **3**, 306 ⟨73JCS(P1)1196⟩; **5**, 763, 833 ⟨54JCS4251⟩, 1111 ⟨80JCS(P1)1029⟩, 1112 ⟨71JCS(C)3262⟩
Clarke, L., **4**, 28 ⟨14JA1899⟩
Clarke, L. F., **3**, 558, 560 ⟨92JOC362⟩, 566, 568 ⟨92JOC1940⟩
Clarke, M. F., **2**, 203 ⟨49JCS315⟩
Clarke, R. L., **2**, 29 ⟨42JA1966⟩
Clarke, T., **1**, 700, 703 ⟨87TL5755⟩; **4**, 286 ⟨89T7161⟩; **5**, 652 ⟨64MI 506-01⟩
Clarkson, S.,**1**, 546 ⟨93TL157⟩
Clasby, M. C., **2**, 1034, 1035 ⟨92SL825⟩
Classon, B., **1**, 587 ⟨90JOC4273⟩; **2**, 7 ⟨88JOC6126⟩
Claude, J., **6**, 623 ⟨82MI 620-01⟩
Claudi, F., **4**, 206 ⟨81JMC1181⟩
Claudia, P., **4**, 154 ⟨83SC129⟩
Claudio, M. R., **3**, 272 ⟨90SC1095⟩
Claunch, R. T., **4**, 472 ⟨68JA4495⟩; **5**, 403, 404 ⟨68JOC3090⟩
Claus, A., **3**, 621, 638 ⟨1878LA(191)33⟩
Claus, C. J., **4**, 193 ⟨51JA5005⟩
Claus, P., **2**, 160 ⟨75T505⟩, 162 ⟨68TL3607⟩
Claus, P. K., **1**, 242 ⟨78M609⟩; **2**, 160 ⟨76CC1003⟩, 160, 164 ⟨74TL3319⟩; **3**, 436, 437 ⟨B-90MI 311-01⟩; **4**, 860 ⟨85M413⟩; **5**, 338 ⟨79JOC2863⟩, 766 ⟨74TL3319⟩
Clausen, K., **5**, 484, 551, 570 ⟨78BSB293⟩, 558, 573, 587 ⟨83T3429⟩, 558, 573, 587, 611 ⟨84JCS(P1)785⟩, 570, 571, 610 ⟨80NJC47⟩, 573 ⟨81TL3635⟩
Clausen, T. P., **1**, 906 ⟨83JOC2146⟩
Clauss, A. D., **4**, 686 ⟨80JOM(201)C31⟩
Clauss, E., **2**, 214 ⟨78CB3159⟩
Clauss, K., **1**, 446 ⟨52LA(577)26, 52LA(578)136⟩; **2**, 873 ⟨52LA(578)136⟩, 873, 879 ⟨52LA(577)26⟩; **4**, 424 ⟨74LA539⟩; **5**, 344 ⟨74LA561⟩
Claussen, K., **5**, 49 ⟨76CB41⟩
Clavel, J. L., **6**, 233, 234 ⟨91CC993⟩
Clawson, L., **1**, 769 ⟨83PAC1733⟩; **3**, 508 ⟨83PAC1733⟩
Clawson, P., **3**, 317 ⟨90JCS(P1)153⟩, 318 ⟨87TL3155, 90JCS(P1)1193⟩
Claxton, E. E., **3**, 203 ⟨88JOC6158⟩
Clay, J. P., **2**, 478 ⟨51JOC892⟩
Clayden, J., **1**, 484 ⟨93CC1682⟩, 836 ⟨93JCS(P1)1849, 93JCS(P1)2913⟩
Clayton, A. F., **1**, 15 ⟨83JCS(P1)87⟩
Clayton, E. M., **6**, 598 ⟨59JA6270⟩
Clayton, F. J., **2**, 910, 911, 931 ⟨76JOC1480⟩
Clayton, J. D., **5**, 199 ⟨69JA3075⟩
Clayton, J. F., **1**, 1092 ⟨76JOC1480⟩
Clayton, J. P., **3**, 559 ⟨77GEP(O)2728884⟩, 708 ⟨74CC278⟩; **6**, 111 ⟨74JCS(P1)22⟩
Cleary, D. G., **1**, 24 ⟨86JOC858⟩, 226, 227 ⟨89T391⟩, 362, 555, 558 ⟨92JA7375⟩
Cleary, M., **1**, 420 ⟨83JOC4272⟩
Cleasson, A., **1**, 513 ⟨78CC621⟩
Cleaver, C. S., **4**, 799 ⟨65JA3716⟩; **6**, 69 ⟨58USP2853531⟩
Cleeburg, F., **5**, 723 ⟨1890JA499⟩
Clef, B. Le, **6**, 105 ⟨83TL1035⟩
Clegg, J. M., **2**, 376 ⟨58JOC1595⟩
Clegg, W., **2**, 473 ⟨91CB2457⟩, 512 ⟨90JOM(399)C21⟩; **3**, 459 ⟨87JCS(D)1071, 87JCS(D)2141⟩; **4**, 173, 237 ⟨91JCS(P1)897⟩, 742 ⟨90JCS(P1)139⟩, 858 ⟨88TL4889, 91JCS(P1)3091⟩, 862 ⟨93JCS(P1)577⟩, 938 ⟨93JCS(P1)343⟩; **5**, 423 ⟨91JCS(P1)897⟩, 663 ⟨83CB724⟩; **6**, 85, 729 ⟨92JOC1696⟩, 98, 99, 101 ⟨91JCS(P1)3091⟩, 126 ⟨91JCS(P1)897⟩, 633 ⟨88CC1175⟩, 698 ⟨93PS(74)279⟩
Cleland, A. J., **6**, 340 ⟨70JOM(24)C61⟩
Cleland, G. H., **5**, 183 ⟨85S681⟩
Cleland, J. H., **5**, 101 ⟨72CJC3741⟩
Clemans, G. B., **1**, 46 ⟨75JOC3686⟩; **2**, 658 ⟨75JOC3686⟩
Clemence, L. W., **2**, 261 ⟨48JA2434⟩
Clemens, D. H., **2**, 399 ⟨65TL1487⟩; **4**, 414 ⟨62JOC3664⟩, 421 ⟨61JOC949⟩, 979 ⟨64JOC2932⟩; **5**, 599

⟨64JOC2932⟩; **6**, 70 ⟨58JA3915⟩, 105, 138, 140 ⟨62JOC3664⟩
Clemens, K. E., **6**, 107 ⟨72JA3201⟩
Clement, A., **3**, 335 ⟨83NJC269⟩, 552 ⟨79NJC149, 82NJC401⟩
Clément, B., **4**, 271 ⟨79TL1737⟩
Clement, B. A., **4**, 740 ⟨76JOC556⟩
Clement, J.-C., **4**, 582 ⟨89JOM(359)205⟩
Clement, K. S., **1**, 422, 784 ⟨84JOC3870⟩
Clement, M., **5**, 244, 247, 254 ⟨52LA(576)61⟩
Clemente, D. A., **4**, 677 ⟨81AG(E)887⟩
Clementi, R., **4**, 316 ⟨54JA3969⟩
Clemmensen, E., **1**, 18 ⟨13CB1837⟩
Clemo, G. R., **3**, 231 ⟨85JCS(P1)2407⟩
Clemo, N. G., **1**, 450 ⟨85JCS(P1)2399⟩
Cleophax, J., **1**, 223 ⟨85TL3121⟩
Clerici, A., **2**, 81 ⟨89JOC3872⟩, 305 ⟨90TL2069⟩; **3**, 311 ⟨74JCS(P2)1699⟩
Clerici, F., **1**, 45 ⟨87S719⟩; **3**, 438 ⟨87S719⟩
Clerin, D., **4**, 979 ⟨76BSF2053⟩
Clery, M., **3**, 684 ⟨91JCS(P1)3389⟩
Clesse, F., **4**, 852, 969 ⟨73BSF581, 68MI 421-01⟩; **5**, 597, 599, 614 ⟨73BSF581⟩
Cleve, A., **1**, 769 ⟨89TL1241⟩, 1070 ⟨93TL5257⟩
Cleve, C., **5**, 635 ⟨86PS(27)289⟩
Clevenger, G. L., **1**, 582 ⟨90CC84⟩
Clezy, P. S., **3**, 106 ⟨72AJC1979⟩, 364 ⟨69AJC239⟩
Clibbens, D. A., **5**, 110 ⟨43OSC(2)557⟩
Clibbens, D. L., **3**, 474 ⟨15JCS1491⟩
Clibze, L. A., **4**, 389 ⟨92MI 408-01⟩
Cliff, G. R., **5**, 1110 ⟨81JCR(S)234⟩
Cliffe, I. A., **2**, 346 ⟨91TL6789⟩
Clifford, A. F., **5**, 1003 ⟨72USP3666784⟩, 1085 ⟨76JFC(7)65, 82IC1616⟩; **6**, 10 ⟨78IJ129⟩, 214 ⟨53JCS2372⟩, 558, 565 ⟨84CB1707⟩, 616, 651 ⟨82JFC(19)411⟩, 617 ⟨83JFC(23)593⟩
Clifford, D. B., **3**, 495 ⟨71JOC3386⟩; **4**, 762, 771 ⟨71JOC3386⟩
Clifford, P. R., **1**, 902 ⟨71JA4219⟩
Clift, S. M., **2**, 681 ⟨83JA640, 84JA8300⟩; **3**, 509 ⟨83JA640⟩
Clifton, G., **6**, 570, 571 ⟨70JMC377⟩
Clin, B., **5**, 373 ⟨87BSF997⟩
Cline, B. L., **3**, 362 ⟨78JOC4910⟩
Cline, R. E., **3**, 362 ⟨50JOC600⟩
Cline, W. K., **2**, 176 ⟨63JOC3537⟩
Clinet, J.-C., **1**, 204, 971, 1053, 1056, 1057, 1058 ⟨81S875⟩
Clinet, J. C., **1**, 597 ⟨87TL5509⟩
Clinet, J.-C., **1**, 971 ⟨77NJC373, 78TL1137, 80TL3987, 82SC739⟩; **2**, 939 ⟨80TL3987, 81S875⟩; **5**, 427 ⟨80TL3987⟩
Clingman, A. L., **4**, 235 ⟨66CAR(2)461⟩
Clinton, R. O., **2**, 128 ⟨45JA594⟩
Clippinger, E., **1**, 905 ⟨65JA376⟩
Clive, D. J. L., **3**, 173, 207 ⟨73CC695⟩
Clive, D. L. J., **1**, 21, 25, 437, 438, 943 ⟨80JA4438⟩, 174, 547, 1131 ⟨84JOC1313⟩, 243 ⟨82JOC1641⟩, 244 ⟨87JOC4943, 90JOC1786, 93CC429⟩, 505, 651 ⟨84CC534⟩, 547 ⟨85CC980⟩, 591, 593, 594, 716 ⟨81JOC231⟩, 592, 601, 712, 713, 716, 780, 781 ⟨78T1049⟩, 592, 608 ⟨74CC100⟩, 594, 928 ⟨89CC332⟩, 601 ⟨73CC695⟩, 609 ⟨78CC770⟩, 688 ⟨91JOC6447⟩, 688, 690 ⟨88JA6914⟩, 814 ⟨82JOC1632⟩, 919 ⟨88PAC1645⟩, 928 ⟨91JOC3801⟩, 942 ⟨78CC41, 91JCS(P1)1433, 93JOC6857⟩, 943 ⟨91JCS(P1)2797⟩, 1131 ⟨85CC1205, 86CC878⟩; **2**, 280, 286, 287, 728 ⟨78T1049⟩, 280 ⟨B-84MI 204-01⟩, 282, 291 ⟨73CC695⟩, 286 ⟨77CC725, 78CC441⟩, 294 ⟨79JOC4279⟩; **3**, 37, 173, 207 ⟨78T1049⟩, 138 ⟨82JOC1632⟩, 141 ⟨82CC828⟩, 144 ⟨84CC534⟩, 618 ⟨70CC1014⟩, 621 ⟨84JOC1313⟩; **4**, 238, 288 ⟨79JOC1883⟩, 286, 779 ⟨87JOC4943, 79JOC4279⟩; **6**, 92 ⟨79JOC1883, 79JOC4279⟩
Clizbe, L. A., **4**, 1014 ⟨81JA2807⟩; **5**, 271, 288 ⟨76JA2352⟩
Cloarec, A.-Y. Le, **5**, 754 ⟨71BCF478⟩, 756 ⟨70BCF200⟩, 759 ⟨72FRP2081556⟩
Clode, D. M., **4**, 180 ⟨79CRV491⟩
Cloke, J. B., **3**, 612 ⟨45JA1249⟩
Clopton, J. C., **1**, 631 ⟨67JA1442, 67JA471⟩
Clos, N., **1**, 76 ⟨89T6901⟩
Closs, G. L., **1**, 848 ⟨65TL287⟩, 907 ⟨66MI 119-09⟩; **3**, 478 ⟨65JA4270⟩, 489 ⟨63JA3796⟩; **4**, 816 ⟨64JA4042⟩; **5**, 844 ⟨65JA4270⟩, 844, 866 ⟨62JA4350⟩
Closs, L. E., **3**, 489 ⟨63JA3796⟩
Closse, A., **6**, 469 ⟨79HCA398⟩
Closson, R. D., **2**, 573 ⟨57JOC598A, 57JOC598B⟩; **5**, 446 ⟨57JOC598⟩; **6**, 523 ⟨58JA6167⟩
Closson, W. D., **2**, 249 ⟨74TL381⟩
Cloudsdale, I. S., **3**, 11 ⟨79JOC4847⟩; **4**, 340 ⟨79JOC4847⟩
Clouet, F. L., **4**, 697, 698 ⟨88JA3231⟩
Clough, A. E., **1**, 776 ⟨90JA228⟩
Clough, J. M., **1**, 617, 704 ⟨91COS(3)861⟩, 754 ⟨78TL4159⟩, 755 ⟨79TL5043⟩; **2**, 613 ⟨90SL761⟩
Clough, R. L., **3**, 518 ⟨76AG(E)543⟩, 522 ⟨76JOM(120)C6⟩
Cloutour, C., **4**, 637, 660 ⟨77JOM(137)157⟩
Cloyd, Jr., J. C., **2**, 853 ⟨75JA53⟩
Cluff, E. F., **2**, 556 ⟨53JA134⟩
Clunz, L. J., **2**, 458 ⟨57JA3570⟩
Clusius, K., **2**, 806 ⟨59HCA739⟩
Clutter, R. J., **2**, 397 ⟨56JOC4003⟩
Clutterbuck, P. W., **5**, 114 ⟨35JCS1467⟩
Coates, B., **1**, 828 ⟨91TL4199⟩
Coates, D. A., **1**, 935 ⟨73JCS(P2)1570⟩
Coates, G. E., **1**, 885 ⟨B-67MI 119-02⟩; **2**, 421 ⟨54JCS22, 54JCS2526, 65JCS1870, 67JCS(A)56⟩, 511 ⟨65JCS3241⟩, 590 ⟨59CI(L)628⟩, 867 ⟨61JCS1000, 61JCS5053⟩; **5**, 1137, 1142, 1143 ⟨63JCS229⟩
Coates, H., **2**, 119 ⟨63BRP917921⟩; **4**, 551 ⟨60BRP842593⟩
Coates, M., **6**, 410 ⟨89EUP310255⟩
Coates, R. M., **1**, 24 ⟨92JOC4598⟩, 214 ⟨82JOC3211⟩, 236 ⟨74TL3955⟩, 591, 693 ⟨72JA4758⟩, 826 ⟨87JA1160⟩, 889 ⟨71JA1027⟩, 1082 ⟨70T3487⟩; **3**, 35 ⟨74TL3955⟩, 435 ⟨86JOC5198⟩; **5**, 327, 636 ⟨86JOC5198⟩
Coats, E., **2**, 379, 388 ⟨64JA2395⟩; **3**, 446 ⟨64JA2395⟩
Coatsworth, L. L., **6**, 247 ⟨92CM979⟩
Cobb, J., **6**, 621, 728 ⟨87CC99⟩
Cobb, R. L., **4**, 28 ⟨58JOC1368⟩
Cobben, P. L. H. M., **6**, 562 ⟨92JA10573⟩
Coblens, K. E., **1**, 5, 9 ⟨82JOC5041⟩
Coburn, C. E., **1**, 209 ⟨83JOC1455⟩
Cocco, M. T., **4**, 972 ⟨88FES103⟩, 986 ⟨92JHC1631⟩; **6**, 95 ⟨75S660⟩
Cochini, F., **1**, 976 ⟨92OM493, 93OM1871⟩
Cochoy, R. E., **4**, 162 ⟨80JOC2874⟩
Cochran, E. L., **1**, 322 ⟨64JCP(40)213⟩
Cochran, J. C., **1**, 455 ⟨82OM586⟩; **2**, 915 ⟨90TL6621⟩; **3**, 91, 97 ⟨90SC609⟩; **4**, 701 ⟨83OM1099⟩
Cock, C. de, **4**, 92 ⟨85T4183⟩
Cockburn, W. F., **6**, 641, 642, 643 ⟨58CJC1541⟩
Cocker, W., **1**, 72 ⟨66JCS(C)41⟩; **3**, 616 ⟨31JCS1391, 31JCS1894⟩
Cockerill, A. F., **1**, 869 ⟨70JCS(B)173, 70JCS(B)179⟩
Cockerill, R. F., **2**, 125 ⟨33JA2932⟩
Cockman, R. W., **6**, 457 ⟨88CC529⟩
Cockroft, R. D., **1**, 826 ⟨69JA2815⟩
Cocks, A. T., **1**, 925 ⟨73HCA1516⟩
Cocolas, G., **5**, 90 ⟨73OSC(5)376⟩
Cocolios, P., **2**, 578 ⟨83JOM(253)65⟩
Cocuzza, A. J., **4**, 216 ⟨77JA4835⟩
Coda, C. A., **5**, 664 ⟨84G133⟩
Codalovi, L. M., **6**, 417 ⟨92EUP486948⟩
Codding, P. W., **2**, 795, 796 ⟨83CJC1562⟩
Codington, J. F., **1**, 45 ⟨46JA1813⟩
Coe, D. E., **2**, 284 ⟨80TL3827⟩
Coe, D. G., **2**, 8 ⟨54JCS2281⟩

Coe, P. L., **1**, 248 ⟨86JCS(P1)475⟩, 1005 ⟨67CC1259, 72JCS(P1)639⟩; **3**, 589 ⟨85JCS(P1)2185⟩; **4**, 3 ⟨73JFC(3)329⟩; **5**, 34, 63, 100 ⟨61JCS808⟩; **6**, 20 ⟨74CC323, 78JCS(P1)202⟩, 43 ⟨75JFC(6)37⟩, 450 ⟨83JFC(22)521⟩
Coelho, F., **1**, 187 ⟨88TL5661⟩
Coenen, H. H., **1**, 449 ⟨87JFC(36)63⟩
Coffen, D. L., **3**, 258 ⟨68JOC2504⟩; **4**, 412 ⟨74JOC167⟩; **6**, 85 ⟨67CC1089, 68JOC2504⟩, 304, 306 ⟨70JHC201⟩
Coffey, R. S., **1**, 668 ⟨68JCS(A)190⟩
Coffield, T. H., **2**, 573 ⟨57JOC598A, 57JOC598B⟩; **5**, 446 ⟨57JOC598⟩
Coffin, B. J., **5**, 62 ⟨B-76MI 502-01⟩
Coffin, M. A., **6**, 85, 729 ⟨92JOC1696⟩, 698 ⟨93PS(74)279⟩
Coffman, D. D., **2**, 169 ⟨60JOC2016⟩, 401 ⟨61JOC4010⟩, 855 ⟨60JOC2015⟩; **3**, 419 ⟨62JA2337⟩; **4**, 5 ⟨49JA490⟩, 130 ⟨60JOC2016⟩, 195 ⟨48JOC223⟩; **5**, 548 ⟨65JA2776⟩, 675, 975 ⟨64JA357⟩, 808 ⟨60JOC2016, 62USP3023208⟩, 808, 989 ⟨62JA2337⟩, 1003 ⟨62JA4275⟩; **6**, 2 ⟨49JOC747⟩, 52, 53, 238, 259, 409, 423, 438 ⟨62JA4275⟩, 233, 235, 236 ⟨59JA3575, 60JOC2016⟩, 242 ⟨62JA2337⟩
Coffman, K. J., **2**, 383 ⟨82SC801⟩; **5**, 774 ⟨83JHC69⟩
Cofler, M., **3**, 268 ⟨84CC1593⟩
Cofré, P., **3**, 113 ⟨90JA1936⟩
Cogen, J. M., **1**, 826 ⟨90TL3241⟩
Coggiola, I. M., **1**, 100 ⟨63MI 103-02⟩
Coghlan, M. J., **1**, 505 ⟨84JOC747⟩, 632, 797, 826 ⟨85JA7352⟩; **2**, 238 ⟨88OS(67)157⟩; **4**, 59 ⟨89TL2033⟩; **6**, 416, 428 ⟨89TL2033⟩
Cohen, A., **1**, 637 ⟨44JCS322⟩; **2**, 890 ⟨32JCS2505, 32JCS593⟩
Cohen, B., **2**, 399, 400 ⟨66JCS(A)1780⟩
Cohen, D., **2**, 693 ⟨64JOC1817⟩
Cohen, D. S., **1**, 1010 ⟨92BMC1619⟩
Cohen, E., **4**, 420 ⟨60JA2731⟩; **6**, 648 ⟨68JMC811⟩
Cohen, H. L., **5**, 293 ⟨53JOC432⟩
Cohen, J., **2**, 118, 124 ⟨93JMC460⟩
Cohen, J. B., **3**, 620 ⟨16PRS232⟩
Cohen, J. F., **1**, 214 ⟨74JA5270⟩
Cohen, J. S., **2**, 410 ⟨93JOC4606⟩
Cohen, L. A., **4**, 223 ⟨58JA3765⟩
Cohen, M. A., **5**, 118 ⟨90TL7223⟩
Cohen, M. C., **4**, 108 ⟨89JOC2869⟩
Cohen, M. J., **4**, 750 ⟨84JOC515⟩
Cohen, M. P., **2**, 317 ⟨89S450⟩; **3**, 251 ⟨72JHC171⟩; **4**, 304 ⟨69JHC429⟩, 307 ⟨70JHC1311⟩
Cohen, M. S., **1**, 168 ⟨63IC1111⟩
Cohen, N., **1**, 514 ⟨76JOC3497⟩, 980 ⟨80JOC2080⟩; **2**, 39 ⟨76JOC3505, 81HCA1158⟩; **4**, 168, 170 ⟨83JA3661⟩; **5**, 159 ⟨80JOC2080⟩
Cohen, S., **1**, 108 ⟨61JCS3452⟩, 248 ⟨87CC381⟩; **3**, 283 ⟨63JOC1379⟩; **4**, 33 ⟨61JCS3452⟩, 751 ⟨57JA4174⟩, 826 ⟨56JOC801⟩
Cohen, S. C., **2**, 959 ⟨77JOM(127)127⟩
Cohen, S. G., **2**, 88 ⟨64JA5326⟩, 403 ⟨55JA2457⟩; **4**, 18, 730 ⟨49JA3439⟩, 743, 750 ⟨50JA3952⟩; **5**, 366 ⟨65JOC1162⟩
Cohen, T., **1**, 235 ⟨82JOC4005⟩, 235, 408, 409, 822 ⟨82TL3019⟩, 236 ⟨86T2803, 87JA4710, 89ACR152, 93JA3855⟩, 302, 303 ⟨87JA7923⟩, 398, 405 ⟨92JA375⟩, 408, 821 ⟨85JOC2965⟩, 468 ⟨75JOC3649⟩, 590 ⟨75JOC812, 76JOC3218, 80OS(59)202⟩, 805 ⟨80JA6900, 83JA520, 83JA6798⟩; **2**, 68 ⟨90JOC4784⟩, 83, 553 ⟨90JOC1528⟩, 422, 551 ⟨80SC311, 89ACR152⟩, 553 ⟨89JOC5657, 90JA6389, 91JA1866, 91JOC5760⟩, 553, 555 ⟨92JOC3680⟩, 647 ⟨77JOC2053⟩, 710 ⟨88OSC(6)737⟩, 999 ⟨84JA1130⟩; **3**, 31, 161 ⟨92JOC2⟩; **5**, 374 ⟨86TL5147⟩; **6**, 641 ⟨92JCS(P1)3179⟩
Cohen, V. I., **2**, 223, 224 ⟨76HCA840⟩; **5**, 516, 522, 626 ⟨77JOC2645⟩, 516, 522, 704 ⟨78JHC237⟩, 522 ⟨77JHC1321, 78JHC1415, 79JHC365⟩, 625 ⟨78S668⟩, 631 ⟨74JOC3043, 76HCA350, 76S728⟩, 642, 704 ⟨78JHC1113⟩; **6**, 596 ⟨80S60⟩, 644 ⟨66BSF73⟩
Cohen, Z., **2**, 86 ⟨75JOC2141, 88OSC(6)43⟩
Cohn, K., **5**, 1129 ⟨70IC1571⟩
Cohn, K. C., **4**, 589 ⟨89OM1118⟩
Cohn, M., **2**, 314 ⟨03M773⟩
Coindard, G., **4**, 145 ⟨72BSF811⟩
Coisne, J. M., **2**, 1100 ⟨82S211⟩
Coke, J. L., **1**, 627 ⟨67JA6701⟩, 628 ⟨68JA5561⟩, 1082, 1083 ⟨77JOC2380⟩
Colapret, J. A., **1**, 812 ⟨85JOC201⟩; **2**, 816 ⟨80JA5866⟩
Colas, Y., **1**, 962 ⟨84TL845⟩
Colau, R., **3**, 269 ⟨79JHC1657⟩
Colberg, H., **3**, 352 ⟨79H(12)1153⟩
Colberg, J. C., **2**, 902, 922 ⟨89JA4873⟩, 906 ⟨93JA6065⟩; **3**, 203 ⟨89JA4873⟩; **4**, 630, 631 ⟨93JA6065⟩, 630, 631, 633, 644 ⟨89JA4873⟩
Colborn, R. E., **4**, 679 ⟨83JCS(D)2661⟩, 680 ⟨83JCS(D)2099⟩, 681 ⟨90JCS(D)761⟩
Colchester, J. E., **5**, 338 ⟨63JCS4918⟩
Colclough, E., **1**, 660, 765 ⟨87CC297⟩
Colclough, T., **2**, 99 ⟨56JCS3006⟩, 144, 173 ⟨61JCS5339⟩
Coldham, I., **1**, 407 ⟨93JCS(P1)1275⟩, 599 ⟨92JCS(P1)2303⟩; **2**, 322 ⟨93JCS(P1)1275⟩; **3**, 35 ⟨91JCS(P1)365⟩
Coldwhite, H., **2**, 852 ⟨82CC691⟩
Cole, L. L., **2**, 25 ⟨75CJC3620⟩
Cole, Q. P., **6**, 261 ⟨44JOC419⟩
Cole, S. J., **1**, 5, 7, 9, 10, 16, 17, 331 ⟨91JCS(P1)103⟩
Cole, S. M., **2**, 138 ⟨77JOC3233⟩
Cole, T. E., **1**, 1034 ⟨86JOC4521⟩; **2**, 516 ⟨87TL3771⟩, 535 ⟨85JA460⟩, 1086 ⟨83OM1316, 86JOC4521⟩; **3**, 83, 102 ⟨77TL781⟩, 135 ⟨87JA5420⟩
Cole-Hamilton, D. J., **2**, 825 ⟨80JCS(D)55⟩; **5**, 220 ⟨85JOM(294)C29⟩
Coleman, B., **3**, 483 ⟨78JA994⟩
Coleman, G. H., **1**, 1144 ⟨36JA2310⟩; **2**, 338 ⟨28JA2739⟩, 779 ⟨47OSC(1)214⟩, 781 ⟨55OSC(3)668⟩; **3**, 459 ⟨55OSC(3)26⟩, 620 ⟨55OSC(3)221⟩; **5**, 25 ⟨55OSC(3)221⟩
Coleman, H. A., **1**, 808 ⟨68JOC4413⟩
Coleman, J. B., **1**, 797 ⟨80JOC900⟩
Coleman, J. P., **1**, 6 ⟨70CC738⟩, 14 ⟨91CC535⟩, 103 ⟨81MI 103-01⟩
Coleman, L. L., **2**, 366 ⟨58JOC1133⟩
Coleman, R. A., **1**, 509, 1101 ⟨79JOC2328⟩
Coleman, R. S., **1**, 523 ⟨87JA2717⟩; **2**, 655 ⟨86JOC5436, 87JA2717, 87TL1027⟩; **3**, 543 ⟨90TL3677⟩
Colens, A., **2**, 758 ⟨79CC1180⟩; **3**, 475 ⟨79CC1180⟩; **4**, 790 ⟨79CC1180⟩, 790, 791 ⟨77NJC369⟩, 889 ⟨77NJC371⟩; **5**, 2 ⟨79CC1180⟩
Coles, J. A., **1**, 422 ⟨52JCS1430⟩
Coles, M. A., **2**, 566 ⟨75IS120⟩
Coles, N., **2**, 927 ⟨92SL143⟩
Coles, R. F., **2**, 491, 496, 498 ⟨47JA927⟩; **4**, 864 ⟨76USP3933914⟩; **5**, 221 ⟨47JA925⟩
Colgan, D., **4**, 647 ⟨84CC1708⟩, 647, 656 ⟨86JCS(D)603⟩; **6**, 190 ⟨84CC1708⟩
Colie, N., **2**, 452 ⟨1888JCS636⟩
Colin, G., **1**, 132 ⟨67CR(C)1204⟩
Colin, J., **2**, 601 ⟨68BSF3385⟩
Colin, J. L., **4**, 199, 314 ⟨83S568⟩
Colla, A., **3**, 246 ⟨91S483⟩
Collado, I. G., **1**, 831 ⟨91TL3209⟩; **2**, 30 ⟨91TL3217, 94T6433⟩
Coll, J., **1**, 527 ⟨85TL6397, 87TL4745⟩, 747 ⟨84T2871⟩; **2**, 624 ⟨88TL5811⟩, 684 ⟨82S186⟩; **3**, 128 ⟨86T4603⟩, 218 ⟨85TL6397, 87TL4745⟩; **4**, 810 ⟨88TL5811⟩; **6**, 695 ⟨83TL3387⟩
Collard, J., **6**, 506 ⟨72AG993⟩
Collard-Charon, C., **5**, 515, 516 ⟨62BSB563⟩, 624, 625 ⟨63BSB304⟩, 1051, 1056 ⟨62BSB531⟩; **6**, 596 ⟨63BSB149⟩, 598 ⟨62BSB541⟩

Collardeau, G., **1**, 414 ⟨50AP(283)36⟩
Collazo, L. R., **1**, 685, 686 ⟨93JOC43⟩
Colle, R., **5**, 735 ⟨80JCS(P1)665⟩
Collela, M. B., **5**, 864 ⟨80ANC967⟩
Collen, G. R., **2**, 530 ⟨83JOC5398⟩
Collet, A., **2**, 374 ⟨87T891⟩; **6**, 33 ⟨85JA6993⟩
Collette, J., **5**, 205 ⟨56JA3819⟩
Colley, A. M., **2**, 711 ⟨91JOC862⟩
Colli, L., **6**, 526 ⟨60G1005⟩
Collienne, R., **5**, 515, 516 ⟨64BSB419⟩
Collier, C. W., **3**, 328 ⟨70SCI87⟩
Collier, IV, L. W., **1**, 74 ⟨83JOC1638⟩
Collier, J. A., **4**, 679 ⟨82OM1350⟩
Collier, M. P., **4**, 656 ⟨71SAP7004922⟩
Collier, M. R., **3**, 458 ⟨72JCS(D)370⟩, 460 ⟨71MI 312-01⟩; **4**, 656 ⟨70JOM(25)C36, 72CC613⟩, 656 ⟨73JCS(D)445⟩; **6**, 396 ⟨71SAP7004922⟩
Collier, T. L., **3**, 404 ⟨86T5157⟩
Collignon, N., **2**, 557 ⟨75JOM(96)139⟩; **3**, 49 ⟨83S634, 84JOM(264)9⟩, 50 ⟨81T1377, 85T427, 92SC219⟩, 639 ⟨90PS(54)171⟩; **4**, 557 ⟨85TL4435, 86JOM(312)283⟩, 557, 576 ⟨84JOM(264)9⟩, 557, 576, 801 ⟨86JOM(304)283⟩, 570 ⟨86S934⟩, 588 ⟨91S31, 93PS(75)99⟩; **6**, 164, 374 ⟨87JOM(323)135⟩, 288 ⟨86JOM(304)283, 87SC1559⟩
Collin, J., **2**, 56 ⟨84JOC2047⟩, 565 ⟨87JOM(328)81⟩; **3**, 7, 119 ⟨84JOC2045⟩; **5**, 439 ⟨91JOC3118⟩
Collin, P. J., **1**, 517 ⟨71AJC2325⟩
Collingnon, N., **2**, 458 ⟨79T1345⟩
Collington, E. W., **1**, 836 ⟨93JCS(P1)1849⟩, 1041 ⟨70JA6676⟩; **2**, 145 ⟨94TL9629, 95SL773⟩, 612 ⟨88TL3365⟩; **3**, 181 ⟨94TL2427⟩, 244 ⟨87CC756⟩, 494 ⟨71T5979⟩; **3**, 90, 97 ⟨70JA6676, 74OS(54)42⟩; **4**, 61, 73, 86 ⟨88TL3365⟩, 306 ⟨88OSC(6)64⟩
Collingwood, S. P., **2**, 8, 32, 36 ⟨87TL4445⟩; **6**, 633 ⟨87TL4445, 88CC1175⟩
Collins, A. M., **1**, 965 ⟨32JA4066⟩
Collins, C. J., **1**, 384 ⟨60QR357, B-66MI 109-01⟩, 385 ⟨55JA5517⟩, 1124 ⟨77JA1669⟩; **3**, 121 ⟨60QR357⟩
Collins, D. J., **1**, 200 ⟨86AJC2095⟩; **4**, 206 ⟨71JCS2166⟩, 454, 471 ⟨83AJC2095⟩; **6**, 72 ⟨89AJC1235⟩
Collins, G. R., **2**, 245 ⟨64JOC1688⟩
Collins, J. B., **6**, 386, 401, 404 ⟨76JA5419⟩
Collins, J. C., **5**, 756 ⟨71CC498⟩
Collins, J. F., **4**, 190 ⟨75TL3489⟩
Collins, J. H., **1**, 157 ⟨66JOC2373⟩; **4**, 575 ⟨66JOC2373⟩
Collins, J. L., **1**, 190, 402 ⟨92TL4735⟩
Collins, P., **4**, 230 ⟨85JCS(P1)587⟩
Collins, P. A., **1**, 774 ⟨79AJC1819⟩
Collins, P. M., **3**, 540 ⟨67JCS(C)1197⟩
Collins, P. W., **1**, 1008 ⟨87JMC193⟩, 1049 ⟨75TL4217⟩, 1049, 1050 ⟨72TL2627, 76TL235⟩
Collins, R. F., **4**, 866 ⟨84GEP3408757A1⟩; **5**, 95 ⟨57CI(L)736⟩
Collins, S., **1**, 486, 545 ⟨87JOC4258⟩, 605 ⟨81TL5111, 83JOC4776⟩; **2**, 76 ⟨88OM2289, 89JOC4154⟩, 238 ⟨80TL2213⟩, 285 ⟨80TL2215, 81JOC3249⟩, 285 ⟨81TL5111, 83JOC4776⟩, 285, 1034, 1035 ⟨83JOC3077⟩, 564 ⟨90JOC3565⟩, 743 ⟨85CJC2313⟩, 949 ⟨85TL1281⟩; **3**, 131 ⟨87TL4391⟩, 171 ⟨83JOC3077⟩, 637 ⟨81JOC3249⟩; **4**, 874 ⟨85CJC2313⟩, 947 ⟨83JOC3077⟩, 953 ⟨87JOC4258⟩
Collins, T. J., **5**, 650 ⟨78JOM(159)73⟩; **6**, 719 ⟨78JOM(159)73⟩, 720, 732 ⟨76JOM(107)C37⟩
Collis, M. J., **1**, 871 ⟨62JA2905⟩
Collman, J. P., **1**, 73 ⟨68IS101⟩, 668 ⟨B-87MI 114-04⟩; **2**, 563 ⟨B-80MI 211-01⟩, 575 ⟨75ACR342⟩, 578 ⟨86ACR209⟩, 956 ⟨B-87MI 219-02⟩, 960 ⟨B-87MI 219-03⟩, 964 ⟨B-87MI 219-05⟩, 984 ⟨B-87MI 219-04⟩, 985 ⟨68IC1298⟩; **3**, 20, 128, 137 ⟨75ACR342⟩, 621 ⟨78JA1119⟩; **5**, 451 ⟨75ACR342⟩, 452 ⟨72JA1788⟩, 460 ⟨68IC27, 68JA2282⟩, 983 ⟨67JA4809⟩
Collonges, R., **1**, 516 ⟨76JOC4089⟩
Collum, D. B., **1**, 68 ⟨83JA6882, 86JA2094⟩, 131 ⟨91JA5751, 91JA9571⟩, 131 ⟨93TL5213⟩, 153 ⟨85JA2078⟩; **2**, 40 ⟨83JA6882, 86JA2094⟩, 554 ⟨92ACR448⟩; **3**, 633 ⟨81TL5011⟩; **4**, 378 ⟨79TL593⟩; **5**, 652 ⟨78JOC4393⟩
Colman, R., **4**, 978 ⟨76JCS(P1)1004⟩
Colobert, F., **4**, 270, 863 ⟨91TL3695⟩
Colombani, D., **2**, 97 ⟨93SC1025⟩
Colombo, D., **6**, 689 ⟨89TL6869⟩
Colombo, L., **2**, 152, 363 ⟨92JA5977⟩; **3**, 37 ⟨82S829⟩; **4**, 265 ⟨78TL3861⟩, 856 ⟨81JCS(P1)1278⟩; **5**, 88 ⟨86JA6394⟩
Colombo, M., **3**, 684, 686 ⟨93JCS(P1)2151⟩
Colomer, E., **5**, 421 ⟨77CC73, 81JOM(205)31⟩
Colon, C., **3**, 264 ⟨80JOC4522, 82SC35⟩
Colón, C. J., **1**, 104 ⟨68CJC2821⟩
Colon, I., **1**, 428 ⟨82JOC2622⟩, 472 ⟨86JOC2627⟩
Colonge, J., **1**, 212 ⟨48BSF818, 48BSF822⟩, 212 ⟨48BSF825, 48BSF827⟩, 475 ⟨48BSF811⟩; **2**, 1016 ⟨47BSF841⟩
Colonna, F. P., **2**, 783 ⟨73T3011, 74T2741⟩
Colonna, M., **2**, 803 ⟨72G527⟩
Colonna, S., **2**, 9, 19 ⟨79JCS(P1)2248⟩, 151 ⟨88CC1451⟩, 151 ⟨92TA921⟩, 151 ⟨92TA95⟩, 393 ⟨77S113⟩; **3**, 634 ⟨81JCS(P1)614⟩; **4**, 74, 75 ⟨72JCS(P2)296, 72S259, 72JCS(P1)1886⟩, 75, 77, 233 ⟨72JCS(P1)1883⟩, 76 ⟨73JA7431, 70CC1441⟩, 233 ⟨76JCS(P2)996⟩, 262 ⟨86JOC891⟩
Colquhoun, H. M., **1**, 212 ⟨B-91MI 105-02⟩; **4**, 799 ⟨80CC879, 88JCS(D)2781⟩; **5**, 387, 388 ⟨86JCS(D)813⟩; **6**, 610 ⟨81JCR(S)276⟩, 610, 619 ⟨80CC879⟩
Colquhoun, I. J., **2**, 437, 828 ⟨82JCS(D)1915⟩; **4**, 546, 562 ⟨78JCR(S)368⟩, 548 ⟨86CC1041⟩, 550, 551, 576 ⟨80IC3195⟩, 704 ⟨88JCS(D)503⟩, 1022 ⟨82JCS(D)1915⟩; **6**, 151 ⟨87PS(38)79⟩, 151, 369 ⟨86IC2699, 82CC930⟩, 370 ⟨85GEP3337939⟩
Colson, J. G., **6**, 57 ⟨68JOC1084⟩
Colson, P.-J., **5**, 162 ⟨94JOC4972⟩
Colter, A. K., **1**, 612 ⟨71JOC1898⟩
Colter, M. A., **2**, 280 ⟨78JOC4885⟩
Colton, C. D., **4**, 97 ⟨79JOC3117⟩
Colton, R., **6**, 332, 351 ⟨83IC3700⟩
Columbo, L., **2**, 375 ⟨86JA6394⟩
Colville, R. J., **4**, 330 ⟨84CJC586⟩
Colvin, E., **2**, 921 ⟨71JA2080, 74JA3684⟩; **3**, 18 ⟨71JA2080⟩; **4**, 646 ⟨B-88MI 414-01⟩
Colvin, E. W., **1**, 216 ⟨B-81MI 105-01⟩, 295 ⟨B-81MI 107-01⟩, 447, 450 ⟨B-81MI 110-01⟩, 648 ⟨78CSR15, B-81MI 114-02, B-88MI 114-02⟩, 758, 759 ⟨B-88MI 116-01⟩, 1078 ⟨73CC151, 77JCS(P1)869⟩; **2**, ⟨B-87MI 210-02⟩, 394 ⟨81HCA2264⟩, 535 ⟨B-81MI 210-01, B-88MI 210-02⟩, 537, 540 ⟨91COS(7)641⟩, 783 ⟨79C1⟩, 909 ⟨B-88MI 218-03⟩; **3**, 292 ⟨B-81MI 306-01⟩, 440 ⟨88T4157⟩, 456 ⟨85CC539⟩; **4**, 354, 357 ⟨B-81MI 408-01⟩, 490 ⟨82TL3835⟩, 831 ⟨B-88MI 419-01⟩, 1044 ⟨B-81MI 423-01⟩; **5**, 24 ⟨76AR(B)199⟩, 377 ⟨88T4157⟩; **6**, 172, 174, 189 ⟨B-81MI 606-01⟩
Colwell, C. E., **4**, 430, 431, 432 ⟨62JOC3598⟩
Comar, D., **6**, 414 ⟨83MI 614-01⟩
Comasseto, J., **2**, 279 ⟨82JCR(S)212⟩
Comasseto, J. V., **1**, 243, 437 ⟨89TL1209⟩, 243, 595 ⟨91S793⟩, 243, 595, 611, 623 ⟨91S897⟩, 245, 441 ⟨89TL7353⟩, 440 ⟨83JOM(253)131, 90SC1681⟩, 595, 611, 716 ⟨86S1⟩, 999 ⟨86SC283, 88OM84⟩, 999, 1035 ⟨90SC2181, 86CC1067⟩; **2**, 728 ⟨76JOM(114)281⟩, 730 ⟨89OM1661⟩, 768 ⟨86S1⟩, 1036, 1037 ⟨81JOM(216)287⟩, 1036, 1037 ⟨88OM84⟩, 1036, 1037 ⟨88SC1979⟩, 1036, 1038 ⟨84S240, 86CC1067, 86SC283⟩; **3**, 176 ⟨91S785⟩
Combellas, C., **1**, 474 ⟨88CC203⟩
Combes, C. E. J., **6**, 402, 404 ⟨81JOM(221)257⟩
Combret, J. C., **2**, 741 ⟨83BSF(2)28, 84S493⟩; **4**, 406 ⟨88BSF1009⟩, 738, 739, 745 ⟨71TL1035⟩

Combrink, K. D., **1**, 810 ⟨92HCA1755⟩
Combs, P., **3**, 197 ⟨87TL611⟩, 198 ⟨89JMC1886⟩; **5**, 370 ⟨89JMC1886⟩
Combs, R. C., **1**, 199 ⟨94TL363⟩
Come, J., **6**, 598 ⟨91TL4823⟩
Comer, F., **5**, 278 ⟨71JCS(C)3540⟩
Comer, W. T., **5**, 767 ⟨74JMC549, 74JOC3080⟩; **6**, 508 ⟨73JMC901⟩
Comfort, D. R., **4**, 412 ⟨69JA1860⟩
Comini, A., **5**, 295 ⟨80S385⟩
Comins, D., **3**, 58, 90, 101 ⟨78S403⟩
Comins, D. L., **1**, 455 ⟨84TL4867⟩; **2**, 420 ⟨84JOC1078⟩, 938 ⟨93JA8851, 93JOC5035⟩, 942 ⟨90JOC292⟩; **3**, 64 ⟨90JOC292⟩, 219 ⟨91TL5697⟩; **4**, 255 ⟨79JA4732, 79JA7104⟩, 299 ⟨92SL615⟩; **5**, 557 ⟨78TL4657, 79JA4732, 79JA7104, 83JA5015⟩; **6**, 120 ⟨78S309⟩, 134 ⟨79JA4732⟩
Comisarow, M. B., **1**, 893, 897 ⟨65JA2997⟩, 894 ⟨67JA5687⟩, 900 ⟨64JA4195⟩, 902 ⟨64JA5682⟩, 915 ⟨67JA5259, 69JA1458⟩; **2**, 1006 ⟨69JA2955⟩
Comisso, G., **5**, 1078 ⟨85S509⟩
Commaseto, J. V., **1**, 437, 438, 441 ⟨92PS(67)183⟩
Commenil, M.-G., **1**, 622 ⟨93SL839⟩
Commerçon, A., **1**, 466 ⟨77JOM(128)1⟩, 1017, 1018 ⟨80T1215, 75TL1465⟩; **2**, 212 ⟨85TL4093⟩, 679 ⟨75TL3833⟩, 682 ⟨75JOM(93)415⟩
Commercon, M., **4**, 405, 408 ⟨91SL111⟩
Commerçon-Bourgain, M., **1**, 312 ⟨84PAC91⟩
Commeyras, A., **1**, 906 ⟨70JA4627⟩, 1130 ⟨82CC433⟩; **4**, 812, 820 ⟨82JFC(20)699, 76CC174⟩; **5**, 58 ⟨76CC885⟩
Commons, T. J., **3**, 468 ⟨78JOC2203⟩; **4**, 70 ⟨78JOC2203⟩
Comninellis, C., **2**, 177 ⟨74S887⟩
Comotti, A., **2**, 698 ⟨92JOC5173⟩
Comoy, P., **1**, 82 ⟨63APO(1)155⟩
Compagnini, A., **5**, 734 ⟨82SC865, 84JCR(S)404⟩, 734, 760 ⟨84JCS(P1)897⟩
Compagnon, P.-L., **3**, 273 ⟨75BSF779⟩, 675 ⟨76S448⟩
Compere, Jr., E. L., **5**, 67 ⟨68JOC2565, 78JOC2702⟩
Compton, H., **4**, 39 ⟨23JCS576⟩
Comstock, M. C., **3**, 702, 703, 712 ⟨90JOC4950⟩
Comstock, W. J., **5**, 723 ⟨1890JA499⟩
Comte, M.-T., **3**, 165 ⟨82TL5031⟩
Comuzzi, C., **5**, 322 ⟨94JA916⟩
Conant, J. B., **1**, 848, 869 ⟨32JA1212⟩; **2**, 456 ⟨20JA830⟩; **3**, 449 ⟨32JA2881⟩
Conant, R., **1**, 882 ⟨72JCS(P1)1535, 73JCS(P1)1858⟩
Conaway, R., **2**, 34, 629, 631 ⟨88JOC4477⟩
Conboy, R. J., **4**, 298 ⟨78JOC3838⟩
Concellon, J. M., **1**, 681 ⟨86CC1665⟩; **2**, 936 ⟨88S234⟩; **3**, 152 ⟨91JCS(P1)2890⟩, 152, 154 ⟨90S1003⟩; **4**, 741, 747, 756, 758 ⟨89JCS(P1)691⟩; **5**, 958 ⟨95TL3937⟩
Concepcion, A. B., **3**, 26 ⟨93JA2059⟩
Concepción, J. I., **1**, 929 ⟨84TL1953⟩; **2**, 13 ⟨86JOC402⟩
Concialini, V., **3**, 310 ⟨91JOC1692⟩
Conde, S., **3**, 305 ⟨81JHC1345⟩; **4**, 938 ⟨91TL5159⟩; **5**, 652 ⟨74T195⟩
Condeiu, C., **4**, 152 ⟨92RRC393⟩
Condert, G., **3**, 240 ⟨86S337⟩
Condo, K., **4**, 347 ⟨74JCS(P1)1279⟩
Condom, R., **2**, 10 ⟨78TL4507⟩
Condon, F. E., **2**, 376 ⟨73OPP219⟩, 377 ⟨72JOC3615⟩, 379 ⟨73OPP225⟩, 380, 381, 382 ⟨73OPP233⟩
Condon, M. E., **3**, 212 ⟨82JMC250⟩
Condon, S., **3**, 229 ⟨89TL1033⟩
Condon, S. L., **2**, 118, 124 ⟨93JMC460⟩; **4**, 279 ⟨92TL1573⟩
Conduché, A., **5**, 837 ⟨08AC(P)84⟩
Condulis, N., **4**, 414 ⟨86JOC2228⟩
Cone, J. L., **5**, 214 ⟨87JOC4933⟩
Confalone, P. N., **1**, 91 ⟨77JOC135⟩; **4**, 268 ⟨83TL5563⟩
Congiu, C., **4**, 972 ⟨88FES103⟩, 986 ⟨92JHC1631⟩
Congreve, M. S., **1**, 829 ⟨93JA5815⟩; **5**, 179 ⟨93JA5815⟩
Congson, L. N., **1**, 382 ⟨84JOC2937⟩
Conia, J., **3**, 138 ⟨85JA1726⟩
Conia, J. M., **1**, 66, 454, 526 ⟨85JA1726⟩, 412 ⟨66TL6115⟩, 422 ⟨68TL2101, 70BSF2981⟩, 526 ⟨75S1⟩, 544 ⟨83S801⟩, 807 ⟨63MI 118-03⟩, 807, 881 ⟨62TL505⟩, 808 ⟨66BSF281, 50BSF533⟩, 881 ⟨50BSF537, 56BSF1040, 61BSF836, 61T45, 63BSF1925, 63BSF1930, 63RCP42⟩; **3**, 140 ⟨81TL649⟩, 154 ⟨76S194⟩, 220 ⟨77TL3505⟩, 224 ⟨73TL2767, 81S289, 81S291⟩, 232 ⟨81TL645⟩, 539 ⟨73CC795, 83T1567⟩; **3**, 9, 123 ⟨78S63⟩, 67, 235 ⟨76S196⟩; **4**, 162 ⟨75AG(E)473⟩, 188 ⟨80SC83⟩
Conklin, R. D., **5**, 675, 678 ⟨81JMC532⟩
Conley, R. A., **4**, 194 ⟨76JA3037⟩
Conley, R. T., **3**, 435 ⟨71MI 311-01⟩
Conlin, R. T., **3**, 424 ⟨87OM 1406⟩; **4**, 627 ⟨92OM3307⟩; **5**, 924 ⟨92OM2317⟩
Conlon, D. A., **1**, 803 ⟨88JA2324⟩
Conlon, L. E., **1**, 414 ⟨57JA2828⟩; **4**, 306 ⟨57JA2825⟩
Conly, J. C., **4**, 27 ⟨53JA1148⟩
Conn, J. J., **3**, 657 ⟨53CJC1211⟩
Conn, M., **2**, 304 ⟨88TL6393⟩
Conn, M. W., **2**, 132 ⟨43OSC(2)547⟩
Conn, R. S. E., **1**, 524, 525 ⟨79JA5283, 77TL2831⟩, 1105 ⟨78JA4624⟩; **2**, 924 ⟨78JA4624⟩; **4**, 1060 ⟨78JA4624⟩
Conneely, J. A., **1**, 803 ⟨79JCS(P2)962⟩
Connell, R., **3**, 193 ⟨81SC947⟩
Connell, R. D., **2**, 50 ⟨89JA9243⟩, 308 ⟨88JOC3845⟩
Connell, S., **2**, 539 ⟨58JOC50⟩
Connelly, C. J., **3**, 109 ⟨91JOC449⟩
Connelly, N. G., **2**, 569 ⟨82COMC-I(3)647⟩; **4**, 676 ⟨89OM412⟩
Conner, W. E., **1**, 503, 504 ⟨83JOC2266⟩
Connet, J. E., **2**, 956 ⟨66JCS(C)106⟩
Connolly, B. A., **5**, 1111 ⟨91TL2505⟩
Connolly, C. B., **5**, 944 ⟨93JOC6952⟩
Connolly, J. M., **5**, 1028 ⟨34JCS822, 35JCS679⟩; **6**, 232 ⟨34JCS822⟩, 297 ⟨37JCS827⟩
Connolly, J. W., **2**, 94 ⟨61JA3985⟩
Connon, N. W., **2**, 397 ⟨69JOC4121⟩
Connor, D., **4**, 221 ⟨72JHC175⟩; **5**, 596 ⟨71JOC1742⟩
Connor, D. T., **4**, 226 ⟨81JHC587⟩; **6**, 558, 564 ⟨93JMC1090⟩, 558, 564 ⟨93JMC1802⟩
Connor, J. A., **5**, 949 ⟨71CC570, 71JCS(A)3368⟩, 953 ⟨67CC1024⟩
Connor, R., **2**, 120 ⟨55OSC(4)363⟩; **4**, 14 ⟨43OSC(2)181⟩, 83, 84 ⟨40JA2596⟩; **5**, 59 ⟨43OSC(2)181⟩
Conole, G., **4**, 675 ⟨88NJC559, 89CC688⟩
Conrad, H., **2**, 407, 420 ⟨81ZN(B)515⟩
Conrad, P., **1**, 821 ⟨87JA5878⟩
Conrad, P. C., **1**, 144 ⟨78JA346⟩
Conrad, W. E., **4**, 210 ⟨75JOC2962⟩
Conradie, W. J., **4**, 56 ⟨65JCS3075⟩
Conrads, M., **6**, 125 ⟨91CB1425⟩
Conran, D. E., **5**, 322 ⟨84JOC106⟩
Conrow, R. B., **2**, 753 ⟨68MI 215-01⟩
Conrow, R. E., **3**, 227, 232 ⟨83JA5679⟩
Conroy, B. K., **4**, 675 ⟨86OM2457⟩, 676 ⟨86OM2384⟩
Conroy, H., **1**, 8 ⟨55JA5960⟩
Considine, J. L., **1**, 83 ⟨73JOM(61)167⟩; **4**, 1068 ⟨74JA1941⟩
Consiglio, C., **1**, 297 ⟨77JA1058⟩
Consiglio, G., **1**, 195, 238 ⟨86T2043⟩; **2**, 583 ⟨91OM2046⟩; **5**, 940 ⟨86OM1976⟩
Consolandi, E., **5**, 233 ⟨91T8767⟩
Consoloni, C., **1**, 198 ⟨88G857⟩, 466 ⟨91JOM(417)289⟩
Consonni, P., **4**, 441 ⟨72JHC827⟩; **5**, 773 ⟨83JMC1187⟩
Constable, A. C., **5**, 451, 937 ⟨83OM1852⟩
Constable, A. G., **2**, 580 ⟨82JA4865⟩
Constable, E. C., **3**, 700 ⟨91JOM(403)C15⟩
Constam, E. J., **5**, 652 ⟨1882LA(214)329⟩
Constantieux, T., **6**, 163 ⟨92TL3903⟩
Constantinescu, D., **3**, 539 ⟨88RRC83⟩
Constantino, M. G., **1**, 516 ⟨86JOC253⟩; **5**, 199 ⟨82JOC3450⟩

Constantinou-Kokotou, V., **2**, 317 ⟨92JCR(S)391⟩
Contardo, N., **1**, 1010 ⟨92BMC1619⟩
Conte, A., **5**, 337 ⟨67CB1248⟩
Conte, V., **2**, 147 ⟨89TL4859⟩
Contelles, J. L. M., **5**, 157 ⟨89CC1160⟩
Contelles, J. M., **6**, 550, 557 ⟨92TL5261⟩
Contento, M., **2**, 21, 23 ⟨80CC1278⟩, 25, 31 ⟨83S306⟩, 28, 30 467 ⟨80S143⟩; **3**, 456, 458 ⟨86TL1695, 93T3809⟩; **5**, 287 ⟨86TL1695⟩
Conti, J., **5**, 1029 ⟨54ACS295⟩
Conti, P. G. M., **1**, 47 ⟨87T3817⟩
Contreras, B., **3**, 63 ⟨90SC2983⟩
Controulis, J., **2**, 260 ⟨48JA1762⟩; **5**, 1125 ⟨46JA2102⟩
Conway, A. C., **2**, 373, 374, 379, 381 ⟨59JA2597⟩
Conway, B. E., **5**, 28 ⟨67CRV623⟩
Conway, J., **1**, 160 ⟨92TL543⟩; **4**, 651 ⟨92TL543⟩
Conway, P., **1**, 821 ⟨71CC1639⟩
Conway, R. J., **5**, 126 ⟨85CA(103)104568c⟩
Conway, W. P., **1**, 599 ⟨74JA7165⟩
Cook, A. A., **2**, 456 ⟨20JA830⟩
Cook, A. F., **2**, 255 ⟨70JA190⟩; **3**, 373 ⟨68JA740⟩
Cook, A. G., **2**, 738, 739, 751 ⟨B-88MI 215-01⟩, 757 ⟨B-69MI 215-02⟩; **3**, 408 ⟨B-88MI 310-01⟩; **4**, 294 ⟨B-88MI 407-01⟩; **6**, 51 ⟨62JOC3686⟩
Cook, A. H., **4**, 318 ⟨B-49MI 407-01⟩, 405 ⟨49JCS2342⟩
Cook, B. R., **2**, 883 ⟨71IC2676⟩
Cook, C. D., **5**, 464 ⟨67CJC301⟩
Cook, C. E., **5**, 131 ⟨87JOC3468⟩
Cook, C. L., **1**, 1069 ⟨52JCS2010⟩
Cook, D., **1**, 918 ⟨72JA6867⟩; **3**, 526 ⟨62CJC2362⟩
Cook, E. S., **2**, 132 ⟨39JA2971⟩
Cook, F., **1**, 630 ⟨67JA5736⟩
Cook, F. B., **1**, 631 ⟨68JA531⟩
Cook, F. L., **3**, 613 ⟨74JOC3416⟩
Cook, G. K., **1**, 1027 ⟨89JOC5828⟩; **4**, 951 ⟨89JOC5828⟩
Cook, G. R., **1**, 835 ⟨92JOC7188⟩; **5**, 572 ⟨94TL1669⟩
Cook, H. G., **2**, 101 ⟨50JCS3125⟩, 409, 410, 412, 413 ⟨49JCS2921⟩
Cook, I. M., **5**, 659, 660 ⟨91JFC(55)47⟩
Cook, J. M., **1**, 776 ⟨90TL3409, 92JOC5121⟩, 810 ⟨93JOC661⟩; **3**, 62 ⟨89JA7504⟩, 310 ⟨89JOC2170⟩; **5**, 322 ⟨81JOC4188⟩
Cook, J. W., **1**, 637 ⟨44JCS322, 49JCS1074⟩
Cook, K. L., **5**, 692 ⟨93TL5983⟩
Cook, L., **1**, 346, 926 ⟨93JOC7640⟩
Cook, M. A., **4**, 1048 ⟨70JOM(24)529⟩; **6**, 60, 191, 289 ⟨71JOM(29)389⟩, 174, 289, 380, 387, 390 ⟨70JOM(24)529⟩, 734 ⟨70CC559⟩
Cook, M., **1**, 929 ⟨86CC1419⟩
Cook, M. I., **5**, 315, 662 ⟨91JOC5256⟩
Cook, M. J., **2**, 347 ⟨71JCS(B)1330⟩; **3**, 3, 114 ⟨79JCS(P1)2493⟩; **5**, 275 ⟨66CC121⟩, 758, 767 ⟨74JCS(P2)546⟩
Cook, P. F. E., **5**, 123 ⟨65JCS4594⟩
Cook, P. M., **1**, 897 ⟨72JA5511⟩
Cook, R. J., **3**, 457 ⟨71JA6703⟩
Cook, R. V., **2**, 491, 495 ⟨19IEC105⟩
Cook, S. L., **2**, 675 ⟨79SC157⟩
Cooke, B. J. A., **1**, 421 ⟨72JCS(P1)2271⟩
Cooke, D. W., **3**, 621 ⟨69JOC3684⟩
Cooke, F., **1**, 422 ⟨79TL1995, 80JOC1046⟩; **3**, 211 ⟨79TL1995⟩; **4**, 355 ⟨77CC513, 77JA4536⟩, 369, 938, 942 ⟨80JOC1046⟩
Cooke, G., **4**, 873 ⟨90CC816⟩; **6**, 698 ⟨93PS(74)279⟩
Cooke, G. E., **4**, 662, 962 ⟨86OM1906⟩
Cooke, Jr., H. G., **5**, 55 ⟨40JA3495⟩
Cooke, J., **4**, 685 ⟨69JCS(A)1872⟩
Cooke, J. M., **4**, 621 ⟨77JA3273⟩
Cooke, M., **4**, 680 ⟨81CC862⟩
Cooke, M. D., **4**, 834 ⟨83TL3373⟩
Cooke, M. J., **5**, 126 ⟨85CA(103)104568c⟩
Cooke, M. P., **2**, 556 ⟨92JOC1495⟩; **3**, 15, 130, 284 ⟨86JOC951⟩
Cooke, Jr., M. P., **1**, 51 ⟨82JOC4963⟩, 214 ⟨85TL3643⟩, 303 ⟨77JA642⟩, 627 ⟨67JA6701⟩; **2**, 525 ⟨87JA931⟩; **4**, 202 ⟨86JOC951⟩, 645, 1057 ⟨87JA931⟩, 652, 655 ⟨93JOC7474⟩
Cooke Jr., M. P., **1**, 233, 709 ⟨81JOC1747⟩; **5**, 451 ⟨70JA6080⟩
Cooke, R., **1**, 112 ⟨74JOC1168⟩
Cooke, R. J., **1**, 797 ⟨87TL5473⟩
Cooke, W. A., **2**, 259, 260 ⟨46JA1860⟩
Cooksey, C. J., **2**, 686 ⟨92JCR(S)202⟩
Cookson, J. P., **5**, 652 ⟨69JOC3204⟩
Cookson, P. G., **2**, 96, 98 ⟨76CC1022⟩
Cookson, R. C., **1**, 207 ⟨85JCS(P1)1589⟩, 537 ⟨67JCS(C)1986⟩, 794, 799, 809 ⟨65TL3955⟩, 799 ⟨71CC107, 71CC108⟩, 807, 809 ⟨60TL29⟩, 862 ⟨61PCS117⟩; **2**, 374 ⟨88OSC(6)936⟩, 858 ⟨65JCS2184⟩, 976 ⟨85JCS(P1)1589⟩; **3**, 20 ⟨80TL1677⟩, 63 ⟨78CC924⟩, 113 ⟨79CC145⟩, 214 ⟨78CC822⟩, 226 ⟨73CC161⟩, 227 ⟨78CC821⟩, 233 ⟨78CC608⟩, 674 ⟨68T1529⟩; **4**, 365 ⟨80TL1677⟩, 957 ⟨76CC990, 78CC821⟩
Cooley, G., **3**, 71 ⟨64TL733⟩
Cooley, J. H., **2**, 303 ⟨89S1⟩; **5**, 1063 ⟨89JOC1048⟩
Cooley, R. A., **2**, 129 ⟨72JMC968⟩
Coolidge, M. B., **6**, 140 ⟨81TL4365⟩
Coombes, R. C., **3**, 362 ⟨83PS(14)139⟩
Coombes, R. G., **2**, 385 ⟨79COC(2)370⟩, 385, 798 ⟨79COC(2)363, 79COC(2)367⟩; **4**, 540 ⟨79COC(2)305⟩
Coombs, M. M., **1**, 15 ⟨83JCS(P1)87⟩
Coombs, R. V., **1**, 130, 889 ⟨65JA275⟩, 961 ⟨74S344⟩; **2**, 696 ⟨65JA275⟩
Coon, C. L., **4**, 298 ⟨90JOC1459⟩; **5**, 763, 815 ⟨66TL3537⟩, 813, 814, 815 ⟨70JOC3093⟩; **6**, 144 ⟨68JCED437⟩, 287 ⟨73S605⟩
Cooney, B. T., **1**, 892 ⟨87AJC1537⟩
Cooney, K. E., **1**, 832 ⟨74HCA771⟩
Cooney, M. J., **1**, 1019 ⟨90JA4054⟩
Coonradt, H. L., **1**, 90, 92 ⟨41JA1563⟩
Cooper, A. B., **3**, 436 ⟨94SL167⟩
Cooper, A. J. L., **5**, 62 ⟨83CRV321⟩
Cooper, B. E., **4**, 566 ⟨70BRP1179242⟩
Cooper, C. M., **1**, 595 ⟨70CC1683⟩; **5**, 278 ⟨71JCS(C)3540⟩
Cooper, C. S., **2**, 377 ⟨83JOC4116⟩
Cooper, D., **1**, 832 ⟨81JCS(P1)2127⟩; **3**, 618 ⟨79TL1725⟩; **4**, 136 ⟨76JCS(D)694⟩
Cooper, D. J., **4**, 20 ⟨66JCS(C)533⟩
Cooper, F. C., **5**, 756 ⟨53JCS255⟩
Cooper, G. D., **1**, 36 ⟨52JA3074⟩, 618 ⟨51JA5187⟩; **2**, 561 ⟨62JOC1493⟩; **4**, 63, 64 ⟨57JA376, 51JA5187⟩, 80 ⟨51JA5184⟩
Cooper, G. F., **1**, 967 ⟨91JOC1083⟩; **5**, 73 ⟨72JOC1947⟩
Cooper, G. R., **4**, 1023, 1028 ⟨85CC614⟩
Cooper, J., **1**, 658 ⟨68IC1014⟩, 831 ⟨87CC1220, 87TL3031⟩, 831 ⟨91JCS(P1)705⟩
Cooper, J. P., **1**, 456 ⟨90JOC3452⟩, 515 ⟨92JOC4786⟩
Cooper, J. W., **1**, 322 ⟨75JA4269⟩; **2**, 1010 ⟨75JA4269⟩
Cooper, K., **1**, 481 ⟨89JA2363⟩; **2**, 345, 347 ⟨88CC1503⟩
Cooper, K. A., **1**, 107 ⟨37JCS1183⟩
Cooper, L., **5**, 274, 294 ⟨80JOC3413⟩
Cooper, M. A., **4**, 344 ⟨69TFS1697⟩
Cooper, M. K., **2**, 828 ⟨78IC880⟩
Cooper, M. S., **2**, 147, 156 ⟨87S1015⟩, 654 ⟨90SL533⟩
Cooper, N. J., **3**, 698 ⟨89OM1538⟩; **4**, 673 ⟨80JCS(D)29⟩; **5**, 444 ⟨93JA4389⟩
Cooper, P., **4**, 111 ⟨71JCS(C)3031⟩, 802 ⟨75JCS(P1)702⟩
Cooper, R. D. G., **2**, 21 ⟨80TL781⟩; **4**, 69 ⟨78JA3933⟩
Cooper, S., **2**, 110 ⟨61JCS5545⟩
Cooper, S. R., **3**, 280 ⟨41OS(21)103⟩
Cooper, T. A., **5**, 1152, 1153, 1159 ⟨80CC333⟩, 1153 ⟨82JST(79)215⟩
Cooper, W. C., **5**, 472 ⟨B-73MI 511-01⟩
Cooper, W. D., **3**, 307 ⟨58JOC1382⟩

Cooper, W. J., **2**, 109 ⟨60JPC682⟩
Coote, S. J., **1**, 194 ⟨93TL3149⟩, 202 ⟨89TL3581⟩
Coover, Jr., H. W., **6**, 118 ⟨56JA4450⟩
Cope, A. C., **1**, 78 ⟨63OSC(4)304⟩, 78, 421 ⟨52JA4867⟩, 104 ⟨63OSC(4)377⟩, 411 ⟨41JA1852, 44JA1684⟩, 411, 806 ⟨40JA441⟩, 411, 807 ⟨41JA1843⟩, 416, 831 ⟨43JA1992⟩, 624 ⟨49JA3929, 63OSC(4)612⟩, 624, 625, 788 ⟨57JA960⟩, 624, 627, 628, 637, 640, 782, 783 ⟨60OR(11)317⟩, 624, 628 ⟨57JA4720⟩, 624, 782, 783 ⟨53JA3212⟩, 625 ⟨59JA2799, 60JA4663⟩, 625, 628 ⟨58JA355⟩, 625, 783 ⟨55JA1628, 57JA4729⟩, 625, 784 ⟨60JA4656⟩, 627 ⟨61JA3861, 63JA1949⟩, 628 ⟨73OSC(5)315⟩, 808 ⟨56JA2547, 56JA2551⟩, 839 ⟨44JA1929⟩, 881 ⟨51JA4158, 55OSC(3)219⟩; **2**, 346 ⟨60OR(11)317⟩, 347 ⟨63OSC(4)612⟩, 756 ⟨49JA3929⟩, 964 ⟨68JA909⟩; **3**, 328 ⟨51OS(31)25⟩, 620 ⟨55OSC(3)219, 57OR(9)107⟩; **4**, 14 ⟨43OSC(2)181⟩, 165 ⟨51JA1668⟩, 880, 883 ⟨45JA1017, 45JA1012⟩; **5**, 46 ⟨57OR(9)107⟩, 59 ⟨43OSC(2)181⟩, 199 ⟨63OSC(4)890⟩, 278 ⟨63OSC(4)339⟩
Cope, O. J., **1**, 658 ⟨64JA1801⟩
Copeland, C., **6**, 544 ⟨82AJC1709⟩
Copeland, C. M., **6**, 589, 593 ⟨88AJC549⟩
Copenhafer, W. C., **5**, 379 ⟨74JA4283⟩
Copenhaver, J. E., **2**, 25 ⟨41OSC(1)142⟩
Copenhaver, J. W., **2**, 478 ⟨54USP2882313⟩; **5**, 829 ⟨54USP2786865⟩
Copéret, C., **2**, 987, 988 ⟨94TL695⟩
Copley, S. D., **1**, 826 ⟨87JA5008⟩
Coppens, P., **4**, 130 ⟨85JCP1517⟩
Copper, J. L., **1**, 1126 ⟨78JA993⟩
Coppert, D. M., **1**, 983 ⟨92JA5998⟩
Coppinger, G. M., **3**, 87 ⟨53JA734⟩; **5**, 299 ⟨54JA1372⟩
Coppola, B. P., **1**, 289 ⟨82JA6879⟩
Coppola, G. M., **1**, 459 ⟨B-84MI 110-02⟩, 954 ⟨B-84MI 120-02⟩; **3**, 9, 123 ⟨84S1021⟩; **4**, 423 ⟨80JHC1163⟩; **5**, 370 ⟨78JHC1169⟩
Coppola, J. C., **2**, 889 ⟨70JCS(C)810⟩
Coq, A. Le, **1**, 958 ⟨83T2185⟩; **3**, 274 ⟨74CR(278)1153⟩
Coquelin, J., **2**, 176 ⟨73BSF210⟩
Corain, B., **3**, 268 ⟨84CC1593, 84JOC4696⟩, 268 ⟨88JCR(S)246⟩
Corallo, M., **6**, 7 ⟨76T2295⟩
Coraor, G. R., **6**, 143 ⟨71JOC2262⟩
Corbally, R. P., **6**, 12 ⟨74JCS(P1)108⟩
Corbeil, J., **5**, 236 ⟨88CHE1297⟩
Corbel, B., **1**, 214 ⟨76JOC3648, 81CJC1415⟩; **2**, 457 ⟨85S1048⟩, 1057 ⟨76TL835⟩; **4**, 923 ⟨73CR(C)395⟩
Corbelin, S., **2**, 1004 ⟨85CB3529⟩
Corbella, A., **5**, 295 ⟨80S385⟩
Corbett, D. F., **3**, 72 ⟨77CC808⟩
Corbett, G. E., **2**, 689 ⟨66JOC153⟩
Corbett, M. D., **2**, 683 ⟨72JOC3944⟩
Corbett, R. L., **5**, 334, 336 ⟨92TL153⟩
Corbett, W. L., **4**, 305 ⟨90JOC2999⟩
Corbier, J., **4**, 884, 891 ⟨70CR(C)1890⟩
Corbin, T. F., **2**, 27 ⟨73OSC(5)328⟩
Corcoran, C., **2**, 478 ⟨69B2557⟩
Corcoran, R. J., **2**, 20 ⟨77JA905⟩, 860 ⟨78HCA2190⟩
Cordella, G., **2**, 225 ⟨74G455⟩
Cordero, F., **2**, 329 ⟨89JOC1748⟩
Cordes, A. W., **5**, 665, 672 ⟨85JOC993⟩
Cordes, E. H., **3**, 449, 450 ⟨62JA826⟩; **6**, 68 ⟨B-69MI 603-01⟩
Cordes, G., **4**, 460 ⟨59MI 410-01⟩; **6**, 159 ⟨59LA(623)103⟩
Cordes, R. E., **2**, 249 ⟨78CJC1183⟩; **4**, 258 ⟨78CJC1183⟩; **5**, 348 ⟨78CJC1183⟩
Cordier, C., **4**, 674 ⟨90OM2993⟩
Cordonnier, G., **3**, 310 ⟨90H(31)427⟩
Cordova, R., **1**, 408 ⟨86T2979⟩; **4**, 370 ⟨86T2979⟩
Cordts, H. P., **2**, 179, 222 ⟨52JA5880⟩
Corelli, F., **4**, 856 ⟨82JHC1493⟩
Corey, E. J., **1**, 5, 9, 10, 936 ⟨75JOC2554⟩, 17 ⟨69JOC3667⟩, 51 ⟨64JA1639⟩, 74 ⟨61JA2957⟩, 108 ⟨54JA6040⟩, 111 ⟨B-89MI 104-01⟩, 116 ⟨92TL6435, 93TL5995⟩, 117, 118 ⟨67JA3911⟩, 117, 118, 466 ⟨68JA5615⟩, 120, 1107 ⟨78JOC3418⟩, 125 ⟨70TL315⟩, 127 ⟨67JA2755⟩, 136 ⟨83TL3165⟩, 140, 693 ⟨66JOC4097⟩, 142, 881 ⟨65JA1345⟩, 153 ⟨76TL3⟩, 210 ⟨65AG(E)1075⟩, 212 ⟨76TL11⟩, 218 ⟨70TL311⟩, 232 ⟨69JOC1233⟩, 304 ⟨65JA1353⟩, 336, 346 ⟨87TL175⟩, 374 ⟨89JA5493⟩, 382 ⟨64JA1652⟩, 389 ⟨64JA478⟩, 417, 831 ⟨91JA4026⟩, 421 ⟨63JA4033⟩, 438 ⟨79TL2317⟩, 450, 662, 1103 ⟨72JA4013⟩, 458 ⟨61TL347⟩, 498 ⟨64JA5570⟩, 504 ⟨90JOC1670⟩, 504, 513 ⟨83TL4883⟩, 523 ⟨73TL4477⟩, 532, 1070 ⟨84JA462⟩, 535 ⟨76TL4705⟩, 535, 1091 ⟨77TL3847⟩, 535, 1091, 1093 ⟨75JOC2265⟩, 540 ⟨73TL1611⟩, 548 ⟨83TL2821⟩, 585 ⟨82TL1979⟩, 601 ⟨87TL4917⟩, 621, 644, 695 ⟨68JA5553⟩, 621, 695 ⟨66JA5656, 68JA5548⟩, 639, 644, 645, 646 ⟨66JA5654⟩, 643, 644 ⟨66JA5652⟩, 643, 644 ⟨66JA5653, 68JA6816, 69JOC3053⟩, 651, 1052, 1053, 1055, 1061 ⟨82TL719⟩, 671 ⟨78JA6294⟩, 695 ⟨67PAC19⟩, 735 ⟨74JOC821⟩, 747 ⟨71TL1821⟩, 752 ⟨88JA5653⟩, 778 ⟨89JOC389⟩, 789 ⟨72TL4651⟩, 801 ⟨72JA8934⟩, 801, 803, 804 ⟨85JA5574⟩, 821 ⟨72JA8932⟩, 826 ⟨82JA6129⟩, 827, 828 ⟨77TL1625⟩, 834 ⟨70JA5522⟩, 839 ⟨82TL3463⟩, 871, 879 ⟨56JA6269⟩, 874 ⟨84TL495⟩, 878 ⟨53E329, 54JA175⟩, 881 ⟨62JA866⟩, 882 ⟨78CB1337⟩, 926 ⟨85TL4291⟩, 977, 1055 ⟨90JA878⟩, 983 ⟨84TL3055⟩, 999, 1052 ⟨68TL5041⟩, 999, 1066 ⟨73TL1495⟩, 1015, 1016 ⟨81JA4618, 82TL1651, 83TL3291, 84TL3059⟩, 1016, 1082, 1083 ⟨84JA2735⟩, 1029 ⟨88JA649⟩, 1048, 1081, 1091 ⟨74JA5581⟩, 1052 ⟨70JA6314⟩, 1063 ⟨87TL4921⟩, 1067 ⟨72TL3769⟩, 1082 ⟨75JOC579⟩, 1131 ⟨88TL57⟩, 1134 ⟨69JA4318, 70TL2245, 78JA4618, 78JA4620⟩; **2**, 25, 32 ⟨72TL4339⟩, 49 ⟨90TL2665, 93JA12579, 93JA3828⟩, 50 ⟨89JA9243⟩, 65 ⟨89TL5207, 89TL6275⟩, 66 ⟨92TL4141⟩, 71 ⟨90JOC784⟩, 72 ⟨87TL5237⟩, 76 ⟨89JA5495⟩, 79, 533 ⟨90JA878⟩, 87 ⟨75TL3183⟩, 119, 124 ⟨92TL4099⟩, 135 ⟨80JA1436, 80JA3663⟩, 135, 136 ⟨80TL3143⟩, 153 ⟨65JA1345⟩, 225 ⟨62JA2938⟩, 351 ⟨83TL2821⟩, 374 ⟨85JOC5391⟩, 390 ⟨87OS(65)166⟩, 393, 396, 783 ⟨78JA6294⟩, 397 ⟨80TL1117, 84JA3682⟩, 420 ⟨84TL491⟩, 478 ⟨84TL2415⟩, 528 ⟨87JA5551, 87JA7925⟩, 528 ⟨92JOC7372, 92TL6907⟩, 533 ⟨89JA5493⟩, 539 ⟨72JA6190⟩, 539, 699 ⟨81TL3455⟩, 540 ⟨82TL4871⟩, 563 ⟨92TL6435⟩, 587 ⟨72JA7210⟩, 587, 1095 ⟨78JOC3418⟩, 588 ⟨86JA7115⟩, 593 ⟨80JA1742⟩, 618, 630, 909 ⟨70TL447⟩, 629 ⟨78JA4618⟩, 629, 970 ⟨67JA4245⟩, 665 ⟨75TL2389⟩, 665, 784, 785 ⟨81TL603⟩, 672 ⟨89JA5472⟩, 675 ⟨73JOC3224⟩, 679 ⟨88JA649⟩, 699 ⟨84TL495⟩, 709 ⟨70JOC777⟩, 778 ⟨75TL3117⟩, 785 ⟨80TL1113⟩, 962 ⟨70JA395⟩, 970 ⟨70JA6314⟩, 970, 971 ⟨74JA5582⟩, 970, 971 ⟨75JOC2265⟩, 999 ⟨83TL3165⟩, 1000 ⟨65JA1353⟩, 1035 ⟨88TL6409⟩, 1045 ⟨70JOC3405⟩; **3**, 4, 115 ⟨79TL399⟩, 10, 18, 124, 133 ⟨76TL3⟩, 11 ⟨80JA1742, 80TL3535⟩, 18, 133 ⟨76TL3667⟩, 30, 33 ⟨91TL5713⟩, 54 ⟨71JA1724⟩, 57 ⟨68JA5616, 75TL2647⟩, 60 ⟨76TL7, 89TL5235⟩, 61 ⟨72TL1815, 78TL5⟩, 62 ⟨80JA1436⟩, 78 ⟨73TL4499⟩, 108 ⟨83TL3291⟩, 116 ⟨85TL5855⟩, 129 ⟨64JA1639⟩, 153 ⟨92TL3435⟩, 162 ⟨83TL3165⟩, 208, 212 ⟨90TL3857, 78TL2461⟩, 214 ⟨75TL925⟩, 215, 220 ⟨78JA6294⟩, 218 ⟨89TL7297, 78TL4597⟩, 222 ⟨87TL5241⟩, 226 ⟨76TL11⟩, 229 ⟨84JA3875⟩, 231 ⟨84JA3384⟩, 286 ⟨62JA4899⟩, 372, 378 ⟨65JA1353⟩, 373 ⟨67TL2325⟩, 494 ⟨70JA397⟩, 495 ⟨72TL3769⟩, 528 ⟨85JA4339⟩, 603 ⟨73TL1495⟩, 621 ⟨75JOC127⟩, 623, 640 ⟨72TL487⟩, 629 ⟨93TL4001⟩, 676 ⟨84JA462⟩, 696 ⟨58JOC1221, 79JA1608⟩, 696 ⟨79TL2745⟩, 701 ⟨87JA287⟩; **4**, 20 ⟨76TL4577⟩, 28 ⟨53JA3297⟩, 80, 83, 84 ⟨69JOC1233⟩, 94 ⟨82JA6816⟩, 171 ⟨80TL3463⟩, 181 ⟨73JOC834⟩, 197, 216 ⟨75TL3267⟩, 199 ⟨76TL809⟩, 217 ⟨84TL3⟩, 219 ⟨75TL3269⟩, 245, 250, 1075

⟨75JOC231⟩, 250 ⟨65AG1134⟩, 250, 391, 392, 957 ⟨66JOC4097⟩, 254 ⟨75TL925⟩, 321 ⟨78TL5⟩, 378, 384, 388 ⟨83TL3165⟩, 380, 389 ⟨83TL3163⟩, 391, 393 ⟨65JA1345⟩, 394 ⟨66JA5656, 68JA5548⟩, 576 ⟨68JA6816⟩, 649 ⟨76TL7, 78CB1362, 78JA2916⟩, 745 ⟨72TL3769⟩, 845 ⟨73JA5829⟩, 850 ⟨67TL3201⟩, 1074 ⟨72JA8932⟩; **5**, 51, 154 ⟨68JA5616⟩, 68, 97 ⟨73JOC4086⟩, 138 ⟨88JA3672⟩, 168 ⟨69JA5675⟩, 168 ⟨70JA397⟩, 171 ⟨70TL311, 74TL2439⟩, 173 ⟨75JA2287, 78JA4618, 79JA7131⟩, 173, 204 ⟨78JA4620⟩, 179 ⟨82JA6818⟩, 223, 374 ⟨72JA6190⟩, 235 ⟨73JA5829⟩, 273, 296 ⟨69JA1233⟩, 339 ⟨72JA7586⟩, 375 ⟨81TL3455⟩, 410 ⟨67JA434⟩, 636 ⟨84TL2639⟩, 652 ⟨57OR(9)388⟩; **6**, 25, 26, 31 ⟨92TL3435⟩, 42 ⟨92JA1906⟩, 71 ⟨83TL5571⟩, 127, 325 ⟨67JA434⟩, 540 ⟨82TL1979⟩, 555 ⟨65JA934⟩, 695 ⟨72TL3769⟩, 705 ⟨82JA4724, 85TL555⟩

Corey, E. R., **3**, 335, 369 ⟨82JA3119⟩; **4**, 258 ⟨82JA3119⟩

Corey, J. Y., **1**, 648 ⟨B-88MI 114-02⟩; **2**, 811 ⟨77JOM(137)56, 79JOM(8)1⟩

Corey, L. D., **1**, 667 ⟨87CJC1821⟩

Corey, R. M., **2**, 741 ⟨73JMC1113⟩

Corfield, P. W. R., **1**, 707 ⟨70JA2581⟩

Corich, M., **4**, 261 ⟨92TL3043⟩

Cork, D. G., **2**, 18, 19 ⟨88CL1877⟩; **5**, 1105 ⟨87JOC681⟩, 1107 ⟨87CC1301⟩

Corkins, H. G., **1**, 705 ⟨78JOC4140⟩; **2**, 778 ⟨80TL2025⟩; **4**, 86 ⟨78JOC4136, 78JOC4140⟩

Corless, P. F., **5**, 1134 ⟨84S1020⟩

Corley, E. G., **2**, 528 ⟨93JOC2791⟩

Corn, J. E., **4**, 30 ⟨73JOC2091⟩

Cornejo, J. J., **3**, 407 ⟨85JOC5382⟩

Cornelis, A., **1**, 600 ⟨78JA3874⟩; **2**, 792 ⟨88TL5657⟩; **3**, 118 ⟨80S849⟩; **4**, 190 ⟨82S162⟩, 431 ⟨85S909⟩

Cornelisse, J., **1**, 364, 365 ⟨93CRV615⟩, 420 ⟨87RTC613⟩; **3**, 667 ⟨70TL4701⟩

Cornell, K. H., **1**, 15 ⟨48JA1490⟩

Corner, J. A., **4**, 58 ⟨93JCS(P1)2303⟩

Cornet, X. B., **4**, 176 ⟨85CAR(144)342⟩

Cornforth, F. J., **1**, 142, 155 ⟨75JA7006⟩

Cornforth, J., **1**, 472 ⟨82JCS(P1)2299⟩; **2**, 843 ⟨82JCS(P1)2289⟩, 844 ⟨82JCS(P1)2299⟩; **3**, 75 ⟨90JCS(P1)1459⟩

Cornforth, J. W., **1**, 458 ⟨69NAT1212⟩; **2**, 631 ⟨42JCS684⟩; **4**, 190 ⟨69JCS(C)1202⟩, 211 ⟨57JCS158⟩

Cornforth, R. H., **2**, 843 ⟨82JCS(P1)2289⟩; **4**, 211 ⟨57JCS158⟩

Cornia, M., **3**, 97 ⟨78JCS(P1)318⟩

Cornils, B., **1**, 202 ⟨B-80MI 105-02⟩; **5**, 24 ⟨B-75MI 502-01⟩

Cornish, C. A., **6**, 121 ⟨85JCS(P1)2585⟩

Cornu, P.-J., **1**, 82 ⟨63BSF798⟩

Coronas, J. M., **5**, 1137 ⟨92ICA173⟩, 1138 ⟨76SRI217⟩, 1139 ⟨77JOM(128)291⟩

Corral, C., **5**, 652 ⟨74T195⟩

Corral, R. A., **4**, 417 ⟨64TL1693⟩, 428 ⟨75JCS(P1)772⟩; **5**, 1126 ⟨85JHC1025⟩; **6**, 221 ⟨52MI 607-02⟩

Corral, S., **6**, 556 ⟨84S172⟩

Correa, A., **3**, 40 ⟨91SC1⟩

Correa, P. E., **2**, 348 ⟨85JOC1563, 88JOC1695⟩

Correa, W. S., **2**, 216 ⟨69BBR(37)593⟩

Corre, E., **3**, 568 ⟨72T4431⟩; **4**, 800 ⟨72T4039⟩, 996 ⟨72T5055⟩

Corre, F., **3**, 61 ⟨81TL625⟩

Corre, G. Le, **6**, 477 ⟨78T3105⟩

Corre, M., **3**, 494 ⟨82S597⟩

Corre, M. L., **2**, 831 ⟨81T2855⟩

Corre, M. Le, **1**, 158 ⟨93TL1011⟩, 640 ⟨89CC1464⟩; **2**, 14 ⟨89CC313⟩; **5**, 694 ⟨89TL3069⟩

Correia, J., **3**, 668 ⟨94S1127⟩

Correia, Y., **2**, 788 ⟨69BRP1167809⟩

Corrie, J. E. T., **5**, 326 ⟨85JCS(P1)883⟩

Corriu, R., **1**, 651 ⟨84JOM(264)207⟩, 759 ⟨70JA7424⟩; **4**, 648, 649 ⟨73JOM(57)C5⟩, 654 ⟨70JA7424⟩

Corriu, R. J. P., **1**, 160 ⟨93JOM(451)C1⟩, 160 ⟨93TL2111⟩, 490 ⟨93CRV1371⟩; **2**, 68 ⟨89T171⟩, 77 ⟨87JOM(328)C17⟩, 264 ⟨80BSB783⟩, 544 ⟨90JOC2878⟩, 546 ⟨80CC168⟩, 547 ⟨93CRV1371, B-91MI 210-03⟩, 600 ⟨80JOM(197)C19⟩, 814 ⟨81JOC3372, 83JOM(255)359⟩, 814 ⟨85OM623⟩, 935 ⟨81TL2985⟩, 946 ⟨72JOM(40)73⟩, 949 ⟨90CRV265⟩; **3**, 14, 82, 95, 102, 104, 107 ⟨87TL3941⟩, 83, 98, 102, 104, 107 ⟨88TL1271⟩, 211 ⟨81TL2985⟩; **4**, 511 ⟨83JOM(259)283⟩, 606 ⟨93JOM(451)C1⟩, 607, 608, 654 ⟨93TL2111⟩, 648 ⟨77JOM(127)281, 78JOM(144)155⟩, 655 ⟨81TL2985⟩; **5**, 410 ⟨70JOM(22)321⟩, 421 ⟨77CC73, 81JOM(205)31⟩, 759, 777 ⟨91S954⟩; **6**, 402, 404 ⟨81JOM(221)257⟩, 576, 577 ⟨91JOM(419)9⟩

Corsaro, A., **3**, 549 ⟨84JCR(S)370⟩; **5**, 734 ⟨82SC865, 84JCR(S)404⟩, 734 ⟨84OMR724⟩, 734, 760 ⟨84JCS(P1)897⟩

Corse, J., **2**, 642, 643 ⟨40JA3250⟩; **6**, 74 ⟨58JA1247⟩

Corse, J. W., **5**, 584, 586 ⟨48JA2837⟩

Corset, J., **2**, 1004 ⟨86PAC1133⟩

Corsico-Coda, A., **2**, 793 ⟨83T689⟩

Corson, B. B., **3**, 618 ⟨44OSC(2)379⟩; **4**, 212 ⟨59JOC1768⟩; **5**, 63 ⟨32OSC(1)336⟩

Corson, F. P., **2**, 607, 624 ⟨72T4883⟩

Cort, A. D., **3**, 155 ⟨91JOC6708⟩

Cortés, D. A., **1**, 865 ⟨82TL795⟩; **3**, 7 ⟨83JA4492⟩, 88 ⟨84JA3374⟩

Cortes, D., **3**, 291 ⟨83OM467⟩

Cortese, N. A., **1**, 79, 85 ⟨78JOC3985⟩; **3**, 12, 127, 621 ⟨78JOC3985⟩, 530 ⟨82JOC3871⟩; **5**, 157 ⟨77JOC3903, 78JOC2952⟩, 282 ⟨82JOC3876, 82JOC3881⟩

Cortez, C., **2**, 659 ⟨79JOC4265⟩

Cortez, R., **6**, 416 ⟨91SC285⟩

Corwin, A. H., **2**, 32 ⟨56JA3135⟩

Cory, R. M., **4**, 99 ⟨73JA7813⟩

Cosack, S., **5**, 365 ⟨88CB833⟩

Cose, R. W. C., **1**, 407, 408 ⟨69CC293⟩

Cosgrove, R. M., **5**, 657 ⟨84JOC546⟩

Coskran, K. J., **4**, 342 ⟨66JA1140⟩; **5**, 1129 ⟨71IC1536⟩

Cosmatos, A., **2**, 274 ⟨65JA4922⟩, 412 ⟨61CB2644⟩

Cosquer, P., **4**, 566 ⟨87PS(30)523⟩

Cossar, B. C., **2**, 121 ⟨62JOC93⟩

Cossement, E., **5**, 233 ⟨80TL3085⟩

Cossement, M., **4**, 842 ⟨83TL2563⟩

Cosset, C., **5**, 941 ⟨91CC881, 95OM1938⟩

Cossio, F. P., **2**, 10 ⟨86JOC4941⟩; **3**, 4, 116 ⟨87T3963⟩, 88 ⟨86CJC225⟩; **5**, 294 ⟨87TL1945⟩

Cossío, F. P., **2**, 398 ⟨90JOC2070⟩; **5**, 17 ⟨86JOC4941⟩

Cossu, S., **4**, 179 ⟨90S313⟩, 770, 779 ⟨90G569⟩; **6**, 556 ⟨91PS(63)51⟩

Cossy, J., **1**, 195, 548 ⟨87TL4547⟩, 287, 362 ⟨90T1859⟩, 1000 ⟨86TL573⟩; **2**, 57 ⟨92TL5045⟩; **3**, 38, 173 ⟨93TL7755⟩, 72 ⟨85TL839⟩, 132 ⟨87S1113⟩; **4**, 274 ⟨88BSF989⟩

Costa, A., **3**, 427 ⟨84TL3493⟩

Costa, A. L., **3**, 31 ⟨93JA7001⟩

Costa, B. de, **5**, 1029 ⟨92JCS(P1)1671⟩

Costa, D. J., **2**, 10 ⟨74T3793⟩

Costa, E., **4**, 151 ⟨92OM3879⟩

Costa, G., **2**, 583 ⟨86ICA(121)27⟩; **5**, 457 ⟨67TL1783⟩

Costa, L., **6**, 50 ⟨74JHC943⟩

Costa, M., **1**, 496 ⟨86JOM(317)373⟩

Costa, T., **5**, 910, 911 ⟨81CB1428⟩

Costanzo, M. J., **6**, 635 ⟨88S460⟩

Costanzo, S. J., **1**, 127, 128, 466 ⟨73JOC3189⟩

Coste, J., **1**, 905 ⟨78JCR(S)62⟩

Costello, G., **5**, 749 ⟨86HCA1224⟩

Costero, A. M., **4**, 749 ⟨85CC296, 86JCR(M)2843⟩; **6**, 695 ⟨85CC296⟩

Costisella, B., **1**, 821 ⟨78JOC2518⟩; **4**, 553 ⟨82EGP154700, 82JPR537⟩, 553 ⟨84EGP206557⟩, 559 ⟨76JPR403⟩, 804 ⟨83T1189⟩, 1002 ⟨81T1227⟩, 1003 ⟨76JPR116⟩, 1004 ⟨82T139, 84T733⟩; **5**, 42 ⟨68AG(E)391⟩, 788

⟨76JPR272⟩; **6**, 43 ⟨67LA(707)35, 68LA(720)146⟩, 48 ⟨77JPR17⟩, 116 ⟨74JPR913⟩, 119 ⟨76TL477, 78JOC2500⟩, 119, 314 ⟨87S44⟩, 120 ⟨78JPR255⟩, 155 ⟨90PS(47)7, 91PS(62)35⟩, 156 ⟨69JPR925⟩, 156, 159 ⟨71LA(750)44⟩, 157 ⟨68AG(E)391, 68AG(E)463, 69JPR577⟩, 159 ⟨72JPR969, 74PS(4)241, 76JPR272⟩, 286 ⟨76JPR116⟩, 367 ⟨72JPR969, 86JPR231⟩, 367, 665 ⟨72JPR87⟩, 661, 733 ⟨78JPR389, 72EGP93766⟩
Cota, D. J., **2**, 308 ⟨69OR(17)213⟩; **5**, 282 ⟨69OR(17)213⟩
Cote, R., **5**, 316 ⟨78JOC3750⟩
Cotgreave, I. A., **2**, 237, 238 ⟨92JA9737⟩
Cotrel, C., **3**, 357 ⟨81TL4507⟩
Cottam, P. D., **2**, 184 ⟨81JCS(P1)2106⟩, 185 ⟨79TL4977⟩
Cott, W. J., **1**, 107 ⟨56JA3484⟩
Cotte, P., **4**, 195 ⟨82JOC2216⟩
Cottens, S., **2**, 612 ⟨84HCA284⟩
Cotter, J. L., **5**, 1013 ⟨66CC336⟩
Cotter, R. J., **1**, 104 ⟨63OSC(4)377⟩; **6**, 421, 426, 460 ⟨64CRV645⟩, 425 ⟨65CI(L)791⟩
Cottet, F., **4**, 170 ⟨87S497⟩
Cottier, L., **4**, 170 ⟨87S497⟩, 1079 ⟨78NJC79, 81T1875⟩, 1079 ⟨81T2515⟩
Cottingham, A. B., **1**, 886 ⟨77JA1881⟩
Cotton, F. A., **1**, 862 ⟨59MI 119-04⟩; **2**, 421 ⟨76JA4477⟩, 468 ⟨71IC2362⟩, 1004 ⟨68IC18⟩; **4**, 576, 577 ⟨63JA2394⟩, 673 ⟨71JA5441, 78IC2093, 78JA802⟩, 674 ⟨78JA5764⟩, 675 ⟨83OM1167⟩, 681 ⟨76JA1774⟩, 683 ⟨69JA6598⟩, 692 ⟨88OM2279, 93JA4145⟩; **5**, 368 ⟨B-80MI 507-01⟩, 943 ⟨71JA2672⟩
Cotton, J. D., **5**, 445 ⟨81ICA165⟩
Cottrell, C. E., **1**, 576 ⟨86JA3739⟩, 613 ⟨87JA3730⟩, 813 ⟨88JA890⟩, 815 ⟨87JA5731⟩; **3**, 217 ⟨88JA890⟩
Cottrell, H. J., **5**, 858 ⟨54JCS2968⟩
Cottrell, I. F., **2**, 906 ⟨88S950⟩; **4**, 509 ⟨91JCS(P1)1091⟩; **5**, 8 ⟨89S598⟩
Couch, D. A., **4**, 52, 59 ⟨72JOC3332⟩; **5**, 940 ⟨77IC677⟩
Coucouvanis, D., **6**, 552 ⟨70JA3342, 70MI 617-01⟩
Coudane, J., **1**, 388 ⟨69CR(C)567⟩
Coudert, G., **1**, 829 ⟨91TL5525⟩; **3**, 90, 97 ⟨93T1421⟩; **4**, 199 ⟨81TL1973⟩
Couffignal, R., **1**, 971 ⟨69BSF898, 70BSF3157⟩; **3**, 131 ⟨78TL3713⟩, 633 ⟨79TL1585⟩; **5**, 223 ⟨85JOM(294)139⟩
Coughlin, A. M., **1**, 570 ⟨92JOC2856⟩
Couladouros, E. A., **5**, 513 ⟨90JA3040⟩
Coulsen, D. R., **6**, 697 ⟨64TL3323⟩
Coulson, C. A., **1**, 911 ⟨B-61MI 119-01⟩
Coulson, D. R., **1**, 214 ⟨74JA5268⟩, 495 ⟨73JOC1483⟩
Coulter, D. E., **1**, 777 ⟨81JOC2804⟩
Coulter, J. M., **1**, 443 ⟨68T4489⟩
Coulter, P. B., **3**, 404 ⟨85JCS(P1)2123⟩
Counsell, R. E., **6**, 606 ⟨87JMC1241⟩
Count, D. J. Le, **5**, 1026 ⟨77S582⟩, 1120 ⟨83JCS(P1)813⟩
Counter, J. A., **1**, 685, 686 ⟨93JOC5900⟩
Courault, K., **3**, 411 ⟨80JPC336⟩, 587 ⟨80ZC211⟩; **5**, 1068 ⟨79ZC59⟩
Courchesne, M., **3**, 89, 96 ⟨93TL3519⟩
Couret, C., **1**, 766 ⟨91OM1771⟩; **2**, 473, 474 ⟨82PS(12)265⟩, 84PS(20)81, 82PS(12)279⟩, 866 ⟨83TL3625⟩, 949 ⟨87JA4411⟩, 1093 ⟨72JOM(34)83⟩; **3**, 501 ⟨94CCR427⟩, 503 ⟨87JA4411, 91JOM(403)93, 92OM3176⟩; **4**, 338 ⟨72IC2274, 73JOM(47)67⟩, 455, 481, 998 ⟨75HCA1316⟩, 485 ⟨91JOM(415)327⟩, 549 ⟨93CC569⟩, 574 ⟨70CR(C)351, 70JOM(24)633⟩, 647 ⟨92JOM(440)233⟩, 662 ⟨87JA4411⟩, 663 ⟨92OM3176⟩; **5**, 720 ⟨75HCA1316⟩; **6**, 131 ⟨77JOM(141)35⟩, 167 ⟨87PS(31)81⟩, 186, 188 ⟨91POL1153⟩, 265, 266 ⟨78BSF(2)361⟩, 373 ⟨82TL4941, 84CC1622⟩, 373, 688 ⟨89CC593⟩, 374, 375 ⟨83TL2769, 83TL3625⟩, 375 ⟨84JAP59193897⟩, 582 ⟨79TL3507⟩, 669 ⟨72JOM(34)83⟩, 709 ⟨94CCR427⟩, 724 ⟨92OM2748⟩
Couret, F., **4**, 338 ⟨72IC2274, 73JOM(47)67⟩, 455, 481, 998 ⟨75HCA1316⟩; **5**, 720 ⟨75HCA1316⟩
Court, A. S., **2**, 709 ⟨72JOC939⟩; **4**, 574 ⟨72JOC939⟩
Court, J. J., **5**, 331 ⟨89TL1773⟩
Courtemanche, G., **2**, 478 ⟨91TL5317⟩, 592 ⟨93SL266⟩, 593 ⟨93TL6053⟩
Courtis, B., **3**, 83 ⟨75JCS(D)2460⟩
Courtneidge, J. L., **1**, 68 ⟨82JCS(P1)1797⟩; **2**, 95 ⟨94TL1003⟩, 98 ⟨82JCS(P1)1807⟩
Courtois, C., **4**, 1065 ⟨77JOM(140)237⟩
Courtois, G., **1**, 543 ⟨75CR(C)999⟩, 544 ⟨78CR(C)265⟩, 973 ⟨81JOM(218)1⟩, 973, 1059 ⟨80JOM(198)1⟩, 974 ⟨86BSF449⟩, 1041 ⟨83BSF(2)21, 81JOM(218)275⟩, 1059 ⟨87BSF93⟩; **2**, 300 ⟨87TL1659⟩, 931 ⟨82BSF395⟩
Courtois, J. E., **5**, 28 ⟨54MI 502-01, 60AQ93⟩
Courtot, C., **3**, 614 ⟨15AC(R)28, 20AC(R)364⟩, 675 ⟨15BSF228⟩
Courtot, P., **1**, 420 ⟨68BSF3401, 69BSF3665⟩, 795 ⟨72BSF4238⟩
Cousineau, T. J., **2**, 675 ⟨79SC157⟩
Cousins, S., **5**, 1110 ⟨80T1839⟩
Cousins, S. J., **6**, 628, 633 ⟨85TL1105⟩
Cousse, H., **4**, 141 ⟨72BSF1361⟩
Cousseau, J., **2**, 19 ⟨89JOC5380⟩, 606, 616 ⟨77JCS(P1)1797⟩, 611 ⟨86BSF910⟩, 623 ⟨80S805⟩; **4**, 768 ⟨89MI 417-01⟩; **5**, 1108 ⟨79CC1026⟩
Coussemant, F., **5**, 710 ⟨66BSF882⟩
Coustard, J. M., **3**, 429 ⟨89TL5763⟩
Coustard, J. M.,**3**, 429 ⟨91BSF730⟩
Coustumer, G. L., **5**, 559 ⟨72CR(C)1215⟩
Coustumer, G. Le, **5**, 584, 586, 614 ⟨73BSF3349⟩
Coutrot, P., **1**, 53 ⟨91COS(8)858⟩, 65, 68, 69 ⟨91COS(8)850⟩, 437, 444, 445, 454 ⟨91COS(8)835⟩; **2**, 618 ⟨75S658⟩; **3**, 197 ⟨78S36⟩, 198 ⟨86S661⟩; **4**, 90, 784 ⟨87S169⟩, 347, 773, 778 ⟨76S107⟩, 473 ⟨89H(28)1179⟩, 577 ⟨86JOM(316)13⟩, 745 ⟨76S197⟩; **5**, 103 ⟨78SC19⟩, 692 ⟨74BSF1965⟩; **6**, 136 ⟨87S169⟩, 266 ⟨75S535⟩
Coutrot, Ph., **4**, 740 ⟨75TL609⟩, 740, 741 ⟨75S535⟩; **5**, 233 ⟨91S23⟩
Coutts, A., **2**, 127 ⟨92JCS(P1)255⟩
Coutts, I. G. C., **2**, 365 ⟨75JCS(P1)2445⟩; **5**, 81, 107 ⟨81S487⟩
Coutts, R. T., **2**, 390 ⟨76T1267⟩, 765 ⟨67JA1537, 72JOC3584⟩; **3**, 310 ⟨72JOC3584⟩
Coutts, S. J., **1**, 816 ⟨90TL4301⟩
Couture, A., **3**, 269 ⟨86CC839⟩, 363 ⟨76JA6218⟩; **4**, 487 ⟨92SC2381⟩; **5**, 516, 522 ⟨87S363⟩, 572, 618 ⟨84JOC714⟩; **6**, 122 ⟨92SC2381⟩
Couture, C., **3**, 661 ⟨85CJC111⟩
Couturier, D., **2**, 28 ⟨81SC669⟩, 742 ⟨87S727⟩; **3**, 659 ⟨85SC473⟩; **5**, 227 ⟨89TL3073⟩
Couty, F., **4**, 202 ⟨93SL349⟩
Couvillio, J. L., **2**, 420 ⟨67JOC510⟩
Covell, A. N., **1**, 1001 ⟨63JCS5889, 63JCS775⟩, 1001 ⟨65JCS894⟩
Coveney, D. J., **1**, 388 ⟨91COS(3)777⟩, 933 ⟨87TL5949⟩; **3**, 136 ⟨90JCS(P1)2721⟩; **5**, 457 ⟨90JCS(P1)2721⟩
Coville, N. J., **4**, 712 ⟨92JOM(427)335⟩
Cowan, D., **4**, 850 ⟨83JOC4713⟩, 876 ⟨87SM647⟩; **5**, 112 ⟨73OSC(5)890⟩
Cowan, D. O., **2**, 67 ⟨62JOC1⟩; **4**, 869 ⟨77CC505⟩, 874 ⟨80CC867⟩, 876 ⟨82CC335, 84JA8303⟩, 876 ⟨88SMB425⟩, 876, 877 ⟨87JA4115⟩; **6**, 552 ⟨77CC660⟩, 591 ⟨75JOC746, 77CC505⟩, 591, 594 ⟨80CC866B⟩, 592 ⟨83CC294⟩
Cowan, J. C., **2**, 778 ⟨65MI 215-02⟩; **5**, 55 ⟨40JA3495⟩
Cowan, R. L., **6**, 516 ⟨87OM2451⟩
Cowap, M. D., **6**, 522, 524 ⟨10JCS798⟩
Coward, J. K., **1**, 1074 ⟨90JOC3147⟩; **3**, 197 ⟨83JOC5001⟩
Cowell, A., **5**, 143 ⟨79TL133⟩, 143, 170 ⟨80JA4193⟩
Cowell, A. B., **6**, 11 ⟨81JFC(17)345⟩
Cowen, K. A., **2**, 310 ⟨90JOC2552⟩; **3**, 44, 188 ⟨90JOC2552⟩
Cowie, D. W., **6**, 90 ⟨33JCS306⟩
Cowie, J. S., **1**, 961 ⟨73JCS(P1)720⟩

Cowie, M., **2**, 1003 ⟨81CJC1463⟩; **4**, 683 ⟨92OM3736⟩; **6**, 332 ⟨87OM283⟩
Cowlagi, B. S., **2**, 753, 760 ⟨76IJC(B)904⟩
Cowle, M., **6**, 332 ⟨83OM1696⟩
Cowles, J. M., **2**, 440, 441 ⟨92JOC511⟩
Cowley, A. A., **2**, 821 ⟨82JA5820⟩
Cowley, A. H., **2**, 430 ⟨83IC547⟩, 434 ⟨69JA2915⟩, 461, 508 ⟨69JA6609⟩, 475 ⟨90AG(E)1150⟩, 476 ⟨73CC813⟩, 488 ⟨71JA2150⟩, 507 ⟨66JA3178⟩, 511 ⟨86CC1543, 88JOM(341)C1⟩, 511 ⟨88POL77, 89AG(E)1208⟩, 511 ⟨90CM221, 91OM1635⟩, 866 ⟨83CC881⟩, 866, 893 ⟨84IC2582, 90AX(C)774⟩; **4**, 114, 802 ⟨69JA1929⟩, 480 ⟨88CC867⟩, 549 ⟨88AG(E)837⟩, 564 ⟨79IC60⟩, 581 ⟨90OM1036, 69JA1934⟩; **5**, 396 ⟨87CC1753⟩, 878, 1155 ⟨87CC980⟩, 884 ⟨88CC171⟩, 889 ⟨88CC753⟩, 890 ⟨88AG(E)838⟩; **6**, 165, 167 ⟨82OM1720⟩, 166 ⟨87CC1092⟩, 167 ⟨83CC528, 84POL389, 84TL3519⟩, 168 ⟨83JA5506, 84IC2582⟩, 169 ⟨83CC881⟩, 172 ⟨86MI 606-01⟩, 173, 174, 176 ⟨90IS239⟩, 191 ⟨81SRI591⟩, 372 ⟨90IS235⟩, 373 ⟨82JA5820, 86JCS(D)1801, 89OM2490⟩, 375 ⟨85JA8211, 85JCS(D)383, 87JCS(D)249⟩, 689 ⟨84JA7015, 85CC1687⟩, 689 ⟨86OM593⟩, 704 ⟨94AG(E)578⟩
Cowling, A. P., **3**, 230 ⟨78CC1006⟩
Cowling, M. P., **2**, 345, 347 ⟨88CC1503⟩
Cox, B. G., **1**, 869, 871 ⟨70JCS(B)194⟩
Cox, D., **3**, 297 ⟨85JMC1832⟩
Cox, D. E., **4**, 668 ⟨90JA2425⟩
Cox, D. G., **1**, 446 ⟨83JA650, 85JA2811⟩; **2**, 444 ⟨88JOC366⟩, 611 ⟨85JA2811⟩, 612 ⟨83JA650⟩; **4**, 117, 118 ⟨88JOC366⟩, 801 ⟨83JA650, 85JA2811⟩; **6**, 265 ⟨88JOC366⟩
Cox, D. P., **2**, 19 ⟨84JOC3216⟩; **4**, 415 ⟨84TL1023⟩, 1081 ⟨85TL1931⟩; **5**, 4 ⟨84JOC3216⟩; **6**, 280 ⟨83JA6513⟩
Cox, G. G., **6**, 119 ⟨92SL975⟩, 628, 636 ⟨94T3195⟩
Cox, J. M., **4**, 226 ⟨67JCS(C)1130⟩
Cox, Jr., J. R., **2**, 694 ⟨58JA5541⟩
Cox, M. E., **2**, 223 ⟨64JCS3832⟩; **4**, 218 ⟨64JCS3832⟩
Cox, M. V.,**5**, 728 ⟨26JA3214⟩
Cox, N. J. G., **1**, 340 ⟨92JCS(P1)1313⟩
Cox, P. B., **2**, 628 ⟨88CC971⟩
Cox, P. J., **1**, 135, 842 ⟨91TA1⟩, 144 ⟨92SL197⟩, 144 ⟨93JCS(P1)3099⟩; **2**, 420 ⟨91TA1⟩; **3**, 124 ⟨91TA1⟩
Cox, R. A., **1**, 871 ⟨68JOC1301⟩, 891 ⟨B-90MI 119-05⟩, 919 ⟨B-89MI 119-06⟩
Cox, S. F., **4**, 405 ⟨49JCS2342⟩
Coxon, J. M., **1**, 185 ⟨90JOC1338⟩, 387 ⟨69T1479, 69T4445⟩, 631 ⟨69T3107⟩; **4**, 182 ⟨70T1311⟩, 182 ⟨74AJC679⟩, 411 ⟨90JOC4777⟩
Coy, D. H., **3**, 600 ⟨70CC456, 73JCS(P1)1561⟩; **4**, 794 ⟨73JCS(P1)1066⟩
Coyle, C., **5**, 8 ⟨88OSC(6)190⟩
Coyle, J. D., **1**, 516 ⟨B-78MI 112-04⟩; **2**, 232 ⟨80S910⟩; **3**, 352 ⟨85T5393⟩; **5**, 24 ⟨78CRV97⟩, 609 ⟨85T5393⟩
Coyle, J. J., **3**, 478 ⟨65JA4270⟩; **5**, 844 ⟨65JA4270⟩, 844, 866 ⟨62JA4350⟩
Coyle, T. D., **1**, 658 ⟨68IC1014⟩; **4**, 632, 644 ⟨68JOM(12)269⟩
Coyne, W. E., **6**, 501 ⟨67JMC541⟩
Coyner, E. C., **3**, 612 ⟨41JA2796⟩
Cozens, A. J., **3**, 433 ⟨83MI 311-01⟩
Cozzi, F., **2**, 362 ⟨77CC723⟩; **3**, 37 ⟨82S805⟩, 431 ⟨89G253⟩, 437 ⟨82JCS(P1)339⟩, 634 ⟨81JCS(P1)614⟩; **4**, 272 ⟨79S535⟩; **5**, 233 ⟨91T8767⟩, 237 ⟨92JOC6339⟩
Cozzi, P. G., **4**, 834 ⟨88JOC4015⟩; **5**, 233 ⟨91T8767⟩
Crab, T. A., **4**, 165 ⟨80JCS(P1)2535⟩
Crabbé, P., **1**, 2, 4, 6 ⟨71LA(748)123⟩, 960 ⟨76CC183⟩, 966 ⟨68JA4733, 75TL4615⟩, 995 ⟨79CC859⟩; **3**, 55 ⟨73CC119⟩, 235 ⟨73JA6655⟩; **4**, 8 ⟨73TL1319⟩
Crabtree, A., **1**, 537 ⟨65JCS4355⟩
Crabtree, H. E., **3**, 542 ⟨68JCS(C)2730⟩
Crabtree, R. H., **1**, 456, 507 ⟨77JOM(141)113⟩, 554 ⟨87JA8025⟩; **2**, 192 ⟨91JOC5503⟩, 915 ⟨90CC1056⟩, 962 ⟨85CRV245⟩; **6**, 369 ⟨90JA7984⟩
Crackett, P. H., **4**, 274 ⟨90JCS(P1)773⟩
Cradock, S., **4**, 360 ⟨67IC1751⟩; **5**, 537 ⟨80M(111)193⟩, 538, 541 ⟨76JCS(D)1661⟩, 539, 541 ⟨72JCS(D)359⟩; **6**, 173 ⟨67IC1751⟩
Cradock, St., **2**, 488, 504 ⟨71ZAAC(381)103⟩
Crafts, J. M., **1**, 462 ⟨1884AC(P)449⟩
Cragg, G. M. L., **2**, 514 ⟨B-73MI 210-01⟩, 900 ⟨B-73MI 218-01⟩
Cragg, R. H., **2**, 214, 215 ⟨73TL655⟩, 260 ⟨70CC425⟩, 267 ⟨68JCS(A)2962, 69CC832⟩, 358 ⟨72JOM(44)37⟩, 1086 ⟨67CB2377⟩; **5**, 1038 ⟨64JCS2108⟩
Cragoe, Jr., E. J., **2**, 491, 496, 498 ⟨47JA927⟩; **5**, 60 ⟨73OSC(5)635⟩, 221 ⟨47JA925⟩; **6**, 506 ⟨68USP3575975⟩
Crahmer, H., **1**, 57 ⟨58CB750⟩
Craig, A. D., **3**, 711 ⟨62JCS548⟩
Craig, A. P., **1**, 595 ⟨67JOC102⟩
Craig, C. A., **3**, 617 ⟨59JA6340⟩
Craig, D., **2**, 1000 ⟨83JHC399⟩; **3**, 36 ⟨91TL6973⟩; **4**, 207 ⟨91TL6973⟩; **5**, 223 ⟨92SL41⟩
Craig, D. C., **1**, 685, 686 ⟨93JOC5900⟩; **2**, 432 ⟨87JOM(323)C1⟩; **4**, 409 ⟨89JOC4771⟩
Craig, J. C., **1**, 1070 ⟨63JCS3712⟩, 1081 ⟨69AG(E)429⟩; **3**, 14 ⟨81S303⟩
Craig, J. W., **2**, 381 ⟨73JMC736⟩
Craig, P. N., **6**, 232 ⟨60JOC60⟩
Craig, R. A., **1**, 896 ⟨50JA163⟩
Craig, R. C., **6**, 578 ⟨80JCS(P1)766⟩
Craig, S. L., **2**, 4, 23, 30 ⟨90JA7431⟩, 4, 23, 30 ⟨93JA3071⟩
Craig, T. A., **3**, 221, 227, 275 ⟨84JA5585⟩
Craig, W. E., **3**, 68 ⟨53OS(33)15⟩; **5**, 1043 ⟨56JA4965⟩
Craig, W. G., **2**, 845 ⟨55JA1864⟩
Crain, D. L., **4**, 743 ⟨72JOC683⟩
Craine, L., **2**, 357, 358, 361, 767 ⟨89CRV689, 85JOC2205⟩; **3**, 436, 437 ⟨89CR689⟩
Craine, L. H., **2**, 767 ⟨84TL1337⟩
Cram, D., **2**, 187 ⟨67JA4099⟩
Cram, D. J., **1**, 19, 43, 624, 782 ⟨62JA1734⟩, 35, 39 ⟨63JA1108⟩, 384 ⟨54JA2753⟩, 473 ⟨87CL19⟩, 503, 505 ⟨56JA2518⟩, 595, 779 ⟨60JA1810⟩, 624 ⟨52JA2137, 54JA5740⟩, 627, 628 ⟨56JA790⟩, 775 ⟨49JA3883⟩, 795 ⟨B-65MI 118-01⟩, 797 ⟨64JA5466, 69JA4459⟩, 844 ⟨73MI 119-18, B-65MI 119-01⟩, 845, 869, 871 ⟨68MI 119-06⟩, 848 ⟨68JA1791⟩, 854 ⟨61JA3688, 64JA5445, 71JA2225, 71JA2231⟩, 854, 855 ⟨63JA3890, B-65MI 119-02⟩, 854, 858, 871, 881 ⟨64JA5457⟩, 858 ⟨68JA2598, 68JA2606, 68JA2612, 71JA2235⟩, 869 ⟨B-65MI 119-04⟩, 871, 881 ⟨61JA3678, 65OS(45)89⟩, 883 ⟨59JA5740, 69JA1009⟩, 886 ⟨B-65MI 119-05⟩, 915 ⟨B-72MI 119-12⟩, 1088 ⟨B-65MI 122-01⟩; **2**, 362, 846 ⟨68JA3869⟩, 843, 845 ⟨79JOC2226⟩; **3**, 234 ⟨47OS(27)9⟩; **4**, 78 ⟨67JA4099⟩
Cram, J. M., **1**, 844 ⟨73MI 119-18⟩
Cramer, C. J., **2**, 356, 784 ⟨86JA1306⟩
Cramer, F., **2**, 106 ⟨58CB1181⟩, 106, 107 ⟨62CB1664⟩
Cramer, R., **1**, 797 ⟨68ACR186⟩; **2**, 401 ⟨61JOC4010⟩; **4**, 765 ⟨40JA1849⟩
Cramer, R. D., **4**, 5 ⟨49JA490⟩
Cramer, R. E., **4**, 581 ⟨86OM2555⟩, 589 ⟨78JA5562, 81IC2466, 84JA5920, 88OM1465, 89OM1192, 90IS177⟩; **5**, 920 ⟨81IC2466⟩, 1096 ⟨80IC2564⟩
Crammer, B., **1**, 801, 802, 803 ⟨88CSR229⟩
Crandall, J., **1**, 955 ⟨93JOC5702⟩
Crandall, J. K., **1**, 248 ⟨86TL2243⟩, 316 ⟨72JOC4236⟩, 495 ⟨92TL5311⟩, 497 ⟨84JOC2125, 91TL3659⟩, 516 ⟨76JOC4089, 79JOC3471⟩, 547 ⟨84JOC4244⟩, 983 ⟨87JA4338, 92JA5998⟩, 1131 ⟨69TL1653⟩; **2**, 85 ⟨83OR(29)345⟩, 389, 390 ⟨92JOC6759⟩, 420 ⟨88OSC(6)948⟩, 716 ⟨85JOC1327⟩, 977 ⟨75JA7171⟩; **3**, 233 ⟨76JOC678⟩, 557 ⟨74JOC489⟩
Crane, P. T., **4**, 742 ⟨74JOC501⟩
Crank, G., **2**, 213 ⟨84AJC2331⟩, 772 ⟨84AJC845⟩; **3**, 541 ⟨70CC206⟩; **5**, 620 ⟨85AJC447⟩, 1122 ⟨84CC53⟩; **6**, 104 ⟨65AJC1967⟩

Crans, D. C., **1**, 149 ⟨92JA8483⟩
Crary, J. W., **3**, 449 ⟨56JA5584⟩
Crass, G., **2**, 89 ⟨75AG(E)762⟩
Crassous, J., **2**, 697 ⟨68BSF4265⟩
Crasto, C., **3**, 618 ⟨93TL1453⟩
Cravador, A., **2**, 278, 294 ⟨85T4793⟩, 281 ⟨81TL2491⟩; **3**, 16, 132 ⟨79S877⟩, 38, 175 ⟨80CC951⟩; **4**, 287, 288 ⟨85T4793, 79S877⟩; **6**, 94 ⟨79JOM(177)1⟩
Craveiro, A. A., **3**, 224 ⟨78JA1267⟩
Cravo, D., **4**, 238 ⟨93SL522⟩
Crawford, G. M., **4**, 685, 722 ⟨76IC2671⟩
Crawford, R., **5**, 205 ⟨56JA3819⟩
Crawford, R. J., **2**, 21 ⟨83JOC1364⟩, 376, 403 ⟨70CJC2745, 72JA7406⟩; **5**, 845 ⟨65CJC126⟩; **6**, 625 ⟨63PCS370⟩
Crawford, S. D., **2**, 4, 23, 30 ⟨90JA7433⟩
Crawford, S. S., **4**, 720 ⟨75JOM(91)C57, 77IC3193⟩
Crawhall, J. C., **2**, 214 ⟨56JCS4066⟩
Crawley, L. C., **2**, 420 ⟨88OSC(6)948⟩; **3**, 557 ⟨74JOC489⟩
Creary, X., **1**, 111, 971, 1056, 1061 ⟨77JA7632⟩, 500 ⟨80JA1611⟩, 905 ⟨83JA7123⟩, 922 ⟨88TL6839⟩; **2**, 28 ⟨87JA2435⟩, 850 ⟨74JOC3612⟩; **3**, 150 ⟨87JOC5026⟩, 465 ⟨86OS(64)207⟩; **5**, 765, 766 ⟨93JA7584⟩
Crease, A. E., **4**, 799 ⟨88JCS(D)2781⟩
Creaser, I., **5**, 386 ⟨75CC974, 81JA3559⟩
Creasy, W. S., **1**, 268 ⟨71JOC2379⟩
Credali, L., **3**, 382 ⟨65JCS807, 65JPS(B)581⟩, 382 ⟨66JPS(A)248, 67JCS(B)117⟩
Creger, P. L., **1**, 881 ⟨67JA2500, 70JA1396⟩, 881 ⟨70JA1397, 70OS(50)58⟩; **5**, 76, 166 ⟨72JOC1907⟩
Cregge, R. J., **1**, 309 ⟨73TL2595⟩, 828 ⟨89TL3377, 89TL6271⟩
Creighton, A. M., **2**, 123 ⟨60JCS1024⟩
Creighton, E. M., **1**, 398 ⟨28JCS3193⟩
Cremer, D., **1**, 909 ⟨78JA4301⟩, 910 ⟨86JA561⟩
Cremer, S. E., **2**, 440, 441 ⟨92JOC511, 67JOC4066⟩; **4**, 114 ⟨81JCS(P2)1138⟩
Cremlyn, R., **2**, 364 ⟨87PS(31)245⟩
Cremlyn, R. J. W., **5**, 712 ⟨55CI(L)1183⟩
Crenshaw, M. D., **2**, 618 ⟨83JOC2782⟩
Crenshaw, R. R., **2**, 166, 173, 232 ⟨65JOC175⟩, 178 ⟨64JOC918⟩
Crépon, E., **1**, 935 ⟨91TL199⟩; **4**, 266 ⟨91T7091⟩
Crescenzi, M., **2**, 21, 28 ⟨88T6525⟩; **4**, 138 ⟨90AG(E)658⟩
Crespan, C. G., **3**, 352 ⟨60JA1515⟩
Cresp, T. M., **1**, 732 ⟨74JCS(P1)2435⟩; **3**, 60 ⟨74JCS(P1)37⟩
Crespi, A. M., **6**, 717, 722 ⟨84OM314⟩, 717, 727 ⟨85OM1830⟩
Crespo, A. M., **3**, 714 ⟨89CC1231⟩
Cressman, E. N. K., **3**, 56 ⟨89JOC4345⟩
Cressman, H. W. J., **5**, 1126 ⟨47OS(27)56⟩
Creswick, M., **4**, 677 ⟨82IC645⟩, 686 ⟨79JOM(165)C17⟩
Creutz, C., **5**, 385 ⟨92IC2318⟩
Crévisy, C., **1**, 1038 ⟨91TL3171⟩
Crich, D., **1**, 172, 940 ⟨85TL757⟩, 231 ⟨86JCS(P1)1613⟩, 320, 330, 332, 342, 344 ⟨B-92MI 108-01⟩, 332, 344 ⟨87MI 108-01⟩, 332, 935 ⟨85TL5943⟩, 332, 935, 1130 ⟨86JCS(P1)39⟩, 595, 599 ⟨87JCR(S)353⟩, 602 ⟨85T4347⟩, 919, 927, 935, 938, 939 ⟨B-92MI 119-17⟩, 935 ⟨83CC939, 84TL1055⟩, 935 ⟨85T3901, 87MI 119-11⟩, 935, 938, 941 ⟨89CRV1413⟩, 940 ⟨84CC774, 86JCS(P1)1603⟩, 1127 ⟨87TL2895⟩, 1129 ⟨87CC786⟩; **2**, 12 ⟨91COS(7)723⟩, 13 ⟨83TL4979⟩, 26 ⟨87S35⟩, 140 ⟨89CRV1413⟩; **3**, 136 ⟨90JCS(P1)2875⟩, 163 ⟨90SL117⟩, 292 ⟨90T2135⟩, 627 ⟨84TL1055⟩; **4**, 174 ⟨91TL2565⟩, 404, 418, 419, 424, 427 ⟨91JCS(P1)1693⟩, 1077 ⟨83CC939, 83TL4979⟩; **5**, 241 ⟨92JA8313⟩, 727 ⟨86JCS(P1)39⟩; **6**, 546, 567 ⟨85TL5943⟩, 550 ⟨86T2329⟩
Criegee, R., **2**, 203 ⟨36CB2753⟩; **3**, 326 ⟨75CB1642⟩
Criegern, T., **2**, 870 ⟨76ZN(B)1058⟩
Crigee, R., **5**, 28 ⟨40AG321⟩
Crim, F. F., **4**, 762, 771 ⟨71JOC3386⟩
Crimmin, M. J., **1**, 674 ⟨77TL441⟩
Crimmin, M. T., **1**, 763 ⟨85JCS(P1)541⟩
Crimmins, M. T., **1**, 356, 357 ⟨91COS(5)123⟩, 482 ⟨92SC1953⟩, 816 ⟨87JA6199⟩, 928 ⟨92TL181⟩; **3**, 222 ⟨84JOC3033⟩, 225 ⟨90JOC4235⟩, 244 ⟨83TL5303, 90JOC5894⟩; **4**, 212 ⟨80JA1198⟩
Crimmins, T. F., **2**, 621 ⟨66CI(L)1635⟩; **4**, 669 ⟨78JOC2170⟩, 706 ⟨78JOM(161)C17⟩
Cripe, K., **1**, 112 ⟨74JOC1168⟩
Cripe, T. H., **1**, 922 ⟨90JOC58⟩
Cripps, H. N., **1**, 498 ⟨58JA751, 59JA2723⟩
Crisan, C., **4**, 383 ⟨59BSF459⟩
Crisco, L. V. T., **3**, 373 ⟨87TL371⟩
Crisp, G. T., **1**, 480 ⟨90OS(68)116⟩, 1027 ⟨84JA4630, 89AJC279⟩, 1029 ⟨84JA7500⟩; **2**, 689 ⟨84JA4630, 85S335⟩, 931 ⟨84JA7500⟩; **3**, 271 ⟨84JA7500, 90SC1665⟩, 273 ⟨89SC1745⟩; **4**, 150 ⟨86AJC1363⟩; **5**, 445 ⟨81ICA165⟩
Crispino, G. A., **2**, 48 ⟨92JOC2768, 94TL543⟩, 49 ⟨92JA7570, 93JOC844⟩
Crist, B. V., **1**, 634 ⟨81JA5314⟩
Crist, D. R., **3**, 434 ⟨89JOC1736⟩
Cristalli, G., **4**, 206 ⟨81JMC1181⟩
Cristante, M., **4**, 552 ⟨86IC712⟩
Cristau, H. J., **1**, 50 ⟨79TL349⟩; **2**, 444 ⟨88S911⟩, 445 ⟨87S648⟩, 446 ⟨79S538⟩, 711 ⟨81S892⟩, 727 ⟨86JOC875⟩, 824 ⟨71LA(751)17⟩, 848 ⟨88S912⟩; **3**, 4, 116 ⟨86TL1775⟩, 48 ⟨B-94MI 301-01⟩, 132 ⟨81SC423⟩, 195 ⟨B-94MI 304-01⟩, 196 ⟨87S648⟩, 493, 496 ⟨B-94MI 313-01⟩; **4**, 875 ⟨86JOC875⟩, 921 ⟨74BSF2263⟩; **5**, 373 ⟨91S382⟩
Cristea, I., **1**, 468 ⟨75JOC3649⟩
Criste, K. O., **2**, 762 ⟨66MI 215-02⟩
Cristobal, B. I., **3**, 577 ⟨89SC1039⟩
Cristol, H., **2**, 445 ⟨87S648⟩, 446 ⟨79S538⟩, 727 ⟨86JOC875⟩; **3**, 132 ⟨81SC7⟩
Cristol, S. J., **1**, 69 ⟨72JOC4341⟩, 628 ⟨60JA4692, 62JOC293⟩, 1065 ⟨54JA4558⟩; **2**, 12, 627 ⟨61JOC280⟩, 26 ⟨82JOC132⟩; **5**, 166 ⟨52JA4083⟩; **6**, 550 ⟨82JOC132⟩
Critcher, D. J., **5**, 340, 347 ⟨93JCS(P1)1581⟩
Crittell, C. M., **1**, 1016 ⟨91JOC3912⟩; **2**, 1020 ⟨90AG(E)287, 90TL4828⟩, 1020 ⟨91JOC3912⟩; **5**, 205, 321 ⟨90TL4821⟩; **6**, 153 ⟨93JOC4105⟩
Crivello, J. V., **2**, 791 ⟨81JOC3056⟩
Crocciani, B., **5**, 801 ⟨82JOM(236)C37⟩
Crocco, G., **5**, 1131 ⟨88AG289⟩; **6**, 673 ⟨88CC1598⟩
Crocco, G. L., **4**, 581 ⟨86CC1154⟩; **5**, 451 ⟨88JA6110⟩
Croce, P. D., **2**, 21 ⟨90S212⟩; **3**, 467 ⟨65CB2715, 77T865⟩
Crochet, K. L., **3**, 363 ⟨70T817⟩
Crociani, B., **3**, 599 ⟨71ICA477⟩; **4**, 1018 ⟨84JOM(269)C15⟩; **5**, 801 ⟨86OM1369⟩, 801, 802 ⟨81JOM(206)C11⟩; **6**, 622 ⟨75JOM(101)C1, 77JCS(D)2015, 79JOM(181)69⟩, 721 ⟨70JOM(25)255⟩
Crocker, M., **4**, 680 ⟨88JA6070⟩
Croft, A. P., **2**, 26 ⟨83JOC3353⟩, 622 ⟨83JOC876⟩
Croft, P. C., **2**, 454 ⟨53JA3379⟩
Crofts, P., **4**, 341 ⟨58QR341⟩
Crofts, P. C., **2**, 448 ⟨70JCS(C)2529⟩, 845 ⟨64JCS1240⟩; **4**, 130 ⟨53JA5738⟩
Crofts, P. S., **2**, 461 ⟨68JCS(B)1416⟩
Croisy, A., **5**, 239 ⟨76JCS(P1)2452⟩
Croker, C., **2**, 435 ⟨83JOM(250)617⟩
Crombie, L., **1**, 460 ⟨78TL3483⟩, 502 ⟨52QR101⟩, 506 ⟨57JCS1622⟩, 606 ⟨92JCS(P1)2683⟩, 734 ⟨69JCS(C)1024⟩, 811 ⟨82TL1405⟩, 815 ⟨88CC539⟩, 816 ⟨76CC684⟩, 957 ⟨55JCS1740⟩; **3**, 78 ⟨57JCS2767⟩, 653 ⟨94TL203⟩
Crombie, L. A., **1**, 533, 536 ⟨87TL4875, 87TL4879⟩
Cromer, R., **2**, 578 ⟨84NJC759⟩

Cromer, R. A., **3**, 141 ⟨82JOC743⟩
Crompton, H., **4**, 761 ⟨20JCS691⟩
Cromwell, N. H., **2**, 30 ⟨55OSC(3)105⟩; **3**, 234 ⟨47OS(27)9⟩, 261, 263 ⟨46CRV83⟩
Cron, M. J., **1**, 15 ⟨48JA1490⟩
Cronheim, G., **2**, 779 ⟨47JOC1⟩
Cronin, D. A., **6**, 611 ⟨64CI(L)1757⟩
Cronin, J., **6**, 623, 627 ⟨72JCS(P2)1050⟩
Cronin, J. P., **4**, 68 ⟨86TL757⟩
Cronje, S., **6**, 723 ⟨94AG(E)672⟩
Cronyn, M. W., **1**, 704 ⟨52JA1225⟩
Crook, S., **5**, 693, 758, 759 ⟨77JCS(P1)1791⟩
Crooker, P. P., **4**, 187 ⟨88S36⟩
Crooks, S. L., **1**, 1053 ⟨87TL527⟩; **3**, 222 ⟨87TL5081⟩
Crooks, III, W. J., **1**, 12 ⟨88JOC432⟩
Crooy, P., **5**, 727 ⟨72S622⟩
Crosbie, K. D., **2**, 428, 429, 432 ⟨69JINC3684⟩
Crosby, D. G., **2**, 654 ⟨61JOC1215⟩; **5**, 95 ⟨73OSC(5)654⟩, 273, 285 ⟨62JOC2205⟩
Crosby, G. A., **1**, 936 ⟨75JOC1966⟩, 1024 ⟨75CC287⟩
Crosby, J., **3**, 685 ⟨83JCS(P1)293⟩, 686 ⟨82TL5453, 85JCS(P1)1517⟩, 686 ⟨86JCR(S)156⟩; **4**, 232 ⟨81JCS(P1)2991⟩, 850 ⟨85CC696⟩; **5**, 652 ⟨76CC240⟩; **6**, 247 ⟨91CC165⟩
Cross, A. D., **1**, 2, 4, 6 ⟨71LA(748)123⟩
Cross, B., **5**, 652 ⟨73USP3954441⟩, 965 ⟨76USP3954441⟩
Cross, B. E., **6**, 472 ⟨81JCS(P1)1293⟩
Cross, G. W., **4**, 136 ⟨76JCS(D)694⟩
Cross, G. W.,**4**, 8 ⟨60PCS81⟩
Cross, R. J., **1**, 668 ⟨B-85MI 114-01⟩; **2**, 465 ⟨77JCS(D)159⟩; **4**, 552 ⟨80JCS(D)871⟩; **5**, 464 ⟨84ACR67⟩, 465 ⟨79JCS(D)1246, 80JCS(D)712, 83JCS(D)359⟩; **6**, 270 ⟨67JOM(7)P20⟩
Crosse, B. C., **2**, 278 ⟨67JCS(A)2014⟩
Crossland, I., **3**, 370, 375 ⟨77ACS(B)890⟩
Crossley, R., **3**, 341 ⟨74JCS(P1)2327⟩
Crossman, J. M., **3**, 483 ⟨73JCS(D)483⟩
Crotti, P., **2**, 318 ⟨90TL4661⟩
Crouch, R. D., **1**, 788 ⟨91JOC6981⟩; **2**, 396 ⟨90T7509⟩; **4**, 764 ⟨91JOC6987⟩
Croudace, M. C., **1**, 530 ⟨81JOC5436⟩; **2**, 985 ⟨81JOC5357, 81JOC5436⟩; **4**, 357 ⟨81JOC5357⟩
Crounse, N. N., **1**, 881 ⟨45JA2106⟩; **3**, 92 ⟨49OR(5)290⟩
Crouse, D. J., **1**, 23 ⟨87JOC2317⟩
Crouse, G. D., **1**, 402, 799 ⟨80JA3972⟩, 603 ⟨83JOC4986⟩; **3**, 221 ⟨83JOC141⟩
Crousse, B., **1**, 1009 ⟨93SL217⟩
Crouzel, C., **6**, 414 ⟨78MI 614-03, 83MI 614-01⟩, 414 ⟨87MI 614-01⟩
Crow, J. P., **3**, 570 ⟨69CC967⟩, 570, 571 ⟨72JCS(D)787⟩
Crow, R., **1**, 207 ⟨90JA4063⟩; **2**, 589 ⟨90JA4063⟩; **6**, 134 ⟨89TL15⟩
Crow, W. D., **2**, 235 ⟨69AJC765, 70T1463⟩; **3**, 552 ⟨70T1493⟩; **5**, 1107 ⟨65JOC2660⟩
Crowe, B. F., **3**, 350 ⟨50JOC81⟩
Crowe, D. F., **1**, 871, 881 ⟨69CC1498⟩, 1082 ⟨67TL3739, 67TL3943⟩
Crowe, W. E., **1**, 199 ⟨86JA3128⟩
Crowley, H., **1**, 1029 ⟨89JMC1820⟩
Crowley, K. J., **1**, 418 ⟨62PCS334, 65T1001⟩, 420 ⟨68JOC3679, 75TL3555⟩, 420 ⟨78T2783⟩
Crowley, P. J., **1**, 227 ⟨93SL277, 93SL279⟩
Crowte, R. J., **2**, 428 ⟨92CC699⟩
Crowther, D. J., **5**, 937 ⟨92OM622⟩
Croxall, W. J., **1**, 1069 ⟨54JA1700⟩; **4**, 188, 208 ⟨48JA2805⟩, 195 ⟨49JA2741⟩, 195, 195 ⟨49JA2736⟩
Crozet, M. P., **1**, 156 ⟨91TL4125⟩
Crozier, R. F., **2**, 354 ⟨75S205⟩
Crudden, C. M., **4**, 661 ⟨92AG(E)232⟩
Cruickshank, F. R., **1**, 1109 ⟨69CRV279⟩
Cruickshank, P. A., **5**, 279 ⟨65JA2492⟩, 279, 1062 ⟨61JOC2525⟩
Crum, J. D., **5**, 62 ⟨58MI 502-01⟩
Crumbie, R. L., **2**, 396 ⟨82JOC4040⟩; **3**, 46, 190 ⟨82JOC4040⟩
Crumbliss, A. L., **4**, 13 ⟨65JA4147⟩; **6**, 62, 244 ⟨65JA4147⟩
Crump, D. R., **1**, 784 ⟨93JHC873⟩
Crump, R. A. N. C., **2**, 546 ⟨92JCS(P1)3277⟩
Crumrine, D. S., **2**, 702 ⟨73JA3310⟩
Cruse, R., **2**, 1034 ⟨60JA3472⟩
Cruse, W. B., **1**, 641, 755 ⟨84JCS(P1)243⟩
Crutchfield, M. M., **2**, 100 ⟨65JOC2692⟩
Crute, T. D., **1**, 226, 227, 751 ⟨88JOC4274⟩; **2**, 86 ⟨88JOC4274⟩
Cruz, A., **3**, 235 ⟨73JA6655⟩
Cruz, R., **3**, 3, 114 ⟨79JCS(P1)2493⟩
Cruz, S. G., **2**, 680 ⟨85JOC1212⟩
Csákvári, B., **6**, 449 ⟨81MI 614-02⟩
Csaky, A. G., **3**, 221 ⟨90TL7669⟩
Csapilla, J., **1**, 1119 ⟨64HCA1590⟩
Cseh, G., **1**, 1111 ⟨64HCA194, 64HCA1590⟩; **2**, 623 ⟨64HCA194⟩
Csirinyi, G., **6**, 558 ⟨85S1149⟩
Csizmadia, I. G., **1**, 879 ⟨75CJC3365⟩, 1109 ⟨71JCP(55)3835, 72CC8⟩, 1109 ⟨74JCS(P2)1808, 77JCS(P2)542⟩, 1109 ⟨78G543⟩, 1143 ⟨73T4199⟩; **3**, 532 ⟨69T2121⟩; **4**, 390 ⟨B-85MI 408-01⟩
Cs Jaszberenyi, J.,**1**, 587 ⟨91TL2569⟩
Csongár, C., **5**, 358, 707 ⟨88ZC24⟩
Csuk, R., **1**, 688 ⟨88JCS(P!)1729⟩; **2**, 678 ⟨86CC1149⟩, 959 ⟨88AOC(28)85⟩; **3**, 30, 161 ⟨91CRV49⟩
Csürös, Z., **6**, 625 ⟨77ACH77⟩
Csutak, J., **6**, 477 ⟨83MI 622-01⟩
Cuadrado, P., **1**, 495 ⟨89T413⟩; **2**, 545 ⟨89SC275⟩, 929 ⟨88TL1825, 91JCS(P1)2811⟩, 929, 930 ⟨89T413, 90CC1030⟩, 930 ⟨92JCS(P1)327⟩; **3**, 203 ⟨89SC275⟩, 356 ⟨88SC1847⟩, 577 ⟨89SC1039⟩; **5**, 417 ⟨89SC275⟩
Cuadro, A. M., **2**, 301 ⟨91SC535⟩
Cu, X.-P., **2**, 89 ⟨88JOC2737⟩
Cubero, I. I., **6**, 550 ⟨93CAR(242)281⟩
Cuculo, J. A., **5**, 807 ⟨52JA710⟩
Cuddy, B. D., **1**, 631 ⟨71JCS(C)3173⟩
Cuenca, T., **2**, 566, 568 ⟨85JOM(295)159⟩
Cueto, O., **4**, 832 ⟨77JOC38⟩
Cuevas, J. C., **1**, 927 ⟨90JA896⟩
Cuevas, J.-C., **4**, 511 ⟨89TL5841⟩, 513 ⟨89TL5837⟩; **5**, 289 ⟨89TL5841⟩; **6**, 162 ⟨89TL5841⟩
Cugnon de Sévricourt, M., **3**, 103 ⟨77BSF142⟩
Cui, S., **4**, 343 ⟨92S258⟩
Cui, W.-F., **5**, 183, 195 ⟨90JCR(S)84⟩, 252 ⟨90MI 505-01⟩, 253 ⟨90JCR(S)230⟩
Culbertson, B. M., **2**, 478 ⟨68JCS(C)922⟩
Culbertson, R., **2**, 418, 419 ⟨71IC900⟩
Culhane, P. J., **5**, 109 ⟨32OSC(1)408⟩
Cullen, E., **2**, 775 ⟨61CJC862⟩
Cullen, E. R., **1**, 686 ⟨82JOC3563⟩; **3**, 391 ⟨81TL4563, 82JOC3563⟩, 392 ⟨85T4843⟩; **4**, 333 ⟨85JCS(P1)107⟩; **5**, 521 ⟨81JA7055⟩, 521, 523 ⟨82JCS(P2)473⟩
Cullen, W. R., **2**, 482 ⟨61CJC1855⟩, 482, 487, 876 ⟨65JOM(4)151⟩, 487 ⟨66JOM(6)364, 70CJC1885, 89CRV713⟩, 499 ⟨63CJC2424, 67CJC379⟩, 503 ⟨59JCS372⟩, 505, 506 ⟨63CJC322⟩, 507 ⟨66JA3178⟩, 512 ⟨64JA1030⟩, 872, 892 ⟨66AOC(4)146⟩, 876 ⟨65JOM(3)406, 73CJC936⟩, 1077, 1078 ⟨66IC1004⟩; **4**, 134 ⟨72JA5702⟩, 137, 141, 153 ⟨66JOM(6)633⟩, 344 ⟨65JOM(4)151⟩, 685 ⟨69JCS(A)1872⟩, 706 ⟨85OM2196⟩, 715 ⟨93JOM(463)C6⟩, 1033 ⟨87JOM(333)269⟩, 1034 ⟨83CJC2354, 86OM1320⟩; **5**, 222 ⟨60CJC472, 92MI 504-01⟩; **6**, 60 ⟨64CJC1123⟩, 65 ⟨70IC1682⟩
Cullis, C. F., **5**, 28 ⟨58CRV403⟩
Cullison, D. A., **3**, 663 ⟨73JA2923⟩
Cullum, T. V., **2**, 212, 219 ⟨51JA3627⟩

Culshaw, D., **4**, 286 ⟨89T7161⟩
Culvenor, C. C. J., **1**, 592 ⟨49JCS282⟩; **2**, 128 ⟨49JCS282⟩, 129 ⟨49JCS278⟩
Culvenor, C. J., **2**, 136 ⟨46JCS1050⟩
Cum, G., **1**, 1063 ⟨85CC1571⟩
Cumbo, C. C., **1**, 658 ⟨64JA4343⟩; **2**, 83 ⟨66JA2201⟩
Cumet, L., **2**, 1016 ⟨47BSF841⟩
Cummerson, D. A., **2**, 147, 156 ⟨87S1015⟩
Cummings, C. A., **5**, 652 ⟨76CC240⟩
Cummins, B., **3**, 242 ⟨60CI(L)348⟩
Cummins, C. C., **4**, 549 ⟨93AG(E)756⟩
Cummins, R. W., **2**, 478 ⟨53JA4520⟩
Cuñat, A. C., **1**, 1082 ⟨92JOC50⟩
Cundari, T. R., **5**, 932 ⟨92JA539⟩
Cun-heng, H., **5**, 355 ⟨85TL5739⟩
Cunico, R. F., **1**, 985 ⟨90TL5607⟩, 1092 ⟨76JOC1480⟩; **2**, 538 ⟨74SRI23⟩, 538, 539 ⟨80JOC4797⟩, 814 ⟨90JOC4634⟩, 907, 908 ⟨76JOM(105)C29, 85JOC5410⟩, 910, 911, 931 ⟨76JOC1480⟩, 921, 922 ⟨77JA7613⟩, 928 ⟨71JOC929⟩, 940 ⟨90TL5607⟩; **3**, 19, 51, 202 ⟨86TL4269⟩, 568 ⟨92JOC1202⟩, 640 ⟨92JOC3331⟩, 684 ⟨83JOC2780⟩; **4**, 354 ⟨90JOC4634, 90TL1945⟩, 359 ⟨92JOC1202, 92JOC6999⟩, 758 ⟨78JOM(162)1⟩, 930 ⟨85JOC5410⟩, 979 ⟨92JOC3331⟩, 1046 ⟨90TL5607⟩; **5**, 426 ⟨92TL6751⟩, 765 ⟨85JOM(281)135⟩; **6**, 24 ⟨91SC2189⟩
Cunkle, G. T., **2**, 25 ⟨86TL6055⟩
Cunneen, J. I., **2**, 117 ⟨47JCS134⟩, 144, 173 ⟨61JCS5339⟩
Cunningham, A. F., **1**, 25 ⟨88JOC1823, 88T6855⟩; **4**, 274, 275 ⟨88T6855⟩
Cunningham, I., **3**, 710 ⟨86CC1196⟩; **5**, 675 ⟨86CC1196⟩
Cunningham, J., **1**, 882 ⟨64TL1191⟩
Cunningham, L. W., **2**, 172 ⟨70JOC3267⟩
Cunningham, W. C., **5**, 723, 827 ⟨71JOC284⟩
Cuomo, J., **2**, 675 ⟨78JOC2073⟩; **5**, 4 ⟨79JOC1016⟩; **6**, 423, 438 ⟨79JOC1016⟩, 462 ⟨80TL819⟩
Cuong, N. K., **5**, 380 ⟨80TL3367⟩
Cupas, C. A., **1**, 380 ⟨64JA2722⟩, 811 ⟨64TL2503⟩, 893, 897 ⟨65JA2997⟩, 894 ⟨67JA5687⟩; **2**, 104 ⟨78S452⟩
Cupertino, D. C., **5**, 220 ⟨85JOM(294)C29⟩
Cupps, T. L., **1**, 1047 ⟨85JOC3972⟩; **3**, 180, 273 ⟨85JOC3972⟩, 633 ⟨83S308⟩
Curci, R., **2**, 146 ⟨76TL4637⟩, 845 ⟨73JOC2703⟩; **3**, 120 ⟨91JA2205⟩; **4**, 188 ⟨89TL257⟩, 237 ⟨74TL4085⟩; **6**, 230 ⟨91JA7654⟩
Curd, F. H. S., **5**, 757 ⟨47JCS160⟩
Curini, M., **3**, 129 ⟨91T7403⟩, 533 ⟨84JOC2052⟩
Curkovic, L., **6**, 444 ⟨77S704⟩
Curner, H. A., **2**, 428 ⟨61JOC5138⟩
Curotto, G., **2**, 918 ⟨93TL3311⟩
Curphey, M., **4**, 979 ⟨74JCS(P1)1611⟩
Curphey, T. J., **2**, 373, 376 ⟨72JOC2259⟩; **3**, 468 ⟨78JOC4666⟩
Curphey, T. L., **5**, 547 ⟨93TL3703⟩
Curran, A. C. W., **3**, 341 ⟨74JCS(P1)2327⟩; **5**, 595 ⟨76JCS(P1)983⟩
Curran, D. P., **1**, 106 ⟨93PAC1153⟩, 172, 230, 320, 334, 336, 344, 546, 919, 935, 946, 948 ⟨88S489⟩, 172, 230, 320, 336, 338, 341, 546, 919, 935, 1126 ⟨88S417⟩, 248 ⟨85JA1488⟩, 320, 326, 327, 328, 919 ⟨91ACR296⟩, 320, 329, 330, 331, 332, 334, 336, 546, 920, 923, 948 ⟨91COS(4)715⟩, 320, 331, 336, 341, 342, 344, 348, 926, 935, 948 ⟨91CRV1237⟩, 320, 336, 339, 340, 341, 342, 344, 347, 348, 546, 919 ⟨91COS(4)779⟩, 320, 919 ⟨91SL63⟩, 324 ⟨93TL4489⟩, 326 ⟨91JA5918⟩, 326, 327, 328 ⟨92JA7007⟩, 326, 328 ⟨90JA6738⟩, 327 ⟨91TL6307, 92JA4436⟩, 327 ⟨93JA8585⟩, 327, 328 ⟨93T4821⟩, 327, 328 ⟨93T4841⟩, 328, 927 ⟨91JOC4335⟩, 334 ⟨89JOC1826⟩, 334, 337, 344, 936 ⟨89JOC3140⟩, 334, 347 ⟨89JA6265⟩, 339, 927 ⟨88JA5900⟩, 339, 927 ⟨93JA6051⟩, 340 ⟨90JA9401⟩, 344 ⟨86TL5821⟩, 344, 831 ⟨90JA5601⟩, 347 ⟨90SL773, 92JA6050⟩, 347, 1132 ⟨85JA1448⟩, 347, 937 ⟨92SL943⟩, 347, 937 ⟨93TL1717⟩, 348 ⟨87JA6558⟩, 348 ⟨89TL2501⟩, 548 ⟨88JA5064⟩, 548, 1132 ⟨85TL4991⟩, 826 ⟨87JA1160⟩, 831 ⟨88T3079⟩, 923 ⟨88ACR206⟩, 927 ⟨90JA896⟩, 1130 ⟨86JA1106, 85T3943⟩; **2**, 14 ⟨91COS(4)779⟩, 14, 34 ⟨87TL2477⟩, 155 ⟨81TL1287⟩, 541 ⟨92TL6931⟩, 565 ⟨90SL773, 92SL943⟩, 587 ⟨92JOC1740⟩; **3**, 142 ⟨86S312⟩, 292 ⟨88S417, 88S489⟩, 627, 637 ⟨92JA4436⟩, 653 ⟨90JA9401⟩, 683 ⟨88MI 320-01⟩; **4**, 274 ⟨81TL1287⟩, 284 ⟨92JA4436⟩, 781 ⟨91T6171⟩, 811 ⟨89JA6265, 89JOC3140⟩; **5**, 432 ⟨92SL631⟩
Curran, W. V., **5**, 639, 640 ⟨88JHC1873⟩
Currie, J. K., **3**, 233 ⟨78CC311, 79TL2665⟩; **5**, 568, 613 ⟨78AG(E)689⟩
Currie, Jr., J. O., **6**, 608 ⟨73JA2982⟩
Currier, H. A., **2**, 428 ⟨59JA1103⟩, 450 ⟨61JOC4626⟩, 454 ⟨61JOC4628⟩
Curry, D. C., **1**, 74, 82, 84 ⟨67JCS(C)1120⟩
Curry, H. M., **2**, 386 ⟨51JA5041⟩
Curry, J. D., **4**, 563 ⟨68JOM(13)199⟩; **6**, 263 ⟨68JOM(13)199⟩
Curry, M. J., **5**, 139 ⟨48JA3781⟩
Curtin, D. Y., **1**, 128, 396 ⟨52JA5381⟩, 397 ⟨54JA494⟩, 847 ⟨60CI(L)262⟩, 1118 ⟨65JA863⟩, 1144 ⟨59JA4714⟩, 1144, 1145 ⟨59JA4719⟩; **2**, 800 ⟨65JA863⟩, 800, 801 ⟨65JA874⟩; **3**, 263 ⟨65JA874⟩, 448 ⟨62JA4887, 62JA4892⟩; **5**, 365 ⟨67JOC1552⟩, 714 ⟨67JA637⟩
Curtis, M. D., **4**, 622 ⟨78IC2324, 78JA6362⟩, 674 ⟨78JOM(161)23, 83OM1172⟩; **5**, 447 ⟨79JOM(172)177⟩
Curtis, N. J., **1**, 21, 25, 437, 438, 943 ⟨80JA4438⟩, 609 ⟨78CC770⟩; **2**, 286 ⟨77CC725⟩; **5**, 386 ⟨84JA625⟩; **6**, 104 ⟨80JOC4038⟩
Curtis, N. R., **1**, 829 ⟨91T7171⟩
Curtis, P. A., **5**, 943, 944 ⟨84CC747⟩
Curtis, R. F., **1**, 1006 ⟨67JCS(C)2011, 67JCS(C)578⟩, 1006 ⟨69JCS(C)2173⟩, 1018 ⟨68TL2919, 71JCS(C)186⟩; **3**, 78 ⟨69JCS2173⟩
Curtis, V. A., **1**, 28, 35 ⟨B-82MI 102-01⟩, 382 ⟨84JOC2937⟩, 636 ⟨81TL199⟩
Curtius, T., **3**, 319 ⟨1885CB2371⟩, 444 ⟨02CB3234⟩, 444, 445 ⟨00CB2460⟩, 462, 468 ⟨1883CB2230, 04CB1261⟩, 468 ⟨06CB1373, 06CB1379⟩, 1884CB953⟩, 468, 476 ⟨04CB1284, 06CB1383, 08CB344⟩, 477 ⟨1894CB59, 1898CB2489⟩; **5**, 989 ⟨1890CB3023, 1894CB778⟩; **6**, 511 ⟨1895JPR(2)472⟩
Curtius, Th., **2**, 1055 ⟨26JPR314⟩
Curts, J., **5**, 579, 580, 623, 624 ⟨60CB1511⟩, 620, 623, 624 ⟨62LA(660)60⟩, 636 ⟨61LA(643)29⟩
Curtui, M., **5**, 222 ⟨92JOM(426)49⟩
Curtze, J., **4**, 416, 423 ⟨75LA2318⟩; **5**, 606 ⟨75LA2318⟩
Cusa, N. W., **4**, 770 ⟨37JCS767⟩
Cusack, N. J., **1**, 74 ⟨76T2157⟩
Cusati, P. F. de, **4**, 59 ⟨90TL1405⟩
Cushman, D. W., **3**, 165 ⟨88JMC2199⟩
Cushman, M., **1**, 583 ⟨90SC1463⟩; **3**, 190 ⟨82S397⟩; **4**, 161 ⟨88JOC5088⟩; **5**, 573, 587 ⟨92T8601⟩
Cusic, J. W., **5**, 43, 62, 114 ⟨43JA262⟩; **6**, 501 ⟨67JMC541⟩
Cussans, N. J., **3**, 132 ⟨80JCS(P1)1654⟩; **5**, 523 ⟨78CC393, 80JCS(P1)1650⟩
Cuthbertson, E., **1**, 243 ⟨75TL1893⟩
Cutler, A. R., **5**, 453 ⟨85OM1247⟩, 457 ⟨87JA5844⟩, 937 ⟨81JOM(213)C31⟩
Cuts, H. W., **1**, 884 ⟨66CC880⟩
Cutting, I., **3**, 55 ⟨81TL2021⟩; **4**, 860 ⟨83CC1209⟩
Cuvigny, T., **1**, 195 ⟨88JOM(339)199⟩, 195, 238 ⟨86JOM(317)383⟩, 212 ⟨75S256⟩, 439, 700 ⟨87T859⟩, 485 ⟨88T111, 83TL4319⟩, 560 ⟨77TL3187⟩, 1000 ⟨65BSF1872⟩; **2**, 420 ⟨75LA719⟩, 421 ⟨69CR(C)834⟩; **3**, 228 ⟨60BSF515⟩, 459 ⟨74BSF1710, 75JOM(87)25⟩; **4**, 272, 275 ⟨82BSF43⟩, 535 ⟨73BSF2989⟩
Cuyas, E., **6**, 329, 330 ⟨92POL2713⟩
Cymerman, J., **1**, 837 ⟨45JCS90⟩
Cymerman-Craig, J., **4**, 413 ⟨55AJC512⟩; **5**, 1035 ⟨54CI(L)785, 56JCS659⟩, 1035 ⟨63OSC(4)700⟩

Cypryk, M., **2**, 467 ⟨85T2471⟩
Cyr, C. R., **1**, 797 ⟨73JA2248⟩; **6**, 719 ⟨73SRI249⟩
Cyr, P. T., **5**, 329 ⟨90TL5297⟩
Cywin, C. L., **1**, 819 ⟨93TL1103⟩
Czaja, R. F., **2**, 137 ⟨69JOC3112⟩
Czarnik, A. W., **1**, 785, 786, 1034 ⟨91COS(5)551⟩
Czarny, A., **4**, 411 ⟨90JOC4777⟩
Czarny, M., **4**, 707 ⟨79JA768⟩
Czech, H., **2**, 393 ⟨78ZC214⟩
Czekanski, T., **4**, 553 ⟨82EGP154700, 82JPR537⟩, 553 ⟨84EGP206557⟩; **5**, 671 ⟨90JPR731⟩
Czempik, H., **1**, 999 ⟨56ZAAC(287)261⟩; **2**, 1058, 1063, 1066, 1069 ⟨56ZAAC(287)261⟩
Czepluch, H., **6**, 445 ⟨80T3543⟩
Czerepinski, R., **6**, 433 ⟨68JA3954⟩
Czernecki, S., **4**, 173 ⟨89JOC610⟩
Czieslik, G., **5**, 383 ⟨72ZAAC(394)26⟩
Czisch, P., **5**, 917 ⟨85OM1310, 85ZN(B)1177⟩
Czombos, J., **1**, 455 ⟨B-84MI 110-01⟩; **3**, 475 ⟨70HCA1057⟩
Czuba, L. J., **1**, 871, 881 ⟨69JOC2324⟩; **2**, 696, 699 ⟨69JOC2324⟩
Czyzewska, E., **1**, 966 ⟨83TL5587⟩
Czyzewski, J., **3**, 356, 377 ⟨79JCS(P1)2340⟩

D

Dabbagh, G., **1**, 118, 119 ⟨91JA631⟩; **2**, 588 ⟨82JA5284⟩; **3**, 287 ⟨82JA5824, 84JOC1119⟩; **4**, 195 ⟨82JOC2216⟩; **5**, 556 ⟨83CC388⟩
Dabbas, N., **3**, 433 ⟨83MI 311-01⟩
Dabby, R. E., **2**, 188 ⟨52JCS4887⟩
Dabdoub, A. M., **3**, 634 ⟨77TL1117⟩
Dabdoub, M. J., **1**, 999 ⟨88OM84⟩, 999, 1035 ⟨90SC2181⟩; **2**, 730 ⟨89OM1661⟩, 1036, 1037 ⟨88OM84, 88SC1979⟩, 1036, 1038 ⟨90SC2181⟩; **4**, 93 ⟨93JOM(460)31⟩; **5**, 241 ⟨92SC1619⟩
Dabdoub, V. M. B., **2**, 730 ⟨89OM1661⟩
Dabestani, S., **4**, 692 ⟨76JA8289⟩
Dabholkar, D. A., **4**, 105 ⟨74JOC1965⟩
Dabkowski, W., **2**, 411 ⟨90AG(E)522⟩
Daboun, H. A., **4**, 420 ⟨83H(20)1615⟩; **5**, 615 ⟨88IJC(B)472⟩
Dabrowska, U., **3**, 355 ⟨71BAP549⟩; **5**, 534 ⟨76CB1779⟩
Dabrowski, J., **3**, 338 ⟨72OMR(4)421, 71BAP549⟩; **4**, 209 ⟨69T4257⟩; **5**, 534 ⟨76CB1779⟩
Dach, R., **2**, 388 ⟨78HCA1622⟩
Dacons, J. C., **3**, 662 ⟨73JOC4363⟩
D'Addieco, A. A., **3**, 328 ⟨51OS(31)25⟩
Daeniker, H. U., **2**, 176, 180 ⟨57HCA2148⟩, 374 ⟨57HCA918⟩
Daeuble, J. F., **1**, 458, 516 ⟨90TL2397⟩; **4**, 224 ⟨89TL2179⟩
Dagani, D., **4**, 153 ⟨82JIC111⟩, 820 ⟨76JOM(104)145⟩
Dagonneau, M., **1**, 495 ⟨82BSF269⟩
Daguer, R., **6**, 114 ⟨64T25⟩
Dahan, F., **3**, 485 ⟨91AG1174, 91AG(E)1154⟩, 682 ⟨92S43⟩; **4**, 547, 579 ⟨93AG(E)1735⟩; **5**, 872 ⟨91AG(E)1154⟩; **6**, 338 ⟨85OM1436⟩, 659 ⟨92CC1274⟩, 669 ⟨91AG(E)1154⟩
Dahan, R., **2**, 612 ⟨84HCA284⟩
Dahill, Jr., R. T., **1**, 747 ⟨65JOC505⟩
Dahl, B. M., **6**, 594, 595 ⟨70ACS959⟩
Dahl, G. H., **4**, 555 ⟨68USP3403176⟩
Dahl, H., **3**, 220 ⟨86S41⟩
Dahl, K., **3**, 684 ⟨68T5251⟩
Dahl, L. F., **1**, 897 ⟨72JA5511⟩; **4**, 148 ⟨86OM630⟩, 689 ⟨69JA7292⟩, 715 ⟨70JA502⟩; **6**, 524 ⟨66JA1821⟩, 525 ⟨79JA6110⟩
Dahl, O., **2**, 108 ⟨91SR670⟩, 469 ⟨86JCR(S)26⟩; **3**, 639 ⟨75ACS(B)863, 76ACS(B)799⟩, 639 ⟨79JCR(S)396, 83ACS(B)639⟩; **4**, 556 ⟨78JCS(P1)947⟩; **6**, 584 ⟨70ACS1094⟩
Dahl, W. E., **4**, 102 ⟨87JHC945⟩; **6**, 445 ⟨87JHC945⟩
Dahlbom, R., **2**, 117 ⟨51ACS690⟩
Dahlenburg, L., **2**, 97 ⟨70T3279⟩, 431 ⟨84IC4170⟩, 439 ⟨86PS(26)143⟩; **4**, 109 ⟨92S710⟩
Dahlhoff, W. V., **4**, 546 ⟨71JCS(A)3495⟩
Dahlman, O., **1**, 4, 6 ⟨83ACS(B)141⟩
Dahlmann, J., **2**, 886 ⟨64LA(678)167⟩, 887 ⟨70JPR10⟩, 888 ⟨67CB1544⟩
Dahm, B., **5**, 1064, 1066 ⟨88CB271⟩
Dahm, D. J., **4**, 850 ⟨74CC751⟩
Dahm, K. H., **1**, 747 ⟨67JA5292⟩
Dahm, S., **3**, 450 ⟨63ACS559⟩
Dahmen, A., **1**, 421 ⟨68TL5215⟩; **3**, 315 ⟨71CC1192⟩
Dahmen, F. J. M., **4**, 825 ⟨79TL2925⟩; **6**, 75 ⟨79TL2925⟩
Dahmlos, J., **6**, 220, 225 ⟨59AG274⟩
Dahms, G., **6**, 252, 274 ⟨74CZ109⟩, 258, 449 ⟨71CB2732⟩
Dahn, H., **2**, 799 ⟨70HCA1637⟩; **3**, 293 ⟨74HCA2201⟩, 479 ⟨73C320⟩; **4**, 162 ⟨70HCA1598⟩
Dahnke, K., **2**, 625 ⟨91JOC6199⟩
Dai, L., **1**, 14 ⟨86TL4343⟩
Dai, L.-X., **2**, 998 ⟨92CC1487⟩
Dai, S. H., **1**, 497 ⟨72JOC950⟩
Dai, W. M., **1**, 1009 ⟨91AG(E)1387⟩
Daich, A., **5**, 367 ⟨92BSF360⟩
Daigle, D., **1**, 999 ⟨66JOM(5)205⟩; **2**, 827, 828 ⟨64JA2299⟩, 841 ⟨65JOC2826⟩
Daigle, D. G., **4**, 113, 118, 120, 122 ⟨67JCED282⟩
Daigle, D. J., **4**, 458 ⟨72JHC715, 75JHC579⟩, 458, 477, 481 ⟨74JHC407⟩, 482 ⟨74JHC1085⟩, 555 ⟨68JCED585, 71USP3607944⟩
Daignault, R. A., **2**, 128 ⟨66JA1828⟩
Daigneault, S., **1**, 594, 928 ⟨89CC332⟩, 688 ⟨91JOC6447, 91JOC3801⟩
Dailey, Jr., O. D., **1**, 20 ⟨87JOC1984⟩; **4**, 260 ⟨76JOC3975⟩, 263 ⟨76JOC3979⟩, 266 ⟨81PS(10)169⟩, 268 ⟨81PS(10)163⟩; **5**, 664 ⟨84JMC1201⟩; **6**, 128 ⟨76JOC3975⟩
Dailey, W. P., **1**, 809 ⟨85JOC3393⟩; **3**, 479 ⟨89TL4197⟩, 480 ⟨88TL6031, 90T7175⟩; **4**, 46 ⟨89JOC5520⟩; **5**, 326 ⟨88TL5719, 89JOC3096⟩, 866 ⟨87TL5801⟩, 868 ⟨90TL7341, 91JOC2258⟩; **6**, 55 ⟨87TL5801⟩, 611 ⟨90JOC3045⟩
Daineko, Z. G., **2**, 891 ⟨74JGU1506⟩
Daines, R. A., **1**, 750 ⟨88JA4685⟩
Dains, F. B., **5**, 1023 ⟨22MI 527-01, 32OSC(1)447⟩, 1049 ⟨34JA719⟩
Dainter, D. S., **6**, 28 ⟨89JCS(P1)283⟩
Dainton, F. S., **6**, 215 ⟨50TFS295⟩
Dajani, E. Z., **1**, 1049 ⟨75TL4217⟩, 1050 ⟨76TL235⟩
Dakin, H. D., **2**, 655 ⟨09JA477⟩; **3**, 620 ⟨16BJ(10)319, 16PRS232⟩, 620 ⟨17BJ(11)79⟩
Dakka, G., **2**, 31 ⟨87TL1223⟩
Dakka, J., **3**, 117 ⟨88JOC3553⟩
Dakternieks, D., **2**, 451, 467, 469 ⟨85PS(24)217⟩; **4**, 660, 700 ⟨93OM2788⟩; **6**, 332, 351 ⟨83IC3700⟩
Dal Bello, G., **1**, 659, 764 ⟨66TL4315⟩; **4**, 664 ⟨65TL3429⟩; **6**, 198 ⟨65TL3429⟩, 198 ⟨66TL4315⟩
Dalbey, T. E., **2**, 496, 498 ⟨57JOC452⟩
Dalcanale, E., **5**, 102 ⟨86JOC567⟩
Daldy, F. G., **4**, 799 ⟨28JCS2756⟩
Dale, A. J., **1**, 737 ⟨71ACS1452⟩; **2**, 886 ⟨75ACS(B)362⟩; **3**, 328 ⟨70ACS3772⟩
Dale, J. A., **5**, 12 ⟨69JOC2543⟩
Dale, J., **1**, 248 ⟨87CC381⟩, 795 ⟨63JCS4091⟩; **2**, 111 ⟨61JCS922⟩, 773 ⟨88ACS(B)354⟩
Dale, J. W., **2**, 288, 291 ⟨58JCS2939⟩, 486, 492, 505 ⟨57JCS3708⟩; **6**, 49 ⟨58JCS2939⟩
Dale, W. J., **3**, 90 ⟨61JOC2225⟩
Dalecki, T. M., **2**, 13 ⟨87CC675⟩
Dalessandro, J., **1**, 4, 7, 9, 17, 28 ⟨78JOC2259⟩
Daley, S. T. A. K., **3**, 372 ⟨87JCS(P1)207⟩
D'Alfonso, G., **4**, 693 ⟨95JOM(488)39⟩
Dalgaard, L., **3**, 347 ⟨74ACS(B)1077⟩; **5**, 547, 555 ⟨74T93⟩, 552, 555 ⟨73T2077⟩
Dalgliesh, C. E., **5**, 24 ⟨B-51MI 502-01, B-52MI 502-01⟩, 62 ⟨B-52MI 502-04, B-52MI 502-08⟩, 84 ⟨B-52MI 502-02⟩
Dalinger, I. L., **6**, 145 ⟨90IZV1816⟩
Daljeet, A. K., **4**, 651 ⟨91JOC1944⟩
Dallacker, F., **5**, 334 ⟨91CZ135⟩, 1047 ⟨58CB1660⟩, 1057 ⟨59M41⟩; **6**, 304 ⟨65LA(689)179⟩, 492 ⟨85ZN(C)344⟩, 596 ⟨59M41⟩
Dalla Cort, A., **2**, 33 ⟨91JOC6708⟩
Dalla Croce, P., **5**, 774, 839, 854, 855 ⟨69G69⟩, 905 ⟨79S299⟩, 1123 ⟨68G511⟩
Dall'Asta, G., **1**, 519 ⟨68JOC3948⟩

Dallemer, F., **5**, 439 ⟨91JOC3118⟩
Dalley, N. K., **4**, 61, 73, 79 ⟨91JOC6878⟩, 218 ⟨94JOC544⟩
Dall'Occo, T., **2**, 942 ⟨84CC258⟩; **3**, 714 ⟨84CC258⟩
Daloze, D., **3**, 694 ⟨89PAC509⟩; **4**, 414 ⟨88TL1691⟩
Dal Piaz, V., **5**, 99 ⟨80JHC621⟩
Dalpozzo, R., **2**, 318 ⟨88TL2251⟩, 346 ⟨90JCS(P1)2133⟩; **3**, 434 ⟨90JOC4456⟩; **4**, 646, 653 ⟨88CC807⟩
Dalton, D. R., **3**, 618 ⟨73CC625⟩
Dalton, H., **4**, 262 ⟨92JA8772⟩
Dalton, J. R., **3**, 662 ⟨79JOC4443⟩
Dalton, L., **3**, 309 ⟨84T3339⟩
Dalton, L. K., **2**, 384 ⟨80AJC1365⟩
Daltrozzo, E., **4**, 987 ⟨79CB484, 81CB2001⟩
Dalverney, G., **4**, 820 ⟨76CC174⟩
Daly, D. G., **1**, 674 ⟨82TL2997⟩
Daly, J. J., **2**, 821 ⟨67HCA1747⟩, 855 ⟨72HCA1991⟩, 1004 ⟨67JCS(A)1913⟩; **6**, 661 ⟨72JCS(D)2267⟩
Daly, N. J., **1**, 777 ⟨73AJC1259⟩
Daly, V. A., **5**, 445 ⟨81ICA165⟩
Daly, W. H., **5**, 979 ⟨74JOC1597⟩; **6**, 416, 430, 486 ⟨88TL5859⟩, 476 ⟨77MI 615-02⟩
Dalziel, J. R., **4**, 455 ⟨81JINC629⟩
Damas, A. M., **4**, 232 ⟨81JCS(P1)2991⟩
Damasevitz, G. A., **1**, 128, 161 ⟨76JOC2214⟩; **2**, 922 ⟨76JOC2214⟩
Damerius, R., **3**, 380 ⟨89ZAAC(578)110⟩
Damiano, J. C., **2**, 551 ⟨80JA7926⟩, 958 ⟨80JA7927⟩
D'Amico, D. C., **3**, 34 ⟨93JA324⟩
D'Amico, J. J., **4**, 14 ⟨51JA4393⟩, 102 ⟨87JHC945⟩; **5**, 1119, 1121 ⟨87PS(29)219, 86PS(26)1⟩; **6**, 304, 305 ⟨67JOC2567, 72USP3652256⟩, 445 ⟨87JHC945⟩, 493, 495 ⟨80PS(8)301⟩, 545 ⟨88JHC1601⟩, 546 ⟨89JHC1245⟩; **6**, 630 ⟨87POL13⟩
Damico, R., **1**, 1144 ⟨63JA2273⟩
Damiens, R., **5**, 751, 753 ⟨56MI 519-02⟩
Damja, R. I., **6**, 380 ⟨85JOM(291)25⟩
Damle, S. B., **1**, 214 ⟨66JA2257⟩
Damm, H., **3**, 275 ⟨86TL505⟩
Damm, W., **1**, 327 ⟨92SL442, 93JA8585⟩, 327 ⟨93TL5885⟩
Dammann, R., **4**, 395 ⟨75CB2368⟩
Dammel, R., **6**, 604 ⟨92ZN(B)351⟩
Damodaran, N. P., **1**, 809, 813 ⟨67TL227⟩
Damon, D. B., **5**, 688 ⟨90JA6439⟩
Damour, D., **1**, 963 ⟨87JOM(319)333⟩, 978 ⟨86T2017, 86T2501⟩, 978 ⟨87TL4689, 88JOM(349)43⟩, 978 ⟨90JOM(396)289⟩
Dampawan, P., **2**, 392 ⟨83S545⟩, 785 ⟨82TL135⟩; **3**, 190 ⟨82JOC1176⟩, 190 ⟨83S545⟩
Damrauer, R., **3**, 695 ⟨66TL189⟩; **4**, 610 ⟨72MI 414-02⟩, 623 ⟨71JA3709⟩; **6**, 60 ⟨67JA1538, 68JA2944⟩, 608 ⟨71JOC1786⟩
Dams, R., **1**, 688 ⟨81BSB1141, 82JOC248⟩, 688 ⟨82MI 115-01⟩, 692 ⟨82BSB149⟩
Dana, D. E., **3**, 464 ⟨75TL1565⟩
Danaha, M. J., **2**, 467 ⟨86JOC3559⟩
Danankov, V. A., **4**, 461 ⟨80IZV1125⟩
Dance, J., **3**, 674 ⟨68T1529⟩
Dancer, R. J., **1**, 486 ⟨90AJC1375⟩
Danchenko, M. N., **2**, 464 ⟨86JGU1567⟩; **4**, 927 ⟨85ZOB838⟩; **5**, 592 ⟨72ZOR528⟩
Danda, H., **2**, 532 ⟨90JOC173⟩
Dandárová, M., **6**, 560 ⟨92CCC1699⟩
Dandiarajan, P. K., **2**, 532, 698 ⟨89JA3441⟩
D'Andrea, S., **5**, 953 ⟨88JA2122⟩
Dané, L. M., **1**, 800 ⟨70TL2755⟩
Danehy, J., **2**, 172 ⟨71JOC2530⟩
Danek, M., **6**, 237 ⟨93CM1321⟩
Danek, S. K., **2**, 404 ⟨87JOC2911⟩
Danen, W. C., **1**, 156 ⟨66JA5663⟩
Daney, M., **2**, 619 ⟨73BSF720⟩
Danforth, R. H., **2**, 652 ⟨70JA3520⟩
Dang, H. P., **1**, 206, 478 ⟨78TL4069⟩, 468, 1097, 1135 ⟨78TL191⟩
Dang, V. A., **4**, 59 ⟨90JOC1847⟩; **6**, 462 ⟨90JOC1⟩
Dang, Y., **1**, 670 ⟨91JOM(405)1⟩
d'Angelo, J., **1**, 129, 872, 873 ⟨76T2979⟩, 558 ⟨90TL879⟩, 869 ⟨75T2979⟩; **2**, 696 ⟨76T2979⟩, 759 ⟨77TL3589⟩, 1050 ⟨74BSF1535⟩; **3**, 60 ⟨87JOC788⟩, 124, 642 ⟨76T2979⟩, 414 ⟨92TA459⟩, 415 ⟨85HOU(E11)233⟩; **4**, 279 ⟨92TA1003⟩; **6**, 106 ⟨70CR(C)468⟩
Danheiser, R. L., **1**, 400 ⟨91COS(5)999⟩, 450, 527, 1061 ⟨83T935⟩, 482 ⟨73JOC1775⟩, 518 ⟨90OS(68)32⟩, 523 ⟨73TL4477⟩, 566, 681, 683 ⟨91JOC1176⟩, 601, 781, 785 ⟨93JOC322⟩, 803 ⟨80JOC1340, 81JA2443, 85JA4579, 91COC(5)999⟩, 805 ⟨81T3943⟩, 812 ⟨82JA7670⟩, 962, 1055 ⟨80JOC3925⟩, 1056, 1058 ⟨86JOC3870⟩, 1060 ⟨85JA7233⟩; **2**, 420 ⟨92OS(71)133⟩, 668 ⟨84JOC1672, 90JA3093⟩, 668, 941, 944 ⟨80JOC4810⟩, 928 ⟨83TL23⟩, 930, 940, 943 ⟨83T935⟩, 930, 944 ⟨89JA389⟩, 1024, 1025 ⟨88TL4917⟩; **3**, 193, 473 ⟨90JOC1959⟩, 211 ⟨85JA8066⟩, 216 ⟨87TL3299⟩, 271 ⟨80JOC4810, 83TL23⟩, 272 ⟨85TL2513⟩; **4**, 358 ⟨85JOC5393⟩, 359 ⟨88OS(66)14⟩, 930, 931 ⟨88T4113⟩, 934 ⟨89JOC2798⟩; **5**, 161 ⟨91JOC1176⟩, 410, 431 ⟨85TL2509⟩, 412 ⟨88OS(66)14⟩, 413, 425 ⟨85JOC5393⟩, 423, 426 ⟨89JOC2798⟩, 423, 430 ⟨88T4113⟩, 424, 429 ⟨85TL2513⟩
Daniel, D., **6**, 472 ⟨92JOC2523⟩
Daniel, D. S., **3**, 468 ⟨78JOC4666⟩
Daniel, H., **6**, 642 ⟨80JMC1217⟩
Daniel, J., **2**, 249 ⟨93SC121⟩
Daniel, J. R., **2**, 289 ⟨79JMC572⟩
Daniel, L. W., **2**, 252 ⟨86JMC2114⟩
Daniel, T., **1**, 898 ⟨89JA7659⟩; **3**, 283, 295, 296, 305 ⟨91S322⟩
Daniel, W., **2**, 101, 102 ⟨28LA(462)186⟩
Danieli, B., **1**, 1121 ⟨73CI(L)1067⟩; **5**, 1045 ⟨90TL7229⟩
Danieli, R., **3**, 253 ⟨83JOC123⟩
Daniels, J. A., **5**, 455 ⟨76CC1032⟩
Daniels, K., **1**, 221 ⟨83JOC409⟩; **4**, 207 ⟨91TL6973⟩
Daniels, P. J. L., **3**, 215 ⟨63JCS3675⟩
Daniels, R., **2**, 382 ⟨79JHC1485⟩
Daniels, R. G., **1**, 633 ⟨82OM1449⟩, 971 ⟨81TL1579⟩; **2**, 940 ⟨81TL1579⟩
Daniels, T. C., **5**, 25, 108 ⟨55JA6682⟩
Daniewski, A. R., **1**, 624, 626 ⟨86JA1039⟩, 697, 698, 700, 703 ⟨85TL5597⟩; **5**, 278 ⟨86JA1039⟩
Daniher, F. A., **5**, 1000 ⟨69JOC2908⟩
Daniil, D., **3**, 470 ⟨78S535⟩
Danikiewicz, W., **3**, 625 ⟨88LA203⟩
Danilenko, V. A., **4**, 874 ⟨79MI 419-01, 80ZOR654⟩, 928 ⟨79ZOB1857, 79ZOB2198⟩, 928 ⟨79ZOB332, 80ZOB573⟩; **5**, 898, 903 ⟨79JGU1740⟩, 920 ⟨81JOU1676⟩
Danilewicz, J. C., **5**, 767 ⟨71JCS(C)1704⟩
Danilova, N. A., **1**, 491 ⟨89S633⟩
Danion, D., **3**, 474 ⟨81TL441⟩, 476 ⟨81AG118, 81AG(E)113⟩, 476, 477 ⟨81BSB615⟩; **4**, 109 ⟨89CJC1125⟩, 328 ⟨91CR(313)517⟩, 489 ⟨70BSF1130⟩, 989 ⟨91CB595⟩, 991, 993, 996 ⟨85BSB475⟩; **5**, 494 ⟨75JOC2990⟩, 633 ⟨91BSF574⟩
Danion-Bougot, R., **1**, 637 ⟨89BSF409⟩; **3**, 474 ⟨81TL441⟩; **4**, 109 ⟨89CJC1125⟩, 993 ⟨87BSF325⟩; **5**, 494 ⟨75JOC2990, 91BSF574⟩
Danishefsky, S., **1**, 99 ⟨75JA380, 75JOC3606⟩, 218, 221 ⟨77JOC394⟩, 243 ⟨83TL1357⟩, 521 ⟨79JA7001⟩, 603 ⟨85JA3891⟩, 831 ⟨80JA6889, 80JA6891⟩, 928 ⟨79ACR66⟩, 1139 ⟨76JOC1669⟩; **2**, 665 ⟨74JA7807, 77JA5810⟩, 665 ⟨81ACR400⟩, 667 ⟨78JOC379⟩, 675 ⟨82TL703⟩; **3**, 38 ⟨78SC211⟩, 186, 228 ⟨80TL805⟩, 207 ⟨80JA2097⟩, 219 ⟨81JA3460⟩, 219, 231 ⟨75JA380⟩, 221 ⟨75JOC538, 69JA2806⟩, 243 ⟨82JA358, 82JA360, 82JOC1981, 82JOC3183⟩, 244 ⟨78JOC379, 84JOC392⟩, 442 ⟨85JOC4650⟩, 464, 470 ⟨61JOC2617⟩; **4**, 174 ⟨82JA358⟩, 174 ⟨87JOC622⟩, 204 ⟨75JOC3807, 76JOC1668, 81OS(60)66⟩, 305 ⟨76JA6715⟩; **5**, 199 ⟨74JA7807⟩, 199 ⟨82OS(61)147⟩

Danishefsky, S. J., **1**, 343 ⟨82JOC2231⟩, 575 ⟨85JA1421⟩, 648, 760 ⟨88JOC3391⟩, 700, 703 ⟨89JOC12⟩, 751 ⟨88JA3929⟩, 780 ⟨87TL4959⟩, 831 ⟨88TL1371, 89JA8231⟩, 835 ⟨84TL3159⟩; **2**, 340 ⟨90JOC1981⟩; **3**, 33, 156 ⟨89JOC4249⟩, 243 ⟨85JA1246, 88JA4368⟩, 243 ⟨88JA7434⟩, 244 ⟨85JA1274, 88JA3929⟩; **4**, 188 ⟨89JA6661⟩; **5**, 336 ⟨90JOC2576⟩, 342 ⟨92JOC7001⟩, 572 ⟨90JOC831⟩, 631 ⟨89TL2743⟩
Dankert, G., **6**, 606 ⟨79GEP2753204⟩
Danklmaier, J., **5**, 694 ⟨89LA665⟩
Dankner, D., **1**, 954 ⟨64T2177⟩
Dankova, E. F., **5**, 567, 571 ⟨87KGS845⟩
Danks, L. J., **2**, 179 ⟨71CJC943⟩
Dankwardt, J. W., **2**, 673 ⟨86JOC271⟩
Dannecker, R., **1**, 527, 944 ⟨85JOC2400⟩
Dannecker-Doerig, I., **1**, 944 ⟨92SL107⟩
Danneels, D., **4**, 231 ⟨75T327⟩
Dannhardt, G., **3**, 360 ⟨89CZ290⟩
Dannley, R. L., **2**, 689 ⟨66JOC153⟩
Dannock, M. C., **1**, 489 ⟨92TL6679⟩
Dan-oh, N., **4**, 704 ⟨92T1449⟩
Danopoulos, A. A., **2**, 437 ⟨90POL2413⟩; **4**, 678 ⟨91CC181⟩
Danplaise, D., **2**, 238 ⟨77JA7743⟩
Dansette, P., **3**, 294 ⟨75TL2557⟩
Dansette, P. M., **2**, 659 ⟨76JOC977⟩
Dantsuji, H., **4**, 900 ⟨93MI 420-01⟩
Danzer, W., **5**, 446, 448 ⟨73AG(E)77⟩
Danzin, C., **2**, 299 ⟨85JCS(P1)2201⟩; **6**, 58, 58 ⟨92BAU370⟩
Danzmann, E., **6**, 218, 220, 224 ⟨75ZAAC(418)109⟩
Daoquan, W., **5**, 1078 ⟨87MI 528-01⟩
Dappen, G. M., **5**, 142 ⟨64JOC2813⟩
Dappen, M. S., **1**, 16 ⟨87JOC3706⟩; **2**, 356, 784 ⟨86JA1306⟩; **3**, 152 ⟨84JOC798⟩
Dapperheld, S., **1**, 428 ⟨90S403⟩
Dapporto, P., **2**, 438 ⟨84JCS(D)575⟩
Darack, F., **2**, 773 ⟨74JA5158⟩
Daraji, R. R., **6**, 547 ⟨85IJC(B)685⟩; **6**, 546 ⟨87JIC194⟩
Daran, J. C., **4**, 681 ⟨92JOM(441)81⟩; **5**, 798 ⟨87JOM(327)339, 89OM2253, 90JOM(381)191⟩, 798, 952 ⟨87JOM(333)245⟩, 953 ⟨92JA8088⟩
Darapsky, A., **3**, 468 ⟨06CB1373⟩, 476 ⟨08CB344⟩
Darby, A. C., **2**, 631 ⟨70JCS(C)1480⟩
Darby, N., **1**, 381 ⟨76JCS(P1)359⟩
Darcel, C., **1**, 962 ⟨94SL457⟩
Darcy, M. G., **2**, 349 ⟨88JOC5856⟩
Dardis, R. E., **2**, 1008 ⟨80JOC524⟩; **3**, 265 ⟨78TL743⟩
Dardoize, F., **4**, 442 ⟨83S951⟩
Da Re, P., **3**, 280 ⟨61JOC3650⟩
Darensbourg, M. Y., **1**, 93 ⟨78JA7425⟩; **3**, 83 ⟨84OM1601⟩
Dario, M. T., **2**, 240 ⟨94TL301⟩
Daris, J.-P., **5**, 236 ⟨88CJC1537⟩
Darkes, P. R., **6**, 647 ⟨78AF1435⟩
Darling, D. L., **2**, 613 ⟨86JOC3242⟩
Darling, G. D., **2**, 560 ⟨87JOC4644⟩
Darling, P., **3**, 4, 115 ⟨81JOC1728⟩
Darling, S. D., **1**, 100 ⟨67JOC2794⟩; **2**, 653 ⟨69OS(49)90⟩; **3**, 530, 531 ⟨68JOC3974⟩
Darlington, W. H., **1**, 315 ⟨80JA4980⟩
Darmadi, A., **2**, 292, 857 ⟨79ZAAC(448)35⟩; **6**, 588 ⟨81ZN(B)1261⟩, 588, 592 ⟨80ZN(B)526⟩
Darmory, F. P., **1**, 506 ⟨75JOC1588⟩
Darnault, G., **5**, 729 ⟨83CI(L)391⟩
Darnbrough, G., **1**, 816 ⟨76CC684⟩
Darnofall, M. E., **4**, 193 ⟨79JOC4825⟩
Daroda, R. J., **4**, 719 ⟨80JCS(D)2315⟩
Da Rold, A., **4**, 720 ⟨89JOM(362)C27, 90IC1817⟩, 720 ⟨92JOM(427)231⟩
Darragh, K. V., **4**, 32 ⟨70JOC1297⟩
Darré, F., **5**, 593 ⟨75BSF829⟩
Darrigo, P., **2**, 39 ⟨94MI 202-03⟩
Darrin, M., **3**, 86 ⟨58CRV1⟩
Dartiguenave, M., **3**, 486 ⟨87OM1822⟩, 487 ⟨90JA5351, 92JOM(438)83⟩; **4**, 548 ⟨90PS(47)327⟩; **5**, 871, 872, 873, 1148 ⟨90JA6277⟩; **6**, 665 ⟨89JOM(372)201⟩, 665, 667, 668, 669, 670 ⟨90JA6277⟩, 673 ⟨88CC1598⟩, 675 ⟨87OM1822, 90JA5351⟩
Dartiguenave, Y., **3**, 486 ⟨87OM1822⟩, 487 ⟨90JA5351, 92JOM(438)83⟩; **4**, 548 ⟨90PS(47)327⟩; **5**, 871, 872, 873, 1148 ⟨90JA6277⟩; **6**, 665 ⟨89JOM(372)201⟩, 665, 667, 668, 669, 670 ⟨90JA6277⟩, 665, 680 ⟨86JA7868⟩, 673 ⟨88CC1598⟩, 675 ⟨87OM1822, 90JA5351⟩
Dartmann, M., **4**, 488 ⟨85CB3227⟩, 611 ⟨89JOM(363)7⟩; **6**, 166 ⟨86AG(E)919⟩, 242, 685 ⟨91EUP450584⟩, 266 ⟨90JOM(399)189⟩, 685, 692 ⟨88CB655⟩, 689 ⟨85CC1687⟩, 689, 692 ⟨85C277⟩
Darzens, G., **2**, 54 ⟨47CR(224)570⟩
Das, A. R., **1**, 14 ⟨92JCS(P1)1881⟩
Das, B., **3**, 273 ⟨90S243⟩
Das, B. C., **5**, 157 ⟨88TL4419⟩
Das, J., **2**, 593 ⟨80JA1742⟩; **3**, 11 ⟨80JA1742⟩, 113 ⟨89JOC5182⟩
Das, K. C., **1**, 811 ⟨69TL3459⟩
Das, M. K., **2**, 474 ⟨86SRI67⟩, 859 ⟨71IC1028⟩; **6**, 492 ⟨92AP(325)267⟩, 514 ⟨90BCJ3658⟩, 515 ⟨76JA5702⟩
Das, N., **2**, 95 ⟨84JA5261⟩
Das, R., **4**, 1029 ⟨91JCS(D)351⟩
Das, S. K., **5**, 1131 ⟨84IJC(B)254⟩
D'Ascoli, R., **3**, 155 ⟨80TL4521⟩, 219 ⟨80JOC4502⟩; **4**, 3 ⟨66JOC916⟩
Dasgupta, F., **5**, 354 ⟨89CC1640⟩
Dasgupta, H. S., **3**, 4, 88, 115 ⟨82S588⟩
Dasgupta, R., **4**, 193 ⟨85TL1581⟩
Das-Gupta, H. N., **2**, 885 ⟨37JIC349⟩
Dash, K. C., **4**, 586 ⟨81CB441⟩; **5**, 1141 ⟨73JA4855⟩
Dashevskaya, T. A., **6**, 233 ⟨79ZOR396, 79ZOR1245⟩
Dashkovskaya, E. V., **5**, 685 ⟨90JOU1408, 90ZOR1629⟩
da Silva, C. C., **5**, 1044 ⟨93TL6717⟩
da Silva, E., **1**, 348 ⟨88CC285⟩
da Silva, G. V. J., **1**, 688, 690 ⟨88JA6914⟩
da Silva Correa, C. M. M., **1**, 949 ⟨90T3061⟩
Da Silva Jardine, P., **2**, 50 ⟨89JA9243⟩; **5**, 138 ⟨88JA3672⟩
das Neves, H. J. C., **2**, 208 ⟨76TL2171⟩
Datcheva, V. K., **3**, 628 ⟨91JA6318⟩
Date, M., **1**, 448 ⟨89CPB36⟩; **2**, 694 ⟨90CPB2637⟩; **4**, 936 ⟨90CPB902⟩
Date, V., **4**, 176 ⟨88JA2662⟩
Dathe, C., **1**, 490 ⟨70JPR150⟩; **2**, 487 ⟨66CB1609⟩, 873 ⟨67MI 217-01⟩; **6**, 60, 267 ⟨64CB1673⟩
Datta, A., **4**, 950 ⟨92SL429, 92TL8035⟩
Datta, A. K., **4**, 1048 ⟨80JCR(M)235⟩
Datta, A. P., **5**, 477 ⟨64AG(E)132⟩
Datta, D., **4**, 950 ⟨92TL8035⟩
Datta, I., **3**, 193 ⟨91SC167⟩
Datta, S. K., **3**, 683 ⟨69JOC2016⟩; **6**, 115 ⟨76JOC813⟩
Datta-Gupta, N., **2**, 260 ⟨66JOC221⟩
Datto, R. L., **2**, 627, 632 ⟨19JA287⟩
Daub, G. H., **1**, 881 ⟨51OR(6)41⟩; **2**, 250 ⟨78JOC4659⟩; **5**, 166 ⟨51OR(6)1⟩
Daub, G. W., **1**, 295 ⟨82JOC161⟩, 828 ⟨81JOC1485⟩; **2**, 606 ⟨85JOC1599⟩, 922 ⟨88T3919⟩; **3**, 141 ⟨82JOC743⟩
Dauben, Jr., H. J., **1**, 865 ⟨63JA3041⟩, 906 ⟨57JA4557, 60JOC1442⟩; **4**, 191 ⟨54JA1359⟩
Dauben, W. G., **1**, 259 ⟨64T1259⟩, 419 ⟨62JOC1910⟩, 420, 421 ⟨73PAC197⟩, 420 ⟨B-80MI 109-02, 73JA3932⟩, 421 ⟨81JOC687, 81T3263⟩, 498 ⟨85TL1429, 85TL989⟩, 632 ⟨77JA3414⟩, 633 ⟨68JA4762⟩, 736 ⟨73TL4425⟩, 794 ⟨64PAC539, 70JA7408⟩, 807, 809 ⟨81TL1583⟩, 826 ⟨90TL3241⟩, 833 ⟨80JOC4413⟩, 876 ⟨85JA2264⟩, 925, 935 ⟨89JOC6101⟩; **2**, 13 ⟨89JOC6101, 89TL2461⟩, 71 ⟨53HCA325⟩, 444 ⟨84JOC4293⟩; **3**, 57 ⟨77JOC682⟩, 234 ⟨88SC1323⟩, 243, 496 ⟨75TL4353⟩, 325

⟨93JOC7635⟩, 494 ⟨73TL4425⟩; **4**, 181 ⟨86JOC4964⟩, 550 ⟨88CB2121⟩; **5**, 346 ⟨82JOC5042⟩, 599 ⟨56JA4135⟩
Daubenspeck, N., **5**, 446 ⟨92CB1583⟩
Daubert, B. F., **4**, 506 ⟨51JA3867⟩
Daudon, M., **2**, 148 ⟨85SUL127⟩
Daufresne, M., **3**, 620 ⟨16PRS232⟩
Daugherty, B. W., **5**, 746 ⟨84TL3651⟩
Daughhetee, Jr., P. H., **5**, 1003 ⟨64JOC246⟩
Daum, H., **6**, 82, 86, 88, 302, 303, 317 ⟨72CB3050⟩
Daumas, M., **3**, 8 ⟨89S64⟩
Dauphin, B., **1**, 961 ⟨85S768⟩
Dauphin, G., **5**, 360 ⟨90CPB2352⟩
Dauplaise, D. L., **1**, 786 ⟨83AG(E)560⟩
D'Auria, M., **2**, 34 ⟨82SC1127⟩; **3**, 4, 115 ⟨82S245⟩, 67 ⟨81TL1041⟩, 155 ⟨80TL4521⟩, 219 ⟨80JOC4502⟩
Daus, K. A., **2**, 4, 23, 30 ⟨90JA7433⟩, ⟨93JA3318⟩ ⟨93JA3071⟩
Dauzonne, D., **2**, 398 ⟨84S1054⟩; **3**, 309 ⟨84JOC4409⟩; **4**, 796 ⟨90SC3339⟩, 798 ⟨87S1020, 90S66⟩
Davalian, D., **3**, 553 ⟨74JOC3906⟩; **4**, 874 ⟨74JOC3906⟩
Davaloy, V. S., **3**, 212 ⟨74S778⟩
Davankov, V. A., **4**, 461 ⟨73IZV955⟩
DaVanzo, J. P., **2**, 103 ⟨60MI 202-03⟩
Dave, H. R., **5**, 305 ⟨70CRV439⟩
Dave, K. G., **1**, 91 ⟨68JOC747⟩
Dave, M. A., **2**, 753, 760 ⟨76IJC(B)904⟩
Dave, M. P., **3**, 363 ⟨76IJC(B)999⟩
Dave, P., **3**, 3, 114 ⟨86SC1343⟩
Dave, P. R., **4**, 429, 430, 431 ⟨93TL6677⟩
Davenport, R. J., **1**, 1031 ⟨89T4293⟩
Daves, Jr., G. D., **5**, 106 ⟨64JOC2006⟩
Davey, D. D., **5**, 275 ⟨87JOC4379⟩
Davey, J. F., **2**, 39 ⟨74MI 202-01⟩
David, G., **4**, 568, 573 ⟨92CB771⟩
David, H. G., **3**, 593 ⟨69JA4761⟩
David, L., **2**, 838 ⟨64JCS4750⟩
David, S., **2**, 684 ⟨83JOC441⟩; **3**, 8, 121 ⟨81TL2885⟩; **6**, 322 ⟨81CC780⟩
Davidkov, K. A., **6**, 502 ⟨73GEP2131789⟩
Davidoff, E. F., **2**, 435 ⟨67JOC781⟩
Davidon, M., **2**, 102 ⟨76BSF833⟩
Davidovich, Y. A., **6**, 68 ⟨86UK1803⟩, 95 ⟨78IZV468⟩
Davidsen, H., **1**, 82 ⟨36CB2106⟩; **4**, 404, 405 ⟨36CB2106⟩
Davidsen, S. K., **4**, 99, 465 ⟨87OS(65)119⟩; **5**, 329 ⟨91JOC5482⟩
Davidsohn, W. E., **2**, 1060, 1061 ⟨68JOM(15)69⟩, 1076, 1080, 1083, 1089, 1092, 1098, 1099, 1101 ⟨67CRV73⟩, 1093 ⟨66JOM(5)29, 67JOM(9)435, 69JOM(16)413⟩
Davidson, A. H., **1**, 215 ⟨81TL1751⟩, 216 ⟨84CC585⟩, 643 ⟨77JCS(P1)1452⟩, 643, 755 ⟨77JCS(P1)550⟩; **4**, 200 ⟨93SL111⟩, 650 ⟨77JCS(P1)550⟩
Davidson, A. J., **1**, 80 ⟨70JOC3273⟩
Davidson, D., **3**, 541 ⟨48JA3426⟩; **4**, 204 ⟨48JA3426⟩
Davidson, F., **2**, 916 ⟨85JOC3666⟩; **4**, 1057 ⟨81JOM(209)1⟩; **5**, 1133 ⟨87JA6532⟩; **6**, 722, 733 ⟨93OM3405⟩
Davidson, G., **2**, 856 ⟨B-92MI 216-01⟩
Davidson, I. M. T., **4**, 614 ⟨93OM289⟩, 617 ⟨73CC323⟩
Davidson, J., **2**, 900 ⟨92JOC3482⟩
Davidson, J. L., **4**, 692 ⟨74JOM(80)C39⟩, 714 ⟨80CC749⟩; **5**, 798 ⟨88JCS(D)27⟩
Davidson, J. M., **2**, 962 ⟨65JCS843⟩
Davidson, P. J., **1**, 663, 668 ⟨76CRV219⟩; **2**, 566 ⟨73JOM(57)269⟩; **4**, 656 ⟨73JOM(57)269⟩; **6**, 174, 186, 190, 191, 195 ⟨76JCS(D)2268⟩
Davidson, R. I., **1**, 108 ⟨82JOC1904⟩
Davidson, R. M., **4**, 563 ⟨80JOC2698⟩
Davidson, R. S., **2**, 460 ⟨67JCS(C)2131⟩, 854 ⟨68JCS(C)2131⟩; **6**, 118 ⟨67JCS(C)1547⟩
Davidson, W. J., **3**, 102 ⟨71AJC1883⟩
Davidson, W. R., **1**, 1109 ⟨76JA6700⟩
Davies, A. G., **1**, 75 ⟨68JCS(C)1474⟩, 657 ⟨72JCS(P2)803⟩; **2**, 95 ⟨56JCS4669, 62JCS4288⟩, 95, 96 ⟨53JCS1541, 54JCS2200⟩, 96, 98 ⟨76CC1022⟩, 96, 99, 109 ⟨65MI 202-02⟩, 99 ⟨57CI(L)1622, 58JCS2372⟩, 112 ⟨59JCS429⟩, 492, 497 ⟨71JCS(C)1660⟩, 501 ⟨70JCS(B)735⟩, 599 ⟨82COMC-I(2)519⟩, 686 ⟨92JCR(S)202⟩, 687 ⟨81JCS(P2)1512⟩, 956 ⟨66JCS(C)106⟩, 1101 ⟨68JCS(C)317⟩; **5**, 716 ⟨67JCS(C)1309⟩, 716, 717 ⟨66JOM(5)590⟩
Davies, A. J., **2**, 576 ⟨B-85MI 211-03⟩
Davies, A. M., **1**, 407, 408 ⟨69CC293⟩
Davies, A. P., **2**, 7 ⟨73CC110⟩, 8, 32, 36 ⟨87TL4445⟩, 62 ⟨86CC831, 88T3761⟩; **6**, 633 ⟨87TL4445⟩
Davies, C., **6**, 448 ⟨92MI 614-01⟩
Davies, C. H., **6**, 721 ⟨74JCS(D)357⟩
Davies, C. W., **5**, 30 ⟨52JCS1607⟩
Davies, D. D., **5**, 792, 794 ⟨70MI 520-02⟩
Davies, D. I., **1**, 346 ⟨94CC41⟩, 628 ⟨62JOC293⟩, 848, 919 ⟨B-78MI 119-05⟩
Davies, D. L., **4**, 679 ⟨83JCS(D)2661, 84JCS(D)2293⟩, 680 ⟨81CC862⟩
Davies, D. T., **6**, 445 ⟨89JAN367⟩
Davies, G., **2**, 484 ⟨32JCS2284⟩, 487, 492, 497, 502 ⟨31JCS185⟩
Davies, G. D., **1**, 494 ⟨87JOC5064⟩
Davies, G. R., **2**, 489, 493, 498 ⟨26PRS523⟩
Davies, H. G., **4**, 171, 176 ⟨91CSR211⟩, 176 ⟨91CSR271⟩; **5**, 262 ⟨B-89MI 506-02⟩
Davies, H. M. L., **1**, 299, 352, 353, 354, 495, 550 ⟨91COS(4)1031⟩, 801 ⟨92TL453⟩, 811 ⟨87T4265, 87TL1853, 88TL975, 93T5203⟩, 812 ⟨91JOC3817, 91JOC5696, 91JOC723, 92TL6935⟩; **3**, 193 ⟨87SC1709⟩, 373 ⟨87TL371⟩; **4**, 349 ⟨86JOC1537⟩, 422 ⟨84JOC113⟩
Davies, I. W., **4**, 268 ⟨91JCS(P1)662⟩, 269 ⟨90TL135⟩
Davies, J., **2**, 184 ⟨81JCS(P1)2106⟩, 185 ⟨79TL4977⟩
Davies, J. A., **5**, 1139 ⟨82OM64⟩
Davies, J. E., **1**, 934 ⟨63JA3419⟩
Davies, J. H., **2**, 439 ⟨66JCS(C)245⟩; **5**, 828 ⟨68JCS(C)431⟩
Davies, J. S., **2**, 540 ⟨87JCS(P1)1107⟩; **5**, 718 ⟨73JCS(P1)2614⟩
Davies, J. W., **4**, 98 ⟨85CC686⟩, 404, 418, 419, 424, 427 ⟨91JCS(P1)1693⟩
Davies, N., **2**, 6 ⟨84JOC3027⟩
Davies, P. B., **4**, 30 ⟨86CI(L)490⟩
Davies, R. A., **6**, 181, 199, 292, 386 ⟨74JOM(69)45⟩
Davies, R. P., **3**, 455 ⟨54JA4931⟩
Davies, R. V., **5**, 593 ⟨76JCS(P1)33, 77JCS(P1)2357⟩
Davies, S. G., **1**, 97 ⟨78JCS(P1)1479⟩, 134 ⟨85CC209, 86CC495⟩, 202 ⟨89TL3581⟩, 202, 203 ⟨93SL323⟩; **2**, 315 ⟨91TA183⟩; **3**, 29 ⟨89TL2971⟩, 66 ⟨77JCS(P1)1346⟩; **5**, 161 ⟨91SL781⟩, 328, 329 ⟨93JCS(P1)459⟩, 329, 330 ⟨88TL6509⟩, 453 ⟨82CC1303, 83CC1202, 83JOM(248)C1, 84CC956, 84JOM207, 84TL2709, 85CC209, 85JOM(285)213, 85TL2125, 86CC495, 86CC607, 86CC609, 86TL3787, 86TL623⟩, 937 ⟨84OM1764⟩, 941 ⟨90AOC(30)1⟩, 943, 944 ⟨84CC747⟩; **6**, 503 ⟨91TL4791⟩
Davies, S. J., **5**, 927 ⟨87JOM(325)185⟩
Davies, T. M., **2**, 393 ⟨67JA725⟩
Davies, V. C., **2**, 354 ⟨75S205⟩
Davies, W., **1**, 592 ⟨49JCS282⟩; **2**, 128 ⟨49JCS282⟩, 129 ⟨49JCS278⟩, 136 ⟨46JCS1050⟩, 183 ⟨32JCS483⟩, 769 ⟨22JCS715⟩; **4**, 11 ⟨22JCS2202⟩
Davies, W. C., **2**, 433, 434, 435 ⟨29JCS1262⟩, 456 ⟨32JCS2880⟩, 483, 872 ⟨30JCS463⟩, 824 ⟨31JCS2109⟩
Davies, W. D., **2**, 1013 ⟨68TL3909⟩
Davies-Fidder, A., **1**, 884 ⟨66T285⟩
d'Avilla, U., **5**, 833 ⟨53JOC64⟩
Davin-Pretelli, E., **6**, 115 ⟨71BSF556⟩
Davis, A. P., **2**, 195 ⟨81CC741⟩, 543 ⟨90TL6721, 90TL6725⟩; **3**, 82 ⟨91COS(8)283⟩; **4**, 519 ⟨81JCS(P1)1934⟩; **5**, 264 ⟨94TL4865⟩, 341 ⟨81T2181⟩
Davis, B. A., **3**, 447 ⟨64CRV149⟩
Davis, C. R., **1**, 990 ⟨91CC566⟩

Davis, C. S., **4**, 47 ⟨67OS(47)23⟩; **5**, 849 ⟨70JPS105⟩
Davis, D. D., **1**, 667 ⟨69JOM(18)P1, 70JOC1303⟩
Davis, D. R., **6**, 264 ⟨87JOC1794⟩
Davis, F. A., **1**, 147 ⟨89TL1613, 90TL1653, 91T173, 93JOC4890⟩, 784 ⟨64JOC3131⟩; **2**, 16 ⟨92TL1153, 93TL3971⟩, 16, 613 ⟨91TL1631⟩, 44 ⟨93PAC633⟩, 150 ⟨82JA5412, 84JOC1465, 88JOC5004, 92JA1428, 92JOC7274⟩, 184 ⟨78TL97, 80JA7967, 86JOC1033⟩, 185 ⟨81JA7016⟩, 251 ⟨83TL3191, 90TL1653⟩, 287 ⟨83TL1213⟩, 342, 344, 349 ⟨89T5703⟩, 357 ⟨73IJS71⟩, 357, 768 ⟨79OPP33⟩, 358 ⟨77JOC967⟩, 361, 792 ⟨76JOC1333⟩, 365 ⟨74JA5000⟩, 652 ⟨87TL5115⟩, 700 ⟨88TL4269⟩, 713 ⟨92TA629⟩, 718 ⟨81JOC3467, 83JOC3071⟩; **3**, 149 ⟨93TL3971⟩, 156 ⟨92CRV919⟩, 171 ⟨90OR(38)655⟩, 183 ⟨88TL4365⟩, 436 ⟨86JOC5024⟩, 437 ⟨92JOC6387, 93PAC633, 93TL6229, 94JOC3243, 94TL9351⟩, 438 ⟨86TL3957, 87TL5115⟩, 634 ⟨92JA1428⟩; **4**, 262 ⟨92JA1428⟩, 312 ⟨93OSC(8)546⟩, 853 ⟨80JOC2597⟩; **5**, 347 ⟨77TL1721⟩
Davis, F. K., **3**, 212 ⟨73JOC3829⟩
Davis, G. A., **3**, 557 ⟨67JA598⟩
Davis, G. W., **2**, 219 ⟨70JOC4148⟩
Davis, H. E., **3**, 345, 346 ⟨67JOC1562⟩
Davis, J., **1**, 1143 ⟨75CC17⟩
Davis, J. A., **2**, 12, 627 ⟨65JOC415⟩
Davis, J. H., **1**, 806, 807 ⟨77JA1499, 77JA2919⟩
Davis, Jr., J. H., **5**, 801 ⟨86ICAL15⟩
Davis, J. T., **1**, 119 ⟨83JOC1767⟩, 204 ⟨83JA5946⟩, 749 ⟨84TL2183⟩; **5**, 155 ⟨84TL2183⟩
Davis, K., **3**, 638 ⟨65JOC3689⟩
Davis, K. E., **1**, 595 ⟨73IJS205⟩; **2**, 153, 184 ⟨67JA718⟩, 608 ⟨64JA427⟩
Davis, K. H., **2**, 129 ⟨65JOC3689⟩
Davis, M., **1**, 82 ⟨65CC71⟩
Davis, M. A., **4**, 554 ⟨65JCED303⟩
Davis, N. M., **3**, 307 ⟨81JOC839⟩
Davis, N. R., **5**, 410, 431 ⟨67JA431⟩; **6**, 127, 132, 325, 326, 520 ⟨67JA431⟩
Davis, P., **5**, 235 ⟨80TL5071⟩
Davis, P. D., **1**, 1015 ⟨84JA6417⟩; **5**, 306 ⟨90TL5201⟩
Davis, R. A., **2**, 200 ⟨68CC9⟩, 200 ⟨71JOC2693⟩; **4**, 31, 34, 44 ⟨63JOC494⟩; **6**, 217 ⟨56USP2744148⟩, 217, 226 ⟨56USP2745886⟩, 223 ⟨53USP2639302, 59USP2871274, 53USP2639301⟩, 226 ⟨53USP2658086⟩, 649 ⟨67JOC1662⟩
Davis, R., **1**, 179 ⟨81JOC2985⟩; **2**, 683 ⟨88SC1379⟩; **3**, 171 ⟨87SC823⟩, 613 ⟨81JOC2985⟩; **6**, 7 ⟨82JCS(D)2281, 84JOM(260)C75, 85JOM(280)397, 90JOM(386)229, 90JOM(397)51⟩
Davis, R. B., **1**, 1045, 1137 ⟨56JA1675⟩; **3**, 273 ⟨56JA1675⟩
Davis, R. C., **3**, 371 ⟨90JCS(P1)2881⟩, 407, 411 ⟨78JOC1467⟩
Davis, R. E., **1**, 898 ⟨71JA6688⟩; **4**, 681 ⟨71CC608, 74JA7108⟩, 687 ⟨90OM2745⟩, 710 ⟨82OM1607⟩
Davis, R. G., **4**, 219 ⟨77TL533⟩
Davis, R. H., **5**, 828 ⟨68JCS(C)431⟩
Davis, S. G., **1**, 539 ⟨88JCS(P1)839⟩
Davis, T., **3**, 579 ⟨88JCS(P2)1071⟩
Davis, W. B., **4**, 683 ⟨92OM3736⟩
Davis, W. D., **2**, 49 ⟨93JOC844⟩
Davis, W. M., **2**, 261, 888 ⟨91JA165⟩, 307 ⟨91JA2321⟩, 927 ⟨93JA10394⟩, 949 ⟨89JA1888⟩; **3**, 512 ⟨90IC1104⟩, 514 ⟨90JA8378⟩, 522 ⟨92JA3367⟩; **4**, 549 ⟨93AG(E)756⟩, 1015 ⟨89OM1371⟩
Davisia, B., **4**, 346 ⟨79T81⟩
Davison, A., **2**, 478 ⟨68JCS1133⟩; **3**, 397 ⟨67CC670⟩; **4**, 683 ⟨69JA6598⟩; **6**, 523 ⟨63JCS1133⟩
Davison, I. G. E., **1**, 949 ⟨91TL49⟩
Davisson, V. J., **6**, 264 ⟨86JOC4768, 87JOC1794⟩
Davy, H., **3**, 339 ⟨72CR(C)623, 75BSF(2)1435⟩; **5**, 551 ⟨82CC457, 85JCR(S)272⟩, 551, 560 ⟨85CI(L)824⟩
Davy, W. A., **2**, 711 ⟨48JCS528⟩
Davydov, A. V., **6**, 282 ⟨68IZV656⟩, 621 ⟨81MI 620-03⟩, 649 ⟨81MI 621-02⟩
Dawber, J. G., **1**, 51 ⟨83JCS(P2)1923⟩
Dawes, C. C., **4**, 190 ⟨75CC432⟩
Dawes, H. M., **6**, 400 ⟨88POL2049⟩
Dawoodi, Z., **4**, 722, 725 ⟨79JOM(170)95⟩
Dawson, A. D., **2**, 160 ⟨77JOC593⟩, 162 ⟨75JOC2759⟩
Dawson, B. T., **1**, 121 ⟨92JA5110, 93JOC2483⟩; **2**, 74, 585 ⟨92JA5110⟩; **3**, 286 ⟨91JA4672⟩
Dawson, D. S., **2**, 876 ⟨65JOM(3)406⟩
Dawson, G. J., **1**, 194 ⟨93TL3149⟩
Dawson, I. M., **1**, 654 ⟨86CC620⟩; **3**, 684 ⟨88CC1339⟩; **4**, 139 ⟨88JCS(P1)2585⟩
Dawson, J. H. J., **1**, 1143 ⟨78JA1928⟩
Dawson, J. K., **6**, 9 ⟨52JCS3490⟩
Dawson, M. I., **1**, 727 ⟨77JOC2783, 77TL4257⟩, 1052 ⟨70TL2247⟩; **3**, 494 ⟨77JOC2783⟩
Dawson, P. A., **6**, 340 ⟨80CC781⟩
Dawson, P. J., **4**, 165 ⟨80JCS(P1)2535⟩
Dawson, R. H., **4**, 1030 ⟨86JOM(309)C56, 87CC826⟩
Dawson, R. L., **3**, 286 ⟨62JA4899⟩
Dawson, T. M., **3**, 242 ⟨71JCS1292⟩
Dawson, T. P., **2**, 496, 498 ⟨57JOC452⟩
Dax, S. L., **1**, 524, 1140 ⟨89TL2627⟩
Day, A. C., **1**, 1120 ⟨67JCS(B)991⟩; **3**, 444 ⟨70OS(50)3⟩, 463 ⟨66JCS(C)467⟩, 543 ⟨73CC247⟩
Day, A. R., **4**, 408 ⟨70JHC355⟩, 414 ⟨60JOC2108⟩; **5**, 194 ⟨50JA3874⟩
Day, C. S., **5**, 469 ⟨78JA7112, 80JA5393⟩; **6**, 516 ⟨81JA2206⟩
Day, J. C., **2**, 27 ⟨74JA5616⟩; **5**, 318 ⟨86JOC4959⟩
Day, M. J., **2**, 765 ⟨75JCS(P1)1764⟩
Day, N. E., **2**, 448 ⟨61JOC4770⟩
Day, R. I., **4**, 319 ⟨55JA3869⟩
Day, R. O., **2**, 856 ⟨79JA3790, 83IC1771⟩; **6**, 661 ⟨86AG(E)457⟩, 686, 688 ⟨87PS(30)495⟩
Day, V. W., **4**, 656 ⟨88OM1828⟩, 692 ⟨76JA8289, 78JA2090⟩; **5**, 459 ⟨82JA1096⟩, 469 ⟨78JA7112, 80JA5393⟩; **6**, 403 ⟨79JA2768⟩, 516 ⟨81JA2206⟩
Dayagi, S., **2**, 751 ⟨B-70MI 215-01⟩; **3**, 404, 405, 407 ⟨B-70MI 310-01⟩, 451 ⟨58JOC2014⟩; **4**, 728 ⟨B-70MI 403-01⟩
Dayal, B., **5**, 294 ⟨74TL2633⟩, 1118 ⟨74TL3135⟩
Dayan, J. E., **5**, 46 ⟨63OSC(4)499⟩
Dayan, M., **4**, 873 ⟨92CC1048⟩
Dayer, F., **3**, 293 ⟨74HCA2201⟩
Dayrit, F. M., **1**, 1105 ⟨81JA4466⟩, 1139 ⟨80JOC3053⟩
Dayrit, M. F., **3**, 147 ⟨80JOC3053⟩
De, A., **3**, 306 ⟨93JCR(S)192⟩; **5**, 580, 615 ⟨78JCS(P1)629⟩
De, B., **1**, 1016, 1082, 1083 ⟨84JA2735⟩
De, K. K., **5**, 1040 ⟨75MI 527-01⟩
Deacon, G. B., **1**, 1038 ⟨81JOM(205)4⟩; **2**, 887 ⟨86ICA(113)43⟩, 888 ⟨84AJC527⟩, 956 ⟨66JCS(C)106⟩; **4**, 697 ⟨74JOM(67)C61⟩
Deadman, J., **4**, 526, 527, 528 ⟨92TL4209⟩
Deady, L. W., **2**, 298 ⟨78AHC(22)71⟩
Deakin, M. R., **5**, 1097 ⟨82CC607⟩
Dean, D. C., **3**, 321 ⟨89CC921, 92JA593⟩, 321 ⟨93JOC4555⟩
Dean, D. L., **3**, 538 ⟨73JA6294⟩
Dean, D. W., **6**, 58, 542 ⟨92TL1839⟩
Dean, F. H., **1**, 2, 6 ⟨65CJC1689⟩; **2**, 18 ⟨73OSC(5)136⟩; **5**, 59 ⟨73OSC(5)580⟩
Dean, F. M., **2**, 999 ⟨82AHC(30)167⟩; **3**, 80 ⟨81JCS(P1)1437⟩, 357, 377 ⟨75JCS(P1)1335⟩, 361 ⟨69JCS(C)2192⟩, 532 ⟨48JCS1674⟩; **4**, 307 ⟨82JCS(P1)1193, 82JCS(P1)2771⟩; **5**, 575 ⟨69JCS(C)2192⟩
Dean, I. C., **5**, 764 ⟨76AJC357⟩
Dean, J. M., **5**, 819 ⟨65JOC2809⟩, 828, 837 ⟨65LA(687)191⟩
Dean, P. D. G., **5**, 78 ⟨65JCS6655⟩
Dean, R. A., **2**, 212, 219 ⟨51JA3627⟩

Dean, W. P., **2**, 76 ⟨88OM2289⟩
Deana, A. A., **1**, 418 ⟨59JA4266⟩; **5**, 95, 306 ⟨73OSC(5)944⟩
De Ancos, B., **6**, 124 ⟨91T3171⟩
De Angelis, F., **6**, 141 ⟨76S798⟩
Deans, F. B., **1**, 447 ⟨57JCS498⟩
Dear, R. E. A., **1**, 914 ⟨62JOC3441⟩; **2**, 624 ⟨70JOC1703⟩; **5**, 1102 ⟨71USP3761501⟩; **6**, 46, 47 ⟨74JFC(4)107⟩, 432, 432 ⟨65JOC1317⟩
Dear, R. E., **2**, 1016 ⟨68JOC819⟩
Deardorff, D. R., **1**, 193 ⟨89JOC2759⟩; **4**, 94 ⟨82JA6816⟩
Dearing, C., **1**, 155 ⟨68CR(C)1162⟩
De, R. L., **6**, 518 ⟨86ZN(B)283⟩
Deatherage, F. E., **2**, 30 ⟨63OSC(4)851⟩
Debad, J. D., **4**, 656 ⟨93OM4572⟩
Debaerdemaeker, T., **3**, 322 ⟨88TL2045⟩; **6**, 268 ⟨90CB945⟩
Debal, A., **1**, 209 ⟨80SC49⟩, 560 ⟨77TL3187⟩, 763 ⟨81CC877, 81TL1595⟩; **3**, 459 ⟨74BSF1710, 75JOM(87)25⟩; **4**, 648 ⟨81CC877⟩
Debattista, F., **2**, 1054 ⟨75JOC124⟩
Debbagh, G., **1**, 205 ⟨86AG(E)760⟩
Debellis, F., **5**, 219, 694 ⟨87JHC325⟩
De Benedetto, L., **4**, 718 ⟨93OM4545⟩
De Beneville, P. L., **5**, 833 ⟨59USP2902514⟩
Debergh, D., **5**, 241 ⟨80SA(A)769⟩
Deberly, A., **5**, 63 ⟨82TL1259⟩
De Bernardinis, S., **1**, 433 ⟨90JCS(P1)428⟩
DeBernardis, J., **1**, 816, 818 ⟨70TL353⟩
De Bernardis, J. F., **5**, 687 ⟨87JHC789⟩
De Bernardo, S., **1**, 755 ⟨88TL2401⟩; **5**, 346 ⟨85JOC3462⟩
DeBlauwe, F., **6**, 79 ⟨80MM252⟩
De Boer, A., **2**, 137 ⟨69JOC3112⟩; **4**, 209, 213 ⟨74JOC77⟩
DeBoer, C. D., **1**, 812 ⟨64JA899⟩
DeBoer, J. A., **1**, 808 ⟨66JA5846⟩
De Boer, Th. J., **4**, 406 ⟨67RTC417⟩
De Boer, T. J., **3**, 467 ⟨54RTC229, 63OSC(4)250⟩
Deboer, T. J., **3**, 432 ⟨79RTC18, 79T147, 80RTC246, 80RTC278⟩
Debon, A., **2**, 617 ⟨75BSF2493⟩
Debono, M., **5**, 48 ⟨69JOC1454⟩
De Botton, M., **3**, 54 ⟨63MI 302-01⟩
De Brabander, J., **6**, 128 ⟨91TL2821⟩
DeBruin, K. E., **1**, 1122 ⟨68JA372⟩; **2**, 440, 441 ⟨69JA7027, 69JA7023⟩, 455 ⟨75JOC1523⟩, 846 ⟨69JA7393, 72JOC2272⟩; **5**, 370 ⟨90PS(50)139⟩, 370, 590, 614, 643 ⟨90JOC6091⟩
DeBrunner, M. R., **4**, 440 ⟨54JA3489⟩
DeBruyn, D. J., **5**, 282 ⟨74JOC2740⟩
De Buyck, L., **2**, 749 ⟨76T3063⟩; **3**, 24 ⟨82TL2853⟩, 27 ⟨88BSB525⟩, 35 ⟨84S46⟩, 184 ⟨82S765⟩, 409 ⟨85TL2709⟩, 409, 563 ⟨82OPP213⟩, 415, 416 ⟨80OPP49, 82S43⟩, 416 ⟨86S192⟩, 417 ⟨82TL789⟩, 418 ⟨83S632, 78JOC2670⟩, 658 ⟨76T3063, 79S741⟩; **4**, 183 ⟨83CB3631⟩; **6**, 509 ⟨77BSB663⟩
Dec, J., **5**, 55 ⟨40JA3495⟩
DeCamp, A., **5**, 325 ⟨89JOC5746⟩
De Carvalho, E., **5**, 361, 362, 711 ⟨85JCR(S)132⟩
De Cesare, J., **2**, 244 ⟨84S1037⟩; **5**, 337 ⟨84S1037⟩
De Checchi, C., **3**, 382 ⟨65JPS(B)581, 66JPS(A)248, 67JCS(B)117⟩
DeChristopher, P. J., **1**, 28 ⟨74JOC3525⟩; **2**, 366 ⟨74JOC3525⟩
Decicco, C. P., **1**, 357 ⟨85TL4707, 87TL2833, 88TL2613⟩
DeCicco, G. J., **1**, 1003 ⟨85JA6546⟩
Deck, K. J., **4**, 709 ⟨92OM2128⟩
Deck, W., **5**, 919 ⟨94AG(E)456⟩
Decker, D. P., **3**, 270 ⟨87JA7495⟩
Decker, O. H. W., **1**, 835 ⟨85JA443, 85JOC5769, 86JOC1377⟩; **3**, 539 ⟨86JOC3067⟩
Decker, Q. W., **2**, 227 ⟨57JOC145⟩
DeClercq, E., **2**, 625 ⟨86JMC213⟩; **6**, 264 ⟨91BMC357⟩, 598 ⟨91JMC3187⟩
Declercq, G., **2**, 662 ⟨75MI 213-01⟩
Declercq, J.-P., **1**, 995 ⟨93TL1779⟩; **2**, 1008 ⟨92POL2351⟩; **3**, 46 ⟨78BS8693⟩, 416, 416 ⟨86JOC3839⟩, 566 ⟨88JOC18⟩, 585 ⟨90CC829, 91JOC4008⟩, 602 ⟨82JA2920⟩; **5**, 601 ⟨85T693⟩, 655 ⟨89CCC3245⟩; **6**, 105 ⟨83TL1035⟩, 373, 688 ⟨89CC593⟩, 653 ⟨90T3897⟩, 688 ⟨90JOC5750⟩
De Clerq, E., **6**, 489 ⟨88JMC1831⟩
Declerq, J. P., **6**, 74 ⟨78BSB693⟩
De Cock, Ch., **4**, 88, 773, 785 ⟨87T4309⟩
De Cock, C. J. C., **1**, 786 ⟨88JOC4859⟩
De Cock, W., **2**, 3 ⟨87S188⟩; **3**, 152 ⟨87S188⟩
Decodts, G., **5**, 132 ⟨85S40⟩
Decora, A. W., **1**, 869 ⟨63T1625⟩
De Corte, B., **2**, 22 ⟨92BSB237⟩; **3**, 563 ⟨87JOC1147⟩; **5**, 697 ⟨92T7345⟩
Decorte, E., **5**, 1078 ⟨85S509⟩
Decorzant, R., **1**, 403, 734 ⟨75HCA1016⟩, 618 ⟨82TL5043⟩, 889 ⟨74HCA1317⟩; **3**, 494 ⟨75HCA1016⟩
De Crauw, T., **4**, 18 ⟨31RTC753⟩
DeCrescenzo, G., **5**, 755 ⟨83SC1083⟩
Decroix, B., **3**, 305 ⟨92URP1710560⟩
Decrouen, J. M., **3**, 339 ⟨74CR(C)259⟩
De Cusati, P. F., **6**, 461 ⟨90TL1405⟩, 462 ⟨90JOC1⟩
Dědek, V., **4**, 33 ⟨84JFC(26)467⟩, 970 ⟨81CCC1389⟩; **6**, 410 ⟨90CZP264358⟩
Dederer, B., **6**, 617 ⟨78CB698⟩
Dedieu, A., **4**, 716 ⟨91CC1274⟩
Dedolph, D., **4**, 898, 907, 908, 911 ⟨85JOC3878⟩
Dedopoulou, D., **1**, 482 ⟨92SC1953⟩
De Duyck, L., **3**, 219 ⟨81JOC3226⟩
Deeb, T. M., **1**, 950 ⟨87JA3163, 90T2317⟩, 950 ⟨90T2329⟩; **2**, 323 ⟨87JA3163⟩
Deeg, A., **2**, 861 ⟨91ZAAC(606)119⟩
Deem, M. F., **1**, 82 ⟨65CC71⟩
Deeming, A. J., **1**, 196 ⟨82COMC-I(4)377⟩; **2**, 582 ⟨78CC344, 84JOM(276)C45⟩; **4**, 682 ⟨82JCS(D)1155⟩, 682, 683 ⟨75JCS(D)1614, 92JCS(D)2423⟩
Deerberg, J., **5**, 342 ⟨94HCA554⟩
Deeter, J., **2**, 726 ⟨93JOC3469⟩
Deev, L. E., **6**, 219 ⟨92RCR75⟩
Defauw, J., **1**, 314 ⟨88JOC50⟩
De Feoand, R. J., **5**, 277 ⟨63JOC2915⟩
Defoin, A., **5**, 330 ⟨89TL7061, 91TA1209⟩, 330, 332 ⟨92HCA109⟩
DeFrees, D. J., **1**, 129, 162 ⟨81JOC1693⟩, 154 ⟨87JA5996⟩, 155 ⟨87JA6004⟩, 906 ⟨89JA1527⟩, 1143 ⟨80JA3334⟩; **2**, 525, 1005 ⟨81JOC1693⟩, 1022 ⟨82JPC2835⟩
DeFrees, S. A., **3**, 351 ⟨90JA3029⟩; **5**, 513 ⟨90JA3029, 90JA3040⟩
Degan, S., **4**, 22 ⟨90JOC5719⟩
Degani, I., **1**, 48 ⟨78JCS(P1)1133⟩; **2**, 134, 143 ⟨83S630⟩, 180 ⟨81S149, 89S957⟩; **3**, 11 ⟨87S79⟩, 16, 132 ⟨81S51⟩, 64 ⟨67G397⟩, 283, 309 ⟨87S311⟩, 306, 307 ⟨93JCR(S)192⟩
Degani, Y., **2**, 751 ⟨B-70MI 215-01⟩; **3**, 404, 405, 407 ⟨B-70MI 310-01⟩; **4**, 728 ⟨B-70MI 403-01⟩; **5**, 1111 ⟨70JA6969, 71JOC2727⟩
Degel, F., **3**, 699, 703 ⟨79AG(E)75⟩, 702 ⟨88CB461⟩; **5**, 1087 ⟨87CB461⟩
Degen, D., **2**, 212 ⟨67AG(E)703⟩
Degen, P., **1**, 465 ⟨71HCA1939⟩; **5**, 151 ⟨71HCA1939⟩
Degener, E., **5**, 652 ⟨68AG(E)856⟩, 654 ⟨68AG942⟩; **6**, 607 ⟨69LA(730)133⟩
Degenhardt, C. R., **1**, 1094 ⟨77TL159⟩; **2**, 911, 946 ⟨77TL159⟩; **4**, 1002 ⟨82SC415⟩; **6**, 123 ⟨90H(30)855⟩, 159 ⟨82SC415⟩
Degering, E. F., **2**, 31 ⟨41OSC(1)36⟩, 102 ⟨38JA2810⟩, 776 ⟨51JA1323⟩; **3**, 426 ⟨39JA3194⟩; **5**, 59, 89 ⟨50JA5137⟩
Degering, E. P., **2**, 366 ⟨39JA3250⟩
Degginger, E. R., **6**, 75 ⟨54JA5736⟩
Deghengi, R., **2**, 618 ⟨62CJC818⟩; **3**, 537 ⟨60JA3201⟩; **6**, 505 ⟨73OSC(5)645⟩

De Giovani, W. F., **3**, 7, 120 ⟨90SC383⟩
Degl'Innocenti, A., **2**, 140 ⟨91JOC7323⟩, 541 ⟨93TL873⟩, 908 ⟨89S647⟩, 918 ⟨91S1201, 92PAC439⟩, 918, 939 ⟨89JOC1473⟩, 935 ⟨92SL883⟩; **3**, 34 ⟨92SL803⟩, 336, 340, 346, 363 ⟨91JOC7323⟩, 346, 353 ⟨93TL873⟩; **4**, 929 ⟨89S647⟩, 930 ⟨87TL4093⟩, 944 ⟨92SL499, 92SL883⟩, 951 ⟨91TL1899⟩, 1018 ⟨92PAC439⟩; **5**, 409 ⟨89S647⟩, 410 ⟨80TL3927⟩, 413 ⟨85AG(E)1068⟩, 416, 423 ⟨87TL4093⟩, 417 ⟨88JOC3612⟩, 419 ⟨85JOC130⟩, 427 ⟨92SL332, 92TL1507⟩, 427, 428 ⟨92SL329⟩, 467 ⟨89JOC2966⟩, 648, 650 ⟨89JOC19⟩; **6**, 325, 335, 520 ⟨86TL5985⟩, 349, 585 ⟨85TL1091⟩, 520, 702 ⟨85AG(E)1068⟩
Degnan, W. M., **5**, 994, 995 ⟨40JA1595⟩
Degner, D., **6**, 77, 78 ⟨86T553⟩
Degny, E., **5**, 633 ⟨91BSF574⟩
de Gonna, V. von, **2**, 472 ⟨91CB1563⟩
De Graaf, P. W. J., **4**, 724 ⟨77JOM(127)391⟩
DeGraff, S. A. G., **3**, 235 ⟨70CC381⟩
De Groot, A., **3**, 54, 253 ⟨82TL4831⟩, 63 ⟨81TL4137, 87RTC549⟩; **4**, 461 ⟨94S23⟩
De Groot, A. E., **3**, 250 ⟨69RTC994⟩
De Guilli, G., **2**, 395 ⟨72S44⟩
Dehaen, W., **5**, 622 ⟨91JCS(P1)607, 93JCS(P1)1719⟩
Dehareng, D., **1**, 570 ⟨93JOC4685⟩
Dehasse-De Lombaert, C. G., **1**, 518 ⟨84TL5043⟩
Dehgani, A., **3**, 219 ⟨91TL5697⟩
Dehlinger, T., **3**, 303 ⟨69JHC141⟩
Dehmer, K., **4**, 900 ⟨67CB2577⟩
Dehmlow, E. V., **1**, 550 ⟨77CB154⟩, 990 ⟨90TL1841⟩, 1038 ⟨82LA1750⟩, 1065 ⟨80LA1, 81T1653⟩, 1065 ⟨86T3569⟩; **3**, 247 ⟨73JOC1451⟩; **4**, 16, 25 ⟨71LA(744)42⟩, 16, 25 ⟨90CB583⟩, 18 ⟨80LA1⟩, 190 ⟨76TL95⟩, 704 ⟨89LA187⟩, 768 ⟨87ZN(B)796⟩, 1080 ⟨84JCR(S)396, 87TL5489, B-84MI 424-02⟩; **5**, 1078 ⟨85IJ219⟩, 1106 ⟨90ZN(B)1091⟩; **6**, 32 ⟨82CB3894, 89LA187⟩, 141 ⟨79LA1456⟩, 462 ⟨81CB1210⟩, 554, 555 ⟨83LA1839⟩
Dehmlow, E. von, **1**, 572 ⟨79S372⟩
Dehmlow, S. S., **1**, 550 ⟨77CB154⟩; **4**, 1080 ⟨B-84MI 424-02⟩
Dehmlow, V., **4**, 1080 ⟨93CC1241⟩
Dehn, R. L., **1**, 1082 ⟨67TL3739, 67TL3943⟩
Dehn, W. H., **2**, 1019 ⟨11JA1598⟩
Dehn, W. M., **2**, 495, 498, 500 ⟨05MI 209-01⟩, 498 ⟨06MI 209-01⟩; **4**, 405, 406 ⟨49JA2271⟩; **5**, 74 ⟨30JA3647, 32OSC(1)89⟩; **6**, 213, 223 ⟨09JA1220⟩, 504 ⟨49JA2297⟩
Dehne, H., **4**, 981 ⟨82JPR915⟩; **6**, 617, 635 ⟨91JPR187⟩, 626 ⟨86ZC204⟩
Dehnert, P., **2**, 293, 499, 500 ⟨79ZN(B)1646⟩, 894 ⟨81ZN(B)48⟩
Dehnicke, K., **5**, 316 ⟨75ZAAC(415)115⟩, 776 ⟨82ZAAC(486)61⟩, 777 ⟨90CZ295⟩, 780 ⟨91ZAAC(596)47⟩, 781 ⟨88ZN(B)1628, 88ZN(B)1679, 89ZAAC(575)10, 90ZN(B)1273⟩, 1016 ⟨71JOM(31)C3⟩, 1143 ⟨75BSF1439⟩; **6**, 733 ⟨66AG(E)841⟩
DeHoff, B., **1**, 177, 178 ⟨89JOC5831⟩
DeHoff, B. S., **1**, 752, 816, 818 ⟨87JOC3883⟩, 816 ⟨87TL723⟩, 816, 818 ⟨86JOC4316, 87JOC3860⟩, 1053 ⟨87TL527⟩; **2**, 9 ⟨87TL723⟩
Deichmann, D., **5**, 1049 ⟨67TL1013⟩
Deinema, M. H., **2**, 251 ⟨52RTC1228⟩
Deinet, A. J., **1**, 45 ⟨46JA1813⟩
Deisenroth, T. W., **6**, 632 ⟨91S86⟩
Deit, H. Le, **2**, 14 ⟨89CC313⟩
Deitchman, D., **6**, 508 ⟨73JMC901⟩
Deiter, R. K., **3**, 217 ⟨85JA4679⟩
De Jeso, B., **2**, 813 ⟨80JOM(193)165⟩, 816 ⟨76JOM(120)217, 77JOM(137)23⟩
DeJeso, B., **2**, 816 ⟨80JOM(186)C9⟩
De Jeso, B., **2**, 816 ⟨85JOC3863⟩; **3**, 406 ⟨80JOM(186)C9⟩
Dejneka, T., **5**, 346 ⟨84H(21)191⟩
De Jong, J., **2**, 753 ⟨72JOC3571⟩
De Jong, R., **3**, 161 ⟨90TL7549⟩
De Jong, R. L. P., **4**, 1017 ⟨87MI 421-01⟩
De Jongh, D. C., **3**, 543 ⟨70JOC4204⟩
Dejonghe, J. P., **3**, 259 ⟨76AG(E)371⟩; **4**, 1016, 1017 ⟨76AG417⟩
Dekant, W., **3**, 550 ⟨91JA5120⟩
deKanter, F. J. J., **2**, 521 ⟨93TL4655⟩
De Keczer, S., **5**, 13 ⟨85MI 501-01⟩
Deker, P. B., **2**, 326 ⟨88JOC2968⟩
Dekerk, J.-P., **2**, 245, 250 ⟨81TL583⟩; **3**, 477 ⟨82TL1103⟩; **5**, 638 ⟨80JOC4366⟩
Deketele, M., **3**, 477 ⟨82TL1103⟩
De Keyser, J.-L., **1**, 786 ⟨88JOC4859⟩
De Kimpe, N., **1**, 115 ⟨B-83MI 104-02⟩; **2**, 22 ⟨79OPP115⟩, 22 ⟨92BSB237⟩, 2, 3, 16, 21, 28 ⟨B-88MI 201-01⟩, 312 ⟨88S587, 91T3407⟩, 322 ⟨88JOC4462⟩, 749 ⟨76T3063⟩, 3 ⟨87S188⟩, 21 ⟨90S595⟩; **3**, 12 ⟨92JOC5761⟩, 24 ⟨82TL2853⟩, 26, 32, 35, 44, 50 ⟨B-88MI 301-01⟩, 35 ⟨84S46⟩, 54 ⟨90BSB41⟩, 148, 152, 155 ⟨B-88MI 304-01⟩, 152 ⟨87S188, 90S595⟩, 162, 165 ⟨B-88MI 304-02⟩, 178, 194 ⟨B-88MI 304-05⟩, 184 ⟨82S765⟩, 219 ⟨81JOC3226⟩, 409 ⟨85AG(E)881, 85TL2709⟩, 409, 563 ⟨82OPP213⟩, 415, 416 ⟨80OPP49, 79OPP115⟩, 416 ⟨82S43, 86S192, 88CC825, 93S1013, 93S89, 86JOC3839, B-88MI 310-02⟩, 417 ⟨82TL789⟩, 418 ⟨83S632, 86TL1707⟩, 563 ⟨78JOC2670, 87JOC1147⟩, 658 ⟨76T3063, 79S741⟩; **4**, 183 ⟨83CB3631⟩; **5**, 348 ⟨94JOC5189⟩, 697 ⟨90T6753, 92T7345⟩; **6**, 509 ⟨77BSB663⟩
De Kimpe, N. G., **1**, 152 ⟨93TL4693⟩
Dekina, T. A., **5**, 1142 ⟨76DOK(230)853⟩
De Kock, C. W., **4**, 703 ⟨78IC625⟩
Dekoker, A., **2**, 758 ⟨80OS(59)26⟩; **4**, 790 ⟨88OSC(6)282⟩; **5**, 2 ⟨79OS(59)26⟩
De Koning, A. J., **2**, 1096 ⟨78JOM(153)C37, 79JOM(174)129⟩
De Koning, J. H., **2**, 675 ⟨81S305⟩
De Koning, L. J., **1**, 590 ⟨88JA2066⟩
Delaby, R., **3**, 612 ⟨36MI 318-01⟩; **5**, 750 ⟨61BCF2067⟩, 751, 753 ⟨56MI 519-01, 56MI 519-02⟩
Delacroix, A., **2**, 726 ⟨81BSF(2)71⟩
del Agua, M. J., **3**, 305 ⟨81JHC1345⟩
DeLaitsch, D. M., **4**, 223 ⟨54JA4962⟩
Delamarche, I., **3**, 303 ⟨93TL2465⟩
DeLaMater, G., **5**, 565, 579 ⟨60CSR45⟩, 591 ⟨61JOC3980⟩, 735 ⟨61CRV45⟩
DeLaMater, M. R., **2**, 652 ⟨77JOC1479⟩
Deland, P. H., **5**, 1124 ⟨75TL3125⟩
Delande, L., **2**, 792 ⟨88TL5657⟩
DeLane, J. C., **1**, 778 ⟨87JOC3688⟩
Delaney, N. G., **3**, 165 ⟨88JMC2199⟩
Delanghe, P. H. M., **2**, 931 ⟨92JOC3270⟩; **4**, 661 ⟨92JOC798⟩
Delaroff, V., **4**, 174 ⟨80T1763⟩
De Lasheras, M. A., **3**, 306 ⟨93JOC5862⟩
De la Torre, J. A., **1**, 837 ⟨94TL15⟩
Delaubenfels, E., **6**, 151 ⟨87PS(38)79⟩
DeLaude, L., **2**, 23 ⟨91TL3705⟩
Delaunay, J., **4**, 950 ⟨77CR(C)281⟩
Delavarenne, S. V., **4**, 764 ⟨69MI 417-01⟩
Delavarenne, S. Y., **1**, 999, 1016 ⟨B-69MI 121-05⟩; **2**, 1011, 1013, 1014, 1017, 1018, 1019 ⟨B-69MI 221-01⟩, 1013, 1047 ⟨70CB1198⟩, 1015 ⟨70CB1216, 70CB1209⟩; **4**, 817 ⟨70CB1198⟩, 819 ⟨73USP3751492⟩, 976, 980 ⟨B-69MI 421-01⟩
Delay, A., **4**, 8 ⟨79TL1913⟩
Delay, F., **5**, 356 ⟨80AG(E)726⟩
Delbecq, F., **1**, 464 ⟨79TL937⟩, 958 ⟨78TL3801, 79NJC321⟩
Delbeke, P., **5**, 622 ⟨91JCS(P1)607⟩
Delbrück, K., **4**, 223 ⟨09CB1476⟩
Del Buttero, P., **2**, 691 ⟨88S344⟩, 692 ⟨87S192⟩; **3**, 430

⟨85CC417⟩; **5**, 497 ⟨73JCS(P1)2540⟩, 497, 499 ⟨74JCS(P2)1483, 74JCS(P2)1676⟩, 676, 677, 906 ⟨85T2503⟩, 905 ⟨79S299, 81G285, 83CC337⟩, 906 ⟨85T3321⟩, 953, 958 ⟨95JOM(486)279⟩, 956 ⟨94TL6377⟩
del Campo, C., **1**, 11, 12 ⟨93T8433⟩
Delduc, P., **3**, 136 ⟨88CC308⟩; **6**, 551 ⟨88CC308⟩
De Leeuw, J. W., **4**, 179 ⟨73RTC1047⟩
Delepeleire, C., **3**, 27 ⟨88BSB525⟩
Delepine, M., **5**, 586 ⟨11CR(153)279⟩, 1023 ⟨07CR1125⟩
De Lera, A. R., **3**, 64 ⟨87TL2921⟩
Deleris, G., **1**, 238 ⟨84TL2135⟩, 978 ⟨75JOM(93)43⟩, 1038, 1138 ⟨76TL2449⟩; **3**, 140 ⟨79JOC3397⟩; **6**, 27 ⟨75JOM(93)43⟩
Delfs, D., **5**, 977 ⟨63BEP631961⟩
Delgado, E., **5**, 1142 ⟨82ICA127⟩
Delgado, J. N., **5**, 215 ⟨65JPS1373⟩
Delgado, M. C., **5**, 45, 71, 97 ⟨82CC1273⟩, 51 ⟨84JOC174⟩
Delgado-Reyes, F., **5**, 952 ⟨91COS(6)485⟩; **6**, 716 ⟨91CRV35⟩
Del Gaudio, J., **4**, 575 ⟨76IS192⟩
Delgiudice, M. R., **5**, 22 ⟨93MI 517-02⟩
DeLiberato, L. A., **4**, 713 ⟨90IC1295⟩
De Lima, C., **4**, 147 ⟨92SL133⟩
Deljac, A., **4**, 259 ⟨73RTC117⟩
Della, E. W., **1**, 900 ⟨87CC833⟩, 923 ⟨90CC1589⟩; **2**, 13 ⟨80JOC4226⟩, 19 ⟨92JOC2850⟩; **4**, 229 ⟨65JOC855⟩
Dellacoletta, B. A., **2**, 622 ⟨82TL371⟩
Dell, C. P., **1**, 831 ⟨89CC1812⟩; **5**, 615 ⟨94JCS(P1)989⟩
Dell'Amico, D. B., **5**, 975, 976 ⟨88CB1891⟩; **6**, 412 ⟨82MI 614-01⟩, 517 ⟨79IC1165⟩
Dellaria, J. F., **2**, 375 ⟨86JA6395a⟩
Dellaria, Jr., J. F., **1**, 36, 486 ⟨77JOC2431⟩, 442 ⟨77JOC3494⟩; **2**, 621, 627 ⟨77JOC2426⟩, 684 ⟨86SC1043⟩
Della Vecchia, L., **4**, 956, 957 ⟨76JA2008⟩
Della Védova, C. O., **6**, 432 ⟨91SA(A)1619, 91ZAAC(600)145⟩, 442 ⟨92ZAAC(609)150⟩, 453 ⟨92SA(A)1179⟩
Dell'Erba, C., **1**, 486 ⟨93T235⟩; **2**, 711 ⟨86T4007⟩; **3**, 664 ⟨87T4625⟩, 664, 667 ⟨90T2205⟩; **4**, 907 ⟨74JCS(P1)2357⟩
Delloria, J. F., **3**, 498 ⟨90TL2661⟩
Delmas, M., **3**, 302 ⟨82SC1121⟩, 373 ⟨87T3945⟩; **4**, 425 ⟨86S643⟩, 701 ⟨69JOM(16)83⟩
Delmastro, M., **2**, 983 ⟨92JOC976⟩
del Mazo, J. M., **6**, 362 ⟨84OPP299⟩
Del Nero, S., **5**, 444 ⟨76JCS(D)203⟩
Deloach, J. A., **3**, 222 ⟨84JOC3033⟩
Delogu, G., **1**, 182, 475 ⟨91JOC183⟩; **4**, 250 ⟨89SL28⟩; **6**, 556 ⟨93JOC1748⟩
De Lombaert, S., **1**, 165 ⟨90JA2264⟩; **3**, 636 ⟨86TL5099⟩, 657, 658 ⟨84TL3475⟩; **5**, 166 ⟨87TL2135⟩
Delorme, D., **1**, 645 ⟨84JA5754⟩; **2**, 459 ⟨90TL6461⟩; **4**, 130 ⟨90TL6461⟩
Deloux, L., **2**, 478 ⟨93CRV763⟩
Delphey, C., **1**, 420 ⟨73T3797⟩
Delpierre, G. R., **2**, 765 ⟨65QR329⟩
Delpon-Lacaze, G., **3**, 503 ⟨92OM3176⟩; **4**, 647 ⟨92JOM(440)233⟩, 663 ⟨92OM3176⟩
Del Pra, A., **5**, 142 ⟨79ICA345⟩
Del Re, G., **6**, 106 ⟨59MI 604-01⟩
Del Rio, I., **5**, 941 ⟨95OM1938⟩
Del'tsova, D. P., **3**, 560, 561, 588 ⟨73BAU1740, 76BAU573⟩, 592 ⟨73DOK(212)628⟩; **4**, 833 ⟨70IZV2140⟩, 979 ⟨86IZV2344⟩; **6**, 106 ⟨767IZV858⟩
Deluca, H. F., **5**, 185 ⟨87MIP8707604⟩
DeLucca, G., **1**, 575 ⟨84JA1518⟩; **2**, 33 ⟨83JOC1678⟩
De Lucchi, O., **1**, 519, 562 ⟨84T2585⟩, 800 ⟨B-83MI 118-01⟩; **2**, 142, 1006 ⟨91OR(40)157⟩, 147 ⟨89TL2575⟩; **3**, 16 ⟨91OR(40)157⟩, 562, 563 ⟨79AG(E)788⟩; **4**, 221, 222, 226, 240, 323 ⟨91OR(40)157⟩, 250 ⟨89SL28⟩, 261 ⟨89TL2575⟩, 272, 854 ⟨91SL565⟩, 277, 854 ⟨92T1485⟩; **6**, 556 ⟨91PS(63)51, 93JOC1748⟩
DeLue, N. R., **2**, 524 ⟨74JA311⟩; **4**, 143 ⟨74JA311⟩
De Lue, N. R., **4**, 143 ⟨77TL3007⟩
Del Valle, L., **2**, 902, 922 ⟨89JA4873⟩; **3**, 203 ⟨89JA4873⟩; **4**, 138, 139 ⟨92OM859⟩, 630, 631, 633, 644 ⟨89JA4873⟩; **6**, 26 ⟨87SC1047⟩
Dema, A. C., **2**, 927 ⟨89JA7615, 90JA7229⟩
DeMahieu, A. F., **2**, 287 ⟨88TL3269⟩; **4**, 399 ⟨88TL3269⟩
Demailly, G., **2**, 7 ⟨91TL3977⟩, 25 ⟨92TL3129, 92TL4901⟩
Demaison, J., **5**, 1160 ⟨94MI 506-04⟩
DeMalde, M., **6**, 6 ⟨56CI(M)371⟩
Demalleray, B. M., **3**, 305 ⟨71HCA353⟩
De March, P., **3**, 231 ⟨81T2391⟩
DeMarco, A. M., **2**, 342, 346 ⟨88JOC1768⟩
Demarco, C., **2**, 211 ⟨63MI 203-01⟩
De Marco, R. A., **4**, 52, 59 ⟨72JOC3332⟩; **5**, 315 ⟨71IC911⟩, 652 ⟨71CC788⟩
DeMarinis, R. M., **6**, 234 ⟨77JOC2024⟩
Demarne, H., **1**, 1037 ⟨65CR(261)1992, 66CR(C)289⟩, 1037 ⟨68BSF205⟩; **2**, 1098 ⟨65CR(261)1992⟩; **6**, 208 ⟨68BSF216⟩
Demartin, F., **4**, 688 ⟨84JOM(267)199⟩
De Mayo, P., **1**, 794, 800 ⟨B-80MI 118-01⟩, 794 ⟨B-63MI 118-01⟩; **3**, 552 ⟨82NJC401⟩
Dembech, P., **1**, 1048 ⟨91TL2169⟩; **2**, 1019 ⟨89S461⟩; **3**, 675 ⟨91TL2013⟩
Dembinskii, I. K., **5**, 694 ⟨76JOU60, 76ZOR64⟩
De Meester, F., **6**, 434 ⟨88EJM561⟩
De Meester, P., **6**, 128 ⟨85TL5001⟩
Demeijere, A., **3**, 709 ⟨91SL87⟩
De Meijere, A., **4**, 109 ⟨89T2957⟩, 309 ⟨91SL87⟩
DeMember, J. R., **1**, 906 ⟨69JA3958, 70JA4627⟩
Demerac, S., **2**, 384 ⟨80AJC1365⟩
Demerseman, P., **2**, 422, 653 ⟨88CC1350⟩, 656 ⟨89S25⟩; **3**, 304 ⟨77JHC861⟩; **4**, 796 ⟨90SC3339⟩, 798 ⟨90S66⟩
De Mesmaeker, A., **1**, 547 ⟨89TL57⟩, 1131 ⟨87TL2637, 87TL2887⟩; **4**, 239 ⟨89TL6311⟩
De Micheli, C., **3**, 678, 681 ⟨79AG(E)721⟩
Demiel, A., **6**, 41 ⟨60JOC993⟩
Demik, N. N., **2**, 471, 472 ⟨91JGU460⟩
Demillequand, M., **2**, 721 ⟨87T5125⟩
Demilo, A. B., **2**, 191 ⟨75S321⟩
Demina, M. M., **1**, 986 ⟨87IZV695⟩
Deming, T. J., **3**, 694 ⟨93JA9101⟩
Demir, A. S., **1**, 560 ⟨92S235⟩; **2**, 385 ⟨92SC2607⟩; **3**, 162 ⟨90SC2279⟩, 162 ⟨92S235⟩
Demirov, G., **5**, 852 ⟨66CB3734⟩
Demjanow, N. J., **1**, 383 ⟨07CB4959⟩
Demko, D. M., **2**, 249 ⟨86TL2099⟩
Demmin, T. R., **2**, 391 ⟨79JOC275⟩; **3**, 192 ⟨79JOC275⟩
Demole, E., **5**, 838 ⟨1874CB790⟩
Demopoulos, B. J., **1**, 19 ⟨83CJC2415⟩
DeMott, D. N., **2**, 27 ⟨63CI(L)1954⟩
Demoulin, A., **1**, 522 ⟨75JA4409⟩
Dempsey, D. F., **4**, 656 ⟨88OM1208⟩
Dempster, C. J., **1**, 811 ⟨70JA703⟩
De Munno, A., **3**, 256 ⟨89H(29)349⟩, 620 ⟨93T8423⟩
Demura, Y., **2**, 444 ⟨88JAP63119491⟩
Demuth, E., **5**, 193 ⟨61CB1762⟩
Demuth, M., **1**, 96 ⟨82AG(E)820⟩, 422 ⟨B-80MI 109-03⟩, 563 ⟨85AG(E)973⟩; **3**, 282 ⟨89SL54⟩; **5**, 199 ⟨87S850⟩
Demuylder, M., **4**, 790, 791 ⟨77NJC369⟩
Demuynck, M., **2**, 130 ⟨62BSF2126, 67BSF1213⟩
Dem'yanets, Z. K., **4**, 975 ⟨89IZV1806⟩
Denarie, M., **5**, 282 ⟨87TL5823⟩
Denat, F., **2**, 942, 949 ⟨92S954⟩
Denenmark, D., **1**, 938 ⟨92JOC6803⟩; **5**, 513 ⟨92JOC6803⟩, 593 ⟨90JOC3442⟩, 739 ⟨89JA1886⟩; **6**, 543, 550, 557 ⟨92JOC6803⟩
Dener, J. M., **1**, 497 ⟨88T7037⟩; **2**, 325 ⟨88JOC6022, 88T7037⟩
Denes, A. S., **1**, 1109 ⟨72CC8⟩
De Net, R. W., **1**, 881 ⟨61JOC644⟩, 1018 ⟨87JMC254⟩; **6**, 104 ⟨64JOC2769⟩

Deneubourg, F., **3**, 694 ⟨89OPP451⟩
Deng, M., **5**, 82 ⟨86MI 502-01⟩, 101 ⟨84TL1797⟩
Deng, M.-Z., **1**, 488 ⟨93JOC1949⟩; **2**, 900 ⟨92JOC4017⟩, 904 ⟨93JOC1949⟩; **5**, 71 ⟨85CC1478⟩, 994 ⟨88T6079⟩
Deng, R. M. K., **2**, 475, 477 ⟨84JCS(D)1911⟩
D'Engenieres, M. D., **1**, 1001 ⟨64BSF2477⟩
Denglazov, N. M., **2**, 1064 ⟨75JGU1832⟩
den Haan, K. H., **2**, 981 ⟨87OM2053⟩; **6**, 194 ⟨87OM1509⟩, 195 ⟨86OM1726⟩
Denham, J. M., **2**, 845 ⟨58JOC1298⟩
Deniau, E., **4**, 487 ⟨92SC2381⟩; **6**, 122 ⟨92SC2381⟩
De Nicola, A., **1**, 166 ⟨92TL6461⟩
deNinno, Jr., F., **5**, 164 ⟨79JA1284⟩
DeNinno, M. P., **1**, 751 ⟨88JA3929⟩; **2**, 66 ⟨90TL7415⟩; **3**, 244 ⟨88JA3929⟩
Denis, J.-M., **1**, 904 ⟨74JA5855⟩; **2**, 337 ⟨85S1131⟩; **3**, 337 ⟨88TL5899⟩, 337, 351 ⟨91T4927⟩, 457 ⟨88TL1287⟩, 563 ⟨87JOC1147⟩, 564 ⟨88T4447⟩, 582 ⟨83CC238⟩, 688 ⟨91AG(E)196⟩, 689 ⟨92CC415⟩, 726 ⟨94CR1413⟩; **4**, 111 ⟨89CC988⟩, 131 ⟨91SC793⟩; **5**, 1152 ⟨91AG(E)196⟩, 1153 ⟨86TL5723⟩, 1158 ⟨92CC415⟩
Denis, J. N., **1**, 440 ⟨82TL3411⟩, 590 ⟨79TL4111, 81TL4009⟩, 592 ⟨83TL1657⟩, 594 ⟨80TL1877⟩, 603 ⟨79TL2697⟩, 713, 1089 ⟨78AG(E)526⟩; **2**, 287 ⟨85CC569⟩, 36 ⟨79TL1801, 83CC229⟩, 281, 284 ⟨79TL2697⟩, 282, 730 ⟨76TL453⟩, 624, 910, 911 ⟨82TL3411⟩, 908 ⟨78AG(E)526⟩; **3**, 39, 175 ⟨76TL453⟩, 40 ⟨91S1216⟩, 74 ⟨82TL3411⟩, 153, 155 ⟨81TL1429⟩, 618, 620 ⟨80CC544, 79TL3995⟩; **4**, 399 ⟨85CC571⟩, 401, 964 ⟨78AG(E)526⟩, 844 ⟨82TL3407⟩; **6**, 88, 94 ⟨82TL3407⟩, 88, 94, 317 ⟨81TL4009⟩
Denis, J. W., **2**, 142 ⟨80CC544⟩
Denis, R. C., **1**, 340 ⟨94TL4531⟩, 340, 1127 ⟨93TL2091⟩
Denise, B., **5**, 952 ⟨91COS(6)485⟩, 953 ⟨95JOM(494)43⟩; **6**, 716 ⟨91CRV35⟩
Denison, J. T., **6**, 212 ⟨58GEP1040011⟩
Denisov, V., **2**, 927, 987, 988 ⟨93TL687⟩
Denisov, V. R., **1**, 1022 ⟨78JOU1095⟩
Denivelle, L., **2**, 94 ⟨54BSF1109⟩; **5**, 1100 ⟨64MI 530-01⟩, 1101, 1103 ⟨65MI 530-01⟩, 1102 ⟨70BSF2637⟩
Denk, R. H., **3**, 445 ⟨62JOC3716⟩
Denk, W., **2**, 456 ⟨56GEP1023033⟩
Denkewalter, R. G., **2**, 223 ⟨68TL3057⟩; **5**, 972 ⟨66JA3163⟩; **6**, 496 ⟨68JA3254⟩
Denkstein, J., **4**, 311 ⟨60CCC2334⟩
Denmark, S. E., **1**, 187 ⟨90JOC1984⟩, 421, 423 ⟨91COS(5)751⟩, 422 ⟨82JA2642, 83HCA2377, 83HCA2397, 86T2821, 88HCA168⟩, 514 ⟨86OS(64)182⟩, 832 ⟨82JA4972, 83JOC3369, 84TL1543, 87JOC5742, 89JA8878, 89TL2469, 91JOC1003, 91JOC5063⟩, 993 ⟨87JOC4031⟩, 1042 ⟨87JA2224⟩; **2**, 78 ⟨88JA984⟩, 356 ⟨90JA311, 90T4857⟩, 356 ⟨91JOC6738⟩, 356, 784 ⟨86JA1306⟩, 398 ⟨88JOC1251⟩, 539 ⟨87JOC165⟩, 785 ⟨93JOC3850⟩, 921 ⟨82JOC4595⟩, 931 ⟨83HCA2377⟩, 998 ⟨91JA723⟩, 1004 ⟨90JA864, 91JA1468⟩; **3**, 41 ⟨93SL359⟩, 55, 229, 271, 272 ⟨88HCA168⟩, 152 ⟨84JOC798⟩, 171 ⟨82JA4972⟩, 190 ⟨88JOC1251⟩, 215 ⟨82JA2642, 83HCA2377⟩; **4**, 152 ⟨92JA2592⟩
Denne, I., **2**, 900 ⟨87TL4169⟩; **5**, 199 ⟨87TL4169⟩
Dennehy, R. D., **3**, 597 ⟨92CC35⟩
Denney, D. B., **1**, 268 ⟨59JA6330, 62JA3944⟩; **2**, 165 ⟨73JA4064⟩, 165 ⟨73JA8191⟩, 434 ⟨67JOC2445⟩, 469 ⟨77JOC782⟩, 674 ⟨66JA1839⟩, 683 ⟨78JOC4672⟩, 830 ⟨69JA5243⟩, 853 ⟨59JA6330⟩, 870 ⟨81JA1785⟩; **3**, 236, 494, 499 ⟨62JOC998⟩, 650 ⟨74JOC2607⟩; **5**, 402 ⟨60ANC1896⟩, 895, 896 ⟨62JOC998⟩
Denney, D. P., **2**, 870 ⟨81JA1785⟩
Denney, D. Z., **2**, 165 ⟨73JA4064⟩, 165 ⟨73JA8191⟩, 830 ⟨69JA5243⟩; **5**, 866 ⟨81JA6164⟩; **6**, 55 ⟨81JA6164⟩, 280, 614 ⟨86TL419⟩
Dennis, J. N., **4**, 964 ⟨82TL3411⟩
Dennis, N., **3**, 228 ⟨76JCS(P1)2289⟩, 669 ⟨72JCS(P1)2054⟩
Dennis, W. E., **3**, 619, 666 ⟨70JOC3253⟩
Dennis, W. H., **2**, 192 ⟨74S295⟩
Denniston, A. D., **4**, 245 ⟨82JOC1145⟩, 252 ⟨75TL4547⟩; **6**, 134 ⟨82JOC1145⟩
Denniston, M. L., **2**, 841 ⟨77IS183⟩
Dennstedt, M., **6**, 81 ⟨1878CB2265⟩
Denny, D. B., **2**, 476 ⟨68TL85⟩
Denny, D. Z., **2**, 476 ⟨68TL85⟩
Denny, M., **1**, 505 ⟨83JA2923⟩
Denny, W. A., **5**, 567 ⟨93JMC2459⟩, 613, 615 ⟨93JCS(P1)1835⟩
Deno, N. C., **1**, 894 ⟨55JA3051, 65JA3000⟩, 894, 900 ⟨62JA2016⟩, 896 ⟨65JA2157⟩, 897, 901, 911 ⟨62JA1498⟩, 897, 901, 911 ⟨63JA2991⟩, 901 ⟨63JA2995⟩, 901, 913 ⟨63JA2998⟩, 902, 912 ⟨64JA1871, B-70MI 119-03⟩, 912 ⟨65JA2153⟩, 912, 913 ⟨65JA4533⟩, 925 ⟨71JA2065⟩; **2**, 20 ⟨71JA438, 79JOC3728⟩; **4**, 193 ⟨47JA2233⟩, 434, 435, 443 ⟨51JA2777⟩
De Noble, J. P., **1**, 514 ⟨76JOC3497⟩
DeNoble, J. P., **4**, 412 ⟨74JOC167⟩
Denot, E., **2**, 18 ⟨60JA4001, 60JA4007⟩
Denson, D. D., **6**, 63 ⟨75JFC(5)475⟩
Dent, B. R., **4**, 742, 748 ⟨89AJC1307⟩
Dent, S. P., **3**, 83 ⟨70CC1703, 75JCS(D)2460⟩, 83 ⟨75JCS(D)2646⟩
Dent, W., **2**, 445 ⟨65JCS1539⟩; **3**, 679 ⟨84JOC3174⟩; **4**, 319 ⟨87JOC235⟩, 509 ⟨85T3529, 89OS(67)133⟩, 510 ⟨85JOC4006⟩
Denurra, T., **1**, 398 ⟨90JOC5532⟩, 432 ⟨89S28, 90JOC5386⟩; **4**, 179 ⟨90S313⟩
Denzel, T., **1**, 988 ⟨74TL1275⟩; **5**, 43 ⟨74TL1275⟩
Denzer, W., **5**, 164, 168 ⟨88JOC4098⟩
Deo, N., **3**, 128 ⟨90TL4105⟩
DePalma, V. M., **1**, 859 ⟨78JA3514⟩
De Pasquale, R. J., **6**, 52 ⟨78JOC1727⟩
Depezay, J.-C., **1**, 486, 1032 ⟨89JOC5324⟩; **3**, 211 ⟨85S766⟩; **4**, 817 ⟨68TL937⟩
De Plato, G., **5**, 652 ⟨03G226⟩
Depner, M., **6**, 69 ⟨35CB2151, 58JPR60⟩
Depraétère, P., **5**, 558 ⟨86MI 513-01⟩
Deprés, J.-D., **1**, 6 ⟨83JA2435⟩
Deprés, J.-P., **1**, 187 ⟨88TL5661⟩, 571 ⟨80JOC2036⟩, 2, 6 ⟨79JA4003⟩; **3**, 234 ⟨80JOC2036⟩; **5**, 49 ⟨89OS(68)41⟩
Deprez, L., **1**, 19 ⟨87TL4759⟩
DePue, R. T., **3**, 515 ⟨90OM2262⟩
DePuy, C. H., **1**, 624, 637, 773, 776, 782, 783 ⟨60CRV431⟩, 627, 628 ⟨56JA790⟩, 632, 633 ⟨60JA634⟩, 870 ⟨68CC1225, 69TL3989⟩; **2**, 346 ⟨60CRV448⟩; **5**, 142 ⟨64JOC2813⟩, 166 ⟨64JOC2810⟩
Deragnaucourt, J., **4**, 229 ⟨87JMC24⟩
Derbesy, G., **2**, 723 ⟨94TL5381⟩; **5**, 247 ⟨92SUL199⟩
der Bruggen, U. Von, **3**, 70 ⟨88CB191⟩; **4**, 188 ⟨88JOC2920⟩
Derbyshire, D. H., **2**, 626 ⟨50JCS573⟩, 631 ⟨50JCS3694⟩
Deregnaucourt, J., **5**, 558 ⟨87JMC24⟩
der Eltz, A. von, **6**, 111 ⟨88LA599⟩
der Emde, H. Von, **1**, 406, 820 ⟨92TL7323⟩
der Emden, W. von, **5**, 586 ⟨60AG268⟩
Dereu, N., **2**, 235 ⟨86EUP165534, 86GEP3443467⟩, 260 ⟨84GEP3229565⟩, 279 ⟨75JHC423⟩; **5**, 334 ⟨92TL3863⟩, 595 ⟨91SC85⟩
Derevyanko, T. Ya., **5**, 826 ⟨67IZV1820⟩
Derge, G. J., **4**, 14 ⟨31JA1594⟩
Dergonov, Y. I., **5**, 968, 969 ⟨73ZOB623⟩, 968, 969 ⟨75ZOB2025⟩
Derguini, F., **2**, 410 ⟨84TL5763⟩
Derguini-Boumechal, F., **1**, 229, 478, 1092 ⟨79TL1503, 76TL3225⟩
Dergunov, Y. I., **2**, 478 ⟨65ZOR1256⟩; **5**, 389, 717 ⟨78ZOB1146⟩, 391 ⟨75ZOB2684, 72ZOB2501⟩, 978 ⟨75ZOB2007⟩
Dergunov, Yu. I., **5**, 716 ⟨76JGU714, 76ZOB714⟩, 717

⟨78JGU1046⟩, 1076, 1077 ⟨87JOM(327)303, 85IZV2824⟩, 1081 ⟨82ZOB921⟩, 1081, 1082 ⟨82ZOB885⟩, 1083 ⟨80ZOB1678⟩, 1085 ⟨89MI 528-01⟩
De Riggi, I., **1**, 949 ⟨92JOC6118⟩
Derijke, D., **3**, 246 ⟨73RTC731⟩
Derkach, G. I., **4**, 130 ⟨68JGU1284⟩, 573 ⟨78URP351438⟩, 574 ⟨72ZOB293⟩; **5**, 652 ⟨59JGU244⟩, 683 ⟨56JGU2239, 56ZOB2009, 58JGU1930, 58ZOB1887, 59ZOB241, 65JGU1019, 65JGU1206, 65ZOB1014, 65ZOB1202⟩, 683, 707 ⟨56JGU2935, 56ZOB2631, 57JGU3284, 57ZOB3248, 57JGU1065⟩, 707 ⟨57ZOB1080, 58JGU2283, 58ZOB2247, 59JGU59, 59ZOB600, 61JGU3192, 68ZOB1867⟩, 707, 861 ⟨61ZOB3424⟩, 707, 861 ⟨68JGU1815⟩, 708 ⟨62JGU167, 62ZOB171, 63JGU1006, 63JGU551, 63ZOB557, 64JGU1152⟩, 708, 709 ⟨65JGU101, 65ZOB1009⟩, 708, 988 ⟨64ZOB1161⟩, 761, 1057 ⟨70UKZ372⟩, 860 ⟨62JGU3689, 62ZOB1873, 63ZOB1017⟩, 861 ⟨62ZOB2245, 63JGU2880, 63JGU2984, 66JGU1101⟩, 967 ⟨69ZOR119⟩, 967, 987 ⟨67AG(E)864⟩, 988 ⟨62ZOB3759, 63ZOB1587, 65ZOB923⟩, 988, 1010, 1017 ⟨69AG(E)421⟩, 997 ⟨69ZOB1712⟩, 1005, 1010 ⟨64ZOB2812⟩, 1008 ⟨66ZOB1639, 66ZOB461⟩, 1052 ⟨69ZOR1285⟩, 1053 ⟨61ZOB2275⟩; **6**, 438 ⟨68ZOR720⟩, 656 ⟨66ZOB461⟩
Derkach, G. J., **2**, 169 ⟨70ZOB1692⟩
Derkach, N. Ya., **2**, 369 ⟨74ZOR2619⟩, 370 ⟨78ZOB1110⟩; **5**, 238 ⟨77ZOR100⟩
Derkach, N. Y., **5**, 341 ⟨74ZOR139, 78ZOR280⟩, 341 ⟨81ZOR529, 86ZOR1189⟩
Derkatsh, G. I., **6**, 445 ⟨69CB2972⟩
der Kimpe, V., **1**, 194 ⟨93TL3309⟩
Dermer, O. C., **2**, 249 ⟨55JA80⟩
Dermott, J. X., **3**, 113 ⟨75JOC3577⟩
Derning, P. H., **5**, 709 ⟨61JOC659⟩
Derocque, J. L., **1**, 1124 ⟨72CB421⟩
Derome, A. E., **3**, 698 ⟨83T2989⟩, 698, 721 ⟨82CC942⟩; **6**, 515 ⟨83T2989⟩
De Roode, W. H., **5**, 781 ⟨77JOM(135)183, 77MI 519-02⟩
De Rosa, M., **2**, 761 ⟨78JOC2639⟩; **5**, 316 ⟨84JOC4784⟩
DeRose, E. F., **3**, 578 ⟨77H(7)247⟩
De Rossi, R. H., **3**, 626 ⟨76JOC3371⟩
der Osten, C. H. von, **3**, 48 ⟨89JA3924⟩
Derouiche, Y., **6**, 515 ⟨91JOM(403)293⟩
Derr, H., **2**, 597 ⟨75IC2809⟩
Dershowitz, S., **3**, 493 ⟨57JOC41⟩
Dert, V., **2**, 10 ⟨79RTC371⟩; **4**, 199 ⟨79RTC371⟩
Derumal, P. J., **2**, 521 ⟨86JOC432⟩
Derunov, V. V., **3**, 597 ⟨89JOM(379)303⟩
DeRussy, D. T., **1**, 813 ⟨88JA890⟩, 815 ⟨88T3139⟩, 1098 ⟨89JOC2278, 90JA5562⟩; **3**, 156 ⟨89JOC4576⟩, 217 ⟨88JA890⟩; **4**, 950, 954 ⟨88T3139⟩
Dervan, P. B., **1**, 650 ⟨76JA1265⟩, 811 ⟨76JA5937⟩; **2**, 379 ⟨79JA2076⟩
Deryckere, A., **5**, 990 ⟨70BSB301⟩
Des Abbayes, H., **3**, 233 ⟨78CC311⟩; **4**, 582 ⟨85JOM(280)C31, 89JOM(359)205⟩; **5**, 452 ⟨86CC1754⟩
Desai, H. S., **3**, 370 ⟨66T7⟩
Desai, J. A., **5**, 760 ⟨79JCS(P1)1255⟩
Desai, M. C., **1**, 1029 ⟨88JA649⟩; **2**, 514, 515, 519 ⟨82JOC5065⟩, 531 ⟨85JA4980⟩, 679 ⟨88JA649⟩; **3**, 135 ⟨82JA6844, 84T1325⟩, 528 ⟨85JA4339⟩; **5**, 377 ⟨94BMC1865⟩
Desai, N. B., **3**, 495 ⟨62JA1745⟩; **4**, 27 ⟨68JOC25⟩, 553 ⟨61JA3539⟩, 745 ⟨62JA1745⟩; **5**, 911, 1093 ⟨61JA3539⟩; **6**, 694 ⟨62JA1745⟩
Desai, S. C., **2**, 386, 798 ⟨65JOC2471⟩
Desai, S. R., **3**, 220 ⟨78JOC4673⟩
Desai, U. R., **5**, 324 ⟨90SC1733⟩
DeSales, J., **3**, 307 ⟨82JOC3668⟩
De Sanctis, Y., **2**, 582 ⟨84JOM(273)C55⟩; **4**, 683 ⟨92JCS(D)2423⟩
De Sarlo, F., **3**, 269 ⟨88JOC2426⟩, 433 ⟨80JCR(S)122⟩, 578 ⟨78JCS(P1)1113⟩, 686 ⟨82S719⟩; **4**, 894 ⟨78JCS(P1)1113⟩
Desauvage, S., **1**, 590 ⟨81TL4009⟩; **4**, 1074, 1075 ⟨84JA3785⟩; **6**, 88, 94, 317 ⟨81TL4009⟩, 92 ⟨84JA3784⟩
Desbene, P.-L., **2**, 626 ⟨89CJC2061⟩
Desbois, M., **3**, 279 ⟨85BSB897⟩
Deschamps, B., **1**, 53 ⟨80JA994⟩, 750, 751 ⟨72T4209, 77TL1137⟩; **3**, 642, 648 ⟨73T2437⟩; **4**, 1026 ⟨92OM1411⟩, 1035 ⟨80JA994⟩
Deschamps, P., **1**, 186 ⟨85S493⟩; **5**, 38 ⟨85S493⟩
Deschler, H., **4**, 416 ⟨69M469⟩
Deschler, U., **4**, 576 ⟨80CB902⟩, 576, 577 ⟨82ZN(B)950⟩, 580, 581, 584, 585 ⟨82CB1379⟩, 583 ⟨81AG(E)586⟩; **5**, 908 ⟨81CB2491, 82ZN(B)950⟩, 908 ⟨83CB1386⟩, 908, 910, 911 ⟨80CB902⟩; **6**, 701 ⟨83CB1386⟩
Deschlker, U., **2**, 166 ⟨87GEP3543567⟩
De Schryver, F. C., **3**, 375 ⟨74JA2964⟩
Descotes, G., **3**, 695, 708 ⟨76TL3427⟩, 695, 708 ⟨79TL163⟩; **4**, 170 ⟨87S497⟩, 174 ⟨75JHC981⟩, 432 ⟨94TL89⟩, 432, 433 ⟨90CC431⟩, 1079 ⟨81CAR(C)1, 81T1875⟩, 1079 ⟨81T2515⟩; **5**, 747 ⟨80TL1459⟩, 1071 ⟨82TL1585⟩; **6**, 309 ⟨76S449, 79CAR(74)127⟩, 309, 310 ⟨78S286⟩, 472 ⟨85JOC118⟩
De Selms, R. C., **3**, 539 ⟨69TL1179⟩
DeSelms, R. H., **5**, 5 ⟨90JA9651⟩
Désert, S., **1**, 834 ⟨92T10315⟩
Desert, S., **5**, 556 ⟨88T3575⟩
Desgranges, C., **2**, 625 ⟨86JMC213⟩
Deshapande, S. S., **2**, 778 ⟨65MI 215-01⟩
Deshayes, C., **2**, 668 ⟨93TL7923⟩; **3**, 453 ⟨80S623⟩
Deshayes, H., **1**, 833 ⟨74BSF2072⟩; **5**, 10 ⟨69CR(C)1325⟩; **6**, 41 ⟨74BSF2072⟩
Deshe, A., **2**, 304 ⟨89S825⟩
Deshmukh, A. R. A. S., **4**, 901, 904 ⟨90JCS(P1)1217⟩
Deshmukh, J. G., **5**, 367 ⟨89IJC(B)186⟩
Deshmukh, M. N., **2**, 147 ⟨84JA8188⟩; **5**, 226 ⟨81CI(L)164⟩
DeShong, F., **3**, 434 ⟨84JOC3421⟩
DeShong, P., **1**, 208 ⟨88JOC4892⟩, 524 ⟨85JOC2309⟩; **2**, 574, 978 ⟨88JA2575⟩; **3**, 231, 303 ⟨85JA5219⟩; **3**, 45 ⟨84JA5598⟩
Deshong, P., **3**, 57 ⟨84TL5355⟩, 218 ⟨87TL2233, 88JOC4892⟩
DeShong, P., **5**, 446 ⟨85JA7788⟩
Deshpande, A. R., **3**, 103 ⟨92IJC(B)526⟩
Deshpande, C. V., **3**, 370 ⟨66T7⟩
Deshpande, P. H., **5**, 735 ⟨69JIC1096⟩
Deshpande, P. K., **1**, 1066 ⟨90T7033⟩; **2**, 85 ⟨90TL4495⟩
Deshpande, R. P., **4**, 203 ⟨83JOC1765⟩
Deshpande, V. H., **6**, 557 ⟨90OPP128⟩
DeSilva, K. G., **2**, 280 ⟨83JCS(D)2143⟩
Desimoni, G., **1**, 826 ⟨93T2093⟩; **4**, 986 ⟨79JPR43⟩
Desio, P. J., **2**, 67, 69 ⟨78CRV491⟩
Desirant, Y., **6**, 222 ⟨58BSB676⟩
Desjardins, C. D., **6**, 49 ⟨73JCS(D)2314, 77CJC3136⟩, 49, 50 ⟨75JCS(D)488⟩
Deslongchamps, P., **1**, 538 ⟨86TL5455, 87TL3457⟩, 878 ⟨B-83MI 119-04⟩, 990 ⟨84CJC1558⟩; **3**, 222 ⟨86TL5455, 87TL3457⟩; **5**, 95 ⟨72CJC2047⟩; **6**, 134 ⟨92CJC2335⟩
Desmaele, D., **1**, 558 ⟨90TL879⟩; **3**, 414 ⟨92TA459⟩; **4**, 279 ⟨92TA1003⟩; **6**, 541 ⟨85TL4941⟩
DesMarteau, D., **3**, 149 ⟨91CC124⟩
DesMarteau, D. D., **2**, 16 ⟨91JOC4925⟩, 613 ⟨87JA7194⟩; **3**, 120 ⟨92TL7245⟩, 419 ⟨84JOC1467⟩; **4**, 104 ⟨92EUP496413⟩, 704 ⟨92JCS(P1)313⟩; **5**, 313 ⟨82POL129, 83JOC771⟩, 315, 662 ⟨91JOC5256⟩, 348, 767 ⟨84HOU(4)883⟩, 659, 660 ⟨91JFC(55)47, 86MI 517-01⟩, 662, 663, 674, 824, 825 ⟨93JCS(P1)505⟩, 662, 808 ⟨84JOC1467⟩, 663 ⟨90JFC(48)367, 91IC535⟩, 807 ⟨81JOC1277⟩, 808, 864 ⟨86MI 521-01⟩, 809 ⟨84JA4266, 88JOC4443⟩; **6**, 232 ⟨80JA2681⟩, 232, 602, 615 ⟨81JFC(18)441⟩, 240 ⟨81JFC(17)463⟩, 251 ⟨71IC2179⟩, 256 ⟨76JA3529, 79IC919⟩, 256 ⟨80IC1330⟩, 256

⟨82JA4034, 83IC805⟩, 256 ⟨93JOC4754⟩, 256, 261 ⟨84JOC3590⟩, 256, 603 ⟨83JOC4844⟩, 257 ⟨86JOC4466⟩, 259, 261, 615 ⟨90JA728⟩, 261, 603, 615 ⟨88JOC4443⟩, 261, 615 ⟨83JOC771⟩, 272 ⟨68IC434, 76JFC(7)501⟩, 279 ⟨89USP4874875, 92EUP496413⟩, 42 ⟨74JOC1298⟩, 42, 231 ⟨75JA13⟩, 53 ⟨79JA7640, 81IC1⟩, 229, 230 ⟨83JOC242⟩, 231, 232 ⟨71JA3882⟩, 428, 430 ⟨82JFC(19)227⟩, 441 ⟨79JFC(14)289⟩, 602, 602 ⟨90EUP353743⟩, 603 ⟨81JOC1277, 84JA4266⟩, 603, 615 ⟨84IC2188⟩, 604 ⟨91JFC(53)181⟩, 614 ⟨93JCS(P1)505⟩, 614, 615 ⟨82POL129⟩
Desmazieres, B., **2**, 248 ⟨76BSF277⟩; **5**, 499 ⟨78JCR(M)5501, 78JCR(S)484⟩
Desmet, M., **5**, 951 ⟨94JOM(479)C12⟩
Desmond, Jr., R. W., **3**, 624 ⟨86JOC1753⟩
Desmond, R., **1**, 940 ⟨88TL281⟩
Desmond, T., **5**, 1097 ⟨83CC457, 84CC75⟩
Desmurs, J. R., **2**, 22 ⟨91SL807⟩; **5**, 315, 316, 318 ⟨93SC2423⟩
Desnuelle, P., **5**, 652 ⟨49BSF251⟩
Desor, D., **4**, 360 ⟨92JOM(433)49⟩; **5**, 432 ⟨90JOM(397)7, 92JOM(433)49⟩
De Souza, J. P., **3**, 170 ⟨75TL921⟩
deSouza, N. D., **3**, 87, 99 ⟨84TL2901⟩
Desper, J. M., **4**, 716 ⟨92ICA557⟩
Despeyroux, B., **1**, 297 ⟨83CC1270⟩; **5**, 142 ⟨84MIP84/1376A1⟩
Despo, A. D., **1**, 978 ⟨80JA5120⟩
Desponds, O., **1**, 113 ⟨93T10175⟩
Desrosiers, P., **3**, 486 ⟨87OM1822⟩; **6**, 675 ⟨87OM1822⟩
Desrosiers, P. J., **4**, 656 ⟨86JA7964⟩
Dess, D. B., **3**, 6, 119 ⟨83JOC4155⟩, 6, 119 ⟨91JA7277⟩; **5**, 204 ⟨91JA7277⟩
Dessaox, O., **3**, 466 ⟨63BSF41⟩
Dessau, R. M., **1**, 336 ⟨74JOC3456⟩, 336, 926 ⟨74JA7977⟩
Dessauer, R., **6**, 143 ⟨71JOC2262⟩
Dessy, R., **1**, 871 ⟨62JA2905⟩
Dessy, R. E., **1**, 539 ⟨66JA3027⟩, 846 ⟨66JA460, 62JA2899⟩, 1040 ⟨56JA1221⟩, 1075 ⟨66JOC624⟩; **5**, 115 ⟨63OSC(4)484⟩
De Staricco, E. R., **6**, 607 ⟨87JFC(37)29⟩
DeStevens, G., **5**, 631 ⟨64JOC1115⟩; **6**, 508 ⟨57JA5710⟩
Destro, R., **1**, 910 ⟨76JA1999, 79AX(B)1846⟩
Desverbne, J.-P., **4**, 621 ⟨89JCR(S)146⟩
Deswarte, S., **3**, 269 ⟨79JHC1657⟩
DeTar, D., **3**, 612 ⟨41JA2796⟩
Detar, D. F., **5**, 206 ⟨55JA6370⟩
DeTar, D. F., **5**, 599 ⟨46JA2029⟩; **6**, 506 ⟨66JA1024⟩
Determann, H., **5**, 84 ⟨63AG(E)358⟩
Detert, H., **1**, 1077 ⟨88LA221⟩
Dethloff, M., **6**, 546 ⟨90MI 617-03⟩
Dethmer, J. K., **2**, 366 ⟨78JMC704⟩
DeTitta, G. T., **3**, 173 ⟨78JOC4552⟩
Detlefsen, W., **1**, 495 ⟨91TL6273⟩
Detre, G., **1**, 430 ⟨73TL447⟩, 1082 ⟨67TL3739⟩
Dettlaf, G., **4**, 681 ⟨76JOM(108)213⟩
Dettlaff, M., **4**, 578 ⟨83JOM(255)305⟩
Dettmeier, U., **1**, 910 ⟨66AG(E)604⟩
Dettmeier, W., **4**, 297, 327 ⟨66CB1944⟩
Detty, M., **2**, 282, 542 ⟨78TL5087⟩, 542, 542 ⟨79TL4189⟩
Detty, M. R., **1**, 611 ⟨80JOC274⟩; **2**, 286, 287 ⟨80JOC274⟩, 541, 542 ⟨82JOC1354, 81JOC1283⟩; **3**, 39, 175 ⟨78TL5087⟩, 39, 175 ⟨79TL4189⟩, 174 ⟨80JOC80⟩; **4**, 93 ⟨85JA6298⟩, 93, 290 ⟨82TL1531, 83TL237⟩; **5**, 728, 732 ⟨80JOC80⟩, 934 ⟨88JOC1203⟩
Deubelly, B., **6**, 164, 371, 373 ⟨88ZN(B)1416⟩
Deuber, T. E., **2**, 204 ⟨71CC1614⟩
Deubzer, B., **2**, 483 ⟨70CB799⟩
Deuchert, K., **3**, 292, 295, 303, 311 ⟨79CB2045⟩
Deus, N., **4**, 1033 ⟨90JOM(384)155⟩
Deuter, J., **1**, 705 ⟨86CB1540⟩, 708 ⟨87CB5⟩
Deutsch, A., **6**, 68 ⟨1879CB115⟩
Deutsch, H., **5**, 1025 ⟨12CB2188⟩
Devani, M. B., **4**, 988 ⟨84TL1291⟩
Devanne, D., **5**, 940, 941 ⟨90CC641⟩
Devant, R., **5**, 66 ⟨86CB2191⟩, 138 ⟨84TL5031⟩
Devaprabhakara, D., **1**, 459 ⟨67JOC4143, 68CJC84, 70JOC347, 70TL4243, 71TL2075, 72TL1391, 75S194⟩, 1106 ⟨79JOM(179)C7⟩
Devaquet, A., **3**, 313 ⟨79JA1095, 79JA1101⟩
Dev, S., **1**, 422 ⟨57JIC169, 68T545⟩, 809, 813 ⟨67TL227⟩; **2**, 679 ⟨74S348⟩; **3**, 4, 115 ⟨79T1789⟩; **4**, 137 ⟨51JA1879⟩; **5**, 52 ⟨68TL2681⟩
Devaud, M., **4**, 700 ⟨84JCR(S)152⟩
Devaux, J. F., **1**, 810 ⟨93JOC2349⟩
Devchand, D. K., **4**, 918, 1075 ⟨86TL4635⟩
Devdhar, R. S., **3**, 370, 376 ⟨74TL3911⟩
Dever, J. L., **4**, 912 ⟨73CA(78)45000b⟩
Devia, A. H., **2**, 619 ⟨87JOC1619⟩
Devia, J. E., **5**, 317 ⟨73SA(A)613⟩
Devillanova, F. A., **6**, 474 ⟨80T1451⟩, 595 ⟨80JHC571⟩
Devincenzis, G., **4**, 415 ⟨83JOC162⟩
Devine, A. M., **4**, 607, 622 ⟨75JCS(D)1434⟩, 617, 623, 656, 661 ⟨75JCS(D)1832⟩, 654, 658 ⟨75JCS(D)1837⟩; **6**, 191 ⟨75JCS(D)1434⟩
Devine, J., **5**, 652 ⟨64MI 506-01⟩
Devitt, F. H., **1**, 1079 ⟨65JOC1015⟩; **3**, 274 ⟨65JOC1015⟩; **4**, 476 ⟨66JCS(B)789⟩
Devlin, J., **6**, 7 ⟨87JCS(P1)1515⟩
De Vohringer, C. M., **6**, 607 ⟨87JFC(37)29⟩
Devos, A., **2**, 8 ⟨89TL3077⟩, 758 ⟨79CC1180⟩; **3**, 475 ⟨79CC1180⟩; **4**, 790 ⟨79CC1180⟩; **5**, 2 ⟨79CC1180⟩
De Vos, M. J., **1**, 240 ⟨85TL6115⟩
De Voss, J. J., **1**, 68 ⟨86CC855⟩
Devotta, S., **6**, 251 ⟨93MI 608-01⟩
De Vries, L., **4**, 979 ⟨71JOC3442⟩
De Waard, E. R., **4**, 179 ⟨73RTC1047⟩
Dewan, J. C., **1**, 397 ⟨89OM1593⟩, 675 ⟨85JA5987⟩; **2**, 306, 927 ⟨89JA4486⟩, 986, 987 ⟨86JA7441⟩; **3**, 512 ⟨86OM2162, 88JA4964⟩, 513 ⟨85JA5999⟩, 516 ⟨86JA2771⟩
Dewar, J. H., **3**, 247 ⟨61JCS3254⟩
Dewar, M. J. S., **1**, 355 ⟨77JA4417⟩, 896 ⟨63JA2245⟩, 905 ⟨77JA377⟩, 906 ⟨59JCS3139⟩, 1109 ⟨69JA5350, 77JA7432⟩; **2**, 378, 381, 382 ⟨73JA1562⟩
Dewar, M. S., **1**, 794 ⟨71AG(E)761⟩; **5**, 112, 114, 116 ⟨66JA3318⟩
Dewey, F. M., **3**, 567 ⟨68TL4207⟩
Dewey, R. S., **6**, 496 ⟨68JA3254⟩
Dewhirst, K. C., **4**, 742 ⟨67JOC1941⟩
Dewidar, A. M., **6**, 68 ⟨64RTC119⟩
De Wilde, H., **1**, 19 ⟨87TL4759⟩
DeWinter, A. J., **1**, 716 ⟨87JOC774⟩
De Witt, C. C., **5**, 85 ⟨43OSC(2)25⟩
De Wolf, W. K., **4**, 38 ⟨85AG(E)585⟩
Dewolfe, R. H., **1**, 837 ⟨B-64MI 118-02⟩; **4**, 177 ⟨B-70MI 404-02⟩, 192 ⟨B-70MI 404-03⟩; **5**, 743 ⟨54JA2411⟩; **6**, 68, 69, 70, 78 ⟨74S153⟩, 68, 69, 80, 81, 91 ⟨B-70MI 603-01⟩, 104, 105 ⟨B-70MI 604-01⟩, 138 ⟨B-70MI 605-03⟩, 296, 297, 307 ⟨B-70MI 610-01⟩, 296 ⟨B-70MI 610-02⟩
Dewsbury, D. J., **5**, 1026 ⟨77S582⟩
Dexheimer, E. M., **2**, 538 ⟨74SRI23⟩, 908 ⟨76JOM(107)229⟩; **4**, 930 ⟨76JOM(107)229⟩; **5**, 416 ⟨76JOM(107)229⟩
Dey, K., **2**, 574 ⟨77JINC153⟩; **3**, 287 ⟨71MI 306-01⟩
Dey, T. L., **2**, 574 ⟨77JINC153⟩
Deycard, S., **1**, 331, 336, 342, 936 ⟨87JOC3509⟩; **4**, 1078 ⟨90JA4284⟩
Deylig, W., **2**, 482, 489 ⟨65ZAAC(336)36⟩, 485, 510 ⟨65CB977⟩
DeYoung, D. J., **6**, 125 ⟨91OM3466⟩
DeYoung, S., **1**, 318 ⟨77JA290⟩
DeYoung, T., **4**, 650 ⟨70JOC1308⟩
Deyrup, J. A., **3**, 595 ⟨75CC34⟩, 602 ⟨78JOC501⟩; **4**, 311 ⟨83CHE(42-1)133⟩

D'Haenens, L., **3**, 532 ⟨84SC163⟩
Dhaher, S. M., **5**, 1144 ⟨88JOM(355)33⟩; **6**, 165, 289, 372, 373 ⟨89JOM(366)39⟩, 396 ⟨87JOM(325)117⟩, 396, 399 ⟨88JOM(355)33, 87CC1183⟩
Dhakte, U. P., **2**, 523 ⟨94JOC2365⟩
Dhaliwal, P. S., **2**, 487 ⟨66JOM(6)364⟩, 499 ⟨67CJC379⟩
Dhanak, D., **2**, 398 ⟨89OS(68)8⟩; **4**, 327 ⟨93OSC(8)550⟩, 906 ⟨90OS(68)8⟩
Dhanoa, D., **1**, 905 ⟨83CJC2403⟩
Dhanoa, D. S., **1**, 120, 465 ⟨87JOC1885⟩; **4**, 396 ⟨82CC358, 87JOC1885⟩
Dhar, D. N., **5**, 107 ⟨86IJC433⟩, 125, 189, 196, 297, 303 ⟨82S506⟩, 344 ⟨86S437⟩
Dhar, P., **2**, 228 ⟨89JOC2998, 92JOC1699⟩
Dhar, R. K., **1**, 831 ⟨89JA3441⟩; **2**, 532 ⟨92JOC2716, 92JOC499⟩, 532, 698 ⟨89JA3441⟩, 532, 698 ⟨93JOC147, 92JOC3767⟩
Dhar, T. G. M., **6**, 629 ⟨92TL1025⟩
Dharmaratne, H. R. W., **1**, 702 ⟨88CJC2860⟩
Dhawan, B., **2**, 694 ⟨85SC411⟩, 694, 845, 846 ⟨86JOC179⟩, 847 ⟨91JOC833⟩; **4**, 470 ⟨88JCR(S)34⟩
Dhawan, K. L., **1**, 213, 478 ⟨80JOC922⟩
Dhawan, S. N., **5**, 1105 ⟨87IJC154⟩
Dhillon, R. S., **1**, 185 ⟨90JOC6356⟩, 488 ⟨86MI 111-01⟩; **2**, 34, 43 ⟨88JOC187⟩, 514 ⟨85TCC23⟩
Dhimane, H., **2**, 313 ⟨89TL4161⟩; **3**, 541 ⟨85TL833, 87TL885⟩, 541 ⟨88TL5919⟩
Dhindsa, A. S., **4**, 870 ⟨90CC470⟩, 873 ⟨90CC816⟩; **6**, 120, 123 ⟨90CC470⟩
Diab, Y., **2**, 364 ⟨76TL1093⟩
Diago-Meseguer, J., **5**, 126, 267, 291 ⟨80S547⟩
D'iakov, V. M., **4**, 342 ⟨79DOK(247)609⟩
Diakur, J., **5**, 234, 239 ⟨76TL4423⟩, 235 ⟨75CJC3693⟩
Dial, C., **2**, 34, 629, 631 ⟨88JOC4477⟩
Diamanti, J., **1**, 881 ⟨67OS(47)20⟩
Diamond, C. J., **6**, 598 ⟨91JMC3187⟩
Diamond, G. B., **1**, 103 ⟨52JA4126⟩
Diamond, J., **6**, 647 ⟨78AF1435⟩
Diamond, L. H., **2**, 374 ⟨54JA4869⟩, 374, 379 ⟨55JA3131⟩
Diamond, S. E., **5**, 386 ⟨74CC622⟩
Dianova, E. N., **4**, 502 ⟨79DOK(246)1130, 82IZV1196, 82IZV2718, 84IZV2023, 84ZOB1511, 85ZOB1471⟩, 558 ⟨81IZV1113⟩; **6**, 678, 691 ⟨91RCR162⟩
Dianova, E. W., **2**, 872 ⟨91RCR162⟩
Diaper, D. G. M., **5**, 24 ⟨59CRV89⟩
Dias, H. V. R., **6**, 722, 733 ⟨93OM3405⟩
Diaz, A., **1**, 918 ⟨72JA6867⟩
Diaz, G. E., **1**, 935 ⟨85JOC2149⟩, 945, 946 ⟨81JA1557⟩
Díaz, I., **5**, 350 ⟨93H(36)1263⟩
Diaz-Ortiz, A., **4**, 826 ⟨92SL893, 93SC1935⟩
DiBattista, F., **1**, 1137, 1141 ⟨75JOC124⟩
Dibble, P. W., **2**, 87 ⟨86JOC3762⟩
Di Bello, C., **5**, 185, 190 ⟨69ZN(B)314⟩
DiBello, N., **3**, 153 ⟨78S227⟩
Di Bianca, F., **2**, 695, 886 ⟨72JOM(38)91⟩
DiBiase, S. A., **3**, 642, 644 ⟨77S629⟩
Dibo, A., **3**, 339, 356, 357 ⟨80BSF(2)530, 80BSF(2)539⟩; **5**, 605, 613 ⟨83BSF(2)277⟩
Dice, D. R., **3**, 335 ⟨74CJC3518⟩
Dice, J. R., **2**, 3 ⟨49JA3107⟩; **3**, 468 ⟨54JA2884, 54JA2887⟩
Dick, C. R., **4**, 12 ⟨62JOC272⟩
Dick, J. H., **2**, 183 ⟨32JCS483⟩
Dick, K. F., **3**, 57 ⟨79JA7131⟩
Dick, T. R., **4**, 546 ⟨71JCS(A)3495⟩
Dicke, R., **1**, 924 ⟨89T951⟩; **2**, 910, 911, 916, 917, 932, 940, 946 ⟨87JOC4868⟩, 916, 930 ⟨85CC354⟩; **3**, 92 ⟨89T951⟩
Dickel, D. F., **2**, 71 ⟨53HCA325⟩
Dicken, C. M., **3**, 45 ⟨84JA5598⟩
Dickens, M. J., **1**, 156 ⟨91TL4709⟩; **2**, 711 ⟨91T8621⟩
Dicker, D. W., **5**, 652 ⟨64MI 506-01⟩
Dickerhoof, D. W., **1**, 897 ⟨72JA5511⟩
Dickerson, J. E., **1**, 494 ⟨78JOC5018, 79JOC4078⟩
Dickert, F. L., **1**, 445 ⟨86JCE930⟩; **2**, 490 ⟨86JCS(D)997⟩
Dickey, J. B., **2**, 36 ⟨43OSC(2)322⟩, 121 ⟨43OSC(2)31⟩, 560 ⟨29JA1576⟩; **4**, 319 ⟨30JA4573⟩
Dickinson, C. L., **4**, 742 ⟨67JOC1941⟩, 829 ⟨59OS(39)13⟩
Dickinson, N., **6**, 241 ⟨82JCS(P1)685⟩
Dickinson, R. P., **1**, 664 ⟨71JCS(C)3447⟩; **3**, 105 ⟨71JCS(C)182⟩
Dickinson, W. C., **2**, 518 ⟨69JA1226⟩
Dickman, D. A., **1**, 149 ⟨87T5095⟩, 149 ⟨89OS(67)60⟩, 150 ⟨85JA7974⟩; **2**, 228 ⟨93S573⟩; **5**, 747 ⟨87T5095⟩
Dickopp, H., **5**, 378 ⟨65TL4007⟩, 718 ⟨63CB1473⟩, 722, 814 ⟨68CB2585⟩
Dickore, K., **2**, 726 ⟨65AG(E)430⟩; **3**, 258 ⟨60CB1068⟩
Dickson, C. A., **4**, 712 ⟨92JOM(427)335⟩
Dickson, G. T., **1**, 637 ⟨49JCS1074⟩
Dickson, H. M., **2**, 274 ⟨65JOC33⟩
Dickson, Jr., J. K., **1**, 828 ⟨89JOC5350⟩
Dickson, L., **1**, 97 ⟨78JOC1007⟩
Dickson, R. S., **2**, 483 ⟨62AJC710⟩; **4**, 687 ⟨69AJC1143, 74AOC323⟩, 688 ⟨67AJC77, 82JOM(224)377⟩, 689 ⟨85OM355⟩, 690 ⟨76AJC2189⟩
Dickstein, J. I., **1**, 536 ⟨B-78MI 112-05⟩, 1016 ⟨B-78MI 121-02⟩; **2**, 1011 ⟨76ACR358, B-78MI 221-02⟩, 1062 ⟨62JA2020, 72JOC2168⟩
Dideberg, O., **5**, 595 ⟨93PS(78)97⟩
Diderrich, G., **6**, 235, 254, 303 ⟨77CB916⟩, 236, 237, 254, 274, 532, 536 ⟨76CB3432⟩, 252, 274 ⟨74CZ109⟩
Didier, E., **1**, 1039 ⟨90SL547⟩
Didiuk, M. T., **1**, 316 ⟨93JA6614, 93JA6997⟩, 316 ⟨94JA3123⟩
Diebert, C. E., **4**, 328 ⟨70JOC1501⟩
Diebold, J. L., **4**, 28 ⟨63OSC(4)254⟩
Dieck, H. A., **1**, 493 ⟨74JA1133, 75JOC1083⟩, 1006, 1008, 1009, 1137 ⟨75JOM(93)253⟩, 1008, 1009, 1137 ⟨75JOM(93)259⟩; **2**, 628, 904 ⟨75JOC1083⟩
Dieck, H., **2**, 813 ⟨86JOM(314)75⟩; **3**, 418 ⟨84CB694⟩
Dieckamnn, W., **5**, 843 ⟨05CB2986⟩
Dieckmann, U., **2**, 572 ⟨72ZAAC(394)293⟩
Dieckmann, W., **3**, 320 ⟨10CB1024⟩
Diederich, F., **1**, 990 ⟨93AG(E)1187⟩, 999 ⟨92AG(E)1240⟩; **2**, 1091 ⟨91JA6943, 92AG(E)1101⟩, 1091 ⟨92AG(E)1241⟩; **4**, 746 ⟨91AG(E)698, 94CC205⟩
Diedrick, C. O., **2**, 298 ⟨76JA269⟩
Diefenbach, H., **3**, 499 ⟨67JPS(B)1039⟩
Dieffenbach, O., **2**, 782 ⟨08GEP1765⟩
Dieffenbacher, A., **3**, 527 ⟨70HCA417⟩
Diego, C. de, **5**, 584, 587 ⟨85H(23)649⟩
Diehl, F., **3**, 359 ⟨89JST(198)307⟩
Diehl, H., **3**, 612 ⟨55OSC(3)372⟩
Diehl, K., **2**, 877 ⟨86CB1953⟩; **5**, 597, 630 ⟨86CB3812⟩
Diehr, H., **2**, 446 ⟨65CB363⟩
Diehr, H. J., **3**, 498 ⟨65CB363⟩
Diekman, J., **3**, 231 ⟨70CC52⟩
Diekmann, J., **3**, 368 ⟨64JA1891⟩, 478 ⟨63JOC2933, 65JOC2272⟩
Diel, P. J., **5**, 404 ⟨91PS(62)15⟩
Dielkus, S., **6**, 184 ⟨92CB1807⟩
Diels, O., **2**, 388 ⟨23CB1933⟩; **3**, 532 ⟨38CB1179⟩
Diemert, K., **2**, 448 ⟨86PS(26)307⟩, 449 ⟨82CB1947⟩, 466 ⟨83PS(15)155⟩, 467 ⟨80PS(15)155⟩, 854 ⟨78CB629⟩; **4**, 550 ⟨82CB1947⟩; **6**, 489 ⟨91PS(60)287, 93PS(83)65⟩
Dieminger, K., **4**, 408 ⟨58CB598⟩
Dienes, M., **6**, 477 ⟨83MIP8303603⟩
Diepers, W., **4**, 296 ⟨66LA(697)171⟩; **6**, 107 ⟨66AG(E)875⟩
Diephouse, T. P., **4**, 1041 ⟨91OM2068⟩
Diephouse, T. R., **2**, 880 ⟨80JOM(202)C95⟩, 880, 893 ⟨83OM1005⟩, 896 ⟨91OM2068⟩; **4**, 1042 ⟨80JOM(202)C95⟩
Diepolder, E., **2**, 780 ⟨1895LA(286)145⟩
Diercks, R., **1**, 519 ⟨83AG(E)778⟩, 1009 ⟨86AG(E)268⟩
Dierdorf, D. S., **2**, 461, 508 ⟨69JA6609⟩
Dierichs, A., **1**, 1065 ⟨27CB950⟩
Dierks, H., **4**, 671 ⟨66AG(E)899⟩

Dieter, J. W., **1**, 196 ⟨87TL5415⟩; **3**, 128 ⟨90TL4105⟩, 251 ⟨83CC1378⟩
Dieter, R. K., **1**, 140, 485 ⟨86T3029, 85JA4679⟩, 485 ⟨86JOC4687⟩; **2**, 315 ⟨93SL407⟩, 588 ⟨87JA2040⟩; **3**, 128 ⟨90TL4105⟩, 188 ⟨92TL5693, 93SL407⟩, 206 ⟨81JA1996⟩, 217 ⟨86JOC4687⟩, 251 ⟨81JOC5031, 83CC1378⟩; **5**, 547, 555 ⟨86T3029⟩, 561 ⟨88S303⟩
Dietl, S., **2**, 865 ⟨88AG(E)709⟩, 880 ⟨88TL535, 88TL539⟩; **3**, 607 ⟨86AG(E)1003⟩; **4**, 549 ⟨88AG(E)389⟩, 565 ⟨88AG(E)709⟩, 1047, 1059 ⟨86AG1020⟩; **6**, 154 ⟨88AG(E)389⟩, 154 ⟨88AG(E)709⟩
Dietrich, A., **5**, 567, 630 ⟨89JMC2493⟩
Dietrich, H., **4**, 671 ⟨66AG(E)899⟩
Dietrich, J., **3**, 418 ⟨84CB694⟩; **4**, 298 ⟨84CB694⟩
Dietrich, K., **5**, 1141 ⟨83JOM(243)245⟩
Dietrich, K.,**6**, 207 ⟨86JOM(301)1⟩
Dietrich, M. A., **4**, 970 ⟨60JA5116⟩; **6**, 15 ⟨60JA6181⟩, 46 ⟨60JA5116⟩
Dietrich, M. W., **2**, 100 ⟨65JOC2692⟩
Dietrich, P., **2**, 686 ⟨63CB3044⟩
Dietrich, R., **3**, 470 ⟨58LA(617)20⟩
Dietrich, W., **5**, 1061, 1078 ⟨81AG855⟩
Dietsch, H., **2**, 471 ⟨76ZN(B)1305⟩
Dietsche, T. J., **1**, 599 ⟨74JA7165⟩
Dietsche, W. H., **2**, 851 ⟨67T3049⟩
Dietschmann, H., **6**, 620, 621, 623 ⟨80S619⟩
Dietz, Jr., A. G., **2**, 647 ⟨77JOC2053⟩
Dietz, E. A., **2**, 448 ⟨74JINC503⟩
Dietz, Jr., E. A., **2**, 434 ⟨73JINC2659⟩; **5**, 1130 ⟨73IC241⟩
Dietz, J., **6**, 624, 625, 633 ⟨87CB339⟩
Dietz, S., **2**, 478 ⟨68JCS(C)922⟩
Dietz, S. D., **3**, 516 ⟨93MI 317-02⟩
Dietze, P. E., **5**, 315 ⟨84JOC5225⟩
Diez-Barra, E., **4**, 826 ⟨92SL893, 93SC1935⟩
Di Fabio, R., **1**, 215 ⟨88TL815⟩
Differding, E., **1**, 518 ⟨84TL5043⟩; **2**, 16 ⟨91SL187, 91TL1779⟩, 16, 365 ⟨88TL6087⟩, 89HCA1248⟩, 938, 944 ⟨87TL397⟩; **3**, 149, 152, 154 ⟨91TL1779⟩, 216 ⟨84TL5043⟩, 424 ⟨87TL397⟩, 557 ⟨85TL1647⟩; **4**, 131 ⟨91SL395⟩; **5**, 314 ⟨92T1595⟩
DiFuria, F., **2**, 146 ⟨76TL4637⟩, 147 ⟨84S325⟩, 147 ⟨86TL6257⟩, 147 ⟨89TL2575⟩; **3**, 275 ⟨90JOC3658⟩; **4**, 237 ⟨74TL4085⟩, 249 ⟨90BSF734⟩, 261 ⟨86TL6257, 89TL2575⟩, 261 ⟨90G165, 92TL3043⟩
Di Gangi, F. E., **5**, 749 ⟨49MI 519-01⟩
Digenis, G. A., **5**, 302 ⟨73TL1627⟩; **6**, 57 ⟨73TL4511⟩
Digenis, G., **6**, 462 ⟨86MI 615-01⟩
Di Giacomo, R., **2**, 451, 467, 469 ⟨85PS(24)217⟩
DiGiulio, A. V., **2**, 377, 403 ⟨58JA6562⟩
Digler, K., **1**, 104 ⟨49HCA489⟩
Dignam, K. J., **3**, 684 ⟨77JCS(P1)1457⟩; **4**, 990 ⟨80JOC4302⟩; **5**, 699 ⟨78JOC388⟩; **6**, 605, 616, 617 ⟨75JCS(P2)1046⟩
Dijkstra, R., **3**, 566 ⟨54RTC575⟩
Dikanskaya, L. D., **6**, 110 ⟨66IZV98⟩
Dike, M., **1**, 345, 946 ⟨80JA3642⟩
Dike, M. S., **3**, 211 ⟨81JOC2400⟩
Dike, S., **1**, 345, 946 ⟨80JA3642⟩
Dike, S. Y., **4**, 304 ⟨78TL3607⟩
Diksic, M., **2**, 8 ⟨93OPP249⟩
Dikusar, E. A., **2**, 1098 ⟨85JGU1691, 86JGU1812⟩; **5**, 1110 ⟨90MI 530-01⟩
Dilgassa, M., **5**, 447 ⟨79JOM(172)177⟩
Dill, J. D., **6**, 386, 401, 404 ⟨76JA5419⟩
Dillard, R. D.,**1**, 506 ⟨61JOC3772⟩
Dillard, R. D., **6**, 509 ⟨64JOC1851⟩
Dillen, J. L. M., **2**, 1000 ⟨91T8611⟩
Diller, W., **6**, 460, 462 ⟨B-91MI 615-01⟩
Dillinger, H. J., **3**, 593 ⟨74T2553⟩
Dillon, Jr., J. L., **1**, 518 ⟨83JOC3382⟩; **4**, 742 ⟨83JOC3382⟩
Dillon, K. B., **2**, 475, 477 ⟨84JCS(D)1911⟩, 476 ⟨88JCS(D)2077, 91CC234⟩, 836, 837 ⟨82POL123⟩; **4**, 133 ⟨91CC234⟩; **5**, 652 ⟨90JCS(D)2593⟩, 872, 1148 ⟨91OM3205⟩, 1129 ⟨88JCS(D)2077⟩; **6**, 659, 667 ⟨91CB1739⟩, 670, 675 ⟨91OM3205⟩
Dillon, M. P., **1**, 573 ⟨92JA9673⟩
Dillon, P. J., **6**, 124 ⟨69CJC4347⟩
Dillon, R. L., **1**, 155, 850, 855, 869 ⟨53JA2439⟩
Dillon, T. E., **2**, 33 ⟨42JA1128⟩
Dills, C. E., **1**, 905 ⟨59LA(623)217⟩
Diluzio, J. W., **1**, 73 ⟨62JA679⟩
Dilworth, B. M., **1**, 704 ⟨86T3731⟩; **4**, 62, 66, 68 ⟨86T3731, 86TL757⟩
DiMaggio, III, A., **2**, 614 ⟨71JOC631⟩
DiMaio, A.-J., **2**, 496 ⟨86OM393, 91OM3764⟩, 500 ⟨90IC798⟩
DiMarco, J. P., **5**, 1121 ⟨81JAP39050⟩
Di Mare, A., **1**, 45 ⟨87S719⟩; **3**, 438 ⟨87S719⟩
DiMare, M., **1**, 703 ⟨86JA2476⟩; **3**, 513 ⟨90JA3875⟩
DiMaria, F., **4**, 166 ⟨72HCA249⟩
Dimas, P. A., **4**, 686 ⟨80JOM(201)C31⟩
Dimchev, Z., **6**, 558 ⟨89MI 617-01⟩
Dime, D. S., **1**, 422 ⟨79TL1999, 80JOC3017⟩; **2**, 911 ⟨80JOC3017⟩; **3**, 211 ⟨80JOC3017⟩
DiMenna, W. S., **3**, 306 ⟨80TL2129⟩
Dimes, R. W., **1**, 563 ⟨81JOC1747⟩
Dimetrenko, A. A., **2**, 356, 357 ⟨91IZV1069⟩
Di Michele, L. M., **1**, 558 ⟨88JA3318⟩
Dimitriadis, E., **1**, 17 ⟨82AJC1895⟩
Dimitriev, V. I., **2**, 434 ⟨78URP635102⟩
Dimitrijevich, S. D., **2**, 364 ⟨88JHC1857⟩
Dimitrikova, L. V., **2**, 756 ⟨88JOU2222⟩
Dimitrova, M. G., **5**, 770 ⟨92MI 519-01, 92MI 519-02⟩
Dimmel, D., **2**, 398 ⟨76JOC3328⟩
Dimmel, D. R., **4**, 619 ⟨72JOC2662⟩; **6**, 208, 405 ⟨75OMS(10)18⟩, 267, 379 ⟨72JOC2662⟩
Dimmerling, A., **3**, 717 ⟨86AP(319)473⟩
Dimmig, D. A., **5**, 185 ⟨55OSC(3)165⟩
Dimock, S. H., **1**, 207 ⟨90JA4063⟩; **2**, 555 ⟨92S191⟩, 589 ⟨90JA4063⟩
Dimroth, K., **1**, 443 ⟨77S339⟩; **2**, 403 ⟨77S339⟩, 1008 ⟨73OSC(5)1135⟩, 1010 ⟨75JA5526⟩; **5**, 366 ⟨77S339⟩, 876, 882 ⟨64AG(E)384, 66CB1325⟩; **6**, 73 ⟨65AG(E)873⟩
Dimroth, O., **2**, 387 ⟨05CB670, 06CB3905⟩; **4**, 304 ⟨02CB984⟩
Dimroth, P., **3**, 562 ⟨64GEP1166771, 65MI 317-01⟩
Dimukhametov, M. N., **2**, 465 ⟨84JGU1704⟩; **5**, 371 ⟨85ZOB2483⟩
Din, L. B., **2**, 753 ⟨78S23⟩
Dinan, F. J., **5**, 313, 360, 361 ⟨B-70MI 507-01⟩
Din, Z. U., **1**, 815 ⟨66CC775⟩, 1143 ⟨73JOC547⟩
d'Incan, E., **3**, 146 ⟨86TL4175⟩
Dinculescu, A., **2**, 1008 ⟨82AHC(2)1⟩
Dines, M., **2**, 166, 169 ⟨69JHC627⟩
Dines, M. B., **2**, 208 ⟨69CC1205⟩
Dinesh, C. U., **4**, 186 ⟨93S1069⟩
Ding, C. Z., **2**, 309 ⟨93JMC3606⟩; **4**, 416 ⟨93SC1467⟩
Ding, M., **3**, 363 ⟨90JOC2421⟩
Ding, S., **2**, 86 ⟨93JOC804⟩
Ding, W., **1**, 54 ⟨86MI 102-01⟩, 1079 ⟨81TL5283⟩; **4**, 598 ⟨84TL4425⟩
Ding, Y., **4**, 343 ⟨92S258⟩, 801 ⟨91S854⟩
Dinges, J., **3**, 325 ⟨93JOC7216⟩
Dinges, W., **2**, 307 ⟨92CC853⟩
Dingle, T. W., **1**, 233 ⟨82JA2560⟩
Dingwall, J. D., **4**, 464 ⟨84JCS(P1)2845⟩
Dingwall, J. G., **2**, 451 ⟨84JCS(P1)2845, 89T3787⟩; **3**, 73, 255 ⟨70JCS(C)2412⟩, 73, 376 ⟨68JCS(C)2543⟩, 341, 376 ⟨69JCS(C)913⟩, 357 ⟨68CC863⟩; **6**, 116 ⟨89T3787⟩
DiNinno, F., **3**, 348, 363 ⟨79JOC3271⟩; **4**, 847 ⟨82TL3535⟩
DiNinno, Jr., F., **1**, 538 ⟨79JA1284⟩
Dinizo, S. E., **3**, 627, 634, 635, 636, 639, 651 ⟨76JOC2846, 77JA182⟩; **5**, 272, 283, 284 ⟨77JA182⟩; **6**, 119 ⟨77S181⟩
Dinkes, L., **6**, 445 ⟨72S81⟩

Dinné, E., **1**, 419, 421 ⟨65TL391⟩, 812 ⟨63AG(E)739⟩
Dinnocenzo, J. P., **1**, 803 ⟨88JA2324⟩
Dinsmore, C. J., **1**, 550 ⟨91TL3755⟩
Dinsmore, R., **2**, 31 ⟨43OSC(2)358⟩
DiNunno, L., **1**, 1123 ⟨65TL4405⟩; **2**, 56 ⟨87T2181⟩
Dionisotti, S., **5**, 22 ⟨93MI 517-02⟩
Dionne, G., **1**, 470 ⟨78CJC419⟩
Di Pardo, R. M., **4**, 417 ⟨86JOC3718, 87JOC3232⟩
DiPasquale, F., **1**, 133 ⟨79TL845⟩
Dipietro, H. R., **3**, 613 ⟨64JOC1970⟩
Di Pietro, J., **1**, 777 ⟨77JOC3899⟩
DiPietro, R. A., **1**, 633 ⟨81JOC1315⟩; **3**, 58 ⟨81JOC1315⟩
DiPol, J., **4**, 412 ⟨74JOC1349⟩, 444 ⟨76JOC3063⟩
Dirania, M. K. M., **4**, 177 ⟨71JCS(C)1213⟩
Di Rienzo, B.,**5**, 726 ⟨83SC753⟩
Dirk, C. W., **4**, 291 ⟨85JA675⟩
Dirks, J. E., **5**, 698, 813 ⟨66JOC3708⟩
Dirksen, H.-W., **6**, 502 ⟨63AG1059⟩
Dirlam, J. P., **1**, 918 ⟨72JA6867⟩
Dirnberger, T., **6**, 426 ⟨86S330⟩
Dirnens, V. V., **4**, 514 ⟨84JOM(268)C29⟩, 515 ⟨86JOM(316)249⟩
Dirnens, W., **5**, 720 ⟨77JOM(134)21⟩
Dirstine, P. H., **4**, 23 ⟨46JOC55⟩
Disanayaka, B. W., **1**, 690 ⟨85CC1282, 87JOC2905⟩, 797 ⟨83S952⟩
Discordia, R. P., **2**, 44 ⟨90TL5603⟩
Disher, J. M., **1**, 1076 ⟨72JOC4467⟩
Dishong, D. M., **4**, 235 ⟨80JOC3634⟩
Di Simone, B., **5**, 377 ⟨94S805⟩
Dismaële, D., **1**, 980 ⟨92TL4447⟩
Disselhoff, H., **3**, 485 ⟨34LA(512)250⟩
Disselnkötter, H., **1**, 1002 ⟨78LA658⟩; **6**, 448 ⟨70GEP2008115⟩
Di Stéfano, C., **4**, 432 ⟨94TL89⟩
DiStefano, F. V., **6**, 405 ⟨76JOM(122)171⟩
Disteldorf, W., **2**, 1066 ⟨76CB546⟩; **5**, 871 ⟨76CB546⟩
Distler, H., **1**, 708 ⟨77CB2978⟩; **2**, 122 ⟨67AG(E)544⟩
Ditrich, K., **2**, 530 ⟨89CB1783, 89LA883⟩, 652 ⟨83S107⟩; **4**, 1056 ⟨87LA977⟩; **5**, 173 ⟨86AG(E)1028⟩
Dittami, J. P., **1**, 802 ⟨84TL1255⟩
Ditterich, E., **3**, 543 ⟨90CB2031⟩
Dittler-Klingemann, A., **3**, 723 ⟨91CB1683⟩
Dittmar, W., **6**, 533 ⟨82EUP62834⟩
Dittmer, D. C., **1**, 242 ⟨84CHEC(7)131⟩, 784 ⟨64JOC3131⟩; **2**, 44 ⟨90TL5603⟩, 196 ⟨B-90MI 203-09⟩, 366 ⟨B-90MI 206-04⟩; **3**, 365 ⟨84JOC4752⟩, 366 ⟨67JA2793, 70JOC4224⟩; **4**, 234 ⟨88JOC5750⟩; **5**, 1113 ⟨84JOC2057⟩
Dittmer, G., **6**, 418 ⟨82MI 614-02⟩
Dittrich, G., **6**, 498 ⟨82TL2833⟩
Dittrich, R., **4**, 802, 803 ⟨81PS(10)127⟩
Dittrich, V., **6**, 619 ⟨80CP1076132⟩
Dittus, G., **1**, 1016, 1108 ⟨43CB80⟩; **2**, 1014 ⟨43CB80⟩
Ditzel, E. J., **6**, 402 ⟨92JOM(437)323⟩
Di Vaira, M., **6**, 552 ⟨93POL13⟩
Divakar, K. J., **2**, 127 ⟨90JCS(P1)969⟩
Dive, G., **1**, 570 ⟨93JOC4685⟩
Diveley, W. R., **5**, 1043 ⟨69JOC616⟩
Diversi, P., **2**, 581 ⟨80JCS(D)1633⟩
Divis, J., **2**, 777 ⟨73MI 215-01⟩
Divisia, B., **3**, 480, 481 ⟨79T181⟩; **6**, 666 ⟨75MI 621-02, 75OMS(10)1021⟩, 666 ⟨79T181⟩
Dix, J. S., **2**, 130 ⟨67JOC282⟩
Dixit, V., **1**, 1070 ⟨85S962⟩; **3**, 273 ⟨85S962⟩
Dixit, V. M., **6**, 264 ⟨87JOC1794⟩
Dixneuf, P., **5**, 940, 941 ⟨95JOM(488)C9⟩; **6**, 721 ⟨74JCS(D)1827⟩
Dixneuf, P. H., **1**, 962 ⟨94SL457⟩; **2**, 676 ⟨91SL755, 93CC850⟩; **4**, 711 ⟨90OM2234⟩; **5**, 400 ⟨88NJC547⟩, 940 ⟨91OM2768⟩, 940, 941 ⟨90CC641, 89AOC(29)163⟩, 941 ⟨91CC881, 92CC1220, 92OM809⟩, 944 ⟨95OM1095⟩
Dixon, A. E., **5**, 1047 ⟨1895JCS1041, 1895JCS565⟩, 1047 ⟨1896JCS1593, 1896JCS855⟩
Dixon, C. E., **3**, 421 ⟨90JOC1086⟩
Dixon, D. A., **4**, 42 ⟨85JA4565⟩, 45 ⟨B-92MI 402-01⟩; **5**, 1133 ⟨87JA6532⟩; **6**, 229 ⟨85JA4565⟩
Dixon, H. B. F., **4**, 565 ⟨77MI 412-01, 78BJ(169)239⟩
Dixon, N. J., **1**, 481 ⟨92JCS(P1)3431⟩; **4**, 952, 953 ⟨89SL52⟩
Dixon, P. S., **2**, 590 ⟨59CI(L)628⟩
Djahanbini, D., **1**, 193 ⟨88S983⟩; **4**, 535 ⟨88S983⟩
Djakovitch, L., **1**, 164 ⟨92SL57⟩
Djamei, S. M., **2**, 537 ⟨86BSF413⟩; **4**, 140 ⟨86BSF413⟩
Djamgaryan, S. M., **5**, 385 ⟨88T5507⟩
Djeghaba, Z., **2**, 1034, 1035 ⟨86JOM(304)115⟩
Djerassi, C., **1**, 23 ⟨51JA1528⟩, 113 ⟨89JOC2117⟩, 439, 545 ⟨86TL2187⟩, 513 ⟨74JOC968⟩, 592 ⟨62CI(L)1236⟩, 630 ⟨65CB3236⟩, 635 ⟨76JA2275⟩, 871 ⟨64JA2832⟩; **2**, 27 ⟨48CRV271⟩, 34 ⟨53JA3493, 55JA3826⟩, 123 ⟨55JA568⟩, 126 ⟨57JA2553⟩, 130 ⟨62JOC1041⟩, 203 ⟨48JCS849⟩, 760 ⟨53JA3632⟩, 900 ⟨79JOC3374⟩; **3**, 7, 119 ⟨51OR(6)207⟩, 208 ⟨70JOC753⟩, 227 ⟨54JA5672⟩; **5**, 116 ⟨58JA1236⟩, 288 ⟨56JOC520⟩, 320 ⟨53JA3493⟩
Djerourou, A. H., **4**, 352 ⟨91TL6325⟩
Djokar, K., **3**, 586 ⟨87LA183⟩, 592 ⟨81CB1976⟩, 592, 593 ⟨85LA2305⟩; **4**, 980, 981, 982 ⟨81CB1976⟩; **5**, 634 ⟨85LA2305⟩
Djuric, S., **1**, 655 ⟨80JA6885⟩; **2**, 543, 811 ⟨81TL1787⟩, 907 ⟨82COMC-I(7)515⟩; **6**, 172 ⟨82COMC-I(7)515⟩
Djuric, S. W., **5**, 327, 328 ⟨92SC871⟩
Dlubala, A., **5**, 550 ⟨87S835⟩
Dluzniewski, T., **4**, 698 ⟨80JOM(201)97⟩
Dmitrichenko, M. Y., **6**, 619 ⟨90ZOB706⟩, 620 ⟨91ZOB643, 92ZOB306⟩
Dmitrichenko, M. Yu., **4**, 766 ⟨89ZOB997⟩
Dmitrichenko, Yu. M., **5**, 656 ⟨87JGU1710, 87ZOB1912⟩
Dmitrienko, G. I., **5**, 1125 ⟨90TL3681⟩
Dmitrienko, G. L., **2**, 767 ⟨82TL821⟩
Dmitrieva, G. V., **4**, 124, 125, 128 ⟨71IZV1159⟩; **6**, 565 ⟨86ZOR2145⟩
Dmitrieva, N. V., **2**, 459 ⟨74ZOB1261⟩
Dmowski, W., **1**, 429 ⟨85JFC(29)273⟩; **6**, 10 ⟨93JFC(60)233⟩, 19 ⟨78PJC71⟩, 39 ⟨83JFC(22)207⟩, 52 ⟨82PJC1369⟩, 233 ⟨88JCS(P1)1179⟩, 233, 234 ⟨87JCS(P1)2119⟩, 440 ⟨83JFC(23)207⟩
Doad, G. J. S., **2**, 629 ⟨89JCR(S)313⟩
Do, J. Y., **1**, 932 ⟨93JA3328⟩
Doak, G. O., **1**, 88, 89 ⟨55JA4262⟩, 271 ⟨88JOM(25)351⟩, 445 ⟨74S328⟩, 646 ⟨B-70MI 114-01⟩, 647 ⟨B-70MI 114-02⟩; **2**, 481, 872 ⟨82CRV15⟩, 481 ⟨B-70MI 209-01⟩, 492 ⟨64JA209, 67ICA34, 67IS92⟩, 509 ⟨74S328⟩, 820, 821 ⟨52JA3414⟩, 844 ⟨53JA1379, 53JA683⟩, 844, 847 ⟨51JA5658, 52JA753⟩, 872, 891 ⟨B-70MI 217-01⟩, 872 ⟨B-89MI 217-01⟩, 884 ⟨40JA167, 46JA1987⟩, 886, 887 ⟨73JOM(55)143⟩; **4**, 595 ⟨82CRV15⟩; **6**, 707 ⟨B-70MI 622-01⟩
Doane, W. M., **5**, 317 ⟨80CAR(79)205⟩; **6**, 276 ⟨69JOC1642, 71CAR(16)145⟩
Do, Q. T., **3**, 352 ⟨90BSF427⟩
Do, Y., **6**, 340 ⟨93IC5883⟩
Dobashi, A., **4**, 70 ⟨80TL3579, 81JA2757⟩
Dobbie, R. C., **2**, 472, 508 ⟨68JCS(A)1406⟩, 482 ⟨67JCS(A)1308⟩; **5**, 1129 ⟨75JCS(D)2368⟩; **6**, 240 ⟨66JCS(A)367⟩, 607 ⟨66JCS(A)933⟩
Döbereiner, U. J., **6**, 502 ⟨83GEP3135111⟩
Doboszewski, B., **1**, 585 ⟨89JOC2217⟩; **2**, 9 ⟨87CJC412⟩; **4**, 45 ⟨84CL1751⟩
Dobrev, A., **2**, 312 ⟨89S963⟩
Dobrotenko, V. V., **2**, 569 ⟨77JOM(131)43, 79DOK(245)198⟩, 569 ⟨82JOM(239)335⟩
Dobrovinskaya, E. K., **5**, 1006 ⟨75ZOB477, 76ZOB1531⟩
Dobrusin, E. M., **5**, 567 ⟨93JMC2459⟩
Dobrzynski, E. D., **6**, 719, 732 ⟨75IC1513⟩
Dobson, J. E., **6**, 181, 269 ⟨69JCS(A)1882⟩

Dobson, N. A., **1**, 504 ⟨61T16⟩
Dodd, D., **1**, 66 ⟨69JCS(B)1071⟩, 676 ⟨68JCS(A)34⟩
Dodd, D. S., **6**, 487 ⟨94TL977⟩
Dodd, J. H., **4**, 346 ⟨80S72⟩
Dodd, M., **5**, 222 ⟨92MI 504-02⟩
Doddema, L. L., **3**, 54, 253 ⟨82TL4831⟩
Dodge, J. A., **2**, 660 ⟨88TL4827⟩
Dodge, R. A., **5**, 63 ⟨32OSC(1)336⟩
Dodonov, V. A., **2**, 879 ⟨88JGU629⟩, 886 ⟨81JGU1056, 85JGU63⟩, 886 ⟨88JGU1934, 89MI 217-02⟩; **5**, 222 ⟨88ZOB1934⟩
Dodson, R. M., **1**, 234 ⟨63CI(L)1203⟩, 417, 829 ⟨49JA2439⟩; **2**, 125 ⟨48JA1176⟩, 153 ⟨67CC1189, 69CC1159⟩, 188 ⟨71JOC2698⟩, 200 ⟨68CC9, 71JOC2693, 71JOC2703⟩, 221, 225 ⟨72JOC2367⟩
Doecke, C. W., **1**, 811 ⟨79JA7599⟩; **5**, 328 ⟨94TL6021⟩
Doedens, R. J., **2**, 1005 ⟨87JA2541⟩; **3**, 422 ⟨83JOC3393⟩
Doel, G. R., **4**, 675 ⟨86CC542⟩
Doering, E., **6**, 82, 83, 91 ⟨55JA509⟩
Doering, F., **6**, 240 ⟨76CA(85)142668⟩, 606 ⟨79GEP2753204, 85GEP3337939⟩
Doering, W. von E., **1**, 299 ⟨56JA5447⟩, 411, 806 ⟨63AG(E)115⟩, 412 ⟨62T67⟩, 803 ⟨63T715⟩, 906 ⟨54JA3203⟩, 912, 914 ⟨58T178⟩; **2**, 193 ⟨49JA2221⟩; **3**, 57 ⟨67JA5505⟩; **4**, 16 ⟨54JA6162⟩, 429, 430, 431 ⟨46JA2252⟩
Doerjer, G., **1**, 1000 ⟨75S423⟩; **4**, 177 ⟨79CB3603⟩
Doerr, I. L., **2**, 212 ⟨61JOC3401⟩; **4**, 54, 57 ⟨73JOC3878⟩
Does, T. V. D., **5**, 1089 ⟨90JOM(388)47⟩
Doetschman, D., **3**, 490 ⟨71S481⟩
Doetzer, R., **6**, 121 ⟨91JOC1816⟩
Dogadina, A. V., **2**, 448 ⟨76JA1290⟩, 1028 ⟨85JGU1771⟩; **3**, 591 ⟨83JGU205, 85JGU26⟩; **4**, 885 ⟨80ZOB984⟩, 914 ⟨84ZOB457⟩, 976 ⟨79ZOR1824⟩
Dogan, O., **1**, 370 ⟨94JOC4⟩
Doherty, A. M., **4**, 284 ⟨92JA3910⟩
Doherty, A. M.,**5**, 327 ⟨95TL197⟩
Doherty, D. G., **4**, 427 ⟨53JA3469⟩
Doherty, G. M., **5**, 1111 ⟨92B7922⟩
Doherty, N. M., **4**, 673 ⟨85CC170⟩
Döhler, H. V., **5**, 1083 ⟨71ZAAC(386)197⟩
Dohmaru, T., **1**, 933 ⟨90JOM(381)155⟩
Dohno, R., **5**, 70 ⟨82NKK811⟩
Doi, J. T., **2**, 7 ⟨87JOC4999⟩, 441, 444 ⟨91PS(56)27⟩
Doi, K., **1**, 143 ⟨86BCJ2801⟩; **6**, 561 ⟨88BCJ3205⟩
Doi, M., **3**, 59 ⟨88JOC1623⟩
Doi, T., **1**, 818 ⟨91JOC3465⟩; **3**, 213 ⟨89JOC4273⟩, 430 ⟨90BCJ1843⟩
Doi, Y., **4**, 238 ⟨91CL415⟩
Doifode, K. B., **2**, 28 ⟨64JOC2025⟩; **3**, 241 ⟨64JOC2025⟩
Doig, S. J., **4**, 687 ⟨90OM2745⟩
Dojo, H., **2**, 623, 628, 902 ⟨83TL731⟩
Doken, K., **3**, 618 ⟨88CL285⟩
Do-Khen, S., **2**, 466 ⟨52IZV894⟩
Doksopulo, T. P., **2**, 1093 ⟨79MI 223-01⟩
Dokuchaeva, I. S., **2**, 1074 ⟨91ZOB611⟩
Dolak, L. A., **5**, 361 ⟨66JOC3038⟩
Dolak, T. M., **1**, 141 ⟨77TL1961⟩; **3**, 169 ⟨77TL1961⟩; **4**, 365, 391, 392 ⟨77TL1961⟩
Dolan, L. A., **6**, 508 ⟨45JA2079, 47JA3150⟩
Dolan, S. C., **1**, 940 ⟨85CC1588⟩
Dolata, D. P., **2**, 67 ⟨90TL4565⟩
Dolbier, Jr., W. H., **1**, 497 ⟨72JOC950⟩
Dolbier, Jr., W. R., **1**, 498 ⟨90JOC711⟩, 510 ⟨85JA3626⟩, 801 ⟨72JA2544, 82JA2498⟩; **4**, 5 ⟨87JOC1872⟩, 7 ⟨82JA2494⟩, 32, 1082 ⟨90JOC589⟩; **6**, 212 ⟨B-88MI 607-01⟩, 227 ⟨84JOC205⟩
Dolbier, W. R., **4**, 1082 ⟨88TL6749⟩, 1083 ⟨90JOC5420⟩
Dolbilkin, K. V., **6**, 145 ⟨76KFZ35⟩
Dolby, L. J., **1**, 613 ⟨77JOC1349⟩; **3**, 231 ⟨71TL3803⟩, 310 ⟨66JA1049⟩
Dolby, P. A., **4**, 1029 ⟨91JCS(D)351⟩
Doldouras, G. A., **1**, 35 ⟨78JA341⟩; **2**, 15, 613 ⟨70JA7494⟩
Dolence, E. K., **3**, 29 ⟨84JOC1378⟩
Dolenko, E. V., **3**, 341 ⟨87KGS1694⟩
Doleschall, G., **2**, 342 ⟨87TL2993⟩; **5**, 199 ⟨86JCS(P1)927⟩
Dolfini, J. E., **4**, 227 ⟨78JOC79⟩; **5**, 575 ⟨75JA5628⟩
Dolgikh, A. N., **6**, 83 ⟨60IZV1901⟩
Dolgoplosk, B. A., **2**, 566 ⟨77JOM(142)145⟩
Dolgopol'skii, I. M., **5**, 694 ⟨76JOU1789, 76ZOR1826⟩
Dolgov, B. N., **1**, 463 ⟨50JGU450⟩
Dolgushina, I. Yu., **4**, 104 ⟨73JOU418⟩
Dolgushin, G. V., **4**, 772 ⟨91ZOB1084⟩, 804 ⟨91JGU983⟩; **5**, 656 ⟨87JGU1710, 87ZOB1912⟩; **6**, 619 ⟨90ZOB706⟩, 620 ⟨91ZOB643, 92ZOB306⟩
Dolgushin, S. V., **4**, 766 ⟨89ZOB997⟩
Dolhaine, H., **4**, 804 ⟨77PS(3)47⟩
Dolinski, R. J., **5**, 998 ⟨65JOC48⟩
Döll, G., **2**, 429 ⟨61CB2664⟩; **4**, 545 ⟨63ZAAC(324)259⟩
Doll, R. J., **2**, 20 ⟨77JA905⟩
Dollase, W. A., **6**, 158 ⟨91PS(56)117⟩
Dollat, J.-M., **1**, 960 ⟨76CC183⟩, 966 ⟨75TL4615⟩
Dolle, R. E., **1**, 581 ⟨85JA1691⟩; **2**, 163 ⟨93TL133⟩; **4**, 45 ⟨84JA4189⟩; **5**, 277 ⟨85JA1695⟩
Doller, D., **2**, 2 ⟨92T9195, 92TL3413⟩, 2 ⟨93TL1681⟩, 108 ⟨91TL4671⟩; **3**, 113 ⟨91MI 303-01⟩
Doller, U., **3**, 227 ⟨88HCA1719⟩
Dolling, U., **3**, 131 ⟨90SC2683⟩
Dolling, U.-H., **1**, 135 ⟨87JOC4745⟩, 558 ⟨88JA3318⟩
Dolling, W., **4**, 900 ⟨78ZC345⟩
Dölling, W., **6**, 545, 546 ⟨89JPR439⟩, 564 ⟨89M871⟩
Dollinger, H., **1**, 148 ⟨84TL1353⟩; **2**, 1001 ⟨84TL1353⟩; **4**, 528 ⟨84TL1353⟩
Dolman, H., **6**, 642 ⟨64RTC1305⟩
Dolmazon, R., **3**, 246 ⟨87BSF873⟩, 527 ⟨70TL3657⟩
Dolnakov, Y. P., **6**, 36 ⟨74ZOR2031⟩
Dol'nikova, T. Yu., **5**, 644 ⟨87ZOB685⟩
Dolphin, D., **2**, 578 ⟨B-82MI 211-01⟩, 583 ⟨87CC1110⟩, 584 ⟨86CC787⟩
Dolphin, J. M., **1**, 821 ⟨83JA127⟩; **2**, 678 ⟨79JA249⟩; **3**, 334 ⟨80JOC2601⟩; **5**, 554 ⟨80JOC2601⟩
Dols, P. P. M. A., **1**, 582 ⟨91TL3739⟩
Dolson, M. G., **3**, 335 ⟨85JOC3216⟩
Dolyuk, V. G., **5**, 345 ⟨74ZOR667⟩
Domagala, J. M., **1**, 387 ⟨78JA1605⟩
Domagala, M., **2**, 887 ⟨86ICA(113)43⟩; **5**, 222 ⟨89ZAAC(574)130, 90ZAAC(582)37⟩
Domagk, G., **2**, 361 ⟨56GEP948330⟩
Domagk, G. F., **1**, 736 ⟨55CB1415⟩
Domaschke, L., **4**, 979 ⟨62CB2095⟩; **6**, 39 ⟨59CB3170⟩
Domascke, L., **6**, 79 ⟨58CB650⟩
Dombek, B. D., **2**, 439 ⟨78JOC3408⟩
Dombroski, M. A., **1**, 347 ⟨90JA2759⟩
Dombrovskii, A. V., **2**, 190 ⟨55ZOB1943⟩, 445 ⟨84ZOB1192⟩; **4**, 555 ⟨83ZOB594⟩, 570 ⟨77ZOB1715⟩; **5**, 597 ⟨69ZOB1282⟩, 904 ⟨71JGU2019⟩; **6**, 208 ⟨81ZOB2241⟩
Dombrovskii, V. A., **2**, 444 ⟨75ZOB1228⟩
Dombrowski, E., **3**, 370 ⟨92IC723⟩
Domeier, L. A., **5**, 1126 ⟨91SC2025⟩
Domenick, R. L., **1**, 905 ⟨87JA7263⟩
Dominguez, C., **3**, 221 ⟨90TL7669⟩; **5**, 1029 ⟨92JCS(P1)1671⟩
Dominguez, D., **1**, 569 ⟨82JOC409⟩, 626 ⟨88H(27)2367⟩
Dominguez, E., **1**, 955 ⟨78TL691⟩; **2**, 89 ⟨88T6677⟩; **4**, 271, 272 ⟨90T7197⟩, 858, 938, 961 ⟨91TL1385⟩
Dominguez, G., **5**, 696 ⟨91OM11, 93JOC3886⟩
Dominguez, R., **4**, 706 ⟨88OM2566⟩
Do-Minh, T., **3**, 314 ⟨70JA1402⟩, 488 ⟨68TL5237⟩, 527 ⟨70JA1766⟩; **4**, 298 ⟨77JOC4217⟩; **5**, 874 ⟨68TL5237⟩
Domling, A., **3**, 694 ⟨93AG(E)563⟩
Dömling, A., **4**, 313 ⟨93T9495, 95T755⟩
Dommisse, R., **4**, 461 ⟨94S23⟩
Domínguez, D., **5**, 162 ⟨82TL5513⟩
Domínguez, E., **4**, 936, 938 ⟨93JOC1596, 91TL5159⟩
Domogatskaya, E. A., **6**, 456 ⟨84MI 614-01⟩

Domont, W., 2, 281 ⟨75AG(E)350⟩
Domoto, T., 3, 91 ⟨88CPB974⟩
Domrachev, G. A., 4, 719 ⟨77JOM(141)313⟩
Domsch, D., 2, 539 ⟨82S1⟩; 6, 43 ⟨82S1⟩
Domschke, G., 2, 361, 792 ⟨78TL4003⟩, 398 ⟨78ZC323⟩; 3, 695, 721 ⟨77S275⟩; 4, 307 ⟨66CB930, 66JPR144⟩
Donadelli, A., 3, 32 ⟨93CC984⟩
Donahue, P. E., 6, 444 ⟨88JOC6145⟩
Donaldson, P. B., 4, 689 ⟨78IC1995⟩
Donalski, M. S., 2, 561 ⟨81JOC3773⟩
Donaruma, L. G., 5, 215, 275, 282, 288, 292, 298 ⟨60OR(11)1⟩
Donate, P. M., 1, 516 ⟨86JOC253⟩
Donath, H., 2, 504 ⟨85ZN(B)1320⟩
Donati, D., 5, 99 ⟨80JHC621⟩
Donaubauer, A., 4, 1026, 1030 ⟨75TL3171⟩
Donaubauer, J., 1, 700, 703 ⟨86JA4603⟩
Donche, A., 3, 650 ⟨66BSF1315⟩
Dondoni, A., 2, 247 ⟨81CC350⟩, 942 ⟨84CC258, 86S757⟩; 3, 29 ⟨92TL7259, B-92MI 301-01⟩, 32 ⟨89TL6063⟩, 43 ⟨92TL4221⟩, 109 ⟨87S998⟩, 311 ⟨84JOC3478, 84JOC590, 91CC1313⟩, 312 ⟨81CC655, 86G133⟩, 368 ⟨71JCS(B)1547⟩, 558 ⟨80JOC3766, 82JOC3998⟩, 560 ⟨84JOC2200⟩, 707, 714 ⟨87S693, 84CC258⟩; 5, 612 ⟨85ZC324⟩, 657 ⟨72JCS(P2)1908⟩, 665 ⟨72JCS(P2)1911⟩, 1038 ⟨71S378⟩, 1073 ⟨77JOC3372⟩; 6, 634 ⟨87JOC3413, 87S693⟩, 637 ⟨90PAC643⟩
Dondy, B., 4, 353 ⟨92S995⟩
Donecke, J., 6, 114 ⟨80LA873⟩
Donegan, G., 2, 158 ⟨89JCS(P1)1550⟩
Donetti, A., 6, 425 ⟨90JMC2101⟩
Dongala, E. B., 6, 7 ⟨77TL4315⟩
Dönges, R., 1, 789 ⟨73AG(E)337⟩
Donker, I. O., 3, 69 ⟨90JHC765⟩
Donleavy, J. J., 5, 93 ⟨43OSC(2)422⟩
Donmaux, A. R., 2, 1566 ⟨63JA3992⟩
Donnelly, D. M. X., 1, 435 ⟨90JCS(P1)2851⟩, 780 ⟨93T7967⟩; 3, 242 ⟨60CI(L)348⟩
Donnelly, K. D., 2, 24 ⟨84TL607⟩
Donnelly, S. J., 5, 282 ⟨74JOC2740⟩
Donnoli, D., 5, 597 ⟨91FES1033⟩
D'Onofrio, F., 2, 34 ⟨82SC1127⟩
Donohoe, T. J., 1, 202, 203 ⟨93SL323⟩
Donohue, J., 2, 855 ⟨71JA3792⟩
Donotrio, F., 1, 578 ⟨85S1159⟩
Donovan, D. J., 1, 894, 899 ⟨77JA5026⟩; 3, 593 ⟨73JOC1319⟩; 4, 42 ⟨81JOC571⟩
Donovan, J., 3, 710 ⟨70TL4079⟩; 6, 623, 627 ⟨72JCS(P2)1050⟩
Donovan, R. J., 2, 920 ⟨92JA6580⟩
Donskikh, I. V., 5, 656 ⟨87JGU1710⟩
Donskikh, V. I., 2, 448 ⟨91ZOB2161⟩, 458 ⟨83JGU1059⟩; 4, 766 ⟨89ZOB997⟩, 926 ⟨82ZOB2375⟩; 5, 656 ⟨87ZOB1912⟩; 6, 619 ⟨90ZOB706⟩, 620 ⟨91ZOB643, 92ZOB306⟩
Don Tilley, T., 4, 656, 657 ⟨93JA5527⟩
Dontsova, N. E., 4, 854 ⟨78IZV2654, 82IZV2327⟩, 861 ⟨84IZV194⟩, 866 ⟨84IZV611⟩
Donzova, N. E., 4, 770, 779 ⟨90G235⟩
Doolittle, R. E., 1, 512 ⟨84S730⟩
Doomes, E., 1, 707 ⟨70JA2581⟩, 709 ⟨76TL1243⟩; 4, 83 ⟨74JOC2526⟩
Doonan, D. J., 6, 721 ⟨73JA4769, 74IC921⟩
Doong, S.-L., 4, 231 ⟨92JOC2217⟩
Doorakian, G. A., 4, 165 ⟨64T2181⟩
Doorenbos, N. J., 5, 199 ⟨65CI(L)648⟩
Doornbos, T., 2, 1030 ⟨56RTC1459, 56RTC481⟩
Döpfer, Kl.-P., 4, 461 ⟨78ZAAC(444)249⟩, 463 ⟨83PS(14)171⟩
Dopico, M., 1, 433 ⟨90JOC991⟩
Dopico, P. G., 1, 991 ⟨93AG(E)554, 93JOC1298⟩; 6, 706 ⟨93AG(E)554⟩
Döpp, D., 3, 61 ⟨89CB585⟩; 5, 313, 322, 350, 384 ⟨85HOU(E5)1135⟩, 1011 ⟨74TL2445⟩
Döpp, H., 5, 313, 322, 350, 384 ⟨85HOU(E5)1135⟩
Doran, M. A., 1, 850 ⟨66JA1272⟩
Doran, R. E., 5, 1047 ⟨1895JCS565⟩
Doran, W. J.,1, 506 ⟨61JOC3772⟩
Dorchak, J., 1, 443 ⟨93TL8051⟩, 938, 946 ⟨91T8969⟩, 939 ⟨92T7435⟩
d'Orchymont, H., 1, 590 ⟨79JA3283⟩; 4, 320 ⟨93S961⟩; 5, 328 ⟨92TL5055⟩
Dordor, I. M., 5, 453 ⟨84CC956, 84TL2709⟩
Dordor-Hedgecock, I. M., 5, 453 ⟨85JOM(285)213, 85TL2125⟩, 453 ⟨86CC607, 86TL3787⟩
Dorff, P., 1, 1078 ⟨92TL3715⟩
Dorfman, A. Ya., 4, 336 ⟨84MI 407-03⟩
Dorfman, L., 3, 233 ⟨63JOC3132⟩; 5, 631 ⟨64JOC1115⟩
Dorfman, L. M., 1, 847 ⟨74JA5708⟩
Dorfman, Ya A., 2, 108 ⟨91JGU999⟩
Dorfmeister, G., 4, 549 ⟨87TL1093⟩
Dorgan, B. J., 2, 543 ⟨90TL6721, 90TL6725⟩
Dörges, C., 4, 484 ⟨90TL6999⟩
Dorhofer, K., 3, 324 ⟨75AG(E)422⟩
Dorigo, A. E., 1, 927 ⟨87JA2195⟩
Döring, H., 6, 431 ⟨28AP(266)277⟩
Dörken, C., 2, 864 ⟨1888CB1505⟩
Dorlars, A., 2, 399 ⟨58HOU(11/2)641, 58HOU(11/2)738⟩
Dorman, J. H., 4, 262 ⟨92JCS(P1)1105⟩
Dorme, R., 4, 187, 208 ⟨83SC629⟩
Dormidonov, I. A., 4, 340 ⟨72ZOB479⟩
Dormond, A., 2, 566 ⟨80CC214⟩
Dormoy, J. R., 3, 309 ⟨93T2885⟩; 5, 191 ⟨72TL4747⟩, 267 ⟨75TL1219⟩
Dorn, C. R., 3, 208 ⟨72JOC268⟩
Dorn, H., 2, 377 ⟨72ZC129⟩, 380 ⟨65CB3377, 67ZC151⟩
Dörner, W., 2, 591 ⟨93JA7027⟩
Dornfeld, C. A., 3, 459 ⟨55OSC(3)26⟩
Dörnhöfer, C., 5, 241 ⟨89JOM(377)281⟩
Dornow, A., 1, 57 ⟨55LA177, 57CB1774⟩; 2, 419 ⟨58ZAAC(294)81⟩; 3, 258 ⟨57LA(602)14⟩, 426 ⟨60CB32⟩; 4, 900 ⟨67CB2577⟩; 5, 822 ⟨55LA(594)185⟩
Dornow, R., 1, 325, 920 ⟨87JOC205⟩
Dornyei, G., 2, 379 ⟨85JPR445⟩
Dorofeenko, G. N., 3, 420 ⟨82RCR469⟩
Dorofeenko, G. N., 5, 688 ⟨81JOU1986, 81ZOR2225⟩
Dorofeeva, O. V., 2, 28 ⟨87IZV2395⟩
Dorofeev, I. A., 3, 361 ⟨80ZOR1842⟩
Dorokhova, E. M., 6, 656 ⟨75ZOR762⟩
Dorokhov, V. A., 1, 657 ⟨65MI 114-01, 69JCS(A)433⟩; 2, 112 ⟨60MI 202-02⟩, 811 ⟨76BAU1374⟩; 3, 458 ⟨68CC250A, 69JCS(A)433⟩; 4, 884 ⟨72IZV1895⟩, 975 ⟨89IZV1806⟩, 979 ⟨90IZV401⟩; 5, 381 ⟨71ZOB2501⟩, 381, 382 ⟨70DOK(194)595⟩, 778 ⟨71DOK(199)1328, 74BAU1593, 78BAU2304, 87IZV165, 89IZV2353⟩; 6, 113 ⟨66IZV364⟩, 504 ⟨91IZV2593⟩, 662 ⟨79DOK(245)121, 82IZV2370, 87IZV1139⟩
Dorokhov, V. I., 5, 1068, 1069 ⟨89ZOR759⟩, 1076 ⟨86ZOR1099⟩
Doronzo, S., 1, 52 ⟨76MI 102-01⟩; 2, 435 ⟨78IS169⟩, 437 ⟨76IS155⟩, 822, 823 ⟨76IS161⟩, 824 ⟨78IS170, 78IS172⟩, 824, 825 ⟨78IS171⟩, 827 ⟨76IS157, 76IS158, 76IS159, 76IS160⟩
Doroshenko, V. V., 2, 169 ⟨70ZOB1692⟩; 5, 787 ⟨73JGU10⟩; 6, 622, 634, 658 ⟨71ZOB2155⟩
Doroshkina, G. M., 5, 679 ⟨91JGU2372, 91ZOB2257⟩
Dorow, R. L., 1, 299, 353 ⟨84OM44, 82TL2261⟩, 832 ⟨91JOC5063⟩; 2, 29 ⟨87TL1123⟩, 375 ⟨86JA6395a⟩, 1004 ⟨90JA864⟩
Dorozhkina, G. M., 4, 489 ⟨82ZOB2444⟩
Dörr, H., 1, 707, 708 ⟨80LA1540⟩
Dorrenbach, F., 2, 472 ⟨90JOM(399)C4⟩
Dörrenbach, F., 6, 243 ⟨89JOM(378)125⟩
Dorritty, M. R. J., 2, 245 ⟨88T4941⟩
Dorsch, D., 1, 134 ⟨83TL3213⟩
Dorsch, H., 1, 104 ⟨68LA(711)46⟩
Dorsey, B. D., 1, 829 ⟨93TL1851⟩

Dorsey, D. A., **3**, 539 ⟨86JOC2814⟩
Dorsky, J., **5**, 66 ⟨55OSC(3)538⟩
Dorwald, F. Z., **5**, 1062 ⟨88LA787⟩
Dorweiler, C., **2**, 1003 ⟨88CB843⟩
Dörzbach, C., **2**, 447 ⟨84ZN(B)618⟩
Doser, K., **2**, 347 ⟨90HOU(E16/1)404⟩
Dos Santos Filho, P. F., **1**, 637 ⟨90JMOCL1⟩
Doss, R. C., **6**, 282, 283 ⟨68USP3419625⟩
Dossena, A., **2**, 24, 31, 35 ⟨85G29⟩, 28 ⟨84T2035⟩; **3**, 27, 154 ⟨84T2035⟩
Dostal, K., **2**, 370 ⟨77CCC1466⟩
Dostal, W., **1**, 1010 ⟨88M751⟩
Dostalek, R., **4**, 571 ⟨92CB2081, 92S787⟩; **5**, 896 ⟨89TL5261⟩; **6**, 520 ⟨92AG(E)1064⟩, 702 ⟨92S787⟩
Dost, N., **4**, 83 ⟨48RTC451⟩
Dostovalova, V. I., **6**, 5 ⟨87IZV808⟩
Dostrovalova, V. I., **4**, 137 ⟨87IZV1087⟩
Dotan, I., **1**, 1143 ⟨76JCP(64)4544⟩
Dotse, A. K., **4**, 212 ⟨91TL2413⟩
Dotson, R., **1**, 1010 ⟨92BMC1619⟩
Doty, J. K., **4**, 764 ⟨78JOC43⟩
Dötz, K. H., **1**, 767 ⟨84AG(E)587⟩, 885 ⟨B-91MI 119-02⟩; **2**, 669 ⟨75AG(E)644⟩, 670 ⟨83JOM(252)C78⟩, 670 ⟨84AG(E)587⟩, 670 ⟨93CC1811⟩; **3**, 508 ⟨B-83MI 315-01⟩, 545 ⟨79AG(E)954⟩; **5**, 444, 931, 932 ⟨84AG(E)587⟩, 652 ⟨69JOM(16)29⟩, 931 ⟨90NJC433⟩, 931 ⟨91SL381⟩, 932 ⟨88JA8413, 90T1235⟩, 934 ⟨95CB645⟩, 935 ⟨94OM4189⟩, 944 ⟨87OM1424, 95CB157⟩, 953 ⟨91CB1635, 92CB2137⟩, 953 ⟨93JOM(459)169⟩; **6**, 716, 719, 722 ⟨91SL381⟩, 716, 719, 722 ⟨B-84MI 622-01⟩
Dotzer, R., **2**, 444 ⟨89S204⟩
Dötzer, R., **2**, 597 ⟨63ZAAC(321)120⟩; **4**, 571 ⟨85TL2769⟩
Dou, D., **4**, 484 ⟨92JA9691⟩
Doubleday, C., **3**, 145 ⟨86TL4671⟩
Doubleday, W., **1**, 236 ⟨87JA4710⟩
Doubleday, W. W., **2**, 553, 555 ⟨92JOC3680⟩
Doucet, H., **2**, 676 ⟨93CC850⟩
Doucoure, A., **2**, 74 ⟨82JOM(236)139⟩
Doudy, B., **3**, 237 ⟨91TL83⟩
Douek, M., **1**, 851, 865 ⟨68JA2698⟩
Dougharty, C. M., **2**, 420 ⟨73JA581⟩
Dougherty, C. M., **2**, 696 ⟨73JA582⟩
Dougherty, G., **2**, 177, 178, 179 ⟨42JOC81⟩, 227 ⟨42JA149, 50JOC865⟩; **4**, 317 ⟨33JA4588⟩
Dougherty, J. T., **3**, 546, 547 ⟨68JOC2738⟩; **5**, 511 ⟨68JOC2738⟩
Doughty, H. W., **4**, 14 ⟨31JA1594⟩
Douglas, A. W., **1**, 558 ⟨88JA3318⟩; **2**, 66 ⟨93TL3243⟩, 528 ⟨93JOC2880⟩
Douglas, C. M., **2**, 836, 837 ⟨64IC428⟩, 904 ⟨59ZN(B)809⟩
Douglas, G. H., **6**, 647 ⟨78AF1435⟩
Douglas, P., **5**, 619 ⟨1892CB1311⟩
Douglass, D., **3**, 273 ⟨90JCS(P1)1507⟩
Douglass, I. B., **2**, 166, 173 ⟨60OS(40)62⟩, 166, 173, 174 ⟨61JOC1996⟩, 166, 173, 232 ⟨59JOC973⟩, 168 ⟨49JOC272⟩, 168 ⟨50JOC795⟩, 168 ⟨53JA4582⟩, 172 ⟨64JOC951⟩, 172, 173, 174 ⟨68JOC2104⟩, 174, 187 ⟨58JOC330⟩, 174, 688 ⟨57JOC536⟩, 174, 717 ⟨70OPP235⟩, 177 ⟨38JA1486⟩, 177, 178 ⟨39JA2548⟩, 178, 187, 196 ⟨65JOC633⟩, 185 ⟨59JOC2004⟩, 194 ⟨62JOC1398⟩, 213 ⟨B-61MI 203-03⟩, 232 ⟨72JOC333⟩, 358 ⟨76JPR1019⟩, 478 ⟨67JOC3645⟩; **4**, 71 ⟨50JOC795⟩, 78 ⟨68JOC2104, 70OPP235⟩, 85 ⟨41JA1571⟩; **5**, 1049 ⟨34JA1609, 34JA719⟩, 1057 ⟨37JA740⟩; **6**, 229 ⟨55JA1899⟩, 229, 255 ⟨56JA6070, 53JA4582⟩
Douglass, III, J. G., **1**, 524 ⟨83TL1675⟩
Douglass, J. E., **4**, 281 ⟨90JHC1433⟩
Doukas, P. H., **6**, 51 ⟨79H(12)637⟩
Doule, M. D., **3**, 539 ⟨89CC303⟩
Doulut, S., **2**, 305 ⟨91TL923⟩
Doumani, T. F., **1**, 56 ⟨58IEC1677⟩
Doumaux, Jr., A. R., **5**, 307 ⟨69JA3992⟩
Douraghi-Zadeh, K., **5**, 849, 1061 ⟨67CRV107⟩; **6**, 506 ⟨67CRV107⟩
Dourtoglou, V.,**5**, 279 ⟨78TL1269, 84S572⟩
Doussal, B. Le, **1**, 958 ⟨83T2185⟩
Dousse, G., **1**, 656 ⟨83JOM(243)281⟩; **6**, 577 ⟨75JOM(88)C35⟩
Doussot, J., **2**, 317 ⟨88S900⟩
Doussot, P., **4**, 353 ⟨92S995⟩
Doutheau, A., **1**, 225 ⟨79S508⟩, 808 ⟨78TL1803⟩, 1033 ⟨82SC557⟩; **2**, 668 ⟨93TL7923⟩; **3**, 230 ⟨85TL4923, 87T2721⟩
Douthit, C. E., **2**, 492 ⟨68IC834⟩
Doviken, L., **4**, 675 ⟨91IC156⟩
Dovlatyan, V. V., **5**, 1120 ⟨80CA(93)95215⟩
Dow, A. W., **3**, 483 ⟨68JA1080, 72JOM(44)279⟩; **4**, 513 ⟨68JA1080⟩
Dow, R. L., **3**, 118 ⟨80SC881⟩; **5**, 329 ⟨86TL4957⟩
Dowd, P., **1**, 245 ⟨89TL6129⟩, 340 ⟨92TL7307⟩, 340, 919, 928, 929, 931 ⟨93CRV2091, 91T4847⟩; **3**, 126 ⟨89T77, 91TL565⟩, 613 ⟨93SC2323⟩; **4**, 162 ⟨85TL2279⟩, 162 ⟨85TL2283⟩
Dowd, S. R., **3**, 124 ⟨63JA2178⟩; **4**, 16 ⟨65JA4259⟩
Dowden, B. F., **2**, 841 ⟨65JOM(4)291⟩
Dowden, J., **4**, 269 ⟨92JOC6390⟩
Dowell, Jr., A. M., **6**, 33, 81, 228 ⟨56JA479⟩
Doweyko, A. M., **4**, 81 ⟨82TL2539⟩
Dowlatshahi, H. A., **1**, 934 ⟨80CC732⟩
Dowling, M., **2**, 245 ⟨88T4941⟩
Downer, J., **2**, 923 ⟨90TL5829⟩; **4**, 661, 702 ⟨90TL5829⟩
Downer, J. D., **2**, 439 ⟨66JCS(C)245⟩
Downes, J. M., **2**, 828 ⟨78IC880⟩
Downes, L. M., **6**, 72 ⟨89AJC1235⟩
Downie, I. M., **1**, 185, 200 ⟨67JOC855⟩; **2**, 7 ⟨68CC1350⟩, 841, 842 ⟨65JCS5771⟩, 845 ⟨64JCS1240⟩; **3**, 92, 99, 101 ⟨93T4015⟩, 613 ⟨67JOC855⟩
Downing, D. C., **4**, 434, 440 ⟨49MI 409-01⟩
Downs, A. J., **2**, 169 ⟨62JCS4361⟩, 492 ⟨85POL1625, 88JCS(D)451⟩; **5**, 1093 ⟨67SA(A)681⟩; **6**, 232 ⟨60JCS3516⟩, 236 ⟨62JCS4361⟩
Downs, A. W., **2**, 290 ⟨68CC1290⟩
Dox, A. W., **5**, 748 ⟨41OSC(1)5⟩
Doxsee, K. M., **1**, 473 ⟨87CL19⟩
Doyama, K., **3**, 64 ⟨88BCJ4353⟩
Doyama, T., **3**, 64 ⟨86TL4497⟩
Doyle, D. L., **3**, 126 ⟨83T345⟩
Doyle, F. P., **5**, 9 ⟨63JCS491⟩
Doyle, J. E., **2**, 346 ⟨78AJC2317⟩
Doyle, K., **6**, 467 ⟨89OPP771⟩
Doyle, K. M., **5**, 640 ⟨86JCS(P1)1873⟩, 652 ⟨74CI(L)803⟩, 769 ⟨74S583⟩, 769 ⟨76T1031⟩
Doyle, M. J., **6**, 716 ⟨73CSR99⟩, 720 ⟨74JCS(D)1494⟩
Doyle, M. P., **1**, 36, 486 ⟨77JOC2431⟩, 234, 299, 352 ⟨86CRV919⟩, 299, 353 ⟨84OM44⟩, 352, 400, 821, 823 ⟨86ACR348, 82TL2261⟩, 353, 354 ⟨91RTC305⟩, 353, 354 ⟨94T4519⟩, 400, 408, 409, 822, 823, 825 ⟨81JOC5094⟩, 400, 823 ⟨84JOC1917⟩, 442 ⟨77JOC3494⟩, 820 ⟨88TL5119⟩; **2**, 11 ⟨78JOC4120⟩, 104 ⟨83JOC3379⟩, 539 ⟨90JOC6082⟩, 619 ⟨87JOC1619⟩, 621, 627 ⟨77JOC2426⟩, 805 ⟨72JOC1597⟩; **3**, 5, 294 ⟨79JOC2955⟩, 118 ⟨80SC881⟩, 203 ⟨88JOC5558⟩, 206 ⟨84JOC1196⟩; **4**, 19 ⟨76CC433⟩, 19, 27 ⟨76JA1627⟩, 138, 154 ⟨88TL5863⟩; **5**, 282 ⟨74JOC2740⟩
Doyle, R. J., **4**, 224 ⟨89TL2179⟩
Doyle, R. R., **3**, 494 ⟨67CC300⟩
Doyle, T. J., **4**, 962 ⟨90TL2541⟩
Doyle, T. R., **6**, 33 ⟨89JA8510⟩
Doyle, T. W., **2**, 128 ⟨66JA1828, 70JOC2716⟩; **6**, 611 ⟨84TL487⟩
Doyler, M. J., **6**, 721 ⟨72CC927⟩
Doyon, J., **2**, 625 ⟨91JOC6199⟩
Drabek, J., **5**, 390 ⟨81EUP54516⟩, 775 ⟨85GEP3501452⟩

Drabowicz, J., **2**, 144 ⟨81SC1025⟩, 144 ⟨B-88MI 203-02⟩, 145 ⟨78S758⟩, 148 ⟨79S39⟩, 151 ⟨92CC1500⟩, 156, 721 ⟨88T6687⟩, 165 ⟨B-90MI 203-03⟩, 184 ⟨89AG(E)97⟩, 184, 194, 239 ⟨B-90MI 203-06⟩, 197 ⟨74CC547, 74S124⟩, 198 ⟨85CC1670⟩, 212 ⟨78S138⟩, 213 ⟨80S32⟩, 217 ⟨76CC220⟩, 218, 724 ⟨85TL5703⟩, 357, 358 ⟨B-90MI 206-01⟩, 450, 842 ⟨75CC382⟩; **3**, 134 ⟨80S125⟩; **4**, 74, 76 ⟨86S831⟩

Drach, B. S., **2**, 407 ⟨64ZOB1441, 64ZOB3055⟩, 407 ⟨65ZOB344⟩; **3**, 418 ⟨78ZOR508⟩; **4**, 99 ⟨88JGU2192⟩, 104 ⟨73JOU418⟩, 106 ⟨68JGU1276⟩, 304 ⟨69ZOR2181⟩, 344 ⟨74ZOB1712⟩, 487 ⟨69ZOB2192⟩, 771 ⟨91ZOB874⟩, 794 ⟨75JOU119, 76JOU2252, 80JOU1762⟩, 966 ⟨73ZOB211⟩, 998 ⟨92ZOB1084, 92ZOB1423⟩, 999 ⟨80ZOB2248, 83ZOB2015, 84ZOB288⟩, 999, 1000 ⟨75ZOB12, 86ZOB321⟩, 999, 1001, 1002, 1007, 1011 ⟨84ZOB2186⟩, 1000, 1001 ⟨78ZOB1994⟩, 1002 ⟨77ZOB1994⟩, 1005 ⟨92ZOB707⟩, 1006, 1010, 1012 ⟨82ZOB1122⟩; **5**, 655 ⟨78JOU881, 78ZOR943⟩, 655 ⟨86JGU2483, 86ZOB2802⟩, 655 ⟨88JGU2192, 88ZOB2465, 89JOU2329⟩, 655, 1053 ⟨89ZOR2597⟩, 655, 656 ⟨84JGU255, 84ZOB288⟩, 655, 657, 659 ⟨76JOU2252, 76ZOR2319⟩, 655, 659 ⟨77JOU1471, 77ZOR1594⟩, 656 ⟨82JGU980, 82ZOB1122, 91UKZ530⟩, 657 ⟨76JOU663, 76ZOR673⟩, 657, 659 ⟨79JOU32, 79ZOR38⟩, 685, 694 ⟨75JOU1926, 75ZOR1913, 77JOU1474, 77ZOR1597⟩, 787, 788 ⟨77JGU441⟩, 965 ⟨70ZOB1933⟩

Dragalov, V. V., **6**, 476 ⟨88PS(39)257⟩

Dräger, M., **1**, 766 ⟨91OM1771⟩; **4**, 570, 585, 660, 704 ⟨92JOM(423)339⟩; **6**, 169 ⟨92OM145⟩, 590 ⟨68AG(E)868⟩, 592 ⟨71CB1429⟩

Draghici, C., **1**, 518 ⟨73TL9⟩

Dragisich, V., **2**, 671 ⟨91JA9873⟩; **5**, 696 ⟨88JA307⟩

Dragisich, W., **3**, 680 ⟨90OM2867⟩

Draguet, C., **5**, 238 ⟨62BSB260⟩

Draheim, S. E., **3**, 639 ⟨90TL2661⟩

Drahnak, T. J., **3**, 502 ⟨81JA1845⟩

Drahowzal, F., **2**, 225 ⟨51M970⟩

Drake, A. F., **4**, 262 ⟨92JCS(P1)1105⟩

Drake, C. B., **2**, 478 ⟨77AJC2479⟩

Drake, G. L., **4**, 555 ⟨68JCED585, 71USP3607944⟩

Drake, J. E., **2**, 239 ⟨93CJC42⟩, 278, 294 ⟨73JCS(D)1716⟩, 292, 294 ⟨80IC1879⟩, 293 ⟨75JIN233⟩, 474 ⟨69JCS(A)2704⟩, 857 ⟨75INC233⟩; **4**, 377 ⟨86IC611⟩, 377, 401 ⟨86ICA51⟩; **5**, 540 ⟨76MI 512-01⟩, 1081 ⟨76JCS(D)366⟩

Drake, Jr., G. L., **4**, 113, 118, 120, 122 ⟨67JCED282⟩

Drake, N. L., **2**, 753 ⟨42OR(1)105⟩

Drake, R. A., **1**, 999 ⟨88TL4181⟩

Drake Jr., G. L., **4**, 453, 456, 458, 475 ⟨72JOC2752⟩

Drakesmith, F. G., **1**, 428 ⟨65JCS5045⟩, 570 ⟨94JCS(P1)3115⟩; **4**, 806 ⟨68JOC472⟩, 817 ⟨68JOC286⟩

Draney, B. W., **1**, 827 ⟨94TL1661⟩

Dransch, G., **2**, 166, 167 ⟨60CB782⟩

Draper, P. M., **2**, 4366 ⟨70TL1687⟩

Drašar, P., **5**, 567 ⟨93SC829⟩

Drauz, K., **3**, 40 ⟨93JOC3568⟩; **4**, 100 ⟨87CZ149⟩

Drauz, K.,**5**, 964 ⟨81CB173⟩

Drauz, K., **6**, 445 ⟨86JOC3494⟩

Drawert, F., **6**, 545 ⟨60CB3056⟩

Dre'an, P., **5**, 1160 ⟨94MI 531-01⟩

Drechsel-Grau, E., **1**, 325, 920 ⟨87JOC205⟩

Drechsler, H.-J., **5**, 655, 660 ⟨78TL1429⟩

Drechsler, H. J., **5**, 685 ⟨78ZN(B)636⟩

Drechsler, K., **2**, 26 ⟨83S314⟩

Dreef, C. E., **2**, 106 ⟨91TL6021⟩, 409 ⟨84TL1307⟩

Drees, F., **5**, 279 ⟨66CB110⟩

Dreeskamp, H., **4**, 583 ⟨72IJ293⟩

Dreet, C. E., **2**, 467 ⟨92TL2357⟩

Drefahl, G., **1**, 57 ⟨58CB750⟩, 745 ⟨64JPR143⟩; **2**, 745 ⟨69JPR162⟩, 982 ⟨58AG166⟩; **3**, 494 ⟨65CB604⟩; **5**, 984 ⟨60CB519⟩

Dreger, E. E., **2**, 102 ⟨44IEC610⟩; **4**, 23 ⟨32OSC(1)357⟩; **5**, 109 ⟨32OSC(1)14⟩, 898 ⟨32OSC(1)23⟩; **6**, 68 ⟨51OSC(1)258⟩

Dreger, L. H., **5**, 164 ⟨61JA606⟩

Dregval, G. F., **5**, 861 ⟨63JGU2880⟩

Dreiding, A. S., **1**, 812 ⟨82HCA451⟩, 826 ⟨65HCA1985, 71HCA1589⟩; **2**, 376 ⟨85HCA220⟩, 831 ⟨76HCA695⟩; **3**, 216 ⟨76HCA1226⟩, 223 ⟨81HCA1123⟩, 223, 273 ⟨85HCA338⟩, 237 ⟨87HCA321, 86HCA560⟩, 352 ⟨76HCA747⟩, 355 ⟨81HCA2369⟩, 526 ⟨68HCA1466, 70HCA120⟩, 527 ⟨70HCA2159, 70HCA417⟩; **4**, 238, 241 ⟨87T4875⟩, 768 ⟨68HCA1466⟩, 979 ⟨77HCA1781⟩; **6**, 425 ⟨86HCA1442⟩

Dreier, F., **4**, 410, 539 ⟨80HCA1190⟩; **6**, 170 ⟨80HCA1190⟩

Dreikorn, B. A., **5**, 1029 ⟨89JHC1735⟩

Dreizler, H., **3**, 694, 716 ⟨91AG(E)1644⟩

Drenth, ,W., **2**, 970, 971 ⟨67JOM(9)295⟩

Drenth, W., **1**, 1114 ⟨78JOC763⟩, 1122 ⟨60RTC1002, 61RTC1285⟩, 1122 ⟨61RTC797, 63RTC375⟩, 1122 ⟨63RTC394, 63RTC405⟩, 1122 ⟨63RTC410, 74JA6692⟩; **2**, 1032 ⟨63RTC375⟩

Dréos Garlatti, R., **2**, 583 ⟨86ICA(121)27⟩

Dresdener, R. D., **6**, 238, 239, 240, 242 ⟨58JA3604⟩

Dresdner, R., **6**, 239 ⟨57JA69⟩

Dresdner, R. D., **5**, 989 ⟨59JA1587⟩; **6**, 253 ⟨59JA574⟩, 257 ⟨58JA1889⟩, 440 ⟨58JOC1576⟩

Dresel, W. H., **6**, 267, 268 ⟨78ZAAC(439)161⟩

Dresely, S., **2**, 75 ⟨88CB1501, 88CB2225⟩, 900 ⟨88S103⟩; **4**, 363 ⟨89CB903⟩

Dressaire, G., **1**, 253 ⟨80TL67⟩

Dressel, J., **1**, 248 ⟨87TL4965⟩, 704, 708 ⟨87JA2857, 88JA5479⟩

Dressler, H., **5**, 1023 ⟨30CB888⟩

Dressler, M., **1**, 490 ⟨70JPR150⟩

Dreux, J., **3**, 232 ⟨72BSF2510⟩

Dreux, M., **2**, 618 ⟨75S658⟩, 1057 ⟨76TL835⟩; **4**, 740 ⟨75TL609⟩, 740, 741 ⟨75S535⟩; **6**, 266 ⟨75S535⟩

Drevin, H., **2**, 219 ⟨72BJ723⟩

Drew, H. D. K., **5**, 1117 ⟨25JCS2307⟩

Drew, J., **3**, 60 ⟨87JOC4047⟩

Drew, K. N., **3**, 708 ⟨91JOC509⟩

Drew, R. A. I., **1**, 68 ⟨86CC855⟩

Drewes, M. W., **3**, 40 ⟨87AG(E)1141⟩, 616 ⟨88TL3295⟩

Drewes, S. E., **1**, 827 ⟨90SC1437⟩; **3**, 273 ⟨90JCS(P1)1507⟩; **4**, 118 ⟨90SC1671⟩

Drewitt, J. G. N., **6**, 653 ⟨36JCS1693⟩

Drewniak, M., **1**, 260 ⟨88JOC5854⟩; **2**, 320 ⟨88JCS(P1)2339⟩; **4**, 449 ⟨88JCS(P1)2339⟩; **5**, 606 ⟨88JOC5854, 88TL1755⟩

Drewniak-Deyrup, M., **6**, 370 ⟨89JHC829⟩

Drews, R., **3**, 553 ⟨84JOM(270)93, 84JOM(272)C40⟩

Drews, S. E., **3**, 498 ⟨88T4653⟩

Drey, C. N. C., **5**, 62 ⟨B-76MI 502-01⟩

Dreyer, G., **1**, 957, 973 ⟨93TL449⟩

Dreyer, G. B., **6**, 487 ⟨93MIP9305026⟩

Dreyfus, H., **3**, 535 ⟨25BRP273622⟩

Driess, M., **4**, 567 ⟨93PS(76)57⟩, 665 ⟨92CB1341⟩; **5**, 892 ⟨92AG(E)316⟩; **6**, 691 ⟨93CC1585⟩

Driggs, R. J., **1**, 128 ⟨78JOM(156)45⟩

Driguez, H., **3**, 43 ⟨78CJC119⟩; **5**, 316, 318 ⟨74JOC3136⟩

Drischel, W., **1**, 1145 ⟨65AG95⟩; **3**, 495 ⟨66CB689⟩; **6**, 25 ⟨64AG(E)513⟩

Driscoll, J. S., **3**, 604 ⟨66CC736⟩; **4**, 420 ⟨72JHC1145⟩, 554 ⟨64JOC2427⟩; **5**, 1093 ⟨62JA4349, 64JOC2427⟩

Drisko, R. W., **4**, 235 ⟨48JA2435⟩

Driss, C., **4**, 190 ⟨90SC2527⟩

Driver, J. E., **6**, 69 ⟨24JA2090⟩

Drivon, G., **6**, 32 ⟨92EUP515258⟩

Drobnica, L., **5**, 1022, 1038 ⟨B-77MI 527-01⟩

Drobnick, J. A., **5**, 864 ⟨71GEP2044574⟩

Droege, M. W., **6**, 246, 247 ⟨89OM1498⟩

Droescher, H., **3**, 470 ⟨68HCA643⟩

Drohne, D., **2**, 890 ⟨70JOM(21)231⟩; **4**, 495, 496 ⟨70JOM(21)131⟩, 495, 497 ⟨73EGP99803⟩, 496, 498, 499 ⟨73JOM(60)95⟩
Drokina, M. I., **6**, 501 ⟨82JOU19582220⟩
Dromzee, Y., **1**, 439 ⟨86T2475⟩; **4**, 720 ⟨90IC1817⟩
Drone, F. J., **2**, 880 ⟨84OM495⟩; **4**, 1039 ⟨85OM1478⟩
Dronkina, M. I., **6**, 238 ⟨66ZOB1309, 71ZOR2000⟩, 619 ⟨84ZOB1110⟩
Dronov, V. I., **4**, 864 ⟨75ZOR415⟩
Droste, H., **3**, 616 ⟨80CB710, 80LA212⟩
Drouin, J., **1**, 66, 454, 526 ⟨85JA1726, 87BSF1006⟩, 526 ⟨87BSF1015⟩, 526, 999 ⟨87TL3923⟩, 544 ⟨83S801⟩; **3**, 138 ⟨85JA1726⟩
Drover, J. C. G., **1**, 230 ⟨87JA4649⟩
Drovneva, R. P., **6**, 656 ⟨78ZOB811, 81ZOB880⟩
Drozd, G. I., **2**, 476 ⟨68ZOB1906⟩
Drozd, V. N., **3**, 361 ⟨77ZOR2012, 80ZOR443⟩, 361 ⟨80ZOR883⟩, 361, 376 ⟨78ZOR2459⟩; **5**, 639 ⟨93S463⟩, 736 ⟨90ZOR214⟩
Drozdova, T. I., **2**, 338, 365 ⟨89RCR250⟩
Drtina, G. J., **1**, 19 ⟨84JA6690⟩, 816 ⟨87JOC2960⟩; **3**, 245 ⟨84TL4467⟩
Drucker, A., **2**, 123 ⟨64JOC360⟩
Drucker, G. E., **1**, 142, 155 ⟨75JA7006⟩, 861 ⟨81JOC632⟩
Druckrey, E., **1**, 521 ⟨66AG(E)1039⟩, 522 ⟨66AG1057⟩
Drueckhammer, D. G., **3**, 159 ⟨91S499⟩; **5**, 235 ⟨92JA7287⟩
Druelinger, M., **1**, 152 ⟨81JA3081⟩; **3**, 124 ⟨81JA3081⟩
Druellinger, M., **3**, 407 ⟨76JA3032⟩
Druey, J., **1**, 398 ⟨62AG(E)155⟩, 513 ⟨59HCA401⟩; **2**, 176, 180 ⟨57HCA2148⟩, 374 ⟨57HCA918⟩; **5**, 769, 852 ⟨55HCA1560⟩
Drukker, A. F., **2**, 373, 374, 379, 381 ⟨59JA2805⟩
Druliner, J. D., **3**, 614 ⟨B-85MI 303-01⟩; **5**, 365 ⟨67JOC1552⟩
Drummond, D. K., **5**, 439 ⟨86JA1722⟩
Drummond, F. E., **5**, 974 ⟨62USP3070618⟩
Drummond, J. T., **5**, 342 ⟨88TL1653⟩; **6**, 58 ⟨89TL7013⟩
Drumright, R. E., **2**, 395 ⟨87OPP471⟩
Dryanska, V., **3**, 408 ⟨84S1038⟩
Dryden, Jr., H. L., **2**, 660 ⟨64JA742⟩
Dryden, N. H., **4**, 656 ⟨91OM2857⟩; **6**, 247 ⟨92CM979⟩
Drygala, P. F., **4**, 454, 471 ⟨83AJC2095⟩
Drysdale, M. J.,**5**, 900 ⟨94CC805⟩
Dschneider, D., **4**, 621 ⟨93JA6625⟩
D'Silva, T. D. J., **6**, 439 ⟨77USP4058549⟩, 448 ⟨85EUP136147, 86JOC3781⟩
Du, C. J. F., **2**, 974 ⟨85JOC3104, 85JOC5524⟩
Du, M. T., **4**, 161 ⟨85JOC5022⟩
Dua, P. R., **5**, 685 ⟨93MI 517-01⟩
Dua, S. K., **1**, 146 ⟨85JOC3674⟩; **3**, 224 ⟨85JOC3674⟩; **4**, 393 ⟨85JOM(285)121⟩
Duah, F. K., **3**, 69 ⟨90JHC765⟩
Duan, J. J.-W., **1**, 573, 581, 584 ⟨92JA9419⟩; **6**, 469 ⟨93JOC3703⟩
Duan, J.-X., **6**, 219, 227 ⟨92CC807⟩, 232, 234 ⟨93CC918⟩
Duan-Hu, J., **1**, 272 ⟨92JOC774⟩
Duar, Y., **1**, 955 ⟨76TL3181⟩
Dubac, J., **1**, 777 ⟨83TL3521⟩; **2**, 22, 28 ⟨91SC11⟩, 942, 949 ⟨92S954, 85TL1315⟩; **3**, 159 ⟨93JOC1835⟩; **6**, 582 ⟨79TL3507⟩
Duballet, P., **3**, 109 ⟨76BSF789⟩
Duban, F., **4**, 420, 421, 426 ⟨55JA745⟩
Dubansky, A., **2**, 782 ⟨59CCC1158⟩
Dubbert, ,R. A., **2**, 985 ⟨92OM4096⟩
Dube, D., **1**, 615, 698, 703 ⟨86JA2776⟩
Dube, M., **6**, 509 ⟨78JOC61⟩
Dube, M. F., **4**, 441 ⟨80T1753⟩
Dubeck, M., **2**, 964 ⟨60JA1939⟩; **4**, 692 ⟨60JA502⟩
Dubenko, L. G., **6**, 609 ⟨78JGU1512⟩
Dubenko, R. G., **5**, 593 ⟨74ZOR1992, 76ZOR681⟩, 593 ⟨76ZOR69⟩, 595 ⟨71ZOR2352, 73ZOR347⟩, 843 ⟨63ZOB3917⟩, 1049 ⟨68ZOR658⟩; **6**, 277 ⟨68ZOR324⟩, 564 ⟨81ZOR191⟩, 653 ⟨57ZOB2134⟩
Düber, E.-O., **2**, 812 ⟨80S630, 82S273⟩
DuBeshter, B., **5**, 379 ⟨74JA4283⟩
Dubey, D. K., **4**, 773 ⟨93TL7645⟩
Dubey, P. K., **6**, 90, 91, 307 ⟨82CC1183⟩
Dubey, S. K., **6**, 443 ⟨87IJC(B)748⟩
Dubiez, R., **5**, 572, 618 ⟨84JOC714⟩
Dubina, T. N., **5**, 703 ⟨85JOU1166⟩
Dubina, V. L., **2**, 756 ⟨88JOU2222⟩; **5**, 638 ⟨66URP181085⟩, 652 ⟨66ZOR1841⟩, 674 ⟨66JOU1809, 76URP534448⟩, 694 ⟨76JOU60, 76ZOR64⟩, 703 ⟨69JOU252, 69ZOR263⟩, 712, 714 ⟨75JOU1719, 75ZOR1724⟩; **6**, 629 ⟨75ZOR1628⟩
Dubini, R., **1**, 816 ⟨74HCA2084⟩
Dubinina, T. N., **3**, 363 ⟨72ZOR1822⟩; **5**, 317, 349 ⟨78ZOR1846⟩, 703 ⟨85ZOR1281, 87JOU75, 87ZOR86⟩, 1032 ⟨72ZOR1822⟩; **6**, 625 ⟨85ZOR1281⟩
DuBois, D. A., **4**, 572 ⟨89IC899⟩; **5**, 887 ⟨87PS(30)463⟩
DuBois, G. E., **2**, 366 ⟨82USP4290975⟩, 367 ⟨80JOC5373⟩; **6**, 606, 645 ⟨92SC1191⟩
Dubois, J.-E., **1**, 318 ⟨75JA1977, 75TL4555, 77JOC2689⟩, 539 ⟨74TL2203⟩, 871 ⟨69CC292⟩; **2**, 697 ⟨78T1349⟩; **3**, 23 ⟨76BSF1875⟩, 526 ⟨81T4189⟩, 567, 620 ⟨80T775⟩
Dubois, R., **2**, 602 ⟨76ACR217⟩; **5**, 29, 34, 43 ⟨67BSF2011⟩, 32 ⟨79T2077⟩
Du Boisson, R. A., **2**, 16, 336 ⟨88JCS(P1)2805⟩; **4**, 108 ⟨86JFC(32)461⟩
Duboudin, F., **1**, 656 ⟨84JOM(267)93⟩; **2**, 699 ⟨87T2075⟩; **4**, 514 ⟨78JOM(156)C25⟩, 624 ⟨88JOM(341)125⟩; **5**, 1134 ⟨82S212⟩
Duboudin, J.-G., **1**, 539 ⟨75JOM(96)C47⟩; **2**, 1034, 1035 ⟨86JOM(304)115⟩
Duboudin, J. G., **2**, 973 ⟨79JOM(168)227, 79JOM(168)233⟩; **4**, 386 ⟨87JOM(331)181⟩
Duboudin, J.-P., **4**, 1065 ⟨79SC53⟩
Dubourg, A., **6**, 373, 688 ⟨89CC593, 90JOC5750⟩
DuBov, S. S., **2**, 166, 170 ⟨65ZOB1412⟩; **4**, 805, 812, 813 ⟨61MI 418-01⟩; **5**, 822 ⟨60ZOB2409⟩; **6**, 258, 259 ⟨65ZOB1412⟩
Dubovik, Y. G., **1**, 103 ⟨75IZV888⟩
Dubovszky, I., **2**, 444 ⟨B-84MI 213-01⟩
Dubrinina, T. N., **2**, 400 ⟨78ZOR1846⟩
Dubs, P., **1**, 595 ⟨78HCA998⟩; **2**, 824 ⟨71HCA710⟩; **3**, 268 ⟨71HCA710⟩; **4**, 256 ⟨78HCA2351⟩; **5**, 247 ⟨76HCA1307⟩
Dubuis, R.,**1**, 318 ⟨73OSC(4)880⟩
Dubuis, R., **1**, 502, 503 ⟨66OS(46)89⟩, 502, 503, 504 ⟨73OSC(5)880⟩
Duc, C. L., **3**, 241 ⟨79JHC821⟩
Duc, Le Van, **6**, 242 ⟨91ZN(B)978⟩
Duc, M., **5**, 1013, 1101 ⟨65ACS438⟩
Ducep, J.-B., **1**, 107 ⟨82OR(27)1⟩; **2**, 143 ⟨68TL5629⟩, 143 ⟨69TL3707⟩, 143 ⟨71T5861⟩, 997 ⟨82OR(27)1⟩, 381 ⟨82OR(27)1⟩, 417 ⟨72CJC2902⟩
Duceppe, J. S., **5**, 1024 ⟨84JHC1685⟩
Duchatsch, W., **1**, 248 ⟨86AG(E)720⟩
Duchek, I., **4**, 551, 570, 588 ⟨79JOM(182)465⟩; **5**, 390 ⟨79JOM(182)465⟩
Duchelle, E. F., **2**, 553, 555 ⟨92JOC3680⟩
Duchene, A., **4**, 388 ⟨87CC29⟩, 388, 389 ⟨85BSF787⟩
Duchenet, V., **2**, 234 ⟨93CC806, 93TL4925⟩
Duchert, R., **5**, 105 ⟨B-45MI 502-02⟩
Duchstein, H. J., **5**, 357 ⟨88AP(321)25⟩
Duck, E. W., **1**, 665 ⟨60LA(629)172⟩
Ducker, J. W., **2**, 1016 ⟨62JCS4995⟩; **3**, 71 ⟨64TL733⟩; **5**, 846 ⟨76JCS(P1)38⟩
Duckworth, A. C., **3**, 534 ⟨62JOC3146⟩; **5**, 199 ⟨62JOC3146⟩
Duckworth, P. S., **2**, 366 ⟨56JCS1755⟩
Duc Le Van, **5**, 1152 ⟨93PS(76)1⟩; **6**, 242, 685 ⟨91HAC385⟩, 685, 692 ⟨88CB655⟩
Ducruix, A., **4**, 693 ⟨75CC615⟩
Dudchenko, T. N., **5**, 776 ⟨85ZOB776⟩
Duddeck, H., **5**, 1079 ⟨87CAR(166)101⟩

Duddey, J. E., **1**, 1110 ⟨66JA4990⟩
Dudek, C. M., **1**, 928 ⟨92TL181⟩
Dudek, H., **2**, 484, 489, 492 ⟨28CB1911⟩, 507 ⟨28CB1906⟩
Dudfield, P., **2**, 22 ⟨85TL5037⟩; **5**, 63 ⟨85HCA216⟩
Dudfield, P. J., **2**, 649 ⟨91COS(7)329⟩
Dudman, C. C., **2**, 1003 ⟨82S419⟩; **3**, 465 ⟨82S419⟩
Dudukina, O. V., **3**, 534 ⟨69ZOB467⟩; **4**, 723 ⟨76DOK(226)1092⟩
Dudzinska-Usarewicz, J., **6**, 546 ⟨85MI 617-03⟩
Dudzinski, B., **2**, 151 ⟨92CC1500⟩
Due, M., **6**, 593 ⟨66ACS2091⟩
Dueber, T. E., **1**, 1111 ⟨74JCS(P2)843⟩, 1111, 1114, 1116 ⟨74JA1100⟩, 1114 ⟨71TL2519, 70JA3802⟩; **2**, 689 ⟨70AG(E)521, 74JA1100, 74OS(54)79, 76JOC4099⟩
Duesler, E. N., **3**, 156 ⟨86JOC130⟩; **4**, 484 ⟨92JA9691⟩
Duetsch, M., **4**, 940, 1014 ⟨92CB2051⟩; **5**, 934 ⟨91AG(E)1658⟩, 946 ⟨93SL486⟩, 958 ⟨92CB2051⟩
Dufaud, V., **4**, 722 ⟨91OM4005⟩
Dufaux, R., **1**, 38 ⟨66HCA1145⟩
Duff, A. W., **4**, 694 ⟨83JA5479⟩
Duff, J. C., **3**, 91 ⟨41JCS547⟩
Duff, J. M., **1**, 652 ⟨76JOM(121)293⟩, 738 ⟨76JOM(113)C11⟩, 759 ⟨70CJC561⟩; **2**, 914, 921, 922 ⟨76JOM(121)293⟩; **3**, 271 ⟨73CJC2024⟩; **4**, 137, 138, 141, 811 ⟨76JOM(122)31⟩, 138, 141, 646, 651, 652, 654, 663 ⟨70CJC561⟩, 506 ⟨77CJC2589⟩, 647, 652, 663 ⟨70JA7567⟩; **5**, 410, 431 ⟨67JA431⟩; **6**, 127, 132, 325, 326, 520 ⟨67JA431⟩, 192 ⟨70CJC561⟩
Duffaut, N., **4**, 355 ⟨75JOM(87)151, 77JOM(142)C35⟩, 933 ⟨72JOM(43)157, 74JOM(80)C25⟩, 1014 ⟨72JOM(43)139, 84OM1660⟩; **5**, 297 ⟨74BSF(2)1949⟩, 416 ⟨72JOM(43)157, 74JOM(80)C25⟩, 417 ⟨78JOM(161)C23⟩, 419 ⟨69JOM(19)P33, 79JIC545⟩, 419, 420 ⟨71JOC3835, 75JOM(93)51⟩, 536 ⟨66MI 512-01⟩, 1076, 1077 ⟨78JOM(148)225⟩; **6**, 124 ⟨71JOM(26)183, 79JOC420⟩, 125, 162 ⟨79JOM(177)137⟩, 162 ⟨71JOM(32)79⟩
Duffin, G. F., **3**, 372 ⟨56JCS3189⟩
Duffner, C. R., **2**, 26 ⟨78T1251⟩
Duffner, P., **1**, 727 ⟨58LA(619)10⟩; **3**, 493 ⟨58LA(619)10⟩
Duffy, D. E., **2**, 206 ⟨93TL3667⟩
Duffy, D. N., **4**, 674 ⟨83OM1172⟩
Duffy, J. J., **2**, 776 ⟨67JOC1124⟩
Duffy, J. L., **2**, 393 ⟨93TL1259⟩
Duffy, N. W., **4**, 721 ⟨94OM511⟩; **6**, 402 ⟨92JOM(437)323⟩
Dufour, M.-N., **3**, 180 ⟨86JCS(P1)1895⟩; **5**, 333 ⟨88TL2661, 89JOC617⟩
Dufour, N., **6**, 685 ⟨89TL4813⟩, 689 ⟨89TL6869⟩
Dufresne, C., **1**, 13, 18, 20 ⟨86JOC3038⟩; **2**, 612 ⟨88CCC2116⟩
Dufresne, R. F., **2**, 787 ⟨74JOC3351⟩
Duggan, A. J., **4**, 166 ⟨72GEP(O)2060217⟩
Duggan, M. E., **2**, 454 ⟨83TL2935⟩
Dugger, H. A., **3**, 352 ⟨76HCA747⟩
Dugger, R. W., **1**, 748, 749 ⟨80SC509⟩
Dughera, S., **3**, 306, 307 ⟨93JCS(P1)273⟩
Duguay, G., **3**, 73 ⟨72BSF4576⟩, 210 ⟨88S564⟩, 414 ⟨86OS(64)164⟩; **4**, 99 ⟨90T5263⟩, 101 ⟨91TA157⟩; **5**, 580 ⟨91BSF760⟩, 633 ⟨86PS(27)327, 88T139⟩, 760 ⟨80JHC767, 84CC157⟩
Dugue, B., **1**, 1059 ⟨87BSF93⟩
Duh, J.-M., **1**, 620 ⟨89JCR(S)275⟩
Duhamel, I., **2**, 22 ⟨91SL807⟩
Duhamel, L., **1**, 180 ⟨88T5495⟩, 453 ⟨77JA8356⟩, 573 ⟨93CC116⟩; **2**, 28 ⟨73BSF1465⟩, 759 ⟨77JA8356, 79JOC3585, 82BSF(2)297, 82JCR(M)621, 82JCR(S)48, 83JCR(M)2101, 83JCR(S)222, 91S649⟩; **3**, 35 ⟨80TL4171⟩, 43 ⟨74BSF331⟩, 62 ⟨81JOC3741⟩, 65 ⟨90TL3129⟩, 76, 262 ⟨83JCR(S)222⟩, 80 ⟨75BSF329⟩, 442 ⟨82TL1605⟩; **4**, 15 ⟨73TL4237⟩, 404, 405, 406, 408, 410, 411, 412 ⟨B-82MI 409-02⟩, 405, 406 ⟨74BSF331⟩; **5**, 62 ⟨63AC(P)315⟩, 315, 316, 318 ⟨93SC2423⟩
Duhamel, P., **1**, 180 ⟨88T5495, 93CC116⟩; **2**, 28 ⟨73BSF1465⟩, 759 ⟨91S649⟩; **3**, 43 ⟨74BSF331⟩; **4**, 15 ⟨73TL4237⟩, 405, 406 ⟨74BSF331⟩; **5**, 283 ⟨87TL2595⟩
Duhemel, L., **4**, 749 ⟨93S1071⟩
Duhl-Emswiler, B. A., **1**, 730 ⟨85JA217⟩; **2**, 152 ⟨77TL4009⟩
Duismann, W., **2**, 404 ⟨74TL265⟩
Dujardin, R., **1**, 68, 69, 454 ⟨84AG(E)224⟩; **2**, 1006 ⟨84AG(E)224⟩; **4**, 1064 ⟨84AG(E)224, 85AG(E)571⟩
Duke, A. J., **1**, 1109 ⟨74JCS(P2)1808⟩
Duke, R. E., **3**, 327 ⟨73JA7287⟩
Duke, R. K., **1**, 520, 1134 ⟨84JOC1898, 82OS(60)81⟩
Dukes, M. D., **6**, 717, 727 ⟨78JOM(153)67⟩
Dulcère, J.-P., **1**, 573 ⟨93CC1038⟩, 955 ⟨88CC237, 93CC270, 93JOC5702, 93JOC5709⟩, 959 ⟨72TL4465, 74CR(C)149⟩, 959 ⟨79TL1845⟩, 959, 960 ⟨74BSF1119⟩, 959, 970 ⟨72CC866⟩, 1131 ⟨87TL2009⟩; **2**, 21 ⟨91SL477⟩, 940 ⟨81TL3179⟩; **3**, 21 ⟨93S399⟩; **4**, 1046 ⟨81TL3179⟩
Dull, D. L., **5**, 12 ⟨69JOC2543⟩
Dull, M. F., **3**, 535, 541 ⟨32JA2432⟩, 536, 541 ⟨35JA774⟩
Dul'nev, P. G., **6**, 562 ⟨81KGS907⟩
Dulou, R., **1**, 546 ⟨62CR(254)1817⟩
du Manoir, J. R., **1**, 232 ⟨72CC541⟩; **2**, 365 ⟨75JA2566⟩
Dumas, D., **2**, 314 ⟨90TL2277⟩; **3**, 180 ⟨90TL2213⟩
Dumas, D. J., **5**, 746 ⟨80JOC2485⟩
Dumas, D. P., **1**, 6 ⟨92JA741⟩; **3**, 48 ⟨93TL403⟩
Dumas, S., **3**, 602 ⟨85JA2192⟩
DuMent, W. W., **2**, 465 ⟨83PS(18)73⟩
Dumler, V. A., **2**, 949 ⟨78ZOB639⟩
Dumont, J. A., **6**, 487 ⟨91JHC1013⟩
Dumont, P., **1**, 786 ⟨88JOC4859⟩
Dumont, W., **1**, 246 ⟨87TL1549⟩, 590 ⟨79TL4111⟩, 592, 593, 713, 716 ⟨76TL1385⟩, 601, 716 ⟨78TL1141⟩, 712, 713 ⟨89H(28)1203⟩, 713 ⟨89T2005, 89T2023, 78TL1145⟩, 716 ⟨90BSF681⟩, 759 ⟨76AG(E)161⟩; **2**, 216 ⟨90TL5007⟩, 278, 294 ⟨85T4793⟩, 279 ⟨92S933⟩, 281 ⟨74AG(E)804, 75AG(E)700⟩, 282, 730 ⟨76TL453⟩, 287 ⟨85CC569, 88TL3265⟩, 287 ⟨88TL3269⟩, 288, 1001 ⟨74AG(E)274⟩, 554 ⟨92TL3381⟩, 728 ⟨77TL3835⟩, 728, 936 ⟨79TL485⟩; **3**, 16, 132 ⟨79S877⟩, 39, 175 ⟨76TL453⟩, 125 ⟨90BSF681⟩; **4**, 88, 91, 288, 400 ⟨77AG(E)541⟩, 238, 239, 283, 375 ⟨75TL1613⟩, 239, 288 ⟨77AG(E)540⟩, 284 ⟨89T2023⟩, 287 ⟨75AG(E)700⟩, 287, 288 ⟨85T4793, 79S877⟩, 289 ⟨92SL638⟩, 375 ⟨79TL485⟩, 375, 400 ⟨81TL4737⟩, 375, 652 ⟨76AG(E)161⟩, 399 ⟨85CC571⟩, 399 ⟨88TL3269, 75AG(E)350⟩, 400 ⟨89T2005⟩, 917 ⟨84TL2883⟩; **6**, 136 ⟨78JOM(149)C10⟩
DuMont, W.-W., **2**, 294 ⟨B-87MI 204-03⟩, 499 ⟨85ZN(B)848, 88OCS587⟩, 729 ⟨91CB1315⟩, 732 ⟨87AG(E)780⟩; **4**, 88, 288 ⟨91CB1315⟩; **5**, 244 ⟨85AG(E)1041⟩, 255, 542 ⟨87CC820⟩, 543 ⟨85AG1051⟩; **6**, 334, 355, 356 ⟨90CB2325⟩, 349 ⟨93CB1355⟩, 355, 356 ⟨88CB2109⟩, 355, 356 ⟨93ZAAC(619)1693⟩
Dumpis, Y. Y., **2**, 377 ⟨69ZOR1735⟩
Dunach, E., **1**, 427, 428 ⟨91SC2377⟩; **2**, 147 ⟨84JA8188⟩, 147 ⟨85PAC1911⟩, 362 ⟨86NJC761⟩; **3**, 634 ⟨85NJC1⟩
Dunbar, B. I., **2**, 393 ⟨75TL71⟩; **3**, 613 ⟨75TL71⟩; **5**, 1134 ⟨75TL71⟩
Dunbar, F. X., **3**, 382, 389 ⟨68CC459⟩
Dunbar, J. E., **2**, 358 ⟨66JOC842, 70JOC279⟩; **6**, 636 ⟨88JOC3120⟩
Dunbar, R. E., **6**, 596 ⟨47JA1833⟩
Duncan, D. M., **3**, 621 ⟨74JOC3050⟩
Duncan, D. P., **2**, 267 ⟨79JOM(164)305⟩; **4**, 571 ⟨78JOM(159)137⟩; **5**, 914 ⟨78JOM(159)137⟩
Duncan, J. A., **1**, 575, 576 ⟨90JA8433⟩, 808 ⟨88JOC196⟩
Duncan, J. H., **1**, 631 ⟨67JA1442, 67JA471⟩, 633 ⟨68JA4762⟩

Duncan, K., **4**, 528 ⟨89B3541⟩
Duncan, M. P., **1**, 74 ⟨87SC703⟩; **3**, 162 ⟨87JOC150⟩
Duncan, S. M., **1**, 525, 826 ⟨86TL4945⟩
Duncan, W. G., **1**, 571 ⟨88OSC(6)210⟩; **2**, 1042 ⟨65JOC1027⟩; **3**, 497 ⟨65JOC1027, 65JOC2543⟩, 539 ⟨76OS(55)32⟩; **4**, 732 ⟨65JOC1027, 65JOC2543⟩, 732 ⟨65TL521⟩; **4**, 8, 732 ⟨64TL1461⟩
Duncan, W. G., **4**, 9 ⟨73OSC(5)390⟩; **6**, 695 ⟨65JOC1027⟩
Duncan, W. P., **2**, 619 ⟨70JOC719⟩
Duncanson, F. D., **2**, 1004 ⟨75JCS(P2)1030⟩
Duncia, J. V., **4**, 780 ⟨91JMC2525⟩
Dundon, C. V., **6**, 522 ⟨47JA1723, 50IS156⟩
Düner, G., **5**, 148 ⟨80AG(E)1029⟩
Dunford, J. A., **1**, 1108 ⟨77TL3545⟩; **2**, 607, 617 ⟨77TL3545⟩
Dung, N., **4**, 989 ⟨82JPR309⟩
Dungoues, J., **2**, 190 ⟨76TL1197⟩
Dunin, A. A., **4**, 576 ⟨27CB291⟩
Dunitz, J. D., **1**, 873 ⟨81HCA1023, 85HCA1373⟩, 873 ⟨85JA5403⟩; **2**, 999 ⟨84HCA224⟩
Dunitz, J. P., **1**, 862 ⟨53NAT121⟩
Dunkelblum, E., **3**, 527 ⟨70HCA2159⟩
Dunken, H., **5**, 560 ⟨88ZC269⟩
Dunker, K.-H., **2**, 936 ⟨86CB535⟩; **4**, 568 ⟨86CB535⟩; **6**, 693 ⟨84ZC384⟩, 693 ⟨86CB535⟩
Dunkerton, L. V., **1**, 263 ⟨75JOC2009⟩; **4**, 224 ⟨88JOC845⟩; **5**, 540 ⟨88JOC845⟩; **6**, 566 ⟨87JHC749⟩
Dunkin, I., **3**, 677 ⟨77ACS(B)848⟩
Dunkin, I. R., **1**, 921 ⟨80CSR1⟩; **3**, 564 ⟨89T259⟩; **5**, 1073 ⟨89T259⟩
Dunlap, R. B., **3**, 707 ⟨91JCS(P1)2495⟩; **5**, 238 ⟨85JOC811⟩, 1054 ⟨91JCS(P1)2495⟩; **6**, 595 ⟨91JCS(P1)2495, 91JOC6733⟩
Dunlap, S. E., **1**, 224 ⟨89HCA1301⟩
Dunlop, R., **4**, 262 ⟨92JA8772, 92JCS(P1)1105⟩
Dunn, A. D., **2**, 123 ⟨B-89MI 203-01⟩; **3**, 617 ⟨82OPP396⟩; **5**, 275 ⟨82OPP396⟩
Dunn, J., **3**, 699 ⟨75JCS(P1)1743⟩
Dunn, J. H., **6**, 522, 523 ⟨60JA1325⟩
Dunn, J. L., **1**, 398 ⟨32JCS1926⟩
Dunn, M. J., **1**, 968 ⟨92CC319, 93SL499⟩
Dunn, M. S., **5**, 84 ⟨47MI 502-01⟩, 90 ⟨63OSC(4)55⟩
Dunn, P. J., **3**, 372 ⟨89JCS(P1)2489⟩
Dunnavant, W. R., **1**, 881 ⟨59OS(39)73, 60OS(40)38⟩, 881 ⟨62JOC2737⟩; **5**, 76 ⟨73OSC(5)564⟩
Dunne, B. A., **4**, 695, 717 ⟨89JOM(372)53, 89AG(E)737⟩
Dunnigan, D. A., **1**, 881 ⟨61JOC644⟩; **3**, 625 ⟨73T1931⟩; **4**, 211 ⟨50JA514⟩; **6**, 641 ⟨63JMC275⟩
Dunning, F., **2**, 261 ⟨27JA1017⟩
Dunning, K. W., **4**, 97 ⟨50JCS2925⟩, 435 ⟨50JCS2920⟩
Dunning, W. J., **4**, 97 ⟨50JCS2925, 50JCS2920⟩
Dunogués, J., **1**, 1038, 1138 ⟨76TL2449⟩, 238 ⟨84TL2135⟩, 284 ⟨76S737⟩, 447, 447 ⟨93SL171⟩, 448 ⟨90TL6179⟩, 450, 451 ⟨89OR(37)57⟩, 963 ⟨81CC1094⟩, 963, 978 ⟨81TL3401, 75JOM(93)43⟩; **2**, 608 ⟨77AG(E)488, 79AG(E)402⟩, 699 ⟨87T2075⟩, 916, 923, 940 ⟨75JOM(87)169⟩, 73, 907 ⟨89OR(37)57⟩, 537 ⟨86BSF413⟩; **2**, 929 ⟨78TL57⟩, 938, 941 ⟨86T669⟩, 938, 941 ⟨92JOC4705⟩, 940 ⟨81TL3401⟩, 945 ⟨79JOC4275, 90TL6179⟩, 1090 ⟨72CR(C)1857, 76MI 223-01⟩; **3**, 140 ⟨79JOC3397⟩, 140 ⟨81TL2883⟩, 213 ⟨76TL1871⟩, 230 ⟨81TL3401⟩; **4**, 140 ⟨86BSF413⟩, 355 ⟨75JOM(87)151, 77JOM(142)C35⟩, 608 ⟨74JOM(71)377⟩, 624 ⟨88JOM(341)125⟩, 933 ⟨72JOM(43)157, 74JOM(80)C25⟩, 1014 ⟨72JOM(43)139, 84OM1660⟩; **5**, 1076, 1077 ⟨78JOM(148)225⟩; **5**, 373 ⟨87BSF997⟩, 416 ⟨72JOM(43)157, 74JOM(80)C25⟩, 417 ⟨78JOM(161)C23⟩, 419 ⟨69JOM(19)P33, 79JOC420⟩, 419, 420 ⟨71JOM(26)183, 75JOM(93)51⟩; **6**, 27 ⟨75JOM(93)43⟩, 124 ⟨71JOM(26)183, 79JOC420⟩, 125 ⟨72JOM(39)C49⟩, 125, 162 ⟨79JOM(177)137⟩, 162 ⟨71JOM(32)79⟩, 162, 163 ⟨91JOM(419)C1⟩, 163 ⟨92TL3903⟩, 379 ⟨88JOM(358)31⟩
Dunstan, I., **4**, 440 ⟨66JCS(C)867, 66JCS(C)870⟩, 440, 442 ⟨66JCS(C)862⟩, 442, 443 ⟨67JCS(C)562⟩
Du Nunno, L., **2**, 1034 ⟨66JCS(B)1188⟩
Duong, K. N. V., **2**, 579 ⟨73JOM(55)375⟩
Duong, T. T., **6**, 417 ⟨91TL4251, 92TL2781⟩, 632 ⟨91S86⟩
Dupas, G., **2**, 398 ⟨87BCJ4492⟩
du Penhoat, C. H., **4**, 272, 275 ⟨82BSF43⟩
Dupin, J. P., **4**, 420 ⟨92PHA754⟩
Dupius, J., **1**, 323, 920 ⟨83TL703⟩
Duplan, J. C., **2**, 364 ⟨76TL1093⟩
Duplantier, A. J., **5**, 586, 609 ⟨89JOC1212⟩
Dupont, A. C., **2**, 13 ⟨90SC1011⟩
Dupont, C., **5**, 24 ⟨B-47MI 502-01, B-47MI 502-02⟩, 24 ⟨B-48MI 502-01⟩, 62 ⟨B-45MI 502-04, B-45MI 502-05⟩, 105 ⟨B-49MI 502-01⟩, 62
Dupont, G., **5**, 24 ⟨58BSF17⟩, 83 ⟨B-41MI 502-01, B-45MI 502-02⟩
Dupont, J., **5**, 801 ⟨90JCS(D)3193⟩
Dupont, L., **5**, 595 ⟨93PS(78)97⟩
Dupont, V., **5**, 327, 351 ⟨93JA8898⟩; **6**, 488 ⟨93JA8898⟩
Dupont-Fenfau, J., **5**, 655 ⟨89CCC3245⟩
Duppa, B. F., **2**, 520 ⟨1859PRS568⟩
Dupre, B., **1**, 16 ⟨87JOC3706⟩
Dupriest, M. T., **3**, 310 ⟨86JOC2021⟩
Dupuis, J., **1**, 325 ⟨85JA4332⟩, 946 ⟨87CB1197⟩; **3**, 621 ⟨85JA4332⟩; **5**, 152 ⟨87OS(65)236⟩
Dupuis, M., **5**, 1020 ⟨90JA4183⟩
Dupuy, C., **2**, 591 ⟨86TL3149⟩; **3**, 12, 128 ⟨85JOC5761⟩
Duraisamy, M., **4**, 816, 818 ⟨93JOC546⟩
Durand, J. F., **2**, 1096 ⟨23CR(176)992, 24BSF166⟩; **6**, 216 ⟨27BSF1251⟩
Durand, M., **2**, 495 ⟨74JOM(77)225⟩; **3**, 466 ⟨63BSF41⟩
Durand, T., **1**, 940 ⟨93TL8245⟩
Duranleau, R., **4**, 1078 ⟨87JOC3243⟩
Duranleau, R. G., **6**, 467 ⟨91JMOC389⟩
Durant, G. J., **6**, 647 ⟨77JMC901⟩
Duranti, E., **3**, 273 ⟨74S357⟩
Dürckheimer, W., **5**, 226 ⟨85AG(E)180⟩
Durden, J. A., **6**, 448 ⟨85EUP136147⟩
Duréault, A., **1**, 486, 1032 ⟨89JOC5324⟩; **2**, 1048 ⟨67CR(C)1496⟩, 1048 ⟨71CR(C)289⟩, 1050 ⟨74BSF1535⟩; **4**, 15 ⟨74BSF1533⟩
Dureja, P., **4**, 743 ⟨82TL5003⟩
Durfee, L. D., **6**, 516, 519 ⟨88CRV1059⟩
Durgaprasad, A., **3**, 116, 158, 161 ⟨90TL5785⟩
Durga Prasad, B. M., **3**, 59 ⟨90TL7521⟩
Durgar'yan, S. G., **4**, 966 ⟨80IZV1147⟩
Durham, L. J., **1**, 871 ⟨64JA2832⟩; **5**, 32 ⟨63OSC(4)635⟩
Durig, J. R., **4**, 130 ⟨85JCP1517⟩
Durkin, K. A., **6**, 255 ⟨91CJC2127⟩
Durkin, K., **1**, 327 ⟨92JA4912⟩
Dürner, G., **1**, 558 ⟨91TL3357⟩; **3**, 540 ⟨77CB3582⟩; **5**, 174 ⟨87AG(E)362⟩
Dürr, H., **2**, 1003 ⟨88CB843⟩
Dürr, M., **4**, 297 ⟨74CB1488⟩
Durrani, A. A., **5**, 110 ⟨73TL4839, 79JCS(P1)2079⟩, 110 ⟨80JCS(P1)1658⟩
Durrant, J. L. A., **6**, 7 ⟨84JOM(260)C75, 85JOM(280)397⟩, 7 ⟨90JOM(386)229⟩
Durrant, M. C., **4**, 568 ⟨87JSP(121)304⟩; **5**, 1130 ⟨87JSP304⟩, 1154 ⟨85JSP(109)8⟩
Durrell, W. S., **5**, 989 ⟨59JA1587⟩
Durst, H. D., **2**, 150 ⟨88JOC5004⟩, 393 ⟨75TL71⟩, 763 ⟨71JA2897⟩; **3**, 613 ⟨75TL71⟩; **5**, 1134 ⟨75TL71⟩
Durst, T., **1**, 142 ⟨74CJC761⟩, 143 ⟨71JA3077⟩, 213, 478 ⟨80JOC922⟩, 214 ⟨76JOC3648, 81CJC1415⟩, 232 ⟨78CJC512, 78S579, 81JCS(P1)1846⟩, 418 ⟨91COS(5)675⟩, 600 ⟨73JA3420⟩, 621, 644, 695 ⟨68JA5553⟩, 621, 695 ⟨66JA5656, 68JA5548⟩, 704 ⟨79COC(3)187⟩, 823 ⟨72JA9261⟩, 844, 918 ⟨B-87MI 119-01⟩, 844 ⟨B-80MI 119-01, B-84MI 119-01⟩, 881 ⟨69MI 119-01⟩; **2**, 142, 144, 153 ⟨79COC(3)121⟩, 149, 200 ⟨76CJC3012⟩, 151 ⟨74CJC761⟩, 200 ⟨73CC4,

73TL2863, 74JA935⟩, 200, 272 ⟨84TL5287⟩, 201 ⟨78TL2353⟩, 202, 207 ⟨70CJC845⟩, 715 ⟨79COC(3)171⟩; **3**, 87 ⟨74JA935⟩, 373 ⟨79TL203⟩; **4**, 74, 75, 76 ⟨70TL4643⟩, 74, 75, 77, 82, 233 ⟨69JA1034⟩, 75, 77 ⟨73CC4⟩, 390 ⟨79COC(3)121, 79COC(3)171⟩, 391 ⟨74CJC761⟩, 393 ⟨73JA3420⟩, 394 ⟨66JA5656, 68JA5548⟩, 923 ⟨81TL191⟩, 1075 ⟨84MI 424-01⟩; **5**, 488, 489 ⟨63TL585⟩, 488, 489 ⟨66CJC819, 63JA2676⟩
Dusart, G., **5**, 772 ⟨83EJM413⟩
Duschinsky, R., **2**, 212 ⟨51JA4464⟩; **6**, 508 ⟨45JA2079, 47JA3150⟩
Dusemund, J., **5**, 652 ⟨88AP(321)41⟩
Dusold, L. R., **1**, 624, 782 ⟨73JOC1742⟩
Dussault, P., **2**, 676 ⟨88JA6276⟩; **5**, 206 ⟨92JOC1009⟩
Dussault, P. H., **1**, 838 ⟨90JA1266⟩
Dussourd, D. E., **1**, 503, 504 ⟨83JOC2266⟩
Düster, D., **4**, 545 ⟨87ZAAC(544)87⟩
Dustman, C. K., **1**, 891 ⟨81MI 119-01⟩
Dusza, J. P., **3**, 494 ⟨60JOC93⟩
Duthaler, R. O., **1**, 166 ⟨94T1539⟩; **2**, 44, 67, 568 ⟨92CRV807⟩, 74 ⟨92JA2321⟩, 80 ⟨89AG(E)494⟩, 657 ⟨84HCA1406⟩; **4**, 131 ⟨91SL395⟩
Dutra, G. A., **1**, 476 ⟨73JA7777⟩
Dutron-Woitrin, F., **5**, 601 ⟨85S79, 85T693⟩, 601, 612 ⟨85S77⟩
Dutson, S. M., **5**, 835 ⟨92JOC4648⟩
Dutt, P. K., **2**, 181 ⟨24JCS1463⟩
Dutta, ,P. C., **3**, 239 ⟨57JA4487⟩
Dutta, A. S., **5**, 351 ⟨86JCS(P1)1655⟩
Dutta, D. K., **2**, 34 ⟨91CI(L)175⟩; **3**, 417 ⟨85SC335⟩
Dutta, K. K., **5**, 367 ⟨90JIC172⟩
Dutto, G., **4**, 718 ⟨93OM4545⟩
Dutton, F. E., **6**, 439 ⟨84USP4473580⟩
Dutton, G. R., **2**, 107 ⟨43OSC(2)109⟩
Dützmann, A., **1**, 898 ⟨21JPR1⟩
Dutzmann, S., **4**, 141 ⟨92JOM(438)45⟩
Duus, F., **2**, 140 ⟨79COC(3)373⟩; **3**, 330, 344, 368, 369 ⟨79COC(3)373⟩, 344 ⟨68T5323, 69T5703, 72T5923, 74T3753, 77ACS(B)40, 79JCS(P2)1532, 82JA5922, 83JCS(P2)1321⟩, 344, 350 ⟨77JOC3123⟩, 344, 350 ⟨86JA630⟩, 350, 364 ⟨77S256⟩, 350, 364 ⟨85S672⟩; **5**, 472 ⟨73SST200, 75SST219⟩, 472, 510, 546 ⟨79COC(3)373⟩, 565 ⟨73SST228, 75SST254, 79COC(3)440⟩, 624, 1022 ⟨79COC(3)461⟩; **6**, 569 ⟨79COC(3)373⟩
Duval, D., **1**, 888 ⟨B-89MI 119-05⟩
Du Vigneaud, V., **1**, 772 ⟨43OSC(2)93⟩; **2**, 126 ⟨30JA4500⟩, 126, 214 ⟨57MI 203-01⟩; **5**, 33 ⟨43OSC(2)93⟩, 89 ⟨32OSC(1)48⟩
Düwel, H., **1**, 910 ⟨70TL1941⟩
Dux III, F., **1**, 4, 7, 9, 17, 28 ⟨78JOC2259⟩
Dvinskikh, V. V., **3**, 436 ⟨85ZOB2077⟩
Dvorak, D., **3**, 71 ⟨88TL2861⟩; **5**, 953 ⟨95OM570⟩
Dvorak, V., **3**, 542 ⟨72TL3443⟩
Dvoretzky, I., **4**, 728 ⟨50JOC1285⟩
Dvornikova, K. V., **4**, 737 ⟨85JFC(28)99⟩
Dwek, R. A., **1**, 870 ⟨66JCS(B)161⟩
Dwivedi, R., **6**, 564 ⟨89IJC(B)439⟩
Dworak, G., **2**, 400 ⟨69M959⟩; **5**, 340 ⟨69M959⟩
Dwyer, S. G., **1**, 727 ⟨70MI 116-04⟩
D'yachenco, O. A., **5**, 685 ⟨90ZOR1629⟩
D'Yachenko, O. A., **5**, 685 ⟨90JOU1408⟩
D'yachkova, L. Y., **5**, 316 ⟨77ZOB968⟩
D'yachkovskaya, O. S., **5**, 390 ⟨77ZOB2294, 80ZOB886⟩, 390 ⟨81ZOB1318⟩, 390, 391 ⟨88IZV859⟩, 1082 ⟨88IZV739⟩
D'yachova, L. Ya., **5**, 810 ⟨75ZOR1981⟩
Dyachuk, A. A., **5**, 1044 ⟨91UKZ89⟩
Dyadchenko, V. P., **4**, 152 ⟨77IZV2417⟩, 723 ⟨72IZV2594⟩; **5**, 385 ⟨88ZOB62⟩
D'yakonova, N. I., **4**, 121, 126 ⟨71JGU2228⟩; **5**, 403 ⟨69JGU597⟩
D'yakonov, I. A., **1**, 907 ⟨67RCR557⟩
D'yakov, V. M., **4**, 140 ⟨75IZV1576⟩, 140 ⟨76IZV2200⟩, 140 ⟨76JGU2076⟩, 569 ⟨75ZOB2010⟩
Dyall, L. K., **2**, 761 ⟨76AJC367⟩; **5**, 214 ⟨89AJC2275⟩
Dyatkina, M. E., **1**, 866 ⟨57DOK(117)1021⟩
Dyatkin, B. L., **2**, 640 ⟨74CSR443, 75ZOR1370⟩, 782 ⟨65IZV1715⟩; **4**, 152 ⟨73JOM(57)423⟩, 730 ⟨75ZOR961⟩, 761, 763 ⟨67ZOR1006⟩, 926 ⟨69CA(70)57950j⟩; **5**, 663, 685, 698, 699, 813 ⟨71T51⟩, 671 ⟨66JGU1340, 66ZOB1326⟩, 811 ⟨67IZV1060, 68IZV1134, 68IZV657, 69IZV958⟩; **6**, 62 ⟨72GEP2101107⟩, 62 ⟨72JOM(34)299, 73JOM(57)423⟩, 65 ⟨75DOK(221)1331, 76ZOR1377⟩, 69 ⟨60IZV231⟩, 219 ⟨65IZV1873⟩
Dyatlova, N. M., **4**, 451, 469 ⟨68RCR7⟩
Dyatlovitskaya, E. V., **1**, 725 ⟨64IZV2003⟩
Dybowski, P., **2**, 258 ⟨90S609⟩
Dybvig, D. H., **5**, 807 ⟨66IC1795⟩, 808 ⟨73USP3751312⟩, 814 ⟨73JOC1065⟩, 815, 817, 818 ⟨73USP3726903⟩; **6**, 279, 281 ⟨73JOC1065⟩, 613 ⟨66IC1795⟩
Dychenko, A. I., **4**, 995 ⟨79ZOR2356⟩; **6**, 654 ⟨72ZOR39⟩
Dyck, B. P., **1**, 609 ⟨92TL4725⟩; **2**, 338 ⟨91CJC1482⟩
Dyckes, D. F., **5**, 185 ⟨85JOC1112⟩
Dyckhoff, F., **5**, 934 ⟨93CB2573⟩
Dye, J. L., **1**, 101 ⟨79AG(E)587⟩
Dye, S. L., **2**, 240 ⟨75TL1739⟩; **3**, 686 ⟨79JOC510⟩; **5**, 201 ⟨75TL1739⟩
Dyer, D. W., **6**, 245 ⟨90JOM(390)C73⟩
Dyer, E., **5**, 990 ⟨72JHC1267⟩; **6**, 143 ⟨68JOC3931⟩, 494 ⟨57JA366⟩
Dyer, G., **2**, 825 ⟨76IS168⟩
Dyer, J. R., **3**, 62 ⟨75JHC1325⟩
Dyer, R. D., **1**, 1018 ⟨87JMC254⟩; **6**, 558, 564 ⟨93JFC(61)1⟩, 558, 564 ⟨93JMC1802⟩
Dyer, S. F., **3**, 65 ⟨79JA1303⟩
Dyer, U. C., **1**, 614 ⟨88JCS(P1)53, 88TL6357⟩, 1099 ⟨88JOC3383⟩
Dygos, J. H., **3**, 140 ⟨90JA7441⟩
Dyke, A. F., **4**, 679 ⟨82JCS(D)1297, 83JCS(D)1417⟩, 679 ⟨83JCS(D)2661⟩, 680 ⟨83JCS(D)2099⟩, 681 ⟨90JCS(D)761, 92CC310⟩
Dyke, S. F., **1**, 493 ⟨81T173⟩
Dyke, W. J. C., **2**, 483, 872 ⟨30JCS463⟩, 484 ⟨32JCS2284⟩, 484, 492, 497, 888 ⟨32JCS1815⟩, 487, 492, 497, 502 ⟨31JCS185⟩, 492 ⟨30JCS2426⟩, 492, 497 ⟨30JCS1920⟩
Dykstra, C. E., **1**, 1143 ⟨78JA1378⟩
Dykstra, R. R., **6**, 135 ⟨93AG(E)1469⟩, 327, 336 ⟨93AG1489⟩
Dykstra, S., **2**, 95 ⟨57JA3474⟩
Dymova, S. F., **5**, 774 ⟨78ZOB2465⟩
Dymowski, K., **5**, 822 ⟨48BSF597⟩
Dynak, J., **1**, 1139 ⟨76JOC1487⟩
Dyong, I., **3**, 43 ⟨79AG(E)933⟩
Dyrbusch, M., **1**, 463 ⟨87AG(E)560⟩; **4**, 536 ⟨86S737, 87AG(E)480⟩, 536 ⟨88AG(E)433⟩
Dysart, M. R., **2**, 838 ⟨62JOC1012⟩
Dyson, G. M., **5**, 1027 ⟨24JCS1702, 26JCS3041⟩, 1027 ⟨27JCS436, 32OSC(1)165⟩, 1027 ⟨42JCS374⟩, 1028 ⟨34JCS822, 35JCS679⟩; **6**, 232 ⟨34JCS822⟩, 297 ⟨37JCS827⟩, 411, 412 ⟨27CRV109⟩
Dyson, P., **6**, 28 ⟨39JCS781⟩
Dyszlewski, A. D., **1**, 1089, 1090 ⟨87TL4127⟩; **2**, 928 ⟨87JOC4760⟩; **4**, 938 ⟨91JOC3556⟩
Dytham, R. A., **1**, 882 ⟨60T246⟩
Dzehemilev, U. M., **2**, 561 ⟨85JOM(285)43⟩
Dzembovska, T., **5**, 595 ⟨86KGS1405⟩
Dzhafarov, I. A., **6**, 558 ⟨90MI 617-02⟩
Dzhagatspanyan, I. A., **5**, 690 ⟨77KFZ38⟩
Dzhakiyaev, G. M., **2**, 1076 ⟨82IZV933⟩
Dzharov, D., **5**, 745 ⟨77MI 519-01⟩
Dzhemilev, U. M., **1**, 316 ⟨85JOM(285)43⟩; **2**, 441 ⟨80IZV734⟩, 561 ⟨83IZV2053, 83IZV218⟩; **3**, 336 ⟨91URP1680699, 92IZV160, 92URP1710560⟩; **4**, 669 ⟨89IZV2562⟩; **5**, 603 ⟨89IZV1319⟩

Dzhumakaev, K. Kh., **5**, 691 <92ZPK1430>
Dzhurinskaya, N. G., **1**, 656 <63MI 114-01>; **6**, 62 <60DOK(131)98>
Dziadulewicz, E., **4**, 842, 845, 851 <89JCS(P1)1793>; **6**, 127 <88JCS(P1)3367>
Dzingeleski, G. D., **1**, 877 <90JOC1019>
Dzintara, M., **2**, 1068 <74JGU1461>
Dziwok, K., **4**, 489 <90CB423>
Dzurilla, M., **5**, 1049 <76CCC1388>; **6**, 558 <87CCC1764>, 565 <90MI 617-01>, 566 <92MI 617-03>
Dzyuba, V. I., **2**, 407 <80ZOB2621>, 408 <81ZOB331>

E

Eaborn, C., **1**, 64 ⟨74JCS(P2)596⟩, 99 ⟨75JCS(P1)470⟩, 318 ⟨79JOM(177)129⟩, 447 ⟨57JCS498, 60JCS179, 72JCS(P2)172, 75JOM(100)43⟩, 448 ⟨64JCS4804⟩, 448, 449 ⟨60JCS1566, 76JCS(P2)925⟩, 452 ⟨73JCS(P1)137⟩, 455 ⟨80JOM(201)233⟩, 999, 1000 ⟨65JOM(4)217, 64JOM(1)420⟩, 1122 ⟨65JCS384⟩; **2**, 176 ⟨68JCS(C)1265⟩, 436 ⟨83JCS(D)905⟩, 536, 538 ⟨B-68MI 210-01⟩, 608 ⟨72CC840⟩, 841 ⟨65JOM(4)291⟩, 937 ⟨75JCS(P1)475⟩, 937, 938 ⟨74JCS(P1)2055, 75JCS(P1)470⟩, 1089 ⟨64JOM(2)97⟩; **3**, 83 ⟨70CC1703, 75JCS(D)2460, 75JCS(D)2646⟩, 287 ⟨71MI 306-01⟩, 664 ⟨72JOM(46)267⟩; **6**, 60, 191, 289 ⟨71JOM(29)389⟩, 165, 289, 372, 373 ⟨89JOM(366)39⟩, 165, 372 ⟨83JCS(D)905⟩, 172 ⟨82JOM(239)93⟩, 172, 189 ⟨B-85MI 606-01⟩, 173, 381, 391 ⟨85JCS(P2)729⟩, 173, 394 ⟨93JOM(462)45⟩, 174 ⟨83JOM(252)281⟩, 174, 176, 381 ⟨92JCS(D)1015⟩, 174, 176, 381, 393 ⟨86CC1043⟩, 174, 289, 380, 387, 390 ⟨70JOM(24)529⟩, 174, 289, 390 ⟨83CC827⟩, 174, 390 ⟨84JOM(269)217⟩, 174, 391 ⟨86CC969⟩, 176, 289, 391 ⟨83CC1390⟩, 176, 381 ⟨85JCS(P2)1687⟩, 176, 394 ⟨91JOM(405)149⟩, 177, 709, 710 ⟨78JOM(157)C50⟩, 177, 709, 710 ⟨80JOM(191)355⟩, 179, 709, 710 ⟨80JOM(186)309⟩, 290 ⟨86JOM(315)C5, 89CC595⟩, 290, 381, 392 ⟨87JCS(P2)1047⟩, 290, 391 ⟨87CC1461⟩, 378 ⟨85JOM(294)305⟩, 380 ⟨81JOM(221)13, 85JOM(291)25, 86JCS(P2)1289, 90JOM(398)59⟩, 381 ⟨86JCS(P2)1363, 93JCS(P2)391, 93JCS(P2)59, 93JOM(451)45⟩, 381, 392 ⟨87JCS(P2)891⟩, 381, 392 ⟨94JOM(466)35⟩, 387 ⟨80JOM(202)157, 82JOM(235)265, 84JOM(272)1, 87JCS(P2)775, 89JCS(D)447⟩, 387, 388, 397 ⟨87JCS(P2)779⟩, 390 ⟨93JCS(D)3259⟩, 392 ⟨87JCS(P2)381⟩, 392, 394 ⟨92JOM(437)41⟩, 393 ⟨80JOM(190)101, 89CC273, 94AG(E)1268, 94JOM(469)129, 94JOM(480)199, 94OM753⟩, 393, 394, 396 ⟨88JCS(D)381⟩, 394 ⟨88CC1389, 90CC1471⟩, 394 ⟨95UP 613-01, 86CC908, 87JOM(320)137, 83JOM(249)23⟩, 395 ⟨87JCS(D)747⟩, 396 ⟨84CC870, 84JOM(263)C23, 85CC1380, 85CC534, 87JOM(325)117⟩, 396, 399 ⟨88JOM(355)33, 87CC1183⟩, 734 ⟨70CC559, 91CC1608⟩
Eachus, R. S., **4**, 93 ⟨85JA6298⟩
Eager, V. W., **3**, 444 ⟨46JA1983⟩
Eagle, M. V., **2**, 941 ⟨53JA2821⟩
Eakin, R. E., **2**, 412 ⟨55JA1866⟩
Eapen, K. C., **2**, 825 ⟨80JFC(15)239⟩; **4**, 768 ⟨86JFC(31)405⟩
Eardley, S., **3**, 208 ⟨65JCS130⟩
Earhart, H. W., **1**, 912, 914 ⟨58T178⟩
Earl, G. W., **2**, 393 ⟨67JA725⟩
Earl, R. A., **4**, 528 ⟨92MI 411-02⟩; **5**, 386 ⟨84JOC4786⟩
Earle, M. J., **2**, 316 ⟨90TL4229⟩; **3**, 92 ⟨92SL77⟩
Earley, W. G., **3**, 422 ⟨88TL3781⟩
Early, J. V., **3**, 310 ⟨85SC1271⟩
Earlywine, A. D., **1**, 637 ⟨84JOC670⟩
Earnshaw, A., **6**, 523, 525 ⟨B-84MI 616-01⟩
Earnshaw, C., **1**, 643 ⟨77JCS(P1)1452⟩; **2**, 680 ⟨79JCS(P2)3099⟩; **3**, 11 ⟨79JCS(P1)3099⟩; **4**, 343 ⟨77CC314⟩, 949 ⟨79JCS(P1)3099⟩
Easdon, J. C., **6**, 63 ⟨88DIS(B)2973⟩
East, M. B., **4**, 212 ⟨89CC178⟩, 662, 962 ⟨86OM1906⟩; **6**, 135 ⟨86JOC3983⟩
Eastep, S. J., **6**, 435 ⟨84MI 614-02⟩, 548 ⟨90JOC1475⟩
Eastham, J. F., **1**, 111 ⟨63JA2171⟩, 384 ⟨B-66MI 109-01⟩, 889 ⟨59JA3652⟩
Eastman, D., **4**, 154 ⟨69JOC2728⟩
Eastman, R. H., **3**, 242 ⟨54JA4118⟩, 247 ⟨46JA2229⟩
Easton, C. J., **1**, 337 ⟨83AJC545⟩; **2**, 394 ⟨88AJC1063⟩; **4**, 65 ⟨83CC1349⟩; **5**, 334 ⟨85JCS(P1)153⟩
Easton, D. B. J., **3**, 356, 361 ⟨72JCS(P1)41⟩; **4**, 850 ⟨72JCS(P1)41⟩
Easton, N. R., **1**, 506 ⟨61JOC3772⟩; **2**, 1018 ⟨65USP3172912⟩; **6**, 509 ⟨64JOC1851⟩
Eastwood, F. W., **1**, 988 ⟨87AJC1675⟩; **3**, 527, 541 ⟨B-80MI 316-03, 70CC206⟩, 541 ⟨74AJC2373, 76JA7421⟩, 541 ⟨77AJC179, 77AJC459⟩; **6**, 104 ⟨65AJC1967⟩
Easty, D. M., **2**, 103 ⟨52JCS1193⟩
Eaton, B., **1**, 451 ⟨86JA1359⟩; **6**, 402 ⟨86OM394⟩
Eaton, P., **1**, 807, 809 ⟨61T13⟩
Eaton, P. E., **1**, 498 ⟨64TL3695⟩, 900 ⟨90JA3225⟩; **2**, 25 ⟨86TL6055⟩, 397 ⟨88JOC5353⟩, 555 ⟨85JA724⟩; **3**, 212 ⟨73JOC4071⟩, 274 ⟨67JA5722⟩; **5**, 73 ⟨72JOC1947⟩, 282 ⟨73JOC4071⟩, 460 ⟨70JA3515⟩
Eatough, N. L., **2**, 403 ⟨62JA2922⟩
Ebata, K., **4**, 1051, 1052 ⟨88CL965⟩
Ebata, T., **1**, 651 ⟨78S746⟩; **3**, 246 ⟨81T709⟩; **4**, 648 ⟨78S746⟩
Ebbinghaus, C. F., **3**, 479 ⟨79JOC4697⟩
Ebel, H. F., **1**, 104 ⟨69FCF387⟩; **2**, 556 ⟨70HOU(13/1)261⟩
Ebeling, E., **4**, 856 ⟨84T3677⟩
Ebeling, H., **5**, 1051 ⟨71ZC420⟩, 1052 ⟨74ZC356⟩
Ebeneder, F., **2**, 175 ⟨42CB34⟩, 178 ⟨42CB344⟩
Ebenezer, W. J., **3**, 60 ⟨92TL4053⟩
Ebenhöch, J., **4**, 136, 140 ⟨86ZN(B)1527⟩, 1052 ⟨87ZN(B)142⟩; **6**, 173, 378 ⟨86ZN(B)1527⟩
Ebenroth, A., **2**, 778 ⟨65JPR123⟩
Eber, B., **4**, 681 ⟨92JOM(441)81⟩
Eberbach, W., **1**, 999 ⟨86SC983⟩; **3**, 315 ⟨80AG(E)47, 80TL4909, 82TL4665, 83CB2383, 84CB2157, 84TL1137⟩, 318 ⟨76TL3299, 76TL3303, 78TL4875, 79TL4049, 81CB2979, 81TL4953, 85CB4035⟩
Eberhard, J. K., **5**, 209 ⟨83CB1183⟩
Eberl, K., **3**, 556 ⟨81OMR(17)180⟩; **6**, 700, 733 ⟨78CB2451⟩
Eberle, D., **1**, 822 ⟨63LA(662)188⟩
Eberle, H.-J., **4**, 581 ⟨81CB413, 83CB1209⟩
Eberle, M., **2**, 394 ⟨90OS(69)19⟩
Eberle, M. K., **4**, 415 ⟨73T4049⟩
Eberlein, J., **4**, 577, 586 ⟨77ZAAC(434)145⟩; **5**, 920, 922 ⟨77CB677⟩; **6**, 702, 707, 708 ⟨77CB677⟩
Eberlein, T. H., **1**, 400 ⟨92JOC3479⟩; **3**, 331 ⟨82JA1445, 86JOC1556⟩, 331, 335 ⟨88JOC2220⟩
Eberlin, J., **4**, 574 ⟨77CB677⟩
Eberly, K., **1**, 871 ⟨40JA113⟩
Eberson, L., **1**, 653 ⟨56ACS633⟩; **5**, 24, 28 ⟨B-78MI 502-02⟩, 26 ⟨71JA5821⟩, 1102 ⟨64ACS1567⟩; **6**, 143 ⟨88ACS(B)666⟩
Eberstein, K., **3**, 160 ⟨76CB3907⟩, 234 ⟨74TL4377⟩
Ebert, G. W., **1**, 121 ⟨84JOC5280, 88JOC4482⟩; **2**, 958, 959 ⟨88JOC4482, 84JOC5280⟩; **3**, 286 ⟨88JOC4482⟩
Ebert, H.-D., **5**, 397 ⟨77AG(E)269⟩
Ebert, K., **6**, 169 ⟨92OM145⟩
Ebetino, F. H., **6**, 123 ⟨90H(30)855⟩, 157 ⟨89EUP298553⟩, 158 ⟨90PS(51/52)23⟩
Ebina, S., **6**, 470 ⟨92OM171⟩
Ebling, J., **2**, 468 ⟨68CB3883, 72ZAAC(394)171⟩

Ebner, C. B., **3**, 195, 220 ⟨87S1055⟩
Ebner, M., **5**, 277 ⟨87JOC2033⟩; **6**, 335 ⟨86JOM(311)63⟩, 335, 336 ⟨85ZN(B)1351⟩
Ebner, W., **4**, 417 ⟨70CB3459⟩; **5**, 831 ⟨70CB82⟩
Ebsworth, E. A. V., **5**, 369, 713 ⟨80JCS(D)1768⟩, 535 ⟨84JCS(D)2301⟩, 538, 541 ⟨76JCS(D)1661⟩, 539 ⟨82JCS(D)1131⟩, 539, 541 ⟨72JCS(D)359⟩, 913 ⟨80CB1637⟩; **6**, 49 ⟨62JCS2290⟩, 165, 167 ⟨82OM1720⟩, 232 ⟨60JCS3516⟩, 457 ⟨88CC529, 93JCS(D)1031⟩
Eby, C. J., **5**, 275 ⟨57JA725⟩
Eby, L. T., **1**, 456, 508 ⟨41JA216⟩
Echagüen, C. O. de, **1**, 579 ⟨93T6717⟩
Echavarren, A. M., **1**, 228, 479 ⟨89T979⟩, 482, 1027 ⟨87JA5478⟩, 1030 ⟨88JA1557⟩; **2**, 86 ⟨88JA4039⟩, 316 ⟨91JOC452⟩, 931 ⟨87JA5478⟩; **3**, 290, 295, 296 ⟨88JA1557⟩; **5**, 455 ⟨90JOM(390)C57⟩
Echegoyen, L., **1**, 890 ⟨92JA3978⟩
Echigo, E., **5**, 233, 506, 509 ⟨87NKK1430⟩
Echigo, Y., **2**, 26 ⟨77CL1013⟩; **3**, 131 ⟨78CL49⟩, 239 ⟨78CL465⟩, 696 ⟨77CL697⟩; **5**, 979 ⟨77CL1345⟩
Echigoya, E., **1**, 674 ⟨76CL1019⟩
Echsler, K-J., **4**, 599, 600, 663 ⟨78TL4391⟩; **6**, 170, 210 ⟨79TL501⟩, 203, 204, 210 ⟨80TL2807⟩
Eck, C. R., **1**, 381 ⟨74JCS(P1)1938, 76JCS(P1)359⟩; **4**, 20 ⟨75JCS(P1)251⟩; **5**, 138 ⟨74TL3135⟩
Eck, J. C., **2**, 29 ⟨43OSC(2)74⟩; **5**, 57 ⟨43OSC(2)74⟩, 86 ⟨43OSC(2)28⟩, 88 ⟨43OSC(2)374⟩
Ecke, A., **2**, 256 ⟨73M503⟩
Ecke, G. G., **6**, 523 ⟨58JA6167⟩
Eckert, H., **2**, 468 ⟨72ZN(B)769⟩; **3**, 696, 699 ⟨87AG(E)894⟩; **4**, 20 ⟨79LA278⟩; **5**, 972 ⟨87AG(E)894⟩; **6**, 416, 447, 462, 471, 483, 486, 487 ⟨87AG(E)894⟩, 425, 487 ⟨78AG(E)361⟩, 485, 487 ⟨79LA278, 76AG(E)681⟩, 487 ⟨91ZN(B)339⟩, 488 ⟨90ZN(B)1715⟩
Eckert, P., **5**, 548 ⟨80CB3605⟩
Eckes, H., **3**, 480 ⟨80CB3303⟩
Eckes, H.-L., **3**, 359 ⟨87AG(E)1046⟩
Eckes, L., **1**, 1110 ⟨78CB1253⟩; **2**, 18, 610 ⟨78S217⟩
Eckhardt, E., **3**, 631 ⟨91LA47⟩; **5**, 135 ⟨91LA47⟩
Eckhardt, H. E., **3**, 582 ⟨78AG(E)465, 79TL2125⟩
Eckmann, M.,**3**, 695 ⟨27M71⟩
Eckoldt, H., **2**, 175 ⟨55HOU(9)3⟩
Eckrich, T. M., **1**, 136 ⟨83TL3165⟩; **2**, 478 ⟨84TL2415⟩, 999 ⟨83TL3165⟩; **3**, 162 ⟨83TL3165⟩; **4**, 378, 384, 388 ⟨83TL3165⟩, 380, 389 ⟨83TL3163⟩
Eckstein, F., **2**, 105 ⟨67AG(E)362⟩
Eckstein, U., **3**, 478, 482 ⟨79LA1002⟩, 480 ⟨71LA(748)207⟩; **5**, 871, 874 ⟨71LA(748)207⟩; **6**, 623, 660, 668 ⟨79LA1002⟩, 670 ⟨71LA(748)207⟩
Eckstein, Z., **4**, 302 ⟨63AHC(2)311, 78AHC(23)1⟩; **5**, 671 ⟨81PJC655⟩, 828 ⟨72MI 521-01⟩
Eckwert, J., **3**, 710 ⟨92AG(E)1218⟩
Edafiogho, I. O., **5**, 332, 333 ⟨91JMC387⟩
Eddine, J. J., **1**, 180 ⟨88T5495⟩
Eddy, M. M., **4**, 668 ⟨90JA2425⟩
Edelmann, F., **3**, 552 ⟨76JOM(118)C41⟩, 553 ⟨82CB1332, 84JOM(270)93⟩; **4**, 589 ⟨89OM1192⟩; **5**, 782 ⟨88JOM(345)C1⟩
Edelmann, F. T., **2**, 872 ⟨89JOM(366)73⟩; **4**, 719 ⟨88ZN(B)1461⟩; **5**, 781 ⟨91ZN(B)1328, 90JOM(338)21⟩, 782 ⟨92CB2171⟩, 893 ⟨89AG(E)397⟩
Edelmann, N. K., **4**, 975 ⟨67JOC3293⟩
Edelmann, T., **2**, 489 ⟨89CB1247⟩
Edelstein, H., **3**, 459 ⟨66JOC3375⟩
Eden, P., **4**, 416 ⟨75JMC177, 75JMC182⟩
Edenberger, D. M., **1**, 1023 ⟨89TL2153⟩
Edenborough, M. S., **3**, 694, 712 ⟨88MI 321-01⟩, 714, 718 ⟨85CC816⟩
Edens, C. O., **3**, 67 ⟨55OSC(3)731⟩, 67, 79 ⟨45OS(25)92⟩
Eder, R., **1**, 530 ⟨88JOM(355)449⟩
Ederer, B., **4**, 635 ⟨92CB2213, 93CB2003⟩
Ederle, H., **4**, 635 ⟨92CB2213⟩
Edgar, A. R., **1**, 43 ⟨87JCS(P1)2371⟩
Edgar, D. E., **6**, 215 ⟨32JA2025⟩
Edgar, K. J., **1**, 1094 ⟨81JOC4600⟩; **3**, 284 ⟨82JOC1585⟩, 626 ⟨81JOC4600⟩
Edinger, A., **3**, 362 ⟨01JPR182⟩, 398 ⟨03JPR72⟩
Edman, J. R., **2**, 651 ⟨84JOC645⟩
Edman, P., **5**, 83 ⟨59MI 502-01⟩
Edmison, M. T., **2**, 249 ⟨55JA80⟩
Edmonds, A. C. F., **1**, 595, 779 ⟨76JCS(P1)459, 70JCS(C)833⟩
Edmonds, M. K., **2**, 27 ⟨90TL7509⟩
Edmonson, Jr., L. J., **2**, 499 ⟨71JA5677⟩
Edmundson, R. S., **4**, 336 ⟨B-87MI 407-03⟩, 451 ⟨B-88MI 410-01⟩
Edo, K., **1**, 1009, 1010 ⟨78H(9)271, 78CPB3843⟩, 1010 ⟨79CPB193⟩
Edsall, J. T., **5**, 83 ⟨47MI 502-02⟩
Edserg, R. L., **2**, 181 ⟨48MI 203-01⟩
Edstrom, E. D., **1**, 246 ⟨91TL5709⟩
Eduok, E. E., **6**, 317 ⟨93T3035⟩
Edward, J. T., **2**, 366 ⟨72CJC3502⟩; **4**, 418 ⟨87CJC282⟩
Edwards, B. E., **3**, 246 ⟨69JOC2767⟩, 343 ⟨62JOC3760⟩; **4**, 245 ⟨59JA3148⟩
Edwards, D. A. V., **3**, 653 ⟨94TL445⟩
Edwards, D. S., **3**, 521 ⟨83OM1505⟩
Edwards, G. L., **1**, 133 ⟨91CC1399⟩; **2**, 640 ⟨91CC1399⟩
Edwards, H. G. M., **4**, 152 ⟨86JOM(314)13⟩
Edwards, J. A., **6**, 113 ⟨76JOC1112⟩
Edwards, J. D., **2**, 110 ⟨55JCS1470, 57JCS348⟩, 111, 112 ⟨58JCS1451⟩
Edwards, J. O., **2**, 845 ⟨73JOC2703⟩
Edwards, J. P., **2**, 998 ⟨91JA723⟩; **4**, 152 ⟨92JA2592⟩
Edwards, Jr., W. R., **5**, 272, 293, 304 ⟨38JA540⟩
Edwards, L. M., **5**, 1140 ⟨71JCS(A)1663⟩
Edwards, M., **1**, 440 ⟨88JOC3815⟩; **5**, 81, 107 ⟨81S487⟩, 1126 ⟨76JCS(P1)716⟩
Edwards, M. L., **1**, 439 ⟨90TL5449⟩; **2**, 309 ⟨90TL3417⟩; **4**, 61, 79, 780, 782 ⟨90TL5449⟩, 73 ⟨89JA1127⟩, 119, 734 ⟨90TL5571⟩; **6**, 266 ⟨90TL5571⟩, 487 ⟨91JMC569⟩
Edwards, M. P., **1**, 654, 697, 700, 703 ⟨83CC630⟩; **3**, 308 ⟨83CC630, 84JOC3503⟩
Edwards, O. E., **2**, 337 ⟨72CJC1167⟩; **3**, 184 ⟨69CJC3489⟩, 468 ⟨59JOC2071⟩
Edwards, O. J. K., **2**, 286, 732 ⟨28JCS2293⟩
Edwards, P. D., **1**, 149 ⟨84JA3270⟩, 1136 ⟨85JOC1019⟩; **2**, 1002 ⟨85JOC1019⟩; **3**, 182 ⟨85JOC1019⟩; **4**, 530 ⟨85JOC1019⟩
Edwards, P. G., **2**, 428 ⟨92CC699⟩, 437 ⟨90POL2413⟩; **4**, 696 ⟨82JA2072⟩
Edwards, P. J., **2**, 642 ⟨92CC1256⟩
Edwards, R. L., **3**, 64 ⟨89JCS(P1)1939⟩
Edwards, W. M., **2**, 801 ⟨54JA1840⟩
Edwards, W. R., **1**, 912, 914 ⟨58T178⟩
Edwin, J., **4**, 632 ⟨83JA2582⟩, 683 ⟨90JA7113⟩
Eeuwhorst, H. G., **4**, 957 ⟨77CC801⟩
Effenberger, F., **1**, 182 ⟨88CB421⟩, 182, 475 ⟨87AG(E)142⟩, 448 ⟨78S755⟩; **2**, 29 ⟨62CB803⟩, 245 ⟨67AG(E)951⟩, 316 ⟨91LA269⟩, 321 ⟨90S575⟩, 322 ⟨91AG(E)873⟩, 656 ⟨73AG(E)775⟩, 665 ⟨87CB1347⟩, 689 ⟨76JOC4099⟩, 943 ⟨84JOC4687⟩, 1006, 1009 ⟨87JA882, 88JOC4379⟩; **3**, 33 ⟨92TL5157⟩, 152, 154 ⟨79CB1670⟩, 246 ⟨82CB2766⟩, 279, 295, 298 ⟨72AG(E)300⟩, 280 ⟨88CB117⟩, 280, 296, 298 ⟨81CB926⟩, 295, 296, 298 ⟨83CB1195⟩, 561 ⟨62AG(E)512⟩, 628 ⟨87AG(E)458, 90S575⟩, 628 ⟨90TL1249, 91TL2605⟩, 628, 631 ⟨93CB779, 91LA47⟩; **4**, 312 ⟨66CB3903⟩, 512 ⟨79LA842⟩, 881 ⟨61AG493, 63CB1350⟩, 881, 882, 972 ⟨64CB3081⟩, 979

⟨66CB3892⟩; **5**, 92 ⟨86LA334⟩, 135 ⟨91LA47⟩, 209 ⟨83CB1183⟩, 236 ⟨80CB2100⟩, 301 ⟨68CB502⟩, 964 ⟨81CB173⟩, 1106 ⟨86CB1594⟩; **6**, 55 ⟨65CB1078⟩, 106 ⟨62AG(E)331, 66CB3892⟩, 138 ⟨66AG(E)132, 68CB1885⟩, 140 ⟨60BRP823519⟩, 140 ⟨63CB1505⟩
Effenberger, P., **2**, 1009 ⟨68TL4265⟩
Effland, R. C., **3**, 307 ⟨88JHC517⟩
Efimova, L. F., **2**, 464 ⟨59ZOB1671⟩
Efner, H. F., **6**, 529, 535 ⟨74IC1778⟩
Efraty, A., **1**, 517 ⟨77CRV691⟩; **3**, 78 ⟨77IC3124⟩, 80 ⟨80JOC4059⟩; **4**, 692 ⟨73JOM(47)145⟩, 706 ⟨77CRV691⟩; **6**, 415 ⟨80JOC4059⟩
Efremov, A. I., **2**, 833, 837, 838, 839 ⟨78ZOB319⟩
Efremov, D. I., **4**, 768 ⟨92ZOR522⟩
Efremov, V. G., **6**, 620 ⟨91ZOB643⟩
Efremov, Yu. Ya., **2**, 894 ⟨83IZV1152⟩; **4**, 106 ⟨90JGU2208⟩, 463 ⟨88ZOB2456⟩, 569 ⟨90PS(53)1⟩, 592 ⟨90ZOB2291⟩; **5**, 406 ⟨88JGU1548⟩, 535 ⟨89JGU1548⟩; **6**, 160 ⟨91PS(61)361⟩, 691 ⟨92PS(66)257⟩
Egan, C., **3**, 684 ⟨91JCS(P2)249⟩; **5**, 838 ⟨91JCS(P2)249⟩
Egan, C. P., **2**, 172 ⟨71JOC2530⟩
Egan, D. A., **2**, 118, 124 ⟨93JMC460⟩
Egan, R., **4**, 743 ⟨55OSC(3)270⟩; **5**, 38 ⟨55OSC(3)270⟩
Egan, W., **1**, 53 ⟨71JA6205⟩; **6**, 489 ⟨93MI 615-02⟩, 491 ⟨89TL7141⟩
Egan Jr., J. W., **4**, 657, 659 ⟨87OM2007⟩
Egashira, H., **3**, 119 ⟨87BCJ4143⟩
Egawa, Y., **5**, 635, 636 ⟨71MI 515-01⟩
Egberink, R. J. M., **1**, 518 ⟨86JOC2004⟩; **6**, 562 ⟨92JA10573⟩
Egbert, E., **4**, 536 ⟨87AG(E)480⟩
Egbertson, M., **6**, 111 ⟨84H(22)281⟩
Ege, G., **3**, 593 ⟨79AG(E)67, 79CB3166⟩, 668 ⟨74S22⟩; **4**, 432 ⟨70CZ215⟩; **6**, 57, 105 ⟨72TL4217⟩
Eger, K., **4**, 979 ⟨92AP(325)551⟩
Egert, E., **1**, 463 ⟨87AG(E)560⟩, 974 ⟨87T1969⟩; **4**, 462 ⟨85AG(E)1067⟩, 536 ⟨86S737, 88AG(E)1194⟩, 536 ⟨88AG(E)433⟩
Egg, H., **5**, 344 ⟨86AP(319)682⟩, 359 ⟨92SC1199⟩
Eggen, M., **1**, 178 ⟨93JOC6174⟩
Eggenspiller, A., **6**, 58, 58 ⟨92BMC407⟩
Egger, K. W., **1**, 925 ⟨73HCA1516⟩
Eggerer, H., **1**, 458 ⟨69NAT1212⟩
Eggerichs, T., **1**, 834 ⟨77JOC172⟩
Eggers, A., **1**, 1029 ⟨93TL4177⟩
Eggers, H., **1**, 727 ⟨58LA(619)10⟩; **3**, 493 ⟨58LA(619)10⟩
Eggers, M. D., **2**, 28 ⟨87JA2435⟩
Eggers, W., **5**, 1064, 1073 ⟨86CB3737⟩
Eggersdorfer, M., **3**, 680 ⟨87S924⟩; **5**, 79 ⟨79LA1547⟩
Eggersmann, G., **4**, 234 ⟨84AP(317)15⟩
Eggert, U., **1**, 364 ⟨88T3899⟩
Eggleston, D. S., **4**, 277 ⟨90SL359⟩; **6**, 632 ⟨87JHC275⟩
Eggnauer, U., **1**, 835 ⟨91TL7233⟩
Egidi, V., **5**, 822 ⟨13G58⟩
Egler, R. S., **2**, 672 ⟨87TL6109⟩
Egli, C., **2**, 663 ⟨75HCA104⟩
Egli, M., **2**, 376 ⟨85HCA220⟩; **6**, 425 ⟨86HCA1442⟩
Eglington, G., **1**, 504 ⟨61T16⟩, 1066 ⟨53OS(33)68⟩, 1103 ⟨65OS(45)39⟩
Eglinton, B., **2**, 1018, 1097 ⟨63JCS2295⟩
Eglinton, G., **1**, 80 ⟨54JCS3197⟩, 1064 ⟨60JCS3614, 64JCS1151⟩; **5**, 189 ⟨54JCS1860⟩
Egocochkin, A. N., **5**, 1081 ⟨88MI 528-01⟩
Egolf, T., **4**, 683 ⟨86OM2413⟩
Egorochkin, A. N., **4**, 141 ⟨81JOM(205)311⟩; **6**, 324, 335 ⟨63BAU1384⟩
Egorova, L. P., **6**, 242 ⟨46ZOB1521⟩
Egorov, M. P., **2**, 947, 949 ⟨B-91MI 218-01⟩
Egorov, Yu. P., **5**, 838 ⟨58ZOB2305⟩
Egron, M., **3**, 4, 115 ⟨82JCS(P1)1967⟩
Egsgaard, H., **2**, 195 ⟨81CC742⟩; **4**, 1081 ⟨83JCS(P2)1687⟩; **5**, 539 ⟨80JCS(P2)1557⟩; **6**, 556, 567 ⟨87CB987⟩
Eguchi, C., **6**, 655 ⟨74BCJ935⟩
Eguchi, H., **5**, 318 ⟨93BCJ2426⟩
Eguchi, M., **2**, 87, 118, 124 ⟨71BCJ3427⟩
Eguchi, S., **1**, 518 ⟨75JA4413⟩, 569 ⟨93JOC5163⟩; **2**, 468 ⟨92OPP209⟩; **3**, 273 ⟨87S1092⟩, 696 ⟨74JOC1239⟩; **4**, 165 ⟨74H(2)177⟩, 232 ⟨84JCS(P1)2641⟩; **5**, 1066 ⟨87NKK1280⟩
Eguren, L., **4**, 581 ⟨89JOM(371)329⟩
Ehara, Y., **5**, 384, 387 ⟨80BCJ1299⟩
Ehle, M., **2**, 859, 865 ⟨89AG(E)225⟩; **4**, 110 ⟨89AG(E)225, 91CB1207⟩
Ehler, D. F., **1**, 21 ⟨73JOC3660⟩; **2**, 914 ⟨74JOM(69)193⟩
Ehler, K. W., **2**, 235 ⟨89JOC1092⟩; **3**, 337 ⟨89JOC1092⟩
Ehlers, J., **3**, 552 ⟨79LA1734⟩; **4**, 851 ⟨80S907⟩; **5**, 552 ⟨80S907⟩, 604 ⟨80CB3010⟩, 604, 623 ⟨79CB1000⟩, 605 ⟨78AG(E)455⟩
Ehlinger, E., **1**, 651, 762 ⟨80JA5004⟩; **2**, 935 ⟨80JA5004⟩; **4**, 355 ⟨77JA4536⟩, 648 ⟨77CC772⟩, 648, 655 ⟨80TL11⟩; **6**, 124 ⟨82TL309⟩
Ehlis, T., **1**, 619 ⟨89TL3131⟩; **4**, 331 ⟨90CB1563⟩; **5**, 489 ⟨90CB1563, 90CB1989⟩
Eholzer, U., **3**, 694, 696, 699, 700, 721, 722, 724, 725 ⟨65AG(E)472⟩, 709 ⟨64AG(E)647⟩, 710 ⟨62AG499⟩, 712, 716 ⟨62AG215, 65CB193⟩
Ehrbar, U., **5**, 1055 ⟨76S519⟩
Ehrenfeld, D., **3**, 514 ⟨87CC129⟩
Ehrenfreund, J., **6**, 116 ⟨89T3787⟩
Ehrenreich, W., **2**, 891 ⟨84CB1999⟩
Ehrhard, A., **6**, 58, 58 ⟨92BMC407⟩
Ehrhardt, H., **1**, 1123 ⟨68TL4617⟩; **3**, 355 ⟨77CB2506⟩
Ehrhardt, K., **2**, 227 ⟨54HCA375⟩
Ehrhart, W. A., **5**, 743, 746, 753 ⟨63JOC1108⟩
Ehrig, V., **2**, 393 ⟨75CB1924, 75CB1961⟩
Ehrler, R., **2**, 392 ⟨86S535⟩; **3**, 46 ⟨86S535⟩, 683 ⟨85MI 320-01⟩
Ehrlich, P., **5**, 109 ⟨43OSC(2)457⟩
Ehrlich, P. P., **3**, 285, 295 ⟨87JOC2615⟩
Ehrlich, R., **2**, 421 ⟨64JA5359, 70IC146⟩
Ehrlich, S., **1**, 318 ⟨77JA290⟩
Ehrlichmann, W., **4**, 410 ⟨82JPR832⟩
Eibach, F., **2**, 945 ⟨78AG(E)278⟩
Eiband, G. R., **1**, 811 ⟨87T4265⟩
Eibisch, H., **2**, 245, 250 ⟨68CB3567⟩
Eichel, W., **1**, 325, 920 ⟨87JOC205⟩; **2**, 39 ⟨76JOC3505⟩
Eichelbaum, G., **5**, 833 ⟨1889CB2973⟩
Eichenauer, H., **1**, 153 ⟨84T1345⟩
Eicher, T., **2**, 67 ⟨B-66MI 202-01⟩, 744 ⟨74CB2186⟩; **3**, 357 ⟨74T4025⟩; **5**, 8 ⟨88S525⟩, 11 ⟨91S1173⟩, 122 ⟨85HOU(E5)659⟩, 182 ⟨89HOU(5)633⟩, 366 ⟨69CB2889⟩, 692, 699, 700 ⟨85HOU(E5/1)812⟩
Eichhorn, B., **2**, 695 ⟨73CB2738⟩; **4**, 114, 128 ⟨73CB2733⟩, 135 ⟨73CB2742⟩
Eichhorn, B. W., **4**, 675 ⟨89OM80, 91IC156⟩, 676 ⟨89OM67⟩
Eichhorn, I., **1**, 482 ⟨89S477⟩; **4**, 163, 176 ⟨B-67MI 404-01⟩
Eichinger, P. H., **2**, 348 ⟨83AJC719⟩
Eichinger, T., **4**, 890 ⟨92HCA124⟩
Eichler, B., **4**, 576, 587 ⟨73EGP94992⟩
Eichler, D., **4**, 412 ⟨67CB2131⟩
Eichler, J., **2**, 751 ⟨74S672⟩; **4**, 530 ⟨74S672⟩
Eichler, S., **1**, 725 ⟨67CB1144⟩
Eichninger, K., **1**, 1004, 1005 ⟨83JCR(M)167⟩
Eickemeyer, D. B., **1**, 379 ⟨58JA3625⟩
Eicken, K., **4**, 830 ⟨74GEP2331675⟩
Eickhoff, D. J., **1**, 218 ⟨85JOC3923⟩; **2**, 1098 ⟨85JOC3923⟩
Eid, Jr., C. N., **3**, 160 ⟨91T975⟩
Eiden, F., **2**, 808 ⟨74AP(307)73⟩; **3**, 308 ⟨87AP(320)1020⟩; **4**, 405 ⟨56CB2873⟩

Eidenschink, R., **1**, 363 ⟨74AG(E)468⟩; **4**, 421 ⟨77CB651⟩
Eidt, S. H., **2**, 595 ⟨78USP4118409⟩
Eidus, Y. T., **6**, 464 ⟨72IZV1635, 72IZV2733⟩, 464 ⟨73IZV562, 73IZV807⟩
Eierdanz, H., **1**, 981 ⟨82AG(E)924⟩
Eigen, M., **1**, 844, 870 ⟨64AG(E)1⟩
Eigen, P., **4**, 838, 1081 ⟨89T6667⟩
Eigenbrot, C., **5**, 951 ⟨87OM988⟩
Eigendorf, U., **4**, 642 ⟨93CB1397⟩
Eijsinga, H., **1**, 496 ⟨76TL947⟩
Eikema Hommes, N. J. R. Van, **6**, 325, 379, 388, 398 ⟨88TL5237⟩
Eiki, T., **2**, 148 ⟨80CL1063⟩
Eilerman, R. G., **3**, 538 ⟨69TL4517⟩
Eilers, E., **3**, 696 ⟨76JPR127⟩
Eilers, K. L., **5**, 142 ⟨64JOC2813⟩, 166 ⟨64JOC2810⟩
Eilerts, N. W., **3**, 516 ⟨93OM2565⟩; **4**, 676 ⟨92AG(E)66⟩
Eilingsfeld, H., **1**, 1026 ⟨60AG836⟩; **4**, 901, 904 ⟨65CB1293⟩; **5**, 577, 608, 609, 610, 728 ⟨63CB2671⟩, 580 ⟨69AG(E)750⟩, 609, 611 ⟨60AG836⟩, 640, 641 ⟨65CB1308⟩; **6**, 51 ⟨60AG836⟩, 70 ⟨63CB2671⟩, 313, 321 ⟨64CB1232⟩
E.-I. Negishi, **1**, 544 ⟨84JA6105⟩; **4**, 655, 658, 659 ⟨88JA646⟩
Einhellig, E., **4**, 493 ⟨73CB3421⟩
Einhellig, K., **5**, 789 ⟨76CZ340⟩
Einholz, W., **5**, 1085 ⟨86ZN(B)1367⟩
Einhorn, A., **1**, 785 ⟨1883CB2208⟩; **3**, 618 ⟨02CB3647⟩; **4**, 296 ⟨05LA(343)207, 08LA(361)113⟩, 297 ⟨08CB24⟩; **6**, 484 ⟨1898LA(300)145⟩
Einhorn, C., **2**, 74 ⟨87JOM(322)177⟩, 551, 958 ⟨89S787⟩; **3**, 178 ⟨91SL37⟩; **4**, 146 ⟨88CC333⟩; **5**, 327 ⟨90SC1105⟩
Einhorn, J., **1**, 166 ⟨92TL6461⟩; **2**, 422 ⟨87JOC4124⟩, 422, 653 ⟨88CC1350⟩, 551, 958 ⟨89S787⟩; **3**, 90 ⟨86TL1791⟩, 178 ⟨91SL37⟩; **5**, 272, 301 ⟨86TL501, 90SC1105⟩
Einstein, F. W. B., **2**, 897 ⟨75JCS(D)172⟩; **4**, 291 ⟨87OM2164⟩, 552 ⟨90IC808⟩, 713 ⟨82CC371⟩, 715 ⟨93JOM(463)C6⟩; **6**, 102 ⟨86CJC732⟩
Eis, M. J., **1**, 204, 478, 1023 ⟨84JA3693⟩, 263 ⟨85TL1153⟩; **2**, 319 ⟨85TL1153⟩
Eisch, J. J., **1**, 13, 20 ⟨92JOC2143⟩, 14 ⟨92JOC1618⟩, 128, 161 ⟨76JOC2214⟩, 144, 1090 ⟨79JOC3277⟩, 145 ⟨80JOC4534⟩, 145, 310, 1090 ⟨79JOC3279⟩, 146 ⟨85JOC3674⟩, 436 ⟨86JA7763⟩, 485, 1032 ⟨84TL4851⟩, 511 ⟨68JOM(11)P7, 68JOM(12)P33⟩, 511 ⟨91COS(8)733⟩, 650 ⟨76JOC2615⟩, 680 ⟨88JOC2829⟩; **2**, 561 ⟨65JA4194⟩, 597 ⟨62JA3605⟩, 815 ⟨82COMC-I(1)599⟩, 816 ⟨64JA4221⟩, 919 ⟨75JOM(86)143⟩, 922 ⟨76JOC2214⟩, 925 ⟨78JOM(159)C13⟩, 948, 968 ⟨71JOC3520⟩, 967 ⟨82COMC-I(1)555, 91COS(8)733⟩, 970 ⟨68JOM(12)P33⟩, 974 ⟨74JA6815⟩, 989 ⟨69JA4575⟩, 1006 ⟨83TL2043⟩, 1083 ⟨90JA1847⟩, 1099 ⟨62JA3820⟩; **3**, 224 ⟨85JOC3674⟩, 404, 408 ⟨86JOC1848⟩, 508 ⟨83TL2043⟩; **4**, 387 ⟨82JOC5051⟩, 393 ⟨85JOM(285)121⟩, 602 ⟨88JOM(341)293⟩, 696 ⟨83TL2043⟩, 698 ⟨88JOC2829⟩, 960 ⟨79JOC3279⟩, 1061 ⟨85JA7219⟩, 1063 ⟨71JOC3520⟩, 1068 ⟨74JA1941, 75LA565⟩, 1068 ⟨82COMC-I(2)641⟩; **5**, 413 ⟨82JOC5051⟩; **6**, 131, 133 ⟨88JOM(341)293⟩
Eisen, N., **1**, 814 ⟨86AG(E)1026⟩; **2**, 31 ⟨87JOC5560⟩; **3**, 90 ⟨86AG(E)1026⟩
Eisen, O., **1**, 552 ⟨89CA(111)23052⟩
Eisenbarth, P., **3**, 640 ⟨87JOC2427⟩; **4**, 512 ⟨87JOC2427⟩
Eisenbraun, E. J., **1**, 7, 19 ⟨82JOC2590⟩, 43 ⟨71JOC737⟩; **3**, 210 ⟨55JA1599⟩; **5**, 66, 74 ⟨73OSC(5)8⟩, 122 ⟨68CI(L)1568⟩
Eisenhart, E. K., **1**, 971, 985 ⟨86JA7791⟩; **5**, 413 ⟨84JOC5282⟩
Eisenhut, M., **2**, 856 ⟨73CC144⟩
Eisenschmidt, V., **6**, 624 ⟨73JPR640⟩
Eisenstein, O., **2**, 580 ⟨82JA4865⟩, 893 ⟨87OM1185⟩; **5**, 932 ⟨86JA2173⟩
Eisentraut, K. J., **4**, 768 ⟨86JFC(31)405⟩
Eisert, M. A., **3**, 494 ⟨64JOM(2)101⟩; **4**, 152 ⟨69JA5027⟩
Eisfeld, W., **1**, 839 ⟨58LA(618)202⟩
Eismont, M. Y., **1**, 353, 354 ⟨94T4519⟩
Eisner, T., **1**, 503, 504 ⟨83JOC2266⟩
Eisner, U., **2**, 229 ⟨71T5753⟩
Eison, M. S., **5**, 334 ⟨86JMC359⟩
Eiss, R., **2**, 1004 ⟨68IC18⟩
Eissenfels, M., **3**, 73 ⟨72T5197⟩
Eissenstat, M. A., **1**, 889 ⟨78JA292⟩; **3**, 616, 621 ⟨88OSC(6)334⟩
Eistert, B., **1**, 394 ⟨35CB200⟩; **2**, 617 ⟨66CB1414⟩, 765 ⟨76CB3462⟩; **3**, 260 ⟨62LA(659)64⟩; **4**, 204 ⟨61CB929⟩; **5**, 855 ⟨63CB3120⟩
Eiter, K., **1**, 573 ⟨72S591⟩, 734 ⟨62LA(658)65⟩, 1002 ⟨78LA658⟩, 1134 ⟨72AG(E)60⟩; **4**, 312 ⟨72LA(765)55⟩
Ejiri, E., **1**, 690 ⟨86BCJ2209⟩
Ejiri, S., **1**, 360 ⟨92JA8707⟩
Ekberg, T., **6**, 550 ⟨93CAR(246)119⟩
Ekogha, C. B. B., **1**, 1089 ⟨83TL4825⟩; **3**, 54, 210 ⟨83TL4825⟩, 55 ⟨83TL4829⟩; **4**, 938 ⟨83TL4825, 85BSF1250⟩
Ekouya, A., **4**, 355 ⟨75JOM(87)151, 77JOM(142)C35⟩; **5**, 420 ⟨75JOM(93)51⟩, 1076, 1077 ⟨78JOM(148)225⟩; **6**, 125, 162 ⟨79JOM(177)137⟩
Eksanov, S. A., **5**, 314 ⟨87IZV2857⟩
Ekundayo, O., **3**, 294 ⟨73MI 306-01⟩
Ekwuribe, N. N., **3**, 14 ⟨81S303⟩
El-Abadelah, M. M., **6**, 611 ⟨76CI(L)164⟩
Elad, D., **1**, 925 ⟨65CC7, 68T4509⟩; **5**, 276 ⟨64JOC1855, 65JOC3361⟩, 286 ⟨62CI(L)362⟩, 298 ⟨63TL77⟩
Elagamey, A. G. A., **4**, 986 ⟨91JPR333⟩
Elahihi, T., **3**, 696, 722 ⟨86ZN(B)132⟩
El-Ahmad, Y., **5**, 615 ⟨93H(36)1979⟩
El Alami, N., **5**, 296 ⟨85SC1233⟩
Elam, E. U., **3**, 345, 346 ⟨67JOC1562⟩, 534 ⟨61CP618722⟩, 546, 547 ⟨68JOC2738⟩; **4**, 438 ⟨61JOC1822⟩; **5**, 511 ⟨68JOC2738⟩
El Amin, B., **1**, 74 ⟨79JOC3442⟩
El Amine, M., **4**, 362 ⟨88SRI317⟩; 627 ⟨88JOM(345)39⟩
El Amouri, H., **4**, 674 ⟨90OM2993, 94OM2244⟩
Elango, V., **3**, 231, 303 ⟨85JA5219⟩; **5**, 446 ⟨85JA7788⟩
El-Ansary, A. L., **5**, 532 ⟨89MI 512-01⟩
El-Antably, S., **2**, 683 ⟨72JOC3944⟩
El Ayoubi, S., **1**, 160 ⟨93TL2111⟩; **4**, 607, 608, 654 ⟨93TL2111⟩
El Ayouki, S., **4**, 606 ⟨93JOM(451)C1⟩
El-Badawi, M., **3**, 270 ⟨85M77⟩
El Badih, A., **6**, 476 ⟨86CI(L)140⟩
Elbahaii, S. A., **5**, 532 ⟨78ZN(B)75⟩
El Bakoush, M., **2**, 445 ⟨68JOC2590⟩
El-Barbary, A. A., **2**, 738, 751 ⟨82T405⟩; **5**, 484 ⟨80ACS(B)597⟩
El-Bassiouny, F. A. A., **4**, 303 ⟨90G677⟩; **5**, 652 ⟨90ACH423, 91ACH35⟩
Elbe, H.-L., **1**, 491 ⟨74CB1654⟩
El-Behairy, M., **4**, 699, 701 ⟨79JOM(172)293⟩; **6**, 405 ⟨79JOM(172)293⟩
El-Berembally, K., **4**, 237 ⟨90SUL157⟩
El-Berins, R., **2**, 748 ⟨71CB3983⟩
Elberling, J. A., **6**, 447 ⟨92JMC3641⟩
Elbers, W., **4**, 134 ⟨92CB801⟩
El-Borai, M., **3**, 557 ⟨80MI 317-03⟩
Elbs, K., **2**, 651 ⟨1893JPR179⟩
Elby, L. T., **2**, 133 ⟨49USP2472471⟩
El-Chaharvi, M., **5**, 27 ⟨78CZ1⟩
El Cherif, S., **2**, 744 ⟨88S138⟩; **3**, 264 ⟨88S138⟩
Elder, E. S., **3**, 369 ⟨84CHEC(7)449⟩
Elder, J. S., **3**, 495 ⟨88TL3003⟩; **4**, 738 ⟨88TL3003⟩
Elder, P. A., **4**, 688 ⟨76JOM(120)C13⟩
Elder, W., **5**, 772 ⟨55CB1284⟩

Elderfield, R. C., **5**, 1126 ⟨49JOC605⟩
El Dik, M., **5**, 373 ⟨70ZOB1949⟩
Eldikov, V. N., **4**, 588 ⟨77ZOB1664⟩
Eldin, S., **1**, 877 ⟨88JA1981, 89JA6419⟩, 877 ⟨91JA1344⟩, 878 ⟨93JOC7100⟩
Eldridge, J. M., **5**, 52 ⟨77JOC3749⟩
El-Dusouqui, O. M. E., **1**, 782 ⟨89JCS(P2)579⟩
El Ebrachi, N. M. A., **3**, 242 ⟨74IJC1052⟩
Elebring, T., **2**, 652 ⟨88JMC1080⟩
Eleev, A. F., **4**, 867 ⟨80IZV892⟩; **6**, 42 ⟨86ZOR1842⟩, 42, 46 ⟨86ZOR1834⟩
Eleftheriou, M.-E., **5**, 575 ⟨91CC116⟩
Eleuterio, H. S., **6**, 42 ⟨72MI 602-01⟩
Elfahham, H. A., **3**, 641, 669 ⟨83H(20)519⟩; **5**, 771 ⟨82JCS(P1)2663, 84BCJ1650⟩
El-Farargy, A. F., **6**, 117 ⟨91PS(57)211⟩
El-Fatatry, H., **4**, 237 ⟨90SUL157⟩
Elfehail, F. E., **3**, 190 ⟨81JOC5151⟩
Elferink, V. H. M., **5**, 605, 614 ⟨81RTC414⟩
Elfert, K., **5**, 29 ⟨74S36⟩
Elgadi, A., **4**, 473 ⟨89H(28)1179⟩
Elgemeie, G. E. H., **3**, 641, 669 ⟨83H(20)519⟩; **5**, 615 ⟨90JCS(P1)1267⟩, 771 ⟨82JCS(P1)2663, 84BCJ1650⟩
El Gendy, A. M., **3**, 66 ⟨80S898⟩
Elgendy, S., **1**, 765 ⟨89TL5643, 89TL5647⟩; **2**, 525, 905 ⟨89TL5643⟩, 525, 905 ⟨89TL5647⟩, 530 ⟨92TL4209⟩; **3**, 127 ⟨89TL5643⟩; **4**, 526, 527, 528 ⟨92TL4209⟩
Elghandour, A. H. H., **4**, 883 ⟨89S775⟩, 984 ⟨86M201⟩; **5**, 677 ⟨91PS(60)119⟩
El Gharbi, R., **4**, 425 ⟨86S643⟩
Elguero, J., **3**, 280 ⟨91TL6709⟩, 451 ⟨84CHEC(5)167⟩, 452 ⟨70BSF4119⟩; **4**, 984 ⟨71BSF283⟩; **5**, 327 ⟨87T2557⟩, 1068 ⟨89JA355⟩, 1122 ⟨84TL5427⟩; **6**, 362 ⟨84OPP299⟩
Elhalim, M. S., **5**, 201 ⟨82IJC(B)589⟩
El-Hamid Ismail, A., **6**, 618, 621 ⟨92T8271⟩
El-Hamouly, W. S., **5**, 215 ⟨86IJC(B)373⟩
El-Hewehi, Z., **6**, 254 ⟨63JPR1⟩
Eliacin, M., **5**, 706 ⟨85JHC1435⟩
Eliadis, J., **6**, 621 ⟨87CZ339⟩
Elian, M., **2**, 483 ⟨70CB799⟩
Elias, W. E., **5**, 727 ⟨70CJC3662⟩
Eliasson, K. M., **5**, 48, 49, 83, 103, 114 ⟨83CC392⟩
Eliazyan, K. A., **5**, 1120 ⟨80CA(93)95215⟩
Elichev, A. A., **4**, 267 ⟨90ZOR1259⟩
Elie, C. J. J., **4**, 176, 199, 352 ⟨90TL2197⟩
Eliel, E. L., **1**, 241 ⟨76JOC1052⟩, 256 ⟨53JA3589⟩, 257 ⟨49JA663⟩, 859 ⟨77JA8262⟩; **2**, 25 ⟨63OSC(4)169⟩, 128 ⟨66JA1828, 70JOC2716⟩, 128 ⟨70JOC3641⟩, 224, 225 ⟨75JOC524⟩, 302 ⟨53OR(7)99⟩, 532 ⟨B-82MI 209-01⟩; **4**, 229 ⟨62TL103, 65JOC855⟩, 230 ⟨85JOC3402, 87OS(65)215⟩, 231 ⟨84JA2943, 84T1333⟩; **5**, 55 ⟨63OSC(4)169⟩, 63, 66, 275 ⟨63OSC(4)58⟩, 998 ⟨57JA2544⟩; **6**, 79 ⟨70JA584⟩
Eliers, E., **3**, 713 ⟨76LA969⟩
Eling, W., **2**, 260 ⟨84GEP3229565⟩
Elinger, M. Y., **2**, 521 ⟨85TL2797⟩
Elinson, M. N., **4**, 197 ⟨88TL1603⟩
Eliseenkov, V. N., **4**, 560 ⟨86ZOB2000⟩
Elisseou, E. M., **2**, 228 ⟨93S490⟩
Elissondo, B., **4**, 300, 533 ⟨84S495⟩, 388 ⟨81JOM(212)C31, 83JOC1559, 88JOM(339)267⟩, 533 ⟨86USP4617409⟩; **5**, 468 ⟨83JOM(251)175⟩
Elix, J. A., **1**, 731 ⟨72AJC545⟩; **2**, 654 ⟨74JCS(P1)1353⟩; **3**, 91 ⟨74JCS(P1)1353⟩, 102 ⟨71AJC1883⟩
El-Jazouli, M., **3**, 562 ⟨88BSF883⟩
El-Kady, M. Y., **5**, 201 ⟨82IJC(B)589⟩
Elkaim, L., **1**, 1076 ⟨91TL5321⟩
Elkaschef, M., **3**, 357 ⟨58JA6312⟩
El Kashef, H. S., **5**, 676, 677 ⟨82JCED103⟩
El-Kateb, A. A., **5**, 610 ⟨83PJC647⟩
El-Kersh, M., **4**, 237 ⟨90SUL157⟩
El-Khamry, A. M. A., **5**, 652 ⟨90ACH423⟩
El Kharraf, Z., **4**, 432, 433 ⟨90CC431⟩
El-Kheli, M. N. A., **6**, 387 ⟨84JOM(272)1, 89JCS(D)447⟩, 393, 394, 396 ⟨88JCS(D)381⟩, **6**, 395 ⟨83JOM(249)23⟩
El-Kholy, I. E.-S., **3**, 357 ⟨59JCS2588, 62JCS1857⟩
El-Khrisky, E.-E. A. M., **5**, 560 ⟨82SUL69⟩
Elkik, E., **1**, 108 ⟨70BSF991⟩; **2**, 757 ⟨68CR(C)623⟩; **3**, 236 ⟨70BSF991, 87BSF861⟩
Elkind, E., **2**, 506 ⟨16CB239⟩
Elkind, V. T., **5**, 580 ⟨56JOC772⟩
Ellenson, J. L., **2**, 402 ⟨68JA7173⟩
Ellerman, J., **1**, 445 ⟨86JCS(D)997⟩; **2**, 432 ⟨67AG(E)355⟩, 490 ⟨86JCS(D)997⟩, 490, 507 ⟨82AG(E)375⟩, 490, 507 ⟨85JOM(290)307⟩, 491 ⟨86ZAAC(539)50, 78ACS(B)646⟩; **5**, 1058 ⟨69AG(E)203⟩
Ellert, O. G., **6**, 341 ⟨90JOM(384)279⟩
Ellery, E., **2**, 783 ⟨47JCS1497⟩
Ellgen, P. C., **4**, 692 ⟨72IC2279⟩
Ellingboe, E. K., **5**, 232 ⟨63OSC(4)928⟩; **6**, 231 ⟨65USP3202718⟩
Elliot, A., **5**, 972 ⟨B-56MI 526-01⟩
Elliot, J., **1**, 836 ⟨93JCS(P1)1849⟩
Elliot, J. D., **1**, 1026 ⟨83JA2904⟩; **3**, 188 ⟨78JCS(P1)1088⟩
Elliot, M., **1**, 724 ⟨74JCS(P1)2470⟩
Elliot, R., **1**, 1026 ⟨83JA2904⟩
Elliot, R. L., **1**, 548 ⟨88JA5064⟩
Elliott, A. J., **5**, 358 ⟨73CJC4115⟩
Elliott, D. F., **2**, 214 ⟨56JCS4066⟩; **5**, 85 ⟨B-55MI 502-03⟩
Elliott, E. D., **3**, 304 ⟨51JA754⟩
Elliott, G. P., **5**, 927 ⟨87JOM(325)185⟩
Elliott, J., **1**, 756 ⟨86TL645, 89TL601⟩
Elliott, J. D., **2**, 299 ⟨77CC758, 78JCS(P1)1088⟩; **3**, 631 ⟨83JOC2294, 84TL591⟩; **6**, 641, 643 ⟨82JCS(P1)2085⟩
Elliott, M., **4**, 746 ⟨74JCS(P1)2470⟩
Elliott, M. E., **3**, 234 ⟨80JOC1109⟩
Elliott, R., **4**, 303 ⟨73CC601⟩
Elliott, R. C., **1**, 36, 486 ⟨77JOC2431⟩, 643 ⟨82JA7041, 82TL5005⟩
Elliott, R. L., **1**, 831 ⟨88T3079⟩; **4**, 208 ⟨94SL152⟩
Ellis, A. J., **2**, 338 ⟨54JCS1750⟩
Ellis, B. A., **5**, 46 ⟨32OSC(1)18⟩
Ellis, B., **2**, 1016 ⟨62JCS4288⟩; **3**, 71 ⟨64TL733⟩
Ellis, C., **4**, 32 ⟨84JFC(26)29⟩
Ellis, D., **1**, 637 ⟨49JCS1074⟩, 968 ⟨93SL499⟩
Ellis, D. E., **1**, 1114 ⟨73JA7916⟩
Ellis, E. W., **4**, 855 ⟨81USP4348529A⟩
Ellis, G. P., **3**, 661, 662, 669 ⟨87CRV779⟩; **5**, 226 ⟨81CHE315⟩
Ellis, I. A., **2**, 416 ⟨70CI(L)499⟩
Ellis, J. E., **2**, 572 ⟨77IC1357⟩; **6**, 522 ⟨81JA6100⟩
Ellis, J. F., **6**, 220 ⟨53JCS1063⟩
Ellis, Jr., H. B., **1**, 154 ⟨87JA5996⟩, 155 ⟨87JA6004⟩
Ellis, K., **2**, 860 ⟨72JCS(P1)1184⟩
Ellis, L. M., **2**, 119 ⟨32JA1674⟩
Ellis, P. S., **5**, 656 ⟨77JHC765⟩
Ellis, R. J., **1**, 795, 801 ⟨64JCS5578, 64JCS959⟩, 805 ⟨63MI 118-02⟩; **4**, 653 ⟨84TL2705⟩
Ellis, W. D., **2**, 619, 626 ⟨65JOC304⟩
Ellison, G. B., **1**, 154 ⟨87JA5996⟩, 155 ⟨87JA6004⟩
Ellison, R. A., **6**, 86 ⟨72JOC2757⟩
Ellman, J. A., **2**, 29 ⟨87TL1123⟩
Ellsworth, E. L., **1**, 207 ⟨88TL893, 90JA4063⟩, 453 ⟨90TL7257⟩; **2**, 589 ⟨90JA4063⟩, 992 ⟨90JA7440⟩, 969 ⟨90TL7257⟩; **3**, 140 ⟨90JA7440⟩
Ellwanger, H., **4**, 71 ⟨54CB300⟩
Ellwanger, R. E., **4**, 209, 213 ⟨74JOC77⟩
Ellzey Jr., S. E., **5**, 592, 609 ⟨63JOC1600⟩
Elmaleln, D. R., **3**, 305 ⟨84JOC2322⟩
Elmasmodi, A., **2**, 742 ⟨87S727⟩
Elmes, B. C., **2**, 384 ⟨80AJC1365⟩
Elmes, P. S., **2**, 507 ⟨70AJC1559⟩; **3**, 550, 552 ⟨87CC573⟩, 614 ⟨79JA6128, 82AJC2041⟩

Elmoghayar, M. R. H., **4**, 984 ⟨86M201⟩; **5**, 532 ⟨84JHC253⟩
Elmore, D., **6**, 598 ⟨56JA4458⟩
Elmore, D. T., **5**, 1048 ⟨58JCS1141⟩
Elmore, S. W., **1**, 810 ⟨91TL319, 92HCA1755⟩
Elmorsy, S. S., **3**, 412 ⟨91TL5421⟩; **5**, 282 ⟨91TL1825⟩
El-Mowafy, A. M., **1**, 30 ⟨84JCS(P1)1671⟩, 629 ⟨81CC96, 82JOC3506⟩
El Murr, N., **2**, 566 ⟨80CC214⟩
Elnagar, H. Y., **1**, 993 ⟨89JA1770⟩
Elnagdi, M. H., **3**, 250 ⟨73IJC128⟩, 641, 669 ⟨83H(20)519, 86H(24)2023⟩, 669,672 ⟨87H(26)497⟩; **4**, 883 ⟨89S775, 77G555⟩; **5**, 676, 677 ⟨82JCED103⟩, 771 ⟨82JCS(P1)2663, 84BCJ1650⟩
El-Nagdy, S., **5**, 652 ⟨90ACH423, 91ACH35⟩
El'natanov, Yu. I., **4**, 481 ⟨82IZV1611⟩; **5**, 397 ⟨75BAU816⟩, 1130 ⟨77IZV249, 82IZV1611⟩
El-Nigumi, Y. O., **2**, 504 ⟨68JINC1715⟩; **4**, 116 ⟨68JINC1715⟩
El'nitskiĭ, A. P., **5**, 208 ⟨59DOK(128)1201⟩
Elofson, R. M., **3**, 363 ⟨64JOC1355⟩
Elokhina, V. N., **3**, 253 ⟨77JOU479⟩; **4**, 942 ⟨90ZOB526⟩; **6**, 565 ⟨90ZOB602⟩
El Ouatib, R., **5**, 1092 ⟨89TL177⟩; **6**, 680 ⟨89TL177⟩
Eloy, F., **3**, 684 ⟨63BSB719⟩; **5**, 219, 833 ⟨62HCA441⟩, 322, 763, 765, 833, 864 ⟨62CRV155⟩, 990 ⟨70BSB301⟩
Elpern, B., **2**, 491, 496, 498 ⟨47JA927⟩; **5**, 221 ⟨47JA875⟩
El Quatib, R., **5**, 886 ⟨93PS(76)41⟩
El-Razak, A., **3**, 482 ⟨81LA1865⟩
El-Sadr, M. M., **3**, 464 ⟨55JA1670⟩
El-Saidi, M., **3**, 324 ⟨92JA8751⟩
Elsäier, A., **2**, 155 ⟨85CB116⟩
Elsässer, A., **6**, 436 ⟨85CB4553⟩
El-Sayad, H. A., **4**, 193 ⟨79JOC4825⟩
El-sayed, G. H., **4**, 105 ⟨71MI 403-01⟩
El Sayed, M., **2**, 945 ⟨93OM2065⟩
El-Sayed El-Kholy, I., **2**, 121 ⟨70JCS(C)1578⟩
Elschenbroich, C., **1**, 885 ⟨B-89MI 119-03⟩; **2**, 880 ⟨87AG(E)229⟩; **4**, 673 ⟨83AG(E)330, 85CC170⟩, 672 ⟨83JA2905⟩, 1039 ⟨87AG(E)229⟩, 1041 ⟨86AG(E)571⟩
El-Seedi, H. R., **5**, 700 ⟨93ACS1004⟩
Elser, W., **4**, 902 ⟨64TL1971, 68OS(48)97⟩
Elsevier, C. E., **4**, 716 ⟨93OM315⟩
Elsevier, C. J., **1**, 466 ⟨85JOC3042⟩, 966 ⟨89JOC3726⟩, 967 ⟨82JOM(240)329⟩, 975 ⟨83JOC1103, 84TL5571⟩, 975, 1004, 1009 ⟨81TL1451⟩; **3**, 599 ⟨93OM3864⟩; **4**, 716 ⟨92ICA689, 93OM3187⟩
Elsey, G. M., **1**, 923 ⟨90CC1589⟩
Elshafie, S. M. M., **5**, 632 ⟨79JPR1034⟩
El-Shamy, H. K., **6**, 214 ⟨53JCS2372⟩
El-Sharabasy, S. A., **5**, 615 ⟨88IJC(B)472⟩
El Sheikh, E., **5**, 924 ⟨79JA83⟩
El Sheikh, M., **3**, 502 ⟨79JA2491⟩; **4**, 930 ⟨79JA83⟩; **5**, 322 ⟨81JOC4188⟩
Elsheimer, S., **4**, 5 ⟨91JOC4322⟩; **6**, 227 ⟨84JOC205, 85DIS(B)840⟩
El-Shekeil, A., **6**, 446 ⟨80JOC5325⟩
Elsinger, F., **5**, 78 ⟨60HCA113⟩
Elslager, E. F., **2**, 373, 378, 381 ⟨64JMC493⟩; **3**, 362 ⟨62JOC4346⟩
Elsmere, S. A., **4**, 208 ⟨94SL152⟩
Elsner, B. B., **1**, 503 ⟨53JCS3156⟩; **2**, 1018 ⟨51JCS893⟩
Elsner, F. H., **4**, 353 ⟨88JA313⟩; **5**, 422 ⟨88JA313⟩
Elsner, G., **2**, 426, 427, 441 ⟨82HOU106⟩, 749 ⟨80GEP2936210⟩
Elsom, L. F., **1**, 491 ⟨76OS(55)48⟩; **2**, 598 ⟨68JOM(15)500⟩
Elston, C. T., **2**, 757 ⟨53JA2014⟩; **5**, 25, 275 ⟨54JA3039⟩
Elsworth, J. F., **5**, 325 ⟨66JCS(C)1477⟩
El-Taweel, F. M. A., **4**, 986 ⟨91JPR333⟩
El-Telbany, F., **2**, 422 ⟨84JOC2438⟩
Elter, G., **2**, 1085 ⟨87CB1437⟩; **6**, 184 ⟨88JOM(347)11⟩
Eltoum, A. O. A., **5**, 681 ⟨93JFC(65)157⟩, 681, 703 ⟨93JFC(65)101⟩
El'tsov, A. V., **1**, 837 ⟨68ZOR1114⟩; **4**, 1017 ⟨86ZOR582⟩; **5**, 601 ⟨70ZOR610⟩; **6**, 599 ⟨69ZOB941⟩
Elvidge, J. A., **4**, 970, 977, 978 ⟨83JCS(P1)1741⟩
Elving, P. J., **1**, 6 ⟨57JA1546⟩, 103 ⟨77CJC3392⟩
Elwahab, L. M., **5**, 201 ⟨82IJC(B)589⟩
El-Wahib, M. F. A., **3**, 464 ⟨55JA1670⟩
El-Wakil, H., **1**, 452 ⟨81SC281⟩; **2**, 630 ⟨87JOC3687⟩
Elwan, N. M., **5**, 675 ⟨90H(31)247⟩, 704 ⟨82H(19)1477⟩
El Wassimy, M. T., **6**, 476 ⟨86CI(L)140⟩, 502 ⟨83T1729⟩
Ely, N., **2**, 602 ⟨76ACR217⟩; **4**, 719, 726 ⟨75IC78⟩
Ely, S. R., **4**, 703 ⟨78IC625⟩
Elyusufi, A. A., **4**, 1013 ⟨90JOM(332)13⟩, 1014 ⟨84OM1660⟩; **5**, 420, 794 ⟨90JOM(391)13⟩
Elz, S., **6**, 632 ⟨88AF7⟩
Elzinga, J., **6**, 7 ⟨80JOC3957⟩
Ema, K., **6**, 636 ⟨84CHEC(3)91⟩
Eman, A., **4**, 238, 239, 283, 375 ⟨75TL1613⟩
Emara, S. A., **5**, 355 ⟨91PS(63)373⟩
Embleton, H. W., **2**, 320 ⟨63OSC(4)795⟩
Emde, H., **2**, 539 ⟨82S1⟩; **3**, 640 ⟨77S636⟩; **4**, 831 ⟨77S867, 83LA816⟩; **6**, 43 ⟨82S1⟩
Emdee, J. L., **1**, 1127 ⟨92JPC2151⟩
Emeléus, H. J., **2**, 165 ⟨46JCS1126⟩, 239 ⟨60JCS1103⟩, 288, 291 ⟨58JCS2939⟩, 292 ⟨63JCS1268⟩, 335 ⟨89MI 206-02⟩, 360 ⟨64JCS5892⟩, 481, 482, 486, 491, 492, 498 ⟨53JCS1552⟩, 486, 492, 505 ⟨57JCS3708⟩, 486, 505 ⟨52JCS2552, 54JCS881⟩, 499, 500 ⟨62JCS2529⟩, 503 ⟨59JCS372⟩, 1014, 1017 ⟨38JCS1358⟩; **3**, 458 ⟨60JCS2614⟩; **4**, 7 ⟨B-69MI 401-01⟩; **5**, 1018 ⟨63JCS1272⟩; **6**, 2 ⟨49JCS2948⟩, 49 ⟨58JCS2939, 62JCS2290⟩, 49 ⟨63JCS1268⟩, 212, 214, 217, 218, 219, 220, 222, 225, 226, 246 ⟨48JCS2188⟩, 212, 244 ⟨B-69MI 607-02⟩, 214 ⟨53JCS2372⟩, 218, 219, 220, 226 ⟨58JCS4245⟩, 233 ⟨52JCS2198⟩, 233, 236, 237 ⟨61JCS2597⟩, 240 ⟨66JCS(A)367, 69JCS(A)431⟩, 241 ⟨66JINC1823⟩, 242 ⟨54JCS3598⟩, 246 ⟨49JCS2953⟩, 409, 419, 422 ⟨48JCS2183⟩, 604 ⟨86JFC(32)415⟩, 607 ⟨66JCS(A)933⟩
Emelius, A. J., **2**, 872 ⟨B-75MI 217-01⟩
Emel'yanova, V. S., **5**, 703 ⟨69JOU252, 69ZOR263⟩
Emer, D., **5**, 566, 618 ⟨57CCA319⟩
Emerson, C. D., **4**, 777 ⟨71USP3592896⟩
Emerson, D. W., **1**, 595 ⟨67JOC102⟩; **5**, 411 ⟨71TL3445⟩, 1040 ⟨60JA3076⟩
Emerson, G. F., **4**, 724 ⟨65JA3254⟩
Emerson, O. H., **5**, 305 ⟨55OSC(3)151⟩
Emerson, T. R., **3**, 406 ⟨65JOC4508⟩
Emerson, W. S., **1**, 629 ⟨63OSC(4)980⟩
Emery, S. E., **1**, 19 ⟨78S763⟩
Emery, T., **2**, 390 ⟨60JA4903⟩, 782 ⟨62JOC1075⟩
Emiliozzi, R., **2**, 126 ⟨59BSF1544⟩, 232 ⟨59BSF1887⟩
Emmerich, Ch., **2**, 896 ⟨93JOM(448)83⟩
Emmerich, G., **3**, 231 ⟨70TL4069⟩
Emmett, J. C., **6**, 647 ⟨77JMC901⟩
Emmick, T. L., **2**, 450 ⟨68JA2459⟩, 840, 841, 845 ⟨68JA3459⟩
Emmons, W. D., **1**, 639, 987 ⟨61JA1733⟩; **2**, 385 ⟨55JA4387⟩, 386 ⟨54JA3468, 55JA6061⟩, 386 ⟨58JOC311⟩, 390 ⟨57JA6522⟩, 397 ⟨55JA4557⟩, 764 ⟨57JA5739⟩; **3**, 565 ⟨62T79⟩, 587 ⟨62JA1316, 64JOC2816⟩; **4**, 312 ⟨73OSC(5)191⟩, 414 ⟨62JOC3664⟩, 421 ⟨61JOC949⟩; **5**, 103 ⟨73OSC(5)547⟩, 205 ⟨73OSC(5)367⟩, 272, 283 ⟨57JA5739⟩, 975 ⟨62JA1316, 67JOC1279⟩, 975, 1031, 1069 ⟨64JOC2816⟩; **6**, 105, 138, 140 ⟨62JOC3664⟩, 144, 145, 369 ⟨63T(S)177⟩, 150 ⟨62T79⟩
Emokpae, T. A., **5**, 1140 ⟨86ICAL17⟩
Emons, H. H., **5**, 560 ⟨88ZC269⟩

Emoto, S., **3**, 101 ⟨74BCJ1687⟩
Emoto, T., **2**, 851 ⟨73BCJ898⟩
Emovon, E. U., **1**, 774 ⟨63JCS1246, 69JCS(B)187⟩
Empfield, J. R., **1**, 147 ⟨91Txi⟩, 573, 581, 584 ⟨92JA9419⟩; **2**, 163 ⟨86JOC829⟩
Emri, J., **5**, 652 ⟨88MI 520-01⟩, 795 ⟨88JOM(344)29⟩; **6**, 491 ⟨88JOM(344)29⟩
Emschwiller, G., **4**, 669 ⟨26CR(183)665⟩
Emsley, J., **2**, 416 ⟨67JCS(A)1918, 67JCS(A)652⟩; **4**, 42 ⟨82JCS(P2)881, 88JCS(P2)1107⟩
Emslie, N. D., **1**, 827 ⟨90SC1437⟩; **4**, 118 ⟨90SC1671⟩
Emziane, M., **2**, 85 ⟨88JOM(346)C7⟩; **3**, 632 ⟨88JMOCL23⟩
Enanoza, R. M., **2**, 205 ⟨75CJC3656⟩
Enas, J. D., **1**, 795 ⟨91JA1355⟩, 1029 ⟨91JA3873⟩
Enayat, E. I., **4**, 303 ⟨90G677⟩
Enda, J., **2**, 932 ⟨81TL3079⟩; **3**, 557 ⟨79CC900, 80JOC5377⟩
Enders, D., **1**, 106 ⟨93CC424⟩, 148 ⟨75AG(E)15⟩, 153 ⟨76TL3⟩, 153 ⟨84T1345⟩, 153 ⟨B-84MI 104-02⟩, 186 ⟨93S1092⟩, 212 ⟨76TL11⟩, 882 ⟨78CB1337⟩; **2**, 331 ⟨93SL401⟩, 380 ⟨87OS(65)173⟩, 385 ⟨74JMC1225, 88OSC(6)542⟩, 385, 420 ⟨75CB1293⟩, 388 ⟨78HCA1622⟩, 759 ⟨91S649⟩; **3**, 10 ⟨B-84MI 301-01⟩, 10, 18, 124, 133 ⟨76TL3⟩, 18, 134 ⟨90SL725⟩, 32, 156 ⟨88TL2437⟩, 34, 163 ⟨93S1013⟩, 35, 35 ⟨94T3349⟩, 44 ⟨93SL226⟩, 51, 202 ⟨87AG(E)351⟩, 60 ⟨76TL7⟩, 124 ⟨B-84MI 304-02⟩, 156 ⟨89HCA980⟩, 172 ⟨93T1821⟩, 183 ⟨82TL639⟩, 204 ⟨93AG(E)598⟩, 226 ⟨76TL11⟩; **4**, 511 ⟨74JMC1225⟩, 530 ⟨72AG(E)301, 75AG(E)15⟩, 649 ⟨76TL7, 78CB1362⟩; **5**, 566, 584, 586, 590, 597, 608, 609 ⟨76CB1309⟩, 597 ⟨73AG(E)1014⟩
Enders, E., **2**, 373 ⟨67HOU(10/2)169⟩; **5**, 842 ⟨67HOU(10/2)378⟩, 843 ⟨65HOU(10/3)538⟩, 844 ⟨67HOU(10/2)467⟩
Endesfelder, A., **2**, 529 ⟨87LA215⟩; **4**, 679 ⟨83JCS(D)2661⟩
Endo, I., **2**, 590 ⟨88CL63⟩
Endo, M., **5**, 1066 ⟨86CL135⟩
Endo, R., **1**, 1044 ⟨72BCJ1846⟩; **6**, 553 ⟨87H(25)123⟩
Endo, S., **3**, 469 ⟨68BCJ2707⟩
Endo, T., **1**, 514 ⟨85CL1851⟩; **2**, 210 ⟨86TL215⟩, 213 ⟨75CL813⟩, 214 ⟨84JOC1434⟩, 215 ⟨67BCJ2388⟩; **3**, 7 ⟨85JOC1332⟩, 684 ⟨94JCS(P1)413⟩; **4**, 181 ⟨90CL2019⟩, 704 ⟨91S368⟩, 825 ⟨88TL2327⟩; **5**, 250 ⟨89S929⟩; **6**, 80 ⟨85JAP(K)60208983⟩, 82 ⟨75BCJ2496⟩, 106 ⟨88TL2327⟩, 298, 299, 300 ⟨84S837⟩, 298, 300 ⟨92MM3829⟩, 561 ⟨88BCJ2203⟩
Endo, Y., **1**, 899 ⟨90TL2623⟩; **2**, 316 ⟨91SL649⟩, 691 ⟨80S461⟩; **3**, 565 ⟨93MI 317-02⟩; **6**, 506 ⟨83S471⟩
Endoh, H., **4**, 419 ⟨90BCJ1160⟩
Endres, G., **5**, 821, 822 ⟨32CB65⟩; **6**, 612 ⟨32CB65⟩
Endrich, K., **4**, 580 ⟨88ZN(B)1293⟩
Enev, V., **1**, 985 ⟨94TL1161⟩
Eng, K. K., **1**, 802 ⟨84TL1255⟩; **2**, 337 ⟨86JOC5045⟩
Engberts, J. B. F. N., **2**, 250, 364, 365 ⟨72TL181⟩, 691 ⟨74JOC2641⟩; **3**, 426 ⟨74S49⟩; **4**, 234 ⟨64TL547⟩, 235 ⟨72TL2477, 77JOC2792⟩, 331 ⟨72RTC209, 74RTC11⟩, 333 ⟨74JOC3215⟩
Engel, C. R., **1**, 470 ⟨78CJC419⟩
Engel, E., **2**, 387 ⟨90HOU(E16a/2)1182⟩
Engel, H., **2**, 274 ⟨64JOC3561⟩
Engel, M., **4**, 700 ⟨84JCR(S)152⟩
Engel, N., **2**, 747 ⟨81TL4263⟩; **5**, 290 ⟨81TL4263⟩
Engel, P., **4**, 852 ⟨89HCA1506⟩
Engel, P. F., **4**, 716 ⟨91CC1274⟩
Engel, P. S., **1**, 330, 924, 925, 944 ⟨80CRV99⟩, 576 ⟨85JA4964⟩, 924 ⟨86JA7935⟩; **6**, 604 ⟨90JOC1503⟩
Engel, R., **2**, 458 ⟨88SC285⟩, 619 ⟨81SC405⟩; **3**, 3, 114 ⟨86SC1343⟩; **4**, 132 ⟨91SC1039⟩, 451 ⟨88OR(36)176⟩
Engelardt, V. A., **4**, 8 ⟨60JA543⟩
Engel, W., **4**, 105 ⟨92AP(325)411⟩
Engelhard, F. J. W., **6**, 4 ⟨35RTC307⟩
Engelhardt, E. L., **2**, 160, 162 ⟨84JMC1634⟩, 613 ⟨81JMC93⟩; **4**, 97 ⟨79JOC3117⟩
Engelhardt, G., **4**, 580 ⟨75JOM393⟩; **5**, 720 ⟨77JOM(134)21⟩; **6**, 43 ⟨67LA(707)35⟩, 159 ⟨72JPR969⟩, 367 ⟨72JPR969⟩
Engelhardt, H. E., **4**, 581 ⟨89JOM(362)117⟩
Engelhardt, L. M., **2**, 1003, 1005 ⟨89JCS(D)105, 84JCS(D)321⟩, 1005 ⟨85JCS(D)337⟩; **4**, 602, 650 ⟨82CC1323⟩, 603, 647 ⟨84JCS(D)311⟩, 604, 646, 647 ⟨84JCS(D)321⟩, 604, 647 ⟨88JCS(D)2403⟩, 626, 662 ⟨87JCS(D)1647⟩, 646 ⟨89JCS(D)105⟩, 646, 647 ⟨85JCS(D)337⟩, 647, 656 ⟨86JCS(D)603⟩, 660 ⟨84JCS(D)331⟩; **5**, 1127 ⟨85JA3910⟩; **6**, 168 ⟨91CC1560⟩, 174, 190 ⟨82CC1323⟩, 196 ⟨87AG(E)681⟩, 196 ⟨88CC336⟩, 196 ⟨93JCS(D)2653⟩
Engelhardt, R., **5**, 848 ⟨1896JPR54⟩
Engelhardt, V. A., **4**, 829, 880, 883, 889, 890, 971, 972 ⟨58JA2788⟩; **5**, 5 ⟨60JA543⟩, 845 ⟨58JA2829⟩; **6**, 19, 39, 52 ⟨60JA543⟩, 245 ⟨58JOC1565⟩
Engels, H. D., **3**, 220 ⟨62CB1495⟩
Engelter, C., **4**, 584, 585 ⟨86JOM(315)255, 82JOM(232)C78⟩
Engemyr, L. B., **3**, 695 ⟨74ACS(A)255⟩
Engen, D. V., **3**, 516 ⟨87OM433⟩
Engenito, Jr., J. S., **4**, 566 ⟨83IC575⟩; **5**, 294 ⟨81JOC4304⟩
England, D. A., **6**, 264 ⟨81CC930⟩
England, D. C., **3**, 545 ⟨81JOC147, 81JOC153⟩, 589 ⟨81JFC(17)265⟩, 590 ⟨73JFC(3)91⟩, 614 ⟨54JA5364⟩; **4**, 5 ⟨60OSC(5)239⟩, 32 ⟨73OSC(5)239⟩, 43, 164 ⟨61JA4670⟩, 44, 86 ⟨81JA5598⟩, 160 ⟨61JA2205⟩, 737 ⟨66JA5582⟩, 761 ⟨68JOC816⟩, 970 ⟨60JA5116⟩; **5**, 56 ⟨73OSC(5)239⟩, 373 ⟨58JA6442⟩; **6**, 15 ⟨60JA6181⟩, 46 ⟨60JA5116⟩, 58 ⟨59JA4801⟩
England, R. C., **4**, 731 ⟨73JFC(3)63⟩
England, W. P., **1**, 807 ⟨92JA2635⟩
Engle, Ch. R., **3**, 537 ⟨60JA3201⟩
Engler, A., **5**, 62 ⟨87OS(66)37⟩
Engler, D. A., **3**, 156 ⟨78JOC188⟩
Engler, E. M., **2**, 68 ⟨72JA4628⟩; **4**, 868 ⟨75CC671⟩, 869 ⟨76TL423⟩, 869, 874 ⟨77CC835⟩, 869, 874 ⟨77JA5909⟩, 874, 875 ⟨76CC148⟩; **6**, 93, 101, 729 ⟨75TL1259⟩, 590 ⟨76CC148, 77JA5909⟩, 590 ⟨83CC235⟩
Engler, J. R., **5**, 546 ⟨72ZAAC(389)145⟩
Engler, R., **1**, 446 ⟨79CB28⟩; **4**, 843 ⟨79CB28⟩
Engler, T. A., **1**, 827 ⟨94TL1661⟩; **3**, 528 ⟨85JA4339⟩; **4**, 647 ⟨84TL4645⟩
Engler, V. R., **5**, 506, 507 ⟨72ZAAC(388)78⟩
Englert, L. E., **3**, 49 ⟨69JOC755⟩
Englert, R. D., **5**, 273, 285 ⟨52JA1010⟩
Englert, U., **3**, 531 ⟨93CB73⟩
Englharat, L. M., **3**, 545 ⟨84CC220⟩
Englin, M. A., **2**, 166, 170 ⟨65ZOB1412⟩; **5**, 807 ⟨68ZOB1408⟩; **6**, 240 ⟨69ZOB199⟩, 242 ⟨61CRV313⟩, 258, 259 ⟨65ZOB1412⟩, 614 ⟨68JGU37⟩, 614, 627 ⟨69JGU183⟩, 242
English, J., **3**, 59 ⟨49JA3310⟩; **5**, 46 ⟨63OSC(4)499⟩; **6**, 522 ⟨47JA1723, 50IS156⟩
English, P. D., **5**, 334 ⟨85JMC1661⟩
English, W. D., **2**, 419 ⟨61JA2122⟩
Englund, B., **4**, 32 ⟨63OSC(4)184⟩
Engman, L., **1**, 595 ⟨81TL1919, 83JOC3918⟩, 610 ⟨87CS515, 87TL1463⟩, 620 ⟨82CC164⟩, 1072 ⟨79TL1509⟩; **2**, 237, 238 ⟨92JA9737⟩, 289 ⟨81JA2715, 81TL1919⟩, 290 ⟨81JOC4194⟩, 616 ⟨87JOC4086⟩, 729 ⟨85JOM(296)357⟩, 735 ⟨90JOM(388)71⟩, 768 ⟨85ACR274⟩; **3**, 175 ⟨89TL2665⟩, 176, 177 ⟨86OM427⟩, 177 ⟨85TL6385⟩; **4**, 88, 89, 93 ⟨89TL2665⟩, 89 ⟨90JOM(391)165⟩, 872 ⟨82JOC3946⟩, 877 ⟨81TL4199⟩; **5**, 242 ⟨80ZN(B)217⟩, 242, 1117 ⟨81JOC4194⟩, 244 ⟨78OPP289⟩, 334 ⟨89JOC2964⟩

Enholm, E. J., **1**, 173, 321, 334 ⟨85T4079⟩, 173, 703 ⟨85JOC4317⟩, 179 ⟨92H(34)2247⟩, 225 ⟨90JOC324⟩, 344, 1130 ⟨85TL3311⟩; **2**, 78 ⟨84TL3927⟩, 81 ⟨89TL1063, 89TL4939⟩, 86 ⟨90JOC324⟩, 311 ⟨90SC981⟩; **3**, 56 ⟨89JOC4345⟩; **6**, 541, 550, 557 ⟨85T4079⟩
Enikeev, K. M., **4**, 132 ⟨91JGU1459⟩, 486 ⟨83IZV432⟩
Enk, E., **3**, 536 ⟨60GEP1081455⟩
Enkelman, V., **4**, 276 ⟨91CB1827⟩
Ennen, J., **6**, 170, 210 ⟨79TL501⟩
Ennis, B. C., **6**, 81 ⟨67JCS(C)30⟩
Ennis, M. D., **1**, 134 ⟨82JA1737⟩, 878 ⟨81PAC1109⟩; **3**, 32 ⟨92JOC1067⟩, 337 ⟨78JA7436⟩; **5**, 148 ⟨81PAC1109⟩
Enokiya, M., **3**, 146, 212, 288, 296 ⟨81CL1135⟩
Enomoto, K., **1**, 813 ⟨85BCJ146⟩; **3**, 338, 339 ⟨89JCS(P1)1241⟩
Enomoto, M., **1**, 954 ⟨86TL4599⟩
Ens, L., **2**, 478 ⟨69JOC940⟩
Ensinger, M. W., **4**, 139 ⟨89JA7199⟩
Ensley, H., **3**, 174, 207 ⟨81TL3043⟩
Ensminger, M., **5**, 1079 ⟨93OM1023⟩
Entenmann, G., **5**, 595 ⟨77CZ508⟩
Enteroth, M., **1**, 212 ⟨83LA226⟩
Entezari-Moghadam, M., **3**, 88, 118 ⟨86SC211⟩, 102 ⟨88BCJ2185⟩, 118 ⟨86SC723⟩
Entezartmoghaday, M., **2**, 215 ⟨88BCJ2185⟩
Entmayr, P., **4**, 741, 747 ⟨76CB2175⟩
Entwhistle, I. D., **1**, 142, 595 ⟨65CC29⟩
Entwistle, D. A., **4**, 236 ⟨93TL5649⟩
Entwistle, D. W., **2**, 107 ⟨77JA1277⟩
Entwistle, E., **1**, 1035 ⟨54JCS147⟩
Entwistle, I. D., **1**, 74, 441 ⟨77JCS(P1)443⟩, 74, 455, 457 ⟨85CRV129⟩; **2**, 291 ⟨76CC61⟩, 379 ⟨82T419⟩; **3**, 13, 83, 93 ⟨80JCS(P1)27⟩
Epa, W. R., **3**, 163 ⟨89TL667⟩; **5**, 142 ⟨94JA7455⟩
Ephritikhine, M., **2**, 675, 676 ⟨78CC694⟩
Epishina, T. A., **5**, 652 ⟨88IZV1691⟩, 669 ⟨87BAU204, 87IZV226⟩, 669 ⟨87JGU1478, 87JGU1479, 87ZOB1658, 87ZOB1659, 88BAU1506, 89BAU672, 89IZV2849, 89IZV745, 89KFZ1317, 90BAU415, 90IZV477⟩, 669, 672 ⟨87BAU1000, 87IZV1086, 89BAU2612, 90BAU1007, 90IZV1122, 89BAU2201⟩, 672 ⟨89IZV2390, 91KFZ33⟩, 672 ⟨92BAU236, 92IZV315⟩, 683 ⟨87BAU1314, 87IZV1422⟩; **6**, 430 ⟨89IZV2849⟩
Epishina, T., **5**, 672 ⟨91JGU1152, 91ZOB1268⟩
Epling, G. A., **1**, 890 ⟨86TL1469⟩; **2**, 83 ⟨92CC1133⟩
Epple, G., **3**, 279, 295, 298 ⟨72AG(E)300⟩, 295, 296, 298 ⟨83CB1195⟩; **5**, 209 ⟨83CB1183⟩
Epple, R., **1**, 58 ⟨55AG710⟩
Eppner, J. B., **3**, 62 ⟨73TL2417⟩
Epprecht, A., **3**, 57 ⟨41HCA1039⟩
Epps, J. L., **1**, 816, 819 ⟨93TL749⟩
Epshtein, G. Y., **2**, 167 ⟨64ZOB1948⟩
Epstein, A. J., **3**, 398 ⟨77CC177⟩
Epstein, E. F., **4**, 715 ⟨70JA502⟩
Epstein, M., **2**, 440 ⟨62T1231⟩, 440, 450 ⟨62T1221, 63USP3116334⟩, 471 ⟨63JA2462⟩
Epstein, W. W., **2**, 133 ⟨91JA3176⟩, 446 ⟨88PS(35)349⟩; **3**, 88 ⟨67CRV247⟩
Epsztajn, J., **5**, 1002 ⟨76TL2691⟩
Epsztein, R., **1**, 999 ⟨73T883⟩, 1041 ⟨85TL3203⟩, 1048 ⟨55CR(240)989⟩
Er, E., **2**, 133 ⟨92HCA2265⟩
Erashko, V. I., **4**, 335 ⟨82IZV161⟩; **6**, 116, 283, 355 ⟨82IZV161⟩, 144 ⟨66UK1740⟩, 145 ⟨68IZV447, 77IZV136⟩, 161, 283, 369 ⟨85IZV439⟩, 281 ⟨75IZV2750⟩, 281, 348 ⟨73IZV350⟩, 283 ⟨77IZV375⟩, 283, 349 ⟨74IZV1350⟩, 283, 349, 355 ⟨89IZV2106⟩, 348, 349, 653 ⟨75IZV1669⟩, 368 ⟨67IZV2708, 73IZV819⟩, 653 ⟨74MI 621-02⟩, 679 ⟨81IZV161⟩
Erasmuson, A., **6**, 472 ⟨81JCS(P1)1293⟩
Erastov, O. A., **2**, 894 ⟨83IZV1152⟩; **4**, 341 ⟨79DOK(244)610, 79IZV866⟩, 341 ⟨80IZV1626⟩, 342 ⟨83IZV1374, 85IZV1102, 86IZV2502, 88IZV159, 89IZV946⟩, 343 ⟨79IZV1863, 86IZV640⟩, 87IZV418⟩, 453, 456 ⟨80IZV2417⟩, 457 ⟨80IZV1438, 80IZV952⟩, 478 ⟨83IZV1379⟩
Erata, T., **2**, 279 ⟨89CC1789⟩, 369 ⟨93JA9826⟩
Erb, W., **5**, 793 ⟨76ZN(B)281⟩
Erbea, R., **5**, 606 ⟨77S250⟩
Erbelding, G., **5**, 1093 ⟨78TL2689⟩
Erber, G., **2**, 561 ⟨85AOC(24)1⟩
Ercolani, C., **5**, 1097 ⟨88CC46⟩
Ercoli, R., **5**, 293 ⟨52JA4496⟩; **6**, 522 ⟨57JA3611, 59G809⟩, 524 ⟨55CI(M)6⟩, 525 ⟨59CI(M)132⟩
Erdbrügger, C. F., **4**, 586 ⟨88OM997⟩
Erdelmeier, I., **1**, 146, 711 ⟨86AG(E)935⟩, 483 ⟨91JA1442, 89JA1125⟩, 763 ⟨85TL4359⟩; **2**, 1000 ⟨86AG(E)938⟩
Erdenmoser, A., **2**, 749 ⟨65CR(261)1992⟩
Erdick, E., **2**, 1095 ⟨92T9577⟩
Erdik, E., **1**, 116, 539, 1094 ⟨84T641⟩, 467 ⟨92T9577⟩; **2**, 305 ⟨89CR1947⟩, 590 ⟨87T2203⟩; **3**, 12, 128 ⟨84T641⟩, 181 ⟨89CRV1947⟩
Erdman, A. A., **2**, 594 ⟨80JOM(201)39⟩
Erdman, J., **1**, 879 ⟨83JOC2957⟩
Erdmann, H., **6**, 412 ⟨1893CB1990⟩
Erdmann, H.-M., **1**, 736 ⟨55CB1415⟩
Erdogan, H., **6**, 561 ⟨92ACR90⟩
Eremeev, A., **5**, 668 ⟨92SC453⟩
Eremeev, A. V., **4**, 793 ⟨84CHE1231⟩; **5**, 661 ⟨91KGS1205, 92CHE579⟩; **6**, 625 ⟨77MI 620-02⟩
Eremenko, I. L., **6**, 341 ⟨90JOM(384)279⟩
Eremenko, L. T., **4**, 108 ⟨89IZV1107⟩, 795 ⟨89BAU635⟩; **5**, 314 ⟨87IZV2857⟩; **6**, 55 ⟨68IZV2307, 68IZV431, 69IZV1331, 73IZV1424⟩, 56 ⟨68IZV912⟩, 145 ⟨69IZV2566⟩, 149 ⟨69IZV1845⟩, 150, 364 ⟨71IZV1594⟩, 250 ⟨93IZV2513⟩, 250, 252 ⟨89IZV113⟩, 261 ⟨68IZV429⟩, 282 ⟨68IZV656, 68IZV676⟩, 284, 364 ⟨74IZV915⟩
Eremin, S. A., **4**, 719 ⟨74IZV2645⟩
Eremov, V. G., **6**, 620 ⟨92ZOB306⟩
Eren, D., **2**, 706 ⟨86JOC3165⟩
Erfort, U., **4**, 385 ⟨83CB1240⟩
Erfurt, G., **4**, 576, 587 ⟨73EGP94992⟩
Ergashev, M. S., **2**, 1056 ⟨87UKZ34⟩
Ergezinger, C., **5**, 781 ⟨88ZN(B)1628, 88ZN(B)1679⟩
Erhardt, J., **4**, 212 ⟨77JOC3458⟩
Erhardt, P. W., **2**, 323 ⟨87SC469⟩
Erhart, M., **6**, 440 ⟨91CB2411⟩
Erian, A. W., **3**, 656, 657 ⟨93CRV1991⟩
Erickson, B. W., **1**, 735 ⟨74JOC821⟩; **3**, 54 ⟨71JA1724⟩
Erickson, D., **2**, 25 ⟨76S389⟩
Erickson, E. H., **5**, 748, 757 ⟨67JMC1123⟩
Erickson, E. R., **6**, 68, 79 ⟨33JA3851⟩
Erickson, G. W., **1**, 152 ⟨81JA3081⟩, 152 ⟨81JA3088⟩; **3**, 124 ⟨81JA3081⟩
Erickson, J. G., **6**, 70 ⟨55JOC1573⟩
Erickson, K. C., **5**, 122 ⟨70TL4011⟩
Erickson, K. L., **1**, 1144 ⟨65JA1142⟩; **2**, 615 ⟨71JOC1024⟩, 622, 625 ⟨65JOC2208⟩
Erickson, R. E., **1**, 561 ⟨61JA927⟩; **2**, 782 ⟨60JA1801⟩; **5**, 115 ⟨73OSC(5)493⟩
Erickson, S. D., **1**, 584 ⟨92JA4128⟩
Erickson, W. F., **1**, 399 ⟨70CC576⟩
Erickson, W. R., **2**, 561 ⟨71TL2183, 81JOC3773⟩
Ericsson, E., **4**, 969, 970, 973 ⟨70ACS3102, 75JST(24)373⟩
Erikova, M. V., **4**, 1056, 1059, 1063 ⟨87ZOB1741⟩
Eriks, K., **1**, 897 ⟨65AX437, 71AX(B)1405⟩
Eriksson, M., **1**, 311 ⟨93JOC7238⟩
Eriyama, Y., **6**, 174 ⟨87AG(E)584⟩
Erk, P., **6**, 304 ⟨88EUP273458⟩
Erker, G., **2**, 858 ⟨88JA624⟩, 968 ⟨83OM1555⟩, 986, 987

⟨79JA3659⟩; **3**, 385 ⟨88JA624⟩, 389 ⟨89AG(E)179⟩, 394 ⟨90AG(E)1067⟩; **4**, 671 ⟨93AG(E)1623⟩, 725 ⟨90JA9620, 91OM3791⟩, 725 ⟨92CB1953, 92JA8531⟩, 725 ⟨92JOM(427)C21, 92OM3517⟩, 725 ⟨93OM4979⟩; **5**, 443 ⟨84ACR103⟩, 797 ⟨85JOM(280)355, 91OM1201⟩, 917 ⟨85OM1310, 85ZN(B)1177⟩, 931 ⟨89AG(E)397⟩
Erle, H.-E., **6**, 620 ⟨79CB640⟩, 620, 621, 623 ⟨80S619⟩
Erlenbach, L., **1**, 43 ⟨58CB2383⟩
Erlenmeyer, H., **4**, 21 ⟨54HCA90⟩; **5**, 365 ⟨56HCA1253⟩
Ermakova, I. V., **6**, 240 ⟨69ZOB199⟩, 614 ⟨68JGU37⟩, 614, 627 ⟨69JGU183⟩
Ermakova, M. I., **5**, 641 ⟨59ZOB1333⟩; **6**, 653 ⟨64ZOB2855⟩
Erman, M. B., **2**, 1093 ⟨71JGU122⟩
Erman, W. F., **1**, 725 ⟨70TL41⟩, 725, 1001 ⟨67TL401⟩
Ermer, O., **1**, 248 ⟨87AG(E)928, 88AG(E)411⟩
Ermer, S., **4**, 683 ⟨80JOM(187)81⟩
Ermili, A., **3**, 361 ⟨81JHC863⟩; **4**, 891 ⟨71G269⟩, 979 ⟨82FES387⟩
Ermolenko, M. S., **3**, 130 ⟨83TL4355⟩, 211 ⟨83TL4359⟩; **5**, 239 ⟨83TL4355⟩
Ermolenko, N. V., **6**, 12 ⟨69IZV196⟩
Ermolov, A. F., **6**, 42 ⟨86ZOR1842⟩
Ermolova, T. I., **2**, 1101 ⟨67DOK(175)1079⟩
Erner, W. E., **5**, 1009, 1037 ⟨64JOC2091⟩
Ernest, I., **3**, 65 ⟨82TL167⟩; **5**, 905 ⟨78JA8214⟩
Ernst, B., **1**, 547 ⟨89TL57⟩, 834 ⟨89TL3517, 91AG(E)1465⟩, 837 ⟨93S729⟩; **3**, 602 ⟨85JA2192⟩; **4**, 239 ⟨89TL6311⟩, 440 ⟨67ZC421⟩
Ernst, B. P., **1**, 275 ⟨92S461⟩
Ernst, F., **2**, 112 ⟨59CB2716⟩
Ernst, M., **2**, 441, 442 ⟨70CB318⟩
Ernst, R. D., **2**, 564, 602 ⟨82COMC-I(3)173⟩; **4**, 678 ⟨92OM3617⟩, 681 ⟨94OM2423⟩
Ernst, T. D., **2**, 650 ⟨89JOC1204⟩
Erokhina, T. S., **4**, 451, 469, 472, 473, 475, 493 ⟨74RCR984⟩, ⟨77JOC376⟩ ⟨77JGU2501⟩
Erokina, V. M., **2**, 454 ⟨62MI 208-01⟩
Erol, K., **6**, 561 ⟨92ACR90⟩
Errede, L. A., **1**, 849 ⟨60JA3652⟩
Errington, K. D., **1**, 78 ⟨37JCS1136⟩
Errington, R. J., **2**, 512 ⟨90JOM(399)C21⟩
Ershova, M. M., **5**, 781 ⟨87MI 519-02⟩
Erskine, G. J., **2**, 566 ⟨84JOM(267)265⟩, 874 ⟨68ICA379⟩
Erson, A. D., **6**, 131 ⟨88JOM(348)317⟩
Ertas, M., **2**, 699 ⟨85HCA961⟩; **5**, 690 ⟨85LA1910⟩
Erte, H., **3**, 480 ⟨61TL9⟩
Ertel, H., **4**, 490 ⟨61TL9⟩, 575, 576, 577 ⟨61CB1987, 62CB581⟩
Ertel, K., **6**, 156 ⟨78MIP87868⟩
Esch, P. M., **1**, 978 ⟨92T3445⟩
Eschenmoser, A., **1**, 414, 415, 827 ⟨64HCA2425⟩, 415 ⟨69HCA1030⟩, 566 ⟨75HCA153⟩, 683 ⟨75HCA1450⟩, 827, 828 ⟨79CT494⟩, 906 ⟨55CI(L)415⟩, 1082 ⟨67HCA2101, 67HCA2108, 71HCA2896, 72HCA1276, 76OS(55)114⟩; **2**, 380 ⟨88OSC(6)56⟩; **3**, 25, 147 ⟨67HCA708⟩, 26, 147 ⟨72HCA1276⟩, 228 ⟨53HCA482⟩, 268 ⟨71HCA353⟩, 659 ⟨69AG(E)343⟩; **4**, 882 ⟨64GEP1183513⟩; **5**, 78 ⟨60HCA113⟩, 131 ⟨63AG(E)211⟩, 749 ⟨86HCA1224⟩; **6**, 104 ⟨63AG296⟩
Escher, S. D., **1**, 618 ⟨82TL5043⟩
Eschonmoss, A., **2**, 824 ⟨71HCA710⟩
Eschwey, M., **6**, 253, 254, 530 ⟨83CB1623⟩
Escobar, G., **3**, 417 ⟨84JCR(S)144⟩; **4**, 304 ⟨84JCR(M)144⟩
Escoula, B., **2**, 19 ⟨86TL1499, 89BSF256⟩
Escudero, J., **3**, 602 ⟨81AG(E)879⟩
Escudié, J., **1**, 656 ⟨90CRV283⟩, 766 ⟨91OM1771⟩; **2**, 473, 474 ⟨82PS(12)265⟩, 473, 474 ⟨84PS(20)81⟩, 474 ⟨82PS(12)279⟩, 866 ⟨83TL3625⟩, 1093 ⟨72JOM(34)83⟩; **3**, 501 ⟨90CRV283, 94CCR427⟩, 503 ⟨87JA4411, 91JOC7356⟩; **4**, 455, 481, 998 ⟨75HCA1316⟩, 485 ⟨91JOM(415)327⟩, 549 ⟨93CC569⟩, 574 ⟨70CR(C)351, 70JOM(24)633⟩, 647 ⟨92JOM(440)233⟩, 662 ⟨87JA4411⟩, 663 ⟨92OM3176⟩; **5**, 720 ⟨75HCA1316⟩, 927 ⟨90CRV283⟩; **6**, 167 ⟨87PS(31)81⟩, 186, 188 ⟨91POL1153⟩, 187, 709, 724 ⟨90CRV283⟩, 265, 266 ⟨78BSF(2)361⟩, 373 ⟨82TL4941, 84CC1622⟩, 373, 688 ⟨89CC593⟩, 374, 375 ⟨83TL2769, 83TL3625⟩, 375 ⟨84JAP59193897⟩, 582 ⟨79TL3507⟩, 669 ⟨72JOM(34)83⟩, 709 ⟨94CCR427⟩, 724 ⟨92OM2748⟩
Escudie, N., **1**, 651 ⟨84JOM(264)207⟩
Escudieoé, J., **2**, 949 ⟨87JA4411⟩
Escudier, M.-C., **5**, 595 ⟨93PS(78)97⟩
Eshelby, J. J., **1**, 227 ⟨93SL277, 93SL279⟩; **2**, 936 ⟨93SL277, 93SL279⟩
Eshelman, J. E., **1**, 816, 819 ⟨93TL749⟩
Eshenmoser, A., **6**, 104 ⟨65HCA1746⟩
Eshtiagh-Hosseini, H., **5**, 1152 ⟨79CC653⟩; **6**, 679 ⟨79CC653⟩
Esipenko, A. A., **5**, 669 ⟨89UKZ1071, 90JOU287⟩, 669 ⟨91UKZ202⟩, 669, 672 ⟨90ZOR339⟩
Esipenko, A. N., **4**, 272 ⟨91ZOR216⟩, 275, 277 ⟨89UKZ1216⟩, 865 ⟨89ZOR1102⟩
Esker, J. L., **1**, 927, 950 ⟨92TL5913, 93JOC4933⟩, 950 ⟨93TL6877⟩; **4**, 1077 ⟨91TL1035⟩; **5**, 334 ⟨93TL6877⟩, 655, 712 ⟨91TL1035⟩
Esko, P., **4**, 555, 557 ⟨91MI 412-01⟩
Esmay, D. L., **6**, 284 ⟨69USP3450762⟩, 362 ⟨72USP3663621⟩
Esmond, R., **5**, 179 ⟨82JOC3358⟩
Espil, L., **1**, 87 ⟨14BSF228⟩
Espinet, P., **2**, 579 ⟨82COMC-I(6)279⟩; **4**, 584 ⟨85JOM(290)125⟩; **5**, 1138 ⟨84ICA215⟩
Espinola, J. M., **2**, 236 ⟨91JOC3520⟩
Espinosa, A., **5**, 350 ⟨93H(36)1263⟩; **6**, 560 ⟨89S923⟩, 563 ⟨88S690⟩
Ess, R. J., **1**, 942 ⟨56JA6414⟩
Essawi, M., **3**, 638 ⟨80SC495⟩
Es-Sayed, M., **2**, 309 ⟨93TL4193⟩
Essenfeld, A. P., **1**, 749 ⟨84TL2183⟩; **5**, 155 ⟨84TL2183⟩
Esser, H., **2**, 403 ⟨56CB685⟩
Esser, W. H., **4**, 177 ⟨77RTC44⟩
Esslinger, M. A., **5**, 586, 609 ⟨89JOC1212⟩
Esteban-Calderon, C., **2**, 1002 ⟨82AX(B)1124⟩
Estermann, H., **2**, 315 ⟨87TL3103⟩
Esteve, Jr., R. M., **1**, 883 ⟨61JA1951⟩
Estham, J. F., **2**, 964 ⟨70C109⟩
Estreicher, H., **1**, 671 ⟨78JA6294⟩; **2**, 393, 396, 783 ⟨78JA6294⟩, 397 ⟨80TL1117⟩, 665, 784, 785 ⟨81TL603, 80TL1113⟩; **3**, 215, 220 ⟨78JA6294⟩
Eswarakrishnan, S., **1**, 415, 827, 828 ⟨85JOC5909⟩, 828, 834 ⟨87JA6716⟩, 876 ⟨B-88MI 119-01⟩; **6**, 9 ⟨B-91MI 601-03⟩
Eswarakrishnan, V., **1**, 619, 707, 949 ⟨86JA4568⟩, 707 ⟨83JA6165, 84TL5469⟩; **2**, 187 ⟨81JOC2025⟩; **4**, 82, 83, 85 ⟨86JA4568⟩, 223 ⟨86JOC3428, 89JA658⟩
Eswaran, S. V., **2**, 163 ⟨85JHC701⟩
Eszenyi, T., **2**, 624 ⟨91TL827⟩
Et, A. L., **3**, 71 ⟨64TL733⟩, 76 ⟨84JOC1288⟩
Etayer, C., **3**, 73 ⟨72BSF4576⟩
Etemad, S., **4**, 869, 874 ⟨77JA5909⟩; **6**, 590 ⟨77JA5909⟩
Etemad-Moghadam, G., **1**, 68, 454 ⟨79S891⟩, 640 ⟨84T5153⟩, 747 ⟨85BSF(2)448⟩; **3**, 608 ⟨91TL3687⟩; **4**, 463 ⟨94S51⟩, 547 ⟨86TL1145, 87T1793⟩, 1025 ⟨91TL3687⟩; **5**, 886 ⟨93PS(76)41⟩, 1092 ⟨89TL177⟩; **6**, 264 ⟨87T1793, 91TL5965⟩, 680 ⟨89TL177⟩, 688 ⟨90JOC5750⟩
Etheridge, S. J., **1**, 603 ⟨85JA3891⟩, 1023 ⟨68CC634⟩
Etienne, A., **2**, 175 ⟨76BSF483⟩, 176 ⟨73BSF210⟩, 248

⟨76BSF277⟩, 788 ⟨69BRP1167809⟩, 887 ⟨45CR(221)562⟩; **5**, 499 ⟨78JCR(M)5501, 78JCR(S)484⟩; **6**, 142 ⟨75CR(C)563⟩, 143 ⟨77CR(C)321⟩
Etienne, M., **5**, 1097 ⟨91JA2324⟩
Etienne, R., **4**, 30 ⟨78HCA2482⟩
Etling, H., **3**, 703, 714 ⟨61AG26⟩, 716 ⟨65CB193⟩
Étlis, V. S., **4**, 891 ⟨71ZOB482⟩; **6**, 324, 335 ⟨63BAU1384⟩, 493 ⟨68ZOR1661⟩, 628, 629 ⟨64JGU4149⟩
Etschenberg, E., **2**, 260 ⟨84GEP3229565⟩
Ettel, M. L., **6**, 514 ⟨91IC2433⟩
Ettel, V., **2**, 440 ⟨60CCC2191⟩
Ettenhuber, E., **2**, 813 ⟨68CB743⟩; **4**, 515 ⟨68CB743⟩
Ettlinger, M., **3**, 603 ⟨82CB161⟩
Ettlinger, M. G., **5**, 636, 831 ⟨57JA1764⟩, 1024 ⟨55JA1831, 56JOC204⟩
Ettouati, L., **2**, 642 ⟨91T9823⟩
Etzel, W. A., **6**, 414 ⟨90MI 614-01⟩
Eubanks, J. R. I., **1**, 629 ⟨91JA8821⟩
Eue, L., **6**, 234 ⟨71GEP2003143⟩
Euerby, M. R., **1**, 436 ⟨86SC779⟩
Eugster, C. H., **3**, 103 ⟨78HCA430⟩, 245 ⟨57HCA2462⟩
Eujen, R., **6**, 51 ⟨80JFC(17)65⟩, 62 ⟨78JA1722⟩, 65 ⟨94JOM(465)153⟩, 245 ⟨92JOM(433)63, 92JOM(434)159⟩, 245 ⟨92JOM(438)57⟩, 717 ⟨86IS52⟩
Eul, W., **6**, 553 ⟨87ZAAC(545)125⟩, 561 ⟨82ZAAC(485)203⟩
Euske, J. M., **4**, 224 ⟨88JOC845⟩; **5**, 540 ⟨88JOC845⟩
Eustace, K. A., **3**, 292 ⟨90T1791⟩
Eustache, J., **3**, 334 ⟨80JOC2601⟩; **5**, 554 ⟨80JOC2601⟩
Eustache, J. M., **5**, 554 ⟨82JOC4384⟩
Evain, E. J., **2**, 303 ⟨89S1⟩; **5**, 1063 ⟨89JOC1048⟩
Evana, D. A., **2**, 680 ⟨80JA3270⟩
Evangelidou-Tsolis, E., **4**, 336 ⟨74PS(4)109⟩
Evans, A. R., **5**, 325 ⟨87JCS(P1)1635⟩
Evans, B. E., **1**, 13 ⟨77S172⟩; **4**, 97 ⟨79JOC3117⟩, 417 ⟨87JOC3232⟩
Evans, C. A., **4**, 229 ⟨92TL4625⟩
Evans, C., **1**, 4, 9 ⟨91JCS(P1)656⟩; **4**, 650 ⟨91JCS(P1)2276⟩
Evans, C. J., **1**, 331, 342 ⟨B-85MI 108-01⟩
Evans, D. A., **1**, 84 ⟨84JA3866⟩, 98 ⟨91JOC741⟩, 129 ⟨B-84MI 104-01⟩, 134 ⟨82JA1737⟩, 136 ⟨74JA5560⟩, 140, 143, 609, 788 ⟨74ACR147⟩, 278, 487, 662, 1103 ⟨76JOC3947⟩, 299 ⟨93CRV1307⟩, 353 ⟨91JA726⟩, 413, 810 ⟨75JA4765⟩, 414 ⟨80JA774⟩, 651 ⟨78JA1548⟩, 703 ⟨86JA2476⟩, 767, 769 ⟨80JA3270⟩, 815 ⟨78JA8170⟩, 822 ⟨72TL5121, 77JA5453⟩, 873, 876 ⟨82TS1⟩, 875 ⟨79JA6120, 80TL3975⟩, 876 ⟨80TL4233⟩, 878 ⟨81JA2127, 81PAC1109, 82MI 119-01, 82TL807, B-84MI 119-02⟩; **2**, 29 ⟨87TL1123⟩, 42 ⟨88JA6917, 90JOC2280⟩, 43 ⟨92JA4042, 93JOC5307⟩, 46 ⟨90JOC1698⟩, 56 ⟨93JA9800⟩, 62 ⟨88JA3560⟩, 63 ⟨90JOC5190⟩, 152, 363 ⟨92JA5977⟩, 154, 196 ⟨74ACR147⟩, 341 ⟨86TL4961⟩, 375 ⟨86JA6395a⟩, 527 ⟨91JA4042, 92JA6671⟩, 532 ⟨92T2127⟩, 532, 698 ⟨79JA6120, 81JA3099, B-82MI 210-01⟩, 541 ⟨77JA5009⟩, 554, 593 ⟨88JA2506⟩, 696 ⟨B-84MI 213-01⟩, 702 ⟨90JA866⟩, 703 ⟨80TL3975⟩; **3**, 11 ⟨79JA371⟩, 32 ⟨92JA9434, 92JOC1067⟩, 36 ⟨74JA5560⟩, 124 ⟨B-84MI 304-01⟩, 127 ⟨90JOC5678⟩, 182 ⟨78CC753⟩, 474 ⟨76JOC3335⟩, 496 ⟨78JA1548⟩, 615 ⟨73CC55, 74JOC914⟩, 615, 616 ⟨73TL4929⟩; **4**, 216 ⟨77JA5009⟩, 339 ⟨78JA3467⟩, 832 ⟨81JA3099⟩, 917 ⟨79JA371⟩; **5**, 80 ⟨89OS(68)83⟩, 148 ⟨81PAC1109⟩, 229 ⟨86TL5683⟩, 277 ⟨69TL1573⟩, 329 ⟨86TL3119, 86TL4957, 86TL4961⟩, 553 ⟨77JA5453⟩, 1134 ⟨74JOC914⟩
Evans, D., **2**, 19 ⟨88JCS(P1)549⟩; **6**, 457 ⟨68IS99⟩
Evans, D. E., **3**, 493 ⟨90JA7001⟩
Evans, D. F., **2**, 564 ⟨71JCS(A)1931⟩
Evans, E. A., **1**, 727 ⟨56JCS4691, 57CI(L)1596⟩
Evans, E. L., **4**, 412 ⟨74JOC167⟩
Evans, E. R., **2**, 707 ⟨66JOC410⟩
Evans, F. J., **1**, 450 ⟨54JA1613⟩; **5**, 421 ⟨54JA1372⟩
Evans, G., **4**, 269 ⟨92JOC6390⟩
Evans, G. L., **2**, 130 ⟨52JA3982⟩
Evans, J. A., **5**, 465 ⟨73CC158⟩
Evans, J., **1**, 669 ⟨74JOM(81)C37⟩; **6**, 343 ⟨89CC1029, 89OM1270⟩, 344 ⟨91JCS(D)1017⟩
Evans, J. C., **1**, 893, 899 ⟨64JA1360⟩, 1037 ⟨92JOC2941⟩
Evans, J. S. A., **2**, 830 ⟨85JA5210, 86JOC5490⟩
Evans, M. L., **5**, 863 ⟨70GEP2011937⟩
Evans, M. M., **4**, 75 ⟨74JOC643⟩
Evans, P. B., **2**, 101 ⟨38JA538⟩
Evans, R., **5**, 840 ⟨25JCS1304⟩
Evans, R. D., **2**, 18 ⟨87S551⟩; **3**, 155 ⟨86S727⟩
Evans, R. M., **3**, 215 ⟨59QR61⟩
Evans, S. A., **1**, 74 ⟨83JOC1638⟩; **2**, 215 ⟨80JOC710⟩
Evans, S., **2**, 478 ⟨69JOC940⟩; **3**, 436 ⟨86JOC5024⟩
Evans, T. E., **5**, 496, 497 ⟨72JOC2080⟩
Evans, T. L., **2**, 150 ⟨86SC1207⟩
Evans, T. W., **5**, 74 ⟨30JA3647⟩
Evans, W. J., **1**, 890 ⟨81JOC3925⟩; **2**, 564 ⟨81CC292⟩, 969 ⟨84JA1291⟩; **3**, 545 ⟨85JA3728⟩; **4**, 703 ⟨93OM2618⟩, 704 ⟨93OM3998⟩, 726 ⟨93JOM(462)141⟩; **5**, 439 ⟨86JA1722⟩, 440 ⟨81CC706⟩
Evdakov, V. P., **2**, 411 ⟨62ZOB3070⟩, 451 ⟨62ZOB1974, 63ZOB1246⟩, 461 ⟨62ZOB3062⟩
Even, L., **2**, 118, 124 ⟨90BSF758⟩
Everett, J. H., **5**, 263 ⟨93JCS(P1)451⟩
Everett, W., **2**, 166 ⟨46JOC469⟩; **6**, 304 ⟨83MI 610-01⟩
Everhardus, H., **2**, 271 ⟨78S359⟩
Everhardus, R. H., **2**, 907 ⟨78RTC69⟩; **4**, 956, 957 ⟨78RTC69, 77CC801⟩
Everitt, G. F., **3**, 570 ⟨72JOM(35)403⟩; **5**, 465 ⟨73CC158⟩
Evers, E. C., **5**, 1136 ⟨59JA5106⟩
Evers, M., **5**, 239 ⟨76JCS(P1)2452⟩
Evers, M. J., **2**, 632 ⟨85JOC1779⟩; **5**, 334 ⟨92TL3863⟩
Evertz, K., **3**, 320 ⟨88TL1677⟩; **4**, 936, 950, 951 ⟨87CB213⟩
Evgenios, D. M., **3**, 351 ⟨73JCS(P1)2866⟩
Evin, G., **5**, 267 ⟨75TL1219⟩
Evitt, E. R., **2**, 978 ⟨80JA7003⟩
Evleth, J. E. M., **2**, 821 ⟨62JOC2192⟩
Evnin, A. B., **3**, 314, 318 ⟨70PAC523⟩
Evrard, G., **2**, 287 ⟨85CC569⟩, 554 ⟨92TL3381⟩; **4**, 1074, 1075 ⟨84JA3785⟩; **6**, 92 ⟨84JA3784⟩
Evstafeeva, N. E., **2**, 949 ⟨78ZOB639⟩
Evstaf'ev, G. I., **5**, 652 ⟨74JGU2371⟩, 737 ⟨73DOK(211)113⟩
Evstigneeva, R. P., **2**, 617 ⟨79S533⟩
Evtikov, Zh. L., **2**, 454 ⟨68DOK(181)877, 69ZOB2367⟩
Eweiss, N. F., **2**, 11 ⟨77S634⟩, 11, 632 ⟨79JCS(P1)433⟩; **5**, 358 ⟨75BCJ365⟩
Ewers, C. L. J., **3**, 422 ⟨91COS(2)1007⟩
Ewing, D. F., **5**, 580, 615 ⟨78JCS(P1)629⟩, 690 ⟨93T1053⟩
Ewing, J. C., **1**, 543 ⟨85JOC2124⟩
Exall, A. M., **2**, 19 ⟨88JCS(P1)549⟩
Exner, O., **5**, 332 ⟨74AG(E)419⟩, 655 ⟨89CCC3245⟩, 657 ⟨72JCS(P2)1908⟩, 665 ⟨72JCS(P2)1911⟩, 665, 672 ⟨85JOC993⟩, 765 ⟨83CCC596⟩, 828, 832 ⟨68CJC1873⟩, 1000 ⟨74AG(E)376⟩
Exon, C., **2**, 985 ⟨83JA2477⟩; **3**, 270 ⟨83JA2477⟩
Extine, M., **2**, 421 ⟨76JA4477⟩
Eyber, G., **6**, 523 ⟨29MI 616-01⟩
Eyer, M., **1**, 156 ⟨86CB575⟩; **2**, 394 ⟨85JA3601⟩; **3**, 191 ⟨85JA3601⟩
Eylander, C., **1**, 999 ⟨74RTC240⟩

Eyles, C. T., **3**, 494 ⟨66JCS67⟩
Eyley, S. C., **1**, 1130 ⟨88CC283⟩; **2**, 1006 ⟨91COS(2)707⟩; **3**, 186 ⟨90SL619⟩, 282, 687 ⟨85TL4649⟩
Eyman, D. P., **3**, 14, 82, 100, 102 ⟨84JOC2279⟩
Eymann, W., **1**, 1111 ⟨76TL3507⟩, 1114, 1115, 1124 ⟨74JA6686⟩
Eynde, J.-J. V., **2**, 310 ⟨89JCS(P1)639⟩
Ezekiel, A. D., **3**, 355 ⟨82BRP2081705⟩
Ezhova, G. I., **2**, 22 ⟨92BAU356⟩
Ezhova, M. B., **2**, 886 ⟨88JGU1934⟩; **5**, 222 ⟨88ZOB1934⟩
Ezmirly, S. T., **1**, 401, 414, 800 ⟨77TL2559⟩
Ezquerra, J., **1**, 826 ⟨90TL6979⟩
Ezzell, B. R., **1**, 53 ⟨74MI 102-01⟩

F

Faas, U., **6**, 105 ⟨58JPR70⟩
Fabbri, D., **4**, 272, 854 ⟨91SL565⟩, 277, 854 ⟨92T1485⟩; **6**, 556 ⟨93JOC1748⟩
Faber, E. M., **2**, 132 ⟨43OSC(2)576⟩
Faber, K., **1**, 19 ⟨83CJC2415⟩; **5**, 118 ⟨91TL341⟩
Faber, P. M., **3**, 712 ⟨94TL9351⟩
Faber, T., **1**, 929 ⟨93JOC1215⟩
Fabian, J., **2**, 130 ⟨62CB1764⟩; **3**, 342 ⟨64AG(E)277⟩; **4**, 245 ⟨62CB1764⟩; **5**, 559 ⟨86JPR(328)812⟩
Fabian, W., **1**, 396 ⟨61LA(642)1⟩
Fabian, W. M. F., **5**, 685 ⟨91MI 517-02⟩
Fabiano, E., **2**, 309 ⟨87S190⟩
Fabina, N. G., **5**, 718 ⟨84JGU426, 84ZOB477⟩
Fabio, P. F., **4**, 161 ⟨85JOC5022⟩
Fabisch, B., **4**, 152 ⟨84JOM(269)219⟩, 656, 663, 700, 724 ⟨84JOM(269)249⟩
Fabre, G., **3**, 50 ⟨81T1377⟩
Fabre, J.-L., **1**, 439, 700 ⟨83TL4311⟩, 485 ⟨85BSF762, 85BSF772⟩, 485, 700 ⟨82TL2469, 83TL4319⟩; **4**, 847 ⟨87BSF823⟩; **6**, 592 ⟨89CC1520⟩
Fabrias, G., **1**, 747 ⟨84T2871⟩; **2**, 1092 ⟨90TL2739⟩; **6**, 695 ⟨83TL3387⟩
Fabricius, D. M., **1**, 808 ⟨79JA1793⟩
Fabrissin, S., **4**, 273, 274 ⟨91JOC3530⟩
Fabry, L., **4**, 1047 ⟨78JOM(161)299⟩
Faburada, A. L., **1**, 879 ⟨83JOC2957⟩
Facchin, G., **3**, 712 ⟨85JOM(294)C21⟩, 723 ⟨84IC961⟩; **5**, 801, 802 ⟨81JOM(206)C11⟩
Fachinetti, G., **5**, 441 ⟨74JOM(71)C5, 77JCS(D)2297⟩, 442 ⟨76CC522⟩, 443 ⟨77JCS(D)1946⟩, 444 ⟨74CC516, 76JCS(D)203⟩; **6**, 341 ⟨79CC396, 79CC397, 80AG411, 81AG215, 81AG94, 87AG681⟩
Fackler, J. P., **2**, 434 ⟨90IS305⟩; **4**, 586 ⟨87OM1992, 88IC836⟩, 587 ⟨88OM2415, 91OM3392⟩, 152 ⟨85CC1278⟩; **6**, 332 ⟨87OM283⟩
Fadal, R., **1**, 84 ⟨87BSF462⟩
Fadeev, M. A., **6**, 250 ⟨93IZV2513⟩, 250, 252 ⟨89IZV113⟩
Fadel, A., **3**, 123 ⟨87S37⟩, 539 ⟨83T1567⟩; **4**, 302 ⟨85HCA1243⟩
Fadel, H. A., **5**, 351 ⟨73IJC735⟩
Fagan, P. F., **2**, 949, 992 ⟨88JA2310⟩
Fagan, P. J., **2**, 600 ⟨78JA4834⟩, 602 ⟨81JA6650⟩, 988 ⟨88JA7128⟩; **3**, 598 ⟨86JA56⟩; **4**, 679 ⟨82JA1134⟩, 680 ⟨88JA6070⟩, 1038 ⟨88JA2310⟩; **5**, 469 ⟨78JA7112, 80JA5393⟩; **6**, 516 ⟨81JA2206⟩
Faggi, C., **5**, 417 ⟨88JOC3612⟩, 467 ⟨89JOC2966⟩
Faggiani, R., **2**, 1007 ⟨86JA1692, 86JA1694⟩, 1008 ⟨83JOC3043, 86JA3613⟩
Fagone, F. A., **1**, 1143 ⟨72JA7862⟩
Fahey, D. R., **5**, 461 ⟨77JA2501⟩
Fahey, J. L., **5**, 274, 286 ⟨73JOC3437⟩; **6**, 113 ⟨73JHC791⟩
Fahey, R. C., **1**, 892 ⟨58JA2326⟩, 896 ⟨63JA2245⟩
Fahmi, A. A., **5**, 676 ⟨84JHC797⟩; **6**, 654 ⟨85JHC813⟩
Fahmy, A. A., **6**, 117 ⟨91PS(57)211⟩
Fahmy, A. F. M., **5**, 351 ⟨73IJC735⟩
Fahmy, M. A. H., **2**, 252 ⟨84USP4450158⟩
Fahmy, S. M., **3**, 653 ⟨83S478⟩; **4**, 984 ⟨77G555⟩; **5**, 365 ⟨83S478⟩
Fahr, E., **3**, 475 ⟨64LA(675)47⟩; **6**, 448 ⟨70AG(E)372⟩, 506 ⟨66AG376⟩
Fahrenholtz, K. E., **3**, 234 ⟨70JOC4141⟩
Fahrmann, U., **2**, 610 ⟨81CB4005⟩
Fahrni, H.-P., **2**, 1049 ⟨78HCA1609⟩; **5**, 280 ⟨78HCA2428, 78HCA2437⟩, 280 ⟨85OS(63)175⟩
Fahse, R., **2**, 201 ⟨76S829⟩
Fain, D., **1**, 883 ⟨90JOC1052⟩
Fainberg, A. H., **1**, 1114, 1116 ⟨57JA4146⟩; **4**, 34 ⟨57JA4170⟩, 755 ⟨57JA4164⟩; **6**, 2 ⟨58JOC322, 61JA2495⟩, 5 ⟨58JA851⟩, 64, 225, 226, 245 ⟨57JA4159⟩
Faingor, B. A., **5**, 1141 ⟨68IZV289⟩
Fainzilberg, A. A., **2**, 334, 365 ⟨91T7447⟩; **4**, 335 ⟨82IZV161⟩, 995 ⟨91IZV1849⟩, 1083 ⟨89IZV356⟩; **5**, 314, 384 ⟨91T7447⟩; **6**, 261, 283 ⟨70IZV2553⟩, 281 ⟨75IZV2750⟩, 281, 348 ⟨73IZV350, 71IZV1073⟩, 283 ⟨72IZV1721, 77IZV2530⟩, 283 ⟨77IZV375⟩, 283, 349 ⟨74IZV1350⟩, 287 ⟨67DOK(176)1086⟩, 323 ⟨75IZV2621, 76IZV2640, 78IZV1091⟩, 368 ⟨67IZV2708⟩, 653 ⟨74MI 621-02⟩, 56 ⟨70IZV387⟩, 56, 282 ⟨71IZV1487⟩, 56, 323 ⟨77IZV139⟩, 116, 283, 355 ⟨82IZV161⟩, 134 ⟨81IC3734⟩, 144 ⟨66UK1740⟩, 145 ⟨68IZV447, 77IZV136, 90IZV1816⟩, 149 ⟨68IZV621, 72IZV1717⟩, 261, 281, 323 ⟨78IZV2348⟩
Fair, C. K., **2**, 602 ⟨86JA40⟩; **4**, 656, 658 ⟨86JA40⟩
Fair, T. W., **2**, 163 ⟨51JA459⟩
Fairchild, C., **1**, 939 ⟨93JOC5028⟩
Fairfax, D. J., **3**, 321 ⟨93JOC4646⟩
Fairfull, A. E. S., **5**, 579 ⟨52JCS742⟩
Fairhurst, R. A., **2**, 316 ⟨90TL4229⟩; **3**, 186 ⟨90SL619⟩
Fairlie, D. P., **5**, 385 ⟨90IC20, 90ICA203⟩, 385, 386 ⟨91IC1564⟩, 780 ⟨90IC140⟩
Faita, G., **1**, 826 ⟨93T2093⟩
Faitondjieva, K. B., **5**, 770 ⟨92MI 519-01, 92MI 519-02⟩
Fajardo-Pinzon, B., **6**, 74 ⟨45JA650⟩
Fajkos, J., **3**, 234 ⟨58CCC2155⟩
Fakhrutdinova, R. A., **4**, 463 ⟨88ZOB2456⟩
Fakley, M. E., **5**, 1140 ⟨77JCS(D)1444⟩
Falardeau, E. R., **6**, 256 ⟨76JA3529⟩
Falb, E., **6**, 486 ⟨93SC2839⟩
Falbe, J., **1**, 296, 297 ⟨B-80MI 107-01⟩; **5**, 16 ⟨B-80MI 501-01⟩; **6**, 36, 43, 51, 57 ⟨85HOU(E5)⟩
Falck, J. R., **1**, 179 ⟨85SC663⟩, 179, 183 ⟨93JOC5892⟩, 183 ⟨92JOC3757⟩, 183 ⟨92TL6723⟩, 580 ⟨86TL267⟩, 590 ⟨75JOC812, 76JOC3218, 80OS(59)202⟩, 1026 ⟨89TL6027⟩, 1134 ⟨78JA4618, 78JA4620⟩; **2**, 6 ⟨85SC663⟩, 446 ⟨84TL5903⟩, 629 ⟨78JA4618⟩; **4**, 175 ⟨94TL1825⟩; **5**, 173 ⟨78JA4618, 79JA7131⟩, 173, 204 ⟨78JA4620⟩, 368 ⟨88SC545⟩; **6**, 511 ⟨88SC545⟩
Falcone, S. J., **5**, 1117 ⟨80JOC1044⟩
Fales, H. M., **2**, 326 ⟨88JOC2968⟩
Faleschini, P., **4**, 683 ⟨92OM154⟩, 711 ⟨94OM1999⟩
Falius, H., **2**, 852 ⟨81ZAAC(480)90⟩
Falk, J., **3**, 598 ⟨87JA2049⟩; **5**, 422 ⟨87JA2049⟩; **6**, 521 ⟨87JA2049⟩
Falk, R. A., **5**, 1011 ⟨69JOC4088⟩
Falkenberg, J., **5**, 605, 616, 634 ⟨68CB1428⟩
Falkof, M. M., **2**, 30 ⟨55OSC(3)350⟩
Faller, A., **1**, 654, 713, 716 ⟨87CC106⟩; **2**, 916 ⟨94SL77⟩
Faller, J. W., **2**, 76 ⟨89JA1937⟩; **3**, 244 ⟨89TL1189⟩, 615 ⟨93TL2275⟩
Faller, P., **3**, 119 ⟨82JOC364⟩
Fallert, M., **4**, 984 ⟨86AP(319)1098⟩
Falling, S. N.,**4**, 29 ⟨75SC33⟩
Fallis, A. G., **1**, 774 ⟨90JA4609, 93JOC2186⟩, 810 ⟨91CJC77⟩; **2**, 93, 138 ⟨87TL523⟩, 141 ⟨88TL897⟩
Fallon, G. D., **1**, 988 ⟨87AJC1675⟩; **3**, 660 ⟨85CC4⟩; **4**, 689 ⟨85OM355⟩
Falmagne, J.-B., **1**, 280 ⟨85CC1168⟩; **3**, 602 ⟨81AG(E)879⟩

Falorni, M., **2**, 9 ⟨86JOC5291⟩, 56 ⟨89JOC2383⟩
Falou, S., **2**, 759 ⟨77TL3589⟩
Fält-Hansen, B., **2**, 359 ⟨92AHC(55)1⟩; **3**, 342 ⟨91S609⟩
Falvello, L. R., **4**, 692 ⟨88OM2279, 93JA4145⟩
Famulok, M., **5**, 212 ⟨89AG(E)337, 89AG(E)468⟩
Fan, J. Y., **2**, 200 ⟨71JOC2703⟩
Fan, R., **1**, 806, 811 ⟨92OR(41)1⟩
Fan, W.-Q., **1**, 260 ⟨91JOC4439⟩, 261 ⟨90JOC3209⟩, 261 ⟨90S1173⟩, 262 ⟨93JOC4376⟩; **3**, 42 ⟨90S1173⟩, 257 ⟨88JHC1287⟩, 260 ⟨91SC1307⟩, 709, 710 ⟨93S45, 91JOC4439⟩, 710 ⟨91S868⟩; **4**, 309 ⟨91JOC4439, 92JOC547⟩, 309 ⟨94S445⟩, 309, 323 ⟨93S229, 91S1147⟩, 446 ⟨93S45⟩; **5**, 625 ⟨94SC1761⟩
Fanaii, G., **5**, 1057 ⟨84S512⟩
Fañanás, F. J., **1**, 1090 ⟨88CC1135, 88TL4859, 89JCR(S)200, 90CC1521, 90CC587, 91CC1001, 92TL831⟩; **2**, 553 ⟨88TL2859⟩
Fanchiang, Y.-T., **2**, 584 ⟨79JA1442⟩
Fandos, R., **4**, 579 ⟨89OM1604⟩; **5**, 797 ⟨91OM2665⟩
Fanelli, J., **2**, 398 ⟨76JOC3328⟩
Fang, F. G., **5**, 572 ⟨90JOC831⟩, 631 ⟨89TL2743⟩
Fang, J.-M., **1**, 620 ⟨89JCR(S)275⟩; **3**, 636 ⟨87TL2853⟩, 638 ⟨90JCS(P1)3365⟩; **4**, 263 ⟨90JOC5515⟩, 849 ⟨82JA7174, 87JOC855⟩; **6**, 127 ⟨87JOC855⟩, 134 ⟨92MI 604-01⟩
Fang, Q., **2**, 927, 945 ⟨89JOC2793⟩, 969 ⟨88TL3445⟩
Fang, Y., **5**, 385 ⟨89JOC1584⟩
Fang, Z., **3**, 436 ⟨94MI 311-01⟩
Fanghaenel, E., **4**, 870 ⟨83S840, 84EGP211356A1⟩, 874 ⟨79ZC192⟩
Fanghanel, E., **3**, 255, 356 ⟨75JPR137⟩, 351 ⟨76JPR127⟩; **4**, 850 ⟨75JPR137⟩; **6**, 113, 304, 306, 316 ⟨76JPR127⟩, 301, 303, 309, 310 ⟨64ZC384, 65ZC386⟩, 500 ⟨79JPR827⟩, 630, 731 ⟨84S667⟩, 729 ⟨82JPR669⟩
Fanizzi, F. P., **5**, 386 ⟨93JA5123⟩, 696 ⟨89JCS(D)947⟩
Fanming, M., **4**, 488 ⟨B-93MI 402-02⟩
Fannes, C. C., **5**, 661 ⟨92TL2049⟩
Fannes, C. I., **5**, 661 ⟨92S705⟩
Fanshawe, W. J., **6**, 655 ⟨67JMC273⟩
Fanta, A. D., **6**, 125 ⟨91OM3466⟩
Fanta, P. E., **1**, 637 ⟨46JA502⟩; **2**, 398 ⟨63OSC(4)844⟩; **5**, 106 ⟨55OSC(3)267⟩, 652 ⟨55OSC(3)661⟩
Fantin, G., **2**, 942 ⟨84CC258⟩; **3**, 32 ⟨89TL6063⟩, 311 ⟨84JOC3478, 84JOC590⟩, 707, 714 ⟨87S693, 84CC258⟩; **6**, 634 ⟨87JOC3413, 87S693⟩, 637 ⟨90H(31)1213, 93H(36)473⟩
Fanwick, P. E., **4**, 583 ⟨92OM1392⟩, 709 ⟨92OM2128⟩, 709, 712 ⟨90OM1345, 91OM3003⟩, 713 ⟨90IC1295⟩, 714 ⟨91IC4710⟩
Fany, J. M., **1**, 828 ⟨82JA7174⟩
Faraci, W. S., **4**, 528 ⟨89B3541⟩
Farag, A. M., **3**, 681 ⟨87JHC577⟩; **5**, 676, 703 ⟨88OPP521, 87JHC577⟩
Faragher, R., **2**, 779 ⟨77JCS(P1)1196⟩; **3**, 379 ⟨77JCS(P1)1196⟩; **4**, 307, 799 ⟨79JCS(P1)249⟩
Faraglia, G., **2**, 695, 886 ⟨72JOM(38)91⟩
Farah, B., **4**, 32 ⟨63JOC2494⟩
Farah, B. S., **2**, 166, 173 ⟨60OS(40)62⟩, 166, 173, 174 ⟨61JOC1996⟩, 166, 173, 232 ⟨59JOC973⟩, 174, 187 ⟨58JOC330⟩, 688 ⟨63JOC2807⟩; **4**, 18 ⟨65JOC1241⟩
Farah, D., **4**, 700 ⟨92JOM(429)311⟩
Farakawa, S., **4**, 888, 900 ⟨69YZ203⟩
Faraone, A., **3**, 234 ⟨70JOC3045⟩
Farardeau, E. R., **2**, 449 ⟨75IC571⟩
Farazi, V., **2**, 439 ⟨90JOM(393)143⟩
Farazmand, I., **2**, 182, 364 ⟨57JCS2574⟩
Farben, I. G., **2**, 748 ⟨39BRP511430⟩
Farbenfabrik Bayer, **2**, 803 ⟨07GEP206455⟩; **6**, 484 ⟨1899GEP114396⟩
Farbenindustrie, (I. G., **6**, 232 ⟨37FRP820795, 38BRP479774⟩
Farbman, M. D., **2**, 21 ⟨55OSC(3)188⟩
Farcasiu, D., **1**, 914 ⟨82ACR46⟩
Farenhorst, E., **2**, 403 ⟨53RTC993⟩
Farge, D., **4**, 229 ⟨87JMC24⟩; **5**, 558 ⟨87JMC24⟩
Farid, P., **1**, 1076 ⟨87CC711⟩
Farid, S., **1**, 181 ⟨63JCS4522⟩
Fariña, F., **3**, 68 ⟨77S642⟩, 658 ⟨88S136⟩; **6**, 124 ⟨91T3171⟩
Farina, J., **2**, 8 ⟨86JOC2637⟩
Farina, J. S., **1**, 1136 ⟨76JOC3213⟩
Farina, P. R., **2**, 381 ⟨75JOC1070⟩
Farina, R., **3**, 291 ⟨83OM467⟩
Farina, V., **1**, 21, 25, 437, 438, 943 ⟨80JA4438⟩, 496 ⟨92TL3563⟩, 939 ⟨93JOC5028⟩, 1004 ⟨90JOC5833⟩
Farines, M., **4**, 213 ⟨70BSF332⟩
Farinholt, L. H., **3**, 451 ⟨30JCS96⟩; **5**, 841 ⟨30JCS94⟩
Farkas, E., **3**, 226 ⟨60JA2402⟩
Farkas, I., **6**, 71 ⟨90LA499⟩
Farkas, M., **5**, 609 ⟨82LA1582⟩
Farley, C. E., **2**, 460, 463 ⟨67JOC236⟩, 840, 841, 854 ⟨67T1065⟩
Farlow, M. W., **4**, 14 ⟨43OSC(2)312⟩; **6**, 212 ⟨55USP2709186, 55USP2709187, 55USP2709188, 55USP2709190, 58GEP1040011⟩, 408, 410 ⟨60IS155⟩, 414 ⟨63OSC(4)524⟩
Farmer, D. M., **5**, 1110 ⟨80T1839⟩
Farmer, H. H., **5**, 37 ⟨63OSC(4)441⟩
Farmer, J. B., **3**, 459 ⟨72JCS(D)1501, 76CC160⟩
Farmer, L., **3**, 37 ⟨90T8267⟩
Farmer, M. L., **2**, 800, 801 ⟨65JA874⟩; **3**, 263 ⟨65JA874⟩
Farmery, K., **3**, 460 ⟨70JCS(A)2279⟩
Farney, R. F., **4**, 392 ⟨78JA5428⟩
Farnham, T., **1**, 613 ⟨77JOC1349⟩
Farnham, W. B., **1**, 906 ⟨69JOC2314⟩, 918 ⟨75JOC700⟩; **2**, 619 ⟨82JA4680⟩, 845, 846 ⟨71CC146⟩, 855 ⟨71JA3792⟩; **4**, 42 ⟨85JA4565⟩; **6**, 28 ⟨B-82MI 601-02⟩, 229 ⟨85JA4565⟩
Farnia, S. M. F., **4**, 697, 698 ⟨88JA3231⟩
Farnier, M., **3**, 76 ⟨89JCS(P1)2095⟩, 77 ⟨88SC1475⟩
Farnoux, C. C., **5**, 742, 748, 751, 754, 755 ⟨B-75MI 519-01⟩
Farnow, H., **1**, 812 ⟨59LA(625)206⟩
Farnsworth, D. W., **3**, 116 ⟨89SC215⟩
Farnum, D. G., **1**, 894, 895 ⟨75APO(11)123⟩, 896 ⟨64JA934⟩, 906 ⟨78JA3847, 83JOC2146⟩, 907 ⟨60JA2651, 67JA5048b⟩; **2**, 152 ⟨77TL4009⟩, 404 ⟨71JA3816⟩; **3**, 446 ⟨63JOC870⟩
Farnung, W., **4**, 913, 918, 925 ⟨84CB2622⟩, 1003, 1004 ⟨84CB1, 77S336⟩
Faro, H. P., **2**, 13 ⟨65JCS2438⟩
Faron, K. L., **1**, 551 ⟨87JOM(334)9⟩; **2**, 607 ⟨86JOC277⟩
Farona, M. F., **5**, 389 ⟨69ICA495⟩
Farooq, O., **1**, 693 ⟨89JOC1375⟩; **2**, 907 ⟨90JOC46⟩, 941 ⟨89S568⟩; **4**, 697, 698 ⟨88JA3231⟩; **5**, 8 ⟨89S566⟩, 992 ⟨90JOC4282⟩; **6**, 45 ⟨89S463⟩
Farooqui, F., **2**, 235 ⟨89JOC1092⟩; **3**, 337 ⟨89JOC1092⟩
Farooqui, T. A., **3**, 300 ⟨76TL3217⟩
Farr, F., **2**, 688 ⟨12LA(391)55⟩
Farr, F. R., **2**, 440, 441 ⟨92JOC511⟩
Farr, J. P., **4**, 691 ⟨81JA3764⟩
Farr, R. A., **1**, 253 ⟨81JOC3293⟩
Farrall, M. J., **3**, 4, 115 ⟨78JOC2618⟩, 4, 115 ⟨81JOC1728⟩, 373 ⟨79TL203⟩
Farrar, V. A., **5**, 332, 333 ⟨91JMC387⟩
Farrar, W. V., **3**, 723 ⟨60JCS3058⟩; **4**, 85, 86 ⟨60JCS3058⟩; **6**, 254, 255 ⟨56JCS508⟩
Farreira, J. T. B., **1**, 1035 ⟨86CC1067⟩
Farrow, M. D., **3**, 723 ⟨24JCS(125)2543⟩
Farrugia, L. J., **6**, 339, 341 ⟨87OM819⟩, 339, 346, 347 ⟨86OM1103⟩, 343 ⟨83JOM(249)273, 86JOM(310)67⟩, 347 ⟨87CC147⟩

Farschtschi, N., **2**, 410 ⟨88TL6843⟩
Farthing, A. C., **5**, 972 ⟨50JCS3213⟩
Farwaha, R., **4**, 150 ⟨89JCS(D)761⟩, 151 ⟨88JCS(D)1773⟩
Fassbender, F.-J., **5**, 1066 ⟨86CB2114⟩
Fasseur, D., **5**, 227 ⟨89TL3073⟩
Fastabend, U., **4**, 1080 ⟨93CC1241⟩
Fataftah, Z. A., **1**, 873, 874, 876 ⟨80JA3959⟩; **2**, 696 ⟨80JA3959⟩
Fatiadi, A. J., **2**, 155 ⟨87S85⟩; **3**, 87 ⟨76S65⟩, 641, 642, 647, 656, 657, 658, 661, 662, 669 ⟨B-83MI 319-01⟩, 656 ⟨86S249⟩; **4**, 168 ⟨87S85⟩; **5**, 28 ⟨74S229⟩, 48 ⟨76S133, 87S85⟩
Fatiadi, J., **3**, 57, 215 ⟨76S133⟩
Fatmi, N., **5**, 324 ⟨92BSB257⟩
Fatome, M., **5**, 388, 389 ⟨88EJM257⟩
Fattorusso, E., **6**, 107 ⟨74JHC529⟩
Fatutta, S., **2**, 783 ⟨66TL1821⟩; **5**, 355 ⟨91G401⟩
Faucher, ,O., **2**, 732 ⟨71IJS(A)18⟩
Faul, D., **2**, 1053 ⟨88TL5355⟩
Faul, M. M., **1**, 353 ⟨91JA726⟩; **2**, 152, 363 ⟨92JA5977⟩
Faulkner, D. J., **1**, 414, 416, 827 ⟨70JA741⟩; **2**, 35 ⟨68JA6225⟩; **3**, 217 ⟨69TL3243⟩, 699 ⟨77JA7367⟩; **5**, 159 ⟨70JA741⟩; **6**, 604 ⟨77JA7367, 78TL1391⟩, 604 ⟨78TL1395⟩
Faulla, W., **3**, 269 ⟨79JCR(S)240⟩
Faull, A. W., **5**, 1035 ⟨79JCR(S)148⟩; **6**, 574 ⟨79JCR(S)240⟩
Faulques, M., **2**, 787, 789 ⟨82S260⟩
Faulstich, H., **2**, 7 ⟨77LA806⟩
Fauq, A. H., **1**, 221, 478 ⟨83JOC4131⟩; **5**, 1052 ⟨91SL654⟩
Fauran, C., **5**, 754 ⟨71BCF478⟩, 756 ⟨70BCF200⟩
Faure, A., **6**, 309 ⟨76S449, 79CAR(74)127⟩, 309, 310 ⟨78S286⟩
Faure, M., **2**, 1071 ⟨69CR(268)358⟩
Faure, R., **1**, 955 ⟨93CC270, 93JOC5702⟩; **4**, 432 ⟨94TL89⟩
Fauss, R., **5**, 1142 ⟨84GEP3430019⟩; **6**, 502 ⟨83GEP3135111⟩
Faust, J., **2**, 1094, 1095 ⟨B-83MI 223-01, B-83MI 223-02, B-86MI 223-01, B-87MI 223-01⟩, 1094 ⟨B-87MI 223-02, B-87MI 223-03, B-87MI 223-04⟩; **3**, 361 ⟨75TL2683⟩; **5**, 532 ⟨76ZC270⟩
Faust, Y., **2**, 19 ⟨79TL1823⟩
Fausto, R., **5**, 506, 507 ⟨89JCS(F2)1945⟩, 506, 508 ⟨90JST67, 87JST381⟩
Fautin, G., **3**, 109 ⟨87S998⟩
Fauvarque, J.-F., **1**, 155 ⟨68CR(C)1162⟩, 467 ⟨81JOM(209)109⟩, 471 ⟨79JOM(177)273⟩
Fava, A., **1**, 821 ⟨78JOC4826, 79JOC4128⟩, 821 ⟨81JOC3315⟩; **2**, 143 ⟨81JOC3315⟩, 156 ⟨88JOC5689⟩, 195 ⟨70JA3035⟩, 362 ⟨67CC985⟩, 688 ⟨73JA2918⟩; **5**, 1038 ⟨61JA2729, 69JOC4032⟩, 1038, 1105 ⟨65JA4791⟩
Favara, D., **1**, 503, 505 ⟨82JOC4254⟩
Favier, R., **2**, 40 ⟨71CJC2586⟩
Favini, G., **1**, 1144 ⟨63G757⟩
Favorskaya, I. A., **1**, 1022 ⟨78JOU1095⟩
Favorskii, A. E., **1**, 391 ⟨1894MI 109-01⟩
Favre, A., **5**, 277 ⟨80JA7578⟩
Favre, B., **4**, 194 ⟨76JA3037⟩
Favre, E., **1**, 1055 ⟨74JOM(76)297, 74JOM(76)305⟩
Favre, H., **2**, 675 ⟨64CJC212⟩
Favrel, F., **5**, 843 ⟨27BSF1603⟩
Favrel, G., **5**, 843 ⟨02CR1312, 27BSF1494⟩
Favret, G., **2**, 326 ⟨91S447⟩
Fawcett, F. S., **5**, 1003 ⟨62JA4275⟩; **6**, 52, 53, 238, 259, 409, 423, 438 ⟨62JA4275⟩
Fawcett, J., **4**, 960 ⟨93JCS(P1)67⟩
Fawcett, J. P., **4**, 687 ⟨74JCS(D)1856⟩
Fawwaz, R. A., **5**, 1063 ⟨72JOC3372⟩
Fawzi, R., **1**, 445 ⟨86JCE930⟩; **2**, 490 ⟨86JCS(D)997⟩; **4**, 112 ⟨86OM2030⟩; **5**, 943 ⟨92JOM(441)63⟩; **6**, 154 ⟨93AG(E)985⟩
Faye, M., **3**, 601 ⟨77JOC4261⟩
Fayed, S., **3**, 302 ⟨82SC1121⟩
Fayos, J., **1**, 613 ⟨77JOC1349⟩
Fazakerly, F. G., **2**, 564 ⟨71JCS(A)1931⟩
Fazely, F., **6**, 489, 490 ⟨90BBR(172)288⟩
Fazio, M. J., **1**, 232 ⟨83JOC494⟩
Feageson, E., **2**, 807 ⟨58JA2686⟩
Feakes, D. A., **6**, 491 ⟨92S380⟩, 514 ⟨91IC2433⟩
Fearnley, S. P., **1**, 791 ⟨94TL1071⟩
Fearon, F. W. G., **2**, 941 ⟨B-69MI 218-02⟩; **4**, 650 ⟨70JOC1308⟩
Feasey, N. D., **4**, 675 ⟨86CC542⟩
Feast, W. J., **3**, 514 ⟨90JA8378⟩
Feaster, J. E., **5**, 354 ⟨92JA6266⟩
Feather, J. A., **1**, 869, 871 ⟨65JCS(B)1752⟩
Feather, P., **2**, 1016 ⟨62JCS4288⟩
Fechner-Schramm, H., **2**, 387 ⟨69LA(725)1⟩
Fedder, J. E., **5**, 995 ⟨65JOC4306⟩; **6**, 57 ⟨65JOC4303⟩
Fedechkina, V. A., **4**, 1000, 1001 ⟨72ZOB1727⟩
Federov, B. P., **6**, 83 ⟨60IZV1828⟩, 111 ⟨70BSF2013⟩
Fedin, E. I., **1**, 870 ⟨67TL4525⟩; **2**, 876 ⟨75IZV897⟩; **4**, 551 ⟨80TL2931⟩; **5**, 385 ⟨72JOM(36)227⟩
Fedoreyev, S. A., **6**, 138 ⟨89T3487⟩
Fedorov, B. P., **6**, 111 ⟨60MI 604-01⟩, 111, 140 ⟨65IZV2179⟩
Fedorov, B. S., **6**, 150, 364 ⟨71IZV1594⟩, 284, 364 ⟨74IZV915⟩
Fedorov, L. A., **4**, 152 ⟨73JOM(57)423⟩; **6**, 64 ⟨73JOM(57)423⟩
Fedorov, V. E., **4**, 706 ⟨70IZV2133⟩
Fedorova, A. V., **1**, 953 ⟨64RCR1⟩
Fedorova, A. V. J., **1**, 1125 ⟨63JGU3508⟩
Fedorova, G. B., **5**, 694 ⟨76JOU1789, 76ZOR1826⟩
Fedorova, G. K., **2**, 463 ⟨82JGU195⟩, 847 ⟨74ZOB85⟩; **5**, 708 ⟨63JGU1006⟩, 860 ⟨63ZOB1017⟩
Fedorova, T. M., **5**, 168 ⟨73CC693⟩
Fedorovich, A. D., **1**, 248 ⟨83RCR61⟩, 794 ⟨81RCR666⟩
Fedorynski, F., **6**, 233 ⟨74S274⟩
Fedorynski, M., **6**, 55, 280, 633 ⟨81JOC5048⟩, 280, 614 ⟨86TL419⟩
Fedot'eva, B. V., **3**, 483 ⟨77JOM(142)155⟩, 533 ⟨75IZV199, 76ZOB930⟩; **4**, 628 ⟨76IZV1887⟩; **6**, 185 ⟨77JOM(142)155⟩
Fedot'eva, I. B., **3**, 483 ⟨77JOM(142)155⟩, 533 ⟨75IZV199, 76ZOB930⟩; **4**, 628 ⟨76IZV1887⟩; **6**, 185 ⟨77JOM(142)155⟩
Fedotova, V. V., **4**, 561 ⟨64ZOB2897⟩
Fedyuk, G. S., **2**, 399, 400 ⟨72ZOR2057⟩
Feely, W., **3**, 89 ⟨57JOC1135⟩
Feger, H., **2**, 539 ⟨82S1⟩
Feger, H.,**3**, 436 ⟨86LA428⟩
Feger, H., **6**, 43 ⟨82S1⟩
Feger, W., **4**, 719 ⟨92JOM(436)333⟩; **6**, 338 ⟨90ZN(B)447⟩
Feher, F., **2**, 212 ⟨67AG(E)703⟩, 360 ⟨58CB2528⟩, 361 ⟨63CB1131⟩
Fehér, F., **2**, 853 ⟨68ZN1030⟩
Feher, F. J., **2**, 962 ⟨84JA1650⟩
Feher, M., **2**, 473 ⟨83AG(E)775⟩; **4**, 1025 ⟨84CB1542⟩
Fehlhaber, H. W., **3**, 433 ⟨83SC649⟩
Fehlhammer, W., **6**, 732 ⟨74AG(E)599⟩
Fehlhammer, W. F., **3**, 680 ⟨87CB2031⟩
Fehlhammer, W. P., **3**, 519, 707 ⟨74AG(E)599⟩, 694 ⟨93CRV1243⟩, 696, 710 ⟨80AG(E)480⟩, 699 ⟨75AG(E)369⟩, 699, 703 ⟨79AG(E)75⟩, 702 ⟨88CB461, 91TL5539⟩, 709 ⟨77AG(E)707⟩, 716 ⟨89JOM(379)97, 93TL8457⟩, 726 ⟨82CC184⟩; **4**, 492 ⟨85AG(E)979⟩, 1017 ⟨75JOM(97)61, 82ZN(B)1044⟩, 1019 ⟨82JOM(224)153⟩; **5**, 658 ⟨89ZN(B)1414⟩, 799 ⟨82JOM(224)153⟩, 904, 906 ⟨85AG(E)979⟩, 983

⟨67AG(E)169⟩, 1087 ⟨87CB461, 88AG(E)1344, 89CB1907⟩, 1088, 1089 ⟨88JOM(347)101, 82CC184⟩, 1089 ⟨87JOM(331)193⟩, 1147 ⟨89ZN(B)419⟩; **6**, 606, 619, 721 ⟨88AG(E)1344, 89CB1907⟩, 675 ⟨82COMC-I(4)523⟩
Fehlner, T. P., **4**, 726 ⟨91OM2726⟩
Fehn, J., **4**, 311, 493 ⟨73AG(E)502, 71CB1826⟩
Fehnel, E. A., **2**, 133, 135 ⟨49JA84⟩
Fehr, C., **1**, 185 ⟨89HCA1537⟩, 582 ⟨92HCA1023⟩, 1082 ⟨79HCA2655⟩; **3**, 130 ⟨87HCA1745⟩, 147 ⟨79HCA2655⟩, 218 ⟨86HCA228⟩; **4**, 166 ⟨81HCA1247⟩
Fehr, L., **5**, 1026 ⟨59LA(621)1⟩
Fehsenfeld, F. C., **1**, 1143 ⟨75JCP(63)3238⟩
Feiccabrino, J. A., **5**, 390 ⟨74JOM(73)319, 75JOM(93)325⟩
Feichtinger, H., **2**, 175 ⟨63CB3068⟩
Feigel, M., **1**, 1088 ⟨88JA6033⟩
Feigelson, G. B., **1**, 648, 760 ⟨88JOC3391⟩; **5**, 631 ⟨89TL2743⟩
Feigerle, C. S., **1**, 1143 ⟨83CPL124⟩
Feighery, W. G., **6**, 343 ⟨91OM2384⟩
Feigl, F., **5**, 864 ⟨34MI 521-01, B-37MI 521-01⟩
Feiken, N., **4**, 723 ⟨93OM2227⟩
Feiler, L. A., **4**, 829 ⟨74CC405⟩
Fein, M. M., **1**, 168 ⟨63IC1111⟩
Feinauer, R., **6**, 104 ⟨71S16⟩, 106 ⟨73AG(E)996⟩, 107 ⟨66AG(E)848, 66AG(E)894⟩, 107 ⟨66LA(698)174, 71AG(E)810⟩, 110 ⟨70LA(742)128⟩, 629 ⟨66CB1912⟩
Feingold, M. H., **2**, 386 ⟨79JOC1881⟩
Feinstein, I., **3**, 80 ⟨80JOC4059⟩; **6**, 415 ⟨80JOC4059⟩
Feiring, A. E., **2**, 15 ⟨77JFC(10)375, 79JOC1252⟩, 28 ⟨88OSC(6)991⟩; **5**, 592, 593 ⟨76JOC148⟩; **6**, 229, 230 ⟨79JOC2907⟩
Feiser, L. F., **4**, 209 ⟨51JA5252⟩
Feisst, J., **2**, 170 ⟨66HCA2344⟩
Feist, F., **6**, 252 ⟨22AG489⟩
Feith, B., **4**, 975 ⟨84SC1073⟩
Feitler, D., **5**, 869 ⟨74TL1391⟩
Fekete, F., **4**, 569 ⟨60USP2920094, 62USP3019248⟩
Felbert, G., **5**, 1054 ⟨66ACS278⟩; **6**, 597 ⟨66ACS281⟩, 598 ⟨66ACS278⟩
Felcht, U., **2**, 861 ⟨77LA1309⟩; **3**, 480 ⟨75CB2040⟩
Felcht, U.-H., **2**, 406 ⟨82HOU(E2)798⟩
Feld, R., **2**, 95 ⟨56JCS4669⟩
Fel'dblyum, V. S., **1**, 294 ⟨68RCR789⟩
Felder, P. W., **2**, 887 ⟨86ICA(113)43⟩
Feldhues, M., **1**, 428 ⟨90S403⟩, 934 ⟨85T4213⟩; **5**, 207 ⟨85T4195, 85T4213⟩, 208 ⟨86T1285⟩
Fel'dman, D. P., **5**, 224 ⟨87ZOB686⟩
Feldman, I. K., **3**, 247 ⟨52JGU278⟩
Feldman, J., **1**, 675 ⟨87MM1169⟩; **3**, 508 ⟨91MI 315-01⟩, 515 ⟨90OM2262⟩
Feldman, K. S., **1**, 231, 949 ⟨90TL823⟩, 928 ⟨88JA3300, 93JOC6851⟩, 949 ⟨93SL827⟩, 1077 ⟨93JA3846⟩; **3**, 315 ⟨83TL5585⟩; **5**, 572 ⟨94JOC1129⟩
Feldman, P. L., **3**, 419 ⟨86S735⟩
Feldmann, R., **1**, 910 ⟨70TL1941⟩
Feldstein, G., **1**, 979 ⟨77JOC353⟩
Feldthus, A., **2**, 685 ⟨91JCR(S)300⟩
Felicetti, M., **5**, 616 ⟨94TL1977⟩
Felici, M., **1**, 546 ⟨84TL3137⟩
Felix, D., **1**, 414, 415, 827 ⟨64HCA2425⟩, 415 ⟨69HCA1030⟩, 1082 ⟨67HCA2101, 71HCA2896⟩, 1082 ⟨72HCA1276, 76OS(55)114⟩; **2**, 380 ⟨88OSC(6)56⟩; **3**, 25, 147 ⟨67HCA708⟩, 26, 147 ⟨72HCA1276⟩; **4**, 882 ⟨64HCA2425⟩
Felix, G., **2**, 608 ⟨77AG(E)488, 79AG(E)402⟩
Félix, G., **2**, 945 ⟨79JOC4275⟩
Felix, R. A., **1**, 490, 1106 ⟨71TL4701⟩
Felker, D., **2**, 538, 540 ⟨80S861⟩
Felkih, A., **3**, 408 ⟨85TL4603⟩
Felkin, H., **1**, 316 ⟨69TL707⟩, 389 ⟨65JCS2513⟩, 397 ⟨69TL821, 77TL3485⟩, 477 ⟨77JOM(127)371⟩, 554 ⟨84TL1279⟩, 1094 ⟨75T2735⟩; **2**, 314 ⟨50CR(230)304⟩, 561 ⟨66TL875, 75CC242⟩, 562 ⟨72TL2285⟩, 675, 676 ⟨78CC694⟩; **4**, 693 ⟨75CC615, 77JOM(129)429⟩; **5**, 453 ⟨82CC1303, 84JOM(262)49⟩
Fell, B., **2**, 449, 466 ⟨74S119⟩
Fell, E., **1**, 654 ⟨78CC358⟩
Fell, S. C. M., **3**, 495 ⟨88TL3003⟩; **4**, 738 ⟨88TL3003⟩
Fellenberger, K., **1**, 800, 816 ⟨66LA(698)80, 70LA(734)106⟩; **6**, 63 ⟨72JA125⟩
Felletschin, G., **1**, 399 ⟨48LA(560)116⟩
Fellinger, L. L., **5**, 298 ⟨38JA579⟩
Fellmann, J., **3**, 517 ⟨80JMOC73⟩
Fellmann, J. D., **3**, 510 ⟨78JA5964⟩, 510, 511 ⟨78JA3359⟩, 510, 511, 512 ⟨80JA6236, 80JA6608⟩, 512 ⟨81JA1440⟩
Fellmann, P., **2**, 697 ⟨78T1349⟩
Fellows, C. A., **1**, 493 ⟨84JA5274⟩, 1108 ⟨78JOC131⟩
Felluga, F., **2**, 396 ⟨89T2099⟩
Feltham, R. D., **2**, 482 ⟨67IS160⟩, 874 ⟨68IC2003⟩, 874, 876 ⟨71JOM(33)347⟩
Felton, E. G., **1**, 663 ⟨65JOC1258⟩
Felton, H. R., **4**, 316 ⟨54JA3969⟩
Felzenstein, A., **5**, 344 ⟨76TL451⟩
Felzmann, J. H., **2**, 617 ⟨66CB1414⟩
Fendesak, G., **4**, 718 ⟨91ZN(B)1169⟩
Fendrick, C. M., **4**, 589, 656, 658 ⟨86JA425⟩, 656 ⟨88OM1828⟩
Feneau-Dupont, J., **3**, 585 ⟨90CC829, 91JOC4008⟩
Feng, H., **4**, 343 ⟨92S258⟩
Feng, X., **3**, 221 ⟨91TL5713⟩; **4**, 692 ⟨93JA4145⟩
Fengl, R. W., **1**, 134 ⟨85TL3075⟩, 480, 1020 ⟨90JOC5359⟩
Fengler, G., **3**, 593 ⟨74T2553⟩; **6**, 471 ⟨91AG(E)1598⟩
Fenical, W., **1**, 418 ⟨67TL4901⟩; **2**, 686 ⟨74JA5580⟩
Fenn, M. D., **5**, 385, 386 ⟨91IC1564⟩
Fenner, B. R., **1**, 596 ⟨91SL873⟩
Fenor, M., **3**, 699, 721 ⟨92TL4779⟩
Fenske, D., **2**, 828, 866 ⟨73CB177⟩; **5**, 400 ⟨76CB359⟩, 780 ⟨91ZAAC(596)47⟩, 781 ⟨89ZAAC(575)10, 90ZN(B)1273⟩
Fenske, R. F., **5**, 932 ⟨76JA441, 82JA3879⟩; **6**, 334 ⟨84OM1038⟩
Fenter, Chr., **3**, 430 ⟨76MI 311-01⟩
Fenton, D. E., **3**, 362 ⟨83PS(14)139⟩
Fenton, D. M., **1**, 1114 ⟨65JOC1808⟩
Fenton, G. F., **2**, 442 ⟨33JCS989⟩
Fenton, G. W., **1**, 49, 640 ⟨29JCS2342⟩, 612 ⟨30JCS705⟩; **2**, 842 ⟨29JCS2342⟩
Fenton, H. S. H., **2**, 650 ⟨1894JCS899⟩
Fenton, S. W., **4**, 165 ⟨51JA1668⟩
Fentress, J., **3**, 612 ⟨44JA545⟩
Fenude, E., **1**, 432 ⟨89S28⟩; **2**, 553 ⟨94SC591⟩
Fenwick, A., **5**, 344, 1005 ⟨85JOC169⟩
Fenzl, W., **2**, 520 ⟨74IS134⟩, 532 ⟨68AG(E)735⟩, 698 ⟨75LA1322, 75LA2201⟩
Feoktistov, A. E., **5**, 323 ⟨88ZOB1056, 88ZOB1299⟩, 722 ⟨83JGU1944⟩, 722, 997 ⟨83ZOB2155⟩, 1001 ⟨84ZOB876⟩
Feoktistov, L. G., **6**, 149 ⟨72IZV1717⟩, 149 ⟨72IZV2603⟩
Ferao, A., **5**, 220, 862 ⟨86S843⟩, 1107 ⟨82S472⟩
Ferdinand, G., **2**, 742 ⟨76S406⟩; **3**, 37 ⟨76S406⟩; **4**, 838 ⟨75LA1484, 76S404⟩, 838 ⟨76S406, 76S408⟩; **6**, 480 ⟨76S408⟩
Fere, A., **4**, 301, 309, 979 ⟨68JCS(C)2721⟩
Férézou, J. P., **1**, 634 ⟨87TL1997⟩, 1071 ⟨91SL611⟩
Ferguson, A. H., **4**, 30 ⟨86CI(L)490⟩
Ferguson, C. P., **4**, 771 ⟨93JFC(63)253⟩; **6**, 528 ⟨92JFC(59)91, 93JFC(60)85⟩
Ferguson, E. E., **1**, 1143 ⟨75JCP(63)3238⟩
Ferguson, G., **1**, 648, 738, 743 ⟨88JCS(P2)1829⟩; **2**, 1004 ⟨75JCS(D)1284⟩, 1007 ⟨91JCS(P2)1563⟩; **3**, 363 ⟨90JCS(D)3697⟩, 500 ⟨88JCS(P2)1829⟩, 614

⟨88OM1761⟩; **4**, 692 ⟨77IC172⟩; **5**, 1097 ⟨83CC457, 84CC75⟩; **6**, 620 ⟨80JCS(P2)867⟩, 708, 709 ⟨88JCS(P2)1829⟩
Ferguson, G. S., **1**, 543 ⟨85JOC2124⟩
Ferguson, H. C., **6**, 508 ⟨73JMC901⟩
Ferguson, I. E. G., **3**, 333, 335 ⟨83CC423⟩, 333, 335 ⟨85JCS(P1)1541⟩
Ferguson, I. J., **4**, 409 ⟨76JCS(P2)1564⟩
Ferguson, L. N., **3**, 83, 96 ⟨46JA2502⟩
Ferguson, M. D., **5**, 606 ⟨93TL5397⟩
Ferguson, R. R., **2**, 192 ⟨91JOC5503⟩
Feringa, B., **1**, 690 ⟨77JA602⟩
Feringa, B. L., **3**, 105 ⟨88S316⟩, 221 ⟨90TL7201⟩, 406 ⟨88S184⟩, 413 ⟨86TL507⟩; **6**, 88 ⟨89TL5481⟩, 124 ⟨90TL7201⟩
Ferland, J. M., **6**, 429 ⟨89TL1901⟩
Fermin, M. C., **3**, 544 ⟨89JA8738⟩
Fernancez-Baeza, J., **4**, 586 ⟨87JOM(331)409, 87JOM(333)129⟩
Fernandes, J. B., **1**, 195 ⟨83S701⟩
Fernandes, V. P., **2**, 168 ⟨61JOC2478⟩
Fernandex, F. J., **3**, 427 ⟨83OPP41⟩
Fernandez, A. H., **1**, 433 ⟨84S481⟩
Fernández, B., **1**, 606 ⟨92JOC3173⟩
Fernández, I., **2**, 36 ⟨93SL489⟩, 152 ⟨92JOC6789⟩,; **3**, 427, 430 ⟨87T551⟩
Fernandez, J. E., **4**, 300 ⟨63JCED600⟩
Fernandez, J. M., **2**, 264 ⟨80BSB783⟩
Fernandez, J. M. G., **5**, 1028 ⟨89CAR(188)35⟩
Fernandez, J. R., **3**, 58 ⟨85JCS(P1)447⟩
Fernández-Alvarez, E., **4**, 897 ⟨85JHC1445⟩; **5**, 334 ⟨88JHC23⟩, 87 ⟨85S773⟩, 199 ⟨88AQ120⟩, 334 ⟨85JHC1353⟩
Fernandez, M., **1**, 837 ⟨94TL15⟩; **3**, 275 ⟨87JOC2537⟩; **4**, 304 ⟨85JOC4231⟩
Fernández, M., **5**, 606 ⟨92H(34)1573⟩
Fernández, M. J., **4**, 685 ⟨90JCS(D)2587⟩
Fernandez, M. L. De S., **4**, 444 ⟨83JCR(S)108⟩
Fernández, R., **2**, 347 ⟨93TL141⟩, 394 ⟨91TL3225⟩; **3**, 46 ⟨92TL3691⟩
Fernández, S., **2**, 329 ⟨88SC2055⟩; 3, 431 ⟨91S713⟩
Fernandez, V. P., **4**, 68 ⟨61JOC2478⟩
Fernandez-Baeza, J., **4**, 584 ⟨89JOM(364)407⟩
Fernández-Lizarbe, J. R., **5**, 126, 267, 291 ⟨80S547⟩
Fernández-Mayoralas, A., **4**, 235 ⟨89CAR(188)81⟩, 938 ⟨91TL5159⟩
Fernandez-Megia, E., **2**, 44 ⟨92TL4637⟩
Fernández-Moro, R., **5**, 164 ⟨91T9929⟩
Fernandez-Picot, I., **1**, 935 ⟨87T2733⟩
Fernández-Resa, P., **5**, 1039 ⟨84S509⟩
Fernandez-Simon, J. L., **1**, 681 ⟨86CC1665⟩; **2**, 936 ⟨88S234⟩; **4**, 741, 747, 756, 758 ⟨89JCS(P1)691⟩
Fernando, Q., **2**, 278 ⟨64ANC1768⟩
Fernekess, H., **2**, 388 ⟨61CB712⟩
Fernelius, W. C., **2**, 491 ⟨60HOU(5/4)1, 60IS116⟩, 872 ⟨60JINC213⟩
Fernholz, H., **1**, 181 ⟨69AG496⟩, 906 ⟨50LA(568)63⟩
Fernholz, M. E., **6**, 213 ⟨B-72MI 607-02⟩
Ferraboschi, P., **2**, 653 ⟨82CC1157⟩; **3**, 3, 114 ⟨86SC43⟩, 88 ⟨80SC75⟩; **4**, 168 ⟨87JCS(P1)2301⟩
Ferraccioli, R., **4**, 1076 ⟨86CC1191⟩; **6**, 25 ⟨90S327⟩
Ferrand, E. F., **1**, 130, 869, 871, 881 ⟨68JOC2222⟩
Ferrara, S., **3**, 661 ⟨B-74MI 319-01⟩
Ferrari, G. F., **1**, 79 ⟨74JOM(70)89⟩
Ferrari, M., **3**, 59 ⟨88S697⟩; **4**, 462 ⟨92SC107⟩
Ferrari, R. P., **4**, 682 ⟨75G939, 75ICA155⟩
Ferrarotti, B., **4**, 429 ⟨87JOC3442⟩
Ferraz, H. M. C., **1**, 243, 437 ⟨89TL1209⟩
Ferré, G., **6**, 57 ⟨69TL2161⟩
Ferreira, D., **2**, 284 ⟨78JCS(P1)1682⟩; **3**, 24, 146 ⟨92TL2383⟩
Ferreira, J. T. B., **1**, 999 ⟨86SC283⟩; **2**, 279 ⟨82JCR(S)212⟩, 1036, 1037 ⟨81JOM(216)287⟩, 1036, 1038 ⟨86CC1067, 86SC283⟩; **4**, 88 ⟨82JCR(S)212⟩, 90 ⟨81JOM(216)287⟩, 93, 291 ⟨84JOM(277)261⟩, 284 ⟨82SC595⟩
Ferreira, L., **2**, 544 ⟨79SC451⟩
Ferreira, L. M., **5**, 212 ⟨91CC1127⟩
Ferreira, T. W., **1**, 436, 438, 484 ⟨82CC840, 79CC637⟩
Ferrer, J. C., **3**, 226 ⟨89JOC3984⟩
Ferrer, P., **1**, 1081 ⟨90JOC3808⟩; **4**, 749 ⟨85CC296, 86JCR(M)2843⟩; **6**, 695 ⟨85CC296⟩
Ferreri, C., **1**, 17 ⟨93JOC249⟩, 885 ⟨91COS(1)139⟩; **2**, 565, 566 ⟨91COS(1)139⟩; **3**, 164 ⟨89S464⟩, 170 ⟨93SC1455⟩
Ferrero, M., **3**, 414 ⟨88TL4863⟩; **5**, 371 ⟨90JCS(P1)2193, 90TL3497⟩, 1066 ⟨88TL4863⟩
Ferretti, A., **2**, 273 ⟨73OSC(5)107⟩
Ferri, F., **1**, 469 ⟨93TL6403⟩
Ferrier, R. J., **1**, 68 ⟨84JCS(P1)1689⟩, 173 ⟨91CC690⟩, 341 ⟨93CRV2779⟩, 769 ⟨87CC1008⟩; **3**, 621 ⟨87CC1007⟩; **4**, 224 ⟨76CAR(52)63⟩, 235 ⟨77CAR(58)397⟩
Ferris, A. F., **6**, 570, 571 ⟨63JOC3140⟩
Ferris, J. P., **2**, 106 ⟨68SCI3836⟩; **3**, 577 ⟨76JOC13⟩, 638 ⟨73OSC(5)32⟩
Ferro, M. P., **4**, 346 ⟨81JOC1828⟩
Ferro, P. G., **1**, 1076 ⟨91TL5321⟩
Ferrosan, **5**, 639, 642 ⟨55MI 515-01⟩
Ferroud, D., **1**, 194 ⟨86TL4573⟩
Ferruti, P., **4**, 301, 309, 979 ⟨68JCS(C)2721⟩
Ferwanah, A., **3**, 75, 76 ⟨84LA649, 73TL3979⟩
Feschenko, N. G., **2**, 449 ⟨69ZOB2139⟩
Feser, R., **4**, 582 ⟨81JOM(219)C29, 85CAR(144)342, 85JOM(281)317⟩; **5**, 458 ⟨79CB834⟩
Feshchenko, N. C., **2**, 449 ⟨78JGU329⟩, 453 ⟨66ZOB564, 78BRP1517876⟩, 463 ⟨82JGU195, 82JGU202⟩, 466 ⟨74ZOB2357⟩, 836 ⟨79ZOB546⟩, 836, 837 ⟨76ZOB252⟩
Feshin, V. P., **5**, 656 ⟨87JGU1710, 87ZOB1912⟩
Feske, E., **2**, 782 ⟨26CB818⟩
Fessenbecker, A., **2**, 903 ⟨90CB2273⟩
Fessenden, J. S., **2**, 543, 811 ⟨61CRV361⟩; **4**, 831 ⟨67JOC3535⟩
Fessenden, R., **2**, 543, 811 ⟨61CRV361⟩; **3**, 289 ⟨60JOC477⟩
Fessenden, R. J., **1**, 127, 128 ⟨57JOC1326⟩; **4**, 831 ⟨67JOC3535⟩
Fessenden, R. W., **1**, 923 ⟨65JCP(43)2704⟩
Fessner, W. D., **1**, 564 ⟨87JA4626⟩, 900 ⟨88JA8599⟩, 918 ⟨83TL5857, 89JA746⟩
Fest, C., **6**, 630 ⟨90GEP3841185⟩, 632 ⟨88GEP(O)3634929⟩
Festal, D., **3**, 254 ⟨70TL1259⟩, 255 ⟨73BSF3339⟩
Fetell, A. I., **2**, 788 ⟨78JOC1238, 78JOC497⟩
Fétizon, M., **1**, 727, 739 ⟨70MI 116-03⟩, 1039 ⟨81CC363⟩; **2**, 693 ⟨69CC112⟩; **3**, 157 ⟨89H(28)521⟩, 165 ⟨87S503⟩, 240, 245 ⟨90S755⟩, 614 ⟨69CC112⟩; **5**, 171 ⟨75T171⟩
Fetizou, M., **3**, 102, 108 ⟨76JHC525⟩
Fetter, N. R., **2**, 421 ⟨62CJC342, 65CJC1884⟩
Fetyukhin, V. N., **4**, 314, 329 ⟨74JOU1565, 74JOU244⟩, 434, 447 ⟨77ZOR723⟩, 436, 441 ⟨77ZOR2449⟩, 992 ⟨78ZOR2624⟩; **5**, 226 ⟨81ZOR2452, 90JOU287⟩, 669 ⟨89UKZ1071, 91UKZ202⟩, 669, 672 ⟨90ZOR339⟩, 1063 ⟨79S738, 81ZOR1420⟩, 1068 ⟨76ZOR2472⟩, 1083 ⟨80ZOB1678⟩
Fetzer, H., **2**, 399 ⟨80CB1950⟩; **5**, 381, 382 ⟨77M1279⟩
Fetzer, I., **2**, 489 ⟨83ZAAC(505)134⟩
Fetzer, U., **3**, 694, 696, 699, 700, 721, 722, 724, 725 ⟨65AG(E)472⟩, 695 ⟨59AG386⟩, 696 ⟨61CB2814⟩; **5**, 654, 655, 657 ⟨62CB126⟩
Feuer, H., **2**, 314 ⟨69JOC1817⟩, 341, 764 ⟨65JOC2880⟩, 377, 382 ⟨70JOC1468⟩, 391 ⟨B-90MI 207-01⟩, 392 ⟨86JHC825⟩, 393 ⟨63OSC(4)368⟩, 396 ⟨61JOC1357⟩, 763 ⟨65JOC2877⟩, 788 ⟨78JOC1238, 78JOC49,

79JOC3410⟩; **3**, 190 ⟨66JOC3152⟩; **4**, 332 ⟨69JOC3104, 70JOC2551⟩, 430, 431, 432 ⟨62JOC3598⟩; **5**, 350 ⟨68IJ587⟩, 772 ⟨77JOC2091⟩, 1003 ⟨58JOC1107⟩; **6**, 145, 147 ⟨60JOC2069⟩, 147 ⟨61JOC391, 62JOC1455⟩
Feuerbach, D. J., **1**, 860 ⟨74JOC2006⟩
Feuerherd, K.-H., **6**, 448, 452, 454 ⟨80AG(E)201, 79EUP7498⟩
Feugeas, C., **6**, 95 ⟨68CR(C)1506⟩, 95, 98 ⟨69BSF325, 70MI 603-02⟩, 111 ⟨69BSF332⟩
Feussner, G., **2**, 529 ⟨80JOM(187)321⟩
Feustel, M., **2**, 1044, 1048 ⟨81LA1907⟩; **3**, 273 ⟨81LA1907⟩
Feutrill, G. I., **1**, 776 ⟨90JOC4051⟩; **5**, 106 ⟨72AJC1731⟩
Fevig, J. M., **1**, 831 ⟨85TL1163⟩; **5**, 331 ⟨91JA5085⟩
Fevig, T. L., **1**, 320, 331, 336, 341, 342, 344, 348, 926, 935, 948 ⟨91CRV1237⟩, 347 ⟨90SL773⟩, 347, 937 ⟨92SL943⟩, 548 ⟨88JA5064⟩
Fiakpui, C. Y., **1**, 1112 ⟨82CJC735⟩; **3**, 310 ⟨87CJC1158⟩
Fialkov, Y. A., **5**, 58 ⟨85ZOR769⟩, 112 ⟨85ZOR2407⟩
Fiandaca, P., **3**, 549 ⟨76JCS(P1)692⟩
Fiandanese, V., **1**, 466 ⟨86JOM(312)343⟩, 467 ⟨85JCS(P1)1115⟩, 470 ⟨92JOC1718⟩, 484 ⟨89TL243⟩; **2**, 740 ⟨91S225⟩, 931 ⟨88TL3705⟩; **3**, 130, 285, 306 ⟨87TL2053⟩, 230 ⟨91CC237⟩, 285 ⟨84TL4805, 85TL3595, 88TL3587⟩; **4**, 476 ⟨91S225⟩, 773 ⟨87S1034⟩, 776 ⟨90PAC1987, 92TL5121⟩
Fiandor, J., **1**, 1089 ⟨90TL597⟩
Fiaschi, R., **2**, 624 ⟨86S122⟩; **4**, 26 ⟨86S122⟩
Fiato, R. A., **2**, 573 ⟨79JOM(172)405⟩
Fiaud, J.-C., **2**, 413 ⟨87JOC5320⟩; **3**, 139 ⟨81CC1159⟩
Fibiger, R., **4**, 313, 314 ⟨88JOC22, 84JOC4237⟩; **5**, 697 ⟨88JOC22⟩
Ficeri, V., **6**, 565 ⟨90MI 615-02⟩
Fichter, F., **1**, 103 ⟨49HCA1414⟩; **2**, 506 ⟨16CB239⟩
Fichter, K. C., **2**, 974 ⟨74JA6815⟩
Fichtner, K., **4**, 442 ⟨83PAC1799⟩
Fichtner, M. W., **2**, 123 ⟨81JOC3141⟩; **4**, 290 ⟨83TL237⟩; **6**, 472 ⟨81JOC3141⟩, 552 ⟨80JOC175⟩
Ficini, J., **1**, 416, 828 ⟨66TL6425⟩, 518 ⟨75TL3151⟩; **2**, 740 ⟨79TL1499⟩, 759 ⟨77TL3589⟩, 1039 ⟨76T1448⟩, 1047 ⟨64BSF871, 65BSF2787⟩, 1048 ⟨67CR(C)1496, 71CR(C)289⟩, 1050 ⟨74BSF1535⟩; **3**, 218 ⟨79TL1499⟩, 566, 602 ⟨76T1449⟩, 601 ⟨72TL1139⟩; **4**, 15 ⟨74BSF1533⟩, 817 ⟨68TL937⟩, 827 ⟨70TL885⟩; **6**, 27 ⟨56BSF1441⟩, 106 ⟨70CR(C)468⟩
Fick, H. H., **1**, 681 ⟨78JOC4574⟩
Fidler, F. A., **2**, 212, 219 ⟨51JA3627⟩
Fiecchi, A., **2**, 653 ⟨82CC1157⟩
Fiechter, A., **5**, 117 ⟨B-84MI 502-01⟩
Fiederling, K., **5**, 911 ⟨82JOM(236)C1⟩
Fiedorek, F. T., **5**, 1106 ⟨52JA1323⟩
Fiegenbaum, P., **4**, 673 ⟨92CB143, 93CB79⟩, 673 ⟨93CB89⟩
Field, D. J., **1**, 403 ⟨80JCS(P1)714⟩
Field, G. F., **4**, 412 ⟨74JOC167⟩
Field, G. I., **3**, 190 ⟨79S295⟩
Field, J. S., **2**, 897 ⟨75JCS(D)172⟩
Field, K., **4**, 14 ⟨71JOC3566⟩
Field, K. W., **2**, 20, 337, 338, 339 ⟨70CRV639⟩, 23 ⟨69S135⟩, 298, 305, 761 ⟨70CR639⟩
Field, L., **1**, 144 ⟨59JA2572⟩, 146 ⟨56JA92⟩, 778, 779 ⟨78S713⟩; **2**, 156 ⟨63OSC(4)674⟩, 166, 173, 232 ⟨65JOC175⟩, 172 ⟨70JOC3267, 73PNA(70)328⟩, 178 ⟨64JOC918⟩, 187 ⟨51JOC964⟩, 187 ⟨81JOC2025⟩, 198 ⟨61JA1256⟩, 211 ⟨B-77MI 203-04⟩, 212, 218 ⟨72JOC2710⟩, 213, 215 ⟨69JOC36, 58JA838⟩, 215 ⟨62JA847, 70JOC749⟩, 216, 218 ⟨69JOC1792, 71JMC995⟩, 218 ⟨71JOC309⟩, 220 ⟨79JOC2654⟩, 358 ⟨70JOC3012⟩, 708 ⟨51JOC946⟩, 1018 ⟨52JA398⟩; **4**, 53 ⟨51JA5870⟩, 216 ⟨69JOC1799⟩, 326 ⟨80JFC(15)29⟩, 395 ⟨52JA3919⟩; **5**, 244 ⟨85JOC4164⟩, 247 ⟨67JOC3467, 71JOC2735⟩, 247 ⟨81JOC1374⟩, 276, 298 ⟨61JA1983⟩, 610, 623, 624 ⟨86PS(26)169⟩, 995 ⟨52JA394⟩
Field, L. D., **1**, 808 ⟨56JA2547⟩; **2**, 35 ⟨81JOC3727⟩, 36 ⟨83JOC2766⟩; **3**, 426 ⟨83JOC2766⟩, 712 ⟨85T1931⟩; **4**, 603, 649 ⟨91OM3167⟩
Field, M., **2**, 176 ⟨76CZ391⟩
Field, S. J., **1**, 394 ⟨83JCS(P1)2387⟩
Fielden, R., **5**, 1108 ⟨73JCS(P1)705⟩
Fieldhouse, J. W., **2**, 204 ⟨69JOC3097⟩
Fieldhouse, S. A., **4**, 687 ⟨74JCS(D)1856⟩; **6**, 340 ⟨70JOM(24)C61⟩
Fielding, H. C., **4**, 737 ⟨78CC475⟩; **5**, 1016 ⟨63BRP923581⟩, 1017 ⟨62BRP907029⟩; **6**, 3 ⟨B-79MI 601-01⟩
Fields, D. B., **4**, 31 ⟨64JOC1591⟩
Fields, D. L., **2**, 121 ⟨62JOC93⟩
Fields, Jr., D. L., **4**, 138 ⟨86JOC3369⟩
Fields, E. K., **1**, 117 ⟨41JA2316⟩; **4**, 463, 469, 492 ⟨52JA1528⟩; **6**, 51 ⟨59CI(L)1216, 62JOC3686⟩
Fields, K. W., **1**, 482 ⟨81JOC197⟩; **2**, 1023 ⟨82JA321, 85JA1429⟩; **3**, 202 ⟨80JA5411⟩, 239 ⟨81JOC197⟩
Fields, R., **4**, 111 ⟨66JCS(C)2075, 70JCS(C)744⟩, 111 ⟨71JCS(C)3031⟩, 736 ⟨78JCS(P1)422⟩, 737 ⟨80JCS(P1)487⟩, 802 ⟨75JCS(P1)702⟩, 821 ⟨71TL1879⟩, 1048 ⟨80JCR(M)235⟩
Fields, T. F., **5**, 639, 640 ⟨88JHC1873⟩
Fields, T. L., **5**, 631 ⟨84S946⟩
Fields, T. R., **1**, 794 ⟨76JA840⟩
Fienemann, H., **1**, 524 ⟨79JOC2802⟩
Fierz-David, H. E., **2**, 803 ⟨B-52MI 215-01⟩
Fieser, L., **2**, 90 ⟨B-89MI 202-06⟩; **3**, 452 ⟨36JA2050⟩
Fieser, L. F., **1**, 503 ⟨B-67MI 112-01⟩, 795 ⟨53JA5421⟩, 885, 886 ⟨B-67MI 119-03⟩, 898 ⟨B-67MI 119-06, B-72MI 119-09⟩, 898 ⟨B-74MI 119-06, B-86MI 119-02, B-90MI 119-06⟩; **2**, 13 ⟨40JA3098⟩, 30 ⟨63OSC(4)195⟩, 657 ⟨B-67MI 213-01⟩; **3**, 101 ⟨55OSC(3)98⟩, 295 ⟨48JA3352⟩; **4**, 429, 430, 431 ⟨46JA2252⟩; **5**, 199 ⟨43OSC(2)560⟩, 290 ⟨55OSC(3)590⟩, 1119 ⟨48JA3212⟩; **6**, 199 ⟨B-77MI 606-02⟩
Fieser, M., **1**, 503 ⟨B-67MI 112-01⟩, 885, 886 ⟨B-67MI 119-03⟩, 898 ⟨B-67MI 119-06, B-72MI 119-09, B-74MI 119-06, B-86MI 119-02, B-90MI 119-06⟩; **2**, 90 ⟨B-89MI 202-06, B-67MI 213-01⟩; **6**, 199 ⟨B-77MI 606-02⟩
Fiesselmann, H., **3**, 251 ⟨54CB841⟩
Fife, T. H., **4**, 189 ⟨71JA1701⟩, 189 ⟨71JOC2357⟩, 218 ⟨70JA5464⟩
Fife, W. K., **3**, 671 ⟨83JOC1375⟩; **5**, 184 ⟨86JOC3744⟩, 184 ⟨86TL4937⟩, 184, 198 ⟨86TL4933⟩
Fifolt, M. J., **2**, 613 ⟨85JOC4576⟩
Figadère, B., **1**, 1029 ⟨91JMC2452⟩
Figeys, H. P., **1**, 503 ⟨70TL5139⟩; **2**, 1100 ⟨82S211⟩; **3**, 449 ⟨66BSB601⟩
Figueredo, M., **1**, 209 ⟨87TL1881⟩
Figueroa, J. J., **4**, 181 ⟨93JOC7274⟩
Filatov, A. S., **5**, 807 ⟨68ZOB1408⟩; **6**, 242 ⟨61CRV313, 62DOK(142)354⟩, 242 ⟨68ZOB709⟩
Filatova, I. M., **2**, 470 ⟨69JGU2161⟩
Filbey, A. H., **5**, 448 ⟨60JA4204⟩
Filby, W. G., **2**, 186 ⟨73JOC4070⟩
Fil'chikov, A. A., **4**, 766 ⟨91ZOB1014⟩, 884 ⟨91ZOB1600⟩
Fild, M., **2**, 428 ⟨90ZAAC(589)187⟩, 431 ⟨90CZ211⟩, 448 ⟨70JCS(A)2359⟩, 855 ⟨72JCS(P2)380⟩; **4**, 282 ⟨76ZN(B)153⟩, 282, 563 ⟨73ZN(B)98⟩, 550 ⟨77CZ259⟩, 550, 553, 563 ⟨88ZAAC(561)157⟩, 551, 553, 564 ⟨88ZAAC(566)90⟩, 553 ⟨79PS(5)337⟩, 553, 562, 563, 564 ⟨87ZAAC(555)109⟩, 563, 564 ⟨81CB1082, 75CC468⟩, 866 ⟨88CZ107⟩; **5**, 1007, 1018 ⟨67CB1082⟩; **6**, 255 ⟨76CZ391⟩, 265 ⟨87CZ176⟩, 442 ⟨67CB1082⟩
Filho, P., **3**, 679 ⟨84JOC3174⟩
Filiberti, M., **2**, 707 ⟨89T7411⟩; **5**, 1111 ⟨89T7411⟩
Filiminov, V. D., **4**, 310 ⟨78ZOR1723⟩

Filimonova, N. P., **6**, 522 ⟨59MI 616-02⟩
Filimonov, V. D., **1**, 1010 ⟨88ZOR650⟩
Filipkowski, M. A., **1**, 980 ⟨93JA3776⟩
Filipovic, L., **2**, 12 ⟨53JOC868⟩
Filippini, L., **3**, 428 ⟨86S463⟩; **5**, 301 ⟨86T885⟩
Filippo, Jr., J. S., **2**, 9 ⟨75JOC1514⟩, 24 ⟨75JA1599⟩
Filippone, P., **6**, 472 ⟨81JCS(P1)1293⟩
Filippou, A. C., **3**, 692 ⟨90JOM(373)179⟩; **6**, 720 ⟨88JOM(355)437⟩
Filippov, M. V., **5**, 398 ⟨91JGU114⟩, 399 ⟨91JGU108⟩
Filippov, O. F., **2**, 449 ⟨68DOK(182)337⟩
Filippova, A. Kh., **4**, 765 ⟨71IZV162⟩, 766 ⟨83IZV2153⟩, 933 ⟨81IZV654⟩, 940 ⟨77IZV596⟩
Fillaut, J.-L., **1**, 164 ⟨92SL57⟩
Fillebeen-Khan, T., **1**, 178 ⟨90SC125⟩, 554 ⟨84TL1279⟩; **3**, 613 ⟨80SC279⟩
Filler, R., **1**, 921 ⟨93JOC6067⟩; **2**, 15, 17, 18 ⟨78IJ71⟩, 16 ⟨85T1111⟩, 613 ⟨70JA6498, 71JOC2917⟩, 75JOC526⟩; **3**, 148 ⟨85T1111⟩; **5**, 55, 89 ⟨62PCS117⟩; **6**, 9 ⟨B-82MI 601-01⟩, 25 ⟨67JOC1217⟩
Fillion, G., **2**, 767 ⟨85TL4443⟩
Fillion, H., **1**, 995 ⟨79CC859⟩; **4**, 837 ⟨92SL493⟩
Fillion, M. P., **2**, 767 ⟨85TL4443⟩
Filonenko, L. P., **2**, 408 ⟨77ZOB56⟩
Filosa, M. P., **1**, 811 ⟨76JOC3490⟩
Filyakova, V. I., **3**, 356 ⟨89ZOR1865⟩
Fimml, W., **2**, 239 ⟨91CB1131⟩; **5**, 1117 ⟨91CB1131⟩; **6**, 357, 358 ⟨91CB1131⟩
Fina, C., **2**, 553 ⟨94SC591⟩
Finan, J. M., **1**, 14 ⟨82TL2719⟩
Finan, P. A., **5**, 295 ⟨62JCS2824⟩
Finch, A., **2**, 110 ⟨61JOC3250⟩, 268 ⟨64T173⟩, 811 ⟨69MI 215-03⟩, 832, 833 ⟨66JCS(B)1162⟩; **5**, 1015 ⟨63SA1621⟩
Finch, H., **2**, 612 ⟨88TL3365⟩; **4**, 61, 73, 86 ⟨88TL3365⟩
Finch, H. D., **4**, 237 ⟨62JOC649⟩
Finch, N., **3**, 302 ⟨71JOC3191⟩; **4**, 413 ⟨73JOC437⟩
Finch, S. C., **5**, 345 ⟨89TL3219⟩
Finck, G., **2**, 656 ⟨08CB4271⟩
Finckenor, L., **2**, 24 ⟨59JA2191⟩
Findeis, M. A., **1**, 700, 703 ⟨87JOC2838⟩
Findeisen, K., **5**, 189, 196 ⟨83GEP3144791⟩, 658 ⟨78S38⟩, 962 ⟨83HOU(E4)738⟩, 1061, 1078 ⟨81AG855⟩, 1134 ⟨81GEP3018821⟩, 1142 ⟨84GEP3430019⟩; **6**, 239 ⟨70GEP1926659⟩, 240 ⟨72CA(77)125952⟩, 297, 299 ⟨72S599⟩, 502 ⟨83GEP3135111⟩, 603, 613 ⟨72GEP2101107⟩
Findeiss, W., **2**, 1093 ⟨67JOM(9)435⟩
Finding, R., **1**, 1045 ⟨71M214⟩; **6**, 77 ⟨70AG(E)456⟩
Findlay, A., **5**, 24 ⟨05JCS747, 09JCS1004⟩
Findlay, J. A., **3**, 494 ⟨69CC733⟩
Findlay, J. D., **1**, 232 ⟨78S579⟩
Findlay, P., **2**, 1022 ⟨87CJC441⟩
Findlay, S. P., **3**, 446 ⟨56JOC644⟩
Fine, L. W., **6**, 25 ⟨64JOC1148⟩
Finelli, A. F., **5**, 35, 199 ⟨63OSC(4)790⟩
Finer-Moore, J., **3**, 187 ⟨80TL809⟩
Finet, J., **3**, 113 ⟨86T5627⟩
Finet, J.-P., **1**, 435 ⟨90JCS(P1)2851⟩, 780 ⟨93T7967⟩; **2**, 675 ⟨86TL3619⟩, 684 ⟨86TL3615⟩
Finet, J. P., **3**, 145 ⟨85JCS(P1)2667⟩
Finger, G. C., **2**, 609 ⟨56JA6034⟩
Finholt, A. E., **2**, 109 ⟨53JA205⟩
Finian, J. M., **2**, 83 ⟨82TL2719⟩
Finiels, A., **3**, 280 ⟨86JOC2128⟩
Fink, D. M., **1**, 450, 527, 1061 ⟨83T935⟩; **2**, 930, 940, 943 ⟨83T935⟩; **3**, 272 ⟨85TL2513⟩; **4**, 358 ⟨85JOC5393⟩, 359 ⟨88OS(66)14⟩; **5**, 410, 431 ⟨85TL2509⟩, 412 ⟨88OS(66)14⟩, 413, 425 ⟨85JOC5393⟩, 424, 429 ⟨85TL2513⟩
Fink, J., **5**, 1155 ⟨87PS(30)479⟩
Fink, W., **4**, 308, 511 ⟨73HCA1117⟩, 508 ⟨74HCA1042⟩
Finkbeiner, H., **5**, 375, 378, 718 ⟨66JA3390⟩
Finkbeiner, H. L., **2**, 561 ⟨62JOC1493⟩; **5**, 70, 95 ⟨63JA616⟩
Finke, P. E., **6**, 485 ⟨92JMC3731⟩
Finke, R. G., **1**, 668 ⟨87MI 114-01⟩; **2**, 956 ⟨B-87MI 219-01⟩, 960 ⟨B-87MI 219-03⟩, 964 ⟨B-87MI 219-05⟩, 984 ⟨B-87MI 219-04⟩; **3**, 621 ⟨78JA1119⟩
Finke, U., **6**, 178, 290, 393 ⟨77ZAAC(430)121⟩, 385, 392 ⟨77ZAAC(430)137⟩
Finkele, C. E., **3**, 369 ⟨91PS(59)419⟩; **5**, 645 ⟨91PS(59)125⟩
Finkelstein, B. L., **1**, 829 ⟨88JOC1922⟩; **5**, 159 ⟨88JOC1922⟩
Finkelstein, H., **1**, 107 ⟨10CB1528⟩
Finkelstein, J. A., **1**, 521 ⟨80TL3323⟩
Finkelstein, M., **6**, 143 ⟨88ACS(B)666⟩
Finkenbine, J. R., **2**, 287 ⟨74TL2091⟩
Finkler, S. H., **4**, 173 ⟨85TL6329⟩
Finley, K. T., **1**, 888 ⟨B-74MI 119-02, B-88MI 119-03⟩; **3**, 418 ⟨B-70MI 310-04⟩
Finn, J., **2**, 196 ⟨86JA5908⟩
Finn, M. G., **1**, 991 ⟨93AG(E)547, 93JOC1298⟩; **5**, 936 ⟨92JA8735, 92OM1759⟩
Finn, M. G.,**5**, 936 ⟨94JA10921⟩
Finn, M. G., **6**, 706 ⟨93AG(E)554⟩
Finnegan, R. A., **1**, 663 ⟨62CI(L)895, 63TL851⟩
Finnegan, W. G., **6**, 16 ⟨49JA298⟩
Finney, N. S., **1**, 962 ⟨89TL5747⟩
Finnimore, S. R., **4**, 674 ⟨82JCS(D)1783⟩
Finocchiaro, P., **1**, 689 ⟨82JCS(P2)77⟩
Finocchio, A. L., **4**, 234 ⟨88JOC5750⟩
Finsen, L., **3**, 66 ⟨81T2375⟩
Finzel, R. B., **2**, 70 ⟨87JA7838⟩
Finzenhagen, M., **2**, 824 ⟨76LA2126⟩
Finzi, P. V., **3**, 477 ⟨63TL1839, 65CB623⟩
Fiorani, T., **1**, 471, 483 ⟨91JOC261⟩
Fioravanti, S., **3**, 181 ⟨94T3829⟩
Fiorentino, M., **3**, 120 ⟨91JA2205⟩
Fiorenza, M., **1**, 242 ⟨83TL2311⟩; **2**, 265 ⟨83TL2311⟩; **5**, 99 ⟨80JHC621⟩, 413 ⟨85AG(E)1068⟩, 419 ⟨85JOC130⟩, 793, 803 ⟨81JCS(P1)3284⟩; **6**, 349, 585 ⟨85TL1091⟩, 520, 702 ⟨85AG(E)1068⟩
Fioretti, M., **2**, 498 ⟨63AC(R)1869⟩
Fiori, G., **5**, 1045 ⟨90TL7229⟩
Firestein, G., **4**, 720 ⟨75JOM(91)C57⟩
Firestone, R. A., **4**, 346 ⟨68GEP1924135⟩
Firl, J., **3**, 556 ⟨75CL51⟩, 556, 564 ⟨81CL527⟩, 680 ⟨80CB2699⟩
Firnau, G., **3**, 308 ⟨91S1079⟩
Firouzabadi, H., **2**, 150 ⟨92BCJ1131⟩, 215 ⟨83SC611, 88BCJ2185⟩, 215 ⟨89S378⟩; **3**, 4 ⟨86S285, 91SC2077⟩, 5 ⟨83BCJ914, 90T6869⟩, 5, 116 ⟨86T719⟩, 88, 118 ⟨86SC211⟩, 100 ⟨78TL839⟩, 102 ⟨88BCJ2185, 86SC723⟩; **5**, 336 ⟨71JCS(C)3550⟩
Firsan, S. J., **3**, 435 ⟨86JOC5198⟩; **5**, 327, 636 ⟨86JOC5198⟩
Firth, B. E., **4**, 183 ⟨72TL5133⟩, 183 ⟨74JOC3145⟩
Firth, Jr., W. C., **2**, 12, 627 ⟨61JOC280⟩, 835, 836 ⟨65IC764⟩, 837 ⟨65IC765⟩; **5**, 995 ⟨68JOC441, 70USP3541120⟩; **6**, 261, 285 ⟨68JOC3489⟩, 284, 362 ⟨73USP3755404, 72USP3699094⟩, 363 ⟨73JOC1080, 73JOC1083⟩, 363 ⟨73JOC1088, 73USP3733360⟩
Fisch, A., **2**, 1100 ⟨82S211⟩
Fischer, A., **1**, 258 ⟨60JOC463⟩, 318 ⟨79JOM(177)129⟩; **2**, 476 ⟨92CB801⟩; **4**, 134 ⟨92CB801⟩, 564 ⟨92CB1325⟩
Fischer, B., **4**, 949 ⟨78JOM(157)41⟩; **5**, 355 ⟨89T3535⟩
Fischer, D. R., **1**, 655 ⟨88JA6579⟩; **2**, 1036 ⟨93JOC5235⟩; **5**, 1108 ⟨92SL535⟩
Fischer, E., **2**, 400 ⟨85HOU(E11)510, 85HOU(E11)532⟩, 782 ⟨1896CB2662⟩; **3**, 450 ⟨1884CB579, 1887CB821⟩; **4**, 223 ⟨09CB1476⟩, 244 ⟨1894CB673⟩, 570, 585, 660, 704 ⟨92JOM(423)339⟩; **5**, 24 ⟨01CB445⟩, 340

⟨85SUL11⟩; **6**, 360, 361 ⟨78JPR335⟩, 622 ⟨82KGS121⟩, 650 ⟨73ZC58⟩, 651 ⟨88JPR900⟩
Fischer, E. O., **1**, 862 ⟨52ZN(B)377, 59MI 119-05⟩; **2**, 483 ⟨70CB799⟩, 882 ⟨60CB1417⟩; **3**, 513 ⟨77CB656⟩, 518 ⟨76AG(E)543⟩, 522 ⟨76JOM(120)C6, 78CB3740⟩, 597 ⟨68CB954⟩, 690 ⟨73AG(E)564⟩, 692 ⟨90JOM(373)179⟩; **4**, 706 ⟨60JA76⟩; **5**, 226 ⟨80CB1010⟩, 444, 445, 446, 936 ⟨67CB2445⟩, 516, 520 ⟨74CB915, 75GEP(O)2410157⟩, 652 ⟨69JOM(16)29, 71JOM(28)237, 79IS164⟩, 897 ⟨73CB1062⟩, 919 ⟨77CB805⟩, 931 ⟨74AG651⟩, 932 ⟨68JOM(13)209, 69AG(E)761⟩, 932, 933, 934, 948 ⟨64AG(E)580⟩, 936 ⟨72JOM(40)159, 72JOM(46)C15, 77CB2574⟩, 942 ⟨77CB1140, 73JOM(56)C37⟩, 949 ⟨74CB3554, 76CB1120, 76CB1128, 77CB2947⟩, 950, 951 ⟨72CB150, 81AG(E)1055⟩, 951, 953 ⟨67JA7141⟩, 952 ⟨67AG(E)961, 68JOM(14)P9, 73CB3910, 73JOM(51)C11, 67CC1024⟩, 953 ⟨68JOM(13)387, 71JOM(28)359, 71JOM(31)C68, 73CB1277, 4CB183, 74CB3326, 76CB1868⟩, 1088 ⟨74JOM(65)53, 75AG(E)760⟩; **6**, 341 ⟨83M851⟩, 716 ⟨64AG(E)580⟩, 716, 719 ⟨76AOC(14)1, 68JOM(12)P1⟩, 719 ⟨69CB1495, 77CB2574⟩, 720 ⟨70AG(E)309, 72CB2558, 76JOM(118)C33⟩, 720, 722 ⟨76JOM(113)C31⟩, 720, 722, 723 ⟨77CB3467⟩, 721 ⟨69JOM(16)275, 77JOM(129)197⟩, 723 ⟨76AG(E)616, 77JOM(128)C49, 78CB3542⟩
Fischer, G., **6**, 290, 291, 393, 709, 710 ⟨89CB409⟩, 382, 709, 710 ⟨87ZN(B)1055⟩, 709 ⟨87CB1605⟩, 709, 710 ⟨88ZN(B)1468, 91CB1981⟩
Fischer, G. W., **2**, 1008 ⟨82AG(E)555⟩; **3**, 72, 73 ⟨77JPR1042⟩, 247 ⟨71CB210⟩, 263 ⟨74JPR469⟩, 366 ⟨88JPR35⟩; **4**, 852 ⟨84CHEC(6)890⟩
Fischer, H., **1**, 325 ⟨B-89MI 108-01⟩, 844 ⟨68MI 119-05⟩, 849 ⟨68JA2471⟩, 866 ⟨62LA(654)64, 62LA(654)64⟩, 921 ⟨B-89MI 119-08⟩, 923 ⟨85MI 119-07⟩, 927 ⟨B-89MI 119-07⟩, 960 ⟨61CB3060⟩, 986 ⟨59CB1849⟩, 991 ⟨92CB2667⟩; **2**, 235 ⟨86EUP165534⟩, 637 ⟨76AG(E)435⟩, 936 ⟨93SL863⟩, 1009 ⟨91T205⟩; **3**, 72, 242 ⟨90CB399⟩, 386 ⟨84AG(E)726, 85JOM(289)C21, 87CB1905, 88JOM(358)229, 89CC667, 89JOM(364)155, 89JOM(377)105, 90JOM(384)305, 92CC563⟩, 386, 397 ⟨87CC559⟩, 390 ⟨91JOM(415)211, 83ZN(B)1365⟩, 400 ⟨83JOM(252)C63⟩, 401 ⟨86JOM(299)C7, 88JOM(348)C1⟩, 508 ⟨B-83MI 315-01⟩, 690 ⟨B-88MI 320-03⟩, 692 ⟨86JOM(317)187⟩; **4**, 63, 69, 80, 220 ⟨49LA(563)54⟩, 580 ⟨84CB3340, 90JOM(385)351⟩, 581 ⟨86JOM(306)203, 87CB583, 91AG(E)998⟩, 600 ⟨82CB645⟩, 830 ⟨78HCA1784⟩, 992 ⟨79CB2369, 80CB1226⟩, 1018 ⟨91OM389⟩; **5**, 108, 887 ⟨85JOM(287)⟩, 605, 614, 627 ⟨87CC559⟩, 605, 614, 627 ⟨88CB2095, 87JOM(332)105⟩, 652 ⟨79IS164⟩, 660 ⟨84JOM(267)277⟩, 744 ⟨72SZP563109⟩, 772, 854, 855, 859 ⟨51LA(574)85⟩, 857 ⟨56CB563⟩, 942 ⟨95CB373⟩, 954 ⟨93JOM(454)133⟩, 1088 ⟨86CB2080⟩; **6**, 610 ⟨86OM2187⟩, 621 ⟨89S875⟩, 716, 719, 722 ⟨B-84MI 622-01⟩, 723 ⟨77JOM(128)C49⟩
Fischer, H. P., **1**, 664 ⟨68C338⟩; **3**, 622, 638 ⟨63HCA1190⟩
Fischer, J., **1**, 53 ⟨80JA994⟩; **4**, 647, 656 ⟨85NJC249⟩, 716 ⟨91CC1274⟩, 1035 ⟨80JA994⟩, 1036 ⟨86OM877⟩; **6**, 237 ⟨89ZAAC(576)225, 92ZAAC(608)69⟩, 669 ⟨89JOC4272⟩
Fischer, Jr., W. F., **1**, 118, 119, 469, 473 ⟨69JA4871⟩, 311 ⟨69JOC3615⟩, 1048 ⟨69JOC3616⟩; **3**, 287 ⟨69JA4871⟩
Fischer, J. W., **3**, 233 ⟨83TL447⟩
Fischer, K., **2**, 999 ⟨87JOC564⟩; **3**, 646 ⟨86CB2590, 86CB3344⟩; **5**, 1083 ⟨85SM347⟩
Fischer, M., **1**, 630 ⟨65CB3236⟩, 785 ⟨68CB2669⟩; **4**, 990 ⟨90CB115⟩, 991, 993, 996 ⟨85BSB475⟩; **5**, 292 ⟨69TL2281⟩
Fischer, O., **2**, 780 ⟨1895LA(286)145⟩
Fischer, P. A., **2**, 19 ⟨90JOC5291⟩
Fischer, P., **5**, 692 ⟨89S256⟩
Fischer, P. M., **1**, 665 ⟨87JCS(P1)475⟩
Fischer, R. A., **2**, 261, 888 ⟨91JA165⟩
Fischer, R., **2**, 412 ⟨90AG(E)782⟩, 478 ⟨77JCR(S)272⟩, 867 ⟨77JCR(M)3101⟩; **4**, 550, 553, 563 ⟨88ZAAC(561)157⟩; **5**, 1074 ⟨89ZAAC(568)35⟩
Fischer, R. D., **5**, 652 ⟨71JOM(28)237⟩
Fischer, R. H., **2**, 391, 392 ⟨80S261⟩; **3**, 190 ⟨79LA612⟩, 190 ⟨80S261⟩
Fischer, R. P., **1**, 487 ⟨75S376⟩
Fischer, R. W., **2**, 585 ⟨93CB45⟩
Fischer, S., **3**, 680 ⟨86AG(E)85⟩, 681 ⟨85AG(E)56⟩
Fischer, U., **4**, 553 ⟨85ZN(B)1152, 86GEP3444678⟩
Fischer, W., **3**, 464 ⟨67JA5312⟩
Fischler, I., **6**, 524 ⟨B-78MI 616-01⟩
Fisera, L., **3**, 77 ⟨86CCC2167⟩, 621 ⟨84JOC276⟩; **5**, 350 ⟨86CCC2167⟩
Fish, P. V., **1**, 216, 979 ⟨93TL7849⟩
Fish, R. H., **2**, 327 ⟨88OM2250⟩, 527 ⟨73JOC158⟩
Fish, R. W., **6**, 29 ⟨61JA4460⟩
Fishbein, J. C., **1**, 590 ⟨88JA5075⟩; **2**, 649 ⟨87JA2824⟩
Fishbein, L., **2**, 30 ⟨56JA1218⟩
Fishbein, P. L., **3**, 539, 543 ⟨84JOC2190⟩
Fishbein, R., **1**, 925 ⟨71JA2065⟩; **2**, 20 ⟨71JA438⟩
Fishcer, H., **1**, 500 ⟨90CC858⟩
Fishel, D. L., **3**, 445 ⟨66JOC677⟩; **5**, 15 ⟨83OS(61)8⟩
Fisher, A., **3**, 619, 666 ⟨70TL1963⟩
Fisher, A. M., **5**, 198 ⟨32JA3738⟩, 437 ⟨88JA957⟩
Fisher, C. L., **2**, 138 ⟨85JOC4390⟩
Fisher, D. R., **2**, 1033 ⟨92SL535⟩
Fisher, G. A., **2**, 473 ⟨91CB2457⟩, 512 ⟨90JOM(399)C21⟩
Fisher, G., **3**, 163 ⟨92TL7647⟩
Fisher, G. B., **2**, 60 ⟨92TL4533⟩, 519 ⟨93TL7693⟩; **4**, 524 ⟨91JOC5691, 93JA440⟩
Fisher, G. J., **3**, 534 ⟨53JOC1055⟩
Fisher, H., **4**, 437 ⟨80AG(E)724⟩
Fisher, H. P., **1**, 858 ⟨71JA2235⟩
Fisher, J., **6**, 549 ⟨83JA7187⟩
Fisher, J. R., **4**, 691 ⟨79IC2808⟩
Fisher, J. W., **5**, 363 ⟨87TL285⟩
Fisher, Jr., W., **2**, 587 ⟨69JOC3615⟩
Fisher, K. J., **4**, 566 ⟨88SRI163⟩
Fisher, L. E., **3**, 39, 40, 178, 183, 184 ⟨90OPP399⟩
Fisher, M., **2**, 976 ⟨83JOC1556⟩
Fisher, R. A., **2**, 873 ⟨90AG(E)771⟩
Fisher, R. P., **1**, 74, 82 ⟨75JOC3599⟩, 163 ⟨73S37⟩, 278, 487, 662 ⟨71JA6309⟩, 278, 531, 662, 1103 ⟨72JA6560⟩, 449 ⟨74S339⟩; **2**, 904 ⟨75S376⟩, 1005 ⟨73S37⟩; **4**, 631 ⟨73S37⟩, 756 ⟨82S127⟩; **5**, 1136 ⟨72JA6560⟩; **6**, 198 ⟨73S37⟩
Fisher, R. W., **2**, 585 ⟨91AG(E)185⟩
Fisher, T. E., **3**, 40 ⟨90S453⟩
Fisher III, A. P., **3**, 398 ⟨77CC177⟩
Fishman, D., **3**, 55 ⟨81S137⟩; **4**, 188 ⟨81S137⟩
Fishman, J., **1**, 871 ⟨66JOC520⟩; **3**, 627 ⟨79JOC67⟩
Fishpaugh, J. R., **1**, 485 ⟨85JA4679⟩; **3**, 217 ⟨85JA4679⟩
Fishwick, B. R., **4**, 798 ⟨86JCS(P1)1171⟩
Fishwick, C. W. G., **4**, 277 ⟨90SL359⟩, 901 ⟨89TL4449, 91TL405⟩; **5**, 566, 613 ⟨89T6771, 89TL4449⟩, 613 ⟨91TL405⟩, 685, 726 ⟨89TL4443⟩
Fisk, M. T., **2**, 25 ⟨63OSC(4)169⟩; **5**, 55 ⟨63OSC(4)169⟩
Fisk, T. E., **1**, 988 ⟨79JA4772⟩, 1145 ⟨80JA6813⟩; **3**, 273 ⟨79S438⟩
Fisyuk, A. S., **6**, 566 ⟨91KGS416⟩
Fiszer, B., **6**, 512 ⟨69RZC1443⟩
Fitch, S. J., **4**, 555 ⟨66GEP1211200⟩
Fitjer, L., **1**, 248 ⟨84TL2329⟩, 713 ⟨86CB1162⟩; **3**, 256 ⟨72CB919⟩
Fitt, J. J., **2**, 33 ⟨81JOC3349⟩; **3**, 302 ⟨71JOC3191⟩; **5**, 607 ⟨76JOC4029⟩, 657 ⟨82JOC3658⟩, 746 ⟨77JOC2639⟩
Fittak, H., **4**, 176 ⟨89JCR(S)152⟩

Fittig, R., **1**, 384 ⟨1860LA(114)54⟩
Fittkau, S., **5**, 729 ⟨85S172⟩
Fitton, A. O., **5**, 692 ⟨89JCR(S)337⟩
Fitton, A. O. S., **4**, 312 ⟨B-68MI 407-02⟩
Fitton, P., **3**, 661 ⟨71JOM(28)287⟩; **5**, 463 ⟨68CC6⟩
Fitz, T., **2**, 631 ⟨85JOC5384⟩; **5**, 306 ⟨86TL919⟩
Fitzgerald, A., **2**, 1003 ⟨78AX(B)828⟩
Fitzgerald, J. A., **2**, 787 ⟨69JOC3279⟩
Fitzgerald, P. H., **2**, 40 ⟨71JA4907⟩
Fitzgerald, R., **2**, 607, 624 ⟨72T4883⟩
Fitzgibbon, M., **2**, 1014, 1098 ⟨38JCS1218⟩
Fitzi, R., **4**, 416 ⟨86AG(E)345⟩
Fitzmaurice, N. J., **2**, 928 ⟨87S1032⟩; **3**, 77 ⟨87S1033⟩, 660 ⟨85CC4⟩
Fitzpatrick, B. M., **1**, 780 ⟨93T7967⟩
Fitzpatrick, G. J., **1**, 1114 ⟨77JA3769⟩
Fitzpatrick, J. D., **4**, 724 ⟨65JA3254⟩
Fitzpatrick, L., **3**, 305 ⟨92JHC247⟩
Fitzsimmons, B. J., **1**, 218, 831 ⟨83JA1988⟩; **2**, 7 ⟨86JOC789⟩; **6**, 626 ⟨89JA2995⟩
Fiyita, T., **2**, 308 ⟨69BCJ3606⟩
Fizet, C., **3**, 617 ⟨76HCA2786⟩
Fjare, K. L., **6**, 522 ⟨81JA6100⟩
Fjeldberg, T., **6**, 186, 192, 195 ⟨86JCS(D)1551⟩, 186, 195 ⟨82CC1407⟩
Fkhitetouani, S., **2**, 39 ⟨94TL4985⟩
Flachskam, Jr., R. L., **1**, 870 ⟨69JOC4083⟩
Flagg, E. E., **5**, 226 ⟨79USP4216316⟩, 1102 ⟨79USP4250105⟩
Flagg, E. M., **5**, 727 ⟨73JOC2242⟩
Flammang, R., **3**, 579 ⟨88JCS(P2)1071⟩, 582 ⟨88JA1337⟩, 584 ⟨92CC1571⟩, 686 ⟨91TL1487⟩; **5**, 698 ⟨92CC1571⟩, 1086, 1087, 1092 ⟨83JA7194⟩
Flamming, R., **3**, 681 ⟨85AG(E)56⟩
Flanagan, P. W., **4**, 796 ⟨76JOC2112⟩
Flanagan, P. W. K., **1**, 869, 871 ⟨69JA2797⟩, 869, 871 ⟨70JA4689⟩
Flanders, E. D., **2**, 133 ⟨76JOC134⟩
Flann, C., **2**, 328, 329 ⟨87JA6097⟩
Flann, C. J., **1**, 67 ⟨83TL4923⟩, 826 ⟨85JA4359⟩; **2**, 331 ⟨87JA6115⟩; **4**, 188 ⟨83TL4923⟩
Flaskamp, E., **2**, 382 ⟨73CB1052⟩
Flatt, S. J., **3**, 5, 115 ⟨79S815⟩
Flechtner, W., **6**, 598 ⟨56JA4458⟩
Fleck, T., **5**, 618, 619 ⟨75LA275⟩
Fleck, T. J., **1**, 721 ⟨89JA5861⟩
Fleet, G. W. J., **1**, 41 ⟨80TL4031⟩, 828 ⟨86TL3053⟩, 828, 832 ⟨82TL109⟩; **3**, 5, 115 ⟨79S815⟩, 5, 116 ⟨77TL3749⟩, 13, 83, 96 ⟨78TL1437⟩, 78 ⟨73TL4499⟩
Flegler, K.-H., **5**, 1141 ⟨76ZAAC(426)288⟩
Fleig, H., **5**, 690 ⟨80LA1216⟩; **6**, 69 ⟨80LA1216⟩
Fleischer, U., **1**, 910 ⟨89JA1147⟩; **3**, 688 ⟨91CC302⟩; **4**, 551, 552 ⟨91CC302⟩, 572 ⟨93AG(E)1359⟩; **5**, 887 ⟨91CC302⟩
Fleischhacker, W., **3**, 91 ⟨89M765⟩
Fleischmann, C., **1**, 821 ⟨78T317⟩
Fleischmann, R., **1**, 73 ⟨72JOM(42)163⟩
Fleming, F. A., **5**, 814 ⟨73JOC1065⟩; **6**, 279, 281 ⟨73JOC1065⟩
Fleming, F. F., **1**, 831 ⟨89CC1665⟩
Fleming, G.-H., **3**, 612 ⟨33JA4161⟩
Fleming, G. L., **6**, 54 ⟨71JCS(C)3833⟩
Fleming, I., **1**, 45 ⟨66JCS(C)425⟩, 130 ⟨80C265⟩, 189, 450, 535 ⟨79S761⟩, 213 ⟨79S446⟩, 301 ⟨B-76MI 107-01⟩, 323, 350, 351 ⟨B-76MI 108-01⟩, 450, 451 ⟨89OR(37)57⟩, 495 ⟨89T413⟩, 519 ⟨76TL881⟩, 556 ⟨79S736⟩, 565 ⟨92JCS(P1)3351⟩, 566 ⟨87JCS(P1)2259⟩, 599 ⟨79TL2179, 79TL993⟩, 599 ⟨79TL995⟩, 643, 755 ⟨77JCS(P1)550⟩, 648, 649 ⟨79COC(3)541⟩, 651, 943 ⟨90CC1575⟩, 653 ⟨78CC176, 78CC177, 78CC178, 80JCS(P1)1493⟩, 654 ⟨83TL4151, 83TL4153, 84JOM(271)281⟩, 655, 1080 ⟨82JCS(P1)1563⟩, 794 ⟨B-76MI 118-01⟩, 815 ⟨84TL5103⟩, 920 ⟨B-76MI 119-02⟩, 1057 ⟨93TL2383⟩, 1081 ⟨69AG(E)429, 71JCS(C)2013⟩; **2**, 73, 907 ⟨89OR(37)57⟩, 392 ⟨88OSC(6)799⟩, 416, 535 ⟨79COC(3)541⟩, 539 ⟨77JCR(S)6⟩, 546 ⟨92JCS(P1)3277, 92JCS(P1)3295⟩, 546 ⟨92JCS(P1)3363⟩, 546, 917, 977 ⟨84JCS(P1)1805⟩, 907, 943 ⟨79S761⟩, 917 ⟨80CC276⟩, 917, 918, 940, 977 ⟨81JCS(P1)2527⟩, 917, 938, 977 ⟨85S899⟩, 921 ⟨81JCS(P1)2415⟩, 929 ⟨86CC1010, 88TL1825, 91JCS(P1)2811⟩, 929, 930 ⟨89T413, 90CC1030⟩, 930 ⟨92JCS(P1)327⟩, 936 ⟨80JCS(P1)2485⟩, 977 ⟨85S898⟩; **3**, 18 ⟨84JCS(P1)119⟩, 36, 56, 168 ⟨79TL2179⟩, 52, 203 ⟨84CC28⟩, 124 ⟨85CC149⟩, 160 ⟨84CC29, 84JCS(P1)1805⟩, 211 ⟨80JCS(P1)2485⟩, 227 ⟨79TL3205⟩, 271 ⟨80CC276, 81T4027⟩, 274 ⟨82JCS(P1)1563⟩, 444 ⟨61JCS5560⟩, 492 ⟨91COS⟩, 639 ⟨81JCS(P1)256⟩, 640 ⟨85JOM(285)173⟩; **4**, 67 ⟨83TL327⟩, 365 ⟨89JCS(P1)115⟩, 620, 1048, 1049 ⟨81JCS(P1)969⟩, 649 ⟨81JCS(P1)2415⟩, 650 ⟨77JCS(P1)550⟩, 966 ⟨79TL2179⟩, 1046 ⟨87JCS(P1)2269⟩, 1059 ⟨81JCS(P1)2527⟩; **6**, 175 ⟨81JCS(P1)969⟩
Fleming, J. A., **1**, 594, 616 ⟨85JOC3622⟩
Fleming, J. S., **2**, 1006 ⟨62TL983⟩; **3**, 446 ⟨63JA2784⟩
Fleming, M. P., **1**, 173, 587, 688 ⟨78JOC3249⟩, 687, 688, 689 ⟨74JA4708, 76JOC896⟩, 688, 689, 690 ⟨78JOC3255⟩; **2**, 12, 627 ⟨79JOC3405⟩; **3**, 49, 60 ⟨78JOC3788⟩; **4**, 255 ⟨79JA4732⟩; **5**, 557 ⟨79JA4732, 83JA5015⟩; **6**, 134 ⟨79JA4732⟩
Fleming, N. B., **3**, 92 ⟨78OPP201⟩
Fleming, S. A., **1**, 357 ⟨86JA306⟩
Flemons, G. F., **3**, 328 ⟨48JCS2138⟩
Flentke, G. R., **4**, 527, 528 ⟨90JBC(265)3738⟩
Fletcher, A. V., **1**, 1075 ⟨75TL1513⟩
Fletcher, D., **1**, 106 ⟨86TL4853⟩
Fletcher, G. L., **3**, 309 ⟨89JOC3264⟩
Fletcher, Jr., H. G., **4**, 184 ⟨73CAR(29)209⟩; **5**, 476, 477 ⟨65JOC1282⟩
Fletcher, M. T., **1**, 68 ⟨86CC855⟩
Fletcher, R. S., **4**, 163 ⟨51JA212⟩
Fletcher, T. L., **2**, 366 ⟨75S39⟩
Fletcher, V. R., **2**, 1006 ⟨93JA497⟩
Fleurant, A., **4**, 796 ⟨90SC3339⟩
Fleury, J., **4**, 979 ⟨76BSF2053⟩
Fleury, J.-P., **3**, 567 ⟨63CR(256)2419⟩; **4**, 423, 427 ⟨55BSF669⟩; **5**, 826 ⟨74T999⟩
Fleury, M.-B., **2**, 308 ⟨93SC817⟩
Flick, W., **2**, 470 ⟨75AG(E)363⟩
Fliege, W., **3**, 709 ⟨83CB3027⟩
Fligge, M., **4**, 181 ⟨67JSP(23)32⟩
Flindt, E. P., **2**, 470 ⟨78ZAAC(447)97⟩
Flippen-Anderson, J., **2**, 384 ⟨92JOC1585⟩; **4**, 298 ⟨90JOC1459⟩, 428 ⟨87JOC4781⟩, 795 ⟨91JOC537⟩, 970, 993 ⟨92JOC235⟩; **6**, 150 ⟨92JOC3026⟩
Fliri, A., **5**, 749 ⟨86HCA1224⟩
Flisak, J., **6**, 365 ⟨90MI 612-01⟩
Fliszár, S., **1**, 895 ⟨76CJC2839⟩
Flitsch, W., **3**, 268 ⟨84LA1878, 87TL4395⟩, 268 ⟨88LA387, 90LA397⟩, 659 ⟨71CB2847⟩; **4**, 299 ⟨70CB3205⟩, 925 ⟨84CB1424⟩; **5**, 753 ⟨52LA(577)234⟩, 934 ⟨89TL1633⟩
Flo, C., **3**, 307 ⟨89JHC1563⟩; **6**, 729 ⟨87CSR75, 89SC2307⟩
Floch, P. Le, **5**, 1157, 1158 ⟨92CC1799⟩; **6**, 679 ⟨89TL817⟩
Floch, Y. Le, **3**, 494 ⟨82S597⟩; **4**, 249 ⟨78BSF(2)595⟩, 317, 322 ⟨69CR(C)1718, 73BSF3499⟩
Flock, F. H., **6**, 213 ⟨B-72MI 607-02⟩
Flogaus, R., **3**, 152 ⟨84S734⟩
Flood, S. H., **3**, 282 ⟨64JA2203⟩
Flood, T., **1**, 963, 978 ⟨80JOC5006, 80JA5120⟩

Flood, T. C., **1**, 692 ⟨72JA6538⟩; **3**, 483 ⟨68JA1080, 72JOM(44)279⟩; **4**, 513 ⟨68JA1080⟩, 656 ⟨86JA7964⟩; **5**, 453 ⟨84JA2853⟩, 1148 ⟨71JOM(29)C25⟩
Florea, C., **3**, 539 ⟨88RRC83⟩
Florencio, F., **4**, 405 ⟨87CJC687⟩
Florent, J., **4**, 172 ⟨81CC556⟩
Florent, J.-C., **3**, 244 ⟨81CC556⟩; **6**, 550 ⟨88CAR(181)253⟩
Florentin, D., **2**, 118, 124 ⟨90BSF758⟩
Flores, A., **4**, 685, 722 ⟨76IC2567⟩
Flores, H. J., **3**, 620 ⟨83SC35⟩
Flores-Vela, A., **6**, 119 ⟨89JPO349⟩
Florez, J., **1**, 466 ⟨93JOC5976⟩; **2**, 553 ⟨93JOC5976⟩; **5**, 71 ⟨83JCS(P1)3019⟩
Flórez, J., **5**, 946, 958 ⟨93CC1068⟩
Florian, W., **1**, 415 ⟨61LA(641)1⟩; **2**, 788 ⟨61LA(641)1⟩; **3**, 720 ⟨61LA(641)1⟩; **4**, 882, 974 ⟨61LA(641)1⟩; **5**, 852 ⟨61LA(641)1⟩; **6**, 104, 105, 313, 314, 318, 321, 730, 731 ⟨61LA(641)1⟩
Floriani, C., **2**, 703 ⟨85IC654⟩; **4**, 696 ⟨83CC1087, 83CC1156⟩; **5**, 441 ⟨74JOM(71)C5, 77JCS(D)2297⟩, 442 ⟨76CC522⟩, 443 ⟨77JCS(D)1946⟩, 444 ⟨74CC516, 76JCS(D)203⟩; **6**, 342 ⟨86AG283, 88NJC621⟩
Florio, E., **1**, 890 ⟨86TL1469⟩
Florio, S., **1**, 1061 ⟨86S638⟩, 1110 ⟨73JCS(P2)2078⟩; **4**, 321 ⟨86S638⟩
Floris, C., **4**, 347 ⟨84PS(19)61⟩, 391 ⟨86TL4625⟩
Florkowski, M. P., **1**, 1040 ⟨48JA3868⟩
Floruss, A., **2**, 472 ⟨91CB1153, 92ZAAC(611)47⟩
Flory, K., **2**, 998 ⟨64TL1137⟩; **3**, 495 ⟨66CB689⟩; **6**, 25 ⟨64AG(E)513⟩
Floss, H. G., **5**, 451 ⟨87JA4837⟩
Flowers, M. C., **1**, 801 ⟨61JCS3547⟩
Flowers, W. T., **3**, 588 ⟨74CC134⟩
Floyd, A. J., **3**, 669, 670 ⟨83T3881⟩
Floyd, C. D., **4**, 620, 1048, 1049 ⟨81JCS(P1)969⟩; **6**, 175 ⟨81JCS(P1)969⟩
Floyd, D., **1**, 120, 1107 ⟨78JOC3418⟩; **2**, 587, 1095 ⟨78JOC3418⟩; **4**, 649 ⟨78JA2916⟩
Floyd, D. M., **1**, 118, 465 ⟨81JA7672⟩; **2**, 1095 ⟨81JA7672⟩; **4**, 905 ⟨90JMC1510⟩; **5**, 342 ⟨82JOC176⟩, 346 ⟨82JOC179, 82JOC5160⟩
Floyd, J. C., **3**, 210 ⟨74TL2877⟩
Floyd, M. B., **1**, 549 ⟨71JA3828⟩; **3**, 140, 156 ⟨79JOC1438⟩; **5**, 759 ⟨68TL3277⟩
Fluchaire, M., **1**, 414 ⟨50BSF121⟩
Fluck, E., **2**, 429 ⟨82ZN(B)965⟩, 450, 842 ⟨75CC382⟩, 864 ⟨63CB3091⟩, 889, 891 ⟨69ZAAC(369)178⟩, 1062 ⟨76MI 222-02, 76ZN(B)172⟩, 1073 ⟨87ZAAC(548)63⟩, 1074 ⟨88ZAAC(567)39⟩; **4**, 308 ⟨73CB69⟩, 437 ⟨74LA1851⟩, 552 ⟨91ZAAC(596)139⟩, 562 ⟨91ZAAC(601)65, 93PS(77)258⟩, 564 ⟨88SRI727⟩, 574 ⟨76ZAAC(421)1⟩, 575 ⟨88ZAAC(567)23⟩; **6**, 490 ⟨81ZN(B)910⟩
Fluckiger, E., **3**, 128 ⟨93T965⟩
Flückiger, R., **2**, 1021 ⟨93JA11702⟩
Fludzinski, P., **1**, 999 ⟨82TL2373⟩, 1016, 1108 ⟨84JA3551⟩; **3**, 146 ⟨82TL2373⟩
Flygare, J. A., **1**, 648 ⟨91SL764⟩; **4**, 139, 652, 655 ⟨91JOC638⟩, 906 ⟨89JOC4723⟩; **5**, 238 ⟨89JOC4723⟩
Flynn, D. L., **3**, 114 ⟨93TL2071⟩; **5**, 282 ⟨84JA6414⟩
Flynn, E. W., **1**, 1144 ⟨59JA4714⟩
Flynn, G. A., **3**, 623, 633 ⟨75JOC1162⟩; **4**, 1004 ⟨86TL1757⟩
Flynn, K. E., **1**, 690 ⟨84JA723⟩
Flynn, R. M., **2**, 849 ⟨79S615⟩; **3**, 495 ⟨83JOC3616⟩; **6**, 264 ⟨80JFC(15)263, 82JFC(20)121⟩, 264 ⟨82JFC(20)617⟩
Foa, M., **3**, 661 ⟨79JOM(173)335⟩
Foá, M., **3**, 661 ⟨B-74MI 319-01⟩
Foa, M., **5**, 143 ⟨85JOM(285)293⟩
Fobare, W. F., **1**, 504 ⟨83JOC5221⟩
Foces-Foces, M. C., **5**, 1066 ⟨89CB307, 89T4263⟩, 1068 ⟨89JA355⟩
Fochi, G., **5**, 443 ⟨77JCS(D)1946⟩; **6**, 341 ⟨87AG681⟩
Fochi, R., **1**, 48 ⟨78JCS(P1)1133⟩; **2**, 134, 143 ⟨83S630⟩, 180 ⟨81S149⟩, 180 ⟨89S957⟩; **3**, 11 ⟨87S79⟩, 16, 132 ⟨81S51⟩, 64 ⟨67G397⟩, 283, 309 ⟨87S311⟩, 306, 307 ⟨93JCS(P1)273⟩; **4**, 249, 250 ⟨81TL1821⟩, 846, 854 ⟨87G227⟩, 1073 ⟨76JCS(P1)323⟩; **6**, 50 ⟨74JHC943⟩, 97 ⟨79S223⟩, 302, 554, 729 ⟨88S22⟩, 472 ⟨80S375, 81S149⟩, 545 ⟨77S873⟩, 553, 554 ⟨86S894⟩
Fochler, M., **6**, 289 ⟨68CB2815⟩
Fockenberg, F., **6**, 276, 537 ⟨86ZN(B)413⟩
Fodor, G., **1**, 726 ⟨61TL579⟩; **5**, 90 ⟨72S464⟩, 655 ⟨80JHC1457⟩, 685 ⟨B-75MI 517-01⟩, 710, 716 ⟨81H(15)165⟩; **6**, 417, 418 ⟨73T3309⟩
Fodor, L., **4**, 273 ⟨89MRC760⟩; **6**, 111 ⟨91JHC481⟩
Fodor, P., **5**, 611 ⟨66CR(C)788⟩
Foerst, D. L., **5**, 234 ⟨76SC453⟩
Foester, R., **5**, 1129 ⟨70IC1571⟩
Fogagnolo, M., **2**, 942 ⟨84CC258⟩; **3**, 32 ⟨89TL6063⟩, 109 ⟨87S998⟩, 311 ⟨84JOC3478, 84JOC590⟩, 312 ⟨86G133⟩, 707, 714 ⟨87S693, 84CC258⟩; **6**, 634 ⟨87JOC3413, 87S693⟩, 637 ⟨90H(31)1213, 93H(36)473⟩
Fogel, D. L., **4**, 648 ⟨75CC536⟩
Fogel, E. R., **1**, 590 ⟨76JOC3218, 80OS(59)202⟩
Fogel, J., **3**, 494 ⟨66JOM(6)205⟩
Fohlisch, B., **3**, 709 ⟨62AG(E)334, 65LA(686)92⟩
Fokeer, A. P., **2**, 569 ⟨B-88MI 210-02⟩
Fokin, A. F., **5**, 699, 710 ⟨81JFC(18)553⟩
Fokin, A. V., **2**, 165 ⟨89IZV197⟩, 167 ⟨85ZOK1835⟩, 170 ⟨88IZV2367⟩; **4**, 104 ⟨86IZV810⟩, 119 ⟨86JGU2394⟩, 135 ⟨71JGU481⟩, 556 ⟨87ZOB54⟩; **5**, 474, 484 ⟨67BAU324⟩, 645 ⟨63ZOB3293⟩, 698 ⟨82IZV937⟩, 698, 762, 813 ⟨78IZV2802⟩, 710 ⟨81IZV2370⟩, 762 ⟨73IZV2159, 80IZV2623⟩; **6**, 42 ⟨85IZV659⟩, 54 ⟨67ZOB152, 79IZV1911⟩, 56, 282 ⟨76IZV489⟩, 236 ⟨92RCR517⟩, 251 ⟨74IZV335⟩, 259 ⟨78IZV486, 68IZV656⟩
Fokt, I., **4**, 176 ⟨89JCR(S)152⟩
Foland, L. D., **3**, 539 ⟨85JA3392, 86JOC3067⟩
Foldi, V. S., **6**, 643 ⟨62JA3673⟩
Földi, Z., **2**, 117 ⟨48JCS1683⟩
Folest, J. C., **6**, 24 ⟨88TL1699⟩
Foley, H. G., **3**, 618 ⟨73CC625⟩
Foley, K. M., **1**, 21 ⟨70JA3232⟩
Foli, C. J., **4**, 5 ⟨91JOC4322⟩
Folin, O., **5**, 976 ⟨1897MI 526-01⟩
Folkers, K., **2**, 177 ⟨41JA3530⟩
Folkins, P. L., **2**, 232 ⟨91JOC904⟩
Folkmann, M., **4**, 59, 60 ⟨90S1159⟩; **6**, 427, 428, 476 ⟨90S1159⟩
Folli, U., **2**, 147 ⟨68JCS(C)1317⟩, 600 ⟨73JCS(P2)638⟩
Fölling, P., **3**, 607 ⟨84AG(E)619⟩; **5**, 1058 ⟨84AG(E)970⟩, 1092 ⟨84AG(E)619⟩; **6**, 680, 687 ⟨84AG(E)619⟩, 680, 687 ⟨84AG(E)970, 86TL1661⟩, 687 ⟨88ZAAC(556)7⟩
Follman, R., **4**, 538 ⟨76CB3047, 76CB3062⟩
Follmann, R., **4**, 392 ⟨80LA1765⟩, 938, 957 ⟨80AG(E)303⟩; **6**, 496 ⟨77AG(E)462⟩
Folsom, H. E., **2**, 148 ⟨92SC1799⟩
Folsom, T. K., **5**, 122 ⟨70TL4011⟩
Folting, K., **4**, 675 ⟨84JA6794, 86OM2171⟩, 675 ⟨86OM2457, 91OM535⟩, 676 ⟨86OM602⟩; **6**, 342 ⟨89JA7283, 90JOM(394)265⟩
Foltynowicz, Z., **2**, 938 ⟨91JMOC113, 92JMOC307⟩, 938 ⟨93PJC1361⟩
Folz, S. D., **5**, 675, 678 ⟨81JMC532⟩
Fomakhin, E. V., **5**, 644 ⟨85ZOB1526⟩, 713 ⟨85JGU1526, 85ZOB1718⟩
Fomenko, M. G., **5**, 708 ⟨61JGU473, 61ZOB516⟩
Fomum, Z. T., **3**, 655, 656 ⟨84T2141⟩; **4**, 979 ⟨88JCS(P1)961⟩
Fonbelo, F., **2**, 553 ⟨88TL2859⟩
Fones, W. S., **3**, 465, 474 ⟨49JA1629⟩; **5**, 285 ⟨49JOC1099⟩

Fong, D.-W., **5**, 528 ⟨67JOC3709⟩
Fong, F. K., **1**, 906 ⟨74JA7638⟩
Fong, H. O., **4**, 267 ⟨80CJC878⟩
Fong, R. H., **6**, 339, 347 ⟨87JA2843, 88OM794⟩
Fonken, G. J., **1**, 418 ⟨65JA3996⟩, 419 ⟨66JA1073⟩, 420 ⟨62TL549⟩
Fonken, G. S., **1**, 881 ⟨52JA831, 63OSC(4)261⟩
Fonong, T., **6**, 264 ⟨82JFC(20)617⟩
Fontaine, X. L. R., **4**, 1023, 1028 ⟨85CC1635⟩
Font, J., **1**, 209 ⟨87TL1881⟩, 217, 225 ⟨88JOC139⟩, 579 ⟨93T6717⟩; **3**, 206 ⟨70CC721⟩, 231 ⟨81T2391⟩
Font-Altaba, M., **4**, 711 ⟨86JOM(314)13⟩; **6**, 331 ⟨87CC472⟩
Fontana, A., **2**, 229 ⟨75CC976⟩
Fontana, F., **1**, 934 ⟨89H(28)489, 90JHC79⟩, 934 ⟨91JOC2866⟩; **3**, 311 ⟨89H(28)489, 91JOC2866⟩
Fontana, S., **5**, 936 ⟨72JOM(40)159⟩; **6**, 250 ⟨92EUP499157, 92JFC(58)290⟩
Fontanella, L., **5**, 566 ⟨68FES642⟩
Fontanelli, R., **6**, 251 ⟨72GEP2111696⟩
Fontanillas Val, J. A., **4**, 93, 291 ⟨84JOM(277)261⟩
Font-Bardía, M., **4**, 713 ⟨90JOM(395)305⟩
Fookes, C. J. R., **3**, 106 ⟨72AJC1979, 81CC797⟩
Fooladi, J., **4**, 353 ⟨91JOM(403)29⟩
Foos, J. S., **3**, 212 ⟨72JA8637⟩
Foote, C. S., **2**, 687 ⟨79TL4363⟩
Foote, J. L., **2**, 752 ⟨61JA743⟩
Footner, H. B., **5**, 1111 ⟨25JCS2887⟩
Forbare, W. F., **1**, 243 ⟨82JOC3348⟩
Forbes, E. J., **5**, 275 ⟨66CC121⟩
Forbes, G. S., **5**, 1015, 1016, 1017 ⟨40JA761⟩, 1016 ⟨47JA1241⟩
Forbes, G. S.,**5**, 1016 ⟨47JA3048, 48JA1043, 48JA1222⟩
Forbes, J. E., **1**, 938 ⟨90TL2565⟩, 939, 946 ⟨90JA2034⟩; **3**, 713 ⟨88TL3343⟩; **6**, 547 ⟨89TL4367, 90JA2034⟩
Forbes, W. F., **3**, 451 ⟨51JCS1762⟩
Forbus, Jr., T. R., **3**, 282, 301 ⟨79JOC313⟩
Forbus, T. R., **1**, 582 ⟨90OSC(7)506⟩
Forchiassin, A., **2**, 783 ⟨74JCS(P1)660⟩
Forchiassin, M., **2**, 783 ⟨66TL1821, 66TL6331⟩, 783 ⟨68T1889⟩; **5**, 354 ⟨79TL4761, 83JHC305, 85G663⟩, 355 ⟨91G401⟩
Ford, F. E., **2**, 600 ⟨63JOC237⟩
Ford, G. H., **4**, 546 ⟨71JCS(A)3495⟩
Ford, Jr., H., **5**, 404 ⟨92TL4137⟩
Ford, J. H., **5**, 85 ⟨55OSC(3)34⟩
Ford, M. E., **2**, 103 ⟨74T2467⟩; **3**, 89, 95 ⟨74SC45⟩; **4**, 194 ⟨76JA3037⟩
Ford, M. J., **1**, 614, 697, 700, 703 ⟨90TL5525⟩; **4**, 200 ⟨90SL255⟩
Ford, P. C., **4**, 674 ⟨77JOM(124)29⟩; **5**, 386 ⟨75IC42⟩; **6**, 516 ⟨88AOC(28)139⟩, 517 ⟨85JA2355, 87IC526⟩
Ford, P. W., **1**, 812 ⟨72JA5910⟩
Ford, R. A., **1**, 869 ⟨67JOC1994⟩
Ford, R. P., **4**, 570 ⟨86MM2089⟩
Ford, R. R., **2**, 466, 467 ⟨84IC2062⟩; **4**, 567 ⟨85IC1993⟩, 572 ⟨84IC2063⟩; **5**, 418, 924, 925 ⟨89OM693⟩, 893 ⟨85IC1993⟩, 924 ⟨87OM2128⟩
Ford, S. G., **5**, 89, 99 ⟨43OSC(2)382, 43OSC(2)140⟩, 99 ⟨43OSC(2)368⟩
Ford, S. H., **1**, 1023 ⟨69TL1379⟩
Ford, T. A., **6**, 5 ⟨50JA2213⟩, 522 ⟨76IC1666⟩
Ford, T. M., **1**, 657 ⟨81JOC647⟩; **2**, 522 ⟨80JOC4067⟩
Ford, W. T., **1**, 858 ⟨68JA2598, 68JA2606⟩, 858 ⟨68JA2612⟩
Förderreuther, M., **2**, 100 ⟨60CB1249⟩; **4**, 177, 178 ⟨60CB1249⟩
Fordham, W. D., **3**, 229, 275 ⟨52JCS3945⟩
Ford-Moore, A. H., **2**, 163 ⟨49JCS2126⟩; **4**, 555 ⟨47JCS1465⟩
Foreman, E. L., **2**, 274 ⟨50JA3595⟩
Foreman, M., **1**, 74, 82 ⟨75JOC3599⟩
Foreman, M. I., **1**, 528 ⟨73JCS(P1)977⟩; **2**, 985 ⟨73JCS(P1)977⟩
Forero-Kelly, Y., **5**, 11 ⟨91S18⟩
Foresti, E., **1**, 834 ⟨81JOC2228⟩; **2**, 245 ⟨81JCS(P1)2322⟩; **4**, 232 ⟨85TL2131⟩; **5**, 354 ⟨87T4249⟩, 648 ⟨88JCS(P1)1499⟩; **6**, 127, 309 ⟨85TL2131⟩
Foret, C. J., **6**, 492, 515 ⟨81JINC457⟩
Forgioue, P. S., **2**, 388 ⟨66JA1078⟩
Forlani, L., **4**, 405 ⟨86G229⟩
Forman, L. E., **5**, 1049 ⟨34JA1609⟩
Forman, M. A., **1**, 799 ⟨86TL4107⟩
Fornaroli, M., **2**, 190 ⟨74JCS(P1)771⟩
Fornasier, R., **1**, 85 ⟨74JOC3195⟩; **4**, 75 ⟨72JCS(P1)1886⟩, 76 ⟨73JA7431⟩
Fornefeld, E. J., **3**, 226 ⟨60JA2402⟩
Forner, K., **2**, 452 ⟨74ZC477⟩; **4**, 554 ⟨66ZC314⟩
Forner, T., **4**, 556 ⟨73ZC55⟩
Forni, A., **2**, 337 ⟨83JOC2640⟩
Forniés, J., **4**, 584 ⟨85JOM(290)125⟩, 692 ⟨88OM2279, 89JCS(D)169, 93JA4145⟩; **5**, 1138 ⟨84ICA215⟩
Foroughi, K., **2**, 477 ⟨74CB639⟩
Forrenbaugh, R. B., **2**, 271, 273 ⟨59JA4810⟩
Forrest, A. K., **4**, 426 ⟨84TL1769⟩
Forrest, D., **4**, 850 ⟨78JA3868⟩; **5**, 516, 526 ⟨77JPC915⟩
Forrest, J., **5**, 293 ⟨46JCS454⟩
Forrester, A. R., **1**, 922 ⟨B-68MI 119-04⟩; **2**, 390, 397 ⟨88OSC(6)803⟩, 764 ⟨72OS(52)77⟩
Forrester, J. L., **2**, 24 ⟨60JOC24⟩
Forsberg, G., **2**, 93 ⟨54ACS135⟩
Forsch, R. A., **5**, 597 ⟨76SC387⟩
Forsche, E., **5**, 821, 839 ⟨62HOU(5/3)1⟩
Forschner, T. C., **5**, 453 ⟨85OM1247⟩
Forshaw, T. P., **2**, 383 ⟨71JCS(C)2404⟩
Forster, A. M., **2**, 492 ⟨85POL1625⟩
Forster, B., **3**, 696, 699 ⟨87AG(E)894⟩; **5**, 972 ⟨87AG(E)894⟩; **6**, 416, 447, 462, 471, 483, 486, 487 ⟨87AG(E)894, 91ZN(B)339⟩
Forster, C., **4**, 557 ⟨90T6645⟩
Förster, H., **1**, 999 ⟨64TL983⟩; **2**, 470 ⟨69M311⟩, 1062 ⟨64TL983⟩; **4**, 481, 487 ⟨68M380⟩; **6**, 665, 686 ⟨82AG(E)448⟩
Forster, M. O., **2**, 769 ⟨20JCS1157⟩; **3**, 463 ⟨10JCS2156, 15JCS260⟩
Forster, S., **3**, 628 ⟨87AG(E)458, 90TL1249⟩
Förster, S.,**5**, 964 ⟨81CB173⟩
Forster, W.-R., **3**, 548 ⟨83CB509, 84AG(E)439⟩
Forsyth, D. A., **1**, 905 ⟨87JA7263⟩, 914 ⟨78JA6299⟩, 915 ⟨76JA6284⟩
Fort, R., **2**, 94 ⟨54BSF1109⟩
Fort, Jr., R. C., **1**, 847, 853 ⟨66MI 119-07⟩, 900 ⟨64JA4194, 64JA4195⟩, 900 ⟨B-73MI 119-07⟩
Fort, Y., **1**, 437 ⟨88TL2963⟩, 437, 438, 439 ⟨90JOC6194⟩, 472 ⟨89JOC4840⟩; **5**, 1105 ⟨92SC617⟩
Forte, P. A., **3**, 618 ⟨71CJC1321⟩; **5**, 689 ⟨76CJC647⟩
Fortenbaugh, R. B., **2**, 621 ⟨58JA1716⟩
Fortes, C. C., **3**, 634 ⟨90SC1943⟩; **4**, 65 ⟨82CC857⟩
Fortes, H. C., **4**, 65 ⟨82CC857⟩
Fortgens, H. P., **1**, 978 ⟨84TL3115⟩; **2**, 324, 326 ⟨87TL3285⟩; **4**, 298 ⟨87TL3285⟩
Forth, B., **1**, 842 ⟨73TL3459⟩
Forth, H., **4**, 18 ⟨24LA(435)219⟩
Fortier, S., **3**, 173 ⟨78JOC4552⟩
Fortin, R., **2**, 135 ⟨83JOC1357⟩, 539 ⟨84TL4717⟩
Fortoul R., C., **4**, 660 ⟨93JOM(448)69⟩
Fortt, S. M., **1**, 1127 ⟨87TL2895⟩; **2**, 26 ⟨87S35⟩; **3**, 136 ⟨90JCS(P1)2875⟩, 292 ⟨90T2135⟩
Fortunak, J., **2**, 392 ⟨88OSC(6)799⟩
Fortunato, J. M., **3**, 127 ⟨76JOC2194⟩, 311 ⟨84JOC5003⟩
Fosbinder, R. J., **4**, 69 ⟨45JA655, 45JA657⟩
Foscolos, G., **3**, 131, 405, 459 ⟨80TL155⟩

Fosker, G. R., **5**, 219 ⟨71JCS(C)1917⟩
Foss, L. E., **5**, 990, 1037 ⟨77JOC1508⟩
Foss, O., **2**, 216 ⟨47ACS307⟩, 237 ⟨47JA2236⟩, 238 ⟨51ACS115, 53ACS227⟩, 289 ⟨59ACS2155, 63ACS1806, 66ACS123⟩, 289, 732 ⟨66ACS132⟩, 369 ⟨80JCS(D)620⟩; **5**, 1116 ⟨47JA2236⟩
Foss, V., **2**, 471 ⟨83JGU2244⟩
Foss, V. L., **2**, 414 ⟨80ZOB1236⟩, 420, 494, 495, 695 ⟨74JOM(78)107⟩, 471 ⟨79JGU2143⟩, 694 ⟨69ZOB2479, 73ZOB1264⟩, 695 ⟨74JOM(78)115⟩, 695, 700 ⟨69ZOB1174⟩; **5**, 395, 396, 399 ⟨89JGU1817⟩, 395, 396, 399, 400 ⟨88JGU201⟩, 395, 396, 399, 400 ⟨88JGU2372⟩, 396, 399 ⟨87DOK114⟩, 398 ⟨91JGU114⟩, 398, 400 ⟨90JGU1810⟩, 399 ⟨91JGU108, 86JGU636⟩, 400 ⟨90JGU1817, 90JGU1822⟩; **6**, 668 ⟨82ZOB2187⟩
Fosse, R., **3**, 455 ⟨24CR(178)811⟩
Fossett, L. A., **4**, 565 ⟨85JA5272⟩
Foster, A., **3**, 307 ⟨80CJC2527⟩
Foster, A. L., **2**, 455 ⟨91JOC3137⟩
Foster, A. M., **3**, 212 ⟨73JOC3829⟩
Foster, B., **4**, 19, 27 ⟨81JCS(P1)78⟩
Foster, C. H., **2**, 215 ⟨70JOC749⟩
Foster, D. F., **5**, 453 ⟨86TL623⟩
Foster, D. J., **5**, 27 ⟨57CRV867⟩, 388 ⟨62ZN(B)135⟩
Foster, G. P., **3**, 411 ⟨87JA5047⟩
Foster, H. M., **2**, 120 ⟨63OSC(4)638⟩
Foster, N. I., **2**, 632 ⟨80S572⟩
Foster, R., **4**, 824 ⟨72JCS(P1)1225⟩
Foster, R. E., **2**, 166 ⟨46JOC499⟩, 645 ⟨45JA1937⟩
Foster, R. F., **1**, 883 ⟨88JA2714⟩
Foster, R. V., **2**, 95, 96 ⟨53JCS1541⟩, 95, 96 ⟨54JCS2200⟩
Foster, T. T., **1**, 624 ⟨49JA3929⟩; **2**, 756 ⟨49JA3929⟩
Foster, W. E., **2**, 626 ⟨58JOC280⟩, 1084 ⟨64JOC3225⟩
Fothergill, G. A., **5**, 295 ⟨62JCS2824⟩
Fotiadu, F., **1**, 499 ⟨90TL4859⟩
Fotsch, C. A., **2**, 556 ⟨88JA4788⟩
Fotsch, C. H., **1**, 316 ⟨88JA4788⟩
Foubelo, F., **1**, 1090 ⟨88CC1135, 88TL4859⟩, 1090 ⟨89JCR(S)200, 90CC1521⟩, 1090 ⟨90CC587, 91CC1001⟩
Foubister, A. J., **1**, 47 ⟨82S1036⟩
Foucard, A., **2**, 466 ⟨90T6715⟩
Foucaud, A., **2**, 394 ⟨86TL493⟩, 744 ⟨82TL2023⟩; **3**, 191 ⟨86TL493⟩, 495 ⟨76BSF2068⟩, 557, 573 ⟨77TL3353⟩, 567 ⟨75CR(C)51, 85JOC771⟩, 567, 574 ⟨82T527⟩, 568 ⟨69TL509⟩, 568 ⟨72T4431⟩, 570 ⟨72CR(C)1309⟩, 573 ⟨77TL3349, 78CC274⟩, 574 ⟨78TL3719⟩; **4**, 183 ⟨71TL231⟩, 800 ⟨69TL2441, 72T4039⟩, 800 ⟨72T5149⟩, 891 ⟨83JOC481⟩, 894 ⟨77TL3349⟩, 996 ⟨72T5055⟩; **5**, 364 ⟨76BCF1119⟩, 728 ⟨82TL2023, 84T1075⟩, 728 ⟨85JOC771⟩; **6**, 571 ⟨90JOC1721⟩, 617 ⟨84T1075, 85JOC771⟩, 629 ⟨78JOC3732⟩, 635 ⟨86JOC4043, 89JOC1185⟩
Fouch, G., **2**, 39 ⟨93CC119⟩
Foulds, C. D., **4**, 285 ⟨84JCS(P1)21⟩
Foulon, J. P., **1**, 312 ⟨80T2305, ⟨84PAC91⟩, 491 ⟨86T1399⟩; **4**, 655 ⟨86T1389⟩
Fouquay, S., **1**, 180 ⟨88T5495⟩
Fouque, D., **3**, 50 ⟨92SC219⟩
Fouquet, E., **1**, 931 ⟨91JA1055⟩
Fouquet, G., **1**, 177 ⟨74AG50⟩, 465 ⟨74AG(E)82⟩
Four, P., **3**, 13, 83, 95, 98, 107 ⟨81JOC4439⟩, 83 ⟨80CC432⟩
Fouré, M., **3**, 590 ⟨80CB811⟩
Fourneron, J. D., **1**, 240 ⟨82T693⟩; **2**, 51 ⟨89JOC4686⟩, 593 ⟨75BSF770⟩; **5**, 135 ⟨90TL4875⟩
Fournet, G., **2**, 11 ⟨86TL1907⟩
Fournier, F., **2**, 29, 30 ⟨84BSB157⟩
Fournier, J. O., **2**, 121 ⟨62JOC93⟩
Fournier, L., **2**, 483 ⟨72SRI53⟩; **4**, 215 ⟨75T809⟩
Fournier, M., **2**, 29, 30 ⟨84BSB157⟩
Fourrey, J.-L., **1**, 1075 ⟨88T147⟩; **2**, 410 ⟨84TL4511⟩, 630 ⟨83TL1605, 88T147⟩
Fourtinon, M., **2**, 813 ⟨80JOM(193)165⟩
Fowler, F. W., **1**, 102 ⟨72JOC1321⟩; **5**, 8 ⟨85JOC2719⟩, 214 ⟨86JOC167⟩
Fowler, J. P., **5**, 10 ⟨69CR(C)1325⟩
Fowler, J. S., **2**, 156 ⟨68JOC3422⟩, 631 ⟨69TL2427, 71JA4841⟩; **4**, 300 ⟨63JCED600⟩
Fowler, K. W., **1**, 473 ⟨80JA790⟩
Fowler, R., **3**, 184, 185 ⟨79T613⟩
Fowles, G. W. A., **2**, 421 ⟨57JCS2078, 62JCS1330⟩, 570 ⟨72JCS(D)2313, 73JCS(D)961⟩
Fox, B. A., **1**, 428 ⟨73OSC(5)346⟩
Fox, B. L., **1**, 881 ⟨66JOC982⟩
Fox, C. J., **4**, 23, 33 ⟨54JA479⟩
Fox, C. M. J., **1**, 1140 ⟨88TL2923⟩; **2**, 271 ⟨88OS(67)108⟩
Fox, D. E., **3**, 310 ⟨92JOC761⟩; **5**, 322 ⟨92SC3067⟩
Fox, D. P., **1**, 1071, 1143 ⟨74JA4562, 75CC17⟩; **6**, 127 ⟨77JOC1667⟩
Fox, G. J., **4**, 24 ⟨76OS(55)20⟩
Fox, H. H., **2**, 888 ⟨48JOC438⟩
Fox, I. S., **2**, 461 ⟨68JCS(B)1416⟩
Fox, J. J., **2**, 212 ⟨61JOC3401⟩
Fox, K. M., **5**, 560 ⟨93JOC3042⟩
Fox, M., **2**, 689 ⟨94JOC6683⟩
Fox, R. B., **2**, 448 ⟨50JA4147⟩
Fox, S. W., **5**, 84 ⟨45MI 502-02⟩
Fox, W. B., **2**, 94 ⟨69JA1310⟩, 98 ⟨70CC216⟩, 99 ⟨67JA4313⟩; **6**, 42, 43 ⟨70JAP7026611⟩, 231 ⟨71CC784⟩, 240 ⟨70CC395⟩, 428 ⟨69TL723⟩
Fox, W. R., **5**, 1111 ⟨80JCS(P1)1029⟩
Foxman, B. M., **1**, 346, 926 ⟨93JOC7640⟩, 929 ⟨93JOC7228⟩; **2**, 873 ⟨90AG(E)771⟩; **3**, 170 ⟨91JOC328⟩
Foxton, M. W., **1**, 511 ⟨68JOM(11)P7, 68JOM(12)P33⟩; **2**, 948, 968 ⟨71JOC3520⟩, 970 ⟨68JOM(12)P33⟩; **4**, 1063 ⟨71JOC3520⟩
Foye, W. O., **2**, 214 ⟨67JPS292⟩
Fozard, A., **3**, 357 ⟨64JCS2760⟩
Fráater, G., **1**, 565, 681, 682, 683 ⟨84T1269⟩
Frach, R., **6**, 608 ⟨83JPR787⟩
Fraenkel, G., **2**, 997, 1003 ⟨89OM2308⟩; **4**, 603, 648, 649 ⟨90JA1382⟩, 603, 649 ⟨90OM1314⟩, 648 ⟨90JA2582, 93JA1551⟩
Fraenkel-Conrat, H., **5**, 83 ⟨56MI 502-01⟩
Fraile, A. G., **2**, 36 ⟨87TL6441⟩, 630 ⟨86S222⟩; **3**, 92, 101 ⟨90CC1571⟩; **4**, 214 ⟨87S49⟩
Frainier, L., **1**, 869, 871 ⟨70JA4689⟩
Frainnet, É., **3**, 80 ⟨73CR(C)1803⟩; **4**, 832 ⟨71JOM(30)C64⟩; **5**, 720 ⟨74CR(C)213⟩
Frajerman, C., **1**, 397 ⟨77TL3485⟩, 477 ⟨77JOM(127)371⟩; **2**, 561 1016, 1018 ⟨66TL875⟩
Frampton, C. S., **4**, 724 ⟨86JOM(307)231⟩
Français, G., **2**, 387 ⟨64CR(259)3737⟩
Francalanchi, F., **5**, 143 ⟨85JOM(285)293⟩
Francalanci, F., **1**, 433 ⟨90JCS(P1)428⟩
France, C. J., **6**, 543, 550 ⟨91T6381⟩
France, D. J., **4**, 166 ⟨70JOC468⟩
Francesconi, L., **5**, 652 ⟨03G226⟩
Francetic, D., **2**, 147 ⟨60TL20⟩, 478 ⟨61CI(L)469⟩
Franchetti, P., **4**, 206 ⟨81JMC1181⟩
Franchimont, A. P. N., **4**, 442 ⟨10RTC296⟩
Franchimont, E., **1**, 999 ⟨62CB319⟩; **2**, 1012, 1013, 1062 ⟨62CB319⟩
Francis, F. E., **2**, 104 ⟨06JCS1⟩
Francis, M. D., **6**, 263 ⟨78JCE760⟩
Francisco, C. G., **1**, 927 ⟨88JCS(P1)3255⟩, 929 ⟨84TL1953⟩, 931 ⟨87TL3397, 92AG(E)772⟩; **2**, 13 ⟨86JOC402⟩; **5**, 1118 ⟨90JCS(P1)2417⟩
Franck, A., **4**, 583, 585, 586, 587 ⟨77CB3536⟩
Franck, R. W., **1**, 882 ⟨65JOC2513⟩; **4**, 187 ⟨92JOC2084⟩

Francke, R., **4**, 338 ⟨88CZ146, 89CZ320⟩
Franckenstein, G.-H., **4**, 408 ⟨77LA956⟩
Franck-Neumann, M., **1**, 70 ⟨78AG(E)690⟩, 954 ⟨79AG(E)688⟩; **3**, 65, 543 ⟨73TL2875⟩, 471 ⟨75BSF247⟩, 473 ⟨69TL2659⟩; **4**, 1076 ⟨86JOM(315)59⟩
Franco, J. O., **4**, 148 ⟨89OM1114⟩
Francoise, G. A., **3**, 414 ⟨92TA459⟩
Francotte, E., **3**, 46 ⟨78BSB693⟩; **4**, 798 ⟨78AG(E)936, 81HCA1208⟩; **6**, 74 ⟨78BSB693⟩
Frandsen, E. G., **3**, 254, 354, 378 ⟨77CC851⟩; **5**, 593 ⟨76TL3347, 77T341⟩
Franenkel, G., **2**, 553, 555 ⟨92JOC3680⟩
Frangin, Y., **1**, 544 ⟨79JOM(166)1⟩; **4**, 1066, 1067 ⟨77JOM(142)9⟩
Frank, A., **3**, 513 ⟨77CB656⟩, 522 ⟨78CB3740⟩; **4**, 695 ⟨83JOM(246)141⟩; **5**, 1088 ⟨75AG(E)502⟩
Frank, A. W., **2**, 837 ⟨61JOC850⟩; **3**, 66 ⟨81JHC549⟩; **4**, 130 ⟨66CJC2593⟩, 453, 456, 458, 475 ⟨72JOC2752⟩, 456 ⟨81PS(10)147⟩, 456, 475, 477 ⟨81CJC27, 81PS(10)207⟩
Frank, C. E., **5**, 42 ⟨61JOC307⟩
Frank, D. S., **3**, 645 ⟨62JOC422⟩
Frank, F. J., **4**, 281 ⟨63JOC1420⟩
Frank, G. A., **1**, 733 ⟨64JOC3327⟩
Frank, H. R., **1**, 637 ⟨46JA502⟩
Frank, L.-R., **2**, 875, 896 ⟨87JOM(336)337⟩
Frank, R., **3**, 406, 416 ⟨80LA1814⟩, 561, 582 ⟨79LA83⟩; **4**, 63, 69, 80, 220 ⟨49LA(563)54⟩; **5**, 657 ⟨79LA74⟩
Frank, R. L., **2**, 119, 120 ⟨46JA2103⟩; **5**, 251 ⟨55OSC(3)116⟩; **6**, 572 ⟨55OSC(3)773⟩
Frank, S., **2**, 835, 836 ⟨65IC764⟩, 837 ⟨65IC765⟩; **6**, 261 ⟨66IC1455⟩, 284, 362 ⟨73USP3755404, 72USP3699094⟩, 363 ⟨73JOC1080, 73JOC1083⟩, 363 ⟨73JOC1088⟩
Frank, W., **2**, 489 ⟨90JOM(386)177⟩
Frank, W. C., **1**, 760 ⟨82JOC3548⟩
Franke, A., **3**, 612 ⟨13M1893⟩
Franke, H.-J., **2**, 686 ⟨75S42⟩
Franke, K., **6**, 141 ⟨79LA1456⟩
Franke, L. A., **1**, 841 ⟨88JA3601, 88JOC2706⟩
Franke, P., **4**, 454, 495, 496, 501, 502 ⟨74JOM(81)187⟩; **6**, 533, 538 ⟨88EGP273832⟩
Franke, R., **4**, 455 ⟨86MI 410-01⟩, 586 ⟨73AG(E)416, 75ICA(13)79⟩, 586, 656 ⟨75CB1321⟩
Franke, U., **6**, 661, 733 ⟨78JPR389⟩
Franke, W., **1**, 1063 ⟨60AG391⟩, 1122 ⟨81JA2770⟩; **5**, 289 ⟨57CB2291⟩
Frankel, E. N., **5**, 142 ⟨73MI 503-02⟩
Frankel, M. B., **2**, 386 ⟨56JA5428⟩; **4**, 442, 443 ⟨61JOC4709⟩, 444 ⟨82JCED94, 83JOC611⟩, 994 ⟨58JOC813⟩; **6**, 56 ⟨72JOC152⟩, 151 ⟨84BRP2123829⟩, 151 ⟨90MI 605-01⟩
Franken, E., **1**, 522 ⟨81TL1675⟩
Franken, P. W., **4**, 47, 48 ⟨77CI(L)127⟩; **6**, 229, 230 ⟨77CI(L)127⟩
Franken, S., **2**, 31 ⟨87JOC5560⟩
Frankforter, G. B., **4**, 743 ⟨26JA3144⟩
Frankhauser, P., **4**, 665 ⟨92CB1341⟩
Frankland, E., **2**, 68, 590 ⟨1849LA(79)171⟩, 520 ⟨1859PRS568⟩
Franklin, R. J., **4**, 268 ⟨91JCS(P1)662⟩, 269 ⟨91TL7743⟩
Franklin, W. J., **5**, 1054, 1057 ⟨65TL3003⟩
Frank-Neuman, M., **1**, 576 ⟨87T853⟩
Frank-Neumann, M., **3**, 30 ⟨90SL641⟩
Franks, S., **2**, 433 ⟨79JCS(P1)3029⟩
Frantz, Jr., A. M., **1**, 901 ⟨62JA4165⟩
Franz, J. A., **1**, 576 ⟨77OS(57)22⟩; **2**, 401 ⟨75JA583, 75JA6137⟩; **5**, 689 ⟨73JA2017⟩
Franz, J. E., **2**, 344 ⟨91TL2711⟩, 391 ⟨65JOC1488⟩, 726 ⟨64JOC2592⟩; **3**, 51, 200 ⟨88JOC4069⟩, 498 ⟨64JOC3721⟩, 686 ⟨78JOC3736⟩; **5**, 235 ⟨91S920⟩, 996 ⟨64JOC2592⟩
Franz, K. D., **4**, 580 ⟨72JOM(42)C35⟩
Franz, L., **5**, 874 ⟨85TL1845⟩
Franz, M., **2**, 943 ⟨67CB2681, 72CB1759⟩, 943 ⟨72CB470⟩
Franz, R., **2**, 19 ⟨80JFC(15)423⟩; **5**, 215 ⟨86MI 504-01⟩; **6**, 409 ⟨80JFC(15)423⟩
Franz, R. G., **3**, 60 ⟨88SC1913⟩
Franze, K.-D., **6**, 511, 581 ⟨73JPR471⟩
Franzen, H., **3**, 444 ⟨02CB3234⟩, 447 ⟨11JPR122⟩
Franzen, V., **1**, 623 ⟨60CB2819⟩; **2**, 632 ⟨54CB1148⟩; **3**, 474 ⟨57LA(602)199⟩
Franzheld, M., **4**, 880 ⟨85JPR297⟩
Franzone, J. S., **4**, 259 ⟨89CZ349⟩
Frappier, F., **2**, 127 ⟨93JA2139⟩
Frasch, M., **6**, 128 ⟨91CB1805⟩, 128, 132 ⟨93CB537⟩
Fraser, G. W., **2**, 370 ⟨71JCS(A)1125⟩; **6**, 257 ⟨67IC1711⟩
Fraser, M. S., **3**, 570 ⟨70JA3489, 72JOM(35)403⟩
Fraser, P. J., **4**, 687 ⟨74AOC323⟩
Fraser, P. S., **3**, 351 ⟨74JOC2509⟩
Fraser, R. R., **1**, 879 ⟨76CJC3809, 78JA657⟩; **2**, 22 ⟨88SC1071⟩; **3**, 10 ⟨79TL3929⟩; **4**, 391 ⟨72JA8795⟩
Fraser, T. E., **4**, 583 ⟨79ZN(B)1218⟩
Fraser, W. A., **5**, 73 ⟨32OSC(1)475⟩
Fraser-Reid, B., **1**, 828 ⟨89JOC5350⟩; **3**, 57 ⟨79CJC645⟩; **4**, 176 ⟨88JA2662⟩, 184 ⟨94CC749⟩, 199 ⟨88TL6549⟩; **5**, 157 ⟨89CC1160⟩, 179 ⟨82JOC3358⟩, 226 ⟨86CAR(150)111⟩
Frasnelli, H., **3**, 474, 484, 485, 486 ⟨74LA1767⟩, 474, 486 ⟨70AG291⟩, 474, 486 ⟨70AG(E)301⟩, 478 ⟨68LA(716)204⟩; **5**, 844, 866, 874 ⟨68LA(716)204⟩, 872 ⟨74LA1767⟩, 873, 874 ⟨70AG(E)301⟩
Frass, W., **6**, 18 ⟨61JA1767⟩
Fráter, G., **3**, 222 ⟨89HCA1515⟩; **5**, 148 ⟨79HCA2825⟩, 148 ⟨79HCA2829⟩, 148 ⟨80HCA1383⟩
Fratzscher, H., **2**, 1060 ⟨64N213⟩
Fraunberger, F., **4**, 297, 327 ⟨66CB1944⟩, 309 ⟨66CB1932⟩
Fravel, H. G., **1**, 794 ⟨76JA840⟩
Fray, G. I., **1**, 564 ⟨94JCS(P1)793⟩, 815 ⟨71JCS(C)970, 77T1527⟩, 1000 ⟨61T18⟩
Frazão, C. M., **1**, 446 ⟨89CB259⟩; **2**, 447 ⟨86JA976⟩, 622 ⟨89CB2213⟩
Frazee, W. J., **4**, 231 ⟨84T1333⟩
Frazer, M. G., **2**, 626 ⟨76S621⟩
Frazer, M. J., **2**, 110 ⟨57JCS739, 61JCS5545⟩
Frazer, M. M., **4**, 427 ⟨55JCS4280⟩
Frazier, C. C., **5**, 445, 453 ⟨78JA1687⟩
Frazier, H. W., **4**, 161 ⟨66JA865⟩
Frazier, K., **1**, 1005 ⟨78TL3195⟩
Frazza, E. J., **2**, 749 ⟨64JOC1800⟩; **3**, 658 ⟨64JOC1800⟩
Frebel, M., **5**, 400 ⟨86CB1857⟩, 1092 ⟨90CB733⟩; **6**, 690 ⟨86CB1857⟩
Fréchard-Ortuno, M., **1**, 1064 ⟨92JOC651⟩
Freche, A., **2**, 226 ⟨71CJC4047⟩, 227 ⟨73TL1401⟩
Fréchet, J. M. J., **2**, 71 ⟨90JOC304⟩; **3**, 4, 115 ⟨78JOC2618⟩, 4, 115 ⟨81JOC1728⟩, 14 ⟨83TL1143⟩, 373 ⟨79TL203⟩; **6**, 477 ⟨85CJC153⟩
Fréchet, J. M. T., **2**, 560 ⟨87JOC4644⟩
Frechette, R., **3**, 267 ⟨89JOC6012⟩
Frechette, R. F., **1**, 20 ⟨87TL2937⟩
Frecht, J. M. J., **2**, 753 ⟨87S383⟩
Fredenhagen, K., **2**, 959 ⟨26ZAAC246⟩
Frederick, J. L., **5**, 368 ⟨62JOC1383⟩
Frederick, M. R., **5**, 1106 ⟨52JA1323⟩
Frederick, R. C., **2**, 761 ⟨72JA3884⟩
Frederiksen, E., **5**, 1051 ⟨36ZAAC(230)31⟩
Frederiksen, P. A. A., **6**, 592, 593 ⟨70ACS2061⟩
Fredga, A., **2**, 278 ⟨61ACS938⟩; **5**, 1113 ⟨70ACS3751⟩
Fredin, L., **5**, 1096 ⟨83JAP5813530⟩
Freeburger, M. E., **5**, 24 ⟨72MI 502-02⟩
Freed, M. E., **4**, 414 ⟨60JOC2108⟩
Freedman, B., **2**, 632 ⟨58JOC1161⟩; **6**, 435 ⟨64FRP1372971⟩

Freedman, H. H., **1**, 898 ⟨B-73MI 119-06⟩, 901 ⟨62JA4165⟩, 909 ⟨64JA734⟩; **3**, 119 ⟨76TL1641⟩; **4**, 165 ⟨64T2181⟩
Freedman, J., **3**, 280 ⟨89JHC1547⟩
Freedman, L., **1**, 91 ⟨61JOC818⟩; **6**, 640 ⟨59JA3728⟩
Freedman, L. D., **1**, 53 ⟨74MI 102-01⟩, 271 ⟨88JOM(25)351⟩, 445 ⟨74S328⟩, 646 ⟨B-70MI 114-01⟩, 647 ⟨B-70MI 114-02⟩; **2**, 481, 872 ⟨82CRV15⟩, 481 ⟨B-70MI 209-01⟩, 492 ⟨64JA209⟩, 509 ⟨74S328⟩, 820, 821 ⟨52JA3414, 62JOC2192⟩, 844 ⟨53JA1379, 53JA683⟩, 844, 847 ⟨51JA5658, 52JA753⟩, 872 ⟨92JOM(442)1, 92JOM(442)61⟩, 872, 891 ⟨B-70MI 217-01⟩, 872 ⟨B-89MI 217-01⟩; **4**, 595 ⟨82CRV15⟩; **6**, 707 ⟨B-70MI 622-01⟩
Freedman, M. B., **4**, 36 ⟨61JA4105⟩
Freedman, R., **5**, 410 ⟨67JA434⟩; **6**, 127, 325 ⟨67JA434⟩
Freeguard, G. F., **2**, 32, 36 ⟨65CI(L)223, 64CI(L)1582⟩
Freek, J. A., **1**, 45 ⟨46JA1813⟩
Freeland, B. H., **4**, 713 ⟨80IC693, 82CC371⟩; **6**, 340 ⟨70JOM(24)C61⟩
Freeman, B. H., **1**, 647, 648, 737, 738, 743 ⟨72T343⟩; **3**, 328 ⟨72CC912⟩, 500 ⟨72T343, 74T2257⟩
Freeman, F., **2**, 221 ⟨83S913, 85S500⟩, 723 ⟨84CRV117⟩; **4**, 203 ⟨77JCED355⟩, 883 ⟨89S698, 91JOC657⟩, 883 ⟨91MI 420-01⟩; **5**, 46 ⟨69CRV591⟩, 247 ⟨94S699⟩
Freeman, J. P., **1**, 636 ⟨67JA1761⟩; **2**, 335 ⟨67JA716⟩, 335, 336 ⟨67JOC4034⟩, 385 ⟨55JA4387⟩, 386 ⟨55JA6061, 61JOC4190⟩, 404, 405 ⟨63JOC2508⟩; **4**, 104 ⟨67JOC4034⟩; **5**, 63, 66, 275 ⟨63OSC(4)58⟩, 314 ⟨61JA3912, 69JA4778⟩; **6**, 104 ⟨67JA716⟩
Freeman, L. D., **1**, 88, 89 ⟨55JA4262⟩
Freeman, L. V., **2**, 430 ⟨67JA1102⟩
Freeman, M. J., **6**, 343 ⟨83JOM(249)273⟩
Freeman, N. K., **5**, 1063 ⟨72JOC3372⟩
Freeman, P. F. H., **5**, 532 ⟨65JCS3312, 70GEP(O)2024393⟩
Freeman, P. K., **2**, 551 ⟨80JOC1924⟩; **3**, 134 ⟨77JOC3205⟩, 532 ⟨65CI(L)424⟩
Freeman, R. C., **2**, 8 ⟨60JA909⟩; **3**, 561 ⟨65JOC3718⟩; **4**, 743, 791 ⟨60JA903, 73OSC(5)387⟩; **5**, 76 ⟨73OSC(5)387⟩, 770, 849 ⟨65JOC3718⟩
Freeman, S., **2**, 862 ⟨91JCS(P2)2081⟩; **4**, 58 ⟨93JCS(P1)2303⟩
Freeman, W. A., **5**, 897 ⟨84JA6082⟩
Freeman, W. P., **5**, 422, 431 ⟨92OM2198⟩
Freenor, F. J., **2**, 25 ⟨76S398⟩, 32 ⟨64JA5037, 72JOC626⟩
Freer, A. A., **2**, 1004 ⟨75JCS(P2)1030⟩; **5**, 330 ⟨91JCS(P1)1001⟩; **6**, 528 ⟨92TL261⟩
Freerks, R. L., **4**, 1014 ⟨81JA2807⟩
Freerksen, R. W., **3**, 134 ⟨83JOC4087⟩, 627, 634, 635, 636, 639, 651 ⟨76JOC2846, 77JA182⟩; **5**, 272, 283, 284 ⟨77JA182⟩
Frei, B., **1**, 812 ⟨84HCA774⟩; **3**, 275 ⟨81JOC1512⟩, 319 ⟨79HCA1645, 82HCA2363⟩; **4**, 250 ⟨81JOC1513⟩, 359 ⟨85HCA44⟩; **5**, 223, 412, 429 ⟨86HCA44⟩, 410, 424 ⟨86HCA1378⟩, 430 ⟨89HCA264⟩; **6**, 127 ⟨86HCA1378, 86HCA44⟩, 128 ⟨84HCA1734⟩, 134 ⟨81JOC1512⟩
Frei, J., **5**, 855 ⟨02CB1084⟩
Freiberg, J., **2**, 1008 ⟨69AP(302)354⟩; **6**, 43 ⟨67LA(707)35⟩
Freidinger, R. M., **1**, 248 ⟨85TL5923⟩, 706, 707 ⟨74JA3332⟩; **2**, 159 ⟨76JA3332⟩
Freidlina, R. K., **5**, 38, 89 ⟨57JGU2481⟩, 55, 60, 89 ⟨62T69⟩
Freidlina, R. Kh., **4**, 13 ⟨83S773⟩, 899, 902 ⟨81CA96556w⟩; **6**, 5 ⟨87IZV808⟩, 323 ⟨82IZV460⟩
Freidlin, G. N., **5**, 105 ⟨80MI 502-01⟩
Freidrich, E. C., **2**, 33 ⟨83JOC1678⟩; **4**, 29 ⟨75SC33⟩
Freifelder, M., **1**, 71, 87 ⟨B-78MI 103-01⟩, 88 ⟨58JA5270, 61JOC1835⟩, 89, 90 ⟨B-71MI 103-01⟩, 91 ⟨62JOC4046⟩, 502 ⟨B-78MI 112-01⟩, 548 ⟨B-71MI 112-01⟩
Freiman, H. S., **3**, 527 ⟨76JOC3303⟩
Freimuth, F., **2**, 212 ⟨54CB1661⟩
Freire, R., **1**, 931 ⟨86TL383, 87TL3397⟩; **3**, 409 ⟨86S1055⟩; **5**, 1125 ⟨86MI 530-01⟩; **6**, 633, 634 ⟨89CC452⟩
Freireich, S., **5**, 318 ⟨72JOM(35)303⟩
Freis, R. E., **5**, 359 ⟨70OPP275⟩
Freiser, H., **2**, 278 ⟨64ANC1768⟩, 279 ⟨89T1611⟩, 941 ⟨53JA2821⟩
Freitag, D., **6**, 471 ⟨91AG(E)1598⟩
Freitag, S., **5**, 1086 ⟨92AG(E)437⟩; **6**, 196 ⟨93CB2247⟩
Freitag, W. O., **5**, 1136 ⟨59JA5106⟩
Frejd, T., **1**, 216 ⟨87TL2753⟩, 218 ⟨90S285⟩, 318 ⟨76ACS(B)485⟩, 465 ⟨83OM598, 87S40⟩, 1072 ⟨76ACS(B)341⟩; **3**, 68, 69 ⟨83JOC1921⟩, 625 ⟨87S40⟩; **4**, 176 ⟨86TL753⟩, 785 ⟨76ACS(B)439, 76CS133⟩, 872 ⟨73ACS2242⟩
Frelek, J., **3**, 157, 161 ⟨91T3753⟩
Fremont, S. L., **4**, 846 ⟨84TL5729⟩
French, A. F., **6**, 653 ⟨63JMC283⟩
French, D., **2**, 94 ⟨93JA1379⟩
French, H. E., **4**, 53 ⟨21JA651⟩; **6**, 504 ⟨26JA1736⟩
French, J. C., **3**, 558, 560, 561 ⟨54JA4398, 53JA657⟩
Frenette, R., **2**, 135 ⟨83JOC1357⟩; **3**, 307 ⟨83JOC3214⟩
Frenette, R. L., **2**, 16 ⟨91JOC5962⟩; **3**, 149 ⟨91JOC5962⟩
Freni, M., **5**, 386 ⟨74JOM(76)355⟩
Frenkel, A. S., **4**, 724 ⟨72ZOB2450⟩
Frenking, G., **2**, 999, 1005 ⟨91CB543⟩, 1003 ⟨91CB861⟩; **3**, 358 ⟨92CB265⟩; **4**, 646 ⟨91CB543⟩; **5**, 928 ⟨84CPL521⟩
Freppel, C., **2**, 40 ⟨71CJC2586⟩
Frère, J.-M., **6**, 434 ⟨88EJM561⟩
Frese, E., **1**, 1040 ⟨65CB3311⟩; **3**, 362, 365 ⟨68CB716⟩, 549 ⟨62CB3077⟩, 550 ⟨70CB949⟩; **6**, 87 ⟨65CB3303⟩
Fresneda, P. M., **5**, 1066 ⟨87S45, 88TL379⟩, 1066 ⟨89CB307⟩, 1107 ⟨82S472⟩
Fressel, H., **2**, 388 ⟨12LA(392)133⟩
Fretwell, P., **1**, 494 ⟨94T359⟩
Fretz, E. R., **1**, 214 ⟨82JOC3211⟩; **2**, 358 ⟨77JOC875⟩
Freudenberg, K., **1**, 21 ⟨28CB1750⟩
Freudenberger, J. H., **2**, 81 ⟨89JA8014⟩, 681 ⟨86OM398⟩; **3**, 516 ⟨85OM1937⟩, 521 ⟨87OM893⟩
Freudenreich, B., **5**, 657 ⟨79LA74⟩
Freund, H., **3**, 628 ⟨65AG(E)1079⟩
Freund, M., **5**, 32 ⟨1884CB780⟩
Freund, W., **3**, 344 ⟨87JOC2154⟩
Frew, A. A., **4**, 585, 691 ⟨82CC614⟩
Frey, A., **4**, 885, 976, 977 ⟨73HCA944⟩
Frey, A. J., **5**, 164 ⟨58T1⟩; **6**, 104 ⟨63GEP1146892, 63T1661⟩, 104 ⟨63T1675⟩
Frey, B., **3**, 224 ⟨90CB363⟩
Frey, F., **2**, 441 ⟨85USP4507503, 85USP4507504⟩
Frey, F. W., **2**, 1101 ⟨B-68MI 223-02⟩; **5**, 1144 ⟨B-68MI 530-03⟩
Frey, G., **5**, 397, 398 ⟨80CB3268⟩, 403 ⟨79CB763, 80CB2769⟩
Frey, H., **1**, 1016 ⟨80LA2061⟩; **2**, 301 ⟨90S931⟩; **4**, 966 ⟨80LA2061⟩
Frey, H. J., **5**, 1111 ⟨64AG861⟩
Frey, H. M., **1**, 402, 794, 795 ⟨69CRV103⟩, 799 ⟨69CC1069, 70JCS(B)1410⟩, 801 ⟨66JCS(A)1342⟩, 805 ⟨63MI 118-02, 66MI 118-03⟩, 807 ⟨68MI 118-01⟩; **3**, 541 ⟨70JCS(B)830⟩
Frey, H. O., **6**, 57, 105 ⟨72TL4217⟩
Frey, M., **3**, 517 ⟨93JA8167⟩, 683 ⟨85MI 320-01⟩; **4**, 979 ⟨92AP(325)551⟩
Frey, P., **2**, 380 ⟨87OS(65)173⟩
Frey, R. M., **1**, 795, 801 ⟨64JCS5578, 61JCS3547⟩, 801 ⟨64JCS959⟩
Freyen, P.,**5**, 367 ⟨94PS(89)57⟩
Freyer, A. J., **2**, 245 ⟨84JA7861⟩
Freyer, W., **4**, 566, 656, 660 ⟨61JOC2604⟩

Freytag, H., **2**, 347, 348 ⟨58HOU(11)190⟩, 361 ⟨56GEP948330⟩
Friary, R., **1**, 91 ⟨81T3615⟩, 1082 ⟨86JOC3214⟩
Friary, R. J., **3**, 322 ⟨67JA5497⟩
Frick, Jr., J. G., **2**, 413 ⟨59JOC1419⟩
Frick, N., **2**, 760 ⟨53JA3632⟩
Frick, U., **2**, 539 ⟨82S1⟩; **6**, 43 ⟨82S1⟩
Fricke, H. J., **6**, 488 ⟨92GEP4219418⟩
Fridinger, T. L., **6**, 77 ⟨67JA5724⟩
Fridkin, M., **2**, 910, 931 ⟨63CJC2977⟩; **4**, 807 ⟨63CJC2977⟩
Fridland, S. V., **2**, 448 ⟨76JGU1213⟩, 459 ⟨74ZOB1261⟩, 833, 837, 838, 839 ⟨78ZOB319⟩; **3**, 50 ⟨87JGU203⟩
Fridman, A. L., **2**, 385 ⟨69RCR640⟩; **4**, 995 ⟨72ZOR2457⟩; **6**, 56, 282 ⟨69IZV2617⟩, 56, 282 ⟨77IZV2058⟩, 145 ⟨70ZOR189⟩, 145 ⟨71ZOR1126⟩, 145 ⟨74ZOR2003⟩, 145 ⟨76KFZ35⟩, 145 ⟨77ZOR1559⟩, 283 ⟨69ZOR1317⟩, 368 ⟨68IZV2839, 69IZV2304⟩
Friebe, B., **5**, 559 ⟨86JPR(328)812⟩
Friebe, T. L., **5**, 329 ⟨93TL7371⟩
Fried, H. E., **1**, 861 ⟨81JOC632⟩; **6**, 254 ⟨75JOC1278⟩
Fried, J. A., **1**, 209 ⟨93JOC7304⟩
Fried, J., **1**, 99 ⟨67JA1044⟩, 1023 ⟨68CC634, 69TL1379⟩, 1023 ⟨70TL2695⟩, 1023, 1024 ⟨73TL3899⟩; **2**, 274 ⟨64JOC3561⟩; **3**, 226 ⟨57JA1130⟩; **4**, 161, 212 ⟨83S891⟩, 734 ⟨84TL4329⟩
Fried, J. H., **1**, 458 ⟨72JA9256⟩, 540 ⟨69JA1853⟩, 881 ⟨60JA1684, 60JA5704⟩; **2**, 1016 ⟨61JA4663⟩; **4**, 36 ⟨61JA4105⟩, 187, 199 ⟨72JA7827⟩, 755 ⟨59JA2078, 60JA3091⟩; **6**, 18 ⟨60JA3091⟩, 18, 225 ⟨59JA2078⟩
Friedel, C., **1**, 462 ⟨1884AC(P)449⟩; **2**, 108 ⟨1867LA(143)118⟩
Friedel, R. A., **6**, 524 ⟨55JA3951⟩
Friederich, C., **4**, 715 ⟨94JOM(471)249⟩
Friederich, W., **5**, 596 ⟨77AP(310)614⟩
Friedheim, E. A., **2**, 237 ⟨71USP3544593⟩
Friedinger, R. M., **1**, 1082 ⟨70T3487⟩; **4**, 417 ⟨87JOC2442⟩
Friedlander, B., **2**, 363 ⟨77TL963⟩
Friedlander, B. T., **2**, 228 ⟨78JOC3481⟩; **5**, 1111 ⟨79S181⟩
Friedlander, P., **3**, 251 ⟨06CB1060⟩
Friedl, T., **1**, 9 ⟨87JOC4299⟩, 880 ⟨86AG(E)349⟩
Friedli, F. E., **4**, 37 ⟨91JOC1663⟩
Friedlina, R. Kh., **1**, 395 ⟨B-65MI 109-01⟩
Friedman, A. J., **2**, 358 ⟨77JOC967⟩, 365 ⟨74JA5000⟩
Friedman, A. R., **2**, 742 ⟨72JOC1902⟩, 743 ⟨74JOC1432⟩; **4**, 864 ⟨72JOC1902⟩
Friedman, B. S., **1**, 462 ⟨58JA5867⟩; **2**, 478 ⟨39JA71⟩
Friedman, H. L., **2**, 274 ⟨50JA3595⟩
Friedman, H. S., **2**, 880 ⟨79CC880⟩
Friedman, L., **1**, 39 ⟨65JA935⟩, 119, 464 ⟨74JA7101⟩, 630 ⟨67JA5736⟩, 631 ⟨65JA505, 68JA531⟩; **2**, 632 ⟨68JOC1636⟩, 726 ⟨60OS(40)93⟩; **5**, 62 ⟨55OSC(3)510⟩, 115 ⟨73OSC(5)810⟩
Friedman, L. J., **3**, 59 ⟨53JA384⟩
Friedman, N., **1**, 897, 901, 911 ⟨63JA2991⟩, 901 ⟨63JA2995⟩, 901, 913 ⟨63JA2998, 65TH⟩
Friedman, O. M., **2**, 414 ⟨54JA655⟩; **3**, 617 ⟨65CI(L)988⟩
Friedmann, R. C., **3**, 310 ⟨92JOC761⟩
Friedrich, A., **1**, 400, 799 ⟨69TL2607⟩
Friedrich, D., **1**, 826 ⟨91JOC3841⟩; **2**, 625 ⟨91JOC6199⟩
Friedrich, E. C., **1**, 352 ⟨82JOC1615⟩, 917 ⟨65JA3267⟩; **2**, 35 ⟨82JOM(226)143⟩, 964 ⟨68JA909⟩
Friedrich, F., **5**, 1027 ⟨65AP(298)51⟩, 1111 ⟨67AP583⟩
Friedrich, H., **2**, 829 ⟨34LA(509)138⟩
Friedrich, H. B., **4**, 148 ⟨89JOM(366)175⟩
Friedrich, H. H., **5**, 1145 ⟨68SAP6903273⟩
Friedrich, H.-J., **5**, 344 ⟨74LA561⟩
Friedrich, J. D., **3**, 635 ⟨87JOC2442⟩; **4**, 257 ⟨87JOC2442⟩; **5**, 630 ⟨87JOC2442⟩
Friedrich, K., **1**, 862 ⟨66AG(E)420⟩, 998 ⟨B-83MI 121-03⟩; **3**, 349 ⟨81TL2971⟩, 373 ⟨76LA641⟩, 493 ⟨59CB2756⟩, 564 ⟨67AG(E)959, 70CB1982⟩, 611, 627 ⟨B-70MI 318-01⟩, 641, 642, 661 ⟨B-70MI 319-01⟩, 641, 661, 662, 663, 664, 665, 666, 673 ⟨B-83MI 319-02⟩; **4**, 792, 894 ⟨67AG(E)959, 70CB1982⟩; **5**, 592, 630 ⟨77TL2139⟩; **6**, 111 ⟨79CB1873⟩
Friedrich, M. E. P., **2**, 483, 487 ⟨30JA376⟩
Friedrich, P., **6**, 720, 722, 723 ⟨77CB3467, 77JOM(128)C49⟩
Friedrichsen, W., **3**, 322 ⟨88TL2045⟩
Friend, R. H., **1**, 623 ⟨92JCS(P1)3225⟩
Friere, R., **2**, 13 ⟨86JOC402⟩
Fries, K., **2**, 656 ⟨08CB4271⟩
Fries, R. W., **5**, 463 ⟨74JA4983⟩
Fries, W., **2**, 1098 ⟨78JOM(159)373⟩
Friese, C., **1**, 653 ⟨89T3163⟩; **6**, 327, 351 ⟨89TL7033⟩
Friesen, R. W., **2**, 767 ⟨82TL821⟩; **5**, 692 ⟨93TL5983⟩
Friess, B., **1**, 496 ⟨92BSF273⟩; **4**, 535 ⟨89TL3963⟩
Friess, R., **5**, 561 ⟨70LA(737)179⟩
Friessen, R. W., **4**, 651 ⟨91JOC1944⟩
Frigo, T. B., **2**, 383 ⟨84JA3366, 86T1769⟩
Friis, P., **3**, 617 ⟨74ACS(B)317⟩
Frijns, J. H. G., **4**, 545 ⟨90OM1222⟩
Frimer, A. A., **2**, 660 ⟨94JOC1831⟩; **4**, 188 ⟨77S578⟩, 205 ⟨89JOC4866⟩
Frimer, A., **2**, 648 ⟨76TL2809⟩
Frimpong-Manso, S., **6**, 558 ⟨90MRC1045⟩
Fringuelli, F., **1**, 601 ⟨B-86MI 114-03⟩; **2**, 146 ⟨93SC3157⟩; **3**, 106 ⟨71CC1441⟩; **4**, 873 ⟨77JCS(P2)775⟩
Fringuelli, R., **1**, 44 ⟨79CC959, 85JCS(P1)493⟩
Friour, G., **3**, 130 ⟨84T683⟩, 152 ⟨84S37⟩, 211, 273 ⟨85S50⟩
Frisch, K. C., **6**, 488 ⟨B-89MI 615-01⟩
Frisch, M. A., **5**, 164 ⟨61JA606⟩
Frisch, M. J., **1**, 1143 ⟨81CPL408⟩
Frischleder, H., **1**, 981 ⟨70TL4315⟩
Frisell, C., **2**, 684 ⟨61OS(41)91⟩
Frisque-Hesbain, A.-M., **2**, 758 ⟨79CC1180⟩; **3**, 475 ⟨79CC1180⟩, 529 ⟨81JA4616⟩, 602 ⟨82JA2920⟩; **4**, 790 ⟨79CC1180⟩; **5**, 2 ⟨79CC1180⟩
Fristad, W. E., **1**, 422 ⟨79TL1999, 80JOC3017⟩, 451 ⟨79JA4420⟩; **2**, 24 ⟨84TL607⟩, 911 ⟨78JOC1620, 80JOC3017⟩; **3**, 126 ⟨80JOC3028⟩, 211 ⟨80JOC3017⟩; **4**, 357 ⟨78JOC1620, 80JOC3028⟩
Fritch, J. R., **1**, 991 ⟨85IJ131⟩; **6**, 403 ⟨79JA2768, 83JA1384⟩
Fritsch, E., **5**, 944 ⟨93JOM(460)C25⟩
Fritsch, F., **1**, 432 ⟨84JMC1575⟩
Fritsch, N., **1**, 521 ⟨79JA7001⟩
Fritsch, P., **1**, 1144 ⟨1894LA(279)319⟩
Fritsche, K., **4**, 506, 513 ⟨91JOM(415)39⟩; **6**, 127 ⟨90JOM(388)47⟩
Fritz, A. W., **5**, 342 ⟨82JOC176⟩, 346 ⟨82JOC5160⟩
Fritz, C., **2**, 473, 474 ⟨70ZAAC(372)171⟩
Fritz, G., **2**, 471 ⟨83ZAAC(504)47, 90ZAAC(589)23⟩, 471 ⟨91ZAAC(549)36⟩, 473 ⟨86ZAAC(535)106⟩, 1092 ⟨70ZAAC(377)48⟩; **4**, 545 ⟨77ZAAC(431)76⟩, 602 ⟨65MI 414-01⟩, 605, 621 ⟨70ZAAC(372)21⟩, 609 ⟨83ZAAC(497)134, 88JOM(341)109⟩, 609, 621 ⟨81ZAAC(473)59⟩, 617 ⟨62ZAAC(315)157, 84ZAAC(512)103, 70ZAAC(372)59⟩, 621 ⟨72ZAAC(391)219, 76ZAAC(426)28, 77ZAAC(433)61, 81ZAAC(478)9, 88ZAAC(556)23⟩, 622 ⟨80ZAAC(464)107⟩, 652 ⟨70ZAAC(377)37⟩, 728 ⟨67AG(E)677⟩, 1050, 1052 ⟨84JOM(271)107⟩; **5**, 1155 ⟨76ZAAC(422)104, 77ZAAC(428)222⟩; **6**, 165 ⟨84ZAAC(511)108⟩, 173 ⟨67AG(E)677⟩, 173, 383 ⟨93ZAAC(619)1494⟩, 178 ⟨72ZAAC(390)157⟩, 178 ⟨72ZAAC(390)191, 81ZAAC(475)87⟩, 178, 190, 191,

385, 391 ⟨83ZAAC(497)21⟩, 178, 290 ⟨84JOM(271)107⟩, 178, 290, 393 ⟨77ZAAC(430)121⟩, 178, 382 ⟨74TCC43⟩, 192 ⟨79ZAAC(448)40⟩, 267 ⟨71AG(E)510, 71ZN(B)480⟩, 267 ⟨72ZAAC(391)219, 74ZAAC(404)103, 76ZAAC(419)213, 78ZAAC(441)125, 79ZAAC(448)55⟩, 267, 268 ⟨71ZAAC(382)217, 71ZAAC(382)9, 78ZAAC(439)161, B-86MI 608-01⟩, 267, 383 ⟨87AG(E)1111, 60ZAAC(303)85⟩, 268 ⟨70ZAAC(372)21, 73ZAAC(395)159, 73ZAAC(399)280, 75ZN(B)965, 79ZAAC(458)37, 83AG(E)730⟩, 290 ⟨79JOM(178)11⟩, 379 ⟨76ZAAC(419)2⟩, 381, 385, 391 ⟨83ZAAC(497)119⟩, 382 ⟨81ZAAC(481)60⟩, 383 ⟨65MI 613-01, 73AG(E)654⟩, 383 ⟨74ZAAC(404)1, 84ZAAC(512)131, 84ZAAC(512)93⟩, 385 ⟨83ZAAC(497)134⟩, 385, 392 ⟨77ZAAC(430)137⟩, 391 ⟨88JOM(341)109⟩, 704 ⟨84ZAAC(511)95, 84ZAAC(518)14⟩
Fritz, H., **1**, 651 ⟨86JOC558⟩; **3**, 315 ⟨80AG(E)47⟩, 617 ⟨76HCA2786⟩; **4**, 177 ⟨79CB3603⟩, 265 ⟨89CB1757⟩; **5**, 375 ⟨86JOC558⟩, 484 ⟨80ACS(B)313⟩, 488, 492 ⟨89CB1757, 90JST171, 85TL5505⟩, 570, 608, 610 ⟨78BSB525⟩, 1124 ⟨85HCA1155⟩
Fritz, H. P., **1**, 862 ⟨59MI 119-05⟩; **2**, 483 ⟨70CB799⟩; **4**, 7 ⟨81ZN(B)1375⟩
Fritz, M., **3**, 694 ⟨93CPB1166⟩
Fritz, P., **3**, 218 ⟨68LA(712)79⟩
Fritz, P. M., **5**, 936 ⟨92JOM(426)343⟩
Fritzberg, A. R., **2**, 309 ⟨89JOC1940⟩
Fritzche, H., **2**, 440 ⟨65CB171⟩; **3**, 391, 586 ⟨61CB2477⟩
Fritzen, E., **1**, 966 ⟨78JOC1389⟩
Fritz-Langhals, E., **2**, 9 ⟨93TL293, 94TL1851⟩
Fritzsche, B., **2**, 363 ⟨70JPR129⟩
Fritzsche, H., **2**, 381 ⟨66JPR96⟩, 428, 431 ⟨65CB1681⟩, 821, 827, 828 ⟨64CB1988⟩; **4**, 69 ⟨60CB2340⟩
Fritzsche, K., **1**, 637, 638 ⟨88TL2805, 88C106⟩
Frobel, E., **3**, 566 ⟨36LA(521)95⟩
Froböse, R., **2**, 1094, 1095 ⟨B-83MI 223-01⟩, 1094, 1095 ⟨B-83MI 223-02⟩, 1094, 1095 ⟨B-86MI 223-01⟩, 1094, 1095 ⟨B-87MI 223-01⟩, 1094 ⟨B-87MI 223-02, B-87MI 223-03⟩, 1094 ⟨B-87MI 223-04⟩
Froch, S., **4**, 1056 ⟨87LA977⟩
Froech, S., **2**, 698 ⟨85TL1643⟩
Froemsdorf, D. H., **1**, 632, 633 ⟨60JA634⟩
Froen, D., **1**, 530 ⟨89S472⟩
Frohberger, P. E., **6**, 439 ⟨82EUP46557⟩
Frohler, M., **5**, 789 ⟨76CZ340⟩
Fröhlich, H., **1**, 296 ⟨74CA(81)120222t⟩; **6**, 267, 268 ⟨71ZAAC(382)9, 71ZAAC(382)217⟩
Fröhlich, H. O., **2**, 569 ⟨79ZAAC(458)89⟩
Fröhlich, R., **2**, 207 ⟨92SL985⟩; **5**, 953 ⟨95OM2447⟩
Fröhling, A., **6**, 84, 86, 303 ⟨62RTC1009⟩
Frohn, H. J., **2**, 492 ⟨86JFC(34)129⟩
Frölisch, B., **3**, 152 ⟨84S734⟩
Fromagest, C., **5**, 83 ⟨53MI 502-02⟩
Fromberg, W., **5**, 797 ⟨85JOM(280)355⟩
Fromeld, H. D., **3**, 10 ⟨63AG(E)683⟩
Fromm, E., **2**, 1034 ⟨12LA(394)325, 23CB2290⟩; **4**, 772 ⟨12LA(394)325, 22CB1014⟩
Frommeld, H.-D., **1**, 881 ⟨64CB3548⟩; **5**, 637, 819, 832, 837 ⟨66JOC157⟩, 828, 837 ⟨65LA(687)191⟩
Fromuth, T. E., **4**, 268 ⟨81PS(10)163⟩
Fronczek, F. R., **1**, 690 ⟨84JOC1627⟩; **4**, 545 ⟨89PS(42)97⟩, 550 ⟨88JA5585, 89IC1872⟩, 562 ⟨86IC1290⟩
Fronza, G., **2**, 359 ⟨81TL5073⟩; **4**, 170 ⟨89JOC5171⟩, 584 ⟨76JOM(111)361⟩
Fronzaglia, A., **5**, 449 ⟨66JOM(5)341⟩
Frosch, F., **3**, 653 ⟨71ZN(B)1124⟩
Frosina, K. V., **6**, 54 ⟨67ZOB152⟩
Frosin, N., **5**, 821 ⟨59DOK(127)337⟩
Frost, A. A., **6**, 386 ⟨81TCA285⟩
Frost, C. G., **1**, 192, 194 ⟨92TA1089⟩, 194 ⟨93TL3149⟩
Frost, II, K. A., **2**, 607, 624 ⟨72T4883⟩
Frost, L. W., **6**, 220, 225 ⟨47IEC404⟩
Frostick, F. C., **1**, 881 ⟨55JOC1531⟩
Frostick Jr., F. C., **5**, 205 ⟨57JA5982⟩
Fröstl, W., **5**, 288 ⟨75HCA587⟩
Frøyen, P., **1**, 737 ⟨71ACS1452⟩, 740, 741 ⟨71ACS2541⟩; **2**, 809 ⟨69ACS2935⟩, 809, 892 ⟨71ACS983⟩, 886 ⟨75ACS(B)362⟩; **3**, 194 ⟨91PS(60)81⟩, 328 ⟨70ACS3772⟩, 591, 592 ⟨74ACS(B)586⟩; **5**, 367, 373 ⟨93PS(78)161⟩
Fruchey, O. S., **2**, 8 ⟨77TL2999, 79JOC359⟩
Frueh, A., **3**, 390 ⟨91JOM(410)C9⟩
Früh, A., **1**, 991 ⟨92CB2667⟩
Fruh, A., **4**, 1018 ⟨91OM389⟩
Frühauf, H.-W., **5**, 849 ⟨72LA(760)102⟩
Fruit, Jr., R. E., **1**, 894 ⟨65JA3000⟩
Fruttero, R., **4**, 429 ⟨87JOC3442⟩
Fry, A., **1**, 629 ⟨91JA8821⟩, 783 ⟨83JA3714⟩; **5**, 998 ⟨68MI 526-01⟩
Fry, A. J., **1**, 87 ⟨B-72MI 103-01⟩, 906 ⟨69JOC2314⟩; **3**, 149 ⟨79TL3357⟩; **4**, 203 ⟨91JOC3283⟩
Fry, D. F., **2**, 319 ⟨88SC29⟩
Fry, D. W., **5**, 567 ⟨93JMC2459⟩
Fry, J. L., **1**, 13, 18 ⟨79TL849⟩, 20 ⟨81OS(60)108⟩, 104 ⟨70JA2542⟩, 136 ⟨90TL6155⟩, 1114 ⟨70JA2538, 70JA2540⟩; **2**, 68 ⟨72JA4628⟩; **3**, 13, 86 ⟨81JOC602⟩, 86, 100 ⟨74CC45⟩
Frydendahl, H., **3**, 342 ⟨91S609⟩
Frydman, B., **1**, 259 ⟨71JA2738⟩; **3**, 106 ⟨74JOC2872⟩
Frydman, N., **1**, 1111 ⟨69CC1044⟩; **4**, 742 ⟨70JA3203⟩
Frydman, R. B., **1**, 259 ⟨71JA2738⟩
Frye, C. L., **2**, 547 ⟨61JA996⟩; **4**, 650 ⟨70JOC1308⟩
Frye, S. V., **4**, 230 ⟨85JOC3402, 87OS(65)215⟩; **5**, 330 ⟨91JOC3750⟩
Fryer, R. I., **3**, 261 ⟨74JHC219⟩, 310 ⟨93SC985⟩; **4**, 412 ⟨74JOC167⟩; **5**, 327 ⟨77JOC3301⟩
Fryermuth, H. B., **2**, 156 ⟨63JOC1140⟩
Fryxell, G. E., **3**, 320 ⟨89JOC817⟩, 321 ⟨90JA3100⟩; **4**, 520 ⟨89JOC644⟩
Fryzuk, M. D., **2**, 436 ⟨88JOC4425⟩, 992 ⟨86TL1537⟩; **3**, 509 ⟨93JA5336⟩; **4**, 566 ⟨85IC642⟩
Fryzuk, M. D.,**5**, 380 ⟨82OM1540, 84JA6993⟩
Fryzuk, M. D., **6**, 516 ⟨89CCR1⟩
Fu, C., **1**, 261 ⟨90JOC3209⟩
Fu, C. C., **3**, 14 ⟨81S303⟩
Fu, G. C., **1**, 299 ⟨93CRV1307⟩; **2**, 42 ⟨88JA6917⟩, 43 ⟨92JA4042⟩, 527 ⟨91JA4042, 92JA6671⟩; **3**, 127 ⟨90JOC5678⟩
Fu, L. H., **3**, 617 ⟨93SC1061⟩; **5**, 666 ⟨93SC1061⟩
Fu, P. P., **1**, 554 ⟨78CRV317⟩; **2**, 659 ⟨78CRV317, 79JOC4265⟩
Fu, W., **6**, 221, 222, 226 ⟨92MI 607-03⟩
Fu, W. Y., **1**, 693 ⟨78JA2140⟩
Fu, X., **1**, 776 ⟨90TL3409, 92JOC5121⟩, 810 ⟨93JOC661⟩, 1038, 1140 ⟨91TL1707⟩
Fuccello, A., **4**, 228 ⟨79CAR(70)217⟩, 238, 287 ⟨77CAR(59)351⟩
Fuchigami, T., **2**, 160 ⟨81JOC5215⟩; **3**, 637 ⟨92TL3161⟩; **4**, 62 ⟨92JOC3755⟩, 62, 63 ⟨90JOC6074⟩, 62, 63 ⟨91CC1027⟩, 62, 63 ⟨93JOC4200⟩, 87, 92 ⟨92TL3161⟩, 221 ⟨86TL3869, 87CL1095⟩; **5**, 194 ⟨87JAP6289642⟩, 758, 762 ⟨76BCJ3607⟩, 766 ⟨74CL247⟩
Fuchikami, T., **1**, 1130 ⟨84TL303⟩; **4**, 613, 614 ⟨78JOM(149)37⟩, 736 ⟨86TL3173⟩, 810 ⟨84TL303⟩, 1053 ⟨77JOM(142)C45⟩; **5**, 194 ⟨88TL4435⟩, 224 ⟨89JOC4511⟩; **6**, 21 ⟨84TL303⟩, 22 ⟨91TL91⟩, 178 ⟨81JA2324⟩, 178, 383 ⟨79JA1348⟩
Fuchita, T., **3**, 618 ⟨78S905⟩
Fuchita, Y., **2**, 582 ⟨85JCS(D)873⟩; **5**, 801 ⟨80BCJ1171⟩
Fuchs, B., **1**, 790 ⟨67TL4945⟩; **2**, 626 ⟨82CC778⟩; **5**, 199 ⟨87S850⟩

Fuchs, C. F., **2**, 117 ⟨48JOC97⟩

Fuchs, E., **4**, 483 ⟨91S1099⟩; **5**, 395 ⟨91CB2843, 91S1099⟩; **6**, 512, 582 ⟨91S1099⟩

Fuchs, E. P. O., **3**, 689 ⟨87JOM(338)329⟩; **4**, 549 ⟨87AG(E)1011⟩; **5**, 1152, 1159 ⟨87JOM(338)329⟩

Fuchs, G., **2**, 117, 124 ⟨65ACS1490⟩; **4**, 327 ⟨70CB2775⟩

Fuchs, H., **6**, 523 ⟨41ZAAC(248)256⟩

Fuchs, J., **6**, 590, 592, 593 ⟨92JOM(427)213⟩

Fuchs, O., **2**, 59, 60 ⟨54NAT125⟩, 59, 60 ⟨55NAT346⟩; **4**, 881 ⟨56CB2060⟩; **6**, 73, 97, 300 ⟨56CB2060⟩

Fuchs, P., **2**, 217 ⟨88SC227⟩

Fuchs, P. L., **1**, 144 ⟨78JA346⟩, 464 ⟨75JA7372⟩, 486 ⟨85JA6137⟩, 572 ⟨78JOC1208⟩, 580, 781 ⟨94JOC311⟩, 618 ⟨90JA8084⟩, 640, 641 ⟨71JA4070⟩, 640, 641 ⟨73JA822⟩, 652 ⟨87SC621⟩, 736 ⟨73JOC3625⟩, 748 ⟨87JA4755⟩, 1067 ⟨72TL3769⟩; **2**, 158 ⟨82JOC3121⟩, 444 ⟨85JOC4252⟩, 446 ⟨77S628⟩, 679 ⟨94JOC311⟩; **3**, 18, 133 ⟨76S456⟩, 55 ⟨71JOC366⟩, 172 ⟨83JOC914⟩, 197 ⟨86SC1745⟩, 220 ⟨85JOC4252⟩, 495 ⟨72TL3769⟩, 496 ⟨74JA1607, 74TL4055⟩; **4**, 174 ⟨86JOC902⟩, 273 ⟨90JA8084⟩, 273, 275 ⟨92TL745⟩, 284 ⟨91JA9864⟩, 365, 393, 394 ⟨91SC(21)1675⟩, 384 ⟨87JA4930⟩, 745 ⟨72TL3769⟩, 948 ⟨88JOC3647⟩; **5**, 330 ⟨88JA2341, 90JA9601⟩, 576 ⟨94JOC348⟩; **6**, 695 ⟨72TL3769⟩

Fuchs, R., **1**, 521 ⟨68AG(E)67⟩; **2**, 25 ⟨75CJC3620⟩; **4**, 894 ⟨67CB2604⟩, 972 ⟨68CB1232⟩; **6**, 426 ⟨86S330⟩, 632 ⟨67CB2604⟩

Fueno, T., **1**, 459 ⟨80JOC1604⟩, 707 ⟨76T1705⟩, 1125 ⟨73JA4090, 74JOC2255⟩; **4**, 1074 ⟨86BCJ453⟩, 1075 ⟨90BCJ3056, 91BCJ2751⟩; **6**, 96, 98 ⟨87JOC2657⟩

Fuentes, L., **5**, 606 ⟨85S86⟩

Fuentes, L. M., **1**, 149 ⟨83JA117⟩, 149 ⟨83JA2082⟩, 149 ⟨84T1361⟩; **4**, 532 ⟨84T1361⟩, 831 ⟨88JOC633⟩

Fugami, K., **1**, 195 ⟨93TL7619⟩, 330, 925, 946 ⟨89BCJ143⟩, 333 ⟨89TL3155⟩, 480 ⟨87CL2203⟩, 801 ⟨88TL1543⟩; **2**, 918, 919, 939 ⟨88T4277⟩; **4**, 188 ⟨89BCJ2050⟩, 808, 811, 813 ⟨89TL3155⟩, 1045 ⟨88T4277⟩

Fuganti, C., **1**, 78 ⟨88JCS(P1)3061⟩; **2**, 359 ⟨81TL5073⟩

Fügedi, P., **4**, 223 ⟨87MI 405-01⟩

Fugger, J., **2**, 374 ⟨55JA1843⟩

Fuggle, J. C., **2**, 421 ⟨72JCS(D)1766⟩

Fugiwara, Y., **2**, 478 ⟨90CPB673⟩

Fügner, A., **4**, 586 ⟨78ZN(B)1325⟩

Fuhr, K. H., **1**, 486 ⟨78JOC4248⟩

Fuhrer, W., **3**, 268 ⟨85HCA1235⟩

Fuhrhop, J., **1**, 885 ⟨B-86MI 119-01⟩

Führling, G., **4**, 792 ⟨70TL1937⟩

Fuhrmann, F., **6**, 444 ⟨83JHC1477⟩

Fuhrmann, K.-D., **4**, 153 ⟨80ZN(B)1376⟩

Fuji, K., **1**, 120 ⟨93TL5127⟩, 134 ⟨91TL7281⟩, 141 ⟨85JOC657⟩, 241 ⟨87CC449⟩, 428 ⟨84JOC3618⟩; **2**, 50 ⟨92TL4021⟩, 658 ⟨80JOC4275⟩, 784 ⟨92S1119⟩, 1062 ⟨87CC1708⟩; **3**, 47, 192 ⟨87S729⟩, 161 ⟨92S852⟩, 162 ⟨89JOC4038⟩, 221 ⟨91TA178⟩; **4**, 235, 236 ⟨86CL1655⟩, 808 ⟨88CC1076⟩; **5**, 83 ⟨81JOC5039⟩, 171 ⟨85JA6404⟩; **6**, 98, 101, 132, 134 ⟨81TL2005⟩, 101 ⟨85JOC657⟩, 119, 126, 132 ⟨85JOC662⟩

Fuji, M., **3**, 47, 192 ⟨88BCJ4029⟩

Fuji, S., **2**, 357, 691 ⟨73JOC1239⟩

Fujiami, T., **2**, 564 ⟨86CC475⟩

Fujihara, H., **1**, 483 ⟨86H(24)3337⟩; **2**, 25 ⟨81TL4409⟩, 212 ⟨91CL141⟩, 271 ⟨86H(24)3019⟩, 279 ⟨89CC1789⟩, 369 ⟨92JCS(P1)2247, 93JA9826⟩; **3**, 310 ⟨86TL3899⟩; **4**, 94 ⟨85CC1266⟩; **5**, 210 ⟨92CL1437⟩, 1106 ⟨81TL4409⟩

Fujii, A., **2**, 849, 850 ⟨71JA3694⟩; **4**, 1034 ⟨91CC1390⟩

Fujii, H., **1**, 836 ⟨91COS(8)895⟩; **6**, 542 ⟨91CPB1659, 92T9433⟩

Fujii, K., **2**, 618 ⟨79BCJ1197⟩; **3**, 495 ⟨79BCJ1197⟩; **6**, 550 ⟨82YGK352⟩

Fujii, M., **1**, 37 ⟨83TL3477⟩, 431 ⟨85JOC3283⟩; **2**, 398 ⟨88BCJ4029⟩; **3**, 46, 190 ⟨87S532⟩, 129 ⟨86CL87⟩; **5**, 403 ⟨84JOC2139, 87T3395⟩

Fujii, M. Y., **2**, 312 ⟨89BCJ3845⟩

Fujii, N., **1**, 188 ⟨90AG(E)609⟩, 188 ⟨91JOC4370⟩, 189 ⟨93JOC1207⟩, 189 ⟨93T9479⟩; **2**, 227 ⟨91SL345⟩; **4**, 218 ⟨89CPB526⟩

Fujii, S., **1**, 571 ⟨88OSC(6)327⟩; **2**, 52 ⟨91CC795⟩; **3**, 224 ⟨76JOC2073, 80OS(59)113⟩

Fujii, T., **1**, 290 ⟨93TL5601⟩, 752 ⟨88H(27)2077⟩, 931 ⟨91JOC2264⟩; **3**, 170 ⟨90BCJ1266⟩, 207 ⟨80TL3779⟩; **4**, 696 ⟨81JOC192⟩, 743 ⟨94JCS(P1)3⟩; **5**, 31 ⟨66BCJ852⟩, 378 ⟨81S451⟩, 560 ⟨84CC1010⟩, 597 ⟨88H(27)2077⟩

Fujii, Y., **1**, 76 ⟨B-56MI 102-01⟩

Fujikawa, K., **6**, 9 ⟨80GEP3008081, 91MI 601-02⟩

Fujikawa, M., **2**, 626 ⟨87BCJ4187, 88BCJ597⟩; **4**, 660 ⟨88JAP267291⟩

Fujikura, S., **1**, 160, 451 ⟨84TL1999⟩; **2**, 972 ⟨84TL1999⟩; **4**, 654 ⟨85TL2101⟩, 1060 ⟨84TL1999⟩

Fujikura, Y., **5**, 143 ⟨75BCJ2075⟩, 282 ⟨77S632⟩

Fujimori, M., **3**, 137 ⟨84BCJ1863⟩

Fujimori, T., **2**, 987 ⟨90CC182⟩; **4**, 858 ⟨90TL4621⟩

Fujimoto, A., **5**, 134 ⟨86CL871⟩

Fujimoto, E., **1**, 5, 8 ⟨86S570⟩; **3**, 129 ⟨86S570⟩

Fujimoto, K., **1**, 208 ⟨86BCJ1363⟩, 405, 816 ⟨83TL513, 93CL1397⟩, 816, 819 ⟨84TL6011⟩, 816, 819 ⟨86TL4185, 91SL629⟩

Fujimoto, K., **2**, 923 ⟨93AG(E)1473⟩; **3**, 59 ⟨81TL1021⟩, 197 ⟨92CC190⟩; **4**, 349 ⟨92CC190⟩

Fujimoto, M., **1**, 428 ⟨84JOC3641⟩; **5**, 1120 ⟨92CC1379⟩

Fujimoto, T., **2**, 314, 352 ⟨91TL525⟩

Fujimoto, Y., **1**, 961 ⟨75CC362⟩; **3**, 152, 154, 155 ⟨82S1021⟩

Fujimura, D., **2**, 74, 572 ⟨89JOC4732⟩

Fujimura, H., **1**, 181 ⟨75CPB1917⟩

Fujimura, M., **5**, 516, 520 ⟨92CL1843⟩

Fujimura, O., **1**, 228 ⟨90JOC1705⟩; **2**, 86 ⟨90JOC1705⟩

Fujimura, S., **6**, 413 ⟨79JOC3573⟩

Fujimura, T., **1**, 483 ⟨88JA3296⟩; **5**, 953, 956 ⟨92JA2985⟩, 958 ⟨93JA9015⟩

Fujinama, T., **3**, 225 ⟨89BCJ2348⟩; **5**, 1078 ⟨76CL811⟩

Fujinami, T., **1**, 364 ⟨89BCJ2348⟩; **2**, 279 ⟨90CL927⟩, 399 ⟨75JOC3291⟩; **5**, 557 ⟨82CL123⟩, 1024 ⟨74JOC1970, 75BCJ2981⟩, 1031 ⟨78NKK265⟩, 1035 ⟨77BCJ425⟩, 1078 ⟨77S889⟩; **6**, 73 ⟨76CL891⟩, 463 ⟨81CL749⟩, 554 ⟨82BCJ1174⟩

Fujinara, Y., **2**, 227 ⟨91CC167⟩

Fujino, H., **6**, 488 ⟨92JAP05339226⟩

Fujino, M., **2**, 274 ⟨78CPB1576⟩

Fujino, S., **4**, 549 ⟨80JOC4683⟩

Fujioka, A., **1**, 466, 470 ⟨76BCJ1958⟩

Fujioka, H., **1**, 827, 828 ⟨89CPB2047⟩, 1092 ⟨83TL1463⟩

Fujisaki, H., **6**, 636 ⟨84CPB4893⟩

Fujisaki, S., **2**, 21 ⟨88S545, 88TL5783⟩, 23 ⟨91CE185⟩, 24 ⟨90BCJ3033⟩, 28 ⟨87BCJ1159⟩, 149 ⟨89BCJ3376⟩, 320 ⟨89CL463⟩, 626 ⟨89BCJ439⟩, 631 ⟨87CL2109, 88BCJ600⟩, 631 ⟨88CL795⟩; **3**, 119 ⟨89BCJ2585⟩; **5**, 48, 49 ⟨85S674⟩, 318 ⟨84CL2045, 85BCJ769⟩, 318 ⟨89JCS(P1)1702, 93BCJ2426⟩

Fujisawa, F., **3**, 303 ⟨82CL1559⟩

Fujisawa, T., **1**, 21, 23 ⟨92TL5567⟩, 183 ⟨90CL593⟩, 183 ⟨92BCJ3487⟩, 230 ⟨80TL2181⟩, 230, 1107 ⟨82TL3587⟩, 352 ⟨92CL61⟩, 476 ⟨81CL1307, 82TL3583⟩, 476 ⟨83TL5745⟩, 481 ⟨82BCJ3555⟩, 970 ⟨84TL4007⟩, 980 ⟨85CL1457⟩; **2**, 18 ⟨91BCJ2596⟩, 25 ⟨84CL1173⟩, 226 ⟨72CL935⟩, 712 ⟨68TL5071⟩, 758 ⟨83CL1791⟩, 837 ⟨64JOC2572⟩, 859 ⟨67BCJ147⟩; **3**, 82, 102, 107 ⟨83TL1543⟩, 123 ⟨89CL1623⟩, 146, 212, 288, 296 ⟨81CL1135⟩, 216 ⟨81CL55⟩, 260 ⟨84CL2055⟩, 286 ⟨82TL5059⟩, 427 ⟨83CL1537⟩, 628 ⟨90CL1615⟩, 632

⟨89TL3811⟩, 633 ⟨89BCJ3851⟩, 634 ⟨73CL287⟩; **4**, 187 ⟨84TL5155⟩; **5**, 14 ⟨83CL835, 87OS(66)121⟩, 131 ⟨82CL1891⟩, 248 ⟨71JOC3546⟩, 553 ⟨83CL1901⟩, 557 ⟨86TL5405, 87NKK1414⟩; **6**, 434 ⟨73CL1315⟩
Fujise, Y., **1**, 816, 818 ⟨86JOC4315⟩
Fujishima, H., **1**, 1050 ⟨92CL695⟩; **3**, 147 ⟨92CL695⟩
Fujita, A., **2**, 27 ⟨93TL5725⟩; **3**, 308 ⟨86T6465⟩; **5**, 194 ⟨88TL4435⟩
Fujita, E., **1**, 141 ⟨85JOC657⟩, 426, 432 ⟨82TL689⟩, 428 ⟨84JOC3618⟩
Fujita, E.,**1**, 431 ⟨88T4095⟩
Fujita, E., **1**, 440 ⟨83CL1457⟩, 440, 616, 668 ⟨83CC619⟩, 490 ⟨81TL4491⟩, 592, 616, 668 ⟨82TL2205⟩, 600 ⟨82CC281⟩, 1080 ⟨85CC697⟩; **2**, 150 ⟨77TL1345⟩, 658 ⟨80JOC4275⟩, 1020, 1063 ⟨85TL4501⟩, 1062 ⟨87CC1708⟩; **3**, 36, 172 ⟨79TL3167⟩, 57 ⟨83TL777⟩, 85 ⟨78CC330⟩, 143 ⟨84CC1007⟩, 143, 225 ⟨81TL1115⟩; **4**, 198 ⟨78TL4115⟩, 392 ⟨84CPB1829⟩; **5**, 40, 106 ⟨78TL5211⟩, 83 ⟨81JOC5039⟩, 223 ⟨85JCS(P1)2361⟩, 277, 295 ⟨80TL841⟩, 329 ⟨86JOC2391⟩, 329, 330 ⟨85CC1418⟩; **6**, 98, 101, 132, 134 ⟨81TL2005⟩, 101 ⟨85JOC657⟩, 116 ⟨86JAP61158986, 86JOC4737⟩, 116 ⟨87NKK1447⟩, 119, 126, 132 ⟨85JOC662⟩
Fujita, H., **1**, 199 ⟨89TL851⟩, 811 ⟨81CC786⟩, 1044 ⟨72BCJ1846⟩
Fujita, J., **2**, 437 ⟨85BCJ2793, 90BCJ2051⟩, 437 ⟨91BCJ2942⟩; **4**, 721 ⟨93JOM(463)169⟩
Fujita, K., **5**, 366 ⟨88JA1638⟩
Fujita, M., **1**, 429 ⟨88T4135⟩, 693 ⟨82CL1697⟩; **2**, 18, 19 ⟨88CL1877⟩, 70 ⟨88T4134⟩, 177, 179 ⟨92CL891⟩, 179 ⟨91CL1391⟩, 305 ⟨91JOC2920⟩, 1000 ⟨82CL1697⟩; **3**, 106 ⟨83TL4573⟩, 108 ⟨82CPB1257⟩, 220 ⟨84TL4959⟩, 294 ⟨88CL1387⟩; **4**, 73 ⟨91TL1463⟩, 262 ⟨80TL2233⟩, 395 ⟨82CL1697⟩, 736 ⟨88T4135⟩, 819 ⟨85JA4085⟩, 858 ⟨90TL3931⟩; **5**, 213 ⟨88JOC5415⟩, 224 ⟨89JOC4511⟩, 1039 ⟨90CC1213⟩, 1105 ⟨87JOC681⟩, 1107 ⟨87CC1301⟩; **6**, 26 ⟨85JA4085, 88T4135⟩, 63 ⟨88BCJ3321⟩
Fujita, R., **1**, 595, 597 ⟨88H(27)261⟩
Fujita, S., **2**, 657 ⟨75TL1695⟩; **6**, 107 ⟨67T45⟩
Fujita, T., **2**, 713 ⟨83CL249⟩; **4**, 197 ⟨89TL5309⟩, 213 ⟨78JOC1248⟩, 736 ⟨93JFC(62)201⟩; **5**, 281 ⟨81S287⟩, 570, 613 ⟨92JOC2419⟩, 630 ⟨88JCS(P1)2203⟩, 1003 ⟨73BCJ1275⟩, 1106 ⟨72CPB427⟩; **6**, 618 ⟨73BCJ303⟩
Fujita, T. S., **4**, 553, 562 ⟨87JA5544⟩
Fujita, Y., **1**, 156 ⟨85SC1123⟩, 187 ⟨83S804⟩, 404, 405, 816, 818 ⟨81JA6492⟩, 413 ⟨78CC972⟩, 469 ⟨86AG(E)935⟩, 613 ⟨86JOC3834⟩, 808 ⟨78CC651⟩, 809 ⟨80TL1347⟩; **3**, 57 ⟨86BCJ3287⟩, 217 ⟨80TL1347⟩; **4**, 183 ⟨86TL69⟩, 225, 238 ⟨92TL239⟩; **5**, 30, 35 ⟨86SC499⟩, 357 ⟨82CL1607⟩; **6**, 70 ⟨76JAP(K)76108012⟩
Fujitsuka, H., **4**, 947 ⟨92JOC5610⟩
Fujiwa, T., **1**, 479 ⟨81S1001⟩
Fujiwara, A., **3**, 704 ⟨82ABC1803⟩
Fujiwara, H., **1**, 439, 450 ⟨82JOC4713⟩
Fujiwara, J., **1**, 1041 ⟨84TL2367⟩; **3**, 269, 437 ⟨83JA6312, 83JA2831⟩
Fujiwara, K., **1**, 552 ⟨91SL651⟩; **4**, 528 ⟨91TL1975⟩; **6**, 113 ⟨76BCJ3567, 80JOC2024⟩
Fujiwara, M., **2**, 318 ⟨89TL739⟩, 699 ⟨89CL1247⟩; **3**, 355 ⟨89MI 308-02⟩, 704 ⟨82ABC1803⟩; **4**, 829 ⟨88JOC5974⟩
Fujiwara, S., **1**, 1045 ⟨86S421⟩; **3**, 303 ⟨88TL1173⟩; **4**, 239, 334, 660 ⟨90OM1355⟩, 334, 909 ⟨90CL2053⟩, 387 ⟨85TL5827⟩; **5**, 1036 ⟨91TL3503⟩
Fujiwara, S.-I., **6**, 498 ⟨90SC703⟩
Fujiwara, T., **1**, 236 ⟨85CL1149⟩, 1053 ⟨87CL1345⟩; **3**, 137 ⟨84BCJ613⟩, 168 ⟨86TL3029⟩, 380 ⟨78CC197⟩, 634 ⟨87CL1963⟩; **4**, 369 ⟨87CL1963⟩, 398 ⟨92CL819⟩, 774 ⟨91TL6563⟩, 938, 958 ⟨84BCJ1863⟩
Fujiwara, W., **4**, 1074 ⟨86BCJ453⟩
Fujiwara, Y., **1**, 462 ⟨86CL357⟩, 692 ⟨78JOC2477⟩, 1004, 1107 ⟨79JOC55, 74TL1067⟩; **2**, 312 ⟨90CL203⟩, 616 ⟨74TL1067⟩; **3**, 699, 721 ⟨92CL1143⟩; **5**, 143 ⟨68TL633⟩, 197 ⟨82CC132⟩
Fujiyama, R., **3**, 136, 203 ⟨90JOC5562⟩
Fujiyama, Y., **5**, 774 ⟨92NKK1344⟩
Fukada, N., **4**, 842 ⟨75JCS(P1)1277⟩, 852 ⟨77JCS(P1)1273⟩; **5**, 547 ⟨73JCS(P1)705, 73JCS(P1)1009, 77JCS(P1)1273, 79JCR(S)140, 80S566, 82YGK123⟩, 555 ⟨77JOC3383⟩, 580 ⟨70JOC2438⟩, 652 ⟨78JCS(P1)558⟩; **6**, 324 ⟨73JCS(P1)1009⟩, 627 ⟨79S554⟩, 632 ⟨85BCJ3379⟩, 636 ⟨93BCJ148⟩
Fukami, H., **2**, 12 ⟨91CE751⟩
Fukami, J., **1**, 747 ⟨71ABC1116⟩
Fukami, S., **3**, 149 ⟨90JA8563⟩
Fukata, F., **2**, 263 ⟨83TL1273⟩
Fukawa, I.,**3**, 439 ⟨76BCJ550⟩
Fukaya, H., **6**, 219 ⟨90CL813⟩, 230 ⟨90JFC(50)173⟩, 257 ⟨90JFC(48)257⟩
Fukazawa, Y., **1**, 818 ⟨93TL8453⟩
Fuks, R., **2**, 1042 ⟨70CB564⟩; **3**, 566 ⟨88JOC18⟩, 687 ⟨80BCB433⟩; **4**, 885 ⟨64AG571⟩; **5**, 751 ⟨73T2147, 78BSB717⟩, 963 ⟨73T297⟩; **6**, 238 ⟨73BSB23⟩, 607, 610 ⟨73T297⟩
Fukuda, H., **4**, 281 ⟨63JOC1420⟩, 825 ⟨88TL2327⟩; **5**, 291 ⟨78BCJ1866⟩; **6**, 106 ⟨88TL2327⟩
Fukuda, K., **3**, 373 ⟨88S178⟩
Fukuda, M., **2**, 399 ⟨73BCJ2156⟩; **4**, 1003, 1004 ⟨75BCJ2103⟩; **5**, 252 ⟨81S637⟩; **6**, 158 ⟨75BCJ1030⟩
Fukuda, N., **3**, 256 ⟨76JCS(P1)1706⟩, 337, 340 ⟨84TL849, 82CC1187⟩; **4**, 852 ⟨76JCS(P1)1706⟩
Fukuda, T., **3**, 633 ⟨89BCJ3851⟩
Fukuda, Y., **1**, 205, 1024 ⟨84TL5423⟩, 1048 ⟨90SL701⟩; **3**, 114, 162 ⟨91JOC3729⟩, 410 ⟨91JOC5812⟩; **4**, 940 ⟨90SL701⟩
Fukugawa, Y., **3**, 684 ⟨84CC1513⟩
Fukuhara, N., **4**, 3 ⟨40JA1171⟩
Fukuhara, T., **1**, 462 ⟨79CL1003⟩, 1009 ⟨90BCJ2124⟩; **2**, 17 ⟨83CL1135, 84CL1241⟩, 19 ⟨85CL1693, 87CL1675⟩, 615 ⟨87SC685⟩; **6**, 552, 554 ⟨87NKK1408⟩
Fukui, K., **1**, 417 ⟨71ACR57⟩; **2**, 279, 288 ⟨67NEP6614945, 68NKK83⟩; **3**, 563 ⟨67JOC3564⟩; **4**, 892 ⟨73CL111, 76JCS(P1)783⟩; **5**, 1039 ⟨62MI 527-01⟩
Fukui, M., **4**, 950 ⟨87JA1269⟩; **5**, 645 ⟨74JOM(78)339⟩, 719, 720 ⟨72JCS(P2)1043⟩; **6**, 684 ⟨69JCS(C)2002⟩
Fukui, Y., **5**, 279 ⟨88CC419⟩
Fukumasa, M., **1**, 217 ⟨91TL1059⟩; **2**, 85 ⟨90TL7031⟩
Fukumoto, K., **1**, 831 ⟨90CC1001⟩, 1082 ⟨90TL6205⟩; **2**, 16 ⟨92JCS(P1)221⟩, 328 ⟨89H(28)63⟩; **3**, 220 ⟨90JOC4497⟩, 294 ⟨87CC1360⟩, 620 ⟨86JOC2833⟩; **4**, 154 ⟨85JCS(P1)1185⟩, 387 ⟨83TL4257⟩; **5**, 131 ⟨82CL1891⟩, 179 ⟨85CPB532⟩, 227 ⟨85H(23)549⟩, 276 ⟨69JCS(C)874⟩, 285 ⟨83JA7345⟩
Fukumoto, S., **2**, 240 ⟨91EUP412699⟩
Fukumoto, T., **5**, 123 ⟨81BCJ1267⟩, 1143 ⟨69MI 530-01⟩, 1144 ⟨70JOM(22)627⟩
Fukumoto, Y., **1**, 212 ⟨93JOC4187⟩
Fukumoto, Y.,**3**, 33 ⟨93JOC4187⟩
Fukumura, M., **2**, 292 ⟨81PS153⟩; **5**, 349 ⟨81PS(10)153⟩
Fukumura, T., **2**, 614 ⟨85CPB1301⟩
Fukunaga, J. Y., **1**, 794 ⟨69TL993⟩
Fukunaga, K., **1**, 638 ⟨79BCJ1107⟩, 1069 ⟨81S879⟩; **2**, 665 ⟨82JOC4376⟩
Fukunaga, T., **1**, 338 ⟨89JA1759⟩, 868 ⟨65JA916⟩; **3**, 526, 527, 546 ⟨60JA2498⟩, 567 ⟨60JA873⟩; **5**, 47, 275 ⟨60JA2498⟩
Fukuoka, D., **1**, 485 ⟨74BCJ503⟩
Fukuoka, S., **5**, 125 ⟨87TL4711⟩, 273, 286, 296 ⟨71JOC2721⟩; **6**, 465 ⟨92JAP04054156⟩, 467 ⟨88JAP63238043⟩

Fukushima, A., **4**, 738 ⟨81TL759⟩
Fukushima, D., **2**, 214 ⟨79CL1077⟩, 214, 220 ⟨77CC407⟩, 214, 220 ⟨77CL893⟩
Fukushima, D. K., **4**, 209 ⟨54JA2943⟩
Fukushima, K., **3**, 309 ⟨90CPB3261⟩, 679 ⟨92CL2173⟩
Fukushima, M., **1**, 364 ⟨89BCJ2348⟩, 1097 ⟨82JA180⟩; **2**, 437 ⟨83JOC2195⟩; **3**, 7, 120 ⟨91BCJ796⟩, 225 ⟨89BCJ2348⟩; **4**, 1033 ⟨80BCJ1138, 82JA180⟩
Fukushima, S., **1**, 490 ⟨89CL1711⟩; **6**, 580 ⟨55MI 618-01⟩
Fukushima, T., **1**, 674 ⟨90AC59⟩; **2**, 28 ⟨85BCJ1599⟩; **4**, 56 ⟨89CC1559⟩; **5**, 510, 561, 563 ⟨90BCJ638⟩; **6**, 21 ⟨89CC1559⟩
Fukutani, Y., **1**, 1130 ⟨85CL1689⟩
Fukuyama, J., **3**, 419 ⟨89JA371⟩
Fukuyama, M., **1**, 869, 871 ⟨70JA4689⟩
Fukuyama, N., **2**, 631 ⟨87BCJ4363⟩
Fukuyama, S., **5**, 898 ⟨68TL3787, 70BCJ2870⟩
Fukuyama, T., **1**, 263 ⟨75JOC2009⟩, 827 ⟨94TL2145⟩; **3**, 13, 40, 85 ⟨90JA7050⟩, 712, 714, 718 ⟨81TL3759⟩; **4**, 218 ⟨74TL1549⟩, 948 ⟨92TL5701⟩; **5**, 155 ⟨79JA259⟩, 681 ⟨83BCJ545⟩
Fukuyama, Y., **1**, 818 ⟨93TL8453⟩; **5**, 68 ⟨74S443⟩, 177 ⟨77JA646⟩
Fukuzawa, S., **1**, 246 ⟨86JCS(P1)1983⟩, 364 ⟨89BCJ2348⟩, 437, 545 ⟨82TL1181⟩, 611 ⟨82TL1177, 83JOM(250)203⟩, 611 ⟨85TL895, 85TL921⟩, 611 ⟨90CE261⟩, 611, 620, 716 ⟨83TL4347⟩, 611, 716 ⟨83JA2748⟩, 620 ⟨85JCS(P1)471⟩; **2**, 24 ⟨80BCJ1390⟩, 279 ⟨90CL927⟩, 287 ⟨83JA2748⟩, 287, 288 ⟨85JCS(P1)471⟩
Fukuzawa, S. I., **2**, 287 ⟨83CC1501, 84CC426⟩, 287, 288 ⟨86JCS(P1)1983⟩, 564 ⟨86CC475⟩
Fukuzawa, S.-i., **3**, 176 ⟨86JCS(P1)1983⟩, 225 ⟨89BCJ2348⟩
Fukuzumi, K., **1**, 444 ⟨80JOM(193)37⟩
Fulihara, H., **2**, 369 ⟨91JA6337⟩
Fulka, C., **1**, 367 ⟨84BSB511⟩
Full, R., **2**, 294 ⟨75ZNB642⟩, 1083, 1094 ⟨75ZAAC(418)273⟩; **5**, 1136 ⟨75ZAAC(418)273⟩
Fuller, A. E., **3**, 131 ⟨82JOC3006⟩
Fuller, A. T., **2**, 103 ⟨47JCS963⟩
Fuller, C. E., **2**, 622 ⟨91JOC4066⟩
Fuller, C. J., **3**, 13, 83, 96 ⟨78TL1437⟩
Fuller, D. J., **1**, 131 ⟨91JA5751⟩
Fuller, G. B., **2**, 284 ⟨80TL3827⟩
Füller, H.-J., **4**, 577, 588 ⟨77CB3528⟩, 579 ⟨77ZN(B)858⟩, 582 ⟨78JOM(160)41⟩, 583 ⟨79ZN(B)1218⟩, 583, 585, 586, 587 ⟨77CB3536⟩, 586 ⟨77IJ149⟩, 587 ⟨75JOM(99)353⟩, 587, 588 ⟨76AG(E)541, 74CB3674⟩; **5**, 1096 ⟨77ZN(B)858⟩; **6**, 706 ⟨77ZN(B)858⟩
Fuller, J. C., **2**, 59 ⟨94JFC(66)123⟩, 60 ⟨92TL4533, 93TL257⟩, 60 ⟨94TL1515⟩
Fuller, N. A., **5**, 26 ⟨63JCS2918⟩
Fuller, S. E., **2**, 338, 339 ⟨81JCS(P2)545⟩
Fuller, W. D., **5**, 972 ⟨76MI 526-01⟩
Fullhart, L., **2**, 125, 135 ⟨49JA1478⟩
Fülling, G., **5**, 179 ⟨88TL5583⟩
Fulmor, W., **6**, 650 ⟨68JA6846⟩
Fülöp, F., **4**, 416, 417 ⟨85S1148⟩; **5**, 363 ⟨92T531⟩; **6**, 557 ⟨83T1829⟩, 558 ⟨82H(19)1191, 83JHC1181⟩, 558 ⟨84JHC1373, 85S1149⟩, 558 ⟨85T1353⟩
Fulpius, B. W., **6**, 496, 546 ⟨78JOC2930⟩
Fulton, D., **5**, 350 ⟨58JA1895⟩
Fultz, W. C., **4**, 582 ⟨87CC1543⟩; **5**, 385 ⟨86OM336⟩; **6**, 516 ⟨85OM939⟩
Fumarola, M. J., **2**, 28 ⟨76S736, 82JOC1588⟩; **5**, 318 ⟨76S736⟩
Fumeaux, J.-P., **5**, 90 ⟨72S464⟩
Fumio, O., **1**, 243 ⟨80SC595⟩
Funabashi, M., **5**, 700, 701 ⟨92JCS(P1)2127⟩
Funabashi, Y., **4**, 231, 940 ⟨88TL4773⟩, 937 ⟨86T2863⟩, 950 ⟨84TL2021⟩; **5**, 152 ⟨86T2863⟩; **6**, 130 ⟨84TL2021⟩
Funabiki, K., **2**, 18 ⟨90TL3167⟩
Funabiki, T., **1**, 73, 74, 79, 80 ⟨72BCJ2723⟩; **2**, 650 ⟨89CL1267⟩; **3**, 662 ⟨83JOM(243)95⟩
Funahashi, M., **1**, 929 ⟨93BCJ819, 93CL545⟩; **3**, 137 ⟨93CL545⟩; **5**, 430 ⟨93BCJ819⟩
Funaioli, T., **6**, 341 ⟨87AG681⟩
Funaki, I., **1**, 240 ⟨88JA7185⟩
Funaki, K., **1**, 651 ⟨78S746⟩; **4**, 648 ⟨78S746⟩
Funaki, Y., **1**, 1134 ⟨85T2369⟩
Funakoshi, K., **3**, 113 ⟨84TL961, 92TL6331⟩
Funakoshi, S., **4**, 218 ⟨89CPB526⟩
Funamizu, M., **3**, 352 ⟨64BCJ1897⟩
Funasaka, W., **4**, 750 ⟨67BCJ1275⟩, 752 ⟨67TL1123, 68BCJ756⟩
Fung, A. P., **1**, 5, 8, 9 ⟨79S59⟩; **2**, 36 ⟨83JOC2766, 79S536⟩, 650 ⟨81JOC4305⟩; **3**, 8, 121, 294 ⟨80S897⟩, 216 ⟨81S473⟩, 426 ⟨83JOC2766⟩, 618 ⟨80S657⟩; **5**, 282 ⟨79S537⟩
Fung, A. P.,**5**, 282 ⟨85OS(63)188⟩
Fung, S., **2**, 976 ⟨80JA6580⟩
Fung, S. H., **5**, 285 ⟨79SC757⟩
Funhoff, A. S., **4**, 550 ⟨88CB2121⟩
Funk, A. H., **2**, 40 ⟨71TL3361⟩
Funk, M. O., **1**, 949 ⟨75JOC3614⟩
Funk, R., **3**, 44 ⟨93SL226⟩
Funk, R. L., **1**, 64 ⟨80JA5253⟩, 448 ⟨77JA5483⟩, 510 ⟨84TL1655⟩, 830, 831 ⟨88JA3298, 80JA6889⟩, 831 ⟨82JA4030, 85JOC707⟩, 831 ⟨86JOC3247, 86T2831, 93JA8847, 93JOC5873⟩; **2**, 632 ⟨76CC833⟩, 944 ⟨75JA5600, 77JA5483⟩; **3**, 536 ⟨88TL1493⟩, 540 ⟨89SL36⟩; **5**, 227 ⟨85JOC707⟩
Funke, B., **5**, 586, 614, 616 ⟨71CB2709⟩; **6**, 296, 297, 298, 300, 313, 320, 321 ⟨77S73⟩
Funke, W., **6**, 107, 109 ⟨69LA(725)15⟩
Fuqua, S. A., **2**, 1042 ⟨65JOC1027⟩; **3**, 497 ⟨65JOC1027, 65JOC2543⟩; **4**, 732 ⟨65JOC1027, 65JOC2543⟩, 732 ⟨65TL521⟩; **4**, 8, 732 ⟨64TL1461⟩
Fuqua, S. A.,**4**, 9 ⟨73OSC(5)390⟩
Fuqua, S. A., **6**, 695 ⟨65JOC1027⟩
Furakawa, N., **2**, 160 ⟨76T2763⟩
Furakoshi, S., **1**, 188 ⟨90AG(E)801⟩
Furan, C., **5**, 759 ⟨72FRP2081556⟩
Furber, M., **1**, 541 ⟨85TL3285⟩, 769 ⟨88TL3339⟩; **3**, 65 ⟨85CC782, 89JCS(P1)683⟩
Furberg, S., **3**, 450 ⟨63ACS559⟩
Furcht, G., **6**, 409, 419 ⟨80EGP139936⟩
Furet, C., **6**, 26, 31 ⟨75JOM(102)423⟩
Furet, Y., **6**, 7 ⟨90JFC(47)95⟩
Furey, W., **2**, 1009 ⟨87AHC(41)275⟩
Furia, F. Di., **3**, 5, 117 ⟨87JOC5467⟩
Furita, H., **2**, 312 ⟨88TL3811⟩
Furlani, C., **2**, 1100 ⟨88G1⟩
Furlani, D., **2**, 79 ⟨88JOM(341)345⟩
Furlei, I. I., **2**, 641 ⟨75DOK(220)146⟩
Furlong, M. T., **1**, 123 ⟨91JOC5974⟩; **2**, 478 ⟨91JOC5974⟩
Furmanova, M. V., **5**, 401 ⟨89JGU1977⟩
Furneaux, R. H., **4**, 224 ⟨76CAR(52)63⟩, 235 ⟨77CAR(58)397⟩
Furrow, Jr., C., **5**, 199 ⟨63OSC(4)242⟩
Furst, A., **2**, 820 ⟨52JA562⟩; **4**, 229 ⟨64JOC724⟩
Furst, G., **5**, 528 ⟨84JA6084⟩
Furst, G. T., **6**, 134 ⟨87JA2504⟩
Furstenburg, G., **2**, 471 ⟨76ZN(B)1305, 83ZAAC(498)57⟩
Fürstner, ,A., **1**, 688 ⟨87S1071, 88JCS(P1)1729⟩; **2**, 678 ⟨86CC1149⟩, 702 ⟨89S571⟩, 958, 959 ⟨93AG(E)164⟩, 959 ⟨88AOC(28)85⟩; **4**, 655 ⟨91JOM(414)295⟩
Furstoss, R., **2**, 51 ⟨89JOC4686⟩; **5**, 316 ⟨78JOC3750⟩

Furubayashi, T., **1**, 274 ⟨73TL787⟩
Furue, M., **6**, 182, 183, 200 ⟨74JOM(69)63⟩
Furuhashi, K., **1**, 217 ⟨91TL1059⟩; **2**, 85 ⟨90TL7031⟩
Furuhata, T., **1**, 242 ⟨84TL4011⟩; **3**, 351 ⟨83JA6151⟩; **4**, 836 ⟨84TL4011⟩; **5**, 234, 242 ⟨82CL885⟩
Furukawa, H., **3**, 137 ⟨84BCJ1863⟩; **4**, 938, 958 ⟨84BCJ1863⟩
Furukawa, I., **3**, 412 ⟨89NKK1748⟩; **5**, 134 ⟨80TL2857⟩, 652 ⟨88CE215⟩, 1025 ⟨89NKK822⟩
Furukawa, J., **1**, 122 ⟨74AOC83⟩, 300 ⟨69T2647⟩; **2**, 67 ⟨74AOC(12)83⟩
Furukawa, M., **2**, 197 ⟨80CPB134⟩, 321 ⟨90CC1385⟩, 363 ⟨76S339, 82CPB1646⟩, 724 ⟨80S937⟩, 725 ⟨82CL2007⟩; **5**, 351 ⟨88H(27)1881⟩, 353 ⟨87S183⟩, 630 ⟨85CPB3479⟩, 1111 ⟨74CL881⟩
Furukawa, N., **1**, 483 ⟨86H(24)3337⟩, 621 ⟨71T4921, 71T6101⟩; **2**, 25 ⟨81TL4409⟩, 32 ⟨73CC212⟩, 212 ⟨91CL141⟩, 271 ⟨86H(24)3019⟩, 279 ⟨89CC1789⟩, 369 ⟨91JA6337, 92JCS(P1)2247⟩, 369 ⟨93JA9826⟩, 401, 745 ⟨76S30⟩, 647 ⟨66BCJ1212⟩; **3**, 310 ⟨86TL3899⟩, 378 ⟨77TL3633, 79T317⟩, 684 ⟨91BCJ318⟩; **4**, 94 ⟨85CC1266⟩, 161 ⟨77JCS(P1)372⟩, 221 ⟨78JCS(P2)1302⟩; **5**, 209 ⟨81T3115⟩, 210 ⟨92CL1437⟩, 349 ⟨81PS(10)153⟩, 1106 ⟨81TL4409⟩
Furukawa, S., **1**, 408 ⟨80JA2125⟩, 448 ⟨89CPB36⟩; **2**, 694 ⟨90CPB2637⟩, 1000 ⟨90CPB329⟩; **3**, 310 ⟨87T5281⟩; **4**, 23, 27 ⟨67BCJ594⟩, 856, 857 ⟨74CL659⟩, 936 ⟨90CPB902⟩
Furukawa, T., **3**, 220 ⟨89JA8277⟩
Furukawa, Y., **1**, 187 ⟨84JA1079⟩; **4**, 426 ⟨82TL3181⟩; **5**, 607 ⟨81TL3413, 82JA4018⟩
Furuno, E., **3**, 360 ⟨92JCS(P1)2135⟩
Furusaki, A., **4**, 673 ⟨69BCJ545⟩
Furusawa, K., **2**, 540 ⟨83MI 210-02, 85MI 210-02⟩
Furusawa, O., **2**, 614 ⟨89CL2213⟩
Furuta, K., **1**, 184 ⟨85JA3343⟩, 927 ⟨88TL2215⟩, 1056, 1057 ⟨84BCJ2768⟩; **2**, 528 ⟨93JA11490⟩; **3**, 639 ⟨84BCJ2768⟩; **4**, 208 ⟨92SL565⟩, 1061 ⟨84BCJ2768⟩; **5**, 30 ⟨88OS(67)76⟩
Furuta, S., **4**, 229 ⟨64JOC724⟩
Furuta, T., **4**, 390 ⟨92JOC2981⟩
Furuune, M., **1**, 1006 ⟨92TL5363⟩
Furuya, M., **5**, 236 ⟨91CL989⟩
Furuya, S., **5**, 1126 ⟨88H(27)2609⟩
Furuya, Y., **6**, 501 ⟨68T2367, 72CL971⟩
Fusco, C., **3**, 120 ⟨91JA2205⟩
Fusco, R., **3**, 467 ⟨63CB802⟩; **5**, 354 ⟨87JHC773⟩, 774, 839, 854, 855 ⟨69G69⟩, 858 ⟨48G332⟩, 1123 ⟨68G511⟩
Fuse, M., **3**, 250, 256 ⟨79CL785⟩; **4**, 956, 957 ⟨79CL785⟩
Fusetani, N., **3**, 701, 704 ⟨92TL6823⟩
Fuson, R. C., **2**, 13 ⟨42OR(1)63⟩, 30 ⟨55OSC(3)209⟩, 138 ⟨46JOC475⟩, 166 ⟨46JOC469, 46JOC499⟩, 494, 506 ⟨47JA559⟩, 632 ⟨58JOC1161⟩, 642, 643 ⟨40JA3250, 44JA1873⟩, 643 ⟨45JA1939, 45JA386⟩, 643, 645 ⟨43JA992⟩, 643, 645, 646 ⟨44JA1109, 44JA1272⟩, 645 ⟨44JA1274, 45JA1937⟩, 645, 646 ⟨44JA1464, 46JA389, 48JA602⟩; **4**, 20 ⟨26JA1093⟩; **5**, 34 ⟨55OSC(3)557⟩, 38 ⟨34CRV275⟩, 76 ⟨43OSC(2)260⟩
Füssel, J., **2**, 1094, 1095 ⟨B-86MI 223-01⟩, 1094, 1095 ⟨B-87MI 223-01⟩, 1094 ⟨B-87MI 223-02, B-87MI 223-03⟩, 1094 ⟨B-87MI 223-04⟩
Fusser, G., **3**, 476, 477 ⟨81BSB615⟩
Fust, K. J., **1**, 57 ⟨57CB1774⟩
Fustero, S., **2**, 322 ⟨89CC1132⟩; **3**, 188 ⟨83JOC2255⟩, 414 ⟨85CB3652, 86CC1179⟩; **4**, 408 ⟨88CC410⟩, 421 ⟨91CC1704⟩
Futamura, S., **2**, 27 ⟨92BCJ345⟩; **3**, 319 ⟨82CC1223⟩
Futterer, E., **6**, 48, 729 ⟨66AHC(7)39⟩
Fuzesi, L., **1**, 820 ⟨76JA7115⟩
Fuzimura, S., **6**, 413 ⟨78GEP2811310⟩
Fuzshenokova, A. V., **4**, 922 ⟨92HAC127⟩
Fyfe, C. A., **1**, 914 ⟨79JA4426⟩; **6**, 109 ⟨68TL659⟩
Fyles, T. M., **2**, 1010 ⟨75JA5526⟩

G

Gaasbeek, C. J., **1**, 901 ⟨68RTC319⟩
Gaasch, J. F., **1**, 668 ⟨72JA5258⟩
Gäb, S., **6**, 231, 232 ⟨85AG48⟩
Gabard, J., **2**, 374 ⟨87T891⟩
Gabbaï, F. P., **6**, 704 ⟨94AG(E)578⟩
Gabbert, G., **5**, 927 ⟨92AG(E)1603⟩
Gabbott, R. E., **1**, 777 ⟨73JCS(P2)1293, 75JCS(P2)317⟩
Gabbutt, C. D., **3**, 172 ⟨94T827⟩
Gabe, E. J., **4**, 1061 ⟨85JA7219⟩
Gabel, R. A., **3**, 108 ⟨78TL227⟩
Gabhe, S. Y., **2**, 652 ⟨77JOC1479⟩
Gabitov, F. A., **4**, 995 ⟨72ZOR2457⟩; **6**, 145 ⟨71ZOR1126, 74ZOR2003, 77ZOR1559⟩
Gabnitskaya, E. S., **2**, 469 ⟨78ZOB2624⟩
Gabor, B., **2**, 984 ⟨91CB2165⟩
Gábor, V., **2**, 59, 60 ⟨54NAT125⟩, 59, 60 ⟨55NAT346⟩
Gabriel, J., **1**, 816, 817 ⟨85JOC1143⟩; **2**, 57 ⟨89JOC159⟩; **4**, 289 ⟨84HCA1070⟩, 816 ⟨83HCA308⟩; **6**, 86, 92 ⟨84HCA1070⟩, 204 ⟨80HCA2046⟩, 317 ⟨84HCA1083⟩
Gabriel, S., **2**, 299 ⟨1887CB224⟩; **6**, 81 ⟨1877CB185⟩
Gabrio, T., **5**, 476, 477, 486, 1050 ⟨68ZC142⟩
Gadallah, F. F., **3**, 363 ⟨64JOC1355⟩
Gadamasetti, K. G., **1**, 806, 811 ⟨92OR(41)1⟩
Gade, L. H., **6**, 174 ⟨93IC2308⟩
Gäde, W., **5**, 1097 ⟨81JOM(213)451⟩
Gadecki, F. A., **1**, 906 ⟨57JA4557⟩
Gadekar, S. M., **5**, 368 ⟨62JOC1383⟩; **6**, 648 ⟨68JMC811⟩
Gadgil, V. R., **3**, 709 ⟨90TL6209⟩
Gadol, S. M., **4**, 687 ⟨90OM2745⟩, 710 ⟨82OM1607⟩
Gadras, A., **1**, 238 ⟨84TL2135⟩
Gadreau, C., **3**, 495 ⟨76BSF2068⟩
Gadwood, R. C., **1**, 713, 715, 716 ⟨83JOC2098, 87JOC774⟩, 812 ⟨82JOC2268, 89OS(67)210⟩; **3**, 142 ⟨82JOC2268⟩, 217 ⟨84JA3869⟩; **4**, 50, 379, 381, 382, 951, 952 ⟨85JOC3255⟩, 399 ⟨83JOC2098⟩
Gadzhieva, P. I., **4**, 290 ⟨91HAC307⟩
Gaede, B., **3**, 78 ⟨79T621⟩
Gaede, P. E., **4**, 715 ⟨94JOM(471)249⟩
Gaeta, K. K., **1**, 243, 437 ⟨89TL1209⟩
Gaetano, K., **2**, 34, 629, 631 ⟨88JOC4477⟩
Gaetzi, K., **5**, 744 ⟨72SZP563109⟩
Gaffney, B. L., **1**, 650 ⟨75SC15⟩
Gafurov, E. K., **4**, 343 ⟨79MI 407-01⟩
Gafurov, R. G., **6**, 150, 364 ⟨71IZV1594⟩, 284, 364 ⟨74IZV915⟩
Gagan, J. M. F., **3**, 69 ⟨70JCS(C)2484⟩
Gage, J. R., **3**, 32 ⟨92JA9434⟩; **5**, 80 ⟨89OS(68)83⟩
Gaggero, N., **2**, 151 ⟨88CC1451⟩, 151 ⟨92TA921⟩
Gagne, M. R., **2**, 56 ⟨93JA9800⟩
Gagnon, J. G., **6**, 261 ⟨67JOC1944, 67USP3355492⟩
Gagnon, P. E., **3**, 453 ⟨53CJC1025⟩
Gagnon, S. D., **3**, 319 ⟨83JA663⟩
Gagoll, A., **2**, 392 ⟨90TL4199⟩
Gahman, T. C., **6**, 79 ⟨88TL2023⟩
Gaiani, G., **5**, 348 ⟨85G265⟩
Gai, Y.-Z., **2**, 341 ⟨87JHC767⟩
Gaidamaka, S. N., **1**, 612 ⟨89ZOR654⟩
Gaidamak, V. A., **2**, 440 ⟨61ZOB3411⟩; **4**, 337 ⟨61ZOB3411⟩
Gaikwad, B. V., **6**, 647 ⟨82PIA441⟩
Gaillot, J.-M., **1**, 706, 707 ⟨79CJC1958⟩
Gaillot, P., **2**, 494 ⟨27CR(185)956⟩, 494, 505, 506 ⟨27CR(185)779⟩
Gainer, G. C., **2**, 706 ⟨49JA1747⟩
Gainer, H., **2**, 212 ⟨51JA4464⟩
Gainsford, G. J., **1**, 784 ⟨93JHC873⟩; **4**, 604, 647 ⟨86JA5949⟩, 742, 748 ⟨89AJC1307⟩
Gainutdinov, R. M., **6**, 659 ⟨90ZOB2694⟩
Gairns, R. S., **4**, 856 ⟨86JCS(P1)501⟩
Gais, H.-J., **1**, 146 ⟨88TL1529, 89AG(E)1025, 92JA4439⟩, 146, 711 ⟨86AG(E)935⟩, 483 ⟨91JA1442⟩, 485 ⟨92TL461⟩, 711 ⟨89JA1125⟩, 763 ⟨85TL4359⟩; **2**, 997, 1000 ⟨91JA4002, 85AG(E)859⟩, 1000 ⟨86AG(E)938, 86AG(E)939, 88TL1259, 89AG(E)1025, ⟨90AG(E)100⟩, 1049 ⟨69HCA2641, 84TL273⟩; **3**, 77 ⟨69HCA2641⟩; **4**, 656 ⟨93JA7215⟩, 885 ⟨69HCA2641⟩, 936, 938 ⟨85AG(E)696⟩; **5**, 135, 138 ⟨84AG(E)142⟩, 154, 174 ⟨84AG(E)143⟩, 174 ⟨84AG(E)145⟩, 236, 239 ⟨78AG(E)267, 77AG(E)244⟩
Gaitanopoulos, D. E., **4**, 102 ⟨85JHC957⟩
Gait, S. F., **2**, 777 ⟨75JCS(P1)26⟩
Gaiter, A., **5**, 847 ⟨14G72⟩
Gaitzsch, E., **2**, 936 ⟨86CB535⟩; **4**, 568 ⟨86CB535⟩; **6**, 693 ⟨84ZC384, 86CB535⟩
Gaitzsch, T., **4**, 146 ⟨86CB1977⟩; **6**, 166, 693 ⟨86CB1977, 84ZC345⟩, 693 ⟨85AG(E)419⟩
Gaj, B. J., **1**, 881 ⟨63JOC1725⟩
Gajda, C., **1**, 191 ⟨93TL4591⟩
Gajda, G. J., **3**, 508 ⟨81JA7358⟩
Gajda, T., **2**, 308 ⟨92SC1929⟩, 412 ⟨74TL3383⟩; **4**, 132 ⟨90S717⟩, 133 ⟨90MI 403-01⟩, 491 ⟨92S367⟩; **5**, 285 ⟨79S549⟩, 300 ⟨81S1005⟩
Gajek, K., **1**, 811 ⟨61LA(644)172⟩
Gajewski, J. J., **1**, 402 ⟨B-71MI 109-01⟩, 412 ⟨87JOC5198⟩, 419 ⟨B-81MI 109-01⟩, 784 ⟨B-81MI 117-02⟩, 794 ⟨71MI 118-01⟩, 800, 825 ⟨B-81MI 118-01⟩, 805 ⟨B-81MI 118-02⟩, 806 ⟨74JA3705⟩, 826 ⟨87JA1170, 89JOC515⟩, 865 ⟨67JA4383⟩
Gajsler, L., **3**, 684 ⟨80JPS(A)635⟩
Gajurel, C. L., **3**, 648 ⟨86CC1779⟩
Gakh, A. A., **2**, 334, 365 ⟨91T7447⟩; **5**, 314, 384 ⟨91T7447⟩; **6**, 145 ⟨83IZV2655⟩, 145 ⟨90TL7379⟩
Gal, A., **1**, 1110, 1112 ⟨73JCS(P2)301⟩, 1110, 1112, 1114 ⟨72JOC1174, 70TL3233⟩, 1112, 1114 ⟨69JA5246⟩; **2**, 627 ⟨45HCA628⟩
Gal, C., **2**, 15 ⟨87JOC2769, 87JOC4928⟩, 15 ⟨88JOC2803⟩
Gal, G., **1**, 80 ⟨70JOC3510⟩; **5**, 762, 816 ⟨65JOC259⟩
Gal, J., **6**, 417, 418 ⟨73T3309⟩
Gal, M. H., **6**, 452 ⟨1866BSF435⟩
Gala, D., **6**, 550 ⟨86JOC4488⟩
Galabov, A. S., **6**, 502 ⟨73GEP2241471⟩
Galafeeva, M. F., **5**, 709 ⟨87MI 517-02⟩
Galakhov, I. V., **4**, 833 ⟨71KGS1645⟩
Galakhov, V. S., **5**, 762 ⟨73IZV2159⟩; **6**, 282 ⟨68IZV656⟩
Galakov, M. V., **4**, 802, 803 ⟨88BAU1686⟩
Galamb, V., **4**, 148 ⟨87OM861⟩
Galambos, G., **2**, 286 ⟨82TL4371⟩; **4**, 240 ⟨82TL4371⟩
Galan, A. A., **2**, 316 ⟨91JOC452⟩; **5**, 223 ⟨86TL4995⟩
Galangay, E., **1**, 961 ⟨74S344⟩; **2**, 274 ⟨64JOC3561⟩
Galas, A. M. R., **4**, 656 ⟨82POL31⟩, 680 ⟨80JCS(D)1771⟩; **5**, 797, 801 ⟨81JCS(D)2088⟩; **6**, 338 ⟨83CC1557, 81CC189⟩, 339 ⟨81CC2496⟩
Galat, A., **5**, 298 ⟨48JA3945⟩
Galatsis, P., **1**, 929 ⟨91TL7493, 93JOC1215⟩
Galbraith, A. R., **1**, 1064 ⟨60JCS3614⟩
Galbraith, H. W., **2**, 776 ⟨51JA1323⟩
Gale, D. M., **3**, 645 ⟨73JOC475⟩
Gale, L. H., **6**, 478 ⟨88JHC193⟩

Galeazzi, E., **3**, 179, 194 ⟨86JOC3374⟩, 235 ⟨73JA6655⟩
Galeeva, I. Z., **4**, 549 ⟨82IZV1196, 82IZV2718⟩, 549 ⟨84IZV2023, 84ZOB1511⟩
Galezowski, W., **4**, 731 ⟨85CJC576⟩
Galimberti, P., **4**, 317 ⟨47G375⟩
Galindo, A., **6**, 327 ⟨92JCS(D)2307⟩, 327, 330 ⟨93IC5569⟩
Galindo, J., **1**, 185 ⟨89HCA1537⟩; **3**, 218 ⟨86HCA228⟩; **4**, 166 ⟨81HCA1247⟩
Galishev, V. A., **6**, 116 ⟨90ZOR1129⟩
Galitskova, N. P., **2**, 436 ⟨67ZOB2080⟩
Galka, R., **4**, 455 ⟨81JINC629⟩
Galkina, M. A., **5**, 223 ⟨90ZOB1928⟩
Galkina, N. E., **4**, 588 ⟨78ZOB1422, 79ZOB1772⟩
Gall, M., **1**, 871, 881 ⟨69JOC2324⟩, 872, 873 ⟨71JOC2361⟩; **2**, 675 ⟨73JOC514⟩, 696, 699 ⟨69JOC2324⟩
Gall, R., **4**, 21 ⟨54HCA90⟩
Gall, T. Y. Le, **5**, 103 ⟨87OS(65)47⟩
Gallaghan, J. A., **2**, 30 ⟨56JA1218⟩
Gallagher, A., **2**, 214 ⟨61JOC3780⟩; **6**, 305 ⟨62JOC4068⟩, 474 ⟨64JOC2442⟩
Gallagher, J., **1**, 111, 585 ⟨93TL4121⟩
Gallagher, M. J., **2**, 414 ⟨66JCS(C)2176⟩, 432 ⟨87JOM(323)C1⟩, 559 ⟨88CC289⟩; **4**, 119 ⟨68AJC1197⟩; **6**, 116 ⟨77TL2987, 80AJC287⟩
Gallagher, M. K., **6**, 332 ⟨87OM283⟩
Gallagher, P. T., **1**, 254 ⟨92SC351⟩, 831 ⟨87CC1220, 87TL3031⟩, 831 ⟨91JCS(P1)705⟩; **5**, 593 ⟨77JCS(P1)2357⟩
Gallagher, T., **2**, 765 ⟨68CC1017⟩; **3**, 72 ⟨89JCS(P1)1793⟩; **4**, 842, 845, 851 ⟨89JCS(P1)1793⟩, 860 ⟨88TL6475, 92T7551⟩; **6**, 127 ⟨88JCS(P1)3367⟩
Gallagher, T. F., **1**, 16 ⟨55JA4367⟩; **4**, 209 ⟨54JA2943⟩
Gallardo, I., **1**, 107 ⟨92JA9576⟩
Galle, J. E., **1**, 144, 1090 ⟨79JOC3277⟩, 145 ⟨80JOC4534⟩, 145, 310, 1090 ⟨79JOC3279⟩, 485, 1032 ⟨84TL4851⟩, 650 ⟨76JOC2615⟩; **4**, 387 ⟨82JOC5051⟩, 602 ⟨88JOM(341)293⟩, 960 ⟨79JOC3279⟩; **5**, 413 ⟨82JOC5051⟩; **6**, 131, 133 ⟨88JOM(341)293⟩
Galledou, B. S., **3**, 238 ⟨83BSF(27)89⟩
Gallego, C. H., **2**, 680 ⟨85JOC1212⟩
Gallego, M. G., **1**, 800 ⟨90CC934⟩; **3**, 418 ⟨87OPP181⟩
Galler, H., **2**, 541 ⟨63CB3289⟩
Galli, C., **3**, 280, 296, 298, 305 ⟨79S303⟩
Galliani, G., **2**, 649 ⟨81T2313⟩; **5**, 773 ⟨83JMC1187⟩
Galli, R., **1**, 934 ⟨71T3575⟩; **2**, 798 ⟨64MI 215-01⟩; **6**, 494 ⟨82JCS(P1)2813⟩, 695 ⟨84T1523⟩
Galliaud, A., **3**, 241 ⟨74BSF1043⟩
Galliaud, M. A., **3**, 246 ⟨72CR(275)897⟩
Gallic, Y. Le, **3**, 65 ⟨90TL3129⟩; **4**, 749 ⟨93S1071⟩
Gallina, C., **3**, 152 ⟨89S466⟩; **6**, 25 ⟨89S466, 90S327⟩
Gallindo, J., **3**, 130 ⟨87HCA1745⟩
Gallis, D. E., **3**, 434 ⟨89JOC1736⟩
Gallivan, Jr., R. M., **1**, 658, 659 ⟨68JA2906⟩
Gallmeier, H.-J., **3**, 349 ⟨81TL2971⟩
Gallo, A. A., **4**, 326 ⟨80JFC(15)29⟩
Gallo, C., **1**, 496 ⟨86JOM(317)373⟩
Gallo, C. J., **3**, 313 ⟨80JA1504⟩
Gallo, D. G., **2**, 381 ⟨73JMC736⟩
Gallo, G. G., **4**, 441 ⟨72JHC827⟩
Gallo, R., **1**, 186 ⟨85S493⟩, 1063 ⟨85CC1571⟩; **3**, 279 ⟨85BSB897⟩; **5**, 38 ⟨85S493⟩
Gallop, M. A., **3**, 486 ⟨84CC1002⟩, 690 ⟨86AOC(25)121⟩; **5**, 942 ⟨86AOC(25)121⟩; **6**, 716, 717, 721 ⟨86AOC(25)121⟩
Gallop, P. M., **2**, 1021 ⟨93JA11702⟩
Gallos, J., **2**, 619, 626, 631 ⟨82JCR(S)150⟩
Gallos, J. K., **1**, 261, 1041 ⟨89S31⟩; **2**, 306 ⟨89S31⟩
Galloy, J., **2**, 150 ⟨82JA5412⟩, 1000 ⟨83JHC399⟩; **5**, 481 ⟨75AG(E)53⟩
Gallucci, C. R., **1**, 841 ⟨88JOC2711⟩
Gallucci, J., **5**, 451 ⟨90JOM(383)C1⟩
Gallucci, J. C., **1**, 1098 ⟨89JOC2278⟩; **5**, 943 ⟨93JA9986⟩
Gallucci, R. A., **5**, 873 ⟨77JA6995⟩
Gallucci, R. R., **3**, 532 ⟨85JOC4404⟩; **4**, 208 ⟨82JOC3517⟩
Gallyamov, M. R., **3**, 197 ⟨87JGU205⟩
Galm, W., **4**, 690 ⟨90AG(E)686⟩, 691 ⟨93CC1459⟩
Galons, H., **2**, 31 ⟨89BSF247⟩
Galoyan, A. M., **1**, 1040 ⟨86AKZ306⟩
Galpin, J. D., **1**, 810 ⟨93TL247⟩
Galppini, E., **2**, 393 ⟨90T7497⟩
Galt, R. H. B., **2**, 753 ⟨60JOC214⟩
Galton, S. A., **2**, 1019 ⟨65JOC1930⟩; **6**, 73 ⟨67JOC2630⟩
Galun, A. B., **4**, 743 ⟨62JOC1426⟩
Galushko, S. V., **3**, 265 ⟨91S207⟩
Galvez, E., **4**, 405 ⟨87CJC687⟩
Gálvez, J., **6**, 624 ⟨94TL2365⟩
Galvin, G., **2**, 590 ⟨59CI(L)628⟩
Galy, J., **4**, 1039 ⟨80AX(B)1344⟩
Galy, J.-P., **3**, 280 ⟨91TL6709⟩
Galyer, A. L., **4**, 966 ⟨76AG(E)609⟩
Galyer, L., **2**, 592 ⟨79IS253⟩
Gamaleya, V. F., **5**, 707 ⟨68ZOB1867⟩, 707, 861 ⟨68JGU1815⟩
Gamayurova, V. S., **2**, 695 ⟨72MI 213-01⟩, 886 ⟨72MI 217-01, 73JGU1921⟩, 891 ⟨74JGU1506⟩
Gambacorta, A., **6**, 141 ⟨76S798⟩
Gambale, R. J., **5**, 174 ⟨81T4059⟩
Gambaretto, G. P., **4**, 31 ⟨76JFC(7)569⟩
Gambaro, M., **2**, 51 ⟨89S295⟩; **3**, 117, 161 ⟨91JOC5812⟩
Gambarotta, S., **2**, 703 ⟨85IC654⟩; **4**, 696 ⟨83CC1087, 83CC1156⟩; **6**, 342 ⟨86AG283⟩
Gambaryan, N. P., **3**, 556 ⟨76RCR630⟩, 560, 561, 588 ⟨73AG(E)675, 65BAU731⟩, 588 ⟨75BAU1274⟩, 592 ⟨73DOK(212)628⟩; **4**, 297, 304 ⟨66IZV1108, 65IZV2046⟩, 894 ⟨75IZV1274⟩, 979 ⟨86IZV2344⟩; **6**, 106 ⟨76IZV858⟩
Gambino, O., **4**, 682 ⟨75ICA155⟩, 683 ⟨72JOM(39)169⟩, 687 ⟨67JOM(8)149⟩, 692 ⟨74ICA5⟩
Gambino, S., **1**, 615 ⟨78MI 114-01⟩
Gambke, B., **3**, 556, 557, 560, 585, 586 ⟨81CB3751⟩
Gamble, A. S., **3**, 515 ⟨93OM2814⟩
Game, C. H., **6**, 719, 732 ⟨74JCS(D)351⟩, 721 ⟨74JCS(D)357⟩
Gamlova, G. V., **5**, 1142, 1145 ⟨79JOM(177)101⟩
Gammal, C. A., **2**, 94 ⟨25JA395⟩
Gammel, F. J., **5**, 226 ⟨80CB1010⟩
Gammill, B., **3**, 266 ⟨79S901⟩
Gammill, R. B., **1**, 521 ⟨79JA7001⟩; **2**, 667 ⟨78JOC379⟩; **3**, 87 ⟨84TL2953, 86JOC3116⟩, 244 ⟨78JOC379⟩, 265 ⟨78TL743⟩; **4**, 253 ⟨80JA3095, 81JOC3555⟩, 302 ⟨81JOC3340⟩
Gamoh, K., **5**, 11 ⟨90MI 501-01⟩
Gamper, N. M., **2**, 464 ⟨59ZOB1671⟩
Gamper, S., **6**, 167, 371 ⟨92CB1333⟩
Gan, W., **4**, 969, 970 ⟨90JPR1035⟩
Ganboa, I., **2**, 28, 30 ⟨84SC939⟩; **3**, 154 ⟨84SC939⟩; **5**, 294 ⟨87TL1945⟩, 299 ⟨82BSF(2)167⟩
Gancarz, R. A., **2**, 238 ⟨80TL1697⟩, 285 ⟨80TL4155, 81JOC4899⟩
Gancarz, R., **4**, 463 ⟨80JPR213⟩, 464 ⟨77S625⟩
Gancher, E., **5**, 559 ⟨66JOC3877⟩
Gande, M. E., **1**, 826 ⟨87JA1170⟩
Gander-Coquoz, M., **4**, 302 ⟨88HCA224⟩
Gandhi, C. S., **3**, 634 ⟨79SC569⟩
Gandhi, R. P., **1**, 521 ⟨68CC552⟩
Gandin, J. M., **6**, 461 ⟨87T5315⟩
Gandini, A., **1**, 896 ⟨65JCS6019⟩; **2**, 204 ⟨67M1518⟩
Gandler, J., **1**, 570 ⟨B-89MI 113-01⟩
Gandler, J. R., **5**, 955 ⟨89OM2282⟩
Gandolfi, R., **3**, 678, 681 ⟨79AG(E)721⟩
Gandolfi, V., **3**, 268 ⟨88JCR(S)246⟩

Gandour, R. D., **1**, 690 ⟨84JOC1627⟩
Gandour, R. W., **3**, 313 ⟨80JA1504⟩
Gandurina, I. A., **4**, 472 ⟨79IZV1110, 79IZV2118⟩
Ganellin, C. R., **1**, 906 ⟨59JCS3139⟩; **6**, 647 ⟨77JMC901⟩
Ganem, B., **1**, 5, 9 ⟨82JOC5041⟩, 72, 75 ⟨86CRV763⟩, 204, 478, 1023 ⟨84JA3693⟩, 263 ⟨85TL1153⟩, 307 ⟨73JA6152⟩, 434, 635 ⟨84JA3353⟩, 451 ⟨83JOC3761⟩, 514 ⟨87JOC5044⟩, 573 ⟨78JA6483⟩, 637 ⟨81TL4163⟩, 637, 1077 ⟨90JA3717⟩, 789 ⟨78JA352⟩, 826 ⟨87JA1170⟩, 1061 ⟨74TL4467⟩; **2**, 146 ⟨80TL689⟩, 319 ⟨85TL1153⟩, 340, 344, 347 ⟨83S537⟩, 931 ⟨73JA6152⟩; **3**, 127 ⟨76JOC2194⟩, 215 ⟨85JOC5897⟩; **4**, 90 ⟨78JOC1607⟩, 807, 1060 ⟨88OSC(6)1033⟩, 913 ⟨92JOC178⟩; **5**, 213 ⟨83S537⟩, 285 ⟨82JA3233⟩, 324 ⟨76TL1951⟩, 330 ⟨91JA5089, 94JA562⟩, 652 ⟨78JOC4393⟩; **6**, 118, 666 ⟨92JOC178⟩
Ganem, B. E., **5**, 51, 154 ⟨68JA5616⟩
Ganesan, K., **2**, 532 ⟨92JOC2716, 92JOC499⟩, 532, 698 ⟨93JOC7162⟩, 532 ⟨94JOC2336⟩
Gangadhar, A., **6**, 450 ⟨88S994⟩
Gangaly, A. K., **4**, 938 ⟨91JOC3556⟩
Gange, D., **1**, 971 ⟨80JA2134⟩
Gangee, A., **3**, 69 ⟨90JHC765⟩
Gangkofner, S., **4**, 890 ⟨93LA207⟩
Gangloff, A. R., **5**, 329 ⟨93TL7371⟩
Ganguly, A., **1**, 91 ⟨81T3615⟩
Ganguly, A. K., **1**, 214 ⟨85TL6301⟩, 1089, 1090 ⟨87TL4127, 87TL4123⟩, 1131 ⟨89H(28)47⟩; **2**, 608, 612 ⟨68CC806⟩; **4**, 961 ⟨85TL6301, 87TL4123⟩; **6**, 553 ⟨81TL3485⟩
Ganguly, R., **1**, 821 ⟨84TL3555⟩
Gani, D., **5**, 207 ⟨87T5341⟩
Gani, P., **3**, 620 ⟨91JCS(P1)954⟩
Ganiev, F. F., **5**, 679 ⟨91JGU2372, 91ZOB2257⟩
Ganis, P., **4**, 628, 701 ⟨88OM210⟩
Ganja, E. A., **2**, 483 ⟨88IC4535⟩, 483, 486 ⟨90IC33⟩
Gannett, P. M., **1**, 700, 703 ⟨86JA4603, 88JOC1064⟩
Gano, J. E., **4**, 352 ⟨89CC1256⟩
Gans, W., **2**, 430, 437, 438 ⟨82ZAAC(491)163⟩
Ganschow, S., **6**, 109 ⟨72JPR145⟩
Gant, T. G., **4**, 306 ⟨94T2297⟩
Gante, J., **4**, 900 ⟨75CB174⟩; **5**, 354 ⟨91CZ215⟩; **6**, 651 ⟨67AG(E)862⟩
Ganter, C., **2**, 124 ⟨72HCA481⟩
Gantert, G. E., **5**, 1043 ⟨56JA4965⟩, 1044 ⟨56JOC1430⟩
Ganther, H. E., **2**, 236 ⟨68B2898⟩, 478 ⟨69B2557⟩
Ganushchak, N. I., **2**, 256 ⟨80ZOB2459⟩; **5**, 1044 ⟨88ZOB616, 91UKZ89⟩
Ganzera, I., **5**, 344 ⟨86AP(319)682⟩
Gao, J., **2**, 597 ⟨94JOC3714⟩
Gao, Q., **2**, 528 ⟨93JA11490⟩
Gao, S., **2**, 683 ⟨93H(36)13⟩
Gao, Y., **1**, 17, 185 ⟨88JA7538⟩; **5**, 1106 ⟨88JA7538⟩
Gao, Z., **1**, 928 ⟨90CJC1698⟩
Gaoni, Y., **1**, 184, 215 ⟨82JOC2564⟩; **4**, 287 ⟨89T2819⟩
Gapinski, R. E., **4**, 845 ⟨79JOC4744⟩; **5**, 235 ⟨78TL4319⟩
Gapski, G., **1**, 690 ⟨78CL803⟩
Gar, T. K., **1**, 656 ⟨B-67MI 114-02⟩; **4**, 98, 99 ⟨87JGU334⟩, 142 ⟨85JGU1079⟩, 608, 638 ⟨69ZOB2281⟩, 609 ⟨72ZOB1361⟩, 621 ⟨75ZOB2672⟩, 626, 637, 640 ⟨69ZOB2601⟩, 628, 639, 640 ⟨72ZOB1521⟩, 638, 640 ⟨69DOK(188)120⟩, 639 ⟨70ZOB2601⟩, 639, 640 ⟨73ZOB625, 76ZOB837⟩, 641 ⟨80ZOB1764⟩, 660, 700 ⟨73ZOB801⟩, 1058 ⟨86ZOB1535⟩; **5**, 969 ⟨75ZOB2185⟩; **6**, 188 ⟨72ZOB1521⟩
Garad, M. V., **2**, 905 ⟨83TL637⟩; **4**, 630, 632, 634 ⟨82JCR(S)132⟩, 642, 645, 1077 ⟨83TL637⟩, 665 ⟨93T2979⟩; **6**, 199 ⟨82JCR(S)132⟩
Garanti, L., **1**, 524 ⟨77JHC947⟩; **4**, 440 ⟨79S380⟩; **5**, 371 ⟨86S772⟩, 676, 677 ⟨84JCS(P1)2535, 81JCS(P1)2245⟩, 677 ⟨93JHC559⟩, 862 ⟨86JCS(P1)2177⟩, 905 ⟨84T2165⟩
Garbalinskaya, N. S., **5**, 385 ⟨85CC171⟩, 652 ⟨88JCS(P1)305⟩
Garbarino, G., **2**, 707 ⟨89T7411⟩, 711 ⟨86T4007⟩; **3**, 664 ⟨87T4625⟩, 664, 667 ⟨90T2205⟩; **5**, 1111 ⟨89T7411⟩
Garbarino, J. A., **2**, 309 ⟨74JCS(P1)2101⟩
Garbauskas, M. F., **2**, 163 ⟨86JOC829⟩
Garber, M., **2**, 835, 836 ⟨65IC764⟩, 837 ⟨65IC765⟩
Garbers, C. F., **1**, 733 ⟨68JCS(C)1982⟩; **4**, 56 ⟨65JCS3075⟩
Garbesi, A., **2**, 362 ⟨67CC985⟩
Garbisch, Jr., E. W., **1**, 95 ⟨68JA3590⟩, 870 ⟨65JA505, 67TL29⟩, 870, 871 ⟨63JA1696⟩
Garbuglio, C., **3**, 255 ⟨58G1226⟩, 399 ⟨64G48⟩
Garbuzova, I. A., **1**, 890 ⟨89JOM(371)279⟩
García, G., **5**, 386 ⟨92JCS(D)53⟩
García, J. M., **2**, 398 ⟨90JOC2070⟩
García Alonso, F. J., **4**, 678 ⟨92OM370⟩
García, M. E., **4**, 674 ⟨93JA3786⟩
García, M. L., **5**, 606 ⟨92H(34)1573⟩
García-López, M. T., **5**, 1039 ⟨84S509⟩
García-Martin, J. C., **4**, 152, 153 ⟨79S893⟩
Garcia, A., **2**, 89 ⟨88T6681⟩; **3**, 671 ⟨84JOC5195⟩
Garcia, B. A., **2**, 31 ⟨76BCJ1989⟩
Garcia, E., **1**, 473 ⟨93JA9276⟩
Garcia, E. E., **3**, 261 ⟨74JHC219⟩
Garciá, H., **3**, 297 ⟨86H(24)2511⟩
Garcia, J., **2**, 521 ⟨87JOC4831⟩; **3**, 673 ⟨85TL1827⟩; **5**, 353, 354 ⟨87TL341⟩, 361 ⟨84JOC3322, 91JOC7038⟩
Garcia, J. C., **2**, 84 ⟨91TL5273⟩
Garcia, J. L., **1**, 474 ⟨81T3957⟩; **2**, 301 ⟨91SC535⟩; **3**, 291 ⟨83OM467⟩, 625 ⟨82JOM(240)C5⟩
Garcia, J. M., **3**, 410 ⟨91CC524⟩; **4**, 514 ⟨91CC524⟩; **5**, 294 ⟨87TL1945⟩
Garcia, M., **1**, 11, 12 ⟨93T8433⟩
Garcia, M. E., **4**, 712 ⟨88JCS(D)2431⟩
Garcia-Blanco, S., **4**, 405 ⟨87CJC687⟩
Garcia-Granda, S., **2**, 1003 ⟨91T5277⟩; **6**, 327 ⟨92AG81⟩, 328 ⟨91JOM(420)C12, 93OM1394⟩, 329, 330 ⟨92POL2713⟩, 329, 330 ⟨93OM2888⟩, 331 ⟨91OM3005, 91OM384⟩
Garcia-Grande, S., **6**, 329, 330 ⟨94OM1336⟩
Garcia-Luna, A., **2**, 452 ⟨83JOC1760⟩; **3**, 618 ⟨80S659⟩, 618, 662 ⟨79S227⟩; **5**, 226 ⟨80S661, 81S790⟩
Garcia-Muñoz, G., **4**, 314 ⟨69JHC921⟩
Garcia Navio, J. L., **2**, 249 ⟨92TL3677⟩; **3**, 171 ⟨92TL3677⟩; **4**, 843 ⟨89H(29)1877⟩; **5**, 606 ⟨85S86⟩
Garcia-Ochoa, S., **1**, 187 ⟨90JOC2294⟩, 579 ⟨92JOC1613⟩
Garcia-Raso, A., **1**, 433 ⟨90JOC991⟩
Garcia Ruano, J. L., **3**, 695, 709 ⟨85TL3381⟩; **4**, 270, 863 ⟨91TL3695⟩
Gard, G. L., **4**, 868 ⟨88IC4329, 89IC3766⟩; **5**, 246 ⟨69JOC4173⟩; **6**, 251 ⟨91JFC(55)313⟩, 252 ⟨90IC4588⟩
Gardano, A., **5**, 143 ⟨85JOM(285)293⟩
Gardein, T., **2**, 1004 ⟨85CB3529⟩
Garden, W. D., **4**, 569 ⟨59BRP815231, 59USP2889349⟩
Gardent, J., **3**, 75 ⟨71BSF3985⟩
Gardette, M., **1**, 468 ⟨84T2741⟩
Gardiner, D. M., **2**, 336, 339 ⟨67JOC1115⟩
Gardiner, J. E., **2**, 413 ⟨50JCS1769⟩
Gardiner, J. M., **1**, 582 ⟨90CC84⟩
Gardiner, M. G., **4**, 603, 649 ⟨91OM3167⟩
Gardlik, J. M., **3**, 284 ⟨83JA7644⟩
Gardlund, S. L., **2**, 972 ⟨63JA3897⟩
Gardlund, Z. G., **2**, 972 ⟨63JA3897⟩
Gardner, D. V., **1**, 637 ⟨70CJC2110⟩
Gardner, J. H., **5**, 74 ⟨43OSC(2)523⟩
Gardner, J. N., **3**, 156 ⟨68JOC3294⟩
Gardner, J. O., **6**, 611 ⟨92TL3113⟩
Gardner, P. D., **1**, 890 ⟨67JA3365⟩
Gardner, P. J., **2**, 832, 833 ⟨66JCS(B)1162⟩
Gardner, R. J., **2**, 377 ⟨70JCS(C)1714⟩
Gardner, S. A., **4**, 689, 690 ⟨73IC2396⟩; **5**, 241 ⟨80JOM(197)111⟩

Gardner, T. G., **4**, 656 ⟨93JA8493⟩
Gardner, T. S., **3**, 107 ⟨51JOC1121⟩; **4**, 424 ⟨57JOC984⟩; **5**, 10 ⟨61JOC1514⟩
Gardrat, C., **1**, 926 ⟨84SC1191⟩; **5**, 206 ⟨85BSB29, 85T4047⟩
Gareau, Y., **1**, 550 ⟨91JOC2523⟩; **2**, 221 ⟨84JA759⟩
Gareev, G. A., **4**, 106 ⟨85JOU806, 88JOU2003⟩
Gareev, R. D., **4**, 343 ⟨72ZOB1861⟩, 488 ⟨79ZOB493⟩, 489 ⟨72ZOB1227, 82ZOB2444⟩; **5**, 709 ⟨84JGU1969, 84ZOB2203⟩
Garegg, P. J., **2**, 7 ⟨79CC978, 80JCS(P1)2866⟩, 7 ⟨82JCS(P1)681, 84S168⟩; **4**, 223 ⟨87MI 405-01⟩; **5**, 354 ⟨89CC1640⟩
Gareh, M. T., **4**, 262 ⟨93TA2139⟩
Garg, C. P., **1**, 556 ⟨92JCS(P1)2565⟩; **2**, 311 ⟨91SL668⟩; **3**, 86 ⟨59TL9⟩, 86, 94, 458 ⟨64JA1085⟩, 126 ⟨61JA2951⟩, 215 ⟨71JOC387⟩
Garg, H. G., **2**, 171 ⟨63JOC2992⟩
Garg, N., **1**, 946 ⟨93T5189⟩
Garibina, V. A., **2**, 1028 ⟨85JGU1771⟩
Gariboldi, P., **5**, 295 ⟨80S385⟩
Garieva, F. R., **4**, 592 ⟨90ZOB2291⟩
Garigipati, R. S., **1**, 408 ⟨86T2979⟩; **2**, 245 ⟨83JA4499, 84JA7861⟩; **4**, 370 ⟨86T2979⟩; **5**, 749 ⟨90TL1969⟩
Garin, J., **5**, 1026 ⟨89BSB289⟩; **6**, 563 ⟨81S961, 91S147⟩
Garino, M., **6**, 420 ⟨26G847⟩
Garito, A. F.,**4**, 869 ⟨75CC867⟩
Gar'kin, V. P., **5**, 349 ⟨81ZOR401⟩
Garkusha, O. G., **1**, 890 ⟨89JOM(371)279⟩
Garkusha-Bozhko, V. S., **5**, 848 ⟨72UKZ171⟩
Garland, R., **5**, 79 ⟨84JA1146⟩
Garland, T. R., **1**, 869 ⟨63T1625⟩
Garlaschelli, L., **4**, 246 ⟨90TL5815⟩
Garmaise, D. L., **2**, 102 ⟨60CJC343⟩; **5**, 1024 ⟨58JA3332⟩; **6**, 533, 540 ⟨62JOC4509⟩
Garmish, J. F., **2**, 403 ⟨68CJC1149⟩
Garín, J., **6**, 605 ⟨84S520⟩, 627 ⟨82S590, 83S375⟩
Garner, A. Y., **4**, 24 ⟨56JA5430⟩, 114, 117 ⟨64USP3161607⟩
Garner, C. M., **2**, 268 ⟨85JA7167⟩
Garner, J., **5**, 943 ⟨90OM2836⟩
Garner, P., **1**, 370 ⟨94JOC4⟩, 373 ⟨83TL1897⟩; **3**, 424 ⟨90JOC412⟩; **4**, 168 ⟨88JOC2979⟩
Garner, S. E., **4**, 681 ⟨92CC310⟩
Garny, F., **5**, 833 ⟨1891CB3426⟩
Garratt, D. G., **1**, 1110 ⟨73JA160⟩, 1123 ⟨76CJC3045, 76CS76⟩, 1123 ⟨76JOC2241⟩; **2**, 4, 5 ⟨B-77MI 201-01⟩, 280 ⟨77JOC1776, 78T2869⟩, 280 ⟨80CJC1030⟩; **5**, 1107 ⟨79CJC2145⟩, 1108 ⟨79CJC2180⟩
Garratt, P. J., **1**, 111 ⟨91COS(3)271⟩, 184 ⟨80S389⟩, 861 ⟨79COC(2)361⟩, 865 ⟨79COC(2)391⟩, 890 ⟨79COC(2)398⟩, 906 ⟨79COC(2)390⟩, 917 ⟨79COC(2)460, B-86MI 119-03⟩, 955 ⟨78TL691⟩, 990 ⟨73JOC864⟩; **2**, 929, 938 ⟨88TL1833, 86TL2761⟩; **5**, 199 ⟨86TL2761⟩; **6**, 632 ⟨89JOC1062⟩
Garraussiam, M., **2**, 256 ⟨87JA1227⟩
Garreau, Y., **1**, 550 ⟨90JOC414⟩; **3**, 322 ⟨90JOC414⟩
Garrels, J. I., **4**, 38 ⟨70JA7359⟩
Garrett, A. B., **1**, 896 ⟨50JA163⟩
Garrett, B. R., **4**, 316 ⟨54JA3969⟩
Garrett, E. R., **4**, 209 ⟨54JA743⟩
Garrett, K., **4**, 676 ⟨88IC3248⟩
Garrigues, B., **4**, 463 ⟨94S51⟩
Garrison, P. J., **3**, 60 ⟨77TL3875⟩; **4**, 924 ⟨82S394⟩
Garritsen, J. W., **6**, 383 ⟨61GEP1157210⟩
Garro-Helion, F., **2**, 304 ⟨93JOC6109⟩
Garrossian, M., **2**, 446 ⟨88PS(35)349⟩
Garrou, P. E., **1**, 444 ⟨85CRV171⟩
Garry, R., **1**, 1124 ⟨72TL2983⟩
Garson, L. R., **5**, 745 ⟨65CJC2640⟩
Garst, M. E., **1**, 241 ⟨80TL4811, 83JOC1362⟩, 524 ⟨83TL1675⟩, 1009 ⟨89JOC249⟩, 1014, 1029 ⟨90JA9330⟩; **3**, 179 ⟨80JOC2307⟩, 253 ⟨73JA250⟩; **4**, 553 ⟨79SC261⟩
Gartman, G. A., **4**, 239 ⟨80JOU1377, 82JOU876⟩, 462 ⟨76ZOB1012⟩; **6**, 100 ⟨86ZOR107⟩
Gartner, A., **5**, 1122 ⟨84MI 530-01⟩
Garufi, M., **5**, 762 ⟨78TL2175⟩
Garver, L. C., **4**, 108 ⟨89JOC2869⟩, 428 ⟨85JOC1699⟩
Garves, K., **1**, 494 ⟨70JOC3273⟩, 667 ⟨68JOC2975⟩
Garvey, D. S., **5**, 173 ⟨82JA5521⟩
Garvey, P. M., **1**, 64, 522 ⟨73JA3078⟩; **3**, 503 ⟨73JA3078⟩; **4**, 627 ⟨73JA3078⟩
Garwood, R. F., **5**, 46 ⟨B-56MI 502-03⟩, 62 ⟨B-56MI 502-02⟩, 105 ⟨B-56MI 502-01⟩
Garza, D. G., **2**, 581 ⟨90OM16, 91OM902⟩
Garzon, G., **4**, 586 ⟨87OM1992, 88IC836⟩
Gasanov, R. G., **6**, 323 ⟨82IZV460⟩, 456 ⟨84MI 614-01⟩
Gasanov, V. S., **6**, 558 ⟨83ZOR92, 84ZOR2158⟩, 558 ⟨85MI 617-01, 90MI 617-02⟩
Gasc, M. B., **2**, 306 ⟨83T703⟩
Gasco, A., **4**, 429 ⟨87JOC3442⟩
Gasdaska, J. R., **3**, 679, 680 ⟨86JA6739⟩; **4**, 520 ⟨89JOC644⟩; **5**, 728 ⟨85JOC4415⟩
Gaset, A., **2**, 22, 28 ⟨91SC11, 92SC1923⟩; **3**, 302 ⟨82SC1121⟩, 373 ⟨87T3945⟩; **4**, 425 ⟨86S643⟩
Gasiecki, A. F., **1**, 1008 ⟨87JMC193⟩
Gasking, D. I., **5**, 431 ⟨85JCS(P1)409⟩
Gasol, V., **1**, 179 ⟨88SC445⟩; **2**, 8 ⟨87S511⟩, 18 ⟨89JOC4294⟩, 19 ⟨87TL4733⟩
Gaspar, C., **5**, 350 ⟨93H(36)1263⟩
Gaspar, P., **6**, 32, 219 ⟨67JOC833⟩
Gaspar, P. P., **1**, 648 ⟨B-88MI 114-02⟩; **2**, 547 ⟨B-91MI 210-03⟩, 947, 949 ⟨B-91MI 218-01⟩; **4**, 748 ⟨80JA6615⟩
Gasparis-Ebeling, T., **6**, 405 ⟨81JMR(44)54⟩
Gasparrini, F., **3**, 133 ⟨79S207⟩; **4**, 260 ⟨90JOC1323⟩
Gasparyan, G. T., **2**, 446 ⟨86AKZ463⟩; **4**, 914 ⟨83CA(98)4614z⟩
Gasselin, V., **2**, 110 ⟨1894MI 202-01⟩
Gasser, O., **4**, 553 ⟨77CB3501⟩, 583, 584, 585, 586, 588 ⟨77CB3517⟩, 586 ⟨77IJ149⟩, 586, 588 ⟨76AG(E)542⟩; **5**, 911, 1093, 1094 ⟨77CB3501⟩, 1093 ⟨75JA6281⟩
Gassman, P. G., **1**, 6 ⟨68JA7271⟩, 430 ⟨68OS(48)68⟩, 499 ⟨91TL6473⟩, 521 ⟨88TL3407⟩, 579 ⟨88JOC5576⟩, 704, 709 ⟨86JOC2397, 83JA667⟩, 709 ⟨88TL4803⟩, 869 ⟨91JA9667⟩, 881 ⟨66JOC982⟩, 885 ⟨67JA946⟩, 904 ⟨80JA1214, 80JA4138⟩, 904 ⟨80JA7613⟩; **2**, 138 ⟨77JOC3233⟩, 139, 140 ⟨77JOC3236⟩, 168 ⟨74JA5508⟩, 238 ⟨82JOC951⟩, 761 ⟨72JA3884, 74JA551, 77TL2235⟩, 762 ⟨72T2749⟩; **3**, 97 ⟨74TL3463⟩, 463 ⟨59JA4751⟩, 615 ⟨78TL3773⟩, 698, 707 ⟨82JA5849, 84TL3259, 85TL4971, 90OSC(7)294⟩; **4**, 67 ⟨83JA667⟩, 186 ⟨88JOC5574⟩; **5**, 338 ⟨77TL2235⟩; **6**, 63 ⟨85TL5243⟩
Gassner, T., **3**, 494 ⟨86CB1350⟩
Gastaldi, S., **1**, 949 ⟨92JOC6118⟩
Gastambide, B., **5**, 191, 196 ⟨85TL1503⟩
Gasteiger, J., **2**, 456 ⟨80TL2687⟩; **3**, 50, 197 ⟨82CB601⟩
Gaston, R. D., **3**, 71 ⟨87S908⟩; **6**, 134 ⟨90JA4552⟩
Gateau-Olesker, A., **1**, 223 ⟨85TL3121⟩
Gatehouse, B. M., **2**, 1008 ⟨77AX(B)3782⟩
Gates, J. N., **2**, 120 ⟨55OSC(4)363⟩
Gates, Jr., J. W., **1**, 432 ⟨66JA4271⟩
Gatilov, Yu. F., **2**, 885 ⟨72JGU536⟩, 886 ⟨73JGU1123⟩, 886, 888 ⟨72JGU538, 72JGU1952⟩, 890 ⟨68JGU370⟩, 891 ⟨70JGU127⟩; **5**, 1132 ⟨74MI 530-01, 74MI 530-02⟩
Gatilov, Y. V., **5**, 212 ⟨85ZOR196⟩
Gatrone, R. C., **6**, 118 ⟨89JOC4272⟩
Gatta, F., **5**, 22 ⟨93MI 517-02⟩
Gattermann, L., **3**, 364 ⟨1895CB2869⟩; **5**, 298 ⟨1888LA(249)244⟩; **6**, 448 ⟨1899CB1116⟩
Gatti, A., **3**, 283, 309 ⟨87S311⟩, 306, 307 ⟨93JCS(P1)273⟩; **6**, 126 ⟨92T2501⟩, 302, 554, 729 ⟨88S22⟩, 553, 554 ⟨86S894⟩

Gatti Comini, S., **1**, 826 ⟨93T2093⟩
Gattow, G., **5**, 506, 507 ⟨72ZAAC(388)78⟩, 546 ⟨72ZAAC(389)145⟩, 556 ⟨76CB3243⟩; **6**, 434, 534, 535 ⟨84ZAAC(508)136⟩, 481 ⟨84ZAAC(513)183⟩, 482 ⟨84ZAAC(514)120⟩, 534 ⟨84ZAAC(509)67, 83ZAAC(502)7⟩, 536 ⟨84ZAAC(509)61⟩, 546 ⟨84ZAAC(515)182⟩, 548 ⟨87ZAAC(552)181⟩, 548, 553 ⟨87ZAAC(550)109⟩, 552 ⟨63ZAAC(321)143, 85ZAAC(530)101⟩, 553 ⟨87ZAAC(545)125, 87ZAAC(554)172⟩, 555, 559 ⟨81ZAAC(481)126⟩, 560 ⟨85ZAAC(522)145, 85ZAAC(531)101, 85ZAAC(531)82, B-77MI 617-01⟩, 561 ⟨82ZAAC(485)203, 82ZAAC(488)94, 85ZAAC(527)130⟩, 561, 564 ⟨85ZAAC(525)112⟩, 563 ⟨85ZAAC(524)111, 85ZAAC(524)117, 85ZAAC(527)125, 87ZAAC(551)191, ⟨88ZAAC(556)141⟩, 564 ⟨85ZAAC(528)168⟩, 565 ⟨84ZAAC(512)231⟩, 590 ⟨68AG(E)868⟩, 592 ⟨71CB1429⟩, 627 ⟨77ZAAC(434)115, 86ZAAC(533)99⟩, 628 ⟨77ZAAC(434)110⟩, 630 ⟨83ZAAC(501)157, 83ZAAC(501)169, 83ZAAC(501)176⟩
Gauchet-Prunet, J. A., **1**, 98 ⟨91JOC741⟩
Gaude, D., **2**, 791 ⟨86SC63⟩
Gaudemar, M., **1**, 544 ⟨79JOM(166)1⟩, 958 ⟨67CR(C)1492⟩, 967 ⟨76JOM(108)159⟩, 971 ⟨69BSF898, 70BSF3157⟩, 1055 ⟨70BSF2175, 74JOM(76)297⟩, 1055 ⟨74JOM(76)305⟩, 1059 ⟨71BSF3071⟩; **2**, 422 ⟨79S463⟩, 590, 1096 ⟨62BSF974⟩; **3**, 461 ⟨74JOM(81)139⟩; **4**, 442 ⟨83S926⟩, 709 ⟨71CR(C)1669⟩, 832 ⟨84JOM(263)C21⟩, 1066, 1067 ⟨77JOM(142)9⟩; **5**, 607 ⟨83TL4001⟩
Gaudemar-Bardone, F., **2**, 422 ⟨79S463⟩; **4**, 394 ⟨91MI 408-01⟩
Gaudemer, A., **2**, 579 ⟨73JOM(55)375⟩
Gaudenzi, M. L., **1**, 472, 1013 ⟨82T313⟩
Gauder, S., **2**, 314 ⟨86OM739⟩
Gaudiano, G., **5**, 821 ⟨59T24⟩, 832 ⟨67G1664⟩
Gaudin, J.-M., **1**, 193 ⟨88S983⟩, 195 ⟨87HCA1477⟩, 195 ⟨88TL4705⟩, 195 ⟨88TL4709⟩, 504 ⟨90TL5749⟩; **4**, 535 ⟨88S983⟩
Gaudino, J. J., **1**, 174, 341 ⟨86JA3102⟩; **5**, 155 ⟨88TL6823⟩
Gaudry, M., **2**, 28 ⟨88OSC(6)193⟩, 29 ⟨69BSF4169⟩; **4**, 405 ⟨88OSC(4)474⟩
Gaudry, R., **2**, 618 ⟨62CJC818⟩; **3**, 615 ⟨55OSC(3)436⟩; **5**, 34, 275, 305 ⟨63OSC(4)496⟩
Gaudy, E. T., **4**, 472 ⟨68JOC3090⟩; **5**, 403, 404 ⟨68JOC3090⟩
Gauffeny, F., **6**, 549 ⟨91T5149⟩
Gaufryau, M.-H., **1**, 656 ⟨84JOM(267)93⟩
Gaughan, E. J., **1**, 1004 ⟨66JOC4071⟩
Gaukhman, A., **3**, 430 ⟨89SC3129⟩; **4**, 929 ⟨84H(22)987⟩, 1044 ⟨88JOM(346)297⟩
Gaukhman, A. P., **2**, 914 ⟨86ZOR140⟩; **6**, 124 ⟨84JOM(265)237⟩
Gaul, J. M., **1**, 21 ⟨70JA3232⟩
Gaule, J., **3**, 463 ⟨16CB1897⟩
Gault, F. G., **1**, 430 ⟨69BSF842⟩
Gault, H., **5**, 62 ⟨B-45MI 502-04, B-45MI 502-05⟩
Gault, R., **1**, 89 ⟨71OS(51)103⟩
Gault, Y., **1**, 554 ⟨84TL1279⟩, 665 ⟨69JOM(16)7⟩
Gaumont, A. C., **3**, 337, 351 ⟨91T4927⟩, 726 ⟨94CR1413⟩
Gaunt, J. C., **4**, 720 ⟨75JOM(102)511⟩
Gauntlett, J. T., **4**, 580 ⟨91JCS(D)1117⟩; **5**, 446 ⟨84CC420, 88JOM(342)C5⟩
Gaur, D. P., **2**, 112 ⟨B-78MI 202-01⟩; **5**, 389 ⟨73JOM(63)221⟩
Gaus, P. L., **2**, 753 ⟨88TL5083⟩; **3**, 83 ⟨84OM1601⟩
Gautam, V. K., **2**, 34, 43 ⟨88JOC187⟩
Gautheron, B., **3**, 636 ⟨86CC424⟩; **4**, 720 ⟨89JOM(362)C27, 90IC1817⟩, 720 ⟨92JOM(427)231⟩
Gautheron, C. M., **3**, 159 ⟨91S499⟩
Gauthier, C., **3**, 280 ⟨86JOC2128⟩
Gauthier, J., **5**, 1024 ⟨84JHC1685⟩; **6**, 429 ⟨89TL1901⟩
Gauthier, J. Y., **1**, 532, 662 ⟨86TL539⟩; **4**, 247 ⟨88TL6729⟩, 248 ⟨88TL6733⟩
Gauthier, M. D., **2**, 1003 ⟨81CJC1463⟩
Gauthier, S., **2**, 753 ⟨87S383⟩
Gautier, A., **1**, 456, 507 ⟨77JOM(141)113⟩; **3**, 694 ⟨1869LA(146)119⟩
Gautier, H., **1**, 474 ⟨88CC203⟩
Gautier, J. A., **1**, 1001 ⟨64BSF2477⟩; **5**, 742, 748, 751, 754, 755 ⟨B-75MI 519-01⟩, 754 ⟨71BCF478⟩, 756 ⟨70BCF200⟩, 759 ⟨72FRP2081556⟩
Gautschi, F., **1**, 1082 ⟨67HCA2101⟩
Gavars, M., **4**, 929 ⟨84H(22)987⟩; **5**, 224 ⟨87ZOB686⟩
Gavars, M. P., **6**, 124 ⟨84JOM(265)237⟩
Gavaskar, K., **2**, 556 ⟨91JA5720⟩
Gavaskar, K. V., **1**, 316 ⟨92JA8053, 94T5957⟩; **2**, 556 ⟨92JA8053⟩
Gavina, F., **4**, 749 ⟨86JCR(M)2843⟩; **6**, 695 ⟨85CC296⟩
Gaviña, P., **4**, 749 ⟨85CC296⟩
Gavin, G., **2**, 627 ⟨90JHC2151⟩
Gavney, J. A., **5**, 450 ⟨90JA4083⟩
Gavrilenko, B. B., **4**, 984 ⟨75JOC2720⟩
Gavrilenko, V. V., **2**, 1093 ⟨67JGU515⟩, 1099 ⟨63IZV1146, 64IZV1989⟩, 1099 ⟨65JGU1677⟩; **3**, 84 ⟨63TL2087⟩, 85 ⟨69T5555⟩; **4**, 698 ⟨75CA(83)10238⟩; **5**, 44 ⟨78IZV2094⟩; **6**, 208 ⟨68IZV910⟩, 208, 210 ⟨83IZV636⟩, 523 ⟨58IZV100⟩
Gavrilov, G. I., **4**, 154, 155 ⟨52JGU1827⟩
Gavrilov, V. I., **2**, 1076 ⟨72JGU1957⟩; **4**, 592 ⟨90ZOB2291⟩
Gavrilov, V. M., **2**, 886, 888 ⟨72JGU538⟩
Gavrilova, G. M., **4**, 777 ⟨90ZOR2056⟩
Gawad, S. M. A., **5**, 615 ⟨88IJC(B)472⟩
Gawley, R. E., **1**, 152 ⟨89JOC175⟩, 154 ⟨84TL263⟩, 306 ⟨76S777⟩; **2**, 321 ⟨89JOC3002⟩, 1003 ⟨91COS(1)459⟩; **3**, 133 ⟨82SC15⟩, 220 ⟨76S777⟩, 425, 431 ⟨88OR(35)1⟩, 427 ⟨84TL263⟩; **4**, 526 ⟨91COS(3)65⟩, 531 ⟨89JOC3002⟩
Gawley, R. E.,**5**, 270, 282, 288 ⟨88OR(35)1⟩
Gawron, M., **2**, 1007 ⟨79PJC861⟩
Gawronska, K., **1**, 634 ⟨81JA5314⟩; **2**, 998 ⟨93JOC546⟩; **4**, 33, 35, 37 ⟨91JOC2193⟩, 816, 818 ⟨93JOC546⟩; **5**, 135, 138 ⟨83HCA2501⟩
Gawronski, J., **2**, 998 ⟨93JOC546⟩; **4**, 33, 35, 37 ⟨91JOC2193⟩, 816, 818 ⟨93JOC546⟩; **5**, 135, 138 ⟨83HCA2501⟩
Gawronski, J. K., **1**, 634 ⟨81JA5314⟩
Gay, R. L., **2**, 621 ⟨66CI(L)1635⟩
Gaydou, E. M., **4**, 187 ⟨72TL4055⟩
Gaydoul, K.-R., **2**, 436, 873 ⟨85TL4067⟩
Gayle, J. B., **2**, 175 ⟨55JOC1165⟩, 273 ⟨54JOC1165⟩
Gaylord, N. G., **1**, 28, 32 ⟨B-56MI 102-01⟩; **4**, 306 ⟨56JA2167⟩
Gaythwaite, W. R., **2**, 286, 732 ⟨28JCS2293⟩
Gazdar, M., **2**, 144 ⟨08JCS1833⟩
Gazieva, G. B., **2**, 288 ⟨59ZOB942⟩; **6**, 49 ⟨59ZOB942⟩
Gazikasheva, A. A., **6**, 121 ⟨92HAC115⟩
Gazis, E., **2**, 358 ⟨64JA3660⟩
Gazit, A., **3**, 564 ⟨88JOC679⟩
Gazizov, M. B., **4**, 340 ⟨71ZOB2575, 72ZOB1647⟩; **5**, 403 ⟨69JGU2541⟩, 404 ⟨67JGU2606⟩; **6**, 116 ⟨90CC1133⟩
Gazizov, T. K., **2**, 254 ⟨87ZOB1429⟩; **4**, 486 ⟨83IZV432⟩; **6**, 583 ⟨90ZOB563⟩
Gazizov, T. Kh., **4**, 915 ⟨68CA(68)49691u, 70ZOB27⟩, 915 ⟨88CA(109)211147t⟩, 921 ⟨90ZOB1940⟩; **5**, 406 ⟨90JGU1940⟩
Gazurova, E., **4**, 764 ⟨88MI 417-01⟩
Gazuzov, T. Kh., **4**, 915 ⟨68ZOB139⟩
Gazzard, E. G., **5**, 966 ⟨69BRP1152877⟩
Gdaniec, M., **6**, 547 ⟨92JCS(P1)47⟩
Gdula, A., **2**, 11 ⟨79S37⟩

Ge, L., **6**, 216 ⟨86MI 607-01⟩
Geanangel, R. A., **6**, 390 ⟨91MI 613-01⟩, 396 ⟨93JCS(D)2891⟩, 491 ⟨84IC3063⟩
Gebala, A., **4**, 719, 726 ⟨75IC78⟩
Gebbing, H., **5**, 339 ⟨79S97⟩
Gebel, I. A., **4**, 140 ⟨91IZV1039⟩
Gebel, W., **5**, 1074 ⟨89ZAAC(568)35⟩, 1074, 1078 ⟨89ZC145⟩
Gebelein, C. G., **5**, 985 ⟨69JOC1677⟩
Gebhardt, B., **4**, 850 ⟨64CB1298⟩
Gebhardt, H., **2**, 99 ⟨35CB1011⟩, 99, 172 ⟨36CB937⟩
Gebhart, M. A., **4**, 281 ⟨90JHC1433⟩
Gebler, W., **5**, 1128 ⟨80AG767⟩
Gebreyes, K., **1**, 619, 707, 949 ⟨86JA4568⟩, 707 ⟨84TL5469⟩; **4**, 82, 83, 85 ⟨86JA4568⟩, 700 ⟨87OM834⟩
Gedge, S., **6**, 80 ⟨81JOC886⟩
Gedridge, Jr., R. W., **2**, 484 ⟨92OM967⟩
Gedymin, V. V., **3**, 531 ⟨71T1317⟩
Gee, S. K., **1**, 812 ⟨82JA7670⟩; **2**, 420 ⟨92OS(71)133⟩, 668 ⟨84JOC1672⟩
Gee, W., **2**, 444 ⟨65JCS3354⟩
Geen, G. R., **1**, 564 ⟨94JCS(P1)793⟩
Geer, S. M., **5**, 774 ⟨83JHC69⟩
Geerts, R., **2**, 511 ⟨86CC1543⟩
Gees, T., **3**, 707 ⟨88CB507⟩; **5**, 658 ⟨88CB507⟩; **6**, 622 ⟨88CB507⟩
Geevers, J., **5**, 662 ⟨71RTC463⟩, 662, 809, 810 ⟨71RTC458, 70JCS(C)875⟩; **6**, 284 ⟨72RTC331⟩, 609 ⟨74TL1687, 74TL1691⟩, 621 ⟨70JCS(C)875⟩
Geffken, D., **4**, 137 ⟨90JOM(381)315⟩, 160 ⟨89TL5243⟩; **5**, 211 ⟨86CB744, 86CZ377⟩, 323 ⟨88AP(321)311, 90AP(323)181⟩, 327 ⟨88AP(321)235⟩
Gefter, E. L., **2**, 451 ⟨61ZOB949⟩
Gegnas, L. D., **1**, 1079 ⟨92JOC5071⟩
Gehlhaus, J., **6**, 106 ⟨71CB873⟩
Gehring, R., **2**, 254 ⟨81PS(11)349⟩
Gehrke, C. W., **5**, 226 ⟨77JC355⟩
Gehrke, M., **4**, 223 ⟨31CB2696⟩
Gehrman, K., **2**, 448 ⟨83GEP(O)3149937⟩
Gehrt, H. H., **2**, 419 ⟨58ZAAC(294)81⟩
Geib, G. D., **2**, 680 ⟨85JOC1212⟩
Geib, J. R., **2**, 1025 ⟨72OPP89⟩
Geib, S. J., **5**, 444 ⟨89JA149⟩; **6**, 338 ⟨89JA149⟩
Geib, S. V., **1**, 326, 327, 328 ⟨92JA7007⟩, 327 ⟨93JA8585⟩
Geibel, K., **4**, 697 ⟨80JOM(191)7⟩; **6**, 206, 207 ⟨80JOM(191)7⟩
Geiger, C. C., **1**, 905 ⟨83JA7123⟩
Geiger, G. A. P., **1**, 1088 ⟨85CB1504⟩
Geiger, R., **5**, 279 ⟨70CB2024, 70CB788⟩, 285 ⟨70CB2041⟩
Geiger, W., **3**, 359 ⟨82LA14⟩, 448 ⟨57AC(R)1244⟩
Geiger, W. E., **4**, 632 ⟨83JA2582⟩, 676 ⟨89OM412⟩, 683 ⟨90JA7113⟩
Geigy, J. R., **2**, 427, 430, 437 ⟨65FRP1401930⟩
Geii, K. H., **6**, 302 ⟨72CB3905⟩
Geike, W., **4**, 581, 582, 583 ⟨83OM257⟩
Gein, L. F., **3**, 475 ⟨78ZOR2234⟩
Gein, V. L., **3**, 475 ⟨78ZOR2234⟩
Geirsson, J. K. F., **3**, 220 ⟨90S993⟩
Geisberger, G., **5**, 381, 382, 1085 ⟨90CB455⟩, 382 ⟨90CB63⟩
Geise, H. J., **1**, 670 ⟨91JOM(405)1⟩, 688 ⟨81BSB1141, 82MI 115-01⟩
Geise, H. Y., **1**, 688 ⟨82JOC248⟩
Geisel, M., **2**, 13 ⟨73S493⟩; **5**, 662, 809 ⟨85CB4459⟩; **6**, 260, 261 ⟨86CB107⟩
Geiseler, G., **2**, 175 ⟨58CB2130⟩
Geisler, J.-P., **2**, 310 ⟨90LA795⟩, 331 ⟨89S608⟩; **3**, 41 ⟨89S608⟩
Geisler, K., **4**, 547, 564 ⟨79CB648⟩; **5**, 624, 625 ⟨84ZC99⟩; **6**, 164, 703 ⟨79CB648⟩, 596 ⟨91EGP291752⟩
Geismann, C., **1**, 1079 ⟨77LA282⟩
Geiss, F., **4**, 204 ⟨61CB929⟩
Geiss, K., **1**, 140, 141 ⟨74AG(E)479⟩
Geiss, K.-H., **1**, 140 ⟨74AG(E)202, 77CB1833⟩; **2**, 272 ⟨74AG(E)202⟩; **4**, 367, 392 ⟨77CB1833⟩; 392 ⟨74AG(E)202, 74AG(E)479, 79CB1420⟩, 395 ⟨76AG(E)437⟩; **6**, 82, 86, 88, 302, 303, 317 ⟨72CB3280⟩
Geissler, B., **4**, 549 ⟨93PS(76)1⟩; **5**, 1152 ⟨92AG(E)758⟩; **6**, 153 ⟨92AG(E)758⟩
Geissler, G., **1**, 638 ⟨53LA(580)44⟩; **3**, 493 ⟨53LA(580)44⟩
Geissler, H., **1**, 282 ⟨94SL511⟩
Geissman, T. A., **5**, 48 ⟨44OR(2)94⟩
Gekhman, A. E., **6**, 410 ⟨89BAU1570⟩
Gelas, J., **4**, 186 ⟨81H(16)1587⟩, 226 ⟨83CJC2103⟩
Gelas-Mialhe, Y., **1**, 706, 707 ⟨79CJC1958⟩; **4**, 779 ⟨77CJC3190⟩
Gelbard, G., **2**, 9, 19 ⟨79JCS(P1)2248⟩, 393 ⟨77S113⟩; **3**, 4 ⟨80TL4653⟩, 4, 115 ⟨86JOC4016⟩
Gelbecke, M., **1**, 503 ⟨70TL5139⟩
Gelbert, M., **3**, 699 ⟨91CB2555⟩
Gelblum, S., **2**, 102 ⟨60CJC343⟩
Geldard, J. F., **5**, 848 ⟨65JOC318⟩
Gelin, R., **3**, 241 ⟨74BSF1043⟩, 527 ⟨70TL3657⟩
Gelin, S., **3**, 212 ⟨78S142, 74BSF1043⟩, 246 ⟨72CR(275)897, 78S291⟩, 246 ⟨84JHC13⟩, 453 ⟨80S623⟩, 527 ⟨70TL3657⟩
Gell, K. I., **3**, 509 ⟨80JOM(184)C1⟩; **4**, 579 ⟨80IC3207⟩; **5**, 917 ⟨80IC3207⟩
Geller, L. E., **2**, 391 ⟨60JA2640⟩, 760 ⟨53JA3632⟩
Gellerman, B. J., **2**, 24 ⟨84TL607⟩
Gellerson, H. E., **1**, 881 ⟨50JA3704⟩
Gellert, H.-G., **1**, 663 ⟨50LA(567)179⟩, 665 ⟨60LA(629)172, 60LA(629)1⟩; **2**, 555 ⟨50LA(567)195⟩, 595 ⟨70HOU(13/4)9⟩, 596 ⟨54LA(589)91, 60LA(629)121⟩, 596 ⟨60LA(629)1⟩, 967 ⟨70HOU(13/4)23⟩; **4**, 699 ⟨60LA(629)1⟩
Gellert, R. W., **4**, 682 ⟨89JOM(365)163⟩, 696 ⟨82JA2072⟩
Gelles, E., **1**, 892 ⟨54JCS2918⟩
Gelling, I. R., **4**, 413 ⟨71JCS(C)780⟩
Gelling, O. J., **3**, 221 ⟨90TL7201⟩; **6**, 124 ⟨90TL7201⟩
Gellrich, M., **1**, 57 ⟨55LA177⟩
Gelmi, M. L., **1**, 45 ⟨87S719⟩; **3**, 438 ⟨87S719⟩
Gelotte, K. O., **2**, 619 ⟨90JHC1549⟩
Geluk, H. W., **6**, 643 ⟨69JMC712⟩
Gemal, A. L., **1**, 5, 9, 10 ⟨80TL3195⟩, 1093 ⟨82JOC3805⟩; **2**, 59 ⟨79JA5848⟩; **3**, 14 ⟨82TL3361⟩, 31 ⟨81TL4077⟩, 249 ⟨75S437⟩
Gemenden, C. W., **4**, 413 ⟨73JOC437⟩
Gemeseus, A., **5**, 826 ⟨1896CB1161⟩
Gemmler, A., **2**, 903 ⟨80CB3352⟩
Genco, N. A., **3**, 666 ⟨73JOC4365⟩
Gendin, D. V., **4**, 656 ⟨84ZOB2025⟩; **5**, 432 ⟨83IZV1675⟩
Gendler, P. L., **5**, 742, 758 ⟨81JOC2455⟩
Gendreau, Y., **1**, 238, 483 ⟨77JOM(142)1⟩; **2**, 679 ⟨79BSF(2)305⟩
Geneste, J. M., **2**, 95 ⟨63BSF470⟩
Geneste, P., **3**, 280 ⟨86JOC2128⟩
Genêt, J.-P., **1**, 193 ⟨88S983⟩, 194 ⟨86TL4573⟩, 469, 494, 1013 ⟨92SL715⟩; **2**, 301 ⟨83TL2745⟩, 304 ⟨93SL680⟩, 437, 444 ⟨90TL6357⟩, 740 ⟨79TL1499⟩; **3**, 181 ⟨91TL2359⟩, 218 ⟨79TL1499⟩; **4**, 465, 478 ⟨92TL77⟩, 535 ⟨88S983⟩, 557, 570, 575 ⟨93PS(77)199⟩; **6**, 461 ⟨87T5315⟩
Genetti, R. A., **4**, 706 ⟨70JOC3888⟩
Geng, L., **3**, 83, 95 ⟨89JOM(376)41⟩
Gengnagel, K., **6**, 504 ⟨81GEP2855882⟩
Genkina, G. K., **4**, 116 ⟨69IZV181⟩
Gennari, C., **1**, 751, 752 ⟨83TL4405⟩; **2**, 698 ⟨92JOC5173⟩; **3**, 37 ⟨82S829⟩, 60 ⟨84JA260⟩; **4**, 265 ⟨78TL3861⟩, 834 ⟨88JOC4015⟩, 856 ⟨81JCS(P1)1278⟩; **5**, 88 ⟨86JA6394⟩
Genoch, K. H., **2**, 148 ⟨66JA3874⟩
Genoni, F., **1**, 79 ⟨74JOM(70)89⟩

Genrari, C., **2**, 375 ⟨86JA6394⟩
Gensler, W. J., **1**, 453, 512 ⟨63JOC1254⟩, 549 ⟨71JA3828⟩, 954 ⟨58JA1376⟩, 1023 ⟨58JA4593⟩; **5**, 28 ⟨57CRV191⟩
Gent, P. A., **1**, 882 ⟨72JCS(P1)1535, 73JCS(P1)1858⟩
Gentes, F. H., **1**, 113 ⟨50JA5127⟩
Gentile, A., **1**, 926 ⟨82TL5587⟩
Gentile, B., **1**, 725 ⟨79HCA2091⟩; **5**, 155 ⟨79HCA2091⟩
Gentilucci, L., **2**, 23 ⟨94CC735⟩
Gentner, R., **5**, 184 ⟨25CB1418⟩
Gentz, F., **5**, 579 ⟨63AG(E)397, 65M1474⟩
Gentzkow, W. von, **4**, 905 ⟨73LA247⟩
Geoffroy, G. L., **4**, 676 ⟨88IC3248, 88OM289⟩, 678 ⟨84JOM(273)333⟩, 679 ⟨81JA1278⟩, 718 ⟨87JOM(334)117⟩; **5**, 449 ⟨89PAC1723⟩, 451 ⟨87JA1584⟩, 946 ⟨90JA3220, 92JA6735⟩, 946, 957 ⟨93JA3806⟩, 955 ⟨94OM4676⟩, 988 ⟨92OM1242⟩; **6**, 344 ⟨86OM2561⟩
Geoffroy, M., **2**, 341, 344 ⟨90TL531⟩; **4**, 598 ⟨81HCA329⟩; **5**, 211 ⟨87MI 504-01⟩, 212, 213 ⟨85CAR375⟩, 214 ⟨85HCA1893⟩
Geoffroy, P., **1**, 539 ⟨88T127⟩
Geoghan, Jr., P., **2**, 40 ⟨67JA1522⟩
Geokjian, P. G., **1**, 1099 ⟨87JOC4819, 87JOC4823⟩, 1099 ⟨87JOC4825⟩
George, A. V. E., **2**, 942 ⟨84CC257⟩; **5**, 559 ⟨78TL4839, 83RTC83⟩
George, B., **5**, 631 ⟨76JOC3233⟩
George, C., **2**, 384 ⟨92JOC1585⟩, 483 ⟨92OM2163⟩, 489, 492 ⟨89OM482⟩, 1053 ⟨90JOC1916⟩; **4**, 428 ⟨87JOC4781, 90JOC2920⟩, 795 ⟨91JOC537⟩, 970, 993 ⟨92JOC235⟩; **5**, 1029 ⟨92JCS(P1)1671⟩; **6**, 150 ⟨92JOC3026⟩
George, C. F., **1**, 19 ⟨85JA5732⟩, 596 ⟨86JA3385⟩; **4**, 298 ⟨90JOC1459⟩; **5**, 277 ⟨85JA5732⟩
George, H. J., **5**, 1027 ⟨24JCS1702, 26JCS3041⟩, 1027 ⟨27JCS436⟩
George, I. A., **6**, 533 ⟨90CL83⟩
George, J. K., **3**, 327 ⟨73JA7287⟩
George, J. W., **2**, 288 ⟨69JA1649⟩, 292 ⟨68JOM(12)401⟩
George, M. V., **2**, 131 ⟨62JCS1967⟩, 388 ⟨68JOC2407⟩, 545 ⟨60JA403⟩, 989 ⟨73JOM(47)225⟩; **5**, 28 ⟨75CRV491⟩
George, P., **2**, 1046 ⟨76C387⟩; **6**, 619, 728 ⟨75C209⟩
George, T. A., **2**, 415, 416, 418, 419 ⟨69JCS(A)992⟩, 920 ⟨68JOM(14)327⟩; **5**, 277 ⟨69JCS(A)992, 69JCS(C)2565⟩; **6**, 61 ⟨65JCS2157⟩, 664 ⟨68JOM(14)327⟩
George, T., **4**, 105 ⟨74JOC1965⟩
George, T. J., **1**, 624, 779, 783 ⟨78JA3927⟩, 834 ⟨77JOC172⟩
George, W. O., **1**, 870 ⟨68JCS(B)132⟩
Georges, B., **4**, 1074, 1075 ⟨84JA3785⟩; **6**, 92 ⟨84JA3784⟩
Georges, M., **4**, 323 ⟨79CJC1153⟩
Georghiou, P. E., **1**, 722 ⟨75JA3512⟩
Georgiadis, M. P., **2**, 311 ⟨89S616⟩; **3**, 222 ⟨91JHC697⟩
Georgiou, K., **3**, 551 ⟨74CC739⟩
Georgopoulos, A., **1**, 512 ⟨86JMC112⟩
Georgoulis, C., **2**, 7 ⟨75BSF607⟩
Georg von Schnering, H., **1**, 785 ⟨91JOC5782⟩
Geraci, L. S., **3**, 31 ⟨93JA8467⟩
Geraghty, N., **1**, 178 ⟨90SC125⟩; **3**, 613 ⟨80SC279⟩
Gerard, F., **1**, 800 ⟨74BSF1924⟩
Gerasimova, E. A., **2**, 838, 839 ⟨50MI 216-01⟩
Gerasimova, E. S., **5**, 270, 283 ⟨71RCR835⟩
Gerasimova, S. S., **6**, 512 ⟨73JGU1019⟩
Gerasimova, T. N., **3**, 536 ⟨72ZOR654⟩
Gerasimova, V. V., **5**, 1025 ⟨88ZOR1771⟩, 1026 ⟨90ZOR2059⟩; **6**, 558 ⟨90ZOR1422⟩
Gerats, I., **4**, 576, 587 ⟨73EGP94992⟩
Geraty, R. A., **5**, 322 ⟨85SC1159⟩
Gerber, H., **1**, 1047 ⟨89TL1073⟩, 1047, 1048 ⟨91LA903⟩; **2**, 1052 ⟨30CB1707⟩; **3**, 351 ⟨92LA927⟩; **5**, 513, 549, 560 ⟨89TL1073⟩, 994 ⟨58AG271⟩, 994, 1018 ⟨58CB1200⟩, 1100, 1101 ⟨60AG1000⟩
Gerber, H. D., **5**, 560, 572, 597, 617 ⟨91LA903⟩
Gerbing, U., **3**, 386 ⟨87CB1905, 89CC667⟩, 386 ⟨89JOM(364)155, 89JOM(377)105⟩, 386, 397 ⟨87CC559⟩, 401 ⟨86JOM(299)C7⟩; **5**, 605, 614, 627 ⟨87CC559⟩, 605, 614, 627 ⟨88CB2095, 87JOM(332)105⟩
Gercek, Z., **2**, 385 ⟨92SC2607⟩
Gerchman, L. L., **4**, 361 ⟨75IC1618⟩; **6**, 21 ⟨79CC128⟩, 62 ⟨78JA1722⟩
Gerdes, H. M., **1**, 166 ⟨79TL3375⟩; **4**, 235 ⟨80JOC3634⟩, 236 ⟨79TL3375⟩; **6**, 126, 135 ⟨79TL3375⟩
Gerdes, J. M., **2**, 444 ⟨84JOC4293⟩; **4**, 181 ⟨86JOC4964⟩, 220 ⟨78TL731⟩, 235 ⟨79TL689⟩
Gerding, H., **6**, 302 ⟨72RTC1117⟩
Gere, J. A., **1**, 841 ⟨88JA3601⟩
Gerecht, B., **3**, 686 ⟨79AG(E)467⟩, 703 ⟨81JOC1046⟩
Gerega, V. F., **5**, 389, 717 ⟨78ZOB1146⟩, 716 ⟨76JGU714⟩, 716 ⟨76ZOB714, 78JGU1046⟩
Gerenstein, D. G., **2**, 410 ⟨88TL6843⟩
Geresh, S., **5**, 118 ⟨93T10099⟩
Gerfin, T., **4**, 566 ⟨91HCA983⟩
Gerhard, V., **2**, 238 ⟨90ZN(B)1625⟩
Gerhardt, C., **5**, 183 ⟨1853MI 504-01⟩
Gerhardt, G. E., **4**, 44 ⟨81JCS(P1)1321⟩; **6**, 36 ⟨77CC259⟩, 36, 251 ⟨81JCS(P1)1321⟩
Gerhardt, H. L., **2**, 102 ⟨43OSC(2)111⟩
Gerhardt, R., **3**, 689 ⟨88AG(E)1534⟩; **4**, 868 ⟨88AG(E)1534, 89CB463⟩, 868 ⟨92CB557⟩
Gerhardt, W., **6**, 504 ⟨65MI 616-01⟩
Gerhart, F., **3**, 478 ⟨67AG50, 67AG(E)74⟩, 478 ⟨68LA(716)204⟩, 696 ⟨76LA183⟩, 697 ⟨68AG(E)805⟩, 704 ⟨70AG(E)458⟩, 713, 717 ⟨72LA(766)116, 69AG(E)672⟩; **4**, 142 ⟨67AG(E)74⟩; **5**, 844, 866, 874 ⟨68LA(716)204⟩; **6**, 719, 732 ⟨67AG(E)560⟩
Gerhart, H. L., **2**, 100 ⟨43OSC(2)112⟩
Géribaldi, S., **1**, 888 ⟨B-89MI 119-05⟩; **2**, 744 ⟨81S880⟩; **3**, 263 ⟨81S880⟩
Gerigk, U., **1**, 331, 342 ⟨90S448⟩
Gerjovich, H. J., **5**, 318 ⟨52JA2189⟩
Gerke, R., **3**, 348 ⟨83TL5861⟩
Gerkin, R. M., **1**, 212 ⟨67JA5850⟩, 386 ⟨71JA1693⟩
Gerlach, H., **1**, 654 ⟨77HCA3039⟩, 1134 ⟨80HCA2426⟩; **5**, 66 ⟨92OS(71)48⟩
Gerlach, K., **2**, 1003 ⟨62CB1108⟩
Gerlach, M., **1**, 331, 342 ⟨90S448⟩, 936 ⟨91JOC5971⟩
Gerlach, R. F., **4**, 674 ⟨83OM1172⟩
Gerlei, A., **4**, 120 ⟨86ACH145⟩
Gerlich, W., **4**, 495, 496 ⟨76ZC64⟩
Gerlits, J. F., **2**, 337 ⟨88JOC2144⟩
Gerlt, J. A., **1**, 869 ⟨91JA9667⟩
Germa, H., **2**, 476 ⟨68CR(C)270⟩; **5**, 705, 706 ⟨77JHC749⟩; **6**, 692 ⟨86PS(26)327⟩
Germain, G., **3**, 46 ⟨78BSB693⟩, 602 ⟨82JA2920⟩, 680 ⟨88HCA1177⟩; **5**, 595 ⟨93PS(78)97⟩, 601 ⟨85T693⟩; **6**, 74 ⟨78BSB693⟩, 105 ⟨83TL1035⟩
German, L., **2**, 15 ⟨B-89MI 201-03⟩, 334 ⟨B-89MI 206-01⟩; **6**, 16 ⟨B-89MI 601-01⟩, 212 ⟨B-89MI 607-01⟩
German, L. S., **2**, 876, 881 ⟨72IZV213⟩; **4**, 134 ⟨72IZV215⟩, 697 ⟨71IZV620⟩, 730 ⟨81IZV1920⟩; **6**, 22 ⟨90IZV491⟩, 54 ⟨63IZV1946⟩, 62 ⟨72IZV85⟩, 69 ⟨60IZV231⟩, 537 ⟨81URP804634⟩
Germanaud, L., **4**, 127 ⟨88BSF699⟩
Germash, A. V., **4**, 1078 ⟨86MI 424-01⟩
Germon, C., **1**, 533, 1107 ⟨80TL3763⟩
Germroth, T. C., **1**, 1114 ⟨76JOC681⟩
Gerner, R., **6**, 560 ⟨85ZAAC(522)145, 85ZAAC(531)101⟩, 561 ⟨82ZAAC(488)94, 85ZAAC(527)130⟩, 561, 564 ⟨85ZAAC(525)112⟩, 563 ⟨85ZAAC(524)111⟩, 563 ⟨85ZAAC(524)117⟩, 563 ⟨85ZAAC(527)125, 85ZAAC(528)168⟩

Gernon, M., **4**, 223 ⟨89JA658⟩
Gero, A., **1**, 869 ⟨54JOC1960, 61JOC3156⟩
Géro, S. D., **1**, 223 ⟨85TL3121⟩, 828 ⟨93CC960⟩, 944 ⟨92T9261⟩; **3**, 59 ⟨81HCA1140⟩, 635 ⟨91TL4713⟩; **6**, 641, 643 ⟨82JCS(P1)2085⟩
Gerold, C., **1**, 503 ⟨51JA5073⟩; **4**, 178 ⟨54JA6113⟩; **5**, 318 ⟨63OSC(4)104⟩
Gerosa, A. Z., **5**, 946, 959 ⟨94ICA233⟩
Gerrans, G. C., **3**, 262 ⟨80TL1373⟩
Gerrard, W., **1**, 447 ⟨58JCS1451, B-61MI 110-01⟩; **2**, 99 ⟨56JCS3006⟩, 100 ⟨39JCS99⟩, 101 ⟨36JCS688, 50JCS2088, 55MI 202-01⟩, 104, 105 ⟨B-73MI 202-01⟩, 110 ⟨55JCS1470, 57JCS348, 57JCS739, 61JCS5545⟩, 111 ⟨56JCS824⟩, 111, 112 ⟨58JCS1451⟩, 111, 267 ⟨59JCS1529⟩, 418, 419 ⟨62JCS1786⟩, 520 ⟨63JCS1582⟩, 900 ⟨B-61MI 218-01⟩; **5**, 184 ⟨53JCS2117⟩, 228 ⟨58JCS3648⟩
Gerrliz, S. W., **2**, 753 ⟨88TL5083⟩
Gershanov, F. B., **4**, 560 ⟨86ZOB2000⟩
Gershheim, L. L., **2**, 117, 138 ⟨47JA2328⟩; **3**, 636 ⟨47JA2328, 55OSC(3)458⟩
Gershon, H., **4**, 3 ⟨66JOC916⟩
Gershtein, L. M., **2**, 403 ⟨69ZOR268⟩
Gerstein, B. C., **1**, 655 ⟨88JA6579⟩
Gerstenberger, M. R. C., **2**, 15 ⟨81AG(E)647⟩; **6**, 51, 54 ⟨81AG(E)647⟩, 274 ⟨83CB3325⟩
Gerster, G., **2**, 454, 457 ⟨83CB2936⟩
Gerster, J. F., **5**, 619 ⟨70JHC863⟩
Gerstl, R., **4**, 32 ⟨67T2549⟩; **6**, 412 ⟨1870CB730⟩
Gerstmans, A., **2**, 792 ⟨88TL5657⟩; **4**, 429, 431 ⟨90JOC6198⟩
Gerstner, F., **5**, 780 ⟨78ZN(B)24, 79JOM(175)33⟩
Gerth, D. B., **1**, 330, 341 ⟨86JOC3726⟩
Gerti, N., **5**, 28 ⟨79MI 502-05⟩
Gertner, D., **2**, 190 ⟨68TL4479⟩; **5**, 318 ⟨72JOM(35)303⟩
Gertsyuk, M. N., **4**, 441 ⟨76ZOR2103, 76ZOR231⟩; **5**, 1068, 1069 ⟨84ZOR2543⟩
Gerus, I. I., **2**, 15 ⟨91RCR1050⟩; **3**, 588 ⟨86JOU1640⟩; **5**, 657 ⟨81JOU1527, 81ZOR1720⟩
Gervais, D., **1**, 456 ⟨86CC1317⟩; **2**, 421 ⟨78ICA163⟩
Gerval, J., **2**, 943 ⟨85CR(301)1289⟩; **5**, 419 ⟨79JIC545⟩; **6**, 124 ⟨79JOC420⟩
Gerval, P., **1**, 656 ⟨84JOM(267)93⟩
Gervasio, G., **4**, 688 ⟨93JOM(444)203⟩; **6**, 352 ⟨82IC3781, 83CC1613⟩
Gervits, L. L., **4**, 640 ⟨76ZOB594⟩; **5**, 376, 389 ⟨76ZOB594⟩
Gerwarth, U. W., **6**, 572 ⟨76JOM(110)15⟩
Gerwin, M., **5**, 618, 619 ⟨75LA275⟩
Gerzina, R. J., **1**, 247 ⟨88JOM(348)199⟩
Geschwend, G., **4**, 209 ⟨88SC2337⟩; **6**, 538 ⟨85DIS(B)840⟩, 650 ⟨76MI 621-01⟩
Geschwinder, P. M., **5**, 19 ⟨82S237⟩
Gesing, E. R. F., **2**, 926 ⟨80AG(E)1023⟩; **3**, 270 ⟨80AG(E)1023⟩; **6**, 402 ⟨81JOM(217)105, 82JOC3192⟩
Gessels, H., **2**, 624 ⟨71C59⟩
Gething, B., **5**, 116 ⟨62JCS186⟩
Getman, D., **1**, 802 ⟨77JA2344⟩
Getman, K. M., **2**, 344 ⟨91TL2711⟩
Geuder, T., **2**, 310 ⟨90LA795⟩
Geue, R. J., **4**, 499 ⟨89CC1648⟩
Geux, W., **1**, 104 ⟨49HCA489⟩
Gever, G., **2**, 374 ⟨49JOC813⟩
Gevorgyan, V., **4**, 704 ⟨91JOM(418)C21⟩, 929 ⟨84H(22)987⟩
Gevorgyan, V. N., **6**, 124 ⟨84JOM(265)237⟩
Gevorkyan, A. A., **1**, 629 ⟨87ZOR1660⟩; **5**, 671 ⟨66JGU1340, 66ZOB1326⟩; **6**, 219 ⟨65IZV1873⟩
Gewald, K., **3**, 80 ⟨80JPR149⟩, 282 ⟨84LA1390⟩; **4**, 833 ⟨75JPR337⟩, 983, 986 ⟨77JPR149⟩, 995 ⟨80JPR87⟩; **5**, 593 ⟨76JPR343⟩, 599 ⟨67AG(E)294⟩; **6**, 628, 631 ⟨71JPR804⟩
Gewert, J. A., **1**, 282 ⟨94SL511⟩
Geyer, S. R., **5**, 560 ⟨88ZC269⟩
Geze, C., **3**, 703 ⟨92ICA73⟩
Gfesser, G. A., **1**, 827 ⟨94TL1661⟩
G. F. Whitfield, **2**, 121 ⟨72JOC1121⟩
Ghaderi, E., **3**, 100 ⟨78TL839⟩
Ghadiri, M. R., **1**, 237, 484, 1031 ⟨86JA1098⟩, 239, 484 ⟨84JA7260⟩
Ghafouripour, A., **5**, 665, 672 ⟨85JOC993⟩, 765, 824 ⟨86JHC1861⟩, 835 ⟨92JOC4648⟩
Ghali, N. I., **2**, 374 ⟨81JOC5413⟩
Ghandehari, M. H., **3**, 553 ⟨74JOC3906⟩; **4**, 874 ⟨74JOC3906⟩
Ghandi, R. P., **1**, 499 ⟨90JCS(P1)2185⟩
Ghanimi, A., **1**, 158 ⟨93TL1011⟩
Ghannam, A., **4**, 353 ⟨91JOC5213⟩, 358, 359 ⟨90JA2392⟩, 387 ⟨88JOC2878⟩
Gharpure, M., **3**, 308 ⟨91S1079⟩
Ghatak, U., **3**, 239 ⟨57JA4487⟩
Ghatak, U. R., **3**, 206 ⟨82JOC5353⟩, 533 ⟨86TL3913⟩; **4**, 170 ⟨87T4433⟩, 193 ⟨85TL1581⟩; **5**, 144 ⟨81IJC(B)411⟩
Ghatge, B. B., **2**, 660 ⟨62JOC1430⟩
Ghattas, A. B. A. G., **5**, 478, 494, 582 ⟨87SUL153⟩, 573 ⟨83T3419⟩; **6**, 503 ⟨82T1163⟩
Ghattas, A.-B., **5**, 560 ⟨82SUL69⟩, 572 ⟨84BSF(2)204⟩
Ghattas, A. G., **5**, 560 ⟨82SUL69⟩; **6**, 476 ⟨86CI(L)140⟩
Ghazal, N. B., **5**, 951 ⟨89JOC4481⟩
Gheerbrant, E., **2**, 14 ⟨89CC313⟩
Ghelfi, F., **2**, 22 ⟨86JCR(S)426, 92T4579⟩, 24 ⟨89JCR(S)108, 89JCR(S)360⟩, 24 ⟨91SC489⟩, 28 ⟨86JCR(S)428⟩; **3**, 27 ⟨86JCR(S)426⟩, 234 ⟨88JCR(S)200⟩; **4**, 249 ⟨93T199⟩
Gheorghiu, M., **4**, 152 ⟨92RRC393⟩
Gheorghiu, M. D., **1**, 518 ⟨73TL9⟩; **3**, 539 ⟨81CSR289, 88RRC83⟩
Ghera, E., **1**, 360 ⟨93JOC6716⟩, 861 ⟨81JOC2059⟩; **3**, 89 ⟨84S504⟩
Ghilardi, C. A., **6**, 327 ⟨82OM778⟩, 332 ⟨83JOM(246)C13, 84JOM(270)251⟩
Ghiringhelli, D., **3**, 16, 132 ⟨82S580⟩
Ghiro, E., **1**, 340, 1127 ⟨93TL2091⟩
Ghobrial, N. M., **6**, 25 ⟨89JCS(P2)251⟩
Ghodoussi, V., **1**, 325, 920 ⟨86JOC5007⟩
Ghosal, S., **1**, 491, 1035 ⟨87JOC4296⟩
Ghose, B. N., **1**, 61 ⟨80JPR503⟩; **2**, 435 ⟨68JCS(A)1909⟩; **4**, 114, 123 ⟨78JIC1254⟩
Ghosez, L., **1**, 210, 225, 613 ⟨88TL2059⟩, 375 ⟨82JA1428⟩, 518 ⟨84TL5043⟩, 522 ⟨75JA4409⟩; **2**, 8 ⟨89TL3061⟩, 721 ⟨87T5125⟩, 748 ⟨75JA2276⟩, 758 ⟨69AG468, 79CC1180, 80OS(59)26, B-76MI 215-01⟩, 759 ⟨74AG272⟩, 938, 944 ⟨87TL397⟩; **3**, 259 ⟨76AG(E)371⟩, 424 ⟨87TL397⟩, 475 ⟨79CC1180⟩, 529 ⟨81JA4616⟩, 557 ⟨85TL1647⟩, 558 ⟨73JA5417⟩, 558, 560 ⟨80TL3081⟩, 601 ⟨72JA2870, 74AG(E)267, B-76MI 317-01⟩, 602 ⟨81AG(E)879, 82JA292, 85JA2192, 90TL4429⟩, 636 ⟨86TL5099⟩, 657, 658 ⟨84TL3475⟩; **4**, 790 ⟨79CC1180, 80TL223, 88OSC(6)282⟩, 790, 791 ⟨77NJC369, B-76MI 418-01, 81H(15)1179⟩, 833 ⟨83T2493⟩, 842 ⟨83TL2563⟩, 889 ⟨77NJC371⟩, 891, 904, 905 ⟨69AG(E)454⟩, 891 ⟨B-76MI 420-01⟩, 897 ⟨86RTC456, B-76MI 420-02⟩, 971, 990 ⟨76MI 421-01, 80CC940⟩, 1016, 1017 ⟨76AG417⟩; **5**, 2 ⟨79CC1180, 79OS(59)26⟩, 166 ⟨87TL2135⟩, 233 ⟨80TL3085⟩, 718 ⟨82JA1428⟩
Ghosh, A., **2**, 156 ⟨81JCS(D)447⟩; **3**, 528 ⟨88CC1421⟩
Ghosh, A. K., **1**, 336, 346 ⟨87TL175⟩, 1029 ⟨88JA649⟩; **2**, 679 ⟨88JA649⟩; **6**, 417 ⟨91TL4251, 92TL2781⟩
Ghosh, J., **6**, 589, 593 ⟨88AJC549⟩
Ghosh, S., **3**, 193 ⟨91SC191⟩, 435 ⟨71MI 311-01⟩
Ghosh, S. K., **1**, 593 ⟨87TL2061, 90T1885⟩; **3**, 62, 221 ⟨89IJC(B)3⟩

Ghoudikian, M., **3**, 552 ⟨75LA1513⟩
Ghozlan, S. A. S., **5**, 677 ⟨87AP(320)850⟩, 771 ⟨84BCJ1650⟩
Ghribi, A., **1**, 479 ⟨84TL3075⟩; **3**, 198 ⟨86S661⟩; **4**, 176 ⟨88PAC49⟩, 192 ⟨84TL3075⟩
Ghuman, M. A., **6**, 390 ⟨91MI 613-01⟩
Giacobbe, T. J., **5**, 1073 ⟨71JOC2142⟩
Giacomelli, G., **1**, 1019, 1141 ⟨78TL2831⟩; **2**, 9 ⟨86JOC5291⟩
Giacomini, D., **2**, 544 ⟨90PAC595⟩; **3**, 440, 456 ⟨87TL5369, 91TL2967⟩, 456, 457, 458 ⟨90PAC605⟩, 456, 458 ⟨86TL1695, 93T3809⟩; **5**, 287 ⟨86TL1695, 87TL5369⟩, 377 ⟨94S805⟩
Giagante, N., **1**, 950 ⟨88JOM(340)317⟩
Giam, C. S., **1**, 486 ⟨81JOC4885⟩; **5**, 1119 ⟨78CJC1026, 79JHC409⟩
Giamalva, D. H., **1**, 495 ⟨83JOC5251⟩
Gianferrara, T., **4**, 273, 274 ⟨91JOC3530⟩
Giang, Y. F., **3**, 529 ⟨86JOC2145⟩
Giangiordano, M. A., **3**, 438 ⟨86TL3957⟩
Giannangeli, M., **5**, 681, 771 ⟨83JHC225⟩
Giannella, M., **4**, 232 ⟨87JMC1934⟩
Giannetto, A., **4**, 151 ⟨90JOM(396)115⟩
Gianni, F. L., **1**, 519 ⟨76TL881⟩
Gianni, M. H., **5**, 314 ⟨69JOC1154⟩
Giannini, U., **4**, 725 ⟨71JOM(26)357⟩
Giannis, A., **3**, 40 ⟨92AG(E)1391⟩
Giannoccaro, P., **6**, 428 ⟨93JOM(451)243⟩
Giansiracusa, J. J., **1**, 906 ⟨82JOC2089⟩
Gibb, V. G., **2**, 598 ⟨78CAR(63)1⟩
Gibbels, U., **1**, 364 ⟨88T3899⟩
Gibbon, G. A., **4**, 360 ⟨67IC1751, 67IC1989⟩, 638 ⟨70MI 414-01⟩, 638, 641 ⟨72IC408⟩; **6**, 173 ⟨67IC1751⟩
Gibbons, E. G., **1**, 1043 ⟨75TL981⟩; **5**, 272, 299, 304 ⟨75TL981⟩
Gibbons, L. K., **4**, 973, 977 ⟨78USP4075001, 76USP4075001⟩
Gibbs, C. G., **4**, 742 ⟨74JOC501⟩
Gibbs, D. E., **2**, 314 ⟨90TL5555⟩
Gibbs, R. A., **1**, 498 ⟨89JA3717⟩, 795 ⟨87TL6021⟩, 993 ⟨88JA4062⟩, 1029 ⟨88JOC1790⟩, 1039 ⟨89JOC4072⟩
Gibian, M. J., **1**, 924, 925 ⟨65JA3413⟩; **2**, 216 ⟨69TL3901⟩
Giblin, G. M. P., **1**, 482 ⟨87CC207⟩
Gibs, G. J., **1**, 100 ⟨71JOC279⟩
Gibson, C. S., **2**, 887 ⟨24JCS2275⟩
Gibson, D., **5**, 22 ⟨91PS(60)239⟩, 371, 403 ⟨90JCS(P1)3263⟩
Gibson, D. H., **4**, 148 ⟨89OM1114⟩; **6**, 517 ⟨84JA1125⟩
Gibson, D. T., **2**, 39 ⟨70B1626, 74MI 202-01⟩; **6**, 90 ⟨33JCS306⟩, 91 ⟨31JCS2637⟩
Gibson, F., **1**, 826 ⟨63NAT388⟩
Gibson, F. S., **5**, 11 ⟨91OS(70)101⟩
Gibson, G. W., **1**, 111 ⟨63JA2171⟩
Gibson, H. W., **4**, 426 ⟨91JOC865⟩; **5**, 24 ⟨69CRV673⟩
Gibson, J. D., **6**, 46 ⟨50JA3642⟩
Gibson, J. F., **2**, 574 ⟨76JCS(D)2204⟩; **4**, 656, 677 ⟨76JCS(D)2204⟩
Gibson, J. L.,**4**, 869 ⟨76JA3916⟩
Gibson, L. L., **3**, 141 ⟨82JOC743⟩
Gibson, M. S., **2**, 299 ⟨68AG986, 68AG(E)919⟩; **3**, 179 ⟨68AG(E)919⟩, 360 ⟨81JCS(P1)2952⟩; **4**, 281, 282 ⟨65T2743⟩; **5**, 356 ⟨65JCS2999⟩, 358 ⟨73CJC4115⟩, 641 ⟨70JCS(C)2106, 74CJC879⟩, 641 ⟨75CJC3211⟩
Gibson, R. E., **3**, 88 ⟨50JA2992⟩
Gibson, R. S., **3**, 59 ⟨85SC371⟩
Gibson, R. W., **5**, 340 ⟨91JCS(P1)3383⟩
Gibson, T., **2**, 4, 34 ⟨81JOC1821⟩
Gibson, V. C., **3**, 511 ⟨86JA5347⟩, 514 ⟨90JA8378⟩
Gidaspov, B., **2**, 386 ⟨78ZOR240⟩
Gidaspov, B. V., **2**, 790 ⟨83RCR284⟩; **5**, 361 ⟨77ZOR1144⟩
Giddings, P. J., **1**, 235, 246 ⟨80TL395⟩, 410, 822 ⟨82JCS(P1)2757, 80TL399⟩; **4**, 90 ⟨80TL395, 82JCS(P1)2757⟩
Giddings, S., **3**, 117 ⟨88JOC1103⟩
Gidley, G. C., **2**, 23 ⟨68JOC38⟩, 33, 95 ⟨68JA808, 84JA5261⟩
Gieiler, W., **5**, 379 ⟨63AG93⟩
Gielen, M., **1**, 924 ⟨B-91MI 119-03⟩; **4**, 156 ⟨81JOM(214)191⟩, 700 ⟨82JOM(236)69, 84JOM(273)303⟩, 700 ⟨84JOM(277)335, 85BSB299⟩
Giella, M., **2**, 138 ⟨48JA1450⟩
Gier, T. E., **3**, 688 ⟨61JA1769⟩; **5**, 1152 ⟨61JA1769⟩
Gieren, A., **4**, 719 ⟨86ZN(B)1431⟩; **6**, 355, 357 ⟨85CC1800⟩, 617 ⟨78CB698⟩
Giering, W. P., **1**, 936 ⟨73JOC3055⟩
Giesbrecht, E., **2**, 279 ⟨56JOC1217⟩, 287 ⟨46JA2671⟩, 732 ⟨59CB40⟩, 735 ⟨55CB1⟩
Giese, B., **1**, 5, 7, 9, 10, 17, 936, 943, 945 ⟨91JOC678⟩, 172, 230, 243 ⟨B-86MI 105-01⟩, 172, 325, 330 ⟨84AG(E)69⟩, 320 ⟨89HOU(E19a)⟩, 320, 323, 324, 920 ⟨83AG(E)753⟩, 320, 323, 329, 330, 331, 332, 341, 342 ⟨B-86MI 108-01⟩, 320, 325, 331 ⟨89AG(E)969⟩, 320, 326, 327, 328, 919 ⟨91ACR296⟩, 320, 330, 331, 332, 341, 343, 935, 948, 1126 ⟨85AG(E)553⟩, 323, 920 ⟨83TL703⟩, 325 ⟨85CB1345, 85JA4332⟩, 325, 326 ⟨84CB2132⟩, 325, 331, 920 ⟨86CB444⟩, 325, 921 ⟨88CB2063, 89JA8311⟩, 326 ⟨91JA1791⟩, 326, 327 ⟨93TL2445⟩, 326, 948 ⟨84TL2743, 91SL423⟩, 327 ⟨91SL425, 92SL442⟩, 327 ⟨92TL2673, 93HCA441⟩, 327 ⟨93JA8585, 93TL5885⟩, 327, 950 ⟨92JOC3994⟩, 330 ⟨89T935⟩, 330, 341 ⟨86JOC3726⟩, 331, 342, 936 ⟨89TL681⟩, 332 ⟨81AG(E)965, 84CB3160⟩, 546 ⟨82AG(E)768, B-86MI 112-01⟩, 919 ⟨89HOU(1)1, 89HOU(2)1⟩, 919, 946, 948 ⟨B-86MI 119-04⟩, 923 ⟨78AG(E)594, 90OS(69)66⟩, 923 ⟨93JOC1083⟩, 931 ⟨91CB387⟩, 937 ⟨89TL2907, 91HCA1135, 91HCA1143⟩, 945 ⟨90TL6013⟩, 946 ⟨87CB1197, 90SL166⟩, 948 ⟨77CB2588, 82CB2526⟩, 948 ⟨84CB3175⟩, 1126 ⟨B-86MI 122-01⟩; **2**, 578, 579 ⟨89AG(E)325⟩, 913 ⟨92JOC3994⟩; **3**, 292 ⟨B-86MI 306-01⟩, 621 ⟨84AG(E)69, 85JA4332⟩, 621 ⟨85T4025⟩; **4**, 385 ⟨83CB1240⟩; **5**, 35 ⟨83S733⟩, 152 ⟨87OS(65)236⟩, 1135 ⟨91JOC528⟩
Giese, F., **6**, 213 ⟨B-72MI 607-02⟩
Giese, H. J., **1**, 692 ⟨82BSB149⟩
Giese, M., **6**, 441 ⟨36CB684⟩
Giese, R. W., **5**, 196 ⟨86JOC3390⟩
Giesel, K., **1**, 364 ⟨88T3899⟩; **3**, 652 ⟨86S548⟩
Giesemann, G., **3**, 696 ⟨82JCR(S)79⟩; **6**, 417 ⟨82JCR(S)79⟩
Giesen, K., **6**, 618 ⟨76CB810⟩
Giessler, W., **5**, 718 ⟨63AG96⟩
Giffard, M., **3**, 337 ⟨90JCR(S)320⟩; **5**, 1108 ⟨79CC1026⟩
Giger, R., **3**, 526 ⟨68HCA1466, 70HCA120⟩; **4**, 768 ⟨68HCA1466⟩
Gigg, J., **1**, 882 ⟨66JCS(C)82⟩
Gigg, R., **1**, 882 ⟨64TL1191, 66JCS(C)82⟩, 882 ⟨68JCS(C)1903, 72JCS(P1)1535⟩, 882 ⟨73JCS(P1)1858⟩; **4**, 176 ⟨80MI 404-01⟩
Gigou, A., **2**, 47 ⟨89AG(E)755⟩
Giguere, R. J., **1**, 525, 826 ⟨86TL4945⟩
Giinsburg, A., **2**, 259, 260 ⟨46JA1860⟩
Giitsidis, O., **4**, 409 ⟨89JOC4771⟩
Gil, G., **1**, 498 ⟨80JCR(S)98, 85T2759⟩, 805 ⟨77TL1779, 80JCR1551⟩, 959 ⟨74CR(C)149, 79TL1845⟩; **3**, 216 ⟨75T849⟩, 658 ⟨86TL3505, 89TL171⟩; **4**, 188 ⟨84TL3805⟩
Gil, J. F., **3**, 31, 161 ⟨93T4923, 94T3437⟩
Gil, S., **1**, 565 ⟨92JCS(P1)3351⟩
Gilak, A., **5**, 1093 ⟨74CB2169⟩
Gilardi, R., **2**, 384 ⟨92JOC1585⟩, 1053 ⟨90JOC1916⟩; **4**, 428 ⟨87JOC4781, 90JOC2920⟩, 428, 429 ⟨93JOC763⟩,

795 ⟨91JOC537⟩, 970, 993 ⟨92JOC235⟩; **6**, 150 ⟨92JOC3026⟩
Gilardi, R. D., **4**, 298 ⟨90JOC1459⟩
Gil-Av, E., **1**, 795 ⟨65JA4107⟩
Gilban, T., **2**, 403 ⟨69CJC2371⟩
Gilbert, A., **4**, 1079 ⟨B-91MI 424-01⟩
Gilbert, A. R., **4**, 573 ⟨58USP2835651⟩
Gilbert, B. A., **1**, 656 ⟨87JOM(332)53⟩; **3**, 624 ⟨93CC669⟩
Gilbert, D. P., **2**, 138 ⟨77JOC3233⟩
Gilbert, E. E., **2**, 157 ⟨63JOC1945⟩, 175, 176, 717 ⟨69S3⟩, 175, 191 ⟨B-65MI 203-02⟩, 688 ⟨63JOC2807⟩, 720 ⟨62CRV549, B-65MI 214-01⟩, 797 ⟨B-65MI 215-03⟩, 1016 ⟨68JOC819⟩; **4**, 18 ⟨65JOC1241⟩, 424 ⟨72S30⟩, 775 ⟨60USP2954316⟩; **6**, 46, 47 ⟨74JFC(4)107⟩
Gilbert, F. L., **2**, 288 ⟨28JCS3179⟩, 289 ⟨29JCS2076⟩
Gilbert, H. E., **3**, 494 ⟨69JA723⟩
Gilbert, H. F., **5**, 699 ⟨82JA6769⟩
Gilbert, J. A., **4**, 419, 427 ⟨93JMC305⟩
Gilbert, J. C., **1**, 495 ⟨83JOC5251⟩, 795 ⟨72TL125⟩, 817, 835 ⟨84TL2303⟩, 831 ⟨86JOC4485, 93JOC6255⟩, 831 ⟨94T1651⟩, 832 ⟨88T7587⟩, 869 ⟨63T1625⟩, 1078 ⟨82JOC1837⟩, 1143, 1144 ⟨70JA7493⟩; **2**, 680, 739 ⟨83JOC448⟩; **3**, 11 ⟨83JOC448⟩
Gilbert, J. M., **1**, 869 ⟨65JA382⟩
Gilbert, J. R., **1**, 869 ⟨70JCS(B)179⟩
Gilbert, K., **3**, 593 ⟨79AG(E)67, 79CB3166⟩
Gilbert, K. E., **2**, 389, 397 ⟨79JOC659⟩, 403, 404, 405 ⟨75JOC1409⟩
Gilbert, L., **1**, 786 ⟨86TL3511⟩
Gilbert, R., **4**, 569 ⟨56USP2768193⟩
Gilbert, R. P., **1**, 1114 ⟨72JA5061⟩
Gilbertson, S. R., **2**, 607 ⟨86JOC277⟩; **3**, 270 ⟨85JA503⟩; **4**, 955 ⟨85JA503⟩; **5**, 934 ⟨86JA520⟩, 956 ⟨85JA503⟩
Gilboa, Y., **5**, 118 ⟨93T10099⟩
Gilbreath, S. G., **5**, 330 ⟨88JA6172⟩
Gilchrist, A. B., **6**, 517 ⟨85OM478⟩
Gilchrist, J. H., **1**, 131 ⟨91JA5751, 91JA9571⟩, 131 ⟨93TL5213⟩
Gilchrist, M., **2**, 807 ⟨64JA1410, 65JA3199⟩
Gilchrist, T. L., **1**, 811 ⟨81JCS(P1)3225⟩, 860, 861 ⟨B-92MI 119-03⟩, 1066, 1081 ⟨B-83MI 121-01⟩; **2**, 159, 401 ⟨77CRV409⟩, 160 ⟨76JCS(P1)2166⟩, 334, 337, 338, 339, 347, 350 ⟨91COS(7)735⟩, 772 ⟨91COS(7)738⟩, 775 ⟨91COS(7)736⟩, 777 ⟨83CSR53⟩, 779 ⟨77JCS(P1)1196⟩; **3**, 104, 107 ⟨B-85MI 303-01⟩, 379 ⟨77JCS(P1)1196⟩; **4**, 307, 799 ⟨79JCS(P1)249⟩, 798 ⟨83CSR53⟩, 1080 ⟨B-69MI 424-01⟩; **5**, 226 ⟨83JCS(P1)49⟩, 351, 363 ⟨91T10053⟩, 362, 367 ⟨87JCS(P1)2511⟩, 764 ⟨76JCS(P1)2166⟩, 766 ⟨75JCS(P1)1964⟩, 1126 ⟨76JCS(P1)716⟩; **6**, 515 ⟨91T10053⟩
Gilday, J. P., **1**, 883 ⟨89JOC5044⟩, 885 ⟨88JOC4972, 90OPP167⟩
Gilde, H. G., **1**, 6 ⟨70CC738⟩
Gildenhorn, H. L., **5**, 1002 ⟨48JA317⟩
Gildersleeve, D. L., **5**, 203 ⟨88JMC362⟩
Giles, D. E., **5**, 1040 ⟨73AJC273⟩
Giles, H. G., **3**, 335, 337, 351 ⟨76CJC537⟩, 337, 351 ⟨74CC409⟩
Giles, M., **3**, 72 ⟨89JCS(P1)1793⟩; **4**, 842, 845, 851 ⟨89JCS(P1)1793⟩
Giles, M. B., **5**, 351 ⟨86JCS(P1)1655⟩
Giles, P. M., **2**, 218 ⟨71JOC309⟩
Giles, P. R., **2**, 202 ⟨93CC1195⟩
Giles, R. G. F., **1**, 732 ⟨74JCS(P1)2435⟩; **3**, 282, 687 ⟨85TL4649⟩, 297 ⟨84JCS(P1)1339⟩
Gilge, U., **3**, 312 ⟨83JHC1011⟩, 722 ⟨83JOM(246)159⟩
Gilgen, P., **3**, 678 ⟨77H(6)143⟩, 679, 680 ⟨74HCA1382⟩
Gilham, P. T., **2**, 105, 106 ⟨59CI(L)542, 58JA6212⟩
Gilheany, D. G., **2**, 48 ⟨90TL3817⟩, 426 ⟨B-90MI 208-01⟩
Gilicinski, A. G., **1**, 570 ⟨92JOC2856⟩
Gilinsky-Sharon, P., **2**, 660 ⟨94JOC1831⟩; **4**, 205 ⟨89JOC4866⟩
Gilje, J. W., **2**, 476 ⟨73CC813⟩; **4**, 581 ⟨86OM2555⟩, 589 ⟨78JA5562, 81IC2466⟩, 589 ⟨84JA5920, 88OM1465⟩, 589 ⟨89OM1118, 90IS177⟩; **5**, 920 ⟨81IC2466⟩, 1096 ⟨80IC2564⟩
Gill, D. S., **1**, 93 ⟨78JCS(D)617⟩
Gill, G. B., **1**, 393 ⟨91COS(3)887⟩, 400, 402, 403 ⟨68QR338⟩, 980 ⟨92JCS(P1)2355⟩; **3**, 193, 530, 531 ⟨91COS(3)887⟩; **6**, 447 ⟨89TL3229⟩
Gill, H. S., **3**, 321 ⟨82TL5099, 83JOC1051⟩, 323 ⟨83JOC1898⟩
Gill, M., **1**, 1091 ⟨81TL1437⟩; **3**, 58, 273 ⟨81TL1437⟩, 157 ⟨86TL1933⟩, 304 ⟨84T621⟩
Gillard, M., **4**, 827 ⟨82TL603⟩
Gillbro, T., **1**, 445 ⟨74JA5032⟩
Gillespie, D. G., **1**, 1069 ⟨83JCS(P1)1689⟩
Gillespie, H. B., **5**, 87 ⟨43OSC(2)489⟩
Gillespie, R. J., **1**, 897 ⟨64CJC502⟩; **3**, 373 ⟨78CC83, 79JCS(P1)2624⟩
Gillessen, D.,**5**, 279, 281 ⟨89TL1927⟩
Gillet, J. P., **3**, 237 ⟨85TL3999⟩; **5**, 58 ⟨82S297⟩
Gillette, G. R., **4**, 567, 1082 ⟨90AG(E)1429⟩; **5**, 1020 ⟨87OM2617⟩; **6**, 125, 668, 669 ⟨91NJC393⟩
Gilliam, W. F., **2**, 538 ⟨46JA1161⟩
Gilliard, J. W., **2**, 539 ⟨84TL4717⟩
Gilliatt, V., **1**, 935 ⟨88OM2220⟩
Gillibrand, M. I., **2**, 386 ⟨49JCS1883⟩
Gillies, C. W., **6**, 410 ⟨80JA7572⟩
Gillies, D. G., **4**, 659 ⟨80JOM(193)21⟩
Gilligan, W. H., **5**, 366 ⟨82JCED97⟩; **6**, 144 ⟨68JOC1247⟩, 145 ⟨86JHC81⟩, 272 ⟨83JEM95, 83USP467715⟩, 308 ⟨83JOC3354⟩, 427, 477 ⟨79S600, 81USP224776⟩, 541, 543 ⟨83JCED131⟩, 449
Gilligau, W., **6**, 449, 492 ⟨85JOC5879⟩
Gilliom, L. R., **1**, 669 ⟨86JA733⟩, 669, 670 ⟨86OM721⟩; **3**, 508 ⟨86OM721⟩
Gillis, B. T., **2**, 771 ⟨65JA4576⟩; **5**, 356 ⟨62JOC413⟩, 363 ⟨68JOC762⟩
Gillis, H. R., **4**, 179 ⟨80JOC4283⟩
Gillman, E. G., **4**, 730 ⟨54JA5423⟩
Gillmann, T., **1**, 497 ⟨93TL607⟩
Gillon, A., **1**, 498 ⟨82JOC3297⟩; **3**, 320 ⟨82T1477⟩
Gillon, D. W., **5**, 1125 ⟨82CC444, 84JCS(P1)147⟩
Gillyard, L., **6**, 415 ⟨89JOC3231⟩
Gilman, H., **1**, 110 ⟨44JA1515⟩, 117 ⟨52JOC1630⟩, 127 ⟨36RTC518⟩, 446 ⟨51JA2880⟩, 649, 763 ⟨62JOC3647⟩, 663 ⟨38JA2333⟩, 881 ⟨45JA2106, 63JOC1725⟩, 1000, 1001 ⟨74SC199⟩, 1043 ⟨31JA4192⟩; **2**, 125, 135 ⟨49JA1478⟩, 129 ⟨45JA1844, 25JA1449⟩, 274 ⟨48JA550⟩, 420 ⟨48JA3945⟩, 421 ⟨56JA4285⟩, 489 ⟨39JOC162⟩, 545 ⟨60JA403⟩, 553 ⟨58JOC2044⟩, 554, 961 ⟨51OR(6)339⟩, 560 ⟨29JA1576⟩, 586 ⟨52JOC1630⟩, 598 ⟨40JA2357, 50JA1760⟩, 600 ⟨53JA2507⟩, 601 ⟨54CRV101⟩, 632 ⟨74JIC1018⟩, 706 ⟨49JA1747⟩, 752 ⟨54OR(8)258⟩, 882 ⟨41JA207⟩, 886 ⟨51JA4470⟩, 889 ⟨51JA2880⟩, 941 ⟨B-69MI 215-04⟩, 942 ⟨46JA1675, 50JA4280⟩, 942 ⟨55JA6380, 57JA4560⟩, 1017 ⟨61CI(L)1619⟩; **3**, 452 ⟨51JA1074⟩, 603 ⟨62JOC3647⟩; **4**, 319 ⟨30JA4573⟩, 354 ⟨54JA2502, 58JA2680⟩, 608 ⟨64JOM(2)277⟩, 647 ⟨51JA5878⟩; **5**, 43 ⟨32OSC(1)361⟩, 112 ⟨43OSC(2)425⟩, 114, 115 ⟨56JA1689⟩, 222 ⟨51JA4470⟩, 254 ⟨60JOC1064⟩, 417 ⟨58JA4529⟩, 595 ⟨33JA1262⟩, 595, 620, 1016 ⟨50JA3045⟩, 595, 729 ⟨24JA493⟩, 652 ⟨54OR(8)258⟩; **6**, 69 ⟨55JA6644⟩
Gilman, J. W., **5**, 994 ⟨93SC335⟩
Gilman, N. W., **1**, 3, 4, 6 ⟨71CC465⟩, 747 ⟨71TL1821⟩, 806, 807 ⟨67CC619⟩; **3**, 57 ⟨68JA5616⟩, 310 ⟨85SC1271⟩; **5**, 51, 154 ⟨68JA5616⟩, 295 ⟨71CC733⟩
Gilman, S., **1**, 604 ⟨80JOC3537⟩, 781 ⟨76JOC1485⟩; **2**, 279 ⟨76JOC1485⟩
Gilmore, W. F., **4**, 127 ⟨87PS(29)287⟩
Gilmour, D. W., **4**, 866 ⟨86EUP174832A2⟩; **5**, 690 ⟨92JCS(P1)2789⟩

Gilon, C., **4**, 446 ⟨83JMC129⟩
Gilow, H. M., **2**, 288 ⟨67JOC2580⟩; **3**, 308 ⟨83OS(62)111⟩
Gilpin, J. R., **2**, 456 ⟨53CJC976⟩
Gilpin, M. L., **4**, 824 ⟨79TL391⟩
Gilyarov, V. A., **4**, 116 ⟨69IZV181⟩, 563 ⟨72IZV2612⟩
Gilyazov, M. M., **4**, 125 ⟨85JGU2497⟩, 126, 127 ⟨79IZV2073⟩
Gimbert, Y., **1**, 570 ⟨93JOC4685⟩
Gimeno, M. C., **4**, 586 ⟨88SRI69⟩
Gimi, R., **2**, 325 ⟨91SL573⟩
Gimi, R. H., **1**, 828 ⟨91TL4267⟩
Gin, D. Y., **4**, 239 ⟨94JA4697⟩
Ginak, A. I., **2**, 5 ⟨81RCR151⟩
Gindelberger, D. E., **6**, 349, 355, 357, 358 ⟨93IC5126⟩
Ginebreda, A., **2**, 299 ⟨78CC742, 81JCS(P1)574⟩
Giner, J.-L., **1**, 113 ⟨89JOC2117⟩
Giner-Sorolla, A., **3**, 450 ⟨59JA2515⟩
Gingras, M., **2**, 87 ⟨89TL279⟩, 89 ⟨88JA7737⟩, 131 ⟨90TL1397⟩; **5**, 242, 337 ⟨87S1122⟩
Gingrich, D. M., **3**, 355 ⟨92MI 308-02⟩
Ginguene, A., **3**, 310 ⟨81JOM(216)139⟩
Giniyatullina, M. A., **2**, 466 ⟨76MI 208-01⟩, 469 ⟨77IZV1437⟩
Ginley, D. S., **4**, 668 ⟨78JOM(157)41⟩, 674 ⟨77ICA85⟩
Ginos, J. Z., **1**, 90 ⟨75JOC1191⟩; **5**, 62 ⟨83CRV321⟩
Ginsberg, D., **1**, 888 ⟨59OR(10)179⟩; **4**, 540 ⟨59OR(10)179⟩; **6**, 25 ⟨50JA5012⟩
Ginsberg, H. F.,**1**, 503 ⟨54JA4446⟩
Ginsburg, D., **3**, 226 ⟨53JA5489⟩; **5**, 46 ⟨59OR(10)179⟩, 66 ⟨50JA5012⟩
Ginsburg, V. A., **4**, 114 ⟨52JGU1575⟩, 140 ⟨52ZOB1783⟩, 555, 569 ⟨58ZOB728⟩; **5**, 807 ⟨67ZOB1413⟩, 822 ⟨60ZOB2409⟩; **6**, 242 ⟨61DOK(141)357, 62DOK(142)354⟩, 613 ⟨67JGU1343, 69JGU1301⟩, 613 ⟨71ZOR2267⟩
Ginzel, K.-D., **6**, 117 ⟨89T1691⟩
Gioacomelli, G., **2**, 56 ⟨89JOC2383⟩
Giocă, B., **5**, 762, 816 ⟨78JCS(P1)309⟩
Gioeli, C., **1**, 244 ⟨91T2237⟩; **4**, 276 ⟨91T3431⟩
Gioia, B., **5**, 762 ⟨76TL1839⟩
Giordano, C., **1**, 391 ⟨84AG(E)413, 85S505⟩, 934 ⟨83ACR27⟩; **2**, 29 ⟨86AG(E)259, 87JOC3018⟩, 29 ⟨87S1039, 89T4243⟩, 29 ⟨90JOC5400⟩; **3**, 152 ⟨89S466⟩, 294 ⟨88TL5561⟩, 301 ⟨90SC383⟩; **4**, 297 ⟨75S789⟩; **5**, 28 ⟨84AG(E)413⟩, 580 ⟨78S917⟩, 606 ⟨72S34, 77S250⟩; **6**, 25 ⟨89S466, 90S327⟩
Giordano, F., **5**, 263 ⟨93JCS(P1)451⟩
Giordano, G., **1**, 456, 507 ⟨77JOM(141)113⟩
Giorgianni, P., **2**, 247 ⟨81CC350⟩; **3**, 368 ⟨71JCS(B)1547⟩; **3**, 369 ⟨86CC964, 90JOC3744⟩, 558 ⟨80JOC3766, 82JOC3998⟩, 560 ⟨84JOC2200⟩; **5**, 647, 648 ⟨86JCS(P1)381, 90JOC3744⟩; **6**, 127 ⟨87JCS(P1)2647⟩
Giovagnoli, D., **3**, 458 ⟨93T3809⟩
Giovani, E., **2**, 135 ⟨92SL303⟩
Giovannoli, M., **4**, 260 ⟨90JOC1323⟩
Gipe, A., **2**, 30 ⟨78JOC2793⟩
Gipe, R. K., **2**, 18 ⟨83JOC3195⟩
Gipstein, E., **1**, 608 ⟨80JOC1486⟩; **4**, 286 ⟨80JOC1486⟩
Giral, L.,**4**, 847 ⟨87BSF823⟩
Giraldo, E., **6**, 425 ⟨90JMC2101⟩
Giralt, E., **5**, 259 ⟨93T11065⟩
Girard, C., **1**, 587 ⟨92TL573⟩; **3**, 224 ⟨73TL2767⟩
Girard, J. P., **1**, 940 ⟨93TL8245⟩
Girard, M., **2**, 50 ⟨93JOC1991⟩
Girard, P., **1**, 3, 6, 8, 10, 17, 587, 680 ⟨80JA2693⟩; **2**, 27 ⟨82BSF(2)327⟩, 565 ⟨80JA2693⟩
Girard, Y., **2**, 525 ⟨88PAC1627⟩; **4**, 217 ⟨90CJC897⟩
Giraud, L., **1**, 156 ⟨91TL4125⟩
Giraudeau, P., **6**, 143 ⟨77CR(C)321⟩
Girault, P.,**5**, 771, 854 ⟨72GEP2146192⟩
Girdwood, R. H., **5**, 28 ⟨60MI 502-01⟩
Girfanova, Y. N., **6**, 582 ⟨76IZV455⟩
Girija, K., **3**, 144 ⟨93TL6631⟩
Girijavallabhan, V. M., **1**, 1131 ⟨89H(28)47⟩; **6**, 553 ⟨81TL3485⟩
Girolami, G. S., **4**, 680 ⟨93CC284, 94OM2309⟩, 696 ⟨93JOM(449)203⟩
Giron, Y., **5**, 118 ⟨93T10099⟩
Girotra, N. N., **4**, 130 ⟨69TL4647⟩; **5**, 173 ⟨68T2443⟩
Giroux, A., **5**, 692 ⟨93TL5983⟩
Giroux, E. L., **4**, 1004 ⟨86TL1757⟩
Girreser, U., **1**, 508 ⟨86AG(E)167⟩
Girsh, D., **1**, 938 ⟨92JOC6803⟩; **5**, 513 ⟨92JOC6662⟩; **6**, 543, 550, 557 ⟨92JOC6803⟩
Girshovich, M. Z., **6**, 599 ⟨69ZOB941⟩
Giselbrecht, K., **6**, 355, 356, 357, 358 ⟨89CB2279⟩, 357, 358 ⟨89CB1255⟩
Gisi, U., **5**, 179 ⟨89LA1233⟩, 179 ⟨89LA1233⟩
Gisler, M., **5**, 275, 298 ⟨76JOC3769⟩
Gist, R. P., **3**, 222 ⟨89JA7634⟩
Gisvold, O., **5**, 748 ⟨55MI 519-01⟩, 749 ⟨49MI 519-01⟩
Gitlitz, M. N., **2**, 409, 421 ⟨69JCS(A)980⟩
Gitliz, M. H., **6**, 5 ⟨67CJC2659⟩
Gitschthaler, K., **4**, 742 ⟨56M323⟩, 763 ⟨56M319⟩; **6**, 39 ⟨56M323⟩
Gittos, M. W., **5**, 593 ⟨76JCS(P1)33, 77JCS(P1)2357⟩
Giua, M., **6**, 596 ⟨59G693⟩
Giudicelli, M. B., **2**, 9, 19 ⟨90TL6527⟩
Giuffre, L., **2**, 190 ⟨74JCS(P1)771⟩
Giuffrida, D., **1**, 571 ⟨92JA6330⟩
Giuliano, C., **1**, 10 ⟨92TL7429⟩
Giuliano, R. M., **6**, 632 ⟨91S86⟩
Giullemin, J. C., **3**, 688 ⟨91AG(E)196⟩
Giumanini, A. B., **1**, 148 ⟨66JOC2047⟩
Giumanini, A. G., **1**, 148 ⟨66JOC2047⟩, 407 ⟨B-71MI 109-02⟩; **3**, 405 ⟨88SC1501⟩
Givens, R. S., **1**, 421 ⟨82JA6874⟩
Gizycki, U. v., **5**, 973 ⟨71AG(E)402, 71AG(E)403⟩, 974 ⟨68AG(E)381⟩
Gjøystdal, A. K., **2**, 1009 ⟨77ACS(B)56⟩
Gladchenko, A. F., **4**, 660 ⟨74ZOB806⟩
Gladfelter, E. J., **2**, 20 ⟨79JOC3728⟩
Gladiali, S., **3**, 11 ⟨77CI(M)839⟩
Gladkova, G. E., **4**, 768 ⟨92ZOR522⟩
Gladshtein, B. M., **4**, 554 ⟨67ZOB2055⟩, 561 ⟨64ZOB2897, 87ZOB2294⟩; **5**, 1131 ⟨67MI 530-01⟩; **6**, 366 ⟨67ZOB2513⟩
Gladstone, B. G., **3**, 544 ⟨93JOC538⟩
Gladstone, P. L., **2**, 338 ⟨91CJC1482⟩
Gladstone, W. A. F., **3**, 681 ⟨69JCS(C)2587⟩; **4**, 314 ⟨67JCS(C)735⟩; **5**, 356 ⟨69JCS(C)2587⟩
Gladych, J. M. Z., **4**, 966 ⟨79COC(2)61⟩
Gladycz, J. A., **2**, 580 ⟨83JA5804⟩
Gladyshev, E. N., **2**, 569 ⟨88MI 211-02⟩; **4**, 656 ⟨76DOK(231)1138⟩
Gladysz, J. A., **1**, 13, 18, 21 ⟨74JA5290⟩, 579 ⟨82TL631⟩; **2**, 580 ⟨79JA1589, 82JA4865⟩, 580 ⟨90TL4417⟩, 679 ⟨82TL631⟩; **4**, 150 ⟨86JA1455⟩, 581 ⟨79JA5440, 79JA6433⟩, 581 ⟨83OM1846, 86CC1154⟩; **5**, 242, 252 ⟨78CC838⟩, 242, 252 ⟨79T2329⟩, 449 ⟨86JA1462⟩, 451 ⟨87JA4837, 88JA6110⟩, 451, 937 ⟨83OM1852⟩, 1097 ⟨92JA5890, 93JA3824⟩; **6**, 475 ⟨78CC838, 79T2329⟩
Gladysz, L. A., **2**, 580 ⟨80JA3299⟩
Glaeske, G., **6**, 562 ⟨81LA1388⟩, 564 ⟨81AP(314)587, 82AP(315)103⟩
Gläiel, W., **2**, 406, 407 ⟨77CB3874⟩
Glamkowski, E. J., **1**, 80 ⟨70JOC3510⟩
Glanville, W. K., **4**, 114 ⟨76JFC(7)153⟩
Glanz, D., **5**, 1074 ⟨89ZAAC(577)165⟩
Glänzer, B. I., **2**, 678 ⟨86CC1149⟩, 959 ⟨88AOC(28)85⟩
Glase, S. A., **3**, 220 ⟨87JOC1440⟩
Glasebrook, A. L., **3**, 389 ⟨34JA2381⟩

Glasel, V. I., **2**, 259 ⟨81ZN(B)447⟩
Glaser, B., **3**, 504 ⟨85AG(E)416⟩; **4**, 642 ⟨86CB3253⟩; **5**, 381, 534 ⟨93CB1119⟩
Glaser, C., **1**, 104 ⟨1870LA(154)137⟩, 1142 ⟨1869CB422⟩
Glaser, R., **1**, 1141 ⟨87JA4237⟩; **5**, 118 ⟨93T10099⟩
Glass, D. S., **1**, 421 ⟨65TL377⟩
Glass, R. S., **2**, 11 ⟨78JOC2291⟩, 132, 202 ⟨B-90MI 203-01⟩, 235 ⟨89JOC1092⟩, 250, 362, 366 ⟨77S798⟩, 263, 541 ⟨73JOM(61)83⟩; **3**, 337 ⟨89JOC1092⟩; **5**, 991 ⟨86JOC5123⟩
Glass, T. E., **1**, 784 ⟨94TL2675⟩; **3**, 224 ⟨71TL2575⟩
Glässel, W., **2**, 407 ⟨75AG(E)629⟩
Glassman, I., **1**, 1127 ⟨92JPC2151⟩
Glatt, H., **5**, 1126 ⟨76JCS(P2)524⟩
Glatz, B., **4**, 179 ⟨79JA2171⟩; **5**, 209 ⟨77AG(E)728⟩
Glauert, R. H., **1**, 41 ⟨52JCS2127⟩
Glavis, F. J., **4**, 188, 208 ⟨48JA2805⟩
Glaze, W. H., **1**, 110 ⟨69JOC641⟩, 663 ⟨65JOC1258⟩; **4**, 698 ⟨70JOM(25)C6⟩
Glazebrook, R. W., **2**, 180 ⟨58JCS2834⟩
Glazer, E. S., **4**, 9 ⟨69JA1386⟩
Glazer, J., **2**, 780 ⟨50JCS2657⟩
Glazier, E. R., **2**, 735 ⟨62JOC4397⟩
Gleason, J. G., **2**, 3 ⟨75JOC3420⟩, 143 ⟨71JA2437⟩, 199 ⟨69TL1447⟩, 199 ⟨71JOC322⟩, 207, 213 ⟨71JOC1314⟩, 212 ⟨70JOC3259⟩, 217 ⟨70TL3551⟩, 411 ⟨78TL3989⟩; **5**, 8 ⟨88OSC(4)727⟩; **6**, 487 ⟨93MIP9305026⟩
Gleason, J. L., **1**, 1079 ⟨92JOC5071⟩
Gleason, M. M., **5**, 944 ⟨93JOC6952, 95AG(E)350⟩
Gleason, T. G., **1**, 244 ⟨91CC354⟩
Gleason, W. B., **4**, 694 ⟨87OM1583⟩
Gleave, T. L., **1**, 891 ⟨35JCS255⟩
Glebova, Z. I., **5**, 394, 404 ⟨80RCR843⟩
Gledhill, A. P., **1**, 442 ⟨86JCS(P1)873⟩; **6**, 493, 494 ⟨89JOC2940⟩
Gleicher, G. J., **1**, 325, 920 ⟨86JOC5007, 89JOC308⟩
Gleim, R. D., **2**, 98 ⟨76JOC3983⟩, 651 ⟨78CC242⟩
Gleiter, R., **1**, 451 ⟨79JA4420⟩, 634 ⟨87JOC2644, 87TL1985⟩, 634 ⟨87TL6159⟩, 690 ⟨84JA5018, 86JA2932⟩, 814 ⟨82AG(E)686⟩, 894 ⟨71JA3927⟩, 1001 ⟨93S558⟩; **3**, 371 ⟨78JOC3893⟩, 372 ⟨79CB260⟩, 530 ⟨91JOC7048⟩; **4**, 162 ⟨86AG(E)999⟩, 312 ⟨66CB3903⟩, 979 ⟨66CB3892⟩; **5**, 301 ⟨68CB502⟩; **6**, 106 ⟨66CB3892⟩, 242, 685 ⟨91EUP450584⟩, 299, 305, 362 ⟨86JOC370⟩
Gleixner, R., **4**, 611 ⟨90JOM(393)33⟩
Glemser, O., **2**, 400 ⟨71CB1247⟩, 401 ⟨79ZAAC(452)103⟩; **5**, 383 ⟨72ZAAC(394)26⟩, 718 ⟨70CB553⟩, 1007, 1018 ⟨67CB1082⟩; **6**, 260 ⟨67MI 608-01, 68ZN(B)743⟩, 409, 410 ⟨67CB1184⟩, 409, 419 ⟨67CB2484, 66CB371⟩, 414 ⟨B-63MI 614-02⟩, 442 ⟨67CB1082⟩
Glenn, J. F., **6**, 494 ⟨57JA366⟩
Glenneberg, J., **3**, 227 ⟨88HCA1719⟩; **5**, 174 ⟨87AG(E)362⟩
Glennon, J. D., **5**, 322 ⟨85SC1159⟩
Gless, R. D., **2**, 252 ⟨92PS(68)147⟩
Gleu, K., **3**, 398 ⟨39CB1246⟩
Glick, G. D., **5**, 943 ⟨83CC880⟩
Glick, M. D., **1**, 66, 69, 454 ⟨82JOC3707⟩; **4**, 206 ⟨82JOC3707⟩, 698 ⟨71JA1035⟩
Glickman, G. A., **4**, 880, 883 ⟨45JA1017⟩
Glickman, S. A., **3**, 328 ⟨51LA(573)163⟩; **4**, 883 ⟨45JA1012⟩
Glidewell, C., **1**, 54 ⟨86T3887⟩, 648 ⟨88S319⟩, 648, 738, 743 ⟨88JCS(P2)1829⟩; **2**, 293 ⟨77JCS(D)527⟩; **3**, 196, 500 ⟨86T3887⟩, 196, 500 ⟨88S319⟩, 363 ⟨90JCS(D)3697, 90JOM(385)369⟩, 363 ⟨91JA7782⟩, 500 ⟨88JCS(P2)1829⟩; **4**, 697 ⟨77JOM(136)7⟩; **6**, 708, 709 ⟨86T3887, 88JCS(P2)1829, 88S319⟩
Gliksmans, G., **5**, 710 ⟨66BSF1376⟩
Glinskii, Yu. D., **2**, 469 ⟨75ZOB2554⟩
Glocker, M. O., **2**, 1009 ⟨91T205⟩; **5**, 1147 ⟨90CB1537⟩
Glockling, F., **2**, 421 ⟨54JCS22, 54JCS2526⟩, 814 ⟨B-69MI 215-04⟩, 867 ⟨65JCS4283⟩, 1092, 1093 ⟨B-88MI 223-01⟩, 1092 ⟨B-69MI 223-01, B-89MI 223-01, B-90MI 223-02⟩; **4**, 141 ⟨74JCS(D)2537⟩; **6**, 380, 387 ⟨79JCR(S)12⟩, 394 ⟨77JCR(S)116⟩, 395 ⟨79JCS(D)767⟩, 396 ⟨78JCR(S)170, 82ICA(58)149⟩, 656 ⟨77JCR(S)116, 79JCS(D)767⟩
Glockner, P., **1**, 517 ⟨64CB2451⟩
Gloede, J., **2**, 617 ⟨63CB1387⟩, 1008 ⟨69AP(302)354⟩; **3**, 618 ⟨63CB1387⟩; **4**, 553 ⟨82CSR493, 84EGP206557⟩; **5**, 710, 712, 714 ⟨82JPR(324)667, 68ZC219⟩, 976 ⟨63CB1387⟩; **6**, 39 ⟨63CB1387⟩, 104 ⟨69ZC201⟩
Gloer, K. B., **3**, 198 ⟨90JOC2842⟩, 245 ⟨88JOC2011⟩
Glokhnov, D. B., **4**, 572 ⟨87ZOB2640⟩
Glosky, C. R., **2**, 599 ⟨57USP2805234⟩
Gloth, R. E., **4**, 707 ⟨78JOM(153)59⟩
Glotova, T. E., **4**, 940 ⟨91ZOB1893⟩
Glotov, E. N., **5**, 1076 ⟨85IZV700⟩; **6**, 615 ⟨85IZV700⟩
Glotter, E., **5**, 164 ⟨72T1275⟩
Glover, D. J., **6**, 170 ⟨61JOC4734⟩
Glover, E. E., **5**, 849 ⟨72JCS(P1)2927⟩
Glover, G. M., **1**, 412, 810 ⟨67JA3462⟩
Glover, Jr., S. A., **1**, 339 ⟨93JOC2115⟩
Glover, P. T., **2**, 529 ⟨90TL7563⟩
Glover, S. A., **5**, 317 ⟨90T7247⟩, 317, 698, 722 ⟨87AJC701⟩, 687, 698 ⟨93AJC1213⟩
Glowinski, R., **1**, 492 ⟨76JOC2661⟩
Gloyna, D., **4**, 556 ⟨75JPR840⟩, 1023 ⟨66AG(E)847, 66ZC28⟩
Glozbach, E., **3**, 482, 483, 484 ⟨80JOM(191)371⟩; **5**, 871, 873, 874 ⟨80JOM(191)371⟩; **6**, 668, 670, 671 ⟨77JOM(132)359⟩, 670, 675 ⟨80JOM(191)371⟩
Gluchowski, C., **3**, 442 ⟨84JOC2512⟩; **5**, 274, 294 ⟨80JOC3413⟩
Glue, S. E. J., **3**, 269 ⟨77S607⟩; **5**, 592 ⟨77S607⟩; **6**, 445 ⟨86TL113⟩, 629 ⟨73JCS(P1)2644⟩
Glukhikh, V. I., **4**, 342 ⟨79DOK(247)609⟩; **5**, 656 ⟨81JGU1015, 81ZOB1201⟩
Glukhovtsev, V. G., **3**, 452 ⟨61CA(55)17526⟩
Glüsenkamp, K.-H., **2**, 412 ⟨90AG(E)782⟩; **3**, 71 ⟨82TL1147⟩
Glushkova, E. N., **3**, 436 ⟨85ZOB2077⟩
Glushkova, M. A., **5**, 781 ⟨87MI 519-02⟩
Glushkova, V. P., **2**, 882 ⟨53MI 217-01⟩
Glushkov, R. G., **2**, 788 ⟨60ZOB1855, 71ZOR1146⟩; **4**, 881 ⟨69MI 420-01⟩; **6**, 104 ⟨77UK685⟩
Glusker, J. P., **2**, 1008 ⟨85JA2067⟩
Gmeiner, P., **2**, 310 ⟨92LA273⟩
Gmünder, J. O., **2**, 383 ⟨84JCS(P2)411⟩
Gnad, J., **5**, 597, 599 ⟨65CB1531⟩
Gnauk, Th., **4**, 1003 ⟨76JPR116⟩; **6**, 286 ⟨76JPR116⟩
Gnoj, O., **3**, 156 ⟨68JOC3294⟩
Gnuchev, N. V., **5**, 826 ⟨67IZV1820⟩
Gnüchtel, A., **2**, 181 ⟨38CB712⟩
Goaller, R. Le, **2**, 791 ⟨86SC63⟩
Goasdoue, C., **3**, 633 ⟨79TL1585⟩; **5**, 607 ⟨83TL4001⟩
Goasdoue, N., **5**, 607 ⟨83TL4001⟩
Göbel, R., **4**, 555 ⟨92PS(73)67⟩
Göbel, T., **1**, 327 ⟨92TL2673⟩
Gobetto, R., **4**, 682 ⟨89JOM(365)163⟩
Gockeln, M., **1**, 561 ⟨93CB1623⟩
Godar, D. E., **5**, 321 ⟨79JOC1447⟩
Godard, A., **3**, 109 ⟨76BSF789⟩
Goddard, III, W. A., **1**, 1143 ⟨77JA2919⟩; **5**, 932 ⟨86JA2180, 86JA4746⟩
Goddard, N., **2**, 474 ⟨69JCS(A)2704⟩
Goddard, R., **4**, 586 ⟨77IJ149⟩, 676 ⟨92ICA741⟩, 686 ⟨77JOM(140)73⟩, 694 ⟨83JA5479⟩, 695 ⟨91AG(E)1124⟩; **5**, 890 ⟨90CB1617⟩
Godefroi, E. F., **3**, 311 ⟨78S675⟩; **4**, 306 ⟨61JA2574⟩

Godefroi, E. M., **1**, 24 ⟨84RTC220⟩
Godfrey, A., **4**, 384 ⟨87TL3911, 88JA6249⟩, 384 ⟨89T495⟩
Godfrey, A. G., **1**, 637, 1077 ⟨90JA3717⟩
Godfrey, C. R. A., **1**, 16 ⟨81JCS(P1)1501⟩; **3**, 132 ⟨82TL957⟩, 171 ⟨93CC1434⟩
Godfrey, I. M., **2**, 654 ⟨74JCS(P1)1353⟩; **3**, 91 ⟨74JCS(P1)1353⟩
Godfrey, J. C., **5**, 9 ⟨63JA643, 68CJC2549⟩, 105 ⟨62CHE347⟩
Godfrey, J. D., **1**, 831 ⟨81JA2446⟩; **3**, 165 ⟨88JMC2199⟩
Godfrey, K. L., **1**, 95 ⟨37JA596⟩
Godfrey, M., **3**, 674 ⟨68T1529⟩
Godfrey, P. D., **3**, 342 ⟨82AJC1747⟩, 550, 552 ⟨87CC573⟩, 582 ⟨84CPL447⟩
Godleski, S. A., **1**, 194 ⟨91COS(4)585⟩
Godovikova, T. I., **6**, 149 ⟨62IZV272⟩, 150 ⟨90IZV1620⟩, 150 ⟨90IZV1623⟩, 368 ⟨60IZV505⟩, 369 ⟨70IZV2108⟩
Godovikov, N. N., **2**, 1033 ⟨83IZV1662, 91RCR885⟩, 1076 ⟨82IZV933⟩; **4**, 123 ⟨86DOK(286)19⟩, 461 ⟨80IZV1125⟩
Godoy, J., **3**, 5 ⟨83HCA1790⟩, 5, 117 ⟨81TL2361⟩
Godschalx, J. P., **3**, 218, 303 ⟨84JA4833⟩
Godson, D. H., **6**, 445 ⟨75JCS(P1)1836⟩
Godt, Jr., H. C., **5**, 765 ⟨56JA1461⟩, 838 ⟨58USP2851489⟩; **6**, 552, 553 ⟨61JOC4047⟩
Godtfredsen, S. E., **5**, 118 ⟨B-88MI 502-01⟩
Godtfredsen, W. O., **3**, 704 ⟨80T515⟩; **4**, 743, 748 ⟨60T149⟩
Godts, F., **3**, 577 ⟨83CC748⟩
Godunova, L. F., **5**, 215 ⟨86IZV2032⟩
Goebel, N., **2**, 1055 ⟨58JOC1051⟩
Goeckner, N. A., **3**, 667 ⟨73JOC481⟩
Goedde, H. W., **5**, 81 ⟨65AG(E)846⟩
Goede, S. J., **3**, 608 ⟨93BSF185⟩, 689, 726 ⟨91CB2677⟩, 710 ⟨91JA6104⟩; **5**, 878, 889, 890 ⟨91CB2677⟩
Goede, S. J., **5**, 890 ⟨92OM3844, 93PS(77)252⟩, 1089 ⟨90JOM(390)C1⟩; **6**, 610 ⟨93JOC6930⟩, 682, 690 ⟨91CB2677⟩
Goedecke, E., **1**, 789 ⟨73AG(E)337⟩
Goedken, V., **2**, 998 ⟨93JOC546⟩; **4**, 816, 818 ⟨93JOC546⟩
Goedken, V. L., **5**, 1097 ⟨82CC607, 88CC46⟩
Goehl, J. E., **6**, 255 ⟨91CJC2127⟩
Goehring, M., **2**, 99 ⟨47CB219⟩, 399 ⟨56AG678⟩; **5**, 338 ⟨52ZN(B)634⟩
Goekjian, P. G., **1**, 1099 ⟨88JOC4151, 88JOC5580⟩
Goel, A. B., **1**, 458, 515 ⟨78JOC757⟩, 884 ⟨82JA6788⟩; **2**, 421 ⟨78IC1862⟩, 969 ⟨78JOC757⟩
Goel, B., **2**, 558 ⟨77CC169⟩
Goel, O. P., **2**, 3 ⟨83TL163⟩; **3**, 188 ⟨91JOC457⟩; **5**, 327 ⟨89OS(67)69⟩, 569, 570 ⟨87S162⟩
Goel, R. G., **2**, 504 ⟨71MI 209-01⟩, 582 ⟨82OM654⟩, 882 ⟨71JCS(A)562⟩, 886 ⟨71CJC2529, 72JOM(36)323⟩, 891 ⟨73JOM(50)129, 74IC1252⟩; **5**, 1133 ⟨73JOM(50)129, 79JOM(166)365⟩
Goel, S. C., **1**, 886 ⟨89JOC1671⟩
Goeldner, M., **6**, 443 ⟨90S1065⟩
Goeppel, D., **1**, 816, 818 ⟨94T3687⟩
Goerdeler, J., **2**, 389 ⟨90HOU(E16a/2)1035⟩, 726 ⟨61CB1067⟩; **3**, 361, 579 ⟨86CB162⟩, 366, 543 ⟨76TL2961⟩, 475 ⟨73CB1496⟩, 552 ⟨85CB851⟩, 552, 579 ⟨74CB502⟩, 558 ⟨72TL1519⟩, 562 ⟨80CB2509, 81CB536⟩; **4**, 412 ⟨63CB1630⟩, 979 ⟨86CB162⟩; **5**, 566, 592, 593, 610 ⟨75CB3066⟩, 581, 597, 599 ⟨63CB1551⟩, 593, 597 ⟨64CB2209⟩, 595, 597, 599 ⟨77CB285⟩, 595, 597, 599, 614 ⟨64CB3106⟩, 597, 599 ⟨63CB526, 65CB1531⟩, 616 ⟨85CB851⟩, 616, 1050 ⟨74CB502⟩, 630 ⟨60CB671, 62CB627⟩, 630 ⟨65CB1556⟩, 630, 631 ⟨60CB663⟩, 630, 633 ⟨61CB2856⟩, 657, 1064 ⟨77CB2996⟩, 762, 815 ⟨53CB400⟩, 763 ⟨53CB890, 60CB397⟩, 765, 766 ⟨57CB1638, 59CB2563⟩, 996 ⟨65CB2954, 66CB3572⟩, 1050 ⟨64TL799, 72TL1513⟩, 1052 ⟨68CB3475, 75CB3071⟩, 1064 ⟨71TL2765, 80CB79⟩, 1064, 1066 ⟨80CB1095⟩, 1064, 1073 ⟨86CB3737⟩; **6**, 443 ⟨85CB2294⟩, 504 ⟨80CB2499⟩, 536 ⟨73CB1487⟩, 559 ⟨82CB1252, 82CB1259⟩, 620 ⟨80CB1095⟩, 651 ⟨59CB2563⟩
Goerdelier, J., **2**, 168 ⟨61CB1682⟩
Goering, H. L., **1**, 187 ⟨86JOC2884⟩, 187 ⟨87JOC897⟩, 187 ⟨89JOC3239⟩; **2**, 117 ⟨56JA348⟩; **4**, 23 ⟨57JA6270, 59JA5937⟩, 23, 38 ⟨55JA3465⟩
Goerlach, A., **4**, 182 ⟨88LA559⟩
Goethals, E. J., **2**, 177 ⟨65BSB21⟩, 232 ⟨69BSB191⟩
Goety, H., **2**, 824 ⟨63LA(665)1⟩
Goetz, H., **2**, 825 ⟨68LA(715)1⟩
Goetze, A. M., **5**, 214 ⟨88JMC1960⟩
Goff, D. A., **4**, 199 ⟨86JOC4711⟩
Goff, D. L., **3**, 436 ⟨86JOC5024⟩
Goff, E. Le, **1**, 537 ⟨64JOC423⟩
Goff, H. M., **2**, 578 ⟨88JA5228⟩
Goff, M. Le, **4**, 969 ⟨68MI 421-01⟩
Goff, N. Le, **1**, 1041 ⟨85TL3203⟩
Goff-Hays, O. Le, **1**, 420 ⟨68BSF3401⟩
Goffic, F. Le, **3**, 8 ⟨89S64⟩, 72 ⟨83TL5209⟩; **4**, 487 ⟨91SC1847⟩
Goffin, E., **2**, 1046 ⟨77JCR(S)105⟩
Goffinet, B., **1**, 419 ⟨55BSF1341⟩
Gofstein, R., **5**, 857 ⟨50MI 521-01⟩
Gofstein, R. M., **3**, 444 ⟨46JA1983⟩
Gogan, N. J., **3**, 307 ⟨81TL4899⟩
Gogerty, J. H., **4**, 416 ⟨75JMC177, 75JMC182⟩
Goggin, P., **2**, 650 ⟨59JCS1804⟩
Gogoberidze, I. T., **4**, 312 ⟨84ZOR1502⟩
Gogoladze, D. D., **5**, 215 ⟨86IZV2032⟩
Gogte, V. N., **1**, 794, 799, 809 ⟨65TL3955⟩; **2**, 394 ⟨87SC1421⟩; **3**, 370, 376 ⟨74TL3911⟩
Goh, J. B., **2**, 931 ⟨92JOC3270⟩
Goh, S. H., **1**, 426 ⟨83CC907⟩
Gohda, N., **3**, 249 ⟨79TL1129⟩; **5**, 1107 ⟨79TL1129, 81JCS(P1)1577⟩
Goheen, D. W., **1**, 393 ⟨59OS(39)37⟩; **2**, 180 ⟨63JOC2485⟩, 278 ⟨55JOC1019⟩; **5**, 146 ⟨59OS(39)37⟩
Going, R. C., **4**, 208 ⟨82JOC3517⟩
Goji, H., **5**, 249 ⟨87NKK1479⟩, 253 ⟨87NKK1475⟩
Gokdale, P. D., **3**, 212 ⟨74S778⟩
Gokel, G., **2**, 133 ⟨B-77MI 203-02⟩
Gokel, G. W., **1**, 166 ⟨79TL3375⟩; **2**, 632 ⟨77TL3519⟩; **3**, 642, 644 ⟨77S629⟩, 695, 696, 700 ⟨88OSC(6)232⟩, 695, 721 ⟨72TL1637, 72AG(E)530⟩; **4**, 235 ⟨80JOC3634⟩, 236 ⟨79TL3375⟩; **6**, 25 ⟨79TL1473⟩, 126, 135 ⟨79TL3375⟩
Gokhale, U., **1**, 605 ⟨83JOC4776⟩; **2**, 285 ⟨83JOC4776⟩
Göknur, O., **4**, 581 ⟨89JOM(371)329⟩
Gokou, C. T., **3**, 414 ⟨86PS(27)327⟩; **5**, 633 ⟨85JCS(P1)1875, 86PS(27)327⟩
Gol, F., **2**, 466 ⟨86CB2767⟩; **4**, 550 ⟨88ZN(B)31⟩
Golan, D. R., **4**, 1080 ⟨88MI 424-01⟩
Golborn, P., **2**, 675 ⟨73JCS(P1)2870⟩; **4**, 912 ⟨73CA(78)45000b⟩, 912, 913 ⟨73S547⟩, 920, 921 ⟨73CA(78)85856s⟩; **6**, 116 ⟨73S547⟩
Gold, A., **1**, 11, 19 ⟨87JOC3205⟩
Gold, B., **6**, ⟨92MI 615-02⟩, 417, 426 ⟨93S103⟩
Goldacre, R. J., **1**, 894 ⟨49JCS1724⟩
Gold, H., **3**, 293 ⟨74HCA2201⟩; **6**, 55 ⟨61CB2594⟩, 105 ⟨60AG956, 63GEP1146892⟩
Gold, J., **1**, 648 ⟨47JA1551⟩; **2**, 537 ⟨47JA1976⟩; **4**, 137 ⟨47JA1976⟩
Gold, M., **5**, 22 ⟨92S1145⟩
Gold, M. H., **4**, 430, 431, 432, 994 ⟨57JOC1665⟩, 440 ⟨51JOC1507⟩
Gold, P. M., **3**, 256 ⟨79TL4433, 81JA3112⟩; **4**, 197 ⟨81JA3112⟩, 253 ⟨80JA3095, 81JOC3555⟩; **6**, 134 ⟨78MI 604-01⟩

Gold, V., **1**, 869, 871 ⟨65JCS(B)1752⟩, 891 ⟨B-67MI 119-04⟩, 895 ⟨52JCS2167⟩, 895, 896 ⟨52JCS2172, 63JCS1930⟩, 896 ⟨65JCS6718, B-67MI 119-05⟩, 1109 ⟨B-67MI 122-01⟩; **4**, 42 ⟨82JCS(P2)881⟩, 1072 ⟨B-65MI 424-01⟩; **5**, 26, 37 ⟨53JCS1409⟩; **6**, 25, 31 ⟨83CC283⟩, 25, 31 ⟨84JCS(P2)1247⟩, 30 ⟨83CC140⟩, 30 ⟨84JCS(P2)1239, 86JCS(P2)471⟩

Goldbeck, O., **5**, 834 ⟨1891CB3658⟩

Goldberg, A. A., **2**, 191 ⟨43JCS4⟩; **5**, 622 ⟨48JCS1919⟩

Goldberg, D. E., **6**, 186 ⟨76CC261⟩

Goldberg, G. M., **1**, 450 ⟨54JA1613⟩; **5**, 421 ⟨54JA1613⟩

Goldberg, K. I., **5**, 450, 935 ⟨89JA1285⟩

Goldberg, M. W., **5**, 566 ⟨33HCA1323⟩; **6**, 506 ⟨65JOC2849⟩

Goldberg, N. N., **1**, 860 ⟨51JA4301⟩

Goldberg, S. I., **1**, 779 ⟨67JOC2059⟩

Goldberg, S. Z., **1**, 890 ⟨74JA1348⟩

Goldberg, Y., **4**, 929 ⟨84H(22)987⟩

Goldberg, Y. S., **4**, 514 ⟨84JOM(268)C29⟩, 515 ⟨86JOM(316)249⟩; **6**, 124 ⟨84JOM(265)237⟩

Goldberg, Yu. Sh., **4**, 1080 ⟨87CC31⟩

Goldblatt, L. A., **5**, 189 ⟨63JOC1905⟩, 189 ⟨63MI 504-01⟩

Goldblum, A., **5**, 403, 404 ⟨88JCS(P2)2029⟩, 403, 404 ⟨89JCS(P1)765⟩

Goldeerg, Y., **3**, 430 ⟨89SC3129⟩

Golden, D. M., **1**, 921, 925 ⟨82ARP493⟩, 1109 ⟨69CRV279⟩

Goldey, R. N., **1**, 112 ⟨53JA3975⟩

Gol'dfarb, E. I., **2**, 259 ⟨90ZOB476⟩

Goldhill, J., **1**, 653 ⟨78CC176, 80JCS(P1)1493⟩, 655, 1080 ⟨82JCS(P1)1563⟩; **3**, 227 ⟨79TL3205⟩, 274 ⟨82JCS(P1)1563⟩

Gol'din, G. S., **2**, 383 ⟨71ZOR258⟩; **5**, 850 ⟨69ZOR1404, 73JGU317⟩, 854 ⟨69ZOR1411⟩, 863 ⟨72ZOB868⟩

Golding, B., **3**, 659 ⟨69AG(E)343⟩

Golding, B. T., **1**, 411, 811 ⟨73CC319⟩; **2**, 8, 32, 36 ⟨87TL4445⟩, 225 ⟨83CC1051, 88JCS(P1)9⟩, 309 ⟨87S190⟩; **6**, 79 ⟨80CC207⟩, 105 ⟨75JCS(P1)1302⟩, 633 ⟨87TL4445, 88CC1175⟩

Gol'ding, I. R., **2**, 1077 ⟨72IZV529⟩

Golding, P., **2**, 104 ⟨93T7037⟩, 386 ⟨88TL2735⟩

Goldish, D. M., **5**, 367 ⟨B-83MI 507-01⟩

Goldman, A., **3**, 80 ⟨80JOC4059⟩; **6**, 415 ⟨80JOC4059⟩

Goldman, A. S., **1**, 554 ⟨92JA9492⟩

Goldman, B. E., **3**, 119 ⟨86T2951⟩

Goldman, E. W.,**5**, 937 ⟨91JA927⟩

Goldman, G., **2**, 964 ⟨62JA1650⟩

Goldman, N., **1**, 130, 889 ⟨65JA275⟩; **2**, 696 ⟨65JA275⟩

Goldmann, F. L., **2**, 864 ⟨63CB3091⟩

Goldner, W., **2**, 471 ⟨85CB3268⟩

Goldschmidt, S., **2**, 404 ⟨58LA(618)173⟩; **3**, 444 ⟨58CB502⟩; **4**, 995 ⟨22CB644⟩; **5**, 208 ⟨52LA(577)153⟩, 1023 ⟨38BRP495510⟩

Goldschmidt, Z., **1**, 801, 802, 803 ⟨88CSR229⟩, 1143 ⟨70JA7606⟩

Goldsmith, B., **1**, 4, 7, 9, 17, 28 ⟨78JOC2259⟩

Goldsmith, D., **5**, 702 ⟨91TL2457⟩

Goldsmith, D. J., **5**, 702 ⟨86TL4701⟩

Goldstein, A., **1**, 773, 776 ⟨55JA1001⟩

Goldstein, D., **3**, 355 ⟨92MI 308-02⟩

Goldstein, J. A., **3**, 482 ⟨76JA7327⟩

Goldstein, J., **3**, 463 ⟨16CB1923⟩

Goldstein, M. J., **1**, 890 ⟨84CC1654⟩

Goldstein, S. A., **4**, 436 ⟨81CJC679⟩

Goldstein, S., **3**, 529 ⟨81JA4616⟩

Goldstein, T. B., **2**, 380, 381, 382 ⟨73OPP233⟩

Goldsworthy, L. J., **2**, 128 ⟨48JCS2177⟩; **4**, 309 ⟨48JCS2174⟩

Goldwhite, H., **2**, 391 ⟨B-65MI 207-01⟩, 407, 862 ⟨75JCS(D)12⟩, 430 ⟨67JA1102⟩, 430, 438 ⟨69T1097⟩, 487, 875 ⟨68JOM(12)133⟩; **4**, 32 ⟨61JCS3825⟩, 111 ⟨66JCS(C)2075⟩, 116 ⟨70OMR(2)81⟩, 123, 125 ⟨66JA3572⟩, 281, 282 ⟨65T2743⟩, 597, 598 ⟨80JCS(D)2428, 76CC623⟩, 755 ⟨60JA3091⟩, 805 ⟨68JOM(12)133⟩; **6**, 18 ⟨60JA3091⟩, 58 ⟨63JCS1083⟩, 64 ⟨61JCS3825⟩, 679 ⟨65JCS6875⟩

Golebiowski, A., **3**, 39 ⟨87T3063, 89CRV149⟩, 40 ⟨92T4231⟩

Golec, Jr., F. A., **5**, 206 ⟨88USP4762945⟩

Golembeski, N. M., **5**, 798, 799, 800 ⟨79JOM(172)251⟩

Golfier, M., **5**, 171 ⟨75T171⟩, 704 ⟨79S946⟩

Golgolab, H., **1**, 1077 ⟨73JHC655, 75JHC801⟩; **6**, 653 ⟨70JHC689⟩

Goli, M. B., **4**, 1022, 1076 ⟨91TL3631⟩

Golino, C., **4**, 1014 ⟨78CJC2286⟩; **5**, 792, 796 ⟨78CJC2286⟩

Golino, C. M., **3**, 501 ⟨74JA7105⟩; **4**, 610, 611 ⟨75JA7371, 74JA7105⟩

Golinski, J., **3**, 192 ⟨85HCA162⟩

Golinski, M., **3**, 615 ⟨93JOC159⟩, 634 ⟨89LA203⟩

Gölitz, P., **6**, 20 ⟨77AG(E)854⟩

Goller, E. J., **2**, 69 ⟨71JOC3311⟩

Gollins, D. W., **3**, 171 ⟨93CC1434⟩; **5**, 329 ⟨90CC720⟩, 376 ⟨90T4733⟩

Gollnick, K., **1**, 521 ⟨73CB674⟩; **3**, 424 ⟨92JOC229⟩, 699 ⟨88TL1127⟩

Gol'Mov, V. P., **5**, 199 ⟨61ZOB665⟩

Golob, A. M., **1**, 413, 810 ⟨75JA4765⟩, 815 ⟨78JA8170⟩

Golobish, T. D., **2**, 163 ⟨86JOC829⟩

Golod, E. L., **2**, 790 ⟨83RCR284⟩

Golodova, K. G., **1**, 225 ⟨72JOU2530⟩

Golokhov, D. B., **4**, 116 ⟨89JGU2194⟩, 116, 133 ⟨89TL2445⟩, 116, 133 ⟨90JGU1536⟩, 116, 133 ⟨91JGU627⟩; **6**, 697 ⟨88ZOB491⟩

Golokhov, D. V., **5**, 895 ⟨88ZOB491⟩

Gololbov, Y. G., **5**, 223 ⟨90ZOB1928⟩

Gololobov, Y. G., **1**, 304 ⟨87T2609⟩; **2**, 468 ⟨92T1353⟩, 1000 ⟨87T2609⟩; **3**, 378 ⟨87T2609⟩, 563 ⟨77JOU1989⟩; **5**, 373, 862 ⟨81T437⟩; **6**, 635 ⟨92IZV425⟩, 656 ⟨74ZOB538⟩

Gololobov, Yu. G., **2**, 467 ⟨79JGU34⟩; **4**, 483 ⟨77ZOB1422, 78ZOR739⟩, 486 ⟨82ZOB930⟩

Golosov, S. N., **4**, 561 ⟨87ZOB2138⟩

Golovanov, A. V., **4**, 125 ⟨86JGU2246⟩, 126 ⟨86JGU2144, 86JGU2242⟩, 128 ⟨87JGU622⟩; **6**, 242 ⟨88ZOB1525⟩

Golovchenko, L. S., **5**, 385 ⟨72JOM(36)227⟩

Golovinskii, E., **5**, 770, 852 ⟨62JGU3330⟩

Golovinsky, E., **5**, 852 ⟨66CB3734⟩

Golovnya, R. V., **4**, 724 ⟨54DOK(97)459⟩

Golstein, S. B., **2**, 566 ⟨77JOM(142)145⟩

Golub, N. M., **2**, 265 ⟨87ZOR1981⟩; **6**, 89 ⟨88URP1421744⟩

Golubeva, E. I., **2**, 464 ⟨59ZOB1671⟩

Golubeva, G. A., **3**, 341 ⟨90KGS1285⟩

Golubev, V. K., **5**, 44 ⟨78IZV2094⟩

Golz, G., **2**, 240 ⟨73LA1644⟩

Gölz, H., **5**, 209 ⟨77AG(E)728⟩

Golz, K., **4**, 569, 570 ⟨74ZAAC(404)204⟩; **6**, 164 ⟨74ZAAC(404)204⟩

Gomberg, M., **1**, 890, 898 ⟨02CB2397⟩, 919 ⟨00CB3150, 00JA757⟩

Gombler, W., **2**, 166 ⟨67AG(E)706⟩, 166 ⟨69AG(E)773⟩, 166 ⟨76JFC(7)115⟩, 166 ⟨76ZN(B)727⟩, 478 ⟨78ZAAC(439)193⟩; **6**, 233 ⟨87JFC(34)475⟩, 252, 274, 275 ⟨92ZN(B)369⟩

Gombos, J., **1**, 91 ⟨76CB2628⟩

Gomelya, N. D., **2**, 463 ⟨82JGU202⟩

Gomes, M. A., **1**, 949 ⟨90T3061⟩

Gomes de Mesquita, A. H., **1**, 897 ⟨65AX437⟩

Gómez, A. M., **1**, 187 ⟨90JOC2294⟩, 579 ⟨92JOC1613⟩, 942 ⟨92CC613⟩

Gomez, E., **5**, 584, 587 ⟨85H(23)649⟩

Gomez, E. E., **4**, 264 ⟨92JA5946⟩
Gomez, F. A., **1**, 168 ⟨91JA5915⟩, 168 ⟨92JOC1384⟩
Gómez, H., **6**, 573 ⟨70JMC334⟩
Gómez, M., **4**, 579 ⟨89OM1604⟩; **6**, 331 ⟨91OM1683⟩
Gómez-Contreras, F., **5**, 366 ⟨91JCS(P1)273⟩
Gomez-Galeno, J., **3**, 55, 59 ⟨86TL2571⟩
Gomez-Parra, F., **3**, 102, 108 ⟨76JHC525⟩
Gomez-Sal, P., **4**, 656 ⟨89JOM(375)59⟩
Gómez-Sánchez, A., **2**, 394 ⟨91TL3225⟩
Gomoll, A. W., **5**, 767 ⟨74JMC549⟩
Gompper, K., **4**, 621 ⟨81ZAAC(478)94⟩
Gompper, R., **1**, 563 ⟨80AG(E)463⟩, 982 ⟨73AG(E)566⟩, 986 ⟨79LA1388⟩, 995 ⟨78TL4263⟩; **2**, 674 ⟨64AG(E)560⟩; **3**, 247 ⟨62LA(659)90⟩, 547 ⟨62CB2861, 62CB2871⟩, 612 ⟨64AG(E)560⟩; **4**, 728 ⟨62AG251, 67CB591⟩, 743, 970 ⟨66CB2900⟩, 843, 852, 899 ⟨62CB2861⟩, 852 ⟨65LA(684)37⟩, 870 ⟨70TL481⟩, 892 ⟨71AG(E)67⟩, 900, 970 ⟨62CB2871⟩, 901 ⟨62ZN851, 65CB1385, 67AG147, 68AG38⟩, 902 ⟨64TL1971, 68OS(48)97⟩, 979 ⟨65CB1391⟩, 998 ⟨82AG559⟩; **5**, 552 ⟨80TL917⟩, 744, 745, 751 ⟨59CB837⟩, 753 ⟨57CB1837⟩; **6**, 139, 140 ⟨59CB329⟩, 140 ⟨60CB1398⟩, 314, 317 ⟨70TL481⟩, 627, 651 ⟨66CB2885⟩
Gomushchak, N. I., **2**, 256 ⟨87ZOB1078⟩
Goncalves, D. C. R. G., **4**, 65 ⟨82CC857⟩
Gonce, F., **6**, 689 ⟨89TL6869⟩
Goncharova, E. L., **5**, 680 ⟨91MI 517-01⟩
Goncharova, L. V., **2**, 841 ⟨76ZOB1701⟩
Gönczi, C., **5**, 220 ⟨89CB1107⟩
Gonda, J., **1**, 834 ⟨91AG(E)1465⟩, 837 ⟨93S729⟩; **5**, 1035 ⟨84CCC295, 86CCC112⟩, 1057 ⟨83CCC3567⟩; **6**, 629 ⟨84CCC295⟩
Gonella, N. C., **4**, 847 ⟨78JOC369⟩; **6**, 698 ⟨78JOC369⟩
Gong, B., **5**, 199 ⟨90JOC4763⟩
Gong, B. Q., **2**, 58 ⟨94TA1075⟩
Gong, J. K., **4**, 583 ⟨92OM1392⟩
Gong, L., **2**, 930 ⟨90JOC3634⟩
Gonis, G., **4**, 174 ⟨66JOC3032⟩
Gönne, S., **4**, 136 ⟨87LA51⟩
Gonnella, N. C., **3**, 254 ⟨79JOC930⟩; **4**, 847, 848 ⟨79JOC930⟩; **5**, 561 ⟨79SC17⟩; **6**, 121, 698 ⟨79JOC930⟩
Gonnermann, J., **1**, 156 ⟨79CB234⟩, 156, 157 ⟨77TL1161⟩; **3**, 191 ⟨79HCA2258⟩
Gonschorrek, C., **1**, 974 ⟨87T2457⟩; **2**, 1029 ⟨87TL785⟩
Gontar, A. F., **5**, 1076 ⟨85IZV700, 87IZV1177⟩; **6**, 238, 242 ⟨75IZV2279⟩, 239, 440 ⟨76IZV209⟩, 240, 615 ⟨80JFC(15)169⟩, 614 ⟨76IZV2381, 76IZV212⟩, 615 ⟨83IZV694, 83MI 620-02, 85IZV700⟩, 616 ⟨77IZV2379, 81IZV2632⟩, 643 ⟨78MI 612-01⟩
Gonther, H. E., **2**, 236 ⟨71B4089⟩
Gonzalez, A., **2**, 320 ⟨90S153, 90SC1149⟩; **3**, 252 ⟨90SC2537⟩
Gonzalez, A. M., **2**, 929 ⟨88TL1825, 91JCS(P1)2811⟩, 930 ⟨90CC1030, 92JCS(P1)327⟩; **3**, 203 ⟨89SC275⟩, 356 ⟨88SC1847⟩, 577 ⟨89SC1039⟩, 653 ⟨86TL2027⟩; **5**, 417 ⟨89SC275⟩
González, B., **2**, 545 ⟨89SC275⟩; **3**, 203 ⟨89SC275, 86TL2027⟩; **5**, 417 ⟨89SC275⟩
Gonzalez, C., **4**, 410 ⟨85H(23)1127⟩
Gonzalez, D., **1**, 525 ⟨80JOC2550⟩
González, E., **1**, 630 ⟨69CC1395⟩
Gonzalez, F. J., **3**, 414 ⟨86CC1179⟩; **4**, 421 ⟨91CC1704⟩
Gonzalez, J., **4**, 5 ⟨91JOC4322⟩; **5**, 201 ⟨91JA8972⟩, 361 ⟨84JOC3322⟩
Gonzalez, J. A., **3**, 621 ⟨84AG(E)69⟩
Gonzalez, J. M., **2**, 18, 24, 31, 34 ⟨85AG(E)319⟩, 18, 610 ⟨91JOC2234⟩, 631 ⟨84JCS(P1)2623⟩, 1019 ⟨87S661⟩; **4**, 749 ⟨93AG(E)893⟩
Gonzalez, M. A., **1**, 150 ⟨91TL5501⟩, 150 ⟨91TL5505⟩, 150 ⟨91TL5509⟩
Gonzalez, O. J., **3**, 434 ⟨89JOC1736⟩
González, R., **1**, 1090 ⟨90CC1521, 90CC587⟩, 1090 ⟨91CC1001, 92TL831⟩
González-Gómez, J. A., **1**, 172, 325, 330 ⟨84AG(E)69⟩
González-Nogal, A. M., **1**, 495 ⟨89T413⟩; **2**, 929, 930 ⟨89T349⟩
Gonzalez-Nuñez, E., **4**, 300 ⟨88JCS(P1)1631⟩, 412 ⟨88CB1813⟩
González Nuñez, M. E., **6**, 230 ⟨91JA7654⟩
Gooch, III, E. E., **2**, 4 ⟨81JOC2582⟩, 34 ⟨80JOC3578, 81SC521⟩, 628 ⟨81SC247⟩
Good, J. J., **3**, 212 ⟨73JOC3829⟩, 623 ⟨79JOC300⟩
Good, R., **3**, 245 ⟨63HCA1259⟩
Goodall, D. C., **2**, 278, 279 ⟨69JCS(A)890⟩
Goodbrand, H. B., **4**, 912 ⟨74S862⟩
Goodburn, T. G., **1**, 1001 ⟨63JCS5889, 65JCS894⟩
Goodchild, J., **3**, 357, 377 ⟨75JCS(P1)1335⟩, 361 ⟨69JCS(C)2192⟩; **5**, 575 ⟨69JCS(C)2192⟩
Goode, M. J., **2**, 492 ⟨88JCS(D)451⟩
Goodfellow, R. J., **4**, 344 ⟨69TFS1697⟩
Goodgame, M. M., **5**, 1124 ⟨75TL3125⟩
Goodhue, C. T., **5**, 735 ⟨67JOC392⟩
Gooding, O. W., **1**, 967 ⟨91JOC1083⟩
Goodlett, V. W., **3**, 546, 547 ⟨68JOC2738⟩; **5**, 511 ⟨68JOC2565⟩
Goodman, F. J., **6**, 555 ⟨76JOC626⟩
Goodman, J. M., **2**, 533, 698 ⟨90T4663, 92JOC5173⟩
Goodman, L., **2**, 129 ⟨58JA1680⟩; **5**, 476, 477 ⟨68JOC1783⟩, 476, 477 ⟨68JOC3727⟩; **6**, 502 ⟨66JMC444⟩
Goodman, M., **4**, 422 ⟨82CC280⟩, 427 ⟨84JOC821⟩, 446 ⟨83MI 409-01⟩; **5**, 84 ⟨57MI 502-02⟩, 140, 279 ⟨56JA1367⟩, 972 ⟨76MI 526-01⟩
Goodman, M. A., **2**, 466, 467 ⟨84IC2062⟩; **4**, 572 ⟨84IC2063⟩
Goodman, M. M., **2**, 279 ⟨83OM1106⟩; **3**, 305 ⟨84JOC2322⟩
Goodrich, B. S., **3**, 234 ⟨61JCS1583⟩
Goodrow, M. H., **2**, 212, 679 ⟨82TL323⟩, 215 ⟨81S457⟩
Goodson, L. H., **4**, 69 ⟨45JA655, 45JA657⟩
Goodwin, C. A., **2**, 530 ⟨92TL4209⟩; **4**, 526, 527, 528 ⟨92TL4209⟩
Goodwin, C. J., **2**, 315 ⟨93JCS(P1)1375, 93SL731⟩
Goodwin, H. P., **5**, 872, 1148 ⟨91OM3205⟩; **6**, 670, 675 ⟨91OM3205⟩
Goodwin, J. A., **4**, 843 ⟨90S991⟩
Goodwin, T. E., **4**, 179 ⟨77SC409⟩
Goodwin Jr., J. T., **4**, 609 ⟨47JA2247, 49USP2483972⟩, 609 ⟨52USP2607791⟩
Goon, D. J. W., **5**, 344 ⟨85JOC4991⟩
Goosen, A., **6**, 431 ⟨71JCS(C)3766⟩
Goossens, D., **3**, 558, 560 ⟨80TL3081⟩
Goossens, H. J. M., **6**, 77, 78 ⟨78S283⟩
Gopalan, A., **1**, 813 ⟨80JA1756⟩
Gopalan, A. S., **2**, 39 ⟨90TL5575⟩; **3**, 275 ⟨81JOC1512⟩; **4**, 250 ⟨81JOC1513⟩; **5**, 322 ⟨94TL5157⟩; **6**, 134 ⟨81JOC1512⟩
Gopalan, R., **3**, 63 ⟨78CC924⟩, 233 ⟨78CC608⟩
Gopichand, S., **6**, 251 ⟨93MI 608-01⟩
Gopichand, Y., **3**, 550 ⟨76TL3093⟩
Gopinath, K. W., **1**, 523 ⟨66JOC2376⟩
Gor, J., **2**, 104 ⟨91BSF222⟩
Goralski, C. T., **1**, 60 ⟨87JOC4014⟩, 443 ⟨89JA384, 91JOC1543⟩, 1037 ⟨92JOC2941⟩; **2**, 59 ⟨94JFC(66)123⟩, 60 ⟨92TL4533, 93TL257⟩, 60 ⟨94TL5201⟩, 157 ⟨70JOC4217⟩, 514 ⟨87JOC4014⟩, 740 ⟨89JA384⟩, 752 ⟨66JA100⟩; **4**, 524 ⟨91JOC5691, 93JA440⟩; **5**, 496, 497 ⟨72JOC2080⟩
Gorbatenko, V. I., **4**, 314, 329 ⟨74JOU1565, 74JOU244⟩, 434, 447 ⟨77ZOR723⟩, 436, 441 ⟨77ZOR2449, 76ZOR2103⟩, 441 ⟨76ZOR231⟩, 992 ⟨78ZOR2624⟩; **5**, 658 ⟨89JOU1365⟩, 658, 1068 ⟨89ZOR1572⟩, 1008

⟨76ZOR1963⟩, 1008 ⟨76ZOR2473, 76ZOR2472⟩, 1068, 1069 ⟨84ZOR2543, 80S85⟩, 1069 ⟨88ZOR986⟩, 1075 ⟨87ZOR2385, 87ZOR2390⟩; **6**, 445 ⟨80S85⟩, 450 ⟨77ZOR290⟩, 451, 606 ⟨93T3227⟩, 453 ⟨71ZOR2229⟩, 609 ⟨78JGU1512⟩, 617 ⟨81ZOR398⟩, 620 ⟨92ZOB324⟩, 657 ⟨78ZOB1425⟩
Gorbatenko, V. J., **5**, 1008 ⟨75ZOR2227⟩
Gorbatenko, W. I., **5**, 1032 ⟨73AG(E)842⟩
Gorbatenko, Zh. K., **2**, 466 ⟨74ZOB2357⟩
Gorbovoi, P. M., **5**, 1044 ⟨88ZOB616, 91UKZ89⟩
Gorbunova, M. G., **3**, 265 ⟨91S207⟩
Gorcian, H., **4**, 553 ⟨64JCS513⟩
Gordeev, A. D., **4**, 794 ⟨80JOU1762⟩; **5**, 655, 659 ⟨77JOU1471⟩, 655, 659 ⟨77ZOR1594⟩, 657, 659 ⟨79JOU32⟩, 657, 659 ⟨79ZOR38⟩
Gordeev, M. F., **1**, 956 ⟨93JOC3038⟩; **4**, 975 ⟨89IZV1806⟩, 979 ⟨90IZV401⟩
Gordetsov, A. S., **5**, 388 ⟨85ZOB1071⟩, 1076, 1077 ⟨87JOM(327)303, 85IZV2824⟩, 1081 ⟨87ZOB405, 88MI 528-01⟩, 1081, 1082 ⟨82ZOB885⟩, 1085 ⟨89MI 528-01⟩
Gordillo, B., **4**, 230 ⟨85TL1927⟩; **6**, 120 ⟨86T1963⟩
Gordon, A., **6**, 266 ⟨90JOM(399)189⟩
Gordon, A. B., **1**, 398 ⟨28JCS3193⟩
Gordon, D. C., **3**, 527 ⟨77JOC2111⟩; **6**, 237 ⟨92OM2947, 93CM1321⟩
Gordon, D. M., **2**, 597 ⟨94JOC3714⟩
Gordon, D. W., **4**, 211 ⟨85SC225⟩
Gordon, E. M., **3**, 165 ⟨88JMC2199⟩, 713 ⟨79JOC1218⟩; **5**, 346 ⟨84H(21)191⟩, 727 ⟨84SC1159⟩
Gordon, E. M. J., **2**, 257 ⟨92JA1521⟩
Gordon, H. B., **2**, 566 ⟨74JOM(74)85⟩
Gordon, I., **4**, 130 ⟨66CJC2593⟩
Gordon, J. C., **4**, 709 ⟨92OM2128⟩, 714 ⟨91IC4710⟩
Gordon, J. R., **2**, 608 ⟨50JA2454⟩
Gordon, K. M., **1**, 601, 610, 611, 621, 646 ⟨75CS9⟩, 716 ⟨75CS9⟩
Gordon, M., **3**, 286, 459 ⟨72JOC3369⟩; **6**, 232 ⟨60JOC60⟩
Gordon, M. D., **2**, 879 ⟨72JA7596⟩
Gordon, M. S., **1**, 849 ⟨68JA2471⟩; **5**, 932 ⟨92JA539⟩
Gordon, P. F., **2**, 662 ⟨84CSR441⟩
Gordon, R. D., **5**, 321 ⟨75USP3914252⟩
Gordon, S. H., **5**, 317 ⟨80CAR(79)205⟩
Gordon, W., **1**, 1142 ⟨67CB1193⟩
Gordy, W., **3**, 694 ⟨42JA2952, 50MI 321-01⟩
Gore, J., **1**, 197 ⟨81JCR(S)278⟩, 225 ⟨79S508⟩, 413, 810 ⟨83TL1157⟩, 459 ⟨80TL51⟩, 495 ⟨78JCR(S)416⟩, 496 ⟨92BSF273⟩, 498 ⟨72TL1189⟩, 515, 1025 ⟨74S573⟩, 660 ⟨71BSF3978⟩, 808 ⟨78TL1803⟩, 809 ⟨84T3277, 86T1333⟩, 810 ⟨76TL677⟩, 955, 971 ⟨93SL105⟩, 958 ⟨78TL3801, 79NJC321⟩, 958 ⟨81TL103⟩, 959 ⟨72TL4465, 76S755⟩, 959, 960 ⟨74BSF1119⟩, 959, 970 ⟨72CC866, 74TL1593⟩, 962 ⟨75BSF2159, 75BSF2166⟩, 962 ⟨84TL845⟩, 967 ⟨79TCC1⟩, 974 ⟨81T1359⟩, 975 ⟨93TL3129⟩, 9 ⟨73CC821, 76BSF169⟩, 11 ⟨86TL1907⟩, 43 ⟨71BSF1649⟩, 314 ⟨90TL2277⟩; **3**, 141 ⟨82TL4263⟩, 144, 217 ⟨86T1333⟩, 180 ⟨90TL2277⟩, 230 ⟨87T2721⟩; **4**, 535 ⟨88S983⟩, 535 ⟨89TL3963⟩
Gore, M. G., **3**, 40 ⟨79CC875⟩
Gore, P. H., **3**, 101 ⟨72JCS(P1)892⟩
Gorelenko, S. V., **4**, 556 ⟨69ZOB839⟩
Gorelic, L., **5**, 55, 89 ⟨62PCS117⟩
Gorelik, V. P., **4**, 429 ⟨77IZV2384⟩
Goren, Z., **2**, 25 ⟨91JOC7186⟩
Gorenstein, D. G., **5**, 402 ⟨84T3215⟩
Göres, E., **4**, 442 ⟨83PAC1799⟩
Goreva, T. V., **5**, 669 ⟨87BAU204, 87IZV226⟩, 669, 672 ⟨90BAU1007⟩, 669, 672 ⟨90IZV1122⟩
Gorgues, A., **1**, 958 ⟨83T2185⟩; **3**, 61 ⟨81TL625⟩, 274 ⟨74CR(278)1153⟩; **6**, 592 ⟨89CC1520⟩
Gorgues, A. H., **4**, 768 ⟨89MI 417-01⟩
Gorham, B. J., **5**, 567, 630 ⟨89JMC2493⟩
Gorissen, H., **1**, 522 ⟨75JA4409⟩
Gorissen, J., **4**, 794 ⟨78BSB391⟩; **6**, 619 ⟨78BSB391⟩
Gorl, U., **2**, 228 ⟨87AG(E)887⟩
Görls, H., **4**, 685 ⟨93JOM(445)133⟩
Gorman, M., **2**, 123 ⟨55JA568⟩
Görmar, G., **5**, 773 ⟨70JPR601⟩, 856 ⟨70JPR610⟩
Gornowicz, G. A., **2**, 1058 ⟨71JA1714⟩; **3**, 565, 711 ⟨70JOM(25)385⟩, 568 ⟨71JA1714⟩; **4**, 516 ⟨70JOM(25)385⟩, 604, 648 ⟨71JA1714⟩; **5**, 254, 536, 539 ⟨66JOC3439⟩; **6**, 162 ⟨70JOM(25)385⟩
Gorodetsky, M., **1**, 870 ⟨67JA1183⟩
Gorrichon-Guignon, L., **2**, 247, 250 ⟨82S281⟩
Gorrod, T. W., **2**, 782 ⟨68TL6155⟩
Görs, S., **4**, 637 ⟨93JA6025⟩
Gorshkova, G. N., **2**, 1018, 1019 ⟨65ZOR415⟩, 1054 ⟨66JOU1440⟩
Gorshkov, U. N., **1**, 1005 ⟨68JOU21⟩
Gorshunov, I. Y., **2**, 820 ⟨84ZOB1917⟩; **6**, 581 ⟨88ZOB1468⟩
Gorshunov, I. Yu., **2**, 461 ⟨88JGU2161⟩
Gorsi, B. L., **5**, 1144 ⟨77IJC(A)1099⟩
Gorsich, R. D., **2**, 942 ⟨55JA6380⟩
Gorska-Poczopko, J., **6**, 546 ⟨90MI 617-03⟩
Gorski, R. A., **5**, 237 ⟨76TL2577⟩
Gorsmann, C., **3**, 550 ⟨91JA5120⟩
Gorth, H., **4**, 699 ⟨67JOM(9)117⟩
Gortner, R. A., **3**, 616 ⟨42JOC164⟩
Gorys, V., **4**, 217 ⟨90CJC897⟩, 831 ⟨79TL4029⟩
Gorzynski, J. D., **1**, 1095, 1099 ⟨72JA9234⟩
Gorzynski Smith, J., **2**, 82 ⟨84S629⟩
Gosavi, R. K., **3**, 527 ⟨70JA1768⟩
Goscinski, O., **1**, 918 ⟨83JCS(P2)305⟩
Gosden, A. F., **1**, 815 ⟨67TL4537⟩
Gosh, S., **5**, 129 ⟨94JA9009⟩
Goshorn, R. H., **2**, 31 ⟨41OSC(1)36⟩
Gosink, T., **1**, 506 ⟨72JOC2989⟩
Gösl, R., **2**, 374, 379, 384 ⟨73OSC(5)43⟩, 380, 383, 384 ⟨59CB2521⟩, 387 ⟨62AG(E)405⟩
Gosling, K., **2**, 435 ⟨68JCS(A)1909⟩, 461 ⟨65JA2113⟩
Gosling, P. D., **5**, 1129 ⟨75JCS(D)2368⟩
Gosmini, C., **4**, 45 ⟨93JFC(63)157⟩
Gosmini, R., **4**, 405, 408 ⟨91SL111⟩
Gosney, I., **1**, 646, 647, 648 ⟨87CSR45⟩, 648, 738, 743 ⟨88JCS(P2)1829⟩, 737, 738 ⟨73T1697⟩, 737 ⟨B-94MI 116-01, 83T206⟩, 739, 740, 742 ⟨77AG(E)487⟩, 790 ⟨93PHC1⟩, 811 ⟨82CC1164⟩; **2**, 235 ⟨69AJC765, 70T1463⟩, 809, 892 ⟨74JCS(P1)460⟩, 837 ⟨83CC857, 74JCS(P1)466⟩; **3**, 49, 196, 492 ⟨87CSR45⟩, 196, 500, 639 ⟨73T1697⟩, 492, 493 ⟨B-79MI 313-01⟩, 500 ⟨83T205, 88JCS(P2)1829⟩, 552 ⟨70T1493⟩; **6**, 707 ⟨87CSR45⟩, 708 ⟨73T1697⟩, 708, 709 ⟨88JCS(P2)1829⟩
Gospodova, T. S., **5**, 607 ⟨94T7193⟩
Gosselck, G., **5**, 900 ⟨67AG(E)249⟩, 1113 ⟨62CB1237⟩
Gosselck, J., **1**, 241 ⟨68AG(E)456, 78AG(E)274⟩, 1072 ⟨65AG(E)1080⟩; **2**, 288 ⟨58CB2345⟩, 877 ⟨75T2121⟩, 1001 ⟨73T917⟩; **3**, 256 ⟨62CB1237⟩, 493 ⟨67AG(E)249⟩
Gosselin, P., **1**, 881 ⟨61BSF836⟩; **4**, 255 ⟨79JOC2807, 80CR(C)183⟩, 256 ⟨78TL2717⟩; **5**, 82 ⟨84S876⟩, 553, 559, 729 ⟨78TL2715⟩, 558, 559 ⟨80TL2421⟩, 561 ⟨79JOC2807⟩; **6**, 627 ⟨87S267⟩
Gösseln, H. J. von, **6**, 445, 540 ⟨91AP(324)917, 91LA405⟩
Gosser, L., **1**, 854 ⟨64JA5445⟩, 854, 855 ⟨63JA3890⟩, 854, 858, 871, 881 ⟨64JA5457⟩, 858 ⟨68JA2598⟩
Gossick, G. J., **5**, 232 ⟨79TL703⟩
Gössinger, E., **2**, 345 ⟨80TL2229⟩
Gosteli, J., **5**, 905 ⟨78JA8214⟩
Gostevskaya, V. I., **4**, 777 ⟨90ZOR2056⟩
Gostevskii, B. A., **5**, 718 ⟨88BAU378, 88IZV460⟩, 719 ⟨83JGU1851, 83JGU200⟩, 719 ⟨83ZOB229⟩, 719, 720 ⟨83BAU1373, 83IZV1515, 83ZOB2051⟩
Goswami, A., **5**, 370 ⟨91CI(L)353⟩

Goswami, H. C., **6**, 422 ⟨33JIC537⟩
Goswami, R., **1**, 303 ⟨77JA642⟩
Goszczyńska, Z., **6**, 635 ⟨77JOC3629⟩
Goth, H., **3**, 680 ⟨80CB2699⟩
Goti, A., **2**, 320 ⟨90TL3351⟩; **3**, 269 ⟨88JOC2426⟩, 431 ⟨91G285⟩
Gotlieb, H. G., **4**, 176 ⟨87CAR(162)145⟩
Goto, E., **4**, 168 ⟨86CC156⟩
Goto, G., **1**, 734 ⟨75CL103⟩; **2**, 135 ⟨80JA1436, 80JA3663⟩, 135, 136 ⟨80TL3143⟩, 240 ⟨91EUP412699⟩; **3**, 299 ⟨92JCS(P1)3401⟩; **4**, 171 ⟨80TL3463⟩; **5**, 332, 333 ⟨92CPB1177⟩
Goto, H., **1**, 481, 1028 ⟨83CPB4346⟩
Goto, J., **3**, 310 ⟨81JA6990, 87CPB4700⟩, 427 ⟨84JAN532⟩
Goto, K., **1**, 577 ⟨94JOC406⟩, 827 ⟨93CC793⟩; **3**, 71 ⟨90CL1705⟩, 144 ⟨82CL939⟩, 431 ⟨87TL4557⟩; **5**, 370 ⟨90JMC2192⟩
Goto, M., **1**, 242 ⟨89JOC4660⟩; **2**, 913 ⟨93OM988⟩, 941 ⟨93T1189⟩; **3**, 208 ⟨90S677⟩, 353 ⟨89JOC4660, 89TL4825⟩, 366 ⟨90TL7641⟩, 551 ⟨89TL1249⟩; **4**, 837 ⟨89TL1249⟩; **5**, 337, 584 ⟨85CB1668⟩, 510, 551, 583, 620 ⟨82LA1229⟩, 547 ⟨82S457, 81LA1798⟩; **6**, 575 ⟨90JCS(P1)1218⟩
Goto, N., **5**, 1122 ⟨87JCS(P1)2531⟩
Goto, T., **1**, 310 ⟨80CL331⟩, 779, 788 ⟨78CL541⟩, 1009 ⟨91CL1271⟩, 1048 ⟨90SL701, 90TL5499⟩; **2**, 233 ⟨86CL349⟩, 234 ⟨89CL2111⟩, 657 ⟨87H(26)65⟩, 702 ⟨77CL679⟩; **3**, 56 ⟨82JA4997⟩, 140 ⟨79CL1483⟩; **4**, 231, 940 ⟨88TL4773⟩, 367, 937 ⟨80CL331⟩, 419 ⟨90BCJ1160⟩, 419, 420 ⟨84CL1599⟩, 936 ⟨82JA4997, 79TL3465⟩, 937 ⟨81CL457, 81TL239, 81TL4287, 84JA3252, 84JOC3517, 86T2863, 87T4759⟩, 940 ⟨90SL701, 91T3727⟩, 950 ⟨84TL2021⟩; **5**, 152 ⟨86T2863⟩; **6**, 68 ⟨65T2059⟩, 130 ⟨81TL4287, 84TL2021⟩
Goto, Y., **1**, 195 ⟨93TL7619⟩; **2**, 769 ⟨89BCJ128⟩; **5**, 215 ⟨86CPB4984⟩
Gotoh, H., **3**, 557 ⟨80JOC5377⟩, 557, 593 ⟨81JCS(P1)2727⟩; **4**, 365 ⟨84JCS(P1)435⟩; **5**, 303 ⟨74JOC3924⟩
Gotoh, Y., **1**, 108 ⟨87BCJ3307, 91BCJ2008⟩, 109 ⟨87BCJ3823⟩, 481 ⟨82BCJ3555⟩; **2**, 16 ⟨91CC179⟩; **3**, 149 ⟨91CC179⟩; **5**, 205 ⟨87BCJ3307⟩
Gotor, V., **2**, 920 ⟨87CC1195⟩; **3**, 414 ⟨85CB3652, 86CC1179⟩, 431 ⟨91S713, 93S45⟩; **5**, 217 ⟨90SL690⟩, 263 ⟨93T4007⟩, 353 ⟨91S350, 92IJC(B)906⟩, 359 ⟨93S287⟩; **6**, 461 ⟨93T10725⟩, 575 ⟨79CC675⟩
Gots, J. S., **5**, 748 ⟨59JOC1309⟩
Götschi, E., **2**, 824 ⟨71HCA710⟩; **3**, 268 ⟨71HCA710⟩
Gott, G. A., **2**, 881 ⟨87AG(E)264⟩
Gott, P. G., **4**, 836, 887 ⟨71JOC2205, 71JOC2211⟩, 979 ⟨64JOC2513⟩; **5**, 481 ⟨81JOC3911⟩
Gottardi, W., **5**, 320, 321 ⟨73M421⟩, 320, 321 ⟨75M1019, 74M611⟩, 1015 ⟨66M115, 70M655⟩, 1015 ⟨71AG(E)416, 71M264⟩; **6**, 451 ⟨69M1860⟩
Gotteland, J.-P., **1**, 194, 288 ⟨89TL1803⟩, 362 ⟨93JOC4298⟩
Gottfried, R., **5**, 1029 ⟨66AG(E)963⟩
Gotthardt, H., **3**, 255 ⟨85TL1627⟩, 324 ⟨75AG(E)422⟩, 372 ⟨79CB260⟩, 687 ⟨81LA1025⟩; **4**, 835 ⟨85CC1154⟩, 850 ⟨84CHEC(6)848⟩; **5**, 550, 560 ⟨87CB411⟩, 551 ⟨79CB1650, 76CB740⟩, 561 ⟨87CB61, 88CB313⟩, 605 ⟨87CB751⟩, 605, 623 ⟨87LA189⟩; **6**, 114 ⟨76TL3563, 80LA873⟩
Gottlieb, H., **2**, 304 ⟨89S825⟩
Gottlieb, H. E., **3**, 369 ⟨94TL953⟩, 564 ⟨86JOC3176⟩
Gotts, R. A., **6**, 261 ⟨24JCS442⟩
Gottschalk, A., **5**, 62 ⟨B-60MI 502-03⟩
Gottschalk, K., **5**, 384 ⟨79AP822⟩
Gottschalk, P., **1**, 925 ⟨85JOC3498⟩
Gottstein, J., **3**, 367 ⟨81JA7032⟩
Götz, A., **2**, 539 ⟨82S1⟩; **6**, 43 ⟨82S1⟩
Götz, J., **2**, 695 ⟨73CB2738⟩
Götz, V., **3**, 458 ⟨76JOM(118)C3⟩
Götzmann, W., **1**, 986 ⟨86AG(E)340⟩
Gouasmia, A.,**4**, 847 ⟨87BSF823⟩
Goubeau, J., **1**, 58 ⟨55AG710, 57CB2733⟩; **2**, 110 ⟨55ZAAC(282)86⟩; **4**, 143 ⟨57LA(604)168⟩, 603, 632, 635 ⟨71CB1347⟩; **5**, 1015 ⟨60CB1379⟩
Gouda, N., **2**, 83 ⟨90JA1286⟩
Goudarzian, N., **3**, 4 ⟨91SC2077⟩
Goudet, H. J., **5**, 105 ⟨B-45MI 502-03⟩
Goudgaon, N. M., **2**, 306 ⟨89SC2409⟩, 307 ⟨88SC1363⟩, 398 ⟨89SC805⟩; **3**, 426 ⟨88SC693⟩
Gough, A., **1**, 668 ⟨68JCS(A)190⟩
Gough, S. T. D., **1**, 1079 ⟨62JCS2333, 64JCS543⟩; **2**, 252 ⟨83USP4383991⟩, 442 ⟨61JCS4263⟩; **3**, 674 ⟨62JCS2333, 64JCS543⟩
Gougou, M., **4**, 1025 ⟨91TL3687⟩
Gougoutas, J., **4**, 905 ⟨90JMC1510⟩
Gougoutas, J. Z., **1**, 200 ⟨90H(31)1359⟩; **2**, 170 ⟨67JA5838⟩; **4**, 71 ⟨67JA5838⟩, 227 ⟨78JOC79⟩; **5**, 321 ⟨75JOC2129⟩, 360 ⟨91JOC3744⟩
Gouin, L., **1**, 539 ⟨77TL167⟩; **2**, 27 ⟨77S786⟩, 606, 616 ⟨77JCS(P1)1797⟩
Goulaouic, P., **3**, 157 ⟨89H(28)521⟩, 165 ⟨87S503⟩
Gould, D., **2**, 24 ⟨59JA2191⟩
Gould, D. E., **2**, 94 ⟨69JA1310⟩, 98 ⟨70CC216⟩; **6**, 42, 43 ⟨70JOC3730⟩, 428 ⟨69TL723⟩
Gould, E. S., **1**, 437 ⟨54JA1706⟩, 852 ⟨B-59MI 119-01⟩, 853 ⟨B-59MI 119-02⟩; **2**, 287 ⟨51JA3196⟩, 729 ⟨56JA4028⟩, 773 ⟨74JOC407⟩
Gould, I. R., **3**, 679 ⟨84JOC3174⟩, 679, 680 ⟨86JA6739⟩; **5**, 728 ⟨85JOC4415⟩
Gould, K. J., **1**, 126 ⟨75TL1633⟩; **2**, 904 ⟨76JCS(P1)2428⟩
Gould, L. D., **1**, 816 ⟨87JA6199⟩
Gould, R. O., **3**, 686 ⟨86JCR(S)156⟩; **4**, 232 ⟨81JCS(P1)2991⟩; **5**, 781 ⟨87JCS(D)2687⟩
Gould, S. J., **3**, 187 ⟨71JA2492⟩
Goulet, M. T., **1**, 397 ⟨87TL1043⟩
Goument, B., **5**, 283 ⟨87TL2595⟩
Goumont, R., **5**, 952 ⟨91COS(6)485⟩, 953 ⟨92JA8088⟩; **6**, 716 ⟨91CRV35⟩
Gourcy, J. G., **6**, 302 ⟨74CC634⟩
Gourdel, Y., **1**, 158 ⟨93TL1011⟩
Gourdie, A. C., **6**, 131 ⟨88JOM(348)317⟩
Goure, W. F., **1**, 1015 ⟨84JA6417⟩; **4**, 614 ⟨78JA6236⟩
Goure (Monsanto), W. F., **6**, 28 ⟨86USP4575565⟩
Gousson, T., **5**, 753 ⟨64CR4067⟩
Gouverneur, V., **4**, 897 ⟨86RTC456⟩
Gouws, I. D., **1**, 828 ⟨91JCS(P1)561⟩
Gouygou, M., **3**, 608 ⟨91TL3687⟩; **5**, 1092 ⟨89TL177⟩; **6**, 373, 688 ⟨89CC593⟩, 680 ⟨89TL177, 90JOC5750⟩
Gouyon, T., **1**, 653 ⟨90JOM(394)37⟩; **4**, 751 ⟨90JOM(394)37⟩
Gouzoles, F., **4**, 536 ⟨90JCS(P1)2251⟩
Gouzoules, F. H., **1**, 67 ⟨86JOC2024⟩
Govea, L. S., **1**, 879 ⟨89JOC773⟩
Gover, T. A., **2**, 33 ⟨60JA3816⟩
Govindachari, T. R., **4**, 94 ⟨84H(22)585⟩
Govindan, C. K., **3**, 414 ⟨83JOC5348⟩
Govindan, S. V., **1**, 580, 781 ⟨94JOC311⟩, 795 ⟨83JOC3581⟩
Govindaraj, N., **5**, 318 ⟨86JOC4959⟩
Govorov, N. N., **2**, 811 ⟨76IZV1823⟩
Gow, R. S., **5**, 199 ⟨49JCS1555⟩
Gowal, H., **4**, 162 ⟨70HCA1598⟩
Gowan, J. E., **3**, 241 ⟨55JCS862⟩
Gowda, N. M. N., **6**, 396 ⟨82ICA(58)149⟩
Gowenlock, B. G., **1**, 452 ⟨78JCS(D)657⟩; **2**, 391 ⟨79CC392⟩, 563 ⟨78JCS(D)657⟩; **4**, 610 ⟨81CC806⟩

Gowland, B. D., **1**, 213, 478 ⟨80JOC922⟩
Gowland, F. W., **3**, 210 ⟨79JOC3448⟩
Gozdz, A. S., **1**, 949 ⟨91JOC2179⟩
Gozzo, F., **4**, 7 ⟨70JCS(C)178⟩, 737 ⟨71JCS(C)925⟩; **6**, 251 ⟨68CI(M)197⟩, 494 ⟨82JCS(P1)2813⟩, 695 ⟨84T1523⟩
Graab, G., **3**, 616 ⟨71CB3594⟩
Graaf, W. de, **1**, 1018 ⟨88T6699⟩
Grab, L. A., **4**, 173 ⟨84TL5009⟩; **5**, 705 ⟨81TL2731⟩
Grabenko, A. F., **5**, 592 ⟨72ZOR528⟩
Graber, D. R., **2**, 218 ⟨87JOC4620⟩, 742 ⟨72JOC1902⟩, 743 ⟨74JOC1432⟩; **4**, 864 ⟨72JOC1902⟩
Grabetz, K., **6**, 405 ⟨79ZN(B)390⟩
Grabiak, R. C., **6**, 28 ⟨86S224⟩
Grabley, F.-F., **2**, 941 ⟨77TL4307, 80TL4251⟩; **3**, 547 ⟨79LA1715⟩, 548 ⟨77TL4307, 79LA1746⟩, 548 ⟨80CB3024⟩, 553 ⟨80TL4251⟩; **4**, 340 ⟨77LA88⟩, 842 ⟨79LA1715⟩; **5**, 589, 593 ⟨79LA1715⟩, 589, 604 ⟨79LA1702⟩, 589, 604, 605, 614 ⟨80CB3024⟩, 589, 614 ⟨77TL4307⟩, 589, 614 ⟨79LA1746, 80CB3010⟩, 626 ⟨80TL4251⟩
Grabley, S., **6**, 337 ⟨81LA264⟩
Gräbner, H., **5**, 1015 ⟨60CB1379⟩
Graboski, G. B., **2**, 393, 396, 398, 783 ⟨86CRV751⟩, 398 ⟨89OS(68)8⟩
Graboski, G. G., **3**, 192 ⟨86CRV751⟩; **4**, 327 ⟨93OSC(8)550⟩, 906 ⟨85JOC2603, 87JOC4693⟩, 906 ⟨90OS(68)8⟩
Graboski, G. S., **4**, 906 ⟨86JOC1012⟩
Grabowski, E. J. J., **1**, 135 ⟨87JOC4745⟩, 558 ⟨88JA3318⟩; **2**, 528 ⟨91JOC751, 91JOC763⟩, 528 ⟨93JOC2791⟩; **3**, 131 ⟨90SC2683⟩; **4**, 247 ⟨89JOC3718⟩; **6**, 446 ⟨91JOC751⟩
Grabowski, J. J., **4**, 271 ⟨90JA607⟩
Grace, D., **1**, 248 ⟨87CC381⟩
Gracey, B. P., **4**, 681 ⟨90JCS(D)761⟩
Grachev, A. A., **4**, 660 ⟨89MI 414-01⟩, 662 ⟨91MI 414-01⟩
Grachev, M. K., **2**, 410 ⟨76ZOB477, 80ZOB1702⟩; **5**, 775 ⟨91ZOB1581⟩
Gracheva, R. A., **2**, 478 ⟨61ZOB217⟩
Gracia, J., **3**, 310 ⟨91CC1687⟩
Gracza, T., **2**, 318 ⟨91S769⟩
Graczyk, P., **4**, 346 ⟨87PS(30)225⟩; **6**, 116 ⟨88JOC3609⟩, 121 ⟨91TL7329⟩, 124 ⟨91TL4189⟩
Grade, M. M., **2**, 150 ⟨86SC1207⟩
Gradel, J., **5**, 476, 492 ⟨92CB1889⟩
Graden, D. W., **1**, 722 ⟨89TL1361⟩
Gradishar, F. J., **6**, 218 ⟨59USP2875254⟩
Gradova, E. V., **4**, 1045 ⟨86ZOB2743⟩
Gradowska, W., **6**, 111 ⟨80PJC145⟩
Grady, B. T., **4**, 44 ⟨94MIP9322265⟩
Grady, G. L., **1**, 9 ⟨72S483⟩; **2**, 89 ⟨72S483⟩
Graefe, A. F., **2**, 388 ⟨58JOC1230⟩
Graefe, J., **1**, 795 ⟨85AG(E)320⟩
Graf, B., **6**, 82, 86, 88, 302, 303, 317 ⟨72CB3050⟩
Graf, E., **2**, 235 ⟨86EUP165534, 86GEP3443467⟩
Graf, H.-D., **2**, 9 ⟨75AG(E)824⟩
Graf, R., **4**, 310 ⟨68AG(E)172⟩; **5**, 298 ⟨68AG(E)172⟩
Graf, W., **1**, 933, 939 ⟨80HCA2328⟩; **4**, 577, 587 ⟨85OM1233⟩
Grafen, P., **2**, 214 ⟨59CB1177⟩; **5**, 81 ⟨65AG(E)846⟩
Graff, M., **6**, 144 ⟨68JOC1247⟩
Graff, W. L. de, **1**, 795 ⟨68CC35⟩
Gräfing, R., **1**, 141 ⟨78CC596⟩, 1072 ⟨76RTC264⟩; **2**, 907 ⟨78RTC69⟩; **4**, 956, 957 ⟨78RTC69⟩
Grafje, H., **2**, 214 ⟨57AG713⟩
Gräfje, H., **2**, 214, 215 ⟨58LA(614)66⟩, 219 ⟨58LA(617)71⟩
Graf-Poncet, J., **5**, 214 ⟨85HCA1893⟩
Gragg, B. R., **4**, 534 ⟨76IC1988⟩
Graham, A. E., **4**, 250 ⟨92T7265⟩
Graham, D., **1**, 652 ⟨69JCS(B)652⟩
Graham, D. P., **2**, 122 ⟨66JOC957⟩
Graham, D. W., **4**, 923 ⟨71CA(75)88759m⟩
Graham, J. C., **3**, 618 ⟨72CJC3857⟩, 619 ⟨73TL3825⟩
Graham, J. D., **3**, 446 ⟨62JA753⟩
Graham, P. J., **6**, 220 ⟨47JA947⟩
Graham, S. L., **3**, 305 ⟨91JOC4260⟩
Graham, W. A. G., **6**, 520 ⟨76JA4678⟩
Graham, W., **1**, 637 ⟨44JCS322⟩
Graham, W. H., **1**, 636 ⟨67JA1761⟩; **2**, 335 ⟨67JA716⟩; **5**, 763, 866 ⟨65JA4396⟩; **6**, 55 ⟨67JA182⟩, 55, 280 ⟨65JA4396⟩, 77 ⟨69TL2223⟩, 104 ⟨67JA716⟩
Graham, W. R. M., **5**, 1096 ⟨91JCP(95)5612⟩
Grahauskas, V., **2**, 335 ⟨69JA1679, 70JOC1545⟩
Grahn, W., **1**, 981 ⟨82AG(E)924⟩
Graille, J., **3**, 280 ⟨86JOC2128⟩
Grakauskas, V., **2**, 13 ⟨69JOC2446⟩, 608 ⟨70JOC723⟩; **4**, 428 ⟨85JOC1699⟩; **5**, 313, 314, 989 ⟨70JOC1545⟩; **6**, 55 ⟨78JOC3485⟩, 56, 282 ⟨68JOC3080⟩, 149 ⟨81JHC1477⟩, 440 ⟨70JA2096⟩, 649 ⟨69JOC2840⟩
Gramatica, P., **1**, 77 ⟨87T4481, 88T1299⟩; **3**, 18, 134 ⟨86SC803⟩, 58 ⟨87T4481⟩
Grambal, F., **5**, 864 ⟨B-75MI 521-03⟩
Gramlich, V., **4**, 566 ⟨91HCA983⟩
Grammaticakis, M. P., **3**, 448 ⟨47BSF438⟩
Gramstadt, T., **2**, 181, 182 ⟨57JCS2640⟩
Gramza, J., **5**, 1113 ⟨84JOC2057⟩
Gramze, R., **2**, 255 ⟨71ZN(B)386⟩
Granados, R., **3**, 666 ⟨83JOC3105⟩
Gran, H. J., **4**, 64 ⟨53LA(581)133⟩; **6**, 44, 91 ⟨53LA(581)133⟩
Granata, A., **2**, 723 ⟨76TL3001⟩; **6**, 478 ⟨80JOC271⟩
Granberg, K. L., **1**, 193 ⟨89TL617⟩
Grandberg, K. I., **2**, 590 ⟨94JOM(474)1⟩; **4**, 152 ⟨77IZV2417⟩, 696 ⟨74JOM(65)131⟩, 723 ⟨69IZV2030, 69IZV2032⟩, 723 ⟨72IZV2594⟩; **5**, 385 ⟨88ZOB62⟩
Grandclaudon, P., **3**, 269 ⟨86CC839⟩; **4**, 487 ⟨92SC2381⟩; **5**, 516, 522 ⟨87S363⟩; **6**, 122 ⟨92SC2381⟩
Grande, M., **4**, 266 ⟨89AQ22⟩
Grandi, R., **2**, 22 ⟨86JCR(S)426, 92T4579⟩, 28 ⟨86JCR(S)428⟩; **3**, 27 ⟨86JCR(S)426⟩, 234 ⟨88JCR(S)200⟩, 489 ⟨77JOC1352⟩
Grandjean, D., **1**, 431 ⟨93TL1155⟩, 1018 ⟨92TL5355⟩
Grandmaison, J.-L., **2**, 667 ⟨78JOC1435⟩
Granger, R., **2**, 697 ⟨68BSF4265⟩; **5**, 142 ⟨60BSF1350⟩
Grangette, H., **5**, 38 ⟨85S493⟩
Granier, M., **5**, 871, 872, 873, 1148 ⟨90JA6277⟩; **6**, 659, 667 ⟨91CB1739⟩, 665, 667, 668, 669, 670 ⟨90JA6277, 88AG(E)1350⟩
Granik, V. G., **2**, 788 ⟨71ZOR1146⟩; **4**, 987 ⟨87KGS1477⟩; **5**, 742 ⟨83RCR377⟩; **6**, 104 ⟨77UK685⟩
Granitzer, W., **1**, 512 ⟨79TL3145, 86JMC112⟩
Granja, J. R., **1**, 606 ⟨92JOC3173⟩, 816, 817 ⟨85TL4959⟩; **3**, 636 ⟨91JA233⟩; **4**, 280 ⟨91TL2193⟩
Granoth, I., **1**, 80 ⟨76JOC3682⟩, 445 ⟨80PS(8)243⟩; **2**, 824 ⟨78JA5130⟩
Grant, A. S., **6**, 646 ⟨91AG(E)1356⟩
Grant, B., **1**, 187 ⟨90JA6615⟩, 513 ⟨74JOC968⟩
Grant, B. E., **2**, 912 ⟨93JA2151⟩; **4**, 709 ⟨92OM2128⟩, 709, 712 ⟨90OM1345, 91OM3003⟩, 713 ⟨90IC1295⟩
Grant, D., **1**, 631 ⟨71JCS(C)3173⟩, 664 ⟨76JCS(P1)234⟩
Grant, D. W., **2**, 879 ⟨80JOM(188)345⟩
Grant, E. B., **3**, 543 ⟨90TL3677⟩
Grant, G. A., **2**, 257 ⟨56CJC1819⟩
Grant, Jr., L. R., **2**, 481, 491, 510 ⟨59JA1⟩; **6**, 283 ⟨78USP4120904⟩
Grant, R., **3**, 685 ⟨84AJC1231⟩
Grant, R. D., **4**, 107 ⟨85AJC1505⟩
Grant, R. W., **2**, 355 ⟨73JOC1813⟩
Grapov, A. F., **2**, 468, 469, 470 ⟨71ZOB1644⟩; **4**, 131 ⟨68JGU2563⟩; **5**, 644 ⟨87ZOB685⟩; **6**, 635 ⟨90ZOB798⟩
Gras, J.-L., **1**, 498 ⟨72TL1189⟩, 499 ⟨82JCR(S)300⟩; **3**, 216 ⟨75T849⟩, 219 ⟨78TL2955⟩, 238 ⟨83BSF(27)89⟩; **4**, 185 ⟨87TL6601⟩, 199 ⟨76TL809⟩

Grashey, R., **3**, 641, 642, 649, 661, 662, 664, 671 ⟨91COS(6)225⟩; **4**, 436 ⟨65JOC74⟩; **6**, 573 ⟨72TL2939⟩
Graske, K., **3**, 703, 722 ⟨88CB1445⟩; **6**, 611 ⟨88CB1445⟩
Graske, K.-D., **3**, 270 ⟨88LA891⟩; **5**, 138 ⟨90MI 503-01⟩, 159 ⟨88LA891⟩
Grass, F., **2**, 3 ⟨87TL5505⟩
Grassberger, M. A., **1**, 727 ⟨70LA(739)211⟩; **2**, 526 ⟨68LA(719)169⟩, 1083 ⟨84HOU(13/3c)1⟩; **3**, 494 ⟨70LA(739)211⟩; **5**, 1141 ⟨70LA(739)211⟩
Grassberger, M., **6**, 183 ⟨84HOU(13/3c)162⟩
Grasselli, J. G., **5**, 389 ⟨69ICA495⟩
Grasselli, P., **1**, 18, 19 ⟨64CI(L)153⟩, 78 ⟨88JCS(P1)3061⟩, 630 ⟨64G537⟩, 679 ⟨70T1281⟩, 680 ⟨67TL1581⟩; **2**, 359 ⟨81TL5073⟩, 1006 ⟨70T1281⟩; **4**, 146 ⟨71T6109⟩, 669 ⟨70T1281⟩
Grassl, Y. N., **1**, 209 ⟨93JOC7304⟩
Grasso, C. P., **5**, 652 ⟨73USP395444;, 965 ⟨76USP3954441⟩
Grate, J. H., **4**, 1016 ⟨76AG(E)171⟩
Grate, J. W., **3**, 545 ⟨85JA3728⟩
Grattan, T. J., **1**, 826 ⟨90JCS(P1)11⟩; **2**, 622 ⟨90JCS(P1)11⟩; **4**, 14 ⟨90JCS(P1)3317⟩
Graul, O., **5**, 838 ⟨1898CB2854⟩
Graul, S. T., **1**, 846 ⟨88JA607⟩
Graule, T., **1**, 508 ⟨86AG(E)167⟩
Graumann, J., **3**, 434 ⟨84LA1545⟩; **5**, 92 ⟨82CZ345⟩
Grauvogl, G., **6**, 371 ⟨93OM2757⟩
Gravel, D., **1**, 340 ⟨94TL4531⟩, 340, 1127 ⟨93TL2091⟩; **3**, 37 ⟨90TL63⟩
Gravens, P. D., **2**, 565 ⟨82COMC-I(3)281, 82COMC-I(3)433⟩, 565, 575 ⟨82COMC-I(3)331⟩
Graves, D. M., **3**, 148 ⟨89JOC661⟩
Gravestock, M. B., **2**, 307, 345, 346, 347, 349 ⟨93CC169⟩; **3**, 219 ⟨71JA4332⟩
Gravier, D., **4**, 420 ⟨92PHA754⟩
Gray, C. E., **1**, 667 ⟨69JOM(18)P1, 70JOC1303⟩; **5**, 792, 794 ⟨70MI 520-02⟩
Gray, D., **2**, 34, 628 ⟨87TL4497⟩, 34, 629, 631 ⟨88JOC4477⟩
Gray, D. N., **4**, 757, 759, 820 ⟨58JOC1661⟩; **5**, 361 ⟨71JOC350⟩
Gray, G. R., **4**, 253 ⟨77TL1617, 78JA3548⟩
Gray, G. W., **3**, 293 ⟨55JCS2686⟩; **5**, 108 ⟨66JCS(C)840⟩
Gray, H. L., **2**, 134 ⟨20JA856⟩
Gray, R. T., **2**, 843 ⟨82JCS(P1)2289⟩; **3**, 208 ⟨70JOC753⟩
Gray, R. W., **3**, 355 ⟨81HCA2369⟩
Gray, S. K., **1**, 1143 ⟨81JA1904⟩
Gray, Y. M., **2**, 904 ⟨59ZN(B)809⟩
Graybill, B. M., **6**, 144, 145, 369 ⟨63T(S)177⟩
Grayson, E. J., **4**, 139 ⟨88T4087⟩
Grayson, I., **4**, 650 ⟨77JCS(P1)550⟩
Grayson, J. I., **1**, 520, 523 ⟨81S753⟩, 643 ⟨77JCS(P1)1452⟩, 643, 755 ⟨77JCS(P1)550⟩; **3**, 35 ⟨77JCS(P1)2263⟩; **4**, 346, 347 ⟨77JCS(P1)2263⟩, 923, 924 ⟨87JCS(P1)967⟩; **5**, 199 ⟨81S753⟩; **6**, 121 ⟨77JCS(P1)2263⟩
Grayson, M., **2**, 442 ⟨59JA4803, 63JA79⟩, 442 ⟨66USP3243450⟩, 444 ⟨60JA3919⟩, 448 ⟨61JOC4770⟩, 460, 463 ⟨67JOC236⟩, 840, 841, 854 ⟨67T1065⟩; **3**, 494 ⟨60JA3919⟩; **4**, 451 ⟨B-64MI 410-01⟩; **5**, 943 ⟨90OM2621⟩
Grayston, M. W., **1**, 173, 587, 688 ⟨78JOC3249⟩; **5**, 866 ⟨76JA1278⟩
Graziani, M., **3**, 92 ⟨80JOM(194)221⟩, 314 ⟨72JOM(46)C29⟩
Graziani, R., **6**, 519 ⟨87JCS(D)2039⟩
Graziano, M. L., **3**, 221 ⟨83S125⟩, 327 ⟨94JCS(P1)147⟩; **6**, 76, 77 ⟨85S762⟩, 107 ⟨68G681, 72JHC1087⟩, 107 ⟨74JHC529, 76JHC205⟩, 107 ⟨87JCR(S)362⟩, 145 ⟨86JHC553⟩
Grazioli, A., **6**, 251 ⟨72GEP2111696⟩
Grdenić, D., **4**, 553 ⟨64JCS513⟩, 697 ⟨91JOM(405)59⟩; **6**, 207 ⟨82CSC1571, 82JOM(238)327⟩, 207 ⟨83JOM(253)283, 84JOM(276)1⟩, 207 ⟨86JOM(306)1, 87JOM(319)1⟩, 207 ⟨91JOM(411)19⟩, 405 ⟨74CC646, 78JOM(153)1⟩
Grdinic, M., **2**, 232, 724 ⟨73IJS307⟩
Gream, G. E., **1**, 905 ⟨66PAC25⟩
Greasley, P. M., **1**, 448 ⟨64JCS4804⟩
Greaves, P. M., **1**, 965 ⟨66JCS(C)1223⟩, 1056 ⟨75T3073⟩; **3**, 649 ⟨68JCS(C)291⟩
Grebenyuk, A. D., **2**, 392 ⟨90JOU583⟩
Greber, G., **5**, 977 ⟨68AG(E)941⟩, 1012 ⟨71CB3131⟩
Grébille, Y., **1**, 628 ⟨92S383⟩
Grechkin, N. P., **2**, 410 ⟨67IZV1606⟩
Greci, L., **2**, 803 ⟨72G527⟩
Greck, C., **1**, 697, 700, 703 ⟨87TL5759⟩; **3**, 181 ⟨91TL2359⟩; **4**, 286 ⟨89T7161⟩
Greco, A., **1**, 519 ⟨68JOC3948⟩; **4**, 617 ⟨87TL3719⟩
Greco, C. C., **2**, 123 ⟨68JOC1275⟩
Greco, M. N., **1**, 83 ⟨80JA1165⟩
Gree, M., **5**, 948, 949 ⟨73JCS(D)975⟩
Grée, R., **1**, 963 ⟨92BSF151⟩, 966 ⟨92AG(E)224, 93T9775⟩; **2**, 47 ⟨89AG(E)755⟩; **3**, 62 ⟨86TL4983⟩
Greeley, A. C., **1**, 811 ⟨86JA5901⟩
Green, A., **6**, 426 ⟨90JAP2209879⟩
Green, A. G., **3**, 358 ⟨03JCS1201⟩
Green, J. H. S., **2**, 956 ⟨66JCS(C)106⟩
Green, D., **4**, 526, 527, 528 ⟨92TL4209⟩
Green, D. C., **4**, 874, 875 ⟨76CC148⟩, 957 ⟨79JOC1476⟩; **6**, 590 ⟨76CC148⟩
Green, D. E., **5**, 28 ⟨B-63MI 502-01⟩
Green, D. L. C., **1**, 165, 879 ⟨91T4223⟩
Green, D. M., **4**, 56, 57 ⟨64JCS766⟩
Green, G., **4**, 165 ⟨84JCS(P1)681⟩
Green, G. F. H., **3**, 208, 234 ⟨61JCS2532⟩
Green, III, F. R., **1**, 1073 ⟨78JA4852⟩; **5**, 174 ⟨78JA4858⟩
Green, I. R., **3**, 297 ⟨84JCS(P1)1339⟩
Green, J. A., **3**, 596 ⟨69TL5093⟩
Green, J., **2**, 103 ⟨60JCS225⟩
Green, J. R., **1**, 196 ⟨93TL4497⟩
Green, J. V., **5**, 1044 ⟨94JMC170⟩
Green, J. W., **5**, 24 ⟨B-57MI 502-01⟩
Green, K. A., **4**, 713 ⟨88OM2450⟩
Green, K., **3**, 200 ⟨89TL4807⟩
Green, M., **2**, 448 ⟨65JCS541⟩, 859 ⟨66JCS(C)1592⟩, 883, 890 ⟨68JCS(A)483⟩; **3**, 552 ⟨70JCS(A)944⟩; **4**, 347 ⟨63JCS1324⟩, 676 ⟨84JCS(D)2455, 86JCS(D)657⟩, 676 ⟨88JCS(D)1843⟩, 685 ⟨69JCS(A)1872⟩, 686 ⟨82JCS(D)1309⟩, 689, 710 ⟨85JCS(D)2483⟩, 691 ⟨75CC451⟩, 692 ⟨80CC1281, 80JCS(D)2182⟩, 712 ⟨84JCS(D)2553⟩, 716 ⟨85JCS(D)2009⟩; **5**, 452 ⟨71JCS(A)367⟩, 465 ⟨74JCS(D)421⟩, 952 ⟨80JA7811⟩, 1139 ⟨70JCS(A)1847⟩; **6**, 340 ⟨82CC51⟩, 343 ⟨82CC773, 83JCS(D)2599⟩, 343 ⟨83JOM(249)273⟩, 603 ⟨65JCS5774⟩, 664 ⟨72JCS(D)1303⟩, 719 ⟨74JCS(D)1189⟩, 719, 732 ⟨74JCS(D)351⟩, 721 ⟨74JCS(D)357⟩, 721, 732 ⟨75JCS(D)939⟩
Green, M. E., **2**, 473 ⟨91CB2457⟩
Green, M. J., **1**, 779 ⟨67JCS(C)532⟩; **6**, 80 ⟨77USP4021459⟩
Green, M. L. H., **1**, 885 ⟨B-67MI 119-02⟩; **2**, 572 ⟨74JCS(D)1324⟩, 597 ⟨90POL2763⟩, 962 ⟨72CC1114⟩; **4**, 673 ⟨80JCS(D)29⟩, 675 ⟨89CC418⟩, 676 ⟨90OM886⟩; **5**, 937 ⟨71JCS(A)794⟩
Green, M. M., **1**, 927 ⟨86JA2381⟩
Green, N. J., **4**, 513 ⟨92JOC6037⟩
Green, R., **4**, 316 ⟨54JA3969⟩
Green, R. H., **3**, 194 ⟨93CC94⟩; **4**, 171, 176 ⟨91CSR211⟩, 176 ⟨91CSR271⟩; **5**, 262 ⟨B-89MI 506-02⟩
Green, S. M., **6**, 642 ⟨65JMC446⟩
Green, T. W., **2**, 658, 683 ⟨B-91MI 213-01⟩

Green, W. H., **4**, 5 ⟨51JA1329⟩
Greenberg, A., **1**, 876 ⟨B-88MI 119-01⟩; **6**, 212 ⟨B-88MI 607-01⟩
Greenberg, B., **2**, 729 ⟨56JA4028⟩
Greenberg, D. M., **5**, 83 ⟨B-51MI 502-02⟩
Greenberg, H., **4**, 316 ⟨54JA3969⟩
Greenberg, S., **6**, 71 ⟨73JA4016⟩
Greenblatt, J., **1**, 1111 ⟨79JA1343, 79JA3967⟩
Greenburg, R. B., **1**, 391 ⟨40JA2590⟩
Greene, A. E., **1**, 2, 6 ⟨79JA4003⟩, 2, 6 ⟨85JOC3957, 83JA2435⟩, 187 ⟨88TL5661⟩, 314 ⟨84JOC931⟩, 530 ⟨90JA9388⟩, 571 ⟨80JOC2036⟩; **2**, 676, 1026 ⟨87JOC2919⟩, 682 ⟨87JOC2303⟩, 1027 ⟨90TL7597⟩; **3**, 40 ⟨91SC1⟩, 145 ⟨83JOC3837⟩, 234 ⟨80JOC2036⟩; **4**, 763 ⟨94JA2153⟩; **5**, 49 ⟨89OS(68)41⟩
Greene, F. D., **1**, 627, 628 ⟨56JA790⟩; **2**, 378, 381 ⟨69JOC2254⟩, 403 ⟨69JOC2263⟩, 403, 404 ⟨70JOC2482⟩, 403, 404, 405 ⟨75JOC1409⟩, 764 ⟨69JOC2269⟩; **5**, 208 ⟨63JOC2168⟩, 354, 359 ⟨73TL4091⟩; **6**, 445 ⟨78JOC4530⟩, 450, 509 ⟨69JOC2254, 69JOC2263⟩, 509 ⟨78JOC922⟩
Greene, J., **4**, 622 ⟨78JA6362⟩
Greene, Jr., J. L., **1**, 321, 591 ⟨76JA951⟩, 600 ⟨78JA3874⟩
Greene, J. M., **4**, 321 ⟨66JOC556⟩
Greene, M. G., **6**, 206 ⟨74JA5640⟩
Greene, R. M. E., **1**, 674 ⟨85CC874, 87JA899⟩
Greene, R. N., **1**, 795 ⟨67JA1404⟩, 870 ⟨68CC1225⟩
Greene, T. W., **2**, 539 ⟨B-91MI 210-02⟩; **3**, 9, 16 ⟨B-91MI 301-01⟩, 123, 132 ⟨B-91MI 304-02⟩; **4**, 176, 180, 184, 185, 186, 198 ⟨B-91MI 404-01⟩, 176 ⟨B-91MI 404-02⟩, 203 ⟨B-91MI 404-03⟩, 209 ⟨B-91MI 404-04⟩; **5**, 22 ⟨B-91MI 507-01⟩, 223 ⟨B-91MI 504-01⟩, 260 ⟨B-91MI 506-02⟩; **6**, 487 ⟨B-91MI 615-03⟩
Greengrass, C. W., **4**, 274 ⟨90JCS(P1)773⟩
Greengrass, G. W., **5**, 905 ⟨78JA8214⟩
Greenhalgh, R., **2**, 108 ⟨67CJC495⟩; **6**, 646 ⟨61CJC1017⟩
Greenhill, J. V., **1**, 260, 263 ⟨91HCA1924⟩; **2**, 744 ⟨71JCS(C)2699⟩; **3**, 261 ⟨77CSR277⟩, 263, 266 ⟨69JCS(B)299⟩, 451 ⟨84CHEC(5)305⟩; **4**, 318, 322 ⟨91HCA1924⟩, 329 ⟨94S107⟩; **5**, 597 ⟨90JCS(P1)1869⟩; **6**, 641 ⟨B-93MI 621-01⟩
Greenhouse, R., **3**, 307 ⟨82JOC3668⟩, 673 ⟨85TL1827⟩; **4**, 410 ⟨85H(23)1127⟩
Greenlee, K. W., **1**, 508 ⟨43JA2020⟩; **2**, 606, 615 ⟨51JA3329⟩
Greenlee, W. J., **3**, 197 ⟨87TL611⟩, 198 ⟨89JMC1886⟩, 615 ⟨83TL4559⟩, 617 ⟨84JOC2632⟩; **5**, 36 ⟨81JOC5351⟩, 370 ⟨89JMC1886⟩
Greenlimb, G. E., **4**, 36 ⟨75JOC2796⟩
Greenlimb, P. E., **2**, 611 ⟨73JFC(3)447⟩; **3**, 495 ⟨75JOC2796⟩; **5**, 894 ⟨75JOC2796⟩; **6**, 695 ⟨75JOC2796⟩
Greenshields, J. N., **5**, 966 ⟨69BRP1152877⟩
Greenspoon, N., **3**, 129 ⟨87JOC5570⟩, 621 ⟨86JA7314⟩
Greenstein, J. P., **5**, 83 ⟨44MI 502-01⟩, 84 ⟨54MI 502-04, 57ANY(69)94⟩
Greenstein, T. P., **5**, 83 ⟨B-61MI 502-01⟩
Greenwood, F. L., **2**, 27 ⟨63OSC(4)108⟩; **5**, 1039 ⟨51JA4495⟩
Greenwood, N. N., **2**, 278 ⟨66CI(L)379⟩, 421 ⟨66JCS(A)706⟩; **6**, 523, 525 ⟨B-84MI 616-01⟩
Greenwood, T. D., **5**, 342 ⟨92JOC5328⟩
Greer, C. M., **5**, 303 ⟨53JA5905⟩
Greer, D., **2**, 530 ⟨92TL4209⟩
Greer, F., **5**, 748 ⟨59JOC1309⟩
Greer, S., **2**, 517 ⟨79JA2352⟩
Greeve, A. E., **3**, 236 ⟨87JA4752⟩
Greeves, N., **1**, 643 ⟨84TL357, 91JCS(P1)1485⟩, 643 ⟨92JCS(P1)3407⟩, 757 ⟨87JCS(P1)2569⟩; **2**, 56, 60 ⟨91COS(8)1⟩, 62, 82 ⟨91COS(8)9⟩; **3**, 197 ⟨87JCS(P1)2569⟩; **6**, 461 ⟨93S1109⟩
Gref, A., **2**, 212, 225 ⟨92T1883⟩
Gregg, D. C., **2**, 193 ⟨53JOC36⟩
Grego, A. M., **5**, 769 ⟨89MI 519-01⟩
Gregor, I. K., **4**, 564 ⟨69ZN(B)1081⟩
Gregor, V., **2**, 782 ⟨59CCC1158⟩
Gregorcic, A., **2**, 18 ⟨87BCJ3083⟩; **4**, 751 ⟨87JFC(37)313⟩
Gregory, J. A., **1**, 654 ⟨86CC620⟩; **4**, 139 ⟨88JCS(P1)2585⟩
Gregory, J. T., **5**, 1106 ⟨52JA1323⟩
Gregory, S., **4**, 656 ⟨88OM1208⟩
Grehn, L., **2**, 299 ⟨87S275, 91ACR285⟩; **5**, 689 ⟨93JCS(P2)655⟩, 977 ⟨88S992⟩
Greiber, D., **3**, 260 ⟨62LA(659)64⟩
Greibrokk, T., **4**, 25 ⟨73TL1367⟩; **5**, 985 ⟨73ACS3368⟩
Greiciute, D., **2**, 712 ⟨77RCR548⟩
Greidanus, J. W., **2**, 1030 ⟨56RTC1459⟩; **3**, 343, 344, 345, 349 ⟨70CJC3530⟩, 343, 349 ⟨69CJC3715⟩
Greif, D., **3**, 354 ⟨88ZC345⟩
Greig, D. G. T., **1**, 595 ⟨70CC1683⟩; **5**, 278 ⟨71JCS(C)3540⟩
Greig, D. J., **3**, 686 ⟨82TL5453⟩; **4**, 850 ⟨85CC696⟩
Greig, M. E., **2**, 103 ⟨60MI 202-03⟩
Grein, F., **1**, 538 ⟨86TL5455⟩
Greiner, I., **5**, 567 ⟨93JOC6076⟩
Greiner, U., **4**, 975 ⟨90LA965⟩; **6**, 321 ⟨90LA965⟩, 728 ⟨79S339⟩
Greisinger, R., **1**, 799 ⟨69JA1578⟩
Greissinger, D., **2**, 512 ⟨82JOM(228)C29⟩
Grekov, L. I., **2**, 440 ⟨85URP1145022⟩
Grelbig, T., **3**, 689 ⟨88AG(E)1534⟩; **4**, 868 ⟨88AG(E)1534⟩
Grelier, S., **1**, 474 ⟨90BSF401⟩; **6**, 162, 163 ⟨91JOM(419)C1⟩
Gremban, R. S., **3**, 698, 707 ⟨84TL3259⟩
Grenan, M. M., **2**, 129 ⟨72JMC968⟩
Grenda, V. J., **5**, 762, 816 ⟨65JOC259⟩
Grendze, M. P., **2**, 1007 ⟨87JA4749⟩
Grenier, M. F., **4**, 1079 ⟨81T2515, B-78MI 408-01⟩
Grenier-Loustalot, M., **3**, 234 ⟨76S33⟩
Grenouillat, D., **5**, 282 ⟨87TL5823⟩; **6**, 643, 645 ⟨90T1839⟩
Grese, T. A., **1**, 289 ⟨91JA7363⟩, 289, 362 ⟨91JA7350⟩, 289, 362 ⟨92JOC686⟩, 1039 ⟨93JOC2468⟩
Gresham, T. L., **5**, 1106 ⟨52JA1323⟩
Gresham, W. F., **6**, 524 ⟨49USP2473993⟩
Gresser, G., **3**, 688 ⟨81ZN(B)16⟩; **5**, 1154 ⟨81ZN(B)16⟩; **6**, 684 ⟨80ZAAC(463)144⟩
Gressier, J.-C., **5**, 546, 550 ⟨85BSF891⟩, 547, 550 ⟨76BSF1145⟩, 561 ⟨75MI 513-01, 81S963⟩
Grethe, G., **4**, 170 ⟨83JOC5315⟩
Greulich, P., **6**, 118 ⟨93HCA2407⟩
Greuter, H., **1**, 831 ⟨88TL3291⟩; **3**, 602 ⟨85JA2192⟩; **4**, 830 ⟨78HCA1784⟩
Greve, B., **1**, 991 ⟨94OM94⟩; **4**, 723 ⟨94OM94⟩
Greve, H., **2**, 188 ⟨77LA1457⟩
Grevels, F.-W., **6**, 338 ⟨88OM812⟩, 524 ⟨B-78MI 616-01⟩
Grewal, G., **4**, 187 ⟨92JOC2084⟩
Grewe, F., **6**, 236 ⟨64AG807⟩, 273, 274 ⟨70GEP1908680⟩, 316 ⟨68BRP1097237⟩
Grewe, R., **3**, 61 ⟨58CB2452⟩, 613 ⟨52LA(575)1⟩
Grey, R. A., **3**, 288 ⟨84JOC2288⟩; **4**, 582 ⟨77IC3187⟩
Grezzo, L. A., **2**, 388 ⟨81JOC2402⟩
Gribasova, N. V., **2**, 1090 ⟨87JGU2482⟩
Gribble, G. W., **1**, 13 ⟨77S172⟩, 19 ⟨78S763, 85JOC5451⟩, 102 ⟨77S859⟩, 475 ⟨85S756⟩; **2**, 300, 310 ⟨87S709⟩; **3**, 309 ⟨89JOC3264, 89JOC4350⟩, 619 ⟨83OPP297⟩; **4**, 165 ⟨84JOC4518⟩
Gribbs, R. H., **3**, 545 ⟨86JA3318⟩
Gribov, B. G., **6**, 149 ⟨62HOU(5/3)735⟩
Grice, N., **1**, 898 ⟨71JA6688⟩; **6**, 517 ⟨79JA1627⟩
Grice, P., **1**, 697, 700, 703 ⟨87TL5759⟩, 697, 700, 703 ⟨87TL5763⟩; **4**, 286 ⟨89T7161⟩
Gridnev, I. D., **2**, 901, 902 ⟨93JOC5351⟩
Griebenow, W., **2**, 745 ⟨74CB1545⟩

Grieco, P. A., **1**, 2, 6 ⟨72JOC2363⟩, 17 ⟨86TL4813⟩, 25, 630 ⟨87JOC3346⟩, 190 ⟨93CC510⟩, 190, 402 ⟨91JA5488⟩, 190, 402 ⟨92TL4735⟩, 209, 651 ⟨75CC537⟩, 210 ⟨74JOC732⟩, 254 ⟨92TL4401⟩, 373 ⟨83TL1897⟩, 409, 821 ⟨73JOC2572⟩, 598, 601 ⟨74TL1097⟩, 601, 602, 607 ⟨74JOC120⟩, 601, 604 ⟨75JA1597⟩, 602, 607 ⟨75JOC542, 76JA1612⟩, 604 ⟨80JOC2247, 80JOC3537⟩, 606 ⟨84CC1345⟩, 607 ⟨76CC582⟩, 608 ⟨77JA5210⟩, 781 ⟨76JOC1485⟩, 821 ⟨72CC702⟩, 826 ⟨89JOC515, 89JOC5849⟩, 836 ⟨80JA7587⟩; **2**, 139 ⟨74TL1097⟩, 196 ⟨82JA5781⟩, 196 ⟨86JA5908⟩, 279 ⟨76JOC1485, 81JOC1215⟩, 629 ⟨79JA4749⟩, 660 ⟨78TL2545⟩; **3**, 173 ⟨78JOC4552⟩, 215 ⟨87JOC3346⟩, 221 ⟨89JOC6008⟩, 308 ⟨81JOC3760⟩, 644, 656 ⟨77JA5210⟩; **5**, 68 ⟨73CC500⟩, 239 ⟨78JOC1283⟩, 239, 281 ⟨81JOC1215⟩, 282 ⟨84JA6414⟩, 652 ⟨79JOC2945⟩
Griedel, B. D., **4**, 385 ⟨90JOC5428, 91JOC5491⟩
Griengl, H., **3**, 628 ⟨93TL4623⟩; **4**, 415 ⟨80LA1573⟩; **5**, 118 ⟨91TL341⟩; **6**, 489 ⟨88JMC1831⟩
Grier, M. C., **3**, 31 ⟨93JOC6543⟩
Grierson, D., **2**, 357 ⟨90OR(39)85⟩
Grierson, D. S., **1**, 238 ⟨80JA1064⟩, 368 ⟨88H(27)1575⟩, 818 ⟨92TL4563⟩, 833 ⟨87TL6461⟩, 834 ⟨90TL523⟩; **2**, 328 ⟨88H(27)1575⟩, 331 ⟨88TL3311⟩; **3**, 355 ⟨87TL6461⟩, 637 ⟨80JA1064⟩, 638 ⟨87TL6179⟩
Griesbaum, K., **1**, 9 ⟨83TL1147⟩, 948 ⟨70AG(E)273⟩; **2**, 137 ⟨63JOC1952⟩, 137, 710, 712 ⟨70AG(E)273⟩; **3**, 148 ⟨81CB1858⟩, 148 ⟨83CB2641⟩, 215 ⟨88CB1795⟩, 326 ⟨68CB463⟩; **4**, 15 ⟨67JOC2651⟩, 15, 24, 29 ⟨65JA3151⟩; **5**, 141 ⟨78AG(E)939⟩
Griesbeck, A., **1**, 230 ⟨87HCA1326⟩; **5**, 161 ⟨87HCA1320⟩; **6**, 79 ⟨87HCA1320⟩
Grieshaber, P., **6**, 104 ⟨68BRP1097237⟩
Griesinger, A., **2**, 757 ⟨63LA(665)101⟩; **4**, 462, 478 ⟨63LA(665)91⟩
Grieson, J. R., **3**, 239 ⟨82CJC210⟩
Griesser, H., **3**, 179, 184 ⟨85LA785⟩; **5**, 15 ⟨89S832⟩
Griessmann, K.-H., **4**, 580, 582 ⟨83ZN(B)711⟩
Grifantini, M., **4**, 206 ⟨81JMC1181⟩
Griffel, F., **2**, 110 ⟨55ZAAC(282)86⟩
Griffin, C. E., **2**, 778 ⟨60JCS1398⟩, 783 ⟨63JA2683⟩, 841 ⟨65JOC97⟩, 849 ⟨66JOC2455, 66TL5049⟩, 849 ⟨68JOC632⟩; **4**, 36, 798 ⟨60JCS1398⟩, 473 ⟨65JA4757⟩
Griffin, D. M., **6**, 36 ⟨52JA2292⟩
Griffin, G. W., **1**, 365 ⟨B-76MI 108-02⟩; **3**, 313 ⟨80JA1504⟩, 314 ⟨70JA1402⟩, 538 ⟨62JA4148⟩
Griffin, J. H., **1**, 299, 353 ⟨84OM44⟩, 400, 823 ⟨84JOC1917⟩
Griffin, P. A., **4**, 607, 622 ⟨75JCS(D)1434⟩; **6**, 191 ⟨75JCS(D)1434⟩
Griffin, R. N., **1**, 883 ⟨61JA1951⟩
Griffin, T. S., **2**, 279, 292 ⟨73JA197⟩
Griffin, W., **2**, 177 ⟨52JCS3334⟩
Griffith, D. L., **4**, 9 ⟨69JA1386⟩
Griffith, E. A. H., **2**, 954 ⟨83OM101⟩
Griffith, E. J., **4**, 451 ⟨B-64MI 410-01⟩
Griffith, J. Q., **5**, 83 ⟨B-55MI 502-02⟩
Griffith, M. G., **1**, 95 ⟨68JA3590⟩
Griffith, R., **3**, 421 ⟨90JOC1086⟩
Griffith, W. P., **3**, 5 ⟨90MI 301-01⟩, 5, 117 ⟨87CC1625⟩, 89 ⟨92SC1967, 90MI 304-01⟩; **4**, 165 ⟨84JCS(P1)681⟩
Griffiths, D. C., **2**, 484 ⟨32JCS2284⟩
Griffiths, D. V., **3**, 498 ⟨89JCS(P1)2425⟩; **4**, 558 ⟨86CC871, 92JCS(P1)479⟩, 922 ⟨81JCS(P1)1363⟩
Griffiths, E. J., **2**, 452 ⟨62JA3442⟩
Griffiths, G., **4**, 861 ⟨87JCS(P2)1253⟩
Griffiths, G. J., **4**, 213 ⟨84JCS(P1)1531⟩; **5**, 79 ⟨84JCS(P1)1531⟩
Griffiths, J., **3**, 540 ⟨68JA5296⟩
Griffiths, J. E., **5**, 1016 ⟨62CJC739⟩; **6**, 246 ⟨60JA5759⟩
Griffiths, P. A., **3**, 498 ⟨89JCS(P1)2425⟩; **4**, 558 ⟨92JCS(P1)479⟩
Griffiths, R. W., **5**, 419 ⟨82JOM(225)331⟩
Grigat, E., **3**, 722 ⟨66CB2361⟩; **5**, 1005 ⟨69AG(E)607⟩, 1100 ⟨67AG(E)206, 72AG(E)949⟩, 1102 ⟨64CB3012⟩; **6**, 616, 618 ⟨72AG(E)949⟩, 629 ⟨64CB3022⟩, 632 ⟨64CB3027⟩
Grigat, H., **5**, 211, 322 ⟨86AP(319)1037⟩
Grigg, R., **1**, 369, 524 ⟨85T3547⟩, 468 ⟨85CC971⟩, 494 ⟨91TL687, 94T359⟩, 546 ⟨93TL153⟩
Grigg, R.,**1**, 546 ⟨93TL157⟩
Grigg, R., **1**, 803 ⟨71CC1248⟩, 809 ⟨73JCS(P1)1913⟩, 837 ⟨93T5177⟩, 839 ⟨91TL279⟩, 1015 ⟨88TL5565⟩; **2**, 158 ⟨89JCS(P1)1550⟩, 158 ⟨91T8297⟩, 245 ⟨88T4941⟩, 352 ⟨91T4007, 92T6929⟩; **3**, 414 ⟨83S1009⟩, 625 ⟨81TL4107⟩; **4**, 193 ⟨84JCS(P1)1643⟩, 311 ⟨85CC1183⟩; **5**, 603, 621 ⟨88T7271⟩; **6**, 7 ⟨87JCS(P1)1515⟩
Grigna-Dubois, M., **1**, 284 ⟨76S737⟩
Grignard, V., **2**, 557 ⟨00CR(130)1322⟩, 1019 ⟨26AC(P)3⟩; **3**, 614 ⟨11MI 318-01, 12MI 318-01⟩, 614 ⟨14MI 318-01, 15AC(R)28⟩, 614 ⟨20AC(R)364, 26BSF1589⟩, 675 ⟨15BSF228, 26AC(P)5⟩; **4**, 59 ⟨19CR(169)1143⟩; **6**, 412, 413 ⟨19CR(169)17⟩
Grignon, J., **1**, 128 ⟨75JOM(96)225⟩
Grignon-Dubois, M., **2**, 190 ⟨76TL1197⟩; **3**, 140 ⟨81TL2883⟩; **4**, 605 ⟨85JOM(280)313⟩
Grigor'ev, A. A., **1**, 612 ⟨89ZOR654⟩
Grigor'eva, I. K., **2**, 832 ⟨69IZV2234⟩
Grigor'eva, N. D., **4**, 955 ⟨81ZOB1857⟩
Grigor'eva, O. V., **4**, 719 ⟨74IZV2645⟩
Grigoryan, D. V., **1**, 1040 ⟨86AKZ306⟩
Grigoryan, N. Y., **2**, 439 ⟨86AKZ258⟩, 446 ⟨90AKZ610⟩
Grigos, V. I., **6**, 145 ⟨69IZV2566⟩
Grigsby, Jr., R. A., **2**, 279 ⟨83JOM(253)31⟩
Grigsby, R. A., **1**, 438 ⟨75JOM(88)175⟩; **2**, 279 ⟨75JOM(88)175⟩
Grill, H., **2**, 244 ⟨70TL1427⟩, 399 ⟨69TL4117⟩
Grill, W., **3**, 687 ⟨55LA(594)21⟩
Griller, D., **1**, 5, 7, 9, 10, 17, 25 ⟨92JOC2427⟩, 5, 7, 9, 10, 17, 936, 943, 945 ⟨91JOC678⟩, 5, 7, 9, 10, 331, 936 ⟨88JOC3641⟩, 5, 9, 10, 936 ⟨89TL2733⟩, 322 ⟨75JA4269⟩, 331, 936 ⟨87JA5267⟩, 919, 922 ⟨76ACR13⟩, 921 ⟨82JA3655⟩, 925 ⟨83JA4701, 89JOC2492⟩, 946 ⟨90JOC3321⟩, 1127 ⟨81JOC2327⟩; **2**, 141, 265 ⟨90JOC6248⟩, 536 ⟨90JOC5413⟩, 1010 ⟨74JA6202, 74JA6203⟩, 1010 ⟨75JA4269, 75JA5526⟩, 1010 ⟨76ACR13⟩; **3**, 440 ⟨88SC1975⟩; **4**, 512, 518 ⟨88SC1975⟩; **5**, 1135 ⟨91JOC528⟩; **6**, 726, 734 ⟨76ACR13⟩
Grillot, G. F., **4**, 316 ⟨54JA3969⟩, 317 ⟨67JOC2891⟩, 333 ⟨66JOC3514, 67JOC3101⟩
Grillott, G. F., **4**, 316 ⟨59JOC1035⟩
Grimaldi, J., **1**, 959 ⟨76S755⟩, 979 ⟨92TL1057⟩; **2**, 351 ⟨92TL1057⟩, 940 ⟨81TL3179⟩; **4**, 1046 ⟨81TL3179⟩
Grimaldo-Morön, J. T., **2**, 444 ⟨91T403⟩
Grim, S., **2**, 447 ⟨59CI(L)849⟩
Grim, S. O., **2**, 432, 475 ⟨74MI 208-01⟩, 435 ⟨67JOC781⟩, 524 ⟨61JA1613⟩, 822 ⟨77JOC1236⟩, 824, 825, 826 ⟨77PS(3)191⟩, 827 ⟨75JOM(94)327⟩; **3**, 495, 497 ⟨66JOC4252, 61JA1617⟩; **4**, 112 ⟨75JOM(94)327⟩, 546, 550, 551 ⟨77IC1770⟩, 550, 551, 576 ⟨80IC3195, 74SRI221⟩, 551 ⟨75IC656, 80ZN(B)832⟩, 551, 576 ⟨76IS195⟩, 552, 563 ⟨80IC1982⟩, 553 ⟨75CC634⟩, 569, 570, 574, 587, 588 ⟨59CI(L)849⟩, 569, 570, 574, 587, 588 ⟨61JA1610, 61JA1613⟩, 575 ⟨76IS192⟩, 1022, 1076 ⟨91TL3631⟩; **5**, 896 ⟨66JOM(5)267⟩; **6**, 151 ⟨80PS(9)123, 87PS(38)79, 91CC979⟩, 151, 369 ⟨86IC2699, 82CC930⟩, 370 ⟨85IC2889⟩
Grime, R. W., **5**, 944 ⟨94JCS(D)1671⟩
Grimes, R. N., **4**, 629 ⟨92CRV175⟩; **6**, 169 ⟨B-70MI 605-01⟩, 406 ⟨92CRV251⟩

Grimley, J., **2**, 110 ⟨54JCS1212⟩
Grimm, A., **3**, 366, 371, 378 ⟨65LA(682)188⟩
Grimm, B., **5**, 215 ⟨86AP(319)325⟩
Grimm, D., **4**, 424 ⟨74LA539⟩
Grimm, E., **3**, 224 ⟨90CB363⟩
Grimm, H. G., **6**, 597 ⟨36CB1356⟩
Grimm, J. W., **6**, 207 ⟨74JPR557⟩
Grimm, K. G., **2**, 541 ⟨77JA5009⟩; **3**, 474 ⟨76JOC3335⟩; **4**, 216 ⟨77JA5009⟩
Grimm, L. F., **5**, 1075 ⟨70CB1664⟩
Grimme, W., **1**, 419, 421 ⟨65TL391⟩, 785 ⟨80TL2799⟩, 795 ⟨65CB756⟩, 812 ⟨63AG(E)739⟩, 910 ⟨65AG(E)354, 66AG(E)604⟩, 910 ⟨68HCA225⟩
Grimmeiss, A. M. H., **2**, 1009 ⟨92AX(C)912⟩
Grimmer, S. S., **3**, 311 ⟨92JA5888⟩
Grimová, J., **5**, 655 ⟨84CCC1505⟩
Grimshaw, J., **5**, 105 ⟨B-76MI 502-03⟩
Grimwood, B. E., **5**, 985 ⟨67JOC3665⟩
Grinberg, S., **5**, 219 ⟨76JCS(P1)1708⟩, 1000 ⟨74TL1965⟩
Grinblat, M. P., **2**, 468 ⟨71JGU475⟩
Grindley, T. B., **1**, 209 ⟨81JOC2831⟩
Grinev, V. M., **2**, 356, 357 ⟨91IZV1069⟩
Grinfeld, A. A., **5**, 113 ⟨84IZV1567⟩, 116 ⟨84TL4989⟩
Grinham, A. R., **3**, 468 ⟨71JCS(C)1384⟩
Grinshtein, E. I., **2**, 430 ⟨66ZOB484⟩, 440 ⟨63ZOB1919⟩
Grinstead, R. R., **2**, 650 ⟨60JA3472⟩
Grinstein, D., **3**, 369 ⟨94TL953⟩
Grinstein, E. I., **2**, 440 ⟨61DOK(139)1359⟩
Grinsteins, V., **5**, 615 ⟨63MI 514-01⟩
Grinter, T. J., **3**, 371 ⟨90JCS(P1)2881⟩
Griot, R., **6**, 104 ⟨63HCA2306⟩
Griot, R. G., **6**, 104 ⟨63T1661⟩
Gripp, V. E., **5**, 199 ⟨52JCS1363⟩
Grisenti, P., **4**, 168 ⟨87JCS(P1)2301⟩
Grishchuk, B. D., **5**, 1044 ⟨88ZOB616, 91UKZ89⟩
Grishin, K. Y., **4**, 1057 ⟨85ZOB2801⟩
Grishin, Y. K., **3**, 540 ⟨84ZOB1817⟩; **4**, 1057 ⟨84ZOB1817⟩
Grishina, O. N., **2**, 838 ⟨80ZOB803⟩, 859 ⟨79ZOB2230⟩
Grisley, D. W., **4**, 554 ⟨64JOC2427, 65T5⟩
Grisley, J. D. W., **2**, 837, 839 ⟨60JA423⟩; **5**, 361 ⟨61JA1191⟩, 1093 ⟨64JOC2427⟩
Grison, C., **4**, 90, 784 ⟨87S169⟩, 473 ⟨89H(28)1179⟩, 577 ⟨86JOM(316)13⟩; **5**, 233 ⟨91S23⟩; **6**, 136 ⟨87S169⟩
Grissom, J. W., **1**, 1009 ⟨92TL2315⟩; **2**, 351 ⟨93JOC6559⟩
Grist, N. J., **4**, 1023, 1032 ⟨88CC673⟩
Grivas, J. C., **5**, 335 ⟨75JOC2029⟩, 749 ⟨61CJC761⟩
Grob, C. A., **1**, 812 ⟨63HCA483⟩, 905 ⟨83ACR426⟩, 1081 ⟨67AG(E)1⟩, 1110, 1111 ⟨71HCA2060, 64HCA194⟩, 1113 ⟨69TL1439, 70HCA2130⟩, 1113, 1114 ⟨70HCA2119⟩, 1119 ⟨64HCA1590, 76TL4191⟩, 1119 ⟨76TL4195⟩; **2**, 13 ⟨73S493⟩, 23 ⟨73S789⟩, 383 ⟨84JCS(P2)411⟩, 623 ⟨64HCA194⟩; **3**, 622, 638 ⟨63HCA1190⟩
Grob, J., **2**, 293, 499, 500 ⟨79ZN(B)1646⟩, 473, 508 ⟨92ZAAC(608)43⟩
Grob, P., **4**, 184 ⟨79LA522⟩
Grobe, J., **2**, 439 ⟨91ZAAC(592)121⟩, 894 ⟨81ZN(B)48⟩; **3**, 501 ⟨80JOM(188)151, 80JOM(197)13⟩, 689 ⟨90CB2317⟩; **4**, 112 ⟨88JOM(344)61, 89ZN(B)175⟩, 112, 116 ⟨90ZN(B)148⟩, 112, 116, 117 ⟨86ZN(B)974, 91ZN(B)978⟩, 142, 519, 568 ⟨77JOM(132)77⟩, 566 ⟨73JOM(61)133, 77ZAAC(433)157⟩, 567, 622, 623 ⟨83ZAAC(500)132⟩, 602 ⟨65MI 414-01⟩, 608, 610, 621 ⟨80JOM(188)151⟩, 611 ⟨89JOM(363)7⟩, 617 ⟨62ZAAC(315)157⟩; **5**, 786, 1087 ⟨91ZN(B)978⟩, 1152 ⟨93PS(76)1⟩; **6**, 49 ⟨90JOM(386)321⟩, 242 ⟨91ZN(B)978⟩, 242, 683 ⟨93ZN(B)58⟩, 242, 685 ⟨91HAC385⟩, 258, 679 ⟨90ZN(B)148⟩, 258, 680, 683 ⟨86ZN(B)149⟩, 264 ⟨85PS(21)349⟩, 265 ⟨90ZN(B)299, 92ZN(B)321⟩, 265, 266, 679, 680, 691 ⟨84AG(E)710⟩, 266 ⟨90JOM(399)189⟩, 383 ⟨65MI 613-01⟩, 678 ⟨86PS(28)239⟩, 679 ⟨87PS(30)401⟩, 684 ⟨93PS(76)265⟩, 685 ⟨90CB2317⟩, 685, 692 ⟨88CB655⟩, 700 ⟨89NJC363, 92CB567⟩
Gröbel, B.-T., **2**, 910, 911, 931 ⟨77CB867⟩, 931, 936 ⟨74AG(E)83, 77CB852⟩; **3**, 16, 20, 132, 137, 291, 295 ⟨77S357⟩, 214 ⟨73AG(E)69⟩, 214 ⟨74TL3171⟩; **4**, 197, 244, 1075 ⟨77S357⟩, 254 ⟨74TL3171⟩, 620 ⟨74AG(E)83⟩, 620, 937, 962 ⟨77CB852⟩, 849 ⟨73CB2277⟩, 964 ⟨77CB867⟩; **6**, 175, 190, 192 ⟨74AG(E)83⟩, 175, 190, 334, 351, 390 ⟨77CB852⟩, 327 ⟨73CB2277⟩
Grobpietsch, T., **5**, 786, 1087 ⟨91ZN(B)978⟩
Grochowski, E., **2**, 831 ⟨82JA6876⟩
Groebel, W., **5**, 337 ⟨59CB2887⟩
Groeger, F. A., **5**, 846 ⟨68TL2463⟩
Groen, S. H., **2**, 1031 ⟨72BCJ509⟩
Groenen, L. C., **6**, 556 ⟨90CC1432⟩
Groenewegen, P., **1**, 873 ⟨78TL491⟩; **2**, 34 ⟨79TL2817⟩; **3**, 21 ⟨78TL491⟩, 27, 35 ⟨79TL2817⟩
Groenweghe, L. C. D., **4**, 125, 128 ⟨61JA1811⟩
Groesmann, H., **2**, 471 ⟨91ZAAC(549)36⟩
Groger, R., **3**, 468 ⟨30CB702⟩
Groh, B. L., **1**, 468 ⟨87JA813⟩; **2**, 932 ⟨87JA813⟩, 1023 ⟨87JA7568⟩
Groh, G., **2**, 566 ⟨63ZAAC(319)230⟩
Grohe, K., **4**, 829, 970 ⟨82CB1733⟩; **6**, 51, 238 ⟨69LA(730)140⟩, 69 ⟨67CB2946⟩, 607 ⟨69LA(730)133⟩
Grohmann, A., **6**, 406 ⟨88AG(E)1544⟩
Groi, H., **5**, 714 ⟨68ZC219⟩
Groipietsch, T., **4**, 116 ⟨91ZN(B)978⟩
Grol, C. J., **3**, 362 ⟨74T3621⟩
Gromek, J. M., **2**, 997, 1002 ⟨85JOM(285)1⟩; **4**, 531 ⟨85JOM(285)1⟩
Gromelski, S. J., **1**, 1144 ⟨72JOC3220⟩; **2**, 802 ⟨72JOC3220⟩
Gromkova, R. A., **4**, 942 ⟨90ZOB526⟩; **6**, 565 ⟨90ZOB602⟩
Gromov, A. V., **4**, 112, 114, 115 ⟨84JGU2250, 84JGU1354⟩, 116 ⟨85JGU1065⟩; **5**, 876, 877 ⟨84ZOB1520⟩; **6**, 680, 681 ⟨84ZOB1520⟩, 696 ⟨85ZOB1194⟩
Gromova, E. M., **4**, 814 ⟨89JGU2040⟩
Gromstadt, T., **2**, 181, 182, 183, 194 ⟨56JCS173⟩
Grønbrk Hazell, R., **6**, 555 ⟨86ACS(B)593⟩
Grondin, R., **5**, 351 ⟨92TL5717⟩
Grönen, J., **1**, 991 ⟨89CB1237⟩
Gronert, S., **5**, 436 ⟨87JA2553⟩
Gröninger, K., **1**, 326, 948 ⟨84TL2743⟩
Gröninger, K. S., **1**, 923 ⟨90OS(69)66⟩
Gronowitz, S., **1**, 318 ⟨76ACS(B)485⟩, 428 ⟨73OSC(5)149⟩, 480 ⟨93JOM(460)127⟩, 1072 ⟨76ACS(B)341, 84JOC2018⟩; **4**, 99 ⟨79S810⟩, 327 ⟨80S322⟩, 785 ⟨76ACS(B)439⟩, 814 ⟨75ACS(B)461⟩, 856 ⟨76CS90⟩, 872 ⟨73ACS2242⟩; **5**, 595 ⟨93S483⟩; **6**, 566 ⟨93S483⟩
Gronwall, S., **1**, 881 ⟨62OS(42)28⟩
Groome, P., **4**, 610 ⟨81CC806⟩
Groot, A. de, **1**, 584 ⟨94JOC374⟩, 600 ⟨81TL4137⟩; **3**, 35 ⟨81TL887, 87RTC489⟩, 122 ⟨85JOC2650⟩
Gros, E. G., **6**, 461, 476 ⟨88ZN(B)1038⟩, 477 ⟨89ZN(B)983⟩
Gros, H., **6**, 104 ⟨69ZC201⟩
Grosa, G., **4**, 259 ⟨89EJM635⟩
Grosjean, B., **3**, 273 ⟨75BSF779⟩, 675 ⟨76S448⟩
Gross, A., **5**, 90 ⟨55LA(591)117⟩
Gross, A. W., **1**, 874 ⟨84TL495⟩, 926 ⟨85TL4291⟩; **2**, 374 ⟨85JOC5391⟩, 390 ⟨87OS(65)166⟩, 420 ⟨84TL491⟩, 699 ⟨84TL495⟩; **3**, 532 ⟨51LA(573)17⟩
Gross, B., **3**, 118 ⟨76S811⟩; **5**, 141 ⟨76S811⟩
Gross, B.,**5**, 279 ⟨78TL1269, 84S572⟩

Gross, D. C., **6**, 517 ⟨85JA2355⟩
Gross, F. C., **2**, 434 ⟨67JOC2445⟩
Gross, G., **3**, 366 ⟨84AG(E)800⟩, 552 ⟨83AG(E)543⟩
Gross, H., **2**, 182, 617 ⟨59CB83, 63CB1387⟩, 1008 ⟨69AP(302)354⟩; **3**, 92, 101 ⟨63CB308⟩, 99, 101 ⟨60CB88⟩, 618 ⟨63CB1387⟩; **4**, 46 ⟨64ZC401⟩, 47, 48 ⟨73OSC(5)221⟩, 553 ⟨82CSR493, 82JPR537, 84EGP206557⟩, 559 ⟨75JPR890, 76JPR403⟩, 1002 ⟨81T1227⟩, 1004 ⟨82T139, 84T733⟩; **5**, 42 ⟨68AG(E)391⟩, 976 ⟨63CB1387⟩, 1023 ⟨65CB2425⟩; **6**, 39 ⟨63CB1387, 67OS(47)47⟩, 43 ⟨66CB2625, 67LA(707)35⟩, 43 ⟨68LA(720)146⟩, 43, 251, 429 ⟨61CB544⟩, 48 ⟨77JPR17⟩, 69, 82 ⟨61CB538⟩, 116 ⟨74JPR913⟩, 120 ⟨78JPR255⟩, 155 ⟨90PS(47)7⟩, 155 ⟨91PS(62)35⟩, 156 ⟨69JPR925⟩, 156, 159 ⟨71LA(750)44⟩, 157 ⟨68AG(E)391, 68AG(E)463, 69JPR577⟩, 159 ⟨72JPR969, 74PS(4)241, 76JPR272⟩, 229 ⟨58CT515⟩, 281 ⟨70JPR475⟩, 286 ⟨76JPR116⟩, 298, 300, 312, 321, 646 ⟨64LA(675)142⟩, 360 ⟨73CB2315⟩, 367 ⟨72JPR969, 86JPR231⟩, 367, 665 ⟨72JPR87⟩, 650 ⟨72CB1709⟩, 661, 733 ⟨78JPR389, 72EGP93766⟩
Gross, M., **6**, 651 ⟨74JPR434⟩
Gross, M. F.,**5**, 936 ⟨94JA10921⟩
Gross, M. L., **4**, 642, 655 ⟨82CC970⟩
Gross, P. M., **3**, 306 ⟨80TL2129⟩
Gross, R. S., **4**, 755, 782 ⟨92JA360⟩, 755, 803 ⟨93TL7197⟩, 812 ⟨94TL1027⟩
Gross, V. H., **5**, 788 ⟨76JPR272⟩
Gross, Von. H., **4**, 1003 ⟨76JPR116⟩
Gross, Z., **4**, 993 ⟨92CJC1022⟩
Grosse, A. V., **2**, 595 ⟨40JOC106⟩; **6**, 32 ⟨52JA1974, 52JA848⟩, 32 ⟨52JA849⟩, 218 ⟨52JA1347⟩, 219 ⟨51JA2461⟩
Grosse, D., **5**, 17, 19 ⟨81S216⟩
Grosse, S., **2**, ⟨76MI 206-01⟩ ⟨76JPR1019⟩
Grosser, F., **3**, 241 ⟨42JA382⟩
Grossert, J. S., **1**, 144 ⟨B-88MI 104-02⟩, 702 ⟨88CJC2860⟩, 706 ⟨74TL2805⟩; **2**, 178, 179, 180 ⟨77CJC407⟩, 249 ⟨70CC305⟩; **4**, 63, 78 ⟨73CC49⟩, 74, 75, 78 ⟨77CJC421⟩, 85 ⟨73CC50⟩, 258 ⟨78CJC1183⟩; **5**, 348 ⟨78CJC1183⟩; **6**, 90, 91, 307 ⟨82CC1183⟩
Grossi, M., **5**, 354 ⟨87T4249⟩, 367 ⟨85OPP385⟩
Grossman, H., **5**, 389 ⟨69ICA495⟩
Grossman, J., **2**, 34 ⟨55JA3826⟩, 126 ⟨57JA2553⟩
Grossman, R. B., **2**, 307 ⟨91JA2321⟩
Grossmann, A., **5**, 851 ⟨08JPR489⟩
Grosso, J. A., **6**, 577 ⟨80JMC1261⟩
Grosspietsch, T., **6**, 242 ⟨91ZN(B)978⟩
Grote, A., **4**, 522 ⟨72LA(755)67⟩; **6**, 637 ⟨72LA(755)67⟩
Grote, C. W., **3**, 201 ⟨91TL5313⟩; **4**, 561 ⟨93PS(75)23⟩
Grote, J., **1**, 374 ⟨84JOC5277⟩
Grotemeier, G., **5**, 148 ⟨81AG(E)207⟩
Groten, B., **2**, 1034 ⟨60JA3799⟩
Grotewold, J., **1**, 657 ⟨68JCS(B)1044⟩
Groth, H., **3**, 309 ⟨41LA(549)238⟩
Groth, P., **3**, 370 ⟨75JCS(P1)2099⟩
Groth, R. H., **5**, 808 ⟨59JA3599⟩; **6**, 257 ⟨59JA3599⟩
Groth, S., **5**, 1101 ⟨92AG(E)866⟩; **6**, 436, 533, 534 ⟨92AG(E)866⟩
Groth, U., **3**, 706 ⟨93CB427, 93LA427⟩; **4**, 535 ⟨79AG(E)863⟩, 536 ⟨83LA1133⟩
Grotjahn, D. B., **2**, 944 ⟨93S579⟩; **5**, 931 ⟨91SL381⟩; **6**, 716, 719, 722 ⟨91SL381⟩
Grotsch, M. M., **3**, 543 ⟨90CB2031⟩
Grous, P. P., **6**, 598 ⟨91JMC3187⟩
Groutas, W. C., **2**, 538, 540 ⟨80S861⟩; **3**, 638 ⟨80SC495⟩
Grove, D. D., **5**, 826, 829, 830 ⟨87JOC4570⟩
Grove, D. M., **1**, 127 ⟨82JA6609, 82OM1366⟩; **2**, 684 ⟨89T5565⟩; **3**, 512 ⟨90RTC446⟩; **4**, 656 ⟨93OM3955⟩, 723 ⟨93OM2227⟩; **6**, 719 ⟨74JCS(D)1189⟩
Grovenstein, E., **1**, 34 ⟨59JA4850⟩, 378 ⟨57JA4985⟩
Grovenstein, Jr., E., **1**, 378, 886 ⟨61JA412⟩, 885 ⟨B-87MI 119-04, 61JA2537⟩, 886 ⟨76JA7835, 77JA1881⟩, 886 ⟨78AG(E)313, 82JA6681⟩, 886 ⟨82JOC2928, 89JOC1671⟩
Grovenstein Jr., E., **1**, 562 ⟨61JA1705⟩
Grover, R., **6**, 448 ⟨85EUP136147⟩
Grover, S. K., **3**, 282 ⟨88MI 306-01⟩
Groves, C. E., **6**, 215 ⟨1870LA(156)60⟩
Groves, I. F., **6**, 7 ⟨82JCS(D)2281⟩
Groves, J. T., **1**, 907 ⟨67JA5048a, 70JA984⟩, 936 ⟨77TL4291⟩; **3**, 108 ⟨84JA630⟩, 224 ⟨74TL909⟩; **4**, 42, 57 ⟨74TL909⟩; **6**, 509 ⟨78JOC922⟩
Groves, W., **6**, 570, 645 ⟨75JMC90⟩
Groving, N., **5**, 652 ⟨65ACS1768⟩
Groziak, M. P., **3**, 109 ⟨91JOC4296⟩; **5**, 275 ⟨86JOC1065⟩, 1063, 1083 ⟨86JOC1277⟩
Grubbs, E. J., **1**, 384, 388 ⟨59JA4733⟩
Grubbs, R. B., **3**, 508 ⟨81JA7358⟩
Grubbs, R. H., **1**, 669 ⟨86JA733⟩, 669, 670 ⟨86OM721⟩, 674 ⟨91COS(5)1115⟩, 674, 675 ⟨82COMC-I(8)499, 87MI 119-11⟩, 675 ⟨88JA7542, 88JA960⟩, 675 ⟨89SCI907⟩, 767 ⟨85JOC2386⟩, 767, 769 ⟨80JA3270, 83PAC1733⟩, 991 ⟨83JA5490⟩; **2**, 680 ⟨80JA3270⟩, 681 ⟨85JOC2386⟩, 702 ⟨83JA1664, 85JOC2316⟩, 979 ⟨82JA7491, 83JA5490⟩; **3**, 508 ⟨83PAC1733, 84PAC59⟩, 508 ⟨86OM721, 89OM583⟩, 517 ⟨93JA7898⟩, 520, 522 ⟨92JA3974⟩; **4**, 670 ⟨81JA5922⟩; **5**, 441 ⟨86JA3318⟩; **6**, 140 ⟨88BCJ171⟩
Gruber, J. M., **1**, 434 ⟨85JOC2668⟩; **3**, 31, 54 ⟨83T861⟩
Gruber, K., **2**, 433 ⟨63CB2186⟩; **4**, 546, 553 ⟨63CB2186⟩
Gruber, L., **3**, 239 ⟨75S708⟩
Gruber, P., **1**, 537 ⟨71AG(E)653⟩; **3**, 680 ⟨71CB3816⟩
Gruber, W., **1**, 383 ⟨73CB1376⟩
Grubert, H., **4**, 706 ⟨60JA76⟩
Grubert, L., **5**, 358, 707 ⟨88ZC24⟩
Grubmuller, P., **1**, 48 ⟨79AG(E)939⟩
Grudoski, D. A., **3**, 179 ⟨80JOC2307⟩
Grudzinskas, C. V., **1**, 535 ⟨78JOC3450⟩; **5**, 15 ⟨78JOC3972⟩
Gruenewald, R., **2**, 446 ⟨61CB186⟩
Gruetzmacher, G., **2**, 761 ⟨74JA5512⟩
Gruetzmacher, H., **6**, 669 ⟨89JOC4426⟩
Gruhzit, O. M., **2**, 260 ⟨48JA1762⟩
Grummitt, O., **2**, 13 ⟨55OSC(3)195⟩; **3**, 444 ⟨46JA914⟩; **4**, 47, 50 ⟨63OSC(4)748⟩, 743 ⟨55OSC(3)270⟩; **5**, 38 ⟨55OSC(3)270⟩, 815 ⟨52JA4876⟩
Grunanager, P., **2**, ⟨83TL1041⟩ ⟨83T689⟩, 354 ⟨90JOC3427⟩; **3**, 477 ⟨63TL1839, 65CB623⟩, 681, 683 ⟨B-84MI 320-03, 91CHE391⟩, 685, 686 ⟨B-71MI 320-01⟩; **5**, 819 ⟨B-71MI 521-01⟩
Grund, H., **1**, 664 ⟨79JCR(S)54⟩
Gründemann, C., **5**, 365 ⟨67CB142⟩
Gründemann, E., **4**, 420 ⟨84JPR279⟩, 889, 972 ⟨66CB2302⟩
Grundevik, I., **6**, 426 ⟨90JAP2209879⟩
Grundke, G., **3**, 27 ⟨85CB4288⟩
Grundke, J., **2**, 581 ⟨83CB159, 83CB176⟩
Grundmann, C., **1**, 912 ⟨38CB442⟩; **3**, 474 ⟨57JA944⟩, 683 ⟨69JOC2016⟩, 684 ⟨67JOC2303⟩, 685, 686 ⟨B-71MI 320-01⟩, 687 ⟨85HOU(E5)1572⟩, 694, 703 ⟨85HOU(E5)611⟩, 722 ⟨58CB1380⟩; **5**, 637, 819, 832, 837 ⟨66JOC157⟩, 656 ⟨62JOC2433⟩, 699 ⟨85HOU(E5/2)1591⟩, 747 ⟨55JA6559⟩, 770, 855 ⟨63AG(E)309⟩, 819 ⟨65JOC2809, B-71MI 521-01⟩, 826, 828 ⟨66FCF62⟩, 828, 837 ⟨65LA(687)191⟩, 833 ⟨B-70MI 521-02⟩; **6**, 603 ⟨62JOC2589⟩
Grundon, M. F., **3**, 242 ⟨75CC772⟩
Grundy, K. R., **6**, 719 ⟨76JOM(113)C45⟩, 720, 732 ⟨75JOM(90)C34⟩, 720, 732 ⟨76JOM(107)C37⟩, 721 ⟨75JOM(91)C61⟩

Grundy, S. L., **1**, 94 ⟨84JCS(D)1747⟩; **5**, 222 ⟨92MI 504-02⟩
Grundy, W., **5**, 1026 ⟨77S582⟩
Grünefeld, J., **5**, 212 ⟨85AP(318)977⟩; **6**, 622 ⟨85AP(318)1057, 85AP(318)992⟩
Grunert, B., **4**, 609, 621 ⟨81ZAAC(473)59⟩
Grung, K. E., **5**, 1117 ⟨89ACS518⟩
Grüning, R., **3**, 484 ⟨74JOM(78)221⟩
Grunner, C., **2**, 471 ⟨77ZN(B)1244⟩
Grunner, T. A., **3**, 686 ⟨78JOC3736⟩
Gruntz, U., **2**, 11 ⟨79JCS(P1)436⟩; **5**, 1107 ⟨78CC133⟩
Grünwald, C., **5**, 944 ⟨95CB49⟩
Grunwald, E., **1**, 844 ⟨65MI 119-07⟩, 1114, 1116 ⟨48JA846⟩, 1114, 1116 ⟨51JA2700⟩, 1114, 1116 ⟨57JA4146⟩
Grunwald, F. A., **2**, 187 ⟨51JOC964⟩, 708 ⟨51JOC946⟩
Grunwell, J. R., **2**, 240 ⟨75TL1739⟩; **3**, 686 ⟨79JOC510, 80JOC3753⟩; **5**, 201 ⟨75TL1739⟩, 234 ⟨76SC453⟩
Grunze, H., **2**, 105 ⟨59CB850⟩
Grusdt, U., **3**, 308 ⟨87AP(320)1020⟩
Gruseck, U., **3**, 221 ⟨87TL2681⟩; **4**, 414 ⟨87TL2681⟩
Gruselle, M., **4**, 674 ⟨90OM2993, 94OM2244⟩
Grushin, V. V., **1**, 427 ⟨91OM1620⟩, 1002 ⟨92JOC2188⟩
Gruszecka, E., **4**, 460 ⟨79PJC2327⟩, 470 ⟨81TL3537⟩
Gruszecka, M., **5**, 716 ⟨87LA483⟩
Gruszecki, W., **5**, 716 ⟨87LA483⟩
Gruter, H. F. M., **4**, 699 ⟨76JOM(117)329⟩
Gruttner, G., **2**, 489 ⟨15CB1473⟩
Grüttner, S., **1**, 463 ⟨87AG(E)683⟩
Grütze, J., **1**, 232 ⟨75AG(E)559⟩
Grützmacher, G., **6**, 691, 733 ⟨93PS(76)21⟩
Grutzmacher, H., **2**, 1067 ⟨88JA6463⟩; **3**, 689 ⟨88JA6463, 90PS(49)301⟩; **4**, 551, 552 ⟨91CC302⟩, 567 ⟨88JA6463⟩, 572 ⟨91AG(E)709, 93AG(E)1359⟩, 588 ⟨91OM938⟩; **5**, 871, 872, 873 ⟨88JA6463⟩, 886 ⟨89AG(E)740⟩, 887 ⟨91CC302⟩, 888, 893 ⟨93CC673, 93PS(76)21⟩, 919 ⟨94AG(E)456⟩, 1086 ⟨92AG(E)437⟩; **6**, 165, 691, 733 ⟨91AG(E)709⟩, 691 ⟨93CC673⟩, 691, 734 ⟨89JA6853⟩, 702 ⟨91CB329⟩, 704 ⟨88JA6463⟩
Grutzmacher, H.-F., **2**, 506, 852 ⟨68CB1381⟩
Grutzner, J. B., **1**, 847 ⟨80JA4709⟩, 865 ⟨73JA6449⟩, 886 ⟨68JA6562⟩
Gruyter, P. de, **5**, 852 ⟨1893CB2385⟩, 858 ⟨01JPR222⟩
Gruz, B. E., **4**, 826 ⟨65ZOB1639⟩
Gryaznov, P. I., **4**, 106 ⟨90JGU2208⟩
Gryaznov, S. M., **2**, 106 ⟨92MI 202-01⟩
Grynkiewicz, G., **4**, 45 ⟨84CL1751⟩, 174 ⟨91TL3313⟩, 176 ⟨89JCR(S)152⟩
Gryszkiewicz-Trochimowski, E., **2**, 465 ⟨60BSF1794⟩; **4**, 129, 130, 131 ⟨67BSF4289⟩; **5**, 822 ⟨48BSF597⟩
Gryszkiewicz-Trochimowski, O., **2**, 465 ⟨60BRP916751⟩
Grzejszczak, S., **1**, 608 ⟨81TL3097⟩, 821 ⟨78JOC2518⟩; **2**, 713 ⟨78JOC473⟩; **3**, 166 ⟨83S332⟩; **4**, 346 ⟨79JOC2967⟩, 804 ⟨83T1189⟩, 835, 847, 848 ⟨78T3081⟩, 847, 848 ⟨76TL2731⟩, 923 ⟨83S332⟩, 924 ⟨81TL3097⟩; **6**, 48 ⟨77JPR17⟩, 119 ⟨76TL477, 78JOC2518, 79JOC2967⟩, 120, 121, 123 ⟨80S127, 81TL3097⟩
Grzeskowiak, N. E., **4**, 540 ⟨82JCS(P1)143⟩
Grzywa, E., **6**, 156 ⟨78MIP87868⟩
Gschneidner, D., **2**, 938 ⟨93JA6625⟩
Gschwend, A. W., **5**, 657 ⟨82JOC3658⟩
Gschwend, H. W., **2**, 33 ⟨81JOC3349⟩, 554, 653, 907, 908, 942, 965, 997 ⟨79OR(26)1⟩; **3**, 90 ⟨79OR(26)1⟩, 633 ⟨78TL4221⟩; **4**, 539 ⟨79OR(26)1⟩, 948 ⟨76OR(26)1⟩; **5**, 607 ⟨76JOC4029⟩, 746 ⟨77JOC2639⟩
Gschwend-Steen, K., **1**, 415 ⟨69HCA1030⟩
Gstach, H., **4**, 329, 444 ⟨85M1329⟩, 440 ⟨90S1048, 90S803⟩, 441 ⟨85T5525⟩, 448 ⟨85M1051, 87M851⟩
Gu, D., **2**, 186 ⟨93TL67⟩
Gu, J.-H., **3**, 615 ⟨92CL1169⟩
Gu, J.-M., **1**, 835 ⟨83JOC4262, 84JOC2708⟩
Gu, X., **4**, 187 ⟨88S36⟩
Gu, Y., **5**, 142 ⟨94JA7455⟩
Gu, Y. G., **2**, 904, 939 ⟨92JOC794⟩
Gu, Y.-Z., **3**, 260 ⟨91SC1307⟩
Gu, Z., **4**, 598 ⟨84TL4425⟩; **6**, 641 ⟨92TL5933⟩
Gu, Z. P., **1**, 585 ⟨89JOC2217⟩
Guaciaro, M. A., **3**, 271 ⟨78TL4661⟩
Gualtieri, F., **3**, 469 ⟨65G983⟩; **4**, 232 ⟨87JMC1934⟩
Guanci, D. F., **4**, 298 ⟨78JOC3838⟩
Guan-lin, H., **3**, 344, 345 ⟨92CJC974⟩
Guanti, G., **2**, 194 ⟨85JOC5592⟩; **4**, 856 ⟨81JCS(P1)1278⟩, 907 ⟨74JCS(P1)2357⟩
Guaraldi, G., **2**, 171, 172, 716 ⟨70JOC2006⟩
Guarna, A., **2**, 320 ⟨90TL3351⟩; **3**, 269 ⟨88JOC2426⟩, 686 ⟨82S719⟩
Guarneri, M., **4**, 475 ⟨82S653⟩
Guasch, J., **5**, 864 ⟨86POL1277⟩
Guastini, C., **2**, 703 ⟨85IC654⟩; **4**, 696 ⟨83CC1087, 83CC1156⟩; **5**, 906 ⟨83CC337⟩; **6**, 342 ⟨86AG283, 88MI 622-02⟩
Guazzaroni, M. E., **2**, 74 ⟨89JOC3087⟩
Gubaidullin, L. Y., **2**, 441 ⟨80IZV734⟩
Gubaidullin, M. G., **5**, 404, 405 ⟨67JGU826⟩
Gubelt, G. B., **4**, 314 ⟨69CJC3983⟩
Gubermator, K., **1**, 814 ⟨82AG(E)686⟩
Gubernick, S., **2**, 652 ⟨87TL5115⟩; **3**, 438 ⟨87TL5115⟩
Gubin, S. P., **4**, 706, 723, 724 ⟨72IZV1823⟩
Gubnitskaya, E. S., **5**, 683 ⟨65JGU1019, 65ZOB1014⟩, 708 ⟨63JGU1006, 63JGU551, 63ZOB557⟩, 708, 709 ⟨65JGU1014, 65ZOB1009⟩, 860 ⟨62ZOB1873, 63ZOB1017⟩, 861 ⟨63JGU2984⟩
Gubsch, P., **2**, 304 ⟨70JOC2826⟩
Gudat, D., **6**, 689 ⟨85CC1687, 86OM593, 87CC10⟩, 689, 690 ⟨87ZAAC(545)7⟩, 689, 692 ⟨85C277, 89JOM(368)C29⟩
Gude, C., **1**, 1010 ⟨92BMC1619⟩
Gudima, A., **6**, 660 ⟨91TL2775⟩
Gudima, A. O., **6**, 680, 681, 689 ⟨92IC3493⟩
Gudina, I. V., **4**, 125 ⟨86JGU2246⟩, 126 ⟨86JGU2242⟩
Gudmundsdottir, A. D., **3**, 220 ⟨90S993⟩
Gueddari, S., **5**, 640 ⟨93JMC1157⟩
Guedin-Vuong, D., **4**, 235 ⟨84JA2954⟩
Guédin-Vuony, D., **5**, 173 ⟨84JA2954⟩
Guedj, R., **2**, 10 ⟨78TL4507, 81JFC(17)127⟩; **6**, 52, 54 ⟨81JFC(18)93⟩
Gueiffier, A., **5**, 360 ⟨90CPB2352⟩
Guella, C., **3**, 63 ⟨87HCA1400⟩
Guemas, J.-P., **3**, 339 ⟨87BSF517⟩; **5**, 606 ⟨82PS(12)325⟩, 760 ⟨80JHC767⟩
Guennec, M. Le, **4**, 111 ⟨89CC988⟩
Guennouni, N., **2**, 182 ⟨92SL581⟩
Guenot, P., **3**, 337 ⟨88TL5899⟩, 688 ⟨91AG(E)196⟩
Guenther, D., **5**, 1006 ⟨69GEP1289526⟩
Guenther, E., **4**, 106 ⟨79EGP137839⟩
Guerbet, M., **2**, 87 ⟨1899CR(128)511⟩
Guerchais, J. E., **4**, 680 ⟨81CC862⟩
Guerchais, V., **4**, 581 ⟨90OM1036⟩, 681 ⟨92CC310⟩
Guerin, C., **1**, 651 ⟨84JOM(264)207⟩, 759 ⟨70JA7424⟩; **2**, 264 ⟨80BSB783⟩, 546 ⟨80CC168⟩, 600 ⟨80JOM(197)C19⟩, 935 ⟨81TL2985⟩; **3**, 211 ⟨81TL2985⟩; **4**, 654 ⟨70JA7424⟩, 655 ⟨81TL2985⟩, 674 ⟨90OM2993⟩
Guerin, P., **5**, 1097 ⟨77CC648⟩; **6**, 717 ⟨77CC648⟩
Guerra, M. A., **6**, 245 ⟨90JOM(390)C73⟩, 246 ⟨86JA4103, 86JOM(307)C58⟩
Guerra, M., **1**, 5, 7, 9, 10, 17, 25 ⟨92JOC2427⟩
Guerrera, F., **5**, 615 ⟨70MI 514-01⟩
Guerrero, A., **1**, 179 ⟨88SC445⟩, 503, 504 ⟨83JOC2266⟩, 747 ⟨84T2805⟩; **2**, 8 ⟨87S511⟩, 18 ⟨89JOC4294⟩, 19 ⟨87TL4733⟩, 1092 ⟨90TL2739⟩; **6**, 695 ⟨83TL3387⟩

Guerrero, A. F., **3**, 151 ⟨88TL6707⟩
Guerrero, Jr., P. G., **4**, 93 ⟨93JOM(460)31⟩
Guerrieri, F., **5**, 27 ⟨73S509⟩
Guerrini, A., **1**, 5, 7, 9, 10, 17, 25 ⟨92JOC2427⟩, 5, 7, 9, 10, 17, 25 ⟨92JOC3405⟩, 936, 945 ⟨90SL219⟩; **2**, 1019 ⟨89S461⟩; **4**, 358 ⟨90T2999⟩
Guertin, J. P., **2**, 762 ⟨66MI 215-02⟩
Guertin, K. R., **2**, 44 ⟨91TL715⟩, 155 ⟨93TL5369⟩; **3**, 156 ⟨91TL715⟩
Guervenou, J., **4**, 560 ⟨91S661, 94PS(88)1⟩
Guest, A. M., **6**, 55 ⟨78JOC3485⟩, 57 ⟨79S311⟩
Guetin, C., **2**, 578 ⟨90JA1874⟩
Guette, C., **2**, 626 ⟨89CJC2061⟩
Guette, J.-P., **2**, 69 ⟨73T3659⟩, 620 ⟨82T2339⟩
Guette, M., **2**, 69 ⟨73T3659⟩
Gueugnot, S., **1**, 121, 1002 ⟨92TL5757⟩, 975, 1004, 1025 ⟨93TL3853⟩
Guevel, R., **1**, 813, 815 ⟨92TL923⟩
Gugel, H., **1**, 1077 ⟨78S536⟩
Gugelchuck, M. M., **3**, 659 ⟨81JOC3671⟩
Guggenberger, L. J., **3**, 510 ⟨78JA3793⟩; **4**, 672 ⟨73IC294⟩, 677 ⟨77JOM(139)C34⟩; **6**, 340 ⟨68IC2265⟩
Guggenheim, T. L., **2**, 543 ⟨84TL1253⟩; **3**, 698, 707 ⟨82JA5849, 90OSC(7)294⟩
Gugger, P., **2**, 487 ⟨90CC1169⟩
Guggisberg, D., **4**, 852 ⟨89HCA1506⟩
Guggolz, E., **2**, 585 ⟨84AG(E)515⟩; **4**, 686 ⟨82CB878⟩
Guglielmetti, R., **4**, 307 ⟨B-90MI 407-01⟩; **6**, 111 ⟨72HCA1782⟩, 115 ⟨71BSF556⟩
Guha, M., **5**, 76 ⟨73OSC(5)384⟩
Guha, P. C., **5**, 103 ⟨55OSC(3)623⟩
Guha, S. K., **3**, 247 ⟨47JIC473, 53JIC379⟩
Guhl, D., **5**, 386 ⟨88JOM(339)213, 88JOM(344)C35⟩, 386 ⟨89JOM(375)245, 89JOM(378)279⟩
Guibe, F., **2**, 304 ⟨93JOC6109⟩; **3**, 13, 83, 95, 98, 107 ⟨81JOC4439⟩, 83 ⟨80CC432⟩
Guibé, F., **4**, 813 ⟨90JOC1857⟩
Guibé-Jampel, E., **4**, 541 ⟨87JOM(320)171⟩; **6**, 477 ⟨78T3105⟩
Guiborel, C., **3**, 474 ⟨81TL441⟩
Guidry, C., **2**, 1000 ⟨83JHC399⟩
Guidu Rao, C., **2**, 516 ⟨80S151⟩
Guigné, A., **5**, 556 ⟨90BSF446⟩
Guijarro, D., **1**, 136 ⟨93T1327⟩; **2**, 311 ⟨93T7761⟩
Guilard, R., **2**, 578 ⟨83JOM(253)65, 88CRV1121⟩; **3**, 76 ⟨89JCS(P1)2095⟩, 77 ⟨88SC1475⟩; **6**, 519 ⟨88CRV1121⟩
Guildford, A. J., **2**, 667 ⟨83CC466⟩
Guiles, J., **1**, 149 ⟨89H(28)295⟩, 150 ⟨91TL5501⟩, 150 ⟨91TL5505⟩, 150 ⟨91TL5509⟩; **2**, 328 ⟨89H(28)295⟩
Guiles, J. W., **1**, 149 ⟨91JOC6873⟩, 150 ⟨92JOC3819⟩
Guilhem, J., **1**, 344 ⟨87TL1413⟩; **6**, 322 ⟨81CC780⟩
Guillanton, G. Le, **3**, 352 ⟨90BSF427⟩, 652, 654 ⟨83T1551⟩; **4**, 277 ⟨90TL3149⟩; **5**, 75 ⟨69BSF2871⟩
Guillaume, D., **1**, 224 ⟨93SL599⟩; **3**, 638 ⟨93SL599⟩
Guillaume, M. R., **2**, 632 ⟨85JOC1779⟩
Guillaumet, G., **1**, 829 ⟨91TL5525⟩; **3**, 90, 97 ⟨93T1421⟩, 240 ⟨86S337⟩; **4**, 199 ⟨81TL1973⟩
Guillemin, J. C., **2**, 337 ⟨85S1131⟩; **3**, 337 ⟨88TL5899⟩, 457 ⟨88TL1287⟩, 564 ⟨88T4447⟩, 582 ⟨83CC238⟩, 689 ⟨92CC415⟩; **4**, 111 ⟨89CC988⟩, 131 ⟨91SC793⟩; **5**, 1152 ⟨91AG(E)196⟩, 1158 ⟨92CC415⟩, 1160 ⟨94MI 531-01⟩
Guillemonat, A., **1**, 1125 ⟨68BSF2120⟩; **2**, 361 ⟨61CR(252)4007⟩, 1071 ⟨69CR(268)358⟩
Guillerm, D., **1**, 1009 ⟨85TL3811, 86TL5857⟩, 1064 ⟨92JOC651⟩; **2**, 297 ⟨78JOM(149)C7⟩, 436 ⟨84JOM(269)107⟩; **5**, 1105 ⟨92TL5047⟩
Guillerm, G., **1**, 973 ⟨73JOM(54)153⟩; **2**, 127 ⟨93JA2139⟩; **5**, 1105 ⟨92TL5047⟩
Guillet, E., **3**, 227 ⟨78TL1155⟩
Guillochon, D., **5**, 727 ⟨92TL5117⟩
Guillot, N., **3**, 409 ⟨89SC2825⟩; **6**, 610, 727 ⟨89SC2825⟩, 653 ⟨90T3897⟩
Guimanini, A. G., **2**, 326 ⟨91S447⟩
Guinamant, J. L., **5**, 25, 56 ⟨86T2275⟩
Guindi, L. H. M., **2**, 398 ⟨90T7443⟩
Guindon, Y., **1**, 327 ⟨91JA9701, 92JA4912⟩, 327 ⟨93TL5885⟩; **2**, 135 ⟨83JOC1357⟩, 142 ⟨84JOC4538⟩, 525 ⟨88PAC1705⟩, 539 ⟨84TL4717⟩; **4**, 199 ⟨87TL2225⟩, 199, 217, 225 ⟨85JOC5379⟩, 217 ⟨90CJC897⟩, 224 ⟨80CAR(86)C3⟩
Guingant, A., **3**, 60 ⟨87JOC4788⟩, 415 ⟨85JA273⟩; **4**, 279 ⟨92TA1003⟩
Guinn, D. E., **3**, 222 ⟨87JOC5528⟩
Guinot, F., **2**, 1008 ⟨71BSF541⟩
Guir, C., **4**, 176 ⟨88PAC49⟩
Guirado, A., **3**, 699, 721 ⟨92TL4779⟩; **6**, 624 ⟨94TL2365⟩
Guiry, P. J., **1**, 435 ⟨90JCS(P1)2851⟩
Guitart, J., **3**, 128 ⟨86T4603⟩
Guitiérez-Puebla, E., **6**, 327 ⟨92JCS(D)2307⟩, 327, 330 ⟨93IC5569⟩
Guitiérrez-Puebla, E., **6**, 327 ⟨89IC2120⟩
Guittet, E., **1**, 225 ⟨81SC723⟩; **2**, 554 ⟨92TL3381⟩, 556 ⟨94T7177⟩; **4**, 249 ⟨93TL1141⟩
Guivisdalsky, P. N., **2**, 84 ⟨89JA3077⟩
Guixer, J., **2**, 299 ⟨78CC742, 81JCS(P1)574⟩
Guizard, C. G., **6**, 410 ⟨79JFC(13)175⟩
Gukasyan, A. O., **6**, 25 ⟨91RCR1318⟩
Guk, Y. V., **2**, 790 ⟨83RCR284⟩
Gukov, A. M., **5**, 698, 762, 813 ⟨78IZV2802⟩; **6**, 259 ⟨78IZV486⟩
Gula, M. J., **1**, 72 ⟨80JOC805⟩
Gulbins, E., **6**, 510 ⟨66LA(698)180⟩, 629 ⟨66CB1912⟩
Gulbins, K., **6**, 510 ⟨61CB3287⟩
Guldner, K., **2**, 891 ⟨84CB1999⟩; **4**, 594 ⟨87ZN(B)118⟩
Gulec, S., **2**, 489, 490 ⟨89JOM(364)67⟩, 501 ⟨88ZN(B)998⟩, 893 ⟨88POL2601⟩, 894 ⟨89PS(44)129⟩; **6**, 169 ⟨92OM145⟩
Gulevich, Y. V., **5**, 467 ⟨85JOM(282)421⟩
Gulevskaya, V. I., **6**, 149 ⟨68IZV621⟩
Gulewitsch, W., **3**, 616 ⟨06CB1181⟩
Gulick, L. G., **3**, 303 ⟨69JHC141⟩
Guliev, A. N., **6**, 518 ⟨80MI 616-01⟩
Gull, M.-R., **4**, 536 ⟨83LA1133⟩
Gull, P., **5**, 1125 ⟨86HCA246⟩
Gull, R., **1**, 183, 209 ⟨86AG(E)754⟩; **3**, 704, 706, 707 ⟨86AG(E)754⟩
Güllich, F., **1**, 42 ⟨59CB916⟩
Gullikson, G. W., **1**, 1008 ⟨87JMC193⟩
Gulliver, D. J., **4**, 288 ⟨84JCS(P2)429⟩
Gullotti, M., **2**, 151 ⟨88CC1451⟩
Gulman, D., **1**, 1127 ⟨86JPC402⟩
Gul'tyai, V. P., **1**, 103 ⟨75IZV888⟩
Gulube, A., **6**, 617, 635 ⟨91JPR187⟩
Gulubov, B., **6**, 558 ⟨89MI 617-01⟩
Gümbel, H., **6**, 668 ⟨85JOM(290)33⟩
Gumby, W. L., **1**, 859, 881 ⟨60JA1386⟩, 869, 871, 881 ⟨59JOC709⟩
Gumenyuk, A. V., **6**, 285 ⟨77ZOB2766⟩, 286, 666 ⟨75ZOB1965⟩
Gumulka, M., **2**, 662 ⟨74CC362⟩
Gunaratne, H. Q. N., **4**, 311 ⟨85CC1183⟩
Gunatilaka, A. A. L., **2**, 225 ⟨90JCR160⟩
Gund, T. M., **1**, 900 ⟨85JA1421⟩
Gundel, J., **3**, 246 ⟨90CB1175⟩
Gundermann, E., **5**, 276, 286 ⟨61AG621⟩
Gundermann, K.-D., **3**, 655 ⟨56CB1263, 74LA1661⟩; **6**, 232 ⟨85HOU(E11)158⟩, 234 ⟨85HOU(E11)665⟩
Gundersen, L.-L., **4**, 202 ⟨89ACS706⟩
Guneratne, R. D., **6**, 58, 264 ⟨94JOC2393⟩, 258 ⟨89JA1773⟩, 266 ⟨90JFC(49)75⟩
Gunertne, R. D., **4**, 802 ⟨90JFC(49)75⟩

Günes, Z. E., **6**, 564 ⟨86AP(319)1064⟩
Gung, W. Y., **4**, 386 ⟨88TL1657, 89T1043⟩, 386 ⟨89TL2183⟩; **5**, 467 ⟨88TL1657⟩
Gunji, H., **5**, 199 ⟨90TL709⟩
Gunn, B. P., **3**, 62, 221 ⟨85TL2869⟩, 209 ⟨79JOC450⟩; **5**, 214 ⟨88JMC1960⟩
Gunn, D. M., **4**, 832 ⟨69JA6195⟩
Gunn, V. E., **4**, 161 ⟨77JOC754⟩
Gunning, H. E., **3**, 337 ⟨65CJC717⟩, 488 ⟨68TL5237⟩, 530 ⟨80PAC1623⟩, 552 ⟨79NJC149⟩; **5**, 874 ⟨68TL5237⟩
Gunstone, F. D., **2**, 44 ⟨60MI 202-01⟩; **5**, 28 ⟨B-57MI 502-02⟩, 652 ⟨53QR175⟩
Günter, O., **3**, 333, 350, 654 ⟨73LA1637⟩
Günter, O. von, **4**, 462, 478 ⟨63LA(665)91⟩
Gunter, W. H. H., **2**, 735 ⟨76CC552⟩
Guntern, O., **1**, 582 ⟨92HCA1023⟩
Günther, D., **2**, 249 ⟨85S66⟩; **3**, 171 ⟨85S66⟩, 534, 541 ⟨63AG(E)608⟩
Gunther, E., **4**, 837 ⟨84T381⟩
Günther, H., **1**, 876 ⟨87AG(E)1212⟩, 146 ⟨89AG(E)1025⟩; **2**, 1000 ⟨89AG(E)1025⟩; **5**, 403, 404 ⟨72CB3357⟩, 485 ⟨80AG(E)563, 81PS(10)213⟩, 554 ⟨78S849⟩, 837 ⟨64AG440⟩
Günther, K., **2**, 186 ⟨73JOC4070⟩; **6**, 606, 607, 618 ⟨69JPR15⟩
Gunther, P., **1**, 705 ⟨86CB1540⟩
Günther, T., **6**, 267 ⟨71ZN(B)480⟩
Günther, W., **1**, 565, 681, 682, 683 ⟨84T1269⟩; **5**, 1042 ⟨90AG(E)1050⟩
Günther, W. H. H., **2**, 127, 278, 292 ⟨66JOC1202⟩, 278 ⟨63MI 204-01⟩, 278, 279, 292 ⟨65JMC845⟩, 279, 292 ⟨64JMC229⟩; **3**, 381, 382, 391 ⟨B-73MI 306-02⟩; **5**, 239 ⟨66JOC1202, 67JOC3929⟩; **6**, 590 ⟨B-73MI 619-04⟩
Güntner, A., **3**, 355, 356 ⟨84AG(E)729⟩
Guntrum, E., **1**, 797 ⟨86S921⟩; **2**, 396 ⟨91S629⟩; **3**, 258 ⟨91S629⟩
Guo, B.-S., **1**, 236 ⟨86T2803, 87JA4710⟩; **2**, 553, 555 ⟨92JOC3680⟩
Guo, B.-Z., **2**, 636, 638 ⟨88ACR135⟩, 639 ⟨87PAC1577⟩
Guo, C. Y., **4**, 803 ⟨90JA3152⟩
Guo, D., **1**, 67, 948 ⟨84TL5239⟩
Guo, G., **1**, 14 ⟨86TL4343⟩
Guo, H., **1**, 1071 ⟨91JOC2045⟩; **2**, 843 ⟨84S781, 86S691⟩
Guo, L., **1**, 109 ⟨93TL4129⟩
Guo, M., **3**, 65, 232 ⟨87HCA1429⟩
Guo, M.-J., **6**, 242 ⟨93TL149⟩
Guo, W., **5**, 866 ⟨81JA6164⟩; **6**, 55 ⟨81JA6164⟩, 55, 280, 633 ⟨81JOC5048⟩, 280, 632 ⟨82JOC4177⟩
Gupta, A., **2**, 458 ⟨88SC285⟩
Gupta, A. K., **1**, 573 ⟨93TA879⟩; **2**, 42 ⟨88BCJ93⟩, 519 ⟨86JOC4296⟩, 519, 531 ⟨87JOC310⟩, 1085 ⟨88JOC1391⟩; **3**, 276 ⟨88JOC1391⟩
Gupta, B. G., **2**, 452 ⟨83JOC1760⟩
Gupta, B. G. B., **2**, 15 ⟨83JOC3356⟩, 19 ⟨83S713⟩, 24, 30 ⟨77S676⟩, 32 ⟨80JOC1638⟩, 36 ⟨79AG(E)612, 79JOC1247⟩, 142 ⟨78JOC4503⟩, 142 ⟨78S137⟩, 150 ⟨79JA5317⟩, 156 ⟨83JOC3585⟩, 537 ⟨77S677⟩; **3**, 8, 121, 294 ⟨80S897⟩, 618 ⟨80S657⟩, 620 ⟨79S36⟩; **5**, 226 ⟨79JOC1247⟩, 270, 282, 300 ⟨79S274⟩
Gupta, D. N., **1**, 357 ⟨68JA6171⟩
Gupta, K. B., **4**, 187 ⟨89SC901⟩
Gupta, K. C., **1**, 733 ⟨76IJC(B)419⟩, 738 ⟨78IJC(B)623, 79IJC(B)637⟩, 741 ⟨80SC457, 80ZN(B)95⟩, 745 ⟨75JIC538⟩; **3**, 254 ⟨78SC315⟩, 500 ⟨78IJC(B)623, 79IJC(B)637⟩
Gupta, K. D., **6**, 253 ⟨86IC275⟩
Gupta, K. K. S., **2**, 832, 833 ⟨66JCS(B)1162⟩
Gupta, O. D., **4**, 564 ⟨90IC573, 91JFC(52)1⟩; **6**, 438 ⟨84CC416⟩
Gupta, O. M., **2**, 334 ⟨90JA2383⟩
Gupta, P. K., **3**, 244 ⟨75JOC1420⟩; **4**, 306 ⟨84IJC(B)1286⟩
Gupta, R., **2**, 695 ⟨90MI 213-02⟩
Gupta, R. P., **2**, 247, 364 ⟨79PS(7)325⟩
Gupta, S., **2**, 112, 558 ⟨93JOM(452)1⟩
Gupta, S. H., **6**, 174, 176, 381 ⟨92JCS(D)1015⟩, 174, 176, 381, 393 ⟨86CC1043⟩
Gupta, S. K., **1**, 80 ⟨69CC1365⟩, 510, 1101 ⟨72JA4370⟩, 1095 ⟨72JA4371, 75JA5249⟩; **2**, 365 ⟨77S39⟩, 522 ⟨71JA4062⟩, 527, 528 ⟨71JA1816⟩, 527, 900 ⟨72JA4370⟩, 527, 900, 901 ⟨75JA5249⟩, 531 ⟨70JA6983, 71JA2802⟩; **3**, 472 ⟨74S577⟩, 621 ⟨69JOC3684⟩; **4**, 442 ⟨70JHC439⟩; **5**, 295 ⟨70JOC825⟩
Gupta, S. N., **1**, 925 ⟨79JOC4123⟩
Gupta, S. P., **4**, 982 ⟨66JIC650⟩
Gupta, S. R., **5**, 107 ⟨86IJC433⟩
Gupta, S. S., **2**, 374 ⟨88OSC(6)936⟩
Gupta, U. C., **1**, 738 ⟨76JOM(112)279⟩
Gupta, V. K., **2**, 507, 894 ⟨87IC1638⟩, 893 ⟨88ICA(150)51⟩
Gupton, J. T., **1**, 253 ⟨89SC2415⟩; **2**, 746 ⟨82SC939⟩; **3**, 162 ⟨89SC2595⟩, 264 ⟨80JOC4522, 82SC35⟩; **5**, 755 ⟨83SC1083⟩
Gurak, J. A., **4**, 668 ⟨82JA2637, 84IC3717⟩; **6**, 204, 205, 404 ⟨83JOM(249)1⟩, 205 ⟨82JA7345⟩, 206 ⟨81JA5951, 82JA2637⟩
Guram, A., **4**, 947 ⟨88JOC3632⟩; **5**, 1115 ⟨88JOC3632⟩
Guram, A. S., **2**, 927 ⟨93JOC5595⟩; **5**, 1115 ⟨93SL259⟩
Gur'ev, N. I., **2**, 492, 509 ⟨78JOM(162)289⟩
Gurevich, P. A., **4**, 465 ⟨84JGU421⟩
Gurfinkel, E., **1**, 1066, 1084 ⟨70T2127⟩
Gurien, H., **1**, 636 ⟨63JOC878⟩
Gurjar, V. G., **5**, 1045 ⟨92SC2741⟩
Gurkova, S. N., **4**, 142 ⟨91MI 403-02⟩
Guro, T., **4**, 448 ⟨90H(31)2029⟩
Gurria, G. M., **2**, 610 ⟨77JOC3173⟩
Gurskaya, G., **4**, 222 ⟨91TL7593⟩
Gurski, A., **2**, 669 ⟨92JA1412, 93JA6101⟩
Gurskii, M. E., **1**, 58 ⟨84JOM(260)17, 84JOM(270)9⟩; **2**, 447 ⟨91MI 208-01⟩, 904 ⟨76IZV1580, 76IZV2062⟩, 1027 ⟨77IZV1456⟩
Gurun, C., **4**, 770 ⟨92SUL251⟩, 770, 776 ⟨92PS(72)225⟩
Gurusamy, N., **1**, 446 ⟨85JA2811⟩; **2**, 611 ⟨85JA2811⟩; **4**, 801 ⟨85JA2811⟩
Gur'yanova, I. V., **2**, 454 ⟨69ZOB2418⟩
Gurylev, É. A., **5**, 369 ⟨78IZV2412⟩
Guryn, R., **5**, 688 ⟨89PJC265⟩; **6**, 620, 621 ⟨80CB2583⟩
Gusar, N. I., **6**, 656 ⟨74ZOB538⟩
Gusarova, N. K., **1**, 441 ⟨91SR1⟩; **2**, 708 ⟨83IZV230⟩
Guseinova, E. S., **6**, 558 ⟨85MI 617-01⟩
Guseinov, F. I., **3**, 197 ⟨87JGU205⟩
Guseinov, M. M., **6**, 412 ⟨72URP345094⟩
Gusel'nikov, L. E., **1**, 63 ⟨81JOM(214)145⟩; **2**, 266 ⟨81JOM(214)415⟩; **3**, 501 ⟨79CRV529⟩; **4**, 610, 611, 614, 627 ⟨79CRV529⟩
Gusev, A. I., **4**, 142 ⟨91MI 403-02⟩, 640 ⟨76ZOB837⟩, 672 ⟨68CC1365⟩
Guseva, F. F., **5**, 576 ⟨91IZV1605⟩
Gush, D. P., **5**, 276 ⟨66JOC3829⟩
Gushchin, A. V., **2**, 879 ⟨88JGU629⟩, 886 ⟨85JGU63⟩
Gushchin, S. I., **6**, 145 ⟨74ZOR2003⟩
Gushchin, V. A., **2**, 886 ⟨88JGU1934⟩
Gushkin, A. V., **5**, 222 ⟨88ZOB1934⟩
Gus'kova, T. A., **5**, 567, 581 ⟨76KFZ41⟩
Gustafson, D. H., **1**, 725, 1001 ⟨67TL401⟩
Gustafson, G., **1**, 879 ⟨83JOC2957⟩
Gustafsson, B., **1**, 311 ⟨77ACS(B)667⟩
Gustavson, L. M., **1**, 816 ⟨86TL1557⟩; **4**, 202 ⟨86TL1557⟩
Gustorf, E. A. K. Von, **6**, 524 ⟨B-78MI 616-01⟩
Gustorf, E. A. Von, **5**, 781 ⟨77JOM(135)183⟩
Gut, R., **1**, 827, 828 ⟨79HCA1922⟩
Gut, S., **1**, 582 ⟨92JOC2270⟩
Gutbrod, H., **4**, 184 ⟨79LA1362⟩, 184 ⟨79LA522⟩

Gutbrod, H.-D., **6**, 70 ⟨80LA1677⟩
Gutch, C. J. W., **2**, 782 ⟨64PCS230⟩
Gutekunst, B., **5**, 923, 924 ⟨81CC191, 82JA5667⟩
Gutekunst, G., **2**, 508, 509 ⟨80ZAAC(462)113⟩; **4**, 594 ⟨80ZAAC(470)144, 80ZAAC(470)157⟩; **5**, 399 ⟨77AG(E)463⟩, 418, 924 ⟨82OM994⟩, 891 ⟨80ZAAC(470)131, 80ZAAC(470)144⟩, 923, 924 ⟨81CC191⟩, 923, 924 ⟨85AX(C)1632, 82JA5667⟩
Gutekunst, G. O., **2**, 134 ⟨20JA856⟩
Guthrie, D. J. S., **4**, 688 ⟨76JOM(120)C13⟩
Guthrie, J. L., **5**, 275 ⟨63OSC(4)513⟩
Guthrie, R. D., **3**, 448 ⟨65JCS2788, 65JCS3528⟩
Guthrie, R. W., **1**, 1029 ⟨89JMC1820⟩
Gutierrez, A. J., **5**, 688, 694 ⟨89JOC4335⟩
Gutierrez, C. G., **1**, 942 ⟨84JOC5206⟩; **4**, 230 ⟨88JOC5179⟩
Gutiérrez, E., **5**, 948 ⟨92JA7288, 94JA791⟩
Gutlérrez-Puebla, E., **4**, 583 ⟨84OM1438⟩; **6**, 332 ⟨88OM2577⟩
Gutman, A. L., **5**, 45 ⟨86JCS(P1)521⟩, 652 ⟨89JOC4263⟩
Gutman, D., **1**, 321, 921 ⟨90ACR375⟩, 1127 ⟨84JA4356⟩
Gutman, G., **1**, 81 ⟨67JOC893⟩
Gutmann, H., **1**, 956 ⟨59HCA719⟩
Gutowski, P., **5**, 561 ⟨88CB313⟩
Gutowsky, H. S., **1**, 922 ⟨67JA2054⟩
Gutsche, C. D., **3**, 125 ⟨54OR(8)364⟩, 538 ⟨62TL1297⟩; **5**, 199 ⟨58JA3711⟩
Gutschow, C., **1**, 458 ⟨B-69MI 122-01⟩
Gutterer, B., **2**, 316 ⟨91LA269⟩; **3**, 631 ⟨91KGS1432⟩
Guttieri, M. J., **1**, 32, 48 ⟨84JOC2875⟩
Guvanova, L. I., **4**, 140 ⟨75IZV1576⟩
Guy, A., **2**, 317 ⟨88S900⟩, 620 ⟨82T2339⟩
Guy, J., **5**, 755 ⟨47BCF406⟩
Guy, R. G., **5**, 1044 ⟨69CI(L)1702⟩, 1105, 1106, 1107, 1108, 1109, 1110 ⟨B-77MI 530-02⟩, 1105, 1110 ⟨61JCS2428⟩, 1105, 1110 ⟨61JCS2436, 78T541⟩, 1110 ⟨80T1839⟩
Guyader, M. Le, **2**, 764 ⟨66BSF1867⟩
Guyaro, D., **2**, 553 ⟨92T4593, 92TL5597⟩
Guyer, J. W., **4**, 342 ⟨70JOC2310⟩
Guyot, J., **2**, 102 ⟨19CR(164)795⟩
Guy Valadon, L. R., **1**, 725 ⟨73JCS(P1)1416⟩
Guziec, Jr., F. S., **1**, 231, 243, 620, 685, 704 ⟨88T6241⟩, 503 ⟨82JOC1221⟩, 684, 685, 686 ⟨74JCS(P1)1794⟩, 685, 686 ⟨75CC539, 93JOC43, 76JCS(P1)2079⟩, 686 ⟨82JOC3563⟩; **2**, 12 ⟨92JOC3772⟩, 858 ⟨75CC539, 76JCS(P1)2079⟩; **3**,, 4, 115 ⟨80S691⟩, 88 ⟨88OPP533⟩, 88, 215 ⟨82JOC1787⟩, 343, 348, 464 ⟨74JCS(P1)1794⟩, 381, 382 ⟨B-87MI 309-02⟩, 381 ⟨B-87MI 309-01⟩, 388 ⟨B-87MI 309-03⟩, 391 ⟨75CC539, 80JOC2890, 81TL4563, 82JOC3563⟩, 391, 527 ⟨76JCS(P1)2079⟩, 392 ⟨85T4843, 84JOC189⟩, 439 ⟨84S479⟩; **4**, 27 ⟨88S547⟩, 333 ⟨85JCS(P1)107⟩, 333 ⟨B-87MI 407-02⟩, 1051 ⟨B-87MI 527-01⟩, 1053 ⟨76JCS(P1)2079⟩; **5**, 226 ⟨83JCS(P2)1471⟩, 521 ⟨81JA7055⟩, 521, 523 ⟨82JCS(P2)473⟩, 575 ⟨89JCR(S)155⟩, 624 ⟨B-87MI 514-01⟩, 631 ⟨B-87MI 515-01⟩; **6**, 587 ⟨B-87MI 619-01, B-87MI 619-02⟩, 589 ⟨89JCS(P1)1068⟩, 598 ⟨94JOC4691⟩
Guziec, L., **6**, 598 ⟨94JOC4691⟩
Guzman, A., **3**, 178 ⟨81JOC1224⟩, 179 ⟨91CJC2059⟩, 179, 194 ⟨86JOC3374⟩
Guzman, L. Sh., **2**, 566 ⟨77JOM(142)145⟩
Guzman, M. C., **1**, 166 ⟨93TL63⟩
Guzzo, P. R., **5**, 315 ⟨94T11091⟩, 329 ⟨94JOC4862⟩
Gverdtsiteli, I. I., **5**, 969 ⟨75ZOB2185⟩
Gverdtsiteli, I. M., **2**, 1093 ⟨79MI 223-01⟩
Gverdtsiteli, M. G., **2**, 904 ⟨76IZV1580, 76IZV2062⟩, 1027 ⟨77IZV1456⟩
Gybin, A. S., **1**, 529 ⟨89IZV495⟩, 530 ⟨86TL1241, 89S472⟩
Gynane, M. J. S., **2**, 597 ⟨74JOM(81)329⟩, 949 ⟨80JOM(202)5⟩; **4**, 597, 598 ⟨80JCS(D)2428, 76CC623⟩; **6**, 165, 167, 168 ⟨80JCS(D)2428⟩, 195 ⟨80IC3637⟩
Györ, M., **6**, 546 ⟨90JCS(P1)2009⟩
Györi, B., **5**, 652 ⟨88MI 520-01⟩, 795 ⟨88JOM(344)29⟩; **6**, 491 ⟨88JOM(344)29⟩
Gysegem, P., **2**, 407, 862 ⟨75JCS(D)12⟩
Gysling, H. J., **4**, 377 ⟨79IC2696⟩; **5**, 241 ⟨80JOM(197)111⟩

H

Ha, D.-C., **2**, 544 ⟨89CRV1447⟩; **5**, 377 ⟨89CRV1447⟩
Ha, H.-J., **4**, 467 ⟨90TL1567⟩
Haack, R. A., **1**, 600, 788 ⟨82JOC4801⟩
Haada, K., **2**, 312 ⟨B-70MI 205-01⟩
Haaf, K., **3**, 696, 707 ⟨90CB635⟩
Haaf, W., **5**, 282 ⟨64CB3234⟩
Haag, A., **1**, 657 ⟨73TL401⟩, 986 ⟨63CB1535⟩; **4**, 522 ⟨72LA(755)67⟩; **5**, 1147, 1148 ⟨70LA(737)80⟩; **6**, 637 ⟨72LA(755)67⟩
Haag, J., **3**, 552, 579 ⟨74CB502⟩; **5**, 616, 1050 ⟨74CB502⟩
Haag, R., **3**, 603 ⟨85CB1720⟩
Haag, W., **1**, 640, 721, 733 ⟨55CB1654⟩; **3**, 494 ⟨55CB1654⟩
Haag, W. G., **2**, 254, 255 ⟨84EUP118109⟩
Haag-Zeino, B., **1**, 831 ⟨90TL4433⟩
Haaima, G., **1**, 1131 ⟨88TL1085⟩; **4**, 811 ⟨93T4229⟩
Haak, P., **1**, 1071, 1143 ⟨74JA4562⟩
Haake, M., **2**, 164 ⟨70TL4449⟩, 360 ⟨76S308⟩, 401 ⟨85HOU(E11)887⟩, 402 ⟨85HOU(E11)1299, 85HOU(E11)1321⟩; **3**, 374 ⟨76S310⟩; **4**, 478, 487 ⟨72CB2233⟩; **5**, 339 ⟨76S308, 79S97⟩; **6**, 625, 632 ⟨91S753⟩
Haake, P., **2**, 807 ⟨73JA8066⟩
Haaland, A., **1**, 862 ⟨79ACR415⟩; **6**, 186, 192, 195 ⟨86JCS(D)1551⟩, 186, 195 ⟨82CC1407⟩
Haan, J. W. de, **1**, 403, 799 ⟨70CC1168⟩, 403, 800 ⟨68RTC298, 70TL2755⟩
Haarman, H. F., **4**, 723 ⟨93OM2227⟩
Haas, A., **2**, 15 ⟨81AG(E)647, 85C134⟩, 292, 857 ⟨79ZAAC(448)35⟩, 361 ⟨75CB1365⟩; **3**, 395 ⟨91CB423⟩; **4**, 867 ⟨92CB571⟩, 872 ⟨84T4963⟩; **5**, 246 ⟨73JFC(3)383⟩, 561 ⟨87CB429⟩, 784 ⟨86JFC(30)439⟩, 812 ⟨B-81MI 521-01⟩, 995, 996, 1018 ⟨63AG(E)552⟩, 1005 ⟨73GEP(O)2153795, 63JCS1272⟩, 1018 ⟨67CB480, 69CB2718⟩, 1018 ⟨71CB2588⟩, 1141 ⟨76ZAAC(426)288⟩; **6**, 51, 54 ⟨81AG(E)647⟩, 80, 237 ⟨92CB571⟩, 94, 307, 536, 734 ⟨84T4963⟩, 228 ⟨82CZ239, 84MI 607-01⟩, 233 ⟨78JFC(11)509, 88JCS(P1)1179⟩, 233, 234 ⟨87JCS(P1)2119⟩, 233, 432 ⟨77CB67⟩, 235 ⟨66AG906⟩, 235, 253, 275, 726 ⟨87CB429⟩, 235, 254, 303 ⟨77CB916⟩, 235, 273, 534 ⟨68CB2609⟩, 236, 237, 254, 274, 532, 536 ⟨76CB3432⟩, 236, 252 ⟨68CB2617⟩, 237 ⟨85CB943, 86JFC(32)415⟩, 237, 256 ⟨92C78⟩, 237, 436, 477, 483 ⟨78CB2891⟩, 252, 274 ⟨74CZ109, 91CB51⟩, 256 ⟨93JCS(D)2547⟩, 256, 536, 537 ⟨76ZAAC(427)114⟩, 258, 449 ⟨71CB2732⟩, 273, 274 ⟨70GEP1908680⟩, 273, 607 ⟨66AG(E)845, 83CB3325⟩, 274, 275, 307, 534 ⟨76CB1976⟩, 276 ⟨74CZ511⟩, 276, 537 ⟨86ZN(B)413⟩, 432 ⟨67AG(E)705, 73JFC(3)383⟩, 432 ⟨91ZAAC(600)145⟩, 432, 435 ⟨69CB2718⟩, 433 ⟨76CB2475⟩, 440 ⟨83CB1257⟩, 450 ⟨66CB3103⟩, 529, 532 ⟨81CB829⟩, 534 ⟨72CB820⟩, 588 ⟨81ZN(B)1261, 91CC1378⟩, 588, 592 ⟨80ZN(B)526⟩, 602, 604 ⟨B-79MI 620-01⟩, 602 ⟨B-80MI 620-01, B-81MI 620-01⟩, 729 ⟨94CB597⟩, 734 ⟨79JA6282⟩
Haas, C. K., **2**, 267 ⟨79JOM(164)305⟩; **4**, 153 ⟨82JIC111⟩
Haas, G., **4**, 322 ⟨57CB444⟩, 322, 417, 424, 427 ⟨57CB50⟩, 417 ⟨57CB53⟩; **5**, 209 ⟨77AG(E)728⟩
Haas, H., **5**, 208 ⟨52LA(577)153⟩
Haas, J., **5**, 885 ⟨91AG(E)310⟩
Haas, P., **2**, 854 ⟨78CB629⟩
Haase, C., **6**, 154 ⟨93AG(E)1424⟩
Haase, K., **5**, 19 ⟨81S715, 82S237⟩
Haase, L., **6**, 157 ⟨69JPR577⟩
Haase, M., **6**, 372 ⟨84ZN(B)1500⟩, 387 ⟨85AG(E)324⟩, 388 ⟨86CB1117⟩
HaÅser, H., **6**, 106 ⟨71CB873⟩
Haba, A., **2**, 370 ⟨77CCC1466⟩
Habashi, A., **5**, 631 ⟨87S228⟩
Habashita, H., **1**, 188 ⟨90AG(E)801⟩, 188 ⟨91JOC4370⟩, 189 ⟨93JOC1207⟩, 189 ⟨93T9479⟩
Habashy, M. M., **5**, 652 ⟨90ACH423⟩
Habaue, S., **1**, 112 ⟨91JA5893⟩, 116 ⟨91JA8955⟩; **2**, 73, 563 ⟨91JA8955⟩, 74 ⟨89JOC5198, 92JOC6386⟩
Habeeb, T. D., **3**, 613 ⟨89JOC4476⟩
Habel, D., **6**, 268 ⟨60ZAAC(303)85⟩
Habendorff, R., **5**, 1040 ⟨57CB1590⟩
Habenicht, G., **1**, 999 ⟨62ZAAC(317)54⟩; **2**, 1082, 1083 ⟨62ZAAC(317)54⟩
Haber, K., **2**, 364 ⟨76JA7617⟩
Haber, M., **4**, 277 ⟨91T1925⟩
Haber, S., **5**, 1157, 1158 ⟨92CC1799⟩
Haber, S. B., **4**, 69 ⟨75JA5957⟩
Haber, W., **1**, 502, 837 ⟨47HCA1911⟩
Haberfield, P., **1**, 871, 881 ⟨61JA3678⟩
Haberland, G. G., **2**, 538 ⟨60JA1883⟩
Häberlein, H., **1**, 157 ⟨62ZN(B)787⟩
Haberman, L. M., **3**, 698, 707 ⟨85TL4971⟩
Habermas, K. L., **1**, 422 ⟨86T2821, 88HCA168⟩; **2**, 539 ⟨87JOC165⟩; **3**, 55, 229, 271, 272 ⟨88HCA168⟩
Habersaat, K., **3**, 404 ⟨73AG584⟩
Habib, M., **2**, 439 ⟨85IC2344⟩
Habib, M. J. A., **1**, 529 ⟨78JCR(M)4401⟩
Habib, R., **3**, 428 ⟨84IJC(B)801⟩
Habich, D., **1**, 208 ⟨88S943⟩, 448 ⟨78S755⟩; **4**, 512 ⟨79LA842⟩; **5**, 975 ⟨92S358⟩
Habicher, W. D., **5**, 1104 ⟨68AG(E)67⟩
Habisch, D., **2**, 374 ⟨62CB680⟩; **5**, 365 ⟨67CB142⟩
Habison, G., **5**, 354 ⟨77S487⟩
Habtemariam, A., **6**, 173, 394 ⟨93JOM(462)45, 91JOM(405)149⟩, 392, 394 ⟨92JOM(437)41, 90CC1471⟩
Habu, Y., **5**, 726 ⟨92CPB602⟩
Haché, B., **2**, 117 ⟨94TL1837⟩
Hache, K., **3**, 540 ⟨77CB3582⟩
Hachem, K., **3**, 122 ⟨80JOC2269⟩
Hachihama, Y., **4**, 211 ⟨64JOC1371⟩
Hachiken, H., **4**, 519 ⟨84H(22)701⟩
Hachiro, Y., **1**, 243 ⟨80SC595⟩
Hachisuka, C., **2**, 313, 544 ⟨91JCS(P1)1341⟩, 316 ⟨91JCS(P1)1767⟩; **3**, 458 ⟨91JCS(P1)1767⟩
Hachiya, I., **3**, 159 ⟨92TL1593⟩
Hacini, S., **1**, 422 ⟨79TL4553⟩; **3**, 212 ⟨79TL4553⟩
Hackenberg, J., **3**, 635 ⟨91CC470, 91S1205⟩; **4**, 1081 ⟨91CC470⟩; **6**, 254 ⟨91CC470⟩
Hacker, R., **3**, 179 ⟨68JPR12⟩; **4**, 647 ⟨86JOM(316)C4⟩
Hackett, S., **4**, 216, 218, 834, 836 ⟨86JOC879⟩, 834 ⟨84TL3539⟩; **6**, 336 ⟨86JOC879⟩
Hackler, L., **6**, 489 ⟨91JCS(P1)37⟩
Hackler, R. E., **1**, 399 ⟨70CC576⟩; **6**, 555 ⟨73JOC2106⟩
Hackley, Jr., B. E., **3**, 108 ⟨70JOC841⟩
Hackmann, J. Th., **5**, 662 ⟨71RTC463⟩, 662, 809, 810 ⟨71RTC458⟩, 809 ⟨70TL3763, 70JCS(C)875⟩, 1125 ⟨72RTC343⟩; **6**, 284 ⟨72RTC331⟩, 621 ⟨70JCS(C)875⟩
Hackney, M. L. J., **2**, 466 ⟨86CC850⟩
Haddad, M., **3**, 541 ⟨85TL833, 87TL881⟩
Haddadin, M. J., **4**, 829 ⟨72TL1353⟩
Hadden, S. K., **5**, 664 ⟨92JOC6662⟩
Haddock, E., **2**, 803 ⟨70JCS(C)2514⟩

Haddon, R. C., **1**, 826 ⟨79AJC1921⟩, 873 ⟨73JA3687⟩, 905 ⟨77JA377⟩, 917 ⟨88JA1108⟩; **3**, 363 ⟨85S639⟩
Hadegorn, I., **3**, 430 ⟨76MI 311-01⟩
Hadel, L. M., **6**, 280, 632, 731 ⟨87TL1969⟩
Hadfield, . S. T., **1**, 15 ⟨83JCS(P1)87⟩
Hadi, M. A., **4**, 138 ⟨89CC957⟩
Hädicke, E., **3**, 318 ⟨81TL4953⟩; **5**, 1044 ⟨78LA227⟩
Hadjiarapoglan, L., **4**, 209 ⟨89TL4223⟩
Hadjiarapoglou, L., **3**, 156 ⟨89TL6497⟩, 424 ⟨92CB571⟩, 500 ⟨88S913, 89TL6673⟩; **4**, 188 ⟨89TL6497⟩, 1081 ⟨88JCS(P1)2839⟩; **5**, 318, 321, 322 ⟨83S207⟩; **6**, 254 ⟨88JCR(S)306⟩, 278, 699, 708 ⟨88S913⟩
Hadji-Walassis, B., **6**, 105 ⟨58JPR70⟩
Hadler, H. I., **3**, 100 ⟨60JOC1896⟩
Hadley, C. R., **1**, 829 ⟨88JOC1922⟩; **5**, 159 ⟨88JOC1922⟩
Hadwick, T., **1**, 384 ⟨58JCS403⟩
Haeberle, K., **6**, 488 ⟨92GEP4219418⟩
Haebich, D., **5**, 1142 ⟨84GEP3430019⟩
Haeckl, F. W., **6**, 2 ⟨41JA3476⟩
Haefliger, W., **3**, 266 ⟨80S236⟩
Haelters, J. P., **2**, 457 ⟨85S1048⟩
Haemers, A., **4**, 461 ⟨94S23⟩
Haenel, J., **2**, 471 ⟨92ZAAC(607)161⟩
Haessig, R., **5**, 1035 ⟨91USP4997967⟩
Haeusler, G., **6**, 367 ⟨74GEP2237879⟩
Hafecz, F., **5**, 1125 ⟨88IJC583⟩
Häfele, B., **3**, 683 ⟨85MI 320-01⟩; **5**, 223 ⟨87S801⟩
Hafer, K., **1**, 901, 913 ⟨63JA2998⟩
Haferburg, D., **2**, 438 ⟨65ZN(B)916⟩
Hafez, E. A. A., **5**, 677 ⟨87AP(320)850⟩
Hafez, T. S., **4**, 320 ⟨92PS(66)87⟩; **5**, 731 ⟨92PS(66)87⟩; **6**, 117 ⟨91PS(57)211⟩
Haffmanns, G., **1**, 524 ⟨84JOC3314⟩; **4**, 513 ⟨84JOC3314⟩, 517 ⟨83TL4303⟩
Haffner, C. D., **1**, 807 ⟨92JA2635⟩
Häffner, J., **1**, 1123 ⟨65TL875⟩, 1125 ⟨64TL2191⟩
Hafner, A., **2**, 44, 67, 568 ⟨92CRV807⟩, 74 ⟨92JA2321⟩, 446 ⟨83HCA1774⟩; **5**, 932 ⟨88JA8413⟩
Hafner, K., **1**, 789 ⟨73AG(E)337⟩, 899 ⟨61LA(650)62⟩, 902 ⟨61AG342⟩, 982 ⟨86AG(E)466⟩, 995 ⟨90PAC531⟩; **2**, 1049 ⟨69HCA2641⟩; **3**, 77 ⟨68AG(E)460, 69HCA2641⟩, 274 ⟨68AG443⟩; **4**, 885 ⟨69HCA2641⟩
Hafner, L., **2**, 157 ⟨71LA(744)15⟩
Hafner, M., **3**, 606, 609 ⟨93S1247⟩
Hafner, T., **4**, 108 ⟨88LA1169⟩
Hafner, W., **6**, 75 ⟨83CB3482, 88T5095⟩
Haga, N., **3**, 634 ⟨73SC265⟩; **4**, 1075 ⟨90BCJ3056, 91BCJ2751⟩
Haga, T., **2**, 701 ⟨84T1381⟩; **6**, 9 ⟨91MI 601-02⟩
Haga, Y., **5**, 285 ⟨83JA7345⟩
Hagaman, E., **2**, 562 ⟨72TL2285⟩
Hagan, C. P., **3**, 623, 633 ⟨75JOC1162⟩
Hagan, J. P., **1**, 410, 821 ⟨75JA6878⟩
Hagarty, J. D., **2**, 771 ⟨65JA4576⟩
Hagedorn, I., **3**, 703, 714 ⟨61AG26⟩, 709 ⟨64AG(E)647⟩, 710 ⟨62AG499⟩, 710, 716 ⟨63AG305⟩, 712, 716 ⟨62AG215, 65CB193⟩, 721 ⟨65AG(E)702⟩; **5**, 852 ⟨66CB850⟩
Hagedorn, P., **6**, 69 ⟨54LA(587)226⟩
Hagel, I., **6**, 213 ⟨B-72MI 607-02⟩
Hägele, B., **4**, 465 ⟨88SC425⟩
Hägele, G., **2**, 454 ⟨88SC425⟩; **4**, 553 ⟨82GEP3111152, 85ZN(B)1152⟩, 553 ⟨86GEP3444678, 92ZN(B)725⟩, 802, 803 ⟨81PS(10)127⟩, 804 ⟨77PS(3)47⟩; **6**, 158 ⟨91PS(56)117⟩
Hagele, K., **2**, 214 ⟨55AG304⟩
Hagele, L. A., **2**, 1088 ⟨76S118⟩
Hägele, W., **6**, 627, 651 ⟨66CB2885⟩
Hagelee, L. A., **4**, 1054 ⟨75JOM(93)21⟩; **6**, 181, 182, 292 ⟨75JOM(93)21⟩, 181, 199, 292, 386 ⟨74JOM(69)45⟩, 182, 183, 199, 292 ⟨73JA5096⟩
Hageman, H. A., **2**, 33 ⟨53OR(7)198⟩; **5**, 1126 ⟨49JOC605⟩
Hagemann, H., **4**, 163 ⟨91HOU(E14a/1)600⟩, 176 ⟨91HOU(E14a/1)1⟩, 178 ⟨91HOU(E14a/1)136⟩, 186 ⟨91HOU(E14a/1)323⟩, 202 ⟨91HOU(E14a/1)412, 91HOU(E14a/1)431⟩, 203, 207 ⟨91HOU(E14a/1)683⟩; **5**, 809 ⟨69AG572⟩, 1004 ⟨73AG(E)999⟩, 1007, 1008 ⟨77AG(E)743, 72AG(E)1011⟩; **6**, 238 ⟨82GEP3044216⟩, 240 ⟨76CA(85)142668⟩, 408, 411, 419 ⟨83HOU(E4)1⟩, 437 ⟨83HOU(E4)1019, 83HOU(E4)1170⟩, 437 ⟨83HOU(E4)1234⟩, 438, 439 ⟨80EUP52842⟩, 445 ⟨69BRP1169158⟩, 448 ⟨71GEP1932830⟩, 451 ⟨73AG(E)999, 77AG(E)743⟩, 606 ⟨85GEP3337939⟩, 607 ⟨69AG(E)606⟩
Hagemann, M., **6**, 258, 679 ⟨90ZN(B)148⟩
Hagen, A., **4**, 442 ⟨83PHA449⟩
Hagen, E. L., **1**, 899 ⟨73ACR53⟩, 900 ⟨68JA6882⟩
Hagen, G. P., **2**, 572 ⟨77IC1357⟩
Hagen, H., **2**, 829 ⟨65CB2431⟩; **5**, 601 ⟨69CZ474⟩, 690 ⟨80LA1216⟩; **6**, 69 ⟨80LA1216⟩, 115 ⟨88LA605⟩, 140 ⟨79GEP2824028⟩, 143 ⟨82CB1721⟩, 643 ⟨84LA108⟩
Hagen, J. P., **1**, 141, 234, 821 ⟨78JOC1185⟩; **2**, 213 ⟨79JOC3230⟩; **4**, 199, 218 ⟨87JOC782⟩, 226 ⟨93JOC506⟩
Hagen, T., **6**, 243 ⟨93AG429, 93JOM(456)19⟩, 243 ⟨93ZN(B)935⟩
Hagen, T. J., **3**, 310 ⟨89JOC2170⟩
Hagen, V., **4**, 442 ⟨83PHA449⟩
Hagen, W., **3**, 109 ⟨88TL3407⟩
Hagenbruch, B., **1**, 248 ⟨84LA340⟩
Hagenmaier, H., **3**, 465 ⟨60ZN(B)751, 63CB1712⟩
Hagens, W., **2**, 1066 ⟨69RTC71, 73RTC762⟩
Hager, D. C., **2**, 827 ⟨78JOC947⟩
Hager, F. D., **5**, 76 ⟨32OSC(1)248⟩
Hager, R., **6**, 173 ⟨89CB2115⟩, 173, 378, 379 ⟨90AG(E)201⟩
Haggerty, B. S., **5**, 946 ⟨92JA6735⟩, 988 ⟨92OM1242⟩
Hagi, T., **4**, 79, 781 ⟨87CPB3959⟩, 102 ⟨91JCS(P1)49⟩
Hagihara, M., **2**, 456 ⟨85BCJ3104, 85BCJ3104⟩
Hagihara, N., **1**, 456 ⟨66BCJ1178⟩, 472, 1008, 1009, 1012, 1137 ⟨75TL4467⟩, 1010, 1012, 1137 ⟨80S627⟩, 1045, 1048 ⟨77S777⟩; **3**, 273 ⟨77S777⟩; **4**, 686 ⟨72CC993⟩; **5**, 304 ⟨77S777⟩
Hagio, K., **3**, 374 ⟨74BCJ909⟩
Hagiwana, D., **6**, 429 ⟨77BCJ718⟩
Hagiwara, D., **5**, 125 ⟨75S456⟩
Hagiwara, H., **3**, 54 ⟨90JCS(P1)1901⟩, 244 ⟨90SC3063⟩, 250, 256 ⟨79CL785, 81CC714⟩; **4**, 956, 957 ⟨79CL785⟩
Hagiwara, I., **3**, 138 ⟨83CL839⟩, 144 ⟨82CL939⟩, 144 ⟨84BCJ242⟩
Hagiwara, K., **5**, 1020 ⟨87OM2270, 88OM558⟩
Hagiwara, Y., **5**, 329 ⟨86JOC2391⟩
Haglid, F., **1**, 91 ⟨68JOC747⟩
Hagmann, W. K., **4**, 212 ⟨77JOC3458⟩
Hague, M. E., **6**, 541 ⟨86CPB430⟩
Hague, M. S., **5**, 413 ⟨82JOC747⟩
Hahl, R., **3**, 229 ⟨84JA3875⟩
Hahn, B., **3**, 349, 365 ⟨81LA10⟩
Hahn, C. H., **3**, 6, 118 ⟨84CC762⟩
Hahn, C.-S., **2**, 145 ⟨90TL2893⟩; **3**, 88 ⟨84S866⟩, 116 ⟨89TL2559, 86TL2875⟩, 215 ⟨85TL2875⟩; **5**, 723 ⟨85JOC1405⟩
Hahn, E., **2**, 564 ⟨86OM1296⟩; **4**, 578 ⟨84AG(E)522⟩
Hahn, F., **4**, 896 ⟨88AP(321)873⟩
Hahn, F. E., **3**, 694, 700, 723 ⟨93AG(E)650⟩, 700 ⟨91JOM(410)C9, 92AG(E)1212⟩, 711, 723 ⟨91AG(E)203, 91CB1683⟩, 723 ⟨92OM84⟩
Hahn, G., **1**, 2, 7, 8, 13, 15, 16, 17, 23, 24 ⟨86JOC1135⟩, 14 ⟨86JOC2596⟩, 977 ⟨82JOC3364⟩, 1052, 1055, 1056, 1058 ⟨83S883⟩, 1056 ⟨84JOC4565⟩; **3**, 129 ⟨86JOC1135⟩, 160 ⟨86JOC2596⟩; **4**, 1061 ⟨84JOC4565⟩

Hahn, G. R., **1**, 178 ⟨87JOC5484⟩
Hahn, J., **2**, 471 ⟨76ZN(B)1305, 83ZAAC(498)57⟩, 855 ⟨86ZAAC(543)7⟩
Hahn, J.-T., **3**, 294 ⟨90JOC6161⟩
Hahn, R. C., **2**, 15 ⟨90TL937⟩, 27 ⟨73OSC(5)328⟩; **4**, 39 ⟨88JOC1331⟩, 214 ⟨88JOC5783⟩
Hahn, T., **6**, 489 ⟨91PS(60)287, 93PS(83)65⟩
Hahn, V., **5**, 566, 618 ⟨57CCA319⟩, 630 ⟨66C175⟩
Hahne, W. F., **1**, 827, 828 ⟨79JOC882⟩
Hahnfeld, J. L., **4**, 7, 32, 35 ⟨77FCR119⟩, 755 ⟨75TL773⟩
Hähnke, M., **2**, 1092 ⟨70ZAAC(377)48⟩; **6**, 178 ⟨72ZAAC(390)157⟩, 178 ⟨72ZAAC(390)191⟩
Haiba, E.-A. I., **2**, 117 ⟨66JOC776⟩
Haider, H. M., **5**, 657 ⟨82JOC3658⟩
Haider, N., **4**, 416 ⟨88JHC119⟩
Haider, S. I., **5**, 1125 ⟨88IJC583⟩
Haiduc, I., **2**, 891 ⟨83JOM(252)295⟩, 941 ⟨B-69MI 215-04⟩; **4**, 692 ⟨73JOM(47)145⟩; **5**, 222 ⟨92JOM(426)49⟩
Haidukewych, D., **1**, 620, 1074 ⟨71CC22⟩
Haigh, D., **1**, 16 ⟨93TA625⟩; **2**, 683 ⟨94T3177⟩; **5**, 374 ⟨86TL5147⟩
Haight, A. R., **1**, 471 ⟨92JA6556⟩
Hailer, E., **3**, 448 ⟨02CB915⟩
Haimova, M., **3**, 310 ⟨86S886⟩
Hain, K., **5**, 606 ⟨77AP(310)1001⟩
Hain, U., **3**, 282 ⟨84LA1390⟩
Haines, A. H., **2**, 649 ⟨B-85MI 213-01⟩; **5**, 325 ⟨B-88MI 507-02⟩
Haines, S. R., **1**, 68 ⟨84JCS(P1)1689⟩
Haire, M. J., **3**, 439 ⟨77JOC3446⟩
Haiss, H., **3**, 462 ⟨60CB1541⟩, 465 ⟨60ZN(B)751, 63CB1712⟩
Hajdu, J., **2**, 144 ⟨93CC70⟩
Hajdu, R. A., **3**, 553 ⟨76JA7872⟩
Hájek, M., **4**, 17 ⟨83CCC1710⟩
Hajek, M., **6**, 7 ⟨80CCC3488, 80CCC3502⟩, 7 ⟨91MI 601-01, 92CCC1291⟩, 7 ⟨92JMOC51⟩, 58 ⟨94TL3537⟩
Hajjatie, M., **2**, 279 ⟨89T1611⟩
Hajjem, B., **5**, 690 ⟨92BSB445⟩
Hajo, M., **3**, 253 ⟨86S137⟩
Hajos, G., **6**, 114 ⟨87JOC2015⟩
Hajos, Z. G., **1**, 578 ⟨90OSC(7)363⟩; **6**, 506 ⟨65JOC2849⟩
Hak, K. L., **2**, 327 ⟨88JHC1383⟩
Hakanson, C. L., **5**, 1120 ⟨81USP4304935⟩
Håkansson, M., **4**, 671 ⟨93OM2268⟩
Hakimelahi, G. H., **2**, 104 ⟨84HCA906⟩; **4**, 12 ⟨79TL3643⟩, 491 ⟨80SC429⟩, 847, 848 ⟨80CJC2780⟩; **6**, 122 ⟨82TL913⟩, 540 ⟨83TL865⟩
Hakushi, T., **5**, 70 ⟨82NKK811⟩
Halamandaris, A., **6**, 508 ⟨57JA5710⟩
Halasz, S. P. von, **2**, 400 ⟨71CB1247⟩; **5**, 718 ⟨70CB553⟩; **6**, 260 ⟨68ZN(B)743⟩
Halazy, S., **1**, 245 ⟨81TL4341⟩, 248 ⟨82TL4385⟩, 623, 713, 716 ⟨79TL4233⟩, 623, 713, 716 ⟨80TL1997⟩, 712, 713 ⟨89H(28)1203⟩, 713, 716 ⟨79CC1136, 81TL1833⟩; **2**, 288 ⟨79TL4233⟩; **4**, 289 ⟨78TL3971⟩, 375, 400 ⟨81TL4737⟩; **6**, 58, 58 ⟨92BMC407⟩, 94 ⟨78TL3971⟩
Halbert, T. R., **4**, 687 ⟨80JA5101⟩
Halcomb, R. L., **1**, 9 ⟨92JA10138⟩; **4**, 188 ⟨89JA6661⟩
Halcour, K., **4**, 328 ⟨63AG1050⟩; **5**, 599 ⟨64AG(E)19⟩
Haldon, B., **4**, 604, 647 ⟨84JA6108⟩
Hale, K. J., **1**, 217 ⟨89TL6963⟩; **2**, 48 ⟨94TL425⟩
Hale, W. F., **1**, 637, 773, 782 ⟨59JA651⟩
Hales, J. L., **4**, 58 ⟨57JCS618⟩
Hales, N. J., **4**, 860 ⟨88TL6475, 92T7551⟩
Haley, E., **4**, 514 ⟨85TL5001⟩; **6**, 128 ⟨85TL5001⟩
Haley, G. J., **1**, 690 ⟨84JA5018, 86JA2932⟩, 690 ⟨86JA515⟩
Haley, M. F., **6**, 255 ⟨83CJC610⟩
Haley, N. F., **2**, 123 ⟨81JOC3141⟩; **4**, 93 ⟨85JA6298⟩, 93, 290 ⟨82TL1531, 83TL237⟩, 877 ⟨81TL1495⟩; **6**, 472 ⟨81JOC3141⟩, 552 ⟨80JOC175⟩, 729 ⟨78JOC678⟩
Haley, T. J., **2**, 614 ⟨71JOC631⟩
Halfor, S. E., **3**, 544 ⟨89JA8738⟩
Halkes, S. J., **4**, 748 ⟨65RTC904⟩
Hall, B., **3**, 456 ⟨78JCS(D)986⟩
Hall, C. D., **2**, 447 ⟨81JCS(P1)3059⟩
Hall, C. R., **2**, 257 ⟨80TL4959⟩, 463 ⟨79TL2729⟩, 466 ⟨76TL3645⟩
Hall, D., **1**, 756 ⟨89TL601⟩
Hall, D. M., **4**, 3, 37 ⟨71JCS(C)279⟩
Hall, D. R., **6**, 105 ⟨75JCS(P1)1302⟩
Hall, D. T., **5**, 101 ⟨72CJC3741⟩
Hall, G. E., **4**, 47, 48 ⟨50JOC715⟩
Hall, H. K., **4**, 827 ⟨82JOC2080⟩; **6**, 79 ⟨80MI 622-01⟩
Hall, I. H., **6**, 492 ⟨92AP(325)267⟩, 492, 515 ⟨91T6915⟩, 514 ⟨89EUP336772, 90EJM301⟩, 514 ⟨92MI 616-01⟩
Hall, Jr., H. K., **1**, 91 ⟨58JA6412⟩; **4**, 827 ⟨82TL603⟩, 993 ⟨90JOC1768⟩; **6**, 68 ⟨85MI 603-01⟩, 446 ⟨80JOC5325⟩, 509 ⟨58JA6409⟩
Hall, J. A., **2**, 547 ⟨61JA996⟩; **5**, 170 ⟨88JOC2371⟩
Hall, J. H., **3**, 684 ⟨80JPS(A)635⟩; **5**, 275, 298 ⟨76JOC3769⟩, 1011 ⟨66JOC1079⟩
Hall, J. K. A., **1**, 881 ⟨67JCS(C)1860⟩
Hall, J. R., **1**, 901 ⟨62JA3199⟩; **5**, 385 ⟨89IC815⟩; **6**, 247 ⟨88JOM(342)399, 93JOM(453)299⟩, 247 ⟨93JOM(453)307⟩
Hall, L. A. R., **4**, 728 ⟨52JA3868⟩
Hall, L. D., **2**, 4 ⟨83HCA1018⟩, 598 ⟨78CAR(63)1⟩, 627 ⟨82CC625⟩; **4**, 134 ⟨72JA5702⟩
Hall, M. B., **5**, 932 ⟨84JA1576⟩
Halland, H. L., **2**, 151 ⟨88CRV473⟩
Hall, P., **2**, 475 ⟨79ZAAC(458)249⟩
Hall, P. L., **1**, 131 ⟨91JA9571⟩
Hall, R. E., **1**, 1114 ⟨70JA2538⟩
Hall, R. G., **2**, 451 ⟨89T3787⟩, 849 ⟨82TL2603⟩; **6**, 116 ⟨89T3787⟩
Hall, R. H., **2**, 138 ⟨49JCS2723⟩
Hallas, G., **4**, 24 ⟨76OS(55)20⟩
Hall, S. S., **1**, 13 ⟨75JOC3151⟩, 97 ⟨71JOC2588⟩; **3**, 131, 405 ⟨87JOC3901⟩
Hall, S. W., **4**, 480 ⟨88CC867⟩, 549 ⟨88AG(E)837⟩; **5**, 396 ⟨87CC1753⟩, 878, 1155 ⟨87CC980⟩, 884 ⟨88CC171⟩, 889 ⟨88CC753⟩, 890 ⟨88AG(E)838⟩
Hall, T. L., **2**, 582 ⟨80JCS(D)1448⟩
Hall, T. N., **6**, 287 ⟨64JOC3587⟩
Hall, W. R., **3**, 372 ⟨74JA1817⟩; **5**, 690 ⟨86JHC1411⟩
Hallberg, A., **1**, 490 ⟨86JOC5286⟩, 1072 ⟨83JHC729⟩; **2**, 682 ⟨89JOC1502⟩, 932 ⟨86JOC5286⟩; **3**, 245 ⟨87TL4215, 88JOC4257⟩, 293 ⟨81JOC5414⟩; **5**, 334 ⟨89JOC2964⟩
Hallenbeck, L. E., **1**, 436 ⟨86JA7763⟩
Hallensleben, M. L., **3**, 619 ⟨72TL2057⟩
Haller, A., **3**, 614 ⟨1889AC(R)222⟩; **4**, 883 ⟨1887BSF23⟩
Haller, F., **2**, 997, 999 ⟨92CB2265⟩, 1003 ⟨91CB861⟩
Haller, H. L., **2**, 178 ⟨41JA1764⟩
Haller, K. J., **3**, 545 ⟨85JA3172⟩; **6**, 334 ⟨84OM1038⟩
Haller, R., **1**, 869 ⟨70T2027, 70T2035⟩
Haller, W. S., **2**, 292 ⟨72JOM(32)97⟩
Halleux, A., **2**, 1046, 1049 ⟨66AG(E)584, 70TL3141⟩; **3**, 596 ⟨64AG(E)752⟩; **4**, 979 ⟨68JCS(C)1726, 70JCS(C)881⟩
Hallgren, J. E., **6**, 341, 401 ⟨73JOM(50)265⟩
Halliday, D. A., **1**, 623 ⟨92JCS(P1)3225⟩
Halliday, R. P., **5**, 849 ⟨70JPS105⟩
Hallinan, E. A., **4**, 734 ⟨84TL4329⟩
Halling, K., **6**, 608, 611 ⟨89LA985⟩
Hallnemo, G., **2**, 652 ⟨88JMC1080⟩
Hallock, J. S., **1**, 68 ⟨83JA6882⟩; **2**, 40 ⟨83JA6882⟩
Halloran, D. J., **1**, 778 ⟨89JOC389⟩
Hallut, A., **1**, 82 ⟨63APO(1)155⟩
Halmann, M., **6**, 116 ⟨86MI 604-01⟩
Halnemer, G., **5**, 151 ⟨90T46⟩

Halova, O., **5**, 900 ⟨90MI 524-01⟩
Halpern, B., **4**, 206 ⟨71JCS2166⟩
Halpern, D. F., **4**, 49 ⟨93JOC7382⟩
Halpern, E., **4**, 763 ⟨52JA4104⟩
Halpern, J., **1**, 77 ⟨87JA1746⟩; **2**, 442 ⟨B-85MI 208-01⟩; **5**, 460 ⟨70JA3515⟩
Halpern, M., **5**, 436 ⟨87JA2553⟩
Halpern, Y., **1**, 901 ⟨73JA4960⟩, 902 ⟨71JA4219⟩, 911 ⟨71JOC2354⟩
Halsall, T. G., **2**, 675 ⟨64JCS1029⟩
Halshimoto, S., **2**, 753 ⟨87CE551⟩
Halsrud, D. A., **6**, 434, 435, 474, 534, 550, 551, 567 ⟨83JOC3966⟩
Halstenberg, M., **2**, 835 ⟨B-79MI 216-01⟩; **6**, 686 ⟨82CB1617⟩, 687 ⟨82CB2371⟩
Halterman, R. L., **1**, 448 ⟨85JA1379⟩, 1085 ⟨81TL4171⟩; **2**, 75, 529, 557 ⟨90JA6339⟩, 524 ⟨79JA248⟩
Haltermann, R. L., **2**, 529 ⟨86JA294⟩
Haltern, B. W. v., **2**, 613 ⟨84CC655⟩
Haltern, B. W. V., **2**, 613 ⟨86JOC1886⟩
Haltinanger, R. C., **2**, 268 ⟨85JA7167⟩
Haltiwanger, R. C., **3**, 368 ⟨84CC502⟩; **4**, 365 ⟨84JOC1691⟩
Halton, B., **1**, 549 ⟨B-87MI 112-01⟩; **4**, 604, 647 ⟨86JA5949⟩
Halver, J. E., **5**, 87 ⟨57JA637⟩
Halverstadt, I. F., **4**, 184, 189 ⟨66JMC127⟩
Halvey, N., **2**, 931 ⟨78OPP181⟩; **4**, 807, 1060 ⟨88OSC(6)1033⟩
Ham, B. M., **5**, 586, 609 ⟨89JOC1212⟩
Hama, M., **2**, 931 ⟨90TL2917⟩
Hamada, A., **1**, 556 ⟨83TL555⟩; **3**, 540, 543 ⟨72TL1849⟩, 540, 543 ⟨75CPB2933⟩
Hamada, M., **3**, 136, 203 ⟨90JOC5562⟩; **5**, 247 ⟨75BCJ2993, 75CL1231⟩
Hamada, S., **4**, 736 ⟨92TL337⟩
Hamada, T., **2**, 717 ⟨86S852⟩
Hamada, Y., **1**, 185, 200 ⟨80S1007⟩; **2**, 266 ⟨83TL4033⟩; **3**, 575 ⟨87JSP(123)476⟩, 613 ⟨80S1007⟩, 616 ⟨79TL4663⟩, 647 ⟨84TL427⟩; **4**, 210 ⟨84TL5409⟩, 1033 ⟨80BCJ1138⟩; **5**, 42 ⟨84OS(62)191⟩, 293 ⟨87CPB2698⟩, 775 ⟨83CPB3139⟩, 1127 ⟨78YZ1361, 79H(12)164⟩
Hamad, M. M., **6**, 117 ⟨91PS(57)211⟩
Ham, G. E., **5**, 1010 ⟨70JOC47⟩
Hamaguchi, F., **4**, 211 ⟨72CPB2123⟩; **5**, 631, 632 ⟨89H(29)435⟩
Hamaguchi, H., **3**, 426 ⟨83JOC2103⟩; **4**, 303 ⟨75JA4264⟩; **5**, 246 ⟨86CL935⟩
Hamaguchi, M., **3**, 324 ⟨75CL499⟩; **4**, 109 ⟨89JOC1135⟩, 277 ⟨92TL1321⟩
Hamaguchi, S., **5**, 118 ⟨81MI 502-06⟩
Hamamot, I., **2**, 394 ⟨87S258⟩
Hamamoto, I., **1**, 37 ⟨85JOC3692, 86JOC2832⟩, 37, 330, 443, 548, 946, 1131 ⟨85T4013⟩, 250 ⟨82CC821, 86JCS(P1)1439⟩, 486 ⟨84CC274⟩, 617, 638 ⟨81CL1139⟩; **3**, 190 ⟨86JOC2832⟩, 192 ⟨87S258⟩
Hamamoto, S., **2**, 410, 411 ⟨86CL1401⟩
Hamamoto, T., **2**, 352 ⟨91TL2555⟩; **5**, 129 ⟨94JOC4749⟩
Hamamura, E. K., **4**, 554 ⟨68TL5731⟩; **5**, 910 ⟨68TL5731⟩
Hamamura, K., **1**, 635 ⟨88BCJ3211⟩; **4**, 721 ⟨93JOM(463)169⟩
Hamamura, T., **3**, 54, 62, 252 ⟨79TL2809⟩
Hamana, H., **3**, 632 ⟨82CL1401⟩; **4**, 549 ⟨80JOC4683⟩
Hamana, M., **2**, 633 ⟨83JOC3297⟩
Hamanaka, E. S., **1**, 4, 9 ⟨79TL4631⟩; **3**, 559 ⟨74TL1515⟩
Hamanaka, N., **2**, 610 ⟨79CL1499⟩; **3**, 55, 252, 275 ⟨81JA4597⟩
Hamanaka, S., **1**, 86 ⟨89BCJ1682⟩, 815 ⟨76BCJ2017⟩, 916 ⟨92PNA(89)915⟩; **4**, 238 ⟨91CL1775⟩; **5**, 240, 243, 253 ⟨91JOC3776, 89CL1825⟩
Hamann, B., **1**, 446 ⟨66ZAAC(343)196⟩; **2**, 481, 486, 487, 502, 505, 510 ⟨65ZAAC(339)289⟩, 481, 486, 489, 505, 506, 510 ⟨65ZAAC(339)298⟩, 481, 486, 505, 510 ⟨64ZAAC(332)179⟩
Hamann, K., **6**, 510 ⟨61CB3287, 66LA(698)180⟩, 629 ⟨66CB1912⟩
Hamann, P. R., **3**, 172 ⟨83JOC914⟩
Hamaoka, T., **1**, 660, 1103 ⟨73JA6456⟩, 661, 1103 ⟨73JA5786⟩; **2**, 606, 628 ⟨73JA5786⟩, 623 ⟨73JA6456⟩, 623, 628 ⟨89JOC6068⟩, 900, 901 ⟨89JOC6075⟩
Hamashima, Y., **5**, 655 ⟨79TL4943⟩; **6**, 107 ⟨82H(18)201⟩
Hamatani, T., **1**, 570 ⟨88T2855⟩
Hamatani, T., **2**, 19 ⟨88T2875⟩
Ham, W. H., **1**, 25, 829 ⟨87JA3025⟩
Hamberger, H., **1**, 918 ⟨72JA6867⟩; **3**, 315 ⟨71CC1190, 71CC1192⟩; **4**, 983 ⟨77TL3619⟩
Hambly, G. F., **2**, 22 ⟨86TL2563⟩
Hamblyn, M., **2**, 365 ⟨75JCS(P1)2445⟩
Hamburger, A., **4**, 297 ⟨08CB24⟩
Hamdan, A., **5**, 657 ⟨82JOC3658⟩
Hamdani, M., **1**, 474 ⟨90BSF401⟩
Hamdouchi, C., **4**, 270, 863 ⟨91TL3695⟩
Hamed, A., **4**, 109 ⟨92S710⟩, 434 ⟨87S745⟩; **5**, 661 ⟨92T8271⟩; **6**, 618, 621 ⟨92T8271⟩, 619 ⟨86T6645, 89S918⟩
Hamedi-Sangsari, F., **3**, 40 ⟨88BSF95⟩
Hamel, E., **3**, 353 ⟨90JMC567⟩
Hamel, E. E., **4**, 430, 431, 432, 994 ⟨57JOC1665⟩
Hamel, L., **3**, 601 ⟨77JOC4261⟩
Hamel, N., **1**, 298 ⟨90JA2803⟩
Hamel, N. N., **6**, 251 ⟨91JFC(55)313⟩
Hamel, P., **3**, 307 ⟨81TL4901, 83JOC3214⟩
Hamelin, J., **2**, 489 ⟨90TL1147⟩, 489, 879 ⟨89TL349⟩; **3**, 639 ⟨86TL5611, 91TL2657⟩, 684 ⟨94JCR(S)116⟩; **4**, 444 ⟨82CJC285⟩, 502 ⟨90TL7607⟩, 566 ⟨87PS(30)523⟩, 567 ⟨86TL5611⟩, 597, 598 ⟨89TL349⟩; **5**, 494 ⟨75JOC2990⟩
Hamell, M., **2**, 631 ⟨50JOC162⟩
Hamer, D., **2**, 491 ⟨61JCS1398⟩
Hamer, J., **2**, 765 ⟨64CRV473⟩; **3**, 438 ⟨B-67MI 311-01⟩
Hamer, N. K., **2**, 807 ⟨66JCS404⟩; **3**, 243 ⟨83JCS(P1)61⟩
Hamer, R. R. L., **3**, 307 ⟨88JHC517⟩
Hamersma, J. W., **2**, 14 ⟨65JOC3985⟩
Hames, B. W., **6**, 332 ⟨83OM1696⟩
Hamid, A. M., **5**, 900 ⟨68JCS(C)1612⟩
Hamill, B. J., **1**, 811 ⟨82CC1164⟩, 1078 ⟨73CC151, 77JCS(P1)869⟩
Hamilton, A. D., **5**, 11 ⟨90JA9586⟩
Hamilton, C. E., **2**, 176 ⟨63JOC3537⟩
Hamilton, C. S., **2**, 491, 496, 498 ⟨47JA927⟩, 736 ⟨40JA1859⟩, 883 ⟨35JA913⟩, 884 ⟨36JA1573⟩, 884, 888 ⟨40JA3142⟩; **3**, 638 ⟨42JA1309⟩; **5**, 221 ⟨47JA875⟩, 842 ⟨46JA588⟩
Hamilton, F. G., **2**, 594 ⟨48OSC(1)519⟩
Hamilton, F. H., **2**, 187 ⟨48OSC(1)492⟩
Hamilton, L. A., **2**, 434 ⟨68USP3392180⟩, 450 ⟨52JA5418⟩, 451 ⟨60USP2957931⟩, 454 ⟨55JA3411⟩
Hamilton, R., **3**, 17, 404 ⟨79TL3201, 85JCS(P1)2123⟩
Hamilton, T., **5**, 1020 ⟨90JA4183⟩
Hamilton, T. P., **5**, 1020 ⟨89JCP1031⟩
Hamilton, W. C., **4**, 493 ⟨75JA38⟩
Hamlet, A. B., **4**, 923 ⟨81TL191⟩
Hamlet, P., **2**, 1013 ⟨65JCP(42)2255⟩
Hamlet, Z., **1**, 986 ⟨70S543⟩
Hamley, P., **2**, 327 ⟨91SL29⟩
Hamlin, R., **4**, 46 ⟨90JA6715⟩
Hamm, K. L., **2**, 379 ⟨58JOC529⟩, 388 ⟨59JA3294⟩
Hammann, I., **5**, 774 ⟨78GEP2701309⟩; **6**, 513 ⟨82EUP43978⟩
Hamm, P. C., **2**, 176, 177 ⟨56JA5580⟩
Hammar, G. W., **6**, 596 ⟨53JOC292⟩
Hamm, R., **2**, 213 ⟨48CB210⟩; **5**, 1111 ⟨48CB210⟩

Hammarström, S., **2**, 135 ⟨80GEP2936210⟩, 135 ⟨80JA3663⟩
Hammen, G., **1**, 1114 ⟨74CB1676⟩
Hammen, P. D., **2**, 188 ⟨71JOC2698⟩, 200 ⟨68CC9⟩, 200 ⟨71JOC2693⟩, 200 ⟨71JOC2703⟩
Hammen, R. F., **1**, 821 ⟨73JA962⟩
Hammer, C. F., **5**, 313, 314 ⟨84JA1543⟩
Hammer, G. G., **6**, 88 ⟨62JA1751⟩
Hammer, R., **4**, 582 ⟨76AG(E)42⟩
Hammer, R. P., **2**, 539 ⟨87JOC165⟩
Hammerich, O., **4**, 221 ⟨84SUL1⟩
Hammerschimdt, F., **3**, 270 ⟨79LA492⟩
Hammerschmidt, E., **2**, 132 ⟨78CB2445⟩
Hammerschmidt, F., **3**, 639 ⟨91JCS(P1)365⟩; **4**, 561 ⟨79LA492⟩
Hammersum, S., **4**, 437 ⟨72ACS1258⟩
Hammerum, S., **4**, 436 ⟨73ACS779⟩, 436, 437 ⟨72TL949⟩; **5**, 641 ⟨74CJC879⟩; **6**, 111, 114, 115 ⟨76ACS(B)837⟩
Hammett, L. P., **1**, 895 ⟨53JA1253, 53JA3955⟩; **5**, 31, 106 ⟨33JA4079⟩
Hammick, D. L., **3**, 362 ⟨48JCS1181⟩, 694 ⟨30JCS1876⟩; **4**, 13, 23 ⟨51JCS1145⟩; **6**, 28 ⟨23JCS2882, 39JCS781⟩, 28 ⟨51JCS1145⟩
Hammond, C. E., **4**, 855 ⟨81USP4348529A⟩; **6**, 342 ⟨89JA7283, 90JOM(394)265⟩
Hammond, G. B., **2**, 457 ⟨86JOC4342⟩; **3**, 198 ⟨86JOC4342⟩, 198 ⟨86TL4265⟩, 198 ⟨87JOC4185⟩
Hammond, G. P., **5**, 687, 698 ⟨93AJC1213⟩
Hammond, G. S., **1**, 299 ⟨62JA1015⟩, 665 ⟨75JOM(85)1⟩, 812 ⟨64JA899⟩, 924 ⟨64JA1911⟩; **3**, 541 ⟨66JA1242⟩, 573 ⟨59JA4878, 60JA5386⟩, 573 ⟨60JA5394⟩; **6**, 32, 219 ⟨67JOC833⟩, 144, 145, 369 ⟨63T(S)177⟩
Hammond, J. J., **4**, 19, 27 ⟨76JA1627⟩
Hammond, J. M., **5**, 772 ⟨77JOC1786⟩
Hammond, P. J., **2**, 870 ⟨81JA1785⟩
Hammond, W. B., **4**, 162 ⟨66JA3672⟩, 406 ⟨67TL3085⟩
Hammons, J. H., **1**, 844, 848 ⟨65MI 119-06, 67JA63⟩, 869 ⟨65JA384⟩
Hammoud, A., **1**, 468, 533, 1009 ⟨87TL1649⟩
Hamner, E. R., **5**, 465 ⟨74CC841⟩
Hamon, D. P. G., **1**, 4, 9 ⟨83AJC2243⟩, 328 ⟨93T6419⟩; **3**, 489 ⟨75CC459⟩; **4**, 102 ⟨92T5163⟩; **5**, 73 ⟨68CC206⟩
Hamon, J.-R., **1**, 165 ⟨82JA7549⟩
Hamon, L., **5**, 953 ⟨92JA8088⟩
Hampden-Smith, M. J., **4**, 151 ⟨88JCS(D)1773⟩; **6**, 342 ⟨89JA7283⟩
Hampel, B., **3**, 234 ⟨61CB1225⟩; **6**, 187 ⟨86OM1944⟩, 187 ⟨86OM730⟩
Hampel, K., **3**, 268 ⟨87TL4395, 88LA387⟩
Hampel, W., **6**, 575 ⟨71ZC12⟩
Hamper, B. C., **3**, 195 ⟨88JOC5558⟩
Hamper, M., **2**, 577 ⟨76JPR1019⟩
Hampp, A., **2**, 437 ⟨90ZN(B)1548⟩
Hamprecht, G., **2**, 368 ⟨81AG(E)151⟩; **3**, 247 ⟨69JOC2339⟩; **5**, 758, 760 ⟨83GEP3134144⟩, 1126 ⟨80MIP-DE283404⟩; **6**, 71 ⟨67AG(E)949⟩, 438 ⟨78GEP2706683⟩, 445 ⟨78GEP2828969⟩
Hamprecht, R., **6**, 573 ⟨72TL2939⟩
Hampton, K. G., **1**, 154 ⟨69JA676⟩, 871 ⟨67JA2664⟩, 879 ⟨65JOC2934⟩, 881 ⟨63JOC1946, 66JOC1035⟩, 881 ⟨67OS(47)92⟩
Hamraz, N., **5**, 631 ⟨76S728⟩
Hamrick, P. J., **1**, 881 ⟨68OS(48)80⟩
Hamsen, A., **2**, 571 ⟨82AG(E)144⟩, 577 ⟨92CB157⟩; **4**, 599, 600, 663 ⟨78TL4391⟩; **6**, 170, 210 ⟨79TL501⟩
Hamson, A. W., **3**, 220 ⟨85JA7967⟩
Hamura, S., **5**, 318 ⟨93BCJ2426⟩
Han, A.-L., **1**, 487 ⟨88CC975⟩
Han, B. H., **2**, 558 ⟨85MI 211-02⟩
Han, C. R., **3**, 215 ⟨85JCS(P1)267⟩
Han, D. I., **3**, 640, 654 ⟨88SC2111⟩; **4**, 370 ⟨89SC2213⟩, 943 ⟨90SC267⟩
Han, F., **5**, 572 ⟨94TL1135⟩
Han, G., **2**, 142 ⟨82SC903⟩
Han, G. Y., **6**, 487 ⟨70JA5748⟩
Han, I.-S., **3**, 121 ⟨85SC759⟩; **5**, 47 ⟨85SC759⟩
Han, J. S., **2**, 139 ⟨91BCJ2524⟩, 142 ⟨90CL2239⟩; **3**, 159 ⟨91CL949⟩, 281, 296, 298 ⟨92CL435⟩; **4**, 609 ⟨93OM2360⟩
Han, K. I., **1**, 436 ⟨86JA7763⟩
Han, L., **2**, 1076, 1080 ⟨92OM3993⟩; **3**, 177, 224 ⟨91TL229⟩
Han, L.-B., **1**, 247 ⟨93OM473⟩; **2**, 730 ⟨92JA7591⟩; **3**, 177 ⟨93CL561⟩; **4**, 948 ⟨92JA7591⟩
Han, N. F., **4**, 1033 ⟨87JOM(333)269⟩
Han, S., **5**, 933 ⟨83JA426⟩
Han, S.-H., **6**, 344 ⟨86OM2561⟩
Han, W., **2**, 44 ⟨93PAC633⟩, 150 ⟨92JA1428⟩; **3**, 437 ⟨93PAC633⟩, 634 ⟨92JA1428⟩; **4**, 262 ⟨92JA1428⟩
Han, Y.-H., **4**, 724 ⟨92OM3169⟩
Han, Y.-K., **2**, 907, 908 ⟨76JOM(105)C29⟩; **4**, 758 ⟨78JOM(162)1⟩
Han, Y. X., **2**, 753 ⟨85JOC1334⟩; **3**, 626 ⟨85JOC1334⟩
Hanack, M., **1**, 104 ⟨67TL3871⟩, 177, 224, 433, 1021 ⟨82S85⟩, 204, 1023 ⟨78S26⟩, 204, 1023 ⟨81S490⟩, 1091 ⟨84TL3437⟩, 1109, 1110, 1117 ⟨B-79MI 122-01⟩, 1110 ⟨70JA4985, 78CB1253⟩, 1110, 1117 ⟨71JA1513⟩, 1111 ⟨76TL3507⟩, 1111, 1113 ⟨78LA1894⟩, 1111, 1117 ⟨72AG(E)714⟩, 1111, 1117, 1124 ⟨77JOC174, 73CB3216⟩, 1113 ⟨78JA7066⟩, 1113, 1114 ⟨80TL503, 69JA2117⟩, 1114 ⟨71TL2171, 74CB1676⟩, 1114, 1115, 1124 ⟨74JA6686⟩, 1115 ⟨75JOC1994, 79CB2453⟩, 1116 ⟨70JA3802, 74TL1637, 77CB199⟩, 1117 ⟨72CB3789, 79AG(E)870, 80CB2015⟩, 1123 ⟨61LA(746)211, 65TL875⟩, 1123 ⟨66TL3847, 68TL4617⟩, 1123 ⟨70LA(733)5⟩, 1124 ⟨72CB421, 74TL2457, 77JA1669, 77TL4377⟩, 1125 ⟨64TL2191⟩, 1141 ⟨85TL1493⟩; **2**, 9 ⟨91S353⟩, 18, 610 ⟨78S217⟩, 36 ⟨87TL6441⟩, 157 ⟨81GEP2926671⟩, 689 ⟨70AG(E)273, 74S293⟩, 689 ⟨76JOC4099, 82S85⟩, 1089 ⟨81JOM(204)315⟩; **3**, 92, 101 ⟨90CC1571⟩, 285 ⟨85TL1285⟩, 492 ⟨89S685⟩, 635 ⟨91CC302⟩; **4**, 214 ⟨87S49⟩, 799 ⟨85CB1008, 89CB331⟩, 1081 ⟨91CC470⟩; **5**, 675 ⟨85CB1008⟩, 1083 ⟨85SM347⟩; **6**, 233 ⟨81TL3047⟩, 254 ⟨91CC470⟩
Hanaee, J., **5**, 597 ⟨90JCS(P1)1869⟩
Hanafusa, F., **3**, 634 ⟨79CL723⟩
Hanafusa, T., **2**, 18 ⟨90TL3167⟩, 19 ⟨86CC793, 92CL1161⟩, 920, 946 ⟨88JOC3539, 90JOC3393⟩; **3**, 116 ⟨83CL379⟩, 613 ⟨84CL725⟩, 616 ⟨87CL687⟩, 660 ⟨85CC838, 86TL1841⟩, 661 ⟨86JOC4714⟩, 668 ⟨86TL2411⟩, 671 ⟨86TL4201⟩; **4**, 426 ⟨82TL2665⟩; **5**, 285 ⟨81CL1143⟩, 1055, 1113 ⟨79S705⟩; **6**, 423, 438 ⟨86CC793⟩
Hanai, K., **3**, 635 ⟨88CPB1698⟩
Hanaizi, J., **4**, 809 ⟨92JOC6903⟩
Hanakawa, M., **3**, 670 ⟨86CPB4545⟩; **5**, 347 ⟨86CPB4545⟩
Hanamoto, T., **1**, 404, 406, 816, 819 ⟨88TL4763⟩, 444 ⟨83JOC2569⟩, 816 ⟨86TL4577⟩; **2**, 843 ⟨83JOC2569⟩; **3**, 189 ⟨91TL3555⟩, 496 ⟨83JOC2569⟩, 628 ⟨90CL1615⟩; **4**, 488, 489 ⟨85JOC1278⟩
Hanaoka, H., **3**, 158 ⟨93CL1767⟩, 158 ⟨93JOC2929⟩
Hanaya, K., **1**, 442 ⟨79JCS(P1)2409⟩
Hanaya, T., **2**, 431 ⟨91BCJ2398⟩, 769 ⟨89BCJ128⟩; **4**, 122 ⟨92CL407⟩
Hanazaki, Y., **2**, 293 ⟨85CL679, 86CL549⟩; **3**, 438 ⟨85CL679⟩
Hanby, W. E., **5**, 972 ⟨B-56MI 526-01⟩
Hance, C. R., **1**, 999 ⟨51JA5864⟩
Hancock, E. M., **1**, 881 ⟨55OSC(3)219⟩; **3**, 620 ⟨55OSC(3)219⟩
Hancock, K. G., **1**, 509 ⟨71JA1308⟩
Handa, K., **3**, 174 ⟨80TL1863⟩
Handa, S., **1**, 473 ⟨86BCJ2044⟩

Hand, J. J., **4**, 166 ⟨70JOC468⟩
Handford, B. O., **4**, 218 ⟨64JCS3832⟩
Handke, G., **1**, 960 ⟨91TL7229⟩, 970 ⟨93CB251⟩; **2**, 940 ⟨93JOC5849⟩
Handke, W., **4**, 553 ⟨79PS(5)337⟩, 553, 562, 563, 564 ⟨87ZAAC(555)109⟩, 866 ⟨88CZ107⟩
Handord, B. O., **2**, 223 ⟨64JCS3832⟩
Handrick, R. G., **4**, 407 ⟨42OS(22)65⟩
Hands, A. R., **1**, 647 ⟨67JCS(C)1099⟩; **3**, 494 ⟨68JCS(C)2448⟩
Hands, D., **4**, 509 ⟨91JCS(P1)1091⟩; **5**, 8 ⟨89S598⟩
Handwerker, B. M., **3**, 170 ⟨88JOC906⟩
Haneda, A., **1**, 3, 6 ⟨81TL2583⟩
Haneda, H., **4**, 660 ⟨88JAP267291⟩
Hanefeld, W., **2**, 249 ⟨81AP(314)799⟩, 389 ⟨90HOU(E16a/2)950⟩; **4**, 901, 904 ⟨89AP(322)593⟩, 901, 904 ⟨90AP(323)619⟩; **6**, 445, 540 ⟨91AP(324)917⟩, 494 ⟨80AP(313)995, 81AP(314)315, 91LA405⟩, 562 ⟨81LA1388⟩, 564 ⟨81AP(314)587, 82AP(315)103, 84AP(317)297, 85AP(318)848, 86AP(319)1064⟩, 566 ⟨92AP(325)173⟩
Hanefield, W., **2**, 157 ⟨81S229⟩
Hanek, M., **3**, 246 ⟨90CB1175⟩
Hanekamp, J., **1**, 464 ⟨90SC2349⟩
Haner, R., **3**, 537 ⟨85JA5396⟩
Hanessian, S., **1**, 177, 178 ⟨89JOC5831⟩, 587 ⟨92TL573⟩, 608 ⟨86TL2949⟩, 615, 698, 703 ⟨86JA2776⟩, 645 ⟨84JA5754⟩; **2**, 8 ⟨82TL4411⟩, 9 ⟨81TL3579⟩, 25 ⟨69JOC2163⟩, 50 ⟨93JOC1991⟩, 459 ⟨90TL6461⟩, 538 ⟨75CJC2975⟩, 539 ⟨79ACR159⟩; **3**, 6, 119 ⟨81S394⟩, 32 ⟨B-83MI 301-02⟩, 228 ⟨86TL2949⟩; **4**, 130 ⟨90TL6461⟩, 171 ⟨83JOC4427⟩, 198 ⟨72JA8929⟩, 224 ⟨80CAR(80)C17, 80CAR(86)C3⟩, 473 ⟨93SL35⟩, 492 ⟨90TL6465⟩; **5**, 277 ⟨80JA7578⟩
Hanewald, K., **5**, 556 ⟨76CB3243⟩
Hanford, W. E., **3**, 526, 538 ⟨46OR(3)108⟩; **6**, 2 ⟨49JOC747⟩, 5 ⟨50JA2213⟩, 80 ⟨48JA2268⟩
Hangauer, D. G., **3**, 207 ⟨80JA3964⟩
Hangauer, Jr., D. G., **5**, 164 ⟨80JA3964⟩
Hangauser, D. G., **3**, 615 ⟨83TL4559⟩
Hangedorn, M., **2**, 468 ⟨89AG(E)1675⟩
Hani, R., **5**, 887 ⟨87PS(30)463⟩
Hanika, G., **6**, 169 ⟨86JOM(312)C1⟩
Haning, H., **1**, 133 ⟨93TL7395⟩; **3**, 124 ⟨93TL7395⟩
Hanisch, H., **2**, 789 ⟨73S109⟩; **3**, 658 ⟨73S109⟩
Hanka, L. J., **1**, 241 ⟨71TL3791⟩
Hanke, D., **2**, 471 ⟨91ZAAC(549)36⟩
Hanke, H. E., **5**, 739 ⟨75JOM(97)39⟩, 1121 ⟨81JCED105⟩
Hanke, J., **6**, 237 ⟨92ZAAC(608)69⟩
Hanke, W., **4**, 317, 499, 500 ⟨81ZC403⟩, 455 ⟨87JOM(326)C9⟩
Hanko, R., **3**, 36 ⟨81AG(E)127⟩; **4**, 382 ⟨81AG(E)127, 85CB2822⟩
Hanks, D. R., **5**, 849 ⟨68JCS(C)1375⟩
Hanlan, A. J. L., **6**, 517 ⟨85OM478⟩
Hanley, I., **2**, 421 ⟨73JCS(D)1876⟩
Hanley, W. S., **2**, 218 ⟨71JMC995⟩; **5**, 247 ⟨71JOC2735⟩
Hanlon, B., **4**, 90 ⟨86JCS(P1)2207⟩; **5**, 360 ⟨86JCS(P1)2207⟩
Hann, M. M., **3**, 39 ⟨82JCS(P1)307⟩
Hanna, C., **2**, 379 ⟨52JA3693⟩
Hanna, I., **1**, 810 ⟨93JOC2349⟩; **3**, 157 ⟨89H(28)521⟩, 165 ⟨87S503⟩, 240, 245 ⟨90S755⟩
Hanna, J. A., **3**, 570 ⟨70JOM(21)P21⟩
Hanna, J., **1**, 687 ⟨73CL1041⟩
Hanna, Jr., ; J. M., **1**, 436 ⟨85JOC1125⟩
Hanna, M. A., **5**, 22 ⟨91PS(60)159⟩
Hanna, N. B., **2**, 364 ⟨88JHC1857⟩
Hannah, D. J., **3**, 16, 132 ⟨79SC301⟩
Hannah, J., **2**, 1016 ⟨61JA4663⟩
Hannahm, D. J., **4**, 655 ⟨81AJC181⟩
Hannan, F. J., **2**, 71 ⟨90JOC784⟩
Hanneman, E. J. M., **2**, 29 ⟨80SC821⟩
Hannick, S. M., **3**, 461 ⟨83JOC3833⟩
Hannig, E., **5**, 652 ⟨78PHA375⟩
Hannigan, T. J., **2**, 367 ⟨82JCR(S)84⟩
Hanninen, K., **5**, 1028 ⟨91JHC1003⟩
Hannon, F. J., **2**, 72 ⟨87TL5237⟩, 588 ⟨86JA7115⟩
Hannum, C. W., **4**, 38, 39 ⟨34JA712⟩
Hanold, N., **1**, 1077 ⟨86T1711⟩
Hanriot, M., **2**, 105 ⟨1879BSF(2)551⟩
Hans, M., **6**, 617, 635 ⟨91JPR187⟩, 626 ⟨86ZC204⟩
Hänsel, E., **2**, 944 ⟨81CB3154⟩
Hänsel, W., **1**, 869 ⟨70T2027, 70T2035⟩
Hansell, D. P., **6**, 558 ⟨85CC1405⟩
Hansen, A. E., **1**, 634 ⟨81JA5314⟩
Hansen, B., **4**, 553 ⟨61JA3539⟩; **5**, 911, 1093 ⟨61JA3539⟩
Hansen, E. T., **6**, 647 ⟨84SC1275⟩
Hansen, H., **5**, 366 ⟨69CB2889⟩
Hansen, H. C., **5**, 336 ⟨85T5145⟩, 483, 590 ⟨86T739⟩; **6**, 254, 275 ⟨87SUL59⟩, 273, 274, 275, 535, 553, 554 ⟨85T5145⟩, 274 ⟨86T739⟩, 274, 275, 554 ⟨84SUL85⟩, 533, 536 ⟨85SUL61⟩
Hansen, H.-J., **1**, 403, 795 ⟨70HCA173⟩, 415 ⟨68HCA1457, 74T1959⟩, 415 ⟨75HCA1293⟩, 420 ⟨72HCA3005⟩, 708 ⟨84HCA434⟩, 734, 795 ⟨72HCA1828, 72HCA1385⟩, 809 ⟨71HCA937, 73HCA75⟩, 827 ⟨63HCA2444, 71MI 118-02⟩, 835 ⟨73HCA105, 77HCA978⟩, 986 ⟨80HCA438, 85HCA2244⟩, 986 ⟨85HCA2249⟩, 993 ⟨73HCA875⟩; **2**, 656 ⟨63HCA2444, 73HCA14⟩, 660 ⟨73HCA75⟩; **3**, 678 ⟨77H(6)143, B-84MI 320-02⟩; **5**, 103 ⟨84OS(62)202⟩, 696 ⟨81HCA1322⟩
Hansen, J., **2**, 997, 1002 ⟨85JOM(285)1⟩; **4**, 531 ⟨85JOM(285)1⟩
Hansen, K., **4**, 561 ⟨93PAC617⟩
Hansen, K. C., **4**, 455, 477 ⟨67JOC2383⟩
Hansen, M. M., **2**, 330 ⟨88JA8734⟩, 702 ⟨87OM2069⟩
Hansen, M. R., **2**, 280 ⟨78JOC4885⟩
Hansen, N. C., **2**, 368 ⟨63ACS2141⟩
Hansen, P., **5**, 554 ⟨82ZC223⟩
Hansen, P.-E., **4**, 340 ⟨80JCS(P1)1627⟩, 841 ⟨77JCS(P1)1723⟩; **5**, 516, 521, 522 ⟨77JCS(P1)1723⟩, 516, 521, 522 ⟨80JCS(P1)1627⟩
Hansen, R. T., **1**, 1049, 1139 ⟨78JA2244, 80JOC3053⟩; **3**, 147 ⟨80JOC3053⟩
Hansen, S. W., **1**, 990 ⟨91CC566⟩; **4**, 821 ⟨86JA4229⟩
Hansert, B., **4**, 683 ⟨93CB2017⟩
Hans Fischer, **1**, 866 ⟨65LA(689)1⟩
Hanskin, A., **6**, 635 ⟨88S460⟩
Hansmersmann, U., **4**, 673 ⟨93CB89⟩
Hanson, A.-T., **2**, 783 ⟨82T389⟩
Hanson, A. W.,**1**, 318 ⟨81JC288⟩
Hanson, C., **2**, 798 ⟨B-76MI 215-02⟩
Hanson, E. M., **6**, 173, 195, 267, 270, 293 ⟨70JOM(24)647⟩, 267, 270, 293, 399 ⟨70JOM(23)361⟩, 270 ⟨71JOM(27)19⟩
Hanson, E. S., **6**, 225, 227 ⟨56USP2755314⟩
Hanson, J. C., **4**, 419 ⟨91TL2683⟩
Hanson, J. R., **1**, 380 ⟨91COS(3)705⟩, 382 ⟨67T793⟩; **2**, 657 ⟨69JCS(C)2349⟩; **3**, 426 ⟨70JCS(C)1182, 74S1⟩
Hanson, R. M., **1**, 220 ⟨91CRV437⟩
Hanson, R. N., **2**, 630 ⟨87JOC3687⟩, 632 ⟨80JOM(186)C33⟩; **6**, 651 ⟨73JOC1591, 79JOC4536⟩
Hanss, I., **3**, 709 ⟨91SL87⟩; **4**, 309 ⟨91SL87⟩
Hanssen, H. W., **3**, 303 ⟨71T3323⟩
Hänssgen, D., **2**, 384 ⟨70CB3733⟩, 400 ⟨81ZAAC(473)80⟩; 474 ⟨90CB1837⟩
Hansske, F., **1**, 939 ⟨83JA4059⟩; **6**, 535 ⟨83JA4059⟩
Hanssle, P., **4**, 950 ⟨72LA(763)208⟩
Hansson, T., **1**, 1031 ⟨90SL48⟩
Hansuld, M. K., **2**, 320 ⟨63OSC(4)795⟩

Hantke, K., **3**, 696 ⟨76JPR127⟩, 706, 707 ⟨73LA1571⟩, 711 ⟨75LA1531⟩, 713 ⟨76CB3964, 76LA969⟩
Hantzsch, A., **3**, 465, 466 ⟨02CB897⟩, 614 ⟨1899CB643⟩; **4**, 880, 883 ⟨1899CB641⟩; **5**, 822 ⟨1892CB705⟩, 838 ⟨1898CB2854, 1899CB3137⟩
Hanuise, J., **2**, 624 ⟨71C59⟩
Hanus, H. D., **3**, 542 ⟨64M1053⟩
Hanyu, Y., **1**, 242 ⟨84TL4011⟩, 983 ⟨86T1989⟩; **3**, 351 ⟨83JA6151⟩; **4**, 836 ⟨84TL4011⟩
Hanzawa, T., **3**, 412 ⟨79BCJ3469⟩
Hanzawa, Y., **1**, 197 ⟨92TL1295⟩, 197 ⟨93JA8835⟩, 197, 1057 ⟨92TL3769⟩, 517 ⟨83JOC1262⟩; **2**, 70 ⟨92TL4469⟩, 949 ⟨82JA6136⟩; **4**, 549 ⟨80JOC4683⟩, 736 ⟨90CPB1104⟩
Hanzel, R. S., **1**, 942 ⟨56JA6414⟩
Hanzlik, R. P., **4**, 177, 181 ⟨78JOC438⟩
Happe, W., **1**, 396 ⟨47LA(557)205⟩
Happel, J., **3**, 674 ⟨61USP3006948⟩
Happer, D. A. R., **1**, 445 ⟨67JOC2701⟩, 892 ⟨87AJC1537⟩; **6**, 177, 709, 710 ⟨78JOM(157)C50⟩, 177, 709, 710 ⟨80JOM(191)355⟩, 179, 709, 710 ⟨80JOM(186)309⟩, 734 ⟨91CC1608⟩
Hapsel-Hentrich, F., **4**, 140 ⟨86IC376⟩
Haque, M. S., **2**, 941, 1023, 1024 ⟨86JA7127⟩, 1023 ⟨85JA1429⟩; **3**, 15 ⟨86JA1325⟩, 154 ⟨85JOC5140⟩
Haque, N., **4**, 983 ⟨92JPR190⟩
Haquin, C., **6**, 635 ⟨86JOC4043⟩
Hara, A., **3**, 59 ⟨72CPB309⟩
Hara, D., **1**, 431 ⟨93JOC4897⟩; **4**, 758, 821 ⟨93JOC4897⟩
Hara, H., **3**, 160 ⟨87TL5891⟩
Hara, K., **1**, 613 ⟨83TL4993⟩; **2**, 141 ⟨90CL1097⟩, 678 ⟨83TL4993⟩; **4**, 197 ⟨83TL4993⟩
Hara, M., **2**, 546 ⟨77CC534⟩; **3**, 145 ⟨85CC1545⟩
Hara, N., **5**, 516, 519, 522, 523, 626 ⟨90BCJ835⟩; **6**, 246 ⟨70CPB784⟩
Hara, R., **2**, 927 ⟨93CC1042⟩, 927, 987, 988 ⟨93TL687⟩
Hara, S., **1**, 185 ⟨90JOC6356⟩, 278 ⟨79BCJ1865⟩, 281 ⟨91TL6749⟩, 489 ⟨88CL809⟩, 542, 1008 ⟨88TL1811⟩, 565 ⟨75TL1545⟩, 1050 ⟨92CL695⟩, 1091, 1094 ⟨86CL459⟩, 1104 ⟨89TL5153⟩; **2**, 76 ⟨91TL6749⟩, 623, 628, 902 ⟨83TL731⟩, 624, 902 ⟨83S1005, 83TL735⟩, 902 ⟨85S406, 89TL5153⟩, 902 ⟨90TL247⟩, 902, 927, 939 ⟨88TL1811⟩; **3**, 27 ⟨85S406⟩, 147 ⟨92CL695⟩, 233 ⟨89CL1959⟩, 370 ⟨77TL2939⟩; **4**, 70 ⟨80TL3579, 81JA2757⟩, 71 ⟨76TL2783⟩, 71, 72 ⟨77TL695⟩, 765 ⟨89TL5153⟩, 954 ⟨89CL1959, 92BCJ2303⟩; **5**, 485, 583 ⟨76TL2783⟩; **6**, 502 ⟨78JHC1231⟩
Hara, Y., **2**, 769 ⟨89BCJ128⟩
Harabe, T., **1**, 1021 ⟨88CPB4337⟩
Harada, F., **5**, 251 ⟨80SC305⟩
Harada, H., **1**, 438 ⟨93JOC6541⟩; **2**, 32, 36 ⟨83CPB4189⟩; **3**, 618 ⟨84CPB2560⟩; **5**, 199 ⟨90JOC1679⟩
Harada, K., **1**, 621 ⟨71T6101⟩; **2**, 52, 53 ⟨91COS(8)139⟩, 974 ⟨85JOC3104, 85JOC5524⟩; **3**, 684 ⟨78CPB3254⟩; **5**, 197 ⟨88JCS(P1)529⟩, 385 ⟨86BCJ2203⟩
Harada, M., **1**, 274 ⟨71JA1508⟩; **2**, 34 ⟨71JA1508⟩
Harada, N., **5**, 251 ⟨80SC305⟩
Harada, R., **5**, 351 ⟨68BCJ2521⟩
Harada, S., **1**, 690 ⟨88JA3578⟩; **3**, 142 ⟨87CL1371⟩, 227 ⟨86S847⟩, 311 ⟨86CPB4916⟩
Harada, T., **1**, 166 ⟨94SL121⟩, 198, 975 ⟨93JOC6166⟩, 431 ⟨92JOC5805, 93JOC4897⟩, 690 ⟨83JA7171, 88JA3578⟩, 807, 834 ⟨80JA2392⟩, 973 ⟨81CL621⟩; **2**, 21 ⟨75JOC2960⟩, 41 ⟨89CC1429⟩, 52 ⟨82CL315⟩; **3**, 281, 298 ⟨91S1216⟩; **4**, 17 ⟨83JA2771⟩, 758, 819, 821 ⟨92JOC5805⟩, 758, 821 ⟨93JOC4897⟩, 816 ⟨88JOC3089⟩; **5**, 63 ⟨82BCJ2186⟩, 607 ⟨78JA1923, 80JA7806⟩, 607, 616 ⟨78TL2167⟩
Haraguchi, K., **1**, 243 ⟨90TL227⟩; **2**, 281 ⟨89S434⟩
Harakal, M. E., **2**, 150 ⟨84JOC1465⟩
Harama, M., **1**, 973 ⟨81JOM(218)1⟩, 973, 1059 ⟨80JOM(198)1, 81JOM(218)275⟩
Haran, G., **6**, 46 ⟨72JCS(P1)34⟩
Harano, K., **1**, 836 ⟨91TL7557⟩; **2**, 123 ⟨87H(26)2583⟩; **6**, 472 ⟨72CPB2348, 73CPB604⟩, 472 ⟨74TL4479, 78CPB3807⟩, 472 ⟨87H(26)2583, 88H(27)2327⟩, 472 ⟨89CPB576, 92OPP200⟩, 473 ⟨91TL7557⟩
Haraoubia, R., **5**, 561 ⟨75MI 513-01⟩
Harasawa, K., **2**, 16 ⟨90JA8563⟩; **3**, 149 ⟨90JA8563⟩
Harata, O., **1**, 450 ⟨81JOM(217)43⟩
Harayama, T., **2**, 665 ⟨77JA5810⟩; **3**, 59 ⟨81CPB766⟩, 87 ⟨93CPB1166⟩
Harbach, J., **2**, 1001, 1002 ⟨93CB1887⟩; **4**, 660 ⟨92LA725⟩
Harbuck, J. W., **3**, 278, 308 ⟨72JOC3618⟩
Harch, P. G., **1**, 895 ⟨75JA2902⟩
Harchelroad Jr., F., **6**, 491, 515 ⟨79JINC1223⟩
Harcourt, D. A., **1**, 1014, 1029 ⟨90JA9330⟩
Hardcastle, I. R., **2**, 657 ⟨94TL1749⟩
Hardegger, E., **2**, 627 ⟨45HCA628⟩, 663 ⟨75HCA104⟩; **5**, 262 ⟨55HCA312⟩
Harden, A., **2**, 50 ⟨94TL4631⟩
Harden, D. B., **4**, 411 ⟨90JOC4777⟩
Harden, W. C., **2**, 261 ⟨27JA1017⟩
Harder, R., **3**, 578 ⟨80AG(E)564⟩
Harder, R. J., **6**, 238 ⟨61JA3422⟩
Harder, S., **2**, 999, 1000 ⟨89OM1688⟩
Hardie, B. A., **3**, 282 ⟨64JA2203⟩
Hardie, W. R., **4**, 184, 189 ⟨66JMC127⟩
Hardin, C. V., **6**, 231 ⟨71CC784⟩
Harding, C. E., **1**, 1110, 1117 ⟨71JA1513⟩, 1124 ⟨72CB421⟩; **2**, 689 ⟨70AG(E)273⟩
Harding, D. R. K., **1**, 834 ⟨76JA3312⟩; **2**, 205 ⟨75CJC3656⟩; **3**, 375 ⟨76CJC2652⟩; **5**, 488, 489 ⟨76CJC2652⟩
Harding, G. F., **2**, 128 ⟨48JCS2177⟩; **4**, 309 ⟨48JCS2174⟩
Harding, K. E., **1**, 68 ⟨84JOC2838⟩, 422, 784 ⟨84JOC3870⟩, 1126 ⟨78JA993⟩; **2**, 93, 354 ⟨91COS(4)363⟩; **3**, 57 ⟨79JA7131⟩, 527 ⟨77JOC4157⟩, 531 ⟨88JOC4877⟩
Harding, L. B., **1**, 1143 ⟨77JA2919⟩
Harding, M. M., **2**, 134 ⟨88TL4477⟩; **3**, 685 ⟨83JCS(P1)293⟩; **4**, 232 ⟨81JCS(P1)2991⟩, 1029 ⟨91JCS(D)351⟩; **5**, 220 ⟨85JOM(294)C29⟩
Harding, P. J. C., **1**, 41 ⟨80TL4031⟩; **3**, 13, 83, 96 ⟨78TL1437⟩
Hardinger, S. A., **2**, 394 ⟨82JOC365⟩; **4**, 333 ⟨81JOC765⟩
Hardstaff, W. R., **4**, 74, 75, 78 ⟨77CJC421⟩, 85 ⟨73CC50⟩; **6**, 254, 255 ⟨75CJC2664⟩
Hardstone, J. D., **5**, 690 ⟨88JMC1031⟩
Hardt, T., **5**, 652 ⟨88AP(321)41⟩
Hardtke, I., **1**, 686 ⟨84CB277⟩; **3**, 348 ⟨84CB277⟩
Hardy, D. G., **5**, 1126 ⟨67JA3281⟩
Hardy, E. E., **6**, 411 ⟨82KO(17)416⟩, 425 ⟨51JA3796⟩, 443 ⟨50JA1888⟩
Hardy, E. M., **1**, 411 ⟨41JA1852⟩, 411, 806 ⟨40JA441⟩; **2**, 122 ⟨55JOC475⟩
Hardy, F. E., **2**, 145 ⟨69JCS(C)2334⟩; **4**, 74, 75 ⟨69JCS(C)2334⟩; **5**, 709 ⟨67JCS(B)1151⟩
Hardy, G., **2**, 348 ⟨88JOC1695⟩
Hardy, Jr., R. A., **6**, 508 ⟨66JMC852⟩
Hardy, J. V. E., **6**, 524 ⟨49USP2473993⟩
Hardy, K. D., **5**, 219 ⟨71JCS(C)1917⟩
Hardy, P. M., **4**, 417, 424 ⟨77JCS(P1)1954⟩, 446 ⟨83MI 409-02⟩
Hardy, W. B., **2**, 621 ⟨58JA1716⟩, 720 ⟨65CI(L)81⟩; **5**, 984 ⟨67TL961, 68JA3295⟩
Hardy-Houis, M., **4**, 229 ⟨87JMC24⟩; **5**, 558 ⟨87JMC24⟩
Hareau, G., **2**, 363 ⟨91T6655⟩
Harel, Z., **1**, 498 ⟨82JOC3297⟩
Haremsa, S., **1**, 708 ⟨87LA679⟩
Haren, Jr., A. C., **1**, 78, 421 ⟨52JA4867⟩
Harendza, M., **1**, 925 ⟨93SL286⟩, 936 ⟨93SL283⟩
Harer, J., **2**, 471 ⟨83ZAAC(504)47⟩

Härer, J., **6**, 582, 684 ⟨85ZAAC(520)120⟩
Hares, O., **1**, 929 ⟨86CC1419⟩
Harger, M. J. P., **2**, 840, 841, 843 ⟨80JCS(P2)154⟩, 843, 846 ⟨82T1511, 80JCS(P2)1505⟩, 862 ⟨82T3073⟩; **4**, 94 ⟨86JCS(P1)1681⟩, 130, 131 ⟨89JCS(P1)563⟩; **5**, 360 ⟨81JCS(P1)3284⟩
Hargis, D. C., **2**, 320 ⟨90TL2991⟩
Hargrave, K. D., **6**, 515 ⟨76JA5702⟩
Hargrave, K. R., **2**, 146 ⟨54PRS(A)389⟩
Hargreaves, G., **5**, 49 ⟨55TFS1105⟩
Hargreaves, M. K., **4**, 3, 37 ⟨71JCS(C)279⟩; **5**, 305 ⟨70CRV439⟩
Hargrove, R. J., **1**, 981 ⟨75JOC657⟩, 1070 ⟨74JOC581⟩, 1111 ⟨74JCS(P2)843⟩, 1114 ⟨71TL2519⟩, 1117 ⟨76T37⟩; **2**, 204 ⟨71CC1614⟩
Hargrove, W. W., **2**, 608, 626 ⟨58JOC1412⟩
Harhash, A. H., **3**, 250 ⟨73IJC128⟩; **6**, 114 ⟨64T25⟩
Harigaya, Y., **3**, 145 ⟨92TL951⟩
Hariharan, P. C., **1**, 1109 ⟨72CPL385⟩
Harimoto, T., **2**, 352 ⟨94TL2205⟩
Häring, M., **5**, 201 ⟨57HCA852⟩; **6**, 629, 630 ⟨58HCA377⟩
Haring, W., **2**, 176, 177 ⟨62AP(295)615⟩
Harirchian, B., **1**, 248 ⟨87ACR371⟩; **2**, 908 ⟨77CC522⟩; **4**, 55 ⟨77CC522⟩
Hark, R. R., **4**, 226 ⟨90JA6263⟩; **5**, 513 ⟨90JA6263⟩
Harkema, S., **1**, 518 ⟨86JOC2004⟩; **3**, 428 ⟨86RTC103⟩; **4**, 1074, 1075 ⟨84JA3785⟩; **6**, 92 ⟨84JA3784⟩
Harkin, I. M., **4**, 580 ⟨91JCS(D)1117⟩
Harkin, S. A., **5**, 1125 ⟨84JCS(P1)147⟩
Harland, P. A., **2**, 299 ⟨84S941⟩; **3**, 188 ⟨84S941⟩
Härle, H., **6**, 605, 619 ⟨82CB3587⟩
Harley-Mason, J., **1**, 45 ⟨66JCS(C)425⟩, 1081 ⟨69AG(E)429⟩; **3**, 444 ⟨61JCS5560⟩
Harling, J. D., **1**, 928 ⟨88CC1380, 92T8031⟩; **6**, 557 ⟨92T8031⟩
Harlow, R. L., **2**, 438 ⟨90OM2653⟩, 926, 927, 988 ⟨87JA2788⟩, 979 ⟨80JA6149⟩; **4**, 582 ⟨87CC1543⟩; **6**, 623, 728 ⟨88JCS(P1)899⟩
Harman, D. G., **5**, 687, 698 ⟨93AJC1213⟩
Harman, L. D., **2**, 175 ⟨68IC391⟩
Harman, M.,**3**, 72 ⟨89JCS(P1)1793⟩
Harman, M., **4**, 842, 845, 851 ⟨89JCS(P1)1793⟩
Harman, P., **2**, 451 ⟨53USP2659714⟩
Harman, W. D., **1**, 94 ⟨88JA7906⟩; **5**, 201 ⟨91JA8972⟩
Harmata, M. A., **1**, 146 ⟨91TL6497⟩, 622 ⟨91T8855⟩, 706 ⟨88TL909⟩, 832 ⟨82JA4972, 83JOC3369⟩, 832 ⟨84TL1543, 89JA8878⟩, 993 ⟨87JOC4031⟩; **3**, 171 ⟨82JA4972⟩; **4**, 655 ⟨93S202⟩
Harmon, A., **5**, 322 ⟨81JOC4188⟩
Harmon, C., **1**, 890 ⟨74JA1348⟩
Harmon, J., **6**, 5 ⟨50JA2213⟩
Harmon, K. M., **1**, 906 ⟨57JA4557, 60JOC1442⟩, 906 ⟨B-73MI 119-08⟩
Harmon, R. E., **1**, 80 ⟨69CC1365⟩; **3**, 472 ⟨74S577⟩, 621 ⟨69JOC3684⟩; **4**, 442 ⟨70JHC439⟩; **5**, 295 ⟨70JOC825⟩, 1040 ⟨75MI 527-01⟩
Harms, K., **1**, 552 ⟨90CB555⟩, 816, 817 ⟨90CB917⟩, 858 ⟨89AG(E)73⟩, 1029 ⟨92CB271⟩; **2**, 997, 998 ⟨93AG(E)164⟩, 997, 999 ⟨92CB2265⟩, 999, 1005 ⟨91CB543⟩, 1000 ⟨85AG(E)573, 86AG(E)1011, 88AG(E)846⟩, 1001 ⟨90CB1853⟩, 1001, 1002 ⟨93CB1887, 92AG(E)79⟩, 1003 ⟨91CB861⟩; **3**, 71 ⟨82TL1147⟩, 566 ⟨86AG(E)373, 89AG(E)1392⟩, 616 ⟨88TL3295⟩, 704 ⟨92AG(E)79⟩; **4**, 646 ⟨91CB543⟩, 816 ⟨93AG(E)1023⟩; **5**, 932 ⟨90T1235⟩, 953 ⟨92CB2137, 93JOM(459)169⟩; **6**, 726 ⟨85CB5007⟩
Harms, R., **1**, 155 ⟨77LA40⟩; **3**, 696 ⟨76JPR127⟩, 696, 702, 704 ⟨77JPR1042, 76LA2105⟩, 704, 717 ⟨73LA611⟩
Harn, N. K., **1**, 820 ⟨88TL5119⟩
Harnden, M. R., **5**, 764 ⟨89JCS(P1)2207⟩
Harnest, G. H., **1**, 45 ⟨46JA1813⟩
Harnisch, H., **1**, 325 ⟨85CB1345⟩; **5**, 35 ⟨83S733⟩
Harns, C. J., **5**, 1126 ⟨76JCS(P1)716⟩
Haromy, T. P., **2**, 1003 ⟨87CAR(168)7⟩
Haroutounian, S. A., **2**, 311 ⟨89S616⟩
Harper, D. J., **3**, 684 ⟨80JPS(A)635⟩
Harper, D. P., **2**, 824 ⟨81JOM(205)161⟩
Harper, J. J., **1**, 1114 ⟨70JA3789⟩
Harper, J. L., **5**, 985 ⟨63CI(L)1280, 66FRP1440591⟩
Harper, J. R., **2**, 496 ⟨90JA4037⟩
Harper, S. D., **4**, 478 ⟨82JA2497⟩
Harpp, D. N., **1**, 111 ⟨74TL3511⟩, 232 ⟨83CJC757⟩, 568 ⟨90TL6291⟩, 651, 760, 981 ⟨78JOC1526⟩; **2**, 3 ⟨75JOC3420⟩, 89 ⟨88JA7737⟩, 120 ⟨86TL3975⟩, 131 ⟨90TL1397⟩, 143 ⟨71JA2437⟩, 143 ⟨78OS(58)138⟩, 151, 688 ⟨76JOC3987⟩, 195 ⟨81CC742⟩, 199 ⟨69TL1447⟩, 199 ⟨71JOC322⟩, 207, 213 ⟨71JOC1314⟩, 212 ⟨70JOC3259⟩, 212, 215, 231 ⟨86TL441⟩, 217 ⟨70TL3551, 71JOC3828⟩, 220 ⟨65JA3965⟩, 221 ⟨70JA7224, 84JA759⟩, 228 ⟨78JOC3481⟩, 231, 723 ⟨83TL5173⟩, 232 ⟨91JOC904⟩, 233 ⟨80JOC5155⟩, 236 ⟨78JA1222⟩, 241 ⟨82JOC2785⟩, 241, 358 ⟨71TL4953⟩, 244 ⟨84S1037⟩, 363 ⟨72TL5313, 77TL963⟩, 363, 688 ⟨73JOC4328⟩, 411 ⟨78TL3989⟩, 541 ⟨76S721⟩, 709 ⟨85TL1795⟩, 723 ⟨76TL3001⟩, 725 ⟨79JOC4135⟩, 931, 932 ⟨76JOM(107)C1⟩, 4366 ⟨70TL1687⟩; **3**, 176 ⟨90TL6291⟩; **4**, 831 ⟨79TL4029⟩
Harpp, D. N., **5**, 338 ⟨82JOC2785⟩
Harpp, D. N., **5**, 342 ⟨72TL5313, 73JOC4328⟩, 605, 622 ⟨83TL4927⟩, 605, 622 ⟨85CJC951⟩, 1111 ⟨79S181⟩; **5**, 8 ⟨88OSC(6)190⟩, 242, 337 ⟨87S1122⟩, 247 ⟨92SUL199⟩, 336 ⟨94JOC806, 84S1037⟩, 338 ⟨78JA1222⟩; **6**, 114 ⟨80JOC3713⟩, 478 ⟨80JOC271⟩, 577, 578 ⟨78JOC337⟩
Harran, P. G., **1**, 929, 935 ⟨93TL4505⟩
Harre, M., **5**, 79 ⟨84JA1146⟩
Harrell, B., **2**, 32 ⟨56JA3135⟩
Harrelson, Jr., J. A., **1**, 922 ⟨90CJC1714⟩
Harri, Jr., J. F., **2**, 4, 29, 392, 623 ⟨63OR(13)150⟩
Harriman, B. R., **3**, 616 ⟨46JA847⟩
Harrington, C. K., **4**, 232 ⟨72JA5115⟩
Harrington, F. P., **6**, 445 ⟨89JAN367⟩
Harrington, K. J., **3**, 541 ⟨74AJC2373, 77AJC459⟩
Harrington, P. M., **1**, 1038 ⟨91JA694⟩
Harrington, T., **5**, 1027 ⟨42JCS374⟩
Harriott, P., **6**, 380, 387 ⟨79JCR(S)12⟩
Harris, A. R., **2**, 19 ⟨86JCR(S)300⟩; **3**, 129 ⟨87SC1587⟩
Harris, B. E., **1**, 799 ⟨90CJC1961⟩
Harris, B. W., **1**, 925 ⟨62JCS4468⟩
Harris, C., **4**, 281, 282 ⟨65T2743⟩
Harris, C. E., **3**, 234 ⟨47OS(27)9⟩
Harris, C. J., **2**, 160 ⟨76JCS(P1)2166⟩; **5**, 342 ⟨89JCS(P1)1815⟩, 764 ⟨76JCS(P1)2166⟩
Harris, C. M., **1**, 879 ⟨69OR(17)155⟩; **2**, 662 ⟨74CC362⟩; **5**, 330 ⟨88JA6172⟩
Harris, C. W., **2**, 941 ⟨55JA6227⟩
Harris, D. H., **2**, 543 ⟨76MI 210-01⟩; **6**, 174, 186, 190, 191, 195 ⟨76JCS(D)2268⟩, 186 ⟨76CC261⟩
Harris, D. J., **2**, 530 ⟨86JA3422⟩; **5**, 152 ⟨89JA2984⟩
Harris, D. L., **1**, 918 ⟨72JA7063⟩
Harris, D. M.,**1**, 449 ⟨50JA5566⟩
Harris, E. E., **4**, 409 ⟨78JOC4276⟩, 743 ⟨26JA3144⟩
Harris, F. W., **5**, 842 ⟨66JHC155⟩
Harris, Jr., G. D., **2**, 309 ⟨89TL5709⟩
Harris, G. H., **3**, 616 ⟨46JA847⟩
Harris, G. S., **1**, 738 ⟨83T297⟩; **2**, 835 ⟨66JCS(A)520⟩; **3**, 328 ⟨72CC912⟩, 500 ⟨83T297⟩
Harris, H. P., **2**, 19 ⟨74JA2250⟩; **5**, 4 ⟨74JA2250⟩
Harris, J., **2**, 700 ⟨75JOM(90)C6⟩; **3**, 311 ⟨92JA5888⟩
Harris, J. C., **2**, 278 ⟨70JOC3086⟩
Harris, J. E., **4**, 553 ⟨63USP3098878⟩, 554 ⟨64JOC2427⟩; **5**, 1093 ⟨62JA4349, 64JOC2427⟩
Harris, J. F., **2**, 166 ⟨63JA749⟩, 688 ⟨64JOC898⟩, 1030 ⟨67JOC2063⟩; **4**, 215 ⟨65JOC2190⟩

Harris, Jr., J. F., **1**, 808 ⟨65TL1359⟩, 919, 948, 950 ⟨63OR(13)150⟩; **3**, 337 ⟨63JA963⟩, 654 ⟨72JOC1340⟩; **5**, 474, 475 ⟨63JA749⟩; **6**, 46, 233, 234 ⟨61JA840⟩, 46, 47, 233, 234 ⟨62JA3148⟩, 81, 233, 234, 302 ⟨67JOC2063⟩, 233 ⟨64JOC898⟩
Harris, J. J., **4**, 199, 218 ⟨87JOC782⟩
Harris, J. M., **1**, 1114 ⟨70JA2538, 70JA2540⟩
Harris, J. W., **4**, 930 ⟨79JA83⟩; **5**, 421 ⟨75JOM(99)379⟩, 924 ⟨76JA3381, 79JA83⟩
Harris, M., **1**, 238 ⟨80JA1064⟩; **2**, 170 ⟨66HCA2344⟩; **3**, 637 ⟨80JA1064⟩; **4**, 836, 887 ⟨71JOC2205⟩
Harris, N. J., **4**, 850 ⟨87S837⟩
Harris, N. V., **6**, 621, 728 ⟨90JMC434⟩
Harris, P. A., **1**, 260 ⟨90T987⟩; **2**, 302 ⟨91TL4247⟩, 319 ⟨90TL3999⟩; **3**, 186, 187 ⟨90T987⟩; **4**, 445 ⟨91CJC1153⟩; **5**, 603 ⟨89SUL117, 89TL3189⟩
Harris, P. G., **2**, 696 ⟨77SC137⟩
Harris, P. J., **4**, 681 ⟨76JCS(D)377⟩, 683 ⟨78JCS(D)403⟩
Harris, R. F., **1**, 1037 ⟨64JOC1872⟩; **2**, 420 ⟨63JOC2775⟩
Harris, R. K., **2**, 855 ⟨72JCS(D)1590, 72JCS(P2)380⟩
Harris, R. L., **2**, 431 ⟨83OM1877⟩; **3**, 376 ⟨74JA6119⟩, 380 ⟨71JA2471⟩
Harris, R. L. N., **4**, 704 ⟨89AJC301⟩; **5**, 731 ⟨74AJC2635⟩
Harris, R. M., **1**, 53 ⟨74MI 102-01⟩
Harris, R. N., **4**, 199 ⟨86JOC4711⟩
Harris, R. O., **6**, 720, 732 ⟨75JOM(90)C34⟩
Harris, S. A., **5**, 63 ⟨32OSC(1)336⟩
Harris, S., **5**, 1079 ⟨93OM1023⟩
Harris, S. J., **2**, 941, 1032 ⟨76CC1008⟩; **3**, 548 ⟨76CC1008, 77JOM(127)C1⟩; **5**, 511, 589 ⟨76CC1008⟩
Harris, S. M., **1**, 988 ⟨87JOC443⟩
Harris, T. M., **1**, 879 ⟨69OR(17)155⟩, 879, 881 ⟨58JA6360, 59JA1154⟩, 881 ⟨59JA1160, 63JOC1946⟩, 881 ⟨65JA82, 66JOC1035⟩, 881 ⟨67OS(47)92, 68OS(48)40⟩; **2**, 662 ⟨74CC362⟩; **4**, 167 ⟨77JA1631⟩, 167 ⟨81JOC2260⟩; **5**, 330 ⟨88JA6172⟩, 374 ⟨84JOC2015⟩
Harris, T. V., **2**, 430 ⟨85IC4437⟩
Harris, W. D., **2**, 101 ⟨51USP2529493⟩
Harris, W. E., **2**, 172 ⟨49MI 203-01⟩
Harrison, A. T., **1**, 131 ⟨91JA5751⟩
Harrison, C. R., **1**, 126 ⟨73CC544, 73TL4491⟩, 126 ⟨75TL1633⟩; **2**, 904 ⟨73CC544, 76JCS(P1)2419⟩, 904 ⟨76JCS(P1)2428, 76JCS(P1)2435⟩; **3**, 264 ⟨80JOC4522⟩, 618 ⟨77S41⟩
Harrison, D. G., **4**, 685 ⟨86CC285⟩
Harrison, D. J., **1**, 946 ⟨90JOC3321⟩
Harrison, G. C., **3**, 612 ⟨55OSC(3)372⟩
Harrison, H. R., **5**, 122 ⟨68CI(L)1568⟩
Harrison, I. T., **1**, 735 ⟨58JCS843⟩; **5**, 259 ⟨B-84MI 506-01⟩
Harrison, J., **2**, 60 ⟨94TL5201⟩
Harrison, J. J., **1**, 474 ⟨81T3957⟩; **2**, 626 ⟨81JOC2169⟩
Harrison, J. O., **3**, 68, 69 ⟨83JOC1921⟩
Harrison, K. N., **2**, 43 ⟨92JA9220⟩
Harrison, L. W., **1**, 66 ⟨84JA4218⟩
Harrison, M. J., **4**, 314 ⟨67JCS(C)735⟩
Harrison, P. G., **1**, 1079 ⟨74JOM(77)C22⟩; **2**, 601 ⟨82COMC-I(2)629⟩, 816 ⟨82COMC-I(2)662⟩, 1100 ⟨B-88MI 223-02⟩; **3**, 603 ⟨74JOM(77)C22⟩; **4**, 726 ⟨71IC522⟩; **6**, 571 ⟨72JCS(P1)130⟩
Harrison, P. J., **1**, 575 ⟨85JA1421⟩; **3**, 106 ⟨81CC797⟩
Harrison, R. G., **1**, 827 ⟨71JCS(C)2950⟩; **5**, 168 ⟨71JCS(C)2944⟩
Harrison, S., **5**, 259 ⟨B-84MI 506-01⟩
Harrison, W. H., **4**, 668 ⟨90JA2425⟩
Harrisson, R. J., **5**, 74 ⟨63OSC(4)493⟩
Harrit, N., **3**, 677 ⟨77ACS(B)848⟩; **5**, 626 ⟨85JCS(P1)907⟩
Harrod, J. F., **1**, 669 ⟨64JA1776⟩; **2**, 913 ⟨65JA16, 67JA1640⟩; **6**, 296, 297 ⟨74TL4409⟩
Harrold, S. J., **2**, 81 ⟨89JA1211⟩
Harrowfield, J. M. B., **5**, 386 ⟨81JA3559⟩
Harrowven, D. C., **1**, 345 ⟨91TL243⟩, 347 ⟨92TL2851⟩; **2**, 84 ⟨91TL243⟩
Harryvan, E., **2**, 1043 ⟨64RTC1211⟩
Hart, A. J., **1**, 859 ⟨79JA633⟩; **3**, 455 ⟨54JA4931⟩
Hart, D. G., **3**, 243, 496 ⟨75TL4353⟩
Hart, D. H., **3**, 494 ⟨73TL4425⟩
Hart, D. J., **1**, 243, 327 ⟨91SL412⟩, 320, 919 ⟨84SCI883⟩, 327, 328 ⟨92JOC4457⟩, 341 ⟨84JA8209⟩, 497 ⟨88T7037⟩, 547, 1131 ⟨85T3959⟩, 651 ⟨78JA1548⟩, 736 ⟨73TL4425⟩, 828 ⟨90JOC6236⟩, 924 ⟨88JA1631, 93TL7819⟩; **2**, 325 ⟨88JOC6022, 88T7037⟩, 544 ⟨89CRV1447⟩; **3**, 439, 440, 456 ⟨83JOC289⟩, 496 ⟨78JA1548⟩, 659 ⟨81JOC3671⟩, 696 ⟨92JOC2018⟩; **4**, 936 ⟨83TL4387⟩, 937 ⟨84JA8209⟩; **5**, 152 ⟨88JA1631⟩, 274, 294 ⟨86JA6054⟩, 277 ⟨87JOC4665⟩, 335 ⟨86JOC1929⟩, 377 ⟨89CRV1447⟩
Hart, D. W., **1**, 669 ⟨74JA8115⟩, 1105 ⟨75JA679⟩; **2**, 4 ⟨74JA8115⟩, 616, 623, 629, 968 ⟨75JA679⟩
Hart, G. C., **1**, 152 ⟨89JOC175⟩
Hart, H., **2**, 636 ⟨79CRV515⟩, 650 ⟨64JOC2397, 71CRV337⟩, 974 ⟨85JOC3104, 85JOC5524⟩, 1006 ⟨62TL983⟩; **3**, 271 ⟨88TL885⟩, 467 ⟨69JA706⟩, 538 ⟨68TL4995, 73JA6294⟩, 538 ⟨77JA4554⟩, 540 ⟨67JCS(C)1197, 68JA5296⟩; **4**, 24 ⟨67JOC2669⟩; **6**, 29 ⟨61JA4460, 70JOC3637⟩, 68 ⟨80JOC2096⟩
Hart, H. H., **2**, 451 ⟨33JCS776⟩
Hart, R., **5**, 989 ⟨56BSB291⟩
Hart, S. A., **6**, 143 ⟨87ZPK1829⟩
Hart, T. W., **5**, 727 ⟨92TL5117⟩; **6**, 239 ⟨80JCS(P1)1544⟩
Harteminck, W. A., **6**, 607, 610 ⟨73T297⟩
Hartemink, M. A., **6**, 238 ⟨73BSB23⟩
Hartemink, M., **5**, 963 ⟨73T297⟩
Hartenhauer, H., **4**, 256 ⟨88ZC269⟩
Harter, W. G., **4**, 537 ⟨88TL761⟩
Hartford, W. H., **3**, 86 ⟨58CRV1⟩
Hartgerink, J., **2**, 577 ⟨80JOC291⟩
Hartgraves, G. A., **1**, 967 ⟨90TL3699⟩; **6**, 64 ⟨88JFC(39)425⟩
Harting, J., **2**, 578, 579 ⟨89AG(E)325⟩
Hartke, K., **1**, 1047 ⟨89TL1073⟩, 1047, 1048 ⟨91LA903⟩; **2**, 140 ⟨91PS(58)223⟩, 240 ⟨73LA1644⟩, 1008 ⟨60CB1305⟩; **3**, 333, 350, 654 ⟨73LA1637⟩, 351 ⟨92LA927⟩, 561, 699, 721 ⟨66CB3163⟩; **4**, 96 ⟨63CB600⟩, 99 ⟨63CB595⟩, 229 ⟨80LA1665⟩, 281 ⟨91LA243⟩, 305 ⟨63CB604⟩, 775 ⟨88AP(321)903⟩, 834 ⟨88LA933⟩, 895 ⟨88AP(321)863⟩, 896 ⟨88AP(321)873, 88AP(321)879⟩, 901 ⟨64TL2737⟩, 901, 904 ⟨64AG781⟩, 973, 974 ⟨70CB133, 70CB147⟩, 984 ⟨86AP(319)1098⟩; **5**, 513, 549, 560 ⟨89TL1073⟩, 513, 559 ⟨85S960⟩, 548, 559 ⟨91TL4679⟩, 548, 559, 584, 586, 614 ⟨80CB919⟩, 554 ⟨76ZC318, 80CB1898⟩, 555 ⟨80LA483, 88AP(321)903⟩, 557, 559 ⟨89LA321⟩, 558 ⟨80AP(313)1029, 83SUL191⟩, 558 ⟨92LA595⟩, 558, 586, 727 ⟨92S1199⟩, 560, 572, 597, 617 ⟨91LA903⟩, 584, 586, 612 ⟨92LA921⟩, 639 ⟨88CZ239⟩, 685 ⟨86TL2743, 89CB657⟩, 685 ⟨89CB669⟩, 776 ⟨80CB1394, 82ZAAC(486)61⟩, 1064 ⟨64AG781⟩; **6**, 538 ⟨80CB1898⟩, 546, 554, 555, 567 ⟨93H(35)77⟩, 622 ⟨66CB3155⟩, 627 ⟨66CB3268⟩, 630 ⟨70AP(303)625⟩
Hartkopf, U., **6**, 380 ⟨82AG(E)443⟩
Hartl, A., **4**, 135, 597 ⟨83CB348⟩
Hartl, H., **5**, 138 ⟨93TA457⟩, 151, 164 ⟨88JA4640⟩
Hartle, R. J., **4**, 569, 574 ⟨66JOC4288⟩
Hartley, D., **4**, 966 ⟨79COC(2)61⟩; **5**, 64 ⟨62JCS4722⟩
Hartley, F. R., **1**, 669 ⟨69CRV799⟩; **2**, 67 ⟨B-87MI 202-01⟩, 433 ⟨79JCS(P1)3029⟩, 579 ⟨82COMC-I(6)471⟩; **4**, 336 ⟨B-90MI 407-03, B-92MI 407-01⟩, 451 ⟨B-90MI 410-01⟩
Hartley, J. B., **3**, 293 ⟨55JCS2686⟩
Hartley, R. C., **1**, 162 ⟨93TL1449⟩
Härtling, S., **3**, 670 ⟨83TL2841⟩

Hartman, B. C., **1**, 217 ⟨73JOC4346⟩
Hartman, G. D., **1**, 618, 706, 707 ⟨82S504⟩; **2**, 160, 162 ⟨84JMC1634⟩, 162 ⟨83TL1011⟩; **3**, 696, 699, 704 ⟨88OSC(6)620⟩; **4**, 81 ⟨82S504⟩
Hartman, R. D., **1**, 618, 706, 707 ⟨82S504⟩; **2**, 160, 162 ⟨84JMC1634⟩, 162 ⟨83TL1011⟩; **4**, 81 ⟨82S504⟩
Hartman, R. J., **5**, 34 ⟨63OSC(4)93⟩
Hartman, W. W., **2**, 36 ⟨43OSC(2)322⟩; **4**, 23 ⟨32OSC(1)357⟩; **5**, 115 ⟨32OSC(1)543⟩, 140 ⟨55OSC(3)650⟩
Hartmann, A., **3**, 481 ⟨79CB2509⟩, 481, 489 ⟨74TL1825, 78CB3068⟩; **5**, 403, 404 ⟨72CB3357⟩, 1022 ⟨83HOU(E4)834⟩; **6**, 430 ⟨57GEP1018054⟩
Hartmann, A. G., **2**, 232 ⟨78JOC3374⟩
Hartmann, E., **5**, 781 ⟨89ZAAC(575)10⟩
Hartmann, G., **6**, 604 ⟨86JFC(34)46⟩
Hartmann, H., **1**, 445, 999 ⟨57ZAAC(290)348, 56ZAAC(287)261⟩, 999 ⟨61ZAAC(312)186, 62ZAAC(317)54⟩, 999 ⟨63N403⟩; **2**, 436, 825, 826 ⟨68LA(714)1⟩, 489, 1081, 1082 ⟨61ZAAC(312)186⟩, 1058 ⟨61AG173⟩, 1058, 1063, 1066, 1069 ⟨56ZAAC(287)261⟩, 1060 ⟨64N213⟩, 1066 ⟨63N403⟩, 1076, 1079 ⟨57ZAAC(290)348⟩, 1076, 1081 ⟨59N321⟩, 1077 ⟨65N59⟩, 1082, 1083 ⟨62ZAAC(317)54⟩; **5**, 559 ⟨64CB1886⟩, 626 ⟨71ZC60⟩, 633 ⟨81ZC271⟩; **6**, 51 ⟨79S241⟩, 572 ⟨73JPR144⟩, 630 ⟨89EGP270530⟩
Hartmann, J., **1**, 113 ⟨76HCA453⟩; **2**, 964 ⟨76HCA453, 76JA4674⟩; **4**, 950 ⟨74S888⟩
Hartmann, M., **6**, 538 ⟨85EGP256705⟩, 650 ⟨76MI 621-01⟩
Hartmann, P., **5**, 92 ⟨92OS(71)200⟩
Hartmann, R. D., **2**, 613 ⟨81JMC93⟩
Hartmann, W., **3**, 556 ⟨75CL51⟩; **4**, 553 ⟨82EGP154700, 84EGP206557⟩; **5**, 10 ⟨89S272⟩, 353 ⟨88S807⟩, 1107 ⟨79GEP2817455⟩; **6**, 108 ⟨79TL513⟩
Hartmanns, J., **1**, 925 ⟨81TL1891⟩
Hartnedy, R. C., **3**, 365 ⟨84JOC4752⟩
Hartner, Jr., F. W., **2**, 681 ⟨83JA640⟩; **3**, 509 ⟨83JA640⟩
Harto, S., **3**, 546 ⟨79CB2698⟩
Hartog, F. A., **2**, 73 ⟨77S18⟩
Hartog, F., **1**, 88 ⟨63MI 103-01⟩
Hartough, H. D., **1**, 42 ⟨51JA4033⟩; **3**, 305 ⟨48JA867, 55OSC(3)14⟩
Hartshorn, A. J., **6**, 721 ⟨75CC929, 78JCS(D)348⟩
Hartshorn, M. P., **1**, 387 ⟨69T1479, 69T4445⟩, 631 ⟨69T3107⟩; **4**, 182 ⟨70T1311⟩, 182 ⟨74AJC679⟩
Hartter, D. R., **4**, 438 ⟨72JOC4136⟩; **5**, 818 ⟨72JOC4136⟩
Hartung, H., **1**, 446 ⟨79CB28⟩; **2**, 258 ⟨93PS(79)131⟩; **4**, 575 ⟨82ZAAC(494)55⟩, 843 ⟨79CB28⟩
Hartung, J., **1**, 325 ⟨85JA4332⟩, 931 ⟨91CB387⟩, 946 ⟨93CB1187⟩; **2**, 48 ⟨92JOC2768⟩, 49 ⟨93JOC844⟩; **3**, 621 ⟨85JA4332⟩; **5**, 655 ⟨88TL3475⟩, 657 ⟨84T405⟩, 657, 1053 ⟨88JPR241⟩
Hartung, Jr., J. B., **2**, 672, 988, 989 ⟨89JA5468⟩
Hartung, W. H., **1**, 11, 12, 18, 19, 21 ⟨53OR(7)263⟩, 860 ⟨45JOC21⟩; **2**, 29 ⟨27JA2517⟩, 609 ⟨46JOC444⟩; **5**, 90 ⟨73OSC(5)376⟩, 823 ⟨29JA2262, 42JOC408⟩
Hartwell, J. L., **3**, 101 ⟨55OSC(3)98⟩
Hartwig, J. F., **4**, 1083 ⟨86TL5907⟩
Hartwig, R., **6**, 626 ⟨86ZC204⟩
Hartwig, U., **3**, 375 ⟨89AG(E)221, 90CB595⟩; **5**, 502 ⟨88CB1435⟩, 652 ⟨89AG(E)221⟩; **6**, 258 ⟨89AG(E)221⟩
Hartwig, W., **1**, 16, 938 ⟨83T2609⟩, 940 ⟨82TL2019⟩, 945 ⟨82CC447⟩; **4**, 535 ⟨79AG(E)863⟩
Hartzell, S. L., **4**, 648 ⟨74TL1403⟩; **6**, 190 ⟨76TL2737⟩
Hartzfeld, H., **4**, 647 ⟨51JA5878⟩
Hartzler, H. D., **1**, 988 ⟨61JA4990, 71JA4527⟩, 1144 ⟨64JA526⟩; **4**, 893, 979, 985 ⟨62JA3387⟩; **6**, 698 ⟨71JA4961⟩
Haruki, E., **5**, 31 ⟨66BCJ852⟩
Harusawa, S., **1**, 836 ⟨91CPB1659, 93T7657⟩, 993 ⟨94TL1255⟩; **3**, 616 ⟨79TL4663⟩, 647 ⟨84TL427⟩, 659, 660 ⟨86CPB4620⟩, 670 ⟨86CPB4545⟩; **5**, 347 ⟨86CPB4545⟩, 1111, 1116 ⟨87TL4189⟩, 1112 ⟨82TL447⟩; **6**, 542 ⟨91CPB1659, 92T9433⟩, 542 ⟨92TL2543⟩, 544 ⟨92CPB2279⟩
Haruta, J., **1**, 1061, 1062 ⟨90JOC4853⟩; **2**, 263, 812 ⟨79TL4311⟩; **3**, 207 ⟨80TL3779⟩; **4**, 188 ⟨82S1089⟩; **5**, 226, 375, 378 ⟨79TL4311⟩, 294 ⟨80S887, 81S451⟩; **6**, 484 ⟨78TL3737⟩
Haruta, J.-I., **3**, 158 ⟨85TL3837⟩, 635 ⟨78CPB784⟩
Haruta, R., **1**, 1055 ⟨82JA7667⟩, 1056, 1057 ⟨84BCJ2768⟩; **3**, 639 ⟨84BCJ2768⟩; **4**, 1061 ⟨84BCJ2768⟩
Haruta, Y., **4**, 193 ⟨89CL737⟩
Harvey, A. B., **2**, 278 ⟨69JCS(D)1023, 72JCP(57)4434⟩
Harvey, B. F., **6**, 220 ⟨56USP2739989⟩
Harvey, D. F., **3**, 244 ⟨84JOC392⟩; **5**, 935 ⟨90TL2529, 91TL5223⟩, 935, 936 ⟨90JA7806⟩; **6**, 719 ⟨90TL2529⟩
Harvey, G. R., **3**, 476 ⟨70TL5225⟩, 476, 477 ⟨66JOC1587⟩, 564 ⟨66JOC3907⟩
Harvey, M. C., **2**, 35 ⟨64CJC2710⟩
Harvey, M. T., **6**, 508 ⟨52USP2592565⟩
Harvey, R. G., **1**, 4, 7 ⟨86JOC3502⟩, 95, 889 ⟨70S161⟩, 432 ⟨79JOC4813⟩, 554 ⟨78CRV317⟩; **2**, 256 ⟨55JA6064⟩, 657 ⟨76JCS(P1)836⟩, 659 ⟨78CRV317, 79JOC4265⟩, 682 ⟨70S161⟩, 855 ⟨63JA1618⟩, 945 ⟨75JOC3097⟩; **3**, 100, 293 ⟨88JOC4587⟩, 294 ⟨90JOC6161⟩, 667 ⟨83JOC5134⟩
Harvey, S., **2**, 559 ⟨88CC289⟩; **4**, 654 ⟨87OM2110, 88CC652⟩, 654 ⟨88JOC3134⟩; **6**, 393 ⟨88JOC3134⟩
Harvie, I., **5**, 466 ⟨70CC198⟩
Harville, R., **2**, 148 ⟨68JOC3976⟩
Harwood, H. J., **2**, 431 ⟨62USP3031509⟩, 461 ⟨63JOC3430⟩, 837, 839 ⟨60JA423⟩, 854 ⟨⟨62JOC4444⟩; **5**, 264 ⟨62CRV99⟩
Harwood, L. M., **1**, 240 ⟨82T693⟩, 377 ⟨B-92MI 109-01⟩, 891 ⟨B-92MI 119-14⟩, 1140 ⟨88TL5017⟩; **2**, 346 ⟨84T4363⟩; **3**, 712, 718, 719 ⟨89SL9⟩
Harzfeld, G., **2**, 437, 1060, 1069 ⟨62CB268⟩, 440 ⟨64CB3430, 67ZAAC(351)18⟩
Hasan, C., **5**, 1056 ⟨35JCS1762⟩
Hase, T. A., **2**, 228 ⟨82SC947⟩; **3**, 20 ⟨81MI 301-01⟩, 136 ⟨82MI 304-01⟩, 182 ⟨B-87MI 304-08⟩
Hase, T., **2**, 238 ⟨91PS281⟩; **4**, 197, 216 ⟨75TL3267⟩
Hasebe, K., **2**, 662 ⟨86TL2401⟩
Hasebe, M., **2**, 13 ⟨88TL6287⟩; **5**, 215 ⟨86TL3239⟩
Hasegama, M., **1**, 189 ⟨93BCJ848⟩
Hasegawa, A., **4**, 184 ⟨73CAR(29)209⟩
Hasegawa, E., **1**, 347, 937 ⟨93TL1717⟩; **4**, 260 ⟨92TL5085⟩, 510 ⟨87JA4421⟩
Hasegawa, H., **4**, 60 ⟨92TL5399⟩, 226 ⟨81BCJ817⟩
Hasegawa, I., **4**, 698 ⟨65JPC3418⟩
Hasegawa, J., **5**, 118 ⟨81MI 502-06⟩
Hasegawa, K., **3**, 246, 543 ⟨75AG(E)636⟩; **6**, 553 ⟨87BCJ435⟩
Hasegawa, M., **1**, 189 ⟨91TL4311⟩, 936 ⟨93SL143⟩; **5**, 344 ⟨90CL905⟩
Hasegawa, N., **2**, 939 ⟨90JOC2978⟩; **3**, 707, 708 ⟨89TL1257⟩
Hasegawa, S., **1**, 826 ⟨90TL4041⟩; **2**, 797 ⟨73BCJ1254⟩
Hasegawa, T., **2**, 14 ⟨93JOC6529⟩, 35 ⟨79BCJ1653⟩, 70 ⟨92TL5965⟩, 212 ⟨84JA5271⟩; **3**, 113 ⟨85CL1385⟩, 352 ⟨87TL203, 90CL563⟩
Hasegawa, Y., **1**, 135 ⟨93TL1963⟩; **5**, 551, 560, 574, 575 ⟨84S827⟩
Hasek, R. H., **3**, 269 ⟨61JOC4775⟩, 534 ⟨61CP618722⟩; **4**, 438 ⟨61JOC1822⟩, 979 ⟨64JOC2513⟩
Hasek, W. R., **3**, 73 ⟨53JA1647⟩; **4**, 8 ⟨60JA543⟩; **5**, 5 ⟨60JA543⟩; **6**, 19, 39, 52 ⟨60JA543⟩
Hasel, W., **1**, 187, 1025 ⟨88CB1461⟩; **3**, 216 ⟨88CB1461⟩
Haseltine, J. N., **2**, 139, 236 ⟨90JOC2576⟩; **5**, 336 ⟨90JOC2576⟩

Hasemann, L., **1**, 785 ⟨91JOC5782⟩
Hasenbach, J., **2**, 472 ⟨91CB1153⟩
Hasenfratz, H., **4**, 206 ⟨81CB1938⟩, 838 ⟨73CB1107⟩
Hasenkamp, R., **3**, 247, 251 ⟨80CB1708⟩
Hasgawy, M., **2**, 141 ⟨90CL1097⟩
Hasha, D. L., **1**, 443 ⟨91JOC1543⟩, 1037 ⟨92JOC2941⟩
Hashida, T., **3**, 605 ⟨87PAC1005⟩, 609 ⟨86JOM(311)C63⟩
Hashigaki, K., **2**, 328 ⟨88TL6949⟩; **3**, 422 ⟨88TL6949⟩
Hashiguchi, S., **1**, 674 ⟨80TL2955⟩; **5**, 351 ⟨91JCS(P1)2435⟩
Hashimoto, C., **2**, 328 ⟨88TL4563⟩
Hashimoto, H., **1**, 465 ⟨72BCJ2582⟩, 473 ⟨86BCJ2044⟩, 483 ⟨86CL407⟩, 494 ⟨90JCS(P1)2207⟩, 648 ⟨85BCJ2812⟩, 676, 678 ⟨84BCJ2335⟩; **2**, 14 ⟨80BCJ770⟩, 398 ⟨88BCJ3020⟩, 618 ⟨78CC446, 79BCJ1197⟩, 859 ⟨80CL729⟩; **3**, 45 ⟨80CL1213, 82BCJ534⟩, 139 ⟨84CL1227⟩, 140 ⟨79CL245⟩, 495 ⟨79BCJ1197⟩; **4**, 39, 40 ⟨71BCJ2864⟩, 250 ⟨75S720⟩, 696 ⟨73NKK381⟩; **5**, 419 ⟨79JOM(175)C1⟩, 726 ⟨85BCJ2721⟩; **6**, 72 ⟨81CL375⟩, 464 ⟨87BCJ1204⟩
Hashimoto, I., **5**, 134 ⟨77JAP(K)77/102237⟩
Hashimoto, K., **1**, 876 ⟨78TL2079⟩, 1038 ⟨83T975⟩; **2**, 263 ⟨81BCJ805⟩; **3**, 311 ⟨93TL115⟩; **5**, 157 ⟨89TL3799⟩, 329 ⟨86JOC2391⟩; **6**, 636 ⟨84CHEC(3)91⟩
Hashimoto, M., **1**, 210 ⟨88CL645⟩, 217 ⟨86TL2015⟩, 505 ⟨83JA5373⟩; **2**, 12 ⟨91CE751⟩; **3**, 696, 708 ⟨84TL5075⟩; **4**, 259 ⟨90JCS(P1)2035⟩; **6**, 447 ⟨92TL3483⟩
Hashimoto, N., **1**, 134 ⟨94TL721⟩; **3**, 125 ⟨82CPB119⟩, 674 ⟨70JOC675⟩; **4**, 166 ⟨92JOC742⟩, 762 ⟨70JOC828⟩; **5**, 132, 873 ⟨81CPB1475, 80TL4619⟩, 873 ⟨81H(15)975, 82CPB119⟩
Hashimoto, S., **2**, 650 ⟨70BCJ293⟩; **3**, 124 ⟨79CPB2760⟩, 129 ⟨92TL2709⟩, 412 ⟨B-89MI 314-01⟩; **4**, 45 ⟨84CL1747⟩, 410 ⟨84H(22)1417⟩; **5**, 134 ⟨80TL2857⟩, 138 ⟨88JA3670⟩, 334 ⟨83OPP315⟩, 652 ⟨88CE215⟩, 1025 ⟨89NKK822⟩
Hashimoto, S.-I., **3**, 62 ⟨80JA1436⟩
Hashimoto, T., **1**, 252 ⟨89JOC4998⟩, 447 ⟨61CA(55)14340⟩; **2**, 279, 288 ⟨67NKK991, 68NKK83⟩, 393 ⟨90JOC2437⟩; **3**, 168 ⟨87CC947⟩; **4**, 842 ⟨82TL1075⟩; **6**, 229 ⟨87JA3708⟩
Hashimoto, Y., **1**, 189 ⟨91TL4311⟩, 189 ⟨93BCJ848⟩, 239 ⟨87CL1691, 87CL1695⟩; **2**, 141 ⟨90CL1097⟩; **4**, 850 ⟨86CL1623⟩; **5**, 1122 ⟨85JCS(P1)2333, 85TL1321⟩, 1122 ⟨87JCS(P1)2531⟩
Hashiyama, T., **2**, 49 ⟨92JOC5067⟩; **3**, 156 ⟨92JOC5067⟩
Hashizuma, A., **3**, 66 ⟨56CB2224⟩
Hashizume, A., **3**, 293 ⟨78TL363⟩; **4**, 339 ⟨77CL485⟩; **6**, 119 ⟨78TL363⟩, 125 ⟨82BCJ224⟩
Hashizume, K., **4**, 854, 857 ⟨83CL767⟩
Hasiak, B., **2**, 742 ⟨87S727⟩
Hasimoto, S., **2**, 579 ⟨92CC1728⟩
Haskelberg, L., **2**, 656 ⟨39JCS398⟩
Haskins, J. T., **5**, 334 ⟨89JMC1024⟩
Haskins, W. C. K., **5**, 1019 ⟨88JA6240⟩
Haslam, E., **5**, 122 ⟨80T2409⟩
Haslanger, M. F., **1**, 1134 ⟨78JA4618⟩; **2**, 778 ⟨75TL3117⟩; **5**, 173 ⟨78JA4618⟩, 488 ⟨81S801⟩
Haslberger, G., **2**, 488 ⟨78CB2702⟩
Haslinger, E., **1**, 91 ⟨76CB2628⟩; **3**, 694 ⟨78M749⟩
Hasma, H., **2**, 749 ⟨76T2979⟩; **3**, 658 ⟨76T1449⟩
Haspel-Hentrich, F., **6**, 229, 230 ⟨86IC376⟩
Hass, A., **3**, 536 ⟨92CB571⟩; **6**, 432 ⟨70AG(E)466⟩
Hass, D., **2**, 416 ⟨70ZC33⟩
Hass, H. B., **1**, 6 ⟨49JA1767⟩, 156 ⟨54JA3972⟩; **3**, 88 ⟨49JA1767, 63OSC(4)932⟩, 612 ⟨31IEC352⟩; **6**, 220, 225 ⟨47IEC404⟩
Hassa, J., **4**, 989 ⟨91CB595⟩
Hassairi, M., **3**, 570 ⟨72CR(C)1309⟩
Hassall, C. H., **4**, 303 ⟨82JCS(P1)2227⟩; **5**, 718 ⟨73JCS(P1)2614⟩
Hassan, D., **1**, 983 ⟨83TL4691⟩
Hassan, F., **4**, 1023, 1028 ⟨85CC614⟩
Hassan, F. S. M., **4**, 1023, 1028 ⟨85CC1635⟩
Hassanali, A., **5**, 999 ⟨65TL4039⟩
Hassan, M. E., **3**, 339, 361 ⟨85T1885, 73CJC3081⟩
Hassan, M. I., **4**, 809 ⟨84JOC725⟩
Hassaneen, H. M., **3**, 681 ⟨87JHC577⟩; **5**, 358 ⟨72T5903, 75BCJ365⟩, 358 ⟨78BCJ512⟩, 675 ⟨90H(31)247⟩, 676 ⟨82H(19)319, 82H(19)57⟩, 676 ⟨84JHC797⟩, 676, 677 ⟨82JHC73, 83JHC719⟩, 703 ⟨87JHC577, 88H(27)695⟩, 704 ⟨80JHC1745, 82H(19)1477⟩
Hassanein, S. M., **2**, 100 ⟨88JOU1473⟩
Hassan-Gonzales, D., **1**, 790 ⟨86T4975⟩
Hassanzadeh, P., **5**, 1096 ⟨92JA9239⟩
Hassdenteufel, J. R., **1**, 1113, 1114 ⟨80TL503⟩
Hasse, L., **6**, 104 ⟨69ZC201⟩
Hasselaar, M., **1**, 1025 ⟨79S292⟩
Hassel, P., **2**, 742 ⟨83S785⟩
Hassel, T., **6**, 604 ⟨87AG(E)359⟩
Hassel, V. T., **5**, 1070 ⟨87AG(E)359⟩
Hasselkuss, G., **4**, 550 ⟨88ZN(B)31⟩
Hasselmann, J., **5**, 192 ⟨61GEP1100612⟩
Hasselström, T., **5**, 652 ⟨30MI 511-01⟩
Hasserodt, M., **2**, 170 ⟨67CB1482⟩
Hasserodt, U., **2**, 253, 462 ⟨63T1563⟩, 428, 431 ⟨65CB1681⟩, 440 ⟨65CB171⟩, 821, 827, 828 ⟨64CB1988⟩; **6**, 449 ⟨68CB113⟩, 539 ⟨89CB113⟩
Hassett, J. W., **4**, 93 ⟨85JA6298⟩
Hassfeld, M., **6**, 478 ⟨77ZC411⟩
Hässig, R., **2**, 998 ⟨83HCA308⟩; **3**, 655 ⟨85TL2395⟩; **4**, 746 ⟨84CB1877⟩, 773 ⟨84SC1285⟩, 816 ⟨83HCA308⟩, 945 ⟨84TL5351⟩; **6**, 204 ⟨80HCA1958⟩, 317 ⟨84HCA1083⟩
Hasskerl, T., **1**, 323, 920 ⟨83TL703⟩
Hasslberger, G., **3**, 500 ⟨78CB2702⟩; **4**, 599 ⟨78CB2702⟩; **5**, 922 ⟨78CB2702⟩
Hass-Lisewska, I., **6**, 414 ⟨78MI 614-02⟩
Hassner, A., **1**, 360 ⟨93JOC6716⟩, 518 ⟨83JOC3382⟩, 777 ⟨81JOC2804⟩; **2**, 22, 28 ⟨74JOC1785⟩, 392 ⟨64JOC1350⟩, 946 ⟨80JOC541⟩, 8022 ⟨74JOC553⟩; **3**, 26 ⟨74JOC1785⟩, 31, 156 ⟨75JOC3427⟩, 47, 192 ⟨74JOC2558⟩, 157, 291 ⟨91COS(1)541⟩, 259 ⟨71JA981, 71JOC258⟩, 269 ⟨74JOC3076⟩, 530, 531 ⟨78JOC2879⟩, 564 ⟨86JOC3176⟩, 633 ⟨73S682⟩, 684 ⟨89S57, 92OPP91⟩; **4**, 313, 314 ⟨88JOC22, 71JA5469⟩, 314 ⟨84JOC4237⟩, 353, 643 ⟨78JOM(156)C12⟩, 614 ⟨80TL429⟩, 742 ⟨83JOC3382, 86TL6389⟩, 829 ⟨72TL1353⟩, 929 ⟨88TL1899⟩, 929, 930, 950, 951, 952 ⟨80JA1577⟩, 930 ⟨80JOC541⟩; **5**, 124 ⟨78TL4475⟩, 332 ⟨63JA2181⟩, 344 ⟨76CRV389⟩, 348 ⟨73TL2783⟩, 355 ⟨89T3535⟩, 384 ⟨79JOC1999⟩, 417, 425 ⟨77JOM(131)C1⟩, 425, 431, 468 ⟨80JA1577⟩, 697 ⟨88JOC22⟩, 905 ⟨76JOC2102⟩, 984 ⟨63AG(E)213, 64TL1125⟩, 984 ⟨65AG(E)370, 65AG(E)449⟩, 984 ⟨65JOC1748, 70JA1326⟩, 985 ⟨67JOC540⟩; **6**, 486 ⟨93SC2839⟩, 608 ⟨73JA2982⟩
Hasso, S., **4**, 682, 683 ⟨75JCS(D)1614⟩
Hasson, H., **2**, 304 ⟨89S825⟩
Hastings, J. S., **3**, 241 ⟨72JCS(P1)2128⟩
Hastings, S. G., **2**, 741 ⟨73JMC1113⟩
Haszeldine, R. H., **4**, 111 ⟨66JCS(C)2075⟩
Haszeldine, R. N., **1**, 431 ⟨53JCS922⟩, 652 ⟨69JCS(B)652⟩, 1130 ⟨51JCS588, 53JCS1199⟩, 1130 ⟨54JCS1634⟩; **2**, 166, 167, 172 ⟨53JCS3219⟩, 181, 182 ⟨57JCS2640⟩, 181, 182, 183, 194 ⟨56JCS173⟩, 183, 194 ⟨66JCS(C)1171⟩, 193 ⟨73JCS(P1)676⟩, 194 ⟨55JCS2901⟩, 288, 291 ⟨58JCS2939⟩, 481, 482, 486, 491, 492, 498 ⟨53JCS1552⟩, 486 ⟨56JCS3631⟩, 486, 492, 505 ⟨57JCS3708⟩, 486, 505 ⟨52JCS2552, 54JCS881⟩, 778 ⟨60JCS1398⟩, 859 ⟨66JCS(C)1592⟩, 1013 ⟨68TL3909⟩; **3**, 483 ⟨73JCS(D)483⟩, 588 ⟨74CC134⟩, 600 ⟨70CC456,

73JCS(P1)1561⟩; **4**, 3 ⟨50JCS2689⟩, 5 ⟨56JCS61⟩, 8 ⟨60PCS81⟩, 31 ⟨54JCS3747⟩, 32 ⟨60JCS4503, 61JCS3825⟩, 32, 33, 35, 37 ⟨52JCS4259⟩, 34 ⟨76JCS(P1)2349⟩, 36, 798 ⟨60JCS1398⟩, 46 ⟨80CC158⟩, 111 ⟨70JCS(C)744, 71JCS(C)3031⟩, 116 ⟨78JFC(11)441⟩, 117 ⟨77JFC(10)27⟩, 136 ⟨76JCS(D)694⟩, 137 ⟨76JCS(P1)513⟩, 140 ⟨75JCS(D)2177⟩, 607, 622 ⟨75JCS(D)1434⟩, 617, 623, 656, 661 ⟨75JCS(D)1832⟩, 654, 658 ⟨75JCS(D)1837⟩, 731 ⟨53JCS3565⟩, 736 ⟨70TL5215, 78JCS(P1)422⟩, 737 ⟨80JCS(P1)487⟩, 794 ⟨73JCS(P1)1066⟩, 802 ⟨75JCS(P1)702⟩, 821 ⟨71TL1879⟩, 979 ⟨68JCS(C)796⟩, 1048 ⟨80JCR(M)235⟩; **5**, 313 ⟨66JCS(C)1514⟩, 988 ⟨56JCS3428⟩; **6**, 2 ⟨49JCS2948, 52JCS4423⟩, 2 ⟨53JCS1548, 54JCS923⟩, 5 ⟨52JCS3490, 53JCS3761⟩, 5, 221 ⟨53JCS922⟩, 32 ⟨57BRP757893, 65BRP1014252⟩, 32, 216, 218, 219 ⟨51JCS584⟩, 45 ⟨72JCS(P1)155, 72JCS(P1)159⟩, 45 ⟨72JCS(P1)2438⟩, 45, 219 ⟨72JCS(P1)1506⟩, 46, 47 ⟨76JCS(P1)1178⟩, 49 ⟨58JCS2939⟩, 54 ⟨65JCS6149, 67JCS(C)2263⟩, 54 ⟨71JCS(C)3833⟩, 58 ⟨63JCS1083⟩, 61 ⟨65JCS2101⟩, 64 ⟨53JCS3607, 61JCS3825⟩, 191 ⟨75JCS(D)1434⟩, 212, 214, 217, 218, 219, 220, 222, 225, 226, 246 ⟨48JCS2188⟩, 213, 216, 220, 223, 224, 225, 226, 227, 228 ⟨52JCS4259⟩, 214 ⟨53JCS2075⟩, 217, 225, 409, 419 ⟨57JA5801⟩, 218 ⟨57JCS30⟩, 219 ⟨49JCS2856, 74JCS(P1)1706⟩, 222 ⟨59JCS13⟩, 232, 236 ⟨53JCS3219⟩, 233 ⟨52JCS2198, 72JCS(P1)2180⟩, 238 ⟨55JCS2532, 80JCS(P1)2254⟩, 239 ⟨65JCS6141, 67JCS(C)1241⟩, 239 ⟨68JCS(C)796, 80JCS(P1)1544⟩, 240 ⟨57JCS1741⟩, 241 ⟨53JCS2075, 54JCS912, 73JCS(P1)1092, 75JCS(D)2225, 75JCS(P1)2033, 80JCS(P1)1960, 80JFC(16)391, 81JCS(P1)455, 81JFC(17)331, 81JFC(17)85⟩, 242 ⟨54JCS3598⟩, 246 ⟨49JCS2953⟩, 312, 436 ⟨84JFC(24)485⟩, 441 ⟨54JCS919⟩, 534 ⟨55JCS3871⟩, 604 ⟨55JCS1881, 56JCS3416⟩, 604 ⟨64JCS1351, 64JCS4066⟩, 679 ⟨65JCS6875⟩

Hata, E., **1**, 1042 ⟨87JOC2094⟩, 1043 ⟨88TL1785⟩

Hata, G., **1**, 813, 814 ⟨80BCJ2958⟩; **3**, 142 ⟨80BCJ2958⟩

Hata, K., **1**, 48 ⟨59BCJ861⟩; **3**, 88 ⟨68JOC3277⟩; **4**, 797 ⟨69BCJ3270⟩

Hata, N., **1**, 439, 450 ⟨82JOC4713⟩, 813 ⟨86TL219⟩; **3**, 217 ⟨86TL219⟩

Hata, T., **2**, 7 ⟨92CL1505⟩, 107 ⟨63JOC481⟩, 135 ⟨75TL1409⟩, 1003 ⟨79AOC(17)407⟩; **3**, 293 ⟨78TL363⟩, 618 ⟨60BCJ1382, 61BCJ99⟩; **4**, 339 ⟨77CL485⟩, 918 ⟨79TL2047⟩; **5**, 403 ⟨78CC285, 84JOC2139⟩, 403 ⟨87T3395⟩, 403, 404 ⟨80JOC4162⟩; **6**, 80 ⟨85S408⟩, 118 ⟨78CC285⟩, 119 ⟨78TL363⟩, 125 ⟨79TL4475, 81CC969⟩, 125 ⟨82BCJ224⟩, 553 ⟨87EUP232012⟩, 582 ⟨79TL3013⟩

Hata, Y., **1**, 592 ⟨75JA2553⟩, 636 ⟨72TL3827, 72TL4659⟩; **2**, 244, 795 ⟨80JOC1691⟩; **6**, 608 ⟨72TL3827⟩

Hatada, A., **4**, 512, 518 ⟨86BCJ2537⟩

Hatajima, T., **2**, 69 ⟨89JA4392⟩

Hatakeyama, S., **1**, 216, 748 ⟨89TL4845⟩, 248 ⟨85TL1333⟩, 596, 597 ⟨76JA3022⟩; **4**, 448 ⟨90H(31)2029⟩; **5**, 155 ⟨89TL4845⟩

Hatam, M., **3**, 4 ⟨91SC2077⟩

Hatanaka, A., **3**, 631 ⟨92BCJ111⟩

Hatanaka, H., **1**, 47 ⟨87S846⟩, 490, 1007 ⟨88JOC918⟩; **4**, 872 ⟨86BCJ2909⟩, 900 ⟨85CC279⟩; **5**, 551, 560, 574, 575 ⟨84S827⟩; **6**, 627, 730 ⟨85S891⟩

Hatanaka, K., **4**, 250 ⟨81TL3243⟩; **6**, 552, 555 ⟨81TL5195⟩

Hatanaka, M., **3**, 108 ⟨85CL1803⟩; **5**, 208 ⟨91CL61⟩; **6**, 77 ⟨73JMC978⟩

Hatanaka, N., **1**, 802 ⟨77JA2344⟩; **3**, 528 ⟨81CC344⟩; **5**, 274, 286 ⟨87JA1798⟩

Hatanaka, T., **2**, 331 ⟨92CC1097⟩

Hatanaka, Y., **1**, 490 ⟨88TL97, 89CL1711⟩, 490 ⟨89JOC268, 91SL845⟩, 490, 1028 ⟨90TL2719⟩, 1007 ⟨89TL2403⟩, 1098, 1140 ⟨84JOC3904⟩; **2**, 69, 564 ⟨84JOC3904⟩, 912 ⟨87TL4715⟩, 912, 931, 945 ⟨91SL845⟩, 931, 932, 939 ⟨89JOC268⟩, 931, 932, 939 ⟨90TL2719⟩; **3**, 88 ⟨83TL2399⟩, 290, 300, 305 ⟨92T2113⟩, 290, 305 ⟨89CL2049⟩

Hatano, C., **6**, 135 ⟨92CC870⟩

Hatano, K., **4**, 506 ⟨92JOC6711⟩

Hatate, M., **4**, 220 ⟨80TL2557⟩

Hatayama, Y., **1**, 186, 225, 229 ⟨89JA7938⟩, 556 ⟨78TL3455⟩; **2**, 281 ⟨88S175⟩

Hatch, III, C. E., **6**, 439 ⟨78JOC3953⟩

Hatch, L. F., **4**, 14 ⟨51JA4393⟩

Hatchard, W. R., **1**, 1142 ⟨63JOC2163⟩; **2**, 166 ⟨46JOC469⟩; **3**, 527 ⟨57JA6261⟩

Hatch, R. P., **5**, 235 ⟨77JOC3960⟩

Hatch III, C. E., **5**, 354 ⟨75CC725⟩

Hatcher, A. S., **3**, 245 ⟨75TL1741⟩

Hatem, J., **1**, 979 ⟨92TL1057⟩; **2**, 351 ⟨92TL1057⟩

Hatfield, G. L., **1**, 805 ⟨83TL2931⟩; **2**, 32 ⟨78TL4483⟩

Hatfield, L. D., **5**, 325, 333 ⟨67JOC1800⟩

Hathaway, B. J., **5**, 703 ⟨77JCS(P2)1136⟩

Hatjiarapoglou, L., **4**, 866 ⟨89JCS(P1)379⟩, 867 ⟨85JA7178, 91PS(58)207⟩

Hatlelid, E. B., **2**, 491, 496, 498 ⟨47JA927⟩

Hatori, K., **4**, 816 ⟨88JOC3089⟩

Hatta, K., **3**, 634 ⟨73CL287⟩

Hatta, T., **1**, 232 ⟨91JHC289⟩; **3**, 189 ⟨B-89MI 304-03⟩

Hatt, H. H., **2**, 373 ⟨43OSC(2)395⟩, 376 ⟨43OSC(2)208⟩, 379, 385 ⟨43OSC(2)211⟩; **3**, 444 ⟨43OSC(2)395⟩

Hattingh, J. T. Z., **5**, 651 ⟨88JOM(355)337⟩

Hatton, K. S., **2**, 677 ⟨86JOC3577⟩

Hatton, W. G., **6**, 508 ⟨53JA1120⟩

Hattori, E., **5**, 246 ⟨82AG(E)150⟩

Hattori, I., **3**, 440 ⟨82S461⟩

Hattori, K., **1**, 180 ⟨86BCJ3617⟩, 246 ⟨86JCS(P1)1983⟩, 830 ⟨94T3099⟩, 1044, 1059 ⟨82TL3395⟩; **2**, 287 ⟨84CC426⟩; **3**, 176 ⟨86JCS(P1)1983⟩, 407, 438 ⟨82TL3395⟩; **5**, 731 ⟨83JA2831⟩

Hattori, N., **5**, 9 ⟨85T3625⟩

Hattori, R., **5**, 337 ⟨81S746⟩, 337, 584 ⟨85CB1668⟩

Hattori, S., **2**, 418 ⟨68LA(720)1⟩

Hattori, T., **1**, 69 ⟨81TL4495⟩

Hatz, E., **2**, 749 ⟨65HCA1746⟩; **5**, 131 ⟨63AG(E)212⟩; **6**, 104 ⟨63AG296, 65HCA1746⟩

Hatzenbühler, D. A., **6**, 244 ⟨75JA187⟩

Hatzigrigoriou, E., **2**, 998 ⟨89LA167, 90AX(C)1300⟩; **3**, 500 ⟨87JCR(S)374⟩

Hatzmann, G., **4**, 441 ⟨68AG(E)293⟩; **5**, 987 ⟨68AG(E)293⟩

Haubein, A. H., **1**, 110 ⟨44JA1515⟩

Haubenstock, H., **5**, 203 ⟨64JOC2993⟩; **6**, 425 ⟨86JOC1180⟩

Haubold, W., **2**, 903 ⟨80CB3352⟩; **5**, 1085 ⟨86ZN(B)1367⟩; **6**, 183 ⟨89JOM(367)19⟩

Haubrich, A., **2**, 940 ⟨93JOC5849⟩

Haubrich, G., **6**, 164, 373 ⟨84CB2063⟩

Haubrich, H., **5**, 763 ⟨60CB397⟩

Haubrichs, R., **1**, 185 ⟨89HCA1537⟩

Hauck, H. F., **3**, 219 ⟨81JOC2428⟩

Hauck, Jr., F. P., **1**, 55, 91 ⟨57JA5279⟩

Hauck, M., **1**, 634 ⟨87JOC2627⟩

Hauck, P. R., **5**, 282 ⟨82JOC3787⟩

Hauck, S. I., **1**, 1004 ⟨90JOC5833⟩

Haufe, G., **1**, 780 ⟨92JOC714⟩; **2**, 18 ⟨87S562⟩; **4**, 787 ⟨90TL2127⟩

Haug, E., **2**, 812 ⟨80S460⟩; **3**, 266 ⟨83LA290⟩; **4**, 973, 974 ⟨83LA290⟩; **5**, 609 ⟨82LA1582⟩, 744 ⟨83S35⟩; **6**, 115 ⟨88LA605⟩, 139 ⟨90CZ176⟩, 143 ⟨82CB1721⟩, 296, 297, 298, 300, 313, 320, 321 ⟨77S73⟩, 643 ⟨84LA108⟩

Haug, Y. K., **5**, 665, 672 ⟨85JOC993⟩

Hauge, R. H., **5**, 1096 ⟨83JCP(87)797⟩; **6**, 247 ⟨90IC3252⟩

Hauge, S., **2**, 289 ⟨59ACS2155⟩
Haugen, G. R., **1**, 1109 ⟨69CRV279⟩
Haugen, R. D., **1**, 688, 690 ⟨88JA6914⟩
Haugwitz, R. D., **6**, 575 ⟨70LA(731)171⟩
Haulser, J., **3**, 418 ⟨81LA1073⟩
Haulton, H. G., **2**, 186 ⟨38JA544⟩
Hauman, C. L., **5**, 720 ⟨78JOM(161)313⟩
Haumeier, L., **4**, 719 ⟨86ZN(B)1431⟩
Haunsøe, N. K., **6**, 238, 540 ⟨72CB2854⟩
Haupe, G., **2**, 18 ⟨86TL4449⟩
Hauptmann, H., **1**, 21, 23, 24, 435, 438 ⟨62CRV347⟩, 813 ⟨86TL1315⟩, 1045 ⟨78S307⟩; **2**, 212 ⟨55JA704⟩, 881, 893 ⟨83JOM(248)269⟩, 888 ⟨72AG(E)441⟩, 1016 ⟨30CB1886⟩; **3**, 79 ⟨78S307⟩; **4**, 134 ⟨74TL303, 83JOM(249)335⟩; **5**, 856 ⟨56CB1081⟩
Hauptmann, S., **1**, 981 ⟨70TL4315⟩; **3**, 328 ⟨69JPR604⟩; **4**, 432 ⟨80ZC437⟩, 880 ⟨85JPR297⟩; **5**, 324 ⟨85S331⟩
Hauptschein, M., **2**, 166 ⟨57JA6248⟩
Hauptschein, M.,**4**, 39 ⟨61JA2383⟩
Hauptschein, M., **6**, 2 ⟨58JOC322, 61JA2495⟩, 3 ⟨61JA2383⟩, 5 ⟨58JA851⟩, 32 ⟨52JA1974, 52JA848⟩, 32 ⟨52JA849⟩, 218 ⟨52JA1347⟩, 219 ⟨51JA2461⟩
Hausberg, H. H., **3**, 711 ⟨75LA1531⟩
Hauschild, L., **4**, 654 ⟨84CB383⟩
Hausen, H.-D., **5**, 1082 ⟨86ZN(B)1223⟩, 1084 ⟨89JOM(359)1⟩
Hausen, H. D., **5**, 780 ⟨79JOM(175)33⟩
Hausen, J., **2**, 108 ⟨24CB795⟩; **4**, 178 ⟨24CB795⟩
Hauser, C. F., **1**, 726 ⟨63JOC372⟩; **3**, 494 ⟨63JOC372⟩
Hauser, C. M., **3**, 620 ⟨68JOC3402⟩
Hauser, C. R., **1**, 144 ⟨73JOM(59)53⟩, 154 ⟨69JA676⟩, 396, 397, 404, 816 ⟨51JA1437⟩, 399 ⟨55JOC1119, 63OSC(4)585⟩, 399 ⟨67JOC1155⟩, 560 ⟨56JA82⟩, 824 ⟨51JA4122⟩, 879 ⟨65JOC2934⟩, 879, 881 ⟨58JA6360, 43OSC(2)268⟩, 881 ⟨54JOC1490, 55JOC1531⟩, 881 ⟨59JA1154, 59JA1160⟩, 881 ⟨59OS(39)73, 60JOC158⟩, 881 ⟨60OS(40)38, 62JOC2737⟩, 881 ⟨63JOC1946, 65JA82⟩, 881 ⟨65JOC2937, 65JOC4135⟩, 881 ⟨66JOC1035, 67OS(47)72⟩, 881 ⟨67OS(47)92, 68OS(48)40⟩, 881 ⟨68OS(48)80⟩, 884 ⟨37JA1823⟩, 1143, 1145 ⟨57JOC1248⟩; **2**, 421 ⟨49JOC310⟩, 621 ⟨66CI(L)1635⟩, 675 ⟨71OS(51)90⟩, 1005 ⟨68JA7287⟩; **3**, 263 ⟨67JOC3817⟩, 616 ⟨73OSC(5)437⟩, 626 ⟨73OSC(5)578⟩, 657 ⟨60JA1786⟩; **4**, 300, 405 ⟨55JA1098⟩, 393 ⟨73JOM(59)53⟩, 706 ⟨65JA1241⟩; **5**, 46 ⟨42OR(1)266⟩, 65 ⟨36JA1304⟩, 76 ⟨73OSC(5)564⟩, 275 ⟨55JOC1448, 57JA725⟩, 998 ⟨37JA121, 50JA4284⟩
Hauser, D., **3**, 266 ⟨80S236⟩, 468, 471 ⟨67HCA1327⟩; **6**, 469 ⟨79HCA398⟩
Hauser, D. J., **1**, 1078 ⟨79TL2961⟩
Hauser, E., **3**, 584 ⟨21HCA887⟩; **5**, 974 ⟨21HCA861⟩
Hauser, F. M., **2**, 393 ⟨88JOC2872⟩, 664 ⟨78JOC178, 79JA1628⟩; **3**, 219 ⟨83JA5688⟩; **4**, 346 ⟨80S72⟩
Häuser, H., **6**, 74 ⟨66CB1892⟩, 106 ⟨64AG(E)380⟩
Hauser, L., **5**, 581 ⟨59CB910⟩
Hauser, W., **2**, 1009 ⟨59LA(626)60⟩; **3**, 695 ⟨27M71⟩
Hausermann, M., **3**, 60 ⟨51HCA1482⟩
Hausigk, D., **5**, 114 ⟨71CB2637⟩
Hauske, J. R., **1**, 885 ⟨91COS(1)77, 92TL3715⟩; **2**, 595 ⟨91COS(1)77⟩
Häusler, S., **2**, 438 ⟨61CB113⟩
Hausmann, H., **4**, 545 ⟨76ZN(B)790⟩
Hausmann, I., **4**, 695, 717 ⟨89AG(E)319⟩
Hausner, C. R., **1**, 999 ⟨51JA5864⟩
Haussmann, W., **5**, 658 ⟨66CB239⟩, 1071 ⟨66CB1252⟩; **6**, 448 ⟨65TL2423, 66CB239⟩, 629 ⟨67AP(300)553⟩, 666 ⟨66CB1252⟩
Hautala, J. A., **1**, 155 ⟨78JOC3095⟩
Hauthal, H. G., **4**, 318 ⟨59LA(622)83⟩
Hauzinger, J., **5**, 359 ⟨92SC1199⟩
Haveaux, B., **2**, 758 ⟨69AG468, 80OS(59)26⟩; **4**, 790 ⟨88OSC(6)282⟩, 891, 904, 905 ⟨69AG(E)454⟩; **5**, 2 ⟨79OS(59)26⟩
Havel, J. J., **2**, 961 ⟨73ACR97⟩
Havelka, F., **6**, 503 ⟨78IZV220⟩
Haven, Jr., A. C., **5**, 1017, 1018 ⟨56JA842⟩
Havill, M. E., **2**, 108 ⟨48JOC280⟩
Havinga, E., **1**, 403 ⟨57RTC689⟩, 403, 795 ⟨64RTC1173⟩, 419 ⟨61T146⟩, 420 ⟨73T1393, 75TL3741⟩; **2**, 756 ⟨66RTC56⟩; **3**, 667 ⟨70TL4701⟩
Haviv, F., **1**, 1018 ⟨87JMC254⟩
Havlik, A. J., **2**, 216 ⟨66QRS285⟩; **5**, 208 ⟨56JA1207⟩
Havlin, R., **1**, 1005 ⟨69JA6464⟩; **2**, 621 ⟨69JA6464⟩; **3**, 273 ⟨69JA6464⟩
Hawari, J. A., **1**, 331, 936 ⟨87JA5267⟩
Hawarth, R. D., **1**, 45 ⟨54JCS3617⟩
Hawes, B. W. V., **1**, 895 ⟨52JCS2167⟩
Hawker, D. W., **4**, 660, 663 ⟨85OM821⟩; **6**, 208, 405 ⟨84JOM(266)37⟩
Hawkins, B., **5**, 932 ⟨88JA8413⟩
Hawkins, B. L., **4**, 25 ⟨71JA4472⟩
Hawkins, C. M., **1**, 412 ⟨87JOC5198⟩
Hawkins, D. G., **2**, 160 ⟨76JCS(P1)2166⟩; **5**, 764 ⟨76JCS(P1)2166⟩
Hawkins, D. W., **1**, 438 ⟨73JCS(P1)654⟩; **3**, 358 ⟨75CC756, 77JCS(P1)515⟩
Hawkins, I., **5**, 800 ⟨87JOM(332)371⟩
Hawkins, J. M., **1**, 374 ⟨91JA7794⟩; **2**, 367 ⟨84JOC3861⟩, 526 ⟨91JA7794⟩
Hawkins, L. D., **3**, 32, 33 ⟨83JOC5180⟩, 238 ⟨86TL4759⟩
Hawkins, M., **4**, 42 ⟨85IC3285⟩
Hawkins, S., **6**, 471 ⟨B-92MI 615-01⟩
Hawks, G. H., **3**, 527 ⟨72OS(52)36⟩; **5**, 8 ⟨88OSC(6)549⟩
Haworth, R. D., **2**, 752 ⟨39JCS1299⟩, 764 ⟨21JCS768⟩
Hawson, A., **2**, 179 ⟨71CJC943⟩
Hawthorne, M. F., **1**, 168 ⟨91JA5915⟩, 168 ⟨92JOC1384⟩, 273 ⟨65JA4585⟩; **2**, 269 ⟨60JA748⟩, 269, 522 ⟨59JA5836⟩, 515 ⟨61JA831⟩, 515, 522 ⟨61JA2541⟩, 524 ⟨61JA367⟩, 653 ⟨57JOC1001⟩; **4**, 574 ⟨58JA3480, 61JA367⟩; **6**, 144, 145, 369 ⟨63T(S)177⟩
Hawthorne, M. J., **2**, 267 ⟨61JA1345⟩
Hay, A. S., **1**, 1034 ⟨60JOC1275, 62JOC3320⟩
Hay, B. P., **1**, 931 ⟨88JA4415, 89JA230⟩
Hay, D., **1**, 148 ⟨83JA2080⟩
Hay, D. R., **1**, 148 ⟨88JA8145⟩
Hay, G. W., **2**, 9 ⟨87CJC412⟩; **3**, 635 ⟨75JCS(P1)180⟩; **4**, 45 ⟨84CL1751⟩, 271 ⟨75JCS(P1)180⟩
Hay, J. M., **1**, 919 ⟨B-74MI 119-07⟩, 922 ⟨B-68MI 119-04⟩
Hay, J. N., **3**, 192 ⟨85HCA162⟩
Hay, J. V., **4**, 167 ⟨77JA1631⟩
Hayakawa, A., **1**, 1010 ⟨90JHC503⟩
Hayakawa, H., **1**, 983 ⟨87TL1803⟩
Hayakawa, I., **3**, 108 ⟨84CPB4914⟩
Hayakawa, I.,**6**, 447 ⟨92TL3483⟩
Hayakawa, K., **1**, 235 ⟨84JOC1985⟩, 666 ⟨73JOC1130⟩; **4**, 171 ⟨86TL947⟩, 892 ⟨76JCS(P1)783⟩; **6**, 207 ⟨71MI 606-01⟩
Hayakawa, N., **3**, 245 ⟨73JA4103⟩
Hayakawa, S., **1**, 929 ⟨91CL1193, 93BCJ819⟩, 929 ⟨93CL545⟩; **3**, 137 ⟨93CL545⟩, 311 ⟨86CPB4916⟩, 628 ⟨88BCJ4379⟩; **5**, 430 ⟨93BCJ819⟩
Hayakawa, T., **3**, 87 ⟨90TL2607⟩
Hayakawa, Y., **1**, 2 ⟨78JA1765, 83OR163⟩, 2, 3, 8 ⟨78JA1786⟩, 363, 364 ⟨78JA1791⟩; **2**, 28 ⟨88OSC(6)520⟩, 106 ⟨86TL4191⟩; **3**, 129 ⟨83OR(29)163, 85T5879⟩, 225 ⟨72JA1629⟩, 245 ⟨73JA4103⟩
Hayama, N., **2**, 632 ⟨80BCJ3691⟩; **3**, 662 ⟨75BCJ3298⟩; **5**, 1110 ⟨92CL509⟩
Hayama, T., **1**, 180 ⟨86TL3403⟩, 603 ⟨82TL4733⟩; **2**, 284 ⟨84JOC3235⟩, 284, 392, 785 ⟨82TL4733, 82CL1109⟩; **4**, 376 ⟨84JOC3235⟩

Hayama, Y., **5**, 438 ⟨90JA7061⟩
Hayami, H., **2**, 918, 931 ⟨83JA4491⟩
Hayami, J., **1**, 617 ⟨78TL763⟩
Hayami, J.-I., **3**, 635, 645 ⟨78TL763⟩
Hayami, K., **1**, 469, 1005 ⟨89SC2199⟩; **2**, 727 ⟨89CL769⟩; **5**, 289, 302 ⟨91CL1443⟩
Hayao, S., **3**, 541 ⟨54JA5563⟩
Hayase, M., **3**, 60 ⟨68TL4359⟩
Hayase, Y., **3**, 49 ⟨69JCS(C)460⟩; **5**, 161 ⟨76JA7874⟩, 1112 ⟨89H(29)841⟩; **6**, 556, 566, 652 ⟨91CPB1939⟩, 636 ⟨88CPB4755, 92MI 621-02⟩
Hayashi, A., **1**, 1038 ⟨91JOC4091⟩, 1050 ⟨92CL2479⟩, 1051 ⟨93JA3362⟩; **2**, 1100 ⟨92CL2479⟩; **4**, 1069 ⟨93CL1881⟩; **5**, 419 ⟨87OM974⟩, 563 ⟨93CL1469⟩
Hayashi, E., **3**, 292 ⟨83CPB3951⟩; **4**, 794 ⟨88CL1887, 89CL905⟩; **5**, 769 ⟨82CPB4521⟩; **6**, 37 ⟨79JFC(15)353⟩, 39 ⟨83JFC(23)123⟩, 219 ⟨90CL813⟩, 230 ⟨90JFC(50)173⟩, 257 ⟨90JFC(48)257⟩, 440 ⟨89JFC(45)293⟩
Hayashi, H., **1**, 449 ⟨88OM1869⟩; **3**, 25 ⟨82CL1637⟩, 628 ⟨86BCJ893⟩
Hayashi, H.,**3**, 628 ⟨86CL931⟩
Hayashi, H., **5**, 645 ⟨74JOM(78)339⟩; **6**, 539 ⟨89JAP03020256⟩
Hayashi, J., **3**, 7, 120 ⟨79TL165⟩, 7, 120 ⟨79TL3861⟩, 7, 120 ⟨80TL1867⟩; **4**, 181 ⟨84JOC2808⟩, 689 ⟨72JOM(39)365⟩
Hayashi, K., **1**, 250 ⟨86JOC4375⟩, 832 ⟨80JA862⟩; **3**, 574 ⟨81CL1197⟩, 717 ⟨90S781⟩, 717, 720 ⟨88T5467⟩; **4**, 517, 538, 1015 ⟨82BCJ1163⟩; **5**, 1023 ⟨92OPP346⟩; **6**, 9 ⟨91MI 601-02⟩, 162 ⟨82BCJ1163⟩
Hayashi, M., **1**, 19 ⟨72BCJ264⟩, 33, 34 ⟨81T2797⟩, 427, 428, 430 ⟨73CL291⟩, 1026 ⟨87CL1975, 88BCJ4037⟩, 1077 ⟨69TL4447⟩; **2**, 143 ⟨72CL259⟩, 348 ⟨90CL547⟩, 610 ⟨79CL1499⟩; **3**, 30, 33 ⟨91SL479⟩, 30, 33 ⟨94T2821⟩, 55, 252, 275 ⟨81JA4597⟩, 140 ⟨87CL289⟩, 628 ⟨90CC1364, 92JCS(P1)3135⟩, 629 ⟨91CC1752, 93JOC1515⟩, 632 ⟨92SL663⟩, 708 ⟨91JOC7356⟩; **4**, 45 ⟨84CL1747⟩, 188 ⟨88CL1101⟩, 192 ⟨74CL15⟩; **5**, 558 ⟨84JCS(P1)85⟩; **6**, 632 ⟨85BCJ3379⟩
Hayashi, N., **2**, 619 ⟨85TL1651⟩; **5**, 550 ⟨74JOM(82)223⟩
Hayashi, R., **1**, 195 ⟨88CC44⟩; **2**, 80 ⟨88CC44⟩
Hayashi, S., **1**, 1078 ⟨79TL2961⟩; **2**, 612 ⟨79CL983⟩; **4**, 231 ⟨87CC1442⟩, 734 ⟨79CL983⟩; **5**, 38 ⟨80CL651⟩, 1111 ⟨74CL881⟩; **6**, 695 ⟨79CL983⟩
Hayashi, T., **1**, 86 ⟨86JOC537⟩, 193 ⟨87TL4849⟩, 194 ⟨88PAC7⟩, 194 ⟨89TL375⟩, 279, 466 ⟨84JA158⟩, 315 ⟨80JA4973⟩, 415, 826 ⟨89SC2109⟩, 473 ⟨80TL845, 88JA8153⟩, 477 ⟨81CC313⟩, 479 ⟨81S1001⟩, 480 ⟨80TL3915⟩, 482 ⟨81TL4449⟩, 826 ⟨89BCJ3835, 91JCR(S)172⟩, 836 ⟨90TL4155⟩, 1097 ⟨82JA180⟩; **2**, 43, 528 ⟨89JA3426⟩, 43, 528 ⟨91TL3387⟩, 51 ⟨88JA5579⟩, 78 ⟨88TL5667⟩, 81 ⟨89JOC6001⟩, 301 ⟨90TL1743⟩, 331 ⟨89JA6301⟩, 437 ⟨83JOC2195⟩, 528 ⟨91TA601⟩, 537 ⟨91JA9887, 92TL7185⟩, 915 ⟨76JOM(114)C19⟩; **3**, 54 ⟨75S727⟩, 138 ⟨80JA4973⟩, 350 ⟨73TL2461, 75JAP(K)7549251⟩, 357 ⟨92JHC841⟩, 637 ⟨92TL3003⟩, 707 ⟨87TL4849, 88TL6321⟩; **4**, 87, 92 ⟨92TL3161⟩, 197 ⟨74S705⟩, 437, 445 ⟨77S647⟩, 651 ⟨90TL2925⟩, 796 ⟨85JOC1547⟩, 1032, 1033 ⟨74TL4405, 76JA3718⟩, 1033 ⟨80BCJ1138, 82JA180⟩, 1033 ⟨86JOC3772, 88JA8153⟩, 1033 ⟨90TA151⟩; **5**, 614 ⟨75JA1608⟩; **6**, 9 ⟨94TA937⟩, 553 ⟨81CPB3158⟩
Hayashi, V., **3**, 684 ⟨84BCJ2531⟩
Hayashi, Y., **1**, 16 ⟨87JOC4878⟩, 195 ⟨89JOC469⟩, 202 ⟨85TL767, 86JOM(299)119⟩, 202 ⟨86S386, 86TL2479⟩, 202 ⟨86TL967⟩, 202, 203 ⟨86JOC2859⟩, 358, 498 ⟨92JA8869⟩, 399 ⟨68TL5381⟩, 536 ⟨81TL2629⟩, 598 ⟨92TL7181⟩, 1140 ⟨90CL1411⟩; **2**, 720 ⟨91BCJ702⟩, 941 ⟨90JOC1108⟩; **3**, 170 ⟨92TL6823⟩, 173 ⟨80TL1497⟩, 374 ⟨74TL1071⟩, 549 ⟨71T1781⟩, 623 ⟨87CL1811⟩, 684 ⟨86BCJ2827⟩, 685 ⟨87BCJ1948⟩; **4**, 777 ⟨93BCJ1866⟩; **5**, 427 ⟨91CL1413⟩, 652 ⟨90TL3603⟩; **6**, 127, 325, 327 ⟨90CL1411⟩
Hayashibe, S., **3**, 360 ⟨91JCS(P1)2281⟩
Hayashida, H., **2**, 77 ⟨89JOC3254⟩, 582 ⟨85JCS(D)873⟩
Hayashida, N., **3**, 87 ⟨93CPB1166⟩
Hayashimatsu, M., **4**, 420 ⟨85JOC4227⟩
Hayashizaki, K., **1**, 473 ⟨88JA8153⟩; **4**, 1033 ⟨88JA8153⟩
Hayasi, Y., **1**, 11, 109 ⟨70BCJ2506⟩, 1024 ⟨86TCC3⟩, 1105, 1139 ⟨80TL1497⟩; **4**, 860 ⟨72BCJ198⟩
Hayden, P. M., **3**, 241 ⟨55JCS862⟩
Hayden, W., **6**, 489 ⟨88JMC1831⟩
Haydock, D. B., **4**, 824 ⟨72JCS(P1)1225⟩
Hayduk, U., **2**, 415 ⟨64ZAAC(331)35⟩
Hayes, B. T., **2**, 379 ⟨70JCS(C)1088⟩
Hayes, K., **2**, 374 ⟨49JOC813⟩
Hayes, L. J., **5**, 723, 827 ⟨71JOC284⟩
Hayes, R., **1**, 803 ⟨71CC1248⟩
Hayes, T. G., **6**, 522 ⟨81JA6100⟩
Hayes, T. K., **4**, 12 ⟨88JA5533⟩
Haygood, J. D., **2**, 193 ⟨51JOC767⟩
Hayler, R. G., **2**, 437 ⟨61JCS896⟩
Haymaker, A., **6**, 54 ⟨66JOC789⟩
Hay-Motherwell, R. S., **1**, 929, 936 ⟨81JCS(P1)2363⟩, 945 ⟨80JCS(P1)2657⟩; **5**, 1054 ⟨80JCS(P1)2657⟩
Haynes, L. J., **5**, 62 ⟨60QR292, B-52MI 502-05⟩, 84 ⟨B-52MI 502-02⟩
Haynes, L. W., **2**, 403 ⟨68CJC1149⟩, 738, 739, 751 ⟨B-88MI 215-01⟩; **4**, 294 ⟨B-88MI 407-01⟩
Haynes, N. B., **3**, 215 ⟨63JCS2420⟩
Haynes, P., **1**, 402, 794 ⟨65T2329⟩
Haynes, R., **2**, 21 ⟨80TL177⟩
Haynes, R. K., **1**, 486 ⟨90AJC1375⟩; **2**, 7 ⟨82AJC517⟩, 999 ⟨88HCA299⟩
Haynes, W. M., **5**, 122 ⟨68CI(L)1568⟩
Haynie, E. C., **1**, 1143 ⟨72JOC1553⟩
Hays, G. R., **6**, 195 ⟨86OM1726⟩
Hays, H. R., **1**, 157 ⟨65JOC1939⟩; **2**, 427, 429, 432, 444 ⟨66JOC3817⟩, 428 ⟨66JOC3391⟩, 450, 452 ⟨68JOC3690, 67JOC1060⟩; **4**, 550 ⟨66JOC3391⟩, 550, 562, 563 ⟨69USP3445522⟩, 575 ⟨65JOC1939⟩
Hays, S. J., **3**, 422 ⟨91JOC4084⟩
Hayter, R. G., **2**, 512 ⟨64JA1030⟩
Hayward, R. C., **2**, 12 ⟨81JCS(P1)2608⟩, 34 ⟨78JCS(P1)126⟩, 45 ⟨74JCS(P1)1858⟩; **3**, 155 ⟨78JCS(P1)126⟩; **4**, 27 ⟨81JCS(P1)2608⟩; **5**, 168 ⟨74JCS(P1)1864⟩
Hayward, R. J., **6**, 507, 577 ⟨75JCS(P1)212⟩
Hayward, W. D., **1**, 688 ⟨91JOC6447⟩
Haywood, D. J., **2**, 675 ⟨80CC980⟩
Hazato, A., **3**, 192 ⟨87T813⟩
Hazawa, Y., **3**, 284 ⟨83JA7644⟩
Hazebroucq, G., **3**, 75 ⟨71BSF3985⟩, 281 ⟨66LA(701)221⟩
Hazeldine, R. N., **2**, 339 ⟨68JCS(C)796⟩
Hazell, R. G., **5**, 336 ⟨85T5145⟩; **6**, 273, 274, 275, 535, 553, 554 ⟨85T5145⟩
Hazelton, K., **2**, 193 ⟨53JOC36⟩
Hazen, G. C., **3**, 193 ⟨81SC947⟩
Hazen, G. G., **6**, 25 ⟨87JOC944⟩
Hazen, K. H., **1**, 1071 ⟨91JOC2045⟩
Hazenberg, J. F. A., **4**, 770, 776 ⟨53RTC813⟩
Hazin, P. N., **4**, 716 ⟨92ICA557⟩
Hazlet, S. E., **2**, 173 ⟨35JA2172⟩
Hazlett, R. N., **2**, 1025 ⟨88S287⟩
He, D. W., **2**, 559, 959 ⟨83AG(E)728⟩
He, H. M., **5**, 342, 345 ⟨91TL283⟩
He, J., **1**, 325, 921 ⟨88CB2063⟩; **2**, 578, 579 ⟨89AG(E)325⟩
He, J.-G., **4**, 507 ⟨84JOM(268)31⟩
He, W., **1**, 826 ⟨90TL6979⟩; **2**, 625 ⟨91JOC6199⟩
He, Y., **6**, 63 ⟨87JFC(37)171⟩
He, Y.-B., **1**, 433 ⟨88S896⟩, 483, 1028 ⟨87TL2387⟩; **2**, 1096 ⟨87TL2387⟩

He, Z.-M., **1**, 373 ⟨83TL1897⟩
Heacock, D. J., **2**, 212 ⟨78S50⟩, 213 ⟨79S8⟩; **5**, 251 ⟨78S50⟩
Head, N. J., **1**, 923 ⟨90CC1589⟩; **2**, 19 ⟨92JOC2850⟩
Head, R. A., **5**, 143 ⟨83EUP127276A2⟩
Headford, C. E., **3**, 387 ⟨83JOM(244)C53⟩
Headford, C. E. L., **4**, 150 ⟨79JA503⟩; **6**, 593 ⟨83JOM(244)C53⟩
Heakin, P., **2**, 570 ⟨74JA5288⟩
Heal, H. G., **2**, 165 ⟨46JCS1126⟩
Healy, M. E., **2**, 816 ⟨64JA4221⟩
Heaney, F., **2**, 158 ⟨91T8297⟩, 312 ⟨91T4007⟩
Heaney, H., **2**, 90 ⟨93MI 202-02⟩, 147, 156 ⟨87S1015⟩, 316 ⟨88TL2377, 90TL4229⟩, 348 ⟨93MI 206-01⟩, 491 ⟨60IS113, 60IS116⟩, 609, 632 ⟨60OS(40)105⟩, 647, 753 ⟨62CRV81⟩, 882 ⟨65JCS5132⟩, 1006 ⟨91COS(2)733⟩, 1006, 1009 ⟨91COS(2)953⟩; **3**, 92 ⟨92SL77⟩, 92, 99, 101 ⟨93T4015⟩, 184 ⟨91COS(2)953⟩, 186 ⟨90SL619⟩, 278, 297 ⟨91COS(2)733⟩, 282, 687 ⟨85TL4649⟩; **4**, 301, 316 ⟨91COS(2)953⟩
Heaney, S. M., **4**, 140 ⟨85SRI321⟩; **5**, 224 ⟨85SRI321⟩
Heap, N., **1**, 797 ⟨66JCS(B)164⟩
Heap, R., **2**, 413 ⟨48JCS1313⟩
Hearn, M. J., **4**, 162 ⟨80JOC2874⟩
Hearn, M. T. W., **5**, 714, 723 ⟨69AJC161⟩
Hearsey, C. H., **5**, 980 ⟨73GEP(O)2225365⟩
Hearst, P. J., **2**, 182 ⟨48JA3955⟩, 182 ⟨50OS(30)58⟩
Heasley, G. E., **2**, 18 ⟨83JOC3195⟩, 30 ⟨78JOC2793⟩, 606 ⟨80JOC4649⟩
Heasley, L. E., **2**, 606 ⟨80JOC4649⟩
Heasley, V. L., **2**, 18 ⟨83JOC3195⟩, 30 ⟨78JOC2793, 80JOC4649⟩; **6**, 54 ⟨76JOC1285⟩
Heath, G. A., **6**, 375 ⟨87JCS(D)249⟩
Heath, M. J., **1**, 630, 631, 632, 633 ⟨67JA5734⟩
Heath, N. J., **2**, 128 ⟨49JCS282⟩, 129 ⟨49JCS278⟩
Heath, N. S., **1**, 592 ⟨49JCS282⟩
Heath, R. L., **2**, 117 ⟨47JCS1477⟩; **3**, 638 ⟨47JCS1500⟩
Heath-Brown, B., **1**, 100 ⟨69CI(L)1595⟩; **3**, 448 ⟨65JCS7185⟩
Heathcock, C., **2**, 392 ⟨64JOC1350⟩; **5**, 984 ⟨63AG(E)213, 65AG(E)449⟩, 984 ⟨65JOC1748, 70JA1326⟩, 985 ⟨67JOC540⟩
Heathcock, C. C., **5**, 984 ⟨64TL1125⟩
Heathcock, C. H., **1**, 130, 131 ⟨B-92MI 104-03⟩, 248 ⟨87H(25)75⟩, 369 ⟨92JOC7056⟩, 469 ⟨76JOC636⟩, 505 ⟨90OSC(7)226⟩, 601, 1093 ⟨76JOC1396⟩, 607, 608 ⟨75JOC2165⟩, 608, 781 ⟨74TL2279⟩, 748, 749 ⟨80SC509⟩, 802 ⟨65JA5264⟩, 813 ⟨81JOC9⟩, 829 ⟨88JOC1922⟩, 873 ⟨B-84MI 119-03⟩, 874 ⟨80JOC1066⟩, 878 ⟨B-84MI 119-05⟩, 879 ⟨91COS(2)181⟩, 883 ⟨91COS(2)133⟩, 884 ⟨91COS(2)234⟩, 1000 ⟨73SC245⟩, 1024, 1139 ⟨76TL1769⟩, 1089 ⟨75JA3822⟩, 1093 ⟨75JOC2970, 76JOC3144⟩; **2**, 330 ⟨88JA8734⟩, 532 ⟨90JOC173⟩, 696 ⟨91COS(2)181⟩, 702 ⟨87OM2069⟩; **3**, 159 ⟨B-92MI 304-01⟩, 168 ⟨90JOC5966⟩, 239 ⟨74S47⟩, 239, 264 ⟨76JOC636⟩, 303 ⟨69JHC141⟩, 643, 644 ⟨74TL2279⟩; **4**, 171 ⟨84JOC3994⟩, 349 ⟨78JOC1256⟩, 953 ⟨75JA3822⟩; **5**, 103 ⟨85OS(64)108⟩, 159 ⟨88JOC1922⟩, 159 ⟨88JOC4736⟩, 277 ⟨87JOC5745⟩, 437 ⟨81TL1881⟩, 572 ⟨94TL3651⟩
Heathcock, C. T., **1**, 875 ⟨77JA247⟩
Heathcock, H.,C., **1**, 873 ⟨B-84MI 119-04⟩
Heathock, C. H., **1**, 979 ⟨88JOC4736⟩
Heatley, P., **2**, 824 ⟨76JCS(P1)2529⟩
Heatlie, J. W. M., **5**, 768, 769, 770, 847, 849, 850, 855 ⟨70CRV151⟩
Heaton, A., **6**, 14 ⟨89JFC(45)86⟩
Heaton, D. E., **2**, 512 ⟨87IC4065⟩
Heaven, M. C., **1**, 1127 ⟨84JA4356⟩
Heavers, A. D., **1**, 606 ⟨92JCS(P1)2683⟩
Hebb, A. M., **2**, 214 ⟨67JPS292⟩
Hebbelynck, M. F., **5**, 15 ⟨51BSB54⟩
Hebborn, P., **2**, 260 ⟨66JOC221⟩
Hebeisen, P., **4**, 440 ⟨84S315⟩, 441 ⟨85T5525⟩
Hebel, D., **2**, 613 ⟨84JOC806⟩; **4**, 44 ⟨84JOC806⟩; **5**, 202 ⟨84JOC806, 90JOC2621⟩; **6**, 264 ⟨91BMC357⟩
Hebenbrock, K.-F., **4**, 312 ⟨72LA(765)55⟩
Heber, D.,**1**, 455 ⟨77T457⟩
Hebert, N. C., **1**, 58 ⟨83JOM(255)135⟩
Hebold, M., **1**, 686 ⟨84CB277⟩; **3**, 348 ⟨84CB277⟩
Hebri, H., **1**, 427, 428 ⟨91SC2377⟩
Hechenbleikner, I., **2**, 428 ⟨59JA1103⟩
Hecht, C., **3**, 401 ⟨84CC686⟩
Hecht, S. M., **2**, 683 ⟨73TL1397⟩; **3**, 466 ⟨72TL5147, 73JOC1821⟩, 618 ⟨93TL1581⟩
Hecht, S. S., **2**, 403, 404 ⟨70JOC2482⟩
Hechtl, W., **1**, 421 ⟨68TL5215⟩
Heck, F. R., **1**, 1006, 1008, 1009, 1137 ⟨75JOM(93)253⟩, 1008, 1009, 1137 ⟨75JOM(93)259⟩
Heck, J., **4**, 672 ⟨83JA2905⟩, 672, 673 ⟨90ZN(B)221, 83AG(E)330⟩, 673 ⟨88JOM(342)45⟩
Heck, J. V., **4**, 59 ⟨84JA1809⟩; **6**, 129 ⟨89TL2873⟩
Heck, R. F., **1**, 32, 457, 507, 1009 ⟨80JOC4926⟩, 79, 85 ⟨78JOC3985⟩, 128 ⟨68JA5531⟩, 192 ⟨B-85MI 105-01⟩, 318 ⟨76JOC265⟩, 493 ⟨68JA5518, 69JA6707⟩, 493 ⟨74JA1133, 75JOC1083⟩, 493 ⟨85MI 111-02, 91COS(4)833⟩, 494 ⟨78JOC3898, 78JOC5018⟩, 494 ⟨79JOC4078⟩, 494, 1096 ⟨78JOC2941⟩, 797 ⟨B-74MI 118-01⟩, 1009 ⟨90JOC63⟩, 1021 ⟨82OR(27)345, 81JOC1061⟩, 1096 ⟨81JOC1067, 82JOC1267⟩; **2**, 628, 904 ⟨75JOC1083⟩, 983 ⟨B-85MI 219-01⟩; **3**, 12, 127, 621 ⟨78JOC3985⟩; **5**, 143 ⟨74JOC3318⟩, 157 ⟨77JOC390, 78JOC2952⟩, 273, 296, 301, 304 ⟨74JOC3327⟩, 456 ⟨61JA4023, 62JA2499⟩, 464 ⟨75JOC2667⟩
Heck, R. K., **2**, 579 ⟨B-85MI 211-04⟩; **3**, 90 ⟨74JA7761⟩
Heckendorn, D. K., **1**, 618, 707, 709 ⟨91JA9682⟩
Heckendorn, R., **4**, 448 ⟨90HCA1700⟩
Hecker, E., **2**, 660 ⟨64CB1940⟩
Hecker, H., **4**, 576, 577 ⟨60CB1339⟩
Hecker, L. R., **6**, 635 ⟨88S460⟩
Heckert, R. E., **2**, 194 ⟨52JA2006⟩
Heckl, B., **5**, 652 ⟨69JOM(16)29⟩, 953 ⟨71JOM(28)359⟩
Heckmann, G., **4**, 552 ⟨91ZAAC(596)139⟩, 562 ⟨91ZAAC(601)65⟩, 574 ⟨76ZAAC(421)1⟩, 575 ⟨88ZAAC(567)23⟩; **5**, 776 ⟨86ZAAC(533)55⟩
Heckmann, R. A., **1**, 726 ⟨68JOC3566⟩
Hedaya, E., **3**, 498 ⟨68T2241⟩
Hedayatullah, M., **2**, 367 ⟨84PS(20)371, 86PS(27)293⟩; **4**, 889 ⟨75BSF2089⟩; **5**, 1100 ⟨64MI 530-01, 68BSF1472⟩, 1101 ⟨67BSF416⟩, 1101, 1103 ⟨65MI 530-01⟩, 1101, 1103 ⟨67BSF422⟩, 1102 ⟨70BSF2637⟩
Hedberg, A., **4**, 905 ⟨90JMC1510⟩
Hedberg, F. L., **4**, 706 ⟨69TL4011⟩
Hedden, D., **5**, 385 ⟨86IC9, 86OM336⟩
Hedgecock, C. J. R., **5**, 161 ⟨91SL781⟩
Hedgecock, Jr., H. C., **2**, 34 ⟨76JA1290⟩, 40 ⟨75JOC1776⟩
Hedges, R. M., **2**, 500, 502 ⟨63JCS(C)798⟩
Hedgley, E. J., **5**, 476, 477 ⟨65JOC1282⟩, 510 ⟨81BRP1589128⟩
Hedhli, A., **2**, 675 ⟨93BSF664⟩
Hedtmann, U., **3**, 157, 161 ⟨91T3753⟩
Hee, L., **6**, 324 ⟨85CB3151⟩
Heeg, M. J., **2**, 25 ⟨91JOC7186⟩; **4**, 724 ⟨92OM3169⟩
Heep, U., **5**, 869 ⟨73LA578⟩
Heer, J. P., **2**, 145 ⟨95SL773⟩, 145, 156 ⟨93JCS(P2)2161⟩
Heerding, D. A., **2**, 922 ⟨88T3919⟩
Heerding, J. M., **2**, 669 ⟨91JOC4048⟩
Heeren, H. K., **3**, 495, 497 ⟨66JOM(5)267⟩
Heeren, J. K., **3**, 494 ⟨64JOM(2)101, 65JA3467⟩; **4**, 551, 570, 588 ⟨64JA1100⟩, 551, 576, 588 ⟨63JA642⟩; **5**, 896 ⟨66JOM(5)267⟩

Heeres, G. J., **1**, 731 ⟨70T1291⟩; **3**, 494 ⟨70T1291⟩
Heeres, H. J., **4**, 703 ⟨93OM2609⟩; **6**, 195 ⟨88CC962⟩
Heerze, L. D., **1**, 688, 690 ⟨88JA6914⟩
Hees, U., **4**, 100 ⟨88AG(E)304⟩, 479, 480 ⟨87CB1645⟩; **5**, 883 ⟨88AG(E)272⟩, 1155 ⟨87CB1645⟩
Heesing, A., **6**, 642, 650 ⟨67ZN(B)820⟩, 649 ⟨65ZN(B)1165, 67TL3851⟩
Heffe, W., **1**, 737 ⟨64MI 116-01⟩
Heffer, J. P., **5**, 229 ⟨64CI(L)544⟩
Heffernan, G. D., **1**, 816 ⟨92CC350⟩
Hefferon, G. J., **1**, 517 ⟨83JOC1262⟩
Heffman, J. W., **2**, 696 ⟨77SC137⟩
Heffner, R. J., **4**, 160 ⟨91SC1055⟩, 160 ⟨91SC2231⟩
Heffner, T. A., **1**, 326, 328 ⟨90JA6738⟩
Hegarty, A. F., **3**, 530 ⟨93S606⟩, 558, 560 ⟨92JOC317⟩, 681 ⟨72JCS(P2)44⟩, 684 ⟨77JCS(P1)1457, 91JCS(P2)249⟩, 687 ⟨80ACR448, 80JA4537⟩, 710 ⟨86CC1196⟩; **4**, 990 ⟨80JOC4302⟩; **5**, 226 ⟨80JA4537⟩, 356 ⟨68JOC753⟩, 358 ⟨73JCS(P2)1422⟩, 672, 821, 825 ⟨86JCS(P2)995⟩, 675 ⟨86CC1196⟩, 675, 703 ⟨84JOC3083⟩, 699 ⟨78JOC388, 77JCS(P2)1136⟩, 703, 704 ⟨71CC689⟩, 714 ⟨77JCS(P2)1085⟩, 714, 723 ⟨77JCS(P2)1080⟩, 716 ⟨80JCS(P2)121⟩, 748 ⟨80CC130, 80TL885⟩, 838 ⟨91JCS(P2)249⟩, 840 ⟨65TL3801⟩, 846 ⟨68TL2463⟩; **6**, 506 ⟨79COC(2)1094⟩, 605, 616, 617 ⟨75JCS(P2)1046⟩, 620 ⟨80JCS(P2)867⟩, 623, 627 ⟨72JCS(P2)1050⟩
Hegarty, F., **1**, 442 ⟨B-78MI 110-01⟩
Hégazi, E., **5**, 592, 593 ⟨33HCA999⟩
Hegde, S. G., **2**, 21 ⟨80TL441, 81TL5019⟩
Hege, B. P., **1**, 726 ⟨68JOC3566⟩
Hegedus, L. S., **1**, 165 ⟨90JA2264⟩, 165 ⟨93JA87⟩, 250 ⟨82JA3727⟩, 315 ⟨80JA4973, 80JA4980⟩, 493 ⟨84T2415⟩, 668 ⟨B-87MI 114-04⟩, 987 ⟨92JA4079⟩; **2**, 563 ⟨B-80MI 211-01⟩, 681 ⟨83JA943⟩, 956 ⟨B-87MI 219-02⟩, 960 ⟨B-87MI 219-03⟩, 964 ⟨B-87MI 219-05⟩, 984 ⟨B-87MI 219-04⟩; **3**, 138 ⟨80JA4973⟩, 214, 293 ⟨83JA943⟩, 289 ⟨73JA3040, 75JA5448⟩, 544 ⟨92JA2991⟩, 635 ⟨85JOC2240⟩; **5**, 162 ⟨94JOC4972⟩, 259 ⟨B-84MI 506-01⟩, 273, 296 ⟨69JA1233⟩, 356 ⟨89JA9194⟩, 375 ⟨89JOC4649⟩, 464 ⟨75JA1184⟩, 696 ⟨85OM1749⟩, 932 ⟨88JA8413⟩, 953 ⟨88JA2122, 88OM702⟩, 953 ⟨90JA6255, 95OM2522⟩
Hegeman, A. D., **3**, 623 ⟨93TL1453⟩
Hegemann, M., **3**, 689 ⟨90CB2317⟩; **4**, 112, 116 ⟨90ZN(B)148⟩; **5**, 1152 ⟨93PS(76)1⟩; **6**, 684 ⟨93PS(76)265⟩, 685 ⟨90CB2317⟩
Heggs, R. P., **2**, 146 ⟨80TL689⟩
Hegmann, J., **3**, 543 ⟨90CB2031⟩
Heh, J. C. K., **2**, 25 ⟨79JOC1961⟩
Hehemann, D. G., **5**, 593 ⟨93JHC887⟩
Hehl, S., **3**, 687 ⟨90S1059⟩; **5**, 215 ⟨90S1129⟩
Hehre, W. J., **1**, 129, 162 ⟨81JOC1693⟩, 893 ⟨75ACR369⟩, 895 ⟨75JA2902⟩, 1109 ⟨70JA2141, 71JA808⟩, 1143 ⟨80JA3334⟩; **2**, 525, 1005 ⟨81JOC1693⟩
Heiba, E. I., **1**, 336 ⟨74JOC3456⟩, 336, 926 ⟨74JA7977⟩
Heibl, C., **3**, 552 ⟨75AG(E)248, 77CB1225⟩; **5**, 511 ⟨77CB1225⟩, 589 ⟨75AG(E)248⟩
Heicklen, J., **1**, 785 ⟨77CRV473⟩; **2**, 1013 ⟨65JPC2484⟩; **3**, 432 ⟨79OPP249⟩
Heid, P. F., **1**, 498 ⟨82JA3676⟩
Heide, W., **2**, 256 ⟨87JA1227⟩
Heidelberger, M., **5**, 275 ⟨41OSC(1)153⟩
Heidenreich, K., **6**, 511 ⟨1895JPR(2)472⟩
Heider, L., **5**, 301 282 ⟨68CB502⟩
Heider, K. J., **3**, 204 ⟨93AG(E)563⟩
Heider, R., **1**, 39 ⟨88CB1039⟩
Heidt, P. C., **3**, 411 ⟨90TL5541⟩
Heikenwälder, C.-R., **4**, 651 ⟨93OM4135⟩
Heikes, J., **5**, 488 ⟨81S801⟩
Heikki, N., **4**, 555, 557 ⟨91MI 412-01⟩
Heil, B., **2**, 330 ⟨91CC1684⟩, 438 ⟨83JOM(253)249⟩; **4**, 510 ⟨91JOM(408)297⟩
Heil, W., **3**, 417 ⟨86LA132⟩
Heilbron, I., **1**, 506 ⟨49JCS1827, 49JCS2028⟩, 506 ⟨49JCS737⟩; **3**, 54, 231 ⟨49JCS737⟩, 273 ⟨49JCS1827⟩; **5**, 83 ⟨49PCS2099⟩
Heilbron, I. M., **1**, 837 ⟨45JCS90⟩; **3**, 238 ⟨49JCS1430⟩, 275 ⟨46JCS39⟩; **4**, 318 ⟨B-49MI 407-01⟩
Heilbronner, E., **1**, 906 ⟨55CI(L)415, 58HCA414⟩, 910 ⟨68HCA225⟩, 1064 ⟨87CB187⟩; **3**, 330 ⟨83HCA801⟩
Heilemann, W., **6**, 252 ⟨92CC1017⟩
Heilig, G., **2**, 1003 ⟨91JOC1453⟩
Heilig, G. C., **2**, 595 ⟨78USP4118409⟩
Heiliger, L., **6**, 726 ⟨89JA8020⟩
Heilmann, S. M., **2**, 69 ⟨85TL981⟩; **3**, 157 ⟨83TL4075⟩, 615 ⟨78S219⟩, 651 ⟨86SC617⟩; **5**, 554 ⟨74JOC511⟩, 1134 ⟨79S523⟩
Heim, B., **4**, 681 ⟨92JOM(441)81⟩
Heim, N., **3**, 227 ⟨88HCA1719⟩; **5**, 174 ⟨87AG(E)362⟩
Heimann, M., **4**, 569 ⟨74ZN(B)485⟩, 570 ⟨78CB2696⟩, 585 ⟨74CB3697⟩, 624 ⟨76AG(E)367⟩; **5**, 914 ⟨78CB2696⟩
Heim, P., **3**, 82 ⟨72JOC2942⟩
Heim, S., **1**, 1023 ⟨68CC634⟩
Heim, U., **4**, 572 ⟨93AG(E)1359⟩; **5**, 888, 893 ⟨93CC673, 93PS(76)21⟩; **6**, 691 ⟨93CC673⟩, 691, 733 ⟨93PS(76)21⟩
Heimbach, H., **1**, 189, 200 ⟨84CB322⟩; **3**, 209 ⟨83CB3702⟩; **4**, 202 ⟨84TL511⟩
Heimbach, P., **1**, 413 ⟨73JOM(49)477⟩, 809 ⟨67AG(E)800, 73JOM(49)483⟩, 812 ⟨69LA(727)194⟩; **2**, 96, 97 ⟨62LA(652)7⟩
Heimer, N. E., **2**, 358 ⟨70JOC3012⟩; **5**, 244 ⟨85JOC4164⟩, 247 ⟨81JOC1374⟩
Heimgartner, H., **1**, 403, 795 ⟨70HCA173⟩, 420 ⟨72HCA3005, 72HCA1385⟩, 814 ⟨79HCA2630⟩; **3**, 678 ⟨77H(6)143, B-84MI 320-02⟩, 679, 680 ⟨74HCA1382, 88HCA1177⟩; **5**, 575 ⟨87HCA1001⟩, 605 ⟨86HCA174⟩, 611 ⟨93T7215⟩, 622 ⟨84HCA534⟩; **6**, 305, 309, 310 ⟨91HCA1500⟩
Hein, F., **2**, 441 ⟨56ZAAC(287)208⟩; **4**, 576, 577 ⟨60CB1339⟩
Hein, M., **4**, 757, 759, 820 ⟨58JOC1661⟩
Heincke, B., **5**, 84 ⟨57AG362⟩
Heindel, N. D., **1**, 384 ⟨63JA1548⟩; **2**, 632 ⟨80S572⟩
Heine, A., **6**, 184 ⟨90CC741⟩
Heine, H.-G., **3**, 601 ⟨93HOU(E15/2)1634⟩; **5**, 1107 ⟨79GEP2817455⟩; **6**, 108 ⟨79TL513⟩
Heine, H. W., **1**, 266 ⟨63JA2743⟩, 636 ⟨70AG(E)374⟩; **2**, 163 ⟨86JOC829⟩; **5**, 1010 ⟨63JA2743⟩
Heine, J., **4**, 338 ⟨88CZ146⟩
Heine, P., **5**, 1074 ⟨89ZAAC(568)35⟩
Heineke, D., **4**, 683 ⟨93CB373⟩, 685 ⟨93AG(E)1048⟩
Heinekey, D. M., **2**, 491 ⟨60IS113, 60IS116⟩
Heinemann, H., **1**, 1145 ⟨67T565⟩
Heinen, H., **2**, 384 ⟨66CB3118⟩; **3**, 597 ⟨85CB952, 86CB3141⟩, 597 ⟨88CB1085, 89CB1139⟩; **5**, 946, 957 ⟨87CB537⟩, 953 ⟨86CB3801⟩, 954 ⟨87CB1297⟩
Heinert, D. H., **1**, 870, 871 ⟨69JOC2817⟩; **5**, 1015 ⟨69USP3470228⟩
Heinicke, J., **2**, 872 ⟨80JOM(186)39⟩; **4**, 495, 496 ⟨76ZC64⟩, 495, 496, 497 ⟨73JPR65⟩, 496, 498, 499 ⟨73JOM(60)95⟩, 497, 498, 499, 500 ⟨85ZC369⟩; **5**, 790 ⟨76JPR409, 89ZC71⟩, 791 ⟨78ZC452, 80JOM(186)39⟩, 791, 892 ⟨79JOM(166)175⟩, 796 ⟨89JOM(364)C17⟩, 880 ⟨91CB493⟩; **6**, 119 ⟨83TL5481⟩, 161 ⟨78ZC452⟩
Heininger, H.-U., **3**, 680 ⟨71CB3816⟩
Heinisch, G., **1**, 934 ⟨78TL619⟩, 1010 ⟨88M253⟩; **4**, 416 ⟨88JHC119⟩
Heinish, G., **1**, 1010 ⟨88M751⟩
Heinlein, T., **4**, 1025 ⟨84CB1542⟩
Heinola, M., **1**, 1123 ⟨68JA3466⟩
Heinrich, P., **6**, 73 ⟨65AG(E)873⟩

Heins, A., **3**, 463 ⟨57LA(604)133⟩
Heintz, A., **4**, 988 ⟨80AG208⟩
Heintz, M., **3**, 620 ⟨93T5091⟩; **5**, 44 ⟨86JCE1013⟩
Heintzelman, R. W., **2**, 163 ⟨71JOC2207⟩, 163 ⟨76S731⟩, 793 ⟨76JOC2207⟩
Heiny, D. K., **5**, 179 ⟨89LA1233⟩
Heinz, G., **4**, 32 ⟨71CB1921⟩, 755 ⟨69CB1944⟩
Heinz, W., **1**, 880 ⟨89TL159⟩, 890 ⟨86CC947⟩
Heinze, H., **4**, 48 ⟨32LA(493)191⟩
Heinze, J., **1**, 103 ⟨89JA2325⟩; **4**, 550 ⟨77CZ259⟩, 649 ⟨83ZN(B)953⟩
Heinze, M., **3**, 363 ⟨02CB375⟩
Heinze, P. L., **1**, 429 ⟨89JFC(44)1637⟩
Heinzelmann, R. V., **2**, 103 ⟨60MI 202-03⟩
Heise, H., **6**, 140 ⟨60AG956⟩
Heiser, B., **5**, 800 ⟨85CB1517⟩; **6**, 324 ⟨85CB3151⟩
Heisig, G. B., **5**, 33 ⟨55OSC(3)213⟩
Heiss, H., **4**, 546, 577 ⟨92GEP4134772⟩
Heissler, D., **2**, 540 ⟨83JOC3252⟩
Heiszwolf, G. J., **1**, 873 ⟨70RTC1153⟩; **2**, 674 ⟨66CC51, 70RTC1153⟩
Heitkamp, S., **2**, 428, 430 ⟨84CB3400⟩, 437 ⟨86CB2921⟩, 440 ⟨83IC3624, 84IC3304⟩
Heitman, H., **3**, 493 ⟨61MI 313-01⟩
Heitmann, P., **2**, 698 ⟨86TL4721⟩
Heitmann, W. R., **1**, 516 ⟨79JOC3471⟩
Heitz, A., **2**, 305 ⟨91TL923⟩
Heitz, M.-P., **2**, 1018 ⟨90TL3141⟩; **4**, 1076 ⟨86JOM(315)59⟩
Heitzer, H., **6**, 51, 238 ⟨69LA(730)140⟩, 606 ⟨79GEP2754604⟩, 607 ⟨69LA(730)133⟩, 617 ⟨76GEP2451635⟩
Heitzmann, M., **3**, 528 ⟨77AG(E)469⟩
Hejchman, E., **3**, 634 ⟨87S452⟩
Hekkert, G. L., **1**, 1122 ⟨61RTC1285, 63RTC405⟩
Helali, S. E., **2**, 663 ⟨75HCA104⟩
Helbert, M., **5**, 577, 608, 609 ⟨79T1087⟩
Helbig, W., **1**, 222 ⟨84LA1165⟩
Helbing, C. H., **1**, 1040 ⟨48JA3868⟩
Helbing, R., **2**, 105 ⟨90ZAAC(580)217⟩
Helboe, P., **3**, 617 ⟨74ACS(B)317⟩
Held, A., **3**, 614 ⟨1889AC(R)222⟩
Held, P., **6**, 651 ⟨74JPR434⟩
Held, W., **3**, 513 ⟨77CB656⟩; **4**, 580 ⟨75JOM(86)C10, 77CB799⟩
Heldroon, G. M., **4**, 1080 ⟨88MI 424-01⟩
Heldt, W. Z., **1**, 484 ⟨65JOC3897⟩; **3**, 695 ⟨61JOC3226⟩; **5**, 215, 275, 282, 288, 292, 298 ⟨60OR(11)1⟩
Helferich, B., **2**, 108 ⟨24CB795⟩, 181 ⟨38CB712⟩, 249, 365 ⟨62LA(657)86⟩, 797 ⟨60LA(635)91⟩; **4**, 178 ⟨24CB795⟩; **5**, 279 ⟨59CB2813⟩
Helfert, H., **6**, 448 ⟨70AG(E)372⟩
Helfrerich, B., **6**, 71 ⟨56CB314⟩
Helfrich, O. B., **2**, 165 ⟨21JA591⟩
Helgesson, G., **4**, 671 ⟨93OM2268⟩
Helgstrand, A. J. E., **6**, 512 ⟨79EUP3008⟩
Helinski, J., **2**, 467 ⟨91TL4981⟩
Helitzer, R., **2**, 336, 339 ⟨67JOC1115⟩
Hell, C., **2**, 28 ⟨1881CB891⟩
Hell, P., **5**, 623 ⟨65AG(E)1095⟩
Hell, P.-M., **5**, 623, 735 ⟨69LA(727)50⟩, 638, 735 ⟨69LA(727)22⟩
Helland, A. C., **1**, 837 ⟨93S729⟩
Hellberg, J. S. E., **2**, 729 ⟨85JOM(296)357⟩
Hellberg, K. H., **6**, 219 ⟨77BEP856233⟩
Hellberg, L. H., **2**, 8 ⟨93SC711⟩
Heller, C. H., **2**, 783 ⟨06CB2339⟩
Heller, E., **2**, 84 ⟨91SL248⟩
Heller, F., **5**, 709 ⟨77GEP751220⟩
Heller, G., **2**, 783 ⟨06CB2334⟩
Heller, H., **3**, 689 ⟨86TL1903⟩
Heller, H. G., **3**, 241 ⟨72JCS(P1)2128⟩
Heller, M. S., **2**, 175 ⟨55JOC1165⟩, 249 ⟨70S545⟩, 273 ⟨54JOC1165⟩; **5**, 348 ⟨70JCED351⟩
Hellermann, L., **3**, 616 ⟨44JA1652⟩
Hellin, M., **5**, 710 ⟨66BSF1376⟩
Helling, J. F., **4**, 706, 709 ⟨62IC227, 61JOC1034⟩
Helliwell, M., **4**, 303 ⟨90JCS(P1)83⟩
Hellman, H., **1**, 822 ⟨63LA(662)188⟩; **3**, 492 ⟨61LA(640)79⟩; **4**, 96 ⟨B-60MI 403-01⟩; **6**, 107 ⟨66AG(E)848⟩
Hellman, S., **2**, 632 ⟨80JOM(186)C33⟩
Hellmann, G., **1**, 146 ⟨89AG(E)1025⟩, 146 ⟨92JA4439⟩, 922 ⟨78CB72⟩; **2**, 1000 ⟨88TL1259, 89AG(E)1025⟩, 1000 ⟨90AG(E)100⟩
Hellmann, H., **2**, 447 ⟨61LA(640)79⟩; **4**, 112 ⟨62LA(659)49⟩, 295 ⟨56CB81⟩, 322 ⟨54CB1690, 57CB444⟩, 322, 417, 424, 427 ⟨57CB50⟩, 416 ⟨B-66MI 409-01⟩, 417 ⟨57CB53⟩, 417, 423 ⟨58CB2432⟩, 442 ⟨56CB1134⟩; **5**, 24 ⟨75JOM(99)C5⟩, 26, 38 ⟨66AG(E)870⟩, 854 ⟨61CB1868⟩, 901 ⟨61TL724⟩
Hellmann, H. J., **2**, 440, 442, 446 ⟨62LA(659)49⟩
Hellmann, J., **2**, 471 ⟨77ZN(B)1244⟩
Hellmann, M., **6**, 222 ⟨57JA5654⟩
Hellwig, G., **1**, 637 ⟨89CB389⟩; **5**, 360 ⟨89CB389⟩
Hellwig, L. C., **4**, 54, 56 ⟨81CJC3055⟩
Hellwig, V., **6**, 433 ⟨76CB2475⟩
Hellwinkel, D., **1**, 445 ⟨64CB769⟩, 922 ⟨78TL4871⟩; **2**, 830, 869 ⟨65CB576⟩, 831, 840, 843 ⟨78CB13, 65AG(E)356⟩, 873, 881 ⟨64CB789⟩, 879 ⟨68CB121, 71CB1761⟩
Hellwinkel, H., **2**, 810 ⟨62AG(E)53⟩
Helm, D. Van der, **2**, 564 ⟨86OM1296⟩
Helm, S., **5**, 927 ⟨88AG(E)1331⟩
Helm, S. W., **5**, 381, 534 ⟨93CB1119, 92CB73⟩
Helmbrecht, J., **5**, 246 ⟨73JFC(3)383⟩; **6**, 432 ⟨73JFC(3)383⟩
Helmchen, G., **1**, 134 ⟨83TL3213⟩, 194 ⟨93TL1769⟩, 786 ⟨87TL183⟩; **4**, 179 ⟨79JA2171⟩; **5**, 32 ⟨85TL6051⟩, 148 ⟨81AG(E)207⟩, 209 ⟨77AG(E)728⟩
Helmholdt, R. B., **4**, 331, 538 ⟨72TL2367⟩
Helmick, L. S., **4**, 392 ⟨78JA5428⟩, 413 ⟨73JOC1949, 76JOC1303⟩
Helmkamp, G. K., **1**, 592 ⟨64JOC3258⟩, 636 ⟨64JOC1316⟩; **2**, 172 ⟨65JOC933⟩
Helmlinger, D., **3**, 222 ⟨89HCA1515⟩
Helmreich, R. F., **5**, 166 ⟨52JA4083⟩
Helmy, A. A. A., **1**, 981 ⟨61JCS2687⟩
Helmy, E., **1**, 779 ⟨70JCS(C)833⟩
Helps, I. M.,**1**, 529 ⟨88JOM(354)233⟩
Helquist, P., **1**, 140 ⟨85JOC3676⟩, 298 ⟨91COS(4)951⟩, 468, 472 ⟨81JA6460⟩, 523 ⟨85TL5393⟩, 541 ⟨86OS(64)1⟩, 541, 1107 ⟨79JOC3888⟩, 781 ⟨89JOC3988, 77JA253⟩; **2**, 308 ⟨88JOC3845⟩; **3**, 65 ⟨88ACS569⟩, 212 ⟨89JOC3988⟩, 217 ⟨78TL1363⟩, 350 ⟨85JOC3676⟩, 519 ⟨79JA6473, 81JA1862⟩, 519 ⟨85JOC5898⟩, 520 ⟨82JA6119⟩; **4**, 264 ⟨79JA1057⟩; **5**, 100 ⟨85OS(64)1⟩, 329 ⟨93TL7371⟩, 951 ⟨87OM988⟩
Helquist, P. M., **1**, 479, 1107 ⟨78TL2465⟩, 1094 ⟨77TL4145⟩, 1095, 1099 ⟨72JA9234, 78TL1363⟩, 1107 ⟨78TL2973, 79JOC1345⟩; **4**, 250 ⟨69TL173⟩; **6**, 134 ⟨69TL173⟩
Helvenston, M. C., **4**, 706 ⟨88OM2566⟩
Helwig, R., **4**, 799 ⟨85CB1008⟩; **5**, 675 ⟨85CB1008⟩
Hemblade, P. R., **1**, 777 ⟨82JCS(P2)1175⟩
Hembre, R. T., **3**, 544 ⟨88JA7868⟩; **4**, 679 ⟨88JA7868⟩
Hemens, C. M., **4**, 444 ⟨84CJC749⟩
Hemer, I., **4**, 33 ⟨84JFC(26)467⟩
Hemesley, P., **1**, 734 ⟨69JCS(C)1024⟩
Hemetsberger, H., **3**, 258 ⟨70M157⟩
Hemingway, M., **4**, 118 ⟨90SC1671⟩
Hemling, T. C., **1**, 988 ⟨85IJ136, 86JA343⟩

Hemmann, S., **6**, 159 ⟨88GEP3611522⟩
Hemmer, R., **2**, 347 ⟨90HOU(E16/1)404⟩
Hemmings, R. T., **2**, 293 ⟨75JIN233⟩, 857 ⟨75INC233⟩; **5**, 540 ⟨76MI 512-01⟩, 1081 ⟨76JCS(D)366⟩
Hemmings, T., **2**, 292, 294 ⟨80IC1879⟩
Hemond, R. C., **4**, 687 ⟨90OM2745⟩
Hempel, D., **6**, 254 ⟨63JPR1, 64ZC457⟩
Hempel, H.-U., **6**, 158 ⟨79ZAAC(457)214⟩
Hempel, J., **6**, 229, 230 ⟨72GEP2129200⟩
Henbest, H. B., **2**, 151 ⟨68JCS(C)2374⟩; **3**, 613 ⟨62JCS954⟩, 634 ⟨67JCS(C)2465⟩; **4**, 207 ⟨64JCS1217⟩; **5**, 168 ⟨59JCS227⟩, 293 ⟨57JCS3032⟩
Hendershot, D. G., **2**, 483 ⟨92OM2163⟩
Henderson, A., **5**, 773 ⟨82TL3357⟩
Henderson, A. D., **5**, 1110 ⟨80T1839⟩
Henderson, B. S., **1**, 797 ⟨90CR1203⟩
Henderson, E., **5**, 1081 ⟨76JCS(D)366⟩
Henderson, H. E., **5**, 540 ⟨76MI 512-01⟩
Henderson, J. W., **1**, 893 ⟨73CSR397⟩
Henderson, M. A., **1**, 979 ⟨88JOC4736⟩; **5**, 159 ⟨88JOC4736⟩
Henderson, M. J., **6**, 190, 196 ⟨86CC672⟩
Henderson, W. A., **2**, 433, 444 ⟨60JA5794⟩, 448 ⟨61JOC4770⟩, 471 ⟨63JA2462⟩; **5**, 316 ⟨86OPP269⟩
Henderson, W., **5**, 802 ⟨90JCS(D)781⟩
Henderson, W. W., **5**, 1100, 1102 ⟨64JA4732⟩
Hendges, S. K., **4**, 650, 656 ⟨93OM338⟩
Hendrick, M. E., **1**, 549 ⟨73JA6286⟩
Hendricks, H. P. G., **1**, 1030 ⟨68RTC69⟩
Hendricks, R. T., **1**, 575, 576 ⟨90JA8433⟩, 808 ⟨88JOC196⟩
Hendrickson, A. R., **3**, 350 ⟨72AJC257⟩
Hendrickson, J. B., **1**, 318 ⟨73TL1513⟩, 442 ⟨61JA1251⟩, 618, 704, 705, 709 ⟨86JA2358⟩, 619, 704, 705, 708 ⟨85JOC2110⟩, 704, 705 ⟨84TL4617⟩, 1071 ⟨89S217⟩; **2**, 157 ⟨86JA2358⟩, 300, 303 ⟨73TL3839⟩, 619 ⟨69JOC3434⟩; **3**, 193, 472, 473 ⟨68JOC3610⟩, 210 ⟨85JOC2110⟩, 618 ⟨75TL277, 76TL603⟩, 662 ⟨87JOC4137⟩; **5**, 123, 280, 291 ⟨75TL277⟩, 196 ⟨89JOC1144⟩, 363 ⟨75JOC3450⟩; **6**, 235 ⟨77JOC3875⟩
Hendriks, A. H. M., **1**, 24 ⟨84RTC220⟩
Hendriksen, L., **6**, 594 ⟨72IJS(A)133⟩
Hendrikson, J. B., **2**, 156 ⟨84S320⟩
Hendrikx, G., **5**, 1066 ⟨86CB2114⟩
Hendriock, J., **4**, 142, 519, 568 ⟨77JOM(132)77⟩
Hendrix, J., **3**, 285, 295 ⟨87JOC2615⟩
Hendry, J. A., **3**, 474 ⟨58JCS1134⟩
Henecka, H., **3**, 448 ⟨57CB1060⟩
Hener, L., **2**, 821, 822 ⟨92CB2187⟩
Henery-Logan, K. R., **1**, 785 ⟨63JA3524⟩; **5**, 81 ⟨59JA3089⟩; **6**, 77 ⟨67JA5724⟩
Heng, R., **5**, 360 ⟨85TL215⟩
Hengartner, U., **1**, 17, 434 ⟨72JA5098⟩; **2**, 693 ⟨72JA5098⟩, 822, 841 ⟨79JOC3741⟩
Hengelsberg, H., **4**, 506, 513 ⟨91JOM(415)39⟩, 602, 621, 929 ⟨92CB607⟩; **5**, 416 ⟨92CB607⟩
Hengesbach, J., **4**, 695 ⟨83JOM(246)141⟩
Hengeveld, J. E., **4**, 67 ⟨83TL139⟩
Hengge, E., **2**, 544 ⟨76AG(E)370⟩
Hengl, L., **3**, 350 ⟨48BSF1120⟩
Hengstberger, H., **5**, 898 ⟨69LA(721)121⟩
Henien, T., **4**, 247 ⟨88TL6729⟩
Henigschmidt-Grossich, R., **2**, 508, 509, 894 ⟨68JCS(A)2623⟩
Henin, F., **2**, 637 ⟨91AG(E)416⟩; **5**, 296 ⟨86TL6339⟩
Henke, B. R., **1**, 369 ⟨92JOC7056⟩
Henke, H., **4**, 353 ⟨92JOM(424)273⟩, 602, 621, 929 ⟨92CB607⟩; **5**, 416 ⟨92CB607⟩
Henkel, G., **2**, 207, 721 ⟨89AG(E)202⟩
Henkel, S., **6**, 393 ⟨94JOM(469)135⟩
Henkel, U., **4**, 943 ⟨77CB2880⟩; **6**, 317, 734 ⟨77CB2880⟩
Henkelmann, J., **3**, 504 ⟨85AG(E)1065⟩; **6**, 713 ⟨85AG(E)1065⟩
Henklein, P., **6**, 533, 538 ⟨88EGP273832⟩
Henkler, T., **3**, 447 ⟨65CB912⟩
Henmi, T., **5**, 333 ⟨94OPP111⟩
Henn, D., **6**, 451 ⟨69M1860⟩
Henn, L., **2**, 876 ⟨84TL1357⟩, 877 ⟨86CB1953⟩, 1027 ⟨84LA1358⟩; **3**, 273 ⟨80JOM(184)317, 81TL2637⟩, 477 ⟨81ZN(B)218⟩
Henn, R., **5**, 476 ⟨87CB1499⟩; **6**, 253 ⟨89JFC(45)225⟩, 254, 530, 532 ⟨88JFC(39)329⟩, 254, 532 ⟨87CB1499⟩, 531 ⟨86SA(A)1281⟩
Henne, A., **2**, 637 ⟨76AG(E)435⟩
Henne, A. L., **1**, 508 ⟨43JA2020⟩; **2**, 18 ⟨44OR(2)49⟩; **3**, 212 ⟨53JCS3628⟩; **4**, 2, 3, 4 ⟨44OR(2)49⟩, 3, 30, 33 ⟨36JA889, 38JA864⟩, 5, 32 ⟨47JA279⟩, 6 ⟨47JA281⟩, 18 ⟨39JA938⟩, 23, 33 ⟨54JA479⟩, 30 ⟨37JA1400, 46JA496⟩, 44 ⟨50JA4378⟩; **6**, 2 ⟨41JA3476⟩, 16 ⟨41JA3478, 47JA1820⟩, 16 ⟨49JA298⟩, 212, 218 ⟨37JA1200⟩
Henneke, K.-W., **1**, 155 ⟨77LA40⟩; **3**, 696, 702, 704 ⟨77LA40⟩
Hennen, W. J., **3**, 159 ⟨91S483⟩; **5**, 315 ⟨82JOC4605⟩, 516, 522 ⟨85JOC1741⟩
Henner, B. J. L., **6**, 402, 404 ⟨81JOM(221)257⟩
Hennessy, B. M., **1**, 576, 1021 ⟨86JOC3098⟩
Hennessy, M. J., **3**, 193, 472 ⟨86JOC4077⟩; **5**, 869 ⟨86JOC4077⟩
Hennig, H., **5**, 852 ⟨68CB3947⟩
Hennig, H.-W., **5**, 220, 403 ⟨86ZN(B)859⟩; 221 ⟨83CZ257, 83CZ363⟩, 221 ⟨85JFC(27)157, 87CZ61⟩
Henning, H. G., **2**, 452 ⟨74ZC477⟩; **4**, 554 ⟨66ZC314⟩, 554, 556 ⟨65ZC419, 73ZC55⟩, 556 ⟨75JPR840⟩, 1023 ⟨66AG(E)847, 66ZC28⟩
Henning, H. J., **3**, 493 ⟨59CB2756⟩
Henning, R., **1**, 156 ⟨76HCA2213⟩, 156 ⟨79CB234⟩, 156, 157 ⟨77TL1161⟩, 325, 326 ⟨83TL5343⟩; **3**, 57 ⟨78AG(E)458⟩, 160 ⟨84CC29⟩; **4**, 255 ⟨79JA4732, 79JA7104⟩; **5**, 557 ⟨79JA4732, 79JA7104⟩, 557 ⟨83JA5015⟩; **6**, 134 ⟨79JA4732⟩
Henning, R. W., **3**, 516 ⟨93OM2565⟩
Henninger, M., **2**, 657, 753 ⟨76S815⟩
Hennion, G. F., **1**, 399 ⟨70JA1769⟩, 506 ⟨61JOC1073⟩, 958, 1125 ⟨49JA1964⟩, 966 ⟨78JOC1389⟩, 970 ⟨76JOC3496, 78JOC1385⟩, 1000 ⟨37JOC1⟩; **3**, 453 ⟨41JA1151⟩; **4**, 16 ⟨39JA1460⟩, 195 ⟨36JA80⟩, 195 ⟨36JA892⟩; **5**, 304 ⟨41JA1151⟩
Hennrichs, W., **2**, 339 ⟨85ZAAC(526)186, 87ZAAC(547)233⟩
Hennson, G. F., **2**, 521 ⟨57JA5190, 58JA617⟩
Henoch, F. E., **1**, 154 ⟨69JA676⟩; **2**, 1005 ⟨68JA7287⟩
Henretta, J. P., **1**, 811 ⟨88TL975⟩
Henrick, C. A., **1**, 456, 502, 504, 505, 506, 734, 1000, 1002 ⟨77T1845⟩, 727 ⟨75JA4327⟩; **3**, 494 ⟨75JA4327, 77T1845⟩
Henrick, K., **2**, 362, 363 ⟨91JCS(P1)2169⟩; **4**, 659 ⟨80JOM(193)21⟩, 675 ⟨88NJC559⟩
Henrick, P., **1**, 89 ⟨53CB167⟩
Henriet, M., **4**, 790 ⟨80TL223⟩
Henriet-Bernard, C., **1**, 979 ⟨92TL1057⟩
Henriksen, L., **3**, 39 ⟨84CC89⟩; **4**, 875 ⟨70ACS3213⟩; **5**, 233, 240, 243, 506, 508 ⟨72ACS1465⟩, 626 ⟨74S501⟩, 627, 642, 726 ⟨79CC1102⟩, 1055 ⟨73IJS389, 76S519⟩; **6**, 591 ⟨67ACS1981, 75JOC746⟩, 591 ⟨84CC89⟩, 592, 593 ⟨70ACS2061⟩, 593, 597 ⟨72ANY(192)101⟩, 594, 597 ⟨82S771⟩
Henriksen, U., **3**, 639 ⟨83ACS(B)639⟩
Henrio, G., **3**, 105, 106 ⟨76BSF265⟩
Henri-Rousseau, O., **1**, 516, 549 ⟨B-78MI 112-03⟩
Henrot, S., **1**, 479 ⟨87TL1781⟩
Henry, H. L., **1**, 1029 ⟨91JMC2452⟩

Henry, J. P., **4**, 31 ⟨70JOC4201⟩
Henry, J. R., **2**, 309 ⟨89TL5709⟩; **3**, 294 ⟨93JOC4745⟩
Henry, Jr., K. J., **1**, 190, 402 ⟨92TL4735⟩, 190 ⟨93CC510⟩
Henry, L., **4**, 405 ⟨1895BSF157⟩; **5**, 570 ⟨1869CB305, 1869CB494⟩
Henry, M. C., **1**, 271, 647, 738, 743 ⟨60JA563⟩; **2**, 1060, 1061 ⟨68JOM(15)69⟩, 1076, 1080, 1083, 1089, 1092, 1098, 1099, 1101 ⟨67CRV73⟩, 1093 ⟨66JOM(5)29, 67JOM(9)435, 69JOM(16)413⟩; **4**, 699 ⟨67JOM(9)117⟩
Henry, P. M., **1**, 1096 ⟨71JOC1886⟩; **2**, 983 ⟨B-80MI 219-01⟩
Henry, R. A., **2**, 781 ⟨58JOC648⟩; **4**, 405, 406 ⟨49JA2271⟩; **6**, 504 ⟨49JA2297⟩
Henry, W., **3**, 617 ⟨82OPP396⟩; **5**, 275 ⟨82OPP396⟩
Hensel, H. R., **3**, 464 ⟨55CB527⟩; **4**, 406 ⟨66CB868⟩
Hensel, R., **2**, 294 ⟨B-87MI 204-03⟩, 465 ⟨83PS(18)73⟩
Hensin, P. D., **2**, 822 ⟨74JOC2296⟩
Hensley, M., **2**, 39 ⟨70B1626⟩
Henson, P. D., **2**, 441 ⟨69JA5645⟩
Henssen, G., **4**, 229 ⟨80LA1665⟩
Henstock, J. B., **1**, 537 ⟨67JCS(C)1986⟩
Hentchoya Hémo, J., **2**, 310 ⟨77TL3469⟩
Henton, D. E., **1**, 841 ⟨73CC223⟩, 1023 ⟨69CC674⟩
Hentschel, M., **5**, 593 ⟨76JPR343⟩
Hentschel, W., **6**, 415 ⟨1887JPR99⟩
Henze, H. R., **3**, 612 ⟨39JA1574⟩; **4**, 48 ⟨39JOC234⟩
Henzen, A. V., **4**, 829 ⟨83TL1303⟩; **6**, 116 ⟨83TL1303⟩
Hepburn, D. D., **2**, 25, 31 ⟨74CI(L)664⟩
Hepburn, D. R., **2**, 25, 31 ⟨76JCS(P1)754⟩
Hepburn, S. P., **2**, 390, 397 ⟨88OSC(6)803⟩, 764 ⟨72OS(52)77⟩
Hepfinger, N. F., **2**, 783 ⟨63JA2683⟩
Hepp, P., **5**, 285 ⟨1877CB327⟩
Hepp, W., **6**, 674, 723 ⟨90JOM(385)221, 86OM173⟩
Heppert, J. A., **3**, 516 ⟨93OM2565⟩; **4**, 673, 722 ⟨89OM1199⟩, 676 ⟨92AG(E)66, 88OM1715⟩
Heppke, G., **5**, 238 ⟨77AG(E)318⟩
Hepworth, J. D., **3**, 172 ⟨94T335⟩
Herald, Jr., D. L., **5**, 51 ⟨71CC10⟩
Heravi, M. M., **1**, 1015 ⟨91TL5243⟩
Herber, J. F., **2**, 391 ⟨65JOC1488⟩
Herber-Brunschweiger, E.,**1**, 455 ⟨77T457⟩
Herberg, A. K., **4**, 489 ⟨65AG1138⟩
Herberg, K., **3**, 480 ⟨65AG1138, 65AG(E)1078⟩
Herberhold, M., **2**, 891 ⟨84CB1999⟩, 963 ⟨86JOM(310)27⟩; **4**, 290 ⟨87JOM(336)153⟩, 594 ⟨87ZN(B)118⟩, 719 ⟨86ZN(B)1431, 87JOM(334)347⟩, 719 ⟨89JOM(371)205, 92JOM(436)333⟩; **5**, 241 ⟨89JOM(377)281⟩; **6**, 247 ⟨90JOM(395)327⟩, 338 ⟨90ZN(B)447⟩
Herberich, G. E., **3**, 531 ⟨93CB297⟩; **4**, 642 ⟨93CB1397⟩, 643 ⟨83ZN(B)1388⟩, 695 ⟨83JOM(246)141, 86JOM(308)153, 86JOM(312)13, 87JOM(319)9, 93JOM(459)1⟩, 695, 717 ⟨89AG(E)319⟩, 695, 717 ⟨89JOM(372)53, 86AG(E)165⟩, 717 ⟨89AG(E)737⟩
Herbert, A. L., **2**, 614 ⟨71JOC631⟩
Herbert, E., **2**, 10 ⟨79JOC3406⟩
Herbert, J., **5**, 944 ⟨94TL3801, 95UP 525-01⟩
Herbert, J. O., **3**, 54 ⟨67CRV153⟩
Herbert, M., **2**, 126 ⟨59BRP823483⟩, 291 ⟨61T1⟩
Herbert, R. B., **1**, 452 ⟨73TL1375⟩, 654 ⟨86CC620⟩; **3**, 694, 712 ⟨88MI 321-01⟩; **4**, 139 ⟨88JCS(P1)2585⟩
Herbert, R. W., **4**, 137 ⟨90JOC4782⟩
Herbert Fischer, **1**, 866 ⟨65LA(689)1⟩
Herbet, J. M., **3**, 65 ⟨89JCS(P1)683⟩
Herbst, G., **5**, 49 ⟨69JOC116⟩
Herbst, H., **3**, 297 ⟨56AG618⟩
Herbst, R. M., **1**, 80 ⟨43OSC(2)491⟩; **5**, 66 ⟨43OSC(2)519⟩, 90 ⟨43OSC(2)491⟩, 652 ⟨43OSC(2)11⟩
Herbst-Irmer, R., **4**, 940, 1014 ⟨92CB2051⟩; **5**, 958 ⟨92CB2051⟩, 1086 ⟨92AG(E)437⟩; **6**, 184 ⟨92CB1807⟩, 196 ⟨93CB2247⟩, 252 ⟨92CC1017⟩
Herbstman, S., **2**, 485, 508, 510 ⟨64JOC986⟩; **5**, 222 ⟨65JOC1259⟩
Hercouet, A., **2**, 320 ⟨91SC1579⟩, 444 ⟨87PS(29)111, 90PS(48)279⟩, 831 ⟨81T2855⟩; **3**, 494 ⟨82S597⟩
Herczegh, P., **3**, 634 ⟨78OPP211⟩; **6**, 88 ⟨88T2063⟩
Herdeis, C., **3**, 717 ⟨86AP(319)473⟩; **4**, 105 ⟨92AP(325)411⟩
Herdewijn, P., **2**, 625 ⟨86JMC213⟩
Herdtweck, E., **3**, 172 ⟨93T1821⟩, 502 ⟨91AG(E)444⟩; **4**, 603 ⟨91CB1575⟩, 611 ⟨92ZN(B)1377⟩, 651 ⟨92JOM(426)1⟩; **5**, 1097 ⟨85AG(E)1062, 89CB445⟩; **6**, 720 ⟨88JOM(355)437⟩
Herfert, R. E., **6**, 63 ⟨57JA3351⟩
Hergenrother, P. M., **5**, 847 ⟨69JPS(A)945⟩
Hergenrother, W. L., **1**, 881 ⟨64JOC221⟩
Herges, R., **1**, 982 ⟨91S1151, 92AG(E)1611⟩
Herget, C., **6**, 523 ⟨61AG579, 65ZN(B)1159⟩
Hergott, H. H., **2**, 539 ⟨82S1⟩; **6**, 43 ⟨82S1⟩
Hergrueter, C. A., **1**, 1094 ⟨77TL4145⟩
Hérion, H., **1**, 1079 ⟨93TL5621⟩
Heritage, G. L., **1**, 310 ⟨05MI 107-01⟩
Herkes, F. E., **4**, 732 ⟨65TL1883, 67JOC1311⟩, 732 ⟨68JOC1854⟩; **6**, 694 ⟨67JOC1311, 68JOC1854⟩
Herlem, M., **1**, 901 ⟨77PAC107⟩
Herling, J., **1**, 795 ⟨65JA4107⟩
Herman, D. F., **6**, 419 ⟨46JA1672⟩
Herman, G., **1**, 590 ⟨75JOC812⟩
Herman, L. G., **5**, 447 ⟨68JOM(15)397⟩
Herman, R. A., **5**, 1126 ⟨89USP4857550⟩
Hermanek, S., **4**, 629 ⟨92CRV175⟩; **5**, 199 ⟨59CCC2748⟩
Hermann, C. K. F., **5**, 342 ⟨92JOC5328⟩
Hermann, E. C., **5**, 285 ⟨71JMC535⟩
Hermann, H. D., **4**, 331 ⟨55CB41⟩
Hermann, P., **2**, 223 ⟨74JPR719⟩
Hermann, R., **2**, 173, 717 ⟨87S72⟩
Hermans, B., **2**, 286 ⟨92TL4629⟩
Hermans, P. H., **2**, 111 ⟨25ZAAC(142)83⟩
Hermawan, I., **2**, 34 ⟨92BCJ1731⟩
Hermecz, I., **3**, 310 ⟨77JCS(P1)789⟩; **4**, 429, 431 ⟨90JOC6198⟩
Hermeling, D., **1**, 561 ⟨88CB1151⟩; **4**, 197 ⟨93GEP4122315⟩
Hermen, M. S., **4**, 650, 656 ⟨93OM338⟩
Hermes, M. E., **3**, 476 ⟨67JA4760⟩; **5**, 1119 ⟨67JA4760⟩
Hermesdorf, M., **3**, 689 ⟨87JOM(338)329⟩; **5**, 1152, 1159 ⟨87JOM(338)329⟩, 1153 ⟨89PS(29)31⟩, 1154, 1155 ⟨89MI 531-01, 87PS(30)479⟩
Hermkens, P. H. H., **3**, 433 ⟨86JOC309⟩
Hermoso, J. A., **5**, 386 ⟨92JCS(D)53⟩
Hernandes, L., **3**, 543 ⟨76JA3728⟩
Hernández-Galán, R., **1**, 831 ⟨91TL3209⟩
Hernández, A., **1**, 659 ⟨75JOM(102)123⟩, 763 ⟨81TL1575⟩; **4**, 188 ⟨73JOC3935⟩
Hernández, A. E., **3**, 128 ⟨93T965⟩
Hernández, D., **1**, 650 ⟨82TL1035⟩; **3**, 130, 202 ⟨85JOC5260⟩
Hernández, E., **1**, 434 ⟨75TL4005⟩; **5**, 386 ⟨85AG(E)961, 85JOM(295)C21⟩, 386 ⟨86JOM(315)245, 87JOM(327)429⟩, 386 ⟨87JOM(328)403, 88JOM(339)213⟩, 386 ⟨88JOM(344)C35⟩
Hernández, J. E., **1**, 1073, 1085 ⟨90SC2733⟩; **2**, 329 ⟨88SC2055⟩; **3**, 206 ⟨90SC131⟩
Hernandez, Jr., L., **3**, 543 ⟨78TL929⟩
Hernandez, O., **2**, 539 ⟨79TL99⟩, 659 ⟨76JOC977⟩; **4**, 260 ⟨76JOC3975⟩, 263 ⟨76JOC3979⟩; **5**, 374 ⟨79TL99⟩; **6**, 128 ⟨76JOC3975⟩
Hernández, R., **1**, 927 ⟨88JCS(P1)3255⟩, 929 ⟨84TL1953⟩, 931 ⟨93TL4865⟩; **2**, 13 ⟨86JOC402⟩; **3**, 409 ⟨86S1055⟩; **5**, 1118 ⟨90JCS(P1)2417⟩, 1119 ⟨87JCS(P1)937⟩, 1125 ⟨86MI 530-01⟩
Hernandez, T. J., **6**, 513 ⟨77GEP2638754⟩

Herndon, J. W., **2**, 927 ⟨93SL363⟩
Herndon, W. C., **1**, 785 ⟨78JOC518⟩
Herold, J., **6**, 409 ⟨20JPR79⟩
Herold, T., **1**, 345, 946 ⟨80JA3642⟩; **2**, 75 ⟨81CB375⟩
Herold, U., **3**, 608 ⟨88TL2935⟩
Heron, B. M., **3**, 172 ⟨94T827⟩; **4**, 298 ⟨94PC 407-01⟩
Herout, V., **1**, 812 ⟨59CCC2371⟩
Herr, B. C., **2**, 452, 841 ⟨52JA4526⟩
Herr, F., **2**, 27 ⟨71JMC982⟩
Herr, H. E., **2**, 31 ⟨50JA1753⟩
Herr, R., **2**, 444 ⟨83CB3567⟩, 446 ⟨85CB3105⟩; **4**, 94 ⟨85OM1208⟩, 551, 560, 1023 ⟨85CB3105⟩, 554 ⟨84CB3374, 84AG(E)247⟩
Herr, R. R., **1**, 108, 464 ⟨59JOC843⟩
Herr, R. W., **1**, 217, 226 ⟨73JOC4263, 70JA4979⟩
Herradon, B., **1**, 311 ⟨87TL3791⟩; **4**, 213 ⟨87TL3791, 89HCA690⟩
Herrera, M., **6**, 547 ⟨80AQ183⟩
Herrick, E. C., **1**, 78 ⟨63OSC(4)304⟩; **5**, 199 ⟨63OSC(4)890⟩
Herrick, R. S., **4**, 675 ⟨83JA2599⟩
Herring, A. M., **4**, 548, 1023, 1028 ⟨86CC882⟩
Herring, D. L., **2**, 836, 837 ⟨64IC428⟩
Herring, K. H., **6**, 76 ⟨77JOC679⟩
Herrinton, P. M., **1**, 216 ⟨83JOC4572⟩
Herriott, A. W., **5**, 47, 52, 115 ⟨74TL1511⟩
Herrman, J., **2**, 478 ⟨69JOC940⟩
Herrmann, A. T., **2**, 984 ⟨91CB2165⟩; **5**, 885 ⟨91AG(E)310⟩, 890 ⟨90CB1617, 90PS(49)337⟩, 890 ⟨90TH29⟩
Herrmann, D., **5**, 1146 ⟨81LA1198⟩; **6**, 537 ⟨75LA1025⟩, 618 ⟨78LA1704⟩
Herrmann, E., **4**, 133, 556 ⟨86ZAAC(536)187⟩, 555 ⟨82ZC307, 86EGP238612⟩; **5**, 1131 ⟨92EGP300936A7⟩; **6**, 581 ⟨92PS(73)81⟩
Herrmann, G., **6**, 143 ⟨85CB3959⟩
Herrmann, J., **6**, 616 ⟨77ZC172⟩
Herrmann, J. L., **1**, 309 ⟨73TL2595⟩; **4**, 855 ⟨73TL4711⟩, 857 ⟨73TL4707⟩
Herrmann, K., **5**, 1144 ⟨79S204⟩
Herrmann, M., **5**, 11 ⟨91S1173⟩
Herrmann, R., **2**, 173 ⟨86TL1493⟩; **3**, 35 ⟨87S159⟩, 696, 699, 721, 722 ⟨85S400⟩, 696, 722 ⟨86ZN(B)132⟩; **4**, 273 ⟨90JCS(P1)2919⟩, 1033 ⟨90JOM(384)155⟩
Herrmann, W. A., **2**, 512 ⟨84AG(E)812, 85CB2472⟩, 585 ⟨84AG(E)515, 91AG(E)185⟩, 585 ⟨93CB45⟩; **3**, 387, 390 ⟨83AG(E)314⟩, 401 ⟨84CC686⟩, 519 ⟨84JOM(264)327⟩, 544 ⟨79JA3133⟩; **4**, 669, 677, 709 ⟨82AOC(20)159⟩, 676 ⟨75JOM(97)245⟩, 677 ⟨81AG(E)887, 82HOU(E1/E2)⟩, 679 ⟨81JOM(204)C21⟩, 686 ⟨77JOM(140)73, 78CB1077⟩, 686 ⟨78ZN(B)911, 79JOM(165)C17⟩, 686 ⟨82CB878, 82JOM(226)C59⟩, 686 ⟨82JOM(236)C18, 83JOM(258)81⟩, 689 ⟨83JOM(256)147⟩, 709 ⟨87JOM(321)257⟩; **5**, 943 ⟨74AG(E)599⟩, 1097 ⟨82AG(E)117, 85AG(E)1062⟩, 1097 ⟨85CB340, 89CB445⟩
Herron, B. F., **1**, 622 ⟨91T8855⟩; **4**, 655 ⟨93S202⟩
Herscheid, J. D. M., **2**, 340 ⟨86CRV697⟩, 613 ⟨84CC655, 86JOC1886⟩; **4**, 186 ⟨73S169⟩, 304 ⟨80JOC1880⟩
Herscovici, J., **3**, 4, 115 ⟨82JCS(P1)1967⟩, 231 ⟨91TL1791⟩
Herse, R. H., **2**, 365 ⟨78IJ60⟩
Hersh, W. H., **6**, 339, 347 ⟨87JA2843, 88OM794⟩
Hershberg, E. B., **1**, 503 ⟨51JA5073⟩; **3**, 100 ⟨55OSC(3)627⟩; **4**, 178 ⟨54JA6113⟩; **5**, 59 ⟨32OSC(1)168⟩
Hershberger, J., **1**, 331, 936 ⟨88T6295, 85TL6289⟩; **2**, 394 ⟨82JOM(225)43⟩
Hershberger, J. W., **1**, 935 ⟨88OM2220⟩
Hershberger, S., **1**, 331, 936 ⟨88T6295, 85TL6289⟩
Hershberger, S. S., **1**, 493 ⟨81TL5231⟩, 1096 ⟨80JOC3840⟩; **3**, 290 ⟨80JOC3840⟩
Hersheid, J. D. M., **5**, 84 ⟨86CRV697⟩
Hershenson, F. M., **5**, 306 ⟨87JMC498, 90JOC560⟩
Hershkowitz, R. L., **2**, 7, 627 ⟨64JA964⟩
Herslof, M., **4**, 856 ⟨76CS90⟩
Hertel, B., **1**, 946 ⟨93CB1187⟩
Hertenstein, U., **3**, 213 ⟨80CB3783⟩, 214 ⟨86CB722⟩, 292, 295, 303, 311 ⟨79CB2045⟩
Herter, W. G., **3**, 404 ⟨88TL761⟩
Herterich, I., **1**, 104 ⟨67TL3871⟩, 1123 ⟨65TL875, 66TL3847⟩, 1123 ⟨70LA(733)5⟩
Hertkorn, N., **4**, 528, 533, 1020 ⟨87CB2081⟩
Hertler, W. R., **3**, 696 ⟨58JOC1221⟩; **4**, 893, 979, 985 ⟨62JA3387⟩; **5**, 1133 ⟨87JA6532⟩
Herunsalee, A., **1**, 1048 ⟨90SL701⟩; **4**, 940 ⟨90SL701, 91T3727⟩
Herunsalee, K., **4**, 742 ⟨83TL527⟩
Hervé, Y., **1**, 344 ⟨87TL1413⟩, 602 ⟨85T4347⟩, 935 ⟨84CC1298, 88T5479⟩; **3**, 635 ⟨87T4297⟩; **4**, 473 ⟨93SL35⟩; **5**, 1115 ⟨86T4983⟩
Herve du Penhoat, C., **1**, 439, 700 ⟨87T859⟩, 485 ⟨88T111⟩, 612 ⟨86T4807⟩, 700 ⟨83TL4319⟩
Herweh, J. E., **2**, 717 ⟨63JOC2766⟩
Herwig, J., **2**, 192 ⟨93CC1592⟩
Herwig, W., **4**, 556 ⟨76GEP2523145⟩
Herynk, J., **2**, 12, 627 ⟨65JOC415⟩
Herz, A. H., **6**, 83, 84 ⟨53JA1668⟩
Herz, J. E., **1**, 630 ⟨69CC1395⟩
Herz, W., **3**, 57 ⟨77JOC813⟩
Herzberger, S., **3**, 556, 557, 560, 585, 586 ⟨81CB3751⟩, 687 ⟨90S1128⟩; **5**, 215 ⟨90S1129⟩
Herzig, C., **2**, 456 ⟨80TL2687⟩; **3**, 50, 197 ⟨82CB601⟩
Herzig, J., **3**, 654 ⟨85JOC3558⟩; **4**, 176 ⟨87CAR(162)145⟩
Herzog, F., **2**, 484 ⟨75JOM(93)191⟩
Herzog, G., **5**, 720 ⟨77JOM(134)21⟩
Herzog, H. L., **2**, 9 ⟨63OSC(4)753⟩
Herzog, W., **5**, 560 ⟨88ZC269⟩
Hes, J., **2**, 105 ⟨74JOC3767⟩
Hesbain-Frisque, A.-M., **1**, 375 ⟨82JA1428⟩, 522 ⟨75JA4409⟩; **3**, 602 ⟨85JA2192⟩; **5**, 718 ⟨82JA1428⟩
Heshmati, P., **1**, 199 ⟨93SL769⟩
Heslinga, L., **5**, 14 ⟨57RTC969⟩; **6**, 41 ⟨57RTC969⟩
Hesp, B., **1**, 517 ⟨67JCS(C)1625⟩
Hess, A., **5**, 237 ⟨80JOC2766⟩
Hess, Jr., B. A., **3**, 337 ⟨91CB2609⟩
Hess, G. P., **5**, 140, 279 ⟨56JA1367, 55JA1067⟩
Hess, N., **1**, 1064 ⟨87CB187⟩
Hess, R. W., **6**, 517 ⟨71JA5424⟩
Hess, T. C., **3**, 530 ⟨85JA7597⟩
Hess, U., **3**, 702 ⟨91TL5539⟩; **6**, 477 ⟨93GEP4140446⟩
Hess, W., **5**, 1145 ⟨68SAP6903273⟩
Hesse, A., **1**, 881 ⟨70OS(50)66⟩; **2**, 420 ⟨71LA(746)149⟩
Hesse, C., **3**, 413 ⟨91SL654⟩
Hesse, G., **2**, 100 ⟨60CB1249⟩; **3**, 107 ⟨56AG438⟩, 680 ⟨68LA(713)1⟩; **4**, 177, 178 ⟨60CB1249⟩, 522 ⟨65LA(687)1, 72LA(755)67⟩, 523 ⟨69LA(722)21⟩; **5**, 823, 824 ⟨55CB130⟩, 1147, 1148 ⟨70LA(737)80⟩; **6**, 637 ⟨72LA(755)67⟩
Hesse, J. E., **2**, 213 ⟨65CI(L)680⟩
Hesse, M., **1**, 993 ⟨73HCA875⟩; **3**, 190 ⟨86HCA1614⟩, 210 ⟨87HCA1393⟩; **4**, 649 ⟨83ZN(B)953⟩; **5**, 272, 283 ⟨82HCA2299⟩, 272, 284 ⟨85HCA484⟩, 307 ⟨89HCA1522, 92HCA62⟩, 887 ⟨83ZAAC(501)57⟩, 924 ⟨87OM2128⟩
Hesse, P., **5**, 597, 599 ⟨03LA(329)341⟩
Hesse, R. H., **1**, 871, 881 ⟨69CC1497⟩; **2**, 15 ⟨76JA3034, 76JA3036⟩, 16 ⟨72CC122⟩, 139 ⟨91JOC6697⟩, 240, 360 ⟨91JOC6702⟩, 335 ⟨74JCS(P1)732, 75CC97⟩, 608, 612 ⟨68CC806⟩, 765 ⟨75JCS(P1)1764⟩; **3**, 352 ⟨75CC557, 79JCS(P1)1166⟩; **5**, 313, 314, 316 ⟨74JCS(P1)732⟩, 735 ⟨91JOC6697⟩; **6**, 231 ⟨74JCS(P1)732⟩
Hesse, Von B., **4**, 215 ⟨74JPR304⟩

Hessell, E. T., **2**, 353 ⟨88JOC5776⟩, 711 ⟨85JOC4812⟩
Hessen, B., **2**, 569 ⟨87OM2354⟩; **3**, 510 ⟨89JA5977⟩; **4**, 671 ⟨93OM2268⟩
Hessner, B., **4**, 695 ⟨86JOM(308)153, 86JOM(312)13⟩, 695 ⟨87JOM(319)9⟩, 695, 717 ⟨89JOM(372)183, 86AG(E)165⟩, 717 ⟨89AG(E)737⟩
Hesson, D., **4**, 69 ⟨75JA5957⟩
Hester, D. M., **4**, 582 ⟨91OM369⟩
Hester, W. F., **3**, 621 ⟨45JA601⟩
Hestmann, K., **4**, 119 ⟨75GEP(O)2412800⟩
Hett, R., **5**, 329 ⟨93TL7371⟩
Hettich, B., **3**, 689 ⟨85JA980⟩
Hettler, H., **2**, 106 ⟨58CB1181⟩
Hettrich, G., **1**, 654 ⟨78CC358⟩
Hetzel, F. W., **6**, 555, 575 ⟨69JOC3604⟩
Hetzl, E., **6**, 444 ⟨83JHC1477⟩
Heubach, G., **2**, 444 ⟨85GEP3332716⟩; **5**, 771 ⟨80LA1376⟩
Heuberger, C., **1**, 933, 939 ⟨80HCA2328⟩; **2**, 657 ⟨84HCA1406⟩
Heublein, A., **3**, 406, 416 ⟨80LA1814⟩, 561, 582 ⟨79LA83⟩
Heublein, G., **2**, 745 ⟨69JPR162⟩, 982 ⟨58AG166⟩
Heuck, C., **6**, 302 ⟨73GEP1694210⟩
Heuck, K., **1**, 325, 326 ⟨84CB2132⟩
Heuer, W., **6**, 506 ⟨79CZ175⟩
Heukelbach, E., **5**, 1063, 1064 ⟨65TL149⟩, 1071 ⟨66CB1252⟩; **6**, 651, 652 ⟨67AP(300)567⟩, 653 ⟨66AP709⟩, 666 ⟨66CB1252⟩
Heuring, D. L., **2**, 182 ⟨74JOC238⟩, 182 ⟨74JOC245⟩
Heuschmann, M., **2**, 377 ⟨81LA943⟩; **3**, 221 ⟨87TL2681⟩; **4**, 122 ⟨82CB3384⟩, 414 ⟨87TL2681⟩; **5**, 557 ⟨83JA5015⟩
Heuser, W., **6**, 218, 220, 224 ⟨75ZAAC(418)109⟩
Heusinger, G., **4**, 232 ⟨87LA451⟩
Heus-Kloos, Y. A., **1**, 1045, 1046 ⟨88JOM(338)289⟩; **3**, 273 ⟨88JOM(338)289⟩; **4**, 138 ⟨87JOM(336)C41⟩, 843 ⟨85S958⟩
Heusler, K., **1**, 927 ⟨64AG(E)525⟩; **2**, 94 ⟨71S501⟩, 95 ⟨61E475⟩
Heusner, A., **6**, 444 ⟨75LA2227⟩
Heusser, H., **3**, 229 ⟨47HCA905⟩
Heveling, J., **1**, 80 ⟨83CC365⟩, 437 ⟨83TL5329⟩
Hevesi, L., **1**, 590 ⟨81TL4009⟩, 601, 716 ⟨78TL1141, 78TL1145⟩; **2**, 278, 294 ⟨85T4793⟩, 281, 290 ⟨78TL3967⟩, 286 ⟨92TL4629⟩, 287 ⟨B-88MI 204-01⟩, 1008 ⟨84TL4441, 92PS(67)155⟩; **3**, 16, 132 ⟨79S877⟩, 36, 39, 168, 176 ⟨85TL6513⟩, 176 ⟨84TL4441⟩; **4**, 287, 288 ⟨85T4793, 79S877⟩, 288 ⟨92TL269⟩, 289 ⟨85TL6513⟩, 844, 875 ⟨85CC1000⟩, 1074, 1075 ⟨84JA3785⟩; **6**, 88, 94, 317 ⟨81TL4009⟩, 91 ⟨84MI 603-01⟩, 92 ⟨84JA3784⟩
Hewerton, W., **2**, 465 ⟨64JCS1020⟩
Hewertson, W., **4**, 545 ⟨62JCS1490⟩
Heweston, G. M., **1**, 777 ⟨73AJC1259⟩
Hewett, W. A., **2**, 117 ⟨57JA3493⟩, 118 ⟨58JOC636⟩, 176, 180 ⟨57JOC980⟩
Hewitson, B., **6**, 46, 47 ⟨76JCS(P1)1178⟩, 219 ⟨74JCS(P1)1706⟩
Hewitt, B. D., **1**, 43 ⟨87JCS(P1)2371⟩
Hewitt, C. D., **6**, 262 ⟨85JCS(P1)1191⟩
Hewitt, D., **4**, 451 ⟨88AHC(43)2⟩
Hewitt, D. G., **4**, 473 ⟨84AJC205⟩
Hewitt, G., **1**, 595 ⟨70CC1683⟩; **5**, 278 ⟨71JCS(C)3540⟩
Hewitt, G. M., **2**, 683 ⟨93H(36)13⟩
Hewkin, C. T., **4**, 237, 783 ⟨91JCS(P1)3103⟩, 858 ⟨88TL4889, 91JCS(P1)3091⟩, 862 ⟨93JCS(P1)577⟩; **5**, 423 ⟨91JCS(P1)3103⟩; **6**, 98, 99, 101 ⟨91JCS(P1)3091⟩, 126 ⟨90TL1877⟩
Hewlett, A. P., **1**, 1043 ⟨31JA4192⟩
Hewson, A. T., **3**, 254 ⟨82TL561⟩, 496 ⟨83JCS(P1)2979⟩; **4**, 856 ⟨85CC618, 90JCS(P1)2967⟩, 925 ⟨78TL3267, 83JCS(P1)2979⟩, 925 ⟨85JCS(P1)2625⟩
Hey, D. H., **1**, 925 ⟨62JCS4468⟩; **2**, 177 ⟨52JCS3334⟩, 250 ⟨48JCS147⟩; **3**, 281 ⟨59JCS1563⟩; **5**, 116 ⟨73OSC(5)51⟩
Hey, E., **5**, 781 ⟨88ZN(B)1679⟩
Heycock, C. T., **6**, 411, 412 ⟨20JCS1410⟩
Heyd, W. E., **1**, 1114, 1115, 1124 ⟨74JA6686⟩
Heydenreich, F., **4**, 583 ⟨72IJ293⟩
Heydkamp, W. R., **2**, 305 ⟨64JA3565⟩
Heydt, A., **2**, 1068 ⟨79AG(E)531⟩; **3**, 481 ⟨80LA590⟩, 482 ⟨80LA305⟩
Heydt, H., **2**, 1068 ⟨77LA1766⟩; **3**, 481 ⟨79S805, 81LA1865⟩, 689 ⟨87JOM(338)329⟩; **4**, 483 ⟨91S1099⟩, 489 ⟨85PS(25)201⟩, 558 ⟨81LA1044⟩; **5**, 395 ⟨91CB2843, 91S1099⟩, 871 ⟨85T825, 86T1815⟩, 871 ⟨89CB565⟩, 1152, 1159 ⟨87JOM(338)329⟩, 1153 ⟨89PS(29)31⟩, 1155 ⟨87PS(30)479⟩; **6**, 512, 582 ⟨91S1099⟩
Heyer, E. W., **1**, 795 ⟨67JA1404⟩
Heyer, G., **4**, 566 ⟨73JOM(61)133⟩
Heyl, D., **1**, 411, 807 ⟨41JA1843⟩; **3**, 108 ⟨48JA3434⟩
Heyl, F. W., **2**, 31 ⟨50JA1753⟩
Heyman, M. L., **2**, 377, 378, 403 ⟨73TL2859⟩
Heymann, H., **4**, 209 ⟨51JA5252⟩
Heymann, M., **5**, 400 ⟨76CB359⟩
Heymann, M. L., **2**, 390 ⟨75JOC1395⟩
Heymans, A., **4**, 790 ⟨29CB409⟩
Heymes, A., **3**, 309 ⟨93T2011⟩
Heyn, A. S., **2**, 561 ⟨71TL2183, 81JOC3773⟩
Heyn, R. H., **4**, 656, 657 ⟨93JA5527⟩
Heyn, W., **2**, 1016, 1017 ⟨30CB1863⟩; **4**, 745 ⟨30CB1868⟩
Heyne, H.-U., **2**, 379 ⟨77JPR195⟩, 403 ⟨88OSC(6)78⟩; **6**, 533, 538 ⟨88EGP273832⟩
Heyne, T. R., **3**, 490 ⟨75JOC756⟩
Heyns, K., **3**, 463 ⟨57LA(604)133⟩, 467 ⟨53CB278, 53CB76⟩; **5**, 28 ⟨57AG600⟩, 48 ⟨60T67⟩
Heys, J. R., **3**, 219 ⟨89JOC4702⟩
Heywang, G., **6**, 421, 422, 423, 424, 425, 428, 429, 432, 434, 436, 437 ⟨83HOU(E4)9, 83HOU(E4)36⟩, 513 ⟨82EUP43978⟩
Heywood, D. L., **4**, 203 ⟨60JOC1699⟩
Hiatt, R., **1**, 37 ⟨68JOC1421⟩
Hibbert, F., **4**, 42 ⟨88JCS(P2)1107⟩
Hiberty, P. C., **5**, 1004 ⟨80NJC127⟩
Hibi, S., **5**, 197 ⟨88JCS(P1)529⟩
Hibino, J., **1**, 678 ⟨85TL5579⟩, 678, 769 ⟨85TL5581⟩; **2**, 977 ⟨84TL2151⟩, 997, 1002 ⟨85JOM(285)163⟩; **4**, 1045 ⟨88T4277⟩
Hibino, S., **1**, 41 ⟨78CC414⟩, 1077 ⟨78S305⟩; **3**, 309 ⟨81JOC2979⟩, 311 ⟨82JA536⟩, 654 ⟨78CPB1874⟩
Hickey, D. M. B., **5**, 205 ⟨88JCS(P1)3103⟩
Hickey, E. R., **1**, 810 ⟨92HCA1755⟩
Hickey, J. P., **4**, 681 ⟨75JOM(99)281⟩
Hickinbottom, W. J., **5**, 46 ⟨B-56MI 502-03⟩, 62 ⟨B-56MI 502-02⟩, 105 ⟨B-54MI 502-01, B-56MI 502-01⟩, 315 ⟨73MI 507-01⟩
Hickman, J., **1**, 658 ⟨68JA4445⟩; **2**, 83 ⟨66JA2201⟩; **4**, 144 ⟨68JA6259⟩
Hickman, R., **5**, 999 ⟨68AJC2756⟩
Hickmans, E. M., **5**, 24 ⟨09JCS1004⟩
Hickmott, P. W., **1**, 129 ⟨82T1975⟩; **2**, 738, 751 ⟨82T3363⟩, 740 ⟨82T1975⟩, 757 ⟨70CI(L)1188⟩, 759 ⟨84T2989⟩; **3**, 262 ⟨76JCS(P1)904⟩, 404 ⟨82T1975, 82T3363⟩; **4**, 294 ⟨82T1975⟩
Hickson, C. L., **1**, 784 ⟨86T2135⟩
Hida, T., **1**, 710 ⟨84JA4547⟩
Hidai, M., **5**, 143 ⟨75BCJ2075⟩, 463 ⟨73BCJ2820, 88CL377⟩; **6**, 456 ⟨74NKK732⟩
Hidaki, S., **4**, 901 ⟨87JHC519⟩
Hidalgo, J., **4**, 184, 189 ⟨66JMC127⟩
Hidebrandt, W., **2**, 293, 499, 500 ⟨79ZOB546, 79ZN(B)1646⟩
Hidenori, O., **3**, 357 ⟨91EUP409164⟩
Hideshima, C., **1**, 202 ⟨91JOC2089⟩
Hideshima, M., **3**, 309 ⟨90CPB3261⟩

Hideyuki, N., **5**, 57 ⟨82CL1255⟩
Hieber, W., **2**, 573 ⟨58LA(618)24⟩; **3**, 568 ⟨66ZAAC(344)285⟩; **4**, 687 ⟨57CB1259⟩; **5**, 449 ⟨60CB901⟩; **6**, 522 ⟨35ZAAC(221)321⟩, 523 ⟨41ZAAC(248)256, 57ZAAC(289)324⟩, 523 ⟨61AG579, 65ZN(B)1159⟩, 524 ⟨39ZAAC(240)261, 43MI 616-01⟩, 524, 525 ⟨40ZAAC(245)321⟩
Hiebl, J., **3**, 696, 708 ⟨88LA765⟩
Hiegel, G. A., **2**, 22 ⟨85SC385⟩; **3**, 6, 119, 161 ⟨92SC1589⟩, 210 ⟨73JOC3637⟩
Hiemstra, H., **1**, 774 ⟨89JA2588⟩, 978 ⟨83TL1407, 84JOC1149⟩, 978 ⟨84TL3115, 86TL1411⟩, 978 ⟨92T3445⟩; **2**, 324, 326 ⟨87TL3285⟩, 1009 ⟨91COS(2)1047⟩; **3**, 422 ⟨88JOC3882, 89JA2588⟩, 422 ⟨90T3849⟩; **4**, 106 ⟨88TL6975⟩, 220 ⟨86T3323⟩, 298 ⟨87TL3285⟩, 327 ⟨94JOC3721⟩; **5**, 374 ⟨86TL1411⟩
Hiers, G. S., **2**, 872 ⟨41OSC(1)550⟩
Hiesk, G. R., **2**, 561 ⟨65JA4194⟩
Hiester, E., **5**, 275 ⟨83CB2037⟩
Hietkamp, S., **2**, 435 ⟨85JOM(296)411⟩, 466 ⟨86CB2767⟩; **4**, 546 ⟨82AG(E)376, 89IS120⟩, 550 ⟨83PS(18)279, 84CB3400⟩, 550 ⟨85JOM(296)411⟩
Hieulle, A., **3**, 455 ⟨24CR(178)811⟩
Higa, K. T., **2**, 508, 511 ⟨86OM1266, 87JOM(325)C7⟩, 511 ⟨90OM275⟩
Higa, M. T., **5**, 404 ⟨77TL155⟩
Higa, T., **2**, 170 ⟨68TL5149⟩; **3**, 249 ⟨70JA5258⟩, 701 ⟨92TL1593⟩; **4**, 71 ⟨75JOC3037⟩
Higaki, M., **1**, 143 ⟨87JOC1414⟩; **4**, 391 ⟨87JOC1414⟩
Higashi, F., **6**, 575 ⟨75S384⟩
Higashi, K., **1**, 976 ⟨91SL697⟩
Higashi, T., **5**, 193 ⟨86CZ377⟩
Higashida, S., **2**, 230 ⟨82CL921⟩
Higashimura, K., **4**, 607, 646 ⟨86JPS(A)1839⟩
Higashino, T., **1**, 1010 ⟨88CPB1935⟩; **2**, 688 ⟨80TL2067⟩; **3**, 292 ⟨83CPB3951⟩, 722 ⟨61CPB635⟩; **5**, 769 ⟨82CPB4521⟩
Higashiura, M., **6**, 300 ⟨92JAP(K)04164085⟩
Higashiura, S., **1**, 192 ⟨93JOC1538⟩
Higashiyama, K., **2**, 313 ⟨92TL235⟩
Higashizaki, S., **4**, 118 ⟨85JCR(S)38⟩
Higginbottom, B., **6**, 45 ⟨72JCS(P1)155⟩
Higgins, C. E., **1**, 778 ⟨61JOC846⟩
Higgins, J., **2**, 627 ⟨67JOC1607⟩
Higgins, J. G., **2**, 7, 837 ⟨73OSC(5)249, 73OSC(5)142⟩
Higgins, Jr., T. D., **6**, 78 ⟨55JA3801⟩
Higgins, S. J., **4**, 548, 1023, 1028 ⟨86CC882⟩, 1023, 1028 ⟨85CC1635⟩
Higgins, W. A., **2**, 845 ⟨55JA1864⟩
Higgins, W., **1**, 779 ⟨70JCS(C)81⟩
Higgs, L. A., **1**, 804 ⟨84TL2447⟩
Highet, P. F., **3**, 87 ⟨68JOC3096⟩
Highet, R. J., **3**, 87 ⟨68JOC3096⟩
Highsmith, T. K., **1**, 1079 ⟨92JOC5071⟩
Higo, M., **3**, 374 ⟨70TL5297⟩
Higson, H. G., **2**, 859 ⟨66JCS(C)1592⟩
Higuchi, C., **6**, 529 ⟨85JAP62113712, 86JAP62176910⟩, 533 ⟨85JAP62120358⟩
Higuchi, H., **1**, 243 ⟨80SC595, 81CL627⟩, 243 ⟨82BCJ182, 87BCJ4027⟩, 594, 611, 621 ⟨81CL447⟩; **2**, 289 ⟨80SC595⟩; **4**, 402 ⟨81CL447⟩; **5**, 1113 ⟨80SC595⟩
Higuchi, K., **2**, 82 ⟨91TL4741⟩, 546 ⟨77CC534⟩, 758 ⟨83CL1791⟩, 915 ⟨80JOM(186)51⟩; **3**, 615 ⟨89CL1393⟩, 615, 627, 634 ⟨93BCJ2016⟩
Higuchi, N., **2**, 50 ⟨88JA3712⟩; **3**, 707 ⟨88TL5151⟩
Higuchi, T., **1**, 279, 466 ⟨84JA158⟩; **2**, 148 ⟨66JA3874⟩, 338 ⟨69JCS(C)626⟩; **3**, 605 ⟨85PS(25)237⟩; **4**, 1051, 1053 ⟨85OM2040, 82OM1473⟩; **5**, 226 ⟨81TL1257⟩, 1086, 1087 ⟨88POL2213⟩; **6**, 178 ⟨81JA2324⟩, 178, 383 ⟨79JA1348⟩
Higuchi, Y., **3**, 412 ⟨91CL1093⟩
Higushi, T., **2**, 579 ⟨93CPB292⟩
Hii, P. S., **2**, 95 ⟨83JOC3126⟩
Hiiro, T., **1**, 245, 437 ⟨87AG(E)1187⟩; **2**, 554 ⟨87AG(E)1187⟩; **3**, 136 ⟨90JA455⟩, 176 ⟨87S1096⟩; **4**, 239, 334, 660 ⟨90OM1355⟩, 652 ⟨87AG(E)1187⟩; **5**, 437 ⟨90JA455⟩; **6**, 498 ⟨90SC703⟩
Hikage, S., **5**, 625 ⟨90CL1403⟩; **6**, 597 ⟨90CL1403⟩
Hikida, T., **3**, 337 ⟨65CJC717⟩
Hikita, T., **5**, 143 ⟨75BCJ2075⟩
Hikota, M., **5**, 173, 199 ⟨87TL4569⟩
Hila, J. E., **2**, 1007 ⟨89AX(C)1520⟩
Hilaire, L., **1**, 430 ⟨69BSF842⟩
Hilard, R. L., **1**, 64 ⟨77AG(E)399⟩
Hilbert, G. E., **3**, 493 ⟨47JA723⟩
Hilbert, M., **5**, 595 ⟨91SC85⟩
Hilbert, P., **3**, 480, 482 ⟨71JOC1379⟩; **4**, 489 ⟨71JOC1379⟩; **5**, 868 ⟨67JA4811, 71JOC1379⟩; **6**, 670 ⟨71JOC1379⟩
Hildahl, G. T., **3**, 232 ⟨56JA4405⟩
Hild, W., **4**, 397 ⟨84CB413⟩
Hildebrand, U., **5**, 244, 246 ⟨86T5969, 85TL4349⟩, 246 ⟨85ZN(B)1563⟩
Hildebrandt, A., **3**, 547 ⟨74ZC92⟩; **6**, 545, 546 ⟨89JPR439⟩
Hildebrandt, B., **2**, 75 ⟨88CB2225⟩
Hildebrandt, K., **3**, 322 ⟨88TL2045⟩
Hildenbrand, M., **5**, 928 ⟨84AG(E)371⟩; **6**, 389, 398, 401 ⟨86AG(E)1112⟩
Hildenbrand, T., **6**, 195 ⟨89JOM(364)289⟩, 195 ⟨92CB1547⟩
Hildenbrand, U., **4**, 568 ⟨87PS(30)349⟩
Hilderbrand, M., **1**, 910 ⟨84AG(E)371⟩
Hilditch, T. P., **2**, 172 ⟨08CB4115⟩, 172 ⟨10JCS2585⟩; **5**, 97 ⟨48PCS243⟩
Hileman, D. L., **3**, 115 ⟨90SC3359⟩
Hileman, J. C., **6**, 523 ⟨61JA2953, B-64MI 616-01⟩
Hilgetag, E., **2**, 380 ⟨65CB3377⟩
Hilgetag, G., **2**, 1099 ⟨63CB1462⟩; **5**, 595 ⟨67JPR302⟩, 773, 849 ⟨68CB2028⟩
Hill, A. F., **3**, 387, 522 ⟨83JA5939⟩; **4**, 150, 722 ⟨83JA5939⟩; **6**, 247 ⟨90JOM(395)327⟩
Hill, A. J., **4**, 18 ⟨34JA2730⟩; **5**, 728 ⟨26JA3214⟩, 744, 745 ⟨54JA3978⟩, 748 ⟨26JA732⟩, 757 ⟨54JA920⟩, 994, 995 ⟨40JA1595⟩
Hill, A. W., **3**, 357, 377 ⟨75JCS(P1)1335⟩, 361 ⟨69JCS(C)2192⟩; **5**, 575 ⟨69JCS(C)2192⟩
Hill, C., **2**, 86 ⟨B-89MI 202-03⟩
Hill, C. L., **1**, 67 ⟨74JA870⟩
Hill, D. A. W., **2**, 108 ⟨63JCS2101⟩
Hill, D. R., **1**, 595 ⟨77JCS(P1)1574⟩; **2**, 153 ⟨77JCS(P1)1574⟩; **4**, 901 ⟨89TL4449, 91TL405⟩; **5**, 566, 613 ⟨89T6771, 89TL4449⟩, 613 ⟨91TL405⟩
Hill, D. T., **4**, 35 ⟨69TL1957⟩
Hill, E. A., **1**, 316 ⟨77AOC(16)131⟩; **2**, 556 ⟨75JOM(91)123⟩; **3**, 532 ⟨74JOC3355⟩
Hill, H. E., **2**, 1030, 1034 ⟨56JA2760⟩
Hill, J., **4**, 177 ⟨71JCS(C)1213⟩; **5**, 692 ⟨89JCR(S)337⟩
Hill, J. E., **2**, 915 ⟨77JOM(137)293⟩
Hill, J. H., **1**, 1016, 1108 ⟨84JA3551⟩
Hill, J. H. M., **2**, 546 ⟨92JCS(P1)3277⟩
Hill, J. M., **1**, 648 ⟨91SL764⟩; **4**, 138, 355 ⟨92JOC386⟩, 660 ⟨91TL3285⟩
Hill, J. O., **6**, 560 ⟨79JINC1277⟩
Hill, J. W., **5**, 52 ⟨43OSC(2)53⟩
Hill, K. A., **2**, 3 ⟨72TL4067⟩, 28, 29 ⟨73JOC2576⟩; **4**, 675 ⟨89CC688⟩
Hill, K., **1**, 800 ⟨B-83MI 118-01⟩
Hill, L., **1**, 987 ⟨91CC1290⟩
Hill, M., **3**, 203 ⟨92JOC386⟩
Hill, M. B., **5**, 812 ⟨72JA8776⟩
Hill, M. E., **2**, 762 ⟨66MI 215-01⟩; **4**, 170 ⟨67JOC2595⟩, 178, 179 ⟨66JOC853⟩, 440, 442, 443 ⟨71JOC3846⟩;

5, 763, 815 ⟨66TL3537⟩, 813, 814, 815 ⟨70JOC3093⟩; **6**, 69, 296, 297 ⟨65JOC411⟩, 144 ⟨68JCED437, 67USP3306939⟩, 297 ⟨68USP3388147⟩
Hill, M. G., **3**, 702, 703, 712 ⟨90JOC4950⟩
Hill, P., **2**, 428 ⟨59JOC356⟩
Hill, R. C., **6**, 469 ⟨79EUP7498⟩
Hill, R. K., **1**, 377 ⟨B-84MI 109-01⟩, 411 ⟨91COS(5)785⟩, 806, 807 ⟨67CC619⟩, 806, 811 ⟨91COC(5)785⟩, 806 ⟨B-84MI 118-03⟩, 823 ⟨B-84MI 118-04⟩, 835 ⟨78TL4337⟩; **3**, 663 ⟨73JA2923⟩; **5**, 201 ⟨64JA965⟩
Hill, R. M., **5**, 89 ⟨32OSC(1)300⟩
Hill, W., **4**, 956, 957 ⟨84AG(E)723⟩
Hill, W. E., **2**, 440 ⟨88ICA(146)111⟩
Hillard, III, R. L., **2**, 944 ⟨75JA5600⟩
Hillebrad, H., **5**, 857 ⟨53LA(581)44⟩
Hillenbrand, H., **5**, 1033 ⟨54LA(590)123⟩
Hillenbrand, O. G., **6**, 537 ⟨64LA(590)123⟩
Hiller, G., **2**, 130 ⟨63AG(E)370⟩; **3**, 344, 363 ⟨63AG(E)370⟩
Hiller, W., **1**, 445 ⟨86JCE930⟩; **2**, 490 ⟨86JCS(D)997⟩, 1003 ⟨82ZN(B)1146⟩; **4**, 112 ⟨86OM2030⟩, 581 ⟨83CB1209⟩, 583 ⟨82ZN(B)1146⟩; **5**, 781 ⟨88ZN(B)1628⟩; **6**, 195 ⟨89JOM(368)139⟩, 391, 394 ⟨91AG(E)324⟩, 395 ⟨92AG(E)1364⟩, 584 ⟨85CB227⟩
Hillers, S., **3**, 163 ⟨93JOC3169⟩
Hillery, P., **2**, 33 ⟨87JOC5466⟩
Hillgärtner, H., **1**, 924 ⟨89T951⟩; **3**, 92 ⟨89T951⟩; **4**, 141, 154, 608, 616 ⟨80JOM(201)197⟩
Hillhouse, G. L., **3**, 545 ⟨88JA4855⟩; **5**, 1096 ⟨88JA4855, 88JA6926⟩
Hillhouse, J. H., **1**, 707 ⟨85CJC1⟩; **5**, 610, 623, 624 ⟨86PS(26)169⟩
Hillier, G. R., **1**, 869, 871 ⟨67JCS(B)827⟩
Hilling, W., **4**, 38 ⟨52JA3895⟩
Hillman, G. R., **5**, 516, 520, 522 ⟨72JMC760⟩
Hillman, M. E. D., **4**, 364 ⟨62JA4715⟩
Hillyard, Jr., R. W., **4**, 513 ⟨78JOM(153)369⟩
Hilp, M., **4**, 96 ⟨70CB104⟩
Hiltbrunner, K., **6**, 204 ⟨79AG(E)785⟩
Hilton, B. D., **2**, 831 ⟨82JA6876⟩
Hilton, C., **1**, 420 ⟨83JOC4272⟩
Hilty, T. K., **4**, 831 ⟨90JOC2972⟩
Hilvert, D., **1**, 826 ⟨88JA5593⟩
Himbert, G., **1**, 1044 ⟨84LA85⟩; **2**, 876 ⟨84TL1357⟩, 877 ⟨86CB1953⟩, 1027 ⟨84LA1358⟩, 1044 ⟨78TL1951, 79LA1828, 81LA1844, 83LA1185⟩, 1044, 1048 ⟨81LA1907⟩, 1047 ⟨72CB2963, 91SL117⟩, 1048, 1068, 1069, 1070 ⟨74CB2513⟩, 1053 ⟨88TL5355⟩, 1093 ⟨79AG(E)405⟩; **3**, 273 ⟨80JOM(184)317, 81LA1907⟩, 273 ⟨81TL2637, 88CB2163⟩, 369 ⟨88CB833, 90TL5579⟩, 369 ⟨91PS(59)419⟩, 467, 473 ⟨70LA(734)70⟩, 477 ⟨72CB2963, 72CB2975⟩, 477 ⟨72S571, 73LA1505⟩, 477 ⟨74CB2513, 79LA1828⟩, 477 ⟨81ZN(B)218⟩, 602 ⟨93HOU(E15/3)3404⟩; **4**, 763 ⟨92S495⟩; **5**, 304 ⟨81LA1844⟩, 365 ⟨88CB833⟩, 596, 617 ⟨85LA2371⟩, 596, 618, 739 ⟨78TL1951⟩, 597, 630 ⟨86CB3812⟩, 645 ⟨91PS(59)125, 85LA1669⟩, 827 ⟨72CB2975⟩; **6**, 620, 651 ⟨82LA2105⟩
Himbert, G. Z., **2**, 1031 ⟨85ZN(B)645⟩
Himbert, J., **3**, 350 ⟨48BSF1120⟩
Himdi-Kabbab, S., **2**, 489 ⟨90TL1147⟩, 489, 879 ⟨89TL349⟩; **3**, 639 ⟨91TL2755⟩; **4**, 502 ⟨90TL7607⟩, 597, 598 ⟨89TL349⟩
Himizu, J., **1**, 1023 ⟨68CC634⟩
Himmel, W., **1**, 296 ⟨74CA(81)120222t, 80AG(E)178⟩; **6**, 192 ⟨79ZAAC(448)40⟩, 267 ⟨79ZAAC(448)55⟩
Himmelsbach, R. J., **1**, 483 ⟨85JA682⟩
Himoe, A., **2**, 608 ⟨61JA4605⟩
Hinckley, D. F., **5**, 25, 34 ⟨50JA4091⟩
Hinde, A. L., **1**, 96, 890 ⟨80JA3370⟩, 96, 890 ⟨80JA4074⟩, 96, 890 ⟨80JA6430⟩, 96, 890 ⟨81JA284⟩
Hinderling, R., **4**, 21 ⟨54HCA90⟩
Hindley, N. C., **1**, 104 ⟨49HCA489⟩
Hindley, R. M., **4**, 308 ⟨75JCS(P1)1670⟩
Hine, J., **1**, 797 ⟨82JOC4758⟩, 797, 882 ⟨82JOC4766⟩, 797, 882 ⟨83JOC584⟩, 870 ⟨69JOC4083⟩, 871 ⟨67JA2664⟩; **4**, 18 ⟨60JA6115⟩, 761 ⟨55JA3886⟩; **6**, 33 ⟨58JA4282⟩, 33, 81, 228 ⟨56JA479⟩, 38 ⟨58JA3002⟩, 45, 69 ⟨57JA5493, 60JA1398⟩, 81 ⟨60JA6118⟩, 88 ⟨62JA1751⟩
Hine, K. E., **3**, 229 ⟨73JA3289⟩
Hiner, R. N., **1**, 1140 ⟨88TL2923⟩
Hines, L. F., **1**, 298 ⟨76JA1806⟩
Hinkamp, J. B., **6**, 216, 217 ⟨45JA1642⟩
Hinkamp, L., **2**, 21 ⟨92LA559⟩
Hinke, A., **2**, 832, 835 ⟨83PS(15)93⟩; **4**, 115 ⟨83PS(15)93⟩
Hinkle, R. J., **2**, 445 ⟨90JOC5033⟩; **3**, 496 ⟨89JOC2783, 90JOC5033⟩; **5**, 944 ⟨93OM3510⟩
Hinkley, D. F., **1**, 795 ⟨55JA5166⟩
Hinkley, J. M., **1**, 879 ⟨83JOC2957⟩
Hinman, M. M., **1**, 353 ⟨91JA726⟩
Hinman, R. L., **2**, 377, 381, 382 ⟨56JA1645⟩, 379 ⟨58JOC529⟩, 388 ⟨59JA3294⟩; **3**, 106 ⟨63JOC3052⟩, 444, 446 ⟨60JOC1775⟩; **5**, 350 ⟨58JA1895⟩
Hinney, H. R., **4**, 333 ⟨81JOC765⟩
Hino, C., **6**, 233, 234 ⟨81TL1997⟩
Hino, F., **5**, 346 ⟨81S610⟩
Hino, J. B., **6**, 504 ⟨61JOC5238⟩
Hino, K., **1**, 809 ⟨80TL1347⟩; **3**, 217 ⟨80TL1347⟩
Hino, T., **2**, 313 ⟨93CPB287⟩, 330 ⟨92TA227⟩; **3**, 238 ⟨86TL4767⟩; **4**, 193 ⟨89JHC1405⟩, 418, 419 ⟨81T1487⟩, 618 ⟨88TL6939⟩; **5**, 567 ⟨69CPB550⟩
Hinoand, T., **3**, 442 ⟨86TL4763⟩
Hinrichs, E. von, **3**, 696 ⟨82JCR(S)79⟩, 696, 721, 722 ⟨72AG(E)929⟩
Hinrichs, G. Von, **6**, 417 ⟨82JCR(S)79⟩
Hinsberg, III, W. D., **2**, 820 ⟨82JA766⟩
Hinsberg, O., **2**, 101 ⟨1890LA(260)40⟩
Hinshaw, B. C., **5**, 516, 522 ⟨85JOC1741⟩
Hinshaw, J. C., **2**, 226 ⟨72TL3567⟩
Hinsken, W., **1**, 563 ⟨85AG(E)973⟩
Hinterding, P., **5**, 957 ⟨89CB365, 90CB2047⟩, 957 ⟨90CB611, 92CB2765⟩
Hintermaier, H., **6**, 503 ⟨79GEP2742158⟩
Hinton, J. F., **5**, 179 ⟨89LA1233⟩
Hinton, S., **2**, 461 ⟨59JCS2835⟩
Hintze, B., **6**, 650, 653 ⟨70CC806⟩, 651 ⟨70BAP569⟩
Hintze, F., **1**, 138 ⟨90AG(E)1422⟩; **6**, 447 ⟨92S1216⟩
Hintze, M., **4**, 1025 ⟨84CB1542⟩
Hintze, M. J., **1**, 873 ⟨87JA5539⟩
Hintzsche, E., **5**, 616 ⟨94TL1977⟩
Hinz, H., **3**, 231 ⟨43CB676⟩
Hinz, J., **6**, 509 ⟨72LA(765)94⟩
Hioki, H., **4**, 858 ⟨88TL3125⟩
Hioki, T., **2**, 437 ⟨83JOC2195⟩; **5**, 607 ⟨80JA7806, 84TL5793⟩
Hippich, S., **6**, 143 ⟨82CB1721⟩
Hippmann, G., **6**, 73 ⟨82MI 603-01⟩
Hipsher, H., **2**, 21 ⟨55OSC(3)188⟩
Hirabayashi, T., **3**, 335, 369 ⟨82JA3119⟩, 337 ⟨82JA312⟩, 337, 351 ⟨82CB492⟩, 358 ⟨83CB273⟩, 363 ⟨71MI 308-01⟩, 527 ⟨77AG(E)105⟩; **4**, 258 ⟨82JA3119⟩; **5**, 235 ⟨72JOM(42)33⟩, 262 ⟨70JOM(25)33⟩, 578, 608, 618 ⟨75JOM(92)139⟩
Hirabayashi, Y., **2**, 858 ⟨77BCJ3007⟩; **4**, 288 ⟨76CL203⟩; **5**, 232 ⟨65BCJ320⟩, 233, 506, 509 ⟨87NKK1430⟩, 240 ⟨78CL1007⟩, 240, 244, 252 ⟨77BCJ3007⟩, 240, 252, 541 ⟨76CL203⟩, 249 ⟨87NKK1479⟩, 253 ⟨87NKK1475⟩
Hiraga, K., **1**, 734 ⟨75CL103⟩
Hirahara, Y., **3**, 565 ⟨93MI 317-02⟩
Hirai, A., **5**, 438 ⟨90JA7061⟩
Hirai, H., **1**, 196 ⟨87BCJ1525⟩; **3**, 696, 708 ⟨84TL5075⟩
Hirai, K., **1**, 208 ⟨86BCJ1363⟩; **2**, 224 ⟨70TL441,

71CPB1022⟩; **3**, 57 ⟨72TL2743⟩, 59 ⟨81TL1021⟩, 255 ⟨71CPB2194, 71TL1137⟩; **4**, 835 ⟨71CPB2194, 71TL1137⟩; **5**, 1048 ⟨63BCJ1214⟩; **6**, 85 ⟨69CPB1931, 77JOC1543⟩, 113 ⟨69CPB1924⟩, 114 ⟨71CPB2222⟩, 543, 557, 563, 578 ⟨88JOC2263⟩
Hirai, S., **3**, 634 ⟨72JA4635⟩; **5**, 81 ⟨78GEP2725519⟩, 1142 ⟨72JA4635⟩
Hirai, T., **1**, 217 ⟨91TL1059⟩; **2**, 85 ⟨90TL7031⟩
Hirai, Y., **1**, 611, 716 ⟨85CC1037⟩; **2**, 327 ⟨87TL77⟩; **5**, 324 ⟨89BCJ2968⟩, 693 ⟨81JOC1769⟩
Hiraiwa, H., **5**, 226 ⟨80CC1029⟩
Hirakawa, K., **3**, 593 ⟨86JHC597⟩
Hirakawa, T., **2**, 626 ⟨87CL627⟩
Hiraki, K., **2**, 582 ⟨85JCS(D)873⟩; **4**, 656 ⟨93BCJ1849⟩; **5**, 801 ⟨80BCJ1171⟩
Hirakura, M., **1**, 881 ⟨58JOC841⟩
Hirama, M., **1**, 552 ⟨91SL651⟩, 702, 1070 ⟨88TL1197⟩, 1050 ⟨92CL2479⟩, 1051 ⟨93JA3362⟩; **2**, 659 ⟨92TL4603⟩, 1100 ⟨92CL2479⟩; **4**, 858 ⟨88TL3125⟩, 1069 ⟨93CL1881⟩; **5**, 152 ⟨87CC1523⟩, 173 ⟨81JA1568⟩, 607 ⟨76JA7084⟩, 607, 616 ⟨82TL881, 76CC776⟩, 616 ⟨82CL529⟩
Hiramatsu, K., **3**, 717 ⟨90S781⟩, 717, 720 ⟨88T5467⟩
Hirano, A., **6**, 138 ⟨80CPB2987⟩
Hirano, J., **5**, 329 ⟨94TL1247⟩
Hirano, M., **2**, 149 ⟨92BCJ1744⟩, 155 ⟨91CL523⟩; **3**, 88, 99 ⟨90BCJ2433⟩, 115 ⟨91BCJ2857⟩, 119 ⟨87BCJ4143, 88JCS(P1)2423⟩, 120 ⟨91BCJ1046⟩; **6**, 115 ⟨91CL1873⟩
Hirano, S., **1**, 422 ⟨80BCJ169⟩; **2**, 571 ⟨77JA3179⟩; **3**, 209 ⟨80BCJ169⟩; **4**, 432 ⟨93TL5097⟩; **5**, 183, 192 ⟨60JOC1265⟩
Hirao, A., **2**, 65, 528 ⟨87BCJ395⟩; **3**, 440 ⟨82S461⟩
Hirao, I., **1**, 204, 1023 ⟨83TL391⟩, 204, 1024 ⟨83TL5121⟩, 223, 1023 ⟨84CC202⟩, 229 ⟨84TL1159, 84TL4549, 84T4261⟩, 1045 ⟨86S421⟩, 1046 ⟨83CL35⟩, 1047 ⟨83TL1719⟩; **2**, 106 ⟨91T8717⟩, 447 ⟨85JAP60152493⟩, 843 ⟨83JOC2569⟩; **3**, 219 ⟨85CL1145⟩, 496 ⟨83JOC2569⟩; **4**, 488, 489 ⟨85JOC1278⟩, 925 ⟨83TL767, 84BCJ2127⟩; **6**, 122 ⟨85CL1099⟩
Hirao, T., **1**, 290 ⟨93TL5601⟩, 558 ⟨78JOC1011⟩, 931 ⟨91JOC2264⟩, 1038 ⟨86TL933⟩; **2**, 14 ⟨83BCJ1881, 93JOC6529⟩, 456 ⟨85BCJ3104⟩, 850 ⟨82BCJ909⟩, 932 ⟨81TL3079⟩; **3**, 62 ⟨82CL1997⟩, 127 ⟨85JA7179⟩, 130, 285, 295 ⟨86TL929⟩, 206 ⟨78JOC1011⟩, 225 ⟨79JA494⟩, 273 ⟨86TL933⟩, 574 ⟨81CL1197⟩, 599 ⟨77TL1009, 80SC233⟩; **4**, 13 ⟨83BCJ1881⟩, 517, 538, 1015 ⟨82BCJ1163⟩, 538 ⟨81AG(E)126⟩, 743 ⟨94JCS(P1)3⟩; **5**, 1072 ⟨75JOC2981⟩; **6**, 7 ⟨90SL217⟩, 162 ⟨82BCJ1163⟩
Hiraoka, K., **1**, 905 ⟨85JA3747⟩
Hiraoka, T., **1**, 1042 ⟨80CC736⟩; **2**, 249 ⟨75BCJ480⟩; **3**, 59 ⟨81TL1021⟩, 559 ⟨76CC516⟩, 581 ⟨76TL1307⟩
Hirashima, T., **2**, 753 ⟨80AG(E)1008⟩; **3**, 621 ⟨87S631⟩; **5**, 596 ⟨93SC2139⟩; **6**, 471 ⟨90CL811⟩, 479 ⟨88TL4767, 90TL4773⟩, 494 ⟨91TL6867⟩
Hirata, F., **1**, 811 ⟨81CC786⟩
Hirata, K., **4**, 802 ⟨88CL211⟩; **5**, 173 ⟨79BCJ1989⟩
Hirata, T., **1**, 619 ⟨88CL2009⟩; **2**, 34 ⟨92TL1299⟩; **4**, 420 ⟨72JHC1145⟩
Hirata, Y., **1**, 19 ⟨72BCJ264⟩; **3**, 63 ⟨90JCS(P1)197⟩; **4**, 222 ⟨76TL65⟩; **6**, 68 ⟨65T2059⟩
Hiratake, J., **3**, 631 ⟨92BCJ111, 92JOC5620⟩
Hiratani, K., **6**, 730 ⟨74BCJ398⟩
Hirayama, A., **3**, 428 ⟨91CL541⟩
Hirayama, M., **2**, 313 ⟨93CPB287⟩
Hirayama, N., **3**, 26 ⟨93JA3943⟩, 159 ⟨90JA7422⟩
Hirayama, T., **6**, 623 ⟨76CPB26⟩
Hiroa, I., **1**, 444 ⟨83JOC2569⟩
Hiroaki, O., **4**, 165 ⟨74H(2)177⟩
Hirobe, M., **2**, 579 ⟨93CPB292⟩; **5**, 226 ⟨81TL1257⟩
Hiroe, Y., **5**, 463 ⟨88CL377⟩
Hiroi, K., **1**, 235 ⟨82CL1871⟩, 409, 821 ⟨73JOC2572⟩, 595, 597, 598, 601, 621 ⟨76JA4887⟩; **2**, 139 ⟨76JA4887⟩, 159 ⟨83CC1470⟩, 212 ⟨80JOC1839⟩, 368 ⟨85S635⟩, 721 ⟨83S1040⟩; **3**, 139 ⟨86CC469⟩, 170 ⟨86CL65⟩, 219 ⟨75JA6911⟩, 559 ⟨85CPB2331, 85CPB4691⟩, 634 ⟨93T1831⟩; **4**, 365 ⟨81CC377⟩, 392 ⟨87CPB1413⟩; **5**, 68 ⟨73CC500⟩
Hiroi, Y., **4**, 872 ⟨85S490⟩; **5**, 281 ⟨83CL449⟩; **6**, 45, 49 ⟨85S497⟩
Hirokami, S., **5**, 693 ⟨81JOC1769⟩
Hirokane, K., **6**, 533 ⟨90JAP03220173⟩
Hirosawa, C., **3**, 169 ⟨92CJC974⟩
Hirose, H., **3**, 498 ⟨68BCJ2815⟩
Hirose, N., **3**, 340 ⟨88MI 308-01⟩
Hirose, T., **1**, 164 ⟨92SL891⟩; **4**, 155, 700 ⟨92SL891⟩
Hirose, Y., **1**, 1048 ⟨90TL5499⟩; **5**, 547 ⟨79JAP7946762⟩
Hirota, H., **1**, 813 ⟨85BCJ146, 87CC1196⟩; **3**, 217 ⟨88BCJ1207⟩, 701, 704 ⟨92TL6823⟩
Hirota, K., **1**, 482 ⟨89JCS(P1)2513⟩; **5**, 747 ⟨69BCJ3310⟩, 792 ⟨67JA2240⟩; **6**, 111 ⟨84H(22)13⟩
Hirota, M., **5**, 1127 ⟨78YZ1361⟩
Hirota, N., **1**, 1029 ⟨87H(26)355⟩; **2**, 1010 ⟨64JA2538⟩
Hirotsu, K., **1**, 195 ⟨89JOC469⟩, 279, 466 ⟨84JA158⟩, 704, 708 ⟨77JA2815⟩; **3**, 605 ⟨85PS(25)237⟩, 607 ⟨91CC124⟩, 609 ⟨90HAC339, 93JOM(456)161⟩; **4**, 559 ⟨90CL643, 91CL2213⟩, 1051, 1053 ⟨85OM1690, 82OM1473⟩; **5**, 1086, 1087 ⟨88POL2213⟩
Hirowatari, N., **3**, 697 ⟨74JOC604⟩; **5**, 88 ⟨74JOC604⟩; **6**, 533 ⟨91JAP03218336⟩
Hiroya, K., **3**, 294 ⟨87CC1360⟩, 620 ⟨86JOC2833⟩
Hirs, C. H. W., **5**, 84 ⟨64MI 502-01⟩
Hirsch, E., **2**, 689 ⟨82CB3687, 82CB399⟩; **3**, 170 ⟨82CB3687⟩
Hirsch, H., **3**, 537 ⟨70JCS(D)699⟩
Hirsch, H. von, **4**, 405 ⟨67CB1289⟩
Hirsch, J. A., **1**, 19, 24 ⟨85JOC2607⟩
Hirsch, J., **2**, 38 ⟨21MI 202-01⟩
Hirschmann, P., **4**, 1019 ⟨82JOM(224)153⟩; **5**, 799 ⟨82JOM(224)153⟩
Hirschmann, R., **1**, 166 ⟨93TL63⟩; **2**, 223 ⟨68TL3057⟩; **4**, 296 ⟨88OSC(6)5⟩; **5**, 972 ⟨66JA3163⟩; **6**, 496 ⟨68JA3254⟩
Hirsekorn, F. J., **1**, 93 ⟨74JA4063, 75JA237⟩, 93 ⟨76IC2379, 78JA2405⟩
Hirsenkorn, R., **1**, 520 ⟨84TL4357⟩, 831 ⟨90TL4433⟩
Hirsh, D. H., **3**, 55 ⟨49JA3468⟩; **4**, 188 ⟨58JA1083⟩
Hirshfeld, A., **5**, 164 ⟨72T1275⟩
Hirshfield, J., **4**, 417 ⟨87JOC2442⟩
Hirshfield, J. M., **6**, 55, 280, 633 ⟨81JOC5048⟩
Hirst, E. L., **5**, 316 ⟨22JCS2169⟩
Hirst, G. C., **1**, 391 ⟨89JA1514⟩
Hirst, G. J. B., **2**, 566 ⟨84JOM(267)265⟩
Hirst, S. C., **3**, 59 ⟨89TL5361⟩
Hirth, C., **6**, 443 ⟨89MI 614-03, 90S1065⟩
Hirtz, H., **6**, 522, 524 ⟨10JCS798⟩
Hirwe, N. W., **3**, 612 ⟨40JIC481⟩
Hirzel, H., **3**, 474, 475 ⟨16CB1978⟩
Hisada, H., **3**, 639 ⟨85TL5835⟩
Hisada, Y., **2**, 563 ⟨92JOC6386⟩
Hisamoto, T., **6**, 215 ⟨81MI 607-01⟩
Hisano, T., **1**, 836 ⟨91TL7557⟩; **2**, 123 ⟨87H(26)2583⟩; **3**, 189 ⟨B-89MI 304-03⟩; **6**, 472 ⟨87H(26)2583, 88H(27)2327⟩, 472 ⟨89CPB576, 92OPP200⟩, 473 ⟨91TL7557⟩
Hiscock, A. K., **2**, 1016 ⟨62JCS4288⟩
Hiscock, S. M., **5**, 238 ⟨70CC1310⟩
Hishmat, O. H., **3**, 242 ⟨74IJC1052⟩
Hiskey, R., **2**, 275 ⟨66JOC2179⟩
Hiskey, R. G., **2**, 219 ⟨61JOC1152, 67JOC97⟩, 219 ⟨70JOC4148, 71JOC488⟩, 219, 220 ⟨69JA7525, 65JA3965⟩, 220 ⟨68JA2677, 70JOC1118⟩, 220 ⟨75JOC250⟩; **4**, 218 ⟨70JOC215⟩, 223 ⟨62JA4789⟩

Hitch, E. F., **2**, 776 ⟨51JA1323⟩
Hitchcock, P., **1**, 63 ⟨76JOM(113)C11⟩; **4**, 101 ⟨91TA157⟩
Hitchcock, P. B., **2**, 569 ⟨81JCS(D)80⟩, 852 ⟨82CC691⟩; **3**, 690 ⟨93AG(E)103⟩; **4**, 547 ⟨88JOM(338)C31⟩, 548 ⟨90CC1307, 93PS(77)274⟩, 549 ⟨94AG(E)2202⟩; **5**, 886 ⟨93CC267⟩, 1160 ⟨93AG(E)103⟩; **6**, 154 ⟨89CC1046, 92JOM(433)C11, 93CC311⟩, 155 ⟨93AG(E)103⟩, 165, 289, 372, 373 ⟨89JOM(366)39⟩, 167 ⟨84CC1669⟩, 169 ⟨87JOM(320)C27⟩, 173, 394 ⟨93JOM(462)45⟩, 174 ⟨83JOM(252)281⟩, 174, 176, 381 ⟨92JCS(D)1015⟩, 174, 176, 381, 393 ⟨86CC1043⟩, 174, 289, 390 ⟨83CC827⟩, 174, 391 ⟨86CC969⟩, 176, 289, 391 ⟨83CC1390⟩, 176, 394 ⟨91JOM(405)149⟩, 179, 709, 710 ⟨80JOM(186)309⟩, 186 ⟨84CC480⟩, 194, 195 ⟨88CC1007, 90JOM(394)57⟩, 349 ⟨85CC1776⟩, 380 ⟨81JOM(221)13, 90JOM(398)59⟩, 381, 392 ⟨87JCS(P2)891⟩, 387 ⟨82JOM(235)265, 84JOM(272)1, 89JCS(D)447⟩, 390 ⟨93JCS(D)3259⟩, 392, 394 ⟨92JOM(437)41⟩, 393 ⟨89CC273, 94AG(E)1268, 94JOM(469)129, 94JOM(480)199, 94OM753⟩, 393, 394, 396 ⟨88JCS(D)381⟩, 394 ⟨88CC1389, 95UP 613-01⟩, 394, 395 ⟨86CC908, 87JOM(320)137, 87JCS(D)747⟩, 396 ⟨84CC870, 84JOM(263)C23, 85CC1380, 85CC534, 87JOM(325)117⟩, 515 ⟨91JOM(403)293⟩, 692 ⟨89JOM(368)C29⟩, 721 ⟨76CC644, 77JCS(D)2160⟩, 721 ⟨78JCS(D)826, 79JCS(D)1929⟩
Hitchcock, S. A., **1**, 343 ⟨90TL3641⟩, 347 ⟨92TL4843⟩
Hitchcock, S. R., **3**, 160 ⟨90S1059⟩
Hitchings, G. J., **3**, 42 ⟨91T2683⟩; **4**, 303 ⟨90JCS(P1)83⟩, 323, 406, 410, 445 ⟨91T2683⟩
Hite, G. A., **1**, 422 ⟨86T2821, 88HCA168⟩; **3**, 55, 229, 271, 272 ⟨88HCA168⟩
Hites, R. D., **5**, 1046 ⟨57CJC832⟩
Hitomi, K., **3**, 660 ⟨83CC170⟩
Hitoshi, T., **2**, 724 ⟨80S937⟩
Hittich, R., **1**, 9 ⟨83TL1147⟩; **6**, 421, 422 ⟨92GEP4023107⟩
Hixon, R. M., **1**, 91 ⟨43OSC(2)566⟩
Hixon, S. C., **1**, 1114 ⟨70JA1082, 71JA6989⟩
Hixson, S. S., **1**, 841 ⟨79MI 118-01, 88JA3601⟩, 841 ⟨88JOC2706, 88JOC2711⟩
Hiyama, H., **4**, 17 ⟨74T2661⟩
Hiyama, T., **1**, 164 ⟨77BCJ3288⟩, 195, 477 ⟨85TL3259⟩, 200 ⟨86S689⟩, 225 ⟨93TL8263⟩, 422 ⟨79JA1599, 80BCJ169⟩, 422 ⟨81BCJ2747⟩, 429 ⟨88T4135⟩, 490 ⟨88TL97, 89CL1711, 89JOC268, 91SL845⟩, 490, 1007 ⟨88JOC918⟩, 490, 1028 ⟨90TL2719⟩, 574 ⟨91TL1215⟩, 677, 1099 ⟨83TL5281, 89TL2403⟩; **2**, 7 ⟨86TL1607⟩, 18 ⟨91SL185⟩, 70 ⟨88T4134⟩, 73, 571 ⟨82BCJ561⟩, 90 ⟨92TL4173, 77JA3179⟩, 912 ⟨87TL4715⟩, 912, 931, 945 ⟨91SL845⟩, 923, 940 ⟨90BCJ3103⟩, 931, 932, 939 ⟨89JOC268⟩, 931, 932, 939 ⟨90TL2719⟩, 939, 940 ⟨92BCJ1280⟩; **3**, 62 ⟨78TL3051⟩, 62, 72, 271 ⟨78TL3047⟩, 209 ⟨80BCJ169⟩, 216 ⟨79JA1599⟩, 220 ⟨84TL4959⟩, 224 ⟨74JA3713, 74TL3297⟩, 224 ⟨78TL771, 80BCJ1010⟩, 244 ⟨83BCJ3088⟩, 290, 300, 305 ⟨92T2113⟩, 290, 305 ⟨89CL2049⟩, 620 ⟨83TL3509⟩, 628 ⟨90CL1615⟩; **4**, 441 ⟨82TL2103⟩, 619 ⟨87TL1811⟩, 652 ⟨82TL1279⟩, 654, 655 ⟨92CL2047⟩, 736 ⟨83TL4113, 88T4135⟩, 806, 817 ⟨84CL1765⟩, 809 ⟨90BCJ1947⟩, 819 ⟨85JA4085⟩, 1045 ⟨85CL1405, 87TL1807⟩; **5**, 213 ⟨88JOC5415⟩; **6**, 10 ⟨92CL827⟩, 26 ⟨85JA4085, 88T4135⟩, 63 ⟨88BCJ3321⟩, 229, 230 ⟨92TL4173⟩, 251 ⟨94SL251⟩
Hiyoshi, T., **2**, 753 ⟨84BCJ2440⟩
Hizatate, S., **2**, 316 ⟨91SL649⟩
Hjeds, H., **2**, 746 ⟨73ACS2802⟩
Hlasta, D. J., **5**, 331 ⟨89TL1773⟩
Hlavacek, R. J., **6**, 596 ⟨51JA1864⟩
Hlavka, J. J., **5**, 279 ⟨56JOC439⟩
Hlawatschek, H., **6**, 630 ⟨83ZAAC(501)157, 83ZAAC(501)169⟩, 630 ⟨83ZAAC(501)176⟩
Hlock, B., **2**, 195 ⟨71S590⟩
Hna, D. H., **2**, 539, 699 ⟨81TL3455⟩
Hnizda, V., **1**, 114 ⟨44JA1389⟩; **2**, 494 ⟨42JA392⟩; **6**, 523 ⟨58USP2822247⟩
Ho, A. J., **3**, 642 ⟨71JOC2026⟩
Ho, C. D., **1**, 977 ⟨83JOC5376⟩; **2**, 939, 940 ⟨83JOC5376⟩
Ho, C. Y., **6**, 598 ⟨91JMC3187⟩
Ho, D., **1**, 752 ⟨87TL3075⟩; **4**, 675 ⟨89OM1626⟩
Ho, D. M., **4**, 148, 150, 581 ⟨90JOM(397)313⟩; **5**, 799 ⟨89OM1523, 89OM2589⟩
Ho, D. S., **5**, 87 ⟨57JA637⟩
Ho, E., **4**, 518 ⟨88T7013⟩
Ho, G.-D., **2**, 356 ⟨90T4857⟩
Ho, G. J., **6**, 280 ⟨89TL2473⟩
Ho, J., **2**, 942 ⟨93JOC6182⟩
Ho, K., **3**, 363 ⟨76JA6218⟩
Ho, K. K., **1**, 185 ⟨93JOC3767⟩
Ho, K. M., **1**, 22, 23, 24, 436 ⟨89JOC4474, 88JOC4466⟩
Ho, L. L., **1**, 618 ⟨73TL1843⟩, 706 ⟨B-77MI 115-01⟩; **2**, 717, 719 ⟨72TL4319⟩
Ho, P.-T., **2**, 6 ⟨84JOC3027⟩; **3**, 40 ⟨93JOC2313⟩
Ho, S., **1**, 769 ⟨83PAC1733⟩; **3**, 508 ⟨83PAC1733⟩
Ho, T., **2**, 1020 ⟨90CC118⟩; **3**, 16, 132 ⟨72S561⟩, 121 ⟨76S609⟩, 129 ⟨79SC241, 79SC665⟩, 133 ⟨76S610⟩
Ho, T.-L., **1**, 347 ⟨B-92MI 108-02⟩, 832 ⟨81SC237⟩; **2**, 24, 30 ⟨77S676⟩, 150 ⟨79SC237⟩, 214 ⟨73SC317, 77SC363⟩, 537 ⟨77S677⟩, 657, 753 ⟨76S815⟩; **3**, 8, 121 ⟨77JOC3097⟩, 208 ⟨81SC7⟩, 219 ⟨82SC995⟩, 222 ⟨83SC207⟩, 618 ⟨75SC299, 75SC423⟩; **4**, 220 ⟨73CC224⟩; **5**, 225 ⟨76AG(E)774⟩ 28 ⟨73S347⟩, 39 ⟨74S715, 75S510⟩, 95 ⟨72CJC2047⟩, 191 ⟨73SC63⟩; **6**, 606, 645 ⟨92SC1191⟩
Ho, W.-B., **3**, 424 ⟨90JOC412⟩
Ho, Y.-B., **1**, 433 ⟨86CC1452⟩
Hoa, H. T., **2**, 585 ⟨93CB45⟩
Hoaglin, R. I., **3**, 55 ⟨49JA3468⟩; **4**, 188 ⟨58JA1083⟩
Hoan, N., **3**, 309 ⟨49RTC5⟩
Hoang, H., **1**, 769 ⟨91S165⟩
Hoare, R. J., **5**, 950 ⟨72JCS(D)653⟩
Hoashi, K., **1**, 253, 486 ⟨87JOC2947⟩; **4**, 231 ⟨87CC1442⟩
Hobbold, M., **5**, 955 ⟨94CC2457⟩
Hobbs, A. J. W., **1**, 533, 536 ⟨87TL4875⟩
Hobbs, C. F., **2**, 580, 822 ⟨81JOC4422⟩; **6**, 140 ⟨71JOC2885⟩
Hobbs, C. J., **6**, 632 ⟨89JOC1062⟩
Hobbs, J. B., **1**, 218 ⟨85TL97⟩
Hobbs, Jr., F. W., **1**, 1010 ⟨89JOC3420⟩; **5**, 151 ⟨86JOC3376⟩
Hobbs, S., **3**, 119 ⟨92JOC1600⟩
Hobbs, S. J., **1**, 826 ⟨87JA1160⟩; **4**, 248 ⟨79JOC656⟩
Hobe, M., **2**, 281 ⟨92SL320⟩
Hoberg, H., **1**, 545 ⟨84JOM(266)203, 84JOM(270)C15⟩; **2**, 988 ⟨82JOM(234)C35, 82S324⟩, 989 ⟨77JOM(127)C29⟩; **3**, 456 ⟨76JOM(118)C1⟩, 458 ⟨76JOM(118)C3⟩; **4**, 153 ⟨67LA(703)1⟩, 693 ⟨77AG(E)183⟩; **5**, 101 ⟨84JOM(266)313⟩, 386 ⟨83JOM(252)359, 85AG(E)961⟩, 386 ⟨85JOM(288)237, 85JOM(295)C21⟩, 386 ⟨86JOM(315)245, 87JOM(327)429⟩, 386 ⟨87JOM(328)403, 88JOM(339)213⟩, 386 ⟨88JOM(344)C35, 89JOM(375)245⟩, 386 ⟨89JOM(378)279, 90JOM(382)C6⟩, 386 ⟨91JOM(403)401, 91JOM(412)225⟩, 803 ⟨75LA1124⟩
Hoberg, J. O., **2**, 969 ⟨92JOC3266⟩
Hobert, K., **3**, 157, 161 ⟨91T3753⟩
Hoblitt, R. P., **5**, 984 ⟨70JA1326⟩
Hobson, J., **3**, 539 ⟨68CC764⟩
Hobson, J. D., **2**, 303 ⟨67JCS(C)2015⟩
Hochsfetter, A., **2**, 314 ⟨03M773⟩
Hochstein, F. A., **5**, 47 ⟨48JA3484⟩
Höchstetter, H., **1**, 708 ⟨81AG(E)677⟩

Hochstrasser, R., **2**, 1022 ⟨89AG(E)181, 89JA2355⟩
Hock, H., **2**, 97 ⟨55CB1544⟩, 112 ⟨59CB2716⟩
Höck, N., **2**, 439 ⟨86PS(26)143⟩
Hock, R., **2**, 858 ⟨88JA624⟩; **3**, 385 ⟨88JA624⟩, 389 ⟨89AG(E)179⟩, 394 ⟨90AG(E)1067⟩
Hocker, J., **3**, 475 ⟨68OS(48)36, 69OPP99⟩; **5**, 872 ⟨73OSC(5)179⟩; **6**, 141 ⟨72AG(E)964, 75CB215⟩, 143 ⟨71LA(748)1⟩, 143 ⟨71LA(751)145⟩, 156 ⟨78LA16⟩, 228, 230 ⟨83HOU(E4)1368⟩, 296, 297 ⟨83HOU(E4)694⟩, 301 ⟨83HOU(E4)705⟩, 308 ⟨83HOU(E4)703⟩, 309 ⟨83HOU(E4)704⟩, 313 ⟨83HOU(E4)710⟩, 730 ⟨83HOU(E4)677⟩, 731 ⟨83HOU(E4)690⟩
Hocking, M. B., **2**, 858 ⟨65JCS2184⟩
Hockless, D. C. R., **2**, 473 ⟨91CB2457⟩
Hockley, M. H., **6**, 308 ⟨90MI 610-01⟩
Hockman, S. K., **6**, 179, 190, 202, 203, 672 ⟨80JA1584⟩
Hockstra, H. R., **2**, 109 ⟨53JA199⟩
Hockswender, Jr., T. R., **1**, 1110 ⟨74JA3574⟩
Hodali, H. A., **5**, 776 ⟨90SRI1413⟩; **6**, 338 ⟨79IC1236⟩
Hodder, O. J. R., **3**, 543 ⟨73CC247⟩
Hodek, W., **6**, 608 ⟨83JPR787⟩
Hodes, H. D., **1**, 199 ⟨78TL4349⟩
Hodge, C. N., **1**, 690 ⟨84JA6450⟩, 919 ⟨89JA8867⟩
Hodge, J. D., **1**, 894, 900 ⟨62JA2016⟩, 897, 901, 911 ⟨62JA1498⟩, 897, 901, 911 ⟨63JA2991⟩, 901 ⟨63JA2995⟩, 901, 913 ⟨63JA2998⟩
Hodge, M. L., **2**, 403 ⟨72S632⟩
Hodge, P., **2**, 7 ⟨75CC622, 84JCS(P1)195⟩, 299 ⟨84S941⟩; **3**, 188 ⟨84S941⟩, 618 ⟨77S41⟩; **5**, 13 ⟨75CC622⟩
Hodges, J. C., **1**, 690 ⟨86JA3513⟩; **3**, 109 ⟨91JOC449⟩
Hodges, K. C., **4**, 1019, 1020 ⟨75JA2702⟩
Hodges, P. J., **1**, 218 ⟨85TL4111⟩, 608 ⟨86TL2949⟩, 615, 698, 703 ⟨86JA2776⟩; **3**, 228 ⟨86TL2949⟩
Hodges, R. S., **2**, 217 ⟨81JBC(256)5064⟩
Hodges, S. L., **3**, 210 ⟨87JOC1603⟩
Hodgins, T., **4**, 743 ⟨72JOC683⟩; **6**, 229 ⟨72JOC2651⟩
Hodgkins, J. E., **5**, 1024 ⟨55JA1831, 56JOC204⟩, 1024 ⟨64JOC3098⟩
Hodgson, D., **6**, 127 ⟨88JCS(P1)3367⟩
Hodgson, G. L., **1**, 381 ⟨74JCS(P1)1938⟩
Hodgson, H. H., **1**, 442 ⟨40JCS207, 42JCS748⟩, 442 ⟨43JCS86⟩
Hodgson, M., **3**, 614 ⟨88OM1761⟩
Hodgson, S. M., **6**, 633 ⟨88CC1175⟩
Hodgson, S. T., **3**, 307 ⟨83JCS(P1)93⟩
Hodosan, F., **5**, 735 ⟨57BSF633, 77RRC1111⟩
Hodson, D., **5**, 867 ⟨68JSC(C)1602⟩, 872 ⟨68JCS(C)2201⟩
Hoederath, W., **5**, 513, 559 ⟨85S960⟩, 558 ⟨83SUL191⟩
Hoefner, A., **5**, 928, 930 ⟨92ZN(B)983⟩
Hoefs, C. A. M., **1**, 776, 778 ⟨89JOC5811⟩
Hoeft, E., **4**, 46 ⟨64ZC401⟩
Hoeg, D. F., **4**, 13 ⟨65JA4147⟩; **6**, 62, 244 ⟨65JA4147⟩
Hoeger, C. A., **3**, 177 ⟨85T4771⟩
Høeg-Jensen, T., **5**, 589 ⟨94JOC1257⟩
Hoehne, S., **2**, 836 ⟨82CB2181⟩
Hoekman, S. K., **3**, 483, 483, 503 ⟨80JA1584⟩; **4**, 602, 612, 619, 626, 627, 628, 637, 640, 652, 663 ⟨80JA1584⟩; **5**, 926, 927 ⟨80JA1584⟩
Hoekstra, A., **2**, 613 ⟨84CC655, 86JOC1886⟩
Hoekstra, J. J., **5**, 815 ⟨68USP3361815⟩; **6**, 649 ⟨68JOC2522⟩
Hoekstra, W., **3**, 229 ⟨87TL1069⟩
Hoekstra, W. P. M., **1**, 1018 ⟨88T6699⟩
Hoelzel, C. B., **2**, 198 ⟨61JA1256⟩, 215 ⟨62JA847⟩
Hoem, A. B., **1**, 835 ⟨91TL179⟩; **3**, 623 ⟨91TL179⟩
Hoerger, F. D., **2**, 203 ⟨54JA5357⟩
Hoerhold, H. H., **6**, 29 ⟨87JPR1131⟩
Hoesch, K., **2**, 76 ⟨21CB206⟩
Hoesch, L., **2**, 376 ⟨85HCA220⟩
Hoess, E., **5**, 631 ⟨88CZ349⟩
Hoeve, W. T., **3**, 182 ⟨80JA7125⟩, 186 ⟨94SC899⟩; **4**, 530 ⟨80JA7125⟩
Hoey, M. D., **4**, 234 ⟨88JOC5750⟩
Hofbauer, G., **2**, 364 ⟨53LA(581)182⟩
Hofbauer, P., **2**, 160 ⟨75T505⟩, 160, 164 ⟨74TL3319⟩; **5**, 766 ⟨74TL3319⟩
Hofelich, T. C., **1**, 891 ⟨87MI 119-10⟩, 905 ⟨85MI 119-06⟩
Höfelmeier, R., **5**, 711 ⟨82AG375, 85TL5281⟩
Höfer, J., **2**, 676 ⟨93CC850⟩
Hofer, O., **1**, 498 ⟨88ACS(B)259⟩
Hofer, P., **2**, 389 ⟨66JOC524⟩
Höfer, R., **5**, 446, 448 ⟨73AG(E)77⟩
Hoff, A., **2**, 368 ⟨65CB2429⟩
Hoff, C. D., **2**, 439 ⟨79JOC3095⟩
Hoff, E. F., **3**, 527 ⟨69JA5679⟩; **4**, 768 ⟨70JOC1501⟩
Hoff, M. C., **2**, 428 ⟨59JOC356⟩, 606, 615 ⟨51JA3329⟩
Hoff, S., **1**, 955 ⟨68RTC916⟩, 971 ⟨68RTC1179⟩; **2**, 939 ⟨68RTC916⟩; **4**, 836 ⟨68RTC1179⟩; **5**, 423 ⟨68MI 509-01⟩
Hoff, W. S., **2**, 155 ⟨80JOC2887⟩
Hoffenberg, D. S., **1**, 881 ⟨55JOC1531⟩; **5**, 275 ⟨55JOC1448⟩
Hoffer, R. K., **3**, 128 ⟨88JOC4133⟩
Hoffer, R. M., **2**, 753 ⟨66JMC249⟩
Hofferth, B., **5**, 595, 620, 1016 ⟨50JA3045⟩
Hoffman, A., **4**, 113, 118, 121 ⟨30JA2995⟩
Hoffman, D. M., **3**, 521 ⟨88CC1489⟩; **4**, 675 ⟨84JA6794, 86OM2457⟩, 676 ⟨84JA6806⟩
Hoffman, E., **1**, 108 ⟨61JCS3452⟩; **4**, 33 ⟨61JCS3452⟩
Hoffman, F. W., **5**, 122 ⟨57JA4759⟩
Hoffman, H., **1**, 639, 752, 753 ⟨58CB61⟩, 639, 752, 753 ⟨59CB2499⟩
Hoffman, J. H., **1**, 102 ⟨77S859⟩
Hoffman, Jr., J. M., **3**, 372 ⟨69JA3953⟩
Hoffman, K., **6**, 43 ⟨82S1⟩
Hoffman, K. P., **6**, 215 ⟨75USP3884985⟩
Hoffman, L. K., **1**, 806 ⟨74JA3705⟩
Hoffman, P. G., **3**, 417 ⟨80JOC161⟩
Hoffman, R., **1**, 398 ⟨92CB1957⟩
Hoffman, R. V., **1**, 179 ⟨93TL2051⟩, 840 ⟨91JOC1014⟩, 1077 ⟨78JA7927⟩; **2**, 23 ⟨92JA6262, 92JOC5700⟩, 357 ⟨76JA6702, 83S34⟩, 357 ⟨85JA4731, 91T1109⟩; **3**, 17, 132 ⟨84JOC4011⟩, 65 ⟨74JOC2939, 78JA7934⟩, 65, 232 ⟨78JA7927⟩, 156 ⟨86JOC130⟩, 163 ⟨86TL5811⟩, 163 ⟨91T1109⟩, 413 ⟨85JOC1859⟩, 431 ⟨87S831⟩; **4**, 183 ⟨86JOC130⟩; **5**, 219 ⟨87S831⟩
Hoffman, R. W., **2**, 75, 529 ⟨88PAC123⟩, 154 ⟨79AG(E)563, 84T2219⟩, 529 ⟨85LA2246, 87LA215⟩, 529 ⟨87LA881⟩, 530 ⟨83TL3209, 89LA883⟩; **6**, 106 ⟨64AG(E)380, 71CB873⟩
Hoffman, T. D., **1**, 883 ⟨69JA1009⟩
Hoffman, W., **3**, 62 ⟨75S781⟩
Hoffman, W. H., **2**, 683 ⟨75JCS(P1)461⟩
Hoffmann, A. K., **4**, 16 ⟨54JA6162⟩, 576 ⟨59JA5519⟩
Hoffmann, C., **1**, 417, 829 ⟨72SC27⟩
Hoffmann, E. G., **4**, 583 ⟨72IJ293⟩
Hoffmann, F. W., **1**, 942 ⟨56JA6414⟩; **2**, 458 ⟨57JA3570⟩, 463 ⟨58JA3945, 59JA148⟩; **6**, 257 ⟨56JA5637⟩
Hoffmann, H., **1**, 910 ⟨65AG(E)354⟩, 999 ⟨64TL983⟩; **2**, 428, 430, 431, 441, 822, 827 ⟨58CB1583⟩, 433, 434, 462, 824, 825 ⟨66CB1134⟩, 434 ⟨67CB692⟩, 446 ⟨61CB1331, 61CB186⟩, 446 ⟨65CB363⟩, 454, 844 ⟨58CB64⟩, 462, 821, 827 ⟨59CB2088⟩, 470 ⟨69M311⟩, 1062 ⟨64TL983⟩; **3**, 363 ⟨67TL2963⟩, 480 ⟨61TL9⟩, 493, 494 ⟨65JCS7140, 62CB2563⟩, 498 ⟨61CB1331, 65CB363⟩; **4**, 462, 490 ⟨62CB2563⟩, 481, 487 ⟨68M380, 61TL9⟩, 575, 576, 577 ⟨59CB2499⟩, 575, 576, 577 ⟨61CB1987⟩, 576 ⟨58CB61, 62CB581⟩; **5**, 774 ⟨78GEP2701309⟩; **6**, 104 ⟨68BRP1097237⟩, 111, 115 ⟨73CB3725⟩, 138, 140 ⟨68CB51⟩
Hoffmann, H. M. R., **1**, 187, 1025 ⟨88CB1461⟩, 364

⟨88T3899⟩, 524 ⟨79JOC2802⟩, 524, 772 ⟨69AG(E)556⟩; **2**, 29 ⟨88OSC(6)512⟩, 639 ⟨72JA7832⟩, 658 ⟨90TL6629⟩; **3**, 8 ⟨85CC812⟩, 216 ⟨88CB1461⟩, 250 ⟨91JOC4084⟩, 652 ⟨82AG(E)859, 86S548⟩; **4**, 99 ⟨89T6113⟩; **5**, 19 ⟨81S715, 82S237⟩; **6**, 107 ⟨72JA3201⟩, 431 ⟨84JOC1174⟩

Hoffmann, J., **1**, 985 ⟨92SL635⟩; **2**, 449 ⟨91AG(E)207⟩; **4**, 110 ⟨91CB1207⟩, 483 ⟨91S1099⟩; **5**, 395 ⟨91S1099⟩; **6**, 153 ⟨91AG(E)207⟩, 512, 582 ⟨91S1099⟩

Hoffmann, K., **1**, 537 ⟨53CB657⟩

Hoffmann, M., **4**, 460 ⟨76C187⟩

Hoffmann, P., **1**, 547 ⟨89TL57⟩; **4**, 239 ⟨89TL6311⟩; **5**, 876, 882 ⟨64AG(E)384, 66CB1325⟩

Hoffmann, R., **1**, 349, 400, 417, 418, 419, 422 ⟨69AG(E)781⟩, 398 ⟨93TL297⟩, 400 ⟨65JA2511⟩, 405, 816 ⟨92AG(E)647⟩, 417, 418, 419, 422 ⟨65JA395⟩, 417, 422 ⟨B-70MI 109-01⟩, 794 ⟨B-70MI 118-01⟩, 1109 ⟨64JCP(40)2480⟩; **2**, 249, 365 ⟨62LA(657)86⟩; **3**, 356 ⟨90ZC247⟩; **4**, 217 ⟨91TL467⟩, 359 ⟨92CB2731⟩, 973 ⟨77IJC(B)297⟩; **5**, 441 ⟨85JA4440⟩, 548, 559, 584, 586, 614 ⟨80CB919⟩, 555 ⟨80LA483⟩, 1097 ⟨80CC509, 81JA3328⟩

Hoffmann, R. W., **1**, 404, 407, 816 ⟨79AG(E)563⟩, 519 ⟨B-67MI 112-02⟩, 885 ⟨B-91MI 119-02⟩, 1081 ⟨B-67MI 121-01⟩; **2**, 74, 529, 530 ⟨81JOC1309⟩, 75 ⟨81CB375, 88CB1501⟩, 75 ⟨88CB2225⟩, 529 ⟨80AG(E)218, 80JOM(187)321⟩, 529 ⟨82AG(E)555, 82CB2357⟩, 530 ⟨89CB1783⟩, 652 ⟨83S107⟩, 698 ⟨85TL1643⟩, 900 ⟨87TL4169, 88S103⟩; **3**, 541 ⟨75CB844⟩; **4**, 25 ⟨90S657⟩, 146 ⟨88JOM(353)C30⟩, 146 ⟨91CB1253⟩, 146 ⟨91CB1259⟩, 146 ⟨91LA811⟩, 288 ⟨92LA643⟩, 363 ⟨89CB903⟩, 363, 373 ⟨86CB1039, 83TL3209⟩, 400, 401 ⟨90TL7419⟩, 660 ⟨92LA725⟩, 1049 ⟨85CB2493⟩, 1056 ⟨87LA977⟩, 1081 ⟨83JCS(P2)1687⟩; **5**, 173 ⟨86AG(E)1028⟩, 199 ⟨87TL4169⟩; **6**, 74 ⟨66CB1892⟩, 107 ⟨67TL3501, 68CB3851⟩, 390 ⟨85CB2493⟩, 630 ⟨77CB37⟩

Hoffmann, U., **1**, 326, 327 ⟨93TL2445⟩

Hoffmann, W., **1**, 422 ⟨76TL3585⟩, 525, 832 ⟨71LA(747)60⟩, 801 ⟨82CB2508, 69LA(729)52⟩

Hoffmann, III, W. A., **6**, 464 ⟨82JOC5209⟩

Hoffmeister, E. H., **6**, 730 ⟨68T6485⟩

Hoffmeister, H., **3**, 690 ⟨91AJC135⟩; **5**, 942 ⟨91AOC(32)227⟩, 1147 ⟨89ZN(B)419⟩; **6**, 723 ⟨91AOC(32)227⟩

Hoffschmidt, R., **5**, 858 ⟨53LA(581)23⟩

Hoffsommer, J. C., **6**, 27 ⟨68CJC2233⟩

Hoffsommer, R. D., **5**, 173 ⟨68T2443⟩

Hofle, G., **1**, 16, 17 ⟨90SL705⟩

Höfle, G., **1**, 537 ⟨71AG(E)653⟩, 985 ⟨75AG(E)58⟩; **2**, 223 ⟨71JA2870⟩; **3**, 699, 700, 723 ⟨76AG(E)113⟩, 702 ⟨90OSC(7)27⟩, 703 ⟨74AG(E)676⟩, 703, 710 ⟨77AG(E)727⟩, 704 ⟨77AG(E)262⟩; **4**, 56 ⟨74AG(E)676⟩; **5**, 129 ⟨78AG(E)569⟩

Höfle, G. A., **1**, 1089 ⟨74JA7125⟩; **2**, 681, 908, 999 ⟨74JA7125⟩; **3**, 157 ⟨74JA7125⟩; **4**, 949 ⟨74JA7125⟩

Höfler, M., **4**, 545 ⟨76ZN(B)790⟩; **6**, 526 ⟨64CB2289⟩

Hofman, E., **2**, 429 ⟨63BRP921463⟩

Hofman, J. E., **1**, 475 ⟨64MI 111-01⟩

Hofmann, A., **6**, 104 ⟨63HCA2306, 63T1675⟩

Hofmann, A. A., **3**, 103 ⟨78HCA430⟩

Hofmann, A. W., **1**, 627 ⟨1851LA(78)253⟩; **2**, 127 ⟨1868CB177⟩, 427 ⟨1871CB430, 1871CB605⟩, 427 ⟨1873CB292⟩, 461, 853 ⟨1857LA(101)1⟩; **3**, 620 ⟨1884CB1404, 1884CB1920⟩, 620 ⟨1886CB1433, 1886CB1822⟩, 693 ⟨1867LA(144)114⟩, 698, 699, 721 ⟨1870CB766⟩; **5**, 566 ⟨1878CB338⟩, 609 ⟨1877CB1095⟩, 1026 ⟨1869CB120⟩, 1035 ⟨1882CB985⟩

Hofmann, C. M., **1**, 411 ⟨41JA1852⟩

Hofmann, H., **5**, 606 ⟨85ZN(B)124⟩

Hofmann, H. P., **4**, 706, 709 ⟨62IC227⟩

Hofmann, J., **1**, 500 ⟨90CC858⟩; **2**, 1009 ⟨91T205⟩; **3**, 386 ⟨88JOM(358)229, 89CC667⟩, 386 ⟨90JOM(384)305⟩; **4**, 581 ⟨91AG(E)998⟩; **5**, 605, 614, 627 ⟨88CB2095⟩; **6**, 621 ⟨89S875⟩

Hofmann, J. E., **1**, 590 ⟨65CI(L)1768⟩, 600, 612 ⟨64JA1561, 63JA2739⟩

Hofmann, K., **2**, 209 ⟨84LA39⟩, 539 ⟨82S1⟩; **4**, 25 ⟨59JA992⟩

Hofmann, K. A., **2**, 1014 ⟨08CB314, 09CB4232⟩, 1098 ⟨08CB316, 09CB4233⟩

Hofmann, L., **4**, 148 ⟨85JOM(289)141⟩, 149, 583 ⟨88OM1106⟩, 582 ⟨81JOM(219)C29⟩, 582 ⟨85JOM(281)317, 82JOM(236)C65⟩

Hofmann, M., **4**, 549 ⟨94AG(E)2202⟩

Hofmann, P., **3**, 508 ⟨B-83MI 315-01⟩, 530, 531 ⟨86S43⟩, 690 ⟨B-88MI 320-03⟩; **4**, 546, 577 ⟨92GEP4134772⟩; **5**, 441 ⟨85JA4440⟩; **6**, 716, 719, 722 ⟨B-84MI 622-01⟩

Hofmann, P. W., **5**, 927 ⟨92CB73⟩

Hofmeister, H., **1**, 5, 9 ⟨83LA705⟩, 558 ⟨89LA2151⟩, 1142 ⟨67CB1200⟩; **2**, 1018 ⟨84AG(E)727⟩; **3**, 227 ⟨82S34⟩

Hofmeister, P., **2**, 879 ⟨76TL3125⟩

Hofmockel, U., **6**, 689 ⟨85CC1687⟩

Höfner, A., **6**, 389, 714 ⟨91AG(E)594⟩

Hofstra, G., **3**, 340 ⟨84TL873⟩

Höft, E., **2**, 96 ⟨65AG(E)524⟩; **3**, 99, 101 ⟨60CB88⟩; **6**, 39 ⟨67OS(47)47⟩, 43, 251, 429 ⟨61CB544⟩, 109 ⟨72JPR145⟩

Hoga, T., **6**, 429 ⟨59JAP594264⟩

Hogale, M. B., **5**, 355 ⟨91IJC(B)717⟩

Hogan, H. P., **1**, 1076 ⟨72JOC4466⟩

Hogan, J. C., **1**, 451, 1105 ⟨79JOC155⟩; **2**, 923 ⟨79JOC155⟩; **4**, 357 ⟨79JOC155⟩

Hogarth, G., **4**, 1023, 1032 ⟨88CC673⟩

Högberg, H. E., **1**, 78 ⟨88JCS(P1)3061⟩

Högberg, T., **4**, 189 ⟨89JOC4549⟩; **5**, 277 ⟨87JOC2033⟩

Hoge, R., **3**, 273 ⟨80JOM(184)317⟩

Högel, J., **6**, 156 ⟨83CB549⟩

Hogenbirk, M., **4**, 618 ⟨89TL6195⟩; **6**, 399 ⟨89TL6195, 92JA7302⟩

Hogen-Esch, T. E., **1**, 858 ⟨77APO(15)153⟩, 859 ⟨66JA307, 66JA318⟩, 859 ⟨69JA4580⟩

Hogeveen, H., **1**, 901 ⟨67CC635, 68RTC319⟩, 901 ⟨69RTC371, 72MI 119-13⟩, 1122 ⟨60RTC1002, 63RTC375⟩, 1122 ⟨63RTC410⟩; **2**, 147 ⟨63JCS4864⟩, 1032 ⟨63RTC375⟩; **3**, 371 ⟨73TL3929⟩; **6**, 7 ⟨80JOC3957⟩

Hogg, A. M., **2**, 431 ⟨91BCJ2398⟩

Hogg, D. R., **2**, 169 ⟨68JCS(C)2713⟩, 184, 718, 719, 724 ⟨79COC(3)261⟩, 718 ⟨79COC(3)311⟩; **5**, 472 ⟨77SST136, 79SST133, 81SST160⟩, 1022 ⟨79COC(3)305⟩

Hogg, J. A., **4**, 209 ⟨54JA743⟩

Hoggarth, E., **2**, 203 ⟨45JCS14⟩; **5**, 62 ⟨B-52MI 502-03⟩, 1036 ⟨50JCS1582⟩

Hoggett, J. G., **3**, 627, 634 ⟨71JCS(C)2151⟩

Hohage, H., **6**, 536 ⟨73CB1487⟩

Hoh, K., **3**, 12, 13, 83, 98, 129 ⟨86JOC5400⟩

Hohenberger, K., **4**, 622 ⟨80ZAAC(464)107⟩

Hohenhorst, M., **3**, 268 ⟨87TL4395, 90LA397⟩

Hohenlohe-Oehringen, K., **5**, 342 ⟨68M1289⟩

Höhle, A., **6**, 620 ⟨76CB2921⟩

Hohlein, U., **1**, 45 ⟨90SL465⟩

Hohlneicher, E., **1**, 910 ⟨68HCA225⟩

Hohmann, F., **4**, 580 ⟨72JOM(42)C35⟩

Hohmann, S., **4**, 578 ⟨76CZ336⟩

Höhn, A., **4**, 499 ⟨89CC1648⟩

Höhne, G., **1**, 95 ⟨87JOC2804⟩; **3**, 349 ⟨75CB673, 78ZN(B)417⟩

Hohne, R., **2**, 201 ⟨76S829⟩

Höhne, S., **4**, 581 ⟨81CB413⟩

Hohner, L., **3**, 492 ⟨65LA(685)1⟩

Hohorst, F. A., **6**, 42 ⟨74JOC1298⟩, 231, 232 ⟨71JA3882⟩, 250 ⟨67JA1809, 68IC624⟩
Hoi, B., **3**, 309 ⟨49RTC5⟩
Hoiberg, C. P., **2**, 102 ⟨69JA4273⟩
Hoiness, C. M., **1**, 299 ⟨73MI 119-19⟩
Hoinowski, A. M., **4**, 48 ⟨75TL3979⟩
Hojatti, M., **2**, 637 ⟨87JA4000⟩, 1022 ⟨87CJC441⟩
Hojo, F., **2**, 1089 ⟨93OM803⟩
Hojo, K., **1**, 811 ⟨70JA6641⟩; **2**, 8, 36 ⟨76CL619⟩, 124, 125 ⟨77CL437⟩; **5**, 125 ⟨76CL267⟩
Hojo, M., **1**, 192 ⟨87TL3497⟩, 286 ⟨92CL1073⟩, 366 ⟨93TL5943⟩, 438 ⟨90CC1284, 93JOC6541⟩; **2**, 139 ⟨91SL539⟩, 140 ⟨92TA1563⟩; **3**, 187 ⟨91TL2371⟩, 230 ⟨86TL353⟩, 232 ⟨90CL113⟩, 246 ⟨86S1013⟩, 246, 261 ⟨76CL499⟩, 253 ⟨75TL1009⟩, 265 ⟨89TL6173, 90S195⟩, 326 ⟨93TL5943⟩; **4**, 74, 75 ⟨76TL613⟩, 74, 80 ⟨68JA4496⟩, 78 ⟨76S697⟩, 226 ⟨82S312⟩, 247 ⟨86JOC1427⟩, 368 ⟨91SL557, 92TL85⟩, 889 ⟨86S1013⟩, 889, 900, 970, 971, 972 ⟨90S195⟩, 900 ⟨90JOC5308⟩; **5**, 170 ⟨85TL3207⟩, 170 ⟨87TL325⟩; **6**, 84, 89 ⟨75JOC963⟩, 634 ⟨91TL5987⟩
Hokari, H., **2**, 14 ⟨80BCJ770⟩; **3**, 45 ⟨80CL1213, 82BCJ534⟩
Hoke, D., **1**, 4, 7 ⟨71JOC1568⟩
Holah, D. G., **4**, 1031 ⟨85JHC513⟩
Holan, G., **6**, 81 ⟨67JCS(C)30⟩
Höland, R., **6**, 213, 227 ⟨1887LA(240)225⟩
Holbert, G. W., **1**, 573 ⟨78JA6483⟩, 789 ⟨78JA352⟩
Holcomb, R. C., **1**, 166 ⟨93TL63⟩, 523 ⟨89TL7157⟩
Holcomb, W. D., **2**, 365 ⟨75JA676⟩
Holcomb, W. F., **3**, 638 ⟨42JA1309⟩; **5**, 1125 ⟨46JA2102⟩
Holcombe, J. L., **2**, 684 ⟨86JOC111⟩
Holde, D., **5**, 184 ⟨25CB1418⟩
Holden, D. A., **5**, 366 ⟨84CJC574⟩; **6**, 489 ⟨84CJC574⟩
Holden, J. R., **6**, 370 ⟨85GEP3337939⟩
Holden, M., **2**, 7 ⟨82AJC517⟩
Holden, M. E., **2**, 556 ⟨47JA1675⟩
Holder, G. M., **2**, 659 ⟨76JOC977⟩
Holder, R. W., **3**, 527 ⟨76JOC3303⟩
Holderbaum, M., **2**, 1003 ⟨88CB843⟩
Hölderich, W., **4**, 545 ⟨77ZAAC(431)76⟩; **5**, 1154 ⟨75CB2484⟩, 1155 ⟨76ZAAC(422)104, 77ZAAC(428)222⟩
Holdrege, C. T., **5**, 9 ⟨63JA643, 68CJC2549⟩
Holick, W., **5**, 905 ⟨78JA8214⟩
Holiday, E. R., **5**, 84 ⟨52MI 502-03⟩
Hölken, A., **3**, 80 ⟨04LA(331)245⟩
Holladay, J. E., **2**, 938 ⟨93JA6625⟩; **4**, 621 ⟨93JA6625⟩
Holland, B., **1**, 1000, 1001 ⟨74SC199⟩
Holland, D. G., **1**, 426 ⟨64JOC3042⟩; **3**, 179 ⟨64RTC1047⟩
Holland, G. F., **4**, 223 ⟨58JA3765⟩
Holland, G. W., **1**, 274, 546 ⟨70JA3503⟩, 277 ⟨71JA2792⟩
Holland, H. L., **2**, 304 ⟨88TL6393⟩; **4**, 190 ⟨76TL3361⟩
Holland, K. D., **4**, 687 ⟨90OM2745⟩
Holland, W., **2**, 29 ⟨53JOC1112⟩
Hollander, Ch. S., **5**, 833 ⟨59USP2902514⟩
Hollander, J., **2**, 772 ⟨56JOC1051⟩; **5**, 967 ⟨67PP1149⟩
Holl, P., **2**, 475 ⟨77AG(E)722⟩, 751 ⟨74S816⟩; **4**, 493 ⟨74S816⟩, 571 ⟨78JOM(159)137⟩; **5**, 914 ⟨78JOM(159)137⟩
Hölle, H. J., **5**, 979 ⟨74JOC1597⟩
Hollebone, B. R., **5**, 1083 ⟨71JCS(A)332⟩
Hollenstein, S., **1**, 899 ⟨93JA7240⟩; **2**, 1007 ⟨93JA1731⟩, 1008 ⟨90AG(E)188⟩
Holler, H. V., **3**, 532 ⟨58JOC1166⟩
Hollfelder, H., **6**, 720, 722 ⟨76JOM(113)C31⟩, 720, 722, 723 ⟨77CB3467⟩
Holliday, A. K., **2**, 110 ⟨52JCS11, 54JCS1212⟩; **6**, 183 ⟨73CC532⟩, 183 ⟨75JCS(D)148⟩
Holliman, F. G., **2**, 288 ⟨45JCS37⟩, 488 ⟨45JCS45⟩
Hollingsworth, C. A., **1**, 1040 ⟨56JA1221⟩
Hollins, R. A., **2**, 629 ⟨79JOC3931⟩
Hollinshead, D. M., **1**, 16 ⟨81JCS(P1)1501⟩; **4**, 165 ⟨83JCS(P1)1579⟩, 165 ⟨84JCS(P1)681⟩
Hollinshead, S. P., **5**, 263 ⟨93JCS(P1)451⟩
Hollis, L. S., **2**, 1009 ⟨89JA1299⟩
Hollis, R., **1**, 340 ⟨92JOC4284⟩
Hollitzer, O., **6**, 502 ⟨79CB727⟩
Holloway, C. E., **6**, 5 ⟨67CJC2659⟩
Holloway, J. H., **6**, 457 ⟨88CC529⟩
Hollstein, W., **2**, 1000 ⟨88AG(E)846⟩
Holm, A., **3**, 365 ⟨77ACS(B)679⟩, 553 ⟨79CC99⟩, 687 ⟨78JCS(P1)1445⟩; **5**, 323 ⟨90S1008⟩, 589 ⟨94JOC1257⟩, 626 ⟨85JCS(P1)907⟩, 652 ⟨65ACS1768⟩, 1013 ⟨70ACS1852⟩, 1013, 1101 ⟨65ACS438, 70ACS1512⟩, 1100, 1102 ⟨B-77MI 530-01, 64ACS825⟩, 1101 ⟨64ACS826, 66ACS2107⟩, 1101 ⟨69ACS1567⟩, 1103 ⟨64ACS2417⟩, 1104 ⟨66ACS2123, 68ACS2019⟩; **6**, 593 ⟨64ACS2417, 66ACS2091⟩
Holm, B., **3**, 293 ⟨81JOC5414⟩
Holm, K. H., **1**, 831 ⟨85JOC5179⟩
Holm, R. H., **3**, 339 ⟨73HCA1671, 73JA613⟩
Holm, S., **2**, 400 ⟨76CB1069⟩; **4**, 862 ⟨73TL2389⟩; **6**, 274, 275, 309 ⟨76CB1069⟩
Holman, D. J., **2**, 435 ⟨68JCS(A)1909⟩
Holman, R. J., **2**, 389 ⟨70JCS(C)2195⟩
Holmberg, B., **6**, 83, 84 ⟨07LA(353)131⟩, 83, 84, 91 ⟨07CB1740⟩
Holmes, A. B., **1**, 177, 178 ⟨92JCS(P1)95⟩, 623 ⟨92JCS(P1)3225⟩, 769 ⟨88TL4333⟩, 829 ⟨91T7171, 93JA10400⟩, 829 ⟨93JA5815⟩, 1021 ⟨86TL6133⟩, 1023 ⟨76TL1539⟩; **2**, 327 ⟨91SL29⟩; **3**, 367 ⟨85JCS(P1)1161⟩; **5**, 179 ⟨93JA5815⟩
Holmes, D., **2**, 43, 528 ⟨93CC1673⟩
Holmes, D. R., **3**, 685 ⟨83JCS(P1)293⟩
Holmes, D. S., **1**, 614 ⟨88TL6357⟩
Holmes, H. L., **3**, 100 ⟨47JA1998⟩, 620 ⟨57OR(9)107⟩; **5**, 46 ⟨57OR(9)107⟩, 114 ⟨43OSC(2)428⟩
Holmes, J., **1**, 397 ⟨60JCS3521⟩
Holmes, J. L., **1**, 921 ⟨88JA7339, 88JA7343⟩, 1141 ⟨84JA521⟩; **2**, 129 ⟨72JMC968⟩, 636 ⟨82JA2648⟩
Holmes, J. M., **1**, 1014, 1029 ⟨90JA9330⟩; **2**, 856 ⟨83IC1771⟩; **6**, 661 ⟨86AG(E)457⟩, 686, 688 ⟨87PS(30)495⟩
Holmes, P. E., **4**, 357 ⟨88TL6395⟩
Holmes, R. R., **1**, 1102 ⟨56JA2173⟩; **2**, 856 ⟨79JA3790, 83IC1771⟩; **6**, 661 ⟨86AG(E)457⟩, 686, 688 ⟨87PS(30)495⟩
Holmes, S. J., **1**, 536 ⟨85JA2568⟩; **2**, 927, 986, 988 ⟨89JA3336⟩; **3**, 270 ⟨85JA2568, 89JA3336⟩; **4**, 581 ⟨84OM476⟩, 809 ⟨89JA3336⟩, 1061 ⟨85JA2568⟩
Holmes, T. F., **6**, 10 ⟨87JOM(325)13⟩, 12 ⟨74JCS(P1)108⟩, 14 ⟨88JCS(P1)921⟩
Holmes, T. J., **3**, 398 ⟨77CC177⟩
Holmes, W. S., **2**, 407 ⟨61JCS1825⟩
Holmes-Smith, R., **1**, 554 ⟨84TL1279⟩
Holmes-Smith, R. D., **2**, 439 ⟨83JCS(P1)861⟩; **4**, 566 ⟨83JCS(P1)861⟩
Holmquist, C. R., **3**, 171, 200 ⟨92TL1131⟩
Holoboski, M. A., **1**, 556 ⟨93TL3021⟩
Holoch, J., **6**, 530, 531 ⟨86CB269⟩
Holsboer, D. H., **6**, 39, 40 ⟨71RTC556⟩, 46, 48 ⟨72RTC349⟩
Holschneider, F., **2**, 720 ⟨25CB409⟩
Holsing, M., **5**, 354 ⟨53JA1933⟩
Holst, W. H., **1**, 1144 ⟨36JA2310⟩
Holsten, J. R., **1**, 144 ⟨59JA2572⟩; **3**, 293 ⟨61JOC4151⟩
Holsworth, D. D., **5**, 993 ⟨89JOC4019⟩
Holt, A., **1**, 652 ⟨69JCS(B)852⟩
Holt, D., **4**, 850 ⟨83JOC4713⟩
Holt, D. A., **1**, 464 ⟨81TL2243⟩; **2**, 22 ⟨85JA7771⟩; **3**, 217 ⟨83JA3348⟩

Holt, E. M., **5**, 407 ⟨91IC1174⟩, 445, 447 ⟨80JA5093⟩; **6**, 338 ⟨85JA8136⟩, 340 ⟨80CC778, 80CC780⟩, 340 ⟨82JA5621⟩, 518 ⟨85IC4449, 87JA7764, 90JOM(383)295⟩
Holt, G., **5**, 867 ⟨68JSC(C)1602⟩, 872 ⟨68JCS(C)2201⟩
Holtan, R. C., **2**, 938, 939 ⟨83T949⟩; **4**, 285 ⟨90JA5609⟩; **5**, 423, 427 ⟨83T949⟩; **6**, 135 ⟨90JA5609⟩
Holterman, H. A. J., **4**, 235 ⟨77JOC2792⟩
Holton, A. G., **6**, 450 ⟨83JFC(22)521⟩
Holton, J., **2**, 565 ⟨79JCS(D)54⟩, 566 ⟨77JA6645⟩; **4**, 584 ⟨81CC698, 82JOM(228)C71⟩, 656 ⟨78CC140⟩, 702 ⟨79JCS(D)54⟩, 725, 726 ⟨79JCS(D)45⟩; **6**, 195 ⟨77JA6645⟩, 195 ⟨78CC140⟩
Holton, R. A., **1**, 23 ⟨87JOC2317⟩, 130 ⟨83TL1345⟩, 842 ⟨84JA5731, 84TL4455⟩, 842 ⟨94JA1597, 94JA1599⟩; **2**, 697, 699 ⟨83TL1345⟩; **4**, 169 ⟨84TL4455⟩, 191 ⟨87JA1597⟩, 219 ⟨77TL533, 80SC911⟩, 959 ⟨86TL2191⟩
Holtschmidt, H., **3**, 699, 721 ⟨62AG848⟩; **5**, 652 ⟨68AG(E)856⟩, 654 ⟨68AG942⟩, 964 ⟨64FRP1469481⟩, 968, 970, 1009 ⟨62AG(E)617⟩, 1005 ⟨62AG(E)632⟩; **6**, 51, 238 ⟨69LA(730)140⟩, 239 ⟨63GEP1222917, 70GEP1926659⟩, 240 ⟨72CA(77)125952⟩, 297, 299 ⟨72S599⟩, 316 ⟨63BEP627486, 68BRP1097237⟩, 603, 613 ⟨72GEP2101107⟩, 607 ⟨62AG(E)632, 69LA(730)133⟩, 617 ⟨76GEP2451635⟩
Holub, D. P., **2**, 311 ⟨90SC981⟩
Holý, A., **1**, 936 ⟨78MI 119-03⟩; **4**, 130 ⟨85CCC1507⟩; **6**, 54 ⟨61CCC3059, 62CCC2886⟩, 80 ⟨67CCC3064⟩, 104 ⟨63CCC2040⟩
Holy, N., **3**, 184, 185 ⟨79T613⟩; **4**, 580, 581, 584, 585 ⟨82CB1379⟩
Holy, N. C., **2**, 393 ⟨67JA725⟩
Holy, N. L., **1**, 93 ⟨79JOC239⟩, 890 ⟨74CRV243⟩; **3**, 45 ⟨77JA944⟩
Holzapfel, C. W., **3**, 620 ⟨85JOC2323⟩; **4**, 190 ⟨91H(32)1445⟩, 228 ⟨90TL2759⟩; **5**, 573 ⟨90SC2235⟩
Holzapfel, W., **4**, 173 ⟨85TL6329⟩
Holze, G., **2**, 282 ⟨58HCA593⟩
Holzer, G., **2**, 765 ⟨76CB3462⟩
Holzer, W., **1**, 1010 ⟨88M253⟩
Holzinger, H., **1**, 978 ⟨88JOC2450⟩
Holzkamp, E., **1**, 665 ⟨55AG541, 60LA(629)172⟩; **2**, 596 ⟨60LA(629)121⟩; **4**, 709 ⟨57LA(605)93⟩
Hölzl, F., **3**, 695 ⟨27M71⟩
Hölzl, W., **4**, 567 ⟨87TL2693⟩; **6**, 129 ⟨89TL4501⟩
Holzmann, G., **5**, 357 ⟨88AP(321)25⟩
Homan, D. F. J., **3**, 309 ⟨89JOC4350⟩
Homberg, O., **5**, 42 ⟨61JOC307⟩
Hombo, T., **3**, 380 ⟨78CC197⟩
Homburg, F., **3**, 538 ⟨64CB1799⟩
Homer, R. B., **5**, 723 ⟨B-70MI 517-03⟩
Homeyer, B., **4**, 94 ⟨88GEP3639877A1⟩; **5**, 774 ⟨78GEP2701309⟩; **6**, 513 ⟨82CSR435⟩
Homma, H., **1**, 1010 ⟨92JMC241⟩
Hommes, H., **2**, 422 ⟨77RTC160⟩, 621 ⟨81TL2495⟩
Homsany, R., **4**, 240 ⟨73JOC3172⟩; **5**, 210 ⟨73JOC3172⟩
Hon, M.-Y., **1**, 801, 802 ⟨89CRV165⟩
Hon, P. M., **3**, 103 ⟨92T1039⟩
Hon, Y.-S., **1**, 900 ⟨85JOC4605⟩
Hona, S. C., **4**, 540 ⟨87JOM(330)357⟩
Honan, M. C., **1**, 776 ⟨85TL6393⟩
Honda, H., **3**, 134 ⟨90SC3265⟩
Honda, K., **1**, 408 ⟨92JOC428⟩, 408, 823 ⟨90JA1999⟩, 839 ⟨90BCJ1328⟩; **2**, 399 ⟨66BCJ1778⟩; **4**, 507 ⟨90JA1999⟩; **5**, 146 ⟨87CC83⟩, 925 ⟨85TL2337⟩
Honda, M., **2**, 20 ⟨73CPB867⟩, 769 ⟨89BCJ128⟩
Honda, O., **5**, 681 ⟨87JHC1391⟩
Honda, T., **1**, 558 ⟨93TA21⟩, 941 ⟨78HCA690⟩; **2**, 858 ⟨87TL6649⟩, 1000 ⟨87JA3010⟩; **3**, 9, 123 ⟨89CL901⟩, 206 ⟨91JCS(P1)954⟩, 222 ⟨90JCS(P1)1733, 90JCS(P1)639⟩, 303 ⟨90JCS(P1)2677⟩, 384, 394 ⟨87TL6649⟩; **5**, 223 ⟨90CPB1836⟩
Honda, Y., **2**, 478 ⟨84T4553⟩; **4**, 908 ⟨93CPB217⟩
Hondo, I., **1**, 456 ⟨92CL1725⟩
Hondo, K., **3**, 145 ⟨87BCJ3421⟩
Hondrelis, J., **5**, 223 ⟨89LA387⟩
Honegger, E., **1**, 1064 ⟨87CB187⟩
Honegger, H., **6**, 116 ⟨77TL2987, 80AJC287⟩
Höneise, W., **5**, 776 ⟨86ZAAC(533)55⟩
Honek, J. F., **2**, 921 ⟨83TL4295⟩
Honert, D., **1**, 574, 584 ⟨91AG(E)1372⟩
Hong, B.-C., **4**, 849 ⟨87JOC855⟩; **6**, 127 ⟨87JOC855⟩
Hong, C. Y., **1**, 4, 7, 9 ⟨83AG(E)562⟩
Hong, F.-E., **4**, 726 ⟨91OM2726⟩
Hong, J. S., **2**, 129 ⟨72JMC968⟩
Hong, P., **4**, 686 ⟨72CC993⟩
Hong, R., **3**, 291 ⟨83OM467⟩
Hong, S., **3**, 197 ⟨89TL3307⟩
Hong, S.-I., **5**, 378 ⟨94S909⟩
Hong, W.-P., **5**, 277 ⟨87JOC4665⟩
Hong, Y., **2**, 76 ⟨89JOC4154⟩; **3**, 131 ⟨87TL4391⟩
Hong, Y.-L. V., **3**, 232 ⟨89JOC5126⟩
Hong Chan, W., **1**, 624, 626 ⟨86JA1039⟩
Hongo, H., **5**, 199 ⟨69TL2465⟩
Hongo, T., **3**, 683 ⟨68BCJ2954⟩
Hongwen, H., **2**, 320 ⟨90SC3563⟩; **3**, 179 ⟨90S615⟩
Honicke-Schmidt, P., **5**, 118 ⟨90CC648⟩
Hönig, H., **5**, 694 ⟨89LA665⟩
Honjo, N., **5**, 215 ⟨86CPB4984⟩
Hönle, W., **6**, 178 ⟨81ZAAC(475)87⟩, 704 ⟨84USP4473580⟩
Honma, T., **3**, 268 ⟨77H(6)1985⟩
Honna, T., **4**, 837 ⟨85CPB2256⟩
Honnen, L. R., **1**, 906 ⟨60JOC1442⟩
Honnens, J., **2**, 469 ⟨86JCR(S)26⟩
Honold, B., **2**, 963 ⟨86JOM(308)289⟩
Honold, J., **4**, 621 ⟨88ZAAC(556)23⟩
Höntsch, R., **5**, 984 ⟨79JPR928⟩
Honwad, V. K., **1**, 1005 ⟨69JA6464⟩; **2**, 621 ⟨69JA6464⟩; **3**, 273 ⟨69JA6464⟩
Honwen, H., **3**, 24 ⟨91S539⟩
Honzatko, R., **6**, 79 ⟨88JA4144⟩
Honzl, J., **4**, 894 ⟨70TL2357, 71CCC3314⟩
Hoock, C., **1**, 982 ⟨91S1151, 92AG(E)1611⟩
Hood, A. R., **6**, 635 ⟨88S460⟩
Hood, H. P., **6**, 415 ⟨19JPC498⟩
Hood, J. N. C., **3**, 196, 500 ⟨82T3355⟩; **6**, 708 ⟨82T3355⟩
Hoogeboom, T. J., **2**, 554 ⟨68JOM(11)615⟩
Hoogenboom, B. E., **3**, 696, 709 ⟨88OSC(6)987, 72TL2369⟩; **4**, 331 ⟨88OSC(6)987⟩
Hoogsteen, K., **1**, 390 ⟨92JA5432⟩; **2**, 528 ⟨91JOC751, 93JOC2791⟩; **3**, 680 ⟨90OM2867⟩; **4**, 509 ⟨91JCS(P1)1091⟩; **5**, 949 ⟨92JA10665⟩; **6**, 446 ⟨91JOC751⟩
Hook, J. M., **1**, 87 ⟨86NPR35⟩
Hook, S. C. W., **2**, 492, 497 ⟨71JCS(C)1660⟩, 501 ⟨70JCS(B)735⟩
Hooks, Jr., H., **5**, 1037 ⟨66AG(E)725⟩; **6**, 448 ⟨66AG(E)672⟩, 494 ⟨66AG210⟩
Hoong, L. K., **2**, 529 ⟨85JA8186⟩
Hooper, D. L., **4**, 444 ⟨83CC276⟩
Hooper, K. C., **2**, 214 ⟨56JCS4066⟩
Hoorfar, A., **4**, 748 ⟨91JCS(P1)2575⟩
Hoornaer, G., **5**, 661 ⟨92S705⟩
Hoornaer, G. J., **5**, 661 ⟨92TL2049⟩
Hoornaert, C., **1**, 547 ⟨82TL2505⟩
Hoornaert, G., **3**, 238 ⟨70TL1821, 73T4241⟩, 238 ⟨75T177⟩
Hoornaert, G. J., **2**, 23 ⟨93T3193⟩
Hoos, W. R., **2**, 476 ⟨68TL5333⟩, 837 ⟨70M776⟩
Hooton, J. R., **2**, 496, 498 ⟨57JOC452⟩

Hooton, K. A., **2**, 867 ⟨65JCS4283⟩
Hoots, J. E., **2**, 827 ⟨81IS178, 82IS175⟩
Hoover, D. J., **1**, 839 ⟨82TL3463⟩; **5**, 688 ⟨90JA6439⟩
Hoover, F. W., **4**, 102, 313 ⟨63JOC1825⟩, 314 ⟨63JOC2082⟩; **5**, 985 ⟨63JOC2082, 64JOC143⟩, 986 ⟨63JOC1825⟩
Hoover, G. J., **2**, 564 ⟨90JOC3565⟩
Hoover, J. F., **3**, 71 ⟨87S908⟩; **4**, 151 ⟨89OM2973⟩, 585 ⟨88OM2082⟩; **5**, 944 ⟨90JA464⟩
Hooz, J., **1**, 314, 1048, 1049, 1139 ⟨71JA7320⟩, 486 ⟨73JOC2574⟩, 1033 ⟨72CJC1105⟩, 1053 ⟨85TL271⟩; **2**, 7 ⟨88OSC(6)634⟩, 533, 697 ⟨87JOC1347, 68JA5936⟩, 1083 ⟨72CJC1005⟩; **3**, 48 ⟨71OPP227⟩, 207 ⟨80SC667⟩, 490 ⟨71OPP47⟩, 622 ⟨68JA6891⟩; **4**, 832 ⟨69JA6195⟩
Hope, D. B., **2**, 188 ⟨70JCS(C)270⟩, 214 ⟨73E389⟩
Hope, E. G., **4**, 288 ⟨84JCS(P2)429⟩
Hope, H., **1**, 530 ⟨85JOC5215⟩, 835 ⟨85JA443⟩; **2**, 475 ⟨91JA3597⟩; **4**, 453 ⟨84IC2550⟩; **6**, 168 ⟨83IC3421⟩, 194, 195 ⟨85JA4337⟩
Hope, P., **5**, 334 ⟨64JCS5679⟩
Hopf, B., **5**, 554 ⟨78ZC90⟩
Hopf, G., **2**, 369 ⟨73ZAAC(401)179⟩
Hopf, H., **1**, 496 ⟨91CB2075⟩, 953 ⟨B-80MI 120-01⟩, 955, 958 ⟨87TL2697⟩, 1069 ⟨89CB377⟩; **2**, 940 ⟨88TL609⟩
Hopff, H., **4**, 30 ⟨63HCA1818⟩; **5**, 298 ⟨60HCA1473⟩; **6**, 443 ⟨49AG183⟩
Höpfner, U., **3**, 352 ⟨79H(12)1153⟩
Hopkins, A. R., **2**, 194 ⟨85JOC5592⟩
Hopkins, K. H., **5**, 718 ⟨73JCS(P1)2614⟩
Hopkins, P. B., **1**, 572 ⟨78JOC1208⟩, 585 ⟨82TL1979⟩, 697, 698, 703 ⟨86TL2095⟩, 697, 698, 703 ⟨87JOC3759⟩, 833 ⟨88TL4229⟩; **2**, 540 ⟨82TL4871⟩, 642 ⟨87JA5553⟩; **5**, 173 ⟨79JA7131⟩; **6**, 540 ⟨82TL1979⟩
Hopkins, R. B., **3**, 113 ⟨88TL2885⟩
Hopkins, R. G., **1**, 799 ⟨69CC1069, 70JCS(B)1410⟩
Hopkins, R. T., **2**, 258 ⟨56JA4447⟩
Hopkins, T. E., **3**, 194 ⟨88T1603⟩; **4**, 206 ⟨89TL3019⟩, 703 ⟨78IC625⟩
Hopkins, T. R., **2**, 255 ⟨65JA4447⟩
Hopkinson, A. C., **1**, 106 ⟨86TL4853⟩, 1109 ⟨71JCP(55)3835⟩, 1143 ⟨73TCA183⟩
Hopkinson, M. J., **5**, 1152 ⟨76CPL(42)460⟩; **6**, 272 ⟨76JFC(7)501⟩, 679 ⟨76CC513⟩
Hoppe, B., **5**, 1012, 1015 ⟨75TL695⟩
Hoppe, D., **1**, 138 ⟨90AG(E)1422, 90AG(E)1424⟩, 138 ⟨92AG(E)1505, 92TL5323⟩, 974 ⟨87T2457⟩; **2**, 997 ⟨84AG(E)932⟩, 997, 999 ⟨89AG(E)69, 91AG(E)321⟩, 1029 ⟨87TL785⟩; **3**, 36 ⟨81AG(E)127⟩, 213 ⟨89TL2919⟩, 484 ⟨69LA(730)1⟩, 697, 704, 713 ⟨74AG(E)789⟩, 704 ⟨71AG(E)331⟩, 704, 707, 719 ⟨75CB1580⟩, 704, 717 ⟨73LA611⟩, 713, 717 ⟨72LA(766)116⟩; **4**, 360, 382 ⟨90AG(E)1424, 92SL764⟩, 381 ⟨84AG(E)932, 81AG(E)127⟩, 382 ⟨82S1045, 85CB2822⟩, 382 ⟨86AG(E)160, 92T8377⟩, 382, 384 ⟨87TL5149, 84AG(E)239⟩, 384 ⟨85TL411⟩, 392 ⟨80LA1765⟩, 538 ⟨76CB3047, 76CB3062, 76TL609⟩, 841 ⟨89TL2915⟩, 938, 957 ⟨80AG(E)303⟩; **5**, 592, 609 ⟨77TL2145⟩, 634 ⟨76TL609⟩, 1046 ⟨73AG(E)923⟩; **6**, 326 ⟨75AG449, 80LA1751⟩, 447 ⟨92S1216⟩, 496 ⟨77AG(E)462⟩
Hoppe, H. W., **5**, 223 ⟨87T2045⟩
Hoppe, I., **1**, 683 ⟨75AG(E)765⟩; **2**, 1023 ⟨79LA219⟩; **3**, 543 ⟨75AG(E)765⟩, 580, 582 ⟨74LA1655⟩, 696 ⟨76LA183⟩, 700 ⟨81LA103⟩, 704 ⟨84LA608⟩, 706, 714 ⟨75JPR137⟩, 706, 714, 717 ⟨84LA600⟩; **4**, 462 ⟨85AG(E)1067⟩, 538 ⟨76TL609⟩, 1004 ⟨85LA555⟩; **5**, 162 ⟨75AG(E)765⟩, 162 ⟨79LA219⟩, 634 ⟨76TL609⟩; **6**, 158 ⟨85LA555⟩
Hoppe, M., **2**, 175 ⟨58CB2130⟩
Hoppe, W., **3**, 465 ⟨63CB1712⟩
Hopper, S. P., **2**, 622 ⟨78JOC3076⟩; **3**, 323 ⟨74JOM(67)341⟩; **4**, 35, 137 ⟨73JOM(51)77⟩, 138 ⟨70JOM(23)99⟩; **6**, 38 ⟨74JOM(67)341⟩, 179, 709, 710 ⟨80JOM(186)309⟩
Hopton, D., **5**, 264 ⟨85JCS(P1)461⟩, 280 ⟨84TL4825⟩
Hopton, F., **4**, 979 ⟨71JHC241⟩
Hopwood, Jr., S. L., **4**, 761, 763 ⟨56JA1685⟩
Horaguchi, T., **4**, 186 ⟨92BCJ304⟩
Horák, J., **2**, 440 ⟨60CCC2191⟩
Horak, M., **1**, 812 ⟨59CCC2371⟩
Horak, R. M., **2**, 39 ⟨93CC119⟩
Horan, N. R., **1**, 316 ⟨93JA6614⟩; **2**, 562 ⟨93JA6614⟩
Horchler v. Locquenghien, K., **3**, 485 ⟨91AG1174, 91AG(E)1154⟩; **6**, 669 ⟨91AG(E)1154⟩
Hordvik, A., **2**, 1008 ⟨66ACS1874⟩; **4**, 863 ⟨80ACS(A)577⟩
Horeau, A., **2**, 561 ⟨71BSF3571⟩; **5**, 290 ⟨65BSF1454⟩
Horeczy, J., **3**, 459 ⟨48JA2012⟩
Horensky, S., **4**, 32 ⟨63JOC2494⟩
Horgan, A. G., **1**, 781 ⟨81JA1232⟩, 834 ⟨81JA1757⟩
Horhold, H.-H., **2**, 245, 250 ⟨68CB3567⟩
Hori, A., **5**, 550 ⟨74JOM(82)223⟩, 609 ⟨77AG(E)787⟩
Hori, I., **3**, 54 ⟨75S727⟩; **4**, 197 ⟨74S705⟩
Hori, K., **2**, 393 ⟨90JOC2437⟩; **4**, 52, 101, 109 ⟨88JCS(P1)1149⟩; **5**, 301 ⟨87JA3402⟩; **6**, 43, 47, 54, 57 ⟨86JOC955⟩, 563, 565 ⟨82JMC557⟩
Hori, M., **1**, 108, 115 ⟨86CPB3599⟩, 181 ⟨75CPB1917⟩, 824 ⟨88CC370⟩, 943 ⟨90JCS(P1)3017⟩; **2**, 143 ⟨90TL3027⟩, 278, 279, 292 ⟨94JOC1011⟩, 288 ⟨87JOC1397⟩, 288 ⟨92JCR(S)393⟩; **3**, 356 ⟨74CL1101⟩, 376 ⟨78TL251, 80JOC2468⟩, 376 ⟨87YGK232⟩, 635 ⟨88CPB1698⟩, 637 ⟨93SC2481⟩; **4**, 232 ⟨84TL757⟩, 288 ⟨91TL105⟩; **5**, 1122 ⟨85JCS(P1)2333, 85TL1321, 87JCS(P1)2531⟩
Hori, S., **3**, 304 ⟨70BCJ3496⟩
Hori, T., **1**, 601, 608, 781 ⟨78JOC1689⟩; **2**, 21 ⟨79JOC4204, 79JOC4208⟩, 281 ⟨78JOC1689⟩, 298 ⟨76JA269⟩, 694 ⟨90CPB2637⟩
Hori, Y., **1**, 1036 ⟨85JOC565, 86BCJ3201⟩; **2**, 676 ⟨86TL2125⟩; **3**, 162 ⟨87JOC2230⟩; **4**, 21 ⟨78CL73⟩; **6**, 552, 554 ⟨86CE311⟩, 552, 554 ⟨87NKK1408⟩
Horibe, I., **1**, 428, 429 ⟨89JOC5308⟩, 813 ⟨70JCS(C)2697, 70JCS(C)2704⟩
Horie, Y., **3**, 626 ⟨89OPP354⟩; **4**, 79, 781 ⟨87CPB3959⟩
Horigome, M., **5**, 1127 ⟨90TL1287⟩
Horiguchi, H., **4**, 205 ⟨76CPB102⟩
Horiguchi, Y., **1**, 86 ⟨89JA6257⟩, 313 ⟨86TL4029⟩; **2**, 702 ⟨83TL3341⟩; **3**, 156 ⟨89TL3323⟩, 220 ⟨89JA8277⟩; **4**, 1064 ⟨87TL1299⟩
Horii, T., **5**, 247 ⟨71JOC3677, 75BCJ2993⟩, 540 ⟨81BCJ790⟩, 1111 ⟨83MI 511-01⟩
Horii, Z.-I., **4**, 303 ⟨61JOC2273⟩
Horiie, T., **3**, 616, 627, 633 ⟨83T967⟩
Horikawa, H., **4**, 310 ⟨79JOC391⟩, 311, 327 ⟨79BCJ826⟩, 424 ⟨82JOC1706⟩
Horikawa, M., **3**, 311 ⟨93TL331⟩; **5**, 251 ⟨80SC305⟩
Horike, A., **4**, 752 ⟨67TL1123⟩
Horiki, K., **4**, 416 ⟨76H(5)203⟩; **5**, 235 ⟨77SC251⟩, 1111 ⟨90H(30)367⟩
Horino, H., **2**, 964 ⟨81JOC4416⟩
Horisaki, M., **2**, 457 ⟨77S179⟩
Horita, A., **2**, 373, 374, 379, 381 ⟨59JA2597⟩
Horita, K., **2**, 90 ⟨88TL4139⟩
Horito, S., **6**, 72 ⟨80BCJ3687, 81CL375⟩
Horiuchi, C. A., **2**, 33 ⟨81S312⟩, 33, 34 ⟨84CL1509, 88CL31⟩, 34 ⟨94BCJ271⟩, 631 ⟨84BCJ2691⟩; **3**, 155 ⟨81S312⟩, 155 ⟨88CL31⟩
Horiuchi, H., **4**, 420 ⟨71BCJ3445⟩
Horiuchi, N., **2**, 920, 946 ⟨88JOC3539⟩
Horiuti, I., **1**, 71 ⟨34MI 103-01⟩
Hörlein, R., **4**, 709 ⟨87JOM(321)257⟩; **5**, 1097 ⟨85CB340⟩
Horler, H., **1**, 325, 331, 920 ⟨86CB444⟩

Horler, M., **3**, 621 ⟨85T4025⟩
Hormi, O., **5**, 55, 99 ⟨87OS(66)173⟩
Hormi, O. E. O., **2**, 672 ⟨88JOC880⟩, 673 ⟨93TL6463⟩; **4**, 553 ⟨90SC1865⟩
Hormuth, P. B., **6**, 622 ⟨68AG(E)299⟩
Horn, D. E., **1**, 1114 ⟨70JA3789⟩
Horn, E., **4**, 716 ⟨93OM315⟩, 723 ⟨93OM2227⟩
Horn, H., **5**, 581, 597, 599 ⟨63CB1551⟩
Horn, H. G., **4**, 802 ⟨78ZN(B)1422⟩
Horn, K. A., **1**, 248 ⟨82TL263⟩
Horn, K., **6**, 362 ⟨79CB1189⟩
Horn, M., **5**, 924 ⟨83TL1477⟩
Horn, P., **3**, 637 ⟨65AG(E)951⟩; **6**, 104 ⟨68CB41⟩
Horn, U., **1**, 1082 ⟨72HCA1276⟩; **3**, 26, 147 ⟨72HCA1276⟩, 108 ⟨76HCA211⟩
Hornaman, E. C., **1**, 632 ⟨74JOC2302⟩
Hornbach, P., **2**, 903 ⟨89ZN(B)1179⟩; **6**, 188, 389, 401 ⟨89CB1881⟩, 389, 398, 401 ⟨86AG(E)1112⟩
Hornback, J. M., **2**, 302 ⟨89TL5853⟩
Hornback, W. J., **1**, 1027 ⟨89JOC5828⟩; **4**, 951 ⟨89JOC5828⟩
Hornbuckle, S. F., **1**, 400, 820 ⟨91CRV263⟩; **3**, 319 ⟨91CRV49⟩, 320 ⟨89JOC817⟩, 323 ⟨91JOC3271⟩, 678, 679 ⟨91CRV263⟩
Horne, K., **4**, 384 ⟨87TL3911, 89T495⟩
Horne, S., **1**, 1042 ⟨88T5451⟩
Horne, W., **4**, 114 ⟨81JCS(P2)1138⟩
Horner, G., **1**, 327 ⟨91JA9701⟩
Horner, J. H., **1**, 950 ⟨93TL5523⟩
Horner, L., **1**, 2, 6, 8, 10 ⟨59CB1700⟩, 76 ⟨68AG(E)942⟩, 395 ⟨55CB934⟩, 639, 752, 753 ⟨58CB61⟩, 639, 752, 753 ⟨59CB2499, 64TL2467⟩; **2**, 27 ⟨59AG349⟩, 188 ⟨65CB1715⟩, 193 ⟨58AC(P)108⟩, 254 ⟨81PS(11)349⟩, 388 ⟨61CB712⟩, 407, 851 ⟨80PS(8)221⟩, 428, 430, 431, 441, 822, 827 ⟨58CB1583⟩, 436 ⟨84PS(19)65⟩, 441 ⟨61GEP1114190, 61LA(646)65⟩, 441 ⟨64PAC225, 65TL1157, 80PS(8)215⟩, 441, 442 ⟨70CB318⟩, 442, 829 ⟨80PS(8)225⟩, 454 ⟨61CB1317⟩, 454, 844 ⟨58CB64⟩, 461 ⟨64TL175⟩, 462, 821, 827 ⟨59CB2088⟩, 490 ⟨75PS(5)265⟩, 764 ⟨57LA(606)24⟩, 769 ⟨53CB1073⟩, 822, 823 ⟨68CB2899⟩, 824, 825 ⟨83PS(15)165⟩, 834, 835 ⟨61CB2122⟩, 860 ⟨59LA(627)142⟩, 881 ⟨75MI 217-02⟩, 886 ⟨64TL3271⟩; **3**, 328 ⟨59LA(627)142⟩, 475 ⟨71LA(747)21⟩, 480 ⟨61TL9⟩, 499, 500 ⟨58CB437⟩, 532 ⟨51LA(573)17, 52CB225, 55CB934⟩, 551 ⟨58LA(614)4⟩; **4**, 220 ⟨57LA(602)135⟩, 343 ⟨70CB2984⟩, 427 ⟨78LA1505⟩, 490 ⟨61TL9⟩, 575 ⟨66TL3315⟩, 575, 576, 577 ⟨59CB2499, 61CB1987⟩, 576 ⟨58CB61, 60CB1371, 61CB1323, 62CB581⟩, 852 ⟨58LA(614)4⟩; **5**, 90 ⟨55LA(591)117⟩, 366 ⟨54LA(587)93⟩, 921 ⟨58CB437⟩, 975 ⟨59LA(627)142⟩, 1128 ⟨85CB676⟩; **6**, 698, 708 ⟨58CB437, 59LA(627)142⟩
Hörnfeldt, A.-B., **1**, 480 ⟨93JOM(460)127⟩; **5**, 595 ⟨93S483⟩; **6**, 566 ⟨93S483⟩
Horng, A., **1**, 163 ⟨73S37⟩, 658 ⟨74JA316⟩, 658, 963 ⟨70JA1427⟩; **2**, 900 ⟨70JA1427, 73S672⟩, 1005 ⟨73S37⟩; **4**, 631 ⟨73S37⟩; **6**, 198 ⟨73S37⟩
Hörnig, H., **6**, 723 ⟨86CB2900, 88JOM(355)243⟩
Hornig, P., **2**, 415, 416, 489, 503 ⟨66JOM(6)259⟩; **5**, 777, 780 ⟨68CB2533⟩
Horning, C. H., **1**, 881 ⟨45JA2110⟩
Horning, D. E., **2**, 647 ⟨73CJC2347⟩
Horning, E. C., **3**, 627 ⟨63OSC(4)144⟩; **5**, 35, 199 ⟨63OSC(4)790⟩, 185 ⟨55OSC(3)165⟩
Horning, M. G., **5**, 185 ⟨55OSC(3)165⟩
Hornung, K.-H., **2**, 26 ⟨62CB2976⟩
Horny, C., **2**, 387 ⟨64CR(259)3737⟩
Hornyak, F. M., **1**, 155 ⟨55JA6026⟩
Horovitch, S., **5**, 8 ⟨88OSC(4)727⟩
Horowitz, N., **3**, 584, 586 ⟨77JA2622⟩
Horrom, B. W., **1**, 881 ⟨60JA2895⟩
Horsch, B., **3**, 628 ⟨90S575, 91TL2605⟩
Horsham, M. A., **1**, 533, 536 ⟨87TL4875, 87TL4879⟩
Horsley, D. B., **2**, 672 ⟨87TL6109⟩
Horspool, W. H., **3**, 413 ⟨86JCR(S)46⟩
Horspool, W. M., **1**, 800 ⟨90CC934⟩, 815 ⟨66CC775⟩; **3**, 542 ⟨71JCS(C)3328⟩; **4**, 1079 ⟨B-92MI 424-02⟩; **5**, 215 ⟨91JCS(P1)223⟩
Horstmann, H., **3**, 34, 37, 164, 169 ⟨77HOU(7/2C)2317⟩; **4**, 880 ⟨76LA1762⟩, 972 ⟨77LA1895⟩; **5**, 630 ⟨60CB671⟩, 630, 631 ⟨60CB663⟩
Horstschäfer, H.-J., **2**, 1077 ⟨68LA(717)1⟩; **6**, 182, 183, 386 ⟨75LA1339⟩
Hortelano, E., **2**, 767 ⟨85JOC2145⟩
Hortmann, A. G., **1**, 421 ⟨63JA4033⟩; **2**, 344 ⟨78JOC2289⟩; **3**, 359 ⟨78JOC3374⟩, 376 ⟨74JA6119⟩, 380 ⟨71JA2471⟩, 638 ⟨88JA4829⟩
Horton, A. D., **3**, 521 ⟨87OM893, 88POL1841⟩; **4**, 675 ⟨88NJC559⟩
Horton, D., **2**, 35 ⟨69JOC3519⟩, 171 ⟨63JOC2992⟩, 171 ⟨69CAR(9)201⟩; **4**, 186 ⟨81H(16)1587⟩, 224, 227 ⟨63MI 405-01⟩
Horton, W. J., **4**, 435, 443 ⟨51JA2769⟩; **5**, 599 ⟨48JA2425⟩
Horvat, R. J., **2**, 820 ⟨52JA562⟩
Horvath, A., **3**, 310 ⟨77JCS(P1)789⟩
Horvath, G., **3**, 310 ⟨77JCS(P1)789⟩
Horváth, I. T., **4**, 688 ⟨81JOM(206)119⟩
Horváth, K., **1**, 30 ⟨79CC300, 80JCS(P1)2554⟩; **2**, 11 ⟨79S437⟩; **5**, 219, 833 ⟨88CB723⟩
Horváth, L., **6**, 449 ⟨81MI 614-02⟩
Horvath, R. F., **1**, 161, 162 ⟨89JOC317⟩; **4**, 520 ⟨89JOC317⟩
Horwell, D. C., **2**, 747 ⟨75JCS(P1)1237⟩; **4**, 404, 418, 419, 424, 427 ⟨91JCS(P1)1693⟩; **5**, 289 ⟨75JCS(P1)1237⟩
Horwitz, C. P., **6**, 338 ⟨85JA8136, 85JA8147⟩, 521 ⟨84AOC(23)219⟩
Horwitz, J. P., **3**, 218 ⟨53JA3268⟩
Hosaka, K., **4**, 408 ⟨89H(29)865⟩
Hosaka, S., **5**, 158 ⟨64TL605⟩
Hosaka, T., **3**, 363 ⟨90H(30)905⟩, 365 ⟨89CPB1999⟩
Hosay, T., **5**, 337 ⟨90TL873⟩
Höschele, G., **4**, 235 ⟨54CB784⟩
Hose, D. R. J., **2**, 362 ⟨93TA2159⟩
Hosekama, S., **5**, 129 ⟨94JOC5532⟩
Hoshi, K., **1**, 578 ⟨87CC490⟩
Hoshi, M., **1**, 1018 ⟨83BCJ2855, 84CL1029⟩; **2**, 20 ⟨79CL961⟩, 43 ⟨91CC205⟩, 273 ⟨85CC715⟩, 922 ⟨86BCJ659, 92BCJ685⟩; **3**, 416, 645 ⟨91CC568⟩, 622 ⟨84CC398⟩, 622, 645 ⟨89CC266⟩; **4**, 815 ⟨88BSJ3764⟩, 1056, 1061 ⟨86BCJ659⟩
Hoshino, M., **1**, 228 ⟨91JOC6483⟩, 899 ⟨90TL2623⟩; **2**, 478 ⟨84T4553⟩, 713 ⟨83CL249⟩; **3**, 164 ⟨85TL1981⟩, 177 ⟨88TL1399⟩, 251 ⟨87CL939⟩, 255 ⟨77CL287⟩, 352, 366 ⟨93TL115⟩, 363 ⟨76JA6218, 90JOC2421⟩, 363 ⟨92CC7⟩, 392 ⟨92CL2289⟩, 547 ⟨81BCJ2845⟩, 668 ⟨80CC791⟩; **4**, 850 ⟨77CL287, 80BCJ1661⟩, 850 ⟨81CC565⟩; **5**, 554, 563, 605 ⟨93BCJ623, 93TL115⟩; **6**, 113 ⟨76BCJ3567, 80JOC2024⟩, 113, 316 ⟨81JCS(P1)618⟩
Hoshino, O., **1**, 584 ⟨92JOC7285⟩; **6**, 449 ⟨87SC1887⟩
Hoshino, S., **3**, 228 ⟨80TL3209⟩; **4**, 400 ⟨80TL3209⟩
Hoshino, T., **3**, 684 ⟨60JA5339⟩
Hoshino, Y., **1**, 554 ⟨87BCJ4468⟩; **4**, 1044 ⟨82CL1663⟩
Hosi, K., **6**, 564 ⟨81S908⟩
Hosking, J. W., **5**, 983 ⟨67JA4809⟩
Hoskins, B. F., **6**, 332, 351 ⟨83IC3700⟩
Hoskins, E. W., **4**, 54, 56 ⟨81CJC3055⟩
Hoskins, S. V., **6**, 717, 718, 722 ⟨82JOM(234)C9⟩, 718 ⟨83CC719⟩, 718, 727 ⟨84JOM(269)C55⟩
Hosmane, N. S., **4**, 656 ⟨77JCR(S)116⟩; **5**, 690 ⟨88JOC5309⟩; **6**, 394 ⟨77JCR(S)116⟩, 406 ⟨90AOC(30)99⟩

Hosmane, R. S., **3**, 673 ⟨85JOC5892⟩; **5**, 351 ⟨90S1095, 83JOC3⟩, 690 ⟨84JOC1212, 88JOC382⟩, 690 ⟨88JOC5309, 88S242⟩
Hosoi, K., **2**, 709 ⟨68BCJ2742⟩
Hosoi, S., **4**, 90 ⟨85CPB1745, 91CPB1365⟩
Hosokawa, H., **5**, 221 ⟨88BCJ4467⟩
Hosokawa, T., **3**, 487 ⟨79CC450, 86OM356⟩; **4**, 193 ⟨90ACR49⟩, 193 ⟨90BCJ166⟩
Hosomi, A., **1**, 253, 486 ⟨87JOC2947⟩, 286 ⟨92CL1073⟩, 314 ⟨77JA1673, 78TL3043⟩, 366 ⟨93TL5943⟩, 438 ⟨93JOC6541⟩, 468 ⟨86TL2881⟩; **2**, 77 ⟨87CC1517, 89JOC3254⟩, 78 ⟨90JOC2415⟩, 139 ⟨91SL557⟩, 300 ⟨88CPB2342⟩, 941 ⟨79S740⟩; **3**, 140 ⟨77JA1673⟩, 140 ⟨79CL245⟩, 186 ⟨82TL547⟩, 187 ⟨91TL2371⟩, 213 ⟨79TL429⟩, 326 ⟨93TL5943⟩, 540 ⟨79S724⟩; **4**, 181 ⟨84JOC2808⟩, 231 ⟨87CC1442⟩, 368 ⟨91SL557⟩, 368 ⟨92TL85⟩, 616 ⟨68CC930⟩, 650 ⟨87CPB1663⟩, 742 ⟨83TL4727⟩, 900 ⟨88JHC959, 90JOC5308⟩; **5**, 378 ⟨66BCJ2263⟩, 419 ⟨79JOM(175)C1⟩; **6**, 634 ⟨91TL5987⟩
Hosomi, K., **4**, 559 ⟨91CL2213⟩
Hosono, K., **5**, 652 ⟨90TL3895⟩
Hosoya, K., **1**, 178 ⟨86BCJ3097⟩
Hosoya, T., **1**, 1071 ⟨91TL6735⟩
Höss, E., **5**, 22 ⟨92S1145⟩
Hossain, M. E., **2**, 564 ⟨86OM1296⟩
Hossain, M. M., **5**, 942 ⟨93JOM(445)143⟩
Hossain, N., **1**, 946 ⟨93T9079⟩
Hosse, W.-H. J., **3**, 430 ⟨76MI 311-01⟩
Hosseini, S., **3**, 27 ⟨90TL179⟩
Hossfeld, R. L., **3**, 616 ⟨42JOC164⟩
Hossner, F., **2**, 684 ⟨88JOM(347)365⟩
Hostetler, M. J., **4**, 710 ⟨92ICA377, 93JA2743⟩, 710 ⟨93OM65⟩
Hota, N. K., **2**, 989 ⟨69JA4575⟩, 1060, 1061, 1076 ⟨71CJC2706⟩; **6**, 60 ⟨64CJC1123⟩
Hotaka, T., **5**, 785, 1130 ⟨70JOC4238⟩
Hotema, M. R., **5**, 835 ⟨92JOC4648⟩
Hotlz, J., **2**, 555, 556 ⟨17CB262⟩
Hotta, Y., **1**, 670, 678, 767 ⟨78TL2417⟩, 678 ⟨80BCJ1698⟩
Hotten, T. M., **3**, 618 ⟨72CC1226⟩
Hotzel, H. H., **4**, 297 ⟨67AP(300)241⟩; **5**, 606 ⟨67AP241⟩
Hou, G., **4**, 420 ⟨92PHA754⟩
Hou, K.-C., **4**, 184 ⟨86OS(64)138⟩, 315 ⟨84S683⟩
Houalla, D., **2**, 411 ⟨65BSF2368⟩, 693 ⟨58MI 213-01⟩
Hou, S.-S., **1**, 950 ⟨91JCR(S)216⟩
Hou, W., **2**, 43 ⟨88TL2967⟩
Houben, J., **3**, 686 ⟨13CB4001⟩; **5**, 546 ⟨07CB1303⟩, 698, 813 ⟨13CB3616⟩, 830 ⟨36CB2352⟩; **6**, 84 ⟨11CB3235, 12CB2942⟩, 625 ⟨13CB2447, 36CB2358⟩
Houchigai, H., **1**, 189 ⟨91TL4311⟩, 189 ⟨93BCJ848⟩
Houdier, S., **4**, 224 ⟨93CAR(248)377⟩
Houge, C., **3**, 529 ⟨81JA4616⟩, 602 ⟨82JA2920⟩
Hough, L., **6**, 561 ⟨91CAR(216)271⟩
Hough, R. R., **2**, 208 ⟨57JA5000⟩
Houghten, R. A., **6**, 651 ⟨79JOC4536⟩
Houghton, E., **3**, 357 ⟨76JOC818⟩; **6**, 115 ⟨76JOC813⟩
Houghton, P. G., **2**, 783 ⟨88S379⟩; **5**, 1073 ⟨85JCS(P1)1471⟩
Houghton, R. P., **2**, 421 ⟨68JCS(C)1704⟩; **3**, 14 ⟨80JCS(P1)756⟩; **6**, 79 ⟨80JCS(P1)756⟩
Houk, A. L., **2**, 560 ⟨32JA3714⟩
Houk, K. N., **1**, 130 ⟨86JA2659⟩, 135, 159 ⟨84JA6467⟩, 323 ⟨86JOC2874⟩, 336 ⟨87JOC959⟩, 415 ⟨88JA2314⟩, 816, 817 ⟨90JOC1421⟩, 927 ⟨87JA2195⟩; **2**, 41 ⟨84T2257⟩, 58, 521 ⟨86JA7404⟩, 999, 1006 ⟨84JA6467⟩; **3**, 313 ⟨80JA1504⟩, 327 ⟨73JA7287⟩; **4**, 378, 391, 816 ⟨84JA6467⟩; **5**, 866 ⟨81JA6164⟩; **6**, 55 ⟨81JA6164⟩
Houlden, S. A., **3**, 532 ⟨69T2121⟩
Houlihan, F., **6**, 477 ⟨85CJC153⟩
Houlihan, W. J., **2**, 382, 384 ⟨69JOC2720⟩; **4**, 415 ⟨69JOC2720⟩, 416 ⟨75JMC177, 75JMC182⟩
Hoult, D. A., **4**, 802 ⟨93JOC1531⟩
Hounshell, W. D., **2**, 1060 ⟨76MI 222-01⟩; **3**, 376 ⟨76JA3615⟩
Houpis, I. N., **1**, 214 ⟨85TL3643⟩, 1029 ⟨88JA649⟩; **2**, 679 ⟨88JA649⟩
Houri, A. F., **1**, 316 ⟨91JA8950, 92JA6692⟩, 316 ⟨93JA6614⟩; **2**, 561 ⟨91JA8518⟩, 562 ⟨92JA6692, 93JA6614⟩
House, H. O., **1**, 104 ⟨67RCP99⟩, 118, 119, 469, 473 ⟨69JA4871⟩, 130 ⟨78JOC700⟩, 130, 871, 872, 881 ⟨65JOC2502⟩, 131 ⟨78JOC2153⟩, 275 ⟨72JA5495⟩, 307 ⟨66JOC646⟩, 311 ⟨66JOC3128, 69JOC3615⟩, 313 ⟨76ACR59⟩, 384, 388 ⟨59JA4733⟩, 390 ⟨55JA5083⟩, 457, 508 ⟨74JOC747⟩, 725 ⟨61JOC4278⟩, 733 ⟨64JOC3327⟩, 871, 872, 881 ⟨65JOC1341⟩, 871, 881 ⟨63JOC3362⟩, 871, 881 ⟨68JOC935⟩, 871, 881 ⟨68JOC943⟩, 871, 881 ⟨69JOC2324⟩, 871 ⟨B-72MI 119-01⟩, 872, 873 ⟨71JOC2361, 62JOC4146⟩, 881 ⟨B-72MI 119-02⟩, 882 ⟨65JOC2513⟩, 884 ⟨B-72MI 119-03⟩, 887 ⟨B-72MI 119-04⟩, 888 ⟨B-72MI 119-06⟩, 889 ⟨B-72MI 119-05⟩, 1048 ⟨69JOC3616⟩, 1093 ⟨75JOC1460⟩, 1142 ⟨73JOC3893⟩; **2**, 420 ⟨78JOC700⟩, 587 ⟨69JOC3615, 73JOC3893⟩, 675 ⟨73JOC514⟩, 696 ⟨65JOC2502⟩, 696, 699 ⟨69JOC2324⟩, 702 ⟨73JA3310⟩, 763 ⟨76JOC863⟩, 771 ⟨B-72MI 215-01⟩; **3**, 124 ⟨78JOC700⟩, 208 ⟨65JOC2942⟩, 287 ⟨69JA4871⟩, 471 ⟨68JOC53⟩, 537 ⟨65JOC2502⟩, 620 ⟨57OR(9)107⟩; **4**, 824 ⟨B-72MI 419-01⟩; **5**, 46 ⟨57OR(9)107⟩, 98 ⟨B-72MI 502-01⟩
House, S., **4**, 960 ⟨93JCS(P1)67⟩
Housecroft, C. E., **5**, 943 ⟨B-90MI 523-02⟩
Houser, J. J., **1**, 894, 900 ⟨62JA2016⟩, 897, 901, 911 ⟨63JA2991⟩, 901 ⟨63JA2995⟩, 901, 913 ⟨63JA2998⟩
Houser, R. W., **1**, 618, 704 ⟨69JA3870⟩, 707 ⟨71JOC1015⟩; **2**, 158 ⟨70JOC905⟩; **4**, 80 ⟨69JA3870, 71JOC1015⟩
Housley, J. R., **6**, 308 ⟨90MI 610-01⟩
Houston, A. H. J., **1**, 398 ⟨47JCS93⟩
Houtekie, M., **4**, 790 ⟨80TL223⟩
Houtteville, M.-C., **5**, 557 ⟨89T4445⟩
Houwen, O. H., **4**, 410 ⟨78TL4995⟩
Houwen-Claassen, A. A. M., **2**, 743 ⟨83S628⟩; **3**, 37 ⟨83S628⟩
Houwing, H. A., **5**, 566, 618 ⟨76TL143⟩; **6**, 605, 624 ⟨81JHC1127⟩
Hovakeemian, G. H., **1**, 448 ⟨83CC502⟩; **2**, 944 ⟨85JA5670⟩
Höver, H., **1**, 907 ⟨60JA2644⟩, 907, 908, 909 ⟨62JA3168⟩
Höver, W., **4**, 310 ⟨70CB3918⟩, 406 ⟨71CB3354⟩, 979 ⟨71LA(748)59⟩; **5**, 599 ⟨71LA(748)59⟩
Hoveyda, A. D., **2**, 527 ⟨92JA6671⟩
Hoveyda, A. H., **1**, 299 ⟨93CRV1307⟩, 316 ⟨91JA5079, 91JA8950⟩, 316 ⟨92JA6692, 93JA6614⟩, 316 ⟨93JA6997, 93JOC4237⟩, 316 ⟨94JA3123⟩; **2**, 42 ⟨88JA6917⟩, 561 ⟨91JA5079, 91JA8950⟩, 562 ⟨92JA6692, 93JA6614⟩, 562 ⟨93JA6997⟩
Hovius, K., **2**, 250, 364, 365 ⟨72TL181⟩, 691 ⟨74JOC2641⟩; **4**, 235 ⟨72TL2477⟩
Howard, A. B., **4**, 617 ⟨73CC323⟩
Howard, A. S., **3**, 262 ⟨80TL1373⟩
Howard, C. C., **1**, 1024, 1134 ⟨80JCS(P1)852, 79JCS(P1)2954⟩
Howard, E., **3**, 685 ⟨1800MI 320-01⟩
Howard, E. G., **3**, 348 ⟨61JA2589⟩, 348, 349, 351 ⟨65JOC1375⟩, 548 ⟨62USP3035030⟩; **5**, 473, 474, 479, 480, 494 ⟨65JOC1375⟩, 474 ⟨61JA2589⟩; **6**, 49, 253, 528, 529, 532 ⟨65JOC1375⟩, 528 ⟨61JA2589⟩
Howard, H. T., **2**, 105 ⟨45JCS382⟩
Howard, J. A. K., **4**, 675 ⟨85JCS(D)905⟩, 681 ⟨76JCS(D)377⟩, 683 ⟨78JCS(D)403⟩, 691 ⟨75CC451⟩,

692 ⟨80JCS(D)2182⟩, 712 ⟨84JCS(D)2553⟩, 716 ⟨85JCS(D)2009⟩, 717 ⟨86AG(E)165⟩; **5**, 927 ⟨87JOM(325)185⟩; **6**, 195 ⟨78JCS(D)734⟩, 343 ⟨82CC773⟩, 698 ⟨93PS(74)279⟩
Howard, J. C., **5**, 1028 ⟨62JOC3701⟩; **6**, 425 ⟨81JOC1720⟩
Howard, J. W., **5**, 73 ⟨32OSC(1)475⟩
Howard, K. J., **4**, 413 ⟨93TL6329⟩, 414 ⟨93JCS(P1)2391⟩
Howard, K. L., **3**, 444 ⟨55OSC(3)351⟩
Howard, M. H., **3**, 207 ⟨89JOC4654⟩
Howard, P. N., **1**, 391 ⟨89JA1514⟩
Howard, T. J., **1**, 72 ⟨75JOC3073⟩
Howard, T. R., **3**, 508 ⟨81JA7358⟩
Howard, W. L., **4**, 178, 184 ⟨60JOC521⟩, 184 ⟨60JOC525⟩, 184 ⟨64USP3127450⟩, 186 ⟨62USP3024284⟩
Howarth, J., **1**, 192, 194 ⟨92TA1089⟩
Howarth, M., **2**, 520 ⟨63JCS1582⟩, 1016, 1018 ⟨66T2829⟩
Howarth, T. T., **3**, 72 ⟨77CC808⟩
Howd, R. A., **5**, 664 ⟨84JMC1201⟩, 668, 830 ⟨86JMC1689, 81JMC1124⟩
Howden, M. E. H., **1**, 665 ⟨87JCS(P1)475⟩
Howe, B. K., **2**, 138 ⟨49JCS2723⟩
Howe, E. E., **5**, 90 ⟨73OSC(5)373⟩
Howe, R., **1**, 868 ⟨65JA914, 65JA915⟩, 905 ⟨65JA376⟩; **6**, 506 ⟨B-65MI 616-02⟩
Howe, R. K., **1**, 735 ⟨71JA3457⟩; **3**, 335 ⟨85JOC3216⟩, 686 ⟨78JOC3736⟩; **5**, 664, 821 ⟨80JOC3916⟩
Howell, A. R., **1**, 335, 346, 495, 946 ⟨90JCS(P1)2709⟩; **2**, 579 ⟨90JCS(P1)2715⟩
Howell, C. F., **1**, 625 ⟨60JA4663⟩
Howell, J. A. S., **5**, 455 ⟨80JCS(D)1845⟩
Howell, J. L., **5**, 1085 ⟨82IC1616⟩; **6**, 10 ⟨78IJ129⟩, 558, 565 ⟨84CB1707⟩
Howell, R. G., **2**, 35 ⟨64CJC2710⟩
Howell, W. C., **1**, 107 ⟨56JA3484⟩
Howells, R. D., **1**, 178 ⟨77CRV69⟩; **4**, 802 ⟨79JA3689, 80JFC(15)543⟩
Howes, A. J., **6**, 187 ⟨86OM1944⟩
Howes, D. A., **2**, 225 ⟨83CC1051⟩; **6**, 79 ⟨80CC207⟩
Howie, R. A., **3**, 363 ⟨90JOM(385)369⟩; **4**, 138, 369, 372 ⟨92JOM(437)111⟩
Howk, B. W., **1**, 1026 ⟨58JA4607⟩; **3**, 79 ⟨63OSC(4)801⟩; **4**, 192 ⟨58JA4607⟩, 192 ⟨63OSC(4)801⟩
Howse, Jr., R. M., **6**, 635 ⟨88S460⟩
Howson, B., **2**, 319 ⟨93SL675⟩
Howton, D. R., **3**, 638 ⟨45JOC277⟩
Hoxmeier, R. J., **4**, 721 ⟨79IC3453⟩
Hoye, P. A. T., **2**, 119 ⟨63BRP917921⟩; **4**, 551 ⟨60BRP842593⟩
Hoye, Th. R., **6**, 719 ⟨93OM2806⟩
Hoye, T. R., **1**, 550 ⟨91TL3755⟩, 590, 591 ⟨80JOC3549⟩; **2**, 985 ⟨93JA1154⟩; **4**, 204 ⟨81JA5618⟩, 213 ⟨87JOC1351⟩
Hoyer, E., **5**, 633 ⟨81ZAAC(480)21⟩, 657 ⟨89ZAAC(576)203⟩; **6**, 565 ⟨81ZAAC(476)7⟩, 594 ⟨68ZC230⟩
Hoyer, J., **2**, 300, 303 ⟨B-75MI 205-01⟩
Hoyle, J., **2**, 175, 248 ⟨B-91MI 203-01⟩, 189 ⟨B-90MI 203-08⟩, 364, 365 ⟨B-90MI 206-05⟩; **5**, 342 ⟨B-91MI 507-02⟩
Hoyle, K. E., **1**, 411, 807 ⟨41JA1843⟩
Hoyle, W., **5**, 772, 851 ⟨67JCS(C)690⟩
Hoyle, Jr., V. A., **4**, 887 ⟨65TL3589, 71JOC2211⟩
Hoyt, C. H., **6**, 236 ⟨61USP3014071⟩
Hoyt, Jr., E. B., **1**, 620 ⟨68JA429⟩
Hoyte, R. M., **3**, 650 ⟨74JOC2607⟩
Hoz, T., **1**, 173, 587, 688 ⟨78JOC3249⟩
Hozumi, M., **5**, 993 ⟨83JAP53/35194⟩; **6**, 558 ⟨89CPB1249⟩
Hrabák, F., **2**, 376 ⟨69CCC4010⟩
Hrabie, J. A., **4**, 175 ⟨82JOC824⟩
Hrabie, J., **1**, 84 ⟨79TL1641⟩
Hrdina, M., **5**, 1106 ⟨72CCC3144⟩
Hrdlovič, P., **3**, 297 ⟨67CRV599⟩
Hretani, M., **3**, 90, 97 ⟨93T1421⟩
Hrib, N. J., **1**, 820 ⟨76JA7115⟩
Hrinyo, T., **2**, 21 ⟨80TL177⟩
Hrivnak, J., **5**, 1039 ⟨69CZP131547⟩
Hrnjez, B. J., **5**, 157 ⟨78JOC2952⟩
Hromatka, O., **3**, 638 ⟨42CB131⟩; **4**, 416 ⟨69M469⟩
Hronec, M., **5**, 24, 118 ⟨82MI 502-02⟩
Hrovat, D. A., **1**, 900 ⟨90JA3227⟩; **3**, 679 ⟨84JOC3174⟩
Hrubiec, R. T., **2**, 7, 8 ⟨84JOC431⟩
Hruby, V. J., **1**, 723 ⟨64JA5041⟩; **2**, 29 ⟨93TA2315, 93TL2561⟩, 29 ⟨94T2391⟩; **3**, 493 ⟨67JA4996⟩; **5**, 300 ⟨71JOC1305⟩
Hrutfiord, B. F., **1**, 885 ⟨62JOC4152⟩
Hsia, D. Y., **3**, 613 ⟨78JOC1017⟩
Hsi, J. D., **1**, 738, 740, 742 ⟨89JOC3229⟩
Hsiao, C.-N., **1**, 615, 700 ⟨82TL1963⟩, 616, 654 ⟨84TL1219⟩, 616, 654 ⟨88JOC2688⟩, 653 ⟨82TL3455⟩; **5**, 329 ⟨85TL4855⟩; **6**, 446 ⟨92TL2629⟩
Hsiao, S.-N., **4**, 306 ⟨93OSC(8)3⟩
Hsiao, Y., **1**, 77 ⟨86JA7117⟩
Hsieh, H.-H., **4**, 74, 75, 76, 78 ⟨73JOC17⟩
Hsieh, S. L., **3**, 424 ⟨89JOC3820⟩
Hsing, N. T., **1**, 738 ⟨81MI 116-01⟩
Hsu, A. C. T., **6**, 611 ⟨87EUP232012⟩
Hsu, D. P., **2**, 927 ⟨93JA10394⟩
Hsu, E.-P. T., **4**, 12 ⟨58JA6233⟩
Hsu, E. T., **1**, 1121 ⟨71TL1385⟩
Hsu, G. J.-H., **4**, 929, 930, 949, 951, 952 ⟨82OM830⟩; **5**, 416 ⟨82OM830⟩
Hsu, H.-B., **1**, 399 ⟨79JOC2348⟩
Hsu, H. C., **2**, 4, 30 ⟨81JOC3113⟩, 628 ⟨81SC247⟩
Hsu, I. H. C., **3**, 302 ⟨71JOC3191⟩
Hsu, J., **1**, 967 ⟨90TL3699⟩
Hsu, L., **4**, 37 ⟨91JOC1663⟩
Hsu, L.-Y., **5**, 277 ⟨87JOC4665⟩; **6**, 345 ⟨84CC392⟩
Hsü, S. K., **1**, 853 ⟨36JCS623, 38JCS78⟩
Hsu, S.-Y., **2**, 91 ⟨86JOC2505⟩, 684 ⟨86JOC2137⟩; **3**, 215 ⟨85JCS(P1)267⟩
Hsu, W.-J., **6**, 450 ⟨89CJC2153⟩
Hsu, W.-L., **6**, 345 ⟨84CC392⟩
Hsu, Y. F., **2**, 165 ⟨73JA4064⟩, 165 ⟨73JA8191⟩
Hsueh, C. M., **1**, 772 ⟨43OSC(2)416⟩; **5**, 33 ⟨43OSC(2)416⟩
Hsu Lee, D.,**1**, 523 ⟨66JOC2376⟩
Hu, B., **1**, 801 ⟨92TL453⟩; **2**, 2 ⟨93TL1681⟩
Hu, C., **2**, 767 ⟨84TL1337, 85JOC2205⟩; **4**, 507 ⟨84JOM(268)31⟩
Hu, C.-M., **4**, 763 ⟨89JFC(42)69⟩
Hu, D., **4**, 549 ⟨93PS(77)5⟩; **6**, 154 ⟨89AG929⟩
Hu, H., **3**, 4, 89, 99, 116 ⟨92SC1217⟩; **4**, 198 ⟨82TL1537⟩
Hu, J., **2**, 43 ⟨88TL2967⟩; **5**, 1020 ⟨90JA4183⟩
Hu, J.-S., **4**, 117, 118 ⟨87TL6317⟩
Hu, L., **1**, 816 ⟨87TL4993⟩
Hu, N. X., **1**, 594, 611, 621 ⟨89JOC4391⟩, 611 ⟨89JCS(P1)1775⟩; **3**, 114 ⟨86TL6099⟩; **5**, 210 ⟨87TL1281, 89JOC4391⟩, 1078 ⟨86TL6099⟩
Hu, S., **2**, 897 ⟨82IC3731⟩
Hu, S. S., **4**, 587 ⟨86JOC5498⟩; **5**, 924 ⟨91OM2752, 91OM2758⟩
Hu, Y., **3**, 4, 89, 99, 116 ⟨92SC1491⟩; **5**, 252 ⟨90MI 505-01⟩
Hu, Y.-L., **5**, 183, 195 ⟨90JCR(S)84⟩, 253 ⟨90JCR(S)230⟩
Hu, Z., **2**, 55 ⟨88TL4113⟩; **6**, 565 ⟨86MI 617-05⟩
Hua, D. H., **1**, 17, 530 ⟨86JA3835⟩, 618, 704 ⟨80JOC1719⟩, 694 ⟨90JOC2128⟩; **2**, 362 ⟨91JOC4⟩, 743 ⟨91S970⟩; **3**, 437 ⟨91JOC4⟩; **4**, 94 ⟨82JA6816⟩, 217 ⟨84TL3⟩; **5**, 179 ⟨82JA6818⟩, 375 ⟨81TL3455⟩, 720 ⟨90JOC3682⟩, 867 ⟨80JOC1717⟩
Huan, Z., **3**, 323 ⟨83TL2829⟩

Huang, B., **6**, 32 ⟨84MI 601-01⟩
Huang, B.-C., **1**, 991 ⟨93OM3971⟩
Huang, B.-G., **5**, 247 ⟨94S699⟩
Huang, C. G., **4**, 49 ⟨93JOC7382⟩
Huang, E. P. Y., **5**, 34 ⟨63OSC(4)93⟩
Huang, F., **1**, 670 ⟨74JA1518⟩; **5**, 458 ⟨74JA1518⟩
Huang, F.-C., **5**, 138 ⟨75JA4144⟩
Huang, F.-J., **3**, 221 ⟨91CC770⟩
Huang, G. T., **1**, 829 ⟨91CC1579⟩
Huang, H.-C., **3**, 60 ⟨89TL5235⟩, 232 ⟨89JOC5126⟩
Huang, H.-N., **2**, 613 ⟨87JA7194⟩; **4**, 44 ⟨88JOC78⟩
Huang, H.-Y., **5**, 386 ⟨91IC1862⟩
Huang, J., **2**, 182 ⟨92TL2657⟩; **5**, 235 ⟨81JA861⟩
Huang, J. C., **1**, 232 ⟨78CJC512⟩
Huang, L., **4**, 318 ⟨86S139⟩
Huang, M. B., **1**, 918 ⟨83JCS(P2)305, 85T6055⟩
Huang, M. G., **4**, 64, 71, 80 ⟨70JOC3002⟩, 229 ⟨68JOC2133⟩
Huang, N. Z., **1**, 482 ⟨82SC795⟩; **3**, 209 ⟨82SC795⟩, 349, 358 ⟨87JOC169⟩; **5**, 336 ⟨87JOC169⟩
Huang, P. Q., **2**, 325 ⟨88T2457⟩; **3**, 422 ⟨88T2457⟩, 616 ⟨87TL547⟩
Huang, P.-T., **2**, 378 ⟨63OSC(4)274⟩, 403 ⟨63OSC(4)66⟩
Huang, R., **3**, 649 ⟨89SC2639⟩
Huang, S., **1**, 939 ⟨93JOC5028⟩
Huang, S. J., **1**, 1121 ⟨71TL1385⟩, 1122 ⟨68JA2432⟩, 1137, 1141 ⟨75JOC124⟩; **2**, 759 ⟨70JOC1204⟩, 1054 ⟨75JOC124⟩
Huang, S. L., **2**, 162 ⟨78JOC4537⟩, 164 ⟨79JOC2510⟩; **3**, 6, 118 ⟨78JOC2480⟩
Huang, S. P., **5**, 199 ⟨89H(28)565⟩
Huang, W., **1**, 504 ⟨83JOC2270⟩; **6**, 32 ⟨84MI 601-01⟩
Huang, W. Y., **4**, 735 ⟨93JCS(P1)795⟩; **6**, 21 ⟨B-94MI 601-01⟩
Huang, X., **1**, 439 ⟨89S42⟩, 623 ⟨87TL801, 88JOC4862⟩; **3**, 4, 88, 116 ⟨82S1091⟩, 176, 647 ⟨88JOC4862⟩, 635 ⟨90SC2291⟩; **4**, 843 ⟨86S967⟩, 900 ⟨91SC1213⟩; **5**, 226 ⟨82SC709⟩
Huang, Y., **1**, 497, 1057 ⟨93JOC1651⟩, 646 ⟨82AOC(20)115⟩, 646, 647 ⟨B-90MI 114-01, 86TL4583⟩, 647 ⟨87JOC3558, 87TL2159, 89JOC2027⟩, 647, 648, 743 ⟨92ACR182⟩, 647, 738 ⟨85TL6447⟩, 647, 741 ⟨87TL2155, 88TL3949⟩, 647, 743 ⟨90JCR(S)388⟩, 648 ⟨86TL2903, 89SC501, 89SC83⟩, 648, 743 ⟨89JOM(366)87⟩, 1047 ⟨91S235⟩; **2**, 843 ⟨84S781⟩; **3**, 79, 274 ⟨85S159⟩, 269, 274 ⟨84S35⟩, 639 ⟨91JOC1381⟩; **4**, 198 ⟨92ACR182⟩, 598 ⟨84TL4425⟩; **5**, 944 ⟨92JOM(431)247⟩
Huang, Y. C. J., **1**, 774 ⟨86JOC113⟩
Huang, Y. T., **1**, 738 ⟨65MI 116-01⟩
Huang, Y.-Z., **1**, 270 ⟨91TL3999⟩, 272 ⟨92JOC774⟩, 272, 743 ⟨90CC85⟩, 488 ⟨93JOC1949⟩, 647 ⟨88LA377⟩, 738 ⟨81MI 116-01⟩, 1079 ⟨81TL5283⟩; **2**, 55 ⟨90TL1185⟩, 73 ⟨88JCS(P1)2855, 88TL1395⟩, 487 ⟨90CC85, 91JOC1381⟩, 487 ⟨92JOC774⟩, 487, 488 ⟨89JOM(366)87⟩, 487, 488 ⟨92ACR182⟩, 900 ⟨92JOC4017⟩, 904 ⟨93JOC1949⟩; **3**, 48 ⟨85TL6447, 88TL3949⟩, 48, 65 ⟨86TL4583⟩, 49 ⟨B-90MI 301-02⟩, 60 ⟨87JOC3558⟩, 196 ⟨88S975⟩, 196, 231 ⟨87TL2131⟩, 196 ⟨B-90MI 304-02⟩, 637 ⟨91JCS(P1)1931⟩, 644 ⟨89SC83⟩, 649 ⟨89SC2639⟩; **4**, 598 ⟨82AOC(20)115⟩
Huang, Z., **4**, 969, 970 ⟨90JPR1035⟩, 972 ⟨87S357⟩, 973 ⟨86CB2208⟩
Huang, Z.-H., **3**, 208 ⟨90S677⟩
Huang, Z.-T., **3**, 265 ⟨90S162⟩; **4**, 899 ⟨90CB541⟩, 900 ⟨90S162⟩
Huang, Z. Z., **2**, 756 ⟨69SC1347⟩
Huang-Minlon, **1**, 11, 18, 19, 41 ⟨49JA3301⟩, 11, 18, 19, 41, 45 ⟨46JA2487⟩
Huang Yao-tseng, **1**, 738 ⟨83MI 116-01⟩, 740 ⟨80MI 116-01⟩, 740, 741 ⟨80MI 116-02⟩
Huba, F., **2**, 1014 ⟨60JOC1626⟩
Hubbard, D. M., **6**, 653 ⟨43JA2390⟩
Hubbard, J. L., **2**, 522 ⟨78S676, 80JOC4067⟩; **3**, 19 ⟨79S701⟩; **4**, 147, 579 ⟨90OM2683⟩, 148, 149 ⟨92JOM(429)369⟩, 364 ⟨78S676, 79JOC467⟩, 364 ⟨79S701, 92HAC223⟩
Hubbard, W. N., **5**, 164 ⟨61JA606⟩
Hubel, W., **2**, 465, 989 ⟨61JA4406⟩; **4**, 681 ⟨59JINC250, 62CB1155⟩, 681 ⟨B-68MI 415-01⟩, 687, 689 ⟨61CB2829⟩
Hubener, G., **4**, 1033 ⟨90JOM(384)155⟩
Huber, B., **3**, 386, 397 ⟨87CC559⟩, 576 ⟨88AG(E)1542⟩; **4**, 550 ⟨86ZN(B)444⟩; **5**, 605, 614, 627 ⟨87CC559⟩, 605, 614, 627 ⟨88CB2095⟩; **6**, 406 ⟨88AG(E)1544⟩
Huber, D., **2**, 753 ⟨88TL635⟩
Huber, E. W., **1**, 440 ⟨91JA7439⟩; **4**, 812 ⟨91JA7439⟩; **5**, 22 ⟨93H(35)315⟩
Huber, F., **2**, 501 ⟨81ZN(B)70⟩, 885 ⟨82ZN(B)815⟩, 887 ⟨86ICA(113)43⟩; **4**, 153 ⟨80ZN(B)1376⟩; **5**, 222 ⟨89ZAAC(574)130, 90ZAAC(582)37⟩
Huber, G., **3**, 374 ⟨73TL4033⟩
Huber, H., **1**, 524 ⟨67JA1753⟩
Huber, J. H.-A., **1**, 841 ⟨73CC223⟩
Huber, J. W., **4**, 467 ⟨77S883⟩
Huber, L. K., **4**, 465 ⟨63JOC2898⟩, 556 ⟨67USP3314900⟩
Huber, M., **5**, 702 ⟨80LA1055⟩; **6**, 522 ⟨76IC1666⟩
Huber, M. K., **4**, 979 ⟨77HCA1781⟩
Huber, P., **1**, 985 ⟨94TL1161⟩; **5**, 214 ⟨86HCA236⟩
Huber, R., **3**, 465 ⟨63CB1712⟩; **4**, 485, 489 ⟨85HCA1730⟩, 621 ⟨72ZAAC(391)219⟩; **5**, 212 ⟨85HCA1730⟩; **6**, 267 ⟨72ZAAC(391)219⟩
Huber, S., **5**, 1041, 1108 ⟨85CC533⟩; **6**, 558 ⟨86HCA1898⟩
Huber, W., **1**, 104 ⟨49HCA489⟩; **2**, 128 ⟨45JA594⟩; **3**, 62 ⟨47HCA1911⟩, 494 ⟨70JA397⟩; **5**, 168 ⟨69JA5675⟩, 168 ⟨70JA397⟩
Huber-Patz, U., **1**, 634 ⟨87TL1985⟩, 705 ⟨84CB966, 86CB1540⟩, 708 ⟨84CB1868⟩, 782 ⟨93LA1133⟩; **6**, 255, 275, 304, 306 ⟨91CB2025⟩
Hubert, A. J., **1**, 549 ⟨78TL1239⟩, 657 ⟨65JCS6669⟩, 795 ⟨63JCS4091⟩, 795, 882 ⟨69S97⟩, 795, 882 ⟨70S405⟩, 956 ⟨68JCS(C)228, 68JCS(C)606⟩
Hubert, C., **4**, 463 ⟨94S51⟩
Hubert, J. C., **4**, 303 ⟨71SC103, 75T1437⟩
Hubert, T. D., **3**, 14, 82, 100, 102 ⟨84JOC2279⟩
Hubert-Habart, M., **4**, 979 ⟨91JHC485⟩; **6**, 620 ⟨84TL1557⟩
Hubertus, G., **1**, 705 ⟨84CB966⟩
Hübler, G., **5**, 1076, 1077, 1083 ⟨78ZN(B)575⟩; **6**, 608 ⟨76ZN(B)1317⟩, 617 ⟨77ZN(B)1003⟩
Hübner, F., **1**, 189, 200 ⟨84CB322⟩
Hübner, J., **5**, 244, 246 ⟨86T5969⟩
Hübner, K., **5**, 597, 599, 614 ⟨62CB926⟩
Hubner, R., **2**, 782 ⟨02CB731⟩
Hübner, T., **6**, 355, 357 ⟨85CC1800⟩
Hubner, W., **5**, 633 ⟨81ZC271⟩
Huboux, A. H., **2**, 923 ⟨90TL3105, 90TL5829⟩; **4**, 661, 702 ⟨90TL5829⟩, 1062 ⟨90TL3105⟩
Hubsch, T., **3**, 246 ⟨89CB1179⟩
Huby, N. J. S., **1**, 812 ⟨92TL6935⟩
Huché, M., **1**, 954 ⟨80T331⟩
Huck, G., **6**, 312 ⟨73ZC465⟩
Hückel, E., **1**, 906 ⟨37MI 119-01, 37MI 119-02⟩
Hückel, W., **1**, 57 ⟨47CB202⟩, 98 ⟨56CB2105⟩
Hucker, H. B., **3**, 295 ⟨53JOC244⟩
Huckestein, M., **3**, 59 ⟨87HCA445⟩
Hückstädt, H., **1**, 993 ⟨92AG(E)90⟩; **5**, 1041 ⟨92AG(E)90⟩
Hudak, E. S., **1**, 1040 ⟨54JOC1580⟩
Hudec, J., **1**, 537 ⟨67JCS(C)1986⟩, 794, 799, 809 ⟨65TL3955⟩, 799 ⟨71CC107, 71CC108⟩, 807, 809 ⟨60TL29⟩, 862 ⟨61PCS117⟩; **2**, 87 ⟨69JCS(C)1605⟩; **4**, 681 ⟨67JCS(C)1862⟩

Hudhomme, P., **4**, 99 ⟨90T5263⟩, 101 ⟨91TA157⟩; **5**, 580 ⟨91BSF760⟩
Hudlický, M., **1**, 107 ⟨B-92MI 104-02⟩, 425, 427, 429, 430 ⟨91COS(8)895⟩, 425, 429 ⟨B-84MI 110-03⟩, 426, 441 ⟨B-76MI 110-01⟩; **1**, 6 ⟨B-76MI 101-01⟩, 21 ⟨B-84MI 101-01⟩; **2**, 2 ⟨78OPP181⟩, 2, 6 ⟨B-83MI 201-01⟩, 10 ⟨87JFC(36)373⟩, 19, 362 ⟨88OR(35)513⟩, 606 ⟨B-83MI 212-01⟩; **3**, 223 ⟨B-84MI 305-01⟩, 467 ⟨80JOC5377⟩; **4**, 706 ⟨88OR(35)513⟩; **6**, 5 ⟨B-76MI 601-01⟩, 9 ⟨B-76MI 601-02⟩, 10 ⟨88OR(35)513⟩, 16 ⟨B-76MI 601-03⟩, 212, 229 ⟨B-92MI 607-01⟩, 225 ⟨B-92MI 607-04⟩
Hudlicky, T., **1**, 86 ⟨87TL5287⟩, 402 ⟨91COS(5)899⟩, 402, 795, 801, 802 ⟨85OR(33)247⟩, 795 ⟨80TL2487, 83JOC3581⟩, 801 ⟨91COC(5)899⟩, 801, 802 ⟨89CRV165⟩, 803, 804 ⟨80JOC5020, 80JA6351⟩, 804 ⟨83JOC4453, 84TL2447, 87JOC4641⟩, 806, 811 ⟨92OR(41)1⟩; **2**, 2, 6 ⟨B-83MI 201-01⟩, 39 ⟨90JA9439⟩, 325 ⟨88JOC2094⟩, 326 ⟨87SC1155⟩, 606 ⟨B-83MI 212-01⟩; **3**, 62 ⟨88JA4735⟩, 212 ⟨81TL3351⟩
Hudlik, A. M., **6**, 125 ⟨89JOC5613⟩
Hudnall, P. M., **2**, 558, 958 ⟨72JA7178, 73CC269⟩
Hudrlik, A. M., **1**, 62 ⟨82JA6809⟩, 509, 511, 531 ⟨B-78MI 112-02⟩, 650 ⟨80JA6894⟩; **2**, 623, 682, 747 ⟨77JA1993⟩, 676 ⟨73JOC4254⟩, 909, 910 ⟨83T877⟩; **4**, 355 ⟨92JOC6552⟩, 357 ⟨85JA4260, 88TL6395⟩, 508 ⟨85TL139⟩, 619, 620 ⟨89JOC5613⟩
Hudrlik, P., **1**, 229 ⟨75JOC2963⟩
Hudrlik, P. F., **1**, 62 ⟨82JA6809⟩, 130 ⟨68JA4462, 68JA4464⟩, 451, 1105 ⟨79JOC155⟩, 509, 511, 531 ⟨B-78MI 112-02⟩, 648, 650 ⟨75JOC2263⟩, 648 ⟨B-88MI 114-03⟩, 649, 650, 761 ⟨75JA1464, 80JA6894⟩, 650, 759 ⟨74TL1133⟩, 651 ⟨75SC359, 76MI 116-01⟩, 759, 761 ⟨75JOC2264⟩; **2**, 539, 696 ⟨68JA4462⟩, 617 ⟨85T1179⟩, 623, 682, 747 ⟨77JA1993⟩, 676 ⟨73JOC4254⟩, 816 ⟨75JOC2963⟩, 909, 910 ⟨83T877⟩, 923 ⟨79JOC155⟩; **4**, 355 ⟨92JOC6552⟩, 357 ⟨77TL591, 79JOC155⟩, 357 ⟨85JA4260, 88TL6395⟩, 508 ⟨85TL139⟩, 619, 620 ⟨89JOC5613⟩; **6**, 125 ⟨89JOC5613⟩
Hudson, A., **4**, 597, 598 ⟨80JCS(D)2428, 76CC623⟩; **6**, 734 ⟨70CC559, 93JA8444⟩
Hudson, A. T., **2**, 225 ⟨83CC1051, 88JCS(P1)9⟩
Hudson, Jr., B. E., **5**, 46 ⟨42OR(1)266⟩
Hudson, C. S., **4**, 223 ⟨43JA1477⟩
Hudson, F. F., **2**, 357 ⟨86JCS(P2)151⟩
Hudson, H. R., **2**, 6 ⟨69S112⟩, 25 ⟨76JCS(P1)104⟩, 25, 31 ⟨74CI(L)664, 76JCS(P1)754⟩, 104, 105 ⟨B-73MI 202-01⟩; **3**, 493 ⟨B-90MI 313-01⟩; **4**, 129 ⟨88PS(40)183⟩
Hudson, R. F., **2**, 807 ⟨B-65MI 215-04⟩; **3**, 494 ⟨63JOC2446⟩, 498 ⟨63HCA2178⟩; **4**, 476 ⟨66JCS(B)789⟩; **5**, 130 ⟨53JCS3352⟩, 347 ⟨86JCS(P2)1211⟩, 405 ⟨73PS287⟩, 531 ⟨78JCS(P2)1167⟩; **6**, 558 ⟨85CC1405⟩
Hudspeth, J. D., **5**, 557 ⟨83JA5015⟩
Hudspeth, J. P., **1**, 760 ⟨80JA2463⟩; **5**, 557 ⟨81TL3925⟩
Huebner, C. F., **3**, 233 ⟨63JOC3132⟩; **5**, 298 ⟨55JA469⟩
Huemer, H., **2**, 268 ⟨64M33⟩
Huesmann, P. L., **1**, 835 ⟨83TL1021⟩
Hueso-Rodriguez, J., **1**, 195 ⟨88TL4709⟩
Huestis, L. D., **4**, 170, 209 ⟨62JA813⟩
Huet, F., **1**, 613 ⟨86TL3733⟩; **3**, 9, 123 ⟨78S63⟩, 210 ⟨88T4463⟩, 220 ⟨77TL3505⟩; **4**, 188 ⟨80SC83⟩, 196 ⟨85S496⟩, 541 ⟨87JOM(320)171⟩; **5**, 199 ⟨92H(34)739⟩
Huet, J., **1**, 1059 ⟨64BSF952⟩; **2**, 312 ⟨64BSF952⟩
Huether, A., **2**, 260 ⟨84GEP3229565⟩
Huff, B., **1**, 1042 ⟨86TL4241⟩; **3**, 109 ⟨88TL3407⟩
Huff, B. J. L., **1**, 871 ⟨69JOC3070⟩
Huff, G. F., **2**, 110 ⟨61USP2992072⟩
Huffman, C. W., **5**, 84 ⟨63CRV625⟩
Huffman, J. C., **2**, 569 ⟨87OM2354⟩, 986, 987 ⟨86JA7411⟩, 986, 987 ⟨87JA2544⟩, 1007 ⟨87JA4749⟩; **3**, 211 ⟨82OM1240⟩, 512 ⟨86JA1502⟩; **4**, 675 ⟨84JA6794, 85OM1312⟩, 675 ⟨86OM2171, 86OM2457⟩, 675 ⟨89OM1626, 89OM80⟩, 675 ⟨91OM535⟩, 676 ⟨84JA6806, 86OM2384⟩, 676 ⟨86OM602, 88POL919⟩, 676 ⟨89OM67⟩, 1080 ⟨87CC469⟩; **5**, 696 ⟨88JA307⟩, 1147 ⟨89CC900⟩; **6**, 345 ⟨84CC392⟩, 515 ⟨89CC900⟩
Huffman, J. W., **4**, 706 ⟨65JOC1600⟩
Huffman, M. A., **2**, 672 ⟨91JA2771, 92OM255⟩
Huffmann, J. C., **4**, 139 ⟨89JA7199⟩
Huffmann, J. W., **2**, 54 ⟨91COS(8)107⟩
Hufnal, J. M., **1**, 166 ⟨79TL3375⟩; **4**, 236 ⟨79TL3375⟩; **6**, 126, 135 ⟨79TL3375⟩
Hug, E., **4**, 427 ⟨34BSF990⟩
Hug, P., **4**, 239 ⟨89TL6311⟩; **5**, 570, 608, 610 ⟨78BSB525⟩
Hug, R., **1**, 734, 795 ⟨72HCA1828⟩
Hügel, H., **4**, 529 ⟨78CB2630⟩
Hügel, R., **2**, 1004 ⟨68IC18⟩
Huggett, A. C., **5**, 214 ⟨87JOC4933⟩
Huggins, D. K., **6**, 523 ⟨61JA2953⟩
Huggins, J. M., **1**, 453, 545 ⟨81JA3002⟩; **4**, 686 ⟨82JOM(226)C59, 83JOM(258)81⟩
Hughes, A. B., **1**, 829 ⟨93JA5815⟩; **5**, 179 ⟨93JA5815⟩
Hughes, A. K., **4**, 656 ⟨93OM1936⟩
Hughes, A. N., **2**, 858 ⟨65JCS2184⟩; **4**, 1031 ⟨70JHC1, 85JHC513⟩
Hughes, B., **6**, 258 ⟨35CB1513⟩
Hughes, C. T., **1**, 734 ⟨66JOC2907⟩
Hughes, D. L., **1**, 135 ⟨87JOC4745⟩; **2**, 6 ⟨92OR(42)335⟩; **5**, 123, 131 ⟨92OR(42)335⟩
Hughes, D. W., **5**, 924 ⟨87OM1246⟩
Hughes, E. C., **5**, 758 ⟨11JCS1493, 11JCS1507⟩
Hughes, E. D., **1**, 107 ⟨37JCS1177, 37JCS1183⟩, 891 ⟨33JCS1571, 33JCS526⟩, 891 ⟨35JCS255⟩, 892 ⟨54JCS2918⟩; **2**, 25 ⟨55JCS3173⟩, 649 ⟨52QR34⟩, 780 ⟨50JCS2657⟩
Hughes, G., **4**, 304 ⟨85JOC4231⟩
Hughes, G. J., **4**, 519 ⟨81JCS(P1)1934⟩
Hughes, J. L., **5**, 217 ⟨67JCED151⟩
Hughes, K. A., **1**, 991 ⟨93AG(E)554, 93JOC1298⟩; **6**, 706 ⟨93AG(E)554⟩
Hughes, L. A., **2**, 570, 573 ⟨82ZN(B)380⟩; **3**, 545 ⟨85JA3728⟩; **5**, 439 ⟨86JA1722⟩
Hughes, L., **1**, 340 ⟨92JOC4284⟩
Hughes, L. R., **1**, 215 ⟨81TL1751⟩
Hughes, M. J., **4**, 57 ⟨93JCS(P1)1493⟩
Hughes, M. T., **1**, 826 ⟨68CC587⟩
Hughes, N. A., **2**, 38, 52, 86, 89 ⟨B-65MI 202-01⟩; **4**, 227 ⟨86CAR(148)25⟩
Hughes, N., **2**, 558, 560 ⟨91JOC698⟩; **6**, 28 ⟨89JCS(P1)283⟩
Hughes, P., **2**, 627 ⟨90JHC2151⟩
Hughes, R., **1**, 1035, 1103 ⟨76TL4385⟩; **2**, 1086 ⟨76TL4385⟩
Hughes, R. A., **6**, 241 ⟨81JCS(P1)455⟩
Hughes, R. J., **4**, 373 ⟨74CC863⟩
Hughes, R. L., **1**, 886 ⟨89JOC1671⟩
Hughes, R. P., **2**, 579 ⟨82COMC-I(5)277⟩; **4**, 687 ⟨90OM2745⟩; **5**, 950 ⟨82OM628, 82OM635⟩
Hughes, S., **4**, 861 ⟨87JCS(P2)1253⟩
Hughes, W. B., **3**, 494 ⟨65JA3467⟩, 495, 497 ⟨66JOC4252⟩; **4**, 212 ⟨54JA5161⟩; **5**, 896 ⟨66JOM(5)267⟩
Hughes, W. L., **5**, 83 ⟨54MI 502-03⟩
Hughmark, P. B., **5**, 276, 298 ⟨61JA1983⟩
Huglin, J. A., **3**, 113 ⟨93CC954⟩
Hugo, S., **4**, 793, 970, 971 ⟨82S199⟩
Hugo, V. I., **3**, 297 ⟨84JCS(P1)1339⟩
Huguenin, R., **6**, 469 ⟨79EUP7498⟩
Hugueny, J. C., **2**, 367 ⟨84PS(20)371⟩
Huguerre, E., **5**, 516, 522 ⟨87S363⟩
Huguet, J., **4**, 238, 241 ⟨87T4875⟩
Huguet, J. L., **1**, 601 ⟨67MI 114-02⟩

Huhn, G. F., **3**, 684, 686 ⟨84JOC4595⟩
Huhn, M., **3**, 613 ⟨89JOC4476⟩
Hui, R. A. H. F., **3**, 88 ⟨78CC952⟩; **4**, 212 ⟨85JOC5167⟩
Hui, R. C., **3**, 136 ⟨85JOC1985, 86TL1473⟩, 158 ⟨92JOC5620⟩; **5**, 238 ⟨84OM327⟩, 437 ⟨84TL1651⟩, 596, 719, 729 ⟨84TL5251⟩
Huie, E. M., **1**, 693 ⟨82JA2923⟩; **4**, 268 ⟨83TL5563⟩
Huie, R., **3**, 433 ⟨85JOC1531⟩
Huis, R., **6**, 195 ⟨86OM1726⟩
Huisgen, R., **1**, 367 ⟨83TL4181, 84BSB511⟩, 421 ⟨68TL5215⟩, 524 ⟨67JA1753, 69AG(E)604⟩, 841 ⟨85TL1057⟩; **2**, 243 ⟨93JOC82⟩, 375 ⟨54LA(590)1⟩, 766 ⟨65TL783⟩, 804 ⟨60AG303⟩, 805 ⟨51LA(573)163⟩; **3**, 315 ⟨71CC1190, 71CC1192⟩, 315 ⟨76TL4643, 77CC439⟩, 315 ⟨77CC440⟩, 320 ⟨82JA4952, 82JA4953⟩, 367 ⟨81JA7032, 83TL4181⟩, 367 ⟨89TL7041, 89TL7045⟩, 370 ⟨87JA902, 89PS(43)63⟩, 407 ⟨58CB12⟩, 452 ⟨62T3⟩, 466 ⟨51LA(573)163, 52LA(575)174⟩, 467 ⟨65CB1138⟩, 478 ⟨66CB475⟩, 572 ⟨89H(29)2069, 90JOC1412⟩, 600 ⟨69CB1833⟩, 677, 678, 680 ⟨80AG(E)947⟩, 678 ⟨62AG(E)50⟩, 679 ⟨72CB1258, 59JOC892⟩, 685 ⟨73CB3291⟩; **4**, 328 ⟨85H(23)2207⟩, 436 ⟨65JOC74⟩, 829 ⟨74CC405⟩; **5**, 201, 842 ⟨62T3⟩, 360 ⟨73OSC(5)650⟩, 611, 710 ⟨58CB972⟩, 657 ⟨72CB1258⟩, 677 ⟨55LA(591)218, 55LA(591)226, 58CB1129⟩, 832 ⟨61AG656⟩, 842 ⟨67CB71⟩, 844 ⟨55LA(591)190⟩, 1046 ⟨72CB2815⟩, 1068 ⟨69CB1848⟩; **6**, 304, 305 ⟨61E566⟩, 309 ⟨61AG656, 72CB2815⟩, 417 ⟨61AG265⟩
Huisman, H. O., **1**, 619, 707 ⟨81T417⟩, 708 ⟨77TL1173, 80H(14)1739⟩, 747 ⟨78JCS(P1)304⟩, 884 ⟨66T285⟩; **3**, 493 ⟨61MI 313-01⟩; **4**, 179 ⟨73RTC1047⟩, 303 ⟨71SC103⟩
Huizinga, W. B., **1**, 786 ⟨87TL357⟩
Hulce, M., **1**, 304 ⟨90OR(38)225, 91COS(4)237⟩; **4**, 960, 964 ⟨81PAC2307⟩
Hulett, L., **5**, 703 ⟨77JCS(P2)1136⟩
Hulkenberg, A., **3**, 618 ⟨82TL1505⟩, 622 ⟨76TL255⟩
Hull, D. C., **5**, 3 ⟨49USP2475966⟩
Hull, K., **1**, 812 ⟨88TL2773⟩; **3**, 229 ⟨89TL1033⟩
Hull, R., **3**, 66 ⟨74JCS(P1)1541, 83CC74⟩, 269 ⟨79JCR(S)240⟩; **5**, 1033 ⟨68JCS(C)1777⟩, 1035 ⟨74JCS(P1)1541, 75JCS(P1)922⟩, 1035 ⟨79JCR(S)148, 79SC477⟩, 1035 ⟨80SC489, 83CC74⟩; **6**, 574 ⟨79JCR(S)240⟩
Hüllmann, M., **2**, 566 ⟨86T2931⟩
Hullot, P., **4**, 535 ⟨73BSF2989⟩
Hulme, A. N., **1**, 831 ⟨90TL7513, 91TL7601⟩; **5**, 161 ⟨90TL7513⟩
Huls, R., **6**, 598 ⟨62BSB541⟩
Hulst, R., **3**, 105 ⟨88S316⟩
Hultin, P. G., **5**, 138 ⟨85CC1563⟩
Humber, D. C., **4**, 229 ⟨92TL4625⟩
Humber, L. G., **2**, 27 ⟨71JMC982⟩
Humble, H., **5**, 840 ⟨25JCS1304⟩
Hümke, K., **6**, 232 ⟨85HOU(E11)158⟩, 234 ⟨85HOU(E11)665⟩
Hummel, C., **1**, 808 ⟨80CB3679⟩
Hummel, K., **1**, 1110, 1117 ⟨71JA1513⟩, 1123 ⟨61LA(746)211, 70LA(733)5⟩; **2**, 689 ⟨70AG(E)273⟩
Hummer, W., **2**, 318 ⟨89TL1517⟩, 318 ⟨91S769⟩; **3**, 120 ⟨91JA2205⟩
Humora, M., **1**, 876 ⟨78JOC2705⟩
Humphlett, W. J., **2**, 117 ⟨66CJC2315⟩
Humphrey, G. R., **5**, 8 ⟨89S598⟩
Humphrey, I., **2**, 40 ⟨18JA822⟩
Humphrey, J. J. S., **2**, 833, 844 ⟨61JA4124⟩
Humphrey, R. E., **2**, 127 ⟨65ANC164, 65MI 203-01⟩
Humphries, J. E., **5**, 840 ⟨25JCS1304⟩
Humphries, R. E., **2**, 959 ⟨77JOM(127)127⟩
Hunadi, R. J., **1**, 1130 ⟨82JOC2251⟩; **4**, 196, 810 ⟨82JOC2251⟩; **5**, 967 ⟨82JOC2251⟩
Hünerbein, J., **5**, 397, 880 ⟨87AG(E)779⟩
Hung, M., **5**, 958 ⟨91OM737⟩
Hung, M.-H., **1**, 196 ⟨84JA6837⟩, 196 ⟨87JA2176⟩, 967 ⟨90TL3703⟩
Hung, P. L. K., **6**, 341, 401 ⟨73JOM(50)265⟩
Hung, R. R., **3**, 48 ⟨91JOC3849⟩
Hung, S. C., **1**, 53 ⟨85JOC4329⟩, 445 ⟨86JOM(312)53⟩; **2**, 374 ⟨81JOC5413⟩, 432 ⟨85JOC4329, 86JOM(312)53⟩
Hung, S.-H., **1**, 24 ⟨91TL3551⟩
Hungate, R., **3**, 243 ⟨88JA7434⟩
Hungate, R. W., **1**, 212 ⟨81JOC1410⟩
Hung, W. M., **5**, 294 ⟨73JOC4073⟩
Hunger, B., **5**, 589, 603, 611 ⟨67JPR294⟩
Hunger, K., **2**, 253, 462 ⟨63T1563⟩
Hungerbühler, E., **4**, 197 ⟨80HCA1960⟩
Hünig, S., **1**, 74 ⟨61TL353⟩, 248 ⟨84LA340⟩, 627 ⟨79LA1925⟩, 688 ⟨84LA1905⟩; **2**, 144 ⟨53LA(579)23⟩, 374, 385, 404 ⟨71CB1088, 71CB1104⟩, 677 ⟨58CB380⟩, 689 ⟨82CB3687, 82CB399⟩, 739 ⟨73OSC(5)808⟩, 999 ⟨87JOC564⟩; **3**, 170 ⟨82CB3687⟩, 213 ⟨80CB3783, 89CB2131⟩, 213 ⟨90CB107⟩, 214 ⟨86CB722⟩, 261 ⟨57CB2833⟩, 292 ⟨80CB302⟩, 292, 295, 303, 311 ⟨79CB2045⟩, 344 ⟨87JOC2154⟩, 355 ⟨77CB2506⟩, 448, 449 ⟨53LA(579)28⟩, 564 ⟨68AG(E)304⟩, 646 ⟨86CB2590, 86CB3344⟩; **4**, 313 ⟨71CB1104⟩; **5**, 366 ⟨69CB2889⟩, 597, 599, 614 ⟨62CB926⟩, 857, 858 ⟨53LA(579)28⟩, 1134 ⟨79S522⟩; **6**, 304 ⟨88EUP293749⟩, 653 ⟨62LA(651)89⟩
Hunkeler, F., **2**, 104 ⟨64HCA1647⟩, 456 ⟨61HCA1812⟩
Hunold, R., **4**, 633 ⟨89AG(E)784⟩, 634 ⟨88AG(E)1370⟩; **5**, 928 ⟨88AG(E)961, 90AG(E)398⟩; **6**, 184, 389, 398, 714 ⟨89AG(E)781⟩, 201, 400 ⟨88AG(E)1370⟩, 389, 398, 401, 713 ⟨89AG(E)784⟩, 389, 714 ⟨91AG(E)594⟩, 397, 398 ⟨90AG(E)1032⟩, 397, 714 ⟨88AG(E)961⟩, 713, 714 ⟨90AG(E)398, 90ZN(B)290⟩
Hunsberger, I. M., **2**, 374 ⟨55JA1843⟩
Hunsdiecker, C., **2**, 627 ⟨42CB291⟩
Hunsdiecker, H., **2**, 627 ⟨42CB291⟩
Hunt, A. R., **1**, 489 ⟨93TL3599⟩
Hunt, B. B., **2**, 845, 860 ⟨57JCS2413⟩
Hunt, C. T., **4**, 691 ⟨81JA3764⟩
Hunt, D. A., **1**, 1094 ⟨81JOC4608⟩
Hunt, D. F., **4**, 193 ⟨72TL3595⟩
Hunt, E., **1**, 604 ⟨78CC518⟩, 821 ⟨72CC757⟩; **4**, 824 ⟨79TL391⟩
Hunt, G., **2**, 457 ⟨72JOC4396⟩, 1071, 1072 ⟨60CI(L)47⟩
Hunt, G. E., **4**, 476 ⟨72JOC4397⟩
Hunt, J. D., **2**, 631 ⟨69TL2427, 71JA4841⟩, 631, 964 ⟨71JA4845⟩; **4**, 194 ⟨73JA3635⟩
Hunt, J. E., **6**, 87 ⟨87TL4153⟩
Hunt, J. T., **1**, 200 ⟨90H(31)1359⟩
Hunt, M. M., **5**, 1140 ⟨79JOM(175)303⟩
Hunt, P. G., **2**, 764 ⟨78JCS(P1)1117⟩
Hunt, R., **4**, 135 ⟨25MI 403-01⟩, 687 ⟨64PCS401, 65IC1270⟩
Hunter, A. D., **4**, 715 ⟨89OM1118, 90CJC41⟩, 715 ⟨92OM864⟩
Hunter, B. K., **2**, 566 ⟨84JOM(267)265⟩
Hunter, D. H., **1**, 886 ⟨B-80MI 119-03⟩
Hunter, D. L., **2**, 904 ⟨66JOC2766⟩
Hunter, G. A., **2**, 1032 ⟨93S1067⟩
Hunter, G. D., **5**, 99 ⟨52BJ163⟩
Hunter, J. E., **4**, 392 ⟨87JA5403⟩
Hunter, L., **6**, 261 ⟨24JCS442⟩
Hunter, R., **1**, 254 ⟨88TL2257⟩, 280 ⟨89TL2013⟩; **2**, 89 ⟨89TL2013⟩; **3**, 268 ⟨89TL4879⟩
Hunter, R. F., **5**, 1027 ⟨26JCS3041, 27JCS436⟩, 1056 ⟨35JCS1762⟩
Hunter, W. E., **2**, 512 ⟨87IC4065⟩, 564 ⟨81CC292⟩, 565 ⟨79JCS(D)54⟩, 566 ⟨77JA6645⟩, 866, 893 ⟨84IC2582⟩, 968 ⟨83OM1555⟩, 969 ⟨84JA1291⟩; **4**, 656 ⟨78CC1081,

78CC140⟩, 656, 663, 709, 724 ⟨81JCS(D)1593⟩, 702 ⟨79JCS(D)54⟩, 725, 726 ⟨79JCS(D)45⟩; **5**, 440 ⟨81CC706⟩; **6**, 168 ⟨84IC2582⟩, 195 ⟨77JA6645⟩, 195 ⟨78CC140⟩, 195 ⟨81JAP(K)56125326⟩, 689 ⟨84JA7015⟩
Hunter, W. F., **5**, 443 ⟨84JCS(D)399⟩
Hunter, W. H., **2**, 181 ⟨32JA3364⟩; **6**, 215 ⟨32JA2025⟩
Huntley, C. M., **5**, 535 ⟨84JCS(D)2301⟩, 539 ⟨82JCS(D)1131⟩
Huntress, E. H., **4**, 14 ⟨48JA2813⟩; **5**, 34, 43 ⟨48JA2824⟩, 59 ⟨32OSC(1)168⟩, 109 ⟨43OSC(2)457⟩; **6**, 214 ⟨B-48MI 607-01⟩
Huntsman, W. D., **1**, 808 ⟨66JA5846⟩; **5**, 325, 333 ⟨67JOC1800⟩
Hupfeld, B., **1**, 183, 209 ⟨86AG(E)754⟩; **3**, 704, 706, 707 ⟨86AG(E)754⟩; **4**, 536 ⟨88AG(E)433⟩
Huppertz, M., **4**, 112 ⟨79ZAAC(459)7, 83CB114⟩; **6**, 288 ⟨83CB114⟩
Huppmann, P., **5**, 348, 767 ⟨84IC2217⟩; **6**, 430 ⟨81ZAAC(472)26⟩
Huq, F., **4**, 656 ⟨71CC1079, 71CC1477⟩
Hur, C. U., **1**, 455 ⟨90TL5039⟩; **3**, 223 ⟨90TL5003⟩
Hürbin, M., **5**, 566 ⟨33HCA1323⟩
Hurd, C. D., **1**, 414 ⟨38JA1905⟩, 835 ⟨57JOC1418⟩; **2**, 117, 138 ⟨47JA2328⟩, 376 ⟨67JA1417⟩, 732 ⟨71IJS(A)18⟩; **3**, 490 ⟨35JA2656⟩, 535 ⟨23JA2167, 23JA3095⟩, 535 ⟨25JA1779, 29JA3614⟩, 535 ⟨40JOC122⟩, 535, 541 ⟨32JA2432⟩, 536, 541 ⟨35JA774, 54JA5563⟩, 615 ⟨45JOC441⟩, 636 ⟨47JA2328, 55OSC(3)458⟩; **4**, 47, 50, 51 ⟨49JA4007⟩, 163 ⟨66JCE527⟩; **5**, 66 ⟨38JA235⟩, 187 ⟨59JOC388⟩, 214 ⟨53JA285⟩, 820 ⟨50JA4697⟩, 1108 ⟨51JA613⟩
Hurd, C. T., **1**, 808 ⟨34JA447⟩
Hurd, D. T., **2**, 40 ⟨48JA2053⟩
Hurd, R. N., **5**, 565, 579 ⟨60CSR45⟩, 591 ⟨61JOC3980⟩, 735 ⟨61CRV45⟩
Hurd, T., **4**, 353 ⟨89CC89⟩
Hurley, M. D., **4**, 42 ⟨93JPC1576⟩
Hurnaus, R., **4**, 743 ⟨72JOC683⟩
Hurni, B., **5**, 280 ⟨78HCA2428⟩
Hurosawa, H., **5**, 1143 ⟨69MI 530-01⟩
Hurst, K. M., **3**, 11 ⟨79JA371⟩, 528 ⟨82JA747⟩; **4**, 339 ⟨78JA3467⟩, 917 ⟨79JA371⟩
Hursthouse, M. B., **3**, 72 ⟨89JCS(P1)1793⟩, 355 ⟨91AG(E)871⟩; **4**, 656 ⟨82POL31⟩, 672 ⟨78JCS(D)1314⟩, 680 ⟨79JA4128, 80JCS(D)1771⟩, 719 ⟨80JCS(D)2315⟩, 842, 845, 851 ⟨89JCS(P1)1793⟩, 873 ⟨90CC816⟩, 993 ⟨92H(34)1005⟩; **5**, 444 ⟨88JCS(D)53⟩, 781 ⟨93POL563⟩, 797, 801 ⟨81JCS(D)2088⟩, 798 ⟨92JCS(D)19⟩; **6**, 187 ⟨86OM1944⟩, 338 ⟨83CC1557⟩, 339 ⟨81CC189, 81CC2496⟩, 400 ⟨88POL2049⟩
Hurtado, F., **5**, 1066 ⟨87S45⟩
Hurtley, W. R. H., **6**, 83 ⟨26JCS2263⟩
Hurwitz, M. J., **1**, 128, 396 ⟨52JA5381⟩
Huryn, D. M., **2**, 67 ⟨91COS(1)49⟩
Hürzeler, M., **4**, 226 ⟨93HCA995⟩, 229 ⟨93HCA1779⟩; **6**, 549 ⟨93HCA995⟩
Husain, A., **1**, 5, 9 ⟨83JOC3667⟩; **2**, 36, 142 ⟨83JOC3667⟩; **5**, 39 ⟨83JOC3667⟩
Husain, M., **3**, 428 ⟨84IJC(B)801, 84IJC(B)801⟩; **6**, 444 ⟨84IJC(B)986, 84IJC(B)986⟩
Husain, S., **3**, 311 ⟨71JOC3368⟩
Husband, J. P. N., **2**, 358 ⟨72JOM(44)37⟩
Husband, L., **3**, 119 ⟨92JOC1600⟩
Husbands, G. E. M., **3**, 371, 372 ⟨69JA3952⟩
Husebye, K., **2**, 289, 732 ⟨66ACS132⟩
Husebye, S., **2**, 289 ⟨63ACS1806⟩
Huseland, D. E., **3**, 12, 128 ⟨88TL3749⟩
Huser, D. L., **1**, 96 ⟨81JOC1593⟩
Husigen, R., **3**, 709 ⟨83CB3027⟩
Husinec, S., **3**, 713 ⟨88TL3343⟩
Husk, G. R., **3**, 519 ⟨83JA258⟩; **5**, 937 ⟨83JA258, 85OM943⟩
Hussain, A., **2**, 338 ⟨69JCS(C)626⟩
Hussain, F. H. S., **4**, 100 ⟨86TL3199⟩
Hussain, H. H., **3**, 271 ⟨86TL5143⟩; **4**, 742 ⟨85JCR(S)182⟩, 742, 747 ⟨88JCR(S)292⟩, 775 ⟨93JCS(P1)1945⟩; **5**, 873 ⟨87T215⟩; **6**, 124 ⟨86TL5143⟩
Hussain, M. S., **4**, 552, 558 ⟨76ZN(B)721⟩, 553 ⟨77CB3501⟩, 585 ⟨77MI 412-02⟩; **5**, 911, 1093, 1094 ⟨77CB3501⟩, 921, 1094 ⟨85OM344⟩; **6**, 370 ⟨76ZN(B)721⟩
Hussain, S. A., **6**, 445 ⟨78JCS(P1)1066⟩
Hussain, S. M., **5**, 615 ⟨88IJC(B)472⟩
Hussain-Bates, B., **5**, 798 ⟨92JCS(D)19⟩
Huss, O. M., **5**, 551 ⟨79CB1650⟩
Huss, R., **5**, 841 ⟨17CB1478⟩
Hussein, A. Q., **4**, 314, 329, 434 ⟨79CB1956⟩; **5**, 220 ⟨87H(26)163⟩, 776 ⟨90SRI1413⟩, 1046 ⟨78CB3750⟩; **6**, 571 ⟨79CB1956⟩
Hussein, F. A., **5**, 770, 850 ⟨68JIC729⟩
Hussein, F. H., **3**, 7, 120 ⟨84TL3363⟩
Husseini, H., **3**, 261 ⟨70S587⟩
Hussey, A. S., **1**, 108, 464 ⟨59JOC843⟩
Hussmann, G. P., **2**, 147, 155 ⟨82JOC3773⟩, 1091 ⟨83JA6316⟩
Hussoin, M. S., **1**, 1071 ⟨89S217⟩; **3**, 662 ⟨87JOC4137⟩; **5**, 196 ⟨89JOC1144⟩
Husson, A., **3**, 419 ⟨76BSF1222⟩
Husson, H.- P., **1**, 104 ⟨94T3139⟩, 224 ⟨93SL599⟩, 238 ⟨80JA1064⟩, 833 ⟨87TL6461⟩, 834 ⟨90TL523⟩; **2**, 325 ⟨88T2457⟩, 328 ⟨88TL4563⟩, 331 ⟨88TL3311⟩; **3**, 355 ⟨87TL6461⟩, 419 ⟨76BSF1222⟩, 422 ⟨88T2457⟩, 616 ⟨87TL547⟩, 637 ⟨80JA1064⟩, 638 ⟨87TL6179, 93SL599⟩; **5**, 212 ⟨73CR(C)1039⟩
Hussong, R., **4**, 489 ⟨85PS(25)201⟩
Husstedt, U., **2**, 30 ⟨79S964, 79S966⟩, 30 ⟨81TL623⟩
Husted, D. R., **5**, 811, 815 ⟨67JOC3859, 66JOC4232⟩, 815, 817, 818 ⟨73USP3726903⟩; **6**, 218 ⟨54JA5141⟩, 259 ⟨66JOC4232⟩, 284, 616, 649 ⟨67JOC3859⟩
Huston, B. L., **2**, 179 ⟨71CJC943⟩
Huston, R., **1**, 812 ⟨82HCA451⟩
Huszar, C., **6**, 645 ⟨91MIP54636⟩
Huszthy, P., **6**, 546 ⟨90JCS(P1)2009⟩
Hutak, A., **2**, 444 ⟨84MIP30022⟩
Hutchings, D. S., **6**, 168 ⟨91CC1560⟩
Hutchings, M. G., **2**, 748 ⟨74CC346⟩; **4**, 296 ⟨91SL485⟩, 523 ⟨75JCS(P1)129, 75JCS(P1)138⟩
Hutchings, R. H., **1**, 149 ⟨93T1807⟩
Hutchins, M. K., **1**, 887 ⟨91COS(8)327⟩
Hutchins, R., **1**, 635 ⟨80OS(59)42⟩
Hutchins, R. O., **1**, 4, 6, 7, 9, 10, 11, 17 ⟨77JOC82⟩, 4, 7 ⟨71JOC1568⟩, 4, 7, 9, 17, 28 ⟨78JOC2259⟩, 14 ⟨81JOC5214⟩, 15, 17, 23, 24, 25 ⟨82JOC4380⟩, 241 ⟨76JOC1052⟩, 635 ⟨75JOC923, 78JOC2299⟩, 635 ⟨79OPP201⟩, 887 ⟨91COS(8)327⟩; **2**, 31 ⟨76JOC1071⟩, 253 ⟨81JA4432⟩, 341, 350 ⟨91COS(8)60⟩, 398 ⟨76JOC3328⟩, 422 ⟨84JOC2438⟩; **3**, 13, 83 ⟨80TL813⟩, 439 ⟨87JOC702, 87TL5619⟩
Hutchins, R. R., **4**, 366, 394, 945, 1060 ⟨84JA3245⟩
Hutchinson, C. R., **4**, 853 ⟨78JOC3922⟩
Hutchinson, D. K., **1**, 486 ⟨85JA6137⟩, 748 ⟨87JA4755⟩; **4**, 384 ⟨87JA4930⟩
Hutchinson, D. W., **2**, 82 ⟨71COC(5)123⟩; **4**, 451, 493 ⟨B-68MI 410-01⟩, 560 ⟨88JOM(346)341⟩; **6**, 123 ⟨88PS(40)1⟩, 263, 288 ⟨85JOM(291)145⟩, 264 ⟨88JOM(340)93, 86JOM(309)C7⟩
Hutchinson, J., **1**, 619, 707, 949 ⟨86JA4568⟩; **4**, 82, 83, 85 ⟨86JA4568⟩
Hutchinson, J. H., **1**, 815 ⟨88CC1281⟩
Hutchinson, J. P., **4**, 699, 700 ⟨83OM106⟩, 701 ⟨84OM1687⟩
Hutchinson, K. D., **1**, 1039 ⟨93JOC2468⟩
Hutchinson, L. L., **2**, 551 ⟨80JOC1924⟩

Hutchinson, R. M., **1**, 435 ⟨90JCS(P1)2851⟩
Hüter, O., **2**, 578, 579 ⟨89AG(E)277⟩
Huth, W., **4**, 561 ⟨84EGP214610⟩
Hutley, B. G., **2**, 824 ⟨76JCS(P1)2529⟩, 844, 846 ⟨77JCS(P2)1705⟩, 847, 848 ⟨77JCS(P2)789⟩
Hutson, D. H., **4**, 224, 227 ⟨63MI 405-01⟩
Hutson, G. V., **2**, 278 ⟨67JCS(A)2014⟩
Hutson, J. L., **6**, 415 ⟨48JA1968⟩
Hüttenhain, S., **3**, 139, 145 ⟨79TL4971⟩
Hüttenhain, S. H., **1**, 189, 200 ⟨84CB322⟩
Huttner, D., **2**, 512 ⟨85ZN(B)1430⟩
Huttner, G., **2**, 422 ⟨84T2211, 84JOM(260)281⟩, 694 ⟨81CB1407⟩, 875, 896 ⟨87JOM(336)337⟩, 1009 ⟨84CB1900⟩; **3**, 513 ⟨77CB656⟩, 556, 557, 560, 585, 586 ⟨81CB3751⟩, 556, 557, 560, 586 ⟨84T893⟩, 602 ⟨82CB172⟩, 687 ⟨84CB1900⟩, 690 ⟨73AG(E)564⟩; **4**, 112 ⟨84CB1695⟩, 563 ⟨77CB1748⟩, 583, 585, 586, 587 ⟨77CB3536⟩, 681 ⟨92JOM(441)81⟩, 695 ⟨83JOM(246)141⟩, 936, 950, 951 ⟨87CB213⟩; **5**, 146 ⟨85TL1631⟩, 166 ⟨84T2211⟩, 1088 ⟨75AG(E)760⟩, 1146 ⟨81CB3655⟩, 1146, 1149 ⟨84CB1161⟩; **6**, 375 ⟨86ZN(B)191⟩, 643 ⟨84CB1161⟩, 720, 722, 723 ⟨77CB3467⟩, 723 ⟨77JOM(128)C49⟩
Hutton, A. T., **6**, 653 ⟨87POL13⟩
Huuhtanen, T. T., **3**, 628 ⟨92TA1223⟩
Huu-Phuoc, T., **6**, 621 ⟨89S918⟩
Huval, C. C., **1**, 470 ⟨93TL3041⟩
Hux, J. E., **4**, 713 ⟨80IC693⟩
Huyck, C. L., **5**, 105 ⟨B-61MI 502-03⟩
Huyer, G., **5**, 85 ⟨91OS(70)1⟩
Huy-Giao, D., **2**, 948 ⟨61CR(253)673⟩
Huylebroeck, J., **2**, 232 ⟨69BSB191⟩
Huynh, C., **1**, 229, 478, 1092 ⟨79TL1503⟩, 503 ⟨65JOC1752⟩, 820 ⟨77SC103⟩, 821 ⟨72TL3587⟩, 971 ⟨83CC1133⟩, 1012, 1013 ⟨88T6337⟩, 1092, 1105 ⟨79TL1073⟩; **2**, 9311 ⟨79TL1073⟩; **3**, 23 ⟨76BSF1445⟩; **4**, 323 ⟨76BSF1482⟩, 853 ⟨78JCR(S)68⟩; **5**, 145 ⟨67BSF985⟩, 554 ⟨75CR(C)1327⟩
Huynh, V., **2**, 814 ⟨83JOM(255)359⟩; **4**, 511 ⟨83JOM(259)283⟩
Huynh Ba Tuong, **6**, 705 ⟨83C10⟩
Huynh-Dinh, T., **2**, 767 ⟨85TL4443⟩
Huys, F., **6**, 52 ⟨79AG(E)615⟩
Huyser, E. S., **1**, 919, 924, 948 ⟨63OR(13)91⟩, 919 ⟨B-70MI 119-06⟩; **2**, 2 ⟨B-73MI 201-03⟩, 21, 27 ⟨70S7, 63CI(L)1954⟩, 104 ⟨58JA4546⟩, 183, 272 ⟨65JOC38⟩; **6**, 5 ⟨63OR(13)91⟩
Huyser, J. J., **4**, 190 ⟨91H(32)1445⟩
Huysmans, W. G. B., **4**, 894 ⟨72JOC1960⟩
Hvoslef, J., **6**, 168 ⟨83IC3421⟩
Hwa, J. C. H., **1**, 628 ⟨73OSC(5)608⟩
Hwan, L., **4**, 396 ⟨86JOM(315)135⟩
Hwang, C.- K., **2**, 9 ⟨84TL2295⟩; **3**, 351 ⟨90JA3029⟩; **5**, 513 ⟨90JA3029, 90JA3040⟩
Hwang, D., **2**, 354 ⟨68OSC(6)670⟩
Hwang, D.-R., **2**, 374 ⟨76JOC3599⟩
Hwang, H., **1**, 905 ⟨80JCR(S)152⟩; **2**, 440, 441 ⟨92JOC511⟩
Hwang, H. J., **2**, 145 ⟨90TL2893⟩
Hwang, J. P., **3**, 18, 133 ⟨90TL6677⟩, 18, 134 ⟨92SC2425⟩
Hwang, J. S., **5**, 723, 827 ⟨71JOC284⟩
Hwang, J.-T., **1**, 649 ⟨73JOC2664⟩; **2**, 81 ⟨90TL4765⟩, 936 ⟨73JOC2664⟩
Hwang, K.-J., **1**, 828 ⟨79TL3057⟩
Hwang, Y. C., **5**, 8 ⟨85JOC2719⟩
Hwang, Y. S., **4**, 432 ⟨93S1218⟩
Hwang Lee, S.-J., **2**, 58 ⟨88T4403⟩
Hwu, J. R., **1**, 656 ⟨87JOM(332)53⟩, 834 ⟨91JCS(P1)3199⟩; **2**, 388 ⟨89JOC1070⟩, 393 ⟨88SC21⟩; **3**, 13, 85 ⟨90JCS(P1)757⟩, 16, 132 ⟨83JOC4432⟩, 369 ⟨91JCS(P1)3199⟩, 426 ⟨90T7413⟩, 433 ⟨85JOC5913⟩, 620 ⟨91JCS(P1)1493⟩, 624 ⟨93CC669⟩; **4**, 182 ⟨85JOC3946⟩, 182 ⟨87JOC188⟩, 234 ⟨83JOC4432⟩; **5**, 582, 611 ⟨93JOC4742⟩, 637, 829 ⟨90T7413⟩, 1134 ⟨84S1020⟩
Hyatt, J. A., **2**, 715 ⟨84S215⟩, 717 ⟨79JOC3847⟩
Hyde, A. J., **2**, 19 ⟨86CC791⟩; **5**, 4 ⟨86CC791⟩
Hyde, C. L., **5**, 772, 850, 852, 855 ⟨71JOC1155⟩; **6**, 55 ⟨71JOC1155⟩
Hyde, J. R., **4**, 699, 700 ⟨83OM106, 82OM404⟩
Hyden, S., **5**, 990, 991 ⟨67CI(L)1406⟩
Hydro, W. R., **6**, 608 ⟨85USP4558160⟩
Hylarides, M. D., **2**, 4, 30 ⟨81JOC3056⟩, 616 ⟨85TL279⟩
Hyldahl, C., **1**, 551 ⟨87JOM(334)9⟩
Hyman, H. H., **2**, 613 ⟨70JA6498, 71JOC2917⟩
Hyne, J. B., **2**, 722, 725 ⟨73CJC3403⟩
Hynes, J. B., **6**, 53 ⟨66IC488⟩, 262 ⟨62JA2751⟩, 284 ⟨68IC1647⟩, 284, 363, 616 ⟨72IC418⟩
Hyodo, C., **1**, 367 ⟨86TL4039⟩
Hyodo, K., **5**, 1065 ⟨81MAC3411⟩
Hyon, M. H., **5**, 274, 294 ⟨86JA6054⟩
Hyre, J. E., **4**, 829 ⟨67OS(47)78⟩
Hyslop, W. I., **5**, 1111 ⟨56JCS96⟩
Hyuga, S., **1**, 489 ⟨88CL809⟩, 1091, 1094 ⟨86CL459⟩; **2**, 902 ⟨90TL247⟩; **3**, 233 ⟨89CL1959⟩; **4**, 954 ⟨89CL1959, 92BCJ2303⟩
Hyun, J. L., **1**, 1003 ⟨85JA6546⟩
Hyyppa, J., **1**, 229 ⟨73ACS3944⟩

I

Iacobelli, J. A., **1**, 576, 1021 ⟨86JOC3098⟩
Iancu, A., **4**, 152 ⟨92RRC393⟩
Ianelli, S., **1**, 539 ⟨88T127⟩
Iarossi, D., **2**, 147 ⟨68JCS(C)1317⟩, 600 ⟨73JCS(P2)638⟩
Iarovenko, N. N., **6**, 229 ⟨58JGU2539⟩
Iavarone, C., **3**, 219 ⟨80JOC4502⟩
Ibaceta-Lizana, J. S. L., **1**, 184 ⟨87JCS(P2)1221⟩
Ibáñaz, P. L., **1**, 145, 1090 ⟨93TL2003⟩
Ibañez, F., **5**, 586 ⟨93JOC4908⟩
Ibanez, L. C., **2**, 18 ⟨60JA4001⟩
Ibarra, C. A., **3**, 170 ⟨89JOC5620⟩; **4**, 325 ⟨92JCS(P1)291⟩
Ibata, T., **2**, 753 ⟨87CL1187⟩; **3**, 324 ⟨75CL499, 86TL4383⟩, 324 ⟨87CL2135, 87TL989⟩, 679 ⟨92CL2197⟩
Ibbotson, A., **3**, 293 ⟨55JCS2686⟩; **5**, 143 ⟨83EUP127276A2⟩, 425 ⟨92T7629⟩
Ibe, K., **5**, 250 ⟨90JOM(386)313⟩
Ibe, M., **3**, 227 ⟨88S210⟩
Ibers, J. A., **1**, 456, 507 ⟨83JA6273⟩; **2**, 432 ⟨83OM979⟩, 438 ⟨73JA4194⟩, 442 ⟨77JA7876⟩; **3**, 570 ⟨70JA3489, 71JA2391⟩; **6**, 332 ⟨83IC411⟩, 517 ⟨85OM590⟩
Ibi, K., **5**, 249 ⟨92ZN(B)558⟩
Ibis, C., **1**, 982 ⟨84LA1873⟩; **4**, 770 ⟨92SUL251⟩, 770, 776 ⟨92PS(72)225⟩, 776, 854 ⟨81CB684⟩
Ibne-Rasa, K. M., **2**, 780 ⟨62JA4962⟩
Iborra, S., **3**, 297 ⟨86H(24)2511⟩
Ibragimov, A. A., **3**, 177, 637 ⟨72JOM(42)399⟩
Ibragimov, A. G., **4**, 669 ⟨89IZV2562⟩
Ibragimov, I. I., **1**, 530 ⟨89S472⟩
Ibragimov, M. A., **3**, 169 ⟨83TL961⟩
Ibrahim, I. T., **5**, 1061 ⟨81CRV589⟩
Ibrahim, J., **3**, 419 ⟨86T4517⟩
Ibrahim, M. K. A., **5**, 677 ⟨91PS(60)119⟩
Ibrahim, M. R., **5**, 433 ⟨92MI 509-01⟩
Ibrahim, N. S., **5**, 631 ⟨87S228⟩
Ibrahim, S., **2**, 848 ⟨88S912⟩
Ibrahim, T., **4**, 14 ⟨90JCS(P1)3317⟩
Ibrakhimov, I., **2**, 1056 ⟨87UKZ34⟩
Ibuka, T., **1**, 188 ⟨85CC1452, 89JA4864, 89JOC4055, 90AG(E)801, 91JOC4370, 92JOC1024⟩, 188, 227 ⟨92SL769⟩, 189 ⟨93JOC1207, 93T9479, 89T435⟩, 305 ⟨87CC1572⟩; **2**, 702 ⟨83CPB128⟩; **5**, 158 ⟨89JA4864⟩
Ichibori, K., **1**, 408, 409, 410, 820, 821, 822 ⟨72JA3870⟩
Ichida, H., **1**, 1140 ⟨90CL1411⟩; **2**, 949 ⟨91TL6877⟩; **6**, 127, 325, 327 ⟨90CL1411⟩
Ichihara, A., **1**, 786 ⟨87S207⟩, 789 ⟨87T5245⟩; **3**, 221 ⟨87S207⟩; **5**, 129 ⟨94JOC4749⟩
Ichihara, J., **2**, 18 ⟨90TL3167⟩, 19 ⟨86CC793, 92CL1161⟩; **3**, 613 ⟨84CL725⟩; **6**, 423, 438 ⟨86CC793⟩
Ichihara, M., **4**, 426 ⟨82TL3181⟩
Ichihara, R., **4**, 862 ⟨76H(4)963⟩
Ichihara, T., **3**, 616 ⟨87CL687⟩
Ichihashi, Y., **2**, 619 ⟨80CL1423⟩
Ichii, T., **1**, 1044 ⟨72BCJ1846⟩
Ichijima, S., **5**, 1032 ⟨74JOC922⟩
Ichikawa, E., **6**, 503 ⟨74YGK936⟩
Ichikawa, J., **1**, 429 ⟨92SL833⟩, 570 ⟨89TL1641⟩; **2**, 825, 853 ⟨92SL833⟩, 829 ⟨92CL547⟩, 906 ⟨89TL1641, 89TL5437⟩, 906 ⟨89TL6379, 92TL337⟩, 906 ⟨92TL3779⟩; **4**, 736 ⟨89TL1641, 89TL6379⟩, 736 ⟨91CL961, 92SL739⟩, 736 ⟨92TL337, 92TL3779⟩, 736 ⟨93JFC(63)281⟩; **5**, 282 ⟨83CL683⟩
Ichikawa, K., **1**, 84 ⟨76BCJ1437⟩, 143 ⟨86BCJ2801⟩, 212, 475 ⟨78BCJ174⟩, 454 ⟨74BCJ2663, 81JCS(P1)991⟩, 493 ⟨80BCJ553⟩, 671 ⟨65JOC1593⟩; **3**, 238 ⟨80JHC289⟩; **4**, 1068 ⟨81JCS(P1)991⟩; **5**, 1113 ⟨80BCJ553⟩, 1113, 1115 ⟨75BCJ1925⟩; **6**, 546, 561 ⟨87S349⟩
Ichikawa, K. I., **3**, 664 ⟨72T3025⟩
Ichikawa, M., **4**, 448 ⟨90H(31)2029⟩; **5**, 706 ⟨85H(23)281⟩
Ichikawa, S., **4**, 970 ⟨91EUP413343⟩
Ichikawa, Y., **1**, 9 ⟨92JA10138⟩, 837 ⟨91SL238⟩; **2**, 159 ⟨92JCS(P1)1497⟩; **3**, 696 ⟨92JCS(P1)2135⟩, 696, 700 ⟨91SL715⟩; **4**, 937 ⟨81CL457, 81TL4287, 84JA3252, 84JOC3517, 86T2863, 87T4759⟩, 950 ⟨84TL2021⟩; **5**, 152 ⟨86T2863⟩; **6**, 130 ⟨81TL4287, 84TL2021⟩, 472 ⟨88TL4957⟩
Ichimoto, I., **3**, 706 ⟨85ABC1761⟩
Ichimura, K., **3**, 561 ⟨70BCJ1874⟩, 574 ⟨66TL807, 67BCJ1933⟩
Ichinari, M., **5**, 1112 ⟨89H(29)841⟩
Ichinine, M., **4**, 618 ⟨93CL267⟩
Ichino, M., **2**, 31 ⟨71TL87⟩
Ichinohe, M., **6**, 193 ⟨93CL267⟩
Ichinose, H., **3**, 553 ⟨92TL7865⟩; **5**, 516, 520 ⟨92TL7865⟩
Ichinose, I., **2**, 31 ⟨76CL283, 80JCS(P1)1051⟩
Ichinose, K., **4**, 766 ⟨92CPB3261⟩
Ichinose, Y., **1**, 330, 925, 946 ⟨89BCJ143⟩, 333 ⟨89TL3155⟩, 1130 ⟨88BCJ2693⟩, 1131 ⟨89TL3159⟩; **2**, 710 ⟨87CL1647⟩, 930, 946 ⟨88BCJ2693, 87BCJ3468⟩, 946 ⟨87TL3709⟩; **4**, 639 ⟨88BCJ2693⟩, 808, 811, 813 ⟨89TL3155⟩; **6**, 21 ⟨89TL3159⟩
Ida, K., **2**, 566 ⟨81CC1140⟩; **6**, 465 ⟨92JAP04356446⟩
Idacavage, M. J., **1**, 133 ⟨79TL845⟩; **2**, 4, 514, 900 ⟨85OR(33)1⟩
Iddon, B., **1**, 664 ⟨71JCS(C)3447⟩; **3**, 105 ⟨71JCS(C)182⟩; **5**, 593 ⟨76JCS(P1)33, 77JCS(P1)2357⟩, 1111 ⟨70JCS(C)1530⟩
Ide, H., **1**, 10 ⟨86JOC5400⟩; **3**, 12, 13, 83, 98, 129 ⟨86JOC5400⟩
Ide, J., **1**, 822 ⟨66TL1787, 68CPB793⟩; **3**, 378 ⟨66TL1787⟩
Ide, W. S., **5**, 51 ⟨43OSC(2)333⟩, 275 ⟨43OSC(2)44⟩
Idei, M., **2**, 631 ⟨82CL1481⟩
Idemoto, T., **1**, 1033 ⟨87JCS(P1)1511⟩
Idogaki, Y., **1**, 827 ⟨88TL655⟩, 827, 829 ⟨87JOC1201⟩
Idoux, J. P., **5**, 755 ⟨83SC1083⟩
Idris, M. S. Hj., **1**, 980 ⟨92JCS(P1)2355⟩
Iedema, A. J. W., **5**, 898, 901, 902 ⟨72RTC37⟩; **6**, 699 ⟨72RTC37⟩
Iemma, F., **3**, 620 ⟨93T8423⟩
Iesce, M. R., **3**, 327 ⟨94JCS(P1)147⟩; **6**, 76, 77 ⟨85S762⟩, 107 ⟨87JCR(S)362⟩, 145 ⟨86JHC553⟩
Iescs, M. R., **3**, 221 ⟨83S125⟩
Ife, R. J., **4**, 900 ⟨89T1801⟩
Iffland, D. C., **1**, 442 ⟨49JA2137⟩; **2**, 403 ⟨61JA747⟩; **4**, 314 ⟨61JA747⟩
Igami, M., **3**, 84 ⟨75BCJ2490, 75TL1063⟩
Igano, K., **5**, 48, 114 ⟨64CPB403⟩
Igarashi, J., **2**, 1000 ⟨90CPB329⟩
Igarashi, M., **5**, 692 ⟨88BCJ3731⟩
Igarashi, S., **4**, 193 ⟨89CL737⟩
Igarashi, T., **1**, 405 ⟨92TL5795⟩; **4**, 379 ⟨92TL5795⟩, 708 ⟨86TL4339⟩
Igau, A., **2**, 1067 ⟨88JA6463⟩; **3**, 596 ⟨89AG(E)621⟩, 689 ⟨88JA6463, 90PS(49)301⟩; **4**, 547, 579 ⟨93AG(E)1735⟩, 567 ⟨88JA6463⟩, 567, 1082 ⟨89AG(E)621⟩; **5**, 871, 872, 873 ⟨88JA6463⟩; **6**, 125 ⟨89AG617⟩, 125, 668, 669 ⟨91NJC393⟩, 665, 680 ⟨86JA7868⟩, 691, 734 ⟨89JA6853⟩, 704 ⟨88JA6463⟩
Igeta, H., **1**, 1010 ⟨78H(9)1397, 80BCJ3273⟩
Ignart'ev, N. V., **6**, 233 ⟨79ZOR1245⟩
Ignasiak, B., **3**, 710 ⟨75JCS(P1)2122⟩

Ignasiak, T., **3**, 710 ⟨75JCS(P1)2122⟩
Ignatenko, A. V., **4**, 639 ⟨70IZV2280⟩
Ignateva, S. N., **3**, 689 ⟨91HAC577⟩; **4**, 342 ⟨85IZV1102, 88IZV159⟩, 343 ⟨79IZV1863, 86IZV640⟩, 343 ⟨87IZV418⟩, 569 ⟨90PS(53)1⟩; **5**, 395 ⟨90BAU2424⟩
Ignatev, N. V., **5**, 58 ⟨85ZOR769⟩
Ignat'ev, V. M., **4**, 802 ⟨67JGU2024⟩
Ignatev, Yu. A., **2**, 845 ⟨82ZOB2721⟩, 881 ⟨85JGU1332⟩
Ignatov, S. M., **4**, 481 ⟨82IZV1611⟩; **5**, 1130 ⟨82IZV1611⟩
Ignatov, V. A., **2**, 358 ⟨79ZOB841⟩, 359 ⟨81ZOB609⟩
Igolen, J., **2**, 767 ⟨85TL4443⟩
Igritskaya, A. V., **6**, 369 ⟨70IZV2108⟩
Iguchi, S., **2**, 362 ⟨91JOC4⟩; **3**, 437 ⟨91JOC4⟩
Ihara, M., **2**, 16 ⟨92JCS(P1)221⟩, 328 ⟨89H(28)63⟩; **3**, 220 ⟨90JOC4497⟩, 665 ⟨77S245⟩; **4**, 154 ⟨85JCS(P1)1185⟩, 387 ⟨83TL4257⟩; **5**, 285 ⟨83JA7345⟩
Ihlo, B., **3**, 451 ⟨63CB658⟩
Ihmels, H., **4**, 162 ⟨91TL6215⟩, 1052 ⟨93TL6541⟩
Ihn, W., **5**, 1118 ⟨70TL1125⟩
Ihrig, K., **1**, 786 ⟨87TL183⟩
Iida, A., **4**, 122 ⟨92CL407⟩
Iida, H., **1**, 693 ⟨82CL1697⟩; **2**, 299 ⟨87JOC1956⟩, 748 ⟨83JOC3531, 84S1041⟩, 1000 ⟨82CL1697⟩; **3**, 31, 157 ⟨86TL3665⟩, 146 ⟨86CL1597⟩, 417, 658 ⟨78S892⟩, 671 ⟨85CL1487⟩; **4**, 262 ⟨80TL2233⟩, 271 ⟨81TL4499⟩, 395 ⟨82CL1697⟩, 854, 857 ⟨82JOC5404, 83CL767, 86JOC326⟩, 857 ⟨86JOC700⟩; **5**, 322, 330 ⟨89JOC4088, 85JOC1818⟩
Iida, I., **5**, 627 ⟨94CL77⟩
Iida, J., **5**, 11 ⟨90MI 501-01⟩
Iida, K., **1**, 651 ⟨83CC921⟩; **2**, 936 ⟨83CC921⟩
Iida, S., **1**, 476 ⟨83TL5745⟩, 970 ⟨84TL4007⟩; **2**, 25 ⟨84CL1173⟩
Iida, T., **5**, 281 ⟨81S287⟩
Iihama, T., **3**, 90, 98 ⟨89JOC24⟩; **4**, 857 ⟨86JOC700⟩
Iijima, I., **6**, 556, 564 ⟨87CPB2840⟩
Iijima, S., **3**, 186 ⟨82TL547⟩
Iimori, T., **2**, 788 ⟨79TL2525⟩; **5**, 135, 138 ⟨84TL2557⟩
Iino, K., **3**, 559 ⟨76CC516⟩, 581 ⟨76TL1307⟩; **4**, 842 ⟨82TL1075⟩
Iio, H., **1**, 180 ⟨87CC358⟩, 779, 788 ⟨78CL541⟩; **4**, 1034 ⟨91CC1390⟩
Iitaka, Y., **1**, 143 ⟨86BCJ2801⟩, 584 ⟨92JOC7285⟩; **3**, 704 ⟨76CPB832⟩; **5**, 671 ⟨84JHC725⟩
Iizuka, H., **2**, 235 ⟨86BCJ2179⟩
Iizuka, K., **3**, 618 ⟨84CPB2560⟩; **5**, 199 ⟨90JMC2621⟩
Iizuka, Y., **5**, 972 ⟨85JOC715⟩; **6**, 413, 415, 425 ⟨85JOC715⟩
Ijadi-Maghsoodi, S., **4**, 607, 621 ⟨88JA2611⟩
Ijaz, A. S., **4**, 21 ⟨89JCS(P1)2009⟩
Ijima, S., **2**, 566 ⟨81CC1140⟩
Ikai, K., **4**, 122 ⟨92CL407⟩
Ikan, R., **2**, 614 ⟨56JA6037⟩
Ikari, K., **5**, 603, 622, 627 ⟨92CL131⟩
Ikariya, T., **3**, 508 ⟨81JA7358⟩
Ikaruja, T., **1**, 674 ⟨90AC59⟩
Ikawa, T., **3**, 59 ⟨76CL29⟩
Ike, Y., **5**, 972 ⟨77BCJ2406⟩
Ikebe, M., **1**, 666 ⟨92JOC17⟩; **2**, 555 ⟨91AG(E)177⟩
Ikeda, H., **3**, 59 ⟨81CPB766⟩, 212 ⟨83JOC2503⟩
Ikeda, I., **2**, 14 ⟨93JOC6529⟩, 89 ⟨88JOC2737⟩; **3**, 617 ⟨78S301⟩
Ikeda, K., **4**, 197 ⟨89TL5309⟩; **5**, 329 ⟨84CL369, 89CPB1179⟩
Ikeda, M., **1**, 348 ⟨88JA1288⟩, 362 ⟨90CC1436⟩; **2**, 35 ⟨79BCJ1653⟩, 92 ⟨81JA6237⟩, 160, 161 ⟨74S277⟩, 161 ⟨72TL4137⟩, 357, 379, 384 ⟨77S1⟩, 357, 691 ⟨73JOC1239⟩, 401 ⟨76T431⟩, 684 ⟨94SL49⟩, 1000 ⟨81JCS(P1)212⟩; **3**, 61 ⟨87JCR(S)296⟩, 227 ⟨86S847⟩, 267 ⟨73JOC4324⟩, 380 ⟨74JOC3519, 78CC197⟩, 635 ⟨78CPB784, 86JCS(P1)1763⟩; **4**, 67 ⟨86JHC1163, 89JCS(P1)879, 93JOC2360⟩, 519 ⟨84H(22)701⟩, 924 ⟨87JCS(P1)1095⟩, 943 ⟨87JCS(P1)589⟩
Ikeda, N., **1**, 651 ⟨81JA5568⟩, 973 ⟨82CL1029⟩, 974, 1056 ⟨82JOC2225⟩, 1055 ⟨82JA7667, 86JA483, 86TL1175⟩, 1056, 1057 ⟨84BCJ2768⟩; **2**, 905 ⟨86JA483⟩; **3**, 639 ⟨84BCJ2768⟩; **4**, 1059, 1060 ⟨81JA5568⟩, 1061 ⟨84BCJ2768⟩
Ikeda, R. M., **5**, 168 ⟨68JOC2819⟩
Ikeda, S., **2**, 252 ⟨79BCJ1431⟩; **4**, 211 ⟨64JOC1371⟩, 931 ⟨92JA9710⟩; **5**, 221 ⟨88BCJ3653⟩, 416 ⟨92JA9710⟩, 561, 563, 588 ⟨90ZC67⟩
Ikeda, S.-I., **3**, 163 ⟨92JOC2⟩
Ikeda, T., **1**, 1112 ⟨76CC612⟩; **2**, 53 ⟨89JA9134⟩; **6**, 510 ⟨71JA6344⟩
Ikeda, Y., **2**, 24 ⟨84BCJ2111, 86BCJ2547⟩, 24 ⟨86BCJ3519⟩, 786 ⟨82BCJ635⟩, 936 ⟨86BCJ657⟩; **3**, 158 ⟨91BCJ1582⟩, 628 ⟨89CL2119⟩, 664 ⟨72T3025⟩
Ikegami, S., **1**, 180 ⟨86TL3403⟩, 404, 406, 816, 819 ⟨88TL4763⟩, 404, 406, 819 ⟨92BCJ1841⟩, 816 ⟨91TL5137⟩, 905 ⟨85JOC1165⟩, 1081 ⟨82TL4607⟩; **3**, 129 ⟨92TL2709⟩, 218, 225 ⟨86JA2090⟩; **4**, 154, 155 ⟨81TL2397⟩, 155 ⟨83CL1303⟩, 321 ⟨86S375⟩; **5**, 138 ⟨88JA3670⟩
Ikegami, Y., **3**, 80 ⟨77H(6)1599⟩
Ikegawa, S., **5**, 43, 112 ⟨81NKK1333⟩
Ikehara, N., **2**, 409, 467 ⟨86TL199⟩
Ikehira, H., **2**, 243, 358 ⟨83S716⟩
Ikehira, T., **1**, 444 ⟨83JOC2569⟩; **2**, 843 ⟨83JOC2569⟩; **3**, 496 ⟨83JOC2569⟩
Ikejiri, T., **2**, 21 ⟨75JOC2960⟩
Ikekawa, N., **1**, 961 ⟨75CC362⟩; **2**, 19 ⟨82JCS(P1)85⟩
Ikemi, Y., **4**, 410 ⟨84H(22)1417⟩
Ikemori, M., **2**, 369 ⟨91JA6337⟩; **3**, 635 ⟨88CPB1698⟩
Ikemoto, I., **2**, 997, 1001 ⟨92JCS(P1)1721⟩; **6**, 594 ⟨86BCJ1741⟩
Ikemoto, K., **4**, 656 ⟨93BCJ1849⟩
Ikemoto, Y., **2**, 312 ⟨88TL3811⟩
Ikenaga, K., **1**, 452, 490 ⟨90CL185, 86JCS(P1)1959⟩, 490 ⟨88CL873⟩; **2**, 916 ⟨93JOC5045⟩, 932 ⟨84TL5789, 87JOC1276⟩; **5**, 220, 222 ⟨84TL5789⟩
Ikeura, C., **4**, 736 ⟨92SL739, 93JFC(63)281⟩
Ikeya, T., **6**, 21 ⟨88JOC2350⟩
Ikizler, A. A., **2**, 11 ⟨79JCS(P1)436⟩
Ikoma, K., **4**, 259 ⟨90JCS(P1)2035⟩
Ikota, N., **1**, 637 ⟨81TL4163⟩; **3**, 415 ⟨86CPB1050⟩; **4**, 90 ⟨78JOC1607⟩
Iksanova, S. V., **2**, 408 ⟨81ZAAC(472)75⟩; **4**, 486 ⟨82ZOB930⟩
Ikura, K., **2**, 187 ⟨66BCJ1309⟩, 208 ⟨68JA7378⟩
Ikuta, S., **1**, 1142 ⟨75MI 122-01, 77JCP(66)4671⟩; **2**, 784 ⟨82S261⟩; **4**, 335 ⟨82S261⟩
Ila, H., **1**, 485 ⟨85JCS(P1)1289⟩, 1076 ⟨82S1100⟩; **4**, 253 ⟨90T2195, 92S1075⟩, 704 ⟨88T1667⟩, 899, 900 ⟨76CC592, 76CC593⟩, 900 ⟨80S748⟩, 903 ⟨84S250⟩; **5**, 555 ⟨82S693, 85JCS(P1)641⟩, 555 ⟨90T5423⟩, 731 ⟨87S547, 88JCS(P1)169⟩, 731 ⟨88T1667⟩
Ila, I., **2**, 74 ⟨93TL3145⟩
Ilardi, L. M., **6**, 428 ⟨91EUP415473⟩
Ilavsky, J., **5**, 24, 118 ⟨82MI 502-02⟩
Ilchmann, G., **4**, 139, 141, 146, 156 ⟨85CB391⟩
Ilett, J. D., **2**, 101 ⟨50JCS3125⟩, 409, 410, 412, 413 ⟨49JCS2921⟩
Iley, D. E., **3**, 57 ⟨79CJC645⟩
Iley, J., **4**, 312 ⟨84JCR(S)108⟩; **5**, 361 ⟨89JCR(S)260⟩, 361, 362, 711 ⟨85JCR(S)132⟩, 362, 657, 711 ⟨92JCS(P2)281, 88CC1249⟩
Iley, J. N., **5**, 370 ⟨87JCS(P2)1489⟩
Iliceto, A., **5**, 1038 ⟨61JA2729⟩, 1038, 1105 ⟨65JA4791⟩
Ilien, B., **6**, 443 ⟨89MI 614-03⟩
Il'ina, L. F., **4**, 493 ⟨88JGU876⟩
Illger, W., **4**, 558 ⟨78CB705⟩
Illingworth, S. M., **2**, 508, 509, 894 ⟨68JCS(A)2623⟩, 544 ⟨70MI 210-01⟩; **4**, 344 ⟨69JOM(17)161⟩
Ilosvay von Nagyilosva, L., **2**, 1094 ⟨01ZAAC(40)123, 1899CB2697⟩

Iloughmane, H., **1**, 777 ⟨83TL3521⟩; **2**, 949 ⟨85TL1315⟩
Ilvespää, A. O., **2**, 691, 692 ⟨63HCA2009⟩; **5**, 1112 ⟨68HCA1723⟩
Il'yasov, A. V., **2**, 894 ⟨83IZV1152⟩; **4**, 476 ⟨79IZV2783⟩, 549 ⟨84ZOB1511⟩; **5**, 373 ⟨73ZOB1451⟩, 398 ⟨88JGU1965⟩
Ilyushin, V. A., **4**, 1055, 1058 ⟨80ZOB692⟩
Ilyusin, M. A., **2**, 790 ⟨83RCR284⟩
Imada, I., **5**, 206 ⟨88CPB178⟩
Imada, M., **4**, 829 ⟨88JOC5974⟩
Imada, N., **2**, 684 ⟨94SL49⟩
Imada, Y., **1**, 192 ⟨93JOC1538⟩; **2**, 308, 344, 351 ⟨88TL2973⟩, 327 ⟨87TL77⟩; **5**, 324 ⟨89BCJ2968⟩
Imaeda, H., **3**, 363 ⟨71MI 308-01⟩; **5**, 235 ⟨72JOM(42)33⟩
Imafuku, K., **2**, 769 ⟨89BCJ128⟩; **3**, 241 ⟨91JHC811⟩, 357, 358 ⟨78BCJ179⟩
Imagawa, M. T., **3**, 356 ⟨90JCS(P1)3003⟩
Imagawa, T., **1**, 769 ⟨80JOC2005⟩; **3**, 498, 651 ⟨84SC1267⟩; **6**, 426 ⟨90MI 614-03⟩, 462 ⟨90MI 615-02⟩
Imai, E., **1**, 108, 115 ⟨86CPB3599⟩; **5**, 1122 ⟨85JCS(P1)2333, 85TL1321⟩, 1122 ⟨87JCS(P1)2531⟩
Imai, H., **3**, 418 ⟨84TL3091⟩; **5**, 796, 802 ⟨84TL3091⟩
Imai, I., **1**, 233, 400 ⟨72JOC1721⟩, 408, 409, 410, 820, 821, 822 ⟨72JA3870⟩; **3**, 373, 378 ⟨72JOC1721⟩; **4**, 815 ⟨84OM1392⟩
Imai, K., **2**, 239 ⟨88CL857⟩
Imai, M., **3**, 303 ⟨90JCS(P1)2677⟩
Imai, N., **1**, 367 ⟨85TL3011⟩; **3**, 91 ⟨88CPB974⟩; **4**, 369 ⟨85TL3011, 87CPB1734⟩; **6**, 129, 282 ⟨87CPB1734⟩
Imai, S., **3**, 353 ⟨89TL4825⟩
Imai, T., **2**, 521 ⟨86JA7402⟩, 531 ⟨85JA4980⟩, 901 ⟨84OM1392⟩, 901, 906 ⟨86JOC5277⟩, 904 ⟨90JOC4986⟩; **4**, 364 ⟨83JA6285⟩
Imai, Y., **4**, 247 ⟨87S164⟩; **5**, 278, 291, 303 ⟨81S991, 82S933⟩
Imaida, M., **5**, 63 ⟨82BCJ2186⟩
Imaizumi, S., **2**, 398 ⟨88BCJ3020⟩; **3**, 244 ⟨88TL5941, 90SC3063⟩
Imajoh-Ohmi, S., **6**, 487 ⟨93CL1401⟩
Imamiya, E., **6**, 558 ⟨89CPB1249⟩
Imamoto, T., **1**, 158 ⟨85JA5301⟩, 158 ⟨87JOC2170⟩, 158 ⟨90JA5244⟩, 202 ⟨82BCJ643⟩, 426, 427 ⟨85CL1491⟩, 515 ⟨87JOC5695⟩, 636 ⟨74CL165⟩, 885 ⟨91COS(1)231⟩, 1038, 1098, 1140 ⟨84TL4233⟩, 1098, 1140 ⟨82CC1042⟩, 1098, 1140 ⟨84JOC3904⟩; **2**, 36 ⟨83S460⟩, 69 ⟨85TL4763, 89JA4392⟩, 69, 564 ⟨84JOC3904⟩, 444 ⟨90JA5244, 91TL3371⟩, 453 ⟨85TL783, 87JOC2170⟩, 474 ⟨85JA3501, 91COS(1)231⟩, 564 ⟨B-94MI 211-01⟩; **3**, 88 ⟨83TL2399⟩, 617 ⟨83S142⟩, 618 ⟨82S591⟩; **4**, 122 ⟨85S982⟩, 570, 575 ⟨90JA5244, 85JA5301⟩, 900 ⟨82CL1933⟩, 954 ⟨84TL4233⟩, 988 ⟨82JCS(P1)1059⟩; **5**, 205 ⟨86CL967⟩, 318, 998 ⟨71BCJ1632⟩, 546 ⟨84S797⟩, 547 ⟨69JOC730⟩, 551, 560, 574, 575 ⟨84S827⟩, 558 ⟨84JCS(P1)85, 71BCJ1639⟩, 998 ⟨71BCJ1644⟩; **6**, 564 ⟨81JCS(P1)2499, 81S908⟩
Imamura, K., **1**, 166 ⟨94SL1⟩
Imamura, M., **1**, 1142 ⟨77JCP(66)4671⟩
Imanaka, T., **1**, 455 ⟨79MI 103-01⟩, 1004, 1107 ⟨79JOC55, 74TL1067⟩; **2**, 314 ⟨91BCJ602⟩, 314, 318 ⟨88CL1759⟩, 616 ⟨74TL1067, 76CL449⟩, 654 ⟨94CC797⟩; **3**, 121 ⟨76CL231, 88CL285⟩, 620 ⟨85CL1339⟩
Iman, M., **3**, 274 ⟨90S631⟩
Imaniech, H., **4**, 651 ⟨92TL543⟩
Imanieh, H., **1**, 160 ⟨92TL543⟩
Imanishi, T., **2**, 444 ⟨82MI 208-01⟩; **3**, 59 ⟨88JOC1623⟩; **5**, 1120 ⟨92CC1379⟩
Imao, S., **5**, 248, 249 ⟨87NKK1442⟩
Imaoka, M., **1**, 590, 591 ⟨78CL413⟩
Imashev, U. B., **6**, 479 ⟨90ZOB1826⟩
Imazaki, H., **3**, 151 ⟨91CC86⟩
Imbach, J. L., **6**, 506 ⟨82JMC178⟩
Imbeaux-Oudotte, M., **3**, 236 ⟨87BSF861⟩
Imbroisi, D. O., **3**, 221 ⟨89TL4309⟩
Imes, R. H., **1**, 869 ⟨67JA3905⟩
Imfeld, P., **5**, 700 ⟨69HCA569⟩
Imhof, W., **1**, 956 ⟨93CB1077⟩; **2**, 875, 896 ⟨93JOM(444)53⟩, 1036, 1037 ⟨90CB417⟩; **4**, 681 ⟨92JOM(441)81⟩
Imhoff, M. A., **1**, 1116 ⟨70JA3802⟩; **2**, 689 ⟨70AG(E)273, 76JOC4099⟩
Imi, K., **1**, 208 ⟨87JOC1013⟩; **2**, 922, 923 ⟨88T4009⟩; **3**, 632, 698 ⟨87JOC1013⟩; **4**, 654 ⟨85TL2101⟩
Immenkeppel, M., **5**, 876, 877, 878 ⟨87ZAAC(553)7⟩, 877, 878 ⟨85TL3551⟩; **6**, 680, 682, 683 ⟨87ZAAC(553)7⟩, 681, 682, 689, 690 ⟨85TL3551⟩, 683 ⟨84AG(E)895⟩
Imoto, E., **5**, 31 ⟨66BCJ852⟩
Imoto, H., **1**, 303 ⟨94JA4131⟩
Imoto, M., **3**, 716, 717 ⟨87JAN1207⟩; **5**, 541 ⟨79NKK1224⟩
Imoto, S., **3**, 659 ⟨76CPB3011⟩
Imoto, T., **3**, 361 ⟨89CC505⟩
Impastato, F. J., **1**, 851 ⟨64JA3283⟩, 885 ⟨62JA4838⟩; **2**, 554 ⟨64JA3283⟩
Imperiali, B., **2**, 532 ⟨81JA1566⟩
Impicciatore, M., **5**, 347 ⟨84FES788⟩
Imrich, J., **5**, 1043 ⟨88ZC141⟩
Imura, M., **4**, 850 ⟨80BCJ1661⟩
Imuta, M., **3**, 357 ⟨91CPB672⟩
Imwinkelried, R., **1**, 165 ⟨90JA2264⟩, 374 ⟨89JA5493⟩, 1026 ⟨87HCA448⟩; **2**, 533 ⟨89JA5493⟩; **4**, 213 ⟨86AG(E)178, 87HCA448⟩; **5**, 953 ⟨88OM702⟩
Inaba, A., **5**, 631 ⟨75H(3)1081⟩
Inaba, M., **2**, 300 ⟨88CPB2342⟩; **4**, 742 ⟨83TL4727⟩; **6**, 316 ⟨78JOC595⟩
Inaba, S., **3**, 146, 289, 303 ⟨83TL2451⟩, 289 ⟨85JOC1373⟩, 616 ⟨75CL331⟩, 617 ⟨75CL737⟩; **5**, 759, 777 ⟨77JOM(140)97, 75JOM(99)C5⟩, 777 ⟨74JOM(72)C11⟩; **6**, 515 ⟨77JOM(140)97⟩
Inaba, T., **1**, 826 ⟨89BCJ3835, 91JCR(S)172⟩
Inada, A., **2**, 55 ⟨86JOC240⟩; **3**, 57 ⟨81JOC240⟩
Inaga, J., **2**, 196 ⟨82JA5781⟩
Inagaki, I., **2**, 311 ⟨87CL1275⟩
Inagaki, M., **2**, 249, 401 ⟨79BCJ1102⟩; **3**, 631 ⟨92BCJ111, 92JOC5643⟩
Inagaki, S., **5**, 1122 ⟨87JCS(P1)2531⟩
Inagaki, T., **5**, 138 ⟨87ABC1345⟩
Inagawa, Y., **6**, 434 ⟨85JHC1643⟩
Inaishi, M., **2**, 21 ⟨79BCJ255, 88OSC(6)90⟩
Inaki, H., **1**, 102 ⟨80H(14)1441⟩
Inamasu, T., **4**, 247 ⟨91TL2039⟩; **5**, 242, 253 ⟨90JCR(S)56⟩
Inamatsu, T., **5**, 665 ⟨90CL559⟩
Inamdar, S. P., **5**, 184 ⟨66JIC615⟩
Inami, H., **1**, 194 ⟨93CC1595⟩
Inami, K., **5**, 368 ⟨85BCJ352⟩
Inamoto, N., **1**, 754 ⟨84CL1997⟩, 836 ⟨86CL501⟩; **2**, 22 ⟨91BCJ3749⟩, 185 ⟨88CC40⟩, 460, 462 ⟨89CL849⟩, 793, 794, 795 ⟨89CL2087⟩, 794, 795 ⟨92JA1830⟩, 840 ⟨69BCJ2975⟩, 851 ⟨73BCJ898, 75BCJ546⟩, 853 ⟨72BCJ506⟩, 854 ⟨71CC1186⟩, 857 ⟨84CL603⟩; **3**, 318 ⟨82CC1055⟩, 332 ⟨87JA279⟩, 337, 340 ⟨84TL849, 82CC1187⟩, 341 ⟨87BCJ1037⟩, 347, 363 ⟨79TL3673⟩, 347, 363 ⟨81BCJ3541⟩, 358 ⟨76TL4283⟩, 359 ⟨74CL1439, 76TL3815⟩, 359 ⟨79AG(E)954, 82AOC(21)241⟩, 359, 361 ⟨79BCJ3640, 72CL9⟩, 361 ⟨75JCS(P1)270⟩, 383 ⟨86CC71, 89JA5949⟩, 392 ⟨83CC1429, 88BCJ861⟩, 439 ⟨76BCJ1913, 76BCJ550, 77H(7)1131, 78BCJ535⟩, 480 ⟨79TL2415⟩, 605 ⟨84TL1809, 85PS(25)237, 87PAC1005, 89TL839⟩, 609 ⟨86JOM(311)C63⟩, 609 ⟨90HAC339⟩, 634 ⟨79CL723⟩, 689 ⟨92CL1053⟩; **4**, 346 ⟨79TL2415⟩, 376 ⟨89JA5949⟩, 608 ⟨87CL2293⟩, 979 ⟨79BCJ3640⟩, 1049 ⟨87JA279⟩; **5**, 516, 523, 524 ⟨86BCJ2529, 86CC71⟩, 524 ⟨87BCJ1037⟩, 560 ⟨84CC1010⟩, 634 ⟨75JCS(P1)270, 76CC562⟩, 634 ⟨78BCJ301⟩, 635 ⟨87H(25)183⟩, 877 ⟨88CL1733⟩, 879 ⟨83CL1653⟩, 884 ⟨88TH17⟩, 1086

⟨90BCJ2736⟩, 1086, 1087 ⟨84TL1809⟩, 1086, 1087 ⟨88POL2213⟩, 1092 ⟨84CC689, 89CC1732⟩, 1092 ⟨89TL187, 89TL5433⟩, 1092 ⟨89TL839⟩, 1115 ⟨89JA5949⟩; **6**, 120, 698 ⟨76TL3695⟩, 176 ⟨87JA279⟩, 304 ⟨75JCS(P1)270⟩, 314 ⟨78BCJ301⟩, 496 ⟨73CC13⟩, 581 ⟨69BCJ2975⟩, 655 ⟨74BCJ935⟩, 680, 690 ⟨88CL1733⟩

Inamoto, T., **5**, 286 ⟨82BCJ643⟩

Inamoto, Y., **5**, 282 ⟨77S632⟩

Inanaga, I., **5**, 173 ⟨79CL1019⟩, 173 ⟨79CL1021⟩

Inanaga, J., **1**, 14 ⟨87TL4437⟩, 16 ⟨91TL4963⟩, 195 ⟨86TL1195⟩, 458, 516 ⟨91TL5559⟩, 606 ⟨84CC1345⟩, 680 ⟨86TL3891, 87CL2101⟩, 962 ⟨86TL5237⟩, 975 ⟨87CL2275⟩, 1045 ⟨81CL1415⟩; **3**, 3, 89, 114 ⟨81BCJ2221⟩, 189 ⟨91TL3039⟩, 223, 275 ⟨79CL1021⟩; **4**, 219 ⟨79CL1277⟩; **5**, 173 ⟨79BCJ1989⟩

Inanici, Y., **5**, 1037 ⟨80CZ365⟩

Inayama, S., **1**, 561 ⟨80CPB282⟩

Inazu, T., **3**, 709 ⟨82TL5335⟩; **5**, 221 ⟨88BCJ4467⟩

Inbal, Z., **4**, 98 ⟨73JOC2251⟩

Inbasekaran, M., **4**, 61, 62, 73, 79, 780 ⟨85JA735⟩, 79, 780 ⟨85CC678⟩

Inch, T. D., **2**, 463 ⟨79TL2729⟩, 466 ⟨76TL3645⟩

Incremona, J. H., **1**, 928 ⟨72JA301, 76JOC523⟩

Indelicato, J. M., **1**, 1110 ⟨68JA6515⟩, 1110, 1115 ⟨69JA6194⟩

Indo, M., **3**, 494 ⟨70BRP1202353⟩

Indzhikyan, M. G., **2**, 437 ⟨90ZOB313⟩, 439 ⟨86AKZ258⟩, 446 ⟨80AKZ408, 86AKZ463⟩, 446 ⟨90AKZ610⟩, 447 ⟨86AKZ457⟩; **4**, 553 ⟨90AKZ537⟩, 914 ⟨80CA186471c, 80CA6622a⟩, 914 ⟨83CA(98)4614z, 86CA(104)19636z⟩

Inenaga, M., **4**, 808 ⟨89TL6701⟩

Ingall, A. H., **3**, 371, 376 ⟨84CHEC(3)885⟩

Ingemann, S., **5**, 437 ⟨85RTC74⟩

Ingendoh, A., **3**, 412 ⟨79CB1297⟩; **4**, 438, 445 ⟨78LA1928⟩

Ingham, R. K., **2**, 12 ⟨56CRV219⟩, 627 ⟨57CRV219⟩, 845 ⟨58JOC1298⟩; **4**, 27, 35, 37 ⟨56CRV219⟩; **5**, 114, 115 ⟨56JA1689⟩

Inghardt, T., **1**, 218 ⟨90S285⟩; **2**, 1099 ⟨90S285⟩

Ingles, D. L., **4**, 237 ⟨59AJC97⟩

Ingleson, D., **2**, 847 ⟨78CC870, 80JCS(P1)1150⟩; **5**, 404 ⟨78CC870, 80JCS(P1)1150⟩

Inglis, F., **2**, 835 ⟨66JCS(A)520⟩

Inglis, J. K. H., **3**, 612 ⟨41OSC(1)254⟩; **5**, 95 ⟨32OSC(1)254⟩

Inglis, T., **5**, 781 ⟨75JCS(D)930⟩

Ingold, C. F., **5**, 347 ⟨83JA7755⟩

Ingold, C. H., **3**, 723 ⟨24JCS(125)2543, 24JCS(125)87⟩

Ingold, C. K., **1**, 49, 640 ⟨29JCS2342⟩, 612 ⟨30JCS705⟩, 853 ⟨38JCS78⟩, 891 ⟨33JCS1571, 33JCS526⟩, 891 ⟨35JCS255⟩, 892 ⟨54JCS2918⟩; **2**, 25 ⟨55JCS3173⟩, 442 ⟨33JCS989⟩, 649 ⟨52QR34⟩, 780 ⟨50JCS2657⟩, 842 ⟨29JCS2342⟩; **6**, 69 ⟨21JCS1222⟩

Ingold, K. U., **1**, 5 ⟨B-83MI 101-01⟩, 322 ⟨75JA4269⟩, 329 ⟨B-73MI 108-02⟩, 330, 341, 936 ⟨84JA343⟩, 331 ⟨85JA4594⟩, 331, 336, 342, 936 ⟨87JOC3509⟩, 340 ⟨92JOC4284⟩, 919, 922 ⟨76ACR13⟩, 919, 927 ⟨B-80MI 119-05⟩, 919 ⟨B-71MI 119-05⟩, 923 ⟨84JA1793⟩, 924, 934 ⟨88JA2877⟩, 925 ⟨93ACR219⟩, 935 ⟨B-83MI 119-01⟩, 936 ⟨83JA3578, 85JA4595⟩, 937 ⟨B-83MI 119-02⟩; **2**, 21 ⟨90ACR219⟩, 388 ⟨73JA3228⟩, 1010 ⟨74JA6202, 74JA6203⟩, 1010 ⟨75JA4269, 75JA5526⟩, 1010 ⟨76ACR13⟩; **3**, 391 ⟨76CC205⟩, 391 ⟨76JPC1901⟩; **4**, 850 ⟨78JA3868⟩, 1078 ⟨90JA4284⟩; **5**, 516, 526 ⟨77JPC915⟩; **6**, 726, 734 ⟨76ACR13⟩

Ingram, A. S., **3**, 376, 377 ⟨74JCS(P1)722, 74JCS(P1)242⟩

Ingram, D. D., **1**, 28 ⟨75JA6873⟩

Ingram, G., **4**, 706 ⟨93JOM(463)163⟩

Ingrosso, G., **2**, 581 ⟨80JCS(D)1633⟩

Ingvorsen, K., **5**, 118 ⟨B-88MI 502-01⟩

Ingwall, J. G., **3**, 73 ⟨69CC466⟩

Inhoffen, H. H., **1**, 736 ⟨55CB1415⟩, 1023 ⟨56CB853⟩; **3**, 55 ⟨59CB1564⟩

Inman, C. E., **4**, 3 ⟨58JA6533⟩

Inman, K. C., **1**, 339, 1130 ⟨86TL2833⟩

Inners, R. R., **1**, 600 ⟨78JA3874⟩

Innocenti, P., **2**, 438 ⟨84JCS(D)575⟩

Ino, A., **1**, 1009 ⟨91CL1271⟩

Inokawa, H., **2**, 457 ⟨77S179, 79BCJ466⟩; **4**, 854 ⟨79BCJ466⟩

Inokawa, S., **2**, 429 ⟨83BCJ219, 77S179⟩, 457 ⟨79BCJ466⟩, 632 ⟨80BCJ3691⟩; **3**, 332, 364 ⟨75BCJ2907⟩, 364, 378 ⟨76BCJ3128⟩, 493 ⟨77S626⟩, 687 ⟨76BCJ3124⟩; **4**, 834 ⟨76BCJ3128⟩, 847 ⟨76S552, 79BCJ466⟩; **5**, 403, 404 ⟨80BCJ1625⟩, 728, 729 ⟨77S626⟩, 775 ⟨79BCJ1541⟩; **6**, 87, 99 ⟨76S551⟩

Inokuchi, T., **1**, 946 ⟨85JOC5875⟩; **2**, 28 ⟨80JOC2731⟩, 83 ⟨90JOC1548⟩; **3**, 7, 120 ⟨91BCJ796⟩, 7, 31 ⟨90JOC462⟩, 46 ⟨86JOC254, 86JOC155⟩, 129 ⟨92CL1895⟩, 152, 154, 155 ⟨80JOC2731⟩; **4**, 181 ⟨83CL1349⟩

Inomata, K., **1**, 333 ⟨73JA967⟩, 503 ⟨83BCJ2539⟩, 619 ⟨88CL2009⟩; **2**, 121 ⟨81CL1457⟩, 743 ⟨92BCJ1379⟩; **3**, 60 ⟨89CL1283⟩, 221 ⟨89CC271⟩; **4**, 193 ⟨89CL737⟩, 412 ⟨91CC462⟩; **5**, 291 ⟨78BCJ1866⟩

Inomata, M., **2**, 57 ⟨89JOC3482⟩

Inoue, A., **1**, 1010 ⟨86CPB1447⟩

Inoue, H., **2**, 638 ⟨76BCJ3673⟩; **4**, 325 ⟨78GEP2705994⟩; **5**, 70, 73 ⟨80CC327⟩; **6**, 503 ⟨90JHC2215⟩

Inoue, I., **1**, 731 ⟨74TL2587⟩; **3**, 494 ⟨74TL2587⟩; **4**, 424 ⟨82JOC1706⟩

Inoue, J., **3**, 634, 635 ⟨87CL887⟩

Inoue, K., **2**, 313 ⟨90TL3023⟩, 626 ⟨88BCJ597⟩; **3**, 5, 118 ⟨83JOC1286⟩, 43 ⟨86TL6083⟩, 223, 275 ⟨79CL1021⟩, 347, 363 ⟨79TL3673⟩, 347, 363 ⟨81BCJ3541⟩, 432 ⟨86JOC549⟩, 618, 697, 722 ⟨61JOC2202⟩; **4**, 300, 311 ⟨82JA5753⟩, 319 ⟨89BCJ1215⟩; **5**, 173 ⟨79CJC2180⟩, 173 ⟨79CL1021⟩, 578, 608, 618 ⟨75JOM(92)139⟩; **6**, 74 ⟨61JOC2202⟩, 558 ⟨89CPB1249⟩

Inoue, M., **1**, 160, 451 ⟨84TL1999⟩, 200 ⟨86S689⟩, 212, 475 ⟨78BCJ174⟩, 1053 ⟨87CL1345⟩; **2**, 972 ⟨84TL1999⟩; **3**, 130, 285 ⟨79TL4303⟩, 616, 627, 633 ⟨83T967⟩, 627 ⟨81TL4279, 80TL389⟩; **4**, 1060 ⟨84TL1999⟩; **5**, 671 ⟨84JHC725⟩

Inoue, N., **2**, 619 ⟨85TL1651⟩, 964 ⟨81JOC4416⟩

Inoue, S., **1**, 34 ⟨75BCJ1892⟩, 77 ⟨87JA1596, 87JA4129⟩, 263 ⟨75JOC2009⟩, 290 ⟨88JOM(25)348, 88OM739⟩, 291 ⟨90OM1325⟩, 408 ⟨92JOC428⟩, 408, 823 ⟨90JA1999⟩, 592 ⟨75JA2553⟩, 730 ⟨70JOC565⟩, 835 ⟨71TL4661⟩, 839 ⟨90BCJ1328⟩; **2**, 223 ⟨75BCJ1665⟩, 700 ⟨87OM2568⟩; **3**, 57 ⟨86CL2035⟩, 157 ⟨85JOC603⟩, 238 ⟨80JHC289⟩, 563 ⟨85CC1773⟩, 616 ⟨90CL1171, 93CL375⟩, 628 ⟨79MAC1089, 81CC229, 82MAC579, 85MAC1755, 86BCJ89, 86CL931, 89CL2119, 90JOC181⟩, 629 ⟨91SL563, 91TL4333, 92CL2443, 92JA7969, 92JOC6778⟩, 632, 638 ⟨93CL975⟩; **4**, 167 ⟨81JCS(P1)1015⟩, 361 ⟨90SC3245⟩, 507 ⟨90JA1999⟩, 522 ⟨91S169⟩, 604, 625, 638 ⟨88JOM(354)155⟩, 609, 638 ⟨89OM1237⟩, 625 ⟨90OM1325⟩, 625, 637 ⟨86OM1197⟩, 625, 637, 1055, 1058 ⟨88OM739⟩, 625, 662 ⟨87OM2568⟩, 626, 662 ⟨87SC1273⟩, 638, 648, 662 ⟨91JOC347⟩, 638, 663 ⟨88JOM(348)25⟩; **5**, 117 ⟨84MI 502-02⟩, 431 ⟨90SC3245⟩; **6**, 185 ⟨87SC1273⟩, 185 ⟨88JOM(354)155⟩, 185 ⟨90OM1325⟩, 185, 186 ⟨88OM739⟩

Inoue, S.-I., **2**, 816 ⟨90JA4897⟩

Inoue, T., **2**, 32 ⟨73CC212⟩, 34 ⟨92TL1299⟩, 362 ⟨77TL2413⟩, 532 ⟨77CL153⟩, 532, 698 ⟨76CL559⟩, 532, 698 ⟨80BCJ174⟩; **3**, 136 ⟨90JA455, 91CC1752⟩, 629 ⟨93JOC1515⟩; **5**, 329 ⟨86JOC2391⟩, 344 ⟨89CL2181⟩, 437 ⟨90JA455⟩

Inoue, Y., **1**, 268 ⟨64T1695⟩, 357 ⟨92CRV741⟩, 483 ⟨86CL407⟩, 676, 678 ⟨84BCJ2335⟩; **2**, 52 ⟨91CC795⟩, 247 ⟨86BCJ3897⟩, 398 ⟨88BCJ3020, 88TL4769⟩, 398 ⟨90T7471⟩; **3**, 139 ⟨84CL1227⟩, 161 ⟨90T7471⟩, 244 ⟨88TL5941, 90SC3063⟩, 264 ⟨65TL2261⟩; **5**, 223 ⟨85JCS(P1)2361⟩, 726 ⟨85BCJ2721⟩; **6**, 116 ⟨86JOC4737, 87NKK1447⟩, 464 ⟨87BCJ1204⟩

Inouye, M., **1**, 1035, 1047 ⟨88TL5379⟩; **3**, 707, 708 ⟨89TL1257⟩; **5**, 728 ⟨89CL1261⟩, 802 ⟨88BCJ3649⟩; **6**, 498 ⟨85CL1671⟩
Inouye, S., **3**, 357 ⟨90JAN1150⟩
Inouye, Y., **1**, 839 ⟨73CC848⟩; **2**, 355 ⟨76JOC300⟩; **3**, 616 ⟨75ABC571⟩; **5**, 183, 192 ⟨60JOC1265⟩
Inove, M., **1**, 749 ⟨86JOC959⟩
Insalaco, M. J., **6**, 489 ⟨88OSC(6)207⟩
Inshakova, V. T., **6**, 243 ⟨89ZOB2628⟩
Interrante, L. V., **4**, 606, 654 ⟨91OM1336⟩
Intini, F. P., **5**, 386 ⟨93JA5123⟩, 696 ⟨89JCS(D)947⟩
Inubushi, A., **1**, 1026 ⟨87CL1975, 88BCJ4037⟩
Inubushi, T., **1**, 925 ⟨89JA6849⟩; **2**, 87 ⟨91JA8981⟩
Inubushi, Y., **3**, 59 ⟨81CPB766⟩, 80 ⟨78BCJ1186⟩
Inui, N., **2**, 51 ⟨90TL7333⟩
Inui, T., **2**, 219 ⟨70JOC4148⟩; **6**, 434 ⟨71BCJ2515⟩
Inukai, K., **6**, 46, 47 ⟨66BCJ2191⟩
Inukai, N., **2**, 3 ⟨76CPB820, 67BCJ2913⟩
Inukai, T., **1**, 373 ⟨67JOC869⟩; **5**, 146 ⟨67JOC869⟩
Invergo, B. J., **3**, 83 ⟨81TL11⟩
Invernizzi, A. G., **5**, 588 ⟨93JHC1079⟩
Ioffe, B. V., **2**, 377 ⟨69ZOR1735⟩, 380 ⟨59ZOB3154⟩, 403 ⟨69ZOR268⟩
Ioffe, S. L., **2**, 394 ⟨86BAU2442⟩; **4**, 524 ⟨74MI 411-01⟩; **5**, 362 ⟨91IZV1108, 91MI 507-06⟩, 380 ⟨83IZV1413⟩, 680 ⟨91MI 517-01⟩; **6**, 65 ⟨76IZV2547⟩, 145 ⟨73IZV122⟩, 147 ⟨85IZV2635⟩, 148 ⟨72IZV490⟩, 148 ⟨73IZV1149⟩, 149 ⟨72IZV2603⟩, 149, 653 ⟨73ZOR896⟩, 169 ⟨74MI 605-01⟩, 368 ⟨72MI 612-01⟩
Ioffe, S. T., **1**, 114 ⟨B-67MI 104-01⟩; **2**, 557 ⟨B-67MI 211-01⟩
Iokubaitite, S. P., **6**, 576 ⟨73RCR587⟩
Ionin, B. E., **2**, 849 ⟨63ZOB2863⟩
Ionin, B. I., **2**, 448 ⟨76JA1290, 90ZOB2394⟩, 848 ⟨65ZOB1577⟩, 1028 ⟨85JGU1771⟩, 1071 ⟨62JGU2355, 63JGU2791, 65JGU1582, 65JGU1910, 67JGU898⟩; **3**, 591 ⟨83JGU205, 85JGU26⟩; **4**, 488 ⟨68ZOB1248⟩, 786 ⟨85ZOR244⟩, 802 ⟨67JGU2024, 87JGU844⟩, 914 ⟨84ZOB457⟩
Ionkin, A. S., **3**, 689 ⟨91HAC577⟩; **4**, 343 ⟨87IZV418⟩, 569 ⟨90PS(53)1⟩; **5**, 395 ⟨90BAU2424⟩, 395, 396 ⟨91PS(55)133⟩; **6**, 160 ⟨91PS(61)361⟩, 691 ⟨92PS(66)257⟩
Ionov, L. B., **2**, 886, 888 ⟨72JGU538, 72JGU1952⟩, 890 ⟨68JGU370, 75JGU1476⟩, 891 ⟨70JGU127, 88JGU1040⟩; **5**, 1132 ⟨74MI 530-01, 74MI 530-02⟩; **6**, 476 ⟨76JGU2441⟩
Iordache, F., **4**, 152 ⟨92RRC393⟩
Iori, M., **4**, 850 ⟨80JCS(P1)2693⟩
Iorio, E. J., **1**, 412, 810 ⟨67JA3462, 70JOC856⟩
Iotschich, Z., **2**, 1018 ⟨05BSF184⟩
Iotsitch, J. I., **6**, 77 ⟨11BSF1308⟩
Iovel, I. G., **4**, 1080 ⟨87CC31⟩
Iovu, M., **4**, 411 ⟨84RRC333⟩
Ipach, I., **2**, 789 ⟨78CB692⟩, 789, 790 ⟨76CB3541, 78S592⟩, 790 ⟨79S981⟩
Ip, D. P., **2**, 613 ⟨81JA1964⟩
Ipaktschi, J., **1**, 736 ⟨73TL4425⟩
Ipaktschl, J., **3**, 494 ⟨73TL4425⟩
Ipatieff, V., **2**, 408 ⟨29CB598⟩
Ipatieff, V. N., **2**, 478 ⟨39JA71⟩
Ippen, J., **1**, 598 ⟨77JA8116⟩; **4**, 851 ⟨80AG(E)66⟩
Ipsale, S., **1**, 1063 ⟨85CC1571⟩
Iqbal, J., **1**, 320, 336, 345, 346, 925, 926, 937, 942, 946 ⟨94CRV519⟩; **2**, 79 ⟨89TL2421⟩, 318 ⟨90TL575⟩; **3**, 139 ⟨93TL6301⟩, 214 ⟨88CC692⟩; **4**, 67 ⟨83TL327⟩, 187 ⟨89SC901⟩; **5**, 264, 275 ⟨87CC114⟩, 329 ⟨93TL7371⟩
Iqbal, M., **4**, 411 ⟨84RRC333⟩
Iqbal, M. Z., **5**, 445, 453 ⟨78JA1687⟩
Iqbal, S. M., **2**, 123, 124 ⟨60JCS1030⟩, 128 ⟨66JCS(C)415⟩
Irani, R. R., **4**, 470 ⟨66JOC1603⟩
Iranmanesh, H., **2**, 675 ⟨78JOC1244⟩; **3**, 323 ⟨78JOC1244⟩
Iranpoor, N., **2**, 84 ⟨90TL735⟩, 135 ⟨91T9861⟩, 222 ⟨91T149⟩, 223 ⟨90SC1047⟩; **3**, 5, 116 ⟨86T719⟩
Iranpour, M., **2**, 945 ⟨93JA1586⟩
Iranshahi, L., **6**, 431 ⟨84JOC1174⟩
Iratcabal, P., **3**, 234 ⟨76S33⟩
Irelan, J. R. S., **2**, 435, 435, 1060, 1064 ⟨69JOC2684⟩
Ireland, R. E., **1**, 17, 434 ⟨72JA5098⟩, 89 ⟨63JOC6⟩, 131, 416, 829, 874, 876 ⟨76JA2868⟩, 218, 831 ⟨83JA1988⟩, 414, 416, 829, 830 ⟨72JA5897⟩, 416, 417, 829, 830, 876 ⟨75TL3975⟩, 417, 795, 830 ⟨91JOC3572⟩, 417, 830, 876 ⟨91JOC650⟩, 434 ⟨69TL2145⟩, 450 ⟨84JA3668⟩, 831 ⟨79CJC1743, 80JA1155⟩, 831 ⟨81JA2446, 85JA3285⟩, 831 ⟨86JOC635, 88JA854⟩, 881 ⟨61OS(41)5⟩, 1052 ⟨70TL2247⟩, 1079 ⟨92JOC5071⟩; **2**, 25 ⟨85JA3285⟩, 420, 696 ⟨76JA2868⟩, 693 ⟨69TL2145, 72JA5098⟩, 693 ⟨83OS(61)116⟩; **3**, 6, 119, 148 ⟨93JOC2899⟩, 140 ⟨90JOC1425⟩, 220 ⟨85JOC2198⟩, 253 ⟨62JOC1615⟩; **4**, 831 ⟨72JA5897, 91JOC650⟩; **5**, 227 ⟨75TL3975, 76JA2868⟩, 239 ⟨85JA3285⟩, 412, 413, 422 ⟨85JOC2198⟩
Irgolic, K., **2**, 414, 504 ⟨68JOM(12)377⟩, 465, 858 ⟨65JOM(4)320⟩, 498, 504 ⟨66JOM(6)17⟩
Irgolic, K. J., **1**, 438 ⟨75JOM(88)175⟩; **2**, 261 ⟨80JCS(P1)2287⟩, 279 ⟨75JOM(88)175, 83JOM(253)31⟩, 292 ⟨72JOM(32)97⟩, 494, 500 ⟨72JOM(39)301⟩, 498, 1080 ⟨71JINC3177⟩, 499 ⟨71JA5677⟩, 768 ⟨B-74MI 215-02⟩, 872 ⟨83JOM(257)67⟩; **4**, 507 ⟨84JOM(268)31⟩; **6**, 308 ⟨90HOU(E12b)1⟩
Iriarte, J., **3**, 235 ⟨73JA6655⟩
Irick, G., **3**, 446 ⟨59JOC1925⟩
Irie, H., **5**, 329 ⟨93TA1573⟩
Irie, M., **3**, 668 ⟨86TL2411⟩; **4**, 924 ⟨87JCS(P1)1095⟩; **5**, 700, 701 ⟨92JCS(P1)2127⟩
Irie, S., **4**, 208 ⟨89CL221⟩
Iriepa, I., **4**, 405 ⟨87CJC687⟩
Irifune, S., **2**, 34 ⟨88S366⟩, 36 ⟨89S283⟩
Iriguchi, J., **4**, 931 ⟨84JA2440⟩; **5**, 415 ⟨84JA2440⟩
Iritani, K., **2**, 983 ⟨86JA2753⟩
Iriuchijima, S., **1**, 210 ⟨72TL4605⟩; **4**, 74 ⟨70BCJ2271, 71S89⟩, 74, 75 ⟨69TL5259⟩, 77 ⟨70S588⟩, 221 ⟨74JA4280⟩
Irmer, E., **2**, 446 ⟨92S577⟩
Irmisch-Pielartzik, B., **5**, 122 ⟨85HOU(E5)659⟩, 182 ⟨89HOU(5)633⟩, 692, 699, 700 ⟨85HOU(E5/1)812⟩
Irngartinger, H., **1**, 634 ⟨87TL1985⟩, 705 ⟨84CB966, 86CB1540⟩, 708 ⟨84CB1868, 87CB5⟩, 782 ⟨93LA521⟩, 986 ⟨86AG(E)340⟩; **3**, 616 ⟨85LA566⟩; **6**, 255, 275, 304, 306 ⟨91CB2025⟩
Ironside, M. D., **4**, 918 ⟨90JCS(P1)2811⟩
Irvin, C. M., **6**, 241 ⟨81JFC(17)591⟩
Irvin, K. J., **6**, 215 ⟨50TFS295⟩
Irvin, R. S., **2**, 361 ⟨60JCS5079⟩; **6**, 607, 609 ⟨58JCS764⟩
Irvine, J., **5**, 62 ⟨55OSC(3)326⟩
Irvine, J. C., **2**, 89 ⟨03JCS1021⟩
Irvine, J. L., **1**, 827 ⟨73JA3348⟩
Irving, A., **5**, 943 ⟨90OM2836⟩
Irving, H., **5**, 841 ⟨32JA263⟩
Irwin, D. C., **3**, 658 ⟨78S894⟩
Irwin, J. G., **3**, 487 ⟨74CC466⟩; **5**, 874 ⟨74CC466⟩
Irwin, K. C., **1**, 37 ⟨68JOC1421⟩
Irwin, R. M., **4**, 621 ⟨77JA3273⟩
Irwin, R. S., **5**, 1105 ⟨61JCS2447⟩, 1105, 1110 ⟨61JCS2436⟩
Irwin, W. J., **4**, 58 ⟨93JCS(P1)2303⟩
Isaa, B., **1**, 827 ⟨91JCS(P1)2089⟩
Isaaca, N. S., **1**, 51 ⟨86TL1209⟩
Isaacs, L. D., **5**, 956 ⟨89TL4061⟩
Isaacs, N. S., **1**, 571 ⟨92JA6330⟩, 844 ⟨B-73MI 119-01⟩, 891 ⟨B-74MI 119-04⟩, 1140 ⟨88TL5017⟩; **2**, 26 ⟨72TL3869⟩, 617 ⟨72CC443⟩; **3**, 541 ⟨70JCS(B)830⟩; **4**, 738 ⟨72CC443⟩, 1072 ⟨B-74MI 424-01⟩; **5**, 652 ⟨76CSR181⟩
Isaacson, E. I., **5**, 215 ⟨65JPS1373⟩
Isabelle, L. M., **2**, 1010 ⟨81JOC179⟩

Isaeva, L. S., **2**, 838, 839 ⟨57DOK(117)996⟩, 876 ⟨58DOK(122)614⟩
Isaev, V. L., **4**, 894 ⟨74MI 420-01⟩
Isagawa, K., **2**, 30 ⟨77CL1117⟩
Isagulyants, G. V., **6**, 145 ⟨73IZV122⟩
Isaka, I., **4**, 205 ⟨76CPB102⟩
Isaka, M., **2**, 533 ⟨89TL7121⟩, 589 ⟨88JA1297, 90JA7428⟩
Isak, H., **6**, 143 ⟨82CB1721⟩
Isaki, Y., **3**, 73 ⟨89TL5289⟩
Isakov, V. V., **4**, 339 ⟨91ZOB909⟩
Isaksson, R., **1**, 248 ⟨87CC381⟩
Isayama, K., **6**, 464 ⟨68TL831⟩
Isayama, S., **2**, 478 ⟨89CL569⟩
Isbell, A. F., **2**, 434 ⟨62JOC2573⟩, 457 ⟨72JOC4396⟩; **3**, 49 ⟨69JOC755⟩; **4**, 476 ⟨72JOC4397⟩
Isbell, I. R., **4**, 467 ⟨64JOC832⟩
Isbell, T. A., **4**, 214 ⟨91JA8168⟩
Ise, S., **4**, 265, 853, 854 ⟨90CPB3242⟩
Isegawa, Y., **6**, 448 ⟨71JOC858⟩
Iseka, K., **4**, 811 ⟨94TA961⟩
Iseki, K., **1**, 1024 ⟨86TL87⟩
Isele, G. L., **5**, 285, 292 ⟨71S266⟩
Iseli, R., **4**, 53 ⟨77HCA1061⟩, 53, 54, 56 ⟨78HCA2047⟩
Iselin, B., **2**, 171 ⟨61HCA61⟩; **6**, 489 ⟨68HCA622⟩
Iselin, M., **1**, 469 ⟨16CB25⟩
Isenberg, N., **2**, 232, 724 ⟨73IJS307⟩
Isenberg, W., **5**, 663 ⟨83CB724⟩
Isensee, G., **6**, 505 ⟨74AP(307)7⟩
Iserson, H., **6**, 217, 225, 409, 419 ⟨57JA5801⟩
Ishar, M. P. S., **1**, 499 ⟨90JCS(P1)2185⟩
Ishchenko, O. S., **5**, 540 ⟨75JGU2421⟩
Ishchuk, Y. N., **5**, 703 ⟨69JOU252, 69ZOR263⟩
Ishi, H., **3**, 87 ⟨93CPB1166⟩
Ishi, Y., **2**, 469 ⟨70BCJ3232⟩; **3**, 639 ⟨79JCS(P1)26⟩; **4**, 90 ⟨91COS(7)762⟩; **5**, 1070 ⟨70OCS23⟩
Ishiba, T., **2**, 224 ⟨70TL441, 71CPB1022⟩; **3**, 255 ⟨71CPB2194⟩; **4**, 835 ⟨71CPB2194⟩; **6**, 85 ⟨77JOC1543⟩, 114 ⟨71CPB2222⟩
Ishibaba, M., **3**, 29, 157 ⟨88BCJ3637⟩
Ishibashi, H., **1**, 362 ⟨90CC1436⟩; **2**, 684 ⟨94SL49⟩, 694 ⟨90CPB3331⟩; **3**, 61 ⟨87JCR(S)296⟩, 227 ⟨86S847⟩, 249 ⟨85S643⟩, 626 ⟨82CPB3574⟩, 635 ⟨86JCS(P1)1763⟩; **4**, 67 ⟨81S324, 86JHC1163, 89JCS(P1)879, 93JOC2360⟩, 924 ⟨87JCS(P1)1095⟩, 943 ⟨87JCS(P1)589⟩
Ishibashi, M., **1**, 210 ⟨72TL4605⟩
Ishibashi, N., **1**, 366 ⟨93TL5943⟩; **3**, 326 ⟨93TL5943⟩
Ishibashi, Y., **2**, 631 ⟨82CL1481⟩
Ishichi, Y., **2**, 950 ⟨92SL843⟩; **4**, 361 ⟨92SL843⟩; **5**, 417, 421, 423 ⟨91JOC1307⟩; **6**, 131 ⟨92SL843⟩
Ishida, A., **1**, 688, 690 ⟨76CL1127⟩; **3**, 63 ⟨75CL1201, 77BCJ1161⟩; **6**, 556, 564 ⟨87CPB2840⟩
Ishida, H., **3**, 106 ⟨80JOC3350⟩
Ishida, M., **1**, 470 ⟨92JOC1161⟩; **3**, 216 ⟨78CC12⟩, 527 ⟨79JOC2067⟩; **4**, 66 ⟨82CL587⟩; **5**, 157 ⟨89TL3799⟩, 242, 250, 506, 509 ⟨87CL475⟩, 244 ⟨86ZC179⟩, 245, 246 ⟨83S552⟩, 245, 246 ⟨83S814⟩, 245, 252 ⟨82S1013, 82AG(E)117⟩, 252 ⟨87NKK1457⟩, 254 ⟨81S370⟩, 274, 286 ⟨80S933⟩, 337 ⟨81S746⟩, 337, 584 ⟨85CB1668⟩, 472, 510 ⟨86OPP369⟩, 547 ⟨82S457⟩, 550, 560 ⟨88SR155, 85PS(22)135⟩, 560 ⟨86BCJ1403⟩, 584 ⟨81LA1798⟩; **6**, 546, 561 ⟨87S349⟩
Ishida, N., **1**, 160 ⟨83JOC2120⟩, 1042 ⟨80CC736⟩; **2**, 923, 924 ⟨82JOM(226)C9⟩; **4**, 654 ⟨84TL4245⟩, 654, 655 ⟨84TL4249⟩
Ishida, T., **2**, 185 ⟨88CC40⟩; **5**, 570, 613 ⟨92JOC2419⟩, 671 ⟨84JHC725⟩
Ishida, Y., **3**, 437 ⟨83TL3255⟩; **5**, 375 ⟨85TL2217⟩, 661, 728 ⟨83TL3255⟩
Ishido, Y., **1**, 925, 946 ⟨93TL1317⟩; **2**, 107 ⟨83CPB2157⟩; **3**, 708 ⟨85MI 321-01⟩; **4**, 910 ⟨83S920⟩
Ishidoya, M., **6**, 306 ⟨89JAP(K)1182311⟩
Ishifune, M., **6**, 24 ⟨90TL7181⟩
Ishigami, T., **2**, 213 ⟨75CL813⟩; **3**, 621 ⟨72JOC1542⟩
Ishigaro, M., **5**, 30, 35 ⟨86SC499⟩
Ishige, O., **4**, 210 ⟨86JOC546⟩; **5**, 132 ⟨86JOC546⟩; **6**, 24 ⟨90TL7181⟩
Ishiguro, H., **2**, 902 ⟨83TL735⟩
Ishiguro, M., **1**, 651 ⟨81JA5568⟩, 973 ⟨82CL1029⟩, 974, 1056 ⟨82JOC2225⟩, 1055 ⟨82JA7667⟩, 1056, 1057 ⟨84BCJ2768⟩; **3**, 265 ⟨90JCS(P1)1891⟩, 625 ⟨84CL1511⟩, 639 ⟨84BCJ2768⟩, 696 ⟨79JA1608, 79TL2745⟩; **4**, 1059, 1060 ⟨81JA5568⟩, 1061 ⟨84BCJ2768⟩
Ishiguro, R., **3**, 369 ⟨90JST(238)71⟩
Ishiguro, T., **3**, 618, 619 ⟨73TL2121⟩
Ishiguro, Y., **3**, 113 ⟨84TL961⟩
Ishihara, A., **6**, 339 ⟨89JOM(368)199⟩
Ishihara, H., **1**, 246 ⟨78CL1007⟩; **2**, 78 ⟨84T2239⟩, 858 ⟨77BCJ3007⟩; **4**, 288 ⟨76CL203⟩; **5**, 233, 506, 509 ⟨87NKK1430⟩, 240 ⟨78CL1007, 86S128⟩, 240, 244, 252 ⟨77BCJ3007⟩, 240, 252, 541 ⟨76CL203⟩, 248, 249 ⟨85S519, 85S520, 87NKK1479⟩, 249 ⟨92ZN(B)558⟩, 249, 563, 640 ⟨88BCJ3931⟩, 250 ⟨87S371, 89S929⟩, 250 ⟨90JOM(386)313⟩, 253 ⟨87NKK1475⟩, 255, 541 ⟨72TL3751⟩, 510, 561, 563 ⟨90BCJ638⟩, 516, 519, 522, 523, 626 ⟨90BCJ835⟩, 541, 542 ⟨92CB417, 93CC277⟩, 561, 563, 588 ⟨90ZC67⟩, 563, 626 ⟨93JA3000⟩
Ishihara, J., **1**, 828 ⟨89JOC1223⟩
Ishihara, K., **2**, 39 ⟨94TL4569⟩, 528 ⟨93JA11490⟩, 679 ⟨90JOC5814⟩; **5**, 652 ⟨90TL3603⟩
Ishihara, K.-I., **2**, 730 ⟨92JA7591⟩; **4**, 948 ⟨92JA7591⟩
Ishihara, M., **1**, 580 ⟨92T10265⟩; **3**, 255 ⟨77CL287⟩; **4**, 850 ⟨77CL287⟩
Ishihara, S., **1**, 822 ⟨72TL2923⟩; **3**, 151 ⟨90TL3571⟩; **6**, 13, 237 ⟨93JA2156⟩, 22, 233, 234 ⟨90TL3579⟩
Ishihara, T., **1**, 434 ⟨87JOC300⟩, 1079 ⟨84TL1377⟩, 1130 ⟨86CL1895⟩; **2**, 77 ⟨91TL1069⟩; **4**, 735 ⟨83TL507, 83TL5657⟩, 737 ⟨88JFC(41)435⟩, 810 ⟨86CL1895⟩; **5**, 346 ⟨90JA9646⟩; **6**, 264 ⟨82JFC(20)121, 82JFC(20)617⟩
Ishihara, Y., **1**, 312 ⟨82JFC(20)313⟩; **2**, 702 ⟨82JOC119⟩; **3**, 299 ⟨92JCS(P1)3401⟩; **5**, 332, 333 ⟨92CPB1177⟩
Ishii, A., **1**, 899 ⟨90TL2623⟩; **3**, 177 ⟨88TL1399⟩, 332 ⟨87JA279⟩, 337, 340 ⟨84TL849, 82CC1187⟩, 341 ⟨87BCJ1037⟩, 363 ⟨90JOC2421, 92CC7⟩, 383 ⟨86CC71⟩, 392 ⟨83CC1429, 88BCJ861⟩, 392 ⟨92CL2289⟩; **4**, 1049 ⟨87JA279⟩; **5**, 516, 523, 524 ⟨86BCJ2529, 86CC71⟩, 524 ⟨87BCJ1037⟩; **6**, 176 ⟨87JA279⟩
Ishii, F., **3**, 358 ⟨76TL4283⟩, 361 ⟨72CL9, 75JCS(P1)270⟩; **5**, 634 ⟨75JCS(P1)270⟩; **6**, 304 ⟨75JCS(P1)270⟩
Ishii, H., **1**, 561 ⟨94T487⟩; **2**, 3 ⟨89CL577⟩, 279, 282 ⟨79CL873⟩
Ishii, M., **4**, 808 ⟨93CC1309⟩, 1034 ⟨91CC1390⟩
Ishii, N., **2**, 929 ⟨92CC1358⟩
Ishii, S., **3**, 108 ⟨85CL1803⟩
Ishii, T., **2**, 363 ⟨94JA2390⟩; **3**, 256 ⟨76JCS(P1)1706⟩; **4**, 570, 573 ⟨87BCJ1831⟩, 852 ⟨76JCS(P1)1706⟩
Ishii, Y., **1**, 209 ⟨79JCS(P1)26⟩, 481 ⟨77CL103⟩; **2**, 3 ⟨76CPB820⟩, 34 ⟨88S366⟩, 36 ⟨89S283⟩, 55 ⟨86JOC240, 88JOC3752⟩; **3**, 7, 215 ⟨86JOC240⟩, 7, 30 ⟨86S774⟩, 57 ⟨81JOC240⟩, 117 ⟨88JOC3587, 88SC869⟩, 352 ⟨87TL203⟩, 363 ⟨71MI 308-01⟩, 635 ⟨85JOC2240⟩; **4**, 405 ⟨71JCS(C)511⟩, 588 ⟨73JOM(51)167⟩; **5**, 235 ⟨72JOM(42)33⟩, 254, 537, 539, 540 ⟨73BCJ244⟩, 262 ⟨70JOM(24)595⟩, 377, 388, 390, 1083 ⟨74JOM(69)353⟩, 387 ⟨75CL197, 75CL641⟩, 390, 721 ⟨71JCS(C)1870⟩, 463 ⟨88CL377⟩, 578, 608, 618 ⟨75JOM(92)139⟩, 645 ⟨74JOM(78)339⟩, 719, 720 ⟨72JCS(P2)1043⟩, 1070 ⟨69TL1421⟩; **6**, 300 ⟨71JOC1176, 72JOC4198⟩, 313 ⟨74JOM(72)103⟩, 322, 337 ⟨B-70MI 605-03⟩, 571 ⟨72JCS(P1)1678⟩, 576 ⟨B-70MI 615-01⟩, 624 ⟨77BCJ3271⟩, 645 ⟨69TL1421⟩, 684 ⟨69JCS(C)2002⟩
Ishikawa, H., **2**, 629 ⟨81CC718⟩
Ishikawa, J., **6**, 464 ⟨87BCJ1204⟩
Ishikawa, K., **1**, 818 ⟨93SL527⟩; **3**, 439 ⟨78BCJ535⟩; **5**, 972 ⟨77BCJ2406⟩; **6**, 120, 698 ⟨76TL3695⟩

Ishikawa, M., **1**, 279 ⟨86TL6369⟩, 450 ⟨81JOM(217)43⟩, 489 ⟨89JA314⟩, 813 ⟨64JCS4578⟩; **2**, 106 ⟨91T8717⟩; **3**, 246 ⟨85TL4727⟩, 408 ⟨87TL1293⟩, 418 ⟨86CC980⟩; **4**, 613, 614 ⟨78JOM(149)37⟩, 613, 617 ⟨91OM2695, 58JOC292⟩, 614 ⟨66JOM(5)226, 93OM2536⟩, 615 ⟨83JOM(250)109, 83OM351⟩, 616 ⟨69CC207, 91JOM(407)157⟩, 1047 ⟨78JOM(161)299⟩, 1048 ⟨79JOM(179)377, 80JOM(194)147⟩, 1051, 1053 ⟨85OM2040⟩, 1052 ⟨81AOC51, 77JOM(142)C45⟩, 1053 ⟨82OM1473⟩; **6**, 178 ⟨81JA2324⟩, 178, 383 ⟨79JA1348⟩, 195 ⟨66JOM(6)451, 82JOM(236)7⟩, 671 ⟨87TL1293⟩, 674 ⟨86CC980⟩
Ishikawa, N., **1**, 825 ⟨81CL1289⟩, 1130 ⟨85JA5186⟩; **2**, 19 ⟨79BCJ3377, 81CL761⟩, 612 ⟨79CL983⟩, 631 ⟨74BCJ1680⟩, 830 ⟨80JOC5052⟩, 1031 ⟨73JOC1361⟩, 1042 ⟨77BCJ3069⟩; **3**, 274 ⟨80CL1327⟩, 351 ⟨73BCJ3285⟩, 661 ⟨75CL277⟩; **4**, 264 ⟨76BCJ502⟩, 415 ⟨74BCJ785⟩, 731 ⟨76BCJ2491, 85JFC(29)445⟩, 734 ⟨79CL983⟩, 735 ⟨78TL4809⟩, 831 ⟨84TL3987, 84TL3991⟩, 937 ⟨85JFC(30)357, 86BSF937⟩, 949 ⟨77CL1379⟩; **5**, 4 ⟨78BCJ1267⟩, 6 ⟨79BCJ3377⟩, 38 ⟨80CL651⟩, 57 ⟨79MI 502-06⟩, 59 ⟨84EUP115085⟩, 234 ⟨77CL483⟩, 280 ⟨80JOC5052⟩; **6**, 22 ⟨82CL1453⟩, 65 ⟨81CL1337⟩, 76, 77 ⟨73JOC1361⟩, 695 ⟨79CL983⟩
Ishikawa, R., **1**, 692 ⟨78JOC2477⟩
Ishikawa, S., **2**, 328 ⟨88TL6949⟩; **3**, 422 ⟨88TL6949⟩
Ishikawa, T., **2**, 84 ⟨91TL667⟩
Ishikawa, Y., **4**, 325 ⟨81CPB2496⟩, 559 ⟨91CL2213⟩
Ishikura, K., **3**, 357 ⟨91EUP409164⟩; **5**, 655 ⟨79TL4943⟩
Ishikura, M., **5**, 301 ⟨82CC741⟩
Ishikura, T., **3**, 684 ⟨84CC1513⟩
Ishimaru, T., **6**, 77 ⟨73JMC978⟩
Ishimaru, Y., **2**, 579 ⟨92CL377⟩
Ishimi, K., **6**, 456 ⟨74NKK732⟩
Ishimoto, N., **4**, 319 ⟨89BCJ1215⟩, 321 ⟨85H(23)2509⟩
Ishimoto, Y., **2**, 326 ⟨93JMC3526⟩
Ishirara, K., **4**, 174 ⟨90JOC5814⟩
Ishiwada, K., **4**, 697 ⟨83JOM(243)245⟩; **5**, 1141 ⟨83JOM(243)245⟩; **6**, 206, 207 ⟨83JOM(243)245⟩
Ishiwari, H., **1**, 769 ⟨80JOC2005⟩
Ishiwata, H., **5**, 199 ⟨86TL4609⟩
Ishiwata, S., **4**, 296 ⟨70CPB1245⟩
Ishiyama, T., **1**, 279 ⟨86TL6369, 87CL25⟩, 279 ⟨91SL687⟩, 489 ⟨89JA314⟩; **2**, 902 ⟨93JA7219⟩; **3**, 228 ⟨87BCJ3471⟩, 290 ⟨91BCJ1999⟩, 290, 295 ⟨91TL6923⟩, 290, 296, 301, 306, 311 ⟨93TL7595⟩, 290, 305, 311 ⟨92TL4465⟩; **5**, 199 ⟨90TL709⟩
Ishizaki, M., **1**, 584 ⟨92JOC7285⟩; **6**, 449 ⟨87SC1887⟩
Ishizaki, Y., **2**, 590 ⟨88CL63⟩
Ishizawa, S., **4**, 736 ⟨90CPB1104⟩
Ishizone, H., **3**, 9, 123 ⟨89CL901⟩
Ishizu, H., **5**, 134 ⟨86CL871⟩
Ishizuka, H., **3**, 351 ⟨89TL2955⟩
Ishizuka, T., **1**, 134 ⟨94TL721⟩
Ishizumi, K., **1**, 573 ⟨78JA6483⟩
Ishmaeva, É. A., **4**, 119 ⟨86JGU501⟩, 489 ⟨82ZOB2444⟩; **6**, 685, 686, 689 ⟨87ZOB901⟩
Isida, T., **1**, 884 ⟨64JOC2781⟩
Isidor, J. L., **4**, 184 ⟨73JOC554⟩
Iskander, Y., **2**, 26 ⟨68CC976⟩
Islam, A. M., **1**, 422 ⟨53JCS2247⟩
Islam, M. A., **5**, 330 ⟨91JCS(P1)1001⟩
Islam, Md. N., **1**, 499 ⟨90JA6416⟩
Islam, M. Q., **2**, 440 ⟨88ICA(146)111⟩
Islam, Q., **3**, 239 ⟨88JCR(S)208⟩; **4**, 742 ⟨90JCS(P1)139⟩
Islamov, R. G., **4**, 345 ⟨83ZOB46⟩; **6**, 583 ⟨93ZOB633⟩, 657 ⟨90ZOB778⟩
Isler, O., **1**, 104 ⟨49HCA489⟩, 156 ⟨57HCA1250⟩, 502, 837 ⟨47HCA1911⟩, 956 ⟨59HCA719⟩; **3**, 62 ⟨47HCA1911, 56HCA249⟩, 71 ⟨59HCA844⟩
Islip, P. J., **5**, 214 ⟨88JMC499⟩
Islyamova, E. V., **4**, 56, 57, 91 ⟨88ZOR1945⟩
Ismaev, I. E., **2**, 894 ⟨83IZV1152⟩
Ismagilova, N. M., **5**, 403 ⟨71JGU2513⟩
Ismagilov, R. K., **4**, 559 ⟨85ZOB1865, 89ZOB1686⟩
Ismail, A. A., **3**, 76 ⟨71ZOR2120⟩
Ismail, A. E.-H., **4**, 109 ⟨92S710⟩; **5**, 661 ⟨92T8271⟩
Ismail, A. F. A., **3**, 498 ⟨40JCS1374⟩
Ismail, M. F., **4**, 303 ⟨90G677⟩; **5**, 355 ⟨91PS(63)373⟩
Ismail, Z. M., **3**, 652 ⟨82AG(E)859, 86S548⟩
Ismailov, A. G., **5**, 63 ⟨84MI 502-01⟩
Ismailov, V. M., **3**, 197 ⟨87JGU205⟩; **4**, 884, 913 ⟨77CA135599v, 76ZOB1652⟩; **5**, 484 ⟨74IZV55⟩; **6**, 518 ⟨80MI 616-01⟩
Isobe, E., **4**, 607, 646 ⟨86JPS(A)1839⟩
Isobe, K., **1**, 202 ⟨85TL767, 86TL967⟩, 202, 203 ⟨86JOC2859⟩, 929 ⟨91CL1193, 93BCJ819⟩; **2**, 585 ⟨81CC808⟩, 999 ⟨79TL833⟩; **3**, 250, 256 ⟨79CL785, 81CC714⟩; **4**, 687 ⟨83JCS(D)1441⟩, 956, 957 ⟨79CL785⟩; **5**, 430 ⟨93BCJ819⟩
Isobe, M., **1**, 310 ⟨80CL331⟩, 779, 788 ⟨78CL541⟩, 1009 ⟨91CL1271⟩, 1026 ⟨93SL413⟩, 1048 ⟨90SL701, 90TL5499⟩; **2**, 702 ⟨77CL679⟩; **3**, 56 ⟨82JA4997⟩; **4**, 199 ⟨75JA6260⟩, 231, 940 ⟨88TL4773⟩, 367, 937 ⟨80CL331⟩, 936 ⟨82JA4997, 79TL3465⟩, 937 ⟨81CL457, 81TL239, 81TL4287, 84JA3252, 84JOC3517, 86T2863, 87T4759⟩, 940 ⟨90SL701, 91T3727⟩, 950 ⟨84TL2021⟩; **5**, 152 ⟨86T2863⟩; **6**, 130 ⟨81TL4287, 84TL2021⟩
Isobe, Y., **1**, 482 ⟨89JCS(P1)2513⟩, 1029 ⟨87H(26)355⟩
Isoda, M., **2**, 363 ⟨82CPB1646⟩, 724 ⟨80S937⟩
Isoe, S., **1**, 651 ⟨87TL211⟩, 929 ⟨84TL5335⟩; **2**, 950 ⟨92SL843⟩; **3**, 32 ⟨92JOC1321⟩, 36, 165 ⟨88CC1468⟩, 36, 165 ⟨93T2011⟩, 143 ⟨84TL5335⟩, 303 ⟨88TL1173⟩, 703, 709 ⟨94CC549⟩; **4**, 197 ⟨91T615, 92JOC1321⟩, 361 ⟨92SL843⟩, 387 ⟨85TL5827⟩, 937 ⟨93T2011⟩; **5**, 343 ⟨92JOC4877⟩, 417, 421, 423 ⟨91JOC1307⟩; **6**, 124 ⟨90JA1962⟩, 129 ⟨91T549⟩, 131 ⟨92SL843⟩, 133 ⟨89TL219⟩
Isogami, Y., **2**, 753 ⟨87CL1187⟩
Isola, M., **2**, 123 ⟨76S326⟩
Isomura, Y., **2**, 205 ⟨80CPB1131⟩; **4**, 563 ⟨88JAP63185993⟩
Isono, K., **5**, 342 ⟨86TL3907⟩
Isowa, Y., **6**, 434 ⟨73BCJ1269⟩
Israel, M., **1**, 10 ⟨81JOC2596⟩; **3**, 174 ⟨81JOC2596⟩; **5**, 999 ⟨73JMC1⟩
Israel, R., **1**, 983 ⟨80JOC3522⟩
Israéli, Y. J., **2**, 1048 ⟨65BSF2187⟩
Israelstam, S. S., **5**, 316 ⟨62JOC4532⟩
Isshiki, K., **3**, 716, 717 ⟨87JAN1207, 87JAN1199⟩, 717 ⟨87JAN1202⟩
Issidorides, C. H., **2**, 213 ⟨66JOC615⟩
Issleib, K., **1**, 446 ⟨66ZAAC(343)196, 75ZAAC(418)158⟩, 737 ⟨67LA(707)120⟩; **2**, 427 ⟨67CB3331, 76ZAAC(425)209⟩, 427, 430, 462 ⟨68CB3612⟩, 427, 437 ⟨59CB1118⟩, 427, 472 ⟨85ZAAC(523)7⟩, 428 ⟨84ZC261⟩, 429 ⟨61CB2664⟩, 429, 461 ⟨61CB2656⟩, 430 ⟨65AG172, 67ZN(B)1238⟩, 430, 437, 438 ⟨82ZAAC(491)163⟩, 430, 454 ⟨67CB2685⟩, 431 ⟨59CB704, 60CB1852, 77PS(3)203⟩, 431, 454, 822, 854 ⟨68CB2197⟩, 432 ⟨61CB392⟩, 433 ⟨54ZAAC(277)258, 63CB2186⟩, 437 ⟨59CB3175⟩, 437, 1060, 1069 ⟨62CB268⟩, 438 ⟨61CB113, 65CB2091⟩, 438 ⟨65ZN(B)916, 66ZAAC(346)241⟩, 439 ⟨65ZN(B)181, 67CB412, 83PS(17)73, 87ZAAC(553)136⟩, 440 ⟨64CB3430, 65ZAAC(336)234, 67ZAAC(351)18, 77ZAAC(436)20⟩, 441 ⟨56ZAAC(287)208⟩, 444 ⟨64ZAAC(328)21⟩, 452 ⟨56ZAAC(288)201⟩, 462 ⟨66ZAAC(343)268⟩, 467 ⟨86ZAAC(535)47⟩, 472 ⟨78PS(4)137⟩, 473 ⟨85PS(21)367⟩, 481, 486, 487, 502, 505, 510 ⟨65ZAAC(339)289⟩, 481, 486, 489, 505, 506, 510 ⟨65ZAAC(339)298⟩, 481, 486, 505, 510 ⟨64ZAAC(332)179⟩, 486, 506, 510 ⟨61AG26⟩, 820, 821 ⟨81TL4475⟩, 826 ⟨62ZC163⟩, 4333 ⟨61CB2244⟩; **4**, 337 ⟨72JPR66, 72T2587⟩, 453 ⟨68CB3612, 85ZC172⟩, 453 ⟨88JOM(355)71⟩, 453, 454 ⟨73ZC139⟩, 453, 454, 477

⟨68CB3619⟩, 453, 454, 477, 478, 481 ⟨67CB2685⟩, 453, 454, 477, 481 ⟨68CB4032⟩, 453, 481 ⟨67TL1489⟩, 454 ⟨71MI 410-01, 73MI 410-01, 74SRI191⟩, 454, 481 ⟨76ZAAC(424)97⟩, 455 ⟨65ZAAC(336)234, 74SRI453, 74ZC243⟩, 458 ⟨83PS(17)73⟩, 458, 477, 481 ⟨85ZAAC(523)7⟩, 461 ⟨78ZAAC(444)249⟩, 463 ⟨80ZAAC(469)109, 83PS(14)171⟩, 545 ⟨59CB3175, 63ZAAC(324)259⟩, 546, 553 ⟨63CB2186⟩, 548 ⟨90PS(49/50)355⟩, 550 ⟨84ZC261, 91ZAAC(594)66⟩, 551 ⟨66LA(699)40, 67LA(707)120⟩, 551 ⟨70JPR135⟩, 551, 552, 564 ⟨70JPR456⟩, 551, 557 ⟨69JPR857⟩, 551, 558 ⟨68LA(713)12⟩, 552 ⟨60MI 412-01, 67LA(707)112⟩, 566 ⟨86ZAAC(535)47⟩, 566, 575 ⟨76ZAAC(422)237⟩, 568 ⟨82ZAAC(488)75⟩, 570, 575, 576 ⟨78ZAAC(447)53, 81ZC341⟩, 575 ⟨82ZAAC(494)55⟩, 997, 998 ⟨79ZAAC(459)131⟩; **5**, 395, 397, 784 ⟨66ZAAC(346)241⟩, 396 ⟨67ZN(B)784⟩, 402 ⟨59CB3183⟩, 784, 881 ⟨78JOM(160)47⟩, 883 ⟨83ZC99⟩, 908 ⟨66LA(699)40⟩, 910 ⟨67LA(707)112⟩; **6**, 151, 701 ⟨70JPR456⟩, 372 ⟨80ZC419⟩, 372, 374 ⟨80ZC153⟩, 372, 688, 689 ⟨81ZAAC(473)85⟩, 490 ⟨87ZAAC(546)142⟩, 491 ⟨85ZAAC(530)16, 86ZAAC(542)37⟩, 511 ⟨70JPR366⟩, 511, 581 ⟨73JPR471⟩, 512 ⟨87EGP242812⟩, 658, 687 ⟨81SRI279⟩, 680, 688 ⟨81ZC357⟩, 686 ⟨81TL4475, 83ZC99, 80JOM(192)33⟩, 687 ⟨85ZAAC(529)216⟩, 701 ⟨66LA(699)40⟩

Itabashi, K., **2**, 122 ⟨83YGK776⟩, 711 ⟨89S955⟩; **3**, 357 ⟨92JHC841⟩, 618 ⟨73CL577⟩

Itagaki, T., **5**, 864 ⟨B-87MI 521-01⟩

Itai, A., **1**, 143 ⟨86BCJ2801⟩; **3**, 704 ⟨76CPB832⟩; **5**, 671 ⟨84JHC725⟩

Itami, A., **3**, 551 ⟨89TL1249⟩; **4**, 837 ⟨89TL1249⟩

Itani, H., **2**, 88 ⟨83JA2908⟩; **3**, 357 ⟨91CPB672⟩

Itano, H. A., **1**, 318 ⟨73T3497⟩

Itaya, T., **6**, 443 ⟨88MI 614-01⟩

Iten, P. X., **3**, 103 ⟨78HCA430⟩

Itenberg, A. M., **4**, 137 ⟨37JGU2495⟩

Ito, A., **2**, 326 ⟨89TL6395⟩

Ito, E., **4**, 506 ⟨92JOC6711⟩, 734 ⟨82TL2323⟩; **6**, 58, 264, 266 ⟨82TL2323⟩

Ito, H., **1**, 197 ⟨92TL1295⟩, 197 ⟨93JA8835⟩, 197, 1057 ⟨92TL3769⟩; **2**, 70 ⟨92TL4469⟩, 73 ⟨88JOC1831, 88JOM(347)5⟩, 73, 597 ⟨89JOM(369)291⟩, 782 ⟨64CPB326⟩; **3**, 158 ⟨93JOC1458⟩, 245 ⟨86JHC1199⟩, 418, 694 ⟨88JOC4158⟩, 723 ⟨93JOC6766⟩; **5**, 802 ⟨88JOC4158⟩, 802, 803 ⟨89CL1603⟩

Ito, I., **4**, 1033 ⟨88JA8153⟩

Ito, K., **2**, 59 ⟨88S995⟩, 65, 528 ⟨87BCJ395⟩, 71 ⟨90JOC304⟩, 313, 544 ⟨91JCS(P1)483⟩, 314 ⟨89JCS(P1)1544, 90JCS(P1)1859⟩, 316 ⟨91JCS(P1)1767, 92TL627⟩, 527 ⟨75CL1095⟩; **3**, 457, 458 ⟨92TL627, 91JCS(P1)1767⟩, 465 ⟨76CPB369, 79CPB682⟩; **4**, 317 ⟨77CPB3385⟩, 322, 323 ⟨73CPB2257⟩, 325 ⟨79CPB1691⟩; **5**, 171 ⟨87JOC4319⟩, 281 ⟨81S287⟩

Ito, M., **1**, 613 ⟨93JCS(P1)987⟩; **2**, 1086 ⟨86JOC4507⟩; **3**, 63 ⟨90JCS(P1)197⟩; **4**, 924 ⟨87JCS(P1)1095⟩

Ito, M. M., **3**, 414 ⟨83BCJ641⟩

Ito, N., **2**, 205 ⟨80CPB1131⟩; **3**, 628 ⟨79MAC1089, 82MAC579⟩

Itô, S., **1**, 215 ⟨84TL5781⟩, 702, 1070 ⟨88TL1197, 835 ⟨90TL727⟩; **2**, 626 ⟨67BCJ418⟩; **3**, 64 ⟨81CC907⟩, 309 ⟨90CPB3261⟩; **4**, 528 ⟨91TL1975⟩, 858 ⟨88TL3125⟩; **5**, 129 ⟨94JOC5532⟩, 152 ⟨87CC1523⟩, 681 ⟨83BCJ545, 84BCJ539⟩; **6**, 560 ⟨87CPB156⟩

Ito, T., **1**, 1016, 1145 ⟨90CC118⟩; **2**, 694 ⟨90CPB3331⟩, 923 ⟨90TL6399⟩; **3**, 664 ⟨80CC1253⟩, 685 ⟨87BCJ1948⟩; **5**, 70 ⟨74CL1035⟩, 143, 304 ⟨75BCJ2091⟩

Ito, W., **1**, 1059 ⟨84CC1004⟩; **4**, 1057 ⟨86JOC886⟩

Ito, Y., **1**, 38, 945 ⟨68JA4182⟩, 193 ⟨87TL4849⟩, 194 ⟨89TL375⟩, 230 ⟨80TL2181⟩, 415, 826 ⟨89SC2109⟩, 473 ⟨88JA8153⟩, 490 ⟨89TL6051⟩, 558 ⟨78JOC1011⟩, 571 ⟨88OSC(6)327⟩, 655 ⟨82JA7609⟩, 1035, 1047 ⟨88TL5379⟩, 1094 ⟨87JOC4412⟩; **2**, 19 ⟨92CL1161⟩, 22 ⟨80JOC2022⟩, 43, 528 ⟨89JA3426⟩, 50 ⟨88JA3712⟩, 51 ⟨88JA5579, 90TL7333⟩, 78 ⟨88TL5667⟩, 240, 359, 768 ⟨70JOC2979⟩, 301 ⟨90TL1743⟩, 311 ⟨90JOC5935⟩, 331 ⟨89JA6301⟩, 528 ⟨91TA601⟩, 545, 546 ⟨92JA3989⟩, 916 ⟨91JOC1948⟩, 920 ⟨89JA6478⟩, 923 ⟨93AG(E)1473⟩, 926 ⟨89JOC3517⟩, 935 ⟨88T3997⟩, 939 ⟨92JA5867⟩; **3**, 32 ⟨89BCJ3038⟩, 77 ⟨73TL2283, 75T1373⟩, 80 ⟨78BCJ1186⟩, 152 ⟨80JOC2022⟩, 158 ⟨93JOC1458⟩, 160 ⟨94T335⟩, 206 ⟨78JOC1011⟩, 224 ⟨76JOC2073, 80OS(59)113⟩, 225 ⟨79JA494⟩, 408 ⟨87TL1293⟩, 418 ⟨84TL3091, 86CC980⟩, 418 ⟨87JA7888, 88TL355⟩, 418 ⟨90SL245⟩, 418, 694 ⟨88JOC4158⟩, 599 ⟨77TL1009, 80SC233, 88PAC583⟩, 694 ⟨90JA2437, 92AG(E)1509⟩, 707 ⟨87TL4849, 88TL5151⟩, 707 ⟨88TL6321⟩, 707, 708 ⟨89TL1257, 91JOC7356⟩, 714, 715 ⟨71T3795⟩, 723 ⟨79JOC2030, 79TL1039⟩, 723 ⟨88S714, 93JOC6766⟩, 724 ⟨86JOM(303)301⟩; **4**, 187 ⟨87TL2723⟩, 651 ⟨90TL2925⟩, 740 ⟨64MI 417-01, 64TL7⟩, 740 ⟨66T2615⟩, 856 ⟨79BCJ2013, 80S736⟩, 1033 ⟨90TA151⟩, 1054 ⟨91JOC1948⟩; **5**, 138 ⟨84T145⟩, 146 ⟨74JOC3273⟩, 146, 692 ⟨75S291⟩, 356 ⟨79TL2253⟩, 552 ⟨80S736⟩, 584, 612 ⟨75T1245⟩, 595, 620 ⟨79TL1039⟩, 635, 636 ⟨71MI 515-01⟩, 728 ⟨89CL1261⟩, 747 ⟨69BCJ3310⟩, 760 ⟨71T3795⟩, 785, 1130 ⟨70JOC4238⟩, 785, 792 ⟨68TL935, 67JA2240⟩, 793 ⟨84TL5139⟩, 793, 796 ⟨87JA7888⟩, 796, 802 ⟨84TL3091, 88BCJ3649⟩, 802 ⟨88JOC4158⟩, 802, 803 ⟨89CL1603⟩, 803, 804 ⟨90JA2437⟩, 902 ⟨67T2137⟩, 1037 ⟨70JOC2118⟩, 1071 ⟨70JOC3995⟩, 1072 ⟨75JOC2981, 93JOC5118⟩; **6**, 127 ⟨85S698⟩, 163, 671, 674 ⟨88JA3692⟩, 384, 671 ⟨91JA8899⟩, 506 ⟨63BCJ1314⟩, 550 ⟨89CC1494, 87TL1293⟩, 672, 674 ⟨87JA7888, 86CC980⟩, 674 ⟨88TL355⟩

Itoh, A., **3**, 221 ⟨91TA178⟩; **4**, 393 ⟨79TL1783⟩

Itoh, E., **5**, 252, 254 ⟨80S721⟩

Itoh, H., **2**, 398 ⟨88BCJ3020⟩

Itoh, K., **1**, 10 ⟨86JOC5400⟩, 63 ⟨85OM1911⟩, 132 ⟨84TL223, 84TL2487⟩, 206, 289, 1092 ⟨82TL1267⟩, 299 ⟨94JA1597⟩, 481 ⟨77CL103⟩, 494 ⟨90JCS(P1)2207⟩; **2**, 398 ⟨85SC527⟩, 540 ⟨84JOC2298⟩; **3**, 29, 157 ⟨88BCJ3637⟩, 128 ⟨86BCJ1747⟩, 363 ⟨71MI 308-01⟩, 426 ⟨90SL477⟩, 663 ⟨88T1827⟩; **4**, 319 ⟨89BCJ1215⟩, 321 ⟨85H(23)2509, 86S375⟩, 588 ⟨73JOM(51)167⟩; **5**, 215 ⟨84TL223⟩, 235 ⟨72JOM(42)33⟩, 245, 246 ⟨83S814⟩, 262 ⟨70JOM(25)33⟩, 325 ⟨85JA5310, 90TL6547⟩, 377, 388, 390, 1083 ⟨74JOM(69)353⟩, 387 ⟨75CL197, 75CL641⟩, 390, 721 ⟨71JCS(C)1870⟩, 645 ⟨74JOM(78)339⟩, 719, 720 ⟨72JCS(P2)1043⟩, 1070 ⟨69TL1421, 70OCS23⟩; **6**, 322, 337 ⟨B-70MI 605-03⟩, 571 ⟨72JCS(P1)1678⟩, 576 ⟨B-70MI 615-01⟩, 645 ⟨69TL1421⟩, 684 ⟨69JCS(C)2002⟩

Itoh, M., **1**, 125 ⟨77BCJ2199⟩, 273 ⟨67JA5708⟩, 274 ⟨71JA1508⟩, 274, 546 ⟨70JA3503⟩, 276 ⟨75CL523, 76TL255⟩, 276 ⟨78BCJ339⟩, 277 ⟨71JA2792⟩, 277, 977 ⟨73S305⟩, 279, 1033 ⟨73JA3080⟩, 469 ⟨77TL3369⟩, 657 ⟨72JA6549⟩, 1033, 1034 ⟨86JOC4507, 75TL1961⟩, 1133 ⟨76CC606⟩; **2**, 34 ⟨71JA1508⟩, 904 ⟨74TL2961⟩, 950 ⟨92SL843⟩; **3**, 209 ⟨87BCJ1839⟩, 622 ⟨75CL523⟩, 703, 709 ⟨94CC85⟩; **4**, 361 ⟨92SL843⟩; **5**, 125 ⟨75S456⟩, 343 ⟨92JOC4877⟩, 561, 563, 588 ⟨90ZC67⟩; **6**, 57 ⟨70CPB784⟩, 131 ⟨92SL843⟩, 429 ⟨77BCJ718⟩

Itoh, N., **1**, 33 ⟨78CC1089⟩; **3**, 153, 154 ⟨92BCJ2800⟩; **4**, 74, 77 ⟨92TL1483⟩; **5**, 360 ⟨91JA6318⟩

Itoh, O., **1**, 493 ⟨80BCJ553⟩; **3**, 290 ⟨92CC453⟩, 634, 635 ⟨83BCJ257⟩; **4**, 223 ⟨83BCJ257, 83BCJ266⟩, 227 ⟨83BCJ270⟩; **5**, 1113 ⟨80BCJ553⟩

Itoh, S., **1**, 1010 ⟨92BCJ2992⟩; **5**, 9 ⟨85T3625⟩, 359 ⟨92JOC7359⟩; **6**, 489 ⟨92JOC7359⟩

Itoh, T., **2**, 91 ⟨79JA159⟩; **3**, 244 ⟨88JA310⟩, 275 ⟨82CPB63⟩, 303 ⟨82CL1559⟩, 632 ⟨89CL1505, 89TL3811⟩, 633 ⟨89BCJ3851⟩; **4**, 60 ⟨92TL5399⟩, 389

⟨90TL6407⟩, 893 ⟨80H(14)619, 82CPB63⟩; **5**, 84 ⟨78AG(E)176⟩, 553 ⟨83CL1901⟩, 557 ⟨86TL5405, 87NKK1414⟩, 996 ⟨68T2583⟩
Itoh, Y., **1**, 217 ⟨86TL2015⟩, 299 ⟨94JA2223⟩; **4**, 311 ⟨87CL1575⟩; **5**, 252 ⟨87NKK1457⟩
Itoho, K., **6**, 501 ⟨68T2367, 72CL971⟩
Itoi, K., **6**, 70 ⟨76JAP7630034⟩
Itok, M., **2**, 305 ⟨87T4079⟩
Itooka, T., **4**, 165 ⟨84SC147⟩; **5**, 199 ⟨87CPB3453⟩
Itotani, K., **3**, 289 ⟨93JOM(443)253⟩
Itsumo, S., **2**, 59 ⟨88S995⟩
Itsuno, S., **2**, 65, 528 ⟨87BCJ395⟩, 71 ⟨90JOC304⟩, 313, 544 ⟨91JCS(P1)1341⟩, 314 ⟨89JCS(P1)1548, 90JCS(P1)1859⟩, 316 ⟨91JCS(P1)1767, 92TL627⟩, 560 ⟨87JOC4644⟩; **3**, 457, 458 ⟨92TL627, 91JCS(P1)1767⟩
Itsushiki, K., **6**, 288 ⟨88JAP63227595⟩
Iturburu, M., **3**, 528 ⟨92TL4823⟩
Ivanchenko, V. I., **6**, 685,686 ⟨90ZOB2238⟩
Ivanov, A. N., **5**, 669 ⟨87BAU204, 87IZV226⟩, 669 ⟨87JGU1478, 87JGU1479⟩, 669 ⟨87ZOB1658, 87ZOB1659⟩, 669, 672 ⟨87BAU1000⟩, 669, 672 ⟨87IZV1086⟩, 669, 672 ⟨90BAU1007⟩, 669, 672 ⟨90IZV1122, 91JGU1152⟩, 672 ⟨91ZOB1268⟩, 683 ⟨87BAU1314, 87IZV1422⟩
Ivanov, A. P., **3**, 434 ⟨84ZOR988⟩
Ivanov, A. Yu., **4**, 569 ⟨83ZOB1831⟩
Ivanov, B. E., **4**, 127 ⟨67JGU1768⟩, 476 ⟨79IZV2783⟩, 478 ⟨69IZV1757⟩, 560 ⟨86ZOB2000⟩; **5**, 369 ⟨85IZV173⟩
Ivanov, C., **3**, 408 ⟨84S1038⟩
Ivanov, D., **5**, 27 ⟨70S615⟩, 98 ⟨76MI 502-01⟩
Ivanov, E. B., **5**, 221 ⟨86IZV2535⟩
Ivanov, I. C., **2**, 309 ⟨89AP(322)181⟩; **4**, 972, 979 ⟨80LA372⟩, 981, 982 ⟨87LA1101⟩; **5**, 599 ⟨87LA1101⟩
Ivanov, L. L., **2**, 1093 ⟨67JGU515⟩, 1099 ⟨64IZV1989, 65JGU1677⟩
Ivanov, S. V., **4**, 807 ⟨90JGU967⟩
Ivanov, V. T., **6**, 104 ⟨64TL47⟩
Ivanov, Yu. Ya., **5**, 669 ⟨89KFZ1317⟩, 672 ⟨91KFZ33⟩
Ivanova, I. A., **6**, 111 ⟨70BSF2013⟩, 111, 140 ⟨65IZV2179⟩
Ivanova, I. S., **6**, 147 ⟨68IZV2379⟩
Ivanova, L. F., **6**, 425 ⟨71JOU1943⟩
Ivanova, L. L., **4**, 698 ⟨75CA(83)10238⟩
Ivanova, M., **4**, 147 ⟨80MI 403-01⟩
Ivanova, N. I., **6**, 565 ⟨86ZOR2145, 90ZOR1364⟩
Ivanova, N. M., **2**, 11 ⟨90OPP215⟩
Ivanova, N. P., **2**, 1093 ⟨81IZV1280⟩
Ivanova, N. R., **2**, 1076 ⟨88JGU2136⟩, 1079 ⟨87JGU2497⟩
Ivanova, R. G., **4**, 911 ⟨68ZOB1523⟩
Ivanova, T. I., **6**, 170 ⟨71IZV1505⟩
Ivanova, V. Kh., **5**, 681 ⟨77IZV393⟩
Ivanova, V. M., **5**, 13 ⟨79MI 501-01⟩
Ivanova, V. N., **6**, 623 ⟨89ZOB714⟩, 659 ⟨86ZOB1427, 87ZOB2467⟩
Ivanova, Zh. M., **4**, 123 ⟨60JGU3986⟩, 130 ⟨68JGU1284⟩
Ivanova, Z. M., **2**, 848 ⟨59ZOB3766⟩; **6**, 656 ⟨74ZOB538⟩
Ivanovics, G., **2**, 177 ⟨44ZPC(281)156⟩
Ivashchenko, D. A., **4**, 658 ⟨85ZOB1629⟩, 1058 ⟨86ZOB1535⟩
Ivasyuk, N. V., **4**, 124 ⟨75JGU1950⟩, 346 ⟨68IZV1625⟩; **6**, 583 ⟨77IZV1177⟩
Iversen, P. V., **3**, 109 ⟨66ACS2649⟩
Iversen, T., **2**, 90 ⟨81CC1240⟩
Iverson, P. E., **2**, 379 ⟨71ACS2337⟩; **4**, 436 ⟨74ACS539⟩
Ivin, K. J., **1**, 674 ⟨85CC874, 87JA899⟩
Ivin, S. Z., **2**, 167 ⟨64ZOB1948⟩, 476 ⟨68ZOB1906⟩, 478 ⟨58JGU2963⟩; **4**, 923 ⟨67ZOB454⟩
Ivshin, V. P., **2**, 385 ⟨69RCR640⟩; **6**, 56, 282 ⟨69IZV2617⟩, 283 ⟨69ZOR1317⟩
Ivshina, T. N., **6**, 56, 282 ⟨69IZV2617⟩, 368 ⟨68IZV2839, 69IZV2304⟩, 368 ⟨72MI 612-01⟩
Iwabuchi, Y., **1**, 776 ⟨89H(28)1861⟩; **2**, 323 ⟨89H(29)1861⟩
Iwadare, T., **1**, 1077 ⟨69TL4447⟩
Iwagami, H., **2**, 241 ⟨85TL3891⟩; **5**, 334 ⟨85TL3891⟩, 336 ⟨86TL3095⟩
Iwagami, Y., **3**, 374 ⟨74TL1071⟩
Iwahara, T., **1**, 160 ⟨84TL1909⟩
Iwai, I., **1**, 1041 ⟨63CPB1049⟩
Iwai, K., **1**, 209 ⟨74CL385⟩, 599 ⟨76SC357⟩
Iwakawa, T., **6**, 556, 566, 652 ⟨91CPB1939⟩, 636 ⟨88CPB4755, 92MI 621-02⟩
Iwaki, D., **1**, 788 ⟨91AG(E)1173⟩
Iwaki, S., **1**, 931 ⟨87CL133⟩; **2**, 46 ⟨88JA4606⟩
Iwakiri, H., **2**, 19 ⟨79BCJ3377⟩; **5**, 6 ⟨79BCJ3377⟩
Iwakuma, T., **1**, 33 ⟨78CC1089⟩; **5**, 152 ⟨87CC1523⟩
Iwakura, H., **1**, 1042 ⟨88TL5391⟩
Iwakura, Y., **3**, 557 ⟨77MI 317-01⟩, 683 ⟨68BCJ2954⟩, 696 ⟨70OS(50)195⟩; **5**, 753 ⟨77JPS2163⟩, 970 ⟨65JOC1158, 66JOC142⟩, 972 ⟨73CL1143, 80OS(59)195⟩, 990 ⟨72MI 526-01⟩; **6**, 230 ⟨80OS(59)195⟩, 415, 428 ⟨79OS(59)195⟩, 462 ⟨76JOC2070⟩
Iwama, N., **2**, 547 ⟨92JA7598⟩
Iwama, T., **2**, 143 ⟨90TL3027⟩, 288 ⟨92JCR(S)393⟩
Iwamatsu, K., **3**, 357 ⟨90JAN1150⟩
Iwamoto, H., **2**, 3 ⟨76CPB820⟩; **3**, 140 ⟨91CL1013⟩
Iwamoto, N., **5**, 974 ⟨66CC828⟩
Iwamura, H., **2**, 794, 795 ⟨92JA1830⟩; **3**, 645 ⟨86TL4177⟩; **4**, 221 ⟨87CC122⟩; **6**, 82 ⟨74BCJ2457⟩, 99 ⟨74BCJ1496⟩, 304 ⟨82BCJ1106⟩
Iwamura, M., **6**, 304 ⟨82BCJ1106⟩
Iwamura, T., **1**, 108, 115 ⟨86CPB3599⟩; **2**, 286, 288, 292 ⟨94CPB811⟩
Iwanaga, K., **1**, 184 ⟨85JA3343⟩; **4**, 900 ⟨93MI 420-01⟩, 973, 980, 983, 984, 986 ⟨88EUP302389⟩; **5**, 30 ⟨88OS(67)76⟩
Iwanaga, Y., **6**, 524 ⟨56JAP10924⟩
Iwanami, S., **2**, 249 ⟨70BCJ3543, 71BCJ2750⟩
Iwano, Y., **1**, 208 ⟨86BCJ1363⟩; **3**, 59 ⟨81TL1021⟩, 559 ⟨76CC516⟩, 581 ⟨76TL1307⟩
Iwanowicz, E., **6**, 641 ⟨92TL5933⟩
Iwanowicz, E. J., **2**, 331 ⟨87TL2083⟩; **6**, 317 ⟨92BMC1607⟩
Iwao, M., **3**, 90, 98 ⟨89JOC24⟩, 310 ⟨87T813⟩
Iwaoka, M., **2**, 281 ⟨88CC1283⟩
Iwaoka, T., **3**, 543 ⟨92S977⟩
Iwasa, A., **3**, 618 ⟨78S905⟩
Iwasa, K., **5**, 212 ⟨85CPB998⟩
Iwasa, S., **1**, 938 ⟨89CC1265, 91JCS(P1)1173⟩, 1090 ⟨89JOC1757⟩; **4**, 938 ⟨89JOC1757⟩; **5**, 513 ⟨89CC1265, 91JCS(P1)1173⟩, 513 ⟨91JOC2849⟩; **6**, 550 ⟨89CC1265, 91JOC2849⟩
Iwasaki, G., **5**, 410, 424 ⟨86HCA1378⟩; **6**, 127 ⟨86HCA1378⟩
Iwasaki, M., **5**, 463 ⟨88CL377⟩
Iwasaki, S., **4**, 607, 654 ⟨67JOM(10)111⟩
Iwasaki, T., **1**, 34 ⟨73JOC2731⟩; **4**, 13 ⟨75TL997⟩, 310 ⟨79JOC391⟩, 311, 327 ⟨79BCJ826⟩, 424 ⟨82JOC1706⟩
Iwasawa, H., **5**, 284 ⟨87H(26)359⟩, 570, 613 ⟨87H(26)359⟩
Iwasawa, N., **1**, 375 ⟨91TA1305⟩, 929 ⟨91CL1193, 93BCJ819⟩, 929 ⟨93CL545⟩; **2**, 701 ⟨82CL353, 84T1381⟩; **3**, 137 ⟨93CL545⟩, 164 ⟨86CL1809⟩, 628 ⟨88BCJ4379⟩; **5**, 430 ⟨93BCJ819⟩, 557 ⟨88CL1021, 88CL1025⟩
Iwase, M., **6**, 456 ⟨74NKK732⟩
Iwase, N., **1**, 839 ⟨90BCJ1328⟩; **3**, 57 ⟨86CL2035⟩
Iwase, S., **1**, 815 ⟨76BCJ2017⟩
Iwashita, T., **1**, 141 ⟨85JOC657⟩; **6**, 101 ⟨85JOC657⟩
Iwata, C., **3**, 59 ⟨88JOC1623⟩; **4**, 303 ⟨61JOC2273⟩; **5**, 1120 ⟨92CC1379⟩
Iwata, I., **4**, 199 ⟨89CL659⟩
Iwata, K., **6**, 502 ⟨78JHC1231⟩
Iwata, M., **3**, 101 ⟨74BCJ1687⟩; **5**, 344 ⟨85BCJ3395⟩
Iwata, N., **1**, 1035 ⟨72JOC3749⟩; **2**, 743 ⟨92BCJ1379⟩
Iwata, S., **1**, 1142 ⟨77JCP(66)4671⟩
Iwayanagi, T., **1**, 66 ⟨77JOM(135)1⟩
Iyengar, R., **2**, 784, 785 ⟨80JOC734⟩

Iyer, P. S., **1**, 904 ⟨86JA1575⟩; **4**, 161, 212 ⟨86S513⟩; **5**, 367, 992 ⟨83JOC3358⟩
Iyer, R., **1**, 619, 707, 949 ⟨86JA4568⟩; **4**, 82, 83, 85 ⟨86JA4568⟩
Iyer, R. N., **6**, 443 ⟨87IJC(B)748⟩, 577 ⟨75JMC913⟩
Iyer, R. P., **6**, 489 ⟨93MI 615-02⟩, 491 ⟨89TL7141⟩
Iyer, R. R., **1**, 487 ⟨91TA277⟩; **2**, 906 ⟨91TL3655⟩; **4**, 815 ⟨91TA277⟩
Iyer, R. S., **1**, 523 ⟨85TL5393⟩, 541 ⟨86OS(64)1⟩; **3**, 519 ⟨85JOC5898⟩; **5**, 100 ⟨85OS(64)1⟩
Iyer, S., **5**, 457 ⟨84JA6451, 87JA2759⟩
Iyer, S. R., **2**, 439 ⟨87JOM(320)339⟩
Iyer, V. S., **2**, 438 ⟨88JOC3016⟩; **3**, 310 ⟨93TL3519⟩
Iyobe, A., **5**, 199 ⟨90JOC1679⟩
Iyoda, J., **3**, 93 ⟨92JOC2677⟩; **4**, 615 ⟨83OM351⟩
Iyoda, M., **1**, 986 ⟨84CL131⟩, 990 ⟨86JA5371⟩, 991 ⟨84CL2005, 89CC1690⟩, 992 ⟨89JA3761, 90CL2149⟩; **3**, 684 ⟨94JCS(P1)413⟩; **4**, 820 ⟨86CC1794⟩; **5**, 936 ⟨94CL1537⟩, 946, 958 ⟨95TL3699⟩, 953 ⟨94CL777⟩
Izatt, R. M., **2**, 133 ⟨76JOC134⟩; **5**, 511 ⟨83JOC2635⟩
Izawa, K., **1**, 1125 ⟨73JA4090, 74JOC2255⟩
Izawa, T., **2**, 788 ⟨79TL2525⟩; **3**, 13, 85 ⟨79BCJ555⟩; **4**, 74 ⟨85BCJ1983⟩
Izawa, Y., **1**, 442 ⟨87JCS(P1)1951⟩, 505, 507 ⟨83BCJ353⟩, 515 ⟨85T2387⟩; **2**, 619 ⟨85TL1651⟩, 621, 627 ⟨84CC1523⟩; **3**, 664 ⟨80CC1096⟩
Izhboldina, L. P., **4**, 766 ⟨89ZOB997⟩
Izhdanov, G. S., **2**, 882 ⟨53MI 217-01⟩
Izmailov, V. M., **4**, 894 ⟨74MI 420-01⟩
Iznoskova, M. G., **4**, 556 ⟨87ZOB54⟩
Izod, K., **6**, 393 ⟨94AG(E)1268, 94OM753⟩
Izquierdo, M. L., **4**, 843 ⟨89H(29)1877⟩
Izsó, G., **6**, 546 ⟨90JCS(P1)2009⟩
Izukawa, H., **4**, 206 ⟨88CC560⟩
Izumi, M., **1**, 143 ⟨86BCJ2801⟩
Izumi, T., **1**, 493 ⟨80BCJ553⟩, 690 ⟨82BCJ2434, 83BCJ192⟩, 690 ⟨83BCJ2023⟩; **3**, 233 ⟨88S604⟩; **4**, 193 ⟨89JHC1405⟩, 193 ⟨90JHC1419⟩, 193 ⟨92JHC1625⟩; **5**, 142 ⟨77BCJ1639⟩, 1113 ⟨80BCJ553⟩
Izumi, Y., **1**, 76 ⟨56MI 103-01⟩, 208 ⟨90CL481⟩, 209 ⟨79BCJ2389⟩; **2**, 82 ⟨91TL4741⟩, 618 ⟨78CC446, 79BCJ1197⟩; **3**, 51, 202 ⟨85TL4229⟩, 77 ⟨89JA2332, 88JOM(344)71⟩, 279 ⟨92CL1987⟩, 495 ⟨79BCJ1197⟩, 615 ⟨89CL1393⟩, 615, 627, 634 ⟨93BCJ2016⟩, 632 ⟨90CL481⟩; **5**, 84 ⟨78AG(E)176⟩, 419 ⟨88JOM(344)71⟩
Izuni, Y., **2**, 84 ⟨90TL7467⟩
Izuoka, A., **2**, 235 ⟨86BCJ2179⟩, 244 ⟨85H(23)1897⟩

J

Jaarsveld, K., **4**, 489 ⟨90JOC4063⟩
Jaber, A., **2**, 52 ⟨89CC1916⟩
Jablonski, C. R., **1**, 870 ⟨69TL3989⟩
Jabre, I., **5**, 589, 612 ⟨90JCR(S)106⟩, 589, 612 ⟨91PS(56)283⟩
Jabre, S., **3**, 659 ⟨85SC473⟩
Jabri, N., **1**, 312 ⟨84PAC91⟩, 468 ⟨81TL959, 83BSF332, 84T2741⟩, 471, 1097, 1107 ⟨81TL3851⟩; **3**, 212 ⟨83TL5081⟩, 212, 228 ⟨86T1369⟩
Jabry, Z., **3**, 582 ⟨86JCR(S)188⟩
Jachiet, D., **1**, 479 ⟨86T5607⟩, 651 ⟨89T381⟩, 1024 ⟨89T6197⟩
Jachmann, J., **5**, 1146 ⟨81LA1198⟩
Jack, T. R., **4**, 692 ⟨77JA4707⟩
Jäckel, K.-P., **1**, 1116 ⟨74TL1637, 77CB199⟩
Jäckh, C., **5**, 1015 ⟨73CB1752⟩; **6**, 419, 442, 451, 452, 453 ⟨73CB1752⟩
Jacklin, A. G., **1**, 506 ⟨57JCS1622⟩, 957 ⟨55JCS1740⟩
Jackman, D. E., **5**, 905 ⟨78JA8214⟩
Jackman, L. M., **1**, 129, 130, 858, 873 ⟨77T2737, 73JA3687⟩, 873 ⟨81JA4494, 77JA4954⟩
Jackman, M., **2**, 128 ⟨45JA594⟩
Jackman, W. F. H., **2**, 127 ⟨37JA2473⟩
Jackson, A., **4**, 306 ⟨61JA2574⟩; **5**, 603 ⟨89SUL117, 89TL3189⟩
Jackson, A. E., **1**, 74, 441 ⟨77JCS(P1)443⟩
Jackson, A. H., **1**, 184 ⟨87JCS(P2)1221⟩; **3**, 308 ⟨66T(S)241⟩
Jackson, B. G., **4**, 881 ⟨85EUP230127⟩; **5**, 31 ⟨58JA217⟩
Jackson, B. L. J., **1**, 387 ⟨69T1479⟩
Jackson, C. B., **1**, 690 ⟨85TL3393⟩
Jackson, D. Y., **1**, 967 ⟨91JOC1083⟩
Jackson, E. L., **5**, 28, 46 ⟨44OR(2)341⟩
Jackson, G., **1**, 739 ⟨76JOM(110)315⟩
Jackson, H., **3**, 294 ⟨90JOC6161⟩
Jackson, H. L., **2**, 478 ⟨51USP2722538⟩; **3**, 695 ⟨63OSC(4)438⟩, 696, 700 ⟨73OSC(5)773⟩
Jackson, I. K., **2**, 824 ⟨31JCS2109⟩
Jackson, J. A., **3**, 198 ⟨90JOC2842⟩
Jackson, J. E., **4**, 138, 154 ⟨88TL5863⟩
Jackson, R. A., **1**, 99 ⟨75JCS(P1)470⟩, 933 ⟨79JCS(P1)1137⟩; **2**, 937 ⟨75JCS(P1)475⟩, 937, 938 ⟨74JCS(P1)2055, 75JCS(P1)470⟩; **6**, 734 ⟨70CC559⟩
Jackson, R. F. W., **1**, 622 ⟨91TL3119⟩, 968 ⟨92CC319, 93SL219⟩, 968 ⟨93SL499⟩; **2**, 90 ⟨88TL2483⟩, 393 ⟨88S229⟩, 591 ⟨92JOC3397⟩; **3**, 130, 634 ⟨90SL735⟩, 190 ⟨88S229⟩, 273 ⟨84JCS(P1)535⟩, 288 ⟨89CC644⟩; **4**, 173, 237 ⟨91JCS(P1)897⟩, 237, 783 ⟨91JCS(P1)3103⟩, 858 ⟨88TL4889, 91JCS(P1)3091⟩, 862 ⟨93JCS(P1)577⟩, 938 ⟨93JCS(P1)343⟩; **5**, 238 ⟨88CC282, 90T7429⟩, 423 ⟨91JCS(P1)3103, 91JCS(P1)897⟩; **6**, 98, 99, 101 ⟨91JCS(P1)3091⟩, 126 ⟨88CC645, 90TL1877⟩, 126 ⟨91JCS(P1)897⟩
Jackson, R. W., **1**, 623 ⟨92JCS(P1)3225⟩, 831 ⟨94TL1317⟩
Jackson, S. A., **4**, 127 ⟨88PS(39)27⟩
Jackson, T., **6**, 28 ⟨89JCS(P1)283⟩
Jackson, T. C., **2**, 60 ⟨94TL1515⟩
Jackson, W., **3**, 61 ⟨80HCA1665⟩
Jackson, W. G., **5**, 385 ⟨90IC20, 90ICA203⟩, 385 ⟨93IC5285⟩, 385, 386 ⟨91IC1564⟩, 387 ⟨91IC4806⟩, 780 ⟨90IC140⟩
Jackson, W. P., **1**, 526 ⟨81JCS(P1)1516⟩; **5**, 173, 235 ⟨82JA5523⟩, 214 ⟨88JMC499⟩, 342 ⟨89JCS(P1)1815⟩
Jackson, W. R., **1**, 545 ⟨82CC1231⟩; **2**, 317 ⟨90TL1447⟩, 888 ⟨84AJC527⟩, 928 ⟨87S1032⟩; **3**, 77 ⟨87S1033⟩, 613 ⟨62JCS954⟩, 614 ⟨79JA6128, 82AJC2041⟩, 614 ⟨82AJC2053, 82TL1621⟩, 628 ⟨88AJC1697⟩, 634 ⟨67JCS(C)2465⟩, 660 ⟨83AJC1975, 85CC4⟩
Jackson, Y. A., **3**, 339 ⟨88JOC3529⟩
Jacksties, W., **2**, 467 ⟨85CB2030⟩
Jacman, L. M., **1**, 826 ⟨63NAT388⟩
Jacob, D., **3**, 359 ⟨86TL5703⟩
Jacob, F., **4**, 555 ⟨60USP2937207⟩
Jacob, III, P., **1**, 277, 977 ⟨73S305⟩, 531, 1103 ⟨77JOC579⟩, 657 ⟨72JA6549⟩
Jacob, K., **3**, 306, 308 ⟨70LA(733)27⟩
Jacob, L. A., **2**, 943 ⟨93S611⟩; **4**, 161 ⟨85JOC5022⟩
Jacob, P., **1**, 946 ⟨78JOM(156)101⟩; **3**, 140 ⟨76JA7832⟩
Jacob, R. A., **6**, 21 ⟨79CC128⟩
Jacob, T. A., **5**, 972 ⟨66JA3163⟩
Jacob, T. M., **2**, 143 ⟨64JA1630⟩
Jacobi, M. A., **4**, 901 ⟨86CB1745, 88CB805⟩
Jacob III, P., **2**, 900 ⟨76JA7832, 77JOC579⟩
Jacobi, P. A., **1**, 20 ⟨87TL2937⟩, 199 ⟨93TL2581⟩, 523 ⟨81JOC2065⟩; **2**, 672 ⟨88TL6873⟩; **3**, 221, 227, 275 ⟨84JA5585⟩, 276 ⟨90JOC202⟩, 707, 714 ⟨79JOC2042⟩; **5**, 329 ⟨94JOC5292⟩; **6**, 111 ⟨84H(22)281⟩
Jacobi, S. A., **4**, 617 ⟨80JA7979⟩
Jacobi, V., **3**, 484 ⟨69LA(730)1⟩
Jacobine, A. F., **6**, 120 ⟨78S309⟩
Jacobs, G. A., **5**, 370 ⟨83JA3743⟩
Jacobs, H. A., **1**, 200 ⟨86AJC2095⟩; **2**, 317 ⟨90TL1447⟩; **3**, 628 ⟨88AJC1697⟩
Jacobs, H. J. C., **1**, 420 ⟨75TL3741, 87RTC613⟩
Jacobs, H. K., **2**, 39 ⟨90TL5575⟩; **5**, 322 ⟨94TL5157⟩
Jacobs, P. B., **1**, 831 ⟨88T3079⟩
Jacobs, R. T., **1**, 776 ⟨90JOC4051⟩
Jacobs, S. A., **3**, 667 ⟨83JOC5134⟩
Jacobs, T. L., **1**, 954 ⟨64T2177⟩, 1114 ⟨65JOC1808⟩, 1122 ⟨44JA686⟩, 1125 ⟨60JA6397⟩, 1125, 1126 ⟨69JA4824⟩; **2**, 89 ⟨50JA2735⟩; **4**, 761, 765 ⟨42JA2635, 40JA1849⟩
Jacobs, W. A., **5**, 275 ⟨41OSC(1)153⟩
Jacobsen, A. E., **2**, 33 ⟨87JOC5466⟩
Jacobsen, C., **4**, 77 ⟨68ACS3256⟩; **6**, 238, 540 ⟨72CB2854⟩, 553 ⟨71CC314⟩, 590 ⟨83CC295⟩
Jacobsen, E. J., **1**, 806, 809 ⟨82JA7225⟩; **3**, 422 ⟨83JOC3393, 88TL3781⟩; **6**, 473 ⟨82JA7225⟩
Jacobsen, E. N., **2**, 47 ⟨88JA1968⟩, 48 ⟨89JOC2263⟩
Jacobsen, G. B., **4**, 548, 1023, 1028 ⟨86CC882⟩, 1023, 1028 ⟨85CC1635⟩
Jacobsen, G. E., **2**, 1003, 1005 ⟨89JCS(D)105⟩; **3**, 545 ⟨84CC220⟩; **4**, 646 ⟨89JCS(D)105⟩
Jacobsen, N., **3**, 335, 346, 351 ⟨78NJC331⟩
Jacobsen, N. E., **2**, 469 ⟨83JA1613⟩
Jacobsen, P., **6**, 118 ⟨B-18MI 607-01⟩
Jacobsen, R. A., **3**, 723 ⟨84JOM(260)347⟩
Jacobsen, R. M., **3**, 220 ⟨78JOC4650⟩
Jacobsen, R. P., **3**, 293 ⟨51JA3463⟩
Jacobsen, S., **5**, 692 ⟨88ACS(B)605⟩
Jacobson, A. E., **1**, 432 ⟨84JMC1575⟩; **3**, 242 ⟨84JOC1051⟩; **4**, 179 ⟨88S233⟩
Jacobson, B. M., **1**, 107 ⟨90JA3554⟩
Jacobson, H. I., **2**, 256 ⟨55JA6064⟩, 855 ⟨63JA1618⟩
Jacobson, I. C., **3**, 88 ⟨89TL2037⟩
Jacobson, M., **1**, 1001 ⟨68JMC373⟩
Jacobson, R. A., **2**, 1016 ⟨33JA4667⟩; **4**, 575 ⟨82ZAAC(494)55⟩; **6**, 79 ⟨88JA4144⟩, 334 ⟨85OM1226⟩, 517 ⟨84OM1523⟩
Jacobson, R., **4**, 171 ⟨82JOC3140⟩
Jacobson, R. M., **3**, 212, 213 ⟨80TL1205⟩, 244 ⟨81TL3565⟩, 638 ⟨79TL771⟩; **4**, 199 ⟨79SC57⟩
Jacobson, S., **6**, 386 ⟨81TCA285⟩

Jacobson, S. E., **3**, 117 ⟨79JOC921⟩; **4**, 267 ⟨73JA6962⟩
Jacobsson, U., **3**, 39 ⟨87T3063⟩
Jacobus, J., **1**, 4 ⟨70CC338⟩, 936 ⟨79JOC449⟩
Jacoby, D., **3**, 496 ⟨90S301⟩
Jacox, M. E., **1**, 923 ⟨67JCP(47)5146⟩; **3**, 555 ⟨63JA278⟩, 565 ⟨79CPH157⟩
Jacquemin, P. V., **5**, 334 ⟨92TL3863⟩
Jacques, J., **2**, 24 ⟨59TL24⟩, 28 ⟨88OSC(6)175, 65BSF71⟩; **3**, 208 ⟨62BSF90⟩
Jacquesy, J.-C., **2**, 626 ⟨80CC110⟩, 650 ⟨86BSF625⟩; **3**, 429 ⟨89TL5763, 91BSF730⟩
Jacquier, R., **1**, 391 ⟨50BSF35⟩, 422 ⟨56BSF1653⟩; **4**, 984 ⟨71BSF283⟩; **5**, 652 ⟨50BSF35⟩
Jacquignon, P., **5**, 239 ⟨76JCS(P1)2452⟩
Jacquot, C., **2**, 176 ⟨73BSF210⟩
Jadhav, K. P., **1**, 142, 1053 ⟨81JOC5041⟩; **4**, 392 ⟨81JOC5041⟩
Jadhav, P. K., **1**, 795 ⟨87OS(65)224⟩; **2**, 42 ⟨B-83MI 202-01⟩, 75, 521 ⟨88JA1535⟩, 423 ⟨87OS(65)224⟩, 514, 515, 519 ⟨82JOC5065⟩, 516 ⟨80JOC3543, 81T3547⟩, 520 ⟨81JOC2988, 85JA2564⟩, 520 ⟨91JCS(P1)2633⟩, 521 ⟨86JOC432⟩, 523 ⟨82JOC5069⟩; **3**, 135 ⟨82JA6844, 84T1325⟩
Jaeger, C. D., **1**, 934 ⟨78JA4903⟩
Jaeger, D. A., **2**, 311 ⟨82TL4651⟩
Jaeger, R., **6**, 633 ⟨91S465⟩
Jaeger, R. H., **1**, 1000 ⟨61T18⟩
Jaeggi, K. A., **6**, 160 ⟨90PS(54)197⟩
Jaehme, J., **5**, 29 ⟨82GEP3116474⟩
Jaén, J. C., **1**, 479 ⟨82JA3165⟩
Jaenicke, L., **1**, 5, 9 ⟨77CB1823⟩, 212 ⟨83LA226⟩; **5**, 92 ⟨61AG449⟩
Jafri, I. A., **1**, 503 ⟨83JCS(P1)387⟩
Jagannatha, S. S., **5**, 272 ⟨84MI 507-02⟩
Jagdmann, Jr., E., **1**, 2, 6 ⟨85JOC3408⟩; **3**, 621 ⟨85JOC3408⟩
Jagdmann, Jr., G. E., **1**, 1136 ⟨85JOC1019⟩; **2**, 1002 ⟨85JOC1019⟩; **3**, 182 ⟨85JOC1019⟩; **4**, 511 ⟨82JA877⟩, 530 ⟨85JOC1019⟩
Jager, A., **6**, 113 ⟨57CB2460⟩
Jäger, H., **6**, 236 ⟨85HOU(E11)1073⟩
Jager, J., **4**, 838, 1081 ⟨89T6667⟩
Jager, L., **4**, 977, 978 ⟨92ZAAC(611)68⟩; **5**, 690, 723 1074 ⟨89ZAAC(568)35, 89ZAAC(577)165⟩, ⟨90ZAAC(588)59⟩, 1070 ⟨85ZC305⟩, 1074 ⟨84ZC379, ⟨89ZAAC(568)35, 89ZAAC(577)165, 1074, 1078 ⟨89ZC145⟩, 1084 ⟨88ZAAC(564)85⟩
Jäger, U., **5**, 340 ⟨86CB1127, 86CB3405⟩
Jäger, V., **1**, 664 ⟨79JCR(S)54⟩, 797 ⟨86S921⟩; **2**, 318 ⟨89TL1517, 91S769⟩, 392 ⟨86S535⟩, 396 ⟨91S629⟩, 1011 ⟨77HOU(5/2a)110⟩, 1052 ⟨69AG(E)273, 75C516⟩; **3**, 46 ⟨84T2199, 86S535⟩, 120 ⟨91JA2205⟩, 258 ⟨91S629⟩, 424 ⟨92CB2361⟩; **4**, 209 ⟨89TL4223⟩; **5**, 223 ⟨87S801⟩; **6**, 417 ⟨80LA122⟩
JägerI. Müller, V., **3**, 683 ⟨85MI 320-01⟩
Jagger, A. J., **6**, 393 ⟨94JOM(469)129, 94OM753⟩
Jagner, S., **2**, 569 ⟨87OM2354⟩; **4**, 671 ⟨93OM2268⟩
Jagodzinski, T., **5**, 631 ⟨90OPP755⟩
Jagoe, C. T., **1**, 190, 402 ⟨91JA5488⟩
Jagow, R. H., **1**, 892 ⟨58JA2326⟩
Jahangir, **1**, 1041 ⟨88CC439⟩; **2**, 313 ⟨88SC893⟩; **5**, 301 ⟨87JOC5378⟩
Jahav, P. K., **2**, 514 ⟨B-86MI 210-01⟩
Jähde, E., **2**, 412 ⟨90AG(E)782⟩
Jahelka, J., **2**, 805 ⟨73CCC706⟩
Jahn, N., **6**, 65 ⟨94JOM(465)153⟩, 245 ⟨92JOM(434)159⟩
Jahngen, E. G. E., **3**, 131 ⟨77JOC1189⟩
Jahngen Jr., E. G. E., **1**, 681, 682 ⟨74JOC1650⟩, 683 ⟨74JOC1322⟩
Jahnke, D., **2**, 187 ⟨66CB1718⟩; **3**, 458 ⟨64CB2661⟩; **5**, 595 ⟨64CB2661, 66CB11⟩
Jahreis, G., **3**, 251 ⟨77S472⟩; **4**, 899 ⟨79JPR699⟩
Jaime, C., **1**, 217, 225 ⟨88JOC139⟩
Jain, A. U., **4**, 536 ⟨86T4223⟩
Jain, B. D., **2**, 565 ⟨73JINC311⟩
Jain, C., **3**, 101 ⟨56MI 303-01⟩
Jain, N., **5**, 205 ⟨88SC2123⟩
Jain, S., **2**, 909, 910 ⟨83T877⟩
Jain, S. C., **1**, 503 ⟨83JOC2274⟩, 503, 504 ⟨83JOC2266⟩
Jain, S. K., **3**, 370 ⟨66T7⟩
Jain, T. C., **1**, 2 ⟨79T885⟩
Jais, M. J. B., **4**, 42 ⟨82JCS(P2)881⟩
Jaiswal, D. K., **5**, 826 ⟨83IJC(B)51⟩
Jakita, K., **4**, 937 ⟨85JFC(30)357⟩
Jakle, H., **3**, 272 ⟨71LA(754)64⟩
Jakob, F., **2**, 375 ⟨54LA(590)1⟩
Jakob, K., **2**, 566 ⟨68ZAAC(356)195⟩
Jakobi, H., **3**, 351 ⟨91CB1747⟩
Jakobi, U., **1**, 282 ⟨94SL511⟩
Jakobovits, J., **3**, 91 ⟨50JCS767⟩
Jakobsen, H. J., **5**, 241 ⟨79JOM(166)373⟩
Jakobsen, M. H., **5**, 323 ⟨90S1008⟩
Jakobsen, P., **3**, 344 ⟨68T5323⟩, 695 ⟨76ACS(B)995⟩
Jakobsen, P.,**5**, 652 ⟨77T3137⟩
Jakobson, G., **2**, 889, 891 ⟨69ZAAC(369)178⟩
Jakubke, H.-D., **5**, 262 ⟨B-89MI 506-04⟩
Jakubke, H. D., **6**, 486 ⟨B-82MI 615-02⟩
Jakubowski, A. A., **1**, 503 ⟨82JOC1221⟩
Jakupovic, J., **1**, 1134 ⟨84T2537⟩
Jakus, V., **5**, 984 ⟨92CCC1505⟩
Jalal, R., **2**, 20 ⟨89SC1697⟩
Jalander, L., **1**, 469 ⟨83ACS(B)173⟩, 482 ⟨88SC343, 90ACS842⟩
Jallageas, J. C., **1**, 1023 ⟨67CR(C)839⟩
Jamart-Gregoire, B., **1**, 539 ⟨88T127⟩
Jambotkar, D., **1**, 732 ⟨62JOC4666⟩
Jamerson, J. D., **4**, 681 ⟨76JA1774⟩
James, A. T., **2**, 780 ⟨50JCS2657⟩; **5**, 97 ⟨63MI 502-01⟩
James, B. G., **1**, 729 ⟨76JCS(P1)1476⟩; **3**, 494 ⟨76JCS(P1)1466⟩
James, B. R., **2**, 583 ⟨87CC1110⟩, 584 ⟨86CC787⟩; **3**, 614 ⟨B-82MI 318-01⟩; **4**, 545 ⟨86OM2220⟩
James, C., **1**, 548 ⟨62CR(255)959⟩; **4**, 229 ⟨87JMC24⟩; **5**, 558 ⟨87JMC24⟩
James, D., **2**, 245 ⟨85CC1449⟩
James, D. E., **1**, 298 ⟨76JA1806, 76JA1810⟩
James, K., **1**, 828 ⟨86TL3053⟩; **2**, 591 ⟨92JOC3397⟩; **3**, 200 ⟨92TL813⟩, 288 ⟨89CC644⟩; **5**, 993 ⟨89JOC4019⟩
James, L. M., **6**, 174, 175, 176, 289, 380, 391, 393, 395 ⟨92OM2938⟩
James, M. R., **2**, 197 ⟨92JCS(P1)2371⟩; **4**, 281 ⟨92JCS(P1)2371⟩
James, P. N., **3**, 107 ⟨63OSC(4)539⟩
James, S. R., **1**, 107 ⟨79COC(1)493⟩, 852 ⟨79COC(1)525⟩; **2**, 2 ⟨79COC(1)493⟩
James, W. J., **5**, 1039 ⟨51JA4495⟩
Jamieson, L. A., **6**, 123 ⟨90H(30)855⟩, 158 ⟨90PS(51/52)23⟩
Jamil, Z., **3**, 5, 118 ⟨85TL6257⟩
Jamison, J. D., **3**, 100 ⟨61JA193⟩
Jamison, S. E., **2**, 104 ⟨58JA4546⟩
Jan, D.-Y., **6**, 345 ⟨84CC392⟩
Jan, G., **2**, 615 ⟨79HCA635⟩
Janardhanam, S., **1**, 810 ⟨92JCS(P1)2727, 93JOC5482⟩, 810 ⟨93SC311⟩
Janarthanan, N., **6**, 580 ⟨93TL6447⟩
Janata, J., **2**, 176 ⟨75JA1492⟩
Janati, T., **5**, 1152 ⟨91AG(E)196⟩, 1158 ⟨92CC415⟩
Jänchen, G., **2**, 726 ⟨69ZC305⟩
Jancis, E. H., **2**, 188 ⟨71JOC2698⟩
Jancke, H., **4**, 602, 607, 637, 660 ⟨90ZAAC(583)195⟩; **5**, 720 ⟨77JOM(134)21⟩
Janda, L., **5**, 655 ⟨84CCC1505⟩
Jander, J., **2**, 339 ⟨85ZAAC(526)186, 87ZAAC(547)233⟩; **6**, 241 ⟨54JCS912⟩, 441 ⟨54JCS919⟩

Jandik, P., **4**, 695 ⟨82AG(E)73⟩
Jane, D. E., **5**, 692 ⟨89JCR(S)337⟩
Janecke, J., **2**, 769 ⟨42CB1096⟩
Janecki, T., **3**, 158 ⟨84S1025⟩; **5**, 12 ⟨89S506⟩
Janes, N. F., **1**, 724 ⟨74JCS(P1)2470⟩; **4**, 746 ⟨74JCS(P1)2470⟩
Jang, D. O., **1**, 17, 938 ⟨91TL7187⟩, 17, 938, 946 ⟨90TL4681⟩, 587 ⟨91TL2569⟩, 937, 938 ⟨92TL5709⟩, 937, 938, 945 ⟨93JOC6838, 91SL435⟩, 938, 946 ⟨90TL3991⟩, 938, 946 ⟨91T8969⟩, 939 ⟨92TL6629⟩; **2**, 536 ⟨91SL435⟩
Jang, G. S., **1**, 630 ⟨92MI 114-01⟩
Jang, Y. M., **1**, 183 ⟨88CC760⟩
Jani, M. K., **6**, 561 ⟨89JIC60⟩
Janickis, V., **2**, 369 ⟨80JCS(D)620⟩
Janik, T. S., **5**, 461 ⟨87JA5145⟩
Janke, N., **2**, 558 ⟨90CB1517⟩, 559, 560, 959 ⟨90CB1507⟩
Jankowski, B. C., **3**, 156 ⟨86JOC130⟩; **4**, 183 ⟨86JOC130⟩
Jankowski, P., **4**, 393 ⟨91CC297⟩
Janku, J., **2**, 617 ⟨88CCC110⟩
Janny, A., **2**, 103 ⟨1883CB175⟩
Janosi, A., **1**, 1079 ⟨93TL5621⟩
Janousek, J., **2**, 1039, 1050 ⟨86TCC89⟩
Janousek, Z., **1**, 919, 922 ⟨85ACR148⟩, 919, 922 ⟨88PAC1635⟩, 982 ⟨73AG(E)566⟩, 995 ⟨93TL1779⟩; **2**, 1042 ⟨82S1083⟩, 1046 ⟨76C387⟩; **3**, 48, 473 ⟨70CCC3618⟩, 409 ⟨89SC2825⟩; **4**, 91, 287, 858, 861 ⟨85T2527⟩, 92 ⟨85T4183⟩; **6**, 7 ⟨92CCC1291⟩, 52 ⟨79AG(E)615⟩, 238 ⟨73AG837⟩, 506 ⟨72AG993⟩, 609, 610, 614, 619, 620 ⟨B-76MI 620-01⟩, 610, 727 ⟨89SC2825⟩, 619 ⟨79AG(E)333, 71AG(E)573⟩, 727, 728, 729 ⟨B-76MI 623-01⟩
Janout, V., **3**, 114 ⟨82JOC3331⟩
Janowicz, A. H., **2**, 962 ⟨82JA352, 83JA3929⟩
Janse, L. C., **2**, 757 ⟨21RTC285⟩
Jansen, B. J. M., **1**, 600 ⟨81TL4137⟩; **3**, 35 ⟨81TL887, 87RTC489⟩, 63 ⟨81TL4137⟩, 253 ⟨87RTC549⟩; **6**, 336 ⟨83SC985⟩
Jansen, J. E., **5**, 1106 ⟨52JA1323⟩
Jansen, J. F. G. A., **3**, 406 ⟨88S184⟩, 413 ⟨86TL507⟩; **6**, 88 ⟨89TL5481⟩
Jansen, M., **1**, 370, 371 ⟨94AG(E)683⟩; **2**, 388 ⟨78HCA1622⟩
Jansons, E., **5**, 545 ⟨76RCR1035⟩
Janssen, C. G. M., **1**, 24 ⟨84RTC220⟩
Janssen, E., **4**, 698 ⟨87JOM(333)155⟩
Janssen, H., **5**, 675 ⟨85CB1632⟩
Janssen, J., **1**, 690 ⟨82CB1234⟩
Janssen, J. F., **6**, 29 ⟨70JOC3637⟩
Janssen, M. A. C., **6**, 598 ⟨91JMC3187⟩
Janssen, M. J., **1**, 731 ⟨70T1291⟩; **3**, 494 ⟨70T1291⟩; **5**, 472 ⟨B-69MI 511-01⟩
Janssen, P. A. J., **6**, 598 ⟨91JMC3187⟩
Janssen, R. H. A. M., **1**, 784 ⟨84JCS(P1)1701⟩
Janssen, S. J., **5**, 384 ⟨91JOM(420)387⟩, 1107 ⟨84S1017⟩
Janssen, W. A. J., **3**, 368 ⟨73S617⟩
Janssen, W., **2**, 503 ⟨69JOM(16)P69⟩
Janssens, F., **3**, 238 ⟨75T177⟩
Jansson, K., **4**, 176 ⟨86TL753⟩
Janulis, E. P., **3**, 679 ⟨84TL405⟩
Janulis, Jr., E. P., **3**, 314 ⟨83JA5929⟩
Januszkiewicz, K. R., **1**, 94 ⟨83OM1055⟩
Janzen, A. F., **2**, 839 ⟨79CJC1903⟩; **4**, 61 ⟨83JFC(22)557⟩, 61, 62, 63 ⟨77CJC3031⟩, 323 ⟨79JHC415⟩, 337, 344 ⟨73CJC1136⟩, 455 ⟨81JINC629⟩; **5**, 378 ⟨71CJC1011⟩
Jaouen, G., **1**, 202 ⟨87JOM(336)143⟩; **2**, 309 ⟨81JOC78⟩; **3**, 699 ⟨78JA4312⟩
Jaouen, G., **4**, 674 ⟨90OM2993⟩
Jaouen, G., **4**, 674 ⟨94OM2244⟩, 713 ⟨82OM225, B-85MI 408-01⟩, 718 ⟨93OM4545⟩
Jappy, J., **3**, 580 ⟨70TL1157⟩; **5**, 794, 803 ⟨71MI 520-02⟩
Jaquelin, S., **5**, 25, 56 ⟨86T2275⟩
Jaquier, R., **3**, 452 ⟨70BSF4119⟩
Jarboe, C. J., **4**, 53, 54 ⟨67JOM(8)361⟩
Jarczewski, A., **4**, 731 ⟨85CJC576⟩
Jardine, F. H., **1**, 73 ⟨65CC131, 65CI(L)560⟩, 73, 79, 80 ⟨67JCS(A)1574⟩, 85 ⟨67JCS(C)270⟩, 456 ⟨81MI 110-01⟩
Jardine, I., **1**, 73, 74 ⟨71CC840⟩
Jardine, P. D., **2**, 976 ⟨83JOC1556⟩; 1035 ⟨88TL6409⟩; **3**, 218 ⟨89TL7297⟩
Jaret, R. S., **6**, 550 ⟨86JOC4488⟩
Jarmer, M., **2**, 471 ⟨90ZAAC(589)23⟩
Jarowicki, K., **3**, 240 ⟨90TL3433⟩
Jarrar, A., **2**, 213 ⟨66JOC615⟩
Jarrett, S. R. M., **6**, 641 ⟨92JCS(P1)3179⟩
Jarstfer, M. B., **2**, 527 ⟨91JA6139⟩
Jarvest, R. L., **5**, 764 ⟨89JCS(P1)2207⟩
Jarvi, E. T., **1**, 439 ⟨90TL5449⟩, 829 ⟨88JOC1922⟩; **2**, 611 ⟨93TL3057⟩; **3**, 150 ⟨85TL2861⟩; **4**, 61, 79, 780, 782 ⟨90TL5449⟩, 73 ⟨89JA1127⟩, 119, 734 ⟨90TL5571⟩, 731, 949 ⟨85TL2861⟩, 736 ⟨93TL3057⟩, 755, 782 ⟨92JA360⟩; **5**, 159 ⟨88JOC1922⟩; **6**, 266 ⟨90TL5571⟩
Jarvie, A. W., **4**, 140, 607 ⟨73JOM(57)261⟩
Jarvie, A. W. P., **1**, 652 ⟨69JCS(B)852⟩
Jarvis, B. B., **1**, 620 ⟨68JA429⟩; **4**, 75 ⟨74JOC643⟩; **5**, 179 ⟨82JOC3358⟩; **6**, 254 ⟨75JOC1278⟩
Jarvis, J. A. J., **4**, 656, 695 ⟨77JCS(D)999⟩
Jarvis, W. F., **4**, 234 ⟨88JOC5750⟩
Jas, G., **4**, 355 ⟨83TL45⟩
Jasche, K., **2**, 439 ⟨67CB412⟩
Jasim, H. A., **6**, 349 ⟨85CC1776⟩
Jasiobedzki, W., **1**, 995 ⟨90PJC123⟩
Jasko, M., **6**, 489 ⟨93MI 615-03⟩
Jason, D. M., **2**, 344 ⟨90JA6696⟩
Jasor, Y., **4**, 405 ⟨88OSC(4)474⟩
Jaspar, B., **5**, 946, 957 ⟨94OM3510⟩
Jasper, B., **5**, 953 ⟨95OM2447⟩, 958 ⟨94CB2475, 94OM3502⟩
Jasperse, C. P., **1**, 320, 331, 336, 341, 342, 344, 348, 926, 935, 948 ⟨91CRV1237⟩, 344, 831 ⟨90JA5601⟩, 347, 937 ⟨92SL943⟩; **2**, 368 ⟨88JOC2389⟩, 565 ⟨92SL943⟩; **3**, 135 ⟨87JA5549⟩; **4**, 71, 88, 288 ⟨86JOC2981⟩; **5**, 334 ⟨87JA5549⟩
Jass, P. A., **6**, 134 ⟨90JA4552⟩
Jastrebski, J. T. B. H., **4**, 724 ⟨80JOM(186)427⟩
Jastrzebski, J. T. B. H., **4**, 660 ⟨86OM1551⟩, 704 ⟨91NJC301⟩
Jasys, V. J., **2**, 344 ⟨90JA6696⟩; **6**, 553 ⟨92JOC4352⟩
Jászay, Z. M., **5**, 1065 ⟨87S520⟩
Jaszay, Z. M., **5**, 1069 ⟨88S397⟩
Jaszberenyi, J. C., **1**, 17, 938 ⟨91TL7187⟩, 17, 938, 946 ⟨90TL4681⟩, 333 ⟨92TL5017, 93TL6505⟩, 443 ⟨93TL8051⟩, 615, 700, 939 ⟨91TL2703⟩, 935 ⟨92T7121⟩, 937, 938 ⟨92TL5709⟩, 937, 938, 945 ⟨93JOC6838, 91SL435⟩, 938, 946 ⟨90TL3991⟩, 938, 946 ⟨91T8969⟩, 939 ⟨89TL2619, 92T7435⟩, 939 ⟨92TL6629⟩, 944 ⟨92JA5904, 92T9261⟩, 944 ⟨93JA8050⟩; **2**, 536 ⟨91SL435, 92TL6629⟩; ; **3**, 635 ⟨92TL5017⟩; **5**, 516, 526 ⟨91TL2703⟩, 1110 ⟨92T2613⟩; **6**, 541, 550 ⟨91TL2703⟩
Jath, G., **2**, 512 ⟨80CB1180⟩
Jaturonrusmee, W., **5**, 1127 ⟨91AJC135⟩
Jaud, J., **3**, 159 ⟨93JOC1835⟩; **4**, 722 ⟨91OM4005⟩; **6**, 688 ⟨90JOC5750⟩
Jauhal, G. S., **5**, 464 ⟨67CJC301⟩
Jauner, T., **3**, 282 ⟨75HCA74⟩
Jaurand, G., **4**, 950, 952 ⟨86TL6201⟩
Jaus, H., **4**, 894, 973, 974 ⟨79LA2096⟩; **6**, 321, 337 ⟨79LA2089⟩, 647 ⟨79LA2096⟩
Jautelat, M., **3**, 373 ⟨67TL2325⟩; **4**, 385 ⟨67LA(702)24⟩; **5**, 28 ⟨81AG(E)703⟩
Javet, P., **2**, 177 ⟨74S887⟩
Jawad, H., **2**, 491 ⟨83JOM(243)417⟩, 493 ⟨82ZN(B)1104⟩, 890, 891 ⟨84JOM(277)257⟩; **6**, 581 ⟨86ZN(B)1142⟩
Jaw, J. Y., **1**, 25, 630 ⟨87JOC3346⟩; **2**, 279 ⟨81JOC1215⟩; **5**, 239, 281 ⟨81JOC1215⟩

Jawdosiuk, M., **1**, 1033 ⟨79CC488⟩
Jaworski, K., **1**, 978 ⟨82JOM(236)177, 85OM333⟩
Jaxa-Chamiec, A. A., **2**, 664 ⟨78CC118⟩; **4**, 285 ⟨84JCS(P1)21⟩
Jayalakshmi, S., **1**, 664 ⟨78CI(L)234⟩
Jay, R., **3**, 477 ⟨1894CB59⟩
Jayaram, C., **4**, 539 ⟨83T2023⟩
Jayasinghe, L. R., **1**, 582 ⟨92JOC2270⟩
Jayasuriya, N., **2**, 219 ⟨92TL451⟩
Jayathirtharao, V., **2**, 202 ⟨84JOC615⟩
Jayatilake, G. S., **2**, 317 ⟨90TL1447⟩; **3**, 628 ⟨88AJC1697⟩
Jayawant, M., **2**, 304 ⟨70JOC2826⟩
Jazouli, M. E., **5**, 551 ⟨84SUL147⟩
Jean, T., **3**, 155 ⟨87JOC3919⟩
Jean, T.-S., **2**, 34 ⟨87JOC3919⟩
Jeanneret-Gris, G., **5**, 12 ⟨90TL75⟩
Jeannin, Y., **1**, 439 ⟨86T2475⟩; **4**, 674 ⟨93JA3786⟩, 681 ⟨92JOM(441)81⟩; **5**, 799 ⟨88JCS(D)1591⟩; **6**, 331 ⟨91OM3005⟩
Jebaratnam, D. J., **1**, 525 ⟨87TL1611⟩; **3**, 620 ⟨87TL1611⟩
Jeczalik, J., **4**, 864 ⟨86MI 419-02⟩
Jedlovská, E., **5**, 665 ⟨76CCC3085⟩
Jefferies, I., **4**, 748 ⟨91JCS(P1)2575⟩
Jefferies, P. R., **1**, 45 ⟨54JCS3617⟩
Jefferson, E. G., **5**, 26, 37 ⟨53JCS1409⟩
Jefferson, R., **5**, 779 ⟨66JCS(A)1584⟩
Jeffery, E. A., **2**, 478 ⟨74JOM(74)365⟩, 1096 ⟨68JOM(11)393⟩; **4**, 699 ⟨70AJC715⟩
Jeffery, E., **2**, 595 ⟨B-72MI 211-01⟩
Jeffery, G. H., **1**, 860 ⟨60JCS4454⟩
Jeffery, J., **4**, 656 ⟨78CC1081⟩; **6**, 194 ⟨79JOM(181)25⟩
Jeffery, J. C., **4**, 656, 663, 709, 724 ⟨81JCS(D)1593⟩, 712 ⟨81CC1255, 81CC867⟩, 712 ⟨84JCS(D)1563, 84JCS(D)1581⟩, 712 ⟨88JCS(D)2431⟩, 712, 716 ⟨82OM1597⟩; **6**, 403 ⟨88JOM(355)231⟩
Jeffery, S. M., **1**, 145 ⟨93JCS(P2)1617⟩, 145 ⟨93JCS(P2)2163⟩
Jeffery, T., **1**, 121, 1002 ⟨89TL2225⟩, 121, 1002 ⟨92TL5757⟩, 494 ⟨84CC1287⟩, 1021 ⟨87S70⟩; **2**, 1018 ⟨88CC909⟩; **3**, 233 ⟨87S70⟩
Jeffery-Luong, T., **1**, 469, 472, 975, 1009, 1137 ⟨83S32⟩, 970, 975 ⟨80TL5019⟩, 971 ⟨82S738⟩
Jefford, C., **4**, 35 ⟨69TL1957⟩
Jefford, C. W., **1**, 47 ⟨62JA3744⟩, 120 ⟨93TL1111, 93TL2911⟩, 120 ⟨93TL7557⟩, 133 ⟨92TL1855⟩; **2**, 52 ⟨89CC1916⟩; **3**, 124 ⟨92TL1855⟩, 307 ⟨90TL995⟩, 701 ⟨92TL1593⟩; **4**, 8 ⟨79TL1913⟩, 214 ⟨83CC1064⟩
Jeffrey, D., **1**, 706 ⟨88TL909⟩
Jeffrey, G. A., **4**, 25 ⟨59JA992⟩
Jeffrey, H., **3**, 305 ⟨80JHC87⟩
Jeffrey, J. G., **6**, 247 ⟨90JOM(388)391⟩
Jeffreys, J. A. D., **4**, 688 ⟨76JOM(120)C13, 80JCS(D)435⟩
Jeffries, P. M., **2**, 753 ⟨88TL5083⟩; **4**, 696 ⟨93JOM(449)203⟩
Jeffs, P. W., **3**, 530 ⟨82JOC3871⟩; **5**, 282 ⟨82JOC3876, 82JOC3881⟩
Jefson, M., **2**, 282 ⟨81TL3561⟩; **3**, 37 ⟨81TL3561⟩
Jeganathan, A., **1**, 560 ⟨92S235⟩; **3**, 162 ⟨92S235⟩
Jeger, O., **1**, 794 ⟨63HCA678⟩; **2**, 71 ⟨53HCA325⟩, 662 ⟨47HCA675⟩; **3**, 319 ⟨79HCA1645, 82HCA2363⟩
Jegham, S., **5**, 157 ⟨88TL4419⟩
Jegou, D., **4**, 411 ⟨65CR(261)2232⟩
Jehangir, M., **3**, 101 ⟨72JCS(P1)892⟩
Jelenc, P., **4**, 884, 891 ⟨70CR(C)1890⟩
Jelinek, A., **2**, 191 ⟨45JA1578⟩
Jelinek, A. G., **6**, 513 ⟨71GEP2040367⟩
Jelinski, L. W., **5**, 556 ⟨83CC388⟩
Jellal, A., **1**, 978, 1061 ⟨86JOC1199⟩; **3**, 652 ⟨83TL2847⟩
Jellison, K. M., **3**, 540 ⟨89SL36⟩
Jeminet, G., **6**, 81, 90, 91 ⟨64CR(259)4051⟩, 302 ⟨67BSF3233, 74CC634⟩
Jemison, R. W., **1**, 401 ⟨80JCS(P1)1458⟩, 407, 823 ⟨80JCS(P1)1436⟩, 408 ⟨80JCS(P1)1450, 69CC294⟩
Jemmis, E. D., **6**, 322 ⟨93OM4267⟩, 386, 401, 404 ⟨76JA5419⟩
Jen, K.-Y., **2**, 478 ⟨80JOC406⟩
Jen, T., **6**, 570, 645 ⟨75JMC90⟩
Jencks, W. P., **1**, 590 ⟨88JA5075⟩, 623 ⟨90JA6950⟩; **2**, 807 ⟨64JA1410, 65JA3199⟩; **3**, 426 ⟨59JA473, 64MI 311-01⟩, 449 ⟨59JA475⟩, 449, 450 ⟨62JA826⟩; **4**, 216 ⟨79JA1476⟩, 294 ⟨72CRV705⟩; **5**, 31 ⟨64JA837⟩, 699 ⟨82JA6769⟩
Jenco, B., **6**, 632 ⟨85OPP256⟩
Jendralla, H., **1**, 575 ⟨84JA1518⟩, 576 ⟨86JA3739⟩; **2**, 802 ⟨79CB2120⟩
Jenevein, R. M., **1**, 6 ⟨68CC934⟩
Jenkins, A., **3**, 444 ⟨46JA914⟩
Jenkins, A. D., **5**, 85 ⟨B-55MI 502-03⟩
Jenkins, C. L. D., **4**, 704 ⟨89AJC301⟩
Jenkins, D. M., **5**, 66 ⟨55OSC(3)538⟩
Jenkins, G. L., **5**, 1125 ⟨47JA365⟩
Jenkins, I. D., **1**, 64 ⟨74JCS(P2)596⟩; **2**, 414 ⟨66JCS(C)2176⟩
Jenkins, J. A., **1**, 811 ⟨76JA5937⟩
Jenkins, Jr., R. H., **2**, 718 ⟨81JOC3467⟩
Jenkins, L. A., **2**, 184 ⟨86JOC1033, 83JOC3071⟩
Jenkins, L. H., **5**, 1017 ⟨59USP2873171⟩
Jenkins, M. J., **6**, 121 ⟨90TL4359⟩
Jenkins, P. R., **1**, 338 ⟨92TL3543⟩, 593, 713 ⟨87CC1540⟩, 633 ⟨82TL3733⟩, 827, 828 ⟨79HCA1922⟩; **2**, 345, 347 ⟨88CC1503⟩; **3**, 202 ⟨91JCS(P1)1893⟩; **4**, 960 ⟨93JCS(P1)67⟩; **5**, 24 ⟨83GSM98, 85GSM96⟩
Jenkins, R. H., **2**, 150 ⟨82JA5412⟩, 184 ⟨80JA7967⟩
Jenkins, R. L., **6**, 443 ⟨50JA1888⟩
Jenkins, R. N., **1**, 53 ⟨74MI 102-01⟩
Jenkins, T. C., **6**, 445 ⟨78JCS(P1)1066⟩
Jenkins, T. E., **1**, 523 ⟨89JA6432⟩
Jenkins, W. W., **1**, 835 ⟨57JOC1418⟩
Jenkins Jr., R. H., **2**, 718 ⟨83JOC3071⟩
Jenko, B., **2**, 322 ⟨91OPP721⟩; **6**, 632 ⟨91OPP721⟩
Jenne, H., **2**, 361 ⟨59CB1149⟩
Jenner, E. L., **4**, 434, 435 ⟨51JA2773⟩, 435, 443 ⟨51JA2769⟩
Jenneskens, L. W., **1**, 774 ⟨88MI 117-01⟩, 776, 778 ⟨89JOC5811⟩; **4**, 38 ⟨85AG(E)585⟩
Jennings, H. J., **5**, 368 ⟨87TL5091⟩
Jennings, J. R., **3**, 458 ⟨65JCS5083, 67JCS(A)1222⟩, 458 ⟨68CC250B⟩; **5**, 383 ⟨68JCS(A)2535⟩
Jennings, L. D., **5**, 135 ⟨90JA7434⟩
Jennings, L. J. A., **5**, 334 ⟨85JMC1661⟩
Jennings, M. N., **5**, 631 ⟨84S946⟩
Jennings, P. W., **3**, 62 ⟨89JOC5186⟩
Jennings, W. B., **2**, 378, 381, 382 ⟨73JA1562⟩; **3**, 439 ⟨85CC582, 88TL3725⟩, 439 ⟨91T5561⟩
Jennings-White, C., **3**, 180 ⟨82TL2533⟩; **4**, 359 ⟨89JOC1789⟩
Jennische, P., **2**, 570 ⟨80JA6744⟩; **3**, 511 ⟨80JA6744⟩; **5**, 918 ⟨80JA6744⟩
Jenny, C., **5**, 575 ⟨87HCA1001⟩, 605 ⟨86HCA174⟩
Jenny, C.-J., **1**, 487 ⟨88CC975⟩
Jenny, E. F., **1**, 398 ⟨62AG(E)155⟩, 513 ⟨59HCA401⟩, 1000 ⟨59AG245⟩; **3**, 470 ⟨68HCA643⟩
Jenny, T., **6**, 558 ⟨86HCA1898⟩, 705 ⟨84TL4097⟩
Jenny, W., **1**, 104 ⟨73C277⟩; **2**, 282 ⟨58HCA593⟩; **5**, 210 ⟨52HCA1429⟩
Jensen, B., **2**, 478 ⟨86ACS(B)609⟩; **5**, 478, 494 ⟨85SUL177⟩, 478, 494, 582 ⟨87SUL153⟩, 483, 590 ⟨86T739⟩; **6**, 254, 275 ⟨87SUL59⟩, 274 ⟨86T739⟩, 533, 536 ⟨85SUL61⟩, 535 ⟨86ACS(B)609, 86SUL203⟩
Jensen, B. L., **2**, 758 ⟨77S848⟩; **4**, 1059 ⟨84TL5711⟩; **5**, 295 ⟨70JOC825⟩
Jensen, E. V., **2**, 256 ⟨55JA6064⟩, 448, 489 ⟨49JOC429⟩, 855 ⟨63JA1618⟩; **6**, 5 ⟨47JA1100⟩, 421 ⟨55JA5033⟩, 422 ⟨55JA3099⟩
Jensen, F., **1**, 415 ⟨88JA2314⟩
Jensen, F. K., **3**, 639 ⟨75ACS(B)863⟩

Jensen, F. R., **1**, 68 ⟨79JOC228⟩; **2**, 26 ⟨75JA2281⟩
Jensen, H., **1**, 178 ⟨74LA369⟩; **5**, 344 ⟨74LA561⟩
Jensen, H. M., **5**, 700 ⟨93ACS1004⟩
Jensen, H. P., **2**, 292 ⟨73JA7917⟩
Jensen, J. A., **3**, 458 ⟨93JOM(456)161⟩; **5**, 1126 ⟨87AP(320)1088⟩
Jensen, J. L., **4**, 216 ⟨79JA1476⟩
Jensen, K. A., **2**, 288 ⟨36BSF2310⟩, 890 ⟨63ACS1875⟩; **4**, 875 ⟨70ACS3213⟩; **5**, 233, 240, 243, 506, 508 ⟨72ACS1465⟩, 472 ⟨B-73MI 511-02⟩, 549 ⟨70MI 513-01⟩, 563 ⟨B-73MI 513-01⟩, 584, 586, 587, 624, 625, 726, 732 ⟨66ACS597⟩, 624 ⟨81MI 514-01⟩, 634 ⟨61ACS1104⟩, 635, 636 ⟨67ACS1936, 61ACS1087⟩, 639, 640 ⟨61ACS1097, 61ACS1109⟩, 641 ⟨61ACS1124, 79ACS137⟩, 1013, 1101 ⟨65ACS438⟩, 1051 ⟨36ZAAC(230)31⟩, 1054 ⟨66ACS278⟩, 1100, 1102 ⟨B-77MI 530-01, 64ACS825⟩, 1101 ⟨64ACS826, 66ACS2107, 69ACS1567⟩, 1103 ⟨64ACS2417⟩; **6**, 587 ⟨B-73MI 619-02⟩, 590, 592 ⟨B-73MI 619-03⟩, 591 ⟨B-73MI 619-05, 70ACS2055⟩, 592, 593 ⟨70ACS2061, 64ACS2417⟩, 593 ⟨66ACS2091⟩, 594 ⟨67ACS2904⟩, 596 ⟨63ACS549, 70QRS45⟩, 597 ⟨66ACS281⟩, 598 ⟨66ACS278⟩
Jensen, K., **4**, 437 ⟨72ACS1258⟩
Jensen, K. F., **6**, 237 ⟨93CM1321⟩
Jensen, K. M., **3**, 651 ⟨86SC617⟩
Jensen, K.-P., **4**, 697 ⟨81ZN(B)188⟩
Jensen, L., **5**, 547, 555 ⟨74T93⟩
Jensen, L. H., **1**, 909 ⟨63JA3302, 66JA198⟩; **2**, 1003 ⟨78AX(B)828⟩
Jensen, M., **3**, 410 ⟨92JA4436⟩
Jensen, M. S., **1**, 216 ⟨93JA2970⟩; **3**, 496 ⟨92TL577⟩
Jensen, O. E., **5**, 573 ⟨85T5595⟩
Jensen, U., **1**, 514 ⟨81CB292⟩
Jen-Sen Dung, **2**, 282 ⟨80JA7950⟩
Jensen-Korte, U., **2**, 373 ⟨90HOU(E16a/1)425⟩
Jenson, F. R., **1**, 69 ⟨72JOC4341⟩
Jenson, T. M., **1**, 752, 816, 818 ⟨87JOC3883⟩, 816 ⟨84JOC1707, 87TL723⟩, 816, 818 ⟨86JOC4316⟩, 816, 818 ⟨87JOC3860⟩; **2**, 9 ⟨87TL723⟩
Jentsch, R., **1**, 209 ⟨73AG(E)323⟩, 242 ⟨79LA451⟩; **2**, 909 ⟨77LA1167⟩; **3**, 704 ⟨71AG(E)331, 76LA2105⟩, 704, 707, 719 ⟨75CB1580⟩, 714, 715 ⟨77LA1167⟩
Jentzsch, A., **1**, 934 ⟨78TL619⟩
Jentzsch, J., **2**, 130 ⟨62CB1764, 63AG(E)370⟩; **3**, 344, 363 ⟨63ACS2325⟩; **4**, 245 ⟨62CB1764⟩
Jentzsch, W.,**5**, 744 ⟨64CB1361⟩
Jentzsch, W., **6**, 52 ⟨64CB1361⟩
Jeon, J.-H., **6**, 468 ⟨94SC305⟩
Jeong, G. H., **5**, 1096 ⟨90JA3332⟩
Jeong, I. H., **4**, 775 ⟨92MI 417-01⟩
Jeong, J. H., **4**, 581 ⟨86OM2555⟩, 589 ⟨90IS177⟩
Jeong, J. U., **1**, 238 ⟨90TL2713⟩; **5**, 418, 428 ⟨87TL3261⟩
Jeong, K.-S., **2**, 48 ⟨92JOC2768⟩, 49 ⟨93JOC844⟩
Jeong, L. S., **4**, 231 ⟨92JOC2217⟩
Jeong, N., **3**, 623 ⟨89JA6474⟩
Jeong, Y., **5**, 173 ⟨94JA5511⟩
Jephcote, V. J., **1**, 603 ⟨91JCS(P1)429⟩; **2**, 79 ⟨89JCS(P1)1529⟩
Jera, C. N., **2**, 85 ⟨89JOC1491⟩
Jerchel, D., **5**, 772 ⟨55CB1284⟩, 772, 854, 855, 859 ⟨51LA(574)85⟩, 855 ⟨50LA(568)185, 57LA(605)191⟩, 856 ⟨42CB75⟩, 857 ⟨56CB563⟩
Jerina, D. M., **2**, 659 ⟨76JOC977⟩; **3**, 294 ⟨75TL2557⟩
Jerkunica, J. M., **1**, 666, 671 ⟨71JA6278⟩
Jermann, F., **3**, 268 ⟨84LA1878⟩
Jernberg, K. M., **1**, 401, 414, 800 ⟨77TL2559⟩
Jernstedt, K. K., **6**, 468 ⟨82JOC4013⟩
Jeroschewski, P., **5**, 554 ⟨82ZC223⟩
Jerris, P. J., **1**, 606 ⟨81JA1501⟩
Jeschke, R., **1**, 834 ⟨89TL3517⟩
Jeschkeit, H., **6**, 486 ⟨B-82MI 615-02⟩
Jeske, G., **2**, 969 ⟨85JA8103, 85JA8111⟩; **6**, 194 ⟨85JA8103⟩, 195 ⟨85JA8091⟩
Jeske, J., **5**, 955 ⟨94CC2457⟩; **6**, 349 ⟨93CB1355⟩, 355, 356 ⟨93ZAAC(619)1693⟩
Jeskey, H., **5**, 105 ⟨57CRV583⟩
Jeso, B. de, **3**, 407 ⟨82SC495⟩
Jespersen, **4**, 830 ⟨87JA3811⟩
Jessep, H. F., **5**, 539, 541 ⟨72JCS(D)359⟩
Jesson, J. P., **4**, 551 ⟨75IC656⟩
Jessup, P. J., **5**, 288 ⟨76TL3089⟩, 991 ⟨78JOC2164⟩
Jester, M. A., **3**, 349 ⟨82JOC3319⟩
Jesthi, P. K., **1**, 163 ⟨76JOM(114)1⟩; **2**, 905 ⟨75JA5608⟩; **4**, 526, 527 ⟨84OM1284⟩, 633 ⟨76JOM(110)25⟩; **6**, 181, 186, 199, 200 ⟨76JOM(110)25⟩, 181, 199 ⟨75JA5608⟩, 199, 201 ⟨76JOM(114)1⟩
Jesthi, R. K., **4**, 665 ⟨76JOM(114)1⟩
Jesus, M. A. de, **4**, 356 ⟨77JOM(132)301⟩; **6**, 26 ⟨87SC1047⟩
Jesus Villa, M., **2**, 89 ⟨88T6681⟩
Jetha, N. K., **4**, 676 ⟨88JCS(D)1843⟩
Jetz, W., **6**, 517 ⟨72JA3799⟩
Jeuell, C. L., **1**, 894 ⟨72JA2044⟩
Jeunge, E. C., **2**, 619 ⟨70JOC719⟩
Jeuring, H. J., **4**, 331 ⟨81JOC2069⟩
Jevnik, M. A., **3**, 451 ⟨55JA1003⟩
Jew, S., **5**, 999 ⟨90TL1559⟩
Jewel, Jr., C. F., **3**, 544 ⟨85JA6715⟩
Jewers, K., **3**, 404 ⟨58JCS2209⟩
Jewson, F. T., **5**, 755 ⟨03JCS766⟩
Jeyaraman, R., **4**, 272, 277 ⟨81S995⟩
Jhingan, A. K., **1**, 530 ⟨87JOC1161⟩
Jhingran, A. G., **6**, 72 ⟨89AJC1235⟩
Jhurry, D., **1**, 474 ⟨90BSF401⟩
Jia, X., **2**, 231 ⟨94TL8833⟩; **5**, 253 ⟨93SC1403⟩
Ji, H., **2**, 945 ⟨93JA1586⟩
Ji, H.-L., **3**, 597 ⟨93JA9846⟩
Jiang, B. A., **4**, 735 ⟨92TL1221⟩
Jiang, B., **3**, 220 ⟨89CC612⟩
Jiang, H.-J., **3**, 260 ⟨91SC1307⟩
Jiang, H.-X., **1**, 272 ⟨92JOC774⟩; **2**, 487 ⟨92JOC774⟩
Jiang, J., **1**, 262, 1041 ⟨89TL3303⟩; **4**, 329 ⟨90S565, 94S107⟩; **5**, 272, 293 ⟨87TL3815⟩, 1031, 1067 ⟨90S565⟩
Jiang, J. L., **4**, 445 ⟨91CJC1153⟩
Jiang, S., **2**, 73 ⟨91TL4639⟩
Jiang, S. J., **1**, 179 ⟨92H(34)2247⟩
Jiang, W., **4**, 587 ⟨86JOC5498⟩
Jiang, X., **6**, 221, 222, 226 ⟨92MI 607-03⟩
Jiang, X. K., **1**, 922 ⟨89JOC5648⟩
Jiang, Y., **3**, 436 ⟨B-94MI 304-02⟩
Jiang, Y. Z., **1**, 135 ⟨93TA1957⟩; **2**, 359, 363 ⟨94JOC914⟩; **3**, 415 ⟨90SC15⟩
Jianhua Zheng, **1**, 739 ⟨85S57⟩
Jian-qi, W., **3**, 330 ⟨83HCA801⟩
Jiao, X.-Y., **4**, 461 ⟨94S23⟩
Ji, T., **2**, 631 ⟨85JOC5384, 85TL2043⟩
Jibodu, K. O., **3**, 687 ⟨83JCS(P1)1067⟩; **5**, 728 ⟨83JCS(P1)1067⟩
Jibril, I., **2**, 422 ⟨84T2211⟩, 875, 896 ⟨87JOM(336)337⟩, 1009 ⟨84CB1900⟩; **3**, 687 ⟨84CB1900⟩; **4**, 112 ⟨84CB1695⟩; **5**, 166 ⟨84T2211⟩; **6**, 375 ⟨86ZN(B)191⟩
Jibril, J., **5**, 146 ⟨85TL1631⟩
Jick, B. S., **5**, 242, 252 ⟨78CC838⟩, 242, 252 ⟨79T2329⟩; **6**, 475 ⟨78CC838, 79T2329⟩
Jigajinni, V. B., **2**, 4 ⟨81JCR(S)376⟩
Jilek, J. O., **5**, 749 ⟨54CLY1210⟩
Jimenco, M. L., **1**, 175 ⟨92TL5261⟩
Jiménez, C., **1**, 948 ⟨84JCS(P1)721⟩; **5**, 284 ⟨81CC1178, 81CC670⟩
Jimenez, H. N., **4**, 307 ⟨90CC1304⟩
Jiménez-Vázquez, H. A., **1**, 891, 906, 918, 919 ⟨91CRV375⟩
Jimeno, M. L., **6**, 550, 557 ⟨92TL5261⟩
Jimonet, P., **3**, 638 ⟨87TL6179⟩
Jinachitra, S., **2**, 190 ⟨79T1315⟩
Jin, F. Q., **4**, 735 ⟨92TL1221, 93JCS(P1)795⟩

Jin, G. Y., **4**, 455 ⟨80TL1845⟩
Jin, G. Yu., **4**, 456 ⟨80TL3467⟩
Jin, H., **1**, 677, 1038, 1099 ⟨86JA5644⟩; **3**, 238 ⟨86TL4759⟩
Jin, H. L., **4**, 229 ⟨92TL4625⟩
Jin, N., **5**, 516, 520 ⟨92CL1843⟩
Jinaraj, V. K., **1**, 661 ⟨88MI 114-04⟩
Jin, S. J., **2**, 597 ⟨91JOC2538⟩
Jin, Y.-Y., **6**, 562 ⟨87JOC4117⟩
Jin, Z., **2**, 41 ⟨94AG(E)851⟩
Jin, Z.-T., **3**, 241 ⟨91JHC811⟩
Jinbo, T., **2**, 55 ⟨80TL2171⟩; **3**, 82 ⟨81S871⟩
Jinbo, Y., **1**, 193 ⟨84TL5921⟩; **5**, 346 ⟨83H(20)1531⟩
Jinbu, Y., **1**, 907 ⟨88JA633⟩
Jindal, S. L., **2**, 232 ⟨72JOC3516⟩
Jiracek, V., **4**, 239 ⟨63CI(L)1397⟩
Jirieny, J., **3**, 18, 133 ⟨78S919⟩
Jitsukawa, K., **2**, 91 ⟨79JA159⟩; **3**, 117 ⟨83TL5009⟩
Jo, S., **2**, 445 ⟨81S53⟩; **4**, 847 ⟨81S53⟩; **6**, 120 ⟨81S53⟩
Joanny, M., **1**, 656 ⟨84JOM(267)93⟩
Job, A., **6**, 522 ⟨27BSF1041, 28CR(187)564⟩
Job, K., **2**, 420 ⟨93OS(72)62⟩
Jobson, R. B., **6**, 25 ⟨86TL3845⟩
Jocachim, H., **4**, 495, 497 ⟨73EGP99803⟩
Jocelyn, P. C., **2**, 127 ⟨B-72MI 203-01⟩
Jochem, G., **5**, 907, 908 ⟨92TL471⟩, 907, 908 ⟨93AG(E)1089⟩
Jochen, G., **4**, 545, 558 ⟨93PS(76)13⟩
Jochims, J. C., **2**, 1009 ⟨84CB1900, 91T205⟩; **3**, 194 ⟨76CB154⟩, 556, 557, 560, 585, 586 ⟨81CB3751⟩, 556, 557, 560, 586 ⟨84T893⟩, 602 ⟨82CB172⟩, 687 ⟨84CB1900, 90S1128⟩, 687 ⟨93S426⟩; **4**, 109 ⟨92S710⟩, 314, 329, 434 ⟨79CB1956, 82CB860⟩, 434 ⟨87S745⟩; **5**, 215 ⟨90S1129⟩, 661 ⟨92T8271⟩, 746 ⟨65CB2128⟩, 1023 ⟨67AG(E)174, 68CB1746⟩, 1046 ⟨76CB154, 78CB3750⟩, 1146 ⟨81CB3655, 83CB1834⟩, 1146, 1149 ⟨84CB1161⟩, 1147 ⟨90CB1537⟩; **6**, 240, 453, 611 ⟨82CB860⟩, 274, 275, 559, 565 ⟨81CB1132⟩, 452, 453 ⟨81CB1746⟩, 559, 565, 605 ⟨81CB2064⟩, 560, 565 ⟨81CB2075⟩, 571 ⟨79CB1956⟩, 605, 619 ⟨82CB3587⟩, 618, 621 ⟨92T8271, 86T6645⟩, 621 ⟨89S918⟩, 643 ⟨84CB1161⟩, 643, 733 ⟨84CB502⟩
Joe, G. H., **1**, 932 ⟨93JA3328⟩
Joesten, M. D., **4**, 563 ⟨69IC1775⟩
Joffe, I., **2**, 108 ⟨48JOC280⟩
Joglar, J., **2**, 322 ⟨89CC1132⟩; **3**, 414 ⟨85CB3652⟩
Jogun, K.-H., **2**, 1006, 1009 ⟨87JA882⟩
Joh, T., **3**, 64 ⟨86TL4497, 88BCJ4353⟩, 151 ⟨91CC86⟩, 621 ⟨73CL379⟩; **5**, 801 ⟨88CC561, 91JCS(D)1531⟩, 801 ⟨92AG(E)851⟩
Johannesen, R. B., **4**, 163 ⟨51JA212⟩
Johannsen, B., **4**, 553 ⟨82COMC-I(1), 84EGP206557⟩
Johannsen, F., **5**, 690 ⟨86CS347⟩
Johannsen, I., **6**, 590 ⟨83CC295⟩, 591 ⟨84CC89⟩
Johannson, I., **3**, 39 ⟨84CC89⟩
Johansen, J. E., **3**, 419 ⟨81JOC4914⟩
Johansen, O. H., **3**, 247 ⟨79ACS(B)669⟩
Johanson, R. G., **1**, 902 ⟨71JA4219⟩
Johansson, K. N. G., **6**, 512 ⟨79EUP3008⟩
Johansson, R., **2**, 7 ⟨82JCS(P1)681, 84S168⟩
Johar, G. S., **5**, 1026 ⟨70IJC759⟩
Johar, P. S., **1**, 992 ⟨93BCJ987⟩; **2**, 597 ⟨91JOC2538⟩, 940 ⟨93BCJ987⟩
Johary, N. S., **2**, 203 ⟨55JCS1307⟩
Johl, R. G., **5**, 83 ⟨B-61MI 502-02⟩
John, D. I., **1**, 38 ⟨79CC345⟩, 235, 246 ⟨80TL395⟩, 410, 822 ⟨82JCS(P1)2757, 80TL399⟩, 945 ⟨81CC901⟩; **4**, 90 ⟨80TL395, 82JCS(P1)2757⟩, 90 ⟨86JCS(P1)2207⟩; **5**, 360 ⟨86JCS(P1)2207⟩
John, E. D., **5**, 684 ⟨92JFC(56)335⟩
John, E. O., **5**, 675 ⟨93IC287⟩, 679, 703 ⟨92ICA527⟩, 692 ⟨88IC3100⟩; **6**, 259, 614, 624 ⟨92JFC(57)293⟩, 260, 261 ⟨92IC329⟩, 262 ⟨92HCA907⟩, 436 ⟨87JFC(36)429⟩
John, G. R., **6**, 455, 517 ⟨86JOM(312)C21⟩
John, J. P., **2**, 435, 435, 1060, 1064 ⟨69JOC2446⟩
John, K. C., **5**, 1135 ⟨78JOC2280⟩
John, P., **4**, 610 ⟨81CC806⟩
John, R. A., **3**, 139 ⟨83JOC4098⟩
Johne, S., **3**, 670 ⟨83TL2841⟩
Johnels, D., **1**, 1019 ⟨90JA4054⟩
Johns, A., **1**, 927, 929 ⟨88TL837, 86CC1419⟩, 929 ⟨87CC1238, 89T7835⟩
Johns, I. B., **3**, 613 ⟨64JOC1970⟩
Johns, R. B., **2**, 107 ⟨88S142⟩
Johns, S. O., **2**, 116, 117, 137, 138 ⟨38JA2452⟩
Johnson, A., **4**, 979 ⟨88JCS(P1)975⟩
Johnson, A. E., **3**, 237 ⟨60JA3102⟩
Johnson, ,F., **2**, 989 ⟨59JA3163⟩
Johnson, A. L., **4**, 9 ⟨71JOC575⟩, 298 ⟨77JOC4217⟩, 780 ⟨91JMC2525⟩
Johnson, A. N., **5**, 24 ⟨B-52MI 502-01⟩
Johnson, A. T., **1**, 241 ⟨80TL4811⟩
Johnson, A. W., **1**, 104 ⟨60T130⟩, 537 ⟨65JCS4355⟩, 646 ⟨61JA417⟩, 646, 737, 738 ⟨60JOC183⟩, 732 ⟨66JOC334⟩, 737 ⟨68CJC461, 70JOC2678⟩, 738, 741 ⟨65CI(L)1726⟩; **2**, 803 ⟨70JCS(C)2514⟩, 877 ⟨70JOC2678⟩, 1000 ⟨69JOC1240⟩; **3**, 374 ⟨B-66MI 308-02⟩, 491 ⟨B-93MI 313-01⟩, 492 ⟨66JA1953⟩, 493 ⟨B-93MI 313-02⟩, 497 ⟨B-93MI 313-03⟩, 498 ⟨61JCS2126⟩, 499 ⟨B-93MI 313-04⟩, 500 ⟨60JOC183⟩; **5**, 24 ⟨B-51MI 502-01⟩, 62 ⟨B-52MI 502-04, B-52MI 502-08⟩, 84 ⟨B-52MI 502-02⟩, 894 ⟨B-66MI 523-01, B-93MI 523-01⟩
Johnson, B. D., **4**, 161 ⟨85JOC5022⟩
Johnson, B. F. G., **3**, 700 ⟨91JOM(403)C15⟩; **4**, 682 ⟨77JCS(D)1328, 87JOM(319)C51⟩, 682 ⟨88JCS(D)1421⟩; **6**, 340 ⟨80CC781⟩, 345 ⟨84CC1089⟩, 352 ⟨80CC812⟩
Johnson, C. D., **5**, 723 ⟨B-70MI 517-03⟩
Johnson, C. G., **4**, 730 ⟨90JOC4448⟩
Johnson, C. K., **5**, 1010 ⟨67JOC1508⟩
Johnson, C. R., **1**, 24 ⟨82JOC1193⟩, 120, 465 ⟨87JOC1885⟩, 158 ⟨87JOC2170⟩, 217, 226 ⟨73JOC4263, 70JA4979⟩, 226 ⟨71JA3047⟩, 304 ⟨73JA7418⟩, 400, 410, 820 ⟨86JA6062⟩, 476 ⟨73JA7777⟩, 622, 710 ⟨73JA6462⟩, 622, 710 ⟨79JA3602⟩, 643 ⟨82JA7041, 82TL5005⟩, 648, 649, 759 ⟨87JOC281⟩, 668 ⟨90JA6729⟩, 705 ⟨78JOC4140, 73JA7424⟩, 711 ⟨81JA7667, 85PS(24)531⟩, 826 ⟨85JA7792, 88JOC3195⟩; **2**, 144, 149 ⟨65JA1109⟩, 148 ⟨69CC365⟩, 149 ⟨62JOC282⟩, 149 ⟨73OSC(5)791⟩, 151 ⟨73S485⟩, 164, 401, 402 ⟨79JOC2055⟩, 362 ⟨71JA5305⟩, 363 ⟨73ACR341⟩, 365 ⟨70JA3825⟩, 399 ⟨79COC(3)233⟩, 399, 401 ⟨79COC(3)241, 79COC(3)215⟩, 401, 402 ⟨79JOC2278, 83JOC1, 79COC(3)223⟩, 402 ⟨79COC(3)237, 79JOC2061⟩, 420 ⟨92OS(71)133⟩, 453 ⟨85TL783, 87JOC2170⟩, 773 ⟨79COC(2)235⟩; **3**, 161 ⟨90TL7549⟩, 170 ⟨82S283⟩, 374 ⟨79COC(3)247⟩, 375 ⟨70JA3815⟩, 378 ⟨71JA5303, 73JA7424⟩, 379 ⟨73ACR341, 73JA4287⟩, 379 ⟨85MI 308-01⟩; **4**, 86 ⟨78JOC4136, 78JOC4140⟩, 387 ⟨88JOC4131⟩, 392 ⟨76S413⟩, 394 ⟨85MI 408-02⟩, 396 ⟨82CC358, 87JOC1885⟩; **5**, 138 ⟨86JA5655⟩
Johnson, D., **2**, 12, 627 ⟨65JOC415⟩
Johnson, D. A., **1**, 4, 9 ⟨79TL4631⟩
Johnson, D. D., **1**, 1067 ⟨92JOC5060⟩
Johnson, D. E., **1**, 778 ⟨87JOC3688⟩
Johnson, D. H., **2**, 287 ⟨73JA4634⟩
Johnson, D. L., **4**, 581 ⟨79JA6433, 83OM1846⟩
Johnson, D. M., **3**, 698, 707 ⟨83TL655⟩
Johnson, D. R.,**5**, 327 ⟨95TL197⟩
Johnson, D. S., **5**, 772, 850, 852, 855 ⟨71JOC1155⟩, 772, 850, 855 ⟨73JOC1344⟩; **6**, 55 ⟨71JOC1155⟩
Johnson, D. W., **1**, 812 ⟨73JOC4117⟩
Johnson, F., **1**, 89 ⟨64JA118⟩, 238 ⟨84JOC1460⟩, 503, 505 ⟨82JOC4254⟩, 871 ⟨65JA5492, 65JA5493⟩, 871 ⟨65JA5513⟩, 871, 876 ⟨68CRV375⟩; **2**, 308 ⟨66AHC(6)95⟩, 657 ⟨85JOC5430⟩; **3**, 229 ⟨54JCS1302⟩,

632 ⟨73OSC(5)614⟩, 687 ⟨66AHC(6)95⟩; **5**, 70 ⟨74JOC1676⟩
Johnson, F. A., **2**, 335 ⟨67JOC416⟩; **5**, 659 ⟨67JOC416⟩; **6**, 208 ⟨89OM2646⟩
Johnson, G., **3**, 422 ⟨91JOC4084⟩; **4**, 32 ⟨84JFC(26)29⟩, 421 ⟨81CC1269⟩, 426 ⟨78CRV65⟩; **5**, 342 ⟨88TL1653⟩; **6**, 58 ⟨89TL7013⟩
Johnson, G. A., **2**, 442 ⟨59JA4803⟩
Johnson, G. J., **2**, 650 ⟨71JOC3184⟩
Johnson, G. M., **1**, 760 ⟨88TL4521⟩
Johnson, H. B., **5**, 34 ⟨63OSC(4)804⟩
Johnson, H. E., **5**, 95 ⟨73OSC(5)654⟩, 273, 285 ⟨62JOC2205⟩
Johnson, J. A., **3**, 495 ⟨71JOC3386⟩, 690 ⟨93AG(E)103⟩; **5**, 1160 ⟨93AG(E)103⟩; **6**, 155 ⟨93AG(E)103⟩
Johnson, J. E., **5**, 665, 672 ⟨85JOC993⟩, 723, 827 ⟨71JOC284⟩, 765, 824 ⟨86JHC1861, 76JOC252⟩, 835 ⟨92JOC4648⟩
Johnson, J. H., **3**, 704 ⟨88JAN454⟩
Johnson, Jr., C. S., **1**, 922 ⟨67JA2054⟩
Johnson, Jr., H. W., **5**, 1003 ⟨64JOC246⟩, 1004 ⟨67JOC1939⟩
Johnson, J. R., **1**, 58 ⟨38JA115⟩, 447 ⟨38JA121⟩; **2**, 112 ⟨38JA121⟩, 906 ⟨38JA105⟩, 1018, 1098 ⟨26JA469⟩; **4**, 318 ⟨B-49MI 407-01⟩; **5**, 98 ⟨42OR(1)210⟩, 290 ⟨41OSC(1)111⟩
Johnson, J. S., **6**, 174, 390, 392 ⟨73JOM(55)209⟩
Johnson, J. W., **1**, 776 ⟨87JA553⟩
Johnson, K., **3**, 426 ⟨39JA3194⟩
Johnson, K. A., **4**, 1080 ⟨88MI 424-01⟩
Johnson, K. K., **2**, 822, 841 ⟨79JOC3741⟩
Johnson, L.,**1**, 503 ⟨51JA4707⟩
Johnson, L. K., **3**, 517 ⟨93JA8130, 93JA8167⟩, 520, 522 ⟨92JA3910⟩
Johnson, M., **5**, 1079 ⟨93OM1023⟩
Johnson, M. C., **5**, 292 ⟨77S31, 77S33⟩, 330 ⟨91JOC3750⟩
Johnson, M. D., **1**, 66 ⟨69JCS(B)1071⟩, 321, 334, 946 ⟨83ACR343⟩, 676 ⟨68JCS(A)34⟩; **2**, 176 ⟨82JCS(P1)849⟩, 551 ⟨82COMC-I(4)331⟩, 597 ⟨70CC1037⟩
Johnson, M. P., **5**, 463 ⟨68CC6⟩; **6**, 516 ⟨91OM2781⟩
Johnson, M. R., **1**, 232 ⟨83JOC494⟩, 478 ⟨79TL4343⟩; **2**, 394 ⟨91OS(70)68⟩
Johnson, M. W., **3**, 156 ⟨81TL607⟩
Johnson, N., **4**, 764 ⟨91JOC6987⟩
Johnson, N. A., **1**, 834 ⟨77JOC172⟩; **2**, 773 ⟨74JOC407⟩
Johnson, O. H.,**1**, 449 ⟨50JA5566⟩; **2**, 412 ⟨55JA2871⟩, 949 ⟨46JA1720⟩
Johnson, P. C., **2**, 381 ⟨73JMC736⟩
Johnson, P. D., **3**, 220 ⟨85JA7967⟩
Johnson, P. Y., **5**, 354 ⟨75CC725⟩
Johnson, R. A., **2**, 92 ⟨91COC(7)389, 91COC(7)411⟩, 98 ⟨78JA7960⟩; **6**, 118 ⟨93JOC4159⟩
Johnson, R. C., **5**, 73 ⟨72JOC1947⟩
Johnson, R. D., **4**, 130 ⟨85JCP1517⟩
Johnson, R. G., **2**, 12 ⟨56CRV219⟩, 627 ⟨57CRV219⟩; **4**, 27, 35, 37 ⟨56CRV219⟩
Johnson, R. L., **1**, 429 ⟨64JA5361⟩
Johnson, R. N., **1**, 1125 ⟨60JA6397⟩; **4**, 688 ⟨76IC90, 76IC97⟩, 689 ⟨76IC107⟩
Johnson, R. P., **1**, 499 ⟨90JA6416⟩, 522 ⟨83JOC273⟩, 954 ⟨89CRV1111⟩, 957 ⟨85JA6615⟩, 990 ⟨84JOC2880, 86TL4679⟩, 1080 ⟨90JA8578⟩
Johnson, R. S., **2**, 187, 273 ⟨48JOC822⟩
Johnson, R. W., **4**, 764 ⟨78JOC43⟩
Johnson, S., **1**, 690 ⟨86JA3513⟩
Johnson, S. A., **1**, 906 ⟨88JA4112⟩
Johnson, S. E., **1**, 74, 82 ⟨75JOC3599⟩, 168 ⟨91JA5915⟩
Johnson, S. H., **4**, 690 ⟨76AJC2189⟩
Johnson, T., **3**, 684 ⟨88CC1339⟩
Johnson, T. B., **2**, 120, 125 ⟨36JA1348⟩, 120, 125 ⟨37JA1837⟩, 177 ⟨37JA2439⟩, 177 ⟨38JA1486⟩, 177, 178 ⟨39JA2548⟩; **4**, 85 ⟨41JA1571⟩; **5**, 1048 ⟨39JA632⟩
Johnson, U., **3**, 234 ⟨61CB1225⟩
Johnson, V. B., **6**, 140 ⟨81TL4365⟩
Johnson, W. C., **2**, 482, 483 ⟨37JA2068⟩
Johnson, W. D., **4**, 914 ⟨68JA5924, 72JA3058⟩
Johnson, W. L., **2**, 8 ⟨77TL2999, 79JOC359⟩
Johnson, W. S., **1**, 82 ⟨63JOC421⟩, 216, 979 ⟨93TL7849⟩, 295 ⟨76AG(E)9, 82JOC161⟩, 295 ⟨91Txi⟩, 414, 416, 827 ⟨70JA741⟩, 881 ⟨51OR(6)41, 52JA831⟩, 881 ⟨63OSC(4)132, 63OSC(4)261⟩, 884 ⟨50JA2395⟩, 1026 ⟨83JA2904⟩, 1122 ⟨73JA4416⟩, 1122, 1138 ⟨78TL2549⟩; **2**, 21 ⟨63OSC(4)162⟩, 35 ⟨68JA6225⟩, 89 ⟨58JA2584, 59T36⟩, 89 ⟨73OSC(2)245⟩, 606 ⟨85JOC1599⟩, 1006 ⟨93JA497⟩; **3**, 2, 113 ⟨56JOC478⟩, 208 ⟨63OSC(4)162⟩, 219 ⟨71JA4332⟩, 275 ⟨81JOC1512⟩, 612 ⟨44JA218⟩, 618, 633 ⟨45JA1745⟩, 631 ⟨83JOC2294, 84TL591⟩; **4**, 211 ⟨50JA514⟩, 250 ⟨81JOC1513⟩; **5**, 39 ⟨70TL4459⟩, 159 ⟨70JA741⟩, 164 ⟨61JA606⟩, 166 ⟨51OR(6)1⟩; **6**, 134 ⟨81JOC1512⟩
Johnston, B., **6**, 100, 102 ⟨89H(28)389⟩
Johnston, B. D., **6**, 101 ⟨88JOC5668⟩, 101, 102 ⟨88JOC3766⟩
Johnston, B. H., **3**, 308 ⟨87JA7553⟩; **4**, 952 ⟨87JA7553⟩
Johnston, C. R., **2**, 164 ⟨79JOC2280⟩
Johnston, G. A. R., **2**, 299 ⟨83JCS(P1)2983⟩
Johnston, G. J., **6**, 31 ⟨86JCS(P2)471⟩
Johnston, H., **2**, 171 ⟨55USP2821554⟩
Johnston, J., **2**, 542 ⟨78TL5091⟩; **3**, 39, 175 ⟨78TL5091⟩; **4**, 239 ⟨78TL5091⟩
Johnston, J. A., **4**, 762, 771 ⟨71JOC3386⟩
Johnston, J. V., **5**, 757 ⟨54JA920⟩
Johnston, L. H., **1**, 331 ⟨85JA4594⟩
Johnston, L. J., **1**, 936 ⟨85JA4595⟩
Johnston, M. I., **1**, 805 ⟨88JOC1848⟩
Johnston, S. M., **4**, 721 ⟨94OM511⟩
Johnston, T. E., **1**, 115 ⟨68JOC2203⟩
Johnston, T. P., **2**, 214 ⟨61JOC3780, 67JOC1261⟩, 255 ⟨79JMC631⟩; **6**, 305 ⟨62JOC4068⟩, 474 ⟨64JOC2442⟩, 504, 506 ⟨63JMC669⟩
Johnston, V. J., **6**, 342 ⟨90JOM(394)265⟩
Johnstone, R. A. W., **1**, 74, 441 ⟨77JCS(P1)443⟩, 74, 455, 457 ⟨85CRV129⟩, 142, 595 ⟨65CC29⟩, 482 ⟨88TL5553⟩; **2**, 56, 60 ⟨91COS(8)259⟩, 291 ⟨76CC61⟩, 379 ⟨82T419⟩; **3**, 13, 83, 93 ⟨80JCS(P1)27⟩, 82 ⟨91COS(8)259⟩; **5**, 273, 285, 300 ⟨79T2169⟩; **6**, 606 ⟨87JCS(P1)1069⟩
Johri, K. K., **2**, 13 ⟨83JOC4158, 86CJC138⟩; **6**, 229, 230 ⟨83JOC242⟩, 428, 430 ⟨82JFC(19)227⟩
Jois, Y. H. R., **4**, 426 ⟨91JOC865⟩
Joklik, J., **2**, 478 ⟨68AG(E)112⟩
Jolidon, S., **1**, 835 ⟨77HCA978⟩
Jolles, Z. E., **6**, 212 ⟨B-66MI 607-01⟩
Jolly, B. S., **6**, 196 ⟨88CC336⟩, 196 ⟨93JCS(D)2653⟩
Jolly, P. W., **1**, 195 ⟨82COMC-I(8)713⟩, 195 ⟨82COMC-I(8)773⟩; **3**, 519 ⟨66JA5044⟩; **4**, 148 ⟨66JA5044⟩
Jolly, R. S., **1**, 523 ⟨90JA4965⟩; **2**, 325 ⟨88JA7536⟩
Jolly, W. L., **2**, 427 ⟨68IS124⟩, 430 ⟨68IS126⟩
Joly, R., **3**, 234 ⟨61BSF559⟩
Jommi, G., **1**, 812 ⟨91TL6969⟩; **2**, 9 ⟨87S920⟩; **3**, 59 ⟨88S697⟩, 169 ⟨87S1126⟩; **4**, 462 ⟨92SC107⟩
Jonas, F., **6**, 730 ⟨83HOU(E4)677⟩, 731 ⟨83HOU(E4)690⟩
Jonas, K., **2**, 575 ⟨76AG(E)621⟩, 811 ⟨77CB2783⟩; **4**, 671 ⟨86AG(E)925⟩, 694 ⟨83JA5479⟩; **5**, 996 ⟨66CB3572⟩; **6**, 154 ⟨87AG(E)908⟩
Jonas, V., **3**, 358 ⟨92CB265⟩
Jonassen, H. B., **1**, 413, 809 ⟨66JOM(6)412⟩
Jonczyic, A., **3**, 634 ⟨89LA203⟩
Jonczyk, A., **2**, 308 ⟨78S882⟩, 749 ⟨86S297⟩
Jończyk, A., **3**, 465 ⟨77BSB739, 78SC569⟩; **3**, 634, 638 ⟨87BSB(96)303⟩; **4**, 82, 83, 237 ⟨75JOC266⟩, 84 ⟨78S883⟩; **5**, 1118 ⟨78S882⟩
Jones, A., **2**, 447 ⟨65AJC903⟩; **3**, 335 ⟨67JPC4111⟩; **5**, 234 ⟨77JCS(P1)1672⟩
Jones, A. B., **1**, 697, 700, 703 ⟨87TL5759⟩, 700, 703 ⟨87TL5755⟩; **2**, 824, 834, 835 ⟨78JA7304⟩; **3**, 424 ⟨91COS(7)151⟩; **4**, 286 ⟨89T7161⟩

Jones, A. D., **4**, 277 ⟨90SL359⟩
Jones, A. M., **1**, 998 ⟨B-91MI 121-01⟩
Jones, A. R., **1**, 884 ⟨50JA2395⟩
Jones, A. S., **2**, 625 ⟨86JMC213⟩
Jones, B. A., **3**, 714 ⟨86JA3153⟩; **5**, 510, 511, 512 ⟨84CRV17⟩, 511 ⟨83JOC2635⟩
Jones, C., **6**, 168 ⟨91CC1560⟩
Jones, C. E., **5**, 1129 ⟨71IC1536⟩
Jones, C. G. P., **6**, 18 ⟨79CC964⟩
Jones, C. H. W., **2**, 279 ⟨83JOM(255)61⟩; **4**, 291 ⟨86OM805, 87OM2164⟩
Jones, C. R., **4**, 580 ⟨79JA7282⟩
Jones, C. W., **2**, 632 ⟨87CC1409⟩; **6**, 233, 234 ⟨90JFC(48)249⟩
Jones, D. A., **2**, 214 ⟨73JOC2965⟩
Jones, D. K., **2**, 66 ⟨93JOC799⟩
Jones, D. M., **1**, 1006 ⟨69JCS(C)2173⟩; **3**, 78 ⟨69JCS2173⟩
Jones, D. N., **1**, 234 ⟨82CC589⟩, 595 ⟨77JCS(P1)1574⟩, 595, 779 ⟨76JCS(P1)459⟩, 601 ⟨70CC86, 67JCS(C)532⟩, 779 ⟨70JCS(C)81, 70JCS(C)833⟩; **2**, 185 ⟨79TL4977⟩; **3**, 169, 172 ⟨83TL405⟩
Jones, D. S., **2**, 309 ⟨89JOC1940⟩; **5**, 278 ⟨60T39⟩
Jones, D. W., **1**, 403 ⟨80JCS(P1)714, 93JCS(P1)681⟩, 403, 800 ⟨84CC990⟩, 841 ⟨86CC1807⟩; **2**, 153 ⟨77JCS(P1)1574⟩; **5**, 364 ⟨72JCS(P1)2728⟩, 365 ⟨76JCS(P1)1150⟩
Jones, E. C. S., **3**, 465 ⟨33JCS363⟩
Jones, E. H. R., **5**, 168 ⟨59JCS1313⟩
Jones, E. M., **5**, 949 ⟨71CC570, 71JCS(A)3368⟩
Jones, E. R. H., **1**, 80 ⟨54JCS3197⟩, 104 ⟨46JCS396⟩, 506 ⟨49JCS1827, 49JCS2028⟩, 506 ⟨49JCS737⟩, 513 ⟨54JCS1854⟩, 519 ⟨46JCS52⟩, 837 ⟨45JCS90⟩, 959 ⟨60JCS341⟩, 1035 ⟨52JCS2014, 54JCS147⟩, 1046 ⟨58JCS1054⟩, 1066 ⟨53OS(33)68⟩, 1069 ⟨52JCS2010⟩; **2**, 875 ⟨58JCS1719⟩; **3**, 54, 231 ⟨49JCS737⟩, 62 ⟨46JCS937⟩, 79 ⟨58JCS1054⟩, 238 ⟨49JCS1430⟩, 252 ⟨46JCS945⟩, 267 ⟨46JCS45⟩, 273 ⟨49JCS1827⟩, 275 ⟨46JCS39⟩; **4**, 429, 430, 431, 439 ⟨65JCS2601⟩; **5**, 74, 102, 103 ⟨58JCS1313⟩, 102 ⟨56CI(L)548⟩, 189 ⟨54JCS1860⟩, 276 ⟨46JCS599⟩
Jones, E. T. T., **6**, 446 ⟨91JOC751⟩
Jones, F. B., **2**, 365 ⟨71JA2692⟩
Jones, F. N., **5**, 476, 487, 582, 613, 614 ⟨70JOC2002⟩; **6**, 540 ⟨69JOC3011⟩
Jones, G., **1**, 747 ⟨68CC543⟩, 1071 ⟨91JOC2045⟩, 1140 ⟨88TL5017⟩; **2**, 394 ⟨67OR(15)204⟩, 504 ⟨68CB1271⟩, 1003 ⟨91T5277⟩; **3**, 308 ⟨83OS(62)111⟩, 357 ⟨64JCS2760⟩; **4**, 19, 27 ⟨81JCS(P1)78⟩, 892 ⟨76JCS(P1)1241⟩, 993 ⟨92H(34)1005⟩
Jones, G. C. H., **2**, 358 ⟨72JOM(44)37⟩
Jones, G. D., **2**, 29 ⟨42JA1966⟩
Jones, G. H., **2**, 12 ⟨67JCS(C)252⟩; **4**, 554 ⟨68TL5731⟩; **5**, 910 ⟨68TL5731⟩
Jones, G. R., **2**, 390 ⟨76T1267⟩
Jones, G. S., **1**, 584 ⟨92JA1491⟩
Jones, G. T., **2**, 780 ⟨50JCS2657⟩
Jones, H., **2**, 183 ⟨55AG32⟩; **5**, 24 ⟨51C169⟩; **6**, 118 ⟨82MI 604-01⟩
Jones, H. W., **1**, 1114, 1116 ⟨51JA2700⟩
Jones, II, G., **1**, 785 ⟨73CC374⟩
Jones, J., **5**, 259, 262 ⟨B-91MI 506-01⟩
Jones, J. B., **2**, 38 ⟨91COS(8)183⟩; **5**, 138 ⟨85CC1563⟩
Jones, J. D., **3**, 73 ⟨54JA1068⟩
Jones, J. E., **3**, 101 ⟨55OSC(3)98⟩; **4**, 298 ⟨77JOC4217⟩; **5**, 290 ⟨55OSC(3)590⟩
Jones, J. G. L., **3**, 244 ⟨75JOC1420⟩
Jones, J. H., **2**, 299 ⟨77CC758, 78JCS(P1)1088⟩; **3**, 188 ⟨78JCS(P1)1088⟩, 310 ⟨59JA1938⟩
Jones, J. I., **4**, 58 ⟨57JCS618⟩; **6**, 429 ⟨57JCS2735⟩
Jones, J. K. N., **2**, 23, 26 ⟨66CJC2339⟩; **3**, 103 ⟨45JCS116⟩
Jones, J. O., **4**, 226 ⟨82JCR(S)116⟩
Jones, J. R., **1**, 871 ⟨67JCS(B)1173⟩; **3**, 426 ⟨44JA352⟩; **4**, 1075 ⟨87AR(C)197⟩
Jones, Jr., M., **1**, 412 ⟨64JA5017⟩, 412, 809, 815 ⟨64JA5019⟩, 549 ⟨73JA6286⟩, 1091 ⟨78TL5193⟩, 1102 ⟨B-78MI 122-01⟩; **3**, 483 ⟨78JA994⟩, 532 ⟨85JOC4404⟩; **4**, 7, 16, 24 ⟨B-85MI 401-01⟩, 1080 ⟨B-73MI 424-02⟩, 1083 ⟨86TL5907, 89JA8491⟩; **5**, 873 ⟨77JA6995⟩
Jones, Jr., W. J., **6**, 314 ⟨67JOC2074⟩
Jones, K., **1**, 345 ⟨92T6875⟩, 346 ⟨89TL5485, 94CC41⟩; **2**, 421 ⟨65JCS1944⟩, 421, 868 ⟨65JOM(3)295⟩, 1005 ⟨91PAC403⟩; **6**, 244 ⟨65JCS2157⟩
Jones, K. D., **1**, 988 ⟨92JCS(P1)747⟩
Jones, K. L., **6**, 290 ⟨89CC595⟩, 381, 392 ⟨94JOM(466)35⟩
Jones, L. B., **1**, 404 ⟨67JA1880⟩
Jones, L. D., **1**, 452 ⟨76JOC1184⟩, 468, 472 ⟨81JA6460⟩
Jones, L. K. N., **5**, 62 ⟨B-52MI 502-06, B-52MI 502-07⟩
Jones, L. W., **2**, 103 ⟨14JA2202⟩
Jones, M., **1**, 591 ⟨74JCS(P1)433⟩, 1143 ⟨B-75MI 122-01⟩; **4**, 49 ⟨84S727⟩; **5**, 873 ⟨92JA1094⟩
Jones, M. D., **1**, 194, 289 ⟨86CC1201⟩
Jones, M. E., **1**, 733 ⟨66JCS(C)1090⟩
Jones, M. E. B., **5**, 823 ⟨61MAC62⟩
Jones, M. F., **4**, 168 ⟨84TL3127⟩, 229 ⟨92TL4625⟩
Jones, N., **1**, 422 ⟨59JCS4017⟩
Jones, N. D., **2**, 184 ⟨81JCS(P1)2106⟩
Jones, N. R., **2**, 160, 162 ⟨84JMC1634⟩
Jones, P. B., **2**, 323 ⟨92JOC1727⟩
Jones, P. C., **6**, 206 ⟨69JA6156⟩
Jones, P. F., **2**, 813 ⟨78JOM(152)C25⟩; **3**, 483 ⟨69CJC4353⟩; **4**, 137 ⟨75CJC332⟩, 609 ⟨68CJC2119⟩, 849 ⟨73JCS(P1)2272⟩; **5**, 410, 431 ⟨67JA431⟩; **6**, 125, 131, 132, 519 ⟨68CJC2119⟩, 127, 132, 325, 326, 520 ⟨67JA431⟩, 128 ⟨73JCS(P1)2272⟩
Jones, P. G., **2**, 100 ⟨91ZN(B)5⟩, 476 ⟨92CB801⟩; **4**, 134 ⟨92CB801⟩, 564 ⟨92CB1325⟩, 585 ⟨88JCS(D)341⟩, 586 ⟨87JOM(331)409, 88OM997⟩, 586 ⟨91JCS(D)2579⟩; **5**, 955 ⟨94CC2457⟩; **6**, 349 ⟨93CB1355⟩, 355, 356 ⟨93ZAAC(619)1693⟩, 581 ⟨92PS(73)81⟩, 726 ⟨85CB5007⟩
Jones, P. H., **1**, 1008 ⟨87JMC193⟩
Jones, P. R., **1**, 127, 128, 466 ⟨73JOC3189⟩, 128 ⟨68JOC1675⟩; **2**, 67, 69 ⟨78CRV491, 71JOC3311⟩; **3**, 502 ⟨80JA4970, 83JA6725⟩, 502 ⟨91OM2529⟩; **4**, 53, 54 ⟨67JOM(8)361⟩, 602 ⟨84OM354⟩, 603, 637 ⟨85OM2206⟩, 606, 611 ⟨77JA8447⟩, 610, 611, 621 ⟨85OM581, 77JA2013⟩, 611 ⟨80JA4970, 82JOM(232)33⟩, 611 ⟨86JA3122⟩, 1048 ⟨87JA913⟩; **5**, 926 ⟨87JA913⟩
Jones, R. A., **2**, 410 ⟨92JOC5535⟩, 473 ⟨92CC726⟩, 475 ⟨90AG(E)1150⟩, 511 ⟨86CC1543, 88JOM(341)C1⟩, 511 ⟨88POL77, 89AG(E)1208⟩, 511 ⟨90CM221, 91OM1635⟩, 512 ⟨86JA1703, 87IC4065⟩, 825 ⟨80JCS(D)55⟩, 895 ⟨92JOM(439)C33⟩; **3**, 106, 308 ⟨82SC231, 64AJC1056⟩, 357 ⟨58JCS3610⟩; **4**, 480 ⟨88CC867⟩, 672 ⟨78JCS(D)446⟩, 680 ⟨79JA4128, 80JCS(D)1771⟩; **5**, 797, 801 ⟨81JCS(D)2088⟩; **6**, 689 ⟨84JA7015⟩
Jones, R., **2**, 36 ⟨69JCS(C)1046⟩, 1000 ⟨88JCS(P1)1023⟩
Jones, R. C. F., **3**, 59 ⟨89TL5361⟩, 247 ⟨83TL4751⟩; **4**, 413 ⟨93TL6329⟩, 414 ⟨82CC282, 93JCS(P1)2391⟩; **5**, 342 ⟨89TL3217⟩, 374 ⟨86TL5147⟩, 374, 375 ⟨86TL5285⟩; **6**, 71 ⟨90JCS(P1)375⟩
Jones, R. E., **5**, 762, 816 ⟨65JOC259⟩
Jones, R. G., **1**, 81 ⟨50JA3539⟩, 117 ⟨52JOC1630⟩; **2**, 421 ⟨56JA4285⟩, 554, 961 ⟨51OR(6)339⟩, 586 ⟨52JOC1630⟩, 598 ⟨40JA2357, 50JA1760⟩; **3**, 247 ⟨51JA5168⟩, 453 ⟨60JOC956⟩; **5**, 76 ⟨47JA2350⟩, 115 ⟨63OSC(4)824⟩, 584, 586 ⟨48JA2837⟩; **6**, 39 ⟨51JA5168⟩
Jones, R. H., **3**, 429 ⟨85T699⟩; **4**, 715 ⟨93JOM(463)C6⟩; **5**, 453 ⟨85JOM(285)213⟩
Jones, R. J., **5**, 264 ⟨87TL5099⟩
Jones, R. L., **3**, 286, 459 ⟨72JOC3369⟩; **5**, 114, 115 ⟨51JCS1208⟩; **6**, 141 ⟨69JCS(C)2251⟩, 448 ⟨86JOC3781⟩

Jones, R. R., **4**, 610, 611 ⟨83JA6725⟩
Jones, R. V. H., **1**, 812 ⟨68CC1229, 70CC892⟩
Jones, R. W., **3**, 245 ⟨75TL1741⟩; **4**, 1029 ⟨91JCS(D)351⟩
Jones, S. R., **5**, 282 ⟨76S32⟩
Jones, T. C., **3**, 486 ⟨84CC1002⟩; **6**, 718 ⟨83CC719⟩
Jones, T. H., **2**, 326 ⟨88JOC2968⟩
Jones, T. K., **1**, 422 ⟨82JA2642, 83HCA2377⟩, 422 ⟨83HCA2397, 86T2821⟩, 514 ⟨86OS(64)182⟩; **2**, 528 ⟨91JOC763⟩, 560 ⟨91JOC751⟩, 921 ⟨82JOC4595⟩, 931 ⟨83HCA2377⟩; **3**, 215 ⟨82JA2642⟩, 272 ⟨83HCA2377⟩, 493 ⟨90JA7001⟩; **6**, 446 ⟨91JOC751⟩
Jones, T. R., **1**, 532, 662 ⟨86TL539⟩
Jones, V. K., **1**, 404 ⟨67JA1880⟩, 733 ⟨64JOC3327⟩
Jones, W., **2**, 1009 ⟨87T1273⟩
Jones, W. A., **5**, 313 ⟨83JOC761⟩
Jones, W. C., **2**, 219 ⟨70JOC4148⟩, 219, 220 ⟨69JA7525⟩
Jones, W. D., **2**, 962 ⟨84JA1650⟩; **3**, 411 ⟨87JA5047⟩; **4**, 690 ⟨92OM2698⟩
Jones, W. E., **2**, 117 ⟨51JCS2123⟩
Jones, W. H., **2**, 484, 492, 497, 888 ⟨32JCS1815⟩; **3**, 426 ⟨59JA2151⟩
Jones, W. J., **2**, 397 ⟨56JOC4003, 73OSC(5)845⟩, 433, 434, 435 ⟨29JCS1262⟩, 483, 872 ⟨30JCS463⟩, 484 ⟨32JCS2284⟩, 484, 492, 497, 888 ⟨32JCS593⟩, 487, 492, 497, 502 ⟨31JCS185⟩, 492 ⟨30JCS2426⟩, 492, 497 ⟨30JCS1920⟩, 824 ⟨31JCS2109⟩; **3**, 88 ⟨59JA4113⟩; **4**, 161 ⟨57JA6562⟩
Jones, W. M., **1**, 1110 ⟨70JA5457⟩, 1119 ⟨67JA1960⟩, 1144 ⟨63JA2273⟩; **2**, 204 ⟨70JA5457⟩, 801 ⟨67JA1960⟩; **3**, 489 ⟨71TL317, 72JOC693⟩; **5**, 937 ⟨85OM944, 86OM178⟩, 937 ⟨92OM622⟩, 938 ⟨92OM1442, 94MIP3494⟩; **6**, 506 ⟨66JA3798⟩
Jones II, G., **1**, 785 ⟨75CC206⟩, 815 ⟨74JA3468⟩
Jong, G. J. de, **6**, 90 ⟨48RTC884⟩
Jong, R. de, **3**, 552 ⟨77S888⟩; **5**, 589, 590, 603, 611, 626 ⟨77S888⟩
Jong, R. L. P. de, **2**, 942 ⟨84CC257⟩; **3**, 548 ⟨82RTC310⟩; **4**, 843 ⟨85S958⟩; **5**, 589 ⟨82RTC310⟩
Jong, T.-T., **2**, 359, 363 ⟨94JOC914⟩; **3**, 242 ⟨90JCS(P1)423⟩
Jonge, C. R. H. I. de, **2**, 772 ⟨B-86MI 215-01⟩
Jonge, C. R. M. I. de, **4**, 894 ⟨72JOC1960⟩
Jongedijk, G., **3**, 122 ⟨85JOC2650⟩
Jongh, H. A. P. de, **4**, 894 ⟨72JOC1960⟩
Jonker, C., **2**, 1032 ⟨64RTC208⟩
Jonkers, F. L., **2**, 10 ⟨79RTC371⟩, 739 ⟨79TL2433, 80TL2671⟩; **3**, 11 ⟨79TL2433, 80TL2671⟩; **4**, 199 ⟨79RTC371⟩, 200, 208 ⟨78JOC3548⟩
Jonsäll, G., **1**, 918 ⟨83JCS(P2)305, 85T6055⟩, 918 ⟨86JA3819⟩
Jonsson, E., **5**, 1113 ⟨70ACS3751⟩
Jonsson, E. U., **2**, 164, 401, 402 ⟨79JOC2055, 79JOC2061⟩; **3**, 375 ⟨70JA3815⟩
Jonsson, E. V., **2**, 365 ⟨70JA3825⟩
Joos, R., **1**, 1082 ⟨72HCA1276⟩; **2**, 380 ⟨88OSC(6)56⟩; **3**, 26, 147 ⟨72HCA1276⟩
Jooss, G., **4**, 432 ⟨70CZ215⟩
Joraschek, N., **1**, 39 ⟨88CB1039⟩
Jordaan, A., **3**, 713, 717 ⟨74MI 321-01⟩
Jordaan, J. H., **1**, 21 ⟨63JOC1395⟩
Jordan, B. K., **5**, 22 ⟨92H(33)657⟩
Jordan, C. L., **1**, 1027 ⟨89JOC5828⟩; **4**, 951 ⟨89JOC5828⟩
Jordan, E., **5**, 65 ⟨36JA1304⟩
Jordan, F., **6**, 365 ⟨90MI 612-01⟩
Jordan, M., **2**, 407, 851 ⟨80PS(8)221⟩, 441 ⟨80PS(8)215⟩, 442, 829 ⟨80PS(8)225⟩; **4**, 427 ⟨78LA1505⟩
Jordan, M. W., **1**, 369, 524 ⟨85T3547⟩
Jordan, R. F., **4**, 656 ⟨93JA8493⟩
Jördens, F., **1**, 936 ⟨91JOC5971⟩
Jore, D., **2**, 297 ⟨78JOM(149)C7⟩, 436 ⟨84JOM(269)107⟩, 437, 443 ⟨79TL1069⟩
Jorg, K., **2**, 512 ⟨84ICAL49⟩
Jørgensen, A., **5**, 690 ⟨86CS347⟩
Jørgensen, E. C., **5**, 279 ⟨71JA6318⟩
Jørgensen, F., **3**, 344 ⟨82JA5922⟩
Jørgensen, F. S., **4**, 433 ⟨84ACS113⟩
Jørgensen, K. A., **2**, 49 ⟨90TL6417⟩, 147 ⟨94JCS(P2)117⟩, 398 ⟨87CC1405⟩; **3**, 411 ⟨80BSB247⟩, 432 ⟨89JOC126⟩; **5**, 572 ⟨84BSF(2)204⟩; **6**, 502 ⟨83T1729⟩, 503 ⟨82T1163⟩
Jørgensen, M., **6**, 561 ⟨91JOC5684⟩
Jørgensen, P. L., **3**, 342 ⟨91S609⟩
Jorgensen, W. L., **1**, 149 ⟨90JA697⟩, 427 ⟨91JOC4718⟩
Jorgenson, M. J., **1**, 802 ⟨65JA5264⟩; **3**, 131, 284 ⟨70OR(18)1⟩; **5**, 103 ⟨73OSC(5)509⟩
Jorgenson, W. L., **1**, 320 ⟨90JCI458, 90JOC9⟩
Jörning, W. P. A., **6**, 549 ⟨92MI 617-05⟩
Jorritsma, R., **4**, 49 ⟨81RTC194⟩
Jorritsmas, R., **1**, 929 ⟨73TL827⟩
Jose, F., **4**, 69 ⟨78JA3933⟩
Joseph-Nathan, P., **4**, 245 ⟨91SC1369⟩, 246 ⟨90SC153⟩
Josey, A. D., **4**, 742 ⟨67JOC1941⟩; **5**, 900 ⟨72USP3647856⟩
Josey, J. A., **2**, 352 ⟨92TL3583⟩
Joshi, B. V., **3**, 709 ⟨88T7243⟩
Joshi, G. C., **1**, 459 ⟨71TL2075⟩
Joshi, Jr., P. C., **6**, 546 ⟨85JIC164⟩
Joshi, K. C., **3**, 350, 364 ⟨77JCS(P1)1127⟩
Joshi, K. K., **4**, 1034 ⟨71BSB651⟩
Joshi, K. M., **5**, 998 ⟨66JIC481⟩
Joshi, N. N., **2**, 72 ⟨89TL5551⟩, 519, 531 ⟨89JA1754⟩, 527 ⟨88JA6246⟩
Joshua, A. V., **1**, 941 ⟨78HCA690⟩
Joshua, C. P., **5**, 652 ⟨72TL4785⟩
Joshua, H., **6**, 496 ⟨68JA3254⟩
Joslin, F. L., **4**, 550 ⟨91POL1713⟩; **6**, 516 ⟨91OM2781⟩
Josse, D., **1**, 474 ⟨88CC203⟩
Jost, E., **3**, 286 ⟨62LA(655)90⟩
Josten, B., **3**, 607 ⟨84AG(E)619⟩; **5**, 1092 ⟨84AG(E)619⟩; **6**, 680, 687 ⟨84AG(E)619, 88ZAAC(556)7⟩
Jostes, F., **4**, 37 ⟨27LA(453)113⟩; **5**, 751 ⟨27LA(451)453⟩
Jouani, A. M., **5**, 991 ⟨92JFC(56)85⟩
Jouannetaud, M.-P., **2**, 626 ⟨80CC110⟩, 650 ⟨86BSF625⟩
Jouben, R., **1**, 271 ⟨83CB473⟩
Joubert, J.-P., **4**, 921 ⟨74BSF2263⟩
Joucla, M. F., **2**, 244, 773, 792, 795 ⟨84CC374⟩
Jouglet, B., **2**, 395 ⟨91SL907⟩
Jouien, R., **1**, 647, 667, 717 ⟨78TL4399⟩
Jouin, P., **3**, 180 ⟨86JCS(P1)1895⟩; **5**, 333 ⟨88TL2661, 89JOC617⟩
Jouitteau, C., **3**, 4 ⟨80TL4653⟩, 4, 115 ⟨86JOC4016⟩
Joukhadar, L., **1**, 16 ⟨81JCS(P1)1501⟩
Jouko, V., **4**, 555, 557 ⟨91MI 412-01⟩
Joule, J. A., **3**, 309 ⟨84T3339⟩, 361 ⟨92JCS(P1)351⟩; **4**, 540 ⟨85CC1021⟩; **5**, 603 ⟨89SUL117, 89TL3189⟩
Joullié, M. M., **1**, 100 ⟨79TL4985⟩; **3**, 92 ⟨86OPP109⟩, 432 ⟨79OPP95⟩, 669 ⟨79H(12)819⟩; **4**, 160 ⟨91SC1055⟩, 160 ⟨91SC2231⟩, 728 ⟨71JOC37⟩; **6**, 111, 114 ⟨69JHC729⟩, 142 ⟨70CC469⟩
Jourdenais, R. A., **2**, 364 ⟨73JA5831⟩
Journet, M., **1**, 340 ⟨94JOC718⟩
Jousseaume, B., **1**, 539 ⟨75JOM(96)C47⟩; **2**, 973 ⟨79JOM(168)227, 79JOM(168)233⟩, 1034, 1035 ⟨86JOM(304)115⟩, 1100 ⟨89T1145⟩; **3**, 303 ⟨91OM366⟩; **4**, 1065 ⟨79SC53⟩; **5**, 467 ⟨85TL6075⟩; **6**, 449 ⟨91OM366⟩
Jousseaume, E., **4**, 608 ⟨74JOM(71)377⟩
Joussef, A. C., **1**, 943 ⟨91JCS(P1)2797⟩
Joussen, R., **2**, 877 ⟨82CB654⟩; **4**, 565, 596, 600 ⟨83CB473⟩, 585 ⟨77CB3930⟩; **6**, 170, 210 ⟨79TL501⟩
Jovanovic, B., **4**, 656 ⟨72CC613⟩
Jovanovic, M. V., **2**, 753 ⟨85JOC1334, 86JOC5157⟩; **3**, 626 ⟨85JOC1334⟩
Jovanovski, G., **6**, 405 ⟨74CC646⟩
Jowett, I. C., **2**, 198, 721 ⟨88S252⟩
Joy, F., **1**, 658 ⟨66JOM(5)506⟩

Joyce, C. J., **3**, 685 ⟨83JCS(P1)293⟩
Joyce, R. M., **6**, 5 ⟨50JA2213⟩
Joyner, H. H., **5**, 547 ⟨93TL3703⟩
Jozsa, A. J., **2**, 1008 ⟨77AX(B)3782⟩
Juan, B., **1**, 865 ⟨82TL795⟩
Juarez-Brambila, J. J., **4**, 524 ⟨93JA440⟩
Juaristi, E., **4**, 230 ⟨85TL1927⟩; **6**, 119 ⟨89JPO349⟩, 120 ⟨86T1963, 91JOC5919⟩
Juaristi, M., **2**, 34 ⟨85T2903⟩; **3**, 4, 115 ⟨85T2903⟩
Jubault, M., **6**, 592 ⟨89CC1520⟩
Jubert, C., **2**, 73 ⟨92JOC6384⟩, 394, 396 ⟨92JOC5431⟩, 590 ⟨92JOC5425⟩, 999, 1002 ⟨93JOC588⟩; **3**, 162 ⟨93JOC588⟩; **4**, 53, 54, 58 ⟨93JOC588⟩
Jucker, E., **2**, 373 ⟨59AG321⟩; **3**, 229 ⟨46HCA704⟩
Judd, G. F., **6**, 244 ⟨54JA474⟩
Jude, S., **2**, 453, 455 ⟨89TL2783⟩
Judelson, D. A., **2**, 156 ⟨84S320⟩
Judge, R. H., **3**, 382 ⟨84JA5406⟩
Judson, P. N., **4**, 970, 977, 978 ⟨80EUP10396⟩, 970, 977, 978 ⟨83JCS(P1)1741⟩
Juengst, C. D., **4**, 570 ⟨87JPS(A)1967⟩
Juenot, P., **5**, 1152 ⟨91AG(E)196⟩
Jufresa, M., **2**, 190, 208 ⟨65BSF787⟩
Juge, S., **1**, 194 ⟨86TL4573⟩; **2**, 437, 444 ⟨90TL6357⟩, 443 ⟨83EUP82057⟩, 478 ⟨85GEP3512781⟩
Jugé, S., **4**, 557, 570, 575 ⟨93PS(77)199⟩
Jugelt, W., **4**, 472 ⟨69T5569⟩
Juhlke, T. J., **6**, 21 ⟨79CC128⟩
Jukes, A. E., **1**, 116 ⟨74AOC215⟩, 318 ⟨B-74MI 121-01⟩; **4**, 1048 ⟨70JOM(24)529⟩; **6**, 174, 289, 380, 387, 390 ⟨70JOM(24)529⟩, 734 ⟨70CC559⟩
Jula, T. F., **1**, 291 ⟨69JOM(16)503⟩; **6**, 60 ⟨68JA2944⟩
Julia, M., **1**, 24 ⟨76BSF513, 76BSF525⟩, 144, 237, 483 ⟨83T3283⟩, 144, 614, 615, 696, 697, 698, 700 ⟨73TL4833⟩, 195 ⟨88JOM(339)199⟩, 195, 238 ⟨86JOM(317)383⟩, 240 ⟨82T693⟩, 337, 548, 919 ⟨71ACR386⟩, 343 ⟨67BSF2641, 69BSF2427⟩, 343 ⟨74PAC553⟩, 439 ⟨86T2475⟩, 439, 700 ⟨82TL3265⟩, 439, 700 ⟨83TL4311⟩, 439, 700 ⟨87T859⟩, 469 ⟨51BSFC13⟩, 483 ⟨87BSF487⟩, 484 ⟨93CC1682⟩, 485 ⟨85BSF762, 85BSF772⟩, 485 ⟨88SCI426⟩, 485, 700 ⟨82TL2469⟩, 548 ⟨62CR(255)959⟩, 612 ⟨73BSF746⟩, 612 ⟨86T4807⟩, 623 ⟨88T2913⟩, 696 ⟨85PAC763, 82TL2457⟩, 700 ⟨82TL2465, 83TL4319⟩, 1071 ⟨91SL611⟩; **2**, 593 ⟨75BSF770⟩, 623 ⟨89JOM(379)201⟩, 628 ⟨90BSF275⟩, 678 ⟨91SL505⟩, 1016 ⟨56BSF1615⟩; **3**, 3 ⟨76BSF1941⟩, 16 ⟨85TL2333⟩, 20, 31 ⟨91SL501⟩, 55 ⟨84TL189⟩, 56 ⟨75BSF1363⟩, 114 ⟨77BSF1941⟩, 238 ⟨49JCS1430⟩, 419 ⟨83T433⟩; **4**, 138 ⟨90SL769⟩, 147 ⟨92SL133⟩, 200 ⟨91SL501⟩, 201 ⟨91SL503⟩, 272, 275 ⟨82BSF43⟩, 742, 746 ⟨89JOM(379)201⟩, 743 ⟨59BSF1828⟩; **5**, 145 ⟨59MI 503-01⟩
Julia, S., **1**, 225 ⟨81SC723⟩, 621 ⟨88TL3255⟩, 622 ⟨93SL839⟩, 816 ⟨74TL2077, 77BSF925⟩, 820 ⟨75CR(C)1231, 77SC103⟩, 821 ⟨72TL3587⟩, 963 ⟨90TL213⟩, 980 ⟨74CC10⟩, 1089 ⟨83TL4825⟩; **2**, 299 ⟨78CC742, 81JCS(P1)574⟩, 201 ⟨88TL3255⟩, 247, 361, 363 ⟨88TL3251⟩, 359 ⟨87BSF181⟩, 359 ⟨87T881, 91T6655⟩; **3**, 23 ⟨76BSF1445⟩, 54, 210 ⟨83TL4825⟩, 55 ⟨83TL4829⟩, 227 ⟨78TL1155⟩, 231 ⟨54BSF780⟩, 228 ⟨53HCA482⟩, 369 ⟨92SL909⟩, 532 ⟨68BSF4913⟩; **4**, 255 ⟨78TL4065⟩, 323 ⟨76BSF1482⟩, 843 ⟨85S958⟩, 853 ⟨78JCR(S)68, 78SC87⟩, 938 ⟨83TL4825, 85BSF1250⟩; **5**, 145 ⟨59MI 503-01⟩, 242 ⟨82SC897⟩, 553 ⟨78TL4065⟩, 554 ⟨75CR(C)1327⟩; **6**, 362 ⟨84OPP299⟩; 475 ⟨82SC897⟩
Julian, P. L., **1**, 10 ⟨50JA362⟩; **3**, 350 ⟨35JA1126⟩
Julin, S., **1**, 1078 ⟨92TL3715⟩
Julius, M., **4**, 146 ⟨88JOM(353)C30⟩, 146 ⟨91LA811⟩, 400, 401 ⟨90TL7419⟩
Juma'a Al-Dulayymi, **2**, 449 ⟨91SL433⟩
Jun, C.-H., **2**, 915 ⟨93JOM(447)177⟩; **5**, 459 ⟨90JOM(390)361⟩
Jun, C. H., **5**, 800 ⟨90MI 520-01⟩
Jun, H., **3**, 689 ⟨93OM4262⟩; **6**, 681, 690 ⟨91JA9379, 93OM4265⟩
Jun, T. X., **1**, 37 ⟨87S821⟩; **3**, 168 ⟨87CC947⟩
Jun, Y. M., **1**, 690 ⟨79JOC4729⟩, 900 ⟨86JOC2969⟩; **4**, 392 ⟨87JA5403⟩
Jund, K., **3**, 472 ⟨80TL15⟩
Junek, H., **1**, 889 ⟨68TL4309⟩; **3**, 653 ⟨71ZN(B)1124⟩; **4**, 55, 57 ⟨80LA1919⟩, 883 ⟨86LA533⟩; **6**, 73 ⟨82MI 603-01⟩

Jung, A., **3**, 631 ⟨85AG(E)989⟩
Junga, H., **1**, 339, 341 ⟨93TL5251⟩
Junga, I. G., **2**, 861 ⟨60JOC666⟩
Junga, J., **5**, 173 ⟨81CL1415⟩
Junga, M., **6**, 275, 278, 279, 307 ⟨72LA(758)132⟩
Jung, F., **1**, 600 ⟨73JA3420⟩; **2**, 200 ⟨73CC4⟩, 200 ⟨73TL2863⟩, 200 ⟨74JA935⟩; **3**, 54 ⟨79S507⟩, 87 ⟨74JA935⟩; **4**, 75, 77 ⟨73CC4⟩, 393 ⟨73JA3420⟩; **5**, 1083 ⟨89TL2379⟩; **6**, 622 ⟨89TL2379⟩
Jung, G. L., **1**, 812 ⟨84TL3959⟩
Jung, I. N., **3**, 502 ⟨91OM2529⟩; **4**, 609 ⟨93OM2360⟩
Jung, J. C., **4**, 201 ⟨93TL3581⟩
Jung, J. H., **3**, 120, 294 ⟨90SC637⟩
Jung, K. W., **1**, 1130, 1132 ⟨89TL827⟩
Jung, K.-Y., **4**, 561 ⟨93PS(75)23⟩
Jung, M., **2**, 299 ⟨85JCS(P1)2201⟩, 437 ⟨90BCJ2051⟩, 1044, 1048 ⟨81LA1907⟩; **3**, 273 ⟨81LA1907⟩; **4**, 214 ⟨94BMC931⟩; **5**, 329 ⟨85TL977⟩
Jung, M. E., **1**, 132, 135 ⟨77TL3791⟩, 154 ⟨76TL1439⟩, 300, 306, 889 ⟨91COS(4)1⟩, 306 ⟨76T3⟩, 556 ⟨77JOC3961⟩, 582 ⟨90CC84⟩, 760 ⟨80JA2463⟩, 805 ⟨83TL2931⟩; **2**, 32 ⟨78TL4483⟩, 36 ⟨77TL2659⟩, 36, 658 ⟨77JOC3761⟩, 444 ⟨78S588⟩, 538 ⟨79OS(59)35⟩, 921 ⟨74JA3684⟩; **3**, 8, 121 ⟨76JOC1479⟩, 9, 123 ⟨77TL4175⟩, 34 ⟨93JA1208⟩, 78 ⟨79T621⟩, 120 ⟨78TL2771⟩, 134 ⟨74JA3683⟩, 187 ⟨91COS(4)1⟩, 206 ⟨77JOC3961⟩, 267 ⟨88JA3965⟩, 502 ⟨76JA7844⟩, 626 ⟨86TL5319⟩, 634 ⟨90H(30)839⟩; **4**, 51 ⟨78S588⟩; **5**, 20, 225 ⟨77JA968⟩, 826, 829, 830 ⟨87JOC4570⟩; **6**, 450 ⟨84CC630⟩, 487 ⟨90H(30)839⟩
Jung, M. H., **6**, 428 ⟨91JAN1083⟩
Jung, S. H., **3**, 498 ⟨86JOC3402⟩; **4**, 652 ⟨88TL25, 93JA10754⟩; **5**, 1080, 1118 ⟨85JA2931, 84TL399⟩
Jung, W. A., **1**, 521 ⟨82CB804⟩; **5**, 14 ⟨82CB804⟩
Jung, Y. S., **4**, 513 ⟨92JOC6037⟩
Jung, Y.-W., **2**, 326 ⟨88JOC5989⟩
Junggebauer, J., **1**, 925 ⟨93SL286⟩
Jungheim, L. N., **1**, 690 ⟨86JA3513⟩; **5**, 363 ⟨87TL285⟩
Jungheim, N., **2**, 212 ⟨80JOC1839⟩
Jungmann, H., **4**, 585 ⟨81ZN(B)1663⟩
Jungmann, P., **3**, 614 ⟨52CB397⟩
Jung-yu, C., **2**, 838, 839, 851 ⟨62ZOB3351⟩
Junino, A., **1**, 498 ⟨80JCR(S)98, 85T2759⟩, 805 ⟨77TL1779, 80JCR1551⟩
Junjappa, H., **1**, 485 ⟨85JCS(P1)1289⟩, 1076 ⟨82S1100⟩; **2**, 74 ⟨93TL3145⟩; **4**, 253 ⟨90T2195, 92S1075⟩, 704 ⟨88T1667⟩, 899, 900 ⟨76CC592, 76CC593⟩, 900 ⟨80S748⟩, 903 ⟨84S250⟩; **5**, 555 ⟨82S693, 85JCS(P1)641⟩, 555 ⟨90T5423⟩, 731 ⟨87S547, 88JCS(P1)169⟩, 731 ⟨88T1667⟩
Junk, P. C., **4**, 654 ⟨87OM2110, 88JOC3134⟩; **6**, 168 ⟨91CC1560⟩, 393 ⟨88JOC3134⟩
Junker, P., **3**, 560 ⟨66LA(700)32⟩
Junkes, P., **4**, 545 ⟨71AG(E)940⟩
Juntunen, S. K., **1**, 237, 238 ⟨87JA6396⟩; **2**, 395 ⟨92TL131⟩
Juraristi, E., **2**, 438 ⟨91JOC5919⟩
Jurášek, A., **5**, 1106 ⟨72CCC3144⟩; **6**, 627 ⟨84CCC2285⟩
Jurayi, J., **1**, 583 ⟨90SC1463⟩
Jurayj, J., **1**, 826 ⟨87JA1170⟩; **5**, 573, 587 ⟨92T8601⟩
Jurch, Jr., G. R., **5**, 244, 546 ⟨79JOC569⟩
Jurczak, J., **3**, 39 ⟨87T3063, 89CRV149⟩, 40 ⟨92T4231⟩, 303 ⟨87JA3981⟩; **4**, 174 ⟨84TL4797⟩; **5**, 344 ⟨85T2441⟩

Jurczyk, S., **4**, 323 ⟨92S1295⟩
Jürgens, B., **5**, 819 ⟨11JPR687⟩, 819, 821, 822 ⟨11JPR453⟩
Jürgens, E., **4**, 220 ⟨57LA(602)135⟩
Jurič, P., **5**, 837 ⟨84CS209⟩
Jurion, M., **2**, 693 ⟨69CC112⟩; **3**, 614 ⟨69CC112⟩
Jurkschat, K., **2**, 253, 260 ⟨82TL1345⟩, 435 ⟨91PS(61)261⟩; **4**, 570, 585, 660, 704 ⟨92JOM(423)339⟩, 660, 700 ⟨93OM2788, 82JOM(236)69⟩, 700 ⟨84JOM(273)303, 84JOM(277)335⟩, 700 ⟨85BSB299⟩
Jurlina, J. L., **2**, 34 ⟨78JCS(P1)126⟩; **3**, 155 ⟨78JCS(P1)126⟩; **4**, 27 ⟨81JCS(P1)2608⟩
Jurlina, J. R., **2**, 12 ⟨81JCS(P1)2608⟩
Jurriens, T. K., **5**, 224, 536 ⟨82JOC3966⟩
Juršíc, B., **5**, 1104 ⟨89JCR(S)104⟩
Jursic, B., **2**, 25 ⟨88S868⟩, 26 ⟨89JCR(S)284⟩, 31, 35 ⟨89JCR(S)104⟩
Juršić, B., **2**, 683 ⟨88T6677⟩; **3**, 119 ⟨88S868⟩
Juskowiak, M., **3**, 426 ⟨89JPR870⟩
Just, C., **3**, 496 ⟨85TL1799⟩
Just, G., **1**, 729 ⟨69JA5364⟩, 1009 ⟨89JOC4453⟩, 1131 ⟨87CJC104⟩; **2**, 633 ⟨88SC1327⟩; **3**, 621 ⟨87JOC3659⟩, 684 ⟨68T5251⟩; **4**, 12 ⟨79TL3643⟩, 491 ⟨80SC429⟩, 847, 848 ⟨80CJC2780⟩
Justin, B., **6**, 241 ⟨81JFC(17)331⟩
Justoni, R., **3**, 612 ⟨39G378, 41G41⟩
Jutand, A., **1**, 467 ⟨81JOM(209)109⟩, 471 ⟨79JOM(177)273⟩, 483 ⟨93SL568⟩
Jutisz, M., **5**, 83 ⟨52BSF821, 53MI 502-02⟩
Jutz, C., **1**, 866 ⟨64AG(E)312⟩; **2**, 790 ⟨72AG(E)315⟩; **3**, 64 ⟨66CB2479⟩, 228 ⟨58CB1867⟩, 231 ⟨75LA874⟩; **4**, 979 ⟨63CB2100⟩; **6**, 51 ⟨B-76MI 602-01⟩
Jutzi, P., **2**, 436 ⟨77CB1269⟩, 484 ⟨74CB1228, 75JOM(93)191⟩, 866 ⟨87JOM(326)C6⟩, 1005 ⟨91OM1282⟩, 4366 ⟨76JOM(118)C8⟩; **3**, 312 ⟨83JHC1011⟩, 722 ⟨83JOM(246)159⟩; **4**, 603 ⟨84CB1885, 91CB2391⟩, 652 ⟨88CB1393⟩, 660, 663 ⟨78JOM(148)247⟩; **6**, 134 ⟨79PS(7)203⟩, 166 ⟨86AG(E)919⟩, 187 ⟨86OM1944⟩, 187 ⟨86OM730⟩, 187 ⟨94OM436⟩, 187, 387 ⟨91OM1647, 91OM3838⟩, 515 ⟨71AG(E)339⟩
Juurik-Hogan, R., **6**, 42, 43 ⟨70JOC3730⟩
Juus, R., **1**, 1082 ⟨76OS(55)114⟩
Juza, R., **1**, 881 ⟨64AG(E)471⟩
Jwili, E. T. P., **6**, 648 ⟨71CJC2315⟩
Jyst, D., **5**, 67 ⟨79MI 502-07⟩

K

Kaack, H., **2**, 824 ⟨73LA1480⟩
Kaae, S., **4**, 77 ⟨68ACS3256⟩
Kaba, R. A., **2**, 1010 ⟨74JA6202⟩
Kabachnick, M. I., **1**, 870 ⟨67TL4525⟩; **2**, 430, 437 ⟨86S198⟩, 449 ⟨68DOK(182)337, 76JGU428⟩, 450 ⟨60DOK(135)323, 63IZV1227⟩, 451 ⟨57DOK(117)817⟩, 454 ⟨59ZOB1450⟩, 462 ⟨61ZOB507, 62ZOB3579⟩, 463 ⟨61IZV604⟩, 464 ⟨59ZOB1671, 60IZV651⟩, 821, 824, 847 ⟨69ZOB841⟩, 838, 839, 851 ⟨62ZOB3351⟩, 841 ⟨65ZOB1574⟩, 848 ⟨68ZOB2285⟩, 1033 ⟨91RCR885⟩; **4**, 113, 120 ⟨70JGU255⟩, 118, 132, 133 ⟨69JGU1490⟩, 119, 120, 121 ⟨80IZV491⟩, 119, 125, 126 ⟨53IZV763, 62DOK(143)211⟩, 123 ⟨86DOK(286)19⟩, 130 ⟨50DOK(75)219, 50IZV39⟩, 339 ⟨73ZOB534, 84PS(21)59⟩, 451, 469 ⟨68RCR7⟩, 463, 469 ⟨52DOK(83)689⟩, 463, 469 ⟨53DOK(92)959, 57IZV1357⟩, 473 ⟨50IZV635, 51IZV95⟩, 546, 553 ⟨86S198⟩, 551 ⟨80TL2931⟩, 551 ⟨92ZOB263⟩, 555 ⟨70ZOB285⟩, 555, 561 ⟨67IZV591⟩, 556 ⟨62IZV2103, 66IZV1954, 67IZV949, 68IZV1417, 68IZV2062⟩, 557 ⟨84ZOB36, 85ZOB1234, 86ZOB1220, 88ZOB1987, 88ZOB1998, 88ZOB2012⟩, 560, 564 ⟨70IZV1326⟩, 563 ⟨72IZV2612, 71IZV2747⟩, 920 ⟨57CA(51)10366g⟩; **6**, 118 ⟨72MI 604-01⟩, 455 ⟨45BAU364, 57IZV48⟩
Kabachnik, M. M., **2**, 471, 472 ⟨91JGU460⟩, 476 ⟨78JGU1088⟩; **4**, 462 ⟨85JGU404⟩, 550 ⟨85ZOB331⟩, 552 ⟨87ZOB949⟩, 552, 562 ⟨79ZOB1446⟩
Kabaki, M., **4**, 361 ⟨90SC3245⟩; **5**, 431 ⟨90SC3245⟩
Kabakoff, D. S., **1**, 894 ⟨70JA3234⟩
Kabakov, A., **5**, 98 ⟨76MI 502-01⟩
Kabalka, G. W., **1**, 19, 41 ⟨75JOC1834, 79SC275⟩, 41, 635 ⟨76JOC574⟩, 125, 280 ⟨68JA818⟩, 126 ⟨69JA2150⟩, 168 ⟨92TL7687⟩, 273 ⟨67JA5710⟩, 274 ⟨68JA4166⟩, 274, 546 ⟨70JA3503⟩, 274, 946 ⟨68JA4165⟩, 275 ⟨71JOM(33)25⟩, 280 ⟨68JA1911⟩, 281 ⟨91TL4619⟩, 510 ⟨76T981⟩, 635 ⟨80OS(59)42⟩, 661 ⟨88MI 114-04⟩, 946 ⟨70JA712, 70JA714⟩, 946 ⟨71JA1024, 86T5523⟩, 962 ⟨78CC726⟩; **2**, 4 ⟨81JOC2582, 84ACR215⟩, 4, 30 ⟨81JOC3113⟩, 34 ⟨76JA1290, 80JOC3578, 81SC521, 82SC101⟩, 34, 628 ⟨87TL4497⟩, 34, 629, 631 ⟨88JOC4477⟩, 40 ⟨75JOC1776⟩, 43 ⟨94SC1019⟩, 58, 526 ⟨90TL4677⟩, 221 ⟨91SL125⟩, 301 ⟨89OM1093, 90SC231⟩, 305 ⟨81JOC4296⟩, 306 ⟨89SC2409⟩, 307 ⟨88SC1363⟩, 341 ⟨85JOC133, 87JHC767, 91COS(8)373⟩, 393, 398, 784 ⟨87OPP283⟩, 398 ⟨84SC1093, 85SC151, 89SC805, 90T7443⟩, 527 ⟨76T981⟩, 608, 632 ⟨82OPP359⟩, 616 ⟨85TL279⟩, 623 ⟨83SC1027⟩, 628 ⟨81SC247⟩, 655 ⟨92TL865, 93TL7667⟩, 698 ⟨70JA712, 70JA714⟩, 772 ⟨91COS(8)364⟩, 776 ⟨91COS(8)366⟩, 907 ⟨83JOM(259)269⟩, 936 ⟨92TL6915⟩; **3**, 133 ⟨92SC2425⟩, 165 ⟨85SC443⟩, 223 ⟨70JA3503⟩, 426 ⟨85CL243, 85JOC133⟩, 426 ⟨85SC1325, 85TL6013⟩, 426 ⟨86SC91, 88SC693⟩, 426 ⟨90SC2453⟩; **4**, 353 ⟨90TL4677⟩, 832 ⟨68JA818⟩; **5**, 45, 71, 97 ⟨82CC1273⟩, 51 ⟨84JOC174⟩, 184, 195, 198 ⟨92TL593⟩, 223 ⟨89OM655⟩, 897 ⟨92TL6915⟩
Kabalker, G. W., **1**, 41 ⟨77S124⟩
Kabat, M. M., **3**, 148 ⟨89JFC(42)435⟩, 200 ⟨93TL8543⟩
Kabayashi, H., **2**, 906 ⟨92TL3779⟩
Kabbara, J., **1**, 979 ⟨92TL8017⟩, 1012 ⟨92SL766⟩
Kabbe, H. J., **2**, 478 ⟨62LA(656)204⟩; **4**, 312 ⟨72LA(765)55⟩
Kabe, Y., **2**, 949 ⟨89JA1888⟩; **3**, 337 ⟨91JA7782⟩, 366 ⟨90TL7641⟩; **4**, 25 ⟨84JA8174⟩, 318 ⟨85CL939⟩; **5**, 413, 430 ⟨79TL871⟩; **6**, 187 ⟨94OM434⟩
Kaberdin, R. V., **4**, 762 ⟨91ZOB56⟩, 771 ⟨93ZOR56⟩, 797 ⟨91JOU48⟩, 970 ⟨82ZOR1835⟩
Kabeta, K., **4**, 1033 ⟨86JOC3772⟩
Kabeya, M., **6**, 675 ⟨85H(23)2371⟩
Kabibulin, I. R., **6**, 555 ⟨91JPR139⟩
Kabi-Satpathy, A., **4**, 657 ⟨89JOM(364)105⟩
Kabitzke, K., **5**, 612 ⟨65CB3819⟩
Kabitzke, K. H., **3**, 699 ⟨64AG(E)641⟩
Kablaoui, M. S., **2**, 658 ⟨74JOC2126⟩
Kabo, A., **2**, 280 ⟨80CJC1030⟩
Kabouche, Z., **2**, 676 ⟨91SL755⟩
Kaburagi, S., **2**, 662 ⟨91CPB1132⟩
Kabusz, S., **4**, 193 ⟨71S312⟩; **6**, 105 ⟨71S312, 72S32⟩, 105 ⟨72S418, 73S423⟩, 105 ⟨75S272⟩
Kabuto, C., **4**, 618 ⟨89JA3748⟩, 1050 ⟨82JA4288⟩, 1051, 1052 ⟨88CL965⟩, 1069 ⟨93CL1881⟩; **6**, 176, 193 ⟨89JA3748⟩, 186 ⟨89AG(E)55⟩, 193 ⟨92CL867⟩
Kabuto, H., **5**, 248, 249 ⟨85S519⟩, 248, 249 ⟨85S520⟩
Kabzinska, K., **4**, 266 ⟨89BAP117⟩
Kachanov, A. V., **3**, 559, 560 ⟨92AG(E)1509⟩
Kachensky, D. F., **1**, 173, 703 ⟨85JOC4317⟩
Kacher, M., **1**, 635 ⟨75JOC923⟩
Kachinski, J., **3**, 80 ⟨77CC53⟩
Kachinsky, J. L. C., **3**, 141 ⟨86JOC1393⟩
Kachkovskaya, L. S., **6**, 685, 686, 689 ⟨87ZOB901, 84ZOB2800⟩, 692 ⟨92ZOB474⟩, 694 ⟨89ZOB2131⟩
Kaczmarek, A. J., **5**, 671 ⟨90JPR731⟩
Kaczmarek, E., **4**, 306 ⟨88S111⟩
Kaczynski, M. A., **2**, 67 ⟨90TL4565⟩
Kaddachi, M. T., **2**, 305 ⟨90TL4735⟩
Kader, A. T., **4**, 218, 235 ⟨62JCS3686⟩
Kadeřábek, V., **4**, 311 ⟨60CCC2334⟩; **6**, 558, 565 ⟨84CCC1577⟩
Kades, E., **5**, 227 ⟨90TL2259⟩
Kadesch, R. G., **5**, 28 ⟨B-63MI 502-02⟩
Kadib-Elban, A., **1**, 311 ⟨83TL5089⟩
Kadibelban, T., **3**, 494 ⟨66AG(E)968⟩
Kadina, M. A., **4**, 639 ⟨70IZV2280⟩; **6**, 62 ⟨67ZOB1040⟩
Kadin, S. B., **1**, 28, 29 ⟨73JOC1348⟩
Kadir, A. A., **5**, 770, 850 ⟨68JIC729⟩
Kadirova, R. F., **6**, 116 ⟨90CC1133⟩
Kadish, K. M., **2**, 578 ⟨88CRV1121⟩; **6**, 519 ⟨88CRV1121⟩
Kado, M., **2**, 855 ⟨67JAP6703120⟩
Kadochnikova, G. D., **5**, 860 ⟨78KGS991⟩
Kadoi, A., **3**, 374 ⟨74TL1071⟩
Kadonaga, J. T., **1**, 803 ⟨81JA2443⟩
Kadorkina, G. K., **2**, 337 ⟨88JA1719⟩
Kadota, I., **1**, 155 ⟨90S401⟩, 177 ⟨89TL3999⟩, 177 ⟨90JOC4417⟩, 1021 ⟨90TL4609⟩; **2**, 79 ⟨92JOC7003⟩, 312 ⟨90S401⟩
Kadyko, M. I., **4**, 112, 114, 115 ⟨84JGU2250⟩
Kadyrov, A. A., **4**, 802, 803 ⟨88BAU1686⟩
Kadyrov, Ch. Sh., **1**, 212 ⟨60MI 105-01⟩
Kadyrov, C. S., **1**, 475 ⟨58DOK31, 60DOK(131)27⟩
Kadyrov, R. A., **4**, 341 ⟨80IZV1626⟩, 342 ⟨83IZV1374⟩
Kadzyauskas, P. P., **3**, 429 ⟨68RCR543⟩
Kaeding, W. W., **4**, 189 ⟨52JA6189⟩
Kaehne, R., **2**, 456 ⟨1898CB1048⟩
Kaesler, R. A., **2**, 607 ⟨86JOC277⟩
Kaesler, R. W., **1**, 521 ⟨82JOC4779⟩; **2**, 670 ⟨85JA1060⟩
Kaestner, G., **6**, 538 ⟨85DIS(B)840⟩
Kaesz, H. D., **1**, 917 ⟨65JA3267⟩; **4**, 681 ⟨60JA4749⟩, 699 ⟨61JA1514⟩, 720 ⟨75JOM(91)C57, 77IC3193⟩, 721 ⟨79IC3453⟩, 812 ⟨60JA6232⟩; **6**, 65, 244 ⟨60JA6228⟩, 517 ⟨82IC1704⟩, 523 ⟨61JA2953⟩

Kaetsu, A., **4**, 699 ⟨91T3281⟩
Kafafi, Z. H., **5**, 1096 ⟨83JCP(87)797⟩
Kafarski, P., **4**, 462 ⟨83CJC2425⟩, 467 ⟨84S577⟩
Kafarski, P. K., **4**, 114 ⟨81JCS(P2)1138⟩
Käfig, F., **4**, ⟨77LA830⟩ ⟨77LA624⟩
Kafka, C. M., **1**, 515 ⟨92JOC4786⟩
Kaftory, M., **1**, 988 ⟨87JA782⟩; **2**, 877 ⟨86CB1953⟩; **3**, 469 ⟨80JCS(P1)2670⟩
Kaga, K., **5**, 560 ⟨86BCJ1403⟩
Kagabu, S., **3**, 626 ⟨89OPP354⟩; **5**, 344 ⟨89CL2181, 91BCJ106⟩, 1112 ⟨88CC1485⟩
Kagan, B., **2**, 463 ⟨59JA148⟩
Kagan, H., **2**, 561 ⟨71BSF3571⟩; **3**, 461 ⟨66BSF1819⟩; **4**, 261 ⟨89JOM(370)43⟩; **5**, 439 ⟨91JOC3118⟩
Kagan, H. B., **1**, 3, 6, 8, 10, 17, 587, 680 ⟨80JA2693⟩, 76 ⟨88BSF846⟩, 208 ⟨87TL5513⟩, 374, 375 ⟨92CRV1007⟩; **2**, 25 ⟨74TL763⟩, 27 ⟨82BSF(2)327⟩, 56 ⟨84JOC2047⟩, 147 ⟨84JA8188⟩, 147 ⟨85PAC1911⟩, 147 ⟨90SL943⟩, 152, 198 ⟨91JOC5991⟩, 362 ⟨86NJC761⟩, 442 ⟨B-90MI 208-02⟩, 564 ⟨86T6573⟩, 565 ⟨80JA2693, 87JOM(328)81⟩, 565 ⟨92T3877⟩, 959 ⟨76PAC177⟩; **3**, 7, 119 ⟨84JOC2045⟩, 615, 632 ⟨87TL5513⟩, 634 ⟨85NJC1⟩; **4**, 261 ⟨90SL643⟩, 706 ⟨90TL3121⟩; **5**, 122 ⟨74JA8113⟩, 290 ⟨65BSF1454⟩
Kagan, J., **4**, 183 ⟨72TL5133⟩, 183 ⟨74JOC3145⟩; **5**, 82 ⟨83JOC4076⟩
Kagarlitskii, A. D., **5**, 105 ⟨79MI 502-09⟩
Kagawa, T., **6**, 529 ⟨88JAP01257116, 89JAP03037110⟩, 529 ⟨89JAP03060417, 89JAP03060418⟩, 529 ⟨89JAP03197310, 91JAP03060419⟩, 533 ⟨88JAP01301654, 88JAP02009858⟩, 533 ⟨90JAP03220173, 91JAP03193759⟩, 533 ⟨93JAP05058990⟩
Kagemoto, T., **5**, 700, 701 ⟨92JCS(P1)2127⟩
Kagen, B. S., **2**, 15, 17 ⟨86CRV(86)997⟩, 334 ⟨86CRV997⟩
Kageyama, H., **5**, 243 ⟨89ZN(B)1050⟩, 244, 250 ⟨94JCS(P1)1083⟩, 249 ⟨92ZN(B)558, 90JOM(386)313⟩, 255, 506, 509, 543 ⟨90TL3587⟩, 541, 542 ⟨92CB417⟩
Kageyama, M., **1**, 697, 700, 703 ⟨90JA7407⟩, 1136 ⟨75JOC1932⟩
Kageyama, T., **3**, 119 ⟨83S815⟩
Kageyama, Y., **1**, 828 ⟨81TL4097⟩
Kägi, B., **6**, 597 ⟨66ACS281⟩
Kagi, D. A., **3**, 226 ⟨73CC161⟩
Kagotami, M., **2**, 157 ⟨78CC367⟩
Kagotani, M., **1**, 1097 ⟨82JA180⟩; **4**, 1033 ⟨80BCJ1138, 82JA180⟩; **5**, 595, 614 ⟨81TL3409⟩, 607 ⟨81TL3413, 82JA4018⟩, 607, 614 ⟨81TL4245⟩
Kagramanov, N. D., **3**, 342 ⟨87IZV2321⟩
Kahana, N., **2**, 304 ⟨89S825⟩
Kaharu, T., **5**, 1089 ⟨92CL1515⟩
Kahil, A. I. M., **3**, 474 ⟨24JA2556⟩
Kahle, G. G., **4**, 415 ⟨73T4049⟩
Kahlen, N., **3**, 709, 711 ⟨60JOC809⟩; **5**, 1136, 1144 ⟨60JOC809⟩
Kahlenberg, H., **5**, 354 ⟨91CZ215⟩
Kahler, F. H., **2**, 491, 496, 498 ⟨47JA927⟩
Kahn, B. E., **1**, 687, 692 ⟨88CRV733⟩
Kahn, F. K., **3**, 700 ⟨91JOM(403)C15⟩
Kahn, H. J., **3**, 363 ⟨45JCS858⟩
Kahn, J., **2**, 471 ⟨B-83MI 202-02⟩
Kahn, M. A., **3**, 169, 172 ⟨83TL405⟩
Kahn, M., **1**, 812 ⟨80TL4547⟩; **2**, 46 ⟨83TL3951⟩, 540 ⟨85JA500⟩, 675 ⟨82TL703⟩; **3**, 207 ⟨80JA2097⟩, 219 ⟨81JA3460⟩; **5**, 353 ⟨94MIP3494⟩, 366 ⟨88JA1638, 89TL2317⟩
Kahne, D., **3**, 633 ⟨81TL5011⟩; **4**, 232 ⟨89JA6881, 93JA1580⟩
Kahne, D. E., **1**, 84 ⟨83JA1072⟩
Kahr, B., **4**, 11 ⟨87JOC3713⟩
Kahr, K., **3**, 427 ⟨60CB132⟩
Kai, M., **4**, 441 ⟨82TL2103⟩
Kai, ,S., **5**, 193 ⟨86EUP268820⟩
Kai, T., **2**, 16 ⟨92JCS(P1)221⟩
Kai, Y., **1**, 250 ⟨86JOC4375⟩, 690 ⟨88JA3578⟩, 992 ⟨90CL2149⟩; **4**, 686 ⟨77BCJ2250⟩, 820 ⟨86CC1794⟩; **5**, 948 ⟨82CC1178⟩
Kaigorodova, L. N., **5**, 215 ⟨86IZV2032⟩
Kaiho, T., **5**, 173 ⟨82JA5521⟩, 173, 235 ⟨82JA5523⟩
Kaila, N., **4**, 187 ⟨92JOC2084⟩
Kaimal, T. N. B., **5**, 18 ⟨86SC1423⟩
Kaim, W., **6**, 173 ⟨78CB3573⟩
Kain, I., **3**, 530, 531 ⟨86S43⟩
Kainer, H., **5**, 857 ⟨53AG442⟩
Kaino, M., **2**, 679 ⟨90JOC5814⟩; **4**, 174 ⟨90JOC5814⟩
Kainosho, M., **1**, 925, 946 ⟨93TL1317⟩
Kaiser, C., **1**, 627 ⟨76OS(55)3⟩; **2**, 191 ⟨64CB2903⟩
Kaiser, E., **6**, 304 ⟨65LA(689)179⟩
Kaiser, E. M., **1**, 101 ⟨88OSC(6)852⟩, 101, 889 ⟨72S391⟩, 144 ⟨73JOM(59)53⟩, 844 ⟨B-73MI 119-02⟩, 879 ⟨77S509⟩, 881 ⟨65JOC2937, 65JOC4135⟩, 881 ⟨67OS(47)72⟩, 885 ⟨75S395⟩; **2**, 205 ⟨75JOC1343⟩, 1005 ⟨68JA7287⟩; **3**, 620 ⟨68JOC3402⟩; **4**, 393 ⟨73JOM(59)53⟩
Kaiser, E. T., **2**, 217 ⟨80MI 203-03, 81CL737⟩
Kaiser, E. W., **2**, 30 ⟨55OSC(3)209⟩
Kaiser, G. V., **1**, 410, 823 ⟨71JA2342⟩
Kaiser, H.-P., **1**, 92 ⟨84JOC4203⟩; **3**, 307 ⟨84JOC4203⟩
Kaiser, L., **2**, 744 ⟨76S535⟩; **6**, 111 ⟨75LA958⟩
Kaiser, M., **5**, 223 ⟨87T2045⟩
Kaiser, R., **1**, 525 ⟨72TL2013⟩, 789 ⟨73AG(E)337⟩
Kaiser, W., **2**, 172 ⟨23CB553⟩; **4**, 313 ⟨67T2781, 75LA266⟩; **5**, 377 ⟨75LA266⟩
Kajfez, F., **2**, 308, 325 ⟨79S161⟩
Kajgaeshi, S., **4**, 421 ⟨75CL607⟩
Kaji, A., **1**, 37 ⟨83CC875, 83TL3477⟩, 37 ⟨85JOC3692, 86JOC2832⟩, 37 ⟨87S821⟩, 37, 330, 443, 548, 946, 1131 ⟨85T4013⟩, 37, 945 ⟨81TL1705⟩, 210 ⟨85TL4747⟩, 235 ⟨81CL315, 82CL351⟩, 240 ⟨82BCJ2965, 87CL809, 88JA7185⟩, 250 ⟨82CC821, 86JCS(P1)1439⟩, 250, 251, 486 ⟨86CC1285⟩, 252 ⟨87JOC4133, 84CC274⟩, 592, 638 ⟨84TL5319⟩, 600, 617, 638 ⟨88JOC251⟩, 617 ⟨78TL763⟩, 617, 638 ⟨81CL1139⟩, 617, 638 ⟨87JOC5111⟩, 635 ⟨88BCJ3211⟩, 946 ⟨86S693⟩; **2**, 394 ⟨85BCJ1069, 87S258⟩, 395 ⟨84S226⟩, 396, 789 ⟨86JOC2139⟩; **3**, 16, 32 ⟨78S881⟩, 46, 190 ⟨87S532⟩, 57 ⟨82JOC5017⟩, 168 ⟨87CC947⟩, 190 ⟨86JOC2832⟩, 192 ⟨87S207⟩, 621 ⟨85CL635⟩, 635, 645 ⟨78TL763⟩, 644 ⟨79S890⟩; **4**, 220 ⟨78BCJ2401⟩, 236 ⟨80BCJ3619⟩, 249 ⟨80S952⟩; **5**, 132 ⟨78BCJ2401⟩; **6**, 495 ⟨71BCJ1393⟩
Kaji, E., **2**, 394 ⟨88OSC(6)503⟩; **3**, 684 ⟨78CPB3254⟩
Kaji, K., **1**, 237 ⟨85JCS(P1)1171⟩, 820 ⟨80CL1061⟩; **4**, 216 ⟨85CL1933⟩; **5**, 678 ⟨84CPB4437⟩, 746 ⟨78CPB2246⟩
Kaji, M., **2**, 104 ⟨82JCS(P1)111⟩
Kaji, R., **1**, 431 ⟨93JOC4897⟩; **4**, 758, 821 ⟨93JOC4897⟩
Kaji, S., **1**, 210 ⟨83JOC5133⟩
Kajigaeshi, S., **2**, 21 ⟨88S545, 88TL5783⟩, 23 ⟨91CE185⟩, 24 ⟨90BCJ3033⟩, 28 ⟨87BCJ1159⟩, 34 ⟨92BCJ1731⟩, 149 ⟨89BCJ3376⟩, 320 ⟨89CL463⟩, 626 ⟨87BCJ4187, 87CL627⟩, 626 ⟨88BCJ597, 89BCJ3373⟩, 626 ⟨89BCJ439⟩, 631 ⟨87CL2109, 88BCJ600⟩, 631 ⟨88CL795, 89BCJ1349⟩, 1031 ⟨72BCJ509, 73JOC1361⟩; **3**, 119 ⟨89BCJ2585⟩; **5**, 48, 49 ⟨85S674⟩, 318 ⟨84CL2045, 85BCJ769⟩, 318 ⟨89JCS(P1)1702⟩; **6**, 76, 77 ⟨73JOC1361⟩
Kajihara, Y., **3**, 28 ⟨91CL1469⟩; **5**, 359 ⟨92JOC7359⟩; **6**, 489 ⟨92JOC7359⟩
Kajikawa, Y., **3**, 163 ⟨92JOC2⟩
Kajiki, T., **4**, 857 ⟨86JOC700⟩
Kajimoto, T., **3**, 470 ⟨73TL4573⟩, 599 ⟨70JOM(23)275⟩
Kajiwara, K., **1**, 141 ⟨85JOC657⟩; **4**, 235, 236 ⟨86CL1655⟩; **6**, 101 ⟨85JOC657⟩
Kajtár-Peredy, M., **5**, 567 ⟨93JOC6076⟩
Kakác, B., **4**, 211 ⟨72MI 404-01⟩
Kakayama, O., **2**, 51 ⟨90TL7333⟩
Kakehi, A., **5**, 681 ⟨83BCJ545, 84BCJ539⟩; **6**, 560 ⟨87CPB156⟩

Kakese, F., **5**, 1078 ⟨69TL2473⟩
Kakeya, N., **6**, 413 ⟨79JOC3573⟩
Kakigano, T., **4**, 681 ⟨92OM989⟩; **5**, 242, 250, 506, 509 ⟨87CL475⟩, 244 ⟨86ZC179⟩
Kakihana, M., **1**, 250 ⟨86JOC4375⟩
Kakihana, T., **3**, 538 ⟨77JA4554⟩
Kakikawa, T., **2**, 330 ⟨92TA227⟩
Kakimoto, M.-A., **4**, 247 ⟨87S164⟩
Kakinami, T., **2**, 21 ⟨88S545, 88TL5783⟩, 23 ⟨91CE185⟩, 24 ⟨90BCJ3033⟩, 28 ⟨87BCJ1159⟩, 149 ⟨89BCJ3376⟩, 320 ⟨89CL463⟩, 626 ⟨87BCJ4187, 87CL627⟩, 626 ⟨88BCJ597, 89BCJ3373⟩, 626 ⟨89BCJ439⟩, 631 ⟨87CL2109, 88BCJ600, 88CL795, 89BCJ1349⟩; **5**, 318 ⟨89JCS(P1)1702⟩
Kakinuma, K., **1**, 552 ⟨89TL4157⟩
Kakinuma, N., **6**, 560 ⟨87CPB156⟩
Kakiuchi, H., **2**, 35 ⟨79BCJ1653⟩
Kakiuchi, K., **1**, 561, 569 ⟨93S460⟩; **3**, 122 ⟨85CL1437⟩
Kakkar, V. V., **2**, 530 ⟨92TL4209⟩; **4**, 526, 527, 528 ⟨92TL4209⟩
Kako, M., **4**, 624 ⟨93OM1514⟩
Kakoi, H., **1**, 263 ⟨75JOC2009⟩
Kakui, T., **2**, 4 ⟨82OM369⟩
Kakui, T., **1**, 490 ⟨79TL1137⟩, 490, 515 ⟨79TL1141⟩, 490, 535 ⟨82OM542⟩; **2**, 4 ⟨78JA290, B-82MI 223-01⟩
Kakurai, T., **3**, 593 ⟨86JHC597⟩
Kakushima, M., **3**, 65 ⟨81TL979⟩
Kalabina, A. V., **2**, 434 ⟨78URP635102⟩
Kalabina, A. V., **5**, 656 ⟨87JGU1710, 87ZOB1912⟩
Kalabina, A. V., **6**, 562 ⟨84ZOR2114⟩
Kalamar, J., **5**, 1039 ⟨69CZP131547⟩
Kalaritis, P., **5**, 55 ⟨90OS(69)10⟩
Kalashnikova, N. A., **5**, 742, 751 ⟨72ZOR1246⟩
Kalashnikov, S. M., **6**, 479 ⟨90ZOB1826⟩
Kalaus, G., **5**, 567 ⟨93JOC6076⟩
Kalb, B., **4**, 806 ⟨73CB1601⟩
Kalbacher, H., **6**, 422 ⟨80CC1265⟩
Kalbag, S., **6**, 491 ⟨84IC3063⟩
Kalb, G. H., **4**, 195 ⟨48JOC223⟩
Kalbas, C., **3**, 386 ⟨92CC488⟩
Kälber, C., **2**, 569 ⟨79ZAAC(458)89⟩
Kalbfus, W., **5**, 936 ⟨72JOM(46)C15⟩, 1088 ⟨74JOM(65)51⟩
Kal'chenko, V. I., **4**, 1004 ⟨86ZOB805⟩; **5**, 863 ⟨80ZOB2413⟩
Kalcheva, V., **5**, 376 ⟨90AF1030⟩; **6**, 558 ⟨89MI 617-01⟩
Kalcheva, V. B., **6**, 502 ⟨73GEP1694210⟩
Kaldor, S. W., **2**, 155 ⟨91TL5043⟩; **3**, 493 ⟨90JA7001⟩
Kalè, V. N., **1**, 591, 593, 594, 716 ⟨81JOC231⟩
Kalenda, N. W., **3**, 711 ⟨58JOC1599⟩
Kaletta, B., **1**, 686 ⟨86T1693⟩
Kaleya, R., **2**, 20 ⟨77JA905⟩
Kalia, K. C., **6**, 650, 655 ⟨80MI 621-01⟩
Kalibabchuk, N. N., **2**, 408 ⟨81ZOB331⟩
Kalikhman, I. D., **2**, 741 ⟨89S446⟩; **3**, 252 ⟨76IZV2040⟩, 430 ⟨85ZOR766⟩, 483 ⟨77JOM(142)155⟩; **4**, 140 ⟨89JOM(361)147⟩, 312 ⟨84ZOR1502⟩, 628 ⟨76IZV1887⟩, 765 ⟨88IZV514⟩, 766 ⟨83IZV2153⟩, 786 ⟨88ZOR509⟩, 940 ⟨77IZV596⟩, 942 ⟨83ZOB1292⟩; **5**, 391, 718 ⟨87ZOB1199⟩, 432 ⟨82IZV949, 83JOM(248)51⟩, 432 ⟨83ZOR1838, 84MI 509-01, 84JGU426⟩, 718 ⟨84ZOB477, 87JGU1073, 88BAU378, 88IZV460⟩, 719 ⟨83JGU1851, 83JGU200, 83ZOB229⟩, 719, 720 ⟨83BAU1373, 83IZV1515, 83ZOB2051⟩, 924 ⟨88IZV1947, 89JOM(368)163⟩; **6**, 131 ⟨83JOM(248)51⟩, 185 ⟨77JOM(142)155⟩, 565 ⟨90ZOB602⟩
Kalikman, I. D., **4**, 942 ⟨90ZOB526⟩
Kalina, D. G., **2**, 602 ⟨82JA1860⟩
Kalinina, G. S., **4**, 637, 641 ⟨75JOM(96)213, 75ZOB1905⟩, 641 ⟨76IZV1342⟩, 660 ⟨72JOM(46)51⟩
Kalinina, O. N., **5**, 385 ⟨88ZOB62⟩
Kalinin, A. V., **2**, 394 ⟨86BAU2442⟩; **5**, 362 ⟨91IZV1108, 91MI 507-06⟩, 380 ⟨83IZV1413⟩, 919 ⟨70DOK(195)98⟩; **6**, 147 ⟨85IZV2635⟩
Kalinin, V. N., **3**, 531 ⟨71T1317⟩, 597 ⟨89JOM(379)303⟩; **4**, 794 ⟨80JOU1762⟩; **5**, 657, 659 ⟨79JOU32⟩, 657, 659 ⟨79ZOR38⟩; **6**, 406 ⟨73JGU848⟩
Kalinnikov, V. T., **4**, 717 ⟨80JOM(201)269⟩
Kalinov, S. M., **2**, 455 ⟨88JGU693⟩
Kalinovskii, I. O., **1**, 492 ⟨84IZV2580⟩, 1004, 1006 ⟨81IZV2836⟩; **3**, 290 ⟨82IZV221⟩
Kalinowski, H.-O., **1**, 210 ⟨77AG(E)264⟩; **2**, 1002 ⟨78HCA3100⟩
Kalinowski, H. O., **2**, 89 ⟨75AG(E)762⟩, 563 ⟨73HOU(13/2a)4, 73HOU(13/2a)533⟩
Kalinowski, H.-O., **4**, 19, 27, 29 ⟨83OMR(21)64⟩; **5**, 177 ⟨77AG(E)264⟩, 597 ⟨77CB3733⟩, 874 ⟨86AG(E)1093⟩; **6**, 731 ⟨71LA(743)1⟩, 732 ⟨69LA(727)228⟩
Kalir, A., **2**, 27 ⟨73OSC(5)825⟩
Kalish, N., **5**, 852 ⟨68JOC1679⟩
Kalish, R., **4**, 843 ⟨71TL2241⟩
Kalish, V. J., **3**, 140 ⟨90JA7441⟩
Kalkote, U. R., **5**, 384 ⟨91SC1889⟩
Kallel, A., **2**, 1007 ⟨89AX(C)1520⟩
Kallenbach, L. R., **2**, 498, 1080 ⟨71JINC3177⟩
Kallenberg, H., **1**, 873 ⟨78TL491⟩; **2**, 34 ⟨79TL2817⟩; **3**, 21 ⟨78TL491⟩, 27, 35 ⟨79TL2817⟩
Kaller, B. F., **1**, 196 ⟨93TL8047⟩, 231 ⟨91TL843⟩
Kaller, R., **6**, 169 ⟨92OM145⟩
Kalli, M., **1**, 1015 ⟨73JCS(P1)1347⟩
Kallio, R. E., **2**, 95 ⟨60JOC621⟩
Kallmerten, J., **1**, 643 ⟨87JOC4303⟩, 816 ⟨90TL4301⟩, 816, 819 ⟨88JOC4631⟩, 816, 819 ⟨93TL749⟩, 816, 819 ⟨93TL753⟩, 816, 831 ⟨88JOC5189, 93TL1103⟩
Kallury, M. R., **5**, 924 ⟨82JA5667⟩
Kallury, R. K., **5**, 923, 924 ⟨81CC191⟩
Kallweit, H., **1**, 692 ⟨92CB149⟩
Kalm, M. J., **2**, 29 ⟨63OSC(4)398, 63OSC(4)608⟩; **5**, 98 ⟨63OSC(4)608⟩, 99 ⟨63OSC(4)616⟩
Kálmán, A., **2**, 413 ⟨85JOC1271⟩, 1003 ⟨86JST141⟩, 1008 ⟨75CSC135⟩; **6**, 557 ⟨83T1829⟩
Kalman, J. R., **1**, 490 ⟨74TL3391⟩
Kalmus, A., **1**, 108 ⟨54JA4137⟩; **5**, 98 ⟨52JCS4521⟩
Kalnins, M. V., **5**, 336 ⟨66CJC2111⟩, 745 ⟨61JOC2563⟩
Kaloustian, M. K., **5**, 512 ⟨79JA2249, 80JA7579⟩, 512 ⟨81TL413⟩, 560 ⟨73JCS(P1)1571⟩, 576 ⟨82SC787⟩; **6**, 96 ⟨80JA7579, 86JA6683⟩
Kalsotra, B. L., **2**, 565 ⟨73JINC311⟩
Kaltenbach, R. F., **1**, 831 ⟨90TL5285⟩
Kaltschmitt, H., **3**, 547 ⟨33CB237⟩
Kaluski, Z., **2**, 1008 ⟨86MI 220-1⟩
Kaluszyner, A., **4**, 743 ⟨62JOC1426⟩, 751 ⟨57JA4174⟩, 826 ⟨56JOC801⟩
Kaluza, L., **5**, 1023 ⟨09M717, 12M363⟩
Kaluza, Z., **4**, 174 ⟨84TL4797⟩; **5**, 344 ⟨85T2441⟩
Kalvin, D., **3**, 704 ⟨85SC267⟩
Kalvin, D. M., **3**, 13 ⟨84T3387⟩
Kalvins, I., **4**, 514 ⟨84JOM(268)C29⟩
Kalvoda, J., **1**, 927 ⟨64AG(E)525⟩, 1082 ⟨67HCA2101⟩; **2**, 94 ⟨71S501⟩, 95 ⟨61E475⟩; **3**, 226 ⟨67HCA269⟩
Kalwinsch, I., **1**, 367 ⟨84BSB511⟩; **3**, 367 ⟨81JA7032⟩
Kalyanam, N., **5**, 275, 295 ⟨82JOC4812⟩
Kalyanaraman, V., **2**, 989 ⟨73JOM(47)225⟩
Kamabuchi, A., **4**, 815 ⟨93SC2851⟩
Kamada, M., **6**, 447 ⟨92TL3483⟩, 623 ⟨76CPB26⟩
Kamada, T., **2**, 323 ⟨90JA2368⟩; **3**, 43 ⟨90JA2368⟩, 618 ⟨89JOC2249⟩; **4**, 197 ⟨89TL5309⟩
Kamai, G., **2**, 414, 415 ⟨55DOK(105)489⟩, 414, 415 ⟨56ZOB126⟩, 436 ⟨62ZOB2848⟩, 442 ⟨56ZOB3426⟩, 461 ⟨29MI 208-01⟩, 496, 497 ⟨63JGU2831⟩, 500 ⟨57DOK(115)305, 57IZV69⟩, 809 ⟨67CA76112w⟩, 838, 839 ⟨50MI 216-01⟩, 881, 883 ⟨72IZV1612⟩, 889 ⟨61JGU3315, 64DOK(157)371⟩; **6**, 242 ⟨46ZOB1521⟩
Kamai, G. H., **2**, 442 ⟨32JGU524⟩
Kamai, G. K., **2**, 501 ⟨73JGU825⟩; **4**, 340 ⟨70ZOB2560⟩
Kamai, G. Kh., **2**, 461 ⟨65URP173230, 66ZOB1436⟩, 889 ⟨66JGU1986, 67JGU1578⟩, 889 ⟨71IZV1726⟩, 890

⟨68JGU370⟩; **4**, 120 ⟨71JGU2183⟩, 993 ⟨68ZOR2116⟩, 993, 994 ⟨71ZOR1670⟩; **5**, 403 ⟨71JGU2513⟩
Kamai, Kh. G., **4**, 911 ⟨68ZOB1523⟩
Kamaike, K., **2**, 107 ⟨83CPB2157⟩
Kamakura, A., **3**, 62, 72, 271 ⟨78TL3047⟩
Kamal, A., **2**, 318 ⟨92TA1361⟩; **3**, 18, 134 ⟨91JCS(P1)2056⟩, 18, 134 ⟨91TL2657⟩; **5**, 344 ⟨87H(26)1051⟩; **6**, 451 ⟨90H(31)1377⟩
Kamal, R., **6**, 558 ⟨87MI 617-02⟩
Kamalov, R. M., **2**, 259 ⟨90ZOB476⟩; **5**, 581 ⟨86ZOB2660⟩, 644 ⟨86ZOB2359⟩, 1122 ⟨91IZV451⟩, 1135 ⟨90ZOB476⟩; **6**, 121 ⟨92HAC115⟩, 538, 539 ⟨89IZV909⟩, 657 ⟨90ZOB778⟩
Kamat, R. J., **1**, 1124 ⟨66JA3152, 69JA4521⟩
Kamata, I., **5**, 706 ⟨85H(23)281⟩
Kamata, K., **2**, 365 ⟨70CC777⟩
Kamata, M., **4**, 260 ⟨92TL5085⟩
Kamata, S., **3**, 634 ⟨73SC265⟩; **5**, 234, 277 ⟨75JA3515⟩, 235 ⟨75CJC3693⟩; **6**, 563 ⟨91ANC1295⟩
Kamata, T., **2**, 326 ⟨93JMC3526⟩
Kamath, K. M., **4**, 54 ⟨71JCS(C)1772⟩, 58 ⟨71JCS(C)3344⟩
Kambe, N., **1**, 245, 437 ⟨87AG(E)1187⟩, 247 ⟨93OM473⟩, 666 ⟨92JOC17⟩; **2**, 281 ⟨88S175⟩, 554 ⟨87AG(E)1187⟩, 555 ⟨91AG(E)177⟩, 710 ⟨92JA5902⟩, 730 ⟨92JA7591⟩; **3**, 38, 174 ⟨82TL4813⟩, 136 ⟨90JA455⟩, 143 ⟨88TL6137⟩, 176 ⟨87S1096⟩, 177, 224 ⟨91TL229⟩, 635 ⟨85JOC2240⟩; **4**, 239, 334, 660 ⟨90OM1355⟩, 334, 909 ⟨90CL2053⟩, 652 ⟨87AG(E)1187⟩, 940 ⟨92JA5902⟩, 940, 947 ⟨91JA9796, 91JOC5721⟩, 948 ⟨92JA7591⟩; **5**, 238, 241 ⟨91JA9796⟩, 437 ⟨90JA455⟩, 625, 626 ⟨85BCJ1448⟩, 1036 ⟨91TL3503⟩; **6**, 479 ⟨88TL4767, 90TL4773⟩, 495, 497 ⟨89AG(E)452⟩, 498 ⟨90SC703⟩
Kambe, S., **4**, 437 ⟨84S259⟩, 437, 438 ⟨84S256⟩, 437, 445 ⟨77S647⟩, 437, 445 ⟨82S1080⟩, 444 ⟨85S100, 81S151⟩; **5**, 606 ⟨85S92⟩
Kambouris, P., **3**, 686 ⟨91TL1487⟩
Kambulova, I. U., **2**, 1076 ⟨88JGU2050⟩
Kameda, K., **1**, 455 ⟨79MI 110-01⟩
Kamei, Y., **1**, 690 ⟨83BCJ192⟩
Kamekawa, K., **5**, 226 ⟨81CL1367⟩
Kamel, M., **6**, 38 ⟨76CB2351⟩
Kamelani, T., **5**, 179 ⟨85CPB532⟩
Kamenar, B., **6**, 405 ⟨74CC646⟩
Kamenasa, S., **3**, 269 ⟨89BCJ860⟩
Kamenecka, T. M., **3**, 198 ⟨91TL2371⟩
Kameswaran, V., **5**, 1126 ⟨89USP4857550⟩
Kametani, S., **2**, 464 ⟨82TL1263⟩; **3**, 352, 353 ⟨88CL717⟩, 353, 360 ⟨82TL1263, 82CL793⟩, 360 ⟨85BCJ667, 92BCJ923⟩
Kametani, T., **1**, 829 ⟨88CC1531⟩; **2**, 1000 ⟨87JA3010⟩; **3**, 9, 123 ⟨89CL901⟩, 222 ⟨90JCS(P1)1733, 90JCS(P1)639⟩, 294 ⟨87CC1360⟩, 620 ⟨86JOC2833⟩, 665 ⟨77S245⟩; **4**, 154 ⟨85JCS(P1)1185⟩, 286 ⟨81T2547⟩, 325 ⟨78H(9)831⟩, 387 ⟨83TL4257⟩; **5**, 159 ⟨88CC1531⟩, 227 ⟨85H(23)549⟩, 276 ⟨69JCS(C)874⟩, 285 ⟨83JA7345⟩; **6**, 511 ⟨70JHC807⟩
Kamey, G., **5**, 827 ⟨70MI 521-01⟩
Kameyama, K., **3**, 91 ⟨88CPB974⟩
Kameyama, M., **1**, 485 ⟨88BCJ1231⟩; **2**, 620 ⟨88BCJ2226⟩, 753 ⟨83CL927⟩; **6**, 7, 24 ⟨87BCJ3687⟩
Kameyama, Y., **1**, 496 ⟨92TL7029⟩
Kamienska-Trela, K., **3**, 338 ⟨72OMR(4)421⟩, 355 ⟨71BAP549⟩
Kamigata, N., **1**, 241 ⟨89JOC2374⟩, 485 ⟨85JOC5045, 88BCJ1231⟩; **2**, 214 ⟨86SUL1⟩, 235 ⟨86BCJ2179, 88BCJ2226⟩, 620 ⟨88CL2017⟩, 997, 1001 ⟨92JCS(P1)1721⟩; **4**, 56 ⟨89CC1559⟩; **5**, 208 ⟨90JFC(46)423⟩, 334 ⟨83OPP315⟩; **6**, 7, 24 ⟨87BCJ3687⟩, 12 ⟨90JFC(46)423⟩, 21 ⟨89CC1559, 83SUL131⟩
Kamijo, T., **2**, 32, 36 ⟨83CPB4189⟩; **3**, 618 ⟨84CPB2560⟩; **5**, 199 ⟨90JOC1679⟩
Kamikawaji, Y., **1**, 235 ⟨84JOC1985⟩
Kamikubo, T., **1**, 596 ⟨92TA853⟩
Kamil, W. A., **4**, 140 ⟨86IC376⟩; **6**, 229, 230 ⟨86IC376⟩
Kamimura, A., **1**, 37 ⟨85JOC3692⟩, 37, 330, 443, 548, 946, 1131 ⟨85T4013⟩, 250 ⟨91S423⟩, 250, 251, 486 ⟨86CC1285⟩, 252 ⟨87JOC4133, 89JOC4998⟩, 592, 638 ⟨84TL5319⟩, 592, 638 ⟨88CC1278⟩, 600, 617, 638 ⟨88JOC251⟩, 617, 638 ⟨87JOC5111⟩, 946 ⟨88BCJ3629⟩; **2**, 393 ⟨90JOC2437, 90S694⟩, 394 ⟨87S258⟩, 395 ⟨84S226⟩, 396, 789 ⟨86JOC2139, 88S921⟩; **3**, 46 ⟨90S694⟩, 192 ⟨87S258⟩
Kamimura, J., **1**, 20 ⟨87S736⟩
Kaminskaya, E. I., **2**, 239 ⟨90ZOR2083⟩; **5**, 337 ⟨86ZOR100⟩
Kaminski, C. W., **2**, 964 ⟨70C109⟩
Kaminski, F. E., **1**, 881 ⟨68OS(48)53⟩; **3**, 626 ⟨73OSC(5)263⟩
Kaminski, J., **5**, 671 ⟨81PJC655⟩
Kaminski, J. J., **4**, 58, 211 ⟨83JOC5280⟩
Kaminski, M., **6**, 39 ⟨83JFC(22)207⟩, 52 ⟨82PJC1369⟩, 440 ⟨83JFC(23)207⟩
Kaminski, Z. J., **4**, 213 ⟨86S649⟩
Kaminsky, D., **2**, 692 ⟨67TL859⟩
Kaminsky, L. S., **5**, 283 ⟨67JCS(C)2128⟩
Kamitori, S., **4**, 1051, 1053 ⟨85OM1690⟩
Kamitori, Y., **3**, 253 ⟨75TL1009⟩; **4**, 247 ⟨86JOC1427⟩
Kamiya, S., **5**, 1118, 1124 ⟨82CPB3125⟩; **6**, 504 ⟨83CPB41⟩
Kamiya, T., **3**, 427 ⟨84JAN532⟩; **6**, 429 ⟨77BCJ718⟩
Kamiya, Y., **2**, 69 ⟨89JA4392⟩; **3**, 319 ⟨82CC1223⟩; **5**, 164 ⟨90CL1549⟩
Kamiyama, K., **5**, 135, 138 ⟨84TL2557⟩
Kamiyama, Y., **2**, 916 ⟨75JA931⟩; **4**, 612, 614 ⟨76JA7424⟩, 1048 ⟨77JA3879⟩
Kamlet, J., **6**, 236 ⟨61USP3014071⟩
Kamlet, M. J., **4**, 297 ⟨61JOC4029⟩; **6**, 56, 261, 279, 287 ⟨68JOC3073⟩, 57 ⟨69JOC45⟩, 144 ⟨62JOC780⟩, 170 ⟨61JOC4734⟩, 297 ⟨68USP3388147⟩
Kamm, O., **2**, 31 ⟨41OSC(1)25⟩; **5**, 106 ⟨32OSC(1)391⟩, 115 ⟨32OSC(1)392⟩
Kammel, G., **2**, 163 ⟨71CB3234⟩
Kammerer, C., **2**, 632 ⟨90JOC4560⟩
Kämmerer, H., **1**, 2, 6, 8, 10 ⟨59CB1700⟩
Kammerl, E., **5**, 1025 ⟨54LA(585)230, 55LA(594)233⟩
Kammeyer, C. W., **4**, 28 ⟨51JA4476⟩
Kamogawa, H., **5**, 1065 ⟨79BCJ533⟩
Kamoriya, A., **5**, 720 ⟨72JA5285⟩
Kampchen, T., **4**, 896 ⟨88AP(321)873⟩; **5**, 685 ⟨89CB669⟩
Kampe, K. D., **5**, 998 ⟨69TL117, 71LA(752)142⟩
Kamper, F., **2**, 20 ⟨76AG(E)306⟩
Kampf, J. W., **2**, 896 ⟨91OM2068⟩; **4**, 1038 ⟨92JA372⟩, 1038, 1041 ⟨92OM1491⟩, 1041 ⟨91OM2068, 92OM2743⟩
Kamphuis, J., **3**, 40 ⟨94TL1581⟩, 340 ⟨84TL873⟩
Kampmeier, J. A., **1**, 1118 ⟨65JA863⟩; **2**, 800 ⟨65JA863⟩, 800, 801 ⟨65JA874⟩, 957 ⟨81JA1847⟩; **3**, 263 ⟨65JA874⟩
Kamyshova, A. A., **4**, 137 ⟨87IZV1087⟩
Kanabe, N., **2**, 311 ⟨87CL1275⟩
Kanabus-Kaminska, J., **1**, 5, 7, 9, 10, 17, 25 ⟨92JOC2427⟩, 331, 936 ⟨87JA5267⟩
Kanai, K., **3**, 222 ⟨90JCS(P1)1733⟩
Kanai, K.-I., **3**, 439, 440, 456 ⟨83JOC289⟩
Kanai, M., **6**, 21 ⟨91TL7425⟩
Kanai, T., **2**, 36 ⟨89S283⟩
Kan, K., **4**, 1003, 1004 ⟨75BCJ2103⟩
Kanakura, A., **3**, 62 ⟨78TL3051⟩; **4**, 652 ⟨82TL1279⟩
Kanal, S., **5**, 1035 ⟨79JCS(P1)2001⟩
Kanamori, F., **4**, 774 ⟨91TL6563⟩
Kanamueller, J. M., **2**, 386 ⟨70JCED356⟩
Kanaoka, Y., **4**, 1077 ⟨72TL4517, 78ACR407⟩; **5**, 572 ⟨92S778⟩, 631, 632 ⟨85JOC1681, 86S768⟩
Kan, R. O., **5**, 37 ⟨73OSC(5)612⟩, 1047 ⟨64JOC2261⟩
Kan, T., **5**, 129 ⟨94JOC5532⟩

Kanatani, R., **1**, 160 ⟨84TL1905⟩, 160 ⟨84TL1909⟩, 160 ⟨84TL1913⟩; **2**, 923, 924 ⟨82JOM(226)C9⟩; **4**, 651, 655 ⟨84TL1913⟩, 654 ⟨84TL1905⟩
Kanazawa, R., **3**, 84 ⟨76S526⟩
Kanazawa, S., **1**, 827 ⟨88JOC3868⟩
Kanazawa, Y., **5**, 353 ⟨87S183⟩
Kanda, T., **3**, 173 ⟨93CC1631⟩; **5**, 242, 250, 506, 509 ⟨87CL475⟩, 242, 506, 509 ⟨89TL1829⟩, 250 ⟨94S282⟩, 255, 506, 509, 543 ⟨90TL3587⟩, 516, 523, 563 ⟨93JA5823⟩, 542 ⟨93CC277, 93CL1469⟩, 563, 626 ⟨93JA3000⟩
Kanda, Y., **1**, 816, 818 ⟨86JOC4315, 87H(25)139⟩; **4**, 948 ⟨92TL5701⟩
Kanda, Z., **4**, 586 ⟨79BCJ2560⟩
Kandasamy, D., **1**, 4, 6, 7, 9, 10, 11, 17 ⟨77JOC82⟩, 4, 7, 9, 17, 28 ⟨78JOC2259⟩
Kandeel, E. M., **5**, 282 ⟨91TL1825⟩
Kandgetcyan, R. A., **3**, 177, 637 ⟨72JOM(42)399⟩
Kandil, A. A., **1**, 787 ⟨87JOC5034⟩; **3**, 639 ⟨87S411⟩; **4**, 362 ⟨86TL3831⟩, 363 ⟨87JOC5121⟩
Kandil, S. A., **1**, 539 ⟨66JA3027⟩
Kandratenko, N. V., **2**, 881 ⟨76JGU2477⟩
Kandror, I. I., **4**, 899, 902 ⟨81CA96556w⟩; **5**, 223 ⟨90ZOB1928⟩, 727 ⟨82IZV2121⟩
Kandul, J. V., **2**, 453 ⟨78BRP1517876⟩
Kane, B. J., **1**, 797 ⟨80JOC900⟩
Kane, J. M., **5**, 351 ⟨92SC1⟩, 557 ⟨83JA5015⟩
Kane, V. V., **2**, 619 ⟨67JA4487⟩; **3**, 126 ⟨83T345⟩; **4**, 82 ⟨67JA4487⟩
Kaneda, A., **5**, 731 ⟨67BCJ2705⟩
Kaneda, K., **1**, 1004, 1107 ⟨79JOC55, 74TL1067⟩; **2**, 91 ⟨79JA159⟩, 314 ⟨91BCJ602⟩, 314, 318 ⟨88CL1759⟩, 616 ⟨74TL1067, 76CL449⟩, 654 ⟨94CC797⟩; **3**, 117 ⟨83TL5009⟩, 618 ⟨88CL285⟩, 620 ⟨85CL1339⟩
Kaneda, M., **3**, 121 ⟨76CL231⟩
Kaneda, Y., **3**, 418 ⟨76JAP7632569⟩
Kanefusa, T., **5**, 248 ⟨85S878⟩
Kanegasaki, S., **6**, 487 ⟨93CL1401⟩
Kanehira, K., **2**, 437 ⟨83JOC2195⟩
Kanehisa, N., **1**, 992 ⟨90CL2149⟩
Kaneko, C., **2**, 16 ⟨92H(33)105⟩, 17 ⟨94CPB459⟩, 39 ⟨90TL7463⟩, 176, 180 ⟨74S876⟩; **3**, 543 ⟨84H(22)2563, 85S184⟩, 543 ⟨90CL1705, 91T5689⟩, 543 ⟨92S977⟩; **4**, 265, 853, 854 ⟨90CPB3242⟩; **5**, 367 ⟨90CPB3184⟩, 652 ⟨88CPB3354, 89CPB2952⟩
Kaneko, H., **6**, 465 ⟨93JAP05255201, 93JAP05310644⟩
Kaneko, I., **2**, 853 ⟨88BCJ4029⟩; **3**, 394 ⟨91CL1053, 92TL1333⟩
Kaneko, K., **2**, 150 ⟨77TL1345⟩; **4**, 198 ⟨78TL4115⟩
Kaneko, T., **1**, 811 ⟨73TL3971, 73TL3975⟩, 811 ⟨74JOC3175⟩; **2**, 658 ⟨71BCJ1986⟩; **3**, 378 ⟨74JOC3175⟩; **5**, 334, 335 ⟨85JA5490⟩, 700, 701 ⟨90MI 517-01⟩
Kaneko, Y., **1**, 515 ⟨85T2387⟩; **2**, 67 ⟨94TL4107⟩, 621, 627 ⟨84CC1523⟩; **3**, 136, 203 ⟨90JOC5562⟩, 168 ⟨86TL3029⟩, 183 ⟨86BCJ457⟩, 634 ⟨87CL1963⟩; **4**, 74 ⟨85BCJ1983⟩, 202 ⟨91JOC2276, 93TL1491⟩, 233 ⟨86BCJ2463⟩, 369 ⟨87CL1963⟩
Kane-Maguire, L. A. P., **6**, 455, 517 ⟨86JOM(312)C21⟩
Kanemasa, S., **1**, 372 ⟨87JOC2523⟩, 552 ⟨89AHC(45)231⟩; **2**, 667 ⟨84CL277⟩, 908 ⟨88BCJ3957⟩, 930, 939 ⟨90BCJ51⟩; **3**, 200 ⟨87BCJ2463⟩, 407, 586 ⟨84JOC2688⟩, 409 ⟨86BCJ1809⟩, 683 ⟨90H(30)719⟩; **4**, 263 ⟨90BCJ466⟩, 302 ⟨87BCJ4079⟩, 367, 936, 938 ⟨88BCJ3957⟩, 367, 946 ⟨90BCJ51⟩, 421 ⟨75CL607⟩, 506, 513 ⟨85BCJ1991⟩, 512, 518 ⟨86BCJ2537⟩, 515 ⟨84H(22)1955⟩, 516 ⟨83CL1131⟩, 517 ⟨84JOC2688⟩, 517, 518 ⟨85H(23)2489⟩, 519 ⟨84CL279⟩, 979 ⟨86JOC1997⟩; **5**, 729 ⟨87JOC2523⟩, 1031 ⟨84JOC2688⟩; **6**, 620 ⟨86JOC1719⟩
Kanemasa, Sh., **5**, 821 ⟨87BCJ2463⟩
Kanematsu, K., **1**, 235 ⟨84JOC1985⟩; **4**, 171 ⟨86TL947⟩
Kanemitsu, K., **6**, 541 ⟨86CPB430⟩
Kanemoto, N., **1**, 290 ⟨88JOM(25)348⟩; **4**, 626, 662 ⟨87SC1273⟩, 638, 663 ⟨88JOM(348)25⟩; **6**, 185 ⟨87SC1273⟩
Kanemoto, S., **1**, 491 ⟨87CL5⟩; **2**, 11 ⟨87TL663, 89BCJ2024⟩, 918, 931 ⟨83JA4491⟩; **3**, 5, 117 ⟨88BCJ3607⟩, 116, 161 ⟨84TL3317⟩, 118, 121 ⟨89BCJ519⟩, 158, 161 ⟨86BCJ105⟩
Kanerva, L. T., **3**, 628 ⟨92TA677⟩
Kaneta, N., **2**, 912, 945 ⟨91JOC3846⟩
Kanetaka, N., **4**, 80 ⟨93BCJ2339⟩; **5**, 272, 283 ⟨92TL1455⟩
Kaneti, J., **3**, 494 ⟨89JCS(P2)1405⟩
Kan-Fan, C., **3**, 419 ⟨76BSF1222⟩
Kanfer, S. J., **1**, 473 ⟨80JA790⟩
Kang, D. K.,**6**, 264 ⟨72CC673⟩
Kang, H.-J., **1**, 826 ⟨90JA3252, 91JA2610⟩
Kang, H.-Y., **1**, 1099 ⟨87JOC4823⟩; **2**, 81 ⟨90TL4765⟩
Kang, J., **1**, 101 ⟨79JOC3737, 83JOC2796⟩, 238 ⟨90TL2713⟩, 630 ⟨92MI 114-01⟩, 1015, 1016 ⟨81JA4618⟩, 1015, 1016 ⟨82TL1651⟩; **2**, 922, 923 ⟨84JOC1838⟩; **4**, 849 ⟨90JOC5555⟩; **5**, 418, 428 ⟨87TL3261⟩; **6**, 705 ⟨82JA4724, 85TL555⟩
Kang, J. W., **2**, 985 ⟨68IC1298⟩
Kang, K. K., **2**, 25 ⟨88JOC3634⟩
Kang, K.-T., **3**, 359 ⟨79BCJ496, 82AOC(21)241⟩, 359, 361 ⟨79BCJ3640⟩, 363 ⟨91MI 308-01⟩; **4**, 139 ⟨92MI 403-01⟩, 979 ⟨79BCJ3640⟩; **5**, 648 ⟨91MI 516-01⟩; **6**, 128 ⟨91MI 604-01⟩
Kang, M., **1**, 1029 ⟨88JA649⟩
Kang, M.-C., **2**, 679 ⟨88JA649⟩
Kang, S., **1**, 1066 ⟨90SL591⟩; **5**, 970 ⟨65JOC1158, 66JOC142⟩
Kangas, P., **2**, 238 ⟨91PS281⟩
Kang, S. H., **1**, 697, 700, 703 ⟨88TL5169⟩
Kang, S.-I., **2**, 281 ⟨88S133⟩, 734 ⟨86JOC295⟩; **5**, 1113 ⟨88S133⟩
Kang, S.-K., **1**, 15, 16 ⟨93JCS(P1)9⟩, 967 ⟨92TA1509⟩; **3**, 31 ⟨91TL367⟩; **6**, 468 ⟨94SC305⟩
Kang, Y., **5**, 215 ⟨87JHC597⟩
Kang, Y. B., **2**, 438 ⟨94CC475⟩
Kang, Y.-H., **2**, 285 ⟨82TL5373, 85T4739⟩
Kanger, T., **1**, 225 ⟨91TA943⟩
Kanghae, W., **3**, 62 ⟨77TL1377⟩, 153 ⟨77TL1225⟩; **4**, 77 ⟨77TL1225⟩
Kania, R., **2**, 133 ⟨91JA3177⟩
Kanig, W., **2**, 489 ⟨82PS(12)149⟩, 507 ⟨78CZ263, 86OCS621⟩; **6**, 168, 192 ⟨83POL291⟩
Kanishchev, M. I., **3**, 75 ⟨88CCC1519⟩
Kanitz, A., **4**, 880 ⟨85JPR297⟩
Kanitz, R., **6**, 455, 517 ⟨86JOM(312)C21⟩
Kanjia, M., **3**, 172 ⟨94T827⟩
Kannan, A., **2**, 1034, 1035 ⟨89CC154⟩
Kannengiesser, G., **4**, 417 ⟨72CJC2902⟩
Kanno, M., **3**, 621 ⟨76BCJ2643⟩
Kanno, T., **1**, 690 ⟨83BCJ2023⟩
Kannonikoff, J., **5**, 652 ⟨1877LA(185)191⟩
Kano, H., **3**, 454 ⟨54CA(48)3342⟩
Kano, K., **1**, 943 ⟨90JCS(P1)3017⟩; **2**, 403 ⟨88BSB461⟩, 579 ⟨92CC1728⟩
Kano, S., **1**, 41 ⟨78CC414⟩, 651 ⟨78S746⟩, 1077 ⟨78JOC4366, 78S305⟩; **3**, 634 ⟨79H(12)681⟩, 654 ⟨78CPB1874⟩; **4**, 648 ⟨78S746⟩; **5**, 284 ⟨87H(26)359⟩, 570, 613 ⟨87H(26)359⟩
Kano, T., **2**, 442 ⟨71JA4301⟩, 715 ⟨84S323⟩; **3**, 309 ⟨81JOC2979⟩
Kano, Y., **3**, 352 ⟨90CL563⟩
Kanofsky, J. R., **3**, 113 ⟨90JA1936⟩
Kanomata, N., **2**, 468 ⟨92TL251⟩
Kanou, K., **1**, 882 ⟨87TL5887⟩
Kanouchi, S., **4**, 612 ⟨74JA5623⟩
Kansal, N. M., **6**, 264 ⟨80IJC(B)610⟩
Kant, J., **1**, 496 ⟨92TL3563⟩, 939 ⟨93JOC5028⟩; **5**, 558 ⟨85JHC1065⟩
Kant, M., **2**, 105 ⟨85EGP231074, 90ZAAC(580)217⟩, 252 ⟨86GEP231075⟩

Kantam, M. L., **3**, 5, 118 ⟨85TL6257⟩; **5**, 1106 ⟨90SC2313⟩
Kantardjiew, I., **3**, 494 ⟨78CB248⟩
Kanter, F. J. J. de, **2**, 925 ⟨92T6087, 92T6105⟩; **3**, 710 ⟨91JA6104⟩
Kanter, J. P., **1**, 629 ⟨91JA6950⟩
Kanters, J. A., **2**, 999, 1000 ⟨89OM1688⟩; **3**, 556 ⟨89AX(C)482⟩
Kanth, J. V. B., **2**, 59 ⟨91JOC5964⟩
Kantlehner, W., **2**, 788 ⟨71CB3475⟩, 812 ⟨80S460⟩; **3**, 266 ⟨83LA290⟩, 419 ⟨91COS(3)485⟩, 627, 637 ⟨71CB924⟩; **4**, 184 ⟨79LA1362⟩, 184 ⟨79LA522⟩, 881 ⟨71CB3475, B-79MI 420-01⟩, 894, 973, 974 ⟨79LA2096⟩, 972, 979 ⟨80LA372⟩, 973, 974 ⟨83LA290⟩, 975 ⟨90LA965⟩; **5**, 609 ⟨82LA1582⟩, 654 ⟨91COS(6)523⟩, 685, 687, 688, 689, 690, 691, 692, 693, 694 ⟨91COS(6)485⟩, 688 ⟨B-79MI 517-01⟩, 726 ⟨91COS(6)536, B-79MI 518-01⟩, 744 ⟨83S35⟩, 747 ⟨80LA344⟩; **6**, 51, 57 ⟨B-79MI 602-03⟩, 51, 57 ⟨B-79MI 602-05⟩, 51 ⟨B-79MI 602-01, B-79MI 602-02⟩, 51 ⟨B-79MI 602-04⟩, 70 ⟨80LA1677⟩, 78 ⟨81S380⟩, 104 ⟨68CB41⟩, 105, 109 ⟨72CB1340⟩, 115 ⟨88LA605⟩, 138 ⟨79S342, B-79MI 605-01⟩, 139 ⟨90CZ176⟩, 140 ⟨79GEP2824028⟩, 140 ⟨83S905⟩, 143 ⟨82CB1721⟩, 296, 297, 298, 300, 313, 320, 321 ⟨77S73⟩, 296, 297, 298, 300, 321 ⟨82LA507, 90LA965⟩, 321, 337 ⟨79LA2089⟩, 643 ⟨84LA108⟩, 647 ⟨79LA2096⟩, 728 ⟨79S339, B-79MI 623-01⟩, 728 ⟨B-79MI 623-02⟩, 729 ⟨81LA70⟩
Kantlehuer, W., **4**, 881 ⟨91COS(6)485⟩
Kantor, E. A., **2**, 265 ⟨87ZOR1981⟩; **6**, 70 ⟨81MI 603-01⟩, 89 ⟨88URP1421744⟩
Kantor, S. W., **1**, 396, 397, 404, 816 ⟨51JA1437⟩, 824 ⟨51JA4122⟩; **5**, 998 ⟨50JA4284⟩
Kanunnikova, E. N., **2**, 886 ⟨81JGU1056⟩
Kanvinde, M. N., **3**, 293 ⟨90SC3259⟩
Kanzaki, S., **1**, 178 ⟨86BCJ3097⟩
Kanzel-Szvoboda, I., **5**, 220 ⟨89CB1107⟩
Kao, L.-C., **1**, 1096 ⟨81JOC1061, 82JOC1267⟩
Kao, L. T. C., **4**, 583, 585 ⟨86JOM(309)225⟩
Kao, S. C., **3**, 83 ⟨84OM1601⟩; **4**, 679 ⟨82OM911⟩; **6**, 517 ⟨79JA1627⟩
Kao, W. L., **1**, 1023 ⟨70TL2695⟩
Kapassakalidis, J. J., **6**, 78 ⟨81S380⟩, 296, 297, 298, 300, 321 ⟨82LA507⟩, 643 ⟨84LA108⟩, 729 ⟨81LA70⟩
Kapaun, G., **4**, 979 ⟨71CB792⟩
Kapicak, L. A., **2**, 875 ⟨67TL4447⟩
Kaplan, E., **1**, 664 ⟨68C338⟩
Kaplan, G. L., **1**, 1029 ⟨89JMC1820⟩
Kaplan, H., **3**, 109 ⟨41JA2654⟩
Kaplan, J., **1**, 334 ⟨89JOC1826⟩
Kaplan, J. K., **1**, 838 ⟨90JA1266⟩
Kaplan, L. A., **2**, 392 ⟨77JOC563⟩; **6**, 144 ⟨62JOC780⟩, 144, 145 ⟨B-70MI 605-02⟩
Kaplan, L., **1**, 321 ⟨B-73MI 108-01⟩; **2**, 1010 ⟨B-73MI 220-1⟩
Kaplan, L. J., **1**, 576 ⟨77OR(25)1⟩; **2**, 843, 845 ⟨79JOC2226⟩
Kaplan, R. B., **4**, 428 ⟨61JA3535⟩; **6**, 150 ⟨61JA3535⟩
Kapnang, H., **2**, 303 ⟨83TL3233⟩, 310 ⟨77TL3469⟩; **4**, 300 ⟨80TL2949⟩
Kapon, M., **3**, 320 ⟨82T1477⟩, 502 ⟨92OM2326⟩; **4**, 267 ⟨90JCS(P2)1987⟩
Kapoor, P. N., **2**, 439 ⟨71JA4158⟩
Kapoor, R. P., **1**, 556 ⟨92JCS(P1)2565⟩; **2**, 311 ⟨91SL668⟩
Kapovits, I., **2**, 1003 ⟨86JST141⟩
Kapp, D. C., **5**, 360 ⟨79S100⟩
Kapp, D. L., **1**, 3, 6 ⟨86JA1265⟩
Kapp, M., **6**, 575 ⟨71ZC12⟩
Kappanna, A. N., **4**, 28 ⟨51JIC675⟩
Kappe, C. O., **3**, 584 ⟨92CC487, 92CC488⟩; **4**, 432 ⟨89JHC1555⟩, 894 ⟨92CC487⟩; **5**, 698 ⟨92CC1571⟩
Kappe, T., **4**, 411 ⟨68M990, 89JHC1555⟩
Kappeler, H., **5**, 32 ⟨61HCA1991⟩
Käppeler, W., **5**, 597, 599, 614 ⟨64LA(673)132⟩
Kappenstein, C., **5**, 1092 ⟨81CB3070⟩
Kappert, M., **5**, 145, 155 ⟨83AG(E)63⟩, 146 ⟨85TL1631⟩
Kappes, D., **5**, 66 ⟨92OS(71)48⟩
Kappos, J. C., **1**, 1063 ⟨87TL4921⟩
Kapteijn, F., **1**, 674 ⟨77CC198⟩
Kaptein, B., **3**, 40 ⟨94TL1181⟩Kapteyn, H., **3**, 78 ⟨73CB2755⟩
Kapur, J. C., **5**, 274, 286 ⟨73JOC3437⟩, 294 ⟨74TL3135⟩
Kapura, A. A., **4**, 923 ⟨83PS(16)345⟩
Kapvits, I., **2**, 160 ⟨77PS(3)9⟩
Karabatsos, G. J., **3**, 446 ⟨63JA2784⟩, 446, 448 ⟨63JA3624⟩
Karabatsos, G. L., **3**, 446 ⟨62JA753⟩
Karabelas, K., **1**, 490 ⟨86JOC5286⟩; **2**, 932 ⟨86JOC5286⟩
Karabinos, J. P., **2**, 214 ⟨63JOC3246⟩
Karady, S., **2**, 342, 346 ⟨88JOC1768⟩; **4**, 48 ⟨75TL3979, 79S970⟩, 302 ⟨84TL4337⟩
Karaeva, O. V., **5**, 369 ⟨90ZOB1183⟩
Karafiat, U., **5**, 640 ⟨92S637⟩
Karafiath, E., **1**, 809 ⟨69JA522⟩
Karaghiosoff, K., **5**, 635 ⟨86PS(27)289⟩
Karagiosov, S. K., **2**, 309 ⟨89AP(322)181⟩
Karakasa, T., **2**, 464 ⟨82TL1263⟩; **3**, 353 ⟨92MI 308-01⟩, 353, 360 ⟨82TL1263, 78JOC4147⟩, 360 ⟨79JOC4151, 80JOC927⟩, 360 ⟨86BCJ3279, 86BCJ335⟩, 360 ⟨89MI 308-03, 90CC1665⟩, 360 ⟨91JCS(P1)2281, 92JCS(P1)2943⟩, 370 ⟨92CC411⟩, 412 ⟨79BCJ3469⟩
Karakhanov, E. A., **1**, 99 ⟨75CC480⟩
Karakhanov, R. A., **5**, 350 ⟨92ZOR1750⟩, 691 ⟨92JOU1399⟩; **6**, 70 ⟨81MI 603-01⟩
Karaman, R., **1**, 136 ⟨90TL6155⟩; **5**, 403, 404 ⟨88JCS(P2)2029⟩, 403, 404 ⟨89JCS(P1)765⟩
Karampatses, P., **6**, 163 ⟨87CB1357⟩, 188, 673, 711 ⟨87CB1203⟩, 673 ⟨87CB1213⟩
Karanewsky, D. S., **2**, 454 ⟨83TL2935⟩, 456 ⟨86TL1751⟩, 656 ⟨76JOC3026⟩; **5**, 360 ⟨91JOC3744⟩
Karapetyan, R. G., **1**, 986 ⟨84AKZ116⟩
Karasaki, Y., **5**, 625 ⟨85JOC384⟩
Karasch, M. S., **2**, 190 ⟨38JOC175⟩, 190 ⟨39JA3093⟩, 958 ⟨B-54MI 219-01⟩
Karasik, A. A., **4**, 342 ⟨89IZV946, 90IZV1133⟩
Karatas, I., **5**, 664, 821 ⟨89OPP517⟩
Karaulov, A. I., **4**, 873 ⟨90CC816⟩
Karavanov, K. V., **2**, 478 ⟨58JGU2963⟩
Karayannis, N. M., **2**, 94 ⟨86ICA179⟩
Karbstein, B., **2**, 1076, 1081 ⟨59N321⟩, 1077 ⟨65N59⟩
Karchefski, E. M., **4**, 203 ⟨77JCED355⟩
Karcher, B. A., **3**, 723 ⟨84JOM(260)347⟩
Kardanov, N. A., **4**, 123 ⟨86DOK(286)19⟩
Kardos, M., **4**, 120 ⟨86ACH145⟩
Karele, B., **2**, 1020 ⟨70JOU889⟩
Karelov, A. A., **4**, 126, 127 ⟨79IZV2073⟩
Karel'skii, V. N., **4**, 622 ⟨69DOK(189)334⟩
Karev, V. N., **2**, 430 ⟨90ZOB1795⟩
Karger, G., **2**, 903 ⟨89ZN(B)1179⟩; **6**, 188, 389, 401 ⟨89CB1881⟩
Karger, L., **2**, 193 ⟨51JOC767⟩
Karger, M. H., **2**, 208 ⟨71JOC532, 71JOC540⟩
Kargin, Y. M., **2**, 845 ⟨82ZOB2721⟩
Kargin, Yu. M., **2**, 881 ⟨85JGU1332⟩
Karikomi, M., **2**, 136 ⟨92CL1655⟩, 318 ⟨93CL1787⟩
Karila, M., **1**, 1055 ⟨69CR(C)342, 73BSF3371⟩
Karim, A., **2**, 753 ⟨69CC578⟩
Karim, M. R., **3**, 217, 245 ⟨88TL6897⟩
Karinskaya, A. S., **2**, 108 ⟨91JGU999⟩
Kariv-Miller, E., **2**, 81 ⟨89JA1211, 89JOC4022⟩
Kariya, Y., **3**, 469 ⟨68AG(E)950⟩
Kariyone, K., **5**, 277 ⟨74TL3995⟩
Karjala, S. A., **5**, 551 ⟨33JA2966⟩
Karl, R., **3**, 694 ⟨91COS(4)1083⟩, 696, 708 ⟨89S718⟩
Karl, W., **2**, 759 ⟨91S649⟩
Karle, B., **2**, 437 ⟨87JOM(335)59⟩

Karlsson, J. O., **1**, 1072 ⟨83JHC729, 84JOC2018⟩; **3**, 539 ⟨85JA3392⟩
Karlsson, S., **4**, 902, 979 ⟨78ACS(B)141⟩; **5**, 592, 593 ⟨72S314⟩
Karlsson, U., **2**, 392 ⟨90TL4199⟩
Karlstedt, N. B., **4**, 344 ⟨72ZOB2418, 73PS(3)55⟩, 766 ⟨89ZOB955⟩, 914 ⟨79ZOB1910⟩, 966 ⟨73MI 420-01⟩
Karmas, G., **1**, 503 ⟨52JA295⟩, 955 ⟨53JA1050⟩; **2**, 421 ⟨56JA4285⟩
Karnatak, K. B., **6**, 546 ⟨85JIC164⟩
Karnaukhova, R. V., **4**, 942 ⟨90ZOB526⟩; **6**, 565 ⟨90ZOB602⟩
Karnes, H. A., **2**, 707 ⟨66JOC3980⟩; **6**, 555, 556 ⟨66JOC3980⟩
Karni, M., **5**, 924 ⟨84CC768⟩
Karnischky, L. A., **3**, 314 ⟨70JA1404⟩, 314, 318 ⟨70PAC523⟩
Karns, T. K. B., **4**, 424 ⟨65JOC2769⟩
Karo, S., **2**, 116, 132 ⟨B-68MI 203-01⟩
Karo, W., **2**, 340, 341, 342, 343, 344, 346, 347, 349, 350, 351, 355, 356 ⟨B-89MI 206-03⟩, 773 ⟨B-68MI 215-03⟩, 776 ⟨B-86MI 215-02⟩; **3**, 425 ⟨B-89MI 311-01⟩, 426 ⟨B-72MI 311-01⟩; **4**, 176 ⟨B-89MI 404-01⟩; **5**, 24 ⟨B-68MI 502-01⟩, 182 ⟨B-86MI 504-01⟩, 322 ⟨B-89MI 507-01⟩, 685, 687, 690 ⟨B-72MI 517-01⟩, 726 ⟨B-72MI 518-01⟩; **6**, 68 ⟨B-86MI 603-01⟩, 500 ⟨B-71MI 616-01⟩
Karodia, N., **1**, 827 ⟨90SC1437⟩
Karol, T. J., **4**, 699, 700 ⟨83OM106, 82OM404⟩, 701 ⟨83OM909⟩
Karpeiskaya, E. I., **5**, 215 ⟨86IZV2032⟩
Karpel, S., **1**, 331, 342 ⟨B-85MI 108-01⟩
Karpel, W. J., **1**, 10 ⟨50JA362⟩
Karpelus, R., **4**, 683 ⟨80JOM(187)81⟩
Karpenko, N. F., **5**, 680 ⟨90BAU294, 90IZV352⟩
Karpf, M., **2**, 668 ⟨82TL4923⟩; **3**, 216 ⟨76HCA1226⟩, 223 ⟨81HCA1123⟩, 223, 273 ⟨85HCA338⟩, 270 ⟨82TL4923, 86HCA560⟩; **4**, 238, 241 ⟨87T4875⟩
Karpha, T. K., **3**, 127 ⟨89SC673⟩
Karpitschka, E. M., **5**, 360 ⟨82S592⟩
Karpitskaya, L. G., **2**, 1019 ⟨79DOK(245)607⟩
Karpova, E. N., **2**, 825 ⟨69ZOB1544⟩
Karpova, N. D., **6**, 55 ⟨68IZV431⟩
Karpyuk, A. D., **1**, 859 ⟨77MI 119-01, 81JOU381⟩
Karra, R., **2**, 889 ⟨90PS(47)357⟩
Karras, M., **1**, 1105 ⟨78JA4624⟩; **2**, 924 ⟨78JA4624⟩, 925, 926 ⟨79JOM(179)C37⟩; **3**, 119 ⟨87TL769⟩; **4**, 1060 ⟨78JA4624⟩
Karrer, P., **1**, 46 ⟨43HCA800⟩, 502 ⟨44HCA1297⟩; **2**, 227 ⟨54HCA375⟩; **3**, 57 ⟨41HCA1039⟩, 229 ⟨46HCA704⟩; **5**, 592 ⟨29HCA554⟩
Karrick, G. L., **1**, 890 ⟨87TL2481⟩
Karsch, H. H., **2**, 435 ⟨79ZN(B)1171, 83ZN(B)1027⟩, 474, 475 ⟨85AG(E)402⟩, 1003 ⟨85OM231⟩; **4**, 112, 546 ⟨82CB823⟩, 545, 546, 547, 575, 582 ⟨77ZN(B)762⟩, 546, 554, 555, 566 ⟨82ZN(B)284⟩, 546, 575 ⟨78GEP2658127⟩, 550 ⟨83ZN(B)1027⟩, 551 ⟨79ZN(B)1178, 84ZN(B)1518⟩, 559 ⟨88PS(37)241⟩, 562 ⟨79ZN(B)31⟩, 566, 588 ⟨83ZN(B)1399⟩, 579 ⟨84JOM(273)195⟩, 582 ⟨74CB3692, 77CB2200⟩, 583 ⟨73AG(E)853, 74CB3684, 74CB93⟩, 587 ⟨85OM231⟩, 588 ⟨86OM1664⟩, 1026 ⟨84TL3687⟩; **5**, 884 ⟨88TH17⟩, 884, 1092 ⟨84TL3687⟩, 887 ⟨86AG(E)454⟩, 908 ⟨79ZN(B)31⟩, 910 ⟨79ZN(B)1171⟩; **6**, 151 ⟨79AG(E)484, 79ZN(B)1171⟩, 151, 369 ⟨82CB818⟩, 152 ⟨84ZN(B)1518⟩, 164, 371, 373 ⟨88ZN(B)1416⟩, 167, 371 ⟨92CB1333⟩, 169 ⟨86JOM(312)C1⟩, 197 ⟨93CB2637⟩, 247 ⟨91CC466⟩, 369 ⟨79ZN(B)1178, 84CC569, 93OM2757⟩, 374 ⟨91OM2884⟩, 394 ⟨90CC1621⟩
Kartashov, A. V., **2**, 15 ⟨89JOU1835⟩, 18 ⟨87JOU230⟩
Kartashov, V. R., **5**, 1044 ⟨91ZOR340⟩
Karten, J. L., **2**, 606 ⟨66T2167⟩
Karthikeyan, M., **3**, 144 ⟨93TL6631⟩
Kartoon, I., **5**, 1000 ⟨74TL1965⟩
Kartsev, V. G., **2**, 1003 ⟨86IZV134⟩
Kartseva, O. S., **5**, 687 ⟨91CHE299, 91KGS369⟩
Kasahara, A., **1**, 690 ⟨82BCJ2434, 83BCJ192⟩, 690 ⟨83BCJ2023⟩; **2**, 816 ⟨83TL511⟩; **3**, 233 ⟨88S604⟩; **4**, 193 ⟨89JHC1405⟩, 193 ⟨90JHC1419⟩; **5**, 142 ⟨77BCJ1639⟩, 296 ⟨75CI(L)745⟩
Kasahara, I., **1**, 77 ⟨87JA1596, 87JA4129⟩
Kasai, N., **1**, 690 ⟨88JA3578⟩, 992 ⟨90CL2149⟩; **2**, 244 ⟨72JOC3615⟩; **3**, 487 ⟨86OM356⟩; **4**, 686 ⟨77BCJ2250⟩, 820 ⟨86CC1794⟩; **5**, 948 ⟨82CC1178⟩
Kasai, Y., **5**, 300 ⟨73TL1595⟩
Kasatkin, A. N., **1**, 469 ⟨89IZV2623⟩, 801 ⟨91TL4591⟩
Kasatkina, T. G., **4**, 715 ⟨88MI 416-01⟩
Kasatochkin, V. I., **2**, 1018, 1019 ⟨65ZOR415⟩
Kaschani-Motlach, M., **2**, 292, 857 ⟨79ZAAC(448)35⟩
Kaschig, J., **5**, 993 ⟨77AG(E)550⟩
Kase, N., **1**, 993 ⟨94TL1255⟩
Kaseda, T., **3**, 419 ⟨89TL4539⟩
Kasei, K., **6**, 488 ⟨92JAP05339542⟩
Kasel, W., **1**, 705 ⟨84CB966⟩, 708 ⟨84CB1868⟩; **5**, 138 ⟨85CC1563⟩
Kasemann, R., **6**, 237 ⟨90JFC(48)207⟩
Kasemsri, P., **1**, 1084 ⟨81CL1241⟩
Kasha, W. C., **2**, 437 ⟨90CB1827⟩
Kashdan, D. S., **3**, 131 ⟨77JOC1189⟩
Kasheva, T. N., **4**, 462 ⟨92ZOB1472⟩; **5**, 683 ⟨76JGU239, 76ZOB243⟩
Kashihara, H., **5**, 630 ⟨85CPB3479⟩
Kashima, C., **2**, 744 ⟨75JOC526⟩; **3**, 357 ⟨84H(22)763⟩; **4**, 899, 979 ⟨84JHC133⟩; **5**, 197 ⟨88JCS(P1)529⟩; **6**, 579 ⟨82H(19)2283⟩
Kashimirov, B. A., **4**, 346 ⟨92ZOB1268⟩
Kashimura, S., **1**, 590 ⟨78TL2807⟩, 615, 698 ⟨78CL69⟩; **2**, 81 ⟨89JOC6001⟩; **3**, 17, 132 ⟨83JOC3338⟩, 426 ⟨83JOC2095⟩; **4**, 210 ⟨86JOC546⟩, 300, 311 ⟨82JA5753⟩; **5**, 132 ⟨86JOC546⟩; **6**, 24 ⟨90TL7181⟩
Kashin, A. N., **1**, 1006 ⟨81JOU18⟩, 1015 ⟨81DOK(261)1141⟩; **6**, 406 ⟨68BAU414⟩
Kashinskii, V. N., **5**, 62 ⟨B-76MI 502-02⟩
Kashirskaya, E. G., **6**, 145 ⟨81IZV2146⟩
Kashiwabara, K., **2**, 437 ⟨85BCJ2793, 90BCJ2051⟩, 437 ⟨91BCJ2942⟩
Kashiwagi, T., **2**, 159 ⟨92JCS(P1)1497⟩; **4**, 205 ⟨76CPB102⟩
Kashtanov, S. A., **6**, 261 ⟨88IZV2639, 88IZV466⟩
Kashutina, M. V., **6**, 368 ⟨67JOM(9)5⟩
Kashuting, M. V., **6**, 149, 653 ⟨73ZOR896⟩
Kashyap, R. P., **3**, 345 ⟨91JOC5932⟩; **4**, 429, 430, 431 ⟨93JOC759⟩; **6**, 317 ⟨93T3035⟩
Kasina, S., **2**, 309 ⟨89JOC1940⟩
Kaska, W. C., **4**, 578 ⟨B-93MI 412-01⟩, 585 ⟨79JOM(182)251⟩; **5**, 386 ⟨70CC100⟩, 917 ⟨80ZN(B)1289⟩, 919 ⟨74JA2847⟩, 1094 ⟨73JA7510⟩; **6**, 705 ⟨83CCR1⟩, 706 ⟨80ZN(B)1289⟩
Kaslow, C. E., **3**, 362 ⟨50JOC600⟩
Kasmai, H., **1**, 865 ⟨75CC201⟩
Kasori, Y., **6**, 465 ⟨92JAP04356446⟩
Kaspar, E., **6**, 25 ⟨58CB2664⟩
Kasper, A. M., **2**, 329 ⟨89JA1517⟩
Kasperovich, V. P., **5**, 655 ⟨78JOU758, 78ZOR820⟩
Kaspers, H., **6**, 316 ⟨68BRP1097237⟩
Kaspi, J., **1**, 1112 ⟨71TL4039, 74JA4518⟩, 1112 ⟨74JA586⟩
Kassab, N. A., **5**, 532 ⟨78ZN(B)75⟩
Kassam, K., **3**, 324 ⟨92JA7969⟩
Kassemann, R., **2**, 881 ⟨86AX(C)1875⟩
Kassinger, R., **4**, 770 ⟨58JA1916, 58JA6450⟩
Kassir, J. M., **1**, 550 ⟨90JOC414⟩; **2**, 668 ⟨91TL5923⟩; **3**, 320 ⟨93TL7755⟩
Kast, M., **4**, 901 ⟨67AG147⟩
Kastner, M. E., **1**, 485 ⟨85JA4679⟩; **3**, 217 ⟨85JA4679⟩
Kasturi, T. R., **2**, 658 ⟨68CJC3625⟩
Kasubick, R. A., **4**, 742 ⟨70JOC1211⟩
Kasuga, H., **5**, 357 ⟨90TL7633⟩

Kasuga, T., **1**, 405, 816, 819 ⟨85CL1729⟩, 816, 819 ⟨84TL6011⟩, 816, 819 ⟨86TL4185⟩
Kasukabe, T., **6**, 558 ⟨89CPB1249⟩
Kasukhin, L. F., **5**, 373, 862 ⟨81T437⟩; **6**, 635 ⟨92IZV425⟩
Kasukhin, L. S., **2**, 467 ⟨79JGU34⟩
Kas'yan, L. I., **5**, 709 ⟨87MI 517-02⟩
Kata, Y., **3**, 635 ⟨85BCJ765⟩
Katada, T., **3**, 696 ⟨74JOC1239⟩; **4**, 232 ⟨84JCS(P1)2641⟩; **5**, 243, 252 ⟨77JOM(129)189⟩, 252, 254 ⟨80S721, 81S370⟩, 510, 551, 583, 620 ⟨82LA1229⟩, 551 ⟨77AG(E)786⟩, 584 ⟨81LA1798⟩
Kataeva, O. N., **6**, 538, 539 ⟨89IZV909⟩
Kataev, E. G., **4**, 861 ⟨78ZOR478⟩
Katagiri, I., **6**, 618 ⟨73BCJ303⟩
Katagiri, M., **1**, 72, 81 ⟨71BCJ166⟩
Katagiri, N., **4**, 265, 853, 854 ⟨90CPB3242⟩; **5**, 367 ⟨90CPB3184⟩, 652 ⟨88CPB3354⟩
Katagiri, S., **5**, 316 ⟨91MI 507-05⟩
Katagiri, T., **2**, 27 ⟨77CL1025⟩; **3**, 57 ⟨82CL1987⟩
Katakai, A., **5**, 864 ⟨86MI 521-02, 87MI 521-03⟩
Katakai, R., **5**, 972 ⟨73CL1143, 85JOC715⟩; **6**, 413, 415, 425 ⟨85JOC715⟩
Katano, K., **3**, 357 ⟨90JAN1150⟩
Kataoka, F.,**1**, 690 ⟨78JOC1612⟩
Kataoka, H., **1**, 690, 806, 807 ⟨88CC354⟩, 704, 707, 709 ⟨88JA7247⟩; **3**, 206 ⟨84CL1133⟩
Kataoka, M., **3**, 54, 62, 252 ⟨79TL2809⟩, 635 ⟨88CPB1698⟩; **4**, 750 ⟨67BCJ1275⟩; **6**, 488 ⟨92JAP05339226⟩
Kataoka, T., **1**, 108, 115 ⟨86CPB3599⟩, 943 ⟨90JCS(P1)3017⟩; **2**, 143 ⟨90TL3027⟩, 278, 279, 292 ⟨94JOC1011⟩, 281 ⟨90H(31)889⟩, 286, 288, 292 ⟨94CPB811⟩, 288 ⟨87JOC1397, 92JCR(S)393⟩; **3**, 356 ⟨74CL1101⟩, 376 ⟨78TL251, 80JOC2468⟩, 376 ⟨87YGK232⟩, 635 ⟨88CPB1698⟩, 637 ⟨93SC2481⟩; **4**, 232 ⟨84TL757⟩, 288 ⟨91TL105⟩; **5**, 81 ⟨78GEP2725519⟩, 1122 ⟨85JCS(P1)2333, 85TL1321⟩, 1122 ⟨87JCS(P1)2531⟩
Kataoka, Y., **1**, 457 ⟨90TL365, 92JOC1615⟩, 676 ⟨87TL1443⟩; **2**, 700 ⟨88TL1065⟩, 927 ⟨90TL365, 90TL369⟩, 927 ⟨92JOC1615⟩, 927, 945 ⟨90JOC5310⟩, 927, 989 ⟨92BCJ1543, 92CL99⟩; **3**, 627 ⟨92BCJ1543⟩; **5**, 344 ⟨91CL1479⟩
Katayama, A., **1**, 1033 ⟨87JCS(P1)1511⟩
Katayama, C.,**4**, 937 ⟨87T4759⟩
Katayama, E., **1**, 389 ⟨83TL4997⟩; **3**, 142, 146 ⟨83TL4997⟩
Katayama, H., **1**, 257 ⟨90JOC408⟩; **3**, 350 ⟨90EUP379979⟩
Katayama, M., **3**, 28 ⟨91CL1469⟩
Katayama, S., **3**, 168 ⟨90CPB3314⟩, 373 ⟨88S178, 89CL973⟩
Katayama, T., **1**, 471 ⟨91CL2073⟩; **5**, 226 ⟨87H(25)79⟩
Katayma, H., **3**, 210 ⟨89TL3685⟩
Katchalski, E., **5**, 84 ⟨51MI 502-01, 58MI 502-04⟩
Katekar, G. F., **3**, 379 ⟨73JA4287⟩
Katerberg, G. J., **5**, 14 ⟨57RTC969⟩; **6**, 41 ⟨57RTC969⟩
Kates, M. R., **1**, 906 ⟨80JA6867, 83JA3571⟩
Kates, S. A., **1**, 336, 346, 926 ⟨88JOC2137⟩, 346 ⟨88TL5209⟩, 347 ⟨90JA2759⟩
Kathawala, F. G., **1**, 767, 769 ⟨89SC2659⟩; **3**, 131, 273, 284 ⟨84TL811⟩
Kätker, H., **1**, 674 ⟨87AG(E)702⟩
Kato, A., **2**, 214 ⟨90JHC1361⟩
Kato, H., **1**, 1109 ⟨68JA1239⟩; **3**, 357 ⟨75JCS(P1)632⟩, 360 ⟨91JCS(P1)2281⟩, 361 ⟨89CC505⟩; **4**, 259 ⟨90JCS(P1)2035⟩, 970 ⟨91EUP413343⟩; **6**, 115 ⟨69BCJ2323⟩, 298, 300 ⟨86JAP(K)61289091⟩
Kato, J.-I., **2**, 701 ⟨83CL1727⟩
Kato, K., **2**, 213 ⟨70JOC4227⟩, 391 ⟨90CL1395, 92CL1137⟩, 831 ⟨92JA4008⟩; **3**, 64 ⟨78CL41⟩, 134 ⟨89CL2233⟩, 192, 428 ⟨91BCJ2948⟩, 561 ⟨71BCJ1362⟩; **4**, 208 ⟨92SL565⟩, 222 ⟨76TL65⟩; **5**, 224 ⟨89JOC4511⟩, 278 ⟨82JCS(P1)2909⟩, 735 ⟨71BCJ1362⟩, 1078 ⟨69TL2473, 70T5731⟩
Katô, M., **1**, 422 ⟨75JA928⟩, 822 ⟨91TL6159, 94CC789⟩, 963 ⟨80TL623⟩, 1136 ⟨75JOC1932⟩; **2**, 21 ⟨78CL223⟩, 638 ⟨76BCJ3673⟩, 223 ⟨80TL623⟩; **4**, 117 ⟨81CC1173⟩, 354 ⟨84BCJ827⟩, 834, 933 ⟨78CC478⟩; **5**, 414 ⟨78CC478⟩, 652 ⟨78JCS(P1)558⟩; **6**, 481 ⟨89CL1009⟩
Kato, N., **1**, 206 ⟨80TL4365⟩, 690 ⟨86BCJ1109⟩, 690, 806, 807 ⟨88CC354⟩; **3**, 243 ⟨85JA1246⟩, 303 ⟨79TL4577⟩; **5**, 775 ⟨83CPB3139⟩
Kato, S., **1**, 186, 225, 229 ⟨89JA7938⟩, 212 ⟨84JA6093⟩; **2**, 370, 768 ⟨91OM1095⟩, 814 ⟨90JOC449⟩; **3**, 173 ⟨93CC1631⟩; **4**, 588 ⟨73JOM(51)167⟩, 709 ⟨74JOM(78)405⟩; **5**, 240 ⟨86S128⟩, 242 ⟨92PS(67)27⟩, 242, 250, 506, 509 ⟨87CL475⟩, 242, 506, 509 ⟨89TL1829⟩, 243 ⟨89ZN(B)1050⟩, 243, 252 ⟨77JOM(129)189⟩, 244 ⟨86ZC179⟩, 244, 250 ⟨94JCS(P1)1083⟩, 245 ⟨86T4593⟩, 245, 246 ⟨83S552⟩, 245, 246 ⟨83S814⟩, 245, 252 ⟨82S1013, 82AG(E)150⟩, 246 ⟨86CL935⟩, 248, 249 ⟨85S519, 85S520, 92ZN(B)558⟩, 249, 563, 640 ⟨88BCJ3931⟩, 250 ⟨89S929, 90JOM(386)313, 94PS(89)57⟩, 252 ⟨87NKK1457⟩, 252, 254 ⟨80S721, 81S370⟩, 254, 537, 539, 540 ⟨73BCJ244⟩, 255, 506, 509, 543 ⟨90TL3587⟩, 255, 541 ⟨72TL3751⟩, 337 ⟨81S746⟩, 337, 584 ⟨85CB1668⟩, 472, 510 ⟨86OPP369⟩, 506, 509 ⟨92JAP(K)4224595⟩, 506, 509 ⟨92MI 512-01⟩, 509, 510, 512 ⟨B-92MI 512-01⟩, 510, 551, 583, 620 ⟨82LA1229⟩, 510, 561, 563 ⟨90BCJ638⟩, 516, 519, 522, 523, 626 ⟨90BCJ825⟩, 516, 523, 563 ⟨93JA5823⟩, 530, 540 ⟨86TL4593⟩, 537 ⟨72IJS279⟩, 541, 542 ⟨92CB417, 93CC277⟩, 546, 550 ⟨72BCJ3492⟩, 547 ⟨82S457⟩, 549 ⟨73IJS359⟩, 549 ⟨74JOM(78)405, 74JOM(82)223⟩, 550, 560 ⟨88SR155⟩, 551 ⟨77AG(E)786, 85PS(22)135⟩, 560 ⟨86BCJ1403⟩, 561, 563, 588 ⟨90ZC67⟩, 563 ⟨93CL1469, 94UP0⟩, 563, 626 ⟨93JA3000⟩, 584 ⟨81LA1798⟩, 584, 620 ⟨72BCJ3653⟩, 609 ⟨77AG(E)787⟩, 691 ⟨91BCJ766, 92BCJ674⟩; **6**, 546, 561 ⟨87S170⟩
Kato, T., **1**, 181 ⟨75CPB1917⟩, 191 ⟨87CL1167⟩, 481, 1028 ⟨83CPB4346⟩, 690 ⟨86BCJ2209⟩, 813, 814 ⟨80BCJ2958⟩; **2**, 31 ⟨76CL283, 80JCS(P1)1051⟩, 453 ⟨87CL1915⟩, 624, 902 ⟨83S1005⟩; **3**, 142 ⟨80BCJ2958⟩, 213 ⟨75TL3299⟩, 231 ⟨78CPB3877⟩, 623 ⟨87CL1167⟩; **4**, 588 ⟨73JOM(51)167⟩; **5**, 215 ⟨86CPB4984⟩
Kato, Y., **4**, 279 ⟨91MI 406-01⟩
Katocs, A., **4**, 273 ⟨89MRC760⟩
Katoh, A., **3**, 357 ⟨84H(22)763⟩; **6**, 579 ⟨82H(19)2283⟩
Katoh, N., **4**, 326, 856, 857 ⟨78TL375⟩
Katoh, S., **2**, 323 ⟨90JA2368⟩, 326 ⟨93JMC3526⟩; **3**, 43 ⟨86TL6083, 90JA2368⟩; **4**, 197 ⟨89TL5309⟩; **5**, 864 ⟨B-87MI 521-01⟩
Katoh, T., **3**, 426 ⟨83CL607⟩; **6**, 447 ⟨92TL3483⟩
Katonak, D. A., **4**, 412 ⟨74JOC167⟩
Katou, T., **1**, 690 ⟨82BCJ2434⟩
Katritzky, A. R., **1**, 30 ⟨79CC300, 79JCS(P1)442, 80JCS(P1)2554, 84AG(E)420, 84JCS(P1)1671⟩, 30, 31 ⟨80T679⟩, 260 ⟨87JCS(P1)805, 88JOC5854⟩, 260 ⟨89JCS(P1)225, 89JCS(P1)2297⟩, 260 ⟨89JOC6022, 90T987⟩, 260 ⟨91JOC4439⟩, 260, 261 ⟨89S323⟩, 260, 261 ⟨89S747⟩, 260, 263 ⟨91HCA1924, 90JOC3209⟩, 261 ⟨90S1173⟩, 261, 1041 ⟨89S31⟩, 262 ⟨91JCS(P1)3295, 93JOC4376⟩, 262, 1041 ⟨89TL3303⟩, 629 ⟨81CC96, 82JCS(P1)2347, 82JOC3506⟩, 845, 859 ⟨88PIA187, B-67MI 119-01⟩, 956 ⟨93JOC3038⟩; **2**, 7 ⟨87CS477⟩, 11 ⟨77S634, 79CC238, 79JCS(P1)436, 79S437, 80JCS(P1)1890, 80JCS(P1)2901, 80T679, 83IJC(B)421, 84AG(E)420⟩, 11, 632 ⟨79JCS(P1)433⟩, 104 ⟨80JOC2515⟩, 121 ⟨85TL469⟩, 301 ⟨91OPP399⟩, 302 ⟨89OPP135, 91TL4247⟩, 304 ⟨90CJC456⟩, 306 ⟨89S31, 89TL3303⟩, 310 ⟨89JCS(P1)225, 89JCS(P1)639⟩, 319 ⟨90TL3999⟩, 320 ⟨88JCS(P1)2339, 93JOC2086⟩, 347 ⟨71JCS(B)1330⟩, 621 ⟨90JHC1737⟩, 710 ⟨91HCA1931⟩ 1008 ⟨92CJC2040⟩; **3**, 3, 114 ⟨79JCS(P1)2493⟩, 29 ⟨91JOC2143⟩, 42 ⟨90S1173, 91T2683⟩, 68, 69

⟨88CB999⟩, 89 ⟨78S619⟩, 157 ⟨87TL1847⟩, 158 ⟨91JOC6917⟩, 186, 187 ⟨90T987⟩, 228 ⟨76JCS(P1)2289⟩, 229 ⟨80JCS(P1)2851⟩, 257 ⟨88JHC1287⟩, 357 ⟨58JCS3610⟩, 406 ⟨87T5171⟩, 433 ⟨83MI 311-01⟩, 579 ⟨88JCS(P2)1071⟩, 669 ⟨72JCS(P1)2054⟩, 709, 710 ⟨93S45, 89TL6657⟩, 710 ⟨90JCS(P1)1847, 91JOC4439, 91S868⟩; **4**, 105 ⟨90JHC139⟩, 296, 300, 310 ⟨87JCS(P1)791⟩, 301 ⟨91S279⟩, 301, 308 ⟨89JOC6022⟩, 306, 310 ⟨90JCS(P1)1717⟩, 308, 445 ⟨87JCS(P1)811⟩, 309 ⟨89JCS(P1)225, 91JOC4439, 92JOC547, 94S445⟩, 309, 323 ⟨93S229⟩, 309, 445 ⟨87JCS(P1)799⟩, 317 ⟨89CS33⟩, 317, 416 ⟨86S804⟩, 318 ⟨91HCA1936⟩, 318, 322 ⟨91HCA1924, 87JCS(P1)781, 89CJC1144⟩, 323 ⟨85JOC1351, 87JCS(P1)775, 91S1147, 92S1295⟩, 323, 406, 410, 445 ⟨91T2683⟩, 329 ⟨90S565, 94S107⟩, 370 ⟨91G471⟩, 383 ⟨87TL1847⟩, 392 ⟨87JOC844⟩, 406, 410 ⟨90JCS(P1)541⟩, 409 ⟨76JCS(P2)1564⟩, 417 ⟨89CC337⟩, 445 ⟨91CJC1153⟩, 446 ⟨93S45⟩, 446, 449 ⟨93JOC2086, 88JCS(P1)2339⟩, 449 ⟨90S663⟩, 520, 540 ⟨87TL5419⟩, 539 ⟨83T2023, 82JCS(P1)143⟩; **5**, 316 ⟨84JOC4784⟩, 596 ⟨91JOC2143⟩, 606 ⟨88JOC5854, 88TL1755⟩, 725 ⟨65JCS5230⟩, 758, 767 ⟨74JCS(P2)546⟩, 1002 ⟨76TL2691, 79JCS(P1)1957, 79JCS(P1)446⟩, 1031, 1067 ⟨90S565⟩, 1107 ⟨78CC133, 80JCS(P1)865⟩; **6**, 143 ⟨77JCS(P1)1162⟩, 143 ⟨90S666⟩, 162 ⟨88RTC641⟩, 170 ⟨91JOC2143⟩, 365 ⟨90JCR(S)330⟩, 370 ⟨89JHC829⟩, 558 ⟨93OPP83⟩, 561 ⟨85H(23)3099⟩

Katrusiak, A., **2**, 11 ⟨84AG(E)420⟩, 1008 ⟨86MI 220-1⟩

Katsoulos, G. A., **2**, 35 ⟨85CC1404⟩

Katsoyannis, P. G., **2**, 412 ⟨55JA5351⟩, 414 ⟨55JA5354⟩

Katsuhara, Y., **6**, 232 ⟨80JA2681⟩, 428, 430 ⟨82JFC(19)227⟩

Katsuhira, T., **1**, 198, 975 ⟨93JOC6166⟩, 431 ⟨92JOC5805, 93JOC4897⟩; **4**, 758, 819, 821 ⟨92JOC5805⟩, 758, 821 ⟨93JOC4897⟩; **5**, 720 ⟨90JOC3682⟩

Katsuka, Y., **1**, 613 ⟨93JCS(P1)987⟩

Katsuki, J., **1**, 16 ⟨91TL4963⟩

Katsuki, M., **1**, 1071 ⟨91TL6735⟩

Katsuki, T., **1**, 180 ⟨86TL3403⟩, 404, 406, 816, 819 ⟨86TL4581⟩, 404, 406, 816, 819 ⟨88TL4763⟩, 404, 406, 819 ⟨92BCJ1841⟩, 816 ⟨86TL4577, 87TL803⟩, 816 ⟨91TL5137⟩, 954 ⟨86TL4599⟩, 1045 ⟨83TL5131⟩; **2**, 82 ⟨83PAC589⟩, 91 ⟨80JA5974⟩, 92 ⟨81JA6237⟩; **3**, 223, 275 ⟨79CL1021⟩, 273 ⟨83TL5131⟩; **5**, 52 ⟨81JOC3936⟩, 125 ⟨87TL4711⟩, 173 ⟨79BCJ1989⟩, 173 ⟨79CJC2180⟩, 173 ⟨79CL1021⟩

Katsumi, I., **3**, 91 ⟨88CPB974⟩

Katsumoto, K., **5**, 1109 ⟨68JOC823⟩

Katsumura, N., **2**, 73, 597 ⟨89JOM(369)291⟩

Katsumura, S., **3**, 303 ⟨88TL1173⟩; **4**, 387 ⟨85TL5827⟩

Katsuno, H., **2**, 969 ⟨83TL1809⟩

Katsura, K., **6**, 533 ⟨87JAP01102058⟩

Katsura, T., **2**, 540 ⟨85MI 210-02⟩

Katsurada, M., **5**, 1120 ⟨92CC1379⟩

Katsuro, Y., **1**, 479 ⟨81S1001⟩, 480 ⟨80TL3915⟩, 482 ⟨81TL4449⟩

Katsuura, A., **5**, 240, 243, 253 ⟨91JOC3776, 89CL1825⟩

Katsuyana, M., **4**, 827 ⟨80H(14)15⟩

Katsyubova, S. A., **2**, 464 ⟨88JGU1131⟩

Kattenberg, J., **1**, 708 ⟨77TL1173⟩

Katti, K. V., **4**, 552 ⟨89IC413, 90IC808⟩

Katyshkina, V. V., **2**, 694 ⟨56ZOB3060⟩

Katz, A. H., **3**, 664 ⟨84TL5473⟩

Katz, J.-J., **1**, 1103 ⟨72S555⟩; **2**, 515 ⟨75JA2791⟩, 530 ⟨73JOC3968⟩; **4**, 363 ⟨75JOC813⟩

Katz, L., **3**, 464 ⟨54JOC718⟩

Katz, R. B., **4**, 900 ⟨89T1801⟩

Katz, R. N., **4**, 1016 ⟨76AG(E)171⟩

Katz, S. A., **3**, 170 ⟨88JOC906⟩

Katz, S., **3**, 212 ⟨72JA8637⟩

Katz, T. J., **1**, 808 ⟨59T19⟩, 811 ⟨70JA6643⟩, 890 ⟨60JA3784, 60JA3785⟩, 901 ⟨62JA3199⟩; **2**, 475 ⟨71JA4065⟩, 670 ⟨85TL2159⟩, 981 ⟨80JA422⟩

Katzenellenbogen, J. A., **1**, 747 ⟨71TL1821⟩, 968 ⟨78JOC555⟩, 1021 ⟨86JMC230⟩, 1052 ⟨70JA6314⟩, 1134 ⟨69JA4318⟩; **2**, 9 ⟨87JOC658⟩, 18 ⟨86JFC(31)99⟩, 629, 970 ⟨67JA4245⟩, 970 ⟨70JA6314⟩, 982 ⟨81JA5459⟩; **4**, 9 ⟨86JOC3508⟩, 527 ⟨85JMC1917⟩; **6**, 134 ⟨90MI 604-01⟩

Katzhendler, J., **5**, 403 ⟨92PS(70)109⟩

Kauck, E. A., **6**, 32, 218 ⟨53USP2647933⟩, 214 ⟨52USP2616927, 55USP2704776⟩

Kauer, J. C., **5**, 100 ⟨73OSC(5)1043⟩, 1100, 1102 ⟨64JA4732⟩

Kauffman, K. C., **1**, 871 ⟨69JA530⟩

Kauffman, T., **1**, 54 ⟨85CB1039⟩; **3**, 404 ⟨73AG584⟩

Kauffman, W. J., **5**, 1013 ⟨74JOC2472⟩

Kauffmann, H., **3**, 686 ⟨13CB4001⟩

Kauffmann, T., **1**, 271 ⟨83CB473⟩, 363 ⟨74AG(E)468⟩, 371, 372 ⟨74AG(E)627⟩, 466 ⟨93CB1447⟩, 647, 667, 717 ⟨78TL4399⟩, 649, 656, 678, 680, 759 ⟨81TL5031⟩, 667 ⟨77AG(E)862⟩, 692 ⟨92CB149⟩, 755 ⟨86CB2150⟩, 1089 ⟨82AG(E)410⟩; **2**, 280, 908 ⟨83CB1001⟩, 421 ⟨67AG(E)633⟩, 436, 873 ⟨85TL4067⟩, 569, 570, 571 ⟨93CB2093, 82AG(E)144⟩, 577 ⟨92CB157⟩, 877 ⟨82CB654⟩, 908, 910, 911 ⟨82AG(E)410⟩; **3**, 447 ⟨64AG206, 64AG(E)342⟩, 447 ⟨65CB912⟩; **4**, 135, 592, 596, 599, 600 ⟨85CB2353⟩, 139, 141, 146, 156 ⟨85CB391⟩, 156, 708 ⟨85CB370⟩, 401 ⟨77AG(E)710⟩, 401, 599, 708, 724 ⟨82CB1810⟩, 421 ⟨77CB651⟩, 535 ⟨70AG(E)163, 77CB2659⟩, 536 ⟨77CB3034⟩, 565 ⟨80TL2803⟩, 565, 596, 600 ⟨83CB473⟩, 577, 578, 579, 581 ⟨86CB2150⟩, 585 ⟨77CB3930⟩, 596, 600 ⟨80AG(E)723⟩, 596, 600 ⟨85CB1039⟩, 599, 600, 663 ⟨78TL4391, 82CB645⟩, 673 ⟨86AG(E)909, 92CB143⟩, 673 ⟨93CB79, 93CB89⟩, 947, 965 ⟨83CB1001⟩; **5**, 769, 848 ⟨64CB3436⟩, 769, 848 ⟨66CB2600⟩, 773 ⟨81CB3684, 63AG(E)217⟩; **6**, 170, 205, 206, 210 ⟨82AG(E)410⟩, 170, 210 ⟨79TL501⟩, 203, 204, 205, 210 ⟨80TCC109⟩, 203, 204, 210 ⟨80TL2807⟩, 209 ⟨85CB380⟩

Kaufhold, M., **1**, 910 ⟨66AG(E)604⟩

Kaufman, B. L., **5**, 404 ⟨68JGU1791⟩

Kaufman, C., **4**, 162 ⟨85TL2279⟩, 162 ⟨85TL2283⟩

Kaufman, D. A., **1**, 1111 ⟨68JA7282⟩

Kaufman, D., **1**, 1110 ⟨69JOC1495⟩, 1122 ⟨74JOC1438, 79JOC3076⟩

Kaufman, E. D., **5**, 213 ⟨81JOC2148⟩

Kaufman, G., **1**, 630 ⟨67JA5736⟩

Kaufman, G. M., **3**, 465 ⟨65JA935⟩

Kaufman, K. D., **3**, 90 ⟨70JCS(C)227⟩

Kaufman, M. H., **4**, 735 ⟨67JOC2749⟩, 736 ⟨68JOC2173⟩

Kaufman, P., **4**, 162 ⟨85TL2279⟩

Kaufman, P. R., **4**, 13 ⟨64JOC1956⟩

Kaufmann, A., **5**, 184 ⟨09CB3484⟩

Kaufmann, E., **5**, 436 ⟨87JA2553⟩

Kaufmann, H., **1**, 1082 ⟨67HCA2101, 67HCA2108⟩

Kaufmann, H. P., **5**, 753 ⟨42CB1585⟩, 1048 ⟨60AP(293)150⟩; **6**, 69 ⟨54LA(587)226⟩, 452 ⟨60AP(293)150⟩

Kaufmann, J., **3**, 80 ⟨77CZ36⟩

Kaufmann, K. D., **3**, 483 ⟨68TL4973⟩; **5**, 873 ⟨68TL4973⟩

Kaufmann, R. J., **5**, 728 ⟨23JA1744⟩

Kaufmann, T., **2**, 577 ⟨92CB163⟩, 589 ⟨92CB2409⟩

Kaufmann, W., **4**, 761 ⟨52M1⟩

Kaufmann, W. E., **6**, 68 ⟨51OSC(1)258⟩

Kaugars, G., **5**, 675, 678 ⟨81JMC532⟩

Kaukorat, T., **2**, 861 ⟨89ZN(B)481⟩

Kaul, F. A., **3**, 516 ⟨93OM2565⟩

Kaul, H.-A., **2**, 512 ⟨82JOM(228)C29⟩

Kaulgud, A. P., **5**, 690 ⟨82IJC(B)989⟩

Kaupp, G., **1**, 418, 420 ⟨78AG(E)150⟩; **4**, 101 ⟨86CB2387⟩, 101, 424 ⟨87CB1897⟩, 411 ⟨90AG(E)933⟩

Kaura, A. C., **4**, 419 ⟨91TL2683⟩

Kaur, G., **3**, 162 ⟨93TL7335⟩; **4**, 192 ⟨93TL7335⟩
Kaur, N. P., **1**, 282 ⟨91JOC3908⟩; **3**, 212 ⟨91CC568⟩
Kaur, V., **4**, 302 ⟨89S677⟩
Kausch, E., **6**, 624, 625, 633 ⟨87CB339⟩, 647 ⟨78LA1543⟩
Kauschka, G., **6**, 223, 228 ⟨76ZC377⟩
Kaushal, R., **4**, 504 ⟨67JIC995⟩
Kaushaner, V. S., **5**, 370 ⟨70ZOB2464⟩
Kaushik, M. P., **2**, 463 ⟨81IJC(B)932⟩, 464 ⟨80JOC2270⟩; **5**, 371 ⟨89CI(L)389⟩
Kaushik, N. K., **6**, 563 ⟨83JIC806⟩
Kavaliunas, A. V., **5**, 1139 ⟨79JOC3069⟩
Kavash, R. W., **3**, 420 ⟨89TL4185⟩
Kaverin, V. V., **4**, 583 ⟨85JOM(288)119⟩
Kavka, M., **1**, 804 ⟨84TL2447⟩
Kawa, H., **2**, 1005 ⟨85JA5313⟩; **4**, 618 ⟨84CC1664⟩, 618, 668 ⟨88POL2023, 90JA2425⟩; **6**, 192, 205, 398 ⟨84CC1664⟩, 192, 398 ⟨88POL2023⟩, 206 ⟨85JA5313⟩
Kawabata, K., **3**, 85 ⟨78CC330⟩; **5**, 277, 295 ⟨80TL841⟩
Kawabata, N., **1**, 115 ⟨73JOC4268⟩, 115, 452 ⟨73T1069⟩, 122 ⟨74AOC83⟩, 300 ⟨69T2647⟩; **2**, 67 ⟨74AOC(12)83⟩, 563 ⟨73T1069⟩; **4**, 669 ⟨76BCJ1177⟩
Kawabata, S., **5**, 245, 252 ⟨82S1013⟩
Kawabata, T., **1**, 25, 630 ⟨87JOC3346⟩, 120 ⟨93TL5127⟩; **2**, 784 ⟨92S1119⟩; **3**, 215 ⟨87JOC3346⟩, 498 ⟨64TL1653⟩
Kawachi, A., **2**, 545, 546 ⟨92JA3989⟩; **4**, 238 ⟨91CL415⟩
Kawachi, H., **5**, 1102 ⟨89JAP3145452⟩, 1125 ⟨86JAP1022060⟩
Kawada, A., **3**, 281, 298, 299 ⟨93CC1157⟩; **4**, 94 ⟨85CC1266⟩; **5**, 351 ⟨91JCS(P1)2435⟩
Kawada, K., **2**, 16 ⟨82TL1165, 86TL4465⟩, 16 ⟨90JA8563, 90OS(69)129⟩; **3**, 149 ⟨82TL1165⟩, 149 ⟨90JA8563⟩; **4**, 8 ⟨85TL2445⟩; **6**, 113, 316 ⟨81JCS(P1)618⟩
Kawada, M., **1**, 613 ⟨83TL4993, 86JOC3896⟩, 613, 700, 1073 ⟨84JA3670⟩, 826 ⟨90TL4041⟩, 962 ⟨93TL2161⟩, 976 ⟨91JOM(417)305, 91SL697⟩, 976 ⟨91TL3397, 91TL7683⟩, 976, 1025 ⟨90TL7179⟩; **2**, 678 ⟨83TL4993⟩; **3**, 60 ⟨83TL1171⟩, 78, 271 ⟨85TL2677⟩, 213 ⟨84CL1259⟩; **4**, 197 ⟨83TL4993⟩, 942 ⟨90T4553⟩; **5**, 346 ⟨91TL7683⟩, 411 ⟨85TL2675, 87JAP6261987⟩; **6**, 126 ⟨85TL2675, 87JAP6261987⟩, 461 ⟨91JOM(417)305⟩
Kawada, Y., **1**, 822 ⟨91TL6159⟩; **6**, 304 ⟨82BCJ1106⟩
Kawagashi, T., **1**, 607 ⟨81TL1809⟩
Kawagishi, T., **4**, 193 ⟨81TL1809⟩
Kawaguchi, T., **5**, 1107 ⟨90CPB1389⟩; **6**, 426 ⟨89MI 614-04⟩
Kawahara, N., **3**, 275 ⟨82CPB63⟩; **4**, 827 ⟨80H(14)15⟩, 893 ⟨80H(14)619, 82CPB63⟩
Kawahara, T., **5**, 223 ⟨88OM366⟩
Kawahara, Y., **5**, 541, 542 ⟨92CB417, 93CC277⟩
Kawaharasaki, N., **5**, 303 ⟨82S933⟩
Kawai, A., **4**, 210 ⟨84TL5409⟩
Kawai, K., **2**, 71 ⟨88PAC1597⟩
Kawai, M., **1**, 209 ⟨74CL385⟩
Kawai, N., **5**, 775 ⟨83CPB2564, 83CPB3139⟩
Kawai, T., **2**, 25 ⟨81TL4409⟩, 393 ⟨90JOC2437⟩; **4**, 94 ⟨85CC1266⟩, 937 ⟨87T4759⟩; **5**, 307 ⟨85CPB5515⟩, 1106 ⟨81TL4409⟩
Kawai, Y., **2**, 40 ⟨91JOC4778, 91TL7075⟩; **3**, 689 ⟨92CL1053⟩
Kawajima, I., **6**, 125 ⟨78BCJ2391⟩
Kawakami, H., **2**, 55 ⟨88CL1633⟩
Kawakami, M., **1**, 1112 ⟨76CC612⟩
Kawakami, N., **3**, 419 ⟨87S824⟩
Kawakami, S., **4**, 1033 ⟨80BCJ1138⟩; **6**, 26 ⟨81TL871⟩
Kawakami, T., **5**, 1120 ⟨92CC1379⟩
Kawakami, Y., **1**, 119, 471, 1092 ⟨83JOC1912⟩; **3**, 307 ⟨83TL3455⟩, 639 ⟨85TL5835⟩
Kawakubo, H., **2**, 326 ⟨93JMC3526⟩
Kawamata, A., **4**, 856 ⟨78CPB685⟩
Kawamori, M., **6**, 504 ⟨82CPB534⟩
Kawamoto, A., **1**, 232 ⟨89BCJ164⟩
Kawamoto, F., **1**, 571 ⟨88OSC(6)327⟩, 1107 ⟨74TL1067⟩; **2**, 616 ⟨74TL1067⟩; **3**, 224 ⟨80OS(59)113⟩
Kawamoto, H., **2**, 431 ⟨91BCJ2398⟩
Kawamoto, I., **3**, 220, 496 ⟨74TL4223⟩; **6**, 553 ⟨87H(25)123⟩
Kawamoto, K., **1**, 816, 817, 818 ⟨85CL1719⟩, 816, 818 ⟨85TL5799⟩, 816, 819 ⟨86TL4899⟩; **2**, 912 ⟨83CC1193, 86JOC3890⟩, 912 ⟨87JOC4864⟩
Kawamoto, T., **1**, 86 ⟨86JOC537⟩, 511, 534, 1106 ⟨87JOC1624⟩; **2**, 922, 923 ⟨88T4009⟩
Kawamukai, H., **3**, 119 ⟨89BCJ2585⟩
Kawamura, E.-i., **4**, 850 ⟨80JCS(P1)2693⟩; **6**, 564 ⟨81JCS(P1)2499⟩
Kawamura, K., **2**, 1000 ⟨87JA3010⟩; **5**, 223 ⟨89TL5821⟩
Kawamura, S., **1**, 187 ⟨84JA1079⟩; **2**, 211 ⟨70BCJ1890⟩; **5**, 170 ⟨85TL3207⟩, 247 ⟨70BCJ1890, 71JOC3677⟩, 247 ⟨75BCJ2993, 75CL1231⟩, 540 ⟨81BCJ790⟩, 1111 ⟨83MI 530-01⟩; **6**, 478 ⟨92JOC1053⟩
Kawamura, T., **3**, 439 ⟨76BCJ1913⟩; **5**, 222 ⟨77BCJ2021⟩, 255, 506, 509, 543 ⟨90TL3587⟩
Kawana, M., **5**, 652 ⟨78JCS(P1)558⟩
Kawanami, H., **3**, 689 ⟨92CL901⟩
Kawanami, Y., **1**, 1045 ⟨81CL1415, 83TL5131⟩; **3**, 273 ⟨83TL5131⟩; **5**, 173 ⟨81CL1415⟩
Kawanishi, K., **2**, 694 ⟨90CPB2637⟩
Kawanishi, M., **2**, 647 ⟨66BCJ1212⟩
Kawanishi, Y., **3**, 117 ⟨83TL5009⟩
Kawanisi, M., **1**, 11, 109 ⟨70BCJ2506⟩, 769 ⟨80JOC2005⟩, 1004 ⟨82CL1485⟩, 1042 ⟨83TL1801, 87JOC1984⟩, 1060 ⟨87CL1519⟩; **3**, 498, 651 ⟨84SC1267⟩; **4**, 156 ⟨76BCJ2837⟩, 810 ⟨82SC1027⟩; **5**, 146 ⟨87CC83⟩, 460 ⟨84JOC4460⟩
Kawano, T., **1**, 571 ⟨91JOC7177⟩; **3**, 694 ⟨90JA2437⟩; **5**, 803, 804 ⟨90JA2437⟩
Kawano, Y., **3**, 674 ⟨70JOC675⟩; **4**, 762 ⟨70JOC828⟩
Kawanobe, W., **5**, 368 ⟨85MAC1803⟩
Kawara, R. O., **4**, 659 ⟨74JOM(70)C21⟩
Kawara, T., **1**, 230 ⟨80TL2181⟩, 481 ⟨82BCJ3555⟩
Kawasaki, A., **1**, 193 ⟨82TL5549⟩; **4**, 437 ⟨64JOC1985⟩
Kawasaki, H., **1**, 135 ⟨93TL1963⟩, 1004 ⟨82CL1485⟩; **3**, 291 ⟨85TL3027, 85TL3031⟩; **4**, 810 ⟨82SC1027⟩; **6**, 128, 327 ⟨85TL3031⟩
Kawasaki, I., **2**, 658 ⟨71BCJ1986⟩; **4**, 448 ⟨90H(31)2029⟩
Kawasaki, T., **2**, 289 ⟨79TL2251⟩; **3**, 249 ⟨79TL1129⟩, 669 ⟨80JCS(P1)1132⟩; **5**, 324 ⟨91H(32)221⟩, 595 ⟨79S887⟩, 1041, 1107 ⟨77TL4417⟩, 1048 ⟨78CI(L)806⟩, 1055, 1114 ⟨79TL2251, 79TL1129⟩, 1107 ⟨81JCS(P1)1577⟩; **6**, 4 ⟨92JAP04164039⟩
Kawasaki, Y., **1**, 186, 225, 229 ⟨89JA7938⟩, 202 ⟨90JOC5923⟩; **2**, 695 ⟨67BCJ428⟩; **5**, 402 ⟨73BCJ1803⟩, 948 ⟨82JCS(D)793⟩
Kawasato, H., **1**, 985 ⟨92CL1357⟩
Kawase, A., **1**, 164 ⟨92SL891⟩; **4**, 155, 700 ⟨92SL891⟩
Kawase, M., **2**, 360 ⟨84JCS(P1)1401⟩; **5**, 317 ⟨89JOC3394⟩, 328 ⟨90SC887⟩, 332 ⟨90TL243⟩
Kawase, T., **1**, 990 ⟨92CC778⟩; **4**, 345 ⟨78JOC1980, 79BCJ3342⟩; **6**, 174 ⟨87AG(E)584⟩, 316 ⟨78JOC595⟩, 698 ⟨83TL3469⟩
Kawase, Y., **2**, 281 ⟨90CZ353⟩; **3**, 304 ⟨70BCJ2884, 70BCJ3496⟩
Kawashima, A., **3**, 8, 122 ⟨79CC303⟩
Kawashima, E., **1**, 925, 946 ⟨93TL1317⟩; **5**, 220 ⟨86JHC1657⟩
Kawashima, H., **3**, 145 ⟨92TL951⟩
Kawashima, K., **3**, 618, 619 ⟨73TL2121⟩
Kawashima, M., **1**, 230 ⟨80TL2181⟩, 476 ⟨81CL1307, 82TL3583⟩, 481 ⟨82BCJ3555⟩; **3**, 428 ⟨91CL763⟩
Kawashima, T., **1**, 754 ⟨84CL1997⟩, 836 ⟨86CL501⟩; **2**, 256 ⟨87JA1227⟩, 460, 462 ⟨89CL849⟩, 547 ⟨92JA7598⟩, 831 ⟨92JA4008⟩; **3**, 400 ⟨93CL1047, 93JA7019, 88BCJ461⟩; **4**, 334 ⟨93CL1047⟩, 570, 573 ⟨87BCJ1831⟩; **6**, 57 ⟨71CPB2629⟩
Kawashima, Y., **5**, 68 ⟨74S443⟩
Kawashiro, K., **5**, 226 ⟨85BCJ1903⟩
Kawashita, T., **6**, 465 ⟨93EUP559212⟩

Kawata, H., **2**, 309 ⟨93BMC2883⟩
Kawata, I., **5**, 197 ⟨82CC132⟩
Kawata, T., **6**, 472 ⟨73CPB604, 78CPB3807⟩
Kawate, T., **2**, 313 ⟨93CPB287⟩, 330 ⟨92TA227⟩; **3**, 613 ⟨84CL725⟩
Kawato, K., **5**, 164 ⟨90CL1549⟩
Kawauchi, T., **5**, 197 ⟨82CC132⟩
Kawazoe, K., **3**, 634 ⟨85H(23)3069⟩; **5**, 1039 ⟨85H(23)3069⟩
Kawazoe, T., **1**, 194 ⟨87CL1237⟩
Kawazoe, Y., **4**, 521 ⟨85TL5547, 86CPB3273⟩; **5**, 551, 560, 574, 575 ⟨84S827⟩
Kawecki, M., **6**, 371 ⟨93OM2757⟩
Kawecki, R., **4**, 266 ⟨89BAP117⟩
Kay, D. J., **4**, 306 ⟨56JA2167⟩
Kay, G., **6**, 68 ⟨1854LA(92)346, 1854PRS135⟩
Kay, I. T., **1**, 769 ⟨84TL2035⟩; **2**, 271 ⟨70JCS(C)2409⟩; **3**, 269 ⟨77S607⟩; **4**, 230 ⟨85JCS(P1)587⟩, 840 ⟨68JCS(C)3011⟩; **5**, 532 ⟨70GEP(O)2024393⟩, 592 ⟨77S607⟩; **6**, 445 ⟨86TL113⟩, 629 ⟨73JCS(P1)2644⟩
Kay, L. T., **2**, 401 ⟨77JCR(S)237⟩
Kay, S. N., **4**, 455 ⟨81JINC629⟩
Kaya, R., **4**, 853 ⟨81JOC196⟩; **5**, 546 ⟨81JOC196⟩
Kayama, M., **5**, 57 ⟨82CL1255⟩
Kayano, T., **6**, 539 ⟨89JAP03020256⟩
Kaydos, J. A., **4**, 67 ⟨83JOC1096⟩
Kaye, J. A., **3**, 279 ⟨53JA745⟩
Kaye, P. T., **5**, 197 ⟨89SC3331⟩
Kayen, A. H. M., **2**, 765 ⟨71RTC548⟩
Kayser, M. M., **5**, 358 ⟨73CJC4115⟩
Kayser, R., **4**, 563 ⟨89ZAAC(576)272⟩
Kazaki, R. O., **2**, 212 ⟨84JA5271⟩
Kazama, Y., **5**, 29 ⟨59JA5817⟩
Kazan, J., **5**, 208 ⟨63JOC2168⟩
Kazanina, T. Yu., **2**, 1074 ⟨91ZN(B)1578⟩
Kazanis, S., **1**, 346, 926 ⟨93JOC7640⟩
Kazankova, M. A., **2**, 1027 ⟨83MI 221-01⟩; **4**, 559, 766 ⟨90ZOB1539⟩, 765 ⟨77ZOR1675, 89ZOB955⟩, 766 ⟨91ZOB1014⟩, 820 ⟨76JGU2367⟩, 884 ⟨91ZOB1600⟩, 914 ⟨84ZOB416⟩, 915 ⟨78ZOB267⟩, 1055, 1058 ⟨80ZOB692⟩
Kazantsev, A. V., **4**, 143 ⟨82IZV1425⟩, 343 ⟨79MI 407-01⟩
Kazembe, T. C. G., **3**, 69 ⟨80T2125⟩
Kazenwadel, W., **2**, 1062 ⟨76MI 222-02, 76ZN(B)172⟩
Kazi, A. B., **1**, 822 ⟨90CL613⟩
Kazimirchuk, E. I., **6**, 457 ⟨87IZV1671⟩
Kazin, V. N., **4**, 979 ⟨91ZOR382⟩
Kazitsyna, L. A., **4**, 587 ⟨68ZOR1685⟩
Kazlauskas, D. A., **5**, 361 ⟨74ZOR2301⟩
Kazlauskas, R., **2**, 299 ⟨83JCS(P1)2983⟩
Kazlauskas, R. J., **1**, 573 ⟨93TA879⟩
Kazmi, S. H.-u-H., **1**, 218 ⟨88SC151⟩
Kazmierczak, F., **1**, 781 ⟨89JOC3988⟩; **3**, 212 ⟨89JOC3988⟩
Kazoura, S. A., **1**, 656 ⟨84OM1340⟩
Kazubski, A., **2**, 58 ⟨92JOC2953⟩
Kazuo, S., **3**, 357 ⟨91EUP409164⟩
Kdalekov, A. M., **2**, 561 ⟨83IZV2053⟩
Ke, B.-W., **4**, 661 ⟨92TL7895⟩
Ke, C. Y., **1**, 185 ⟨93JOC3767⟩
Ke, M., **2**, 583 ⟨87CC1110⟩, 584 ⟨86CC787⟩
Keable, H. R., **3**, 460 ⟨72JCS(D)153⟩
Keable, J., **2**, 815 ⟨76JCS(D)1⟩; **3**, 456 ⟨78JCS(D)986⟩, 457 ⟨76JCS(D)1⟩
Kealy, T. J., **1**, 862 ⟨51NAT1039⟩
Keana, J. F. W., **4**, 191 ⟨B-63MI 404-01⟩
Kearney, J. A., **5**, 358 ⟨73JCS(P2)1422⟩
Kearns, M. M., **2**, 367 ⟨82JCS(P1)677⟩
Kearny, J. A., **5**, 703, 704 ⟨71CC689⟩
Keasey, A., **4**, 693 ⟨78JCS(D)1825, 78JCS(D)1830⟩
Keat, R., **2**, 864 ⟨B-73MI 216-01⟩; **4**, 550, 563 ⟨76JCS(D)1113⟩, 552 ⟨80JCS(D)871⟩
Keay, B. A., **2**, 942 ⟨87TL5965⟩
Keay, J. G., **1**, 87 ⟨86AHC(39)1⟩; **4**, 310 ⟨94H(37)1951⟩
Kebarle, P., **1**, 905 ⟨85JA3747⟩
Kebekbaeva, M. M., **2**, 108 ⟨91JGU999⟩
Keblys, K. A., **5**, 448 ⟨60JA4204⟩
Keck, G. E., **1**, 173, 321, 334 ⟨85T4079⟩, 173, 334 ⟨82JA5829⟩, 173, 703 ⟨85JOC4317⟩, 344, 1130 ⟨85TL3311⟩, 348 ⟨87JOC2958⟩, 961 ⟨82TL3051⟩; **2**, 78 ⟨84TL1879, 84TL3927⟩; **3**, 31 ⟨93JA8467, 93JOC6543⟩, 56 ⟨89JOC4345⟩; **5**, 322, 330 ⟨93JOC6083⟩, 326 ⟨79TL1185, 79TL4627⟩, 326 ⟨81JA3173⟩; **6**, 541, 550, 557 ⟨85T4079⟩
Keck, H., **6**, 139, 140 ⟨59CB329⟩, 140 ⟨60AG956⟩
Keck, R., **2**, 11 ⟨82TL1549⟩
Kecskes, A., **6**, 223 ⟨81MIP68355⟩
Keder, N. L., **5**, 917 ⟨80ZN(B)1289⟩; **6**, 706 ⟨80ZN(B)1289⟩
Kedesdy, E., **4**, 797 ⟨08CB2190⟩
Kee, A., **2**, 327 ⟨91SL29⟩
Kee, M. L., **2**, 174, 717 ⟨70OPP235⟩; **4**, 78 ⟨70OPP235⟩
Keefe, J. R., **1**, 869 ⟨B-90MI 119-02⟩
Keefer, L. K., **2**, 379 ⟨84JOC3470⟩
Keefer, R. M., **5**, 204 ⟨59JA2374⟩
Keeffe, J. R., **1**, 896 ⟨64JA4727⟩; **2**, 637 ⟨87JA4000, 90JA4862⟩
Keegan, D. S., **1**, 816, 818 ⟨91JOC1185⟩
Keehn, P. M., **3**, 119 ⟨82TL35⟩, 618 ⟨76TL603⟩; **4**, 436 ⟨81CJC679⟩
Keek, J., **2**, ⟨B-74MI 206-01⟩
Keeley, D., **1**, 390 ⟨73JA3068⟩
Keeley, D. E., **4**, 216 ⟨77JA4835⟩
Keenan, T. P., **1**, 166 ⟨93TL63⟩
Keene, F. R., **5**, 385 ⟨73JA5649⟩, 386 ⟨81JA3559⟩
Keeping, J. W., **3**, 714, 718 ⟨85CC816⟩
Kees, F., **2**, 176 ⟨74S489⟩, 401 ⟨81CB774⟩; **5**, 364 ⟨81CB802⟩
Kees, K. L., **1**, 688, 689, 690 ⟨78JOC3255, 82TL1777⟩, 690 ⟨87T5489⟩
Keese, R., **1**, 1038 ⟨93TL1467⟩; **3**, 659 ⟨69AG(E)343⟩
Keese, W., **3**, 27 ⟨85CB4288⟩
Keglevich, G., **2**, 450 ⟨86JCS(P2)1029⟩
Kehm, B. B., **6**, 504 ⟨63OSC(4)515⟩
Keidrowski, G. V., **5**, 1062 ⟨88LA787⟩
Keifer, L., **6**, 509 ⟨78JOC61⟩
Keiko, N. A., **2**, 741 ⟨89S446⟩; **3**, 430 ⟨85ZOR766⟩
Keiko, V. V., **1**, 63 ⟨81JOM(214)145⟩; **2**, 266 ⟨81JOM(214)415⟩; **3**, 533 ⟨76ZOB930⟩; **4**, 368 ⟨81JOM(204)13⟩, 507 ⟨87JOM(326)159⟩, 933 ⟨81IZV654⟩, 940 ⟨77IZV596⟩
Keil, R., **1**, 453 ⟨90TL7257⟩; **2**, 588 ⟨92TL5857⟩, 969 ⟨90TL7257⟩
Keilen, G., **2**, 155 ⟨92ACS867⟩
Keim, R., **6**, 217, 220 ⟨31ZAAC(201)245⟩
Keim, W., **6**, 456 ⟨69JOM(19)161⟩
Keinan, E., **1**, 86 ⟨87JOC2576⟩, 194 ⟨80TL2591⟩, 194, 1053 ⟨83JOC5302⟩, 975, 1025 ⟨86JOC4006⟩; **2**, 14 ⟨90JOC2927⟩, 36 ⟨87JOC4846⟩, 49 ⟨92TL6407, 92TL6411⟩, 86 ⟨75JOC2141, 88OSC(6)43⟩, 397 ⟨77JOC844⟩, 706 ⟨86JOC3165⟩; **3**, 123 ⟨90JOC2927⟩, 129 ⟨87JOC5570⟩, 139 ⟨80TL2591⟩, 621 ⟨86JA7314⟩, 654 ⟨85JOC3558⟩; **5**, 20 ⟨90JOC3922⟩
Keinan, E. H., **3**, 128, 621 ⟨87JOC2576⟩
Kein, W., **2**, 192 ⟨93CC1592⟩
Keindl, M. C., **2**, 221 ⟨83S913, 85S500⟩
Keinen, E., **4**, 176 ⟨87CAR(162)145⟩
Keiner, P., **4**, 217 ⟨91TL467⟩
Keino, K., **3**, 303 ⟨90JCS(P1)2677⟩
Keiser, J. E., **2**, 149 ⟨73OSC(5)791⟩
Keister, J. B., **6**, 340, 344 ⟨83OM219⟩, 343 ⟨83OM1044, 91OM2384⟩, 343, 344 ⟨85JOM(287)357⟩
Keitel, I., **4**, 559 ⟨75JPR890, 76JPR403⟩, 1002 ⟨81T1227⟩; **6**, 119, 314 ⟨87S44⟩, 120 ⟨78JPR255⟩, 155 ⟨91PS(62)35⟩
Keiter, R. L., **4**, 588 ⟨76JOM(107)73⟩
Keith, A. N., **4**, 714 ⟨80CC749⟩

Keith, D. D., **5**, 315 ⟨78JOC3713⟩
Keith, D. R., **3**, 61 ⟨79S132⟩
Keith, L. H., **4**, 706 ⟨65JOC1600⟩
Kekulé, A., **2**, 647 ⟨1866MI 213-01⟩
Kelarev, V. I., **5**, 350 ⟨92ZOR1750⟩, 691 ⟨92JOU1399⟩
Kelbaugh, P. R., **6**, 553 ⟨92JOC4352⟩
Kelber, C., **5**, 547 ⟨11CB1693⟩
Keldsen, G. L., **1**, 51 ⟨78JA7312⟩
Kell, D. A., **1**, 524 ⟨85JOC2309⟩
Kelland, J. W., **2**, 565 ⟨82COMC-I(3)433⟩
Kellbaugh, P. R., **2**, 1020 ⟨90JA6696⟩
Kelleghan, W. J., **1**, 1137 ⟨81JOC2280⟩
Kelleher, R. G., **3**, 231 ⟨80CC486⟩
Keller, C. E., **1**, 917 ⟨66JA604, 66JA606⟩
Keller, H., **2**, 414 ⟨58HCA244⟩, 1036, 1037 ⟨90CB417⟩; **4**, 491 ⟨93H(36)1925⟩, 562 ⟨86TL1903⟩; **6**, 27 ⟨56CB2578⟩
Keller, K., **2**, 872 ⟨89JOM(366)73⟩; **4**, 800 ⟨86CB3150⟩
Keller, L., **1**, 468, 472 ⟨81JA6460⟩
Keller, T. H., **1**, 195 ⟨89TL5883⟩, 195 ⟨90TL1265⟩
Keller, W., **6**, 183 ⟨89JOM(367)19⟩
Kellermann, W., **4**, 245 ⟨93S149⟩
Keller-Schierlein, W., **5**, 214 ⟨86HCA236⟩
Kellert, M. D., **2**, 27 ⟨63OSC(4)108⟩
Kelley, A. E., **1**, 36 ⟨52JA3074⟩
Kelley, C., **5**, 449 ⟨89PAC1723⟩, 946 ⟨90JA3220, 92JA6735⟩
Kelley, G. W., **1**, 146 ⟨90JOC3840⟩
Kelley, S. M., **2**, 625 ⟨85HCA813⟩
Kelling, H., **4**, 602, 607, 637, 660 ⟨90ZAAC(583)195⟩
Kellner, D., **1**, 942 ⟨91JCS(P1)1433⟩
Kellner, J., **1**, 1012 ⟨88TL3651⟩
Kellner, K., **2**, 440 ⟨84ZC365⟩; **4**, 317, 499, 500 ⟨81ZC403⟩, 451, 452, 455, 481 ⟨84ZC365⟩, 455 ⟨80JOM(193)307, 87JOM(326)C9⟩, 455, 462, 478, 481, 499, 500, 501 ⟨78JOM(149)167⟩, 477, 481, 503 ⟨80ZC152⟩, 483, 499, 500 ⟨74JPR851⟩; **6**, 161 ⟨78JOM(149)167⟩
Kellogg, K. B., **6**, 231 ⟨48JA3986⟩
Kellogg, M. S., **1**, 4, 9 ⟨79TL4631⟩, 420, 421 ⟨73PAC197⟩
Kellogg, R., **2**, 55 ⟨91COS(8)79⟩; **5**, 1011 ⟨66JOC1079⟩
Kellogg, R. M., **1**, 367 ⟨76T2165⟩, 685 ⟨72JOC4045⟩, 706 ⟨74TL2805⟩; **2**, 117 ⟨B-69MI 203-01⟩, 120 ⟨87OS(66)150⟩, 124 ⟨86JOC3664⟩, 167 ⟨74JA3146⟩; **3**, 250 ⟨80JOC4804⟩, 368 ⟨72JOC4045, 76T2165⟩, 371 ⟨73TL3929, 84CHEC(4)713⟩; **4**, 234 ⟨80JOC4126⟩; **5**, 177 ⟨81JA5183⟩
Kellogg Co., M. W., **2**, 478 ⟨54USP2882313⟩
Kellom, D. B., **1**, 773 ⟨53JA6011⟩; **2**, 757 ⟨53JA2014⟩
Kellom, D. Y., **1**, 773 ⟨53JA6011⟩
Kelly, B. J., **4**, 515 ⟨89TL2703⟩
Kelly, C. J., **5**, 1113 ⟨84TL3387⟩
Kelly, D. P., **1**, 895 ⟨71JA464⟩, 906 ⟨82JOC2089⟩, 914 ⟨72JA2034⟩; **2**, 404 ⟨87JOC2911⟩
Kelly, D. R., **1**, 1024, 1134 ⟨80JCS(P1)852⟩; **3**, 714, 718 ⟨84CC133⟩; **5**, 262 ⟨B-89MI 506-02⟩
Kelly, E. G., **5**, 436 ⟨73JA8118⟩
Kelly, F. W., **1**, 773 ⟨71MI 117-01⟩
Kelly, J., **2**, 768 ⟨72ACS2877⟩
Kelly, J. M., **5**, 567, 630 ⟨89JMC2493⟩
Kelly, J. W., **2**, 830 ⟨85JA5210, 86JOC5490⟩
Kelly, L. F., **5**, 1126 ⟨85TL501⟩
Kelly, M. G., **4**, 47 ⟨68JOC3335⟩
Kelly, M. J., **1**, 173 ⟨90JCS(P1)2352⟩, 971, 985 ⟨86JA7791⟩; **2**, 202 ⟨84TL5889⟩, 938, 939 ⟨83T949, 82JA1119⟩; **5**, 423, 427 ⟨82JA1119⟩, 423, 427 ⟨83T949⟩; **6**, 543 ⟨91JCS(P1)787⟩
Kelly, P., **4**, 77 ⟨68ACR299⟩
Kelly, R. C., **4**, 168 ⟨73JA7156⟩
Kelly, R. D., **4**, 654, 656 ⟨89JOM(361)123⟩
Kelly, R. E., **6**, 349 ⟨85CC1776⟩
Kelly, R. W., **2**, 23 ⟨68JCS(C)416⟩
Kelly, S. E., **1**, 648 ⟨91COS(1)729⟩, 676, 677 ⟨91COS(1)807⟩, 678 ⟨91COS(1)749⟩, 696, 702 ⟨91COS(1)792⟩, 744 ⟨91COS(1)761⟩, 758 ⟨91COS(1)731⟩, 1045 ⟨84TL1757⟩; **2**, 997, 1003 ⟨91COS(1)729⟩
Kelly, T. A., **1**, 831 ⟨86JOC4485⟩, 832 ⟨88T7587⟩
Kelly, W., **5**, 622 ⟨48JCS1919⟩
Kelly, W. J., **1**, 19 ⟨78S763⟩; **4**, 332, 428, 429 ⟨83JOC332⟩
Kelly, W. L., **1**, 869, 871, 881 ⟨68JOC3477⟩
Kelly, W. S. J., **4**, 546 ⟨71JCS(A)3495⟩
Kelmers, A. D., **2**, 374 ⟨56JA3874⟩
Kelsey, D. R., **1**, 472 ⟨86JOC2627⟩, 1114 ⟨71JA1941, 73CC589⟩
Kelsey, R., **4**, 174 ⟨66JOC3032⟩
Kelvin, D. M., **2**, 517 ⟨88T5597⟩
Kem, K. M., **4**, 563 ⟨89USP4818774⟩
Kemal, Ö., **4**, 317 ⟨81JCS(P1)1569⟩
Kemmitt, R. D. W., **1**, 194, 289 ⟨86CC1201⟩; **2**, 575 ⟨82COMC-I(5)1⟩, 824 ⟨68JCS(A)2149⟩; **5**, 465 ⟨73CC158, 74CC841⟩, 466 ⟨70CC198⟩, 802 ⟨89JCS(D)345, 90JCS(D)781⟩
Kemmler, M., **6**, 154 ⟨93AG(E)985⟩
Kemp, B., **6**, 491 ⟨84IC3063⟩
Kemp, D., **5**, 204 ⟨89T3299⟩
Kemp, D. S., **2**, 127 ⟨93JOC2216⟩, 229 ⟨89JOC3640⟩; **3**, 620 ⟨87TL464⟩, 620, 621 ⟨87TL4645⟩
Kemp, R. A., **2**, 430 ⟨83IC547⟩; **6**, 165, 167 ⟨82OM1720⟩, 191 ⟨81SRI591⟩
Kempe, T., **1**, 704, 708 ⟨77JA6946⟩; **2**, 177 ⟨73ACS1451⟩
Kemper, B., **2**, 529 ⟨84T2219, 85LA2246⟩
Kempf, D., **6**, 417 ⟨92EUP486948⟩
Kempf, D. J., **1**, 148 ⟨83JA2080⟩
Kempf, H., **1**, 576 ⟨87T853⟩
Kempf, K. W., **4**, 872 ⟨84T4963⟩; **6**, 94, 307, 536, 734 ⟨84T4963⟩
Kempter, G., **4**, 410 ⟨82JPR832⟩
Kenan-Yaser, H., **2**, 245 ⟨90JA7653⟩
Kendall, E. C., **2**, 30 ⟨41OSC(1)521⟩; **3**, 632 ⟨41OSC(1)256⟩; **5**, 56 ⟨32OSC(1)131⟩, 85 ⟨32OSC(1)21⟩
Kendall, J. D., **3**, 372 ⟨56JCS3189⟩; **6**, 84 ⟨45USP2389153, 48JCS687⟩
Kendall, P. E., **2**, 589 ⟨74JA2806⟩
Kendall, P. M., **3**, 289 ⟨75JA5448⟩
Kendall, R. V., **5**, 626 ⟨70JOC806⟩
Kende, A. S., **1**, 391 ⟨60OR(11)261⟩, 615, 698 ⟨90TL7105⟩, 690 ⟨86JA3513⟩, 797 ⟨82JOC163⟩, 926 ⟨88JA2210⟩, 999 ⟨82TL2373⟩, 1016, 1108 ⟨84JA3551⟩, 1038 ⟨88TL4217⟩, 1065 ⟨88JOC2655⟩, 1143 ⟨70JA7606⟩; **2**, 341 ⟨91TL1699⟩; **3**, 46, 189 ⟨91TL1699⟩, 130, 286 ⟨78SC59⟩, 146 ⟨82TL2373⟩, 208 ⟨82TL1751⟩, 538 ⟨69TL4517⟩, 696, 719 ⟨93TL579⟩; **4**, 208 ⟨85TL5411⟩; **5**, 48 ⟨60OR(11)261⟩, 223 ⟨89TL5821⟩, 555 ⟨82SC829⟩
Kendo, M., **2**, 631 ⟨87CL2109⟩
Kendurkar, P. S., **1**, 268, 738, 739 ⟨76JOM(108)175⟩, 738 ⟨73JOM(60)247, 75JOM(85)173⟩, 738 ⟨75JOM(96)237⟩
Kengerli, M. I., **5**, 433 ⟨80MI 509-01⟩
Ken-ichiro Nishiwaki, **6**, 190, 194, 393 ⟨73TL4193⟩
Kenley, R. A., **5**, 664 ⟨84JMC1201⟩, 830 ⟨81JMC1124⟩
Kennard, C. H., **4**, 603, 649 ⟨91OM3167⟩
Kennard, C. H. L., **1**, 572 ⟨93JOC2501⟩; **2**, 29 ⟨93JOC2501⟩; **6**, 247 ⟨93JOM(453)299⟩
Kennard, O., **1**, 641, 755 ⟨84JCS(P1)243⟩; **2**, 889 ⟨70JCS(C)810⟩
Kennedy, B. W., **3**, 328 ⟨72CC912⟩
Kennedy, D. J., **4**, 509 ⟨91JCS(P1)1091⟩; **5**, 8 ⟨89S598⟩
Kennedy, E. R., **1**, 993 ⟨76JOC3191⟩
Kennedy, J. A., **6**, 558, 564 ⟨93JMC1802⟩
Kennedy, J. D., **1**, 83 ⟨73JOM(61)167⟩
Kennedy, J. P., **1**, 128 ⟨70JOC532⟩, 462 ⟨82MI 111-01⟩
Kennedy, J. R., **6**, 221, 223 ⟨79JCED251⟩, 223, 226, 227, 228 ⟨79JMR559⟩
Kennedy, K. E., **4**, 182 ⟨94TL57⟩

Kennedy, M., **2**, 142 ⟨91COS(7)193⟩; **3**, 232 ⟨90JCS(P1)1047⟩; **4**, 62, 66 ⟨91COS(7)206⟩
Kennedy, R. J., **2**, 192 ⟨60JOC1901⟩
Kennedy, R. M., **1**, 23 ⟨87JOC2317⟩, 842 ⟨84TL4455⟩; **2**, 58, 521 ⟨86JA7404, 86JA7402⟩; **4**, 169 ⟨84TL4455⟩, 191 ⟨87JA1597⟩
Kennedy, S., **6**, 345 ⟨84CC392⟩
Kennedy, V. H., **4**, 1080 ⟨88MI 424-01⟩
Kennedy, V. O., **1**, 1010 ⟨90SL427⟩
Kennemann, A., **4**, 234 ⟨84AP(317)15⟩
Kennepohl, D. K., **2**, 477 ⟨90IC5081⟩; **6**, 622 ⟨90IC5081⟩
Kenner, G. W., **1**, 432 ⟨55JCS522⟩; **2**, 694 ⟨55JCS522⟩; **3**, 308 ⟨66T(S)241⟩; **5**, 84 ⟨57MI 502-02⟩, 264 ⟨85JCS(P1)461⟩, 278 ⟨60T39⟩, 712 ⟨55CI(L)1087, 56JCS1231⟩
Kenner, J., **3**, 451 ⟨36CB341⟩, 465 ⟨33JCS363, 37JCS1551⟩, 465, 490 ⟨35JCS286⟩
Kennewell, P.,**1**, 546 ⟨93TL153⟩
Kennewell, P. D., **2**, 163 ⟨80CSR477⟩, 664 ⟨78CC118⟩; **3**, 39 ⟨82JCS(P1)307⟩; **4**, 330 ⟨75CSR189, 80CSR477⟩; **5**, 1123 ⟨92T7703⟩; **6**, 111 ⟨90TL123⟩
Kenngott, E., **2**, 1055 ⟨26JPR314⟩
Kennon, L., **5**, 847 ⟨67JHC483⟩
Kenny, A., **6**, 241 ⟨81JFC(17)85⟩
Kenny, C., **2**, 81 ⟨89JA8236⟩
Kenny, D. H., **2**, 11 ⟨79JCS(P1)436⟩
Kenny, III, J. W., **1**, 1143 ⟨78JA6930⟩
Keno, Y., **5**, 701 ⟨71CPB1606⟩
Kent, D. E., **4**, 119, 129 ⟨86JCS(P1)913⟩; **6**, 263 ⟨84JCS(P1)1119⟩
Kent, P. W., **1**, 2, 6 ⟨60JCS298⟩
Kent, R. E., **3**, 618 ⟨55OSC(3)493⟩; **4**, 827 ⟨46JA1922⟩; **5**, 275 ⟨55OSC(3)490⟩; **6**, 70, 80 ⟨46JA1922⟩
Kenyon, G. L., **1**, 869 ⟨91JA9667⟩; **4**, 563 ⟨80JOC2698⟩
Kenyon, J., **1**, 398 ⟨47JCS93⟩; **2**, 118 ⟨50JCS3309⟩, 127 ⟨37JA2473⟩, 188 ⟨52JCS4887⟩, 286, 732 ⟨28JCS2293⟩; **3**, 620 ⟨16CB2697⟩; **5**, 214 ⟨46JCS25⟩, 1000 ⟨41JCS263⟩
Kenyon, R. L., **3**, 231 ⟨43OS(23)78⟩
Kenyon, W. G., **1**, 881 ⟨65JOC2937, 65JOC4135⟩, 881 ⟨67OS(47)72⟩
Keogh, J., **5**, 905 ⟨76JOC2102⟩
Keough, P. T., **2**, 442 ⟨59JA4803⟩, 444 ⟨60JA3919⟩; **3**, 494 ⟨60JA3919⟩
Kepler, J. A., **1**, 84 ⟨87JMC1505⟩
Kepler, K. D., **2**, 4, 23, 30 ⟨90JA7431⟩, 4, 23, 30 ⟨93JA3071⟩
Keppens, M. A., **1**, 152 ⟨93TL4693⟩
Keraly, F. X. L., **2**, 387 ⟨64CPB326⟩
Kerbachi, R., **6**, 615, 728 ⟨87JFC(37)259⟩
Kerbage, E., **4**, 837 ⟨92SL493⟩
Kerber, R., **1**, 674 ⟨76MAC1349⟩
Kerber, R. C., **2**, 393 ⟨67JA725⟩; **3**, 519 ⟨81JA1862⟩, 520 ⟨82JA6119⟩
Kerby, M. C., **4**, 675 ⟨91IC156⟩
Kerdeskey, F. A. J., **2**, 532 ⟨81JA1566⟩; **3**, 306 ⟨91TL2003⟩
Kerekes, I., **2**, 6 ⟨73S786⟩, 6, 11, 17, 392, 615 ⟨79JOC3872⟩, 17 ⟨73S779⟩, 18 ⟨73S780⟩, 19 ⟨74JA925⟩; **4**, 10 ⟨74JA925⟩, 32 ⟨79JOC3842⟩; **5**, 4 ⟨79JOC3726, 73S487⟩, 5 ⟨74JA925⟩; **6**, 52 ⟨74JA925⟩, 439 ⟨79JOC3872⟩
Kerekgyarto, J., **4**, 190 ⟨82S421⟩
Kerfanto, M., **4**, 309 ⟨74CR(C)221⟩, 317, 322 ⟨69CR(C)1718, 70CR(C)436⟩, 322 ⟨73BSF3499⟩, 405, 411 ⟨75BSF196, 65CR(261)2232⟩
Kergoat, R., **4**, 674 ⟨92OM721⟩
Kergomard, A., **2**, 95 ⟨63BSF470⟩; **5**, 318 ⟨61BSF2360⟩; **6**, 81, 90, 91 ⟨64CR(259)4051⟩, 302 ⟨67BSF3233⟩
Kerkmann, T., **5**, 729 ⟨80AG(E)466⟩
Kermadec, D. de, **2**, 8 ⟨93TL2757⟩
Kermode, N. J., **4**, 584 ⟨81CC698, 82JOM(228)C71⟩
Kern, H., **6**, 413 ⟨78MI 614-01⟩
Kern, W., **2**, 631 ⟨60LA(634)84⟩
Kerr, J. A., **1**, 321 ⟨68QR549⟩
Kerr, M. A., **1**, 556 ⟨92JA5959⟩, 580 ⟨88SC1905⟩; **3**, 57 ⟨88SC1905⟩
Kerr, R. G., **2**, 285, 1034, 1035 ⟨83JOC3077⟩, 388 ⟨82CJC2711⟩; **3**, 171 ⟨83JOC3077⟩, 174 ⟨85T4759⟩; **4**, 290, 335 ⟨85JOM(286)171⟩; **5**, 335, 338 ⟨86CJC308⟩, 1056, 1108 ⟨85JOM(286)171⟩
Kerrigan, V., **6**, 212, 214, 217, 218, 219, 220, 222, 225, 226, 246 ⟨48JCS2188⟩
Kersalov, A. V., **4**, 794 ⟨75JOU119⟩
Kerscher, T., **5**, 944 ⟨93JOM(460)C25⟩
Kerschl, S., **2**, 903 ⟨85CC1199, 86CC1170⟩, 903 ⟨86CC403⟩; **4**, 665 ⟨85CC1199⟩, 1068 ⟨88JOM(338)195⟩
Kersting, F., **5**, 869 ⟨61LA(647)11, 61LA(647)1⟩
Kerwin, Jr., J. F., **1**, 831 ⟨80JA6889⟩; **3**, 243 ⟨82JA358, 82JA360, 82JOC1981, 82JOC3183⟩; **4**, 174 ⟨82JA358⟩; **5**, 25, 108 ⟨46JA1844⟩, 273, 285 ⟨80JA7026⟩
Kerwood, J. E., **5**, 336 ⟨69JOC51⟩
Kerzhner, B. K., **5**, 566, 567, 610, 657 ⟨76ZOR2213, 76JOU2148⟩; **6**, 52 ⟨76ZOR2213⟩
Keschmann, E., **3**, 478 ⟨75T1817⟩
Keshavamurthy, K. S., **5**, 125, 189, 196, 297, 303 ⟨82S506⟩
Kesheng, F., **4**, 488 ⟨93MI 410-01⟩
Kesler, P., **4**, 203 ⟨86JOC3811⟩
Kesling, H. S., **2**, 612 ⟨79CL983⟩; **4**, 734 ⟨79CL983, 81JFC(18)293⟩; **6**, 213, 222, 223, 224, 226, 227, 228 ⟨82JFC(20)89⟩, 695 ⟨75TL3789, 79CL983⟩
Kesper, K., **3**, 359 ⟨89JST(198)307⟩
Kessar, S. V., **1**, 282 ⟨91JOC3908⟩, 1081 ⟨91COS(4)483⟩; **3**, 212 ⟨91CC568⟩
Kesselmayer, M. A., **4**, 763 ⟨87JA5029⟩; **6**, 280 ⟨86JA99⟩
Kessenikh, A. V., **6**, 62 ⟨67ZOB1040⟩
Kessick, M. A., **1**, 896 ⟨65JCS6718⟩
Kessler, H., **4**, 974 ⟨70CB973⟩; **6**, 731 ⟨71LA(743)1⟩, 732 ⟨69LA(727)228⟩
Kessler, K., **3**, 631 ⟨85AG(E)989⟩
Kessler, M., **3**, 694 ⟨50MI 321-01⟩
Kessler, P., **6**, 443 ⟨90S1065⟩
Kessler, W., **3**, 578 ⟨88JOC4997⟩
Kesten, S., **5**, 327 ⟨89OS(67)69⟩
Kester, J., **1**, 865 ⟨86TL5931⟩
Kesti, M. R., **2**, 912 ⟨92OM1095⟩
Kestner, M. M., **4**, 332 ⟨73JA3356⟩
Ketcha, D. M., **1**, 19 ⟨85JOC5451⟩; **2**, 164 ⟨84SC915⟩; **3**, 309 ⟨89JOC3264, 89JOC4350⟩; **5**, 350 ⟨85JOC2224⟩
Ketcham, R., **1**, 732 ⟨62JOC4666⟩, 784 ⟨82AG(E)225⟩; **3**, 334, 351, 369, 550 ⟨82AG(E)225⟩; **4**, 223, 225 ⟨74JOC2010⟩
Ketley, A. D., **6**, 69 ⟨60JA1398⟩
Ketterman, K. J., **5**, 43 ⟨73JOC1491⟩
Kettle, S. F. A., **1**, 860 ⟨B-65MI 119-03⟩
Kettler, P. B., **6**, 151 ⟨91CC979⟩
Kettler, R., **2**, 385 ⟨84LA1468⟩; **4**, 312 ⟨84LA1468⟩
Kettlewell, B., **3**, 537 ⟨73CC921⟩
Kettner, C. A., **4**, 527, 528 ⟨90JBC(265)3738, 84JBC(259)15106⟩, 528 ⟨85USP4499082, 88EUP293881⟩, 528 ⟨92MI 411-02⟩
Keukeleire, D. de, **1**, 356 ⟨78TL2057⟩
Keul, H., **3**, 148 ⟨81CB1858⟩, 326 ⟨75CB1642⟩; **5**, 141 ⟨78AG(E)939⟩
Keum, T., **3**, 617 ⟨79S112⟩
Keumi, T., **1**, 156 ⟨93T537⟩; **2**, 650 ⟨79S536, 81JOC4305⟩; **3**, 280, 296, 298, 305 ⟨80S139⟩, 280, 305 ⟨88BCJ455⟩, 282, 295, 296 ⟨90CL783⟩
Keus, D., **5**, 707 ⟨84JOC3466⟩
Keuser, U., **5**, 593, 597 ⟨64CB2209⟩, 595, 597, 599, 614 ⟨64CB3106⟩
Kevan, L., **2**, 1013 ⟨65JCP(42)2255⟩
Kevill, D. N., **6**, 421 ⟨B-72MI 614-01⟩
Kevill, P. N., **6**, 431 ⟨67JOC2633⟩
Key, M. E., **2**, 492 ⟨67IS92⟩

Keyer, R. A., **4**, 703 ⟨93OM2618⟩
Keys, B. A., **2**, 256 ⟨87JA1227⟩
Keys, B., **1**, 420 ⟨79JCS(P2)810⟩
Keys, D. E., **1**, 576 ⟨85JA4964⟩
Keys, R. T., **3**, 573 ⟨59JA4878, 60JA5394⟩
Keyton, D. J., **1**, 1131 ⟨69TL1653⟩
Keyzer, H., **2**, 430, 438 ⟨69T1097⟩
Kezar, III, H. S., **3**, 38, 174, 207 ⟨81JOC2920⟩
Kezdi, M., **6**, 558 ⟨85JPR1007⟩
Kezuka, H., **1**, 635, 1075 ⟨83T3279⟩
Khabashesku, V. N., **4**, 1083 ⟨87DOK(296)403⟩
Khac, T. B., **4**, 405 ⟨88OSC(4)474⟩
Khachatryan, R. A., **2**, 437 ⟨90ZOB313⟩, 439 ⟨86AKZ258⟩, 446 ⟨90AKZ610⟩, 447 ⟨86AKZ457⟩; **4**, 553 ⟨90AKZ537⟩
Khachkuruzov, G. A., **6**, 223 ⟨59MI 607-03⟩
Khadam, S. R., **1**, 548 ⟨81JOC2622⟩
Khadzhieva, B., **6**, 558 ⟨89MI 617-01⟩
Khafizova, G. S., **4**, 493 ⟨88JGU876⟩
Khai, B. T., **3**, 86 ⟨89JOC949⟩
Khaikin, L. S., **5**, 395 ⟨86JST(139)255⟩
Khairullina, S. V., **6**, 564 ⟨83ZOB785⟩
Khairullina, V. K., **6**, 564 ⟨83ZOB785⟩
Khairullin, R. A., **6**, 116 ⟨90CC1133⟩
Khairullin, R. S., **4**, 120 ⟨71JGU2183⟩
Khairullin, U. K., **4**, 125, 127, 128 ⟨68JGU2006⟩
Khairullin, V. K., **2**, 854 ⟨69ZOB341⟩; **4**, 124, 125 ⟨69JGU577⟩, 124, 125, 128 ⟨71IZV1159, 78JGU1813⟩, 129 ⟨78JGU1168⟩, 555 ⟨68ZOB2071⟩, 556 ⟨76ZOB266⟩, 561 ⟨66ZOB296⟩; **5**, 534 ⟨73JGU2635⟩; **6**, 538, 539 ⟨89IZV909⟩, 538, 539, 584 ⟨88ZOB1489⟩, 563 ⟨81ZOB530⟩, 564 ⟨81ZOB28⟩
Khalaf, A. A., **1**, 180, 212 ⟨B-84MI 105-01⟩
Khalafi-Nezhad, A., **2**, 104 ⟨84HCA906⟩
Khalaturnik, M. V., **2**, 444 ⟨75ZOB1228⟩
Khalid, M., **5**, 559 ⟨93TL2605⟩, 563 ⟨91CC964⟩
Khalifa, M. A. E., **4**, 984 ⟨77G555⟩
Khalikov, S. K., **6**, 547 ⟨85KPS844⟩
Khalil, A. M., **5**, 22 ⟨91PS(60)159⟩
Khalil, F. Y., **1**, 51 ⟨78CJC1933⟩
Khalil, H., **3**, 34, 60 ⟨77JOC2123⟩
Khalilov, Kh. D., **5**, 139 ⟨80MI 503-01⟩
Khalitov, F. G., **2**, 883 ⟨83JGU1832⟩
Khaliullin, F. A., **6**, 549 ⟨88ZOR117⟩
Khamliche, L., **3**, 650 ⟨86TL5491⟩
Khamsi, J., **2**, 675 ⟨86TL3619⟩, 684 ⟨86TL3615⟩
Khamsitthideth, S., **3**, 567, 574 ⟨82T527, 78CC274⟩
Khan, A. Q., **1**, 218 ⟨88SC151⟩
Khan, A. Z.-Q., **3**, 355 ⟨91JOC1902, 91JOC4919⟩
Khan, G., **5**, 101 ⟨79CC215⟩
Khan, G. M., **5**, 275 ⟨66CC121⟩
Khan, G. R., **5**, 316 ⟨84JOC4784⟩; **6**, 558 ⟨91PJS422⟩
Khan, H. A., **3**, 56 ⟨82TL2399⟩; **4**, 66 ⟨82TL2399⟩, 67 ⟨82TL5083⟩
Khan, H. R., **5**, 620 ⟨85AJC447⟩
Khan, J., **2**, 472 ⟨85AG(E)504⟩
Khan, J. H., **5**, 201 ⟨89T5979⟩
Khan, K. M., **1**, 831 ⟨89CC1812⟩
Khan, M. A., **1**, 595 ⟨89JCR(S)30, 91JCR(S)62⟩; **4**, 153, 588 ⟨91OM2159⟩, 153, 698 ⟨86OM525⟩, 187 ⟨89SC901⟩; **6**, 469 ⟨89JCR(S)15⟩
Khan, M., **1**, 199 ⟨94TL363⟩
Khan, N., **3**, 57 ⟨75JCS(P1)1457⟩, 528 ⟨87CC1728⟩
Khan, N. A., **2**, 30 ⟨63OSC(4)851, 63OSC(4)969⟩
Khan, N. H., **3**, 428 ⟨84IJC(B)801⟩; **4**, 441, 444 ⟨81IJC1003⟩; **5**, 94 ⟨79SC363⟩, 1041 ⟨78SC497⟩; **6**, 444 ⟨84IJC(B)986⟩
Khan, R. A., **1**, 1143 ⟨82CPL493⟩
Khan, R. H., **6**, 627 ⟨91JCR(S)160⟩
Khan, S. A., **2**, 626 ⟨88JOC1799⟩
Khan, S. I., **5**, 826, 829, 830 ⟨87JOC4570⟩
Khan, T., **1**, 456, 507 ⟨77JOM(141)113⟩
Khan, W. A., **1**, 148 ⟨66JOC2047⟩, 148 ⟨66JOC2061⟩; **4**, 914 ⟨72JA3058⟩
Khana, M. S., **1**, 556 ⟨92JCS(P1)2565⟩
Khananashvili, L. M., **5**, 720 ⟨72IJS319, 70IZV1157⟩
Khanapure, S. P., **1**, 940 ⟨93TL8245⟩; **2**, 753 ⟨86JOC5157⟩; **3**, 626 ⟨87JOC1333, 87JOC2619⟩
Khanbabaee, K., **3**, 88 ⟨90CB1357⟩
Khand, I. U., **1**, 528 ⟨73JCS(P1)975, 73JCS(P1)977⟩, 529 ⟨77JCR(M)153, 77JCR(M)168⟩, 529 ⟨78JCR(M)4401, 78JCR(M)4454⟩; **2**, 985 ⟨73JCS(P1)977, 77ANY(295)2⟩
Khandelwal, B. L., **2**, 690 ⟨85JOM(291)185⟩
Khandelwal, G. D., **3**, 542 ⟨71JCS(C)3328⟩
Khandker, M. N. I., **1**, 999 ⟨87IJC(B)77⟩
Khangurov, A. V., **2**, 1036 ⟨90ZOR2635⟩
Khanna, M. S., **2**, 311 ⟨91SL668⟩
Khanna, P. L., **2**, 20 ⟨77JA905⟩
Khanna, R. K., **1**, 68 ⟨83TL1411⟩; **4**, 587 ⟨86JOC5498⟩; **6**, 151, 369 ⟨86IC2699⟩, 370 ⟨85GEP3337939⟩
Khanna, V. V., **2**, 531 ⟨92JOC6608⟩
Khanolkar, A. D., **1**, 316 ⟨92JA8053⟩; **2**, 556 ⟨91JA5720, 92JA8053⟩
Kharasch, H. S., **2**, 23 ⟨39JA3432⟩ 557 ⟨B-54MI 211-01⟩
Kharasch, M. S., **1**, 114 ⟨B-54MI 104-01⟩, 117 ⟨41JA2316⟩, 311 ⟨41JA2308⟩; **2**, 67 ⟨B-54MI 202-01⟩, 96 ⟨58JOC1322⟩, 116 ⟨51JOC524⟩, 117 ⟨48JOC97⟩, 165 ⟨55JA1041⟩, 165 ⟨58JOC381⟩, 176 ⟨48JA3526⟩, 448, 489 ⟨49JOC429⟩, 557 ⟨B-54MI 211-01⟩; **3**, 323 ⟨53JOC1030⟩; **4**, 11 ⟨39JA2142⟩, 24 ⟨35JA2463⟩, 38, 39 ⟨34JA712⟩; **5**, 16 ⟨42JA329⟩; **6**, 5 ⟨47JA1100, 53JOC328⟩, 217 ⟨37JOC76⟩
Kharasch, N., **2**, 126 ⟨59JOC1029⟩, 169 ⟨63JOC1901⟩, 172, 216, 217 ⟨B-61MI 203-01⟩, 194 ⟨58JOC1694⟩, 216 ⟨66QRS285⟩, 357, 688 ⟨46CRV269⟩; **5**, 208 ⟨56JA1207⟩; **6**, 233 ⟨50JA3529⟩
Kharchenko, A. V., **5**, 655 ⟨89JOU2329⟩, 655, 1053 ⟨89ZOR2597⟩, 656 ⟨91UKZ530⟩
Kharchenko, V. I., **6**, 671, 672 ⟨92MI 621-01⟩
Khardin, A. P., **2**, 440 ⟨85URP1145022⟩
Khare, A. B., **6**, 666 ⟨91S405⟩
Kharitonov, A. V., **5**, 669 ⟨87JGU1478, 87ZOB1658⟩
Kharitonov, V. G., **2**, 447 ⟨85IZV1200⟩, 877 ⟨84DOK(274)606, 84IZV700⟩, 877 ⟨86IZV181⟩, 878 ⟨87JOU1417⟩; **3**, 496 ⟨89IZV1182⟩
Kharitonova, N. I., **4**, 124, 125 ⟨69JGU577⟩
Kharrat, A., **1**, 946 ⟨90JOC3321⟩; **5**, 206 ⟨85BSB29, 85T4047⟩
Khasanov, M. K., **2**, 442 ⟨67ZOB2522⟩
Khasapov, B. N., **4**, 524 ⟨74MI 411-01⟩; **6**, 169 ⟨74MIP 614-01⟩
Khaskin, B. A., **2**, 361 ⟨75ZOB1005⟩; **4**, 561 ⟨87ZOB2138⟩
Khaskin, G. I., **6**, 562 ⟨81KGS907⟩
Khathing, D. T., **3**, 4, 88, 115 ⟨82S580⟩
Khatib, A. A., **5**, 354 ⟨90IC309⟩
Khatri, H., **3**, 13 ⟨79CC79⟩
Khatri, H. H., **3**, 697 ⟨78JOC734⟩
Khatri, H. N., **1**, 835 ⟨78TL4337⟩
Khatri, N. A., **3**, 311 ⟨82JA536⟩, 618 ⟨79TL4907⟩
Khatri, S. G., **4**, 103 ⟨85S323⟩
Khattar, R., **4**, 682 ⟨87JOM(319)C51, 88JCS(D)1421⟩
Khatuntsev, G. D., **5**, 977 ⟨69DOK(181)115⟩, 1014 ⟨68DOK(181)115⟩, 1015 ⟨71ZOB2340⟩
Khaustova, T. I., **4**, 142 ⟨88IZV1946⟩
Khayat, M. A. R., **5**, 709, 861 ⟨66T975⟩, 774 ⟨89JPR149⟩
Khazal, N. M. S., **6**, 7 ⟨90JOM(386)229, 90JOM(397)51⟩
Khbeis, S. G., **5**, 871 ⟨85T811⟩
Khedija, H., **4**, 863 ⟨72T3545⟩
Kheifets, N. V., **6**, 413 ⟨68ZPK1380⟩
Khekimov, Y. K., **4**, 775 ⟨90CA230883w⟩; **6**, 88 ⟨88MI 603-01⟩, 478 ⟨87JGU365⟩
Khemani, K. C., **1**, 707 ⟨85CJC1⟩
Khéribet, T.,**4**, 10 ⟨80JOC5333⟩
Khetagurova, S. Sh., **4**, 341 ⟨80IZV1626⟩
Khidekel, M. L., **2**, 310 ⟨80RCR14⟩
Khim, Y. H., **2**, 212, 218 ⟨72JOC2710⟩
Khire, U. R., **1**, 1055 ⟨93TL15⟩; **4**, 260 ⟨89TL4007, 90JCS(P1)3217⟩

Khisamova, Z. L., **2**, 414, 415 ⟨55DOK(105)489⟩, 414, 415 ⟨56ZOB126⟩
Khisamutdinov, G. K., **6**, 261, 283 ⟨70IZV2553⟩, 283 ⟨71IZV1073, 72IZV1721⟩
Khisamutdinova, R. Sh., **2**, 461 ⟨65URP173230⟩
Khismatullina, L. A., **2**, 442 ⟨53DOK(92)69, 56ZOB3426⟩
Khitrova, O. M., **4**, 637 ⟨76MI 414-01⟩
Khlebnicova, T. S., **3**, 239 ⟨85S784⟩
Khlebnikov, A. F., **3**, 596 ⟨90JOU1642⟩; **4**, 1082 ⟨89UK1122⟩; **6**, 51 ⟨73ZOR2346, 77ZOR1857⟩
Khmelevskaya, L. V., **5**, 781 ⟨87MI 519-02⟩
Khmel'nitskii, L. I., **5**, 667, 671 ⟨91BAU438, 91IZV506⟩, 838 ⟨58ZOB2305, 60IZV1783, 61IZV477⟩; **6**, 150 ⟨60IZV1783, 70KGS590, 90IZV1620, 90IZV1623⟩, 369 ⟨70IZV2108⟩
Kho, E., **1**, 635 ⟨85JOC2798⟩
Khodabocus, A., **1**, 942 ⟨91JCS(P1)1433⟩
Khodzhalyev, T. K., **6**, 88 ⟨88MI 601-01⟩
Khoina, E. V., **4**, 865 ⟨78ZOR1947⟩, 866 ⟨78UKZ183⟩
Khokhar, A. R., **5**, 586 ⟨69JCS(C)1117⟩
Khokhlov, P. S., **2**, 252 ⟨84ZOB2545⟩, 449 ⟨84ZOB2703⟩; **4**, 346 ⟨92ZOB1268⟩
Khokhlova, A. M., **5**, 822 ⟨60ZOB2409⟩
Kholodenko, G. É., **6**, 493 ⟨68ZOR1661⟩
Khomenko, L. A., **4**, 771 ⟨90ZOR740⟩
Khomik, L. I., **4**, 719 ⟨73DOK(209)869⟩
Khomutov, R. M., **4**, 464 ⟨78IZV1951⟩, 472 ⟨79IZV1110, 79IZV2118⟩; **5**, 826 ⟨67IZV1820⟩
Khorana, H. G., **2**, 106 ⟨58JA6212⟩, 108 ⟨53JCS2257⟩, 143 ⟨64JA1630⟩, 414 ⟨61JA649⟩; **5**, 185 ⟨63JA1997⟩, 279 ⟨55CI(L)1087⟩
Khor, T. C., **2**, 16 ⟨75JOM(91)C20⟩
Khorkhoyanu, L. V., **2**, 458 ⟨65ZOB732⟩
Khorlina, I. M., **3**, 82 ⟨64ZOB1029⟩, 84 ⟨62TL619, 63TL2087⟩
Khorlin, A. Y., **2**, 32 ⟨70TL2517⟩
Khorrami, A., **2**, 677 ⟨86JOC3577⟩
Khorshev, S. Ya., **4**, 141 ⟨81JOM(205)311⟩
Khosavanna, S., **4**, 962 ⟨93T6483⟩
Khoshdel, E., **2**, 7 ⟨84JCS(P1)195⟩, 19 ⟨81JFC(17)93⟩; **3**, 149 ⟨77JOC3527⟩; **4**, 4, 12, 13, 21, 30, 33 ⟨77JOC3527⟩
Khosravi, E., **3**, 514 ⟨90JA8378⟩
Khosrowshahi, J. S., **2**, 13 ⟨87CC675⟩; **6**, 55 ⟨90MI 602-01⟩
Khotimskii, V. S., **4**, 966 ⟨80IZV1147⟩
Khotinen, A. V., **3**, 200 ⟨90S515⟩
Khotinsky, E., **2**, 906 ⟨09CB3091⟩
Khouri, F., **5**, 512 ⟨79JA2249, 80JA7579⟩, 512 ⟨81TL413⟩
Khouri, F. F., **6**, 96 ⟨80JA7579, 86JA6683⟩
Khourzon, R., **1**, 1071 ⟨91SL611⟩
Khouw, V. T., **5**, 1108 ⟨73CJC1741⟩
Khramtsova, S. Y., **4**, 617 ⟨87IZV1424⟩
Khranovskii, V. A., **2**, 408 ⟨77ZOB56⟩
Khrashcheva, O. V., **4**, 322 ⟨77ZOR1930⟩
Khripach, V. A., **2**, 924, 925 ⟨90TL4937⟩; **3**, 247 ⟨75JGU2572⟩; **4**, 809 ⟨90TL4937⟩
Khromova, N. Y., **4**, 641 ⟨80ZOB1764⟩
Khromova, Z. I., **6**, 613 ⟨69JGU1692⟩
Khrustalev, V. A., **2**, 373 ⟨80JOU244⟩
Khunen, L., **2**, 348 ⟨66CB3384⟩
Khuong-Huu, F., **1**, 1024 ⟨90TL7145, 92SL266⟩
Khurana, J. M., **1**, 568 ⟨91S827⟩; **2**, 4 ⟨88IJC(B)1129⟩
Khurana, J. P. S., **6**, 113 ⟨87S990⟩
Khusainova, N. G., **4**, 561 ⟨63ZOB1045, 64ZOB3938⟩, 926 ⟨84ZOB1758⟩
Khusnullina, G. M., **5**, 371 ⟨90IZV1139⟩
Khusnutdinova, E. A., **5**, 708 ⟨86JGU1096, 86ZOB1245⟩
Khusnutdinova, É. K., **4**, 493 ⟨88JGU1148, 88JGU876⟩
Khutoretskii, V. M., **6**, 56 ⟨69IZV1188, 70IZV387⟩, 56, 282 ⟨71IZV1487⟩, 145 ⟨83IZV2655⟩
Khvostenko, V. I., **2**, 641 ⟨75DOK(220)146⟩
Khwaja, H., **2**, 156 ⟨81JCS(D)447⟩, 156 ⟨83JCS(D)415⟩
Ki, D. K., **2**, 327 ⟨88JHC1383⟩
Kiaeezadeh, F., **3**, 5, 116 ⟨86T719⟩
Kianpour, A., **1**, 6 ⟨56JPC1454⟩
Kibar, R., **3**, 148 ⟨81CB1858⟩
Kibardin, A. M., **4**, 106 ⟨90JGU1541⟩, 486 ⟨83IZV432⟩
Kibardina, L. K., **5**, 369 ⟨78ZOB2468, 92IZV220⟩, 369, 370 ⟨78ZOB695⟩, 402 ⟨78JGU414⟩; **6**, 563 ⟨81ZOB530⟩, 564 ⟨81ZOB28⟩
Kibayashi, C., **1**, 1067 ⟨92JA10653⟩; **2**, 299 ⟨87JOC1956⟩; **3**, 419 ⟨89TL4539⟩; **4**, 191 ⟨89TL4165⟩; **5**, 322, 330 ⟨89JOC4088, 85JOC1818⟩, 330 ⟨91CC1237, 91TL4325⟩, 330 ⟨94JOC1358⟩
Kibayashi, T., **2**, 34 ⟨88S366⟩
Kibbel, H. U., **6**, 580 ⟨81JPR41⟩
Kibirev, V. K., **6**, 104 ⟨66UKZ204⟩
Kibler, C. J., **5**, 184 ⟨55OSC(3)28⟩, 652 ⟨55OSC(3)108⟩
Kice, J. L., **1**, 595 ⟨67JOC1631⟩; **2**, 122 ⟨63JOC957, 66JA5245⟩, 187 ⟨63JOC870⟩, 208 ⟨68JA7378⟩, 236 ⟨91JOC3520⟩, 238 ⟨80TL1697⟩, 285 ⟨80TL4155, 81JOC4899⟩, 285 ⟨82TL5373, 85T4739⟩, 368 ⟨93JOC917⟩, 725 ⟨81JOC4894⟩, 734 ⟨86JOC295⟩, 735 ⟨84JOC3106⟩; **3**, 365 ⟨85JOC32⟩, 368 ⟨91JOC1424, 91JOC1431⟩; **5**, 1107 ⟨92JOC5270⟩
Kida, S., **4**, 407 ⟨79TL5011⟩
Kidd, J. M., **2**, 166, 167, 172 ⟨53JCS3219⟩, 194 ⟨55JCS2901⟩; **6**, 232, 236 ⟨53JCS3219⟩, 534 ⟨55JCS3871⟩
Kidd, K. B., **2**, 511 ⟨86CC1543, 88JOM(341)C1⟩, 512 ⟨87IC4065⟩
Kido, F., **1**, 822 ⟨89TL1575, 90CC418⟩, 822 ⟨90CL613, 90T4887⟩, 822 ⟨91TL6159, 94CC789⟩
Kido, K., **5**, 244, 250 ⟨94JCS(P1)1083⟩
Kido, M., **2**, 1000 ⟨81JCS(P1)212⟩; **5**, 1122 ⟨87JCS(P1)2531⟩
Kidoura, K., **3**, 66 ⟨82TL873⟩
Kidwai, M. M., **5**, 94 ⟨79SC363⟩
Kidwell, R. L., **2**, 653 ⟨69OS(49)90⟩; **3**, 530, 531 ⟨68JOC3974⟩
Kieczykowski, G. R., **1**, 209 ⟨78JOC788⟩; **3**, 623 ⟨75TL4647⟩; **4**, 224 ⟨78JA1938⟩, 855 ⟨73TL4711⟩
Kiedrowski, G. V., **2**, 658 ⟨82S683⟩
Kiefel, M. J., **3**, 157 ⟨86TL1933⟩
Kiefer, E. F., **1**, 498 ⟨68JA4187⟩, 794 ⟨69TL993⟩
Kiefer, G., **4**, 979 ⟨66CB3892⟩; **6**, 106 ⟨66CB3892⟩
Kiefer, H., **5**, 566 ⟨75LA1984⟩; **6**, 445 ⟨72S39⟩
Kiefere, J. M., **2**, 101 ⟨38JA538⟩
Kieffer, R., **2**, 876 ⟨75BSF(2)1185⟩
Kiehlmann, E., **4**, 437 ⟨73CJC2448⟩; **6**, 365 ⟨73CJC2448⟩
Kiehs, K., **6**, 632 ⟨86GEP(O)3504453⟩
Kiel, W. A., **1**, 21, 25, 437, 438, 943 ⟨80JA4438⟩; **2**, 286 ⟨77CC725⟩, 580 ⟨80JA3299, 82JA4865⟩
Kielbasinoki, P., **2**, 144 ⟨B-88MI 203-02⟩, 156, 721 ⟨88T6687⟩; **3**, 166 ⟨83S332⟩; **4**, 923 ⟨83S332⟩; **5**, 243, 540 ⟨76JCS(P1)564⟩, 1061 ⟨81T233⟩; **6**, 635 ⟨77JOC3629⟩
Kiely, D. E., **3**, 117 ⟨87JOC1149⟩
Kiely, J. S., **1**, 1004, 1005, 1006 ⟨77JOC2626⟩
Kiener, V., **5**, 952 ⟨67AG(E)961⟩
Kienitz, L., **4**, 894, 973, 974 ⟨79LA2096⟩; **6**, 296, 297, 298, 300, 313, 320, 321 ⟨77S73⟩, 321, 337 ⟨79LA2089⟩, 647 ⟨79LA2096⟩
Kienlen, J.-C., **1**, 774 ⟨87TL1463⟩
Kiennemann, A., **2**, 876 ⟨75BSF(2)1185⟩
Kienzle, A., **5**, 1077 ⟨93CB2569⟩
Kienzle, F., **1**, 495 ⟨78HCA449⟩; **2**, 398 ⟨65CJC3074⟩, 621 ⟨70JOC528⟩, 631 ⟨71JA4841⟩, 631, 964 ⟨71JA4845⟩; **3**, 702 ⟨73HCA1671⟩; **4**, 194 ⟨73JA3635⟩
Kierstead, R. W., **1**, 1029 ⟨89JMC1820⟩; **3**, 234 ⟨70JOC4141⟩; **5**, 164 ⟨58T1⟩; **6**, 72 ⟨70TL5111⟩
Kiesel, V., **2**, 507 ⟨71JPR259⟩
Kiesewetter, D. O., **2**, 18 ⟨86JFC(31)99⟩
Kiesgen de Richter, R., **2**, 42 ⟨90JOC2855⟩
Kiessel, M., **2**, 677 ⟨58CB380⟩

Kieszkowski, J., **5**, 1107, 1112 ⟨83T4175⟩
Kifer, E. W., **4**, 638 ⟨70MI 414-01⟩, 638, 641 ⟨72IC408⟩
Kigasawa, K., **4**, 325 ⟨78H(9)831⟩
Kigoshi, H., **1**, 710 ⟨84JA4547⟩; **5**, 329 ⟨94TL1247⟩
Kiguchi, T., **2**, 352 ⟨94TL2205⟩
Kihara, K., **5**, 47 ⟨65BCJ893⟩
Kihara, N., **2**, 141 ⟨90CL1097⟩
Kihlberg, J., **4**, 176 ⟨86TL753⟩
Kijama, M., **2**, 214 ⟨84JOC1434⟩
Kiji, J., **1**, 94 ⟨82CL603⟩, 1035 ⟨89JOM(360)39⟩; **2**, 434 ⟨82CL977⟩, 619 ⟨80CL1423⟩; **4**, 14 ⟨64JA4851⟩; **5**, 142 ⟨63TL1437⟩, 158 ⟨64TL605⟩
Kiji, S., **2**, 34 ⟨88CL31⟩; **3**, 155 ⟨88CL31⟩
Kiji, T., **5**, 289, 302 ⟨91CL1443⟩
Kijima, A., **1**, 339 ⟨90TL2315⟩
Kijima, I., **6**, 469 ⟨78BCJ3549⟩
Kijima, M., **1**, 514 ⟨85CL1851⟩
Kikalj, D., **3**, 621 ⟨89S612⟩
Kikani, B. B., **2**, 311 ⟨91S176⟩; **3**, 264 ⟨91S176⟩
Kikelj, D., **3**, 621 ⟨88S981⟩; **4**, 824 ⟨88S981⟩, 883, 905 ⟨88CB1817⟩, 896, 897 ⟨89JHC1335⟩
Kikkawa, I., **3**, 130, 285 ⟨80S877⟩
Kikkawa, S., **2**, 1016, 1033 ⟨63JA1648⟩
Kikuch, T., **2**, 615 ⟨89SC865⟩
Kikuchi, A., **3**, 309 ⟨93CRV2207⟩
Kikuchi, K., **2**, 997, 1001 ⟨92JCS(P1)1721⟩; **5**, 125 ⟨77BCJ1863⟩, 1024 ⟨91JAP(K)3193760⟩; **6**, 594 ⟨86BCJ1741⟩
Kikuchi, N., **2**, 34 ⟨93TL2165⟩
Kikuchi, T., **1**, 667 ⟨84CL1897, 85TL2205⟩; **3**, 419 ⟨89TL4539⟩; **4**, 699 ⟨91T3281⟩; **6**, 541 ⟨86CPB430⟩
Kikugawa, K., **2**, 31 ⟨71TL87⟩; **4**, 856 ⟨78CPB685⟩; **6**, 57 ⟨71CPB2629⟩
Kikugawa, Y., **1**, 102 ⟨69TL3101⟩; **2**, 139 ⟨81CL1157⟩, 360 ⟨84JCS(P1)1401⟩; **3**, 430 ⟨92JOC3245⟩, 430, 666 ⟨91S750⟩, 431 ⟨93S433⟩, 617 ⟨93SC1061⟩; **5**, 317 ⟨89JOC3394⟩, 328 ⟨90SC887, 94JOC929⟩, 332 ⟨90TL243⟩, 333 ⟨94OPP111⟩, 665, 666, 824 ⟨91S441⟩, 665, 672, 824, 825 ⟨92JOC3245⟩, 666 ⟨93SC1061⟩, 745 ⟨69TL699⟩, 746 ⟨78CPB2246⟩
Kikui, J., **6**, 51 ⟨76JOC3794⟩
Kikukawa, K., **1**, 452, 490 ⟨90CL185⟩, 486 ⟨81JOC4885⟩, 486 ⟨81T31⟩, 486 ⟨83JOC1333, 86JCS(P1)1959⟩, 490 ⟨88CL873⟩, 1033 ⟨87JCS(P1)1511⟩; **2**, 932 ⟨84TL5789, 87JOC1276⟩; **3**, 90 ⟨84JOM(270)283⟩, 290 ⟨82CL35⟩; **5**, 197, 198 ⟨80TL2877⟩, 198 ⟨81JOC4413⟩, 220 ⟨82JOM(235)243⟩, 220, 222 ⟨84TL5789, 77BCJ2021⟩
Kiladze, J. K., **6**, 89 ⟨88URP1421744⟩
Kiladze, T. K., **2**, 265 ⟨87ZOR1981⟩; **6**, 70 ⟨81MI 603-01⟩
Kilama, J. J., **5**, 262 ⟨94MI 506-02⟩
Kilbourn, M. R., **2**, 9 ⟨87JOC658⟩, 13 ⟨86CJC138⟩, 18 ⟨86JFC(31)99⟩; **5**, 796 ⟨82MI 520-01⟩
Kilbourne, H. W., **4**, 33 ⟨53JA4091⟩
Kilburn, J., **2**, 546 ⟨92JCS(P1)3295⟩
Kilby, B. A., **2**, 413 ⟨50JCS1769⟩
Kilday, W., **5**, 366 ⟨65JOC64⟩
Kil'disheva, O. V., **2**, 173 ⟨81IZV159⟩, 211 ⟨81IZV1850⟩
Kilduff, J. E., **2**, 821 ⟨82JA5820⟩, 866, 893 ⟨84IC2582⟩; **6**, 167 ⟨83CC528⟩, 168 ⟨84IC2582⟩, 373 ⟨82JA5820, 86JCS(D)1801⟩, 375 ⟨87JCS(D)249⟩
Kilényi, S. N., **2**, 664 ⟨85TL4419⟩
Kilgour, J. A., **4**, 614 ⟨76JA7746⟩; **5**, 873 ⟨77JA6995⟩
Kilian, R., **5**, 289 ⟨69T5745⟩
Kilian, W., **2**, 545 ⟨69JOM(18)299⟩
Killian, D. B., **4**, 195 ⟨36JA80⟩, 195 ⟨36JA892⟩
Killing, H., **2**, 916, 930 ⟨85CC354⟩
Killinger, T. A., **1**, 525 ⟨77TL2831, 78JOC2161⟩; **4**, 827 ⟨79JA6023⟩
Killmer, Jr., L. B., **4**, 825 ⟨83JOC298⟩
Killoran, M., **5**, 151 ⟨87JA2426⟩
Kilmas, M. T., **1**, 199, 200 ⟨87JA5749⟩
Kilner, A. E. H., **1**, 871 ⟨51JCS2430⟩
Kilner, M., **3**, 460 ⟨70CC552, 70JCS(A)2279⟩, 460 ⟨71JCS(A)292, 72JCS(D)153⟩, 460, 461 ⟨74JCS(D)1620⟩; **5**, 781 ⟨75JCS(D)930, 84ICA85⟩, 781 ⟨86JCS(D)1359, 87JCS(D)2687⟩; **6**, 448 ⟨92MI 614-01⟩
Kilthau, G., **2**, 879 ⟨68CB121⟩
Kim, B. H., **1**, 831 ⟨88T3079⟩; **3**, 142 ⟨86S312⟩; **4**, 226 ⟨90JA6263⟩; **5**, 343 ⟨94SL197⟩, 513 ⟨90JA6263⟩
Kim, B. J., **1**, 238 ⟨90TL2713⟩
Kim, B. M., **2**, 48 ⟨90TL3003⟩, 49 ⟨90TL4317⟩, 521 ⟨85JA4549, 87JOC4831⟩, 696 ⟨91COS(2)239⟩, 698 ⟨86JA8279⟩; **5**, 237 ⟨86JA8279⟩
Kim, C., **3**, 131, 284 ⟨83JOC1550⟩; **5**, 142 ⟨94JA7455⟩
Kim, C. J., **1**, 915 ⟨69JA1458⟩
Kim, C. K., **1**, 130 ⟨91JPO449⟩; **5**, 219, 694 ⟨87JHC325⟩; **6**, 163 ⟨87CB1357⟩, 186, 188, 292, 388, 397, 711 ⟨86CB2966⟩, 188, 388, 397, 673, 711 ⟨86CB2980⟩, 188, 673, 711 ⟨87CB1203⟩, 673 ⟨87CB1213⟩
Kim, C. S. Y., **4**, 16 ⟨59JA5008⟩
Kim, C. U., **1**, 722 ⟨75JA3512⟩; **2**, 25, 32 ⟨72TL4339⟩; **5**, 339 ⟨72JA7586⟩
Kim, D., **1**, 183 ⟨86TL943⟩, 183 ⟨88CC760⟩, 334, 347 ⟨89JA6265⟩, 339, 927 ⟨88JA5900⟩, 344 ⟨86TL5821⟩; **3**, 311 ⟨82JA536⟩; **4**, 781 ⟨91T6171⟩, 811 ⟨89JA6265⟩
Kim, D. H., **4**, 849 ⟨90JOC5555⟩
Kim, D. S. H. L., **4**, 883 ⟨89S698, 91JOC657⟩, 883 ⟨91MI 420-01⟩
Kim, D.-W., **2**, 179 ⟨92S1203⟩
Kim, D. Y., **3**, 639 ⟨87SC953⟩; **6**, 120 ⟨86MI 604-02⟩
Kim, H., **1**, 25, 630 ⟨87JOC3346⟩, 606 ⟨84CC1345⟩; **3**, 151 ⟨88TL6707⟩, 215 ⟨87JOC3346⟩; **4**, 191 ⟨87JA1597⟩
Kim, H.-B., **4**, 959 ⟨86TL2191⟩
Kim, H. J., **5**, 361 ⟨86S970⟩
Kim, H. O., **1**, 179 ⟨93TL2051⟩, 840 ⟨91JOC1014⟩; **4**, 231 ⟨92JOC2217⟩; **5**, 237, 277 ⟨87JOC4531⟩
Kim, H. P., **3**, 690 ⟨87AOC(27)51⟩; **5**, 942 ⟨87AOC(27)51⟩
Kim, H. R., **3**, 120, 294 ⟨90SC637⟩, 151 ⟨92CL603⟩; **6**, 280, 632 ⟨90TL4715⟩
Kim, H. S., **2**, 327 ⟨88OM2250⟩
Kim, I. O., **1**, 183 ⟨88CC760⟩; **3**, 628 ⟨94SC103⟩
Kim, J., **1**, 109 ⟨93TL4129⟩
Kim, J. D., **3**, 82 ⟨87TL4575⟩
Kim, J. E., **3**, 13, 82 ⟨87TL2389⟩, 13, 82, 94, 99 ⟨87JOC5030, 87MI 303-01⟩, 82 ⟨87TL4575, 88MI 303-01⟩, 82, 99 ⟨87TL6231⟩, 86 ⟨87MI 303-02⟩
Kim, J. H., **1**, 630 ⟨92MI 114-01⟩, 832 ⟨91JOC5063⟩; **5**, 359 ⟨91TL3449⟩; **6**, 428 ⟨91JAN1083⟩, 480 ⟨92JC241⟩
Kim, J.-I. I., **1**, 1096 ⟨81JOC1061, 81JOC1067⟩
Kim, J. K., **2**, 138 ⟨82JA3231, 85JOC4390⟩
Kim, J. M., **2**, 41 ⟨94TL2⟩; **3**, 3, 114 ⟨93TL5113⟩, 13, 85 ⟨91TL6903⟩, 82 ⟨90MI 303-01, 94MI 303-01⟩
Kim, J. N., **3**, 120, 294 ⟨90SC637⟩, 151 ⟨92CL435⟩, 684 ⟨90H(31)1693, 91H(32)477⟩; **5**, 664 ⟨92JOC6649⟩
Kim, J. S., **4**, 139 ⟨92MI 403-01⟩
Kim, J.-U., **4**, 510 ⟨87JA4421⟩
Kim, K., **1**, 573 ⟨92JOC5557⟩; **2**, 615 ⟨71JOC1024⟩, 626 ⟨86CC1159⟩; **3**, 374 ⟨75JOC3857⟩; **5**, 368 ⟨86TL4749, 93MI 507-01⟩, 637 ⟨91S951⟩, 736 ⟨93JOC7001⟩; **6**, 641 ⟨88TL3183⟩, 651 ⟨92TL4963⟩
Kim, K. D., **1**, 1018 ⟨91TL6085⟩; **2**, 910 ⟨90TL6137⟩; **4**, 772, 939, 948 ⟨90TL6137⟩, 962 ⟨90TL2541, 91TL6085⟩
Kim, K. E., **1**, 86 ⟨88SC89⟩; **3**, 127 ⟨88SC89⟩
Kim, K. J., **4**, 849 ⟨90JOC5555⟩
Kim, K. K., **3**, 151 ⟨92CL603⟩
Kim, K. M., **3**, 151 ⟨92CL603⟩
Kim, K. R., **6**, 480 ⟨92JC241⟩
Kim, K. S., **2**, 145 ⟨90TL2893⟩; **3**, 6, 118 ⟨84CC762⟩, 88 ⟨84S866⟩, 116 ⟨89TL2559, 86TL2875⟩, 215 ⟨85TL2875⟩; **4**, 190 ⟨78S48⟩; **5**, 418, 428 ⟨87TL3261⟩
Kim, M., **2**, 214, 220 ⟨77CL893⟩; **5**, 1126 ⟨70TL4731⟩; **6**, 428 ⟨91JAN1083⟩
Kim, M.-G., **5**, 1120 ⟨74JOC1494⟩
Kim, M. J., **5**, 135 ⟨92JOC1605⟩

Kim, M. W., **4**, 240 ⟨87TL4925⟩; **6**, 111 ⟨89JOC1782⟩, 340 ⟨93IC5883⟩
Kim, N.-J., **1**, 307 ⟨88TL2195⟩; **2**, 324 ⟨88TL2195⟩
Kim, S., **1**, 4, 7, 9 ⟨83AG(E)562⟩, 39 ⟨85JOC1927⟩, 314, 1050 ⟨90TL7627⟩, 446 ⟨90TL2901⟩, 629 ⟨91TL2043⟩, 932 ⟨93JA3328⟩, 939 ⟨86JOC2613⟩, 1134 ⟨78JA4618, 78JA4620⟩; **2**, 11 ⟨88JOC3111⟩, 446 ⟨89MI 208-01⟩, 629 ⟨78JA4618⟩, 639 ⟨88JA8244, 91JCS(D)443⟩; **3**, 83 ⟨87JOC2114, 87PAC1005⟩, 83 ⟨87TL1913⟩, 85 ⟨84JOC1717⟩, 130, 287 ⟨83JOC2608⟩, 140 ⟨91SC25⟩, 147 ⟨90TL7627⟩, 498 ⟨88TL5413⟩, 618, 619, 697, 721 ⟨86TL1925⟩; **4**, 201, 217 ⟨93TL5769⟩, 216 ⟨89TL6697⟩, 257 ⟨87JOC2114⟩, 284 ⟨91JA9864⟩, 918 ⟨90SL115⟩; **5**, 126 ⟨85CC473, 85JOC560⟩, 138 ⟨85JOC1751⟩, 173 ⟨78JA4618⟩, 173 ⟨79JA7131⟩, 173, 204 ⟨78JA4620⟩, 280 ⟨85TL1341, 86CC719⟩, 862 ⟨94S1057⟩, 1030 ⟨86MI 527-01⟩, 1030, 1078 ⟨85TL1661⟩, 1030, 1078 ⟨86JOC2613, 86TL1925⟩; **6**, 340 ⟨93IC5883⟩, 417 ⟨86H(24)1625⟩, 517 ⟨84OM1523⟩
Kim, S. C., **1**, 14 ⟨80JOC1⟩, 485 ⟨83OM779⟩
Kim, S.-G., **1**, 15, 16 ⟨93JCS(P1)9⟩, 967 ⟨92TA1509⟩, 1124 ⟨73TL4767⟩
Kim, S. H., **2**, 179 ⟨92S1203⟩; **5**, 862 ⟨94S1057⟩
Kim, S. M., **3**, 220 ⟨62CB1495⟩
Kim, S. S., **2**, 76 ⟨89JA5495⟩, 446 ⟨89MI 208-01⟩; **3**, 83 ⟨87JOC2114, 87TL1913⟩; **4**, 257 ⟨87JOC2114⟩; **5**, 280 ⟨86CC719⟩; **6**, 79 ⟨88JA4144⟩
Kim, S.-W., **2**, 662 ⟨89CPB1434, 91CPB1132⟩, 662, 664 ⟨89CPB304⟩; **3**, 207 ⟨81CPB2733⟩
Kim, T. H., **4**, 924 ⟨87MI 420-01⟩; **6**, 119 ⟨85TL3479⟩, 121 ⟨88SC1611⟩, 504 ⟨90JOC3699⟩
Kim, T. V., **4**, 1008, 1009 ⟨86ZOB2681, 90ZOB229⟩; **5**, 644 ⟨89ZOB1917⟩, 652 ⟨90JGU1165⟩, 657 ⟨87JGU1104, 87ZOB1233⟩, 659 ⟨86ZOB2681⟩, 659, 788, 789 ⟨86JGU2372⟩, 787 ⟨89JGU1496⟩, 788 ⟨85JGU402⟩, 788 ⟨89JGU1129, 88JGU30⟩, 789 ⟨89JGU1123⟩
Kim, U. T., **2**, 144 ⟨93CC70⟩
Kim, W. J., **1**, 697, 700, 703 ⟨88TL5169⟩; **5**, 138 ⟨85JOC1751⟩; **6**, 428 ⟨91JAN1083⟩, 559 ⟨82MI 617-03⟩
Kim, Y. C., **1**, 446 ⟨90TL2901⟩; **4**, 918 ⟨90SL115⟩; **5**, 126 ⟨85JOC560⟩
Kim, Y. G., **1**, 629 ⟨91TL2043⟩
Kim, Y. H., **2**, 150 ⟨88TL6453⟩, 193 ⟨79TL821⟩, 193 ⟨81T37⟩, 214 ⟨79CL1077⟩, 214, 220 ⟨77CC407⟩, 243 ⟨80BCJ2023⟩, 621, 627 ⟨78TL4519⟩, 621, 627 ⟨80BCJ1065⟩, 717, 718 ⟨92CL1483⟩, 724 ⟨81BCJ1443⟩, 726 ⟨79TL3307, 85TL3821⟩; **4**, 74, 75 ⟨93S209⟩, 201 ⟨93TL3581⟩; **5**, 361 ⟨86S970⟩, 368 ⟨86TL4749, 93MI 507-01⟩
Kim, Y. J., **1**, 39 ⟨85JOC1927⟩; **4**, 260 ⟨91T8091⟩
Kim, Y.-J., **5**, 378 ⟨94S909⟩
Kim, Y. S., **3**, 82, 99 ⟨87TL6231⟩
Kimball, R. K., **3**, 633 ⟨44OSC(2)498⟩
Kimbrough, D. R., **1**, 826 ⟨87JA1170⟩
Kimel, W., **1**, 416, 831 ⟨43JA1992⟩; **3**, 233 ⟨57JOC1611⟩
Kimling, H., **1**, 635, 635 ⟨70TL761⟩; **3**, 352 ⟨79H(12)1153⟩
Kimmel, T., **2**, 422 ⟨84JOC2494⟩; **3**, 636 ⟨86TL5099⟩
Kimpe, N. D., **4**, 189 ⟨94S427⟩
Kimpe, N. de, **4**, 12 ⟨79SC575⟩
Kimpenhaus, W., **6**, 38 ⟨76CB2351⟩
Kimura, F., **6**, 9 ⟨91MI 601-02⟩
Kimura, H., **5**, 322 ⟨93BCJ327⟩
Kimura, J., **3**, 8, 122 ⟨79CC303⟩
Kimura, K., **1**, 578 ⟨87CC490⟩, 677, 1099 ⟨83TL5281⟩, 811 ⟨81CC786⟩; **2**, 73, 571 ⟨82BCJ561⟩, 370, 768 ⟨91OM1095⟩; **3**, 365 ⟨77BCJ2751⟩; **5**, 1122 ⟨85JCS(P1)2333, 85TL1321⟩, 1122 ⟨87JCS(P1)2531⟩
Kimura, M., **1**, 638 ⟨79BCJ1107⟩, 795 ⟨70BCJ1120⟩; **2**, 88 ⟨86JOC3913⟩, 685 ⟨84JCS(P2)451⟩; **3**, 162 ⟨85JOC4655⟩, 645 ⟨86TL4177⟩; **4**, 221 ⟨87CC122⟩, 414 ⟨76S201⟩; **5**, 248, 249 ⟨85S519⟩
Kimura, N., **1**, 558 ⟨93TA21⟩
Kimura, S., **3**, 417, 658 ⟨78S892⟩
Kimura, T., **1**, 558 ⟨86T2971⟩; **2**, 18, 19 ⟨88CL1877⟩, 177, 179 ⟨92CL891⟩, 217 ⟨81CL737⟩, 227 ⟨91CC167⟩, 478 ⟨90CPB673⟩; **3**, 206 ⟨86TL2791⟩, 294 ⟨88CL1387⟩; **4**, 247 ⟨86JOC1427⟩; **5**, 154 ⟨86T2971⟩, 1039 ⟨90CC1213⟩, 1105 ⟨87JOC681⟩, 1107 ⟨87CC1301⟩
Kimura, Y., **1**, 404, 405, 816, 817, 818 ⟨83JOC279⟩, 404, 406, 816 ⟨81T4097, 82TL3931⟩, 1140 ⟨84CL1543, 86CPB1531⟩; **2**, 614 ⟨89CL2213, 90CL769⟩, 677 ⟨79S688⟩; **3**, 22 ⟨82TL3931⟩, 137 ⟨79CL321⟩, 357 ⟨91CPB672⟩, 631 ⟨88TL6957⟩; **6**, 495 ⟨79CL1361⟩
Kinast, G., **2**, 1008 ⟨76AG(E)239⟩
Kinastowski, S., **4**, 306 ⟨83S654, 88S111⟩
Kinder, D. H., **4**, 527 ⟨85JMC1917⟩, 528 ⟨90USP4963655⟩
Kinder, F. R., **3**, 322 ⟨90TL6835, 93JOC21⟩
Kinder, L. L., **5**, 772 ⟨77JOC1862⟩
Kindermann, K., **3**, 179 ⟨68JPR12⟩
Kindler, H., **1**, 57 ⟨47CB202⟩
Kindler, K., **1**, 47 ⟨66CB227⟩, 89 ⟨53CB167⟩; **5**, 566, 579, 625 ⟨23LA(431)187⟩
Kindt, P., **6**, 564 ⟨89M871⟩
King Jr, A. D., **5**, 445, 453 ⟨78JA1687⟩
King, A. O., **1**, 468, 471, 533, 1020, 1097, 1104 ⟨78JA2254⟩, 473 ⟨88OS(66)67⟩, 479, 534, 1106 ⟨76CC17⟩, 1006 ⟨78JOC358⟩, 1006, 1140 ⟨77CC683⟩, 1070 ⟨80JOC2526, 85OS(64)44⟩; **2**, 629, 975, 976 ⟨79S501⟩, 925, 969, 976 ⟨78JA2254, 80JOC2526⟩, 1096 ⟨77CC683, 78JOC358⟩
King, B. J., **1**, 1065 ⟨62JOC3688⟩
King, C., **2**, 162 ⟨60JOC352⟩; **4**, 562 ⟨86IC1290⟩
King, C.,**5**, 652 ⟨60JOC352⟩
King, C. R., **3**, 92, 97 ⟨87TL2473⟩
King, E., **5**, 843 ⟨24LA(439)211⟩
King, F. D., **2**, 753 ⟨74CC256⟩, 769 ⟨75S738⟩, 1055 ⟨78CC351⟩; **5**, 351, 363 ⟨91T10053⟩, 362, 367 ⟨87JCS(P1)2511⟩, 1000 ⟨78CC351⟩
King, F. E., **4**, 309 ⟨48JCS2174⟩
King, G., **5**, 49 ⟨36JCS1788⟩
King, G. S., **5**, 735 ⟨74JCS(P1)1499⟩
King, G. S. D., **5**, 604 ⟨84BSB405⟩; **6**, 652 ⟨81BSB63⟩
King, G. T., **4**, 143 ⟨86IC4309⟩
King, H., **2**, 103 ⟨47JCS963⟩, 890 ⟨32JCS2505⟩; **5**, 85 ⟨32OSC(1)298⟩, 690 ⟨93T11029⟩
King, J. A., **1**, 451 ⟨86JA1359⟩; **5**, 183 ⟨51JA4911⟩, 272, 276 ⟨46JA525⟩, 566, 579, 591, 618, 621 ⟨55JA4062⟩, 566, 608 ⟨47JA1207⟩, 599 ⟨48JA4143⟩
King, Jr., J. A., **4**, 690 ⟨83JA4846⟩; **6**, 443 ⟨92JOC5136⟩, 444 ⟨88JOC6145⟩
King, J. C., **3**, 224 ⟨73JOC304⟩
King, J. D., **5**, 934 ⟨91TL7759⟩
King, J. F., **1**, 232 ⟨72CC541⟩, 707 ⟨85CJC1⟩, 834 ⟨76JA3312⟩; **2**, 9 ⟨79TL3615⟩, 177 ⟨80S285⟩, 179 ⟨71CJC943⟩, 202 ⟨75ACR10⟩, 204 ⟨81TL3573⟩, 205 ⟨75CJC3656, 78JA1637⟩, 365 ⟨75JA2566⟩; **3**, 375 ⟨69CJC4509, 71JA6304⟩, 375 ⟨76CJC2652, B-91MI 308-02⟩; **4**, 78 ⟨77CJC2323⟩; **5**, 472, 495 ⟨75ACR10⟩, 488, 489 ⟨63TL585⟩, 488, 489 ⟨66CJC819⟩, 488, 489 ⟨76CJC2652, 63JA2676⟩, 1106 ⟨78JA1637⟩
King, J. P., **2**, 201 ⟨89TL2763⟩; **4**, 556 ⟨65IC198, 67USP3332986⟩
King, K., **1**, 786 ⟨68CRV415⟩
King, K. D., **1**, 772 ⟨79MI117-01⟩
King, L., **2**, 28 ⟨64JOC3459⟩
King, L. C., **2**, 125 ⟨48JA1176⟩, 212 ⟨51JA2647⟩; **3**, 357 ⟨51JA300⟩; **6**, 596 ⟨51JA1864⟩
King, N., **2**, 768 ⟨89UP-01⟩
King, R. B., **2**, 414 ⟨84JOC1784⟩, 439 ⟨71JA4158, 76JOC972⟩, 439 ⟨79JOC3095⟩, 466 ⟨85IC3136⟩, 576 ⟨70ACR417⟩, 853 ⟨75JA53⟩; **3**, 715 ⟨85T3235⟩; **4**, 681 ⟨60JA4749, 73JOM(60)C57⟩, 692 ⟨73JOM(47)145⟩, 821 ⟨70ACR417, 73JA1811⟩, 1019, 1020 ⟨75JA2702⟩; **5**, 407 ⟨91IC1174⟩, 445, 453 ⟨78JA1687⟩, 448 ⟨64JA1267⟩, 449 ⟨66JOM(5)341⟩, 452 ⟨63JA1918, 64JOM(2)15⟩, 943 ⟨63JA1922⟩; **6**, 247 ⟨61JA3593⟩, 517 ⟨63JA1918⟩, 518 ⟨85IC4449, 87JA7764⟩, 518 ⟨90JOM(383)295⟩

King, R. M., **1**, 1134 ⟨84T2415⟩
King, R. R., **2**, 660 ⟨79JOC4194⟩
King, R. W., **1**, 624, 637, 773, 776, 782, 783 ⟨60CRV431⟩; **2**, 346 ⟨60CRV448⟩; **3**, 451 ⟨64JA732⟩; **4**, 342 ⟨66JA1140⟩
King, S. A., **1**, 194, 289 ⟨90JA408, 89JA5902⟩
King, S. B., **5**, 330 ⟨91JA5089, 94JA562⟩
King, S. M., **2**, 969 ⟨87TL3895⟩
King, S. N., **3**, 539 ⟨86JOC2814⟩
King, T. J., **3**, 704 ⟨80T515⟩; **4**, 656 ⟨77JCR(S)116⟩, 799 ⟨80CC879⟩; **6**, 394 ⟨77JCR(S)116⟩, 610, 619 ⟨80CC879⟩
King, T. P., **1**, 34 ⟨51JA2718⟩; **6**, 496, 546 ⟨78JOC2930⟩
King, W. B., **4**, 313 ⟨75JCS(P1)61, 77JCS(P1)282⟩; **5**, 105 ⟨32OSC(1)252⟩
Kingma, A. J., **1**, 305 ⟨87HCA2201⟩
Kingsbury, C. A., **1**, 595, 779 ⟨60JA1810⟩, 854 ⟨61JA3688⟩, 871, 881 ⟨61JA3678⟩; **2**, 651 ⟨84JOC645⟩
Kingsbury, W. D., **2**, 148 ⟨69CC365⟩; **6**, 487 ⟨91JMC98⟩
Kingston, B. M., **4**, 656 ⟨71SAP7004922⟩; **6**, 396 ⟨71SAP7004922⟩
Kingston, D. G. I., **6**, 550 ⟨93JOC3798⟩
Kini, A., **1**, 505 ⟨83JA2923⟩, 690 ⟨78CL803⟩
Kini, A. M., **6**, 87 ⟨87TL4153⟩
Kinjo, J., **6**, 264 ⟨91BMC357⟩
Kinkead, S. A., **2**, 367 ⟨84JA7496⟩
Kinloch, E. F., **1**, 457, 508 ⟨74JOC747⟩
Kinnear, A. M., **4**, 130 ⟨52JCS3437⟩
Kinney, C. R., **5**, 595, 729 ⟨24JA493⟩
Kinney, W. A., **1**, 603 ⟨83JOC4986⟩, 632, 797, 826 ⟨85JA7352⟩
Kino, Y., **2**, 55 ⟨88JOC3752⟩
Kinokshita, M., **6**, 134 ⟨88BCJ2147⟩
Kinoshita, H., **1**, 503 ⟨83BCJ2539⟩, 619 ⟨88CL2009⟩; **2**, 743 ⟨92BCJ1379⟩; **4**, 193 ⟨89CL737⟩; **5**, 291 ⟨78BCJ1866⟩
Kinoshita, K., **3**, 597 ⟨80JOM(190)C39⟩, 616 ⟨90CL1171⟩, 628 ⟨89CL2119⟩
Kinoshita, M., **1**, 779 ⟨75TL2841⟩; **2**, 361 ⟨84JOC2724⟩, 683 ⟨79JOC638⟩; **4**, 264 ⟨76BCJ256⟩; **5**, 766 ⟨83CL885⟩; **6**, 550 ⟨79JA6116⟩
Kinoshita, N., **6**, 560 ⟨87CPB156⟩
Kinoshita, T., **1**, 100 ⟨75BCJ1865⟩, 867 ⟨90JOC996⟩; **2**, 1000 ⟨90CPB329⟩
Kinoshita, Y., **6**, 550 ⟨82YGK352⟩
Kinowski, A., **3**, 702 ⟨93S1215⟩
Kinoyan, F. S., **1**, 1040 ⟨86AKZ306⟩
Kinsella, M. A., **1**, 362, 555, 558 ⟨92JA7375⟩
Kinsey, A. C., **2**, 358, 361 ⟨71JCS(C)3867⟩
Kinsman, R. G., **3**, 669, 670 ⟨83T3881⟩
Kinsman, R. L., **6**, 32 ⟨52JA849⟩
Kinson, P. L., **3**, 470 ⟨70JA2591⟩
Kinstle, T. H., **1**, 639, 644, 747 ⟨68CC1699⟩
Kinter, C. M., **1**, 788 ⟨90JOC3967, 91JOC6981⟩; **5**, 159 ⟨90JOC3967⟩
Kinyazev, A. P., **5**, 340 ⟨89ZOB611⟩
Kinzelmann, ,H. G., **2**, 559, 560, 959 ⟨90CB1507⟩
Kinzinger, L., **2**, 310 ⟨90LA795⟩
Kinzy, W., **4**, 174 ⟨87TL1981⟩
Kiovsky, T. E., **2**, 1009 ⟨68JA4666⟩
Kipnis, F., **2**, 126, 223 ⟨49JA2270⟩; **4**, 216 ⟨52JA1068⟩, 223 ⟨51JA822⟩
Kipphardt, H., **2**, 380 ⟨87OS(65)173⟩
Kiprianov, A. I., **6**, 232 ⟨52ZOB2216⟩
Kiprof, P., **2**, 585 ⟨91AG(E)185, 93CB45⟩; **5**, 925 ⟨92JA4910⟩, 955 ⟨92CB1141⟩
Kira, M. A., **4**, 882 ⟨83MI 420-01⟩
Kira, M., **1**, 759 ⟨73TL4193⟩, 922 ⟨88JPO197⟩; **2**, 77 ⟨88JA4599⟩, 78 ⟨89JA6429, 89TL1099⟩, 546 ⟨71TL1511⟩, 936 ⟨73TL4193⟩; **3**, 502 ⟨91JA3986⟩; **4**, 395 ⟨67CC889⟩, 613 ⟨88CL1441⟩, 618 ⟨88TL6939⟩, 620, 653 ⟨73TL4193⟩, 1052 ⟨80AG632, 80TL3077⟩; **6**, 190, 194, 393 ⟨73TL4193⟩, 383 ⟨88CL1441⟩
Kirby, A. F., **3**, 451 ⟨52JA4331, 53JA6026⟩
Kirby, A. J., **2**, 807 ⟨B-67MI 215-01⟩; **4**, 852 ⟨66T3001⟩; **5**, 786 ⟨B-67MI 520-01⟩
Kirby, G. W., **2**, 105 ⟨63CI(L)1877⟩, 140 ⟨91JCS(P1)2329⟩, 289, 292 ⟨88JCS(P1)1913⟩, 354 ⟨77CSR1⟩, 414 ⟨58JCS3039⟩; **3**, 333 ⟨91JCS(P1)2329⟩, 333, 335 ⟨83CC1325⟩, 333, 335 ⟨83CC423⟩, 333, 335 ⟨85JCS(P1)1541⟩, 333, 340 ⟨84CC1469⟩, 333, 340 ⟨84CC922⟩, 345, 349, 351, 369 ⟨90JCS(P1)3175⟩, 384 ⟨86CC1152⟩, 636 ⟨88JCS(P1)1913⟩, 657 ⟨67CC218⟩; **4**, 290 ⟨88JCS(P1)1913⟩, 490 ⟨82TL3835⟩; **5**, 326 ⟨77CRV1, 85JCS(P1)883⟩, 330 ⟨88TL6173, 91JCS(P1)1001⟩, 330 ⟨93JCS(P1)1397⟩, 336 ⟨88JCS(P1)1913⟩, 336, 1113 ⟨86CC1152⟩
Kirby, P., **2**, 439 ⟨66JCS(C)245⟩, 803 ⟨70JCS(C)2514⟩; **5**, 828 ⟨68JCS(C)431⟩
Kirby, R. H., **5**, 43 ⟨32OSC(1)361⟩
Kirchen, R. P., **1**, 895 ⟨78JA5134⟩, 918 ⟨81JA588⟩
Kircher, F., **3**, 452 ⟨41LA(547)1⟩
Kircher, H. W., **1**, 797 ⟨87JOC2586⟩
Kirchgässner, U., **5**, 954 ⟨91JA2228⟩
Kirchhoff, R. A., **1**, 622, 710 ⟨73JA6462⟩, 622, 710 ⟨79JA3602⟩; **3**, 379 ⟨73JA4287⟩
Kirchmeier, R. L., **2**, 334 ⟨90JA2383⟩; **4**, 564 ⟨90IC573, 91JFC(52)1⟩, 803 ⟨90JA3152⟩; **5**, 246 ⟨75MI 505-01⟩, 659 ⟨92IC4917⟩, 675 ⟨93IC287⟩, 679 ⟨90JFC(48)395⟩, 679, 703 ⟨90IC4255⟩, 679, 703 ⟨92ICA527⟩, 684 ⟨92JFC(56)335⟩; **6**, 251, 256, 259, 624 ⟨90JFC(48)395⟩, 258 ⟨89JA1773⟩, 258, 278 ⟨89CJC1795⟩, 259 ⟨93IC4802⟩, 259, 614, 624 ⟨92JFC(57)293⟩, 259, 643, 651, 653 ⟨92IC488⟩, 260 ⟨92IC4917⟩, 260, 261 ⟨92IC329⟩, 260, 615 ⟨89IC3345, 75MI 608-01⟩, 262 ⟨90HAC167, 92HCA907⟩, 277 ⟨75IC592⟩, 603 ⟨90IC571⟩
Kirchmeyer, S., **2**, 1003 ⟨91JOC1453⟩; **4**, 315 ⟨83S500⟩
Kirchner, D. G., **2**, 753 ⟨74JOC1758⟩
Kirchner, G., **5**, 135 ⟨85JA7072⟩
Kirchner, I., **1**, 934 ⟨78TL619⟩
Kirchner, J. J., **1**, 833 ⟨88TL4229⟩
Kirchner, R. M., **2**, 438 ⟨73JA4194⟩
Kirchoff, R. A., **2**, 164 ⟨79JOC2280⟩
Kireev, S. L., **1**, 529 ⟨89TL4021⟩
Kirenskaya, L. I., **5**, 476 ⟨57JGU2301⟩; **6**, 239 ⟨59ZOB2157⟩, 440 ⟨57ZOB2243⟩
Kirichenko, T. I., **5**, 1024 ⟨85ZOB2100⟩; **6**, 476 ⟨85ZOB2100⟩, 575 ⟨80TL313⟩
Kirihara, M., **1**, 1140 ⟨87TL4583, 88CPB3897⟩; **5**, 223 ⟨90CPB1836⟩
Kirihata, M., **3**, 706 ⟨85ABC1761⟩
Kirilin, A. D., **4**, 140 ⟨88JGU81⟩; **5**, 978 ⟨75ZOB2007⟩, 1075 ⟨85ZOB1308⟩, 1079 ⟨86ZOB1525⟩; **6**, 626 ⟨88ZOB393, 91ZOB1202⟩
Kirilina, N. I., **4**, 140 ⟨88JGU81⟩
Kirillov, A. I., **6**, 425 ⟨71JOU1943⟩
Kirilov, E. M. G., **3**, 494 ⟨89JCS(P2)1405⟩
Kirilov, M., **1**, 644 ⟨88PS(40)69⟩, 747 ⟨72M1651⟩; **2**, 1004 ⟨85CB3529, 88AX(C)626⟩; **3**, 494 ⟨89JCS(P2)1405⟩
Kirillova, L. P., **1**, 1006 ⟨74JOU1990⟩; **3**, 267 ⟨84JGU28⟩; **4**, 106 ⟨85JOU806, 88JOU2003⟩
Kirilyuk, B. A., **2**, 265 ⟨87ZOR1981⟩; **6**, 89 ⟨88URP1421744⟩
Kirino, O., **5**, 769 ⟨70MI 519-01⟩, 847 ⟨70JPB1696⟩, 848 ⟨72JPJ275⟩
Kirisawa, M., **5**, 280 ⟨85CL123⟩
Kirk, B. E., **1**, 498, 518 ⟨74JCS(P1)1209, 71CC1144⟩, 518 ⟨74JCS(P1)1844⟩; **2**, 19 ⟨88JCS(P1)549⟩
Kirk, D. N., **1**, 631 ⟨69T3107⟩
Kirk, J. M., **2**, 90 ⟨88TL2483⟩; **5**, 238 ⟨90T7429⟩
Kirk, K. L., **6**, 264 ⟨91BMC357⟩
Kirk, P. G., **5**, 1129 ⟨69JCS(A)2190⟩
Kirkemo, C. L., **5**, 177 ⟨77JA646⟩
Kirkiacharian, S., **5**, 827 ⟨83S220⟩
Kirkley, R. K., **1**, 636 ⟨70JA1784⟩

Kirkpatrick, C. M., **5**, 932 ⟨85JA1993⟩
Kirkpatrick, D., **1**, 126 ⟨73CC544, 73TL4491⟩; **2**, 26 ⟨72TL3869⟩, 617 ⟨72CC443⟩, 883, 890 ⟨68JCS(A)483⟩, 904 ⟨73CC544, 76JCS(P1)2435⟩; **4**, 738 ⟨72CC443⟩
Kirkschleger, B., **4**, 146 ⟨88JOM(346)C1⟩
Kirkwood, M. W., **6**, 655 ⟨57CJC527⟩
Kirman, J., **4**, 111 ⟨66JCS(C)2075⟩
Kirmreuther, H., **2**, 1014 ⟨08CB314, 09CB4232⟩, 1098 ⟨08CB316, 09CB4233⟩
Kirmse, R., **6**, 565 ⟨81ZAAC(476)7⟩
Kirmse, W., **1**, 258 ⟨76AG(E)251⟩, 383 ⟨73CB1376⟩, 630 ⟨66LA(691)41⟩, 631 ⟨65CB4022⟩, 637 ⟨89CB389⟩, 799 ⟨72CC716⟩, 801 ⟨82CB2508⟩, 905 ⟨83JA950, 84CB2510⟩, 1145 ⟨64AG(E)651⟩; **2**, 387 ⟨74CB745⟩; **3**, 47 ⟨67CB1491⟩, 466 ⟨67LA(707)44⟩, 531, 532 ⟨B-71MI 316-01⟩, 551 ⟨58LA(614)4⟩; ;**4**, 7, 16, 24 ⟨B-71MI 401-01⟩, 16, 24 ⟨B-64MI 401-01⟩, 619, 621 ⟨89JA4127⟩, 852 ⟨58LA(614)4⟩, 1080 ⟨B-71MI 424-01⟩; **5**, 360 ⟨89CB389⟩
Kirner, W. R., **2**, 133 ⟨29JA3131⟩
Kirollos, K. S., **1**, 980 ⟨92JCS(P1)2355⟩
Kirpichenko, S. V., **1**, 63 ⟨81JOM(214)145⟩; **2**, 266 ⟨81JOM(214)415⟩; **4**, 368 ⟨81JOM(204)13⟩, 507 ⟨87JOM(326)159⟩
Kirpicheva, E., **6**, 250 ⟨93IZV2513⟩
Kirrmann, A., **2**, 757 ⟨68CR(C)623⟩
Kirsanov, A. V., **2**, 399, 400 ⟨67ZOR1282⟩, 407 ⟨65ZOB344⟩, 449 ⟨69ZOB2139⟩, 453 ⟨66ZOB564⟩, 469 ⟨78ZOB2624⟩, 860 ⟨56ZOB907⟩; **3**, 363 ⟨72ZOR1822⟩, 567 ⟨71JOU413⟩; **4**, 104 ⟨73JOU418⟩, 304 ⟨69ZOR2181⟩, 487 ⟨69ZOB2192⟩, 573 ⟨77URP403314⟩, 999, 1000 ⟨75ZOB12⟩; **5**, 373 ⟨69JGU1445⟩, 644 ⟨60ZOB3006, 61ZOB582⟩, 657 ⟨76JOU663, 76ZOR673⟩, 683 ⟨56JGU2239, 56ZOB2009⟩, 683 ⟨58JGU1930, 58ZOB1887⟩, 683, 707 ⟨56JGU2935, 56ZOB2631, 57JGU3284, 57ZOB3248, 57JGU1065⟩, 707 ⟨57ZOB1080, 58JGU2283, 58ZOB2247, 59JGU596, 59ZOB600⟩, 707, 861 ⟨61ZOB3424⟩, 708 ⟨62JGU167, 62ZOB171⟩, 787 ⟨73JGU10⟩, 810 ⟨72ZOR695⟩, 860 ⟨62JGU3689, 62ZOB1873⟩, 861 ⟨62ZOB2245, 68ZOB326, 68ZOB850⟩, 965 ⟨70ZOB1933⟩, 988 ⟨54ZOB1033, 59ZOB2256⟩, 988 ⟨61ZOB1607, 61ZOB3062, 62ZOB3759⟩, 1005, 1010 ⟨64ZOB2812⟩, 1032 ⟨72ZOR1822⟩, 1053 ⟨61ZOB2275⟩; **6**, 238, 610 ⟨71ZOR2084⟩, 285, 367, 666 ⟨73ZOB544⟩, 367 ⟨72ZOB1169⟩, 537 ⟨87UKZ395⟩, 616, 728 ⟨72JGU97⟩
Kirsanov, V. A., **5**, 707 ⟨61JGU3192⟩
Kirsanova, I. A., **5**, 1052 ⟨69ZOR1285⟩
Kirsanova, N. A., **5**, 761, 1057 ⟨70UKZ372⟩; **6**, 658 ⟨73ZOR1815⟩
Kirsch, A., **1**, 950 ⟨93CB1171⟩
Kirsch, C., **3**, 550 ⟨91CB1397⟩
Kirsch, G., **2**, 368 ⟨B-87MI 206-02⟩; **3**, 72 ⟨75CR(281)35⟩; **5**, 333 ⟨B-87MI 507-01⟩
Kirsch, J. K., **5**, 31 ⟨64JA837⟩
Kirschke, K., **2**, 659 ⟨72JPR170⟩; **4**, 989 ⟨86TL4281⟩; **5**, 357 ⟨85JPR(327)893⟩
Kirschleger, B., **1**, 664 ⟨88JOM(346)C1⟩; **3**, 152 ⟨84TL835⟩, 152, 154 ⟨81S68⟩; **4**, 138, 146, 667 ⟨85BSF825⟩, 139 ⟨86BSF470⟩, 146 ⟨84TL835⟩
Kirschleger, R., **2**, 998 ⟨84TL835⟩
Kirschning, A., **4**, 648, 655 ⟨88TL4281⟩, 938 ⟨91JOC717⟩
Kirschstein, G., **6**, 213 ⟨B-74MI 607-02⟩
Kirss, R. U., **6**, 237 ⟨92OM2947, 93CM1321⟩
Kirst, H. A., **1**, 999, 1052 ⟨68TL5041, 70JA6314⟩; **2**, 314 ⟨88SC777⟩, 970 ⟨70JA6314⟩
Kirste, B., **3**, 270 ⟨88LA891⟩; **5**, 159 ⟨88LA891⟩
Kirstein, H., **5**, 131 ⟨87JOC3784⟩
Kirstein, H. M., **5**, 138 ⟨91JA910⟩
Kirtane, J. G., **1**, 842 ⟨83T2323⟩; **2**, 91 ⟨83T2323⟩
Kirtley, S. W., **2**, 571 ⟨82COMC-I(3)1079, 82COMC-I(3)1255⟩, 571 ⟨82COMC-I(3)783⟩
Kirwan, J. N., **1**, 5, 7, 9, 10, 16, 17, 331 ⟨91JCS(P1)103, 90TL5093⟩
Kiryu, Y., **1**, 120 ⟨93TL5127⟩, 241 ⟨87CC449⟩; **3**, 161 ⟨92S852⟩
Kiryushkin, A. A., **6**, 104 ⟨64TL47⟩
Kirzner, V. N., **2**, 448 ⟨76JGU572⟩; **4**, 566 ⟨76ZOB575⟩
Kisalus, J. C., **2**, 441 ⟨83JOC4466, 84JA7021⟩
Kischkel, H., **2**, 440 ⟨84ZN(B)356⟩, 493, 495, 499, 500 ⟨81ZN(B)1105⟩, 507 ⟨83ZAAC(502)175, 86OCS621⟩; **4**, 455 ⟨84ZN(B)356, 86MI 410-01⟩
Kischkewitz, J., **1**, 54 ⟨88ZAAC(560)141⟩; **4**, 704 ⟨90JFC(46)265⟩; **6**, 243 ⟨90JFC(46)265⟩
Kise, H., **2**, 35 ⟨81TL1017⟩, 121 ⟨72JOC1121⟩, 298 ⟨78JOC3345⟩, 478 ⟨71TL1761⟩; **5**, 350 ⟨71TL1761⟩
Kise, M., **2**, 647 ⟨66BCJ1212⟩
Kise, M. A., **5**, 93 ⟨43OSC(2)422⟩
Kise, N., **2**, 81 ⟨90TL1303⟩; **3**, 151 ⟨82TL1609⟩; **6**, 25, 26 ⟨82TL1609, 81TL871⟩, 26 ⟨82TL4801⟩
Kiseki, Y., **1**, 486 ⟨81JOC4885⟩
Kiselev, V. V., **4**, 465 ⟨84JGU421⟩
Kiseleva, E. I., **4**, 1008, 1009 ⟨86ZOB2681, 90ZOB229⟩; **5**, 644 ⟨89ZOB1917⟩, 652 ⟨90JGU1165⟩, 657 ⟨87JGU1104, 87ZOB1233⟩, 659 ⟨86ZOB2681⟩, 659, 788, 789 ⟨86JGU2141⟩, 787 ⟨89JGU1496⟩, 788 ⟨85JGU402⟩, 788 ⟨89JGU1129, 88JGU30⟩, 789 ⟨89JGU1123⟩
Kiselyov, A. S., **4**, 245 ⟨93T2151⟩; **6**, 145 ⟨90TL7379⟩
Kisfaludy, L., **5**, 277 ⟨67JA183, 74TL1785⟩
Kish, F. A., **1**, 896 ⟨65JA2157⟩
Kishi, K., **2**, 301 ⟨90TL1743⟩; **3**, 618 ⟨89JOC2249⟩; **4**, 857 ⟨79S880⟩
Kishi, N., **1**, 404, 405, 816, 817, 818 ⟨83JOC279⟩, 816 ⟨92T4087⟩; **2**, 932 ⟨92SL189⟩; **3**, 211 ⟨83TL795⟩
Kishi, Y., **1**, 14 ⟨82TL2719⟩, 196 ⟨93TL8047⟩, 263 ⟨75JOC2009⟩, 478 ⟨79TL4343⟩, 489 ⟨87JA4756⟩, 677, 1038, 1099 ⟨86JA5644⟩, 748 ⟨81T3873, 82JA1109⟩, 831 ⟨89JA2735⟩, 835 ⟨71TL4661⟩, 978 ⟨87JOC1370⟩, 1092 ⟨83TL1463⟩, 1099 ⟨87CS573, 87JOC4819, 87JOC4823, 87JOC4825, 88JOC3383, 88JOC4151, 88JOC5580, 88TL4909⟩; **2**, 46 ⟨83TL3947⟩, 73 ⟨81T3873⟩, 83 ⟨82TL2719⟩, 656 ⟨76JOC3026⟩; **3**, 238 ⟨86TL4759, 86TL4767⟩, 244 ⟨87TL3463⟩, 442 ⟨86TL4763⟩, 461 ⟨83JOC3833⟩; **4**, 216 ⟨77JA4835⟩, 218 ⟨74TL1549⟩; **5**, 155 ⟨79JA259⟩, 223 ⟨88TL4909⟩; **6**, 68 ⟨65T2059⟩
Kishida, M., **3**, 635 ⟨88CPB1698⟩
Kishida, T., **3**, 122 ⟨85CL1437⟩
Kishida, Y., **1**, 822 ⟨66TL1787, 68CPB793⟩; **3**, 57 ⟨72TL2743⟩, 374 ⟨70CPB490⟩, 378 ⟨66TL1787⟩
Kishigama, Y., **3**, 57 ⟨81JOC240⟩
Kishigami, Y., **2**, 55 ⟨86JOC240⟩
Kishikawa, K., **4**, 206 ⟨92BCJ2366⟩; **5**, 325 ⟨87JOC2327⟩
Kishikawa, T., **1**, 582 ⟨92BCJ2366⟩
Kishimoto, H., **1**, 1140 ⟨86TL195, 87CPB1405⟩, 1140 ⟨87TL4583, 88CPB3897⟩
Kishimoto, K., **2**, 25 ⟨81TL4409⟩, 28 ⟨85BCJ1599⟩; **5**, 1106 ⟨81TL4409⟩
Kishimoto, T., **3**, 418 ⟨76JAP7632569⟩
Kishimura, K., **1**, 185 ⟨90JOC6356⟩
Kisielowski, L., **1**, 248 ⟨83CB2378⟩; **2**, 1051, 1063 ⟨83CB1320⟩
Kisilenko, A. A., **4**, 794 ⟨80JOU1762⟩; **5**, 657, 659 ⟨79JOU32⟩, 657, 659 ⟨79ZOR38⟩; **6**, 609, 619 ⟨83KGS798⟩
Kisin, A. V., **4**, 98, 99 ⟨87JGU334⟩, 140 ⟨83JFC(23)339⟩, 156 ⟨78JGU769⟩, 603, 625, 637, 701 ⟨71JOM(30)C57⟩, 639 ⟨70ZOB2601⟩, 640 ⟨76ZOB837⟩, 660 ⟨77ZOB1751⟩; **5**, 1079 ⟨86ZOB1525⟩; **6**, 626 ⟨88ZOB393, 91ZOB1202⟩
Kiso, Y., **1**, 451 ⟨75JA5603⟩, 466 ⟨73JOM(50)C12⟩, 466, 470 ⟨76BCJ1958⟩, 471 ⟨72JA9268, 75JA4405⟩; **2**, 227 ⟨91CC167⟩, 478 ⟨90CPB673⟩, 915 ⟨74JOM(76)105, 74JOM(76)95⟩
Kiso, Y., **5**, 199 ⟨90JMC2621⟩

Kiso, Y., **5**, 226 ⟨80CC1029⟩
Kispert, L. D., **1**, 909 ⟨74JOC378⟩, 923 ⟨67JCP(46)3193⟩
Kissel, H., **5**, 838 ⟨1899CB3137⟩
Kissel, T., **4**, 229 ⟨80LA1665⟩; **5**, 554 ⟨76ZC318, 80CB1898⟩; **6**, 538 ⟨80CB1898⟩
Kissel, W. J., **4**, 206 ⟨66TL6145⟩
Kissinger, L. W., **4**, 440 ⟨57JOC1662⟩, 994 ⟨66JOC369⟩; **5**, 825 ⟨58JOC1517⟩
Kissman, H. M., **6**, 108 ⟨52CB949⟩
Kistenmacher, T. J., **4**, 876 ⟨84JA8303, 87SM647⟩; **6**, 592 ⟨83CC294⟩
Kisuno, A., **5**, 371 ⟨86BCJ3696⟩
Kita, M., **4**, 721 ⟨93JOM(463)169⟩
Kita, Y., **1**, 822 ⟨73TL2351⟩, 827, 828 ⟨89CPB2047⟩, 1061, 1062 ⟨90JOC4853⟩, 1140 ⟨85TL1549, 86TL195⟩, 1140 ⟨87CPB1405, 87TL4583⟩, 1140 ⟨88CPB3897⟩; **2**, 263 ⟨83TL1273⟩, 263, 812 ⟨79TL4311⟩, 289 ⟨79TL2251⟩, 665 ⟨81TL4283, 82JOC4376⟩, 941 ⟨90JOC1108⟩; **3**, 108 ⟨82CPB1257⟩, 158 ⟨85TL3837⟩, 207 ⟨80TL3779⟩, 249 ⟨79TL1129⟩, 378 ⟨74JCS(P1)1125⟩, 669 ⟨80JCS(P1)1132⟩; **4**, 188 ⟨82S1089⟩, 223 ⟨93TL4063⟩, 418 ⟨91TL2375⟩, 661 ⟨91SL911, 92JCS(P1)2813⟩, 828 ⟨93SL273⟩; **5**, 189, 195, 282 ⟨86JOC4150⟩, 189, 199 ⟨84TL6027⟩, 215 ⟨89S334⟩, 223 ⟨90CPB1836⟩, 226, 375, 378 ⟨79TL4311⟩, 294 ⟨80S887⟩, 377 ⟨89JCS(P1)1862, 81S451⟩, 595 ⟨79S887⟩, 1041, 1107 ⟨77TL4417⟩, 1048 ⟨78CI(L)579⟩, 1055, 1114 ⟨79TL2251, 79TL1129⟩, 1107 ⟨81JCS(P1)1577⟩; **6**, 484 ⟨78TL3737⟩
Kitada, C., **2**, 274 ⟨78CPB1576⟩
Kitade, Y., **1**, 1029 ⟨87H(26)355⟩
Kitaev, Y. P., **2**, 379 ⟨85ZOR124⟩; **3**, 446 ⟨58CA(52)9980⟩, 448 ⟨63BAU408⟩, 681 ⟨72RCR495, 76ZOR1676⟩; **5**, 356 ⟨75ZOR173, 75ZOR348⟩, 652 ⟨76ZOR1557⟩, 860 ⟨74ZOR2200⟩
Kitaev, Yu. P., **5**, 681 ⟨77IZV393⟩
Kitagaki, S., **4**, 661 ⟨91SL911, 92JCS(P1)2813⟩
Kitagawa, K., **2**, 304 ⟨90S735⟩; **4**, 284 ⟨91TL375⟩
Kitagawa, M., **2**, 103 ⟨36MI 202-01⟩
Kitagawa, O., **2**, 34 ⟨92TL1299, 93TL2165⟩; **4**, 734 ⟨85CPB5137⟩
Kitagawa, S., **3**, 366 ⟨78CPB2092⟩
Kitagawa, T., **1**, 867 ⟨90JOC996⟩; **3**, 710 ⟨83CPB723⟩; **4**, 442 ⟨83CPB723⟩, 538 ⟨79CPB2857⟩; **5**, 367 ⟨77CPB1651⟩; **6**, 21 ⟨91TL375⟩
Kitagawa, Y., **2**, 420 ⟨75TL1859⟩
Kitahara, T., **1**, 218, 221 ⟨77JOC394⟩, 649 ⟨88T4339⟩, 1134 ⟨79TL3021⟩, 1139 ⟨76JOC1669⟩; **2**, 665 ⟨74JA7807⟩, 546; **3**, 221 ⟨75JOC538⟩, 634 ⟨84JOC3281⟩; **4**, 305 ⟨76JA6715⟩; **5**, 173 ⟨84T2935⟩, 199 ⟨74JA7807⟩, 199 ⟨82OS(61)147⟩
Kitahara, Y., **2**, 31 ⟨76CL283⟩; **3**, 213 ⟨75TL3299⟩, 352 ⟨64BCJ1897⟩
Kitai, M., **1**, 451 ⟨75TL2825⟩, 1045, 1138 ⟨78TL2301⟩; **3**, 238, 270, 273 ⟨78TL2301⟩
Kitai, T., **1**, 1077 ⟨69TL4447⟩
Kitajima, H., **1**, 156 ⟨93T537⟩; **3**, 280, 296, 298, 305 ⟨80S139⟩, 280, 305 ⟨88BCJ455⟩, 282, 295, 296 ⟨90CL783⟩
Kitajima, M., **5**, 324 ⟨92H(33)121⟩
Kitajima, T., **2**, 540 ⟨84JOC2298⟩
Kitamura, A., **1**, 576 ⟨85JA4964⟩; **5**, 634 ⟨78BCJ301⟩; **6**, 314 ⟨78BCJ301⟩
Kitamura, K., **1**, 193 ⟨84TL5921⟩; **3**, 213 ⟨83TL3489⟩
Kitamura, M., **1**, 77 ⟨86JA7117, 87JOC3174⟩, 310 ⟨80CL331⟩, 779, 788 ⟨78CL541⟩; **2**, 53 ⟨87JA5856⟩, 67 ⟨91AG(E)49⟩, 71 ⟨88PAC123, 89JA4028⟩, 478 ⟨B-89MI 211-03⟩; **3**, 56 ⟨82JA4997⟩; **4**, 367, 937 ⟨80CL331⟩, 936 ⟨82JA4997, 79TL3465⟩, 937 ⟨81CL457, 81TL239, 84JA3252, 84JOC3517⟩
Kitamura, S., **6**, 207 ⟨71MI 606-01⟩
Kitamura, T., **1**, 541, 1018 ⟨87JA7561⟩, 985 ⟨92CL1357, 91JCS(P1)2892⟩, 1120 ⟨79TL1619, 80T3229⟩; **2**, 1020 ⟨88JOC4105⟩, 1028 ⟨89JA2225⟩, 1029 ⟨88JA3272⟩; **3**, 621 ⟨73CL379⟩; **5**, 317 ⟨89JOC3394⟩, 328 ⟨90SC887⟩; **6**, 547 ⟨89BCJ967⟩
Kitani, Y., **3**, 487 ⟨79CC450, 86OM356⟩
Kitano, H., **2**, 279, 288 ⟨67NEP6614945, 68NKK83⟩; **5**, 1039 ⟨62MI 527-01⟩
Kitano, K., **2**, 316 ⟨92TL627⟩; **3**, 457, 458 ⟨92TL627⟩
Kitano, M., **6**, 96, 98 ⟨87JOC2657⟩
Kitano, T., **4**, 507 ⟨91S996⟩
Kitano, Y., **2**, 921 ⟨88T4073⟩
Kitao, O., **4**, 230 ⟨84TL5797⟩
Kitaoka, M., **2**, 713 ⟨85CL805, 87JOC1078⟩
Kitaoka, S., **5**, 183, 192 ⟨60JOC1265⟩
Kitatani, K., **1**, 164 ⟨77BCJ3288⟩; **3**, 224 ⟨74TL3297⟩
Kitaura, K., **5**, 334 ⟨85JMC1772⟩
Kitawaki, R., **5**, 1118 ⟨57NKZ567, 60JOC1043⟩, 1125 ⟨61JOC4122⟩
Kitayama, R., **2**, 159 ⟨83CC1470⟩, 721 ⟨83S1040⟩
Kitayama, T., **2**, 398 ⟨90T7471⟩
Kitayima, K., **2**, 202 ⟨82TL5427⟩
Kitazawa, N., **2**, 16 ⟨92H(33)105⟩
Kitazawa, S., **2**, 226 ⟨92CC600⟩; **3**, 370 ⟨92CC600⟩
Kitazume, T., **1**, 1130 ⟨85JA5186⟩; **2**, 19 ⟨81CL761⟩, 830 ⟨80JOC5052⟩; **3**, 274 ⟨80CL1327⟩, 351 ⟨73BCJ3285⟩; **4**, 415 ⟨74BCJ785⟩, 731 ⟨76BCJ2491, 85JFC(29)445⟩, 736 ⟨88S614, 89JOC5630⟩, 736 ⟨92SL977, 93JFC(62)201⟩; **5**, 280 ⟨80JOC5052⟩; **6**, 21 ⟨88JOC2350⟩, 22 ⟨82CL1453⟩, 65 ⟨81CL1337⟩, 253 ⟨77JA4194, 78IC2173⟩
Kitching, W., **1**, 65 ⟨69PAC1⟩, 67, 69 ⟨81JOC563⟩, 68 ⟨86CC855⟩, 572 ⟨93JOC2501⟩, 846 ⟨66JA460⟩, 895 ⟨79JOC2652⟩; **2**, 29 ⟨93JOC2501⟩, 210 ⟨88OM1196⟩; **6**, 207 ⟨68MI 606-01⟩
Kitigawa, T., **2**, 437 ⟨91BCJ2942⟩; **5**, 189, 196 ⟨87CPB1262⟩
Kitigawa, Y., **1**, 534 ⟨82TL2087⟩
Kitihara, T., **1**, 591 ⟨86TL1343⟩
Kitoh, Y., **3**, 153, 154 ⟨92BCJ2800⟩; **4**, 74, 77 ⟨92TL1483, 93TL2331⟩
Kitschke, B., **5**, 1122 ⟨80AG(E)566⟩
Kitson, F., **6**, 221, 222, 226 ⟨93BSF599⟩
Kitson, F. G., **5**, 1133 ⟨87JA6532⟩
Kittisopikul, S., **1**, 936 ⟨77TL4291⟩; **4**, 734 ⟨84TL4329⟩
Kitzing, R., **1**, 521 ⟨68AG(E)67⟩
Kiuchi, F., **4**, 90 ⟨91COS(7)762⟩
Kiuchi, S., **4**, 857 ⟨86JOC700⟩
Kivinen, A., **5**, 130 ⟨B-72MI 503-03⟩
Kivisikk, R., **5**, 410 ⟨65CJC1175⟩
Kivsanov, A. V., **4**, 10 ⟨73S787⟩
Kiyohara, M., **5**, 371 ⟨79TL3851⟩
Kiyohara, Y., **6**, 322, 337 ⟨70MI 611-01⟩, 576 ⟨70MI 618-02⟩
Kiyoi, T., **1**, 473 ⟨88JA8153⟩; **3**, 214 ⟨90JOC2554⟩; **4**, 1033 ⟨88JA8153⟩; **5**, 1072 ⟨93JOC5118⟩
Kiyokawa, H., **3**, 378 ⟨74JCS(P1)1125⟩
Kiyonaga, H., **1**, 836 ⟨91TL7557⟩; **6**, 472 ⟨88H(27)2327, 92OPP200⟩, 473 ⟨91TL7557⟩
Kiyooka, S., **1**, 290 ⟨85CL1829, 87CL495⟩, 486 ⟨87CL1775⟩; **2**, 67 ⟨94TL4107⟩; **3**, 136, 203 ⟨90JOC5562⟩; **4**, 202 ⟨91JOC2276, 93TL1491⟩; **5**, 432 ⟨85CL1829, 87CL495⟩
Kiyoshima, Y., **2**, 123 ⟨69TL3631⟩
Kiyotaka, H., **5**, 1012 ⟨89NKK1881⟩
Kizaki, H., **4**, 142, 521 ⟨91CL97⟩
Kizim, N. G., **4**, 152 ⟨73JOM(57)423⟩; **6**, 64 ⟨73JOM(57)423⟩
Kjaer, A., **5**, 584, 586, 587 ⟨52ACS327⟩, 828, 831 ⟨70ACS1367⟩, 832 ⟨87ACS(B)29⟩, 1029 ⟨54ACS295⟩; **6**, 595 ⟨89H(28)269⟩
Kjellin, C., **5**, 831 ⟨1897LA(289)117⟩
Kjelsberg, F., **3**, 549 ⟨20HCA853⟩
Kjelsberg, M. A., **3**, 516 ⟨87OM433⟩
Kjersgaard, D., **5**, 828, 831 ⟨70ACS1367⟩

Kjonaas, R. A., **1**, 491 ⟨83JOC1924⟩; **2**, 702 ⟨86JOC3993⟩; **3**, 128 ⟨86JOC3993, 86TL1437⟩, 128 ⟨88JOC4133⟩
Kkiefer, G., **2**, 245 ⟨67AG(E)951⟩
Klaassen, A. A. K., **1**, 890 ⟨74AOC215⟩
Klabunde, K. J., **1**, 1143 ⟨72JA7862⟩; **2**, 582 ⟨85IS1961⟩, 961 ⟨75ACR393⟩; **3**, 355, 356 ⟨84AG(E)729⟩; **4**, 152 ⟨76JFC(7)95⟩; **5**, 550 ⟨87JOC1156⟩, 1096 ⟨90JA3332⟩; **6**, 235, 535 ⟨84JOC3854⟩, 529, 535 ⟨74IC1778, 84JA263⟩, 694 ⟨72JA820⟩
Klabunde, M., **4**, 276 ⟨91CB1827⟩
Klabunde, U., **5**, 951, 953 ⟨67JA7141⟩
Klabunovskii, E. I., **5**, 215 ⟨86IZV2032⟩
Klaebe, A., **4**, 478 ⟨81PS(10)207⟩
Klaeboe, P., **5**, 1117 ⟨77ACS(A)884⟩
Klaeser, K., **6**, 555, 559 ⟨81ZAAC(481)126⟩, 627 ⟨77ZAAC(434)115⟩, 628 ⟨77ZAAC(434)110⟩
Klaff, N., **4**, 695, 717 ⟨89AG(E)319⟩, 695, 717 ⟨89JOM(372)183⟩
Klager, F., **3**, 687 ⟨55LA(594)21⟩
Klager, K., **1**, 530, 890 ⟨48LA(560)1⟩; **2**, 386 ⟨56JA5428⟩; **4**, 430, 431, 432, 994 ⟨57JOC1665⟩
Klages, C.-P., **3**, 345, 346 ⟨80CB2255⟩, 348 ⟨77AG(E)725, 77JCR(S)146⟩; **5**, 510, 516, 523 ⟨79JCR(S)160, 77AG743⟩, 623 ⟨77AG(E)725⟩
Klages, F., **1**, 28 ⟨66LA(691)15⟩; **2**, 380 ⟨41LA(547)1⟩, 383, 384 ⟨59CB1842⟩, 1009 ⟨59LA(626)60⟩; **3**, 452 ⟨41LA(547)1⟩; **4**, 47 ⟨59CB1818⟩; **6**, 628 ⟨64CB735⟩
Klahn-Oliva, A. H., **6**, 517 ⟨85OM478⟩
Klahre, G., **1**, 639, 752, 753 ⟨59CB1950⟩; **3**, 480 ⟨61TL9⟩; **4**, 575, 576, 577 ⟨59CB2499⟩, 575, 576, 577 ⟨61CB1987⟩
Klair, S. S., **2**, 134 ⟨88TL4477⟩, 147 ⟨90SL457⟩, 694 ⟨90CSR147⟩; **4**, 249 ⟨92T5933⟩, 261 ⟨90SL457⟩, 263 ⟨89JCS(P1)2441⟩, 929 ⟨90CSR147⟩; **5**, 409, 411, 412, 417 ⟨90CSR147⟩
Klamann, D., **2**, 9 ⟨52M1398⟩, 176, 181 ⟨55CB201⟩, 225 ⟨51M970⟩, 363, 399 ⟨59CB1910⟩, 364 ⟨53LA(581)182⟩; **4**, 181 ⟨67JSP(23)32⟩, 333 ⟨90HOU(E12b)1⟩; **5**, 300 ⟨54CB1294⟩, 333 ⟨85HOU(E11)1⟩
Klambunde, K. J., **1**, 232 ⟨73JA1547⟩
Klamman, D., **4**, 163 ⟨91HOU(E14a/1)600⟩, 176 ⟨91HOU(E14a/1)1⟩, 178 ⟨91HOU(E14a/1)136⟩, 186 ⟨91HOU(E14a/1)323⟩, 202 ⟨91HOU(E14a/1)412, 91HOU(E14a/1)431⟩, 203, 207 ⟨91HOU(E14a/1)683⟩
Klanberg, F., **6**, 340 ⟨68IC2265⟩
Klanderman, B. H., **3**, 88 ⟨66JOC2618⟩
Klang, J. A., **1**, 692 ⟨89JA2324⟩
Klapotke, T., **2**, 500, 501 ⟨87POL1593⟩
Klarberg, B., **4**, 420 ⟨60JA2731⟩
Klarhe, G., **4**, 490 ⟨61TL9⟩
Klarner, F., **1**, 794 ⟨84MI 118-01⟩
Klärner, F. G., **1**, 799 ⟨80JA2469⟩, 800 ⟨72AG(E)832⟩; **3**, 394 ⟨90AG(E)609⟩
Klas, N., **1**, 54 ⟨85CB1039⟩, 271 ⟨83CB473⟩; **4**, 565, 596, 600 ⟨83CB473⟩, 596, 600 ⟨85CB1039⟩
Klason, P., **2**, 169 ⟨1887CB2376⟩
Klatt, M., **3**, 44 ⟨93SL121⟩
Klauck, W., **6**, 488 ⟨92GEP4215648⟩
Klauke, E., **3**, 699 ⟨69AG(E)20⟩; **6**, 9 ⟨81JFC(18)281⟩, 28 ⟨82S951⟩, 51, 238 ⟨69LA(730)140⟩, 229, 230 ⟨72GEP2129200⟩, 234 ⟨71GEP2003143⟩, 236 ⟨64AG807, 82GEP3044216⟩, 240 ⟨72CA(77)125952⟩, 262 ⟨85JCS(P1)1191⟩, 433 ⟨88JFC(40)365⟩, 438, 439 ⟨80EUP52842, 74GEP2311662⟩, 439 ⟨82JFC(19)553⟩, 453 ⟨73GEP2131789⟩, 529 ⟨65GEP1219455⟩, 602, 615 ⟨66AG(E)848⟩, 603, 613 ⟨72GEP2101107⟩
Klaus, K., **4**, 115 ⟨84JGU1354⟩; **5**, 876, 877 ⟨84ZOB1520⟩; **6**, 680, 681 ⟨84ZOB1520⟩
Klausener, A., **6**, 465 ⟨93EUP558996, 93EUP559001⟩, 465 ⟨93GEP4134688⟩
Klausner, Y. S., **5**, 278 ⟨74S549⟩
Klautke, S., **2**, 472 ⟨83MI 208-02⟩
Klaveness, J., **6**, 134 ⟨86JOM(303)189⟩
Klaver, W. J., **1**, 774 ⟨89JA2588⟩, 978 ⟨84JOC1149, 86TL1411⟩; **3**, 422 ⟨89JA2588⟩; **5**, 374 ⟨86TL1411⟩
Klayman, D. L., **2**, 220 ⟨68QRS189⟩, 237, 278, 291, 292 ⟨65JOC2454⟩, 279, 292 ⟨73JA197⟩, 290 ⟨66JOC3396⟩; **5**, 1056 ⟨81EJM317⟩; **6**, 594, 596 ⟨69JOC3549⟩, 637 ⟨65JOC2454⟩
Klebach, T. C., **2**, 1003 ⟨84T765⟩; **4**, 483 ⟨84T991⟩
Kleb, K. G., **2**, 797 ⟨60LA(635)91⟩
Klebanski, E. O., **2**, 866 ⟨85ZOB2141⟩
Klebanskii, A. L., **2**, 468 ⟨71JGU475⟩
Klebe, G., **2**, 1002, 1003 ⟨87AG(E)78⟩
Klebe, J., **3**, 490 ⟨60JA247⟩
Klebe, J. F., **5**, 375, 378, 718 ⟨66JA3390⟩
Kleeman, H.-W., **4**, 528 ⟨89EUP315574⟩
Kleeman, M., **4**, 282 ⟨66AG(E)594⟩
Kleemann, A., **6**, 445 ⟨86JOC3494⟩
Kleemann, G., **3**, 380 ⟨78AG(E)516, 79AG(E)944⟩, 380 ⟨83CB645, 84CB3255⟩, 380 ⟨87AG(E)799⟩
Kleene, R. D., **4**, 212 ⟨54JA5161⟩
Kleiderer, E. C., **5**, 114 ⟨43OSC(2)538⟩
Kleier, D. A., **6**, 567 ⟨88JOC38⟩
Kleijn, H., **1**, 514 ⟨79S430⟩, 541 ⟨77TL2023⟩, 544 ⟨79TL3327, 81JOM(206)257⟩, 544 ⟨81RTC337⟩, 967 ⟨82JOM(240)329, 85JOC5143⟩, 975 ⟨84TL5571⟩, 975, 1004, 1009 ⟨81TL1451⟩, 1004, 1006, 1025 ⟨82RTC405, 82RTC97⟩; **2**, 589 ⟨79JOM(172)377⟩; **3**, 642 ⟨78S454, 79S430⟩; **4**, 704 ⟨92S432⟩, 940 ⟨84JOM(276)317⟩, 960, 963 ⟨86JOM(302)1⟩; **5**, 548, 550, 559, 560 ⟨79S432⟩
Klein, A., **3**, 40 ⟨92AG(E)1391⟩; **4**, 540 ⟨87JOM(330)357⟩
Klein, B., **3**, 684 ⟨94JA1316⟩
Klein, D. A., **3**, 618 ⟨71JOC3050⟩
Klein, F., **5**, 727 ⟨1878CB1825⟩, 744 ⟨1878CB4⟩, 748 ⟨1877CB1889⟩
Klein, G., **1**, 811 ⟨79JA7599⟩; **5**, 1028 ⟨62JOC3701⟩
Klein, H., **1**, 496 ⟨76TL947⟩; **2**, 1034 ⟨60JA3472⟩
Klein, H. C., **3**, 279 ⟨48JA1647, 53JA745⟩
Klein, H. F., **2**, 575 ⟨84HOU(13/9b)628⟩, 597 ⟨67CB1129⟩; **3**, 487 ⟨90JA5351⟩; **4**, 582 ⟨74CB3692, 76AG(E)42⟩, 582 ⟨77CB2200⟩, 583 ⟨74CB93⟩; **5**, 461 ⟨73AG(E)402⟩, 1137 ⟨88CB1115⟩; **6**, 75 ⟨90JA5351⟩
Klein, J., **1**, 540 ⟨73JCS(P2)1971, 74JOM(67)467⟩, 705 ⟨84CB966, 86CB1540⟩, 708 ⟨87CB5⟩, 1054 ⟨B-78MI 121-01⟩, 1066, 1084 ⟨70T2127⟩; **4**, 668 ⟨75CC877, 78JOM(157)1⟩; **6**, 206 ⟨73JCS(P2)599, 75CC877⟩, 206 ⟨76T1839, 83T2733⟩
Klein, J. H., **6**, 206 ⟨74JA5640⟩
Klein, J. L., **2**, 741 ⟨83BSF(2)28, 84S493⟩; **4**, 406 ⟨88BSF1009⟩
Klein, J. T., **3**, 307 ⟨88JHC517⟩
Klein, K.-M., **1**, 1142 ⟨66CB1642⟩
Klein, K. P., **1**, 399 ⟨67JOC1155⟩; **2**, 391 ⟨79JOC275⟩; **3**, 192 ⟨79JOC275⟩
Klein, P., **1**, 828 ⟨86JOC123⟩; **5**, 568 ⟨81JOC3558⟩, 570, 610, 618 ⟨80TL4061⟩
Klein, R. A., **5**, 142 ⟨64JOC2813⟩
Klein, R. J., **2**, 724 ⟨60CB2736⟩
Klein, R. S., **5**, 690 ⟨86JHC1757⟩
Klein, S., **2**, 556 ⟨94JOC2925⟩; **4**, 45 ⟨93JFC(63)157⟩
Klein, S. A., **1**, 28 ⟨74JOC3525⟩; **2**, 366 ⟨74JOC3525⟩
Klein, W., **1**, 991 ⟨94OM94⟩; **4**, 723 ⟨94OM94⟩
Klein, W. J. de, **4**, 894 ⟨72JOC1960⟩
Klein, W. R., **1**, 121 ⟨93JOC2492⟩, 207 ⟨93TL3063⟩; **3**, 128 ⟨92JOC2249⟩; **5**, 539 ⟨80JCS(P2)1557⟩
Kleinberg, J., **2**, 12 ⟨47CRV381⟩, 183, 272 ⟨65JOC38⟩; **5**, 275 ⟨55OSC(3)516⟩
Kleine, B., **2**, 814 ⟨76TL2173⟩
Kleine, K.-M., **3**, 366 ⟨65CB3081⟩
Kleine, W., **5**, 652 ⟨79IS164⟩, 1088 ⟨75AG(E)760⟩; **6**, 700, 733 ⟨78CB2451⟩, 723 ⟨76AG(E)616, 77JOM(128)C49⟩, 723 ⟨78CB3542⟩
Kleiner, F. G., **2**, 1101 ⟨68LA(716)19⟩
Kleiner, H. J., **2**, 448 ⟨73AG(E)877⟩; **4**, 119, 121 ⟨72GEP(O)2060217, 74LA751⟩, 556 ⟨76GEP2523145⟩; **5**, 264 ⟨80AG(E)133⟩
Kleiner, K., **5**, 161 ⟨93AG(E)1191⟩

Kleinert, H. D., **2**, 118, 124 ⟨93JMC460⟩
Kleinman, E. F., **2**, 1006, 1009 ⟨91COS(2)893⟩; **3**, 184, 186 ⟨91COS(2)893⟩, 436 ⟨91COS(2)994⟩, 440 ⟨91COS(2)934⟩; **4**, 315 ⟨91COS(2)893⟩
Kleinpeter, E., **3**, 354 ⟨88ZC345⟩
Kleinschmidt, R. F., **1**, 839 ⟨44JA1929⟩
Kleinschmit, P., **2**, 166 ⟨87GEP3543567⟩
Kleinstück, R., **3**, 534 ⟨71AG(E)132⟩; 618 ⟨71CB2025⟩, 618, 619 ⟨71CB1030⟩; **4**, 552 ⟨71CB2250⟩; **5**, 1063 ⟨71CB1335⟩
Klemenkova, Z. S., **4**, 557 ⟨85ZOB1234⟩
Klement, I., **2**, 591 ⟨94TL1177⟩, 1096 ⟨93TL4623⟩; **3**, 614, 646, 665, 670, 675 ⟨93TL4623⟩
Klementschitz, W., **4**, 742 ⟨56M323⟩, 761 ⟨52M1⟩, 763 ⟨56M319⟩; **6**, 39 ⟨56M323⟩
Klemka, A. J., **3**, 618 ⟨40JA1432⟩
Klemm, D., **4**, 209 ⟨88SC2337⟩; **6**, 538 ⟨85EGP256705⟩
Klemm, K., **3**, 579 ⟨89JCS(P2)1841⟩; **4**, 987 ⟨79CB484, 81CB2001⟩; **5**, 744, 745, 751 ⟨59CB837⟩; **6**, 139, 140 ⟨59CB329⟩
Klemm, L. H., **1**, 523 ⟨66JOC2376⟩
Klempier, N., **3**, 628 ⟨93TL4769⟩; **5**, 118 ⟨91TL341⟩
Klempner, D., **6**, 488 ⟨B-89MI 615-01⟩
Klenha, J., **4**, 239 ⟨63CI(L)1397⟩
Klenk, E., **5**, 62 ⟨56AG349⟩
Klenk, H., **2**, 656 ⟨73AG(E)775⟩; **3**, 280, 296, 298 ⟨81CB926⟩
Klepa, T. I., **2**, 428 ⟨90ZOB697⟩
Klepel, M., **4**, 588 ⟨80EGP142887⟩
Kleschick, W. A., **1**, 874 ⟨80JOC1066⟩, 875 ⟨77JA247⟩; **4**, 349 ⟨78JOC1256⟩; **6**, 636 ⟨88JOC3120⟩
Klesse, R., **4**, 138, 139 ⟨92OM859⟩, 188 ⟨85JOC3627⟩
Klessing, K., **1**, 708 ⟨74CB1915, 76CB793⟩, 708 ⟨77CB2978⟩
Klett, M. W., **1**, 957 ⟨85JA6615⟩
Kleune, D., **6**, 537 ⟨75LA1025⟩
Klewe, B., **6**, 170 ⟨72ACS1874⟩
Klibanov, A. M., **5**, 118 ⟨84MI 502-04⟩, 135 ⟨84JA2687, 85JA7072⟩
Klich, M., **6**, 511, 605 ⟨86T2677⟩
Klicic, J., **3**, 424 ⟨92CB2361⟩
Klicnar, J., **6**, 558, 565 ⟨84CCC1577⟩
Kliegel, W., **3**, 434 ⟨84LA1545⟩; **4**, 405 ⟨65CB4036, 67CB2515⟩, 408 ⟨77LA956⟩, 435 ⟨66CB2686, 73AP(306)161⟩, 437 ⟨66CB1678⟩; **5**, 92 ⟨82CZ345⟩
Kliegman, J. M., **4**, 298 ⟨70JOC3140, 70T2555⟩
Kligman, F. L., **6**, 366 ⟨81ZOB1035⟩
Klima, W. L., **1**, 468, 536, 1104 ⟨78TL1027⟩, 1070 ⟨80JOC2526⟩; **2**, 969 ⟨78TL1027⟩, 976 ⟨80JOC2526⟩
Klimanov, V. I., **6**, 410 ⟨89BAU1570⟩
Klimars, M., **4**, 856 ⟨84T3677⟩
Klimenko, G. R., **3**, 707 ⟨89ZOR1573⟩
Klimko, V. T., **3**, 69 ⟨64JGU109⟩
Klimkovich, E. A., **6**, 629 ⟨75ZOR1628⟩
Klimova, A. I., **6**, 406 ⟨68BAU414⟩
Klimova, E. I., **4**, 719 ⟨90MI 416-01⟩
Klimova, T. A., **6**, 147 ⟨68IZV2379⟩
Klimovic, J., **1**, 986 ⟨93CM357⟩
Klimstra, P. D., **2**, 95 ⟨60JOC621⟩
Kline, D. N., **3**, 250 ⟨87TL913, 88JOC2238⟩; **4**, 275 ⟨92JOC298⟩; **5**, 425 ⟨87TL913⟩
Kline, E. A., **1**, 64, 522 ⟨73JA3078⟩; **3**, 503 ⟨73JA3078⟩; **4**, 627 ⟨73JA3078⟩
Kline, E., **1**, 522 ⟨72JOM(42)C21⟩; **4**, 610 ⟨72JOM(42)C21⟩
Kline, M. W., **1**, 398 ⟨52JA5179⟩
Klinekole, B. W., **2**, 23 ⟨92JA6262⟩
Kling, K. E., **2**, 167 ⟨50CB87⟩; **4**, 71 ⟨49USP2483972⟩
Klinge, M., **4**, 536 ⟨88AG(E)1194⟩
Klingebiel, U., **2**, 418 ⟨81CB527⟩; **4**, 606 ⟨76AG(E)619⟩, 649 ⟨81ZAAC(482)65, 83ZN(B)953⟩, 1050 ⟨85JOM(291)277⟩; **5**, 887 ⟨83ZAAC(501)57⟩, 1074 ⟨78ZN(B)950⟩; **6**, 290, 379, 380 ⟨84ZAAC(510)169⟩, 372 ⟨84ZN(B)1500⟩, 379 ⟨85JOM(291)277, 84ZAAC(510)175⟩, 380 ⟨85ZN(B)1023, 87JOM(323)1⟩, 387 ⟨85AG(E)324⟩, 388 ⟨86CB1117⟩
Klingenberg, J. J., **5**, 66 ⟨63OSC(4)110⟩
Klingensmith, K. A., **3**, 502 ⟨84JA525⟩
Klingert, B., **6**, 64 ⟨88JOM(339)17⟩
Klinghoffer, V., **2**, 118, 124 ⟨93JMC460⟩
Klingl, H., **2**, 477 ⟨74CB639⟩
Klingler, R. J., **4**, 674 ⟨78JOM(161)23⟩
Klingler, T. C., **4**, 332 ⟨69JOC3104⟩
Klingner, E., **4**, 602, 621, 929 ⟨92CB607⟩; **5**, 416 ⟨92CB607⟩
Klingsberg, E., **2**, 12 ⟨55OSC(3)119⟩; **3**, 73, 254 ⟨66JOC3489⟩, 73, 376, 378 ⟨63JA3244⟩, 256 ⟨72JOC3226⟩, 371 ⟨61JA2934, 68JOC2915⟩, 389 ⟨66JA5045⟩; **4**, 852 ⟨72JOC3226⟩
Klingstedt, T., **1**, 465 ⟨83OM598, 87S40⟩; **3**, 625 ⟨87S40⟩
Klink, F. W., **6**, 253 ⟨86JFC(32)89⟩
Klink, W., **1**, 753 ⟨64TL2467⟩; **4**, 577 ⟨62CB581⟩
Klinkert, G., **1**, 1134 ⟨79JCS(P1)2954⟩
Klinkhammer, K. W., **6**, 195 ⟨92CB1547⟩
Klioze, S. S., **1**, 506 ⟨75JOC1588⟩
Klippe, M., **3**, 72, 242 ⟨90CB399⟩
Klix, R. C., **1**, 388 ⟨85TL985⟩
Klobucar, W. D., **3**, 16 ⟨76SC575⟩
Kloche, E., **1**, 561 ⟨93CB1623⟩
Kloek, J. A., **2**, 366, 368 ⟨76JOC4028⟩, 401 ⟨79JOC305⟩, 745 ⟨78JOC1460⟩; **3**, 309 ⟨69JA2342⟩
Kloesters, W., **2**, 447, 1064 ⟨77AG(E)45⟩
Kloeters, W., **2**, 1064 ⟨78TL3343⟩
Kloetzel, M. C., **5**, 205 ⟨51JA4958⟩, 285 ⟨52JA1321⟩
Kloft, M., **5**, 592, 609 ⟨77TL2145⟩
Kloker, W., **2**, 862 ⟨70ZAAC(375)140⟩
Kloosterman, D. A., **1**, 831 ⟨90TL1977⟩; **3**, 302 ⟨90TL1977⟩
Kloosterziel, H., **1**, 402, 795 ⟨69RTC266⟩, 403, 799 ⟨70CC1168⟩, 403, 800 ⟨68RTC298⟩, 404 ⟨65RTC241⟩, 794 ⟨63RTC741, 70TL2755⟩, 873 ⟨70RTC1153⟩; **2**, 226 ⟨53RTC612⟩, 251 ⟨52RTC1228⟩, 360 ⟨75CC413⟩, 674 ⟨66CC51, 70RTC1153⟩, 723 ⟨54RTC129⟩; **3**, 364 ⟨75RTC1⟩; **4**, 16 ⟨62RTC925⟩
Klopfenstein, C. E., **1**, 523 ⟨66JOC2376⟩
Klopman, G., **1**, 58, 841 ⟨72JA7859⟩, 901 ⟨69JA3261⟩
Klöpping, H. L., **6**, 560 ⟨51RTC917, 51RTC949⟩
Klopsch, A., **5**, 1143 ⟨75CB664⟩
Klosa, J., **5**, 126 ⟨57AG135⟩
Klose, G., **1**, 234 ⟨63CI(L)1203⟩; **2**, 188 ⟨71JOC2698⟩
Klosowski, J. M., **4**, 650 ⟨70JOC1308⟩
Klosse, T. R., **2**, 335 ⟨75CC97⟩
Kloster-Jensen, E., **2**, 1016 ⟨64ACS1629, 66T965⟩
Klöter, G., **5**, 348, 767 ⟨84IC2217⟩; **6**, 238 ⟨79JA347⟩, 275 ⟨80AG(E)942⟩
Kloth, B., **2**, 471 ⟨77ZN(B)1244⟩, 865 ⟨77ZAAC(432)67⟩
Klotz, J. R. M., **5**, 24 ⟨42CRV97⟩
Klotz, K.-P., **3**, 616 ⟨80CB710⟩
Klotz, U. J., **6**, 617 ⟨84CZ404⟩, 635 ⟨85M651⟩
Klotzbücher, W. K., **6**, 522 ⟨76IC1666, 77IC822⟩
Klötzer, D., **2**, 440 ⟨84ZC224⟩
Klötzer, W., **3**, 294 ⟨71JA4321⟩; **5**, 360 ⟨82S592⟩
Klouras, N., **5**, 374, 376, 718 ⟨91ZAAC(603)89⟩
Kluba, K., **2**, 107 ⟨71T3163⟩
Kluba, M., **2**, 377 ⟨81S537⟩
Kluck, D., **2**, 157 ⟨81S229⟩, 249 ⟨81AP(314)799⟩
Kluess, C., **4**, 719 ⟨73JOM(56)269, 76ZAAC(423)112⟩
Klug, J. T., **3**, 55 ⟨81S137⟩; **4**, 188 ⟨81S137⟩
Klug, W., **1**, 248 ⟨85AG(E)1046⟩; **2**, 361 ⟨75CB1365⟩; **5**, 246 ⟨73JFC(3)383⟩; **6**, 235 ⟨66AG906⟩, 235, 273, 534 ⟨68CB2609⟩, 236, 252 ⟨68CB2617⟩, 258, 449 ⟨71CB2732⟩, 273, 274 ⟨70GEP1908680⟩, 273, 607 ⟨66AG(E)845⟩, 432 ⟨73JFC(3)383⟩, 534 ⟨72CB820⟩
Kluge, A. F., **1**, 458 ⟨72JA9256⟩; **2**, 680 ⟨78TL3629⟩; **3**, 11 ⟨78TL3629, 79JOC4847⟩; **4**, 187, 199 ⟨72JA7827⟩, 338 ⟨90OSC(7)160⟩, 340 ⟨79JOC4847⟩, 490 ⟨77JOC376⟩; **6**, 113 ⟨76JOC1112⟩

Kluge, H., **6**, 608 ⟨83JPR787⟩
Kluge, R., **1**, 556 ⟨90T2371⟩; **5**, 365 ⟨87ZC27⟩
Kluger, E. W., **2**, 358 ⟨77JOC967⟩, 365 ⟨74JA5000⟩
Kluiber, R. W., **1**, 881 ⟨63OSC(4)261⟩
Klumann, M., **3**, 269, 530 ⟨62AG32⟩
Klumpp, E., **6**, 352 ⟨68JOM(11)207⟩
Klumpp, G. W., **1**, 194 ⟨88TL3579⟩, 194 ⟨93TL3309⟩, 795 ⟨68CC35, 72TL125⟩, 795 ⟨73RTC605⟩; **2**, 324 ⟨88TL3579⟩, 521 ⟨93TL4655⟩, 554, 556 ⟨86RTC1⟩, 555 ⟨94T5933⟩, 682 ⟨75TL3429⟩, 925 ⟨87TL2889, 92T6087⟩, 925 ⟨92T6105⟩; **4**, 528 ⟨90RTC305, 88TL5237⟩, 618 ⟨89TL6195⟩, 668, 699 ⟨87RTC514⟩, 669 ⟨88RTC393⟩, 1047, 1058, 1062 ⟨83T3073⟩, 1061 ⟨87TL2889⟩; **5**, 424 ⟨83T3073⟩, 424, 468 ⟨83CC239⟩, 427 ⟨82TL3101⟩; **6**, 130, 133 ⟨83CC239⟩, 130, 133 ⟨83T3073⟩, 325, 379, 388, 398 ⟨88TL5237⟩, 399 ⟨89TL6195⟩
Klun, J. A., **1**, 652 ⟨93TL1593⟩
Klunder, A. J. H., **1**, 582 ⟨91TL3739⟩, 786 ⟨87TL357, 94TL2787⟩
Klunder, J. M., **1**, 116, 118 ⟨91COS(3)207⟩, 223 ⟨89JOC1295⟩; **2**, 721 ⟨87JOC2598⟩; **3**, 632 ⟨89JOC1295⟩
Klünenberg, H., **1**, 934 ⟨78AG(E)47⟩
Klusener, P. A. A., **1**, 880 ⟨86AG(E)465⟩
Klusener, P., **1**, 9 ⟨87JOC4299⟩
Klusik, H., **2**, 903 ⟨84ZN(B)1042⟩; **4**, 1048, 1054, 1057 ⟨84ZN(B)19042⟩, 1082 ⟨84AG(E)826⟩; **5**, 927, 928 ⟨83AG(E)877, 84AG(E)369⟩, 928 ⟨84AG(E)372, 84AG(E)826⟩; **6**, 713 ⟨82JOM(234)C17, 84ZN(B)1042⟩
Klussmann, P., **2**, 472 ⟨69ZAAC(370)171⟩
Klutchko, S., **3**, 251 ⟨72JHC171⟩; **5**, 596 ⟨71JOC1742⟩
Kluth, J., **5**, 1120 ⟨87EUP209779⟩
Kluth, J. F., **6**, 232 ⟨90HOU(E16a)214, 90HOU(E16a)271⟩
Klynev, M. V., **2**, 310 ⟨80RCR14⟩
Klyuchinskii, S. A., **4**, 766 ⟨89ZOB2145⟩
Klyuev, N. A., **3**, 362 ⟨89MI 308-04⟩
Kmieciak, A., **2**, 472 ⟨92ZAAC(611)47⟩
Kmiecik-Lawrynowicz, G., **4**, 830 ⟨87JA3811⟩; **6**, 280 ⟨86JOC2168⟩
Kminek, I., **1**, 986 ⟨93CM357⟩
Knaack, D. F., **2**, 23 ⟨60JOC20⟩
Knaak, J., **6**, 410 ⟨66CB371⟩
Knaap, C. T., **4**, 660 ⟨86OM1551⟩
Knab, R., **5**, 433 ⟨92MI 509-01⟩
Knabe, B., **2**, 879 ⟨71CB1761⟩
Knaggs, J. A., **2**, 7 ⟨73CC110⟩
Knapczyk, J. W., **2**, 824, 834, 835 ⟨78JA7304⟩
Knapp, Jr., F. F., **2**, 279 ⟨83JOM(253)31, 83OM1106⟩, 623 ⟨83SC1027⟩, 907 ⟨83JOM(259)269⟩; **3**, 305 ⟨84JOC2322⟩
Knapp, S., **1**, 582 ⟨94JOC281⟩, 782 ⟨92TL1025⟩; **3**, 18, 133 ⟨76TL3667⟩, 165 ⟨80TL4301⟩; **5**, 11 ⟨91OS(70)101⟩, 285, 377 ⟨85TL1803⟩, 378, 379 ⟨88JOC4006, 88JOC4773⟩; **6**, 616 ⟨82TL3539⟩, 629 ⟨92TL1025⟩
Knauer, W., **5**, 1097 ⟨90AG(E)318⟩
Knaup, W., **3**, 553, 554 ⟨91OM3967⟩; **5**, 944 ⟨91CB1121⟩
Knaus, E. E., **2**, 765 ⟨69JHC989, 72JOC1690⟩; **3**, 310 ⟨87CJC1158⟩; **4**, 949 ⟨80CJC130⟩; **5**, 1119 ⟨78CJC1026, 79JHC409⟩
Knaus, G. N., **1**, 664 ⟨73JA3408⟩
Knauss, L., **5**, 953 ⟨71JOM(31)C68⟩
Knausz, D., **6**, 449 ⟨81MI 614-02⟩
Knebel, W. J., **6**, 517 ⟨71JA5424⟩
Knebl, R., **4**, 568 ⟨87PS(30)349⟩; **5**, 879, 880 ⟨85MI 523-01⟩, 1155 ⟨85MI 531-01⟩
Knecht, M., **1**, 677 ⟨93SL837⟩
Kneen, G., **5**, 214 ⟨88JMC499⟩
Kneidl, F., **2**, 879 ⟨73AG(E)931, 74AG(E)668⟩, 879 ⟨76TL3125⟩; **3**, 680 ⟨71CB3816⟩
Kneisley, J. W., **2**, 30 ⟨55OSC(3)209⟩, 643 ⟨45JA386⟩, 645 ⟨44JA1274⟩, 645, 646 ⟨44JA1464⟩
Kneubueler, W., **1**, 104 ⟨73C277⟩
Kneuper, H.-J., **4**, 673 ⟨85CC170⟩; **5**, 1097 ⟨85AG(E)1062, 89CB445⟩
Kneyama, K., **5**, 659 ⟨93JOC32⟩
Kniel, P., **1**, 441 ⟨68HCA371⟩
Knieps (Riedel de Haen), R., **6**, 32 ⟨92GEP4025227⟩
Knierzinger, A., **4**, 485, 489 ⟨85HCA1730⟩; **5**, 212 ⟨85HCA1730⟩
Kniesel, H., **4**, 719 ⟨86ZN(B)1431, 87JOM(334)347⟩, 719 ⟨89JOM(371)205⟩
Kniežo, L., **5**, 612 ⟨85ZC324⟩, 1041, 1049 ⟨90SC509⟩, 1043 ⟨88ZC141⟩
Knifton, J. F., **3**, 426 ⟨73JOC3296⟩; **5**, 142 ⟨76JOC793⟩, 142 ⟨78MI 503-01⟩; **6**, 467 ⟨91JMOC389⟩
Knight, D. W., **1**, 16 ⟨93TA625⟩, 463 ⟨91COS(3)241⟩, 831 ⟨84CC827, 86JCS(P1)161⟩, 831 ⟨87CC1220, 87TL3031⟩, 831 ⟨89CC1812, 91JCS(P1)705⟩; **2**, 307, 345, 346, 347, 349 ⟨93CC169⟩; **4**, 170 ⟨79JCS(P1)62⟩; **5**, 24 ⟨78GSM111, 79GSM67⟩, 24 ⟨80GSM75, 81GSM87⟩, 24 ⟨82GSM100, 86GSM131⟩, 24 ⟨87GSM130, 88GSM75⟩, 24 ⟨89GSM89, 90GSM91⟩, 24 ⟨92GSM79⟩, 227 ⟨84CC827⟩
Knight, E. C., **3**, 451 ⟨36CB341⟩
Knight, G. J., **5**, 1013 ⟨66CC336⟩
Knight, J., **1**, 1127 ⟨87JCS(P1)1237⟩
Knight, J. G., **1**, 196 ⟨90SL224⟩
Knight, J. W., **1**, 629 ⟨79TL3815, 91JA4712⟩
Knight, K. S., **1**, 244 ⟨91CC354⟩
Knight, V., **2**, 1013 ⟨65JPC2484⟩
Knights, E. F., **1**, 82 ⟨68IJ691⟩, 658, 659 ⟨68JA4439⟩; **2**, 516 ⟨68JA5280⟩
Knipe, A. C., **1**, 623 ⟨B-81MI 114-01⟩, 845 ⟨B-91MI 119-01⟩, 879 ⟨B-89MI 119-01⟩
Knipper, R., **6**, 414 ⟨83MI 614-01⟩
Knipprath, W. G., **2**, 622 ⟨66MI 212-01⟩
Knist, J., **1**, 383 ⟨73CB1376⟩
Knittel, D., **3**, 258 ⟨70M157⟩; **5**, 215 ⟨86M(117)491⟩
Knobler, C., **5**, 798 ⟨89OM2253⟩
Knobler, C. B., **1**, 999 ⟨92AG(E)1240⟩; **2**, 580 ⟨83JA5804⟩, 1091 ⟨91JA6943, 92AG(E)1241⟩; **4**, 746 ⟨91AG(E)698⟩
Knobloch, T. P., **2**, 505, 506 ⟨77CZ399⟩
Knoch, F., **1**, 837 ⟨87CB131⟩; **2**, 936 ⟨86CB535⟩; **3**, 605, 607 ⟨86CB2466, 84AG(E)619⟩, 609 ⟨86JOM(307)93⟩; **4**, 146 ⟨86CB1977⟩, 559 ⟨83CB2275⟩, 566 ⟨84CB2693, 85CB814⟩, 567 ⟨85CB4068⟩, 568 ⟨86CB535⟩, 1026 ⟨86CB2466⟩; **5**, 675 ⟨85CB1632⟩, 880 ⟨83CB938⟩, 889 ⟨85CB1352⟩, 1058 ⟨84AG(E)970⟩, 1092 ⟨84AG(E)587⟩, 1092, 1093, 1095 ⟨83CB2275⟩; **6**, 164, 373 ⟨84CB2063⟩, 166 ⟨85CB814⟩, 166, 689 ⟨84CB2693⟩, 166, 693 ⟨86CB1977⟩, 371 ⟨86CB2609⟩, 680, 687 ⟨84AG(E)579⟩, 680, 687 ⟨84AG(E)970⟩, 683 ⟨84AG(E)895⟩, 686 ⟨83CB1873, 82CB2371⟩, 687 ⟨86CB2748, 86TL1661⟩, 687 ⟨88USP4754072⟩, 693 ⟨85AG(E)419, 86CB535⟩, 693 ⟨88PS(36)147⟩
Knochel, P., **1**, 37, 191 ⟨84HCA261⟩, 123 ⟨88TL2395, 91JA5735⟩, 123 ⟨91JOC5974⟩, 138 ⟨92AG(E)1459⟩, 163 ⟨90JA7431⟩, 207, 315, 538 ⟨91COS(4)865⟩, 248 ⟨86TL4427, 86TL4431⟩, 467 ⟨93CRV2117⟩, 468, 471 ⟨93T29⟩, 486 ⟨90JOC5446⟩, 539 ⟨91COS(4)872, 91COS(4)877⟩, 539, 541 ⟨91COS(4)896⟩, 542 ⟨91COS(4)886, 91COS(4)888⟩, 543 ⟨91COS(4)879⟩, 679 ⟨86TL1043⟩, 679, 836 ⟨86TL1039⟩, 885 ⟨91COS(1)211⟩, 974, 1056 ⟨92JOC4070⟩, 991 ⟨91JA9888, 94OM94⟩, 1004 ⟨93JOC2694⟩, 1017 ⟨89TL4799⟩; **2**, 67, 590 ⟨93CRV2117⟩, 69 ⟨88TL3887, 90TL7575⟩, 69, 72, 591 ⟨92JOC1956⟩, 69, 72, 591 ⟨93TL3115⟩, 73 ⟨92JOC6384⟩, 159 ⟨85TL4455⟩, 478 ⟨90OS(70)195, 91JOC5974⟩, 588, 590 ⟨88JOC2390, 91COS(1)211⟩, 590 ⟨92OS(70)195⟩, 591 ⟨93JA7027, 93TL7911⟩, 592 ⟨94SL410⟩, 594 ⟨93JOC2694⟩, 783, 785, 786 ⟨85T4861⟩, 785 ⟨82S1017, 82TL3897⟩, 900 ⟨92JOC3482⟩, 910 ⟨86TL5727⟩, 925 ⟨91JA5735⟩, 936,

945 ⟨84TL4383⟩, 976, 977, 978 ⟨91COS(4)865⟩, 998 ⟨89JOC5202⟩, 999, 1002 ⟨93JOC588⟩, 1005 ⟨90JA7431⟩, 1006 ⟨86TL1039, 86TL4427⟩, 1096 ⟨93TL4623⟩; **3**, 17, 132 ⟨88TL6701⟩, 23 ⟨88TL6697⟩, 157 ⟨92TL3717⟩, 162 ⟨93CRV2117⟩, 162 ⟨93JOC588⟩, 192 ⟨85T4861⟩, 201 ⟨90TL1833⟩, 204 ⟨90JA7431⟩, 287 ⟨88JOC2390⟩, 303 ⟨89JOC5202⟩, 614, 646, 665, 670, 675 ⟨93TL4623⟩, 623 ⟨89JA6474⟩; **4**, 53, 54, 58 ⟨93JOC588⟩, 153 ⟨92JOM(438)11⟩, 335 ⟨85T4861⟩, 357 ⟨84TL4383⟩, 385 ⟨89JOC5202⟩, 396 ⟨92T2025⟩, 398 ⟨86TL1043⟩, 651, 724 ⟨88TL6697⟩, 665 ⟨90JA7431⟩, 701 ⟨86TL4427⟩, 709 ⟨90JOC5446, 93S530⟩, 709, 724 ⟨86TL1039⟩, 723 ⟨94OM94⟩, 772, 963 ⟨91JA5735⟩, 815 ⟨92TL3717⟩; **5**, 937 ⟨95OM3163⟩
Knockel, P., **2**, 74 ⟨90JA6146⟩, 785 ⟨84HCA261⟩
Knoerzer, G., **4**, 631, 635, 635 ⟨90ZN(B)15⟩
Knoesel, F., **5**, 782 ⟨90JOM(338)21⟩
Knoess, H. P., **1**, 123 ⟨91JOC5974⟩; **2**, 478 ⟨91JOC5974⟩
Knofel, W., **3**, 550 ⟨70CB949⟩
Knoflach, J., **5**, 360 ⟨82S592⟩
Knol, A., **2**, 925 ⟨92T6105⟩
Knol, P. G., **2**, 884 ⟨54IZV410⟩
Knölker, H.-J., **1**, 448 ⟨87AG(E)1035⟩
Knoll, A., **5**, 565, 614 ⟨87ZC8⟩, 633 ⟨85JPR(327)455⟩, 760 ⟨84S51⟩
Knoll, F., **2**, 507 ⟨69JA4986⟩; **4**, 111 ⟨B-89MI 403-01⟩, 478, 487 ⟨71CB31⟩; **5**, 876 ⟨89MI 523-01⟩, 876, 1154 ⟨81AG(E)731⟩, 910 ⟨79CB1093⟩, 1093 ⟨76AG(E)702⟩; **6**, 665, 686 ⟨82AG(E)448⟩, 678, 689, 692 ⟨81AG(E)731, 79AG(E)873⟩, 692 ⟨B-89MI 622-01⟩
Knoll, F. M., **1**, 413 ⟨80JA865⟩, 836, 838 ⟨78JA4822⟩
Knoll, L., **4**, 580 ⟨78JOM(148)C25, 79CB2709⟩, 580 ⟨79JOM(182)77⟩, 580, 581 ⟨80JOM(193)47, 77ZN(B)1268⟩
Knolle, J., **1**, 827, 828 ⟨77TL1625⟩
Knollmüller, K. O., **4**, 556 ⟨70USP3534125⟩
Knollmüller, M., **4**, 416 ⟨69M469⟩
Knop, B., **2**, 465 ⟨65AG(E)244, 66CB1663⟩
Knop, H., **6**, 184 ⟨92CB1807⟩
Knoppova, V., **4**, 764 ⟨88MI 417-01⟩
Knorr, L., **6**, 485 ⟨03CB1280⟩
Knorr, R., **2**, 742 ⟨83S785⟩, 961 ⟨77TL3969⟩; **4**, 9 ⟨69JA1034⟩, 949, 950 ⟨84AG(E)366⟩; **5**, 279, 281 ⟨89TL1927⟩
Knorre, R., **3**, 76 ⟨84JOC1288⟩
Knors, C., **5**, 951 ⟨87OM988⟩
Knösal, F., **5**, 782 ⟨92CB2171⟩
Knoth, W., **6**, 548 ⟨87ZAAC(552)181⟩, 548, 553 ⟨87ZAAC(550)109, 87ZAAC(554)172⟩, 563 ⟨87ZAAC(551)191, 88ZAAC(556)141⟩
Knoth, W. H., **3**, 709, 711 ⟨58JA4151⟩; **4**, 607, 608, 621, 622, 654 ⟨58JOC1392⟩; **5**, 1133 ⟨58JA4151⟩
Knott, E. B., **1**, 256 ⟨47JCS1190⟩; **5**, 652 ⟨45JCS686⟩
Knott, S. J., **6**, 445 ⟨89JAN367⟩
Knotz, F., **3**, 446, 450 ⟨58M718⟩
Knotz, H., **5**, 1068 ⟨86M(117)1437⟩
Knouzi, N., **2**, 324 ⟨87TL1757⟩
Know, L. H., **2**, 180 ⟨73OSC(5)196⟩
Knowles, J. R., **1**, 826 ⟨87JA5008⟩
Knowles, P., **4**, 866 ⟨84GEP3408757A1⟩
Knowles, W. S., **1**, 76 ⟨68CC1445, 77JA5946⟩; **2**, 391 ⟨65JOC1488⟩, 580, 822 ⟨81JOC4422⟩
Knox, G. R., **1**, 19, 43, 624, 782 ⟨62JA1734⟩, 528 ⟨73JCS(P1)975, 73JCS(P1)977⟩, 862 ⟨58PCS289, 61JCS4610⟩; **2**, 103 ⟨90OM301⟩, 182 ⟨58JCS692⟩, 985 ⟨73JCS(P1)977⟩; **4**, 688 ⟨76JOM(120)C13⟩
Knox, I., **1**, 511 ⟨77JOC3981⟩
Knox, L. H., **1**, 2, 4, 6 ⟨71LA(748)123⟩, 900 ⟨39JA3184⟩, 906 ⟨54JA3203⟩
Knox, S. A. R., **4**, 673 ⟨85CC170⟩, 674 ⟨80JOM(202)C49, 82JCS(D)1783⟩, 675 ⟨86CC542⟩, 678 ⟨90JOM(400)255⟩, 679 ⟨82JCS(D)1297, 83JCS(D)1417, 83JCS(D)2661, 84JCS(D)2293⟩, 680 ⟨81CC862, 83JCS(D)2099⟩, 681 ⟨76JCS(D)377, 90JCS(D)761, 92CC310⟩, 683 ⟨78JCS(D)403, 79JCS(D)1801⟩, 1023, 1032 ⟨88CC673⟩
Knox, S. D., **1**, 595, 779 ⟨76JCS(P1)459⟩
Knudsen, C. G., **3**, 180, 210 ⟨83JOC2260⟩
Knudsen, M. J., **1**, 530 ⟨85JOC5215⟩
Knudsen, R. D., **2**, 648 ⟨74JOC3343⟩; **5**, 291 ⟨70JOC1219⟩
Knunyanto, I. L., **2**, 167 ⟨66DOK(170)96⟩, 167 ⟨68IZV1889⟩, 173 ⟨81IZV159⟩, 183, 249 ⟨75IZV2530⟩, 211 ⟨81IZV1850⟩, 214 ⟨64IZV644⟩, 227 ⟨80IZV2111⟩, 464, 854 ⟨66ZOB500⟩, 640 ⟨74CSR443, 75ZOR1370⟩, 641 ⟨77DOK(233)591, 78IZV1193⟩, 641 ⟨78IZV493⟩, 712 ⟨77RCR548⟩, 782 ⟨65IZV1715⟩, 876, 881 ⟨72IZV213⟩, 883 ⟨55IZV472⟩; **3**, 363 ⟨29CB3048⟩, 495 ⟨81RCR280, 82RCR1⟩, 527 ⟨67DOK(173)1117, 67FCR107⟩, 531 ⟨68T1341⟩, 560, 561, 588 ⟨73AG(E)675, 65BAU731⟩, 588 ⟨75JFC(6)227, 76BAU573⟩, 592 ⟨73DOK(212)628⟩; **4**, 134 ⟨72IZV215⟩, 297, 304 ⟨66IZV1108, 65IZV2046⟩, 697 ⟨71IZV620, 79IZV1831⟩, 730 ⟨75ZOR961⟩, 731 ⟨75JFC(6)227, 80CA71025t⟩, 736 ⟨76IZV895⟩, 737 ⟨81JFC(17)441⟩, 761, 763 ⟨67ZOR1006⟩, 799 ⟨60MI 418-01, 91BAU1705⟩, 802, 805 ⟨60BAU1851⟩, 803 ⟨76BAU853, 76BAU873⟩, 805, 812, 813 ⟨61MI 418-01⟩, 820 ⟨82BAU2498⟩, 833 ⟨71KGS1645⟩, 867 ⟨80IZV892⟩, 894 ⟨72IZV1347, 74IZV358⟩, 894 ⟨74MI 420-01⟩, 926 ⟨69BSF2712⟩, 966 ⟨74IZV2545⟩, 970 ⟨71DOK(201)1359⟩, 1023 ⟨72ZOB346⟩; **5**, 22 ⟨91BAU1705⟩, 59 ⟨78DOK(239)330⟩, 474, 484 ⟨67BAU324⟩, 475 ⟨77BAU206⟩, 502 ⟨71BAU959, 72AG(E)583⟩, 663, 685, 698, 699, 813 ⟨71T51⟩, 671 ⟨66JGU1340, 66ZOB1326⟩, 679 ⟨91IZV1918⟩, 811 ⟨67IZV1060, 68IZV1134⟩, 811 ⟨68IZV657⟩, 813 ⟨69IZV958⟩, 821 ⟨59DOK(127)337⟩, 1076 ⟨85IZV700⟩; **6**, 54 ⟨63IZV1946, 74IZV2144, 76ZOR1377⟩, 65 ⟨75DOK(221)1331⟩, 69 ⟨60IZV231⟩, 219 ⟨65IZV1873⟩, 238, 242 ⟨75IZV2279⟩, 239, 440 ⟨76IZV209⟩, 240, 615 ⟨80JFC(15)169⟩, 241 ⟨60DOK(132)602⟩, 537 ⟨81URP804634⟩, 614 ⟨76IZV2381, 76IZV212⟩, 615 ⟨83IZV694, 83MI 620-02⟩, 615 ⟨85IZV700⟩, 616 ⟨77IZV2379, 81IZV2632⟩, 643 ⟨78MI 612-01⟩, 694 ⟨82RCR1⟩
Knupfer, H., **3**, 452 ⟨62OS(41)813⟩, 694, 696, 699, 700, 721, 722, 724, 725 ⟨65AG(E)472⟩; **5**, 201, 842 ⟨62T3⟩
Knüppel, P. C., **4**, 550 ⟨88ZN(B)31⟩; **6**, 373 ⟨89OM2490⟩
Knutson, F. J., **1**, 636 ⟨81TL199⟩
Knutson, P. L. A., **1**, 879 ⟨77S509⟩; **2**, 205 ⟨75JOC1343⟩
Knuz, H., **4**, 462 ⟨92S90⟩
Ko, C.-W., **1**, 145 ⟨92T8963⟩
Ko, J. S., **1**, 39 ⟨85JOC1927⟩
Ko, K.-Y., **4**, 231 ⟨84T1333⟩; **6**, 428 ⟨91JAN1083⟩
Ko, O.-H., **1**, 752 ⟨87TL3075⟩
Ko, S. B., **1**, 1130, 1132 ⟨89TL827⟩
Ko, S. S., **1**, 748 ⟨82JA1109⟩
Ko, S. Y., **1**, 14 ⟨85JOC5678⟩; **2**, 52 ⟨94CC1775⟩; **3**, 632 ⟨87JOC667⟩
Ko, Y. H., **5**, 432 ⟨90JOM(397)7⟩
Ko, Y. K., **5**, 126 ⟨85CC473⟩, 280 ⟨85TL1341⟩; **6**, 417 ⟨86H(24)1625⟩
Ko, Y. Y. C. Y. L., **6**, 156 ⟨89NJC891⟩, 165 ⟨90TL3859⟩
Kobayashi, G., **2**, 1003 ⟨79AX(B)2443⟩; **4**, 842 ⟨75CPB2390⟩, 844 ⟨73CPB2770, 75CPB2749⟩, 888, 900 ⟨69YZ203⟩, 979 ⟨73YZ612⟩; **5**, 559 ⟨79H(12)485⟩, 560 ⟨75YZ980⟩
Kobayashi, H., **1**, 570 ⟨89TL1641⟩; **2**, 19 ⟨85CL233⟩, 175 ⟨75CL307⟩, 906 ⟨89TL1641, 89TL5437⟩, 906 ⟨89TL6379, 92TL337⟩, 908 ⟨88BCJ3957⟩; **3**, 140 ⟨79CL245⟩; **4**, 367, 936, 938 ⟨88BCJ3957⟩, 736 ⟨89TL1641, 89TL6379⟩, 736 ⟨91CL961, 92TL337⟩, 736 ⟨92TL3779⟩
Kobayashi, K., **1**, 490 ⟨89TL6051⟩, 929 ⟨93JCS(P1)3007⟩; **2**, 453 ⟨87CL1915⟩, 920 ⟨89JA6478⟩, 926

⟨89JOC3517⟩; **3**, 80 ⟨78BCJ1186⟩, 220 ⟨84TL4959⟩, 429 ⟨88JOC4897⟩, 620 ⟨83TL3509⟩, 648 ⟨86CC1779⟩, 723 ⟨79JOC2030, 79TL1039⟩, 724 ⟨86JOM(303)301⟩; **4**, 68 ⟨70BCJ1223⟩, 156 ⟨76BCJ2837⟩, 156 ⟨77BCJ1353⟩, 748 ⟨78TL4253, 84JA3584⟩; **5**, 595, 620 ⟨79TL1039⟩; **6**, 594 ⟨86BCJ1741⟩

Kobayashi, M., **1**, 64, 449 ⟨86BCJ2807⟩, 64, 449 ⟨86BCJ2807⟩, 241 ⟨89JOC2374⟩, 467 ⟨80JA3298⟩, 485 ⟨85JOC5045⟩, 543 ⟨80JOC5223⟩, 704, 707, 708 ⟨86CL433⟩, 706, 708 ⟨84CL833⟩, 706, 708 ⟨87JOC1703⟩, 1106 ⟨80S1034⟩; **2**, 120 ⟨86TL3975⟩, 214 ⟨86SUL1⟩, 235 ⟨86BCJ2179⟩, 244 ⟨85H(23)1897⟩, 251 ⟨73S667⟩, 285 ⟨82CC438⟩, 360 ⟨75CL991, 78JOC652⟩, 399 ⟨66BCJ1778⟩, 546 ⟨77CC534⟩, 627 ⟨69OPP221⟩, 629, 976 ⟨80JA3298⟩, 797 ⟨79CL149⟩, 915 ⟨80JOM(186)51⟩, 915, 923 ⟨81JOM(216)149, 80JOC5223⟩, 976 ⟨80S1034⟩; **3**, 20 ⟨89TL6371⟩, 620 ⟨85CL1339⟩, 707 ⟨88TL6321⟩; **5**, 334 ⟨83OPP315⟩, 359 ⟨92JOC7359⟩, 451 ⟨87JA4837⟩; **6**, 24 ⟨83SUL131⟩, 489 ⟨92JOC7359⟩

Kobayashi, N., **3**, 708 ⟨85MI 321-01⟩; **5**, 248 ⟨71JOC3546⟩; **6**, 434 ⟨73CL1315⟩

Kobayashi, S., **1**, 38, 945 ⟨68JA4182⟩, 462 ⟨72JA7448⟩, 485 ⟨73CL1097⟩, 1112 ⟨76BCJ2560, 76CC612⟩, 1113, 1114 ⟨80CC103⟩, 1120 ⟨76CC180, 79TL1619⟩, 1120 ⟨80T2531⟩; **2**, 72 ⟨89TL7095⟩, 78 ⟨93TL3453⟩, 81 ⟨89CL1401⟩, 139 ⟨91BCJ2524⟩, 142 ⟨90CL2239⟩, 568 ⟨92T5691⟩, 688 ⟨70TL5115⟩, 788 ⟨79TL2525⟩; **3**, 40 ⟨82TL525⟩, 159 ⟨91CL949⟩, 159 ⟨92TL1625⟩, 159 ⟨93JOC2647⟩, 162 ⟨93S371⟩, 243 ⟨82JA358, 82JA360⟩, 243 ⟨82JOC1981⟩, 281, 296, 298 ⟨92CL435⟩, 281, 298 ⟨91CL1059⟩, 281, 298 ⟨91S1216⟩, 281, 298, 299 ⟨93CC1157⟩, 442 ⟨86TL4763⟩, 543 ⟨90CL1705⟩, 615 ⟨91CL537⟩, 629 ⟨91CL541⟩; **4**, 174 ⟨82JA358⟩, 188 ⟨88CL1101⟩, 354 ⟨88JA4826⟩; **5**, 135, 138 ⟨84TL2557⟩, 236 ⟨86CL1817, 87CL743⟩, 236 ⟨89CL1001, 90CL1019⟩, 236 ⟨91CL1901, 91CL989⟩, 291 ⟨76CL57⟩, 418 ⟨91TL4929⟩, 652 ⟨89CPB2952⟩, 747 ⟨69BCJ3310⟩, 753 ⟨79TL2517⟩, 785, 792 ⟨68TL935, 67JA2240⟩, 1037 ⟨70JOC2118⟩; **6**, 547 ⟨89BCJ967⟩

Kobayashi, T.-A., **3**, 406 ⟨85S967⟩

Kobayashi, T., **1**, 202 ⟨86S386, 86TL2479⟩, 202 ⟨86TL967⟩, 202, 203 ⟨86JOC2859⟩, 900 ⟨88JA8599⟩, 1015 ⟨81CC333⟩, 1042 ⟨80CC736⟩, 1047 ⟨85TL3463⟩; **2**, 28 ⟨80JOC2731⟩, 249 ⟨75BCJ480⟩, 743 ⟨92BCJ1379⟩; **3**, 46 ⟨86JOC254⟩, 77 ⟨84JCS(P1)2103⟩, 152, 154, 155 ⟨80JOC2731⟩, 213 ⟨75TL3299⟩, 269 ⟨85JCS(P1)1401⟩, 276 ⟨81CC333⟩, 290 ⟨85TL3463⟩, 625 ⟨83CL193⟩; **4**, 136, 138, 141 ⟨91OM1960⟩, 248 ⟨88TL3971⟩, 421 ⟨75CL607⟩, 616 ⟨86BCJ1509⟩, 1052 ⟨78JOM(162)C43⟩; **5**, 170 ⟨85TL3207⟩, 226 ⟨82JOM(213)C12⟩

Kobayashi, Y., **1**, 215 ⟨84TL5781⟩, 226 ⟨90JA1661⟩, 228 ⟨88TL343⟩, 468 ⟨93BCJ2446⟩, 651 ⟨83CC921⟩, 1079 ⟨82TL343⟩, 1107 ⟨85CC1636⟩; **2**, 7, 835 ⟨68CPB1009⟩, 19 ⟨82JCS(P1)85⟩, 20 ⟨68CPB1784, 73CPB867⟩, 122 ⟨83YGK776⟩, 631 ⟨77JCR(S)215⟩, 900 ⟨90JOC5324⟩, 936 ⟨83CC921⟩, 1045 ⟨78JOC199⟩; **3**, 32 ⟨89BCJ3038⟩, 162 ⟨92CL2039⟩, 628 ⟨85MAC1755, 86BCJ893, 86CL931⟩; **4**, 183 ⟨86TL69⟩, 549 ⟨80JOC4683⟩, 734 ⟨85CPB5137⟩, 736 ⟨90CPB1104⟩, 811 ⟨94TA961⟩, 31 ⟨74CPB2767⟩, 1060 ⟨84CC1130⟩; **5**, 1125 ⟨86MI 530-02⟩; **6**, 9 ⟨B-82MI 601-01⟩, 14 ⟨80JCS(P1)2755⟩, 22, 246 ⟨79TL4071⟩, 232 ⟨79TL3865, 69TL4095⟩, 246 ⟨70CPB2334⟩, 300 ⟨71JOC1176⟩, 447 ⟨92TL3483⟩, 624 ⟨77BCJ3271⟩

Kobayushi, K., **2**, 202 ⟨82TL5427⟩

Köbbing, S., **5**, 951 ⟨93CB1849⟩

Kobdri, T., **2**, 226 ⟨72CL935⟩

Kobel, M., **6**, 213 ⟨59JA4801⟩

Kober, E., **2**, 166 ⟨59JA4810⟩; **3**, 474 ⟨57JA944⟩; **5**, 656 ⟨62JOC2585⟩, 966, 1013 ⟨69JOC1162⟩, 1033 ⟨69AG(E)760⟩; **6**, 603 ⟨62JOC2589⟩

Kober, F., **2**, 478 ⟨76ZAAC(420)74⟩, 482, 485, 487, 510 ⟨83ZAAC(501)89⟩, 493, 502 ⟨82S173⟩, 494 ⟨80ZAAC(466)179, 81ZAAC(472)75⟩, 496 ⟨79CZ119⟩, 500, 501 ⟨73ZAAC(402)43⟩, 500, 504 ⟨73ZAAC(401)243⟩, 500, 891 ⟨75ZAAC(412)202, 73ZAAC(397)97⟩, 502 ⟨80ZAAC(466)183⟩, 503 ⟨75JOM(94)393⟩, 695 ⟨74ZAAC(406)52⟩; **3**, 80 ⟨77CZ36⟩, 435 ⟨77CZ133⟩; **4**, 593 ⟨75JOM393, 82S173⟩, 593 ⟨91ZAAC(605)151⟩; **5**, 222 ⟨88CZ106, 90CZ71⟩

Kober, F. von, **4**, 593, 594 ⟨75ZAAC202⟩

Kober, R., **1**, 1042 ⟨86S372⟩; **4**, 465 ⟨83MI 410-01⟩

Kober, W., **2**, 539 ⟨82S1⟩; **5**, 718 ⟨76S259⟩; **6**, 43 ⟨82S1⟩

Köberle, K., **2**, 720 ⟨25CB409⟩

Kobinger, W., **6**, 642 ⟨80JMC1217⟩

Kobiro, K., **1**, 561, 569 ⟨93S460⟩

Koblik, A. V., **2**, 1008 ⟨82AG(E)555⟩

Koblitz, W., **6**, 418 ⟨37CB1080⟩

Kobolov, N. A., **6**, 145 ⟨76KFZ35⟩

Kobori, T., **2**, 712 ⟨68TL5071⟩

Kobori, Y., **1**, 279, 466 ⟨84JA158⟩

Koboto, I., **5**, 141 ⟨68JOC2525⟩

Kobrakov, K. I., **5**, 687 ⟨91CHE299, 91KGS369⟩

Köbrich, G., **1**, 164 ⟨72AG(E)473⟩, 491 ⟨74CB1654⟩, 1068, 1069, 1081 ⟨B-69MI 121-02⟩, 1144 ⟨63ZN(B)1125, 65AG(E)49⟩, 1144 ⟨67CB2011⟩, 1145 ⟨65AG95, 67T565⟩; **2**, 618 ⟨72CB1683⟩, 906 ⟨67CB3371⟩, 997 ⟨72AG(E)473⟩, 998 ⟨64TL1137⟩, 1016 ⟨65AG(E)49⟩; **3**, 232 ⟨62LA(654)131⟩, 494, 495 ⟨62AG33, 66CB689⟩; **4**, 145, 741, 816, 817, 820 ⟨72AG(E)473⟩, 145, 741, 817 ⟨67AG(E)41⟩, 145, 817 ⟨63ZN(B)1125⟩, 146 ⟨91CB1253⟩, 741, 747 ⟨76CB2175⟩, 758 ⟨66CB670, 69BSF2712⟩, 806 ⟨73CB1601⟩; **6**, 25 ⟨64AG(E)513⟩, 62 ⟨65TL969⟩, 244 ⟨67AG15⟩, 267, 293, 378 ⟨70TL4693⟩, 293 ⟨67AG(E)41, 72AG(E)473⟩

Kobrzhitskii, V. V., **5**, 226 ⟨81ZOR2452⟩

Kobs, U., **1**, 924 ⟨89T951⟩, 1043 ⟨90CB2191⟩; **3**, 92 ⟨89T951⟩

Kobutani, T., **5**, 691 ⟨92BCJ674⟩

Kobzareva, V. P., **4**, 100 ⟨89JGU200⟩

Kočevar, M., **5**, 824, 834 ⟨82JHC1397⟩, 837 ⟨81H(15)293, 84CS209⟩

Koch, A., **1**, 923 ⟨93JOC1083⟩; **2**, 578, 579 ⟨89AG(E)277⟩

Koch, B., **5**, 246 ⟨73JFC(3)383⟩; **6**, 256, 536, 537 ⟨76ZAAC(427)114⟩, 276 ⟨74CZ511⟩, 432 ⟨73JFC(3)383⟩, 588, 592 ⟨80ZN(B)526⟩

Koch, D., **2**, 472 ⟨78CB3838⟩, 865 ⟨77ZAAC(432)67⟩; **5**, 348 ⟨88ZN(B)1495⟩

Koch, E., **3**, 365 ⟨77ACS(B)679⟩

Koch, E.-W., **1**, 1141 ⟨85TL1493⟩

Koch, G. K., **1**, 74 ⟨69MI 103-01⟩

Koch, H. F., **4**, 36 ⟨61JA4105⟩

Koch, H. J., **5**, 844 ⟨55LA(591)190⟩

Koch, J., **5**, 677 ⟨55LA(591)218, 55LA(591)226⟩; **6**, 489, 490 ⟨90BBR(172)288⟩

Koch, K., **1**, 926 ⟨88JA2210⟩; **2**, 1063, 1065, 1066, 1069, 1070 ⟨72ZAAC(394)74⟩; **5**, 214 ⟨86JOC167⟩

Koch, O., **1**, 364 ⟨88T3899⟩

Koch, P., **1**, 248 ⟨87AG(E)928⟩; **2**, 362 ⟨67CC985⟩; **6**, 495 ⟨75TL2087⟩

Koch, S., **3**, 339 ⟨73IC2589⟩

Koch, S. A., **4**, 673 ⟨78IC2093⟩

Koch, S. A. M., **4**, 54, 58 ⟨83JOC635⟩

Koch, T., **5**, 307 ⟨92HCA62⟩

Koch, T. H., **5**, 1012 ⟨73JA3957⟩

Koch, W., **1**, 906 ⟨89JA1527⟩; **2**, 997, 999 ⟨92CB2265⟩, 999, 1005 ⟨91CB543⟩; **4**, 646 ⟨91CB543⟩; **5**, 338 ⟨52ZN(B)634⟩

Kochanek, P. M., **1**, 841 ⟨77TL763⟩

Kochendoerfer, E., **2**, 830 ⟨64CB741⟩

Kochergin, V. P., **4**, 156 ⟨78JGU769⟩, 588 ⟨77ZOB1664⟩, 660 ⟨75ZOB2448, 77ZOB1751⟩

Kocheshkov, K. A., **2**, 520 ⟨B-67MI 210-01⟩, 882 ⟨53MI 217-01⟩; **6**, 522 ⟨40DOK(26)54⟩

Kochetkova, V. A., **6**, 5 ⟨87IZV808⟩
Kochetkov, B. B., **4**, 941 ⟨71ZOB719, 72ZOB870⟩
Kochetkov, C. A., **5**, 385 ⟨85CC171⟩
Kochetkov, K. A., **5**, 652 ⟨88JCS(P1)305⟩
Kochetkov, N. K., **2**, 982 ⟨49IZV305⟩; **3**, 130 ⟨83TL4355⟩, 239 ⟨58JGU3024⟩; **5**, 239 ⟨83TL4355⟩
Kochhar, K. S., **4**, 203 ⟨83JOC1765⟩
Kochi, ,J. K., **2**, 590, 960 ⟨73JA6487⟩
Kochi, H., **3**, 418 ⟨76JAP7632569⟩
Kochi, J. F., **1**, 117, 119, 465 ⟨71S303⟩
Kochi, J. K., **1**, 117, 1092 ⟨71JA1483⟩, 119 ⟨72JOM(42)205⟩, 321, 923 ⟨74JA2201⟩, 467 ⟨76JOC502⟩, 491 ⟨72BCJ1120⟩, 561 ⟨68T2215⟩, 668 ⟨B-78MI 114-04⟩, 669 ⟨74JA6140⟩, 671 ⟨77JOM(124)327, 77JOM(124)349⟩, 671 ⟨77JOM(124)371⟩, 919 ⟨B-73MI 119-09, B-73MI 119-10⟩, 924, 925 ⟨70JA4395⟩, 929 ⟨B-73MI 119-14⟩, 934 ⟨B-73MI 119-16⟩, 946 ⟨86JA4790⟩, 1099 ⟨79JA7547⟩; **2**, 12 ⟨72OR(19)279⟩, 31 ⟨87CC1013, 89JOC3028⟩, 90 ⟨B-81MI 202-01⟩, 97 ⟨62JA2785⟩, 590 ⟨73JCS(D)2620, 74JA6140⟩, 652 ⟨73JA7114⟩, 960 ⟨72JOM(40)C81⟩; **3**, 190 ⟨93TL1859⟩; **4**, 27 ⟨72OR(19)279⟩, 300 ⟨71JOC3112⟩, 1077 ⟨B-73MI 424-01⟩; **5**, 229 ⟨68JOC75⟩; **6**, 147 ⟨85JOC5245⟩, 147 ⟨86RTC278⟩, 147 ⟨86RTC286⟩, 369 ⟨93TL1859⟩
Kochi, J. R., **1**, 560 ⟨72OR(19)279⟩
Kochkol'da, S. P., **4**, 556 ⟨69ZOB839⟩
Kochlelov, F., **6**, 77 ⟨11BSF1308⟩
Kochmann, W., **2**, 252 ⟨86GEP231075⟩; **4**, 106 ⟨79EGP137839⟩
Kochta, J., **1**, 837 ⟨87CB131⟩
Kocian, O.,**1**, 1137 ⟨93JCS(D)2629⟩
Kocienski, P., **1**, 224 ⟨90T4503⟩, 455 ⟨93SL1133⟩, 481 ⟨89JA2363⟩, 614 ⟨85PS(24)477⟩, 648 ⟨91COS(6)1000⟩, 654, 713, 716 ⟨87CC106⟩, 678 ⟨88TL3357⟩, 1094 ⟨83TL3905, 84CC151⟩; **2**, 561 ⟨89TL6753⟩; **3**, 240 ⟨90TL3129⟩,541 ⟨89TL1833, 93JCS(P1)1549⟩; **4**, 655 ⟨90PAC1933⟩, 765 ⟨89TL1833⟩, 934, 935 ⟨91TL6961⟩, 952, 953 ⟨89SL52⟩, 962, 963, 964 ⟨92SL886⟩; **5**, 173 ⟨85CC1386⟩, 343 ⟨92S582⟩
Kociénski, P. H., **1**, 568 ⟨91COS(6)975⟩
Kocienski, P. J., **1**, 144, 696 ⟨85PS(24)97⟩, 412, 810 ⟨67JA3462⟩, 481 ⟨92JCS(P1)3419, 92JCS(P1)3431⟩, 614, 616, 696, 698, 700 ⟨81CI(L)548⟩, 615, 616 ⟨79JCS(P1)1290⟩, 615, 654, 696, 697, 698, 700 ⟨78JCS(P1)829⟩, 616, 654 ⟨79TL2649⟩, 616, 696, 700 ⟨80JCS(P1)1045, 78JCS(P1)834⟩, 700, 702 ⟨80JCS(P1)1400⟩, 700, 703 ⟨87JCS(P1)2171⟩, 821 ⟨80CC1096⟩, 979 ⟨77JOC353⟩, 1076 ⟨76JOC3650⟩; **2**, 28 ⟨88OSC(6)991⟩; **3**, 20 ⟨80TL1559⟩, 23 ⟨80CC1096⟩; **4**, 369 ⟨80CC1096, 83TL3905⟩, 953 ⟨87JCS(P1)2189⟩
Kockmann, W., **4**, 576, 587 ⟨73EGP94992⟩
Köckritz, A., **3**, 590 ⟨92PS(73)185⟩
Kockritz, A., **5**, 790 ⟨91PS(63)95⟩
Köckritz, A., **6**, 156 ⟨91PS(63)95⟩
Köckritz, P., **3**, 587 ⟨80ZC211⟩
Kockritz, P., **5**, 1068 ⟨79ZC59⟩
Kocor, M., **3**, 294 ⟨73TL3635⟩; **4**, 209 ⟨69T4257⟩
Kocour, C., **3**, 535 ⟨23JA2167⟩
Kocourek, J., **4**, 239 ⟨63CI(L)1397⟩
Kočovský, P., **1**, 194 ⟨89JA4981⟩
Kocovsky, P., **1**, 690 ⟨85TL2171⟩; **2**, 112 ⟨94T529⟩, 314 ⟨90SL677⟩
Kocur, J., **3**, 247 ⟨74CB2115⟩
Kocy, O., **5**, 575 ⟨75JA5628⟩
Koda, A., **4**, 205 ⟨76CPB102⟩
Koda, S., **3**, 617 ⟨66BCJ854⟩
Kodaira, K., **1**, 466 ⟨88BCJ1625⟩; **6**, 37 ⟨79JFC(15)353⟩, 39 ⟨78JFC(12)1⟩, 51 ⟨77JFC(9)279⟩, 410 ⟨76JAP7630034⟩
Kodama, A., **5**, 324 ⟨91H(32)221⟩
Kodama, H., **2**, 23, 596 ⟨77JOM(142)71⟩; **5**, 29 ⟨59JA5817⟩
Kodama, M., **1**, 215 ⟨84TL5781⟩, 818 ⟨93TL8453⟩
Kodama, S., **1**, 471 ⟨75JA4405⟩; **6**, 580 ⟨55MI 618-01⟩
Kodama, S.-I., **1**, 466, 470 ⟨76BCJ1958⟩
Kodama, T., **1**, 264 ⟨90TL6379⟩; **2**, 319 ⟨90TL6379⟩; **3**, 638 ⟨90TL6331⟩
Kodama, Y., **1**, 143 ⟨86BCJ2801⟩
Kodde-Busch, H., **2**, 167 ⟨50CB87⟩
Koddebusch, H., **4**, 71 ⟨50CB87⟩
Kodera, T., **5**, 997 ⟨70JAP7019893⟩
Kodera, Y., **2**, 308, 344, 351 ⟨88TL2973⟩, 853 ⟨87CL357⟩; **3**, 409 ⟨85TL4633⟩; **5**, 196 ⟨82MI 504-01⟩; **6**, 300 ⟨92JAP(K)04164085⟩
Kodess, M. I., **4**, 311 ⟨83IZV2568⟩
Kodian, G. N., **6**, 14 ⟨80ZOB1897⟩
Kodicek, E., **3**, 226 ⟨71JCS859⟩
Kodomari, M., **2**, 4 ⟨88IJC(B)1129⟩, 626 ⟨88JOC2093⟩, 710 ⟨91BCJ3485⟩; **3**, 618 ⟨73CL577⟩; **5**, 252 ⟨81S637⟩
Kodomori, M., **2**, 620 ⟨87CL1901⟩
Koebernick, W., **1**, 587 ⟨77CB2127⟩; **3**, 234 ⟨74TL4377⟩
Koeck, H., **3**, 618 ⟨32BSB196⟩
Koeffler, H. P., **1**, 1029 ⟨91JMC2452⟩
Koegel, R. J., **5**, 84 ⟨57ANY(69)94⟩
Koegler, S., **3**, 424 ⟨92JOC229⟩, 699 ⟨88TL1127⟩
Koehl, W. J., **1**, 847 ⟨60CI(L)262⟩
Koehle, H. J., **4**, 856 ⟨84T3677⟩
Koehler, K. F., **3**, 621 ⟨84JOC276⟩
Koellemann, C., **6**, 357, 358 ⟨91OM2101⟩
Koelliker, U., **3**, 494 ⟨70JA397⟩; **5**, 168 ⟨70JA397⟩
Koelsch, C. F., **3**, 612 ⟨44JA306⟩; **5**, 114 ⟨55OSC(3)791⟩
Koenen, G., **5**, 1047 ⟨58CB1660⟩
Koenig, A. Von, **6**, 360 ⟨76GEP2460330⟩
Koenig, H., **2**, 795, 796 ⟨83CJC1562⟩
Koenig, H. E., **2**, 442 ⟨B-85MI 208-02⟩
Koenig, J., **2**, 796 ⟨83CC73⟩
Koenig, K. E., **1**, 450, 451 ⟨73JA3416⟩, 490, 1106 ⟨71TL4701⟩
Koenig, K. H., **6**, 438 ⟨78GEP2706683⟩
Koenig, M., **3**, 608 ⟨91TL3687⟩; **4**, 463 ⟨94S51⟩, 547 ⟨86TL1145, 87T1793⟩, 1025 ⟨91TL3687⟩; **5**, 886 ⟨93PS(76)41⟩, 1092 ⟨89TL177⟩; **6**, 264 ⟨87T1793, 91TL5965⟩, 373 ⟨84CC1622⟩, 373, 688 ⟨89CC593⟩, 680 ⟨89TL177, 90JOC5750⟩
Koenig, N. H., **2**, 120, 180 ⟨58JOC1525⟩
Koenig, T., **1**, 934 ⟨B-73MI 119-15⟩; **4**, 13 ⟨70JOC508⟩
Koenig, T. M., **3**, 632 ⟨92TL3281⟩
Koenigs, L., **2**, 829 ⟨34LA(509)138⟩
Koenigsberger, R. U., **1**, 72 ⟨75JOC3073⟩
Koepf, H., **6**, 177, 178, 382, 387, 396, 709, 710 ⟨87ZN(B)1062⟩, 382, 709, 710 ⟨87ZN(B)1055⟩
Koeplin, R., **6**, 262 ⟨84JFC(24)457⟩
Koeppe, R. E., **5**, 1111 ⟨53JA4655⟩
Koeppen, J., **4**, 98 ⟨87CZ247⟩
Koerber, K., **2**, 930 ⟨91TL1187⟩
Koerner, E. A., **5**, 781 ⟨77JOM(135)183⟩
Koerner, M., **1**, 120 ⟨88JA2641⟩, 312 ⟨87TL945⟩; **2**, 992 ⟨88JA2641⟩
Koerwitz, F. L., **3**, 245 ⟨86JOC2525⟩
Koestler, R. C., **1**, 862 ⟨58CI(L)1589⟩
Koevoet, A. L., **1**, 403 ⟨57RTC689⟩
Köffer, D. P. J., **4**, 695 ⟨86JOM(308)153⟩, 695, 717 ⟨89JOM(372)183, 86AG(E)165⟩
Kofler, M., **1**, 104 ⟨49HCA489⟩, 502, 837 ⟨47HCA1911⟩; **3**, 62 ⟨47HCA1911⟩
Kofod, H., **2**, 120 ⟨63OSC(4)491⟩
Kofron, W. G., **1**, 154 ⟨76JOC439⟩, 881 ⟨62JOC2737⟩; **3**, 263 ⟨67JOC3817⟩
Koft, E. R., **1**, 516 ⟨82JA5568, 84JA2115⟩; **2**, 617 ⟨82JA2659⟩; **3**, 303 ⟨84JA2115⟩; **4**, 331 ⟨87JOC3466⟩
Koga, G., **2**, 420 ⟨70JOC960⟩; **3**, 721, 722 ⟨70TL3309⟩
Koga, K., **1**, 33, 34 ⟨81T2797⟩, 55 ⟨76TL57, 76TL61⟩, 135 ⟨89CPB1120, 90CC1657⟩, 135 ⟨93TL1963⟩, 199 ⟨89TL851⟩, 473 ⟨92JA8732⟩; **2**, 50 ⟨89CC1921,

90TL1741⟩; **3**, 124 ⟨79CPB2760⟩, 291 ⟨85TL3027, 85TL3031⟩, 415 ⟨86CPB1050⟩, 468 ⟨71TL4495, 75T227⟩, 639 ⟨84TL1599⟩; **6**, 128, 327 ⟨85TL3031⟩
Koga, N., **3**, 721, 722 ⟨70TL3309⟩
Koga, R., **4**, 656 ⟨93BCJ1849⟩
Koga, T., **3**, 293 ⟨88CL1141⟩, 361 ⟨89CC505⟩; **5**, 452 ⟨88CL1141⟩
Kogami, K., **4**, 192, 193 ⟨84BCJ1876⟩
Kogan, T. P., **1**, 234 ⟨82CC589⟩; **3**, 310 ⟨90T6623⟩
Koganty, R. R., **3**, 351 ⟨86JA3811⟩; **6**, 57 ⟨73TL4511⟩
Kogen, H., **3**, 325 ⟨89JA8957⟩; **5**, 157 ⟨87CC311⟩
Koguchi, T., **1**, 1055 ⟨84CL1939⟩
Koguchi, Y., **5**, 199 ⟨90TL709⟩
Kogure, T., **3**, 12 ⟨72TL5035⟩
Koh, D.,**3**, 156 ⟨92CC1571⟩
Koh, L. L., **1**, 897 ⟨71AX(B)1405⟩
994⟩
Koh, Y. J., **3**, 197 ⟨93TL2147⟩
Kohama, H., **1**, 980 ⟨85CL1457⟩; **4**, 187 ⟨84TL5155⟩
Kohama, M., **4**, 942 ⟨90T4553⟩
Kohara, N., **6**, 561 ⟨88BCJ3205⟩
Kohara, T.,**5**, 226 ⟨81BCJ1720⟩
Kohara, T., **5**, 462 ⟨81BCJ2161⟩
Kohda, A., **1**, 1097 ⟨81CC951⟩; **3**, 151 ⟨81JOC509⟩
Kohda, K., **4**, 521 ⟨86CPB3273⟩
Kohei, K., **4**, 193 ⟨90JHC1419⟩
Kohji, K., **3**, 357 ⟨91EUP409164⟩
Kohl, B., **5**, 552 ⟨80TL917⟩
Kohl, K., **5**, 581 ⟨59CB910⟩
Kohl, R. R., **5**, 942 ⟨93JOM(445)143⟩
Köhlbrandt, S., **4**, 109 ⟨92S710⟩
Kohler, E., **3**, 493 ⟨1899CB1566⟩; **4**, 309, 310 ⟨63LA(664)130⟩
Kohler, E. P., **1**, 310 ⟨05MI 107-01⟩; **3**, 612 ⟨33JA4299⟩; **6**, 254 ⟨35JA217⟩
Kohler, F., **4**, 598 ⟨75CB2649⟩
Köhler, F. H., **2**, 1003 ⟨85OM231⟩; **4**, 528, 533, 1020 ⟨87CB2081⟩, 581, 582, 583 ⟨83OM257⟩, 587 ⟨75JOM(99)353, 85OM231⟩, 1019 ⟨76JOM(121)211⟩, 1026 ⟨84TL3687⟩; **5**, 884, 1092 ⟨84TL3687⟩, 936 ⟨79CB708⟩
Köhler, H., **3**, 568 ⟨67ZAAC(354)69⟩; **4**, 556, 557, 699 ⟨90ZAAC(586)25⟩, 576, 587 ⟨73EGP94992⟩, 977, 978 ⟨92ZAAC(611)68⟩; **5**, 690, 723 ⟨90ZAAC(588)59⟩, 1070 ⟨85ZC305⟩, 1074 ⟨84ZC379, 89ZAAC(568)35, 89ZAAC(577)165⟩, 1074, 1078 ⟨89ZC145⟩, 1081, 1085 ⟨72ZC147⟩, 1084 ⟨88ZAAC(564)85⟩
Köhler, H.-J., **1**, 1109 ⟨78JA5297⟩
Köhler, J., **2**, 586 ⟨68JOM(12)225⟩
Kohler, J. J., **6**, 512 ⟨68USP3401214⟩
Kohler, K.-H., **3**, 366, 543 ⟨76TL2961⟩
Köhler, R., **5**, 657 ⟨89ZAAC(576)203⟩
Kohler, R.-D., **5**, 690 ⟨80LA1216⟩; **6**, 69 ⟨80LA1216⟩
Köhler, V. H., **5**, 1083 ⟨71ZAAC(386)197⟩
Kohler, W., **4**, 223 ⟨31CB2696⟩
Kohlhase, A. H., **2**, 181 ⟨32JA2441⟩
Kohlhase, W. L., **6**, 218 ⟨54JA5141⟩
Kohlman, D. T., **1**, 136 ⟨90TL6155⟩
Kohl-Mines, E., **1**, 986 ⟨85HCA2244, 85HCA2249⟩
Kohmoto, S., **1**, 216 ⟨90JCS(P1)1226⟩, 582 ⟨92BCJ2366⟩, 938 ⟨89CC1265, 91JCS(P1)1173⟩, 1090 ⟨89JOC1757⟩; **3**, 191 ⟨89CC110⟩, 684 ⟨92NKK420⟩; **4**, 206 ⟨92BCJ2366⟩, 208 ⟨89CL221⟩, 938 ⟨89JOC1757⟩; **5**, 513 ⟨89CC1265, 91JCS(P1)1173, 91JOC2849⟩; **6**, 550 ⟨89CC1265, 91JOC2849⟩
Kohmoto, Y., **2**, 30 ⟨81JOC3312⟩
Kohn, G. K., **6**, 448 ⟨71GEP2128672⟩
Kohn, H., **3**, 550 ⟨76TL3093⟩; **5**, 855 ⟨72JOC3504⟩, 1080, 1118 ⟨85JA2931, 84TL399⟩; **6**, 479 ⟨83JOC5033⟩
Kohn, J., **6**, 624 ⟨91MM2302⟩
Kohne, B., **5**, 241 ⟨79JOM(166)373⟩
Köhne, R., **4**, 622 ⟨62CB3030⟩
Kohnke, F. H., **1**, 571 ⟨92JA6330⟩
Kohnke, J., **2**, 165, 402 ⟨71CB2023⟩, 399 ⟨70CB2152, 71CB2648, 71CB3875⟩
Kohno, M., **2**, 318 ⟨90BCJ1252⟩, 518 ⟨91TL1479⟩
Kohno, S., **2**, 14 ⟨83BCJ1881⟩; **4**, 13 ⟨83BCJ1881⟩
Kohoda, H., **1**, 408, 409, 410, 820, 821, 822 ⟨72JA3870⟩
Kohra, S., **2**, 77 ⟨87CC1517⟩, 140 ⟨92TA1563⟩, 300 ⟨88CPB2342⟩; **3**, 361, 362 ⟨91JHC1245⟩; **4**, 231 ⟨87CC1442⟩, 368 ⟨92TL85⟩, 650 ⟨87CPB1663⟩, 844 ⟨91JHC1245⟩, 900 ⟨88JHC959⟩; **6**, 634 ⟨91TL5987⟩
Kohricht, K.-H., **2**, 509 ⟨89JOM(373)307⟩
Kohrt, A., **5**, 558 ⟨92LA595⟩
Koida, Y., **5**, 769 ⟨70MI 519-01⟩, 847 ⟨70JPB1696⟩, 848 ⟨72JPJ275⟩
Koidan, G. N., **2**, 407 ⟨89ZOB2135, 89ZOB2136⟩, 407, 408 ⟨90ZOB465⟩; **5**, 379 ⟨85ZOB1080⟩; **6**, 242 ⟨93ZOB1240⟩, 242, 682 ⟨92ZOB948⟩, 267 ⟨90ZOB709⟩, 680, 682 ⟨89ZOB1902⟩, 681 ⟨88ZOB1923, 88ZOB482⟩, 690 ⟨89ZOB2133⟩, 696 ⟨87ZOB2141, 87ZOB2794, 88ZOB1461, 88ZOB2164, 89ZOB1904, 89ZOB959, ⟨90ZOB2814⟩
Koide, K., **1**, 106 ⟨93CC617⟩
Koide, T., **2**, 227 ⟨91SL345⟩; **4**, 218 ⟨89CPB526⟩
Koie, K., **4**, 221 ⟨87CC122⟩
Koike, S., **1**, 674 ⟨90AC59⟩
Koike, T., **3**, 173 ⟨93CC1631⟩, 365 ⟨89CPB1999⟩; **5**, 250 ⟨94S282⟩
Koike, W., **2**, 650 ⟨70BCJ293⟩
Koike, Y., **2**, 546 ⟨77CC534⟩
Koitabashi, T., **3**, 596 ⟨75BCJ2287⟩
Koizumi, S., **1**, 829 ⟨88CC1531⟩; **5**, 159 ⟨88CC1531⟩
Koizumi, T., **1**, 248 ⟨87JOC5063⟩, 767, 769 ⟨88BCJ467⟩; **2**, 455, 846 ⟨79S110⟩, 459 ⟨78TL3741⟩; **3**, 627 ⟨91JA6318⟩; **4**, 52, 101, 109 ⟨88JCS(P1)1149⟩, 79, 781 ⟨87CPB3959⟩, 101 ⟨87JCS(P1)2203⟩, 102 ⟨91JCS(P1)49⟩, 108 ⟨87JOC5061, 91CPB3120⟩, 108, 780 ⟨89JOC5453⟩, 221 ⟨87CL1095⟩, 270, 854, 859, 860 ⟨86SC233⟩, 285 ⟨85TL6205⟩, 831 ⟨74CPB2767⟩; **5**, 360 ⟨91JA6318⟩; **6**, 43, 47, 54, 57 ⟨86JOC955⟩
Kojić-Prodić, B., **6**, 195 ⟨86OM1726⟩
Kojima, A., **2**, 188 ⟨84CL2125⟩; **4**, 976 ⟨70JCS(C)476, 71JCS(C)196⟩
Kojima, K., **3**, 35 ⟨85T4449⟩, 394 ⟨91CL1013⟩
Kojima, M., **2**, 18 ⟨87JFC(34)337⟩, 614 ⟨85CPB1301⟩; **5**, 403, 404 ⟨80BCJ1625⟩; **6**, 554 ⟨80CPB110⟩
Kojima, S., **2**, 460, 462 ⟨89CL849⟩, 469 ⟨70BCJ3232⟩; **6**, 604 ⟨B-91MI 621-01⟩
Kojima, T., **1**, 373 ⟨67JOC869⟩; **3**, 634 ⟨73CL287⟩; **5**, 146 ⟨67JOC869⟩
Kojima, Y., **1**, 206 ⟨80TL4365⟩; **3**, 303 ⟨79TL4577⟩; **6**, 111 ⟨84H(22)13⟩
Kojitani, T., **4**, 72 ⟨75JOC3540⟩
Kok, G. B., **4**, 876, 877 ⟨87JA4115⟩
Kokel, B., **3**, 473 ⟨80AG754, 80AG(E)716⟩; **6**, 620 ⟨84TL1557⟩
Kokel, N., **6**, 488 ⟨92GEP4219418⟩
Koketsu, J., **2**, 469 ⟨70BCJ3232⟩; **4**, 405 ⟨71JCS(C)511⟩
Kokil, P. B., **3**, 163 ⟨88JA2987⟩; **5**, 339 ⟨87TL5431⟩
Kokko, B. J., **2**, 305 ⟨38JGU1825, 82JOC2822⟩
Kokorekina, V. A., **6**, 149 ⟨72IZV1717⟩, 149 ⟨72IZV2603⟩
Kokorev, G. I., **2**, 881 ⟨89JGU2500⟩, 883 ⟨83JGU1832⟩, 885, 890 ⟨87JGU2496⟩, 889 ⟨89JGU2254⟩, 891 ⟨91JGU405⟩; **5**, 369, 373 ⟨88ZOB829⟩, 370 ⟨91ZOB450⟩
Kokorev, K. I., **6**, 564 ⟨89ZOB2520⟩
Kokorina, L. G., **5**, 1132 ⟨74M611, 74MI 530-02⟩
Kokosova, A. S., **5**, 619 ⟨92ZOR1750⟩, 691 ⟨92JOU1399⟩
Kokotos, G., **2**, 317 ⟨92JCR(S)391⟩
Koksal, Y., **4**, 849 ⟨84LA450⟩
Köksal, Y., **5**, 174 ⟨84LA450⟩, 174 ⟨84LA462⟩
Kokubo, M., **4**, 506 ⟨80BCJ789⟩
Kokubo, T., **6**, 552 ⟨82BCJ339⟩
Kokuryo, Y., **3**, 246, 261 ⟨76CL499⟩
Kolaczewska, A., **4**, 651 ⟨91JOC1944⟩

Kol, M., **2**, 94 ⟨91JA2648⟩; **4**, 52 ⟨92JA7643⟩
Kolani, N., **1**, 978 ⟨81TL3609, 82JOM(236)177⟩
Kolar, A. J., **4**, 309 ⟨75TL3579⟩; **5**, 315 ⟨77S457⟩
Kolasa, T., **5**, 322 ⟨85S929⟩; **6**, 446 ⟨92TL2629⟩
Kolbah, D., **2**, 308, 325 ⟨79S161⟩; **3**, 425 ⟨74MI 311-01⟩
Kolb, H. C., **2**, 292 ⟨94CRV2483⟩
Kolb, M., **3**, 214 ⟨73AG(E)69, 74TL3171⟩; **4**, 254 ⟨74TL3171, 77LA811⟩, 254, 618, 853 ⟨77LA830⟩, 849 ⟨73CB2277⟩; **5**, 183 ⟨86TL4437⟩, 225 ⟨81SC763⟩; **6**, 327 ⟨73CB2277⟩
Kolb, O., **3**, 553 ⟨81AG(E)593, 85CB873⟩; **6**, 352 ⟨80CB1654⟩
Kolb, R., **1**, 708 ⟨77CB2978⟩; **5**, 760 ⟨69T5437⟩; **6**, 143 ⟨75CB1142⟩
Kolb, U., **4**, 570, 585, 660, 704 ⟨92JOM(423)339⟩
Kolb, V. M., **1**, 706 ⟨78JOC1985, B-77MI 115-01⟩
Kolbasina, V. D., **2**, 949 ⟨78ZOB639⟩
Kolbe, A., **2**, 467 ⟨91CB2705⟩; **4**, 556, 557, 699 ⟨90ZAAC(586)25⟩
Kolbe, W., **4**, 338 ⟨79ZC417⟩; **6**, 645, 666 ⟨73ZC192⟩
Kolbeck, W., **2**, 766 ⟨65TL783⟩; **3**, 709 ⟨83CB3027⟩
Kolbina, V. E., **4**, 926 ⟨82ZOB2375⟩
Kolc, J., **3**, 502 ⟨76JA7844⟩, 542 ⟨72TL3443, 75JA6586⟩
Koldan, G. N., **4**, 559 ⟨88ZOB1665⟩
Kolditz, L., **5**, 560 ⟨88ZC269⟩; **6**, 223, 228 ⟨76ZC377⟩
Koldobskii, A. B., **6**, 611 ⟨91ZOR533⟩
Koldobskii, G. I., **2**, 319 ⟨78RCR1084⟩; **5**, 270, 283 ⟨71RCR835⟩
Kolenbrander, H. M., **2**, 270 ⟨69CJC3271⟩
Kolenko, I. P., **6**, 42 ⟨73ZOR2013⟩
Kolesnik, N. P., **2**, 438 ⟨90ZOB1243⟩
Kolesnikova, I. V., **5**, 1070 ⟨88JFC(40)217, 89ZOR1523⟩, 1070 ⟨89ZOR1689⟩; **6**, 272 ⟨88ZOR1513⟩, 609 ⟨83JFC(22)439⟩, 616 ⟨84ZOR1197⟩, 619 ⟨88JFC(40)217, 93JPO319⟩
Kolesnikov, A. M., **6**, 367 ⟨75ZOB2093⟩
Kolesnikov, R. V., **6**, 368 ⟨74ZOR1793⟩
Kolesnikova, S. A., **2**, 358 ⟨69ZOR1088⟩
Kolesnikov, S. P., **4**, 142 ⟨71DOK(196)85⟩
Kolesova, M. B., **5**, 601 ⟨70ZOR610⟩
Kolesov, V. S., **4**, 698 ⟨75CA(83)10238⟩
Kolhe, J. N., **1**, 548 ⟨81JOC2622⟩; **4**, 536 ⟨86T4223⟩
Kolind-Andersen, H., **5**, 552, 555 ⟨73T2077⟩
Kolinski, A., **6**, 19 ⟨78PJC71⟩
Kolkmann, F., **4**, 802 ⟨78ZN(B)1422⟩; **6**, 263 ⟨84JCS(P1)1119⟩, 264 ⟨81CC930⟩
Köll, P., **1**, 925 ⟨81TL1891⟩
Kollár, L., **4**, 510 ⟨91JOM(408)297⟩
Kollat, P., **2**, 1008 ⟨85JA2067⟩
Kolle, P., **2**, 474 ⟨88CB871⟩
Kölle, U., **4**, 719 ⟨92JOM(436)333⟩
Kolleck, M., **5**, 49 ⟨76CB41⟩
Kollegger, G., **4**, 655 ⟨91JOM(414)295⟩
Kollek, L., **2**, 1016 ⟨30CB1886⟩, 1016, 1017 ⟨30CB1863⟩; **4**, 745 ⟨30CB1868⟩
Kollenz, G., **3**, 366 ⟨84AG(E)800⟩, 584 ⟨92CC487⟩; **4**, 894 ⟨92CC487⟩; **5**, 217 ⟨87OPP52⟩, 556 ⟨77LA1751⟩
Koller, M., **3**, 273 ⟨86HCA560⟩
Kollmannsberger-von Nell, G., **3**, 556, 557, 560, 585, 586 ⟨81CB3751⟩
Kollmar, H., **1**, 1109 ⟨71TCA65⟩
Kollmeier, H. J., **5**, 652 ⟨71JOM(28)237⟩
Kollmeier, J., **4**, 688 ⟨76JOM(120)C13⟩
Kollmeier, M. J., **6**, 720 ⟨70AG(E)309⟩
Kollmeyer, W. D., **1**, 848 ⟨68JA1791⟩; **4**, 907 ⟨77CA121332u⟩, 983 ⟨75BEP821282⟩
Kollonitsch, J., **1**, 35 ⟨78JA341⟩, 127 ⟨66JCS(A)453⟩; **2**, 15, 613 ⟨70JA7494⟩, 59, 60 ⟨54NAT125⟩, 59, 60 ⟨55NAT346⟩, 111 ⟨61JA1515⟩, 117 ⟨48JCS1683⟩; **5**, 199 ⟨64CI(L)1867⟩
Kolm, H., **3**, 532 ⟨63CB1948⟩
Kolm, H.-G., **5**, 638 ⟨73LA207⟩
Kolmakova, E. I., **6**, 369 ⟨77IZV1965⟩
Kolobova, N. E., **2**, 464 ⟨55IZV823⟩, 478 ⟨56IZV927⟩; **3**, 597 ⟨89JOM(379)303⟩; **4**, 672 ⟨68CC1365, 69IZV100⟩, 706 ⟨68IZV686⟩, 726 ⟨73IZV2815⟩; **6**, 457 ⟨87IZV1671⟩
Kolodiazhnyi, I., **3**, 549 ⟨87TL881⟩
Kolodiazhnyi, O. I., **2**, 453 ⟨81TL1231⟩; **3**, 593 ⟨80TL3983⟩; **4**, 116, 133 ⟨89TL2445⟩; **5**, 894 ⟨89ZC396⟩, 1086 ⟨83PS(18)39⟩, 1086, 1087 ⟨82TL4933⟩, 1086, 1087 ⟨83ZOB1093⟩; **6**, 683 ⟨83ZOB473⟩
Kolodka, T. V., **5**, 655 ⟨87ZOB475⟩, 655, 788, 790 ⟨87JGU415⟩, 713 ⟨90JGU849, 90ZOB963⟩
Kolodyazhnaya, S. N., **2**, 802 ⟨72KGS1690⟩
Kolodyazhnyi, O. I., **2**, 476 ⟨91JGU464⟩; **3**, 585, 586, 587 ⟨80JGU45⟩; **4**, 115 ⟨74JGU1386⟩, 116 ⟨89JGU2194, 90JGU1541⟩, 116, 133 ⟨90JGU1536⟩, 116, 133 ⟨91JGU627, 89JGU285⟩, 570 ⟨90ZOB695, 91MI 412-02⟩, 572 ⟨87ZOB2640⟩, 572, 575 ⟨87ZOB831⟩, 574, 587 ⟨80TL3984⟩; **5**, 397 ⟨86JGU2141⟩, 401 ⟨89JGU1977⟩, 879 ⟨81JGU1883⟩, 895 ⟨88ZOB491⟩, 898, 915 ⟨81JGU2125⟩, 902 ⟨77JGU658⟩, 902, 903 ⟨79JGU88, 75JGU2509⟩, 997 ⟨69ZOB1712⟩; **6**, 278, 699 ⟨77ZOR275⟩, 680, 683 ⟨85ZOB264, 85ZOB2797⟩, 683 ⟨87PS(30)433⟩, 684 ⟨80ZOB233, 81ZOB2189⟩, 689 ⟨85ZOB1862, 85ZOB2214⟩, 694, 699, 700 ⟨83RCR1096⟩, 694, 700, 701, 702 ⟨91RCR391⟩, 696 ⟨89ZOB330⟩, 697 ⟨88ZOB491⟩, 699 ⟨77ZOB2390, 82ZOB1538⟩, 699, 700 ⟨79ZOB104⟩
Kolokol'tseva, I. G., **4**, 488 ⟨68ZOB1248, 70ZOB2618⟩
Kolomeitsev, A. A., **6**, 50 ⟨79ZOR1561⟩, 232, 234, 236, 237 ⟨85S667⟩, 234 ⟨94S145⟩, 235 ⟨90S1151⟩
Kolomiets, A. F., **2**, 167 ⟨85ZOK1835⟩, 448 ⟨66CA-P2297⟩, 886 ⟨67JGU855⟩; **4**, 135 ⟨71JGU481⟩; **5**, 666, 671, 692 ⟨86MI 517-02⟩
Kolomiets, A. K., **6**, 236 ⟨92RCR517⟩
Kolomnikova, G. D., **1**, 906 ⟨67RCR735⟩; **6**, 635 ⟨92IZV425⟩
Kolonits, P., **1**, 747 ⟨66JOC1447⟩; **5**, 220 ⟨89CB1107⟩
Kolonko, K. J., **1**, 630, 632 ⟨75TL1811⟩, 630, 633 ⟨78JOC1404⟩
Kolosova, N. D., **4**, 627 ⟨80IZV2668⟩; **6**, 187, 710 ⟨83IZV959⟩, 387, 395 ⟨79BAU2222⟩
Kolotilo, N. V., **6**, 680 ⟨85ZOB1188⟩
Kolovos, M. G., **2**, 230 ⟨84TL4153⟩
Kolpakova, I. D., **6**, 118 ⟨72MI 604-01⟩
Kolshorn, H., **1**, 1077 ⟨78S536, 86T1711⟩; **3**, 359 ⟨87AG(E)1046⟩
Kolter, T., **3**, 40 ⟨92AG(E)1391⟩
Kolthoff, I. M., **2**, 126 ⟨46MI 203-01⟩, 172 ⟨49MI 203-01⟩
Kol'tsova, A. N., **5**, 850 ⟨73JGU317⟩, 863 ⟨72ZOB868⟩
Kol'Tsov, A. I., **3**, 540 ⟨68DOK(179)892⟩
Koltzenburg, G., **4**, 1078 ⟨80JCS(P2)883, 81JCS(P2)143⟩, 1078 ⟨83AG(E)500⟩
Kolycheva, M. T., **2**, 15 ⟨91RCR1050⟩
Kolyukhina, G. P., **4**, 661 ⟨70ZOB2607⟩
Koma, Y., **5**, 997 ⟨63BCJ95⟩
Komachev, V. V., **4**, 768 ⟨75ZOB2374⟩
Komada, M., **5**, 250 ⟨89S929⟩
Komaki, K., **6**, 472 ⟨78CPB3807⟩
Komaki, R., **4**, 317 ⟨77CPB3385⟩
Komamura, C., **5**, 1078 ⟨85SC171⟩
Komar, D. A., **2**, 925 ⟨78JOM(159)C13⟩
Komarov, E. V., **4**, 802 ⟨87JGU844⟩
Komarov, N. V., **2**, 1090 ⟨63IZV97⟩, 1101 ⟨67DOK(175)1079⟩; **4**, 541 ⟨80ZOB1427⟩
Komarov, V. A., **5**, 698, 762, 813 ⟨78IZV2802⟩; **6**, 54 ⟨67ZOB152⟩, 56, 282 ⟨76IZV489⟩, 259 ⟨78IZV486⟩
Komarov, V. N., **6**, 60 ⟨67IZV693⟩
Komarov, V. Ya., **4**, 786 ⟨85ZOR244⟩
Komarov, V. Y., **3**, 591 ⟨85JGU26⟩
Komarova, L. I., **4**, 57 ⟨86ZOR2327⟩
Komarova, T. N., **4**, 940 ⟨91ZOB1893⟩
Komatsu, H., **3**, 61 ⟨87JCR(S)296⟩, 620 ⟨86JOC3007⟩

Komatsu, K., **1**, 867 ⟨90JOC996⟩, 907 ⟨82BCJ3257, 88JA633⟩, 909 ⟨80TL947⟩; **5**, 404 ⟨92JOC6216⟩
Komatsu, M., **1**, 284 ⟨91SL919⟩, 1010 ⟨92BCJ2992⟩, 1043 ⟨89CL773⟩; **2**, 303 ⟨71S89⟩; **4**, 328 ⟨85S276⟩, 979 ⟨74JOC3198⟩; **5**, 10, 367 ⟨91SL919⟩, 359 ⟨90TL5327, 92JOC7359⟩, 364 ⟨85H(23)677⟩, 989 ⟨85S276⟩, 1032 ⟨74JOC957⟩; **6**, 489 ⟨92JOC7359⟩
Komatsu, N., **1**, 609, 983 ⟨92CC46, 93JOC3697⟩
Komatsu, T., **1**, 674 ⟨87CC388⟩; **4**, 1057 ⟨86JOC886⟩
Komatsu, Y., **5**, 245, 246 ⟨83S552, 86CL935⟩
Komazawa, S., **4**, 319 ⟨89BCJ1215⟩, 321 ⟨85H(23)2509⟩
Komazawa, T., **4**, 669 ⟨76BCJ1177⟩
Komel'kova, V. I., **6**, 565 ⟨87ZC24⟩
Komen, A. E., **4**, 716 ⟨93OM315⟩
Komen, C. M. D., **2**, 925 ⟨92T6105⟩
Komery, J., **2**, 179 ⟨71CJC943⟩
Kometani, T., **2**, 631 ⟨85JOC5384, 85TL2043⟩; **5**, 306 ⟨86TL919⟩
Komissarova, E. V., **4**, 941 ⟨79ZOB812⟩
Komissarova, N. G., **3**, 8 ⟨89S940⟩
Komiya, I., **3**, 357 ⟨90JAN1150⟩
Komiya, S., **2**, 590 ⟨88CL63⟩
Komiya, T., **2**, 853 ⟨87CL357, 88BCJ4029⟩
Komiya, Y., **2**, 853 ⟨87CL357, 88BCJ4029⟩; **3**, 336 ⟨88CL1145⟩, 597 ⟨80JOM(190)C39⟩; **4**, 415 ⟨88CL1145⟩; **5**, 603 ⟨88CL1145⟩, 603, 622 ⟨91BCJ2402⟩
Komizo, K., **2**, 399 ⟨75JOC3291⟩
Komkov, A. V., **4**, 979 ⟨90IZV401⟩
Komkov, I. P., **2**, 478 ⟨58JGU2963⟩
Komnenos, T., **1**, 306 ⟨1883LA(218)145⟩
Komori, O., **2**, 123 ⟨69TL3631⟩
Komori, S., **2**, 162 ⟨72S203⟩, 251 ⟨73S667⟩; **4**, 72 ⟨75JOC3540⟩; **5**, 967 ⟨72JOC3030⟩, 1003 ⟨71JOC3542, 73BCJ1275⟩; **6**, 445 ⟨71BCJ2182⟩, 618 ⟨73BCJ303⟩
Komori, T., **3**, 91 ⟨92CL901⟩
Komornicki, A., **1**, 415 ⟨88JA2314⟩, 905 ⟨77JA377⟩
Komoto, K., **2**, 160 ⟨76T2763⟩
Komoto, R. G., **3**, 621 ⟨78JA1119⟩
Kompa, K. L., **2**, 388 ⟨64ZAAC(332)238⟩, 752 ⟨66CB1347⟩
Kompan, O. E., **5**, 211 ⟨91MI 504-02⟩
Komura, M., **4**, 202 ⟨91JOC2276⟩
Komuro, T., **5**, 563, 626 ⟨93JA3000⟩
Komya, K., **6**, 465 ⟨92JAP04054156⟩
Konaka, R., **5**, 48, 113 ⟨62JOC1597⟩
Konakahara, T., **6**, ⟨92MI 615-02⟩, 417, 426 ⟨93S103⟩
Konarek, J., **6**, 42 ⟨73CCC66⟩
Koncos, R., **1**, 462 ⟨58JA5867⟩
Konda, Y., **6**, 138 ⟨80CPB2987⟩
Kondakov, D. Y., **1**, 197 ⟨93JA8485⟩
Kondo, A., **2**, 632 ⟨85CL411⟩
Kondo, F., **2**, 546 ⟨75JOM(92)C46⟩
Kondo, H., **1**, 464 ⟨73CL667⟩, 813, 814 ⟨80BCJ2958⟩; **3**, 142 ⟨80BCJ2958⟩; **5**, 351 ⟨68BCJ2521⟩
Kondo, I., **1**, 1035 ⟨72JOC3749⟩
Kondo, K., **1**, 144 ⟨75TL1007⟩, 159 ⟨74AG(E)407⟩, 299, 354 ⟨77TL109⟩, 400 ⟨72CC860⟩, 466, 1097 ⟨79JOC2408⟩, 823, 834 ⟨72CC62⟩; **2**, 24 ⟨94TL737⟩, 200, 222 ⟨69TL2743, 72JA5786⟩, 268 ⟨74TL793, 75BCJ3682⟩, 268 ⟨75JOC3644, 75TL2689⟩, 398 ⟨85ABC2331⟩, 753 ⟨80AG(E)1008⟩, 1000 ⟨67T3049⟩; **3**, 46 ⟨86JOC254⟩, 56 ⟨75TL1007⟩, 365 ⟨72CC62⟩, 668 ⟨86TL2411⟩, 734 ⟨82TL2323, 82TL2327⟩; **4**, 734 ⟨82TL2323, 82TL2327⟩; **5**, 795, 796 ⟨75BCJ3682⟩, 795, 796 ⟨75TL2689⟩, 1065 ⟨81MAC3411⟩, 1078 ⟨85SC171⟩; **6**, 14 ⟨81CL1719⟩, 58, 264, 266 ⟨82TL2323⟩, 465 ⟨71TL4885⟩, 465, 481 ⟨75BCJ108⟩, 466 ⟨74TL803⟩, 495, 497 ⟨89AG(E)452⟩, 498 ⟨79S597⟩, 505 ⟨79S735⟩, 510 ⟨71JA6344⟩
Kondo, M., **2**, 626 ⟨88BCJ597⟩; **5**, 193 ⟨86EUP268820⟩
Kondo, N., **5**, 249 ⟨87NKK1479⟩, 253 ⟨87NKK1475⟩
Kondo, S., **1**, 408, 409, 410, 820, 821, 822 ⟨72JA3870⟩, 820 ⟨71JOC1732⟩; **2**, 35 ⟨85CC400, 88S403⟩, 702 ⟨77CL679⟩; **3**, 357 ⟨90JAN1150⟩
Kondo, T., **1**, 182 ⟨79BCJ2377⟩, 196 ⟨85CL1477⟩, 196 ⟨89JOM(369)C51⟩, 483 ⟨86CL407⟩, 518 ⟨94AG(E)580⟩, 676, 678 ⟨84BCJ2335⟩; **3**, 290 ⟨88JOM(345)397⟩; **5**, 286 ⟨86JOM(309)333⟩
Kondo, Y., **1**, 1009 ⟨92S746⟩, 1010 ⟨82CPB2410, 82CPB2417⟩, 1010 ⟨83S312, 88CPB1890⟩, 1010 ⟨88CPB2248⟩, 1034 ⟨88H(27)475⟩; **2**, 677 ⟨84S245⟩, 685 ⟨84JCS(P2)451⟩, 942 ⟨93JOC6182⟩, 1027 ⟨92SL502⟩; **4**, 195 ⟨84S245⟩, 196 ⟨86CPB2362⟩
Kondoh, A., **5**, 223 ⟨88OM366⟩
Kondoh, H., **3**, 9, 123 ⟨89CL901⟩
Kondracki, P. A., **4**, 108 ⟨91JA8807⟩
Kondratenko, N. V., **4**, 771 ⟨90ZOR740⟩; **5**, 204 ⟨77S574⟩; **6**, 50 ⟨79ZOR1561⟩, 82, 302 ⟨69ZOB1755⟩, 232, 233, 234 ⟨75S721⟩, 232, 234, 236, 237 ⟨85S667⟩, 233, 234 ⟨84ZOR115, 67ZOB1770⟩, 235 ⟨90S1151⟩
Kondrateva, G. Y., **5**, 201 ⟨59IZV484⟩
Kondrateva, V. V., **2**, 884 ⟨54ZOB1259⟩
Kondrat'ev, P. V., **4**, 706 ⟨81JOM(209)233⟩
Kondyurina, T. F., **2**, 469 ⟨67ZOB2787⟩
Konecny, V., **6**, 335 ⟨75MI 611-01⟩, 560 ⟨92CCC1699⟩, 562 ⟨84MI 617-02, 84MI 617-04⟩
Konen, D. A., **4**, 842 ⟨76T2507⟩; **5**, 555 ⟨76T2507⟩
Konev, A. N., **5**, 1077 ⟨85IZV2824⟩
Koneva, N. Yu., **4**, 583 ⟨85JOM(288)119⟩
Kong, B. H., **1**, 130 ⟨91JPO449⟩
Kong, F., **2**, 22 ⟨88SC1071⟩, 122, 127 ⟨89SC3143⟩
Kong, W., **1**, 498 ⟨88JOC4807⟩
Kong, Y. K., **5**, 924 ⟨88OM2245⟩
Kongkathip, B., **4**, 193 ⟨84JCS(P1)1643⟩
Kongshaug, P. A., **6**, 516 ⟨87OM372⟩
Konieczny, M., **1**, 432 ⟨79JOC4813⟩; **2**, 684 ⟨72JOC2267⟩; **3**, 618 ⟨78ZN(B)1033⟩
König, B., **6**, 87 ⟨65CB3303⟩, 304 ⟨67CB767⟩
König, G., **3**, 280, 296, 298 ⟨81CB926⟩
König, H., **3**, 378 ⟨65CB3733⟩, 487 ⟨90JA5351⟩, 532 ⟨38CB1179⟩; **6**, 675 ⟨90JA5351⟩
König, H.-B., **5**, 636 ⟨54CB825⟩; **6**, 445 ⟨82MI 614-03⟩
König, J., **1**, 402, 794 ⟨70CB426⟩, 795 ⟨65LA(688)28⟩; **2**, 624 ⟨70TL1629⟩
König, K., **1**, 1038 ⟨67TL495⟩
König, K.-H., **2**, 382, 383 ⟨70CB2052⟩; **4**, 102 ⟨90T7729⟩; **6**, 445 ⟨78GEP2828969⟩, 448, 452, 454 ⟨80AG(E)201, 79EUP7498⟩, 494 ⟨79GEP2921130⟩
König, L., **1**, 117, 1092 ⟨83CB882⟩
König, R., **1**, 649, 656, 678, 680, 759 ⟨81TL5031⟩; **4**, 139, 141, 146, 156 ⟨85CB391⟩, 156, 708 ⟨85CB370⟩
König, W. A., **1**, 166, 183 ⟨93TL1917⟩
König, W., **1**, 537 ⟨71ACS2541⟩; **5**, 279 ⟨70CB2024, 70CB788⟩, 285 ⟨70CB2041⟩
Konigstein, F.-J., **3**, 64 ⟨59LA(622)37⟩, 85 ⟨58AG165⟩
Konigstein, O., **2**, 631 ⟨60LA(634)84⟩
Konijn, M., **2**, 1003 ⟨84T765⟩
Koning, H. de, **1**, 747 ⟨78JCS(P1)304⟩
Köning, K., **5**, 962 ⟨83HOU(E4)738⟩
Koning, K. H., **2**, 368 ⟨81AG(E)151⟩
Konings, M. S., **4**, 680 ⟨88JA6070⟩
Konishi, H., **1**, 94 ⟨82CL603⟩, 1035 ⟨89JOM(364)245⟩; **2**, 434 ⟨82CL977⟩, 619 ⟨80CL1423⟩; **3**, 356 ⟨73JA3043⟩, 399 ⟨75CC359⟩
Konishi, M., **1**, 279, 466 ⟨84JA158⟩, 477 ⟨81CC313⟩, 1097 ⟨82JA180⟩; **2**, 437 ⟨83JOC2195⟩; **4**, 1033 ⟨80BCJ1138, 82JA180⟩, 1033 ⟨86JOC3772⟩
Konishi, S., **2**, 14 ⟨89JOC4989⟩
Konishi, T., **4**, 659 ⟨78BCJ1397⟩
Konkina, T. N., **5**, 391 ⟨87ZOB1643⟩, 1081 ⟨87ZOB405⟩
Konkol, W., **2**, 262 ⟨61CB1263⟩; **5**, 376 ⟨61CB1263⟩
Konno, A., **3**, 637 ⟨92TL3161⟩; **4**, 62 ⟨92JOC3755⟩, 62, 63 ⟨90JOC6074, 91CC1027⟩, 87, 92 ⟨92TL3161⟩
Konno, H., **5**, 322 ⟨93BCJ327⟩
Kono, D. H., **5**, 17 ⟨73TL5121⟩
Kono, H., **2**, 527 ⟨75CL1095⟩; **3**, 490 ⟨71OPP47⟩; **4**, 797 ⟨69BCJ3270⟩

Kono, K., **1**, 486 ⟨83JOC1333⟩, 1033 ⟨87JCS(P1)1511⟩; **3**, 290 ⟨82CL35⟩; **5**, 197, 198 ⟨80TL2877⟩, 198 ⟨81JOC4413⟩
Kono, S., **6**, 4 ⟨91EUP421322, 92EUP473105⟩
Konoike, T., **1**, 521 ⟨87JA5167⟩; **3**, 357 ⟨91CPB672⟩, 427 ⟨73BCJ3198⟩
Kononov, L. I., **6**, 558 ⟨85KGS339⟩
Konopacki, D. B., **2**, 228 ⟨93S573⟩
Konopelski, J. P., **1**, 382 ⟨84JOC4094⟩; **3**, 160 ⟨91T447⟩
Konopski, L., **1**, 626 ⟨90TL1865⟩
Konovalova, I. V., **2**, 454 ⟨60ZOB2348⟩, 476 ⟨91MI 208-03⟩, 1074 ⟨89ZOB1979⟩; **4**, 343, 561 ⟨78ZOB1001⟩, 493 ⟨88JGU1148, 88JGU876⟩; **5**, 708 ⟨86JGU1096, 86ZOB1245⟩, 709 ⟨84JGU1969, 84ZOB2203⟩, 786 ⟨75JGU2512⟩; **6**, 118 ⟨79S81⟩
Konrad, P., **2**, 421 ⟨75ZN(B)681⟩
Konradi, A. W., **2**, 81 ⟨89JA8014, 90JOC4506⟩; **3**, 424 ⟨94JA1316⟩
Kon'shina, L. O., **6**, 145 ⟨76KFZ35⟩
Konstante, G., **5**, 224 ⟨87ZOB686⟩
Konstantinov, Yu. S., **4**, 761, 763 ⟨67ZOR1006⟩
Konstantinovic, S., **2**, 286 ⟨87TL4343⟩
Köntges, R., **4**, 802 ⟨78ZN(B)1422⟩
Konya, N., **1**, 635 ⟨88BCJ3211⟩
Konyushkin, L. D., **2**, 133 ⟨90S841⟩; **3**, 634 ⟨90S841⟩
Konz, M. J., **5**, 533 ⟨79USP4302238⟩
Koo, J.-Y., **2**, 344 ⟨78JOC2289⟩
Koock, S. U., **1**, 1126 ⟨69JA3292, 69JA3299⟩
Kooi, J., **3**, 699 ⟨75JCS(P1)1743⟩; **5**, 658 ⟨76JA1099⟩
Kooijman, H.,**5**, 890 ⟨92OM3844⟩
Kooistra, D. A., **5**, 282 ⟨74JOC2740⟩
Kook, A. M., **6**, 116 ⟨90CC1133⟩
Kool, L. B., **2**, 963 ⟨86JOM(310)27⟩
Koolpe, G. A., **2**, 281 ⟨78JOC4252⟩, 423, 908 ⟨78JOC3794⟩; **4**, 289, 965 ⟨78JOC3794⟩, 375, 400, 964 ⟨78JOC4252⟩
Koop, D. A., **2**, 172 ⟨64JOC951⟩, 194 ⟨62JOC1398⟩
Koop, H., **4**, 564 ⟨75CC468⟩
Koops, R. W., **3**, 31 ⟨90JOC5058⟩
Koos, M., **3**, 627 ⟨93T1541⟩
Koosha, K., **2**, 587 ⟨78BSF(2)575⟩
Kooyman, E. C., **1**, 773 ⟨63RTC1123⟩, 774 ⟨72RTC3⟩; **2**, 127 ⟨60T76⟩, 403 ⟨53RTC993⟩
Kooyman, J. G. A., **1**, 1030 ⟨68RTC69⟩; **2**, 1033 ⟨69RTC179⟩
Kopach, M. E., **5**, 201 ⟨91JA8972⟩
Kopaevich, Yu. L., **2**, 876, 881 ⟨72IZV213⟩; **4**, 134 ⟨72IZV215⟩
Köpe, H., **2**, 868 ⟨64CB2395⟩
Kopecký, J., **5**, 280 ⟨66CI(L)1529⟩
Kopecky, K. R., **1**, 299 ⟨62JA1015⟩, 883 ⟨59JA5740⟩; **2**, 403 ⟨69CJC2371⟩; **6**, 469 ⟨88CJC2234⟩
Kopeiki, V. V., **4**, 979 ⟨91ZOR382⟩
Köpf, H., **6**, 177, 178, 392, 709, 710 ⟨86JOM(315)9⟩, 290, 291 ⟨87CB653⟩
Kopf, J., **2**, 1004 ⟨85CB3529⟩; **4**, 109 ⟨89T2819⟩
Kopinke, F.-D., **1**, 925 ⟨89JOC3571⟩
Kopka, I. E., **1**, 873, 874, 876 ⟨80JA3959⟩; **2**, 696 ⟨80JA3959⟩
Köpke, B., **3**, 330 ⟨83CC238⟩, 346 ⟨82JCR(S)314⟩, 349, 365 ⟨81LA10⟩
Köplin, R., **6**, 615 ⟨83JFC(22)175⟩
Kopola, N., **4**, 535 ⟨89TL3963⟩
Kopp, A., **4**, 513 ⟨84JOM(264)87⟩
Kopp, R., **1**, 1114, 1115, 1124 ⟨74JA6686⟩, 1115 ⟨79CB2453⟩
Koppel, I., **5**, 689 ⟨93JCS(P2)655⟩
Koppel, J., **5**, 689 ⟨93JCS(P2)655⟩
Koppell, G. A., **4**, 69 ⟨78JA3933⟩
Köppelmann, E., **3**, 404 ⟨73AG584⟩; **4**, 535 ⟨70AG(E)163, 77CB2659⟩
Koppenhoefer, B., **2**, 12 ⟨87OS(66)151⟩; **5**, 60 ⟨87OS(66)151⟩
(Koppers Co., Inc., USA)869, **4**, ⟨84ZC261⟩ ⟨84USP4465845A⟩
Koppes, W. M., **5**, 14 ⟨75JOC3026⟩
Kopping, B., **1**, 5, 7, 9, 10, 17, 936, 943, 945 ⟨91JOC678⟩, 327, 950 ⟨92JOC3994⟩, 331, 342, 936 ⟨89TL681, 90TL6013⟩; **2**, 913 ⟨92JOC3994⟩; **5**, 1135 ⟨91JOC528⟩
Koppitz, P., **4**, 829 ⟨74CC405⟩
Kopple, K. D., **5**, 972 ⟨69JOC1631⟩
Koptyug, V. A., **1**, 103 ⟨91ZOR1137⟩, 914 ⟨74BAU1031, 84TCC1⟩
Kopylova, B. V., **5**, 55, 60, 89 ⟨62T69⟩; **6**, 323 ⟨82IZV460⟩
Korataeva, F. Kh., **4**, 923 ⟨83PS(16)345⟩
Korbacz, K., **1**, 608 ⟨81TL3097⟩; **4**, 924 ⟨81TL3097⟩; **6**, 123 ⟨81TL3097⟩
Korbonits, D., **5**, 219, 833 ⟨88CB723⟩, 220 ⟨89CB1107⟩
Korbukh, I. A., **2**, 812 ⟨78JOU1101⟩; **5**, 690 ⟨82KGS1682⟩
Korbukh, J. A., **5**, 690 ⟨82CHE1302⟩
Korchev, O. I., **6**, 302 ⟨86MI 610-01⟩
Korchevin, N. A., **3**, 361 ⟨80ZOR1842⟩
Kordish, R. J., **1**, 16 ⟨81JOC4843⟩
Koreeda, M., **1**, 566 ⟨84JOC2079, 84TL4881⟩, 738, 740, 742 ⟨89JOC3229⟩, 806 ⟨80JOC1172⟩, 816 ⟨89JOC5415, 93TL2597⟩, 816, 819 ⟨86JOC4090⟩, 816, 833 ⟨85JA5572⟩, 832 ⟨83JOC2122⟩; **2**, 540 ⟨90JA8175⟩, 677 ⟨84TL4881⟩; **5**, 209 ⟨83CC1113, 84JOC3875⟩; **6**, 533 ⟨90CL83⟩
Koren, B., **5**, 836 ⟨84H(22)1545⟩, 1122 ⟨84MI 530-01⟩
Korenchenko, O. V., **5**, 672 ⟨89BAU2201, 89IZV2390⟩
Korepin, A. G., **5**, 314 ⟨87IZV2857⟩
Koretskii, A. S., **4**, 329 ⟨74JOU244⟩
Korff, J., **3**, 456 ⟨76JOM(118)C1⟩
Korff, R., **3**, 200 ⟨82AG(E)371⟩
Korhonen, I. O. O., **4**, 54 ⟨82ACS(B)721⟩
Korinth, J., **6**, 236 ⟨67GEP1232954⟩
Koritsanszky, T., **6**, 114 ⟨87JOC2015⟩
Korkor, O., **1**, 811 ⟨87TL1853⟩
Kormachev, V. V., **2**, 461 ⟨66ZOB1436⟩
Kormendy, M., **3**, 542 ⟨62JOC3365⟩
Kormer, A., **2**, 1080 ⟨62ZOB1349⟩
Kormer, M. V., **4**, 885 ⟨91ZOB1192⟩; **5**, 691 ⟨91CHE424, 91KGS532⟩
Kormer, V. A., **1**, 960 ⟨60ZOB1056⟩
Kornberg, B. E.,**5**, 327 ⟨95TL197⟩
Kornblum, N., **1**, 36 ⟨44OR(2)262, 52JA3074⟩, 200, 238 ⟨88JOC1475⟩, 250 ⟨75AG(E)734⟩, 442 ⟨49JA2137⟩, 637, 638 ⟨71JA4316, 77JOC2944⟩; **2**, 392 ⟨62OR(12)101⟩, 393 ⟨63OSC(4)454, 63OSC(4)724⟩, 393 ⟨67JA725⟩, 393, 395 ⟨73JOC1418⟩, 397 ⟨56JOC4003, 73OSC(5)845⟩; **3**, 88 ⟨59JA4113⟩; **4**, 161 ⟨57JA6562⟩, 161 ⟨66JA865⟩, 328 ⟨74JA2580⟩, 332 ⟨73JA3356⟩, 332, 428, 429 ⟨83JOC332⟩, 428 ⟨75AG(E)734, B-82MI 409-01⟩; **5**, 275 ⟨79JOC4727⟩
Korneev, S. M., **3**, 193 ⟨91S195⟩
Kornemann, D., **2**, 73 ⟨92JOC6384⟩
Korner, V. A., **2**, 900 ⟨65ZOB1669⟩
Kornet, M. J., **2**, 377, 381, 383 ⟨68JHC397⟩, 381 ⟨84JHC1709⟩, 382 ⟨79JHC1485⟩
Kornetzky, D., **2**, 302 ⟨88S775⟩; **4**, 530 ⟨88S775, 91S637⟩; **6**, 561, 563 ⟨91S637, 88S775⟩
Kornev, V. I., **2**, 890 ⟨75JGU1476⟩
Kornfeld, E. C., **1**, 81 ⟨50JA3539⟩; **5**, 566, 579, 584, 586, 612, 618 ⟨51JOC131⟩
Kornilov, M., **6**, 104, 105 ⟨64CCC645⟩
Kornilov, M. Y., **5**, 652 ⟨90JGU1165⟩
Kornilova, Y. I., **3**, 275 ⟨50JGU1262⟩
Kornprobst, J. M., **1**, 781 ⟨74TL2203⟩; **5**, 282 ⟨68BSF3657⟩
Kornrumpf, W., **4**, 7 ⟨81ZN(B)1375⟩
Kornuta, P. P., **4**, 800 ⟨76JOU782⟩
Korobitsyna, I. K., **1**, 393 ⟨67RCR260⟩; **3**, 193 ⟨91S181⟩; **5**, 271, 292 ⟨67RCR260⟩
Korobkina, L. I., **5**, 864 ⟨B-69MI 502-03⟩

Korobov, M. S., **3**, 347, 356 ⟨73ZOR1717⟩
Koroleva, T. I., **6**, 635 ⟨90ZOB798⟩
Korolev, V. A., **3**, 342 ⟨87IZV2321, 87IZV2398⟩, 342 ⟨92MC67⟩
Korol'kov, V. V., **4**, 726 ⟨73DOK(209)1113⟩
Koroniak, H., **4**, 1083 ⟨84JFC(24)503⟩
Körösi, J., **5**, 318 ⟨61CB480⟩
Korostova, S. E., **3**, 430 ⟨85ZOR766⟩
Korotaeva, I. M., **2**, 1091 ⟨79JGU536⟩
Korotaeva, N. M., **4**, 267 ⟨90ZOR1259⟩
Korotkii, Yu. V., **5**, 1062 ⟨89ZOR2196⟩
Korotkova, V. P., **2**, 454 ⟨60ZOB2995⟩; **4**, 554 ⟨69ZOB845⟩
Korovyakov, A. P., **6**, 476 ⟨76JGU2441⟩
Korp, J. D., **5**, 1097 ⟨85CB340⟩
Korpar-Čolig, B., **4**, 697 ⟨84CCA689, 91JOM(405)59, 92CSC1116⟩, 697 ⟨93OM4708⟩; **6**, 207 ⟨82JOM(238)327, 84CCA689⟩, 207 ⟨84JOM(276)1, 91JOM(411)19⟩, 405 ⟨74CC646, 78JOM(153)1⟩
Korpela, T., **5**, 701 ⟨77OPP57⟩
Korpics, C. J., **5**, 34 ⟨63OSC(4)93⟩
Korpiun, O., **2**, 441 ⟨67JA4784⟩, 807, 845 ⟨68JA4842⟩
Korpusik, M., **3**, 613 ⟨89JOC4476⟩
Korsakova, I. S., **6**, 147 ⟨69IZV2309⟩
Korsarych, Z., **1**, 235 ⟨82JOC4005⟩, 235, 408, 409, 822 ⟨82TL3019⟩
Korshak, V. V., **2**, 430 ⟨69JOM(17)201⟩, 1018, 1019 ⟨63IZV2213⟩
Korshevets, I. K., **3**, 405 ⟨84S947⟩
Korshunov, A. I., **4**, 1045 ⟨86ZOB2743⟩
Korswagen, R., **4**, 679 ⟨81AG(E)1049⟩
Korswagen, R. P., **4**, 580 ⟨88ZN(B)1293⟩, 581 ⟨89JOM(371)329⟩
Kort, D. A., **4**, 650, 656 ⟨93OM338⟩
Kort, M. E., **1**, 583 ⟨94JOC332⟩; **2**, 932 ⟨91TL855⟩; **4**, 355 ⟨91TL855⟩
Korte, F., **2**, 245 ⟨62T1527⟩, 253, 462 ⟨63T1563⟩, 428, 431 ⟨65CB1681⟩, 440 ⟨65CB171⟩, 821, 827, 828 ⟨64CB1988⟩; **3**, 672 ⟨62CB281⟩, 721 ⟨66TL881⟩; **4**, 413 ⟨66CB724⟩; **5**, 849 ⟨72LA(755)101⟩, 1037 ⟨79ZN(B)31⟩
Korte, W. D., **1**, 112 ⟨70JOC22, 74JOC1168⟩; **5**, 919 ⟨74JA2360⟩
Korth, H. G., **1**, 922 ⟨90APO(25)131⟩, 946 ⟨87CB1197⟩; **5**, 654 ⟨85HOU(E5/1)628, 85HOU(E5/1)632⟩, 664 ⟨85HOU(E5/1)631⟩
Korvenranta, J., **5**, 864 ⟨82ZOR1813⟩
Korver, G. L., **1**, 825 ⟨79JA2746⟩
Korvola, J., **2**, 636 ⟨88JA7984⟩
Korytnyk, W., **2**, 109 ⟨60JA3866⟩; **3**, 294 ⟨64JOC574⟩; **4**, 164 ⟨67JHC625⟩
Korzeniowski, S. H., **2**, 632 ⟨77TL3519⟩
Kos, A. J., **1**, 135, 159 ⟨84JA6467⟩, 852 ⟨83T1141⟩, 1088 ⟨85CB1504⟩; **2**, 999, 1006 ⟨84JA6467⟩; **3**, 497 ⟨86CB1331⟩; **4**, 378, 391, 816 ⟨84JA6467⟩, 667, 668 ⟨83AG(E)733⟩; **6**, 205 ⟨83AG(E)733⟩, 697, 700 ⟨86CB1331⟩
Kosachev, I. P., **2**, 845 ⟨82ZOB2721⟩
Kosacheva, É. M., **5**, 369 ⟨85IZV173⟩
Kosack, S., **2**, 1047 ⟨91SL117⟩; **3**, 273 ⟨88CB2163⟩, 369 ⟨88CB833⟩
Kosak, A., **6**, 632 ⟨85OPP256⟩
Kosak, A. I., **3**, 305 ⟨55OSC(3)14⟩
Kosak, A. J., **3**, 305 ⟨48JA867⟩
Kosaka, T., **5**, 769 ⟨70MI 519-01⟩, 847 ⟨70JPB1696⟩
Kosarev, Yu. S., **2**, 451 ⟨62ZOB1974⟩
Kosarych, Z., **1**, 408, 821 ⟨85JOC2965⟩
Kosba, H., **3**, 297 ⟨62JOC2293⟩
Kosbahn, W., **3**, 556, 564 ⟨81CL527⟩
Koschel, D., **6**, 213 ⟨B-74MI 607-02⟩
Koschmieder, S. U., **2**, 473 ⟨92CC726⟩; **5**, 798 ⟨92JCS(D)19⟩
Koščik, D., **5**, 1057 ⟨83CCC3567⟩; **6**, 558 ⟨87CCC1764⟩, 565 ⟨90MI 617-01⟩, 566 ⟨92MI 617-03⟩
Koseki, K., **1**, 591 ⟨86TL1343⟩
Koser, G. F., **1**, 11 ⟨78JOC2676⟩, 108 ⟨B-83MI 104-01⟩; **2**, 148 ⟨90JA5672⟩, 320 ⟨87JOC476⟩, 1020 ⟨81JOC4324, 84JOC4700⟩, 1020 ⟨90SL365⟩; **3**, 163 ⟨88JA2987⟩, 163 ⟨88JOC210⟩, 163 ⟨93TL779⟩, 163, 178 ⟨90SL365⟩; **5**, 321 ⟨87JOC476⟩, 339 ⟨87TL5431⟩, 999 ⟨86JOC2669, 88JOC5158⟩
Köser, H. G., **1**, 865 ⟨86TL5931⟩
Koshar, R. J., **4**, 864 ⟨76USP3932526, 76USP3984357⟩; **5**, 811, 815 ⟨67JOC3859⟩, 815, 817, 818 ⟨73USP3726903⟩; **6**, 254, 278 ⟨73JOC3358⟩, 259 ⟨66JOC4232⟩, 260, 284 ⟨68USP3387033⟩, 284, 616, 649 ⟨67JOC3859⟩
Koshar, R. S., **5**, 815 ⟨66JOC4232⟩
Koshchii, V. A., **1**, 492 ⟨86UKZ1090⟩
Koshiba, M., **3**, 290, 409 ⟨86BCJ677⟩
Koshikawa, O., **4**, 857 ⟨86JOC700⟩
Koshimies, J. K., **3**, 20 ⟨81MI 301-01⟩, 136 ⟨82MI 304-01⟩
Koshino, H., **2**, 902 ⟨85S406⟩; **3**, 27 ⟨85S406⟩
Koshino, J., **1**, 198 ⟨79SC809⟩; **3**, 19 ⟨83SC1149⟩, 135 ⟨81CL1059⟩; **4**, 935 ⟨81CL1059, 83CL933⟩, 935 ⟨83SC1149⟩
Koshizawa, S., **5**, 547 ⟨73JCS(P1)1009⟩; **6**, 324 ⟨73JCS(P1)1009⟩
Koshti, N. M., **5**, 322 ⟨94TL5157⟩
Koshtoyan, S. D., **4**, 833 ⟨70IZV2140⟩
Kosina, A. N., **4**, 723 ⟨69IZV2030, 69IZV2032⟩
Kosinskaya, I. M., **2**, 407 ⟨67ZOB2693, 67ZOB856⟩; **6**, 621 ⟨78JGU1078⟩, 656 ⟨74ZOB1264⟩
Kosir, I., **2**, 615 ⟨92CC274⟩
Koski, W. S., **1**, 921 ⟨63JA384⟩
Koskinen, A., **3**, 638 ⟨82H(19)2115⟩
Koskinen, A. M. P., **3**, 193, 472 ⟨90CC652⟩, 207 ⟨89JOC4654⟩
Kosley, Jr., R. W., **1**, 980 ⟨82JOC389⟩; **2**, 740 ⟨76JA7104⟩
Koslowski, J. A., **1**, 311, 540 ⟨91COS(4)169⟩
Koslowski, T., **4**, 352, 353 ⟨92CB591⟩
Kosolapoff, G. M., **2**, 450 ⟨68JCS839⟩, 452 ⟨65JCS6638⟩, 454 ⟨49JA369, 51JA5392⟩, 454 ⟨53JA3379⟩, 455 ⟨51JA4101, 51JA5466⟩, 456 ⟨45JA1180⟩, 458 ⟨51JA855⟩, 625 ⟨53JA3596⟩, 837 ⟨50JA4291⟩, 838, 845, 846 ⟨50JA4292⟩, 847 ⟨48JA3465⟩; **4**, 130 ⟨53JA5738⟩, 336 ⟨B-72MI 407-01⟩, 451 ⟨B-72MI 410-01⟩, 469 ⟨53JA5278⟩, 472 ⟨48JA1283⟩, 473 ⟨47JA2112⟩, 550, 562, 563 ⟨69USP3445522⟩, 550 ⟨B-72MI 412-01⟩, 554 ⟨61JCS2423, 53JA1500⟩, 1023 ⟨72CA(77)126770j⟩; **5**, 800 ⟨B-72MI 520-01⟩, 900 ⟨72MI 523-01⟩; **6**, 701 ⟨B-72MI 614-01⟩
Kosower, E., **5**, 62 ⟨55OSC(3)510⟩
Kosower, E. M., **1**, 35 ⟨71ACR193⟩; **2**, 216 ⟨69BBR(37)593⟩, 402 ⟨71ACR193, 71JA1992⟩; **3**, 244 ⟨63JOC687⟩; **5**, 366 ⟨71ACR193⟩
Kosower, N. S., **2**, 216 ⟨69BBR(37)593⟩
Kossa, Jr., W. C., **1**, 495 ⟨69TL2313⟩
Kossanyi, J., **1**, 842 ⟨73TL3459⟩
Kossev, K., **1**, 644 ⟨88PS(40)69⟩
Kossmehl, G., **3**, 105 ⟨86CB3198⟩
Kost, A. N., **5**, 116 ⟨68JIC1109⟩
Kost, V. N., **5**, 38, 89 ⟨57JGU2481⟩, 55, 60, 89 ⟨62T69⟩
Kosta, K., **6**, 156, 157 ⟨90PS(47)261⟩
Kostenko, L. I., **4**, 970 ⟨88ZOR2374⟩
Koster, D. F., **2**, 420, 422 ⟨73JOC489⟩
Koster, G., **2**, 530 ⟨89CB1783⟩
Köster, H., **1**, 885 ⟨77CB717⟩
Koster, J. B., **1**, 522 ⟨71S139⟩
Köster, R., **1**, 60 ⟨83CB1336⟩, 657 ⟨59AG520⟩, 727 ⟨70LA(739)211⟩, 1001 ⟨75LA2130⟩; **2**, 112 ⟨66AG(E)580⟩, 132, 270 ⟨92LA835⟩, 418 ⟨68LA(720)1⟩, 520 ⟨74IS134⟩, 523 ⟨65LA(682)21, 74IS141⟩, 524 ⟨67JA2782⟩, 526 ⟨68LA(719)169⟩, 532 ⟨68AG(E)735⟩, 698 ⟨75LA1322, 75LA2201⟩, 1041 ⟨60LA(639)33⟩, 1077 ⟨68LA(717)1⟩, 1083 ⟨82HOU(13/3b)1, 84HOU(13/3c)1⟩, 1083, 1084, 1087,

1088, 1097 ⟨68LA(717)21⟩, 1085 ⟨82HOU(13/3b)69⟩, 1087 ⟨75LA2130⟩, 1087, 1088 ⟨65TL1901, 75CB395⟩, 1088 ⟨76S118⟩; **3**, 494 ⟨70LA(739)211⟩; **4**, 631 ⟨62AG(E)508⟩, 699 ⟨57LA(608)1⟩; **5**, 228 ⟨74LA112⟩, 380 ⟨83HOU(13)87⟩, 381 ⟨93CB565⟩, 578, 610 ⟨92LA1081⟩, 1141 ⟨70LA(739)211⟩; **6**, 172, 181 ⟨82HOU(13/3a)1⟩, 172, 204 ⟨64AOC(2)257⟩, 182, 183, 386 ⟨75LA1339⟩, 183 ⟨64TL1667⟩, 183 ⟨84HOU(13/3c)162⟩, 183 ⟨85AG(E)326⟩
Koster, W. H., **4**, 227 ⟨78JOC79⟩; **5**, 346 ⟨82JOC179, 84H(21)191⟩, 370 ⟨83JA3743⟩
Kosti, S., **3**, 469 ⟨80JCS(P1)2670⟩
Kostić, N. M., **5**, 932 ⟨82JA3879⟩
Kostikov, R. R., **1**, 907 ⟨67RCR557⟩; **2**, 136 ⟨90KGS1477⟩; **3**, 596 ⟨90JOU1642⟩; **4**, 35 ⟨83ZOR1625⟩, 1082 ⟨89UK1122⟩; **6**, 51 ⟨73ZOR2346, 77ZOR1857⟩
Kostka, K., **5**, 370 ⟨92AP325⟩
Kostlan, C. R., **6**, 558, 564 ⟨93JFC(61)1⟩
Kostova, A. G., **4**, 85 ⟨35MI 402-01⟩
Kostrova, S. M., **2**, 841 ⟨78ZOB91⟩
Kostsova, A. G., **2**, 178 ⟨48JGU729⟩
Kostuba, L. J., **4**, 188 ⟨79JMC1264⟩
Kostyanovskii, R. G., **2**, 337 ⟨88JA1719⟩, 356, 357 ⟨79ZV924, 81T4245, 91IZV1069⟩, 439 ⟨68IZV190⟩, 641 ⟨75DOK(220)146⟩; **3**, 559, 560 ⟨92BAU2039⟩; **4**, 155 ⟨68IZV270⟩, 295 ⟨60DOK(135)853⟩, 297 ⟨61DOK(139)877⟩, 323 ⟨87IZV2399⟩, 481 ⟨82IZV1611⟩, 886 ⟨69DOK(186)835⟩; **5**, 317 ⟨85T575⟩, 395, 397, 402 ⟨68T2995⟩, 397 ⟨75BAU816⟩, 1130 ⟨77IZV249, 82IZV1611⟩
Kostyk, M. D., **1**, 894 ⟨90JA8912⟩
Kostyuk, A. J., **4**, 915 ⟨75ZOB556⟩
Kostyuk, A. S., **3**, 484 ⟨75ZOB819⟩, 534 ⟨69ZOB467, 70ZOB707⟩, 540 ⟨84ZOB1817⟩; **4**, 886 ⟨69JOM(17)241⟩, 1056, 1059, 1063 ⟨87ZOB1741⟩, 1057 ⟨84ZOB1817, 85ZOB2801⟩; **5**, 376 ⟨69JOM(17)241, 73ZOB1753⟩, 376, 389 ⟨74ZOB1753, 73ZOB603⟩, 390 ⟨70ZOB707⟩
Kosugi, H., **1**, 209 ⟨74CL385⟩, 318 ⟨80JA1333⟩, 599 ⟨76SC357⟩; **2**, 713 ⟨85CL805, 87JOC1078⟩; **3**, 210, 256 ⟨82CL815⟩, 250, 256 ⟨79CL785, 81CC714⟩; **4**, 956, 957 ⟨79CL785⟩
Kosugi, K., **1**, 483, 484 ⟨80TL87⟩; **6**, 73 ⟨76CL891⟩
Kosugi, M., **1**, 194 ⟨87CL1237⟩, 471 ⟨77CL301, 84CL1225⟩; **2**, 753 ⟨83CL927⟩; **3**, 138 ⟨83CL839⟩, 140 ⟨77JOM(129)36⟩, 142 ⟨87CL1371⟩, 144 ⟨82CL939⟩, 144 ⟨84BCJ242⟩, 288 ⟨77CL1423⟩, 290, 408 ⟨86CL1197⟩, 290, 409 ⟨86BCJ677⟩, 293 ⟨87BCJ767⟩, 625 ⟨84CL1511⟩; **4**, 14 ⟨83BCJ2480⟩; **5**, 468 ⟨87BCJ3462⟩, 631, 632 ⟨89H(29)435⟩
Kosugi, Y., **1**, 585 ⟨89JOC2217⟩; **5**, 631, 632 ⟨89H(29)435⟩
Kosyan, S. M., **1**, 629 ⟨87ZOR1660⟩
Kosyukov, V. P., **6**, 445 ⟨69ZOB220⟩
Koszmehl, G., **3**, 494 ⟨73AG(E)237⟩
Koszuk, J., **3**, 197, 231 ⟨80TL2287⟩
Koszyk, F. J., **1**, 795 ⟨80TL2487, 83JOC3581⟩
Koszyk, F. U., **1**, 803, 804 ⟨80JOC5020⟩
Kotaka, N., **3**, 363 ⟨90JOC2421⟩
Kotake, H., **1**, 503 ⟨83BCJ2539⟩, 619 ⟨88CL2009⟩; **2**, 121 ⟨81CL1457⟩; **5**, 291 ⟨78BCJ1866⟩
Kotali, A., **3**, 429 ⟨91OPP593⟩
Kotali, E., **6**, 562 ⟨87JCS(P1)2759⟩
Kotani, T., **1**, 1015 ⟨92TL3503⟩
Kotani, Y., **1**, 431 ⟨93JOC4897⟩; **4**, 758, 821 ⟨93JOC4897⟩
Kotel'nikova, G. P., **6**, 242 ⟨62DOK(142)354⟩
Koten, I. A., **4**, 56 ⟨73OSC(5)145⟩
Kotenko, A. A., **4**, 970 ⟨88ZOR2374⟩
Kotera, M., **3**, 359 ⟨76TL3815, 82BCJ243⟩
Kothari, P. J., **6**, 573 ⟨80JHC1369⟩
Kothe, N., **6**, 620 ⟨76CB2921⟩
Kotia, N. K., **2**, 278 ⟨71JOC1700⟩
Kotila, S., **5**, 934 ⟨95CB645⟩
Kotlyarevskii, I. L., **1**, 1006 ⟨70IZV1144, 71IZV1534, 72IZV472⟩, 1010 ⟨80IZV726⟩; **4**, 560 ⟨83IZV2802, 84IZV1122⟩, 560 ⟨87IZV860⟩
Kotnis, A. S., **2**, 658 ⟨91TL3441⟩
Koto, H., **4**, 122 ⟨85S982⟩; **5**, 205 ⟨86CL967⟩
Koto, S., **2**, 398 ⟨88OSC(6)797⟩
Kotora, M., **6**, 7 ⟨91MI 601-01, 92JMOC51⟩
Kotsuji, K., **4**, 392 ⟨87CPB1413⟩
Kotsuji, Y., **3**, 226 ⟨90JCS(P1)1745⟩
Kotsuki, H., **1**, 120 ⟨86T2873⟩, 155 ⟨90S401⟩, 177 ⟨89TL3999⟩, 177 ⟨90JOC4417⟩, 1021 ⟨90TL4609⟩; **2**, 60 ⟨88BCJ2684⟩, 84 ⟨93TL4031⟩, 312 ⟨90S401⟩; **3**, 9, 123 ⟨85TL705⟩; **4**, 176 ⟨92SL97⟩
Kotsuki, M., **2**, 588 ⟨86T2873⟩
Kotte, K., **5**, 1111 ⟨67AP583⟩
Kottirsch, G., **1**, 496, 1051 ⟨90JA2816⟩; **6**, 487 ⟨93EUP560730⟩
Kottler, A., **6**, 655 ⟨57GEP1002355⟩
Kottong, W. G., **6**, 284 ⟨69USP3450762⟩
Kottwitz, B., **2**, 448 ⟨86PS(26)307⟩
Kotun, S. P., **6**, 602, 602 ⟨90EUP353743⟩
Kotyński, A., **4**, 472 ⟨80S1028⟩
Kotynski, A., **6**, 122 ⟨78JCR(S)41⟩
Kotyński, A., **6**, 156, 157 ⟨90PS(47)261⟩
Koujiri, T., **3**, 265 ⟨90JCS(P1)1891⟩
Kouklis, A. J., **1**, 369 ⟨92JOC7056⟩
Koul, V. K., **3**, 438 ⟨83S554⟩
Koulkes, M., **1**, 506 ⟨49CR(228)1501⟩
Koumaglo, K., **2**, 935 ⟨84TL717⟩
Koumbouris, B., **2**, 512 ⟨84AG(E)812, 85CB2472⟩
Koumoto, N., **3**, 123 ⟨89CL1623⟩
Koutchérov, V. F., **5**, 168 ⟨60BSF1412⟩
Koutecký, J., **1**, 866 ⟨64CCC1932⟩
Koutsantonis, G. A., **4**, 711 ⟨91JOM(407)391⟩
Kouwenhoven, A. P., **1**, 498 ⟨75AG(E)709⟩, 674 ⟨81CC1081⟩
Kouwenhoven, C. G., **1**, 518 ⟨72TL5269⟩
Kovac, A., **5**, 1039 ⟨69CZP131547⟩
Kovac, J., **3**, 257 ⟨75CCC1750⟩; **4**, 970 ⟨88ZOR2374⟩; **5**, 665 ⟨76CCC3085⟩, 1106 ⟨72CCC3144⟩; **6**, 627 ⟨84CCC2285⟩
Kovacic, P., **2**, 20, 337, 338, 339 ⟨70CRV639⟩, 23 ⟨69S135⟩, 298, 305, 761 ⟨70CR639⟩, 306 ⟨76JA5678⟩, 338 ⟨68OS(48)4, 69JOC911⟩, 403 ⟨68CJC1149⟩, 405 ⟨74CC217⟩, 608 ⟨64JA427⟩, 752 ⟨61JA221, 61JA743⟩, 752 ⟨66JA100⟩, 773, 775 ⟨76JOC1751⟩; **4**, 14 ⟨71JOC3566⟩; **5**, 317 ⟨75JA1180⟩
Kovacs, I., **5**, 880 ⟨91CB493⟩
Kovacs, J., **5**, 277 ⟨67JA183⟩, 1079 ⟨87CAR(166)101⟩
Kovács, K., **3**, 475 ⟨70HCA974⟩; **6**, 104 ⟨68TL2713⟩
Kovacs, L., **2**, 444 ⟨84MIP30022⟩
Kovács, M., **5**, 745 ⟨80S883⟩
Kovács, O. J., **6**, 71 ⟨63PCS374⟩
Kovács, Ö. K. J., **6**, 73 ⟨67T4181⟩
Kovacs, T., **6**, 264 ⟨91BMC357⟩
Koval, I. V., **2**, 401 ⟨90RCR819⟩; **3**, 374 ⟨91ZOR217⟩; **4**, 326 ⟨82ZOR841⟩, 330 ⟨93SR149⟩; **5**, 334 ⟨81ZOR533⟩
Kovalenko, A. L., **4**, 418 ⟨90JGU2516, 91JGU1072⟩
Kovalenko, L. V., **6**, 497 ⟨88ZOB26⟩
Kovaleva, L. F., **1**, 1010 ⟨88ZOR650⟩
Kovaleva, T. V., **2**, 836 ⟨79ZOB546⟩, 836, 837 ⟨76ZOB252⟩
Kovaleva, V. F:, **5**, 967 ⟨69ZOR1888⟩
Kovalev, I. P., **1**, 1051 ⟨90TL7063⟩
Kovalev, V. A., **4**, 794 ⟨75JOU119, 76JOU2252⟩, 794 ⟨80JOU1762⟩; **5**, 655, 657, 659 ⟨76JOU2252⟩, 655, 657, 659 ⟨76ZOR2319⟩, 655, 659 ⟨77JOU1471⟩, 655, 659 ⟨77ZOR1594⟩, 657 ⟨76JOU663, 76ZOR673⟩, 657, 659 ⟨79JOU32⟩, 657, 659 ⟨79ZOR38⟩, 685, 694 ⟨75JOU1926⟩, 685, 694 ⟨75ZOR1913⟩, 685, 694 ⟨77JOU1474⟩, 685, 694 ⟨77ZOR1597⟩, 787, 788 ⟨77JGU441⟩

Kovaleve, T. V., **2**, 466 ⟨74ZOB2357⟩
Kovar, D., **6**, 178 ⟨81JA2324⟩
Kovar, P., **2**, 118, 124 ⟨93JMC460⟩
Kovar, R. A., **2**, 418, 419 ⟨71IC900⟩, 421 ⟨71IC893⟩, 597 ⟨75IC2809⟩
Kovar, R., **3**, 127 ⟨76JOC1939⟩
Kovelesky, A. C., **3**, 576 ⟨69JA5887⟩
Kovel'man, I. R., **2**, 812 ⟨78JOU1101⟩
Kovenya, V. A., **2**, 407 ⟨75ZOB2394⟩, 408 ⟨78ZOB2686⟩
Köver, A., **3**, 8 ⟨85CC812⟩
Kovesdi, I., **4**, 273 ⟨89MRC760⟩
Kovredov, A. I., **4**, 143 ⟨82IZV1425⟩, 143 ⟨85IZV809⟩
Kovsman, E. P., **5**, 105 ⟨80MI 502-01⟩
Kovtun, E. A., **4**, 541 ⟨80ZOB1427⟩
Kovtun, V. Yu., **4**, 563 ⟨72IZV2612⟩
Kovyrzina, V. P., **2**, 888 ⟨72JGU1952⟩
Kowalczyk, B. A., **1**, 925, 935 ⟨89JOC6101⟩; **2**, 13 ⟨89JOC6101⟩
Kowalczyk, J. J., **2**, 668 ⟨90JA3093⟩
Kowalewska, W., **6**, 546 ⟨90MI 617-03⟩
Kowalik, J., **4**, 461 ⟨78CL1103⟩, 472 ⟨81S57⟩, 472, 473 ⟨81PJC713⟩
Kowalski, C. J., **1**, 482 ⟨81JOC197⟩; **2**, 282 ⟨80JA7950⟩, 941, 1023, 1024 ⟨86JA7127⟩, 1023 ⟨82JA321, 85JA1429⟩, 1023 ⟨87TL2463, 92JOC7194⟩; **3**, 15 ⟨86JA1325⟩, 154 ⟨85JOC5140⟩, 202 ⟨80JA5411⟩, 239 ⟨81JOC197⟩, 247, 539 ⟨88JA3693⟩
Kowalski, J., **2**, 108, 452 ⟨85TL4965⟩; **4**, 566, 573 ⟨88JOM(356)285⟩
Kowalski, M. H., **2**, 445 ⟨90JOC5033⟩; **3**, 496 ⟨89JOC2783, 90JOC5033⟩
Kowalski, P., **4**, 1013 ⟨90JOM(332)13⟩; **5**, 420, 794 ⟨90JOM(391)13⟩
Kow, R., **1**, 162, 163 ⟨73JA2715⟩, 162, 765 ⟨72JA6854⟩; **2**, 525 ⟨72JA6854⟩; **4**, 664 ⟨72JA6854⟩
Kowitt, F. R., **1**, 671 ⟨60JA739⟩
Kowitz, E., **2**, 1018 ⟨60LA(638)33⟩
Kowitz, F., **4**, 976 ⟨60LA(638)33⟩
Koyama, H., **2**, 420 ⟨92OS(71)133⟩
Koyama, I., **1**, 1113, 1114 ⟨80CC103⟩
Koyama, K., **2**, 28 ⟨91CC1418⟩, 83 ⟨90JA1286⟩; **3**, 35 ⟨85T4449⟩; **4**, 1077 ⟨72TL4517⟩
Koyama, M., **2**, 398 ⟨88OSC(6)797⟩
Koyama, S., **6**, 4 ⟨91EUP421322, 92EUP473105⟩
Koyama, T., **6**, 465 ⟨93EUP528498⟩
Koyama, Y., **4**, 955 ⟨80JOM(201)477⟩
Koyano, H., **3**, 49, 198 ⟨89TL4787⟩
Koyano, K., **6**, 445 ⟨73CJC333⟩
Koyasu, Y., **5**, 463 ⟨88CL377⟩
Koyoma, T., **3**, 394 ⟨89TL2095⟩
Koyuncu, D., **2**, 216 ⟨90TL5007⟩
Kozak, J., **3**, 40 ⟨92T4231⟩
Kozarich, J. W., **1**, 869 ⟨91JA9667⟩; **2**, 683 ⟨73TL1397⟩; **3**, 466 ⟨72TL5147, 73JOC1821⟩; **4**, 53 ⟨90JOC3029⟩
Kozawa, K., **2**, 986, 988 ⟨89CL761⟩
Kozawa, Y., **2**, 282 ⟨78JCS(P1)1273⟩; **5**, 1115 ⟨78JCS(P1)1273⟩
Kozerski, L., **2**, 789 ⟨83S489⟩
Kozeschkow, K. A., **1**, 446 ⟨34CB317⟩
Kozheynikova, A. N., **1**, 1006 ⟨70IZV1144⟩
Kozhich, D. T., **2**, 617 ⟨79S533⟩
Kozhina, N. D., **3**, 77 ⟨86CCC2167⟩; **5**, 350 ⟨86CCC2167⟩
Kozhushko, B. N., **4**, 1007, 1009, 1011, 1012, 1013 ⟨89ZOB571⟩, 1007, 1010, 1013 ⟨83ZOB548⟩; **5**, 787 ⟨73JGU10, 90JGU864⟩; **6**, 160 ⟨77ZOB321⟩, 285 ⟨77ZOB2766⟩, 285, 367, 666 ⟨73ZOB544⟩, 286, 606 ⟨88ZOB1516⟩, 286, 666 ⟨75ZOB1965⟩, 367 ⟨75ZOB2093, 89ZOB571⟩, 622, 634, 658 ⟨71ZOB2155⟩
Kozhushkov, S. I., **1**, 500 ⟨91ZOR488⟩; **2**, 445 ⟨87ZOR2120⟩
Koziara, A., **2**, 308, 407 ⟨92S1063⟩, 309 ⟨89JCR(S)296⟩, 317 ⟨87S487⟩, 412 ⟨70T3527, 76T1649⟩, 414 ⟨67T2243⟩; **5**, 273, 285 ⟨79S527, 79S549⟩, 1031 ⟨88JPR473⟩
Kozik, T. A., **5**, 1025 ⟨78KGS1497⟩
Kozikowski, A. P., **1**, 83 ⟨80JA1165⟩, 521 ⟨87JA5167⟩, 534 ⟨82TL2087⟩, 633, 829 ⟨87JOC3541⟩, 693 ⟨82JA2923⟩, 703 ⟨84TL2085⟩, 736 ⟨73TL4395⟩, 815 ⟨79CC106⟩; **2**, 278 ⟨78JOC2735⟩, 665 ⟨75TL2389⟩, 942 ⟨90TL6785⟩, 999 ⟨79TL833⟩; **3**, 106 ⟨80JOC3350⟩, 214 ⟨75TL925⟩, 220 ⟨87JOC1375⟩, 282, 295, 302, 305, 307, 309 ⟨80JA860⟩, 295, 302, 305, 307, 309 ⟨85T4821⟩, 494 ⟨73TL4425⟩, 498 ⟨86JOC3402⟩, 632 ⟨83JOC366⟩, 683 ⟨84ACR410⟩, 725 ⟨91SL654⟩; **4**, 187 ⟨80CC477⟩, 254 ⟨75TL925⟩, 538 ⟨80JOC2548⟩; **5**, 239 ⟨78JOC2735, 85T4821⟩, 666, 671 ⟨83JOC366⟩, 1052 ⟨91SL654⟩; **6**, 98 ⟨80JOC2236⟩, 487 ⟨94TL977⟩
Kozikowski, J., **2**, 573 ⟨57JOC598A, 57JOC598B⟩; **5**, 446 ⟨57JOC598⟩
Kozima, S., **1**, 1060 ⟨87CL1519⟩; **2**, 989 ⟨69JA4575⟩; **4**, 156 ⟨76BCJ2837⟩, 156 ⟨77BCJ1353⟩
Kozina, A. P., **5**, 1076, 1077 ⟨87JOM(327)303⟩, 1081 ⟨87ZOB405⟩
Koziora, A. P., **5**, 1081 ⟨88MI 528-01⟩
Koziski, K. A., **2**, 653 ⟨87JOC674⟩
Kozlikovskii, Y. B., **1**, 492 ⟨86UKZ1090⟩
Kozlov, A. I., **4**, 884 ⟨91ZOB1600⟩
Kozlov, E. S., **4**, 121, 122 ⟨80JGU802⟩, 342 ⟨80ZOB2424⟩, 344 ⟨74ZOB1029⟩; **6**, 609 ⟨78JGU1512⟩
Kozlov, L. M., **5**, 820 ⟨60ZOB1937⟩
Kozlov, N. S., **4**, 462 ⟨76ZOB1012⟩
Kozlov, V. A., **5**, 644 ⟨87ZOB685⟩
Kozlov, V. V., **1**, 36 ⟨63RCR59⟩; **6**, 653 ⟨74MI 621-01⟩
Kozlova, E. V., **4**, 462 ⟨92ZOB1472⟩
Kozlova, G. K., **2**, 464 ⟨60IZV651⟩
Kozlowska-Gramsz, E., **5**, 1071 ⟨82TL1585⟩
Kozlowski, J. A., **1**, 177 ⟨83JOC3334⟩, 207 ⟨84JOC1147⟩, 207, 312 ⟨84TL5959⟩, 207, 312 ⟨85JOM(285)437⟩, 217, 478 ⟨84JOC3928⟩; **2**, 586, 588 ⟨85JA3197⟩, 589 ⟨85JOM(285)437⟩; **3**, 128 ⟨84T5005⟩
Kozlowski, J., **1**, 204, 207 ⟨82JA2305⟩
Kozluk, T., **4**, 1079 ⟨81T1875⟩
Kozuka, S., **2**, 148 ⟨66BCJ364⟩, 688 ⟨80TL2067⟩; **3**, 177 ⟨80BCJ817⟩, 374 ⟨77CL725⟩; **5**, 334, 336 ⟨81CC510⟩, 666 ⟨84BCJ2011⟩, 1144 ⟨86MI 530-03⟩
Kozyukov, V. P., **2**, 394 ⟨86BAU2442⟩; **4**, 295, 313 ⟨81ZOB2382⟩, 295, 313 ⟨81ZOB2382, 82ZOB1386⟩, 313 ⟨82ZOB1386, 83ZOB119⟩; **5**, 323 ⟨88ZOB1056, 88ZOB1299⟩, 721, 722 ⟨78JGU1313⟩, 721, 722 ⟨78ZOB1429, 81JGU2055⟩, 722 ⟨81JGU2055, 81ZOB2382⟩, 722 ⟨81ZOB2382, 82JGU1222⟩, 722 ⟨82JGU1222, 82ZOB1386⟩, 722 ⟨82ZOB1386, 83JGU1944⟩, 722, 863 ⟨71DOK(199)103⟩, 722, 997 ⟨83ZOB2155⟩, 968 ⟨69FRP1563380, 69ZOB2598, 71BRP1235156, 79RCR473⟩, 968, 969 ⟨70DOK(190)110⟩, 968, 969 ⟨73ZOB623⟩, 968, 969 ⟨75ZOB2025, 72ZOB2501⟩, 969 ⟨79ZOB784, 80ZOB620⟩, 977 ⟨69DOK(181)115⟩, 978 ⟨73ZOB2089, 75ZOB2007, 81ZOB1814⟩, 983 ⟨80ZOB960, 81ZOB2017⟩, 1000 ⟨73UK1451⟩, 1001 ⟨84ZOB876⟩, 1006 ⟨75ZOB477, 76ZOB1531⟩, 1006 ⟨77ZOB2259, 83ZOB1096⟩, 1006 ⟨83ZOB1434⟩, 1007 ⟨81ZOB239⟩, 1014 ⟨68DOK(181)115, 74ZOB553⟩, 1016 ⟨68ZOB1179⟩, 1076, 1077 ⟨78ZOB1594⟩, 1076, 1077 ⟨78ZOB1943⟩, 1082 ⟨78ZOB1313⟩, 1085 ⟨89MI 528-01⟩, 1127 ⟨84URP570267⟩; **6**, 147 ⟨85IZV2635⟩, 443 ⟨69ZOB2598⟩
Kpegba, K., **5**, 556 ⟨86TL1505, 89T2041⟩, 557 ⟨90TL1853⟩
Kpoton, A., **2**, 264 ⟨80BSB783⟩
Kraak, A., **3**, 540 ⟨61RTC810⟩
Kraatz, A., **3**, 231 ⟨75LA874⟩
Kraatz, U., **6**, 110 ⟨73TL1219⟩, 437 ⟨83HOU(E4)1023⟩, 528, 529, 532, 533 ⟨83HOU(E4)407⟩, 528, 540, 541, 543, 545, 552 ⟨83HOU(E4)420⟩
Krach, T., **3**, 159 ⟨91S499⟩
Krack, W., **4**, 903 ⟨82LA585⟩; **5**, 566, 610, 726 ⟨77LA1291⟩

Kraebel, C. M., **1**, 1077 ⟨93JA3846⟩
Kraemer, P., **5**, 705, 706 ⟨77JHC749⟩
Kraemer, R., **5**, 883 ⟨83TL5885⟩; **6**, 160 ⟨83PS(18)27⟩, 686 ⟨83TL5885⟩
Kraer, A., **4**, 632 ⟨89ZN(B)96⟩
Kraeutler, B., **1**, 934 ⟨78JA4903⟩
Krafft, G. A., **1**, 617, 704 ⟨82T2857⟩, 942 ⟨85TL135⟩; **2**, 254, 299 ⟨85TL4867⟩, 289 ⟨86JA1314, 88JA8671⟩, 289 ⟨88JA8679⟩, 982 ⟨81JA5459⟩; **3**, 333 ⟨85TL1947⟩, 383 ⟨86JA1314⟩, 394 ⟨87TL3887, 88JA8679⟩, 636 ⟨88JA8671⟩; **4**, 287 ⟨88JA8679⟩, 287, 376, 393 ⟨88JA8671⟩, 365, 376, 570 ⟨87TL5121⟩, 947 ⟨88JOC3632⟩; **5**, 554 ⟨82JOC4384⟩, 1113 ⟨86JA1314⟩, 1115 ⟨88JOC3632⟩; **6**, 130 ⟨87TL5121, 88JA8671⟩
Krafft, M. E., **1**, 12 ⟨88JOC432⟩, 130 ⟨83TL1345⟩, 528, 529 ⟨88TL999⟩; **2**, 697, 699 ⟨83TL1345⟩; **3**, 118 ⟨86JOC5482⟩; **4**, 191 ⟨87JA1597⟩
Kraft, A., **1**, 623 ⟨92JCS(P1)3225⟩
Kraft, F., **3**, 447 ⟨11JPR122⟩
Kraft, M. Ya., **2**, 694 ⟨56ZOB3060⟩
Kraft, M. Y., **2**, 825, 829 ⟨63ZOB3050⟩, 825, 829 ⟨64ZOB1446⟩
Kraft, R., **5**, 773, 849 ⟨68CB2028⟩
Kraft, S., **4**, 595 ⟨92ZAAC(607)157, 92ZAAC(607)164⟩
Krägeloh, K., **2**, 539 ⟨82S1⟩; **4**, 985 ⟨84JA6015⟩; **6**, 43 ⟨82S1⟩
Kragushi, M. N.,**3**, 429 ⟨91T447⟩
Kraihanzel, C. S., **1**, 999 ⟨67JOM(8)239⟩; **4**, 545 ⟨76JOM(117)159⟩; **5**, 447 ⟨68JOM(15)397⟩
Krainev, A. G., **4**, 560 ⟨84DOK(277)371, 85BBA(818)96⟩
Kraits, Z. S., **2**, 693 ⟨59DOK(124)1251⟩
Kraka, E., **1**, 910 ⟨86JA561⟩
Kraker, A. J., **5**, 567 ⟨93JMC2459⟩
Kral, V., **3**, 71 ⟨88TL2861⟩, 75 ⟨88CCC1519⟩
Kralichkina, M. G., **2**, 695 ⟨72MI 213-01⟩, 885 ⟨72JGU536⟩, 886 ⟨72MI 217-01⟩
Kralik, M. S., **4**, 678 ⟨92OM3617⟩
Krallmann, R., **4**, 603 ⟨91CB2391⟩
Kramar, V., **1**, 871, 881 ⟨63JOC3362, 62JOC4146⟩
Kramarova, E. P., **4**, 140 ⟨89JOM(361)147⟩; **5**, 374 ⟨88ZOB1093⟩
Krämer, A., **5**, 696 ⟨85OM1749⟩; **6**, 183 ⟨90AG(E)292⟩, 188, 389, 401 ⟨89CB1881⟩
Kramer, D. L., **4**, 61 ⟨89JMC997⟩
Kramer, D. N., **4**, 186 ⟨47JA2246⟩
Kramer, E. A., **5**, 378 ⟨71CJC1011⟩
Kramer, G. M., **1**, 893 ⟨69JA4819⟩, 906 ⟨75APO(11)177, 89MI 119-09⟩
Kramer, G. W., **1**, 449 ⟨B-75MI 110-01⟩, 838 ⟨77JOM(132)9⟩; **2**, 514, 516, 520 ⟨B-75MI 210-01⟩, 517 ⟨78JOC1058, 79JA2966⟩, 521 ⟨74JOM(73)1⟩, 524 ⟨75JOM(90)C1, 76JA1964⟩, 900 ⟨B-75MI 218-01⟩
Kramer, K. A. W., **3**, 488 ⟨63NAT903⟩
Kramer, L., **4**, 493 ⟨75JA38⟩
Kramer, M. P., **3**, 355, 356 ⟨84AG(E)729⟩; **5**, 550 ⟨87JOC1156⟩; **6**, 235, 535 ⟨84JOC3854, 84JA263⟩
Kramer, R., **1**, 1107 ⟨77JA253⟩
Kramer, R. L., **2**, 118, 135 ⟨21JA880⟩
Kramer, S. W., **1**, 1008 ⟨87JMC193⟩; **3**, 140 ⟨90JA7441⟩
Krämer, T., **1**, 138 ⟨90AG(E)1424⟩
Kramer, T., **4**, 360, 382 ⟨90AG(E)1424, 86AG(E)160⟩, 382 ⟨92T8377⟩, 382, 384 ⟨87TL5149⟩
Kramer, W., **4**, 896, 897 ⟨89JHC1335⟩; **5**, 1028 ⟨90JHC433⟩; **6**, 118 ⟨93HCA2407⟩
Kramm, H., **5**, 856 ⟨64LA(678)132⟩
Kramolowsky, R., **5**, 1058 ⟨69AG(E)20⟩
Krannich, L. A., **2**, 892 ⟨72IC1226⟩
Krannich, L. K., **2**, 434, 485, 872 ⟨90IC3502⟩, 484 ⟨90POL289⟩, 506 ⟨69IC1032⟩, 507, 894 ⟨87IC1638⟩, 893 ⟨88ICA(150)51⟩
Krantz, A., **1**, 958 ⟨87JA3491⟩, 980 ⟨84JA1877⟩, 1042 ⟨88T5451⟩; **3**, 552, 553 ⟨77JA4842, 76JA7872⟩; **5**, 1012, 1015 ⟨75TL695⟩
Kranz, H., **2**, 438 ⟨68CB3438⟩; **6**, 118 ⟨73ZAAC(399)1⟩
Krapcho, A. P., **1**, 95, 96, 889 ⟨59JA3658⟩, 681, 682 ⟨74JOC1650⟩, 683 ⟨74JOC1322⟩, 881 ⟨67OS(47)20, 64JA1096⟩, 1114 ⟨70JA3789⟩; **3**, 124, 131 ⟨82S805, 77JOC1189⟩, 344 ⟨71JOC3885⟩, 721 ⟨62JOC1089⟩; **5**, 52 ⟨77JOC3749⟩, 750 ⟨62JOC1255⟩
Krapivin, A. M., **4**, 966 ⟨80IZV1147⟩
Krapivin, G. D., **6**, 627 ⟨84CCC2285⟩
Krapp, W., **2**, 831, 840, 843 ⟨78CB13⟩
Krasavtsev, I. I., **5**, 1062 ⟨89ZOR2196⟩
Krasik, P., **2**, 330 ⟨92TA1283⟩
Krasilnikova, E. A., **2**, 461 ⟨67ZOB1173⟩, 462 ⟨68ZOB609⟩
Krasil'nikova, E. V., **4**, 656 ⟨76DOK(231)1138⟩
Krasnansky, R., **2**, 1010 ⟨87JOC3062⟩
Krasnec, Jr., L., **5**, 33 ⟨71ZC110⟩
Krasnoperova, E. N., **2**, 891 ⟨88JGU1040⟩
Krasodomska, M., **5**, 597, 607 ⟨93M721⟩
Krass, N., **2**, 650 ⟨91SL39⟩
Krasuski, W., **5**, 548 ⟨82LA1451⟩
Kratsch, H. C., **3**, 497 ⟨68TL71⟩
Kratt, M. J., **2**, 102 ⟨31ZOB190⟩
Kratzer, O., **1**, 723, 733 ⟨62CB1894⟩
Kratzer, R. H., **4**, 481 ⟨85JFC(30)269⟩; **5**, 786 ⟨85JFC(30)269, 85JFC(30)289⟩
Krauo, G. A., **3**, 723 ⟨82IC2178⟩
Kraus, A., **2**, 781 ⟨1881M371⟩
Kraus, C. A., **4**, 699 ⟨30JA695⟩; **6**, 185 ⟨32JA1622⟩, 185 ⟨52JA1418⟩
Kraus, G. A., **1**, 574 ⟨84S885⟩, 664 ⟨78JOC4235⟩, 827 ⟨89JOC3137⟩, 900 ⟨85JOC4605⟩, 949 ⟨87TL4011⟩, 1005 ⟨78TL3195⟩; **2**, 264 ⟨91TL2189⟩; **3**, 65 ⟨78JOC4235⟩, 226 ⟨82SC521⟩, 245 ⟨91JOC4147⟩; **4**, 846 ⟨78JOC4235⟩, 1079 ⟨90TL1819⟩; **6**, 134 ⟨78JOC4235⟩
Kraus, G., **1**, 525 ⟨76JA6747⟩
Kraus, H., **3**, 446 ⟨57JOC204⟩
Kraus, H.-J., **4**, 693 ⟨79CC814, 80CB1072⟩, 694 ⟨83JA5479⟩; **5**, 897 ⟨83CB1955⟩; **6**, 123, 698 ⟨83CB1955⟩
Kraus, K., **5**, 359 ⟨92SC1199⟩
Kraus, R., **6**, 565 ⟨90MI 615-02⟩
Kraus, W., **2**, 9 ⟨75AG(E)824⟩
Krause, J. G., **3**, 431, 437, 618 ⟨75S502⟩
Krause, L. J., **2**, 434 ⟨80CC671⟩; **4**, 7 ⟨81JA2995⟩; **6**, 246 ⟨80CC671, 81JA2995⟩
Krause, M., **2**, 174 ⟨68AG970⟩, 174 ⟨69CB3357⟩, 658 ⟨90TL6629⟩
Krause, N., **1**, 960 ⟨91TL7229⟩, 970 ⟨93CB251, 93CB261⟩, 970 ⟨93LA521⟩, 1024 ⟨88CB1315⟩, 1038, 1039 ⟨87CB1845⟩; **2**, 940 ⟨93JOC5849⟩
Krause, W., **2**, 339 ⟨51CB170⟩; **5**, 601 ⟨78JPR313⟩
Krause-Göing, R., **2**, 989 ⟨77JOM(127)C29⟩; **4**, 693 ⟨77AG(E)183⟩
Krause-Loevenich, D., **5**, 765, 766 ⟨57CB1638⟩
Krauss, D., **4**, 321 ⟨89CB2311⟩
Krauss, J. C., **3**, 633 ⟨83S308⟩
Krauss, P., **3**, 551 ⟨71T5953⟩
Krausse, N., **2**, 940 ⟨90CB2173⟩
Krausz, F., **3**, 297 ⟨63CR(256)5594⟩
Krauth, C. A., **4**, 282 ⟨85EUP146838⟩
Krauthausen, E., **6**, 236 ⟨85HOU(E11)614⟩
Krauze, A., **2**, 22 ⟨92BSB237⟩
Kravchenko, A. L., **4**, 574 ⟨63IZV1563⟩
Kravchenko, V. V., **4**, 970 ⟨88ZOR2374⟩
Kravetz, M., **1**, 325, 920 ⟨86JOC5007⟩
Kravitz, J. I., **2**, 672 ⟨88TL6873⟩; **5**, 329 ⟨94JOC5292⟩
Kravtsov, D. N., **5**, 385 ⟨72JOM(36)227, 88ZOB62⟩, 1141 ⟨68IZV289⟩
Krawiec, M., **6**, 317 ⟨93T3035⟩
Krawiecki, Cz., **2**, 466 ⟨57RZC715⟩
Krawieka, B., **2**, 857 ⟨62MI 216-01⟩
Krayevsky, A., **6**, 489 ⟨93MI 615-03⟩

Krayushkin, M. M., **5**, 363 ⟨91T447⟩, 666 ⟨90BAU1538, 90IZV1692⟩; **6**, 29 ⟨89CS81⟩, 653 ⟨79IZV813⟩
Kreager, A., **1**, 530 ⟨89S472⟩
Krebs, A., **1**, 522 ⟨81TL1675⟩, 570 ⟨91COS(6)949⟩, 635, 635 ⟨70TL761⟩, 686 ⟨81TL4937, 84CB277⟩, 686 ⟨86T1693⟩, 1075 ⟨68TL4511⟩; **2**, 943 ⟨84JOC4687⟩, 947 ⟨83TL4083⟩; **3**, 330 ⟨83HCA801⟩, 348 ⟨84CB277⟩, 352 ⟨79H(12)1153⟩, 355, 356 ⟨84AG(E)729⟩, 464 ⟨81TL4937⟩, 550, 551 ⟨91JOC1317⟩
Krebs, A. W., **1**, 907 ⟨65AG(E)10⟩
Krebs, B., **4**, 488 ⟨85CB3227⟩, 611 ⟨89JOM(363)7⟩; **5**, 946, 957 ⟨94OM3510⟩, 958 ⟨94CB725, 94OM3502⟩; **6**, 166 ⟨86AG(E)919⟩, 242, 685 ⟨91EUP461563⟩, 266 ⟨90JOM(399)189⟩, 552 ⟨63ZAAC(321)143⟩, 685, 692 ⟨88CB655⟩, 689 ⟨85CC1687⟩, 689, 692 ⟨85C277⟩
Krebs, E., **5**, 652 ⟨55OSC(3)649⟩
Krebs, E.-P., **1**, 797 ⟨81HCA1023⟩; **3**, 269 ⟨79HCA497⟩
Krech, F., **2**, 427, 472 ⟨85ZAAC(523)7⟩, 428 ⟨84ZC261⟩, 439 ⟨83PS(17)73, 87ZAAC(553)136⟩, 439 ⟨91ZAAC(594)66⟩, 444 ⟨64ZAAC(328)21⟩, 473 ⟨85PS(21)367⟩, 478 ⟨91ZAAC(600)195⟩; **4**, 458 ⟨83PS(17)73⟩, 458, 477, 481 ⟨85ZAAC(523)7⟩, 550 ⟨84ZC261, 91ZAAC(594)66⟩
Krech, K., **2**, 429, 461 ⟨61CB2656⟩, 433 ⟨63CB2186⟩; **4**, 546, 553 ⟨63CB2186⟩
Krech, V., **2**, 427 ⟨76ZAAC(425)209⟩
Kreeger, R. L., **3**, 502 ⟨76JA7846⟩; **4**, 352 ⟨85TL1115⟩
Kreevoy, M. M., **1**, 671 ⟨60JA739⟩, 892 ⟨55JA1122⟩
Kregesmann, R., **4**, 599, 600, 663 ⟨78TL4391⟩
Krehbiel, G., **5**, 823, 824 ⟨55CB130⟩
Kreher, R., **2**, 388 ⟨73CB3097, 78AG(E)203⟩; **3**, 247 ⟨76AG(E)382⟩, 435 ⟨82CB2679⟩
Kreider, E. M., **1**, 379 ⟨70OR(18)99⟩; **2**, 719 ⟨70OR(18)99⟩
Kreilein, K., **4**, 621 ⟨77ZAAC(433)61⟩
Kreis, G., **3**, 690 ⟨73AG(E)564⟩; **5**, 919 ⟨78CB3283⟩, 948 ⟨73JOM(56)C37⟩, 952 ⟨73CB3910⟩, 1088 ⟨74JOM(65)53⟩; **6**, 723 ⟨78CB3542⟩
Kreisel, N., **4**, 761 ⟨52M1⟩
Kreiser, W., **3**, 59 ⟨87HCA445⟩, 71 ⟨73LA2078⟩
Kreissl, F. R., **2**, 878 ⟨83AG(E)1001⟩, 878, 896 ⟨84CB427⟩, 878, 896 ⟨86JOM(317)175, 83OM1266⟩; **3**, 513 ⟨77CB656⟩, 518 ⟨76AG(E)543⟩, 545 ⟨86AG(E)643, 88JOM(355)267⟩, 690 ⟨B-88MI 320-03⟩; **4**, 580 ⟨75JOM(86)C10, 77CB799⟩; **5**, 919 ⟨77CB3040, 77CB805⟩, 919 ⟨78CB3283⟩, 936 ⟨77CB2574⟩, 942 ⟨77CB1140⟩, 948 ⟨73JOM(56)C37⟩, 949 ⟨77CB2947⟩, 952 ⟨73CB3910, 95CB289⟩, 953 ⟨74CB183⟩, 955 ⟨92CB1141⟩, 1088 ⟨74JOM(65)53⟩; **6**, 700, 733 ⟨78CB2451⟩, 716, 719, 722 ⟨84MI 617-04⟩, 719 ⟨77CB2574⟩, 720 ⟨72CB2558, 76JOM(118)C33⟩, 720, 722 ⟨76JOM(113)C31⟩, 720, 722, 723 ⟨77CB3467, 77JOM(129)197⟩, 723 ⟨76AG(E)616, 77JOM(128)C49⟩, 723 ⟨78CB3542⟩
Kreisz, S., **1**, 987 ⟨86CB1208⟩
Kreiter, C. G., **1**, 917 ⟨65JA3267, 66JA2047⟩; **3**, 386 ⟨89JOM(377)105⟩, 597 ⟨78CB1223, 73AG(E)564⟩; **4**, 582 ⟨74CB3692⟩; **5**, 652 ⟨71JOM(28)237⟩, 932 ⟨69AG(E)761⟩, 949 ⟨74CB3554⟩, 950, 951 ⟨72CB150⟩, 952 ⟨73CB3910⟩, 955 ⟨68AG(E)390⟩; **6**, 153 ⟨89AG(E)1013⟩, 720 ⟨72CB2558⟩
Kreitmeier, P., **3**, 605, 607 ⟨88AG(E)1360⟩, 606 ⟨89TL3939⟩
Kreke, G. W., **2**, 132 ⟨39JA2971⟩
Kremer, K. A. M., **1**, 153 ⟨84T1345⟩; **3**, 519 ⟨81JA1862⟩, 520 ⟨82JA6119⟩
Kremer, M., **6**, 379 ⟨92JOM(429)1⟩
Kremer, P. W., **2**, 440, 441 ⟨92JOC511⟩; **4**, 114 ⟨81JCS(P2)1138⟩
Kremlev, M. M., **4**, 326 ⟨82ZOR841⟩, 817 ⟨87JOU246⟩; **5**, 58 ⟨85ZOR769⟩, 334 ⟨81ZOR533⟩, 345 ⟨69ZOR279, 74ZOR667⟩
Krempp, M., **1**, 448 ⟨90TL6179⟩; **2**, 945 ⟨90TL6179⟩
Krennrich, G., **4**, 162 ⟨86AG(E)999⟩
Krentzien, H., **2**, 33 ⟨83JOM(243)175⟩
Krepinsky, J. J., **4**, 175 ⟨87TL5595⟩; **6**, 88 ⟨91CJC415⟩
Krepski, L. R., **1**, 688, 689, 690 ⟨78JOC3255⟩; **2**, 69 ⟨85TL981⟩; **3**, 157 ⟨83TL4075⟩, 530, 531 ⟨78JOC2879⟩, 651 ⟨86SC617⟩
Krepysheva, N. E., **2**, 476 ⟨89JGU634⟩
Krerke, U., **2**, 110 ⟨53ZN608⟩
Kresge, A. J., **1**, 869 ⟨B-90MI 119-02⟩; **2**, 40 ⟨71JA4907⟩, 636, 637, 1022 ⟨90ACR43⟩, 637 ⟨84JA460, 87JA4000⟩, 637 ⟨90JA4862, 91SCI395⟩, 638 ⟨74JA4494, 86HCA1331⟩, 638 ⟨89CC1548⟩, 1022 ⟨87CJC1719, 87CJC441⟩, 1022 ⟨89JA2355, 89JA8297⟩, 1040 ⟨91AG(E)1356⟩; **6**, 646 ⟨91AG(E)1356⟩
Kresinski, R. A., **4**, 587 ⟨91OM3392⟩
Krespan, C. G., **2**, 1006 ⟨69JA415⟩; **3**, 352 ⟨61JA3434⟩, 459 ⟨65JOC1398⟩, 590 ⟨73JFC(3)91⟩; **4**, 36 ⟨62JOC1813⟩, 44, 86 ⟨81JA5598⟩, 45 ⟨B-92MI 402-01⟩, 97, 444 ⟨65JOC1398⟩, 731 ⟨73JFC(3)63⟩, 737 ⟨66JA5582⟩, 761 ⟨68JOC816⟩, 767 ⟨86JOC326⟩, 799 ⟨65JA3716⟩, 1076 ⟨84JA5544⟩; **5**, 1011, 1102 ⟨69JOC1278⟩; **6**, 32, 219 ⟨58JOC2016, 62JOC3995⟩, 49, 58 ⟨62JOC3584⟩, 64 ⟨60JOC105⟩, 245 ⟨58JOC1565⟩, 645 ⟨91MIP 601-01⟩
Kress, A., **1**, 909 ⟨74JOC378⟩
Kress, J., **1**, 674 ⟨85CC874, 87JA899⟩; **2**, 681 ⟨86CC531⟩; **3**, 514 ⟨87CC129, 89CC1062⟩, 515 ⟨82CC514, 85CC793⟩; **5**, 932 ⟨87JA3953⟩
Kress, J. K., **3**, 515, 518 ⟨87JA3953⟩
Kress, M. H., **1**, 196 ⟨93TL8047⟩
Kress, T. H., **2**, 942 ⟨88S803⟩
Kress, W., **4**, 697 ⟨80JOM(191)7, 81ZN(B)188⟩, 697 ⟨83JOM(243)245, 86JOM(301)1⟩; **5**, 1141 ⟨83JOM(243)245⟩; **6**, 206, 207 ⟨80JOM(191)7⟩, 206, 207 ⟨83JOM(243)245, 83JOM(256)217⟩, 207 ⟨86JOM(300)167⟩
Krestel, M., **6**, 625 ⟨86CB3236, 87CB1271⟩
Kresze, G., **2**, 100 ⟨62AG(E)97⟩, 163 ⟨89LA605⟩, 240 ⟨59CB1048⟩, 244 ⟨70TL1427⟩, 245 ⟨67AG109, 80LA629⟩, 245 ⟨80LA843⟩, 247 ⟨80AG(E)737⟩, 360 ⟨82TL1243⟩, 366 ⟨72LA(762)93⟩, 399 ⟨69TL4117⟩, 793 ⟨62AG(E)89, 83SR215⟩, 796 ⟨66TL1671, 67AG(E)149⟩, 796 ⟨B-68MI 215-05⟩; **3**, 374 ⟨73TL4033⟩; **4**, 304 ⟨64CB483⟩, 304, 311 ⟨64CB490⟩; **5**, 341 ⟨88PS(35)1⟩, 652 ⟨64CB483⟩, 670, 671 ⟨87BSB51⟩, 731 ⟨81LA65⟩
Kretchmer, R. A., **1**, 492 ⟨76JOC2661⟩, 812 ⟨68JOC4258⟩
Kretov, A. E., **6**, 621 ⟨81MI 620-03⟩, 649 ⟨81MI 621-02⟩
Kretschmann, H., **4**, 556, 557, 699 ⟨90ZAAC(586)25⟩
Kretschmann, M., **4**, 977, 978 ⟨92ZAAC(611)68⟩; **5**, 1074 ⟨89ZAAC(568)35, 89ZAAC(577)165⟩
Kretschmar, H. C., **1**, 725 ⟨70TL41⟩
Kretschmer, G., **1**, 784 ⟨75TL1335⟩
Kretsu, G., **5**, 193 ⟨58ZOB1864⟩
Kretzschmar, G., **1**, 332 ⟨84CB3160⟩, 332, 935, 1130 ⟨86JCS(P1)39⟩, 935 ⟨83TL5889, 84TL1055⟩, 948 ⟨84CB3175⟩; **3**, 627 ⟨84TL1055⟩; **5**, 595 ⟨66CB11⟩, 727 ⟨83TL5889, 86JCS(P1)39⟩
Kretzschmar, K., **1**, 332 ⟨81AG(E)965⟩
Kreuder, A., **3**, 452 ⟨25CB1974⟩
Kreutzberger, A., **3**, 94 ⟨69AP(302)828, 71AP(304)362⟩; **5**, 747 ⟨55JA6559, 72T4877⟩
Kreutzer, K. A., **2**, 82 ⟨91JA5093⟩
Kreutzkamp, N., **2**, 427 ⟨54CB919⟩; **3**, 480 ⟨65AG1138, 65AG(E)1078⟩; **4**, 321 ⟨66AP(299)906⟩, 460 ⟨59MI 410-01⟩, 489 ⟨65AG1138⟩, 561 ⟨59CB1695⟩; **6**, 159 ⟨59LA(623)103⟩
Kreutzmann, J., **5**, 587, 614 ⟨84SUL39⟩
Kreuz, K. L., **2**, 398 ⟨71JOC2574⟩
Kreuzer, F.-H., **5**, 638 ⟨73LA207⟩
Krewson, C. F., **5**, 83 ⟨B-55MI 502-02⟩
Krey, P., **4**, 981 ⟨82JPR915⟩
Kreze, G., **2**, 353 ⟨76AG(E)760⟩
Kricheldorf, H. R., **5**, 225 ⟨79AG(E)689⟩, 373 ⟨72S695⟩,

537 ⟨71S435, 73MI 512-01⟩, 538, 540 ⟨72MI 512-01⟩, 977 ⟨68AG(E)941⟩, 978 ⟨70S649, 72AG(E)128⟩, 991 ⟨72S551⟩, 992 ⟨72CB3958, 73CB3753⟩, 992 ⟨76S329⟩, 993 ⟨77AG(E)550⟩, 1012 ⟨71CB3131, 73LA772⟩, 1026 ⟨70AG(E)526⟩, 1027 ⟨71LA(748)101⟩, 1030 ⟨70S539⟩, 1031 ⟨75AG(E)502⟩

Krichevskii, L. A., **4**, 343 ⟨76ZOB783⟩

Kricka, L. J., **1**, 357 ⟨74S539⟩

Kricks, R. J., **4**, 29 ⟨76S313⟩

Kriechbaum, G., **4**, 686 ⟨82CB878⟩

Kriechbaum, G. W., **3**, 519 ⟨84JOM(264)327⟩

Kriecheldorf, H. R., **2**, 860 ⟨72S695⟩

Krief, A., **1**, 21, 25, 437 ⟨76TL2643⟩, 25 ⟨81TL1623⟩, 140, 143, 145, 146, 164 ⟨91COS(3)85⟩, 240 ⟨85TL6115⟩, 245 ⟨81TL4341, 85TL1093⟩, 245 ⟨86CC457, 86TL1723⟩, 246 ⟨87CC1214, 87TL1549⟩, 246 ⟨91TL417⟩, 248 ⟨82TL4385⟩, 301, 712, 713 ⟨80T2531⟩, 390 ⟨84TL2713⟩, 440 ⟨82TL3411⟩, 590 ⟨79TL4111, 81TL4009⟩, 592 ⟨83TL1657⟩, 592, 593, 713, 716 ⟨76TL1385⟩, 592, 593, 713, 716 ⟨76TL3743⟩, 592, 593, 713, 716 ⟨82CC564⟩, 592, 593, 716 ⟨76TL3227⟩, 592, 594, 716 ⟨78TL2693⟩, 594 ⟨80TL1877⟩, 601, 716 ⟨78TL1141⟩, 603 ⟨79TL2697⟩, 623 ⟨78TL187⟩, 623, 713, 716 ⟨79TL4233⟩, 623, 713, 716 ⟨80TL1997⟩, 712, 713 ⟨89H(28)1203⟩, 712, 713, 714, 715, 716, 717 ⟨B-87MI 115-02⟩, 712, 713, 714, 716 ⟨91COS(1)629⟩, 712, 714 ⟨87TCC1⟩, 713 ⟨89T2005, 89T2023⟩, 713, 1089 ⟨78AG(E)526⟩, 713, 714, 716 ⟨84IJ125⟩, 713, 716 ⟨79CC1136, 78TL1145⟩, 716 ⟨81TL1833, 90BSF681⟩, 759 ⟨76AG(E)161⟩; **2**, 32 ⟨80CC656⟩, 36 ⟨79TL1801, 83CC229⟩, 141 ⟨91TL2509⟩, 142 ⟨80CC544⟩, 216 ⟨90TL5007⟩, 278, 294 ⟨85T4793⟩, 279 ⟨84TL3625, 92S933⟩, 281 ⟨74AG(E)804, 75AG(E)350⟩, 281 ⟨75AG(E)700, 81TL2491⟩, 281 ⟨89T2023, 92SL317⟩, 281 ⟨92SL320⟩, 281, 284 ⟨79TL2697⟩, 281, 290 ⟨78TL3967⟩, 282, 730 ⟨76TL453⟩, 287 ⟨85CC569, 88TL3265⟩, 287 ⟨88TL3269, B-88MI 204-01⟩, 288 ⟨79TL4233⟩, 288, 1001 ⟨74AG(E)274⟩, 478 ⟨89T2005⟩, 554 ⟨85TL1093, 86TL1719⟩, 554 ⟨86TL1723, 92TL3381⟩, 556 ⟨94T7177⟩, 624, 910, 911 ⟨82TL3411⟩, 728 ⟨77TL3835⟩, 728, 936 ⟨79TL485⟩, 908 ⟨78AG(E)526⟩, 997, 1001 ⟨80T2531, 91COS(1)629⟩; **3**, 16, 132 ⟨79S877⟩, 38, 175 ⟨80CC951⟩, 39, 175 ⟨76TL453⟩, 125 ⟨90BSF681⟩, 153, 155 ⟨81TL1429⟩, 175 ⟨80T2531⟩, 207 ⟨78TL187⟩, 618, 620 ⟨80CC544, 79TL3995⟩; **4**, 238, 239, 283, 375 ⟨75TL1613⟩, 239, 288 ⟨77AG(E)540⟩, 284 ⟨89T2023⟩, 287 ⟨75AG(E)700⟩, 287, 288 ⟨85T4793, 79S877⟩, 288 ⟨86TL1723⟩, 289 ⟨92SL638⟩, 375 ⟨79TL485⟩, 375, 398 ⟨91COS(1)629⟩, 375, 400 ⟨81TL4737⟩, 375, 652 ⟨76AG(E)161⟩, 377, 390, 394, 395, 398, 400, 527, 646, 948 ⟨80T2531⟩, 390, 398 ⟨91COS(3)85⟩, 396, 401 ⟨84IJ125, B-87MI 408-04⟩, 399 ⟨78TL2693, 84TL3629⟩, 399 ⟨85CC571, 88TL3269⟩, 400 ⟨75AG(E)350, 89T2005⟩, 401, 964 ⟨78AG(E)526⟩, 827 ⟨70TL885⟩, 844 ⟨82TL3407⟩, 917 ⟨84TL2883, 82TL3411⟩; **6**, 88, 94 ⟨82TL3407⟩, 88, 94, 317 ⟨81TL4009⟩, 91 ⟨84MI 603-01, 78TL3971⟩, 94 ⟨79JOM(177)1⟩, 136 ⟨78JOM(149)C10⟩, 698 ⟨86T1209⟩

Krief, P., **5**, 374 ⟨85JOC1712⟩; **6**, 627 ⟨89S132⟩

Krieff, A., **3**, 74 ⟨82TL3411⟩

Krieg, G., **2**, 421 ⟨69ZAAC(364)316⟩

Krieger, A., **4**, 131 ⟨91SL395⟩

Krieger, C., **1**, 233 ⟨85CB1230⟩; **4**, 851 ⟨80AG(E)66⟩, 992 ⟨79CB2369⟩

Krieger, J. K., **1**, 465, 1090 ⟨76OS(55)103⟩, 491, 1105 ⟨71JA1379⟩

Krieger, L., **5**, 1058 ⟨84AG(E)970⟩; **6**, 680, 687 ⟨84AG(E)970⟩, 684 ⟨88MI 622-02⟩

Kriegesmann, R., **1**, 647, 667, 717 ⟨78TL4399⟩, 667 ⟨77AG(E)862⟩; **4**, 135, 592, 596, 599, 600 ⟨85CB1798⟩, 156, 708 ⟨85CB370⟩, 401, 599, 708, 724 ⟨82CB1810⟩

Krimen, L. I., **2**, 308 ⟨69OR(17)213⟩; **5**, 282 ⟨69OR(17)213⟩

Krimm, H., **2**, 390 ⟨58CB1057⟩; **5**, 292 ⟨58CB1057⟩, 652 ⟨67GEP1228249⟩, 966 ⟨66GEP1228249⟩, 1012 ⟨70GEP(O)1913273⟩

Krimmel, J. A., **5**, 361 ⟨71JOC350⟩

Krimmer, H.-P., **3**, 530 ⟨85JA7597⟩

Krimmer, S., **3**, 246 ⟨91S483⟩

Kriner, W. A., **4**, 608, 621 ⟨64JOC1601⟩; **5**, 1136 ⟨59JA5106⟩

Krinitskaya, L. V., **6**, 118 ⟨72MI 602-01⟩

Kris, E. J., **3**, 294 ⟨64JOC574⟩

Krisha, M. V., **1**, 248 ⟨87JOC4265⟩

Krishan, K., **5**, 331 ⟨73TL2711⟩

Krishma, M. V., **4**, 953 ⟨87JOC4258⟩

Krishna, A., **1**, 286 ⟨87TL2649⟩; **3**, 144 ⟨93TL6301⟩

Krishna, M. V., **1**, 440 ⟨88JOC3815⟩, 486, 545 ⟨87JOC4258⟩, 605 ⟨87TL1737, 89JOC4146⟩

Krishna, R., **5**, 924 ⟨82JA5667⟩

Krishnamurthy, D., **3**, 31 ⟨93JOC6476⟩

Krishnamurthy, R., **1**, 243, 327 ⟨91SL412⟩, 327, 328 ⟨92JOC4457⟩, 924 ⟨93TL7819⟩

Krishnamurthy, S., **1**, 3 ⟨80JOC849⟩, 4, 7, 8, 10 ⟨82JOC276⟩, 4, 7, 9, 10, 17 ⟨83JOC3085⟩, 4, 8, 10, 17 ⟨80JOC2550⟩, 14 ⟨80JOC1⟩, 428 ⟨69JOC3918⟩, 485 ⟨83OM779⟩; **2**, 83 ⟨73JA8486, 76JA3383⟩, 534 ⟨86JOC238⟩, 708 ⟨89JOC4458⟩

Krishnamurthy, T., **2**, 229 ⟨71T5753⟩

Krishnamurthy, V. V., **1**, 900 ⟨85JA2764⟩, 904 ⟨86JA1575⟩; **2**, 791 ⟨82JOC596⟩

Krishnamurti, M., **1**, 504 ⟨61T16⟩

Krishnamurti, P., **1**, 900 ⟨90JA6422⟩

Krishnamurti, R., **1**, 180, 212, 462 ⟨91COS(3)293⟩, 904 ⟨90JOC6061⟩; **2**, 36 ⟨B-91MI 201-01⟩, 70 ⟨89JA393⟩; **3**, 707 ⟨90S646⟩; **5**, 1045 ⟨90S646⟩

Krishnan, A. M., **2**, 385 ⟨89JCR(S)12, 91CL569⟩; **6**, 653 ⟨91CL569⟩

Krishnan, K., **1**, 826 ⟨93JOC7751⟩, 828 ⟨89TL6577⟩, 829 ⟨93JCS(P1)667⟩

Krishnan, P., **5**, 1045 ⟨92SC2741⟩

Krishnan, R., **1**, 1143 ⟨81CPL408⟩

Krishnan, V., **6**, 594 ⟨67ACS2904⟩

Krishna-Prasad, Y. S. R., **5**, 84 ⟨61T176⟩

Krishna Rao, G. S., **3**, 91 ⟨87IJC(B)914⟩

Krishtal, V. S., **4**, 1004 ⟨86ZOB805⟩; **5**, 863 ⟨80ZOB2413⟩

Krishtalik, L. I., **1**, 248 ⟨87TCC149⟩

Krist, H. G., **5**, 934 ⟨93CB2573⟩

Kristen, G., **4**, 1080 ⟨88JOC1806⟩

Kristensen, R., **3**, 354 ⟨73ACS2517⟩

Kristián, P., **5**, 612 ⟨85ZC324⟩ 1022, 1038 ⟨B-77MI 527-01⟩, 1035 ⟨84CCC295, 86CCC112⟩, 1040 ⟨69MI 527-01⟩, 1049 ⟨76CCC1388⟩, 1054 ⟨72CCC3066⟩, 1057 ⟨83CCC3567⟩; **6**, 558 ⟨87CCC1764⟩, 629 ⟨84CCC295⟩

Kristiansen, E. S. S., **6**, 593, 597 ⟨72ANY(192)101⟩, 594 ⟨72IJS(A)133⟩

Kristiansen, O., **5**, 775 ⟨85GEP3501452⟩; **6**, 619 ⟨80CP1076132⟩

Kristinsson, H., **5**, 355 ⟨86HCA333⟩, 668 ⟨79S102⟩, 1124 ⟨85HCA1155⟩

Kriukov, V., **4**, 222 ⟨91TL7593⟩

Krivchun, M. N., **2**, 448 ⟨90ZOB2394⟩

Krivdin, L. B., **6**, 562 ⟨84ZOR2114⟩

Krivinka, P., **4**, 894 ⟨70TL2357, 71CCC3314⟩

Krivonogov, V. P., **4**, 864 ⟨75ZOR415⟩

Krivozheiko, K. M., **1**, 837 ⟨68ZOR1114⟩

Kriwetz, G., **5**, 556 ⟨77LA1751⟩

Krizan, T. D., **1**, 1089 ⟨83JA6155⟩

Kïrmse, W., **2**, 802 ⟨79CB2120⟩

Krober, H., **3**, 546 ⟨75ZC91⟩

Kroboth, K. C., **2**, 683 ⟨72JOC3944⟩

Kroeger, J., **3**, 238 ⟨36JOC163⟩

Kroft, E. R., **3**, 479 ⟨87JOC3466⟩

Krogh, J. A., **2**, 359 ⟨79S228⟩, 359, 361 ⟨82LA121⟩; **3**, 617 ⟨79S722⟩, 618 ⟨78S703, 79ZN(B)511⟩; **5**, 334 ⟨79S228⟩

Krogh-Jespersen, K., **1**, 909 ⟨78JA4301, 86JA134⟩; **4**, 830 ⟨87JA3811⟩; **5**, 927 ⟨84AG(E)825⟩, 928 ⟨85JA2773⟩; **6**, 280 ⟨86JOC2168⟩, 280, 632 ⟨82JOC4177⟩
Krogsgaard-Larsen, P., **2**, 746 ⟨73ACS2802⟩
Krohmer, P., **4**, 603, 632, 635 ⟨71CB1347⟩
Krohn, J., **4**, 898 ⟨73LA443⟩; **5**, 606 ⟨69CB3786⟩, 631 ⟨73LA476⟩
Krohn, K., **2**, 546 ⟨94AG(E)99⟩; **3**, 88 ⟨90CB1357⟩, 539 ⟨88LA943⟩; **4**, 856 ⟨84T3677⟩, 1075 ⟨91JOC6038⟩
Krohnke, F., **3**, 258 ⟨60CB1068⟩, 494 ⟨50CB291⟩; **3**, 66 ⟨63AG(E)380⟩
Kröhnke, F., **3**, 88 ⟨39CB440⟩; **5**, 46 ⟨62AG(E)626⟩
Krois, D., **4**, 163 ⟨80T1345⟩
Kroker, J., **4**, 1041 ⟨86AG(E)571⟩
Krokhina, S. S., **4**, 476 ⟨79IZV2783⟩; **5**, 221 ⟨86IZV2535⟩, 369 ⟨85IZV173⟩
Krolevets, A. A., **5**, 369 ⟨88ZOB2628⟩, 1128 ⟨91IZV2847, 91MI 530-02⟩, 1131 ⟨92IZV187⟩
Krolikiewicz, K., **3**, 424 ⟨90T3849⟩, 671 ⟨83S316⟩
Krolikowski, D., **1**, 253 ⟨89SC2415⟩
Kroll, F., **5**, 953 ⟨93JOM(459)169⟩
Kroll, W.-R., **1**, 665 ⟨60LA(629)172⟩, 665, 666 ⟨60LA(629)53⟩; **2**, 596 ⟨60LA(629)121⟩; **4**, 699 ⟨60LA(629)53⟩
Krolls, U., **2**, 3 ⟨83TL163⟩; **5**, 315 ⟨61JA4469⟩, 327 ⟨89OS(67)69⟩, 569, 570 ⟨87S162⟩
Krolski, M. E., **3**, 226 ⟨82SC521⟩
Kromer, K., **3**, 687 ⟨81LA1025⟩
Krommes, P., **2**, 485 ⟨75JOM(88)329, 75JOM(92)181⟩; **3**, 482 ⟨75JOM(93)339, 75JOM(97)59⟩, 482 ⟨76JOM(110)195, 77JOM(132)359⟩, 483 ⟨77JOM(127)19⟩, 485 ⟨76JOM(120)131⟩; **5**, 871 ⟨76JOM(110)195⟩; **6**, 668 ⟨75JOM(93)339⟩
Kron, J., **3**, 545 ⟨89JOM(373)203⟩; **6**, 723 ⟨86CB2900, 88JOM(355)243⟩, 723 ⟨89JOM(373)203⟩
Kron, T. E., **4**, 119, 120, 121 ⟨80IZV491⟩, 339 ⟨84PS(21)59⟩
Kron, V. A., **2**, 458 ⟨83JGU1059⟩
Kronenberg, M. E., **2**, 756 ⟨66RTC56⟩
Kroner, I., **6**, 489 ⟨66CB2039⟩
Kroner, J., **6**, 686, 688 ⟨87PS(30)495⟩
Kroner, M., **3**, 66 ⟨67CB3172⟩; **5**, 15 ⟨89S832⟩
Krönhnke, F., **2**, 1003 ⟨62CB1108⟩
Krook, M. A., **5**, 322 ⟨83T2571⟩
Kroon, J., **4**, 660 ⟨86OM1551⟩; **5**, 377 ⟨91JA5742⟩
Kroon, J. L., **6**, 649 ⟨67JOC1662⟩
Kropf, H., **2**, 97 ⟨55CB1544, 70T3279⟩, 686 ⟨75S42, 88HOU(E13)761⟩, 687 ⟨76LA2331, 78T1231⟩; **4**, 314 ⟨66LA(700)1⟩
Kropp, J. E., **5**, 984 ⟨70JA1326⟩
Kropp, K., **2**, 968 ⟨83OM1555⟩, 986, 987 ⟨79JA3659⟩
Kropp, P. J., **1**, 108 ⟨82JOC1904⟩, 794 ⟨76JA840⟩, 1120 ⟨76JA4319⟩; **2**, 4, 23, 30 ⟨90JA7433⟩, 4, 23, 30 ⟨93JA3071⟩; **4**, 25 ⟨83JOC2084⟩
Kroto, H. W., **2**, 43 ⟨94CC463⟩; **3**, 551 ⟨74CC739⟩, 579 ⟨89JCS(P2)1841⟩, 582 ⟨90JA3779⟩; **4**, 568 ⟨87JSP(121)304⟩; **5**, 1130 ⟨87JSP304⟩, 1152 ⟨76CPL(42)460, 78JA446⟩, 1152 ⟨79CC653, 80JSP(82)185⟩, 1152 ⟨82CL981, 85JSP(111)415⟩, 1152, 1153, 1159 ⟨80CC333⟩, 1153 ⟨79JCS(D)1405, 81JSP(90)507⟩, 1153 ⟨81JSP(90)512, 82JST(79)215⟩, 1154 ⟨82JSP(92)158, 85JSP(109)8⟩; **6**, 678 ⟨82CSR435⟩, 679 ⟨76CC513, 78JA446⟩, 679 ⟨79CC653⟩
Krotovich, I. N., **5**, 710 ⟨81IZV2370⟩
Krotz, L., **4**, 840, 861 ⟨91TL2177⟩
Krouse, S. A., **3**, 691 ⟨89MM2569⟩
Krow, G. R., **1**, 799 ⟨75JA3837⟩; **2**, 654 ⟨93OR(43)251⟩, 674 ⟨74S572⟩; **3**, 556 ⟨71AG(E)435⟩, 560 ⟨72JOC2364⟩; **5**, 141 ⟨81T2697⟩, 283 ⟨81T1283⟩
Krowczynski, A., **2**, 789 ⟨83S489⟩
Krowiak, S. A., **1**, 196, 410 ⟨93TL4385⟩
Krowicki, K., **2**, 725 ⟨81JOC4894⟩
Krsek, G., **6**, 524 ⟨48JA383⟩
Krubsack, A. J., **2**, 170 ⟨68TL5149⟩, 170 ⟨80SC245⟩; **3**, 249 ⟨70JA5258, 75JOC3179⟩; **4**, 71 ⟨75JOC3037⟩
Krück, K.-H., **2**, 680 ⟨62CB2514⟩; **4**, 340 ⟨62CB2514⟩; **6**, 697 ⟨62CB2514⟩
Kruck, T., **6**, 526 ⟨64CB2289⟩
Kruczynski, L. J., **2**, 839 ⟨79CJC1903⟩
Krueger, A. C., **1**, 145 ⟨93SL829⟩
Krueger, C., **6**, 123, 698 ⟨83CB1955⟩
Krueger, D. S., **2**, 665 ⟨77TL611⟩
Krueger, K., **4**, 492 ⟨72CZ691⟩
Krueger, K. E., **3**, 725 ⟨91SL654⟩; **5**, 1052 ⟨91SL654⟩
Krueger, W., **6**, 71 ⟨71CAR(19)139⟩
Kruerke, U., **2**, 481 ⟨B-90MI 209-01⟩
Krüerke, U., **4**, 687, 689 ⟨61CB2829⟩
Krugalya, O. A.,**5**, 874 ⟨80RCR357⟩
Krug, W. P., **6**, 552 ⟨77CC660⟩
Krüger, C., **1**, 545 ⟨84JOM(265)323⟩; **2**, 435 ⟨85JOM(296)411⟩, 561 ⟨85AOC(24)1⟩, 575 ⟨76AG(E)621⟩, 984 ⟨91CB2165⟩, 1003 ⟨88CB843⟩; **3**, 61 ⟨89CB585⟩, 394 ⟨90AG(E)1067⟩, 414 ⟨85CB3652⟩, 456 ⟨63CB2132⟩, 597 ⟨86CB3141⟩, 640 ⟨67JOM(9)125⟩; **4**, 425 ⟨90S951⟩, 550 ⟨85JOM(296)411⟩, 579 ⟨84JOM(273)195⟩, 582, 583 ⟨74CB3706⟩, 583, 584, 585, 586, 588 ⟨77CB3517⟩, 586 ⟨77IJ149⟩, 604, 647 ⟨87CB471⟩, 632 ⟨83JA2582⟩, 659 ⟨85CB4248⟩, 671 ⟨86AG(E)925⟩, 676 ⟨92ICA741⟩, 686 ⟨77JOM(140)73⟩, 693 ⟨77AG(E)183, 77CB1763⟩, 694 ⟨69AG(E)678, 83JA5479⟩, 695 ⟨91AG(E)1124⟩, 709 ⟨87JOM(321)257⟩, 711 ⟨73CSC347⟩, 720 ⟨90ZAAC(581)41⟩, 725 ⟨90JA9620, 91OM3791⟩, 725 ⟨92CB1953, 92JA8531⟩, 725 ⟨92JOM(427)C21, 92OM3517⟩, 725 ⟨93OM4979⟩; **5**, 395 ⟨91CB2843⟩, 721 ⟨63CB2138⟩, 797 ⟨91OM1201⟩, 885 ⟨91AG(E)310⟩, 890 ⟨90CB1617⟩, 897 ⟨83CB1955⟩, 917 ⟨85OM1310⟩, 951 ⟨81AG(E)1055⟩; **6**, 154 ⟨87AG(E)908⟩, 338 ⟨88OM812⟩, 406 ⟨88AG(E)1544⟩, 615 ⟨83JFC(22)175⟩, 701 ⟨84OM38⟩, 729 ⟨94CB597⟩
Krüger, C. R., **1**, 881 ⟨64JOM(1)476⟩; **2**, 696 ⟨63AG(E)617⟩, 699 ⟨64JOM(1)476⟩; **3**, 568 ⟨63AG(E)617⟩
Krüger, E. A., **1**, 107 ⟨26MI 104-01⟩
Krüger, F., **4**, 561 ⟨73GEP2228928⟩
Kruger, G., **1**, 871 ⟨68JOC1750⟩
Kruger, G. J., **5**, 651 ⟨88JOM(355)337, 85OM275⟩, 951 ⟨94JOM(479)C12⟩; **6**, 333 ⟨84JOM(262)69⟩, 540 ⟨92POL893⟩
Krüger, K., **5**, 595 ⟨67JPR302⟩
Krüger, K. E., **6**, 41, 46 ⟨58CB806⟩
Kruger, M., **3**, 694, 716 ⟨91AG(E)1644⟩
Krüger, P., **5**, 825 ⟨1884CB1685, 1885CB727⟩
Kruger, T. L., **2**, 439 ⟨90JOM(393)143⟩
Krüger, U., **3**, 688 ⟨91CC302⟩; **4**, 551, 552 ⟨91CC302⟩; **5**, 698 ⟨77CZ154⟩, 887 ⟨91CC302⟩; **6**, 702 ⟨91CB329⟩
Krüger, W., **4**, 550 ⟨77CZ259⟩
Krügerke, T., **3**, 379, 380 ⟨88CB1977, 87AG(E)799⟩
Kruglaya, O. A., **3**, 483 ⟨77JOM(142)155⟩, 533 ⟨75IZV199, 76ZOB930⟩; **4**, 628 ⟨76IZV1887⟩, 638 ⟨80IZV1451⟩, 660 ⟨72JOM(46)51⟩, 933 ⟨81IZV654⟩, 940 ⟨77IZV596⟩; **5**, 432 ⟨84MI 509-01⟩, 1137 ⟨69ZOB2365⟩; **6**, 62 ⟨72JOM(34)299⟩, 185 ⟨77JOM(142)155⟩
Kruglik, L. I., **3**, 563 ⟨77JOU1989⟩
Kruglya, O. A., **6**, 62 ⟨72IZV85⟩
Kruglyak, Y. L., **5**, 822 ⟨65ZOB248⟩; **6**, 603 ⟨69JGU1694⟩, 613 ⟨69JGU1235, 69JGU1692⟩, 613 ⟨72JGU293, 72JGU799⟩, 613 ⟨72JGU803⟩
Kruithof, K. J. H., **4**, 1047, 1058, 1062 ⟨83T3073⟩; **5**, 424 ⟨83T3073⟩, 424, 468 ⟨83CC239⟩, 427 ⟨82TL3101⟩; **6**, 130, 133 ⟨83CC239⟩, 130, 133 ⟨83T3073⟩
Kruizinga, W. H., **5**, 177 ⟨81JA5183⟩
Krukonis, A. P., **2**, 1010 ⟨71JPC1246⟩
Krumm, A., **4**, 478, 487 ⟨71CB31⟩

Krumm, B., **5**, 382 ⟨93JOM(446)25⟩; **6**, 243 ⟨93JOM(446)25⟩
Krumpe, K. E., **1**, 550 ⟨90JOC414, 91JOC2523⟩, 550 ⟨92T5385⟩; **3**, 321 ⟨89CC921⟩, 322 ⟨90JOC414⟩
Krumpolc, M., **4**, 362 ⟨72CCC1392⟩
Krupadanam, G. L. D., **1**, 827 ⟨91SC1455⟩
Kruppa, V., **2**, 437 ⟨87PS(31)33⟩; **5**, 566, 595, 610 ⟨87PS(31)33⟩
Krupp, F., **1**, 666 ⟨60LA(629)14⟩; **2**, 43, 112 ⟨60LA(629)241⟩; **4**, 699 ⟨60LA(629)14⟩
Krus, K., **6**, 562 ⟨88ZOR2019⟩
Kruse, B., **1**, 816 ⟨90TL4425⟩; **4**, 381 ⟨89CB2023⟩
Kruse, C. G., **2**, 10 ⟨79RTC371⟩, 311 ⟨90TA163⟩, 330 ⟨90T1653⟩; **3**, 628 ⟨90T979⟩, 631 ⟨93T1063⟩; **4**, 64 ⟨77TL885⟩, 199 ⟨79RTC371⟩, 200, 208 ⟨78JOC3548⟩, 210 ⟨76TL1725⟩; **6**, 48 ⟨79JOC1847⟩, 48, 120 ⟨77TL885⟩
Kruse, C. W., **2**, 609 ⟨56JA6034⟩; **4**, 20 ⟨58JA1942⟩
Kruse, H. W., **2**, 388 ⟨57JA572⟩
Kruse, L. I., **1**, 130, 304 ⟨77CC233⟩; **3**, 106, 309 ⟨84JOC3195⟩, 124 ⟨77CC233⟩, 244 ⟨77JOC3846⟩
Kruse, W., **1**, 669 ⟨72JOM(42)C39⟩; **2**, 360 ⟨58CB2528⟩
Krusic, P. J., **1**, 923 ⟨76JA230⟩
Krusteva, R., **3**, 408 ⟨84S1038⟩
Krutak, J. J., **2**, 717 ⟨79JOC3847⟩
Krutikov, V. F., **4**, 469 ⟨92ZOB2708⟩
Krutošíková, A., **5**, 1106 ⟨72CCC3144⟩; **6**, 560 ⟨92CCC1699⟩
Krutskii, L. N., **4**, 340 ⟨70ZOB2560⟩
Krutzsch, H., **5**, 1004 ⟨67JOC1939⟩
Krutzsch, H. C., **4**, 732, 751, 752 ⟨68TL71⟩, 751, 752, 753 ⟨70JOC2125⟩, 762 ⟨71JOC2351⟩
Kryczka, B., **6**, 309 ⟨79CAR(74)127⟩
Kryger, A. C., **3**, 100 ⟨60JOC1896⟩
Krylova, G. V., **5**, 105 ⟨81MI 502-04⟩
Krysan, D. J., **2**, 669 ⟨92JA1412⟩
Kryshtal, G. V., **1**, 888 ⟨84RCR744⟩; **3**, 71 ⟨88TL2861⟩
Krysina, L. S., **5**, 864 ⟨69MI 521-01⟩
Krysina, V. K., **4**, 886 ⟨69JOM(17)241⟩; **5**, 376 ⟨69JOM(17)241⟩
Krzhizhevskii, A. M., **6**, 54 ⟨74IZV2144⟩
Krzyanowski, P., **3**, 426 ⟨89JPR870⟩
Krzyzanowska, B., **3**, 439 ⟨78S521, 82S270⟩; **4**, 1008 ⟨88PJC165⟩
Ktnecs, P. R., **5**, 864 ⟨71GEP2044574⟩
Ku, A. T., **2**, 1009 ⟨72JA1668⟩
Ku, B., **3**, 197 ⟨89TL3307⟩; **4**, 246 ⟨89SC433⟩, 572 ⟨89SC1891⟩
Ku, H., **2**, 632 ⟨81JOC5239⟩; **5**, 657 ⟨78JOC381⟩
Ku, T., **2**, 162 ⟨75JOC2759⟩
Ku, W.-S., **5**, 582 ⟨92S1219⟩, 582, 611 ⟨93JOC4742⟩
Kuan, C.-P., **2**, 908 ⟨85JOC5410⟩; **3**, 568 ⟨92JOC1202⟩; **4**, 354 ⟨90JOC4634, 90TL1945⟩, 359 ⟨92JOC1202, 92JOC6999⟩, 930 ⟨85JOC5410⟩, 979 ⟨92JOC3331⟩
Kuban, R. J., **3**, 238 ⟨85TL4911⟩; **4**, 989 ⟨86TL4281⟩
Kubas, G. J., **2**, 357, 365 ⟨79JOC3867⟩
Kubato, S., **5**, 769 ⟨70MI 519-01⟩
Kübel, B., **1**, 985 ⟨75AG(E)58⟩
Kubel, H., **6**, 571 ⟨79LA263⟩
Kubiak, C. P., **4**, 583 ⟨92OM1392⟩
Kubica, Z., **6**, 156 ⟨78MIP87868⟩
Kubicz, E., **5**, 366 ⟨93PJC711⟩
Kubik, D. A., **4**, 177 ⟨74JOC2815⟩
Kubiniok, S., **2**, 294 ⟨B-87MI 204-03⟩, 732 ⟨87AG(E)780⟩; **4**, 88, 288 ⟨91CB1315⟩
Kubinsky, J., **3**, 530, 531 ⟨09CB4213⟩
Kubler, D. G., **4**, 188 ⟨58JA1083⟩
Kubo, A., **3**, 419 ⟨87S824⟩
Kubo, I., **2**, 655 ⟨81TL3451⟩; **3**, 208 ⟨82TL1751⟩
Kubo, K., **2**, 205 ⟨80CPB1131⟩
Kubo, Y., **1**, 494 ⟨85TL1519⟩; **3**, 644 ⟨79S890⟩
Kuboike, S., **6**, 636 ⟨93AOC(35)211⟩
Kubota, M., **1**, 73 ⟨68IS101⟩; **5**, 384, 387 ⟨80BCJ1299⟩, 385 ⟨78BCJ2909⟩, 460 ⟨71JA1368⟩, 949 ⟨73JCS(D)1267⟩, 983 ⟨67JA4809⟩
Kubota, S., **5**, 346 ⟨83H(20)1531⟩, 847 ⟨70JPB1696⟩, 848 ⟨72JPJ275⟩; **6**, 563, 565 ⟨82JMC557⟩
Kubota, T., **2**, 830 ⟨80JOC5052⟩, 859 ⟨83CL441⟩; **5**, 280 ⟨80JOC5052⟩, 655 ⟨79TL4943⟩; **6**, 107 ⟨82H(18)201⟩
Kubota, Y., **1**, 149 ⟨84T1361⟩; **4**, 532 ⟨84T1361⟩, 618 ⟨88TL6939⟩
Kucera, D. J., **4**, 187, 199 ⟨89JOC5695⟩
Kucera, H. W., **5**, 410, 411 ⟨75JOM(87)263⟩
Kučera, J., **3**, 48, 473 ⟨70CCC3618, 66TL1109⟩
Kucera, T. J., **6**, 145, 147 ⟨60JOC2069⟩
Kucerovy, A., **1**, 139 ⟨88JA842⟩; **2**, 999 ⟨88JA842⟩; **4**, 379 ⟨88JA842⟩
Kucharzewska-Rusek, E., **2**, 802 ⟨70RZC1131⟩
Kuchen, ,W., **2**, 466 ⟨83PS(15)155⟩, 448 ⟨86PS(26)307⟩, 449 ⟨82CB1947⟩, 462 ⟨68CB3454⟩, 465 ⟨65AG(E)244, 66CB1663⟩, 467 ⟨80PS(15)155⟩, 468 ⟨74ZN(B)611⟩, 469 ⟨75ZAAC(413)266⟩, 832, 835 ⟨83PS(15)93⟩, 836, 837 ⟨81CB3485⟩, 854 ⟨78CB629⟩; **4**, 115 ⟨83PS(15)93⟩, 550 ⟨82CB1947⟩; **6**, 489 ⟨91PS(60)287, 93PS(83)65⟩
Kuchen, W. von, **2**, 1063, 1065, 1066, 1069, 1070 ⟨72ZAAC(394)74⟩
Kucher, R. V., **5**, 28 ⟨79MI 502-03⟩, 530 ⟨82MI 512-01⟩
Kucherova, N. F., **2**, 847 ⟨51DOK(78)709⟩
Kucherov, V. F., **1**, 747 ⟨64IZV1341, 65IZV1504⟩; **2**, 748 ⟨68T4677⟩; **5**, 24 ⟨B-69MI 502-02⟩
Kuchin, A. V., **1**, 1022 ⟨85IZV1109⟩; **4**, 698 ⟨88IZV1448⟩, 1068 ⟨88ZOB1567⟩
Kucinski, P., **3**, 215 ⟨73CC56⟩
Kuck, J. A., **2**, 906 ⟨38JA105⟩
Kücken, M., **6**, 580 ⟨81JPR41⟩
Kuckert, E., **3**, 597 ⟨87CB1939⟩
Kuckertz, H., **2**, 697 ⟨63AG(E)47⟩
Kucsman, A., **2**, 148 ⟨82JCS(P2)1075⟩, 160 ⟨77PS(3)9⟩, 1003 ⟨86JST141⟩
Kuczkowski, R. L., **3**, 326 ⟨B-84MI 307-02⟩
Kucznierz, R., **5**, 578, 610 ⟨92LA1081⟩
Kudav, N. A., **2**, 791 ⟨88SC2161⟩; **3**, 280 ⟨83CI(L)479⟩
Kudinov, A. R., **4**, 694, 717 ⟨87JOM(336)187, 87DOK(293)1137⟩
Kudinova, V. V., **2**, 471 ⟨79JGU2143⟩, 694 ⟨69ZOB2479⟩
Kudo, H., **1**, 442 ⟨79JCS(P1)2409⟩
Kudo, I., **6**, 558 ⟨89CPB1249⟩
Kudo, K., **1**, 189 ⟨91TL4311⟩, 189 ⟨93BCJ848⟩; **3**, 90 ⟨76BCJ1681⟩; **5**, 463 ⟨73BCJ2820⟩
Kudo, M., **3**, 629 ⟨91TL4333, 92JA7969⟩; **4**, 507 ⟨91EUP488681⟩, 613 ⟨88CL1441⟩; **6**, 383 ⟨88CL1441⟩
Kudo, S., **5**, 157 ⟨79JCS(P1)771⟩
Kudoh, E., **5**, 516, 520 ⟨92CL1843⟩
Kudou, N., **3**, 233 ⟨88S604⟩
Kudryavtsev, A. A., **2**, 407 ⟨89ZOB2135⟩, 408 ⟨80ZOB951⟩, 427 ⟨91ZOB1021⟩; **6**, 696 ⟨87ZOB2141⟩
Kudryavtseva, L. V., **2**, 876 ⟨75IZV897⟩, 1077, 1081, 1097 ⟨68IZV2287⟩
Kudyakov, N. M., **4**, 342 ⟨79DOK(247)609⟩, 569 ⟨75ZOB2010⟩
Kudzin, Z. H., **4**, 466 ⟨78S469, 81S643⟩, 472 ⟨80S1028⟩
Kudzma, L. V.,**3**, 311 ⟨91S1021⟩
Kuehe, M. E., **5**, 285 ⟨61JA1492⟩
Kuehle, E., **3**, 721 ⟨64CA(60)6795⟩; **5**, 738 ⟨59USP3053876⟩; **6**, 273, 274 ⟨70GEP1908680⟩, 439 ⟨82EUP46557⟩, 513 ⟨82EUP43978⟩
Kuehn, E. D., **1**, 1000 ⟨73JOC3588⟩; **3**, 273 ⟨73JOC3588⟩
Kuehne, M. E., **1**, 97 ⟨73OSC(5)400⟩; **2**, 738, 751 ⟨70S510⟩, 1048 ⟨68JOC4406⟩; **3**, 224 ⟨73JOC304⟩, 633 ⟨59JA5400, 70JOC161⟩
Kuehne, U., **2**, 439 ⟨83PS(17)73⟩
Kuehnling, W. R., **6**, 631 ⟨84JOC2404⟩
Kuemmerle, D., **6**, 353, 354 ⟨88JOM(394)183⟩
Kueng, W., **5**, 97 ⟨83MI 502-01⟩
Kuenzler, A., **6**, 596 ⟨91EGP291752⟩
Kueppers, H., **4**, 381 ⟨67LA(704)120⟩

Küerke, U., **2**, 1077 ⟨70JOM(21)837⟩, 1084 ⟨56ZN(B)676⟩, 1094, 1095 ⟨B-87MI 223-01⟩, 1094 ⟨B-87MI 223-02, B-87MI 223-03⟩, 1094 ⟨B-87MI 223-04⟩
Kuester, E. U., **2**, 437 ⟨84ZN(B)115⟩
Kugel, R. L., **4**, 577 ⟨63JPS(A)3627⟩
Kuhara, M., **2**, 61 ⟨88TL5419⟩
Kuhla, D. E., **3**, 234 ⟨80JOC1109⟩
Kuhl, G., **1**, 999 ⟨61ZAAC(312)186⟩; **2**, 489, 1081, 1082 ⟨61ZAAC(312)186⟩
Kuhl, R., **5**, 73 ⟨50CB390⟩
Kühle, E., **2**, 138, 165 ⟨71S563⟩, 165 ⟨71S617⟩, 165, 716 ⟨70S561⟩, 165 ⟨B-90MI 203-04⟩, 357 ⟨B-73MI 206-01⟩, 365 ⟨57GEP1101407⟩, 688 ⟨B-73MI 213-01⟩, 726 ⟨65AG(E)430⟩; **3**, 699 ⟨69AG(E)20⟩; **5**, 989 ⟨B-71MI 526-03⟩, 1004, 1032, 1070 ⟨69AG(E)20⟩, 1036 ⟨70AG(E)54⟩, 1071 ⟨65AG(E)430⟩; **6**, 234 ⟨71GEP2003143⟩, 236 ⟨64AG807⟩, 316 ⟨63BEP627486, 63BEP632635⟩, 316 ⟨68BRP1097237⟩, 435 ⟨70AG(E)54⟩, 449 ⟨62AG861, 70S542⟩, 452, 614, 616, 617, 643 ⟨69AG(E)20⟩, 453 ⟨68FRP1517379, 73GEP2131789⟩, 453 ⟨75GEP2351556⟩, 529 ⟨65GEP1219455⟩, 602 ⟨67AG(E)649⟩, 602, 604 ⟨83HOU(E4)522⟩, 602, 604, 605, 606 ⟨B-71MI 620-01⟩, 606, 617 ⟨62AG(E)647⟩, 614, 616 ⟨83HOU(E4)543⟩, 623, 624, 628, 629, 630 ⟨83HOU(E4)561⟩, 626, 627, 628 ⟨83HOU(E4)579⟩, 631, 632, 633 ⟨83HOU(E4)587⟩, 634, 635, 636 ⟨83HOU(E4)597⟩, 640, 642, 643 ⟨83HOU(E4)608⟩
Kühlein, K., **1**, 178 ⟨74LA369⟩; **3**, 571 ⟨66TL3415⟩; **5**, 232, 233 ⟨85HOU(E5)832⟩, 629 ⟨85HOU(E5)1266⟩, 635, 636 ⟨85HOU(E5)1279⟩, 639 ⟨85HOU(E5)1291⟩, 699 ⟨85HOU(E5/2)1218⟩, 846 ⟨85HOU(E5)1300⟩; **6**, 664 ⟨66TL3423⟩
Kuhlemann, F., **5**, 857 ⟨1893CB2978⟩
Kuhlman, H., **3**, 621, 633 ⟨88OSC(6)866⟩
Kuhlmann, D., **4**, 535 ⟨77CB2659⟩; **5**, 773 ⟨81CB3684⟩
Kuhlmann, G. E., **3**, 366 ⟨67JA2793, 70JOC4224⟩
Kuhlmann, H., **3**, 211 ⟨77CB1007⟩
Kühltau, H. P., **6**, 507 ⟨60CB2810⟩
Kuhm, P., **2**, 1074 ⟨88ZAAC(567)39⟩
Kühn, A., **4**, 693 ⟨77CB1763, 79JOM(179)421⟩, 693 ⟨79JOM(179)439⟩
Kühn, F., **3**, 585 ⟨92LA15⟩
Kuhn, H., **1**, 936 ⟨91JOC5971⟩
Kuhn, H. J., **1**, 521 ⟨73CB674⟩
Kuhn, M., **2**, 484 ⟨74CB1228, 75JOM(93)191⟩, 511 ⟨74AG(E)84⟩, 511, 512 ⟨80CB3318⟩; **5**, 911 ⟨80CB3318⟩
Kuhn, N., **2**, 406, 408 ⟨74AG(E)811⟩, 406, 408 ⟨74JOM(82)C3⟩, 409, 416 ⟨79ZN(B)888⟩
Kuhn, R., **1**, 866 ⟨62LA(654)64, 65LA(689)1⟩, 867 ⟨67AG(E)635⟩, 912 ⟨38CB442⟩, 954 ⟨65CB2611⟩, 960 ⟨61CB3060⟩, 986 ⟨59CB1849⟩; **3**, 66 ⟨68LA(715)106⟩; **4**, 53, 56 ⟨67CB2120⟩, 183 ⟨61CB2258⟩; **5**, 855 ⟨50LA(568)185⟩, 856 ⟨42CB75⟩, 857 ⟨53AG442⟩
Kuhn, S., **2**, 411 ⟨59LA(625)88⟩; **5**, 4 ⟨56CB862⟩
Kuhn, S. J., **1**, 462 ⟨64JOC2317⟩, 899 ⟨63JA1328⟩, 914 ⟨58JA6535, 58JA6541⟩, 914 ⟨60JA2380⟩; **2**, 182 ⟨62JOC2667⟩; **3**, 91, 93, 94, 99, 103 ⟨B-64MI 303-01⟩, 93 ⟨60JA2380⟩, 282 ⟨63JA1328, 64JA2203⟩; **6**, 410, 420, 422 ⟨56JOC1319⟩, 432, 432 ⟨65JOC1317⟩
Kühn, T., **2**, 943 ⟨81CB2293⟩
Kuhn, W., **1**, 797 ⟨86S921⟩; **2**, 670 ⟨83JOM(252)C78⟩
Kühne, H.-M., **2**, 825 ⟨82PS(13)331⟩
Kühne, U., **2**, 427, 472 ⟨85ZAAC(523)7⟩, 428 ⟨84ZC261⟩, 473 ⟨85PS(21)367⟩; **4**, 458 ⟨83PS(17)73⟩, 458, 477, 481 ⟨85ZAAC(523)7⟩
Kühnemund, K.-H., **6**, 729 ⟨82JPR669⟩
Kuhnen, F., **2**, 13 ⟨73S493⟩
Kuhnen, L., **2**, 156 ⟨66AG(E)937⟩
Kuhner, S., **3**, 628 ⟨91TL2605⟩
Kühnle, A., **6**, 233 ⟨81TL3047⟩
Kühnle, W., **4**, 536 ⟨87AG(E)480⟩
Kühr, H., **1**, 104 ⟨67AG(E)801⟩
Kuimova, M. E., **1**, 657 ⟨68DOK(179)1344⟩, 658 ⟨68IZV548⟩
Kuindersma, K. A., **3**, 371 ⟨73TL3929⟩
Kuipers, J. A. M., **3**, 365 ⟨81S295⟩
Kuipers, P. J., **6**, 558, 564 ⟨93JMC1802⟩
Kuipers, W., **3**, 556 ⟨89AX(C)482⟩
Kuivila, H. G., **1**, 83 ⟨73JOM(61)167⟩, 330, 341, 936 ⟨63JOC2165⟩, 447 ⟨61JA2159⟩, 935 ⟨68ACR299⟩; **2**, 600 ⟨78JOC750⟩, 601 ⟨68JOC802⟩; **3**, 83 ⟨66JA571⟩; **4**, 13 ⟨68ACR299⟩, 699, 700 ⟨83OM106, 82OM404⟩, 700 ⟨87OM834, 92JOM(429)311⟩, 701 ⟨83OM909, 84OM1687⟩; **6**, 405 ⟨76JOM(122)171⟩
Kujath, E., **4**, 998 ⟨82AG559⟩
Kujime, S., **1**, 929 ⟨93JCS(P1)3007⟩
Kujoi, T., **2**, 78 ⟨88TL5667⟩
Kukemnhohner, T., **2**, 568 ⟨86TL5711⟩
Kukhar, V. P., **2**, 15 ⟨91RCR1050, 91RCR850⟩, 467 ⟨91ZN(B)258⟩; **3**, 363 ⟨72ZOR1822⟩, 567 ⟨71JOU413⟩, 618 ⟨74S563⟩; **4**, 110 ⟨79S747⟩, 451, 469, 493 ⟨87RCR859⟩, 462 ⟨92ZOB1472⟩, 555, 1004, 1008 ⟨79ZOB2217⟩, 561 ⟨79ZOB1470⟩, 800 ⟨76JOU782⟩, 800 ⟨78JGU711, 79JGU1947⟩, 865 ⟨78ZOR1947, 79ZOR1477⟩, 866 ⟨83ZOR2417⟩, 971 ⟨73ZOR39⟩, 987, 988 ⟨74ZOR36⟩, 991 ⟨78ZOB778⟩; **5**, 57 ⟨73S545⟩, 385 ⟨93S117⟩, 401 ⟨89JGU1977⟩, 683 ⟨66JGU1276, 66ZOB1260⟩, 683 ⟨76JGU239, 76ZOB243⟩, 683, 707 ⟨87UKZ397⟩, 810 ⟨72ZOR695⟩, 879 ⟨81JGU1883⟩, 902 ⟨77JGU658⟩, 903 ⟨75JGU2509⟩, 1032 ⟨72ZOR1822⟩; **6**, 19 ⟨82ZOR228⟩, 58 ⟨79ZOB1025⟩, 238, 610 ⟨71ZOR2084⟩, 278, 699 ⟨77ZOR275⟩, 286 ⟨79ZOB1470⟩, 367 ⟨72ZOB1169⟩, 450 ⟨77ZOR290⟩, 609, 619 ⟨83KGS798⟩, 610, 727 ⟨73ZOR39⟩, 612, 623 ⟨81ZOR180⟩, 612, 727 ⟨74ZOR449⟩, 612, 727 ⟨79ZOR215, 73ZOR43⟩, 619 ⟨74ZOR36, 75ZOR71⟩, 619 ⟨78ZOR1841, 84ZOB1110⟩, 620 ⟨82UKZ395⟩, 625 ⟨75ZOR2223⟩, 658 ⟨73ZOR1815⟩, 680, 683 ⟨85ZOB264, 83ZOB473⟩, 683 ⟨85ZOB2797, 87PS(30)433⟩, 684 ⟨80ZOB233, 81ZOB2189⟩, 689 ⟨85ZOB1862, 85ZOB2214⟩, 694, 699, 700 ⟨83RCR1096⟩
Kukharev, B. F., **3**, 707 ⟨89ZOR1573⟩
Kukhtin, V. A., **4**, 485 ⟨82JGU392⟩, 768 ⟨75ZOB2374⟩
Kukla, M. J., **3**, 311 ⟨84JOC5003⟩; **6**, 598 ⟨91JMC3187⟩
Kukovinetz, A. G., **2**, 561 ⟨83IZV2053⟩
Kukovitskii, D. M., **6**, 127 ⟨88ZOB1066⟩
Kuksis, A., **5**, 24 ⟨59CRV89⟩
Kula, J., **4**, 203 ⟨86SC833⟩
Kula, M.-R., **3**, 628 ⟨90AG(E)386⟩
Kulagowski, J. J., **6**, 119 ⟨92SL975⟩, 628, 636 ⟨94T3195⟩
Kulak, A. N., **1**, 801 ⟨91TL4591⟩
Kulanthaivel, P., **1**, 4, 7 ⟨87TL3883⟩
Kul'bitskii, G. N., **5**, 742, 751 ⟨72ZOR1246⟩
Kul'chitskii, M. M., **4**, 992 ⟨76UKZ204⟩
Kulcsár, L., **5**, 745 ⟨80S883⟩
Kulenovic, S. T., **2**, 22, 29, 34 ⟨78JOC3687⟩, 558 ⟨77SC223⟩
Kuleshova, N. D., **2**, 214 ⟨64IZV644⟩
Kul'ganek, V. V., **1**, 888 ⟨84RCR744⟩
Kulibaba, N. K., **5**, 861 ⟨68ZOB326, 68ZOB850⟩; **6**, 616, 728 ⟨72JGU97⟩
Kulikov, A. S., **5**, 667, 671 ⟨91BAU438⟩, 667, 671 ⟨91IZV506⟩
Kulikov, Y. M., **6**, 653 ⟨74MI 621-01⟩
Kulishov, V. I., **6**, 562 ⟨81KGS907⟩
Kulka, M., **2**, 124 ⟨56CJC1093⟩; **3**, 279 ⟨54JA5469⟩
Kulkarni, A. K., **1**, 62 ⟨82JA6809⟩; **2**, 617 ⟨85T1179⟩, 909, 910 ⟨83T877⟩; **4**, 355 ⟨92JOC6552⟩, 357 ⟨85JA4260⟩, 508 ⟨85TL139⟩
Kulkarni, D. G., **1**, 801, 802 ⟨93SL875, 91T8155⟩, 802 ⟨91TL1107⟩
Kulkarni, S. A., **2**, 654 ⟨92SC1555⟩; **3**, 293 ⟨90SC3167⟩
Kulkarni, S. B., **1**, 422 ⟨68T545⟩

Kulkarni, S. U., **1**, 487, 488 ⟨88JOC239⟩, 509 ⟨82JOM(225)63⟩, 531 ⟨82JOM(225)C1⟩, 531, 662 ⟨82JOC171⟩; **2**, 10, 658 ⟨83S249⟩, 516 ⟨80S151⟩, 522, 526 ⟨82JOC863⟩, 901 ⟨82JOC3808, 88JOC239⟩, 5266 ⟨80JOC4540⟩; **3**, 13, 82 ⟨79S704⟩, 135 ⟨82JOC3808⟩; **4**, 815 ⟨82JOC3808, 82S193⟩, 815 ⟨82S195⟩
Kulkarni, S. V., **2**, 522, 526 ⟨80JOC4542⟩
Kulkarni, Y. D., **5**, 341 ⟨86IJC(A)57⟩, 734 ⟨86IJC57⟩; **6**, 546 ⟨82JIC882⟩
Kulkarni, Y. S., **1**, 7 ⟨87JOC307⟩, 176 ⟨85TL5675⟩, 805 ⟨88JOC1848⟩; **2**, 22 ⟨87JOC307⟩
Kulkowit, S., **3**, 206 ⟨78CC1069⟩; **4**, 90 ⟨80CC506⟩
Kullman, W., **5**, 262 ⟨B-89MI 506-05⟩
Kullmer, H., **2**, 788 ⟨71CB440⟩
Kullnig, R. K., **2**, 1006 ⟨93HOU(E/19d)713⟩; **3**, 88 ⟨87JOC3905⟩; **5**, 457 ⟨87JA5844⟩
Kul'nevich, V. G., **6**, 627 ⟨84CCC2285⟩
Kulomzina-Pletneva, S. D., **1**, 516 ⟨79T77⟩
Kulter, J., **4**, 550 ⟨82CB1947⟩
Kulumbetova, K. Zh., **4**, 564 ⟨71IZV2747⟩
Kulunbetova, K. Zh., **4**, 560, 564 ⟨70IZV1326⟩
Kumada, M., **1**, 160 ⟨83JOC2120⟩, 160 ⟨84TL1905⟩, 160 ⟨84TL1909⟩, 160 ⟨84TL1913⟩, 279 ⟨80PAC669⟩, 279, 466 ⟨84JA158⟩, 450 ⟨81JOM(217)43⟩, 451 ⟨75JA5603, 73JOM(50)C12⟩, 466, 470 ⟨76BCJ1958⟩, 466, 473 ⟨78OS(58)127⟩, 467 ⟨76CL1237⟩, 471 ⟨72JA9268⟩, 471 ⟨75JA4405⟩, 472 ⟨77TL4089, 80TL845⟩, 477 ⟨81CC313⟩, 479 ⟨81S1001⟩, 480 ⟨80TL3915⟩, 482 ⟨81TL4449⟩, 490 ⟨79TL1137⟩, 490, 515 ⟨79TL1141⟩, 490, 535 ⟨82OM542⟩, 1097 ⟨82JA180⟩, 1106 ⟨78TL2161⟩; **2**, 4 ⟨78JA290, 78TL1809⟩, 4 ⟨82OM369, B-82MI 223-01⟩, 437 ⟨83JOC2195⟩, 537 ⟨66TL5493⟩, 630 ⟨87JOC1100⟩, 915 ⟨74JOM(76)105, 74JOM(76)95⟩, 915 ⟨75JOM(86)C27, 76JOM(114)C19⟩, 923, 924 ⟨82JOM(226)C9⟩, 931 ⟨84TL83⟩; **3**, 18 ⟨84TL321⟩, 145 ⟨76CL1239⟩; **4**, 140 ⟨71JOM(30)349⟩, 395 ⟨67CB1087⟩, 607, 654 ⟨67JOM(10)111⟩, 613, 614 ⟨78JOM(149)37, 58JOC292⟩, 614 ⟨66JOM(5)226⟩, 615 ⟨83JOM(250)109, 83OM351⟩, 616 ⟨68CC930, 69CC207⟩, 651, 655 ⟨84TL1913⟩, 654 ⟨77IJ265, 84TL1905⟩, 1032, 1033 ⟨74TL4405, 76JA3718⟩, 1033 ⟨80BCJ1138, 82JA180⟩, 1033 ⟨86JOC3772⟩, 1047 ⟨78JOM(161)299⟩, 1048 ⟨79JOM(178)157, 80JOM(194)147⟩, 1052 ⟨81AOC51, 83JOM(254)13⟩, 1053 ⟨77JOM(142)C45, 82OM1473⟩; **5**, 378 ⟨66BCJ2263⟩; **6**, 178 ⟨81JA2324⟩, 178, 383 ⟨79JA1348⟩, 195 ⟨66JOM(6)451, 82JOM(236)7⟩
Kumadaki, I., **1**, 1079 ⟨82TL343⟩; **2**, 20 ⟨73CPB867⟩, 631 ⟨77JCR(S)215⟩; **6**, 9 ⟨88JOC3637⟩, 14 ⟨80JCS(P1)2755⟩, 22, 246 ⟨79TL4071⟩, 232 ⟨79TL3865, 69TL4095⟩, 246 ⟨70CPB2334⟩
Kumadaki, J., **4**, 549 ⟨80JOC4683⟩
Kumagai, M., **1**, 651 ⟨74TL4005⟩; **2**, 915 ⟨74JOM(66)C14⟩; **3**, 640 ⟨74TL4005⟩
Kumagai, T., **3**, 66 ⟨82TL873⟩; **5**, 223 ⟨85JCS(P1)2361⟩, 329 ⟨86JOC2391⟩, 329, 330 ⟨85CC1418⟩; **6**, 116 ⟨86JAP61158986, 86JOC4737⟩, 116 ⟨87NKK1447⟩
Kumagami, H., **3**, 696 ⟨90T6351⟩
Kumagawa, T., **3**, 36 ⟨85BCJ2849⟩, 209 ⟨86TL2471⟩
Kumai, S., **4**, 767 ⟨80BP2029827⟩; **6**, 29 ⟨92JAP04149143⟩, 32 ⟨91EUP450584⟩
Kumamoto, T., **1**, 178 ⟨86BCJ3097⟩; **2**, 688 ⟨70TL5115⟩, 709 ⟨68BCJ2742⟩; **3**, 541 ⟨64JOC2242⟩; **4**, 226 ⟨81BCJ817⟩; **5**, 898 ⟨68TL3787, 70BCJ2870⟩, 1125 ⟨86MI 530-02⟩
Kumamoto, Y., **1**, 983 ⟨86T1989⟩; **3**, 351 ⟨87TL4833⟩
Kuman, S. S., **5**, 215 ⟨87IJC(B)620⟩
Kumanotani, J., **3**, 191 ⟨81CL1677⟩, 191 ⟨82CL83⟩
Kumar, A., **3**, 17, 132 ⟨84JOC4011⟩, 413 ⟨85JOC1859⟩; **4**, 37 ⟨91JOC1663⟩, 899, 900 ⟨76CC592, 76CC593⟩, 900 ⟨80S748⟩; **6**, 650, 655 ⟨80MI 621-01⟩
Kumar, B., **6**, 546 ⟨82JIC882⟩
Kumar, B. V., **6**, 561 ⟨85IJC(B)1298⟩
Kumar, C. V., **1**, 1106 ⟨79JOM(179)C7⟩
Kumar, D., **5**, 356 ⟨93JCR(S)244⟩
Kumar, G., **4**, 434 ⟨84H(22)2351⟩
Kumar, K., **1**, 248 ⟨83CB2378⟩; **3**, 347, 363 ⟨92TL5955⟩, 592 ⟨83CB2708⟩
Kumar, M. U., **1**, 802 ⟨91T8155⟩, 949 ⟨91TL45⟩, 950 ⟨90T2345, 90TL1675⟩
Kumar, P., **2**, 145 ⟨92CC84⟩; **4**, 186 ⟨93S1069⟩, 246 ⟨93S67⟩; **5**, 291 ⟨79S726⟩
Kumar, R., **2**, 145 ⟨92CC84⟩, 437 ⟨88JOM(342)263⟩
Kumar, R. C., **2**, 367 ⟨84JA7496⟩
Kumar, S., **4**, 437 ⟨88IJC(B)912⟩; **6**, 111 ⟨90IJC(B)180⟩
Kumaraswamy, G., **1**, 285 ⟨92CC1313⟩, 286 ⟨87TL2649⟩; **3**, 422 ⟨88TL4153⟩
Kumaravel, G., **2**, 34 ⟨84TL233⟩; **3**, 155 ⟨84TL233⟩; **5**, 321 ⟨84TL233⟩
Kumari, L. K., **3**, 293 ⟨83IJC(B)1246⟩
Kumari, N., **1**, 738 ⟨75JOM(96)237⟩
Kumashivo, I., **2**, 192 ⟨67JOC3032⟩
Kumata, K., **4**, 872 ⟨86BCJ2909⟩
Kumawat, K., **3**, 350, 364 ⟨77JCS(P1)1127⟩
Kumazawa, T., **1**, 617, 709 ⟨88JA4866⟩; **4**, 906 ⟨78CC362⟩
Kume, A., **5**, 403 ⟨84JOC2139⟩; **6**, 80 ⟨85S408⟩, 125 ⟨79TL4475, 81CC969⟩, 125 ⟨82BCJ224⟩
Kume, F., **4**, 230 ⟨87OS(65)215⟩
Kume, Y., **1**, 23 ⟨92TL6367⟩
Kumegawa, N., **1**, 992 ⟨91JA9604, 93BCJ987⟩; **2**, 940 ⟨93BCJ987⟩
Kumerth, D. C., **5**, 132 ⟨74JOC1968⟩
Kumiko, N., **3**, 357 ⟨91EUP462531⟩
Kumler, W. D., **6**, 598 ⟨56JA97⟩
Kumli, K., **2**, 443 ⟨59JA248⟩
Kumli, K. F., **2**, 436, 442, 443 ⟨64JA2378⟩
Kumli, K. W., **2**, 842 ⟨59JA3805⟩
Kümmel, R., **2**, 427 ⟨67CB3331⟩, 427, 430, 462 ⟨68CB3612⟩, 430 ⟨65AG172⟩; **4**, 453 ⟨68CB3612⟩, 453, 454, 477 ⟨68CB3619⟩; **5**, 396 ⟨67ZN(B)784⟩
Kummer, D., **2**, 473, 474 ⟨70ZAAC(372)171⟩; **4**, 602 ⟨65MI 414-01⟩; **6**, 268 ⟨90CB945⟩, 383 ⟨65MI 613-01⟩, 656 ⟨88JOM(339)241⟩
Kümmerlen, J., **2**, 1088 ⟨92CB1597⟩
Kummert, K., **4**, 259, 855 ⟨90S271⟩
Kumobayashi, H., **1**, 77 ⟨87JA1596, 87JA4129⟩; **2**, 53 ⟨93JA3318⟩; **3**, 127 ⟨92TL635⟩, 158 ⟨93JOC2929⟩
Kumon, N., **3**, 383 ⟨89JA5949⟩; **4**, 376 ⟨89JA5949⟩; **5**, 1115 ⟨89JA5949⟩
Kunai, A., **2**, 786 ⟨83TL4443⟩; **3**, 430 ⟨90BCJ1843⟩; **4**, 614 ⟨93OM2536⟩
Kunakova, R. V., **3**, 336 ⟨91TL6973⟩
Kunamoto, T., **3**, 227 ⟨81CL1341⟩
Kunath, A., **5**, 161 ⟨93AG(E)1191⟩
Kunau, W.-H., **5**, 97 ⟨76AG(E)61⟩
Kunczik, T., **1**, 212 ⟨83LA226⟩
Kunda, S. A., **2**, 616 ⟨85TL279⟩; **5**, 51 ⟨84JOC174⟩
Kunda, U. S., **5**, 51 ⟨84JOC174⟩
Kündgen, U., **5**, 889 ⟨85CB1352⟩; **6**, 689 ⟨82AG(E)219⟩
Kundhu, N. G., **3**, 273 ⟨90S243⟩
Kündig, E. P., **1**, 25 ⟨88JOC1823, 88T6855⟩; **2**, 558, 560 ⟨81HCA2606, 82TL3901⟩; **4**, 274, 275 ⟨88T6855⟩
Kundiger, D., **4**, 826 ⟨43OS(23)45⟩, 887, 889 ⟨40JA1281⟩
Kundiger, D. G., **4**, 743 ⟨58JA5988⟩, 745 ⟨56JA6101⟩
Kundler, S., **4**, 665 ⟨93CB1361⟩
Kundu, K., **6**, 393 ⟨94JOM(480)199⟩
Kundu, N. G., **4**, 103 ⟨85S323⟩
Kundu, S. K., **4**, 792 ⟨70JOC4250⟩
Kuneda, N., **2**, 187 ⟨68BCJ3015⟩, 187 ⟨B-77MI 203-03⟩
Kunert, D. M., **3**, 543 ⟨78TL929⟩
Kunert, F., **6**, 73 ⟨60LA(632)38⟩
Kunerth, D. C., **2**, 684 ⟨76JOC732⟩
Kunesch, N., **4**, 176 ⟨87TL3569⟩
Kung, F. E., **2**, 100 ⟨47JA2158⟩
Kunieda, N., **1**, 779 ⟨75TL2841⟩; **3**, 651 ⟨86TL5109⟩; **4**, 264 ⟨76BCJ256⟩
Kunieda, T., **1**, 134 ⟨94TL721⟩; **5**, 226 ⟨81TL1257⟩

Kunish, F., **2**, 698 ⟨86TL4721⟩
Kunishi, T., **5**, 766 ⟨83CL885⟩
Kunishige, M., **2**, 986, 988 ⟨89CL761⟩
Kunishima, M., **1**, 431 ⟨88T4095⟩, 1016, 1145 ⟨90CC118⟩; **2**, 1020 ⟨90CC118⟩, 1020, 1063 ⟨85TL4501⟩, 1062 ⟨87CC1708⟩; **3**, 162 ⟨89JOC4038⟩; **4**, 808 ⟨88CC1076⟩
Kunita, K., **3**, 21, 141 ⟨88JOC4489⟩
Kunitomi, H., **1**, 976 ⟨91SL697⟩
Kunitskaya, G. P., **5**, 349 ⟨77ZOR2055⟩
Kunitskaya, L. A., **2**, 890 ⟨75JGU1476⟩
Kuniyasu, H., **1**, 1073 ⟨91CL1805⟩; **2**, 710 ⟨92JA5902⟩; **4**, 940 ⟨92JA5902⟩, 940, 947 ⟨91JA9796⟩; **5**, 238, 241 ⟨91JA9796⟩
Kunkel, G., **2**, 310 ⟨90LA795⟩
Kunkel, W., **2**, 662 ⟨68LA(717)54⟩
Kunkely, H., **4**, 686 ⟨82CB878⟩
Kunng, F.-A., **1**, 835 ⟨83JOC4262, 84JOC2708⟩
Kuno, H., **1**, 456 ⟨92CL1725⟩; **2**, 55 ⟨88CL1633⟩; **3**, 7 ⟨91BCJ312⟩
Kunô, K., **4**, 156 ⟨77BCJ1353⟩
Kuno, S., **2**, 759 ⟨81JOM(218)309⟩
Kuno, Y., **1**, 327 ⟨94CC987⟩
Kunshenko, B. V., **6**, 37 ⟨83JFC(22)105⟩
Kunstmann, P., **2**, 773 ⟨70JA1587⟩
Kunstmann, R., **3**, 600 ⟨69CB1816⟩; **5**, 1058 ⟨69AG(E)203⟩
Kuntsevich, A. D., **5**, 708 ⟨61TL807⟩
Kuntsevich, H. D., **5**, 708 ⟨61JGU473⟩
Kuntz, B. A., **2**, 76 ⟨89JOC4154⟩
Kunugi, A., **4**, 774, 779 ⟨93MI 417-01⟩
Kunugita, E., **3**, 20 ⟨89TL6371⟩
Kunungi, A., **4**, 856 ⟨86MI 419-01⟩
Kunz, D., **1**, 24 ⟨66CB1393⟩; **5**, 515, 516 ⟨66CB1393⟩
Kunz, H., **2**, 203 ⟨91SL163⟩, 595 ⟨91AG(E)694⟩; **4**, 176, 180 ⟨91COS(6)631⟩; **5**, 1042 ⟨90AG(E)1050⟩
Kunz, M., **2**, 1094 ⟨B-87MI 223-02, B-87MI 223-03⟩, 1094 ⟨B-87MI 223-04⟩
Kunz, R. A., **4**, 64, 71 ⟨74JOC2648⟩
Kunz, R., **3**, 716 ⟨94AG(E)330⟩; **4**, 743, 970 ⟨66CB2900⟩, 979 ⟨65CB1391⟩
Kunz, T., **5**, 164 ⟨88AG(E)268⟩
Kunz, W., **2**, 444 ⟨87PS(33)41⟩; **4**, 105 ⟨87PS(33)41⟩
Kunz, Y. K., **5**, 824 ⟨76JOC252⟩
Kunze, H., **6**, 563 ⟨84MI 617-03⟩, 687 ⟨82CB2371⟩
Kunze, O., **5**, 513, 559 ⟨85S960⟩, 557, 559 ⟨89LA321⟩
Kunze, U., **2**, 437 ⟨87PS(31)33⟩, 440 ⟨87PS(29)373⟩; **3**, 549 ⟨82CB3653⟩; **5**, 546 ⟨69CB3347⟩, 566, 595, 610 ⟨87PS(31)33⟩; **6**, 581 ⟨86ZN(B)1142, 87ZN(B)77⟩, 581 ⟨87ZN(B)860⟩, 584 ⟨85CB227⟩
Kunzek, H., **5**, 397 ⟨73JOM(49)149⟩
Künzel, O., **2**, 1096 ⟨62CB2155⟩
Künzel, W., **5**, 397 ⟨52CB239⟩
Künzer, H., **1**, 24, 144 ⟨91TL1949⟩, 555 ⟨89T6409⟩
Kunzer, H., **3**, 613, 627, 698 ⟨83T961⟩
Kunzmann, E., **1**, 204, 1023 ⟨78S26⟩
Kuo, C. H., **1**, 79 ⟨83JOC1991⟩; **5**, 173 ⟨68T2443⟩
Kuo, D. L., **1**, 195 ⟨90TL1265⟩, 815 ⟨88CC1281⟩
Kuo, E., **1**, 502 ⟨87JOC3126⟩
Kuo, E. Y., **1**, 962 ⟨89TL5747⟩, 1038 ⟨91JA694⟩
Kuo, F., **3**, 197 ⟨86SC1745⟩
Kuo, G. F.-H., **5**, 951 ⟨87OM988⟩
Kuo, G.-H., **3**, 519 ⟨85JOC5898⟩, 520 ⟨82JA6119⟩
Kuo, H.-L., **1**, 489 ⟨89JOC868⟩; **2**, 630 ⟨89JOC868⟩
Kuo, M., **4**, 318 ⟨85CL939⟩
Kuo, S.-C., **1**, 1130 ⟨86JA1106⟩
Kuo, S. J., **1**, 473 ⟨80JA790⟩
Kuo, Y.-C., **3**, 531, 532 ⟨82CPB526⟩; **5**, 867 ⟨82CPB899⟩
Kuo, Y.-H., **3**, 62 ⟨86TL2757⟩
Kuo, Y.-N., **4**, 832 ⟨72JOM(46)73⟩
Kupchan, S. M., **1**, 431 ⟨60JOC2217⟩
Kupchik, E. J., **2**, 889 ⟨65IC973⟩, 889, 890 ⟨68JOM(11)627⟩; **5**, 389, 1121 ⟨76SRI345⟩, 390 ⟨74JOM(73)319, 75JOM(93)325⟩, 739 ⟨75JOM(97)39, 81JCED105⟩
Kupczyk-Subotkowska, L., **3**, 368 ⟨91JOC1424⟩
Kuper, D. G., **3**, 532 ⟨65CI(L)424⟩
Küper, S., **4**, 536 ⟨88AG(E)433⟩
Kupfer, O., **3**, 463 ⟨11CB2197⟩
Kupfer, R., **3**, 417 ⟨85CB3089⟩, 422 ⟨86CB857⟩; **5**, 694 ⟨85CB3089⟩; **6**, 625 ⟨86CB3236, 87CB1271⟩, 730 ⟨92CB2487⟩
Kupin, B. S., **2**, 1032 ⟨67ZOR2102⟩; **3**, 548 ⟨62ZOR1759, 71ZOR1120⟩; **5**, 589, 590, 603, 611 ⟨71ZOR1120⟩
Kuplennik, Z. I., **2**, 408 ⟨77ZOB56⟩
Kupper, R., **1**, 1122 ⟨74JOC1438, 79JOC3076⟩; **2**, 385 ⟨80JOC2919⟩; **4**, 335 ⟨79JOC2326⟩, 896 ⟨80JOC2919⟩
Kupper, R. J., **2**, 831 ⟨82JA6876⟩
Kuppers, E., **2**, 1019 ⟨04CB4415⟩
Kuppers, H., **4**, 381 ⟨64TL1503⟩
Küppers, H. J., **2**, 133 ⟨91CC475⟩
Kupprat, G., **2**, 474 ⟨86ZAAC(533)146⟩
Kuptsova, T. S., **4**, 660 ⟨77ZOB1751⟩
Kurafuji, T., **6**, 465 ⟨94EUP581240⟩
Kuraishi, T., **3**, 310 ⟨87T5281⟩
Kuramochi, T., **1**, 934 ⟨91CL1691, 91TL6559⟩, 934 ⟨93JCS(P1)2417⟩
Kuramoto, M., **1**, 102 ⟨69TL3101⟩
Kuramshin, É. M., **6**, 479 ⟨90ZOB1826⟩
Kuranashi, S., **6**, 533 ⟨90JAP03220173, 93JAP05058990⟩
Kuraoka, S., **4**, 519 ⟨84CL279⟩
Kurapov, P. B., **3**, 362 ⟨89MI 308-04⟩
Kurasawa, Y., **5**, 706 ⟨85H(23)281⟩
Kurasoshi, M., **4**, 810 ⟨86CL1895⟩
Kurata, H., **1**, 649 ⟨88T4339⟩, 990 ⟨92CC778⟩, 991 ⟨89CC1690⟩
Kurata, Y., **1**, 590, 693 ⟨72TL737⟩
Kurauchi, M., **6**, 627 ⟨79S554⟩
Kurbanov, D., **4**, 775 ⟨90CA230883w⟩; **6**, 88 ⟨88MI 603-01⟩, 478 ⟨87JGU365⟩
Kurbanov, I., **6**, 88 ⟨88MI 601-01⟩
Kure, N., **5**, 700 ⟨93ACS1004⟩
Kurebayashi, Y., **4**, 325 ⟨81CPB2496⟩
Kurek, A., **4**, 209 ⟨69T4257⟩
Kurek, J. T., **5**, 282 ⟨69JA5647⟩
Kurella, M. G., **6**, 662 ⟨82IZV2370⟩
Kurg, V. V., **4**, 771 ⟨91ZOB874⟩, 998 ⟨92ZOB1084⟩
Kurguzova, A. M., **4**, 342 ⟨71ZOB1964⟩
Kurhajec, G. A., **5**, 814 ⟨73JOC1065⟩; **6**, 279, 281 ⟨73JOC1065⟩
Kuri-Brena, F., **2**, 683 ⟨93H(36)13⟩
Kurihara, A., **1**, 936 ⟨93SL143⟩; **4**, 900 ⟨82CL1933⟩
Kurihara, T., **1**, 179 ⟨76TL2455⟩, 836 ⟨91COS(8)895, 93T7657⟩, 993 ⟨94TL1255⟩; **3**, 621 ⟨76TL2455⟩, 647 ⟨84TL427⟩, 659, 660 ⟨86CPB4620⟩, 670 ⟨86CPB4545⟩; **5**, 347 ⟨86CPB4545⟩, 1111, 1116 ⟨87TL4189⟩; **6**, 542 ⟨91CPB1659, 92T9433⟩, 542 ⟨92TL2543⟩, 544 ⟨92CPB2279⟩
Kuriki, T., **5**, 340 ⟨88CPB877⟩
Kurishima, K., **6**, 383 ⟨91JA7790⟩
Kurisu, M., **4**, 21 ⟨84JHC1157⟩
Kurita, A., **2**, 4 ⟨78JA290, 78TL1809⟩, 4 ⟨82OM369, B-82MI 223-01⟩
Kurita, H., **5**, 753 ⟨77JPS2163⟩
Kurita, J., **1**, 561 ⟨91H(32)2089⟩; **4**, 808 ⟨93CC1309, 93CC1817⟩
Kurita, K., **3**, 557 ⟨77MI 317-01⟩, 696 ⟨70OS(50)195⟩; **5**, 972 ⟨80OS(59)195⟩; **6**, 230 ⟨80OS(59)195⟩, 415, 428 ⟨79OS(59)195⟩, 462 ⟨76JOC2070⟩
Kurita, M., **1**, 422 ⟨68TL3635⟩
Kurita, Y., **3**, 427 ⟨83CL1537⟩
Kuriu, Y., **5**, 205 ⟨86JFC(31)37⟩; **6**, 12 ⟨86JFC(31)37⟩
Kuriyama, N., **2**, 304 ⟨90S735⟩
Kuriyama, T., **3**, 628 ⟨90CL1615⟩
Kurkin, A. N.,**3**, 195 ⟨83JGU610⟩
Kurkin, A. N., **3**, 195 ⟨83JGU656⟩
Kurkjy, R. P., **6**, 421, 426, 460 ⟨64CRV645⟩, 425 ⟨65CI(L)791⟩

Kurkovskaya, L. N., **5**, 516, 521 ⟨75JOU1077⟩
Kurn, N., **5**, 90 ⟨73JOC457⟩
Kurobe, H., **4**, 154 ⟨85JCS(P1)1185⟩, 387 ⟨83TL4257⟩; **5**, 199 ⟨89GSM89⟩
Kuroboshi, M., **1**, 574 ⟨91TL1215⟩, 1130 ⟨86CL1895⟩; **2**, 18 ⟨91SL185, 91TL1215⟩, 77 ⟨91TL1069⟩, 90 ⟨92TL4173⟩; **4**, 737 ⟨88JFC(41)435⟩; **6**, 10 ⟨92CL827⟩, 229, 230 ⟨92TL4173⟩, 251 ⟨94SL251⟩
Kuroda, A., **1**, 797 ⟨84CL1541⟩; **5**, 199 ⟨86TL4609⟩
Kuroda, C., **1**, 813 ⟨85BCJ146⟩
Kuroda, H., **1**, 1079 ⟨82TL343⟩; **3**, 88, 99 ⟨90BCJ2433⟩; **5**, 189, 196 ⟨87CPB1262⟩
Kuroda, K., **1**, 1064 ⟨80TL4021⟩; **2**, 27 ⟨93TL5725⟩, 624 ⟨80TL4021⟩; **3**, 66 ⟨79TL1607⟩; **5**, 185 ⟨78S928⟩; **6**, 488 ⟨92JAP05339226⟩
Kuroda, S., **1**, 404, 406, 816, 819 ⟨88TL4763⟩, 404, 406, 819 ⟨92BCJ1841⟩, 816 ⟨87TL803, 91TL5137⟩; **3**, 64 ⟨78CL41⟩
Kuroda, T., **1**, 677, 1099 ⟨83TL5281⟩, 677, 1099 ⟨86JA6048⟩, 1038, 1140 ⟨85TL5585⟩; **3**, 59 ⟨85TL5585⟩; **5**, 134 ⟨80TL2857⟩
Kuroda, Y., **3**, 66 ⟨81JOC3175⟩; **6**, 300 ⟨72JOC4198⟩
Kuroiwa, K., **6**, 434 ⟨73BCJ1269⟩
Kurokawa, H., **3**, 184 ⟨86TL5759⟩
Kurokawa, M., **2**, 205 ⟨82CPB332⟩
Kurokawa, N., **3**, 40 ⟨84TL1071⟩
Kurokawa, T., **1**, 836 ⟨91CPB1659⟩; **6**, 542 ⟨91CPB1659⟩
Kurokawa, Y., **5**, 332 ⟨89JOC5945⟩
Kuroki, K., **2**, 363 ⟨82CPB1646⟩
Kuroki, T., **5**, 223 ⟨87CPB1741⟩
Kurono, M., **3**, 151 ⟨77CL995⟩
Kurono, Y., **4**, 506 ⟨92JOC6711⟩
Kurosawa, H., **1**, 69 ⟨81TL4495⟩, 186, 225, 229 ⟨89JA7938⟩, 248 ⟨87JOM(334)243⟩; **2**, 394 ⟨82TL2965⟩, 598 ⟨70JOM(25)353, 82OM440⟩; **4**, 659 ⟨74JOM(70)C21, 78BCJ1397⟩; **5**, 1143 ⟨69JAP47019769⟩, 1144 ⟨68JOM(12)241, 69JAP47016293⟩, 1144 ⟨70JOM(22)627⟩; **6**, 517 ⟨75IC2148⟩
Kurosawa, K., **2**, 22 ⟨87BCJ809⟩, 631 ⟨87BCJ4363⟩; **3**, 101 ⟨89BCJ545⟩, 709 ⟨82TL5335⟩; **4**, 660 ⟨90BCJ3036⟩
Kurotaki, A., **5**, 164 ⟨80S403⟩
Kuroyama, Y., **1**, 1010, 1012, 1137 ⟨80S425⟩
Kurozawa, H., **2**, 597 ⟨82COMC-I(1)725⟩
Kurozumi, S., **4**, 279 ⟨91MI 406-01⟩
Kurr, B. G., **1**, 693 ⟨78JA2140⟩
Kurras, E., **4**, 580 ⟨73AG(E)854, 75JOM(84)C13⟩, 580 ⟨78JOM(160)35⟩, 708 ⟨65JOM(4)114⟩
Kursanov, D. N., **1**, 13, 281 ⟨74S633⟩, 76 ⟨67T2235, 80CC748⟩; **4**, 706 ⟨81JOM(209)233⟩, 706, 1036 ⟨79JOM(168)C33⟩, 707 ⟨83JOM(244)C21⟩; **5**, 97 ⟨84JOM(267)309⟩
Kurtev, B., **5**, 62, 84 ⟨B-80MI 502-01⟩
Kurtev, K., **2**, 825 ⟨80JCS(D)55⟩; **4**, 797 ⟨77UKZ1000⟩
Kurth, M. J., **1**, 236 ⟨85JOC3846⟩, 590, 591 ⟨80JOC3549⟩, 821 ⟨91TL335⟩, 831 ⟨93JOC4589⟩, 835 ⟨85JA443, 85JOC5769⟩, 835 ⟨86JOC1377, 87TL1031⟩, 835 ⟨88S362⟩; **2**, 133 ⟨90JOC2286⟩, 393 ⟨93TL1259⟩, 701 ⟨85JOC3846⟩
Kurts, A. L., **2**, 7 ⟨89JOU587, 89JOU642⟩; **3**, 430 ⟨B-83MI 311-01⟩
Kurtz, A. N., **2**, 131 ⟨69JA5415⟩
Kurtz, D. W., **3**, 577 ⟨66CC689⟩
Kurtz, P., **3**, 622 ⟨51LA(572)49⟩
Kurtz, W., **2**, 689 ⟨76JOC4099⟩
Kurusu, Y., **1**, 37 ⟨84TL4417⟩, 195 ⟨88CC44⟩, 195 ⟨88JA4473⟩, 196 ⟨87BCJ1525⟩, 238, 246 ⟨89BCJ2913⟩; **2**, 73 ⟨89CL1647⟩, 74, 80 ⟨88TL3563, 88CC44⟩, 80 ⟨88JA4473⟩; **3**, 117, 161 ⟨84TL4417⟩, 162 ⟨92CL2039⟩
Kuryama, K., **6**, 317 ⟨93JAP(K)05062690⟩
Kuryla, W. C., **4**, 829 ⟨67OS(47)78⟩; **6**, 74, 75 ⟨64JOC2773⟩
Kurz, D., **2**, 853 ⟨68ZN1030⟩
Kurz, J., **2**, 258 ⟨71CB3740⟩
Kurz, J. L., **2**, 278 ⟨70JOC3086⟩
Kurz, K., **5**, 376 ⟨83JOC4111⟩
Kurz, M. E., **1**, 336, 346 ⟨83JOC1373⟩, 926 ⟨84JOC1603⟩; **2**, 650 ⟨71JOC3184⟩
Kurze, R., **4**, 561 ⟨84EGP214610⟩
Kurzer, F., **1**, 78 ⟨78T1261⟩; **2**, 26 ⟨78T1251⟩, 172 ⟨53JCS549⟩, 688 ⟨63OSC(4)937⟩; **5**, 344 ⟨49JCS3029⟩, 635 ⟨65AHC(5)123⟩, 640 ⟨86JCS(P1)1873⟩, 652 ⟨74CI(L)803⟩, 769 ⟨74S583⟩
Kurzer, F.,**5**, 769 ⟨76T1031⟩; **5**, 83 ⟨73OSC(5)1046⟩
Kurzer, F., **5**, 849 ⟨68JCS(C)1375⟩, 849, 1061 ⟨67CRV107⟩, 1120 ⟨63OSC(4)213⟩; **6**, 506 ⟨67CRV107⟩
Kürzinger, A., **2**, 64 ⟨88JOM(346)413⟩
Kurzmeier, H., **5**, 1029 ⟨90JOC3749⟩
Kusaba, T., **5**, 184 ⟨87BCJ4325⟩
Kusabayashi, S., **3**, 326 ⟨94JCS(P1)643⟩, 368 ⟨91JCS(P1)3043⟩
Kusaji, M., **3**, 680 ⟨86ACR342⟩
Kusakabe, M., **1**, 451 ⟨86CL1473⟩, 1107 ⟨85CC1636⟩; **4**, 1060 ⟨84CC1130⟩
Kusama, H., **5**, 427 ⟨91CL1413⟩
Kusama, Y., **3**, 169 ⟨90JOC595⟩
Kusamran, K., **3**, 154 ⟨79CL209⟩; **4**, 77 ⟨79CL209⟩
Kusano, K., **3**, 90 ⟨90TL6835⟩
Kusano, T., **2**, 106 ⟨91PS(55)195⟩, 107 ⟨88BCJ3359⟩
Kusano, Y., **2**, 7 ⟨86TL1607⟩
Kusayama, Y., **3**, 557 ⟨77MI 317-01⟩; **5**, 753 ⟨77JPS2163⟩
Kusche, A., **1**, 816, 817 ⟨90CB917⟩; **4**, 217 ⟨91TL467⟩
Kuschel, R., **3**, 380 ⟨91JA233⟩; **4**, 868 ⟨92CB557⟩
Kushin, A. V., **2**, 595 ⟨83ZOB2519⟩
Kushkova, N. P., **4**, 133 ⟨36JGU283⟩
Kushner, A. S., **1**, 904 ⟨65JA4017⟩
Kushner, S., **5**, 1125 ⟨48JOC924⟩
Kushnir, V. N., **2**, 444 ⟨75ZOB1228⟩; **4**, 570 ⟨77ZOB1715⟩
Kusogi, H., **3**, 140 ⟨80JA1333⟩
Kusomoto, T., **2**, 69, 564 ⟨84JOC3904⟩
Kussner, C. L., **3**, 357, 362 ⟨76JOC3784⟩
Küsters, W., **3**, 365 ⟨73JA2383⟩
Kusters, W., **3**, 365 ⟨74JA3502⟩
Küsthardt, N., **2**, 585 ⟨84AG(E)515⟩
Kusuda, S., **1**, 594, 668 ⟨92JOC3145⟩
Kusuma, K., **4**, 333 ⟨80S565⟩
Kusume, K., **1**, 469, 1005 ⟨89SC2199⟩; **3**, 634, 635 ⟨87CL887⟩
Kusumi, T., **2**, 655 ⟨81TL3451⟩
Kusumoto, K., **6**, 80 ⟨85JAP(K)60208983⟩
Kusumoto, S., **6**, 487 ⟨93CL1401⟩
Kusumoto, T., **1**, 158 ⟨85JA5301⟩, 158 ⟨90JA5244⟩, 202 ⟨82BCJ643⟩, 426, 427 ⟨85CL1491⟩, 1098, 1140 ⟨82CC1042⟩, 1098, 1140 ⟨84JOC3904⟩; **2**, 36 ⟨83S460⟩, 444 ⟨90JA5244⟩, 474 ⟨85JA3501⟩, 923, 940 ⟨90BCJ3103⟩, 939, 940 ⟨92BCJ1280⟩; **3**, 628 ⟨90CL1615⟩; **4**, 570, 575 ⟨90JA5244, 85JA5301⟩, 619 ⟨87TL1811⟩, 654, 655 ⟨92CL2047⟩, 809 ⟨90BCJ1947⟩, 1045 ⟨85CL1405, 87TL1807⟩; **5**, 286 ⟨82BCJ643⟩
Kusunose, S., **3**, 319 ⟨82CC1223⟩
Kusyakov, R. A., **5**, 354 ⟨81ZOR2153⟩
Kuta, G. S., **3**, 595 ⟨75CC34⟩, 602 ⟨78JOC501⟩
Kutateladze, A. G., **2**, 232 ⟨92MI 203-02⟩
Kutateladze, T. G., **5**, 1107 ⟨92JOC5270⟩
Kutchan, T. M., **1**, 402, 795, 801, 802 ⟨85OR(33)247⟩, 803, 804 ⟨80JOC5020, 80JA6351⟩
Kutepov, D. F., **6**, 500 ⟨57JGU2590, 57JGU2882⟩
Kutepow, N. V., **5**, 27 ⟨68AG(E)329⟩, 142 ⟨68AG352⟩
Kutin, S. V., **2**, 883 ⟨83JGU1832⟩
Kutner, A., **1**, 1076, 1118 ⟨51JA4199⟩; **2**, 801 ⟨51JA4199⟩; **5**, 185 ⟨87MIP8707604⟩
Kutnetsova, M. G., **6**, 444 ⟨80ZOB875⟩
Kutney, G. W., **4**, 70 ⟨81JOC4911⟩, 330 ⟨84CJC586⟩
Kutney, J. P., **1**, 928 ⟨90CJC1698⟩, 941 ⟨78HCA690⟩;

2, 683 ⟨93H(36)13⟩; **3**, 303 ⟨71T3323⟩; **5**, 199 ⟨89H(28)565⟩; **6**, 416 ⟨75SC47⟩
Kutschy, K., **5**, 1043 ⟨88ZC141⟩
Kutschy, P., **6**, 558 ⟨87CCC1764⟩, 565 ⟨90MI 617-01⟩, 566 ⟨92MI 617-03⟩
Kutter, E., **3**, 247 ⟨62LA(659)90⟩; **4**, 728 ⟨62AG251⟩, 852 ⟨65LA(684)37⟩, 901 ⟨67AG147⟩
Kutter, J., **2**, 449 ⟨82CB1947⟩
Kutyrev, A. A., **1**, 888 ⟨91RCR72⟩; **2**, 259 ⟨82URP891678⟩
Kutyrev, G. A., **2**, 259 ⟨82URP891678⟩; **4**, 923 ⟨83PS(16)345⟩; **6**, 583 ⟨93ZOB633⟩
Kutzelnigg, W., **1**, 1109 ⟨73CPL309⟩
Kuushenko, B. V., **6**, 19 ⟨B-89MI 601-02⟩
Kuvaldin, K. V., **4**, 766 ⟨89ZOB2145⟩
Kuwabara, T., **3**, 113 ⟨93TL1299⟩
Kuwae, A., **3**, 635 ⟨88CPB1698⟩
Kuwahara, K., **1**, 1136 ⟨75JOC1932⟩
Kuwajima, I., **1**, 86 ⟨89JA6257⟩, 122 ⟨87JA8056, 87TL337⟩, 133 ⟨82JA1025, 85ACR181⟩, 281 ⟨75JA3257⟩, 283 ⟨76TL1699⟩, 313 ⟨86TL4029⟩, 471 ⟨82JA6831⟩, 483 ⟨88JA3296⟩, 485, 1083 ⟨84JOC1230⟩, 590, 693 ⟨72TL737⟩, 600 ⟨72TL649⟩, 620 ⟨81JOC5246⟩, 876 ⟨78TL2079⟩, 881 ⟨76TL1817⟩, 963 ⟨80TL623⟩, 980 ⟨79JA2208⟩, 1038 ⟨76AG(E)498, 83T975⟩, 1127 ⟨86TL1355⟩; **2**, 69 ⟨87JA8056⟩, 263 ⟨81BCJ805⟩, 265 ⟨78BCJ2183⟩, 537 ⟨84JOC1140⟩, 702 ⟨83TL3341⟩, 921 ⟨77TL2001⟩, 962 ⟨70JA395⟩, 1091 ⟨77CC76⟩; **3**, 7, 120 ⟨82JOC837⟩, 10, 123 ⟨83JA5703⟩, 38, 175 ⟨81BCJ3100⟩, 38, 175 ⟨81BCJ3510⟩, 74 ⟨79TL3461, 81TL2183⟩, 144 ⟨82JA6831⟩, 147 ⟨81JOC5246⟩, 156 ⟨89TL3323⟩, 159 ⟨93TL6281⟩, 175 ⟨78TL1277⟩, 187 ⟨92JOC7009⟩, 220 ⟨89JA8277⟩, 223 ⟨80TL623⟩, 228 ⟨80TL3209⟩, 230 ⟨79JA2208⟩, 623, 640 ⟨72TL487⟩, 636 ⟨78S602⟩; **4**, 354 ⟨84BCJ827⟩, 400 ⟨80TL3209⟩, 834, 933 ⟨78CC478⟩, 932 ⟨77JA4181⟩, 1064 ⟨87TL1299⟩; **5**, 12, 96 ⟨87OS(66)87⟩, 190 ⟨64JOC1385⟩, 412, 422 ⟨76CL993⟩, 413 ⟨77JA5827, 78TL259⟩, 414 ⟨76TL1591, 77JA4181⟩, 414 ⟨78CC478⟩, 414, 425 ⟨78JOM(145)C1⟩, 420 ⟨78BCJ2391⟩, 423 ⟨84BCJ2167⟩; **6**, 125 ⟨76TL1591⟩
Kuwamura, T., **3**, 632 ⟨69BCJ1345⟩
Kuwata, S., **5**, 279 ⟨88CC419⟩
Kuwatani, Y., **1**, 992 ⟨89JA3761⟩
Kuwatsuka, S., **4**, 979 ⟨86MI 421-01⟩
Kuwayama, S., **4**, 270, 854, 859, 860 ⟨86SC233⟩, 285 ⟨85TL6205⟩
Kuylenstjevna, K. G., **5**, 831 ⟨1897LA(289)117⟩
Kuyper, J., **2**, 399 ⟨74JOM(74)289⟩
Kuyper, L. F., **1**, 859 ⟨77JA8262⟩
Kuze, S., **6**, 471 ⟨93EUP561363⟩
Kuzina, V. I., **2**, 887 ⟨86JGU287⟩; **5**, 390 ⟨80ZOB886⟩
Kuzma, P. C., **5**, 374 ⟨84JOC2015⟩
Kuz'menko, L. S., **4**, 800 ⟨76JOU782⟩
Kuzmenok, N. M., **3**, 274 ⟨81JGU1053⟩
Kuzmich, D., **3**, 310 ⟨86JOC2021⟩
Kuz'micheva, V. I., **5**, 863 ⟨72ZOB868⟩
Kuzmierkiewicz, W., **1**, 260, 263 ⟨91HCA1924⟩; **2**, 710 ⟨91HCA1931⟩; **4**, 308, 445 ⟨87JCS(P1)811⟩, 318 ⟨91HCA1936⟩, 318, 322 ⟨91HCA1924⟩, 392 ⟨87JOC844⟩
Kuz'mina, G. N., **6**, 565 ⟨86MI 617-03⟩
Kuz'mina, L. G., **4**, 583 ⟨85JOM(288)119⟩, 721 ⟨79IZV1528⟩; **5**, 385 ⟨88ZOB62⟩
Kuz'mina, L. V., **4**, 485 ⟨82JGU392⟩
Kuz'mina, N. A., **4**, 13 ⟨83S773⟩; **5**, 385 ⟨93S117⟩
Kuz'min, K. I., **2**, 876 ⟨82MI 217-01⟩, 879 ⟨88JGU1173⟩, 1076 ⟨88JGU2136⟩, 1078 ⟨68JGU1295, 70JGU1100⟩, 1079 ⟨69JGU1068, 69JGU1797⟩, 1079 ⟨87JGU2497⟩
Kuz'min, O. V., **4**, 140 ⟨69DOK(185)311⟩
Kuz'min, V. K., **2**, 886 ⟨73JGU1921⟩
Kuz'min, V. S., **6**, 284 ⟨92MC52⟩
Kuznetsov, A. L., **2**, 1093 ⟨81IZV1280⟩; **4**, 812 ⟨79JGU625⟩
Kuznetsov, E. V., **2**, 442 ⟨63ZOB2631, 67ZOB2522⟩, 1008 ⟨90AHC(50)157⟩; **6**, 366 ⟨81ZOB1035⟩, 512 ⟨73JGU1019⟩
Kuznetsov, O. M., **3**, 8 ⟨89S940⟩
Kuznetsova, A. I., **1**, 421 ⟨42IZV392⟩
Kuznetsova, L. D., **4**, 135 ⟨71JGU481⟩; **6**, 56, 282 ⟨76IZV489⟩, 251 ⟨74IZV335⟩
Kuznetsova, T. S., **1**, 500 ⟨91ZOR488⟩; **2**, 445 ⟨87ZOR2120⟩
Kuzovleva, R. G., **4**, 561 ⟨65ZOB354⟩
Kuzuhara, H., **5**, 344 ⟨85BCJ3395⟩
Kuzuna, S., **5**, 601 ⟨79CPB1352⟩
Kvasha, Z. N., **2**, 448 ⟨66CA-P2297, 68CA-P106877⟩
Kvasov, B. A., **5**, 385 ⟨72JOM(36)227⟩
Kvitko, I. Ya., **2**, 779 ⟨64ZOB2094⟩; **4**, 1017 ⟨86ZOR582⟩; **5**, 516, 521 ⟨75JOU1077⟩
Kvitko, S. M., **6**, 653 ⟨72ZOR332⟩
Kwa, T. L., **1**, 464 ⟨69T5771⟩
Kwag, C.-Y., **4**, 597 ⟨91MI 413-01, 91POL319⟩
Kwaitkowski, G. T., **4**, 576 ⟨68JA6816⟩
Kwak, K. H., **3**, 18, 133 ⟨90TL6677⟩, 18, 134 ⟨92SC2425⟩; **5**, 368 ⟨92TL3165⟩
Kwan, M. H., **2**, 229 ⟨75CJC455⟩
Kwan, M. M. Y., **1**, 353, 354 ⟨94T4519⟩
Kwart, H., **1**, 624, 779, 783 ⟨78JA3927⟩, 624, 783 ⟨81JA4650⟩, 781 ⟨81JA1232⟩, 786 ⟨68CRV415⟩, 834 ⟨74JOC1575, 77JOC172⟩, 834 ⟨81JA1757⟩, 836 ⟨73JOC3658⟩, 838 ⟨72JA2515, 73JA8678⟩; **2**, 707 ⟨66JOC410⟩
Kwart, L. D., **1**, 781 ⟨81JA1232⟩, 834 ⟨81JA1757⟩; **2**, 561 ⟨75CC242⟩; **3**, 62 ⟨88JA4735⟩
Kwasigroch, C. A., **1**, 1056, 1058 ⟨86JOC3870⟩, 1060 ⟨85JA7233⟩
Kwasnik, W., **6**, 217 ⟨53USP2658927⟩, 409, 410, 418, 419, 420 ⟨B-63MI 614-01⟩, 409, 419, 420 ⟨B-75MI 614-01⟩
Kwetkat, K., **5**, 467 ⟨90S1001⟩
Kwiatek, J., **2**, 478 ⟨54USP2882313⟩
Kwiatkowska, S., **6**, 650, 653 ⟨70CC806⟩, 651 ⟨70BAP569⟩
Kwiatkowski, G. T., **1**, 639, 644, 645, 646 ⟨66JA5654⟩, 643, 644 ⟨66JA5652⟩, 643, 644 ⟨66JA5653⟩, 643, 644 ⟨68JA6816⟩
Kwiatkowski, S., **3**, 684 ⟨86M1091⟩
Kwik, W. L., **6**, 242 ⟨93JOM(448)219⟩, 243 ⟨92JST(268)389, 93JCS(D)663⟩
Kwok, F. C., **2**, 636, 638 ⟨88ACR135⟩, 639 ⟨86TL3275, 87PAC1577⟩
Kwok, I. M. Y., **4**, 843 ⟨90S991⟩
Kwok, R., **5**, 291 ⟨68JA4706⟩
Kwon, C.-H., **5**, 344 ⟨85JOC4993⟩, 1120 ⟨87SC1677, 90JOC3403⟩
Kwon, H., **3**, 713 ⟨93CC1760⟩; **5**, 415 ⟨92JOM(437)C13⟩
Kwon, H. B., **3**, 243 ⟨90JOC3114⟩
Kwon, O. O., **2**, 41 ⟨94TL2⟩; **3**, 3, 114 ⟨93TL5113⟩, 13, 85 ⟨91TL6903⟩, 82 ⟨B-94MI 301-01⟩
Kwon, S., **3**, 635 ⟨78CPB784⟩
Kwon, T., **1**, 929 ⟨92JOC2399⟩
Kwon, Y. C., **1**, 406, 816, 817 ⟨85TL5013⟩, 816, 817 ⟨85TL5017⟩, 816, 817 ⟨85TL5021⟩; **3**, 622 ⟨82TL2077⟩, 623 ⟨81JOC229⟩
Kwong, H.-L., **2**, 48 ⟨90TL2990, 92JOC2768⟩
Kwong, K. S., **1**, 575, 576 ⟨90JA8433⟩, 808 ⟨88JOC196⟩
Kyandzhetsian, R. A., **5**, 900, 904 ⟨79ZOR2396⟩, 903, 904 ⟨74JGU1677⟩, 920 ⟨81JOU1676⟩
Kyaw, M., **4**, 229 ⟨65JCS1298⟩
Kyba, E. P., **1**, 157 ⟨76JA4805⟩; **2**, 431 ⟨83OM1877, 85IC1613⟩, 1071 ⟨81TL1875⟩; **5**, 789 ⟨83TL1679⟩
Kybart, C., **5**, 928, 929 ⟨92AG(E)1238, 92AG(E)1384⟩; **6**, 188 ⟨92AG(E)1384⟩, 201, 715 ⟨92AG(E)1238⟩
Kybett, A. P., **3**, 293 ⟨89CC1355⟩; **6**, 233, 234 ⟨90JFC(48)249⟩
Kyburz, R., **4**, 53, 54, 56 ⟨78HCA2047⟩, 54, 56 ⟨71HCA1037⟩

Kyi, Z.-Y., **3**, 452 ⟨52JCS1321⟩
Kyle, R. H., **2**, 420 ⟨48JA3945⟩
Kyler, K., **6**, 705 ⟨85TL555⟩
Kyler, K. S., **1**, 155 ⟨84OR(31)1⟩, 491, 1035 ⟨87JOC4296⟩; **2**, 92 ⟨88JA4818⟩; **3**, 134 ⟨83JOC4087⟩, 292 ⟨84OR47⟩, 620 ⟨84OR1⟩, 628 ⟨91JA6992⟩; **4**, 94 ⟨84JOC1084⟩, 220 ⟨88TL3773⟩, 365, 938 ⟨81JOC5182, 83JOC34⟩, 938 ⟨83JOC383⟩, 938, 939 ⟨83JA619⟩
Kyllingstad, V. L., **1**, 732 ⟨66JOC334⟩; **2**, 824, 834, 835 ⟨78JA7304⟩
Kynast, U., **6**, 196 ⟨87AG(E)681⟩
Kynaston, W., **4**, 58 ⟨57JCS618⟩
Kyogoku, Y., **1**, 925, 946 ⟨93TL1317⟩
Kyono, K., **5**, 356 ⟨79TL2253⟩
Kyrides, L. P., **5**, 15 ⟨37JA206⟩
Kyrtopoulos, S. A., **2**, 385 ⟨76CC877⟩
Kyte, C. T., **1**, 860 ⟨60JCS4454⟩
Kyung, S. H., **2**, 566 ⟨86T2931⟩
Kyushin, S., **1**, 922 ⟨88JPO197⟩
Kyutoku, H., **1**, 590 ⟨78TL2807⟩
Kyz'mina, L. G., **1**, 530 ⟨86TL1241⟩

L

Laabassi, M., **1**, 963 ⟨92BSF151⟩
Laakmann, K., **4**, 604, 647 ⟨87CB471⟩
Laakso, L. M., **1**, 217 ⟨89TL6963⟩
Laakso, T. M., **6**, 504 ⟨57JA5717⟩
Laali, K., **1**, 247 ⟨88JOM(348)199⟩; **2**, 907 ⟨90JOC46⟩
Laali, K. K., **6**, 153 ⟨93JOC4105⟩
Laane, J., **2**, 497 ⟨75JOM(91)295⟩; **4**, 621 ⟨77JA3273⟩
Laï, R., **4**, 580 ⟨86CC1208⟩
Laarhoven, W. H., **1**, 841 ⟨83JOC2202⟩
Laasch, P., **2**, 1009 ⟨56CB209⟩; **3**, 687 ⟨56CB209⟩
Laatsch, H., **1**, 275 ⟨92S461⟩
Labadie, J. W., **3**, 211, 288 ⟨83JOC4634⟩
Labadie, S. S., **1**, 1015 ⟨84JA6417⟩
Labahn, D., **2**, 880 ⟨91JOM(402)55⟩
Labar, D., **1**, 592, 593, 713, 716 ⟨82CC564⟩, 601, 716 ⟨78TL1141⟩, 712, 713 ⟨89H(28)1203, 78TL1145⟩, 716 ⟨90BSF681⟩; **2**, 281, 290 ⟨78TL3967⟩; **3**, 125 ⟨90BSF681⟩
Labaree, D., **1**, 835 ⟨89JOC2893, 90JA7682⟩
Labastida, V., **6**, 119 ⟨89JPO349⟩
Labaudinière, L.,**1**, 444 ⟨83PS(15)359⟩; **2**, 594 ⟨93TL2303⟩; **4**, 809 ⟨92JOC6903⟩
Labaudiniere, R., **2**, 727 ⟨86JOC875⟩; **3**, 132 ⟨81SC423⟩; **4**, 875 ⟨86JOC875⟩
Labaw, C. S., **6**, 251, 625 ⟨82JHC1205⟩, 632 ⟨87JHC275⟩
l'Abbé, G., **2**, 245, 250 ⟨81TL583⟩; **3**, 259 ⟨71JA981, 71JOC258⟩, 477 ⟨82TL1103⟩, 558 ⟨82TL2909⟩, 577 ⟨83CC748⟩; **5**, 1014 ⟨73JOC675⟩, 604 ⟨84BSB405⟩, 622 ⟨90BSB281, 91JCS(P1)607⟩, 622 ⟨93JCS(P1)1719⟩, 638 ⟨80JOC4366⟩, 962, 1022 ⟨87S525⟩, 1040 ⟨90JHC1059⟩, 1123 ⟨79TL49, 81JOC4478⟩; **6**, 573 ⟨92T7505⟩, 629 ⟨65T1537⟩, 652 ⟨81BSB63⟩
LaBel, N. A., **2**, 354 ⟨68OSC(6)670⟩
LaBelle, B. E., **1**, 530 ⟨85JOC5215⟩; **4**, 566, 568 ⟨81IC3200⟩
LaBerge, J. M., **4**, 295, 407, 412 ⟨73JOC3288⟩, 444 ⟨76JOC3063⟩
LaBerge, S., **4**, 995 ⟨70JMC1248⟩
Labhart, M. P., **1**, 618, 707, 708 ⟨83HCA1090⟩
Labiad, B., **4**, 246 ⟨89SC31⟩; **5**, 199 ⟨93SC419⟩; **6**, 82 ⟨89SC31⟩
Labidalle, S., **2**, 31 ⟨89BSF247⟩
Labinger, J. A., **1**, 515 ⟨91COS(8)667⟩; **2**, 4, 568, 969 ⟨76AG(E)333⟩, 569 ⟨82COMC-I(3)705⟩
Lablache-Combier, A., **5**, 572, 618 ⟨84JOC714⟩
Laborde, E., **1**, 14, 226, 227 ⟨87JOC4898⟩, 285 ⟨92TL6607⟩
Laboue, B., **1**, 1045 ⟨89TL3545⟩; **3**, 130 ⟨92TL4221⟩
Labounty, L., **3**, 311 ⟨92JA5888⟩
Labourer, J., **3**, 125 ⟨90BSF681⟩
Laboureur, J. L., **1**, 246 ⟨87TL1549⟩, 390 ⟨84TL2713⟩, 712, 713 ⟨89H(28)1203⟩, 716 ⟨90BSF681⟩; **2**, 287 ⟨88TL3265⟩
Labrie, F., **2**, 31 ⟨94SC733⟩
Lacadie, J. A., **2**, 232 ⟨72JOC333⟩, 358 ⟨76JPR1019⟩
la Cal, M. T. de, **3**, 577 ⟨89SC1039⟩
Lacerda, M. D. S., **3**, 241 ⟨79JHC821⟩
Lacey, E., **6**, 415 ⟨87MI 614-02⟩
Lacey, M. J., **3**, 263 ⟨70AJC841⟩
Lacey, R. N., **3**, 526 ⟨B-64MI 316-01⟩
Lach, B., **3**, 618 ⟨1884CB1571⟩
Lach, D., **1**, 982 ⟨73AG(E)566⟩
Lachaise, I., **6**, 7 ⟨92TL213⟩
Lachance, A., **1**, 881 ⟨70TL3903⟩
Lachance, P., **2**, 749 ⟨88TL2295⟩
Lachen, B., **1**, 340 ⟨89JA8309⟩
Lacher, B., **1**, 326 ⟨89JA8311⟩; **2**, 226 ⟨88T1153⟩, 627 ⟨87T4321⟩
Lacher, J. R., **1**, 6 ⟨56JPC1454⟩; **4**, 5 ⟨51JA824⟩, 31, 34 ⟨51JA711⟩, 32 ⟨48JA1283⟩, 36, 751 ⟨56JA59⟩, 757, 759, 820 ⟨58JOC1661⟩, 761, 763 ⟨56JA1685, 52JA4104⟩; **5**, 318 ⟨52JA2189⟩; **6**, 36 ⟨52JA2292⟩
Lachhein, S., **1**, 546 ⟨82AG(E)768⟩; **5**, 35 ⟨83S733⟩
Lachmann, J., **4**, 489 ⟨90CB423⟩, 562 ⟨91CB275⟩; **6**, 167, 371 ⟨92CB1333⟩
Lachmann, U., **4**, 556 ⟨75JPR840⟩
Lachowicz, A., **2**, 569, 570 ⟨77ZAAC(431)88⟩
Lackmann, R., **4**, 940, 1014 ⟨92CB2051⟩; **5**, 934 ⟨91AG(E)1658⟩, 958 ⟨92CB2051⟩
Lacombe, J. M., **4**, 176 ⟨88CAR(181)246⟩, 225 ⟨88TL4293⟩
Lacombe, S., **3**, 335 ⟨90JOC2596⟩, 369 ⟨91NJC533⟩, 582 ⟨89JCS(F2)741⟩
la Concepcion Foces-Foces, M. de, **4**, 408 ⟨88CC410⟩
LaCoste, W., **2**, 465 ⟨1885CB2109⟩
LaCount, R. B., **1**, 104 ⟨60T130⟩, 537 ⟨64JOC423⟩, 646 ⟨61JA417⟩
Lacour, J., **1**, 263 ⟨92JA3993⟩; **3**, 194 ⟨92JA741⟩; **4**, 443 ⟨93JA9347⟩
Lacour, M., **3**, 273 ⟨75BSF779⟩
Lacquin, R., **5**, 24 ⟨48MI 502-01, B-47MI 502-01⟩, 24 ⟨B-47MI 502-02⟩, 62 ⟨45MI 502-01, B-45MI 502-04⟩, 62 ⟨B-45MI 502-05⟩, 105 ⟨45MI 502-02, B-45MI 502-02⟩, 105 ⟨B-49MI 502-01⟩
Lacrampe, G., **2**, 815 ⟨83RTC65⟩
Lacroix, L., **1**, 582 ⟨91TL3739⟩
Ladbury, J. W., **5**, 28 ⟨58CRV403⟩
Ladd, J. A., **6**, 381 ⟨92JOM(427)9⟩
Ladd, M. F. C., **3**, 358 ⟨75CC756⟩
Ladduwahetty, T., **2**, 327 ⟨91SL29⟩
Ladeishchikova, E. V., **4**, 1055, 1058 ⟨80ZOB692⟩
Ladenberger, V., **4**, 756 ⟨67CB3893⟩
Ladenburg, A., **2**, 108 ⟨1867LA(143)118⟩
Ladhar, F., **4**, 425 ⟨86S643⟩
Ladika, M., **1**, 988 ⟨81JA6437, 87JA782⟩, 1070 ⟨81CC459⟩; **2**, 1094 ⟨81JA6437⟩; **3**, 273 ⟨81S29⟩
Ladjama, D., **3**, 54 ⟨79S507⟩
Ladlow, M., **1**, 345 ⟨84TL4317⟩; **5**, 574 ⟨86CC1756⟩
Ladner, W., **2**, 529 ⟨82CB2357⟩, 530 ⟨89LA883⟩; **5**, 150 ⟨83CB3413⟩
Ladner, W. E., **5**, 138 ⟨84JA7250⟩
Ladouceur, B., **6**, 77 ⟨49MI 603-1⟩
Ladouceur, G., **1**, 826 ⟨89JOC4278⟩
Ladoulis, S. J., **1**, 196 ⟨85JOM(285)C13⟩
Ladow, R. J., **6**, 204, 379 ⟨72CC1078⟩
Ladsooksathit, S., **2**, 619 ⟨81SC405⟩
Laduranty, J., **4**, 321 ⟨87JOM(335)283⟩
Ladurée, D., **5**, 550 ⟨77RTC118⟩, 558 ⟨77RTC254⟩
Ladyzhnikova, T. D., **6**, 149 ⟨86ZOR2618⟩, 149 ⟨87ZOR2624⟩, 149 ⟨88ZOR644⟩
Laemmle, J. T., **1**, 1039 ⟨75CRV521⟩
Lafargue, D., **4**, 637, 660 ⟨77JOM(137)157⟩
Lafferty, J. J., **5**, 92 ⟨83EUP66456⟩
Laffitte, J.-A., **1**, 619, 707, 949 ⟨86JA4568⟩; **2**, 437, 444 ⟨90TL6357⟩; **4**, 82, 83, 85 ⟨86JA4568⟩, 223 ⟨86JOC3428⟩
LaFlamme, P., **1**, 299 ⟨56JA5447⟩
LaForge, R. A., **5**, 25, 108 ⟨46JA1844⟩

Lafrance, R. J., **4**, 444 ⟨83CC721, 84CJC749⟩
la Fuente Blanco, J. A. de, **3**, 192 ⟨90JOC1349⟩
Lagain, D., **2**, 229 ⟨83T3839⟩; **3**, 343 ⟨80JOC2517⟩, 343, 351 ⟨79TL2677⟩
Lagally, H., **6**, 524, 525 ⟨40ZAAC(245)321, 43ZAAC(251)96⟩
Laganis, E. D., **1**, 862 ⟨80JA6633⟩; **2**, 916 ⟨85JOC3666⟩; **3**, 353 ⟨80JA6634⟩
Läge, M., **5**, 946, 957 ⟨94OM3510⟩, 958 ⟨94CB2475, 94OM3502⟩
Lage, N., **3**, 562 ⟨88BSF883, 91JCS(P1)3389⟩, 562 ⟨91PS(59)141⟩; **5**, 731 ⟨91JCS(P1)3389, 91PS(59)141⟩
Lagh, M., **4**, 698, 708 ⟨91CB1511⟩
Laghmari, M., **2**, 330 ⟨91CC1684⟩
Lago, J. M., **5**, 367 ⟨83SC289⟩
Lagodzinskaya, G. V., **4**, 795 ⟨89BAU635⟩
Lagow, R., **4**, 668 ⟨90JA2425⟩; **6**, 206 ⟨85JA5313⟩
Lagow, R. J., **2**, 1005 ⟨85JA5313⟩; **4**, 44 ⟨81JCS(P1)1321, 88JOC78⟩, 52 ⟨74JA7588⟩, 153, 154 ⟨78IC618⟩, 618 ⟨84CC1664⟩, 618, 668 ⟨88POL2023, 82JA2637⟩, 668 ⟨84IC3717, 90JA9415⟩; **6**, 21 ⟨79CC128⟩, 36 ⟨73JOC3617, 77CC259⟩, 36, 251 ⟨81JCS(P1)1321⟩, 37 ⟨75JOC3271⟩, 46, 252 ⟨77IC2974⟩, 62 ⟨78JA1722⟩, 192, 205, 398 ⟨84CC1664⟩, 192, 398 ⟨88POL2023⟩, 204, 205, 404 ⟨83JOM(249)1⟩, 205 ⟨73JA1343, 75CC302⟩, 205 ⟨79JA2214, 79JOC2311⟩, 205 ⟨82JA7345⟩, 206 ⟨81JA5951, 82JA2637⟩, 245 ⟨90JOM(390)C73⟩, 246 ⟨86JA4103, 86JOM(307)C58⟩, 297, 301 ⟨89JOC1990⟩
Lagowski, J. J., **6**, 243 ⟨59QR233⟩
Lagowski, J. M., **1**, 859 ⟨B-67MI 119-01⟩
Lagrange, G., **2**, 578 ⟨83JOM(253)65⟩
Lagrone, C. B., **6**, 340 ⟨86IC2472⟩
Lagu, B., **3**, 128 ⟨90TL4105⟩
Laguerre, M., **2**, 929 ⟨78TL57⟩, 938, 941 ⟨86T669⟩, 945 ⟨79JOC4275⟩
Laguna, A., **4**, 585 ⟨88JCS(D)341⟩, 586 ⟨83ICA63, 88SRI69⟩, 691 ⟨75CC451⟩; **5**, 1138 ⟨73SRI237⟩, 1141 ⟨73MI 530-01⟩
Laguna, M. A., **3**, 653 ⟨86TL2027⟩
Laguna, M., **4**, 586 ⟨83ICA63, 88SRI69⟩
Lagunas, M.-C., **4**, 586 ⟨91JCS(D)2579⟩
Lahav, M., **2**, 98 ⟨72IJ101⟩
Lahlil, K., **1**, 570 ⟨93JOC4685⟩
Lahm, G. P., **3**, 220 ⟨78JOC4650⟩
Lahoti, R. J., **3**, 283, 295, 296, 305 ⟨91S320⟩; **5**, 384 ⟨91SC1889⟩
Lahournère, J.-C., **2**, 600 ⟨71JOM(33)C4⟩, 601 ⟨71JOM(33)C7⟩
Lahousse, F., **1**, 592 ⟨81AG(E)585⟩; **4**, 71 ⟨81AG(E)585⟩, 92 ⟨85T4183⟩; **6**, 44 ⟨81AG(E)585⟩
la Hoz, A. de, **4**, 826 ⟨92SL893, 93SC1935⟩
Lahoz, F. J., **4**, 682 ⟨87JOM(319)C51⟩, 685 ⟨90JCS(D)2587⟩
Lai, B., **2**, 683 ⟨75JCS(P1)461⟩
Lai, C., **3**, 696 ⟨92JOC2018⟩
Lai, C. H., **1**, 2, 8 ⟨79JOC641⟩; **3**, 129 ⟨79JOC641⟩
Lai, E. K. Y., **5**, 316 ⟨91H(32)481⟩
Lai, H. K., **3**, 124 ⟨79TL4121⟩
Lai, J. T., **5**, 67 ⟨82USP4347379⟩
Lai, L.-L., **5**, 582, 611 ⟨93JOC4742⟩, 625 ⟨93S870⟩
Lai, R., **2**, 33 ⟨83JOM(243)175⟩; **5**, 918 ⟨85CC1151⟩
Lai, Y.-H., **1**, 11, 18, 19 ⟨80TL2637⟩, 233 ⟨88JOC4472⟩, 687 ⟨80OPP361⟩
Lai, Y. H., **2**, 558, 958 ⟨81S585⟩
Lai, Y-H., **2**, 944 ⟨85JA5670⟩
Laib, H., **3**, 500 ⟨53LA(560)57⟩
Laidig, K., **1**, 928 ⟨86TL1553⟩
Laifa, E. A., **5**, 478, 494, 582 ⟨87SUL153⟩
Laikhter, A. L., **4**, 995 ⟨91IZV1849⟩
Laimann, E., **2**, 399, 400 ⟨76CB2442⟩
Laine, R. M., **4**, 674 ⟨77JOM(124)29⟩
Laing, S. B., **3**, 87 ⟨68JCS(C)2915⟩
Laird, E. R., **1**, 320 ⟨90JCI458, 90JOC9⟩
Laird, T., **1**, 401 ⟨80JCS(P1)1458⟩, 407, 823 ⟨80JCS(P1)1436⟩, 408 ⟨80JCS(P1)1450⟩; **3**, 82 ⟨79COC(5)1105⟩
Laishev, V. Z., **4**, 874 ⟨81JOU578⟩, 877 ⟨79ZOR2596⟩; **5**, 627 ⟨81ZOR2064⟩
Laitinen, H. A., **2**, 126 ⟨46MI 203-01⟩
Laitinen, H., **1**, 103 ⟨42JA2365⟩
Lajis, N. H., **4**, 199 ⟨78JOC3964⟩; **5**, 141 ⟨78JOC371⟩
Lajoie, G., **5**, 219, 574 ⟨85CJC3089, 83TL3815⟩, 574 ⟨86JA182⟩
Lajunen, M., **1**, 905 ⟨85ACR254⟩
Lake, C. H., **6**, 343 ⟨91OM2384⟩
Lake, D. E., **6**, 215 ⟨51USP2553518⟩
Lakhanpal, S., **5**, 710, 714 ⟨92CCC540⟩, 710, 714 ⟨93CCC2150⟩
Lakhrissi, M., **3**, 495 ⟨91SL583⟩
Lakhtin, V. G., **4**, 1045 ⟨86ZOB2743⟩
Lakhvich, F. A., **3**, 239 ⟨85S784⟩, 247 ⟨75JGU2572⟩
Lakin, D., **4**, 199, 218 ⟨87JOC782⟩
Lakouraj, M. M., **3**, 4 ⟨91SC2077⟩
Lakritz, J., **2**, 376 ⟨58JOC1595⟩; **3**, 407 ⟨B-63MI 310-01⟩
Lakshmaiah, G., **1**, 285 ⟨92CC1313⟩, 370 ⟨93TL4861⟩
Lakshmanan, S., **6**, 27 ⟨80CJC485⟩
Lakshmi Kantam, M., **4**, 185 ⟨93SL429⟩
Lakshmikantham, M. V., **2**, 290 ⟨77JA1470⟩; **3**, 339 ⟨88JOC3529⟩, 349, 358 ⟨87JOC169, 86JOC411⟩, 358 ⟨87JOC1874⟩, 365 ⟨85JOC1550⟩, 371, 372 ⟨84CHEC(4)1037, 75ACR139⟩; **4**, 847, 848, 870 ⟨80H(14)271⟩, 850, 869 ⟨78ANY355⟩, 855 ⟨78JOC82⟩, 869 ⟨75CC867⟩, 877 ⟨81TL4199⟩; **5**, 336 ⟨87JOC169⟩, 528 ⟨84JA6084⟩, 528, 529 ⟨89JOC4746⟩, 561 ⟨79SC17⟩, 1113 ⟨80JOC2632⟩, 1117 ⟨77JA1470⟩; **6**, 101 ⟨80H(14)271⟩
Lakshminarayana, G., **1**, 1019 ⟨90TL7227⟩
Lakshmi Santhi, P., **4**, 185 ⟨93SL429⟩
Lal, A., **6**, 502 ⟨78BCJ1245⟩
Lal, B., **2**, 8 ⟨73TL3937⟩; **3**, 87, 99 ⟨84TL2901⟩; **5**, 294 ⟨74TL2633⟩
Lal, G. S., **1**, 147 ⟨89TL1613⟩, 147 ⟨90TL1653⟩; **2**, 15 ⟨92CC595⟩, 15, 17 ⟨93JOC2791⟩, 251 ⟨90TL1653⟩, 941, 1023, 1024 ⟨86JA7127⟩, 1023 ⟨87TL2463⟩; **3**, 149 ⟨93JOC2791⟩, 171 ⟨90OS(69)158⟩, 247, 539 ⟨88JA3693⟩
Lal, S. G., **2**, 150 ⟨88JOC5004⟩
LaLancette, E. A., **1**, 863 ⟨65JA1941⟩
Lalancette, J. M., **1**, 881 ⟨70TL3903⟩; **2**, 226 ⟨71CJC4047⟩, 227 ⟨73TL1401⟩, 533 ⟨64CJC2903⟩; **3**, 620 ⟨71CJC2990⟩
Lalande, J., **1**, 165 ⟨85CC83⟩
Lalande, R., **1**, 925 ⟨71BSF461⟩
Lalezari, I., **1**, 1077 ⟨70AG(E)464, 71JOC2836⟩, 1077 ⟨73JHC655, 75JHC801⟩, 1077 ⟨77JHC567⟩; **3**, 553 ⟨73JOC338⟩; **4**, 333 ⟨84CHEC(6)346⟩, 852 ⟨73JHC11⟩, 874 ⟨73JOC338, 77S765⟩, 874 ⟨79JHC1405⟩; **6**, 653 ⟨70JHC689⟩
Lalima, Jr., N. J., **1**, 278 ⟨78JOC1279⟩
Lalitha, S., **3**, 530 ⟨91JOC7048⟩
Lallemand, J. Y., **1**, 214 ⟨75TL585⟩, 223, 224 ⟨87T2701⟩, 763 ⟨81TL1595⟩, 810 ⟨93JOC2349⟩; **4**, 238, 241 ⟨88TL2179⟩
Lally, D. A., **3**, 157 ⟨86TL1933⟩
Lalonde, J. J., **2**, 395 ⟨87JOC1601⟩
Lalonde, M., **1**, 654 ⟨85S817⟩; **2**, 539 ⟨85S817⟩
Lalor, F. J., **5**, 1097 ⟨83CC457, 84CC75⟩
Laltes, A., **2**, 306 ⟨83T703⟩
Lalu, J. P., **5**, 313 ⟨66JCS(C)1514⟩
Lam, B., **1**, 906 ⟨78JA3847⟩
Lam, C. H., **1**, 22, 23, 24, 436 ⟨89JOC4474⟩
Lam, C. T., **5**, 652 ⟨77IS95⟩, 950, 951 ⟨74JOM(70)273⟩

Lam, H. Y. P., **2**, 656 ⟨79CJC3304⟩
Lam, J. N., **3**, 29 ⟨91JOC2143⟩, 158 ⟨91JOC6917, 579 ⟨88JCS(P2)1071⟩; **4**, 322 ⟨89CJC1144⟩, 323 ⟨87JCS(P1)775⟩; **5**, 596 ⟨91JOC2143⟩; **6**, 143 ⟨90S666⟩, 170 ⟨91JOC2143⟩
Lam, L. K. M., **1**, 104 ⟨70JA2542⟩, 1114 ⟨70JA2538⟩
Lam, W. Y., **6**, 232, 602, 615 ⟨81JFC(18)441⟩, 256 ⟨82JA4034⟩, 257 ⟨86JOC4466⟩
LaMaire, S. J., **1**, 536 ⟨87TL295⟩; **2**, 969 ⟨87TL295, 87TL3895⟩; **3**, 614 ⟨87TL295⟩
Lamal, D. M., **5**, 866 ⟨76JA1278⟩
Lamanec, T. R., **2**, 342, 346 ⟨88JOC1768⟩
Lamanna, W. M., **4**, 694 ⟨86JA2096, 87OM1583⟩
Lamanova, I. A., **2**, 809 ⟨67CA76112w⟩, 891 ⟨74JGU1506⟩
la Mare, P. B. D. de, **2**, 5 ⟨B-76MI 201-01⟩
la Mare, P. B. O. de, **1**, 837 ⟨B-63MI 118-04⟩
Lamartine, R., **2**, 619 ⟨74JOC1744⟩
LaMattina, J. L., **3**, 184 ⟨80S329⟩, 246, 620 ⟨83TL2059⟩; **5**, 746 ⟨77JOC1523⟩
Lamaty, F., **1**, 1015 ⟨90TL4393⟩; **2**, 980 ⟨90TL4393⟩, 988 ⟨89TL5105⟩
Lamaty, G., **2**, 1008 ⟨71BSF541⟩; **4**, 215 ⟨75T809⟩
Lamazouère, A. M., **4**, 854, 855 ⟨84T2951⟩; **5**, 593 ⟨67CR(C)99, 75BSF829⟩, 595 ⟨93PS(78)97⟩
Lambeck, R., **4**, 314 ⟨66LA(700)1⟩
Lambert, A., **2**, 117 ⟨47JCS1477⟩; **5**, 820 ⟨49JCS42⟩
Lambert, B. F., **1**, 97 ⟨73OSC(5)400⟩
Lambert, C., **1**, 205, 1024 ⟨84TL5423⟩, 571 ⟨85T3331⟩; **2**, 983 ⟨86JA2753⟩; **5**, 22 ⟨91T9053⟩, 132 ⟨85TL4439⟩, 577, 617 ⟨85T3331⟩, 595 ⟨91SC85⟩
Lambert, C. E., **4**, 262 ⟨90MI 406-01⟩
Lambert, F. L., **2**, 619, 626 ⟨65JOC304⟩
Lambert, J. B., **1**, 14 ⟨85JA7978⟩, 808 ⟨79JA1793⟩, 905 ⟨78JA2501⟩; **2**, 70 ⟨87JA7838⟩, 161 ⟨72JA208⟩, 287 ⟨73JA4634⟩, 431 ⟨71T4245⟩; **4**, 748 ⟨78TL4253, 80JA6615⟩, 748 ⟨84JA3584⟩
Lambert, J. L., **1**, 868 ⟨66JA1905⟩
Lambert, Jr., R. L., **4**, 152, 155, 156 ⟨75JOM(88)255⟩, 155 ⟨75JOM(88)287⟩, 806 ⟨77JOM(142)39⟩; **6**, 173, 195, 267, 270, 293 ⟨70JOM(24)647⟩, 380, 391 ⟨77JOM(142)39⟩
Lambert, S. J., **1**, 661 ⟨88MI 114-04⟩
Lamberth, C., **1**, 923 ⟨93JOC1083⟩; **4**, 176 ⟨91TL7369⟩
Lamberton, A. H., **2**, 385 ⟨51QR75⟩, 386 ⟨49JCS1883⟩, 386, 405 ⟨55JCS3997, 69JCS(C)397⟩
Lamberts, J. J. M., **1**, 841 ⟨83JOC2202⟩; **5**, 1028 ⟨76MI 527-01⟩
Lambeth, D. O., **2**, 160 ⟨79JOC2632⟩
Lambrecht, G., **6**, 444 ⟨89AF539⟩
Lambrecht, J., **3**, 556, 557, 560, 585, 586 ⟨81CB3751⟩, 556, 557, 560, 586 ⟨84T893⟩, 602 ⟨82CB172⟩; **5**, 1146 ⟨81CB3655⟩
Lamchen, M., **1**, 41 ⟨55JCS2044⟩; **2**, 765 ⟨65QR329⟩; **5**, 283 ⟨67JCS(C)2128⟩, 318 ⟨50JCS747⟩, 325 ⟨66JCS(C)1477⟩
Lam-Chi, Q., **3**, 43 ⟨79AG(E)933⟩
Lamidey, A.-M., **2**, 97 ⟨90SC239⟩
Lamirand, J., **1**, 199 ⟨94TL363⟩
Lamm, B., **1**, 896 ⟨64JA4727, 66JA120⟩; **2**, 804 ⟨69ACS2983⟩; **5**, 991 ⟨74ACS(B)699⟩
Lammerink, B. H. M., **2**, 743 ⟨83S628⟩; **3**, 37 ⟨83S628⟩, 365 ⟨81S295⟩, 368 ⟨78TL811⟩; **4**, 570, 926 ⟨84JOC263⟩; **5**, 562 ⟨78TL811, 83RTC400⟩, 562, 645 ⟨84JOC263⟩
Lammers, R., **4**, 188 ⟨88JOC2920⟩
Lammertsma, K., **1**, 552 ⟨89AG(E)1321⟩; **2**, 757 ⟨94JA642⟩; **4**, 697, 698 ⟨88JA3231⟩
La Monica, G., **5**, 386 ⟨74JOM(76)355, 75JOM(88)237⟩, 386 ⟨76JCS(D)1777⟩
LaMontagne, M. P., **5**, 363 ⟨68JOC762⟩
Lamothe, M., **4**, 365, 393, 394 ⟨91SC(21)1675⟩
Lamothe, P. J., **6**, 208, 405 ⟨75OMS(10)18⟩
Lamothe, S., **1**, 162 ⟨93TL1449⟩
Lamotte, G., **1**, 38, 945 ⟨79TL2291, 80JCS(P1)2657⟩; **5**, 1054 ⟨80JCS(P1)2657⟩
Lamoureux, G. V., **2**, 124 ⟨93JOC633⟩; **3**, 239 ⟨86CJC520⟩
la Moya Cerero, S. de, **3**, 92, 101 ⟨90CC1571⟩
Lampa, E., **5**, 597 ⟨91FES1033⟩
Lamparter, E., **1**, 1113 ⟨73CB3216⟩, 1117 ⟨72CB3789⟩
Lampe, F. W., **2**, 1006 ⟨65JA2490⟩
Lampe, J., **1**, 874 ⟨80JOC1066⟩
Lampham, D. J., **2**, 291 ⟨83JCS(P1)471⟩
Lampilas, M., **1**, 223, 224 ⟨87T2701⟩
Lampin, J. P., **1**, 751 ⟨77TL1137⟩
Lampman, G. M., **2**, 176 ⟨82JCS(P1)849⟩
Lamprecht, H., **3**, 451 ⟨63CB651⟩
Lamza, L., **2**, 825 ⟨61CB2514⟩, 825, 841 ⟨64JPR294⟩
Lan, X., **2**, 320 ⟨93JOC2086⟩; **4**, 309 ⟨94S445⟩; **6**, 370 ⟨89JHC829⟩
Lan, X. F., **4**, 446, 449 ⟨93JOC2086⟩
Lancaster, M., **1**, 232 ⟨81JCS(P1)1846⟩
Lancaster, R., **1**, 42 ⟨56JOC918⟩
Lancelin, J., **4**, 172 ⟨83TL4833⟩
Lancelin, J.-M., **1**, 513 ⟨83TL4833⟩
Lancelot, C. J., **1**, 104 ⟨70JA2542⟩, 915 ⟨B-72MI 119-12⟩, 1114 ⟨70JA2538⟩
Lancelot, J.-C., **5**, 706 ⟨84JHC91⟩
Lancombe, L., **6**, 33 ⟨85JA6993⟩
Land, A. H., **4**, 237 ⟨41JOC888⟩
Landais, P., **6**, 414 ⟨87MI 614-01⟩
Landau, A., **3**, 356 ⟨91CB665⟩
Landau, E. F., **2**, 717 ⟨47JA1219⟩
Landauer, S. R., **2**, 8 ⟨53JCS2224, 54JCS2281⟩
Lander, G. D., **5**, 755 ⟨03JCS766⟩
Landesman, H., **1**, 307 ⟨63JA207⟩; **2**, 738, 758 ⟨63JA207⟩
Landgrebe, J. A., **2**, 675 ⟨78JOC1244⟩; **3**, 321 ⟨82TL5099, 83JOC1051⟩, 323 ⟨78JOC1071, 78JOC1244⟩, 323 ⟨83JOC1898, 83TL2829⟩, 323 ⟨89TL4089⟩
Landgrebe, K., **1**, 574 ⟨84S885⟩
Landheimer, F., **3**, 54, 231 ⟨49JCS737⟩
Landi, Jr., J. J., **3**, 97 ⟨91SC167⟩
Landick, R. C., **4**, 346 ⟨75TL4531⟩
Landini, D., **2**, 4, 23, 30 ⟨80JOC3527⟩, 9 ⟨75CC950, 75S430⟩, 9, 35 ⟨74CC879⟩, 19 ⟨74S428⟩, 25 ⟨74S37⟩, 25, 31, 35 ⟨92JCS(P1)2309⟩, 32 ⟨78S771⟩, 2999 ⟨76S389⟩; **3**, 3, 114 ⟨79CI(L)213, 79S134⟩, 238 ⟨69JCS(C)1204⟩, 427, 430 ⟨90S333⟩, 643 ⟨90SC965⟩; **4**, 74, 75 ⟨72JCS(P2)296⟩, 233 ⟨72CC734, 76JCS(P2)996⟩; **5**, 29, 106 ⟨82JOC154⟩, 499 ⟨74JCS(P2)1676⟩
Landis, C. R., **1**, 77 ⟨87JA1746⟩; **5**, 384 ⟨92OM2024⟩
Landis, II, R. T., **2**, 382 ⟨73JA2719⟩, 403 ⟨74JA1788⟩
Landis, M. E., **5**, 366 ⟨79JOC2288⟩
Landis, M. S., **1**, 835 ⟨92JOC7188⟩
Landman, B., **2**, 530 ⟨83TL3209⟩
Landmann, B., **2**, 101 ⟨1887LA(241)150⟩; **4**, 363, 373 ⟨86CB1039, 83TL3209⟩
Landmann, H., **2**, 1034 ⟨23CB2290⟩
Landolt, R. G., **1**, 795 ⟨67JA1404⟩
Landon, M. J., **2**, 405 ⟨72JA4392⟩
Landon, P., **3**, 293 ⟨89CC1355⟩
Landor, P. D., **1**, 961 ⟨73JCS(P1)720⟩, 1005 ⟨75JCS(P1)1628⟩, 1005, 1006 ⟨73TL1019⟩, 1015 ⟨73JCS(P1)1347⟩; **3**, 655, 656 ⟨84T2141⟩; **4**, 979 ⟨88JCS(P1)975⟩
Landor, S. R., **1**, 458 ⟨B-82MI 110-01⟩, 808 ⟨65JCS6784⟩, 954 ⟨B-82MI 120-01⟩, 961 ⟨73JCS(P1)720⟩, 965 ⟨66JCS(C)1223⟩, 990 ⟨67JCS(C)194⟩, 1005 ⟨75JCS(P1)1628⟩, 1005, 1006 ⟨73TL1019⟩, 1015 ⟨73JCS(P1)1347⟩, 1056 ⟨75T3073⟩; **2**, 677 ⟨65JCS5225⟩; **3**, 649 ⟨68JCS(C)291⟩, 655, 656 ⟨84T2141⟩; **4**, 979 ⟨88JCS(P1)975⟩

Landquist, J. K., **5**, 472 ⟨81SST165⟩, 565 ⟨81SST177⟩
Landro, F. J., **6**, 204, 205, 404 ⟨83JOM(249)1⟩, 205 ⟨82JA7345⟩
Landsberg, B. M., **3**, 551 ⟨74CC739⟩
Landscheidt, A., **3**, 211 ⟨77CB1007⟩
Landscheidt, H., **6**, 465 ⟨93EUP558996, 93EUP559001⟩, 465 ⟨93GEP4134688⟩
Lane, A. G., **3**, 69 ⟨70JCS(C)2484⟩
Lane, A. P., **2**, 509 ⟨67JA1040⟩, 859 ⟨67JCS1028⟩, 862 ⟨67JCS(A)1492⟩
Lane, C. F., **1**, 510 ⟨76T981⟩; **2**, 30 ⟨70JA6660, 88T2763⟩, 30 ⟨88T2773⟩, 310, 763 ⟨75S135⟩, 527 ⟨76T981⟩; **6**, 61 ⟨70JA7212⟩
Lane, G. C., **1**, 243 ⟨90TL5697⟩
Lane, J. F., **1**, 113 ⟨50JA5127⟩, 394 ⟨41JA1674⟩; **2**, 320, 756 ⟨46OR(3)267⟩; **3**, 612 ⟨44JA545⟩; **5**, 998 ⟨46OR(3)267⟩, 1000 ⟨41JA1674⟩
Lanears, R., **3**, 684 ⟨63BSB719⟩
Lane, R. W., **3**, 339 ⟨73HCA1671⟩
Lane, S., **3**, 256 ⟨84TL1039⟩
Laneman, S. A., **4**, 545 ⟨89PS(42)97⟩, 550 ⟨88JA5585, 89IC1872⟩
Lanfredi, A. M. M., **4**, 683 ⟨80JOM(187)81⟩, 713 ⟨81JOM(221)93⟩
Langa, F., **5**, 215 ⟨91JCS(P1)223⟩
Lang, E. S., **6**, 580 ⟨88SC301⟩
Lang, F., **3**, 649 ⟨73JOC479⟩; **4**, 739 ⟨73JOC479⟩
Lang, G., **4**, 133, 556 ⟨86ZAAC(536)187, 86EGP238612⟩
Lang, H., **1**, 956 ⟨93CB1077⟩; **2**, 875, 896 ⟨92JOM(441)389⟩, 875, 896 ⟨93JOM(444)53⟩, 1036, 1037 ⟨90CB417⟩, 1079 ⟨90ZN(B)1529⟩; **4**, 620 ⟨91OM551⟩, 714 ⟨93JOM(447)C1⟩; **6**, 190, 381, 392 ⟨91OM551⟩, 375 ⟨86ZN(B)191⟩
Lang, Jr., S. A., **3**, 360 ⟨80JOC4857⟩; **5**, 631 ⟨84S946⟩, 759 ⟨79JOC4160⟩, 760 ⟨80JOC3750⟩
Láng, K. L., **6**, 104 ⟨68TL2713⟩
Láng, L. K., **6**, 71 ⟨63PCS374⟩, 73 ⟨67CC13, 67T4181⟩
Lang, P., **1**, 259 ⟨64T1259⟩
Lang, R. W., **1**, 831 ⟨88TL3291⟩, 986 ⟨80HCA438, 85HCA2249⟩; **2**, 16 ⟨91TL1779⟩, 16, 365 ⟨88TL6087⟩, 16, 365 ⟨89HCA1248⟩; **3**, 149, 152, 154 ⟨91TL1779⟩, 494 ⟨80HCA438⟩; **4**, 14 ⟨86HCA881⟩; **5**, 103 ⟨84OS(62)202⟩; **6**, 64 ⟨88JOM(339)17⟩
Lang, W., **3**, 69 ⟨60CB2743⟩
Langdale-Smith, R. A., **1**, 808 ⟨68JOC4413⟩
Langdon, S. P., **5**, 770 ⟨84JCS(P1)993⟩
Lange, B., **3**, 699, 700, 723 ⟨76AG(E)113⟩, 702 ⟨90OSC(7)27⟩, 703, 710 ⟨77AG(E)727⟩, 704 ⟨77AG(E)262⟩
Lange, B. C., **1**, 129, 130, 858, 873 ⟨77T2737, 81JA4494⟩
Lange, C., **2**, 30 ⟨91TL4135⟩
Lange, G., **3**, 488 ⟨73JOM(54)23⟩; **6**, 242, 683 ⟨93ZN(B)58⟩, 265 ⟨90ZN(B)299⟩
Lange, G. L., **1**, 357 ⟨85TL4707, 85TL6163⟩, 357 ⟨87TL2833, 88TL2613⟩; **2**, 7 ⟨90SC1473⟩; **3**, 208, 209, 241 ⟨79JOC3066⟩
Lange, H., **4**, 136 ⟨82LA1946⟩, 136, 368 ⟨90JOM(388)57⟩, 152 ⟨84JFC(26)1⟩, 153 ⟨85JFC(27)309⟩; **6**, 64 ⟨84JFC(26)435, 88JFC(41)185⟩, 65, 66 ⟨85JFC(27)309⟩
Lange, H. G., **6**, 423 ⟨85JFC(28)219⟩
Lange, K., **4**, 562 ⟨93PS(77)258⟩
Lange, L., **2**, 294 ⟨B-87MI 204-03⟩, 729 ⟨91CB1315⟩; **4**, 88, 288 ⟨91CB1315⟩
Lange, M., **3**, 699, 700 ⟨78TL3027⟩; **5**, 1097 ⟨77CC648, 78AG(E)781⟩; **6**, 717 ⟨77CC648⟩
Lange, U., **5**, 1081, 1085 ⟨72ZC147⟩
Lange, W., **2**, 414, 482, 503 ⟨64ZAAC(326)280⟩
Langemann, A., **1**, 883 ⟨59JA5740⟩
Langen, D. Von, **3**, 165 ⟨88JMC2199⟩
Langendries, R., **3**, 375 ⟨74JA2964⟩
Langendries, R. F. J., **2**, 205 ⟨74TL4781⟩
Langenfeld, B., **2**, 431 ⟨60USP2959621⟩
Langensee, P., **1**, 890 ⟨86CC947⟩
Langer, C., **2**, 460 ⟨83ZAAC(504)147⟩
Langer, E., **2**, 575 ⟨84HOU(13/9b)1⟩, 579 ⟨84HOU(13/9b)287, 84HOU(13/9b)464⟩, 601 ⟨75HOU(13/7)1⟩, 828, 866 ⟨73CB177⟩, 1101 ⟨75HOU(13/7)73⟩; **3**, 272 ⟨71LA(754)64⟩; **4**, 163 ⟨80T1345⟩; **5**, 400 ⟨76CB359⟩; **6**, 172 ⟨75HOU(13/7)1⟩, 518 ⟨73AG(E)842⟩
Langer, F., **2**, 592 ⟨94SL410⟩
Langer, H. G., **6**, 410 ⟨67USP3322823⟩
Langer, H. J., **2**, 722, 725 ⟨73CJC3403⟩
Langer, J., **2**, 183 ⟨55AG32⟩; **5**, 4 ⟨58CB2553⟩
Langer, Jr., A. W., **5**, 43 ⟨65MI 502-01⟩
Langer, M., **5**, 448 ⟨75JOM(87)223⟩, 476, 477 ⟨43HCA1722⟩
Langer, S. H., **2**, 539 ⟨58JOC50⟩
Langerbeins, K., **2**, 472 ⟨85ZN(B)1274⟩
Langford, R. B., **2**, 169 ⟨63JOC1901⟩, 194 ⟨58JOC1694⟩
Lang-Fugmann, S., **2**, 388 ⟨90HOU(E16a/2)1227⟩; **4**, 294 ⟨92HOU(E14a/3)545⟩, 315, 319 ⟨92HOU(E14a/3)483⟩
Langhals, E., **3**, 572 ⟨89H(29)2069, 90JOC1412⟩
Langhans, K. P., **2**, 427, 430 ⟨90ZN(B)203⟩, 429 ⟨87CB1707⟩; **4**, 112, 546 ⟨90CB995⟩, 546 ⟨87CB1707⟩; **6**, 165 ⟨87CB1707⟩
Langhauser, F., **2**, 984 ⟨91CB2165⟩; **4**, 725 ⟨92OM3517⟩; **5**, 885 ⟨91AG(E)310⟩, 890 ⟨90CB1617⟩
Langheck, M., **2**, 167 ⟨53CB557⟩, 167 ⟨54CB325⟩
Langheck, S. M., **6**, 233 ⟨53CB557⟩
Langkamerer, C. M., **2**, 783 ⟨36JA225⟩
Langkammerer, C. M., **2**, 396 ⟨73OSC(5)445⟩; **4**, 86 ⟨35JA2360⟩; **6**, 49, 58 ⟨62JOC3584⟩
Langkammerev, C. M., **2**, 190 ⟨34JA1628⟩
Langler, R. F., **2**, 178, 179, 180 ⟨77CJC407⟩, 179 ⟨76CJC498⟩; **4**, 63, 78 ⟨73CC49⟩, 74, 75, 78 ⟨77CJC421⟩, 85 ⟨73CC50⟩, 258 ⟨78CJC1183⟩, 267 ⟨80CJC878⟩; **5**, 348 ⟨78CJC1183⟩; **6**, 254, 255 ⟨75CJC2664⟩, 254, 255 ⟨80CJC1996, 83CJC610⟩, 255 ⟨91CJC2127⟩
Langley, B. W., **2**, 403, 404 ⟨52JCS4191⟩
Langley, W. D., **2**, 782 ⟨03CB3803, 55OSC(3)334⟩; **5**, 115 ⟨55OSC(3)334⟩
Langlois, B., **6**, 229, 230 ⟨86BSF925⟩, 233, 234 ⟨91CC993⟩
Langlois, B. R., **6**, 11 ⟨91TL7525⟩, 38, 45 ⟨88JFC(41)247⟩
Langlois, D. P., **4**, 211 ⟨48JA2624⟩
Langlois, N., **3**, 638 ⟨93T77⟩
Langlois, Y., **1**, 253 ⟨80TL67⟩, 626 ⟨90TL1865⟩, 626, 630, 655, 789 ⟨82JA7666⟩; **5**, 214 ⟨81TL4405⟩, 351 ⟨93TL457⟩
Langone, J. J., **1**, 549 ⟨71JA3828⟩
Langrand, G., **5**, 226 ⟨85TL1857⟩
Langsch, D., **2**, 861 ⟨91ZAAC(606)119⟩
Langsdorf, Jr., P., **6**, 513 ⟨71GEP2040367⟩
Langsdorf, Jr., W. P., **6**, 497, 513 ⟨74USP3849102⟩
Langston, S. P., **6**, 264 ⟨91TL6425⟩
Langwald, M., **3**, 684 ⟨86M1091⟩
Lanigan, D., **5**, 1044 ⟨69CI(L)1702⟩
Lankamp, H., **1**, 922 ⟨68TL249⟩
Lankelma, H. P., **2**, 253 ⟨56JA976⟩
Lanneau, G., **4**, 648 ⟨77JOM(127)281⟩
Lanneau, G. F., **2**, 876 ⟨75JOM(85)179⟩; **3**, 14, 82, 95, 102, 104, 107 ⟨87TL3941⟩, 83, 98, 102, 104, 107 ⟨88TL1271⟩; **4**, 648 ⟨78JOM(144)155⟩; **5**, 759, 777 ⟨91S954⟩; **6**, 576, 577 ⟨91JOM(419)9⟩
Lannert, K. P., **4**, 563 ⟨69IC1775⟩
Lansard, J.-P., **1**, 314 ⟨84JOC931⟩; **3**, 145 ⟨83JOC3837⟩
Lansbury, P. T., **1**, 102 ⟨63JA2236⟩, 311 ⟨82TL2627⟩, 396, 397 ⟨66JA78, 62JOC1933⟩; **2**, 478 ⟨62JOC1933⟩; **3**, 207 ⟨80JA3964⟩, 209 ⟨82TL2623⟩, 295 ⟨61JOC670⟩; **4**, 961 ⟨85TL169⟩; **5**, 26 ⟨73TL1569⟩, 164 ⟨80JA3964⟩
Lansinger, J. M., **1**, 1136 ⟨79CC124⟩; **4**, 26 ⟨79SC341⟩

Lantenois, M., **6**, 216 ⟨15CR(156)652⟩
Lantseva, L. T., **4**, 761, 763 ⟨67ZOR1006⟩; **5**, 475 ⟨77BAU206⟩
Lantz, A., **6**, 7 ⟨89JFC(45)215⟩
Lantzsch, R., **5**, 28 ⟨81AG(E)703⟩
Lanza, P., **2**, 213 ⟨50G180⟩
Lanza, S., **4**, 151 ⟨90JOM(396)115⟩
Lanza, T. J., **4**, 196 ⟨86TL1653⟩
Lanz, J. W., **2**, 75 ⟨88CB1501⟩
Lanzendorfer, U., **3**, 543 ⟨90CB2031⟩
Lanznaster, N., **4**, 441 ⟨85T5525⟩
Laonigro, G., **2**, 45 ⟨76TL3661⟩
Laos, I., **3**, 227 ⟨57JA6308⟩
Laoutidis, J., **5**, 690 ⟨88LA1107⟩
Lapalme, R., **2**, 91 ⟨83JA7719⟩; **6**, 327 ⟨83JA7719⟩
Lapenue, M., **2**, 619 ⟨73BSF720⟩
Lapidus, A. L., **6**, 465 ⟨77IZV363⟩
Lapin, A. A., **2**, 446 ⟨87ZOB296⟩, 877 ⟨84IZV719⟩; **4**, 343, 561 ⟨78ZOB1001⟩
Lapitajs, G., **4**, 540 ⟨87JOM(330)357⟩
Lapkin, I. I., **2**, 949 ⟨78ZOB639⟩; **3**, 436 ⟨85ZOB2077⟩; **4**, 239 ⟨80JOU1377, 82JOU876⟩; **5**, 62 ⟨B-76MI 502-02⟩, 438 ⟨66JGU1945⟩; **6**, 82 ⟨71ZOR1887⟩, 84 ⟨62ZOB745⟩, 92 ⟨71ZOR473⟩, 100 ⟨86ZOR107⟩, 656 ⟨78ZN849, 81ZOB880⟩
LaPoint, R. C., **3**, 545 ⟨89JA9057⟩
Laporte, O., **4**, 514 ⟨78JOM(156)C25⟩; **5**, 1134 ⟨82S212⟩
Laporterie, A., **1**, 777 ⟨83TL3521⟩; **2**, 949 ⟨85TL1315, 90CRV215⟩
Lapouyade, P., **5**, 416 ⟨74JOM(80)349⟩, 419 ⟨79JIC545⟩; **6**, 124 ⟨79JOC420⟩
Lapouyade, R., **2**, 619 ⟨73BSF720⟩
Lappert, F., **2**, 421 ⟨69CC1064⟩
Lappert, M. F., **1**, 447 ⟨58JCS1451⟩, 657 ⟨69JCS(A)433⟩, 658 ⟨66JOM(5)506⟩, 663, 668 ⟨76CRV219⟩; **2**, 99 ⟨56JCS3006⟩, 110 ⟨55JCS1470, 57JCS348, 57JCS739⟩, 111 ⟨56CRV959, 56JCS824⟩, 111, 112 ⟨58JCS1451⟩, 111, 267 ⟨59JCS1529⟩, 414, 423 ⟨B-80MI 207-01⟩, 415, 416, 418, 419 ⟨69JCS(A)992⟩, 418, 419, 1085 ⟨57JCS381, 65JCS4256⟩, 419, 533 ⟨59JCS2927⟩, 421 ⟨65JCS1944⟩, 421, 868 ⟨65JOM(3)295⟩, 520 ⟨B-67MI 210-02, 62JCS4088⟩, 543 ⟨76MI 210-01, B-80MI 210-01⟩, 565 ⟨79JCS(D)54, 82COMC-I(3)559⟩, 565 ⟨82COMC-I(3)635⟩, 566 ⟨73JOM(57)269, 77JA6645⟩, 569 ⟨81JCS(D)80⟩, 582 ⟨76AOC345, 80JCS(D)1448⟩, 815 ⟨73JCS(D)157, B-80MI 215-02⟩, 852 ⟨82CC691⟩, 920 ⟨68JOM(14)327⟩, 949 ⟨80JOM(202)5⟩, 1005 ⟨82CC14⟩, 1083 ⟨64JOM(1)384⟩, 1086 ⟨67CB2377⟩; **3**, 456, 457 ⟨73JCS(D)157⟩, 457, 460, 461 ⟨73JCS(D)151⟩, 458 ⟨68CC250A, 69JCS(A)433⟩, 458 ⟨72JCS(D)370⟩, 460 ⟨71MI 312-01⟩, 460, 461 ⟨71CC215⟩, 482 ⟨67CC836⟩, 483, 484 ⟨70JCS(A)2954⟩; **4**, 584 ⟨81CC698⟩, 584 ⟨82JOM(228)C71⟩, 602, 650 ⟨82CC1323⟩, 604, 646, 647 ⟨82CC14⟩, 647, 657 ⟨80CC1284⟩, 647, 660 ⟨92JCS(D)775⟩, 656 ⟨70JOM(25)C36, 71SAP7004922⟩, 656 ⟨72CC613, 73JCS(D)445⟩, 656 ⟨73JOM(57)269, 78CC1081⟩, 656 ⟨78CC140, 78PAC703⟩, 656, 663, 709, 724 ⟨81JCS(D)1593⟩, 656, 695 ⟨77JCS(D)999⟩, 660 ⟨82JOM(233)C28⟩, 702 ⟨79JCS(D)54⟩, 725, 726 ⟨79JCS(D)45⟩, 849 ⟨73JCS(P1)2272⟩; **5**, 228 ⟨58JCS3648⟩, 277 ⟨69JCS(A)992, 69JCS(C)2565⟩, 443 ⟨84JCS(D)399⟩, 779 ⟨66JCS(A)1584⟩, 874 ⟨67CC836⟩, 931, 940 ⟨72CRV545⟩, 1016 ⟨65JCS4256⟩, 1038 ⟨64JCS2108⟩, 1133 ⟨66MI 530-01⟩, 1144 ⟨66MI 530-02⟩; **6**, 61 ⟨65JCS2157⟩, 128 ⟨73JCS(P1)2272⟩, 165, 167, 168 ⟨80JCS(D)2428⟩, 167 ⟨84CC1669⟩, 169 ⟨87JOM(320)C27⟩, 174, 186, 190, 191, 195 ⟨76JCS(D)2268⟩, 174, 190 ⟨82CC1323⟩, 186 ⟨76CC261⟩, 186 ⟨84CC480⟩, 186, 192, 195 ⟨86JCS(D)1551⟩, 186, 195 ⟨82CC1407⟩, 187 ⟨81JOM(212)C4⟩, 190, 196 ⟨86CC672⟩, 194 ⟨79JOM(181)25⟩, 194, 195 ⟨88CC1007, 90JOM(394)57⟩, 195 ⟨74JOM(76)C45, 77JA6645, 78JCS(D)734, 80IC3637, 81JCS(D)814⟩, 196 ⟨88CC336, 93JCS(D)2653⟩, 396 ⟨71SAP7004922⟩, 599 ⟨80CC635⟩, 664 ⟨68JOM(14)327⟩, 672 ⟨70JCS(A)2954⟩, 716 ⟨73CSR99⟩, 716, 721 ⟨72CRV545⟩, 720 ⟨74JCS(D)1494⟩, 720, 721 ⟨77JCS(D)1272, 71CC400⟩, 721 ⟨72CC851, 72CC927⟩, 721 ⟨74JCS(D)1591, 74JCS(D)1827⟩, 721 ⟨75CC929, 75JOM(100)139⟩, 721 ⟨76CC644, 77JCS(D)1283⟩, 721 ⟨77JCS(D)2160, 77JCS(D)2172⟩, 721 ⟨77MI 622-01, 78JCS(D)348⟩, 721 ⟨78JCS(D)826, 78JCS(D)837⟩, 721 ⟨79JCS(D)1929⟩
Lappert, M. J., **4**, 597, 598 ⟨80JCS(D)2428, 76CC623⟩
Lappert, M. L., **6**, 349 ⟨85CC1776⟩
la Pradilla, R. F. de, **1**, 14, 226, 227 ⟨87JOC4898⟩, 600, 983 ⟨92TL4985⟩
Lapsley, R. W., **1**, 637 ⟨44JCS322⟩
Lapteva, L. I., **3**, 50 ⟨87JGU203⟩
Lapteva, V. L., **4**, 322 ⟨77ZOR1930⟩
Lapworth, A., **2**, 764 ⟨21JCS768⟩; **3**, 612 ⟨41OSC(1)181⟩, 616 ⟨31JCS1391, 31JCS1894⟩, 626, 634 ⟨22JCS1699⟩; **5**, 35 ⟨32OSC(1)451⟩, 95 ⟨32OSC(1)181⟩
Laqua, A., **3**, 562 ⟨80CB2509⟩
Larbig, W., **1**, 665, 666 ⟨60LA(629)53⟩; **2**, 523 ⟨65LA(682)21⟩; **4**, 699 ⟨60LA(629)53⟩
Larchar, A. W., **2**, 130 ⟨52JA3982⟩; **5**, 579, 580 ⟨53JOC748⟩
Larchevêgue, M., **1**, 209 ⟨80SC49⟩, 212 ⟨75S256⟩, 479 ⟨87TL1781⟩, 560 ⟨77TL3187⟩, 763 ⟨81CC877, 81TL1595⟩, 837 ⟨92TL4313⟩, 1009 ⟨93SL217⟩, 8 ⟨92TL7337⟩, 165 ⟨85CC83⟩, 179 ⟨90TL2149⟩; **2**, 420 ⟨75LA719⟩; **3**, 459 ⟨74BSF1710⟩, 459 ⟨75JOM(87)25⟩, 652 ⟨83S297⟩; **4**, 648 ⟨81CC877⟩; **5**, 152 ⟨89CC31⟩
Lardicci, L., **1**, 198 ⟨88G857⟩, 466 ⟨91JOM(417)289⟩, 1016 ⟨87TL6073⟩, 1019, 1141 ⟨78TL2831⟩; **2**, 9 ⟨86JOC5291⟩, 56 ⟨89JOC2383⟩
Lardy, I. A., **1**, 1064 ⟨64JCS1151⟩
Larin, M. F., **4**, 656 ⟨84ZOB2025⟩; **5**, 718 ⟨84JGU426, 84ZOB477⟩, 719, 720 ⟨83BAU1373⟩, 719, 720 ⟨83IZV1515⟩
Larin, N. A., **1**, 463 ⟨50JGU450⟩
Larionov, V. I., **6**, 127 ⟨88ZOB1066⟩
Larizza, A., **5**, 746 ⟨64JOC3697⟩
Larkin, D. R., **1**, 889 ⟨59JA3652⟩
Larkin, J., **5**, 332 ⟨63JA2181⟩
Larkin, J. M., **2**, 398 ⟨71JOC2574⟩
Larock, R., **1**, 494 ⟨88TL905⟩
Larock, R. C., **1**, 16, 21 ⟨B-89MI 101-01⟩, 66 ⟨84JA4218⟩, 67 ⟨B-85MI 102-02⟩, 128 ⟨78JOM(156)45⟩, 128, 1095 ⟨78AG(E)27⟩, 128, 1095 ⟨82T1713⟩, 230 ⟨88TL5069, 89TL3487⟩, 230 ⟨90SL341⟩, 454 ⟨84JOC3411, B-85MI 110-02⟩, 491 ⟨86MI 111-02⟩, 491, 1095 ⟨76JOC2241⟩, 492 ⟨78JOC1468⟩, 492, 1095 ⟨77JOC1680⟩, 493 ⟨81TL5231, 84JA5274⟩, 541 ⟨B-89MI 112-01⟩, 687 ⟨B-89MI 115-01⟩, 882 ⟨B-89MI 119-02⟩, 948 ⟨B-86MI 119-06⟩, 957, 974, 1060 ⟨86JOC2623⟩, 1095 ⟨72JA4371, 72JOM(36)1⟩, 1095 ⟨B-85MI 122-01⟩, 1096 ⟨80JOC3840⟩, 1108 ⟨78JOC131⟩; **2**, 2, 6, 15, 18, 28 ⟨B-89MI 201-01⟩, 4 ⟨91COS(4)269⟩, 38, 93 ⟨B-89MI 202-01⟩, 40 ⟨B-86MI 202-01⟩, 73 ⟨B-89MI 202-02⟩, 86 ⟨89TL3487, 90JOC6244⟩, 86 ⟨93JOC804, 93SL145⟩, 93 ⟨B-89MI 202-04⟩, 316 ⟨B-89MI 205-02⟩, 318 ⟨B-89MI 205-01⟩, 520 ⟨70JA2467⟩, 592 ⟨78AG(E)27⟩, 752 ⟨B-89MI 215-01⟩, 753 ⟨B-89MI 215-02⟩, 757 ⟨B-89MI 215-03⟩, 928 ⟨93JOC4579⟩, 982 ⟨B-86MI 219-01⟩; **3**, 25 ⟨89TL6629⟩, 57 ⟨B-89MI 302-01⟩, 85, 87 ⟨B-89MI 303-01⟩, 140 ⟨93JOC2847⟩, 158 ⟨B-89MI 304-01⟩, 161 ⟨B-89MI 304-02⟩, 211 ⟨78JOC710⟩, 212 ⟨76TL3097⟩, 215, 223 ⟨B-89MI 305-01⟩, 290 ⟨80JOC3840⟩; **4**, 171, 188, 194 ⟨B-86MI 404-01⟩, 697 ⟨73JOM(61)27⟩; **5**, 142 ⟨75JOC3227⟩, 142 ⟨78AG(E)27⟩, 259 ⟨B-89MI 506-01⟩

Laronze, J.-Y., **1**, 628 ⟨92S383⟩
Laroyer, C. B., **6**, 537 ⟨63CJC1643⟩
Larpent, C., **1**, 446 ⟨88TL4577, 93TL4331⟩; **2**, 446 ⟨90T6381⟩; **3**, 639 ⟨88T6107⟩
Larrabee, C. E., **1**, 866 ⟨50JA1245⟩
Larrabee, R. B., **1**, 848 ⟨65TL287⟩
Larraza, M. I., **3**, 620 ⟨83SC35⟩
Larscheid, M. E., **2**, 822, 841 ⟨79JOC3741⟩
Larsen, B. D., **5**, 626 ⟨85JCS(P1)907⟩
Larsen, C. A., **2**, 879 ⟨91CM96⟩
Larsen, C., **5**, 605, 622 ⟨85CJC951⟩; **6**, 114 ⟨80JOC3713⟩, 570, 571 ⟨69ACS324⟩, 574 ⟨68ACS1898⟩, 577, 578 ⟨78JOC337⟩, 594, 595 ⟨70ACS959⟩
Larsen, D. S., **3**, 245 ⟨90JCS(P1)1339⟩; **5**, 1107, 1109 ⟨81JCS(P1)58⟩
Larsen, D. W., **2**, 117 ⟨56JA348⟩
Larsen, E., **5**, 641 ⟨79ACS137⟩
Larsen, E. H., **5**, 76 ⟨73OSC(5)379⟩
Larsen, E. R., **3**, 532 ⟨58JOC1166⟩
Larsen, J. D., **5**, 703 ⟨88JMC2066⟩
Larsen, J. R., **5**, 52 ⟨77JOC3749⟩
Larsen, J. W., **6**, 43 ⟨72CRV357⟩
Larsen, K., **3**, 323 ⟨89TL4089⟩
Larsen, K. E., **3**, 683 ⟨84T2985⟩
Larsen, O., **6**, 584 ⟨70ACS1094⟩
Larsen, P. O., **3**, 617 ⟨74ACS(B)317⟩
Larsen, R. O., **1**, 751 ⟨83PS(16)339⟩
Larsen, S., **3**, 639 ⟨79JCR(S)396⟩
Larsen, S. D., **1**, 234 ⟨88JA5932⟩; **5**, 177 ⟨87JOC3937⟩
Larsne, C., **2**, 228 ⟨78JOC3481⟩
Larson, A. C., **2**, 357, 365 ⟨79JOC3867⟩
Larson, D. W., **4**, 23 ⟨59JA5937⟩
Larson, E., **3**, 243 ⟨85JA1246⟩, 244 ⟨85JA1274⟩
Larson, E. L., **4**, 1060 ⟨88TL1355⟩
Larson, G. L., **1**, 434 ⟨75TL4005⟩, 457, 508 ⟨78JOC4555⟩, 648 ⟨89JOM(360)39⟩, 649 ⟨88T3781⟩, 650 ⟨82TL1035⟩, 654 ⟨69JA514⟩, 659 ⟨75JOM(102)123⟩, 759 ⟨89OS(67)125⟩, 763 ⟨81TL1575⟩, 981 ⟨88TL1355⟩; **2**, 420 ⟨88SC253⟩, 907 ⟨B-89MI 218-02⟩, 1005 ⟨89JOM(360)39⟩; **3**, 130, 202 ⟨85JOC5260⟩, 135 ⟨80SC813⟩, 225 ⟨86TL2223⟩, 272 ⟨90SC1095⟩; **4**, 138, 139 ⟨92OM859⟩, 188 ⟨73JOC3935⟩, 188 ⟨85JOC3627⟩, 357 ⟨78JOM(146)C8, 80JOM(198)15⟩, 652 ⟨88T3781⟩, 831 ⟨88JOC633⟩, 1048, 1058, 1059, 1060, 1062 ⟨86OM2274⟩; **6**, 26 ⟨87SC1047⟩, 183, 200, 396 ⟨73JOM(57)225⟩, 183, 200, 396, 400 ⟨69JA6541⟩, 405 ⟨70JA231⟩
Larson, H. O., **2**, 391, 392 ⟨B-69MI 207-02⟩; **4**, 161 ⟨57JA6562⟩; **5**, 332 ⟨63JA827⟩
Larson, K. D., **3**, 407 ⟨85JOC5382⟩
Larson, S., **2**, 34, 629, 631 ⟨88JOC4477⟩
Larson, S. B., **2**, 364 ⟨88JHC1857⟩
Larsson, E., **2**, 126 ⟨28CB1439⟩
Larsson, F. C. V., **4**, 903 ⟨74T1283⟩; **5**, 547 ⟨72T5341⟩, 556 ⟨74RTC258⟩; **6**, 303, 306 ⟨75RTC1⟩
Lasaletta, J.-M., **2**, 347 ⟨93TL141⟩
Lasalle, L., **5**, 1160 ⟨94MI 531-01⟩
Lasalle, P. de, **2**, 422 ⟨84T2211⟩; **5**, 166 ⟨84T2211⟩
la Salud, R. de, **2**, 36 ⟨93SL489⟩
Lasch, J. G., **2**, 866 ⟨83CC881⟩, 866, 893 ⟨84IC2582⟩; **6**, 168 ⟨83JA5506⟩, 168 ⟨84IC2582⟩, 169 ⟨83CC881⟩, 689 ⟨84JA7015⟩
Laschat, S., **2**, 459 ⟨92S90⟩; **4**, 462 ⟨92S90⟩
Laschi, F., **1**, 507 ⟨88IC4429⟩
las Heras, F. G. de, **5**, 1039 ⟨84S509⟩
Laskin, S., **4**, 47 ⟨72CEN55, 72CEN62⟩
Laskorin, B. N., **5**, 403, 404 ⟨74JGU1685, 72JGU1256⟩
Laskovics, F. M., **2**, 22 ⟨77JA6672⟩; **3**, 152 ⟨77JA6672⟩
Laskowski, S. C., **2**, 128 ⟨45JA594⟩
Laslett, R. L., **3**, 308 ⟨64AJC1056⟩
Lasne, M.-C., **1**, 786, 787 ⟨85S121⟩; **2**, 36 ⟨90SC41⟩; **3**, 233 ⟨85S121⟩, 564 ⟨88T4447⟩, 582 ⟨83CC238, 86JCR(S)188⟩
Lasouris, U. I., **6**, 277 ⟨75IC592⟩
Lassaletta, J.-M., **3**, 46 ⟨92TL3691⟩
Lassalle, G., **5**, 76 ⟨90OS(69)173⟩
Lassila, J. D., **3**, 538, 539 ⟨68JA2449⟩
Lassmann, E., **2**, 400 ⟨71ZN(B)73⟩
Lassnigg, D., **4**, 416 ⟨88JHC119⟩
Last, L. A., **1**, 214 ⟨82JOC3211⟩
Last, W.-D., **2**, 174 ⟨68AG970, 69CB3357⟩
Lastovskii, R. P., **6**, 118 ⟨72MI 602-01⟩, 500 ⟨46MI 616-01⟩
Lastra, E., **1**, 165 ⟨93JA87⟩; **3**, 544 ⟨92JA1428⟩
Laszlo, P., **2**, 792 ⟨86ACR121, 88TL5657⟩; **3**, 118 ⟨80S849⟩; **4**, 190 ⟨82S162⟩, 431 ⟨85S909⟩; **5**, 368 ⟨84TL3701⟩
Laszlo, S. E. de, **1**, 614, 697, 700, 703 ⟨90TL5525⟩; **5**, 278 ⟨86CC344⟩
Latesky, S. L., **5**, 1097 ⟨87JA4731⟩
Lathan, W. A., **1**, 1109 ⟨70JA2141, 71JA808⟩, 1109 ⟨72CPL385⟩
Lathbury, D., **2**, 765 ⟨68CC1017⟩
Lathbury, D. C., **1**, 248 ⟨88CC81⟩
Latif, F., **2**, 9 ⟨87LA617⟩
Latimer, L. H., **3**, 249 ⟨78JOC1031⟩
Latimer, P. H., **2**, 191 ⟨40JOC24⟩
La Torre, F., **3**, 4, 57, 115 ⟨79S356⟩; **5**, 275 ⟨80S243⟩
Latowski, T., **6**, 609 ⟨89ZN(B)1589⟩
Latrofa, A., **5**, 282 ⟨83S1013⟩
Latscha, H. P., **6**, 622 ⟨68AG(E)299⟩
Lattes, A., **1**, 68, 454 ⟨79S891⟩, 531, 662 ⟨82T1713⟩, 674 ⟨77CC502, 77RTC(M)91⟩; **2**, 19 ⟨86TL1499, 89BSF256⟩, 444 ⟨82PS(13)97⟩; **3**, 406 ⟨74S812⟩
Lattke, E., **2**, 961 ⟨77TL3969⟩
Lattre, J. de, **4**, 218 ⟨12BSB323⟩
Lattrell, R., **5**, 226 ⟨85AG(E)180⟩, 1005 ⟨72CB2800⟩, 1106 ⟨74LA901⟩
Lattuada, L., **5**, 946 ⟨91OM807⟩; **6**, 723 ⟨93OM2994⟩
Latxague, L., **1**, 474 ⟨90BSF401⟩
Latyaeva, V. N., **2**, 566 ⟨72SRI33⟩, 569 ⟨77JOM(131)43, 79DOK(245)198⟩; **4**, 656 ⟨76DOK(231)1138⟩; **5**, 388 ⟨85ZOB1071⟩, 1081 ⟨87ZOB405⟩, 1081, 1082 ⟨82ZOB885⟩
Lau, C., **6**, 49 ⟨76JFC(7)261⟩
Lau, C. K., **1**, 13, 18, 20 ⟨86JOC3038⟩; **2**, 612 ⟨88CJC1479⟩; **3**, 239 ⟨82CJC210⟩
Lau, H. H., **1**, 493 ⟨84JA5274⟩; **3**, 720 ⟨79LA1444⟩
Lau, J., **3**, 48 ⟨91JOC3591⟩
Lau, K. S. Y., **1**, 471 ⟨77ACR434⟩, 1137 ⟨81JOC2280⟩; **2**, 582, 960 ⟨77ACR434⟩; **4**, 818 ⟨78JOC1595⟩; **5**, 143 ⟨76JA5832⟩, 463 ⟨74JA4983⟩
Lau, P. W. K., **1**, 535 ⟨77TL3317⟩, 648, 651, 762 ⟨78TL2383⟩; **2**, 624, 630 ⟨77TL3317⟩, 935 ⟨78TL2383⟩; **3**, 58 ⟨77TL3317⟩; **4**, 655 ⟨78TL2383⟩, 831 ⟨79TL4029⟩
Laub, R., **6**, 434 ⟨88EJM561⟩
Laub, R. J., **1**, 126 ⟨75TL1633⟩; **2**, 904 ⟨76JCS(P1)2419⟩
Laube, T., **1**, 873 ⟨85CB764, 85HCA1373⟩, 873 ⟨85JA5403⟩, 899 ⟨93JA7240⟩, 900 ⟨86AG(E)349⟩, 906 ⟨87AG(E)560, 89JA9224⟩; **2**, 999 ⟨84HCA224⟩, 1007 ⟨93JA1731⟩, 1008 ⟨90AG(E)188⟩; **3**, 192 ⟨85HCA162⟩, 537 ⟨85JA5396⟩
Laubach, B., **5**, 1086, 1092 ⟨80TL2497⟩; **6**, 665, 686 ⟨82AG(E)448⟩, 680, 684 ⟨84TL4447, 83CB1873⟩
Laubach, G. D., **3**, 226 ⟨60JA4293⟩
Laubengayer, A. W., **5**, 1016 ⟨43JA1783⟩
Lauble, S., **6**, 173, 383 ⟨93ZAAC(619)1494⟩
Laucher, D., **1**, 310 ⟨87TL5279⟩
Laudenfeld, B., **2**, 431, 461 ⟨62CB64⟩
Lauderdale, S. C., **5**, 730 ⟨73JOC3951⟩
Lauener, H., **6**, 646, 647 ⟨75AF1477⟩
Lauer, R. F., **1**, 601, 602, 603, 607, 608, 781 ⟨73TL1979⟩,

601, 604, 605, 606, 607 ⟨73JA6137⟩, 601, 608, 781 ⟨73JA2697⟩, 601, 610, 611, 621, 646 ⟨75CS9⟩, 603, 608, 781 ⟨74JOC429⟩, 608 ⟨72JA7154⟩, 716 ⟨75CS9⟩; **2**, 85 ⟨73JA2697⟩, 282, 287 ⟨73JA6137⟩, 286 ⟨73TL1979⟩; **3**, 37, 173, 207 ⟨73JA6137⟩, 175 ⟨73TL1979⟩
Lauer, W. M., **2**, 783 ⟨36JA225⟩; **4**, 86 ⟨35JA2360⟩
Laufer, D. A., **6**, 416, 417 ⟨91JOC891⟩, 467 ⟨89OPP771⟩
Laufer, S., **3**, 360 ⟨89CZ290⟩
Laugal, J. A., **1**, 101 ⟨84TL2089⟩
Laughlin, R. G., **2**, 164 ⟨67JA2435⟩, 402 ⟨68JA2651⟩, 452 ⟨67JOC1060⟩
Laugraud, S., **3**, 60 ⟨87JOC4788⟩
Lauher, J. W., **5**, 951 ⟨87OM988⟩; **6**, 174 ⟨93IC2308⟩
Lauke, H., **2**, 565 ⟨84OM69⟩, 602 ⟨74JOM(67)61, 84JA6841⟩, 969 ⟨85JA8111⟩; **6**, 195 ⟨85JA8091⟩
Launay, M., **1**, 439, 700 ⟨82TL3265, 82TL2465⟩
Laurence, K. A., **5**, 246 ⟨78JFC(11)327⟩
Laurenco, C., **4**, 347, 773, 778 ⟨76S107⟩, 478 ⟨74MI 410-01, 76T2089⟩; **5**, 132, 139 ⟨76CR(C)483⟩
Laureni, J., **3**, 552, 553 ⟨77JA4842, 76JA7872⟩
Laurent, A., **1**, 495 ⟨78JCR(S)416⟩, 780 ⟨92JOC714⟩; **2**, 18 ⟨86TL4449, 87S562⟩, 364 ⟨76TL1093⟩; **3**, 356 ⟨90ZC247⟩; **5**, 282 ⟨68BSF3657⟩
Laurent, E., **2**, 16 ⟨83TL903, 86BSF955⟩, 16 ⟨87TL2359, 89T4431⟩; **3**, 149 ⟨87TL2359⟩; **4**, 62 ⟨90TL2287⟩; **6**, 11 ⟨91TL7525⟩
Laurent, H., **1**, 5, 9 ⟨83LA705⟩; **2**, 36 ⟨85TL5445⟩, 38 ⟨83AG(E)406⟩, 103 ⟨69T761⟩, 1018 ⟨84AG(E)727⟩; **3**, 64 ⟨66CB3057⟩, 227 ⟨82S34⟩
Laurent, J.-P., **2**, 495 ⟨74JOM(77)225⟩
Laurent, M. A., **4**, 437 ⟨1837LA(21)130⟩
Laurent, P. A., **4**, 728 ⟨67BSF571⟩
Laurent, P., **4**, 302 ⟨78BSF(2)83⟩, 582 ⟨89JOM(359)205⟩; **5**, 452 ⟨86CC1754⟩
Laurie, J. C. V., **4**, 712 ⟨84JCS(D)1563⟩
Lauritzen, S. E., **1**, 706, 707 ⟨81ACS(B)263⟩
Lauron, H., **1**, 439 ⟨86T2475⟩
Lautenberger, W. J., **4**, 979 ⟨70JPC2722⟩
Lautens, M., **1**, 191 ⟨92TL2617⟩, 191 ⟨93TL4591⟩, 196 ⟨87T4817⟩, 525 ⟨85JA1781, 85TL4887⟩; **2**, 923 ⟨90TL3105, 90TL5829⟩, 931 ⟨92JOC3270⟩; **4**, 661 ⟨92AG(E)232, 92JOC798⟩, 661, 702 ⟨90TL5829⟩, 1062 ⟨90TL3105⟩
Lautenschlaeger, F., **3**, 700 ⟨62CJC863⟩
Lautenschlaeger, F. K., **2**, 221, 226 ⟨69JOC3991⟩; **4**, 979 ⟨71JHC241⟩
Lautenschlager, H.-H., **4**, 618 ⟨92JOM(428)1⟩
Lauterbach, E. H., **1**, 839 ⟨91TL4481⟩
Lauterwein, J., **5**, 934 ⟨89TL1633⟩
Lauwers, M., **2**, 36 ⟨79TL1801⟩
Lauzon, G. de, **1**, 53 ⟨80JA994⟩; **2**, 472 ⟨83JA6871⟩; **4**, 1035 ⟨80JA994⟩
Laval, J.-P., **1**, 674 ⟨77CC502⟩
Laval, J. P., **1**, 674 ⟨77RTC(M)91⟩; **2**, 444 ⟨82PS(13)97⟩
Lavallée, J. F., **1**, 327 ⟨91JA9701, 92JA4912, 538 ⟨86TL5455, 87TL3457⟩; **3**, 89, 96 ⟨93TL3519⟩, 222 ⟨86TL5455, 87TL3457⟩
Lavallee, P., **2**, 8 ⟨82TL4411⟩, 538 ⟨75CJC2975⟩; **4**, 198 ⟨72JA8929⟩
Lavanish, J. M., **3**, 526 ⟨68TL6003⟩; **4**, 768 ⟨68TL6003⟩
Lavastre, O., **5**, 940, 941 ⟨95JOM(488)C9⟩
Lavayssiere, H., **1**, 656 ⟨83JOM(243)281⟩; **6**, 577 ⟨75JOM(88)C35⟩
Lavender, M. L., **5**, 400 ⟨92JOM(426)C28⟩
Laver, W. M., **2**, 190 ⟨34JA1628⟩
Laverne, D. S., **2**, 212, 232 ⟨47JA1710⟩
Laves, E., **6**, 90 ⟨1892CB361⟩, 90, 91 ⟨1892CB347, 1890CB1414⟩
Lavi, Y., **2**, 235 ⟨90H(32)2099⟩
Lavie, D., **5**, 66 ⟨50JA5012⟩; **6**, 25 ⟨50JA5012⟩
Lavielle, G., **2**, 680 ⟨78MI 213-01⟩; **4**, 340 ⟨73TL173⟩, 738, 739, 745 ⟨71TL1035⟩
Lavielle, S., **1**, 143 ⟨78JA1558⟩
Lavigne, G., **6**, 155 ⟨82OM1114⟩
Lavigne, J. B., **1**, 34 ⟨51JA2718⟩
Lavin, E., **5**, 595 ⟨39JA104⟩
Lavin, K. D., **4**, 706 ⟨84JA2337⟩
Laviron, E., **4**, 720 ⟨89JOM(362)C27⟩
Lavoie, A. C., **1**, 439 ⟨83JA5075⟩; **2**, 709 ⟨83JA5075⟩
LaVoie, E. J., **4**, 167 ⟨77JOC1045⟩
Lavrent'ev, A. N., **4**, 114 ⟨75JGU1668⟩, 125 ⟨86JGU2246⟩, 126 ⟨86JGU2144, 86JGU2242⟩, 128 ⟨87JGU622⟩; **6**, 242 ⟨88ZOB1525⟩
Lavrent'yev, V. I., **4**, 507 ⟨87JOM(326)159⟩
Lavrenyuk, T. Ya., **5**, 685, 694 ⟨75JOU1926⟩, 685, 694 ⟨75ZOR1913⟩
Lavrinovich, L. I., **2**, 606 ⟨84JOM(268)5⟩; **5**, 381 ⟨71ZOB2501⟩, 778 ⟨78BAU2304⟩; **6**, 662 ⟨79DOK(245)121⟩
Lavrov, V. E., **6**, 476 ⟨88PS(39)257⟩
Lavrov, V. I., **2**, 674 ⟨68RCR907⟩
Lavrova, E. E., **2**, 428 ⟨88ZOB33⟩; **4**, 337 ⟨86JGC425⟩
Lavrukhin, B. D., **5**, 223 ⟨90ZOB1928⟩
Law, C. K., **1**, 804 ⟨83CJC3227⟩
Law, J. Y., **3**, 215 ⟨87JOC3346⟩
Law, K.-W., **1**, 486, 545 ⟨87JOC4258⟩, 605 ⟨83JOC4776⟩; **2**, 285 ⟨83JOC4776⟩, 743 ⟨85CJC2313⟩; **4**, 322 ⟨87JCS(P1)781⟩, 874 ⟨85CJC2313⟩, 953 ⟨87JOC4258⟩
Law, W. C., **3**, 164 ⟨85JA4175⟩
Lawesson, S.-O., **1**, 881 ⟨62OS(42)28⟩; **2**, 758 ⟨76BSB805⟩, 89, 652 ⟨59JA4230⟩, 256 ⟨81ACS(B)239⟩, 684 ⟨61OS(41)91⟩, 694 ⟨74RTC252⟩, 738, 751 ⟨82T405⟩; **3**, 344 ⟨68T5323, 69T5703⟩, 346, 353 ⟨82T993⟩, 346, 363 ⟨78BSB223⟩, 347 ⟨74ACS(B)1077⟩, 358 ⟨86BSF283⟩, 360 ⟨81T197⟩, 411 ⟨80BSB247⟩; **4**, 207 ⟨63AK(20)225⟩, 903 ⟨74T1283⟩; **5**, 484, 551, 570 ⟨78BSB293⟩, 547 ⟨72T5341⟩, 547, 555 ⟨74T93⟩, 551 ⟨77ACS(B)460, 84T2663⟩, 552, 555 ⟨73T2077⟩, 556 ⟨74RTC258⟩, 558, 573, 587 ⟨83T3429⟩, 558, 573, 587, 611 ⟨84JCS(P1)785⟩, 560 ⟨82SUL69⟩, 570, 571, 608, 610, 618 ⟨78BSB229⟩, 570, 571, 610 ⟨80NJC47⟩, 570, 571, 613 ⟨81ACS(B)239⟩, 570, 575, 608, 610, 618 ⟨77BSB321⟩, 570, 575, 608, 610, 618 ⟨77BSB693⟩, 570, 608, 610 ⟨78BSB525⟩, 572 ⟨84BSF(2)204⟩, 573 ⟨81TL3635, 83T3419⟩, 573 ⟨85T5595⟩, 581, 611, 643, 644 ⟨85PS(25)297⟩, 600, 621 ⟨75ACS(B)538⟩, 600, 621 ⟨75ACS(B)604⟩, 618 ⟨78BSB299⟩, 76 ⟨73OSC(5)379⟩, 484 ⟨80ACS(B)597⟩; **6**, 233 ⟨60ACS2230, 62ACS117⟩, 303, 306 ⟨75RTC1⟩, 502 ⟨83T1729⟩, 503 ⟨82T1163⟩
Lawhead, J., **2**, 491, 496, 498 ⟨47JA927⟩
Lawless, E. W., **2**, 175 ⟨68IC391⟩
Lawless, G. A., **6**, 390 ⟨93JCS(D)3259⟩
Lawlor, F. E., **2**, 166 ⟨57JA6248⟩
Lawrance, Jr., W. W., **5**, 122 ⟨71TL3453⟩
Lawrence, C. A., **1**, 637 ⟨44JCS322⟩
Lawrence, K. A., **6**, 253 ⟨79JA5949⟩
Lawrence, L. M., **2**, 958 ⟨80JA2493⟩
Lawrence, N. J., **3**, 202 ⟨91JCS(P1)1893⟩; **4**, 960 ⟨93JCS(P1)67⟩
Lawrence, R., **6**, 111 ⟨89JOC1782⟩
Lawrence, R. F., **1**, 828 ⟨80TL4335, 83T3731⟩
Lawrence, T., **1**, 337 ⟨83AJC545⟩, 547 ⟨79JCS(P2)1535⟩
Lawrynowicz, W., **2**, 19 ⟨84JOC3216⟩; **4**, 1083 ⟨86TL5907⟩; **5**, 4 ⟨84JOC3216⟩
Laws, D. R. J., **3**, 649 ⟨68JCS(C)291⟩
Lawson, A., **4**, 86 ⟨69JCS(C)652⟩; **5**, 83 ⟨73OSC(5)1046⟩, 767 ⟨70JCS(C)1429⟩; **6**, 652 ⟨70JCS(C)1429⟩
Lawson, E. N., **1**, 572 ⟨93JOC2501⟩; **2**, 29 ⟨93JOC2501⟩
Lawson, I. W., **2**, 463 ⟨79TL2729⟩
Lawson, J. E., **2**, 198 ⟨61JA1256⟩, 215 ⟨58JA838⟩
Lawson, J. P., **3**, 496 ⟨81JOC3119⟩
Lawson, K. R., **4**, 319 ⟨84JCS(P1)865⟩, 843

⟨84JCS(P1)859⟩; **5**, 552, 559 ⟨84JCS(P1)865⟩, 558, 559 ⟨84JCS(P1)859⟩
Lawson, T., **1**, 700, 703 ⟨88JOC1064⟩
Lawston, I. W., **3**, 487 ⟨74CC466⟩, 537 ⟨74CC513⟩; **5**, 874 ⟨74CC466⟩
Lawter, L. M. L., **1**, 1026 ⟨86JOC2607⟩
Lawton, G., **5**, 351 ⟨85TL3171⟩, 354 ⟨85TL3167⟩
Lawyer, C. B., **2**, 396 ⟨61JOC1357⟩
Lay, J. O., **4**, 642, 655 ⟨82CC970⟩
Laycock, D. E., **1**, 669 ⟨78TL3307⟩; **2**, 171 ⟨81JOC289⟩, 577 ⟨80JOC291⟩
Layer, R. W., **3**, 404, 406 ⟨63CRV489⟩
Layh, M., **2**, 665 ⟨87CB1347⟩; **4**, 659, 698 ⟨91JOM(415)181, 90POL277⟩; **6**, 195 ⟨89JOM(364)289⟩, 195 ⟨89JOM(368)139⟩, 195 ⟨90POL277⟩, 195 ⟨92CB1547⟩, 391, 394 ⟨91AG(E)324⟩, 395 ⟨92AG(E)1364⟩
Layrolle, P., **6**, 264 ⟨91TL5965⟩
Layton, R. B., **1**, 314, 1048, 1049, 1139 ⟨71JA7320⟩, 1033 ⟨72CJC1105⟩; **2**, 1083 ⟨72CJC1005⟩
Layzell, T. P., **2**, 43, 528 ⟨93CC1673⟩; **5**, 652 ⟨88MI 520-01⟩, 795 ⟨88JOM(344)29⟩
Lazar, I., **5**, 652 ⟨88MI 520-01⟩, 795 ⟨88JOM(344)29⟩; **6**, 491 ⟨88JOM(344)29⟩
Lazar, J. G., **5**, 1134 ⟨84S1020⟩
Lazareva, M. I., **2**, 365 ⟨85T4279⟩
Lazarev, A. N., **3**, 540 ⟨65DOK(164)357, 68DOK(179)892⟩
Lazaridis, N. V., **2**, 16 ⟨86TL2715, 87JOC4307⟩; **3**, 26, 148 ⟨86TL2715⟩
Lazaro, E., **4**, 264 ⟨92RTC163⟩
Lazbin, I. M., **2**, 320 ⟨87JOC476⟩; **5**, 321 ⟨87JOC476⟩, 999 ⟨86JOC2669⟩
Lazerte, J. D., **4**, 737 ⟨53JA2698⟩; **6**, 32, 218 ⟨53USP2647933⟩, 218 ⟨55USP2704776⟩
Lazhko, E. I., **4**, 115 ⟨84JGU1354⟩, 116 ⟨85JGU1065⟩, 559, 766 ⟨90ZOB1539⟩; **5**, 876, 877 ⟨84ZOB1520⟩; **6**, 680, 681 ⟨84ZOB1520⟩, 696 ⟨85ZOB1194⟩
Lazier, W. A., **5**, 299 ⟨41OSC(1)99⟩
Lazovenko, A. N., **2**, 358 ⟨79ZOB841⟩
Lazraq, M., **1**, 766 ⟨91OM1771⟩; **2**, 949 ⟨87JA4411⟩; **3**, 503 ⟨87JA4411, 91POL1153⟩; **4**, 549 ⟨93CC569⟩, 662 ⟨87JA4411⟩; **6**, 186, 188 ⟨91POL1153⟩
Lazukina, I. A., **4**, 110 ⟨79S747⟩
Lazukina, L. A., **4**, 866 ⟨83ZOR2417⟩; **6**, 450 ⟨77ZOR290⟩, 619 ⟨84ZOB1110⟩
Lazzari, D., **2**, 918 ⟨93TL3311⟩, 918, 940 ⟨92SL981⟩
Le, N. A., **4**, 1083 ⟨89JA8491⟩
Le, P. H., **1**, 85 ⟨83JOC3370⟩
Le, T. Q., **1**, 712, 713 ⟨89H(28)1203⟩
Lea, K. R., **4**, 32 ⟨48JA1550⟩
Lea, R. E., **1**, 67 ⟨76JA5973⟩
Leach, J. B., **2**, 278 ⟨66CI(L)379⟩, 811 ⟨69MI 215-03⟩
Leach, M., **6**, 650 ⟨74JOC1166⟩
Leader, H., **1**, 80 ⟨76JOC3682⟩; **5**, 371, 403 ⟨90JCS(P1)3263⟩, 403, 404 ⟨88JCS(P2)2029⟩, 403, 404 ⟨89JCS(P1)765⟩
Leahy, D. E., **5**, 316 ⟨84JOC4784⟩
Leahy, J., **6**, 254, 255 ⟨75CJC2664⟩
Leake, N. H., **1**, 45 ⟨46JA1813⟩
Leake, P. H., **5**, 98 ⟨56CRV27⟩
Leal, G., **1**, 904 ⟨59JA3160⟩
Leamer, R. A., **2**, 342, 346 ⟨88JOC1768⟩
Leandri, G., **4**, 907 ⟨74JCS(P1)2357⟩
Leanna, M. R., **2**, 942 ⟨88S803⟩; **3**, 153 ⟨93TL4481⟩
Learmonth, E. K., **2**, 688, 688 ⟨36JCS327⟩
Learn, D. B., **2**, 386, 798 ⟨65JOC2471⟩
Learn, K., **1**, 15, 17, 23, 24, 25 ⟨82JOC4380⟩; **2**, 422 ⟨84JOC2438⟩; **5**, 639, 640 ⟨88JHC1873⟩
Learn, K. S., **1**, 815 ⟨86JA7873⟩, 815, 1098 ⟨88JA879⟩
Learned, A. E., **1**, 988 ⟨88CC301, 88JOC3122⟩
Leary, T. S., **3**, 612 ⟨45JA1249⟩
Leatham, M. J., **2**, 653 ⟨71CC1475⟩
Léauté, I., **3**, 337 ⟨90JCR(S)320⟩
Leavell, K. H., **1**, 801 ⟨73JA8096⟩
Leaver, D., **3**, 254, 371, 376 ⟨70JCS(C)1202⟩, 255, 360, 361 ⟨65JCS32⟩, 255, 376 ⟨64CI(L)461⟩, 356, 361 ⟨72JCS(P1)41⟩, 371 ⟨62JCS5104, 90JCS(P1)2881⟩; **4**, 850 ⟨72JCS(P1)41⟩; **6**, 113 ⟨65JCS32⟩
Leavitt, F. C., **2**, 989 ⟨59JA3163⟩
Leazer, J. L., **4**, 247 ⟨89JOC3718⟩
Leban, I., **5**, 834 ⟨79JOC1695⟩
Lebaud, J., **1**, 833, 834 ⟨79PS(7)69⟩; **5**, 552 ⟨79PS(7)69⟩
Lebedeva, A. B., **5**, 1075 ⟨85ZOB1308⟩, 1079 ⟨86ZOB1525⟩
Lebedeva, A. D., **2**, 1090 ⟨63IZV97⟩
Lebedeva, G. P., **2**, 886 ⟨81JGU1056⟩
Lebedev, A. T., **5**, 567, 571 ⟨87KGS845⟩
Lebedev, A. V., **5**, 1079 ⟨86ZOB1525⟩; **6**, 626 ⟨88ZOB393, 91ZOB1202⟩
Lebedev, B. A., **2**, 386 ⟨78ZOR240⟩; **5**, 361 ⟨77ZOR1144⟩
Lebedev, E. P., **3**, 336 ⟨79ZOB1084⟩
Lebedev, O. V., **5**, 838 ⟨58ZOB2305, 60IZV1783⟩, 838 ⟨61IZV477⟩; **6**, 150 ⟨60IZV1783⟩, 150 ⟨70KGS590⟩
Lebedev, S. A., **2**, 1093 ⟨71JGU122⟩; **4**, 619, 628 ⟨70ZOB939⟩, 640 ⟨76ZOB594⟩; **5**, 376, 389 ⟨76ZOB594⟩
Lebedev, V. B., **2**, 1032 ⟨64ZOB1899⟩; **4**, 125 ⟨86JGU2246⟩
Lebedev, V. N., **6**, 621 ⟨81MI 620-02⟩, 646 ⟨81MI 621-01⟩
Lebedev, V. V., **4**, 126 ⟨86JGU2242⟩
LeBel, N. A., **1**, 624, 628 ⟨57JA4720⟩, 625 ⟨59JA2799⟩, 625, 784 ⟨60JA4656⟩, 627 ⟨61JA3861⟩
Lebel, N. A., **2**, 137 ⟨69JOC3112⟩
LeBel, N. A., **3**, 433 ⟨85JOC3938, 85TL4331⟩
LeBelle, M. J., **1**, 142 ⟨74CJC761⟩, 823 ⟨72JA9261⟩; **2**, 151 ⟨74CJC761⟩; **4**, 391 ⟨74CJC761⟩
Leber, P. A., **1**, 784 ⟨94TL2675⟩, 799 ⟨86TL4107⟩
Leberre, A., **2**, 248 ⟨76BSF277⟩
Lebeveda, A. B., **6**, 626 ⟨88ZOB393, 91ZOB1202⟩
Lebioda, L., **1**, 545 ⟨86JA1940⟩, 633, 1053 ⟨84JA6006⟩; **2**, 581 ⟨91OM902⟩
Leblanc, J. C., **3**, 83 ⟨85JOM(292)225⟩
Leblanc, J.-P., **2**, 310 ⟨92TL6295⟩
Leblanc, R., **3**, 568 ⟨69TL509, 72T4431⟩; **4**, 800 ⟨69TL2441, 72T4039⟩
Leblanc, Y., **1**, 645 ⟨84JA5731⟩; **2**, 7 ⟨86JOC789⟩, 8 ⟨82TL4411⟩; **5**, 351 ⟨92TL5717⟩; **6**, 626 ⟨89JA2995⟩
Lebonc, A., **4**, 950 ⟨77CR(C)281⟩
Leborgne, F., **6**, 480 ⟨90MI 615-01⟩
LeBot, S., **4**, 580 ⟨86CC1208⟩
Lebreton, J., **1**, 752, 816, 818 ⟨87JOC3883⟩, 816 ⟨87TL723⟩, 816, 818 ⟨87TL3323, 88JA2925⟩, 818 ⟨88JOC4108, 88TL3547⟩, 818 ⟨90JOC227⟩; **2**, 9 ⟨87TL723⟩; **3**, 222 ⟨90JOC227⟩
Lebreton, L., **6**, 444 ⟨89JHC231⟩
Lebrun, H., **1**, 787 ⟨80T2497⟩; **3**, 582 ⟨78TL463⟩
Lecea, B., **2**, 14 ⟨85T4657, 86CJC2342⟩, 34 ⟨85T2903⟩; **3**, 4, 115 ⟨85T2903⟩, 632 ⟨89JCS(P1)1692⟩; **5**, 294 ⟨87TL1945⟩, 329 ⟨87JCS(P1)845⟩
Lecher, H., **2**, 127 ⟨22CB2423⟩, 133 ⟨15CB524⟩, 219 ⟨22CB1424⟩, 720 ⟨25CB409⟩
Lecher, H. Z., **2**, 122 ⟨55JOC475⟩
Lechevallier, A., **2**, 398 ⟨86S826⟩; **3**, 9, 123 ⟨78S63⟩; **4**, 108 ⟨86S826, 86S828⟩, 188 ⟨80SC83⟩
Lechler, R., **5**, 1084 ⟨89JOM(359)1⟩
Leckey, R. G., **2**, 491 ⟨61JCS1398⟩
Leckta, T., **1**, 919 ⟨89JA8867⟩
Leclerc, G., **2**, 753 ⟨88TL635⟩; **3**, 471 ⟨75BSF247⟩; **6**, 611 ⟨79BSF(2)520⟩, 619 ⟨74TL3765⟩
Lecolier, S., **4**, 211 ⟨87SC1467⟩; **5**, 994 ⟨88T6079⟩, 1105 ⟨88T1619⟩; **6**, 297 ⟨75FRP2235188⟩

Lecomte, J., **6**, 221, 222, 227 ⟨37CR(204)1927⟩
Lecomte, J.-P., **5**, 1097 ⟨81IC3119⟩
Lecomte, P., **2**, 726 ⟨81BCJ2847⟩
LeConte, J. N., **5**, 1035 ⟨49JA2240⟩
Leconte, M., **1**, 674 ⟨85CC1816⟩
Lecoq, A., **5**, 22 ⟨91TL2765⟩, 327, 351 ⟨93JA8898⟩; **6**, 488 ⟨93JA8898⟩
le Corre, M., **2**, 320 ⟨91SC1579⟩, 444 ⟨87PS(29)111⟩
Lectka, T., **1**, 688 ⟨89JOC3748⟩
L'Ecuyer, Ph., **2**, 775 ⟨61CJC862⟩
Lê Dao, H., **3**, 293 ⟨74HCA2201⟩
Leddy, B. P., **2**, 11 ⟨79JCS(P1)436⟩
Leder, J., **1**, 1092 ⟨83TL1463⟩
Ledercq, D., **4**, 648 ⟨78JOM(144)155⟩
Lederer, M., **4**, 187 ⟨63CB77⟩
Ledermann, M., **4**, 100 ⟨88AG(E)304⟩; **5**, 883 ⟨88AG(E)272⟩
Ledford, T. G., **3**, 657 ⟨60JA1786⟩
Ledig, B., **2**, 1001, 1002 ⟨93CB1887, 92AG(E)79⟩; **3**, 704 ⟨92AG(E)79⟩
Ledig, W. O., **3**, 208 ⟨71JOC4124⟩
Leditschke, H., **2**, 1097 ⟨74HOU(13/2b)26, 74HOU(13/2b)9⟩; **6**, 172, 204, 207 ⟨74HOU(13/2b)1⟩
Lednar, P. W., **2**, 582 ⟨80JCS(D)1448⟩
Lednicer, D., **1**, 1143, 1145 ⟨57JOC1248⟩; **2**, 28 ⟨87JOC5624⟩; **3**, 626 ⟨73OSC(5)578⟩
Lednor, P. W., **2**, 582 ⟨76AOC345⟩
Ledon, H., **3**, 475 ⟨74S347⟩
Ledoux, I., **1**, 474 ⟨88CC203⟩
Ledwith, A., **1**, 357 ⟨74S539⟩, 881 ⟨64PCS108⟩; **6**, 488 ⟨89MI 614-04⟩
Lee, A., **3**, 90, 96 ⟨92TL7499⟩
Lee, A. G., **2**, 597 ⟨B-71MI 211-01⟩
Lee, A. M. W., **2**, 82 ⟨83PAC589⟩
Lee, A. O., **4**, 956, 957 ⟨76JA2008⟩
Lee, A. W. M., **6**, 550 ⟨89SC547⟩, 554 ⟨88SC1531⟩
Lee, B., **1**, 148 ⟨89JOC458⟩; **2**, 639 ⟨91JCS(D)443⟩; **5**, 655 ⟨85JHC1121⟩
Lee, B. C., **1**, 130 ⟨91JPO449⟩
Lee, B. H., **5**, 328 ⟨83JOC24⟩, 572 ⟨94TL1135⟩
Lee, C., **2**, 197 ⟨89PS(45)35⟩, 718 ⟨90S391⟩; **3**, 116 ⟨79JOC3446⟩, 133 ⟨93SL427⟩; **5**, 47, 49, 52, 113 ⟨81TL1655⟩
Lee, C. C., **1**, 383 ⟨59JA4390⟩, 1112 ⟨82CJC735⟩; **3**, 100 ⟨47JA1998⟩
Lee, C.-D. S., **6**, 476 ⟨77MI 615-02⟩
Lee, C. H., **1**, 177 ⟨87H(26)1291⟩; **3**, 707 ⟨92JA4089⟩
Lee, C.-K., **1**, 173 ⟨91CC690⟩
Lee, C.-L., **4**, 545 ⟨86OM2220⟩, 691 ⟨81JA3764⟩
Lee, C. M., **1**, 769 ⟨90CAR(201)185⟩; **3**, 103 ⟨92T1039⟩; **5**, 625 ⟨62JOC3671⟩
Lee, C.-S., **2**, 397 ⟨92SL337⟩; **5**, 274, 294 ⟨86JA6054⟩
Lee, C. V., **1**, 1114 ⟨71TL2519⟩
Lee, C. W., **2**, 398 ⟨92JCS(P1)601⟩
Lee, C.-W., **2**, 730 ⟨94JCS(P1)717⟩
Lee, C.-Y., **4**, 616 ⟨87OM1861⟩
Lee, D., **1**, 1066 ⟨90SL591⟩
Lee, D. C., **2**, 928 ⟨89JOC2271⟩, 980 ⟨88JA7255⟩
Lee, D. G., **3**, 116 ⟨82JOC2790, 83JA3188⟩; **5**, 101 ⟨72CJC3741⟩
Lee, D. H., **1**, 417, 831 ⟨91JA4026⟩, 977, 1055 ⟨90JA878⟩; **2**, 79, 533 ⟨90JA878⟩
Lee, D. P., **2**, 7 ⟨75JOC1669⟩
Lee, D. S., **1**, 4, 10 ⟨92TL6673⟩
Lee, D.-S., **2**, 359, 363 ⟨94JOC914⟩
Lee, D. Y., **2**, 942 ⟨83JOC4156⟩
Lee, E., **1**, 4, 10 ⟨92TL6673⟩, 24, 439 ⟨93TL4541⟩, 455 ⟨90TL5039⟩, 1130, 1132 ⟨89TL827⟩; **3**, 223 ⟨90TL5039⟩
Lee, E. K. C., **1**, 784 ⟨72JA7235⟩
Lee, F. K. C., **5**, 173 ⟨82JA4686⟩
Lee, F. L., **4**, 1061 ⟨85JA7219⟩
Lee, G. A., **3**, 119 ⟨76TL1641⟩, 684 ⟨82S5508⟩
Lee, G. C. M., **1**, 1014, 1029 ⟨90JA9330⟩
Lee, G. E., **3**, 433 ⟨80JA373⟩
Lee, G.-H., **1**, 24, 439 ⟨93TL4541⟩; **4**, 263 ⟨90JOC5515⟩, 583, 585 ⟨86JOM(309)225⟩, 683 ⟨93JOM(461)207⟩, 711 ⟨93OM108⟩; **6**, 348 ⟨90JCS(D)3033⟩
Lee, G. M., **1**, 937 ⟨88T4671⟩; **6**, 23, 25 ⟨90JOC1281⟩
Lee, G. R., **3**, 698 ⟨89OM1538⟩; **5**, 444, 937 ⟨87OM1839⟩
Lee, G. T. S., **4**, 777 ⟨75CC107⟩
Lee, H., **1**, 4, 7 ⟨86JOC3502⟩, 611, 620 ⟨81CC277⟩; **2**, 287, 288 ⟨81CC277⟩; **3**, 100, 293 ⟨88JOC4587⟩; **5**, 736 ⟨93JOC7001⟩; **6**, 651 ⟨92TL4963⟩
Lee, H. D., **2**, 522, 526 ⟨80JOC4542⟩, 522, 526 ⟨82JOC863⟩, 5266 ⟨80JOC4540⟩; **4**, 815 ⟨82S193, 82S195⟩
Lee, H.-H., **1**, 624, 628 ⟨57JA4720⟩, 959 ⟨60JCS341⟩; **5**, 1044 ⟨79JCS(P1)757⟩
Lee, H.-K., **4**, 229 ⟨69JOC2080⟩; **6**, 195 ⟨93JOM(443)C39⟩
Lee, H. L., **1**, 697, 700, 703 ⟨88JOC4282⟩, 749, 828 ⟨87JOC1372⟩
Lee, H. M., **2**, 921, 922 ⟨77JA7613⟩
Lee, Ho. H., **5**, 1115 ⟨78S152⟩
Lee, H. P. C., **2**, 170 ⟨65USP3454631⟩
Lee, H. S., **3**, 13, 85 ⟨91TL6903⟩, 82 ⟨92OPP327⟩, 84 ⟨92OPP335⟩, 86 ⟨92OPP331⟩
Lee, H.-W., **6**, 468 ⟨94SC305⟩
Lee, H. Y., **3**, 325 ⟨89JA8954, 89JA8957⟩
Lee, I., **1**, 130 ⟨91JPO449⟩; **5**, 444 ⟨93JA4389⟩
Lee, J., **1**, 429, 1089 ⟨93TL415⟩, 1066 ⟨90SL591⟩; **3**, 107 ⟨51JOC1121⟩; **4**, 424 ⟨57JOC984⟩, 841 ⟨93TL415⟩; **5**, 10 ⟨61JOC1514⟩
Lee, J. B., **1**, 185, 200 ⟨67JOC855⟩; **2**, 7 ⟨68CC1350⟩, 979 ⟨82JA7491⟩; **3**, 508 ⟨81JA7358⟩, 613 ⟨67JOC855⟩; **5**, 13 ⟨66JA3440⟩
Lee, J. C., **3**, 13, 82 ⟨87TL2389⟩, 13, 85 ⟨91TL6731⟩, 82 ⟨88MI 303-01, 90MI 303-01⟩, 82 ⟨92OPP91⟩, 84 ⟨92OPP335⟩
Lee, J.-G., **1**, 28 ⟨75JA6873⟩, 799 ⟨80JA4841⟩; **2**, 25 ⟨88JOC3634⟩, 621, 627, 632 ⟨92TL3167⟩; **3**, 18, 133 ⟨90TL6677⟩, 18, 134 ⟨92SC2425⟩; **5**, 368 ⟨92TL3165⟩
Lee, J. H., **1**, 238 ⟨90TL2713⟩; **2**, 1054 ⟨69CC1393, 74BCJ946⟩; **4**, 849 ⟨90JOC5555⟩; **5**, 418, 428 ⟨87TL3261⟩
Lee, J. I., **3**, 130, 287 ⟨83JOC2608⟩; **5**, 126 ⟨85JOC560⟩
Lee, J. M., **1**, 314, 1050 ⟨90TL7627⟩; **3**, 140 ⟨91SC25⟩, 147 ⟨90TL7627⟩; **4**, 201, 217 ⟨93TL5769⟩
Lee, J.-S., **1**, 15, 16 ⟨93JCS(P1)9⟩; **5**, 924 ⟨79JA6750⟩
Lee, J.-T., **1**, 13 ⟨90TL4101⟩; **3**, 212 ⟨73JOC4071⟩; **5**, 282 ⟨73JOC4071⟩
Lee, J. W., **2**, 1034, 1035 ⟨90SL290⟩; **4**, 347 ⟨89SC2209⟩, 347, 781, 783, 840 ⟨90SC273⟩
Lee, J. Y., **2**, 441 ⟨85USP4507503, 85USP4507504⟩; **5**, 343 ⟨94SL197⟩
Lee, K., **2**, 394 ⟨89SC3055⟩, 457 ⟨91JOC5556⟩; **3**, 50, 198 ⟨91JOC5556⟩, 197 ⟨91SC279⟩, 231 ⟨91S213⟩; **4**, 260 ⟨91T8091⟩, 603, 649 ⟨83OM21⟩, 927, 929 ⟨91SC279⟩; **6**, 123 ⟨92TL5375⟩
Lee, K. E., **2**, 1092 ⟨93JOM(449)53⟩
Lee, K.-H., **1**, 4, 10 ⟨92TL6673⟩; **4**, 281 ⟨91LA243⟩, 1078 ⟨83JCS(P2)1071⟩
Lee, K.-I., **6**, 428 ⟨91JAN1083⟩
Lee, K.-J., **2**, 446 ⟨82JOC2768⟩; **5**, 862 ⟨94S1057⟩
Lee, K. S., **3**, 516 ⟨87OM433⟩
Lee, K.-W., **2**, 600 ⟨82OM1496⟩; **3**, 13, 82 ⟨87TL2389⟩, 13, 82, 94, 99 ⟨87JOC5030, 88H(27)1595⟩, 82 ⟨88MI 303-01⟩, 584, 586 ⟨74CC962, 74JOC3780, 77JA2622⟩
Lee, L., **2**, 404 ⟨71JA3816⟩
Lee, L. F., **1**, 1076 ⟨72JOC4468⟩; **2**, 763 ⟨76JOC863⟩; **3**, 335 ⟨85JOC3216⟩
Lee, L. F. H., **5**, 138 ⟨75JA4144⟩

Lee, L.-H., **6**, 429 ⟨65JA3943⟩
Lee, L. T. C., **2**, 524 ⟨66JA4757⟩; **4**, 832 ⟨67JA6804⟩
Lee, M., **1**, 357 ⟨85TL6163, 87TL2833⟩; **6**, 592 ⟨83CC294⟩
Lee, M. E., **3**, 502 ⟨83JA6725, 91OM2529⟩; **4**, 610, 611 ⟨83JA6725⟩, 610, 611, 621 ⟨85OM581, 82JOM(232)33⟩
Lee, M. H., **6**, 559 ⟨82MI 617-03⟩
Lee, M. J. L., **6**, 447 ⟨92JMC3641⟩
Lee, M. W., **1**, 785 ⟨78JOC518⟩
Lee, N. H., **3**, 118 ⟨86TL2875⟩, 215 ⟨85TL2875⟩
Lee, P., **4**, 370 ⟨91G471⟩
Lee, P. H., **2**, 446 ⟨89MI 208-01⟩; **3**, 498 ⟨88TL5413⟩
Lee, R. A., **4**, 656 ⟨91OM2191⟩
Lee, R. C.-Y., **4**, 564 ⟨79IC60⟩
Lee, R. E., **1**, 866 ⟨86JA5078⟩
Lee, R. W. K., **1**, 498 ⟨89JA3717⟩
Lee, S., **1**, 238 ⟨90TL2713⟩; **4**, 181 ⟨90CL2019⟩, 216 ⟨89TL6697⟩; **6**, 68 ⟨80JOC2096⟩
Lee, S.-B., **4**, 704 ⟨91S368⟩
Lee, S.-D., **4**, 704 ⟨91S368⟩
Lee, S. E., **3**, 13, 85 ⟨91TL6903⟩, 82 ⟨92OPP327⟩, 84 ⟨92OPP335⟩, 86 ⟨92OPP331⟩
Lee, S.-H., **1**, 307 ⟨88TL2195⟩; **2**, 324 ⟨88TL2195⟩, 365, 607, 624 ⟨86JA2445⟩, 517 ⟨85JOC1582⟩; **4**, 154 ⟨83SC129⟩, 536 ⟨90JCS(P1)2251⟩
Lee, S.-J., **1**, 24 ⟨91TL3551⟩; **2**, 981 ⟨80JA422⟩
Lee, S. J. H., **4**, 356, 643, 645 ⟨88T4033⟩
Lee, S.-M., **1**, 233 ⟨88JOC4472⟩
Lee, S. P., **2**, 25 ⟨78CL923⟩; **5**, 280 ⟨79SC31⟩
Lee, S. W., **2**, 66 ⟨88TL4453⟩, 177, 178, 179 ⟨42JOC81⟩; **3**, 492 ⟨66JA1953⟩; **4**, 336 ⟨90JOC2644⟩; **5**, 576 ⟨94JOC348⟩
Lee, S. Y., **1**, 636 ⟨69TL4001⟩, 805 ⟨88JOC1848⟩; **2**, 639 ⟨88JA8244⟩
Lee, T., **3**, 62 ⟨79TL2297⟩
Lee, T. B. K., **2**, 773 ⟨73JCS(P1)2778⟩; **4**, 48 ⟨75TL3979⟩
Lee, T. V., **5**, 199 ⟨89SL30⟩, 223 ⟨86TL4995⟩
Lee, V., **2**, 681 ⟨90OS(69)72⟩, 802 ⟨73JOC2435⟩; **3**, 531, 532 ⟨70OS(50)77⟩; **5**, 144 ⟨70OS77⟩
Lee, V. J., **1**, 300 ⟨91COS(4)69⟩, 312 ⟨91COS(4)139⟩; **5**, 639, 640 ⟨88JHC1873⟩
Lee, V. Y., **6**, 590 ⟨83CC235⟩
Lee, W., **1**, 1010 ⟨92BMC1619⟩
Lee, W. C., **1**, 790 ⟨87CC943⟩; **3**, 222 ⟨89JA7634⟩
Lee, W. J., **1**, 816 ⟨87TL4993⟩
Lee, W.-K., **1**, 151 ⟨89TL1197, 90JOC2578, 93JOC1109⟩; **2**, 922, 923 ⟨84JOC1838⟩; **3**, 182 ⟨89TL1197⟩; **4**, 511, 530 ⟨89TL1197⟩
Lee, W.-S., **4**, 260 ⟨91T8091⟩, 417 ⟨72CJC2902⟩, 689, 690 ⟨77JOM(127)93⟩
Lee, Y. H., **4**, 218 ⟨94JOC544⟩
Lee, Y.-J., **2**, 249 ⟨74TL381⟩
Lee, Y. K., **1**, 303 ⟨76JOC4044⟩
Lee, Y. W., **1**, 84 ⟨87JMC1505⟩; **2**, 504 ⟨93JCS(D)663⟩; **6**, 243 ⟨92JST(268)389, 93JCS(D)663⟩
Leebrick, J. R., **2**, 695 ⟨65FRP1397139⟩; **5**, 42 ⟨61JOC307⟩
Leeder, W. R., **2**, 1077, 1078 ⟨66IC1004⟩
Leedham, K., **1**, 1130 ⟨54JCS1634⟩; **6**, 2 ⟨53JCS1548⟩
Leeds, J. P., **2**, 314 ⟨88SC777⟩; **5**, 457 ⟨84JA6451⟩
Leeds, N. S., **4**, 209 ⟨54JA2943⟩
Leeming, S. A., **1**, 1140 ⟨88TL5017⟩
Leenson, I. A., **6**, 147 ⟨80MI 605-01⟩, 148 ⟨73IZV1149⟩
Leeper, F. J., **1**, 1131 ⟨89TL5017⟩
Leeper, R. W., **2**, 601 ⟨54CRV101⟩
Leermakers, P. A., **1**, 299 ⟨62JA1015⟩
Lee-Ruff, E., **1**, 106 ⟨86TL4853⟩
Lees, M., **5**, 606 ⟨82PS(12)325⟩, 760 ⟨84CC157⟩
Lees, W. J., **2**, 127 ⟨91JOC7328⟩; **3**, 159 ⟨93JOC1887⟩
Leese, R. M., **1**, 13 ⟨77S172⟩
Leeson, L. J., **2**, 111 ⟨61JPS193⟩
Leeson, P. D., **2**, 328 ⟨88TL6985⟩; **5**, 205 ⟨88JCS(P1)3103⟩
Lefebre, Y., **5**, 652 ⟨75JMC581⟩
Lefebvre, C., **3**, 55 ⟨84TL189⟩
Lefebvre, G., **1**, 750, 751 ⟨72T4209⟩; **3**, 642, 648 ⟨73T2437⟩
Leffek, K. T., **4**, 731 ⟨85CJC576⟩
Leffers, W., **2**, 1005 ⟨91OM1282⟩
Lefferts, J. L., **4**, 806 ⟨77JOM(142)39⟩; **6**, 293, 399 ⟨74JA6237⟩, 380, 381, 388, 399 ⟨76JOM(116)257⟩, 380, 391 ⟨77JOM(142)39⟩
Leffler, J. E., **2**, 687 ⟨72JOC3096⟩; **5**, 52 ⟨49CRV385⟩
Leffler, M. T., **2**, 261 ⟨48JA2434⟩; **5**, 10 ⟨55OSC(3)547⟩, 1119 ⟨48JA3212⟩
Lefker, B. A., **1**, 1130 ⟨87T5663⟩
Lefkowitz, S. L., **5**, 377 ⟨94BMC1865⟩
Lefloch, P., **5**, 1158 ⟨93CC1295⟩
LeFloc'h, Y., **4**, 322 ⟨70CR(C)436⟩
Lefort, D., **6**, 5 ⟨85T4503⟩
Lefrançois, J.-M., **1**, 836 ⟨91TL5969⟩; **2**, 594 ⟨92PAC1857⟩
Lefrançois, M., **1**, 665 ⟨69JOM(16)7⟩
Leftin, M. H., **1**, 436 ⟨85JOC1125⟩, 484 ⟨84CC617⟩, 970 ⟨85JOC1122⟩
Leftwick, A. P., **2**, 1016 ⟨62JCS4288⟩
Legedz, S., **2**, 198 ⟨85CC1670⟩
Léger, F., **2**, 759 ⟨91S649⟩
Léger, J. M., **2**, 938, 941 ⟨86T669⟩; **3**, 346 ⟨91TL1787⟩
Léger, L., **4**, 255 ⟨74CR(C)695, 75BSF657⟩
Leger, S., **1**, 532, 662 ⟨86TL539⟩
Léger-Lambert, M.-P., **2**, 938, 941 ⟨92JOC4705⟩
Leggetter, B. E., **2**, 128 ⟨63CJC2671⟩
Legido, M., **3**, 410 ⟨91CC524⟩; **4**, 514 ⟨91CC524⟩; **6**, 163 ⟨92TL3903⟩
Legin, G. Ya., **6**, 287 ⟨67DOK(176)1086⟩
Leginus, J. M., **3**, 45 ⟨84JA5598⟩, 434 ⟨84JOC3421⟩
Legiuns, J. M., **3**, 57 ⟨84TL5355⟩
Legler, J., **2**, 421 ⟨67AG(E)544⟩
le Goaller, R., **2**, 299 ⟨80T1223⟩
LeGoff, E., **1**, 521 ⟨82JOC4779⟩; **3**, 303 ⟨89SC787⟩
LeGoff, N., **1**, 999 ⟨73T883⟩
LeGoffic, F., **4**, 465 ⟨85S62⟩
Legrand, L., **5**, 586 ⟨66BSF3828⟩
Legrand, N., **3**, 703 ⟨92ICA73⟩
Legrand, Y., **2**, 1046 ⟨77JCR(S)105⟩
Legras, Y., **2**, 478 ⟨85GEP3512781⟩
Legraverend, M., **5**, 199 ⟨92H(34)739⟩
Legrel, P., **5**, 652 ⟨87S306⟩
Legreneur, S., **6**, 560 ⟨84JCS(P1)1005⟩
LeGrow, G. E., **4**, 137, 138, 141, 811 ⟨76JOM(122)31⟩; **5**, 410 ⟨65CJC1175⟩
Legters, J., **5**, 229 ⟨92RCT211⟩
Legueut, C., **1**, 763 ⟨81TL1595⟩
Legzdins, P., **2**, 573 ⟨88OM1877⟩; **4**, 656 ⟨93OM4572⟩
Lehm, H.-U., **4**, 410 ⟨82JPR832⟩
Lehman, H. A., **4**, 550 ⟨83EGP201449⟩
Lehman, H., **1**, 58 ⟨55AG710⟩
Lehman, P. G., **1**, 99 ⟨74TL2395⟩
Lehman, T. von, **1**, 1114 ⟨75JA1531⟩
Lehmann, B., **1**, 522 ⟨75LA1113⟩
Lehmann, C., **4**, 745 ⟨88JOC5554⟩; **5**, 138 ⟨91JA910⟩
Lehmann, G., **6**, 412, 418 ⟨30CB1221⟩, 508 ⟨47JA3150⟩
Lehmann, H. A., **4**, 561 ⟨84EGP214610⟩; **5**, 560 ⟨88ZC269⟩
Lehmann, H., **6**, 139 ⟨90CZ176⟩
Lehmann, H.-D., **4**, 901 ⟨68AG38⟩
Lehmann, L., **3**, 706 ⟨93CB427, 93JOM(461)81⟩
Lehmann, M., **3**, 465, 466 ⟨02CB897⟩
Lehmann, R., **1**, 113 ⟨93T10175⟩; **2**, 557, 907, 999 ⟨84TL745⟩; **4**, 378 ⟨84TL745⟩
Lehmann, W. J., **2**, 109 ⟨59JCP(30)1222⟩
Lehmeier, T., **2**, 529 ⟨85LA2246⟩
Lehmkhul, H., **2**, 595 ⟨70HOU(13/4)9⟩

Lehmkuhel, H., **4**, 659 ⟨85CB4248⟩
Lehmkuhl, H., **1**, 495 ⟨72JOM(34)1⟩, 665 ⟨60LA(629)1⟩; **2**, 561, 562 ⟨81BSF(2)87⟩, 577 ⟨85CB2407⟩, 581 ⟨83CB159, 83CB176⟩, 595 ⟨64AG(E)640, 68LA(719)40⟩, 596 ⟨60LA(629)1⟩, 967 ⟨70HOU(13/4)23⟩, 1041 ⟨60LA(639)33⟩, 1098 ⟨70HOU(13/4)154⟩; **4**, 654 ⟨84CB383⟩, 698 ⟨66TL2315, 73LA708⟩, 698 ⟨87JOM(333)155⟩, 699 ⟨60LA(629)1, 73LA715⟩, 720 ⟨90ZAAC(581)41⟩; **6**, 204, 208 ⟨70HOU(13/4)1⟩
Lehn, J., **2**, 860 ⟨78HCA2190⟩
Lehn, J.-M., **5**, 114 ⟨76JA1743⟩; 360 ⟨85TH86⟩
Lehn, W. L., **3**, 89 ⟨57JOC1135⟩
Lehnen, H., **3**, 557 ⟨91TL4041⟩; **4**, 559 ⟨91TL4279⟩; **5**, 691 ⟨91TL4279⟩
Lehner, H., **4**, 163 ⟨80T1345⟩
Lehner, J., **5**, 588, 632 ⟨85ZC219⟩; **6**, 559 ⟨85ZC219⟩
Lehnert, W., **3**, 618 ⟨71TL1501⟩
Lehnig, M., **1**, 922 ⟨89JA5845⟩
Lehong, N., **4**, 224 ⟨80CAR(80)C17⟩
Lehr, F., **1**, 156 ⟨76HCA2213⟩, 156, 157 ⟨77TL1161⟩, 157 ⟨76AG(E)505⟩; **2**, 783 ⟨79C1⟩; **3**, 57 ⟨78AG(E)458⟩, 191 ⟨79HCA2258⟩; **4**, 540 ⟨76AG(E)505⟩
Lehr, W., **5**, 341 ⟨81CB80⟩
Lehrich, F., **1**, 955, 958 ⟨87TL2697⟩
Lei, B., **1**, 774 ⟨90JA4609, 93JOC2186⟩
Lei, G., **2**, 879 ⟨B-85MI 211-04⟩
Lei, K. L., **1**, 2, 8 ⟨79JOC641⟩; **3**, 129 ⟨79JOC641⟩
Lei, X., **3**, 145 ⟨86TL4671⟩
Lei, X.-H., **5**, 600, 621 ⟨91OPP435⟩
Leibetseder, H., **5**, 344 ⟨86AP(319)682⟩
Leibner, J. E., **1**, 936 ⟨79JOC449⟩
Leicester, H. M., **2**, 729 ⟨43OSC(2)238⟩; **4**, 4 ⟨38JA864⟩, 18 ⟨39JA938⟩
Leicester, J., **2**, 19 ⟨63OSC(4)525⟩
Leicher, W., **5**, 208 ⟨52LA(577)153⟩
Leigh, A. J., **5**, 199 ⟨89SL30⟩
Leigh, G. J., **2**, 579 ⟨82COMC-I(5)541⟩
Leight, W. J., **1**, 799 ⟨90CJC1961⟩
Leighton, J. L., **3**, 32 ⟨92JA9434⟩
Leighton, J. P., **1**, 1005 ⟨75JCS(P1)1628⟩, 1005, 1006 ⟨73TL1019⟩
Leiiring, E., **4**, 453, 454, 477 ⟨68CB3619⟩, 453, 454, 477, 481 ⟨68CB4032⟩, 455 ⟨74SRI453⟩
Leij, M. van der, **5**, 562 ⟨78TL811⟩
Leimeister, H., **5**, 316 ⟨75ZAAC(415)115⟩, 1016 ⟨71JOM(31)C3⟩
Leimer, K. R., **5**, 226 ⟨77JC355⟩
Leimgruber, W., **2**, 749 ⟨70USP3542848⟩
Lein, Jr., G. H., **2**, 600 ⟨78JOC750⟩
Leinberger, P., **6**, 617 ⟨78CB698⟩
Leinwetter, M., **4**, 177, 181 ⟨78JOC438⟩
Leir, C. M., **2**, 166 ⟨72JOC887⟩
Leis, D. G., **6**, 74, 75 ⟨64JOC2773⟩
Leise, M., **1**, 956 ⟨93CB1077⟩; **4**, 714 ⟨93JOM(447)C1⟩
Leising, M., **1**, 325, 331, 920 ⟨86CB444⟩; **3**, 621 ⟨85JA4332⟩
Leissring, E., **2**, 820, 821 ⟨81TL4475⟩; **4**, 337 ⟨72JPR66, 72T2587⟩, 337 ⟨79ZC57⟩, 453 ⟨85ZC172, 88JOM(355)71⟩, 482 ⟨78PS(4)59⟩, 548 ⟨79ZC416, 81T753⟩, 548 ⟨90PS(49/50)355⟩; **5**, 883 ⟨80TL1141, 83ZC99⟩; **6**, 686 ⟨80TL1141, 81TL4475⟩, 686 ⟨81ZC407, 83ZC99⟩
Leistner, D., **5**, 365 ⟨87ZC27⟩
Leistner, S., **5**, 587 ⟨68ZC376⟩
Leitch, L. C., **2**, 377 ⟨54CJC545⟩, 1012, 1015 ⟨54CJC500⟩; **4**, 311 ⟨68CJC385⟩
Leites, L. A., **4**, 629 ⟨92CRV175⟩; **6**, 170 ⟨71IZV1505⟩
Leith, B., **5**, 1152 ⟨93PS(76)1⟩
Leitner, P., **4**, 290 ⟨87JOM(336)153⟩; **5**, 241 ⟨89JOM(377)281⟩
Leito, I., **5**, 689 ⟨93JCS(P2)655⟩
Leitz, H. F., **2**, 393 ⟨75CB1924⟩
Lejczak, B., **4**, 462 ⟨83CJC2425⟩, 467 ⟨84S577⟩
Lejeune, G., **3**, 91, 98 ⟨55BSF1594⟩
Lekar, M. G., **5**, 595 ⟨71ZOR2352⟩
Lelandais, D., **3**, 210 ⟨81CJC945⟩
Lelandais, P., **1**, 133 ⟨92TL1855⟩; **3**, 124 ⟨92TL1855⟩
Leland, J. K., **6**, 375 ⟨87JCS(D)249⟩
Lele, S. V.,**3**, 280 ⟨83CI(L)479⟩
Lelieveld, P., **4**, 264 ⟨92RTC163⟩
Lellouche, I., **2**, 64 ⟨89TL5141⟩, 212, 225 ⟨92T1883⟩
Lellouche, J.-P., **2**, 47 ⟨89AG(E)755⟩, 753 ⟨86TL6193⟩; **6**, 480 ⟨90MI 615-01⟩
LeMahieu, R. A., **1**, 1029 ⟨89JMC1820⟩; **2**, 659 ⟨76JOC977⟩; **3**, 234 ⟨70JOC3045⟩; **6**, 72 ⟨70TL5111⟩
LeMahieu, R., **1**, 498 ⟨64JA5570⟩
Lemaire, B., **2**, 624 ⟨71C59⟩
Lemaire, H., **2**, 191 ⟨55JA6231⟩
Lemaire, M., **2**, 317 ⟨88S868⟩, 620 ⟨82T2339⟩
Lemaire-Audoire, S., **2**, 304 ⟨93SL680⟩
Lemal, D. M., **1**, 790 ⟨66JA1335, 66JA2858⟩, 862 ⟨80JA6633⟩; **2**, 153 ⟨73JA920⟩, 153 ⟨73JA922⟩, 379, 388 ⟨64JA2395⟩; **3**, 353 ⟨80JA6634⟩, 446 ⟨64JA2395⟩, 453 ⟨65JA393⟩, 497 ⟨64TL245⟩; **4**, 46 ⟨89JOC5520, 90JA6715⟩, 160, 755 ⟨89JOC5502⟩; **6**, 697 ⟨64TL245⟩
Lemal, M., **1**, 61 ⟨B-70MI 102-01⟩
Lemanach, L., **6**, 7 ⟨90JFC(47)95⟩
Lemarié, M., **4**, 870 ⟨92TL6131⟩; **5**, 527 ⟨91TL7411⟩, 563 ⟨92TL6131⟩
LeMasters, W. L., **2**, 358 ⟨77JOC875⟩
Lemay, G., **3**, 131, 405, 459 ⟨80TL155⟩
Lemenovskii, D. A., **4**, 696 ⟨74JOM(65)131⟩, 723 ⟨69IZV2030, 72IZV2594⟩
Lemeune, S., **2**, 304 ⟨93SL680⟩
Lemieux, R., **4**, 217 ⟨90CJC897⟩
Lemieux, R. U., **1**, 24 ⟨49JA2870⟩
Lemieux, R. U.,**3**, 2, 113 ⟨56JOC478⟩
Lemieux, R. U., **4**, 224 ⟨51CJC1079⟩
Lemin, A. J., **3**, 227 ⟨54JA5672⟩
Lemire, A. E., **4**, 61 ⟨83JFC(22)557⟩
Lemke, T. L., **2**, 163 ⟨83JHC899⟩
Lemmen, P., **3**, 696, 708 ⟨89S718⟩
Lemmen, T. H., **2**, 569 ⟨87OM2354⟩
Lemmer, K. L., **2**, 366 ⟨78JMC704⟩
Lemmer, L. J., **2**, 366 ⟨78JMC704⟩
Lemmerman, L. F., **1**, 881 ⟨55OSC(3)44⟩
Lemmon, D. H., **2**, 1093 ⟨67SA(A)1169⟩
LeMoing-Orliac, M. A., **3**, 567, 574 ⟨82T527⟩
Lemor, A., **2**, 317 ⟨88S900⟩
Lempert, K., **5**, 1118 ⟨74TL2939⟩; **6**, 546 ⟨90JCS(P1)2009⟩, 560 ⟨83JCS(P1)2011⟩
Lempert-Sréter, M., **6**, 560 ⟨83JCS(P1)2011⟩
Lenaers, R., **5**, 219, 833 ⟨62HCA441⟩, 322, 763, 765, 833, 864 ⟨62CRV155⟩
Lenarda, M., **3**, 314 ⟨72JOM(46)C29⟩, 568 ⟨72JOM(39)217⟩
Lencu, V., **6**, 223 ⟨81MIP68355⟩
Lender, E., **4**, 411 ⟨68M990⟩
Lenfant, M., **2**, 331 ⟨93BMC681⟩
Lengfeld, F., **2**, 100 ⟨1895CB449⟩; **6**, 452 ⟨1895MI 614-01⟩
Lenglet, L., **1**, 634 ⟨87TL1997⟩
Lengyel, I., **3**, 722 ⟨64JA1356⟩; **4**, 297, 327 ⟨66CB1944⟩, 309 ⟨66CB1932⟩
Lenhard, R. H., **4**, 179 ⟨64JA2183⟩
Lenhardt, H., **1**, 325, 326 ⟨84CB2132⟩
Lenk, C. T., **2**, 34 ⟨53JA3493⟩; **5**, 320 ⟨53JA3493⟩
Lennartz, H.-W., **1**, 801 ⟨82CB2508⟩
Lennartz, J., **4**, 166 ⟨77LA1807⟩
Lennon, J., **4**, 930 ⟨79JA83⟩; **5**, 924 ⟨79JA83⟩
Lennon, J. M., **4**, 137 ⟨75CJC332⟩
Lennon, M., **3**, 294 ⟨75JCS(P1)622⟩

Lennon, P., **2**, 577 ⟨76JOM(108)93, 77JA8426⟩
Le Noble, W. J., **1**, 891 ⟨78MI 119-01⟩, 905 ⟨84JA810⟩; **2**, 674 ⟨70S1⟩; **3**, 345 ⟨91JOC5932⟩, 369 ⟨92JOC3120⟩
Lenoir, D., **1**, 687 ⟨89S883⟩, 688, 689 ⟨77S553, 80JCR(S)396, 88JOC661⟩, 906 ⟨80JA683⟩; **5**, 244 ⟨85AG(E)1041⟩, 543 ⟨85AG1051⟩
Lensch, C., **6**, 380 ⟨85ZC309⟩
Lensen, N., **2**, 315 ⟨91SL625⟩; **3**, 42 ⟨92JOC4563⟩
Lensink, C., **2**, 330 ⟨93TA215⟩
Lenthe, M., **6**, 606 ⟨85GEP3337939⟩
Lentsch, S. E., **3**, 120 ⟨80TL4695⟩
Lentz, C. M., **1**, 304 ⟨77TL3215⟩
Lentz, D., **3**, 380 ⟨79AG(E)944⟩, 694, 703 ⟨94AG(E)1315⟩, 694, 716 ⟨91AG(E)1531⟩, 699, 703 ⟨84JFC(24)523⟩, 703, 722 ⟨88CB1445, 92CC1523⟩, 716 ⟨93AG(E)1456⟩; **4**, 683 ⟨88CB1413, 90IC4396⟩, 683 ⟨92OM2916⟩, 821 ⟨90AG(E)315⟩; **6**, 373 ⟨92ZAAC(617)53⟩, 430 ⟨81ZAAC(472)26⟩, 611 ⟨84JFC(24)523, 88CB1445⟩, 613, 623 ⟨85IC4665⟩, 721 ⟨90CB751⟩
Lenz, B. G., **2**, 400 ⟨85HOU(E11)943⟩; **3**, 171, 368 ⟨84TL5947, 84CC502⟩; **5**, 472, 488 ⟨85HOU(E11/1)911⟩, 472, 495 ⟨85HOU(E11/2)1326⟩, 562 ⟨85JOC2930⟩, 734, 736 ⟨84RTC342⟩
Lenz, F., **1**, 1123 ⟨76JOC2331⟩
Lenz, G., **4**, 146 ⟨86CB1977⟩; **6**, 166, 693 ⟨86CB1977⟩
Lenz, G. R., **5**, 288 ⟨81JHC691⟩, 288, 290, 297, 302 ⟨87S421⟩, 288, 297, 302 ⟨78S489⟩
Lenz, R., **1**, 327 ⟨93HCA441⟩
Leo, G. C., **3**, 51, 200 ⟨88JOC4069⟩
Leon, C., **2**, 561 ⟨66TL875⟩
León, E. I., **5**, 1118 ⟨90JCS(P1)2417⟩; **6**, 633, 634 ⟨89CC452⟩
Leon, G., **4**, 843 ⟨91S481⟩
Leonard, E. K., **2**, 413 ⟨59JOC1419⟩
Leonard, J., **1**, 305, 306 ⟨94MI 107-01⟩, 791 ⟨94TL1071⟩
Leonard, J. D., **5**, 306 ⟨87JMC498⟩
Leonard, J. E., **3**, 120 ⟨80TL4695⟩
Leonard, Jr., W. R., **6**, 107 ⟨91TL597⟩
Leonard, M. J., **2**, 346 ⟨84T4363⟩
Leonard, N. J., **1**, 55, 91 ⟨57JA5279⟩, 82 ⟨57JA6210⟩, 884 ⟨52JA1704⟩; **2**, 33 ⟨46JA2402⟩, 129 ⟨67JOC677⟩, 149 ⟨62JOC282⟩, 943 ⟨82JA2613⟩, 1008 ⟨63JOC3021⟩; **3**, 422 ⟨63JOC3021⟩; **4**, 305 ⟨67JA4456⟩; **5**, 690 ⟨83JOC3⟩, 1107 ⟨65JOC2660⟩
Leonard, R., **5**, 755 ⟨83SC1083⟩
Leonard, W. R., **1**, 244 ⟨85TL6431⟩; **3**, 175 ⟨85TL6431⟩
Léonard-Coppens, A. M., **1**, 592, 593, 716 ⟨76TL3227⟩
Leonardo, C. L., **5**, 1068 ⟨89JA355⟩
Leone, C. L., **1**, 43 ⟨91TL1691⟩
Leone, J. T., **1**, 6 ⟨57JA1546⟩
Leone-Bay, A., **2**, 7, 621 ⟨90JOC3415, 88JOC2858⟩; **3**, 129 ⟨86JOC2378⟩
Leonesi, D., **5**, 1054 ⟨63AC(R)1113⟩
Leong, A. Y. W., **3**, 20, 55 ⟨78TL5175⟩
Leong, W., **1**, 449 ⟨89TL6487⟩
Leong, W. W.-H., **3**, 240 ⟨83TL2361⟩; **4**, 951 ⟨83TL2361⟩
Leonhard, K., **6**, 352 ⟨79AG663, 80CB1654⟩
Leonhardt, J., **1**, 574 ⟨94AG(E)117⟩
Leonhardt, W., **6**, 366 ⟨83AG(E)632⟩
Leoni, P., **4**, 693 ⟨95JOM(488)39⟩
Leonov, A. A., **2**, 1028 ⟨85JGU1771⟩; **3**, 591 ⟨83JGU205, 85JGU26⟩
Leonowicz, M. E., **4**, 687 ⟨80JA5101⟩
Leon-Ruaud, P., **2**, 133 ⟨91T5185⟩
Leonte, M., **3**, 446 ⟨62CA(56)4645⟩
Leonteva, I. N., **5**, 737 ⟨77JGU916⟩
Leont'eva, I. V., **4**, 551 ⟨80TL2931, 92ZOB263⟩, 557 ⟨84ZOB36, 86ZOB1220⟩, 557 ⟨88ZOB1987, 88ZOB1998⟩, 557 ⟨88ZOB2012⟩
Leont'eva, L. I., **4**, 719 ⟨73DOK(209)869, 74IZV2645⟩
Leont'eva, L. M., **4**, 524 ⟨74MI 411-01⟩; **6**, 65 ⟨76IZV2547⟩, 169 ⟨74MI 605-01⟩
Leopold, D., **3**, 380 ⟨89ZAAC(578)110⟩
Leopold, E. J., **3**, 57 ⟨86OS(64)164⟩
Lepage, Y., **3**, 66 ⟨80S898⟩
Lepesa, A. M., **5**, 707 ⟨61JGU3192⟩, 707, 861 ⟨61ZOB3424⟩, 708 ⟨62JGU167, 62ZOB171⟩
Lépine, F., **5**, 574 ⟨83TL3815⟩
Lepisto, M., **6**, 143 ⟨88ACS(B)666⟩
L'Eplattenier, F. A., **2**, 746 ⟨76S543⟩
L'Eplattenier, F., **5**, 455 ⟨70HCA1091⟩
Leplawy, M. T., **4**, 213 ⟨86S649⟩; **5**, 278 ⟨60T39⟩
Lepley, A. R., **1**, 148 ⟨66JOC2047⟩, 148 ⟨66JOC2061⟩, 399 ⟨65JOC3888, 69JA1237⟩, 407 ⟨B-71MI 109-02⟩
Leplyanin, G. V., **4**, 267 ⟨90ZOR1259⟩; **6**, 302 ⟨86MI 610-01⟩
Lepoire, D. M., **2**, 104 ⟨83JOC3379⟩
Lepone, G. E., **6**, 630 ⟨84JHC61⟩
Leppard, D. G., **2**, 831 ⟨76HCA695⟩
Leppert, E., **5**, 537 ⟨71S435, 73MI 512-01⟩, 538, 540 ⟨72MI 512-01⟩, 992 ⟨76S329⟩
Leppik, R. A., **3**, 102 ⟨71AJC1883⟩
le Quan, M., **2**, 1093 ⟨67BSF777⟩
Lequan, M., **4**, 534 ⟨76JOM(113)C13⟩
Lequitte, M., **2**, 457 ⟨86TL3515⟩; **4**, 338 ⟨86TL3515⟩
Lera, A. R. de, **1**, 419 ⟨91COS(5)699⟩
Leray, N., **1**, 923 ⟨77JA8348⟩
Lerch, A., **4**, 49, 86 ⟨63CB2266⟩
Lerch, I., **6**, 655 ⟨57GEP1002355⟩
Lerch, U., **2**, 162 ⟨71JOC3687⟩, 162 ⟨71JOC3861⟩
Lerche, H., **3**, 72, 242 ⟨90CB399⟩
Lerdal, D. A., **2**, 348 ⟨78JOC2662⟩
Lerer, M., **4**, 212 ⟨67BSF4172⟩
Leriverend, C., **5**, 248 ⟨94S761⟩, 587, 611 ⟨93TL6741⟩
Leriverend, M.-L., **1**, 387 ⟨75CR(C)791⟩, 422 ⟨68TL2101, 70BSF2981⟩
Leriverend, P., **1**, 387 ⟨75CR(C)791⟩, 412 ⟨66TL6115⟩; **3**, 343 ⟨78RTC121⟩
Lerman, O., **2**, 16 ⟨83JOC724⟩, 613 ⟨81JOC4629, 84JOC806⟩; **4**, 44 ⟨84JOC806⟩; **5**, 202 ⟨80JOC672, 81JOC4629⟩, 202 ⟨84JOC806⟩
Lermontov, S. A., **2**, 280 ⟨93TL703⟩
Lerner, L. R., **4**, 24 ⟨67JOC2669⟩
Lerner, M., **5**, 595, 617 ⟨39JA105⟩
Lerouge, P., **1**, 985 ⟨84TL1987⟩; **2**, 282 ⟨82TL1557, 84TL1983⟩; **3**, 38 ⟨82TL1557⟩, 74 ⟨84TL1987⟩; **6**, 93 ⟨82TL1557, 85BSF1219⟩
Leroux, Y., **2**, 939 ⟨71JOM(30)295, 71TL591⟩, 939 ⟨71TL593, 73TL2585⟩; **3**, 214 ⟨73TL2585⟩; **5**, 427 ⟨71TL591, 73TL2585⟩
Leroy, J., **2**, 10 ⟨79JOC3406⟩, 19 ⟨79TL353⟩; **4**, 731 ⟨87JOC290⟩; **5**, 659 ⟨73TL2771⟩
Lerpinière, J., **3**, 541 ⟨93JCS(P1)1549⟩
Lerschinky, K. L., **2**, 366, 368 ⟨76JOC4028⟩
Lerstrup, K. A., **4**, 876 ⟨88SMB425⟩, 876, 877 ⟨87JA4115⟩; **5**, 627, 642, 726 ⟨79CC1102⟩; **6**, 561 ⟨91JOC5684⟩
Lerstrup, K., **3**, 39 ⟨84CC89⟩; **4**, 876 ⟨82CC336, 84JA8303⟩, 876 ⟨87SM647⟩; **6**, 591 ⟨84CC89⟩, 592 ⟨83CC294⟩
Lesage, M., **1**, 5, 7, 9, 10, 331, 936 ⟨88JOC3641⟩, 5, 9, 10, 936 ⟨89TL2733, 89JOC2492⟩; **2**, 536 ⟨90JOC5413⟩; **3**, 468 ⟨59JOC2071⟩
Lesage, M. D., **2**, 141, 265 ⟨90JOC6248⟩
Lesbre, M., **2**, 814, 815 ⟨B-71MI 215-02⟩, 946, 948, 949 ⟨B-71MI 218-01⟩, 948 ⟨61CR(253)673⟩; **4**, 521 ⟨72JOM(34)C18⟩
Leschinsky, K. L., **2**, 401 ⟨79JOC305⟩, 745 ⟨78JOC1460⟩; **5**, 632 ⟨75JOC66⟩; **6**, 28 ⟨86S224⟩
Leshina, T. V., **4**, 656 ⟨84ZOB2025⟩
Lesiak, K., **5**, 1057 ⟨75S785⟩; **6**, 122 ⟨75S785⟩, 264 ⟨91BMC357⟩

Lesiak, T., **5**, 990 ⟨93JPR294⟩, 991 ⟨88JPR825⟩
Lesiecki, H., **5**, 397 ⟨77AG(E)269⟩, 403 ⟨79CB763, 79CB793⟩; **6**, 161 ⟨78ZN849⟩
Lesimple, P., **4**, 950, 952 ⟨86TL6201⟩
Leslie, E. J., **2**, 293 ⟨77JCS(D)527⟩
Leslie, V. J., **1**, 815 ⟨67TL4537⟩
Lesma, G., **1**, 601 ⟨84JCS(P1)1593⟩; **5**, 1045 ⟨90TL7229⟩
Lespagnol, C., **3**, 619 ⟨86TL347⟩; **5**, 223, 374 ⟨88JHC59, 88JHC49⟩, 718 ⟨86JHC183⟩
Lessard, J., **1**, 785 ⟨73CJC981⟩; **3**, 43 ⟨78CJC119⟩; **5**, 316 ⟨78JOC3750⟩, 316, 318 ⟨74JOC3136⟩
Lessard, M. V., **1**, 1122 ⟨68JA2432⟩; **2**, 759 ⟨70JOC1204⟩
Lessen, T. A., **2**, 249 ⟨86TL2099⟩
Lesser, R., **2**, 734 ⟨12CB1835⟩, 768 ⟨24CB1077⟩; **3**, 256 ⟨12CB1835⟩
Lessmann, K., **1**, 925 ⟨93SL286⟩, 936 ⟨93SL283⟩
Lester, C. T., **3**, 449 ⟨56JA5584⟩
Lester, D., **4**, 67 ⟨83TL139⟩
Lester, D. J., **2**, 390 ⟨80JCS(P1)1212⟩; **3**, 18, 134 ⟨80JCS(P1)1212⟩, 88 ⟨78CC952⟩, 118, 464 ⟨79CC705⟩; **5**, 366 ⟨80JCS(P1)1212⟩
Leston, G., **2**, 393 ⟨63OSC(4)368⟩; **4**, 430, 431, 432 ⟨62JOC3598⟩
LeSuer, W. M., **2**, 462, 842 ⟨55JA3526⟩; **5**, 83 ⟨63OSC(4)919⟩
Lesuisse, D., **1**, 505 ⟨83JA5373⟩
Lesur, B. M., **1**, 305 ⟨83TL2231⟩; **5**, 152 ⟨83TL2231⟩, 152 ⟨89JA2984⟩
Letchford, B. R., **4**, 730 ⟨62CI(L)1472⟩
Lete, E., **2**, 89 ⟨88T6681⟩
Letellier, M., **3**, 60 ⟨87JOC4047⟩, 440 ⟨88SC1975⟩; **4**, 512, 518 ⟨88SC1975⟩
Leth, T. D., **6**, 641 ⟨63JMC275⟩
Leto, J. R., **1**, 530 ⟨61JA2944⟩
Leto, M. F., **1**, 530 ⟨61JA2944⟩
Letourneau, F., **6**, 417, 418 ⟨73T3309⟩
LeTourneau, M. E., **4**, 79, 780 ⟨85CC678⟩
Letourneux, Y., **5**, 296 ⟨85SC1233⟩
Letsch, G., **2**, 381 ⟨66JPR96⟩
Letschert, H.-P., **1**, 370, 371 ⟨94AG(E)683⟩
Letsinger, R. L., **1**, 113, 114 ⟨50JA849⟩, 846 ⟨50JA4842⟩; **2**, 450 ⟨68JA2459⟩, 556 ⟨47JA950⟩, 840, 841, 845 ⟨68JA3459⟩; **3**, 100 ⟨61JA193⟩, 664 ⟨66JA2884⟩; **4**, 28 ⟨51JA4476⟩
Lett, R., **1**, 1134 ⟨78JA4618, 78JA4620⟩; **2**, 629 ⟨78JA4618⟩; **3**, 419, 420 ⟨88TL6541⟩; **5**, 173 ⟨78JA4618⟩, 173, 204 ⟨78JA4620⟩
Lett, R. M., **1**, 812 ⟨82JOC2268⟩; **2**, 931 ⟨88TL6541⟩; **3**, 142 ⟨82JOC2268⟩, 217 ⟨84JA3869⟩
Lettieri, G., **5**, 746 ⟨64JOC3697⟩
Lettre, H., **3**, 614 ⟨52CB397⟩
Letulle, M., **3**, 359 ⟨91TL2013⟩
Leu, L., **4**, 182 ⟨87JOC188⟩
Leu, S.-L., **3**, 242 ⟨90JCS(P1)423⟩
Leube, H., **3**, 481 ⟨59CB1345⟩
Leuchs, H., **6**, 486 ⟨06CB857⟩
Leuchs, K., **2**, 103 ⟨1890CB203⟩
Leuck, D. J., **2**, 310 ⟨90JOC2552⟩; **3**, 44, 188 ⟨90JOC2552⟩
Leue, C., **6**, 187 ⟨94OM436⟩, 387 ⟨91OM3838⟩
Leuenberger, H., **5**, 214 ⟨86HCA236⟩
Leuenberger, H. G. W., **1**, 77 ⟨76HCA1832⟩
Leuer, M., **4**, 567 ⟨83CC1171⟩, 568 ⟨89CB453⟩, 568, 573 ⟨92CB771⟩; **6**, 168 ⟨83CC1171⟩, 288 ⟨89CB453⟩
Leukel, G., **5**, 224, 1007 ⟨88S990⟩, 1006, 1007 ⟨89JPS(A)4341⟩
Leumann, C., **1**, 797 ⟨93HCA418⟩
Leung, H.-W., **1**, 1110 ⟨73JA160⟩
Leung, K. K., **1**, 598 ⟨75TL4197⟩
Leung, P. H., **2**, 438 ⟨86IC3396⟩
Leung, T., **1**, 449, 977 ⟨74JA5620⟩, 977, 1056 ⟨78JA5561⟩; **2**, 901 ⟨77JA5192⟩; **4**, 814 ⟨77JA5192⟩
Leung, W., **3**, 25 ⟨89TL6629⟩
Leung, W-P., **2**, 1003, 1005 ⟨89JCS(D)105, 84JCS(D)1801⟩, 1005 ⟨84JCS(D)321, 85JCS(D)337⟩; **4**, 604, 646, 647 ⟨84JCS(D)1801⟩, 604, 646, 647 ⟨84JCS(D)321⟩, 604, 647 ⟨88JCS(D)2403⟩, 646 ⟨89JCS(D)105⟩, 646, 647 ⟨85JCS(D)337⟩, 647, 656 ⟨86JCS(D)603⟩, 647, 660 ⟨92JCS(D)775⟩, 660 ⟨82JOM(233)C28, 84JCS(D)331⟩; **6**, 190 ⟨84JCS(D)1801⟩, 194, 195 ⟨90JOM(394)57⟩, 195 ⟨93JOM(443)C39⟩
Leung, W. Y., **3**, 303 ⟨89SC787⟩
Leung-Toung, R., **3**, 584 ⟨92CC487, 92CC488⟩; **4**, 894 ⟨92CC487⟩
Leupold, M., **5**, 950, 951 ⟨72CB150⟩
Leurs, S., **6**, 573 ⟨92T7505⟩
Leuser, W. M., **2**, 255, 256 ⟨52JA161⟩
Leusink, A. J., **2**, 676, 970, 971 ⟨67JOM(9)285⟩, 970, 971 ⟨67JOM(9)295⟩; **4**, 695 ⟨75JOM(84)117⟩, 701 ⟨69JOM(16)91, 83OM1099⟩, 708 ⟨73JOM(55)419⟩
Leuski, U., **6**, 630 ⟨89EGP270530⟩
Leutert, T., **2**, 445, 445 ⟨78HCA1708⟩
Lévai, A., **3**, 361 ⟨91BSF976, 92JCR(S)163⟩
Levaillant, R., **2**, 102 ⟨36AC(P)560⟩
Levand, O., **4**, 161 ⟨57JA6562⟩
Levan'kov, S. V., **4**, 339 ⟨91ZOB1315⟩
Levas, E., **4**, 739, 746 ⟨70CR(C)1467, 72CR(C)1357⟩
Levas, M., **4**, 743 ⟨63BSF2147⟩, 746 ⟨72CR(C)1357⟩
Levashova, G. B., **5**, 709 ⟨87MI 517-02⟩
Levason, W., **2**, 872, 874 ⟨75JOM(88)171⟩, 872, 896 ⟨B-79MI 217-02, 76JCS(D)1429⟩; **4**, 288 ⟨84JCS(P2)429⟩
Levasseur, L., **1**, 412 ⟨65JA1150⟩
Levchenko, E. S., **2**, 399, 400 ⟨67ZOR1282⟩, 399, 400 ⟨72ZOR2057, 78ZOR1846⟩, 407 ⟨67ZOB852⟩; **3**, 363 ⟨72ZOR1822⟩; **5**, 317, 349 ⟨78ZOR1846⟩, 337 ⟨86ZOR100⟩, 341 ⟨74ZOR139⟩, 341 ⟨78ZOR280, B-71MI 507-02⟩, 703 ⟨85JOU1166, 85ZOR1281⟩, 703 ⟨87JOU75, 87ZOR86⟩, 718 ⟨84JOU1963, 84ZOR2153⟩, 775 ⟨84ZOR1784⟩, 1032 ⟨72ZOR1822⟩; **6**, 625 ⟨85ZOR1281⟩, 656 ⟨75ZOR762⟩
Levchenko, N. I., **6**, 629 ⟨75ZOR1628⟩
Levedev, E. P., **5**, 538, 539 ⟨79JGU940⟩
Levenberg, B., **3**, 617 ⟨68JBC(243)6009⟩
Levenberg, P. A., **1**, 516 ⟨86S184⟩, 606 ⟨81JA1501⟩; **3**, 115 ⟨81S567⟩, 223, 275 ⟨86S184⟩
Levene, R., **1**, 540 ⟨73JCS(P2)1971, 74JOM(67)467⟩
Levens, E., **2**, 906 ⟨63OSC(4)68⟩
Levenson, C. H., **2**, 163, 237 ⟨84JMC228⟩
Lever, D. C., **2**, 133 ⟨91JA3176⟩
Lever, J. R., **1**, 210 ⟨82TL2379⟩; **3**, 224 ⟨82TL2379⟩
Lever, Jr., O. W., **1**, 1089 ⟨74JA7125⟩; **2**, 681, 908, 999 ⟨74JA7125⟩; **4**, 917 ⟨81TL3789⟩, 948 ⟨76JOC2312, 76JOC2874⟩, 949 ⟨74JA7125⟩, 953 ⟨75CC519⟩
Lever, O. W., **3**, 20, 136 ⟨76T1943⟩
Levering, D. R., **4**, 428 ⟨63T(S)3⟩
Lever Jr., O. W., **3**, 157 ⟨74JA7125⟩
Leveson, L. L., **3**, 532 ⟨66T209⟩
Levesque, G., **5**, 546 ⟨78TL3847⟩, 546, 550 ⟨85BSF891⟩, 547, 550 ⟨76BSF1145, 87S835⟩, 558 ⟨86MI 513-01⟩, 561 ⟨75MI 513-01, 81S963⟩
Levi, B. A., **1**, 129, 162 ⟨81JOC1693⟩; **2**, 525, 1005 ⟨81JOC1693⟩
Levi, G., **2**, 135 ⟨25G973⟩
Levi, R., **3**, 616 ⟨08CB2925⟩
Levi, S., **3**, 294 ⟨88TL5561⟩
Levi, T. G., **5**, 548 ⟨24G389⟩
Levillain, J., **6**, 734 ⟨93JA8444⟩
Levin, A. A., **6**, 149, 653 ⟨73ZOR896⟩
Levin, B. B., **2**, 887 ⟨53MI 217-02⟩
Levin, D., **1**, 641, 643 ⟨88JCS(P1)1799, 84TL357⟩, 643 ⟨85TL505, 86TL2265⟩, 643 ⟨92JCS(P1)2155, 92JCS(P1)3407⟩; **3**, 197 ⟨88JCS(P1)1799⟩

Levin, J., **2**, 164 ⟨78TL503⟩
Levin, J. I., **5**, 30 ⟨82TL2347⟩, 329 ⟨82SC989⟩
Levin, N., **5**, 823 ⟨42JOC408⟩
Levin, V. P., **6**, 223, 226, 227, 228 ⟨72MI 607-01⟩
Levin, Ya. A., **4**, 125 ⟨85JGU2497⟩
Levine, B. H., **1**, 1015 ⟨91TL5243⟩
Levine, H. H., **4**, 440 ⟨51JOC1507⟩
Levine, R., **1**, 860 ⟨51JA4301, 56JA1723⟩; **2**, 422 ⟨60JA472⟩, 631 ⟨50JOC162⟩; **4**, 731 ⟨55JA4168⟩
Levine, S. G., **1**, 812, 813 ⟨81JOC2199⟩; **2**, 680 ⟨58JA6150⟩; **3**, 493 ⟨58JA6150⟩, 620 ⟨89TL4767⟩
Levinson, A. S., **5**, 39 ⟨63JOC2184⟩
Levinson, M., **3**, 358 ⟨86JOC411⟩, 365 ⟨85JOC1550⟩
Levinson, M. I., **3**, 346, 355 ⟨85T5061⟩; **4**, 108 ⟨91USP5043489⟩; **5**, 512, 570 ⟨85T5061⟩, 528 ⟨84JA6084⟩
Levinson, S. H., **4**, 825 ⟨83JOC298⟩
Levisalles, J., **2**, 799 ⟨69CC1515⟩
Levisky, J. A., **2**, 752 ⟨66JA100⟩
Levita, G., **2**, 688 ⟨73JA2918⟩
Levitina, E. S., **5**, 215 ⟨86IZV2032⟩
Levitt, T. E., **5**, 27 ⟨70T1539⟩, 234 ⟨77JCS(P1)1672⟩
Levitz, R., **3**, 638 ⟨79TL771⟩
Levkov, Y. L., **2**, 609, 621 ⟨60DOK(133)855⟩
Levkovskaya, G. G., **2**, 338, 365 ⟨89RCR250⟩; **3**, 252 ⟨76IZV2040⟩; **4**, 312 ⟨84ZOR1502⟩
Levorse, A. T., **5**, 285, 377 ⟨85TL1803⟩, 378, 379 ⟨88JOC4006, 88JOC4773⟩
Levskaya, G. S., **2**, 883 ⟨65JGU1253⟩, 886 ⟨67JGU855⟩
Levy, A., **2**, 1034 ⟨60JA3472⟩
Levy, A. B., **1**, 278 ⟨78JOC1279, 78JOC4684⟩, 278 ⟨79TL2313⟩, 449 ⟨B-75MI 110-01⟩, 486 ⟨73JOC2574⟩, 487 ⟨80S945⟩, 1108 ⟨77TL3545⟩; **2**, 305 ⟨87T4079⟩, 514, 516, 520 ⟨B-75MI 210-01⟩, 526 ⟨72JOM(44)233⟩, 607, 617 ⟨77TL3545⟩, 900 ⟨B-75MI 218-01⟩, 906 ⟨76TL2201⟩; **4**, 314 ⟨71JA5469⟩, 934, 949 ⟨76TL2201⟩, 935, 949 ⟨78JOM(156)123⟩
Levy, D., **5**, 191 ⟨69TL4563⟩
Levy, E. S., **3**, 106 ⟨74JOC2872⟩
Levy, G. C., **3**, 366 ⟨67JA2793⟩
Levy, H., **1**, 411 ⟨44JA1684⟩
Levy, I., **2**, 126, 223 ⟨49JA2270⟩
Lévy, J., **1**, 628 ⟨92S248⟩
Levy, J. B., **1**, 895 ⟨53JA1253, 53JA3955⟩
Levy, J. N., **4**, 129 ⟨89B8270⟩; **6**, 117 ⟨89CC246⟩
Levy, L. K., **6**, 82, 83, 91 ⟨55JA509⟩
Levy, M., **5**, 83 ⟨50MI 502-01⟩
Levy, N., **2**, 393 ⟨47QR358⟩
Levy, S., **3**, 498 ⟨58JOC2036⟩
Lew, G., **1**, 488, 983, 1033, 1034 ⟨91T343⟩, 532, 662 ⟨73CC874⟩, 532, 662 ⟨74JOC2321⟩; **2**, 901, 904 ⟨91T343⟩; **4**, 815 ⟨91SM2535⟩
Lewandos, G. S., **2**, 558, 560 ⟨83ICAL241⟩
Lewars, E. G., **2**, 205 ⟨75CJC3656⟩; **3**, 530 ⟨83CRV519, 85JOC135⟩
Lew, W., **5**, 142 ⟨94JA7455⟩
Lewellyn, M. E., **6**, 558 ⟨90JOC5230⟩
Lewicki, J. W., **2**, 735 ⟨76CC552⟩
Lewin, A. H., **3**, 92 ⟨78OPP201⟩
Lewis, A. D., **5**, 566, 579, 591, 618, 621 ⟨55JA4062⟩
Lewis, A. J., **1**, 387 ⟨69T4445⟩; **4**, 182 ⟨70T1311⟩
Lewis, C. N., **3**, 212 ⟨86S944⟩
Lewis, D. B., **4**, 675 ⟨85JCS(D)905⟩; **5**, 927 ⟨87JOM(325)185⟩
Lewis, D. F., **6**, 552 ⟨72JA1563⟩
Lewis, D. T., **5**, 528 ⟨40JCS831⟩
Lewis, E. O., **4**, 771 ⟨93JFC(63)253⟩
Lewis, E. S., **1**, 107 ⟨90JA3554⟩, 801 ⟨73JA8096⟩, 853, 898 ⟨50JA1005⟩, 892 ⟨54JA791⟩; **2**, 25 ⟨52JA308⟩; **6**, 116 ⟨90CC1133⟩
Lewis, F. D., **3**, 407 ⟨B-70MI 310-03⟩
Lewis, G. E., **4**, 675 ⟨85JCS(D)905⟩
Lewis, H. H., **3**, 474 ⟨25JA1728⟩
Lewis, J. A., **1**, 68 ⟨86CC855⟩; **5**, 342 ⟨92JOC5328⟩
Lewis, J., **1**, 30 ⟨79JCS(P1)442⟩; **3**, 700 ⟨91JOM(403)C15⟩; **4**, 682 ⟨77JCS(D)1328, 87JOM(319)C51⟩, 682 ⟨88JCS(D)1421⟩; **5**, 1002 ⟨76TL2691, 79JCS(P1)446⟩; **6**, 340 ⟨80CC781⟩, 345 ⟨84CC1089⟩, 352 ⟨80CC812⟩
Lewis, J. J., **3**, 124 ⟨85CC149⟩; **6**, 632 ⟨87JHC275⟩
Lewis, J. W., **1**, 443 ⟨63PCS19, 68T4489⟩, 443, 660 ⟨64TL2039⟩, 659 ⟨69JCS(B)863⟩
Lewis, K. E., **1**, 419 ⟨64JCS3080⟩
Lewis, L. L., **2**, 614 ⟨47JA2022⟩
Lewis, L. R., **4**, 306 ⟨61JA2574⟩
Lewis, M. D., **1**, 534 ⟨87TL5129⟩
Lewis, N., **1**, 16 ⟨93TA625⟩
Lewis, N. G., **1**, 941 ⟨78HCA690⟩
Lewis, N. J., **2**, 652 ⟨77JOC1479⟩
Lewis, P. H., **4**, 698 ⟨53JCP(21)986⟩
Lewis, P. J., **2**, 539 ⟨90JOC6082⟩
Lewis, P. L., **6**, 76 ⟨77JOC679⟩
Lewis, R. A., **2**, 807, 845 ⟨68JA4842⟩
Lewis, R. G., **1**, 91 ⟨68JOC747⟩, 725, 1001 ⟨67TL401⟩
Lewis, R. N., **2**, 237 ⟨48JA1115⟩
Lewis, R. T., **5**, 868 ⟨92T1465⟩
Lewis, S. C., **6**, 504 ⟨90H(31)1393⟩
Lewis, T. A., **5**, 652 ⟨54CI(L)1154⟩
Lewis, T. B., **3**, 82 ⟨63JOC2918⟩
Lewis, T. P., **1**, 849 ⟨69T4117⟩
Lewis, W., **1**, 451 ⟨81S999⟩, 511 ⟨78JOC2739⟩; **2**, 616 ⟨79JA5101⟩, 922, 931 ⟨78JOC2739⟩, 922, 931 ⟨81S999⟩; **3**, 247 ⟨82TL353⟩; **4**, 206 ⟨82TL353⟩, 758, 816 ⟨79JA5101⟩, 808, 1059 ⟨78JOC2739⟩
Lewton, D. A., **1**, 595 ⟨77JCS(P1)1574⟩; **2**, 153 ⟨77JCS(P1)1574⟩
Lex, J., **1**, 248 ⟨85AG(E)1046, 86AG(E)508⟩, 248 ⟨87AG(E)928, 88AG(E)411⟩; **5**, 867 ⟨84AG(E)717⟩
Lexa, D., **2**, 578 ⟨90JA1874⟩
Ley, H., **5**, 838, 855 ⟨14CB2938⟩
Ley, K., **5**, 1008 ⟨72AG(E)1011⟩; **6**, 448 ⟨71GEP1932830⟩
Ley, S. V., **1**, 196 ⟨90SL224⟩, 226 ⟨81JCS(P1)270⟩, 526 ⟨81JCS(P1)1516⟩, 596, 780 ⟨87TL221⟩, 614, 697, 700, 703 ⟨90TL5525⟩, 654, 697, 700, 703 ⟨83CC630⟩, 697, 700, 703 ⟨87TL5759⟩, 697, 700, 703 ⟨87TL5763⟩, 697, 703 ⟨89T7161⟩, 700, 703 ⟨87TL5755⟩, 757 ⟨85T3825⟩, 780 ⟨91T6813⟩, 1023 ⟨92SL291⟩, 1031 ⟨89T4017, 90SL48⟩; **2**, 239 ⟨79JCS(P1)3013⟩, 271 ⟨88OS(67)108⟩, 357 ⟨75CC855⟩, 390 ⟨80JCS(P1)1212⟩, 551 ⟨B-89MI 211-01⟩, 690 ⟨81JCS(P1)1473⟩; **3**, 5 ⟨90MI 301-01⟩, 5, 117 ⟨87CC1625⟩, 18, 134 ⟨80JCS(P1)1212⟩, 88 ⟨78CC952⟩, 89 ⟨92SC1967, 90MI 304-01⟩, 132 ⟨80JCS(P1)1654⟩, 197 ⟨87TL345⟩, 234 ⟨81TL3031⟩, 308 ⟨83CC630, 84JOC3503⟩; **4**, 165 ⟨83JCS(P1)1579⟩, 165 ⟨84JCS(P1)681⟩, 200 ⟨90SL255⟩, 200 ⟨91T1329⟩, 236 ⟨86T4333, 93TL5649⟩, 286 ⟨89T7161⟩; **5**, 164 ⟨91T9929⟩, 277 ⟨87TL3019⟩, 278 ⟨86CC344⟩, 366 ⟨80JCS(P1)1212⟩, 523 ⟨78CC393, 79CC755⟩, 523 ⟨80JCS(P1)1650, 81T(S)213⟩
Leyck, S., **2**, 235 ⟨86EUP165534, 86GEP3443467⟩, 235 ⟨88GEP3626554⟩
Leyendecker, F., **3**, 165 ⟨82TL5031⟩
Leyerer, H., **2**, 435 ⟨87JOM(334)369⟩
Leyshon, W. M., **2**, 274 ⟨75JCS(P1)1925⟩
Leyzdins, P., **4**, 656 ⟨91OM2857⟩
Lezzi, A., **1**, 1006, 1013 ⟨84T2773⟩, 1010 ⟨85T621⟩, 1010, 1013 ⟨84S571⟩
L'Haridon, P., **4**, 674 ⟨92OM721⟩
Lheureux, M., **2**, 949 ⟨90CRV265⟩
Lhommet, G., **3**, 411 ⟨87TL2353⟩, 496 ⟨90S301⟩, 541 ⟨85TL833, 87TL881⟩
Lhoste, J. M., **3**, 442 ⟨83T1777⟩
Lhoste, P., **2**, 330 ⟨91CC1684⟩; **3**, 632 ⟨88JMOCL23⟩

Li, A., **6**, 699 ⟨93JFC(60)175⟩
Li, B. L., **2**, 467, 468 ⟨86IC361⟩
Li, B.-L., **4**, 566 ⟨83IC575⟩, 567 ⟨85IC1993⟩
Li, B. L., **5**, 893 ⟨85IC1993⟩
Li, C., **4**, 190 ⟨93HCA211⟩, 343 ⟨92S258⟩, 801 ⟨91S854⟩; **6**, 118 ⟨92PS(69)75⟩, 624 ⟨91MM2302⟩
Li, C. J., **1**, 568 ⟨90TL6291⟩, 826 ⟨93CRV2023⟩, 1050 ⟨93TL2271⟩; **2**, 73 ⟨91TL7017⟩, 233 ⟨93TL903⟩; **3**, 176 ⟨90TL6291⟩
Li, C.-S., **1**, 345 ⟨87CC166⟩, 633, 829 ⟨87JOC3541⟩; **2**, 632 ⟨87JOC691⟩; **4**, 1083 ⟨84JCS(P1)1561⟩
Li, G., **2**, 29 ⟨93TA2315, 93TL2561⟩
Li, G. M., **3**, 336, 340, 385 ⟨93CE53⟩, 553 ⟨92TL7865⟩; **5**, 516, 520 ⟨92TL7865⟩
Li, G. S., **1**, 21 ⟨70JA3222⟩
Li, H., **3**, 345 ⟨91JOC5932⟩
Li, H. J., **2**, 577 ⟨91JOM(419)245⟩
Li, H.-Y., **1**, 552 ⟨89TL4157⟩
Li, J., **1**, 647 ⟨86TL4583⟩, 956 ⟨93JOC3038⟩; **2**, 730 ⟨92S281⟩; **3**, 48, 65 ⟨86TL4583⟩; **5**, 873 ⟨92JA1094⟩
Li, J. C., **4**, 60 ⟨93SL195⟩
Li, J.-S., **1**, 651 ⟨82CC969⟩; **2**, 78 ⟨89JOC5768⟩
Li, L., **1**, 828 ⟨90SC423⟩; **3**, 13, 40, 85 ⟨90JA7050⟩; **4**, 969, 970 ⟨90JPR1035⟩
Li, L.-H., **1**, 161 ⟨91TL2879⟩
Li, M., **1**, 647 ⟨88LA377⟩
Li, M. K., **1**, 436 ⟨85CC1610⟩
Li, M. P., **1**, 111 ⟨74TL3511⟩
Li, M. Y., **1**, 65, 66 ⟨84JA3703⟩
Li, M.-Y., **1**, 663 ⟨83OM554⟩
Li, N.-S., **1**, 488 ⟨93JOC1949⟩; **2**, 904 ⟨93JOC1949⟩
Li, P., **5**, 38 ⟨86JOC4354⟩
Li, P. T. J., **2**, 590 ⟨81JOC4323⟩
Li, Q.-L., **3**, 257 ⟨88JHC1287⟩
Li, S., **1**, 646, 647 ⟨90JOM(400)121⟩
Li, S.-C., **6**, 409 ⟨44MI 614-01⟩
Li, S. O., **2**, 412 ⟨55JA1866⟩
Li, S.-W., **3**, 49 ⟨B-90MI 301-02⟩, 196 ⟨88S975⟩, 196 ⟨B-90MI 304-02⟩, **3**, 649 ⟨89SC2639⟩
Li, T., **1**, 414, 416, 827 ⟨70JA741⟩, 456, 502, 504, 505 ⟨73S457, 94TL2145⟩; **2**, 35 ⟨68JA6225⟩; **5**, 159 ⟨70JA741⟩
Li, T. H., **1**, 506 ⟨72JOC2989⟩
Li, W., **2**, 330 ⟨89JOC265⟩, 569, 570, 571 ⟨93CB2093⟩
Liaw, S.-J., **3**, 613 ⟨79JOC2029⟩
Li, W. K., **4**, 805, 812, 813 ⟨61MI 418-01⟩
Li, W. S., **2**, 9 ⟨84TL2295⟩; **3**, 87, 88, 95, 293 ⟨89S293⟩
Li, X., **1**, 367 ⟨84BSB511⟩; **2**, 31 ⟨94SC733⟩, 104 ⟨93S207⟩, 256 ⟨93JCS(P1)1091⟩, 1076, 1080 ⟨92OM3993⟩; **6**, 221, 222, 226 ⟨92MI 607-03⟩
Li, X.-Y., **1**, 922 ⟨89JOC5648⟩; **4**, 117, 118 ⟨87TL6317⟩
Li, Y., **3**, 196 ⟨90TL5897⟩, 436 ⟨94MI 311-01⟩; **4**, 264 ⟨92JA5946⟩
Li, Z., **1**, 196 ⟨87TL5415⟩, 212 ⟨93TL3539⟩; **2**, 843 ⟨84S781⟩, 846 ⟨86S240⟩
Li, Z.-Q., **1**, 270 ⟨91TL3999⟩
Li, Z.-W., **5**, 948 ⟨94JA11584⟩
Liaan-Jensen, S., **1**, 747, 748 ⟨73ACS1401⟩
Liang, C. K., **2**, 403 ⟨68CJC1149, 71S204⟩
Liang, D. G., **1**, 747, 748 ⟨69AJC2145⟩
Liang, G., **1**, 900 ⟨85JA2764⟩, 902 ⟨72JA3544, 72JA6434⟩, 906 ⟨73JA8698, 74JA195⟩, 906 ⟨77JA5683, 80JA683⟩, 914 ⟨78JA6299⟩, 918 ⟨75JOC700⟩; **4**, 42 ⟨81JOC571⟩
Liang, H. T., **1**, 785 ⟨58JA3585⟩
Liang, L., **2**, 1027 ⟨93T2159⟩
Liang, M., **5**, 958 ⟨91OM737⟩
Liang, P., **6**, 365 ⟨55OSC(3)803⟩
Liang, W. C., **2**, 802 ⟨73JOC2438⟩
Liang, Y., **2**, 307 ⟨88SC1363⟩
Liangzhu, L., **5**, 1078 ⟨87MI 528-01⟩
Lianza, F., **4**, 566 ⟨91HCA983⟩
Liao, H., **1**, 272 ⟨92JOC774⟩
Liao, L.-F., **4**, 849 ⟨87JOC855⟩; **6**, 127 ⟨87JOC855⟩
Liao, Q., **1**, 270 ⟨88S321⟩; **3**, 133 ⟨93SL407⟩, 500 ⟨88S321⟩
Liao, S., **2**, 558 ⟨90CB1517⟩
Liao, Y., **1**, 647, 743 ⟨90JCR(S)388⟩; **2**, 487 ⟨90CC85, 91JOC1381⟩, 487 ⟨92JOC774⟩, 531 ⟨92JOC6608⟩; **3**, 639 ⟨91JOC1381⟩
Libbey, A. J., **1**, 906 ⟨57JOC1243⟩
Libby, P. R., **4**, 61 ⟨89JMC997⟩
Libermann, D., **2**, 100 ⟨33BSF1070⟩; **3**, 350 ⟨48BSF1120⟩
Libers, R., **4**, 439 ⟨63T(S)23⟩
Libertini, E., **1**, 689 ⟨82JCS(P2)77⟩
Libertini, L. J., **2**, 379 ⟨74JMC948⟩
Li Bin-quan, **1**, 738 ⟨83MI 116-01⟩
Libis, B., **3**, 567 ⟨63CR(256)2419⟩
Libman, D. D., **5**, 769, 847, 848 ⟨56JCS2253⟩
Libman, J., **1**, 811 ⟨72JA9228⟩; **4**, 204 ⟨69T1679⟩
Licandro, E., **2**, 553 ⟨92CC1623⟩, 691 ⟨88S344⟩, 692 ⟨87S192⟩; **3**, 430 ⟨85CC417⟩; **5**, 676, 677, 906 ⟨85T2503⟩, 905 ⟨79S299⟩, 905 ⟨81G285⟩, 905 ⟨84T2165, 83CC337⟩, 906 ⟨85T3321⟩, 943 ⟨92CC1623⟩, 946 ⟨91OM807⟩, 946, 959 ⟨94ICA233⟩, 953, 958 ⟨95JOM(486)279⟩, 956 ⟨94TL6377⟩; **6**, 723 ⟨93OM2994⟩
Licastro, S. A. de, **4**, 338 ⟨72T6013⟩
Licchelli, M., **4**, 617 ⟨87TL3719⟩
Lichstein, B. M., **6**, 409 ⟨71GEP2105907⟩
Licht, U., **6**, 488 ⟨92GEP4219418⟩
Lichtenberg, F., **4**, 382 ⟨85CB2822⟩, 384 ⟨84AG(E)239⟩
Lichtenberger, J., **4**, 423, 427 ⟨55BSF669⟩; **5**, 105 ⟨B-49MI 502-01⟩
Lichtenthaler, F. W., **2**, 693 ⟨61CRV607⟩; **3**, 32 ⟨B-92MI 301-02⟩; **4**, 920 ⟨61CRV607⟩
Lichtenthäler, L., **2**, 746 ⟨75S426⟩
Lichtenthaler, R. N., **4**, 988 ⟨80AG208⟩
Lichtenwalter, G. D., **4**, 354 ⟨58JA2680⟩; **5**, 254 ⟨60JOC1064⟩
Lichty, J. G., **5**, 965 ⟨44USP2362648⟩
Licini, G., **2**, 147 ⟨86TL6257⟩, 147 ⟨89TL4859⟩; **4**, 249 ⟨90BSF734⟩, 261 ⟨86TL6257, 89TL2575⟩, 261 ⟨90G165, 92TL3043⟩, 861 ⟨88MI 419-02⟩
Lickiss, P. D., **4**, 656 ⟨92CSR271⟩; **5**, 418, 924, 925 ⟨89OM693, 85JA4338⟩, 925 ⟨89OM709⟩; **6**, 174 ⟨83JOM(252)281⟩, 290 ⟨86JOM(315)C5, 89CC595⟩, 290, 391 ⟨87CC1461⟩, 378 ⟨85JOM(294)305⟩, 380 ⟨81JOM(221)13, 90JOM(398)59⟩, 381 ⟨92JOM(427)9, 93JCS(P2)391⟩, 381 ⟨93JCS(P2)59⟩, 381, 392 ⟨87JCS(P2)891⟩, 381, 392 ⟨94JOM(466)35⟩, 387 ⟨86JOM(308)261, 89JCS(D)447⟩, 390 ⟨93JCS(D)3259⟩, 394 ⟨95UP 613-01⟩, 515 ⟨91JOM(403)293⟩
Lidaks, M., **5**, 224 ⟨87ZOB686⟩
Liddell, D. A., **5**, 293 ⟨46JCS454⟩
Lidert, Z., **4**, 99 ⟨79S810⟩, 327 ⟨80S322⟩
Lidon, M. J., **5**, 1066 ⟨90JCS(P1)1727⟩, 1107 ⟨82S472⟩
Lidor, R., **2**, 626 ⟨88JOC5545⟩; **3**, 431 ⟨89IJ187⟩
Lidy, W., **3**, 440, 457 ⟨76CB1491⟩, 615, 628 ⟨73CB587⟩, 628, 632 ⟨73TL1449⟩; **5**, 721, 722 ⟨76CB2542⟩, 994 ⟨76CB1491⟩
Lie, K., **3**, 279, 292 ⟨70MI 306-01⟩
Lieb, F., **1**, 1002 ⟨78LA658⟩; **4**, 170 ⟨87LA607⟩; **5**, 882 ⟨76TL3489⟩; **6**, 437 ⟨83HOU(E4)1275⟩
Lieb, M., **2**, 15 ⟨85C134⟩; **6**, 237, 436, 477, 483 ⟨78CB2891⟩
Liebein, F., **3**, 710 ⟨81AP(314)459⟩
Liebenow, H., **2**, 831 ⟨80AG740⟩
Lieber, E., **5**, 1046 ⟨57CJC832, 63JOC194⟩; **6**, 537 ⟨63CJC1643⟩
Lieberknecht, A., **3**, 179, 184 ⟨85LA785⟩; **4**, 465 ⟨82AG(E)776⟩; **5**, 351 ⟨92S482⟩, 692 ⟨89S256⟩
Liebeskind, L. S., **1**, 134 ⟨85TL3075⟩, 480, 1020

⟨90JOC5359⟩, 836 ⟨90JOC4614⟩, 978 ⟨88TL4253⟩, 1047 ⟨87TL1857⟩; **2**, 669 ⟨92JA1412⟩, 672 ⟨87JOC2631, 91JA2771⟩; **3**, 273 ⟨87TL1857⟩, 544 ⟨85JA6715⟩; **5**, 457 ⟨84JA6451, 87JA2759⟩, 457 ⟨90TL3723⟩
Liebig, H., **5**, 54 ⟨76CZ3⟩, 599 ⟨72LA(757)147⟩
Liebig, J., **5**, 19 ⟨1832MI 501-01⟩
Liebig, R., **2**, 380 ⟨69ZC105⟩
Liebman, A. A., **3**, 92 ⟨79JHC799⟩; **4**, 168, 170 ⟨83JA3661⟩
Liebman, J. F., **6**, 212 ⟨B-88MI 607-01⟩
Liebscher, J., **5**, 565, 614 ⟨87ZC8, 94CC1437⟩, 631 ⟨88ZC182⟩, 633 ⟨85JPR(327)455⟩, 660 ⟨84JPR(326)311⟩, 760 ⟨84S51⟩; **6**, 51 ⟨79S241⟩, 610, 613, 617, 618, 621 ⟨88S655⟩
Lied, T., **5**, 174 ⟨84AG(E)145⟩, 236, 239 ⟨78AG(E)267⟩
Liedhegener, A., **3**, 472 ⟨70LA(739)174⟩, 475 ⟨68OS(48)36, 69OPP99⟩, 480 ⟨71LA(748)207⟩, 498 ⟨67T2701⟩; **5**, 595 ⟨67LA(710)118⟩, 869 ⟨68CB3734⟩, 871, 874 ⟨71LA(748)207⟩, 872 ⟨73OSC(5)73⟩; **6**, 143 ⟨71LA(746)92⟩, 666 ⟨68CB3734, 68TL3171⟩, 670 ⟨71LA(748)207⟩
Liedholm, B., **3**, 99 ⟨92JCS(P1)2235⟩
Liedle, S., **5**, 488, 492 ⟨90JST171⟩; **6**, 243 ⟨92JST(268)389⟩
Liedtke, A., **2**, 510 ⟨89IC1627⟩
Liedtke, J. D., **1**, 660 ⟨65JA1526⟩; **2**, 904 ⟨65JA1526⟩
Lieke, W., **3**, 693 ⟨1859LA(112)316⟩
Lie Ken Jie, M. S. F., **1**, 986 ⟨92CPL243⟩; **6**, 481, 482 ⟨91JCS(P2)501⟩
Liem, B. J., **5**, 848 ⟨71ZN(B)296⟩
Lien, A. P., **1**, 897 ⟨50IEC2103⟩; **2**, 126 ⟨50JA731⟩
Lienemann, A., **6**, 105 ⟨75S272⟩
Lienert, J., **3**, 571 ⟨68LA(718)24⟩; **5**, 1078 ⟨68LA(718)24⟩
Lienhard, C., **6**, 623 ⟨70GEP1915668⟩
Lienhard, G. E., **1**, 870 ⟨69JA1146⟩; **4**, 525, 526, 527 ⟨81JA5241⟩, 899 ⟨68JA3781⟩
Lienhard, U., **2**, 1049 ⟨78HCA1609⟩; **5**, 280 ⟨78HCA2428, 78HCA2437⟩, 280 ⟨85OS(63)175⟩
Lienig, D., **2**, 291 ⟨65ZAAC(335)289⟩
Liepa, A. J., **3**, 106 ⟨72AJC1979⟩; **5**, 1126 ⟨85TL501⟩
Liepert, T. K., **2**, 68 ⟨89TL3901⟩
Liepins, E., **4**, 515 ⟨86JOM(316)249⟩; **5**, 224 ⟨87ZOB686⟩
Liepinsh, E., **4**, 514 ⟨84JOM(268)C29⟩
Liepin'sh, E. E., **4**, 807, 811, 813 ⟨91JGU1295⟩, 810 ⟨91JGU1306⟩
Lies, R., **1**, 364 ⟨88T3899⟩; **3**, 652 ⟨86S548⟩
Liesching, D., **2**, 748 ⟨77S495⟩, 812 ⟨76S746⟩; **3**, 657 ⟨77S495⟩
Liese, T., **1**, 1069 ⟨86CB2995⟩
Lieser, T., **3**, 467 ⟨50CB137⟩
Liétge, S., **4**, 557 ⟨85TL4435⟩, 557, 576, 801 ⟨86JOM(304)283⟩
Lietge, S., **6**, 288 ⟨86JOM(304)283⟩
Liétjé, S., **4**, 570 ⟨86S934⟩
Lieurance, B. A., **3**, 309 ⟨89JOC4350⟩
Liewen, M. B., **1**, 571 ⟨88OSC(6)210⟩; **3**, 539 ⟨76OS(55)32⟩
Liggero, S. H., **1**, 1114 ⟨70JA3789⟩
Light, J., **1**, 5, 936 ⟨90TL2957, 93OS(72)199⟩
Light, K., **2**, 445 ⟨68JOC2590⟩
Light, K. K., **1**, 90 ⟨66JOC870⟩
Lightner, D. A., **1**, 592 ⟨62CI(L)1236⟩, 634 ⟨81JA5314⟩; **2**, 187 ⟨67JA4099⟩; **4**, 78 ⟨67JA4099⟩
Lightowler, M., **4**, 863 ⟨92SL730⟩
Ligtvoet, G. J., **4**, 184 ⟨87RTC545⟩
Liguori, A., **2**, 350 ⟨83JHC1207⟩; **3**, 618 ⟨87S168⟩
Liisberg, S., **4**, 743, 748 ⟨60T149⟩
Lijewski, L., **2**, 66 ⟨90TL7415⟩
Li-Lan, S., **1**, 766 ⟨93T6821⟩
Lilan Shi, **1**, 269, 738 ⟨88TL5295⟩, 739 ⟨87S370⟩
Lilga, M. A., **4**, 545 ⟨86OM2220⟩; **6**, 517 ⟨85OM590⟩
Lilienfeld, W. M., **1**, 881 ⟨45JA1273⟩
Liljefors, T., **1**, 1000 ⟨88S250⟩
Lille, Ü., **1**, 225 ⟨91TA843⟩
Lillevik, H. A., **3**, 616 ⟨42JOC164⟩
Lilley, G. L., **1**, 475 ⟨81JOC2709⟩
Lilley, K. J., **1**, 475 ⟨81JOC2709⟩
Lillford, P. L., **3**, 527 ⟨67JCS(B)360⟩
Lillie, B. M., **4**, 182 ⟨94TL969⟩
Lillie, C., **6**, 642 ⟨80JMC1217⟩
Lillie, T. J., **1**, 739, 740, 742 ⟨77AG(E)487⟩
Lillocci, C., **1**, 629 ⟨88JOC1733⟩
Lilly, F.,**5**, 750 ⟨61BCF2067⟩
Lilly, R. L., **2**, 691 ⟨62CI(L)559⟩
Lilyquist, M. R., **4**, 731 ⟨55JA2783⟩
Lim, B. B., **5**, 351 ⟨90S1095⟩, 690 ⟨88JOC382, 88JOC5309⟩, 690 ⟨88S242⟩
Lim, H., **1**, 690 ⟨82JOC5229⟩
Lim, H. M., **2**, 176 ⟨87JOC2162⟩
Lim, L. B. L., **1**, 595, 599 ⟨87JCR(S)353⟩; **3**, 163 ⟨90SL117⟩; **4**, 174 ⟨91TL2565⟩
Lim, R. M., **2**, 444 ⟨78S588⟩; **4**, 51 ⟨78S588⟩
Lim, S., **1**, 207 ⟨93AG(E)1068⟩
Lim, S. T., **3**, 83 ⟨87JOC2114⟩
Lim, T. F. O., **3**, 502 ⟨80JA4970⟩; **4**, 606, 611 ⟨77JA8447, 77JA2013⟩, 611 ⟨80JA4970⟩, 735 ⟨81JOM(205)301⟩
Limbach, H.-H., **1**, 574, 584 ⟨91AG(E)1372⟩
Limberg, C., **3**, 395 ⟨91CB285⟩; **6**, 237, 256 ⟨92C78, 93JCS(D)2547⟩, 588 ⟨91CC1378⟩
Limburg, W. W., **5**, 379 ⟨62RTC430⟩
Limich, V. V., **5**, 1024 ⟨85ZOB2100⟩; **6**, 476 ⟨85ZOB2100⟩
Limmer, J., **5**, 384, 387 ⟨70CB2388⟩
Limmer, S., **4**, 548 ⟨91JOM(413)55⟩
Limori, T., **1**, 584 ⟨92JA4128⟩
Lin, C., **1**, 78 ⟨83JOC3162⟩, 808 ⟨78JA877⟩, 1023 ⟨69TL1379⟩
Lin, C.-E., **3**, 343 ⟨91JMC1292⟩
Lin, C.-H., **1**, 9 ⟨92JA10138⟩, 1023 ⟨68CC634, 70TL2695⟩; **3**, 273 ⟨76SC503⟩; **4**, 171 ⟨82JOC615⟩, 616 ⟨87OM1861⟩, 661 ⟨92TL7895⟩
Lin, C. M., **3**, 353 ⟨90JMC567⟩
Lin, C. T., **2**, 876 ⟨75JOM(85)179⟩; **6**, 450 ⟨89CJC2153⟩
Lin, D. C. T., **1**, 971 ⟨85JA6046⟩
Lin, F., **1**, 983 ⟨92JA5998⟩
Lin, F.-T., **4**, 652 ⟨93JA10754⟩; **5**, 151 ⟨87JA2426⟩
Lin, G.-J., **2**, 580 ⟨82JA4865⟩
Lin, G.-Y., **2**, 580 ⟨80JA3299⟩; **5**, 451, 937 ⟨83OM1852⟩
Lin, H.-S., **1**, 434, 815 ⟨87TL31⟩, 558, 575, 584 ⟨90JA9284⟩, 815, 1098 ⟨88JA879⟩; **2**, 238 ⟨88OS(67)157⟩; **3**, 206 ⟨87TL31⟩
Lin, I. J. B., **4**, 396 ⟨86JOM(315)135⟩, 583 ⟨90JCS(D)2509⟩, 583, 585 ⟨86JOM(309)225⟩
Lin, J., **1**, 663 ⟨65JOC1258⟩; **2**, 939 ⟨93JA3794⟩, 1028 ⟨86JA7832⟩; **4**, 202 ⟨93SL349⟩; **6**, 317 ⟨92BMC1607⟩, 641 ⟨92TL5933⟩
Lin, J. J., **1**, 4, 7, 8, 10, 17 ⟨78JOC1263⟩, 75, 457, 514 ⟨78JOC2567⟩, 458, 515 ⟨78JOC757, 77TL4481⟩; **2**, 969 ⟨78JOC757⟩; **3**, 127 ⟨76JOC1939, 76TL3865⟩; **4**, 1078 ⟨87JOC3243⟩
Lin, J. M., **5**, 214 ⟨86JOC167⟩
Lin, K.-C., **2**, 517 ⟨89JOC5814⟩; **3**, 419 ⟨90TL7549⟩; **4**, 22 ⟨90JOC5719⟩
Lin, K.-K. G., **3**, 460 ⟨75IC1675⟩
Lin, K. T., **2**, 149 ⟨78JOC2717⟩
Lin, K.-Y., **4**, 216 ⟨91JA7767⟩
Lin, L., **3**, 6, 119, 148 ⟨93JOC2847⟩
Lin, L. C., **1**, 791 ⟨87JOC244⟩
Lin, M.-T., **2**, 999 ⟨84JA1130⟩
Lin, N.-H., **2**, 196 ⟨82JA5781⟩
Lin, P., **4**, 649 ⟨85TL2391⟩
Lin, P.-Y., **5**, 582 ⟨92S1219⟩, 582, 611 ⟨93JOC4742⟩
Lin, R. I.-S., **5**, 247 ⟨94S699⟩

Lin, S., **1**, 1047 ⟨91S235⟩
Lin, S.-C., **3**, 13, 40, 85 ⟨90JA7050⟩
Lin, S. S., **2**, 1091 ⟨91JA6943⟩
Lin, S. T., **2**, 582 ⟨85IS1961⟩
Lin, T. S., **1**, 585 ⟨90TL3829⟩
Lin, W., **4**, 680 ⟨93CC284, 94OM2309⟩
Lin, W.-H., **6**, 297, 301 ⟨89JOC1990⟩
Lin, X., **2**, 529 ⟨90TL7563⟩; **4**, 936 ⟨89CC1872⟩
Lin, Y., **2**, 307 ⟨88JHC1035⟩; **3**, 118, 218 ⟨87TL3115⟩, 229 ⟨88TL1045⟩, 360 ⟨80JOC4857⟩; **5**, 631 ⟨84S946⟩, 759 ⟨79JOC4160⟩, 760 ⟨80JOC3750⟩
Lin, Y. C., **1**, 991 ⟨93OM3971⟩; **4**, 678 ⟨83JA1679⟩; **6**, 517 ⟨82IC1704⟩
Lin, Y.-i., **1**, 906 ⟨69JA6861⟩
Lin, Y. T., **2**, 595 ⟨86JOM(307)273⟩
Lin, Y.-T., **6**, 641 ⟨88TL3183⟩
Lin, Y.-Z., **4**, 275 ⟨90JOC247⟩
Lin, Z., **3**, 190 ⟨93TL1859⟩; **6**, 369 ⟨93TL1859⟩
Lincoln, D. N., **1**, 894 ⟨65JA3000⟩, 912, 913 ⟨65JA4533⟩
Lincoln, F. H., **1**, 729 ⟨69CC303, 69JA5364⟩; **3**, 494 ⟨69CC303⟩; **4**, 209 ⟨54JA743⟩
Lind, H., **5**, 780 ⟨65LA(688)14⟩; **6**, 506 ⟨66AG376⟩
Lindbeck, A. C., **3**, 182 ⟨89JOC5651⟩; **4**, 530 ⟨89JOC5651⟩, 531 ⟨91JA8546⟩
Lindberg, P., **4**, 299 ⟨77JCS(P1)684⟩
Lindberg, W. O., **5**, 28 ⟨B-55MI 502-01⟩
Lindblad, C. G., **5**, 652 ⟨69ACR65⟩
Linde, II, R. G., **1**, 193 ⟨89JOC2759⟩
Lindel, H., **2**, 254 ⟨81PS(11)349⟩; **5**, 1128 ⟨85CB676⟩
Lindell, S. D., **4**, 554 ⟨90TL5381⟩, 863 ⟨92SL730⟩
Lindemann, H., **3**, 468 ⟨30CB702⟩, 694 ⟨30CB1650⟩; **4**, 18 ⟨24LA(435)219⟩, 25 ⟨23LA(431)270⟩
Linden, A., **6**, 305, 309, 310 ⟨91HCA1500⟩
Linden, S.-M., **1**, 797, 882 ⟨83JOC584⟩
Linder, J., **3**, 107 ⟨85TL2155⟩
Linder, S. M., **1**, 883 ⟨87HCA1858, 87HCA1879⟩; **3**, 143 ⟨87HCA1858⟩
Linderman, R. J., **1**, 479 ⟨83JOC4621⟩, 540 ⟨81JOC3696⟩; **3**, 148 ⟨89JOC661⟩, 162 ⟨88TL3911⟩, 496 ⟨83TL3043⟩; **4**, 353 ⟨91JOC5213⟩, 354 ⟨88JOC1569⟩, 358, 359 ⟨90JA2392⟩, 361, 379 ⟨92JOC5078⟩, 384 ⟨87TL3911, 88JA6249⟩, 385 ⟨90JOC5428, 91JOC5491⟩, 387 ⟨88JOC2878⟩; **5**, 412, 413, 425 ⟨88JOC1569⟩
Lindermann, R. J., **4**, 384 ⟨89T495⟩
Lindert, A., **2**, 29, 34 ⟨71TL3995⟩, 420, 696 ⟨71JA2318⟩
Lindgren, A., **2**, 278 ⟨61ACS938⟩
Lindgren, B., **5**, 1114 ⟨73ACS726, 76ACS(B)941⟩, 1114 ⟨77ACS(B)1⟩
Lindgren, P. T., **5**, 1120 ⟨81USP4304935⟩
Lindhard, M., **5**, 984 ⟨30CB2528, 30CB2544⟩
Lindhardt, J. C., **2**, 423 ⟨76TL2565⟩
Lindinger, W., **1**, 1143 ⟨75JCP(63)3238, 76JCP(64)4544⟩
Lindlar, H., **1**, 156 ⟨57HCA1250⟩, 318 ⟨73OSC(4)880⟩, 502, 503 ⟨66OS(46)89⟩, 502, 503, 504 ⟨73OSC(5)880⟩; **3**, 71 ⟨59HCA844⟩
Lindley, A., **4**, 737 ⟨78CC475⟩
Lindley, J., **1**, 1005 ⟨84T1433⟩
Lindley, J. M., **2**, 753 ⟨78S23⟩
Lindley, P. F., **6**, 621 ⟨84CC1105⟩, 621, 728 ⟨87CC99⟩
Lindley, W. A., **1**, 436 ⟨83JOC4419⟩
Lindner, C., **3**, 552, 579 ⟨74CB502⟩, 558 ⟨72TL1519⟩, 562 ⟨80CB2509, 81CB536⟩; **5**, 616, 1050 ⟨74CB502⟩; **6**, 504 ⟨80CB2499⟩
Lindner, E., **1**, 445 ⟨86JCS(D)997⟩; **2**, 437 ⟨84ZN(B)115, 87JOM(335)59⟩, 437 ⟨92ZN(B)693⟩, 438 ⟨68CB3438⟩, 442 ⟨86ZN(B)10⟩, 490 ⟨86JCS(D)997⟩, 836 ⟨82CB2181⟩, 1003 ⟨82ZN(B)1146⟩; **4**, 112 ⟨86OM2030⟩, 563 ⟨72CB3261⟩, 581 ⟨81CB413, 83CB1070⟩, 581 ⟨83CB1209⟩, 583 ⟨82ZN(B)1146⟩, 921 ⟨83CB3141, 83ZN(B)726⟩; **5**, 220, 221 ⟨81CB2272⟩, 397 ⟨77AG(E)269, 78ZN(B)1457⟩, 397, 398 ⟨80CB3268⟩, 403 ⟨79CB1456, 79CB763⟩, 403 ⟨79CB793, 80CB2769⟩, 449 ⟨74ZN(B)275⟩, 546 ⟨69CB3347⟩, 943 ⟨92JOM(441)63⟩; **6**, 154 ⟨93AG(E)1424⟩, 327, 330 ⟨75JOM(94)229⟩, 656 ⟨78ZN849⟩
Lindner, H. J., **1**, 146 ⟨89AG(E)277⟩, 146, 711 ⟨86AG(E)935⟩, 326 ⟨89JA8311⟩; **2**, 997, 1000 ⟨91JA4002, 85AG(E)859⟩, 1000 ⟨86AG(E)938, 86AG(E)939⟩, 1000 ⟨88TL1153, 89AG(E)1025⟩, 1000 ⟨90AG(E)100⟩; **5**, 1122 ⟨80AG(E)566⟩
Lindner, L., **6**, 414 ⟨78MI 614-01⟩
Lindner, R., **1**, 737 ⟨67LA(707)120⟩; **4**, 551 ⟨66LA(699)40, 67LA(707)120⟩, 551, 558 ⟨68LA(713)12⟩, 557 ⟨67LA(707)112⟩; **5**, 908 ⟨66LA(699)40⟩, 910 ⟨67LA(707)112⟩; **6**, 701 ⟨66LA(699)40⟩
Lindquist, R. N., **4**, 525 ⟨77JA6435⟩
Lindsay, A. J., **6**, 517 ⟨84OM1523⟩
Lindsay, C. M., **4**, 846 ⟨84TL995⟩; **6**, 317 ⟨84TL991⟩
Lindsay, D. A., **1**, 331, 336, 342, 936 ⟨87JOC3509, 83JA3578⟩
Lindsay, D. G., **3**, 68 ⟨64T2091⟩
Lindsay, R. J., **2**, 751, 752 ⟨79COC(2)131⟩
Lindsay Smith, J. R., **2**, 338, 339 ⟨81JCS(P2)545⟩; **3**, 419 ⟨71JCS(C)3060⟩
Lindsell, W. E., **2**, 557 ⟨82COMC-I(2)155⟩
Lindsey, A. S., **5**, 105 ⟨57CRV583⟩
Lindsey, J. J., **2**, 787 ⟨74JOC3351⟩
Lindsey, Jr., R. V., **2**, 427 ⟨59JA6273⟩, 440 ⟨68JOC3088⟩; **4**, 970 ⟨60JA5116⟩; **5**, 839 ⟨58JA5793⟩, 1133 ⟨58JA4151⟩; **6**, 15 ⟨60JA6181⟩, 46 ⟨60JA5116⟩, 58 ⟨59JA4801⟩
Lindsey, R. V., **3**, 709, 711 ⟨58JA4151⟩
Lindsey Jr., R. V., **4**, 607, 608, 621, 622, 654 ⟨58JOC1392⟩; **5**, 373 ⟨58JA6442⟩
Lindstaedt, J., **3**, 351, 548 ⟨83CB66, 83CB509⟩
Lindstedt, E. I., **5**, 652 ⟨87JOM(334)255⟩
Lindstedt, E.-L., **5**, 151 ⟨86ACS(B)466⟩
Lindstrom, E. G., **3**, 449 ⟨48JA1381⟩
Lindstrom, M. J., **2**, 27 ⟨74JA5616⟩
Lindswell, W. E., **1**, 452 ⟨78JCS(D)657⟩; **2**, 563 ⟨78JCS(D)657⟩
Linebarrier, D. L., **2**, 76 ⟨89JA1937⟩
Lineberger, W. C., **1**, 1143 ⟨83CPL124⟩
Linek, E. V., **3**, 348, 363 ⟨79JOC3271⟩
Lineva, A. N., **4**, 656 ⟨76DOK(231)1138⟩
Linford, L., **5**, 651 ⟨88JOM(355)337⟩, 951 ⟨85OM275⟩; **6**, 333 ⟨84JOM(262)69⟩, 540 ⟨92POL893⟩
Ling, D., **2**, 131 ⟨72JA4784⟩; **4**, 368 ⟨72JA4784⟩
Ling, F. L., **1**, 738 ⟨B-81MI 105-01⟩
Ling Chwang, T., **4**, 605 ⟨76JA8412⟩; **6**, 185, 206 ⟨73JA3324⟩
Lingens, F., **4**, 322 ⟨54CB1690⟩; **5**, 83 ⟨68AG(E)350⟩
Ling Fang-le, **1**, 738 ⟨B-83MI 114-02⟩
Lingham, I. N., **4**, 446 ⟨83MI 409-02⟩
Lingier, W. R. F., **5**, 773 ⟨71CB639⟩
Lingo, S. P., **3**, 612 ⟨39JA1574⟩
Linhard, M., **6**, 439, 452 ⟨40CB177⟩
Linhart, F., **5**, 1126 ⟨80MIP-DE283404⟩
Link, H., **1**, 812 ⟨63HCA483⟩; **3**, 622, 638 ⟨63HCA1190⟩
Link, J. O., **6**, 25, 26, 31 ⟨92TL3435⟩, 42 ⟨92JA1906⟩
Link, K.-H., **2**, 387 ⟨71AG(E)813⟩
Link, M., **6**, 290, 291, 393, 709, 710 ⟨89CB409⟩
Link, P., **6**, 487 ⟨93ZN(B)1000⟩
Link, R., **1**, 574, 584 ⟨91AG(E)1372⟩
Linke, K.-H., **2**, 381, 382 ⟨73ZN(B)829, 73CB1052⟩; **5**, 848 ⟨71ZN(B)296⟩
Linke, S., **2**, 697 ⟨68JA5936⟩; **3**, 622 ⟨68JA6891⟩; **5**, 282, 300 ⟨78S303⟩
Linker, K. H., **5**, 1134 ⟨81GEP3018821⟩
Linker, T., **1**, 330 ⟨89T935⟩
Lin'kova, M. G., **2**, 167 ⟨66DOK(170)96⟩, 167 ⟨68IZV1889⟩, 173 ⟨81IZV159⟩, 211 ⟨81IZV1850⟩, 214 ⟨64IZV644⟩, 227 ⟨80IZV2111⟩

Linn, C. B., **1**, 117 ⟨36JA816⟩
Linn, W. J., **3**, 317 ⟨65JA3657⟩
Linn, W. S., **1**, 1125 ⟨68JA6741⟩
Linneman, E., **2**, 54 ⟨1872LA(161)178⟩
Linney, I. D., **2**, 52 ⟨94CC1775⟩; **5**, 340 ⟨91JCS(P1)3383⟩, 340, 347 ⟨93JCS(P1)1581⟩
Linoh, H., **4**, 136 ⟨87LA51⟩
Linoh, K., **5**, 348 ⟨88ZN(B)1495⟩
Linsker, F., **2**, 888 ⟨44JA191⟩
Linstead, R. P., **1**, 78 ⟨37JCS1136⟩; **3**, 275 ⟨56JCS3070⟩; **5**, 199 ⟨54JCS3722⟩
Linstrumelle, G., **1**, 111, 464, 971, 1052, 1056 ⟨75CC561⟩, 121, 1002 ⟨92TL5757⟩, 204, 971, 1053, 1056, 1057, 1058 ⟨81S875⟩, 206, 478 ⟨78TL4069⟩, 229, 478, 1092 ⟨79TL1503⟩, 464, 1090 ⟨75S434⟩, 465, 1090 ⟨76OS(55)103⟩, 467, 1009 ⟨81TL315⟩, 468 ⟨84TL6001⟩, 468, 1097, 1135 ⟨78TL191⟩, 468, 533, 1009 ⟨87TL1649⟩, 469 ⟨93TL6403⟩, 469, 472, 975, 1009, 1137 ⟨83S32⟩, 970, 975 ⟨80TL5019⟩, 971 ⟨77NJC373, 78TL1137⟩, 971 ⟨80TL3987, 82S738⟩, 971 ⟨82SC739, 83CC1133⟩, 974 ⟨76TL275⟩, 975, 1004, 1025 ⟨93TL3853⟩, 980 ⟨74CC10⟩, 1009 ⟨85TL3811, 86TL5857⟩, 1009 ⟨93SL217⟩, 1012 ⟨91TL6109⟩, 1012, 1013 ⟨88T6337⟩, 1064 ⟨92JOC651⟩, 1066 ⟨92TL2681⟩, 1092 ⟨76TL3225⟩, 1092, 1105 ⟨79TL1073⟩; **2**, 410 ⟨84TL5763⟩, 939 ⟨80TL3987, 81S875⟩, 9311 ⟨79TL1073⟩; **3**, 532 ⟨68BSF4913⟩; **5**, 427 ⟨80TL3987⟩
Linti, G., **2**, 474 ⟨88CB871⟩; **5**, 927 ⟨90AG(E)199, 92CB73⟩
Linyorva, A. N., **2**, 569 ⟨77JOM(131)43⟩
Liogon'kii, B. I., **5**, 24 ⟨73RCR939⟩
Lion, C., **3**, 23 ⟨76BSF1875⟩, 526 ⟨81T4189⟩; **5**, 32 ⟨79T2077⟩, 388, 389 ⟨88EJM257⟩
Lions, F., **5**, 848 ⟨65JOC318⟩
Liorber, B. G., **2**, 465 ⟨60DOK(135)1150⟩; **3**, 49 ⟨73RCR538⟩; **5**, 820 ⟨60ZOB1937⟩; **6**, 659 ⟨86ZOB1427, 87ZOB2467⟩
Liotta, C. L., **1**, 873 ⟨85TL1599⟩; **2**, 19 ⟨74JA2250⟩; **3**, 613 ⟨74JOC3416⟩, 634 ⟨77TL1117⟩; **5**, 4 ⟨74JA2250⟩, 395, 396 ⟨84TL1249⟩; **6**, 27 ⟨66CJC575⟩
Liotta, D., **2**, 155 ⟨80JOC2887⟩, 234 ⟨B-87MI 203-01⟩, 280 ⟨78TL4977⟩, 282 ⟨80TL3643, 81JOC2605⟩, 282 ⟨84ACR28⟩, 542 ⟨78TL5091⟩; **3**, 36, 37 ⟨B-87MI 301-02⟩, 37, 173 ⟨85T4727⟩, 38, 174, 207 ⟨81JOC2920⟩, 39, 175 ⟨78TL5091⟩, 173, 174, 183 ⟨B-87MI 304-03⟩, 174 ⟨80TL3643⟩, 174 ⟨85T4881⟩, 174, 207 ⟨81TL3043, 82JOC1598⟩, 229 ⟨87TL1069⟩, 256 ⟨79SC697⟩; **4**, 239 ⟨78TL5091⟩, 333 ⟨B-87MI 407-01⟩, 401 ⟨80JOC2736, 81JOC4301⟩; **5**, 457 ⟨90TL3723⟩
Liotta, D. C., **1**, 327 ⟨92JA4912⟩; **2**, 30 ⟨93H(35)701⟩, 66 ⟨93JOC799⟩, 688 ⟨92JOC1948⟩; **3**, 164 ⟨92JOC1940⟩; **4**, 230 ⟨91JA9377⟩; **5**, 334 ⟨92JOC1948⟩
Liotta, R., **1**, 510, 1101, 1103 ⟨79JA96⟩; **2**, 517 ⟨76JA5297, 77JA3427⟩, 517 ⟨78JOC1058, 79JA2966⟩, 900, 901, 906 ⟨79JA96⟩; **4**, 631 ⟨79JA96⟩
Liou, K. F., **2**, 595 ⟨86JOM(307)273⟩
Lipatova, I. P., **2**, 879 ⟨88JGU1173⟩, 1076 ⟨88JGU2050⟩
Lipczynska-Kochany, E., **1**, 935 ⟨91CRV477⟩
Lipfert, G., **2**, 233 ⟨59LA(627)195⟩
Lipina, E. S., **6**, 54, 261 ⟨89ZOR2490⟩
Lipinski, C. A., **1**, 1052 ⟨70TL2247⟩
Lipinski, M., **3**, 696, 700 ⟨73OSC(5)300⟩, 721 ⟨61OS(41)13⟩
Lipisko, B., **3**, 186, 228 ⟨80TL805⟩
Lipka, H., **1**, 496 ⟨91CB2075⟩
Lipkind, M. B., **2**, 615 ⟨89IZV1935⟩
Lipkowitz, K. B., **3**, 338, 345 ⟨79JOC486⟩, 345, 352 ⟨77TL3417⟩
Lipnitskii, V. F., **5**, 627 ⟨81ZOB1842⟩, 642 ⟨85ZOB2318⟩; **6**, 115 ⟨81ZOB1842⟩
Lipowitz, J., **2**, 657 ⟨73JOC162⟩
Lipowska, M., **5**, 597, 607 ⟨93M721⟩
Lipp, M., **5**, 1047 ⟨58CB1660⟩, 1057 ⟨59M41⟩; **6**, 596 ⟨59M41⟩
Lipp, P., **3**, 638 ⟨43CB1275⟩
Lippard, S. J., **2**, 1009 ⟨89JA1299⟩; **3**, 356 ⟨91TL4713⟩; **6**, 552 ⟨70JA3342, 72JA1563⟩
Lippert, A., **2**, 261 ⟨38JA2370⟩
Lippert, B. J., **1**, 440 ⟨91JA7439⟩; **4**, 812 ⟨91JA7439⟩
Lippmaa, E., **2**, 407 ⟨80ZOB2621⟩
Lippmaa, E. T., **6**, 368 ⟨72MI 612-01⟩
Lippold, U., **5**, 928, 930 ⟨88AG(E)960⟩
Lippsmeier, B., **4**, 119 ⟨75GEP(O)2412800⟩; **6**, 232 ⟨69CB2150⟩
Lipschutz, B. H., **1**, 453 ⟨90TL7257⟩
Lipschutz, S., **6**, 232 ⟨60JOC60⟩
Lipscomb, R. D., **2**, 166 ⟨46JOC499⟩
Lipshutz, B. H., **1**, 120 ⟨88JA2641⟩, 120, 1107 ⟨78JOC3418⟩, 124 ⟨92T5709⟩, 177 ⟨83JOC3334⟩, 204, 207 ⟨82JA2305⟩, 206 ⟨90SL119, 84JOC1147⟩, 207 ⟨88TL893, 90JA4063⟩, 207, 312 ⟨84TL5959⟩, 207, 312 ⟨85JOM(285)437⟩, 212 ⟨81JOC1410⟩, 217, 478 ⟨84JOC3928, 87TL945⟩, 473 ⟨93JA9276⟩, 654 ⟨80TL3343⟩, 773, 778, 779, 780, 781, 782 ⟨91COS(1)107⟩, 1042 ⟨86TL4241⟩; **1**, 86 ⟨89SL64⟩, 116, 120, 206, 465 ⟨87S325⟩, 116, 206, 311, 465, 476 ⟨92OR(41)135⟩, 118, 465 ⟨81JA7672⟩, 119 ⟨82JA4696⟩, 120 ⟨86T2873⟩; **2**, 555 ⟨92S191⟩, 585 ⟨91COS(1)107⟩, 585, 702, 910, 918, 919, 924, 931 ⟨92OR(41)135⟩, 586, 588 ⟨85JA3197⟩, 587, 1095 ⟨78JOC3418, 86T2873⟩, 588 ⟨90JA4032, 92TL5857⟩, 588, 1095 ⟨90SL119⟩, 588, 917 ⟨87S325⟩, 589 ⟨85JOM(285)437, 90JA4063⟩, 667 ⟨86CRV795⟩, 969 ⟨90TL7257⟩, 992 ⟨90JA7440⟩, 1094 ⟨89T349⟩, 1095 ⟨81JA7672⟩; **3**, 9, 123 ⟨85TL705⟩, 12, 128 ⟨92OR(41)135⟩, 109 ⟨88TL3407, 84T5005⟩, 140 ⟨90JA7440⟩, 204 ⟨89TL4617⟩; **4**, 176 ⟨81TL4603⟩, 184 ⟨81JOC2419⟩, 199 ⟨80TL3343⟩; **6**, 134 ⟨89TL15⟩
Lipsky, S. D., **1**, 97 ⟨71JOC2588⟩
Liptak, A., **4**, 190 ⟨82S421⟩
Liptak, D., **4**, 698 ⟨82OM1307⟩
Lipton, M., **5**, 277 ⟨77TL4171⟩
Lipton, M. F., **1**, 630, 632 ⟨75TL1811⟩, 633 ⟨78JOC1409, 80JOM(186)155⟩; **5**, 263 ⟨80OS(59)49⟩
Liptuga, N. I., **4**, 573 ⟨77URP403314, 78URP351438⟩, 573 ⟨84ZOB59⟩, 574 ⟨72ZOB293⟩; **5**, 1005, 1010 ⟨64ZOB2812, 66ZOB461⟩; **6**, 656 ⟨66ZOB461⟩
Liras, S., **1**, 827 ⟨89JOC3137⟩; **6**, 472 ⟨92JOC2523⟩
Lis, R., **2**, 196 ⟨86JA5908⟩
Lischewski, M., **4**, 455 ⟨74ZC243⟩, 551 ⟨70JPR135⟩, 551, 557 ⟨69JPR857⟩
Lischka, H., **1**, 1109 ⟨78JA5297⟩
Liska, F., **4**, 970 ⟨81CCC1389⟩
Liskamp, R. M. J., **1**, 406 ⟨92T6425⟩, 817 ⟨91TL5401⟩; **2**, 199 ⟨81JOC5408⟩; **3**, 634 ⟨81JOC5408⟩; **4**, 264 ⟨81JOC3273⟩
Lisko, J. R., **5**, 937 ⟨85OM944⟩
Lissavetzky, J., **6**, 556 ⟨84S172⟩
Lissel, M., **1**, 572 ⟨79S372⟩, 1065 ⟨80LA1, 81T1653⟩; **2**, 26 ⟨83S314⟩; **4**, 18 ⟨80LA1⟩; **6**, 462 ⟨81CB1210⟩
Lissi, E. A., **1**, 657 ⟨68JCS(B)1044⟩
List, A. K., **3**, 545 ⟨88JA4855⟩; **5**, 1096 ⟨88JA4855, 88JA6926⟩
List, D., **2**, 886 ⟨64LA(678)167⟩
Lister, M. A., **2**, 533 ⟨88TL585⟩, 533, 698 ⟨90T4663⟩
Lister, S. G., **1**, 654, 697, 700, 703 ⟨83CC630⟩; **3**, 308 ⟨83CC630, 84JOC3503⟩
Lister-James, J., **2**, 15 ⟨B-80MI 218-01⟩
Listl, M., **6**, 425, 487 ⟨78AG(E)361⟩
Liston, A. J., **2**, 675 ⟨66JOC2105, 70CC111⟩
Listov, A. A., **4**, 894 ⟨74MI 420-01⟩
L'Italien, Y. J., **5**, 306 ⟨87JMC498⟩
Litano, K., **5**, 639 ⟨91MI 515-01⟩

Litchel, K. E., **3**, 721 ⟨65AG(E)702⟩
Lithgow, A., **3**, 701 ⟨92TL1593⟩
Litkei, G., **3**, 259 ⟨79LA174⟩
Litscher, G., **2**, 544 ⟨76AG(E)370⟩
Litterer, H., **1**, 428 ⟨90S403⟩
Littger, R., **4**, 632 ⟨93AG(E)295⟩
Little, D. A., **1**, 890 ⟨81JOC3909⟩
Little, G. M., **4**, 840, 961 ⟨90JCS(P1)2775⟩
Little, G. R., **4**, 182 ⟨70T1311⟩
Little, R. D., **1**, 359 ⟨85JA2495, 91COS(5)239⟩, 359, 360 ⟨86CRV875⟩, 359, 944 ⟨81JA2744⟩, 527, 944 ⟨85JOC2400, 79JA7129⟩, 944 ⟨81TL4389, 85JOC3940⟩, 944 ⟨85TL3417, 92SL107⟩; **2**, 403 ⟨78JOC2921⟩; **3**, 132 ⟨80TL3339⟩; **4**, 234 ⟨80TL3339⟩
Little, R. L., **2**, 726 ⟨60OS(40)93⟩; **3**, 446, 470 ⟨58JA2257⟩
Little, W., **3**, 5, 116 ⟨77TL3749⟩
Little, W. F., **2**, 985 ⟨68IC1298⟩
Little, W. K., **1**, 862 ⟨58CI(L)1589⟩
Littlefield, L. B., **2**, 435 ⟨78JOC3508⟩
Littlewood, D. M., **2**, 764 ⟨78JCS(P1)1117⟩
Littlewood, P. S., **3**, 242 ⟨71JCS1292⟩
Littmann, E. R., **5**, 24 ⟨42CRV97⟩
Litvinenko, L. M., **2**, 717 ⟨64MI 214-01⟩
Litvinova, K. D., **6**, 39 ⟨76ZOR1798⟩
Litvinova, O. V., **5**, 1016 ⟨66ZOB952⟩
Litvinov, I. A., **3**, 689 ⟨91HAC577⟩; **4**, 342 ⟨88IZV159⟩, 549 ⟨82IZV2718, 84IZV2023⟩; **5**, 370 ⟨91ZOB450⟩; **6**, 538, 539 ⟨89IZV909⟩
Litvinov, V. P., **2**, 133 ⟨90S841⟩; **3**, 634 ⟨90S841⟩, 656 ⟨85S98⟩; **5**, 615 ⟨89ZOR1323⟩
Litvyak, I. G., **5**, 389 ⟨76ZOB75⟩
Liu, A., **1**, 1038 ⟨92JA9279⟩
Liu, A. H., **3**, 512 ⟨88JA4964⟩
Liu, B., **1**, 906 ⟨89JA1527⟩
Liu, C.-C., **3**, 221 ⟨91CC770⟩; **6**, 253 ⟨86JFC(32)89⟩
Liu, C. Q., **3**, 637 ⟨91MI 318-01⟩
Liu, C.-S., **4**, 616 ⟨87OM1861⟩
Liu, C. W., **4**, 583 ⟨90JCS(D)2509⟩
Liu, D., **5**, 746 ⟨84TL3651⟩
Liu, E. K. S., **4**, 153, 154 ⟨78IC618⟩
Liu, F.-Z., **4**, 521 ⟨85TL5547, 86CPB3273⟩
Liu, G., **1**, 599 ⟨87SC1839⟩; **3**, 89, 100 ⟨81JOC4617⟩; **6**, 565 ⟨86MI 617-05⟩
Liu, G. L., **3**, 415 ⟨90SC15⟩
Liu, H., **1**, 106 ⟨93OSC(8)210⟩, 181 ⟨90JOC3961⟩, 328, 927 ⟨91JOC4335, 90JA896⟩; **3**, 123 ⟨86SC1357⟩
Liu, H.,**3**, 300 ⟨90JOC3961⟩
Liu, H.-H., **4**, 230 ⟨90TL257⟩, 232 ⟨90TA143⟩
Liu, H.-J., **2**, 25 ⟨78CL923⟩, 36 ⟨88SC119⟩, 69 ⟨91TL5473⟩; **3**, 121 ⟨85SC759⟩, 161 ⟨82CJC94⟩, 239 ⟨86CJC520⟩; **5**, 131 ⟨78TL4461⟩, 234 ⟨91SC2097⟩, 280 ⟨79SC31⟩; **5**, 47 ⟨85SC759⟩
Liu, H. T., **1**, 339, 927 ⟨88JA5900⟩
Liu, J., **5**, 335 ⟨86JOC1929⟩
Liu, J. C., **1**, 681, 682, 785 ⟨72JA2000⟩; **5**, 161 ⟨72JA2000⟩
Liu, J.-H., **4**, 825 ⟨83JOC298⟩
Liu, J. S., **1**, 894 ⟨65JA3000⟩, 894, 900 ⟨62JA2016⟩, 912, 913 ⟨65JA4533⟩
Liu, J. Z., **4**, 657 ⟨84OM825⟩
Liu, K., **1**, 905 ⟨80JCR(S)152⟩; **2**, 476 ⟨90JA8575⟩; **4**, 716 ⟨93OM1213⟩
Liu, K. C., **5**, 664, 821 ⟨80JOC3916⟩
Liu, K. K.-C., **3**, 32 ⟨90JOC4897⟩, 631 ⟨89TL1917⟩
Liu, K.-T., **4**, 194 ⟨76JA3037⟩
Liu, L., **5**, 336 ⟨86JOC5332⟩
Liu, L. H., **1**, 635 ⟨53JA2153⟩
Liu, L.-K., **2**, 478 ⟨80JOC406⟩; **3**, 87, 88, 95, 293 ⟨89S293⟩; **4**, 583 ⟨90JCS(D)2509⟩; **5**, 384 ⟨90OM560⟩; **6**, 348 ⟨91OM2485⟩
Liu, M.-C., **1**, 585 ⟨90TL3829⟩
Liu, M. T. H., **3**, 324 ⟨86TL4383, 87CL2135⟩, 324 ⟨87TL1011, 90JA744⟩; **4**, 138, 154 ⟨88TL5863⟩
Liu, N. H., **2**, 573 ⟨86OM2452⟩
Liu, Q., **2**, 663 ⟨91MI 213-01⟩
Liu, R.-S., **1**, 419 ⟨67JA112⟩, 420 ⟨71JA1532⟩, 505 ⟨83JA2923⟩, 613 ⟨86TL3337⟩, 690 ⟨78CL803⟩, 752 ⟨88TL419⟩; **3**, 231 ⟨85TL2869⟩; **4**, 711 ⟨93OM108⟩
Liu, S.-H., **3**, 219 ⟨82SC995⟩
Liu, S. K., **4**, 555 ⟨66GEP1211200⟩
Liu, S.-L., **6**, 378 ⟨64MI 613-01, 70JOM(24)89⟩
Liu, S.-T., **2**, 431 ⟨83OM1877, 85IC1613⟩
Liu, T. M. H.,**5**, 745 ⟨61JOC2563⟩
Liu, W., **3**, 4, 115 ⟨80S223⟩
Liu, Y., **3**, 519 ⟨B-89MI 315-02⟩, 520 ⟨B-89MI 315-01⟩; **4**, 168, 170 ⟨83JA3661⟩, 347 ⟨74JCS(P1)1279⟩
Liu, Y.,**5**, 937 ⟨91JA927⟩
Liu, Y., **6**, 478 ⟨91JMC2049⟩
Liu, Y.-C., **5**, 135 ⟨90JCS(P1)2559⟩
Liu, Y. S., **3**, 699, 721 ⟨92CL1143⟩
Liu, Y.-Y., **3**, 92 ⟨79JHC799⟩
Liu, Z., **1**, 587 ⟨90JOC4273⟩; **2**, 7 ⟨88JOC6126⟩; **3**, 241 ⟨91JHC697⟩; **4**, 972 ⟨87S357⟩; **6**, 216 ⟨86MI 607-01⟩
Liu, Z.-D., **2**, 1033 ⟨93JOC1924⟩, 1036 ⟨92SC1997⟩
Liu, Z.-R., **1**, 13, 20 ⟨92JOC2143⟩, 14 ⟨92JOC1618⟩
Livanstov, M. V., **4**, 458 ⟨91ZOB1478⟩, 459 ⟨91URP1618747⟩, 459, 484 ⟨91ZOB1016⟩
Livantov, M. V., **2**, 433 ⟨87URP1294811⟩
Livantsova, L. I., **3**, 270 ⟨81JGU1105⟩; **4**, 827 ⟨84ZOB1925⟩; **5**, 702 ⟨91JGU1268, 91ZOB1389⟩, 723 ⟨89JGU2055, 89ZOB2293⟩
Livantsov, M. V., **2**, 433, 473 ⟨91ZN(B)1578⟩, 434 ⟨86ZOB2794, 87ZOB884⟩, 451 ⟨84JGU2237, 89JGU2130⟩; **4**, 126 ⟨86JGU1258⟩, 458 ⟨92ZOB946⟩, 562 ⟨90ZOB1420⟩, 914 ⟨79ZOB1910, 87ZOB1406⟩; **5**, 401 ⟨87JGU1254⟩; **6**, 117 ⟨86ZOB2427⟩, 117, 490 ⟨88PS(35)329⟩, 281 ⟨84ZOB2504⟩, 313 ⟨88ZOB2168⟩, 314 ⟨88ZOB1447, 88ZOB1927, 87JGU1260⟩, 490 ⟨87JGU1484⟩
Liverton, N. J., **1**, 207 ⟨85JCS(P1)1589⟩; **2**, 976 ⟨85JCS(P1)1589⟩; **5**, 179 ⟨86JA2451⟩
Livezey, M.,**1**, 506 ⟨61JOC3156⟩
Livinghouse, T., **1**, 217 ⟨73JOC4346⟩, 244 ⟨85TL6431⟩, 523 ⟨90JA4965⟩, 524 ⟨86JOC1159⟩, 1134 ⟨78CC754⟩; **2**, 325 ⟨88JA7536⟩, 684 ⟨86JOC111⟩; **3**, 175 ⟨85TL6431⟩, 410 ⟨92JA5459, 87JA590⟩, 707 ⟨92JA4089, 92T2209⟩, 708 ⟨87S391⟩, 711 ⟨84SC639⟩; **4**, 216, 218, 834, 836 ⟨86JOC879⟩, 516, 537 ⟨84SC639⟩, 518 ⟨83CC210, 85T3559⟩, 834 ⟨84TL3539⟩; **5**, 728 ⟨83CC210⟩; **6**, 336 ⟨86JOC879⟩
Livingston, J. R., **2**, 112 ⟨68MI 202-02⟩
Livingstone, J. G., **2**, 867 ⟨61JCS1000, 61JCS5053⟩
Livingstone, S. E., **2**, 128 ⟨56JCS437⟩; **6**, 563 ⟨72ICA11⟩
Livshits, V. A., **4**, 979 ⟨86IZV2344⟩
Liwschitz, Y., **5**, 199 ⟨57JCS4397⟩
Liz, R., **1**, 833 ⟨84CC1427⟩; **2**, 740 ⟨86S960⟩, 744 ⟨84S304⟩, 1097 ⟨83JCS(P1)1087, 88CC121⟩; **3**, 75 ⟨86CC1180⟩, 413 ⟨85CC1375⟩, 417 ⟨83JCS(P1)1093⟩; **4**, 1067 ⟨83JCS(P1)1087⟩
Lizuka, K., **2**, 32, 36 ⟨83CPB4189⟩
Lizzi, M. J., **2**, 746 ⟨82SC939⟩; **3**, 264 ⟨80JOC4522⟩
Ljutkina, F. W., **2**, 102 ⟨31ZOB190⟩
Llabrès, G., **5**, 241 ⟨80SA(A)769⟩
Llama, E. F., **1**, 11, 12 ⟨93T8433⟩
Llamas, K., **3**, 577 ⟨86JA5543⟩; **4**, 892, 979 ⟨86JA5543⟩
Llamas Marcos, A., **6**, 70 ⟨84MIP521372⟩
Llavona, L., **3**, 152, 154 ⟨90S1003⟩
Llebaria, A., **1**, 106 ⟨93JOM(451)1⟩; **2**, 624 ⟨88TL5811⟩; **4**, 810 ⟨88TL5811⟩
Llera, J.-M., **1**, 828 ⟨89JOC5350⟩; **2**, 152 ⟨92JOC6789⟩, 347 ⟨93TL141⟩
Llewellyn, D. R., **1**, 384 ⟨58JCS403⟩; **5**, 652 ⟨54CI(L)1154⟩
Llewellyn, F. J., **2**, 1002 ⟨54AX145⟩

Llinares, A., **4**, 598 ⟨81HCA329⟩
Llinas-Brunet, M., **1**, 327 ⟨91JA9701⟩; **2**, 36 ⟨86H(24)3043⟩; **3**, 239 ⟨86CJC520⟩, 701 ⟨89JOC1483⟩
Llobera, A., **3**, 713 ⟨88TL3343⟩
Llonch, J.-P., **3**, 80 ⟨73CR(C)1803⟩
Llorente, I., **4**, 749 ⟨93AG(E)893⟩, 772 ⟨90TL7375⟩; **5**, 1108 ⟨90TL7375⟩
Lloyd, A. C., **1**, 321 ⟨68QR549⟩
Lloyd, B. R., **4**, 585, 691 ⟨82CC614⟩, 1023, 1032 ⟨88CC673⟩
Lloyd, C. L., **6**, 151 ⟨92OM26⟩
Lloyd, D., **1**, 54 ⟨67CI(L)787, 86T3887⟩, 646, 647, 648 ⟨87CSR45⟩, 647, 648, 737, 738, 743 ⟨72T343, 67CC1042⟩, 648 ⟨88S319⟩, 648, 738, 743 ⟨88JCS(P2)1829⟩, 737, 738 ⟨73T1697⟩, 737, 738, 743 ⟨72T353⟩, 737 ⟨B-94MI 116-01, 68CI(L)1277⟩, 738 ⟨83T297⟩, 738, 742 ⟨69JCS(C)2464⟩, 739, 740, 742 ⟨77AG(E)487⟩, 845, 891 ⟨B-66MI 119-01⟩, 861 ⟨B-66MI 119-02⟩, 906 ⟨B-66MI 119-03⟩, 907 ⟨B-66MI 119-04⟩, 916 ⟨B-66MI 119-05⟩; **2**, 1004 ⟨67CC1042⟩, 1007 ⟨91JCS(P2)1563⟩; **3**, 49, 196, 492 ⟨87CSR45⟩, 69 ⟨70JCS(C)2484⟩, 196, 500 ⟨82T3355⟩, 196, 500 ⟨86T3887⟩, 196, 500 ⟨88S319⟩, 196, 500, 639 ⟨73T1697⟩, 328 ⟨72CC912⟩, 373 ⟨67CI(L)118⟩, 498 ⟨67CI324⟩, 498 ⟨72T353, 71JCS(C)2941⟩, 500 ⟨72T343, 74T2257⟩, 500 ⟨83T297, 88JCS(P2)1829⟩; **6**, 707 ⟨87CSR45⟩, 708 ⟨73T1697, 82T3355⟩, 708, 709 ⟨86T3887, 88JCS(P2)1829, 88S319⟩
Lloyd, H. A., **2**, 326 ⟨88JOC2968⟩
Lloyd, J. E., **3**, 457 ⟨64JCS1649⟩, 458 ⟨65JCS2662, 65JCS5083⟩
Lloyd, J. M., **1**, 629 ⟨82JCS(P1)2347⟩
Lloyd, Jr., J. C., **2**, 439 ⟨76JOC972⟩
Lloyd, W. G., **4**, 193 ⟨69JOC3949⟩
Lloyd-Williams, P., **5**, 259 ⟨93T11065⟩
Llusar, R., **4**, 692 ⟨88OM2279⟩
Llyin, S. G., **6**, 138 ⟨89T3487⟩
Lím, D., **1**, 130 ⟨73JOM(50)9⟩
Lo, C. P., **4**, 327 ⟨61JOC3591⟩
Lo, K. M., **1**, 693 ⟨78JA2140⟩
Lo, L., **4**, 247 ⟨88TL6729⟩
Lo, P. Y., **4**, 1044 ⟨90JOM(391)19, 90JOM(396)299⟩
Lo, S. M., **3**, 289 ⟨73JA3040, 75JA5448⟩
Lo, Y. S., **2**, 322 ⟨88SC205⟩; **3**, 468 ⟨74JOC1444⟩
Loacher, J. R., **3**, 532 ⟨58JOC1166⟩
Loader, C. E., **1**, 19 ⟨83CJC2415⟩; **3**, 106 ⟨66CJC1007⟩, 307 ⟨80CJC2527, 81TL4899⟩, 307 ⟨85S353⟩, 669, 670 ⟨81CJC2673⟩
Loan, L. D., **3**, 326 ⟨65JA737⟩
Lobanov, D. I., **2**, 821, 824, 847 ⟨69ZOB841⟩, 848 ⟨68ZOB2285⟩
Lobanov, O. P., **4**, 999 ⟨80ZOB2248, 83ZOB2015⟩, 999 ⟨84ZOB288⟩, 999, 1001, 1002, 1007, 1011 ⟨84ZOB2186⟩, 1000, 1001 ⟨78ZOB1994⟩, 1002 ⟨77ZOB1994⟩, 1006, 1010, 1012 ⟨82ZOB1122⟩; **5**, 655, 656 ⟨84JGU255⟩, 655, 656 ⟨84ZOB288, 82JGU980⟩, 656 ⟨82ZOB1122⟩
Löbberding, A., **6**, 550 ⟨86T2329⟩
Lobeck, W. G., **5**, 334 ⟨86JMC359⟩
Lober, F., **2**, 361 ⟨56GEP948330⟩
Lobering, H. G., **3**, 231 ⟨75LA874⟩
Lobo, A. M., **5**, 212 ⟨82TL1391, 91CC1127⟩, 214, 635 ⟨84S829⟩
Lobo, A. P., **1**, 884 ⟨65JOC181⟩
Lobo, M. J. S., **2**, 208 ⟨76TL2171⟩
Locatell, Jr., L., **5**, 98 ⟨55CRV659⟩
Locati, L., **3**, 222, 303 ⟨93TL7089⟩
Lochead, A., **3**, 333, 335 ⟨83CC1325⟩, 333, 335 ⟨83CC423⟩, 333, 335 ⟨85JCS(P1)1541⟩
Lochead, A. W., **3**, 333, 340 ⟨84CC1469⟩, 333, 340 ⟨84CC922⟩
Löchel, I., **4**, 581 ⟨89JOM(371)329⟩
Löchel, W., **4**, 404, 405, 408, 410 ⟨53CB1463⟩
Locher, J., **5**, 838 ⟨1874CB670⟩
Locher, R., **1**, 1048 ⟨81AG(E)569⟩
Lochmann, L., **1**, 130 ⟨73JOM(50)9⟩
Lochschmidt, S., **4**, 550 ⟨86AG(E)253, 86ZN(B)444⟩; **6**, 660, 686 ⟨83AG(E)545⟩
Lochte, H. L., **1**, 104 ⟨41JA2975⟩; **2**, 388, 403 ⟨22JA2556⟩; **3**, 459 ⟨48JA2012⟩
Lociuro, S., **3**, 181 ⟨83TL593⟩
Lock, C. J. L., **1**, 894 ⟨90JA8912⟩; **2**, 1007 ⟨86JA1692, 86JA1694⟩, 1007 ⟨90JA8912⟩, 1008 ⟨83JOC3043, 86JA3613⟩, 1009 ⟨89JA5424⟩
Lock, G., **1**, 42 ⟨43CB1252⟩
Lock, J. D., **2**, 9 ⟨79TL3615⟩, 205 ⟨78JA1637⟩; **5**, 1106 ⟨78JA1637⟩
Lock, R. L., **3**, 533 ⟨77JOC3165⟩
Lockard, J. P., **2**, 151 ⟨73S485⟩
Locke, J. M., **2**, 198 ⟨61JA1256⟩, 215 ⟨62JA847⟩
Lockhart, I. M., **3**, 267 ⟨65JCS3610⟩
Lockhart, J., **1**, 907 ⟨61JA2375⟩
Lockhart, J. C., **2**, 110 ⟨61JOC3250⟩
Lockhart, J. L. B., **2**, 832, 833 ⟨63OSC(4)784⟩
Lockley, W. J. S., **1**, 725 ⟨73JCS(P1)1416⟩
Locquenghien, K. H., **5**, 872 ⟨91AG(E)1154⟩
Locquenghien, K. H. Von, **4**, 665 ⟨91ZN(B)1207⟩
Locquin, R., **5**, 83 ⟨B-41MI 502-01⟩
Lodaya, J. S., **3**, 163 ⟨88JA2987⟩, 163 ⟨88JOC210⟩
Loebel, J., **1**, 146 ⟨90JOC3840⟩; **2**, 564 ⟨86OM1296⟩
Loeber, D. E., **3**, 57 ⟨75JCS(P1)1457⟩
Loehr, T. M., **2**, 488 ⟨71JA2150⟩
Loeppky, R. N., **4**, 74 ⟨68TL5415⟩
Loerzer, T., **3**, 348 ⟨83TL5861⟩
Loeschorn, C. A., **5**, 1113 ⟨84TL3387⟩
Loev, B., **2**, 413 ⟨57JOC1186⟩; **3**, 542 ⟨62JOC3365⟩; **5**, 29, 30, 92 ⟨64CI(L)193⟩; **6**, 570, 645 ⟨75JMC90⟩
Loevenich, D., **5**, 763 ⟨53CB890⟩
Loevenich, J., **1**, 1065 ⟨27CB950⟩; **2**, 1052 ⟨30CB1707⟩
Loevy, H., **5**, 595 ⟨03CB585⟩, 652 ⟨04CB874⟩
Loew, O., **6**, 221 ⟨1869ZC624⟩
Loewe, L., **3**, 565 ⟨38CB1627⟩
Loewe, M. F., **1**, 149, 594, 716 ⟨86JOC3108⟩; **4**, 532 ⟨86JOC3108⟩
Loewenthal, H. J. E., **1**, 622 ⟨92T2515⟩; **4**, 191 ⟨59T269⟩
Loffet, A., **4**, 58, 59 ⟨86S627⟩; **6**, 480 ⟨93MI 615-01⟩
Löffler, A., **2**, 307 ⟨90S125⟩, 1026 ⟨92S495⟩; **4**, 763 ⟨92S495⟩
Löffler, W., **6**, 296, 297, 298, 300, 321 ⟨82LA507⟩
Loftfield, R. B., **1**, 392 ⟨51JA4707⟩, 393 ⟨54JA35⟩
Lofthouse, T. J., **3**, 640 ⟨89IC492⟩
Loftus, F., **5**, 1083 ⟨89TL2379⟩; **6**, 622 ⟨89TL2379⟩
Logan, M. E., **4**, 975 ⟨89EUP297872⟩
Logan, T. J., **1**, 1063 ⟨67JOC105⟩; **2**, 428 ⟨66JOC3391⟩, 470 ⟨70CA88061⟩; **4**, 550 ⟨66JOC3391⟩; **6**, 246 ⟨63JOC1129⟩
Loganathan, V., **1**, 494 ⟨91TL687⟩
Logemann, E., **1**, 1000 ⟨75S423⟩; **3**, 280 ⟨75CB1570⟩; **4**, 177 ⟨79CB3603⟩; **5**, 570, 608, 610 ⟨78BSB525⟩
Lögers, M., **2**, 412 ⟨90AG(E)782⟩
Loginova, E. I., **4**, 478 ⟨69IZV1757⟩
Loginova, G. M., **4**, 488 ⟨79ZOB493⟩
Logothetis, A. L., **2**, 7622 ⟨67PC-01⟩; **4**, 104 ⟨68JOC2330⟩; **5**, 698, 761, 812, 814, 817, 819 ⟨66JOC3689⟩, 811 ⟨68JOC2330⟩
Logue, M. W., **1**, 1006, 1045 ⟨75JOC131⟩, 1135 ⟨82JOC2549⟩; **3**, 273 ⟨75JOC131, 82JOC2549⟩
Logusch, E.,**4**, 225 ⟨81JA3215⟩
Logusch, E. W., **2**, 344 ⟨91TL2711⟩; **3**, 51, 200 ⟨88JOC4069⟩; **4**, 210 ⟨84TL4195⟩
Loh, J.-P., **1**, 693 ⟨81JOC4072⟩; **3**, 220 ⟨87JOC1440⟩
Loh, R., **3**, 231 ⟨85TL2873⟩

Loh, T.-P., **2**, 70 ⟨90JA6737⟩
Lohani, H. C., **6**, 563 ⟨82MI 617-01⟩
Lohaus, G., **3**, 326 ⟨54CB1708⟩, 618 ⟨70OS(50)18⟩, 618, 666 ⟨67CB2719⟩, 669 ⟨70OS(50)52⟩; **5**, 348 ⟨70TL127⟩, 1005 ⟨72CB2791, 72CB2800⟩, 1106 ⟨74LA901⟩
Lohberger, S., **3**, 694 ⟨91COS(4)1083⟩
Löher, H. J., **5**, 152 ⟨81HCA2808⟩
Lohmann, H., **5**, 657, 1064 ⟨77CB2996, 71TL2765⟩
Löhmann, L., **1**, 396, 397 ⟨42LA(550)260⟩
Lohmer, K., **6**, 360 ⟨76GEP2460330⟩
Lohner, W., **3**, 256 ⟨81JOM(208)35⟩; **5**, 241 ⟨79JOM(166)373, 81JOM(205)167⟩
Lohr, A. D., **5**, 1043 ⟨69JOC616⟩
Lohr, L. L., **6**, 691 ⟨84JPC1981⟩
Löhr, T., **5**, 605 ⟨87CB751⟩, 605, 623 ⟨87LA189⟩
Lohray, B. B., **1**, 185 ⟨92S1035⟩, 186 ⟨93S1092⟩; **2**, 47 ⟨92TA1317⟩, 100 ⟨92S1035⟩, 322 ⟨89TL2623⟩; **3**, 34, 163 ⟨93S1092⟩, 51, 202 ⟨87AG(E)351⟩, 156 ⟨89HCA980⟩
Lohrenz, J. C. W., **2**, 997, 999 ⟨92CB2265⟩, 1001, 1002 ⟨93CB1887⟩
Lohrenz, J. W. C., **2**, 999 ⟨93CB1873⟩
Lohse, C., **2**, 792 ⟨78ACS(B)625⟩; **3**, 366 ⟨72TL5213⟩
Lohse, O., **4**, 229 ⟨93TA1547⟩
Loibner, H., **1**, 179 ⟨76HCA2100⟩; **3**, 613 ⟨76HCA2100⟩; **5**, 1107 ⟨76HCA2100⟩
Loim, N. M., **1**, 13, 281 ⟨74S633⟩, 76 ⟨67T793⟩; **4**, 706 ⟨81JOM(209)233⟩, 706, 1036 ⟨79JOM(168)C33⟩
Lois, T. A., **1**, 881 ⟨68JA4505⟩
Loiselle, P. M., **1**, 499 ⟨90JA6416⟩
Loizou, G., **1**, 827 ⟨90SC1437⟩
Lojanapiwatna, V., **4**, 856 ⟨80AJC1073⟩
Lok, C. M., **3**, 667 ⟨70TL4701⟩
Lokanatha Rai, K. M., **3**, 157, 291 ⟨91COS(1)541⟩
Löken, B., **3**, 218 ⟨56JA876⟩; **4**, 191 ⟨54JA1359⟩
Lokensgard, J. P., **3**, 532 ⟨74JOC3355⟩
Lokshin, B. V., **1**, 890 ⟨89JOM(371)279⟩; **4**, 557 ⟨85ZOB1234⟩
Loktionova, R. A., **5**, 713 ⟨86ZOB1262⟩
Loktionsova, R. A., **5**, 713, 788 ⟨86JGU1113⟩
Loleit, H., **2**, 491, 505 ⟨35JCS366⟩
Löliger, J., **3**, 468 ⟨74JOC1444⟩
Löliger, P., **3**, 659 ⟨69AG(E)343⟩
Lollo, C. P., **5**, 723 ⟨85JOC1405⟩
Lomakina, A. V., **5**, 787 ⟨90JGU864⟩
Lomakina, S. I., **1**, 469 ⟨89IZV2623⟩, 1022 ⟨85IZV1109⟩; **4**, 1068 ⟨88ZOB1567⟩
Lomanova, N. F., **5**, 567, 581 ⟨76KFZ41⟩
Lomas, J. S., **1**, 883 ⟨90JOC1052⟩, 923 ⟨87JOC2627⟩
Lombard, A., **6**, 540 ⟨92POL893⟩
Lombard, R., **3**, 232 ⟨58BSF1458⟩; **5**, 208 ⟨64BSF542⟩
Lombardo, F., **2**, 81 ⟨89JOC4022⟩
Lombardo, L., **1**, 678 ⟨82TL4293⟩, 678, 767 ⟨87OS(65)81⟩; **3**, 193 ⟨80S368⟩, 419 ⟨89JA371⟩
Lombardo, P., **4**, 164, 209 ⟨65JOC3834⟩
Lomölder, R., **5**, 207 ⟨87AG(E)1253⟩
Lonchabon, G., **2**, 175 ⟨76BSF483⟩
Lonchambon, G., **2**, 726 ⟨81BSF(2)71⟩; **6**, 142 ⟨75CR(C)563⟩, 143 ⟨77CR(C)321⟩
Lonergan, G., **5**, 223 ⟨89LA387⟩
Long, A., **1**, 50 ⟨79TL349⟩
Long, A. G., **3**, 208 ⟨65JCS130⟩, 208, 234 ⟨61JCS2532⟩; **4**, 56, 57 ⟨64JCS766⟩; **5**, 62 ⟨B-52MI 502-04⟩, 570 ⟨64JCS1190⟩
Long, D. A., **2**, 13 ⟨83TL1933⟩
Long, F. A., **1**, 916 ⟨64JA322, 64JA327⟩, 916 ⟨64JA331⟩
Long, G. C., **6**, 589 ⟨89JCS(P1)1068⟩
Long, G. G., **2**, 492 ⟨64JA209, 67ICA34⟩, 492 ⟨67IS92, 68IC834⟩
Long, G. W., **2**, 681 ⟨84TL395⟩
Long, J. W., **2**, 24 ⟨51JA998⟩
Long, K. P., **3**, 482 ⟨77JA1267, 82JOC1284⟩; **6**, 118, 119, 636, 666 ⟨82JOC1284, 77JA1267⟩
Long, L. H., **2**, 32, 36 ⟨65CI(L)223, 64CI(L)1582⟩
Long, M. B., **5**, 990 ⟨72JHC1267⟩
Long, M. de, **1**, 364, 365 ⟨90PAC1597⟩
Long, N. R., **2**, 941, 1023 ⟨78JOC376⟩; **3**, 537 ⟨78JOC376⟩
Long, Q.-H., **4**, 309, 323 ⟨93S229⟩; **6**, 558 ⟨93OPP83⟩
Long, R. A., **5**, 619 ⟨70JHC863⟩
Long, T. V., **2**, 488 ⟨71JA2150⟩; **4**, 726 ⟨71IC522⟩
Long, X., **2**, 573 ⟨86OM2452⟩
Longobardi, M., **2**, 204 ⟨83JHC839⟩; **5**, 501 ⟨83JHC1549⟩
Longobardo, L., **2**, 688 ⟨93SC1515⟩; **3**, 159 ⟨92T1299⟩, 170 ⟨93SC1515⟩
Longone, D. T., **3**, 494 ⟨67CC300⟩
Longoni, G., **5**, 461 ⟨75CC267⟩; **6**, 525 ⟨79JA6110⟩
Longuet-Higgins, H. C., **1**, 853 ⟨46JCS636⟩
Loninova, I. V., **2**, 476 ⟨91MI 208-03⟩
Lönn, H., **4**, 223 ⟨87MI 405-01⟩, 224 ⟨85CAR(139)115⟩; **6**, 546 ⟨91TL7453⟩
Lönnberg, H., **4**, 222 ⟨91TL7593⟩
Loo, S. N., **2**, 608, 612 ⟨68CC806⟩
Look, G. C., **4**, 181 ⟨86JOC4964⟩
Look, M., **4**, 912 ⟨81MI 420-01⟩
Looker, J. H., **2**, 651 ⟨84JOC645⟩; **3**, 475 ⟨67CJC1727⟩
Loomis, G. L., **4**, 745, 746 ⟨75TL1373⟩
Looney, M. G., **1**, 829 ⟨91T7171, 93JA5815⟩; **5**, 179 ⟨93JA5815⟩
Loong, W. A., **3**, 249 ⟨75JOC3179⟩
Loontjes, J. A., **1**, 704 ⟨80RTC39⟩; **4**, 270, 272 ⟨80RTC39⟩
Loos, W. T., **3**, 628 ⟨90SL735⟩
Loosli, H.-R., **4**, 197 ⟨80HCA1960⟩
Loosmore, S. M., **2**, 205 ⟨78JA1637⟩; **5**, 1106 ⟨78JA1637⟩
Loots, M. J., **1**, 318 ⟨80JA1333⟩, 491, 536 ⟨77TL1303⟩, 536, 1105 ⟨77JA8045, 78TL4381⟩; **2**, 969 ⟨78TL4381, 80JA1333⟩, 969, 992 ⟨77JA8045⟩; **3**, 140 ⟨80JA1333⟩
Lopes, A., **6**, 448 ⟨86JOC3781⟩
Lopez, A. F., **3**, 626 ⟨76JOC3371⟩
Lopez, E. M., **6**, 329 ⟨94JOM(467)231⟩
Lopez, F., **1**, 146 ⟨89AG(E)277⟩; **2**, 1000 ⟨89AG(E)1025⟩; **4**, 772 ⟨90TL7375⟩
López, F., **5**, 1108 ⟨90TL7375⟩; **6**, 121 ⟨88S562⟩
Lopez, F. J., **5**, 199 ⟨89H(28)565⟩
López, G., **2**, 148 ⟨71IJS(A)130⟩; **5**, 386 ⟨92JCS(D)53⟩
López, J. A., **4**, 685 ⟨90JCS(D)2587⟩
López, J. C., **1**, 579 ⟨92JOC1613⟩, 942 ⟨92CC613⟩; **2**, 9 ⟨91S353⟩
Lopez, L., **1**, 187 ⟨78CC1085⟩, 238 ⟨83JCS(P1)2953⟩, 592 ⟨76S200⟩, 1025 ⟨79TL3873⟩; **2**, 28 ⟨77JCS(P1)501⟩, 626 ⟨71JCS(C)3652⟩; **3**, 153 ⟨78S139⟩, 154 ⟨77JCS(P1)501⟩, 319 ⟨83JA663⟩; **4**, 188 ⟨89TL257⟩; **5**, 775 ⟨77JCS(P1)1806⟩
López, L. A., **4**, 421 ⟨89TL4573⟩; **6**, 571 ⟨89S228⟩
López, M. C., **2**, 398 ⟨90JOC2070⟩
Lopez, M. C., **3**, 4, 116 ⟨87T3963⟩, 632 ⟨89JCS(P1)1692, 90TL2205⟩
Lopez, R., **4**, 864 ⟨91TA93⟩
Lopez, R. C. G., **3**, 334 ⟨82CC1029⟩, 334, 335 ⟨83T1487⟩
Lopez-Calahorra, F., **3**, 182 ⟨88S314⟩
Lopez-Cepero, I. M. de, **3**, 130, 202 ⟨85JOC5260⟩
López-Espinosa, M. T. P., **6**, 550 ⟨93CAR(242)281⟩
Lopez-Lazaro, A. L., **5**, 1066 ⟨94TL3817⟩
Lopez-Leonardo, C., **1**, 835 ⟨91TL4041⟩; **3**, 623 ⟨91TL4041, 93T5153⟩
López-Prado, J., **1**, 68 ⟨83T2863⟩; **5**, 152 ⟨83T2863⟩
Lopez-Rodriguez, M. L., **1**, 30 ⟨84JCS(P1)1671⟩
López-Sánchez, M. I., **4**, 325 ⟨92JCS(P1)291⟩
Lopez Sastre, J. A., **4**, 858 ⟨88CJC2975, 93JCR(S)60⟩
Lopp, M., **1**, 225 ⟨91TA943⟩
Lopresti, R., **2**, 39 ⟨76JOC3505⟩
Lopresti, R. J., **2**, 39 ⟨81HCA1158⟩; **4**, 448 ⟨58JA4921⟩

Lopresti, R. L., **4**, 168, 170 ⟨83JA3661⟩
Lopusinski, A., **2**, 256 ⟨82CC1362⟩, 475 ⟨82JA290⟩
Lopyrev, T. T., **5**, 718 ⟨88BAU378⟩
Lopyrev, V. A., **5**, 718 ⟨84JGU426, 84ZOB477⟩, 718 ⟨88IZV460⟩; **6**, 413 ⟨68ZPK1380⟩
Loran, J. S., **2**, 847 ⟨76JCS(P2)1444⟩
Lorand, J. P., **2**, 355 ⟨73JOC1813⟩
Lora-Tamayo, M., **5**, 366 ⟨91JCS(P1)273⟩
Lorber, M., **5**, 984 ⟨70JA1326⟩
Lorber, M. E., **1**, 633 ⟨68JA4762⟩; **5**, 985 ⟨67JOC540⟩
Lorberth, J., **2**, 485 ⟨75JOM(88)329, 75JOM(92)181⟩, 504 ⟨85ZN(B)1320⟩; **3**, 482 ⟨67CC836, 75JOM(93)339⟩, 482 ⟨75JOM(97)59, 76JOM(110)195⟩, 482 ⟨77JOM(132)359⟩, 482, 483, 484 ⟨80JOM(191)371⟩, 483 ⟨77JOM(127)19⟩, 483, 484 ⟨70JCS(A)2954⟩, 483, 484, 488 ⟨71JOM(27)303, 68JOM(15)251⟩, 484 ⟨74JOM(78)221⟩, 485 ⟨76JOM(120)131⟩, 488 ⟨73JOM(54)23⟩; **5**, 871 ⟨76JOM(110)195⟩, 871, 873, 874 ⟨80JOM(191)371, 67CC836⟩; **6**, 668 ⟨75JOM(93)339⟩, 668, 670, 671 ⟨77JOM(132)359⟩, 670, 675 ⟨80JOM(191)371⟩, 672 ⟨70JCS(A)2954⟩
Lorcy, D., **4**, 870 ⟨90CC470⟩, 873 ⟨90CC816⟩
Lorcy, D., **6**, 120, 123 ⟨90CC470⟩
Lord, W. A., **5**, 846 ⟨76JCS(P1)38⟩
Lore, A. M., **5**, 580 ⟨70JOC2102⟩, 832 ⟨70CI(L)238⟩
Loren, S., **1**, 374 ⟨91JA7794⟩; **2**, 526 ⟨91JA7794⟩
Lorenčak, P., **1**, 835 ⟨91JOC970⟩; **3**, 541 ⟨86CC369⟩, 582 ⟨88JA1337⟩; **4**, 900 ⟨91JOC970⟩
Lorente, A., **4**, 405 ⟨87CJC687⟩; **5**, 606 ⟨85S86, 92H(34)1573⟩
Lorenz, B., **6**, 594 ⟨68ZC230⟩
Lorenz, D., **1**, 745 ⟨64JPR143⟩
Lorenz, D. H., **3**, 698, 721 ⟨63JOC1707⟩
Lorenz, G., **3**, 621, 633 ⟨88OSC(6)866⟩
Lorenz, H., **3**, 690 ⟨73AG(E)564⟩
Lorenz, I., **6**, 655, 733 ⟨79AG(E)693⟩
Lorenz, J., **3**, 547 ⟨81LA1361⟩
Lorenz, K. T., **1**, 248 ⟨87ACR65⟩
Lorenz, R., **3**, 184, 185 ⟨79T613⟩, 448 ⟨57CB1060⟩
Lorenz, W., **2**, 461 ⟨62BEP609076⟩, 464 ⟨58MIP1067011⟩; **4**, 743 ⟨48CB422⟩
Lorenzi, G. P.,**1**, 434 ⟨66JOC321⟩
Lorenzi, G. P., **1**, 999 ⟨66JOC329⟩
Lorenzo, A., **2**, 11 ⟨80S853⟩; **5**, 680 ⟨88H(27)1935, 89S843⟩, 1066 ⟨94TL3817⟩, 1122 ⟨84TL5427⟩; **6**, 504 ⟨92S297⟩
Lorenzo, M. A., **5**, 680 ⟨89JCR(S)262⟩
Loreto, M. A., **3**, 181 ⟨94T3829⟩
Lorette, N. B., **4**, 178, 184 ⟨60JOC521⟩, 184 ⟨60JOC525⟩, 184 ⟨64USP3127450⟩, 186 ⟨62USP3024284⟩
Lorey, O., **6**, 134 ⟨79PS(7)203⟩
Lorimer, J. P., **1**, 925 ⟨87CSR239⟩; **2**, 19 ⟨86JCR(S)300⟩
Loring, H. S., **2**, 126 ⟨30JA4500⟩
Lorne, R., **1**, 206, 478 ⟨78TL4069⟩, 464, 1090 ⟨75S434⟩; **2**, 359 ⟨87BSF181⟩, 363 ⟨91T6655⟩; **3**, 23 ⟨76BSF1445⟩; **4**, 323 ⟨76BSF1482⟩, 938 ⟨85BSF1250⟩
Lorz, E., **5**, 756 ⟨48JA1904⟩
Los, M., **4**, 166 ⟨70JOC468⟩
Los, M. G., **4**, 640 ⟨76ZOB837⟩
Losanitsch, S. M., **5**, 1023 ⟨1891CB3021⟩
Losch, R., **3**, 552, 579 ⟨74CB502⟩; **5**, 616, 1050 ⟨74CB502⟩, 1064 ⟨71TL2765, 80CB79⟩
Löschmann, I., **4**, 322 ⟨54CB1690⟩
Losee, K., **5**, 354 ⟨53JA1933⟩
Lösel, W., **6**, 504 ⟨86CB2553⟩
Losen, J., **1**, 1065 ⟨27CB950⟩
Lösner, A., **2**, 887 ⟨1894CB263⟩
Loss, A., **2**, 504 ⟨85ZN(B)1320⟩
Losse, G., **5**, 193 ⟨61CB1762⟩
Lossen, H., **5**, 999 ⟨1872LA(161)347, 1877LA(186)1⟩, 999 ⟨1889LA(252)170⟩
Lossen, W., **5**, 826, 999 ⟨1895LA(281)169⟩, 835 ⟨1884CB1587, 1889LA(252)214⟩
Lossing, F. P., **1**, 894 ⟨B-63MI 119-02⟩, 921 ⟨88JA7339, 88JA7343⟩, 925 ⟨83JA4701⟩, 1109 ⟨70CJC955, 71CJC357⟩, 1109 ⟨72CJC3973⟩; **2**, 636 ⟨82JA2648⟩; **3**, 335 ⟨67JPC4111⟩
Lothrop, W. C., **3**, 452 ⟨36JA2050⟩
Lotsch, W., **1**, 922 ⟨70T5343⟩
Lotsh, G., **1**, 1010 ⟨88M751⟩
Lotter, H., **3**, 183 ⟨82TL639⟩
Lottes, A. C., **1**, 499 ⟨91TL6473⟩; **3**, 323 ⟨89TL4089⟩
Lotto, G., **2**, 49 ⟨90TL2665⟩
Lotts, K. D., **4**, 60 ⟨76TL3381⟩; **5**, 1122 ⟨79TL3131⟩
Lotz, S., **4**, 720 ⟨93AG(E)710⟩; **6**, 560 ⟨85ZAAC(531)82⟩, 627 ⟨86ZAAC(533)99⟩
Lotz, W. W., **2**, 1001 ⟨73T917⟩
Lötzsch, K., **3**, 447 ⟨65CB912⟩
Lou, B., **1**, 14 ⟨86TL4343⟩
Lou, J.-D., **3**, 4 ⟨92SC767⟩, 57 ⟨87S179⟩
Lou, M.-L., **1**, 439 ⟨87JOC2224⟩
Lou, W. X., **3**, 4 ⟨92SC767⟩
Lou, W.-X., **3**, 57 ⟨87S179⟩
Loubinoux, B., **1**, 458, 515 ⟨77TL1069⟩, 1039 ⟨90SL547⟩; **4**, 199 ⟨81TL1973⟩, 199, 314 ⟨83S568⟩
Loudas, B. L., **5**, 1102 ⟨68USP3733349⟩
Loudon, A. G., **1**, 627 ⟨67JCS(B)509⟩
Loudon, G. M., **2**, 692 ⟨75JOC126⟩; **3**, 617 ⟨75JOC126⟩; **4**, 422 ⟨84JOC4272⟩; **5**, 999 ⟨84JOC4272, 84JOC4277⟩
Loudon, J. D., **1**, 637 ⟨49JCS1074⟩
Lough, A. J., **4**, 613 ⟨91OM3292⟩; **5**, 923, 924 ⟨92OM3088, 91OM2752⟩, 924 ⟨91OM2758⟩
Lougheed, G. S., **4**, 47 ⟨67OS(47)23⟩
Loughlin, W. A., **1**, 486 ⟨90AJC1375⟩
Louis, D., **1**, 975 ⟨93TL3129⟩
Louis, E., **2**, 5411 ⟨68IC1253⟩; **5**, 952 ⟨68JOM(14)P9⟩
Louis, J.-M., **3**, 102, 108 ⟨76JHC525⟩; **5**, 171 ⟨75T171⟩
Loukakou, E., **4**, 891 ⟨83JOC481⟩
Lounasmaa, M., **3**, 638 ⟨82H(19)2115⟩
Loupy, A., **2**, 19 ⟨88SC1661⟩, 25, 31 ⟨86CC1250, 88SC1275⟩, 31 ⟨89BSF247⟩, 157 ⟨89SC2939⟩; **3**, 634 ⟨87S56⟩, 642 ⟨77S126⟩; **5**, 132 ⟨85S40⟩, 199 ⟨93SC419⟩
Lourak, M., **1**, 472 ⟨89JOC4840⟩
Lourens, R., **4**, 483 ⟨84T991⟩
Louw, R., **1**, 624, 779, 783 ⟨78JA3927⟩, 774 ⟨72RTC3⟩; **2**, 150 ⟨76CC496⟩; **4**, 47, 48 ⟨77CI(L)127⟩; **5**, 438 ⟨74TL3377, 76C199⟩, 438 ⟨76JCS(P2)1895⟩; **6**, 229, 230 ⟨77CI(L)127⟩
Louwen, J. N., **1**, 127 ⟨82JA6609⟩
Love, C. J., **1**, 74 ⟨73JCS(P1)2509⟩
Love, E. M., **2**, 405, 773 ⟨73JA3070⟩
Love, G. M., **2**, 773 ⟨76S726⟩
Lovejoy, R. W., **2**, 1093 ⟨67JCP(46)658⟩
Lovel, C. G., **1**, 545 ⟨82CC1231⟩; **3**, 614 ⟨82AJC2053, 82TL1621⟩, 660 ⟨83AJC1975⟩
Lovelace, A. M., **4**, 731 ⟨55JA2783⟩; **6**, 5 ⟨55JA768⟩
Lovelace, T. C., **2**, 325 ⟨88JOC2094⟩; **5**, 329 ⟨94JOC1231⟩
Lovell, M. F., **5**, 759 ⟨79JOC4160⟩
Lovely, C. J., **3**, 439 ⟨88TL3725, 91T5561⟩
Loveridge, E. L., **5**, 291 ⟨70JOC1219⟩
Lovett, J. R., **6**, 369 ⟨71USP3562309⟩
Lovey, A. J., **3**, 131 ⟨77JOC1189⟩
Lovey, R. G., **3**, 436 ⟨94SL167⟩
Lovich, S. F., **2**, 160 ⟨84JOC2629⟩
Lovie, J. C., **4**, 610 ⟨66JCS(C)1137⟩
Loving, B. A., **1**, 890 ⟨67JA3365⟩
Low, C. M. R., **2**, 551 ⟨B-89MI 211-01⟩
Löw, D., **4**, 574 ⟨91CB199⟩
Low, G. H., **3**, 694 ⟨71T3069⟩
Low, J. N., **4**, 138, 369, 372 ⟨92JOM(437)111⟩
Low, M., **5**, 277 ⟨74TL1785⟩
Löw, O., **2**, 438 ⟨66ZAAC(346)241⟩; **5**, 395, 397, 784 ⟨66ZAAC(346)241⟩

Löw, P., **2**, 742 ⟨83S785⟩
Low, P. J., **4**, 678 ⟨94JOM(464)191⟩
Löw, R., **3**, 477, 479 ⟨78HCA97, 66TL5821⟩
Lowe, C., **1**, 942 ⟨91JCS(P1)1433⟩
Lowe, J. A., **1**, 154 ⟨76TL1439⟩; **3**, 502 ⟨76JA7844⟩
Lowe, J. L., **5**, 579 ⟨52JCS742⟩
Löwe, U., **3**, 139, 145 ⟨79TL4971⟩
Löwe, W., **4**, 234 ⟨84AP(317)15⟩
Lowen, G. T., **3**, 626 ⟨86TL5319⟩
Lowenthal, R. E., **1**, 353 ⟨91TL7373⟩
Lowery, J. A., **2**, 111 ⟨61JPS193⟩
Lowery, M. K., **2**, 20, 337, 338, 339 ⟨70CRV639⟩, 298, 305, 761 ⟨70CR639⟩, 338 ⟨69JOC911⟩
Lowinger, T. B., **1**, 942 ⟨90CJC1636⟩
Lown, E. M., **3**, 530 ⟨80PAC1623⟩
Lown, J. W., **1**, 11, 19 ⟨82JOC4304⟩; **2**, 290 ⟨66JOC3396⟩; **3**, 317 ⟨71CJC3443⟩, 351 ⟨86JA3811⟩, 368 ⟨70JOC1716⟩, 584 ⟨73TL3727⟩
Lowndes, P. R., **4**, 519 ⟨81JCS(P1)1934⟩
Lowry, T. H., **1**, 107 ⟨B-87MI 104-01⟩, 845 ⟨B-87MI 119-02⟩, 886, 893 ⟨B-87MI 119-05⟩, 891 ⟨B-87MI 119-06⟩, 904 ⟨B-87MI 119-07⟩, 915 ⟨B-87MI 119-09⟩
Lowry, T. M., **2**, 288 ⟨28JCS3179⟩, 289 ⟨29JCS2076⟩; **5**, 652 ⟨25JCS279⟩
Lowy, A., **4**, 297 ⟨21JA341⟩
Loy, G., **1**, 988 ⟨87G625⟩
Lozac'h, N., **3**, 255 ⟨64BSF3254, 74BSF1196⟩, 339, 356, 357 ⟨80BSF(2)530⟩, 354, 376 ⟨84CHEC(6)1049, 80BSF(2)539⟩, 363 ⟨69BSF827⟩; **5**, 586 ⟨66BSF3828⟩, 605, 613 ⟨83BSF(2)277⟩; **6**, 48 ⟨80AHC(27)151⟩
Lozinskii, M. O., **4**, 272 ⟨91ZOR216⟩, 275, 277 ⟨89UKZ1216⟩, 864 ⟨90UKZ1310⟩, 865 ⟨89ZOR1102⟩; **5**, 685 ⟨90JOU1408, 90ZOR1629⟩, 1022 ⟨68RCR363⟩, 1049 ⟨65ZOR1415⟩, 1062 ⟨89ZOR2196⟩; **6**, 254 ⟨88ZOR1327⟩, 654 ⟨63ZOB113⟩
Lozinsky, M. O., **5**, 843 ⟨61ZOB1621, 64UKZ68⟩
Lrmmle, J. T., **2**, 67 ⟨75CRV521⟩
Lu, D. A., **5**, 71 ⟨85CC1478⟩
Lu, L. D. L., **2**, 348 ⟨83JOC3608⟩; **5**, 173, 235 ⟨82JA5523⟩
Lu, M. C., **2**, 129 ⟨70JOC4209⟩
Lu, P., **1**, 886 ⟨76JA7835⟩
Lu, P.-C., **1**, 886 ⟨82JA6681, 82JOC2928⟩
Lu, P. P. Y., **4**, 679 ⟨82OM911⟩
Lü, Q.-H., **1**, 527 ⟨85AG(E)316⟩
Lu, S.-B., **1**, 444 ⟨85JA1424⟩
Lu, S.-L., **4**, 57 ⟨76JA6752⟩
Lu, S.-T., **1**, 627 ⟨88H(27)751⟩, 784 ⟨85H(23)3085⟩
Lu, X., **2**, 879 ⟨85MI 217-01⟩; **3**, 83, 95 ⟨89JOM(376)41⟩, 118, 218 ⟨87TL3115⟩, 229 ⟨88TL1045, 89JOC1105⟩; **6**, 7 ⟨90JCS(P1)2031⟩
Lu, Y., **6**, 556 ⟨88MI 617-02⟩
Lu, Z.-H., **1**, 120 ⟨93TL7557⟩; **3**, 222 ⟨91TL1467⟩
Lub, J., **3**, 430 ⟨86RTC161⟩, 430, 433 ⟨86RTC22⟩, 434 ⟨84RTC328⟩
Lübbers, T., **1**, 346 ⟨92SL743⟩
Lubell, W. D., **3**, 40, 184 ⟨87JA236⟩, 179, 184 ⟨88JA7447⟩
Lubenets, E. G., **3**, 536 ⟨72ZOR654⟩
Luber, J., **2**, 476 ⟨76CB3581⟩
Luberoff, B. J., **4**, 193 ⟨69JOC3949⟩
Lubineau, A., **1**, 826 ⟨90TL4147⟩; **3**, 192 ⟨92TL8073⟩
Lubisch, W., **4**, 925 ⟨84CB1424⟩
Lublin, A., **3**, 444, 445 ⟨00CB2460⟩
Lubosch, W., **1**, 137 ⟨78HCA512⟩, 210 ⟨77AG(E)264⟩; **2**, 1002 ⟨78HCA3100⟩; **3**, 182 ⟨76AG(E)313⟩; **4**, 530 ⟨80HCA102⟩; **5**, 177 ⟨77AG(E)244⟩, 566, 584, 586, 590, 597, 608, 609 ⟨76CB1309⟩, 597 ⟨77CB3733⟩, 606 ⟨76AG(E)313⟩
Lubovich, A. A., **4**, 706, 723, 724 ⟨72IZV1823⟩
Luc, V. M., **6**, 652 ⟨81BSB63⟩
Lucarini, M., **1**, 5, 7, 9, 10, 17, 25 ⟨92JOC3405⟩, 17 ⟨93JOC249⟩
Lucas, E. A., **4**, 675 ⟨91OM535⟩
Lucas, H. J., **2**, 410 ⟨50JA5491⟩
Lucas, J. M., **1**, 1065 ⟨62JOC3688⟩
Lucas, R., **5**, 298 ⟨55JA469⟩
Lucca, G. de, **2**, 35 ⟨82JOM(226)143⟩
Lucchese, R. R., **2**, 364 ⟨76JA7617⟩
Lucchesini, F., **3**, 256 ⟨89H(29)349⟩, 620 ⟨93T8423⟩; **4**, 875 ⟨88HCA1242⟩; **6**, 100, 124 ⟨88HCA1242⟩
Lucchetti, J., **1**, 25 ⟨81TL1623⟩, 592, 594, 716 ⟨78TL2693⟩; **4**, 287, 288 ⟨85T4793⟩, 289 ⟨78TL3971⟩, 399 ⟨78TL2693⟩; **6**, 94 ⟨78TL3971⟩
Lucchi, O. de, **1**, 520 ⟨87CB531⟩, 617 ⟨84JOC596⟩; **4**, 770, 779 ⟨90G569⟩, 840, 861 ⟨91TL2177, 88MI 419-02⟩
Lucchini, V., **1**, 617 ⟨84JOC596⟩, 1109 ⟨74JCS(P2)1808, 77JCS(P2)542⟩, 1109 ⟨78G543⟩; **4**, 272, 854 ⟨91SL565⟩, 277, 854 ⟨92T1485⟩
Lucci, R. D., **1**, 693 ⟨78JA2140⟩; **4**, 167 ⟨75JOC1371⟩
Lucente, G., **1**, 595 ⟨70CC1683⟩
Lucero, M. J., **1**, 949 ⟨90TL5551⟩
Luche, J., **3**, 127 ⟨87TL2347⟩
Luche, J.-L., **1**, 1093 ⟨82JOC3805⟩, 156 ⟨91TL4709⟩, 166 ⟨92TL6461⟩, 314 ⟨84JOC931⟩, 960 ⟨76CC183⟩, 966 ⟨75TL4615⟩, 5, 9, 10 ⟨80TL3195⟩; **2**, 25 ⟨74TL763⟩; **2**, 551 ⟨80JA7926⟩, 551, 958 ⟨89S787⟩, 591 ⟨86TL3149⟩, 59 ⟨79JA5848⟩, 74 ⟨87JOM(322)177⟩, 422 ⟨87JOC4124⟩, 422, 653 ⟨88CC1350⟩, 958 ⟨80JA7927⟩; **3**, 12 ⟨85TL829⟩, 12, 128 ⟨85JOC5761⟩, 14 ⟨82TL3361⟩, 31 ⟨81TL4077⟩, 90 ⟨86TL1791⟩, 145 ⟨83JOC3837⟩, 178 ⟨91SL37⟩; **4**, 146 ⟨88CC333⟩; **5**, 122 ⟨74JA8113⟩, 152 ⟨88TL5369⟩, 272, 301 ⟨86TL501⟩, 327 ⟨90SC1105⟩
Luche, M. J., **1**, 143 ⟨78JA1558⟩
Luche, M.-J., **1**, 2, 6 ⟨85JOC3957, 83JA2435⟩
Lucherini, A., **2**, 581 ⟨80JCS(D)1633⟩
Luchetti, J., **2**, 278, 294 ⟨85T4793⟩
Luchter, K. M., **2**, 405 ⟨72JA4392⟩
Lucien, H. W., **4**, 50 ⟨49JA258⟩
Luck, E., **2**, 405 ⟨74CC217⟩
Luck, J. M., **3**, 620 ⟨61JBC(236)715⟩
Lucke, A. J., **2**, 210 ⟨91TL807⟩
Lücke, E., **2**, 739 ⟨73OSC(5)808⟩; **3**, 261 ⟨57CB2833⟩; **4**, 580 ⟨85ZN(B)968⟩
Luckenbach, G., **5**, 756 ⟨1884CB1428⟩
Luckenbach, R., **1**, 51 ⟨B-73MI 102-01⟩, 53 ⟨75CB803⟩; **2**, 441 ⟨71TL2177, 79TL3081⟩, 441, 442 ⟨73MI 208-01⟩, 822, 823 ⟨68CB2899⟩, 853 ⟨77ZN(B)584⟩, 888 ⟨73S307⟩
Luckenbaugh, R., **4**, 195 ⟨49JA2741⟩, 195, 195 ⟨49JA2736⟩
Luckner, M., **1**, 35 ⟨B-72MI 102-01⟩
Luczak, J., **6**, 509 ⟨78JOC2132⟩
Ludde, H., **5**, 1107 ⟨48M316⟩
Luders, L., **1**, 498 ⟨85TL1429⟩
Ludsteck, D., **3**, 485 ⟨54CB1887⟩
Ludwig, III, E. G., **2**, 499 ⟨86JOM(308)289⟩, 508 ⟨86JOM(303)197⟩
Ludwig, E., **5**, 633 ⟨81ZAAC(480)21⟩
Ludwig, Jr., E. G., **2**, 491, 505, 893 ⟨83OM1859, 83OM1573⟩
Ludwig, J. W., **2**, 770 ⟨80JOC4666⟩
Ludwig, P., **2**, 498 ⟨64MI 209-01⟩
Ludwig, R. A., **6**, 560, 561 ⟨B-62MI 617-02⟩
Ludwig, R., **5**, 161 ⟨93AG(E)1191⟩
Ludwig, U., **1**, 398, 399 ⟨69TL3415⟩
Ludwikow, M., **3**, 620, 627 ⟨74AG(E)665⟩, 634 ⟨72TL2391⟩
Ludwlg, , IIIE. G., **2**, 505 ⟨82OM1408⟩
Lue, N. R. de, **2**, 23 ⟨88T2785⟩, 30 ⟨88T2773⟩, 34 ⟨76JA1290, 76S114⟩, 34 ⟨88T2751⟩
Lue, P., **1**, 260 ⟨88JOC5854, 89JCS(P1)225⟩, 260, 263 ⟨91HCA1924⟩; **2**, 310 ⟨89JCS(P1)225⟩; **3**, 710 ⟨89TL6657⟩; **4**, 309 ⟨89JCS(P1)225⟩, 318, 322

⟨91HCA1924⟩; **5**, 606 ⟨88JOC5854⟩; **6**, 641 ⟨B-93MI 621-01⟩
Luedtke, A. E., **1**, 900 ⟨86JOC2969⟩
Luedtke, G., **1**, 523 ⟨90JA4965⟩; **3**, 707 ⟨92T2209⟩
Luegno, J. I., **1**, 566 ⟨84JOC2079, 84TL4881⟩, 816 ⟨89JOC5415⟩, 816, 833 ⟨85JA5572⟩; **2**, 677 ⟨84TL4881⟩
Lueoend, R., **2**, 398 ⟨88JOC1251⟩; **3**, 190 ⟨88JOC1251⟩
Luerssen, H., **3**, 680 ⟨86ACR342⟩
Luethens, M. L., **2**, 434 ⟨90IS305⟩
Luft, G., **5**, 192 ⟨86GEP3511050⟩, 192 ⟨89GEP3811344⟩
Luftmann, H., **2**, 20 ⟨76AG(E)306⟩, 21 ⟨92LA559⟩
Lugade, A. G., **5**, 561 ⟨88S303⟩
Lugan, N.,**5**, 946 ⟨90JA3220⟩
Lugan, N., **5**, 946 ⟨92JA6735⟩, 955 ⟨94OM4676⟩
Luger, P., **3**, 380 ⟨87AG(E)799, 91JA233⟩, 689 ⟨88AG(E)1534⟩; **4**, 745 ⟨88JOC5554⟩, 868 ⟨88AG(E)1534⟩; **5**, 138 ⟨91JA910⟩, 151, 164 ⟨88JA4640⟩
Lugli, G., **2**, 602 ⟨73ICA319⟩
Lugo, N., **3**, 342 ⟨83PS(17)47⟩
Lugtenburg, J., **1**, 420 ⟨73T1393⟩, 957 ⟨76RTC80⟩; **3**, 667 ⟨70TL4701⟩; **5**, 15 ⟨89RTC51⟩
Luh, T.-Y., **1**, 2, 8 ⟨79JOC641⟩, 22, 23, 238, 436 ⟨90S89⟩, 22, 23, 24, 436 ⟨89JOC4474⟩, 22, 23, 437 ⟨85JOC5413, 85CC1610⟩, 436 ⟨88JOC4466⟩, 991 ⟨87CC981, 90JOC1874⟩, 1127 ⟨93JOC5574⟩; **2**, 936 ⟨90JA9356⟩, 939 ⟨93S349⟩; **3**, 365 ⟨90JOC1874⟩; **4**, 244 ⟨91ACR257⟩; **5**, 285 ⟨79SC757⟩
Luh, Y., **3**, 634 ⟨76TL3799⟩
Luheshi, A. B. N., **5**, 688 ⟨90TL123⟩
Luheshi, A.-B. N., **6**, 111 ⟨90TL123⟩
Luhmann, U., **3**, 247, 251 ⟨80CB1708⟩
Lührs, K., **1**, 47 ⟨66CB227⟩
Lui, C. Y., **1**, 906 ⟨69JA3958, 70JA4627⟩
Lui, H., **3**, 124 ⟨79TL4121⟩
Lui, H.-J., **2**, 36 ⟨86H(24)3043⟩
Lui, K.-T., **3**, 215 ⟨71JOC387⟩
Lui, S. C., **3**, 490 ⟨35JA2656⟩
Luibrand, R. T., **1**, 563 ⟨81JOC1874⟩
Luidhardt, T., **2**, 34, 629, 631 ⟨88JOC4477⟩
Luijten, J. G. A., **2**, 1100 ⟨64RTC295⟩
Luinstra, G. A., **1**, 670 ⟨87CC849⟩; **6**, 194 ⟨87OM1509⟩
Luis, A., **1**, 826 ⟨88JOC3195⟩
Luis, S. V., **1**, 1081 ⟨90JOC3808⟩; **4**, 749 ⟨85CC296, 86JCR(M)2843⟩; **6**, 695 ⟨85CC296⟩
Luitjen, J. G. A., **2**, 478 ⟨50OS(36)86⟩, 601 ⟨56CI(L)352⟩
Lukacovic, M. F., **2**, 212, 214 ⟨74JOC253⟩
Lukacs, A., **5**, 448 ⟨82OM1730⟩
Luk, K.-C., **5**, 345, 346 ⟨88JOC5507⟩; **6**, 543 ⟨88S226⟩
Lukanov, L. K., **4**, 488 ⟨91SC1951, 92S263⟩
Lukas, J., **1**, 893, 901 ⟨67JA4739⟩, 895 ⟨68JA933⟩
Lukas, J. H., **1**, 498 ⟨75AG(E)709⟩
Lukas, K. L., **4**, 936, 938 ⟨85AG(E)696⟩; **5**, 135, 138 ⟨84AG(E)142⟩
Lukasch, H., **5**, 928, 930 ⟨88AG(E)960⟩
Lukashev, N. V., **2**, 414 ⟨80ZOB1236⟩, 1027 ⟨90PS(49/50)179⟩; **4**, 559, 766 ⟨90ZOB1539, 91ZOB1014⟩, 884 ⟨91ZOB1600⟩; **6**, 668 ⟨82ZOB2187⟩
Lukashina, S. G., **2**, 336 ⟨83JOU1203⟩
Lukaszewski, H., **5**, 1050 ⟨64TL1087, 64TL855⟩, 1052 ⟨66JOC722⟩
Luk, T., **3**, 129 ⟨79JOC641⟩
Luke, G. P., **1**, 491, 1035 ⟨87JOC4296⟩
Lüke, H.-W., **2**, 814 ⟨77AG(E)535⟩; **5**, 645 ⟨75AG(E)427⟩, 645, 739 ⟨75LA1808⟩; **6**, 571 ⟨79LA263⟩
Lukehart, C. M., **2**, 927 ⟨89JA7615, 90JA7229⟩, 927 ⟨92SL681⟩; **5**, 450 ⟨76JA2365⟩, 801 ⟨78IC2369, 81JA1467⟩, 801 ⟨82IC1278, 83IC3954⟩, 801 ⟨86ICA67, 86ICAL15⟩, 943 ⟨71JA2672⟩
Lukes, G. E., **6**, 254 ⟨68USP3376314⟩
Lukevics, E., **4**, 514 ⟨84JOM(268)C29⟩, 515 ⟨86JOM(316)249⟩, 704 ⟨91JOM(418)C21⟩, 929 ⟨84H(22)987⟩, 1044 ⟨88JOM(346)297⟩, 1080 ⟨87CC31⟩; **6**, 124 ⟨84JOM(265)237⟩, 558 ⟨85KGS339⟩
Lukevits, E., **2**, 914 ⟨86ZOR140⟩, 1068 ⟨74JGU1461⟩
Lukina, Yu. A., **4**, 140 ⟨76IZV2200⟩, 140 ⟨76JGU2076⟩, 569 ⟨75ZOB2010⟩
Lukin, K. A., **1**, 500 ⟨87ZOR2545, 91ZOR488⟩
Luknitskii, F. I., **3**, 526 ⟨69RCR487⟩
Luksza, M., **2**, 512 ⟨82JOM(228)C29, 84ICAL49⟩; **5**, 911 ⟨82JOM(236)C1⟩
Lukszo, J., **4**, 461 ⟨78CL1103, 78PJC321⟩, 461 ⟨78PJC959⟩, 461, 463 ⟨77S239⟩
Lukyanenko, N. G., **4**, 418 ⟨89ZOB2435⟩
Luk'yanenko, N. G., **5**, 1024 ⟨85ZOB2100⟩; **6**, 476 ⟨85ZOB2100⟩
Lukyanenko, N. G., **6**, 575 ⟨80TL313⟩
Luk'yanenko, S. N., **4**, 1009 ⟨90ZOB1187⟩; **5**, 369, 713 ⟨90ZOB1187, 90JGU1057⟩, 713 ⟨90ZOB1186⟩, 713, 788 ⟨90JGU1059⟩
Luk'yanov, O. A., **2**, 391 ⟨89RCR470⟩; **4**, 429 ⟨77IZV2384⟩, 995 ⟨75IZV2342⟩; **5**, 680 ⟨90BAU294, 90IZV352⟩; **6**, 145 ⟨81IZV2146⟩, 149 ⟨68ZOR231⟩, 284 ⟨92MC52⟩
Luloff, J. S., **2**, 362, 363 ⟨69JOC2397⟩
Lum, K. K., **1**, 773 ⟨69MI 117-01⟩
Lum, R. C., **4**, 271 ⟨90JA607⟩
Lum, R. T., **2**, 986, 987 ⟨86JA7441⟩
Lumb, J. T., **1**, 885 ⟨67JA946⟩; **5**, 709 ⟨64JCS1189⟩
Lumbroso, ,H., **1**, 648, 738, 743 ⟨88JCS(P2)1829⟩
Lumbroso, H., **3**, 500 ⟨88JCS(P2)1829⟩; **4**, 873 ⟨77JCS(P2)775⟩; **6**, 708, 709 ⟨88JCS(P2)1829⟩
Lumin, S., **5**, 368 ⟨88SC545⟩
Lumin, S.,**6**, 511 ⟨88SC545⟩
Lumma, Jr., W. C., **2**, 613 ⟨81JMC93⟩
Lumma, W. C., **4**, 48 ⟨75TL3979⟩
Luna, D., **3**, 671 ⟨84JOC5195⟩
Luna, H., **3**, 62 ⟨88JA4735⟩
Lunazzi, L., **2**, 386 ⟨81JCS(P2)1449⟩
Lund, F. J., **4**, 59, 60 ⟨90S1159⟩; **6**, 427, 428, 476 ⟨90S1159⟩, 477 ⟨91MIP9110639⟩
Lund, H., **1**, 427 ⟨B-83MI 110-01⟩; **3**, 107 ⟨63ACS2325⟩, 109 ⟨66ACS2649⟩; **6**, 555 ⟨86ACS(B)593⟩
Lund, P. R., **3**, 323 ⟨78JOC1071⟩
Lundberg, K. L., **4**, 455 ⟨69IC1336⟩
Lundberg, K. R., **4**, 566 ⟨69IC862⟩
Lundberg, W. O., **5**, 28 ⟨57MAC25, 63MI 502-01⟩, 28 ⟨B-63MI 502-02⟩
Lundeen, A. J., **2**, 112 ⟨71USP3600417⟩; **5**, 636, 831 ⟨57JA1764⟩
Lundehn, J.-R., **2**, 399, 400 ⟨76CB2442⟩
Lundell, J., **5**, 701 ⟨77OPP57⟩
Lundgren, H. P., **5**, 83 ⟨49MI 502-01⟩
Lundmark, B. R., **5**, 356 ⟨89JA9194⟩
Lundquist, J. A., **1**, 934 ⟨64JOC921⟩
Lundstedt, T., **5**, 599 ⟨86ACS(B)534, 86ACS(B)694⟩, 599 ⟨87ACS(B)157, 87ACS(B)164⟩
Lundt, I., **1**, 2, 4, 8, 9 ⟨84CAR(130)125⟩
Luneva, L. K., **2**, 1093 ⟨66JGU570⟩
Lunin, V. V., **6**, 611 ⟨91ZOR533⟩
Lüning, U., **1**, 325 ⟨85CB1345⟩, 325, 326 ⟨84CB2132⟩, 950 ⟨93CB1171⟩
Luning, U., **3**, 699 ⟨91CB2555⟩
Lunn, G., **2**, 327 ⟨87JOC1043⟩, 379 ⟨84JOC3470⟩
Lunn, P. M., **3**, 317 ⟨90JCS(P1)153⟩
Lunn, R. J., **1**, 828 ⟨86TL3053⟩
Lunn, W. H. W., **5**, 340 ⟨92T8615⟩
Lunsford, C. D., **4**, 302 ⟨51JA2596⟩
Luo, F.-T., **1**, 542 ⟨82TL27⟩, 1045, 1047 ⟨83TL5181⟩, 1048 ⟨93TL5911⟩; **2**, 160 ⟨84JOC2629⟩, 975 ⟨82TL27⟩; **3**, 130, 288 ⟨83TL5181⟩, 675 ⟨93TL5911⟩; **4**, 954, 1062 ⟨83JOC1560⟩

Luo, H., **6**, 230 ⟨92JOC4749⟩
Luo, J., **1**, 776 ⟨90JA228⟩
Luo, W., **5**, 234 ⟨91SC2097⟩
Luo, X.-G., **2**, 392 ⟨86JHC825⟩
Luong-Thi, N. T., **4**, 656 ⟨78CC1081⟩, 656, 663, 709, 724 ⟨81JCS(D)1593⟩
Lupi, A., **1**, 382 ⟨82HCA371⟩; **6**, 550 ⟨82HCA371⟩
Luppold, E., **3**, 64 ⟨76ZN(B)1654⟩, 272 ⟨75CB237⟩
Luputiu, G., **4**, 98 ⟨86AP(319)954⟩
Lur'e, É. P., **2**, 640 ⟨74DOK(217)1320⟩, 641 ⟨75DOK(220)146⟩; **3**, 588 ⟨75BAU1274⟩; **4**, 736 ⟨76IZV895⟩, 803 ⟨76BAU853, 76BAU873⟩, 894 ⟨75IZV1274⟩; **5**, 475 ⟨77BAU206⟩
Lur'e, L. F., **4**, 314, 329 ⟨74JOU1565⟩; **5**, 1008 ⟨76ZOR1963, 76ZOR2473⟩; **6**, 450 ⟨77ZOR290⟩, 617 ⟨81ZOR398⟩, 655, 685 ⟨91ZOB401⟩
Lurz, C., **4**, 989 ⟨91CB595⟩, 991 ⟨91JHC1863⟩
Lusby, W. R., **1**, 652 ⟨93TL1593⟩
Lusch, K. J., **1**, 304 ⟨82T2939⟩
Lusch, M. J., **1**, 470 ⟨74JOC3174⟩
Lusenkova, M. A., **3**, 597 ⟨89JOM(379)303⟩
Lushnikova, T. V., **3**, 588 ⟨75JFC(6)227⟩; **4**, 731 ⟨75JFC(6)227⟩
Lusinchi, X., **1**, 46 ⟨88TL2571⟩; **3**, 408 ⟨85TL4603⟩; **5**, 789 ⟨79T43⟩
Lusk, D. I., **4**, 13 ⟨65JA4147⟩; **6**, 62, 244 ⟨65JA4147⟩
Luskin, L. S., **5**, 1043 ⟨56JA4965⟩, 1044 ⟨56JOC1430⟩
Luss, H. R., **4**, 93 ⟨85JA6298⟩
Lusser, M., **6**, 372 ⟨84OM1916, 86JOM(301)269⟩
Lustig, M., **2**, 94 ⟨66JA4531⟩; **6**, 250 ⟨67CC870⟩, 251 ⟨65IC1828⟩
Lusztky, J., **1**, 925 ⟨93ACR219⟩
Lusztyk, J., **1**, 330, 341, 936 ⟨84JA343⟩, 331 ⟨85JA4594⟩, 331, 336, 342, 936 ⟨87JOC3509⟩, 923 ⟨84JA1793⟩, 924, 934 ⟨88JA2877, 83JA3578⟩, 936 ⟨85JA4595⟩; **2**, 21 ⟨90ACR219⟩; **4**, 1078 ⟨90JA4284⟩
Luterbacher, A., **5**, 184 ⟨09CB3484⟩
Lüth, B., **3**, 689 ⟨90CB2317⟩; **6**, 684 ⟨93PS(76)265⟩, 685 ⟨90CB2317⟩
Luthardt, H. J., **6**, 107 ⟨67TL3501, 68CB3851⟩
Luthe, H., **1**, 325, 920 ⟨87JOC205⟩
Lüthe, H., **4**, 695 ⟨86JOM(310)151⟩
Luther, I. G. W., **2**, 826, 866 ⟨77IS186⟩
Lüthje, K., **5**, 1048 ⟨60AP(293)150⟩
Luthman, K., **5**, 132 ⟨87JOC3777⟩
Luthra, N. P., **5**, 238 ⟨85JOC811⟩
Lütjens, H., **2**, 69, 72, 591 ⟨93TL3115⟩
Lutjens, H., **3**, 550, 551 ⟨91JOC1317⟩
Lutomsky, K. A., **5**, 172 ⟨B-84MI 503-01⟩
Lutsenko, A. F., **5**, 401 ⟨87JGU1254⟩
Lutsenko, A. I., **2**, 876 ⟨75IZV897⟩
Lutsenko, I. F., **1**, 62 ⟨70MI 102-02⟩; **2**, 414 ⟨80ZOB1236⟩, 420, 494, 495, 695 ⟨74JOM(78)107⟩, 433 ⟨87URP1294811⟩, 434 ⟨86ZOB2794, 87ZOB884⟩, 436 ⟨67ZOB2080, 68MI 208-01⟩, 448 ⟨77JGU707⟩, 449 ⟨76JGU428⟩, 451 ⟨84JGU2237, 89JGU2130⟩, 467 ⟨79JGU616⟩, 471 ⟨79JGU2143, 83JGU2244⟩, 476 ⟨89JGU87⟩, 693 ⟨59DOK(124)1251⟩, 694 ⟨69ZOB2479⟩, 695 ⟨73ZOB1264, 74JOM(78)115⟩, 695, 700 ⟨69ZOB1174, 68JOM(14)229⟩, 700, 701 ⟨72PAC409, 70JOM(24)359⟩, 1027 ⟨83MI 221-01, 87MI 221-01⟩, 1093 ⟨71JGU122⟩; **3**, 195 ⟨72CA(78)136370⟩, 195 ⟨83JGU610⟩, 195 ⟨83JGU656⟩, 270 ⟨81JGU1105⟩, 484 ⟨75ZOB819⟩, 534 ⟨69USP3462489, 70ZOB707⟩; **4**, 115 ⟨84JGU1354⟩, 116 ⟨85JGU1065⟩, 126 ⟨86JGU1258⟩, 344 ⟨72ZOB2418, 73PS(3)55⟩, 462 ⟨85JGU404⟩, 546 ⟨76URP539892, 81ZOB484⟩, 546, 550, 562 ⟨77ZOB775⟩, 547 ⟨76ZOB571, 83ZOB699⟩, 548 ⟨87ZOB2606⟩, 550 ⟨79ZOB712, 80ZOB1881⟩, 550 ⟨80ZOB949, 85ZOB331⟩, 551 ⟨78ZOB757, 80ZOB989⟩, 552 ⟨76ZOB568, 77ZOB2689⟩, 552 ⟨87ZOB831⟩, 552, 562 ⟨74ZOB276⟩, 552, 562 ⟨79ZOB1446⟩, 555 ⟨77ZOB2636, 83ZOB1673⟩, 563 ⟨87ZOB525⟩, 564 ⟨89ZOB101⟩, 566 ⟨76ZOB575⟩, 566, 568, 576 ⟨76ZOB1218⟩, 569 ⟨72ZOB112⟩, 574, 588 ⟨75ZOB1486⟩, 616 ⟨79ZOB2785⟩, 637 ⟨76MI 414-01⟩, 640 ⟨76ZOB594⟩, 661 ⟨70ZOB2607⟩, 765 ⟨77ZOR1675⟩, 820 ⟨76JGU2367⟩, 886 ⟨69JOM(17)241⟩, 913 ⟨76ZOB565⟩, 914 ⟨84ZOB416, 87ZOB1406⟩, 915 ⟨75ZOB556, 78ZOB267⟩, 966 ⟨73MI 420-01⟩, 1055, 1058 ⟨80ZOB692⟩; **5**, 376 ⟨69JOM(17)241, 73ZOB1753⟩, 376, 389 ⟨74ZOB1753⟩, 376, 389 ⟨76ZOB594, 73ZOB603⟩, 390 ⟨70USP3509199⟩, 395, 396, 399 ⟨89JGU1817⟩, 395, 396, 399, 400 ⟨88JGU201⟩, 395, 396, 399, 400 ⟨88JGU2372⟩, 396, 399 ⟨87DOK114⟩, 876, 877 ⟨84ZOB1520⟩, 884 ⟨81DOK(256)1401, 81ZOB482⟩, 1092 ⟨87JGU2350⟩; **6**, 117 ⟨86ZOB2427⟩, 117, 490 ⟨88PS(35)329⟩, 265, 679 ⟨81ZOB2630⟩, 281 ⟨84ZOB2504⟩, 313 ⟨88ZOB2168⟩, 314 ⟨88ZOB1447, 88ZOB1927, 87JGU1260⟩, 490 ⟨87JGU1484⟩, 668 ⟨82ZOB2187⟩, 678 ⟨84ZC345⟩, 679 ⟨81DOK(256)1401⟩, 680, 681 ⟨84ZC345⟩, 696 ⟨85ZOB1194⟩
Lutsenko, I. P., **2**, 471 ⟨78JGU207⟩, 476 ⟨78JGU1088⟩; **4**, 112, 114, 115 ⟨84JGU2250⟩
Lutsenko, L. N., **4**, 864 ⟨69MI 419-01⟩
Lutskaya, N. V., **4**, 765 ⟨88IZV514⟩
Luttikhede, H. J. G., **2**, 569 ⟨87OM2354⟩
Lüttke, W., **1**, 690 ⟨82CB1234⟩; **2**, 403, 404 ⟨65LA(687)236⟩
Luttke, W., **3**, 247, 251 ⟨80CB1708⟩, 256 ⟨72CB919⟩
Lüttke, W., **3**, 348 ⟨83TL5861⟩, 657 ⟨69AG(E)458⟩
Luttmann, K., **3**, 231 ⟨89CB1653⟩
Lüttringhaus, A., **2**, 214 ⟨55AG304⟩; **5**, 285, 292 ⟨71S266⟩; **6**, 502 ⟨63AG1059⟩
Lutz, G., **1**, 918 ⟨89JA746⟩
Lutz, H. D., **4**, 264 ⟨91CB2449⟩
Lutz, R. E., **1**, 45 ⟨46JA1813⟩; **2**, 175 ⟨55JOC1165⟩, 273 ⟨54JCS3659⟩; **3**, 274 ⟨51JOC51⟩; **5**, 25, 34 ⟨50JA4091⟩
Lutz, R. P., **1**, 411, 413, 414, 415, 806, 825, 826, 827, 836 ⟨84CRV205⟩, 809 ⟨76JOC2048⟩
Lutz, W., **5**, 721 ⟨79CB2158⟩; **6**, 240, 439 ⟨79CB2158⟩
Lutz, W. B., **2**, 223 ⟨56JA2469⟩; **5**, 332 ⟨71JOC3835⟩
Luu-Duc, C., **5**, 749 ⟨87MI 519-01⟩
Lux, F., **6**, 523 ⟨65ZN(B)1159⟩
Lux, R., **2**, 163 ⟨89LA605⟩; **5**, 341 ⟨88PS(35)1⟩
Luxembourg, K., **2**, 414, 416, 418 ⟨1896CB710⟩
Luxen, A., **2**, 613 ⟨88TL1501⟩, 731 ⟨82TL3905⟩
Luzanova, M. B., **2**, 442 ⟨62ZOB553⟩
Luz, Z., **1**, 870 ⟨67JA1183⟩
Luzgina, G. M., **3**, 267 ⟨84JGU28⟩; **4**, 106 ⟨85JOU806⟩
Luzikova, E. V., **4**, 559, 766 ⟨90ZOB1539⟩
Luzikov, Y. N., **4**, 603, 625, 637, 701 ⟨71JOM(30)C57⟩, 627 ⟨80IZV2668⟩; **5**, 876, 877 ⟨84ZOB1520⟩; **6**, 696 ⟨85ZOB1194⟩
Luzikov, Yu. N., **4**, 115 ⟨84JGU1354⟩, 116 ⟨85JGU1065⟩, 884 ⟨91ZOB1600⟩, 914 ⟨84ZOB416⟩; **6**, 680, 681 ⟨84ZOB1520⟩
Luzzio, F. A., **2**, 397 ⟨84JA3682⟩; **3**, 4, 115 ⟨80S691⟩, 4, 115 ⟨89JOC5387⟩, 88 ⟨88OPP533⟩, 88, 215 ⟨82JOC1787⟩
L'vova, M. S., **6**, 261, 283 ⟨70IZV2553⟩, 283 ⟨71IC2179⟩
Lwanga, M. M., **1**, 1056 ⟨75T3073⟩
Lwowski, W., **2**, 298 ⟨63JA1200, 69JA5107⟩, 298 ⟨B-84MI 205-01⟩, 793 ⟨65JA4359⟩; **3**, 363 ⟨69TL3611⟩; **5**, 367 ⟨B-71MI 507-01⟩, 845 ⟨73CC932⟩; **6**, 620 ⟨86JOC1719⟩
Lyalin, B. V., **5**, 317 ⟨92MI 507-02⟩; **6**, 56 ⟨87IZV306⟩
Lyalin, V. V., **6**, 61 ⟨68ZOB2813⟩
Lyapina, T. V., **5**, 341 ⟨74ZOR139, 78ZOR280⟩, 341 ⟨81ZOR529⟩

Lyashenko, G. S., **4**, 765 ⟨71IZV162⟩, 766 ⟨83IZV2153⟩, 940 ⟨77IZV596⟩
Lyashenko, Yu., **5**, 672 ⟨90JGU1718, 90ZOB1923⟩
Lyashenko, Yu. E., **5**, 22 ⟨93PS(72)153⟩, 669, 672 ⟨89BAU2628⟩, 669, 672 ⟨89IZV2865⟩, 669, 672 ⟨91PS(60)85⟩
Lycan, W. R., **4**, 31, 34 ⟨51JA711⟩; **5**, 318 ⟨52JA2189⟩
Lyerla, J. R., **1**, 914 ⟨79JA4770⟩
Lygo, B., **1**, 757 ⟨85T3825⟩; **2**, 319 ⟨93SL764⟩; **3**, 171 ⟨90SL282⟩
Lygo, B.,**3**, 188 ⟨93SL764⟩
Lygo, B., **4**, 236 ⟨86T4333⟩, 286 ⟨89T7161⟩
Lyle, G. G., **2**, 385, 422, 798 ⟨76S462⟩
Lyle, M. A., **4**, 141 ⟨74JCS(D)2537⟩
Lyle, P. A., **2**, 657 ⟨84HCA1406⟩
Lyle, R. E., **2**, 385, 422, 798 ⟨76S462⟩
Lyle, T. A., **1**, 295 ⟨82JOC161⟩, 812 ⟨84HCA774⟩; **4**, 97 ⟨88EUP264183⟩
Lynch, A. T., **2**, 902 ⟨87JA5867⟩
Lynch, G. J., **1**, 67 ⟨81JOC531⟩
Lynch, J., **5**, 146 ⟨86JA284⟩
Lynch, J. E., **4**, 230 ⟨87OS(65)215⟩, 231 ⟨84JA2943⟩
Lynch, L. E., **2**, 69 ⟨85TL981⟩; **3**, 651 ⟨86SC617⟩; **4**, 875, 877 ⟨84JOC1653⟩
Lynch, M., **4**, 527, 528 ⟨90JBC(265)3738⟩
Lynch, M. J., **4**, 811 ⟨93T4229⟩
Lynch, P. P., **1**, 443 ⟨63PCS19, 68T4489⟩
Lynch, T. J., **4**, 706 ⟨88OM2566⟩
Lynch, T. R., **3**, 256 ⟨71CJC1477⟩, 547 ⟨71CJC1456⟩; **4**, 852 ⟨71CJC1477⟩; **5**, 354 ⟨71CJC1598⟩
Lynch, T.-Y., **1**, 919, 922 ⟨89JA7558⟩, 922 ⟨90JOC58⟩
Lynch-Hart, U. E., **6**, 147 ⟨61JOC391⟩
Lynch Jr., M. A., **6**, 523 ⟨54JA3831⟩
Lynd, R. A., **1**, 465 ⟨76S625⟩, 514 ⟨77S52⟩, 533 ⟨76S816⟩, 534 ⟨74S658⟩
Lynd, R., **6**, 208 ⟨71CC1593⟩
Lynn, D. G., **5**, 199 ⟨90JOC4763⟩
Lynn, J. W., **5**, 847 ⟨59CRV841⟩, 997 ⟨65JOC1260⟩
Lynshenko, G. S., **4**, 933 ⟨81IZV654⟩
Lyon, D. R., **2**, 488 ⟨45JCS30⟩
Lyon, G. D., **1**, 28 ⟨74JOC3525⟩; **2**, 366 ⟨74JOC3525⟩
Lyons, D. E.,**4**, 29 ⟨75SC33⟩
Lyons, J. E., **1**, 428 ⟨69JOC638⟩
Lyons, R. E., **3**, 400 ⟨27CB824, 31CB530⟩
Lypka, G. N., **4**, 323 ⟨79JHC415⟩
Lysenko, S. N., **1**, 1022 ⟨78JOU1095⟩
Lysenko, V. P., **1**, 304 ⟨87T2609⟩; **2**, 1000 ⟨87T2609⟩; **3**, 378 ⟨87T2609⟩
Lysenko, Z., **2**, 21 ⟨85SC977⟩; **3**, 27 ⟨85SC977⟩
Lyssikatos, J. P., **4**, 202 ⟨91COS(2)661⟩
Lyster, M. A., **2**, 36, 658 ⟨77JOC3761⟩, 538 ⟨79OS(59)35⟩; **5**, 20, 225 ⟨77JA968⟩
Lysy, R., **5**, 1116 ⟨72BSF3559⟩
Lyszak, E. L., **4**, 650, 656 ⟨93OM338⟩
Lythgoe, B., **1**, 590, 591, 615, 700 ⟨77TL4223⟩, 615, 616 ⟨79JCS(P1)1290⟩, 615, 654, 696, 697, 698, 700 ⟨78JCS(P1)829⟩, 616, 696, 700 ⟨80JCS(P1)1045⟩, 698, 1073 ⟨78TL2625, 78JCS(P1)834⟩, 700, 702 ⟨80JCS(P1)1400⟩, 735 ⟨58JCS843⟩, 754 ⟨76JCS(P1)2386, 78JCS(P1)590⟩, 821 ⟨72CC757⟩, 827 ⟨71JCS(C)2950⟩, 1073 ⟨79JCS(P1)2429⟩; **2**, 403, 404 ⟨52JCS4191⟩; **3**, 242 ⟨71JCS1292⟩; **5**, 168 ⟨71JCS(C)2944⟩
Lythgoe, D. J., **5**, 690 ⟨92JCS(P1)2789⟩
Lyttle, D. A., **2**, 392 ⟨49JA3079⟩
Lyttle, M. H., **2**, 409, 410 ⟨91JOC4608⟩
Lytwyn, E., **5**, 380 ⟨80TL3367⟩
Lyzwa, P., **2**, 184 ⟨89AG(E)97⟩

M

Ma, C., **5**, 329 ⟨91TL2577⟩
Ma, D., **3**, 118, 218 ⟨87TL3115⟩, 229 ⟨88TL1045, 89JOC1105⟩
Ma, J. C. N., **3**, 87 ⟨68JOC3096⟩
Ma, K. W., **3**, 224 ⟨74TL909⟩; **4**, 42, 57 ⟨74TL909⟩
Ma, M.-M., **6**, 378 ⟨70JOM(24)89⟩
Ma, P., **1**, 14 ⟨82JOC1378⟩
Ma, S., **2**, 43 ⟨88TL2967⟩; **4**, 186 ⟨93TL5269⟩; **6**, 7 ⟨90JCS(P1)2031⟩
Ma, T. S., **6**, 68 ⟨32JA2964⟩
Ma, Y., **2**, 307 ⟨92CC853⟩
Ma, Y.-X., **2**, 253 ⟨89JA2321, 90TL1953⟩
Maack, A., **3**, 607 ⟨92TL1981⟩
Maartmann-Moe, K., **2**, 290 ⟨81ACS(A)151⟩; **5**, 657 ⟨89ZAAC(576)203⟩, 1117 ⟨81ACS(A)151⟩
Maas, G., **1**, 258 ⟨B-86MI 106-01⟩, 352, 637 ⟨87TCC75, B-86MI 114-04⟩, 808 ⟨80CB3679⟩, 990 ⟨85CC1812⟩; **2**, 449 ⟨91AG(E)207⟩, 1008 ⟨83JOC3038⟩, 1024, 1025 ⟨85JOC2801⟩; **3**, 48 ⟨B-86MI 301-02⟩, 193 ⟨B-86MI 304-05⟩, 193 ⟨B-87MI 304-09⟩, 482 ⟨76CB2039⟩, 483 ⟨87CB635⟩, 532 ⟨85JOC2801⟩, 689 ⟨86TL1903⟩; **4**, 321 ⟨89CB2311⟩, 549 ⟨92AG(E)1055⟩, 557 ⟨76CB2039⟩, 558 ⟨78CB705⟩, 562 ⟨86TL1903⟩, 975 ⟨84SC1073⟩; **5**, 865, 868, 874 ⟨B-86MI 522-01⟩, 868 ⟨B-86MI 522-04⟩, 869, 871 ⟨76CB2039⟩, 870 ⟨B-86MI 522-02⟩, 870 ⟨B-86MI 522-03, 85T811⟩, 871 ⟨85T825⟩, 885 ⟨92AG(E)1055⟩; **6**, 153 ⟨91AG(E)207⟩, 154 ⟨92AG(E)1055⟩, 692 ⟨81TCC71⟩
Maasböl, A., **5**, 444, 445, 446, 936 ⟨67CB2445⟩, 932, 933, 934, 948 ⟨64AG(E)580⟩; **6**, 716 ⟨64AG(E)580⟩
Maaske, B., **4**, 726 ⟨94JOM(468)235⟩
Maassarani, F., **4**, 647, 656 ⟨85NJC249⟩
Maassen, J. A., **2**, 390 ⟨71RTC373⟩
Maatta, E. A., **2**, 602 ⟨81JA6650⟩
Maberly, T. R., **5**, 937 ⟨84OM1764⟩
Mabon, G., **3**, 652, 654 ⟨83T1551⟩
Mabry, R. J., **2**, 39 ⟨70B1626⟩
Mabry, T. J., **6**, 480 ⟨92JC241⟩
Mac, Y. C., **1**, 178 ⟨66AJC521⟩
MacAlpine, G. A., **1**, 422 ⟨76JCS(P1)410⟩, 1082 ⟨78CJC308⟩
Macaluso, A., **2**, 765 ⟨64CRV473⟩
Macaulay, S. R., **2**, 423 ⟨80JOC734⟩
Macbeth, A. K., **5**, 316 ⟨22JCS2169⟩
MacBride, J. A. H., **3**, 66 ⟨83CC74⟩; **5**, 1035 ⟨83CC74⟩
Maccagnani, G., **2**, 209 ⟨77JCS(P1)1469⟩, 245 ⟨81JCS(P1)2322⟩; **3**, 368 ⟨71JCS(B)1547, 76CC431⟩, 368 ⟨79JCS(P1)1720⟩, 369 ⟨86CC964, 90JOC3744⟩; **4**, 232 ⟨85TL2131⟩, 371, 944 ⟨89JCS(P1)2083⟩; **5**, 647 ⟨81CC822⟩, 647, 648 ⟨86JCS(P1)381, 88CC365⟩, 648 ⟨88JCS(P1)1499, 90JOC3744⟩, 648, 650 ⟨89JCS(P1)2083⟩; **6**, 127 ⟨81CC822⟩, 127, 128 ⟨90H(31)47⟩, 127, 309 ⟨85TL2131⟩
Maccarone, E., **2**, 288 ⟨83JCS(P2)1605⟩; **3**, 492 ⟨82G25⟩
Macchia, F., **2**, 135 ⟨92SL303⟩, 318 ⟨90TL4661⟩
Macchioni, A., **5**, 799 ⟨91G101⟩
Macciantelli, D., **2**, 386 ⟨81JCS(P2)1449⟩; **5**, 647, 648 ⟨86JCS(P1)381⟩
Maccioni, A., **1**, 738 ⟨64MI 116-02⟩; **2**, 147 ⟨61TL607⟩; **4**, 231 ⟨76S797⟩, 972 ⟨88FES103⟩, 986 ⟨92JHC1631⟩
Maccoll, A., **1**, 921 ⟨88JA7339⟩
MacCormack, G. K., **6**, 255 ⟨91CJC2127⟩
Maccoss, M., **6**, 487 ⟨93EUP547699⟩
MacDiarmid, A. G., **2**, 110 ⟨61JA3384⟩, 399, 400 ⟨66JCS(A)1780⟩, 508 ⟨66AG(E)418⟩, 512 ⟨67MI 209-01⟩; **3**, 711 ⟨62JCS548⟩; **5**, 1136 ⟨59JA5106⟩
MacDonald, A. L., **2**, 361 ⟨73CJC2504⟩
Macdonald, D., **1**, 810 ⟨93TL247⟩
MacDonald, D. L., **2**, 105 ⟨68CAR(6)253⟩
Macdonald, J. E., **1**, 828 ⟨79TL3057, 80TL4335⟩; **2**, 22 ⟨87OS(66)194⟩, 942 ⟨87TL1851⟩; **4**, 651 ⟨87TL1851⟩
MacDonald, J. G., **5**, 605, 622 ⟨83TL4927⟩, 605, 622 ⟨85CJC951⟩; **6**, 126 ⟨83JOC3189⟩
Macdonald, J. N., **3**, 337 ⟨77CC483⟩
Macdonald, P. L., **3**, 231 ⟨70CC52⟩
MacDonald, S. A., **4**, 427 ⟨84JOC821⟩
Macdonald, T. L., **1**, 108 ⟨80JA7760⟩, 136 ⟨74JA5561⟩, 138, 139 ⟨84JA3376⟩, 139 ⟨88JA842⟩, 197 ⟨78CC876⟩, 197 ⟨80JOC4740⟩, 413 ⟨86JOC1124⟩; **2**, 940 ⟨80JOC4740⟩, 999 ⟨88JA842⟩; **3**, 141 ⟨86JOC1124⟩; **4**, 379 ⟨88JA842⟩
MacDonald, W. A., **1**, 738 ⟨83T297⟩; **3**, 196, 500 ⟨82T3355, 83T297⟩; **6**, 708 ⟨82T3355⟩
MacDonell, D. L., **2**, 183 ⟨72JOC3549⟩
Macdonell, G. D., **2**, 214 ⟨81BEP886595⟩
MacDowell, D. W., **1**, 808 ⟨56JA2547, 56JA2551⟩
MacDowell, D. W. H., **1**, 436 ⟨83JOC4419⟩, 808 ⟨86JOC183⟩
MacDuffie, D. E., **6**, 233, 236, 237 ⟨61JCS2597⟩
Macey, W. A. T., **2**, 19 ⟨63OSC(4)525⟩
MacGillavry, C. H., **1**, 897 ⟨65AX437⟩
MacGregor, N. S., **2**, 180 ⟨63JOC2485⟩
Machackova, O., **2**, 805 ⟨72CCC3467, 73CCC706⟩
Machat, E., **1**, 1004, 1005 ⟨83JCR(M)167⟩
Machelett, B., **4**, 685 ⟨76ZC116⟩
Machell, G., **2**, 101 ⟨55MI 202-01⟩
Machi, D., **2**, 87 ⟨91JA8981⟩
Machida, H., **1**, 1010 ⟨85CPB1761⟩; **3**, 164 ⟨85TL1981⟩; **5**, 1054 ⟨90MI 524-01⟩
Machida, K., **5**, 171 ⟨85JA3285⟩
Machida, M., **3**, 65 ⟨81TL4487⟩, 427 ⟨89S206⟩; **4**, 1077 ⟨82S1078⟩; **5**, 572 ⟨92S778⟩, 631, 632 ⟨85JOC1681, 86S768⟩
Machida, Y., **2**, 87 ⟨75TL3183⟩, 1072 ⟨79JOC865⟩; **4**, 506 ⟨92JOC6711⟩, 508 ⟨91S117⟩
Machiguchi, T., **3**, 352 ⟨87TL203, 90CL563⟩
Machii, D., **1**, 925 ⟨89JA6849⟩
Machii, Y., **3**, 617 ⟨78S301⟩
Machinaga, N., **4**, 191 ⟨89TL4165⟩
Macho, V., **1**, 906 ⟨82BSB422, 82JA907⟩
Machon, Z., **5**, 528, 529 ⟨86S142⟩
Macicek, J., **2**, 1004 ⟨88AX(C)626⟩; **3**, 494 ⟨89JCS(P2)1405⟩
Maciej, T. W., **6**, 282 ⟨85JCS(P1)1935⟩
Maciejewska, U., **4**, 731 ⟨85CJC576⟩
Maciejewsk, L., **1**, 816, 818 ⟨89TL2549⟩; **5**, 990 ⟨93JPR294⟩
Mack, C. H., **2**, 1055 ⟨58JOC1051⟩; **5**, 592, 609 ⟨63JOC1600⟩
Mack, H. G., **5**, 675 ⟨93IC287⟩
Mack, J., **2**, 417, 418 ⟨69IC279⟩
Mack, J. P. G., **1**, 877 ⟨89JA6419⟩
Mack, K.-E., **2**, 1009 ⟨88JOC4379⟩

Mack, P. O. L., **3**, 542 ⟨72CC451⟩
Mack, W., **5**, 832 ⟨61AG656⟩, 1046 ⟨72CB2815⟩; **6**, 309 ⟨61AG656, 72CB2815⟩, 417 ⟨61AG265⟩
MacKay, D., **2**, 379, 388 ⟨66JCS(C)813⟩; **5**, 366 ⟨80CJC1316⟩; **6**, 626 ⟨83CJC1213⟩
Mackay, G. I., **1**, 107 ⟨74JA4027, 76CJC1643⟩
MacKay, M. F., **3**, 113 ⟨93CC954⟩
Mackay, W. D., **3**, 494 ⟨69CC733⟩
MacKenzie, A. R., **3**, 36 ⟨91TL6973⟩; **4**, 207 ⟨91TL6973⟩
Mackenzie, C. A., **4**, 177, 179 ⟨55JOC1695⟩
Mackenzie, G., **2**, 32 ⟨76JCS(P1)1446⟩
Mackenzie, K., **1**, 723 ⟨B-70MI 116-02⟩, 795 ⟨B-64MI 118-01⟩, 797 ⟨B-70MI 118-02⟩, 1075 ⟨75TL1513⟩; **2**, 402 ⟨B-75MI 207-02⟩
Mackenzie, P. B., **3**, 508 ⟨84PAC59⟩
MacKenzie, T. E., **4**, 283 ⟨87JCS(D)757⟩
Mackie, A. G., **1**, 446 ⟨B-94MI 110-02⟩; **2**, 881 ⟨87AG(E)264⟩
Mackie, J. D. H., **2**, 25 ⟨55JCS3173⟩
Mackie, R. K., **2**, 7 ⟨B-79MI 201-01⟩, 835 ⟨B-79MI 216-02⟩; **3**, 341 ⟨73JCS(P1)657⟩; **5**, 110 ⟨77JFC(10)437⟩
Mackiewicz, P., **5**, 316 ⟨78JOC3750⟩
Mackinnon, J. W. M., **5**, 326 ⟨85JCS(P1)883⟩
Macklin, J. W., **4**, 697 ⟨75JOM(99)31, 76SA(A)1459⟩
Macklin, P. D., **4**, 718 ⟨87JOM(334)117⟩
Mackor, A., **2**, 389 ⟨70RTC164⟩, 391 ⟨70RTC151, 70RTC159⟩
Mackor, E. L., **1**, 893 ⟨64PCS147⟩, 914 ⟨61MP241, B-70MI 119-05⟩
Mackowiak, A., **6**, 546 ⟨88MI 617-01⟩
MacLachlan, F. N., **5**, 354 ⟨71CJC1598⟩
MacLaren, J. A., **6**, 507 ⟨77AJC455⟩
Maclean, A. F., **3**, 534 ⟨53JOC1055⟩
MacLean, C., **1**, 914 ⟨61MP241, B-70MI 119-05⟩, 922 ⟨68TL249⟩
MacLean, D. B., **1**, 1041 ⟨88CC439⟩; **4**, 54, 56 ⟨81CJC3055⟩
MacLeod, J. K., **1**, 1130 ⟨88TL391⟩
MacLow, C., **6**, 489 ⟨93MI 615-03⟩
MacMillan, A., **5**, 924 ⟨88JOM(341)C9⟩
MacMillan, J., **1**, 382 ⟨72JCS(P1)1274⟩, 940 ⟨85CC1588⟩
MacMillan, J. H., **5**, 992 ⟨73JOC2982⟩
MacMullin, R. B., **2**, 94 ⟨25JA395⟩
MacNaughton, N. W., **4**, 435, 443 ⟨51JA2769⟩
MacNeil, P. A.,**5**, 380 ⟨82OM1540, 84JA6993⟩
MacNicol, D. D., **1**, 243 ⟨75TL1893⟩
MacNicol, M., **1**, 398 ⟨28JCS3193⟩
MaCollough, J. F., **2**, 458 ⟨51JA855⟩
Macomber, D. W., **4**, 934 ⟨91OM2121⟩; **5**, 955 ⟨89OM1275⟩, 958 ⟨91OM2121, 91OM737⟩
Macomber, R. S., **1**, 988 ⟨85IJ136, 86JA343⟩, 993 ⟨76JOC3191⟩, 1114 ⟨75JA1531⟩, 1125, 1126 ⟨69JA4824⟩; **2**, 28 ⟨75JOC1990⟩
Macor, J. E., **1**, 200 ⟨91TL7195⟩, 200 ⟨93H(35)349⟩
MacPhee, J.-A., **3**, 567, 620 ⟨80T775⟩; **5**, 32 ⟨79T2077⟩
MacPherson, A. J., **2**, 441 ⟨84JCR(S)32⟩
MacPherson, D. T., **1**, 194, 288, 546 ⟨87JA3483⟩; **4**, 925 ⟨85JCS(P1)2625⟩
MacPherson, E. J., **6**, 507 ⟨71JOC2516⟩
MacPhillamy, H. B., **5**, 298 ⟨55JA469⟩
Macquarrie, D. J., **3**, 293 ⟨89CC1355⟩
MacRory, P. P., **2**, 881 ⟨87AG(E)264⟩
MacSweeney, D. F., **1**, 381 ⟨74JCS(P1)1938⟩
Madaj, Jr., E. J., **1**, 379 ⟨86JA3466⟩; **2**, 478 ⟨86JA3466⟩
Madan, O. P., **3**, 498 ⟨65TL201, 66T567⟩
Madappat, K. V., **3**, 703 ⟨89AG(E)1256⟩
Madaule, Y., **6**, 374, 375 ⟨83TL2769, 84JCR(S)274⟩
Madawinata, K., **1**, 155 ⟨77LA40⟩, 242 ⟨79LA451⟩; **3**, 696, 702, 704 ⟨77LA40⟩, 704 ⟨76LA2105⟩
Madden, M. J., **5**, 772, 850, 855 ⟨73JOC1316⟩
Madding, G. D., **5**, 655 ⟨85JHC1121⟩; **6**, 546 ⟨87JHC581⟩
Maddock, J., **4**, 268 ⟨91JCS(P1)662⟩, 269 ⟨90TL135⟩
Maddocks, P. J., **1**, 460 ⟨78TL3483⟩
Maddox, M. L., **2**, 935 ⟨85TL5375⟩
Madeja, R., **2**, 656 ⟨73HCA14⟩
Madelung, W., **3**, 722 ⟨15CB953⟩
Mader, H., **1**, 524 ⟨69AG(E)604⟩
Mader, M., **1**, 1045 ⟨78S307⟩; **3**, 79 ⟨78S307⟩
Mäder, T., **4**, 229 ⟨93HCA1779⟩
Madesclaire, M., **2**, 144 ⟨86T5459⟩, 713 ⟨92TA629⟩; **5**, 360 ⟨90CPB2352⟩
Madhav, R., **2**, 787 ⟨69JOC3279⟩
Madhavarao, M., **2**, 577 ⟨76JOM(108)93⟩
Madhukar, P., **4**, 934 ⟨91OM2121⟩; **5**, 955 ⟨89OM1275⟩, 958 ⟨91OM2121, 91OM737⟩
Mäding, P., **2**, 440 ⟨84ZC224⟩; **5**, 784, 785, 791 ⟨86ZC407⟩
Madison, S. A., **6**, 428 ⟨91EUP415473⟩
Madjdabadi, F. A., **2**, 13 ⟨83TL1933⟩
Madonik, A. M., **3**, 527 ⟨77JOC2111⟩
Mador, I. L., **5**, 16 ⟨B-76MI 501-01⟩; **6**, 469 ⟨78MI 615-01⟩
Madroñero, R., **2**, 308 ⟨66AHC(6)95⟩; **4**, 314 ⟨69JHC921⟩; **5**, 652 ⟨74T195⟩, 655 ⟨87S628⟩
Madsen, J. O., **2**, 758 ⟨76BSB805⟩
Madsen, R., **4**, 184 ⟨94CC749⟩
Madson, M. A., **2**, 263 ⟨82JOC3336⟩; **5**, 719 ⟨82JOC3336⟩
Maduakor, E. C., **3**, 90, 96 ⟨92TL7499⟩
Maeda, A., **2**, 150 ⟨77TL1345⟩, 539 ⟨94TL8413⟩
Maeda, H., **2**, 81 ⟨89JOC4022⟩, 665 ⟨82JOC4376⟩; **3**, 196 ⟨88CPB613⟩, 424 ⟨94CPB1041⟩; **4**, 828 ⟨93SL273⟩; **5**, 1107 ⟨90CPB1389⟩
Maeda, K., **1**, 108, 115 ⟨86CL433⟩, 532 ⟨82JOC2117⟩; **2**, 630 ⟨87JOC1100⟩, 914 ⟨88TL6955⟩; **3**, 18 ⟨84TL321, 82JOC2117⟩, 158 ⟨86TL65⟩; **4**, 92 ⟨93TL1311⟩; **5**, 659 ⟨93JOC32⟩; **6**, 256 ⟨93TL1311⟩
Maeda, M., **1**, 990 ⟨82JOC1435⟩; **2**, 18 ⟨87JFC(34)337⟩, 614 ⟨85CPB1301⟩
Maeda, N., **2**, 104 ⟨82JCS(P1)111⟩
Maeda, R., **3**, 108 ⟨68CPB897⟩
Maeda, T., **1**, 816 ⟨92T4087⟩, 817 ⟨86TL4189⟩; **2**, 855 ⟨67JAP6703120⟩; **3**, 309 ⟨90CPB3261⟩; **5**, 81 ⟨78GEP2725519⟩
Maeda, Y., **1**, 824 ⟨94JCS(P1)393⟩; **5**, 171 ⟨87JOC4319⟩
Maehara, M., **5**, 1112 ⟨88CC1485⟩
Maehata, E., **1**, 980 ⟨85CL1457⟩
Maehr, H., **6**, 650 ⟨74JOC1166⟩
Maejima, K., **1**, 198, 975 ⟨93JOC6166⟩, 431 ⟨93JOC4897⟩; **4**, 758, 821 ⟨93JOC4897⟩
Maekawa, E., **1**, 243 ⟨88JA4815⟩, 603 ⟨88JCS(P1)575⟩; **2**, 285 ⟨88JCS(P1)575, 89JCS(P1)1927⟩; **5**, 248 ⟨85S878⟩, 248, 249 ⟨87NKK1442, 88BCJ2205⟩, 249 ⟨89JCS(P1)1927⟩
Maekawa, S., **2**, 669 ⟨94TL605⟩
Maekawa, T., **1**, 434 ⟨87JOC300⟩, 1079 ⟨84TL1377⟩; **4**, 197 ⟨92JOC1321⟩; **5**, 417, 421, 423 ⟨91JOC1307⟩
Maekawa, T.-i., **6**, 124 ⟨90JA1962⟩
Maemura, K., **1**, 486 ⟨81JOC4885⟩
Maennle, F., **1**, 574, 584 ⟨91AG(E)1372⟩
Maeno, K., **2**, 21 ⟨88S545⟩
Maercker, A., **1**, 68, 69, 454 ⟨84AG(E)224⟩, 229 ⟨87CB1695⟩, 441 ⟨92JOM(441)1⟩, 508 ⟨86AG(E)167⟩, 638 ⟨65OR(14)270⟩; **2**, 1004 ⟨65OR(14)270⟩, 1006 ⟨84AG(E)224⟩; **4**, 264 ⟨91CB2489⟩, 620, 646, 653 ⟨87CB1695⟩, 647 ⟨84JOM(269)C40⟩, 650 ⟨84JOM(273)C57⟩, 667, 668 ⟨83AG(E)733⟩, 737 ⟨65OR(14)270⟩, 816, 819 ⟨93AG(E)1023⟩, 1064 ⟨84AG(E)224, 85AG(E)571⟩; **6**, 172, 204, 205 ⟨87TCC(138)1, 83AG(E)733⟩, 205 ⟨86OS378, 92AG(E)584⟩, 405 ⟨84AG(E)995⟩

Maerten, G., **5**, 590, 609 ⟨63LA(669)66⟩, 590, 609, 623 ⟨68LA(712)58⟩, 623, 624 ⟨68LA(712)46⟩
Maeshima, T., **5**, 694 ⟨87NKK1304⟩
Maestro, M. A., **1**, 576 ⟨92JOC5208⟩
Maestro, M. C., **3**, 658 ⟨88S136⟩; **6**, 124 ⟨91T3171⟩
Maeta, H., **2**, 70 ⟨92TL5965⟩
Maetzke, T., **2**, 999 ⟨88HCA299⟩
Maffei, S., **4**, 416 ⟨76S748⟩
Maffre, P., **2**, 50 ⟨93JOC1991⟩
Magar, S. S., **4**, 273, 275 ⟨92TL745⟩
Magari, H., **1**, 813 ⟨87CC1196⟩
Magdanyi, L., **2**, 444 ⟨84MIP30022⟩
Magdasieva, N. N., **5**, 900, 904 ⟨79ZOR2396⟩
Magdasieva, T. V., **5**, 920 ⟨81JOU1676⟩
Magdeeva, R. K., **2**, 464 ⟨80JGU2159⟩
Magdesieva, N. M., **4**, 874 ⟨80ZOR654⟩
Magdesieva, N. N., **3**, 177, 637 ⟨72JOM(42)399⟩; **4**, 928 ⟨79ZOB1857, 79ZOB2198⟩, 928 ⟨79ZOB332, 80ZOB573⟩; **5**, 898, 903 ⟨79JGU1740⟩, 903, 904 ⟨74JGU1677⟩
Magedov, I. V., **5**, 639 ⟨93S463⟩
MaGee, D. I., **2**, 1027 ⟨93T2159⟩
Magee, R. J., **6**, 560 ⟨79JINC1277⟩
Magennis, S. A., **1**, 669 ⟨74JA6140⟩; **2**, 590 ⟨74JA6140⟩, 590, 960 ⟨73JA6487⟩
Magerlein, B. G., **5**, 46 ⟨49OR(5)413⟩
Mägerlein, H., **5**, 584, 589 ⟨75S47⟩; **6**, 236 ⟨71S478⟩, 529 ⟨74S26⟩
Mageswaran, S., **1**, 400, 408, 410, 821, 823 ⟨81JCS(P1)1953⟩, 681, 682 ⟨76JCS(P1)884⟩
Maggini, M., **4**, 162 ⟨91TL6215⟩
Maggiulli, C. A., **5**, 219, 694 ⟨87JHC325⟩
Magi, M., **1**, 18, 39 ⟨63T1127⟩, 630 ⟨62TL1261⟩
Magi, M. Y., **6**, 368 ⟨72MI 612-01⟩
Magid, R. M., **1**, 186 ⟨80T1901⟩; **2**, 6 ⟨80T1901⟩, 8 ⟨77TL2999, 79JOC359⟩, 8 ⟨81JOC824⟩
Magidson, O. Yu., **2**, 788 ⟨60ZOB1855⟩
Magin, R. W., **1**, 882 ⟨65JOC2513⟩
Maginn, S. J., **2**, 134 ⟨88TL4477, 92T5933⟩; **4**, 249 ⟨92T5933⟩
Maglioli, P., **4**, 250 ⟨89SL28⟩
Magnien, E., **6**, 118 ⟨82MI 604-01⟩
Magnin, D. R., **1**, 340 ⟨89JA8309⟩, 340, 341 ⟨86JA2787⟩; **4**, 563 ⟨92USP5103036⟩
Magnotta, F. A., **2**, 717 ⟨63JOC2766⟩
Magnus, P., **1**, 263 ⟨92JA3993⟩, 422 ⟨79TL1995, 80JOC1046⟩, 583 ⟨94T1397⟩, 651, 762 ⟨80JA5004⟩, 655 ⟨80JA6885⟩, 813 ⟨80JA1756⟩, 971 ⟨80JA2134⟩, 1009 ⟨88JA1626⟩; **2**, 680 ⟨79CC822⟩, 908 ⟨77CC522⟩, 926 ⟨87JA7495, 87JOC1483⟩, 935 ⟨80JA5004⟩, 985 ⟨83JA2477⟩; **3**, 11 ⟨79CC822⟩, 164, 173 ⟨92TL6111⟩, 181 ⟨90JA462⟩, 194 ⟨92JA767⟩, 194 ⟨92TL2777⟩, 211 ⟨79TL1995, 82OM1240⟩, 211, 253 ⟨85JOC1621⟩, 218 ⟨89JOC5148⟩, 270 ⟨83JA2477, 85TL4851⟩, 270 ⟨87JA7495⟩; **4**, 355 ⟨77CC522⟩, 369, 938, 942 ⟨80JOC1046⟩, 443 ⟨93JA9347⟩, 648, 655 ⟨80TL11⟩, 650 ⟨82OM553⟩, 936 ⟨82OM1243, 85JOC1621⟩; **5**, 344 ⟨90JA462⟩, 574 ⟨86CC1756⟩; **6**, 124 ⟨82TL309⟩
Magnus, P. D., **1**, 140, 146, 436, 704 ⟨77T2019⟩; **2**, 309 ⟨74JCS(P1)2101⟩, 357 ⟨75CC855⟩, 543, 811 ⟨81TL1787⟩, 727, 729, 730, 731, 734 ⟨79COC(3)491⟩, 907 ⟨82COMC-I(7)515⟩; **3**, 366 ⟨72CC632, 73JCS(P1)1580⟩; **4**, 190 ⟨75CC432⟩, 355 ⟨77CC513, 77JA4536⟩, 355 ⟨78CC297⟩, 390 ⟨77T2019⟩, 1079 ⟨72JCS(P1)1103⟩; **5**, 560 ⟨73JCS(P1)1571⟩, 735 ⟨74JCS(P1)853⟩, 846 ⟨76JCS(P1)38⟩, 1051 ⟨79COC(3)522⟩; **6**, 172 ⟨82COMC-I(7)515⟩
Magnuson, H. J., **2**, 844 ⟨53JA1379, 53JA683⟩
Magnusson, B., **2**, 130 ⟨62ACS1536, 62ACS772⟩, 130 ⟨63ACS273⟩
Magnusson, G., **1**, 216 ⟨87TL2753⟩; **3**, 62 ⟨73JOC1380⟩; **4**, 176 ⟨86TL753⟩; **6**, 550 ⟨93CAR(246)119⟩
Mago-Karacsony, E., **1**, 85 ⟨79MIP15154⟩
Magolda, R. L., **1**, 591, 594, 716 ⟨79CC83⟩, 603 ⟨81JA3486⟩; **2**, 282 ⟨79S982, 80JA1404⟩; **3**, 38, 174 ⟨79S982⟩, 308 ⟨81JA6969⟩
Magomedov, G. K. I., **4**, 724 ⟨72ZOB2450⟩
Magoos, L., **2**, 1097 ⟨68MI 223-01⟩; **4**, 656 ⟨77JCR(S)116⟩; **6**, 394 ⟨77JCR(S)116⟩
Magosch, K-H., **6**, 110 ⟨70LA(742)128⟩, 107 ⟨71AG(E)810⟩
Magraby, M. A., **3**, 339, 361 ⟨85T1885⟩
Magriotis, P. A., **1**, 238 ⟨84JOC1460⟩, 1018 ⟨91TL6085⟩; **2**, 910 ⟨90TL6137⟩, 915 ⟨91TL5047⟩; **3**, 116 ⟨85TL5855⟩, 540 ⟨93TL2071⟩, 701 ⟨87JA287⟩; **4**, 772, 939, 948 ⟨90TL6137⟩, 962 ⟨90TL2541, 91TL6085⟩, 962, 963 ⟨91TL5047⟩
Magrum, G. R., **2**, 1023 ⟨87JA7568⟩
Mague, J. T., **2**, 478 ⟨67JCS(A)1347⟩, 874 ⟨68JOC1681⟩; **4**, 455, 477 ⟨67JOC2383⟩, 550 ⟨91POL1713⟩; **6**, 151 ⟨92OM26⟩, 516 ⟨91OM2781⟩
Maguet, L., **4**, 212 ⟨67BSF4172⟩
Maguire, A. R., **2**, 18 ⟨93JCS(P1)1795⟩; **3**, 232 ⟨90JCS(P1)1047⟩
Maguire, J. A., **1**, 554 ⟨92JA9492⟩; **6**, 406 ⟨90AOC(30)99⟩
Maguire, M. H., **2**, 451 ⟨55JCS2039⟩
Mah, H., **2**, 748 ⟨71CB3983⟩
Mah, T., **1**, 519 ⟨76TL881⟩; **5**, 63 ⟨78CC961⟩
Mahaderan, V., **2**, 39 ⟨74MI 202-01⟩
Mahaffy, C. A. L., **1**, 529 ⟨78JCR(M)4454⟩
Mahajan, A. R., **3**, 536 ⟨91S1027⟩
Mahajan, M. P., **2**, 249 ⟨87TL2641⟩, 744 ⟨89S631⟩; **5**, 330 ⟨91JCS(P1)1001⟩
Mahajan, S. W., **2**, 32, 36 ⟨85TL3863⟩
Mahajna, M., **5**, 403 ⟨91JOC4791⟩
Mahalanabis, K. K., **1**, 599, 779 ⟨77HCA2388⟩; **3**, 90, 98 ⟨89JOC24⟩
Mahale, V. B., **4**, 656 ⟨77JCR(S)116, 79JCS(D)767⟩; **6**, 394 ⟨77JCR(S)116⟩, 395 ⟨79JCS(D)767⟩, 396 ⟨78JCR(S)170⟩
Mahan, J. E., **5**, 461 ⟨77JA2501⟩
Mahapatro, S. N., **2**, 619 ⟨87JOC1619⟩
Mah, R. W. H., **2**, 907 ⟨63JOC2171⟩; **4**, 144 ⟨63JOC2171⟩
Mahasneh, A. S., **3**, 162 ⟨90SC2279⟩
Mahato, S. B., **4**, 745 ⟨88JOC5554⟩
Mahendran, M., **1**, 894 ⟨90JA8912⟩; **2**, 1007 ⟨86JA1692, 90JA8912⟩, 1008 ⟨86JA3613⟩
Maher, J. P., **4**, 681 ⟨92CC310⟩
Mahindroo, V. H., **1**, 487 ⟨91TA277⟩; **4**, 815 ⟨91TA277⟩
Mahindroo, V. K., **1**, 1034 ⟨91JOC1500⟩
Mahiou, B., **1**, 920 ⟨89JOC308⟩
Mahjoub, A., **5**, 546 ⟨78TL3847⟩, 546, 550 ⟨85BSF891⟩
Mahler, E. de, **2**, 1076 ⟨21BSF1024⟩
Mahler, H., **4**, 758 ⟨87TL5145⟩, 758, 818 ⟨91CB1379⟩
Mahler, J. E., **1**, 917 ⟨62JA2842⟩
Mähler, L., **6**, 213 ⟨B-72MI 607-02⟩
Mahler, W., **2**, 477, 869 ⟨65IC119⟩
Mahli, B. S., **4**, 742 ⟨90JCS(P1)1881⟩
Mahmood, S., **2**, 880 ⟨87AG(E)229⟩
Mahmood, T., **5**, 1125 ⟨88IJC583⟩
Mahmoud Ali, S., **5**, 1110 ⟨81JCR(S)234⟩
Mahmoud, S., **4**, 1039 ⟨87AG(E)229⟩
Mahmoudi, M., **1**, 816, 818 ⟨89TL2549⟩
Mahni, F., **4**, 322 ⟨87JCS(P1)781⟩
Mahon, M., **4**, 165 ⟨83JCS(P1)1131⟩

Mahon, M. F., **1**, 522 ⟨91JCS(P1)1255⟩; **4**, 268 ⟨91JCS(P1)662⟩, 269 ⟨90TL135⟩, 860 ⟨92T7551⟩; **5**, 340, 347 ⟨93JCS(P1)1581⟩; **6**, 395 ⟨90JCS(D)2643⟩
Mahoney, W. S., **1**, 86 ⟨88JA291⟩; **3**, 12, 128 ⟨88JA291⟩
Mahoud, M. M., **5**, 781 ⟨86JCS(D)1359⟩
Mahoungou, J. R., **1**, 974 ⟨86BSF449⟩
Mahr, N., **1**, 991 ⟨93AG(E)1315⟩
Mahran, M. R. H., **4**, 464 ⟨89JPR906⟩
Mahrwald, R., **3**, 220 ⟨87S1012⟩
Mai, I. I., **5**, 691 ⟨92ZPK1430⟩
Mai, K., **3**, 426 ⟨85OPP183⟩, 615, 616 ⟨84TL4583, 85SC157⟩, 618 ⟨86S1037, 86TL2203⟩; **5**, 226, 1134 ⟨86JOC3545⟩, 1076 ⟨87JOC275⟩, 1122 ⟨86SC1823⟩; **6**, 380 ⟨86JOC3545⟩
Maia, A., **4**, 233 ⟨72CC734⟩
Maia, A. M., **2**, 9 ⟨75CC950⟩; **4**, 233 ⟨76JCS(P2)996⟩
Maia, J. A., **5**, 765, 824 ⟨86JHC1861⟩
Maidanovich, N. K., **4**, 486 ⟨82ZOB930⟩, 1004 ⟨86ZOB805⟩, 1009 ⟨89ZOB2492⟩; **5**, 713 ⟨89ZOB2492⟩, 713, 788 ⟨89JGU2228⟩
Maidment, M. S., **4**, 56 ⟨85JCS(P1)1567⟩
Maier, A., **2**, 1009 ⟨88JOC4379⟩
Maier, G., **1**, 420 ⟨67AG(E)402⟩, 521 ⟨82CB804⟩, 789 ⟨74AG(E)425⟩; **3**, 233 ⟨69TL4987⟩, 337 ⟨91CB2609⟩, 358 ⟨83CB273, 92CB265⟩, 455 ⟨68CB1354⟩, 504 ⟨85AG(E)1065⟩, 686 ⟨87AG(E)155⟩, 703 ⟨88AG(E)938⟩, 710 ⟨92AG(E)1218⟩; **4**, 19, 27, 29 ⟨83OMR(21)64⟩, 610 ⟨81AG(E)597⟩; **5**, 14 ⟨82CB804⟩, 874 ⟨85TL1845, 86AG(E)1093⟩, 1154 ⟨81AG(E)197⟩; **6**, 713 ⟨85AG(E)1065⟩
Maier, L., **1**, 444 ⟨B-72MI 110-01⟩; **2**, 426, 428, 440 ⟨B-72MI 208-01⟩, 426, 428, 440 ⟨B-72MI 208-01⟩, 427, 444 ⟨B-72MI 208-03⟩, 428 ⟨59AG574, 63MI 208-01⟩, 428, 440 ⟨66HCA842⟩, 431 ⟨61CB3051, 61CB3056⟩, 431 ⟨B-72MI 208-02, B-72MI 208-02⟩, 441 ⟨64HCA2137⟩, 444 ⟨66HCA2488, 87PS(33)41⟩, 444, 824, 882 ⟨57JA5884⟩, 449 ⟨62JINC275⟩, 451 ⟨63HCA2667⟩, 452 ⟨68HCA405⟩, 458 ⟨90PS(47)465⟩, 459, 460, 462, 851 ⟨66HCA1249⟩, 461 ⟨64HCA27⟩, 462 ⟨71PS(1)71⟩, 463 ⟨64HCA1448⟩, 465, 858 ⟨66HCA1000⟩, 475 ⟨B-72MI 208-04, B-72MI 208-04⟩, 820, 821 ⟨74MI 216-01⟩, 820 ⟨B-72MI 216-01, 67HCA1747⟩, 854 ⟨62JINC1073, 64HCA120⟩, 855 ⟨72HCA1991⟩, 872 ⟨60JINC213⟩; **4**, 105 ⟨87PS(33)41⟩, 112, 115, 121, 123, 127, 128 ⟨71HCA1651⟩, 113, 121, 556 ⟨69HCA858⟩, 120 ⟨70HCA2069⟩, 120, 122 ⟨72MI 403-01⟩, 125, 473 ⟨79JOM(178)157⟩, 127, 556 ⟨69HCA827⟩, 336 ⟨B-72MI 407-01⟩, 343 ⟨92S258⟩, 344 ⟨72PS(1)237⟩, 451 ⟨B-72MI 410-01⟩, 452 ⟨67HCA1723⟩, 452, 466 ⟨65AG(E)527⟩, 453, 550 ⟨66HCA842⟩, 463 ⟨83PS(14)295⟩, 463, 466 ⟨81PS(11)139, 68HCA1608⟩, 466 ⟨81PS(11)149⟩, 470 ⟨67HCA1742, 80PS(8)67⟩, 472 ⟨93S955⟩, 546 ⟨65HCA1034, 66USP3253033⟩, 550 ⟨70USP3496231, 70USP3518312, B-72MI 412-01⟩, 556 ⟨68AG(E)384, 68AG(E)385, 69HCA845, 70HCA1940, 73USP3734954, 74USP3845169⟩, 563 ⟨65HCA133, 72USP3632634⟩, 803 ⟨73MI 418-01⟩; **5**, 404 ⟨91PS(62)15⟩, 800 ⟨B-72MI 520-01⟩, 900 ⟨72MI 523-01⟩; **6**, 157, 159 ⟨81PS(11)311⟩, 701 ⟨B-72MI 614-01⟩
Maier, M., **4**, 176 ⟨85TL2065⟩
Maier, M. E., **1**, 554 ⟨91TL3679⟩, 1021 ⟨91TL53⟩, 1038 ⟨92TL7511⟩; **2**, 936 ⟨93SL863⟩; **3**, 320 ⟨88TL1677⟩; **5**, 572 ⟨90JOC831⟩
Maier, N. A., **2**, 594 ⟨80JOM(201)39⟩
Maier, R., **3**, 246 ⟨82CB2766⟩
Maier, S., **5**, 781 ⟨88ZN(B)1628⟩
Maier, T., **6**, 78 ⟨81S380⟩, 138 ⟨79S342⟩, 296, 297, 298, 300, 313, 320, 321 ⟨77S73⟩, 296, 297, 298, 300, 321 ⟨82LA507⟩, 643 ⟨84LA108⟩, 729 ⟨81LA70⟩
Maier, W., **6**, 488 ⟨92GEP4215648⟩
Maier, W. F., **1**, 32, 48 ⟨84JOC2875, 79AG(E)939⟩, 502 ⟨87JOC3126⟩, 530 ⟨87JOC1161⟩; **2**, 422 ⟨80LA1471⟩; **3**, 83 ⟨86JA2608⟩
Maierhofer, A., **4**, 297, 327 ⟨66CB1944⟩
Maignan, C., **5**, 82 ⟨84S876⟩
Maignan, J., **3**, 344 ⟨73BSF(2)1973⟩
Maigrot, N., **4**, 545, 565 ⟨93PS(77)254⟩
Maihara, A., **6**, 461 ⟨90TL5629⟩
Maikap, G. C., **1**, 568 ⟨91S827⟩
Maikova, A. I., **4**, 912 ⟨67ZOB1623⟩; **5**, 707 ⟨68JGU2501, 68ZOB2586⟩; **6**, 314 ⟨67ZOB2137⟩
Mailhe, A., **2**, 125 ⟨09BSF405⟩
Maillard, B., **1**, 331, 336, 342, 936 ⟨87JOC3509, 83JA3578⟩; **2**, 97 ⟨90SC239, 93SC1025⟩, 98 ⟨93T2477⟩; **5**, 206 ⟨85BSB29, 85T4047⟩
Maillet, R., **2**, 441 ⟨80TL2525⟩
Main, A. J., **3**, 65 ⟨82TL167⟩
Mainen, E. L., **5**, 24 ⟨72MI 502-02⟩
Maiocchi, L., **6**, 425 ⟨90JMC2101⟩
Maione, A. M., **3**, 5 ⟨84S955⟩; **4**, 201 ⟨87S250⟩
Maiorana, S., **2**, 691 ⟨88S344⟩, 692 ⟨87S192⟩; **3**, 430 ⟨85CC417⟩; **5**, 497 ⟨73JCS(P1)2540⟩, 497, 499 ⟨74JCS(P2)1483, 74JCS(P2)1676⟩, 905 ⟨79S299, 81EUP54516⟩, 906 ⟨83CC337, 85T3321⟩, 943 ⟨92CC1623⟩, 946, 959 ⟨94ICA233⟩, 953, 958 ⟨95JOM(486)279⟩, 956 ⟨94TL6377⟩; **6**, 723 ⟨93OM2994⟩
Maiorano, S., **2**, 553 ⟨92CC1623⟩
Mairanovskii, S. G., **1**, 103 ⟨75IZV888⟩
Maire, J. C., **4**, 701 ⟨69JOM(16)83⟩
Mairino, M., **3**, 57 ⟨87TL2099⟩
Mais, A., **1**, 8 ⟨59JA3644⟩
Mais, F. J., **6**, 467 ⟨93EUP543234⟩
Maisey, R. F., **1**, 747 ⟨68CC543⟩; **5**, 292 ⟨77S31, 77S33⟩
Maissant, J.-M., **2**, 650 ⟨82JMOC333⟩
Maisto, K. D., **2**, 25 ⟨93SC2199⟩
Maiti, P. K., **6**, 492 ⟨92AP(325)267⟩
Maitland, D. J., **3**, 64 ⟨89JCS(P1)1939⟩
Maitlis, P. M., **1**, 93 ⟨77CC427, 78JCS(D)617⟩, 94 ⟨84JCS(D)1747⟩; **2**, 579 ⟨82COMC-I(6)279⟩, 581 ⟨85JCS(D)573⟩, 585 ⟨81CC808⟩, 980 ⟨76ACR93⟩; **4**, 687 ⟨83JCS(D)1441⟩, 689 ⟨69JA7292, 69JOM(19)169⟩, 693 ⟨78JCS(D)1825, 78JCS(D)1830⟩, 715 ⟨67CJC2011⟩
Maitra, S. K., **4**, 745 ⟨88JOC5554⟩
Maitte, P., **3**, 89 ⟨89BSB221⟩, 219 ⟨87JHC497⟩, 293 ⟨69BSF915⟩; **4**, 181 ⟨80BSB759⟩
Majat, E. E., **2**, 556 ⟨47JA950⟩
Majchrzak, M. W., **2**, 943 ⟨86T1299⟩; **3**, 307 ⟨85TL5035⟩, 323 ⟨89CJC1753⟩
Majenski, J. M., **2**, 221 ⟨81TL3659⟩
Majer, J. R., **6**, 84 ⟨48JCS687⟩
Majeste, R. J., **6**, 509 ⟨78JOC922⟩
Majetich, G., **1**, 314 ⟨83TL1909, 88JOC50⟩, 525, 826 ⟨86TL4945⟩, 812 ⟨88TL2773⟩, 957, 973 ⟨93TL449⟩; **3**, 229 ⟨89TL1033⟩, 624 ⟨86JOC1745, 86JOC1753⟩
Majewski, M., **1**, 215, 576 ⟨82TL1343⟩, 688, 690 ⟨88JA6914⟩, 828 ⟨84JOC2682⟩
Majewski, P., **2**, 453 ⟨89PS(45)151⟩; **6**, 242 ⟨91PS(55)185⟩
Majewski, R. F., **2**, 381 ⟨73JMC736⟩
Majewski, T. E., **6**, 143 ⟨68JOC3931⟩
Majid, A., **2**, 175 ⟨74IC2710⟩, 421 ⟨73JCS(D)1876⟩
Majidi, A., **4**, 604 ⟨93JOC2517⟩
Ma Jing-ji, **1**, 740 ⟨80MI 116-01⟩, 740, 741 ⟨80MI 116-02⟩
Major, A., **5**, 857 ⟨55JA4297⟩
Majoral, J.-P., **2**, 470 ⟨86IC2695⟩; **4**, 459 ⟨83TL1975⟩, 547, 579 ⟨93AG(E)1735⟩, 552 ⟨86IC712⟩; **5**, 705 ⟨78PS(4)317⟩, 705, 706 ⟨77JHC749⟩, 883 ⟨83TL5885⟩;

6, 160 ⟨83PS(18)27⟩, 685 ⟨89TL4813⟩, 686 ⟨83TL5885⟩, 689 ⟨89TL6869⟩
Major, R. T., **2**, 243, 360 ⟨56JA6181⟩
Majumdar, A., **3**, 273 ⟨90S243⟩
Majumdar, D., **2**, 905 ⟨80CC39⟩; **4**, 362 ⟨80JA7588⟩, 362, 373, 526 ⟨83OM1529⟩, 525 ⟨79JOM(170)259⟩, 642, 645 ⟨80CC39⟩, 642, 645 ⟨83OM230⟩; **6**, 203 ⟨80CC39, 80JOM(184)C41⟩, 203 ⟨83OM230⟩
Majumdar, M. K., **2**, 533 ⟨62JCS4088⟩
Mak, C. C., **5**, 934 ⟨94T2003⟩
Mak, C. P., **1**, 554 ⟨81JOC1⟩
Mak, H. J., **2**, 797 ⟨76SC129⟩
Makabe, T., **3**, 340 ⟨88MI 308-01⟩
Makabe, Y., **3**, 21, 141 ⟨88JOC4489⟩, 413 ⟨85TL5563⟩
Makami, K., **2**, 932 ⟨92SL189⟩
Makani, S., **2**, 626 ⟨80CC110⟩
Makarenkova, L. M., **6**, 65 ⟨76IZV2547⟩
Makarova, L. G., **1**, 65 ⟨B-72MI 102-02⟩, 128 ⟨B-67MI 104-02⟩; **2**, 1097 ⟨B-67MI 223-01⟩; **4**, 669, 684, 722 ⟨73IZV2796⟩, 685 ⟨72IZV1600⟩; **6**, 172, 207 ⟨B-67MI 606-01⟩
Makarova, L. L., **2**, 891 ⟨88JGU1040⟩
Makarov, G. M., **6**, 657 ⟨90ZOB778⟩
Makarov, K. N., **4**, 731 ⟨86IZV91⟩, 737 ⟨81JFC(17)441⟩; **5**, 663, 685, 698, 699, 813 ⟨71T51⟩, 811 ⟨68IZV1134⟩, 811 ⟨68IZV657, 69IZV958⟩
Makarov, S. P., **2**, 166, 170 ⟨65ZOB1412⟩; **4**, 135, 136 ⟨52JGU1569⟩, 154, 155 ⟨52JGU1827⟩; **5**, 822 ⟨60ZOB2409⟩; **6**, 240 ⟨69ZOB199⟩, 242 ⟨61DOK(141)357, 62DOK(142)354, 68ZOB709⟩, 258, 259 ⟨65ZOB1412⟩, 614 ⟨68JGU37⟩, 614, 627 ⟨69JGU183⟩
Makarov, Yu. V., **4**, 706 ⟨68IZV686⟩, 726 ⟨73IZV2815⟩
Mak, T. C. W., **1**, 709 ⟨78TL4269⟩, 983 ⟨85TL631⟩, 986 ⟨91BCJ659⟩; **2**, 436 ⟨90POL2901⟩; **6**, 195 ⟨93JOM(443)C39⟩, 247 ⟨93JOM(453)299⟩
Mäke, S., **5**, 183 ⟨68AG(E)465⟩
Makeeva, E. A., **3**, 341 ⟨90KGS1285⟩
Makhaev, V. D., **2**, 442 ⟨83URP1046248, 84ZOB2550⟩
Makhamatkhanov, M. M., **6**, 563 ⟨82MI 617-02⟩
Makheva, G. F., **6**, 613 ⟨88KFZ143⟩
Makhmutova, S. F., **4**, 128 ⟨78JGU1813⟩, 129 ⟨78JGU1168⟩
Makhon'kov, D. I., **4**, 800 ⟨79JOU2207⟩; **5**, 316 ⟨77ZOB968⟩
Makhova, I. V., **4**, 197 ⟨88TL1603⟩
Makhova, N. N., **5**, 667, 671 ⟨91BAU438⟩, 667, 671 ⟨91IZV506⟩
Makhsumov, A. G., **2**, 1018, 1019 ⟨65ZOR415⟩, 1056 ⟨87UKZ34⟩
Maki, H., **3**, 366 ⟨90TL7641⟩
Maki, Y., **1**, 482 ⟨89JCS(P1)2513⟩, 1029 ⟨87H(26)355⟩; **2**, 214 ⟨84H(22)1017⟩; **4**, 413 ⟨76TL1199⟩, 420 ⟨76CPB235⟩; **5**, 278 ⟨82JCS(P1)2909⟩; **6**, 12 ⟨79JAP79283⟩, 111 ⟨84H(22)13⟩
Makin, M. I. H., **2**, 213 ⟨84AJC2331⟩, 772 ⟨84AJC845⟩; **5**, 1122 ⟨84CC53⟩
Makin, S. M., **3**, 57 ⟨59JGU3925⟩, 76 ⟨71ZOR2120⟩
Makino, I., **6**, 543, 557, 563, 578 ⟨88JOC2263⟩
Makino, K., **4**, 101, 105 ⟨87JFC(35)677⟩; **6**, 229 ⟨92JFC(59)417⟩
Makino, N., **3**, 637 ⟨82JCS(P1)1303⟩
Makino, S., **1**, 2 ⟨78JA1765⟩, 2, 3, 8 ⟨78JA1786⟩; **3**, 225 ⟨72JA1772⟩, 245 ⟨73JA4103⟩
Makino, T., **3**, 616 ⟨75ABC571⟩
Makino, Y., **1**, 86 ⟨89BCJ1682⟩
Makinouchi, S., **3**, 293 ⟨88CL1141⟩; **5**, 452 ⟨88CL1141⟩
Makisumi, Y., **1**, 816 ⟨66TL6393⟩; **3**, 362 ⟨66TL6399⟩
Makita, A., **5**, 199 ⟨86JAN1257⟩
Makita, H., **5**, 327 ⟨81JGU2125⟩
Makitra, R. G., **2**, 860 ⟨56ZOB907⟩
Maklyaev, F. L., **4**, 554 ⟨60ZOB1602⟩; **6**, 263 ⟨60ZOB1602⟩
Makomo, H., **5**, 589 ⟨93PS(80)31⟩
Makosza, A., **3**, 465 ⟨77BSB739⟩
Makosza, M., **1**, 613, 622 ⟨87S1142⟩, 1033 ⟨79CC488⟩; **2**, 308 ⟨78S882⟩; **3**, 99, 105 ⟨87TL3021⟩, 620, 627 ⟨74AG(E)665⟩, 625 ⟨88LA203⟩, 634 ⟨72TL2391⟩, 634, 638 ⟨87BSB(96)303⟩, 702 ⟨93S1215⟩; **4**, 82, 83, 237 ⟨75JOC266⟩, 736 ⟨84JFC(24)387⟩; **5**, 1118 ⟨78S882⟩; **6**, 233 ⟨74S274⟩
Makovetskii, Yu. P., **6**, 612, 623 ⟨81ZOR180⟩
Makoveyeva, N. P., **4**, 603, 625, 637, 701 ⟨71JOM(30)C57⟩
Makowka, B., **2**, 472 ⟨85ZN(B)1274⟩
Makriyannis, A., **6**, 506 ⟨73JOC1652⟩
Maksakova, M. V., **5**, 850 ⟨73JGU317⟩, 863 ⟨72ZOB868⟩
Maksimenko, A. A., **5**, 211 ⟨91MI 504-02⟩
Maksimov, Y. V., **6**, 653 ⟨72ZOR332⟩
Maksimova, L. I., **5**, 601 ⟨70ZOR610⟩
Maksudov, N. K., **6**, 561 ⟨81UKZ26⟩, 563 ⟨82MI 617-02⟩
Maksyutov, E. M., **6**, 147 ⟨80MI 605-01⟩
Makushima, M., **3**, 307 ⟨81TL4901, 83JOC3214⟩
Makuta, T., **1**, 228 ⟨93TL2135⟩
Mal, D., **1**, 239 ⟨90IJC(B)390⟩; **2**, 156 ⟨88JCR308⟩; **3**, 219 ⟨83JA5688⟩
Malacria, M., **1**, 194, 288 ⟨89TL1803⟩, 225 ⟨79S508⟩, 340 ⟨94JOC718⟩, 362 ⟨93JOC4298⟩, 413, 810 ⟨83TL1157⟩, 808 ⟨78TL1803⟩, 809 ⟨84T3277, 86T1333⟩, 959 ⟨76S755, 77T2813⟩, 967 ⟨79TL7⟩; **2**, 944 ⟨93JOC4298⟩; **3**, 141 ⟨82TL4263⟩, 144, 217 ⟨86T1333⟩; **4**, 140 ⟨89S687⟩, 837 ⟨92SL493⟩
Malakhova, I. G., **2**, 430, 437 ⟨86S198⟩; **4**, 546, 553 ⟨86S198⟩
Malament, D. S., **2**, 404 ⟨70JA4586⟩
Malaprade, L., **5**, 28 ⟨28BSF683, 28CR(186)382⟩
Malarek, D. H., **1**, 884 ⟨67JOC2257⟩
Malaspina, C., **4**, 986 ⟨79JPR43⟩
Malassa, I., **3**, 417 ⟨85S177⟩
Malatesta, M. C., **5**, 461 ⟨75CC267⟩
Malavaud, C., **4**, 478 ⟨89MI 410-01⟩
Malbec, F., **5**, 705 ⟨84JHC1689⟩
Maldonado, L., **1**, 309 ⟨74JA5272⟩
Maldonado, V. C. de, **4**, 831 ⟨88JOC633⟩
Malecha, J. W., **2**, 75, 521 ⟨91JOC5243⟩
Maleck, G., **6**, 649 ⟨67TL3851⟩
Malecki, F. E., **1**, 28 ⟨66LA(691)15⟩
Maleev, A. V., **2**, 445 ⟨87ZOR2120⟩
Malek, F., **1**, 933 ⟨79JCS(P1)1137⟩
Malek, G.,**5**, 652 ⟨68JA1651⟩
Málek, J., **1**, 18, 21 ⟨69TL1739⟩; **3**, 84 ⟨72S217⟩; **6**, 7 ⟨80CCC3488, 80CCC3502⟩
Maleki, M., **4**, 917 ⟨81TL3789, B-81MI 420-02⟩
Malekin, S. I., **6**, 613 ⟨69JGU1235, 69JGU1692⟩, 613 ⟨72JGU293, 72JGU799⟩, 613 ⟨72JGU803⟩
Malek-Yazdi, F., **5**, 516, 519, 522 ⟨76JOC729⟩, 589, 626 ⟨77S328⟩
Malenfant, E., **2**, 79 ⟨93SL81⟩
Malenki, D. M., **5**, 713, 788 ⟨86JGU1113⟩
Malenko, D. M., **4**, 1009 ⟨90ZOB1187⟩; **5**, 369, 713 ⟨90ZOB1187⟩, 643 ⟨86ZOB1467, 84JGU1925⟩, 713 ⟨84ZOB2154, 86ZOB1262⟩, 713 ⟨90JGU1057, 90ZOB1186⟩, 713, 788 ⟨90JGU1059⟩, 737 ⟨86ZOR1654, 86JGU1467⟩
Maleski, R. J., **2**, 398 ⟨72JOC2810⟩
Malet, C., **3**, 226 ⟨89JOC3984⟩
Maletina, E. A., **5**, 1135 ⟨74URP449912, 82S277⟩
Maletina, I. I., **1**, 109 ⟨88JOU530⟩; **4**, 864 ⟨79ZOR2416⟩; **5**, 204 ⟨77S574⟩, 205 ⟨88ZOR593⟩; **6**, 14 ⟨93RHA232⟩

Maletzko, C., **6**, 255, 275, 304, 306 ⟨91CB2025⟩
Maletzo, C., **5**, 478, 492 ⟨91CB1311⟩
Malevannaya, R. A., **4**, 113, 120 ⟨70JGU255⟩, 118, 132, 133 ⟨69JGU1490⟩, 555 ⟨70ZOB285⟩
Malevannaya, Z. P., **5**, 199 ⟨61ZOB665⟩
Malfroot, T., **2**, 303 ⟨84JOC2081⟩; **4**, 58 ⟨84JOC1700, 90JOC2240⟩, 920, 921 ⟨90JOC5982⟩, 921 ⟨89FP2610926⟩; **5**, 282 ⟨87TL5823⟩; **6**, 428 ⟨84JOC2081⟩, 429 ⟨90JOC2240, 90JOC5982⟩
Malherbe, R., **1**, 7, 833 ⟨83JOC860⟩, 811 ⟨76JA5937, 78HCA3096⟩
Malhotra, N., **4**, 370 ⟨91G471⟩; **6**, 558 ⟨93OPP83⟩, 559 ⟨82JCS(P1)653⟩
Malhotra, R., **2**, 32 ⟨80JOC1638⟩, 36 ⟨79AG(E)563, 79JOC1247⟩, 103 ⟨B-89MI 202-05⟩, 1053 ⟨87JOC2294⟩; **3**, 279, 295, 296 ⟨78S672⟩; **5**, 226 ⟨79JOC1247⟩
Malhotra, R. C., **4**, 773 ⟨93TL7645⟩
Malhotra, S. K., **1**, 795, 871, 881 ⟨65JA3228⟩, 871 ⟨62JA3402, 65JA5492⟩, 871 ⟨65JA5493, 65JA5513⟩, 871, 881 ⟨63JA1538⟩, 871, 881 ⟨64JA1997⟩, 871, 881 ⟨66JA1332⟩, 871, 881, 882 ⟨62TL669⟩
Mali, R. S., **1**, 827 ⟨94CC251⟩; **3**, 90 ⟨79S906⟩, 109 ⟨74T4153⟩; **5**, 110 ⟨76S123⟩
Maliar, F., **3**, 659 ⟨85SC473⟩
Malichenko, B. F., **5**, 967 ⟨69URP241428⟩
Malik, A. A., **2**, 317 ⟨89S450⟩; **5**, 967 ⟨91JOC3043⟩
Malik, A., **1**, 218 ⟨88SC151⟩; **2**, 9 ⟨87LA617⟩; **4**, 413 ⟨82ZN(B)512⟩
Malik, K. M. A., **4**, 672 ⟨78JCS(D)1314⟩, 680 ⟨79JA4128, 80JCS(D)1771⟩, 719 ⟨80JCS(D)2315⟩
Malik, M., **4**, 472 ⟨90TL1759⟩
Malik, W. U., **5**, 1038 ⟨69JCED110⟩; **6**, 641 ⟨68JMC1268⟩
Malinka, W., **5**, 347 ⟨86FES676⟩
Malinovskii, M. S., **6**, 429 ⟨53JGU229⟩, 581 ⟨75ZOB1484⟩
Malinowski, M., **1**, 688 ⟨81BSB1141, 82JOC248⟩, 688 ⟨82MI 115-01⟩, 692 ⟨82BSB149⟩
Maliphant, G. K., **5**, 749, 750 ⟨51JCS485⟩
Malisch, M., **4**, 135, 597 ⟨83CB348⟩
Malisch, W., **2**, 511 ⟨74AG(E)84⟩, 511, 512 ⟨74JOM(76)C7⟩, 511, 512 ⟨80CB3318, 74AG(E)670⟩, 512 ⟨75CB700, 76CB692⟩, 512 ⟨80CB1180, 82JOM(228)C29⟩, 512 ⟨84ICAL49⟩; **4**, 569 ⟨69AG(E)372, 69CB83⟩, 569 ⟨70CB3007, 70CB97⟩, 570, 571 ⟨70CB3448, 74JOM(77)C15⟩, 573 ⟨74AG(E)540⟩, 580, 582 ⟨83ZN(B)711⟩, 581 ⟨83CB690⟩; **5**, 448 ⟨82CB1488⟩, 911 ⟨80CB3318, 82JOM(236)C1⟩, 912 ⟨70CB3007, 70CB3448⟩, 912 ⟨75JOM(99)421⟩, 913 ⟨75CB2554⟩, 914 ⟨74AG(E)540⟩, 914, 1095 ⟨70CB97⟩, 920 ⟨73JOM(61)C15⟩; **6**, 373 ⟨70CB3448⟩, 689 ⟨85CC1687⟩, 700, 703 ⟨70AG(E)77⟩, 700, 703 ⟨71CB150, 74AG(E)540⟩
Malissar, D. G. S., **3**, 273 ⟨90JCS(P1)1507⟩
Maliverney, C., **5**, 601 ⟨90TL6339⟩
Malkeil, J., **6**, 27 ⟨42JA2515⟩
Mal'kevich, N. Yu., **2**, 841 ⟨78ZOB91⟩; **4**, 464 ⟨77ZOB579⟩
Malkiewich, C. D., **5**, 652 ⟨77IS95⟩
Malkin, T., **5**, 28 ⟨57MAC25, 63MI 502-01⟩, 28 ⟨B-55MI 502-01, B-63MI 502-02⟩
Mälkönen, P. J., **5**, 1028 ⟨91JHC1091⟩
Mal'kova, G. Y., **2**, 569 ⟨82JOM(239)335⟩
Mallaiah, B. V., **1**, 555 ⟨83S310⟩
Mallaiah, K., **2**, 74 ⟨93TL3145⟩
Mallamo, J. P., **1**, 23, 236 ⟨81JA2886⟩; **4**, 958 ⟨78TL3995⟩, 960, 964 ⟨81PAC2307⟩
Mallart, S., **3**, 181 ⟨91TL2359⟩
Mallela, S. P., **6**, 396 ⟨93JCS(D)2891⟩
Malleron, J. L., **3**, 139 ⟨81CC1159⟩
Mallet, J.-M., **1**, 623 ⟨88T2913⟩; **4**, 238 ⟨93SL522⟩
Malley, M. F., **1**, 200 ⟨90H(31)1359⟩; **4**, 905 ⟨90JMC1510⟩; **5**, 360 ⟨91JOC3744⟩
Mallick, I. M., **1**, 716 ⟨87JOC774⟩
Mallien, M., **6**, 390 ⟨93JCS(D)3259⟩, 393 ⟨94AG(E)1268⟩
Mallinson, P. R., **6**, 528 ⟨92TL261⟩
Mallory, D., **4**, 295, 407, 412 ⟨73JOC3288⟩, 444 ⟨76JOC3063⟩
Mallouli, A., **3**, 66 ⟨80S898⟩
Malm, J., **1**, 480 ⟨93JOM(460)127⟩
Malmberg, H., **2**, 587 ⟨82TL3823⟩; **5**, 151 ⟨82T1509⟩
Malmberg, W.-D., **3**, 345, 349 ⟨83LA1694⟩; **5**, 510, 516, 523 ⟨79JCR(S)160⟩
Malone, G. R., **3**, 27 ⟨74JOC618⟩, 61 ⟨74JOC623⟩, 286 ⟨69JA5886⟩
Malone, J. F., **1**, 369, 524 ⟨85T3547⟩; **2**, 158 ⟨91T8297⟩, 245 ⟨88T4941⟩; **4**, 262 ⟨92JCS(P1)1105⟩, 698 ⟨72JCS(D)2649⟩
Malone, T. C., **2**, 328, 329 ⟨87JA6097⟩
Maloney, T. W., **3**, 368 ⟨70JOC1716⟩
Malotki, P. von, **4**, 454 ⟨73MI 410-01⟩; **6**, 511 ⟨70JPR366⟩
Malovetskii, Y. P., **2**, 444 ⟨91ZOB1075⟩
Malovichko, N. N., **3**, 362 ⟨89MI 308-04⟩
Malovik, V. V., **2**, 453 ⟨78BRP1517876⟩
Malpass, D. B., **1**, 680 ⟨88JOC2829⟩; **4**, 698 ⟨88JOC2829⟩
Malpass, J. R., **3**, 539 ⟨68CC764⟩; **4**, 98 ⟨85CC686⟩; **5**, 330 ⟨87TL5937, 92TL273⟩
Malrieu, J.-P., **3**, 497 ⟨79JA7169⟩; **6**, 694 ⟨79JA7169⟩
Malsuyama, H., **2**, 997, 1001 ⟨92JCS(P1)1721⟩
Malte, A., **1**, 1005 ⟨69JA6464⟩; **2**, 621 ⟨69JA6464⟩; **3**, 273 ⟨69JA6464⟩
Malte, A. M., **1**, 618, 706 ⟨69JA7510⟩; **4**, 81 ⟨69JA7510⟩
Maltesson, A., **4**, 814 ⟨75ACS(B)461⟩
Malthouse, J. P. G., **5**, 1111 ⟨92B7922⟩
Mal'tsev, A. K., **3**, 342 ⟨87IZV2321, 87IZV2398⟩; **4**, 1083 ⟨87DOK(296)403⟩
Malygina, I. L., **4**, 706 ⟨58DOK(120)1263⟩
Malygin, V., **6**, 613 ⟨88KFZ143⟩
Malysheva, I. P., **5**, 390, 391 ⟨88IZV859⟩, 1082 ⟨88IZV739⟩
Malysheva, S. F., **2**, 1064 ⟨75JGU1832⟩
Malyshev, V. A., **5**, 691 ⟨92JOU1399⟩
Malyutina, I. V., **2**, 428 ⟨90ZOB697⟩
Malz, H., **3**, 721 ⟨64CA(60)6795⟩; **5**, 738 ⟨59USP3053876⟩; **6**, ⟨63BEP632635⟩
Mamada, A., **1**, 236 ⟨85CL1149⟩
Mamaev, V. P., **6**, 110 ⟨66IZV98⟩
Mamakov, K. A., **2**, 501 ⟨73JGU825⟩, 881, 883 ⟨72IZV1612⟩, 889 ⟨64DOK(157)371, 66JGU1986⟩, 889 ⟨71IZV1726⟩
Mamantov, A., **2**, 405 ⟨72JA4392⟩
Mamba, A., **5**, 303 ⟨74JOC3924⟩
Mambu, L., **1**, 1009 ⟨93SL217⟩
Mamdapur, V. R., **3**, 62, 221 ⟨89IJC(B)3⟩
Mamedov, S. A., **5**, 369 ⟨90ZOB1183⟩
Mamer, O. A., **3**, 294 ⟨66JOC2708⟩
Mami, I., **3**, 625 ⟨76JOC2502⟩
Mamishov, A. Kh., **5**, 139 ⟨80MI 503-01⟩
Mammalis, P., **2**, 103 ⟨60JCS225⟩
Mamman, A. S., **5**, 15 ⟨83T1187⟩
Mammarella, R. E., **4**, 648 ⟨78JOM(156)299⟩
Mammi, M., **3**, 255 ⟨58G1226⟩, 382 ⟨65JCS807, 67JCS(B)117⟩, 399 ⟨64G48⟩
Mamolo, M. G., **5**, 769 ⟨89MI 519-01⟩
Mamos, P., **5**, 374, 376, 718 ⟨91ZAAC(603)89⟩
Mamuzić, R. I., **1**, 91 ⟨67T215⟩
Man, E. H., **6**, 233, 235, 236 ⟨59JA3575⟩, 408, 410 ⟨60IS155⟩

Man, M. W. C., **1**, 160 ⟨93JOM(451)C1⟩, 160 ⟨93TL2111⟩
Manabe, H., **6**, 498 ⟨85CL1671⟩
Manabe, K., **4**, 279 ⟨91MI 406-01⟩
Manada, N., **6**, 465 ⟨94EUP581240⟩
Manaeva, T. V., **5**, 652 ⟨90KGS1430⟩
Manandhar, M., **4**, 910 ⟨83S920⟩
Manas, A. R. B., **3**, 283 ⟨87T1847⟩; **6**, 88, 317 ⟨75CC216⟩
Manas, M. M., **2**, 347 ⟨71JCS(B)1330⟩
Manasse, O., **5**, 821 ⟨1893LA(274)95⟩
Manassero, M., **6**, 525 ⟨79JA6110⟩
Man, T. O., **1**, 827 ⟨89JOC3137⟩; **4**, 1083 ⟨84JCS(P1)1561⟩
Mancelle, N., **3**, 43 ⟨74BSF331⟩; **4**, 405, 406 ⟨74BSF331⟩
Mancera, O., **3**, 234 ⟨56JOC1432⟩; **5**, 288 ⟨56JOC520⟩
Manceron, L., **2**, 1022, 1023 ⟨88JPC37⟩
Manchand, P. S., **5**, 40, 107 ⟨71CC667⟩, 345, 346 ⟨88JOC5507⟩
Mancheño, B., **1**, 136 ⟨93T1327⟩, 237 ⟨89TL3837⟩
Manchot, W., **6**, 412, 418 ⟨30CB1221⟩, 524 ⟨36ZAAC(226)385⟩
Manchot, W. J., **6**, 524 ⟨36ZAAC(226)385⟩
Mancinelli, P. A., **4**, 853 ⟨80JOC2597⟩
Mancini, F., **2**, 698 ⟨90CC1680⟩
Mancini, I., **3**, 63 ⟨87HCA1400⟩
Mancini, M. L., **2**, 921 ⟨83TL4295⟩
Mancuso, A. J., **2**, 1055 ⟨81S165⟩; **3**, 6, 118 ⟨78JOC2480⟩, 6, 88, 118 ⟨81S165⟩; **5**, 209 ⟨81S165⟩
Mandai, T., **1**, 613 ⟨83TL4993, 86JOC3896⟩, 613, 700, 1073 ⟨84JA3670⟩, 826 ⟨90TL4041⟩, 962 ⟨93TL2161⟩, 975 ⟨93JA5865, 93JOM(451)15⟩, 976 ⟨91JOM(417)305, 91SL697⟩, 976 ⟨91TL3397, 91TL7683⟩, 976, 1025 ⟨90TL7179⟩; **2**, 678 ⟨83TL4993⟩; **3**, 60 ⟨83TL1171⟩, 78, 271 ⟨85TL2677⟩, 213 ⟨84CL1259⟩, 651 ⟨86TL5109⟩; **4**, 197 ⟨83TL4993⟩, 942 ⟨90T4553⟩; **5**, 346 ⟨91TL7683⟩, 411 ⟨85TL2675⟩; **6**, 126 ⟨85TL2675⟩, 461 ⟨91JOM(417)305⟩
Mandal, A. K., **2**, 32, 36 ⟨85TL3863⟩, 515 ⟨77JOM(135)C10, 78S146⟩, 516 ⟨79TL4777, 80JOC3543⟩, 519 ⟨81T3547⟩, 523 ⟨82JOC5069⟩; **5**, 298 ⟨81S977⟩
Mandal, N. B., **4**, 745 ⟨88JOC5554⟩
Mandal, S. B., **1**, 11, 19 ⟨82JOC4304⟩
Mandal, S. K., **4**, 148 ⟨89OM1114⟩, 148, 150, 581 ⟨90JOM(397)313⟩
Mandanas, B. Y., **1**, 639, 644, 747 ⟨68CC1699⟩
Mandard, X., **1**, 983 ⟨83TL4691⟩
Mandel, B. J., **4**, 412 ⟨74JOC167⟩
Mandel, G. S., **1**, 815 ⟨78JA8170⟩, 831 ⟨85JA3285⟩; **2**, 25 ⟨85JA3285⟩; **5**, 239 ⟨85JA3285⟩
Mandel, H. G., **5**, 744, 745 ⟨54JA3978⟩
Mandel, N., **2**, 855 ⟨71JA3792⟩
Mandel, N. S., **1**, 815 ⟨78JA8170, 85JA2872⟩; **2**, 25 ⟨85JA3285⟩; **5**, 239 ⟨85JA3285⟩
Mander, L. N., **1**, 87 ⟨86NPR35⟩, 769 ⟨88TL3339⟩, 823 ⟨73JOC2915⟩; **2**, 699 ⟨84TL5953⟩; **3**, 193 ⟨80S368⟩
Mandeville, W. H., **1**, 119, 312 ⟨74JOC400⟩
Mandl, J. R., **4**, 586 ⟨78ZN(B)1325⟩
Mandronero, R., **3**, 687 ⟨66AHC(6)95⟩
Mane, R. B., **5**, 367 ⟨89IJC(B)186⟩
Manekar-Tilve, A., **1**, 827 ⟨94CC251⟩
Manescalchi, F., **2**, 19, 25, 31, 35 ⟨76S472⟩, 21, 23 ⟨80CC1278⟩, 25, 31 ⟨83S306⟩, 28, 30 ⟨80S143⟩, 103 ⟨85TL3369⟩; **3**, 20 ⟨78JOC1598⟩
Maness, D. D., **1**, 1110 ⟨70JA5457⟩; **2**, 204 ⟨70JA5457⟩
Manfre, R. J., **2**, 925 ⟨78JOM(159)C13⟩
Manfredi, A., **1**, 1076 ⟨85JOC2372⟩; **2**, 151 ⟨88CC1451⟩; **4**, 540 ⟨87JOM(330)357⟩; **5**, 607 ⟨85T4929⟩
Manfredi, K. P., **3**, 62 ⟨89JOC5186⟩
Manfredini, S., **2**, 617, 624 ⟨90TL5633⟩, 744 ⟨83S902⟩; **3**, 263 ⟨83S902⟩
Manfredy, A., **2**, 1034, 1035 ⟨85JOC2372⟩
Manganey, P., **1**, 670, 967 ⟨90JA8042⟩
Mange, K. C., **2**, 19 ⟨89JFC(43)405⟩
Mangeney, P., **1**, 312 ⟨84PAC91⟩, 496 ⟨92BSF171, 85TL4197⟩, 967 ⟨91T1677⟩; **2**, 86 ⟨89TL2387, 89TL2391⟩, 315 ⟨91SL625⟩, 381 ⟨91TL1171, 92JOC4521⟩; **3**, 42 ⟨91TL1171, 92JOC4563⟩; **4**, 176 ⟨88PAC49⟩, 176 ⟨90TA477⟩, 405, 408 ⟨91SL111⟩; **5**, 152 ⟨88TL4411⟩, 214 ⟨81TL4405⟩
Mangeot, J.-P., **5**, 327, 351 ⟨93JA8898⟩; **6**, 488 ⟨93JA8898⟩
Mangini, A., **4**, 390 ⟨B-85MI 408-01⟩
Manginot, E., **5**, 373 ⟨91S382⟩
Mangold, H. K., **2**, 272 ⟨64JOC3055⟩
Mangold, R., **1**, 399 ⟨48LA(560)116⟩
Mangoni, L., **1**, 587 ⟨81TL3551⟩; **2**, 45 ⟨75G377, 76TL3661⟩
Mangum, M. G., **1**, 550, 1143 ⟨75JA3854⟩, 1071, 1143 ⟨74JA4562, 75JA1459⟩, 1143 ⟨75JA6478⟩
Manhas, M. S., **1**, 725 ⟨69JCS(C)2728⟩; **2**, 683 ⟨75JCS(P1)461⟩; **3**, 527 ⟨67T4769⟩; **4**, 425 ⟨78JHC949⟩; **5**, 273, 285 ⟨81JOC1229⟩, 274, 286 ⟨73JOC3437⟩, 274, 286 ⟨92S358⟩, 294 ⟨74TL2633⟩; **6**, 113 ⟨73JHC791⟩
Mani, R. S., **4**, 382 ⟨89TL5413⟩
Mani, S. R., **3**, 374 ⟨75JOC3857⟩
Manimaran, T., **4**, 447 ⟨88JCS(P1)2137⟩; **5**, 963 ⟨86H(24)2813, 87JCS(P1)2163, 87S907⟩; 964, 1119 ⟨89JCR(S)331⟩
Manisse, N., **1**, 835 ⟨91JOC970⟩; **4**, 900 ⟨91JOC970⟩
Manitto, P., **1**, 77 ⟨87T4481, 88T1299⟩; **3**, 18, 134 ⟨86SC803⟩, 58 ⟨87T4481⟩
Maniwa, K., **4**, 221 ⟨74JA4280⟩
Manjunatha, S. G., **1**, 194 ⟨91CC372⟩; **5**, 611 ⟨90TL1327⟩
Manley, B. C., **2**, 1005 ⟨85JA5313⟩; **4**, 618, 668 ⟨88POL2023⟩; **6**, 192, 398 ⟨88POL2023⟩, 206 ⟨85JA5313⟩
Manley, N. C., **5**, 205 ⟨64JOC2099⟩
Manley, P. W., **5**, 1121 ⟨93JMC2004⟩
Mann, A., **1**, 738 ⟨90AG(E)1454⟩; **5**, 199 ⟨86TL2913⟩
Mann, B. E., **2**, 824 ⟨71JCS(A)2976⟩; **5**, 446 ⟨88JOM(342)C5⟩
Mann, C. D. M., **6**, 340 ⟨70JOM(24)C61⟩
Mann, C. K., **2**, 653 ⟨70JA2042⟩
Mann, E. G., **2**, 508, 865 ⟨72JCS(P1)2548⟩
Mann, F. G., **1**, 41 ⟨52JCS2127⟩, 445 ⟨65JCS4120⟩; **2**, 26 ⟨54JCS2819⟩, 123 ⟨45JCS666⟩, 160 ⟨22JCS1052⟩, 170 ⟨22JCS594⟩, 288 ⟨45JCS37⟩, 437, 821 ⟨52JCS3039⟩, 445 ⟨54JCS2832⟩, 461 ⟨59JCS2835⟩, 488 ⟨45JCS30, 45JCS45⟩, 809, 810, 825 ⟨37JCS527⟩, 810 ⟨32JCS958⟩, 810, 892 ⟨72JCS(P1)2793⟩, 829 ⟨52JCS4453⟩, 872 ⟨B-70MI 217-02⟩, 875 ⟨58JCS1719⟩, 889 ⟨70JCS(C)810⟩; **5**, 26, 94 ⟨45JCS751⟩
Mann, G., **4**, 432 ⟨80ZC437⟩, 880 ⟨85JPR297⟩; **5**, 324 ⟨85S331⟩
Mann, J., **1**, 363, 364 ⟨86T4611⟩, 391 ⟨91COS(3)839⟩; **2**, 15 ⟨87CSR381⟩; **3**, 230 ⟨78CC1006⟩
Mann, J. S., **1**, 745 ⟨75JIC538⟩
Mann, K., **3**, 354 ⟨74S578⟩
Mann, K. R., **3**, 702, 703, 712 ⟨90JOC4950⟩
Mann, W., **3**, 612 ⟨1881CB1645⟩
Manna, S., **1**, 179 ⟨85SC663⟩, 580 ⟨86TL299⟩; **2**, 6 ⟨85SC663⟩
Mannafov, T. G., **4**, 56, 57, 91 ⟨88ZOR1945⟩
Mannen, S., **1**, 318 ⟨73T3497⟩
Mannich, C., **1**, 82 ⟨36CB2106⟩; **3**, 60 ⟨43CB1480⟩; **4**, 404, 405 ⟨36CB2106⟩
Manning, D., **2**, 42 ⟨85AG(E)878⟩

Manning, D. T., **1**, 808 ⟨68JOC4413⟩; **5**, 823 ⟨63JOC1673⟩
Manning, H. W., **1**, 942 ⟨93JOC6857⟩; **4**, 444 ⟨84CJC749⟩
Manning, R. A., **1**, 46 ⟨75JOC3686⟩; **2**, 658 ⟨75JOC3686⟩
Manninga, T. M., **1**, 287 ⟨88JA1602⟩
Manninga, T. N., **1**, 288 ⟨86JA6051⟩
Mannschreck, A., **4**, 806 ⟨73CB1601⟩
Mano, S., **1**, 297 ⟨93JA7033⟩
Manojlovic-Muir, L., **4**, 585, 691 ⟨82CC614⟩, 656 ⟨72CC613⟩, 714 ⟨80CC749⟩; **6**, 721 ⟨71CC400⟩
Manolescu, I., **3**, 539 ⟨88RRC83⟩
Manolopoulo, M., **2**, 647 ⟨62CJC2175⟩
Manoussakis, G., **2**, 504 ⟨68JINC1715⟩; **4**, 116 ⟨68JINC1715⟩
Manriquez, J. M., **2**, 602 ⟨81JA6650⟩; **3**, 598 ⟨86JA56⟩; **5**, 469 ⟨78JA7112, 80JA5393⟩; **6**, 516 ⟨81JA2206⟩
Mansell, H. L., **2**, 346 ⟨91TL6789⟩
Mansell, V. G., **1**, 870 ⟨68JCS(B)132⟩
Mansfield, G. H., **1**, 80 ⟨54JCS3197⟩
Mansfield, J. W., **1**, 869, 871 ⟨67JCS(B)827⟩
Manske, R., **2**, 877 ⟨75T2121⟩
Mansour, A. I., **6**, 173, 381, 391 ⟨85JCS(P2)729⟩
Mansour, E. M. E., **2**, 26 ⟨68CC976⟩
Mansour, E.-S. M. E., **5**, 264 ⟨92JOC6371⟩
Mansour, S. A., **5**, 615 ⟨87SRI525⟩
Mansour, T. S., **2**, 749 ⟨88TL2295⟩; **4**, 229 ⟨92TL4625⟩
Mansukhani, R. I., **5**, 370 ⟨78JHC1169⟩
Mansuy, D., **1**, 63 ⟨76JOM(110)139⟩; **3**, 699, 700 ⟨78TL3027⟩; **5**, 640 ⟨93JMC1157⟩, 1097 ⟨77CC648, 78AG(E)781⟩, 1097 ⟨80PAC681, 81IC3119⟩; **6**, 717 ⟨77CC648⟩
Mantanari, F., **5**, 102 ⟨86JOC567⟩
Mantegani, A., **4**, 416 ⟨64G595⟩
Mantei, R. A., **5**, 236 ⟨91TL6507⟩
Mantell, G. J., **2**, 116 ⟨51JOC524⟩
Manteuffel, R. M., **5**, 858 ⟨34LA(512)97⟩
Manthey, J. W., **1**, 200, 238 ⟨88JOC1475⟩
Manthey, M. K., **2**, 906 ⟨90JOC4581⟩
Manthorpe, P. A., **4**, 176 ⟨80MI 404-01⟩
Mantione, R., **2**, 939 ⟨71JOM(30)295, 71TL591⟩, 939 ⟨71TL593⟩; **5**, 427 ⟨71TL591⟩
Mantlo, N. B., **1**, 455 ⟨84TL4867⟩
Mantovani, A., **5**, 801 ⟨82JOM(236)C37⟩, 801, 802 ⟨81JOM(206)C11⟩
Mantus, E. K., **3**, 636 ⟨93TL1085⟩; **4**, 245 ⟨93TL1085⟩
Mantz, I. B., **1**, 232, 234 ⟨71JA676⟩
Manuel, C. P., **5**, 944 ⟨83JCS(D)2397⟩
Manuel, G., **1**, 656 ⟨84OM1340⟩; **2**, 949 ⟨66BSF327, 90CRV215⟩; **3**, 502 ⟨81T2875⟩
Manuel, T. A., **2**, 989 ⟨59JA3163⟩; **4**, 681 ⟨60JA4749⟩
Manuilova, E. A., **6**, 410 ⟨74URP424809⟩
Manuilov, A. F., **4**, 119 ⟨65JGU2053⟩, 550, 554 ⟨65ZOB1602⟩
Manukina, T. A., **4**, 637 ⟨76MI 414-01⟩, 661 ⟨70ZOB2607⟩
Manukov, E. N., **1**, 626 ⟨88MI 114-01⟩
Manulkin, Z. M., **2**, 882 ⟨51ZOB93⟩
Manwaring, R. S., **5**, 168 ⟨71JCS(C)2944⟩
Manyik, R. M., **1**, 399 ⟨55JOC1119⟩
Manz, G., **3**, 661 ⟨31LA(488)111⟩
Manzer, L. E., **4**, 578, 579 ⟨76IC2567⟩, 677 ⟨77JOM(139)C34⟩; **5**, 940 ⟨72IC503, 73JOM(47)C17⟩; **6**, 8 ⟨93M 601-01⟩
Manzoni, L., **3**, 35 ⟨93JOC3165⟩
Manzoor-i-khuda, M., **3**, 78 ⟨57JCS2767⟩
Manzucchi, A., **2**, 149 ⟨80S563⟩; **3**, 4, 115 ⟨78S534⟩
Mao, C.-L., **2**, 675 ⟨71OS(51)90⟩
Mao, D. T., **1**, 401, 414, 800 ⟨77TL2559⟩, 805 ⟨78CC479⟩; **3**, 22 ⟨77TL2559⟩
Mao, M. K., **5**, 235 ⟨91S920⟩
Mao, S. S. H., **3**, 509 ⟨93JA5336⟩
Maone, T., **6**, 106 ⟨59MI 604-01⟩
Mapelli, C., **5**, 367 ⟨89JOC145⟩
Mapson, L. W., **5**, 98 ⟨55MI 502-01⟩
Maquestiau, A., **3**, 579 ⟨88JCS(P2)1071⟩, 582 ⟨88JA1337⟩; **5**, 773 ⟨87T4185⟩, 1086, 1087, 1092 ⟨83JA7194⟩
Mar, E. K., **2**, 83, 85 ⟨92JOC46⟩; **4**, 389 ⟨91TL5683⟩
Marais, C. F., **5**, 651 ⟨88JOM(355)337⟩
Marais, D., **3**, 72 ⟨83TL5209⟩
Marans, N. S., **5**, 276 ⟨66JOC3829⟩; **6**, 145 ⟨50JA5329⟩
Marascia, F. J., **6**, 229 ⟨55JA1899⟩
Marastani, M., **4**, 422 ⟨85JMC769⟩
Marat, R. K., **4**, 61, 62, 63 ⟨77CJC3031⟩
Marathey, M. G., **2**, 28 ⟨64JOC2025⟩; **3**, 241 ⟨64JOC2025⟩
Maraval, M., **2**, 22, 28 ⟨91SC11⟩
Maraver, J., **1**, 130 ⟨87JCPB765⟩
Marayama, K., **2**, 517 ⟨81JA1969, 81JA3229⟩
Marazza, F., **4**, 649 ⟨80HCA555⟩
Marbet, R., **1**, 156 ⟨57HCA1250⟩, 979 ⟨67HCA1158⟩; **3**, 229, 233 ⟨67HCA1158⟩
Marble, L. K., **1**, 187 ⟨90JOC1984⟩
Marcaccini, S., **3**, 702 ⟨90H(31)1855⟩, 706 ⟨89S641, 90LA935⟩
Marcantoni, E., **2**, 318 ⟨88TL2251⟩, 346 ⟨90JCS(P1)2133, 93CC1373⟩; **3**, 190 ⟨92SL64⟩, 434 ⟨90JOC4456⟩
Marcelo-Curto, M. J., **5**, 212 ⟨91CC1127⟩
Marcewicz-Rojewska, B., **6**, 637 ⟨89MI 620-01⟩
March, J., **1**, 107 ⟨B-92MI 104-01⟩, 352 ⟨B-92MI 108-03⟩, 552 ⟨B-92MI 119-02⟩, 861 ⟨B-92MI 119-04⟩, 872 ⟨B-92MI 119-05⟩, 883 ⟨59JA1617, 63JCE212⟩, 883 ⟨B-92MI 119-06⟩, 884 ⟨B-92MI 119-07, B-92MI 119-08⟩, 884 ⟨B-92MI 119-09⟩, 885 ⟨77JCS(P1)460, B-92MI 119-10⟩, 887 ⟨B-92MI 119-11⟩, 889 ⟨B-92MI 119-12, B-92MI 119-13⟩, 891, 904 ⟨B-92MI 119-15⟩; **2**, 132 ⟨B-92MI 203-01⟩, 608 ⟨B-85MI 212-01⟩, 722 ⟨B-92MI 214-01⟩, 758 ⟨B-85MI 215-01⟩; **3**, 408 ⟨B-92MI 310-01⟩; **4**, 160 ⟨B-92MI 404-01⟩, 214 ⟨B-92MI 404-02⟩, 1072, 1075, 1077, 1080 ⟨B-92MI 424-01⟩; **5**, 294 ⟨81JOC4304⟩
March, J. C., **1**, 32, 34, 36 ⟨B-85MI 102-01⟩
March, L. A., **4**, 856 ⟨85CC618⟩
Marchalin, S., **3**, 305 ⟨93ACS287⟩
Marchand, A., **1**, 656 ⟨84JOM(267)93⟩
Marchand, A. P., **1**, 6 ⟨87JOC4784⟩, 637 ⟨84JOC670⟩; **2**, 32 ⟨69CI(L)200⟩; **4**, 253 ⟨93TL1463⟩, 428 ⟨87JOC4781⟩, 429, 430, 431 ⟨93JOC759⟩; **5**, 112, 114, 116 ⟨66JA3318⟩
Marchand, E., **2**, 744 ⟨82TL2023⟩; **3**, 557, 573 ⟨77TL3353⟩, 567 ⟨75CR(C)51, 85JOC771⟩, 573 ⟨77TL3349⟩, 574 ⟨78TL3719⟩; **4**, 894 ⟨77TL3349⟩; **5**, 728 ⟨82TL2023, 84T1075⟩, 728 ⟨85JOC771⟩; **6**, 571 ⟨90JOC1721⟩, 617 ⟨84T1075, 85JOC771⟩, 617 ⟨92TL6155⟩, 635 ⟨86JOC4043, 89JOC1185⟩
Marchand-Brynaert, J., **2**, 758 ⟨B-76MI 215-01⟩, 759 ⟨74AG272⟩; **3**, 558, 560 ⟨80TL3081⟩, 601 ⟨72JA2870, 74AG(E)267⟩, 601 ⟨B-76MI 317-01⟩; **4**, 790, 791 ⟨B-76MI 418-01⟩, 833 ⟨83T2493⟩, 842 ⟨83TL2563⟩, 891 ⟨B-76MI 420-01⟩, 905 ⟨B-76MI 420-02⟩, 971, 990 ⟨76MI 421-01⟩; **5**, 233 ⟨80TL3085⟩; **6**, 434 ⟨88EJM561⟩
March, P. de, **3**, 315 ⟨91T7713⟩, 320 ⟨82JA4952, 82JA4953⟩
Marchant, N. S., **4**, 675 ⟨89OM1626⟩, 676 ⟨86OM602, 88POL919⟩
Marchelli, R., **2**, 24, 31, 35 ⟨85G29⟩, 28 ⟨84T2035⟩; **3**, 27, 154 ⟨84T2035⟩
Marchenko, A. P., **2**, 407 ⟨89ZOB2135, 89ZOB2136⟩, 407, 408 ⟨90ZOB465, 78ZOB2686⟩, 408 ⟨80ZOB951⟩; **3**, 494

⟨90JGU613⟩; **4**, 559 ⟨88ZOB1665⟩; **5**, 369 ⟨88ZOB706⟩, 379 ⟨85ZOB1080⟩; **6**, 14 ⟨80ZOB1897⟩, 242 ⟨93ZOB1240⟩, 242, 682 ⟨92ZOB948⟩, 267 ⟨90ZOB709⟩, 612, 623 ⟨81ZOR180⟩, 680, 682 ⟨89ZOB1902⟩, 681 ⟨88ZOB1923, 88ZOB482⟩, 690 ⟨89ZOB2133⟩, 696 ⟨87ZOB2141, 87ZOB2794⟩, 696 ⟨88ZOB1461, 88ZOB2164⟩, 696 ⟨89ZOB1904, 89ZOB959⟩, 696 ⟨90ZOB2814⟩
Marchenko, G. A., **6**, 284, 364, 369 ⟨75IZV2838⟩, 364, 368 ⟨77IZV1563⟩
Marchese, G., **1**, 187 ⟨78CC1085⟩, 467 ⟨85JCS(P1)1115⟩, 484 ⟨89TL243⟩, 1025 ⟨79TL3873⟩; **2**, 874 ⟨68MI 217-01⟩, 931 ⟨88TL3705⟩; **3**, 130, 285, 306 ⟨87TL2053⟩, 230 ⟨91CC237⟩, 285 ⟨84TL4805, 85TL3595⟩, 285 ⟨88TL3587⟩; **4**, 773 ⟨87S1034⟩
Marchesini, A., **1**, 1076 ⟨85JOC2372⟩; **2**, 1034, 1035 ⟨85JOC2372⟩; **5**, 354 ⟨87JHC773⟩
Marchetti, F., **1**, 1010 ⟨85T621⟩; **5**, 442 ⟨76CC522⟩
Marchetti, L., **2**, 787 ⟨67AC(R)1266⟩
Marchetti, M., **2**, 553 ⟨94SC591⟩; **3**, 617, 619 ⟨82SC25⟩; **6**, 417 ⟨82SC25⟩
Marchi, M., **5**, 68, 95 ⟨77G499⟩
Marchiori, M. L. P. F. C., **3**, 131 ⟨81JOC3928⟩
Marciniec, B., **2**, 536 ⟨B-92MI 210-01⟩, 938 ⟨91JMOC113, 92JMOC307⟩; **4**, 566 ⟨88SRI695⟩, 568 ⟨88MIP4121⟩
Marcinow, Z., **1**, 87 ⟨92OR(42)1⟩
Marco, G. J., **4**, 737 ⟨60JA1260⟩
Marco, J. A., **4**, 749 ⟨85CC296, 86JCR(M)2843⟩; **6**, 695 ⟨85CC296⟩
Marco, J. L., **5**, 343 ⟨93TL5627⟩
Marco-Contelles, J., **1**, 175 ⟨92TL5261⟩, 175 ⟨93JOC2894⟩; **6**, 473, 544, 550 ⟨93JOC2894⟩
Marcotullio, M. C., **2**, 44 ⟨89TL3175⟩; **3**, 129 ⟨91T7403⟩
Marcuccio, S. M., **1**, 489 ⟨92TL6679⟩
Marcum, V. C., **4**, 63, 69 ⟨67JOC204⟩
Marcus, E., **5**, 217 ⟨67JCED151⟩
Marcus, N. L., **5**, 318 ⟨66JA3051⟩, 1120 ⟨69JOC1808⟩
Marcus, R., **4**, 738 ⟨62JA1312⟩
Marcus, W. Y., **2**, 617 ⟨85JCS(P1)1127⟩
Marcuzzi, F., **1**, 200 ⟨82JOC4577⟩, 318 ⟨75TL2771, 79JOC3022⟩, 1003 ⟨82CC959⟩, 1122 ⟨76JA3295⟩
Marczak, S., **3**, 240 ⟨90TL3433⟩; **4**, 393 ⟨91CC297⟩
Mardamigo, C. L., **3**, 414 ⟨85TL4403⟩
Marder, T. B., **2**, 43 ⟨93CJC930⟩, 527 ⟨92JA9350⟩, 1083 ⟨93JOM(449)27⟩; **4**, 582 ⟨87CC1543⟩, 716 ⟨85JCS(D)2009⟩
Mardones, M. A., **2**, 475 ⟨90AG(E)1150⟩, 511 ⟨91OM1635⟩, 895 ⟨92JOM(439)C33⟩
Marecek, J. F., **2**, 869 ⟨74JA7269⟩; **4**, 493 ⟨75JA38⟩
Maréchal, E., **1**, 462 ⟨82MI 111-01⟩
Mareda, J., **4**, 8 ⟨79TL1913⟩
Marek, I., **1**, 496 ⟨92BSF171⟩, 670, 967 ⟨90JA8042⟩, 836 ⟨91TL5969, 91T1677⟩; **2**, 1006 ⟨92SL633⟩; **4**, 176 ⟨88PAC49⟩
Marekowa, U., **4**, 602, 607, 637, 660 ⟨90ZAAC(583)195⟩
Maren, T. H., **2**, 260 ⟨46JA1864⟩
Marenets, M. S., **5**, 988 ⟨59ZOB2256, 61ZOB1607⟩; **6**, 232 ⟨54JGU885⟩
Mares, F., **3**, 117 ⟨79JOC921⟩
Maresca, L., **5**, 386 ⟨93JA5123⟩
Maretina, I. A., **2**, 1042 ⟨66ZOR1994⟩; **4**, 885 ⟨80ZOB984, 91ZOB1192⟩, 976 ⟨79ZOR1824, 81ZOR1169⟩; **5**, 691 ⟨88JOU1049, 88ZOR1165⟩, 691 ⟨91CHE424, 91KGS532⟩
Marfat, A., **1**, 479, 1107 ⟨78TL2465⟩, 541, 1107 ⟨79JOC3888, 77JA253⟩, 1107 ⟨78TL1363, 78TL2973⟩, 1107 ⟨79JOC1345⟩; **2**, 135 ⟨80JA1436, 80JA3663⟩, 135, 136 ⟨80TL3143⟩; **3**, 217 ⟨78TL1363⟩
Marfella, A., **5**, 597 ⟨83FES101⟩
Margaretha, P., **3**, 413 ⟨83JOC1925⟩
Margaretta, P., **2**, 133 ⟨92HCA2265⟩
Margaryan, A. Kh., **1**, 516 ⟨79T77⟩
Margerum, D. W., **2**, 337 ⟨82IC2545⟩
Margolin, A. L., **5**, 135 ⟨90JA2849⟩
Margolin, Z., **1**, 142, 155 ⟨75JA7006⟩
Margolis, D. S., **2**, 278, 292 ⟨57JCS799⟩; **3**, 391 ⟨57JCS799⟩
Margot, C., **1**, 113 ⟨89JOC2117⟩, 579 ⟨90T2411⟩
Margrave, J. L., **5**, 164 ⟨61JA606⟩, 1096 ⟨83JAP5813530⟩; **6**, 247 ⟨90IC3252⟩
Marguerite, J., **4**, 171 ⟨85OS(63)147⟩
Marhold, A., **6**, 9 ⟨81JFC(18)281⟩, 28 ⟨82S951⟩, 228, 232, 237, 238 ⟨83HOU(E4)625⟩, 228, 437 ⟨83HOU(E4)1203⟩, 610, 614, 727 ⟨83HOU(E4)655⟩
Marhold, J., **2**, 777 ⟨73MI 208-01⟩
Mari, T., **5**, 131 ⟨82CL1891⟩
Maria, P. C., **2**, 10 ⟨78TL4507⟩
Mariano, P. S., **1**, 835 ⟨83JOC4262, 83TL1021⟩, 835 ⟨84JOC2708, 90JA7431⟩, 871 ⟨68JA6091⟩; **3**, 414 ⟨85TL47⟩, 420 ⟨83JA6160, 83T3845⟩, 420 ⟨89TL4185⟩; **4**, 510 ⟨87JA4421⟩, 513 ⟨92JOC6037⟩, 518 ⟨85TL47, 88T7013⟩
Mariano, S.,P., **1**, 835 ⟨89JOC2893⟩
Maricich, T. J., **2**, 298 ⟨63JA1200⟩, 724 ⟨84JOC1931⟩; **4**, 232 ⟨72JA5115⟩
Maricich, T. M., **2**, 364 ⟨73JA5831⟩
Maricq, J., **3**, 62 ⟨58JOC157⟩
Marin, B., **3**, 409 ⟨89SC2825⟩; **6**, 610, 727 ⟨89SC2825⟩
Marin, J. M., **5**, 445 ⟨84JA3214⟩, 800 ⟨89JOM(377)157⟩
Marin, R., **6**, 524 ⟨39ZAAC(240)261⟩
Marinas, J. M., **3**, 671 ⟨84JOC5195⟩
Mar'in, V. P., **5**, 388 ⟨85ZOB1071⟩
Marinelli, E., **1**, 278 ⟨79TL2313⟩
Marinelli, E. R., **1**, 487 ⟨80S945⟩
Marinelli, F., **1**, 546 ⟨85T5121⟩, 1029 ⟨89TL2581⟩; **2**, 983 ⟨92JOC976⟩; **4**, 169 ⟨91T1525⟩
Marinelli, G. P., **3**, 661 ⟨79JOM(173)335⟩
Marinetti, A., **2**, 438 ⟨92S157⟩; **4**, 713 ⟨80JOM(197)335, 82OM225⟩, 713 ⟨85OM1123⟩; **5**, 407 ⟨82OM1488, 84PS(19)311⟩; **6**, 692 ⟨90OM793⟩
Maringgele, W., **2**, 399 ⟨80CB1950⟩, 811 ⟨77JOM(141)249⟩, 1085 ⟨87CB1437⟩; **4**, 145 ⟨71M118⟩, 1015 ⟨82JOM(231)191⟩; **5**, 380 ⟨76ZN(B)1546⟩, 380, 382 ⟨78CB538⟩, 381 ⟨77M1027, 84CB2112⟩, 381, 382 ⟨77M1279⟩, 381, 382 ⟨79M473⟩, 381, 660, 717 ⟨77ZAAC(433)94, 77ZAAC(436)173⟩, 383, 722 ⟨78ZAAC(443)148⟩, 739 ⟨77JOM(141)241⟩, 777, 778 ⟨78ZAAC(445)107⟩, 795, 796 ⟨82JOM(231)191⟩, 863, 864 ⟨71M121⟩; **6**, 184 ⟨88JOM(347)11⟩, 184 ⟨90CC741⟩, 184 ⟨92CB1807⟩, 662 ⟨80ZAAC(468)99, 82JOM(231)191⟩
Marini, F., **1**, 782 ⟨93JOC1349⟩
Marini-Bettòlo, G., **1**, 498 ⟨80T719⟩
Marino, G., **3**, 106 ⟨71CC1441⟩
Marino, J. P., **1**, 14, 226, 227 ⟨87JOC4898⟩, 226 ⟨81JOC5379⟩, 479 ⟨82JA3165, 83JOC4621⟩, 540 ⟨81JOC3696⟩, 600, 983 ⟨92TL4985⟩, 654 ⟨84JOC3671⟩, 811 ⟨73TL3971, 73TL3975⟩, 811 ⟨74JOC3175, 76TL3245⟩, 1136 ⟨76JOC3213⟩; **3**, 378 ⟨74JOC3175⟩; **4**, 240 ⟨87TL4925⟩, 346 ⟨75TL4531, 81JOC1828⟩; **6**, 111 ⟨89JOC1782⟩
Marino, Y. L., **3**, 577 ⟨70T1779⟩
Marinovic, N., **1**, 604 ⟨76JA1612⟩; **2**, 660 ⟨78TL2545⟩
Marinovic, N. N., **1**, 338, 342, 1127 ⟨83TL1871⟩
Mariorana, S., **5**, 946 ⟨91OM807⟩
Maris, P. A., **3**, 206 ⟨81JOC5348⟩, 417 ⟨80JOC4952⟩
Marita, T., **3**, 411 ⟨80BCJ3696⟩
Marji, D., **3**, 419 ⟨86T4517⟩

Mark, H. W., **1**, 905 ⟨78JA2501⟩; **2**, 214 ⟨88USP4721813⟩
Mark, I., **2**, 47 ⟨88JA1968⟩, 48 ⟨89JOC2263⟩
Markandu, J., **1**, 839 ⟨91TL279⟩; **2**, 352 ⟨91T4007⟩
Markarian, G., **1**, 959 ⟨75BSF1407⟩
Mark, V., **2**, 411 ⟨73OSC(5)602⟩; **4**, 476 ⟨66OS(46)31⟩
Markby, R., **2**, 670 ⟨58JA1009⟩
Märkel, R., **5**, 660 ⟨84JOM(267)277⟩
Marker, C., **2**, 150 ⟨1865LA(136)891⟩
Market, J., **1**, 826 ⟨71HCA1589⟩
Markey, F. X., **2**, 123 ⟨55JA568⟩
Markezich, R. L., **1**, 1122 ⟨73JA4416⟩
Markham, J. R. E., **2**, 822 ⟨74JOC2296⟩
Markham, R. T., **2**, 434 ⟨73JINC2659⟩, 448 ⟨74JINC503⟩
Markies, P. R., **6**, 406 ⟨91AOC(32)147⟩
Markiewicz, W., **2**, 282 ⟨81JOC2605⟩
Markiewicz, W. T., **2**, 540 ⟨80TL4523⟩
Märkl, G., **1**, 157 ⟨63AG(E)153⟩, 157 ⟨63ZN(B)84⟩, 731 ⟨73S295⟩, 1012 ⟨88TL3651⟩, 1079 ⟨61CB3005⟩; **2**, 439 ⟨85CB2365⟩, 853 ⟨68TL1755⟩, 865 ⟨88AG(E)709⟩, 878 ⟨90TL6331⟩, 879 ⟨73AG(E)931, 74AG(E)668⟩, 879 ⟨76TL3125⟩, 880 ⟨88TL535, 88TL539⟩, 881, 893 ⟨83JOM(248)269⟩, 888 ⟨72AG(E)441⟩; **3**, 499, 649 ⟨61CB2996⟩, 605, 607 ⟨88AG(E)1360⟩, 606 ⟨89TL3939⟩, 606 ⟨90TL4429, 86AG(E)1003⟩, 607 ⟨92TL1981, 93TL2915⟩, 608 ⟨88TL2935, 88TL463⟩, 609 ⟨90TL6331⟩, 690 ⟨86AG(E)264⟩; **3**, 68, 494 ⟨62CB3003⟩, 445 ⟨61TL811, 73S295⟩; **4**, 1026, 1030 ⟨75TL3171⟩, 1027, 1030 ⟨85CB2365⟩, 1038 ⟨76TL2599⟩, 1047, 1059 ⟨86AG1020⟩ 134 ⟨74TL303, 83JOM(249)335⟩, 456 ⟨80TL3467⟩, 456, 482 ⟨81TL229, 81TL1105⟩, 484 ⟨90TL6999⟩, 490 ⟨61TL809⟩, 502 ⟨86TL2957⟩, 549 ⟨87AG(E)236, 87TL1093⟩, 549 ⟨88AG(E)389⟩, 565 ⟨88AG(E)709⟩, 567 ⟨87TL2693⟩, 594 ⟨90TL6331⟩, 627 ⟨80TL1405⟩, 628, 639 ⟨86TL4015⟩, 850 ⟨67TL3201⟩; **5**, 882 ⟨76TL3489⟩, 895, 896 ⟨61CB2996⟩, 904, 905 ⟨61TL807⟩, 924 ⟨83TL1477⟩, 1155, 1157 ⟨85TL5507⟩, 1155, 1157 ⟨86TL171⟩, 1160 ⟨86AG(E)264⟩; **6**, 129 ⟨89TL4501⟩, 154 ⟨88AG(E)389⟩, 154 ⟨88AG(E)709⟩, 161 ⟨86TL2957⟩, 374 ⟨90TL6331⟩, 659 ⟨62ZN(B)782⟩
Markl, G. Z., **3**, 493 ⟨63NAT1136⟩
Märkl, R., **1**, 1111, 1113 ⟨78LA1894⟩
Märkl, V. G., **4**, 455 ⟨80TL1845⟩
Markley, K. S., **5**, 28 ⟨B-60MI 502-01⟩
Markley, L. D., **2**, 742 ⟨70JOC3275⟩
Markmann, J., **6**, 75 ⟨88T5095⟩
Marko, D. E., **1**, 14 ⟨85JA7978⟩
Marko, I., **2**, 88 ⟨91COS(3)913⟩; **3**, 602 ⟨85JA2192⟩
Markó, I. E., **1**, 233, 398, 407 ⟨91COS(3)913⟩, 821, 823 ⟨91COC(3)913⟩; **2**, 24 ⟨91SL733, 91TL1831⟩, 598 ⟨90JOC3368⟩; **3**, 44 ⟨92SL275⟩, 130 ⟨90JOC3368⟩, 147 ⟨92TL1763⟩, 294 ⟨90SL345⟩; **5**, 18, 276, 293, 299 ⟨90TL7237⟩
Markó, L., **2**, 438 ⟨83JOM(253)249⟩, 439 ⟨79JOC3095⟩; **4**, 455 ⟨80JOM(193)47⟩, 688 ⟨81JOM(206)119, 93JOM(444)203⟩; **6**, 352 ⟨68JOM(11)207⟩
Markoupoulou, O., **2**, 145, 156 ⟨93JCS(P2)2161⟩
Markova, E. I., **6**, 565 ⟨86MI 617-03⟩
Markova, S. A., **4**, 698 ⟨88IZV1448⟩, 1068 ⟨88ZOB1567⟩
Markova, V. U., **2**, 900 ⟨65ZOB1669⟩
Markova, Yu. V., **5**, 827 ⟨67ZOR1207⟩
Markovac, A., **3**, 108 ⟨70JOC841⟩
Markov, P., **2**, 226 ⟨77JPR1036⟩
Markov, V. I., **2**, 756 ⟨88JOU2222⟩; **6**, 608 ⟨73ZOR1759⟩
Markovich, E. A., **5**, 820 ⟨60ZOB1937⟩
Markovki, L. N., **5**, 888 ⟨84PS(19)189⟩
Markovnikoff, W., **2**, 87 ⟨01CB3246⟩
Markovski, N., **5**, 349 ⟨77ZOR2055⟩
Markovskii, L. N., **2**, 183 ⟨75S801⟩, 238 ⟨83ZOR343⟩, 239 ⟨90ZOR2083⟩, 399 ⟨77ZOR727⟩, 399, 400 ⟨67ZOR1282⟩, 399, 400 ⟨72ZOR2057⟩, 406 ⟨79ZOB1908⟩, 407 ⟨67ZOB852, 80ZOB2621⟩, 408 ⟨81ZOB331⟩, 438 ⟨90ZOB1243⟩, 866 ⟨85ZOB2141⟩; **3**, 363 ⟨72ZOR1822⟩, 374 ⟨78ZOR1659⟩, 385 ⟨91ZOR408⟩, 602 ⟨86JGU221⟩; **4**, 10 ⟨73S787⟩, 110 ⟨82JOU2240⟩, 111 ⟨89T6019⟩, 121, 122 ⟨80JGU802⟩, 122 ⟨80S823⟩, 326 ⟨78ZOR1659⟩, 800 ⟨76JOU782⟩, 1011 ⟨80ZOB343⟩; **5**, 5 ⟨75S801⟩, 349 ⟨B-71MI 507-02⟩, 863 ⟨80ZOB2413⟩, 876 ⟨89T6019⟩, 1032 ⟨72ZOR1822⟩; **6**, 621 ⟨73ZOR633, 74ZOR1000⟩, 655, 685 ⟨91ZOB401⟩, 678 ⟨86ZOB253⟩, 678 ⟨89T6019⟩, 678 ⟨B-88MI 622-01⟩, 680 ⟨85ZOB1188⟩, 684 ⟨84ZOB965, 82ZOB1925⟩, 685 ⟨83ZOB1672, 85ZOB1437⟩, 685, 686 ⟨85ZOB221⟩, 685, 686, 689 ⟨87ZOB901⟩, 687 ⟨84ZOB715, 84PS(19)189⟩, 689 ⟨84ZOB2800, 87ZOB1433⟩, 690 ⟨89ZOB2133⟩, 692, 693 ⟨92MI 622-01, 83ZOB693⟩, 693 ⟨91ZN(B)1539⟩, 694 ⟨89ZOB2131⟩, 685,686 ⟨90ZOB2238⟩
Markovskii, V. I., **6**, 284, 364, 369 ⟨75IZV2838⟩, 364, 368 ⟨77IZV1563⟩
Markowitz, M., **3**, 13, 83 ⟨80TL813⟩
Markowska, A., **2**, 465 ⟨60RZC1675⟩
Markowski, V., **3**, 315 ⟨71CC1192, 76TL4643⟩, 315 ⟨77CC439, 77CC440⟩
Markowskii, L., **6**, 660 ⟨91TL2775⟩
Marks, B. S., **2**, 388 ⟨55JA4100⟩
Marks, J., **3**, 179 ⟨80JOC2307⟩
Marks, M. J., **3**, 218, 276, 408 ⟨82JOC52⟩, 418 ⟨81JOC5405⟩; **5**, 796 ⟨82JOC52⟩
Marks, M. W., **4**, 696 ⟨82JA2072⟩
Marks, T. J., **2**, 43 ⟨92JA9220⟩, 564, 602 ⟨82COMC-I(3)173, 74JOM(67)61⟩, 602 ⟨76ACR223, 81JA6650⟩, 602 ⟨82JA1860, 83JA6824⟩, 602 ⟨84JA6841, 86JA40⟩, 969 ⟨85JA8103, 85JA8111⟩; **3**, 598 ⟨86JA56⟩; **4**, 589, 656, 658 ⟨86JA425⟩, 656 ⟨88OM1828⟩, 656, 658 ⟨86JA40, 82JA7357⟩, 658 ⟨86JA6805⟩, 683 ⟨69JA6598⟩, 703 ⟨74JA7586⟩; **5**, 469 ⟨78JA7112, 80JA5393⟩; **6**, 194 ⟨85JA8103⟩, 195 ⟨85JA8091⟩, 516 ⟨81IC3734⟩
Marks, V., **4**, 205 ⟨89JOC4866⟩
Markstein, J., **2**, 615 ⟨71JOC1024⟩
Marktscheffel, F., **2**, 442, 444 ⟨60JOC1996⟩
Markusch, P., **3**, 473, 479, 480 ⟨69LA(722)45, 66TL6199⟩, 480 ⟨69AG577, 69AG(E)612⟩; **5**, 868 ⟨66TL6199⟩; **6**, 664 ⟨71LA(753)143⟩
Markwell, R. E., **2**, 15 ⟨76JA3034, 76JA3036⟩, 15 ⟨B-80MI 218-01⟩
Marley, H., **2**, 906 ⟨88S950⟩; **6**, 443 ⟨89JCS(P1)1727⟩
Marlin, J. E., **1**, 832 ⟨87JOC5742, 89TL2469⟩, 832 ⟨91JOC1003⟩
Marlowe, C. K., **2**, 133 ⟨91JA3177⟩; **4**, 1014 ⟨81JA2807⟩
Marman, J. W., **2**, 676, 970, 971 ⟨67JOM(9)285⟩
Marman, T. H., **1**, 68 ⟨84JOC2838⟩
Marmet, D., **3**, 695, 708 ⟨79TL163⟩; **5**, 747 ⟨80TL1459⟩
Marmo, E., **5**, 597 ⟨83FES101, 85FES120⟩
Marmon, R. J., **1**, 403 ⟨93JCS(P1)681⟩, 950 ⟨92TL4993⟩; **2**, 358 ⟨94T1275, 94T1295⟩
Marmor, R. S., **2**, 658 ⟨72JOC2901⟩; **3**, 480, 482 ⟨70TL2493⟩, 480, 482 ⟨71JOC1379⟩, 489 ⟨71JOC128⟩; **4**, 489 ⟨71JOC1379⟩, 563, 740, 741, 801 ⟨73JOM(59)237⟩; **5**, 868 ⟨67JA4811, 70TL2493⟩, 868 ⟨71JOC1379⟩, 870 ⟨71JOC128⟩; **6**, 266, 288 ⟨73JOM(59)237⟩, 371 ⟨73JOM(59)231⟩, 670 ⟨71JOC1379⟩
Marín, J. M., **4**, 583 ⟨84OM1438⟩
Marnoki, K.,**3**, 437 ⟨83TL943⟩
Marnung, T., **4**, 973 ⟨75JST(24)373⟩

Marolewski, A. E., **1**, 209 ⟨93JOC7304⟩
Marolt, P., **2**, 1008 ⟨80S227⟩
Marongiu, E., **6**, 321 ⟨72G558⟩
Maronglu, G., **2**, 386 ⟨81JCS(P2)1449⟩
Maroshin, Yu. V., **2**, 1090 ⟨63IZV97⟩
Marosvari, A., **2**, 444 ⟨B-84MI 213-01⟩
Marotta, E., **2**, 391, 392 ⟨90OPP707⟩, 395 ⟨86S237⟩; **3**, 46, 191 ⟨90OPP707⟩, 47, 192 ⟨86S237⟩
Maroy, K., **2**, 289 ⟨63ACS1806, 66ACS123⟩
Marples, B. A., **1**, 929 ⟨89TL3343, 92T6883⟩; **6**, 544, 557 ⟨92T6883⟩
Marples, P. K., **4**, 545 ⟨76JOM(117)159⟩
Marquais, S., **1**, 125, 465 ⟨93SL45⟩
Marquarding, D., **4**, 493 ⟨76LA36⟩
Marquardt, A., **2**, 490 ⟨1887CB1516, 1888CB2035⟩, 886 ⟨1889LA(251)323⟩
Marquardt, D. J., **1**, 950 ⟨90T2317, 90T2329⟩, 950 ⟨90T2345⟩
Marquardt, F.-H., **5**, 1029 ⟨66HCA1716⟩
Marquardt, G., **6**, 383 ⟨73AG(E)654, 74ZAAC(404)1⟩
Marques, E. C., **2**, 439 ⟨85JA7423⟩; **4**, 550 ⟨85JA7423⟩
Marques, M. M., **5**, 212 ⟨82TL1391⟩
Marquestiau, A., **3**, 681 ⟨85AG(E)56⟩
Marquet, A., **1**, 143 ⟨78JA1558⟩; **2**, 24 ⟨59TL24⟩, 28 ⟨88OSC(6)175, 88OSC(6)193⟩, 29 ⟨69BSF4169⟩, 118, 124 ⟨90BSF758⟩, 229 ⟨87JOC3154⟩; **3**, 208 ⟨62BSF90⟩
Marquet, B., **2**, 16 ⟨86BSF955, 87TL2359⟩; **3**, 149 ⟨87TL2359⟩
Marquez, A., **5**, 759 ⟨82MI 519-01⟩
Marquez, V. A., **4**, 420 ⟨72JHC1145⟩
Marquis, D. M., **5**, 473 ⟨63USP3097236⟩; **6**, 528 ⟨60USP2962529⟩
Marquis, E. T., **1**, 884 ⟨65JOC181⟩; **4**, 1078 ⟨87JOC3243⟩
Marquis Jr, R. W., **5**, 331 ⟨91JA5085⟩
Marra, A., **4**, 235 ⟨89CAR(188)81⟩; **6**, 549 ⟨91T5149⟩
Marr, D. H., **3**, 618 ⟨72CJC3857⟩
Marr, G., **1**, 155 ⟨73JOM(47)417⟩
Marraud, M., **6**, 488 ⟨93JA8898⟩
Marre, M. R., **4**, 478 ⟨81PS(11)87⟩
Marrero, J. J., **1**, 931 ⟨86TL383⟩
Marrero, R., **3**, 31, 54 ⟨83T861⟩
Marriott, J. G., **5**, 306 ⟨87JMC498⟩
Marron, B. E., **5**, 513 ⟨90JA3040⟩
Marsaioli, A. J., **5**, 1044 ⟨93TL6717⟩
Marsais, F., **2**, 942 ⟨83T2009⟩; **3**, 310 ⟨81JOM(216)139⟩
Marsch, M., **2**, 997, 999 ⟨92CB2265, 91AG(E)321⟩, 1000 ⟨85AG(E)573, 86AG(E)1011⟩, 1000 ⟨88AG(E)846⟩, 1001 ⟨90CB1853⟩, 1002, 1003 ⟨87AG(E)78⟩, 144 ⟨93CB1887⟩; **3**, 566 ⟨86AG(E)373, 89AG(E)1392⟩, 704 ⟨91ZOR408⟩; **4**, 646 ⟨91CB543⟩, 816 ⟨93AG(E)1032⟩
Marschall, R., **6**, 373 ⟨92ZAAC(617)53⟩, 721 ⟨90CB751⟩
Marschall-Weyerstahl, H., **5**, 184 ⟨85LA1935⟩
Marschner, C., **3**, 213 ⟨89CB2131, 90CB107⟩
Marschner, F., **1**, 550 ⟨77CB154⟩; **3**, 349 ⟨75CB673⟩
Marschoff, C. M., **2**, 1008 ⟨87CJC2000⟩
Marsden, E., **1**, 442 ⟨40JCS207⟩
Marsden, J. C., **2**, 223 ⟨64JCS3832⟩; **4**, 218 ⟨64JCS3832⟩
Marsden, K., **4**, 150 ⟨79JA503⟩; **6**, 717, 718 ⟨80JA1206⟩, 722 ⟨88JOM(338)393⟩
Marsel, C. J., **3**, 674 ⟨61USP3006948⟩
Marsella, J. A., **1**, 570 ⟨92JOC2856⟩
Marsella, M., **5**, 135 ⟨90JCS(P1)2559⟩
Marsh, B. K., **1**, 248 ⟨87ACR65⟩
Marsh, C. R., **1**, 1005 ⟨⟨67CC1259, 72JCS(P1)639⟩
Marsh, F. D., **2**, 619 ⟨82JA4680⟩; **3**, 476 ⟨67JA4760⟩; **5**, 1119 ⟨65JA2296, 67JA4760⟩; **6**, 28 ⟨78USP4098831, B-82MI 601-02⟩
Marsh, J. E.,**6**, 207 ⟨27JCS2658⟩
Marshalkina, T. I., **6**, 475 ⟨85ZOR1670⟩
Marshalkin, V. N., **6**, 475 ⟨85ZOR1670⟩
Marshall, A. G., **6**, 345 ⟨84CC392⟩
Marshall, C. W., **3**, 227 ⟨57JA6308⟩
Marshall, D., **3**, 63 ⟨56JCS4082⟩
Marshall, D. R., **1**, 612 ⟨75CC940⟩; **2**, 1007 ⟨91JCS(P2)1563⟩
Marshall, G. L., **4**, 288 ⟨84JCS(P2)429⟩
Marshall, J. A., **1**, 24 ⟨86JOC858⟩, 82 ⟨63JOC421⟩, 196, 233, 396, 404 ⟨91COS(3)975⟩, 226 ⟨89CRV1503⟩, 226, 227 ⟨89T391⟩, 226, 227, 751 ⟨88JOC4274⟩, 374 ⟨84JOC5277⟩, 505 ⟨84JOC747⟩, 552 ⟨89JOC5854⟩, 633, 1053 ⟨84JA6006⟩, 660 ⟨70TL3861⟩, 690 ⟨84JA723⟩, 752, 816, 818 ⟨87JOC3883⟩, 816 ⟨84JOC1707, 87TL723⟩, 816, 817 ⟨90JOC1421⟩, 816, 818 ⟨86JOC4316⟩, 816, 818 ⟨87JOC3860⟩, 816, 818 ⟨87TL3323, 88JA2925⟩, 818 ⟨88JOC4108, 88TL3547⟩, 818 ⟨88TL4913, 90JOC227⟩, 957 ⟨90JOC3450⟩, 966 ⟨93JOC7180⟩, 967 ⟨86TL4845⟩, 1053 ⟨87TL527⟩, 1057 ⟨92JOC1242, 92JOC3387⟩; **2**, 9 ⟨87TL723⟩, 79 ⟨89TL2183⟩, 86 ⟨88JOC4274, 90JOC1540⟩, 88 ⟨91COS(3)975⟩, 118, 124 ⟨93SL209⟩; **3**, 222 ⟨87TL5081, 90JOC227⟩, 227, 232 ⟨83JA5679⟩, 253 ⟨62JOC1615⟩, 623, 633 ⟨75JOC1162⟩; **4**, 203 ⟨77JOC1794⟩, 240 ⟨82JOC693⟩, 386 ⟨88TL1657, 89T1043⟩, 386 ⟨89TL2183⟩, 846 ⟨71TL871⟩; **5**, 207 ⟨86TL5197⟩, 467 ⟨88TL1657, 92SL537⟩
Marshall, J. L., **1**, 430 ⟨68OS(48)68⟩; **4**, 24 ⟨83JA3411⟩; **5**, 122 ⟨70TL4011⟩; **6**, 120 ⟨78S309⟩
Marshall, J. R., **3**, 612 ⟨31IEC352⟩
Marshall, M. D., **2**, 431 ⟨62USP3031509⟩
Marshall, W. S., **5**, 276, 298 ⟨61JA1492⟩
Marshallsay, G. J., **6**, 698 ⟨93PS(74)279⟩
Marsh, R., **6**, 30 ⟨84JCS(P2)1239⟩
Marshman, R. W., **1**, 149 ⟨92JA8483⟩; **5**, 456, 801 ⟨90JA8369⟩
Marsi, K. L., **2**, 441 ⟨69JA4724, 74JOC265⟩, 830 ⟨69JA5243⟩, 848 ⟨56JA3063⟩
Marsi, M., **1**, 481 ⟨84JA7264⟩, 579 ⟨82TL631⟩; **2**, 679 ⟨82TL631⟩; **3**, 138 ⟨84JA7264⟩
Marsich, N., **4**, 696 ⟨79JOM(174)121⟩
Marsman, J. W., **4**, 708 ⟨73JOM(55)419⟩
Marsmann, H., **6**, 534 ⟨72CB820⟩
Marsmann, H. C., **2**, 493, 495, 496 ⟨70JA3969⟩, 507 ⟨69JA4986⟩; **4**, 802 ⟨78ZN(B)1422⟩
Marson, C. M., **1**, 30 ⟨84AG(E)420⟩; **2**, 7 ⟨87CS477⟩, 11 ⟨84AG(E)420⟩, 202 ⟨93CC1195⟩; **3**, 68, 69 ⟨88CB999⟩; **4**, 323 ⟨85JOC1351⟩; **5**, 944 ⟨94TL6717⟩, 955 ⟨95JOM(494)C12⟩
Marsura, A., **5**, 749 ⟨87MI 519-01⟩
Marszak, I., **1**, 506 ⟨49CR(228)1501, 50BSF1305⟩
Marszak-Fleury, A., **1**, 506 ⟨49CR(228)1501, 50BSF1305⟩
Martí, J. M., **5**, 386 ⟨92JCS(D)53⟩
Martel, A., **5**, 236 ⟨88CJC1537⟩
Martel, B., **5**, 536 ⟨66MI 512-01⟩; **6**, 194 ⟨72JOM(40)C53⟩
Martel, J., **1**, 503 ⟨65JOC1752⟩; **4**, 757 ⟨81EUP30887⟩; **5**, 145 ⟨67BSF985⟩
Martel, R., **5**, 588 ⟨93BMC619⟩
Martell, A. E., **4**, 473 ⟨65JA2567⟩; **5**, 115 ⟨73OSC(5)617⟩
Martelli, G., **2**, 103 ⟨85TL3369⟩, 316 ⟨90JOC4199⟩, 544 ⟨90PAC595⟩; **3**, 253 ⟨83JOC123⟩, 440, 456 ⟨87TL5369⟩, 456, 457, 458 ⟨90PAC605⟩, 456, 458 ⟨86TL1695⟩, 457 ⟨91SL243, 90JOC4199⟩; **5**, 287 ⟨86TL1695, 87TL5369⟩, 377 ⟨91SL243⟩
Martelli, S., **4**, 206 ⟨81JMC1181⟩
Marten, D. F., **1**, 482 ⟨74TL925⟩; **5**, 944 ⟨80CC341⟩
Martens, C., **5**, 638 ⟨80JOC4366⟩
Martens, D., **1**, 886 ⟨78CB2480⟩
Martens, H., **3**, 238 ⟨70TL1821, 73T4241⟩, 238 ⟨75T177⟩
Martens, H. F., **2**, 485, 486, 487, 505 ⟨73JOM(51)223⟩, 491, 505 ⟨77JOM(136)173⟩; **4**, 699 ⟨76JOM(117)329⟩
Martens, J., **2**, 64, 528 ⟨92TA1475⟩; **4**, 100 ⟨87CZ149⟩; **5**, 238 ⟨77AG(E)318⟩; **6**, 445 ⟨86JOC3494⟩
Marti, R., **2**, 74 ⟨92JA2321⟩
Martigny, P., **1**, 615 ⟨78MI 114-01⟩

Martin, A. A., **5**, 583 ⟨87ZC90⟩, 588, 632 ⟨85ZC219, 84ZC438⟩; **6**, 559 ⟨81ZC438, 85ZC219⟩, 560 ⟨90ZC90⟩
Martin, A., **3**, 126 ⟨83T345⟩
Martin, A. M., **1**, 193 ⟨89JOC2759⟩
Martin, A. R., **1**, 489 ⟨93ACS221⟩; **4**, 223, 225 ⟨74JOC2010⟩
Martin, B., **1**, 870 ⟨62JA4824⟩; **5**, 615 ⟨93SC2259⟩
Martin, C. A., **2**, 311 ⟨82TL4651⟩
Martin, C. W., **3**, 323 ⟨78JOC1071, 83JOC1898⟩
Martin, D., **3**, 614, 633 ⟨65CB3662⟩; **4**, 74 ⟨70JPR683⟩, 889, 972 ⟨66CB2302⟩; **5**, 1004 ⟨67AG(E)168⟩, 1013, 1101 ⟨64TL2829⟩, 1023 ⟨65CB2425⟩, 1100 ⟨67ZC123, 64AG(E)311⟩, 1103 ⟨79GEP142333⟩; **6**, 300 ⟨65CB3286⟩, 533 ⟨65CB2063⟩, 534 ⟨65CB2059⟩, 624 ⟨65CB3662, 73JPR289⟩
Martin, D. G., **1**, 241 ⟨71TL3791⟩; **2**, 21 ⟨63OSC(4)162⟩; **3**, 208 ⟨63OSC(4)162⟩
Martin, D. J., **1**, 808 ⟨65JA4501⟩; **2**, 123 ⟨68JOC1275⟩
Martin, D. P., **5**, 235 ⟨92JA7287⟩
Martin, D. R., **2**, 434 ⟨73JINC2659⟩, 448 ⟨74JINC503⟩, 841 ⟨77IS183⟩; **4**, 552 ⟨86ICA(115)L29⟩; **5**, 1130 ⟨73IC241⟩; **6**, 492, 515 ⟨81JINC457⟩
Martin, E. L., **1**, 18 ⟨42OR(1)155⟩; **5**, 199 ⟨43OSC(2)560⟩
Martin, F. T., **2**, 168 ⟨50JOC795⟩; **4**, 71 ⟨50JOC795⟩
Martin, G., **1**, 469 ⟨59AC(P)541⟩; **2**, 1000 ⟨83JHC399⟩; **3**, 342 ⟨83PS(17)47⟩
Martin, H., **1**, 665 ⟨55AG541⟩, 666 ⟨60LA(629)14⟩, 1064 ⟨87CB187⟩; **2**, 596 ⟨54LA(589)91⟩; **4**, 698 ⟨85ZN(B)182⟩, 699 ⟨60LA(629)14⟩
Martin, H.-D., **1**, 708 ⟨81AG(E)677⟩
Martin, J., **2**, 852 ⟨76JPC2417⟩; **4**, 586 ⟨87JOM(331)409⟩; **5**, 70 ⟨74JOC1676⟩
Martin, J.-A. F., **3**, 413 ⟨86JCR(S)46⟩
Martin, J. C., **1**, 576 ⟨72JA5003, 77OS(57)22⟩, 582 ⟨90OSC(7)506⟩, 894 ⟨70JA1660⟩, 922 ⟨67JA2054⟩, 923 ⟨B-73MI 119-11⟩, 939 ⟨85SC401⟩, 1089 ⟨83JA6155⟩; **2**, 401 ⟨75JA583, 75JA6137⟩, 547 ⟨79JA1591⟩, 942 ⟨83JOC4156⟩; **3**, 6, 119 ⟨79JA5294⟩, 6, 119 ⟨83JOC4155⟩, 6, 119 ⟨91JA7277⟩, 269 ⟨61JOC4775⟩, 282, 301 ⟨79JOC313⟩; **4**, 438 ⟨61JOC1822⟩, 836, 887 ⟨71JOC2205⟩, 881 ⟨66JOC2966, 65TL3589⟩, 887 ⟨71JOC2211⟩, 979 ⟨64JOC2513, 66USP3227714⟩; **5**, 204 ⟨74JA3319, 91JA7277⟩, 481 ⟨81JOC3911⟩, 689 ⟨73JA2017⟩
Martin, J. F., **3**, 418 ⟨87OPP181⟩
Martin, J. G., **2**, 351 ⟨89JOC1221⟩; **5**, 214 ⟨88JMC1960⟩
Martin, J. L., **2**, 18 ⟨83JOC3195⟩
Martin, J. O., **1**, 738, 741 ⟨65CI(L)1726⟩
Martin, J. W. L., **1**, 53 ⟨88JA4346⟩; **2**, 438 ⟨86IC3396⟩
Martin, K., **3**, 293 ⟨89CC1355⟩; **5**, 1023 ⟨38BRP495510⟩
Martin, K. E., **3**, 535 ⟨29JA3614⟩
Martin, K. R., **2**, 841 ⟨65JOC97⟩
Martin, K. V., **3**, 351 ⟨64JCS2944⟩; **5**, 473, 479 ⟨62USP3048629⟩, 473, 479 ⟨64JCS2944⟩
Martin, L. D., **1**, 1092 ⟨82JOC3630⟩
Martin, M., **1**, 769 ⟨82PAC1915⟩; **3**, 480 ⟨71LA(748)207⟩; **5**, 871, 874 ⟨71LA(748)101⟩; **6**, 670 ⟨71LA(748)207⟩
Martin, M. B., **5**, 214 ⟨88JMC1960⟩
Martin, M. L., **5**, 577, 608, 609 ⟨79T1087⟩
Martin, M. R., **3**, 68 ⟨77S642⟩, 658 ⟨88S136⟩
Martin, M. V., **3**, 68 ⟨77S642⟩
Martin, O. R., **5**, 895, 896 ⟨79HCA1401⟩
Martin, P., **4**, 12 ⟨85T4057⟩, 14 ⟨80HCA1947⟩
Martin, P.,**4**, 830 ⟨78HCA1784⟩
Martin, P. A., **5**, 322 ⟨94TL5157⟩
Martin, R., **2**, 439 ⟨91ZAAC(592)121⟩, 656 ⟨89S25⟩, 657 ⟨92OPP369⟩; **3**, 297 ⟨63CR(256)5594⟩, 482 ⟨85T819⟩; **4**, 979 ⟨77HCA1781⟩; **5**, 325 ⟨87JCS(P1)1635⟩
Martin, R. B., **2**, 773 ⟨B-68MI 215-04⟩
Martin, R. H., **2**, 1006 ⟨65JA2490⟩; **5**, 15 ⟨51BSB54⟩
Martin, R. L., **3**, 350 ⟨72AJC257⟩, 397 ⟨69AJC891⟩
Martin, S., **1**, 1017 ⟨82TL4329⟩; **2**, 1036, 1037 ⟨90CB417⟩; **4**, 806, 819 ⟨84JOM(264)155⟩; **6**, 265 ⟨92ZN(B)321⟩
Martin, S. F., **1**, 16 ⟨87JOC3706⟩, 815 ⟨82JOC3190⟩, 831 ⟨80T419⟩, 860 ⟨74JA8095⟩; **2**, 330 ⟨89JOC265⟩, 352 ⟨92TL3583⟩, 816 ⟨80JA5866⟩; **3**, 60 ⟨77TL3875⟩, 114 ⟨78TL1943⟩, 220 ⟨78JOC4673⟩, 222 ⟨87JOC5528, 89JA7634⟩, 442 ⟨84JOC2512⟩; **4**, 99, 465 ⟨87OS(65)119⟩, 924 ⟨82S394⟩; **5**, 28 ⟨79S633⟩, 769, 849 ⟨72JOC3958⟩; **6**, 58, 542 ⟨92TL1839⟩, 472 ⟨92JOC2523⟩
Martin, S. J., **1**, 836 ⟨93JCS(P1)1275⟩, 1031 ⟨84JA4263⟩; **3**, 166 ⟨82JA3228⟩; **4**, 129 ⟨85JCR(S)92⟩, 131 ⟨87JCS(P1)181⟩; **6**, 121 ⟨87JCS(P1)181⟩
Martin, T. A., **1**, 45 ⟨46JA1813⟩
Martin, T. R., **5**, 443 ⟨84JCS(D)399⟩; **6**, 599 ⟨80CC635⟩
Martin, V. A., **3**, 160 ⟨90S1059⟩; **4**, 736 ⟨92JOC5530⟩
Martin, V. S., **1**, 14 ⟨82JOC1378⟩; **2**, 82 ⟨83PAC589⟩, 92 ⟨81JA6237⟩, 155 ⟨92T3571⟩; **5**, 52 ⟨81JOC3936⟩
Martin, W. B., **4**, 96, 98, 99, 308, 416, 417, 424, 425, 427 ⟨65OR(14)52⟩
Martin, Y. C., **1**, 1018 ⟨87JMC254⟩
Martín, M. R., **6**, 124 ⟨91T3171⟩
Martín, N., **5**, 343 ⟨93TL5627⟩
Martín, V. S., **2**, 84 ⟨88TL2701⟩
Martina, D., **1**, 70 ⟨78AG(E)690⟩; **3**, 30 ⟨90SL641⟩; **4**, 1076 ⟨86JOM(315)59⟩
Martina, D. P., **4**, 280 ⟨92TL717⟩
Martina, V., **3**, 285 ⟨84TL4805⟩
Martinelli, G., **1**, 1078 ⟨92TL3715⟩
Martinelli, L., **1**, 732 ⟨62JOC4666⟩
Martinelli, M. J., **5**, 329 ⟨94JOC5292⟩
Martinet, P., **4**, 806 ⟨89JOM(367)1⟩
Martinetti, G., **3**, 653 ⟨89JCS(P1)105⟩
Martinez, A. G., **1**, 433 ⟨84S481⟩, 1116 ⟨70JA3802⟩; **2**, 9 ⟨91S353⟩, 36 ⟨87TL6441, 88S58⟩, 630 ⟨86S222⟩; **4**, 214 ⟨87S49⟩
Martinez, D., **2**, 170 ⟨80EGP139258⟩, 170 ⟨82JPR631⟩
Martinez, G., **1**, 566, 681, 682 ⟨81JOC3359⟩
Martinez, G. R., **1**, 408, 524, 821, 823 ⟨79JA6452⟩; **2**, 445 ⟨79JA6452⟩; **3**, 308 ⟨81JOC3760⟩, 496 ⟨79JA6452⟩; **4**, 517 ⟨79JA6452, 80JA7993⟩
Martinez, H., **3**, 342 ⟨83PS(17)47⟩
Martinez, J., **2**, 305 ⟨91TL923⟩; **3**, 44 ⟨90TL7319⟩; **6**, 506 ⟨82JMC178⟩
Martinez, J. P., **2**, 904 ⟨90JA7423⟩
Martinez, M. A., **6**, 476, 551 ⟨85PS(25)91⟩, 534 ⟨86S760⟩, 558 ⟨85S975⟩
Martinez-Aragon, A. B., **4**, 858 ⟨93JCR(S)60⟩
Martinez, S. J., **3**, 309 ⟨84T3339⟩
Martinez, V., **6**, 563 ⟨81S961⟩
Martinez Carreras, S., **4**, 656 ⟨89JOM(375)59⟩
Martinez-Davila, C., **1**, 803 ⟨80JOC1340, 81JA2443⟩, 805 ⟨81T3943⟩
Martinez-Gallo, J. J., **4**, 808 ⟨87JCS(P1)1017⟩
Martinez-Gallo, J. M., **2**, 24, 31 ⟨86JCR(S)274, 85CC1422⟩; **3**, 27, 155 ⟨86S678⟩
Martinezi, A. G., **2**, 36 ⟨84S608⟩; **3**, 57 ⟨84S608⟩
Martinez-Merino, V., **5**, 334 ⟨85JHC1353, 88JHC23⟩
Martinez-Ripoll, M., **2**, 1002 ⟨82AX(B)1124⟩
Martinho-Simões, J. A., **1**, 5, 7, 9, 10, 17, 25 ⟨92JOC2427⟩
Martini, C. M., **3**, 252 ⟨51JA2544⟩
Martini, T., **6**, 106 ⟨72TL3957⟩
Martin-Ramos, J. D., **4**, 858 ⟨93JCR(S)60⟩
Martins, J., **5**, 354 ⟨53JA1933⟩
Martins, M. E., **2**, 1008 ⟨87CJC2000⟩
Martinsen, A., **3**, 695 ⟨74ACS(A)255⟩
Martínez, A., **4**, 897 ⟨85JHC1445⟩
Martínez, A. G., **3**, 92, 101 ⟨90CC1571⟩
Martínez-Grau, A., **5**, 343 ⟨93TL5627⟩
Martínez-Ripoll, M., **5**, 386 ⟨92JCS(D)53⟩
Marton, D., **2**, 79 ⟨88JOM(341)345⟩
Marton, M. T., **1**, 785 ⟨73CC374⟩
Martorell, G., **1**, 433 ⟨90JOC991⟩, 482, 1027 ⟨92JOC678⟩
Martre, A. M., **3**, 16, 132 ⟨90TL2599⟩
Martvon, A., **3**, 194 ⟨83CCC3421⟩; **5**, 1028 ⟨83CCC3421⟩

Marty, R. A., **3**, 335, 337, 351 ⟨76CJC537⟩, 337, 351 ⟨74CC409⟩, 375 ⟨71JA6304, 74JA2964⟩
Martymyak, I. E., **5**, 1044 ⟨91UKZ89⟩
Martynov, A. V., **2**, 1036, 1037 ⟨90IZV1865⟩; **4**, 786 ⟨85ZOR2467, 87ZOR60⟩, 786 ⟨88ZOR509, 89ZOR1470⟩, 786 ⟨89ZOR1773⟩, 787 ⟨90IZV1693, 90ZOR978⟩, 787 ⟨91IZV2106, 92ZOB2730⟩, 872 ⟨91MI 419-01⟩
Martynov, B. I., **4**, 152 ⟨73JOM(57)423⟩; **6**, 62 ⟨72GEP2101107, 72JOM(34)299⟩, 64 ⟨73JOM(57)423⟩
Martynov, I. V., **4**, 494 ⟨87DOK(297)1132, 87IZV1680⟩, 494 ⟨88IZV936⟩; **5**, 369 ⟨88ZOB2628⟩, 370 ⟨88ZOB2424⟩, 652 ⟨88IZV1691⟩, 669 ⟨87BAU204, 87IZV226⟩, 669 ⟨87JGU1478, 87JGU1479⟩, 669 ⟨87ZOB1658, 87ZOB1659⟩, 669 ⟨88BAU1506, 89BAU672⟩, 669 ⟨89IZV2849, 89IZV745⟩, 669 ⟨89KFZ1317, 90BAU415⟩, 669 ⟨90IZV477⟩, 669, 672 ⟨87BAU1000⟩, 669, 672 ⟨87IZV1086⟩, 669, 672 ⟨89BAU2612⟩, 669, 672 ⟨89BAU2628⟩, 669, 672 ⟨89IZV2865⟩, 669, 672 ⟨90BAU1007⟩, 669, 672 ⟨90IZV1122⟩, 669, 672 ⟨91PS(60)85, 89BAU2201⟩, 672 ⟨89IZV2390, 90JGU1718⟩, 672 ⟨90ZOB1923, 91JGU1152⟩, 672 ⟨91KFZ33, 91ZOB1268⟩, 672 ⟨92BAU236, 92IZV315⟩, 683 ⟨87BAU1314, 87IZV1422⟩, 1076 ⟨87IZV1177⟩, 1128 ⟨91IZV2847, 91MI 530-02⟩, 1131 ⟨92IZV187⟩; **6**, 261 ⟨88IZV2639, 88IZV466⟩, 430 ⟨89IZV2849⟩, 603 ⟨69JGU1694⟩, 613 ⟨69JGU1235, 69JGU1692⟩, 613 ⟨70JGU540, 72JGU293⟩, 613 ⟨72JGU799, 72JGU803⟩, 613 ⟨89ZOB1455⟩
Martynov, J. V., **5**, 822 ⟨65ZOB248⟩
Martynov, V. F., **4**, 340 ⟨72ZOB479⟩
Martynov, V. I., **4**, 551 ⟨80ZOB989⟩
Martynova, L. G., **4**, 152 ⟨73JOM(57)423⟩; **6**, 64 ⟨73JOM(57)423⟩
Martynova, L. L., **6**, 242 ⟨62DOK(142)354⟩, 613 ⟨71ZOR2267⟩
Martynovskii, A. A., **3**, 362 ⟨89MI 308-04⟩
Martynyuk, A. P., **2**, 407 ⟨67ZOB896⟩; **3**, 418 ⟨78ZOR508⟩; **4**, 999 ⟨80ZOB2248⟩, 999, 1001, 1002, 1007, 1011 ⟨84ZOB2186⟩
Martz, J. T., **2**, 303 ⟨84JOC2081⟩; **4**, 58 ⟨81EUP40153, 83GEP3241568⟩, 58 ⟨84JOC2081⟩; **6**, 25 ⟨79TL1473⟩, 428 ⟨83GEP3241568, 84JOC2081⟩
Marugg, J. E., **2**, 409 ⟨84TL1253⟩, 469 ⟨86JCR(S)26⟩
Marui, S., **4**, 218 ⟨82CL1555⟩
Marullo, N. P., **1**, 777 ⟨66TL6279⟩; **4**, 162 ⟨68JOC2368⟩
Marumo, K., **2**, 944 ⟨91S1163⟩; **3**, 557,558 ⟨91S1163⟩; **4**, 522 ⟨91S169⟩; **5**, 874 ⟨91S1163⟩
Maruoka, K., **1**, 303 ⟨94JA4131⟩, 387 ⟨92T3749⟩, 415, 826 ⟨88JA7922⟩, 415, 826 ⟨90JA316⟩, 825, 826 ⟨90JA7791, 89TL1265⟩, 826 ⟨90TL377⟩, 1041 ⟨84TL2367⟩, 1044, 1059 ⟨82TL3395⟩, 1106 ⟨85AG(E)668⟩, 1130 ⟨85CL1689⟩; **2**, 30, 34 ⟨86JA6036⟩, 596 ⟨94T8983⟩; **3**, 9, 122 ⟨91SL491⟩, 26 ⟨93JA3943⟩, 34 ⟨91T6983⟩, 127 ⟨88BCJ2241⟩, 159 ⟨90JA7422⟩, 184 ⟨91COS(6)763⟩, 244 ⟨88JA310⟩, 269, 437 ⟨83JA6312⟩, 407 ⟨85AG(E)668⟩, 407, 438 ⟨82TL3395⟩, 435 ⟨91COS(6)786, 83JA2831⟩, 437 ⟨83TL3255⟩, 437, 438 ⟨83JA672⟩; **5**, 421 ⟨90JA7791, 91SL793⟩, 421, 431 ⟨91TA647⟩, 661, 728 ⟨83TL3255⟩, 731 ⟨83JA2831⟩, 777 ⟨81JA4186⟩
Maruta, T., **5**, 1102 ⟨89JAP3145452⟩
Maruyama, A., **4**, 448 ⟨90H(31)2029⟩
Maruyama, K., **1**, 119 ⟨80JA2318⟩, 226 ⟨87CL963⟩, 312 ⟨82JFC(20)313⟩, 362 ⟨90CC1436⟩, 450, 510 ⟨78CC702⟩, 488 ⟨77JA5652⟩, 531, 663 ⟨77CC852⟩, 540 ⟨79JOC1744⟩, 713, 714, 716 ⟨84JOC1096⟩, 1001 ⟨80CL669⟩, 1038, 1039 ⟨86CC102⟩, 1059 ⟨84CC1004⟩; **2**, 67 ⟨82H(18)357⟩, 78 ⟨84T2239⟩, 566 ⟨81TL2895⟩, 702 ⟨82JOC119⟩, 703 ⟨80TL4607⟩; **3**, 412 ⟨91CL1093⟩; **4**, 642, 648, 659 ⟨84JOC1096⟩, 1057 ⟨86JOC886⟩
Maruyama, S., **3**, 547 ⟨81BCJ2845⟩
Maruyama, T., **1**, 228 ⟨93TL2135⟩; **2**, 71 ⟨90JOC304⟩, 528 ⟨93JA11490⟩; **3**, 502 ⟨91JA2610⟩
Marvel, C. S., **1**, 503 ⟨51JA1066⟩, 772 ⟨43OSC(2)416, 43OSC(2)93⟩; **2**, 29 ⟨43OSC(2)74, 55OSC(3)523⟩, 29 ⟨55OSC(3)848⟩, 31 ⟨41OSC(1)25⟩, 125 ⟨50JA1033⟩, 187, 273 ⟨48JOC822⟩, 191 ⟨43OSC(2)558⟩, 192 ⟨41OSC(1)84⟩, 483, 487 ⟨30JA376⟩, 609, 621 ⟨32OSC(1)170⟩; **3**, 612 ⟨41OSC(1)536⟩; **4**, 29 ⟨32OSC(1)358⟩, 47 ⟨32OSC(1)369⟩; **5**, 33 ⟨43OSC(2)416, 43OSC(2)93⟩, 33 ⟨55OSC(3)495⟩, 33, 89 ⟨55OSC(3)705⟩, 34 ⟨32OSC(1)289⟩, 55 ⟨40JA3495⟩, 57 ⟨43OSC(2)74⟩, 57, 64 ⟨29JA260⟩, 76 ⟨32OSC(1)248⟩, 85 ⟨43OSC(2)310⟩, 88 ⟨43OSC(2)374⟩, 89 ⟨32OSC(1)48⟩, 105 ⟨32OSC(1)252⟩, 115 ⟨55OSC(3)822⟩, 299 ⟨41OSC(1)99⟩, 551 ⟨55JA5997⟩, 898 ⟨32OSC(1)23⟩
Marvell, E. N., **1**, 401, 800 ⟨79TL2779⟩, 412, 813 ⟨70TL509⟩, 418, 419, 420 ⟨B-80MI 109-01⟩, 419, 420 ⟨65TL385⟩, 419, 420, 794 ⟨73T3781, 73T3797⟩, 420 ⟨83JOC4272⟩, 456, 502, 504, 505 ⟨73S457⟩, 506 ⟨72JOC2989⟩, 808 ⟨78JA877⟩
Marwat, A. G., **4**, 585 ⟨77MI 412-02⟩
Marwuet, A., **2**, 127 ⟨93JA2139⟩
Marx, J., **3**, 490 ⟨54CB1499⟩
Marx, J. A., **6**, 453 ⟨93TL239⟩
Marx, J. N., **2**, 34 ⟨83T1529⟩, 660 ⟨74JA2121⟩
Marx, K. H., **5**, 174 ⟨84LA474⟩
Marx, M., **1**, 307 ⟨78TL2445⟩
Marx, R., **6**, 90 ⟨41CB1667⟩
Marxer, A., **2**, 445, 445 ⟨78HCA1708⟩, 691, 692 ⟨63HCA2009⟩
Marx-Moll, L., **3**, 490 ⟨54CB1499⟩
Marxus, R., **6**, 244 ⟨62JA1312⟩
Maryanoff, B. E., **1**, 4, 6, 7, 9, 10, 11, 17 ⟨77JOC82⟩, 638, 639, 647, 721, 744 ⟨89CRV863⟩, 722 ⟨89TL1361⟩, 730 ⟨85JA217⟩; **2**, 61 ⟨94TL4891⟩, 253 ⟨81JA4432⟩, 462 ⟨73CC273⟩, 1003, 1004 ⟨89CRV863⟩; **3**, 48, 195, 493 ⟨89CRV863⟩, 376 ⟨74JA5648, 74JA5650⟩, 376 ⟨74JA5651, 75JA2718⟩; **6**, 694 ⟨89CRV863⟩
Maryanoff, C. A., **1**, 4, 6, 7, 9, 10, 11, 17 ⟨77JOC82⟩, 4, 7, 9, 17, 28 ⟨78JOC2259⟩; **2**, 31 ⟨76JOC1071⟩, 300, 310 ⟨90TL5595⟩; **3**, 376 ⟨76JA3615⟩; **5**, 657 ⟨84JOC546⟩; **6**, 641, 645 ⟨86JOC1882⟩
Mar'yasova, V. I., **4**, 656 ⟨84ZOB2025⟩
Maryin, V. P., **4**, 723 ⟨92JOM(441)241⟩
Marynick, D. S., **5**, 932 ⟨85JA1993⟩
Marzabaoi, M. R., **3**, 638 ⟨88JA4829⟩
Marzi, W., **2**, 504 ⟨80CB2928⟩
Marzin, C., **3**, 452 ⟨70BSF4119⟩
Marzorati, L., **2**, 104 ⟨80JOC2515⟩
Mas, J.-M., **2**, 31 ⟨92SC2187⟩
Mas, T. de, **5**, 570, 618 ⟨81JCR(S)58⟩
Masada, H., **2**, 684 ⟨78CL57⟩
Masahiro, Y., **5**, 990 ⟨72MI 526-01⟩
Masaki, C., **2**, 617 ⟨81BCJ2843⟩
Masaki, H., **4**, 937 ⟨87T4759⟩
Masaki, M., **3**, 563 ⟨67JOC3564⟩; **6**, 413 ⟨78GEP2811310, 79JOC3573⟩
Masaki, Y., **1**, 237 ⟨85JCS(P1)1171⟩, 601, 604 ⟨75JA1597⟩, 820 ⟨80CL1061⟩, 1016, 1145 ⟨90CC118⟩; **2**, 326 ⟨89TL6395⟩, 784 ⟨92S1119⟩, 1020 ⟨90CC118⟩; **4**, 183 ⟨93CL17⟩, 199 ⟨89CL659⟩, 216 ⟨85CL1933⟩, 808 ⟨89TL6701⟩; **5**, 204 ⟨90JA5677⟩
Masakuni, O., **3**, 357 ⟨91EUP409164⟩
Masamune, H., **3**, 632 ⟨87JOC667⟩
Masamune, S., **1**, 14 ⟨82JOC1378⟩, 353 ⟨91TL7373⟩, 400, 799 ⟨69JA4322⟩, 517 ⟨80T343⟩, 749 ⟨84TL2183⟩, 811 ⟨70JA6641⟩, 884 ⟨66CC880⟩; **2**, 521 ⟨86JA7402⟩, 521 ⟨87JOC4831⟩, 532 ⟨81JA1566, 85AG(E)1⟩, 532, 697 ⟨79TL1665⟩, 532, 698 ⟨79TL2229⟩, 533 ⟨89TL7357⟩, 696 ⟨91COS(2)239, 86JA8279⟩, 947, 949 ⟨88TL3383, 82JA6136⟩, 949 ⟨85TL1281, 89JA1888⟩, 949 ⟨91AG(E)902⟩; **2**, 58, 521 ⟨86JA7404⟩, 77, 521

⟨89JA1892, 85JA4549⟩; **3**, 232 ⟨65CI(L)184⟩; **4**, 628 ⟨88TL3383⟩, 643, 645 ⟨89JA1892⟩; **5**, 74 ⟨86TL4537⟩, 155 ⟨84TL2183⟩, 161 ⟨76JA7874⟩, 173 ⟨81JA1568⟩, 173 ⟨82JA5521⟩, 173, 235 ⟨82JA5523⟩, 234, 239 ⟨76TL4423⟩, 234, 277 ⟨75JA3515, 75CJC3693⟩, 237 ⟨86JA8279⟩; **6**, 174 ⟨87AG(E)584⟩
Masamune, S. E., **1**, 722 ⟨75JA3512⟩
Masamune, T., **2**, 765 ⟨76JCS(P1)2365⟩
Masana, J., **2**, 299 ⟨81JCS(P1)574⟩
Masano, S., **5**, 327 ⟨81JHC555⟩
Masát, J., **6**, 236 ⟨66FRP1437908⟩
Masatsuga, Y., **1**, 580 ⟨92T10265⟩
Masawaki, T., **4**, 947 ⟨91JOC5721⟩
Mascarella, S. W., **3**, 222 ⟨84JOC3033⟩
Mascareñas, J. L., **1**, 816, 817 ⟨87TL2099⟩, 1029 ⟨88TL1203⟩; **3**, 57 ⟨87TL2099⟩
Mascaretti, O. A., **2**, 9 ⟨86JOC3217⟩; **4**, 100 ⟨88T7007⟩
Maschke, A., **2**, 100 ⟨62AG(E)97⟩, 793 ⟨62AG(E)89⟩
Masci, B., **2**, 792 ⟨85JOC4081⟩
Mascitelli, L., **3**, 92 ⟨90SC2565⟩
Mascolo, G., **2**, 931 ⟨88TL3705⟩
Mase, T., **3**, 218, 225 ⟨86JA2090⟩
Mash, E. A., **2**, 67 ⟨90TL4565⟩
Mashchenko, N. V., **4**, 462 ⟨85JGU404⟩
Mashima, K., **2**, 53 ⟨93JA3318⟩
Mashino, T., **2**, 579 ⟨93CPB292⟩
Mashlyakovskii, L. N., **2**, 848 ⟨65ZOB1577⟩, 1071 ⟨65JGU1582⟩
Mashoshina, S. N., **4**, 552, 562 ⟨74ZOB276⟩, 574, 588 ⟨75ZOB1486⟩
Masi, J. F., **2**, 109 ⟨60JPC682⟩
Masias, A., **6**, 562 ⟨88ZOR2019⟩
Masiero, S., **4**, 232 ⟨91TL2971⟩; **6**, 127 ⟨91TL2971, 91TL6801⟩
Masilamani, D., **1**, 4, 6, 7, 9, 10, 11, 17 ⟨77JOC82⟩; **2**, 21, 619 ⟨81JOC4486⟩, 31 ⟨76JOC1071⟩
Masjedizadeh, M. R., **1**, 944 ⟨92SL107⟩
Maskaeva, E. V., **4**, 152 ⟨89MI 403-03⟩, 152 ⟨91MI 403-01⟩
Maslak, P., **1**, 949 ⟨91JOC2179⟩
Maslankiewicz, A., **3**, 362 ⟨92H(34)247⟩
Maslen, H. S., **2**, 1002 ⟨54AX145⟩
Maslennikov, I. G., **4**, 114 ⟨75JGU1668⟩, 125 ⟨86JGU2246⟩, 126 ⟨86JGU2144, 86JGU2242⟩, 128 ⟨87JGU622⟩; **6**, 242 ⟨88ZOB1525⟩, 613 ⟨89ZOB1455⟩
Maslin, D. N., **3**, 84 ⟨63TL2087⟩, 85 ⟨69T5555⟩
Maslov, P. G., **6**, 223, 228 ⟨58MI 607-01⟩, 223, 228 ⟨66ZOB1355⟩
Maslov, Yu. P., **6**, 223, 228 ⟨58MI 607-01⟩, 223, 228 ⟨66ZOB1355⟩
Maslyanskii, G. N., **1**, 56 ⟨71MI 102-01⟩
Masnovi, J. M., **6**, 147 ⟨85JOC5245⟩, 147 ⟨86RTC286⟩
Masnyk, M., **4**, 393 ⟨91CC297⟩
Mason, C. J., **5**, 1056 ⟨81EJM317⟩
Mason, C. T., **4**, 50 ⟨49JA258⟩
Mason, D., **2**, 651 ⟨53JCS3623⟩
Mason, J. C., **2**, 1093 ⟨67BSF777⟩
Mason, J. P., **2**, 386 ⟨51JA5041⟩; **3**, 467 ⟨54JA441⟩; **6**, 27 ⟨42JA2515⟩
Mason, J. S., **5**, 587, 593 ⟨93JMC1604⟩
Mason, M. G., **4**, 93 ⟨85JA6298⟩; **6**, 332 ⟨83IC411⟩
Mason, M. S., **5**, 87 ⟨57JA637⟩
Mason, P. M., **4**, 267 ⟨80CJC878⟩
Mason, R., **1**, 641, 642 ⟨83TL5293⟩, 756 ⟨83TL593⟩, 757 ⟨87JCS(P1)2569⟩; **3**, 197 ⟨87JCS(P1)2569⟩; **4**, 687 ⟨64PCS401, 68JCS(A)1293⟩
Mason, R. E., **2**, 188 ⟨52JCS4887⟩
Mason, T. J., **1**, 925 ⟨87CSR239⟩; **2**, 19 ⟨86JCR(S)300⟩
Mass, G., **3**, 489 ⟨B-86MI 312-01⟩; **4**, 480 ⟨87AG(E)1257⟩
Massa, S., **4**, 856 ⟨82JHC1493⟩
Massa, W., **1**, 858 ⟨89AG(E)73⟩, 981 ⟨82AG(E)924⟩; **2**, 504 ⟨85ZN(B)1320⟩, 1000 ⟨86AG(E)1011⟩; **4**, 281 ⟨91LA243⟩, 633 ⟨89AG(E)784⟩, 672 ⟨83JA2905⟩, 672, 673 ⟨90ZN(B)221, 83AG(E)330⟩, 698, 708 ⟨91CB1511⟩, 896 ⟨88AP(321)873⟩, 1041 ⟨86AG(E)571⟩, 1082 ⟨87AG(E)546⟩; **5**, 685 ⟨86TL2743, 89CB657⟩, 928 ⟨87AG(E)546, 87AG(E)798⟩, 928 ⟨88AG(E)961, 88ZN(B)801⟩, 928 ⟨90AG(E)398⟩, 928, 929 ⟨92AG(E)1238⟩, 928, 930 ⟨92ZN(B)983, 90AG(E)399⟩, 929 ⟨92AG(E)1384⟩; **6**, 160 ⟨82ZAAC(485)23⟩, 184, 389, 398, 714 ⟨89AG(E)781⟩, 188 ⟨92AG(E)1384⟩, 188, 711 ⟨87AG(E)798⟩, 201, 397, 715, 716 ⟨90AG(E)1030⟩, 201, 400 ⟨88AG(E)1370⟩, 201, 715 ⟨92AG(E)1238⟩, 202, 724 ⟨87AG(E)546⟩, 324, 684 ⟨84ZAAC(517)75⟩, 389, 398, 401, 713 ⟨89AG(E)784⟩, 389, 714 ⟨91AG(E)594⟩, 397, 398 ⟨90AG(E)1032⟩, 397, 714 ⟨88AG(E)961, 86AG(E)1111⟩, 400 ⟨90AG(E)399⟩, 582, 684 ⟨85ZAAC(520)139⟩, 711 ⟨87PAC1011⟩, 713, 714 ⟨90AG(E)398, 88ZN(B)75⟩, 714 ⟨90ZN(B)290⟩
Massad, S. K., **3**, 32, 33 ⟨83JOC5180⟩
Massanet, G. M., **1**, 831 ⟨91TL3209⟩; **2**, 30 ⟨94T6433⟩
Massé, G., **2**, 344, 347, 357 ⟨85S807⟩; **3**, 78 ⟨88S907⟩
Masse, G. M., **3**, 34, 164 ⟨75JOC1294⟩
Massé, J., **4**, 508 ⟨76CR(C)473⟩. 646, 654 ⟨75JOM(93)71⟩, 648, 649 ⟨73JOM(57)C5⟩
Masse, J. P., **4**, 648 ⟨77JOM(127)281⟩; **5**, 410 ⟨70JOM(22)321⟩
Massengale, J. T., **2**, 413 ⟨57JOC1186⟩
Massey, A. G., **2**, 872 ⟨85JOM(287)57⟩, 959 ⟨77JOM(127)127⟩
Massey, W. D., **2**, 813 ⟨75JOC1090⟩
Massias, F., **3**, 172 ⟨89TL7389⟩
Massie, Jr., S. P., **1**, 881 ⟨45JA2106⟩
Massi-Mauri, M., **6**, 525 ⟨59CI(M)132⟩
Massin, C. C., **5**, 203 ⟨88JMC362⟩
Massiot, G. S., **2**, 724 ⟨77JA4405⟩; **4**, 220 ⟨77JA4405⟩
Massol, M., **2**, 947, 949 ⟨73JOM(56)1⟩; **4**, 152, 155, 156 ⟨75JOM(88)255⟩
Masson, A., **1**, 544 ⟨78CR(C)265⟩
Masson, J., **3**, 343 ⟨76TL4297, 77T3089⟩; **4**, 255 ⟨76TL4775⟩
Masson, J. C., **1**, 1000 ⟨68BSF1085⟩
Masson, S., **1**, 1058 ⟨77T2949⟩; **2**, 617 ⟨75BSF2493⟩; **3**, 550 ⟨86TL4313⟩, 562 ⟨88BSF883, 91POL1153⟩; **4**, 255 ⟨79JOC2807, 80CR(C)183⟩, 256 ⟨77T2949, 78TL2717⟩, 847 ⟨89TL3415⟩; **5**, 551 ⟨84SUL147⟩, 551, 729, 731 ⟨84T1573⟩, 553, 559, 729 ⟨78TL2715⟩, 557, 559 ⟨82TL4087⟩, 558, 559 ⟨80TL2421⟩, 561 ⟨79JOC2807⟩, 589 ⟨93PS(80)31⟩, 731 ⟨91JCS(P1)3389, 91PS(59)141⟩; **6**, 121, 316 ⟨89TL3415⟩, 121, 316 ⟨92JOC4507⟩, 123 ⟨88PS(40)1, 90TL1151⟩, 734 ⟨93JA8444⟩
Massonne, K., **5**, 11 ⟨91S1173⟩
Massoud, A., **5**, 952 ⟨91COS(6)485⟩; **6**, 716 ⟨91CRV35⟩
Massuda, D., **1**, 318 ⟨75TL3383⟩, 652 ⟨77JA936⟩, 1094 ⟨76S801⟩; **2**, 616 ⟨75TL3383⟩, 911 ⟨76S801⟩
Massy, D. J. R., **4**, 315, 325 ⟨87S589⟩
Massy-Westropp, R. A., **1**, 17 ⟨82AJC1895⟩; **4**, 102 ⟨92T5163⟩
Massy-Westropp, R. J., **1**, 328 ⟨93T6419⟩
Mastagli, P., **2**, 662 ⟨75MI 213-01⟩
Mastalerz, H., **1**, 821 ⟨83JA127⟩; **3**, 360 ⟨81JCS(P1)2952⟩
Mastalerz, P., **4**, 124 ⟨87S477⟩, 345 ⟨B-92MI 407-02⟩, 460 ⟨74RZC1119, 76RZC661⟩, 460 ⟨79PJC2327⟩, 461 ⟨78CL1103⟩, 467 ⟨79S985, 84S577⟩, 472 ⟨81S57⟩, 472, 473 ⟨81PJC713, 78PJC2271⟩, 473 ⟨81PJC411⟩, 476 ⟨73TL5201⟩, 555 ⟨65RZC1129⟩
Mastellari, A., **3**, 311 ⟨84JOC3478⟩
Mastellari, A. R., **2**, 942 ⟨86S757⟩
Masters, A. P., **3**, 530, 531 ⟨86JOC3558⟩; **5**, 447 ⟨90CJC502⟩
Masthoff, R., **2**, 421 ⟨69ZAAC(364)316⟩
Mastin, T. N., **2**, 255, 256 ⟨52JA161⟩
Mastorilli, E., **2**, 624 ⟨86S122⟩; **4**, 26 ⟨86S122⟩
Mastroianni, D., **2**, 302 ⟨92SC2305⟩
Mastryukova, T. A., **2**, 255 ⟨87ZOB786⟩, 454

⟨59ZOB1450⟩, 462 ⟨61ZOB507, 62ZOB3579⟩, 464 ⟨80JFC(15)423⟩, 841 ⟨65ZOB1574⟩, 854 ⟨76PS(1)211⟩; **4**, 463, 469 ⟨53DOK(92)959⟩, 551 ⟨80TL2931, 92ZOB263⟩, 557 ⟨84ZOB36, 85ZOB1234⟩, 557 ⟨86ZOB1220, 88ZOB1987⟩, 557 ⟨88ZOB1998, 88ZOB2012⟩
Masuda, H., **5**, 664 ⟨84BCJ2184⟩
Masuda, K., **4**, 422 ⟨79JOC4543⟩
Masuda, R., **3**, 230 ⟨86TL353⟩, 232 ⟨90CL113⟩, 246 ⟨86S1013⟩, 246, 261 ⟨76CL499⟩, 253 ⟨75TL1009, 86S137⟩, 265 ⟨89TL6173, 90S195⟩; **4**, 74, 75 ⟨76TL613⟩, 78 ⟨76S697⟩, 226 ⟨82S312⟩, 247 ⟨86JOC1427⟩, 889 ⟨86S1013⟩, 889, 900, 970, 971, 972 ⟨90S195⟩; **6**, 84, 89 ⟨75JOC963⟩
Masuda, S., **1**, 182 ⟨79BCJ2377⟩, 463 ⟨80CC86, 83BCJ1089⟩, 475 ⟨82BCJ167⟩; **2**, 453 ⟨87CL1915⟩; **3**, 704 ⟨82ABC1803⟩
Masuda, T., **1**, 181 ⟨75CPB1917⟩, 797 ⟨84CL1541⟩, 979 ⟨82JOC4478⟩; **2**, 138 ⟨92TL947⟩, 748 ⟨84S1043⟩; **4**, 221 ⟨78JCS(P2)1302⟩, 607, 646 ⟨86JPS(A)1839⟩
Masuda, Y., **1**, 238 ⟨90TL2879⟩, 663 ⟨80BCJ1652⟩, 1018 ⟨83BCJ2855, 84CL1029⟩; **2**, 20 ⟨79CL961⟩, 43 ⟨91CC205⟩, 273 ⟨85CC715⟩, 289 ⟨76CL1115, 76CL785⟩, 623 ⟨78CL665⟩, 922 ⟨86BCJ659⟩; **3**, 416, 645 ⟨91CC748⟩, 622 ⟨84CC398⟩, 622, 645 ⟨89CC266, 74BCJ2511⟩; **4**, 815 ⟨88BSJ3764⟩, 1056, 1061 ⟨86BCJ659⟩; **5**, 1114 ⟨76CL1115, 76CL785⟩
Masue, K., **5**, 1036 ⟨78BCJ659⟩
Masui, M., **2**, 108 ⟨79COC(3)1013⟩, 219, 227 ⟨84CC843⟩, 385 ⟨88CPB459⟩, 390 ⟨84CPB2609⟩; **3**, 120 ⟨83CC479⟩, 196 ⟨88CPB613⟩, 496 ⟨85TL2199, 87CPB4960⟩; **5**, 1078 ⟨85CPB373⟩, 1107 ⟨90CPB1389⟩
Masui, Y., **2**, 145 ⟨89JA5002⟩; **5**, 324 ⟨87CC1471, 90JOC1744⟩
Masuko, K., **1**, 178 ⟨86BCJ3097⟩
Masuko, M., **6**, 652 ⟨92MI 621-02⟩
Masumi, N., **3**, 90 ⟨90TL6887⟩
Masumoto, H., **5**, 561, 563, 588 ⟨90ZC67⟩
Masumoto, K., **4**, 247 ⟨91TL2039⟩
Masumoto, T., **5**, 801 ⟨80BCJ1171⟩
Masunaga, T., **2**, 850 ⟨82BCJ909⟩
Masunova, A. Y., **1**, 500 ⟨91ZOR488⟩
Masuo, K., **4**, 142, 521 ⟨91CL97⟩
Masuoka, Y., **1**, 734 ⟨75CL103⟩
Masure, D., **1**, 429 ⟨78S128⟩, 511 ⟨82JOM(226)C55⟩, 833 ⟨74BSF2072⟩, 1090 ⟨76S761⟩; **2**, 626 ⟨89CJC2061⟩; **3**, 67, 237 ⟨78S128⟩, 67, 237 ⟨78S458⟩; **6**, 41 ⟨74BSF2072⟩
Masuya, H., **1**, 734 ⟨75CL103⟩; **5**, 640, 642 ⟨80CPB2116⟩
Masuyama, Y., **1**, 37 ⟨84TL4417⟩, 195 ⟨88CC44⟩, 195 ⟨88JA4473⟩, 196 ⟨87BCJ1525⟩, 238, 246 ⟨89BCJ2913⟩; **2**, 73 ⟨89CL1647⟩, 74, 80 ⟨88TL3563, 88CC44⟩, 80 ⟨88JA4473⟩, 287 ⟨77CL835⟩; **3**, 117 ⟨84TL173⟩, 117, 161 ⟨84TL4417⟩, 162 ⟨92CL2039⟩, 634 ⟨76TL2967⟩, 636, 637 ⟨77CL835⟩
Masuzawa, M., **4**, 259 ⟨90JCS(P1)2035⟩
Matachek, J. R., **6**, 333, 334 ⟨86IC2877, 84OM1038⟩
Matacz, Z., **4**, 428 ⟨79PJC187⟩; **5**, 692 ⟨88BAP139⟩
Mataka, S., **1**, 232 ⟨91JHC289⟩; **2**, 793 ⟨71MI 215-01⟩, 794 ⟨76H1243⟩, 796 ⟨71BCJ1896⟩; **3**, 263 ⟨84S663⟩, 464 ⟨75CC114⟩; **4**, 21 ⟨84JHC1157⟩, 273 ⟨92JCS(P1)1179⟩
Matano, Y., **2**, 1082 ⟨92BCJ3504⟩; **3**, 196 ⟨93TL8457⟩; **4**, 432 ⟨93S1218⟩; **5**, 344 ⟨91CL105⟩
Matassa, V. G., **5**, 342 ⟨90JMC2621⟩
Mateescu, G. D., **1**, 900 ⟨85JA2764⟩, 906 ⟨73JA8698⟩, 909 ⟨70JA1430⟩, 914 ⟨72JA2034, 78JA6299⟩; **4**, 42, 44 ⟨71JA781⟩
Mateev, I. S., **5**, 1157 ⟨74CA121069⟩
Mateev, Yu. I., **5**, 1075 ⟨87ZOR2390⟩
Matelich, M. C., **1**, 287, 361 ⟨91JA9007, 92S151⟩
Mateos, A. F., **3**, 192 ⟨90JOC1349⟩
Matera, C., **5**, 597 ⟨85FES120⟩
Matern, E., **2**, 471 ⟨90ZAAC(589)23⟩; **4**, 621 ⟨76ZAAC(426)28⟩, 1014 ⟨78CJC2286⟩; **5**, 792, 796 ⟨78CJC2286⟩; **6**, 178 ⟨81ZAAC(475)87⟩, 267, 268 ⟨B-86MI 608-01⟩
Materne, C., **4**, 972 ⟨73LA573⟩
Matesich, M. A., **1**, 1122 ⟨65JA2295⟩
Matheny, N. P., **1**, 887 ⟨64JA5503⟩
Mather, A. N., **1**, 216 ⟨84CC585⟩; **2**, 921 ⟨88T3953⟩
Mathes, W., **4**, 237 ⟨51CB648⟩
Mathew, C. P., **2**, 534 ⟨81JOC2712⟩
Mathew, J., **2**, 21 ⟨90JOC3880⟩; **3**, 190 ⟨82S397⟩
Mathew, S. I., **6**, 114 ⟨74JCS(P1)540⟩
Mathews, D. P., **4**, 119, 734 ⟨90TL5571⟩; **6**, 10, 84 ⟨86TL4861⟩
Mathey, F., **1**, 53 ⟨80JA994⟩, 751 ⟨77TL1137⟩; **2**, 19 ⟨79TL353⟩, 430 ⟨85TL1717⟩, 430, 438 ⟨90CRV997, 92S157⟩, 440 ⟨83CJC21⟩, 441 ⟨79JOM(177)255, 80TL2525⟩, 441 ⟨82JOM(231)237⟩, 472 ⟨83JA6871⟩, 894 ⟨93BSF521⟩; **3**, 149 ⟨77TL2797⟩, 197 ⟨78S36⟩; **4**, 10 ⟨71T3965⟩, 112 ⟨78TL2407⟩, 341 ⟨73T707, 82TL1565⟩, 341 ⟨84JOM(266)285⟩, 545 ⟨87JOM(328)349⟩, 545, 565 ⟨93PS(77)254⟩, 569 ⟨78ICAL294⟩, 570, 585, 588 ⟨82S725⟩, 585 ⟨76S766⟩, 1026 ⟨73TL3255, 74T3127⟩, 1026 ⟨76T2395, 92OM1411⟩, 1026, 1031, 1035 ⟨88CRV429⟩, 1034 ⟨77JA3537⟩, 1034 ⟨77JOM(139)77, 80JA994⟩, 1037 ⟨76TL4155, 78JA5748⟩, 1039 ⟨79ICAL67, 80AX(B)1344⟩; **5**, 397 ⟨91T6673⟩, 407 ⟨82OM1488, 84PS(19)311⟩, 1157, 1158 ⟨92CC1799, 93CC1295⟩; **6**, 154 ⟨91CC1305⟩, 229, 230 ⟨73TL2253⟩, 679 ⟨89TL817⟩, 692 ⟨87AG(E)275, 90OM793⟩, 692 ⟨92ACR90⟩
Mathian, B., **2**, 148 ⟨85SUL127⟩
Mathias, J. P., **1**, 571 ⟨92JA6330⟩
Mathias, L. J., **6**, 631 ⟨79S561⟩
Mathias, R., **4**, 36 ⟨73TL611⟩
Mathiason, D. R., **4**, 562 ⟨68IC709⟩
Mathies, D., **4**, 417 ⟨73MI 409-01⟩
Mathieson, M. T., **6**, 247 ⟨93JOM(453)299, 93JOM(453)307⟩
Mathieu, B., **3**, 479 ⟨80CB1632⟩
Mathieu, D., **3**, 279 ⟨85BSB897⟩
Mathieu, J., **2**, 393, 394, 396 ⟨B-73MI 207-01⟩
Mathieu, R., **4**, 711 ⟨86JOM(314)13⟩, 713 ⟨90JOM(395)305⟩, 722 ⟨91OM4005⟩; **5**, 955 ⟨94OM4676⟩; **6**, 338 ⟨85OM1436⟩
Mathis, J., **1**, 1010 ⟨92BMC1619⟩
Mathisen, D., **1**, 108 ⟨76JA2580⟩, 112 ⟨79JA2207⟩
Mathisen, D. E., **1**, 110 ⟨82JOC3521⟩; **6**, 425 ⟨82JOC3521⟩
Mathre, D. J., **1**, 134 ⟨82JA1737⟩, 878 ⟨81PAC1109⟩; **2**, 66 ⟨93JOC799⟩, 528 ⟨91JOC751, 91JOC763⟩, 528 ⟨93JOC2880⟩; **5**, 148 ⟨81PAC1109⟩; **6**, 446 ⟨91JOC751⟩
Mathur, M. A., **2**, 379 ⟨81JINC3389⟩, 435 ⟨74IS129⟩
Mathur, N., **4**, 37 ⟨91JOC1663⟩
Mathur, N. K., **2**, 45 ⟨71TL691⟩; **5**, 652 ⟨77SC57⟩
Mathvink, R. J., **1**, 933 ⟨89JOC1777⟩; **3**, 136 ⟨90JA4008⟩, 292 ⟨88JOC3377, 89JOC1777⟩, 292 ⟨92JOC1321⟩
Matia, M. P., **2**, 301 ⟨91SC535⟩
Matier, W. L., **5**, 767 ⟨74JMC549, 74JOC3080⟩; **6**, 508 ⟨73JMC901⟩
Matkin, D. A., **3**, 80 ⟨81JCS(P1)1437⟩
Matkovic-Čalogović, D., **4**, 697 ⟨92CSC1116, 93OM4708⟩; **6**, 207 ⟨83JOM(253)283, 86JOM(306)1⟩, 207 ⟨87JOM(319)1, 91JOM(411)19⟩
Matl, V. G., **2**, 219 ⟨71JOC488⟩
Matlin, A. R., **1**, 596 ⟨86JA3385⟩
Matlin, S. A., **3**, 65 ⟨81CC10, 90JCS(P1)89⟩, 532 ⟨72JCS(P1)2623⟩; **5**, 325 ⟨72CC1222⟩, 720 ⟨79JCS(P1)2478⟩
Matlock, P. L., **3**, 621 ⟨78JA1119⟩
Matner, M., **1**, 1078 ⟨66TL1663⟩
Matoba, K., **3**, 659 ⟨76CPB3011⟩; **5**, 199 ⟨87CPB3453⟩, 597, 599 ⟨84SC1257⟩

Matolcsy, G., **4**, 120 ⟨86ACH145⟩; **5**, 547 ⟨72JOC1727⟩
Matos, J. R., **5**, 117 ⟨85JOC1992⟩
Matos, R. M., **5**, 886 ⟨93CC267⟩
Matough, M. F. S., **2**, 7 ⟨68CC1350⟩; **3**, 92 ⟨92SL77⟩
Matović, R., **1**, 982 ⟨91CC294⟩
Matrka, M., **2**, 777 ⟨73MI 208-01⟩
Matrosov, E. I., **4**, 556 ⟨68IZV2062⟩
Matsbara, Y., **5**, 639 ⟨91MI 515-01⟩
Matseescu, G. D., **4**, 42 ⟨81JOC571⟩
Matser, H. J., **3**, 250 ⟨74TL4179⟩
Matsnaga, K., **6**, 73 ⟨76CI(L)164⟩
Matsnev, V. V., **4**, 800 ⟨76JOU782⟩
Matsoukas, J., **5**, 223 ⟨89LA387⟩
Matsuba, S., **3**, 282 ⟨89BCJ3371⟩
Matsubara, H., **5**, 746 ⟨78CPB2246⟩
Matsubara, S., **1**, 208 ⟨90TL6209⟩, 264 ⟨90TL6379⟩, 491 ⟨87CL5⟩, 570 ⟨88T2855⟩; **2**, 19 ⟨88T2875⟩, 319 ⟨90TL6379⟩, 918, 919, 939 ⟨88T4277⟩, 922, 923 ⟨88T4009⟩, 977 ⟨84TL2151⟩, 997, 1002 ⟨85JOM(285)163⟩; **3**, 5, 117 ⟨88BCJ3607⟩, 30 ⟨92CL2173⟩, 615 ⟨91CL1447⟩, 632 ⟨90TL6209⟩, 638 ⟨90TL6379⟩, 645 ⟨86TL4177⟩; **4**, 221 ⟨87CC122⟩, 231 ⟨90JA8189, 92CL2173⟩, 1045 ⟨88T4277⟩
Matsubara, Y., **1**, 899 ⟨87BCJ3721, 88NKK157⟩
Matsuda, A., **1**, 1010 ⟨85CPB1761, 92JMC241⟩; **3**, 3 ⟨78BCJ3016⟩; **5**, 27 ⟨77MI 502-01⟩, 1054 ⟨90MI 527-01⟩
Matsuda, F., **5**, 9 ⟨85T3625⟩, 129 ⟨94JOC5532⟩
Matsuda, H., **1**, 570 ⟨88T2855⟩; **2**, 19 ⟨88T2875⟩, 699 ⟨89CL1247⟩, 701 ⟨93CC1067, 93CL1953⟩; **3**, 140 ⟨91CL1181⟩; **4**, 829 ⟨88JOC5783⟩
Matsuda, I., **1**, 206, 1094 ⟨78CL773⟩, 209 ⟨78S221, 79BCJ2389⟩, 209 ⟨79JCS(P1)26⟩; **3**, 51, 202 ⟨85TL4229⟩, 77 ⟨89JA2332, 90JA6120, 88JOM(344)71⟩, 609 ⟨86JOM(311)C63⟩, 639 ⟨79JCS(P1)26⟩; **5**, 346 ⟨91TL7431⟩, 377, 388, 390, 1083 ⟨74JOM(69)353⟩, 387 ⟨75CL197, 75CL641⟩, 390, 721 ⟨71JCS(C)1870⟩, 419 ⟨88JOM(344)71⟩; **6**, 571 ⟨72JCS(P1)1678⟩
Matsuda, K., **1**, 372 ⟨87JOC2523⟩, 1063 ⟨91CC948⟩; **2**, 658 ⟨71BCJ1986⟩; **3**, 407, 586 ⟨84JOC2688⟩; **4**, 512, 518 ⟨86BCJ2537⟩, 515 ⟨84H(22)1955⟩, 516 ⟨83CL1131⟩, 517 ⟨84JOC2688⟩, 517, 518 ⟨85H(23)2489⟩, 979 ⟨86JOC1997⟩; **5**, 729 ⟨87JOC2442⟩, 1031 ⟨84JOC2688⟩
Matsuda, M., **5**, 143 ⟨68TL633⟩
Matsuda, S., **1**, 1035 ⟨72JOC3749⟩, 1061, 1062 ⟨90JOC4853⟩; **4**, 661 ⟨91SL911, 92JCS(P1)2813⟩
Matsuda, T., **1**, 452, 490 ⟨90CL185⟩, 486 ⟨81JOC4885⟩, 486 ⟨81SC895⟩, 486 ⟨83JOC1333, 86JCS(P1)1959⟩, 490 ⟨88CL873⟩, 1033 ⟨87JCS(P1)1511⟩; **2**, 932 ⟨84TL5763, 87JOC1276⟩; **3**, 90 ⟨84JOM(270)283⟩, 290 ⟨82CL35⟩, 628 ⟨90CC1364, 92JCS(P1)3135⟩, 634, 651 ⟨84JA7890⟩; **5**, 197, 198 ⟨80TL2877⟩, 198 ⟨81JOC4413⟩, 220 ⟨82JOM(235)243⟩, 220, 222 ⟨84TL5789, 77BCJ2021⟩, 601, 602 ⟨76PS75, 72IJS89⟩
Matsuda, Y., **1**, 829 ⟨88CC1531⟩; **2**, 41 ⟨89CC1429⟩, 391 ⟨88S805⟩; **3**, 428 ⟨88S805⟩, 456 ⟨89JCS(P1)279⟩, 458 ⟨88JCS(P1)945⟩; **4**, 842 ⟨75CPB2390⟩, 844 ⟨73CPB2770, 75CPB2749⟩, 888, 900 ⟨69YZ203⟩, 901 ⟨87JHC519⟩, 979 ⟨73YZ612⟩, 983, 986 ⟨85JHC937⟩; **5**, 159 ⟨88CC1531⟩, 559 ⟨79H(12)485⟩, 560 ⟨75YZ980⟩
Matsue, H., **3**, 136, 203 ⟨90JOC5562⟩
Matsueda, E. R., **2**, 230 ⟨82CL921⟩
Matsueda, G. R., **2**, 217 ⟨81CL737⟩; **6**, 641 ⟨92JOC2497, 93SC657⟩, 641 ⟨93TL3389⟩
Matsueda, R., **2**, 217 ⟨80MI 203-02, 80MI 203-03⟩, 217 ⟨81CL737⟩, 230 ⟨82CL921⟩; **5**, 652 ⟨70TL1901⟩
Matsuguma, H. J., **2**, 691 ⟨57IS122⟩
Matsui, F., **4**, 205 ⟨76CPB102⟩
Matsui, H., **3**, 428 ⟨88BCJ461⟩; **4**, 774, 779 ⟨93MI 417-01⟩
Matsui, K., **1**, 841 ⟨70TL1175⟩, 1007 ⟨89TL2403⟩; **4**, 734 ⟨82TL2323⟩, 1053 ⟨82OM1473⟩; **5**, 1048 ⟨63BCJ1214⟩; **6**, 14 ⟨81CL1719⟩, 58, 264, 266 ⟨82TL2323⟩
Matsui, M., **1**, 503, 1052 ⟨75T1846⟩, 1134 ⟨79TL3021⟩; **2**, 627 ⟨69OPP221⟩; **4**, 225 ⟨77CAR(54)C17⟩; **6**, 71 ⟨76CAR(51)C13⟩
Matsui, S., **1**, 240 ⟨82BCJ2965⟩; **4**, 23, 27 ⟨67BCJ594⟩, 236 ⟨80BCJ3619⟩; **6**, 550 ⟨87BCJ1853⟩
Matsui, T., **1**, 382 ⟨81JOC3376⟩; **3**, 494 ⟨70BRP1202353⟩; **6**, 629 ⟨73CPB62⟩
Matsukara, T., **1**, 63 ⟨85OM1911⟩
Matsukawa, M., **1**, 680 ⟨87CL2101⟩
Matsukawa, S., **2**, 80 ⟨93CC327⟩
Matsuki, T., **1**, 1042 ⟨87JOC2094⟩
Matsukubo, H., **3**, 357 ⟨75JCS(P1)632⟩
Matsukura, H., **1**, 212 ⟨93CPB55⟩
Matsumaga, S., **4**, 888, 900 ⟨69USP3463835⟩
Matsumara, A., **2**, 563 ⟨73T1069⟩
Matsumara, S., **1**, 956 ⟨93CC1406⟩
Matsumara, Y., **1**, 17 ⟨79TL2157⟩; **2**, 323 ⟨88BCJ3029⟩; **3**, 261 ⟨82JA6697⟩
Matsumaru, N., **4**, 850 ⟨81CC565⟩
Matsumoto, A., **2**, 1054 ⟨69CC1393, 74BCJ946⟩; **3**, 36 ⟨81CC263⟩; **4**, 1033 ⟨80BCJ1138⟩
Matsumoto, H., **1**, 273 ⟨67JA5708⟩, 299 ⟨94JA2223⟩, 490 ⟨79TL1137⟩; **2**, 4 ⟨78JA290, B-82MI 223-01⟩, 21 ⟨78CL223⟩, 83 ⟨90JA1286⟩, 288 ⟨92JCR(S)393⟩, 912 ⟨78AG(E)279⟩; **4**, 407 ⟨79TL5011, 80H(14)97⟩, 1044 ⟨82CL1663⟩; **5**, 578, 608 ⟨78H(11)139⟩; **6**, 7 ⟨73TL5147, 75TL899⟩, 7 ⟨78JOC1734⟩
Matsumoto, K., **1**, 34 ⟨73JOC2731⟩; **2**, 301 ⟨91CC306⟩, 753 ⟨87CE551⟩, 920 ⟨91BCJ1461⟩; **3**, 6, 118 ⟨93CL561⟩, 317 ⟨71CJC3443⟩, 717 ⟨90S781⟩, 717, 720 ⟨88T5467⟩; **4**, 13 ⟨75TL997⟩, 310 ⟨79JOC391⟩, 311, 327 ⟨79BCJ826⟩, 410 ⟨84H(22)1417⟩, 424 ⟨82JOC1706⟩, 811 ⟨92TL7031⟩; **5**, 84 ⟨77NKK109⟩, 423 ⟨84BCJ2167⟩; **6**, 60, 62 ⟨93T8487⟩, 646 ⟨68JOC552⟩
Matsumoto, K. E., **3**, 266 ⟨71JA3299⟩
Matsumoto, M., **1**, 73, 74, 79, 80 ⟨72BCJ2723⟩, 178 ⟨86H(24)3149⟩, 1064 ⟨80TL4021⟩; **2**, 540 ⟨84JOC2298⟩, 624 ⟨80TL4021⟩; **3**, 64 ⟨81CC907, 79TL1607⟩, 634 ⟨87H(26)913⟩, 663 ⟨88T1827⟩; **5**, 223 ⟨87CPB1741⟩, 596 ⟨93SC2139⟩; **6**, 115 ⟨69BCJ2323⟩, 556, 564 ⟨87CPB2840⟩
Matsumoto, S., **1**, 452, 490 ⟨90CL185, 88CL873⟩, 674 ⟨76CL1019⟩; **2**, 916 ⟨93JOC5045⟩; **3**, 7, 120 ⟨91BCJ318⟩, 7, 31 ⟨90JOC462⟩, 309 ⟨90CPB3261⟩
Matsumoto, T., **1**, 382 ⟨81TL4313⟩, 962 ⟨93TL2161⟩, 1071 ⟨91TL6735⟩; **2**, 36 ⟨83S460⟩, 921 ⟨88T4073⟩; **3**, 157 ⟨85JOC603⟩; **4**, 653 ⟨91TL2049⟩; **5**, 9 ⟨85T3625⟩
Matsumoto, Y., **1**, 193 ⟨87TL4849⟩; **2**, 43, 528 ⟨89JA3426⟩, 51 ⟨88JA5579⟩, 78 ⟨88TL5667⟩; **3**, 43 ⟨86TL6083⟩, 101 ⟨89BCJ2736⟩, 160 ⟨94T335⟩, 707 ⟨87TL4849⟩; **4**, 516 ⟨86CL1193⟩, 1033 ⟨90TA151⟩
Matsumura, A., **1**, 115 ⟨73JOC4268⟩, 115, 452 ⟨73T1069⟩
Matsumura, C., **1**, 1035 ⟨89JOM(364)245⟩
Matsumura, H., **3**, 357, 358 ⟨78BCJ179⟩; **6**, 111 ⟨79CC485⟩
Matsumura, K., **3**, 310 ⟨86TL3899⟩, 674 ⟨70JOC675⟩; **4**, 507 ⟨91EUP488681⟩
Matsumura, N., **5**, 70, 73 ⟨80CC327⟩, 70, 73 ⟨83CC1487⟩
Matsumura, T., **6**, 462 ⟨76JOC2070⟩
Matsumura, Y., **1**, 432 ⟨79JOC4508⟩, 590 ⟨78TL2807⟩, 615, 698 ⟨78CL69⟩, 615, 698, 702 ⟨77JOC2036⟩; **2**, 323 ⟨90JA2368⟩, 457 ⟨81TL3249⟩, 486, 510 ⟨69MI 209-01⟩, 502 ⟨67MI 209-02⟩, 502, 889 ⟨68JOM(11)299⟩, 890 ⟨75JOM(96)57⟩; **3**, 7, 120 ⟨79TL165⟩, 7, 120 ⟨79TL3861⟩, 7, 120 ⟨80TL1867⟩, 17, 132 ⟨83JOC3338⟩, 43 ⟨86TL6083, 90JA2368⟩, 73 ⟨89TL5289⟩, 219 ⟨84JOC300⟩, 261 ⟨82TL1201⟩, 269, 437 ⟨83JA6312⟩, 432 ⟨86JOC549, 83JA2831⟩, 437, 438 ⟨83JA672⟩, 618 ⟨89JOC2249⟩; **4**, 136 ⟨68MI 403-01⟩, 201 ⟨69JA2803, 87S1099⟩, 300, 311 ⟨82JA5753⟩, 303 ⟨75JA4264⟩, 595 ⟨71MI 413-01, 72JOM(43)333⟩, 595 ⟨75JOM57⟩; **5**, 731 ⟨83JA2831⟩; **6**, 115 ⟨69BCJ2323⟩

Matsumur, Y., **4**, 197 ⟨89TL5309⟩
Matsunaga, A., **1**, 1077 ⟨69TL4447⟩
Matsunaga, I., **5**, 223 ⟨87CPB1741⟩; **6**, 447 ⟨86JAN230⟩
Matsunaga, K., **2**, 162 ⟨72S203⟩; **3**, 57 ⟨89TL3693⟩; **4**, 317 ⟨77CPB2964⟩
Matsunaga, S., **3**, 701, 704 ⟨92TL6823⟩; **4**, 197 ⟨91T615⟩, 808, 811, 813 ⟨89TL3155⟩; **5**, 343 ⟨92JOC4877⟩, 417, 421, 423 ⟨91JOC1307⟩
Matsunaga, S.-i., **1**, 333 ⟨89TL3155⟩; **6**, 124 ⟨90JA1962⟩, 129 ⟨91T615⟩, 133 ⟨89TL219⟩
Matsunami, N., **5**, 250 ⟨87S371⟩
Matsunami, S., **6**, 655 ⟨74BCJ935⟩
Matsuo, H., **1**, 744 ⟨60CPB819⟩; **4**, 202 ⟨91JOC2276⟩
Matsuo, M., **1**, 66 ⟨77JOM(135)1⟩
Matsuo, S., **3**, 197 ⟨92CC190⟩; **4**, 349 ⟨92CC190⟩
Matsuo, T., **2**, 19 ⟨86CC793⟩; **5**, 640, 642 ⟨80CPB2116⟩; **6**, 423, 438 ⟨86CC793⟩
Matsuo, Y., **3**, 360 ⟨92BCJ923⟩; **4**, 614 ⟨93OM2536⟩
Matsuoka, H., **4**, 708 ⟨86TL4339⟩
Matsuoka, K., **2**, 54 ⟨88JA4475⟩
Matsuoka, M., **3**, 707 ⟨87TL4849⟩
Matsuoka, R., **4**, 1064 ⟨87TL1299⟩
Matsuoka, R. T., **1**, 146 ⟨92SL27⟩
Matsuoka, S., **1**, 1009 ⟨90BCJ2124⟩; **3**, 273 ⟨87S1092⟩
Matsuoka, T., **5**, 131 ⟨87JOC3784⟩; **6**, 472 ⟨88H(27)2327, 89CPB576⟩
Matsuoka, Y., **3**, 289 ⟨81CL531⟩; **6**, 556, 564 ⟨87CPB2840⟩
Matsuoko, M., **1**, 193 ⟨87TL4849⟩
Matsura, T., **3**, 262 ⟨79JHC527⟩; **6**, 384, 671 ⟨91JA8899⟩
Matsushima, H., **2**, 160, 161 ⟨74S277⟩, 401 ⟨76T431⟩; **3**, 267 ⟨73JOC4324⟩
Matsushita, H., **1**, 456 ⟨92CL1725⟩, 476 ⟨81TL3737⟩, 937 ⟨93JCS(P1)2339⟩; **2**, 55 ⟨88BCJ3283, 88CL1633⟩, 757 ⟨77BCJ1513⟩, 975 ⟨84OS(62)31⟩, 975, 976 ⟨81JA2882⟩; **3**, 7 ⟨91BCJ312⟩
Matsushita, M., **4**, 432 ⟨93TL5097⟩
Matsushita, T., **3**, 357 ⟨90CRV879⟩, 360 ⟨91JCS(P1)2281, 92JCS(P1)2135⟩
Matsushita, Y., **1**, 942 ⟨83JCS(P1)403⟩; **3**, 292 ⟨83CPB3951⟩
Matsusita, H., **4**, 774 ⟨91TL6563⟩
Matsuto, M., **1**, 63 ⟨85OM1911⟩
Matsuura, H., **3**, 332, 364 ⟨75BCJ2907⟩; **5**, 1152 ⟨82CL981, 85JSP(111)415⟩
Matsuura, K., **1**, 156 ⟨93T537⟩
Matsuura, S., **3**, 357 ⟨91CPB672⟩
Matsuura, T., **3**, 62 ⟨86TL2757⟩, 77 ⟨73TL2283, 75T1373⟩, 418 ⟨84TL3091, 86CC980⟩, 418 ⟨87JA7888, 88TL355⟩, 427 ⟨73BCJ3198⟩; **4**, 410 ⟨93JOC6949⟩; **5**, 47 ⟨65BCJ893⟩, 299 ⟨79CC970⟩, 356 ⟨79TL2253⟩, 584, 612 ⟨75T1245⟩, 793, 796 ⟨87JA7888⟩, 796, 802 ⟨84TL3091⟩; **6**, 163, 671, 674 ⟨88JA3692⟩, 672, 674 ⟨87JA7888, 86CC980⟩, 674 ⟨88TL355⟩
Matsuura, Y., **2**, 46 ⟨88JA4606⟩
Matsuyama, H., **1**, 241 ⟨89JOC2374⟩, 704, 707, 708 ⟨86CL433⟩, 706, 708 ⟨84CL833⟩, 706, 708 ⟨87JOC1703⟩; **2**, 214 ⟨86SUL1⟩, 244 ⟨85H(23)1897⟩, 620 ⟨88BCJ2226⟩; **3**, 226 ⟨90JCS(P1)1745⟩; **5**, 936 ⟨94CL1537⟩, 946, 958 ⟨95TL3699⟩, 953 ⟨94CL777⟩
Matsuyama, K., **2**, 21 ⟨75JOC2960⟩
Matsuyama, N., **3**, 170 ⟨86CL65⟩; **4**, 618 ⟨88TL6787⟩
Matsuyama, Y., **3**, 419 ⟨87S824⟩
Matsuzaka, H., **5**, 463 ⟨88CL377⟩
Matsuzaki, E., **1**, 409, 821, 822 ⟨72JOC3791⟩
Matsuzaki, K., **1**, 430 ⟨85S676⟩; **2**, 566 ⟨67JOM(10)P9⟩; **4**, 365 ⟨84JCS(P1)435⟩, 773, 781, 783 ⟨85S556⟩; **5**, 235 ⟨93JA5302⟩
Matsuzaki, T., **6**, 465 ⟨93EUP559212⟩
Matsuzawa, S., **1**, 313 ⟨86TL4029⟩; **2**, 589 ⟨88JA1297⟩; **4**, 1051, 1053 ⟨85OM2040⟩
Matsuzawa, Y., **5**, 584 ⟨81LA1798⟩
Mattalia, J.-M., **6**, 550 ⟨92JA4111⟩
Matt, P. von, **1**, 194 ⟨93AG(E)566⟩
Mattay, J., **4**, 181 ⟨81LA1105⟩; **5**, 145 ⟨78CB2206⟩, 951 ⟨93CB1849⟩; **6**, 125 ⟨91CB1425⟩
Mattenberger, A., **1**, 220 ⟨82AG(E)71, 82AG(S)161⟩
Matter, Y. M., **1**, 325, 920 ⟨87JOC205⟩
Mattes, H., **2**, 10 ⟨87TL1697⟩
Mattes, K., **3**, 539 ⟨73JA6134⟩
Matteson, D., **4**, 144, 362, 373, 525 ⟨89CRV1535⟩, 631, 635 ⟨90JOC2274⟩
Matteson, D. S., **1**, 65 ⟨69MI 102-01⟩, 126 ⟨B-87MI 104-04⟩, 138 ⟨89JA4399⟩, 163 ⟨70JOM(24)263⟩, 163 ⟨76JOM(114)1⟩, 163 ⟨77JA3196⟩, 163 ⟨82OM20⟩, 163 ⟨85OM1690⟩, 258 ⟨57JA2217⟩, 447 ⟨B-87MI 110-01⟩, 651, 762 ⟨81TL2751⟩, 660 ⟨65JA1526, 89T1859⟩, 764 ⟨83OM236⟩, 787 ⟨87JOC5034⟩; **2**, 529 ⟨83JA2077, 86S973⟩, 529 ⟨93JOC6545⟩, 529, 530, 935 ⟨81TL2751⟩, 529, 900, 904 ⟨89CRV1535⟩, 529, 900, 904, 905 ⟨89T1859⟩, 530 ⟨87TL4499, 59JA5004⟩, 904 ⟨60JA4228, 65JA1526⟩, 905 ⟨75JA5608, 80CC39⟩, 907 ⟨63JOC2171⟩, 1005 ⟨85OM1690⟩; **3**, 35 ⟨82JOC2479⟩, 167 ⟨82OM280⟩, 712, 714, 715 ⟨68JA3761⟩; **4**, 144 ⟨63JOC2171⟩, 144 ⟨90S200⟩, 144, 146 ⟨91SL631⟩, 144, 373 ⟨68JOC3055⟩, 146 ⟨86TL795⟩, 362 ⟨80JA7588, 80JA7590⟩, 362 ⟨83JA2077, 86TL3831⟩, 362 ⟨87JOC5116⟩, 362, 372, 373 ⟨B-87MI 408-01⟩, 362, 373, 526 ⟨83OM1529⟩, 363 ⟨87JOC5121⟩, 372, 373, 525 ⟨78JA1325⟩, 389, 664, 665 ⟨89JA4399, 79JOM(170)259⟩, 525 ⟨89T1859, 91PAC339⟩, 525, 526, 527 ⟨81JA5241⟩, 525, 632, 635 ⟨66JOM(6)100⟩, 526 ⟨83OM1536, 87TL4499⟩, 526, 527 ⟨84OM1284, 84OM614, 92MI 411-01⟩, 528 ⟨89B3541⟩, 631, 634 ⟨63JA2684⟩, 631, 697 ⟨64JOC2742⟩, 632, 632, 633 ⟨69JOM(20)19⟩, 632, 633 ⟨75S147, 68JA2194⟩, 633 ⟨70ACR186, 70JOM(24)263⟩, 633 ⟨76JOM(110)25⟩, 634 ⟨77JA3196⟩, 634, 664 ⟨82OM20⟩, 642 ⟨83OM236⟩, 642, 645 ⟨80CC39⟩, 642, 645 ⟨83OM230⟩, 664, 665 ⟨85OM1690, 76JOM(114)1⟩, 697 ⟨73JOM(54)35⟩, 750, 1067 ⟨78JOM(152)1⟩, 813, 934 ⟨90HAC135⟩, 814 ⟨63JOC369, 74JOM(69)53⟩, 849 ⟨79JOC1352⟩, 1054 ⟨75JOM(93)21, 78JOC950⟩; **6**, 61 ⟨69JOM(16)33⟩, 124, 135 ⟨78JA1325⟩, 181 ⟨73IC2472⟩, 181, 182 ⟨70JOM(24)263⟩, 181, 182, 198 ⟨75S147⟩, 181, 182, 292 ⟨75JOM(93)21⟩, 181, 186, 199, 200 ⟨76JOM(110)25⟩, 181, 199 ⟨75JA5608⟩, 181, 199, 292, 386 ⟨74JOM(69)45⟩, 181, 386 ⟨68JA2194⟩, 181, 386 ⟨69JOM(20)19⟩, 181 ⟨B-77MI 606-01⟩, 182 ⟨70JOM(21)P6⟩, 182 ⟨78JOM(152)1⟩, 182, 183, 199, 292 ⟨73JA5096⟩, 182, 183, 200 ⟨74JOM(69)63⟩, 182, 199 ⟨78JOC950⟩, 182, 396 ⟨74JOM(69)53⟩, 183, 200, 292, 386, 400 ⟨73JOM(57)231⟩, 183, 200, 396 ⟨73JOM(57)225⟩, 183, 200, 396, 400 ⟨69JA6541⟩, 199 ⟨80JOC1091⟩, 199 ⟨82OM20⟩, 199, 201 ⟨76JOM(114)1, 85OM1690⟩, 203 ⟨80CC39, 80JOM(184)C41⟩, 203 ⟨83OM230⟩, 405 ⟨70JA231⟩
Matteucci, M., **4**, 216 ⟨91JA7767⟩, 219 ⟨90TL2385⟩
Matthes, D., **4**, 1026, 1030 ⟨75TL3171⟩, 1038 ⟨76TL2599⟩
Matthew, B. R., **2**, 317 ⟨90TL1447⟩
Matthews, A. O., **5**, 115 ⟨32OSC(1)392⟩
Matthews, B. R., **3**, 628 ⟨88AJC1697⟩
Matthews, C. N., **3**, 547, 549, 604 ⟨68JA3842⟩, 549, 604 ⟨66TL5707⟩, 602 ⟨69ACR373, 66CC736⟩, 604 ⟨67CC137, 67JOC3554⟩, 604 ⟨68CI(L)653⟩; **4**, 553 ⟨63USP3098878⟩, 554 ⟨64JOC2427, 65T5⟩; **5**, 1093 ⟨62JA4349, 64JOC2427⟩
Matthews, D. P., **1**, 439 ⟨90TL5449⟩, 440 ⟨91JA7439, 93OS(72)216⟩, 967 ⟨87TL2207⟩; **2**, 611 ⟨90TL973, 93TL3057⟩; **3**, 673 ⟨86JOC3228⟩; **4**, 52 ⟨90TL973⟩, 61, 79, 780, 782 ⟨90TL5449⟩, 73 ⟨89JA1127⟩, 79 ⟨90JOC2973⟩, 736 ⟨93TL3057⟩, 812 ⟨91JA7439, 94TL1027⟩; **5**, 1127 ⟨85TL6273⟩; **6**, 266 ⟨90TL5571⟩
Matthews, E. W., **5**, 1133 ⟨87JA6532⟩

Matthews, J. S., **5**, 652 ⟨69JOC3204⟩
Matthews, R. S., **1**, 218 ⟨85JOC3923⟩, 221 ⟨83JOC409⟩; **2**, 1098 ⟨85JOC3923⟩
Matthews, R. W., **4**, 659 ⟨80JOM(193)21⟩
Matthews, V., **6**, 54 ⟨67JCS(C)2263⟩, 604 ⟨64JCS1351⟩
Matthews, W. S., **1**, 142, 155 ⟨75JA7006⟩, 618 ⟨74TL1105⟩, 618, 706 ⟨69JA7510, B-77MI 115-01⟩; **4**, 81 ⟨69JA7510⟩
Matthey, G., **3**, 92, 101 ⟨63CB308⟩
Matthies, D., **3**, 417 ⟨85S177⟩; **4**, 98 ⟨87CZ247⟩, 101 ⟨86CB2387⟩, 101, 424 ⟨87CB1897⟩, 298 ⟨85AP(318)473⟩, 311 ⟨68AP(301)867⟩; **5**, 606 ⟨74AP(307)801, 77AP(310)1001⟩
Matti, G. Y., **3**, 582 ⟨90JA3779⟩
Mattiew, E., **2**, 436 ⟨84JOM(269)107⟩
Mattingly, P. G., **5**, 273, 285 ⟨80JA7026⟩
Mattingly, T. W., **2**, 298 ⟨63JA1200⟩
Mattinson, B. J. H., **6**, 240 ⟨57JCS1741⟩
Mattioda, G., **3**, 40 ⟨88BSF67⟩
Mattocks, K. L., **4**, 466 ⟨82S188⟩
Mattox, W. J., **5**, 549 ⟨30JA332⟩
Mattschei, P. K., **6**, 181 ⟨72MI 606-01⟩, 181 ⟨73IC2472⟩, 182, 183, 386 ⟨75LA1339⟩
Mattson, R. J., **2**, 310 ⟨90JOC2552⟩; **3**, 44, 188 ⟨90JOC2552⟩
Matulenko, M. A., **1**, 583 ⟨94JOC332⟩
Matulianskiene, R., **5**, 690 ⟨87KGS1131, 87CHE907⟩
Matulic-Adamic, J., **6**, 58 ⟨94TL3227⟩
Maturová, E., **5**, 655 ⟨84CCC1505⟩
Matusbara, Y., **5**, 694 ⟨87NKK1304⟩
Matusch, R., **5**, 554 ⟨76ZC318, 80CB1898⟩; **6**, 538 ⟨80CB1898⟩
Matusiak, G., **2**, 1009 ⟨85H(23)1513⟩
Matusiak, M., **4**, 491 ⟨92S367⟩
Matuszak, C. A., **1**, 97 ⟨78JOC1007⟩
Matuszak, M. P., **6**, 429 ⟨34JA2007⟩
Matuszewski, B., **1**, 421 ⟨82JA6874⟩
Matuszko, A. J., **5**, 821 ⟨63JOC2260⟩
Matveenko, V. I., **2**, 170 ⟨88IZV2367⟩
Matveeva, E. D., **2**, 7 ⟨89JOU587, 89JOU642⟩
Matveeva, E. V., **2**, 440 ⟨85URP1145022⟩
Matveeva, L. V., **2**, 451 ⟨67ZOB1692⟩
Matveev, Y. I., **6**, 620 ⟨92ZOB324⟩
Matveev, Yu. I., **5**, 658 ⟨89JOU1365⟩, 658, 1068 ⟨89ZOR1572⟩, 1068, 1069 ⟨84ZOR2543, 88ZOR986⟩, 1075 ⟨87ZOR2385⟩
Matyukhina, E. N., **2**, 883 ⟨65JGU1253⟩
Matyushe, A. G., **2**, 463 ⟨82JGU195, 82JGU202⟩
Matyushecheva, G. I., **4**, 793 ⟨77JOU1918⟩
Matz, J. R., **1**, 690 ⟨82TL1777, 82TL2723⟩, 690 ⟨84JA5018, 86JA2932⟩, 690 ⟨86JA515, 87T5489⟩, 805 ⟨80JA6900, 83JA520⟩; **2**, 422, 551 ⟨80SC311⟩; **4**, 68 ⟨81TL2455⟩, 216, 218, 226, 380 ⟨80JA6900⟩, 366, 394, 945, 1060 ⟨84JA3245⟩; **5**, 414 ⟨80JA6900⟩
Matzcit, A., **1**, 561 ⟨93CB1623⟩
Matzke, T., **4**, 580, 581 ⟨90CB739⟩
Matzner, M., **6**, 421, 426, 460 ⟨64CRV645⟩, 425 ⟨65CI(L)791⟩
Mauermann, H., **2**, 969 ⟨85JA8111⟩; **6**, 195 ⟨85JA8091⟩
Maughan, W., **2**, 299 ⟨84S941⟩; **3**, 188 ⟨84S941⟩
Maugras, I., **5**, 333 ⟨88TL2661⟩
Maujean, A., **3**, 65 ⟨73TL1635⟩
Maul, R., **5**, 1145 ⟨68SAP6903273⟩
Maulding, D. R., **6**, 252 ⟨72JOC1458⟩
Mauleón, D., **3**, 666 ⟨83JOC3105⟩
Maumy, M., **1**, 343 ⟨67BSF2641, 69BSF2427⟩; **2**, 1066 ⟨72BSF1600⟩; **5**, 750 ⟨93TL6395⟩
Maurer, D., **3**, 424 ⟨92JOC229⟩
Maurer, H., **2**, 492 ⟨86JFC(34)129⟩
Maurer, P. J., **3**, 180, 284 ⟨84JA1095⟩
Maurin, R., **1**, 498 ⟨80JCR(S)98, 85T2759⟩, 805 ⟨77TL1779, 80JCR1551⟩, 979 ⟨92TL1057⟩; **2**, 351 ⟨92TL1057⟩; **3**, 216 ⟨75T849⟩; **4**, 791 ⟨81H(15)1179⟩
Maury, P. B., **5**, 194 ⟨54JA5803⟩
Mauthner, F., **6**, 27 ⟨25JPR125⟩
Mautner, H. G., **2**, 278 ⟨63MI 204-01, 68JMC446⟩, 278, 279, 292 ⟨65JMC845⟩, 279, 292 ⟨64JMC229⟩; **5**, 516, 520, 522 ⟨72JMC760⟩, 563 ⟨70JMC214⟩, 625 ⟨62JOC3671⟩; **6**, 598 ⟨56JA97, 59JA6270⟩, 637, 732 ⟨62JOC2899⟩
Mautner, K., **6**, 133 ⟨89JOC4042⟩
Mautz, D. S., **1**, 840 ⟨91JA4895⟩
Maux, P. Le, **3**, 699 ⟨78JA4312⟩
Mauz, E., **4**, 581 ⟨91AG(E)998⟩
Mauz, O., **6**, 302 ⟨73GEP1694210⟩
Mauzé, B., **1**, 242 ⟨85S1069⟩, 544 ⟨72JOM(44)69⟩; **2**, 74 ⟨82JOM(236)139⟩; **3**, 639 ⟨92SC2229⟩, 650 ⟨90SC2251⟩; **4**, 1065 ⟨77JOM(131)321⟩, 13 ⟨77JA5317⟩, 704 ⟨91JOM(411)69⟩
Mavity, J. M., **2**, 595 ⟨40JOC106⟩
Mavlyutov, R. F., **6**, 479 ⟨90ZOB1826⟩
Mavrin, G. V., **6**, 659 ⟨90ZOB1986, 90ZOB2694⟩
Mavrin, V. Yu., **6**, 156, 313 ⟨88ZOB2167⟩
Mavunkel, B., **4**, 905, 911 ⟨85CC845⟩
Maw, G. N., **1**, 614, 697, 700, 703 ⟨90TL5525⟩; **5**, 66 ⟨92OS(71)48⟩; **6**, 628, 633 ⟨85TL1105⟩
Maw, W., **6**, 467 ⟨24JCS(125)2259⟩
Mawby, R. J., **5**, 447 ⟨64JA3994⟩, 455 ⟨76CC1032⟩, 465 ⟨74JCS(D)421⟩
Mawhinney, T. P., **2**, 263 ⟨82JOC3336⟩; **5**, 719 ⟨82JOC3336⟩
Maximov, O. B., **6**, 138 ⟨89T3487⟩
Maxwell, J. R., **1**, 677 ⟨88TL2599⟩
Maxwell, R. D., **1**, 1144 ⟨36JA2310⟩
Maxwell, R. J., **5**, 1109 ⟨78MI 530-01⟩
May, A., **3**, 370 ⟨90CB1475⟩
May, C. J., **4**, 692 ⟨77JA4707⟩
May, D. D., **5**, 203, 205 ⟨83JA3999⟩
May, D. R., **2**, 126 ⟨46MI 203-01⟩
May, E. M., **2**, 190 ⟨38JOC175⟩
May, H. J., **4**, 198 ⟨66AG(E)420⟩; **6**, 645 ⟨80AF1733⟩
May, K. D., **1**, 708 ⟨75TL947⟩
May, L. M., **3**, 57 ⟨79JA7131⟩
May, P., **5**, 566 ⟨13JCS2272⟩
May, P. D., **5**, 329 ⟨91JOC5482⟩
May, P. J., **4**, 56, 57 ⟨64JCS766⟩
May, R. G., **5**, 85 ⟨88JA2237⟩
May, R. W., **1**, 772 ⟨67JCS(B)557⟩
Maya, W., **6**, 231 ⟨69JA2902⟩
Mayatepek, H., **2**, 462 ⟨68CB3454⟩
Maybhate, S. P., **6**, 627 ⟨91S220⟩
Maycock, A. L., **1**, 1135 ⟨76ACR313⟩
Maycock, C. D., **2**, 566 ⟨81HCA1552⟩; **3**, 8, 121 ⟨87JCS(P1)1221⟩; **4**, 396 ⟨81HCA357⟩; **6**, 134 ⟨81HCA357⟩
Maydonovitch, D. J., **4**, 687 ⟨80JA5101⟩
Mayence, A., **1**, 260, 261 ⟨89S323⟩; **4**, 417 ⟨89CC337⟩
Mayer, A., **3**, 516 ⟨87OM433⟩
Mayer, B., **1**, 708 ⟨81AG(E)677⟩; **4**, 1049 ⟨85CB2493⟩; **6**, 390 ⟨85CB2493⟩
Mayer, C., **3**, 348 ⟨75ZN(B)760⟩
Mayer, E. P., **3**, 595 ⟨93CB1551⟩; **5**, 381, 534 ⟨93CB1119⟩
Mayer, G. D., **6**, 314, 565 ⟨81JCS(P1)52⟩
Mayer, H., **3**, 577 ⟨61JA1010⟩
Mayer, H. A., **2**, 437 ⟨90CB1827⟩; **6**, 154 ⟨93AG(E)1424⟩
Mayer, J. M., **3**, 508 ⟨B-88MI 315-01⟩, 517 ⟨87JA7213⟩
Mayer, K. K., **2**, 163 ⟨68TL319⟩; **4**, 549 ⟨87AG(E)236⟩, 686 ⟨82JOM(236)C18⟩; **5**, 1013 ⟨68TL319⟩
Mayer, M. D., **2**, 1006 ⟨81LA1415⟩
Mayer, R., **1**, 24 ⟨66CB1393⟩; **2**, 130 ⟨62CB1764, 63AG(E)370⟩, 361, 792 ⟨78TL4003⟩, 386, 399 ⟨85HOU(E11)558⟩, 398 ⟨78ZC323⟩, 398 ⟨85PS(23)277, 76ZC437⟩, 399 ⟨82ZC237, 85HOU(E11)579⟩, 399 ⟨85HOU(E11)584⟩, 401 ⟨85HOU(E11)1031⟩; **3**, 342 ⟨64AG(E)277⟩, 343 ⟨67CB93⟩, 344, 363 ⟨63AG(E)370⟩, 347 ⟨63CB3096⟩, 349 ⟨66CB1771⟩, 361 ⟨75TL2683⟩,

364 ⟨68CB3517⟩, 546 ⟨75ZC91⟩, 695, 721 ⟨77S275⟩; **4**, 245 ⟨62CB1764⟩, 256 ⟨88ZC269⟩, 850 ⟨64CB1298⟩; **5**, 226 ⟨85HOU(E5/1)615⟩, 476, 477 ⟨75ZC348⟩, 476, 477, 484 ⟨65CB829⟩, 476, 477, 510 ⟨63JPR214⟩, 480 ⟨B-79MI 511-02⟩, 513, 592, 593, 620 ⟨68CB3517⟩, 515, 516 ⟨66CB1393⟩, 546, 549 ⟨85HOU(E5)891⟩, 546, 549 ⟨B-79MI 513-02⟩, 550 ⟨76ZC355⟩, 554 ⟨77ZC366, 78ZC90⟩, 554 ⟨89JPR243⟩, 559 ⟨64CB1886⟩, 560 ⟨88ZC269⟩, 561 ⟨77ZC92⟩, 565 ⟨88SR1⟩, 584, 586, 609 ⟨64ZC457⟩, 586 ⟨72ZC133⟩, 589, 603, 611 ⟨67JPR294⟩, 599 ⟨64AG(E)705, 67AG(E)294⟩, 601 ⟨78JPR313, 88ZC233⟩, 1104 ⟨68AG(E)67⟩, 1111 ⟨64AG861⟩; **6**, 301, 303, 309, 310 ⟨64ZC384⟩, 500 ⟨82JPR227⟩, 628, 631 ⟨71JPR804⟩
Mayer, R. P., **1**, 385 ⟨59JA1497⟩
Mayer, T., **4**, 321 ⟨89CB2311⟩
Mayer, U., **1**, 1084 ⟨63CB342⟩
Mayer, W., **6**, 69 ⟨67CB2946⟩
Mayes, N., **1**, 168 ⟨63IC1111⟩
Mayhew, S. G., **5**, 1111 ⟨92B7922⟩
Maynard, G. D., **1**, 414 ⟨92JA5018⟩, 883 ⟨89JOC5044, 89JOC5054⟩, 1098 ⟨90JA277⟩; **5**, 425 ⟨91JOC5480⟩
Maynard, J. T., **4**, 30 ⟨63JOC112⟩; **6**, 18 ⟨63JOC112⟩
Maynard, R. B., **4**, 589 ⟨78JA5562, 81IC2466⟩, 589 ⟨84JA5920, 90IS177⟩; **5**, 920 ⟨81IC2466⟩, 1096 ⟨80IC2564⟩
Maynard, S. C., **4**, 961 ⟨85TL169⟩
Maynert, E. W., **2**, 643 ⟨45JA1939⟩, 645 ⟨45JA1937⟩
Maynes, G. G., **1**, 928 ⟨73JA856⟩
Mayo, D. T., **1**, 804 ⟨80JA6163⟩
Mayo, F. R., **1**, 923 ⟨67JA2654⟩; **2**, 4, 116, 117, 439 ⟨40CRV351⟩, 190 ⟨38JOC175⟩, 190 ⟨39JA3093⟩; **4**, 23 ⟨40CRV351⟩
Mayo, G. L., **2**, 1031 ⟨59JA3386⟩
Mayo, P. de, **1**, 356 ⟨68CC982, 82CJC425⟩, 377, 380, 382, 395, 414 ⟨B-63MI 109-01⟩; **2**, 228 ⟨73TL1561⟩; **3**, 335, 337, 351 ⟨76CJC537⟩, 335, 346, 351 ⟨78NJC331⟩, 337, 351 ⟨74CC409⟩, 347 ⟨78TL4621⟩, 358 ⟨79JOC1977⟩, 359 ⟨74CC877, 77CJC3763⟩, 362 ⟨74JCS(P1)2559⟩, 363 ⟨76JA6218⟩, 365 ⟨73JA2383, 74JA3502⟩, 375 ⟨69CJC4509, 71JA6304⟩, 375 ⟨74JA2964⟩; **5**, 634 ⟨74TL91⟩
Mayo, S. L., **3**, 363 ⟨85S639⟩
Mayoral, J., **6**, 563 ⟨81S961⟩
Mayr, A., **3**, 690 ⟨91AOC(32)222⟩, 692 ⟨86JA548⟩, 699 ⟨75AG(E)369⟩; **4**, 1019 ⟨82JOM(224)153⟩; **5**, 446 ⟨92CB1583⟩, 799 ⟨82JOM(224)153⟩, 942 ⟨91AOC(32)227⟩; **6**, 517 ⟨82IC1704⟩, 723 ⟨91AOC(32)227⟩
Mayr, H., **1**, 294 ⟨90AG(E)1371⟩; **3**, 70 ⟨88CB191⟩; **4**, 188 ⟨88JOC2920⟩, 192 ⟨89S128⟩, 637 ⟨93JA6025⟩; **6**, 252 ⟨88S961⟩
Mayrhofer, R., **5**, 374 ⟨86AP(319)203⟩
Mays, M., **4**, 876 ⟨87SM647⟩
Mays, M. D., **4**, 876 ⟨88SMB425⟩
Mays, M. J., **4**, 675 ⟨88NJC559, 89CC688⟩
Mays, R. F., **2**, 322 ⟨88SC205⟩
Mazalov, S. A., **6**, 12 ⟨69IZV196⟩
Mazanek, M., **6**, 566 ⟨92MI 617-02⟩
Mazany, A. A., **4**, 586 ⟨88IC836⟩
Mazany, A. M., **6**, 332 ⟨87OM283⟩
Mazdiyasni, H., **5**, 214 ⟨88JMC1960⟩
Mazepa, I. K., **2**, 453 ⟨78BRP1517876⟩
Mazerolles, P., **1**, 656 ⟨84JOM(267)93⟩, 1000 ⟨60BSF856⟩; **2**, 814, 815 ⟨B-71MI 215-02⟩, 946, 948, 949 ⟨B-71MI 218-01⟩, 948 ⟨61CR(253)673, 66BSF327⟩; **3**, 502 ⟨81T2875⟩; **4**, 141 ⟨72BSF1361⟩; **6**, 172 ⟨B-89MI 606-02⟩, 582 ⟨79TL3507⟩
Maziak, L., **5**, 574 ⟨83TL3815, 86JA182⟩
Maziarczyk, H., **6**, 546 ⟨90MI 617-03⟩
Mazid, M., **3**, 355 ⟨91AG(E)871⟩
Mazid, M. A., **5**, 781 ⟨93POL563⟩
Mazières, M. R., **3**, 689 ⟨92TL2981⟩; **5**, 1158 ⟨92TL2981⟩; **6**, 660 ⟨91TL2775⟩, 690 ⟨92TL2981⟩
Mazik, M., **6**, 621 ⟨93H(36)1521⟩
Mazilkina, M. V., **6**, 656 ⟨81ZOB880⟩
Mazouz, A., **3**, 636 ⟨86CC424⟩
Mazouz, F., **5**, 22 ⟨91JHC1511⟩, 640 ⟨93JMC1157⟩
Mazumdar, S. N., **2**, 249 ⟨87TL2641⟩; **6**, 636 ⟨87TL2641⟩
Mazume, H., **3**, 361, 362 ⟨91JHC1245⟩; **4**, 844 ⟨91JHC1245⟩
Mazume, T., **5**, 232 ⟨65BCJ320⟩
Mazur, D. J., **3**, 331 ⟨86JOC1556⟩
Mazur, R. H., **1**, 383 ⟨59JA4390⟩; **2**, 11 ⟨51JA2509⟩, 214 ⟨73JOC2965⟩
Mazur, Y., **1**, 869, 870 ⟨67JOC2162, 67JA1183⟩, 881 ⟨57JA2906, 58JA5220⟩, 881 ⟨58JA6296⟩, 1111 ⟨69CC785⟩; **2**, 26, 33 ⟨63CI(L)609⟩, 86 ⟨75JOC2141, 88OSC(6)43⟩, 208 ⟨71JOC532, 71JOC540⟩, 397 ⟨77JOC844⟩; **4**, 204 ⟨69T1679⟩, 742 ⟨70JA3203⟩
Mazurek, M. A., **2**, 444 ⟨78S588⟩; **4**, 51 ⟨78S588⟩; **6**, 71 ⟨65CJC1918⟩
Mazurkiewicz, W., **5**, 483, 590 ⟨86T739⟩; **6**, 274 ⟨86T739⟩
Mazza, A., **5**, 657 ⟨78JOC381⟩
Mazzanti, G., **2**, 209 ⟨77JCS(P1)1469⟩, 245 ⟨81JCS(P1)2322⟩, 938 ⟨91TL2971⟩; **3**, 368 ⟨71JCS(B)1547, 76CC431⟩, 368 ⟨79JCS(P1)1720⟩, 369 ⟨86CC964⟩; **4**, 232 ⟨85TL2131, 91TL2971⟩, 371, 944 ⟨89JCS(P1)2083⟩, 859 ⟨92TL6383⟩; **5**, 562 ⟨92TL6383⟩, 647 ⟨81CC822⟩, 647, 648 ⟨86JCS(P1)381, 88CC365⟩, 648 ⟨88JCS(P1)1499⟩, 648, 650 ⟨89JCS(P1)2083⟩, 1038 ⟨71S378⟩; **6**, 127 ⟨81CC822, 87JCS(P1)2647⟩, 127 ⟨91TL2971, 91TL6801⟩, 127, 128 ⟨90H(31)47⟩, 127, 309 ⟨85TL2131⟩
Mazzei, A., **2**, 602 ⟨73ICA319⟩
Mazzei, M., **3**, 361 ⟨81JHC863⟩; **4**, 979 ⟨82EUP49618⟩
Mazzocchi, P. H., **3**, 80 ⟨77CC53⟩, 92, 97 ⟨87TL2473⟩; **5**, 325 ⟨89JOC5746⟩
Mazzucato, U., **5**, 1038 ⟨61JA2729⟩
Mbafor, J. T., **3**, 655, 656 ⟨84T2141⟩; **4**, 979 ⟨88JCS(P1)961⟩
M'Boula, J., **2**, 935 ⟨81TL2985⟩; **3**, 211 ⟨81TL2985⟩
M'boungou-M'passi, A., **2**, 637 ⟨91AG(E)416⟩
Mbuyi, M., **5**, 334 ⟨89BSB395⟩
McAdam, D. P., **6**, 589, 593 ⟨88AJC549⟩, 612 ⟨88AJC563⟩
McAdams, III, L. V., **1**, 620 ⟨71JA476⟩; **2**, 29 ⟨88OSC(6)403⟩; **3**, 479 ⟨71JA476⟩; **4**, 83 ⟨88OSC(6)403⟩, 83, 85, 86 ⟨71JA476⟩
McAfee, F., **2**, 735 ⟨84JOC3106⟩
McAfee, M. J., **1**, 811 ⟨88TL975⟩
McAlees, A., **4**, 151 ⟨88JCS(D)1773⟩
McAlees, A. J., **3**, 706 ⟨90JCS(P1)2115⟩; **4**, 150 ⟨89JCS(D)761⟩
McAllan, D. T., **1**, 100 ⟨51JCS2556⟩; **2**, 212, 219 ⟨51JA3627⟩; **5**, 199 ⟨52JCS1363⟩
McAndrew, B. A., **1**, 832 ⟨72JCS(P1)367⟩
McArthur, C. R., **6**, 445 ⟨73CJC333⟩
McArthur, R. E., **6**, 212, 217 ⟨40JA3477⟩, 216 ⟨50IS37⟩
McAuliffe, C. A., **1**, 446 ⟨B-94MI 110-02⟩; **2**, 440 ⟨88ICA(146)111⟩, 872, 874 ⟨75JOM(88)171⟩, 872, 896 ⟨B-79MI 217-02, 76JCS(D)1429⟩, 881 ⟨87AG(E)264⟩, 883 ⟨71IC2676⟩
McBain, D. S., **5**, 318 ⟨86JOC4959⟩
McBee, E. T., **4**, 33 ⟨53JA4091⟩, 64, 65, 80 ⟨52JA3594⟩, 743 ⟨72JOC683⟩, 826 ⟨47IEC412⟩; **5**, 320 ⟨73OSC(5)663⟩; **6**, 16, 44, 45, 234 ⟨52JA3594⟩, 220 ⟨47JA947⟩, 220, 225 ⟨47IEC404⟩, 229 ⟨72JOC2651⟩, 244 ⟨54JA474⟩
McBrady, J. J., **5**, 814 ⟨73JOC1065⟩; **6**, 261 ⟨67JOC1944⟩, 279, 281 ⟨73JOC1065⟩
McBrearty, Jr., C. F., **2**, 414, 504 ⟨68JOM(12)377⟩
McBride, B. J., **1**, 241 ⟨83JOC1362⟩, 524 ⟨83TL1675⟩, 1009 ⟨89JOC249⟩
McBride, J. J., **3**, 711 ⟨52JA5247⟩

McBride, J. M., **1**, 104 ⟨71JA1544⟩, 922 ⟨74T2009⟩; **2**, 404 ⟨70JA4586⟩
McBride, Jr., J. J., **5**, 1133 ⟨52JA5247⟩
McBride, L. J., **2**, 409, 410 ⟨83TL245⟩
McBride, W. R., **2**, 388 ⟨57JA572, 59JA5546⟩, 388, 389 ⟨69JOC2997⟩
McCabe, P. H., **2**, 222 ⟨80CC100⟩
McCaffrey, D. J. A., **2**, 433 ⟨79JCS(P1)3029⟩
McCague, R., **1**, 4, 9 ⟨91JCS(P1)656⟩, 576 ⟨87JCS(P1)1011⟩; **4**, 650 ⟨91JCS(P1)2276⟩
McCaig, A. E., **2**, 354 ⟨93TL3939⟩
McCain, J. H., **3**, 664 ⟨66JA2884⟩
McCaleb, G. S., **6**, 504, 506 ⟨63JMC669⟩
McCall, M. A., **5**, 580, 615, 616 ⟨62JOC2433⟩
McCallum, D., **2**, 373, 376, 403 ⟨84JOC336⟩
McCallum, J. S., **1**, 551 ⟨81JA7677⟩; **2**, 670 ⟨81JA7677⟩
McCallum, K. S., **3**, 565 ⟨62T79⟩; **6**, 150 ⟨62T79⟩
McCallum, R. A., **5**, 84 ⟨57ANY(69)94⟩
McCallum, R. J., **1**, 142, 155 ⟨75JA7006⟩
McCandlish, Jr., C. S., **2**, 25, 31 ⟨75JOC134⟩
McCandlish, L. E., **4**, 697 ⟨75JOM(99)31⟩
McCanna, T. D., **1**, 523 ⟨85TL5393⟩
McCann, P. P., **6**, 487 ⟨91JHC1013⟩
McCann, S., **1**, 826 ⟨89JOC5849⟩
McCants, D., **2**, 144, 149 ⟨65JA1109⟩
McCarney, C. C., **3**, 530 ⟨75JCS(P1)1600⟩
McCarry, B. E., **1**, 1122 ⟨73JA4416⟩; **3**, 219 ⟨71JA4332⟩
McCarthy, A. R., **5**, 641 ⟨68CC499, 74JCS(P1)627⟩
McCarthy, B. A., **1**, 346 ⟨93JOC6217⟩
McCarthy, D. G., **3**, 560 ⟨80JCS(P2)579⟩, 687 ⟨80JA4537⟩; **5**, 226 ⟨80JA4537⟩, 714 ⟨77JCS(P2)1085⟩, 714, 723 ⟨77JCS(P2)1080⟩
McCarthy, G. D., **2**, 257 ⟨56CJC1819⟩
McCarthy, J. R., **1**, 439 ⟨90TL5449⟩, 440 ⟨91JA7439, 93OS(72)216⟩, 967 ⟨87TL2207⟩; **2**, 309 ⟨90TL3417⟩, 611 ⟨90TL973, 93TL3057⟩; **3**, 673 ⟨86JOC3228⟩; **4**, 52 ⟨90TL973⟩, 61, 62, 73, 79, 780 ⟨85JA735⟩, 61, 79, 780, 782 ⟨90TL5449⟩, 73 ⟨89JA1127⟩, 79 ⟨90JOC2973⟩, 79, 780 ⟨85CC678⟩, 119, 734 ⟨90TL5571⟩, 736 ⟨93TL3057⟩, 755, 782 ⟨92JA360⟩, 755, 803 ⟨93TL7197⟩, 812 ⟨91JA7439, 94TL1027⟩, 1080 ⟨87CC469⟩; **5**, 1127 ⟨85TL6273⟩; **6**, 10, 84 ⟨86TL4861⟩, 266 ⟨90TL5571⟩
McCarthy, Jr., J. R., **5**, 751 ⟨76BEP831934, 77BRP1466047⟩
McCarthy, K., **3**, 419 ⟨89JA371⟩
McCarthy, K. E., **1**, 120 ⟨86T2873⟩, 207, 312 ⟨85JOM(285)437⟩; **2**, 588 ⟨86T2873⟩, 589 ⟨85JOM(285)437⟩
McCarthy, P. A., **2**, 91 ⟨82TL4199⟩
McCarthy, W. C., **3**, 104 ⟨63OSC(4)918⟩; **5**, 990, 1037 ⟨77JOC1508⟩
McCarty, C. G., **3**, 435 ⟨B-70MI 311-01⟩
McCarty, F. J., **5**, 652 ⟨68JMC534⟩
McCarty, J. E., **1**, 624 ⟨54JA5740⟩
McCasland, G. E., **4**, 229 ⟨64JOC724⟩
McCaulay, D. A., **1**, 897 ⟨50IEC2103⟩
McCauley, J. P., **2**, 150 ⟨84JOC1465⟩
McClanahan, J. L., **5**, 985 ⟨63CI(L)1280⟩
McClard, R. W., **4**, 127 ⟨88PS(39)27⟩, 553, 562 ⟨87JA5544⟩, 563 ⟨89TL411⟩
McClaugherty, D. C., **5**, 723, 827 ⟨71JOC284⟩
McClean, D. C., **1**, 625, 783 ⟨55JA1628⟩
McCleery, P. P., **5**, 401 ⟨89JCS(P1)1319⟩; **6**, 117 ⟨89JCS(P1)1319⟩
McCleland, C. W., **2**, 24 ⟨B-83MI 201-02⟩
McClellan, M. B., **6**, 254 ⟨68USP3376314⟩
McClelland, A. L., **6**, 231 ⟨65USP3202718⟩
McClelland, R. A., **2**, 649 ⟨87JA2824⟩; **4**, 1072 ⟨89JA4967⟩; **6**, 80 ⟨81JOC886⟩
McClenaghan, I., **5**, 690 ⟨92JCS(P1)2789⟩
McClenaghnan, J., **4**, 613 ⟨91OM3292⟩
McClennen, W. R., **6**, 247 ⟨61JA1598⟩
McClevarty, J. A., **2**, 478 ⟨68JCS1133⟩
McCleverty, J. A., **5**, 387, 388 ⟨86JCS(D)813⟩, 455 ⟨63JCS4096⟩; **6**, 523 ⟨63JCS1133⟩
McClinton, D. A., **6**, 9, 14, 228 ⟨92T6555⟩, 11 ⟨92T6585⟩
McClinton, M. A., **6**, 9, 14, 228 ⟨92T6555, 92T6585⟩, 15 ⟨88CC638⟩, 233, 234 ⟨90JFC(48)249⟩
McClory, M. R., **1**, 143 ⟨71JA3077⟩, 800 ⟨74JA1962⟩
McCloskey, A. L., **1**, 881 ⟨63OSC(4)261⟩; **2**, 418, 419 ⟨60JA6242, 61JA2122⟩; **4**, 211 ⟨50JA514⟩
McCloskey, C. J., **1**, 688, 690 ⟨78JOC3609⟩; **3**, 143 ⟨85JOC175⟩
McCloskey, C. M., **2**, 781 ⟨55OSC(3)668⟩; **3**, 620 ⟨55OSC(3)221⟩; **5**, 25 ⟨55OSC(3)221⟩
McCloskey, J. E., **1**, 2 ⟨79T885⟩
McCloskey, P. J., **2**, 326 ⟨88JOC1380⟩
McClure, C. K., **2**, 533, 698 ⟨90T4663⟩; **3**, 201 ⟨91TL5313⟩, 331 ⟨86JOC1556⟩; **4**, 561 ⟨93PS(75)23⟩
McClure, D. J., **6**, 603 ⟨88IC570⟩
McClure, G. L., **4**, 183 ⟨73TL1599⟩
McClure, J. D., **3**, 498 ⟨67TL2401⟩, 498, 651 ⟨70JOC3045⟩
McClure, R. E., **5**, 831 ⟨62JOC304⟩
McClusker, P. A., **2**, 521 ⟨57JA5190, 58JA617⟩
McCluskey, J. G., **2**, 303 ⟨67JCS(C)2015⟩
McClymont, E. L., **3**, 312 ⟨93TL7595⟩; **5**, 993 ⟨92SC411⟩
McCollum, G. J., **1**, 142, 155 ⟨75JA7006⟩, 618 ⟨73TL1843, 74TL1105⟩; **2**, 156 ⟨73TL289⟩; **4**, 75 ⟨73TL289⟩
McColm, E. M., **3**, 612 ⟨41OSC(1)536⟩
McComb, S., **1**, 1038 ⟨92JA9279⟩
McCombe, K. M., **3**, 405 ⟨86JCS(P1)867⟩
McCombie, H., **2**, 104 ⟨45JCS380⟩; **4**, 77 ⟨31JCS1913⟩, 770 ⟨37JCS767⟩
McCombie, S. M., **4**, 938 ⟨91JOC3556⟩
McCombie, S. W., **1**, 11, 16, 935, 938 ⟨75JCS(P1)1574⟩, 214 ⟨85TL6301⟩, 508 ⟨87TL5395⟩, 1089, 1090 ⟨87TL4127, 87TL4123⟩; **4**, 961 ⟨85TL6301, 87TL4123⟩; **5**, 516, 522, 526, 627 ⟨75JCS(P1)1574⟩; **6**, 553 ⟨81TL3485⟩, 557 ⟨75JCS(P1)1574⟩
McConaghy, Jr., J. S., **1**, 881 ⟨68JOC3425⟩; **4**, 168 ⟨68JCS(C)7⟩
McConnell, R. L., **6**, 118 ⟨56JA4450⟩
McConnell, W. W., **5**, 991 ⟨86JOC5123⟩
McConville, D. H., **1**, 675 ⟨93JA4413⟩
McCormack, M. T., **5**, 703 ⟨77JCS(P2)1136⟩; **6**, 620 ⟨80JCS(P2)867⟩
McCormick, C. J., **4**, 141, 142 ⟨71MI 403-02⟩
McCormick, F. B., **2**, 580 ⟨82JA4865⟩
McCormick, J. E., **4**, 103 ⟨85JCS(P1)93⟩
McCormick, J. P., **2**, 11 ⟨75CC303, 80JOC2566⟩; **3**, 156 ⟨81TL607⟩
McCortney, B. A., **1**, 107 ⟨90JA3554⟩
McCowan, J. D., **1**, 178 ⟨77CRV69⟩; **2**, 566 ⟨84JOM(267)265⟩
McCoy, P. A., **1**, 475 ⟨81JOC2709⟩
McCoy, R. K., **3**, 65 ⟨88SC51⟩
McCrae, D. A., **4**, 245 ⟨82JOC1145⟩; **6**, 134 ⟨82JOC1145⟩
McCrae, W., **2**, 1018, 1097 ⟨63JCS2295⟩
McCrann, P. M., **2**, 311 ⟨82TL4651⟩
McCrary, A. L., **2**, 127 ⟨65MI 203-01⟩
McCready, R. J., **3**, 220 ⟨82TL2355⟩
McCrindle, R., **4**, 150 ⟨89JCS(D)761⟩, 151 ⟨88JCS(D)1773⟩
McCullagh, L., **3**, 579 ⟨68JA1923⟩
McCulloch, A. W., **1**, 318 ⟨81JC288⟩, 522 ⟨69CJC2391⟩, 525 ⟨79TL1963⟩
McCullogch, J. J., **1**, 800 ⟨74JA1962⟩
McCullough, D. W., **4**, 652 ⟨88TL25⟩
McCullough, J. D., **2**, 287 ⟨51JA3196⟩, 729 ⟨45JA1965⟩
McCullough, K. J., **3**, 284 ⟨83JA7644⟩, 326 ⟨94JCS(P1)643⟩
McCullough, L. G., **1**, 675 ⟨85JA5987⟩

McCullough, R., **4**, 876 ⟨87SM647⟩
McCullough, R. D., **4**, 876 ⟨88SMB425⟩, 876, 877 ⟨87JA4115⟩
McCullum, G. W., **2**, 305 ⟨81JOC4296⟩
McCulough, J. F., **2**, 454 ⟨51JA5392⟩
McCurry, Jr., P. M., **1**, 521 ⟨79JA7001⟩
McCusker, P., **2**, 111 ⟨57AG677⟩
McCusker, P. A., **2**, 1016, 1018 ⟨37JA1307⟩
McDaniel, Jr., C. R., **4**, 24 ⟨83JA3411⟩
McDaniel, Jr., R. L., **1**, 812, 813 ⟨81JOC2199⟩
McDaniel, K., **2**, 160 ⟨84JOC2500⟩
McDermott, G. A., **3**, 692 ⟨86JA548⟩; **5**, 446 ⟨92CB1583⟩
McDermott, I. R., **5**, 714 ⟨79JCS(P2)634⟩
McDermott, J. X., **1**, 668, 669 ⟨73JA4451⟩, 668, 669 ⟨76JA6521⟩; **5**, 441 ⟨76JA6529⟩
McDermott, S., **2**, 367 ⟨83S192⟩
McDevitt, J. P., **4**, 9 ⟨71JOC575⟩
McDonald, A. N., **3**, 543 ⟨73CC247⟩
McDonald, C. E., **4**, 182 ⟨94TL57⟩
McDonald, F. E., **2**, 988 ⟨90JA4956⟩; **5**, 342 ⟨92JOC7001⟩, 944 ⟨93JOC6952, 95AG(E)350⟩, 957 ⟨94JA9363⟩
McDonald, G. J., **4**, 440, 442, 443 ⟨71JOC3846⟩
McDonald, III, J. H., **1**, 1027 ⟨89JOC5828⟩; **2**, 68 ⟨80TL1031⟩; **4**, 378 ⟨79TL593⟩, 951 ⟨89JOC5828⟩
McDonald, J. F., **3**, 51, 200 ⟨88JOC4069⟩
McDonald, R., **1**, 1110 ⟨73JA160⟩; **3**, 307 ⟨81TL4899⟩
McDonald, R. N., **1**, 212 ⟨72JOC2957⟩; **3**, 536 ⟨64JA4866⟩; **4**, 50 ⟨63JA4004, 67JA6573⟩, 50 ⟨68JOC2934, 70JA5664⟩
McDonald, R. S., **2**, 40 ⟨71JA4907⟩
McDonald, S. R., **4**, 713 ⟨89OM2077⟩
McDonald, T. L., **4**, 379 ⟨84JA3376⟩
McDonald, W. S., **4**, 698 ⟨72JCS(D)2649⟩, 1028, 1030 ⟨82CC287⟩
McDonough, C. S., **1**, 835 ⟨91TL179⟩; **3**, 623 ⟨93TL1453⟩
McDougal, P. G., **1**, 1089 ⟨87JOC4817⟩; **4**, 165 ⟨89JOC91⟩, 220 ⟨78JA7103⟩
McDougall, D. C., **3**, 333, 335 ⟨83CC423⟩, 333, 335 ⟨85JCS(P1)1541⟩
McDowell, J., **5**, 330 ⟨49JA1691⟩
McElheny, G. C., **5**, 591 ⟨61JOC3980⟩
McElhill, E. A., **1**, 852 ⟨50JA408⟩
McElhinney, R. S., **4**, 103 ⟨85JCS(P1)93⟩, 226 ⟨82JCR(S)116⟩; **5**, 1023 ⟨66JCS(C)950⟩
McElligott, P. J., **2**, 978 ⟨84OM134⟩
McElroy, A. B., **1**, 643 ⟨91JCS(P1)1485⟩, 754 ⟨85TL2119, 85TL5709⟩, 836 ⟨93JCS(P1)1275⟩
McElroy, A. D., **2**, 110 ⟨61USP2992266⟩
McElvain, S. M., **2**, 29 ⟨42JA1963, 42JA1966⟩, 191 ⟨45JA1578⟩, 609, 621 ⟨32OSC(1)170⟩; **3**, 210 ⟨55JA1599⟩, 618 ⟨55OSC(3)493⟩; **4**, 824, 888 ⟨49CRV453⟩, 825 ⟨50JA1661, 55JA5601⟩, 826 ⟨43OS(23)45⟩, 827 ⟨46JA1922⟩, 880, 883 ⟨49JA40⟩, 887, 889 ⟨40JA1281⟩, 888, 889 ⟨45JA202⟩; **5**, 115 ⟨55OSC(3)740⟩, 139 ⟨48JA3781⟩, 275 ⟨55OSC(3)490⟩, 551 ⟨33JA2966⟩, 706, 845 ⟨55JA4571⟩; **6**, 69 ⟨50JA1661⟩, 69, 74, 80 ⟨42JA2525⟩, 70 ⟨51JA2233, 53JA3987⟩, 70 ⟨58JA3915⟩, 70, 77 ⟨42JA1825⟩, 70, 80 ⟨46JA1922⟩, 74 ⟨36JA529, 37JA2266⟩, 74 ⟨45JA650, 51JA3807⟩, 74, 80 ⟨42JA1963⟩, 75 ⟨54JA5736⟩, 75 ⟨55JA5601⟩, 76 ⟨59JA2579, 37JA1273⟩
McElvain, S. S., **1**, 538 ⟨79JA1284⟩; **5**, 164 ⟨79JA489⟩
McEnroe, F. J., **1**, 97 ⟨71JOC2588⟩
McEvoy, F. J., **5**, 25, 64 ⟨73JOC4044⟩
McEwan, W. E., **2**, 436, 442, 443 ⟨64JA2378, 59JA248⟩
McEwen, I., **1**, 777 ⟨82JCS(P2)1179⟩
McEwen, N. E., **1**, 268 ⟨62JA676, 62JA677⟩
McEwen, W. E., **1**, 50 ⟨87JOC4829⟩, 51 ⟨78JA7312⟩; **2**, 824, 834, 835 ⟨78JA7304⟩, 842 ⟨59JA3805⟩, 848 ⟨56JA3063⟩, 876 ⟨75JOM(85)179⟩; **3**, 492 ⟨62JA676⟩; **4**, 994 ⟨63T49⟩
Mcewen, W. K., **1**, 848, 869 ⟨36JA1124⟩
McEwen, W. L., **2**, 1018, 1098 ⟨26JA469⟩; **5**, 52 ⟨43OSC(2)53⟩
McFadden, H. G., **4**, 704 ⟨89AJC301⟩
McFadyen, J. S., **1**, 39 ⟨36JCS584⟩
McFarland, J. W., **2**, 743 ⟨83S628⟩; **3**, 37 ⟨83S628⟩, 680 ⟨59JOC892⟩; **5**, 344 ⟨81SR215⟩, 1018 ⟨72IJS319⟩
McFarland, P. E., **3**, 130 ⟨72JA5106⟩
McFarlane, I. M., **6**, 543, 550 ⟨91T6381⟩
McFarlane, K. L., **3**, 162 ⟨89SC2595⟩
McFarlane, N., **1**, 881 ⟨64PCS108⟩
McFarlane, W., **2**, 435 ⟨67JOC781⟩, 437, 828 ⟨82JCS(D)1915⟩, 466 ⟨77JCS(D)1674⟩; **4**, 546, 562 ⟨78JCR(S)368⟩, 548 ⟨86CC1041, 88POL129⟩, 548 ⟨88POL239, 92JCS(D)2353⟩, 550, 551, 576 ⟨80IC3195⟩, 704 ⟨88JCS(D)503⟩, 1022 ⟨82JCS(D)1915⟩; **6**, 151 ⟨87PS(38)79⟩, 151, 369 ⟨86IC2699, 82CC930⟩, 370 ⟨85GEP3337939⟩
McFarlin, R. F., **3**, 83 ⟨56JA252⟩
McFaull, A. J., **5**, 1044 ⟨56JOC1430⟩
McGaffin, G., **3**, 44, 187 ⟨89S359⟩
McGahren, W. J., **2**, 773 ⟨70JA1587⟩
McGann, P. E., **6**, 509 ⟨78JOC922⟩
McGarrity, J. F., **3**, 479 ⟨73C320⟩
McGarry, D., **5**, 377 ⟨88T4157⟩
McGarry, D. G., **3**, 440 ⟨88T4157⟩, 456 ⟨85CC539⟩; **4**, 226 ⟨90JA6263⟩; **5**, 513 ⟨90JA6263⟩; **6**, 134 ⟨87JA2504, 90JA3696⟩
McGarry, E. J., **5**, 126 ⟨85CA(103)104568c⟩
McGarry, L. W., **5**, 934 ⟨88JOC1203⟩
McGarry, P. F., **5**, 924 ⟨92OM2317⟩
McGarvey, G., **2**, 606 ⟨77SC475⟩, 607, 914, 921, 922 ⟨78JOC4424⟩; **4**, 1059 ⟨79JOC4623⟩
McGarvey, G. J., **1**, 138, 139 ⟨84JA3376⟩, 139 ⟨88JA842⟩, 218, 831 ⟨83JA1988⟩, 981 ⟨88TL1355⟩; **2**, 88 ⟨86JOC3913⟩, 516 ⟨85TL6297⟩, 999 ⟨88JA842⟩; **3**, 162 ⟨85JOC4655⟩; **4**, 379 ⟨84JA3376, 88JA842⟩, 952 ⟨84JOC4091⟩, 1048, 1058, 1059, 1060, 1062 ⟨86OM2274⟩, 1060 ⟨88TL1355⟩
McGeary, C. A., **6**, 165, 289, 372, 373 ⟨89JOM(366)39⟩, 393 ⟨89CC273, 94JOM(480)199⟩
McGeary, M. J., **5**, 952, 953 ⟨85OM2102⟩
McGeary, R. P., **1**, 774 ⟨93AJC1321⟩
McGee, J., **1**, 456 ⟨79JOM(178)43⟩
McGee, J. J., **6**, 572 ⟨73OSC(4)801⟩
McGee, L. R., **1**, 876 ⟨80TL3975⟩, 878 ⟨81PAC1109⟩; **2**, 703 ⟨80TL3975⟩; **5**, 148 ⟨81PAC1109⟩
McGee, T. W., **2**, 24 ⟨60JOC24⟩
McGeehan, G. M., **3**, 707 ⟨86JOC2103⟩
McGettigan, C., **1**, 458, 516 ⟨90TL2397⟩; **3**, 12, 128 ⟨88TL3749⟩
McGhie, J. F., **1**, 16 ⟨81JCS(P1)1501⟩, 438 ⟨73JCS(P1)654⟩; **2**, 747 ⟨75JCS(P1)1237⟩; **3**, 215 ⟨63JCS3675⟩, 358 ⟨75CC756, 77JCS(P1)515⟩; **5**, 289 ⟨75JCS(P1)1237⟩
McGill, J. M., **2**, 310 ⟨93S1089⟩, 354 ⟨89TL1327⟩
McGillivray, G., **2**, 631 ⟨69TL2427, 71JA4841⟩, 652 ⟨70JA3520⟩; **3**, 307 ⟨77JOC4248⟩, 308 ⟨66TL6399⟩
McGimpsey, W. G., **3**, 314 ⟨85JA7204⟩
McGinnis, C., **3**, 444 ⟨50JA2890⟩
McGlinchey, M. J., **1**, 165 ⟨82JA7549⟩; **2**, 961 ⟨73ACR97⟩; **4**, 707, 720 ⟨77JOM(141)85⟩, 713 ⟨82OM225, 85OM1123⟩, 718 ⟨93OM4545⟩; **6**, 265, 269 ⟨70JCS(A)31⟩
McGlothin, R. E., **2**, 461 ⟨61JOC5205⟩, 500, 502 ⟨63JCS(C)798⟩
McGowan, G., **2**, 181 ⟨1884JPR280⟩
McGowan, L. S., **1**, 221 ⟨83JOC409⟩
McGrady, S., **2**, 492 ⟨88JCS(D)451⟩
McGrane, P. L., **3**, 410 ⟨92JA5459⟩
McGrath, J. A., **1**, 970 ⟨78JOC1382⟩
McGrath, T. F., **4**, 731 ⟨55JA4168⟩
McGreer, D., **5**, 205 ⟨56JA3819⟩
McGreer, D. E., **1**, 637 ⟨70CJC2110⟩, 747 ⟨68CJC2225⟩; **2**, 764 ⟨62CJC177⟩

McGregor, A. C., **6**, 487 ⟨57JA6180⟩
McGregor, J. H., **3**, 632 ⟨45JCS535⟩
McGregor, R. R., **4**, 609 ⟨47JA2247⟩
McGregor, S. D., **1**, 790 ⟨66JA1335, 66JA2858⟩
McGregor, W. H., **3**, 620 ⟨62B53⟩
McGregor, W. M., **2**, 140 ⟨93CSR199⟩; **3**, 345, 349, 351, 369 ⟨90JCS(P1)3175⟩
McGuckin, R. M., **3**, 439 ⟨85CC582⟩
McGuigan, C., **2**, 412 ⟨93S311⟩
McGuiness, M., **3**, 193, 472 ⟨90TL4987⟩
McGuire, M. A., **1**, 315 ⟨80JA4973⟩; **2**, 689 ⟨94JOC6683⟩; **3**, 138 ⟨80JA4973⟩
McGuire, R. R., **2**, 17 ⟨77TL363⟩
McGuirk, P. R., **1**, 479, 1107 ⟨78TL2465⟩, 541, 1107 ⟨79JOC3888, 77JA253⟩, 1107 ⟨78TL1363, 78TL2973⟩, 1107 ⟨79JOC1345⟩; **3**, 217 ⟨78TL1363⟩
McHale, D., **2**, 103 ⟨60JCS225⟩
McHendry, L. H., **6**, 28 ⟨83USP4419514⟩
McHenry, B. M., **4**, 366, 394, 945, 1060 ⟨84JA3245⟩
McHenry, W. E., **3**, 556 ⟨B-80MI 317-02⟩
Mchich, M., **4**, 185 ⟨87TL6601⟩
McHugh, G. P., **3**, 446 ⟨22JCS1648⟩
McInnes, A. G., **1**, **1**, 318 ⟨81JC288⟩, 522 ⟨69CJC2391⟩, 525 ⟨79TL1963⟩
McInnis, E. L., **1**, 420 ⟨B-80MI 109-02⟩
McIntosh, C. L., **1**, 522 ⟨72JA6229⟩; **3**, 539 ⟨73JA6134⟩, 542 ⟨70JA7001, 73JA4061⟩, 543 ⟨69JA4309, 71CC770⟩, 543 ⟨73JA244, 73JA247⟩; **6**, 75 ⟨71CC383⟩, 497 ⟨75GEP2435407⟩, 513 ⟨76GEP2553147⟩
McIntosh, D., **6**, 525 ⟨76JA3167⟩
McIntosh, J. C., **3**, 415 ⟨86CJC726⟩
McIntosh, J. M., **3**, 34 ⟨83CJC1872⟩, 34, 164 ⟨75JOC1294⟩, 34, 60 ⟨77JOC2123⟩, 234 ⟨71CJC3045⟩; **4**, 912 ⟨74S862⟩
McIntosh, M. C., **2**, 309 ⟨89TL5709⟩, 921, 922 ⟨93JOC4823⟩
McIntyre, D. D., **5**, 366 ⟨80CJC1316⟩; **6**, 626 ⟨83CJC1213⟩
McIntyre, J. S., **1**, 893, 899 ⟨64JA1360⟩
McIntyre, S., **3**, 35 ⟨91JCS(P1)451⟩
McIsaac, Jr., J. E., **5**, 275 ⟨71JOC3048⟩
McIver, Jr., R. T., **1**, 883 ⟨78JOC815⟩, 1143 ⟨80JA3334⟩
McIvor, R. A., **2**, 257 ⟨56CJC1819⟩
McKay, A. F., **2**, 102 ⟨60CJC343⟩, 456 ⟨53CB1483⟩; **3**, 466 ⟨47JA3028, 48JA1974⟩; **5**, 1024 ⟨58JA3332⟩, 1125 ⟨62CJC1403⟩; **6**, 508 ⟨51JOC1829, 53CJC896⟩, 508 ⟨53JA1120⟩, 533, 540 ⟨62JOC4509⟩
McKay, D., **5**, 134 ⟨78GEP2822472⟩
McKay, G. R., **4**, 825 ⟨55JA5601⟩; **6**, 75 ⟨55JA5601⟩
McKay, W. R., **3**, 438 ⟨81JCS(P1)2443⟩; **4**, 105, 300 ⟨81JCS(P1)2435⟩, 304, 305 ⟨81JCS(P1)2443⟩; **5**, 696 ⟨81JCS(P1)2435⟩
McKean, D. R., **1**, 471 ⟨87JOC422⟩, 579 ⟨82TL323⟩; **2**, 31 ⟨76BCJ1989⟩
McKechnie, J. S., **1**, 898 ⟨71JCS(B)918⟩
McKee, B. H., **3**, 243 ⟨90JOC3114⟩
McKee, D., **2**, 30 ⟨78JOC2793⟩
McKee, G. A., **6**, 27 ⟨80CJC485⟩
McKee, J. R., **2**, 311 ⟨91S176⟩; **3**, 264 ⟨91S115⟩; **6**, 27 ⟨80CJC485⟩
McKee, M. L., **5**, 929 ⟨92AG(E)1384⟩; **6**, 188 ⟨92AG(E)1384⟩
McKee, R., **4**, 305 ⟨76JA6715⟩
McKee, R. L., **5**, 1027 ⟨46JA2506⟩; **6**, 109 ⟨65JOC829⟩
McKee, S. P., **6**, 417 ⟨91TL4251, 92TL2781⟩
McKee, W. E., **2**, 452 ⟨51JA4590⟩
McKeer, L. C., **2**, 620 ⟨83TL3117⟩
McKeever, C. H., **2**, 13 ⟨42OR(1)63⟩, 642, 643 ⟨40JA3250⟩; **6**, 524 ⟨49USP2476263, 49USP2477553⟩, 524 ⟨49USP2477554⟩
McKellar, R., **2**, 1000 ⟨83JHC399⟩
McKellin, W. H., **2**, 143 ⟨51JA2251⟩
McKelvey, J. M., **4**, 93 ⟨85JA6298⟩
McKelvey, R. D., **1**, 421 ⟨73JA3932⟩
McKelvie, N., **4**, 553 ⟨61JA3539⟩, 745 ⟨62JA1745⟩; **5**, 911, 1093 ⟨61JA3539⟩; **6**, 694 ⟨62JA1745⟩
McKenna, C., **3**, 482 ⟨76JA7327⟩
McKenna, C. E., **4**, 129 ⟨89B8270⟩, 130 ⟨79CC739⟩; **5**, 404 ⟨77TL155⟩; **6**, 117 ⟨89CC246⟩, 264 ⟨81JOC4573⟩, 666 ⟨91S405⟩
McKenna, J., **1**, 2, 6 ⟨85JOC3408⟩; **2**, 161 ⟨80JCS(P2)385⟩; **3**, 404 ⟨58JCS2209⟩, 621 ⟨85JOC3408⟩
McKenna, J. M., **2**, 161 ⟨80JCS(P2)154⟩
McKenna, M.-C., **5**, 404 ⟨77TL155⟩
McKenna, P., **5**, 802 ⟨89JCS(D)345⟩
McKennis, J. S., **4**, 681 ⟨74JA7108⟩
McKennon, M. J., **1**, 473 ⟨93S250⟩; **3**, 40 ⟨93JOC3568⟩
McKenny, R. L., **2**, 892 ⟨67IC1178⟩
McKenzie, A., **4**, 161 ⟨88JOC5088⟩, 918 ⟨88JCS(P1)1039⟩
McKenzie, B., **3**, 632 ⟨41OSC(1)256⟩
McKenzie, B. F., **2**, 30 ⟨41OSC(1)521⟩; **5**, 56 ⟨32OSC(1)131⟩, 85 ⟨32OSC(1)21⟩
McKenzie, G., **5**, 690 ⟨93T1053⟩
Mckenzie, J. R., **3**, 162 ⟨88TL3911⟩
McKenzie, S., **3**, 73, 376 ⟨68JCS(C)2543⟩, 341 ⟨70JCS(C)145, 73JCS(P1)657⟩, 341, 363 ⟨66CC401⟩, 388 ⟨79JCS(P1)2334⟩; **5**, 472 ⟨70SST181⟩, 565 ⟨70SST201⟩
McKenzie, T. C., **3**, 220 ⟨77S608⟩
McKeon, J. E., **2**, 1566 ⟨63JA3992⟩; **5**, 307 ⟨69JA3992⟩, 463 ⟨68CC6⟩
McKeon, T. F., **2**, 193 ⟨53JOC36⟩
McKeough, D., **3**, 372 ⟨72JA6215, 74JA4268⟩
McKervey, A., **4**, 90 ⟨80CC506⟩
McKervey, M. A., **1**, 48 ⟨79AG(E)939⟩, 631 ⟨71JCS(C)3173⟩, 664 ⟨76JCS(P1)234⟩, 674 ⟨82TL2997⟩, 704 ⟨86T3731⟩; **3**, 166 ⟨82TL2509⟩, 167 ⟨83TL117⟩, 174, 234, 256 ⟨85JCS(P1)2193⟩, 194 ⟨93CC94⟩, 206 ⟨78CC1069⟩, 231 ⟨80CC486⟩, 232 ⟨90JCS(P1)1047⟩, 508 ⟨94CPB1041⟩; **4**, 62, 66 ⟨91COS(7)206⟩, 62, 66, 68 ⟨86T3731, 86TL757⟩, 69 ⟨83TL117⟩, 90 ⟨85JCS(P1)2193⟩
McKew, J. C., **1**, 831 ⟨93JOC4589⟩
McKie, J. A., **2**, 565 ⟨91JOC4112⟩
McKillip, W., **3**, 447 ⟨62JOC3946⟩
McKillop, A., **1**, 81 ⟨65JA1984⟩, 489 ⟨92T8117⟩, 491 ⟨76OS(55)48⟩, 1076 ⟨72AG(E)48⟩; **2**, 13 ⟨83TL1933⟩, 103 ⟨74T2467⟩, 598 ⟨68JOM(15)500⟩, 626 ⟨69TL1623, 72JOC88⟩, 631 ⟨69TL2427, 71JA4841⟩, 631, 964 ⟨71JA4845⟩, 652 ⟨70JA3520⟩, 776 ⟨70JOC1670⟩; 982 ⟨71JA7331, 73JA1296⟩; **3**, 4, 115 ⟨79S401⟩, 89, 95 ⟨74SC45⟩, 146 ⟨77TL1827⟩, 303 ⟨67T897⟩, 527 ⟨72OS(52)36⟩, 664 ⟨84TL5473⟩; **4**, 194 ⟨73JA3635⟩, 194 ⟨76JA3037⟩, 194 ⟨76JA6750⟩, 194 ⟨82COMC-I(7)465⟩; **5**, 8 ⟨88OSC(6)549⟩, 183 ⟨68JA2422⟩, 204 ⟨89T3299⟩, 773 ⟨82TL3357⟩, 1135 ⟨78JOC2280⟩
McKillop, A. G., **2**, 155 ⟨83TL1505⟩, 216 ⟨90TL5007⟩
McKillop, A. M., **5**, 718 ⟨87SC657⟩
McKinney, R. J., **3**, 614 ⟨85MI 318-01⟩, 626 ⟨85JOC5370⟩; **4**, 683 ⟨78JCS(D)403, 79JCS(D)1801⟩
McKinnie, B. G., **4**, 382 ⟨77JA5213⟩; **5**, 237 ⟨77TL1839⟩
McKinnon, D. M., **3**, 254, 371, 376 ⟨70JCS(C)1202⟩, 255, 360, 361 ⟨65JCS32⟩, 255, 376 ⟨64CI(L)461⟩, 339 ⟨71CJC3299, 69CJC2039⟩, 361 ⟨73CJC3081⟩, 361, 371 ⟨68CJC1855, 62JCS5104⟩, 371 ⟨84CHEC(6)783⟩; **4**, 850 ⟨84CHEC(6)798⟩; **5**, 634 ⟨70CJC1991⟩; **6**, 113 ⟨65JCS32⟩
McKittrick, B., **2**, 627 ⟨90JHC2151⟩
McKittrick, B. A., **3**, 215 ⟨85JOC5897⟩
McKnight, M. V., **3**, 303 ⟨82JCS(P1)1343⟩

McKown, W. D., **2**, 1031 ⟨73JOC1361⟩; **6**, 76, 77 ⟨73JOC1361⟩
McKusick, B. C., **1**, 560 ⟨59OS(39)68⟩; **2**, 130 ⟨52JA3982⟩; **3**, 352 ⟨60JA1515⟩, 695 ⟨63OSC(4)438⟩, 696, 700 ⟨73OSC(5)773⟩; **4**, 742 ⟨67JOC1717⟩; **5**, 579, 580 ⟨53JOC748⟩
McLachlin, D. M., **5**, 184 ⟨55OSC(3)28⟩
McLain, S. J., **2**, 440 ⟨83JA6355⟩; **3**, 511 ⟨79JA3210⟩; **4**, 453 ⟨83JA6355⟩
McLane, C., **2**, 343, 763 ⟨76TL233⟩
McLaren, K. L., **1**, 175 ⟨88JA1633⟩; **2**, 351 ⟨88JA1633⟩
McLaughlin, A. R., **5**, 384 ⟨79JOC1999⟩
McLaughlin, G. M., **2**, 442 ⟨82IC1007⟩; **6**, 599 ⟨80CC635⟩, 720 ⟨74JCS(D)1494⟩, 720, 721 ⟨77JCS(D)1272, 74JCS(D)1591⟩
McLaughlin, K. C., **1**, 81 ⟨50JA3539⟩; **5**, 115 ⟨63OSC(4)824⟩
McLaughlin, K. T., **3**, 140 ⟨90JA7441⟩
McLaughlin, L. M., **1**, 811 ⟨82CC1164⟩
McLaughlin, M. L., **5**, 395, 396 ⟨84TL1249⟩
McLay, G. W., **5**, 183 ⟨68JA2422⟩
McLay, N. R., **5**, 263 ⟨93JCS(P1)451⟩
Mclean, A., **3**, 294 ⟨75JCS(P1)622⟩
McLean, A. D., **1**, 154 ⟨87JA5996⟩, 155 ⟨87JA6004⟩, 1109 ⟨76JA875, 76JCP(64)4159⟩; **2**, 1022 ⟨82JPC2835⟩
McLean, J. W., **2**, 378, 379 ⟨71JMC1052⟩
McLean, M. J., **2**, 625 ⟨86JMC213⟩
McLean, R. A., **6**, 570, 645 ⟨75JMC90⟩
McLean, R. A. N., **2**, 858 ⟨69MI 216-01⟩
McLean, S., **1**, 402, 794 ⟨65T2329, 69CJC1555⟩; **5**, 377 ⟨94BMC1865⟩
McLean, W. N., **1**, 482 ⟨88TL5553⟩
McLeary, C. D., **4**, 211 ⟨41JA1537⟩
McLennan, A. J., **5**, 465 ⟨83JCS(D)359⟩
McLeod, A. J., **2**, 190 ⟨79T1315⟩
McLeod, C. M., **4**, 300, 315 ⟨21JCS1470⟩
McLeod, D. J., **5**, 32 ⟨63OSC(4)635⟩
McLeod, J. K., **3**, 92 ⟨78AJC1533⟩
McLernon, J. L., **4**, 715 ⟨92OM864⟩
McLick, J., **2**, 25 ⟨79JOC1961⟩
McLoughlin, J., **2**, 57 ⟨89JOC159⟩
McLoughlin, J. I., **1**, 816, 818 ⟨91JOC1185⟩
McMahon, H. C., **1**, 506 ⟨67ANY(145)91⟩
McMahon, W. G., **3**, 88 ⟨87JOC3905⟩; **6**, 632 ⟨91S86⟩
McManaman, J. L., **6**, 51 ⟨71JOC3627⟩
McManus, S. P., **1**, 891 ⟨B-73MI 119-05⟩; **6**, 43 ⟨72CRV357⟩
McMaster, D., **1**, 1053 ⟨85TL271⟩
McMath, A. M., **4**, 39 ⟨27JCS538⟩
McMeekin, T. L., **5**, 83 ⟨46MI 502-01⟩
McMeeking, J., **2**, 565 ⟨82COMC-I(3)433⟩, 565, 575 ⟨82COMC-I(3)331⟩; **3**, 457, 460, 461 ⟨73JCS(D)151⟩, 460 ⟨71MI 312-01⟩, 460, 461 ⟨71CC215⟩; **4**, 656 ⟨78CC140⟩; **6**, 195 ⟨78CC140⟩, 720 ⟨74JCS(D)1494⟩
McMichael, K. D., **1**, 825 ⟨79JA2746⟩
McMillan, F. H., **5**, 183 ⟨51JA4911⟩, 272, 276 ⟨46JA525⟩, 566, 608 ⟨47JA1207⟩, 599 ⟨48JA4143⟩, 621 ⟨48JA868⟩
McMillan, R. M., **2**, 788 ⟨79JOC3410⟩
McMillen, D. F., **1**, 921, 925 ⟨82ARP493⟩
McMills, M. C., **1**, 760 ⟨88TL4521⟩, 831 ⟨90TL1977⟩; **3**, 302 ⟨90TL1833⟩
McMordie, R. A. S., **4**, 262 ⟨92JA8772, 92JCS(P1)1105⟩
McMorris, T. C., **1**, 85 ⟨83JOC3370⟩
McMullen, C. H., **2**, 742 ⟨66JCS217⟩; **3**, 267 ⟨66JCS(B)1217⟩
McMullen, G. L., **3**, 541 ⟨74AJC2385, 76JA7421⟩, 541 ⟨77AJC179⟩
McMurray, J. E., **2**, 81 ⟨89TL1169⟩, 395 ⟨88OSC(6)648⟩; **3**, 530, 531 ⟨87JOC4885⟩; **5**, 107 ⟨72SC389⟩
McMurray, J. S., **5**, 185 ⟨85JOC1112⟩
McMurry, J. E., **1**, 156, 309 ⟨73JOC4367⟩, 173, 587, 688 ⟨78JOC3249⟩, 357 ⟨80TL2477⟩, 480 ⟨88ACR47⟩, 480, 1027 ⟨80TL4313⟩, 482 ⟨83TL2723⟩, 687, 688 ⟨83ACR405⟩, 687, 688 ⟨89CRV1513⟩, 687, 688, 689 ⟨74JA4708, 76JOC896⟩, 688 ⟨83JA1660, 89JOC3748⟩, 688, 689, 690 ⟨78JOC3255, 82TL1777⟩, 690 ⟨82TL2723, 84JA5018⟩, 690 ⟨84JA6450, 85TL2167⟩, 690 ⟨85TL2171, 86JA2932⟩, 690 ⟨86JA515, 87JOC4885⟩, 690 ⟨87T5489⟩, 816 ⟨87TL3209⟩, 919 ⟨89JA8867⟩; **2**, 392 ⟨88OSC(6)799⟩, 689 ⟨83TL979⟩, 746 ⟨88OSC(6)592⟩; **3**, 210 ⟨74JOC2217⟩, 215 ⟨69JA3676⟩, 224 ⟨71TL2575⟩, 684 ⟨73OS(53)59⟩; **5**, 164 ⟨71CC488⟩
McNab, H., **1**, 782 ⟨82JCS(P1)1845⟩, 784 ⟨86T2135⟩; **2**, 1007 ⟨91JCS(P2)1563⟩
McNab, J. G., **4**, 24 ⟨35JA2463⟩
McNally, J. J., **3**, 305 ⟨92JHC247⟩
McNally, J. P., **5**, 941 ⟨90AOC(30)1⟩
McNamara, J. M., **4**, 247 ⟨89JOC3718⟩; **5**, 329 ⟨94JOC1231⟩
McNaughton, D., **3**, 550, 552 ⟨87CC573⟩; **5**, 1152 ⟨85JSP(111)415⟩, 1154 ⟨85JSP(109)8⟩
McNeely, S. A., **1**, 1120 ⟨76JA4319⟩
McNeil, R. L., **6**, 254, 255 ⟨80CJC1996⟩
McNeill, A. H., **2**, 79 ⟨90TL6239, 94S322⟩
McNelis, E., **4**, 750 ⟨84JOC515⟩, 760 ⟨92TL2285, 92TL7705⟩
McNich, H. A., **2**, 553 ⟨58JOC2044⟩
McNicol, D. D., **6**, 528 ⟨92TL261⟩
McNierney, E., **4**, 160, 755 ⟨89JOC5502⟩
McNiven, N. L., **1**, 628 ⟨52JCS153⟩
McOmie, J. F. W., **5**, 26, 37 ⟨73S416⟩, 588 ⟨48AR198⟩
McPartlin, M., **4**, 675 ⟨88NJC559, 89CC688⟩; **6**, 345 ⟨84CC1089⟩
McPartlin, M. M., **2**, 362, 363 ⟨91JCS(P1)2169⟩
McPhail, A., **1**, 91 ⟨81T3615⟩
McPhail, A. T., **1**, 326 ⟨90TL1679⟩, 326, 327 ⟨91TL707⟩, 326, 327 ⟨92JA7664⟩, 327, 328 ⟨91JA7002, 90JA6740⟩; **2**, 253 ⟨81JA4432⟩, 482 ⟨86JOM(308)281⟩, 482, 504 ⟨90POL319⟩, 511 ⟨87JOM(325)C7, 88JOM(354)287⟩, 927 ⟨90JA7229⟩; **4**, 597 ⟨91MI 413-01, 91POL319⟩; **6**, 491 ⟨84IC4322, 86S833⟩, 491 ⟨89IS79⟩, 515 ⟨76JA5702⟩
McPhail, D. R., **2**, 927 ⟨90JA7229⟩
McPhall, A. T., **6**, 514 ⟨89EUP336772⟩
McPhee, D. J., **1**, 946 ⟨90JOC3321⟩; **3**, 440 ⟨88SC1975⟩; **4**, 512, 518 ⟨88SC1975⟩
McPhee, J. R., **5**, 85 ⟨59MI 502-02⟩
McPhee, W. D., **2**, 12 ⟨55OSC(3)119⟩
McPherson, A. M., **4**, 1068 ⟨74JA1941⟩
McPherson, C. A., **1**, 1078 ⟨72JA8084, 72SC331⟩; **3**, 274 ⟨72SC331⟩
McPherson, K. A., **4**, 681 ⟨90JCS(D)761⟩
McPherson, M., **4**, 850 ⟨85CC696⟩
McQuillen, F. J., **1**, 73, 74 ⟨71CC840, 73JCS(P1)2509⟩; 506 ⟨59JCS2902⟩, 507 ⟨69CC1503⟩
McQuire, L., **4**, 918 ⟨90JCS(P1)2811⟩
McRae, J. A., **3**, 626, 634 ⟨22JCS1699⟩
McShane, L., **4**, 69 ⟨78JA3933⟩
McSorley, K., **1**, 255 ⟨90CC1532⟩; **2**, 302 ⟨90CC1532⟩
McVeigh, I., **2**, 218 ⟨71JMC995⟩; **5**, 247 ⟨71JOC2735⟩
McVey, S., **4**, 689 ⟨69JOM(19)169⟩
McVicar, W. K., **4**, 147, 579 ⟨90OM2683⟩, 148, 149 ⟨92JOM(429)369⟩
McWhinnie, W. R., **1**, 601 ⟨B-87MI 114-03⟩; **2**, 280 ⟨83JCS(D)2143, 86JCS(D)2003⟩; **2**, 690 ⟨85JOM(291)185⟩; **6**, 563 ⟨88JCS(D)2363⟩
Mead, D., **3**, 231 ⟨85TL2873⟩
Mead, E. J., **2**, 59 ⟨53JA6263⟩, 109, 532 ⟨56JA3613⟩
Mead, K. A., **4**, 679 ⟨84JCS(D)2293⟩, 680 ⟨81CC862, 83JCS(D)2099⟩, 681 ⟨90JCS(D)761⟩, 712 ⟨81CC867⟩; **6**, 340 ⟨82CC51⟩, 343 ⟨82CC773, 83JCS(D)2599⟩

Mead, K. T., **1**, 282 ⟨89TL6829⟩, 1038, 1039 ⟨87TL1019⟩
Meade, C. F., **4**, 706, 706 ⟨73JOM(51)1⟩
Meader, Jr., A. L., **1**, 394 ⟨48JOC763⟩
Meadows, J. D., **6**, 468 ⟨82JOC4013⟩, 469 ⟨84JA5304⟩
Meadows, J. H., **2**, 969 ⟨84JA1291⟩
Meakins, G. D., **3**, 307 ⟨80JCR(S)42⟩; **4**, 429, 430, 431, 439 ⟨65JCS2601⟩; **5**, 655 ⟨87JCS(P1)1153⟩, 1125 ⟨82CC444, 84JCS(P1)147⟩; **6**, 572 ⟨87JCS(P1)1153⟩
Meals, B. N., **2**, 483 ⟨44JOC211⟩
Meals, R. N., **2**, 538 ⟨46JA1161⟩
Meana, M. C., **3**, 55 ⟨73CC119⟩
Means, G. E., **1**, 74 ⟨79JOC3442⟩, 637 ⟨89JOC5424⟩
Meanwell, N. A., **1**, 643 ⟨82TL5005⟩, 711 ⟨81JA7667, 85PS(24)531⟩; **2**, 401, 402 ⟨83JOC1⟩; **3**, 169, 172 ⟨83TL405⟩, 170 ⟨82S283⟩
Mears, R. J., **3**, 204 ⟨93T177⟩
Meathrel, W. G., **4**, 43 ⟨74CJC951⟩
Mebane, A. D., **1**, 503 ⟨52JA295⟩, 955 ⟨53JA1050⟩
Mebane, R., **1**, 241 ⟨76JOC1052⟩
Mechoulam, R., **1**, 268 ⟨58JA4386⟩; **2**, 650 ⟨83S1046⟩; **3**, 328 ⟨58JA4386⟩; **4**, 826 ⟨56JOC801⟩; **5**, 112 ⟨69CC343⟩
Mecke, Jr., R., **6**, 509 ⟨56CB343⟩
Mecke, R., **5**, 506 ⟨56CB1110⟩; **6**, 509 ⟨56CB343⟩
Meckler, H., **2**, 619 ⟨90JHC1549⟩
Meckstroth, W., **6**, 610, 727 ⟨92ZAAC(617)136⟩, 613 ⟨93ZN(B)19⟩
Medcalf, C., **3**, 460 ⟨70JCS(A)2279⟩
Mederer, K., **1**, 1041 ⟨87S848⟩
Medich, J. R., **2**, 420 ⟨92OS(71)133⟩; **4**, 387 ⟨88JOC4131⟩
Medici, A., **2**, 942 ⟨84CC258, 86S757⟩; **3**, 109 ⟨87S998⟩, 311 ⟨84JOC3478, 84JOC590⟩, 312 ⟨81CC655, 86G133⟩, 707, 714 ⟨87S693, 84CC258⟩; **6**, 634 ⟨87JOC3413, 87S693⟩, 637 ⟨90H(31)1213, 93H(36)473⟩
Medina, J. C., **2**, 92 ⟨88JA4818⟩; **4**, 220 ⟨88TL3773⟩
Medinger, L., **2**, 457 ⟨85S1048⟩
Medlik-Balan, A., **4**, 668 ⟨75CC877⟩; **6**, 206 ⟨75CC877, 76T1839⟩
Medved, T. Ya., **2**, 463 ⟨61IZV604⟩, 464 ⟨60IZV651⟩; **4**, 451, 469 ⟨68RCR7⟩, 463, 469 ⟨52DOK(83)689⟩, 463, 469 ⟨53DOK(92)959, 57IZV1357⟩, 473 ⟨50IZV635, 51IZV95⟩, 555, 561 ⟨67IZV591⟩, 556 ⟨62IZV2103, 66IZV1954⟩, 556 ⟨67IZV949, 68IZV1417⟩, 556 ⟨68IZV2062⟩, 560, 564 ⟨70IZV1326, 71IZV2747⟩
Medved, T. Y., **6**, 118 ⟨72MI 604-01⟩
Medvedeva, A. S., **1**, 986 ⟨87IZV695⟩; **4**, 942 ⟨83ZOB1292⟩, 1055 ⟨80ZOB2044⟩; **5**, 226 ⟨84IZV1897⟩; **6**, 516 ⟨84IZV1897⟩
Medvedev, A. E., **4**, 126 ⟨86JGU2144⟩
Medvedeva, E. N., **5**, 718 ⟨84JGU426, 84ZOB477⟩
Medvedeva, M. D., **5**, 369 ⟨78ZOB2468⟩
Medyansteva, N. M., **6**, 429 ⟨53JGU229⟩
Mee, J. D., **3**, 601 ⟨74JA4712⟩
Meek, D. W., **2**, 438 ⟨73JA4194⟩, 439 ⟨84IC778⟩, 465 ⟨66IC1297⟩, 825 ⟨76IS168⟩, 883 ⟨71IC2676⟩
Meek, J. S., **2**, 12 ⟨73OSC(5)126⟩, 156 ⟨68JOC3422⟩; **5**, 273, 285 ⟨52JA1010⟩
Meek, T. D., **6**, 487 ⟨93MIP9305026⟩
Meenakshi, R., **3**, 4, 116 ⟨91SC419⟩
Meen, R. H., **4**, 836, 887 ⟨71JOC2205⟩, 881 ⟨66JOC2874⟩; **5**, 481 ⟨81JOC3911⟩
Meerholtz, C., **1**, 494 ⟨94T359⟩
Meerholtz, K., **1**, 103 ⟨89JA2325⟩
Meerholz, C. A., **3**, 262 ⟨80TL1373⟩; **5**, 523 ⟨79CC755, 81T(S)213⟩
Meerpoel, L., **5**, 661 ⟨92S705⟩
Meerwein, H., **1**, 380 ⟨22CB2500⟩, 415 ⟨61LA(641)1⟩; **2**, 89 ⟨63HOU(6/3)1⟩, 112 ⟨36JPR251⟩, 679 ⟨65HOU(6/3)97⟩, 788 ⟨61LA(641)1⟩, 1009 ⟨56CB209⟩, 1021, 1025, 1027 ⟨65HOU(6/3)116⟩; **3**, 280 ⟨33CB411⟩, 687 ⟨56CB209⟩, 720 ⟨61LA(641)1⟩; **4**, 881 ⟨56CB2060⟩, 882, 974 ⟨61LA(641)1⟩; **5**, 227 ⟨37JPR251⟩, 687 ⟨65HOU(6/3)295⟩, 852 ⟨61LA(641)1⟩; **6**, 71 ⟨55AG374⟩, 73 ⟨60LA(632)38, 65HOU(6)361⟩, 73, 97, 300 ⟨56CB2060⟩, 104, 105, 313, 314, 318, 321, 730, 731 ⟨61LA(641)1⟩
Meese, C. O., **2**, 203 ⟨84S1041⟩; **5**, 506 ⟨77JST25⟩, 624, 625, 732 ⟨76CB956⟩, 736 ⟨71LA(753)169⟩
Meester, P. de, **4**, 514 ⟨85TL5001⟩; **5**, 765, 824 ⟨86JHC1861⟩
Meesters, A. C. M., **5**, 637 ⟨77CC597⟩
Meesters, L., **3**, 221 ⟨90TL7201⟩; **6**, 124 ⟨90TL7201⟩
Meetsma, A., **3**, 510 ⟨89JA5977⟩; **4**, 656 ⟨93OM184, 93OM1936⟩, 671 ⟨93OM2268⟩; **5**, 797 ⟨91OM2665⟩; **6**, 194 ⟨87OM1509⟩, 195 ⟨88CC962⟩
Meeuwissen, H. J., **1**, 445 ⟨83T4225⟩; **2**, 441 ⟨83T4225⟩
Meganem, F., **4**, 534 ⟨76JOM(113)C13⟩
Megera, I. V., **5**, 631 ⟨75ZOB2116⟩
Meguri, H., **1**, 568 ⟨90CC730⟩
Mehdi, S., **4**, 73 ⟨89JA1127⟩, 755, 803 ⟨93TL7197⟩
Mehl, A. F., **2**, 387 ⟨88JOC1467⟩
Mehl, W., **1**, 325, 921 ⟨88CB2063⟩
Mehler, G., **2**, 577 ⟨85CB2407⟩
Mehler, K., **4**, 659 ⟨85CB4248⟩
Mehlhoff, M. A., **6**, 573 ⟨80JHC1369⟩
Mehlhorn, A., **2**, 398 ⟨88CCC2116⟩
Mehltretter, C. L., **5**, 62 ⟨53MI 502-01⟩
Mehmandoust, M., **1**, 837 ⟨92TL4313⟩
Mehr, L., **3**, 452 ⟨55JA984⟩
Mehra, M. M., **1**, 1023 ⟨70TL2695⟩
Mehra, S. C., **5**, 1038 ⟨69JCED110⟩; **6**, 641 ⟨68JMC1268⟩
Mehra, U., **6**, 113 ⟨87S990⟩, 115 ⟨84TL1849⟩
Mehratra, A., **2**, 132 ⟨81SC409⟩
Mehrora, A. K., **5**, 39 ⟨83JOC3667⟩
Mehrotra, A. K., **1**, 5, 9 ⟨83JOC3667⟩; **2**, 36, 142 ⟨83JOC3667⟩, 784, 785 ⟨80JOC734⟩; **4**, 203 ⟨82S962⟩
Mehrotra, K. N., **5**, 362 ⟨84BCJ1667⟩; **6**, 507 ⟨80IJC(B)653⟩, 507, 576 ⟨80S1001⟩
Mehrotra, M. M., **1**, 1131 ⟨88TL57⟩; **3**, 231 ⟨84JA3384⟩
Mehrotra, R. C., **2**, 112 ⟨B-78MI 202-01⟩; **4**, 584, 587 ⟨85IJC(A)398⟩, 588 ⟨86JOM(303)351⟩; **5**, 389 ⟨73JOM(63)221, 77JOM(142)165⟩, 1144 ⟨77IJC(A)1099⟩; **6**, 78 ⟨63PIA(A)45⟩
Mehrotra, S. K., **2**, 866, 893 ⟨84IC2582⟩; **5**, 389 ⟨77JOM(142)165⟩; **6**, 167 ⟨83CC528⟩, 168 ⟨84IC2582⟩, 245 ⟨90JOM(390)C73⟩
Mehrsheikh-Mohammadi, M. E., **1**, 922 ⟨88TL6839⟩; **2**, 28 ⟨87JA2435⟩
Mehta, A. S., **1**, 627 ⟨63JA1949⟩
Mehta, D. V., **4**, 105 ⟨74JOC1965⟩
Mehta, G., **1**, 3, 8 ⟨87JA7230⟩, 524 ⟨79TL2625⟩, 907 ⟨67JA5048b⟩; **2**, 12, 28 ⟨87JA7230⟩, 762 ⟨72T2749⟩; **3**, 142 ⟨93TL8157⟩, 151 ⟨85SC991⟩, 530 ⟨91JOC7048⟩; **6**, 550 ⟨81CC756⟩
Mehta, M. D., **5**, 9 ⟨63JCS491⟩
Mehta, N. B., **5**, 595 ⟨62JOC4412⟩
Mehta, P. G., **4**, 742, 979 ⟨93MM916⟩
Mehta, S., **2**, 657 ⟨69JCS(C)2349⟩; **4**, 238 ⟨91TL4435⟩
Mehta, S. R., **5**, 110 ⟨76S113⟩
Mehta, V. D., **6**, 576, 577 ⟨91JOM(419)9⟩
Meichle, W., **4**, 838 ⟨89AP(322)263⟩
Meichsner, G., **1**, 63 ⟨87CB1049⟩
Meidar, D., **1**, 462 ⟨79NJC269⟩; **3**, 114 ⟨78S671⟩; **4**, 191 ⟨81S282⟩
Meidine, M. F., **6**, 692 ⟨89JOM(368)C29⟩
Meier, G. P., **1**, 821 ⟨87JA5878⟩; **3**, 422 ⟨88TL3781⟩
Meier, H., **1**, 393 ⟨75AG(E)32⟩, 1075 ⟨71S215⟩, 1077 ⟨72S235, 78S536⟩, 1077 ⟨79CB3728, 86T1711⟩, 1077 ⟨88LA221⟩; **2**, 389 ⟨71HOU(10/1)891⟩; **3**, 246 ⟨88S274⟩, 359 ⟨86TL5703, 87AG(E)1046⟩, 359 ⟨90CB1143, 91JHC573⟩, 470 ⟨78S535⟩, 551 ⟨71T5953, 77AG(E)835⟩, 552 ⟨77T449⟩, 553 ⟨72TL445⟩; **4**, 195 ⟨88S274⟩, 232 ⟨90CB1143⟩, 874 ⟨72TL445⟩; **5**, 511 ⟨77T449, 78JHC697⟩, 511 ⟨79JHC1295, 80CB2519⟩, 1003, 1013 ⟨77AG(E)555⟩; **6**, 511 ⟨57LA(607)67⟩

Meier, I. K., **1**, 683 ⟨89JA3069⟩
Meier, K., **3**, 229 ⟨47HCA905⟩
Meier, K. D., **1**, 1000 ⟨59AG245⟩
Meier, L., **5**, 206 ⟨87S517⟩
Meier, M., **1**, 266 ⟨84TL3441⟩; **3**, 621 ⟨87CB1⟩
Meier, M. S., **1**, 335 ⟨88TL167⟩, 946 ⟨89JOC1320⟩
Meier, R., **2**, 420 ⟨53CB1483⟩
Meier, S., **3**, 567, 580, 585, 586 ⟨90CB2339⟩; **6**, 730 ⟨92CB2487⟩
Meier, W. E., **5**, 285 ⟨71JMC535⟩
Meier zu Köcker, I., **5**, 1057 ⟨59M41⟩; **6**, 596 ⟨59M41⟩
Meignan, G., **1**, 446 ⟨93TL4331⟩; **2**, 446 ⟨90T6381⟩
Meiiner, I., **4**, 453 ⟨68CB3612⟩
Meijboom, N., **2**, 439 ⟨86EUP184872⟩
Meijer, E. W., **3**, 91 ⟨82OR(28)1⟩; **4**, 17 ⟨82OR(28)1⟩
Meijer, J., **1**, 496 ⟨76TL947⟩, 536 ⟨82TL2797⟩, 544 ⟨81JOM(206)257, 81RTC337⟩, 967 ⟨82JOM(240)329⟩, 975, 1004, 1009 ⟨81TL1451⟩, 989 ⟨75TL2923⟩, 999 ⟨74RTC240⟩, 1004, 1006, 1025 ⟨82RTC405⟩, 1017, 1018 ⟨81JOM(217)267, 82RTC97⟩, 1058 ⟨81S551⟩, 1094 ⟨77TL1823⟩; **2**, 616, 624 ⟨77RTC168⟩, 924, 925 ⟨77TL1823⟩, 1032 ⟨71RTC1098, 77RTC179⟩; **3**, 337 ⟨71IJS(B)85⟩, 548 ⟨72RTC578, 74RTC26⟩, 548 ⟨77RTC179, 82RTC310⟩; **4**, 245 ⟨74RTC242⟩, 704 ⟨92S432⟩, 809 ⟨77TL1823⟩, 843 ⟨73RTC1067⟩, 843 ⟨74RTC184⟩, 963 ⟨75RTC14, 75TL2923⟩; **5**, 510 ⟨74RTC51⟩, 547, 552 ⟨78S439⟩, 548, 550, 559, 560 ⟨79S432⟩, 550 ⟨73RTC601, 75RTC83⟩, 554 ⟨80TL1771⟩, 589 ⟨82RTC310⟩, 589, 614, 618 ⟨72RTC578⟩, 1105 ⟨78S577⟩
Meijere, A. de, **1**, 166, 183 ⟨93TL1917⟩, 494 ⟨94SL189⟩, 900 ⟨73AG(E)399⟩, 975 ⟨92JA4051⟩, 1069 ⟨86CB2995, 89CB377⟩; **2**, 309 ⟨93TL4193⟩, 932 ⟨90SL365⟩; **3**, 44, 187 ⟨89S359⟩, 113 ⟨76AG(E)761⟩; **4**, 940, 1014 ⟨92CB2051⟩; **5**, 934 ⟨91AG(E)1658⟩, 946 ⟨93SL259⟩, 958 ⟨92CB2051⟩; **6**, 20 ⟨77AG(E)854⟩, 380 ⟨82AG(E)443⟩
Meijs, G. F., **3**, 104 ⟨81CC595⟩
Meiklejohn, R. A., **5**, 815 ⟨66JOC4232⟩; **6**, 259 ⟨66JOC4232⟩
Meilahn, M. K., **6**, 51 ⟨71JOC3627, 78JOC1346⟩
Meili, J. E., **1**, 808 ⟨56JA2551⟩
Meineke, E. W., **5**, 919 ⟨77CB3040⟩, 942 ⟨77CB1140⟩; **6**, 720 ⟨72CB2558⟩
Meinema, H. A., **1**, 72, 88 ⟨90RTC87⟩; **2**, 485, 486, 487, 505 ⟨73JOM(51)223⟩, 488 ⟨70JOM(22)653⟩, 491, 505 ⟨77JOM(136)173⟩, 695, 885, 886, 889 ⟨70JOM(25)139⟩
Meinhardt, D., **1**, 769 ⟨83PAC1733⟩; **3**, 508 ⟨83PAC1733⟩
Meinhardt, N. A., **1**, 993 ⟨62JOC1828⟩; **2**, 1073 ⟨62JOC1828⟩
Meinhart, J. D., **3**, 508 ⟨89OM583⟩, 545 ⟨86JA3318⟩; **5**, 441 ⟨86JA3318⟩; **6**, 140 ⟨88BCJ171⟩
Meinhoffer, J., **5**, 84 ⟨62C385⟩
Meinhold, H. J., **2**, 170 ⟨80EGP139258, 82JPR631⟩
Meinke, P. T., **1**, 942 ⟨85TL135⟩; **2**, 289 ⟨86JA1314, 88JA8671⟩, 289 ⟨88JA8679⟩; **3**, 333 ⟨85TL1947⟩, 383 ⟨86JA1314⟩, 394 ⟨87TL3887, 88JA8679⟩, 636 ⟨88JA8671⟩; **4**, 287 ⟨88JA8679⟩, 287, 376, 393 ⟨88JA8671⟩, 365, 376, 570 ⟨87TL5121⟩, 947 ⟨88JOC3632⟩; **5**, 1113 ⟨86JA1314⟩, 1115 ⟨88JOC3632⟩; **6**, 130 ⟨87TL5121, 88JA8671⟩
Meinwald, J., **1**, 503 ⟨83JOC2274⟩, 503, 504 ⟨83JOC2266, 83JOC2270⟩, 631 ⟨68JA800⟩, 776 ⟨90JOC4051⟩; **2**, 238 ⟨77JA7743⟩, 282 ⟨81TL3561⟩, 350 ⟨59JA5800⟩, 391 ⟨64JA4074⟩; **3**, 37 ⟨81TL3561⟩, 463 ⟨59JA4751⟩, 532 ⟨65CI(L)425⟩; **5**, 235 ⟨81JA861⟩
Meinwald, Y. C., **2**, 391 ⟨64JA4074⟩
Meiramov, M. G., **4**, 143 ⟨82IZV1425⟩
Meise, W., **5**, 655 ⟨88LA833⟩
Meisel, J. L., **2**, 403 ⟨74TL3847⟩
Meisels, A., **1**, 934 ⟨63JA3419⟩
Meisenheimer, J., **4**, 430, 431, 432 ⟨03CB434⟩, 431, 432 ⟨06CB2543⟩
Meiser, P., **4**, 308 ⟨73CB69⟩
Meissner, H., **6**, 418 ⟨37CB1080⟩
Meissner, I., **2**, 427, 430, 462 ⟨68CB3612⟩
Meister, A., **5**, 62 ⟨83CRV321⟩, 81 ⟨58MI 502-03⟩
Meister, H., **1**, 1063 ⟨60AG391⟩; **2**, 403, 404, 405 ⟨64LA(679)83⟩
Meister, J., **1**, 948 ⟨77CB2588⟩
Meister, R., **6**, 213 ⟨B-72MI 607-02⟩
Meisters, A., **1**, 186 ⟨72CC595⟩; **2**, 478 ⟨74JOM(74)365⟩
Meixner, A., **1**, 999 ⟨63N403⟩; **2**, 1066 ⟨63N403⟩
Meixner, H., **6**, 304 ⟨88EUP260979⟩
Meixner, J., **1**, 323, 920 ⟨83TL703⟩
Mejean, A., **6**, 443 ⟨89MI 614-03⟩
Mekata, H., **3**, 129 ⟨86CL87⟩
Mekelburger, H. B., **1**, 873 ⟨91COS(2)99⟩; **2**, 696 ⟨91COS(2)99⟩
Mekhael, K., **5**, 214 ⟨85HCA1893⟩
Mekhalfia, A., **3**, 294 ⟨90SL345⟩; **5**, 18, 276, 293, 299 ⟨90TL7237⟩
Meklati, B., **1**, 155 ⟨68CR(C)1162⟩
Melamed, M., **2**, 906 ⟨09CB3091⟩
Melby, L. R., **1**, 560 ⟨59OS(39)68⟩; **4**, 5 ⟨60OSC(5)239⟩, 32 ⟨73OSC(5)239⟩, 829 ⟨59OS(39)13⟩, 970 ⟨60JA5116⟩; **5**, 56 ⟨73OSC(5)239⟩, 373 ⟨58JA6442⟩; **6**, 46 ⟨60JA5116⟩
Meldal, M., **5**, 323 ⟨90S1008⟩
Melega, W. P., **1**, 1094 ⟨77TL159⟩; **2**, 911, 946 ⟨77TL159⟩
Melenchuk, S. N., **5**, 615 ⟨89ZOR1323⟩
Meléndez, E., **5**, 1026 ⟨89BSB289⟩; **6**, 563 ⟨91S147, 81S961⟩, 605 ⟨84S520⟩, 627 ⟨83S375⟩
Meléndez, F., **6**, 627 ⟨82S590⟩
Meleshonkova, N. N., **4**, 719 ⟨90MI 416-01⟩
Melhado, L. L., **5**, 690 ⟨88JOC3852⟩
Meli, A., **1**, 507 ⟨88IC4429⟩; **4**, 851 ⟨83CC1309⟩; **6**, 325 ⟨80AG1055, 81JOM(218)81⟩, 327 ⟨82OM778⟩, 332 ⟨83JOM(246)C13, 84JOM(270)251⟩
Melikyan, G. G., **1**, 199 ⟨94TL363⟩; **2**, 640 ⟨74DOK(217)1320, 75ZOR1370⟩, 641 ⟨75DOK(220)146⟩; **6**, 65 ⟨75DOK(221)1331, 76ZOR1377⟩
Melis, R., **2**, 683 ⟨78JOC4672⟩
Melis, S., **4**, 347 ⟨84PS(19)61⟩, 391 ⟨86TL4625⟩; **6**, 95 ⟨75S660⟩
Mella, S. L., **6**, 509 ⟨81JMC1089⟩
Meller, A., **2**, 399 ⟨80CB1950⟩, 811 ⟨77JOM(141)249⟩, 1085 ⟨87CB1437⟩; **4**, 145 ⟨71M118⟩, 606 ⟨76AG(E)619⟩, 1015 ⟨74M684, 82JOM(231)191⟩; **5**, 380 ⟨76ZN(B)1546⟩, 380, 382 ⟨78CB538⟩, 381 ⟨77M1027, 84CB2112⟩, 381, 382 ⟨77M1279⟩, 381, 382 ⟨79M473⟩, 381, 660, 717 ⟨77ZAAC(433)94, 77ZAAC(436)173⟩, 383, 722 ⟨78ZAAC(443)148⟩, 739 ⟨77JOM(141)241⟩, 777, 778 ⟨78ZAAC(445)107⟩, 795, 796 ⟨82JOM(231)191⟩, 863, 864 ⟨71M121⟩, 1147 ⟨70M627⟩; **6**, 183 ⟨91CB2645⟩, 184 ⟨88JOM(347)11⟩, 184 ⟨90CC741⟩, 184 ⟨92CB1807⟩, 196 ⟨93CB2247⟩, 622 ⟨69M1823⟩, 662 ⟨82JOM(231)191⟩
Mellerio, G., **5**, 588 ⟨93JHC1079⟩
Mellet, C. O., **5**, 1028 ⟨89CAR(188)35⟩
Mellin, T. N., **3**, 198 ⟨89JMC1886⟩; **5**, 370 ⟨89JMC1886⟩
Mello, R., **3**, 120 ⟨91JA2205⟩; **4**, 188 ⟨89TL257⟩; **6**, 230 ⟨91IZV2593⟩
Melloni, G., **1**, 182, 475 ⟨91JOC183⟩, 200 ⟨82JOC4577⟩, 318 ⟨75TL2771, 79JOC3022⟩, 398 ⟨90JOC5532⟩, 432 ⟨89S28, 90JOC5386⟩, 1111 ⟨70JCS(C)2625⟩, 1112 ⟨68TL4039, 69CC1520⟩, 1112 ⟨71JCS(C)3018⟩, 1122 ⟨76JA3295⟩, 1123 ⟨65TL4399, 65TL4405⟩; **2**, 9 ⟨89TL4555⟩; **4**, 179 ⟨90S313⟩
Mello Périssé, A. C. de, **5**, 856 ⟨56CB1081⟩
Mellor, J. M., **1**, 7 ⟨87TL5339⟩; **2**, 24 ⟨87CC1577⟩, 284 ⟨80TL3827⟩; **5**, 282 ⟨76S32⟩, 1108 ⟨87JCS(P1)2729⟩
Mellor, M. T. J., **2**, 824 ⟨76JCS(P1)2529⟩, 844, 846 ⟨77JCS(P2)1705⟩, 847, 848 ⟨77JCS(P2)789⟩

Mel'nichenko, N. V., **4**, 314, 329 ⟨74JOU1565⟩, 441 ⟨76ZOR2103, 76ZOR231⟩; **6**, 657 ⟨78ZOB1425⟩
Mel'nichenko, N. W., **5**, 1008 ⟨75ZOR2227⟩
Mel'nichuk, E. A., **2**, 449 ⟨69ZOB2139, 78JGU329⟩, 836, 837 ⟨76ZOB252⟩
Mel'nik, Y. I., **2**, 256 ⟨87ZOB1078⟩
Mel'nikov, A. A., **6**, 55, 282 ⟨87ZOR1657⟩, 149 ⟨87ZOR2624⟩
Melnikov, L. L., **2**, 468, 469, 470 ⟨71ZOB1644⟩
Mel'nikov, N. N., **2**, 361 ⟨75ZOB1005⟩; **4**, 131 ⟨68JGU2563⟩; **5**, 644 ⟨87ZOB685⟩, 774 ⟨78ZOB2465⟩, 966 ⟨66ZOB1707⟩; **6**, 230 ⟨30JPR210, 30JPR81⟩, 635 ⟨90ZOB798⟩
Mel'nikova, L. G., **4**, 995 ⟨91IZV1849⟩
Mel'nikova, T. G., **6**, 145 ⟨81IZV2146⟩
Mel'nitskii, I. A., **2**, 265 ⟨87ZOR1981⟩; **6**, 89 ⟨88URP1421744⟩
Mélot, J.-M., **2**, 394 ⟨86TL493⟩; **3**, 191 ⟨86TL493⟩
Melpolder, J. B., **1**, 318 ⟨76JOC265⟩
Melton, J., **1**, 156, 309 ⟨73JOC4367⟩; **2**, 395 ⟨88OSC(6)648⟩
Meltz, C. N., **1**, 119 ⟨83JOC1767⟩, 204 ⟨83JA5946⟩, 1041 ⟨83TL4503⟩
Meltzer, R. I., **2**, 692 ⟨67TL859⟩; **5**, 566, 579, 591, 618, 621 ⟨55JA4062⟩
Melville, M. G., **3**, 91, 97 ⟨90SC609⟩
Melvin, H. W., **5**, 595, 620, 1016 ⟨50JA3045⟩
Melvin, Jr., L. S., **1**, 233 ⟨B-75MI 105-01⟩, 398 ⟨B-75MI 109-01⟩, 1134 ⟨78JA4618, 78JA4620⟩; **2**, 629 ⟨78JA4618⟩, 778 ⟨75TL3117⟩; **3**, 374 ⟨B-75MI 308-01⟩; **5**, 173 ⟨78JA4618⟩, 173, 204 ⟨78JA4620⟩
Melvin, T., **1**, 1129 ⟨87CC786⟩
Melzer, D., **5**, 1097 ⟨84JOM(263)67⟩
Memarian, H. R., **3**, 61 ⟨89CB585⟩
Memmesheimer, H., **1**, 985 ⟨92SL635⟩; **2**, 449 ⟨91SL433⟩
Memura, K., **3**, 498, 651 ⟨84SC1267⟩
Mena, M., **4**, 656 ⟨89JOM(375)59⟩; **5**, 797 ⟨91OM1201⟩
Menachem, Y., **2**, 16 ⟨79CC479⟩
Menachery, M., **3**, 358 ⟨86JOC411⟩
Ménager, L., **2**, 561 ⟨71BSF3571⟩
Menahem, Y., **5**, 202 ⟨80JFC(16)19⟩
Menapace, L. W., **1**, 330, 341, 936 ⟨63JOC2165⟩
Ménard, G., **5**, 743 ⟨76BSF301⟩
Menard, M., **5**, 236 ⟨88CHE1297⟩
Menard, P. R., **4**, 352 ⟨85TL1115⟩
Menashe, N., **3**, 431 ⟨89IJ187⟩
Menasse, R., **3**, 65 ⟨82TL167⟩
Mencarelli, P., **2**, 28 ⟨87G83⟩; **4**, 415 ⟨83JOC162⟩
Menchen, S. M., **1**, 21, 25, 437, 438, 943 ⟨80JA4438⟩, 609 ⟨78CC770⟩; **2**, 294 ⟨79JOC4279⟩; **4**, 238, 288 ⟨79JOC1883, 79JOC4279⟩; **6**, 92 ⟨79JOC1883, 79JOC4279⟩
Mendelowitz, P. L., **2**, 622 ⟨78JOC3076⟩
Mendelsohn, J., **2**, 701 ⟨68JOM(11)97⟩
Mendelsohn, R., **6**, 365 ⟨90MI 612-01⟩
Mendelson, L., **1**, 468, 472 ⟨81JA6460⟩
Mendelson, S. A., **4**, 365, 366 ⟨85TL2965⟩
Mendelson, W. L., **4**, 825 ⟨83JOC298⟩
Mendenhall, G. D., **1**, 925 ⟨85JOC3220⟩; **3**, 407 ⟨85JOC5382⟩
Méndez-Castrillón, P. P., **5**, 1039 ⟨84S509⟩
Mendoza, A., **4**, 750, 1067 ⟨78JOM(152)1⟩, 849 ⟨79JOC1352⟩; **6**, 182 ⟨78JOM(152)1⟩
Mendoza, J. de, **2**, 316 ⟨91JOC452⟩
Mendoza, J. S., **1**, 615, 698 ⟨90TL7105⟩; **3**, 46, 189 ⟨91TL1487⟩
Mendívil, C., **4**, 584 ⟨85JOM(290)125⟩
Menéndez, E.,**3**, 431 ⟨91S629⟩; **5**, 217 ⟨90SL690⟩
Menendez, J. C., **3**, 616 ⟨86TL3285⟩
Menes, R., **1**, 534 ⟨87TL5129⟩
Menezes, R. F., **5**, 199 ⟨89TL3295⟩
Meng, Q.-J., **4**, 438 ⟨88SC2289, 89S307⟩
Meng, X.-J., **1**, 281 ⟨91TL4619⟩
Menge, G. A., **3**, 616 ⟨34JA2197⟩
Menge, M., **4**, 563 ⟨81AP(314)218⟩
Mengel, R., **3**, 359 ⟨90CB1143⟩; **4**, 232 ⟨90CB1143⟩
Menger, F., **2**, 379, 388 ⟨64JA2395⟩; **3**, 446 ⟨64JA2395⟩
Menger, F. M., **3**, 116 ⟨79JOC3446⟩; **5**, 47, 49, 52, 113 ⟨81TL1655⟩
Menges, D., **2**, 577 ⟨92CB163⟩
Menicagli, R., **1**, 1049 ⟨89JOC2391⟩; **2**, 393 ⟨87T5411, 90T7497⟩, 393, 783 ⟨88JOC45, 89JOC2391⟩
Menichi, G., **6**, 620 ⟨84TL1557⟩
Menichincheri, M., **1**, 601, 781, 785 ⟨93JOC322⟩
Menichini, F., **1**, 784 ⟨84JCS(P1)1701⟩
Menkisoglou-Spyroudi, O., **5**, 205 ⟨86JCS(P1)795⟩
Mennenga, H., **4**, 580 ⟨73AG(E)854, 75JOM(84)C13⟩; **6**, 207 ⟨74JPR557⟩
Mennona, F. A., **1**, 1029 ⟨89JMC1820⟩
Menon, B., **1**, 847 ⟨B-80MI 119-02⟩, 859 ⟨79JOC317⟩
Menon, B. C., **1**, 870 ⟨69JOC4083⟩, 871 ⟨67JA2664⟩
Menozzi, G., **5**, 344 ⟨86FES539⟩, 501 ⟨86JHC455, 87JHC633⟩
Mentrup, A., **2**, 441 ⟨61GEP1114190, 61LA(646)65⟩
Mentzel, C., **3**, 264 ⟨73CB450⟩
Mentzen, B., **4**, 713 ⟨B-85MI 408-01⟩
Menu, M. J., **3**, 486 ⟨87OM1822⟩, 487 ⟨90JA5351⟩; **4**, 548 ⟨90PS(47)327⟩; **5**, 871, 872, 873, 1148 ⟨90JA6277⟩; **6**, 665 ⟨89JOM(372)201⟩, 665, 667, 668, 669, 670 ⟨90JA6277⟩, 665, 680 ⟨86JA7868⟩, 673 ⟨88CC1598⟩, 675 ⟨87OM1822, 90JA5351⟩
Menz, F., **3**, 472 ⟨67TL739, 68CB2622⟩, 472 ⟨70LA(739)174⟩
Menz, G., **2**, 903 ⟨77ZN(B)1400⟩
Menzel, D., **2**, 817 ⟨64CB1155⟩
Menzel, H., **3**, 483 ⟨68JA1080, 72JOM(44)279⟩; **4**, 513 ⟨68JA1080⟩
Menzel, I., **1**, 1075 ⟨71S215⟩; **3**, 553 ⟨72TL445⟩; **4**, 874 ⟨72TL445⟩
Menzel, J., **4**, 567 ⟨85CB4068⟩
Menzel, M., **4**, 837, 860 ⟨82S206⟩
Menzer, W., **3**, 711 ⟨58AG656⟩
Meot-Ner, M., **1**, 892 ⟨87JA7947⟩
Meou, A., **3**, 658 ⟨86TL3505⟩
Meïr, M. S., **2**, 579 ⟨88TL167⟩
Merault, G., **1**, 963 ⟨72JOM(39)C44⟩; **2**, 916, 923, 940 ⟨75JOM(87)169⟩, 1090 ⟨72CR(C)1857⟩; **5**, 297 ⟨74BSF(2)1949⟩
Mercer, A. J. H., **1**, 647 ⟨67JCS(C)1099⟩; **2**, 508, 865 ⟨72JCS(P1)2548⟩; **3**, 494 ⟨68JCS(C)2448⟩
Mercer, F., **3**, 543 ⟨78TL929⟩
Mercer, J., **4**, 606, 620, 623, 653 ⟨90OM2677⟩; **6**, 191 ⟨90OM2677⟩
Mercer, R. J., **4**, 676 ⟨88JCS(D)1843⟩
Mercere, N. M. I., **6**, 221 ⟨52MI 607-02⟩
Merchan, F., **5**, 1026 ⟨89BSB289⟩; **6**, 563 ⟨81S961⟩; **6**, 627 ⟨82S590⟩
Merchan, F. L., **3**, 43 ⟨92TL4221⟩; **6**, 563 ⟨91S147⟩, 605 ⟨84S520, 83S375⟩
Merchan Alvarez, F. L., **6**, 605 ⟨83AQ(C)115⟩
Merchant, J. R., **4**, 304 ⟨78TL3607⟩
Mercier, A., **4**, 485 ⟨91TL2125⟩
Mercier, C., **3**, 226 ⟨91TL1433⟩
Mercier, F., **2**, 430 ⟨85TL1717⟩, 441 ⟨79JOM(177)255, 82JOM(231)237⟩; **4**, 545 ⟨87JOM(328)349⟩; **6**, 154 ⟨91CC1305⟩
Mercier, S., **5**, 27 ⟨78MI 502-01⟩
Merck, W., **5**, 222 ⟨1856LA(97)329⟩
Merckling, N., **2**, 227 ⟨54HCA375⟩
Merdès, R., **4**, 557, 570, 575 ⟨93PS(77)199⟩
Merényi, R., **1**, 592 ⟨81AG(E)585⟩, 919, 922 ⟨85ACR148⟩, 919, 922 ⟨88PAC1635⟩; **2**, 1014 ⟨64AG(E)747, 64AG(E)755⟩; **4**, 71 ⟨81AG(E)585⟩, 88, 773, 785 ⟨87T4309⟩, 91, 287, 858, 861 ⟨85T2527⟩, 92 ⟨85T4183⟩, 681 ⟨62CB1155⟩, 798 ⟨78AG(E)936, 81HCA1208⟩ 829 ⟨71CB2786⟩; **5**, 601 ⟨85S79⟩, 601, 612 ⟨85S77⟩; **6**, 44

⟨81AG(E)585⟩, 52 ⟨79AG(E)615⟩, 607 ⟨78CB1619⟩, 619 ⟨79AG(E)333⟩
Mereshkowsky, B. K., **2**, 87 ⟨23LA(431)231⟩
Meresz, O., **3**, 532 ⟨69T2121⟩
Mereyala, H. B., **1**, 812 ⟨84HCA774⟩
Mergardt, B., **2**, 938 ⟨91AG(E)1687⟩
Mergen, W. W., **2**, 812 ⟨80S460⟩; **3**, 266 ⟨83LA290⟩; **4**, 972, 979 ⟨80LA372⟩, 973, 974 ⟨83LA290⟩; **6**, 139 ⟨90CZ176⟩, 643 ⟨84LA108⟩
Merger, F., **2**, 683 ⟨79GEP(O)2807762⟩; **6**, 506 ⟨68CB2419⟩
Merger, R., **1**, 1001 ⟨93S558⟩
Meric de Beffon, C. de, **2**, 585 ⟨91AG(E)185⟩
Meriem, A., **4**, 459 ⟨83TL1975⟩; **6**, 160 ⟨83PS(18)27⟩
Merienne, C., **2**, 579 ⟨73JOM(55)375⟩
Merifield, E., **5**, 342 ⟨93HCA957⟩
Merijanian, A., **2**, 483, 502, 890 ⟨64IC580⟩, 497, 502 ⟨66IC187, 64JOM(1)369⟩
Merino, I., **5**, 953 ⟨95OM2522⟩
Merino, P., **3**, 43 ⟨92TL4221⟩, 311 ⟨91CC1313⟩; **5**, 1026 ⟨89BSB289⟩
Meritt, M. V., **2**, 98 ⟨78JA7960⟩
Merk, W., **4**, 681 ⟨74JA7108⟩
Merkel, C., **1**, 1000 ⟨75S423⟩
Merkel, C. M., **4**, 552 ⟨86ICA(115)L29⟩
Merkel, D., **2**, 618 ⟨72CB1683⟩
Merkel, R., **3**, 549 ⟨82CB3653⟩
Merkel, T. F., **2**, 377, 403 ⟨81JOC442⟩
Merkel, W., **5**, 739 ⟨70CB32⟩; **6**, 616, 617 ⟨75CB2290⟩
Merker, R. L., **2**, 538 ⟨60JA1883⟩; **4**, 603 ⟨65JOM(4)98⟩; **6**, 173 ⟨63JA2243⟩, 173, 378 ⟨64JOC953⟩, 173, 378 ⟨65JOM(4)98⟩
Merkle, H. R., **2**, 906 ⟨67CB3371⟩; **6**, 62 ⟨65TL969⟩
Merkle, U., **3**, 470 ⟨78S535⟩; **5**, 511 ⟨80CB2519⟩
Merkushev, E. B., **2**, 630 ⟨88S923⟩, 1019 ⟨79DOK(245)607⟩
Merle, A., **4**, 174 ⟨75JHC981⟩
Merlic, C. A., **1**, 237 ⟨88JA5216⟩, 238 ⟨90JOC1127⟩; **2**, 670 ⟨91JA7418⟩; **3**, 544 ⟨93JOC538⟩; **5**, 946 ⟨93TL227⟩
Merlin, P., **4**, 414 ⟨88TL1691⟩
Merling, G., **1**, 906 ⟨1891CB3108⟩
Merlino, S., **1**, 1010 ⟨85T621⟩; **5**, 442 ⟨76CC522⟩
Merrer, Y. Le, **3**, 211 ⟨85S766⟩
Merricks, D., **2**, 362, 363 ⟨91JCS(P1)2169⟩
Merrifield, J. H., **5**, 451, 937 ⟨83OM1852⟩
Merrill, R. E., **2**, 961 ⟨74JOC3452⟩
Merritt, A., **1**, 578 ⟨86TL1449⟩
Merritt, J. E., **1**, 340 ⟨91T8663⟩, 346 ⟨88TL5209⟩
Merritt, L., **3**, 212 ⟨90JHC1709⟩
Merritt, R. F., **2**, 335 ⟨67JOC416⟩; **5**, 659 ⟨67JOC416⟩; **6**, 439 ⟨67JOC1633⟩
Merrow, R. T., **2**, 103 ⟨55CRV485⟩
Mersch, R., **2**, 1009 ⟨56CB209⟩; **3**, 687 ⟨56CB209⟩
Merslavič, M., **5**, 836 ⟨86M(117)221⟩
Merten, R., **4**, 310 ⟨62AG872⟩, 426 ⟨62AG(E)592⟩; **5**, 322, 971 ⟨70S589⟩, 979 ⟨65CB1097⟩; **6**, 105 ⟨62AG(E)592⟩, 141 ⟨72AG(E)964⟩, 141 ⟨75CB215⟩, 143 ⟨71LA(751)145⟩, 156 ⟨78LA16⟩
Mertens, A., **4**, 315 ⟨83S500⟩, 1072 ⟨83CB3926⟩
Mertens, H., **4**, 900 ⟨86AP(319)161⟩
Mertes, M. P., **2**, 105 ⟨74JOC3767⟩; **4**, 229 ⟨69JOC2080⟩
Mertin, A., **3**, 709 ⟨91SL87⟩; **4**, 309 ⟨91SL87⟩
Mertis, K., **2**, 575 ⟨75JOM(99)C19⟩
Mertsalov, S. L., **6**, 109 ⟨67KGS1096⟩
Mertz, C., **1**, 623 ⟨60CB2819⟩
Mertz, W., **1**, 181 ⟨72S142⟩
Merwin, L. H., **6**, 158 ⟨91PS(56)117⟩
Merwin, R., **5**, 944 ⟨95OM2102⟩
Mery, J., **5**, 185 ⟨85T5331⟩
Merz, A., **1**, 731 ⟨73S295⟩; **3**, 323 ⟨74S724⟩, 494 ⟨73S295⟩; **4**, 134 ⟨83JOM(249)335⟩
Merz, H., **5**, 1038 ⟨58CB683⟩
Mesch, K. A., **2**, 441 ⟨83JOC4466, 84JA7021⟩
Meseri, J., **6**, 221 ⟨52MI 607-01⟩
Meshcheryakov, A. P., **3**, 452 ⟨61CA(55)17526⟩
Meshram, H. M., **2**, 230 ⟨93TL2521⟩; **3**, 18, 134 ⟨91JCS(P1)2056⟩, 18, 134 ⟨91TL2657⟩
Meske, M., **1**, 980 ⟨93JA3776⟩
Meskens, F. A. J., **4**, 176, 177 ⟨81S501⟩
Meskens, F., **4**, 181 ⟨84MI 404-01⟩, 186 ⟨83MI 404-01⟩
Meske-Schüller, J., **1**, 104 ⟨68LA(711)65⟩
Meslans, M. M., **5**, 4 ⟨1894MI 501-01⟩
Meslin, J.-C., **2**, 744 ⟨74S298⟩; **3**, 363 ⟨91S543⟩; **5**, 633 ⟨85PS(25)279⟩, 760 ⟨80JHC767, 83PS(15)143⟩
Mesnard, D., **1**, 971 ⟨91JOM(403)299⟩, 973 ⟨92JOM(440)277⟩, 974 ⟨86BSF449⟩, 978 ⟨82JOM(236)177⟩, 1059 ⟨87BSF93⟩; **4**, 1047 ⟨82TL4083⟩
Messa, W., **4**, 634 ⟨88AG(E)1370⟩
Messbauer, B., **2**, 460 ⟨83ZAAC(504)147⟩
Messeri, T., **2**, 917, 922 ⟨92G65⟩
Messerle, B. A., **4**, 603, 649 ⟨91OM3167⟩
Messerle, L. W., **2**, 570 ⟨80JA6744⟩; **3**, 510 ⟨78JA3793⟩, 510, 511, 512 ⟨80JA6236⟩, 511 ⟨80JA6744, 81JA1440⟩; **5**, 918 ⟨80JA6744⟩
Messerotti, W., **3**, 489 ⟨77JOC1352⟩
Messier, A., **2**, 759 ⟨82JCR(M)621, 82JCR(S)48⟩; **3**, 35 ⟨80TL4171⟩
Messinger, P., **2**, 188 ⟨77LA1457⟩; **4**, 333 ⟨80S565⟩
Messmer, A., **5**, 641 ⟨53CB697⟩, 1079 ⟨64AG(E)228, 87CAR(166)101⟩; **6**, 114 ⟨87JOC2015⟩
Mestdagh, H., **2**, 944 ⟨85JA5670⟩; **3**, 419 ⟨83T433⟩
Mester, L., **3**, 450 ⟨65AG580, 65AG(E)574⟩, 450 ⟨65JA596⟩; **5**, 641 ⟨53CB697⟩, 855 ⟨62BSF186⟩, 857 ⟨55JA4297, 55JA4301⟩
Mester, T., **3**, 259 ⟨79LA174⟩
Mestre, F., **1**, 231 ⟨89H(28)171⟩; **6**, 551 ⟨89H(28)171⟩
Mestres, R., **5**, 189 ⟨81S218⟩, 300 ⟨82S288⟩, 1063 ⟨81S373⟩, 1079 ⟨81BSF(2)361⟩
Mestroni, G., **5**, 457 ⟨67TL1783⟩
Mesureur, D., **2**, 751 ⟨89S309, 92JOC3807⟩; **3**, 150 ⟨92JOC3807⟩
Meszaro, Z., **3**, 310 ⟨77JCS(P1)789⟩
Meszaros, Z., **2**, 1008 ⟨75CSC135⟩
Mesztìczky, A., **6**, 449 ⟨81MI 614-02⟩
Metais, E., **3**, 303, 306 ⟨92TL5245⟩
Metcalf, B., **2**, 299 ⟨85JCS(P1)2201⟩
Metcalf, B. W., **1**, 1138 ⟨78TL1581⟩; **3**, 150 ⟨85TL2861⟩, 472 ⟨80TL15⟩; **4**, 731, 949 ⟨85TL2861⟩
Metcalfe, J., **1**, 784 ⟨72JA7235⟩
Metcalfe, S., **1**, 54 ⟨86T3887⟩, 648 ⟨88S319⟩, 648, 738, 743 ⟨88JCS(P2)1829⟩; **3**, 196, 500 ⟨86T3887, 88S319, 88JCS(P2)1829⟩; **6**, 708, 709 ⟨86T3887⟩, 708, 709 ⟨88JCS(P2)1829⟩, 708, 709 ⟨88S319⟩
Metcalfe, T. P., **5**, 114, 115 ⟨51JCS1208⟩
Metcalf III, C. A., **1**, 176 ⟨93JA2581⟩
Mete, A., **1**, 218 ⟨85TL97⟩; **4**, 970 ⟨91EUP425030⟩
Meth-Cohn, O., **2**, 39 ⟨93CC119⟩, 244 ⟨84CC1144⟩, 298 ⟨87ACR18⟩, 753 ⟨78S23⟩; **3**, 104 ⟨75TL4705⟩, 105 ⟨80S(59)133⟩; **4**, 237 ⟨88JCS(P1)2663⟩, 938 ⟨93JCS(P1)343⟩; **5**, 381 ⟨B-88MI 507-01⟩, 1108 ⟨73JCS(P1)705⟩; **6**, 507, 577 ⟨75JCS(P1)212⟩
Metha, G., **3**, 669 ⟨78S374⟩
Metha, P., **6**, ⟨92MI 615-02⟩
Methvink, R. J., **3**, 136 ⟨89JOC177⟩
Metivier, P., **1**, 427 ⟨91JOC4718⟩
Metler, T., **1**, 538 ⟨68T4285⟩; **3**, 253 ⟨68T4285⟩
Metlesics, W., **5**, 354 ⟨65JOC1311⟩
Metni, M. R., **2**, 1003 ⟨91T5277⟩
Métra, P., **4**, 444 ⟨82CJC285⟩
Metras, F., **3**, 234 ⟨76S33⟩; **4**, 1079 ⟨81T2515, B-78MI 408-01⟩
Mettegang, H., **3**, 638 ⟨43CB1275⟩
Metternich, H.-J., **4**, 568, 573 ⟨92CB771⟩
Metternich, R., **2**, 529 ⟨85LA2246, 87LA881⟩, 702

⟨90JA866⟩; **4**, 528 ⟨92EUP471651⟩; **6**, 487 ⟨93EUP560730⟩
Mettile, F. J., **1**, 431 ⟨82JFC(20)157⟩
Mettille, F. J., **1**, 426 ⟨68FCR315⟩
Mettler, C., **3**, 618 ⟨02CB3647⟩
Metwally, S. A. M., **1**, 201 ⟨75BCJ2939⟩; **5**, 690 ⟨90S704⟩
Metz, F., **2**, 578 ⟨84NJC759⟩
Metz, J. T., **2**, 41 ⟨84T2257⟩
Metz, P., **1**, 828 ⟨90SL97⟩; **2**, 207 ⟨92SL985, 93SL399⟩, 207, 721 ⟨89AG(E)202⟩; **5**, 685 ⟨88T6841⟩
Metzenthin, T., **6**, 304 ⟨88EUP253527⟩
Metzer, H., **2**, 340, 343, 344, 346, 350, 353, 354, 355, 356 ⟨71HOU(10/1)1091⟩
Metzer, K. G., **6**, 445 ⟨82MI 614-03⟩
Metzger, C., **3**, 527 ⟨68CB1120⟩; **4**, 901, 904 ⟨68CB1131⟩
Metzger, H., **2**, 389 ⟨71HOU(10/1)891⟩; **3**, 378 ⟨65CB3733⟩; **5**, 215 ⟨68HOU(10/4)184⟩, 819 ⟨68HOU(10/4)1⟩; **6**, 597 ⟨36CB1356⟩
Metzger, H. G., **2**, 874 ⟨68IC2003⟩, 874, 876 ⟨71JOM(33)347⟩
Metzger, J., **1**, 925 ⟨81TL1891⟩; **2**, 33 ⟨83JOC5409⟩, 40 ⟨71S527⟩; **3**, 279 ⟨85BSB897⟩; **5**, 918 ⟨85CC1151⟩; **6**, 111 ⟨72HCA1782⟩, 115 ⟨71BSF556⟩
Metzger, J. V., **4**, 318 ⟨84CHEC(6)321⟩, 328 ⟨84CHEC(6)312⟩
Metzler, N., **4**, 635 ⟨93CB2003, 93OM2423⟩
Metzler, O., **2**, 662 ⟨47HCA1883, 47HCA675⟩
Metzner, P., **1**, 834 ⟨92T10315⟩; **2**, 140 ⟨91PS(59)295⟩; **3**, 331 ⟨92S1185⟩, 343 ⟨76TL4297, 77T3089⟩, 343 ⟨78T2289⟩, 347 ⟨92TL6111⟩; **4**, 255 ⟨82CC335, 85BSF881⟩, 257 ⟨80TL4657⟩; **5**, 248 ⟨94S761⟩, 248, 562 ⟨88CC390⟩, 527 ⟨91TL7411⟩, 550 ⟨92S1185⟩, 550, 556 ⟨89RHA152⟩, 551 ⟨85JCR(S)272⟩, 551, 560 ⟨85CI(L)824⟩, 552 ⟨83TL3617, 91(59)PS1⟩, 553 ⟨86T2025⟩, 556 ⟨82CC335, 83TL4203, 85BSF881, 86BSF817, 86TL1505, 87TL409, 88T3575, 89T2041, 90BSF446⟩, 557 ⟨90TL1853⟩, 587, 611 ⟨93TL6741⟩, 616 ⟨85T1289⟩
Meuche, D., **1**, 906 ⟨58HCA414⟩
Meunier, A. C., **6**, 215, 216, 221 ⟨39JA435⟩
Meunier, B., **1**, 429 ⟨81JOM(204)345⟩
Meunier, C. J., **1**, 591 ⟨88JCS(P1)813⟩; **6**, 559 ⟨88JCS(P1)813⟩
Meunier, I., **3**, 76 ⟨89JCS(P1)2095⟩
Meunier, P., **3**, 636 ⟨86CC424⟩
Meunier-Piret, J., **2**, 435 ⟨91PS(61)261⟩; **4**, 700 ⟨84JOM(277)335⟩
Meuret, J., **6**, 174 ⟨93CB2227⟩, 174 ⟨93JOM(462)31⟩, 349 ⟨90JOM(398)65⟩
Meurling, P., **2**, 123 ⟨72ACS279⟩
Meusen, A., **2**, 374, 379, 384 ⟨73OSC(5)43⟩, 380, 383, 384 ⟨59CB2521⟩
Mews, R., **2**, 401 ⟨79ZAAC(452)103⟩, 402 ⟨78AG(E)530⟩; **5**, 662, 663 ⟨82CC839⟩, 662, 809 ⟨85CB4459, 83CB724⟩; **6**, 251 ⟨91JFC(55)313⟩, 252 ⟨83JFC(23)325, 90IC4588⟩, 252 ⟨91CB1353, 92CC1017⟩, 253 ⟨86IC275, 92CB535⟩, 260, 261 ⟨86CB107⟩, 440 ⟨91CB2411⟩, 604 ⟨86JFC(34)46⟩, 726 ⟨85CB4997, 85CB5007⟩
Meuwsen, A., **2**, 99 ⟨35CB1011⟩, 99, 172 ⟨36CB937⟩
Mewshaw, R. E., **3**, 239 ⟨89TL3753⟩
Mewwly, R., **2**, 457 ⟨92T2911⟩
Meye, O., **3**, 452 ⟨23CB1136, 24CB1108⟩
Meyer, A., **1**, 958 ⟨83T2185⟩; **4**, 135, 597 ⟨83CB348⟩
Meyer, A. Y., **6**, 206 ⟨76T1839⟩
Meyer, B., **4**, 979 ⟨76BSF2053⟩
Meyer, C., **2**, 313 ⟨93SL386⟩, 592 ⟨93SL266⟩, 593 ⟨93TL6053⟩
Meyer, E., **1**, 725 ⟨76LA1475⟩; **3**, 693 ⟨1866JPR147⟩
Meyer, E. F., **1**, 459 ⟨63JA2881⟩
Meyer, F., **2**, 693 ⟨52N353⟩
Meyer, F. C., **5**, 652 ⟨00LA(314)200⟩
Meyer, F. E., **1**, 494 ⟨94SL189⟩
Meyer, G., **2**, 497 ⟨1883CB1439⟩; **3**, 231 ⟨43CB676⟩; **6**, 236 ⟨71S478⟩, 529 ⟨74S26⟩
Meyer, G. R., **1**, 884 ⟨82JA6788⟩
Meyer, H., **2**, 439 ⟨87ZAAC(553)136⟩, 820, 821 ⟨81TL4475⟩; **3**, 359 ⟨89JST(198)307⟩; **4**, 634 ⟨85AG(E)788⟩, 880 ⟨76LA1762⟩, 972 ⟨77LA1895⟩, 1082 ⟨87AG(E)546⟩; **5**, 784, 881 ⟨78JOM(160)47⟩, 883 ⟨80TL1141⟩, 927, 928 ⟨85AG(E)788, 87AG(E)546⟩, 928 ⟨87AG(E)798, 88ZN(B)801⟩; **6**, 188, 711 ⟨87AG(E)798⟩, 201 ⟨85AG(E)788⟩, 202, 724 ⟨87AG(E)546⟩, 484 ⟨B-31MI 615-01⟩, 686 ⟨80TL1141, 81TL4475⟩, 686 ⟨81ZC407⟩, 687 ⟨80JOM(192)33⟩, 711 ⟨87PAC1011⟩, 714 ⟨88ZN(B)801⟩
Meyer, H. J., **5**, 581, 611, 643, 644 ⟨85PS(25)297⟩
Meyer, H. S., **2**, 783 ⟨06CB2334⟩
Meyer, H.-W., **4**, 903 ⟨75LA19⟩; **5**, 735, 736 ⟨73LA462, 74LA776⟩
Meyer, J., **2**, 94 ⟨36ZAAC(228)341⟩, 810, 853, 859 ⟨19HCA635⟩; **3**, 498 ⟨19HCA619⟩, 555, 591 ⟨19HCA635⟩, 561, 584, 591 ⟨20CB72⟩
Meyer, K., **5**, 106 ⟨55OSC(3)637⟩
Meyer, M., **4**, 652 ⟨88CB1393⟩
Meyer, M. D., **3**, 106, 309 ⟨84JOC3195⟩
Meyer, N., **1**, 716 ⟨77LA846⟩; **2**, 281 ⟨77LA846⟩, 632 ⟨80CB1304⟩; **4**, 238, 284 ⟨77LA846⟩, 378 ⟨76AG(E)438, 80CB1290⟩; **6**, 632 ⟨86GEP(O)3504453⟩
Meyer, P.-H., **4**, 49, 310 ⟨71S150⟩
Meyer, R., **3**, 474, 484, 485, 486 ⟨74LA1767⟩, 695 ⟨59AG386⟩, 696, 721, 722, 723 ⟨60CB239⟩, 706 ⟨77LA1174⟩; **5**, 872 ⟨74LA1767⟩; **6**, 536 ⟨79ZC289⟩
Meyer, R. B., **1**, 881 ⟨60JOC158⟩; **2**, 163, 237 ⟨84JMC228⟩
Meyer, R. F., **2**, 741 ⟨73JMC1113⟩; **6**, 632 ⟨63JOC2902⟩
Meyer, S., **2**, 437 ⟨87JOM(335)59⟩
Meyer, U., **2**, 866 ⟨87JOM(326)C6⟩; **6**, 166 ⟨86AG(E)919⟩
Meyer, V., **3**, 546 ⟨1890CB1571⟩; **4**, 439 ⟨1875LA(175)88⟩; **5**, 652 ⟨1882LA(214)329⟩, 838 ⟨1873CB1492, 1874CB425⟩, 838 ⟨1874CB670, 1875LA(175)93⟩
Meyer, W. L., **1**, 46 ⟨75JOC3686⟩, 578 ⟨86TL1449⟩, 884 ⟨65JOC181⟩; **2**, 658 ⟨75JOC3686⟩; **5**, 39 ⟨63JOC2184⟩, 179 ⟨89LA1233⟩
Meyer-Bäse, K., **4**, 585 ⟨88JCS(D)341⟩, 586 ⟨88OM997⟩
Meyers, A. I., **1**, 89 ⟨71OS(51)103⟩, 134 ⟨91T9503, 93JOC7507⟩, 134 ⟨93TL7701⟩, 149 ⟨83JA117⟩, 149 ⟨83JA2082⟩, 149 ⟨84JA3270⟩, 149 ⟨84T1361⟩, 149 ⟨87T5095⟩, 149 ⟨88JA4778⟩, 149 ⟨89H(28)295⟩, 149 ⟨89OS(67)60⟩, 149 ⟨91JOC2091⟩, 149 ⟨91JOC6873⟩, 149 ⟨92JA8483⟩, 149 ⟨93T1807⟩, 149, 594, 716 ⟨86JOC3108⟩, 150 ⟨85JA7974, 91TL5501, 91TL5505, 91TL5509, 92JOC3819⟩, 151 ⟨93JA6652, 93JOC6538, 81JA3081, 81JA3088⟩, 310 ⟨87TL5279⟩, 357 ⟨86JA306⟩, 473 ⟨93S250⟩, 473, 539 ⟨85T837⟩, 483 ⟨85JA682, 88JA7178⟩, 664 ⟨70TL4355, 73JA3408⟩, 722 ⟨72JOC2579⟩, 845, 860 ⟨92T2589⟩, 1001, 1134 ⟨90JOC791⟩, 1041 ⟨70JA6676⟩, 1130 ⟨87T5663⟩, 1136 ⟨85JOC1019⟩; **2**, 12, 627 ⟨79JOC3405⟩, 328 ⟨89H(28)295⟩, 398 ⟨67JOC4134⟩, 653 ⟨85T837⟩, 816 ⟨78JOC3245⟩, 974 ⟨83TL4935, 85CC690⟩, 1002 ⟨85JOC1019⟩; **3**, 27 ⟨74JOC618⟩, 40 ⟨93JOC3568⟩, 49, 60 ⟨78JOC3788⟩, 58, 90, 101 ⟨78S403⟩, 61 ⟨73JOC36, 74JOC623⟩, 90, 97 ⟨70JA6676, 74OS(54)42⟩, 108 ⟨78TL227⟩, 124 ⟨81JA3081⟩, 182 ⟨80JA7125⟩, 182 ⟨85JOC1019⟩, 182 ⟨91JOC2143⟩, 211, 576 ⟨73JOC2129⟩, 219 ⟨89JOC4673⟩, 286 ⟨69JA5886⟩, 294 ⟨80JOC2785⟩, 407 ⟨76JA3032⟩, 494 ⟨71T5979⟩, 496 ⟨81JOC3119, 83TL3043⟩, 576 ⟨69JA5887, 70TL4355⟩, 621 ⟨93TL5459⟩, 665 ⟨84TL2941⟩, 707 ⟨69TL5151⟩; **4**, 210 ⟨79JA2501⟩, 255 ⟨79JA4732, 79JA7104⟩, 306 ⟨76AG(E)270, 85T837⟩, 306 ⟨88OSC(6)64, 88OSC(6)905⟩, 306 ⟨94T2297⟩, 321 ⟨66JOC556⟩, 511 ⟨82JA877⟩, 530 ⟨80JA7125, 85JOC1019⟩, 532 ⟨84T1361, 86JOC3108⟩, 532 ⟨86JOC872⟩; **5**, 79 ⟨84JA1146⟩, 80 ⟨80JOC2785⟩, 166 ⟨75JOC1186⟩, 168, 172 ⟨B-84MI 503-02⟩, 172 ⟨B-84MI 503-01⟩, 557 ⟨78TL4657, 79JA4732⟩, 557 ⟨79JA7104, 81TL3925⟩,

557 ⟨83JA5015, 85T5089⟩, 747 ⟨87T5095⟩, 1014 ⟨69TL5151⟩; **6**, 107 ⟨91TL597⟩, 134 ⟨79JA4732⟩
Meyers, C. Y., **1**, 618 ⟨73TL1843, 74TL1105⟩, 618, 704 ⟨80JOC1719⟩, 618, 706 ⟨69JA7510, 78JOC1985⟩, 706 ⟨B-77MI 115-01⟩; **2**, 156 ⟨73TL289⟩, 717, 719 ⟨72TL4319⟩; **4**, 75 ⟨73TL289⟩, 81 ⟨69JA7510⟩; **5**, 86 ⟨63OSC(4)39⟩, 867 ⟨80JOC1717⟩
Meyers, H. V., **1**, 697, 698, 703, 1099 ⟨88JA5198⟩
Meyers, M. D., **6**, 261 ⟨66IC1455⟩, 363 ⟨73JOC1088⟩
Meyers, T. C., **6**, 421 ⟨55JA5033⟩, 422 ⟨55JA3099⟩
Meyer zu Reckendorf, W., **5**, 140 ⟨70CB37⟩
Meyle, E., **6**, 129 ⟨89LA975⟩
Meynaud, A., **4**, 103 ⟨71BSF942⟩
Meyr, R., **3**, 328 ⟨58AG702⟩, 696, 700 ⟨73OSC(5)300⟩, 721 ⟨61OS(41)13⟩
Meystre, C., **2**, 95 ⟨61E475⟩
Mezheritskaya, L. V., **5**, 688 ⟨81JOU1986, 81ZOR2225⟩
Mezheritskii, V. V., **2**, 1008 ⟨82AG(E)555⟩; **5**, 688 ⟨81JOU1986, 81ZOR2225⟩
Mezzina, E., **3**, 456 ⟨91TL2967⟩, 458 ⟨90TL3481, 93T3809⟩; **6**, 121 ⟨90JCS(P1)19⟩
Mhaskar, S. Y., **1**, 1019 ⟨90TL7227⟩
Mhatre, S., **4**, 899, 900 ⟨76CC592⟩; **5**, 110 ⟨77JFC(10)437⟩
Mi, A.-Q., **1**, 135 ⟨93TA1957⟩; **2**, 359, 363 ⟨94JOC914⟩
Miah, M. A. J., **1**, 464 ⟨84JOC737⟩; **3**, 311 ⟨85JOC5436⟩
Miao, G., **3**, 128 ⟨93T965⟩
Miao, S. W., **2**, 362 ⟨91JOC4⟩, 743 ⟨91S970⟩; **3**, 437 ⟨91JMC1292⟩
Miarka, S. V., **5**, 1000 ⟨59JOC1293⟩
Miccoli, G., **1**, 466 ⟨86JOM(312)343⟩
Micetich, R. G., **1**, 1077 ⟨68CJC1057, 71OPP163⟩; **3**, 494 ⟨71CJC2143⟩; **5**, 577 ⟨76TL967⟩, 736 ⟨79CP1051008⟩
Michael, A., **1**, 306 ⟨1887JPR349, 1887MI 107-01⟩; **3**, 634 ⟨37JA744⟩; **4**, 766 ⟨1886CB845⟩; **5**, 593, 595 ⟨1887JPR449⟩
Michael, H., **4**, 683 ⟨88CB1413⟩
Michael, J., **3**, 388, 389 ⟨88CC1494⟩
Michael, J. D., **6**, 625 ⟨85TL4149, 87S170⟩, 628, 633 ⟨85TL1105⟩
Michael, J. P., **3**, 268 ⟨89TL4879⟩
Michael, M. A., **4**, 950 ⟨91T6539⟩
Michaelides, M. R., **3**, 621 ⟨86TL3353⟩; **5**, 152 ⟨89JA2984⟩
Michaelies, A., **2**, 490 ⟨1886LA(233)60⟩
Michaelis, A., **2**, 101 ⟨1887LA(241)150⟩, 386 ⟨1893CB310⟩, 409, 410 ⟨03LA(326)149, 03LA(326)160⟩, 414, 416, 418 ⟨1896CB710⟩, 451 ⟨1874CB1688⟩, 455 ⟨01LA(315)43⟩, 456 ⟨1898CB1048⟩, 465 ⟨1885CB2109⟩, 861 ⟨1896LA(293)193, 1896LA(293)215⟩, 886 ⟨1889LA(251)323⟩, 887 ⟨1894CB263⟩, 890 ⟨02LA(321)141, 1882CB1952⟩, 906 ⟨01LA(315)19, 01LA(315)26⟩, 906 ⟨1882CB180, 1894CB244⟩; **3**, 80 ⟨04LA(331)245⟩, 493 ⟨1899CB1566⟩, 618, 619 ⟨1893LA(274)312⟩
Michaelis, W., **1**, 325, 920 ⟨87JOC205⟩
Michaels, Jr., R. J., **3**, 105 ⟨50JA1422, 63OSC(4)915⟩
Michaels, R. J., **1**, 881 ⟨61JOC644⟩
Michael-Schulz, H., **4**, 683 ⟨90IC4396, 92OM2916⟩
Michaelski, J., **2**, 256 ⟨82CC1362⟩
Michaely, W. J., **1**, 547 ⟨84JOC4244⟩
Michajlyszyn, V., **5**, 749 ⟨54CLY1210⟩
Michalik, M., **6**, 651 ⟨88JPR900⟩
Michalowicz, W., **2**, 800 ⟨58JOC1800⟩
Michalska, M., **2**, 397 ⟨68BAP567⟩
Michalski, J., **2**, 187, 199 ⟨60JCS1665⟩, 411 ⟨90AG(E)522⟩, 412 ⟨57RZC879⟩, 441 ⟨71AG(E)494⟩, 451 ⟨55RZC960⟩, 452 ⟨70CC1495⟩, 461 ⟨62RZC1781⟩, 465 ⟨60RZC1675⟩, 466 ⟨57RZC715⟩, 467 ⟨85T2471, 91TL4981⟩, 475 ⟨82JA290⟩, 846 ⟨51NAT1035⟩, 857 ⟨62MI 216-01⟩; **5**, 1107, 1112 ⟨83T4175⟩
Michalski, T. J., **4**, 585 ⟨80PS(9)121⟩
Michalson, E. T., **5**, 375 ⟨89JOC4649⟩
Micha-Screttas, M., **2**, 551 ⟨78JOC1064⟩
Michaud, C., **6**, 231 ⟨80JFC(15)201⟩, 441 ⟨80JFC(15)37⟩
Michaud, D. P., **2**, 758 ⟨77S848⟩
Michaud, H., **6**, 503 ⟨79GEP2742158⟩, 623 ⟨70GEP1915668⟩
Micheel, F., **5**, 753 ⟨52LA(577)234⟩, 1040 ⟨56CB133, 57CB1590⟩
Michejda, C. J., **2**, 35 ⟨71JA1389⟩, 381 ⟨73JOC2412⟩, 385 ⟨80JOC2919⟩, 387 ⟨80JA3883, 81JA442⟩, 387 ⟨88JOC1467⟩, 387, 388 ⟨83S476⟩, 831 ⟨82JA6876⟩; **4**, 335 ⟨79JOC2326⟩
Michejda, J., **4**, 896 ⟨80JOC2919⟩
Michel, A., **4**, 1074, 1075 ⟨84JA3785⟩; **6**, 92 ⟨84JA3784⟩
Michel, B., **3**, 479 ⟨73C320⟩
Michel, G., **6**, 335 ⟨92C130⟩
Michel, H., **6**, 712, 713 ⟨90AG(E)401⟩
Michel, H.-J., **6**, 538 ⟨85DIS(B)840⟩
Michel, I. M., **6**, 317 ⟨92BMC1607⟩
Michel, R., **2**, 166 ⟨87GEP3543567⟩; **5**, 84 ⟨55FOR349⟩
Michel, S. T., **4**, 50, 379, 381, 382, 951, 952 ⟨85JOC3255⟩
Michel, U., **1**, 1117 ⟨79AG(E)870⟩
Micheletti, R., **6**, 425 ⟨90JMC2101⟩
Michelin, R. A., **3**, 700 ⟨80IC3853⟩, 712 ⟨85JOM(294)C21⟩, 723 ⟨80IC3850, 84IC961⟩; **6**, 247 ⟨89JCS(D)1149, 90OM1449⟩, 663 ⟨90JCS(D)1197⟩, 721 ⟨88IC2809⟩
Michelot, D., **1**, 111, 464, 971, 1052, 1056 ⟨75CC561⟩, 971 ⟨82SC739⟩, 974 ⟨76TL275⟩, 980 ⟨74CC10⟩, 1055 ⟨89SC1705⟩, 1097 ⟨83S130⟩; **3**, 23 ⟨76BSF1445⟩; **4**, 323 ⟨76BSF1482⟩
Michelott, E. L., **1**, 195 ⟨83S701⟩
Michelotti, E. L., **1**, 436 ⟨85JOC1125⟩, 480 ⟨84JOC4894⟩, 482 ⟨79JA2246⟩, 484 ⟨79CC637, 84CC617⟩, 970 ⟨85JOC1122⟩
Michels, G., **4**, 100 ⟨88AG(E)304⟩; **5**, 883 ⟨88AG(E)272⟩, 924 ⟨88CB1363⟩
Michels, J. G., **4**, 437 ⟨63JOC1144⟩
Michels, M., **2**, 472 ⟨85AG(E)504⟩
Michener, E., **2**, 674 ⟨74S572⟩
Michie, J. K., **4**, 53, 54 ⟨81JCS(P1)785⟩, 203 ⟨81S824⟩
Michl, J., **1**, 866 ⟨64CCC1932⟩; **3**, 501, 502 ⟨85CRV419, 81JA1845⟩, 502 ⟨83JA6176, 84JA525⟩, 542 ⟨72TL3443⟩; **5**, 923 ⟨85CRV419⟩; **6**, 177, 709 ⟨85CRV419⟩, 177 ⟨B-89MI 606-01, B-89MI 622-02⟩
Michlot, D., **1**, 821 ⟨72TL3587⟩
Michne, W. F., **3**, 145 ⟨92TL5287⟩
Michnick, T. J., **4**, 144, 146 ⟨91SL631⟩
Michnik, T. J., **4**, 527 ⟨92MI 411-01⟩
Michno, D. M., **1**, 420 ⟨B-80MI 109-02⟩, 421 ⟨81JOC687, 81T3263⟩; **3**, 57 ⟨77JOC682⟩
Micke, W., **5**, 934 ⟨89TL1633⟩
Mickel, S. J., **2**, 1064 ⟨84CC200⟩; **4**, 123 ⟨90EUP402312⟩, 306 ⟨93OSC(8)3⟩, 487 ⟨82T2513⟩
Mickus, D. E., **4**, 380 ⟨89TL3011⟩
Micles, J., **2**, 214 ⟨67JPS292⟩
Mico, I., **3**, 627 ⟨93T4327⟩
Micromastoras, E. D., **5**, 765 ⟨83CCC596⟩
Micucci, D. D., **5**, 652 ⟨68JMC534⟩
Midcalf, C., **3**, 460 ⟨70CC552, 71JCS(A)292⟩, 460, 461 ⟨74JCS(D)1620⟩
Middlebrooks, M., **4**, 467 ⟨77S883⟩
Middlemass, E. D., **2**, 440 ⟨82JOC905⟩
Middlemiss, D., **1**, 202 ⟨89TL3581⟩; **3**, 29 ⟨89TL2971⟩
Middleton, D. S., **1**, 244 ⟨88TL1315, 89TL3865⟩, 942 ⟨91JCS(P1)1433⟩, 943 ⟨90T545⟩; **4**, 187, 826 ⟨89SC21⟩
Middleton, J., **4**, 136 ⟨76JCS(D)694⟩
Middleton, S., **1**, 341 ⟨93CRV2779⟩; **2**, 507 ⟨70AJC1559⟩
Middleton, W. J., **2**, 16 ⟨80JA4845, 85JOC3698⟩, 19 ⟨75JOC574, 88OSC(6)835⟩, 19 ⟨89JFC(43)405⟩, 166, 170 ⟨65JOC1384⟩, 362 ⟨88OSC(6)440⟩, 385, 386, 792 ⟨66JA3842⟩, 1012, 1013 ⟨59JA803, 58USP2831835⟩; **3**, 149 ⟨80JA4845⟩, 330 ⟨65JOC1384⟩, 348 ⟨61JA2589⟩, 348, 349, 351 ⟨65JOC1375⟩, 459 ⟨65JOC1398⟩; **4**, 42 ⟨85JA4565⟩, 97 ⟨73JOC3924⟩, 97, 444 ⟨65JOC1398⟩,

256 ⟨65JOC1384⟩, 297 ⟨88OSC(6)664⟩, 829, 880, 883, 889, 890, 971, 972 ⟨58JA2788⟩, 971, 972 ⟨59USP2883368⟩; **5**, 203 ⟨85JOC3698⟩, 322, 719 ⟨84JOC4541⟩, 473 ⟨75JOC129⟩, 473, 474, 475, 479 ⟨62USP3069395⟩, 473, 474, 479, 480, 494 ⟨65JOC1375⟩, 474 ⟨61JA2589⟩, 476 ⟨64USP3136781⟩, 479, 480 ⟨63USP3113936⟩, 492 ⟨65JOC1390⟩, 665, 672 ⟨84JOC919⟩, 845 ⟨58JA2829⟩, 973 ⟨66NEP6603215⟩; **6**, 10 ⟨75USP3914265, 76USP397691⟩, 49, 253, 528, 529, 532 ⟨65JOC1375⟩, 229 ⟨85JA4565⟩, 260 ⟨84JOC4541⟩, 528 ⟨61JA2589⟩
Midgley, J. M., **1**, 197 ⟨76JCS(P1)1384⟩
Midland, M., **2**, 57 ⟨89CRV1553⟩
Midland, M. M., **1**, 274 ⟨71JA1506, 71JA1508⟩, 279, 1033 ⟨73JA3080⟩, 333, 946 ⟨72AG(E)692⟩, 405, 816, 817, 818 ⟨84JOC1842⟩, 406, 816, 817 ⟨85TL5013⟩, 449 ⟨75JOC2845, B-75MI 110-01⟩, 486 ⟨73JOC2574⟩, 657 ⟨72JA6549, 78JOM(156)203⟩, 816 ⟨85JA3915⟩, 816, 817 ⟨85JOC1143⟩, 816, 817 ⟨85TL5017⟩, 816, 817 ⟨85TL5021⟩, 816, 818 ⟨91JOC1185⟩, 946 ⟨71JA1024, 86T5523⟩, 977 ⟨77JOC2650⟩, 1033, 1034 ⟨86JOC4507⟩, 1037 ⟨75JOC2250⟩, 1085 ⟨81TL4171⟩; **2**, 34 ⟨71JA1508⟩, 34 ⟨85JA3915, 89JOC159⟩, 58 ⟨92JOC2953⟩, 97 ⟨87T4059⟩, 138, 712 ⟨71JA3291⟩, 305 ⟨87T4079⟩, 514, 516, 520 ⟨B-75MI 207-02⟩, 514 ⟨B-83MI 210-01⟩, 517 ⟨79JA2352⟩, 524 ⟨79JA248⟩, 900 ⟨B-75MI 218-01⟩, 1086 ⟨86JOC4507⟩; **3**, 31 ⟨90JOC5058⟩, 622 ⟨82TL2077⟩, 623 ⟨81JOC229⟩
Midlemas, E. D., **2**, 453 ⟨82JA1893⟩
Midollini, S., **6**, 327 ⟨82OM778⟩
Midorikawa, H., **3**, 54 ⟨75S727⟩, 350 ⟨73TL2461, 75JAP(K)7549251⟩; **4**, 197 ⟨74S705⟩
Midura, W., **1**, 821 ⟨78JOC2518⟩; **2**, 148 ⟨79S39⟩, 713 ⟨78JOC473⟩; **3**, 219 ⟨86NJC567⟩; **6**, 119 ⟨78JOC2500⟩, 474 ⟨84TL2489⟩
Midura, W. H., **4**, 346, 924 ⟨92TA1515⟩; **6**, 123, 124 ⟨92TA1515⟩
Mielcarek, J. J., **4**, 547 ⟨88JFC(40)41⟩
Mieloszynski, J. L., **3**, 352 ⟨85RTC9⟩
Miesch, M., **1**, 576 ⟨87T853⟩
Miet, C., **4**, 176 ⟨87TL3569⟩
Miftakhova, R. G., **2**, 496, 497 ⟨63JGU2831⟩
Miftakhov, M. N., **6**, 659 ⟨90ZOB2694⟩
Miftakhov, M. S., **1**, 491 ⟨89S633⟩; **3**, 8 ⟨89S940⟩
Miftakova, R. G., **2**, 885 ⟨72JGU1960⟩
Migdalof, B. H., **3**, 231 ⟨69JA2806⟩
Miginac, L., **1**, 543 ⟨75CR(C)999⟩
Miginiac, L., **1**, 242 ⟨85S1069, 76TL3083⟩, 544 ⟨72JOM(44)69, 78CR(C)265⟩, 963 ⟨87JOM(319)333⟩, 971 ⟨91JOM(403)299⟩, 973 ⟨92JOM(440)277⟩, 973, 1059 ⟨80JOM(198)1⟩, 974 ⟨86BSF449⟩, 978 ⟨82JOM(236)177, 84TL651, 85OM333, 86T2017, 86T2501, 87TL4689, 88JOM(349)43, 90JOM(396)289⟩, 1059 ⟨81JOM(218)275, 87BSF93⟩; **2**, 74 ⟨82JOM(236)139⟩, 300 ⟨87TL1659⟩; **3**, 60 ⟨88TL4717⟩, 639 ⟨92SC2229⟩, 650 ⟨90SC2091⟩; **4**, 321 ⟨87JOM(335)283⟩, 704 ⟨91JOM(411)69⟩, 1047 ⟨82TL4083⟩, 1065 ⟨77JOM(140)237⟩; **5**, 754 ⟨74BCF994⟩
Miginiac, P., **1**, 127 ⟨64BSF2485, 68CR(C)1614⟩, 311 ⟨83TL5089⟩, 318 ⟨77JOM(132)445⟩, 800 ⟨74BSF1924⟩, 961 ⟨85S768⟩, 973 ⟨81JOM(218)1⟩, 1041 ⟨83BSF(2)21⟩; **2**, 931 ⟨82BSF395⟩; **4**, 192 ⟨81JAP(K)56002956⟩, 192 ⟨86CB1725⟩; **6**, 729 ⟨87JOM(328)249⟩
Miginiac-Groizeleau, L., **1**, 127 ⟨64BSF2485⟩
Migita, T., **1**, 194 ⟨87CL1237⟩, 233, 400 ⟨72JOC1721⟩, 408, 409, 410, 820, 821, 822 ⟨72JA3870⟩, 409, 821, 822 ⟨72JOC3791⟩, 471 ⟨77CL301, 84CL1225⟩, 820 ⟨71JOC1732⟩; **2**, 753 ⟨83CL927⟩; **3**, 138 ⟨83CL839⟩, 140 ⟨77JOM(129)36⟩, 142 ⟨87CL1371⟩, 144 ⟨82CL939⟩, 144 ⟨84BCJ242⟩, 288 ⟨77CL1423⟩, 290, 408 ⟨86CL1197⟩, 290, 409 ⟨86BCJ677⟩, 293 ⟨87BCJ767⟩, 373, 378 ⟨72JOC1721⟩, 625 ⟨84CL1511⟩; **4**, 14 ⟨83BCJ2480⟩; **5**, 468 ⟨87BCJ3462⟩
Miglierini, G., **4**, 462 ⟨92SC107⟩
Migliorese, K. G., **2**, 613 ⟨75JOC807⟩; **3**, 252 ⟨74JOC843⟩
Mignani, G., **2**, 3 ⟨87TL5505⟩, 21 ⟨92TL495⟩; **3**, 226 ⟨91TL1171⟩
Mignani, M., **1**, 288 ⟨86JA6051⟩
Mignani, S., **1**, 193, 288 ⟨89JA7487⟩; **2**, 938 ⟨86JOC3435⟩
Mignani, S. M., **1**, 287 ⟨88JA1602⟩
Migniac, P., **2**, 69 ⟨87JOC4796⟩
Migron, Y., **3**, 149 ⟨79TL3357⟩
Miguel, D., **6**, 327 ⟨92AG81⟩, 328 ⟨91JOM(420)C12, 93OM1394⟩, 329 ⟨94JOM(467)231⟩, 329, 330 ⟨92POL2713⟩, 329, 330 ⟨93OM2888⟩, 329, 330 ⟨94OM1336⟩, 331 ⟨87CC472, 91OM1683⟩, 331 ⟨91OM3005, 91OM384⟩
Miguel, J. A., **6**, 331 ⟨87CC472, 91OM1683⟩, 331 ⟨91OM3005, 91OM384⟩
Mihai, G., **3**, 232 ⟨62T257⟩
Mihailov, B. M., **3**, 365 ⟨59MI 308-01⟩
Mihailović, M. L., **1**, 91 ⟨67T215⟩; **3**, 464 ⟨70T1101⟩, 620 ⟨65TL461⟩
Mihailovic, M. Lj., **2**, 286 ⟨87TL4343⟩
Mihailovski, A., **6**, 656 ⟨74OMR(6)494⟩
Mihalyi, E., **5**, 83 ⟨57MI 502-05⟩
Mihelich, E. D., **1**, 221 ⟨83JOC409⟩; **4**, 306 ⟨76AG(E)270⟩; **5**, 166 ⟨75JOC1186⟩
Mihichuk, L., **2**, 876 ⟨73CJC936⟩
Mihm, G., **4**, 610 ⟨81AG(E)597⟩
Mihoubi, M. N., **1**, 955 ⟨88CC237, 93JOC5702, 93JOC5709⟩
Miittendorf, J., **3**, 422 ⟨90T4049⟩
Mijs, W. J., **1**, 789 ⟨72CC347⟩; **2**, 772 ⟨B-86MI 215-01⟩; **4**, 894 ⟨72JOC1960⟩
Mikaelian, G. S., **1**, 530 ⟨86TL1241, 89S472⟩
Mikaeloff, A., **4**, 202 ⟨93SL349⟩
Mikailov, B. M., **2**, 514 ⟨B-84MI 210-01⟩, 533 ⟨61DOK(139)385⟩
Mikama, K., **1**, 405, 816 ⟨83TL513⟩; **4**, 852 ⟨78CL1243⟩
Mikami, H., **3**, 426 ⟨83JOC2103⟩
Mikami, K., **1**, 161 ⟨84TL131⟩, 196, 404, 816 ⟨86CRV885⟩, 374 ⟨91TA643⟩, 404, 405, 816, 817, 818 ⟨83JOC279⟩, 404, 405, 816, 818 ⟨81JA6492⟩, 404, 406, 816 ⟨81TL69⟩, 404, 406, 816, 820 ⟨86TL4511⟩, 404, 407, 816 ⟨91S594⟩, 405, 816, 817 ⟨84T2303⟩, 405, 816, 817 ⟨84TL565⟩, 405, 816, 819 ⟨85CL1729⟩, 415, 826 ⟨87TL5879⟩, 525 ⟨87CL2347⟩, 816 ⟨83CL1379, 92T4087⟩, 816, 817, 818 ⟨85CL1719⟩, 816, 818 ⟨85TL5799⟩, 816, 819 ⟨84TL6011⟩, 816, 819 ⟨86TL4185⟩, 816, 819 ⟨86TL4899⟩, 816, 819 ⟨87CL69⟩, 817 ⟨86TL4189, 91SL629⟩, 826 ⟨90JA4035⟩; **2**, 70 ⟨90JA6737⟩, 80 ⟨89JA1940, 90JA3949⟩, 80 ⟨92JA6566, 93CC327⟩, 88 ⟨86CRV885⟩, 923 ⟨84TL131⟩; **3**, 22 ⟨82CL1349⟩, 211 ⟨83TL795⟩, 615 ⟨92CL1169⟩
Mikana, M., **6**, 124 ⟨91TL4189⟩
Mikaye, H., **1**, 37 ⟨85JOC3692⟩, 37, 330, 443, 548, 946, 1131 ⟨85T4013⟩
Mikazaki, Y., **1**, 290 ⟨86S776⟩
Mikhail, G., **3**, 282 ⟨89SL54⟩
Mikhailenko, F. A., **6**, 104 ⟨66UKZ204⟩
Mikhailik, S. K., **4**, 130 ⟨68JGU1284⟩
Mikhaillyants, S. A., **4**, 626, 637, 640 ⟨69ZOB2601⟩
Mikhailov, B. M., **1**, 58 ⟨84JOM(260)17, 84JOM(270)9⟩, 60 ⟨82JOM(226)115⟩, 657 ⟨60ZOB3615, 65MI 114-01⟩, 657 ⟨68DOK(179)1344⟩, 658 ⟨62IZV1373, 68IZV548⟩, 838 ⟨74PAC505⟩, 1030 ⟨80S904⟩; **2**, 112 ⟨60MI 202-02⟩, 269 ⟨60IZV386⟩, 606 ⟨84JOM(268)5⟩, 811 ⟨76BAU1374, 76IZV1823⟩, 847 ⟨51DOK(78)709⟩, 904 ⟨76IZV1580, 76IZV2062⟩, 1027 ⟨77IZV1456⟩, 1084 ⟨57IZV812⟩; **4**, 884 ⟨72IZV1895⟩; **5**, 381 ⟨71ZOB2501⟩, 381, 382 ⟨70DOK(194)595⟩, 778 ⟨71DOK(199)1328, 74BAU1593⟩, 778 ⟨77PAC749, 78BAU2304⟩, 1136 ⟨59DOK(127)571⟩; **6**, 113 ⟨66IZV364⟩, 662 ⟨79DOK(245)121, 82IZV2370⟩
Mikhailov, V. A., **5**, 1070 ⟨88JFC(40)217, 89ZOR1523⟩, 1070 ⟨89ZOR1689⟩; **6**, 619 ⟨88JFC(40)217, 93JPO319⟩

Mikhailov, V. S., **1**, 103 ⟨75IZV888⟩
Mikhailova, O. B., **2**, 468, 469, 470 ⟨71ZOB1644⟩
Mikhailyants, S. A., **4**, 608, 638 ⟨69ZOB2281⟩, 638, 640 ⟨69DOK(188)120⟩
Mikhailyuchenko, N. K., **4**, 1007, 1010 ⟨77ZOB75⟩, 1007, 1010, 1013 ⟨83ZOB548⟩
Mikhaleva, A. I., **3**, 430 ⟨85ZOR766⟩
Mikhalin, N. V., **4**, 560 ⟨83IZV2802, 84IZV1122⟩, 560 ⟨87IZV860⟩
Mikheer, M. L., **2**, 208 ⟨72IZV2510⟩
Mikheev, E. P., **6**, 522 ⟨59MI 616-02⟩
Mikheichev, G. A., **6**, 145 ⟨71ZOR30⟩
Mikhlina, E. E., **1**, 45 ⟨70KGS495⟩
Miki, H., **2**, 244 ⟨69TL447⟩
Miki, J., **1**, 842 ⟨72BCJ1777⟩
Miki, K., **3**, 487 ⟨86OM356⟩; **5**, 948 ⟨82CC1178⟩
Miki, M., **3**, 659, 660 ⟨86CPB4620⟩
Miki, S., **5**, 129 ⟨94JOC4749⟩
Miki, T., **5**, 377 ⟨89JCS(P1)1862⟩, 640, 642 ⟨80CPB2116⟩; **6**, 9 ⟨88JOC3637⟩
Miki, Y., **1**, 836 ⟨86CL501⟩; **4**, 519 ⟨84H(22)701⟩
Mikio, M., **5**, 57 ⟨82CL1255⟩
Mikiulec, R. A., **2**, 214 ⟨73JOC2965⟩
Mikler, L., **5**, 1035 ⟨86CCC112⟩
Mikoiaczyk, M., **1**, 821 ⟨78JOC2518⟩
Mikol, G. J., **2**, 174 ⟨B-68MI 203-02⟩; **5**, 989 ⟨72JOC724⟩
Mikolajczak, J., **6**, 529, 532 ⟨81CB829⟩
Mikolajczak, K. L., **1**, 955 ⟨65JOC2983⟩
Mikolajczyk, M., **1**, 608 ⟨81TL3097⟩; **2**, 124 ⟨87S1009⟩, 144 ⟨81SC1025, B-88MI 203-02⟩, 145 ⟨78S758⟩, 148 ⟨79S39⟩, 151 ⟨92CC1500⟩, 165 ⟨B-90MI 203-03⟩, 184 ⟨89AG(E)97⟩, 197 ⟨74CC547⟩, 197 ⟨74S124⟩, 198 ⟨85CC1670⟩, 212 ⟨78S138⟩, 213 ⟨80S32⟩, 217 ⟨76CC220⟩, 218, 724 ⟨85TL5703⟩, 450, 842 ⟨75CC382⟩, 713 ⟨78JOC473⟩; **3**, 35 ⟨87S659⟩, 166 ⟨83S332⟩, 197 ⟨84S691⟩, 219 ⟨86NJC567⟩; **4**, 346 ⟨79JOC2967, 87PS(30)225⟩, 346, 924 ⟨92TA1515⟩, 347 ⟨73S669, B-94MI 407-02⟩, 804 ⟨83T1189⟩, 835, 847, 848 ⟨78T3081⟩, 847, 848 ⟨76TL2731⟩, 923 ⟨83S332⟩, 924 ⟨81TL3097⟩, 925 ⟨89S101⟩; **5**, 243, 540 ⟨76JCS(P1)564⟩, 1061 ⟨81T233⟩; **6**, 116 ⟨88JOC3609⟩, 119 ⟨76TL477, 78JOC2518⟩, 119 ⟨79JOC2967⟩, 120, 121, 123 ⟨80S127⟩, 121 ⟨91TL7329, 92T8697, 81TL3097⟩, 123, 124 ⟨92TA1515, 91TL4189⟩, 129, 326, 350, 351 ⟨89S101⟩, 474 ⟨84TL2489⟩, 509 ⟨78JOC2132⟩, 635 ⟨77JOC3629⟩
Mikoshiba, R., **3**, 557 ⟨80JOC5385⟩
Mikroyannidis, J. A., **4**, 338, 339 ⟨85PS(25)39⟩
Mikulaschek, G., **2**, 358 ⟨61CB634⟩
Mikulcik, P., **3**, 700 ⟨91JOM(403)C15⟩
Mikulshina, V. V., **1**, 737 ⟨68JOM(13)263⟩; **3**, 496 ⟨89IZV1182⟩
Mikulshin, V. V., **3**, 328 ⟨68JOM(13)263⟩
Mikulski, C. M., **2**, 94 ⟨86ICA179⟩
Milani, F., **3**, 88 ⟨83S749⟩
Milano, P. D., **2**, 133 ⟨91JA3176⟩
Milanova, R. K., **2**, 683 ⟨93H(36)13⟩
Milas, N. A., **5**, 102 ⟨43OSC(2)302⟩
Milcent, R., **5**, 22 ⟨91JHC1511⟩, 355 ⟨92JHC1081⟩, 640 ⟨93JMC1157⟩, 704 ⟨79S946⟩, 705 ⟨84JHC1689⟩; **6**, 444 ⟨89JHC231⟩
Milch, A., **2**, 723 ⟨1888JPR(2)207⟩
Milchereit, A., **5**, 386 ⟨85JOM(288)237⟩
Milczarck, R., **5**, 890 ⟨90PS(49)337⟩
Milczarek, R., **6**, 154 ⟨87AG(E)908⟩
Mildner, P., **1**, 503 ⟨53JCS3294⟩
Mile, B., **1**, 921 ⟨68AG(E)507⟩
Miles, D. E., **1**, 166 ⟨79TL3375⟩; **4**, 236 ⟨79TL3375⟩; **6**, 126, 135 ⟨79TL3375⟩
Miles, D. H., **5**, 78 ⟨72TL3987⟩
Miles, J. A., **3**, 376 ⟨74JA6119⟩
Miles, L. W. C., **2**, 128 ⟨50JCS2938⟩
Miles, M. G., **4**, 850 ⟨74CC751, 75JOC2577⟩, 972 ⟨68JOC1506⟩; **6**, 120 ⟨75JOC2577⟩
Miles, M. L., **1**, 726 ⟨63JOC372⟩; **3**, 494 ⟨63JOC372⟩
Miles, P. D., **6**, 647 ⟨77JMC901⟩
Miles, S. J., **6**, 186 ⟨84CC480⟩, 187 ⟨81JOM(212)C4⟩, 195 ⟨80IC3637⟩
Miles, W. H., **1**, 834 ⟨81JA1757⟩; **3**, 519 ⟨83JOM(254)333, 85JA2924⟩, 520 ⟨82JA3761⟩; **4**, 679 ⟨82JA1134⟩
Milette, S., **1**, 513 ⟨83TL4899⟩, 514 ⟨84TL2179⟩; **3**, 62 ⟨83TL4899⟩
Milewich, L., **2**, 389 ⟨66JOC524⟩
Milewska, M., **5**, 322 ⟨85S929⟩
Milewska, M. J., **5**, 7, 15 ⟨88S475⟩, 214 ⟨87AJC1919, 87TL1817⟩
Milewski-Mahrla, B., **4**, 583 ⟨81AG(E)586⟩, 586 ⟨81CB441⟩; **5**, 910, 911 ⟨81CB1428⟩, 1092 ⟨81CB3070⟩; **6**, 701 ⟨84OM38⟩
Mil'grom, A. E., **5**, 754 ⟨69URP256779, 69URP256781⟩
Milhailovic, M. L., **2**, 93 ⟨70S209⟩
Milhavet, J.-C., **3**, 218 ⟨79BSF381⟩
Milhavet, J. C., **5**, 360 ⟨90CPB2352⟩
Milicia, A. P., **2**, 358 ⟨77JOC875⟩
Milius, R. A., **1**, 452 ⟨81SC281⟩
Milivojevic, D., **2**, 322 ⟨91OPP721⟩; **6**, 632 ⟨85OPP256, 91OPP721⟩
Miljkovic, D., **2**, 9 ⟨93CCC902⟩; **3**, 659 ⟨69AG(E)343⟩
Milkowski, J. D., **2**, 223 ⟨68TL3057⟩; **4**, 296 ⟨88OSC(6)5⟩
Milks, J. E., **3**, 317 ⟨62JA1315, 64JA3814⟩
Milks, W. N., **6**, 217 ⟨56USP2744147⟩
Millan, S. D., **1**, 929 ⟨91TL7493, 93JOC1215⟩
Millar, I. T., **2**, 437, 821 ⟨52JCS3039⟩, 445 ⟨54JCS2832⟩, 491 ⟨60IS113, 60IS116⟩, 609, 632 ⟨60OS(40)105⟩, 769 ⟨B-66MI 215-04⟩, 829 ⟨52JCS4453⟩, 882 ⟨65JCS5132⟩
Millar, R., **5**, 692 ⟨89JCR(S)337⟩
Millar, R. B., **2**, 328 ⟨88TL6715⟩
Millar, R. W., **2**, 386 ⟨88TL2735⟩
Millard, A. A., **1**, 467 ⟨77JA4833⟩
Millard, B. J., **1**, 398 ⟨63JCS3397⟩
Mill, T., **1**, 37 ⟨68JOC1421⟩; **5**, 812 ⟨72JA8776⟩
Millbanks, C., **1**, 346 ⟨94CC41⟩
Millen, J., **5**, 567, 630 ⟨89JMC2493⟩
Miller, A., **4**, 54 ⟨79JOC4722⟩, 200 ⟨93SL111⟩, 279 ⟨90TL1043⟩; **5**, 326, 330 ⟨89SL32⟩, 664 ⟨84JMC1201⟩
Miller, A. E., **6**, 641 ⟨86S777⟩
Miller, A. E. G., **3**, 86 ⟨59JOC627⟩
Miller, A. G., **3**, 686 ⟨79JOC510⟩
Miller, A. R., **2**, 646 ⟨76JOC3599⟩
Miller, A. S., **4**, 829 ⟨72TL1353⟩
Miller, ,S. R., **2**, 969 ⟨88TL1631⟩
Miller, B., **1**, 808, 809 ⟨75ACR245, 65JA5115⟩; **2**, 660 ⟨63JOC345⟩
Miller, C. B., **4**, 767 ⟨57USP2803665, 57USP2803666⟩; **6**, 217 ⟨56USP2748177⟩
Miller, C. D., **4**, 107 ⟨70IC975⟩
Miller, C. H., **4**, 172 ⟨75JA7182⟩
Miller, C. S., **3**, 617 ⟨55OSC(3)646⟩
Miller, D., **5**, 9 ⟨63JCS491⟩
Miller, D. B., **1**, 128 ⟨66JOC908⟩, 149 ⟨88JA4778⟩; **4**, 796 ⟨76JOC2112⟩
Miller, D. D., **1**, 556 ⟨83TL555⟩, 688 ⟨83JA1660⟩
Miller, D. G., **2**, 656 ⟨76TL1637⟩
Miller, D. J., **5**, 1123 ⟨92T7703⟩; **6**, 628, 636 ⟨94T3195⟩
Miller, D. L., **4**, 642, 655 ⟨82CC970⟩
Miller, E. G., **3**, 463 ⟨59JA4751⟩
Miller, F. A., **2**, 1093 ⟨67SA(A)1169⟩
Miller, F. W., **1**, 1119 ⟨67JA1960⟩; **2**, 801 ⟨67JA1960⟩
Miller, G., **6**, 558 ⟨83JHC1223⟩
Miller, G. E., **2**, 132 ⟨43OSC(2)576⟩
Miller, G. P., **1**, 890 ⟨93CI(L)226⟩
Miller, H. K., **2**, 627, 632 ⟨19JA287⟩
Miller, III, M. C., **6**, 514 ⟨92MI 616-01⟩
Miller, J. A., **1**, 130 ⟨71JOC1112⟩, 207, 465, 511, 512, 533, 534, 542, 543, 1106, 1139 ⟨84OR(32)375⟩, 449 ⟨89TL6487⟩, 512 ⟨83JA1383⟩, 536 ⟨85JA2568⟩, 543 ⟨84TL5863⟩, 669 ⟨84TL3407⟩, 1019 ⟨83S128⟩, 1045,

1047 ⟨83TL5181⟩; **2**, 35 ⟨74TL2691, 76JCS(P1)416⟩, 629 ⟨89JOC998⟩, 629, 974, 975, 976 ⟨84TL5863⟩, 657 ⟨87JOC322⟩, 902 ⟨81S288⟩, 910, 919, 922, 968 ⟨84OR375⟩, 925 ⟨85TL5863⟩, 925, 931, 973 ⟨83JA6761⟩, 927, 986, 988 ⟨89JA3336⟩, 969 ⟨84TL3407⟩; **3**, 130, 288 ⟨83TL5181⟩, 218 ⟨83JA6761⟩, 270 ⟨85JA2568⟩; **4**, 28 ⟨76JCS(P1)416⟩, 53, 54 ⟨81JCS(P1)785⟩, 203 ⟨81S824⟩, 808, 1062, 1068 ⟨84OR(32)375⟩, 809 ⟨83JA6761, 89JA3336⟩, 816 ⟨89JOC998⟩, 917 ⟨81TL3789, B-81MI 420-02⟩, 1061 ⟨85JA2568⟩; **5**, 417, 426 ⟨81S288, 81JA6217⟩; **6**, 118 ⟨77JCS(P1)1898, 77TL1065⟩
Miller, J., **6**, 721 ⟨72JA417⟩
Miller, J. B., **3**, 463 ⟨59JOC560⟩
Miller, J. D., **4**, 213 ⟨87JOC1351⟩
Miller, J. G., **4**, 979 ⟨70JPC2722⟩
Miller, J. J., **1**, 69 ⟨72JOC4341⟩; **5**, 383 ⟨70JOM(24)595⟩
Miller, J. M., **4**, 190 ⟨76TL3361⟩; **6**, 233, 234 ⟨90JFC(48)249⟩
Miller, Jr., R. E., **1**, 612 ⟨71JOC1898⟩
Miller, Jr., W. T., **4**, 16 ⟨59JA5008⟩, 34 ⟨57JA4170⟩, 152 ⟨68JA7367⟩, 755 ⟨57JA4164, 59JA2078, 60JA3091⟩; **6**, 18 ⟨60JA3091, 61JA1767⟩, 18, 225 ⟨59JA2078⟩, 64, 225, 226, 245 ⟨57JA4159⟩
Miller, J. S., **1**, 883 ⟨78JOC815⟩; **6**, 732 ⟨72IC2069⟩
Miller, L. A., **4**, 561 ⟨63USP3093672⟩
Miller, L. E., **5**, 86 ⟨63OSC(4)39⟩
Miller, L. L., **1**, 403, 799 ⟨71JA650, 69JA1578⟩, 1110 ⟨69JOC1495⟩, 1111 ⟨68JA7282⟩; **2**, 657 ⟨78JOC3078⟩; **5**, 714 ⟨67JA637⟩, 1110 ⟨84JOC5198⟩
Miller, M., **1**, 987 ⟨92JA4079⟩
Miller, M. J., **1**, 238, 613 ⟨80JA5979⟩; **2**, 171 ⟨69CAR(9)201⟩, 241 ⟨85TL3891⟩, 692 ⟨75JOC126⟩; **3**, 259 ⟨71JA981⟩, 617 ⟨75JOC126⟩; **4**, 306 ⟨93OSC(8)3⟩; **5**, 273, 285 ⟨80JA7026, 87TL6257⟩, 285, 322 ⟨87JOC4471⟩, 285, 327, 328 ⟨85TL5385⟩, 315 ⟨94T11091, 83T2571⟩, 322 ⟨86ACR49⟩, 328 ⟨83JOC24⟩, 329 ⟨85TL4855, 85TL977, 91TL2577, 94JOC4862⟩, 330 ⟨94JOC4602⟩, 334 ⟨85TL3891⟩, 334, 336 ⟨86JOC3133, 86JOC5332⟩, 336 ⟨86TL3095⟩
Miller, M. L., **1**, 194, 288, 289 ⟨89JA6482⟩, 288 ⟨88JA3687⟩
Miller, N. E., **2**, 487 ⟨65IC1458⟩; **3**, 500 ⟨65IC1458⟩; **4**, 143 ⟨86IC4309⟩, 145 ⟨91IC2228⟩, 455 ⟨69IC1336⟩, 524 ⟨64IC1196, 69IC275, 77JOM(137)131, 81IC1328, 84JOM(269)123, 88JOM(349)11⟩, 562 ⟨68IC709⟩, 566 ⟨69IC862⟩, 569 ⟨65JA390⟩, 571, 574, 597 ⟨65IC1458⟩; **5**, 912, 913, 921, 1093 ⟨65IC1458⟩, 913 ⟨65JA390⟩; **6**, 492 ⟨91IC2228⟩, 708 ⟨65IC1458⟩
Miller, P. C., **2**, 1004 ⟨91JA1468⟩
Miller, P. E., **3**, 573 ⟨62JOC4053⟩
Miller, R., **2**, 396 ⟨61JOC1357⟩; **5**, 370 ⟨74MI 507-01⟩
Miller, R. B., **1**, 653, 654 ⟨74TL543⟩; **2**, 606 ⟨77SC475⟩, 607, 617, 622, 628, 922 ⟨74TL543⟩, 607, 914, 921, 922 ⟨78JOC4424⟩, 922, 931 ⟨83TL2055⟩, 925, 926 ⟨84JOC725⟩; **3**, 210 ⟨73TL5037⟩, 419 ⟨88TL6715⟩; **4**, 809 ⟨84JOC725⟩, 1059 ⟨79JOC4623⟩
Miller, R. C., **2**, 430, 840 ⟨59JOC2013⟩; **6**, 213 ⟨57JA20⟩
Miller, R. D., **1**, 177 ⟨83JOC3334⟩, 579 ⟨82TL323⟩; **2**, 1003 ⟨91JOC1453⟩; **3**, 538 ⟨72JA663⟩, 655 ⟨85TL2395⟩; **4**, 773 ⟨84SC1285⟩, 945 ⟨84TL5351⟩; **6**, 317 ⟨84TL991⟩
Miller, R. F., **1**, 928 ⟨88JA3300⟩; **2**, 668 ⟨90JA3093⟩; **3**, 193, 473 ⟨90JOC1959⟩
Miller, R. G., **6**, 516 ⟨87OM372⟩
Miller, R. L., **1**, 491, 534 ⟨70JA6678⟩
Miller, R. W., **2**, 453 ⟨82JA1893⟩
Miller, S. A., **1**, 812 ⟨89OS(67)210⟩, 862 ⟨52JCS632⟩, 869, 871 ⟨70JA4689⟩; **6**, 215 ⟨67MI 607-01⟩
Miller, S. C., **4**, 736 ⟨93TL3057⟩
Miller, S. I., **1**, 536 ⟨B-78MI 112-05⟩, 538 ⟨68T4285⟩, 1016 ⟨B-78MI 121-02⟩, 1110 ⟨77T1211⟩; **2**, 849, 850 ⟨71JA3694⟩, 1011 ⟨76ACR358, B-78MI 221-02⟩, 1012, 1014 ⟨52JA629⟩, 1016, 1033 ⟨63JA1648⟩, 1017 ⟨73OSC(5)921⟩, 1028 ⟨71T2651, 71TL1753⟩, 1062 ⟨62JA2020, 72JOC2168⟩; **3**, 252 ⟨74JOC843⟩, 253 ⟨68T4285⟩; **4**, 984 ⟨75JOC2720⟩
Miller, S. J., **3**, 32 ⟨92JOC1053⟩, 634 ⟨90H(30)839⟩; **6**, 487 ⟨90H(30)839⟩
Miller, T. G., **6**, 38 ⟨60JOC2009⟩
Miller, T. M., **1**, 668, 669 ⟨86OM1473⟩
Miller, W. H., **1**, 1099 ⟨88JOC5580⟩, 1143 ⟨81JA1904⟩
Miller, W. R., **2**, 778 ⟨65MI 215-02⟩
Miller, W. T., **4**, 36 ⟨61JA4105⟩
Millership, J. S., **1**, 197 ⟨76JCS(P1)1384⟩
Millet, G. H., **3**, 3, 114 ⟨79JCS(P1)2493⟩
Millich, F., **3**, 694 ⟨72CRV1010⟩
Milliet, P., **5**, 789 ⟨79T43⟩
Milligan, B., **2**, 122 ⟨63JOC3608⟩, 172, 220 ⟨62JCS2172⟩, 176, 177 ⟨59JCS2969⟩, 228 ⟨65JCS2901⟩; **5**, 601, 621 ⟨61JCS1194⟩
Milligan, D. E., **1**, 923 ⟨67JCP(47)5146⟩; **3**, 555 ⟨63JA278⟩
Milligan, J. R., **5**, 360 ⟨90JCS(P1)3103⟩
Millington, D., **2**, 465 ⟨77JCS(D)159⟩
Millon, J., **1**, 464, 1090 ⟨75S434⟩
Mills, A., **3**, 117 ⟨88JOC1103⟩
Mills, A. K., **1**, 871 ⟨60HCA1915⟩
Mills, E. J., **2**, 129 ⟨40JA1173⟩
Mills, J. E., **5**, 591, 609 ⟨86S482⟩, 657 ⟨84JOC546⟩; **6**, 641, 645 ⟨86JOC1882⟩
Mills, J. F., **2**, 24 ⟨73MI 201-01⟩
Mills, J. G., **5**, 687, 698 ⟨93AJC1213⟩
Mills, J. L., **2**, 434 ⟨69JA2915⟩, 488 ⟨71JA2150⟩, 506, 894 ⟨75JOM(84)1⟩; **4**, 1042 ⟨78ICAL331⟩
Mills, J. S., **2**, 1016 ⟨62JCS4288⟩; **5**, 116 ⟨58JA1236⟩
Mills, L. S., **1**, 505 ⟨83TL409⟩
Mills, M. J., **3**, 617 ⟨82OPP396⟩; **5**, 275 ⟨82OPP396⟩
Mills, N. S., **1**, 879 ⟨89JOC773⟩
Mills, O. S., **3**, 419 ⟨86JCS(P1)619⟩, 460 ⟨67CC245⟩; **4**, 688 ⟨64PCS187⟩, 692 ⟨59JA4757⟩; **5**, 932 ⟨65AG(E)1082⟩, 950 ⟨72JCS(D)653⟩
Mills, R. J., **1**, 448 ⟨89JOC4372⟩; **2**, 627 ⟨84TL483, 89JOC4372⟩; **4**, 604, 646 ⟨89JOC4372⟩
Mills, R. M., **4**, 686 ⟨82JCS(D)1309⟩; **6**, 340 ⟨82CC51⟩, 343 ⟨82CC773, 83JCS(D)2599⟩
Mills, R. W., **1**, 381 ⟨74JCS(P1)1938⟩; **4**, 20 ⟨75JCS(P1)251⟩
Mills, S., **1**, 148 ⟨83JA2080⟩, 940 ⟨88TL281⟩
Mills, S. D., **1**, 340 ⟨92JCS(P1)1313⟩
Mills, W. J., **5**, 1147 ⟨89CC900, 91IC1046⟩; **6**, 515 ⟨89CC900, 91IC1046⟩, 515 ⟨92IC4911⟩
Milne, C. R. C., **2**, 569 ⟨81JCS(D)80⟩; **4**, 1018 ⟨78JA1164⟩
Milne, D., **1**, 580, 842 ⟨93CC884⟩
Milne, G. M., **1**, 49, 51, 157 ⟨70JA2139⟩; **3**, 493 ⟨70JA2139⟩
Milne, H. B., **5**, 87 ⟨57JA637, 57JA645⟩, 366 ⟨65JOC64⟩
Milner, D. J., **2**, 615 ⟨92SC73⟩
Milner, D. L., **2**, 963 ⟨69JA6983⟩
Milner, J., **3**, 685 ⟨83JCS(P1)293⟩
Milner, P. H., **4**, 419 ⟨91TL2683⟩
Milobendzki, T., **2**, 105 ⟨17MI 202-01⟩
Milolajczuk, M., **2**, 437, 443 ⟨79T1531, 79JA7292⟩, 443 ⟨80PAC959⟩
Milolajczyk, M., **2**, 156, 721 ⟨88T6687⟩
Milone, L., **4**, 683 ⟨72JOM(39)169⟩, 692 ⟨74JOM(64)271⟩, 718 ⟨83POL77⟩
Milot, G., **1**, 151 ⟨93JA6652⟩, 151 ⟨93JOC6538⟩
Milowsky, A. S., **4**, 517 ⟨86JMC2241⟩
Milstein, D., **1**, 471, 473, 1145 ⟨79JA4992⟩; **3**, 90 ⟨89CC1816⟩, 288 ⟨79JOC1613⟩, 288, 296 ⟨78JA3636⟩; **4**, 582 ⟨87CC1543⟩
Milstein, N., **1**, 212 ⟨68JHC337⟩
Milstein, S. R., **3**, 407 ⟨74JOC3923⟩
Miltschitzky, G., **6**, 409 ⟨34ZAAC(221)154⟩
Miltz, W., **2**, 751 ⟨90S750⟩; **4**, 99 ⟨88T5403⟩
Mil'vitskaya, E. M., **1**, 795, 801 ⟨76RCR469⟩

Milvitskaya, J., **5**, 351, 366 ⟨91TL133⟩
Mima, H., **2**, 369 ⟨91JA6337⟩
Mimasu, R., **6**, 122 ⟨85CL1099⟩
Mimouni, N., **4**, 588 ⟨91S31, 93PS(75)99⟩
Mimura, M., **3**, 631 ⟨92BCJ111⟩; **6**, 623 ⟨76CPB26⟩
Mimura, S., **6**, 561 ⟨87BCJ1807⟩
Mimura, T., **1**, 404, 406, 816 ⟨81T4097⟩; **3**, 57 ⟨77TL2425⟩, 126 ⟨79TL531⟩; **6**, 495 ⟨79CL1361⟩
Minabe, M., **5**, 226 ⟨87GSM130⟩
Minach, L., **4**, 440 ⟨84S315⟩
Minai, M., **5**, 193 ⟨86EUP268820⟩
Minakata, S., **1**, 1010 ⟨92BCJ2992⟩, 1043 ⟨89CL773⟩
Minakawa, H., **1**, 929 ⟨93JCS(P1)3007⟩
Minakimi, S., **5**, 636 ⟨68JAP6827842⟩
Minami, I., **1**, 193 ⟨85JOC1523⟩, 196 ⟨84TL5157⟩, 196 ⟨85JOM(296)269⟩, 558 ⟨86T2971⟩, 558, 561 ⟨87S992, 84S1009⟩, 962 ⟨86CC922, 87S603⟩, 975 ⟨86TL731⟩; **3**, 57, 215 ⟨86TL1805⟩, 122 ⟨84TL2791⟩, 139 ⟨83CL1325, 83TL1793, 83TL4713, 84TL5157, 85JOC1523⟩, 206 ⟨83TL1797, 83TL5635, 83TL5639, 84CL1133, 86TL2791⟩, 623 ⟨87JOC2988⟩, 645 ⟨86CC118⟩; **5**, 154 ⟨86T2971⟩
Minami, K., **1**, 240 ⟨87CL809, 88JA7185⟩; **3**, 464 ⟨66CC730⟩, 622 ⟨81BCJ629⟩; **5**, 295, 299 ⟨66CC17⟩; **6**, 107 ⟨82H(18)201⟩
Minami, L., **1**, 558 ⟨83TL5635⟩
Minami, M., **1**, 829 ⟨90JOC2108⟩
Minami, N., **1**, 748 ⟨82JA1109⟩, 881 ⟨76TL1817⟩; **5**, 412, 422 ⟨76CL993⟩, 414 ⟨76TL1591⟩, 414, 425 ⟨78JOM(145)C1⟩, 420 ⟨78BCJ2391⟩; **6**, 125 ⟨76TL1591, 78BCJ2391⟩
Minami, T., **1**, 195 ⟨89JOC469⟩, 202 ⟨86JOM(299)119, 86TL2479, 86TL967⟩, 202, 203 ⟨86JOC2859⟩, 225 ⟨93TL8263⟩, 429 ⟨92SL833⟩, 444 ⟨83JOC2569⟩, 608 ⟨79CB355⟩, 1038 ⟨91JOC4091⟩; **2**, 244 ⟨69TL447, 72JOC3810⟩, 399 ⟨73BCJ2156⟩, 447 ⟨85JAP60152493, 88JOC2937⟩, 478 ⟨92S333⟩, 662 ⟨86TL2401⟩, 825, 853 ⟨92SL833⟩, 829 ⟨92CL547⟩, 843 ⟨79CB355, 83JOC2569⟩; **3**, 197 ⟨92CC190⟩, 216 ⟨78CC12⟩, 496 ⟨83JOC2569, 89JOC974⟩, 496 ⟨91RHA270⟩, 527 ⟨79JOC2067⟩, 593 ⟨70JOC2076⟩; **4**, 66 ⟨82CL587⟩, 326 ⟨74JOC3412⟩, 348 ⟨79CB355⟩, 349 ⟨92CC190⟩, 488, 489 ⟨85JOC1278⟩, 736 ⟨92SL739, 92TL3779, 93JFC(63)281⟩, 925 ⟨83TL767, 84BCJ2127⟩; **5**, 1033 ⟨68TL2651⟩, 1123 ⟨81JOC4478⟩
Minami, T.,**5**, 652 ⟨75BCJ3259⟩
Minami, T., **6**, 122 ⟨85CL1099⟩, 123, 697 ⟨79CB355⟩
Minamida, I., **4**, 900 ⟨93MI 420-01⟩, 973, 980, 983, 984, 986 ⟨88EUP302389⟩
Minamikawa, H., **3**, 628 ⟨87CL2073, 88BCJ4379⟩
Minamikawa, J., **2**, 357, 379, 384 ⟨77S1⟩, 357, 691 ⟨73JOC1239⟩
Min, S., **5**, 245 ⟨86T4593⟩, 530, 540 ⟨86TL4593⟩
Min, S. J., **3**, 3, 114 ⟨93TL5113⟩, 13, 85 ⟨91TL6903⟩
Minasyan, G. G., **4**, 914 ⟨83CA(98)4614z, 86CA(104)19636z⟩
Minasz, R. J., **4**, 16 ⟨65JA4259⟩
Minato, A., **1**, 466, 470 ⟨76BCJ1958⟩, 467 ⟨87JA1257⟩, 473 ⟨80TL845⟩; **2**, 931 ⟨84TL83⟩
Minato, H., **1**, 813 ⟨64JCS4578, 70JCS(C)2697⟩, 813 ⟨70JCS(C)2704⟩; **2**, 175 ⟨75CL307⟩, 360 ⟨75CL991, 78JOC652⟩
Minato, M., **1**, 228 ⟨88TL343⟩; **2**, 45 ⟨90JOC766⟩; **3**, 113 ⟨87TL3683⟩
Mincheva, Z., **5**, 376 ⟨90AF1030⟩
Mincione, E., **2**, 26 ⟨78TL4575⟩; **3**, 206 ⟨77S773⟩
Mincuzzi, A., **1**, 592 ⟨76S200⟩
Mindl, J., **6**, 558, 565 ⟨84CCC1577, 81CCC1970⟩
Mine, N., **1**, 462 ⟨86CL357⟩
Minemura, K., **4**, 154 ⟨85JCS(P1)1185⟩
Mineo, I. C., **6**, 206 ⟨65JA3788⟩
Mineshima, F., **4**, 842 ⟨75JCS(P1)1277⟩
Mineura, K., **5**, 334 ⟨85JMC1772⟩
Ming, H. H., **2**, 331 ⟨88TL6493⟩
Mingaleva, K. S., **1**, 748 ⟨80ZOB76⟩; **2**, 1032 ⟨64ZOB1899⟩; **4**, 841 ⟨80ZOR720⟩
Ming, N. S., **5**, 1125 ⟨90TL3681⟩
Ming, S., **4**, 488 ⟨B-93MI 402-02⟩
Minghetti, G., **4**, 697 ⟨70JOM(22)5⟩; **6**, 721 ⟨70JOM(25)255, 73G373⟩
Ming-Hui, D., **3**, 75 ⟨90JCS(P1)1459⟩
Minguillón, C., **3**, 666 ⟨83JOC3105⟩
Mini, V., **5**, 828, 837 ⟨65LA(687)191⟩
Minielli, J. L., **5**, 334 ⟨86JMC359⟩
Minieri, P. P., **1**, 637 ⟨48JA4045⟩; **2**, 308 ⟨48JA4045⟩; **5**, 281, 282 ⟨48JA4045⟩
Minikawa, J., **2**, 161 ⟨72TL4137⟩
Minisci, F., **1**, 318 ⟨76TCC1⟩, 923 ⟨77JA7960⟩, 925 ⟨73S1⟩, 926 ⟨82TL5587⟩, 934 ⟨71T3575, 83ACR27, 85TL4803, 89H(28)489, 90JHC79, 91JOC2866⟩; **2**, 3 ⟨94TL1427⟩; **3**, 311 ⟨74JCS(P2)1699, 89H(28)489⟩, 311 ⟨91JOC2866⟩; **4**, 439 ⟨69OPP5⟩; **5**, 1110 ⟨59G2428⟩; **6**, 6 ⟨56CI(M)371⟩, 7 ⟨75ACR165⟩
Miniski, F., **2**, 798 ⟨64MI 215-01⟩
Mink, J., **4**, 697 ⟨86JOM(301)1⟩
Mink, J.,**6**, 207 ⟨86JOM(301)1⟩
Mink, R., **4**, 706 ⟨78JA6382⟩
Minkin, V. I., **2**, 731, 732 ⟨79RCR343⟩; **3**, 347, 356 ⟨73ZOR1717⟩; **4**, 290 ⟨91HAC307⟩, 290, 291 ⟨88KGS1144⟩; **5**, 1078 ⟨88ZOR1581⟩; **6**, 50 ⟨79ZOR1561⟩
Minkwitz, R., **2**, 238 ⟨90ZN(B)1625⟩, 510 ⟨89IC1627⟩; **5**, 246, 254 ⟨88ZAAC(566)151⟩; **6**, 610, 727 ⟨92ZAAC(617)136⟩, 613 ⟨93ZN(B)19⟩, 615, 728 ⟨87JFC(37)259⟩
Minnella, A. E., **1**, 13 ⟨75JOC3151⟩
(Minnesota MiningMfg. Co.)864, **4**, ⟨76USP4075001⟩ ⟨76USP3933914⟩
Minobe, M., **1**, 165 ⟨85T2751⟩, 165 ⟨85T3663⟩
Minobe, M., **1**, 187 ⟨84JA1079⟩
Minoguchi, K., **5**, 681 ⟨87JHC1391⟩
Minoguchi, M., **4**, 857 ⟨86JOC508⟩
Minoja, A. P., **4**, 693 ⟨95JOM(488)39⟩
Minokawa, O., **3**, 58 ⟨77S721⟩
Minor, C. A., **5**, 1109 ⟨68JOC823⟩
Minot, C., **2**, 229 ⟨83T3839⟩
Minoura, M., **3**, 400 ⟨93JA7019⟩; **4**, 334 ⟨93CL1047⟩
Minowa, N., **1**, 106 ⟨93CC617⟩, 558, 828 ⟨93CC619⟩
Minssen, M., **2**, 1016 ⟨65BSF71⟩
Minton, M. A., **3**, 474 ⟨77JOC3757⟩
Mintz, E. A., **4**, 658 ⟨85OM1308⟩
Mintz, E., **1**, 545 ⟨86JA1940⟩; **2**, 954 ⟨83OM101⟩, 978 ⟨84OM134⟩
Mintz, M. J., **2**, 338 ⟨73OSC(5)184⟩
Minung, L., **5**, 1125 ⟨86MI 530-02⟩
Minville, R. M., **2**, 782 ⟨65CR(260)2512⟩
Minyaeva, L. G., **5**, 688 ⟨81JOU1986, 81ZOR2225⟩
Mio, S., **4**, 937 ⟨86T2863⟩, 950 ⟨84TL2021⟩; **5**, 152 ⟨86T2863⟩; **6**, 130 ⟨84TL2021⟩
Miocque, M., **1**, 1001 ⟨64BSF2477⟩; **2**, 31 ⟨89BSF247⟩; **5**, 742, 748, 751, 754, 755 ⟨B-75MI 519-01⟩, 754 ⟨71BCF478⟩, 756 ⟨70BCF200⟩, 759 ⟨72FRP2081556⟩
Mion, L., **1**, 343 ⟨67BSF2641⟩, 1023 ⟨67CR(C)839⟩; **3**, 616 ⟨93BSF513⟩
Mion, M., **1**, 1023 ⟨67CR(C)839⟩
Mioskowski, C., **1**, 179 ⟨85SC663⟩, 183 ⟨92JOC3757⟩, 269, 647 ⟨83TL4419⟩, 580 ⟨89TL179⟩, 738 ⟨90AG(E)1454⟩, 739, 740 ⟨83SC1193⟩, 1026 ⟨89TL6027⟩; **2**, 6 ⟨85SC663⟩, 10 ⟨89TL557⟩, 135 ⟨80JA1436, 80JA3663⟩, 446 ⟨84TL5903⟩, 1018 ⟨90TL3141⟩; **3**, 48, 57 ⟨89TL179⟩; **4**, 186 ⟨88TL4583⟩; **5**, 328 ⟨92TL5055⟩
Miotti, U., **2**, 142, 1006 ⟨91OR(40)157⟩
Miotti, U.,**3**, 16 ⟨91OR(40)157⟩
Miotti, U., **3**, 365 ⟨66CC586, 68CC724⟩, 365 ⟨70JCS(B)325⟩; **4**, 221, 222, 226, 240, 323 ⟨91OR(40)157⟩

Miquel, J. F., **5**, 81 ⟨55MI 502-02⟩
Miquel, P., **5**, 1047 ⟨1877MI 527-01⟩
Mir, G. N., **6**, 647 ⟨78AF1435⟩
Mir, I., **4**, 34 ⟨76JCS(P1)2349⟩
Miranda, E. I., **2**, 58, 526 ⟨90TL4677⟩; **4**, 248 ⟨86TL6305⟩, 353 ⟨90TL4677⟩, 354 ⟨92JA10078⟩; **5**, 422 ⟨92JA10078⟩
Miranda, M. A., **3**, 297 ⟨86H(24)2511⟩
Miranda, M., **6**, 598 ⟨91JMC3187⟩
Miranda, R., **4**, 246 ⟨90SC153⟩
Mirando, P., **1**, 556 ⟨80JOC4702⟩; **2**, 28 ⟨80JOC4702⟩
Miranova, A. A., **4**, 864 ⟨79ZOR2416⟩
Miranov, V. F., **4**, 313 ⟨82ZOB1386⟩
Mir, Q.-C., **6**, 253 ⟨79JA5949, 81JA406⟩
Miravitlles, C., **5**, 1137 ⟨84OM996⟩; **6**, 30 ⟨91JOC103⟩
Mirbach, M., **2**, 872 ⟨B-90MI 217-01⟩
Mirejovsky, D., **1**, 1114 ⟨78JOC763⟩
Mirkin, C. A., **4**, 718 ⟨87JOM(334)117⟩
Mironenko, D. A., **4**, 462 ⟨92ZOB1472⟩
Mironiv, V. F., **4**, 660 ⟨77ZOB1751⟩
Mironova, A. A., **1**, 109 ⟨88JOU530⟩; **5**, 205 ⟨88ZOR593⟩; **6**, 14 ⟨93RHA232⟩
Mironova, E. A., **2**, 464 ⟨60IZV651⟩
Mironov, A. F., **2**, 617 ⟨79S533⟩
Mironova, N. V., **5**, 1007 ⟨81ZOB239⟩; **5**, 969 ⟨79ZOB784, 80ZOB620⟩
Mironova, N. V.,**5**, 969 ⟨82ZOB2654⟩
Mironov, A. V., **6**, 583 ⟨93ZOB633⟩
Mironova, Z. N., **4**, 339 ⟨73ZOB534⟩
Mironov, G. S., **4**, 979 ⟨91ZOR382⟩
Mironov, V. A., **1**, 248 ⟨83RCR61⟩, 794 ⟨81RCR666⟩
Mironov, V. D., **5**, 1016 ⟨68ZOB1179⟩
Mironov, V. F., **1**, 656 ⟨63MI 114-01, B-67MI 114-02⟩; **2**, 476 ⟨91MI 208-03⟩; **4**, 98, 99 ⟨87JGU334⟩, 142 ⟨85JGU1079⟩, 142 ⟨91MI 403-02⟩, 142 ⟨92MI 403-02⟩, 156 ⟨78JGU769⟩, 295, 313 ⟨81ZOB2382, 83ZOB119⟩, 574 ⟨63IZV1563⟩, 608, 638 ⟨69ZOB2281⟩, 609 ⟨72ZOB1361⟩, 621 ⟨75ZOB2672⟩, 626, 637, 640 ⟨69ZOB2601⟩, 628, 639, 640 ⟨72ZOB1521⟩, 638, 640 ⟨69DOK(188)120⟩, 639 ⟨70ZOB2601⟩, 639, 640 ⟨73ZOB625, 76ZOB837⟩, 641 ⟨80ZOB1764⟩, 660 ⟨72ZOB631, 74ZOB806, 75ZOB2448, 76ZOB1043⟩, 660, 700 ⟨73ZOB801⟩, 1058 ⟨86ZOB1535⟩; **5**, 323 ⟨88ZOB1056, 88ZOB1299⟩, 719 ⟨83JGU1513, 83ZOB1683⟩, 721, 722 ⟨78JGU1313⟩, 721, 722 ⟨78ZOB1429, 72JGU1859⟩, 722 ⟨81JGU2055, 81ZOB2382⟩, 722 ⟨82JGU1222, 82ZOB1386, 83JGU1944⟩, 722, 863 ⟨71DOK(199)103⟩, 722, 978 ⟨72ZOB1870⟩, 722, 997 ⟨83ZOB2155⟩, 962, 968, 969, 978, 1006, 1012, 1015 ⟨84JOM207⟩, 968 ⟨69FRP1563380, 69ZOB2598⟩, 968 ⟨71BRP1235156, 79RCR473⟩, 968, 969 ⟨70DOK(190)110, 73ZOB1502, 73ZOB623, 75ZOB2025, 72ZOB2501⟩, 969 ⟨75ZOB2185, 79ZOB784⟩
Mironov, V. F.,**5**, 969 ⟨82ZOB2654⟩, 977 ⟨69DOK(181)115⟩
Mironov, V. F., **5**, 978 ⟨73ZOB2089, 75ZOB2007, 81ZOB1814⟩, 983 ⟨80ZOB960, 81ZOB2017⟩, 1000 ⟨73UK1451⟩, 1001 ⟨84ZOB876⟩, 1006 ⟨75ZOB477, 76ZOB1531, 77ZOB2259, 83ZOB1096, 83ZOB1434⟩, 1014 ⟨68DOK(181)115, 74ZOB553⟩, 1015 ⟨71ZOB2340, 72ZOB879, 73ZOB1044⟩, 1076, 1077 ⟨78ZOB1943⟩; **6**, 62 ⟨60DOK(131)98⟩, 188 ⟨72ZOB1521⟩, 267 ⟨57IZV199⟩, 443 ⟨69ZOB2598⟩, 444 ⟨80ZOB875⟩, 445 ⟨69ZOB220, 77ZOB1063⟩
Mironovskaya, N. M., **5**, 774 ⟨78ZOB2465⟩
Miroshnichenko, V. V., **2**, 407 ⟨89ZOB2136⟩, 407, 408 ⟨90ZOB465, 80ZOB951⟩; **6**, 14 ⟨80ZOB1897⟩
Mirrashed, M., **1**, 1077 ⟨77JHC567⟩
Mirrington, R. N., **5**, 106 ⟨72AJC1731⟩
Mirskova, A. N., **2**, 338, 365 ⟨89RCR250⟩, 1036, 1037 ⟨90IZV1865⟩; **4**, 94 ⟨86ZOR1923⟩, 312 ⟨84ZOR1502⟩, 765 ⟨88IZV514⟩, 772 ⟨90ZOB227, 91ZOB1084⟩, 783 ⟨83ZOR1344, 84ZOR972, 85ZOR965⟩, 786 ⟨85ZOR2467, 87ZOR60, 88ZOR509, 89ZOR1470, 89ZOR1773⟩, 787 ⟨90IZV1693, 90ZOR978, 91IZV2106, 92ZOB2730⟩, 804 ⟨91JGU983⟩, 872 ⟨91MI 419-01⟩, 926 ⟨82ZOB2375⟩
Mirskov, R. G., **2**, 1091 ⟨86JGU849⟩, 1093 ⟨81IZV1280⟩; **3**, 548 ⟨78ZOB2137⟩; **4**, 140 ⟨91IZV1039⟩, 617 ⟨87IZV1424⟩, 812 ⟨79JGU625⟩, 813 ⟨73JGU679⟩; **5**, 540 ⟨75JGU2421⟩
Mirviss, S. B., **6**, 74 ⟨51JA3807⟩
Mirza, N. A., **1**, 794, 799, 809 ⟨65TL3955⟩
Mirzabekyants, N. S., **4**, 966 ⟨74IZV2545⟩
Mirzai, H., **5**, 624, 625, 732 ⟨76CB906⟩
Mirzaoglu, R., **5**, 664 ⟨90SRI437, 92SRI851⟩
Misaki, S., **4**, 731, 767 ⟨85JFC(29)471⟩
Misaki, Y., **6**, 555 ⟨92CC1410⟩
Misani, F., **2**, 163 ⟨51JA459⟩
Misawa, H., **1**, 613 ⟨86JOC3834⟩, 613, 700, 1073 ⟨86JOC3830⟩, 1036 ⟨85JOC565⟩; **3**, 57 ⟨86BCJ3287⟩
Misch, M. A., **2**, 89 ⟨65JHC152⟩
Mischke, P., **4**, 523 ⟨69LA(722)21⟩
Mise, T., **1**, 1097 ⟨82JA180⟩; **4**, 1033 ⟨80BCJ1138, 82JA180⟩
Miser, J. R., **2**, 647 ⟨77JOC2053⟩
Mishani, E., **2**, 94 ⟨93JA1379⟩; **4**, 52 ⟨92JA7643⟩
Mishima, T., **3**, 224 ⟨74TL3297⟩; **4**, 734 ⟨85CPB5137⟩
Mishina, E. A., **5**, 1077 ⟨85IZV2824⟩
Mishina, T., **3**, 618 ⟨78S905⟩
Mishin, M. A., **5**, 702 ⟨88JOU2149, 88ZOR2384⟩
Mishra, A. K., **6**, 514 ⟨71IC1896⟩
Mishra, P., **2**, 543 ⟨86TL4391⟩; **3**, 184 ⟨86TL4391⟩, 415 ⟨86CJC726⟩
Mishra, R. S., **6**, 264 ⟨80IJC(B)610⟩
Mishrikey, M. M., **2**, 121 ⟨70JCS(C)1578⟩
Misima, S., **2**, 28 ⟨80JOC2731⟩; **3**, 152, 154, 155 ⟨80JOC2731⟩
Misiorny, A., **6**, 512 ⟨79EUP3008⟩
Misiti, D., **3**, 4, 57, 115 ⟨79S356⟩; **4**, 168 ⟨82JOC2995⟩, 260 ⟨90JOC1323⟩; **5**, 275 ⟨80S243⟩
Misiura, K., **2**, 412 ⟨85JOC1815⟩
Miskevich, G. N., **3**, 418 ⟨78ZOR508⟩
Mis'kevich, G. N., **5**, 655 ⟨78JOU881, 78ZOR943⟩
Mislankar, S. G., **5**, 203 ⟨88JMC362⟩
Mislow, K., **1**, 53 ⟨71JA6205⟩; **2**, 89 ⟨51JA4043⟩, 440 ⟨69JA2788⟩, 440, 441 ⟨69JA7027, 67JA4784⟩, 441 ⟨69JA5645, 69JA7012, 69JA7023⟩, 441, 443 ⟨71FCF61⟩, 462 ⟨73CC273⟩, 807, 845 ⟨68JA4842⟩, 845, 846 ⟨71CC146, 69JA7393⟩, 855 ⟨71JA3792⟩, 949 ⟨86JA6664⟩; **3**, 376 ⟨74JA5648, 74JA5650, 74JA5651, 75JA2718, 76JA3615⟩, 457 ⟨71JA6703⟩; **4**, 11 ⟨87JOC3466⟩
Mison, P., **1**, 906 ⟨80JA683⟩
Misono, A., **1**, 103 ⟨68BCJ2921⟩; **3**, 617 ⟨66BCJ854⟩
Misra, B. N., **5**, 214 ⟨88IJC(B)1159⟩, 328, 332, 710, 714 ⟨89CCC2738⟩, 665 ⟨84IJC(B)728⟩, 710, 714 ⟨92CCC540⟩, 710, 714 ⟨93CCC2150⟩
Misra, N., **2**, 891 ⟨89POL581⟩; **5**, 318, 369 ⟨89POL581⟩
Misra, R. A., **3**, 7, 120 ⟨89S403⟩; **4**, 806 ⟨73CB1601⟩
Misra, R., **5**, 52 ⟨68TL2681⟩
Misra, R. N., **2**, 623, 682, 747 ⟨77JA1993⟩; **4**, 357 ⟨77TL591⟩; **5**, 413 ⟨82JOC747⟩
Misra, S. C., **1**, 16 ⟨81JCS(P1)1501⟩; **3**, 358 ⟨75CC756⟩
Misra, T. N., **5**, 384 ⟨76JIC1230⟩
Missakian, M. G., **4**, 223, 225 ⟨74JOC2010⟩
Missol, U., **2**, 374 ⟨74JPR347⟩; **5**, 357 ⟨74ZC265⟩
Misterkiewicz, B., **2**, 226 ⟨88T1153⟩
Mistry, K. M., **1**, 815 ⟨88CC539⟩
Mistryukov, E. A., **3**, 405 ⟨84S947⟩
Misu, D., **1**, 1038 ⟨86TL933⟩; **3**, 127 ⟨85JA7179⟩, 130, 285, 295 ⟨86TL929⟩, 273 ⟨86TL933⟩
Misumi, A., **1**, 184 ⟨85JA3343⟩
Misumi, S., **1**, 232 ⟨78SC285, 89BCJ164⟩, 243 ⟨80SC595, 81CL627⟩, 243 ⟨82BCJ182, 87BCJ4027⟩, 594, 611, 621 ⟨81CL447⟩, 661 ⟨67JA4531⟩; **2**, 289 ⟨80SC595⟩, 623

⟨67JA4531⟩, 623, 628 ⟨89JOC6068⟩; **3**, 668 ⟨86TL2411⟩; **4**, 402 ⟨81CL447⟩; **5**, 1113 ⟨80SC595⟩
Mita, N., **1**, 466, 1097 ⟨79JOC2408⟩
Mita, T., **1**, 515 ⟨87JOC5695⟩, 1098, 1140 ⟨84JOC3904⟩; **2**, 69, 564 ⟨84JOC3904⟩
Mitamura, S., **2**, 202 ⟨82TL5427⟩; **3**, 281, 298, 299 ⟨93CC1157⟩; **4**, 857 ⟨79S880⟩
Mitani, M., **2**, 28 ⟨91CC1418⟩, 83 ⟨90JA1286⟩; **3**, 497 ⟨68JA2728⟩
Mitani, T., **5**, 549 ⟨73IJS359⟩, 584, 620 ⟨72BCJ3653⟩
Mitarbeit, U., **6**, 51, 238 ⟨69LA(730)140⟩
Mitch, F. A., **4**, 602, 607 ⟨54JA1619⟩
Mitchard, L. C., **5**, 937 ⟨71JCS(A)794⟩
Mitchell, A. G., **4**, 58 ⟨93JCS(P1)2303⟩
Mitchell, D., **3**, 632 ⟨92TL3281⟩
Mitchell, D. K., **5**, 919 ⟨74JA2847⟩
Mitchell, F. W., **2**, 410 ⟨50JA5491⟩
Mitchell, J. A., **5**, 275, 276 ⟨31JA1879⟩
Mitchell, J., **1**, 690 ⟨86JA515⟩
Mitchell, J. C., **5**, 366 ⟨79JOC2288⟩; **6**, 621, 728 ⟨87CC99⟩
Mitchell, J. D., **4**, 546, 550, 551 ⟨77IC1770, 74SRI221⟩, 551 ⟨80ZN(B)832⟩, 551, 576 ⟨76IS195⟩, 553 ⟨75CC634⟩
Mitchell, M. A., **1**, 493 ⟨81TL5231⟩
Mitchell, M. B., **1**, 489 ⟨92T8117⟩; **4**, 277 ⟨90SL359⟩, 900 ⟨89T1801⟩
Mitchell, M. J., **5**, 95, 306 ⟨73OSC(5)944⟩
Mitchell, P. R. K., **3**, 297 ⟨84JCS(P1)1339⟩
Mitchell, R. B., **5**, 1113 ⟨84JOC2057⟩
Mitchell, R. C., **5**, 360 ⟨90JCS(P1)3103⟩
Mitchell, R. D., **1**, 865 ⟨82TL795⟩
Mitchell, R. H., **1**, 11, 18, 19 ⟨80TL2637⟩, 233 ⟨75TL219, 82JA2560⟩, 704 ⟨73TL4395⟩, 713, 714, 716 ⟨74CC990⟩, 1074 ⟨73CC955⟩; **2**, 14 ⟨89SL55⟩; **3**, 310 ⟨93TL3683⟩; **4**, 64 ⟨73TL4395⟩
Mitchell, R. S., **1**, 881 ⟨50JOC850⟩
Mitchell, R. W., **1**, 507 ⟨71JCS(A)3224⟩
Mitchell, T. F., **2**, 373, 374, 379, 381 ⟨59JA2805⟩
Mitchell, T. N., **1**, 468 ⟨92S803⟩; **2**, 814 ⟨76TL2173⟩, 910, 911, 916, 917, 932, 940, 946 ⟨87JOC4868⟩, 916, 930 ⟨85CC354⟩, 1092 ⟨92S803⟩; **3**, 571, 580 ⟨74JOM(71)39⟩; **4**, 152 ⟨84JOM(269)219⟩, 567 ⟨89JOM(368)167⟩, 656, 663, 700, 724 ⟨84JOM(269)249⟩, 661, 701, 1062 ⟨85OM1044⟩, 699, 701 ⟨79JOM(172)293⟩, 700 ⟨91OM936⟩, 746, 813, 1069 ⟨86OM1991⟩, 813, 1066 ⟨83JOM(256)37⟩, 1060, 1062 ⟨85JOM(281)163, 81JOM(210)C17⟩, 1063, 1065, 1069, 1070 ⟨87JOM(322)151⟩, 1065 ⟨87JOM(322)141⟩, 1065, 1066, 1068, 1069 ⟨83JOM(252)47⟩; **5**, 467 ⟨90S1001⟩; **6**, 209 ⟨83JOM(252)47, 85OM1044⟩, 209 ⟨86OM1991⟩, 395, 399, 401, 405 ⟨85JOM(291)179, 79JOM(172)293⟩
Mitchell, T. R. B., **3**, 625 ⟨81TL4107⟩
Mitchell Jr., J., **3**, 618 ⟨45JA161⟩
Mitgau, R., **5**, 304 ⟨55CB301⟩
Mito, N., **6**, 443 ⟨88MI 614-01⟩
Mitra, A., **1**, 405, 406, 816, 817, 820 ⟨78JA1927⟩; **3**, 203, 204 ⟨78TL2659⟩; **4**, 378 ⟨78JA1927, 79TL593⟩
Mitra, R. B., **1**, 389 ⟨64JA478⟩, 498 ⟨64JA5570⟩, 707 ⟨75IJC251⟩; **2**, 225 ⟨62JA2938⟩
Mitsch, R. A., **5**, 808 ⟨65JA328, 72USP3674785⟩, 808 ⟨73USP3751312⟩, 814 ⟨73JOC1065⟩, 816 ⟨84USP4446068⟩, 1080 ⟨67CC59⟩; **6**, 52 ⟨68JOC3675⟩, 55, 261, 285 ⟨68JOC1847⟩, 242, 613 ⟨66JOC3833⟩, 254, 278 ⟨73JOC3358⟩, 261 ⟨64JHC233, 66JHC245⟩, 279, 281 ⟨73JOC1065⟩, 285 ⟨72USP3637663⟩, 604 ⟨67JA3868, 68JHC41, 84USP4446068⟩, 616 ⟨67JHC389⟩
Mitsch, R., **5**, 1080 ⟨67JA5007⟩
Mitschke, K.-H., **2**, 488 ⟨69CB4136, 73CB1226, 73CB3645⟩, 488, 504 ⟨71ZAAC(386)139⟩, 502 ⟨71CB1842⟩, 888 ⟨71ZAAC(386)147⟩
Mitschler, A., **1**, 53 ⟨80JA994⟩; **4**, 1034 ⟨77JA3537⟩, 1035 ⟨80JA994⟩, 1037 ⟨78JA5748⟩
Mitsuda, Y., **3**, 431 ⟨87TL4557⟩
Mitsudera, H., **2**, 214 ⟨90JHC1361⟩
Mitsudo, T.-A., **2**, 676 ⟨86TL2125⟩; **3**, 162 ⟨87JOC2230⟩; **6**, 339 ⟨89JOM(368)199⟩
Mitsudo, T., **1**, 493 ⟨92JOM(423)405⟩, 518 ⟨94AG(E)580⟩, 1036 ⟨85JOC565, 86BCJ3201⟩; **2**, 310 ⟨74TL1879⟩; **3**, 83 ⟨71BCJ2569⟩, 84 ⟨73TL3535, 75BCJ2490⟩, 84 ⟨75TL1063⟩, 597 ⟨80JOM(190)C39⟩; **6**, 457 ⟨72CL483⟩
Mitsugi, C., **6**, 434 ⟨85JHC1643⟩
Mitsuhashi, H., **3**, 54 ⟨66TL4113⟩; **4**, 408 ⟨89H(29)865⟩
Mitsuhashi, K., **5**, 664 ⟨84BCJ2184⟩, 681 ⟨87JHC1391⟩; **6**, 28 ⟨77JHC881, 78JHC893⟩
Mitsuhashi, T., **1**, 265 ⟨77TL2441⟩
Mitsuhira, Y., **4**, 959 ⟨78CL517⟩
Mitsui, H., **3**, 431 ⟨83TL1049⟩, 432 ⟨84CC874, 84H(21)483⟩
Mitsui, K., **5**, 332 ⟨90TL243⟩
Mitsui, T., **1**, 747 ⟨71ABC1116⟩
Mitsunaga, T., **3**, 350 ⟨90EUP379979⟩
Mitsunobu, O., **1**, 179 ⟨72TL1279⟩, 179 ⟨76TL2455⟩, 179 ⟨81S1⟩; **2**, 6, 87, 118, 124, 299, 683 ⟨81S1⟩, 38, 82, 83 ⟨91COS(6)1⟩, 87, 118, 124 ⟨71BCJ3427⟩, 87, 118, 124 ⟨72JA679⟩, 107 ⟨63JOC481⟩, 118, 124 ⟨67BCJ2380⟩, 213 ⟨70JOC4227⟩; **3**, 8, 122 ⟨79CC303⟩, 8, 122 ⟨81TL2295⟩, 179, 188 ⟨91COS(6)65⟩, 561 ⟨71BCJ1362⟩, 621 ⟨76TL2455⟩; **5**, 123, 131, 138 ⟨81S1⟩, 177 ⟨84TL2163⟩, 735 ⟨71BCJ1362⟩, 1078 ⟨69TL2473, 70T5731⟩
Mitsuya, H., **6**, 491 ⟨89TL7141⟩
Mittag, E., **2**, 194, 239 ⟨19CB413⟩
Mittag, J., **6**, 268 ⟨79ZAAC(458)37⟩
Mittakanti, M., **3**, 640 ⟨89IC492⟩; **5**, 688 ⟨90IC3218⟩; **6**, 491 ⟨92S380⟩, 491, 515 ⟨90IC554⟩, 492 ⟨92AP(325)267, 90IC3218⟩
Mittal, P. K., **4**, 283 ⟨87JCS(D)757⟩
Mittal, R. S. D., **2**, 791 ⟨85S70⟩; **5**, 138 ⟨75JA4144⟩
Mitt, T., **1**, 697, 700, 703 ⟨88JOC4282⟩
Mitteil, I., **4**, 177, 178 ⟨38CB1803⟩
Mittelbach, M., **4**, 883 ⟨86LA533⟩
Mittermaier, J., **5**, 62 ⟨41AG51⟩
Mitulla, K., **2**, 1089 ⟨81JOM(204)315⟩
Mitzel, N. W., **6**, 702 ⟨92CB1053⟩
Mitzlaff, M., **4**, 303 ⟨75GEP2503114⟩
Miura, A., **2**, 238 ⟨82JOC951⟩
Miura, H., **2**, 331 ⟨89JA5886⟩; **6**, 467 ⟨94EUP583789⟩
Miura, I., **1**, 141 ⟨85JOC657⟩; **6**, 101 ⟨85JOC657⟩
Miura, K., **1**, 23, 236 ⟨81JA2886⟩, 330, 925, 946 ⟨89BCJ143⟩, 801 ⟨88TL1543⟩; **2**, 106 ⟨91T8717⟩, 913 ⟨93BCJ2348, 93BCJ2356⟩; **4**, 737, 743 ⟨91CL1319⟩, 737, 743 ⟨92BCJ1513⟩, 811 ⟨92TL7031⟩
Miura, M., **1**, 483, 484 ⟨80TL87, 79TL43⟩, 494 ⟨90JCS(P1)2207⟩, 1006 ⟨92TL5363⟩; **2**, 305 ⟨93BCJ1297⟩; **4**, 959 ⟨78CL517, 80CL1209⟩; **5**, 668, 830 ⟨86JMC1689⟩
Miura, R., **1**, 63 ⟨85OM1911⟩
Miura, S., **4**, 683 ⟨80JOM(187)81⟩
Miura, T., **2**, 238 ⟨82JOC951⟩, 285 ⟨82CC438⟩; **4**, 183 ⟨93CL17⟩, 193 ⟨92JHC1625⟩
Miura, Y., **2**, 361 ⟨84JOC2724⟩; **5**, 766 ⟨83CL885⟩
Miwa, A., **1**, 829 ⟨87TL2597⟩
Miwa, H., **3**, 357 ⟨91CPB672⟩
Miwa, T., **1**, 100 ⟨75BCJ1865⟩, 1067 ⟨92JOC5060⟩; **3**, 25 ⟨82CL1637⟩, 304 ⟨70BCJ2884⟩, 419 ⟨87S824⟩, 567 ⟨60JA873⟩; **5**, 68 ⟨74S443⟩, 275 ⟨57JA2530⟩
Miwatashi, S., **2**, 927, 989 ⟨92BCJ1543, 92CL99⟩; **3**, 627 ⟨92BCJ1543⟩
Mixan, C. E., **2**, 161 ⟨72JA208⟩, 287 ⟨73JA4634⟩
Mixato, H., **2**, 797 ⟨79CL149⟩
Miyabo, A., **1**, 867 ⟨90JOC996⟩
Miyachi, N., **1**, 813 ⟨86TL219⟩; **2**, 1036, 1037 ⟨91JCS(P1)2049⟩; **3**, 217 ⟨86TL219⟩
Miyachi, Y., **5**, 199 ⟨86TL4609⟩
Miyadera, T., **4**, 842 ⟨82TL1075⟩

Miyagawa, K., **5**, 245, 246 ⟨83S552⟩, 245, 246 ⟨83S814⟩, 245, 252 ⟨82S1013⟩, 252, 254 ⟨80S721, 81S370⟩, 337 ⟨81S746⟩
Miyagawa, Y., **3**, 597 ⟨80JOM(190)C39⟩
Miyagishima, T., **5**, 635, 636 ⟨71MI 515-01⟩; **6**, 556, 564 ⟨87CPB2840⟩
Miyahara, H., **5**, 657 ⟨88H(27)155⟩
Miyahara, I., **3**, 609 ⟨93JOM(456)161⟩
Miyahara, M., **5**, 699 ⟨81BCJ3863⟩; **6**, 504 ⟨83CPB41⟩
Miyai, J., **2**, 927 ⟨90TL369⟩
Miyaji, H., **4**, 931 ⟨86CL181⟩
Miyaji, K.,**3**, 628 ⟨86CL931⟩
Miyake, ,A., **1**, 464 ⟨73CL667⟩
Miyake, A., **1**, 813, 814 ⟨80BCJ2958⟩; **3**, 142 ⟨80BCJ2958⟩
Miyake, F., **1**, 697, 698, 703 ⟨86TL2095⟩, 697, 698, 703 ⟨87JOC3759⟩
Miyake, H., **1**, 37 ⟨83CC875, 83TL3477⟩, 37, 945 ⟨81TL1705⟩, 252 ⟨86TL3025⟩, 617, 638 ⟨81CL1139⟩; **2**, 610 ⟨79CL1499⟩; **3**, 9 ⟨88S394⟩, 57 ⟨82JOC5017⟩, 621 ⟨85CL635⟩; **4**, 249 ⟨80S952⟩
Miyake, J., **5**, 625 ⟨85JOC384⟩, 625, 626 ⟨85BCJ1448⟩
Miyake, K., **3**, 166 ⟨78BCJ3008⟩
Miyake, M., **1**, 943 ⟨90JCS(P1)3017⟩; **5**, 280 ⟨85CL123⟩; **6**, 618 ⟨73BCJ303⟩
Miyake, N., **1**, 451 ⟨75JA5603⟩
Miyake, S., **6**, 547 ⟨89BCJ967⟩, 552, 554 ⟨86CE311⟩
Miyake, T., **2**, 629 ⟨81CC718⟩; **3**, 127 ⟨92TL627⟩; **4**, 761, 771 ⟨76BCJ1931⟩; **5**, 249 ⟨87NKK1479⟩, 253 ⟨87NKK1475⟩; **6**, 95 ⟨77MI 603-01⟩
Miyake, Y., **2**, 360 ⟨84JCS(P1)1401⟩
Miyaki, F., **1**, 681 ⟨86T1581⟩
Miyakoshi, T., **3**, 191 ⟨81CL1677⟩, 191 ⟨82CL83⟩, 226 ⟨84TL3621⟩
Miyamoto, C., **3**, 704 ⟨82ABC1803⟩
Miyamoto, H., **2**, 50 ⟨92TL4021⟩
Miyamoto, I., **3**, 361 ⟨89CC505⟩
Miyamoto, K., **4**, 193 ⟨89JHC1405⟩
Miyamoto, M., **5**, 332, 333 ⟨92CPB1177⟩
Miyamoto, N., **1**, 485 ⟨74BCJ503⟩
Miyamoto, O., **3**, 57 ⟨86CL2035⟩
Miyamoto, S., **6**, 178 ⟨81JA2324⟩
Miyamoto, S., **6**, 178, 383 ⟨79JA1348⟩
Miyamoto, T., **1**, 822 ⟨73TL2351⟩; **3**, 378 ⟨74JCS(P1)1125⟩, 380 ⟨74JOC3519⟩
Miyamoto, Y., **3**, 629 ⟨93JOC1458⟩
Miyano, K., **1**, 100 ⟨75BCJ1865⟩
Miyano, M., **1**, 603 ⟨81JOC1846⟩; **3**, 208 ⟨72JOC268⟩; **5**, 12 ⟨65JA3958⟩
Miyano, O., **5**, 205 ⟨86JFC(31)37⟩, 360 ⟨83JFC(22)91⟩; **6**, 12 ⟨86JFC(31)37⟩, 233, 234 ⟨82TL3929⟩
Miyano, S., **1**, 473 ⟨86BCJ2044⟩; **2**, 14 ⟨80BCJ770⟩, 348 ⟨83JOC3608⟩, 398 ⟨88BCJ3020⟩, 618 ⟨78CC446, 79BCJ1197⟩, 859 ⟨80CL729⟩; **3**, 45 ⟨80CL1213, 82BCJ534⟩, 252 ⟨76JCS(P1)1146⟩, 495 ⟨79BCJ1197⟩, 625 ⟨66TL1509, 67CPB1811, 71JOC2948⟩; **4**, 39, 40 ⟨71BCJ2864⟩, 696 ⟨73NKK381⟩
Miyano, T., **1**, 942 ⟨82TL443⟩; **2**, 156 ⟨80BCJ3615⟩, 457 ⟨79BCJ466⟩; **4**, 854 ⟨79BCJ466⟩
Miyaoka, H., **1**, 582 ⟨89CPB2882⟩
Miyasaka, A., **4**, 259 ⟨90JCS(P1)2035⟩
Miyasaka, T., **1**, 243 ⟨90TL227⟩, 1010 ⟨85CPB1761, 92JMC241⟩; **2**, 281 ⟨89S434⟩; **3**, 286 ⟨86CL449⟩; **4**, 240 ⟨B-92MI 404-01⟩; **5**, 134 ⟨86CL871⟩, 215 ⟨85CL701⟩, 277, 295 ⟨80TL841⟩
Miyashi, M., **4**, 310 ⟨79JOC391⟩
Miyashi, T., **1**, 815 ⟨74JA3468⟩; **3**, 80 ⟨77H(6)1599⟩, 221 ⟨84CC179⟩
Miyashiro, Y., **4**, 900 ⟨90JOC5308⟩
Miyashita, A., **1**, 76 ⟨84T1245⟩; **6**, 115 ⟨91CL1873, 93CL13⟩
Miyashita, M., **1**, 228 ⟨91JOC6483⟩, 601, 602, 607 ⟨74JOC120⟩; **2**, 783 ⟨82CL1505, 85ACR284⟩; **3**, 39, 175 ⟨80S664⟩, 160 ⟨87TL4293⟩, 192 ⟨89CC841⟩, 435 ⟨90T7569⟩, 637 ⟨82JCS(P1)1303⟩; **4**, 208 ⟨89CL221⟩, 906 ⟨78CC362⟩; **5**, 329 ⟨93TA1573⟩
Miyashita, S., **2**, 830 ⟨80JOC5052⟩; **5**, 280 ⟨80JOC5052⟩
Miyashitam, M., **3**, 410 ⟨90CL239⟩
Miyata, K., **1**, 422 ⟨85CL1137⟩; **3**, 145 ⟨87BCJ1063⟩, 145 ⟨87BCJ3421⟩
Miyata, M., **3**, 246 ⟨85TL4727⟩
Miyata, O., **4**, 908 ⟨93CPB217⟩; **5**, 726 ⟨92CPB602⟩
Miyata, S.-i., **1**, 931 ⟨91JOC2264⟩
Miyauchi, A., **5**, 432 ⟨85CL1829⟩
Miyauchi, H., **4**, 736 ⟨89JOC5630⟩
Miyauchi, T., **5**, 547 ⟨73JCS(P1)1009⟩; **6**, 324 ⟨73JCS(P1)1009⟩
Miyaura, A., **1**, 531 ⟨81CL879⟩
Miyaura, M., **1**, 469 ⟨77TL3369⟩
Miyaura, N., **1**, 125 ⟨77BCJ2199⟩, 228 ⟨82JOM(233)13⟩, 276 ⟨75CL523, 76TL255⟩, 277 ⟨71JA2792⟩, 277, 977 ⟨73S305⟩, 279 ⟨86TL6369, 87CL25, 87S373, 91SL687⟩, 279, 1033 ⟨73JA3080⟩, 279, 532 ⟨81TL127⟩, 472, 489, 531 ⟨81JOM(213)C53⟩, 472, 532, 1104 ⟨79CC866⟩, 473, 489 ⟨81SC513⟩, 480, 482, 489 ⟨93JOC2201, 86TL3745⟩, 489 ⟨89JA314, 90OS(68)130⟩, 489 ⟨92BCJ2863, 92SL207⟩, 489 ⟨93CL1429⟩, 489, 531 ⟨86CL1329⟩, 532 ⟨82JOC2117⟩, 532, 1020 ⟨79TL3437⟩, 532, 1020 ⟨85JA972⟩, 532, 1094 ⟨80BCJ1471⟩, 532, 1104 ⟨80TL2865⟩, 542, 1008 ⟨88TL1811⟩, 657 ⟨72JA6549⟩, 1009 ⟨90BCJ2124⟩, 1033, 1034 ⟨86JOC4507, 75TL1961⟩; **2**, 527, 900, 902 ⟨89TL3789⟩, 528 ⟨90TL231⟩, 901 ⟨87S373, 93JA7219⟩, 902, 927, 939 ⟨88TL1811⟩, 904 ⟨74TL2961⟩, 1086 ⟨86JOC4507⟩; **3**, 24 ⟨82JOC2117⟩, 228 ⟨87BCJ3471⟩, 290 ⟨91BCJ1999⟩, 290, 295 ⟨91TL6923⟩, 290, 296, 301, 306, 311 ⟨93TL7595⟩, 290, 305, 311 ⟨92TL4465⟩, 552 ⟨79CL535⟩; **4**, 815 ⟨93SC2851⟩; **5**, 157 ⟨81CL879⟩
Miyazaki, A., **1**, 599 ⟨76SC357⟩
Miyazaki, H., **5**, 317 ⟨77MI 507-01, 80MI 507-02⟩
Miyazaki, K., **5**, 630 ⟨90CL1651⟩; **6**, 495 ⟨71BCJ1393⟩
Miyazaki, S., **5**, 28 ⟨77MI 502-02⟩
Miyazaki, S.-I., **4**, 940, 947 ⟨91JA9796⟩; **5**, 238, 241 ⟨91JA9796⟩
Miyazaki, T., **1**, 991 ⟨84CL2005⟩; **3**, 592 ⟨89JCS(P1)2140⟩; **5**, 226 ⟨80CC1029⟩, 252 ⟨87NKK1457⟩, 731 ⟨83JA2831⟩
Miyazaki, Y., **2**, 946 ⟨86S776⟩
Miyazawa, E., **4**, 610, 617, 622 ⟨87BCJ2263⟩
Miyazawa, M., **1**, 192 ⟨92SL323⟩; **3**, 160 ⟨87TL3515⟩, 276 ⟨86TL373⟩
Miyazawa, T., **2**, 331 ⟨92CC1097⟩; **3**, 7 ⟨85JOC1332⟩; **5**, 279 ⟨88CC419⟩
Miyazawa, Y., **1**, 704, 707, 708 ⟨86CL433⟩, 706, 708 ⟨84CL833⟩, 706, 708 ⟨87JOC1703⟩
Miyoshi, H., **1**, 66, 454 ⟨80JCS(P1)1098⟩, 493 ⟨80BCJ553⟩; **2**, 624, 629 ⟨79CL1357⟩; **4**, 1068 ⟨81JCS(P1)991⟩; **5**, 1113 ⟨80BCJ553⟩
Miyoshi, K., **2**, 290 ⟨81NKK472⟩; **3**, 15, 90 ⟨82BCJ1663⟩
Miyoshi, M., **1**, 34 ⟨73JOC2731, 81JCS(P1)991⟩; **4**, 13 ⟨75TL997⟩, 311, 327 ⟨79BCJ826⟩, 424 ⟨82JOC1706⟩
Miyoshi, N., **1**, 245, 437 ⟨87AG(E)1187⟩; **2**, 279, 282 ⟨79CL873⟩, 281 ⟨88S175⟩, 311 ⟨87CL1275⟩, 554 ⟨87AG(E)1187⟩; **3**, 38, 174 ⟨82TL4813⟩, 176 ⟨87S1096⟩; **4**, 240 ⟨77TL851⟩, 652 ⟨87AG(E)1187⟩; **5**, 240, 243, 253 ⟨91JOC3776⟩, 1036 ⟨91TL3503⟩; **6**, 505 ⟨79S735⟩
Miyoshi, S., **1**, 577 ⟨94JOC406⟩
Mizhiritskii, M. D., **2**, 540 ⟨88RCR447⟩; **3**, 336 ⟨79ZOB1084⟩; **5**, 538, 539 ⟨79JGU940⟩, 583 ⟨86ZOB1547⟩; **6**, 537 ⟨86ZOB1547⟩
Mizoguchi, K., **3**, 173 ⟨93CC1434⟩
Mizoguchi, M., **3**, 7, 120 ⟨79TL165⟩, 7, 120 ⟨79TL3861⟩, 7, 120 ⟨80TL1867⟩
Mizoguchi, T., **2**, 219 ⟨67JOC97⟩; **3**, 468 ⟨71TL4495, 75T227⟩
Mizokami, T., **2**, 690 ⟨91CL459⟩

Mizoroki, T., **5**, 143, 304 ⟨75BCJ2091⟩
Mizrahi, V., **5**, 370 ⟨82JOC3533⟩
Mizrakh, L. I., **2**, 451 ⟨63ZOB1246⟩, 461 ⟨62ZOB3062⟩
Mizsak, S. A., **6**, 118 ⟨93JOC4159⟩
Mizu, Y., **1**, 598 ⟨92TL7181⟩; **3**, 170 ⟨92TL7181⟩
Mizuguchi, T., **2**, 765 ⟨76JCS(P1)2365⟩
Mizukami, H., **5**, 659 ⟨93JOC32⟩
Mizukami, S., **5**, 636 ⟨66CPB1249⟩, 830 ⟨67CPB61⟩, 830, 831 ⟨66CPB1263⟩
Mizuki, Y., **2**, 219, 227 ⟨84CC843⟩; **5**, 1078 ⟨85CPB373⟩
Mizumura, A., **5**, 627 ⟨94CL77⟩
Mizuno, A., **1**, 185, 200 ⟨80S1007⟩; **3**, 613 ⟨80S1007⟩; **4**, 420 ⟨85JOC4227⟩
Mizuno, H., **3**, 596 ⟨75BCJ2287⟩; **5**, 972 ⟨77BCJ2406⟩
Mizuno, K., **1**, 348 ⟨88JA1288⟩; **2**, 677 ⟨79S688⟩; **6**, 558 ⟨89CPB1249⟩
Mizuno, T., **5**, 596 ⟨93SC2139⟩; **6**, 471 ⟨90CL811⟩, 479 ⟨88TL4767, 90TL4773⟩, 494 ⟨91TL6867⟩, 495, 497 ⟨89AG(E)452⟩
Mizuno, W., **4**, 947 ⟨91JCS(P1)1555⟩
Mizuoka, A., **2**, 14 ⟨89JOC4989⟩
Mizusawa, Y., **6**, 134 ⟨88BCJ2147⟩
Mizusuna, A., **1**, 991 ⟨89CC1690⟩
Mizuta, M., **4**, 588 ⟨73JOM(51)167⟩; **5**, 232 ⟨65BCJ320⟩, 243, 252 ⟨77JOM(129)189⟩, 246 ⟨82AG(E)150⟩, 252, 254 ⟨80S721, 81S370⟩, 254, 537, 539, 540 ⟨73BCJ244⟩, 337 ⟨81S746⟩, 337, 584 ⟨85CB1668⟩, 510, 551, 583, 620 ⟨82LA1229⟩, 537 ⟨72IJS279⟩, 546, 550 ⟨72BCJ3492⟩, 547 ⟨82S457⟩, 549 ⟨73IJS359⟩, 549 ⟨74JOM(78)405, 74JOM(82)223⟩, 551 ⟨77AG(E)743⟩, 584 ⟨81LA1798⟩, 584, 620 ⟨72BCJ3653⟩, 609 ⟨77AG(E)787⟩
Mizutaki, S., **1**, 180 ⟨86BCJ3617⟩, 246 ⟨86JCS(P1)1983⟩; **2**, 287 ⟨84CC426⟩; **3**, 176 ⟨86JCS(P1)1983⟩
Mizutani, M., **1**, 187 ⟨84JA1079⟩; **3**, 626 ⟨82CPB3574⟩; **5**, 607 ⟨80JA7806⟩
Mizutani, T., **5**, 516, 523, 563 ⟨93JA5823⟩
Mizuyama, K.,**4**, 844 ⟨75CPB2749⟩
Mkolajczyk, M., **6**, 48 ⟨77JPR17⟩
Mkrtchyan, A. P., **5**, 690 ⟨77KFZ38, 90CHE1170⟩, 690 ⟨90KGS1403⟩
Mladenova, M., **4**, 394 ⟨91MI 408-01⟩
Mladenov, I., **4**, 797 ⟨77UKZ1000⟩
Mlakar, B.,**5**, 652 ⟨64CB483⟩
Mlaker, B., **4**, 304 ⟨64CB483⟩
Mlekuz, M., **4**, 713 ⟨85OM1123⟩
Mlinarić-Majerski, K., **1**, 704, 709 ⟨86JOC2397⟩, 709 ⟨88TL4803⟩
Mlochowski, J., **2**, 279 ⟨84S439⟩; **3**, 7 ⟨90TL2177⟩, 89, 96, 98, 99, 108 ⟨84S747⟩, 617 ⟨89S223⟩; **4**, 288 ⟨84S439⟩; **5**, 366 ⟨93PJC711⟩; **6**, 92 ⟨84S439⟩
Mloston, G., **1**, 367 ⟨84BSB511⟩; **3**, 367 ⟨89TL7041, 89TL7045⟩, 572 ⟨89H(29)2069⟩; **4**, 328 ⟨85H(23)2207⟩
Mlotkowska, B., **2**, 257 ⟨91LA1361⟩, 373 ⟨78TL4731⟩; **6**, 48 ⟨77JPR17⟩, 116 ⟨74JPR913⟩
Moad, G., **1**, 339 ⟨74CC472⟩
Moal, H. Le, **5**, 64 ⟨78JCR(S)198⟩
Mo, Y., **1**, 902 ⟨72JA3544⟩
Mo, Y. K., **1**, 58, 841 ⟨72JA7859⟩, 901 ⟨73JA4960⟩, 914 ⟨72JA2034⟩
Mobashery, S., **2**, 25 ⟨91JOC7186⟩; **6**, 486 ⟨92JOC2755⟩
Mobbili, G., **3**, 159 ⟨92T1299⟩
Mobbs, R. H., **4**, 1076 ⟨65AFC(4)50⟩; **6**, 41, 58, 60, 65 ⟨65AFC(4)50⟩
Mobele, B. I., **1**, 21, 23 ⟨92TL5567⟩
Möbius, L., **3**, 467 ⟨65CB1138⟩, 478 ⟨66CB475⟩; **4**, 901, 904 ⟨65CB1293⟩
Mochalina, E. P., **2**, 782 ⟨65IZV1715⟩; **6**, 69 ⟨60IZV231⟩
Mochalin, V. B., **1**, 916 ⟨77RCR530⟩
Mochalin, V. B.,**3**, 57 ⟨59JGU3925⟩
Mocheda, K., **2**, 563 ⟨87JOM(332)247⟩
Mochel, W. E., **6**, 80 ⟨48JA2268⟩
Mochida, K., **1**, 449 ⟨88OM1869⟩; **4**, 140, 141 ⟨78JOM(154)353⟩, 637, 641 ⟨84BCJ796⟩; **6**, 244 ⟨88BCJ3002⟩
Mochizuki, D., **2**, 326 ⟨93JMC3526⟩
Mochizuki, H., **2**, 620 ⟨88CL2017⟩
Mochizuki, K., **3**, 2 ⟨79CL443⟩
Mochizuki, S., **6**, 215 ⟨81MI 607-01⟩
Mock, G. V., **5**, 98 ⟨55CRV659⟩
Mock, S., **5**, 954 ⟨93CB657⟩; **6**, 720, 723 ⟨93JOM(459)55⟩
Mock, W. L., **1**, 74 ⟨61JA2957⟩, 458 ⟨61TL347⟩, 620, 790 ⟨66JA2857⟩, 620, 791 ⟨69JA5682⟩; **2**, 245, 250, 366 ⟨75JA6521⟩; **4**, 193 ⟨81JOC2557⟩
Mockel, A., **3**, 602 ⟨82JA2920⟩
Möckel, K., **6**, 560 ⟨80ZC101⟩
Möckel, R., **6**, 49 ⟨94AG(E)323⟩
Moczar, E., **3**, 450 ⟨65JA596⟩; **5**, 855 ⟨62BSF186⟩
Modak, A. B., **5**, 864 ⟨83IJC(A)307, 86SRI113⟩
Modak, A. S., **6**, 58, 264 ⟨94JOC2393⟩, 258 ⟨89JA1773⟩
Modarai, B., **3**, 149 ⟨77JOC3527⟩; **4**, 3, 37 ⟨71JCS(C)279⟩, 4, 12, 13, 21, 30, 33 ⟨77JOC3527⟩
Modena, G., **1**, 200 ⟨82JOC4577⟩, 318 ⟨78JOC1521, 79JOC3022⟩, 519, 562 ⟨84T2585⟩, 617 ⟨84JOC596⟩, 1109 ⟨72CC8, 74JCS(P2)1808⟩, 1109 ⟨77JCS(P2)542, 78G543⟩, 1109, 1110 ⟨71ACR73⟩, 1111 ⟨70JCS(C)2625⟩, 1112 ⟨68CC1363, 68CC1676, 68TL4039, 69CC1520, 71JCS(B)1569, 71JCS(B)374, 71JCS(B)381, 71JCS(C)3018, 72JCS(P1)1136, 72JCS(P1)216, 72JCS(P2)2026⟩, 1123 ⟨65TL4399, 65TL4405⟩; **2**, 142, 1006 ⟨91OR(40)157⟩, 145 ⟨60G383⟩, 146 ⟨76TL4637⟩, 147 ⟨84S325⟩, 147 ⟨86TL6257⟩, 147 ⟨89TL2575⟩, 147 ⟨89TL4859⟩, 192, 212, 216 ⟨B-74MI 203-04⟩, 1034 ⟨66JCS(B)1188⟩
Modena, G.,**3**, 16 ⟨91OR(40)157⟩
Modena, G., **3**, 275 ⟨90JOC3658⟩; **3**, 5, 117 ⟨87JOC5467⟩; **4**, 221, 222, 226, 240, 323 ⟨91OR(40)157⟩, 249 ⟨90BSF734⟩, 261 ⟨86TL6257, 89TL2575⟩, 261 ⟨90G165, 92TL3043⟩, 772 ⟨63CA(59)11313c⟩
Moder, M. P., **2**, 943 ⟨82JA2613⟩
Moder, T. I., **2**, 26 ⟨75JA2281⟩
Moderhack, D., **3**, 560 ⟨86CB3411⟩
Modest, E. J., **2**, 252 ⟨86JMC2114⟩; **5**, 999 ⟨73JMC1⟩, 1118 ⟨66JHC387⟩
Modi, S. P., **4**, 950 ⟨91T6539⟩
Modinos, A. G. J., **4**, 1042 ⟨73CC258⟩
Modonov, V. B., **4**, 765 ⟨88IZV514⟩
Modrego, J., **4**, 685 ⟨90JCS(D)2587⟩
Modrick, M. A., **4**, 579 ⟨83OM154⟩; **6**, 707 ⟨83OM154⟩
Modritzer, K., **2**, 415, 492 ⟨59CB2637⟩, 487 ⟨56ZN(B)750⟩
Mödritzer, K., **2**, 879 ⟨56ZN(B)751, 56ZN(B)753⟩
Modro, A., **1**, 318 ⟨77JOC2021⟩, 1123 ⟨76CJC3045, 76JOC2331⟩
Modro, A. M., **2**, 106 ⟨92OPP57⟩
Modro, J. M., **2**, 187, 199 ⟨60JCS1665⟩
Modro, T. A., **2**, 176 ⟨75JA1492⟩, 1000 ⟨91T8611⟩; **4**, 570 ⟨91PS(55)41⟩; **5**, 370 ⟨82JOC3533⟩, 371, 373, 707 ⟨93HAC11⟩, 714 ⟨92PS(72)201⟩
Mody, N. V., **3**, 187 ⟨80TL809⟩
Mody, P. N., **3**, 363 ⟨76IJC(B)999⟩
Moe, H., **4**, 212 ⟨59JOC1768⟩
Moed, H. D., **6**, 642 ⟨64RTC1305⟩
Moedritzer, K., **2**, 416 ⟨64IC609⟩, 448 ⟨61JA4381⟩; **4**, 123 ⟨67USP3360556⟩, 125, 127, 128 ⟨61JA4381⟩, 463, 470 ⟨72SRI317, 66JOC1603⟩, 470 ⟨73SRI75⟩
Moegelin, W., **6**, 512 ⟨87EGP242812⟩
Moeller, E., **2**, 298, 303 ⟨57HOU(11/1)1⟩, 302 ⟨57HOU(11/1)961⟩
Moeller, K. D., **1**, 289 ⟨89H(28)321⟩, 944 ⟨85TL3417, 92SL107⟩; **3**, 421 ⟨93JA1434⟩; **5**, 126 ⟨93JA11434⟩
Moenius, T., **5**, 223 ⟨86AG(E)994⟩
Moenius, V., **4**, 881 ⟨85S1062⟩
Moennighoff, H., **4**, 865 ⟨73AP(306)389⟩
Möens, L., **2**, 80 ⟨91COS(8)129⟩; **3**, 184 ⟨82S765⟩, 416 ⟨82S43⟩, 416, 416 ⟨86JOC3839⟩, 417 ⟨82TL789⟩
Moens, L. J., **6**, 598 ⟨91JMC3187⟩
Moerck, R., **1**, 422 ⟨80JOC1046⟩; **4**, 369, 938, 942 ⟨80JOC1046⟩

Moerikofer, A. W., **1**, 1100 ⟨63JA2063⟩; **2**, 518 ⟨61JA3417, 63JA2063⟩
Moerlein, S. M., **1**, 449 ⟨87JFC(36)63, 87JOC664⟩; **2**, 949 ⟨87JOC664⟩
Moersch, G. W., **1**, 681 ⟨71JOC1149⟩
Moesch, L., **3**, 44 ⟨90HCA405⟩
Moezzi, B., **4**, 453 ⟨84IC2550⟩
Moffat, J., **3**, 357 ⟨51JA300⟩; **5**, 1108 ⟨51JA613⟩
Moffat, J. G., **2**, 162 ⟨71JOC3687⟩, 162 ⟨71JOC3861⟩, 414 ⟨61JA649⟩
Moffatt, J. G., **2**, 8 ⟨70JOC2319⟩, 105 ⟨64BBR(17)146⟩; **3**, 6, 118 ⟨63JA3027⟩, 373 ⟨68JA740⟩; **4**, 554 ⟨68TL5731⟩; **5**, 910 ⟨68TL5731⟩; **6**, 71 ⟨73JA4016⟩
Moffatt, M. E., **1**, 899 ⟨63JA1328⟩; **3**, 282 ⟨63JA1328⟩
Moffett, R. B., **2**, 395 ⟨63OSC(4)652⟩; **3**, 459 ⟨55OSC(3)562⟩; **4**, 189 ⟨63OSC(4)427⟩
Mogami, T., **6**, 498 ⟨90SC703⟩
Mogel, L., **4**, 989 ⟨82JPR309⟩
Mögelin, W., **4**, 463 ⟨80ZAAC(469)109⟩; **6**, 491 ⟨85ZAAC(530)16⟩
Mogendorf, K.-D., **3**, 578 ⟨88JOC4997⟩
Moggi, G., **2**, 175 ⟨70TL1313⟩
Mogias, J., **6**, 237 ⟨92ZAAC(608)69⟩
Mohacsi, A., **5**, 132 ⟨82SC453⟩
Mohacsi, E., **2**, 684 ⟨83SC827⟩; **4**, 271 ⟨62JA684⟩; **5**, 727 ⟨84SC1159⟩
Mohai, B., **2**, 572 ⟨74ZAAC(412)11⟩
Mohajer, D., **2**, 215 ⟨88BCJ2185⟩; **3**, 88, 118 ⟨86SC211⟩, 102 ⟨88BCJ2185, 86SC723⟩
Mohamadi, F., **1**, 68 ⟨83JA6882, 86JA2094⟩; **2**, 40 ⟨83JA6882, 86JA2094⟩
Mohan, J. J., **2**, 528 ⟨91JOC751⟩, 560 ⟨91JOC763⟩; **6**, 446 ⟨91JOC751⟩
Mohan, L., **2**, 397 ⟨89JOC5783⟩
Mohan, R. M., **1**, 346, 926 ⟨93JOC7640⟩
Mohandas, T. P., **5**, 15 ⟨83T1187⟩
Mohanraj, S., **1**, 482 ⟨83TL2723⟩
Mohanty, S., **3**, 282 ⟨88MI 306-01⟩
Mohareb, R. M., **3**, 653 ⟨83S478⟩, 669,672 ⟨87H(26)497⟩; **5**, 365 ⟨83S478⟩, 631 ⟨87S228⟩
Moharir, Y. E., **5**, 1026 ⟨75JIC148⟩
Mohedano, J. M., **4**, 405 ⟨87CJC687⟩
Mohialdin-Khaffof, S. N., **2**, 15 ⟨92CC595⟩
Möhlau, R., **3**, 363 ⟨02CB375⟩
Mohmand, S., **3**, 335, 369 ⟨82JA3119⟩, 337 ⟨77CC287, 82JA312⟩, 337, 351 ⟨82CB492⟩, 358 ⟨83CB273⟩, 527 ⟨77AG(E)105⟩; **4**, 258 ⟨82JA3119⟩
Mohnke, K., **5**, 753 ⟨42CB1585⟩
Mohr, G., **4**, 900 ⟨75CB174⟩
Mohr, J.-T., **5**, 129 ⟨94JOC1160⟩
Mohr, K., **4**, 453, 454 ⟨73ZC139⟩
Mohr, P., **5**, 135, 138 ⟨83HCA2501⟩
Mohr, R., **4**, 229 ⟨84AG(E)890⟩, 447 ⟨85S973⟩
Mohraz, M., **3**, 330 ⟨83HCA801⟩
Mohri, A., **4**, 478 ⟨82PS(13)85⟩
Mohri, M., **1**, 497 ⟨92SL721⟩
Mohri, S., **5**, 223 ⟨90CPB1836⟩
Mohri, T., **2**, 1035 ⟨88TL6409⟩
Möhrle, H., **4**, 416 ⟨76AP503⟩; **5**, 215 ⟨86AP(319)325⟩, 289 ⟨69T5745⟩; 566, 590, 606, 609, 610 ⟨72T199⟩
Mohrmann, K. H., **6**, 606 ⟨85GEP3337939⟩
Mohtacheni, R., **4**, 708, 726 ⟨84ZN(B)798⟩
Mohtasham, J., **6**, 251 ⟨91JFC(55)313⟩, 252 ⟨90IC4588⟩
Mohyla, J., **6**, 584 ⟨85CB227⟩
Moilliet, J. S., **4**, 737 ⟨78CC475⟩
Moimas, F., **5**, 1078 ⟨85S509⟩
Moine, G., **4**, 959 ⟨84JA6097⟩
Moingeon, M., **3**, 146 ⟨86TL4175⟩
Moir, R. Y., **4**, 391 ⟨65JA5498⟩
Moisak, I. E., **2**, 25, 31 ⟨69JOU971⟩
Moise, C., **3**, 83 ⟨85JOM(292)225⟩
Moiseenkov, A. M., **2**, 11 ⟨90OPP215⟩
Moiseeva, N. I., **6**, 410 ⟨89BAU1570⟩
Moiseeva, O. A., **2**, 834, 835, 840 ⟨78ZOB232⟩
Moiseev, I. I., **6**, 410 ⟨89BAU1570⟩
Moiseev, I. K., **6**, 145 ⟨76KFZ35⟩
Moiseev, S. K., **4**, 723 ⟨80JOM(185)C6, 82JOM(224)C53⟩
Moiseichuk, K. L., **2**, 1098 ⟨85JGU1691, 86JGU1812⟩; **5**, 1110 ⟨90MI 530-01⟩
Moitghinejad, E., **2**, 215 ⟨89S378⟩
Mojarrad, F., **2**, 1003 ⟨91T5277⟩
Mojé, S., **1**, 516 ⟨66JA4964⟩, 1005 ⟨69JA6464⟩
Moje, S., **2**, 621 ⟨69JA6464⟩; **3**, 621 ⟨66JA4964⟩
Mokhtar, R., **2**, 13 ⟨86CJC138⟩
Mokhtary, P. B., **5**, 631 ⟨76S728⟩
Moklyschuck, L. I., **5**, 58 ⟨85ZOR769⟩
Mokrushin, V. S., **5**, 567, 571 ⟨87KGS845⟩
Mokry, L. M., **6**, 704 ⟨94AG(E)578⟩
Mol, J. C., **1**, 674 ⟨77CC198, 81CC1081⟩
Molan, S., **5**, 764 ⟨76JHC487⟩
Molander, A. G., **3**, 228 ⟨77JOC3106⟩
Molander, G. A., **1**, 2, 7, 8, 13, 15, 16, 17, 23, 24 ⟨86JOC1135⟩, 14 ⟨86JOC2596⟩, 216 ⟨89JOC3114⟩, 360 ⟨86JA4683⟩, 449, 510 ⟨86JOC4512⟩, 885 ⟨91COS(1)251⟩, 937 ⟨92CRV29⟩, 1018 ⟨81JOC645⟩, 1037 ⟨85JOC1577⟩, 1043 ⟨79S104⟩, 1050 ⟨77JA954⟩; **2**, 42 ⟨94JOC2676⟩, 69, 564 ⟨90JOC4990⟩, 81 ⟨89JA8236, 91COS(1)251⟩, 564 ⟨92CRV29, B-89MI 211-02⟩, 565 ⟨91JOC4112⟩, 925, 973 ⟨83JOC5409⟩, 969 ⟨92JOC3266⟩; **3**, 129 ⟨86JOC1135⟩, 130 ⟨93JOC7216⟩, 147 ⟨77JA955⟩, 160 ⟨86JOC2596⟩, 163 ⟨93JA5821⟩, 228 ⟨84JOC5024⟩; **6**, 133 ⟨89JOC4042⟩
Molchanova, G. N., **3**, 540 ⟨84ZOB1817⟩; **4**, 1057 ⟨84ZOB1817⟩
Molchanova, N. R., **5**, 334 ⟨81ZOR533⟩
Molchanov, A. P., **4**, 1082 ⟨89UK1122⟩
Moldes, I., **4**, 689 ⟨95JOM(489)C65⟩, 711 ⟨86JOM(315)C22⟩, 713 ⟨90JOM(395)305⟩
Moldowan, J. M., **1**, 630 ⟨85JOC5460⟩
Mole, S. J., **4**, 662, 962 ⟨86OM1906⟩
Mole, T., **1**, 186 ⟨72CC595⟩; **2**, 478 ⟨74JOM(74)365⟩, 595 ⟨B-72MI 211-01⟩, 1096 ⟨68JOM(11)393⟩; **4**, 699 ⟨70AJC715⟩
Molenaar, E., **3**, 574 ⟨76RTC154⟩; **4**, 777 ⟨68RTC49⟩
Molenaar, E. J. F., **1**, 929 ⟨68RTC1301⟩
Molenda, R. P., **2**, 432, 475 ⟨74MI 208-01⟩
Molina, A., **3**, 171 ⟨92TL3435⟩
Molina, G. A., **1**, 36, 442 ⟨73JCS(P1)541⟩
Molina, G., **3**, 530 ⟨82JOC3871⟩; **5**, 282 ⟨82JOC3876⟩
Molina, J., **6**, 469 ⟨88CJC2234⟩
Molina, J. M., **4**, 858 ⟨93JCR(S)60⟩
Molina, M. T., **1**, 827 ⟨89JOC3137⟩
Molina, P., **1**, 835 ⟨91TL4041⟩; **2**, 11 ⟨80S853⟩, 121 ⟨85TL469⟩; **3**, 585 ⟨90CC829, 91JOC4008⟩, 623 ⟨91TL4041, 93T5153⟩; **5**, 220, 862 ⟨86S843⟩, 350 ⟨93H(36)1263⟩, 622, 1067 ⟨89TL2847⟩, 680 ⟨89JCR(S)262, 89S843⟩, 711 ⟨85SC643⟩, 975, 1031 ⟨82S596⟩, 1065 ⟨93JOC5264⟩, 1066 ⟨87S45, 88JOC4654⟩, 1066 ⟨88TL379, 89CB307, 89T4263, 90CC7, 90JCS(P1)1727, 94TL3817⟩, 1067, 1079 ⟨90T1063⟩, 1068 ⟨89JA355, 82SC573⟩, 1107 ⟨82S472⟩, 1122 ⟨84TL5427⟩; **6**, 504 ⟨92S297⟩, 560 ⟨89S923⟩, 563 ⟨88S690⟩
Molina, R., **5**, 680 ⟨88H(27)1935⟩
Molina, Y., **4**, 300 ⟨88JCS(P1)1631⟩, 412 ⟨88CB1813⟩
Molina-Buendía, P., **5**, 1002 ⟨76TL2691, 79JCS(P1)1957⟩
Molin, H., **6**, 491 ⟨89CAR(194)209⟩
Molin, M., **1**, 413 ⟨73JOM(49)477⟩, 809 ⟨73JOM(49)483⟩; **2**, 200 ⟨74JA935⟩; **3**, 87 ⟨74JA935⟩
Molina Molina, J., **4**, 858 ⟨88CJC2975⟩
Molinari, A., **6**, 635 ⟨88S460⟩
Molinari, F., **2**, 698 ⟨92JOC5173⟩
Molinari, H., **5**, 607 ⟨85T4929⟩, 946 ⟨91OM807⟩
Molinero, E. G., **5**, 773 ⟨82TL3357⟩
Molines, H., **4**, 731 ⟨87JOC290⟩, 736 ⟨92JOC5530⟩
Molins, E., **6**, 30 ⟨91JOC103⟩

Molla, E., **6**, 387 ⟨89JCS(D)447⟩
Moll, E., **4**, 493 ⟨71CB1826⟩
Moll, F., **4**, 98 ⟨86AP(319)954⟩
Moll, M., **2**, 491 ⟨86ZAAC(539)50⟩
Moll, R., **4**, 215 ⟨74JPR304⟩; **6**, 651 ⟨79ZC250⟩
Mollbach, A., **4**, 583 ⟨72IJ293⟩
Mollenkopf, M., **5**, 355 ⟨86HCA333⟩
Moller, A., **4**, 989 ⟨86TL4281⟩
Möller, A., **5**, 357 ⟨85JPR(327)893⟩
Moller, A., **5**, 624, 625 ⟨84ZC99⟩
Möller, F., **1**, 573 ⟨72S591⟩; **2**, 347, 348 ⟨58HOU(11)190⟩
Möller, G., **6**, 729 ⟨94CB597⟩
Möller, J., **6**, 560 ⟨83JCS(P1)2011⟩
Moller, K. E., **5**, 27 ⟨63AG(E)719⟩
Möller, M., **3**, 247 ⟨76AG(E)382⟩; **4**, 112 ⟨88JOM(344)61⟩; **5**, 344 ⟨88CB689⟩, 639 ⟨88CB701⟩
Möller, M. H., **3**, 459 ⟨93CB779⟩
Möller, T., **2**, 577 ⟨92CB157⟩
Mollier, Y., **3**, 254 ⟨70BSF2018, 70TL1259⟩, 363 ⟨69BSF827⟩, 366 ⟨71CR(C)278, 74BSF(2)471⟩; **5**, 559 ⟨72CR(C)1215⟩, 584, 586, 614 ⟨73BSF3349⟩; **6**, 637, 732 ⟨86NJC51⟩
Mollin, J., **5**, 630 ⟨87CCC2087⟩
Mollov, N. M., **5**, 345 ⟨90SC2083⟩
Molloy, K.,**1**, 522 ⟨91JCS(P1)1255⟩
Molloy, K. C., **4**, 268 ⟨91JCS(P1)662⟩, 269 ⟨90TL135⟩, 860 ⟨92T7551⟩; **5**, 340, 347 ⟨93JCS(P1)1581⟩; **6**, 395 ⟨90JCS(D)2643⟩
Molloy, R. M., **5**, 48 ⟨69JOC1454⟩
Molodtsov, S. S., **2**, 888 ⟨72JGU1952⟩; **6**, 476 ⟨76JGU2441⟩
Molodykh, Z. V., **6**, 634 ⟨90ZOB222⟩
Moloney, M. G., **1**, 490 ⟨88JCS(P1)2847⟩, 946 ⟨92T9373⟩, 1145 ⟨89JCS(P1)333⟩; **2**, 230 ⟨91T5603⟩
Moloy, K. G., **3**, 598 ⟨86JA56⟩
Moltzen, E. K., **2**, 478 ⟨86ACS(B)609⟩; **3**, 355, 356 ⟨84AG(E)729⟩, 550, 551 ⟨91JOC1317⟩; **5**, 550 ⟨87JOC1156⟩; **6**, 235, 535 ⟨84JOC3854⟩, 435 ⟨86SUL93, 84JA263⟩, 535 ⟨86ACS(B)609, 86SUL203⟩, 555 ⟨86ACS(B)593⟩
Molyneaux, J. M., **4**, 843 ⟨89JHC1771⟩
Molz, T., **1**, 1077 ⟨86T1711⟩
Momchilova, S., **1**, 644 ⟨88PS(40)69⟩
Momenteau, M., **2**, 578 ⟨90JA1874⟩
Mommers, A. A., **1**, 1141 ⟨84JA521⟩
Mommers, S., **1**, 1037 ⟨81S459⟩
Momose, T., **4**, 165 ⟨84SC147⟩; **5**, 199 ⟨87CPB3453⟩
Momot, V. V., **4**, 1004 ⟨86ZOB805⟩; **5**, 1076 ⟨86ZOR1099⟩
Momtchev, M., **4**, 147 ⟨80MI 403-01⟩
Monagle, J. J., **3**, 455 ⟨56JA4470⟩; **6**, 643 ⟨62JA3673⟩
Monahan, L. C., **1**, 1130 ⟨88TL391⟩
Monahan, R., **3**, 229 ⟨87TL1069⟩
Monari, M., **6**, 332 ⟨86G101⟩
Moncur, M. V., **1**, 865 ⟨73JA6449⟩
Mond, L., **6**, 522, 524 ⟨10JCS798⟩
Mondello, C., **1**, 876 ⟨78JOC2705⟩
Monden, M., **1**, 180 ⟨87CC358⟩
Mondeshka, D., **5**, 344 ⟨89SC3113⟩
Mondet, J. C., **6**, 297 ⟨75FRP2235188⟩
Mondon, A., **1**, 721 ⟨57LA(603)115⟩; **3**, 493 ⟨57LA(603)115⟩
Mondovi, B., **2**, 211 ⟨63MI 203-01⟩
Mondt, J. L., **2**, 213 ⟨70TL1203⟩
Monera, O. D., **1**, 637 ⟨89JOC5424⟩
Money, T., **1**, 381 ⟨74JCS(P1)1938, 76JCS(P1)359⟩, 815 ⟨88CC1281⟩; **4**, 20 ⟨75JCS(P1)251⟩
Monforte, J., **3**, 9, 123 ⟨85TL705⟩
Monge, A., **4**, 583 ⟨84OM1438⟩, 897 ⟨85JHC1445⟩; **5**, 199 ⟨88AQ120⟩, 334 ⟨85JHC1353, 88JHC23⟩, 948 ⟨92JA7288, 94JA791⟩; **6**, 327 ⟨89IC2120, 92JCS(D)2307⟩, 327, 330 ⟨93IC5569⟩
Mongelli, N., **5**, 1107 ⟨78CC133⟩
Mongeot, H., **1**, 658 ⟨70BSF2505⟩
Monick, J. A., **2**, 38, 87 ⟨B-68MI 202-01⟩
Monin, E. A., **4**, 548 ⟨87ZOB2637⟩, 550 ⟨85ZOB331⟩, 552 ⟨87ZOB949⟩, 654 ⟨92MI 414-01⟩, 6540 ⟨92MIP173178⟩; **5**, 1092 ⟨87JGU2350⟩
Moniot, J. L., **3**, 200 ⟨84TL4741⟩
Moniotte, P., **5**, 337 ⟨90TL873⟩
Monkiewicz, J., **1**, 159 ⟨84JOC1522⟩; **3**, 197, 231 ⟨80TL2287⟩
Monkovic, I., **2**, 478 ⟨61CI(L)469⟩
Monn, J. A., **4**, 107 ⟨89TL911⟩
Monneret, C., **6**, 550 ⟨88CAR(181)253⟩
Monno, S., **5**, 1125 ⟨86JAP1022060⟩
Monobe, H., **3**, 286 ⟨86CL449⟩
Monrad, M. S., **2**, 398 ⟨84SC1093⟩
Monreal, M. C. F., **3**, 170 ⟨89JOC5620⟩
Monro, A. M., **6**, 642 ⟨65JMC446⟩
Monsef-Mirzai, Z., **2**, 280 ⟨83JCS(D)2143⟩
Monso, J. M., **2**, 1010 ⟨71JA2215⟩
Montagne, M., **5**, 62 ⟨B-45MI 502-01⟩
Montaña, A.-M., **1**, 529 ⟨85T5995⟩
Montanari, F., **1**, 85 ⟨74JOC3195⟩; **2**, 9 ⟨75CC950⟩, 9, 35 ⟨74CC879⟩, 19 ⟨74S428⟩, 25 ⟨74S37⟩, 32 ⟨78S771, 79TL5055⟩, 147 ⟨61TL607, 63JCS4864, 68JCS(C)1317⟩; **3**, 4 ⟨79S134⟩, 7 ⟨87JOC2559⟩, 31 ⟨89JOC2970⟩, 238 ⟨69JCS(C)1204⟩, 661 ⟨79JOM(173)335⟩; **4**, 75 ⟨72JCS(P1)1886⟩, 76 ⟨73JA7431⟩, 233 ⟨70CC1441⟩, 770 ⟨57CA(51)12849g, 57CA(51)5721f⟩, 770, 772 ⟨57G1061⟩
Montanari, S., **3**, 169 ⟨87S1126⟩
Montanari, V., **3**, 120 ⟨92TL7245⟩
Montanucci, M., **2**, 278, 729 ⟨83JOC4289⟩, 709 ⟨83JOC4795⟩, 711 ⟨83S751⟩
Montason, M., **3**, 71 ⟨59HCA844⟩
Montaudon, E., **5**, 206 ⟨85T4047⟩
Montavon, M., **1**, 156 ⟨57HCA1250⟩
Monteagudo, M., **2**, 103 ⟨92SC677⟩
Montecalvo, D. F., **1**, 954 ⟨73TL3181⟩
Monteil, R. L., **5**, 716 ⟨78JCS(P1)1338⟩
Monteiro, H. J., **3**, 170 ⟨75TL921⟩, 193 ⟨87SC983⟩, 249 ⟨75S437, 77JOC2324⟩
Montemayor, R. G., **2**, 409 ⟨73IC2482⟩
Montenarh, M., **6**, 512 ⟨77CB2368⟩
Montero, N., **2**, 104 ⟨91BSF222⟩
Montes de Lopez-Cepero, I., **1**, 759 ⟨89OS(67)125⟩
Montevecchi, P. C., **3**, 166 ⟨88TL2381⟩; **5**, 561 ⟨76JOC2639⟩
Montgomery, A. M., **3**, 698, 707 ⟨83TL655⟩
Montgomery, C. D., **1**, 157 ⟨85IC3680⟩; **6**, 516 ⟨89CCR1⟩
Montgomery, D., **1**, 828 ⟨91TL4199⟩
Montgomery, H. A. C., **2**, 413 ⟨58JCS1963⟩
Montgomery, J. A., **3**, 357, 362 ⟨76JOC3784⟩; **4**, 227 ⟨92JMC533⟩; **6**, 109 ⟨65JOC829⟩, 504, 506 ⟨63JMC669⟩
Montgomery, J., **3**, 169, 172 ⟨93JOC6476⟩; **5**, 953 ⟨90JA6255⟩
Montgomery, L. K., **1**, 321, 923 ⟨74JA2201⟩; **2**, 1007 ⟨87JA4749⟩
Montgomery, R. E., **2**, 833, 834, 840, 841 ⟨63JOC3315⟩
Monti, D., **1**, 77 ⟨87T4481, 88T1299⟩; **3**, 18, 134 ⟨86SC803⟩, 58 ⟨87T4481⟩
Montijn, P. P., **1**, 955, 1125 ⟨65RTC31⟩, 986 ⟨67RTC129⟩, 1030 ⟨68RCR789⟩, 1072 ⟨64RTC456⟩; **2**, 1043 ⟨64RTC1211⟩; **3**, 61 ⟨65RTC31⟩
Montillier, J. P., **2**, 151, 688 ⟨76JOC3987⟩, 221 ⟨70JA7224⟩
Montoya, J., **5**, 455 ⟨90JOM(390)C57⟩
Montserrat, J. M., **1**, 466 ⟨93JOC5976⟩; **5**, 946, 958 ⟨93CC1068⟩
Montury, M., **1**, 459 ⟨80TL51⟩; **2**, 940 ⟨80TL163⟩
Monzu, J.-P., **1**, 842 ⟨73TL3459⟩
Moodie, R. B., **2**, 99 ⟨57CI(L)1622, 58JCS2372⟩
Moody, C. J., **1**, 233, 235 ⟨88TL6005⟩, 400 ⟨90T6501⟩,

825 ⟨87AHC(42)203⟩; **2**, 159, 401 ⟨77CRV409⟩, 160 ⟨76JCS(P1)2166⟩, 1000 ⟨88TL6009⟩; **3**, 106 ⟨87JCS(P1)931⟩, 166 ⟨90T6501⟩; **4**, 539 ⟨90SL413, 91JCS(P1)323⟩, 856 ⟨86JCS(P1)501⟩; **5**, 351 ⟨85TL3171⟩, 354 ⟨85TL3167⟩, 655, 657 ⟨87JCS(P1)1389⟩, 764 ⟨76JCS(P1)2166⟩, 766 ⟨75JCS(P1)1964⟩, 767 ⟨77CRV409⟩; **6**, 119 ⟨92SL975, 82AHC(30)1⟩, 628, 636 ⟨94T3195⟩
Moody, C. M., **1**, 486 ⟨93TL4667⟩
Moody, H. M., **3**, 40 ⟨94TL1777⟩
Moody, R. J., **1**, 163 ⟨77JA3196⟩, 163 ⟨82OM20⟩; **2**, 905 ⟨75JA5608⟩; **4**, 634 ⟨77JA3196⟩, 634, 664 ⟨82OM20⟩; **6**, 181, 199 ⟨75JA5608⟩, 199 ⟨80JOC1091⟩, 199 ⟨82OM20⟩
Mooiweer, H. H., **1**, 975 ⟨84TL5571⟩; **2**, 324, 326 ⟨87TL3285⟩; **4**, 106 ⟨88TL6975⟩, 298 ⟨87TL3285⟩
Mooji, J. J., **1**, 890 ⟨74AX(B)833⟩
Mook, H. W., **4**, 38 ⟨28JCS2125⟩
Mook, Jr., R., **1**, 338, 1127, 1129 ⟨86TL4529⟩, 548 ⟨88OS(66)75⟩, 548, 1127, 1131, 1132 ⟨87JA2829⟩, 548, 1132 ⟨83JA3720⟩; **4**, 175 ⟨83JA3720⟩
Mook, R., **1**, 342 ⟨83JA3741⟩
Mookherjee, B. D., **3**, 208 ⟨71JOC4124⟩
Moon, C.-W., **2**, 926 ⟨92JOC3178⟩
Moon, H. W., **6**, 608, 611, 653 ⟨72JOC2005⟩
Moon, W. M., **2**, 805 ⟨72JOC383⟩
Moon, Y. C., **2**, 356 ⟨90JA311⟩
Mooney, B. A., **2**, 348 ⟨83AJC719⟩
Mooney, E. F., **2**, 418, 419 ⟨62JCS1786⟩, 520 ⟨63JCS1582⟩
Moons, C., **2**, 155 ⟨80BSF411⟩
Mooradian, A., **2**, 692 ⟨67TL407⟩; **3**, 100 ⟨47JA1998⟩
Moore, A. J., **4**, 847, 848 ⟨91S26⟩, 870 ⟨90CC470, 90PAC473⟩, 873 ⟨90CC816⟩; **6**, 120, 123 ⟨90CC470⟩, 120, 123 ⟨91JCS(P1)157⟩, 698 ⟨91S26, 93PS(74)279⟩
Moore, A. T., **5**, 90 ⟨73OSC(5)586⟩
Moore, B. D., **3**, 514 ⟨87CC129⟩
Moore, C., **1**, 883 ⟨89CC161⟩; **4**, 237 ⟨88JCS(P1)2663⟩
Moore, C. J., **1**, 68 ⟨86CC855⟩
Moore, D. R., **3**, 547 ⟨71CJC1456⟩
Moore, D. W., **1**, 543 ⟨85JOC2124⟩; **3**, 429 ⟨85H(50)5123⟩; **4**, 295, 407, 412 ⟨73JOC3288, 74JOC1349⟩, 444 ⟨76JOC3221⟩, 736 ⟨68JOC2173⟩; **6**, 54 ⟨66JOC789⟩
Moore, E. J., **3**, 311 ⟨92JA5888⟩
Moore, G. A., **5**, 264 ⟨85JCS(P1)461⟩
Moore, G. G., **3**, 281 ⟨70JOC2351⟩; **5**, 3 ⟨69JOC2486⟩
Moore, G. G. I., **2**, 179 ⟨79JOC1708⟩; **6**, 45 ⟨79JOC1708⟩
Moore, G. J., **1**, 426 ⟨64JOC3042⟩; **4**, 799 ⟨66JCS(C)2304⟩; **6**, 63 ⟨75JFC(5)475⟩
Moore, G. L., **1**, 1006, 1045 ⟨75JOC131⟩; **3**, 273 ⟨75JOC131⟩; **6**, 74 ⟨73USP3745220⟩
Moore, H. W., **1**, 571 ⟨88OSC(6)210⟩; **2**, 669 ⟨91JOC4048⟩; **3**, 275 ⟨87JOC2537⟩, 539 ⟨76OR(23)1, 81CSR289, 85JA3392, 86JOC2814, 86JOC3067, 91JOC6094⟩, 539, 543 ⟨84JOC2190, 76JA3728⟩, 543 ⟨78TL929⟩, 546 ⟨70JA4132, 71JA2812⟩, 653 ⟨84CC1066⟩; **4**, 304 ⟨85JOC4231⟩; **5**, 367 ⟨B-83MI 507-01⟩
Moore, I., **4**, 712 ⟨81CC1255, 84JCS(D)1563⟩, 712 ⟨84JCS(D)1581⟩
Moore, J. A., **1**, 393 ⟨50JA2884⟩; **3**, 465 ⟨62OS(41)16⟩, 468 ⟨54JA2884, 54JA2887⟩; **4**, 742, 979 ⟨93MM916⟩; **5**, 332, 333 ⟨91JMC387⟩, 359 ⟨91TL3449⟩
Moore, J. L., **1**, 697, 700, 703 ⟨86JOC3916⟩
Moore, J. S., **5**, 314 ⟨90JOC3374⟩
Moore, J. W., **4**, 698 ⟨71JA1035⟩
Moore, K. M., **2**, 148 ⟨77S791⟩
Moore, K. W., **4**, 834 ⟨83TL3373⟩
Moore, L. L., **5**, 652 ⟨73OSC(5)687⟩
Moore, L. O., **4**, 31 ⟨70JOC4201⟩
Moore, M. A., **2**, 381 ⟨61JOC2740⟩
Moore, M. F., **2**, 410 ⟨85JOC2019⟩
Moore, M. L., **2**, 310 ⟨49OR(5)301⟩
Moore, M. W., **1**, 542, 543 ⟨81JOC4093⟩; **2**, 975 ⟨81JOC4093⟩
Moore, P. T., **1**, 627 ⟨61JA3861⟩
Moore, R. E., **1**, 811 ⟨71JA3087⟩
Moore, T., **2**, 94 ⟨86ICA179⟩
Moore, W. H., **2**, 717 ⟨79JOC3847⟩
Moore, W. M., **6**, 143 ⟨87ZPK1829⟩
Moore, W. R., **1**, 624, 628 ⟨57JA4720⟩, 627 ⟨61JA3861⟩; **4**, 24 ⟨70JA5469⟩, 755, 782 ⟨92JA360⟩
Moorehead, A. W., **1**, 907 ⟨B-87MI 119-08⟩
Moores, C. J., **4**, 419 ⟨91TL2683⟩
Moorhoff, C. M., **3**, 231 ⟨87TL559⟩
Moormann, A. E., **2**, 310 ⟨93SC789⟩
Moormeier, L. F., **5**, 42 ⟨61JOC307⟩
Moors, R. G., **1**, 629 ⟨89JA2258⟩
Moorthy, K. E., **1**, 556 ⟨83TL555⟩
Moorthy, S. N., **1**, 459 ⟨75S194⟩
Moos, J., **6**, 44 ⟨58LA(616)1⟩
Moosavipour, H., **3**, 4 ⟨86S285⟩
Moos, W. H., **5**, 306 ⟨90JOC560⟩
Moosmayer, A., **1**, 922 ⟨67TL3877⟩
Moosmuller, F., **5**, 1025 ⟨54LA(585)230⟩
Mootoo, D. R., **4**, 176 ⟨88JA2662⟩, 199 ⟨88TL6549⟩
Mootz, D., **2**, 861 ⟨91ZAAC(606)119⟩; **4**, 553 ⟨92ZN(B)725⟩
Mora, J., **3**, 427 ⟨84TL3493⟩
Moradpour, A., **1**, 570 ⟨93JOC4685⟩; **4**, 843 ⟨87TL3805⟩; **6**, 627 ⟨89S132⟩
Moraes, F., **4**, 291 ⟨85JA675⟩
Morales, C. B., **4**, 1048, 1058, 1059, 1060, 1062 ⟨86OM2274⟩
Morales, O., **1**, 620, 1074 ⟨71CC22⟩
Morales-Rios, M. S., **4**, 245 ⟨91SC1369⟩
Moralev, S. N., **2**, 1033 ⟨91RCR885⟩
Moran, D., **4**, 737 ⟨80JCS(P1)487⟩
Moran, E. J., **3**, 717 ⟨91TL3807⟩
Moran, H. W., **1**, 1114 ⟨70JA1082, 71JA6989⟩; **2**, 168 ⟨84JMC1379⟩
Moran, J. R., **1**, 367 ⟨84BSB511⟩, 841 ⟨85TL1057⟩; **4**, 232 ⟨90T1783⟩, 266 ⟨89AQ22⟩
Morán, J. R., **5**, 554 ⟨90T1057⟩
Moran, S., **1**, 154 ⟨87JA5996⟩, 155 ⟨87JA6004⟩
Moran, T. A., **1**, 754 ⟨76JCS(P1)2386, 78JCS(P1)590⟩
Morand, P., **3**, 4, 116 ⟨86TL1775⟩, 60 ⟨87JOC4047⟩; **4**, 175 ⟨91JOC1788⟩
Morandini, F., **1**, 195, 238 ⟨86T2043⟩; **5**, 940 ⟨86OM1976⟩
Morath, R. J., **4**, 319 ⟨52JA3885, 55JA3869⟩
Moravskii, A. P., **4**, 1080 ⟨83ZOB416⟩
Morbach, W., **3**, 495 ⟨77S679⟩; **5**, 894 ⟨77AG(E)180⟩; **6**, 265 ⟨77S699⟩
Morbitzer, L., **6**, 471 ⟨91AG(E)1598⟩
Morck, H., **3**, 355 ⟨79S361⟩
Mordenti, L., **1**, 458, 515 ⟨77TL1069⟩
Mordini, A., **2**, 140 ⟨91JOC7323⟩, 541 ⟨93TL873⟩, 918 ⟨91S1201⟩, 935 ⟨92SL883⟩; **3**, 34 ⟨92SL803⟩, 336, 340, 346, 363 ⟨91JOC7323⟩, 346, 353 ⟨93TL873⟩; **4**, 944 ⟨92SL499, 92SL883⟩, 951 ⟨91TL1899⟩; **5**, 427 ⟨92SL332, 92TL1507⟩, 427, 428 ⟨92SL329⟩
Mordvintseva, E. A., **6**, 613 ⟨70JGU540⟩
More, K. M., **4**, 61, 63, 73 ⟨77S791⟩
More, P. G., **3**, 647 ⟨89JOM(371)397⟩
Moreau, B., **1**, 143 ⟨78JA1558⟩
Moreau, J., **1**, 1051, 1053, 1054, 1055, 1058, 1060 ⟨B-80MI 121-02⟩, 1055 ⟨70BSF2175⟩, 1059 ⟨71BSF3071⟩; **3**, 131 ⟨78TL3713⟩
Moreau, J. J. E., **1**, 160 ⟨93JOM(451)C1⟩, 160 ⟨93TL2111⟩; **2**, 814 ⟨81JOC3372, 83JOM(255)359⟩, 814 ⟨85OM623⟩, 946 ⟨72JOM(40)73⟩; **4**, 511 ⟨83JOM(259)283⟩, 606 ⟨93JOM(451)C1⟩, 607, 608, 654 ⟨93TL2111⟩
Moreau, J.-L., **1**, 967 ⟨76JOM(108)159⟩; **5**, 223 ⟨85JOM(294)139⟩

Moreau, P., **1**, 829 ⟨91TL5525⟩, 1023 ⟨67CR(C)839⟩, 1130 ⟨82CC433⟩; **4**, 812, 820 ⟨82JFC(20)699, 76CC174⟩; **5**, 58 ⟨76CC885⟩
Moreau, R.-C., **5**, 584, 586, 612 ⟨65BSF3623⟩, 586 ⟨65BSF3628⟩, 611 ⟨66CR(C)788⟩, 753 ⟨64CR4067⟩; **6**, 644 ⟨66BEP672205⟩
Morehouse, S. M., **5**, 229 ⟨64CI(L)544⟩
Morel, A., **5**, 83 ⟨B-41MI 502-01⟩
Morel, C., **1**, 504 ⟨90TL5749⟩
Morel, D., **2**, 3 ⟨87TL5505⟩, 21 ⟨92TL495⟩
Morel, G., **2**, 744 ⟨82TL2023⟩; **3**, 557, 573 ⟨77TL3353⟩, 567 ⟨85JOC771⟩, 567, 574 ⟨82T527, 77TL3349⟩, 574 ⟨78CC274, 78TL3719⟩; **4**, 894 ⟨77TL3349⟩; **5**, 728 ⟨82TL2023, 84T1075⟩, 728 ⟨85JOC771⟩; **6**, 571 ⟨90JOC1721⟩, 617 ⟨84T1075, 85JOC771, 92TL6155⟩, 635 ⟨86JOC4043, 89JOC1185⟩
Morel, J., **3**, 104, 105, 106 ⟨69BSF2511⟩, 105, 106 ⟨76BSF265⟩, 305 ⟨93ACS287⟩
Moreland, C. G., **2**, 492 ⟨68IC834⟩
Moreland, D. W., **1**, 876 ⟨85JA2264⟩
Moreland, S., **4**, 905 ⟨90JMC1510⟩
Morel, S., **3**, 280 ⟨91TL6709⟩
Morel, Y., **1**, 474 ⟨90BSF401⟩
Morella, A. M., **2**, 24 ⟨84TL1197⟩
Morell, W., **6**, 405 ⟨77ZN(B)1022, 79ZN(B)390⟩
Morellet, G., **2**, 650 ⟨86BSF625⟩
Morely, J. O., **2**, 657, 753 ⟨76S528⟩
Moreno, M., **1**, 107 ⟨92JA9576⟩
Moreno Lopez, M. J., **6**, 70 ⟨84MIP521372⟩
Moreno-Mañas, M., **2**, 324 ⟨92SC1249⟩
More O'Ferrall, R. A., **4**, 1075 ⟨87MI 424-01⟩
Morera, E., **1**, 433 ⟨86TL5541⟩, 435 ⟨84TL4821, 90OS(68)138⟩, 435 ⟨90SC1293⟩, 489 ⟨92TL4815⟩, 1029 ⟨86S320⟩, 1030 ⟨91TL6449⟩, 1099 ⟨84TL2271⟩; **5**, 296, 301 ⟨86TL3931⟩, 331 ⟨85TL1109⟩
Morére, A., **5**, 388 ⟨92MI 507-01⟩, 389 ⟨92JOM(423)C5, 92PS(70)75⟩
Moret, E., **1**, 113 ⟨93T10175⟩
Moret, M., **5**, 946, 959 ⟨94ICA233⟩
Moretó, J. M., **1**, 106 ⟨93JOM(451)1⟩, 527 ⟨85TL6397, 87TL4745⟩; **2**, 624 ⟨88TL5811⟩, 684 ⟨82S186⟩; **3**, 218 ⟨85TL6397, 87TL4745⟩; **4**, 810 ⟨88TL5811⟩
Moretti, I., **2**, 337 ⟨83JOC2640⟩; **3**, 406 ⟨70S141⟩
Moretti, R., **1**, 120 ⟨88JA2641⟩; **5**, 342 ⟨89TL6009⟩; **6**, 134 ⟨89TL15⟩
Morey, M. C., **4**, 176 ⟨81TL4603⟩, 184 ⟨81JOC2419⟩
Morf, D., **4**, 444 ⟨58CB660⟩
Morgan, A. D., **3**, 14 ⟨80JCS(P1)756⟩; **6**, 79 ⟨80JCS(P1)756⟩
Morgan, B., **5**, 334 ⟨85JMC1661⟩
Morgan, C. D., **2**, 188 ⟨70JCS(C)270⟩
Morgan, D. R., **2**, 230 ⟨91T5329⟩
Morgan, E. D., **1**, 1000 ⟨61T18⟩; **5**, 559 ⟨67T3785⟩
Morgan, G. H., **4**, 614 ⟨93OM289⟩
Morgan, G. T., **2**, 288 ⟨28JCS321⟩, 489, 493, 498 ⟨26PRS523⟩; **5**, 1117 ⟨25JCS2307⟩
Morgan, H. W., **5**, 1129 ⟨60IS84⟩
Morgan, I. T., **1**, 651, 943 ⟨90CC1575⟩
Morgan, J. F., **2**, 491, 496, 498 ⟨47JA927⟩; **5**, 221 ⟨47JA875⟩
Morgan, Jr., H. S., **5**, 298 ⟨50JOC800⟩
Morgan, K. D., **3**, 59 ⟨76S25⟩
Morgan, L. R., **2**, 1055 ⟨58JOC648⟩
Morgan, M., **2**, 1000 ⟨83JHC399⟩
Morgan, P., **2**, 126 ⟨46MI 203-01⟩
Morgan, P. H., **2**, 341, 342, 350 ⟨75T2595⟩
Morgan, P. W., **2**, 452, 841 ⟨52JA4526⟩
Morgan Jr., T. K., **5**, 384 ⟨79JOC1999⟩
Morgan, V. L., **3**, 131 ⟨82JOC3006⟩
Morgans, Jr., D., **1**, 837 ⟨94TL15⟩
Morge, R. A., **2**, 218 ⟨87JOC4620⟩
Morgenstern, J., **3**, 342 ⟨64AG(E)277⟩
Morgenstern, M. A., **4**, 673, 722 ⟨89OM1199, 88OM1715⟩
Morgenthau, Jr., J. L., **4**, 193 ⟨51JA5005⟩
Mori, A., **3**, 45 ⟨80CL1213⟩, 616 ⟨90CL1171, 91CL145⟩, 616 ⟨93CE53⟩, 628 ⟨89CL2119, 90JOC181⟩, 629 ⟨91SL563, 91TL4333, 92CL2443, 92JA7969, 92JOC6778⟩, 632, 638 ⟨93CL975⟩, 635 ⟨85JOC2240⟩; **4**, 174 ⟨85JOC5444⟩; **5**, 184 ⟨87BCJ4325⟩
Mori, A. L., **4**, 589 ⟨84JA5920⟩
Mori, E., **4**, 420 ⟨71BCJ3445⟩
Mori, F., **6**, 70 ⟨76JAP(K)76108012⟩
Mori, H., **2**, 107 ⟨83CPB2157⟩; **3**, 234 ⟨62CPB429⟩, 430, 666 ⟨91S750⟩; **5**, 303 ⟨92BCJ1636⟩, 665, 666, 824 ⟨91S750⟩, 972 ⟨77BCJ2406⟩
Mori, I., **1**, 415 ⟨81TL3985⟩; **3**, 168 ⟨90JOC5966⟩; **5**, 1043 ⟨84TL4683⟩
Mori, K., **1**, 165 ⟨85T2751⟩, 165 ⟨85T3663⟩, 165 ⟨86T295⟩, 503, 1052 ⟨75T1846⟩, 591 ⟨86TL1343⟩, 649 ⟨88T4339⟩, 747 ⟨71ABC1116⟩, 782 ⟨94LA41⟩, 943 ⟨90JCS(P1)3017⟩, 1134 ⟨79TL3021, 85T2369⟩; **2**, 123 ⟨69JOC4170⟩; **3**, 246 ⟨81T709⟩, 433 ⟨85S279⟩, 634 ⟨84JOC3281⟩, 698 ⟨66JOC4170⟩; **4**, 250 ⟨75S720⟩; **5**, 118 ⟨84MI 502-03⟩, 143, 304 ⟨75BCJ2091⟩, 173 ⟨84T2935⟩; **6**, 465 ⟨93EUP528498⟩
Mori, M., **1**, 494 ⟨85TL1519⟩, 591 ⟨86TL1343⟩; **2**, 123 ⟨69TL3631⟩, 912, 945 ⟨91JOC3846⟩, 927, 988 ⟨92JOC3519⟩; **5**, 157 ⟨77TL1037, 79JCS(P1)771, 79TL1133⟩, 296 ⟨83JOC4058⟩, 301 ⟨78JOC1684, 82CC741, 86CC841⟩
Mori, R., **2**, 440, 829 ⟨91TA667⟩
Moriarty, K. J., **1**, 489 ⟨90TL6509⟩
Moriarty, R. M., **1**, 74 ⟨87SC703⟩, 109 ⟨89JA6443, 90JA1297⟩, 109 ⟨93TL4129⟩, 900 ⟨90JA3228⟩, 1076 ⟨87CC711⟩; **2**, 13 ⟨87CC675⟩, 25 ⟨89TL791⟩, 44 ⟨86ACR244⟩, 405 ⟨90SC2353⟩, 476 ⟨90JA8575⟩, 1020 ⟨90SL365⟩; **3**, 162 ⟨87JOC150⟩, 163 ⟨89TL667⟩, 163, 178 ⟨90SL365⟩, 181 ⟨85SC649⟩, 194 ⟨88T1603⟩, 294 ⟨75TL2557⟩; **4**, 184 ⟨86ACR244⟩, 184 ⟨86OS(64)138⟩, 198 ⟨82TL1537⟩, 206 ⟨89TL3019⟩, 315 ⟨84S683⟩, 1081 ⟨90S431⟩; **5**, 366 ⟨87SC409⟩, 897 ⟨84JA6082⟩
Mori, S., **1**, 173 ⟨90H(31)1189⟩, 1042 ⟨88TL5391⟩; **2**, 532, 697 ⟨79TL1665⟩, 753 ⟨84TL429⟩; **3**, 125 ⟨82CPB3380⟩; **5**, 173 ⟨81JA1568⟩, 873 ⟨82CPB3380, 90O(68)S1⟩
Mori, T., **1**, 841 ⟨70TL1175⟩; **2**, 579 ⟨93CPB292⟩, 758 ⟨83CL1791⟩; **3**, 82, 102, 107 ⟨83TL1543⟩, 286 ⟨82TL5059⟩, 620 ⟨85CL1339⟩; **4**, 238 ⟨91CL415⟩, 877 ⟨86CL311⟩; **5**, 14 ⟨83CL835⟩; **6**, 524 ⟨56JAP10924⟩
Mori, W., **3**, 9, 123 ⟨89CL901⟩
Mori, Y., **1**, 623 ⟨83TL2599⟩; **2**, 4, 23 ⟨78CL833⟩, 19 ⟨91CPB1088⟩, 61 ⟨88TL5419, 89TL4383, 89TL4387⟩, 81 ⟨89JOC6001⟩; **3**, 428 ⟨88BCJ461⟩, 593 ⟨70JOC2076⟩
Moriayma, M., **2**, 440, 441, 453 ⟨83JA4727⟩
Morikawa, A., **3**, 7 ⟨77BCJ2773⟩, 118 ⟨76BCJ1656⟩
Morikawa, I., **4**, 1033 ⟨90TA151⟩
Morikawa, K., **3**, 156 ⟨92JOC5067⟩
Morikawa, M., **4**, 14 ⟨64JA4851⟩; **5**, 142 ⟨63TL1437⟩
Morikawa, T., **4**, 734 ⟨85CPB5137⟩; **6**, 504 ⟨82CPB534⟩
Morimoto, A., **1**, 831 ⟨85JOC5179⟩; **6**, 468 ⟨82JOC4013⟩
Morimoto, H., **4**, 218 ⟨89CPB526⟩; **5**, 206 ⟨88CPB178⟩
Morimoto, K., **4**, 889, 900, 970, 971, 972 ⟨90S195⟩; **6**, 229 ⟨92JFC(59)417⟩
Morimoto, O., **6**, 20 ⟨89TL109⟩
Morimoto, S., **5**, 226 ⟨85BCJ1903⟩
Morimoto, T., **1**, 1041 ⟨84CC794⟩; **2**, 149 ⟨92BCJ1744⟩, 155 ⟨91CL523⟩; **3**, 88, 99 ⟨90BCJ2433⟩, 115 ⟨91BCJ2857⟩, 119 ⟨87BCJ4143, 88JCS(P1)2423⟩, 120 ⟨91BCJ1046⟩, 406 ⟨85CL1371⟩, 713 ⟨85CC1584⟩; **4**, 313 ⟨84CC794⟩, 509 ⟨85CPB4596⟩; **5**, 134 ⟨77JAP(K)77/102237⟩
Morimoto, Y., **2**, 331 ⟨87CPB3845⟩
Morimura, S., **4**, 420 ⟨71BCJ3445⟩
Morin, C., **1**, 530 ⟨90JA9388⟩
Morinaga, K., **2**, 20 ⟨68CPB1784⟩
Morin, Jr., J. M., **1**, 803 ⟨80JOC1340⟩; **3**, 211 ⟨85JA8066⟩
Morinaka, H., **6**, 533 ⟨83JAP60155152⟩

Morinaka, Y., **1**, 818 ⟨93TL8453⟩
Morin, L., **1**, 833, 834 ⟨79PS(7)69⟩; **3**, 351 ⟨77BSF688⟩; **5**, 552 ⟨79PS(7)69⟩
Morin, R. B., **5**, 577 ⟨76TL967⟩, 736 ⟨79CP1051008⟩
Morin, R. D., **2**, 89 ⟨51IEC1596⟩
Morini, G., **5**, 347 ⟨84FES788⟩
Morioka, M., **6**, 134 ⟨88BCJ2147⟩
Morisaki, M., **1**, 961 ⟨75CC362⟩; **2**, 19 ⟨82JCS(P1)85⟩
Morisaki, Y., **1**, 613, 700, 1073 ⟨84JA3670⟩
Morise, X., **4**, 131 ⟨91SC793⟩
Morishima, H., **3**, 40 ⟨82SC25⟩
Morishima, N., **5**, 971 ⟨75JAP89344⟩
Morishita, T., **5**, 209 ⟨81T3115⟩
Morita, A., **6**, 501 ⟨68MI 606-01⟩
Morita, H., **4**, 63, 64 ⟨57JA376⟩; **5**, 671 ⟨84JHC725⟩
Morita, K., **1**, 183 ⟨92BCJ3487⟩; **2**, 10 ⟨68BCJ1724⟩; **3**, 498 ⟨68BCJ2815⟩, 618, 619 ⟨73TL2121⟩, 674 ⟨70JOC675⟩; **4**, 762 ⟨70JOC828⟩
Morita, K.-I., **2**, 26 ⟨67JOC31⟩
Morita, N., **1**, 248 ⟨88BCJ1281⟩
Morita, S., **4**, 549 ⟨80JOC4683⟩
Morita, T., **1**, 13, 15, 17 ⟨81S32⟩, 156 ⟨93T537⟩; **2**, 11, 36 ⟨79S379, 78CC874⟩, 847 ⟨78TL2523⟩; **5**, 370 ⟨90JMC2192⟩
Morita, Y., **1**, 132 ⟨89JOC1785⟩, 1056, 1061 ⟨90JOC441⟩; **2**, 112 ⟨66AG(E)580⟩, 398 ⟨85SC527⟩; **3**, 32 ⟨92JOC1321⟩, 136 ⟨90JA455⟩, 703, 709 ⟨94CC549⟩; **4**, 197 ⟨92JOC1321⟩, 361 ⟨92SL843⟩, 842 ⟨75CPB2390⟩; **5**, 437 ⟨90JA455⟩; **6**, 131 ⟨92SL843⟩
Morita, Z., **2**, 1000 ⟨67T4279⟩; **3**, 373 ⟨67T4279⟩
Moritani, I., **1**, 531 ⟨75JA5606⟩, 729 ⟨71JCS(C)3252⟩, 842 ⟨72BCJ1777⟩; **2**, 268 ⟨74TL793, 75BCJ3682, 75JOC3644, 75TL2689⟩; **5**, 143 ⟨68TL633⟩, 795, 796 ⟨75BCJ3682⟩, 795, 796 ⟨75TL2689⟩
Moritani, T., **1**, 1078 ⟨92CC721⟩; **5**, 582 ⟨67BCJ2209⟩
Moritz, P., **4**, 136 ⟨92JOM(427)293⟩, 140 ⟨93OM4930⟩; **6**, 243 ⟨93OM4930⟩
Moriwake, T., **2**, 84 ⟨90TL221, 91TL667⟩
Moriwaki, H., **3**, 311 ⟨86CPB4916⟩
Moriwaki, M., **2**, 21 ⟨88S545, 88TL5783⟩, 24 ⟨90BCJ3033⟩, 355 ⟨76JOC300⟩, 626 ⟨89BCJ439⟩, 631 ⟨88CL795⟩
Moriya, H., **2**, 566 ⟨81CC1140⟩
Moriya, K., **4**, 94 ⟨88GEP3639877A1⟩
Moriya, M., **1**, 220 ⟨93CC614⟩
Moriya, O., **1**, 936 ⟨82JA5564⟩, 1127, 1130, 1131 ⟨84CL1437⟩; **3**, 684 ⟨94JCS(P1)413⟩; **4**, 187 ⟨86JCS(P1)1351⟩; **6**, 561 ⟨88BCJ2203⟩
Moriya, T., **1**, 489 ⟨93CL1429⟩; **4**, 736 ⟨91CL961⟩, 815 ⟨93SC2851⟩
Moriyama, K., **4**, 902 ⟨87JCS(P1)1501⟩
Moriyama, S., **3**, 360 ⟨89MI 308-03⟩
Moriyama, T., **1**, 613 ⟨86JOC3896⟩
Moriyasu, M., **1**, 1060 ⟨87CL1519⟩
Morizawa, Y., **1**, 248 ⟨82TL2871⟩, 452 ⟨84TL3221⟩, 803 ⟨84MI 118-02⟩; **2**, 624, 946, 948 ⟨84TL3221⟩, 918 ⟨85T3257⟩, 918, 931 ⟨83JA4491⟩, 929 ⟨84TL1163⟩, 947, 977 ⟨84TL3217, 84TL2151⟩, 977 ⟨84TL2221⟩, 997, 1002 ⟨85JOM(285)163⟩; **4**, 141, 811, 812 ⟨84TL3221⟩, 652 ⟨82TL1279⟩
Morken, J. P., **1**, 316 ⟨91JA8950, 92JA6692⟩, 316 ⟨93JA6997, 93JOC4237, 94JA3123⟩; **2**, 561 ⟨91JA8950⟩, 562 ⟨92JA6692, 93JA6997⟩
Morken, P. A., **1**, 990 ⟨91CC566⟩; **4**, 746, 748, 760, 817, 821 ⟨93JA5430⟩, 816, 820 ⟨94T2993⟩
Mørkved, E. H., **5**, 716 ⟨79ACS(B)544, 80ACS(B)313⟩
Morliere, P., **5**, 1097 ⟨77CC648⟩; **6**, 717 ⟨77CC648⟩
Morlock, R., **6**, 510 ⟨66LA(698)180⟩
Mormann, W., **3**, 64 ⟨72CB1815⟩; **5**, 224, 1007 ⟨88S990⟩, 1006, 1007 ⟨89JPS(A)4341⟩
Mornet, R., **1**, 539 ⟨77TL167⟩; **2**, 27 ⟨77S786⟩
Moro, K., **5**, 223 ⟨87CPB1741⟩; **6**, 447 ⟨86JAN230⟩
Moroder, F., **4**, 475 ⟨82S653⟩
Moroder, L., **2**, 218 ⟨82ZPC(363)1461⟩
Moroe, T., **3**, 494 ⟨70BRP1202353⟩
Morokuma, K., **6**, 691 ⟨84JPC1981⟩
Morooka, ,Y., **5**, 223 ⟨88OM366⟩
Morooka, T., **6**, 529 ⟨91JAP03060419⟩
Moro-Oka, Y., **3**, 59 ⟨76CL29⟩, 117 ⟨85TL2107⟩; **4**, 657 ⟨86JOM(299)C35⟩, 674 ⟨93OM2925⟩, 679 ⟨92CC1031⟩, 681 ⟨92OM989⟩
Morozov, A. B., **4**, 669 ⟨89IZV2562⟩
Morretti, R., **2**, 992 ⟨88JA2641⟩
Morrill, T. C., **2**, 4 ⟨93S973⟩; **5**, 1109 ⟨68JOC823⟩
Morris, A., **1**, 2, 6 ⟨60JCS298⟩
Morris, C. J. O. R., **2**, 456 ⟨32JCS2880⟩
Morris, D. G., **1**, 737 ⟨71ACS1452⟩
Morris, D. L., **3**, 639 ⟨88OSC(6)932⟩
Morris, G., **2**, 841, 842 ⟨65JCS5771⟩
Morris, G. E., **6**, 465 ⟨87CC410⟩
Morris, G. F., **4**, 743 ⟨58JA5988⟩
Morris, J., **2**, 554, 593 ⟨88JA2506⟩
Morris, J. H., **2**, 418 ⟨82COMC-I(1)311⟩, 811 ⟨69MI 215-03⟩
Morris, K. B., **2**, 1004 ⟨91JCS(P2)1077⟩
Morris, K. M., **4**, 802 ⟨93JOC1531⟩
Morris, M. J., **4**, 675 ⟨89CC688⟩, 679 ⟨84JCS(D)2293⟩; **5**, 400 ⟨92JOM(426)C28⟩
Morris, P. E., **3**, 117 ⟨87JOC1149⟩
Morris, R. C., **2**, 462 ⟨53USP2642461⟩
Morris, R. E., **2**, 1025 ⟨88S287⟩
Morris, S. A., **6**, 603 ⟨88IC570⟩
Morris, W., **2**, 497 ⟨75JOM(91)295⟩
Morrish, W. N., **5**, 841 ⟨57JOC692⟩
Morris-Natschke, S., **2**, 252 ⟨86JMC2114⟩
Morrison, A. L., **5**, 712 ⟨55CI(L)1183⟩
Morrison, D. C., **2**, 167, 257 ⟨55JA181⟩, 841 ⟨50JA4820⟩, 847, 861 ⟨51JA5896⟩
Morrison, D. E.,**1**, 506 ⟨61JOC3156⟩
Morrison, D. S., **6**, 462 ⟨90JOC1⟩
Morrison, G. A., **5**, 1110 ⟨81JCR(S)234⟩
Morrison, G. F.,**3**, 48 ⟨71OPP227⟩
Morrison, H., **3**, 464, 470 ⟨61JOC2617⟩
Morrison, III, W. H., **3**, 418, 697 ⟨74JOC600, 71OS(51)31⟩; **5**, 796, 797 ⟨71OS(51)31⟩
Morrison, J. A., **2**, 434 ⟨80CC671⟩, 483 ⟨88IC4535⟩, 483, 486 ⟨90IC33⟩, 903 ⟨91CRV35⟩; **4**, 7 ⟨81JA2995⟩, 140 ⟨75IC1614⟩; **6**, 62 ⟨78JA1722⟩, 173 ⟨75JOM(92)163⟩, 181 ⟨91CRV35⟩, 206 ⟨81JA5951⟩, 244, 717 ⟨93AOC(35)211⟩, 246 ⟨80CC671, 81JA2995⟩, 246, 247 ⟨89JOM(376)149, 83MI 622-01⟩
Morrison, M. A., **5**, 273, 285 ⟨80JA7026⟩
Morrison, W. H., **2**, 268 ⟨74JOC600⟩; **3**, 696, 700 ⟨88OSC(6)751⟩, 697 ⟨70JA6675⟩
Morrison, III, W. H. I., **5**, 68, 420, 792, 796, 797 ⟨74JOC600⟩, 792, 796, 797 ⟨70JA6675⟩
Morrissey, C. J., **1**, 462 ⟨58JA5867⟩
Morrissey, M. M., **1**, 84 ⟨84JA3866⟩
Morrissey, P., **3**, 479 ⟨79JOC4697⟩
Morritz, F. L., **1**, 462 ⟨58JA5867⟩
Morrow, C. J., **2**, 435, 435, 1060, 1064 ⟨69JOC2684⟩
Morrow, D. F., **2**, 381 ⟨73JMC736⟩, 753 ⟨66JMC249⟩
Morrow, S. D., **2**, 394 ⟨82JOC365⟩; **4**, 333 ⟨81JOC765⟩
Morsch, K., **5**, 24 ⟨32M61, 32M65⟩, 24 ⟨32M68⟩
Mörsdorf, P., **3**, 586, 587 ⟨83CB741⟩
Morse, A. T., **2**, 1012, 1015 ⟨54CJC500⟩
Morse, J. G., **2**, 414 ⟨73IC2119⟩, 449 ⟨⟨75IC571⟩; **4**, 114 ⟨76JFC(7)153⟩, 547 ⟨88JFC(40)41⟩
Morse, K. W., **2**, 414 ⟨73IC2119, 75IC571⟩; **3**, 640 ⟨89IC492⟩; **4**, 114 ⟨76JFC(7)153⟩; **5**, 688 ⟨90IC3218⟩; **6**, 491 ⟨92S380⟩, 491, 515 ⟨90IC554⟩, 492 ⟨92AP(325)267⟩, 514 ⟨88IC302, 91IC2228, 90IC3218⟩
Morse, S. D., **5**, 246 ⟨78JFC(11)327⟩
Morss, L. R., **2**, 602 ⟨83JA6824⟩
Mortag, M., **6**, 560 ⟨80ZC101⟩

Mortell, T. R., **3**, 642 ⟨72TL669⟩
Mortensen, J., **4**, 649 ⟨83ZN(B)953⟩
Mortensen, K., **6**, 590 ⟨83CC295⟩
Mortezania, R., **2**, 16 ⟨88JOC5153⟩
Mortier, J., **4**, 1019 ⟨88OM2553⟩; **5**, 955 ⟨88OM2553⟩
Mortikov, V. Y., **3**, 656 ⟨85S98⟩
Mortillaro, L., **3**, 382 ⟨65JCS807, 65JPS(B)581⟩, 382 ⟨66JPS(A)248, 67JCS(B)117⟩
Mortimer, C. T., **2**, 491 ⟨52JCS4331⟩
Mortimer, R. J.,**1**, 1137 ⟨93JCS(D)2629⟩
Mortimore, M., **1**, 678 ⟨88TL3357⟩
Mörtl, A., **2**, 445 ⟨80ZN(B)990⟩, 449 ⟨83ZC249⟩
Mortlock, A. A., **6**, 503 ⟨9⟨91TL4791⟩
Mortlock, S. V., **1**, 595, 601, 624 ⟨91COS(6)1011⟩, 602 ⟨87CC880⟩; **2**, 79 ⟨88TL2479⟩
Morton, A. A., **2**, 556 ⟨47JA1675, 47JA950⟩, 556 ⟨53JA134⟩, 557 ⟨55JOC981⟩; **5**, 595 ⟨41JA314⟩
Morton, D. R., **1**, 1122, 1138 ⟨78TL2549⟩; **4**, 248 ⟨79JOC656⟩
Morton, D. V. A., **4**, 1023, 1032 ⟨88CC673⟩
Morton, D. W., **4**, 571 ⟨82OM623⟩
Morton, G. O., **6**, 650 ⟨68JA6846⟩
Morton, H. E., **1**, 811 ⟨78JOC3630, 83CJC1226⟩, 1092 ⟨79JOC3437, 80JOC4263⟩; **2**, 142 ⟨84JOC4538⟩, 525 ⟨88PAC1627⟩, 600 ⟨87CJC78⟩; **3**, 153 ⟨93TL4481⟩, 273 ⟨78CC1033⟩; **4**, 199, 217, 225 ⟨85JOC5379⟩; **5**, 160 ⟨89T363⟩
Morton, I. D., **5**, 62 ⟨B-52MI 502-03⟩
Morton, Jr., G. A., **2**, 421 ⟨56JA4285⟩
Morton, Jr., J. W., **5**, 652 ⟨54OR(8)258⟩
Morton, J. W., **2**, 752 ⟨54OR(8)258⟩
Morton, M. D., **3**, 516 ⟨93OM2565⟩
Morton, P. D., **4**, 465 ⟨63JOC2898⟩
Morton, P. R., **4**, 697, 698 ⟨88JA3231⟩
Morton, S., **2**, 440 ⟨83IC3624⟩
Morton, W. D., **2**, 16, 336 ⟨88JCS(P1)2805⟩
Morton-Blake, D. A., **2**, 859 ⟨67JCS1028⟩, 862 ⟨67JCS(A)1492⟩
Morton Jr., D. R., **1**, 710 ⟨79JOC2880⟩
Mortreux, A., **1**, 692 ⟨84CC341⟩
Morvan, J. M., **4**, 405, 411 ⟨75BSF196⟩
Morvillo, A., **2**, 890, 891, 893 ⟨88IC3730, 84IC1504⟩; **5**, 1138 ⟨84JOM(276)431⟩
Mory, R., **2**, 182 ⟨59HCA1653⟩; **5**, 9 ⟨59HCA1653⟩; **6**, 52 ⟨59HCA1653⟩
Morzycki, J. W., **3**, 132 ⟨82TL957⟩; **5**, 316 ⟨91H(32)481⟩
Mosaka, T., **3**, 363 ⟨91CPB489⟩
Mosandl, A., **4**, 232 ⟨87LA451⟩
Mosandl, T., **3**, 584 ⟨92CC1571⟩; **5**, 698 ⟨92CC1571⟩
Mosbach, K., **5**, 118 ⟨81MI 502-07, 82MI 502-01⟩
Moschi, E., **5**, 99 ⟨80JHC621⟩
Mosedale, A. J., **5**, 288 ⟨88JCS(P1)2185⟩
Moseley, K., **4**, 715 ⟨67CJC2011⟩
Moseley, R. H., **5**, 1012 ⟨73JA3957⟩
Moser, E., **5**, 932 ⟨68JOM(13)209⟩, 952 ⟨68JOM(14)P9, 68JOM(15)147⟩, 953 ⟨68JOM(13)387⟩; **6**, 719 ⟨68JOM(12)P1⟩, 721 ⟨69JOM(16)275⟩
Moser, E. O., **5**, 952 ⟨68JOM(15)147⟩
Moser, G. A., **4**, 706, 706 ⟨73JOM(51)1⟩
Moser, H., **5**, 749 ⟨86HCA1224⟩
Moser, W. R., **4**, 24 ⟨70JA5469⟩
Moses, C. G., **2**, 126 ⟨26JA776⟩
Moses, L. M., **2**, 561 ⟨81JOC3773⟩
Moses, P., **4**, 751 ⟨57JA4174⟩
Moses, P. R., **4**, 846 ⟨74JA945⟩
Moses, R. C., **2**, 326 ⟨87CC509⟩
Mosetlig, E., **5**, 1106 ⟨63MI 530-01⟩
Mosettig, E., **3**, 13, 82, 85, 86 ⟨54OR(8)218⟩, 13, 83 ⟨48OR(4)362⟩
Moshenberg, R., **4**, 105 ⟨86JOC3123, 87JOC101⟩; **5**, 366 ⟨85JOC5604⟩
Moshentseva, L. V., **5**, 116 ⟨68JIC1109⟩
Mosher, C. W., **5**, 830 ⟨81JMC1124⟩; **6**, 502 ⟨66JMC444⟩
Mosher, H. M., **6**, 641 ⟨88TL3183⟩
Mosher, H. S., **2**, 9 ⟨72JA4184⟩, 64 ⟨73JOC1870, 73TL1505⟩, 67 ⟨62JOC1⟩, 95 ⟨54JA2984, 54JA2987⟩, 95 ⟨57JA3474⟩, 396 ⟨82JOC4040⟩; **3**, 46 ⟨85TL6269⟩, 46, 190 ⟨82JOC4040⟩, 638 ⟨44JA725, 45JA1071⟩; **5**, 12 ⟨69JOC2543⟩
Mosher, W. A., **1**, 384 ⟨63JA1548⟩; **3**, 295 ⟨61JOC670⟩, 445 ⟨58JA3048⟩
Moshida, Y., **2**, 19 ⟨81CL761⟩
Moshkina, T. M., **5**, 709 ⟨84JGU1969, 84ZOB2203⟩
Moshnikov, S. A., **6**, 490 ⟨87JGU1484⟩
Moskal, ⟨85RTC177⟩9
Moskal, J., **3**, 214 ⟨84TL2585⟩, 709, 720 ⟨85RTC177⟩; **6**, 122 ⟨85RTC177⟩
Moskaleva, R. N., **4**, 794 ⟨78JOU135⟩; **6**, 504, 508 ⟨79JOU1077⟩
Moskalevskaya, L. S., **2**, 847 ⟨74ZOB85⟩
Moskau, D., **1**, 876 ⟨87AG(E)1212⟩
Moskova, V. V., **4**, 884, 913 ⟨77CA135599v⟩
Moskovits, M., **6**, 522 ⟨76IC1666⟩
Moskowitz, M., **4**, 316 ⟨54JA3969⟩
Moskva, V. V., **1**, 888 ⟨91RCR72⟩; **2**, 459 ⟨74ZOB2621⟩; **3**, 49 ⟨73RCR538⟩, 197 ⟨87JGU205⟩; **4**, 132 ⟨91JGU1459⟩, 465 ⟨84JGU421⟩, 559 ⟨85ZOB1865, 89ZOB1686⟩, 912 ⟨67ZOB1623⟩; **5**, 484 ⟨74IZV55⟩, 707 ⟨68JGU2501, 68ZOB2586⟩; **6**, 156, 313 ⟨88ZOB2167⟩, 314 ⟨67ZOB2137⟩
Mosleh, A., **1**, 483 ⟨93SL568⟩
Moss, F., **6**, 241 ⟨75JCS(P1)2033⟩
Moss, G. I., **3**, 541 ⟨70CC206⟩
Moss, J. H., **2**, 486, 492, 505 ⟨57JCS3708⟩
Moss, J. R., **4**, 148 ⟨86ICA(119)177⟩, 148 ⟨89JOM(366)175⟩, 582 ⟨83JCS(D)1495⟩, 584 ⟨79JOM(182)C20⟩, 584, 585 ⟨86JOM(315)255, 82JOM(232)C78⟩; **5**, 943 ⟨90OM2836, 95AOC(37)169⟩, 948, 949 ⟨73JCS(D)975⟩; **6**, 719 ⟨74JCS(D)1189⟩, 719, 732 ⟨74JCS(D)351⟩
Moss, N., **1**, 243 ⟨82JOC1641⟩, 812 ⟨84TL3959, 85TL2735⟩
Moss, R. A., **1**, 909 ⟨86JA134⟩, 1102 ⟨B-78MI 122-01⟩, 1143 ⟨B-73MI 122-01, B-75MI 122-01⟩; **2**, 376 ⟨75JOC1213⟩, 405 ⟨72JA4392⟩, 405, 773 ⟨73JA3070, 73JCS(P1)2778⟩, 773 ⟨74ACR421, 76S726⟩; **3**, 465 ⟨68T2881⟩, 465, 466 ⟨66JOC1082, 74ACR421⟩; **4**, 7, 16, 24 ⟨B-85MI 401-01⟩, 18 ⟨79S425⟩, 32 ⟨67T2549⟩, 415 ⟨84TL1023⟩, 816 ⟨64JA4042⟩, 830 ⟨87JA3811⟩, 1080 ⟨B-73MI 424-02⟩, 1081 ⟨85TL1931⟩, 1083 ⟨86TL5907⟩; **5**, 866 ⟨81JA6164⟩, 943 ⟨83CC880⟩; **6**, 55 ⟨81JA6164, 89ACR15⟩, 55, 280, 633 ⟨81JOC5048⟩, 280 ⟨83JA6513, 86JOC2168⟩, 280 ⟨89TL2473, 90JPO694⟩, 280, 614 ⟨86TL419⟩, 280, 632 ⟨82JOC4177, 90TL4715⟩, 280, 632, 731 ⟨87TL1969⟩, 488 ⟨92JA9386⟩
Moss, R. E., **2**, 383, 384 ⟨82JCS(P1)603⟩
Moss, V. A., **3**, 657 ⟨67CC218⟩
Moss, W. O., **1**, 471 ⟨92JA6556⟩; **3**, 72 ⟨89JCS(P1)1793⟩; **4**, 842, 845, 851 ⟨89JCS(P1)1793⟩, 860 ⟨88TL6475, 92T7551⟩
Mosseron-Canet, M., **1**, 726 ⟨71BSF(2)4551⟩
Mosset, P., **1**, 580 ⟨86TL299⟩, 966 ⟨92AG(E)224, 93T9775⟩; **3**, 303 ⟨93TL2275⟩
Mossman, C. J., **1**, 510 ⟨84TL1655⟩
Mostafavipoor, Z., **3**, 5 ⟨83BCJ914⟩
Most, Jr., E. E., **2**, 384 ⟨57JOC358⟩
Mosterd, A., **3**, 250 ⟨74TL4179⟩
Mosti, L., **5**, 501 ⟨86JHC455, 87JHC633⟩
Mostovoi, M. V., **1**, 657 ⟨65MI 114-01⟩
Mostowicz, D., **5**, 344 ⟨91CAR283⟩
Mota, J. F., **5**, 1028 ⟨89CAR(188)35⟩
Motavaze, G., **4**, 889 ⟨75BSF2089⟩
Motes, J. M., **1**, 851 ⟨62JA2465⟩, 871 ⟨70JA2445, 70JA3697⟩
Motevalli, M., **5**, 444 ⟨88JCS(D)53⟩
Motherway, R., **5**, 1111 ⟨92B7922⟩

Motherwell, R. S. H., **1**, 38, 945 ⟨79TL2291⟩, 940 ⟨82TL2019⟩; **6**, 550, 557 ⟨81JCS(P1)2363⟩
Motherwell, W. B., **1**, 16, 938 ⟨81S743⟩, 38, 945 ⟨79TL2291⟩, 38, 945 ⟨80S68⟩, 133 ⟨91CC1399, 92TL6187⟩, 320, 330, 332, 342, 344 ⟨B-92MI 108-01⟩, 527 ⟨89TL7107⟩, 546 ⟨B-92MI 112-01⟩, 692 ⟨92TL3899⟩, 919, 927, 935, 938, 939 ⟨B-92MI 119-17⟩, 919, 928 ⟨92MI 119-18, 88CC1380⟩, 928 ⟨92T8031⟩, 929, 936 ⟨81JCS(P1)2363⟩, 934 ⟨80CC732⟩, 935 ⟨83CC939, 85T3901⟩, 935, 938 ⟨81PAC15⟩, 940 ⟨82TL2019⟩, 945 ⟨80JCS(P1)2657, 80JCS(P1)2665⟩, 945 ⟨82CC447⟩; **2**, 13 ⟨83TL4979⟩, 785 ⟨82CC551, 83BSF(2)61⟩; **3**, 8, 121 ⟨81CC1232⟩, 113 ⟨86T5627⟩, 118, 464 ⟨79CC705⟩, 132 ⟨82TL957⟩, 145 ⟨85JCS(P1)2667⟩, 405 ⟨84CC337, 84TL3707⟩, 408 ⟨86JCS(P1)2243⟩; **4**, 1077 ⟨83CC939, 83TL4979⟩; **5**, 868 ⟨92T1465⟩, 1054 ⟨80JCS(P1)2657⟩; **6**, 550, 557 ⟨81JCS(P1)2363, 92T8031⟩, 560 ⟨84JCS(P1)1005⟩
Motherwell, W. D. S., **2**, 889 ⟨70JCS(C)810⟩
Mothes, V., **5**, 551, 729, 731 ⟨84T1573⟩
Motohashi, S., **2**, 3 ⟨82S1021⟩; **3**, 152, 154, 155 ⟨82S1021⟩, 227 ⟨88CPB1169⟩
Motohiro, H., **3**, 357 ⟨91EUP409164⟩
Motoki, S., **2**, 226 ⟨92CC600⟩, 244 ⟨78CL591⟩, 464 ⟨82TL1263⟩; **3**, 352, 353 ⟨88CL717⟩, 353, 360 ⟨82TL1263, 78JOC4147⟩, 360 ⟨79JOC4151, 80JOC927, 82CL793, 85BCJ667, 86BCJ3279, 86BCJ335, 90CC1665, 91JCS(P1)2281, 92BCJ923, 92JCS(P1)2943⟩, 365 ⟨77BCJ2751⟩, 370 ⟨77JOC3922, 84SR33⟩, 370 ⟨92CC411, 92CC600⟩, 412 ⟨79BCJ3469⟩, 592 ⟨80BCJ2582, 89JCS(P1)2140⟩; **5**, 1066 ⟨86CL135⟩
Motoki, Y., **1**, 197 ⟨93JA8835⟩
Motoo, Y., **3**, 628 ⟨86BCJ893⟩
Motornyï, S. P., **5**, 476 ⟨57JGU2301⟩; **6**, 239 ⟨59ZOB2157⟩, 440 ⟨57ZOB2243⟩, 613 ⟨60JGU4029⟩
Motoyama, I., **4**, 797 ⟨69BCJ3270⟩
Motoyama, M., **4**, 227 ⟨91CC1421⟩
Motoyama, Y., **1**, 374 ⟨91TA643⟩
Motoyoshiya, J., **3**, 557 ⟨79CC900, 80JOC5385⟩, 557, 593 ⟨81JCS(P1)2727⟩, 574 ⟨81CL1197, 86JHC597⟩; **4**, 365 ⟨84JCS(P1)435⟩, 802 ⟨88CL211⟩
Motozawa, K., **6**, 564 ⟨81JCS(P1)2499⟩
Motsak, G. V., **5**, 105 ⟨80MI 502-01⟩
Motsarev, G. V., **6**, 243 ⟨89ZOB2628⟩
Mott, A. W., **2**, 233 ⟨84S657⟩; **6**, 275, 535 ⟨84SUL241⟩, 434, 435, 474, 534, 550, 551, 567 ⟨83JOC4750⟩, 435 ⟨84JCS(P1)2615, 84JOC1043⟩, 435 ⟨84MI 614-02⟩
Mottadelli, S., **2**, 25, 31, 35 ⟨92JCS(P1)2309⟩
Mott, F., **2**, 134 ⟨32JPR257⟩
Mott, L., **5**, 17 ⟨66LA(693)132⟩, 1078 ⟨68LA(718)24⟩
Mott, R. C., **2**, 34 ⟨79JOC1731⟩; **3**, 155 ⟨79JOC1731⟩
Mottaghinejad, E., **3**, 5 ⟨90T6869⟩
Motte, J.-C., **2**, 1039, 1050 ⟨86TCC89⟩, 1051, 1052 ⟨75C515, 75C516⟩
Motte-Coppe, G., **5**, 601 ⟨85T693⟩
Motter, R. F., **2**, 1031 ⟨59JA3386⟩
Mottoh, A., **4**, 841 ⟨84CC1028⟩
Mottus, E. H., **4**, 698 ⟨73JOM(50)47⟩
Motyka, L. A., **3**, 303 ⟨86JA1056⟩
Motyl, K. M., **4**, 679 ⟨82JA7325⟩
Motz, D., **4**, 707 ⟨88AG(E)939, 88CB799⟩
Motz, P. L., **5**, 799 ⟨89OM1523, 89OM2589⟩
Motzkus, E., **2**, 134 ⟨32JPR257⟩, 172, 194 ⟨39CB657⟩
Mouk, M. L., **2**, 129 ⟨72JMC968⟩
Mouko-Mpegna, D., **4**, 388, 389 ⟨85BSF787⟩; **5**, 468 ⟨83JOM(251)175⟩
Moulden, H. N., **2**, 694 ⟨61JCS3079⟩
Moulds, C., **4**, 216 ⟨91JA7767⟩
Moulds, L. de. V., **2**, 101 ⟨35JCS901⟩
Moule, D. C., **3**, 382 ⟨84JA5406⟩
Moulines, F., **1**, 164 ⟨89CC614⟩, 164 ⟨92SL57⟩; **2**, 699 ⟨87T2075⟩; **5**, 1134 ⟨82S212⟩
Moulines, H., **3**, 67 ⟨85S754⟩
Moulines, J., **2**, 97 ⟨90SC349⟩
Mouloungi, Z., **2**, 28 ⟨92SC1923⟩
Moulton, C. J., **3**, 195 ⟨80JCS(D)299⟩
Mouna, A., **2**, 331 ⟨93BMC681⟩
Mountfield, B. A., **2**, 111, 267 ⟨59JCS1529⟩
Mountford, P., **4**, 675 ⟨89CC418⟩, 676 ⟨90OM886⟩
Moura Campos, M. de, **4**, 348 ⟨64CI(L)1461⟩
Mourad, M. S., **2**, 341 ⟨85JOC133⟩; **3**, 426 ⟨85JOC133⟩
Moureu, C., **3**, 618 ⟨1894BSF1067⟩; **5**, 59 ⟨32OSC(1)166⟩
Mouri, M., **2**, 528 ⟨93JA11490⟩
Mouriño, A., **1**, 576 ⟨92JOC5208⟩, 606 ⟨92JOC3173⟩, 816, 817 ⟨85TL4959, 87TL2099⟩, 1029 ⟨86TL1523, 88TL1203⟩; **3**, 294 ⟨75H(3)449⟩
Mourning, M. C., **1**, 628 ⟨68JA5561⟩
Mousa, H. A. H., **5**, 703 ⟨88H(27)695⟩
Mouslouhouddine, M., **2**, 741 ⟨84S493⟩
Moussa, A., **1**, 578 ⟨88SC1795⟩
Moussa, F., **1**, 999, 1035 ⟨90SC2181⟩; **2**, 1036, 1038 ⟨90SC2181⟩
Moussebois, C., **5**, 219, 833 ⟨62HCA441⟩
Mousseron, M., **1**, 422 ⟨56BSF1653⟩
Mousseron-Canet, M., **2**, 764 ⟨65CR(260)2851, 67BSF1296⟩
Mousset, G., **1**, 615 ⟨78MI 114-01⟩; **3**, 16, 132 ⟨90TL2591⟩
Moussiaux, B., **3**, 694 ⟨89PAC509⟩
Moustakis, C. A., **3**, 392 ⟨84JOC189, 85T4843⟩; **5**, 226 ⟨83JCS(P2)1471⟩
Moutardier, G., **2**, 662 ⟨75MI 213-01⟩
Moutevelis-Minakakis, P., **2**, 230 ⟨84TL4153⟩; **5**, 367 ⟨85JCS(P1)2277⟩
Movchun, V. N., **6**, 235 ⟨90S1151⟩
Mowart, W., **4**, 656 ⟨70CC1369⟩
Mowat, W., **4**, 656 ⟨71CC1079, 71CC1477⟩; **5**, 461 ⟨71JCS(A)850⟩
Mowlem, T. J., **2**, 711 ⟨91T8621⟩
Mowry, D. T., **2**, 127 ⟨54JA585⟩; **3**, 611, 612, 613, 614, 619, 620, 621, 627, 636, 641 ⟨48CRV189⟩, 621 ⟨47JA573⟩; **4**, 205 ⟨50JA2535⟩, 205, 213 ⟨47JA2358⟩
Moya, E., **4**, 269 ⟨92JOC6390⟩; **5**, 572 ⟨94TL2057⟩
Moyano, A., **1**, 529 ⟨85T5995⟩, 530 ⟨90JA9388⟩, 1085 ⟨88S707⟩; **2**, 84, 318 ⟨91TL6931, 91TL6935⟩, 676, 1026 ⟨87JOC2919⟩, 682 ⟨87JOC2303⟩; **3**, 218 ⟨88T2657⟩, 243 ⟨90TL7505⟩; **4**, 196 ⟨88S707⟩, 763 ⟨94JA2153⟩, 851 ⟨90TL2169⟩
Moya-Portuguez, M., **3**, 558, 560 ⟨80TL3081⟩
Moyer, J. A., **5**, 334 ⟨89JMC1024⟩
Moyer, J. R., **5**, 205 ⟨64JOC2099⟩
Moyle, M., **1**, 1070 ⟨63JCS3712⟩; **5**, 74 ⟨63OSC(4)493⟩, 1035 ⟨54CI(L)785, 56JCS659⟩, 1035 ⟨63OSC(4)700⟩
Moynehan, T. M., **3**, 281 ⟨59JCS1563⟩
Mozaffari, A., **2**, 394 ⟨91OS(70)68⟩
Mozdzen, E. C., **1**, 21 ⟨79JOC2185⟩, 401, 800, 883 ⟨78JA2134⟩
Mozingo, R., **3**, 13, 83 ⟨48OR(4)362⟩
Mozolis, V. V., **6**, 576 ⟨73RCR587⟩
Mozumi, M., **1**, 834 ⟨86H(24)3347⟩; **5**, 284 ⟨87CC1627⟩
Mozzon, M., **6**, 247 ⟨90OM1449⟩, 663 ⟨90JCS(D)1197⟩, 721 ⟨88IC2809⟩
Mpango, G. W. B., **3**, 655, 656 ⟨84T2141⟩
M'Pati, J., **5**, 214 ⟨81TL4405⟩
Mrabet, H., **5**, 700, 701 ⟨92SC1985⟩
Mrani, M., **2**, 541 ⟨82JA3104⟩; **3**, 336, 340, 346, 347, 363, 392 ⟨82JA3104⟩; **5**, 512, 575 ⟨82JA3104⟩
Mratkhuzina, T. A., **6**, 145 ⟨76KFZ35⟩
Mraz, T. J., **1**, 13, 18 ⟨79TL849⟩
Mrogenstern, H., **3**, 435 ⟨82CB2679⟩
Mrotzek, H., **3**, 552 ⟨79LA1734⟩; **5**, 539 ⟨80JCS(P2)1557⟩, 624, 625, 732 ⟨76CB956⟩
Mroz, C. T., **6**, 573 ⟨68JOC851⟩
Mu, J.-D., **2**, 487 ⟨92JOC774⟩

Mualla, M., **1**, 248 ⟨86TL2243⟩
Mubarak, M. S., **1**, 68 ⟨85JOC673⟩
Mucciaro, T. P., **1**, 799 ⟨92JA5878⟩
Muccigrosso, D. A., **3**, 117 ⟨79JOC921⟩
Mucheli, M. V. R.,**3**, 280 ⟨83CI(L)479⟩
Muchmore, C. R., **5**, 951 ⟨89JOC4481⟩
Muchmore, D. C., **1**, 17, 434 ⟨72JA5098⟩; **2**, 693 ⟨72JA5098⟩
Muchowski, J. M., **1**, 92 ⟨84JOC4203⟩, 177 ⟨87H(26)1291⟩, 256 ⟨87SC863⟩, 755 ⟨81JOC459⟩; **2**, 445 ⟨81USP4296042⟩, 647 ⟨73CJC2347⟩, 935 ⟨85TL5375⟩; **3**, 39, 40, 178, 183, 184 ⟨90OPP399⟩, 178 ⟨81JOC1224⟩, 179 ⟨91CJC2059⟩, 179, 194 ⟨86JOC3374⟩, 307 ⟨82JOC3668, 84JOC4203, 85JMC1037⟩, 638 ⟨90CJC1305⟩, 673 ⟨85TL1827⟩
Muci, A. R., **2**, 56 ⟨93JA9800⟩
Muck, D. L., **6**, 506 ⟨66JA3798⟩
Mucke, W., **6**, 534 ⟨65CB2059⟩
Muckensturm, B., **1**, 774 ⟨87TL1519⟩
Mudi, A. K., **5**, 864 ⟨83IJC(A)307⟩
Mudryk, B., **1**, 236 ⟨93JA3855⟩, 1033 ⟨79CC488⟩; **2**, 83, 553 ⟨90JOC1528, 89JOC5657⟩, 553 ⟨90JA6389, 91JA1866⟩, 553 ⟨91JOC5760⟩
Muehlbacher, M., **6**, 264 ⟨86JOC4768⟩
Muehlbauer, E., **5**, 977 ⟨63BEP631961⟩
Mueller, G., **3**, 386 ⟨87CB1905⟩, 386, 397 ⟨87CC559⟩; **4**, 629 ⟨88ZN(B)727⟩, 647 ⟨86JOM(316)C4⟩
Mueller, H., **5**, 793 ⟨76ZN(B)281⟩
Mueller, K. H., **5**, 1120 ⟨87EUP209779⟩
Mueller, M. E., **1**, 922 ⟨90JOC58⟩
Mueller, N., **3**, 639 ⟨91JCS(P1)365⟩
Mueller, P., **4**, 30 ⟨78HCA2482⟩
Mueller, P. H., **4**, 748 ⟨78TL4253, 80JA6615⟩, 748 ⟨84JA3584⟩
Mueller, R., **1**, 490 ⟨70JPR150⟩
Mueller, R. H., **1**, 131, 416, 829, 874, 876 ⟨76JA2868⟩, 414, 416, 829, 830 ⟨72JA5897⟩; **2**, 420, 696 ⟨76JA2868⟩; **4**, 831 ⟨72JA5897⟩; **5**, 73 ⟨72JOC1907⟩, 227 ⟨76JA2868⟩
Mueller, W. H., **2**, 117 ⟨69JOC2955⟩, 166, 169 ⟨69JHC627⟩, 208 ⟨69CC1205⟩, 359 ⟨68JOC2111⟩; **4**, 15 ⟨67JOC2651⟩; **5**, 248 ⟨67JOC2925, 68JOC1533⟩
Mueller-Westerhoff, U. T., **4**, 706 ⟨93JOM(463)163⟩
Muench, A., **4**, 479, 480 ⟨84CC1634⟩; **6**, 117 ⟨84CC1634⟩
Muenstedt, R., **1**, 991 ⟨84JOM(264)135, 84JOM(271)181⟩
Mues, C., **1**, 828 ⟨90SL97⟩; **5**, 685 ⟨88T6841⟩
Muetterties, E., **6**, 212 ⟨58GEP1040011⟩
Muetterties, E. L., **1**, 87, 93 ⟨79ACR324, 74JA4063⟩, 93 ⟨76IC2379, 78JA2405, 78JA7425⟩, 456 ⟨81ICA1⟩; **2**, 166 ⟨62IC756⟩, 477, 869 ⟨65IC119⟩, 964 ⟨59JA2597⟩; **4**, 524 ⟨64IC1196⟩, 692 ⟨76JA8072, 78JA2090⟩; **5**, 675, 975 ⟨64JA357⟩; **6**, 212 ⟨55USP2709186, 55USP2709187, 55USP2709188, 55USP2709190⟩, 217 ⟨56USP2757214⟩, 217, 220, 225 ⟨55USP2709184⟩, 233, 235, 236 ⟨59JA3575⟩
Muffler, H., **6**, 138 ⟨68CB1885⟩
Mufti, A. S., **5**, 1017 ⟨65JCS5055⟩
Mugdan, M., **6**, 215 ⟨37GEP68065⟩
Mügge, C., **2**, 253, 260 ⟨82TL1345⟩, 439 ⟨91ZAAC(594)66⟩,; **4**, 550 ⟨91ZAAC(594)66⟩
Mugnier, Y., **4**, 720 ⟨89JOM(362)C27⟩
Mugno, M., **3**, 536 ⟨63CI(M)1216⟩
Mugrage, B., **3**, 181 ⟨90JA462⟩; **5**, 344 ⟨90JA462⟩
Muhammad, N., **5**, 999 ⟨73JMC1⟩
Mühlbauer, E., **4**, 47 ⟨59CB1818⟩
Mühlbauer, G., **1**, 813 ⟨86TL1315⟩
Mühle, H., **4**, 53 ⟨78S593⟩, 53, 54, 56 ⟨78HCA2047⟩
Muhlemann, C., **5**, 92 ⟨92OS(71)200⟩
Muhlstadt, M., **2**, 170 ⟨80EGP139258⟩
Mühlstädt, M., **2**, 170 ⟨82JPR631⟩
Muhlstadt, M., **5**, 641 ⟨88JCS(P2)821⟩
Muhn, R., **1**, 327 ⟨92TL2673⟩, 330 ⟨89T935⟩
Muhr, W., **2**, 632 ⟨80S572⟩
Mui, J. Y.-P., **4**, 16 ⟨65JA4259⟩; **6**, 266 ⟨66JOM(5)185⟩, 608 ⟨71JOC1786⟩; **6**, 38 ⟨64JA2961⟩, 60 ⟨68JA2944⟩
Muir, C. N., **1**, 387 ⟨69T1479⟩
Muir, K. W., **4**, 585, 691 ⟨82CC614⟩, 656 ⟨72CC613⟩, 714 ⟨80CC749⟩; **6**, 721 ⟨71CC400⟩
Mujata, T., **2**, 753 ⟨80AG(E)1008⟩
Mujsce, A. M., **1**, 118, 119 ⟨91JA631⟩
Mukai, C., **1**, 706 ⟨88TL909⟩; **2**, 1000 ⟨81JCS(P1)212⟩
Mukai, I., **4**, 199 ⟨89CL659⟩
Mukai, K., **1**, 238 ⟨83JOC1293⟩; **2**, 257 ⟨79TL5015, 83JOC1293⟩, 855 ⟨68TL5015⟩; **4**, 485 ⟨88TL663⟩
Mukai, T., **1**, 196 ⟨85CL1477⟩, 196 ⟨89JOM(369)C51⟩, 795 ⟨70BCJ1120⟩, 982 ⟨87AG(E)335⟩; **3**, 80 ⟨77H(6)1599⟩, 221 ⟨84CC179⟩
Mukaiyama, S., **3**, 3, 89, 114 ⟨81BCJ2221⟩
Mukaiyama, T., **1**, 189 ⟨86CL1009⟩, 191 ⟨87CL1167⟩, 239 ⟨87CL1691, 87CL1695⟩, 308 ⟨76AJC1635, 76CL163⟩, 333 ⟨73JA967⟩, 427, 428, 430 ⟨73CL291⟩, 485 ⟨73CL1097⟩, 590 ⟨75CL871⟩, 590, 591 ⟨78CL413⟩, 590, 693 ⟨74CL1523⟩, 687 ⟨73CL1041⟩, 688, 690 ⟨76CL1127⟩, 936 ⟨84CL2069⟩, 973 ⟨81CL621⟩, 1026 ⟨87CL1975, 88BCJ4037⟩, 1039 ⟨79CL447, 80CL255⟩, 1039 ⟨84CL405⟩, 1071 ⟨79CL481⟩, 1134 ⟨77CL959⟩; **2**, 8 ⟨77CL383⟩, 8, 36 ⟨76CL619⟩, 26 ⟨77CL1013⟩, 43 ⟨89CL7573⟩, 71 ⟨79BCJ3371, 79JA1455⟩, 81 ⟨89CL1401⟩, 106 ⟨67JOC1063⟩, 107 ⟨63JOC481⟩, 118, 124, 213, 217 ⟨68TL5907⟩, 124, 125 ⟨77CL437⟩, 139 ⟨91BCJ2524⟩, 142 ⟨90CL2239⟩, 143 ⟨72CL259⟩, 215 ⟨67BCJ2388⟩, 344 ⟨78CL1057⟩, 361 ⟨70TL3411⟩, 391 ⟨90CL1395, 92CL1137⟩, 401 ⟨65BCJ1989⟩, 478 ⟨89CL569⟩, 517 ⟨80CL993⟩, 524 ⟨72BCJ2244⟩, 532 ⟨77CL153, 82OR(28)203⟩, 532, 698 ⟨76CL559, 80BCJ174⟩, 656 ⟨75CL1041⟩, 688 ⟨70TL5115⟩, 701 ⟨82CL353, 83CL1727, 84T1381, 87CL293⟩, 709 ⟨68BCJ2742⟩, 712 ⟨93CL1⟩, 837 ⟨64JOC2572⟩, 859 ⟨67BCJ147⟩; **3**, 7 ⟨77BCJ2773⟩, 13, 85 ⟨79BCJ555⟩, 25 ⟨82CL1637⟩, 54 ⟨83CL1207⟩, 63 ⟨75CL1201, 77BCJ1161⟩, 82, 107 ⟨74CL1447⟩, 84, 107 ⟨75CL215⟩, 118 ⟨76BCJ1656⟩, 120 ⟨77CL757⟩, 131 ⟨78CL49⟩, 134 ⟨89CL2233⟩, 139 ⟨86CL1009⟩, 140 ⟨79CL1483, 87CL289⟩, 156 ⟨91CL281, 93CL1579⟩, 158 ⟨91CL1499⟩, 159 ⟨91CL949⟩, 164 ⟨86CL1809⟩, 168 ⟨84CL1531, 85CL1871⟩, 220 ⟨77AG(E)817, 87S1043⟩, 239 ⟨78CL465⟩, 279, 281, 298 ⟨86CL165⟩, 281, 296, 298 ⟨92CL435⟩, 281, 298 ⟨91CL1059⟩, 281, 298 ⟨91S1216⟩, 281, 298 ⟨92CL1751, 93BCJ3729⟩, 286 ⟨73JA4763, 74BCJ1777⟩, 289 ⟨81CL531⟩, 374 ⟨70TL5297⟩, 541 ⟨64JOC2242⟩, 615 ⟨91CL537⟩, 618 ⟨60BCJ1382, 61BCJ99⟩, 618, 697, 722 ⟨61JOC2202⟩, 623 ⟨87CL1167, 87CL1811⟩, 629 ⟨91CL541⟩, 684 ⟨60JA5339⟩, 696 ⟨77CL697⟩, 698, 721 ⟨62JOC3651, 65BCJ858⟩; **4**, 176 ⟨87S1043⟩, 187 ⟨84CL265⟩, 188 ⟨88CL1101⟩, 192 ⟨74CL15⟩, 193 ⟨76CL769⟩, 218 ⟨82CL1555⟩, 408 ⟨78CL1253, 80CL17⟩, 833 ⟨62BCJ71⟩, 850 ⟨86CL1623⟩, 902 ⟨67BCJ2641⟩; **5**, 125 ⟨76CL267, 77BCJ1863⟩, 190 ⟨64JOC1385⟩, 214, 1001 ⟨61JOC782⟩, 226 ⟨81CL1367, 81CL285⟩, 234 ⟨76CL741⟩, 236 ⟨74CL187, 86CL1013, 86CL1817, 87CL743, 89CL1001, 90CL1019, 91CL1901, 91CL989⟩, 237, 277 ⟨86CL915⟩, 237, 277 ⟨87CL293⟩, 282 ⟨83CL683⟩, 291 ⟨76CL57⟩, 557 ⟨88CL1021, 88CL1025⟩, 652 ⟨70TL1901⟩, 774, 850 ⟨68TL3569⟩, 898 ⟨68TL3787, 70BCJ2870⟩, 979 ⟨77CL1345, 79AG(E)707⟩, 997 ⟨63BCJ95⟩, 1001 ⟨61JOC4381⟩, 1005 ⟨62BCJ1104⟩, 1025 ⟨77CL573⟩, 1078 ⟨77CL575⟩, 1121 ⟨73CL1225⟩; **6**, 74 ⟨61JOC2202⟩, 105, 106 ⟨67BCJ2641⟩, 113 ⟨65BCJ2107, 66BCJ2005⟩
Mukanov, I. P., **5**, 1132 ⟨74MI 530-01, 74MI 530-02⟩
Mukerjee, A. K., **5**, 28 ⟨73S327⟩, 291 ⟨79S726⟩, 1110 ⟨78T1731⟩; **6**, 573 ⟨91CRV1⟩
Mukerji, I., **1**, 205 ⟨86AG(E)760⟩
Mukerji, S. K., **3**, 57 ⟨81ACS(B)643⟩

Mukhacheva, A., **2**, 453 ⟨56ZOB2463⟩
Mukhacheva, O. A., **2**, 466 ⟨52IZV894⟩; **4**, 728 ⟨71ZOB1873⟩
Mukhamadaliev, N. M., **3**, 531 ⟨68T1341⟩
Mukhametshin, F. H., **2**, 167 ⟨85ZOK1835⟩
Mukhametshin, F. M., **6**, 145 ⟨70ZOR189⟩, 145 ⟨74ZOR2003⟩
Mukhametzyanova, É. Kh., **4**, 121, 126 ⟨71JGU2228⟩, 124 ⟨75JGU1950⟩
Mukhametzyanova, E. Kh., **4**, 127 ⟨67IZV1535⟩
Mukhametzyanova, Ekh., **4**, 346 ⟨68IZV1625⟩
Mukherjee, A., **3**, 369 ⟨92JOC3120⟩
Mukherjee, P., **6**, 514 ⟨90BCJ3658⟩
Mukherjee, R., **5**, 358 ⟨71CC1113⟩
Mukherjee, R. N., **4**, 32 ⟨61JCS3825⟩; **5**, 1137, 1142, 1143 ⟨63JCS229⟩; **6**, 64 ⟨61JCS3825⟩
Mukherjee, S., **3**, 306 ⟨93JCR(S)192⟩
Mukherjee, T. K., **4**, 28, 38 ⟨56JA2264⟩
Mukherjhee, D., **3**, 127 ⟨89SC673⟩
Mukhopadhyay, A., **4**, 202 ⟨90SC687⟩; **5**, 943 ⟨82CC215⟩, 944 ⟨83JCS(D)2397⟩
Mukhopadhyay, R., **2**, 398 ⟨85S886⟩
Mukhopadhyay, T., **1**, 927 ⟨86JA2381⟩; **2**, 394 ⟨82HCA1101⟩
Mukhtarov, A. S., **5**, 373 ⟨73ZOB1451⟩
Mukkanti, K., **2**, 753 ⟨89TL251⟩
Mulamba, T., **1**, 1092 ⟨78TL3711⟩
Mularski, C. J., **3**, 246, 620 ⟨83TL2059⟩
Mul, W. P., **4**, 716 ⟨92ICA689⟩
Muldakhmetov, Z. M., **4**, 343 ⟨76ZOB783⟩
Mulder, R. J., **4**, 218, 235 ⟨64TL543⟩, 218, 235 ⟨67TL3057, 67TL3061⟩
Mulders, J., **1**, 870 ⟨69JOC4083⟩
Mulengi, J. K., **5**, 324 ⟨92BSB257⟩
Muler, L. I., **5**, 1018 ⟨64ZOB3609⟩
M-ul-Hague, **4**, 114 ⟨81JCS(P2)1138⟩
Mulholland, D. L., **1**, 917 ⟨83JOC1431⟩
Mulholland, K. G., **5**, 203 ⟨88JMC362⟩
Mulholland, T. P. C., **4**, 824 ⟨72JCS(P1)1225⟩
Mullah, K. B., **4**, 61, 73, 79 ⟨91JOC6878⟩, 218 ⟨94JOC544⟩
Mullane, M., **5**, 672, 821, 825 ⟨86JCS(P2)995⟩
Mullen, G. B., **3**, 434 ⟨90HCA169⟩
Mullen, G. P., **5**, 238 ⟨85JOC811⟩
Müllen, K., **1**, 101 ⟨87AG(E)204⟩, 880 ⟨89TL159⟩, 890 ⟨86CC947⟩
Mullen, K., **2**, 1013 ⟨68TL3909⟩
Müllen, K., **4**, 276 ⟨91CB1827⟩
Müller, A., **2**, 890 ⟨71CB3782⟩, 997, 998 ⟨93AG(E)1032⟩; **3**, 426 ⟨60CB32⟩; **4**, 99 ⟨63CB595⟩, 762 ⟨93JFC(60)61⟩, 771 ⟨93JFC(63)253⟩, 816 ⟨93AG(E)1032⟩
Müller, A.-K., **5**, 589, 603, 611 ⟨67JPR294⟩
Müller, B., **1**, 883 ⟨87HCA1858, 87HCA1879⟩; **2**, 338 ⟨83CB3931⟩; **3**, 143 ⟨87HCA1858⟩; **4**, 442 ⟨58LA(615)34⟩
Müller, C., **4**, 35 ⟨71TL3869⟩
Müller, C. E., **2**, 259 ⟨90TL501⟩; **5**, 337 ⟨89AP(322)343⟩
Muller, C. L., **1**, 470 ⟨92JOC1161⟩; **3**, 220 ⟨90TL5897⟩
Muller, D., **1**, 194, 353 ⟨91HCA232⟩; **2**, 483 ⟨86OCS638⟩, 493 ⟨86ZN(B)1129⟩, 505 ⟨82AG(E)439, 83ZN(B)125⟩, 505 ⟨86OCS636⟩
Müller, D., **2**, 893 ⟨83JOM(253)C21⟩; **5**, 223 ⟨87T2045⟩
Müller, D.-W., **2**, 437 ⟨59CB3175⟩; **4**, 545 ⟨59CB3175⟩
Müller, E., **1**, 922 ⟨67TL3877⟩; **2**, 2, 6 ⟨60HOU(5/4)1⟩, 2, 6, 18 ⟨62HOU(5/3)1⟩, 169, 170 ⟨64CB2614⟩, 170 ⟨63CB3050⟩, 385 ⟨84LA1468⟩, 387, 388 ⟨67HOU(10/2)827⟩, 402, 404 ⟨67HOU(10/2)757⟩; **3**, 64 ⟨66CB2479, 76ZN(B)1654⟩, 272 ⟨71LA(754)64, 75CB237⟩, 275 ⟨73CB39⟩, 462 ⟨60CB1541⟩, 465 ⟨60ZN(B)751, 63CB1712⟩, 468 ⟨04CB1261⟩, 485 ⟨34LA(512)250, 54CB1887⟩, 485 ⟨57CB1302⟩, 551 ⟨71T5953⟩, 570, 571 ⟨68LA(718)1, 66AG(E)514⟩, 680 ⟨73AG(E)155⟩, 722 ⟨61ZN(B)845⟩; **4**, 312 ⟨84LA1468⟩, 434 ⟨87S745⟩; **5**, 350 ⟨67HOU(10/2)123⟩; **6**, 619 ⟨86T6645⟩
Müller, E. P., **2**, 860 ⟨84HCA1238⟩; **5**, 373 ⟨85CC1764⟩
Müller, F., **4**, 313 ⟨67T2781⟩; **5**, 855 ⟨52CB470⟩; **6**, 234 ⟨38USP2108606⟩, 653 ⟨62LA(651)89⟩
Muller, F. J., **4**, 830 ⟨74GEP2331675⟩
Müller, G., **1**, 51, 446 ⟨87CB1403, 89CB123⟩, 634 ⟨87TL1985⟩, 727, 728 ⟨66AG(E)667⟩, 1088 ⟨88JA6033⟩; **2**, 435 ⟨85JOM(296)411⟩, 444 ⟨89AG(E)1700⟩, 446 ⟨85CB3105⟩, 447 ⟨86JA976⟩, 474, 475 ⟨85AG(E)402⟩, 512 ⟨87JOM(325)159⟩, 561 ⟨85AOC(24)1⟩, 1002 ⟨86AG(E)1103⟩, 1003 ⟨85OM231⟩; **3**, 545 ⟨86AG(E)643⟩, 576 ⟨88AG(E)1542⟩; **4**, 94 ⟨85OM1208⟩, 1023 ⟨89CB1857⟩, 1052 ⟨87ZN(B)142⟩, 118 ⟨87OM2489⟩, 397 ⟨80JOM(194)9⟩, 489 ⟨90CB423⟩, 548, 1023 ⟨88CB1241⟩, 550 ⟨85JOM(296)411, 86ZN(B)444⟩, 551 ⟨84ZN(B)1518⟩, 551, 560, 1023 ⟨85CB3105⟩, 562 ⟨91CB275⟩, 570 ⟨87CB789⟩, 574 ⟨80CB1480⟩, 579 ⟨84JOM(273)195⟩, 581, 582, 583 ⟨83OM257⟩, 586 ⟨81CB441⟩, 587 ⟨85OM231⟩, 588 ⟨86OM1664⟩; **5**, 226 ⟨87CB39⟩, 605, 614, 627 ⟨87CC559⟩, 605, 614, 627 ⟨88CB2095⟩, 833 ⟨1886CB1491⟩, 887 ⟨86AG(E)454⟩, 911 ⟨84ZN(B)1456⟩, 932 ⟨90T1235⟩, 979 ⟨65CB1097⟩, 1095 ⟨87CB789⟩, 1097 ⟨88AG(E)583⟩, 1137 ⟨84OM996, 92ICA173⟩, 1138 ⟨76SRI217⟩; **6**, 152 ⟨84ZN(B)1518⟩, 164, 371, 373 ⟨88ZN(B)1416⟩, 167, 371 ⟨92CB1333⟩, 173 ⟨89CB2115⟩, 173, 378, 379 ⟨90AG(E)201⟩, 369 ⟨84CC569⟩, 374 ⟨91OM2884⟩, 383 ⟨89ZAAC(579)75⟩, 392, 710 ⟨84JOM(271)381⟩, 392, 710 ⟨87OM35⟩, 673 ⟨86CC591⟩, 701 ⟨84OM38⟩, 701 ⟨84ZN(B)1456⟩, 706 ⟨86AG(E)574, 85AG(E)229⟩, 710 ⟨87OM32⟩
Muller, G. W., **1**, 359, 944 ⟨81JA2744, 79JA7129⟩
Müller, H., **1**, 511, 534, 542 ⟨60LA(629)222⟩; **2**, 967 ⟨56CB444⟩, 967, 974 ⟨60LA(629)222⟩; **3**, 364 ⟨65AG(E)154⟩; **4**, 685 ⟨93JOM(445)133⟩, 698 ⟨60LA(629)222⟩; **6**, 208 ⟨60LA(629)222⟩
Muller, H.-G., **4**, 895 ⟨88AP(321)863⟩, 896 ⟨88AP(321)873, 88AP(321)879⟩
Müller, H.-J., **4**, 710 ⟨86AG(E)734, 89CB1901⟩
Müller, H. P., **5**, 1070 ⟨87AG(E)359⟩; **6**, 604 ⟨87AG(E)359⟩
Müller, J., **2**, 504 ⟨71ZAAC(381)103, 85ZN(B)1320⟩, 565 ⟨84OM69⟩, 997, 1000 ⟨91JA4002⟩; **3**, 690 ⟨73AG(E)564⟩; **4**, 690 ⟨93JOM(459)325, 93ZN(B)1558⟩, 715 ⟨94JOM(471)249⟩; **5**, 1143 ⟨75CB664⟩; **5**, 652 ⟨69JOM(16)29, 71JOM(28)237⟩, 857 ⟨01JPR199⟩, 950, 951 ⟨72CB150⟩, 952 ⟨68JOM(14)P9, 73CB3910⟩; **6**, 422 ⟨83LA248⟩, 723 ⟨86OM173⟩, 729 ⟨87CSR75⟩
Muller, J.-C., **4**, 883 ⟨83HCA809⟩
Müller, K., **3**, 659 ⟨69AG(E)343⟩
Müller, K.-D., **6**, 572 ⟨76JOM(110)15⟩
Müller, M., **3**, 699 ⟨91CB2555⟩; **6**, 129 ⟨89LA975⟩
Müller, M. C., **2**, 57 ⟨92TL5045⟩
Müller, N., **1**, 422 ⟨76TL3585⟩, 892 ⟨58JA3489⟩; **2**, 373 ⟨90HOU(E16a/1)529⟩, 383 ⟨90HOU(E16a/1)561⟩; **3**, 62 ⟨75S781⟩; **6**, 20 ⟨91T549⟩
Müller, O., **2**, 169 ⟨65CB1455⟩
Müller, P., **1**, 257 ⟨78TL4727⟩; **2**, 984 ⟨89AG(E)610⟩; **3**, 5 ⟨83HCA1790⟩, 5, 117 ⟨81TL2361⟩, 509 ⟨89AG(E)610⟩
Muller, R., **1**, 686 ⟨84CB277⟩, 74 ⟨61TL353⟩; **2**, 487 ⟨66CB1609⟩, 511 ⟨80CB3517⟩, 873 ⟨67MI 217-01⟩, 1096 ⟨62CB2155⟩; **3**, 348 ⟨84CB277⟩, 483 ⟨68ZC262⟩; **4**, 622 ⟨62CB3030⟩; **6**, 60, 267 ⟨64CB1673⟩, 61 ⟨64CB1115, 66CB793⟩, 173 ⟨85ZC309⟩, 173, 267 ⟨58CB22⟩, 383 ⟨59CB1018, 61EGP22169⟩, 383 ⟨63CB2894, 64CB1111⟩
Müller, R.-G., **6**, 714 ⟨90ZN(B)290⟩
Muller, R. J., **1**, 871, 881 ⟨62JA3164⟩
Müller, R. K., **1**, 1082 ⟨72HCA1276, 76OS(55)114⟩; **2**, 380 ⟨88OSC(6)56⟩
Muller, R. K., **3**, 26, 147 ⟨72HCA1276⟩

Müller, S., **1**, 522 ⟨81TL1675⟩; **2**, 111 ⟨61JPS193⟩
Müller, U., **1**, 565, 681, 682, 683 ⟨84T1269⟩; **2**, 504 ⟨85ZN(B)1320⟩; **3**, 495 ⟨90T4417⟩; **6**, 340 ⟨75JOM(102)109⟩, 655, 733 ⟨79AG(E)693, 66AG(E)841⟩, 733 ⟨80ZAAC(468)165⟩
Müller, V., **3**, 477, 479 ⟨78HCA97⟩
Müller, W., **2**, 733 ⟨35CB1540⟩; **3**, 64 ⟨66CB2479⟩; **6**, 61 ⟨64CB1115⟩, 383 ⟨61EGP22169, 63CB2894⟩, 383 ⟨64CB1111⟩
Muller-Starke, H., **3**, 613, 627, 698 ⟨83T961⟩, 634 ⟨84TL3301⟩
Mulley, S., **2**, 134 ⟨92T5933, 88TL4477⟩; **4**, 249 92T5933⟩
Mullican, M. D., **2**, 663 ⟨80JOC5002⟩, 665 ⟨82TL4551, 87OS(65)98⟩; **6**, 558, 564 ⟨93JMC1090⟩
Mulliez, M., **5**, 370 ⟨80PS(8)27, 85BSF1211⟩, 370 ⟨86BSF101⟩
Mulligan, P. J., **4**, 995 ⟨70JMC1248⟩
Mulliken, R. S., **1**, 892 ⟨58JA3489⟩
Mullin, J. G., **1**, 1029 ⟨89JMC1820⟩
Mullin, V. A., **2**, 239 ⟨86ZOB898⟩
Mullins, D., **2**, 363 ⟨77TL963⟩, 411 ⟨78TL3989⟩
Mullins, D. F., **1**, 232 ⟨83CJC757⟩
Mullins, M., **3**, 16 ⟨75TL2017⟩; **4**, 222, 365, 393 ⟨75TL2017⟩
Mullins, M. J., **1**, 234, 821 ⟨78TL519, 87JA5878⟩
Mullins, S. J., **1**, 570 ⟨94JCS(P1)3115⟩
Mullins, S. T., **2**, 613 ⟨86BSF930, 86CC1623⟩
Mullis, J. C., **3**, 632, 633 ⟨82JOC2873⟩
Müllners, W., **6**, 492 ⟨85ZN(C)344⟩
Mulqueen, G. C., **5**, 327 ⟨93T9137⟩
Mulraney, M. J., **4**, 545 ⟨81JOM(217)51⟩
Multani, R. K., **2**, 565 ⟨73JINC311⟩
Mulvaney, J. E., **1**, 539 ⟨69JOC1936⟩; **2**, 972 ⟨63JA3897⟩
Mulvaney, M., **3**, 147 ⟨80JA2508⟩
Mulvey, R. E., **3**, 459 ⟨87JCS(D)1071, 87JCS(D)2141⟩
Mulzer, J., **1**, 681 ⟨81CC891⟩, 681, 683 ⟨79CC52⟩, 829 ⟨93S615⟩; **2**, 324 ⟨89JA7500⟩, 422 ⟨84T2211⟩, 677 ⟨77TL4651, 79CC52⟩; **3**, 40 ⟨86JOC5294⟩, 270 ⟨88LA891⟩; **5**, 62 ⟨81MI 502-02⟩, 122 ⟨91COS(6)323⟩, 129 ⟨94JOC1160⟩, 131 ⟨87JOC3784⟩, 138 ⟨86T5961, 90MI 503-01⟩, 138 ⟨91JA910, 93TA457⟩, 145, 155 ⟨83AG(E)63⟩, 146 ⟨85TL1631⟩, 151, 164 ⟨88JA4640⟩, 159 ⟨88LA891⟩, 161 ⟨79AG(E)622⟩, 164, 168 ⟨88JOC4098⟩, 166 ⟨84T2211⟩, 168 ⟨84MI 503-01⟩, 170 ⟨79AG(E)793⟩, 729 ⟨80AG(E)466⟩
Mumma, R. O., **2**, 102 ⟨66MI 202-02, 69JA4273⟩
Mumm, O., **5**, 1040 ⟨40CB843⟩
Mumzhieva, N. G., **4**, 126, 127 ⟨79IZV2073⟩
Munakata, H., **1**, 582 ⟨92BCJ2366⟩; **4**, 206 ⟨92BCJ2366⟩
Mun, K. R., **5**, 637 ⟨91S951⟩
Munavalli, S., **4**, 771 ⟨93JFC(63)253⟩; **6**, 528 ⟨92JFC(59)91, 93JFC(60)85⟩
Munavu, R. M., **4**, 197 ⟨75TL4543⟩
Munawar, M. A., **2**, 621 ⟨86IJC328⟩, 626 ⟨88JOC1799⟩
Münch, A., **5**, 892 ⟨81ZN(B)1080⟩
Munch, G., **3**, 616 ⟨1896CB62⟩
Munch, J. C., **5**, 823 ⟨29JA2262⟩
Münch, W., **4**, 37 ⟨27LA(453)113⟩; **5**, 131 ⟨87JOC3784⟩, 151, 164 ⟨88JA4640⟩
Munch, W., **5**, 751 ⟨27LA(451)453⟩
Munch-Petersen, J., **5**, 30 ⟨73OSC(5)762⟩
Mundhenke, R. F., **2**, 613 ⟨85JOC4576⟩
Mundles, E., **2**, 212 ⟨54CB1661⟩
Mundt, O., **2**, 428 ⟨78ZAAC(443)42⟩, 507 ⟨82ZAAC(486)70⟩, 509, 511 ⟨89ZAAC(569)62⟩; **5**, 400 ⟨80ZAAC(468)55⟩; **6**, 160 ⟨80ZAAC(462)130⟩, 160 ⟨82ZAAC(485)23⟩, 678 ⟨83PS(14)267⟩, 684 ⟨84ZAAC(517)89⟩, 684, 687 ⟨79ZAAC(459)87⟩
Mundy, B. P., **3**, 338, 345 ⟨79JOC486⟩, 345, 352 ⟨77TL3417⟩; **5**, 146 ⟨84SC227⟩
Mundy, D., **1**, 601 ⟨70CC86⟩
Munehira, S., **4**, 226 ⟨82S312⟩
Munekata, S., **4**, 767 ⟨80BP2029827⟩
Munekata, T., **1**, 657, 658 ⟨67JA567⟩
Muneyuki, R., **5**, 588 ⟨75TL3063⟩
Mungall, W. S., **2**, 47 ⟨88JA1968⟩, 366 ⟨78JMC704⟩
Munger, J. D., **1**, 831 ⟨82JA4030⟩
Munger, Jr., J. D., **1**, 831 ⟨85JOC707, 86T2831⟩; **3**, 325 ⟨89JA8957⟩; **5**, 227 ⟨85JOC707⟩
Münger, K., **1**, 923 ⟨85MI 119-07⟩
Munier, P., **3**, 195 ⟨91T3347⟩
Munjal, R. C., **1**, 909 ⟨86JA134⟩; **4**, 18 ⟨79S425⟩
Munoz, A., **2**, 111 ⟨91CAR(225)113⟩
Munoz, B., **5**, 263 ⟨93TL1247⟩
Muñoz, L., **3**, 193, 472 ⟨90CC652⟩
Muñoz, M. A., **6**, 327 ⟨92JCS(D)2307⟩
Munoz, S., **2**, 36 ⟨93SL489⟩
Munroe, B. M., **1**, 889 ⟨67JOC2778⟩
Munroe, J. E., **1**, 826 ⟨82JA6129⟩, 1027 ⟨89JOC5828⟩; **4**, 951 ⟨89JOC5828⟩
Munsch, H., **2**, 1008 ⟨71BSF541⟩
Munson, J. W., **3**, 530 ⟨B-80MI 316-02⟩; **5**, 322 ⟨B-82MI 507-01⟩
Munster, I., **1**, 816, 818 ⟨94T3687⟩
Münster, I., **4**, 217 ⟨91TL467⟩
Münster, P., **4**, 99 ⟨88T5403⟩, 102 ⟨87S223⟩
Munsterer, H., **2**, 793 ⟨83SR215⟩
Munt, S. P., **1**, 1127 ⟨89CC480⟩
Muntwyler, O., **5**, 145 ⟨24HCA390⟩
Munyemana, F., **2**, 8 ⟨89TL3077⟩
Münze, R., **2**, 440 ⟨84ZC224⟩
Münzenberg, K. J., **4**, 563 ⟨81AP(314)218⟩
Münzmay, T., **2**, 1003 ⟨88CB843⟩
Mura, Jr., A. J., **1**, 590 ⟨75JOC812, 76JOC3218⟩; **4**, 66, 68 ⟨75TL4433⟩
Murabayashi, A., **6**, 556, 566, 652 ⟨91CPB1939, 92MI 621-02⟩
Murafuji, T., **1**, 648 ⟨89CC1749⟩, 743 ⟨88CL849⟩; **3**, 196, 500 ⟨90BCJ950, 88CL847⟩; **6**, 709 ⟨90BCJ950⟩
Muraglia, E., **1**, 926 ⟨93JOC7610⟩
Murahashi, E., **1**, 988 ⟨90TL7469⟩
Murahashi, S., **3**, 21, 141 ⟨88JOC4489⟩, 113 ⟨93TL1299⟩, 294 ⟨93S433⟩, 432 ⟨87TL2383⟩; **4**, 193 ⟨90ACR49, 90BCJ166⟩; **5**, 1139 ⟨86JOC898⟩
Murahashi, S.-I., **1**, 192 ⟨93JOC1538⟩, 193 ⟨82TL5549⟩, 466, 1097 ⟨79JOC2408⟩, 488 ⟨77JA5652⟩, 531, 663 ⟨77CC852⟩, 969 ⟨80JOC4536⟩; **2**, 145 ⟨89JA5002⟩, 308, 344, 351 ⟨88TL2973⟩, 318 ⟨90BCJ1252⟩, 327 ⟨87TL77⟩, 606 ⟨84OS(62)39⟩, 654 ⟨92TL7557⟩, 940 ⟨80JOC4536⟩; **3**, 158 ⟨93CL1767⟩, 158 ⟨93JOC2929⟩, 176 ⟨80JA2456⟩, 407 ⟨85CC613⟩, 409 ⟨85TL4633⟩, 413 ⟨85TL5563⟩, 431 ⟨83TL1049⟩, 432 ⟨84CC874, 84H(21)483⟩, 487 ⟨79CC450, 86OM356⟩, 638 ⟨87TL6469⟩, 642 ⟨77TL4429⟩; **5**, 171 ⟨87JOC4319⟩, 324 ⟨87CC1471, 89BCJ2968⟩, 324 ⟨90JOC1744⟩
Murai, F., **3**, 177 ⟨88TL1399⟩
Murai, H., **3**, 530 ⟨88JA1025⟩
Murai, S., **1**, 186, 225, 229 ⟨89JA7938⟩, 202 ⟨90JOC5923⟩, 212 ⟨84JA6093, 93JOC4187⟩, 245, 437 ⟨87AG(E)1187⟩, 556 ⟨78TL3455⟩, 666 ⟨86JOC2389, 92JOC17⟩; **2**, 279, 282 ⟨79CL873⟩, 281 ⟨88S175, 77S874⟩, 542 ⟨88JA1976⟩, 554 ⟨87AG(E)1187⟩, 753 ⟨80AG(E)1008⟩, 912 ⟨83CC1193, 86JOC3890⟩, 912 ⟨87JOC4864⟩, 939 ⟨91JA7778⟩; **3**, 20 ⟨83JA1370⟩
Murai, S.,**3**, 33 ⟨93JOC4187⟩
Murai, S., **3**, 38, 173 ⟨77S874⟩, 38, 174 ⟨82TL4813⟩, 163 ⟨92JOC2⟩, 176 ⟨87S1096⟩, 246, 543 ⟨75AG(E)636⟩, 293 ⟨77S776⟩, 336, 340, 385 ⟨88JA1976, 88TL6965⟩, 394 ⟨89TL2095⟩; **4**, 240 ⟨77TL851⟩, 652 ⟨87AG(E)1187⟩, 931 ⟨84JA2440, 92JA9710⟩; **5**, 346 ⟨90JA9646⟩, 415 ⟨84JA2440⟩, 416 ⟨92JA9710⟩, 438 ⟨90JA7061⟩, 625 ⟨85JOC384⟩, 625, 626 ⟨85BCJ1448⟩; **6**, 7 ⟨81AG(E)475⟩, 498 ⟨79S597⟩, 505 ⟨79S735⟩

Murai, T., **1**, 186, 225, 229 ⟨89JA7938⟩, 212 ⟨84JA6093⟩; **2**, 281 ⟨88S175⟩, 370, 768 ⟨91OM1095⟩, 814 ⟨90JOC449⟩; **3**, 173 ⟨93CC1631⟩; **5**, 242, 506, 509 ⟨89TL1829⟩, 243 ⟨89ZN(B)1050⟩, 244, 250 ⟨94JCS(P1)1083⟩, 245 ⟨86T4593⟩, 246 ⟨86CL935⟩, 248, 249 ⟨85S519, 85S520, 92ZN(B)558⟩, 249, 563, 640 ⟨88BCJ3931⟩, 250 ⟨90JOM(386)313, 94PS(89)57⟩, 252 ⟨87NKK1457⟩, 255, 506, 509, 543 ⟨90TL3587⟩, 472, 510 ⟨86OPP369⟩, 509, 510, 512 ⟨B-92MI 512-01⟩, 510, 561, 563 ⟨90BCJ638⟩, 516, 523, 563 ⟨93JA5823⟩, 530, 540 ⟨86TL4593⟩, 541, 542 ⟨92CB417⟩, 542 ⟨93CC277⟩, 561, 563, 588 ⟨90ZC67, 93CL1469⟩, 563 ⟨94UP0⟩, 563, 626 ⟨93JA3000⟩

Muraka, K., **3**, 3 ⟨78BCJ3016⟩

Murakami, H., **3**, 418, 694 ⟨88JOC4158⟩

Murakami, K., **1**, 789 ⟨87T5245⟩; **4**, 506 ⟨80BCJ789⟩

Murakami, M., **1**, 189 ⟨86CL1009⟩, 191 ⟨87CL1167⟩, 800 ⟨90JOC1506⟩, 803 ⟨79CL927⟩, 1035, 1047 ⟨88TL5379⟩; **2**, 3 ⟨76CPB820⟩, 205 ⟨80CPB1131⟩, 219 ⟨67BCJ2913⟩, 539 ⟨94TL8413⟩, 916 ⟨91JOC1948⟩, 923 ⟨93AG(E)1473⟩, 986, 988 ⟨89CL761⟩; **3**, 139 ⟨86CL1009⟩, 158 ⟨93JOC1458⟩, 168 ⟨85CL1871⟩, 220 ⟨87S1043⟩, 279, 281, 298 ⟨86CL165⟩, 418 ⟨87JA7888, 88TL355⟩, 616 ⟨72JA4654⟩, 623 ⟨87CL1167⟩, 634 ⟨72JA4644⟩, 694 ⟨90JA2437, 92AG(E)1509⟩, 707 ⟨88TL5151⟩, 707, 708 ⟨89TL1257, 91JOC7356⟩, 723 ⟨88S714, 93JOC6766⟩; **4**, 176 ⟨87S1043⟩, 187 ⟨84CL265⟩, 205 ⟨76CPB102⟩, 1054 ⟨91JOC1948⟩; **5**, 236 ⟨86CL1013⟩, 728 ⟨89CL1261⟩, 793, 796 ⟨87JA7888⟩, 802 ⟨88BCJ3649, 88JOC4158⟩, 802, 803 ⟨89CL1603⟩, 803, 804 ⟨90JA2437⟩, 1065 ⟨81MAC3411⟩, 1078 ⟨85SC171⟩; **6**, 163, 671, 674 ⟨88JA3692⟩, 384, 671 ⟨91JA8899⟩, 465 ⟨94EUP581240⟩, 550 ⟨89CC1494⟩, 636 ⟨84CHEC(3)91⟩, 672, 674 ⟨87JA7888, 88TL355⟩

Murakami, S., **2**, 949 ⟨85TL1281⟩; **4**, 193 ⟨89JHC1405⟩; **6**, 495, 497 ⟨89AG(E)452⟩

Murakami, T., **1**, 983 ⟨93JOC3697⟩

Murakami, Y., **2**, 213 ⟨76CL857⟩, 380 ⟨83CPB423⟩; **3**, 158 ⟨93JOC2929⟩, 307 ⟨88H(27)1855⟩, 309 ⟨80H(14)1939, 88CPB2023⟩, 309 ⟨90CPB3261⟩

Murakata, M., **1**, 135 ⟨90CC1657⟩

Murakawa, K., **2**, 149 ⟨89BCJ3376⟩; **5**, 318 ⟨89JCS(P1)1702⟩

Muraki, M., **1**, 333 ⟨73JA967⟩; **3**, 82, 107 ⟨74CL1447⟩, 84, 107 ⟨75CL215⟩

Muralidharan, K. R., **1**, 605 ⟨87TL1737, 89JOC4146⟩

Muralidharan, S., **4**, 1073 ⟨86CJC1949, 88JA4336⟩

Muralidharan, V. B., **1**, 5, 9 ⟨82JOC5041⟩

Murali Dhar, T. G., **1**, 782 ⟨92TL1025⟩

Muramatsu, ,H., **6**, 46, 47 ⟨66BCJ2191⟩

Muramatsu, S., **3**, 220, 496 ⟨74TL4223⟩

Muramatsu, T., **1**, 442 ⟨79JCS(P1)2409⟩

Muranaka, K., **1**, 568 ⟨87BCJ2917⟩

Murano, K., **1**, 827, 828 ⟨89CPB2047⟩

Muraoka, M., **3**, 256 ⟨76JCS(P1)1688⟩, 338 ⟨82CL101, 85CC1299⟩, 338, 339 ⟨89JCS(P1)1241⟩, 356 ⟨84JHC1445, 90JCS(P1)3003⟩; **4**, 842 ⟨75JCS(P1)1277⟩, 852 ⟨76JCS(P1)1706, 77JCS(P1)1273⟩; **5**, 547 ⟨73JCS(P1)1009, 77JCS(P1)1273⟩, 547 ⟨79JAP7946762, 80S566, 82YGK123⟩, 555 ⟨77JOC3383⟩; **6**, 324 ⟨73JCS(P1)1009⟩

Muraoka, O., **5**, 199 ⟨87CPB3453⟩

Murase, I., **3**, 714, 715 ⟨71T3795⟩; **5**, 760 ⟨71T3795⟩

Murase, M., **2**, 123 ⟨87H(26)2583⟩, 662 ⟨80CC593⟩; **3**, 363 ⟨90H(30)905, 91CPB489⟩, 365 ⟨89CPB1999⟩; **6**, 472 ⟨87H(26)2583⟩

Murase, N., **3**, 26 ⟨93JA3943⟩

Murata, A., **5**, 194 ⟨87JAP6289642⟩

Murata, H., **5**, 1152 ⟨82CL981⟩

Murata, I., **1**, 866 ⟨73MI 119-19⟩; **3**, 367 ⟨77AG(E)866⟩

Murata, K., **6**, 505 ⟨79S735⟩

Murata, M., **1**, 238 ⟨88TL4105⟩; **2**, 4 ⟨82OM369, B-82MI 223-01⟩; **3**, 40 ⟨76JOC3317⟩; **5**, 171 ⟨85JA6404⟩, 293 ⟨87CPB2698⟩

Murata, N., **3**, 311 ⟨92TL5373⟩

Murata, S., **1**, 209 ⟨78S221, 79BCJ2389⟩, 209 ⟨79JCS(P1)26⟩, 813 ⟨83JA625⟩; **2**, 263 ⟨80TL767⟩, 305 ⟨93BCJ1297⟩, 539 ⟨81T3899⟩; **3**, 615 ⟨81T3899⟩, 639 ⟨79JCS(P1)26⟩; **4**, 192 ⟨80JA3248⟩, 386 ⟨83JA625⟩

Murata, T., **1**, 651 ⟨87TL211⟩; **4**, 197 ⟨91T615⟩; **6**, 124 ⟨90JA1962⟩, 129 ⟨91T615⟩

Murata, Y., **1**, 382 ⟨81TL4313⟩, 503 ⟨83BCJ2539⟩, 823 ⟨90CL2069⟩

Muratake, H., **3**, 696 ⟨90T6351⟩, 696, 701 ⟨89TL1815⟩; **5**, 572 ⟨87TL2265⟩

Murat, M., **2**, 125 ⟨09BSF405⟩

Muratov, N. N., **6**, 37 ⟨83JFC(22)105⟩

Murau, L. A., **4**, 86 ⟨74JOC1449⟩

Müraus, A., **1**, 225 ⟨91TA943⟩

Murayama, A., **6**, 43, 47, 54, 57 ⟨86JOC955⟩

Murayama, E., **1**, 667 ⟨84CL1897, 85TL2205⟩; **2**, 540 ⟨83JOC3252⟩; **3**, 209 ⟨83JOC3252⟩, 220 ⟨82TL1047⟩; **4**, 250 ⟨82TL1047⟩, 699 ⟨91T3281⟩, 708 ⟨86TL4339⟩

Murayama, H., **1**, 976 ⟨91JOM(417)305, 91TL3397⟩, 976, 1025 ⟨90TL7179⟩; **6**, 461 ⟨91JOM(417)305⟩

Murayama, K., **3**, 187 ⟨92JOC7009⟩; **4**, 420 ⟨71BCJ3445⟩; **6**, 465 ⟨92JAP04356446, 93JAP05255197⟩

Murayama, Y., **5**, 334 ⟨85JMC1772⟩

Murch, R. M., **4**, 602, 607 ⟨54JA1619⟩

Murdoch, G. C., **4**, 846 ⟨61JOC1987⟩

Murdock, H. R., **6**, 415 ⟨19JPC498⟩

Murdzek, J. C., **1**, 675 ⟨85JA5987⟩

Murdzek, J. S., **1**, 675 ⟨87MM2640⟩; **3**, 513 ⟨87OM1373, 90JA3875⟩

Mure, H., **5**, 1103 ⟨90JSP62⟩

Murillo, C. A., **4**, 673 ⟨78JA802⟩

Murla, M., **6**, 227 ⟨84JOC205⟩

Muro, S., **1**, 990 ⟨92CC778⟩

Murobushi, T., **3**, 636 ⟨78S602⟩

Murohashi, T., **3**, 543 ⟨92S977⟩

Murphey, W. A., **1**, 860, 861 ⟨51JA3308⟩

Murphree, S. S., **1**, 470 ⟨92JOC1161⟩; **3**, 220 ⟨90TL6145⟩; **4**, 275 ⟨92JOC298⟩

Murphy, B. L., **2**, 19 ⟨90JOC5291⟩

Murphy, C. J., **1**, 16 ⟨87JOC3706⟩, 686 ⟨82JOC3563⟩; **3**, 391 ⟨80JOC2890, 81TL4563, 82JOC3563⟩, 392 ⟨85T4843⟩; **4**, 333 ⟨85JCS(P1)107⟩; **5**, 521 ⟨81JA7055⟩, 521, 523 ⟨82JCS(P2)473⟩

Murphy, C. K., **2**, 16 ⟨93TL3971⟩, 44 ⟨90TL5603⟩, 150 ⟨92JOC7274⟩; **3**, 149 ⟨93TL3971⟩

Murphy, C. N., **6**, 479 ⟨73AJC755⟩

Murphy, D. B., **6**, 604 ⟨64JOC1613⟩

Murphy, G. J., **4**, 13 ⟨77JA5317⟩, 14 ⟨77JOM(141)71⟩, 34, 137 ⟨75JOM(92)7⟩, 756 ⟨73JOM(49)117⟩

Murphy, J. A., **1**, 927, 929 ⟨88TL837, 86CC1419⟩, 929 ⟨87CC1238, 88TL955⟩, 929 ⟨89T7835, 93TL867⟩; **2**, 664 ⟨79CC1165⟩

Murphy, J., **1**, 11, 19 ⟨82JOC4304⟩

Murphy, L. C., **3**, 648, 668 ⟨84NAT119⟩

Murphy, M., **2**, 653 ⟨69OS(49)90⟩

Murphy, N. J., **1**, 256 ⟨53JA3589⟩

Murphy, P., **6**, 631 ⟨84JOC2404⟩

Murphy, P. J., **1**, 580, 842 ⟨93CC884⟩; **2**, 914, 915 ⟨90TL1051⟩, 921 ⟨88T3953⟩; **3**, 48, 195 ⟨88CSR1⟩

Murphy, T. J., **3**, 451 ⟨64JA732⟩

Murphy, W. S., **1**, 881 ⟨68OS(48)80⟩; **2**, 305 ⟨64JA3565⟩; **6**, 127 ⟨79TL1827, 80JCS(P1)2678⟩

Murphy, Z. L., **3**, 577 ⟨69JOC3451⟩; **4**, 979 ⟨69JOC3451⟩

Murrary, R. C., **2**, 192 ⟨33JCS739⟩

Murray, A., **3**, 652 ⟨86CC1230⟩

Murray, A. R., **1**, 98 ⟨51JCS1945⟩

Murray, A. W., **4**, 918 ⟨88JCS(P1)1039, 90JCS(P1)2811⟩, 918, 1075 ⟨86TL4635⟩

Murray, B., **6**, 168 ⟨83IC3421⟩
Murray, C. K., **2**, 607 ⟨86JOC277⟩
Murray, III, H. H., **2**, 434 ⟨90IS305⟩; **4**, 152 ⟨85CC1278⟩, 586 ⟨87OM1992, 88IC836⟩
Murray, J. G., **6**, 522 ⟨61JA1287⟩
Murray, K. J., **1**, 58 ⟨61JOC631⟩, 58, 59 ⟨86T5497⟩, 75, 85 ⟨59JA4108⟩
Murray, K. W., **6**, 241 ⟨80JCS(P1)1960⟩
Murray, M., **1**, 953 ⟨77HOU(5)967⟩, 960 ⟨77HOU(5)1030, 77HOU(5)1037⟩; **2**, 835 ⟨70JCS(C)878, 72JCS(D)1590⟩; **4**, 691 ⟨75CC451⟩; **5**, 407 ⟨70JCS(C)878⟩
Murray, N. D., **3**, 652 ⟨86CC1230⟩
Murray, R. A., **6**, 620 ⟨86JOC1719⟩
Murray, R. E., **1**, 514 ⟨77S52⟩, 1048 ⟨80S150⟩; **2**, 436 ⟨87USP4668823⟩; **3**, 675 ⟨80S150⟩; **4**, 808, 811 ⟨81JOC1292⟩
Murray, R. K., **2**, 845, 846 ⟨71CC146⟩, 855 ⟨71JA3792⟩; **3**, 538 ⟨68TL4995⟩
Murray, R. W., **2**, 90 ⟨89CRV1187⟩, 232 ⟨72JOC3516⟩, 343 ⟨89SC3509⟩, 344 ⟨90JOC2954⟩, 377, 379 ⟨60JA2728⟩, 397 ⟨89JOC5783⟩; **3**, 326 ⟨65JA737⟩
Murray, S., **3**, 357, 377 ⟨75JCS(P1)1335⟩
Murray, S. G., **2**, 872, 874 ⟨75JOM(88)171⟩; **4**, 288 ⟨84JCS(P2)429⟩
Murray, T. F., **1**, 546, 1024 ⟨81JA7520⟩, 1023, 1024 ⟨76CC907, 78JOC353⟩; **2**, 980 ⟨79JA4107⟩; **5**, 30 ⟨32OSC(1)523⟩, 170 ⟨76CC907⟩
Murray, T. P., **2**, 662 ⟨74CC362⟩
Murray, W., **5**, 11 ⟨91S18⟩
Murray, W. V., **5**, 664 ⟨92JOC6662⟩
Murray Jr., R. K., **3**, 28 ⟨93JOC5493⟩
Murray-Rust, J., **3**, 373 ⟨78CC83⟩
Murray-Rust, P., **3**, 373 ⟨78CC83⟩
Murrell, J. N., **1**, 860 ⟨B-65MI 119-03⟩
Murrer, B. A., **2**, 764 ⟨78JCS(P1)1117⟩
Murtagh, V., **2**, 334 ⟨91JFC(52)389⟩; **5**, 22 ⟨91JFC(52)389⟩
Murtas, S., **2**, 581 ⟨80JCS(D)1633⟩
Murthy, K. S. K., **1**, 688 ⟨91JOC6447⟩, 688, 690 ⟨88JA6914⟩; **5**, 344 ⟨86S437⟩
Murtiashaw, C. W., **3**, 59 ⟨84JA4558⟩, 211 ⟨81JOC2400⟩, 310 ⟨92JOC761⟩
Murugesan, N., **3**, 419 ⟨89JA371⟩
Muruoku, K., **2**, 306 ⟨80JOC2739⟩
Musallam, H. A., **1**, 84 ⟨87JMC1505⟩; **3**, 162 ⟨87JOC150⟩
Musavirov, R. S., **6**, 127 ⟨88ZOB1066⟩
Muscatella, M. J., **6**, 340, 344 ⟨83OM219⟩
Musco, A., **1**, 471, 483 ⟨91JOC261⟩; **4**, 683 ⟨69JA6598⟩
Musgrave, W. K. R., **1**, 428 ⟨65JCS5045⟩; **2**, 650 ⟨59JCS1804⟩; **4**, 34, 36 ⟨61JCS3779⟩, 146 ⟨62JCS1993⟩; **6**, 2, 3 ⟨61JCS3779⟩, 12 ⟨74JCS(P1)108⟩, 220 ⟨53JCS1063⟩
Musher, J. I., **2**, 1001 ⟨74JA4099⟩
Mushkalo, I. L., **4**, 866 ⟨83ZOR2417⟩
Mushkin, Y. I., **5**, 389, 717 ⟨78ZOB1146⟩, 969 ⟨72ZOB2501⟩
Mushkin, Yu. I., **5**, 717 ⟨78JGU1046⟩, 1085 ⟨89MI 528-01⟩
Musierowicz, S., **4**, 343 ⟨77PS(3)345⟩
Musik, I., **6**, 637 ⟨89MI 620-01⟩
Musin, R. Z., **2**, 461 ⟨88JGU2161⟩, 881 ⟨89JGU2500⟩, 891 ⟨91JGU405⟩, 894 ⟨83IZV1152⟩; **4**, 463 ⟨88ZOB2456⟩, 561 ⟨91ZOB2698⟩, 592 ⟨90ZOB2291⟩; **5**, 406 ⟨88JGU1548, 91BAU661⟩, 535 ⟨89JGU1548⟩; **6**, 691 ⟨92PS(66)257⟩
Musio, R., **2**, 740 ⟨91S225⟩; **4**, 476 ⟨91S225⟩
Musker, W. K., **1**, 654 ⟨69JA514⟩; **2**, 212, 679 ⟨82TL323⟩, 215 ⟨81S457⟩, 247 ⟨90JOC4156⟩, 441, 444 ⟨91PS(56)27⟩, 524 ⟨69IC255⟩; **4**, 522 ⟨67TL995⟩
Musliner, W. J., **1**, 432 ⟨66JA4271⟩
Muslinkin, A. A., **4**, 561 ⟨91ZOB2698⟩
Muslukhov, R. R., **4**, 669 ⟨89IZV2562⟩
Musser, A. K., **2**, 158 ⟨82JOC3121⟩
Musser, J. H., **2**, 392 ⟨88OSC(6)799⟩; **6**, 504 ⟨90H(31)1393⟩
Musser, M. T., **1**, 200, 238 ⟨88JOC1475⟩; **2**, 393 ⟨67JA725⟩
Musso, H., **1**, 504, 623 ⟨88CB2239⟩, 510 ⟨88AG(E)298⟩; **5**, 83 ⟨56AG313⟩
Musstein, P., **2**, 878 ⟨87CB1281⟩
Mustafa, A., **6**, 114 ⟨64T25⟩
Mustafa, O. E. A., **5**, 355 ⟨91PS(63)373⟩
Mustafina, E. M., **3**, 336 ⟨91URP1680699, 92IZV160⟩, 336 ⟨92URP1710560⟩
Mustaphin, A. H., **3**, 200 ⟨90S515⟩
Musytafaev, Sh. A., **6**, 558 ⟨90MI 617-02⟩
Mutchler, J. P., **6**, 27 ⟨66CJC575⟩
Müter, B., **3**, 659 ⟨71CB2847⟩
Muth, B. F., **2**, 175 ⟨55HOU(9)2⟩
Muth, C. L., **1**, 21 ⟨79JOC2185⟩
Muth, F., **2**, 797 ⟨55HOU(9)343⟩
Muth, H., **5**, 353 ⟨80HOU(4)736⟩
Muth, K., **4**, 21 ⟨62JOC757⟩; **5**, 95, 306 ⟨73OSC(5)944⟩
Muthard, D. A., **4**, 847 ⟨82TL3535⟩
Muthukrishnan, R., **4**, 957 ⟨76HCA13⟩
Muthukumara Pillai, S., **1**, 294 ⟨86CRV353⟩
Muthukumaraswamy, N., **2**, 220 ⟨75JOC250⟩
Muthuramu, K., **3**, 346 ⟨80CC243, 83JOC4482⟩, 368 ⟨82JOC127⟩
Muthusamy, S., **5**, 1039 ⟨89OPP228⟩
Mutin, R., **1**, 674 ⟨77CC502, 77RTC(M)91⟩
Muto, S., **5**, 240 ⟨86S128⟩, 250 ⟨89S929⟩
Mutou, T., **5**, 329 ⟨94TL1247⟩
Mutschler, E., **6**, 444 ⟨89ACR15⟩
Mutterer, F., **3**, 108 ⟨76HCA211⟩
Mutterties, E. L., **1**, 93 ⟨75JA237⟩
Muxfeldt, H., **4**, 179 ⟨79JA2171⟩; **5**, 209 ⟨77AG(E)728⟩
Muyanchi, A., **1**, 290 ⟨85CL1829⟩
Muzaffar, A., **3**, 353 ⟨90JMC567, 90SC713⟩
Muzard, M., **4**, 252 ⟨92S965⟩
Muzart, J., **3**, 8, 33, 121, 163 ⟨93S11⟩, 37 ⟨85JOM(280)269⟩, 72 ⟨85TL839⟩, 215 ⟨86BSF65⟩, 275 ⟨88TL2321, 89SC2061⟩, 293 ⟨86TL3139, 87TL2131⟩, 294 ⟨92SC1993⟩; **5**, 28 ⟨92CRV113⟩, 296 ⟨86TL6339⟩
Muzika, K., **3**, 59 ⟨73TL2309⟩
Muzovskaya, E. V., **5**, 1006 ⟨83ZOB1096⟩
Muzyka, P. V., **6**, 693 ⟨91ZN(B)1539⟩
M. van den Broek, L. A. G.,**4**, 264 ⟨92RTC163⟩
M.-Xin Ding, **3**, 392 ⟨92CL2289⟩
Myamoto, S., **6**, 461 ⟨94JAP06041019⟩
Myatt, J., **6**, 297 ⟨72MI 610-01⟩
Mychajlowskij, W., **1**, 63, 451 ⟨74TL3479⟩, 111 ⟨74TL3511⟩, 535 ⟨77TL3317⟩, 648, 652, 982 ⟨74TL171⟩, 651, 760, 981 ⟨78JOC1526⟩; **2**, 624, 630 ⟨77TL3317⟩, 931, 932 ⟨76JOM(107)C1⟩, 931, 932, 935 ⟨76TL4439⟩
Mychajlowsky, W., **3**, 58 ⟨77TL3317⟩
Myers, A. G., **1**, 801, 803, 804 ⟨85JA5574⟩, 962 ⟨89TL5747⟩, 1038 ⟨91JA694⟩; **4**, 239 ⟨94JA4697⟩
Myers, D. K., **2**, 449 ⟨71JOC1285⟩
Myers, G. S., **4**, 294 ⟨48JA3659⟩
Myers, J. D., **1**, 636 ⟨70AG(E)374⟩
Myers, M., **1**, 302, 303 ⟨87JA7923⟩; **4**, 365, 366 ⟨85TL2965⟩; **5**, 142 ⟨73JCS(P1)2796⟩
Myers, M. R., **6**, 128 ⟨89JOC1290⟩
Myers, P. L., **3**, 194 ⟨93CC94⟩; **5**, 1120 ⟨87JMC951⟩
Myers, R. F., **1**, 1122, 1138 ⟨78TL2549⟩
Myers, R. R., **6**, 506 ⟨43OSC(2)462⟩
Myers, R. W., **2**, 1003 ⟨87CAR(168)7⟩
Myers, T. C., **2**, 456 ⟨57JOC972⟩
Myers, W. H., **2**, 435 ⟨74IS129⟩
Myerson, J., **1**, 67, 68 ⟨84T2317⟩

Myhre, M., **4**, 979 ⟨71JHC241⟩
Myhre, P. C., **1**, 906 ⟨82BSB422, 82JA907⟩, 906 ⟨90JA8991⟩
Mykytka, J. P., **2**, 614 ⟨71JOC631⟩
Mynott, R., **2**, 581 ⟨83CB159, 83CB176⟩, 984 ⟨89AG(E)610⟩; **3**, 509 ⟨89AG(E)610⟩; **4**, 583 ⟨85CB298⟩; **5**, 908 ⟨84T3273⟩, 917 ⟨85OM1310, 85ZN(B)1177⟩
Myong, S. O., **3**, 132 ⟨80TL3339⟩; **4**, 234 ⟨80TL3339⟩
Myoung, Y. C., **3**, 64 ⟨90JOC292⟩
Myrboh, B., **1**, 1076 ⟨82S1100⟩
Myrczek, J., **6**, 396 ⟨93JCS(D)2891⟩
Mysin, N. I., **5**, 391 ⟨75ZOB2684⟩, 1081 ⟨82ZOB921⟩
Mysov, E. I., **1**, 493 ⟨74JOM(72)163⟩; **4**, 736 ⟨76IZV895⟩, 803 ⟨76BAU853, 76BAU873⟩; **6**, 22 ⟨90IZV491⟩

N

Naae, D. G., **2**, 612 ⟨79CL983⟩; **3**, 495 ⟨73JA8467, 83JOC3616⟩; **4**, 7 ⟨73JA8467⟩, 34 ⟨79JOC1394⟩, 52 ⟨80JOC1394⟩, 728 ⟨71JFC(1)123⟩, 733 ⟨77JOC1780⟩, 733, 734 ⟨73SC197, 79CL983⟩, 734 ⟨81JFC(18)293⟩, 754 ⟨71JFC(1)381⟩; **5**, 321 ⟨75JOC2129⟩; **6**, 695 ⟨75TL3789⟩
Naae, D. J., **6**, 695 ⟨79CL983⟩
Naal, N. T., **3**, 178 ⟨81JOC1224⟩
Nabetani, M., **4**, 102 ⟨91JCS(P1)49⟩
Nabeya, A., **3**, 61 ⟨73JOC36⟩
Nabi, S. N., **2**, 172 ⟨63JCS2636⟩, 239 ⟨60JCS1103⟩
Nace, H. R., **1**, 775, 776 ⟨62OR(12)57, 53JA2118⟩; **2**, 111 ⟨55JA1578⟩, 125, 127 ⟨53JA2118⟩; **5**, 146 ⟨73JOC1941⟩
Naceur, J. B., **4**, 713 ⟨B-85MI 408-01⟩
Nachbaur, E., **5**, 1015 ⟨66M115, 66M361⟩
Nachod, F. C., **3**, 252 ⟨51JA2544⟩
Naddaka, V. I., **5**, 340 ⟨89ZOB611⟩, 349 ⟨81ZOR401⟩, 1078 ⟨88ZOR1581⟩
Nadeau, R., **4**, 53, 54 ⟨67JOM(8)361⟩
Nadel, M. E., **1**, 889 ⟨64JA1096⟩
Nader, F. W., **1**, 981 ⟨85AG(E)851⟩, 987 ⟨86CB1208⟩, 988 ⟨86AG(E)93⟩; **6**, 79 ⟨70JA584⟩
Naderi, M., **2**, 215 ⟨83SC611⟩
Nadi, A.-I., **3**, 246 ⟨84JHC13⟩
Nadir, U., **3**, 438 ⟨83S554⟩
Nadir, U. K., **2**, 357, 768 ⟨79OPP33⟩; **5**, 347 ⟨77TL1721⟩
Nadj, M. M., **6**, 522 ⟨40DOK(26)54⟩
Nadler, D., **2**, 4366 ⟨76JOM(118)C8⟩
Nadler, E. B., **2**, 643 ⟨90JOC2673⟩, 645 ⟨87JA2112⟩, 930 ⟨90TL555⟩; **5**, 406 ⟨91JOC4241⟩
Naef, R., **2**, 588 ⟨86JA7115⟩, 935 ⟨85TL5375⟩; **4**, 197 ⟨80HCA1960⟩, 213 ⟨84T1313⟩, 302 ⟨83JA5390⟩, 419, 424 ⟨85HCA135⟩
Naegele, W., **2**, 137 ⟨63JOC1945⟩; **4**, 15, 24, 29 ⟨65JA3151⟩
Naegeli, P., **1**, 525 ⟨72TL2013⟩
Naegely, P. C., **4**, 71 ⟨80JOC752⟩
Näf, F., **1**, 403, 734 ⟨75HCA1016⟩, 465 ⟨71HCA1939⟩, 618 ⟨82TL5043⟩, 889 ⟨74HCA1317⟩; **3**, 187 ⟨71JA2492⟩, 494 ⟨75HCA1016, 77HCA1161⟩; **5**, 151 ⟨71HCA1939⟩
Nafikova, A. A., **4**, 549 ⟨84ZOB1511⟩
Nagabhushan, T. L., **1**, 508 ⟨87TL5395⟩
Nagabundi, S., **5**, 710, 716 ⟨81H(15)165⟩
Nagahara, S., **3**, 34 ⟨91T6983⟩
Nagai, Y., **5**, 777 ⟨74JOM(72)C11⟩
Nagai, K., **1**, 77 ⟨87JOC3174⟩; **4**, 14 ⟨83BCJ2480⟩
Nagai, M., **1**, 867 ⟨90JOC996⟩; **2**, 239 ⟨88CL857⟩; **4**, 881 ⟨68TL497⟩; **6**, 106 ⟨68TL497⟩
Nagai, N., **5**, 1078 ⟨85CPB373⟩
Nagai, S., **4**, 365 ⟨84JCS(P1)435⟩; **5**, 226 ⟨80TL1467⟩
Nagai, T., **2**, 207 ⟨78BCJ1459⟩, 249, 401 ⟨79BCJ1102⟩; **4**, 109 ⟨89JOC1135⟩, 198 ⟨80TL3919⟩, 277 ⟨92TL1321⟩, 811 ⟨94TA961⟩
Nagai, W., **6**, 60 ⟨90T3859⟩, 165 ⟨90TL3859⟩
Nagai, Y., **2**, 21 ⟨78CL223⟩, 264 ⟨73JOM(50)26⟩, 540 ⟨73JOM(50)C26⟩, 546 ⟨77CC534⟩, 912 ⟨78AG(E)279⟩, 915 ⟨74JOM(66)C14, 80JOM(186)51⟩, 915, 923 ⟨81JOM(216)149⟩; **3**, 616 ⟨75CL331⟩, 617 ⟨75CL737⟩, 640 ⟨74TL4005⟩; **4**, 1044 ⟨82CL1663⟩; **5**, 57 ⟨82CL1255⟩, 777 ⟨75JOM(99)C5⟩; **6**, 7 ⟨73TL5147, 75TL899, 78JOC1727⟩
Nagai, Z., **3**, 498, 651 ⟨84SC1267⟩
Nagakawa, T., **2**, 397 ⟨71TL2053⟩
Nagaki, D. A., **4**, 148 ⟨86OM630⟩
Nagakura, I., **1**, 469 ⟨82CJC1256⟩, 804 ⟨83CJC3227⟩, 811 ⟨78JOC3630, 83CJC1226⟩; **3**, 217 ⟨82CJC1256⟩, 239 ⟨75SC193, 82CJC210⟩
Nagamitsu, T., **5**, 235 ⟨93JA5302⟩
Nagamo, M., **6**, 629 ⟨73CPB62⟩
Naganawa, H., **3**, 717 ⟨87JAN1199, 87JAN1202⟩
Nagano, H., **1**, 327 ⟨94CC987⟩
Nagano, T., **1**, 1010 ⟨88CPB2248⟩
Nagano, Y., **4**, 21 ⟨78CL73⟩; **5**, 516, 520 ⟨92CL1843⟩; **6**, 552, 554 ⟨86CE311⟩, 552, 554 ⟨87NKK1408⟩
Nagao, K., **3**, 207 ⟨81CPB2733⟩
Nagao, S., **2**, 21 ⟨84CL877⟩; **5**, 678 ⟨84CPB4437⟩
Nagao, Y.,**1**, 431 ⟨88T4095⟩, 431, 1068 ⟨88JA6565⟩, 931 ⟨87CL133⟩, 1016, 1145 ⟨90CC118⟩, 1080 ⟨85CC697⟩; **2**, 46 ⟨88JA4606⟩, 150 ⟨77TL1345⟩, 917, 918 ⟨88JA6565⟩, 1020 ⟨90CC118⟩, 1020, 1063 ⟨85TL4501⟩, 1062 ⟨87CC1708⟩; **3**, 36, 172 ⟨79TL3167⟩, 85 ⟨78CC330⟩, 143 ⟨84CC1007⟩, 162 ⟨89JOC4038⟩; **4**, 198 ⟨78TL4115⟩, 808 ⟨88CC1076, 89TL6701⟩; **5**, 204 ⟨90JA5677⟩, 223 ⟨85JCS(P1)2361⟩, 277, 295 ⟨80TL841⟩, 316 ⟨91MI 507-05⟩, 329 ⟨86JOC2391⟩, 329, 330 ⟨85CC1418⟩; **6**, 116 ⟨86JAP61158986, 86JOC4737⟩, 116 ⟨87NKK1447⟩
Nagaoka, H., **1**, 189 ⟨86CL1009⟩, 582 ⟨89CPB2882⟩, 748 ⟨81T3873⟩; **2**, 73 ⟨81T3873⟩; **3**, 139 ⟨86CL1009⟩, 226 ⟨84TL3621⟩, 279, 281, 298 ⟨86CL165⟩, 528 ⟨82JA747⟩
Nagaoka, T., **3**, 430 ⟨90BCJ1843⟩
Nagarajan, K., **4**, 94 ⟨84H(22)585⟩; **5**, 690 ⟨82IJC(B)989⟩
Nagarajan, M., **1**, 176 ⟨88TL107⟩, 330 ⟨88JOC1432⟩; **4**, 277 ⟨84S757⟩
Nagarajan, S., **1**, 514 ⟨87JOC5044⟩; **6**, 606, 645 ⟨92SC1191⟩
Nagarajan, S. C., **4**, 50, 379, 381, 382, 951, 952 ⟨85JOC3255⟩
Nagaraju, S., **1**, 416 ⟨92JCS(P1)311⟩
Nagasaka, T., **3**, 309 ⟨77CPB3023⟩
Nagasaki, N., **5**, 48, 49 ⟨85S674⟩
Nagasaki, Y., **4**, 660 ⟨90BCJ3036⟩
Nagasawa, F., **3**, 311 ⟨39CB1470⟩
Nagasawa, H., **3**, 47, 192 ⟨87S729⟩; **5**, 171 ⟨85JA6404⟩
Nagasawa, H. T., **5**, 344 ⟨85JOC4993⟩, 1120 ⟨87SC1677, 90JOC3403⟩; **6**, 447 ⟨92JMC3641⟩
Nagasawa, M., **1**, 1001 ⟨74TL423⟩
Nagasawa, N., **2**, 702 ⟨77CL679⟩
Nagasawa, S., **3**, 115 ⟨91BCJ2857⟩
Nagasawa, T., **5**, 118 ⟨90PAC1441⟩; **6**, 434 ⟨73BCJ1269⟩
Nagase, H., **1**, 710 ⟨84JA4547⟩; **3**, 343 ⟨91JMC1292⟩; **4**, 222 ⟨76TL65⟩; **5**, 726 ⟨79TL4671⟩
Nagase, K., **2**, 518 ⟨62JA190⟩
Nagase, S.,**1**, 323 ⟨86JOC2874⟩; **3**, 605 ⟨85PS(25)237⟩; **4**, 3 ⟨66BCJ219⟩, 624 ⟨93OM1514⟩; **5**, 1086, 1087 ⟨88POL2213⟩; **6**, 20 ⟨67FCR77⟩, 36 ⟨76BCJ1888⟩, 37 ⟨79JFC(13)519, 79JFC(15)353⟩, 39 ⟨78JFC(12)1, 78JFC(12)359⟩, 39 ⟨83JFC(23)123⟩, 51 ⟨77JFC(9)279⟩, 410 ⟨76JAP7630034⟩
Nagase, T., **1**, 754 ⟨84CL1997⟩; **5**, 995 ⟨88CL1385⟩; **6**, 409 ⟨70JAP7026611⟩
Nagase, Y., **5**, 223 ⟨85JCS(P1)2361⟩
Nagashima, A., **4**, 264 ⟨76BCJ502⟩
Nagashima, E., **1**, 990 ⟨83CPB3306⟩
Nagashima, H., **1**, 493 ⟨84T2699⟩; **3**, 114 ⟨82TL3085⟩, 118 ⟨85T5645⟩, 175 ⟨78TL799⟩, 521 ⟨89JA2352⟩; **4**, 199

⟨89CL659⟩; **5**, 346 ⟨91TL7431⟩, 678 ⟨84CPB4437⟩, 746 ⟨78CPB2246⟩; **6**, 7 ⟨81CL1169, 85T393⟩
Nagashima, M., **3**, 360 ⟨90CC1665⟩, 370 ⟨92CC411⟩
Nagashima, N., **4**, 1033 ⟨80BCJ1138⟩
Nagashima, S., **2**, 912 ⟨78AG(E)279⟩
Nagashima, T., **3**, 46 ⟨90S694⟩
Nagata, K., **1**, 248 ⟨87JOC5063⟩; **2**, 326 ⟨89TL6395⟩; **4**, 60 ⟨92TL5399⟩, 101 ⟨87JCS(P1)2203⟩, 108 ⟨87JOC5061, 91CPB3120⟩, 108, 780 ⟨89JOC5453⟩; **5**, 636 ⟨66CPB1249, 68JAP6827842⟩, 830 ⟨67CPB61⟩, 830, 831 ⟨66CPB1263, 69CPB653⟩; **6**, 560 ⟨87CPB156⟩
Nagata, M., **5**, 693 ⟨81JOC1769⟩
Nagata, R., **1**, 601 ⟨87TL4917⟩, 988 ⟨90TL7469⟩, 993 ⟨89TL4995⟩
Nagata, S., **2**, 17 ⟨84CL1241⟩, 19 ⟨85CL1693⟩
Nagata, T., **1**, 927 ⟨88TL2215⟩
Nagata, W., **1**, 208 ⟨70JCS2365⟩; **2**, 88 ⟨83JA2908⟩; **3**, 49 ⟨69JCS(C)460⟩, 60 ⟨68TL4359⟩, 614, 616, 633 ⟨77OR(25)255⟩, 616 ⟨72JA4654⟩, 632 ⟨70JCS(C)2365⟩, 634 ⟨72JA4635, 72JA4644, 73SC265⟩; **5**, 81 ⟨78GEP2725519⟩, 655 ⟨79TL4943⟩, 1142 ⟨72JA4635, 72OS5290⟩, 1142, 1143 ⟨65USP3444161⟩; **6**, 107 ⟨82H(18)201⟩, 111 ⟨79CC396⟩
Nagata, Y., **4**, 361 ⟨90SC3245⟩; **5**, 431 ⟨90SC3245⟩
Nagatani, T., **2**, 326 ⟨93JMC3526⟩
Nagatsuka, M., **5**, 731 ⟨67BCJ2705⟩
Nagel, A., **1**, 99 ⟨75JA380⟩; **3**, 219, 231 ⟨75JA380⟩
Nagel, C. J., **5**, 709 ⟨83JOC888⟩
Nagel, D., **3**, 616 ⟨71CB3594⟩
Nagel, D. L., **1**, 700, 703 ⟨88JOC1064⟩
Nagel, D. W., **6**, 138 ⟨76T2625⟩
Nagel, F. J., **3**, 593 ⟨69JA4761⟩
Nagel, K., **1**, 665 ⟨60LA(629)210⟩; **2**, 596 ⟨54LA(589)91⟩, 1009 ⟨68TL4265⟩; **4**, 668 ⟨55ZAAC(282)345⟩; **6**, 206 ⟨55ZAAC(282)345⟩
Nagel, M., **3**, 417 ⟨85CB3089⟩; **4**, 229 ⟨84AG(E)890⟩; **5**, 694 ⟨85CB3089⟩
Nagel, R. von, **6**, 267, 293, 378 ⟨70TL4693⟩
Nagel, U., **1**, 76 ⟨84AG(E)435⟩; **2**, 437 ⟨88CB1123, 92CB1061⟩; **4**, 710 ⟨86AG(E)734⟩
Nagelkerke, R., **6**, 100, 102 ⟨89H(28)389⟩, 101 ⟨88JOC5668⟩
Nagendrappa, G., **1**, 459 ⟨70JOC347, 70TL4243⟩, 574 ⟨90S843⟩, 650 ⟨80JA6894⟩; **4**, 357 ⟨85JOM(280)31⟩
Nageswar, Y. V. D., **6**, 450 ⟨88S994⟩
Naghski, J., **5**, 83 ⟨B-55MI 502-02⟩
Nagira, K., **1**, 486 ⟨81T31⟩; **5**, 197, 198 ⟨80TL2877⟩, 198 ⟨81JOC4413⟩
Nagl, A., **6**, 207 ⟨83JOM(253)283⟩
Nagler, M., **1**, 498 ⟨82JOC3297⟩
Nagler, R. C., **2**, 421 ⟨55JA1796⟩
Nago, H., **4**, 273 ⟨92JCS(P1)1179⟩
Nagubandi, S., **5**, 655 ⟨80JHC1457⟩
Nagura, K., **2**, 650 ⟨74T3021⟩
Nagy, F., **5**, 709 ⟨82MI 517-01⟩
Nagy, M., **3**, 12 ⟨92JOC5761⟩
Nagy, S. M., **3**, 618 ⟨55OSC(3)690⟩
Nagy, T., **1**, 154 ⟨84TL263⟩; **3**, 427 ⟨84TL263⟩
Nagy-Magos, Z., **4**, 455 ⟨80JOM(193)307⟩
Nahabedian, K. V., **1**, 447 ⟨61JA2159⟩
Nahar, K. R., **5**, 226 ⟨80IJC(A)69⟩
Nahm, S., **1**, 1047 ⟨81TL3815⟩; **3**, 13, 85, 273, 285 ⟨81TL3815⟩, 490 ⟨80JOC4065⟩; **5**, 677 ⟨81JOC1402⟩
Nahr, U., **5**, 1073 ⟨85CB526⟩
Naider, F., **2**, 133 ⟨92TL1435⟩
Naidoo, K., **1**, 181 ⟨88T1567⟩
Naidorf, S., **4**, 742 ⟨86TL6389⟩
Naidu, B. N., **1**, 400 ⟨93JA1177⟩
Naidu, M. V., **3**, 87, 96 ⟨79S144⟩
Naihan, H., **2**, 1047 ⟨91SL117⟩
Naik, A. R., **5**, 670, 819 ⟨66CJC297⟩
Naik, R. G., **2**, 530 ⟨85JOC5586⟩
Naik, V. G., **1**, 802 ⟨91T8259⟩
Naimushin, A. I., **6**, 479 ⟨90ZOB1826⟩
Nainan, K. C., **4**, 563 ⟨69IC2671⟩
Naipawer, R. E., **1**, 72 ⟨73JA6379⟩
Nair, H. K., **4**, 151, 152 ⟨92JFC(56)341⟩; **6**, 58, 264 ⟨94JOC2393⟩, 246, 247 ⟨89JOM(376)149⟩
Nair, M., **1**, 1078 ⟨79TL2961⟩
Nair, M. D., **3**, 361 ⟨82T1673⟩; **4**, 94 ⟨82IJC(B)1, 82T1673⟩; **5**, 760 ⟨79JCS(P1)1762⟩
Nair, P., **1**, 493 ⟨81T173⟩
Nair, P. G., **5**, 652 ⟨72TL4785⟩
Nair, P. M., **5**, 15 ⟨83T1187⟩
Nair, V., **1**, 36 ⟨79TL1181⟩, 550 ⟨91COS(4)999⟩; **2**, 368 ⟨87OPP466⟩; **6**, 15 ⟨89JA8502⟩
Naish, P. J., **4**, 679 ⟨82JCS(D)1297, 83JCS(D)1417⟩
Naito, I., **1**, 1120 ⟨80T3229⟩
Naito, K., **3**, 206 ⟨91JCS(P1)954⟩
Naito, T., **2**, 352 ⟨94TL2205⟩; **4**, 908 ⟨93CPB217⟩; **5**, 726 ⟨92CPB602⟩
Naito, Y., **3**, 231 ⟨79JA5070⟩
Najafi, M. R., **2**, 922, 929 ⟨91JOC2468⟩; **4**, 1056 ⟨86JOC1330⟩
Najam, A. A., **2**, 608 ⟨72CC840⟩
Najdi, S. D., **3**, 218 ⟨90JA441⟩
Nájera, C., **1**, 145 ⟨89JOC1491⟩, 145, 1090 ⟨93TL2003⟩, 237 ⟨89TL3837⟩, 486, 1031 ⟨89TL173⟩, 948 ⟨84JCS(P1)721⟩, 1089, 1090 ⟨87TL6709⟩, 1090 ⟨89JCS(P1)1387⟩; **2**, 24, 31 ⟨86JCR(S)274, 85CC1422⟩, 392 ⟨90TL4199⟩; **3**, 27, 155 ⟨86S678⟩, 256 ⟨88JCS(P1)1029, 88JOC4708⟩, 429 ⟨85CC126, 85T699⟩, 627 ⟨93T4015⟩; **4**, 393 ⟨92TL4065⟩, 808 ⟨87JCS(P1)1017⟩, 938, 961 ⟨87TL6709⟩, 938, 961 ⟨88JOC4708⟩; **5**, 284 ⟨81CC1178, 81CC670⟩, 1106 ⟨91T5193⟩
Najim, S. T., **6**, 290 ⟨86JOM(315)C5⟩, 290, 391 ⟨87CC1461⟩, 381 ⟨93JCS(P2)391, 93JCS(P2)59⟩, 381, 392 ⟨87JCS(P2)891⟩
Naka, H., **3**, 142 ⟨87CL1371⟩; **5**, 468 ⟨87BCJ3462⟩
Naka, M., **4**, 669 ⟨76BCJ1177⟩
Nakabayashi, S., **3**, 357 ⟨90JAN1150⟩; **4**, 225 ⟨81CAR(95)308⟩
Nakabayashi, T., **2**, 211 ⟨59JOC807, 72IJS199⟩, 211, 722 ⟨71JOC3677⟩, 723 ⟨61JOC2483⟩; **5**, 247 ⟨75BCJ2993, 75CL1231⟩, 540 ⟨81BCJ790⟩, 1111 ⟨83MI 530-01⟩
Nakada, M., **4**, 354 ⟨88JA4826⟩; **5**, 418 ⟨91TL4929⟩
Nakadaira, Y., **1**, 449 ⟨88OM1869⟩, 922 ⟨88JPO197⟩; **2**, 916 ⟨75JA931⟩; **4**, 612 ⟨74JA5623⟩, 612, 614 ⟨76JA7424⟩, 613 ⟨88CL1441⟩, 616 ⟨86BCJ1509⟩, 640 ⟨82CL1855⟩, 1048 ⟨77JA3879⟩, 1050 ⟨82JA4288⟩, 1051, 1052 ⟨88CL965, 78JOM(162)C43⟩, 1052 ⟨80AG632, 80TL3077⟩; **6**, 380 ⟨86JOM(309)247⟩, 383 ⟨88CL1441⟩
Nakadate, M., **6**, 504 ⟨83CPB41⟩
Nakagawa, H., **3**, 634 ⟨87CL1963⟩; **4**, 369 ⟨87CL1963⟩, 521 ⟨86CPB3273⟩; **6**, 472 ⟨92OPP200⟩
Nakagawa, I., **2**, 135 ⟨75TL1409⟩
Nakagawa, K., **3**, 464 ⟨66CC730⟩, 620 ⟨63CPB296⟩; **4**, 62, 63 ⟨90JOC6074, 91CC1027⟩; **5**, 48, 113 ⟨62JOC1597⟩, 48, 114 ⟨64CPB403⟩, 251 ⟨80SC305⟩, 295, 299 ⟨66CC17⟩
Nakagawa, M., **1**, 476 ⟨81CL1307⟩; **2**, 313 ⟨93CPB287⟩; **3**, 442 ⟨86TL4763⟩; **5**, 567 ⟨69CPB550⟩
Nakagawa, N., **3**, 200 ⟨87BCJ2463⟩; **5**, 821 ⟨87BCJ2463⟩
Nakagawa, R., **4**, 521 ⟨88JOM(339)259⟩
Nakagawa, T., **5**, 48, 49 ⟨85S674⟩, 318 ⟨84CL2045, 85BCJ769⟩
Nakagawa, Y., **1**, 826 ⟨88JOC3195⟩, 1036 ⟨85JOC565⟩; **2**, 50 ⟨88JA3712⟩, 939 ⟨92JA5867⟩; **3**, 76 ⟨72CPB1325⟩; **4**, 221 ⟨86TL3869⟩
Nakahama, S., **2**, 65, 528 ⟨87BCJ395⟩, 71 ⟨90JOC304⟩; **3**, 440 ⟨82S461⟩; **5**, 368 ⟨85MAC1803⟩

Nakahara, J. H., **4**, 481 ⟨85JFC(30)269⟩; **5**, 786 ⟨85JFC(30)269, 85JFC(30)289⟩
Nakahara, T., **6**, 80 ⟨85JAP(K)60208983⟩
Nakahara, Y., **2**, 10, 20 ⟨85TL4207⟩, 18, 280 ⟨89CC1881⟩; **3**, 308 ⟨86T6465⟩
Nakahashi, K., **2**, 320 ⟨91JOC5263⟩; **5**, 224 ⟨89JOC4511⟩, 274, 286 ⟨87JA1798⟩
Nakahata, M., **5**, 63 ⟨82BCJ2186⟩
Nakahira, H., **1**, 666 ⟨92JOC17⟩; **2**, 555 ⟨91AG(E)177⟩; **3**, 177, 224 ⟨91TL229⟩
Nakai, A., **4**, 90 ⟨91CPB1365⟩
Nakai, T., **1**, 816 ⟨91CL43⟩
Nakai, E., **1**, 405, 816, 817 ⟨85CL1723⟩, 405, 816, 817 ⟨88TL4587⟩
Nakai, H., **2**, 610 ⟨79CL1499⟩; **3**, 151 ⟨77CL995⟩; **5**, 345 ⟨84JA1133⟩, 972 ⟨73CL1143⟩
Nakai, K., **1**, 189 ⟨93JOC1207⟩, 189 ⟨93T9479⟩; **6**, 636 ⟨93BCJ148⟩
Nakai, S., **2**, 30, 34 ⟨86JA6036⟩, 108 ⟨79CPB1700⟩
Nakai, T., **1**, 196, 404, 816 ⟨86CRV885⟩, 210 ⟨75TL4027⟩, 374 ⟨91TA643⟩, 404, 405, 816, 817, 818 ⟨83JOC279⟩, 404, 405, 816, 818 ⟨81JA6492⟩, 404, 406, 816 ⟨81TL69⟩, 404, 406, 816, 820 ⟨86TL4511⟩, 404, 407, 816 ⟨91S594⟩, 405 ⟨92TL5795⟩, 405, 816 ⟨83TL513⟩, 405, 816, 817 ⟨84T2303, 84TL565, 85CL1723, 88TL4587⟩, 405, 816, 819 ⟨85CL1729⟩, 415, 826 ⟨87TL5879⟩, 525 ⟨87CL2347⟩, 552 ⟨91JOC5973⟩, 810 ⟨93T1025⟩, 816 ⟨82TL3931, 83CL1379, 92T4087, 93CL1397⟩, 816, 817 ⟨84CL255⟩, 816, 817, 818 ⟨85CL1719⟩, 816, 818 ⟨85TL5799⟩, 816, 819 ⟨84TL6011, 86TL4185, 86TL4899, 87CL69⟩, 817 ⟨86TL4189⟩, 818 ⟨93SL527, 91SL629⟩, 823 ⟨90CL2069⟩, 825 ⟨81CL1289⟩, 826 ⟨90JA4035⟩, 828 ⟨81TL4097⟩; **2**, 21 ⟨81TL2291⟩, 70 ⟨90JA6737⟩, 80 ⟨89JA1940, 90JA3949⟩, 88 ⟨86CRV885⟩, 612 ⟨79CL983⟩, 932 ⟨92SL189⟩, 1042 ⟨77BCJ3069⟩; **3**, 22 ⟨82CL1349, 82TL3931⟩, 57 ⟨74TL3625, 77TL2425⟩, 126 ⟨79TL531⟩, 211 ⟨83TL795⟩, 615 ⟨92CL1169⟩; **4**, 379 ⟨92TL5795⟩, 392 ⟨74TL3625⟩, 734 ⟨79CL983⟩, 735 ⟨78TL4809⟩, 831 ⟨84TL3987, 84TL3991⟩, 852 ⟨78CL1243⟩, 949 ⟨77CL1379⟩; **5**, 38 ⟨80CL651⟩; **6**, 314 ⟨70BCJ3528⟩, 495 ⟨79CL1361⟩, 695 ⟨79CL983⟩, 730 ⟨74BCJ398⟩
Nakai, Y., **3**, 427 ⟨84JAN532⟩; **6**, 244 ⟨88BCJ3002⟩
Nakaido, S., **1**, 233, 400 ⟨72JOC1721⟩, 408, 409, 410, 820, 821, 822 ⟨72JA3870⟩, 820 ⟨71JOC1732⟩; **3**, 373, 378 ⟨72JOC1721⟩
Nakaiida, S., **5**, 242, 506, 509 ⟨89TL1829⟩
Nakajima, I., **1**, 466, 470 ⟨76BCJ1958⟩
Nakajima, J., **5**, 378 ⟨66BCJ2263⟩
Nakajima, J.-I., **4**, 614 ⟨58JOC292⟩
Nakajima, K., **6**, 487 ⟨93CL1401⟩
Nakajima, M., **1**, 135 ⟨90CC1657⟩; **2**, 50 ⟨89CC1921, 90TL1741⟩, 385 ⟨84CC451, 84TL2619⟩; **3**, 66 ⟨56CB2224⟩, 293 ⟨78TL363⟩, 353 ⟨81S36⟩; **4**, 918 ⟨79TL2047⟩; **6**, 119 ⟨78TL363⟩, 125 ⟨79TL4475, 82BCJ224⟩
Nakajima, N., **2**, 40 ⟨91JOC4778⟩, 90 ⟨88TL4139⟩, 1000 ⟨81JCS(P1)212⟩; **5**, 1139 ⟨86JOC898⟩
Nakajima, S., **3**, 88 ⟨78CL1397⟩
Nakajima, T., **1**, 182 ⟨79BCJ2377⟩, 212, 475 ⟨75BCJ960⟩, 463 ⟨80CC86⟩, 463 ⟨83BCJ1089, 82BCJ167⟩, 613 ⟨83TL4993⟩; **2**, 453 ⟨87CL1915⟩, 542 ⟨88JA1976⟩, 678 ⟨83TL4993⟩; **3**, 336 ⟨89SC2431⟩, 336, 340, 385 ⟨88JA1976, 93CE53, 88TL6965⟩, 390, 400 ⟨89JA8749⟩, 394 ⟨89TL2095⟩, 553 ⟨92TL7865⟩; **4**, 197 ⟨83TL4993⟩, 931 ⟨86CL181⟩; **5**, 429 ⟨86CL177⟩, 516, 520 ⟨92TL7865⟩, 597, 599 ⟨84CPB1658⟩, 597, 599 ⟨84SC1257, 87CPB3139⟩; **6**, 481 ⟨89CL1009⟩
Nakajima, Y., **1**, 179 ⟨76TL2455⟩; **3**, 616 ⟨75ABC571⟩, 621 ⟨76TL2455⟩; **5**, 189, 199 ⟨84TL6027⟩
Nakajo, E., **1**, 1094 ⟨87JOC4412⟩; **2**, 935 ⟨88T3997⟩; **5**, 793 ⟨84TL5139⟩; **6**, 127 ⟨85S698⟩
Nakama, A., **5**, 173 ⟨79CL1019⟩, 173 ⟨79CL1021⟩
Nakamichi, T., **6**, 306 ⟨89JAP(K)1182311⟩
Nakamine, T., **1**, 702, 1070 ⟨88TL1197⟩
Nakamizo, A., **5**, 334 ⟨85JMC1772⟩
Nakamizo, Y., **5**, 334 ⟨85JMC1772⟩
Nakamori, S., **4**, 226 ⟨81BCJ817⟩
Nakamoto, Y., **1**, 182 ⟨79BCJ2377⟩, 212, 475 ⟨75BCJ960⟩; **5**, 1120 ⟨92CC1379⟩
Nakamura, A., **1**, 504 ⟨88BCJ2859⟩; **2**, 442 ⟨71JA4301, 77JA7876⟩, 970 ⟨76JMOC285⟩; **4**, 231 ⟨90JA8189⟩, 589 ⟨84JA5920⟩, 603, 649 ⟨83OM21⟩; **5**, 441 ⟨85JA4440⟩, 463, 800, 801 ⟨69JA7196⟩
Nakamura, E., **1**, 86 ⟨89JA6257⟩, 122 ⟨87JA8056, 87TL337⟩, 122 ⟨89JA3091, 91SL539⟩, 124 ⟨92T5709⟩, 133 ⟨82JA1025, 85ACR181⟩, 281 ⟨75JA3257⟩, 283 ⟨76TL1699⟩, 313 ⟨86TL4029⟩, 360 ⟨89JA7285, 90JOC5553⟩, 360 ⟨92JA8707⟩, 483 ⟨88JA3296⟩, 876 ⟨78TL2079⟩, 925 ⟨89JA6849⟩, 925, 946 ⟨93TL1317⟩, 1038 ⟨76AG(E)498, 83T975⟩; **2**, 69 ⟨87JA8056⟩, 87 ⟨91JA8981⟩, 263 ⟨81BCJ805⟩, 587 ⟨92T5709⟩, 589 ⟨88JA1297, 90JA7428⟩, 702 ⟨83TL3341, 91SL539⟩; **3**, 156 ⟨89TL3323⟩, 211 ⟨83CC499⟩, 636 ⟨78S602⟩; **4**, 388 ⟨92JA9715⟩, 830 ⟨89JA7285⟩; **5**, 953, 956 ⟨92JA2985⟩, 958 ⟨92JA9715, 93JA9015⟩
Nakamura, H., **1**, 426, 428 ⟨87OPP442⟩; **2**, 626 ⟨87BCJ4187, 88BCJ597⟩; **3**, 55 ⟨77CL345⟩, 662 ⟨83JOM(243)95⟩; **5**, 251 ⟨80SC305⟩
Nakamura, K., **1**, 143 ⟨87JOC1414⟩, 426, 428 ⟨87OPP442⟩, 431 ⟨85JOC3283⟩; **2**, 40 ⟨91JOC4778, 91TL7075⟩, 69 ⟨85TL4763, 89JA4392⟩, 398 ⟨88TL4769, 90T7471⟩, 1000 ⟨67T3049⟩; **3**, 129 ⟨86CL87⟩, 161 ⟨90T7471⟩, 348 ⟨75BCJ2403⟩, 351 ⟨79JOC2244⟩, 373 ⟨67T4279⟩, 628 ⟨90CL1615⟩; **4**, 391 ⟨87JOC1414⟩, 654, 655 ⟨92CL2047⟩; **5**, 652 ⟨90TL3603⟩
Nakamura, M., **4**, 610, 617, 622 ⟨87BCJ2263⟩
Nakamura, N., **2**, 184 ⟨83JA7172⟩; **3**, 286 ⟨77CL645⟩; **4**, 67 ⟨93JOC2360⟩
Nakamura, R., **1**, 674 ⟨76CL1019⟩
Nakamura, S., **5**, 225 ⟨89BCJ539⟩, 418 ⟨91TL4929⟩, 898 ⟨68YZ1039⟩
Nakamura, T., **1**, 197, 1057 ⟨92TL3769⟩, 250 ⟨86JOC4375⟩, 570 ⟨94JCS(P1)3115⟩, 813 ⟨85BCJ146⟩; **2**, 69 ⟨88CC610⟩, 591 ⟨89OS(67)98⟩; **3**, 159 ⟨93TL6281⟩; **5**, 694 ⟨87NKK1304⟩, 936 ⟨94CL1537⟩
Nakamura, W., **4**, 291 ⟨83MI 406-01⟩
Nakamura, Y., **2**, 123 ⟨69JOC4170⟩, 288 ⟨92JCR(S)393⟩, 997, 1001 ⟨92JCS(P1)1721⟩; **5**, 541 ⟨79NKK1224⟩
Nakane, M., **3**, 592 ⟨89JCS(P1)2140⟩; **4**, 853 ⟨78JOC3922⟩; **5**, 1066 ⟨86CL135⟩
Nakanishi, A., **3**, 698 ⟨81JOC5445⟩; **5**, 546 ⟨72MI 513-01⟩, 1118 ⟨81JOC5445⟩; **6**, 533 ⟨83JAP60155152⟩
Nakanishi, H., **2**, 24 ⟨83CC947⟩; **5**, 560 ⟨85PS(22)135⟩; **6**, 546, 561 ⟨87S170⟩
Nakanishi, K., **1**, 690 ⟨86BCJ1109⟩; **2**, 133 ⟨91CC90⟩, 655 ⟨81TL3451⟩
Nakanishi, S., **2**, 213 ⟨76CL857⟩; **3**, 137 ⟨79CL321⟩; **6**, 421 ⟨55JA5033⟩, 422 ⟨55JA3099⟩
Nakanishi, T., **4**, 219 ⟨89JOC5998, 91CL121⟩, 618 ⟨89JA3748, 93CL267⟩; **6**, 176, 193 ⟨89JA3748⟩, 193 ⟨92CL867⟩, 193 ⟨93CL267⟩
Nakano, K., **2**, 219 ⟨67BCJ2913⟩; **4**, 205 ⟨76CPB102⟩
Nakano, M., **2**, 938 ⟨91CC703⟩; **3**, 640 ⟨82S915⟩, 640, 646, 659 ⟨83S917⟩; **4**, 202 ⟨91JOC2276⟩, 508 ⟨85CC1684⟩; **5**, 597, 599, 614 ⟨83S225⟩; **6**, 14 ⟨80JCS(P1)2755⟩
Nakano, S., **6**, 26 ⟨81TL871⟩, 115 ⟨93CL13⟩
Nakano, T., **2**, 21 ⟨78CL223⟩, 39 ⟨90TL7463⟩, 55 ⟨86JOC240, 88JOC3752⟩, 944 ⟨91S1163⟩; **3**, 7, 215 ⟨86JOC240⟩, 7, 30 ⟨86S774⟩, 57 ⟨81JOC240⟩, 557,558 ⟨91S1079⟩; **5**, 57 ⟨82CL1255⟩, 874 ⟨91S1163⟩; **6**, 7 ⟨73TL5147, 78JOC1734⟩
Nakano, Y., **1**, 666 ⟨73JOC1130⟩; **4**, 109 ⟨89JOC1135⟩; **5**, 692 ⟨88BCJ3731⟩

Nakao, A., **3**, 528 ⟨86TL5471⟩
Nakao, K., **1**, 468 ⟨93BCJ2446⟩
Nakao, R., **1**, 933 ⟨90JOM(381)155⟩; **5**, 123 ⟨81BCJ1267⟩
Nakao, T., **1**, 690 ⟨88JA3578⟩
Nakashima, K., **1**, 1063 ⟨91CC948⟩; **5**, 157 ⟨89TL6529⟩
Nakashima, S., **1**, 182 ⟨79BCJ2377⟩
Nakasuji, K., **3**, 367 ⟨77AG(E)866⟩
Nakata, M., **1**, 815 ⟨76BCJ2017⟩, 956 ⟨93CC1406⟩
Nakata, S., **4**, 238 ⟨91CL1775⟩
Nakata, T., **1**, 478 ⟨79TL4343⟩, 590, 591 ⟨88TL659⟩, 976 ⟨91JOM(417)305, 91TL3397⟩, 976, 1025 ⟨90TL7179⟩; **2**, 61 ⟨83TL2653⟩; **5**, 48, 113 ⟨62JOC1597⟩, 157 ⟨89TL6529⟩; **6**, 461 ⟨91JOM(417)305⟩
Nakatani, H., **3**, 635 ⟨86JCS(P1)1763⟩; **4**, 67 ⟨93JOC2360⟩, 943 ⟨87JCS(P1)589⟩
Nakatani, K., **1**, 929 ⟨84TL5335⟩, 1029 ⟨91TL3405⟩; **2**, 429 ⟨83BCJ219⟩, 669 ⟨94TL605⟩; **3**, 30, 33 ⟨94T2821⟩, 129 ⟨88TL2283⟩, 143 ⟨84TL5335⟩; **6**, 447 ⟨92TL3483⟩
Nakatani, N., **2**, 138 ⟨92TL947⟩
Nakatani, S., **3**, 36, 165 ⟨88CC1468⟩, 36, 165 ⟨93T2011⟩; **4**, 937 ⟨93T2011⟩
Nakatini, F., **2**, 323 ⟨88BCJ3029⟩
Nakatomi, S., **6**, 425 ⟨79JAP7961121⟩
Nakatsu, K., **3**, 597 ⟨80JOM(190)C39⟩
Nakatsubo, F., **4**, 216 ⟨77JA4835⟩
Nakatsugawa, K., **3**, 616 ⟨75CL331⟩
Nakatsuji, H., **1**, 1109 ⟨68JA1239⟩; **5**, 932 ⟨84JA5892⟩, 933 ⟨83JA426⟩
Nakatsuji, S., **1**, 1063 ⟨91CC948⟩; **3**, 54, 62, 252 ⟨79TL2809⟩
Nakatsuka, M., **1**, 571 ⟨88OSC(6)327⟩, 655 ⟨82JA7609⟩, 1079 ⟨89JOC4267⟩; **2**, 22 ⟨80JOC2022⟩; **3**, 152 ⟨80JOC2022⟩, 224 ⟨80OS(59)113⟩
Nakatsuka, N., **1**, 400, 799 ⟨69JA4322⟩
Nakatsuka, S., **4**, 218 ⟨74TL1549⟩
Nakatsuka, S.-I., **2**, 657 ⟨87H(26)65⟩
Nakatsuka, T., **1**, 471 ⟨75JA4405⟩; **3**, 54 ⟨83CL1207⟩; **4**, 218 ⟨82CL1555⟩
Nakatsukasa, J., **2**, 457 ⟨77S179⟩
Nakatsukasa, S., **1**, 1038, 1140 ⟨85TL5585⟩; **2**, 918, 919, 939 ⟨88T4277⟩; **3**, 59 ⟨85TL5585⟩; **4**, 1045 ⟨88T4277⟩
Nakaya, C., **4**, 432 ⟨93S1218⟩; **5**, 344 ⟨91CL105⟩
Nakaya, M., **4**, 349 ⟨92CC190⟩
Nakayama, A., **4**, 398 ⟨92CL819⟩, 654, 655 ⟨92CL2047⟩, 847 ⟨78CC74⟩, 850 ⟨76JA7440⟩
Nakayama, E., **1**, 822 ⟨72TL2923⟩
Nakayama, J., **1**, 442 ⟨70T4609⟩, 899 ⟨90TL2623⟩; **2**, 27 ⟨93TL5725⟩, 478 ⟨84T4553⟩, 713 ⟨83CL249⟩; **3**, 3 ⟨78BCJ3016⟩, 15 ⟨82BCJ2289⟩, 164 ⟨85TL1981⟩, 177 ⟨88TL1399⟩, 251 ⟨87CL939⟩, 255 ⟨77CL287⟩, 352, 366 ⟨93TL115⟩, 363 ⟨90JOC2421, 92CC7⟩, 392 ⟨92CL2289⟩, 547 ⟨81BCJ2845⟩, 668 ⟨80CC791⟩; **4**, 88, 89 ⟨88CL1317, 77CL287⟩, 850 ⟨80BCJ1661, 81CC565⟩; **5**, 548 ⟨92CC1522⟩, 554, 563, 605 ⟨93BCJ623, 93TL115⟩, 627 ⟨94CL77⟩; **6**, 82 ⟨75S436⟩, 99 ⟨74CC166, 75S38⟩, 113 ⟨76BCJ3567, 80JOC2024⟩, 113, 316 ⟨81JCS(P1)618⟩
Nakayama, K., **1**, 408, 409, 410, 820, 821, 822 ⟨72JA3870⟩; **2**, 108 ⟨90JA6936⟩; **5**, 146 ⟨74JOC3273⟩
Nakayama, M., **2**, 447 ⟨88JOC2937⟩; **3**, 197 ⟨92CC190⟩; **5**, 208 ⟨90JFC(46)423⟩; **6**, 12 ⟨90JFC(46)423⟩
Nakayama, N., **1**, 156 ⟨93T537⟩; **2**, 1089 ⟨93OM803⟩
Nakayama, S., **2**, 851 ⟨75BCJ546⟩, 854 ⟨71CC1186⟩; **5**, 205 ⟨86JFC(31)37⟩; **6**, 12 ⟨86JFC(31)37⟩
Nakayama, Y., **1**, 931 ⟨92JCS(P1)1837⟩; **2**, 440, 829 ⟨91TA667⟩; **4**, 731 ⟨85JFC(29)445⟩; **5**, 411 ⟨85TL2675⟩; **6**, 126 ⟨85TL2675⟩, 488 ⟨92JAP05339542⟩, 627 ⟨84JOC1703⟩
Nakayasu, S., **1**, 1015 ⟨92TL3503⟩
Nakayma, K., **1**, 820 ⟨71JOC1732⟩
Nakazaki, M., **1**, 690 ⟨83JA7171, 88JA3578⟩, 990 ⟨82JOC1435⟩
Nakazawa, A., **2**, 398 ⟨85ABC2331⟩
Nakazawa, H., **4**, 581 ⟨83OM1846⟩
Nakazawa, M., **1**, 818 ⟨93TL5923⟩, 1023 ⟨93SL37⟩
Nakazawa, T., **1**, 1035 ⟨72JOC3749⟩; **2**, 711 ⟨89S955⟩
Nakazima, Y., **3**, 348 ⟨75BCJ2403⟩
Nakazono, Y., **1**, 1042 ⟨83TL1801⟩; **3**, 15, 90 ⟨82BCJ1663⟩
Nakhmanovich, A. S., **3**, 253 ⟨77JOU479⟩; **4**, 940 ⟨91ZOB1893⟩, 942 ⟨90ZOB526⟩; **6**, 565 ⟨90ZOB602⟩
Nakonieczna, L., **5**, 322 ⟨85S929⟩, 323 ⟨87S418⟩
Nakova, N. Z., **4**, 826 ⟨81ZOR1550⟩; **6**, 74 ⟨78ZOR2229⟩
Nalbandy, M., **3**, 6, 119, 161 ⟨92SC1589⟩
Nalbandyan, G. K., **5**, 690 ⟨90CHE1170, 90KGS1403⟩
Nalewajek, D., **2**, 16 ⟨91JOC5962⟩; **3**, 149 ⟨91JOC5962⟩; **4**, 291 ⟨85JA675⟩, 874 ⟨80CC866⟩; **6**, 591, 594 ⟨80CC866A⟩, 628, 730 ⟨80CC866⟩, 631 ⟨81JOM(217)329⟩
Nalley, E. A., **5**, 824 ⟨76JOC252⟩
Nally, J., **1**, 327, 328 ⟨91JA7002, 90JA6740⟩
Nam, G.-S., **4**, 467 ⟨90TL1567⟩
Nam, K. D., **4**, 260 ⟨91T8091⟩
Nam, K.-S., **6**, 468 ⟨94SC305⟩
Namane, A., **2**, 767 ⟨85TL4443⟩
Namanworth, E., **1**, 893 ⟨67JA3576⟩, 894 ⟨70JA3234⟩, 915 ⟨67JA5259⟩
Namata, T., **2**, 360 ⟨71IJS215⟩
Namavari, M., **2**, 25, 31, 35 ⟨90TL4973⟩; **4**, 627 ⟨92OM3307⟩
Namba, T., **2**, 298 ⟨78JOC3345⟩; **6**, 428 ⟨73TL2287⟩
Nambiar, K. P.,**4**, 225 ⟨81JA3215⟩; **5**, 173 ⟨79JA7131⟩
Nambiar, S., **4**, 224 ⟨89TL2179⟩
Nambu, H., **1**, 125 ⟨69JA2147, 69JA6854⟩, 126 ⟨70JA5790⟩, 464 ⟨69JA6852⟩; **3**, 541 ⟨64JOC2242⟩, 622 ⟨69JA6854⟩, 622, 627 ⟨70JA5791⟩, 698, 721 ⟨62JOC3651⟩
Nambu, Y., **1**, 514 ⟨85CL1851⟩; **2**, 214 ⟨84JOC1434⟩
Nambudiry, M. E. N., **1**, 754 ⟨76JCS(P1)2386, 78JCS(P1)590⟩
Names, J., **3**, 310 ⟨89JOC2170⟩
Nametkin, N. J., **2**, 266 ⟨81JOM(214)415⟩
Nametkin, N. S., **1**, 63 ⟨81JOM(214)145⟩; **3**, 501 ⟨79CRV529⟩; **4**, 140 ⟨69DOK(185)311⟩, 610, 611, 614, 627 ⟨79CRV529⟩, 622 ⟨69DOK(189)334⟩, 656, 658 ⟨74IZV2861, 75IZV1600⟩
Namigata, F., **4**, 750 ⟨67BCJ1275⟩
Namwindwa, E. S., **2**, 147 ⟨91SL80⟩; **4**, 261 ⟨90SL457, 91SL80⟩
Namy, J. L., **1**, 3, 6, 8, 10, 17, 587, 680 ⟨80JA2693⟩; **2**, 56 ⟨84JOC2047⟩, 564 ⟨86T6573⟩, 565 ⟨80JA2693, 87JOM(328)81⟩; **3**, 7, 119 ⟨84JOC2045⟩; **5**, 439 ⟨91JOC3118⟩
Nanasawa, M., **5**, 1065 ⟨79BCJ533⟩
Nanbu, H., **6**, 20 ⟨89TL109⟩
Nandan, S. R., **1**, 582 ⟨94JOC281⟩
Nandi, K., **5**, 566, 592, 593, 610 ⟨75CB3066⟩
Nandi, K. N., **2**, 291 ⟨41JCS70⟩
Nangia, A., **1**, 176 ⟨93JA2581⟩, 826 ⟨88TL1669⟩; **5**, 189, 196 ⟨84JCR(S)100⟩
Naniwa, Y., **3**, 47, 192 ⟨87S729⟩
Nanjappa, P., **2**, 517 ⟨88T5597⟩
Nanjundiah, B. S., **1**, 802 ⟨91TL1107⟩
Nanjunidiah, B. S., **1**, 801 ⟨92TL1645⟩
Nannessian, S., **6**, 71 ⟨80MI 603-01⟩
Nanninga, T. M., **2**, 555 ⟨84OS(62)58⟩
Nanninga, T. N., **1**, 632 ⟨85JA1293⟩
Nantermet, P. G., **3**, 225 ⟨90JOC4235⟩
Nantz, M. H., **2**, 533 ⟨89TL7357⟩
Naoki, H., **2**, 655 ⟨81TL3451⟩
Naora, H., **1**, 504 ⟨88BCJ2859⟩

Naota, T., **2**, 606 ⟨84OS(62)39⟩, 654 ⟨92TL7557⟩; **3**, 113 ⟨93TL1299⟩, 158 ⟨93CL1767, 93JOC2929⟩; **5**, 171 ⟨87JOC4319⟩, 1139 ⟨86JOC898⟩
Napier, D. R., **3**, 446, 470 ⟨58JA2257⟩
Napier, R. P., **3**, 249 ⟨70JOC1709⟩
Napoletano, M., **6**, 608 ⟨88SC1171⟩
Napoli, J. J., **1**, 808 ⟨79JA1793⟩
Napoli, M., **4**, 31 ⟨76JFC(7)569⟩
Napolitano, E., **2**, 624 ⟨86S122⟩; **4**, 26 ⟨86S122⟩
Nappier, T. E., **2**, 438 ⟨73JA4194⟩
Naqvi, S. M., **1**, 402, 795, 801, 802 ⟨85OR(33)247⟩
Nara, S., **5**, 129 ⟨94JOC5532⟩
Narabu, S., **2**, 136 ⟨92CL1655⟩
Naraguchi, K., **4**, 240 ⟨92MI 405-01⟩
Narain, R. P., **5**, 785 ⟨85POL1151⟩; **6**, 78 ⟨63PIA(A)45⟩
Narang, C. K., **5**, 652 ⟨77SC57⟩
Narang, S. C., **2**, 32 ⟨80JOC1638⟩, 35 ⟨81JOC3727, 79AG(E)612⟩, 36 ⟨79JOC1247, 83JOC2766⟩, 36, 538 ⟨82T2225⟩, 88, 104 ⟨77JCS(P1)1114, 78S452⟩, 142 ⟨78JOC4503⟩, 142 ⟨78S137⟩, 150 ⟨79JA5317⟩, 452 ⟨83JOC1760⟩, 791 ⟨82JOC596⟩; **3**, 279, 295, 296 ⟨78S672⟩, 426 ⟨83JOC2766⟩, 618 ⟨80S657, 80S659⟩; **4**, 191 ⟨81S282⟩; **5**, 226 ⟨79JOC1247, 80S661⟩, 270, 282, 300 ⟨79S274⟩
Narasaka, K., **1**, 178 ⟨88BCJ571⟩, 308 ⟨76BCJ779⟩, 358, 498 ⟨92JA8869⟩, 375 ⟨91TA1305⟩, 427, 428, 430 ⟨73CL291⟩, 582 ⟨93OSC(8)210⟩, 929 ⟨91CL1193, 93CI(L)226⟩, 1134 ⟨77CL959⟩, 1140 ⟨90CL1411⟩; **2**, 143 ⟨72CL259⟩, 394 ⟨91CL423⟩, 656 ⟨75CL1041⟩; **3**, 7 ⟨77BCJ2773⟩, 25 ⟨82CL1637⟩, 120 ⟨77CL757⟩, 137 ⟨93CL545⟩, 628 ⟨87CL2073, 88BCJ4379⟩; **4**, 235 ⟨84JA2954⟩, 252 ⟨92CL1229⟩; **5**, 173 ⟨84JA2954⟩, 427 ⟨91CL1413⟩, 430 ⟨93BCJ819⟩, 557 ⟨88CL1025⟩; **6**, 127, 325, 327 ⟨90CL1411⟩
Narasako, K., **1**, 929 ⟨93BCJ819⟩
Narasimhamurthy, N., **6**, 544 ⟨88TL827⟩
Narasimhan, N., **1**, 108 ⟨80JA7760⟩
Narasimhan, N. S., **3**, 90 ⟨79S906⟩, 109 ⟨74T4153⟩
Narasimhan, S., **1**, 14 ⟨83JOC3091⟩; **2**, 629 ⟨84JOC4822⟩
Narasunhan, S., **2**, 320 ⟨81S441⟩
Narayana, C., **1**, 168 ⟨92TL7687⟩; **2**, 85 ⟨90TL6977⟩, 221 ⟨91SL125⟩, 312 ⟨89SC561⟩, 655 ⟨92TL865, 93TL7667⟩; **3**, 133 ⟨92SC2587⟩; **5**, 27 ⟨85S253⟩
Narayanan, B. A., **1**, 652 ⟨87TL6261⟩; **2**, 89 ⟨90JOC768⟩; **4**, 655, 656 ⟨87TL6261⟩
Narayanan, C. R., **3**, 133 ⟨77CI(L)454⟩
Narayanan, K., **1**, 454 ⟨84JOC3411⟩; **3**, 310 ⟨89JOC2170⟩
Narayanan, K. S., **2**, 444 ⟨91USP4997952⟩
Narayanan, K. V., **3**, 216 ⟨71CRV429⟩
Nardelli, M., **1**, 539 ⟨88T127⟩
Nardi, D., **4**, 852 ⟨74AC(R)305⟩
Nardi, N., **2**, 438 ⟨84JCS(D)575⟩
Nardin, G., **4**, 585 ⟨74JOM(74)C23⟩, 696 ⟨79JOM(174)121⟩, 711 ⟨94OM1999⟩
Nared, K. D., **1**, 826 ⟨88JA5593⟩
Nargund, R. P., **1**, 17 ⟨86TL4813⟩
Narimatsu, S., **1**, 206, 289, 1092 ⟨82TL1267⟩
Narine, B., **3**, 105 ⟨80S(59)133⟩
Narisada, M., **1**, 428, 429 ⟨89JOC5308⟩
Narisano, E., **4**, 856 ⟨81JCS(P1)1278⟩; **5**, 336 ⟨90TA141⟩
Narisawa, S., **2**, 80 ⟨92JA6566⟩
Narita, H., **5**, 385 ⟨86BCJ2203⟩
Narita, K., **3**, 356 ⟨74CL1101⟩, 376 ⟨78TL251⟩; **6**, 434 ⟨73BCJ1269⟩
Narita, M., **3**, 40 ⟨82TL525⟩; **4**, 851 ⟨76S489⟩; **6**, 121 ⟨76BCJ1996⟩, 121, 698 ⟨75CC960⟩
Narizuka, S., **4**, 62 ⟨92JOC3755⟩, 62, 63 ⟨93JOC4200⟩
Narjes, F., **4**, 938 ⟨91JOC717⟩
Narr, B., **3**, 722 ⟨61ZN(B)845⟩
Narsimhan, N. S., **2**, 752 ⟨83TL4733⟩
Naruchi, K., **4**, 23, 27 ⟨67BCJ594⟩; **5**, 325 ⟨82JOC5283⟩
Narukawa, Y., **3**, 544 ⟨92JA2991⟩
Narula, A. K., **2**, 112, 558 ⟨93JOM(452)1⟩
Naruse, H., **1**, 518 ⟨94AG(E)580⟩
Naruse, K., **1**, 230, 1107 ⟨82TL3587⟩, 476 ⟨82TL3583⟩; **3**, 146, 212, 288, 296 ⟨81CL1135⟩
Naruse, M., **1**, 207 ⟨74T3037⟩, 1033 ⟨73TL1847, 74T2159⟩; **5**, 330 ⟨94JOC1358⟩
Naruse, N., **3**, 634 ⟨73OSC(5)266⟩
Naruse, Y., **2**, 679 ⟨86TL1363⟩; **4**, 174 ⟨90JOC5814⟩
Naruta, Y., **1**, 23 ⟨80JOC4097⟩, 226 ⟨87CL963⟩
Narwid, T. A., **1**, 832 ⟨74HCA771⟩
Nasaka, N., **2**, 916 ⟨93JOC5045⟩
Nasakin, O. E., **4**, 768 ⟨75ZOB2374⟩
Nash, J. J., **1**, 778 ⟨89JOC389⟩
Nash, S. A., **3**, 87 ⟨84TL2953, 86JOC3116⟩
Nashed, M. A., **4**, 176 ⟨82CAR(102)99⟩, 176 ⟨82CC1274⟩
Nasibov, Sh. S., **2**, 356, 357 ⟨81T4097⟩
Nasielski, J., **3**, 449 ⟨66BSB601⟩
Nasipuri, D., **1**, 99, 890 ⟨59T148⟩; **5**, 76 ⟨73OSC(5)384⟩
Naslund, L. A., **1**, 797 ⟨68JOC221⟩
Näsman, J. H., **4**, 553 ⟨90SC1865⟩
Naso, F., **1**, 187 ⟨78CC1085⟩, 466 ⟨86JOM(312)343⟩, 467 ⟨85JCS(P1)1115⟩, 470 ⟨92JOC1718⟩, 484 ⟨88PAC79, 89TL243⟩, 1065 ⟨74JCS(P1)340⟩, 1110 ⟨73JCS(P2)2078⟩, 1112 ⟨68CC1363⟩; **2**, 740 ⟨91S225⟩, 874 ⟨68MI 217-01⟩, 931 ⟨88TL3705⟩; **3**, 230 ⟨91CC237⟩, 285 ⟨88TL3587⟩; **4**, 476 ⟨91S225⟩, 773 ⟨87S1034⟩, 776 ⟨92TL5121⟩
Nason, D. M., **6**, 553 ⟨92JOC4352⟩
Nass, R., **6**, 615, 728 ⟨87JFC(37)259⟩
Nassar, M. N., **6**, 462 ⟨86MI 615-01⟩
Nasser, J., **3**, 639 ⟨90PS(54)171⟩
Nassimbeni, L. R., **4**, 584, 585 ⟨86JOM(315)255, 82JOM(232)C78⟩; **6**, 607 ⟨83MI 620-01⟩
Nast, R., **2**, 1096 ⟨62CB2155⟩, 1098 ⟨64CB3317⟩
Nasu, A., **2**, 216 ⟨92BCJ626⟩
Nasutavicus, W. A., **4**, 301 ⟨64JOC407⟩
Nasybullina, S. S., **2**, 809 ⟨67CA76112w⟩
Natale, N., **2**, 398 ⟨76JOC3328⟩
Natale, N. R., **1**, 635 ⟨78JOC2299, 79OPP201⟩, 635 ⟨80OS(59)42⟩; **3**, 4 ⟨83SC817⟩, 219 ⟨82TL5009⟩
Natalie, Jr., K. J., **1**, 413 ⟨86JOC1124⟩; **3**, 141 ⟨86JOC1124⟩
Natalini, B., **1**, 44 ⟨85JCS(P1)493⟩; **3**, 311 ⟨92TL3003⟩
Nataniel, T., **2**, 855 ⟨86ZAAC(543)7⟩
Natarajan, S., **1**, 1082, 1083 ⟨77JOC2380⟩; **2**, 507 ⟨82JOM(231)119⟩; **4**, 94 ⟨84H(22)585⟩
Natat, A., **4**, 215 ⟨75T809⟩
Natchus, M. G., **1**, 86 ⟨87TL5287⟩, 804 ⟨87JOC4641⟩
Natekar, M. V., **1**, 707 ⟨75IJC251⟩
Nathan, A. H., **2**, 136 ⟨41JA2361⟩; **4**, 209 ⟨54JA743⟩
Nathan, E. C., **2**, 28 ⟨88OSC(6)991⟩
Nathan, III, E. C., **1**, 909 ⟨68JA4495⟩
Nathan, W. S., **5**, 199 ⟨52JCS1363⟩
Nath, R., **4**, 773 ⟨93TL7645⟩
Natile, G., **4**, 260 ⟨90JOC1323⟩; **5**, 386 ⟨93JA5123⟩, 696 ⟨89JCS(D)947⟩
Nativ, E., **5**, 403 ⟨68TL4405⟩
Nativi, C., **1**, 667 ⟨87TL2751⟩; **2**, 907 ⟨91JOC1951⟩
Natsibullin, F. Y., **6**, 55 ⟨68IZV431, 69IZV1331⟩, 55 ⟨73IZV1424⟩, 56 ⟨68IZV912⟩, 261 ⟨68IZV429⟩, 282 ⟨68IZV656, 68IZV676⟩
Natsugari, H., **1**, 942 ⟨83JCS(P1)403⟩; **2**, 244, 245 ⟨84JA7867⟩; **5**, 351 ⟨91JCS(P1)2435⟩
Natsuki, R.,**4**, 844 ⟨73CPB2770⟩, 979 ⟨73YZ612⟩
Natsume, M., **1**, 1042 ⟨83H(20)601⟩; **3**, 696 ⟨90T6351⟩, 696, 701 ⟨89TL1815⟩; **5**, 572 ⟨87TL2265⟩
Natta, G., **5**, 293 ⟨52JA4496⟩; **6**, 522 ⟨57JA3611⟩, 524 ⟨55CI(M)6⟩
Natu, A. A., **2**, 394 ⟨87SC1421⟩; **5**, 384 ⟨91SC1889⟩
Nau, P., **5**, 142 ⟨60BSF1350⟩

Naughton, A. B. J., **1**, 782 ⟨92TL1025⟩; **6**, 629 ⟨92TL1025⟩
Naujoks, E., **2**, 940 ⟨88TL609⟩
Naumann, D., **1**, 54 ⟨88ZAAC(560)141⟩; **2**, 483, 873 ⟨87JOM(334)323⟩, 881 ⟨86AX(C)1875⟩; **4**, 152 ⟨84JFC(26)1⟩, 153 ⟨85JFC(27)309⟩, 704 ⟨90JFC(46)265⟩; **5**, 205, 321 ⟨89JFC(45)401⟩; **6**, 49 ⟨94AG(E)323⟩, 64 ⟨88JFC(41)185⟩, 65, 66 ⟨85JFC(27)309⟩, 220 ⟨76JFC(8)177, 89JFC(45)401⟩, 220, 246 ⟨91CJC327⟩, 237 ⟨89ZAAC(576)225, 90JFC(48)207, 92ZAAC(608)69⟩, 243 ⟨87JOM(334)323, 90JFC(46)265⟩, 243 ⟨91JOM(417)C47⟩, 245 ⟨91JOM(407)1⟩, 246 ⟨89JOM(368)131⟩
Naumann, D. J., **6**, 64 ⟨84JFC(26)435⟩
Naumann, K., **2**, 440 ⟨69JA2788⟩, 440, 441 ⟨69JA7027, 69JA5645⟩, 441 ⟨69JA7012, 69JA7023⟩; **4**, 106 ⟨79EGP137839⟩
Naumann, M., **6**, 522 ⟨30GEP531402, 31GEP547025⟩
Naumann, W., **5**, 366 ⟨54LA(587)93⟩
Naumov, A. D., **4**, 660 ⟨74ZOB806⟩; **5**, 381, 382 ⟨70DOK(194)595⟩
Naumov, V. A., **3**, 689 ⟨91HAC577⟩; **5**, 370 ⟨91ZOB450⟩; **6**, 538, 539 ⟨89IZV909⟩
Naumova, L. V., **4**, 926 ⟨84ZOB1758⟩
Nauta, W. T., **1**, 898 ⟨37RTC535⟩
Nauta, W. Th., **1**, 922 ⟨68TL249⟩
Navany, S. C., **5**, 226 ⟨81S790⟩
Navarrete, P., **5**, 759 ⟨82MI 519-01⟩
Navarrini, W., **4**, 104 ⟨92EUP496413⟩; **6**, 250 ⟨92EUP499157, 92EUP499158, 92JFC(58)143, 92JFC(58)290⟩, 279 ⟨89USP4874875, 92EUP496413⟩
Navarro, C., **2**, 97 ⟨93SC1025⟩
Navarro, F. J. G., **3**, 170 ⟨89JOC5620⟩
Navarro, M., **3**, 7, 120 ⟨90SC399⟩
Navarro, R., **4**, 584 ⟨85JOM(290)125⟩, 692 ⟨89JCS(D)169⟩; **5**, 1138 ⟨84ICA215⟩
Navasaki, K., **5**, 148 ⟨86CL81⟩
Nava Salgado, V. O., **1**, 785 ⟨91JOC5782⟩
Navech, J., **4**, 459 ⟨83TL1975⟩; **5**, 705 ⟨78PS(4)317⟩, 705, 706 ⟨77JHC749⟩, 883 ⟨83TL5885⟩; **6**, 160 ⟨83PS(18)27⟩, 686 ⟨83TL5885⟩, 692 ⟨86PS(26)327⟩
Navio, J. L. G., **3**, 306 ⟨93JOC5862⟩
Nawa, M., **2**, 859 ⟨80CL729⟩
Nay, B., **2**, 620 ⟨85S1157⟩
Nayak, U., **1**, 412, 810 ⟨67JA3462⟩
Nayak, U. G., **2**, 129 ⟨69JOC97⟩
Nayak, V. R., **4**, 749, 750 ⟨79IJC(B)500⟩
Nayler, J. H. C., **5**, 9 ⟨63JCS491⟩, 219 ⟨71JCS(C)1917⟩; **6**, 111 ⟨74JCS(P1)22⟩
Naylor, A., **1**, 115 ⟨93SL189⟩, 202 ⟨89TL3581⟩; **2**, 931 ⟨93SL189⟩; **3**, 29 ⟨89TL2971⟩; **4**, 807 ⟨93SL189⟩
Naylor, Jr., C. A., **5**, 74 ⟨43OSC(2)523⟩
Naylor, R. A., **2**, 847 ⟨76JCS(P2)1444⟩
Naylor, R. F., **2**, 117 ⟨47JCS1532⟩, 132 ⟨47JCS1106⟩
Nayyar, N. K., **1**, 320, 336, 345, 346, 925, 926, 937, 942, 946 ⟨94CRV519⟩; **2**, 23 ⟨92JA6262⟩
Nazar, B., **2**, 534 ⟨85JOC549⟩
Nazarenko, T. I., **6**, 219 ⟨92RCR75⟩
Nazaretian, V. P., **6**, 236 ⟨79S972⟩
Nazarov, D. V., **4**, 1081 ⟨88KGS852⟩
Nazarov, I. N., **1**, 79 ⟨56ZOB1175⟩, 421 ⟨42IZV392, 43IZV389, 57JGU693, 60JGU765⟩, 421, 422 ⟨44IZV65, 46IZV633⟩, 422 ⟨47IZV205⟩; **3**, 57 ⟨59JGU3925⟩
Nazarova, D. V.,**3**, 57 ⟨59JGU3925⟩
Nazarova, G. N., **2**, 440 ⟨85URP1145022⟩
Nazeer, M., **5**, 330 ⟨88TL6173, 93JCS(P1)1397⟩
Nazemi, H., **3**, 621 ⟨78CJC41⟩
Nazer, B., **2**, 126, 708 ⟨84S498⟩, 268 ⟨84JA8001⟩, 526 ⟨86JOC5264⟩; **3**, 13, 82 ⟨87JOC5400, 84JA8001⟩, 82 ⟨86JOC5264⟩
Nazeruddin, G. M., **1**, 801 ⟨92TL1645⟩
Nazimuddin, M., **6**, 563 ⟨91POL823⟩
Nazmutdinova, V. N., **5**, 221 ⟨89IZV418⟩, 406 ⟨88JGU1548, 91BAU661⟩, 535 ⟨89JGU1548⟩, 1131 ⟨91IZV237⟩
Nazran, A. S., **2**, 373, 376, 403 ⟨84JOC336⟩
Ncube, S., **4**, 250 ⟨78TL2345⟩
Ndip, G. M., **2**, 904 ⟨92JOC5768⟩
Ndong-Mebah, J. M., **2**, 255 ⟨92PS(73)49⟩
Neal, A. M., **4**, 699 ⟨30JA695⟩
Neal, R. A., **5**, 247 ⟨81JOC1374⟩
Neal, T., **1**, 1122 ⟨79JOC3076⟩
Neal, T. R., **6**, 264 ⟨86JOC4768⟩
Neale, D. W., **6**, 247 ⟨88JOM(342)399, 93JOM(453)299⟩, 247 ⟨93JOM(453)307⟩
Neale, R. S., **5**, 318 ⟨66JA3051⟩, 1120 ⟨69JOC1808⟩
Neale, S. R., **1**, 927 ⟨71S1⟩
Neamati, N., **4**, 174 ⟨91TL3313⟩
Neary, A. P., **1**, 552 ⟨89CC1090⟩
Neber, P., **5**, 841 ⟨12CB3732, 13CB2032⟩, 841 ⟨13CB2370⟩
Neber, P. W., **2**, 780 ⟨42LA(550)182⟩
Nebergall, W. H., **2**, 942 ⟨50JA4702⟩, 949 ⟨46JA1720⟩
Nebuchla, M., **1**, 432 ⟨84JMC1575⟩
Nechaeva, A. I., **4**, 660 ⟨76ZOB1043⟩
Nechitailo, L. G., **5**, 28 ⟨79MI 502-03⟩
Nechvatal, A., **2**, 101 ⟨50JCS2088⟩
Nechvatal, E., **2**, 3 ⟨B-72MI 201-01⟩
Neckers, D. C., **1**, 925 ⟨79JOC4123, 85JOC3498⟩; **3**, 98 ⟨77JOC3514⟩
Nedelec, J.-Y., **2**, 14 ⟨91SL659, 92TL213⟩; **6**, 24 ⟨88TL1699⟩, 5 ⟨85T4503⟩, 7 ⟨92TL213⟩
Nédélec, L., **4**, 174 ⟨80T1763⟩
Nedolya, N. A., **5**, 1025 ⟨88ZOR1771⟩, 1026 ⟨90ZOR2059⟩; **6**, 558 ⟨90ZOR1422⟩, 565 ⟨87ZC24⟩
Nedugov, A. N., **4**, 239 ⟨80JOU1377, 82JOU876⟩; **6**, 100 ⟨86ZOR107⟩
Née, G., **2**, 751 ⟨92JOC3807⟩; **3**, 150 ⟨92JOC3807⟩
Neeb, M. K., **2**, 439 ⟨90JOM(393)143⟩
Needles, H. L., **1**, 881 ⟨66JOC989⟩
Neef, G., **1**, 558 ⟨89LA2151⟩
Neeildlein, R., **4**, 896, 897 ⟨89JHC1335⟩
Neeman, M., **2**, 89 ⟨58JA2584, 59T36⟩, 89 ⟨73OSC(2)245⟩; **4**, 751 ⟨57JA4174⟩
Neemann, J., **2**, 418 ⟨81CB527⟩
Neenan, T. X., **1**, 1010 ⟨88JOC2489⟩
Neergaard, J. R., **4**, 252 ⟨71JOC2731⟩
Neese, H. J., **2**, 512 ⟨70MI 209-02⟩, 566 ⟨69JOM(20)129, 71JOM(32)313⟩
Neese, P., **4**, 112 ⟨86OM2030⟩
Neeser, J.-R., **2**, 4 ⟨83HCA1018⟩
Neeson, S. J., **1**, 823 ⟨88TL3993⟩
Nef, I. U., **3**, 694 ⟨1892LA(270)267⟩
Nef, J. U., **2**, 87 ⟨01LA(318)157⟩; **5**, 747 ⟨1892LA(270)267⟩
Nefedov, B. K., **6**, 464 ⟨72IZV1635, 72IZV2733⟩, 464 ⟨73IZV562, 73IZV807⟩, 503 ⟨78IZV220⟩
Nefedov, M. N., **4**, 707 ⟨83JOM(244)C21⟩
Nefedov, O. M., **1**, 529 ⟨89TL4021⟩; **2**, 11 ⟨90OPP215⟩, 609, 621 ⟨60DOK(133)855⟩, 615 ⟨89IZV1935⟩, 947, 949 ⟨B-91MI 218-01⟩; **3**, 232 ⟨77AG(E)646⟩, 342 ⟨87IZV2398, 92MC67⟩; **4**, 142 ⟨71DOK(196)85⟩, 1083 ⟨87DOK(296)403⟩
Nefedov, S. E., **6**, 341 ⟨90JOM(384)279⟩
Nefedov, V. A., **5**, 1116 ⟨68MI 530-02⟩
Neff, D. L., **3**, 573 ⟨59JA4878⟩
Neff, J. U., **5**, 819 ⟨1894LA(280)305⟩
Nefkens, G. H. L., **5**, 228 ⟨83T2995⟩, 229 ⟨85T6063⟩
Nefkens, S. C. A., **2**, 583 ⟨91OM2046⟩
Neganova, E. G., **5**, 395, 396, 399 ⟨89JGU1817⟩, 395, 396, 399, 400 ⟨88JGU201, 88JGU2372⟩, 396, 399

⟨87DOK114⟩, 398 ⟨91JGU114⟩, 398, 400 ⟨90JGU1810⟩, 399 ⟨91JGU108, 86JGU636⟩, 400 ⟨90JGU1817, 90JGU1822⟩
Negele, M., **4**, 695 ⟨86JOM(308)153⟩
Negeshi, E., **3**, 270 ⟨89JA3336⟩
Negeshi, E.-I., **3**, 270 ⟨85JA2568⟩
Negishi, A., **1**, 159 ⟨74AG(E)407⟩; **2**, 200, 222 ⟨69TL2743, 72JA5786⟩
Negishi, E., **1**, 66, 69, 70 ⟨B-80MI 102-01⟩, 133 ⟨79TL845, 83TL1341⟩, 142, 1053 ⟨81JOC5041⟩, 207 ⟨87ACR65⟩, 277 ⟨71JA2792⟩, 333, 946 ⟨71JA3777⟩, 450, 510 ⟨76JOC3484⟩, 453 ⟨81JA4985⟩, 453, 542 ⟨81PAC2333⟩, 464 ⟨88JA5383⟩, 467 ⟨80JA3298, 91TL4453⟩, 468 ⟨77JA3168, 88OS(66)60, 91TL6683⟩, 468, 471, 533, 1020, 1097, 1104 ⟨78JA2254⟩, 468, 533, 1106 ⟨76JA6729⟩, 468, 533, 536 ⟨87JA2393⟩, 468, 536, 1104 ⟨78TL1027⟩, 468, 542, 1004 ⟨82ACR340⟩, 468, 543 ⟨78JA2252⟩, 471, 533 ⟨76CC596⟩, 473 ⟨88OS(66)67⟩, 476 ⟨81TL3737⟩, 479, 534, 1106 ⟨76CC17⟩, 481 ⟨91TL5903⟩, 488, 983, 1033, 1034 ⟨91T343⟩, 491 ⟨87JOM(334)181⟩, 532, 1100, 1103 ⟨73CC606⟩, 532, 662 ⟨73CC874⟩, 532, 662 ⟨74JOC2321⟩, 534 ⟨76TL1927⟩, 534, 542, 543 ⟨78TL2357⟩, 536 ⟨85JA2568, 86JOC4080⟩, 536 ⟨86TL2829, 87TL917⟩, 536, 542 ⟨85MI 112-04, 82TL27⟩, 542, 543 ⟨81JOC4093, 80JOC5223⟩, 543 ⟨84TL5863⟩, 544 ⟨83OM563, 88S1⟩, 665 ⟨B-80MI 114-01, B-80MI 114-02⟩, 669 ⟨84TL3407⟩, 991 ⟨81JA1276⟩, 999 ⟨74CC762, 83COMC-I337⟩, 1003, 1021 ⟨75JA7385⟩, 1006 ⟨78JOC358⟩, 1006, 1140 ⟨77CC683⟩, 1006, 1035 ⟨87CC477⟩, 1015 ⟨90TL4393⟩, 1036 ⟨91JA1440⟩, 1045, 1047 ⟨83TL5181⟩, 1070 ⟨80JOC2526, 85OS(64)44⟩, 1070, 1094 ⟨87TL5793⟩, 1085 ⟨77CC318, 83JA6344⟩, 1094 ⟨85TL5671⟩, 1106 ⟨80S1034⟩, 1102 ⟨77TL411⟩, 1103 ⟨72S555⟩; **2**, 4, 514, 900 ⟨85OR(33)1⟩, 30 ⟨87TL5793⟩, 134 ⟨B-80MI 203-01⟩, 160 ⟨84JOC2629⟩, 515 ⟨75JA2791⟩, 515, 516, 518, 900 ⟨74S77⟩, 516 ⟨71JOM(28)C1, 73JA6757⟩, 516, 522 ⟨72JA3561⟩, 518 ⟨67JA5478, 69JA1224, 77T2331⟩, 524 ⟨75JOC814⟩, 565, 919, 925, 926, 969, 986 ⟨88S1⟩, 568, 986 ⟨91COS(5)1163⟩, 568, 986, 988 ⟨86TL2829⟩, 596, 970 ⟨80TL1501⟩, 629, 974, 975, 976 ⟨84TL5863⟩, 629, 975, 976 ⟨79S501⟩, 629, 976 ⟨80JA3298⟩, 900 ⟨82COMC-I(7)303, 82COMC-I(7)323⟩, 901, 904 ⟨91T343⟩, 909, 923, 961 ⟨86JA3402⟩, 919, 925 ⟨78JOM(156)213⟩, 923, 969 ⟨91JOC2590, 85TL5863⟩, 925, 931, 973 ⟨83JA6761⟩, 925, 969, 976 ⟨78JA2254⟩, 925, 974, 975 ⟨78JA2252⟩, 925, 974, 975 ⟨85JA6639⟩, 926, 931, 976 ⟨83JA6344⟩, 926, 931, 973, 976 ⟨84JA6105⟩, 927, 986, 988 ⟨89JA3336⟩, 927, 987, 988 ⟨93TL687⟩, 931 ⟨85JA8289⟩, 954 ⟨B-80MI 219-02⟩, 961 ⟨74JOC3452⟩, 968 ⟨B-80MI 219-03⟩, 969 ⟨74CC762, 77JA3168⟩, 969 ⟨84TL3407, 85MI 219-01⟩, 969 ⟨88TL1631⟩, 969 ⟨78TL1027⟩, 972 ⟨90TL493⟩, 973 ⟨83OM563⟩, 973, 976 ⟨88JOC913⟩, 974, 975 ⟨81PAC2333, 86OCS467, 82TL27⟩, 975 ⟨84OS(62)31⟩, 975, 976 ⟨81JA2882, 78TL2357⟩, 975 ⟨81JOC4093⟩, 976 ⟨80JOC2526, 80JOC5223, 80S1034, 81JA4985, 91TL6683⟩, 976, 978 ⟨81JA1276⟩, 980 ⟨89JA3454, 90TL4393⟩, 980 ⟨92PAC323, 93T5471⟩, 980 ⟨90JA8590⟩, 986, 987 ⟨87CL623⟩, 986, 988 ⟨89CL761, 89CC852⟩, 986, 991 ⟨91OM825⟩, 987 ⟨90CC182⟩, 988 ⟨89TL5105⟩, 988 ⟨89CS457, 91JA7424⟩, 991, 992 ⟨82TL3327, 82ACR340⟩, 992 ⟨87JA2393⟩, 1096 ⟨77CC683, 78JOC358⟩; **3**, 130, 288 ⟨83TL5181⟩, 139 ⟨83JOC4098⟩, 211 ⟨81JA1277⟩, 218 ⟨83JA6761⟩, 223 ⟨73CC606⟩, 228 ⟨87CL1007⟩, 270 ⟨86JOC4080⟩, 271 ⟨86TL775⟩, 289 ⟨75JOC1675⟩; **4**, 372 ⟨75JOC814⟩, 392 ⟨81JOC5041⟩, 652 ⟨90JOC5406⟩, 658 ⟨89JA3089⟩, 750, 1066 ⟨81JA1276⟩, 808, 1059 ⟨86JA3402⟩, 809 ⟨83JA6761, 88JA5383, 89JA3336⟩, 815 ⟨91T343⟩, 954, 1062 ⟨83JOC1560⟩, 1048, 1062 ⟨84JA6105⟩, 1061 ⟨85JA2568, 86TL2829⟩, 1061, 1066 ⟨87TL917, 85JA6639⟩; **5**, 464 ⟨89JA8018⟩; **6**, 172, 204 ⟨82COMC-I(7)265⟩, 172 ⟨B-80MI 606-01⟩
Negishi, Y., **3**, 318 ⟨82CC1055⟩, 625 ⟨84CL1511⟩
Negoita, N., **1**, 1132 ⟨83T3943⟩
Negoro, A., **5**, 240, 243, 253 ⟨91JOC3776⟩
Negoro, K., **2**, 478 ⟨80CC1183⟩, 663 ⟨82JOC1200⟩; **4**, 943 ⟨87JOC1256⟩
Negoro, T., **2**, 24 ⟨84BCJ2111, 86BCJ2547⟩, 24 ⟨86BCJ3519⟩
Negrebetskii, V. V., **4**, 557 ⟨88ZOB1987⟩; **5**, 775 ⟨91ZOB1581⟩; **6**, 693 ⟨91ZOB79⟩, 685,686 ⟨90ZOB2238⟩
Negrini, A., **4**, 770 ⟨57CA(51)12849g, 57CA(51)5721f⟩, 770, 772 ⟨57G1061⟩
Negrini, E., **2**, 942 ⟨86S757⟩
Negron, A., **4**, 353 ⟨89JOC4051⟩
Nehashi, T., **5**, 223 ⟨87CPB1741⟩
Neher, H. T., **4**, 188, 208 ⟨48JA2805⟩
Nehl, H., **4**, 698 ⟨73LA708⟩
Nehring, R., **6**, 107 ⟨66AG(E)848⟩
Neidlein, R., **1**, 707, 708 ⟨80LA1540⟩; **2**, 984 ⟨86AG(E)640⟩; **3**, 621 ⟨88S981, 89S612⟩; **4**, 439 ⟨64AG382⟩, 485 ⟨93LA955⟩, 491 ⟨93H(36)1925⟩, 550 ⟨88CB2121⟩, 824 ⟨88S981⟩, 875 ⟨88HCA1242⟩, 883, 905 ⟨88CB1817⟩, 890 ⟨92HCA124⟩, 978 ⟨91HCA579⟩, 981 ⟨90S959⟩; **5**, 341 ⟨81CB80⟩, 596 ⟨77AP(310)614⟩, 658 ⟨66CB239⟩, 996 ⟨67CB698⟩, 1028 ⟨90JHC433⟩, 1063, 1064 ⟨65TL149⟩, 1071 ⟨66CB1252⟩; **6**, 44, 95 ⟨62CB1859⟩, 100, 124 ⟨88HCA1242⟩, 118 ⟨93HCA2407⟩, 448 ⟨65TL2423, 66CB239⟩, 617 ⟨75AP(308)379, 78CB698⟩, 617 ⟨84CZ404⟩, 629 ⟨67AP(300)553⟩, 635 ⟨85M651⟩, 651 ⟨77AP820⟩, 651, 652 ⟨67AP(300)567⟩, 653 ⟨66AP709⟩, 666 ⟨66CB1252⟩
Neier, R., **5**, 1041, 1108 ⟨85CC533⟩; **6**, 558 ⟨86HCA1898⟩
Neikam, W. C., **1**, 901 ⟨62JA3199⟩
Neikova, N. A., **6**, 502 ⟨73GEP2241471⟩
Neiland, O., **2**, 1020 ⟨70JOU889⟩
Neiland, O. J., **6**, 555 ⟨92KGS1117⟩
Neilands, J. B., **2**, 390 ⟨60JA4903⟩, 782 ⟨62JOC1075⟩
Neilands, O., **1**, 11 ⟨77JOU1033⟩
Neilsen, J. B., **1**, 390 ⟨92JA5432⟩
Neilson, D. G., **5**, 685, 687, 688, 690, 691, 692, 694 ⟨B-75MI 517-02⟩, 685, 687, 690, 726, 742, 769, 849 ⟨61CRV179⟩, 685, 687 ⟨B-91MI 517-03⟩, 698 ⟨B-75MI 517-03, B-91MI 517-04⟩, 699 ⟨B-75MI 517-04, B-91MI 517-05⟩, 726 ⟨B-91MI 518-01⟩, 742, 769 ⟨B-75MI 519-02⟩, 768, 769, 770, 847, 849, 850, 855 ⟨70CRV151⟩, 768, 769, 772, 773 ⟨B-75MI 519-03⟩, 826 ⟨B-75MI 521-01⟩, 830, 847 ⟨B-75MI 521-02⟩
Neilson, R. H., **2**, 466 ⟨79IC347, 82IC3568⟩, 466 ⟨83IC814⟩, 466, 467 ⟨84IC2062⟩, 467, 468 ⟨86IC361⟩, 861, 862 ⟨77IC1460⟩; **4**, 129 ⟨87PS(33)147⟩, 566 ⟨81IC1679, 82IC3568⟩, 566 ⟨83IC575⟩, 570 ⟨86MM2089, 87OM959⟩, 571 ⟨77CC308, 82OM623⟩, 572 ⟨84IC2063, 89IC899⟩, 573 ⟨87TL6121⟩; **5**, 887 ⟨81IC1679, 87PS(30)463⟩, 893 ⟨85IC1993⟩; **6**, 129 ⟨87TL6121⟩, 164 ⟨83PS(18)43⟩, 165, 167, 193, 688 ⟨85OM339⟩, 350, 694 ⟨84CC698⟩, 689 ⟨88IC784⟩, 692 ⟨84OM1132⟩
Neiman, L. A., **4**, 312 ⟨73TL1889⟩
Neimysheva, A. A., **2**, 464, 854 ⟨66ZOB500⟩; **5**, 708 ⟨61JGU473, 61ZOB516⟩; **6**, 236 ⟨59ZOB3401⟩, 238 ⟨59ZOB2159⟩, 260 ⟨59ZOB2169⟩, 602 ⟨59JGU2662⟩, 604, 612 ⟨59JGU2131⟩
Neirabeyeh, M. A., **5**, 141 ⟨76S811⟩
Neises, B., **5**, 124 ⟨85OS(63)183⟩, 183 ⟨86TL4437⟩
Neiteler, C., **4**, 673 ⟨93CB89⟩
Neiteler, G., **2**, 577 ⟨92CB163⟩
Neithamer, D. R., **3**, 545 ⟨89JA9057⟩

Nekhoroshkov, V. M., **4**, 569 ⟨90PS(53)1⟩; **6**, 160 ⟨91PS(61)361⟩
Nekrasova, G. V., **6**, 54, 261 ⟨89ZOR2490⟩
Nekrasov, A. S., **5**, 46 ⟨70MI 502-01⟩
Nekrasov, V. V., **6**, 230 ⟨30JPR210, 30JPR81⟩
Nekrasov, Yu. V., **2**, 478 ⟨90K668⟩
Nel, M., **1**, 483 ⟨87BSF487⟩
Nelles, J., **4**, 196 ⟨37USP2091373⟩
Nelsen, S. F., **1**, 3, 6 ⟨86JA1265⟩; **2**, 382 ⟨73JA2719, 73TL2321⟩, 382 ⟨76JA5269⟩, 383 ⟨84JA3366, 86T1769⟩, 388 ⟨81JOC2402⟩, 403 ⟨66JA137, 74JA1788⟩; **3**, 60 ⟨68JOC775⟩
Nelson, A. J., **6**, 28 ⟨89JCS(P1)283⟩
Nelson, D. A., **1**, 630 ⟨85JOC5460⟩; **4**, 545 ⟨86OM2220⟩
Nelson, D. J., **1**, 449 ⟨89JOC6064⟩; **2**, 23 ⟨88JOC5664⟩, 910 ⟨86JOC2381⟩; **3**, 90, 97 ⟨77JOC3308⟩
Nelson, D. L., **2**, 293 ⟨75JIN233⟩, 857 ⟨75INC233⟩
Nelson, G. E., **2**, 441 ⟨85USP4507501, 85USP4507502⟩
Nelson, G. L., **1**, 400, 799, 805 ⟨67JA5503⟩, 799 ⟨70JA1096⟩
Nelson, J. A., **3**, 633 ⟨70JOC161⟩; **5**, 962 ⟨57CRV47⟩
Nelson, J. D., **5**, 295 ⟨70JOC825⟩
Nelson, J. F., **1**, 127 ⟨36RTC518⟩
Nelson, J. S., **5**, 189 ⟨63JOC1905⟩, 189 ⟨63MI 504-01⟩
Nelson, J. V., **1**, 414 ⟨80JA774⟩, 873, 876 ⟨82TS1⟩, 875 ⟨79JA6120⟩; **2**, 532, 698 ⟨79JA6120⟩, 532, 698 ⟨81JA3099⟩, 532 ⟨B-82MI 210-01⟩; **4**, 832 ⟨81JA3099⟩
Nelson, J. W., **6**, 70, 77 ⟨42JA1825⟩
Nelson, L. E., **2**, 913, 914 ⟨61JA4381⟩; **4**, 610 ⟨65JOC2618⟩
Nelson, L. L., **1**, 1004, 1005, 1006 ⟨77JOC2626⟩
Nelson, N. A., **1**, 625, 783 ⟨55JA1628⟩, 1000 ⟨58JA6626⟩
Nelson, N., **4**, 47 ⟨72CEN55, 72CEN62⟩
Nelson, R., **5**, 1016 ⟨68JA6247⟩
Nelson, R. V., **1**, 807 ⟨80TL2125⟩; **4**, 219 ⟨80SC911⟩
Nelson, S. G., **3**, 198 ⟨91TL2581⟩
Nelson, S. J., **1**, 430 ⟨73TL447⟩; **6**, 439 ⟨82USP4333883, 84USP4473580⟩
Nelson, S. M., **4**, 546 ⟨71JCS(A)3495⟩
Nelson, T. D., **1**, 788 ⟨91JOC4339⟩; **2**, 665 ⟨92T9111⟩
Nelson, T. P., **2**, 153 ⟨67CC1189⟩, 153 ⟨69CC1159⟩
Nelson, V., **2**, 773, 775 ⟨76JOC1751⟩, 1031 ⟨73JOC1361⟩; **6**, 76, 77 ⟨73JOC1361⟩
Nelyubin, B. V., **6**, 5 ⟨87IZV808⟩
Nemecek, C., **2**, 147 ⟨85PAC1911⟩, 362 ⟨86NJC761⟩
Nemery, I., **3**, 636 ⟨86TL5099⟩
Nemeth, A., **6**, 645 ⟨91MIP54636⟩
Nemeth, D., **5**, 151 ⟨87JA2426⟩
Németh, S., **5**, 987 ⟨90AG(E)1147⟩
Nemeth, S. T., **5**, 641 ⟨75CJC3211⟩
Nemoto, H., **1**, 189 ⟨93JOC1207⟩, 227 ⟨89T435⟩, 816, 818 ⟨86JOC4315, 87H(25)139⟩, 831 ⟨90CC1001⟩, 1082 ⟨90TL6205⟩; **2**, 54 ⟨88JA4475⟩, 250 ⟨93SL465⟩; **3**, 213 ⟨83TL3489⟩; **4**, 154 ⟨85JCS(P1)1185⟩, 286 ⟨81T2547⟩, 307 ⟨90CC1304⟩, 387 ⟨83TL4257⟩; **5**, 227 ⟨85H(23)549⟩
Nemoto, N., **3**, 8, 122 ⟨79CC303⟩
Nenitzescu, C. D., **1**, 899, 901, 914 ⟨B-68MI 119-03⟩; **3**, 232 ⟨61JCS3566, 62T257⟩, 238 ⟨B-64MI 305-01⟩
Nenninger, H., **6**, 360 ⟨88CZ382⟩
Nenninger, J., **6**, 616 ⟨85CB1371⟩
Neogi, A. N., **5**, 199 ⟨71CI(L)545⟩
Neogi, G., **1**, 74 ⟨84JCS(D)1477⟩
Neoh, S. B., **1**, 955 ⟨78TL691⟩
Neplyuev, V. M., **4**, 275, 277 ⟨89UKZ1216⟩, 332 ⟨77JOU2224⟩, 864 ⟨90UKZ1310⟩, 865 ⟨78ZOR1947, 79ZOR1477, 89ZOR1102⟩, 866 ⟨77ZOR508, 78UKZ183⟩, 866 ⟨83ZOR2417⟩; **5**, 595 ⟨71ZOR2352, 73ZOR347⟩, 1049 ⟨68ZOR658⟩; **6**, 254 ⟨88ZOR1327⟩, 275, 277, 316 ⟨76ZOR2583⟩, 277 ⟨68ZOR324⟩
Neplyuev, V. N., **6**, 275 ⟨76ZOR558⟩
Neplyvev, V. M., **4**, 272 ⟨91ZOR216⟩
Nercesian, D., **1**, 1077 ⟨77JHC567⟩
Nerdel, F., **2**, 824 ⟨63LA(665)1⟩; **4**, 181 ⟨67LA85⟩
Neretin, V. V., **4**, 638 ⟨80IZV1451⟩
Neri, O., **1**, 587 ⟨81TL3551⟩
Nery, R., **6**, 505 ⟨66JCS(C)350⟩
Nes, W. R., **4**, 297 ⟨50JA5409⟩
Nesbit, R. J., **4**, 689 ⟨85OM355⟩
Nesbitt, S., **1**, 1009 ⟨90JOC63⟩
Nesbitt, S. L., **2**, 541 ⟨77JA4836⟩; **4**, 216 ⟨77JA5009⟩
Nesener, E., **5**, 397 ⟨73JOM(49)149⟩
Neset, S. M., **5**, 324 ⟨93ACS1141⟩
Nesi, R., **5**, 532 ⟨67JHC54⟩
Nesmeyanov, A. N., **1**, 114 ⟨B-67MI 104-01⟩, 128 ⟨B-67MI 104-02⟩, 304 ⟨87T2609⟩, 446 ⟨34CB317⟩; **2**,**2**, 446 ⟨75IZV2845⟩, 464 ⟨55IZV823⟩, 520 ⟨B-67MI 210-01⟩, 557 ⟨B-67MI 211-01⟩, 693 ⟨59DOK(124)1251⟩, 838, 839 ⟨57DOK(117)996⟩, 876 ⟨58DOK(122)614, 67DOK(172)1329⟩, 881 ⟨45BAU251, 64IZV1197⟩, 881 ⟨64IZV1202, 71IZV2103⟩, 884 ⟨53DOK(91)1341, 54IZV410⟩, 956 ⟨29CB1010⟩, 971 ⟨67DOK(174)96⟩, 982 ⟨49IZV305, 54IZV1008⟩,1000 ⟨87T2609⟩; **3**, 378 ⟨87T2609⟩; **4**, 152 ⟨77IZV2417⟩, 669, 684, 722 ⟨73IZV2796⟩, 672 ⟨68CC1365, 69IZV100⟩, 685 ⟨72IZV1600⟩, 696 ⟨74JOM(65)131, 77JOM(137)217⟩, 696, 723 ⟨78JOM(153)115⟩, 706 ⟨57DOK(112)439, 58DOK(120)1263, 68IZV686, 70IZV2133⟩, 706, 723, 724 ⟨72IZV1823⟩, 719 ⟨73DOK(209)869, 74IZV2645⟩, 723 ⟨69IZV2030, 69IZV2032, 71DOK(198)590, 72DOK(202)362⟩, 723 ⟨72IZV2594, 76DOK(226)1092, 80JOM(185)C6⟩, 724 ⟨54DOK(97)459, 62IZV1683⟩, 726 ⟨73DOK(209)1113, 73IZV2815⟩; **5**, 38, 89 ⟨57JGU2481⟩, 55, 60, 89 ⟨62T69⟩; **6**, 172, 207 ⟨B-67MI 606-01⟩, 522 ⟨40DOK(26)54, 40DOK(26)57⟩, 522 ⟨59MI 616-02⟩
Nesmeyanova, O. A., **4**, 724 ⟨54DOK(97)459⟩
Nesmeyanov, N. A., **1**, 647 ⟨67ZOR598⟩, 737 ⟨68JOM(13)263⟩, 738 ⟨65BAU1434, 65IZV1474⟩, 741 ⟨72BAU2056, 72IZV2113⟩; **2**, 447 ⟨85IZV1200⟩, 877 ⟨84DOK(274)606, 84IZV700⟩, 877 ⟨86IZV181⟩, 878 ⟨87JOU1417⟩; **3**, 328 ⟨68JOM(13)263⟩, 496 ⟨89IZV1182⟩; **4**, 587 ⟨64IZV772, 65JOM(4)202⟩, 587 ⟨66DOK(171)111, 66ZOR942⟩, 587 ⟨68ZOR1685⟩; **5**, 902 ⟨63DOK(151)856⟩, 919 ⟨65JOM(4)202, 70DOK(195)98⟩
Nesmeyanov, N., **2**, 1097 ⟨B-67MI 223-01⟩
Ness, A. B., **4**, 195 ⟨48JOC223⟩
Nesterenko, G. N., **6**, 149 ⟨69IZV1845⟩
Nesterenko, V. D., **5**, 373 ⟨73ZOB1451⟩, 405 ⟨77JGU505⟩
Nesterov, L. V., **2**, 476 ⟨89JGU634⟩
Nesterov, N. I., **1**, 748 ⟨80ZOB76⟩
Nesterov, V. N., **5**, 615 ⟨89ZOR1323⟩
Nesterova, L. I., **4**, 483 ⟨77ZOB1422, 78ZOR739⟩, 1009 ⟨90ZOB1187⟩; **5**, 369, 713 ⟨90ZOB1187, 90JGU1057⟩, 713 ⟨90ZOB1186⟩, 713, 788 ⟨90JGU1059⟩
Nestle, M. O., **6**, 401, 402, 404 ⟨79JOM(178)227⟩
Nestler, B., **1**, 785 ⟨91JOC5782⟩
Nestrick, T. J., **1**, 85, 457 ⟨74CRV567⟩
Nesynov, E. P., **4**, 992 ⟨79ZOB1087⟩; **5**, 727 ⟨70ZOR805⟩
Neto, C. C., **6**, 489 ⟨90BBR(171)458⟩
Netsch, K.-P., **1**, 835 ⟨91JOC970⟩; **3**, 366 ⟨84AG(E)800⟩, 584 ⟨92CC488⟩; **4**, 900 ⟨91JOC970⟩
Netscher, T., **2**, 173 ⟨87S683⟩, 239 ⟨91T8145⟩
Nettelbeck, C., **4**, 661, 701, 1062 ⟨85OM1044⟩; **6**, 209 ⟨85OM1044⟩
Netter, H., **2**, 414 ⟨58HCA244⟩
Netzel, D. A., **3**, 280 ⟨84JOC4226⟩
Netzel, M. A., **4**, 938 ⟨83JOC383⟩
Neu, P., **5**, 220, 403 ⟨86ZN(B)859⟩, 221 ⟨87CZ61⟩
Neubauer, D., **5**, 27 ⟨68AG(E)329⟩, 142 ⟨68AG352⟩
Neubauer, G., **6**, 313, 321 ⟨64CB1232⟩
Neubauer, R., **4**, 525 ⟨86JOC1610⟩

Neubeck, R., **3**, 200 ⟨84TL4741⟩, 212 ⟨82JMC250⟩
Neuberg, C., **2**, 38 ⟨21MI 202-01⟩
Neuberger, A., **5**, 84 ⟨48MI 502-01⟩
Neubert, M. E., **5**, 15 ⟨83OS(61)8⟩
Neubig, R., **6**, 606 ⟨87JMC1241⟩
Neubold, H. B., **4**, 99 ⟨73JA7813⟩
Neucombe, M., **3**, 10 ⟨B-83MI 301-01⟩, 124 ⟨B-83MI 304-01⟩
Neuenbuger, M., **1**, 327 ⟨93HCA441⟩
Neuenschwander, M., **2**, 1049 ⟨69HCA2641, 78HCA1609⟩; **3**, 77 ⟨68AG(E)460, 69HCA2641⟩, 273 ⟨81C336⟩, 274 ⟨68AG443⟩; **4**, 53 ⟨77HCA1061, 78S593⟩, 53, 54, 56 ⟨78HCA2047⟩, 54, 56 ⟨71HCA1037⟩, 852 ⟨89HCA1506⟩, 885 ⟨69HCA2641, 73HCA1318, 73HCA1331⟩, 885, 976, 977 ⟨73HCA944⟩; **5**, 280 ⟨78HCA2428, 78HCA2437⟩, 280 ⟨85OS(63)175⟩
Neuenschwander, P., **1**, 664 ⟨68C338⟩
Neuffer, J., **2**, 436, 443, 450 ⟨86JOM(301)289⟩
Neuffer, L., **2**, 103 ⟨14JA2202⟩
Neugebauer, D., **4**, 576 ⟨80CB902⟩, 581, 582, 583 ⟨83OM257⟩; **5**, 226 ⟨80CB1010⟩, 908, 910, 911 ⟨80CB902⟩; **6**, 151 ⟨79AG(E)484⟩, 341 ⟨83M851⟩
Neugebauer, F. A., **1**, 866 ⟨62LA(654)64⟩, 922 ⟨78TL4871⟩; **4**, 436 ⟨67LA(706)107⟩, 437 ⟨80AG(E)724⟩, 992 ⟨79CB2369, 80CB1226⟩
Neugebauer, W., **1**, 1088 ⟨85CB1504⟩; **4**, 668 ⟨82CC1184⟩
Neuhaus, R., **4**, 768 ⟨87ZN(B)796⟩
Neuhaus, S., **5**, 1146 ⟨81LA1198⟩
Neuhauser, H., **3**, 302 ⟨91LA1365⟩, 680 ⟨87S924⟩; **5**, 22 ⟨92S1145⟩
Neukirchinger, K., **5**, 381, 382, 1085 ⟨90CB455⟩
Neukom, C., **1**, 1134 ⟨80JA5425⟩; **2**, 39 ⟨76JOC3505, 81HCA1158⟩
Neumaier, H., **4**, 119 ⟨75GEP(O)2412800⟩, 125 ⟨80USP4224241⟩
Neumair, G., **3**, 568 ⟨66ZAAC(344)285⟩
Neuman, Jr., R. C., **2**, 403 ⟨70JOC3401⟩; **4**, 29, 40 ⟨66JOC1857⟩
Neuman, P. N., **1**, 418 ⟨65JA3996⟩
Neumann, B., **1**, 248 ⟨85AG(E)1046⟩; **4**, 603 ⟨91CB2391⟩; **5**, 893 ⟨91CB989⟩; **6**, 187 ⟨94OM436⟩, 187, 387 ⟨91OM1647, 91OM3838⟩
Neumann, E., **2**, 170 ⟨80EGP139258⟩, 170 ⟨82JPR631⟩
Neumann, F. W., **5**, 742, 751, 755 ⟨44CRV351⟩
Neumann, G., **6**, 661 ⟨72JOM(42)293⟩
Neumann, H., **1**, 111, 1090 ⟨76TL4839, 78CB2785⟩; **2**, 188 ⟨65CB1715⟩, 909, 910 ⟨76TL4839⟩, 909, 910 ⟨78CB2785⟩, 961 ⟨74CB847⟩; **5**, 1111 ⟨70JA6969⟩
Neumann, H. C., **5**, 205 ⟨70JMC948⟩
Neumann, H. G., **5**, 215 ⟨86MI 504-01⟩
Neumann, N., **5**, 692 ⟨84CB1935⟩
Neumann, R., **1**, 1065 ⟨86T3029⟩; **3**, 91 ⟨86S569⟩; **5**, 658 ⟨81M(112)959⟩
Neumann, S., **5**, 407 ⟨78CC946⟩
Neumann, T. L., **6**, 28 ⟨86S224⟩
Neumann, W. P., **1**, 4, 5, 9, 16, 172, 230, 330, 341, 428, 919, 935, 938, 942, 946 ⟨87S665⟩, 17, 939 ⟨92SL801⟩, 331, 342 ⟨90S448⟩, 922 ⟨89JA5845⟩, 924 ⟨89T951, B-91MI 119-03⟩, 925 ⟨93SL286⟩, 936 ⟨91JOC5971, 92JCS(P1)3165⟩, 936 ⟨93SL283⟩, 950 ⟨88JOM(340)317⟩, 1038 ⟨67TL495⟩, 1043 ⟨90CB2191⟩; **2**, 478 ⟨76JOM(105)183⟩, 599 ⟨62LA(653)157⟩, 601 ⟨64LA(705)10, 67LA(707)15⟩, 868 ⟨66LA(692)1⟩, 947 ⟨88TL5245, 90SL113, 91SL431⟩, 947, 949 ⟨83JA897⟩, 947, 949 ⟨91CRV311⟩, 1100 ⟨B-70MI 223-02⟩, 1101 ⟨68LA(716)19⟩; **3**, 92 ⟨89T951⟩, 570, 571 ⟨68LA(718)1, 66AG(E)514⟩, 571 ⟨66TL3415⟩; **4**, 141, 154, 608, 616 ⟨80JOM(201)197⟩, 360 ⟨79CB936⟩; **5**, 344, 593 ⟨93JOC7022⟩, 780 ⟨65LA(688)14⟩, 1006 ⟨91CB1997⟩; **6**, 187 ⟨91CRV311⟩, 661 ⟨72JOM(42)293⟩, 664 ⟨66TL3423⟩
Neumeier, R., **3**, 723 ⟨91CB1683⟩
Neumeister, J., **5**, 141 ⟨78AG(E)939⟩
Neümuller, B., **4**, 575 ⟨88ZAAC(567)23⟩
Neumüller, B., **6**, 195 ⟨91ZN(B)1539⟩
Neumuller, O.-A., **1**, 839 ⟨58LA(618)202⟩
Neunhoeffer, H., **1**, 522 ⟨75LA1113⟩, 790 ⟨73LA437, 74LA1190⟩; **4**, 436 ⟨84CHEC(3)531⟩; **5**, 362 ⟨92H(33)893, 92T651⟩, 640 ⟨92S637⟩, 849 ⟨72LA(760)102⟩, 852 ⟨68CB3947⟩, 856 ⟨71LA(750)12⟩; **6**, 109 ⟨84CHEC(3)91⟩
Neunhoeffer, O., **2**, 825 ⟨61CB2514⟩
Neunlist, S., **1**, 1077 ⟨88CC830⟩
Neurath, H., **5**, 28 ⟨B-63MI 502-03⟩, 83 ⟨44MI 502-01⟩
Neureiter, N. P., **1**, 592, 646 ⟨59JA578⟩, 801 ⟨59JOC2044⟩; **2**, 118 ⟨60JA5354⟩; **4**, 83 ⟨65JOC1313⟩
Neuth, J. F., **2**, 340, 342, 350, 353, 355, 356 ⟨90HOU(E16)1⟩
Neutzner, J., **6**, 178, 190, 191, 385, 391 ⟨83ZAAC(497)21⟩
Neuvar, E. W., **5**, 814 ⟨73JOC1065⟩; **6**, 52 ⟨68JOC3675⟩, 616 ⟨67JHC389⟩
Neuy, A., **4**, 138 ⟨89CC957⟩
Nevaleinen, V., **2**, 66 ⟨92TA1441, 92TA1563⟩, 66 ⟨92TA921, 92TA933⟩
Neveux, M., **2**, 676 ⟨91SL755⟩
Nevill, W. A., **3**, 645 ⟨62JOC422⟩
Neville, R. G., **6**, 572 ⟨73OSC(4)801⟩
Nevkina, N. N., **2**, 886 ⟨89MI 217-02⟩
Nevrekar, N. B., **3**, 280 ⟨83CI(L)479⟩
New, J. S., **5**, 334 ⟨86JMC359⟩
Newall, A. R., **2**, 89 ⟨69JCS(B)749⟩
Newallis, P. E., **4**, 164, 209 ⟨65JOC3834⟩, 297 ⟨64JOC3114⟩
New, R. G. A., **3**, 694 ⟨30JCS1876⟩
Newberry, G., **1**, 446 ⟨28JCS3046⟩
Newbold, A. J., **2**, 654 ⟨90SL533⟩
Newbold, B. T., **2**, 776 ⟨61JCS4260⟩
Newbold, G. T., **3**, 229 ⟨54JCS1302⟩
Newbury, G., **5**, 744, 750, 754 ⟨47JCS738⟩
Newcomb, M., **1**, 331, 342 ⟨86JA4132⟩, 334 ⟨89JOC1826⟩, 808 ⟨80JOC4793⟩, 923 ⟨88ACR206⟩, 927, 950 ⟨92TL5913⟩, 949 ⟨91TL45, 87JA3163⟩, 950 ⟨90T2317, 90T2329, 90T2345, 90TL1675, 91JOC1309, 93JOC4933, 93TL5523, 93TL6877⟩; **2**, 323 ⟨87JA3163⟩, 770 ⟨80JOC4666⟩; **3**, 404 ⟨B-83MI 310-01⟩, 411 ⟨80JOC1489⟩, 621 ⟨86JA240, 90TL2975⟩; **4**, 1077 ⟨91TL1035⟩; **5**, 274, 294 ⟨80JOC3413⟩, 334 ⟨93TL6877⟩, 655, 712 ⟨91TL1035⟩
Newington, I. M., **1**, 154 ⟨86CC176⟩
Newirth, T. L., **5**, 436 ⟨73JA8118⟩
Newkirk, A. E., **2**, 611 ⟨46JA2467⟩
Newkirk, J. D., **1**, 392 ⟨51JA3900⟩; **5**, 66 ⟨55OSC(3)538⟩
Newkom, C., **1**, 1134 ⟨83JA3656⟩
Newkome, G. R., **1**, 637, 638 ⟨90JOC5801⟩; **2**, 827 ⟨78JOC947⟩; **3**, 445 ⟨66JOC677⟩; **4**, 183 ⟨73TL1599⟩
Newlands, L. R., **5**, 768, 769, 770, 847, 849, 850, 855 ⟨70CJC3662⟩
Newlands, M. J., **2**, 339 ⟨68JCS(C)796⟩; **3**, 600 ⟨70CC456, 73JCS(P1)1561⟩; **4**, 136 ⟨76JCS(D)694⟩, 607, 622 ⟨75JCS(D)1434⟩, 794 ⟨73JCS(P1)1066⟩, 979 ⟨68JCS(C)796⟩; **6**, 61 ⟨65JCS2101⟩, 191 ⟨75JCS(D)1434⟩, 239 ⟨68JCS(C)796⟩
Newman, B. C., **2**, 128 ⟨66JA1572, 70JOC3641⟩
Newman, H., **1**, 1138 ⟨73JOC2254⟩; **4**, 312 ⟨69CC369⟩
Newman, M. J., **6**, 254, 255 ⟨75CJC2664⟩
Newman, M. S., **1**, 21, 24 ⟨50JA4296⟩, 885 ⟨58T88⟩, 896 ⟨50JA163⟩, 1075 ⟨58JOC665⟩, 1076 ⟨72JOC4468⟩, 1076, 1118 ⟨51JA4199, 56JA4654⟩, 1118 ⟨69JA5677, 70JA7564⟩, 1119, 1143, 1144 ⟨69JOC1220⟩, 1143 ⟨68JA4189⟩, 1143 ⟨69JA6461⟩, 1143 ⟨73JOC547⟩, 1143 ⟨74JOC761, 70JA4309⟩, 1144 ⟨70JA4312, 72JOC3220⟩; **2**, 21 ⟨55OSC(3)188⟩, 223 ⟨56JA2469⟩, 627

⟨41JA2847⟩, 630, 801 ⟨70JA4309⟩, 679 ⟨73JOC2910⟩, 707 ⟨66JOC3980⟩, 753 ⟨60JOC214, 51JA4199⟩, 801 ⟨54JA1840, 69JA6461⟩, 801 ⟨69JOC1220, 70JA4312⟩, 801 ⟨70JA7564, 70JOC2412⟩, 802 ⟨72JOC3220, 73JOC2435⟩, 802 ⟨73JOC2438⟩, 1025 ⟨72OPP89, 77OS(57)65⟩; **3**, 66, 70 ⟨56JOC638⟩, 286 ⟨48JOC592⟩, 426 ⟨73OSC(5)1031⟩, 447 ⟨62JOC3946⟩, 463 ⟨59JOC385⟩, 474 ⟨49JA1506⟩, 489 ⟨68JA4189, 69JA6461⟩, 489 ⟨69JOC1220⟩, 526, 527, 546 ⟨60JA2498⟩, 531, 532 ⟨70OS(50)77⟩, 567 ⟨60JA873⟩, 613 ⟨56JA450, 58JOC797⟩; **4**, 18 ⟨78JOC4367⟩, 178 ⟨55OSC(3)502⟩, 211 ⟨41JA1537⟩, 743 ⟨56JOC638, 62JA995⟩; **5**, 25, 35 ⟨56JA450⟩, 34 ⟨56JA5416⟩, 46 ⟨49OR(5)413⟩, 47, 275 ⟨60JA2498⟩, 93 ⟨73OSC(5)1031⟩, 106 ⟨41JA2431⟩, 110, 115 ⟨61JOC1759⟩, 114 ⟨43OSC(2)428, 63OSC(4)484⟩, 144 ⟨70OS77⟩, 229 ⟨50JA5163⟩, 275 ⟨57JA2530⟩, 294 ⟨73JOC4073⟩, 1002 ⟨48JA317, 57JA213⟩; **6**, 28 ⟨78JOC4367⟩, 79 ⟨73JA278⟩, 555, 556 ⟨66JOC3980⟩, 555, 575 ⟨69JOC3604⟩

Newman, R. H., **5**, 168 ⟨68JOC2819⟩

Newman, R. M., **6**, 205 ⟨82JA7345⟩

Newman, T. H., **2**, 821 ⟨82JA5820⟩; **6**, 373 ⟨82JA5820⟩

Newman-Evans, D. D., **1**, 16 ⟨81JOC4843⟩

Newton, C. G., **6**, 543, 550 ⟨91T6381⟩

Newton, D. J., **1**, 539 ⟨69JOC1936⟩

Newton, Jr., R. J., **1**, 962 ⟨78CC726⟩

Newton, R. F., **1**, 173 ⟨90JCS(P1)2352⟩, 218 ⟨85TL97⟩, 234 ⟨82CC589⟩, 826 ⟨83CC932⟩, 1024, 1134 ⟨80JCS(P1)852, 79JCS(P1)2954⟩

Newton, T. W., **2**, 546, 917, 977 ⟨84JCS(P1)1805⟩, 917, 918, 940, 977 ⟨81JCS(P1)2527⟩; **3**, 18 ⟨84JCS(P1)119⟩, 203 ⟨84JCS(P1)1805⟩; **4**, 1059 ⟨81JCS(P1)2527⟩

Neynaber, C. B., **4**, 298 ⟨78JOC3838⟩

Nezu, Y., **3**, 713 ⟨85CC1584⟩; **4**, 509 ⟨85CPB4596⟩

N'Gabé, D. Y., **2**, 938, 941 ⟨86T669⟩

Ng, C. T., **1**, 436 ⟨88JOC4466⟩

Ng, D. K. P., **2**, 936 ⟨90JA9356⟩

Ng, H.-Y., **2**, 228 ⟨73TL1561⟩; **3**, 359 ⟨74CC877, 77CJC3763⟩

Ng, J. S., **1**, 120 ⟨88JA2641⟩; **2**, 992 ⟨88JA2641⟩; **3**, 140 ⟨90JA7441⟩

Ng, S. C., **5**, 346 ⟨87TL4319⟩

Ng, T. W., **2**, 1060, 1061, 1076 ⟨71CJC2706⟩; **4**, 692 ⟨77IC172⟩

Ng, W.-K., **6**, 380, 387 ⟨79JCR(S)12⟩

Ngochindo, R. I., **3**, 303 ⟨82JCS(P1)1343⟩

Ngoi, T. H. J., **1**, 949 ⟨91JCR(S)1⟩, 950 ⟨91JCR(S)216⟩

Ngooi, T. K., **5**, 179 ⟨88TL5583⟩

N'Guessan, Y. T., **5**, 633 ⟨91BSF574⟩

Nguyen, A. C., **4**, 525 ⟨77JA6435⟩

Nguyen, B. T., **2**, 33 ⟨86JOC2206⟩

Nguyen, D., **3**, 89, 96 ⟨93TL3519⟩

Nguyen, M. C., **5**, 946 ⟨93TL227⟩

Nguyen, N., **1**, 799 ⟨90CJC1961⟩; **6**, 55 ⟨90MI 602-01⟩

Nguyen, N. H., **1**, 448 ⟨85JA1379⟩

Nguyen, N. V., **3**, 275 ⟨87JOC2442⟩, 539 ⟨85JA3392⟩, 653 ⟨84CC1066⟩; **4**, 304 ⟨85JOC4231⟩, 970, 993 ⟨92JOC235⟩, 994 ⟨92TL2949⟩; **6**, 150 ⟨92JOC3026⟩

Nguyen, P.-N., **1**, 926 ⟨84JOC1603⟩

Nguyen, S., **1**, 811 ⟨86JA5901⟩; **2**, 588 ⟨86T2873⟩

Nguyen, S. L., **1**, 120 ⟨86T2873⟩, 207, 312 ⟨84TL5959⟩, 207, 312 ⟨85JOM(285)437⟩; **2**, 589 ⟨85JOM(285)437⟩

Nguyen, S. T., **3**, 520, 522 ⟨92JA3974⟩

Nguyen, T., **1**, 481 ⟨91TL5903⟩; **2**, 88 ⟨91TL5903⟩, 923, 969 ⟨91JOC2590⟩; **3**, 67 ⟨85S754⟩, 237 ⟨89JOC5640⟩; **5**, 481, 487, 582 ⟨87JFC(35)523⟩; **6**, 63 ⟨77JOC565⟩

Nguyen, T. H., **1**, 149 ⟨92JA8483⟩

Nguyen, T. M., **5**, 971 ⟨92JOC7364⟩

Nguyen Thi, K. H., **2**, 744 ⟨82TL2023⟩; **5**, 728 ⟨82TL2023, 84T1075⟩; **6**, 617 ⟨84T1075⟩

Ni, J., **4**, 706 ⟨85OM2196⟩

Ni, Z., **1**, 980 ⟨93JA3776⟩

Ni, Z.-J., **1**, 22, 23, 238, 436 ⟨90S89⟩; **2**, 936 ⟨90JA9356⟩

Niangoran, C. E., **2**, 445 ⟨87S648⟩; **3**, 196 ⟨87S648⟩

Nibbering, N. M. M., **1**, 590 ⟨88JA2066⟩, 1143 ⟨78JA1928⟩; **4**, 50 ⟨79JA7617⟩; **5**, 437 ⟨85RTC74⟩

Nibi, S., **6**, 648 ⟨68JMC811⟩

Niburg, S. C., **4**, 487 ⟨82PS(13)319⟩

Nicaise, O., **3**, 41 ⟨93SL359⟩

Nicasio, M. C., **5**, 948 ⟨92JA7288, 94JA791⟩

Niccolai, G. P., **4**, 679 ⟨94OM2527⟩, 680 ⟨88JA6070⟩

Niccoli, A., **1**, 506 ⟨79T929⟩

Nice, L. E., **4**, 182 ⟨94TL57⟩

Nichino, K., **3**, 367 ⟨77AG(E)866⟩

Nicholas, A. M. de P., **1**, 921 ⟨84CJC1850, 84CJC1860⟩

Nicholas, K., **4**, 681 ⟨71CC608⟩, 724 ⟨70CC161⟩

Nicholas, K. M., **1**, 196 ⟨85JOM(285)C13⟩, 196 ⟨87TL5415⟩, 198 ⟨87ACR207⟩, 199 ⟨78TL4349, 80TL1595, 94TL363⟩, 199, 1025 ⟨83TL2239⟩, 200 ⟨93JA6438⟩; **2**, 307 ⟨92CC853⟩; **3**, 247 ⟨91H(32)477⟩; **4**, 688 ⟨71TL3475, 87ACR207⟩

Nicholls, B., **5**, 168 ⟨59JCS227⟩

Nicholls, D., **4**, 58 ⟨93JCS(P1)2303⟩

Nicholls, S. M., **4**, 681 ⟨92CC310⟩

Nichols, D. E., **4**, 188 ⟨79JMC1264⟩; **6**, 577 ⟨80JMC1261⟩

Nichols, D. I., **2**, 824 ⟨68JCS(A)2149⟩

Nichols, J. M., **2**, 855 ⟨72JCS(P2)380⟩

Nichols, L. D., **4**, 681 ⟨60JA4749⟩

Nichols, M. B., **6**, 577 ⟨80JMC1261⟩

Nichols, P. D., **3**, 347 ⟨72CJC3923⟩

Nicholson, D. A., **4**, 338 ⟨71JOC3843⟩, 563 ⟨68JOM(13)199⟩; **5**, 432 ⟨65IC1747⟩; **6**, 118 ⟨71JOC3843⟩, 263 ⟨68JOM(13)199, 71JOC1835⟩, 263, 518 ⟨67JOC4111⟩

Nicholson, J., **5**, 366 ⟨65JOC1162⟩

Nicholson, J. M., **5**, 332, 333 ⟨91JMC387⟩

Nicholson, J. S., **2**, 691 ⟨62CI(L)1244⟩

Nicholson, T., **6**, 349 ⟨87IC1488⟩

Nickel, W., **1**, 686 ⟨86T1693⟩

Nickel, W.-U., **1**, 686 ⟨81TL4937⟩; **3**, 464 ⟨81TL4937⟩

Nickon, A., **1**, 248 ⟨B-87MI 119-10⟩, 631 ⟨72JA7081⟩, 821 ⟨84TL3555⟩, 862 ⟨B-87MI 119-03⟩, 868 ⟨66JA1905⟩, 905 ⟨86JOC3736⟩, 906 ⟨69JA6861⟩; **2**, 249 ⟨68JA4096⟩, 710 ⟨85T4443⟩; **3**, 465 ⟨81JOC4685⟩

Nickson, T. E., **1**, 311 ⟨82TL2627⟩; **2**, 608 ⟨85S669⟩

Niclas, H., **4**, 903 ⟨66ZC417⟩

Niclas, H. J., **6**, 624 ⟨73JPR289⟩

Nicodem, D. E., **3**, 131 ⟨81JOC3928⟩

Nicola, M., **6**, 425 ⟨90JMC2101⟩

Nicolaides, E. D., **3**, 468 ⟨54JA2884, 54JA2887⟩

Nicolaou, G. A., **6**, 621 ⟨84CC1105⟩, 621, 728 ⟨87CC99⟩

Nicolaou, K. C., **1**, 14 ⟨82CC1292⟩, 25 ⟨81T4097⟩, 106 ⟨93CC617⟩, 558, 828 ⟨93CC619⟩, 581 ⟨85JA1691⟩, 591, 594, 716 ⟨79CC83⟩, 601, 604 ⟨B-84MI 114-01⟩, 603 ⟨81JA3486, 79JA3704⟩, 617, 709 ⟨88JA4866⟩, 704, 707, 709 ⟨88JA7247⟩, 750 ⟨88JA4685⟩, 754 ⟨84CC349⟩, 990 ⟨73JOC864⟩, 1001 ⟨84CC350⟩, 1009 ⟨91AG(E)1387⟩, 1038 ⟨92JA9279⟩, 1134 ⟨78JA4618, 78JA4620, 89S898⟩; **2**, 87 ⟨75TL3183⟩, 279 ⟨81JOC1215⟩, 281 ⟨79JA3704⟩, 282 ⟨79S982, 80JA1404⟩, 286 ⟨81T4097, B-87MI 204-02⟩, 623, 624 ⟨84JA5734⟩, 623, 624 ⟨86S453⟩, 629 ⟨78JA4618⟩; **3**, 38, 174 ⟨79S982⟩, 308 ⟨81JA6969, 81TL4647⟩, 351 ⟨90JA3029⟩; **4**, 45 ⟨84JA4189⟩, 224, 225 ⟨83JA2430⟩, 226 ⟨90JA6263⟩, 333 ⟨B-84MI 407-01⟩; **5**, 171 ⟨74TL2439⟩, 173 ⟨75JA2287, 78JA4618⟩, 173, 204 ⟨78JA4620⟩, 239, 281 ⟨81JOC1215⟩, 239, 336, 337 ⟨85T4835⟩, 277 ⟨85JA1695⟩, 513 ⟨90JA3029, 90JA3040, 90JA6263⟩; **6**, 134 ⟨87JA2504, 90JA3696⟩

Nicolas, K. M., **3**, 147 ⟨80JA2508⟩

Nicolas, M., **1**, 104 ⟨68CR(C)1834⟩

Nicolau, C. P., **5**, 193 ⟨85T5133⟩

Nicolau, K. C., **2**, 9 ⟨84TL2295⟩
Nicolaus, B. J. R., **2**, 103 ⟨62HCA358⟩, 182, 248 ⟨62HCA717⟩
Nicolaus, R. A., **6**, 107 ⟨68G681⟩
Nicolay, J. F., **1**, 104 ⟨94T3139⟩
Nicolet, B. H., **2**, 117 ⟨35JA1098⟩
Nicoletti, R., **6**, 141 ⟨76S798⟩
Nicolini, M., **6**, 622 ⟨75JOM(101)C1, 79JOM(181)69⟩
Nicpon, P., **2**, 465 ⟨66IC1297⟩
Nidy, E. G., **2**, 98 ⟨78JA7960⟩
Nie, P., **5**, 1002 ⟨76TL2691, 79JCS(P1)446⟩
Nie, P.-L., **1**, 30 ⟨79JCS(P1)442⟩; **2**, 11 ⟨77S634⟩, 11, 632 ⟨79JCS(P1)433⟩
Niebergall, H., **2**, 431 ⟨60AG210⟩, 431, 461 ⟨62CB64⟩
Niebergalland, H., **2**, 431 ⟨60USP2959621⟩
Nieberl, S., **6**, 114 ⟨76TL3563⟩
Niecke, E., **2**, 406 ⟨91AG(E)217⟩, 408 ⟨82CB185⟩, 470 ⟨75AG(E)363, 86IC2695⟩, 472 ⟨91CB1563⟩; **4**, 484 ⟨81AG(E)131⟩, 488 ⟨85CB3227⟩, 558 ⟨80CB1549⟩, 567 ⟨83CC1171⟩, 568 ⟨89CB453⟩, 568, 573 ⟨92CB771⟩; **5**, 885 ⟨89TL459⟩, 887 ⟨81AG(E)131⟩, 1074 ⟨67ZN(B)785⟩; **6**, 166 ⟨90CB1245⟩, 168 ⟨81AG(E)131⟩, 168 ⟨83CC1171⟩, 288 ⟨89CB453⟩, 689 ⟨85CC1687, 86OM593, 87CC10⟩, 689, 690 ⟨87ZAAC(545)7⟩, 689, 692 ⟨85C277⟩, 692 ⟨89JOM(368)C29⟩, 694 ⟨81CC72⟩
Niedenzu, K., **1**, 729, 730 ⟨B-76MI 116-01⟩; **4**, 107 ⟨70IC975⟩
Niederhauser, A., **4**, 885 ⟨73HCA1318, 73HCA1331⟩, 885, 976, 977 ⟨73HCA944⟩
Niederhauser, W., **3**, 621 ⟨45JA601⟩
Niederhauser, W. D., **2**, 167 ⟨50USP2580695⟩
Niedermann, H.-P., **3**, 359 ⟨86TL5703, 87AG(E)1046⟩, 359 ⟨90CB1143⟩; **4**, 232 ⟨90CB1143⟩
Niedermeyer, U., **3**, 628 ⟨90AG(E)386⟩
Niederprüm, H., **2**, 689 ⟨73LA20⟩
Nieduzak, T. R., **1**, 521 ⟨87JA5167⟩
Nief, F., **4**, 545 ⟨87JOM(328)349⟩, 1036 ⟨86OM877⟩; **5**, 397 ⟨91T6673⟩
Nieger, M., **2**, 409, 503 ⟨90CB975⟩, 472 ⟨91MI 208-02⟩, 474 ⟨90CB1837⟩; **4**, 462 ⟨85AG(E)1067⟩, 568 ⟨89CB453⟩
Nieger, M.,**5**, 934 ⟨95CB645⟩
Nieger, M., **6**, 288 ⟨89CB453⟩
Nieh, H.-C., **4**, 930 ⟨92JOC7010⟩
Nieh, H. C., **5**, 410, 411 ⟨92JOC7010⟩
Nieh, M. T., **1**, 692 ⟨72JA6538⟩
Nielek, S., **5**, 991 ⟨88JPR825⟩
Nielsen, A. T., **1**, 156 ⟨B-69MI 104-01⟩, 884 ⟨65OS(45)25⟩; **4**, 295, 407, 412 ⟨73JOC3288⟩, 298 ⟨90JOC1459, 74JOC1349⟩, 430, 431 ⟨78JOC2460⟩, 430, 431, 432 ⟨62JOC1813⟩, 444 ⟨76JOC3221⟩; **5**, 690 ⟨79JOC633⟩, 1003 ⟨58JOC1107⟩
Nielsen, C. J., **5**, 1117 ⟨77ACS(A)884⟩
Nielsen, C. K., **6**, 648 ⟨78JMC773⟩
Nielsen, C. M., **3**, 48 ⟨91JOC3591⟩
Nielsen, J., **6**, 535 ⟨91SUL143⟩
Nielsen, J. B., **5**, 675 ⟨90JFC(48)407⟩; **6**, 603 ⟨88IC570⟩
Nielsen, K. E., **5**, 1125 ⟨90TL3681⟩
Nielsen, P. H., **5**, 584, 586, 587, 624, 625, 726, 732 ⟨66ACS597⟩; **6**, 570, 571 ⟨69ACS324⟩, 574 ⟨68ACS1898⟩, 594, 595 ⟨70ACS351⟩, 596 ⟨63ACS549⟩
Nielsen, R. B., **1**, 397 ⟨89OM1593⟩; **2**, 261, 888 ⟨91JA165⟩, 927 ⟨89JA2870⟩, 969 ⟨87TL3895⟩, 986, 987 ⟨88CRV1047⟩; **3**, 370 ⟨88JA3171⟩; **4**, 1061 ⟨89JA2870⟩
Nielsen, R. P., **2**, 387 ⟨62JCS3818⟩
Nielsen, S. B., **6**, 232 ⟨93TL2973⟩
Nielsen-Marsh, S., **2**, 428 ⟨92CC699⟩
Nielson, A. T., **4**, 995 ⟨79JOC633⟩
Nielson, J., **2**, 469 ⟨86JCR(S)26⟩
Nielson, P. H., **2**, 890 ⟨63ACS1875⟩
Nielson, R. H., **4**, 566 ⟨88OM572, 89IC3219⟩, 566, 572 ⟨83OM921⟩, 567 ⟨83PS(18)43⟩
Nielson, T., **2**, 398 ⟨65CJC840⟩
Niemann, B., **2**, 414 ⟨58HCA244⟩
Niemann, J., **2**, 472 ⟨91CB1563⟩
Niemann, U., **6**, 233 ⟨78JFC(11)509⟩, 233, 432 ⟨77CB67⟩, 418 ⟨82MI 614-02⟩
Niemer, B., **4**, 710 ⟨93AG(E)923⟩
Niemeyer, C., **2**, 533 ⟨90TL3863⟩
Niemeyer, H. M., **4**, 175 ⟨91JOC1788⟩; **5**, 325 ⟨91H(32)1687⟩
Niemeyer, M., **6**, 604 ⟨86JFC(32)415⟩
Nientiedt, J., **5**, 1152 ⟨93PS(76)1⟩; **6**, 258, 680, 683 ⟨86ZN(B)149⟩, 679 ⟨87PS(30)401⟩, 684 ⟨93PS(76)265⟩, 685, 692 ⟨88CB655⟩, 700 ⟨89NJC363⟩
Niepel, H., **6**, 51 ⟨80JFC(17)65⟩
Nierenstein, M., **3**, 474 ⟨15JCS1491, 24JA2551⟩, 474 ⟨24JA2556, 25JA1728⟩; **5**, 110 ⟨43OSC(2)557⟩
Niermann, H., **2**, 601 ⟨67LA(707)15⟩
Nie-Sarink, M. J. de, **2**, 614 ⟨84MI 212-01⟩
Niess, R., **2**, 1009 ⟨68TL4265, 88JOC4379⟩; **5**, 301 ⟨68CB502⟩
Niessen, W. von, **4**, 352, 353 ⟨92CB591⟩
Niessner, M., **2**, 1041 ⟨83AG(E)53⟩; **5**, 302 ⟨82TL3255⟩
Nieter-Burgmayer, S. J., **4**, 675 ⟨83JA2599⟩
Nieuwenhuis, J., **3**, 247 ⟨58RTC761⟩
Nieuwland, J. A., **1**, 1000 ⟨37JOC1⟩; **2**, 606, 623 ⟨36JA1806⟩, 1018 ⟨32JA787, 33JCS741⟩, 1018, 1019 ⟨33JA2150⟩; **3**, 238 ⟨36JOC163⟩, 618 ⟨37JA1202⟩; **4**, 24 ⟨36JA1806⟩, 195 ⟨36JA892⟩
Nieuwstad, Th. J., **1**, 96 ⟨72RTC1069⟩
Nieves, I., **1**, 434 ⟨75TL4005⟩
Niewöhner, U., **4**, 170 ⟨87LA607⟩
Nifant'eva, L. V., **5**, 966 ⟨66ZOB1707⟩
Nifantev, E. E., **2**, 255 ⟨87ZOB786⟩, 410 ⟨76ZOB477, 80ZOB1702⟩, 414 ⟨72ZOB1936⟩, 451 ⟨67ZOB1692⟩, 464 ⟨80JGU2159⟩, 469 ⟨71ZOB2372⟩; **4**, 476 ⟨71ZOB2372⟩; **5**, 370 ⟨81PS(10)261⟩, 775 ⟨91ZOB1581⟩
Nifant'ev, N. E., **4**, 550 ⟨80ZOB1881⟩
Nifant'ev, N. F., **5**, 370 ⟨81PS(10)261⟩, 884 ⟨81ZOB482⟩
Nigam, A., **6**, 566 ⟨87JHC749⟩
Nigam, S. C., **2**, 304 ⟨89SC3139⟩
Nigh, W. G., **2**, 28 ⟨B-73MI 201-04⟩
Nightingale, D., **3**, 426 ⟨44JA352⟩
Nightingale, D. V., **3**, 295 ⟨53JOC244⟩
Nigmatullina, R. F., **4**, 864 ⟨75ZOR415⟩
Nihira, T., **4**, 847 ⟨89TL2091⟩
Nihonyana, M., **2**, 264 ⟨73JOM(50)26⟩
Nihonyanagi, M., **2**, 540 ⟨73JOM(50)C26⟩
Nii, Y., **1**, 829 ⟨87TL2597⟩; **2**, 788 ⟨79TL2525⟩; **5**, 753 ⟨79TL2517⟩
Niibo, Y., **6**, 126 ⟨87TL2147, 89JOC5003⟩
Niihata, S., **1**, 358, 498 ⟨92JA8818⟩
Niimi, H., **6**, 313 ⟨74JOM(72)103⟩, 624 ⟨77BCJ3271⟩
Niino, H., **5**, 334 ⟨85JMC1772⟩
Niinomi, Y., **1**, 451 ⟨82CC56⟩; **3**, 660 ⟨82CC56⟩
Niitsu, T., **3**, 609 ⟨86JOM(311)C63⟩, 689 ⟨92CL1053⟩; **5**, 877 ⟨88CL1733⟩, 884 ⟨88TL333⟩, 1086 ⟨90BCJ2736⟩, 1086, 1087 ⟨88POL2213⟩, 1092 ⟨89TL187, 89TL5433⟩; **6**, 680, 690 ⟨88CL1733⟩
Niiyama, K., **1**, 710 ⟨84JA4547⟩
Niizato, H., **5**, 701 ⟨71CPB1606⟩
Niizeki, S., **5**, 1125 ⟨86JAP1280464⟩
Nijhoff, J., **4**, 703 ⟨93OM2609⟩
Nikaido, T., **6**, 7 ⟨75TL899⟩
Nikam, S. S., **1**, 977 ⟨83JOC5376⟩, 1059 ⟨85JOC2193⟩; **2**, 930 ⟨86JOC745⟩, 939, 940 ⟨83JOC5376⟩
Nikam, S. S.,**5**, 327 ⟨95TL197⟩
Nikanorov, V. A., **5**, 1142, 1145 ⟨79JOM(177)101⟩
Nikiforov, T. T., **5**, 1111 ⟨91TL2505⟩
Nikiforova, T. Y., **6**, 242 ⟨68ZOB709⟩
Nikishina, I. S., **4**, 1058 ⟨86ZOB1535⟩; **5**, 1079 ⟨86ZOB1525⟩; **6**, 626 ⟨88ZOB393, 91ZOB1202⟩

Nikishin, G. I., **1**, 331 ⟨61IZV1099⟩, 925 ⟨60DOK(131)580, 61DOK(136)1099⟩, 925 ⟨61IZV2065⟩, 926 ⟨72JOU2515⟩, 1051 ⟨90TL7063⟩; **2**, 365 ⟨85T4279⟩; **3**, 214 ⟨88JOM(348)123⟩; **4**, 197 ⟨88TL1603⟩; **5**, 168 ⟨73CC693⟩
Nikishova, N. G., **2**, 7 ⟨89JOU642⟩
Nikitin, E. V., **2**, 845 ⟨82ZOB2721⟩, 881 ⟨85JGU1332⟩
Nikitin, V. S., **5**, 721, 722 ⟨78JGU1313⟩, 721, 722 ⟨78ZOB1429⟩, 1082 ⟨78ZOB1313⟩
Nikitina, V. I., **3**, 231, 269 ⟨68JGU301⟩; **4**, 561 ⟨68ZOB292⟩; **5**, 737 ⟨77JGU916⟩; **6**, 160 ⟨75ZOB1450⟩
Nikitskaya, E. S., **3**, 247 ⟨52JGU278⟩
Niklaus, A., **3**, 163 ⟨93JOC3165⟩
Nikoforova, T. Y., **6**, 242 ⟨61CRV313⟩
Nikolaeva, A. D., **4**, 993 ⟨68ZOR2116, 75ZOR452⟩, 993, 994 ⟨71ZOR1670⟩; **5**, 827 ⟨70MI 521-01⟩; **6**, 145 ⟨71ZOR1126, 80MI 605-02⟩
Nikolaeva, I. L., **5**, 369 ⟨89ZOB2392⟩
Nikolaeva, N. A., **4**, 570 ⟨62DOK(147)117⟩
Nikolaeva, N. V., **6**, 160 ⟨91PS(61)361⟩, 691 ⟨92PS(66)257⟩
Nikolaeva, S. N., **3**, 540 ⟨84ZOB1817⟩; **4**, 814 ⟨89JGU2040⟩, 1056, 1059, 1063 ⟨87ZOB1741⟩, 1057 ⟨84ZOB1817, 85ZOB2801⟩
Nikolaev, A. V., **4**, 339 ⟨73ZOB534⟩
Nikolaev, N. I., **6**, 223 ⟨77MI 607-01⟩
Nikolaeva, S. V., **6**, 127 ⟨88ZOB1066⟩
Nikolaeva, V. G., **4**, 728 ⟨71ZOB1873⟩
Nikolaus, D., **5**, 548 ⟨82LA1451⟩
Nikoleav, V. A., **3**, 193 ⟨91S195⟩
Nikoleva, S. N., **4**, 626 ⟨91ZOB921⟩
Nikol'skaya, A. N., **3**, 674 ⟨77RCR374⟩
Nikonorov, K. V., **2**, 450 ⟨48ZOB2008⟩; **5**, 369 ⟨78IZV2412⟩
Nikonov, G. N., **4**, 336 ⟨94AHC(61)59⟩, 342 ⟨83IZV1374, 85IZV1102⟩, 342 ⟨86IZV2502, 88IZV159⟩, 342 ⟨89IZV946, 90IZV1133, 92IZV196⟩, 453, 456 ⟨80IZV2417⟩, 457 ⟨80IZV1438, 80IZV952⟩, 478 ⟨83IZV1379⟩
Nile, T. A., **2**, 915 ⟨77JOM(137)293⟩
Nilkantiah, P. M., **4**, 11 ⟨44JIC112⟩
Nilsen, N. O., **1**, 990 ⟨84TL2887⟩
Nilson, M. E., **5**, 820 ⟨50JA4697⟩
Nilsson, A., **1**, 82 ⟨85ACS(B)187⟩
Nilsson, K., **5**, 118 ⟨81MI 502-07, 82MI 502-01⟩
Nilsson, L., **2**, 746 ⟨79ACS(B)547⟩, 758 ⟨76ACS(B)1000⟩
Nilsson, M., **1**, 311 ⟨77ACS(B)667, 93JOC7238⟩; **2**, 587 ⟨82TL3823⟩, 783 ⟨82T389⟩; **5**, 151 ⟨82T1509⟩, 151 ⟨86ACS(B)466⟩, 652 ⟨87JOM(334)255⟩
Nilsson, N. H., **2**, 400 ⟨76CB1069⟩; **3**, 346, 363 ⟨78BSB223⟩; **4**, 862 ⟨74CB2345⟩, 901 ⟨74S433⟩; **5**, 592, 593 ⟨72S314, 74S433⟩; **6**, 238, 540 ⟨72CB2854⟩, 274, 275, 309 ⟨76CB1069⟩, 553 ⟨71CC314⟩
Nilsson, W., **5**, 54 ⟨62JOC629⟩
Nilsson, Y. I. M., **1**, 192 ⟨93JA6609⟩
Nimgirawath, S., **3**, 154 ⟨79CL209⟩; **4**, 77 ⟨79CL209⟩
Nimitz, J. S., **2**, 396 ⟨82JOC4040⟩; **3**, 46, 190 ⟨82JOC4040⟩; **5**, 234 ⟨78TL3523⟩
Nimkar, S., **5**, 457 ⟨90TL3723⟩
Nimmesgern, H., **1**, 342 ⟨85JOC5620⟩, 1127 ⟨85TL957⟩; **2**, 202 ⟨84JOC615⟩; **3**, 320 ⟨88JA2894⟩; **4**, 510 ⟨85JOC4006⟩; **6**, 637 ⟨82CB2516⟩
Nimomiya, K., **5**, 265, 278, 993 ⟨74T2151⟩
Nimura, N., **5**, 1049 ⟨79CPB1137⟩
Nindel, H., **2**, 486, 487, 510 ⟨70ZC118⟩
Nineham, A. W., **4**, 309 ⟨48JCS2174⟩; **5**, 855, 857 ⟨55CRV355⟩
Ning, R. Y., **2**, 129 ⟨67JOC677⟩; **5**, 327 ⟨77JOC3301⟩
Ninomiya, I., **2**, 352 ⟨94TL2205⟩; **4**, 908 ⟨93CPB217⟩; **5**, 726 ⟨92CPB602⟩
Ninomiya, K., **5**, 993 ⟨72JA6203⟩
Ninomiya, S.-I., **4**, 521 ⟨85TL5547, 86CPB3273⟩
Ninomiya, Y., **4**, 263 ⟨90BCJ466⟩
Nir, Z., **2**, 691 ⟨72T3833⟩
Nisar, M., **1**, 14 ⟨89JA6280⟩, 558, 561 ⟨87S992, 84S1009⟩; **3**, 124 ⟨85JOC3416⟩, 623 ⟨87JOC2988⟩
Nisbet, H. B., **3**, 452 ⟨45JCS126⟩
Nisbet, M. P., **2**, 836, 837 ⟨82POL123⟩
Nishi, K., **1**, 1061, 1062 ⟨90JOC4853⟩
Nishi, S., **3**, 139 ⟨80JA6381⟩, 139 ⟨86JOC421⟩; **5**, 607 ⟨80JA7806⟩
Nishi, T., **1**, 1113, 1114 ⟨80CC103⟩; **4**, 603, 649 ⟨83OM21⟩; **5**, 157 ⟨87CC311⟩, 173, 199 ⟨87TL4569⟩
Nishibayashi, Y., **1**, 609, 983 ⟨92CC46, 93JOC3697⟩
Nishida, A., **2**, 1024, 1025 ⟨88TL4917⟩; **3**, 223 ⟨90JA902⟩; **5**, 318 ⟨85BCJ769, 93BCJ2426⟩
Nishida, H., **6**, 80 ⟨85JAP(K)60208983⟩
Nishida, I., **1**, 884 ⟨83JA1598⟩
Nishida, K., **1**, 811 ⟨81CC786⟩
Nishida, K.-i., **4**, 699 ⟨91T3281⟩
Nishida, M., **5**, 243, 252 ⟨77JOM(129)189⟩
Nishida, S., **1**, 564 ⟨91JOC997⟩, 690 ⟨78JOC1612⟩, 729 ⟨71JCS(C)3252⟩, 800 ⟨90JOC1506⟩, 803 ⟨79CL927⟩; **2**, 904 ⟨90JOC4986⟩
Nishida, T., **1**, 413 ⟨78CC972⟩, 808 ⟨78CC651⟩, 809 ⟨80TL1347⟩; **3**, 217 ⟨80TL1347⟩; **5**, 324 ⟨91H(32)221⟩; **6**, 70 ⟨76JAP7630034⟩
Nishida, Y., **6**, 465 ⟨93EUP559212⟩
Nishide, K., **1**, 426, 432 ⟨82TL689⟩; **2**, 658 ⟨80JOC4275⟩, 784 ⟨92S1119⟩; **4**, 806, 817 ⟨84CL1765⟩, 809 ⟨90BCJ1947⟩; **5**, 40, 106 ⟨78TL5211⟩, 83 ⟨81JOC5163⟩
Nishide, Y., **4**, 193 ⟨89CL737⟩
Nishidu, S., **1**, 842 ⟨72BCJ1777⟩
Nishigaki, M., **5**, 248 ⟨85S878⟩
Nishiguchi, I., **3**, 146 ⟨77CL1021⟩, 212 ⟨83JOC2503⟩, 621 ⟨87S631⟩; **5**, 596 ⟨93SC2139⟩; **6**, 479 ⟨88TL4767, 90TL4773⟩, 494 ⟨91TL6867⟩
Nishiguchi, J., **6**, 471 ⟨90CL811⟩
Nishiguchi, S., **2**, 390 ⟨84CPB2609⟩
Nishiguchi, T., **1**, 444 ⟨80JOM(193)37⟩; **3**, 118 ⟨89JOC1531⟩, 121 ⟨89JOC3001⟩
Nishihara, A., **5**, 141 ⟨68JOC2525⟩
Nishihara, H., **3**, 117 ⟨88SC869⟩
Nishihara, T., **3**, 161 ⟨88JOC1343⟩
Nishihara, Y., **3**, 186 ⟨84TL5405⟩; **4**, 307 ⟨67TL3769⟩
Nishihata, K., **1**, 143 ⟨86BCJ2801⟩
Nishihira, K., **6**, 465 ⟨93EUP559212⟩
Nishii, N., **4**, 686 ⟨72CC993⟩
Nishii, S., **1**, 188 ⟨89JA4864⟩, 188 ⟨89JOC4055⟩, 305 ⟨87CC1572⟩, 1026 ⟨86JA7116⟩, 1038, 1039 ⟨86CC102⟩; **5**, 158 ⟨89JA4864⟩
Nishijima, K., **6**, 464 ⟨68TL659⟩
Nishikawa, H., **6**, 555 ⟨92CC1410⟩
Nishikawa, M., **1**, 217 ⟨86TL2015⟩; **2**, 197 ⟨80CPB134⟩
Nishikawa, S., **2**, 440, 829 ⟨91TA667⟩
Nishikawa, T., **1**, 1009 ⟨91CL1271⟩, 1048 ⟨90TL5499⟩; **2**, 84 ⟨90TL221⟩
Nishikaze, N., **4**, 979 ⟨74JOC3198⟩
Nishimoto, M., **4**, 67 ⟨89JCS(P1)879⟩
Nishimoto, Y., **4**, 193 ⟨90JHC1419⟩
Nishimura, A., **1**, 1015 ⟨93TL2139⟩; **3**, 651 ⟨86TL5109⟩
Nishimura, H., **3**, 266 ⟨71JA3299⟩; **5**, 652 ⟨90TL3603⟩
Nishimura, J., **1**, 300 ⟨69T2647⟩, 895 ⟨74JA3548⟩; **4**, 206 ⟨63OSC(4)713⟩
Nishimura, K., **1**, 193 ⟨82TL5549⟩, 671 ⟨65JOC1593⟩; **4**, 615 ⟨83JOM(250)109⟩, 1048 ⟨80JOM(194)147⟩; **6**, 178 ⟨81JA2324⟩, 383 ⟨82JOM(236)7⟩
Nishimura, M., **1**, 352 ⟨92CL61⟩
Nishimura, O., **2**, 274 ⟨78CPB1576⟩
Nishimura, S., **1**, 72, 81 ⟨71BCJ166⟩, 1034 ⟨88H(27)475⟩; **3**, 408 ⟨87TL1293⟩; **5**, 113 ⟨69TL1019⟩; **6**, 9 ⟨80GEP2920386⟩, 671 ⟨87TL1293⟩

Nishimura, T., **1**, 822 ⟨73TL2351⟩; **2**, 34 ⟨92BCJ1731⟩, 139 ⟨91BCJ2524⟩, 142 ⟨90CL2239⟩; **3**, 281, 298 ⟨91CL1059⟩
Nishinaga, A., **3**, 412 ⟨91CL1093⟩; **5**, 299 ⟨79CC970⟩
Nishino, H., **2**, 22 ⟨87BCJ809⟩, 631 ⟨87BCJ4363⟩; **3**, 101 ⟨89BCJ545⟩; **6**, 300 ⟨92JAP(K)04164085⟩
Nishino, M., **1**, 464 ⟨73CL667⟩, 813, 814 ⟨80BCJ2958⟩; **3**, 142 ⟨80BCJ2958⟩
Nishinomiya, T., **5**, 1102 ⟨89JAP3145452⟩
Nishio, K., **3**, 159 ⟨93JOC2647⟩
Nishio, M., **1**, 143 ⟨86BCJ2801⟩; **2**, 701 ⟨93CC1067, 93CL1953⟩
Nishio, O., **1**, 1133 ⟨76CC606⟩
Nishio, T., **2**, 146 ⟨91JCS(P1)1717⟩, 185 ⟨88SC2083⟩, 228 ⟨73BCJ2253⟩; **3**, 247 ⟨80S1013⟩, 343 ⟨73BCJ2253⟩; **5**, 627 ⟨94CL77⟩; **6**, 575 ⟨82JCS(P1)2149⟩
Nishioka, E., **2**, 331 ⟨89JA5886⟩
Nishioka, H., **3**, 87 ⟨93CPB1166⟩
Nishioka, K., **1**, 986 ⟨84CL131⟩
Nishioka, R., **4**, 198 ⟨80TL3919⟩
Nishioka, T., **3**, 631 ⟨92BCJ111, 92JOC5643⟩
Nishioka, Y., **6**, 482 ⟨89BCJ2117⟩
Nishitani, K., **2**, 282 ⟨80TL4417⟩, 625 ⟨81TL3745⟩, 976 ⟨82JA4708⟩; **3**, 38 ⟨80TL4417⟩; **5**, 166 ⟨87TL655⟩, 173 ⟨82JA4708⟩; **6**, 472, 473 ⟨82JA4708⟩
Nishitani, T., **4**, 424 ⟨82JOC1706⟩
Nishiwaki, K., **1**, 759 ⟨73TL4193⟩; **2**, 7, 27 ⟨93JOC4382⟩; **4**, 620, 653 ⟨73TL4193⟩; **5**, 1137 ⟨83JCS(D)1841⟩; **5**, 337, 584 ⟨85CB1668⟩, 948 ⟨82CC1178⟩
Nishiwaki, K.,**5**, 948 ⟨82JCS(D)793⟩
Nishiwaki, K.-I., **2**, 936 ⟨73TL4193⟩
Nishiwaki, N., **1**, 1043 ⟨89CL773⟩
Nishiwaki, T., **2**, 744 ⟨75S20⟩; **4**, 850 ⟨80JCS(P1)2693⟩
Nishiyama, C., **3**, 307 ⟨88H(27)1855⟩
Nishiyama, H., **1**, 63 ⟨85OM1911⟩, 132 ⟨84TL223, 84TL2487⟩, 206, 289, 1092 ⟨82TL1267⟩, 299 ⟨94JA2223⟩; **2**, 540 ⟨84JOC2298⟩; **3**, 663 ⟨88T2413⟩; **5**, 215 ⟨84TL223⟩, 325 ⟨85JA5310⟩, 726 ⟨79TL4671⟩
Nishiyama, K., **2**, 544 ⟨87BCJ2289⟩, 753 ⟨83CC1322⟩; **3**, 665 ⟨84CL773⟩; **4**, 432, 433 ⟨87T693, 88S106⟩, 438 ⟨88BCJ609⟩, 516 ⟨83CC1322⟩; **5**, 1042 ⟨87BCJ2289⟩, 1106 ⟨87BCJ2692⟩; **6**, 558 ⟨93TL3745⟩
Nishiyama, R., **6**, 9 ⟨80GEP3008081⟩
Nishiyama, S., **6**, 80 ⟨85S408⟩
Nishiyama, T., **3**, 7, 31 ⟨90JOC462⟩
Nishiyama, Y., **1**, 86 ⟨89BCJ1682⟩; **4**, 227 ⟨91CC1421⟩, 238 ⟨91CL1775⟩; **5**, 240, 243, 253 ⟨91JOC3776, 89CL1825⟩
Nishizawa, E. E., **6**, 445 ⟨89JMC228⟩
Nishizawa, H., **2**, 84 ⟨93TL4031⟩
Nishizawa, M., **1**, 16 ⟨87JOC4878⟩, 604 ⟨76JA1612⟩, 607 ⟨76CC582⟩, 781 ⟨76JOC1485⟩; **2**, 63 ⟨91COS(8)159⟩, 64 ⟨84JA6709⟩, 279 ⟨76JOC1485⟩; **3**, 225 ⟨81BCJ2233⟩
Nishizawa, S., **1**, 5, 7 ⟨84S949⟩
Nisian-Neilson, P., **2**, 466 ⟨82IC3568⟩
Nisian-Neilson, P., **2**, 466, 467 ⟨84IC2062⟩
Nisiwaki, M., **5**, 329 ⟨94TL1247⟩
Niskova, A. N., **3**, 252 ⟨76IZV2040⟩
Nissan, R. A., **4**, 298 ⟨90JOC1459⟩
Nitikin, V. S., **4**, 658 ⟨85ZOB1629⟩
Nitsch, H., **2**, 353 ⟨76AG(E)760⟩
Nitsche, M., **6**, 317 ⟨78CB3644⟩
Nitsuta, K., **6**, 488 ⟨92JAP05339226⟩
Nitta, H., **3**, 629 ⟨91SL563, 91TL4333⟩, 629 ⟨92JA7969, 92JOC6778⟩
Nitta, K., **1**, 677 ⟨86JA7408⟩; **2**, 73 ⟨88TL5263⟩, 74, 572 ⟨89JOC4732⟩, 618, 625, 630 ⟨86JA7408⟩
Nitta, M., **2**, 468 ⟨92TL251⟩; **3**, 77 ⟨84JCS(P1)2103⟩, 269 ⟨85JCS(P1)1401⟩
Nitta, Y., **2**, 213 ⟨66JA2328, 67JOC727⟩, 227 ⟨88CPB791⟩; **3**, 294 ⟨66JA2328⟩; **4**, 426 ⟨86H(24)25⟩; **5**, 307 ⟨85CPB5515, 87CPB364⟩, 652 ⟨88CPB3125⟩
Nitti, P., **5**, 355 ⟨91G401⟩
Nittolo, S., **6**, 370 ⟨85IC2889⟩
Nitz, S., **5**, 1037 ⟨79ZN(B)768⟩
Nitz, T. J., **4**, 179 ⟨84TL3047⟩
Nitzschke, M., **2**, 130 ⟨63AG(E)370⟩; **3**, 344, 363 ⟨63AG(E)370⟩
Nitzschmann, R. E., **3**, 568 ⟨67JOM(8)547⟩
Niu, C.-I., **2**, 127, 214 ⟨55JA416⟩
Niu, C. R., **3**, 103 ⟨92SL909⟩
Nivard, R. J. F., **3**, 344, 345, 363 ⟨73S149⟩; **4**, 177, 190 ⟨69CC1175⟩, 304 ⟨80JOC1880⟩, 830 ⟨73RTC11, 77JOC3128⟩; **5**, 512, 566, 568, 570, 608 ⟨73S149⟩; **6**, 44 ⟨71RTC1123⟩, 55, 107 ⟨69RTC289⟩, 75 ⟨76JCS(P1)1048, 77JOC3128⟩, 75 ⟨81RTC13⟩
Niven, M. L., **4**, 148 ⟨86ICA(119)177⟩, 584, 585 ⟨86JOM(315)255, 82JOM(232)C78⟩; **6**, 607 ⟨83MI 620-01⟩
Nivert, C., **1**, 544 ⟨72JOM(44)69⟩
Nivert, C. L., **6**, 401, 402, 404 ⟨77JA5209⟩
Nivikova, O. P., **5**, 702 ⟨91JGU1268⟩
Nivorozhkin, L. E., **3**, 347, 356 ⟨73ZOR1717⟩
Niwa, H., **1**, 710 ⟨84JA4547⟩; **3**, 113 ⟨85CL1385⟩, 365 ⟨77BCJ2751⟩; **5**, 199 ⟨86TL4609⟩
Niwa, I., **2**, 282 ⟨77S874⟩; **3**, 38, 173 ⟨77S874⟩
Niwa, S., **1**, 1039 ⟨90JCS(P1)937⟩; **2**, 67 ⟨92CRV833⟩, 71 ⟨88JOC927, 89JCS(P1)109⟩, 1096 ⟨90JCS(P1)937⟩
Nix, M., **5**, 152 ⟨87OS(65)236⟩
Nixdorf, M., **3**, 616 ⟨85LA566⟩
Nixon, J. F., **2**, 478 ⟨91MI 208-04⟩; **3**, 609, 688 ⟨88CRV1327, 93CI(L)404⟩, 690 ⟨93AG(E)103⟩; **4**, 548 ⟨90CC1307, 93PS(77)274⟩, 549 ⟨94AG(E)2202⟩; **5**, 886 ⟨93CC267⟩, 1152 ⟨76CPL(42)460, 78JA446, 79CC653, 80JSP(82)185⟩, 1152, 1153, 1159 ⟨80CC333⟩, 1153 ⟨79JCS(D)1405, 81JSP(90)512, 82JST(79)215⟩, 1154 ⟨82JSP(92)158, 85JSP(109)8⟩, 1160 ⟨93AG(E)103⟩; **6**, 154 ⟨89CC1046⟩, 154 ⟨92JOM(433)C11, 93CC311⟩, 155 ⟨93AG(E)103⟩, 679 ⟨76CC513, 78JA446, 79CC653⟩, 690, 692 ⟨88CRV1327, 89JOM(368)C29⟩
Nixou, J. F., **5**, 1153 ⟨81JSP(90)507⟩
Niyazymbetov, M. E., **3**, 634 ⟨90S841⟩; **4**, 38 ⟨90IZV1802⟩, 1083 ⟨89IZV356, 89UK1105⟩; **6**, 56 ⟨87IZV306⟩, 149 ⟨82IZV2637⟩
Nizamov, I. S., **2**, 254 ⟨93PS(79)179⟩, 464 ⟨88JGU1131⟩; **4**, 345 ⟨86ZOB216, 87DOK(292)360⟩, 921 ⟨84ZOB1324⟩; **5**, 405 ⟨89JGU2388⟩
Niznik, G. E., **1**, 38 ⟨78JOC2396⟩; **2**, 268 ⟨74JOC600⟩; **3**, 418, 697 ⟨69JA7778⟩, 418, 697 ⟨74JOC600⟩, 696, 699, 721, 722 ⟨72JOC187⟩, 696, 700 ⟨88OSC(6)751, 70JA6675⟩, 697 ⟨71TL4965⟩; **5**, 68, 420, 792, 796, 797 ⟨74JOC600⟩, 792, 796, 797 ⟨70JA6675⟩, 796 ⟨69JA7778⟩, 796, 797 ⟨71OS(51)31⟩
Nizrvik, G. E., **3**, 697 ⟨71OS(51)31⟩
Njav, V. C. O., **3**, 234 ⟨83JOC1007⟩
Nkansah, A. N., **6**, 116 ⟨86MI 604-01⟩
Nkengfack, A. E., **4**, 979 ⟨88JCS(P1)961⟩
Nkeng Peh, M. L., **5**, 918 ⟨85CC1151⟩
Nkiliza, J., **3**, 346 ⟨91TL1787⟩
No, B. I., **2**, 430 ⟨90ZOB1795⟩, 448 ⟨91ZOB1906⟩, 469 ⟨75ZOB2554⟩; **5**, 722 ⟨75JGU935, 75ZOB949⟩, 774 ⟨78ZOB78⟩
Noack, K., **6**, 455 ⟨68JOM(13)411⟩
Noack, R., **5**, 932 ⟨90T1235⟩
Noall, W. I., **2**, 24 ⟨77JCS(P1)226⟩
Noata, T., **3**, 407 ⟨85CC613⟩
Nobayashi, Y., **1**, 204, 1024 ⟨83TL5121, 84T4261⟩
Nobel, D., **5**, 962 ⟨89CRV1927⟩
Nober, G., **3**, 452 ⟨41LA(547)1⟩
Nobes, R. H., **2**, 1022 ⟨82JOC1869⟩
Nobis, J. F., **5**, 27 ⟨57CRV641⟩
Noble, Jr., P., **4**, 795 ⟨70JOC4236⟩; **5**, 232 ⟨63OSC(4)924⟩; **6**, 144, 150 ⟨64CRV19⟩

Noble, W. J. Le, **1**, 988 ⟨81JA4638⟩
Nobuhara, M., **3**, 704 ⟨76CPB832⟩
Nobuharu, S., **3**, 357 ⟨91EUP462531⟩
Nobumasa, N., **2**, 627 ⟨69NAT765⟩
Nocera, J., **1**, 556 ⟨80JOC4702⟩; **2**, 28 ⟨80JOC4702⟩
Nocher, A.-M. Le, **3**, 347 ⟨92TL6151⟩, 368 ⟨91TL747⟩
Nocito, V., **4**, 108 ⟨91USP5043489⟩
Noda, I., **5**, 510, 551, 583, 620 ⟨82LA1229⟩, 584 ⟨81LA1798⟩
Noda, Y., **1**, 1015 ⟨90TL4393⟩; **2**, 923, 969 ⟨91JOC2590⟩, 980 ⟨90TL4393⟩; **3**, 255 ⟨81CL805⟩
Node, M., **1**, 134 ⟨91TL7281⟩, 241 ⟨87CC449⟩, 426, 432 ⟨82TL689⟩, 428 ⟨84JOC3641⟩; **2**, 658 ⟨80JOC4275⟩, 784 ⟨92S1119⟩; **3**, 47, 192 ⟨87S729⟩, 161 ⟨92S852⟩, 221 ⟨91TA178⟩; **5**, 40, 106 ⟨78TL5211⟩, 83 ⟨81JOC5163⟩, 171 ⟨85JA6404⟩
Nodiff, E. A., **6**, 218 ⟨52JA1347⟩, 232 ⟨60JOC60⟩
Noding, S. A., **2**, 68 ⟨79JOC4371⟩
Noe, C. R., **5**, 354 ⟨77S487⟩
Noe, E. A., **4**, 9 ⟨72JA2020⟩
Noe, M. C., **1**, 116 ⟨93TL5995⟩; **2**, 49 ⟨93JA12579⟩
Noe, R., **4**, 725 ⟨90JA9620⟩; **5**, 797 ⟨91OM1201⟩
Noël, D., **6**, 309 ⟨67BSF2239⟩
Noël, Y., **2**, 921 ⟨74JA3684⟩
Noels, A. F., **1**, 549 ⟨78TL1239⟩
Noeth, H., **4**, 373 ⟨83CB1479⟩
Noftle, R. E., **4**, 32 ⟨84JFC(26)29⟩
Nogaideli, A. I., **5**, 720 ⟨70IZV1157, 72IZV1150⟩
Nogami, Y., **3**, 361 ⟨89CC505⟩
Nógrádi, M., **2**, 67 ⟨B-86MI 202-03⟩
Noguchi, H., **2**, 381 ⟨81CPB3387⟩
Noguchi, M., **2**, 757 ⟨77BCJ1513⟩; **3**, 372 ⟨77H(6)1173⟩
Noguchi, S., **5**, 215 ⟨85CL701⟩, 601 ⟨79CPB1352⟩
Noguchi, T., **2**, 434 ⟨82CL977⟩
Noguchi, Y., **2**, 197 ⟨80CPB134⟩, 724 ⟨80S937⟩
Nogues, P., **4**, 827 ⟨82TL603⟩
Noguez, J. A., **3**, 60 ⟨84JA260⟩
Noguichi, Y., **2**, 363 ⟨82CPB1646⟩
Nogusa, H., **3**, 426 ⟨83JOC2095⟩
Nohair, K., **2**, 14 ⟨91SL659, 92TL213⟩; **6**, 7 ⟨92TL213⟩
Nohara, A., **5**, 74 ⟨73TL1995⟩, 363, 367, 774 ⟨74TL1187⟩
Nohara, W., **3**, 279 ⟨92CL1987⟩
Nohira, H., **2**, 53 ⟨93TL2351⟩, 816 ⟨83TL511⟩; **4**, 142, 521 ⟨91CL97⟩, 506 ⟨80BCJ789⟩, 610, 617, 622 ⟨87BCJ2263⟩, 833 ⟨62BCJ71⟩; **5**, 214, 1001 ⟨61JOC782, 61JOC4381⟩; **6**, 113 ⟨65BCJ2107⟩, 115 ⟨91CL1873, 93CL13⟩
Nohlen, M., **5**, 386 ⟨90JOM(382)C6, 91JOM(412)225⟩
Noji, Y., **2**, 19 ⟨81BCJ1151⟩; **4**, 52 ⟨77TL2893, 81BCJ1151⟩; **5**, 6 ⟨77TL2337, 77TL2893⟩
Nojima, M., **2**, 6 ⟨73S786⟩, 6, 11, 17, 392, 615 ⟨79JOC3872⟩, 17 ⟨73S779⟩, 18 ⟨73S780⟩, 19 ⟨74JA925⟩, 797 ⟨73BCJ1254⟩; **3**, 326 ⟨94JCS(P1)643⟩, 368 ⟨91JCS(P1)2890⟩; **4**, 10 ⟨74JA925⟩, 32 ⟨79JOC3842⟩; **5**, 4 ⟨79JOC3726⟩, 5 ⟨73S487, 74JA925⟩, 282 ⟨76JCS(P1)2205⟩; **6**, 52 ⟨74JA925⟩, 439 ⟨79JOC3872⟩
Nokaka, Y., **6**, 533 ⟨83JAP60056958⟩
Nokami, J., **1**, 779 ⟨75TL2841, 78TL4903⟩, 1055 ⟨84CL1939⟩; **2**, 21 ⟨81TL3193⟩; **3**, 634, 651 ⟨84JA7890, 86TL5109⟩; **4**, 220 ⟨80TL2557⟩, 264 ⟨76BCJ256⟩; **6**, 461 ⟨90TL5629⟩
Nokikova, Z. S.,**3**, 195 ⟨83JGU610⟩
Nolan, M. C., **2**, 969 ⟨87JA203⟩
Noland, W. E., **2**, 394 ⟨73OSC(5)833⟩; **4**, 439 ⟨63T(S)23⟩
Nolde, K., **2**, 1012, 1013 ⟨62M952⟩
Nolen, E. G., **1**, 255 ⟨90CC1532⟩; **2**, 302 ⟨90CC1532⟩
Nolen, III, H. W., **5**, 668, 830 ⟨86JMC1689⟩
Noli, M., **6**, 412 ⟨28G443⟩
Noll, J. E., **4**, 506 ⟨51JA3867⟩; **5**, 1024 ⟨55JA3149⟩
Noll, K., **3**, 561 ⟨65JOC3718⟩; **5**, 770, 849 ⟨65JOC3718⟩
Noller, C. R., **1**, 117 ⟨36JA816, 64JA1112⟩, 127 ⟨29JA594⟩; **2**, 31 ⟨43OSC(2)358⟩, 107 ⟨43OSC(2)109⟩, 182 ⟨48JA3955, 50OS(30)58⟩; **5**, 298 ⟨43OSC(2)586⟩
Nolte, C., **3**, 356 ⟨91LA111⟩
Nolte, E., **3**, 613 ⟨52LA(575)1⟩
Nolte, J. G., **4**, 708 ⟨73JOM(55)419⟩
Nolte, M., **4**, 725 ⟨92CB1953, 92JOM(427)C21⟩, 725 ⟨93OM4979⟩
Nolte, R., **2**, 858 ⟨88JA624⟩; **3**, 385 ⟨88JA624⟩
Nolte, R. J. M., **3**, 708 ⟨78JOC1972⟩
Nolte, U., **5**, 893 ⟨91CB989⟩
Noltemeyer, M., **2**, 489 ⟨89CB1247⟩, 880 ⟨91JOM(402)55⟩, 1085 ⟨87CB1437⟩; **4**, 536 ⟨88AG(E)1194⟩, 800 ⟨86CB3150⟩; **5**, 381 ⟨84CB2112⟩, 606 ⟨85ZN(B)124⟩, 781 ⟨91ZN(B)1328⟩, 782 ⟨90JOM(338)21⟩, 934 ⟨91AG(E)1658⟩, 946 ⟨93SL486⟩; **6**, 183 ⟨91CB2645⟩, 252 ⟨91CB1353⟩
Noltes, J. G., **1**, 127 ⟨82JA6609⟩; **2**, 485, 486, 487, 505 ⟨73JOM(51)223⟩, 488 ⟨70JOM(22)653⟩, 491, 505 ⟨77JOM(136)173⟩, 585, 589 ⟨82COMC-I(2)709⟩, 601 ⟨56CI(L)352⟩, 695, 885, 886, 889 ⟨70JOM(25)139⟩; **4**, 626 ⟨67TL1443⟩, 695 ⟨75JOM(84)117, 75JOM(84)129⟩, 696 ⟨75JOM(102)551⟩, 701 ⟨69JOM(16)91, 83OM1099⟩, 724 ⟨80JOM(186)427⟩; **6**, 185 ⟨67TL1443⟩
Nome, F., **5**, 652 ⟨87SC1741, 89SC1181⟩
Nomoto, R., **1**, 1010 ⟨92JMC241⟩
Nomoto, T., **1**, 442 ⟨87JCS(P1)1951⟩, 515 ⟨85T2387⟩; **2**, 235 ⟨86H(24)303⟩, 621, 627 ⟨84CC1523⟩
Nomura, E., **2**, 89, 683 ⟨88CL1773⟩
Nomura, H., **6**, 558 ⟨89CPB1249⟩
Nomura, K., **1**, 1063 ⟨91CC948⟩; **4**, 67 ⟨86JHC1163⟩; **5**, 301 ⟨87JA3402⟩; **6**, 555 ⟨92CC1410⟩
Nomura, M., **1**, 494 ⟨90JCS(P1)2207⟩, 1006 ⟨92TL5363⟩; **2**, 305 ⟨93BCJ1297⟩
Nomura, N., **1**, 187 ⟨91SL251⟩
Nomura, T., **2**, 264 ⟨54MI 203-01⟩
Nomura, Y., **1**, 603 ⟨82CC871, 82TL4733⟩; **2**, 130 ⟨60BCJ1743⟩, 282, 284 ⟨82TL1361, 81CL1715⟩, 284 ⟨82CC871, 82CL1733⟩, 284 ⟨84JOC3235⟩, 284, 392, 785 ⟨82TL4733⟩, 368 ⟨85S212, 82CL1109⟩, 1036, 1038 ⟨82CL1244⟩; **3**, 228 ⟨76JCS(P1)2289⟩, 348 ⟨60BCJ1743⟩, 412 ⟨79TL3453⟩, 414 ⟨83BCJ641⟩, 637 ⟨82CC871, 82TL1361⟩; **4**, 376 ⟨84JOC3235⟩, 516 ⟨86CL1193⟩; **5**, 335 ⟨85S212⟩, 993 ⟨83JAP53/35194⟩, 1116 ⟨81CL1069⟩
Nona, S. N., **6**, 241 ⟨80JFC(15)423, 81JCS(P1)455⟩
Nonaka, M., **3**, 63 ⟨58JOC1580⟩
Nonaka, T., **2**, 160 ⟨81JOC5215⟩, 903, 919 ⟨86TL2007⟩, 918, 923 ⟨86T4427⟩; **4**, 221 ⟨86TL3869, 87CL1095⟩
Nongkynrih, I., **2**, 744 ⟨89S631⟩
Nonhebel, D. C., **1**, 870 ⟨67JCS(C)1716, 68JCS(C)676⟩, 870 ⟨68T1869⟩, 919, 928 ⟨93CSR347⟩, 919 ⟨B-79MI 119-04⟩; **4**, 1077 ⟨B-74MI 424-02, B-79MI 424-01⟩
Nonomura, K., **2**, 370, 768 ⟨91OM1095⟩
Nonoshita, K., **1**, 415, 826 ⟨88JA7922⟩, 415, 826 ⟨90JA316, 89TL1265⟩; **3**, 127 ⟨88BCJ2241⟩
Nonoyama, M., **4**, 721 ⟨93JOM(463)169⟩
Nooda, Y., **1**, 980 ⟨84JOC1204⟩
Nooi, J. R., **2**, 1032 ⟨61RTC244, 62RTC533⟩
Noordik, H., **1**, 890 ⟨74AX(B)833⟩
Noori, G. F. M., **3**, 282 ⟨83JCS(P1)1075⟩
Nooy, A. E. J. de, **1**, 185 ⟨91S347⟩; **6**, 317 ⟨91S347⟩
Noravyan, A. S., **5**, 690 ⟨77KFZ38⟩
Norbeck, D. W., **1**, 831 ⟨85JA3285⟩; **2**, 25 ⟨85JA3285⟩; **3**, 220 ⟨85JOC2198⟩; **5**, 239 ⟨85JA3285⟩, 412, 413, 422 ⟨85JOC2198⟩; **6**, 417 ⟨92EUP486948⟩
Norberg, B., **2**, 287 ⟨85CC569⟩
Norberg, T., **4**, 223 ⟨87MI 405-01⟩
Norberto, F., **5**, 361 ⟨89JCR(S)260⟩, 361, 362, 711 ⟨85JCR(S)132⟩, 362, 657, 711 ⟨92JCS(P2)281⟩

Norbury, A., **2**, 900, 901 ⟨91TL6239⟩
Norbury, A. H., **2**, 858 ⟨73ACS1939⟩
Norcross, B. E., **1**, 1076 ⟨76JOC3650⟩
Norcross, R. D., **2**, 533, 698 ⟨90T4663⟩
Nord, F. F., **3**, 350 ⟨50JOC81⟩
Nordahl, J. G., **1**, 811 ⟨88TL975⟩
Nordal, V., **3**, 357 ⟨69ACS1975⟩
Nordeen, C., **2**, 684 ⟨86SC1043⟩
Nordin, I. C., **1**, 104 ⟨61JA198⟩; **4**, 306 ⟨66JHC531⟩
Nordmann, R., **5**, 1125 ⟨86HCA246⟩
Norell, J. B., **2**, 656 ⟨73JOC1924⟩
Norell, J. R., **3**, 297 ⟨73JOC1924⟩; **5**, 496 ⟨63JA3231⟩, 658 ⟨70JOC1619⟩
Norem, N. T., **6**, 247 ⟨90IC3252⟩
Noren, G. H., **5**, 1020 ⟨87OM2617⟩
Noren, J. O., **6**, 491 ⟨89CAR(194)209⟩ , 512 ⟨79EUP3008⟩
Norikova, N. V., **2**, 881 ⟨64IZV1197, 64IZV1202⟩
Norin, T., **1**, 828 ⟨90SC423⟩; **2**, 177 ⟨73ACS1451⟩, 1011 ⟨85SC463⟩
Norio, S., **5**, 1012 ⟨89NKK1881⟩
Noritake, Y., **1**, 187 ⟨91SL251⟩
Nørkjrr, K., **3**, 352, 366 ⟨93CB73⟩
Norley, M. C., **2**, 916 ⟨94SL77⟩
Norlung, L. L., **2**, 366 ⟨78JMC704⟩
Norman, A. D., **2**, 47, 474 ⟨70IC870⟩, 428 ⟨84IC4130⟩, 466 ⟨86CC850⟩, 473 ⟨68JA6556⟩, 473, 474 ⟨85IC1107, 71JOM(28)81⟩
Norman, A. W., **1**, 1029 ⟨91JMC2452⟩, 1039 ⟨89JOC4072⟩
Norman, D. G., **1**, 5, 9 ⟨83S304⟩; **4**, 190 ⟨85S751⟩
Norman, G. R., **2**, 255, 256 ⟨52JA161⟩, 462, 842 ⟨55JA3526⟩
Norman, J. J., **1**, 107 ⟨57JA2311⟩; **2**, 19 ⟨57JA2311⟩
Norman, L. R., **2**, 660 ⟨74JA2121⟩
Norman, N. C., **2**, 473 ⟨91CB2457⟩, 512 ⟨90JOM(399)C21⟩, 866 ⟨83CC881⟩, 866, 893 ⟨84IC2582⟩; **4**, 676 ⟨84JCS(D)2455, 88JCS(D)1843⟩; **6**, 167 ⟨83CC528, 84TL3519⟩, 168 ⟨83JA5506⟩, 168 ⟨84IC2582⟩, 169 ⟨83CC881⟩, 172 ⟨86MI 606-01⟩, 173, 174, 176 ⟨90IS239⟩, 372 ⟨90IS235⟩, 373 ⟨86JCS(D)1801⟩, 375 ⟨85JA8211, 85JCS(D)383, 87JCS(D)249⟩, 689 ⟨84JA7015⟩
Norman, O. L., **1**, 379 ⟨58JA3625⟩
Norman, R. O. C., **2**, 338, 339 ⟨81JCS(P2)545⟩, 388 ⟨72JCS(P1)1692⟩; **3**, 419 ⟨71JCS(C)3060⟩, 681 ⟨69JCS(C)2587⟩; **4**, 314 ⟨67JCS(C)735⟩; **5**, 356 ⟨69JCS(C)2587⟩, 366 ⟨72JCS(P1)1692⟩
Normand, F. L., **4**, 458 ⟨72JHC715⟩
Normand, H., **5**, 132, 139 ⟨76CR(C)483⟩
Normant, H., **1**, 212 ⟨75S256⟩, 1000 ⟨65BSF1872⟩, 1092 ⟨60MI 122-01⟩; **2**, 420 ⟨75LA719⟩, 421 ⟨69CR(C)834⟩, 1042, 1050, 1073 ⟨65CR(260)1984⟩, 1072 ⟨66BSF1707⟩; **3**, 54 ⟨63MI 302-01⟩, 228 ⟨60BSF515⟩; **4**, 383 ⟨59BSF459⟩, 395 ⟨64CR(259)830⟩; **5**, 554 ⟨75CR(C)1327⟩; **6**, 27 ⟨56BSF1441⟩
Normant, J., **1**, 464 ⟨66BSF1888⟩; **2**, 682 ⟨75JOM(93)415⟩, 1031 ⟨63BSF1876⟩; **3**, 130 ⟨84T621⟩; **4**, 18 ⟨⟨63BSF1868⟩, 176 ⟨88PAC49⟩, 761, 770 ⟨63BSF1876⟩, 763, 771 ⟨75T891, 75T897⟩
Normant, J. F., **1**, 116, 453, 539 ⟨78PAC709⟩, 164 ⟨77BSF765⟩, 230, 476 ⟨80TL935⟩, 238, 483 ⟨77JOM(142)1⟩, 248 ⟨86TL4427, 86TL4431⟩, 312 ⟨80T2305, 84PAC91⟩, 429 ⟨78S128⟩, 465 ⟨77TL3263⟩, 465, 1017, 1018 ⟨79S826⟩, 466 ⟨77JOM(128)1⟩, 468 ⟨81TL959, 83BSF332, 84T2741⟩, 471, 1097, 1107 ⟨81TL3851⟩, 477 ⟨77JCR(M)2101⟩, 478, 479 ⟨80T1961, 78S528⟩, 479 ⟨78TL2027, 84TL3075, 86T5607⟩, 490 ⟨90TL2879⟩, 491 ⟨76JOM(113)99, 86T1399⟩, 491, 1108 ⟨74JOM(77)269⟩, 496 ⟨84IJ113, 92BSF171⟩, 511 ⟨82JOM(226)C55⟩, 533, 1107 ⟨80TL3763⟩, 538, 539, 540, 541, 542, 1093, 1106, 1139 ⟨81S841⟩, 540 ⟨71TL2583, 74BSF1656, 75JOM(96)471⟩, 541 ⟨82TL5151, 84OS(62)1⟩, 653 ⟨90JOM(394)37⟩, 670, 967 ⟨90JA8042⟩, 679 ⟨86TL1043⟩, 679, 836 ⟨86TL1039⟩, 833 ⟨74BSF2072, 91TL5969⟩, 967 ⟨85TL4197, 91T1677⟩, 1000 ⟨65BSF859⟩, 1000, 1002, 1004, 1047 ⟨72S63⟩, 1017 ⟨63BSF1876, 63BSF1888, 82TL4329⟩, 1017, 1018 ⟨80T1215⟩, 1017, 1067 ⟨82TL4325, 75TL1465⟩, 1038 ⟨77TL3383⟩, 1045 ⟨70TL2659⟩, 1067 ⟨74BSF(2)1731, 75S458, 79S502⟩, 1090 ⟨76S245, 76S761⟩, 1092 ⟨78TL3711⟩, 1108 ⟨74JOM(77)281⟩; **2**, 86 ⟨89TL2387, 89TL2391⟩, 159 ⟨85TL4455⟩, 186 ⟨69JOM(121)123⟩, 313 ⟨93SL386⟩, 420 ⟨75JOM(97)325⟩, 556 ⟨94JOC2925⟩, 574 ⟨B-83MI 211-02⟩, 592 ⟨93SL266⟩, 593 ⟨93TL6053⟩, 594 ⟨92PAC1857, 93TL2303, 93TL7575⟩, 594, 998 ⟨93SL665⟩, 624 ⟨74S803⟩, 629 ⟨74JOM(77)269⟩, 629, 977 ⟨71TL2583⟩, 676 ⟨77BSF693⟩, 679 ⟨75TL3833, 79BSF(2)305⟩, 903, 909, 910, 917, 919, 924, 925, 976, 977, 978 ⟨81S841⟩, 910 ⟨86TL5727⟩, 919 ⟨79S826⟩, 936, 945 ⟨84TL4383⟩, 976, 977, 978 ⟨76MI 219-01⟩, 977 ⟨82TL5151⟩, 998 ⟨90JOM(400)19⟩, 1006 ⟨86TL1039, 86TL4427⟩, 1006 ⟨92SL633⟩, 1015 ⟨75S458⟩, 1095 ⟨73BSF2137⟩; **3**, 67, 237 ⟨78S128⟩, 67, 237 ⟨78S458⟩, 125 ⟨79S968⟩, 130 ⟨79SC639⟩, 152 ⟨84S37⟩, 211, 273 ⟨85S50⟩, 212 ⟨83TL5081⟩, 212, 228 ⟨86T1369⟩, 222 ⟨80T1961⟩, 236 ⟨78S29⟩, 237 ⟨85TL3999, 70TL2659⟩, 273 ⟨73BSF2137⟩; **4**, 22 ⟨75BSF1797⟩, 45 ⟨93JFC(65)67⟩, 192 ⟨84TL3075⟩, 357 ⟨84TL4383⟩, 398 ⟨86TL1043⟩, 405, 408 ⟨91SL111⟩, 655 ⟨86T1389⟩, 701 ⟨86TL4427⟩, 709, 724 ⟨86TL1039⟩, 751 ⟨90JOM(394)37⟩, 806 ⟨89JOM(367)1⟩, 806, 819 ⟨84JOM(264)155⟩, 809 ⟨92JOC6903⟩, 817, 818, 820 ⟨90JOM(400)19⟩, 963 ⟨77BSF693⟩, 1016, 1017 ⟨76AG417⟩; **5**, 10 ⟨69CR(C)1325⟩, 58 ⟨82S297⟩, 152 ⟨88TL4411⟩; **6**, 41 ⟨74BSF2072⟩
Normant, M. H., **5**, 559 ⟨72CR(C)1215⟩
Noronha, G., **5**, 971 ⟨92JOC7364⟩
Norris, D. J., **5**, 199 ⟨92T4863⟩
Norris, J. F., **2**, 25 ⟨41OSC(1)144⟩, 89 ⟨32JA2088⟩, 495 ⟨19IEC817⟩; **3**, 618 ⟨40JA1432⟩
Norris, R. D., **4**, 31, 34 ⟨73JA182⟩
Norris, R. K., **1**, 37 ⟨82JOC1047⟩, 474 ⟨91COS(4)451⟩; **5**, 735 ⟨74JCS(P1)853⟩
Norris, R. O., **4**, 16 ⟨39JA1460⟩
Norris, T. H., **6**, 415 ⟨48JA1968⟩
Norris, W. L., **2**, 128 ⟨48JCS2177⟩; **4**, 309 ⟨48JCS2174⟩
Norris, W. P., **1**, 1065 ⟨54JA4558⟩; **6**, 105 ⟨72T1965⟩
Norsoph, E. B., **3**, 483 ⟨78JA994⟩
Norstrom, R. J., **1**, 1143 ⟨73TCA183⟩
Nortey, S. O., **6**, 635 ⟨88S460⟩
North, B., **4**, 690 ⟨68JA1060⟩
North, M., **3**, 30, 628 ⟨93SL807⟩, 618, 619 ⟨90T8267⟩; **5**, 259 ⟨94MI 506-04⟩
North, P. C., **1**, 505 ⟨83TL409⟩; **3**, 61 ⟨82TL2811⟩
North, P. R., **6**, 247 ⟨92CM979⟩
Northington, D. J., **3**, 489 ⟨71TL317, 72JOC693⟩
Northrop, J. H., **1**, 886 ⟨89JOC1671⟩
Norton, D. G., **3**, 530, 531 ⟨63OSC(4)348⟩
Norton, J. A., **3**, 621 ⟨42CRV319⟩
Norton, J. R., **1**, 204, 1023, 1139 ⟨75TL51⟩, 546, 1024 ⟨81JA7520⟩, 1023, 1024 ⟨76CC907⟩, 1023, 1024 ⟨78JOC353⟩; **2**, 956 ⟨B-87MI 219-02⟩, 960 ⟨B-87MI 219-03⟩, 964 ⟨B-87MI 219-05⟩, 980 ⟨79JA4107⟩, 984 ⟨B-87MI 219-04⟩; **3**, 544 ⟨88JA7868⟩; **4**, 679 ⟨82JA7325, 88JA7868⟩; **5**, 170 ⟨75TL51, 76CC907⟩
Norton, R. V., **2**, 172, 173, 174 ⟨68JOC2104⟩, 478 ⟨67JOC3645⟩; **4**, 78 ⟨68JOC2104⟩
Norton, T., **3**, 546 ⟨1877CB701⟩
Norton, T. R., **3**, 279 ⟨55USP2711414⟩
Norwood, V. M., **6**, 514 ⟨88IC302⟩
Nose, M., **1**, 986 ⟨84CL131⟩

Nose, S., **6**, 580 ⟨55MI 618-01⟩
Noskov, V. G., **6**, 366 ⟨67ZOB2513⟩
Nosova, V. M., **4**, 140 ⟨83JFC(23)339⟩, 641 ⟨80ZOB1764⟩, 1045 ⟨86ZOB2743⟩, 1058 ⟨86ZOB1535⟩
Nosseir, M., **3**, 357 ⟨58JA6312⟩
Nosyreva, V. V., **4**, 775 ⟨92ZOR1780⟩
Notani, J., **5**, 125 ⟨75S456⟩
Note, H., **3**, 627 ⟨91JA6318⟩; **5**, 360 ⟨91JA6318⟩
Notes, J. G., **4**, 626 ⟨71JOM(29)409⟩
Nöth, H., **2**, 42 ⟨85AG(E)878⟩, 109 ⟨60CB1078⟩, 112 ⟨68ZAAC(358)44⟩, 358 ⟨61CB634⟩, 408, 415 ⟨63AG(E)218⟩, 410 ⟨61CB1505⟩, 414, 415, 891 ⟨64ZAAC(330)233, 64ZAAC(331)35⟩, 421 ⟨75ZN(B)681⟩ , 467 ⟨85CB2030⟩, 474 ⟨88CB871⟩, 474, 509 ⟨72CB1962⟩, 520 ⟨71MI 210-01⟩, 811 ⟨77CB2783⟩, 1048 ⟨78JOM(145)17⟩, 1086 ⟨67CB2377⟩; **3**, 504 ⟨85AG(E)416⟩, 572 ⟨90JOC1412⟩, 595 ⟨93CB1551⟩, 606 ⟨90TL4429⟩; **4**, 484 ⟨90TL6999, 92JA9691⟩, 632 ⟨93AG(E)295⟩, 635 ⟨92CB2213, 93CB2003 93OM2423⟩, 642 ⟨86CB3253⟩; **5**, 381, 382, 1085 ⟨90CB455⟩, 381, 534 ⟨93CB1119⟩, 382 ⟨90CB63⟩, 433 ⟨68CB2502⟩, 907, 908 ⟨93AG(E)1089⟩, 927 ⟨88AG(E)1331, 90AG(E)199⟩, 927 ⟨92CB73⟩, 934 ⟨94TL561⟩; **6**, 405 ⟨81JMR(44)54⟩, 712 ⟨85MI 622-01, 87MI 622-01⟩
Nöth, W., **2**, 421 ⟨76ZN(B)697⟩
Noto, T., **5**, 223 ⟨87CPB1741⟩; **6**, 447 ⟨86JAN230⟩
Notté, P., **4**, 791 ⟨81H(15)1179⟩
Nottke, J. E., **1**, 537 ⟨69JOC4134⟩
Notzumoto, S., **1**, 816 ⟨66TL6393⟩
Nouguier, R., **1**, 674 ⟨77RTC(M)91⟩; **4**, 185 ⟨87TL6601⟩
Nour, M. A., **3**, 412 ⟨91TL5421⟩; **5**, 282 ⟨91TL1825⟩
Noureldin, N. A., **3**, 116 ⟨82JOC2790, 83JA3188⟩
Noureldin, R., **2**, 627 ⟨90JHC2151⟩
Nouri, R., **4**, 1014 ⟨90MI 421-01⟩
Nourrisson, G., **5**, 760 ⟨80JHC767⟩
Novack, V. J., **1**, 269, 738 ⟨81JA1283⟩; **2**, 702 ⟨90JA866⟩; **3**, 196 ⟨81JA1283⟩, 231 ⟨84JA1148⟩; **4**, 165 ⟨84JA1148⟩
Novak, B. M., **1**, 675 ⟨88JA7542, 88JA960⟩; **3**, 694 ⟨93JA8467⟩
Novak, I., **6**, 228 ⟨90BCJ1278⟩
Novak, K., **5**, 447 ⟨66JOM(6)265⟩
Novak, L., **2**, 1015 ⟨92TL683⟩; **4**, 47 ⟨68JOC3335⟩; **6**, 695 ⟨92TL683⟩
Novak, M. A., **2**, 383 ⟨71ZOR258⟩
Novak, P. M., **3**, 536 ⟨88TL1493⟩
Novell, J. R., **2**, 204 ⟨69JOC3097⟩
Novelli, A., **2**, 749 ⟨68T2481⟩
Novelli, R., **5**, 205 ⟨88JCS(P1)3103⟩
Novi, M., **1**, 486 ⟨93T235⟩; **2**, 711 ⟨86T4007⟩; **3**, 664 ⟨87T4625⟩, 664, 667 ⟨90T2205⟩; **5**, 1111 ⟨89T7411⟩
Novichkova, A. S., **6**, 100 ⟨86ZOR107⟩
Novik, L. A., **2**, 386 ⟨78ZOR240⟩
Novikov, A. N., **1**, 1010 ⟨88ZOR650⟩; **2**, 631 ⟨60MI 212-01⟩
Novikova, N. V., **2**, 876 ⟨67DOK(172)1329⟩
Novikova, O. P., **5**, 702 ⟨91ZOB1389⟩, 723 ⟨89JGU2055, 89ZOB2293⟩
Novikova, S. A., **5**, 215 ⟨86IZV2032⟩
Novikova, T. S., **6**, 150 ⟨70KGS590⟩
Novikova, V. M., **4**, 560 ⟨83IZV2802⟩
Novikova, Z. S., **2**, 436 ⟨67ZOB2080, 68MI 208-01⟩, 448 ⟨76JGU572, 77JGU707⟩, 449 ⟨76JGU428⟩, 471, 472 ⟨91JGU460⟩, 476 ⟨78JGU1088, 89JGU87⟩; **3**, 195 ⟨72CA(78)136370⟩, 195 ⟨83JGU656⟩; **4**, 462 ⟨85JGU404⟩, 546 ⟨76URP539892, 81ZOB484⟩, 546, 550, 562 ⟨77ZOB775⟩, 547 ⟨76ZOB571, 83ZOB699⟩, 548 ⟨87ZOB2637⟩, 550 ⟨79ZOB712, 80ZOB1881, 80ZOB949, 85ZOB331⟩, 551 ⟨78ZOB757, 80ZOB989⟩, 552 ⟨77ZOB2689, 87ZOB949⟩, 552, 562 ⟨74ZOB276, 79ZOB1446⟩, 552, 562 ⟨79ZOB470⟩, 555 ⟨77ZOB2636, 83ZOB1673⟩, 563 ⟨87ZOB525⟩, 564 ⟨89ZOB101⟩, 566 ⟨76ZOB575⟩, 566, 568, 576 ⟨76ZOB1218⟩, 569 ⟨72ZOB112⟩, 574, 588 ⟨75ZOB1486⟩; **5**, 884 ⟨81DOK(256)1401, 81ZOB482⟩, 1092 ⟨87JGU2350⟩; **6**, 679 ⟨81DOK(256)1401⟩
Novikov, M. S., **3**, 596 ⟨90JOU1642⟩
Novikov, S. S., **2**, 385 ⟨69RCR640⟩; **5**, 838 ⟨58ZOB2305, 60IZV1783⟩, 838 ⟨61IZV477⟩; **6**, 56, 282 ⟨69IZV2617⟩, 56, 282 ⟨77IZV2058⟩, 144 ⟨59UK484⟩, 145 ⟨73IZV122⟩, 147 ⟨69IZV2309⟩, 148 ⟨72IZV490⟩, 149 ⟨62IZV272, 67ZOB1163, 68IZV621, 68ZOR231, 72IZV2603⟩, 150 ⟨60IZV1783, 70KGS590⟩, 283 ⟨69ZOR1317⟩, 284, 364, 369 ⟨75IZV2838⟩, 364, 368 ⟨77IZV1563⟩, 368 ⟨60IZV505, 68IZV2839, 69IZV2304, 70IZV2108⟩
Novikov, V. M., **4**, 587 ⟨64IZV772, 65JOM(4)202⟩, 587 ⟨66ZOR942⟩; **5**, 919 ⟨65JOM(4)202⟩
Novosad, J., **5**, 575 ⟨91CC116⟩
Novoselova, M. N., **4**, 768 ⟨92ZOR522⟩
Novoselova, R. I., **2**, 440 ⟨63ZOB1919⟩
Novosel'tseva, G. I., **2**, 1019 ⟨79DOK(245)607⟩
Novotorov, Y. N., **4**, 656 ⟨76DOK(231)1138⟩
Novotortsev, V. M., **6**, 341 ⟨90JOM(384)279⟩
Novozhilova, T. I., **4**, 793 ⟨70JOU750⟩
Novruzov, S. A., **5**, 484 ⟨74IZV55⟩
Nowak, F., **3**, 716 ⟨93AG(E)1456⟩
Nowak, G., **1**, 445, 999 ⟨57ZAAC(290)348⟩; **2**, 1076, 1079 ⟨57ZAAC(290)348⟩
Nowakowski, J., **5**, 972 ⟨92JPR187⟩
Nowakowski, L. H., **5**, 675, 678 ⟨81JMC532⟩
Nowakowski, P. M., **2**, 812 ⟨79JOM(8)95⟩
Nowak-Wydra, B., **2**, 7 ⟨87CS477⟩
Nowell, I. W., **4**, 856 ⟨85CC618⟩, 1042 ⟨73CC258⟩
Nowick, J. S., **1**, 566, 681, 683 ⟨91JOC1176⟩; **4**, 930, 931 ⟨88T4113⟩, 934 ⟨89JOC2798⟩; **5**, 161 ⟨91JOC1176⟩, 423, 426 ⟨89JOC2798⟩, 423, 430 ⟨88T4113⟩, 971 ⟨92JOC7364⟩
Noyce, D. S., **1**, 681, 785 ⟨66JOC4043⟩, 1122 ⟨65JA2295, 68JA1020⟩, 1122 ⟨68JA1023, 68JA372⟩
Noyes, P. R., **6**, 78 ⟨55JA3801⟩
Noyes, R. M., **2**, 1012, 1014 ⟨52JA629⟩
Noyes, W. A., **2**, 104 ⟨33JA3888, 43OSC(2)108⟩, 388, 403 ⟨22JA2556⟩; **3**, 612 ⟨04JA1545⟩; **5**, 305 ⟨41OSC(1)457⟩
Noyochi, S., **3**, 286 ⟨86CL449⟩
Noyori, R., **1**, 2 ⟨78JA1765, 83OR163⟩, 2, 3, 8 ⟨78JA1786⟩, 2, 8 ⟨78BCJ2745⟩, 76 ⟨84SC1191⟩, 77 ⟨86JA7117, 87JA1596⟩, 77 ⟨87JA4129, 87JOC3174⟩, 77 ⟨90ACR345⟩, 132 ⟨85JA3348, 88JA4718⟩, 132 ⟨89JOC1785⟩, 218 ⟨70TL311⟩, 363, 364 ⟨78JA1791⟩, 422 ⟨68TL3635, 75JA928⟩, 456 ⟨91COS(8)443⟩, 504 ⟨84AG(E)847⟩, 607 ⟨81TL1809⟩, 884 ⟨83JA1598⟩, 1056, 1061 ⟨90JOC441⟩; **2**, 28 ⟨88OSC(6)520⟩, 53 ⟨87JA5856, 89JA9134⟩, 64 ⟨84JA6709⟩, 67 ⟨91AG(E)49⟩, 71 ⟨88PAC1597, 89JA4028⟩, 263 ⟨80TL767⟩, 478 ⟨B-89MI 211-03⟩, 539 ⟨81T3899⟩, 638 ⟨76BCJ3673⟩, 816 ⟨90JA4897⟩, 841 ⟨88OS(67)20⟩; **3**, 54 ⟨71JA1724⟩, 129 ⟨83OR(29)163, 85T5879⟩, 225 ⟨72JA1772, 81BCJ2233⟩, 245 ⟨73JA4103⟩, 615 ⟨81T3899⟩, 621 ⟨72JOC1542⟩; **4**, 45 ⟨84CL1747⟩, 181, 182 ⟨80TL1357⟩, 192 ⟨80JA3248, 82TL3595⟩; **5**, 171 ⟨70TL311⟩; **6**, 107 ⟨67T45⟩
Noyori, Y., **6**, 124 ⟨80TL1357⟩
Nozaki, H., **1**, 14, 221 ⟨82TL3597⟩, 141 ⟨73JA2694⟩, 141, 834 ⟨73JA2693⟩, 160, 1094 ⟨77TL1805⟩, 160, 451 ⟨84TL1999⟩, 161, 1094 ⟨76JOC2941⟩, 161, 450, 1106 ⟨76JOC2215⟩, 164 ⟨77BCJ3288⟩, 205, 1024 ⟨84TL5423⟩, 206 ⟨74JA5254⟩, 207 ⟨74T3037⟩, 207, 1100 ⟨77T1845⟩, 248 ⟨82TL2871⟩, 274 ⟨73TL787⟩, 415 ⟨81TL3985⟩, 422 ⟨68TL3635, 79JA1599⟩, 422 ⟨80BCJ169, 81BCJ2747⟩, 451 ⟨75TL2825⟩, 452 ⟨84TL3221, 85CL53⟩, 480, 1027 ⟨84BCJ108⟩, 485

⟨74BAU1031⟩, 491 ⟨87CL5⟩, 511 ⟨77TL3641⟩, 539 ⟨B-83MI 112-01⟩, 565 ⟨75TL1545⟩, 650, 761 ⟨76JOC2940⟩, 651 ⟨74BCJ2529, 74JA1620⟩, 670, 678, 767 ⟨78TL2399⟩, 677, 1099 ⟨83TL5281, 86JA6048⟩, 678 ⟨80BCJ1670, 85TL5579⟩, 678, 769 ⟨85TL5581⟩, 761 ⟨77TL1807⟩, 803 ⟨84MI 118-02⟩, 841 ⟨70TL1175⟩, 1027 ⟨80TL2531, 81TL1609⟩, 1033 ⟨73TL1847, 74T2159⟩, 1038, 1140 ⟨85TL5585⟩, 1045, 1138 ⟨78TL2301⟩, 1094 ⟨77T2987⟩; **2**, 73, 571 ⟨82BCJ561⟩, 215 ⟨90TL3591⟩, 420 ⟨75TL1859, 77JA3179⟩, 624, 630 ⟨85CL53⟩, 624, 946, 948 ⟨84TL3221⟩, 699 ⟨79CL379⟩, 918 ⟨85T3257, 85TL4629⟩, 918 ⟨86CB1581⟩; **2**, 918, 919, 939 ⟨88T4277⟩, 918, 923 ⟨86T4427⟩, 918, 931 ⟨83JA4491⟩, 922 ⟨76JOC2215, 76JOC2941⟩, 922 ⟨77T2987⟩, 922, 923 ⟨88T4009⟩, 922, 931 ⟨82TL3387⟩, 924 ⟨77TL1805, 79JOM(177)145⟩, 929 ⟨84TL1163⟩, 947, 977 ⟨84TL3217⟩, 972 ⟨84TL1999, 84TL2151⟩, 977 ⟨84TL2221⟩, 983 ⟨86JA2753⟩, 997, 1002 ⟨85JOM(285)53⟩, 1000 ⟨67T4279⟩, 1088 ⟨76TL3969⟩, 1098 ⟨84BJC108⟩; **3**, 5 ⟨81TL1605⟩, 5, 117 ⟨80TL4843⟩, 5, 117 ⟨88BCJ3607⟩, 21 ⟨73JA2693⟩, 35 ⟨88JA5209⟩, 54 ⟨73TL2465⟩, 55 ⟨89JOC2779⟩, 62 ⟨78TL3051⟩, 62, 72, 271 ⟨78TL3047⟩, 116, 161 ⟨84TL3317⟩, 118 ⟨82TL539⟩, 118, 121 ⟨89BCJ519⟩, 125 ⟨77BCJ1592⟩, 126 ⟨84BCJ108⟩, 158, 161 ⟨86BCJ105⟩, 159 ⟨90JA901⟩, 209 ⟨80BCJ169⟩, 216 ⟨79JA1599⟩, 224 ⟨74JA3713, 74TL3297⟩, 224 ⟨78TL771, 80BCJ1010⟩, 238, 270, 273 ⟨78TL2261⟩, 244 ⟨83BCJ3088⟩, 289 ⟨75TL4239, 85BCJ2425⟩, 373 ⟨67T4279⟩, 616, 627, 633 ⟨83T967⟩, 627 ⟨81TL4279, 80TL389⟩; **4**, 17 ⟨74T2661⟩, 19 ⟨74JA3010⟩, 141, 811, 812 ⟨84TL3221⟩, 216 ⟨89T1209⟩, 225, 238 ⟨92TL239⟩, 248 ⟨88TL3971⟩, 393 ⟨79TL1783⟩, 648 ⟨74JA1620⟩, 652 ⟨82TL1279⟩, 654 ⟨85TL2101⟩, 704 ⟨92T1449⟩, 860 ⟨72BCJ198⟩, 956, 957 ⟨73JA2694⟩, 1045 ⟨88T4277⟩, 1060 ⟨84TL1999⟩, 1061, 1062 ⟨79JOM(177)145⟩; **5**, 29 ⟨59JA5817⟩, 251 ⟨90TL3591⟩, 552 ⟨73JA5803⟩, 1043 ⟨84TL4683⟩; **6**, 25, 31 ⟨74JA3010⟩, 107 ⟨67T45⟩, 126 ⟨89JOC5003⟩
Nozaki, K., **1**, 297 ⟨93JA7033⟩, 330, 925, 946 ⟨89BCJ143⟩, 333 ⟨88TL1041⟩, 925, 938, 946, 1131 ⟨88TL6127⟩, 939, 946, 1130 ⟨88TL6125⟩, 1130, 1132 ⟨87JA2547⟩; **2**, 710 ⟨87CL1647⟩, 903, 919 ⟨86TL2007⟩, 923 ⟨87JA2547⟩, 946 ⟨87TL3709⟩; **3**, 12, 128 ⟨91BCJ2585⟩; **6**, 550 ⟨88TL6125⟩
Nozaki, Y., **4**, 816 ⟨88JOC3089⟩
Nozawa, A., **5**, 1070 ⟨69TL1421⟩; **6**, 645 ⟨69TL1421⟩
Nozawa, K., **1**, 45 ⟨81CPB3489⟩
Nozawa, R., **3**, 309 ⟨90CPB3261⟩
Nozawa, S., **1**, 274 ⟨71JA1508⟩, 274, 546 ⟨70JA3503⟩; **2**, 34 ⟨71JA1508⟩; **3**, 223 ⟨70JA3503⟩
Nozoe, S., **1**, 382 ⟨64JA1652⟩; **2**, 660 ⟨63JOC783⟩
Nozoe, T., **1**, 899 ⟨87BCJ3721, B-88MI 119-02⟩; **2**, 769 ⟨89BCJ128⟩
Nozowa, A., **5**, 1070 ⟨70OCS23⟩
Nozulak, J., **4**, 536 ⟨83LA1133⟩
Nsunda, K. M., **2**, 1008 ⟨84TL4441⟩; **3**, 36, 39, 168, 176 ⟨85TL6513⟩, 176 ⟨84TL4441⟩; **4**, 289 ⟨85TL6513⟩, 844, 875 ⟨85CC1000⟩
Nubbemeyer, U., **1**, 834 ⟨91AG(E)1465⟩
Nubel, P. O., **4**, 677 ⟨82JA4955⟩
Nuber, B., **2**, 865 ⟨88AG(E)709⟩; **3**, 607 ⟨86AG(E)1003⟩; **4**, 549 ⟨88AG(E)389⟩, 565 ⟨88AG(E)709⟩, 581 ⟨89JOM(371)329⟩, 1047, 1059 ⟨86ACR250⟩; **6**, 154 ⟨88AG(E)389, 88AG(E)709⟩
Nuberl, S., **6**, 114 ⟨80LA873⟩
Nübling, C., **2**, 392 ⟨88OSC(6)799⟩; **4**, 266 ⟨90CAR(202)1⟩
Nucciarelli, L., **3**, 155 ⟨80TL4521⟩
Nudelman, A., **2**, 362, 846 ⟨68JA3869⟩; **3**, 654 ⟨85JOC3558⟩; **4**, 176 ⟨87CAR(162)145⟩; **6**, 486 ⟨93SC2839⟩
Nudenberg, W., **2**, 116 ⟨51JOC524⟩; **3**, 323 ⟨53JOC1030⟩; **6**, 5 ⟨53JOC328⟩
Nuechterlein, D., **6**, 42, 43 ⟨70JOC3730⟩
Nuel, D., **1**, 976 ⟨92OM493, 93OM1871⟩; **6**, 338 ⟨85OM1436⟩
Nuenhoeffer, H., **2**, 106, 107 ⟨62CB1664⟩
Nuevar, E. W., **6**, 279, 281 ⟨73JOC1065⟩
Nugent, M. J., **3**, 440 ⟨88T4157⟩; **5**, 377 ⟨88T4157⟩
Nugent, R. A., **6**, 118 ⟨93JOC4159⟩
Nugent, R. M., **2**, 245, 250, 366 ⟨75JA6521⟩
Nugent, W. A., **1**, 587 ⟨90JA6408⟩, 671 ⟨77JOM(124)327, 77JOM(124)349⟩, 671 ⟨77JOM(124)371⟩; **2**, 24 ⟨78TL3427⟩, 85 ⟨89JA4525⟩, 926, 927, 988 ⟨87JA2788⟩, 926, 988 ⟨84JA6422⟩, 949, 992 ⟨88JA2310, 88JA7128⟩, 988 ⟨89JA6435⟩; **3**, 508 ⟨B-88MI 315-01⟩, 626 ⟨85JOC5370⟩; **4**, 1038 ⟨88JA2310⟩; **5**, 151 ⟨86JOC3376⟩
Nugiel, D. A., **2**, 645 ⟨85JA3669⟩
Nuhfer, P. A., **2**, 373, 374, 379, 381 ⟨59JA2597⟩
Nuhn, P., **4**, 239 ⟨64AP(297)461⟩
Nuistein, P., **5**, 911 ⟨84ZN(B)1456⟩, 921, 1094 ⟨87CB1281⟩
Nukui, S., **1**, 495 ⟨94T371⟩
Numaka, S., **4**, 659 ⟨78BCJ1397⟩
Numano, K., **1**, 232 ⟨91JCS(P2)1703⟩
Numata, H., **1**, 216, 748 ⟨89TL4845⟩; **5**, 155 ⟨89TL4845⟩
Numata, M., **2**, 214 ⟨90JHC1361⟩
Numata, S., **4**, 659 ⟨74JOM(70)C21⟩
Numata, T., **1**, 821 ⟨77TL1337⟩; **2**, 145 ⟨81S204⟩; **3**, 634 ⟨81S204⟩, 634, 635 ⟨83BCJ257, 77TL1337⟩; **4**, 221 ⟨78JCS(P2)1302⟩, 223 ⟨83BCJ257, 83BCJ266⟩, 227 ⟨83BCJ270⟩, 233 ⟨72BCJ2794⟩
Numazawa, M., **1**, 578 ⟨87CC490⟩
Nun, E., **4**, 672 ⟨83JA2905⟩
Nunami, K., **3**, 717 ⟨90S781⟩, 717, 720 ⟨88T5467⟩
Nuner, R., **2**, 364 ⟨87PS(31)245⟩
Nunes, F., **1**, 2, 3, 6 ⟨63JA1609⟩
Nunez, M. T., **2**, 84 ⟨88TL2701⟩
Nunez, R., **4**, 138, 139 ⟨92OM859⟩
Nunez, W., **3**, 113 ⟨83HCA177⟩
Nunn, B., **5**, 334 ⟨85JMC1661⟩
Nunn, C. M., **2**, 511 ⟨88JOM(341)C1, 88POL77⟩, 511 ⟨90CM221, 91OM1635⟩, 893 ⟨90AX(C)774⟩; **4**, 480 ⟨88CC867⟩, 581 ⟨90OM1036⟩; **5**, 889 ⟨88CC753⟩, 890 ⟨88AG(E)838⟩; **6**, 373 ⟨89OM2490⟩
Nunn, C. N., **4**, 549 ⟨88AG(E)837⟩
Nunn, K., **1**, 966 ⟨92AG(E)224, 93T9775⟩
Nunn, M. J., **2**, 35 ⟨74TL2691, 76JCS(P1)416⟩; **4**, 28 ⟨76JCS(P1)416⟩
Nunno, L. D., **1**, 1110 ⟨73JCS(P2)2078⟩
Nunokawa, Y., **1**, 1018 ⟨84CL1029⟩
Nunomoto, S., **1**, 119, 471, 1092 ⟨83JOC1912⟩; **3**, 307 ⟨83TL3455⟩
Nupponen, H., **6**, 263 ⟨92JCS(P2)835, 92PS(70)183⟩, 263 ⟨93TL4551⟩
Nuretdinova, O. N., **5**, 576 ⟨91IZV1605⟩
Nuretdinov, I. A., **2**, 410 ⟨67IZV1606⟩, 465 ⟨84JGU1704⟩, 466 ⟨76MI 208-01⟩, 857 ⟨78ZOB2673⟩; **5**, 371 ⟨85ZOB2483, 90IZV1139⟩
Nurgalieva, A. N., **2**, 1076 ⟨82IZV933⟩
Nurmi, T. T., **1**, 316 ⟨87JA2442⟩
Nurmukhamedova, L. V., **2**, 857 ⟨78ZOB2673⟩
Nürnberg, O., **5**, 918, 919 ⟨92CC1631⟩
Nurrenbach, A., **3**, 704 ⟨70AG(E)458⟩
Nurtdinov, S. Kh., **4**, 120 ⟨71JGU2183⟩, 463 ⟨88ZOB2456⟩; **5**, 403 ⟨71JGU2513⟩; **6**, 564 ⟨83ZOB785⟩
Nurula, C. K., **2**, 474 ⟨88CB871⟩
Nurushev, R. A., **2**, 595 ⟨83ZOB2519⟩
Nuss, J. M., **1**, 1015 ⟨91TL5243⟩

Nussbaumer, P., **2**, 83 ⟨92TL7507⟩
Nussbeutel, U., **1**, 924 ⟨89T951⟩; **3**, 92 ⟨89T951⟩
Nussim, M., **2**, 40 ⟨61JOC630⟩
Nusstein, P., **4**, 565 ⟨84ZN(B)1456⟩, 565, 593 ⟨87CB1281⟩; **6**, 701 ⟨84ZN(B)1456, 85OM344⟩, 708 ⟨87CB1281⟩
Nutaitis, C. F., **1**, 475 ⟨85S756⟩
Nutatis, C. F., **2**, 300, 310 ⟨87S709⟩
Nutile, A. N., **1**, 881 ⟨60JA5704⟩
Nutting, H. S., **6**, 185 ⟨32JA1622⟩, 212, 225 ⟨31USP1961622⟩
Nützel, K., **2**, 557 ⟨73HOU(13/2a)47⟩, 591 ⟨73HOU(13/2a)553⟩, 1095 ⟨73HOU(13/2a)608, 73HOU(13/2a)622, 73HOU(13/2a)859⟩
Nuvole, A., **5**, 657 ⟨78JHC1513⟩
Nuzhdina, Y. A., **6**, 621 ⟨74ZOR1000⟩
Nwaji, M. N., **3**, 65, 232 ⟨74TL2255⟩
Nwauka, S. O., **2**, 619 ⟨89SC799⟩
Nwaukwa, S. O., **3**, 119 ⟨82TL35⟩
Nwe, K. T., **4**, 648 ⟨92JOC6107⟩
Nyberg, D. D., **2**, 103 ⟨57JA1222⟩
Nyberg, K., **2**, 748 ⟨76S545⟩; **5**, 24, 28 ⟨B-78MI 502-02⟩, 1102 ⟨64ACS1567⟩
Nyberg, W. H., **6**, 653 ⟨63JMC283⟩
Nybergh, B., **1**, 384 ⟨22CB1960⟩
Nyburg, S. C., **5**, 923, 924 ⟨85AX(C)1632, 79JA6750⟩, 924 ⟨82JA5667⟩
Nychka, H. R., **6**, 16 ⟨64FRP1369784⟩
Nycz, T. J., **5**, 990 ⟨72JHC1267⟩
Nyeki, O., **5**, 277 ⟨74TL1785⟩
Nyholm, R., **2**, 872 ⟨B-75MI 215-01⟩
Nyholm, R. S., **5**, 1083 ⟨71JCS(A)332⟩
Nyns, C., **1**, 592 ⟨81AG(E)585⟩; **4**, 71 ⟨81AG(E)585⟩; **6**, 44 ⟨81AG(E)585⟩
Nystrom, J. E., **1**, 523 ⟨85TL5393⟩; **2**, 301 ⟨83TL2745⟩
Nystrom, R. F., **1**, 20 ⟨58JA2896⟩
Nyulaszi, L., **5**, 880 ⟨91CB493⟩
Nyzam, V., **2**, 530 ⟨93TL6899⟩

O

Oae, S., **2**, 7, 27 ⟨93JOC4382⟩, 25 ⟨81TL4409⟩, 32 ⟨73CC212⟩, 116 ⟨B-77MI 203-01⟩, 145 ⟨81S204⟩, 148 ⟨66BCJ364⟩, 160 ⟨76T2763⟩, 187 ⟨66BCJ1309⟩, 187 ⟨68BCJ3015⟩, 187 ⟨B-77MI 203-03⟩, 193 ⟨79TL821⟩, 193 ⟨81T37⟩, 214 ⟨79CL1077⟩, 214, 220 ⟨77CC407⟩, 214, 220 ⟨77CL893⟩, 223 ⟨75BCJ1665⟩, 235 ⟨74TL1059⟩, 243 ⟨80BCJ2023⟩, 292 ⟨75TL649, 81PS153⟩, 360 ⟨71IJS215⟩, 401, 745 ⟨76S30⟩, 621, 627 ⟨78TL4519⟩, 621, 627 ⟨80BCJ1065⟩, 647 ⟨66BCJ1212⟩, 708 ⟨83BCJ3802, 84BCJ232⟩, 712 ⟨60JA5359⟩, 724 ⟨81BCJ1443⟩, 726 ⟨79TL3307⟩; **3**, 170 ⟨B-88MI 304-03⟩, 310 ⟨86TL3899⟩, 378 ⟨77TL3633, 79T181⟩, 634 ⟨81S204⟩, 634, 635 ⟨83BCJ257, 77TL1337⟩; **4**, 161 ⟨77JCS(P1)372⟩, 221 ⟨78JCS(P2)1302⟩, 223 ⟨83BCJ257, 83BCJ266⟩, 227 ⟨83BCJ270⟩, 233 ⟨72BCJ2794⟩, 390 ⟨B-88MI 408-01⟩; **5**, 209 ⟨81T3115⟩, 349 ⟨75TL649, 81PS(10)153⟩, 546 ⟨72MI 513-01⟩, 550 ⟨72T3203⟩, 1106 ⟨81TL4409⟩; **6**, 82 ⟨64T417⟩, 303, 307 ⟨72BCJ960⟩
Oakdale, U. O., **2**, 889 ⟨32JA2993⟩
Oakes, F. T., **4**, 949, 950 ⟨80JOC4959⟩, 949, 950 ⟨82JOC3094⟩
Oakes, M. L., **2**, 18 ⟨83JOC3195⟩
Oakes, T. R., **3**, 593 ⟨69JA4761, 73JOC1319⟩
Oakley, D., **6**, 465 ⟨87CC410⟩
Oakley, R. T., **2**, 795, 796 ⟨83CJC1562, 83CC73⟩; **5**, 777, 778 ⟨87JOM(331)161⟩
Oakwood, T. S., **5**, 64 ⟨55OSC(3)114⟩
Oayama, H., **2**, 228 ⟨73BCJ2253⟩
Oba, M., **2**, 544 ⟨87BCJ2289⟩; **4**, 432, 433 ⟨87T693⟩, 438 ⟨88BCJ609⟩; **5**, 1042 ⟨87BCJ2289⟩, 1106 ⟨87BCJ2692⟩; **6**, 558 ⟨93TL3745⟩
O'Bannon, P. E., **3**, 479 ⟨89TL4197⟩, 480 ⟨88TL6031, 90T7341⟩; **5**, 326 ⟨88TL5719, 89JOC3096⟩, 868 ⟨90TL7341, 91JOC2258⟩
Obara, H., **3**, 282 ⟨89BCJ3371⟩
Obara, Y., **3**, 293 ⟨87BCJ767⟩
Obase, H., **4**, 970 ⟨91EUP413343⟩
Obata, N., **3**, 470, 593 ⟨69TL3403⟩, 582 ⟨71CC587, 69TL3407⟩, 596 ⟨75BCJ2287⟩
Obata, T., **2**, 106 ⟨67JOC1063⟩
Obayashi, M., **2**, 70 ⟨88T4134⟩, 924 ⟨77TL1805, 79JOM(177)145⟩; **3**, 616, 627, 633 ⟨83T967, 80TL389⟩; **4**, 734 ⟨82TL2323, 82TL2327⟩, 736 ⟨83TL4113, 88T4135⟩, 806, 817 ⟨84CL1765⟩, 1061, 1062 ⟨79JOM(177)145⟩; **5**, 601 ⟨79CPB1352⟩; **6**, 26 ⟨88T4135⟩, 58, 264, 266 ⟨82TL2323⟩
Obayashi, R., **2**, 285 ⟨92JOC111⟩
Obaza-Nutaitis, J. A., **4**, 165 ⟨84JOC4518⟩
Obendorf, D., **6**, 357, 358 ⟨85JOM(295)C1⟩
Oberender, H., **2**, 659 ⟨72JPR170⟩
Oberhammer, H., **3**, 380 ⟨79AG(E)944⟩; **5**, 488, 492 ⟨90JST171⟩, 675 ⟨93IC287⟩; **6**, 243 ⟨92JST(268)389⟩, 253 ⟨86IC275, 92CB535⟩, 604 ⟨86JFC(32)415⟩, 613, 623 ⟨85IC4665⟩
Oberhummer, W., **5**, 769 ⟨31M106, 33M285⟩
Oberle, K., **2**, 445 ⟨68JOC2590⟩
Obermeyer, A., **5**, 1077 ⟨93CB2569⟩
Oberster, A. G., **2**, 1016 ⟨61JA4663⟩
Obertas, J., **5**, 1039 ⟨69CZP131547⟩
Obeyama, J., **4**, 231, 940 ⟨88TL4773⟩
Obinger, M., **6**, 503 ⟨79GEP2742158⟩
O'Blenes, S. B., **6**, 255 ⟨91CJC2127⟩
Obrecht, D., **3**, 275 ⟨89HCA117⟩; **4**, 277 ⟨91HCA27⟩
Obrecht, J.-P., **4**, 485, 489 ⟨85HCA1730⟩; **5** 1, 92 ⟨92OS(71)200⟩, 212 ⟨85HCA1730⟩
Obrecht, R., **3**, 696, 699, 721, 722 ⟨85S400⟩
O'Brien, A. S., **2**, 126 ⟨46MI 203-01⟩
O'Brien, B. A., **3**, 419 ⟨84JOC1467⟩; **5**, 395, 396 ⟨84TL1249⟩, 662 ⟨86MI 517-01⟩, 662, 808 ⟨84JOC1467⟩, 808, 864 ⟨86MI 521-01⟩, 809 ⟨84JA4266⟩; **6**, 257 ⟨86JOC4466⟩, 603 ⟨84JA4266⟩, 603, 615 ⟨84IC2188⟩
O'Brien, C., **3**, 435 ⟨64CRV81⟩
O'Brien, D. H., **2**, 377, 403 ⟨65JOC3278⟩, 499 ⟨71JA5677⟩; **4**, 507 ⟨84JOM(268)31⟩
O'Brien, J. F., **4**, 698 ⟨65JPC3418⟩
O'Brien, J. L., **2**, 399 ⟨65TL1487⟩; **4**, 979 ⟨64JOC2932⟩; **5**, 599 ⟨64JOC2932⟩
O'Brien, J. P., **4**, 650, 656 ⟨93OM338⟩
O'Brien, Jr., H. A., **2**, 685 ⟨82JOC359⟩
O'Brien, M. H., **2**, 492 ⟨67ICA34, 68IC834⟩
O'Brien, M. J., **2**, 701 ⟨85JOC3846⟩
O'Brien, R. E., **2**, 630 ⟨62JCS470⟩; **3**, 464 ⟨62JCS470⟩
O'Brien, R. J., **6**, 340 ⟨70JOM(24)C61⟩
Obrycki, R., **2**, 849 ⟨66JOC2455, 66TL5049⟩, 849 ⟨68JOC632⟩
Obushak, N. D., **2**, 256 ⟨87ZOB1078⟩
Ocando, E., **4**, 552 ⟨86IC712⟩
Ocando-Mavarez, E., **2**, 470 ⟨86IC2695⟩
Ocawa, M., **3**, 7, 215 ⟨86JOC240⟩
Occo, T. D., **3**, 312 ⟨86G133⟩
Ochai, H., **2**, 69 ⟨87JOC4418⟩
Ochal, Z., **2**, 308 ⟨78S882⟩; **5**, 1118 ⟨78S882⟩
Ochiai, E., **3**, 311 ⟨39CB1470⟩
Ochiai, H., **2**, 69 ⟨88CC610⟩, 591 ⟨89OS(67)98⟩; **3**, 130 ⟨85TL5529⟩, 161 ⟨88JOC1343⟩, 290, 295, 296 ⟨83TL3869⟩; **5**, 170 ⟨85TL3207⟩; **6**, 383 ⟨82JOM(236)7⟩
Ochiai, M., **2**, 46 ⟨88JA4606⟩, 150 ⟨77TL1345⟩, 917, 918 ⟨88JA6565⟩, 1020 ⟨90CC118⟩, 1020, 1063 ⟨85TL4501⟩, 1062 ⟨87CC1708⟩; **3**, 57 ⟨83TL777⟩, 143 ⟨84CC1007⟩, 143, 225 ⟨81TL1115⟩, 162 ⟨89JOC4038⟩; **4**, 183 ⟨93CL17⟩, 220 ⟨78JA7103⟩, 392 ⟨84CPB1829⟩, 808 ⟨88CC1076, 89TL6701⟩, 881 ⟨68TL497⟩; **5**, 83 ⟨81JOC5163⟩, 204 ⟨90JA5677⟩, 329 ⟨86JOC2391⟩, 329, 330 ⟨85CC1418⟩; **6**, 106 ⟨68TL497⟩, 113, 316 ⟨81JCS(P1)618⟩, 116 ⟨86JOC4737, 87NKK1447⟩
Ochi, K., **2**, 662 ⟨80CC593⟩
Ochi, M., **2**, 60 ⟨88BCJ2684⟩, 312 ⟨90S401⟩
Ochi, N., **2**, 582 ⟨85JCS(D)873⟩
Ochiuni, M.,**3**, 439 ⟨76BCJ1913⟩
Ocho de Retana, A., **5**, 199 ⟨88AQ120⟩
Ochs, W., **5**, 599 ⟨72LA(757)147⟩
Ochsler, B., **4**, 551, 570, 588 ⟨79JOM(182)465, 80EGP142887⟩; **5**, 390 ⟨79JOM(182)465⟩
Ockrymick, S. B., **2**, 822 ⟨74JOC2296⟩
Ocone, L. R., **2**, 843 ⟨66IS71⟩
O'Connell, E. J., **3**, 538 ⟨62JA4148⟩
O'Connell, E. M., **2**, 355 ⟨73JOC1813⟩
O'Connell, T. J., **3**, 311 ⟨92JA5888⟩
O'Conner, B. R., **4**, 767, 950 ⟨68JOC1991⟩
O'Connor, B., **3**, 496 ⟨85TL1799⟩; **4**, 658 ⟨89JA3089⟩
O'Connor, B. R., **2**, 800 ⟨65JA863⟩; **5**, 476, 487 ⟨72USP3637749⟩, 476, 487, 582, 613, 614 ⟨70JOC2002⟩
O'Connor, E. J., **3**, 520 ⟨82JA6119⟩; **5**, 451 ⟨87JA4837⟩
O'Connor, G. L., **2**, 111 ⟨55JA1578⟩, 125, 127 ⟨53JA2118⟩
O'Connor, J., **2**, 173 ⟨74JA3921⟩, 186 ⟨74JA3929⟩; **3**, 334 ⟨74JA3929⟩
O'Connor, J. M., **2**, 927, 945 ⟨93JA9846, 93JA1586⟩; **3**, 520 ⟨82JA3761⟩, 545 ⟨85JA3172⟩, 597 ⟨93JA9846⟩; **5**, 944 ⟨87JA7578, 89JA4129⟩, 944 ⟨90JA6232, 95OM2102⟩; **6**, 402 ⟨86OM394⟩
O'Connor, N., **3**, 171 ⟨90SL245⟩

O'Connor, T. J., **2**, 1060, 1061, 1076 ⟨71CJC2706⟩
Oda, D., **2**, 394, 784 ⟨86JOC4368⟩; **3**, 169 ⟨90JOC595⟩, 184 ⟨86TL5759⟩
Oda, H., **2**, 326 ⟨89TL6395⟩, 624, 630 ⟨85CL53⟩, 624, 946, 948 ⟨84TL3221⟩, 918 ⟨85T3257⟩, 929 ⟨84TL1163⟩, 946 ⟨87BCJ3468⟩, 947, 977 ⟨84TL3217, 84TL2221⟩; **4**, 141, 811, 812 ⟨84TL3221⟩, 199 ⟨89CL659⟩; **5**, 678 ⟨84CPB4437⟩
Oda, J., **2**, 355 ⟨76JOC300⟩; **3**, 616 ⟨75ABC571⟩, 631 ⟨92BCJ111, 92JOC5643⟩; **5**, 172 ⟨87JA7188⟩
Oda, K., **5**, 631, 632 ⟨85JOC1681, 86S768⟩
Oda, M., **2**, 13 ⟨89CC1636⟩; **3**, 473 ⟨72JA4787⟩, 667 ⟨79CL1427⟩; **4**, 820 ⟨86CC1794⟩
Oda, R., **3**, 498 ⟨64TL1653⟩, 549 ⟨71T1781⟩; **4**, 740 ⟨64MI 417-01, 64TL7⟩, 740 ⟨66T2615⟩; **5**, 902 ⟨67T2137⟩; **6**, 506 ⟨63BCJ1314⟩
Oda, T., **2**, 145 ⟨89JA5002⟩; **5**, 324 ⟨87CC1471, 90JOC1744⟩
Oda, Y., **3**, 113 ⟨93TL1085⟩
Odagaki, Y., **5**, 1086, 1087 ⟨88POL2213⟩
Odaira, Y., **3**, 122 ⟨85CL1437⟩; **5**, 113 ⟨69TL1019⟩
O'Dea, J., **3**, 532 ⟨74JOC3355⟩
Odeh, N., **5**, 350 ⟨90SC1811⟩
Odenthal, J., **3**, 352 ⟨79H(12)1153⟩
Odera, J., **5**, 411 ⟨87JAP6261987⟩; **6**, 126 ⟨87JAP6261987, 92JAP04149153⟩
Odinets, I. L., **2**, 476 ⟨89JGU87⟩; **4**, 547 ⟨83ZOB699⟩, 562 ⟨83ZOB1673⟩, 563 ⟨87ZOB525⟩, 564 ⟨89ZOB101⟩
Odlerova, Z., **5**, 630 ⟨87CCC2087⟩
Odo, K., **5**, 758, 762 ⟨76BCJ3607⟩, 766 ⟨74CL247⟩; **6**, 335 ⟨75YGK136⟩, 503 ⟨74YGK936⟩
Odom, H. C., **3**, 211 ⟨71OS(51)115⟩
Odom, J. D., **2**, 1083 ⟨90JA1847⟩; **3**, 707 ⟨91JCS(P1)2495⟩; **4**, 813 ⟨76JOM(108)1⟩; **5**, 238 ⟨85JOC811⟩, 1054 ⟨91JCS(P1)2495⟩; **6**, 265, 269 ⟨70JCS(A)31⟩, 595 ⟨91JCS(P1)2495, 91JOC6733⟩
O'Donnell, M. J., **3**, 406 ⟨82JOC2663⟩; **4**, 445 ⟨85TL695⟩, 1080 ⟨87CC469⟩; **5**, 746 ⟨84TL3651⟩
O'Donoghue, A. I., **3**, 273 ⟨89SC1745⟩
O'Donohue, A. M., **3**, 18, 134 ⟨79TL4583⟩
O'Donovan, D. G., **6**, 641 ⟨53JA4053⟩
Odorisio, P. A., **2**, 432 ⟨84JOC2906⟩
O'Dowd, M. L., **3**, 202 ⟨80JA5411⟩
Odriozola, J. M., **3**, 6, 118 ⟨91JOC5948⟩
Odubela, A. A., **5**, 282 ⟨74JOC2740⟩
O'Dwyer, J. B., **4**, 68, 71, 81, 83, 84 ⟨74JOC2516⟩
Oediger, H., **2**, 860 ⟨59LA(627)142⟩; **3**, 328 ⟨59LA(627)142⟩, 499, 500 ⟨58CB437⟩; **5**, 921 ⟨58CB437⟩, 975 ⟨59LA(627)142⟩; **6**, 698, 708 ⟨58CB437, 59LA(627)142⟩
Oehlschlager, A. C., **2**, 26, 32 ⟨71T2617⟩, 607 ⟨89JOC5064⟩, 903 ⟨88TL261, 91T1163⟩, 917 ⟨91JOC770⟩
Oehme, G., **4**, 580 ⟨73AG(E)854, 75JOM(84)C13⟩; **6**, 207 ⟨74JPR557⟩
Oehme, H., **2**, 427, 430, 462 ⟨68CB3612⟩, 430, 454 ⟨67CB2685, 66ZAAC(343)268⟩; **4**, 337 ⟨72JPR66, 72T2587⟩, 337 ⟨79ZC57⟩, 453 ⟨68CB3612⟩, 453, 454 ⟨73ZC139⟩, 453, 454, 477 ⟨68CB3619⟩, 453, 454, 477, 478, 481 ⟨67CB2685⟩, 453, 454, 477, 481 ⟨68CB4032⟩, 453, 481 ⟨67TL1489⟩, 454 ⟨71MI 410-01⟩, 455 ⟨74SRI453⟩, 482 ⟨78PS(4)59⟩, 548 ⟨79ZC416, 81T753⟩; **5**, 883 ⟨80TL1141, 83ZC99⟩; **6**, 686 ⟨80TL1141, 81ZC407⟩, 686 ⟨83ZC99⟩
Oehme, V. H., **4**, 453, 454, 481 ⟨73JPR526⟩
Oehmigen, T., **2**, 428 ⟨90ZAAC(589)187⟩, 431 ⟨90CZ211⟩
Oei, H. Y., **4**, 321 ⟨66AP(299)906⟩
Oerple, K., **2**, 262, 537, 539 ⟨85TL5515⟩
Oertel, F., **6**, 556 ⟨92PHA251⟩
Oertel, G., **5**, 964 ⟨64GEP1205087⟩, 968, 970, 1009 ⟨62AG(E)617⟩, 974 ⟨68AG(E)381⟩, 1033 ⟨54LA(590)123⟩
Oertel, M., **2**, 252 ⟨86GEP231075⟩; **4**, 561 ⟨84EGP214610⟩
Oertle, K., **2**, 537, 538, 539 ⟨85TL5511⟩; **5**, 224, 375 ⟨85TL5515⟩
Oeschger, T. R., **5**, 346 ⟨92OPP147⟩
Oeser, H. G., **6**, 454 ⟨79EUP7498⟩
Oesterle, T., **2**, 539 ⟨82S1⟩; **5**, 718 ⟨85S403⟩; **6**, 43 ⟨82S1⟩
Oesterle, Th., **5**, 718 ⟨87LA687⟩
Oesterling, R. E., **4**, 3 ⟨58JA6533⟩
Oestreich, E., **2**, 399 ⟨76ZC437⟩
Oestreich, T. E., **4**, 413 ⟨73JOC1949⟩
Oetker, C., **3**, 568 ⟨73CB2144⟩
Oettmeier, W., **4**, 311 ⟨70CB2314⟩
Öfele, K., **3**, 513 ⟨68AG(E)950⟩; **5**, 934 ⟨93CB2573⟩
O'Ferrall, R. A. M., **4**, 100 ⟨90JOC3244⟩
Offe, H. A., **5**, 636 ⟨54CB825⟩; **6**, 445 ⟨82MI 614-03⟩
Offenhaur, R. D., **3**, 60 ⟨68JOC775⟩
Offerman, R. J., **4**, 105 ⟨90JHC139⟩; **6**, 162 ⟨88RTC641⟩
Offermann, K., **3**, 694, 696, 699, 700, 721, 722, 724, 725 ⟨65AG(E)472⟩, 696 ⟨61CB2814⟩
Offermanns, H., **4**, 328 ⟨67AG(E)907⟩; **5**, 599 ⟨67AG(E)907⟩
Ofitserov, E. N., **4**, 342 ⟨78ZOB2653⟩; **6**, 583 ⟨77IZV1177⟩
Ofner, A., **3**, 62 ⟨58JOC157⟩, 233 ⟨57JOC1611⟩
Ofner, H., **2**, 16 ⟨91SL187⟩
Ofori-Okai, G., **4**, 223 ⟨89JA658⟩
Ofori-Okai, K. G., **4**, 724 ⟨86JOM(307)231⟩
Oftedahl, E., **5**, 1046 ⟨63JOC194⟩
Oftedahl, M. L., **4**, 864 ⟨67USP3335188⟩
Oga, N., **3**, 265 ⟨90JCS(P1)1891⟩
Oganesyan, S. S., **5**, 28 ⟨81MI 502-03⟩
Ogasawara, H., **3**, 543 ⟨84H(22)2563, 85S224⟩
Ogasawara, K., **2**, 85 ⟨89CC1344⟩, 323 ⟨89H(29)1861⟩; **3**, 221 ⟨89CC271⟩; **4**, 168 ⟨86CC156⟩, 346 ⟨87CL1569⟩, 412 ⟨91CC462⟩, 827 ⟨88CC189⟩; **5**, 164 ⟨80S403⟩, 607 ⟨76JA7084⟩, 607, 616 ⟨82TL881, 76CC776⟩, 616 ⟨82CL529⟩
Ogata, K., **2**, 78 ⟨90JOC2415⟩, 139 ⟨91SL557⟩, 140 ⟨92TL85⟩, 1000 ⟨90CPB329⟩; **4**, 73 ⟨91TL1463⟩, 368 ⟨91SL557, 92TL85⟩; **6**, 634 ⟨91TL5987⟩
Ogata, M., **4**, 407 ⟨79TL5011, 80H(14)97⟩, 520 ⟨86JOC3897⟩; **5**, 578, 608 ⟨78H(11)139⟩; **6**, 61 ⟨89T637⟩
Ogata, T., **2**, 457 ⟨77S179, 79BCJ466⟩; **3**, 290, 408 ⟨86CL1197⟩, 332, 364 ⟨75BCJ2907⟩, 353 ⟨81S36⟩, 364, 378 ⟨76BCJ3128⟩, 493 ⟨77S626⟩, 687 ⟨76BCJ3124⟩; **4**, 834 ⟨76BCJ3128⟩, 847 ⟨76S552⟩, 854 ⟨79BCJ466⟩; **5**, 403, 404 ⟨80BCJ1625⟩, 728, 729 ⟨77S626⟩, 775 ⟨79BCJ1541⟩; **6**, 87, 99 ⟨76S551⟩
Ogata, Y., **2**, 21 ⟨75JOC2960, 77CI(L)538⟩, 21 ⟨79BCJ255, 88OSC(6)90⟩, 156 ⟨81CJC718⟩, 348 ⟨B-89MI 206-04⟩, 650 ⟨74T3021⟩, 685 ⟨84JCS(P2)451⟩; **4**, 343 ⟨70JOC597⟩, 437 ⟨64JOC1985⟩; **5**, 205 ⟨67T3327⟩
Ogawa, A., **2**, 285 ⟨92JOC111⟩, 311 ⟨87CL1275⟩, 542 ⟨88JA1976⟩, 554 ⟨87AG(E)1187⟩, 710 ⟨92JA5902⟩, 730 ⟨92JA7591⟩; **3**, 90 ⟨90TL6887⟩, 136 ⟨90JA455⟩, 143 ⟨88TL6137⟩, 176 ⟨87S1096⟩, 336, 340, 385 ⟨88JA1976⟩; **4**, 239, 334, 660 ⟨90OM1355⟩, 334, 909 ⟨90CL2053⟩, 652 ⟨87AG(E)1187⟩, 940 ⟨92JA5902⟩, 940, 947 ⟨91JA9796, 91JOC5721⟩, 948 ⟨92JA7591⟩; **5**, 238, 241 ⟨91JA9796⟩, 240, 243, 253 ⟨91JOC3776, 93MI 505-01⟩, 437 ⟨90JA455⟩, 510 ⟨91COS(6)461⟩, 624 ⟨91COS(6)476⟩, 625 ⟨85JOC384⟩, 625, 626 ⟨85BCJ1448⟩; **6**, 479 ⟨88TL4767, 90TL4773⟩, 495, 497 ⟨89AG(E)452⟩
Ogawa, H., **3**, 361 ⟨89CC505⟩; **5**, 801 ⟨88CC561, 91JCS(D)1531⟩
Ogawa, K., **4**, 837 ⟨85CPB2256⟩
Ogawa, M., **2**, 34 ⟨88S366⟩, 36 ⟨89S283⟩, 55 ⟨86JOC240, 88JOC3752⟩; **3**, 7, 30 ⟨86S774⟩, 117 ⟨88JOC3587, 88SC869⟩, 279 ⟨92CL1895⟩; **6**, 461 ⟨91JOM(417)305⟩
Ogawa, S., **2**, 25 ⟨81TL4409⟩, 816 ⟨83TL511⟩; **5**, 699 ⟨81BCJ3863⟩, 1106 ⟨81TL4409⟩
Ogawa, T., **2**, 630 ⟨88S236⟩, 632 ⟨85CL411⟩, 727 ⟨89CL769⟩; **3**, 165 ⟨89BCJ1358⟩, 308 ⟨86T6465⟩, 500

⟨88CL847⟩, 634, 635 ⟨87CL887⟩; **4**, 187 ⟨87TL2723⟩, 225 ⟨77CAR(54)C17, 81CAR(95)308⟩, 286 ⟨89BCJ1358⟩, 750 ⟨88S236⟩; **5**, 242, 253 ⟨90JCR(S)56⟩, 289, 302 ⟨91CL1443⟩, 329 ⟨94TL1247⟩, 344 ⟨91CL105⟩, 630 ⟨90CL1651⟩; **6**, 71 ⟨76CAR(51)C13⟩, 150 ⟨88BCJ2927⟩
Ogawa, Y., **2**, 518 ⟨91TL1479⟩; **3**, 417, 658 ⟨78S892⟩; **4**, 408 ⟨89H(29)865⟩
Ogawara, R., **5**, 1143 ⟨69JAP47019769⟩
Ogden, P. H., **5**, 1080 ⟨67CC59, 67JA5007⟩; **6**, 239 ⟨67JCS(C)2302⟩, 242, 613 ⟨66JOC3833⟩, 604 ⟨67JA3868, 68JHC41⟩, 616 ⟨67JHC389⟩
Ogg, C. L., **5**, 205 ⟨55JA4037⟩
Ogger, L., **4**, 656 ⟨82ZN(B)957⟩
Ogi, Y., **5**, 249 ⟨87NKK1479⟩, 253 ⟨87NKK1475⟩
Ogihara, K., **5**, 196 ⟨82MI 504-01⟩
Ogilvie, K. K., **2**, 107 ⟨77JA1277⟩; **6**, 540 ⟨83TL865⟩
Ogilvie, W., **4**, 432, 433 ⟨87CJC166⟩
Ogilvie, W. W., **4**, 226 ⟨90JA6263⟩; **5**, 513 ⟨90JA6263⟩
Ogima, M., **3**, 233 ⟨89CL1959⟩; **4**, 954 ⟨89CL1959⟩
Ogini, W. O., **2**, 582 ⟨82OM654⟩
Ogino, H., **3**, 357 ⟨90JAN1150⟩
Ogino, T., **2**, 44 ⟨80TL177⟩; **3**, 2 ⟨79CL443⟩; **5**, 276 ⟨69JCS(C)874⟩
Ogino, Y., **2**, 48 ⟨90TL2990⟩, 49 ⟨93JOC844⟩; **5**, 563 ⟨94UP0⟩
Ogiso, A., **3**, 77 ⟨89JA2332, 90JA6120⟩
Ogle, J., **6**, 598 ⟨56JA4458⟩
Ogle, J. R., **5**, 1048 ⟨58JCS1141⟩
Ogliaruso, M. A., **5**, 24 ⟨B-79MI 502-01, B-91MI 502-01⟩, 122 ⟨B-92MI 503-01⟩
Oglobin, K. A., **4**, 35 ⟨83ZOR1625⟩
Ogloblin, K. A., **2**, 787 ⟨64ZOB1522⟩; **5**, 702 ⟨88JOU2149, 88JOU2154⟩, 702 ⟨88ZOR2384, 88ZOR2389⟩; **6**, 51 ⟨73ZOR2346, 77ZOR1857⟩
Ognjanova, V. H., **5**, 345 ⟨90SC2083⟩
Ognyanov, V. I., **3**, 210 ⟨87HCA1393⟩, 628 ⟨91JA6992⟩; **5**, 307 ⟨89HCA1522, 92HCA62⟩
Ogorevc, B., **5**, 1118, 1122 ⟨83MI 530-02⟩
Ogoshi, H., **2**, 960 ⟨75JA6461⟩; **3**, 356 ⟨73JA3043⟩, 367 ⟨77AG(E)866⟩, 399 ⟨75CC359⟩
Ogric, C., **5**, 952 ⟨95CB289⟩
Ogston, A. G., **5**, 83 ⟨55MI 502-03⟩
Oguchi, Y., **3**, 627 ⟨92BCJ923⟩; **5**, 344 ⟨91CL1479⟩
Oguni, N., **2**, 348 ⟨90CL547⟩; **3**, 30, 33 ⟨91SL479⟩, 30, 33 ⟨94T2821⟩, 628 ⟨90CC1364, 92JCS(P1)3135⟩, 629 ⟨91CC1752, 93JOC1515⟩, 632 ⟨92SL663⟩
Ogura, A., **3**, 158 ⟨91BCJ1389⟩
Ogura, F., **2**, 289 ⟨80SC595⟩, 293 ⟨81CL447⟩, 753 ⟨84CL853⟩; **3**, 39, 176 ⟨89CL607⟩, 89 ⟨84CL891⟩, 114 ⟨86TL6099⟩, 376 ⟨76JA3615⟩; **4**, 238 ⟨91CL415⟩, 402 ⟨81CL447⟩, 850 ⟨88CC822⟩; **5**, 210 ⟨87TL1281, 89JOC4391⟩, 242 ⟨86CL977⟩, 1078 ⟨86TL6099⟩, 1113 ⟨80SC595⟩, 1117 ⟨82SC131⟩
Ogura, H., **3**, 145 ⟨90TL265⟩, 275 ⟨82CPB63⟩; **5**, 226 ⟨80TL1467⟩, 235 ⟨85JOC273⟩, 1025 ⟨89CPB2334⟩, 1049 ⟨79CPB1137⟩; **6**, 507 ⟨82SC213⟩
Ogura, K., **2**, 153, 1001 ⟨91COS(1)505⟩, 305 ⟨91JOC2920⟩, 748 ⟨84S1043⟩; **3**, 31, 157 ⟨86TL3665⟩, 61 ⟨78TL1305⟩, 130, 285 ⟨79TL4303⟩, 146 ⟨86CL1597⟩, 157 ⟨91COS(1)505⟩, 671 ⟨85CL1487⟩; **4**, 73 ⟨91TL1463⟩, 74 ⟨71S89⟩, 74, 77 ⟨71BCJ1726⟩, 80 ⟨93BCJ2339⟩, 197 ⟨71TL3151⟩, 233 ⟨70CC1689, 72BCJ2023⟩, 262 ⟨80TL2233⟩, 271 ⟨81TL4499⟩, 326, 856, 857 ⟨74JA1960⟩, 326, 856, 857 ⟨78TL375⟩, 390, 391 ⟨91COS(1)505⟩, 395 ⟨82CL1697⟩, 854, 857 ⟨82JOC5404, 83CL767⟩, 856 ⟨72TL1383, 75S385, 79BCJ2013, 80S736⟩, 856, 857 ⟨74CL659, 79S880⟩, 857 ⟨86JOC508, 86JOC700⟩, 858 ⟨90TL4621⟩; **5**, 199 ⟨90JMC2621⟩, 552 ⟨80S736⟩
Ogura, M., **4**, 616 ⟨69CC207⟩; **5**, 118 ⟨81MI 502-06⟩
Oguri, T., **3**, 173 ⟨78JOC4552⟩
Oguro, K., **4**, 955 ⟨75CC899, 78JOM(159)417⟩; **5**, 272, 283 ⟨92TL1455⟩, 940 ⟨78JOM(159)417⟩, 1137 ⟨76IC2346⟩; **6**, 517 ⟨76IC2346⟩
Ogurtsov, V. A., **6**, 150 ⟨90IZV1620⟩, 150 ⟨90IZV1623⟩
Oh, C.-H., **6**, 480 ⟨92JC241⟩
Oh, D. Y., **2**, 730 ⟨94JCS(P1)717⟩; **3**, 197 ⟨89TL3307⟩, 197 ⟨91SC279⟩, 197 ⟨93TL2147⟩, 231 ⟨91S207⟩; **4**, 246 ⟨89SC433⟩, 347 ⟨89SC2209⟩, 347, 781, 783, 840 ⟨90SC273⟩, 370 ⟨89SC2213⟩, 572 ⟨89SC1891⟩, 924 ⟨87MI 420-01⟩, 927, 929 ⟨91SC279⟩, 943 ⟨90SC267⟩; **5**, 1018 ⟨67CB480⟩; **6**, 119 ⟨85TL3479⟩, 120 ⟨86MI 604-02⟩, 121 ⟨88SC1611⟩, 123 ⟨92TL5375⟩
Oh, G.-M., **5**, 378 ⟨94S909⟩
Ohakara, M., **4**, 72 ⟨75JOC3540⟩
Oh, K. C., **6**, 651 ⟨92TL4963⟩
O'Halloran, J. K., **3**, 710 ⟨70TL4079⟩; **4**, 413 ⟨73JOC1949, 76JOC1303⟩
Ohannesian, L., **2**, 22 ⟨84JOC2032⟩, 620 ⟨86S868⟩; **3**, 93, 94 ⟨87CRV671⟩, 152 ⟨84JOC2032⟩, 209 ⟨88JOC1110⟩; **6**, 557 ⟨88JOC1110⟩
O'Hara, B., **6**, 489 ⟨93MI 615-03⟩
Ohara, E., **2**, 784, 786 ⟨81CC261⟩
Ohara, M., **2**, 657 ⟨75AG(E)820⟩
Ohara, T., **3**, 209 ⟨87BCJ1839⟩
Oharu, K., **6**, 380 ⟨86JOM(309)247⟩
Ohashi, H., **4**, 615 ⟨83OM351⟩
Ohashi, M., **4**, 850 ⟨85CC1803⟩
Ohashi, O., **5**, 1103 ⟨90JSP62⟩, 1152, 1153, 1159 ⟨80CC333⟩, 1153 ⟨82JST(79)215⟩
Ohashi, Y., **3**, 139 ⟨85JOC1523⟩, 465 ⟨76CPB369⟩
Ohasi, M., **4**, 167 ⟨81JCS(P1)1015⟩
Oh, S. Y., **3**, 13, 82 ⟨87TL2389, 87MI 303-01⟩, 82 ⟨87TL4575, 88H(27)1595⟩, 86 ⟨87MI 303-02⟩
Oh, T., **2**, 88 ⟨91TL4647⟩; **3**, 422 ⟨88TL3781⟩
Oh, Y. I., **4**, 165 ⟨89JOC91⟩
Ohba, M., **5**, 597 ⟨88H(27)2077⟩
Ohba, N., **3**, 156 ⟨81T473⟩
Ohba, T., **5**, 70, 73 ⟨80CC327⟩
Ohe, K., **3**, 146 ⟨90JCS(P1)1697⟩, 163 ⟨92JOC2⟩; **4**, 931 ⟨92JA9710⟩; **5**, 241 ⟨87JOM(334)C45⟩, 242 ⟨87JOM(334)C43, 89JOM(361)63⟩, 346 ⟨90JA9646⟩, 416 ⟨92JA9710⟩, 438 ⟨90JA7061⟩
Ohe, M., **4**, 302 ⟨87BCJ4079⟩
Ohe, T., **3**, 290 ⟨92CC415⟩
Oh-e, T., **3**, 290, 305, 311 ⟨92TL4465⟩
Ohfune, Y., **2**, 629 ⟨79JA4749⟩; **3**, 40 ⟨84TL1071⟩
Ohi, N., **5**, 223 ⟨87CPB1903⟩; **6**, 447 ⟨86JAN230⟩
Ohimane, H., **2**, 312 ⟨88TL3811⟩
Ohira, N., **2**, 753 ⟨84CL853⟩
Ohira, S., **2**, 680 ⟨89SC561⟩
Ohiro, S., **5**, 872 ⟨89SC561⟩
Ohishi, H., **5**, 726 ⟨92CPB602⟩; **6**, 544 ⟨92CPB2279⟩
Ohishi, M., **3**, 157 ⟨85JOC603⟩
Oh-ishi, T., **6**, 556, 564 ⟨87CPB2840⟩
Ohkawa, S., **5**, 278 ⟨82JCS(P1)2909⟩
Ohkawa, T., **4**, 827 ⟨88CC189⟩
Ohkawara, T., **2**, 197 ⟨80CPB134⟩, 724 ⟨80S937⟩
Ohki, E., **2**, 660 ⟨63JOC783⟩; **6**, 553 ⟨81CPB3158⟩
Ohki, H., **2**, 695 ⟨88TL4719⟩
Ohki, S., **3**, 309 ⟨77CPB3023⟩; **4**, 211 ⟨72CPB2123⟩
Ohki, T., **5**, 325 ⟨85JA5310⟩
Ohkubo, K., **6**, 501 ⟨72CL971⟩
Ohkubo, T., **2**, 14 ⟨89JOC4982⟩
Ohkuma, M., **2**, 139 ⟨91SL557⟩; **4**, 368 ⟨91SL557, 92TL85⟩
Ohkuma, T., **2**, 53 ⟨89JA9134⟩; **3**, 276 ⟨86TL373⟩
Ohlendorf, H. W., **2**, 749 ⟨76S741⟩
Öhler, E., **3**, 270 ⟨85M77, 91S357⟩; **4**, 339 ⟨91CB175, 91LA229⟩; **5**, 166 ⟨70AG(E)457⟩
Ohlinger, H., **6**, 443 ⟨49AG183⟩
Ohlmeyer, M., **2**, 528 ⟨91TA613⟩
Ohlmeyer, M. J., **2**, 42 ⟨88JOC5178⟩, 527 ⟨91JA6139, 91JOC1020⟩
Ohloff, G., **3**, 25, 147 ⟨67HCA708⟩, 147 ⟨79HCA2655⟩, 494 ⟨75HCA1016, 77HCA1161⟩; **4**, 166 ⟨81HCA1247⟩

Ohmari, H., **3**, 496 ⟨85TL2199, 87CPB4960⟩
Ohmasa, N., **2**, 731 ⟨81CL1117⟩
Ohme, R., **2**, 373 ⟨68ZC41⟩, 374 ⟨61CB2166, 68JPR257⟩, 374, 378, 380 ⟨68LA(713)74⟩, 374, 380 ⟨70JPR349⟩, 377 ⟨62CB2012⟩, 379 ⟨77JPR195⟩, 403 ⟨63AG(E)157, 88OSC(6)78⟩, 806 ⟨71JPR642⟩; **5**, 353 ⟨65TL1857, 67JPR86⟩; **6**, 70 ⟨68LA(716)207⟩
Ohmizu, H., **3**, 146 ⟨77CL1021⟩, 151 ⟨82TL1609⟩; **4**, 300, 311 ⟨82JA5753⟩; **6**, 25, 26 ⟨82TL1609, 81TL871⟩, 26 ⟨82TL4801⟩
Ohmori, H., **2**, 108 ⟨79CPB1700⟩, 219, 227 ⟨84CC843⟩, 385 ⟨88CPB459⟩; **3**, 196 ⟨88CPB613⟩, 424 ⟨94CPB1041⟩; **5**, 1078 ⟨85CPB373⟩, 1107 ⟨90CPB1389⟩
Ohmori, K., **2**, 26 ⟨84S841⟩
Ohms, G., **6**, 581 ⟨92PS(73)81⟩
Ohmura, H., **3**, 360 ⟨82CL793⟩
Ohnaka, Y., **4**, 420 ⟨85JOC4227⟩
Ohnishi, A., **3**, 723 ⟨88S714⟩
Ohnishi, M., **5**, 972 ⟨69JOC1631⟩
Ohnishi, R., **2**, 650 ⟨78JOC865⟩
Ohnishi, S., **3**, 350 ⟨83CPB1929⟩
Ohnishi, Y., **5**, 292 ⟨81CPB1063⟩
Ohno, A., **2**, 40 ⟨91JOC4778, 91TL7075⟩, 116 ⟨B-77MI 203-01⟩, 398 ⟨88TL4769, 90T7471⟩; **3**, 129 ⟨86CL87⟩, 161 ⟨90T7471⟩, 348 ⟨75BCJ2403⟩, 351 ⟨79JOC2244⟩, 662 ⟨75BCJ3298⟩; **4**, 391 ⟨87JOC1414⟩; **5**, 652 ⟨90TL3603⟩; **6**, 82 ⟨64T417⟩
Ohno, A. Y., **2**, 312 ⟨89BCJ3845⟩
Ohno, H., **3**, 616 ⟨93CL375⟩, 629 ⟨91SL551⟩
Ohno, K., **2**, 622, 627 ⟨66TL4713⟩; **3**, 65 ⟨81TL4487⟩; **5**, 429 ⟨86CL177⟩, 458 ⟨66JA3452⟩, 459 ⟨69S1⟩, 726 ⟨79TL4671⟩, 1152 ⟨82CL981, 85JSP(111)415⟩, 1153 ⟨81JSP(90)507, 81JSP(90)512, 82JST(79)215⟩, 1154 ⟨82JPR(324)667⟩
Ohno, M., **2**, 72 ⟨89TL1657, 89TL7095⟩, 568 ⟨92T5691⟩, 788 ⟨79TL2525⟩; **3**, 40 ⟨82TL525⟩, 273 ⟨87S1092⟩, 433 ⟨85S279⟩, 634 ⟨73OSC(5)266⟩, 698 ⟨81JOC5445⟩; **4**, 354 ⟨88JA4826⟩, 424 ⟨70JA343⟩; **5**, 10, 367 ⟨91SL919⟩, 135, 138 ⟨84TL2557, 84T145⟩, 345 ⟨84JA1133⟩, 418 ⟨91TL4929⟩, 753 ⟨79TL2517⟩, 1118 ⟨81JOC5445⟩
Ohno, S., **3**, 217 ⟨88BCJ1207⟩, 356 ⟨74CL1101⟩, 376 ⟨78TL251, 80JOC2468⟩
Ohno, T., **2**, 81 ⟨89CL1401⟩, 142 ⟨90CL2239⟩; **3**, 159 ⟨91CL949⟩, 281, 298 ⟨91CL1059⟩
Ohno, Y., **2**, 618 ⟨79BCJ1197⟩; **3**, 495 ⟨79BCJ1197⟩
Ohnogi, T., **4**, 736 ⟨88S614, 89JOC5630⟩
Ohnuma, T., **3**, 217 ⟨86TL219⟩
Ohoka, M., **5**, 967 ⟨72JOC3030⟩, 1003 ⟨71JOC3542, 73BCJ1275⟩; **6**, 445 ⟨71BCJ2182⟩, 618 ⟨73BCJ303⟩
Ohorodnik, A., **2**, 448 ⟨83GEP(O)3149937⟩
Öhrlein, R., **2**, 392 ⟨86S535⟩; **3**, 46 ⟨86S535⟩
Ohsaki, H., **3**, 723 ⟨88S714⟩; **4**, 613, 617 ⟨91OM2695⟩
Ohsaki, T., **5**, 340 ⟨88CPB877⟩
Ohsawa, A., **2**, 20 ⟨73CPB867⟩; **4**, 60 ⟨92TL5399⟩
Ohsawa, S., **5**, 179 ⟨89CL1775⟩
Ohsawa, T., **3**, 665 ⟨77S245⟩; **5**, 285 ⟨83JA7345⟩
Ohshima, M., **3**, 54 ⟨83CL1207⟩, 139 ⟨86CL1009⟩, 168 ⟨85CL1871⟩, 279, 281, 298 ⟨86CL165⟩; **4**, 187 ⟨84CL265⟩
Ohshima, T., **2**, 249, 401 ⟨79BCJ1102⟩
Ohshiro, Y., **2**, 14 ⟨83BCJ1881, 93JOC6529⟩, 244 ⟨72JOC3810⟩, 303 ⟨71S89⟩, 456 ⟨85BCJ3104⟩, 843 ⟨79CB355⟩, 850 ⟨82BCJ909⟩, 932 ⟨81TL3079⟩; **3**, 62 ⟨82CL1997⟩, 557 ⟨79CC900, 80JOC5377⟩, 574 ⟨81CL1197⟩, 593 ⟨70JOC2076⟩; **4**, 13 ⟨83BCJ1881⟩, 66 ⟨82CL587⟩, 328 ⟨85S276⟩, 348 ⟨79CB355⟩, 517, 538, 1015 ⟨82BCJ1163⟩, 538 ⟨81AG(E)126⟩, 743 ⟨94JCS(P1)3⟩, 979 ⟨74JOC3198⟩; **5**, 10, 367 ⟨91SL919⟩, 274, 286 ⟨80S933⟩, 359 ⟨90TL5327, 92JOC7359⟩, 364 ⟨85H(23)677⟩, 989 ⟨85S276⟩, 1032 ⟨74JOC957⟩; **6**, 7 ⟨90SL217⟩, 123, 697 ⟨79CB355⟩, 162 ⟨82BCJ1163⟩, 489 ⟨92JOC7359⟩
Ohshita, J., **4**, 613, 617 ⟨91OM2695⟩, 616 ⟨91JOM(407)157⟩
Ohst, H., **4**, 643 ⟨83ZN(B)1388⟩
Ohsuga, Y., **2**, 662 ⟨91CPB1132⟩
Ohsugi, E., **3**, 108 ⟨68CPB897⟩
Ohsugi, M., **5**, 282 ⟨77S632⟩
Ohsuki, S., **4**, 171 ⟨86TL947⟩
Ohsumi, T., **5**, 286 ⟨86JOM(309)333⟩, 362 ⟨92H(33)893, 92T651⟩; **6**, 443 ⟨88MI 614-01⟩
Ohta, A., **3**, 632 ⟨90CL481⟩; **5**, 327 ⟨81JHC555⟩; **6**, 434 ⟨85JHC1643⟩
Ohta, H., **2**, 84, 445 ⟨88BCJ4476⟩, 363 ⟨94JA2390⟩, 398 ⟨85ABC2331, 87CL191⟩, 853 ⟨87CL357, 88BCJ4029⟩, 858 ⟨87TL6649⟩; **3**, 319 ⟨82CC1223⟩, 336 ⟨88CL1145, 91BCJ2402⟩, 384, 394 ⟨87TL6649, 91CL1053⟩, 394 ⟨92TL1333⟩, 631 ⟨88TL6957⟩, 666 ⟨70CC1601⟩; **4**, 415 ⟨88CL1145⟩; **5**, 164 ⟨90CL1549⟩, 179 ⟨89CL1775⟩, 196 ⟨82MI 504-01⟩, 603 ⟨88CL1145⟩, 603, 622 ⟨91BCJ2402⟩, 603, 622, 627 ⟨92CL131⟩
Ohta, K., **2**, 35 ⟨85CC400⟩; **4**, 773, 781, 783 ⟨85S556⟩
Ohta, M., **3**, 561 ⟨70BCJ1874⟩, 563 ⟨67JOC3564⟩, 574 ⟨66TL807, 67BCJ1933⟩; **4**, 976 ⟨71JCS(C)196⟩; **5**, 24 ⟨79MI 502-02⟩, 639, 846 ⟨54JPJ821⟩, 642 ⟨73BCJ2921, 54BCJ624⟩
Ohta, N., **3**, 599 ⟨77TL1009, 80SC233⟩
Ohta, S., **3**, 311 ⟨86CPB4916⟩, 319 ⟨82HCA2363⟩; **4**, 448 ⟨90H(31)2029⟩; **5**, 125, 126 ⟨82S833⟩
Ohta, T., **2**, 753 ⟨84BCJ2440⟩; **3**, 127 ⟨92TL635⟩; **4**, 389 ⟨90TL6407⟩
Ohta, Y., **2**, 949 ⟨89JA1888⟩
Ohtake, T., **4**, 696 ⟨73NKK381⟩
Ohtaki, T., **6**, 187 ⟨94OM434⟩
Ohto, N., **2**, 98 ⟨84MI 202-01⟩
Ohtsubo, A., **5**, 236 ⟨91CL989⟩
Ohtsuka, N., **2**, 712 ⟨68TL5071⟩
Ohtsuka, Y., **6**, 625 ⟨70BCJ187⟩
Ohtsuki, K., **3**, 146 ⟨86CL1597⟩; **4**, 857 ⟨86JOC326⟩
Ohuchi, K., **3**, 244 ⟨88TL5941, 90SC2983⟩
Ohuchi, S., **2**, 7 ⟨92CL1505⟩
Ohuchi, Y., **5**, 30, 35 ⟨86SC499⟩
Ohuchida, S., **3**, 55, 252, 275 ⟨81JA4597⟩, 229 ⟨84JA3875⟩
Ohya, R., **3**, 668 ⟨80CC791⟩
Oi, R., **2**, 684 ⟨88CL899⟩; **3**, 46 ⟨86JOC254⟩
Oiarbide, M., **2**, 398 ⟨90JOC2070⟩; **3**, 162 ⟨87TL5361⟩
Oida, H., **5**, 249 ⟨87NKK1479⟩, 253 ⟨87NKK1475⟩
Oida, S., **5**, 234 ⟨84TL2793⟩, 236 ⟨85TL673⟩, 245 ⟨86T4593⟩, 530, 540 ⟨86TL4593⟩; **6**, 553 ⟨81CPB3158⟩
Oida, T., **2**, 247 ⟨86BCJ3897⟩, 720 ⟨91BCJ702⟩; **3**, 267 ⟨89JOC6012⟩; **6**, 552 ⟨82BCJ339⟩, 552, 555 ⟨81TL5195⟩
Oiji, Y., **4**, 970 ⟨91EUP413343⟩
Oikawa, H., **5**, 126, 280, 300 ⟨85JOC760⟩; **6**, 417 ⟨83S908⟩
Oikawa, I., **3**, 592 ⟨80BCJ2582⟩
Oikawa, M., **5**, 129 ⟨94JOC5532⟩
Oikawa, S., **6**, 498 ⟨92CL1389⟩
Oikawa, Y., **3**, 310 ⟨76H(4)1859⟩; **4**, 201 ⟨82TL889⟩; **5**, 173, 199 ⟨87TL4569⟩
Oine, T., **6**, 143 ⟨81JOC1571⟩
Oiry, J., **6**, 506 ⟨82JMC178⟩
Oishi, M., **3**, 26 ⟨93JA3943⟩; **6**, 106 ⟨68TL497⟩
Oishi, T., **2**, 61 ⟨83TL2653⟩, 365 ⟨70CC777⟩, 402 ⟨78TL3427⟩; **3**, 310 ⟨81JA6990⟩; **4**, 881 ⟨68TL497⟩; **5**, 157 ⟨89TL6529⟩
Oishi, Y., **2**, 684 ⟨78CL57⟩
Ojika, M., **5**, 329 ⟨94TL1247⟩
Ojima, I., **2**, 222 ⟨72JA5786⟩, 264 ⟨73JOM(50)26⟩, 320 ⟨91JOC5263⟩, 540 ⟨73JOM(50)C26⟩, 840 ⟨69BCJ2975⟩, 915 ⟨74JOM(66)C14⟩, 920 ⟨92JA6580⟩; **3**, 12 ⟨72TL5035⟩, 106 ⟨83TL4573⟩, 365 ⟨72CC62⟩, 528 ⟨81CC344⟩, 616 ⟨75CL331⟩, 617 ⟨75CL737⟩, 640 ⟨74TL4005⟩; **4**, 810 ⟨84TL303⟩; **5**, 224 ⟨89JOC4511⟩, 274, 286 ⟨87JA1798⟩, 759, 777 ⟨77JOM(140)97, 74JOM(72)C11⟩, 777 ⟨75JOM(99)C5⟩; **6**, 21 ⟨84TL303⟩, 515 ⟨77JOM(140)97⟩, 581 ⟨69BCJ2975⟩

Ojo, I. A. O., **4**, 129 ⟨88PS(40)183⟩
Oka, H., **3**, 469 ⟨68BCJ2707⟩
Oka, K., **3**, 349, 365 ⟨79JOC1736⟩, 370 ⟨77TL2939⟩; **4**, 70 ⟨80TL3579, 81JA2757⟩, 71 ⟨76TL2783⟩, 71, 72 ⟨77TL695, 79JOC1736⟩; **5**, 123 ⟨81BCJ1267⟩, 485, 583 ⟨76TL2783⟩
Oka, S., **2**, 398 ⟨88TL4769⟩; **3**, 129 ⟨86CL87⟩, 348 ⟨75BCJ2403⟩, 351 ⟨79JOC2244⟩, 429 ⟨84CC289, 88JOC4897⟩, 662 ⟨75BCJ3298⟩; **4**, 391 ⟨87JOC1414⟩
Okabe, E., **4**, 842 ⟨75JCS(P1)1277⟩
Okabe, N., **5**, 375 ⟨85TL2217⟩
Okabe, T., **5**, 550 ⟨72T3203⟩
Okachi, R., **5**, 334 ⟨85JMC1772⟩
Okada, A., **2**, 544 ⟨72JOM(36)C13⟩, 546 ⟨71TL1511⟩
Okada, E., **3**, 230 ⟨86TL353⟩, 232 ⟨90CL113⟩, 246 ⟨86S1013⟩, 265 ⟨89TL6173, 90S195⟩; **4**, 889 ⟨86S1013⟩, 889, 900, 970, 971, 972 ⟨90S195⟩
Okada, H., **2**, 394 ⟨82TL2965⟩, 598 ⟨82OM440⟩; **5**, 199 ⟨86JAN1257⟩
Okada, K., **2**, 13 ⟨89CC1636⟩; **3**, 289 ⟨75TL4239⟩; **4**, 559 ⟨91CL2213⟩, 889, 900, 970, 971, 972 ⟨90S195⟩; **5**, 1036 ⟨91TL3503⟩
Okada, M., **2**, 478 ⟨80CC1183⟩, 663 ⟨82JOC1200⟩; **3**, 139 ⟨86JOC421⟩, 227 ⟨86S847⟩, 635 ⟨86JCS(P1)1763⟩; **4**, 60 ⟨92TL5399⟩, 67 ⟨86JHC1163⟩, 230 ⟨84TL5797⟩
Okada, N., **4**, 873 ⟨86CL1861⟩, 877 ⟨86CL311⟩
Okada, R., **4**, 80 ⟨93BCJ2339⟩; **5**, 272, 283 ⟨92TL1455⟩
Okada, S., **2**, 71 ⟨88PAC123, 89JA4028⟩, 311 ⟨92JOC1504⟩
Okada, T., **3**, 304 ⟨70BCJ2884⟩
Okada, Y., **2**, 214 ⟨90JHC1361⟩, 440, 829 ⟨91TA667, 92CL547⟩; **3**, 496 ⟨89JOC974⟩; **4**, 737 ⟨88JFC(41)435⟩, 810 ⟨86CL1895⟩; **5**, 589, 641 ⟨90JHC567⟩
Okafor, C. O., **3**, 357 ⟨91DP149⟩; **5**, 1109 ⟨73JOC4383⟩
Okahara, M., **2**, 19 ⟨81BCJ1151⟩, 89 ⟨88JOC2737⟩, 162 ⟨72S203⟩, 251 ⟨73S667⟩; **3**, 617 ⟨78S301⟩; **4**, 52 ⟨77TL2893, 81BCJ1151⟩, 72 ⟨75JOC3540⟩; **5**, 6 ⟨77TL2337, 77TL2893⟩, 1003 ⟨73BCJ1275⟩
Okajima, N., **5**, 589, 641 ⟨90JHC567⟩
Okamato, Y., **4**, 874 ⟨81CC669⟩
Okamoto, A., **3**, 607 ⟨91CC124⟩
Okamoto, H., **5**, 691 ⟨91BCJ766, 92BCJ674⟩
Okamoto, J., **2**, 459 ⟨73CL599⟩; **5**, 864 ⟨86MI 521-02, B-87MI 521-01⟩
Okamoto, K., **2**, 13 ⟨89CC1636⟩; **3**, 430 ⟨92JOC3245⟩; **5**, 206 ⟨88CPB178⟩, 665, 672, 824, 825 ⟨92JOC3245⟩
Okamoto, M., **3**, 311 ⟨86CPB4916⟩, 698, 721 ⟨62JOC3651⟩; **5**, 125, 126 ⟨82S833⟩
Okamoto, S., **2**, 855 ⟨74JAP7401534⟩, 923 ⟨90TL6399⟩
Okamoto, T., **2**, 21 ⟨88S545⟩, 24 ⟨90BCJ3033⟩, 28 ⟨87BCJ1159⟩, 34 ⟨92BCJ1731⟩, 320 ⟨89CL463⟩, 626 ⟨87BCJ4187, 87CL627, 88BCJ597, 89BCJ3373, 89BCJ439⟩, 631 ⟨87CL2109, 88BCJ600, 88CL795, 89BCJ1349⟩, 691 ⟨80S461⟩, 753 ⟨87CE551⟩; **3**, 264 ⟨73TL4533⟩, 429 ⟨84CC289, 88JOC4897⟩, 456 ⟨89JCS(P1)279⟩, 458 ⟨88JCS(P1)945⟩, 662 ⟨75BCJ3298⟩, 704 ⟨76CPB832⟩; **5**, 1126 ⟨88H(27)2609⟩; **6**, 506 ⟨83S471⟩
Okamoto, Y., **2**, 11, 36 ⟨79S379⟩, 34 ⟨83S461, 78CC874⟩, 106 ⟨85BCJ3393, 91PS(55)195⟩, 107 ⟨88BCJ3359⟩, 321 ⟨90CC1385⟩, 457 ⟨83S916⟩, 459 ⟨81S811⟩, 847 ⟨78TL2523⟩; **3**, 605 ⟨90TL2311⟩, 640 ⟨82S915⟩, 640, 646, 659 ⟨83S917⟩; **4**, 240 ⟨73JOC3172⟩, 792 ⟨70JOC4250⟩, 869 ⟨84USP4465845A⟩, 979 ⟨91JHC485⟩; **5**, 210 ⟨73JOC3172⟩, 253 ⟨89CL1825⟩, 706 ⟨85GSM96⟩; **6**, 158 ⟨75BCJ1030⟩
Okamura, F., **2**, 348 ⟨90CL547⟩
Okamura, H., **4**, 959 ⟨78CL517, 80CL1209⟩
Okamura, K., **3**, 66 ⟨81JOC3175⟩
Okamura, W. H., **3**, 64 ⟨87TL2921⟩
Okanara, M., **2**, 156 ⟨80BCJ3615⟩
Okano, ,M., **2**, 281 ⟨80CC412⟩
Okano, H., **2**, 79 ⟨92JOC7003⟩
Okano, K., **4**, 358 ⟨85JOC5393⟩, 359 ⟨88OS(66)14⟩; **5**, 412 ⟨88OS(66)14⟩, 413, 425 ⟨85JOC5393⟩, 753 ⟨79TL2517⟩
Okano, M., **2**, 24 ⟨74BCJ143, 74BCJ3121⟩, 24 ⟨74BCJ692, 80BCJ1390⟩, 282 ⟨78JCS(P1)1273, 79JCS(P1)1206⟩, 282 ⟨80JOC1953, 81JOC3021⟩, 282 ⟨82TL2105, 84JOC3796, 85JCS(P1)373⟩, 284 ⟨81JOC4727⟩, 286 ⟨81CC546⟩, 597 ⟨76BCJ2762⟩, 617 ⟨77JCS(P1)676⟩, 617, 629 ⟨75CC925⟩, 624, 629 ⟨79CL1357⟩, 631 ⟨74BCJ147⟩, 633 ⟨83JOC3297⟩; **3**, 37, 174 ⟨84JOC3796⟩, 174 ⟨82TL2105⟩; **4**, 250 ⟨81TL3243⟩, 740 ⟨64MI 417-01, 64TL7, 66T2615⟩, 761, 771 ⟨76BCJ1931⟩, 1068 ⟨81JCS(P1)991⟩; **5**, 251 ⟨77BCJ220⟩, 282 ⟨81CC546⟩, 282, 300 ⟨81JOC4727⟩, 902 ⟨67T2137⟩, 1032, 1056 ⟨77BCJ722⟩, 1036 ⟨77BCJ2785, 78BCJ659⟩, 1042 ⟨75BCJ3205, 79BCJ3611⟩, 1053 ⟨78BCJ1818⟩, 1105 ⟨74BCJ2745⟩, 1113 ⟨80BCJ553⟩, 1113, 1115 ⟨75BCJ1925, 78JCS(P1)1273⟩; **6**, 95 ⟨77MIP53536⟩, 506 ⟨63BCJ1314⟩, 552 ⟨82BCJ339⟩, 552, 555 ⟨81TL5195⟩
Okano, S., **2**, 619 ⟨69JOC3434⟩
Okano, T., **2**, 434 ⟨82CL977, 80CL1423⟩
Okauchi, T., **2**, 394 ⟨91CL423⟩; **4**, 252 ⟨92CL1229⟩, 900 ⟨93MI 420-01⟩, 973, 980, 983, 984, 986 ⟨88EUP302389⟩
Okawa, K., **3**, 361 ⟨75JCS(P1)270⟩; **5**, 634 ⟨75JCS(P1)270⟩; **6**, 304 ⟨75JCS(P1)270⟩
Okawa, T., **6**, 465 ⟨92JAP04108765⟩
Okawara, M., **2**, 131 ⟨84TL2675⟩, 188 ⟨84CL2125⟩, 214 ⟨84JOC1434⟩, 256 ⟨81S547⟩, 287 ⟨77CL835⟩, 362 ⟨77TL2413⟩, 946 ⟨86S776⟩; **3**, 7 ⟨85JOC1107⟩, 57 ⟨74TL3625⟩, 118 ⟨76TL4597⟩, 119 ⟨83S815⟩, 634 ⟨76TL2967⟩, 636, 637 ⟨77CL835⟩; **4**, 187 ⟨86JCS(P1)1351⟩, 392 ⟨74TL3625⟩, 847 ⟨78CC74⟩, 850 ⟨76JA7440⟩; **5**, 560 ⟨77TL4607⟩; **6**, 298, 299, 300 ⟨84S837⟩, 309 ⟨72BCJ1797⟩, 314 ⟨70BCJ3528⟩, 730 ⟨74BCJ398⟩, 731 ⟨79S182⟩
Okawara, R., **2**, 486, 510 ⟨69MI 209-01⟩, 502 ⟨67MI 209-02⟩, 502, 889 ⟨68JOM(11)299⟩, 598 ⟨70JOM(25)353⟩, 695 ⟨67BCJ428⟩; **4**, 136 ⟨68MI 403-01⟩, 153 ⟨72JOM(43)117⟩, 220 ⟨80TL2555⟩, 595 ⟨71MI 413-01, 72JOM(43)333⟩, 659 ⟨78BCJ1397⟩, 955 ⟨75CC899⟩; **5**, 940 ⟨78JOM(159)417⟩, 1143 ⟨69MI 530-01⟩, 1144 ⟨68JOM(12)241, 70JOM(22)627⟩
Okawara, T., **2**, 321 ⟨90CC1385⟩, 363 ⟨76S339⟩, 725 ⟨82CL2007⟩; **5**, 351 ⟨88H(27)1881⟩, 353 ⟨87S183⟩, 630 ⟨85CPB3479⟩
Okawasa, M., **2**, 210 ⟨86TL215⟩
Okawo, M., **4**, 970 ⟨70MI 421-01⟩
Okazaki, H., **2**, 617 ⟨77JCS(P1)676⟩; **3**, 35 ⟨88JA5209⟩; **5**, 223 ⟨87CPB1741⟩; **6**, 447 ⟨86JAN230⟩
Okazaki, K., **5**, 301 ⟨87JA3402⟩
Okazaki, M., **4**, 67 ⟨93JOC2360⟩
Okazaki, M. E., **2**, 543 ⟨86TL4391⟩; **3**, 184 ⟨86TL4391⟩
Okazaki, R., **2**, 22 ⟨91BCJ3749⟩, 185 ⟨88CC40⟩, 547 ⟨92JA7598⟩, 793, 794, 795 ⟨89CL2087⟩, 794, 795 ⟨92JA1830⟩, 831 ⟨92JA4008⟩, 851 ⟨73BCJ898, 75BCJ546⟩, 854 ⟨71CC1186⟩, 949 ⟨91TL6877⟩; **3**, 318 ⟨82CC1055⟩, 331 ⟨88YGK1149⟩, 332 ⟨87JA279⟩, 337, 340 ⟨84TL849, 82CC1187⟩, 341 ⟨87BCJ1037⟩, 347, 363 ⟨79TL3673⟩, 347, 363 ⟨81BCJ3541⟩, 358 ⟨76TL4283⟩, 359 ⟨74CL1439, 76TL3815⟩, 359 ⟨79BCJ496, 82BCJ243⟩, 359, 361 ⟨79BCJ3640, 72CL9⟩, 361 ⟨75JCS(P1)270⟩, 383 ⟨86CC71, 89JA5949⟩, 392 ⟨83CC1429, 88BCJ861⟩, 400 ⟨93CL1047, 93JA7019⟩; **4**, 311 ⟨87BCJ3291, 87CL1575⟩, 334 ⟨93CL1047⟩, 376 ⟨89JA5949⟩, 608 ⟨87CL2293⟩, 653 ⟨91TL2049⟩, 979 ⟨79BCJ3640⟩, 1049 ⟨87JA279⟩; **5**, 246 ⟨86CL935⟩, 516, 523, 524 ⟨86BCJ2529, 86CC71⟩, 524 ⟨87BCJ1037⟩, 560 ⟨84CC1010⟩, 634 ⟨75JCS(P1)270, 76CC562, 78BCJ301⟩, 635 ⟨87H(25)183⟩, 1115 ⟨89JA5949⟩; **6**, 176 ⟨87JA279⟩, 304 ⟨75JCS(P1)270⟩, 314 ⟨78BCJ301⟩
Okazaki, S., **4**, 512 ⟨90S36⟩, 654 ⟨77IJ265⟩
Okazoe, T., **2**, 681 ⟨87JOC4410⟩, 700 ⟨88TL1065⟩; **3**, 161 ⟨87JOC4410⟩
Okecha, S. A., **3**, 638 ⟨77JCS(P1)1811⟩; **4**, 321 ⟨77JCS(P1)1811⟩; **6**, 113 ⟨83CI(L)205⟩

Okeda, Y., **4**, 850 ⟨85CC1803⟩
Oketani, K., **6**, 636 ⟨84CPB4893⟩
Okeya, S., **4**, 687 ⟨83JCS(D)1441⟩
Okhlobystina, L. V., **6**, 56 ⟨69IZV1188, 70IZV387⟩, 56, 282 ⟨71IZV1487⟩, 170 ⟨71IZV1505⟩, 287 ⟨67DOK(176)1086⟩, 469 ⟨82JGU131⟩
Okhlobystin, O. Yu., **2**, 434 ⟨58MI 208-01⟩, 598 ⟨64JOM(2)281⟩, 1096 ⟨65JOM(3)257⟩; **6**, 170 ⟨71IZV1505⟩, 368 ⟨67JOM(9)5⟩, 406 ⟨68BAU414⟩, 523 ⟨58IZV100⟩
Okhunia, T., **2**, 53 ⟨87JA5856⟩
Oki, M., **4**, 66 ⟨76BCJ553⟩, 68 ⟨70BCJ1223⟩; **6**, 48, 82 ⟨76BCJ553, 74BCJ2457⟩, 82 ⟨75BCJ2496⟩, 99 ⟨74BCJ1496⟩
Okide, G. B., **6**, 621 ⟨92JHC1551⟩
Okimoto, M., **5**, 698 ⟨93CL521⟩
Okina, E. A., **3**, 634 ⟨90SC1943⟩
Okinoshima, H., **2**, 915 ⟨75JOM(86)C27⟩
Okmura, T., **3**, 632 ⟨70JCS(C)2365⟩
Okogun, J. I., **4**, 1079 ⟨72JCS(P1)1103⟩
Okonkwo, G. C., **6**, 558, 564 ⟨93JMC1802⟩
Okonogi, T., **3**, 494 ⟨74TL2587⟩
Okoro, U. C., **3**, 357 ⟨91CRV363⟩
Okorodudu, A. O. M., **2**, 801 ⟨69JOC1220⟩; **3**, 489 ⟨68JA4189, 69JOC1220⟩
Okruszek, A., **2**, 441 ⟨71AG(E)494⟩, 452 ⟨70CC1495⟩
Oku, A., **2**, 41 ⟨89CC1429⟩; **3**, 538 ⟨77JA4554⟩; **4**, 17 ⟨83JA2771⟩, 758, 819, 821 ⟨92JOC5805⟩, 758, 821 ⟨93JOC4897⟩, 816 ⟨88JOC3089⟩
Oku, J., **3**, 628 ⟨79MAC1089, 81CC229⟩, 628 ⟨82MAC579⟩
Oku, M., **4**, 779 ⟨72JA1012, 74TL4157⟩
Oku, T., **4**, 679 ⟨92CC1031⟩
Okubo, M., **3**, 412 ⟨80BCJ281⟩
Okuda, H., **3**, 361, 362 ⟨91JHC1245⟩; **4**, 844 ⟨91JHC1245⟩; **5**, 559 ⟨79H(12)485⟩
Okuda, I., **5**, 1025 ⟨78YGK1090⟩
Okuda, J., **4**, 603 ⟨91CB1575⟩
Okuda, K., **4**, 365 ⟨84JCS(P1)435⟩
Okuda, T., **3**, 704 ⟨82ABC1803⟩
Okuda, Y., **2**, 456 ⟨85BJC2425⟩, 918 ⟨85TL4629⟩, 918, 923 ⟨86T4427⟩; **3**, 289 ⟨85BCJ2425⟩
Okude, Y., **2**, 73, 571 ⟨82BCJ561⟩
Okuede, Y., **2**, 571 ⟨77JA3179⟩
Okukada, N., **5**, 452 ⟨88CL1141⟩
Okukado, N., **2**, 160 ⟨84JOC2629⟩, 629, 975, 976 ⟨79S501⟩, 925, 969, 976 ⟨78JA2254⟩, 969 ⟨78TL1027, 78TL2357⟩, 992 ⟨87JA2393⟩, 1096 ⟨77CC683⟩; **3**, 293 ⟨88CL1141⟩; **5**, 173 ⟨78JOM(159)137, 79CL1021⟩
Okuma, K., **2**, 84, 445 ⟨88BCJ4476⟩, 360 ⟨78JOC652⟩, 363 ⟨94JA2390⟩, 853 ⟨87CL357, 88BCJ4323⟩, 858 ⟨87TL6649⟩; **3**, 336 ⟨88CL1145, 91BCJ2402⟩, 336 ⟨92RHA117⟩, 384, 394 ⟨87TL6649, 91CL1053⟩, 394 ⟨92TL1333⟩; **4**, 415 ⟨88CL1145⟩; **5**, 603 ⟨88CL1145⟩, 603, 622 ⟨91BCJ2402⟩, 603, 622, 627 ⟨92CL131⟩
Okumoto, H., **4**, 197 ⟨80CL617⟩
Okumura, F. S., **5**, 582 ⟨67BCJ2209⟩
Okumura, N., **4**, 437 ⟨64JOC1985⟩
Okumura, R., **6**, 471 ⟨93EUP561363⟩
Okumura, T., **4**, 321 ⟨86S375⟩
Okunaka, R., **5**, 223 ⟨90CPB1836⟩
Okuno, K., **4**, 212 ⟨54JA693⟩
Okuno, T., **4**, 828 ⟨93SL273⟩
Okushi, T., **6**, 471 ⟨90CL811⟩, 494 ⟨91TL6867⟩
Okutani, T., **6**, 558 ⟨89CPB1249⟩
Okuyama, S., **6**, 484 ⟨78TL3737⟩
Okuyama, T., **4**, 1074 ⟨82TL2665, 86BCJ453⟩, 1075 ⟨90BCJ3056, 91BCJ2751⟩; **6**, 96, 98 ⟨87JOC2657⟩
Olah, G. A., **2**, 6 ⟨73S786, 74S653⟩, 6, 11, 17, 392, 615 ⟨79JOC3872⟩, 6, 17 ⟨90SL267⟩, 6, 17, 18 ⟨93S693⟩, 11 ⟨74S652, 81HCA2528, 83HCA1028⟩, 12 ⟨74S896⟩, 13 ⟨74S560, 76JOC1627⟩, 13 ⟨B-64MI 201-01⟩, 15 ⟨83JOC3356⟩, 15, 17 ⟨86JFC(33)377, 73S779⟩, 18 ⟨73S780⟩, 19 ⟨74JA925, 83S713⟩, 22 ⟨84JOC2032⟩, 24, 30 ⟨77S676⟩, 32 ⟨80JOC1638⟩, 35 ⟨81JOC3727⟩, 36 ⟨79AG(E)612, 79JOC1247⟩, 36 ⟨83JOC2766⟩, 36, 142 ⟨83JOC3667⟩, 36, 538 ⟨82T2225, B-91MI 201-01⟩, 70 ⟨89JA393⟩, 103 ⟨B-89MI 202-05⟩, 104 ⟨78S452, 93S207⟩, 142 ⟨78JOC4503, 78S137⟩, 144 ⟨90JA3697⟩, 150 ⟨79JA5317⟩, 156 ⟨83JOC3585⟩, 384 ⟨92JOC1585⟩, 392 ⟨87JOC701⟩, 397 ⟨92SL337⟩, 408, 414 ⟨60CJC1428, 60CJC1431⟩, 411 ⟨59LA(625)88⟩, 452 ⟨83JOC1760⟩, 537 ⟨77S677⟩, 620 ⟨86S868⟩, 650 ⟨78JOC865, 79S536, 81JOC4305, 89JOC1204⟩, 650 ⟨91SL39⟩, 657, 753 ⟨76S815⟩, 790 ⟨79MI 215-02⟩, 791 ⟨82JOC596⟩, 907 ⟨90JOC46⟩, 941 ⟨89S568⟩, 1006 ⟨69JA2955, 70SCI1298, 72S307⟩, 1007 ⟨67JA2993, 68JA938⟩, 1007 ⟨69JA5801⟩, 1008 ⟨72S306⟩, 1009 ⟨68JA4666⟩; **3**, 3, 114 ⟨77S419⟩, 8, 121 ⟨77JOC3097, 78JA5396⟩, 8, 121, 294 ⟨80S897⟩, 14, 90, 101 ⟨84S228⟩, 18, 134 ⟨80S220⟩, 90, 101 ⟨81AG(E)878⟩, 91, 93, 94, 99, 103 ⟨B-64MI 303-01⟩, 93 ⟨60JA2380⟩, 93, 94 ⟨87CRV671⟩, 114 ⟨78S671⟩, 121 ⟨76S609⟩, 133 ⟨76S610, 93SL427⟩, 151, 153 ⟨79TL3653⟩, 152 ⟨84JOC2032⟩, 216 ⟨81S473⟩, 278, 282, 295, 296, 297, 299, 300, 301, 302, 305, 306, 309 ⟨B-64MI 306-01⟩, 279, 295, 296 ⟨78S672⟩, 282 ⟨63JA1328, 64JA2203, B-73MI 306-02⟩, 426 ⟨83JOC2766⟩, 617 ⟨79S112⟩, 618 ⟨80S657, 80S659⟩, 618, 662 ⟨79S227⟩, 620 ⟨79S36⟩, 632, 698 ⟨90JOC2016⟩, 707 ⟨90S646⟩; **4**, 10 ⟨74JA925⟩, 32 ⟨79JOC3842⟩, 41 ⟨B-64MI 402-01⟩, 42 ⟨75JA2293, 81JOC571⟩, 42, 44 ⟨71JA781⟩, 161, 212 ⟨86S513⟩, 191 ⟨81S282⟩, 203 ⟨82S962⟩, 216 ⟨90S104⟩, 315 ⟨83S500⟩, 428, 429 ⟨93JOC763⟩, 697, 698 ⟨88JA3231⟩, 1072 ⟨83CB3926, B-68MI 424-01⟩; **5**, 282 ⟨85OS(63)188⟩, 367, 992 ⟨83JOC3358, 90JOC4282⟩, 1045 ⟨90S646⟩; **5**, 4 ⟨79JOC3872⟩, 5 ⟨73S487, 74JA925⟩, 6 ⟨78JA5396⟩, 8 ⟨89S566⟩, 10 ⟨91S1177⟩, 36, 108 ⟨65JOC2386⟩, 39 ⟨83JOC3667⟩, 105 ⟨B-64MI 502-03⟩, 225 ⟨76AG(E)774⟩, 226 ⟨79JOC1247, 80S661, 81S790⟩, 270, 282, 300 ⟨79S274⟩, 282 ⟨79S537⟩; **6**, 410, 420, 422 ⟨56JOC1319⟩, 423 ⟨74S654⟩, 432, 432 ⟨65JOC1317⟩, 439 ⟨79JOC3872⟩, 442 ⟨B-64MI 614-01⟩, 726 ⟨89JA8020⟩, 729 ⟨93AG(E)767⟩; **6**, 45 ⟨89S463⟩, 52 ⟨74JA925⟩, 212 ⟨B-92MI 607-02⟩, 229 ⟨87JA3708⟩
Olah, G., **2**, 182 ⟨62JOC2667⟩, 608, 619 ⟨57JCS1823⟩; **3**, 495 ⟨90S885⟩, 618 ⟨78S702⟩; **5**, 4 ⟨56CB862⟩
Olah, J. A., **2**, 6, 11, 17, 392, 615 ⟨79JOC3872⟩, 13 ⟨74S560, 76JOC1627⟩; **3**, 279, 295, 296 ⟨78S672⟩; **4**, 32 ⟨79JOC3842⟩; **5**, 4 ⟨79JOC3726⟩, 36, 108 ⟨65JOC2386⟩, 105 ⟨B-64MI 502-03⟩
Olah, J. A., **6**, 229 ⟨87JA3708⟩
Olah, J. A., **6**, 439 ⟨79JOC3872⟩, 442 ⟨B-64MI 614-01⟩
Olah, V. A., **4**, 190 ⟨82S421⟩
Olander, C. P., **5**, 1023 ⟨22MI 527-01, 32OSC(1)447⟩
Olano, B., **2**, 322 ⟨89CC1132⟩; **3**, 188 ⟨83JOC2255⟩; **4**, 408 ⟨88CC410⟩
O'Laughlin, J. W., **4**, 554, 563 ⟨61JA1722⟩
Olbrysch, O., **4**, 698 ⟨73LA708⟩, 699 ⟨73LA715⟩
Olczak, R. T., **2**, 613 ⟨85JOC4576⟩
Ol'Dekop, Y. A., **2**, 594 ⟨80JOM(201)39⟩; **4**, 970 ⟨82ZOR1835⟩; **5**, 208 ⟨59DOK(128)1201⟩
Ol'dekop, Yu. A., **2**, 1098 ⟨85JGU1691⟩; **4**, 762 ⟨91ZOB56⟩, 797 ⟨91JOU48⟩; **5**, 1110 ⟨90MI 530-01⟩
Ol'dekop, Yu., **2**, 1098 ⟨86JGU1812⟩
Oldenberg, E. B., **2**, 123 ⟨55JA568⟩
Oldenziel, O. H., **2**, 982 ⟨71JA7331, 73JA1296⟩; **3**, 621 ⟨73TL1357, 77JOC3114⟩, 694 ⟨72SC281⟩, 696, 709 ⟨88OSC(6)987⟩; **4**, 331 ⟨88OSC(6)987⟩
Oldershaw, C. F., **2**, 103 ⟨32JA977⟩
Oleinik, A. V., **6**, 680, 682 ⟨89ZOB1902⟩
Oleinik, V. A., **2**, 407 ⟨89ZOB2136⟩; **4**, 559 ⟨88ZOB1665⟩; **6**, 242, 682 ⟨92ZOB948⟩, 681 ⟨88ZOB1923, 88ZOB482⟩, 696 ⟨87ZOB2141, 87ZOB2794, 88ZOB1461, 88ZOB2164, 89ZOB1904, 89ZOB959⟩
Olejniczak, B., **5**, 1031 ⟨89S300⟩

Oleksyszyn, J., **2**, 491, 505, 893 ⟨83OM1859⟩; **4**, 460 ⟨78C253⟩, 467 ⟨78PJC1949, 79S985⟩, 470 ⟨81TL3537⟩, 487 ⟨77S571⟩
Oles, S. R., **5**, 852 ⟨68JOC1679⟩
Olesen, S. O., **5**, 570, 571, 610 ⟨80NJC47⟩
Olfky, S. J., **5**, 276 ⟨66JOC3829⟩
Olfson, R. A., **6**, 461 ⟨90TL1405⟩
Olgemoller, B., **3**, 699 ⟨75AG(E)369⟩
Olgilvie, K. K., **2**, 467 ⟨86JOC3559⟩
Olijnsma, T., **4**, 331 ⟨72RTC209⟩
Oliva, A., **4**, 357 ⟨78JOM(146)C8⟩, 843 ⟨91S481⟩; **5**, 513 ⟨88SC323⟩
Olivares, A. M., **2**, 504 ⟨77SRI195⟩
Oliver, J. E., **2**, 10 ⟨75JOC2410⟩, 26 ⟨88OSC(6)424⟩, 26, 33 ⟨76JOC3279⟩, 191 ⟨75S321⟩; **6**, 647 ⟨72IJS(A)105⟩
Oliver, J. P., **4**, 698 ⟨71JA1035, 77AOC(15)235⟩, 698 ⟨82OM1307⟩
Oliver, M., **2**, 21 ⟨80TL177⟩
Oliver, W. L., **2**, 431 ⟨71T4245⟩
Oliveri, S., **5**, 769 ⟨89MI 519-01⟩
Olivero, A. G., **4**, 396 ⟨81HCA357⟩; **6**, 134 ⟨81HCA357⟩
Oliveros-Desherces, E., **3**, 406 ⟨74S812⟩
Oliveto, E. P., **2**, 24 ⟨59JA2191⟩; **4**, 178 ⟨54JA6113⟩; **5**, 105 ⟨62CHE179⟩, 318 ⟨63OSC(4)104⟩
Olivier, A., **5**, 1083 ⟨89TL2379⟩; **6**, 622 ⟨89TL2379⟩
Olivier, P. J., **6**, 723 ⟨94AG(E)672⟩
Olivier, S. C. J., **5**, 33, 108 ⟨27RTC600⟩
Olivucci, M., **4**, 610 ⟨93JA3322⟩
Olkhovick, V. K., **2**, 924, 925 ⟨90TL4937⟩; **4**, 809 ⟨90TL4937⟩
Oller, M., **3**, 213 ⟨80CB3783⟩
Ollinger, J., **3**, 35 ⟨74TL3955⟩
Ollis, W. D., **3**, 294 ⟨90SL345⟩, 492 ⟨B-79MI 313-02⟩, 704 ⟨80T515⟩; **5**, 326 ⟨90JCS(P1)1623⟩, 639 ⟨50JCS3389⟩, 641 ⟨68CC499, 74JCS(P1)627⟩; **6**, 611 ⟨79JCS(P1)724⟩
Ollmann, G., **2**, 1008 ⟨69AP(302)354⟩
Ollmann, R. J., **4**, 550 ⟨88CB2121⟩
Olmstead, H. D., **2**, 696, 699 ⟨69JOC2324⟩, 702 ⟨73JA3310⟩
Olmstead, M. M., **2**, 133 ⟨90JOC2286⟩, 212, 679 ⟨82TL323⟩, 1005 ⟨87JA2541, 91OM1282⟩; **4**, 565 ⟨85JA5272⟩, 656 ⟨90OM1720⟩, 691 ⟨81JA3764⟩; **6**, 194, 195 ⟨85JA4337⟩, 349 ⟨85CC1674⟩
Olmsted, A. W., **2**, 25 ⟨41OSC(1)144⟩
Olofson, R. A., **2**, 27 ⟨84JOC2795⟩, 303 ⟨84JOC2081⟩, 420 ⟨73JA581⟩, 675 ⟨78JOC2073⟩, 696 ⟨73JA582⟩; **3**, 432 ⟨79OPP249⟩, 577 ⟨61JA1007, 61JA1010⟩, 577 ⟨66T(S)415, 70T1779⟩; **4**, 58 ⟨81EUP40153, 83GEP3241568⟩, 58 ⟨84JOC2081, 90JOC2240⟩, 59 ⟨90JOC1847, 90TL1405⟩, 60 ⟨76TL3381⟩, 920, 921 ⟨90JOC5982⟩; **5**, 4 ⟨79JOC1016⟩, 626 ⟨70JOC806⟩, 855 ⟨72JOC3504⟩, 1066, 1073 ⟨70TL3453⟩, 1122 ⟨79TL3131⟩; **6**, 25 ⟨79TL1473⟩, 421 ⟨90JOC1851⟩, 423 ⟨90JOC1847⟩, 423, 438 ⟨79JOC1016⟩, 428 ⟨83GEP3241568, 84JOC2081⟩, 429 ⟨78JOC752, 90JOC2240⟩, 429 ⟨90JOC5982⟩, 462 ⟨80TL819, 90JOC1⟩
Olofson, R., **4**, 921 ⟨89FP2610926⟩
Olomucki, M., **5**, 101 ⟨58CR(246)1877, 60AC(P)845⟩, 103 ⟨87OS(65)47⟩
Olschwang, D., **6**, 95 ⟨68CR(C)1506⟩, 95, 98 ⟨69BSF325⟩, 95, 98 ⟨70MI 603-02⟩, 111 ⟨69BSF332, 71BSF3354⟩
Olsen, B. A., **2**, 172 ⟨65JOC933⟩
Olsen, C. E., **5**, 589 ⟨94JOC1257⟩
Olsen, D. K., **6**, 51 ⟨78JOC1346⟩
Olsen, G. E., **5**, 640 ⟨61ACS1109⟩
Olsen, H., **2**, 773 ⟨74JA5158⟩
Olsen, J. D., **4**, 656 ⟨93JA8493⟩
Olsen, L. D., **4**, 262 ⟨90JA3191, 90MI 406-01⟩, 262 ⟨92JA8772⟩
Olsen, R. J., **3**, 369 ⟨76JA1264⟩
Olsen, R. K., **3**, 221 ⟨91TL5721⟩; **4**, 309 ⟨75TL3579⟩; **5**, 237, 277 ⟨87JOC4531⟩, 315 ⟨77S457, 82JOC4605⟩, 1109 ⟨65JOC184⟩
Olsen, R. S., **5**, 70 ⟨85JOC4877⟩
Ol'shevskaya, V. A., **4**, 143 ⟨85IZV809⟩
Olson, A. R., **5**, 595 ⟨41JA314⟩
Olson, E. S., **3**, 468 ⟨77JHC1255⟩
Olson, G. L., **3**, 59 ⟨76S25⟩
Olson, J. T., **5**, 344 ⟨92SC2729⟩
Olson, R. E., **2**, 938, 939 ⟨83T949⟩; **3**, 256 ⟨81JA3112⟩; **4**, 197 ⟨81JA3112⟩; **5**, 410 ⟨79JA2225⟩, 423, 427 ⟨83T949⟩
Olson, R. K., **2**, 223 ⟨78JOC1270⟩
Olson, W. L., **4**, 148 ⟨86OM630⟩
Olsson, K., **6**, 85, 89 ⟨76CS122⟩
Oltmann, K., **4**, 400, 401 ⟨90TL7419⟩
Oltra, J. E., **2**, 617 ⟨91JHC939⟩
Oltvoort, J. J., **2**, 675 ⟨81S305⟩
Oluwadiya, J. O., **6**, 109 ⟨90JHC1953⟩
Oly, M., **4**, 109 ⟨89T2819⟩
Omae, I., **2**, 1100 ⟨89MI 223-02⟩
O'Mahony, M. J., **6**, 472, 473 ⟨85CC1326⟩
O'Mahony, R., **3**, 244 ⟨90JOC5894⟩
Omar, A. H. H., **6**, 28 ⟨89JCS(P1)283⟩
Omelanczuk, J., **2**, 437, 443 ⟨79T1531, 79JA7292⟩, 450, 842 ⟨75CC382⟩, 851 ⟨92CC1718⟩
Omichi, H., **5**, 864 ⟨86MI 521-02, 87MI 521-03⟩
Omietanski, G., **2**, 374 ⟨56JA3874⟩
Omietanski, G. M., **2**, 383 ⟨57CRV1021⟩, 384 ⟨56JA1211⟩
Omo, A., **3**, 129 ⟨86S570⟩
Omodei-Salé, A., **5**, 773 ⟨83JMC1187⟩
Omori, H., **3**, 279, 292 ⟨70MI 306-01⟩; **4**, 681 ⟨92OM989⟩
Omori, K., **4**, 828 ⟨93SL273⟩; **6**, 51 ⟨77JFC(9)279⟩
Omori, Y., **5**, 1111, 1116 ⟨87TL4189⟩
Omote, Y., **3**, 247 ⟨80S1013⟩, 357 ⟨84H(22)763⟩, 366 ⟨84JOC1837⟩; **4**, 899, 979 ⟨84JHC133⟩; **5**, 197 ⟨88JCS(P1)529⟩, 630 ⟨88JCS(P1)2203⟩; **6**, 114 ⟨86TL1335⟩, 575 ⟨82JCS(P1)2149⟩, 579 ⟨82H(19)2283⟩
Omoto, Y., **2**, 185 ⟨88SC2083⟩
Omura, H., **3**, 293 ⟨77S776⟩; **6**, 539 ⟨89JAP03020256⟩
Omura, K., **3**, 6, 118 ⟨78T1651⟩; **5**, 235 ⟨93JA5302⟩
Omura, S., **5**, 235 ⟨93JA5302⟩; **6**, 138 ⟨80CPB2987⟩
Omura, T., **2**, 960 ⟨75JA6461⟩
Omura, Y., **6**, 70 ⟨76JAP(K)76108012⟩
Ona, H., **3**, 357 ⟨91CPB672⟩
Onada, K., **2**, 1088 ⟨76TL3969⟩
On, H. P., **2**, 616 ⟨79JA5101⟩, 922 ⟨80S803⟩, 922, 931 ⟨81S999⟩; **4**, 756 ⟨82S127⟩, 758, 816 ⟨79JA5101⟩, 808, 811 ⟨81JOC1292⟩
Onaka, M., **2**, 82 ⟨91TL4741⟩; **3**, 140 ⟨79CL1483⟩, 289 ⟨81CL531⟩, 615 ⟨89CL1393⟩, 615, 627, 634 ⟨93BCJ2016⟩, 632 ⟨90CL481⟩
Onaka, N., **2**, 84 ⟨90TL7467⟩
Onak, T., **6**, 169 ⟨82COMC-I(1)411⟩, 183 ⟨74JCS(D)665⟩
Onak, T. P., **2**, 520 ⟨B-75MI 210-02⟩
Onami, T., **3**, 632 ⟨89JOC1295⟩
Onda, K., **3**, 153, 154 ⟨92BCJ2800⟩; **4**, 74, 77 ⟨92TL1483, 93TL2331⟩
Onda, K.-I., **3**, 158 ⟨91JOC4129⟩
Onda, M., **6**, 138 ⟨80CPB2987⟩
Ondrus, T. A., **5**, 1119 ⟨78CJC1026, 79JHC409⟩
O'Neal, H. R., **2**, 466 ⟨83IC814⟩
O'Neil, I. A., **2**, 943 ⟨88JOC1815⟩; **3**, 309 ⟨84JOC4409⟩, 696, 704 ⟨90TL2047⟩, 696, 713 ⟨90SL603⟩, 712, 718, 719 ⟨89SL9, 90TL2051⟩
O'Neil, R. M., **3**, 371 ⟨90JCS(P1)2881⟩
O'Neil, T. H., **2**, 693 ⟨83OS(61)116⟩
O'Neill, B. T., **3**, 284 ⟨91COS(1)397⟩
O'Neill, J. J., **2**, 412 ⟨51JA5202⟩
O'Neill, P., **3**, 530 ⟨93S606⟩, 558, 560 ⟨92JOC362⟩
Oney, I., **2**, 807 ⟨67JA6972⟩
Ong, B. S., **2**, 120 ⟨76TL319⟩, 931, 932 ⟨76JOM(107)C1⟩; **4**, 216 ⟨76TL319⟩
Ongania, K.-H., **5**, 586, 609 ⟨82M(113)593⟩

Ong, T.-S., **6**, 517 ⟨84JA1125⟩
Ongena, R., **3**, 206 ⟨81T2085⟩
Ongini, E., **5**, 22 ⟨93MI 517-02⟩
Ong-Lee, A., **3**, 307 ⟨83JHC1565⟩
Onischenko, A. A., **6**, 145 ⟨70IZV948⟩
Onishchenko, A. A., **4**, 995 ⟨75IZV2342⟩
Onishchenko, P. P., **4**, 539 ⟨70MI 411-01⟩
Onishchuk, N. P., **5**, 685 ⟨90JOU1408, 90ZOR1629⟩
Onishi, K., **6**, 4 ⟨92JAP04108765⟩
Onishi, M., **4**, 656 ⟨93BCJ1849⟩
Onishi, T., **3**, 57 ⟨86BCJ3287⟩, 217 ⟨80TL1347⟩
Onishi, Y., **2**, 148 ⟨66BCJ364⟩
Onisko, B. C., **3**, 240 ⟨88H(27)651⟩
Onistschenko, A., **2**, 319 ⟨87T565⟩; **5**, 345 ⟨86CB2678⟩
Onitsuka, K., **5**, 801 ⟨91JCS(D)1531, 92AG(E)851⟩
Onji, J., **2**, 222 ⟨91T149⟩
Onnis, V., **4**, 986 ⟨92JHC1631⟩
Ono, H., **6**, 110 ⟨71JA4075⟩
Ono, K., **2**, 855 ⟨74JAP7401534⟩; **3**, 30, 33 ⟨94T2821⟩, 614 ⟨26BSF1589⟩; **4**, 673 ⟨69APO(7)1⟩
Ono, M., **2**, 30 ⟨81JOC3312⟩; **3**, 59 ⟨81CPB766⟩
Ono, N., **2**, 393 ⟨90JOC2437⟩, 394 ⟨85BCJ1069, 87S258⟩, 395 ⟨84S226⟩, 396, 789 ⟨86JOC2139, 88S921⟩; **3**, 46, 190 ⟨87S532⟩, 57 ⟨82JOC5017⟩, 168 ⟨87CC947⟩, 190 ⟨86JOC2832⟩, 192 ⟨87S258⟩, 621 ⟨85CL635⟩, 635, 645 ⟨78TL763⟩, 644 ⟨79S890⟩; **4**, 220 ⟨78BCJ2401⟩, 249 ⟨80S952⟩, 328 ⟨74JA2580⟩; **5**, 132 ⟨78BCJ2401⟩
Ono, S., **5**, 774, 850 ⟨68TL3569⟩
Ono, S.-I., **6**, 561 ⟨87BCJ1807⟩
Ono, T., **2**, 311 ⟨92JOC1504⟩; **3**, 634 ⟨79H(12)681⟩, 634, 651 ⟨84JA7890⟩, 654 ⟨78CPB1874⟩; **5**, 334 ⟨85JMC1772⟩
Ono, Y., **2**, 108 ⟨93MI 202-01⟩; **5**, 252 ⟨87NKK1457⟩
Onodera, A., **2**, 234 ⟨89CL2111⟩
Onodera, J., **3**, 282 ⟨89BCJ3371⟩
Onodera, K., **5**, 183, 192 ⟨60JOC1265⟩
Onoe, A., **2**, 24 ⟨74BCJ143, 74BCJ3121⟩, 24 ⟨74BCJ692⟩, 617 ⟨77JCS(P1)676⟩, 617, 629 ⟨75CC925⟩, 631 ⟨74BCJ147⟩
Onomura, O., **4**, 201 ⟨87S1099⟩
Onopchenko, A., **2**, 753 ⟨79JOC1233, 79JOC3671⟩; **4**, 993 ⟨76MI 421-02⟩
Onoue, H., **3**, 357 ⟨91EUP409164⟩, 464 ⟨66CC730⟩; **5**, 295, 299 ⟨66CC17⟩
Onoyama, K., **6**, 563 ⟨91ANC1295⟩
Onozawa, T., **2**, 444 ⟨90JA5244⟩; **4**, 570, 575 ⟨90JA5244⟩
Onshiro, Y., **2**, 244 ⟨69T4283⟩
Ontiveros, C. D., **2**, 483 ⟨88IC4535⟩
Ontoria, J. M., **3**, 6, 118 ⟨91JOC5948⟩, 528 ⟨92TL4823⟩
Onuki, H., **3**, 701, 704 ⟨92TL6823⟩
Onyiriuka, S., **3**, 65, 232 ⟨74TL2255⟩
Onys'ko, P. P., **4**, 1008, 1009 ⟨86ZOB2681, 89ZOB2492⟩, 1009 ⟨90ZOB229⟩; **5**, 644 ⟨89ZOB1917, 90ZOB1389⟩, 652 ⟨90JGU1165⟩, 657 ⟨87JGU1104, 87ZOB1233⟩, 659 ⟨86ZOB2681⟩, 659, 788, 789 ⟨86JGU2372⟩, 660 ⟨87ZOB1514⟩, 660, 788 ⟨87JGU1349⟩, 713 ⟨89ZOB2492⟩, 713, 788 ⟨89JGU2228⟩, 738 ⟨90JGU1389⟩, 787 ⟨89JGU1496⟩, 788 ⟨85JGU402, 89JGU1129⟩, 789 ⟨88JGU30, 89JGU1123⟩
Onyszchuk, M., **5**, 388 ⟨92MI 507-01⟩, 389 ⟨92JOM(423)C5, 92PS(70)75⟩, 1016 ⟨62CJC739⟩
Oohara, T., **2**, 85 ⟨88TL2851⟩, 322 ⟨88TL4093⟩; **4**, 76 ⟨88TL313⟩
Ooi, H. C., **6**, 104 ⟨82JCS(P1)2871⟩
Ooi, I.-H., **4**, 767 ⟨89JOC1479⟩
Ooi, N. S., **3**, 433 ⟨80JCR(S)366⟩
Ooi, T., **3**, 9, 122 ⟨91SL491⟩, 34 ⟨91T5689⟩
O-Oka, M., **5**, 634 ⟨76CC562, 78BCJ301⟩
Ooka, M., **6**, 314 ⟨78BCJ301⟩
Ooms, P. H. J., **3**, 344, 345, 363 ⟨73S149⟩; **4**, 830 ⟨77JOC3128⟩; **5**, 512, 566, 568, 570, 608 ⟨73S149⟩; **6**, 75 ⟨76JCS(P1)1048, 77JOC3128⟩
Oon, S.-M., **4**, 138, 154 ⟨88TL5863⟩
Oosawa, Y., **4**, 552 ⟨75CL1259, 76JOM(122)113⟩
Oose, M., **3**, 120 ⟨91BCJ1046⟩
Oosthuizen, H. E., **3**, 694 ⟨83AOC(22)209⟩
Oota, S., **2**, 444 ⟨88JAP63119491⟩
Ooyama, J., **5**, 1138 ⟨76BCJ464⟩
Oozumi, Y., **2**, 537 ⟨91JA9887, 92TL7185⟩, 537 ⟨93TL2335⟩
Oparina, L. A., **3**, 361 ⟨80ZOR1842⟩
op den Brouw, P. M., **6**, 39 ⟨75S527⟩
Opeida, I. A., **5**, 28 ⟨79MI 502-03⟩
Opel, A., **2**, 999 ⟨93CB1873⟩
Openshaw, H. T., **2**, 412 ⟨45JCS660⟩; **5**, 975 ⟨45JCS660⟩
Opitz, A., **4**, 604, 647 ⟨91S1221⟩
Opitz, G., **2**, 204 ⟨67AG(E)107⟩, 757 ⟨63LA(665)101⟩, 817 ⟨64CB1155⟩; **3**, 269, 530 ⟨62AG32⟩, 369 ⟨67AG(E)107⟩; **4**, 282 ⟨66AG(E)594, 66TL5263⟩, 295 ⟨56CB81⟩, 331 ⟨90CB1563⟩, 416 ⟨B-66MI 409-01⟩, 979 ⟨65LA(684)103⟩; **5**, 472, 488, 495, 497, 500 ⟨67AG(E)107⟩, 489 ⟨90CB1563, 90CB1989⟩, 489, 490 ⟨86TL167⟩, 495 ⟨66TL5263⟩, 496, 497, 500 ⟨65LA(684)103⟩
Opitz, R., **2**, 572 ⟨74ZAAC(412)11⟩
Oplinger, J. A., **4**, 224 ⟨88JOC2953⟩
Oppelt, J. C., **2**, 783 ⟨66JCE144⟩
Oppenauer, R.,**3**, 7, 119 ⟨37RTC137⟩
Oppenhauer, R., **5**, 129 ⟨66M62⟩
Oppenheim, A., **3**, 546 ⟨1877CB701⟩
Oppenheim, A. P., **4**, 688 ⟨82JOM(224)377⟩
Oppenheim, W., **2**, 1017 ⟨64IJ121⟩
Oppermann, M., **3**, 255 ⟨85TL1627⟩; **4**, 835 ⟨85CC1154⟩
Oppitz, G., **3**, 269 ⟨63LA(662)178⟩
Oppoltzer, W., **4**, 649 ⟨80HCA555⟩
Oppolzer, W., **2**, 22 ⟨85TL5037⟩, 71 ⟨88TL5645⟩, 202 ⟨84TL5889⟩, 331 ⟨92JA5900⟩, 346 ⟨94JA3139⟩, 346, 349 ⟨79TL4391⟩, 354 ⟨77AG(E)10⟩, 532 ⟨90JA2767⟩, 558, 559 ⟨84TL3305⟩, 560 ⟨82TL3901⟩, 562 ⟨89AG(E)38⟩, 919 ⟨92HCA170⟩; **3**, 189 ⟨92JA5900⟩, 373 ⟨67TL2325⟩, 528 ⟨86TL5471⟩; **4**, 107 ⟨92JA5900⟩; **5**, 63 ⟨85HCA216⟩, 152 ⟨81HCA2808⟩, 288 ⟨75HCA587⟩, 342 ⟨87HCA1666, 89TL6009⟩, 342 ⟨93HCA957, 94HCA554⟩, 347 ⟨90TL4117⟩; **6**, 461 ⟨91HCA465⟩
Oppong, I. V., **3**, 161 ⟨82CJC94⟩
Orabi, M. O. A., **3**, 80 ⟨81JCS(P1)1437⟩
Orahovatz, A., **3**, 365 ⟨85JOC1550⟩
Orama, M., **5**, 864 ⟨83ACS(A)631⟩
Orama, O., **4**, 563 ⟨77CB1748⟩; **6**, 375 ⟨86ZN(B)191⟩
Oram, D. E., **4**, 565 ⟨85JA5272⟩
Orange, C., **5**, 132 ⟨85S40⟩
Oravec, P., **3**, 77 ⟨86CCC2167⟩; **5**, 350 ⟨86CCC2167⟩
Orazi, O. O., **4**, 417 ⟨64TL1693⟩, 428 ⟨75JCS(P1)772⟩; **5**, 1126 ⟨85JHC1025⟩; **6**, 221 ⟨52MI 607-02⟩
Orazov, O. G., **4**, 775 ⟨90CA230883w⟩
Orazsakhatov, B., **6**, 341 ⟨90JOM(384)279⟩
Orban, J., **2**, 699 ⟨84TL5099⟩
Orbe, M., **5**, 132 ⟨87JOC3777⟩
Orbović, N., **3**, 464 ⟨70T1101⟩
Orchin, M., **2**, 735 ⟨62JOC3401⟩; **4**, 148, 150, 581 ⟨90JOM(397)313⟩; **5**, 25, 108 ⟨51JA436⟩
Orda, V. V., **5**, 204 ⟨77S574⟩, 205 ⟨88ZOR593⟩; **6**, 14 ⟨93RHA232⟩, 61 ⟨68ZOB2813⟩, 229 ⟨64ZOB1979⟩, 233, 234, 237 ⟨65ZOB1628⟩
Ordonez, M., **6**, 119 ⟨89JPO349⟩
Oreana, M., **3**, 4, 115 ⟨76JA6737⟩
O'Regan, M., **3**, 513 ⟨90JA3875⟩
O'Regan, M. B., **3**, 514 ⟨90JA8378⟩
O'Reilly, B. C., **6**, 647 ⟨89TL7313⟩
O'Reilly, N. J., **5**, 681 ⟨83JFC(23)293, 93JFC(65)157⟩, 681, 703 ⟨93JFC(65)101⟩; **6**, 63 ⟨85TL5243⟩
Oremek, G., **6**, 540 ⟨80CB750⟩, 620, 621 ⟨80CB2583⟩
Oremus, V., **6**, 305, 309, 310 ⟨91HCA1500⟩
Orena, M., **2**, 87 ⟨81S793⟩; **3**, 4 ⟨76S394⟩, 57, 89 ⟨76CC190, 76TL3985⟩, 158 ⟨84JOC701⟩; **5**, 1125 ⟨89TL1679⟩; **6**, 469 ⟨82JOC4626⟩

Oren, J., **5**, 199 ⟨87S850⟩
Orere, D. M., **3**, 18, 133 ⟨78S919⟩, 621 ⟨77CC280⟩
Oreshko, G. V., **4**, 108 ⟨89IZV1107⟩, 795 ⟨89BAU635⟩; **6**, 55 ⟨68IZV2307, 73IZV1424⟩, 250, 252 ⟨89IZV113⟩
Organ, T. D., **3**, 426 ⟨70JCS(C)1182⟩
Orghedani, R., **3**, 280 ⟨91TL6709⟩
Orgura, K., **2**, 1000 ⟨82CL1697⟩
Ori, K., **3**, 29, 157 ⟨88BCJ3637⟩
Oribe, T.,**4**, 831 ⟨74CPB2767⟩
Orita, A., **5**, 438 ⟨90JA7061⟩
Orita, H., **3**, 87 ⟨90TL2607⟩
Orlandini, A., **6**, 325 ⟨80AG1055, 81JOM(218)81⟩, 327 ⟨82OM778⟩, 332 ⟨83JOM(246)C13, 84JOM(270)251⟩
Orlando Jr., C. M., **5**, 314 ⟨69JOC1154⟩
Orlinski, R., **5**, 152 ⟨83JA7200⟩
Orlova, E. V., **5**, 854 ⟨69ZOR1411⟩
Orlov, A. M., **2**, 167 ⟨66DOK(170)96⟩
Orlova, N. A., **4**, 98, 99 ⟨91JGU1875⟩, 418 ⟨84JGU635⟩; **5**, 374 ⟨88ZOB1093⟩
Orlova, T. I., **3**, 434 ⟨84ZOR988⟩
Orlova, T. N., **4**, 979 ⟨91ZOR382⟩
Orlova, T. Y., **5**, 97 ⟨84JOM(267)309⟩
Orlov, B. A., **2**, 460, 851 ⟨62ZOB4017⟩
Orlov, G. I., **5**, 719 ⟨83JGU1513, 83ZOB1683⟩, 978 ⟨81ZOB1814⟩, 983 ⟨80ZOB960, 81ZOB2017⟩
Orlov, N. F., **4**, 570 ⟨71MI 412-01⟩; **5**, 404 ⟨68JGU1791⟩
Orlov, R. I., **5**, 1001 ⟨84ZOB876⟩
Ormerod, E., **6**, 653 ⟨26LA(446)260⟩
Ormerod, J., **5**, 142 ⟨73JCS(P1)2796⟩
Ormiston, R. A., **3**, 49, 196, 492 ⟨87CSR45⟩; **6**, 707 ⟨87CSR45⟩
Ornaf, R. M., **3**, 165 ⟨80TL4301⟩; **5**, 285, 377 ⟨85TL1803⟩
Ornfelt, J., **2**, 126, 223 ⟨49JA2270⟩; **4**, 216 ⟨52JA1068⟩, 223 ⟨51JA822⟩
Ornstein, P. L., **2**, 36 ⟨77TL2659⟩; **3**, 9, 123 ⟨77TL4175⟩, 32 ⟨92JOC1067⟩
Oro, L. A., **4**, 685 ⟨90JCS(D)2587⟩
Orochena, S. F., **4**, 25 ⟨59JA992⟩
O'Rourke, D., **4**, 60 ⟨93SL195⟩
Orpen, A. G., **2**, 570 ⟨90OM2207⟩; **4**, 151 ⟨92OM3879⟩, 545 ⟨90OM1222⟩, 675 ⟨86CC542⟩, 676 ⟨84JCS(D)2455, 86JCS(D)657⟩, 676 ⟨88JCS(D)1843⟩, 679 ⟨83JCS(D)2661⟩, 681 ⟨90JCS(D)761, 92CC310⟩, 689, 710 ⟨85JCS(D)2483⟩, 1023, 1032 ⟨88CC673⟩; **6**, 343 ⟨83JOM(249)273⟩
Orphanides, G. G., **3**, 65, 232 ⟨78JA7927⟩
Orr, D. E., **2**, 746 ⟨84S618⟩
Orr, G., **3**, 539 ⟨73JA6134⟩, 542 ⟨73JA4061⟩
Orr, J. C., **5**, 285 ⟨72CAR(23)251⟩
Orr, R., **2**, 30 ⟨78JOC2793⟩
Orrell, K. G., **4**, 283 ⟨87JCS(D)757⟩
Ortaggi, G., **2**, 26 ⟨78TL4575⟩; **3**, 206 ⟨77S773⟩
Ort, M. R., **4**, 698 ⟨73JOM(50)47⟩
Ort, O., **2**, 326 ⟨88S985⟩
Ortar, G., **5**, 296, 331 ⟨85TL1109⟩, 301 ⟨86TL3931⟩
Ortega, C., **2**, 7 ⟨82JCS(P1)681⟩
Ortega, T., **6**, 559 ⟨89JOC2978⟩
Orten, J. M., **5**, 89 ⟨32OSC(1)300⟩
Orthner, L., **4**, 555 ⟨60USP2937207⟩
Ortiz, E., **4**, 652 ⟨88T3781⟩
Ortiz, F. L., **4**, 421 ⟨91CC1704⟩
Ortiz, M. J., **3**, 414 ⟨86JCS(P1)2021⟩, 418 ⟨87OPP181⟩
Ortiz de Montellano, P. R., **2**, 578 ⟨B-86MI 211-02⟩; **4**, 734 ⟨83JOC4432⟩
Orwat, M. J., **5**, 223 ⟨89TL5821⟩
Orwig, B. A., **2**, 217 ⟨70TL3551⟩
Orwoll, E. F., **2**, 647 ⟨55OSC(3)130⟩
Orzech, C. E., **2**, 1016, 1033 ⟨63JA1648⟩, 1062 ⟨62JA2020⟩
Osa, T., **3**, 617 ⟨66BCJ854⟩
Osada, A., **2**, 104 ⟨82JCS(P1)1963⟩
Osada, M., **5**, 693 ⟨84CPB4694⟩
Osaka, K., **3**, 60 ⟨83TL1171⟩
Osaka, M., **3**, 616 ⟨90CL1171⟩
Osaka, N., **3**, 663 ⟨88T2413⟩
Osakada, K., **6**, 470 ⟨92OM171⟩
Osaki, H., **6**, 542 ⟨91CPB1659, 92T9433⟩, 542 ⟨92TL2543⟩
Osaki, M., **4**, 193 ⟨76CL769⟩
Osaki, T., **5**, 402 ⟨73BCJ1803⟩
Osanai, K., **5**, 155 ⟨89TL4845⟩
Osanova, N. A., **2**, 832 ⟨69IZV2234⟩, 879 ⟨60JGU3203⟩
Osarych, Z., **3**, 72 ⟨80TL3959⟩
Osato, R. L., **5**, 193 ⟨51JA5553⟩
Osawa, A., **6**, 434 ⟨73CL1315⟩
Osawa, K., **5**, 1065 ⟨79BCJ533⟩
Osawa, M., **3**, 628 ⟨90CL1615⟩; **4**, 654, 655 ⟨92CL2047⟩
Osawa, N., **5**, 681 ⟨83BCJ545⟩
Osawa, T., **2**, 52 ⟨82CL315, 91CC795⟩
Osborn, H. M. I., **2**, 319 ⟨93SL675⟩
Osborn, J. A., **2**, 681 ⟨86CC531⟩; **3**, 514 ⟨87CC129, 89CC1062⟩, 515 ⟨82CC514, 85CC793⟩, 515, 518 ⟨87JA3953⟩; **5**, 932 ⟨87JA3953⟩; **6**, 155 ⟨82OM1114⟩, 457 ⟨68IS99⟩
Osborn, M. E., **3**, 621 ⟨80JOC167⟩
Osborn, R. B. L., **3**, 552 ⟨70JCS(A)944⟩
Osborne, A. G., **4**, 685, 706, 719 ⟨79JOM(181)425⟩
Osborne, B. P., **2**, 421 ⟨62JCS1330⟩
Osborne, C. E., **2**, 168 ⟨53JA4582⟩, 206 ⟨59JA2002⟩; **6**, 255 ⟨53JA4582⟩
Osborne, D. W., **4**, 31, 34, 44 ⟨63JOC494⟩
Osborne, J. A., **2**, 478 ⟨67JCS(A)1347⟩
Osborne, J. E., **4**, 5 ⟨56JCS61⟩
Osborne, S., **3**, 164, 173 ⟨91SL87⟩
Osburn, S. W., **2**, 123 ⟨57JOC561⟩
Osei, R. D., **2**, 439 ⟨83JCS(P1)861⟩; **4**, 566 ⟨83JCS(P1)861⟩
Osei-Twum, E. Y., **2**, 373, 376, 403 ⟨84JOC336⟩
Osella, D., **4**, 682 ⟨89JOM(365)163⟩, 718 ⟨83POL77, 93OM4545⟩
Osgood, E. R., **2**, 778 ⟨80TL2025⟩
O'Shaughnessy, B., **2**, 4 ⟨93S735⟩
O'Shea, P. D., **2**, 632 ⟨87JCS(P1)1039⟩
Oshida, J., **2**, 19 ⟨82JCS(P1)85⟩
Oshika, T., **2**, 444 ⟨91TL3371⟩
Oshikawa, T., **2**, 6 ⟨84BCJ2675⟩, 429 ⟨83BCJ219⟩, 438 ⟨85S291⟩, 459 ⟨90BCJ2728⟩; **4**, 122 ⟨92CL407⟩, 924 ⟨86BCJ3293⟩; **5**, 403, 404 ⟨90BCJ2728⟩; **6**, 119 ⟨86BCJ3293⟩
Oshiki, T., **2**, 444 ⟨90JA5244⟩; **4**, 570, 575 ⟨90JA5244⟩
Oshima, K., **2**, 420 ⟨75TL1859⟩, 624, 630 ⟨85CL53⟩, 624, 946, 948 ⟨84TL3221⟩, 681 ⟨87JOC4410⟩, 699 ⟨79CL379⟩, 710 ⟨87CL1647⟩, 903, 919 ⟨86TL2007⟩, 918 ⟨85T3257, 85TL4629, 86CB1581⟩, 918, 919, 939 ⟨88T4277⟩, 918, 923 ⟨86T4427⟩, 918, 931 ⟨83JA4491⟩, 920 ⟨91BCJ1461⟩, 922, 931 ⟨82TL3387, 87JA2547⟩, 927 ⟨90TL365, 90TL369, 92JOC1615⟩, 929 ⟨84TL1163⟩, 930, 946 ⟨88BCJ2693, 87BCJ3468⟩, 946 ⟨87TL3709⟩, 947, 977 ⟨84TL3217, 84TL2151⟩, 977 ⟨84TL2221⟩, 997, 1002 ⟨85JOM(285)53⟩, 1098 ⟨84BJC108⟩, ⟨85CB97⟩ ⟨85BJC2425⟩; **3**, 5 ⟨81TL1605⟩, 5, 117 ⟨76TL2503, 80TL4843, 88BCJ3607⟩, 6, 118 ⟨93CL571⟩, 12, 128 ⟨91BCJ2402⟩; **4**, 141, 811, 812 ⟨84TL3221⟩, 188 ⟨89BCJ2050⟩, 393 ⟨79TL1783⟩, 639 ⟨88BCJ2693⟩, 648 ⟨74JA1620⟩, 737, 743 ⟨91CL1319, 92BCJ1513⟩, 808, 811, 813 ⟨89TL3155⟩, 811 ⟨92TL7031⟩, 873 ⟨86CL1861⟩, 944 ⟨91TL2783⟩, 956, 957 ⟨73JA2694⟩, 1045 ⟨88T4277⟩; **5**, 552 ⟨73JA5803⟩, 1043 ⟨84TL4683⟩; **6**, 21 ⟨89TL3159⟩, 60, 62 ⟨93T8487⟩, 550 ⟨88TL6125⟩
Oshima, T., **3**, 572 ⟨89H(29)2069⟩; **4**, 198 ⟨80TL3919⟩; **5**, 248, 249 ⟨87NKK1442⟩
Oshimi, M., **6**, 558 ⟨93TL3745⟩
Oshino, H., **2**, 69 ⟨87JA8056⟩
Oshio, A., **2**, 69 ⟨93CC811⟩
Oshio, H., **6**, 443 ⟨88MI 614-01⟩
Oshio, I., **2**, 662 ⟨89CPB1434⟩
Oshry, L., **5**, 62 ⟨82OPP249⟩

Osina, O. M., **6**, 12 ⟨69IZV196⟩
Osipova, M. P., **2**, 889 ⟨67JGU1578⟩; **4**, 485 ⟨82JGU392⟩
Osipova, T. I., **4**, 464 ⟨78IZV1951⟩, 472 ⟨79IZV1110, 79IZV2118⟩
Osipov, O. A., **2**, 834, 835, 840 ⟨78ZOB232⟩, 841 ⟨76ZOB1701⟩
Osipov, V. N., **4**, 346 ⟨92ZOB1268⟩
Osman, A., **5**, 676 ⟨82GSM100⟩
Osowska, K., **2**, 808 ⟨76AG(E)302⟩
Osowska-Pacewicka, K., **2**, 307 ⟨90S505⟩, 308 ⟨92SC1929⟩, 317 ⟨87S487⟩, 373 ⟨87S485⟩
Ospenson, J. N., **6**, 236 ⟨58USP2821554⟩
Ospica, A., **2**, 294 ⟨72CB454⟩
Ospici, A., **2**, 1083, 1094 ⟨75ZAAC(418)273⟩; **5**, 1136 ⟨75ZAAC(418)273⟩
Ossig, G., **6**, 196 ⟨93CB2247⟩
Ossola, F., **3**, 442 ⟨87ICA275⟩; **6**, 519 ⟨87JCS(D)2039⟩
Osswald, G., **3**, 614 ⟨1899CB643⟩; **4**, 880, 883 ⟨1899CB641⟩
Ostanina, L. P., **4**, 340 ⟨71ZOB2575, 72ZOB1647⟩
Oster, B. W., **2**, 988 ⟨82JOM(234)C35, 82S324⟩; **5**, 101 ⟨84JOM(266)313⟩, 386 ⟨83JOM(252)359⟩
Osterberg, A. C., **6**, 508 ⟨66JMC852⟩
Osterburg, J., **3**, 452 ⟨23CB1136, 24CB1108⟩
Ostercamp, D. L., **3**, 406 ⟨87T5171⟩
Östermann, T., **2**, 895 ⟨67JOM(8)105⟩
Ostermeier, J., **5**, 952 ⟨95CB289⟩
Osterroht, C., **2**, 506, 852 ⟨68CB1381⟩; **5**, 612 ⟨65CB3819⟩
Ostertag, R., **6**, 213 ⟨B-59MI 607-01⟩
Östman, B., **2**, 104 ⟨67ACS1257⟩
Ostoja Starzewski, K. A., **4**, 578 ⟨B-93MI 412-01⟩, 580 ⟨72JOM(42)C35⟩, 583 ⟨92OM2701⟩
Ostrander, M., **6**, 489 ⟨93MI 615-03⟩
Ostrikova, V. N., **4**, 717 ⟨80JOM(201)269⟩
Ostrogovich, G., **5**, 1126 ⟨76JCS(P2)524⟩
Ostroumova, N. G., **5**, 827 ⟨67ZOR1207⟩
Ostroumov, I. G., **5**, 691 ⟨88JOU1049, 88ZOR1165⟩
Ostrovskii, V. A., **2**, 319 ⟨78RCR1084⟩
Ostrowski, M., **6**, 349 ⟨93CB1355⟩, 355, 356 ⟨93ZAAC(619)1693⟩
Ostrowski, S., **3**, 702 ⟨93S1215⟩
Ostrum, G. K., **2**, 28 ⟨64JOC3459⟩
Ostveen, E. A., **4**, 893 ⟨77RTC68⟩
Ostwald, R., **2**, 69, 72, 591 ⟨93TL3115⟩
Osuch, C., **2**, 726 ⟨64JOC2592⟩; **3**, 498 ⟨64JOC3721⟩; **5**, 996 ⟨64JOC2592⟩
Osuga, D. T., **2**, 12 ⟨73OSC(5)126⟩
Osuka, A., **2**, 14 ⟨84BCJ303⟩, 142 ⟨80CL143⟩, 708, 723 ⟨84CL139⟩, 731 ⟨81CL1117⟩, 1001 ⟨83TL5109⟩; **3**, 625 ⟨83CL193⟩
O'Sullivan, A. C., **2**, 139 ⟨91JOC6697⟩, 240, 360 ⟨91JOC6702⟩; **4**, 285 ⟨84JCS(P1)21⟩; **5**, 735 ⟨91JOC6697⟩
O'Sullivan, D. A., **5**, 857 ⟨53JA5309⟩
O'Sullivan, M. B., **3**, 194 ⟨93CB2151⟩
Oswald, A. A., **2**, 105 ⟨59CJC1498⟩, 137 ⟨63JOC1952⟩, 213 ⟨B-66MI 203-01⟩, 408, 414 ⟨60CJC1428⟩, 408, 414 ⟨60CJC1431⟩
Oswald, A., **2**, 411 ⟨59LA(625)88⟩
Oswiecimska, M., **4**, 559 ⟨76JPR403⟩
Oszczapowicz, J., **5**, 864 ⟨B-91MI 521-02⟩
Ota, H., **4**, 808 ⟨93CC1817⟩
Ota, K., **3**, 171 ⟨81S661⟩
Otaka, A., **2**, 227 ⟨91SL345⟩; **4**, 218 ⟨89CPB526⟩
Otaka, T., **4**, 731 ⟨76BCJ2491⟩
Otake, K., **2**, 74, 80 ⟨88TL3563, 88CC44⟩
Otani, H., **3**, 352 ⟨87TL203⟩; **4**, 820 ⟨86CC1794⟩
Otani, N., **5**, 1031 ⟨78NKK265⟩, 1078 ⟨76CL811, 77S889⟩
Otani, S., **2**, 301 ⟨91CC306⟩; **4**, 410 ⟨84H(22)1417⟩
Otera, J., **2**, 215 ⟨90TL3591⟩; **3**, 35 ⟨88JA5209⟩, 55 ⟨89JOC2779⟩, 60 ⟨83TL1171⟩, 78, 271 ⟨85TL2677⟩, 159 ⟨90JA901⟩, 213 ⟨84CL1259⟩; **4**, 197 ⟨83TL4993⟩, 197 ⟨88S95⟩, 216 ⟨89T1209⟩, 248 ⟨88TL3971⟩, 704 ⟨92T1449⟩; **5**, 251 ⟨90TL3591⟩, 402 ⟨73BCJ1803⟩, 411 ⟨85TL2675⟩; **6**, 126 ⟨85TL2675, 87TL2147, 89JOC5003⟩
Otero, A., **2**, 881 ⟨78JOM(149)315, 78JOM(154)13⟩
Otero, J., **4**, 225, 238 ⟨92TL239⟩
Oth, J. F. M., **2**, 1014 ⟨64AG(E)747, 64AG(E)755⟩
Othen, D. G., **2**, 815 ⟨76JCS(D)1⟩; **3**, 457 ⟨76JCS(D)1⟩
Othler, E., **3**, 412 ⟨79CB107⟩
Otomatsu, T., **5**, 279 ⟨88CC419⟩
Otonari, Y. A., **5**, 994 ⟨93SC335⟩
Otsubo, T., **2**, 289 ⟨80SC595⟩, 293 ⟨81CL447⟩, 753 ⟨84CL853⟩; **3**, 39, 176 ⟨89CL607⟩, 114 ⟨86TL6099⟩; **4**, 238 ⟨91CL415⟩, 402 ⟨81CL447⟩, 850 ⟨88CC822⟩; **5**, 210 ⟨87TL1281, 89JOC4391⟩, 242 ⟨86CL977⟩, 1078 ⟨86TL6099⟩, 1113 ⟨80SC595⟩, 1117 ⟨82SC131⟩
Otsuji, Y., **2**, 30 ⟨77CL1117⟩, 133 ⟨91CC90⟩, 677 ⟨79S688⟩; **3**, 137 ⟨79CL321⟩
Otsuka, K., **6**, 469 ⟨94CL495⟩
Otsuka, M., **3**, 40 ⟨82TL525⟩
Otsuka, S., **2**, 442 ⟨71JA4301, 77JA7876⟩, 816 ⟨90JA4897⟩, 970 ⟨76JMOC285⟩; **3**, 698 ⟨66JOC4170⟩; **5**, 463, 800, 801 ⟨69JA7196⟩, 801, 802 ⟨76JCS(D)327⟩
Otsuka, T., **3**, 613 ⟨58JOC797⟩; **5**, 357 ⟨87H(26)2081, 90TL7633⟩
Otsuki, T., **2**, 459 ⟨81S811⟩
Otsuto, T., **3**, 89 ⟨84CL891⟩
Ott, D. G., **2**, 873 ⟨B-86MI 217-01⟩
Ott, E., **2**, 1014 ⟨30CB1941, 31CB1324⟩, 1014 ⟨43CB80⟩; **3**, 534, 541 ⟨08CB2208⟩; **4**, 797 ⟨43CB88⟩
Ott, H., **5**, 262 ⟨55HCA312⟩; **6**, 104 ⟨63HCA2306, 63T1675⟩
Ott, K. C., **2**, 979 ⟨82JA7491⟩; **3**, 508 ⟨84PAC59⟩; **4**, 670 ⟨81JA5922⟩
Ott, R. A., **3**, 13, 86 ⟨81JOC602⟩
Ott, W., **3**, 366 ⟨84AG(E)800⟩; **5**, 556 ⟨77LA1751⟩
Otte, R., **6**, 540 ⟨92POL893⟩, 723 ⟨90CC1722⟩
Otten, C., **4**, 203 ⟨70HCA1330⟩
Ottenheijm, H. C. J., **2**, 199 ⟨81JOC5408⟩, 340 ⟨86CRV697⟩; **3**, 433 ⟨86JOC309⟩, 634 ⟨81JOC5408⟩; **4**, 264 ⟨81JOC3273, 92RTC163⟩, 304 ⟨80JOC1880⟩; **5**, 14 ⟨83OS(61)1⟩, 84 ⟨86CRV697⟩
Ottenheym, J. H., **6**, 173 ⟨61GEP1157210⟩
Otter, J. C., **5**, 375, 376 ⟨90OM1557⟩
Otter, P., **5**, 142 ⟨73JCS(P1)2796⟩
Otterbacher, T. J., **5**, 30 ⟨32OSC(1)290⟩
Ottersen, T., **3**, 247 ⟨79ACS(B)669⟩
Ott-Lastic, J., **5**, 241 ⟨89JOM(377)281⟩
Ottlinger, R., **3**, 680 ⟨80CB2699⟩; **4**, 493 ⟨78S44⟩; **5**, 634 ⟨77CB2114⟩
Ottman, G., **6**, 494 ⟨66AG210⟩
Ottmann, G., **5**, 1033 ⟨69AG(E)760⟩, 1037 ⟨66AG(E)725⟩
Ottmeyer, W., **2**, 1014 ⟨30CB1941⟩
Otto, A., **3**, 263 ⟨91S115⟩
Otto, C., **4**, 603 ⟨84CB1885⟩
Otto, H., **6**, 335 ⟨86JOM(311)63⟩
Otto, H.-H., **2**, 248 ⟨82TL5389⟩; **4**, 321 ⟨67AP(300)647⟩; **5**, 374 ⟨86AP(319)203⟩, 384 ⟨86AP(319)216⟩; **6**, 129 ⟨89LA975⟩
Otto, H. W., **4**, 838 ⟨89AP(322)263⟩
Otto, J. A., **4**, 775 ⟨60USP2954316⟩
Otto, J., **4**, 708 ⟨65JOM(4)114⟩
Otto, R., **2**, 723 ⟨1888JPR(2)207⟩
Otto, W. H., **4**, 355 ⟨81AG(E)581⟩; **6**, 125 ⟨81AG(E)581⟩
Ottolenghi, A., **2**, 910, 931 ⟨63CJC2977⟩; **4**, 807 ⟨63CJC2977⟩
Ottomann, G., **6**, 448 ⟨66AG(E)672⟩
Ottow, E., **6**, 469 ⟨84JA5304⟩
Ou, K., **2**, 753 ⟨84TL839⟩
Ouchi, T., **5**, 541 ⟨79NKK1224⟩
Ouchida, S., **5**, 173 ⟨79CJC2180⟩, 173 ⟨79CL1021⟩
Oudenes, J., **2**, 533, 697 ⟨87JOC1347⟩; **3**, 207 ⟨80SC667⟩
Ouellette, T. J., **3**, 456 ⟨71MI 312-02⟩, 456, 458, 459 ⟨72IC242⟩

Ouertani, M., **2**, 27 ⟨82BSF(2)327⟩
Ours, C. W., **3**, 625 ⟨73T1931⟩; **6**, 642 ⟨68JMC1129⟩
Oussaid, B., **4**, 463 ⟨94S51⟩
Ousset, J. B., **2**, 446 ⟨84TL5903⟩; **3**, 48, 57 ⟨89TL179⟩
Out, G. J. J., **2**, 324 ⟨88TL3579⟩
Outurquin, F., **4**, 89 ⟨88TL5893⟩
Ouzounis, D., **4**, 545 ⟨87ZAAC(544)87⟩
Ouzzine, K., **5**, 940 ⟨91OM2768⟩, 941 ⟨92OM809⟩
Ovadia, D., **3**, 320 ⟨82T1477⟩
Ovakimyan, M. Z., **2**, 437 ⟨90ZOB313⟩, 446 ⟨80AKZ408, 86AKZ463⟩
Ovakimyan, M. Zh., **4**, 914 ⟨80CA186471c, 80CA6622a⟩, 914 ⟨83CA(98)4614z, 86CA(104)19636z⟩
Ovaska, T. V., **2**, 556 ⟨91JA5720⟩, 972 ⟨90TL627⟩
Ovchinnikov, Y. A., **6**, 104 ⟨64TL47⟩
Ovechko, G. V., **6**, 250 ⟨93IZV2513⟩
Overberger, C. G., **2**, 123 ⟨64JOC360⟩, 377, 403 ⟨58JA6562⟩, 377, 403 ⟨81JOC442⟩, 378 ⟨63OSC(4)274⟩, 379 ⟨83JHC121⟩, 388 ⟨55JA4100, 63OSC(4)66⟩, 478 ⟨53JA4520⟩; **3**, 455 ⟨56JA4470⟩; **6**, 476 ⟨77MI 615-02⟩
Overheu, W., **4**, 229 ⟨84AG(E)890⟩
Overman, J. D., **2**, 217 ⟨74S59⟩
Overman, L. E., **2**, 217 ⟨74S59⟩, 326 ⟨88JA612⟩, 328, 329 ⟨87JA6097⟩, 331 ⟨87JA6115⟩, 543 ⟨86TL4391⟩, 922 ⟨88T3919⟩, 931 ⟨88TL6541⟩, 1009 ⟨91COS(2)1007⟩; **3**, 184 ⟨86TL4391⟩, 184, 187, 420 ⟨91COS(2)1007⟩, 187 ⟨93JOC4662⟩, 220 ⟨82TL2355⟩, 419 ⟨85HCA745⟩, 419, 420 ⟨88TL6541⟩, 421 ⟨88JA612, 88TL901⟩, 422 ⟨83JOC2685, 83JOC3393, 88TL3781, 89JOC1236, 89TL647⟩; **4**, 187, 199 ⟨89JOC5695⟩, 199 ⟨88JA2248⟩, 410 ⟨93JOC6949⟩, 1014 ⟨81JA2807⟩; **5**, 271, 288 ⟨76JA2352⟩, 287 ⟨91JOC1933, 76TL3089⟩, 331 ⟨91JA5085⟩, 991 ⟨78JOC2164⟩, 1108 ⟨88JA612⟩; **6**, 473 ⟨82JA7225, 84AG(E)579⟩
Ovodova, T. V., **2**, 392 ⟨90JOM(400)19⟩
Owada, H., **2**, 281 ⟨80CC412⟩, 282 ⟨82TL2105, 84JOC3796⟩, 282 ⟨85JCS(P1)373⟩, 284 ⟨81JOC4727⟩, 286 ⟨81CC546⟩; **3**, 37, 174 ⟨84JOC3796⟩, 174 ⟨82TL2105⟩; **5**, 282 ⟨81CC546⟩, 282, 300 ⟨81JOC4727⟩
Owczarczyk, Z., **2**, 749 ⟨86S297⟩; **3**, 99, 105 ⟨87TL3021⟩, 634, 638 ⟨87BSB(96)303⟩
Owen, B. B., **6**, 522 ⟨47JA1723, 50IS156⟩
Owen, C. R., **3**, 588 ⟨74CC134⟩
Owen, D. M., **6**, 20 ⟨74CC323⟩
Owen, D. W., **2**, 600 ⟨83JOM(255)473⟩
Owen, J. J., **2**, 100 ⟨34JA1204⟩
Owen, L. N., **2**, 123 ⟨60JCS1024⟩, 123, 124 ⟨60JCS1030⟩, 128 ⟨50JCS2938, 50JCS579⟩, 128 ⟨66JCS(C)415⟩, 203 ⟨49JCS315, 55JCS1307⟩; **4**, 20 ⟨66JCS(C)533⟩, 226 ⟨67JCS(C)1130⟩, 229 ⟨65JCS1298⟩; **6**, 309 ⟨67JCS(C)807⟩
Owen, N. E. T., **2**, 25 ⟨76S398⟩
Owen, N. L., **3**, 337 ⟨77CC483⟩
Owen, T. C., **2**, 629 ⟨89JCR(S)313⟩
Owen, W. J., **4**, 566 ⟨70BRP1179242⟩
Owens, A. H., **2**, 323 ⟨87SC469⟩
Owens, D. A., **6**, 508 ⟨73JMC901⟩
Owens, K., **2**, 394 ⟨82JOM(225)43⟩; **4**, 148 ⟨89OM1114⟩; **6**, 517 ⟨84JA1125⟩
Owens, M., **3**, 577 ⟨86JA5543⟩; **4**, 892, 979 ⟨86JA5543⟩
Owens, P. W., **2**, 1071 ⟨81TL1875⟩
Owens, R. A., **2**, 941 ⟨76TL1553⟩; **3**, 527 ⟨76TL1553⟩; **4**, 807 ⟨76TL1553, 77DIS(B)1214⟩; **6**, 622 ⟨77JOC3220⟩
Owens, R. M., **3**, 613 ⟨73JA3613⟩
Owens, T. A., **3**, 337 ⟨78JA7436⟩
Owens, W., **2**, 629 ⟨79JA4749⟩
Owens, W. H., **3**, 84 ⟨93JA82⟩, 84, 85 ⟨92TL3717⟩
Owings, F. W., **2**, 689 ⟨94JOC6683⟩
Owston, P. G., **4**, 442, 443 ⟨49JCS1638⟩
Oxenius, R., **5**, 739 ⟨70CB32⟩
Oxenrider, B. C., **2**, 391 ⟨79JOC275⟩; **3**, 192 ⟨79JOC275⟩
Oxley, P., **5**, 689, 710 ⟨48JCS1514, 48JCS1618⟩, 749, 750 ⟨51JCS485⟩, 749, 757 ⟨46JCS147⟩, 750, 751 ⟨47JCS1110⟩
Oya, M., **5**, 972 ⟨73CL1143⟩
Oya, T., **2**, 626 ⟨67BCJ418⟩
Oyama, H., **3**, 337, 340 ⟨84TL849, 82CC1187⟩
Oyamada, H., **2**, 269 ⟨84CL251⟩; **5**, 221 ⟨88BCJ3653⟩
Oyamada, K., **6**, 629 ⟨73CPB62⟩
Oyamada, Y., **5**, 1066 ⟨86CL135⟩
Oyler, A. R., **4**, 907 ⟨80JOC933⟩
Ozaki, A., **5**, 143, 304 ⟨75BCJ2091⟩
Ozaki, H., **5**, 63 ⟨82BCJ2186⟩
Ozaki, J.-I., **6**, 24 ⟨83SUL131⟩
Ozaki, K., **2**, 398 ⟨87CL191⟩; **5**, 403 ⟨87T3395⟩; **6**, 143 ⟨81JOC1571⟩
Ozaki, M., **4**, 979 ⟨86MI 421-01⟩
Ozaki, S., **2**, 106 ⟨92TL1313⟩, 390 ⟨84CPB2609⟩, 590 ⟨88CL63⟩, 725 ⟨70JAP7321096-R⟩; **3**, 120 ⟨83CC479⟩; **5**, 962, 964, 965, 985, 986 ⟨72CRV457⟩, 972 ⟨77BCJ2406⟩, 995 ⟨88CL1385⟩
Ozaki, T., **6**, 417, 426 ⟨93S103⟩
Ozaki, Y., **2**, 662 ⟨89CPB1434, 91CPB1132⟩, 662, 664 ⟨89CPB304⟩; **4**, 17 ⟨74T2661⟩, 310 ⟨79JOC391⟩; **5**, 558 ⟨82CJC199⟩
Ozawa, F., **5**, 27 ⟨85MI 502-01⟩; **6**, 470 ⟨92OM171⟩
Ozawa, K., **3**, 108 ⟨85CL1803⟩, 361 ⟨72CL9, 75JCS(P1)270⟩; **5**, 634 ⟨75JCS(P1)270⟩; **6**, 304 ⟨75JCS(P1)270⟩
Ozawa, M., **4**, 193 ⟨89CL737⟩
Ozawa, S., **2**, 650 ⟨89CL1267⟩, 699 ⟨79CL379⟩
Ozazaki, S., **2**, 305 ⟨90S36⟩
Ozbalik, N., **5**, 728 ⟨88T3501⟩
Ozegowski, S., **6**, 155 ⟨90PS(47)7⟩
Ozeki, K., **4**, 413 ⟨76TL1199⟩
Ozeki, M., **6**, 504 ⟨82CPB534⟩
Ozin, G. A., **6**, 522 ⟨76IC1666, 77IC822⟩, 525 ⟨76JA3167⟩
Ozog, F. J., **3**, 357 ⟨51JA300⟩
Ozorio, A. A., **3**, 20, 55 ⟨78TL5175⟩

P

Paal, B., **4**, 855 ⟨82AJC785⟩
Paal, G., **6**, 111 ⟨69T5995⟩
Paar, J. E., **4**, 332 ⟨69JOC3104⟩
Pabel, M., **5**, 943 ⟨92JOM(441)63⟩
Pabon, H. J. J., **1**, 505 ⟨71RTC961⟩, 506 ⟨82JCS(P1)545⟩; **2**, 423 ⟨76TL2565⟩
Pabon, Jr., R. A., **1**, 248 ⟨87ACR65⟩, 805 ⟨84JA2730⟩
Pabst, E. W., **3**, 627, 634, 635, 636, 639, 651 ⟨76JOC2846⟩
Pabst, P., **2**, 778 ⟨65JPR123⟩
Pabst, W. E., **3**, 651 ⟨77JA182⟩; **5**, 272, 283, 284 ⟨77JA182⟩
Pac, C., **2**, 752 ⟨87JOC753⟩
Pacák, J., **4**, 223, 225 ⟨63M290, 59CCC2566⟩, 226 ⟨61CCC2206⟩
Pacansky, J., **3**, 539 ⟨73JA6134⟩, 542 ⟨73JA4061⟩, 543 ⟨73JA244⟩
Pacansky, T. J., **6**, 476 ⟨B-77MI 615-01⟩
Pacaud, R. A., **2**, 23 ⟨43OSC(2)336⟩
Pace, R. D., **3**, 426 ⟨90SC2453⟩
Pachaly, B., **1**, 936 ⟨81AG(E)1043⟩; **4**, 1058 ⟨84AG(E)454⟩
Pachaly, P., **5**, 384 ⟨79AP822⟩
Pachinger, W., **1**, 195 ⟨90TL1265⟩
Pachler, K. G. R., **6**, 138 ⟨76T2625⟩
Pachter, I. J., **2**, 223 ⟨88OSC(6)1016⟩; **5**, 205 ⟨51JA4958⟩, 285 ⟨52JA1321⟩
Pacifici, J. G., **2**, 403 ⟨69JOC2263⟩; **6**, 509 ⟨69JOC2263⟩
Paciorek, K. J. L., **4**, 481 ⟨85JFC(30)269⟩; **5**, 786 ⟨85JFC(30)269, 85JFC(30)289⟩
Packendorff, K., **2**, 1014 ⟨30CB1941, 31CB1324⟩
Packer, K. J., **2**, 499, 500 ⟨62JCS2529⟩
Pacofsky, G. J., **1**, 504 ⟨83JOC5221⟩; **3**, 222 ⟨89JA7634⟩
Pacreau, A., **5**, 952 ⟨91COS(6)485⟩; **6**, 716 ⟨91CRV35⟩
Padberg, H. J., **4**, 594 ⟨76AG(E)56, 78JCR(S)252⟩
Paddick, K. E., **4**, 676 ⟨86JCS(D)657⟩
Paddison, Jr., J. R., **2**, 1003 ⟨87CAR(168)7⟩
Paddock, N. L., **2**, 407 ⟨61JCS1825⟩
Paddon-Row, M. N., **1**, 130 ⟨86JA2659⟩, 323 ⟨86JOC2874⟩; **2**, 41 ⟨84T2257⟩, 999 ⟨88HCA299⟩
Padegimas, S. J., **5**, 294 ⟨67JA7131⟩
Padeken, H. G., **2**, 391, 392, 394 ⟨71HOU(10/1)12⟩
Padeker, H. H., **2**, 784 ⟨71HOU(10/1)9⟩
Paderes, G. D., **1**, 427 ⟨91JOC4718⟩
Padias, A. B., **4**, 993 ⟨90JOC1768⟩; **6**, 68 ⟨85MI 603-01⟩, 79 ⟨80MM252⟩
Padma, S., **1**, 3, 8 ⟨87JA7230⟩; **2**, 12, 28 ⟨87JA7230⟩
Padmanabhan, P., **1**, 1131 ⟨89TL5017⟩
Padmanabhan, S., **1**, 199, 1025 ⟨83TL2239⟩; **2**, 221 ⟨91SL125⟩, 304 ⟨90TL6977⟩; **3**, 165 ⟨89BCJ1358⟩; **4**, 286 ⟨89BCJ1358⟩
Padolik, L. L., **5**, 451 ⟨90JOM(383)C1⟩, 943 ⟨93JA9986⟩
Padova, G., **1**, 524 ⟨77JHC947⟩
Padova, R., **6**, 653 ⟨26LA(446)260⟩
Padwa, A., **1**, 342 ⟨85JOC5620⟩, 355 ⟨84JOC1353⟩, 365, 366, 368 ⟨91COS(4)1069⟩, 365 ⟨B-76MI 108-02, B-84MI 108-01⟩, 366 ⟨91ACR22⟩, 400, 820 ⟨91CRV263⟩, 470 ⟨92JOC1161⟩, 523 ⟨B-84MI 112-01⟩, 524 ⟨84JOC3314⟩, 550 ⟨90JOC414, 91JOC2523, 92T5385⟩, 571 ⟨81OS(60)53⟩, 736 ⟨74JOC1318⟩, 802 ⟨77JA2344⟩, 808 ⟨80JA2797⟩, 980 ⟨93JA3776⟩, 1089, 1090 ⟨87TL4127⟩, 1127 ⟨85TL957⟩; **2**, 95 ⟨62JOC2976⟩, 354 ⟨76AG(E)123, B-84MI 206-01⟩, 668 ⟨91TL5923⟩, 671 ⟨92JA5881⟩, 927, 945 ⟨93JOC21⟩, 928 ⟨87JOC4760, 90T1145⟩; **3**, 45, 160, 189, 678 ⟨91COS(4)1069⟩, 220 ⟨90TL6145⟩, 250 ⟨87TL913, 88JOC2238⟩, 314 ⟨B-84MI 307-01, 91COS(4)1089⟩, 319 ⟨91CRV213⟩, 320 ⟨88JA2894, 89JOC817, 93TL7853⟩, 321 ⟨89CC921, 89TL1491, 90JA3100, 92JA593, 93JOC4646⟩, 322 ⟨90JOC414, 93JOC21⟩, 323 ⟨91JOC3271⟩, 490 ⟨80JOC4065⟩, 621 ⟨83TL3447, 84JOC276⟩, 640 ⟨87JOC2427⟩, 677, 678 ⟨B-84MI 320-01, 90MI 320-01⟩, 678, 679 ⟨76ACR371, 91CRV263⟩, 679 ⟨84JOC3174⟩, 679, 680 ⟨86JA6739, B-82MI 320-01⟩; **4**, 154 ⟨69JOC2728⟩, 275 ⟨91JOC6386, 92JOC298⟩, 319 ⟨87JOC235⟩, 509 ⟨85T3529, 89OS(67)133⟩, 510 ⟨85JOC4006⟩, 512 ⟨87JOC2427⟩, 513 ⟨84JOC3314⟩, 517 ⟨83TL4303⟩, 520 ⟨89JOC644⟩, 938 ⟨91JOC3556⟩; **5**, 425 ⟨87TL913⟩, 657 ⟨78JOC381⟩, 677 ⟨81JOC1402⟩, 728 ⟨85JOC4415⟩; **6**, 126 ⟨83JOC3189⟩
Paetow, M., **1**, 138 ⟨92TL5323⟩
Paetow, U., **2**, 490 ⟨1886LA(233)60⟩
Paetsch, J. D. H., **2**, 1071 ⟨69JOC1483⟩
Paetzoid, P., **2**, 471, 474 ⟨91CB765⟩, 811 ⟨67FCF437⟩; **3**, 485 ⟨89CB595⟩; **6**, 186, 387, 712 ⟨89CB1057⟩, 187, 188, 292, 388, 389, 673, 712 ⟨89CB595⟩, 187, 387, 712, 713 ⟨90CB747⟩, 188, 388, 391, 712 ⟨87CB1069⟩
Paetzold, R., **2**, 101 ⟨60ZAAC(303)53⟩, 287 ⟨68ZAAC(360)293⟩, 288, 289, 292 ⟨67ZAAC(352)167⟩, 291 ⟨60ZAAC(305)88, 65ZAAC(335)289⟩, 291, 292 ⟨65ZAAC(338)195⟩, 369 ⟨73ZAAC(401)179⟩
Paetzold, V. R., **2**, 369 ⟨65ZAAC(338)22⟩
Pag, D., **2**, 95 ⟨94TL1003⟩
Paganelli, S., **1**, 297, 414 ⟨93MI 107-01⟩
Pagani, G. A., **4**, 1076 ⟨86CC1191⟩
Pagano, A. S., **2**, 386 ⟨58JOC311⟩, 397 ⟨55JA4557⟩; **4**, 312 ⟨73OSC(5)191⟩; **5**, 205 ⟨73OSC(5)367⟩
Page, G. A., **1**, 80 ⟨63OSC(4)136⟩
Page, G., **2**, 195, 688 ⟨70CC1466⟩
Page, P. C. B., **1**, 165 ⟨89T7643⟩, 284 ⟨87TL4455⟩, 1043 ⟨88S621⟩; **2**, 134 ⟨88TL4477, 92T5933⟩, 143 ⟨89T7643⟩, 145 ⟨94TL9629⟩, 145 ⟨95SL773⟩, 145, 156 ⟨93JCS(P2)2161⟩, 145, 156 ⟨93SC1507⟩, 147 ⟨90SL405⟩; **3**, 181 ⟨94TL2427⟩, 20, 29, 31, 137, 157, 160, 169 ⟨89T7643⟩, 43, 182 ⟨87CC775⟩, 244 ⟨87CC756⟩; **4**, 929 ⟨90CSR147⟩, 1076 ⟨85CC742⟩; **5**, 304 ⟨88S621⟩, 409, 411, 412, 417 ⟨90CSR147⟩, 410, 423, 424 ⟨90T2573⟩, 423 ⟨86TL2527, 86TL5421⟩
Paget, W. E., **2**, 4 ⟨81JCR(S)376⟩, 525, 900 ⟨80CC1169⟩
Pagini, G., **2**, 103 ⟨62HCA358⟩
Pagliarin, R., **2**, 9 ⟨87S920⟩; **4**, 462 ⟨92SC107⟩
Paglietti, G., **1**, 102 ⟨76JCS(P1)45⟩
Pagni, A., **5**, 953, 958 ⟨95JOM(486)279⟩
Pagni, R., **5**, 676 ⟨82H(19)319⟩
Pagni, R. M., **1**, 891 ⟨84T4161⟩; **2**, 34, 628 ⟨87TL4497⟩, 34, 629, 631 ⟨88JOC4477⟩
Pagnoni, U. M., **2**, 22 ⟨86JCR(S)426, 92T4579⟩, 24 ⟨89JCR(S)108, 89JCR(S)360⟩, 24 ⟨91G559, 91SC489⟩, 28 ⟨86JCR(S)428⟩, 150 ⟨93SC1759⟩; **3**, 27 ⟨86JCR(S)426⟩, 234 ⟨88JCR(S)200⟩, 489 ⟨77JOC1352⟩; **4**, 249 ⟨93SR149⟩
Pagnori, U. M., **2**, 22 ⟨90JCR(S)188⟩
Paguer, D., **1**, 833, 834 ⟨79PS(7)69⟩
Pah, C. S., **1**, 15, 16 ⟨93JCS(P1)9⟩
Pahde, C., **1**, 649, 656, 678, 680, 759 ⟨81TL5031⟩; **2**, 569, 570, 571 ⟨93CB2093⟩
Pahil, S. S., **4**, 504 ⟨67JIC995⟩
Pai, B. R., **4**, 94 ⟨84H(22)585⟩
Pai, G. G., **2**, 517 ⟨85JOC1384⟩, 531 ⟨85JOC1385⟩
Paiaro, G., **1**, 65 ⟨88G577⟩; **4**, 550, 701 ⟨90G53⟩, 628, 701 ⟨88G577⟩, 628, 701 ⟨86G471, 88OM210⟩
Paige, H. L., **6**, 49, 50 ⟨75JCS(D)488⟩
Paige, J. N., **5**, 1010 ⟨67JHC178⟩

Paik, Y. H., **4**, 162 ⟨85TL2279⟩, 162 ⟨85TL2283⟩
Paikin, D. M., **2**, 464 ⟨59ZOB1671⟩
Pailer, M., **2**, 268 ⟨64M33⟩
Pailthorp, J. R., **6**, 217 ⟨54USP2694739⟩
Pain, D. L., **5**, 858 ⟨54JCS2968⟩
Pain, G. N., **4**, 686 ⟨82JCS(D)1309⟩, 688 ⟨82JOM(224)377⟩, 689 ⟨85OM355⟩
Paine, A. J., **3**, 661 ⟨85CJC111⟩
Paine, R. T., **2**, 474 ⟨88CB871⟩; **4**, 484 ⟨92JA9691⟩
Painter, E. P., **6**, 596 ⟨47JA1833⟩
Paisley, S. D., **1**, 187 ⟨86JOC2884⟩
Pajunen, E. O., **4**, 553 ⟨90SC1865⟩
Pak, C. S., **1**, 24, 439 ⟨93TL4541⟩, 86 ⟨86TL2409⟩; **2**, 66 ⟨88TL4453⟩
Pak, V. D., **4**, 462 ⟨76ZOB1012⟩
Pakinkar, S. K., **1**, 842 ⟨83T2323⟩; **2**, 91 ⟨83T2323⟩
Pakrashi, S. C., **2**, 398 ⟨85S886⟩; **5**, 601, 621 ⟨89IJC(B)673⟩
Pakulski, M., **2**, 821 ⟨82JA5820⟩, 866 ⟨83CC881⟩, 866, 893 ⟨84IC2582⟩; **5**, 1107, 1112 ⟨83T4175⟩; **6**, 167 ⟨83CC528⟩, 168 ⟨83JA5506, 84IC2582⟩, 169 ⟨83CC881⟩, 173, 174, 176 ⟨90IS239⟩, 372 ⟨90IS235⟩, 373 ⟨82JA5820, 86JCS(D)1801⟩, 375 ⟨85JA8211, 85JCS(D)383⟩, 375 ⟨87JCS(D)249⟩
Pal, A., **4**, 120 ⟨86ACH145⟩
Pal, A. K., **4**, 745 ⟨88JOC5554⟩
Pala, G., **4**, 416 ⟨64G595⟩
Palacios, F., **3**, 414 ⟨88TL4863⟩; **4**, 656 ⟨89JOM(375)59⟩; **5**, 371 ⟨90JCS(P1)2193, 90TL3497⟩, 1066 ⟨88TL4863⟩; **6**, 121 ⟨88S562⟩
Palacios, S. M., **2**, 438 ⟨84JOC3387, 84JOC4609⟩
Palamidis, E., **4**, 577 ⟨85JOM(297)C30⟩
Palaska, E., **6**, 561 ⟨92AF1453⟩
Palazón, J. M., **2**, 155 ⟨92T3571⟩
Palazzi, A., **5**, 1088 ⟨91AG(E)847⟩
Palazzo, F. C., **5**, 822 ⟨13G58⟩
Pal'chik, R. I., **3**, 540 ⟨71ZOB240⟩
Pale, P., **1**, 431 ⟨93TL1155⟩, 1018 ⟨92TL5355⟩
Paleček, J., **5**, 324 ⟨91CCC1317⟩
Paleev, B. B., **5**, 754 ⟨69URP256779, 69URP256781⟩
Palei, B. A., **4**, 698 ⟨75CA(83)10238⟩; **6**, 208 ⟨68IZV910⟩
Palenik, G. J., **5**, 937 ⟨92OM622⟩
Palenzuela, J. A., **1**, 795 ⟨91JA1355⟩, 993 ⟨89JA1770⟩
Paleta, O., **4**, 5 ⟨71CCC1867⟩, 970 ⟨81CCC1389⟩; **6**, 4 ⟨72CLY937, 77FCR39⟩, 5 ⟨66CCC3584, 67CCC3888, 71CCC1867, 74CCC1330⟩, 42 ⟨73CCC66⟩
Paleveda, Jr., W. J., **5**, 972 ⟨66JA3163⟩; **6**, 496 ⟨68JA3254⟩
Palfey, B. A., **6**, 611 ⟨90JOC3045⟩
Palfreyman, M. N., **5**, 587, 593 ⟨93JMC1604⟩
Pali, L., **6**, 645 ⟨91MIP54636⟩
Palkowitz, A. D., **1**, 752 ⟨89JOC3009⟩; **2**, 75 ⟨90JA6348⟩, 75, 529, 557 ⟨90JA6339⟩
Palla, O., **6**, 494 ⟨82JCS(P1)2813⟩, 695 ⟨84T1523⟩
Pallai, P., **4**, 422 ⟨82CC280⟩
Pallaud, R., **1**, 104 ⟨68CR(C)1834⟩
Pallavinci, M., **2**, 53 ⟨91TL4381⟩
Pallini, U., **6**, 6 ⟨56CI(M)371⟩
Palma, P., **3**, 618 ⟨83S740⟩; **5**, 800 ⟨89JOM(377)157, 90OM583⟩
Pal Mahajan, M., **6**, 636 ⟨87TL2641⟩
Palme, H.-J., **3**, 220 ⟨87S1012⟩
Palmer, B. D., **1**, 654, 697, 700, 703 ⟨83CC630⟩; **3**, 308 ⟨83CC499, 84JOC3503⟩
Palmer, C., **3**, 612 ⟨1889JA89⟩
Palmer, C. J., **4**, 265 ⟨92MI 406-01⟩
Palmer, C. S., **2**, 507 ⟨28JA536⟩
Palmer, D. E., **2**, 815 ⟨73JCS(D)157⟩; **3**, 456, 457 ⟨73JCS(D)157⟩, 457, 460, 461 ⟨73JCS(D)151⟩
Palmer, J. R., **1**, 1049 ⟨75TL4217⟩, 1050 ⟨76TL235⟩
Palmer, L. C., **2**, 388 ⟨55JA4100⟩
Palmer, M. J., **5**, 690 ⟨88JMC1031⟩
Palmer, P. J., **4**, 821 ⟨71TL1879⟩
Palmer, R., **1**, 382 ⟨84JOC2937⟩; **3**, 436 ⟨86JOC5024⟩
Palmertz, I., **5**, 991 ⟨74ACS(B)699⟩
Palmieri, G., **3**, 262, 263 ⟨90S895⟩; **4**, 260 ⟨90JOC1323⟩
Palmisano, G., **1**, 601 ⟨84JCS(P1)1593⟩; **2**, 931 ⟨93SL771⟩; **3**, 229 ⟨75CC892⟩; **5**, 1045 ⟨90TL7229⟩
Palominos, M. A., **5**, 513 ⟨86CL1251⟩; **6**, 476, 551 ⟨85PS(25)91⟩, 479 ⟨90S825⟩, 546, 547 ⟨83JCS(P1)2641⟩
Palomo, A.-L., **5**, 280, 295 ⟨84S413⟩; **6**, 57 ⟨69TL2161⟩
Palomo, C., **2**, 10 ⟨86JOC4941⟩, 14 ⟨84TL1103, 85T4657, 86CJC2342⟩, 28, 30 ⟨84SC939⟩, 34 ⟨85T2903⟩, 36 ⟨84TL3123⟩, 262 ⟨85TL475⟩, 398 ⟨90JOC2070⟩; **3**, 4, 115 ⟨83TL4367, 85T2903⟩, 4, 116 ⟨87T3963⟩, 6, 118 ⟨91JOC5948⟩, 88 ⟨86CJC225⟩, 154 ⟨84SC939⟩, 162 ⟨87TL5361⟩, 410 ⟨91CC524⟩, 528 ⟨92TL4823⟩, 618 ⟨84TL3365⟩, 632 ⟨89JCS(P1)1692, 90TL2205⟩; **4**, 514 ⟨91CC524⟩; **5**, 17 ⟨82S684, 86JOC4941⟩, 189 ⟨81S218⟩, 223, 374 ⟨85TL475⟩, 294 ⟨87TL1945⟩, 299 ⟨82BSF(2)167⟩, 300 ⟨82S288⟩, 329 ⟨87JCS(P1)845⟩, 367 ⟨83SC289, 84TL3365⟩, 371 ⟨82BSF(2)369⟩, 372 ⟨81S993⟩, 1063 ⟨81S373⟩, 1079 ⟨81BSF(2)361⟩; **6**, 162, 163 ⟨91JOM(419)C1, 92TL3903⟩
Palomo-Coll, A. L., **5**, 126, 267, 291 ⟨80S547⟩, 188, 195, 199 ⟨81S616⟩, 193 ⟨85T5133⟩
Palop, J. A., **4**, 897 ⟨85JHC1445⟩; **5**, 199 ⟨88AQ120⟩
Palopoli, F. P., **5**, 652 ⟨68JMC534⟩
Pálosi, E., **5**, 219, 833 ⟨88CB723⟩
Palou, E., **6**, 622 ⟨66CB3155⟩
Palovich, M., **5**, 432 ⟨92SL631⟩
Palter, A., **2**, 520, 900, 901 ⟨83TL1433⟩
Palumbo, G., **1**, 587 ⟨81TL3551⟩, 885 ⟨91COS(1)139⟩; **2**, 302 ⟨92SC2305⟩, 688 ⟨93SC1515⟩, 724 ⟨81S888⟩; **3**, 164 ⟨89S464⟩, 170 ⟨93SC1515⟩
Palumbo, P. S., **1**, 618, 704, 705, 709 ⟨86JA2358⟩, 619, 704, 705, 708 ⟨85JOC2110⟩, 704, 705 ⟨84TL4617⟩; **2**, 157 ⟨86JA2358⟩; **3**, 210 ⟨85JOC2110⟩
Pályi, G., **4**, 148 ⟨87OM861⟩, 688 ⟨81JOM(206)119⟩; **6**, 517 ⟨85M1103, 85OM1523⟩
Palyulin, V. A., **4**, 970 ⟨79ZOR1099⟩
Pan, B., **4**, 94 ⟨82JA6816⟩
Pan, B.-C., **5**, 179 ⟨82JA6818⟩
Pan, H. L., **2**, 366 ⟨75S39⟩
Pan, S. H., **6**, 25 ⟨86TL3845⟩
Pan, T. P., **2**, 308 ⟨69BCJ3606⟩
Pan, X., **3**, 220 ⟨89CC612⟩
Pan, Y., **1**, 905 ⟨87JA7263⟩; **3**, 633 ⟨78TL4221⟩
Pan, Y.-G., **1**, 556 ⟨77JOC3961⟩
Pan, Y. G., **3**, 206 ⟨77JOC3961⟩
Panasenko, A. I., **3**, 362 ⟨89MI 308-04⟩
Panasenko, T. G., **3**, 374 ⟨91ZOR217⟩
Panchit, U. K., **3**, 90, 95 ⟨79CC117⟩
Pancoast, T. A., **2**, 152 ⟨77TL4009⟩
Pancrazi, A., **1**, 1071 ⟨91SL539⟩
Panda, C. S., **2**, 557 ⟨87S645⟩; **4**, 653 ⟨87S645⟩
Panda, R. K., **1**, 74 ⟨84JCS(D)1477⟩
Pande, C. S., **5**, 205 ⟨88SC2123⟩
Pande, K. C., **1**, 448, 449 ⟨60JCS1566⟩; **2**, 598 ⟨64TL3393⟩; **4**, 561 ⟨71USP3622654⟩
Pande, L. M., **6**, 270 ⟨91SRI401⟩
Pandell, A. J., **1**, 476 ⟨67JA5421⟩
Pandey, A., **2**, 318 ⟨90TL575⟩
Pandey, B., **4**, 186 ⟨93S961⟩, 246 ⟨93S67⟩, 260 ⟨89TL4007, 90JCS(P1)3217⟩
Pandey, G., **1**, 285 ⟨92CC1313⟩, 286 ⟨87TL2649⟩, 370 ⟨93TL4861⟩, 943 ⟨93CC1636⟩; **2**, 303 ⟨90TL1199⟩; **3**, 144 ⟨93TL6631⟩, 422 ⟨88TL4153⟩
Pandey, P. S., **6**, 80, 84 ⟨80JOC740⟩
Pandey, R. C., **5**, 52 ⟨68TL2681⟩
Pandey, V. K., **5**, 685 ⟨93MI 517-01⟩; **6**, 563 ⟨82MI 617-01⟩
Pandhi, S. B., **5**, 355 ⟨90JOC2196⟩; **6**, 113 ⟨87S990⟩
Pandiarajan, P. K., **1**, 831 ⟨89JA3441⟩
Pandit, M. I., **6**, 611 ⟨61MI 620-01⟩
Pandit, U. K., **3**, 91, 107 ⟨92S1221⟩, 235 ⟨70CC381⟩, 310 ⟨76RTC220⟩
Pandley, P. N., **1**, 428 ⟨82S876⟩

Pandolfo, L., **1**, 65 ⟨88G577⟩; **4**, 550, 701 ⟨90G53⟩, 628, 701 ⟨88G577⟩, 628, 701 ⟨88OM210, 86G471⟩
Pandya, A. D., **3**, 363 ⟨76IJC(B)999⟩
Paneccasio, V., **1**, 1137, 1141 ⟨75JOC124⟩; **2**, 1054 ⟨75JOC124⟩
Panek, J. S., **2**, 46 ⟨90JA4873⟩, 935 ⟨90JOC5564, 90TA801⟩, 1005 ⟨91COS(1)579⟩; **3**, 160 ⟨93JA7898⟩; **4**, 354 ⟨90JOC6071⟩, 355 ⟨90JA4873⟩; **5**, 93 ⟨91OS(70)79⟩, 409 ⟨92OPP553⟩
Panella, J. P., **3**, 632 ⟨73OSC(5)614⟩
Paneque, M., **4**, 583 ⟨84OM1438⟩; **5**, 800 ⟨90OM583⟩
Paneth, F. A., **2**, 491, 505 ⟨35JCS366⟩
Panetta, J. A., **4**, 175 ⟨82JOC946⟩
Panfilovich, Z. U., **2**, 879 ⟨88JGU1173⟩, 1076 ⟨88JGU2136⟩, 1078 ⟨68JGU1295⟩, 1079 ⟨87JGU2497⟩
Pang, Z., **5**, 1078 ⟨87MI 528-01⟩
Panicucci, R., **2**, 373, 376, 403 ⟨84JOC336⟩
Panizzi, L., **3**, 239 ⟨47G549⟩
Pankayatselvan, R., **2**, 307 ⟨92CC853⟩
Pankert, D., **4**, 288 ⟨92TL269⟩
Pan'kov, A. K., **6**, 145 ⟨72KGS713⟩
Pankov, V. I., **2**, 440 ⟨85URP1145022⟩
Pankowski, J., **1**, 697, 698, 700, 703 ⟨85TL5597⟩
Pankowski, M., **2**, 436 ⟨85IC533⟩
Pankratz, M., **2**, 1008 ⟨85JOC4553⟩
Pannell, K. H., **4**, 136, 138, 141 ⟨91OM1960⟩, 614, 656 ⟨74JOM(78)C35⟩, 639, 656, 663 ⟨93OM3979⟩, 685, 722 ⟨76IC2671⟩
Panossian, S., **5**, 580 ⟨78S917⟩, 606 ⟨77S250⟩
Panouse, J. J., **1**, 82 ⟨63BSF798⟩
Panova, N. I., **6**, 84 ⟨62ZOB745⟩
Panquin, A. M., **4**, 416 ⟨48AG267⟩
Pansare, S. V., **5**, 85 ⟨91OS(70)1⟩
Panse, P., **1**, 725 ⟨67CB1144⟩
Pansegrau, D., **1**, 539 ⟨88JA7178⟩
Pansegrau, P. D., **1**, 248 ⟨87TL4965⟩; **2**, 974 ⟨83TL4935, 85CC690⟩; **3**, 665 ⟨84TL2941⟩
Panshiri, F. M., **5**, 1052 ⟨75CB3071⟩
Panster, P., **2**, 511, 512 ⟨74AG(E)670, 74JOM(76)C7⟩, 512 ⟨75CB700, 76CB692⟩; **5**, 912 ⟨75JOM(99)421⟩, 913 ⟨75CB2554⟩
Pant, A. K., **6**, 546 ⟨85JIC164⟩
Pant, B. C., **2**, 1076 ⟨69JOM(17)165⟩, 1101 ⟨69JOM(16)413⟩; **5**, 210 ⟨72TL4779⟩, 550 ⟨74JOM(65)51⟩; **6**, 267 ⟨67JCS(C)1470⟩
Pant, B. P., **5**, 826 ⟨83IJC(B)51⟩
Pant, C. M., **4**, 448 ⟨84CC1289⟩
Pant, J., **2**, 1009 ⟨85CB4259⟩
Pantaloni, A., **3**, 180 ⟨86JCS(P1)1895⟩; **5**, 333 ⟨88TL2661, 89JOC617⟩
Panteleeva, A. R., **4**, 127 ⟨67IZV1535, 67JGU1768⟩, 127, 128 ⟨68IZV1557⟩, 556 ⟨67DOK(177)340⟩
Pantev, T. P., **5**, 770 ⟨92MI 519-01, 92MI 519-02⟩
Pantukh, B. I., **4**, 560 ⟨86ZOB2000⟩
Panunzio, M., **2**, 19, 25, 31, 35 ⟨76S472⟩, 103 ⟨85TL3369⟩, 316 ⟨90JOC4199⟩; **3**, 20 ⟨78JOC1598⟩, 440, 456 ⟨87TL5369, 91TL2967⟩, 456, 457, 458 ⟨90PAC605⟩, 456, 458 ⟨86TL1695⟩, 457 ⟨91SL243, 90JOC4199⟩, 458 ⟨90TL3481, 93T3809⟩, 621, 625 ⟨75JCS(P1)1273⟩; **5**, 287 ⟨86TL1695, 87TL5369⟩, 377 ⟨91SL243, 94S805⟩
Panzer, H. P., **1**, 786 ⟨83AG(E)560⟩
Panzica, R. P., **3**, 362 ⟨78JOC4910⟩
Paoli, P., **3**, 702 ⟨90H(31)1855⟩, 706 ⟨91LA843⟩
Paolini, I. de, **5**, 821 ⟨30G700⟩, 822 ⟨30G703⟩
Paolini, J. P., **1**, 440 ⟨93OS(72)216⟩; **5**, 652 ⟨68JMC534⟩
Paolucci, C., **1**, 821 ⟨78JOC4826, 79JOC4128⟩, 821 ⟨81JOC3315⟩; **2**, 143 ⟨81JOC3315⟩, 156 ⟨88JOC5689⟩
Papa, A. J., **2**, 749 ⟨66JOC1426⟩; **6**, 649 ⟨66JOC1426⟩
Papa, D.,**1**, 503 ⟨54JA4446⟩
Papa, D., **5**, 36 ⟨48JA3091⟩
Papadatos, A., **5**, 241 ⟨92JA8313⟩
Papadopoulos, E. P., **2**, 213 ⟨66JOC615⟩; **3**, 98 ⟨77JOC3514⟩; **5**, 592 ⟨73JOC667, 74JOC2540⟩, 592, 593, 620, 631 ⟨76JOC962, 76JOC3233⟩; **6**, 558 ⟨85S423⟩
Papadopoulos, K., **3**, 172 ⟨93T1541⟩
Papadopoulos, M., **4**, 867 ⟨91PS(60)131⟩
Papadopoulos, S., **1**, 1027 ⟨89AJC279⟩
Papageorgiou, C., **6**, 541 ⟨87HCA138⟩
Papageorgiou, G., **2**, 316 ⟨88TL2377, 90TL4229⟩; **3**, 186 ⟨90SL619⟩
Papagni, A., **2**, 553 ⟨92CC1623⟩; **5**, 676, 677, 906 ⟨85T2503, 85T3321⟩, 943 ⟨92CC1623⟩, 946 ⟨91OM807⟩, 946, 959 ⟨94ICA233⟩, 956 ⟨94TL6377⟩; **6**, 723 ⟨93OM2994⟩
Papahatjis, D. P., **3**, 308 ⟨81JA6969, 81TL4647⟩; **4**, 45 ⟨84JA4189⟩, 224, 225 ⟨83JA2430⟩
Papaleo, S., **3**, 706 ⟨91LA843⟩
Papanastassiou, Z. B., **1**, 3, 6 ⟨64JOC2870⟩
Papasergio, R. I., **2**, 1003, 1005 ⟨89JCS(D)105⟩; **4**, 603, 647 ⟨84JCS(D)311⟩, 603, 647, 657 ⟨82CC1023⟩, 604 ⟨83CC1419⟩, 604, 647, 648 ⟨90JCS(D)1161⟩, 626, 662 ⟨87JCS(D)1647⟩, 646 ⟨89JCS(D)105⟩, 647 ⟨84CC1708⟩, 647, 656 ⟨86JCS(D)603, 87JCS(D)3085⟩; **6**, 190 ⟨84CC1708⟩, 190, 195 ⟨83CC1419⟩, 190, 196 ⟨86CC672⟩, 195 ⟨84CC612⟩
Papavassiliou, G. C., **4**, 869 ⟨91SM2535⟩
Pape, C., **3**, 615 ⟨1896CZ90⟩
Papenberg, G., **4**, 889 ⟨80AP(314)65⟩, 983 ⟨81AP65⟩
Papenberg, M., **4**, 673 ⟨92CB143, 93CB79⟩, 673 ⟨93CB89⟩
Papenfuhs, T., **6**, 504 ⟨81GEP2855882⟩
Papies, O., **1**, 785 ⟨80TL2799⟩
Papillon-Jegou, D., **4**, 309 ⟨74CR(C)221⟩
Papini, A., **1**, 242 ⟨83TL2311⟩; **2**, 265 ⟨83TL2311⟩
Papini, P., **5**, 583 ⟨59G526, 66G125⟩
Papoula, M. T. B., **3**, 118, 464 ⟨79CC705⟩, 145 ⟨85JCS(P1)2667⟩
Papouskova, Z., **4**, 352 ⟨73CCC1522⟩
Papp, G. P., **2**, 440 ⟨66JOC588⟩
Pappalardo, J. A., **2**, 298, 320 ⟨B-65MI 205-01⟩; **3**, 238 ⟨63OSC(4)185⟩; **4**, 196 ⟨50JA2613⟩, 196 ⟨63OSC(4)558⟩
Pappalardo, P., **6**, 124 ⟨82TL309⟩
Pappas, B. C., **1**, 517 ⟨69JOC520⟩
Pappas, J. J., **5**, 559 ⟨66JOC3877⟩
Pappas, S. P., **1**, 517 ⟨69JOC520⟩
Pappo, R., **1**, 888 ⟨59OR(10)179⟩, 1049 ⟨75TL4217⟩, 1049, 1050 ⟨72TL2627, 76TL235⟩
Pappo, R.,**3**, 2, 113 ⟨56JOC478⟩
Pappo, R., **4**, 540 ⟨59OR(10)179⟩; **5**, 46 ⟨59OR(10)179⟩
Papsheva, N. P., **5**, 1025 ⟨88ZOR1771⟩
Papulov, Yu. G., **6**, 223, 226, 227, 228 ⟨72MI 607-01⟩
Papworth, T., **4**, 689 ⟨95JOM(489)C65⟩
Paqet, F., **1**, 68 ⟨84JA8313⟩
Paquer, D., **1**, 1058 ⟨77JOM(140)257⟩; **2**, 255 ⟨92PS(73)49⟩; **3**, 339 ⟨74CR(C)259⟩, 342, 352 ⟨72IJS(B)269⟩, 343 ⟨78RTC121⟩, 344 ⟨76RTC172⟩, 351 ⟨77BSF688, 85RTC9⟩, 360 ⟨73IJS173⟩, 361, 376 ⟨76BSF(2)120⟩; **4**, 255 ⟨75BSF1439⟩; **5**, 232 ⟨75BSF1439⟩, 550 ⟨77RTC118⟩, 552 ⟨79PS(7)69⟩, 558 ⟨77RTC254⟩, 634 ⟨76BSF120⟩
Paquette, L. A., **1**, 18, 564 ⟨84JA6702⟩, 19 ⟨84JA6690⟩, 25, 829 ⟨87JA3025⟩, 49 ⟨88JOC477⟩, 96 ⟨73OSC(5)467⟩, 144, 617, 704, 707 ⟨68ACR209⟩, 144, 617, 704, 707 ⟨77OR(25)1⟩, 186 ⟨92T7383⟩, 246 ⟨85JOC5200⟩, 248 ⟨82TL263, 87TL4965⟩, 284 ⟨84JA6442⟩, 362, 555, 558 ⟨92JA7375⟩, 402, 799 ⟨80JA3972⟩, 414 ⟨92JA5018⟩, 422 ⟨79TL1999, 80JOC3017⟩, 434, 815 ⟨87TL31⟩, 451 ⟨79JA4420⟩, 517 ⟨75T2855, 83JOC1262⟩, 558, 575, 584 ⟨90JA9284⟩, 563 ⟨80JOC5378⟩, 575 ⟨84JA1518⟩, 576 ⟨86JA3739⟩, 603 ⟨83JOC4986⟩, 613 ⟨77OS(57)53, 87JA3730⟩, 617 ⟨68MI 114-04⟩, 618, 704 ⟨69JA3870⟩, 632, 797, 826 ⟨85JA7352⟩, 633 ⟨82OM1449⟩, 634 ⟨87JA3174⟩, 653 ⟨84JOC1317⟩, 654 ⟨83JOC4976⟩, 690 ⟨84JOC3610, 84JOC3618⟩, 704, 708 ⟨77JA2815, 77JA6946, 78JOC4527, 87JA2857⟩, 707 ⟨71JOC1015⟩, 707, 708 ⟨74JA5801, 74JA4715⟩, 708 ⟨76JA4936, 86TL1895,

88JA5479, 88JA8197⟩, 709 ⟨64JA4085⟩, 776 ⟨87TL2795, 90JA228⟩, 785 ⟨70JA1763⟩, 806 ⟨90AG(E)609, 90SL67⟩, 810 ⟨91TL319, 92HCA1755, 93JOC4245, 93JOC4952⟩, 811 ⟨79JA7599⟩, 812 ⟨85JOC201⟩, 813 ⟨88JA890⟩, 813, 815 ⟨92TL923, 86JA7873⟩, 815 ⟨87JA5731, 88T3139, 88TL273⟩, 815, 1098 ⟨88JA879⟩, 816 ⟨87JA3017, 87JOC2960⟩, 816, 817 ⟨86JA3841⟩, 826 ⟨89JOC4278, 90JA3252, 90T4487, 90TL6979, 91JA2610, 91JA2762, 91JOC3841, 93TL3523⟩, 865, 917, 918 ⟨78AG(E)106⟩, 883 ⟨89JOC5044, 89JOC5054⟩, 885 ⟨88JOC4972, 88JOC4978, 90OPP167⟩, 900 ⟨88JA8599⟩, 918 ⟨75JOC700⟩, 971 ⟨81TL1579⟩, 1094 ⟨77TL159⟩, 1098 ⟨89JOC2278, 90JA277⟩, 1098 ⟨90JA5562⟩; **2**, 103 ⟨60MI 202-03⟩, 158 ⟨70JOC905⟩, 238 ⟨88OS(67)157⟩, 361 ⟨B-95MI 202-01⟩, 619 ⟨67JA4487⟩, 625 ⟨91JOC6199⟩, 911 ⟨78JOC1620, 80JOC3017, 84CJC2415⟩, 911, 946 ⟨77TL159⟩, 921, 938 ⟨93JOC2714⟩, 940 ⟨81TL1579⟩; **3**, 16 ⟨76SC575⟩, 102, 103 ⟨92JA3910⟩, 126 ⟨80JOC3028⟩, 141 ⟨90AG(E)609⟩, 156 ⟨89JOC4576, G2CEN2⟩, 206 ⟨87TL31⟩, 211 ⟨80JOC3017⟩, 217 ⟨88JA890, 91JA2610⟩, 220, 639 ⟨84JOC3604⟩, 221 ⟨83JOC141⟩, 234 ⟨83JA7358⟩, 284 ⟨83JA7644⟩, 621 ⟨80JOC167⟩; **4**, 64 ⟨64JA4085, 64JA4089⟩, 64, 69, 80 ⟨68JOC1080⟩, 69 ⟨67JA4483⟩, 79 ⟨77OR(25)1, 64JA4383⟩, 80 ⟨69JA3870, 71JOC1015⟩, 82 ⟨67JA4487⟩, 85 ⟨73OSC(5)231⟩, 179 ⟨84TL3047⟩, 224 ⟨88JOC2953⟩, 284 ⟨92JA3910⟩, 307 ⟨66JOC1232⟩, 357 ⟨80JOC3028⟩, 950, 954 ⟨88T3139⟩; **5**, 30 ⟨89OS(68)198⟩, 76 ⟨90OS(69)173⟩, 271, 288 ⟨62JA4987, 63JA3288, 64OS(44)41⟩, 425 ⟨91JOC5480⟩, 496, 497 ⟨64JOC2854⟩

Paquin, A. M., **6**, 484 ⟨46ZN(B)518⟩

Paquin, R. J., **3**, 453 ⟨53CJC1025⟩

Paquot, C., **3**, 617 ⟨46MI 318-01⟩

Paradisi, C., **2**, 240 ⟨94TL301⟩

Paradisi, M. P., **3**, 31 ⟨82T1827⟩

Paradkar, M. V., **3**, 293 ⟨90SC3259⟩

Parady, T. E., **1**, 706 ⟨B-77MI 115-01⟩

Paragamian, V., **5**, 657 ⟨84JOC546⟩

Parahall, G. W., **2**, 431 ⟨68IS157⟩

Parakin, O. V., **2**, 845 ⟨82ZOB2721⟩, 881 ⟨85JGU1332⟩

Paramanov, V. V., **6**, 650 ⟨75MI 621-01⟩

Parameswaran, K. N., **3**, 617 ⟨65CI(L)988⟩

Parashikov, V., **5**, 344 ⟨89SC3113⟩

Paraskewas, S., **5**, 275 ⟨74S574⟩

Paratian, J. M., **6**, 22 ⟨92CC53⟩

Parayre, E., **4**, 508 ⟨76CR(C)473⟩

Parchomenko, N. A., **6**, 687 ⟨84ZOB715⟩

Pardasani, R., **4**, 665 ⟨93T2979⟩, 1077 ⟨86TL5033⟩

Pardhasaradhi, M., **3**, 293 ⟨83IJC(B)1246⟩

Pardini, V. L., **3**, 497 ⟨81JCS(P1)1520⟩

Pardo, C., **2**, 25, 31 ⟨86CC1250, 88SC1275⟩

Pardo, R., **1**, 422 ⟨79TL4553⟩; **3**, 212 ⟨79TL4553⟩

Paredes, R., **5**, 144 ⟨63JOC3512⟩

Parekh, H., **6**, 561 ⟨87MI 617-01⟩

Parekh, S. I., **1**, 16, 940 ⟨93TL2733⟩, 777 ⟨94T639⟩, 919 ⟨B-92MI 119-16⟩; **4**, 352 ⟨89CC1256⟩; **6**, 550 ⟨93TL2733⟩

Parello, J., **3**, 406 ⟨74S812⟩, 450 ⟨65JA596⟩; **5**, 185 ⟨85T5331⟩

Parent, K.,**5**, 745 ⟨61JOC2563⟩

Parent, P., **6**, 480 ⟨90MI 615-01⟩

Parent, R. A., **1**, 214 ⟨66JA2257⟩

Parg, A., **6**, 445 ⟨78GEP2828969⟩

Parham, M. E., **6**, 76, 77 ⟨73JOC1361⟩

Parham, W. E., **1**, 417, 829 ⟨49JA2439⟩, 452 ⟨76JOC1184⟩, 1090 ⟨77JOC260⟩; **2**, 1030, 1031, 1034 ⟨56JA4783⟩, 1031 ⟨59JA3386, 72BCJ509⟩, 1031 ⟨73JOC1361⟩; **3**, 235 ⟨69JOC34⟩, 284 ⟨82ACR300⟩; **4**, 7, 16, 24 ⟨63OR(13)55⟩, 32 ⟨57JOC730⟩, 170, 209 ⟨62JA813⟩, 223 ⟨54JA4962⟩, 979 ⟨66TL2311⟩

Parihar, P., **5**, 678, 679 ⟨81JCED418⟩

Parikh, A. R., **6**, 561 ⟨85MI 617-02, 86MI 617-01⟩, 561 ⟨87MI 617-01⟩

Parikh, D. K., **5**, 389, 1121 ⟨76SRI345⟩

Parikh, J. R., **3**, 57 ⟨67JA5505⟩

Paris, G., **5**, 34, 275, 305 ⟨63OSC(4)496⟩

Paris, G. Y., **2**, 102 ⟨60CJC343⟩

Paris, J., **3**, 40 ⟨94TL1181⟩; **5**, 755 ⟨47BCF406⟩

Paris, J. M., **1**, 144, 614, 615, 696, 697, 698, 700 ⟨73TL4833⟩

Pariser, E. R., **4**, 178 ⟨48JA2827⟩

Parish, E. J., **3**, 115 ⟨90SC3359⟩, 134 ⟨90SC3265⟩; **5**, 78 ⟨72TL3987⟩

Parish, R. C., **5**, 123, 126 ⟨65JOC927⟩

Parish, S., **3**, 134 ⟨90SC3265⟩

Parishchemko, A. A., **2**, 448 ⟨77JGU707⟩

Parisi, G., **2**, 214 ⟨57JA3503⟩

Pariza, R. J., **2**, 444 ⟨85JOC4252⟩; **3**, 220 ⟨85JOC4252⟩

Park, B. K., **2**, 145, 156 ⟨93JCS(P2)2161⟩, 145, 156 ⟨93SC1507⟩; **4**, 250 ⟨92T7265⟩

Park, B.-S., **4**, 1079 ⟨91MI 424-02⟩

Park, C., **3**, 223 ⟨90TL5039⟩

Park, C. H., **3**, 363 ⟨91MI 308-01⟩; **5**, 648 ⟨91MI 516-01⟩; **6**, 128 ⟨91MI 604-01⟩

Park, C.-H., **6**, 468 ⟨94SC305⟩

Park, C. M., **1**, 455 ⟨90TL5039⟩

Park, C. Y., **2**, 47 ⟨91TL1003⟩; **4**, 139 ⟨92MI 403-01⟩

Park, D.-C., **1**, 15, 16 ⟨93JCS(P1)9⟩

Park, D. J., **5**, 215 ⟨90SC2901⟩

Park, H., **1**, 563 ⟨80JOC5378⟩; **5**, 862 ⟨94S1057⟩, 999 ⟨90TL1559⟩

Park, H. G., **5**, 999 ⟨90TL1559⟩

Park, J., **2**, 639 ⟨87CC1213, 88JA8244⟩, 949 ⟨89JA1888⟩

Park, J. D., **1**, 6 ⟨56JPC1454⟩; **3**, 532 ⟨58JOC1166⟩; **4**, 5 ⟨51JA1329⟩, 31, 34 ⟨51JA711⟩, 32 ⟨48JA1550⟩, 36, 751 ⟨56JA59⟩, 757, 759, 820 ⟨58JOC1661⟩, 761, 763 ⟨56JA1685, 52JA4104⟩; **5**, 273, 285 ⟨52JA1010⟩, 318 ⟨52JA2189⟩; **6**, 36 ⟨52JA2292⟩, 214 ⟨44USP2351390⟩

Park, J. G., **2**, 684 ⟨89CC1363⟩

Park, J. H., **1**, 629 ⟨91TL2043⟩; **2**, 11 ⟨88JOC3111⟩; **4**, 201, 217 ⟨93TL5769⟩, 216 ⟨89TL6697⟩

Park, J.-H., **5**, 378 ⟨94S909⟩

Park, J. M., **4**, 168 ⟨88JOC2979⟩

Park, J. S., **4**, 127 ⟨87PS(29)287⟩

Park, J.-Y., **1**, 1127 ⟨84JA4356⟩

Park, K., **1**, 465 ⟨93JOC4866⟩

Park, K. H., **1**, 579, 584 ⟨94JOC394⟩

Park, K. K., **2**, 398 ⟨92JCS(P1)601⟩

Park, K. P., **4**, 467 ⟨90TL1567⟩

Park, M., **5**, 999 ⟨90TL1559⟩

Park, M.-G., **3**, 543 ⟨73JA5412⟩

Park, S. B., **1**, 86 ⟨88SC89⟩; **3**, 127 ⟨88SC89⟩; **4**, 155, 534 ⟨93JOC523⟩, 531, 533 ⟨92JOC2220⟩

Park, S. U., **1**, 331, 342 ⟨86JA4132⟩

Park, S.-U., **3**, 621 ⟨86JA240, 90TL2975⟩

Park, S. W., **1**, 183 ⟨88CC760⟩

Park, S. Z., **3**, 193, 473 ⟨90JOC1959⟩

Park, W.-S., **1**, 884 ⟨83TL1667⟩; **2**, 57 ⟨88JOC1231⟩, 534 ⟨86JOC3278, 86JOC3396⟩, 535 ⟨87JOC4020⟩; **3**, 623 ⟨86JOC398⟩

Park, Y. J., **2**, 717, 718 ⟨92CL1483⟩; **4**, 74, 75 ⟨93S209⟩

Parkanji, L., **2**, 413 ⟨85JOC1271⟩

Parkanyi, C., **3**, 680 ⟨80JHC833⟩; **5**, 771 ⟨80JHC877⟩

Párkányi, L., **2**, 1003 ⟨86JST141⟩; **5**, 199 ⟨86JCS(P1)927⟩

Parker, A. J., **1**, 178 ⟨66AJC521⟩; **2**, 126 ⟨59JOC1029⟩; **5**, 1040 ⟨73AJC273⟩

Parker, C. O., **3**, 565 ⟨62T79⟩; **4**, 439 ⟨72JOC922⟩, 824 ⟨56JA4944⟩; **6**, 144, 145, 369 ⟨63T(S)177⟩, 150 ⟨62T79⟩

Parker, D., **1**, 177 ⟨83JOC3334⟩, 217, 478 ⟨84JOC3928⟩; **2**, 546 ⟨92JCS(P1)3277⟩; **3**, 614 ⟨88OM1761⟩

Parker, D. A., **1**, 120 ⟨86T2873⟩, 207, 312 ⟨84TL5959⟩, 207, 312 ⟨85JOM(285)437, 87TL945⟩; **2**, 588 ⟨86T2873⟩, 589 ⟨85JOM(285)437⟩

Parker, D. G., **6**, 10 ⟨87JOM(325)13⟩, 14 ⟨88JCS(P1)921⟩

Parker, D. M., **2**, 448 ⟨70JCS(C)2529⟩
Parker, E. A., **2**, 103 ⟨33JA766⟩
Parker, K. A., **1**, 339, 1130 ⟨86TL2833⟩, 522 ⟨89JA5948⟩, 980 ⟨82JOC389⟩, 1043 ⟨75TL981⟩; **2**, 653 ⟨87JOC674⟩, 740 ⟨76JA7104⟩; **5**, 272, 299, 304 ⟨75TL981⟩
Parker, M. H., **1**, 583 ⟨94JOC332⟩
Parker, R. E., **1**, 108 ⟨63AFC63⟩
Parker, S. R., **3**, 92 ⟨78OPP201⟩
Parker, V. D., **1**, 117 ⟨64JA1112⟩; **5**, 986 ⟨68TL3341⟩
Parker, W. E., **5**, 205 ⟨55JA4037⟩
Parker, W. L., **3**, 704 ⟨88JAN454⟩
Parkes, G. D., **6**, 653 ⟨36JCS1693⟩
Parkhomenko, N. A., **5**, 1063 ⟨82ZOR668⟩
Parkin, A., **2**, 613 ⟨86BSF930, 86CC1623⟩
Parkin, S., **2**, 475 ⟨91JA3597⟩
Parkin, S. S. P., **6**, 590 ⟨83CC235⟩
Parks, J. E., **5**, 696 ⟨74JOM(71)453⟩
Parlar, H., **5**, 1037 ⟨79ZN(B)768, 80CZ365⟩
Parlier, A., **5**, 798 ⟨87JOM(327)339, 89OM2253⟩, 798 ⟨90JOM(381)191⟩, 798, 952 ⟨87JOM(333)245⟩, 944 ⟨86CC514, 91COS(6)485⟩, 953 ⟨92JA8088, 95JOM(494)43⟩; **6**, 716 ⟨91CRV35⟩
Parlor, S., **3**, 15, 90, 97, 101 ⟨84TL1843⟩
Parmantier, M., **5**, 481 ⟨75AG(E)53⟩
Parmar, S. S., **6**, 573 ⟨80JHC1369⟩
Parmerter, S. M., **3**, 448, 449 ⟨59OR(10)1⟩
Parnell, E. W., **2**, 797 ⟨60JCS4366⟩
Parnes, H., **1**, 5, 9, 935 ⟨79JOC151⟩; **5**, 13 ⟨85MI 501-01⟩, 82 ⟨79MI 502-08⟩
Parnes, Z. N., **1**, 13, 281 ⟨74S633⟩, 76 ⟨67T2235, 80CC748⟩, 906 ⟨67RCR735⟩; **3**, 58, 211 ⟨84S991⟩; **4**, 706 ⟨81JOM(209)233⟩, 706, 1036 ⟨79JOM(168)C33⟩
Parnham, M., **2**, 235 ⟨86EUP165534⟩
Parnham, M. J., **2**, 235 ⟨88GEP3626554⟩
Parnham, W. E., **4**, 440 ⟨50JA3843⟩
Parodi, B., **5**, 344 ⟨86FES539⟩
Parquette, J. R., **1**, 595, 598 ⟨93JOC1579⟩; **3**, 170 ⟨93JOC1579⟩
Parrado, P., **5**, 199 ⟨88AQ120⟩
Parrain, J. L., **1**, 1020 ⟨92TL3647⟩; **4**, 388 ⟨92JOM(437)C19⟩
Parratt, M. J., **4**, 131 ⟨83CC886, 87JCS(P1)181⟩, 801 ⟨82CC1270, 86JCS(P1)1417⟩, 802 ⟨86JCS(P1)1425⟩; **6**, 121 ⟨87JCS(P1)181⟩, 289 ⟨83CC886, 86JCS(P1)1425⟩
Parrett, F. W., **2**, 448 ⟨76SRI115⟩
Parrick, J., **4**, 21 ⟨89JCS(P1)2009⟩
Parries, G. S., **3**, 613 ⟨78JOC1017⟩
Parrilli, M., **2**, 45 ⟨75G377, 76TL3661⟩
Parrinello, G., **1**, 471 ⟨87JOC422⟩; **2**, 583 ⟨91OM1183⟩; **3**, 3 ⟨87JA7122⟩
Parrino, V. A., **6**, 640 ⟨59JA3728⟩
Parris, C. L., **5**, 116 ⟨73OSC(5)73⟩
Parrish, D. R., **1**, 578 ⟨90OSC(7)363⟩; **6**, 506 ⟨65JOC2849⟩
Parrott, M. J., **1**, 848, 919 ⟨B-78MI 119-05⟩; **5**, 264 ⟨85JCS(P1)461⟩, 280 ⟨84TL4825⟩
Parry, F. H., **3**, 527 ⟨69JA5679, 71JOC1486⟩
Parry, III, F. H., **4**, 768 ⟨70JOC1515⟩
Parry, R. J., **2**, 619, 626 ⟨65JOC304⟩; **3**, 161 ⟨90TL7549⟩; **4**, 264 ⟨92JA5946⟩
Parry, R. W., **2**, 409 ⟨73IC2482⟩; **4**, 115 ⟨76JINC55⟩, 341 ⟨76JFC(7)1⟩, 549 ⟨76MI 412-01⟩
Parshall, G. W., **1**, 663, 668 ⟨76CRV243⟩, 669 ⟨71JA3793⟩, 669, 766, 767, 769 ⟨78JA3611⟩, 670 ⟨B-80MI 114-03⟩; **2**, 427 ⟨59JA6273⟩, 680 ⟨78JA3611⟩; **3**, 508 ⟨78JA3611⟩; **4**, 672 ⟨71JA3793⟩, 725 ⟨78JA3611⟩; **5**, 1138, 1139, 1140 ⟨74JA2360⟩; **6**, 58 ⟨59JA4801⟩
Parshina, L. G., **5**, 694 ⟨76JOU1789, 76ZOR1826⟩
Parshina, V. A., **2**, 440 ⟨61ZOB3411, 61ZOB3417⟩, 460, 851 ⟨62ZOB4017⟩; **4**, 119 ⟨65JGU2053⟩, 121 ⟨61JGU3189⟩, 336 ⟨61ZOB3417⟩, 337 ⟨61ZOB3411⟩, 554 ⟨69ZOB1247⟩
Parshin, V. A., **2**, 442 ⟨62ZOB553⟩
Parsons, A. F., **1**, 946 ⟨92T9373⟩
Parsons, A. S., **3**, 268 ⟨89TL4879⟩
Parsons, I. W., **6**, 37 ⟨71T4533⟩
Parsons, J. L., **1**, 80 ⟨69CC1365⟩; **3**, 621 ⟨69JOC3684⟩
Parsons, M. E., **6**, 647 ⟨77JMC901⟩
Parsons, P., **1**, 1127 ⟨87JCS(P1)1237⟩
Parsons, P. J., **1**, 115 ⟨93SL189⟩, 227 ⟨93SL277, 93SL279⟩, 248 ⟨88CC81⟩, 347 ⟨90CC1543⟩, 552 ⟨89CC1090⟩, 816 ⟨92CC350⟩, 993 ⟨93SL931⟩, 1130 ⟨88CC283⟩; **2**, 936 ⟨93SL277, 93SL279⟩; **3**, 55 ⟨81TL2021⟩, 214 ⟨78CC822⟩, 227 ⟨78CC821⟩; **4**, 807 ⟨93SL189⟩, 860 ⟨83CC1209⟩, 957 ⟨76CC990, 78CC821⟩
Parsons, T. D., **6**, 243 ⟨67JA3446⟩
Parsons, T. F., **2**, 178 ⟨64JOC918⟩
Parsons, W. H., **3**, 198 ⟨89JMC1886⟩; **5**, 370 ⟨89JMC1886⟩
Partch, R. E., **2**, 86 ⟨67JA3662⟩
Partington, S. M., **1**, 883 ⟨88JCS(P2)983⟩
Partis, R. A., **3**, 666 ⟨73JOC4365⟩
Parton, S. K., **3**, 228 ⟨76JCS(P1)2289⟩
Partos, R. D., **1**, 988 ⟨69JA6112⟩; **3**, 568 ⟨65JA5068⟩, 591, 592, 603 ⟨69JA6112⟩; **5**, 860 ⟨65JA5068, 66JA4996⟩
Partovi, M. H., **3**, 553 ⟨74JOC3906⟩; **4**, 874 ⟨74JOC3906⟩
Partridge, B., **4**, 59 ⟨84JA1809⟩
Partridge, J. J., **2**, 519 ⟨73JA7171⟩
Partridge, M. W., **5**, 748 ⟨53MI 519-01⟩, 750 ⟨47JCS390⟩, 750, 751 ⟨47JCS1110⟩, 750, 751, 753 ⟨73JCS(P1)453⟩, 751, 753 ⟨49JCS1308⟩, 756 ⟨53JCS255⟩, 757 ⟨49JCS3043⟩
Partridge, S. M., **5**, 84 ⟨50CI(L)383⟩
Parveen, S., **6**, 558 ⟨91PJS422⟩
Parvez, M., **1**, 408 ⟨86T2979⟩, 830, 831 ⟨88JA3298, 93JOC5873⟩, 937 ⟨88T4671⟩, 949 ⟨93JOC6851⟩; **4**, 370 ⟨86T2979⟩; **5**, 1097 ⟨83CC457, 84CC75⟩; **6**, 93 ⟨92CC1539, 91IC1776⟩
Parvin, K., **2**, 437 ⟨59CI(L)541⟩
Parvizi, B., **4**, 978 ⟨76JCS(P1)1004⟩
Parvulescu, L., **3**, 539 ⟨88RRC83⟩
Pasanen, P., **4**, 229 ⟨B-80MI 405-01⟩; **5**, 701 ⟨77OPP57⟩
Pascali, V., **1**, 439 ⟨73CC351⟩
Pascard, C., **2**, 363 ⟨91T6655⟩; **4**, 693 ⟨75CC615⟩; **6**, 322 ⟨81CC780⟩
Paschalidis, C., **2**, 444 ⟨89AG(E)1700⟩; **4**, 548, 1023 ⟨88CB1241, 89CB1857⟩
Pasche, W., **4**, 979 ⟨69AP(302)81⟩; **5**, 965 ⟨69AP335, 69AP617⟩
Pascual, A., **1**, 812 ⟨84HCA774⟩; **5**, 1064 ⟨89HCA556⟩
Pascual, S., **5**, 595 ⟨93PS(78)97⟩
Pasedach, H., **1**, 525, 832 ⟨69LA(729)52, 71LA(747)60⟩
Pasenok, S., **6**, 243 ⟨91JOM(417)C47⟩
Pasetti, A., **6**, 42 ⟨66JOC2312⟩
Pasevina, K. I., **5**, 813 ⟨80IZV2623⟩
Pashalidis, I., **3**, 401 ⟨88JOM(348)C1⟩
Pashinin, V. A., **5**, 386 ⟨74ZOR1428⟩
Pashinkin, A. P., **4**, 915 ⟨70ZOB27⟩
Pashinnik, V. E., **2**, 183 ⟨75S801⟩; **4**, 10 ⟨73S787⟩; **5**, 5 ⟨75S801⟩
Pashkevich, K. I., **2**, 227 ⟨88ZOR1088⟩; **3**, 356 ⟨89ZOR1865⟩; **4**, 311 ⟨83IZV2568⟩; **6**, 219 ⟨92RCR75⟩
Paskins, K. N., **4**, 24 ⟨76OS(55)20⟩
Paskovich, D., **6**, 32, 219 ⟨67JOC833⟩
Paskucz, L., **2**, 361 ⟨69CB1489⟩
Pasmanyuk, S. V., **4**, 478 ⟨69IZV1757⟩
Pasquali, M., **4**, 693 ⟨95JOM(488)39⟩
Pasquato, L., **1**, 520 ⟨87CB531⟩, 617 ⟨84JOC596⟩; **4**, 840, 861 ⟨91TL2177⟩
Pasquet, G., **3**, 194 ⟨80TL931⟩
Pasquini, M. A., **2**, 299 ⟨80T1223⟩
Passacantilli, P., **3**, 92 ⟨90SC2565⟩
Pass, F., **2**, 427 ⟨62M230⟩, 428 ⟨59M148⟩, 429, 822, 823 ⟨59M792⟩
Passalacqua, V., **2**, 787 ⟨67AC(R)1266⟩

Passet, B. V., **5**, 742, 751 ⟨72ZOR1246⟩
Passmore, J., **6**, 49 ⟨73JCS(D)2314, 76JFC(7)261⟩, 49 ⟨77CJC3136⟩, 49, 50 ⟨75JCS(D)488⟩
Pasta, P., **2**, 151 ⟨92T9577⟩
Pasterczyk, J. W., **2**, 484 ⟨90PS(48)157⟩
Pasternack, G., **2**, 887 ⟨67JOC3707⟩
Pasternak, V. I., **3**, 618 ⟨74S563⟩; **4**, 971 ⟨73ZOR39⟩; **6**, 238, 610 ⟨71ZOR2084⟩, 367 ⟨72ZOB1169⟩, 610, 727 ⟨73ZOR39⟩, 612, 623 ⟨81ZOR180⟩, 612, 727 ⟨74ZOR449⟩, 612, 727 ⟨79ZOR215⟩, 619 ⟨78ZOR1841, 84ZOB1110⟩, 625 ⟨75ZOR2223⟩
Pasternak, Y., **1**, 1056 ⟨62CR(255)1750⟩
Pasto, D. J., **1**, 74 ⟨61JA2957, 91OR(40)91⟩, 458 ⟨61TL347, 91COS(8)471⟩, 459 ⟨73TL2517⟩, 498 ⟨82JA3676, 88JOC4807, 92JOC1146⟩, 499 ⟨80TL4787⟩, 658 ⟨64JA4343, 68JA4445, 73JOM(60)19⟩, 658, 659 ⟨66JOC2777⟩, 667 ⟨68JOC2975⟩, 922 ⟨88JA8164⟩, 954 ⟨84T2805⟩, 966 ⟨78JOC1389⟩, 970 ⟨76JOC3496, 78JOC1382, 78JOC1385⟩; **2**, 83 ⟨66JA2201⟩, 524 ⟨70TL215⟩, 532 ⟨71JOC1790⟩, 1010 ⟨87JOC3062⟩; **4**, 142 ⟨66JOC2773⟩, 144 ⟨68JA6259⟩, 631 ⟨64JA3039⟩, 632 ⟨69T1557⟩, 632, 634 ⟨67TL723, 69T1571⟩
Pastor, A., **6**, 327 ⟨92JCS(D)2307⟩, 327, 330 ⟨93IC5569⟩
Pastor, S. D., **2**, 353 ⟨88JOC5776⟩, 432 ⟨84JOC2906⟩, 457 ⟨92T2911⟩, 711 ⟨85JOC4812⟩; **4**, 1034 ⟨90JOC1649⟩
Pastore, H. D. O., **1**, 637 ⟨90JMOCL1⟩
Pastour, P., **3**, 104, 105, 106 ⟨69BSF2511⟩, 105, 106 ⟨76BSF265⟩, 107 ⟨68BSF4117⟩, 109 ⟨76BSF789⟩; **5**, 756 ⟨78JCR(S)132⟩
Pastukhova, I. V., **4**, 478, 493 ⟨76ZOB1246⟩, 873 ⟨77JGU2501⟩
Pastushenko, E. V., **6**, 478 ⟨85ZOR2206, 86JOU270⟩
Pastushok, V. N., **4**, 418 ⟨89ZOB2435⟩
Pastusznak, J. J., **2**, 229 ⟨⟨77MI 203-05, 81JOC1868⟩
Pasynkiewicz, S., **4**, 698 ⟨80JOM(201)97⟩
Pasynskii, A. A., **4**, 672 ⟨68CC1365, 69IZV100⟩, 717 ⟨80JOM(201)269⟩; **6**, 341 ⟨90JOM(384)279⟩
Patachke, H. P., **2**, 793 ⟨62AG(E)89⟩
Patai, S., **1**, 258 ⟨B-68MI 106-01, B-78MI 106-01⟩, 445 ⟨B-94MI 110-01⟩, 447, 450 ⟨B-89MI 110-01⟩, 458 ⟨B-80MI 110-01⟩, 888 ⟨B-64MI 119-01⟩, 889 ⟨62JCS1741, 62JCS396, 70JCS(B)469⟩, 1109 ⟨B-64MI 122-01⟩, 1133 ⟨B-83MI 122-01⟩; **2**, 38 ⟨B-71MI 202-01⟩, 116 ⟨B-74MI 203-01⟩, 142 ⟨B-88MI 203-01⟩, 172 ⟨B-90MI 203-05⟩, 481 ⟨B-94MI 209-01⟩, 547 ⟨B-89MI 210-01⟩, 773 ⟨B-75MI 215-01⟩, 872 ⟨B-94MI 217-01⟩
Patai, S.,**3**, 196 ⟨B-94MI 304-02⟩, 451 ⟨58JOC2014⟩, 501 ⟨B-89MI 314-01⟩; **4**, 330 ⟨B-81MI 407-01⟩, 333 ⟨B-86MI 407-01⟩; **5**, 1, 16, 17 ⟨B-72MI 501-01⟩; **6**, 104, 105 ⟨B-79MI 604-01⟩, 308, 311, 312, 313, 317 ⟨B-86MI 610-02⟩, 308, 311, 312, 313, 317 ⟨B-87MI 610-01⟩
Patalinghug, W. C., **6**, 168 ⟨91CC1560⟩
Patane, M. A., **1**, 826 ⟨90TL6799⟩
Pataud-Sat, M., **2**, 544 ⟨90JOC2878⟩, 814 ⟨81JOC3372, 83JOM(255)359⟩, 814 ⟨85OM623⟩
Patchet, A. A., **1**, 79 ⟨83JOC1991⟩; **3**, 198 ⟨89JMC1886⟩; **4**, 409 ⟨78JOC4276⟩, 979 ⟨76TL2405⟩; **5**, 370 ⟨89JMC1886⟩
Patchornik, A., **5**, 1111 ⟨70JA6969, 71JOC2727⟩
Pate, B. D., **2**, 627 ⟨82CC625⟩
Patel, A. C., **4**, 950 ⟨85JCS(P1)1201⟩
Patel, A. N., **1**, 965 ⟨66JCS(C)1223⟩
Patel, B. A., **1**, 32, 457, 507, 1009 ⟨80JOC4926⟩, 494 ⟨78JOC3898, 78JOC5018⟩, 1096 ⟨81JOC1061, 81JOC1067⟩, 1096 ⟨82JOC1267⟩; **5**, 157 ⟨77JOC3903⟩
Patel, B. P., **3**, 231 ⟨90JOC3⟩
Patel, C. S., **1**, 891 ⟨33JCS526⟩
Patel, D. V., **1**, 700, 703 ⟨86JA4603⟩; **6**, 616 ⟨82TL3539⟩
Patel, G., **2**, 530 ⟨92TL4209⟩; **4**, 526, 527, 528 ⟨92TL4209⟩
Patel, H., **4**, 161 ⟨88JOC5088⟩
Patel, H. A., **2**, 1060, 1061, 1076 ⟨71CJC2706⟩
Patel, M., **5**, 93 ⟨91OS(70)79⟩, 330 ⟨86TL683⟩; **6**, 450 ⟨87JOC2319⟩
Patel, M. S., **3**, 616 ⟨70CJC1881⟩
Patel, N. G., **4**, 268 ⟨83TL5563⟩
Patel, N. R., **5**, 659 ⟨92IC4917⟩, 679 ⟨90JFC(48)395⟩, 679, 703 ⟨92ICA527⟩; **6**, 251, 256, 259, 624 ⟨90JFC(48)395⟩, 259 ⟨93IC4802⟩, 259, 643, 651, 653 ⟨92IC488⟩, 260 ⟨92IC4917⟩, 262 ⟨92ICA527⟩
Patel, P., **4**, 540 ⟨85CC1021⟩; **5**, 361, 362, 711 ⟨85JCR(S)132⟩
Patel, R., **4**, 409 ⟨76JCS(P2)1564⟩
Patel, R. C., **2**, 11 ⟨79CC238, 80JCS(P1)1890⟩, 11 ⟨80JCS(P1)2901⟩; **3**, 433 ⟨83MI 311-01⟩
Patel, R. R., **3**, 208 ⟨71JOC4124⟩
Patel, S. K., **4**, 365 ⟨89JCS(P1)115⟩
Patel, S. T., **1**, 570 ⟨92CC1477⟩; **3**, 150 ⟨92CC1477⟩
Patel, U., **5**, 334 ⟨89JMC1024⟩
Patel, V. F., **1**, 335, 346, 495, 946 ⟨90JCS(P1)2709⟩, 345, 946 ⟨90JCS(P1)2691⟩, 346 ⟨86TL2303⟩, 670 ⟨87CC871⟩, 933 ⟨87TL5949, 88TL707, 90JCS(P1)2703⟩; **2**, 578 ⟨90JCS(P1)2703⟩, 578, 579 ⟨90JCS(P1)2709⟩; **3**, 136 ⟨90JCS(P1)2721⟩; **5**, 457 ⟨90JCS(P1)2721⟩
Patel, V. V., **4**, 868 ⟨75CC671⟩, 869 ⟨76TL423⟩, 869, 874 ⟨77CC835, 77JA5909⟩; **6**, 93, 101, 729 ⟨75TL1259⟩, 590 ⟨77JA5909⟩
Pater, R., **5**, 641, 846 ⟨65JOC3739⟩; **6**, 83 ⟨62JOC2858⟩
Paternò, E., **6**, 411, 412 ⟨1878G233⟩
Paterson, E. S., **2**, 555 ⟨B-85MI 211-01⟩
Paterson, I., **1**, 133 ⟨79TL1519⟩, 213 ⟨79S446⟩, 556 ⟨79S736⟩, 599 ⟨79TL993, 79TL2179⟩, 599 ⟨79TL995⟩, 640 ⟨93SL774⟩, 831 ⟨90TL7513, 91TL7601⟩; **2**, 533 ⟨88TL585, 89TL1293⟩, 533 ⟨89TL7121⟩, 533, 698 ⟨88CI(L)390, 90T4663⟩
Paterson, I., **2**, 698 ⟨92JOC5173⟩, 701 ⟨92TL4233⟩; **3**, 36, 124, 168 ⟨88T4207⟩, 36, 56, 168 ⟨79TL2179⟩, 56 ⟨82TL2399⟩, 139, 145 ⟨79TL1519⟩, 159 ⟨93JOC4182⟩, 164, 173 ⟨91SL145⟩, 227 ⟨79TL3205⟩, 247 ⟨90TL2213⟩, 639 ⟨81JCS(P1)256⟩; **4**, 66 ⟨82TL2399⟩, 67 ⟨82TL5083⟩, 966 ⟨79TL2179⟩; **5**, 161 ⟨90TL7513⟩, 173 ⟨82JA4686⟩
Paterson, T. M., **5**, 326, 330 ⟨89SL32⟩
Pathak, D. D., **2**, 438 ⟨92JCS(D)1751⟩
Pathak, R. K., **5**, 533 ⟨88JCR(S)376⟩
Pathak, V. N., **3**, 350, 364 ⟨77JCS(P1)1127⟩
Patheiger, M., **4**, 668 ⟨55ZAAC(282)345⟩; **6**, 206 ⟨55ZAAC(282)345⟩
Pathirana, H. M. K. K., **2**, 280 ⟨86JCS(D)2003⟩
Patience, J. M., **5**, 342 ⟨89TL3217⟩
Patil, G., **3**, 426 ⟨85OPP183⟩, 615, 616 ⟨84TL4583, 85SC157⟩, 618 ⟨86S1037, 86TL2203⟩; **5**, 226, 1134 ⟨86JOC3545⟩, 1076 ⟨87JOC275⟩, 1122 ⟨86SC1823⟩; **6**, 380 ⟨86JOC3545⟩
Patil, P., **4**, 511 ⟨89TL5841⟩; **5**, 289 ⟨89TL5841⟩; **6**, 162 ⟨89TL5841⟩
Patin, H., **1**, 446 ⟨88TL4577⟩; **2**, 446 ⟨90T6381⟩; **3**, 639 ⟨88T6107⟩
Patnaik, S., **6**, 237 ⟨93CM1321⟩
Patney, H. K., **4**, 246 ⟨93TL7127⟩
Patois, C., **4**, 576 ⟨93PS(75)143⟩
Paton, A. C., **1**, 89 ⟨64JA118⟩
Paton, J. M., **5**, 316, 318 ⟨74JOC3136⟩
Paton, R. M., **3**, 684 ⟨80JPS(A)635, 88CC1339⟩, 685 ⟨83JCS(P1)293⟩, 686 ⟨82TL5453, 85JCS(P1)1517, 86JCR(S)156, 89CSR33⟩; **4**, 232 ⟨81JCS(P1)2991⟩, 850 ⟨85CC696⟩
Patonay, T., **3**, 259 ⟨79LA174⟩
Patorra, A., **6**, 245 ⟨92JOM(438)57⟩
Patricia, J. J., **1**, 316 ⟨87JA2442⟩, 336, 346, 926 ⟨88JOC2137⟩; **2**, 672 ⟨89JOC38⟩
Patrick, C. R., **4**, 730 ⟨62CI(L)1472⟩; **5**, 116 ⟨62JCS186⟩
Patrick, D. W., **1**, 601, 610, 611, 621, 646 ⟨75CS9⟩, 716 ⟨75CS9⟩
Patrick, J. B., **6**, 108 ⟨52CB949, 52JA3855⟩
Patrick, J. E., **1**, 816, 818 ⟨70TL353⟩, 816, 818 ⟨71JA3556⟩; **6**, 608 ⟨68CC968⟩

Patrick, Jr., T. M., **1**, 629 ⟨63OSC(4)980⟩
Patrick, T. B., **1**, 1076 ⟨72JOC4467⟩, 1143 ⟨69JA6461, 72JOC1553⟩, 1144 ⟨70JA4312⟩; **2**, 13 ⟨83JOC4158, 86CJC138⟩, 15, 17 ⟨86CRV(86)997⟩, 16 ⟨88JOC5153⟩, 334 ⟨86CRV997⟩, 613 ⟨86JOC3242⟩, 753 ⟨74JOC1758⟩, 801 ⟨69JA6461, 70JA4312⟩; **3**, 27 ⟨90TL179⟩, 489 ⟨69JA6461⟩; **4**, 10 ⟨81JOC3917⟩, 52 ⟨84TL1019⟩
Patrie, W. J., **3**, 5, 294 ⟨79JOC2955⟩
Patroni, J. J., **6**, 477 ⟨87AJC795⟩
Patsanovskii, I. I., **4**, 119 ⟨86JGU501⟩, 489 ⟨82ZOB2444⟩; **6**, 685, 686, 689 ⟨87ZOB901⟩
Patsch, M., **1**, 398, 399 ⟨69TL3415⟩; **5**, 580 ⟨69AG(E)750⟩
Patsch, M. R., **6**, 63 ⟨69OS(49)86⟩
Patt, W. C., **3**, 109 ⟨91JOC328⟩
Pattenden, G., **1**, 320, 335, 345, 933, 946 ⟨88CSR361⟩, 335 ⟨88CC1110⟩, 335, 346, 495, 946 ⟨90JCS(P1)2709⟩, 340 ⟨92JCS(P1)1313⟩, 343 ⟨90TL3641⟩, 343, 931 ⟨93TL6787⟩, 345 ⟨84TL4317, 91TL243⟩, 345, 946 ⟨90JCS(P1)2691⟩, 346 ⟨86TL2303⟩, 347 ⟨92TL2851, 92TL4843, 94TL2413, 94TL2417⟩, 357 ⟨80CC1195⟩, 450 ⟨85JCS(P1)2399⟩, 460 ⟨78TL3483⟩, 497 ⟨85T4001⟩, 670 ⟨87CC871⟩, 690 ⟨85TL3393, 86TL399, 88JCS(P1)1085⟩, 725, 733 ⟨68JCS(C)1997⟩, 729 ⟨68JCS(C)1984, 76JCS(P1)1476⟩, 734 ⟨69JCS(C)1024⟩, 754 ⟨78TL4159⟩, 755 ⟨79TL5043⟩, 811 ⟨82TL1405⟩, 816 ⟨76CC684⟩, 933 ⟨87TL5949, 88TL707⟩, 946 ⟨90JCS(P1)2703⟩; **2**, 578, 579 ⟨90JCS(P1)2691, 90JCS(P1)2709⟩, 660 ⟨82TL4843⟩; **3**, 7, 120 ⟨84TL3363⟩, 60 ⟨92TL4053⟩, 136 ⟨90JCS(P1)2721⟩, 231 ⟨85JCS(P1)2407⟩, 232 ⟨81JCS(P1)3147⟩, 494 ⟨76JCS(P1)1466⟩; **4**, 170 ⟨79JCS(P1)62⟩; **5**, 304 ⟨93TL2677⟩, 327 ⟨93T9137⟩, 457 ⟨90JCS(P1)2721⟩; **6**, 447 ⟨89TL3229, 91TL259⟩
Patterson, C. W., **1**, 929 ⟨86CC1419, 87CC1238⟩, 929 ⟨88TL955, 93TL867⟩
Patterson, E. L., **2**, 214 ⟨54JA1828⟩
Patterson, J., **5**, 214 ⟨53JA285⟩
Patterson, Jr., J. W., **2**, 683 ⟨88SC1379⟩
Patterson, J. W., **1**, 829 ⟨91CC1579, 93T4789⟩; **5**, 164 ⟨71CC488⟩
Patterson, L. E., **5**, 291 ⟨68JA4706⟩
Patterson, R. G., **3**, 128 ⟨93T965⟩
Patterson, R. T., **4**, 995 ⟨89AX(C)1751⟩
Patterson, S. E., **4**, 411 ⟨90JOC4777⟩
Patterson, W., **1**, 1070 ⟨80JOC2526⟩; **2**, 976 ⟨80JOC2526, 80S1034⟩
Pattison, D. B., **5**, 272, 276 ⟨45JA1783⟩
Pattison, F. L. M., **1**, 2, 6 ⟨65CJC1689⟩, 107 ⟨56JA3484, 57JA2311⟩; **2**, 18 ⟨73OSC(5)136⟩, 19 ⟨57JA2311⟩; **5**, 55, 57, 59, 62 ⟨56JA2255⟩, 59 ⟨73OSC(5)580⟩
Pattison, I., **3**, 458 ⟨68CC250B⟩, 461 ⟨68JCS(A)57⟩
Pattison, J. B., **2**, 36 ⟨69JCS(C)1046⟩
Pattison, V. A., **1**, 396, 397 ⟨62JOC1933, 66JA78⟩; **2**, 478 ⟨62JOC1933⟩; **4**, 736 ⟨70JOC2096⟩; **6**, 57 ⟨68JOC1084⟩
Patton, A., **2**, 580 ⟨83JA5804⟩
Patton, D. S., **1**, 6 ⟨68JA7271⟩
Patttee, L., **3**, 129 ⟨79JOC2568⟩
Patwardhan, S. A., **2**, 679 ⟨74S348⟩
Paty, P. B., **2**, 542 ⟨78TL5091⟩; **3**, 39, 175 ⟨78TL5091⟩; **4**, 239 ⟨78TL5091⟩
Patzak, A., **5**, 344 ⟨86AP(319)682⟩
Patzschke, H. P., **2**, 100 ⟨62AG(E)97⟩
Pauer, F., **3**, 370 ⟨90CB1475⟩; **6**, 440 ⟨91CB2411⟩
Paugam, J.-P., **2**, 14 ⟨92TL213⟩, 1057 ⟨76TL835⟩; **6**, 7 ⟨92TL213⟩
Paukstelis, J. V., **2**, 1008 ⟨63JOC3021⟩; **3**, 422 ⟨63JOC3021⟩; **4**, 96 ⟨B-69MI 403-01⟩; **5**, 1120 ⟨74JOC1494⟩, 1126 ⟨70TL4731⟩; **6**, 42 ⟨74JOC1298⟩
Paul, G., **4**, 555 ⟨82ZC307⟩
Paul, H., **5**, 773, 849 ⟨68CB2028⟩
Paul, I. C., **1**, 898 ⟨71JCS(B)918⟩; **2**, 1001 ⟨74JA4099⟩, 1007 ⟨65JCS1097⟩
Paul, K. G., **1**, 503, 505 ⟨82JOC4254⟩
Paul, K. J., **4**, 509 ⟨91JCS(P1)1091⟩
Paul, L., **4**, 793 ⟨67CB1087⟩
Paul, M., **4**, 832 ⟨71JOM(30)C64⟩; **5**, 720 ⟨74CR(C)213⟩
Paul, N. C., **2**, 104 ⟨93T7037⟩, 386 ⟨88TL2735⟩
Paul, P. F. M., **1**, 503 ⟨53JCS3156⟩; **2**, 1018 ⟨51JCS893⟩
Paul, R., **1**, 414 ⟨50BSF121⟩; **3**, 685 ⟨63BSF140⟩
Paul, R. C., **2**, 498 ⟨54JCS881⟩; **4**, 504 ⟨67JIC995⟩
Paul, S. M., **5**, 1029 ⟨B-79MI 517-01⟩
Paul, W., **2**, 831 ⟨80AG740⟩; **3**, 375, 387 ⟨84AG(E)58⟩, 387, 391 ⟨83AG(E)316⟩; **4**, 149, 583 ⟨88OM1106⟩, 582 ⟨81JOM(219)C29, 82JOM(236)C65, 85JOM(281)317, 85CB261, 85CB3032⟩
Paulen, W., **5**, 1019 ⟨83AG(E)785, 83TL2639⟩; **6**, 454 ⟨83AG(E)785⟩, 454, 686 ⟨83CB109⟩, 454, 686 ⟨83TL2639⟩, 519 ⟨83CB2371⟩
Pauler, C. T., **1**, 777 ⟨81JOC2804⟩
Pauling, L., **1**, 846 ⟨B-60MI 119-01⟩, 852 ⟨B-60MI 119-02⟩, 869 ⟨B-60MI 119-03⟩, 923 ⟨69JCP(51)2767⟩; **3**, 694 ⟨42JA2952⟩
Paulmier, C., **1**, 243 ⟨B-86MI 105-02⟩, 601 ⟨B-86MI 114-01⟩, 712 ⟨B-86MI 115-02⟩, 780 ⟨B-86MI 117-01⟩, 985 ⟨84TL1987⟩; **2**, 280 ⟨B-86MI 204-01⟩, 282 ⟨82TL1557, 84TL1983⟩, 286 ⟨B-86MI 204-02, B-86MI 204-03⟩, 368, 369 ⟨B-86MI 206-01⟩; **3**, 37 ⟨B-86MI 301-01⟩, 38 ⟨82TL1557⟩, 74 ⟨84TL1987⟩, 104, 105, 106 ⟨69BSF2511⟩, 174 ⟨B-86MI 304-02⟩, 175 ⟨B-86MI 304-04⟩, 177 ⟨B-86MI 304-03⟩, 381 ⟨B-86MI 309-01⟩; **4**, 89 ⟨88TL5893⟩, 333 ⟨B-84MI 407-02⟩, 398 ⟨B-86MI 408-01⟩; **5**, 472 ⟨B-86MI 511-01⟩, 624 ⟨B-86MI 514-01⟩, 732 ⟨B-86MI 518-01⟩, 1000 ⟨B-65MI 527-01⟩, 1051 ⟨B-86MI 527-01⟩, 1108 ⟨79BSF237⟩; **6**, 93 ⟨82TL1557, 85BSF1219⟩, 97, 100 ⟨75JOC2911⟩, 308, 311, 312, 313 ⟨B-86MI 610-03⟩, 587 ⟨B-86MI 619-01⟩
Pauls, H. W., **5**, 226 ⟨86CAR(150)111⟩
Pauls, N., **4**, 581 ⟨83CB1070⟩
Paulsen, H., **1**, 4, 7 ⟨76CB3598⟩, 310 ⟨82TL3171⟩, 587 ⟨77CB2127⟩, 876 ⟨66AG(E)899⟩, 940, 941 ⟨85T4253⟩; **2**, 936 ⟨75CB1732⟩; **3**, 160 ⟨76CB3907⟩, 234 ⟨74TL4377⟩; **4**, 912 ⟨75MI 420-01⟩; **5**, 28 ⟨57AG600⟩; **6**, 549 ⟨88LA75⟩
Paulshock, M., **5**, 285 ⟨71JMC535⟩
Paulson, G., **2**, 690 ⟨78MI 213-02⟩
Paulson, G. O., **3**, 304 ⟨63JOC398⟩
Paulson, M. C., **3**, 449 ⟨48JA1381⟩
Paulson, S. E., **1**, 155 ⟨87JA6004⟩
Paulus, E. F., **2**, 1007 ⟨89AX(C)1520⟩; **6**, 129 ⟨89LA975⟩
Paulus, H., **2**, 1000 ⟨88TL1259⟩
Paulus, J., **5**, 105 ⟨B-49MI 502-01⟩
Pauluth, D., **3**, 8 ⟨85CC812⟩
Pauly, H., **6**, 27 ⟨23CB979⟩
Pauly, M., **3**, 619 ⟨86TL347⟩; **5**, 223, 374 ⟨88JHC59, 88JHC49⟩, 718 ⟨86JHC183⟩
Pauptit, R. A., **6**, 718, 727 ⟨84JOM(269)C55⟩
Pausacker, K. H., **2**, 136; **3**, 451 ⟨50JCS1328, 52JCS661⟩; **5**, 204 ⟨53JCS107⟩
Pauson, P. L., **1**, 528 ⟨73JCS(P1)975, 73JCS(P1)977, 85T5855⟩, 529 ⟨77JCR(M)153, 77JCR(M)168, 78JCR(M)4401, 78JCR(M)4454, 88JOM(354)233⟩, 530 ⟨88JOM(355)449⟩, 862 ⟨51NAT1039, 58PCS289⟩, 862 ⟨61JCS4610⟩, 1032 ⟨63JCS2990⟩; **2**, 182 ⟨58JCS692⟩, 684 ⟨75JCS(D)1677⟩, 985 ⟨73JCS(P1)977, 77ANY(295)2⟩; **3**, 218 ⟨85T5855⟩, 492 ⟨63JCS2990⟩, 567 ⟨71TL3367⟩; **4**, 688 ⟨76JOM(120)C13⟩
Paust, J., **4**, 49, 86 ⟨63CB2266⟩; **6**, 63 ⟨69OS(49)86, 72JA125⟩
Pautet, F., **2**, 148 ⟨85SUL127⟩
Pauw, J. E., **1**, 688 ⟨82TL5485⟩
Pauwels, R., **6**, 598 ⟨91JMC3187⟩
Pavel, G. V., **2**, 455 ⟨85JGU2198⟩
Paveni, M., **4**, 418 ⟨87CJC282⟩
Pavez, H., **5**, 759 ⟨82MI 519-01⟩

Pavia, A. A., **4**, 225 ⟨88TL4293⟩
Pavia, A. P., **4**, 176 ⟨88CAR(181)246⟩
Pavia, M. R., **5**, 306 ⟨87JMC498, 90JOC560⟩
Pavkovic, S. F., **5**, 934 ⟨89JOC3249⟩
Pavlath, A., **2**, 608, 619 ⟨57JCS1823⟩
Pavlath, A. E., **2**, 762 ⟨66MI 215-02⟩; **6**, 419 ⟨64JOC3007⟩, 419, 424 ⟨65JOC1639⟩, 421 ⟨65JOC4104⟩
Pavlenko, N. G., **3**, 567 ⟨71JOU413⟩; **4**, 555, 1004, 1008 ⟨79ZOB2217⟩, 800 ⟨76JOU782, 78JGU711, 79JGU1947⟩, 987, 988 ⟨74ZOR36⟩, 991 ⟨78ZOB778⟩; **5**, 810 ⟨72ZOR695⟩; **6**, 612, 727 ⟨74ZOR449⟩, 619 ⟨74ZOR36⟩, 620 ⟨82UKZ395⟩
Pavlenko, N. V., **2**, 476, 614 ⟨89JGU469⟩
Pavlenko, S. A., **6**, 671, 672 ⟨92MI 621-01⟩
Pavlichenko, M. G., **4**, 560 ⟨86ZOB2000⟩
Pavlik, F. J., **6**, 42 ⟨70JOC2054⟩
Pavlis, R. R., **2**, 35 ⟨64JA2956⟩
Pavlova, L. A., **2**, 876 ⟨82MI 217-01⟩, 1078 ⟨70JGU1100⟩, 1079 ⟨69JGU1068, 69JGU1797⟩; **6**, 68 ⟨86UK1803⟩, 95 ⟨78IZV468⟩
Pavlova, N. N., **4**, 239 ⟨80JOU1377, 82JOU876⟩; **6**, 92 ⟨71ZOR473⟩, 100 ⟨86ZOR107⟩
Pavlov, K. V., **4**, 142 ⟨92MI 403-02⟩
Pavlov, V. A., **4**, 132 ⟨91JGU1459⟩; **6**, 622 ⟨90ZOB223⟩, 659 ⟨86ZOB1427, 87ZOB2467⟩
Pavlov, V. M., **4**, 833 ⟨71KGS1645⟩
Pavlow, V. M., **2**, 208 ⟨72IZV2510⟩
Pawelchak, G. A., **5**, 358 ⟨73CJC4115⟩
Pawelke, G., **5**, 382 ⟨93JOM(446)25⟩; **6**, 14 ⟨89JFC(42)429⟩, 15 ⟨91JFC(52)229⟩, 51 ⟨80JFC(17)65⟩, 243 ⟨89JOM(378)125, 93AG429, 93JOM(446)25, 93JOM(456)19, 93ZN(B)935⟩, 262 ⟨84JFC(24)457⟩, 602 ⟨88CC105⟩, 604 ⟨92ZN(B)351⟩, 615 ⟨83JFC(22)175⟩, 615, 632 ⟨84JFC(26)321⟩
Pawellek, D., **4**, 743 ⟨62JA995⟩; **5**, 809 ⟨66AG908⟩
Pawelzik, H., **5**, 620, 623, 624 ⟨62LA(660)60⟩, 636 ⟨61LA(643)29⟩
Pawlak, M., **1**, 726 ⟨73HCA1176⟩; **3**, 494 ⟨77HCA1161⟩, 496 ⟨75JOC100⟩
Pawlak, M. S., **1**, 506 ⟨67ANY(145)91⟩
Pawlenko, S., **2**, 385 ⟨90HOU(E16a/2)1137⟩, 401 ⟨85HOU(E11)1123⟩, 535 ⟨80HOU(13/5)⟩, 599 ⟨78HOU(13/6)181⟩, 1089 ⟨80HOU(13/5)33⟩; **4**, 294 ⟨92HOU(E14a/3)545⟩, 315, 319 ⟨92HOU(E14a/3)483⟩, 831 ⟨93HOU(E15/2)1742⟩; **6**, 172, 189 ⟨80HOU(13/5)27⟩, 172, 204 ⟨78HOU(13/6)181⟩, 236 ⟨85HOU(E11)1055, 85HOU(E11)1067⟩, 236 ⟨85HOU(E11)1084⟩
Pawlowski, G., **2**, 1089 ⟨81JOM(204)315⟩
Pawlowski, N. E., **2**, 122 ⟨66JA5245⟩
Paxton, H. W., **2**, 491, 496, 498 ⟨47JA927⟩
Payard, M., **4**, 547 ⟨86TL1145⟩
Payette, D., **3**, 621 ⟨78CJC41⟩
Payling, D. W., **1**, 75 ⟨71JINC1157⟩
Payne, D. S., **2**, 859 ⟨67JCS1028⟩, 862 ⟨67JCS(A)1492⟩
Payne, G. B., **1**, 223 ⟨62JOC3819⟩; **3**, 374 ⟨68JOC3517⟩, 527 ⟨66JOC718⟩; **4**, 213 ⟨59JOC284⟩; **5**, 51 ⟨57JOC1680⟩, 205 ⟨59JOC1354⟩, 494 ⟨68JOC3517⟩, 709 ⟨61JOC651, 61JOC659⟩
Payne, J. H., **4**, 125, 128 ⟨61JA1811⟩
Payne, J. J., **4**, 229 ⟨92TL4625⟩
Payne, M. W., **6**, 340, 344 ⟨83OM219⟩
Payne, N. C., **1**, 157 ⟨85IC3680⟩; **4**, 713 ⟨80IC693⟩, 1018 ⟨78JA1164⟩
Payne, S. E., **5**, 1019 ⟨88JA6240⟩
Payzant, J. D., **1**, 107 ⟨74JA4027, 76CJC1643⟩
Paz, J. L. de, **5**, 1068 ⟨89JA355⟩
Paz, M. A., **2**, 1021 ⟨93JA11702⟩
Pazhitnova, N. V., **4**, 916 ⟨80ZOB1225⟩
Pazik, J. C., **2**, 489, 492 ⟨89OM482⟩; **4**, 659 ⟨88OM1516⟩
Pazo, Y., **3**, 209 ⟨88JOC1110⟩; **6**, 557 ⟨88JOC1110⟩
Pazos, J. F., **2**, 764 ⟨69JOC2269⟩
Peach, J. M., **3**, 712, 718, 719 ⟨89SL9⟩
Peach, M., **6**, 93 ⟨92CC1539⟩
Peach, M. E., **1**, 897 ⟨62JCS600⟩; **2**, 137 ⟨B-74MI 203-03⟩
Peacock, K., **4**, 814 ⟨63JOC369⟩
Peacock, N. J., **1**, 618, 704 ⟨80JOC1719⟩; **5**, 867 ⟨80JOC1717⟩
Peacock, R. D., **2**, 824 ⟨68JCS(A)2149⟩
Peacock, V., **2**, 382 ⟨76JA5269⟩
Peacock, V. E., **2**, 388 ⟨81JOC2402⟩
Peak, D. A., **2**, 176, 177 ⟨51JCS3292⟩, 691 ⟨62CI(L)1244⟩; **4**, 899 ⟨52JCS4067⟩; **5**, 579 ⟨52JCS742⟩, 710 ⟨48JCS1618⟩, 749, 750 ⟨51JCS485⟩
Peak, S. A., **2**, 48 ⟨94TL425⟩
Peake, C. J., **5**, 664 ⟨86SC763⟩
Peake, S. C., **2**, 476 ⟨70JCS(A)1049⟩, 855 ⟨72JCS(P2)380⟩
Peake, S. L., **1**, 108 ⟨78JA4888⟩
Pearce, A., **1**, 643, 755 ⟨77JCS(P1)550⟩; **2**, 936 ⟨80JCS(P1)2485⟩; **3**, 211 ⟨80JCS(P1)2485⟩, 639 ⟨81JCS(P1)256⟩; **4**, 650 ⟨77JCS(P1)550⟩
Pearce, A. A., **1**, 443, 660 ⟨64TL2039⟩, 659 ⟨69JCS(B)863⟩
Pearce, B. C., **1**, 1134 ⟨85TL2237⟩
Pearce, C. A., **2**, 418, 419, 1085 ⟨57JCS381⟩
Pearce, L. B., **2**, 386 ⟨79JOC1881⟩
Pearce, R., **1**, 99 ⟨75JCS(P1)470⟩, 663, 668 ⟨76CRV219⟩; **2**, 565 ⟨79JCS(D)54⟩, 566 ⟨73JOM(57)269, 77JA6645⟩, 937 ⟨75JCS(P1)475⟩, 937, 938 ⟨74JCS(P1)2055⟩, 937, 938 ⟨75JCS(P1)470⟩; **4**, 656 ⟨73JCS(D)445, 73JOM(57)269⟩, 656 ⟨78CC140⟩, 656, 695 ⟨77JCS(D)999⟩, 702 ⟨79JCS(D)54⟩, 725 ⟨75CC621⟩, 725, 726 ⟨79JCS(D)45⟩; **6**, 195 ⟨77JA6645, 78CC140⟩
Pearl, I. A., **5**, 114 ⟨55OSC(3)745, 63OSC(4)972⟩
Pearlman, B. A., **1**, 594, 616 ⟨85JOC3622⟩
Pearlman, P. S., **3**, 13 ⟨80JOC3449⟩
Pearlson, W. H., **4**, 737 ⟨53JA968⟩; **6**, 32, 218 ⟨46JA968, 53USP2647933⟩, 218 ⟨49JA2499, 55USP2704776⟩, 419 ⟨46JA1672⟩
Pearn, E. J., **5**, 1015 ⟨63SA1621⟩
Pearn, J., **2**, 110 ⟨61JOC3250⟩
Pearsall, M. A., **5**, 1019 ⟨88JA7228⟩; **6**, 345 ⟨84CC1089⟩
Pearse, P. L., **2**, 433, 434, 435 ⟨29JCS1262⟩
Pearson, A. J., **1**, 98 ⟨B-85MI 103-02⟩, 196 ⟨82COMC-I(8)939⟩, 292 ⟨75TL3617⟩, 885 ⟨B-85MI 119-02⟩; **2**, 91 ⟨86JOC2505⟩, 684 ⟨86JOC2137, 89CC1363⟩, 985 ⟨92OM4096⟩; **3**, 215 ⟨85JCS(P1)267⟩; **4**, 536 ⟨90JCS(P1)2251⟩, 856 ⟨92T7527⟩; **5**, 146 ⟨79JCS(P1)1255⟩
Pearson, C. J., **5**, 351 ⟨85TL3171⟩, 354 ⟨85TL3167⟩
Pearson, D. A., **3**, 60 ⟨84JA260⟩
Pearson, D. E., **2**, 38 ⟨B-77MI 202-01⟩, 608, 626 ⟨58JOC1412, 76S621⟩, 626, 685 ⟨67JOC2358⟩, 696 ⟨74CRV45⟩; **3**, 279, 281, 295, 296, 297, 298, 301, 305, 309 ⟨72S533⟩, 286, 459 ⟨72JOC3369⟩, 526 ⟨B-70MI 316-01⟩; **4**, 48 ⟨B-70MI 402-01⟩, 176 ⟨B-70MI 404-01⟩; **5**, 1 ⟨B-77MI 501-01⟩, 24 ⟨B-70MI 502-01, B-77MI 502-01⟩, 112 ⟨73OSC(5)890⟩, 303 ⟨53JA5905⟩
Pearson, D. L., **1**, 906 ⟨57JA4557⟩
Pearson, K., **4**, 4 ⟨79JFC(13)325⟩
Pearson, M. J., **4**, 426 ⟨85TL377⟩, 447 ⟨77JCS(P1)189, 86JCS(P1)1077⟩; **5**, 223 ⟨85JCS(P1)1927⟩, 345 ⟨89TL3219⟩; **6**, 111 ⟨74JCS(P1)22⟩
Pearson, M. M., **2**, 539 ⟨90JOC6082⟩
Pearson, N. R., **1**, 977 ⟨81JOC829, 82JOC3364⟩; **2**, 268, 522, 526, 900 ⟨80JA5919⟩; **3**, 502 ⟨79JA2491⟩
Pearson, P. S., **2**, 292 ⟨70IC106⟩
Pearson, R. G., **1**, 155, 850, 855, 869 ⟨53JA2439⟩, 886 ⟨63JA3533, 67JA1827⟩; **5**, 447 ⟨64JA3994⟩
Pearson, T. G., **3**, 389 ⟨38JCS409⟩
Pearson, W. H., **1**, 190 ⟨92JOC2986⟩, 371 ⟨94TL2641⟩, 372 ⟨86JA2769, 92JA1329⟩; **2**, 325 ⟨89TL6661⟩, 517 ⟨89JOC5814⟩, 753 ⟨81JA2483⟩; **3**, 182 ⟨89JOC5651⟩, 404 ⟨88TL761⟩, 411 ⟨90TL5541⟩, 419 ⟨90TL7571⟩, 457 ⟨92JA1329⟩; **4**, 22 ⟨90JOC5719⟩, 213 ⟨87JOC1353⟩, 329 ⟨81JA2483⟩, 530 ⟨89JOC5651⟩, 531 ⟨91JA8546⟩, 533, 537 ⟨92JOC6354, 88TL761⟩, 537 ⟨92JA1329⟩

Pease, J., **1**, 5, 9, 935 ⟨79JOC151⟩
Peaston, W. C., **3**, 257 ⟨68JCS(C)2481⟩
Pebler, J., **4**, 581, 582, 583 ⟨83OM257⟩, 632 ⟨83JA2582⟩; **5**, 776 ⟨82ZAAC(486)61⟩
Pecchi, S.,**3**, 34 ⟨92SL803⟩
Pech, R., **4**, 979 ⟨91PHA741⟩
Pechacek, J. T., **1**, 1038, 1140 ⟨91TL1707⟩
Pecher, B., **4**, 245 ⟨93S149⟩
Pechet, M. M., **1**, 871, 881 ⟨69CC1497⟩; **2**, 15 ⟨76JA3034, 76JA3036, B-80MI 218-01⟩, 16 ⟨72CC122⟩, 139 ⟨91JOC6697⟩, 240, 360 ⟨91JOC6702⟩, 335 ⟨74JCS(P1)732, 75CC97⟩, 391 ⟨60JA2640⟩, 608, 612 ⟨68CC806⟩, 765 ⟨75JCS(P1)1764⟩; **3**, 352 ⟨75CC557, 79JCS(P1)1166⟩; **5**, 313, 314, 316 ⟨74JCS(P1)732⟩, 735 ⟨91JOC6697⟩; **6**, 231 ⟨74JCS(P1)732⟩
Pechmann, H. von, **5**, 842 ⟨1894CB2121⟩, 842, 859 ⟨1894CB320⟩, 857, 858, 859 ⟨1892CB3175⟩, 858 ⟨1896CB2166⟩
Pechukas, A., **2**, 482, 483 ⟨37JA2068⟩
Pechurina, S. Ya., **2**, 1093 ⟨71JGU122⟩
Pecile, C., **5**, 1053 ⟨62AC(R)79⟩
Peck, C. J., **4**, 46 ⟨80CC158⟩
Peck, D. R., **1**, 826 ⟨87JA1160⟩; **4**, 204 ⟨81JA5618⟩
Peck, E. B., **2**, 270 ⟨76JMC1423⟩
Peck, R. M., **2**, 270 ⟨76JMC1423⟩
Pecquet, P., **5**, 199 ⟨92H(34)739⟩
Pecunioso, A., **1**, 1049 ⟨89JOC2391⟩; **2**, 393 ⟨87T5411, 90T7497⟩, 393, 783 ⟨88JOC45, 89JOC2391⟩
Pedan, V. P., **5**, 694 ⟨76JOU60, 76ZOR64⟩; **6**, 629 ⟨75ZOR1628⟩
Pedatella, S., **2**, 688 ⟨93SC1515⟩; **3**, 170 ⟨93SC1515⟩
Peddle, G. J. D., **4**, 386 ⟨66JOM(5)486⟩, 609 ⟨68CJC2119⟩, 614, 621 ⟨72JA5837⟩; **5**, 410 ⟨63CJC2351⟩, 431 ⟨60JA5102, 66JOM(5)106, 66JOM(5)486⟩; **6**, 125, 131, 132, 519 ⟨68CJC2119⟩, 515 ⟨69CC462⟩
Pedersen, A. B., **5**, 690 ⟨86CS347⟩
Pedersen, B. S., **3**, 346, 363 ⟨78BSB223⟩; **5**, 484, 551, 570 ⟨78BSB293⟩, 551 ⟨77ACS(B)460⟩, 570, 571, 608, 610, 618 ⟨78BSB229⟩, 570, 575, 608, 610, 618 ⟨77BSB693⟩, 570, 608, 610 ⟨78BSB299, 78BSB525⟩
Pedersen, C., **1**, 2, 4, 8, 9 ⟨84CAR(130)125⟩; **5**, 634 ⟨61ACS1104⟩, 636 ⟨61ACS1087⟩, 639, 640 ⟨61ACS1097, 61ACS1109⟩, 641 ⟨61ACS1124⟩
Pedersen, C. J., **5**, 106 ⟨67JA7017⟩
Pedersen, C. L., **2**, 792 ⟨78ACS(B)625⟩; **3**, 677 ⟨77ACS(B)848⟩
Pedersen, C. T., **2**, 388 ⟨64ACS2199⟩; **3**, 356 ⟨74ACS(B)367⟩, 366 ⟨72TL5213, 74AG(E)349⟩; **5**, 1053, 1054 ⟨63ACS1459, 66ACS278⟩; **6**, 598 ⟨66ACS278⟩
Pedersen, E. B., **2**, 694 ⟨74RTC252⟩; **3**, 48 ⟨91JOC3591⟩, 344 ⟨69T5703⟩; **5**, 745 ⟨77ACS(B)261, 79S546⟩, 1126 ⟨87AP(320)1088⟩
Pedersen, F., **3**, 550 ⟨91CB1485⟩
Pedersen, K. O., **5**, 83 ⟨48MI 502-03⟩
Pedersen, S. D., **1**, 990 ⟨91CC566⟩
Pedersen, S. F., **2**, 81 ⟨89JA8014⟩, 672, 988, 989 ⟨89JA5468⟩, 979 ⟨82JA6808⟩; **3**, 514 ⟨82JA7483⟩
Pedersen, U., **5**, 551 ⟨84T2663⟩, 558, 573, 587 ⟨83T3429⟩
Pederson, I., **4**, 863 ⟨80ACS(A)577⟩
Pederson, R. L., **3**, 32 ⟨90JOC4897⟩, 48 ⟨89JA3924⟩, 159 ⟨91S499⟩
Pederson, S. F.,**3**, 424 ⟨94JA1316⟩
Pedlar, J. M., **5**, 532 ⟨70GEP(O)2024393⟩
Pedler, A. E., **6**, 20 ⟨74CC166, 78JCS(P1)202⟩
Pedlow, Jr., G. W., **1**, 301 ⟨41JA758⟩
Pednekar, P. R., **1**, 6 ⟨87JOC4784⟩
Pedoussaut, M., **1**, 388 ⟨69CR(C)567⟩
Pedrini, P., **2**, 942 ⟨84CC258, 86S757⟩; **3**, 109 ⟨87S998⟩, 311 ⟨84JOC590, 84JOC3478⟩, 312 ⟨81CC655, 86G133⟩, 368 ⟨79JCS(P1)1720⟩, 707, 714 ⟨84CC258, 87S693⟩; **6**, 634 ⟨87JOC3413, 87S693⟩, 637 ⟨90H(31)1213, 93H(36)473⟩
Pedro, J. R., **2**, 36 ⟨93SL489⟩
Pedrocchifantoni, D., **2**, 39 ⟨94MI 202-03⟩
Pedrocci-Fantoni, G., **2**, 359 ⟨81TL5073⟩
Pedrosa, R., **2**, 306 ⟨90SL763⟩, 312 ⟨83JOC2255⟩, 320 ⟨90S153, 90SC1149⟩; **3**, 252 ⟨90SC2537⟩
Peek, R. J., **2**, 625 ⟨87TL3857⟩
Peel, J. B., **2**, 171 ⟨29JCS1048⟩
Peel, S. R., **2**, 597 ⟨90POL2763⟩
Peeples, C. J., **2**, 323 ⟨92JOC1727⟩
Peerboom, R., **5**, 437 ⟨85RTC74⟩
Peet, N. P., **2**, 675 ⟨73JOC514⟩; **3**, 240 ⟨89JOC3913⟩, 468 ⟨75JHC1191⟩; **4**, 61, 62, 73, 79, 780 ⟨85JA735⟩, 79, 780 ⟨85CC678⟩; **5**, 22 ⟨93H(35)315⟩, 327 ⟨89JOC3913⟩
Peeters, H., **2**, 749 ⟨80GEP2912344⟩
Pegg, N. A., **1**, 1098 ⟨90JA277⟩; **3**, 156 ⟨89JOC4576⟩
Pegg, W. J., **1**, 10 ⟨81JOC2596⟩; **3**, 174 ⟨81JOC2596⟩
Pegram, J. J., **1**, 654 ⟨80TL3343⟩; **2**, 914, 939 ⟨88TL6719⟩; **4**, 176 ⟨81TL4603⟩, 199 ⟨80TL3343⟩
Pegues, J. F., **3**, 621 ⟨80JOC167⟩
Pehk, T., **1**, 225 ⟨91TA843⟩
Peiffer, G., **2**, 1071 ⟨69CR(268)358⟩
Peisach, D., **1**, 346, 926 ⟨93JOC7640⟩
Peiseler, B., **4**, 235 ⟨88TL5233⟩
Peisker, A., **5**, 334 ⟨91CZ135⟩
Pelah, Z., **1**, 630 ⟨65CB3236⟩
Pelc, B., **3**, 226 ⟨71JCS859⟩
Pelchowicz, Z., **2**, 458 ⟨61JCS238⟩
Pelcman, B., **5**, 199 ⟨87TL4441⟩
Pelczar, F. L., **1**, 83 ⟨73JOM(61)167⟩
Pelczer, I., **4**, 120 ⟨86ACH145⟩
Peleties, N., **2**, 281 ⟨69AG(E)450⟩; **3**, 175 ⟨69AG(E)450⟩, 497 ⟨72CB511⟩; **4**, 289, 398, 400 ⟨69AG(E)450⟩, 399, 400 ⟨72CB511⟩; **6**, 92, 307, 317, 697 ⟨72CB511⟩, 93, 94 ⟨69AG(E)450⟩
Pélichet, C., **5**, 455 ⟨70HCA1091⟩
Pelinski, L., **1**, 816, 818 ⟨89TL2549⟩
Pelizarri, G., **5**, 847 ⟨14G72⟩
Pelizzi, G., **2**, 881, 883, 886, 890 ⟨84JCS(D)2365⟩, 890, 891, 893 ⟨84IC1504, 88IC3730⟩; **6**, 517 ⟨79IC1165⟩
Pel'kis, N. P., **5**, 718 ⟨84JOU1963, 84ZOR2153⟩; **6**, 656 ⟨75ZOR762⟩
Pel'kis, P. S., **4**, 332 ⟨77JOU2224⟩, 865 ⟨78ZOR1947, 79ZOR1477⟩, 995 ⟨79ZOR2356⟩; **5**, 1022 ⟨68RCR363⟩, 1049 ⟨65ZOR1415, 68ZOR658⟩; **5**, 592 ⟨72ZOR528⟩, 593 ⟨74ZOR1992, 76ZOR69, 76ZOR681⟩, 595 ⟨71ZOR2352, 73ZOR347⟩, 727 ⟨70ZOR805⟩, 843 ⟨61ZOB1621, 63ZOB3917⟩; **6**, 275 ⟨76ZOR558⟩, 277 ⟨68ZOR324⟩, 564 ⟨81ZOR191⟩, 573 ⟨61ZOB3726⟩, 653 ⟨57ZOB2134, 65ZOR735⟩, 654 ⟨63ZOB113, 72ZOR39⟩
Pellacani, L., **3**, 153 ⟨78S227⟩, 181 ⟨83TL593⟩, 181 ⟨94T3829⟩
Pellechia, P. J., **2**, 74 ⟨90JOC1868⟩
Pellegata, R., **3**, 229 ⟨75CC892⟩
Pellegrini, J. P., **2**, 626 ⟨81JOC2169⟩
Peller, R. P., **2**, 841 ⟨65JOC97⟩
Pellerin, B., **4**, 111 ⟨89MI 403-02⟩; **5**, 1153 ⟨86TL5723⟩
Pellet, M., **3**, 9, 123 ⟨78S63⟩, 210 ⟨88T4463⟩, 220 ⟨77TL3505⟩
Pelletier, J. D., **2**, 32 ⟨94TL1051⟩
Pelletier, S. W., **1**, 4, 7 ⟨87TL3883⟩; **3**, 187 ⟨80TL809⟩; **5**, 51 ⟨71CC10⟩
Pellicciari, R., **1**, 44 ⟨79CC959, 85JCS(P1)493⟩; **3**, 311 ⟨92TL3003⟩
Pelling, S., **4**, 582 ⟨83JCS(D)1495⟩
Pellissier, H., **3**, 658 ⟨86TL3505, 89TL171⟩
Pellmont, B., **1**, 956 ⟨59HCA719⟩
Pellon, P., **1**, 158 ⟨93TL1011⟩, 162 ⟨89JA8737⟩; **2**, 489 ⟨90TL1147⟩; **3**, 639 ⟨86TL5611⟩; **4**, 137 ⟨89JA8737⟩, 566 ⟨87PS(30)523⟩, 567 ⟨86TL5611⟩, 597, 598 ⟨89TL349⟩
Peloso, C., **5**, 192 ⟨83TL813⟩
Pelster, H., **1**, 899 ⟨61LA(650)62⟩, 902 ⟨61AG342⟩; **5**, 977 ⟨63BEP631961⟩

Pelter, A., **1**, 75 ⟨B-88MI 103-01⟩, 126 ⟨73CC544, 73TL4491⟩, 126 ⟨75TL1633⟩, 163 ⟨83TL623, 83TL627, 83TL631⟩, 204 ⟨82TL5229⟩, 212 ⟨93T3007⟩, 449 ⟨B-88MI 110-01⟩, 509 ⟨91COS(8)703⟩, 509, 510 ⟨83TL1433, B-88MI 112-01⟩, 509 ⟨B-88MI 112-02⟩, 657, 658, 946 ⟨79COC(3)791⟩, 660, 764 ⟨83TL635⟩, 660, 765 ⟨87CC297, 85TL5093⟩, 765 ⟨89TL5643, 89TL5647⟩, 838 ⟨B-80MI 118-02⟩, 988 ⟨92JCS(P1)747⟩, 999 ⟨88TL4181⟩, 1035 ⟨75CC857⟩, 1035, 1103 ⟨76TL4385⟩; **2**, 4, 14, 900 ⟨91COS(7)593⟩, 418 ⟨79COC(3)925⟩, 514, 518, 519, 520, 523, 528, 529, 531, 533 ⟨B-88MI 210-01⟩, 515 ⟨78TL4715⟩, 516 ⟨79TL4777⟩, 525 ⟨83TL631⟩, 525, 1005 ⟨83TL623⟩, 525, 1005 ⟨83TL627⟩, 525, 905 ⟨89TL5643, 89TL5647⟩, 533 ⟨76JCS(D)2087⟩, 748 ⟨74CC346⟩, 900 ⟨79COC(3)687, 79COC(3)689⟩, 900, 1005 ⟨91COS(1)487⟩, 900, 901 ⟨91TL6239⟩, 900, 903, 904, 906 ⟨B-88MI 218-01⟩, 904 ⟨73CC544, 76JCS(P1)2419, 76JCS(P1)2428, 76JCS(P1)2435⟩, 905 ⟨83TL635, 83TL637⟩, 966 ⟨B-88MI 219-01⟩, 1005 ⟨91PAC403⟩, 1086 ⟨75CC857, 76TL4385⟩; **3**, 3, 114 ⟨83TL1433⟩, 19 ⟨83TL631⟩, 127 ⟨89TL5643⟩, 135 ⟨81CC1149⟩, 247 ⟨82TL353⟩, 412 ⟨91TL5353⟩; **4**, 206 ⟨82TL353⟩, 250 ⟨78TL2345⟩, 373 ⟨74CC863, 81CC1149⟩, 523 ⟨75JCS(P1)129, 75JCS(P1)138⟩, 642, 645, 1077 ⟨83TL637⟩, 664 ⟨83TL623, 83TL631⟩, 664 ⟨93T2988⟩, 665 ⟨93T2979⟩, 704 ⟨93T3007⟩, 832 ⟨93HOU(E15/2)1737⟩, 840, 961 ⟨90JCS(P1)2775⟩, 1077 ⟨86TL5033⟩; **5**, 234 ⟨77JCS(P1)1672⟩, 282 ⟨91TL1825⟩, 381 ⟨B-88MI 507-01⟩, 565 ⟨70T1539⟩; **6**, 172 ⟨79COC(3)687⟩, 172, 198, 201 ⟨91COS(1)487⟩, 172, 198 ⟨B-87MI 606-02, B-88MI 606-01⟩, 199 ⟨83TL623⟩
Pelter, M. W., **5**, 30 ⟨89OS(68)198⟩
Pelton, E. A., **1**, 209 ⟨93JOC7304⟩
Peltonen, S., **1**, 229 ⟨73ACS3944⟩
Peltzer, III, E. T., **1**, 636 ⟨70AG(E)374⟩
Peña, M. R., **1**, 494, 1029 ⟨85JOC2302⟩, 1027 ⟨89JA5417⟩
Penacles, S., **5**, 87 ⟨85S773⟩
Penberthy, C., **4**, 416 ⟨75JMC177, 75JMC182⟩
Penco, S., **1**, 433 ⟨90JCS(P1)428⟩; **3**, 284 ⟨84CC530⟩
Pendalwar, S. L., **1**, 456 ⟨90JOC3452⟩, 515 ⟨92JOC4786⟩
Pendergast, W., **5**, 690 ⟨86JHC1411⟩
Pendse, A. D., **2**, 683 ⟨78JOC4672⟩
Pendyala, V. R., **6**, 251 ⟨93MI 608-01⟩
Penenory, A., **1**, 922 ⟨89JA5845⟩
Penev, P., **4**, 490 ⟨71CC490⟩
Peng, C. H., **5**, 87 ⟨57JA645⟩
Peng, C. T., **5**, 25, 108 ⟨55JA6682⟩
Peng, G., **1**, 827 ⟨94TL2145⟩
Peng, J., **3**, 707 ⟨91JCS(P1)2495⟩; **5**, 1054 ⟨91JCS(P1)2495⟩; **6**, 595 ⟨91JCS(P1)2495, 91JOC6733⟩
Peng, M.-L., **1**, 24 ⟨91TL3551⟩
Peng, S., **4**, 713 ⟨85OM1123⟩
Peng, S.-M., **4**, 263 ⟨90JOC5515⟩, 683 ⟨93JOM(461)207⟩, 711 ⟨93OM108⟩; **6**, 348 ⟨90JCS(D)3033⟩
Peng, T. S., **2**, 580 ⟨90TL4417⟩
Peng, W.-S., **2**, 359, 363 ⟨94JOC914⟩
Peng, X., **1**, 54 ⟨86MI 102-01⟩, 1038, 1140 ⟨91TL1707⟩; **3**, 410 ⟨92JA5466⟩
Penke, B., **3**, 475 ⟨70HCA1057⟩
Pen'kovskii, V. V., **6**, 671, 672 ⟨92MI 621-01⟩
Penland, A. D., **2**, 421 ⟨71JCS(A)1237, 72JCS(D)326⟩
Penman, D. R., **3**, 104 ⟨53JCS1742⟩
Penmasta, R., **1**, 900 ⟨90JA3228⟩; **2**, 25 ⟨89TL791⟩; **3**, 163 ⟨89TL667⟩; **5**, 366 ⟨87SC409⟩
Penn, G., **6**, 489 ⟨88JMC1831⟩
Penn, J. H., **3**, 84 ⟨93JA82⟩, 84, 85 ⟨92TL3737⟩
Penn, R. E., **2**, 184 ⟨78JA3622⟩; **3**, 335, 369 ⟨82JA3119⟩, 337 ⟨78JA7436, 76JA1264⟩; **4**, 258 ⟨82JA3119⟩
Pennanen, P., **4**, 553 ⟨90SC1865⟩
Penner, G. H., **1**, 418 ⟨86CJC793⟩
Pennetreau, P., **4**, 429, 431 ⟨90JOC6198⟩
Penney, C., **5**, 568, 571 ⟨93T10489⟩
Penney, T., **4**, 869, 874 ⟨77JA5909⟩; **6**, 590 ⟨77JA5909⟩
Penning, T. D., **5**, 138 ⟨86JA5655⟩
Penninger, S., **4**, 355 ⟨77LA624⟩, 493 ⟨78S526⟩
Pennings, M. L. M., **3**, 435 ⟨83JOC1816⟩
Pennington, W. T., **4**, 563 ⟨B-88MI 410-01⟩; **5**, 665, 672 ⟨85JOC993⟩
Penny, M., **1**, 177 ⟨93CC1111⟩
Pensionerova, G. A., **5**, 656 ⟨81JGU1015, 81ZOB1201⟩
Penso, M., **2**, 25, 31, 35 ⟨92JCS(P1)2309⟩; **3**, 427, 430 ⟨90S333⟩, 643 ⟨90SC965⟩; **5**, 1030 ⟨91S1001⟩
Penton Jr., H. R., **1**, 575 ⟨73JOC26⟩; **3**, 370 ⟨73JA279⟩, 370, 374 ⟨74JOC2885⟩; **5**, 335, 336 ⟨73JA279⟩
Penzenstadler, E., **4**, 611 ⟨92ZN(B)805, 92ZN(B)1377⟩; **6**, 174, 192 ⟨92ZN(B)805⟩
Penzhorn, R. D., **2**, 186 ⟨73JOC4070⟩
Penzlin, G., **1**, 885 ⟨B-86MI 119-01⟩
Penzo, M., **4**, 777 ⟨75CC107⟩
Peoples, P. R., **1**, 847 ⟨80JA4709⟩
Peperkamp, H. A., **6**, 642 ⟨64RTC1305⟩
Peperzak, R. M., **3**, 35 ⟨87RTC489⟩
Pepin, Y., **3**, 621 ⟨78CJC41⟩
Pepino, R., **3**, 702 ⟨90H(31)1855⟩, 706 ⟨89S641, 90LA935⟩, 706 ⟨91LA843⟩
Peppe, C., **4**, 153, 588 ⟨91OM2159⟩, 153, 698 ⟨86OM525⟩
Pepper, J. M., **2**, 647 ⟨62CJC2175⟩
Pepperman, A. B., **4**, 342 ⟨75JOC2056⟩; **5**, 102 ⟨81JOC5039⟩
Pepperman Jr., A. B., **4**, 458 ⟨72JHC715, 75JHC579⟩, 458, 477, 481 ⟨74JHC407⟩, 482 ⟨74JHC1085⟩
Peppin, A., **2**, 183, 194 ⟨66JCS(C)1171⟩
Peppmöller, R., **6**, 642, 650 ⟨67ZN(B)820⟩
Peprzak, R. M., **3**, 253 ⟨87RTC549⟩
Perakyla, H., **2**, 228 ⟨82SC947⟩
Peraldo, M., **6**, 526 ⟨60G1005⟩
Perales, A., **4**, 325 ⟨92JCS(P1)291⟩
Peran, J., **2**, 268 ⟨64T173⟩
Perchais, J., **5**, 826 ⟨74T999⟩
Perchec, P. Le, **1**, 526 ⟨75S1⟩, 808 ⟨66BSF281⟩; **3**, 539 ⟨73CC795⟩; **4**, 127 ⟨88BSF699⟩
Perchinummo, M., **1**, 934 ⟨71T3575⟩
Perchonock, C. D., **1**, 521 ⟨80TL3323⟩
Percival, A., **1**, 653 ⟨78CC178⟩; **2**, 921 ⟨81JCS(P1)2415⟩; **4**, 649 ⟨81JCS(P1)2415⟩, 970, 977, 978 ⟨83JCS(P1)1741⟩
Percy, J. M., **1**, 570 ⟨92CC1477⟩
Perdomo, G. R., **4**, 175 ⟨87TL5595⟩
Pereira, S., **1**, 417 ⟨93MI 109-01⟩, 830 ⟨93MI 118-01⟩, 876 ⟨93MI 119-01⟩
Perekalin, V. V., **4**, 428 ⟨76RCR1052⟩, 451, 490, 493 ⟨92RCR1220⟩, 826 ⟨81ZOR1550⟩, 994 ⟨85ZOR1111⟩; **6**, 54, 261 ⟨89ZOR2490⟩, 147 ⟨69ZOR2246⟩, 149 ⟨66ZOR1902, 69ZOR220, 69ZOR1313, 73ZOR269, 88ZOR644⟩, 653 ⟨72ZOR332⟩
Perekhod'ko, V. S., **5**, 827 ⟨70MI 521-01⟩; **6**, 145 ⟨80MI 605-02⟩
Perelman, M., **3**, 226 ⟨60JA2402⟩
Perelygina, O. P., **4**, 827 ⟨84ZOB1925⟩
Pereshein, V. V., **5**, 388 ⟨85ZOB1071⟩
Pereslegina, N. S., **6**, 565 ⟨86MI 617-03⟩
Peretyazhko, M. Z., **6**, 573 ⟨61ZOB3726⟩
Peretz, M., **1**, 194, 1053 ⟨83JOC5302⟩; **6**, 116 ⟨86MI 604-01⟩
Perevalova, E. G., **4**, 152 ⟨77IZV2417⟩, 696 ⟨74JOM(65)131⟩, 706 ⟨57DOK(112)439, 58DOK(120)1263⟩, 719 ⟨73DOK(209)869, 74IZV2645⟩, 723 ⟨69IZV2030, 69IZV2032, 72IZV2594⟩, 724 ⟨54DOK(97)459⟩; **5**, 385 ⟨88ZOB62⟩, 821 ⟨59DOK(127)337⟩
Perevalov, V. P., **5**, 652 ⟨90KGS1430⟩
Pereye, M., **6**, 26, 31 ⟨75JOM(102)423⟩
Pereyre, M., **1**, 128 ⟨75JOM(96)225, B-87MI 104-05⟩, 132 ⟨67CR(C)1204, 69TL505⟩, 132 ⟨70CR(C)100, 73JOM(55)273⟩, 452 ⟨B-87MI 110-05⟩, 919 ⟨B-86MI 119-05⟩; **2**, 599 ⟨B-87MI 211-07⟩, 601 ⟨68BSF3385,

75JOM(88)79⟩, 701 ⟨67BSF1928, 68JOM(11)97⟩, 1094 ⟨B-87MI 223-05⟩; **3**, 640 ⟨67PNA(265)939⟩; **4**, 358 ⟨88CC503, 90JOM(391)283⟩, 388 ⟨81JOM(212)C31, 83JOC1559⟩, 533 ⟨86USP4617409, 88JOM(339)267⟩; **5**, 415 ⟨92JOM(437)C13⟩; **6**, 172, 204 ⟨B-87MI 606-01⟩
Perez, C., **3**, 8, 122 ⟨86T4133⟩
Perez, D., **1**, 86 ⟨87JOC2576⟩; **2**, 36 ⟨87JOC4846⟩; **3**, 123 ⟨90JOC2927⟩, 128, 621 ⟨87JOC2576⟩, 129 ⟨87JOC5570⟩
Perez, F., **2**, 7 ⟨86JOC789⟩
Pérez, J. A. M., **1**, 606 ⟨92JOC3173⟩
Perez, J. J., **3**, 231, 303 ⟨85JA5219⟩
Perez, L. A., **6**, 280, 632 ⟨82JOC4177⟩
Pérez, M. A., **2**, 360 ⟨82TL1243⟩; **5**, 731 ⟨81LA65⟩
Pérez, P. J., **5**, 948 ⟨92JA7288⟩; **6**, 327 ⟨89IC2120⟩
Perez, S., **3**, 8, 122 ⟨86T4133⟩
Perez-Carreño, E., **2**, 1003 ⟨91T5277⟩
Perez-Cordero, E., **1**, 890 ⟨92JA3978⟩
Perez-Martinez, J. A., **6**, 327 ⟨92AG81⟩, 328 ⟨91JOM(420)C12, 93OM1394⟩, 329 ⟨94JOM(467)231⟩, 329, 330 ⟨92POL2713, 93OM2888, 94OM1336⟩
Perez Moya, L. A., **3**, 530, 531 ⟨86S43⟩
Perez-Ossorio, R., **3**, 413 ⟨86JCR(S)46⟩, 414 ⟨85TL4403, 86JCS(P1)2021⟩, 417 ⟨84JCR(S)144⟩, 418 ⟨87OPP181⟩; **4**, 304 ⟨84JCR(M)144⟩
Perez-Soler, R., **4**, 174 ⟨91TL3313⟩
Perfetti, R. B., **3**, 119 ⟨76TL3499⟩
Perfileva, I. A., **2**, 439 ⟨59DOK(127)1217⟩
Perhavc, M., **5**, 836 ⟨84H(22)1545⟩
Periasamy, M., **1**, 541 ⟨88TL4313⟩, 906 ⟨82JOC2089⟩; **2**, 34 ⟨90TL1919⟩, 300 ⟨93JOC3156⟩, 312 ⟨89SC565⟩, 316, 325 ⟨92T4623⟩, 969 ⟨87CC495⟩; **3**, 694 ⟨B-83MI 321-01⟩; **5**, 27 ⟨85S253⟩
Periasamy, M. P., **1**, 155 ⟨74JA3711⟩, 155 ⟨77JA2631⟩; **3**, 697 ⟨71TL4965, 74JOC611⟩
Pericàs, M. A., **1**, 529 ⟨85T5995⟩, 530 ⟨90JA9388⟩, 1085 ⟨88S707⟩; **2**, 84, 318 ⟨91TL6931, 91TL6935⟩, 1026 ⟨81T1441, 82T1505, 87OS(65)68, 87T2311, 90JA7405⟩, 1027 ⟨88CC942⟩; **3**, 243 ⟨90TL7505⟩, 247, 540 ⟨85S1118, 90JOC395⟩; **4**, 196 ⟨88S707⟩, 767 ⟨77TL4433⟩, 851 ⟨90TL2169⟩
Pericas, M., **4**, 763 ⟨94JA2153⟩
Pericas, N. A., **3**, 540 ⟨77TL4437⟩
Perich, J. W., **2**, 107 ⟨88S142⟩
Perichon, J., **1**, 427, 428 ⟨91SC2377⟩; **3**, 146 ⟨86TL4175⟩, 620 ⟨93T5091⟩; **5**, 44 ⟨85TL1509, 86GEP3522304⟩
Périchon, J., **6**, 24 ⟨88TL1699⟩
Périchon, S., **6**, 22 ⟨92CC53⟩
Perie, J. J., **2**, 306 ⟨83T703⟩
Peries, R., **1**, 1091 ⟨77JA7365⟩; **2**, 676 ⟨77JA7365⟩; **3**, 60 ⟨77JA7365⟩
Peringer, P., **2**, 860 ⟨84HCA1238⟩; **5**, 373 ⟨85CC1764⟩; **6**, 372 ⟨84OM1916, 86JOM(301)269⟩
Periot, P., **3**, 236 ⟨78S29⟩
Perisamy, M., **2**, 59 ⟨91JOC5964⟩
Perjési, P., **6**, 562 ⟨91M1047⟩
Perkin, A. G., **3**, 358 ⟨03JCS1201⟩
Perkin, Jr., W. H., **4**, 11 ⟨22JCS2202⟩, 24, 29 ⟨07JCS816⟩
Perkins, E. G., **5**, 98 ⟨63OSC(4)444⟩
Perkins, M. J., **2**, 389 ⟨70JCS(C)2195⟩; **5**, 116 ⟨73OSC(5)51⟩; **6**, 445 ⟨78JCS(P1)1066⟩
Perkinson, N. A., **5**, 759 ⟨79JOC4160⟩
Perkow, W., **2**, 693 ⟨52N353⟩
Perlier, S., **5**, 558 ⟨86MI 513-01⟩
Perlikowska, W., **2**, 124 ⟨87S1009⟩, 437, 443 ⟨79T1531⟩
Perlin, A. S., **3**, 8 ⟨B-69MI 301-02⟩, 121 ⟨B-69MI 304-02⟩; **5**, 28 ⟨B-69MI 502-03⟩; **6**, 71 ⟨65CJC1918⟩
Perlin, T. J., **2**, 118, 124 ⟨93JMC460⟩
Perlman, M. E., **2**, 413 ⟨88JOC1761⟩
Perl'mutter, B. L., **4**, 142 ⟨71DOK(196)85⟩
Perlmutter, P., **1**, 300 ⟨B-92MI 107-01⟩; **2**, 928 ⟨87S1032⟩; **3**, 615, 633, 634, 638 ⟨B-92MI 318-01⟩, 660 ⟨85CC4⟩
Perlmutter, R., **1**, 537 ⟨B-92MI 112-02⟩
Perlova, E. M., **4**, 152 ⟨88JOM(341)225⟩; **5**, 1142 ⟨88JOM(341)225⟩
Perlstein, M. E., **3**, 218 ⟨60JOC1966⟩
Permutter, P., **3**, 438 ⟨84TL5951⟩
Pernak, J., **1**, 260 ⟨91JOC4439⟩; **3**, 710 ⟨91S868⟩; **4**, 309 ⟨91JOC4439, 92JOC547⟩, 323 ⟨91S1147⟩
Perner, M., **5**, 212 ⟨85CZ389⟩
Perner, R. J., **2**, 66 ⟨90TL7415⟩
Pernet, A. G., **4**, 198 ⟨72JA8929⟩
Perni, R. B., **2**, 22 ⟨87OS(66)194⟩; **3**, 619 ⟨83OPP297⟩; **4**, 247 ⟨89SC2383⟩
Peron, D., **5**, 944 ⟨95OM1095⟩
Perozzi, E. F., **1**, 576 ⟨77OR(25)1⟩; **2**, 547 ⟨79JA1591⟩
Perregaard, J., **2**, 694 ⟨74RTC252⟩; **5**, 551 ⟨77ACS(B)460⟩, 570, 575, 608, 610, 618 ⟨77BSB321⟩, 600, 621 ⟨75ACS(B)538, 75ACS(B)604⟩
Perren, E. A., **4**, 130 ⟨52JCS3437⟩
Perret, C., **2**, 558, 560 ⟨81HCA2606, 82TL3901⟩
Perret, R., **1**, 185 ⟨89HCA1537⟩; **3**, 130 ⟨87HCA1745⟩
Perri, M., **5**, 370 ⟨83JA3743⟩
Perri, S. T., **3**, 539 ⟨86JOC3067⟩; **5**, 351 ⟨90JOC6037⟩
Perrichon, H., **2**, 1019 ⟨26AC(P)3⟩; **3**, 675 ⟨26AC(P)5⟩
Perrier, G., **5**, 727 ⟨92TL5117⟩
Perrier, H., **3**, 90, 98 ⟨89JOC24⟩
Perrier, M., **2**, 735 ⟨50BSF245⟩; **3**, 378 ⟨79BSF(2)199, 79BSF(2)205⟩
Perrin, C. L., **1**, 103 ⟨65MI 103-02⟩; **2**, 1007 ⟨66JA4934⟩; **5**, 723 ⟨85JOC1405⟩
Perrin, D. E., **2**, 455 ⟨75JOC1523⟩
Perrin, L. C., **5**, 847 ⟨B-91MI 521-01⟩
Perrin, P., **2**, 753 ⟨86TL6193⟩
Perrin, R., **2**, 619 ⟨74JOC1744⟩
Perrini, G., **2**, 288 ⟨83JCS(P2)1605⟩; **3**, 492 ⟨82G25⟩; **5**, 734 ⟨84JCR(S)404⟩, 734, 760 ⟨84JCS(P1)897⟩
Perrino, A. C., **1**, 506 ⟨61JOC1073⟩
Perrio, S., **1**, 834 ⟨89CC414, 91T6275⟩; **5**, 557 ⟨89CC414, 91T6275⟩, 557 ⟨92T4135, 93T3131⟩
Perriot, P., **1**, 164 ⟨77BSF765⟩, 1067 ⟨75S458, 79S502⟩; **2**, 1015 ⟨75S458⟩; **3**, 125 ⟨79S968⟩, 652 ⟨83S297⟩; **4**, 740, 741 ⟨75S458⟩
Perrocheau, J., **5**, 1153 ⟨86TL5723⟩
Perron, F., **4**, 171, 176 ⟨89CRV1617⟩; **6**, 79 ⟨88TL2023⟩
Perron, Y. G., **5**, 9 ⟨63JA643, 68CJC2549⟩
Perrone, E., **2**, 6 ⟨85G393⟩
Perronet, J.,**5**, 771, 854 ⟨72GEP2146192⟩
Perrot, M., **3**, 14, 82, 95, 102, 104, 107 ⟨87TL3941⟩, 83, 98, 102, 104, 107 ⟨88TL1271⟩
Perrot-Petta, M., **5**, 759, 777 ⟨91S954⟩
Perry, D. A., **1**, 655, 1080 ⟨82JCS(P1)1563⟩; **3**, 271 ⟨81T4027⟩, 274 ⟨82JCS(P1)1563⟩, 331 ⟨83JA1683, 86JOC1556⟩, 331, 332 ⟨86JA2985⟩
Perry, D. H., **5**, 26, 37 ⟨73S416⟩
Perry, F. M., **4**, 770, 779 ⟨65T2899⟩; **5**, 354 ⟨53JA1933⟩
Perry, M., **2**, 247, 250 ⟨82S281⟩
Perry, M. A., **3**, 88 ⟨50JA2992⟩
Perry, M. W. D., **1**, 154 ⟨83CC1040⟩; **3**, 293 ⟨83CC1040⟩; **4**, 536 ⟨83CC1040, 86T4223⟩
Perry, R., **1**, 1002, 1134 ⟨83TL5185⟩; **3**, 65, 232 ⟨83TL5185⟩
Perry, S. G., **5**, 26 ⟨63JCS2918⟩, 26, 37 ⟨60JCS3070⟩
Perry, W. D., **6**, 208 ⟨89OM2646⟩
Perry, W. L., **2**, 274 ⟨50JA3595⟩
Persad, A., **1**, 144 ⟨92SL197⟩, 144 ⟨93JCS(P1)3099⟩
Persau, C., **2**, 471 ⟨91ZAAC(549)36⟩
Pershin, D. G., **1**, 58 ⟨84JOM(260)17⟩
Pershin, G. N., **5**, 567, 581 ⟨76KFZ41⟩
Persico, D. F., **4**, 44 ⟨88JOC78⟩
Persico, N., **5**, 597 ⟨85FES120⟩
Persmark, U., **1**, 560 ⟨62ACS(B)523⟩
Person, D., **5**, 694 ⟨89TL3069⟩
Person, H., **5**, 364 ⟨76BCF1119⟩
Person, M., **4**, 891 ⟨83JOC481⟩

Persons, P. E., **1**, 752 ⟨87TL3075⟩
Persson, B., **5**, 1114 ⟨67ACS837⟩
Persson, J., **3**, 175 ⟨89TL2665⟩; **4**, 88, 89, 93 ⟨89TL2665⟩
Perst, H., **3**, 556 ⟨93HOU(E15/3)2531⟩, 582 ⟨78AG(E)465, 79TL2125⟩; **6**, 729 ⟨B-71MI 623-01⟩
Perter, A., **4**, 230 ⟨85JCS(P1)587⟩
Perthuis, J., **1**, 69 ⟨85BSF75⟩
Perumal, S., **1**, 260, 263 ⟨91HCA1924⟩; **4**, 318 ⟨91HCA1936⟩, 318, 322 ⟨91HCA1924⟩
Perumattam, J., **3**, 250 ⟨87TL913, 88JOC2238⟩; **5**, 425 ⟨87TL913⟩
Perutz, R. N., **5**, 794 ⟨78CC762⟩
Peruyero, C., **5**, 1139 ⟨77JOM(128)291⟩
Perveen, R., **6**, 558 ⟨87MI 617-02, 91PJS422⟩
Perveev, F. Y., **2**, 427 ⟨60ZOB784⟩
Perzanowski, H. P., **5**, 201 ⟨89T4599⟩
Pesce, G., **1**, 187 ⟨78CC1085⟩, 592 ⟨76S200⟩, 1025 ⟨79TL3873⟩; **2**, 28 ⟨77JCS(P1)501⟩; **3**, 154 ⟨77JCS(P1)501⟩
Peschard, O., **1**, 180 ⟨88T5495⟩
Peschel, R., **4**, 74 ⟨70JPR683⟩
Peseckis, S. M., **1**, 221 ⟨83TL1377⟩, 700, 703 ⟨82TL4879⟩; **2**, 1011 ⟨84JOC3429⟩
Peseke, K., **3**, 357 ⟨91EGP294941⟩, 547 ⟨75ZC19⟩; **4**, 967 ⟨88MI 421-01⟩, 970 ⟨84MI 421-01⟩
Pesel, H., **2**, 812 ⟨78ZN(B)1465⟩; **5**, 1134 ⟨78ZN(B)1465⟩
Pesotskaya, G. V., **4**, 971 ⟨73ZOR39⟩
Pestana, D. C., **2**, 475 ⟨91JA3597⟩
Pestunovich, A. E., **5**, 1076, 1077 ⟨85ZOB621⟩
Pestunovich, V. A., **2**, 1091 ⟨86JGU849⟩; **4**, 137 ⟨76CCC386⟩, 140 ⟨89JOM(361)147⟩
Petasis, N. A., **1**, 601, 604 ⟨B-84MI 114-01⟩, 826 ⟨90TL6799, 93TL1721⟩; **2**, 286 ⟨B-87MI 204-02⟩, 937 ⟨92SL665⟩; **4**, 333 ⟨B-84MI 407-01⟩; **5**, 239, 336, 337 ⟨85T4835⟩
Pete, J. P., **1**, 195, 548 ⟨87TL4547⟩, 287, 362 ⟨90T1859⟩, 630 ⟨67BSF675⟩, 1000 ⟨86TL573⟩; **2**, 637 ⟨91AG(E)416⟩; **3**, 37 ⟨85JOM(280)269⟩, 72 ⟨85TL839⟩, 294 ⟨91TL3993⟩; **4**, 274 ⟨88BSF989⟩; **5**, 296 ⟨86TL6339⟩
Peter, R., **1**, 127 ⟨80CC1202⟩; **2**, 566 ⟨80AG(E)1011⟩, 702 ⟨81TL4691⟩
Petercakova, D., **5**, 1043 ⟨88ZC141⟩
Peterman, K., **5**, 784 ⟨86JFC(30)439⟩
Peterman, K. E., **4**, 107 ⟨76IC743⟩; **5**, 313 ⟨74IC2705⟩; **6**, 51, 52 ⟨75IC1223⟩
Petermann, J., **5**, 577, 590, 608, 609 ⟨75T1209⟩
Peters, D., **2**, 165 ⟨58JOC381⟩
Peters, D. A. V., **1**, 2, 6 ⟨65CJC1689⟩
Peters, D. G., **1**, 68 ⟨85JOC673⟩
Peters, E., **6**, 222 ⟨57JA5654⟩, 580 ⟨81JPR41⟩
Peters, E.-M., **1**, 566 ⟨93CB1481⟩, 785 ⟨91JOC5782⟩; **2**, 1003 ⟨83M1097⟩; **3**, 380 ⟨79AG(E)944⟩, 689 ⟨85JA980⟩; **4**, 609 ⟨88JOM(341)109⟩; **6**, 173, 383 ⟨93ZAAC(619)1494⟩, 381, 385, 391 ⟨83ZAAC(497)119⟩, 391 ⟨88JOM(341)109⟩
Peters, E. N., **1**, 895 ⟨73JA2400, 77JA1712⟩
Peters, F. M., **2**, 421 ⟨65CJC1884⟩
Peters, G. A., **4**, 883 ⟨61JOC412⟩; **5**, 749 ⟨61JOC412⟩
Peters, G., **2**, 460 ⟨62JOC2198⟩
Peters, H., **2**, 440 ⟨83IC3624, 84IC3304⟩
Peters, J. A., **2**, 841 ⟨65JOC97⟩
Peters, J., **2**, 454 ⟨88SC425⟩; **4**, 465 ⟨88SC425⟩, 566 ⟨81ZAAC(475)18⟩; **5**, 888 ⟨81ZAAC(475)18⟩; **6**, 164 ⟨81ZAAC(475)18⟩, 293, 693 ⟨82AG(E)80⟩, 688, 689 ⟨81TL4957⟩
Peters, J. L., **3**, 640 ⟨89IC492⟩
Peters, K., **1**, 566 ⟨93CB1481⟩, 785 ⟨91JOC5782⟩, 966 ⟨92AG(E)224⟩; **2**, 437 ⟨90ZN(B)1548⟩, 732 ⟨87AG(E)780⟩, 1003 ⟨83M1097, 83M783⟩; **3**, 213 ⟨89CB2131⟩, 231 ⟨89CB1653⟩, 553, 554 ⟨91OM3967⟩, 581 ⟨92JOM(423)329⟩; **4**, 609 ⟨88JOM(341)109⟩
Peters, K., **4**, 695, 717 ⟨89JOM(372)183⟩
Peters, K., **4**, 991 ⟨91JHC1863⟩, 991, 993, 996 ⟨85BSB475⟩; **6**, 173, 383 ⟨93ZAAC(619)1494⟩, 381, 385, 391 ⟨83ZAAC(497)119, 88JOM(341)109⟩
Peters, N. K., **1**, 96 ⟨81JOC1593⟩
Peters, S. C., **1**, 1026 ⟨93SL413⟩
Peters, T. B., **4**, 583 ⟨92OM1392⟩
Peters, U., **1**, 690 ⟨87TL3091⟩
Peters, W., **2**, 861 ⟨91ZAAC(606)119⟩
Peterseim, M., **1**, 17, 939 ⟨92SL801⟩, 936 ⟨91JOC5971, 92JCS(P1)3165⟩
Petersen, C. S., **3**, 450 ⟨63ACS559⟩
Petersen, H., **4**, 301 ⟨73S247⟩, 423, 427, 492 ⟨73S243⟩
Petersen, H. J., **6**, 647 ⟨84SC1275⟩, 648 ⟨78JMC773⟩
Petersen, J. L., **2**, 569 ⟨87OM2354⟩; **4**, 659 ⟨89OM2461, 93OM3890⟩
Petersen, J. M., **5**, 698, 813 ⟨66JOC3708⟩
Petersen, J. S., **2**, 58, 521 ⟨86JA7404⟩
Petersen, M. R., **1**, 414, 416, 827 ⟨70JA741⟩
Petersen, S., **6**, 413 ⟨52HOU(8)84⟩, 424 ⟨52HOU(8)101⟩
Petersen, U., **6**, 488 ⟨83HOU(E4)273⟩, 489 ⟨83HOU(E4)274, 83HOU(E4)279, 83HOU(E4)284⟩, 500 ⟨83HOU(E4)334⟩
Peterson, A. C., **2**, 440, 441 ⟨92JOC511⟩
Peterson, B., **1**, 216 ⟨83JOC2499⟩
Peterson, B. H., **4**, 213 ⟨87JOC1351⟩
Peterson, C. J., **5**, 361 ⟨63JOC2294⟩
Peterson, D., **1**, 648, 650 ⟨75JOC2263⟩, 649, 650, 761 ⟨75JA1464⟩, 650, 759 ⟨74TL1133⟩, 651 ⟨75SC359⟩, 759, 761 ⟨75JOC2264⟩
Peterson, D. J., **1**, 140, 693 ⟨67JOC1717⟩, 157 ⟨65JOC1939⟩, 157 ⟨66JOC2373⟩, 648, 649, 758 ⟨68JOC780⟩, 1063 ⟨67JOC105⟩; **2**, 434 ⟨67JOC1717⟩, 545 ⟨60JA403⟩, 709 ⟨68JOC780⟩, 1001 ⟨71JA4027⟩; **4**, 365, 391 ⟨71USP3597463⟩, 365, 391, 392 ⟨67JOC1717⟩, 377, 390, 394, 398 ⟨72MI 408-02⟩, 397 ⟨71GEP2114367, 76USP3987191⟩, 528, 533 ⟨71JA4027⟩, 546, 575 ⟨67JOM(8)199⟩, 566 ⟨65JOC1939, 70USP3511865⟩, 575 ⟨66JOC950, 66JOC2373, 66TL3315⟩, 608, 621, 622, 624, 646 ⟨72MI 414-01⟩, 646, 916 ⟨68JOC780⟩, 649 ⟨67JOM(9)373⟩, 961 ⟨71JOM(26)215⟩
Peterson, E., **3**, 547 ⟨33CB237⟩
Peterson, G. A., **1**, 551 ⟨87JOM(334)9⟩; **2**, 607 ⟨86JOC277⟩, 670 ⟨85JA1060⟩
Peterson, G. E., **3**, 247 ⟨83TL4751⟩
Peterson, H., **2**, 243, 360 ⟨56JA6181⟩, 445 ⟨76GEP1768461⟩
Peterson, H. J., **1**, 896 ⟨65JA2157⟩
Peterson, J. C., **1**, 633, 1053 ⟨84JA6006⟩
Peterson, J. L., **1**, 436 ⟨83JOC4419⟩; **4**, 656 ⟨91OM2191⟩, 657 ⟨89JOM(364)105⟩, 657, 659 ⟨87OM2007⟩
Peterson, J. O., **1**, 102 ⟨63JA2236⟩
Peterson, J. R., **1**, 246 ⟨85JOC5200⟩; **2**, 24 ⟨84TL607⟩, 672 ⟨87TL6109⟩, 846 ⟨72JOC2272⟩
Peterson, J. S., **2**, 521 ⟨85JA4549⟩
Peterson, K., **3**, 323 ⟨83TL2829⟩
Peterson, L. I., **2**, 1051 ⟨68TL5357⟩
Peterson, L. K., **2**, 470 ⟨69CJC4281⟩; **4**, 437 ⟨72CC841, 73CJC2448⟩; **6**, 365 ⟨73CJC2448⟩
Peterson, M. L., **2**, 190 ⟨54JA3945⟩, 191 ⟨54JA3957⟩; **4**, 362 ⟨87JOC5116⟩
Peterson, M. R., **3**, 217 ⟨69TL3243⟩; **5**, 159 ⟨70JA741⟩
Peterson, P. E., **1**, 12 ⟨64JOC2325⟩, 769 ⟨88JOC1903⟩, 963, 978 ⟨80JA5120, 80JOC5006⟩, 1021 ⟨80TL4047⟩, 1110 ⟨66JA4990, 68JA6515⟩, 1110, 1115 ⟨69JA6194⟩, 1122 ⟨65JA2295⟩, 1124 ⟨66JA3152, 69JA4521⟩; **2**, 910 ⟨86JOC2381⟩; **4**, 352 ⟨80S318⟩, 1059 ⟨84TL5711⟩
Peterson, S. S., **2**, 532 ⟨85AG(E)1⟩
Peterson, W. J., **2**, 872, 881 ⟨75SC393⟩
Peterson, W. R., **2**, 416, 417 ⟨69MI 207-03⟩
Peterson, Jr., W. R., **5**, 992 ⟨72JOC1738⟩
Petiniot, N., **1**, 549 ⟨78TL1239⟩
Petisi, J. P., **1**, 34 ⟨58JA1654⟩

Petit, A., **1**, 403 ⟨49BSF501⟩; **2**, 31 ⟨89BSF247⟩; **5**, 132 ⟨85S40⟩
Petit, E. L., **1**, 88, 89 ⟨55JA4262⟩
Petit, F., **1**, 692 ⟨84CC341⟩
Petit, H., **3**, 496 ⟨90S301⟩; **4**, 181 ⟨80BSB759⟩
Petit, M., **1**, 692 ⟨84CC341⟩
Petit, Y., **1**, 179 ⟨90TL2149⟩, 837 ⟨92TL4313⟩; **3**, 652 ⟨83S297⟩; **5**, 152 ⟨89CC31⟩
Petitpierre, J. C., **5**, 816 ⟨73GEP2242772⟩
Petiushina, T. A., **3**, 540 ⟨71ZOB240⟩
Petneházy, I., **5**, 1065 ⟨87S520⟩, 1069 ⟨88S397⟩
Petragnani, N., **1**, 243, 595 ⟨91S793⟩, 243, 595, 611, 623 ⟨91S897⟩, 516 ⟨86JOC253⟩, 595, 611, 716 ⟨86S1⟩, 727 ⟨64CB3293⟩; **2**, 279 ⟨82JCR(S)212⟩, 727 ⟨63CB247⟩, 728 ⟨76JOM(114)281⟩, 768 ⟨86S1⟩, 856 ⟨65CI(L)1076⟩, 857 ⟨68CB3070⟩, 1036, 1037 ⟨81JOM(216)287⟩, 1036, 1038 ⟨84S240⟩; **3**, 176 ⟨91S793, 91S897⟩, 177 ⟨87SC443⟩; **4**, 88 ⟨82JCR(S)212⟩, 88, 90, 348 ⟨76JOM(114)281⟩, 90 ⟨81JOM(216)287⟩, 290 ⟨70CB2271⟩, 291 ⟨83MI 404-01⟩, 333 ⟨91S793, 91S897⟩, 333 ⟨B-94MI 407-01⟩, 348 ⟨64CI(L)1461⟩, 348, 349 ⟨78JOM(152)295⟩; **5**, 28, 46, 148 ⟨82S521⟩, 472 ⟨91S793, 91S897⟩, 903 ⟨76JOM(114)281, 77JOM(124)1⟩; **6**, 697 ⟨76JOM(114)281⟩
Petraitis, J. J., **1**, 841 ⟨81T3967⟩, 980 ⟨82JOC389⟩; **2**, 740 ⟨76JA6702⟩; **3**, 144 ⟨81T3967⟩
Petrakis, K. S., **4**, 161, 212 ⟨83S891⟩
Petranyi, G., **1**, 512 ⟨84JMC1539, 86JMC112⟩
Petrasch, A., **4**, 497, 498, 499, 500 ⟨85ZC369⟩
Petrasiunas, G. L. R., **3**, 347 ⟨78TL4621⟩
Petre, J., **4**, 203 ⟨86JOC3811⟩
Petrenko, V. V., **5**, 530 ⟨B-82MI 507-01⟩
Petres, J., **3**, 475 ⟨70HCA974⟩
Petrie, M. A., **2**, 511 ⟨93JCS(D)1737⟩
Petrie, P. S., **6**, 212, 225 ⟨31USP1961622⟩
Pétrier, C., **1**, 314 ⟨84JOC931⟩, 1093 ⟨82JOC3805⟩; **2**, 591 ⟨86TL3149⟩; **3**, 12 ⟨85TL829⟩, 12, 128 ⟨85JOC5761⟩, 14 ⟨82TL3361⟩, 127 ⟨87TL2347⟩, 145 ⟨83JOC3837⟩
Petrignani, J.-F., **1**, 242 ⟨87OM1625⟩, 976 ⟨92MI 119-18⟩; **2**, 33 ⟨83JOC5409⟩
Petrillo, A., **1**, 554 ⟨92JA9492⟩
Petrillo, G., **1**, 486 ⟨93T235⟩; **2**, 707 ⟨89T7411⟩, 711 ⟨86T4007⟩; **3**, 664 ⟨87T4625⟩, 664, 667 ⟨90T2205⟩; **5**, 1111 ⟨89T7411⟩
Petrini, M., **1**, 37 ⟨85S269⟩, 39 ⟨90JOC5159⟩, 578 ⟨92JOC2160⟩; **2**, 318 ⟨87S713⟩, 346 ⟨90JCS(P1)2133, 93CC1373⟩, 391, 392 ⟨90OPP707⟩, 393, 394 ⟨86S849⟩, 395 ⟨86S237, 86S1024⟩, 395 ⟨87S711⟩, 396 ⟨86SC1781, 90T7531⟩, 397 ⟨92TL4835⟩; **3**, 18, 133 ⟨88JCS(P1)2563⟩, 46 ⟨84T3809⟩, 46, 191 ⟨90OPP707⟩, 47 ⟨86S1024⟩, 47, 192 ⟨86S237⟩, 190 ⟨92SL64⟩, 191 ⟨84S529⟩, 434 ⟨90JOC4456⟩
Petrissans, J., **3**, 234 ⟨76S33⟩
Petriva, T. D., **5**, 1070 ⟨89ZOR1689⟩
Petro, C., **1**, 905 ⟨79JA522⟩
Petrosyan, U. S., **2**, 476 ⟨90JGU398⟩
Petrosyan, V. A., **2**, 133 ⟨90S841⟩; **3**, 634 ⟨90S841⟩; **4**, 38 ⟨90IZV1802⟩, 1083 ⟨89IZV356, 89UK1105⟩; **5**, 317 ⟨92MI 507-02⟩; **6**, 56 ⟨87IZV306⟩, 149 ⟨82IZV2637⟩
Petrosyan, V. P., **6**, 650 ⟨75MI 621-01⟩
Petrosyan, V. S., **2**, 433, 473 ⟨91ZOB1016⟩; **4**, 458 ⟨91ZOB1478, 92ZOB946⟩, 459 ⟨91URP1618747⟩, 459, 484 ⟨91ZOB1016⟩, 562 ⟨90ZOB1420⟩, 626 ⟨91ZOB921⟩; **5**, 702 ⟨91JGU1268, 91ZOB1389⟩
Petrov, A. A., **1**, 953 ⟨64RCR1⟩, 960 ⟨60ZOB1056⟩; **2**, 292 ⟨69ZOB1804⟩, 448 ⟨76JA1290, 90ZOB2394⟩, 454 ⟨68DOK(181)877, 69ZOB2367⟩, 477 ⟨77ZOB1432⟩, 731 ⟨65ZOR2115⟩, 820, 821 ⟨63ZOB3774⟩, 838, 839 ⟨73ZOB2178⟩, 849 ⟨63ZOB2863⟩, 900 ⟨65ZOB1669⟩, 1021, 1025, 1027, 1030, 1036 ⟨89RCR948⟩, 1028 ⟨85JGU1771⟩, 1032 ⟨64ZOB1899, 65ZOR2157⟩, 1032 ⟨67ZOR2102⟩, 1038 ⟨87RCR267⟩, 1042 ⟨66ZOR1994⟩, 1071 ⟨62JGU2355, 63JGU2791⟩, 1071 ⟨65JGU1910, 67JGU898⟩, 1080 ⟨62ZOB1349⟩, 1090 ⟨87JFC(36)373⟩, 1101 ⟨62DOK(143)855⟩; **3**, 548 ⟨62ZOR1759, 71ZOR1120⟩, 591 ⟨83JGU205, 85JGU26⟩; **4**, 488 ⟨68ZOB1248, 70ZOB2618⟩, 766 ⟨89ZOB2145⟩, 786 ⟨74ZOR1986⟩, 802 ⟨67JGU2024⟩, 807, 811, 813 ⟨91JGU1295⟩, 810 ⟨91JGU1306⟩, 834 ⟨80ZOR13⟩, 841 ⟨74ZOR2269⟩, 872 ⟨74ZOR2456⟩, 874 ⟨81JOU578⟩, 877 ⟨79ZOR2596⟩, 885 ⟨80ZOB984⟩, 913 ⟨74ZOB2067, 76CA192821t⟩, 914 ⟨84ZOB457⟩, 941 ⟨71ZOB719, 72ZOB870⟩, 948 ⟨79ZOB829, 80ZOB1053⟩, 955 ⟨81ZOB1857⟩, 976 ⟨79ZOR1824⟩; **5**, 589, 590, 603, 611 ⟨71ZOR1120⟩, 627 ⟨81ZOR2064⟩, 691 ⟨88JOU1049, 88ZOR1165⟩; **6**, 116 ⟨90ZOR1129⟩
Petrov, A., **4**, 471 ⟨76MI 410-02⟩
Petrov, A. D., **1**, 331 ⟨61IZV1099⟩, 925 ⟨60DOK(131)580, 61DOK(136)1099⟩, 925 ⟨61IZV2065⟩; **2**, 609, 621 ⟨60DOK(133)855⟩, 1089 ⟨52DOK(85)1297⟩; **4**, 569 ⟨62DOK(143)840, 63DOK(148)875⟩, 569, 570, 573 ⟨65IZV286⟩, 570 ⟨62DOK(147)117⟩; **6**, 62 ⟨60DOK(131)98⟩
Petrov, B. I., **4**, 641 ⟨75ZOB1905⟩; **5**, 1137 ⟨69ZOB2365⟩; **6**, 62 ⟨72IZV85, 72JOM(34)299⟩
Petrov, G., **1**, 747 ⟨72M1651⟩; **2**, 1004 ⟨85CB3529, 88AX(C)626⟩; **3**, 494 ⟨89JCS(P2)1405⟩
Petrov, I. B., **4**, 660 ⟨72JOM(46)51⟩
Petrov, J. S., **5**, 345 ⟨90SC2083⟩
Petrov, K. A., **2**, 257 ⟨91ZOB1115⟩, 357 ⟨90RCR832⟩, 407 ⟨56ZOB3378⟩, 411 ⟨62ZOB3070⟩, 440 ⟨61ZOB3411, 61ZOB3417⟩, 442 ⟨62ZOB553⟩, 451 ⟨62ZOB1974, 63ZOB1246⟩, 454 ⟨60ZOB2995⟩, 458 ⟨65ZOB732⟩, 460, 851 ⟨62ZOB4017⟩, 461 ⟨62ZOB3062⟩, 841 ⟨78ZOB91⟩; **4**, 53 ⟨77JGU2501⟩, 119 ⟨65JGU2053⟩, 121 ⟨61JGU3189⟩, 336 ⟨61ZOB3417⟩, 337 ⟨61ZOB3411⟩, 451, 469, 472, 473, 475, 493 ⟨74RCR984⟩, 464 ⟨77ZOB579⟩, 478, 493 ⟨76ZOB1246⟩, 490 ⟨76ZOB1250⟩, 490 ⟨76ZOB1495, 76MI 410-03⟩, 550, 554 ⟨65ZOB1602, 60ZOB1602⟩, 554 ⟨69ZOB1247, 69ZOB845⟩, 561 ⟨83ZOB541, 84URP1067006⟩, 916 ⟨80ZOB1225⟩; **5**, 565, 579 ⟨69RCR21⟩, 708 ⟨61JGU473, 61ZOB516⟩; **6**, 80 ⟨77ZOR943⟩, 236 ⟨59ZOB3401⟩, 238 ⟨59ZOB2159⟩, 260 ⟨59ZOB2169⟩, 263 ⟨60ZOB1602⟩, 602 ⟨59JGU2662⟩, 604, 612 ⟨59JGU2131⟩Petrov, K. I., **3**, 76 ⟨71ZOR2120⟩
Petrov, M. L., **2**, 1038 ⟨87RCR267⟩; **3**, 548 ⟨62ZOR1759, 71ZOR1120⟩; **4**, 834 ⟨80ZOR13⟩, 874 ⟨81JOU578⟩, 877 ⟨79ZOR2596⟩, 913 ⟨74ZOB2067, 76CA192821t⟩; **5**, 589, 590, 603, 611 ⟨71ZOR1120⟩, 627 ⟨81ZOR2064⟩; **6**, 116 ⟨90ZOR1129⟩
Petrov, V. A., **3**, 120 ⟨92TL7245⟩; **4**, 104 ⟨92EUP496413⟩, 730 ⟨81IZV1920⟩; **5**, 662, 663, 674, 824, 825 ⟨93JCS(P1)505⟩; **6**, 256 ⟨93JOC4754⟩, 279 ⟨92EUP496413⟩, 614 ⟨93JCS(P1)505⟩, 643 ⟨78MI 612-01⟩
Petrova, G. G., **2**, 336 ⟨83JOU1203⟩
Petrova, G. M., **4**, 554 ⟨69ZOB1247⟩
Petrova, I., **1**, 644 ⟨88PS(40)69⟩
Petrova, J., **2**, 618 ⟨75S658⟩; **4**, 347, 773, 778 ⟨76S107⟩, 740, 741 ⟨75S535⟩; **6**, 266 ⟨75S535⟩
Petrova, L. V., **3**, 452 ⟨61CA(55)17526⟩
Petrova, M., **3**, 678 ⟨B-76MI 320-01⟩
Petrova, S., **1**, 552 ⟨89CA(111)23052⟩
Petrova, T. D., **5**, 654 ⟨87RCR1125⟩, 712 ⟨88RCR234, 88UK405⟩, 1070 ⟨88JFC(40)217, 89ZOR1523⟩; **6**, 272 ⟨88ZOR1513⟩, 609 ⟨77JFC(9)505, 83JFC(22)439, 86ZOR1297⟩, 616 ⟨84ZOR1197⟩, 619 ⟨88JFC(40)217, 93JPO319⟩
Petrovanu, M., **1**, 398 ⟨B-76MI 109-01⟩
Petrovich, J. P., **5**, 476, 477 ⟨64TL3363⟩, 476, 477 ⟨65JOC2882⟩
Petrovich, L. M., **4**, 716 ⟨90JOM(387)C31, 92ICA557⟩
Petrovskaya, E. A., **4**, 583 ⟨85JOM(288)119⟩
Petrovskaya, L. I., **2**, 436 ⟨67ZOB2080⟩

Petrovskii, P. V., **1**, 870 ⟨67TL4525⟩; **2**, 430 ⟨69JOM(17)201⟩, 447 ⟨85IZV1200⟩, 876 ⟨75IZV897⟩, 877 ⟨84DOK(274)606, 86IZV181⟩, 878 ⟨87JOU1417⟩; **3**, 496 ⟨89IZV1182⟩; **4**, 123 ⟨86DOK(286)19⟩, 551 ⟨80TL2931, 92ZOB263⟩, 557 ⟨84ZOB36, 85ZOB1234, 86ZOB1220, 88ZOB1987, 88ZOB1998, 88ZOB2012⟩, 694, 717 ⟨87JOM(336)187⟩, 706 ⟨81JOM(209)233⟩; **5**, 1131 ⟨92IZV187⟩; **6**, 635 ⟨92IZV425⟩

Petrovskii, R. V., **4**, 553 ⟨90AKZ537⟩

Petrovsky, P. V., **3**, 597 ⟨89JOM(379)303⟩

Petrow, V., **2**, 1016 ⟨62JCS4288⟩, 1016, 1018 ⟨66T2829⟩

Petrow, V. A., **3**, 363 ⟨45JCS858⟩

Petruk, G. M., **6**, 238 ⟨82GEP3044216⟩

Petrukhov, L. P., **2**, 1091 ⟨86JGU849⟩

Petrulenis, L., **6**, 558 ⟨85KGS339⟩

Petruneva, R. M., **2**, 448 ⟨91ZOB1906⟩

Petrunin, V. A., **2**, 448 ⟨68ZOB1551⟩

Petry, R. C., **2**, 335, 336 ⟨67JOC4034⟩; **4**, 104 ⟨67JOC4034⟩; **5**, 314 ⟨61JA3912, 69JA4778⟩

Petrzilka, M., **1**, 520, 523 ⟨81S753⟩; **2**, 282 ⟨79HCA1406⟩, 346, 349 ⟨79TL4391⟩; **3**, 38, 174 ⟨79HCA1406⟩; **4**, 51 ⟨78HCA2286, 78HCA3075⟩; **5**, 199 ⟨81S753⟩

Petter, W., **5**, 373 ⟨85CC1764⟩

Pettersson, L., **1**, 216 ⟨87TL2753⟩

Petterson, L. L., **2**, 418, 419 ⟨60JA6242⟩

Petterson, R. C., **2**, 614 ⟨71JOC631⟩

Pettett, M. G., **3**, 367 ⟨85JCS(P1)1161⟩

Petti, M. A., **4**, 170 ⟨86JOC1704⟩

Pettig, D., **4**, 536 ⟨86S737, 88AG(E)1194⟩

Pettit, G. R., **1**, 3, 4, 6 ⟨64CJC572⟩, 21, 23, 435 ⟨62OR(12)356⟩, 438 ⟨66OR(12)356⟩; **3**, 620 ⟨85JOC2323⟩; **5**, 223 ⟨87CJC2390⟩, 745 ⟨61JOC2563, 65CJC2640⟩; **6**, 501, 508 ⟨70JOC843, 65CJC1798⟩

Pettit, R., **1**, 866, 910 ⟨56CI(L)1306⟩, 866, 910 ⟨60JA1972⟩, 898 ⟨71JA6688⟩, 904 ⟨59JA3160⟩, 917 ⟨62JA2842, 66JA604⟩, 917 ⟨66JA606⟩; **3**, 83, 102 ⟨77TL781⟩, 519 ⟨66JA5044⟩; **4**, 148 ⟨66JA5044⟩, 679 ⟨80JA1752, 82OM1350⟩, 679 ⟨82OM911⟩, 681 ⟨71CC608, 74JA7108⟩, 688 ⟨71TL3475⟩, 706 ⟨68PAC235⟩, 724 ⟨65JA3254, 68JA1057, 70CC161⟩; **6**, 517 ⟨79JA1627⟩

Pettit, R. J., **2**, 680 ⟨85JOC1212⟩

Pettitt, D. J., **1**, 592 ⟨64JOC3258⟩; **2**, 172 ⟨65JOC633⟩

Pettus, Jr., J. A., **1**, 811 ⟨71JA3087⟩

Petty, C. B., **5**, 991 ⟨78JOC2164⟩

Petty, H. E., **1**, 212 ⟨72JOC2957⟩

Petty, J. D., **1**, 879 ⟨77S509⟩

Petty, J. T., **5**, 1019 ⟨88JA6240⟩

Petty, M. C., **4**, 873 ⟨90CC816, 90CC816⟩

Petty, S. R., **5**, 760 ⟨80JOC3750⟩

Petukhova, A. S., **4**, 570 ⟨70ZOB609, 71ZOB2186⟩

Petukhova, I. Yu., **4**, 775 ⟨91ZOR1796⟩

Petukhova, N. P., **4**, 770, 779 ⟨90G235⟩, 854 ⟨78IZV2654, 82IZV2327⟩, 861 ⟨84IZV194⟩, 866 ⟨84IZV611⟩

Petukhov, P. A., **3**, 638 ⟨93SL580⟩

Petukhov, S. A., **6**, 145 ⟨74ZOR2003⟩

Petz, W., **5**, 900 ⟨93OM4056⟩; **6**, 337 ⟨76JOM(105)C19⟩, 517 ⟨83OM1044⟩

Petzhold, K., **2**, 38 ⟨83AG(E)406⟩

Petzold, G., **4**, 554, 556 ⟨65ZC419⟩

Pevarello, P., **3**, 684, 686 ⟨93JCS(P1)2151⟩

Pevzner, M. S., **6**, 145 ⟨72KGS713⟩

Pews, R. G., **1**, 861 ⟨81JHC939⟩; **2**, 21 ⟨85SC977⟩, 654 ⟨91SC2335⟩; **3**, 27 ⟨85SC977⟩

Peyreye, M., **2**, 78 ⟨72JOM(35)C20⟩

Peyronel, G., **4**, 687 ⟨67G1327⟩

Peyronel, J.-F., **2**, 413 ⟨87JOC5320⟩

Peyton, A. L., **2**, 377 ⟨83JOC4116⟩

Peyton, D. H., **4**, 868 ⟨89IC3766⟩

Peyton, K. B., **2**, 22 ⟨85SC385⟩

Pez, G. P., **1**, 570 ⟨92JOC2856⟩; **4**, 671 ⟨76JA8072, 79JA6933⟩

Pfab, J., **2**, 391 ⟨79CC392⟩; **4**, 38, 40 ⟨76TL943⟩

Pfab, W. Z., **1**, 862 ⟨52ZN(B)377⟩

Pfaendler, H. R., **1**, 1110, 1111 ⟨71HCA2060⟩, 1113 ⟨70HCA2130⟩; **5**, 905 ⟨78JA8214⟩

Pfaff, K., **2**, 748 ⟨78S897⟩; **3**, 657 ⟨78S897⟩

Pfaltz, A., **1**, 194 ⟨93AG(E)566⟩, 194, 353 ⟨91HCA232⟩, 220 ⟨82AG(E)71, 82AG(S)161⟩; **3**, 128 ⟨93TL7725⟩, 183 ⟨84TL2977⟩

Pfannstiel, K., **2**, 167 ⟨50CB87⟩, 769 ⟨42CB1096⟩; **4**, 71 ⟨50CB87⟩

Pfau, M., **1**, 546 ⟨62CR(254)1817⟩; **3**, 415 ⟨85JA273⟩

Pfeffer, B., **3**, 148 ⟨81CB1858⟩

Pfeffer, M., **4**, 647, 656 ⟨85NJC249⟩, 716 ⟨91CC1274⟩, 723 ⟨93OM2227⟩; **5**, 801 ⟨90JCS(D)3193⟩

Pfeffer, P. E., **2**, 393 ⟨70TL699⟩; **4**, 842 ⟨76T2507⟩; **5**, 555 ⟨76T2507⟩

Pfeifer, C. R., **1**, 958 ⟨55JOC95, 55JOC1337⟩; **2**, 1099 ⟨55JOC1337⟩

Pfeifer, K.-P., **3**, 369 ⟨90TL5725, 91PS(59)419⟩; **5**, 645 ⟨91PS(59)125⟩

Pfeifer, W. D., **1**, 1110, 1117 ⟨71JA1513⟩; **2**, 689 ⟨70AG(E)521⟩

Pfeiffer, H., **2**, 597 ⟨51ZAAC(267)39⟩; **5**, 856 ⟨26CB1162⟩

Pfeiffer, I., **2**, 597 ⟨51USP2722538⟩

Pfeiffer, J. M., **2**, 888 ⟨84AJC527⟩

Pfeil, B., **4**, 676 ⟨92ICA741⟩

Pfeil, E., **3**, 628 ⟨65AG(E)1079⟩

Pfenninger, J., **1**, 933, 939 ⟨80HCA2328⟩

Pfister, G., **1**, 434 ⟨69TL2145⟩; **2**, 693 ⟨69TL2145⟩

Pfister, J. R., **2**, 683 ⟨88SC1379⟩

Pfister, M., **2**, 203 ⟨89MI 203-02⟩

Pfister, T., **5**, 1120 ⟨87EUP209779⟩

Pfisterer, H., **3**, 387, 390 ⟨83AG(E)314⟩, 519 ⟨84JOM(264)327⟩; **4**, 581 ⟨89JOM(371)329⟩, 686 ⟨82JOM(226)C59, 83JOM(258)81⟩

Pfister-Guillou, G., **3**, 255 ⟨64BSF3254⟩

Pfister-Guillouzo, G., **3**, 335 ⟨90JOC2596, 93JA324⟩, 369 ⟨91NJC533⟩, 582 ⟨89JCS(F2)741⟩

Pfitzner, K. E., **2**, 105 ⟨64BBR(17)146⟩; **3**, 6, 118 ⟨63JA3027⟩

Pflästerer, G., **1**, 1001 ⟨93S558⟩

Pflaum, D. J., **2**, 1019 ⟨34JA1106⟩

Pflaum, S., **3**, 607 ⟨92TL1855⟩

Pflaumbaum, W., **5**, 550, 560 ⟨87CB411⟩, 561 ⟨87CB61, 88CB313⟩

Pfleger, R., **6**, 113 ⟨57CB2460⟩

Pfleging, E., **4**, 834 ⟨88LA933⟩; **5**, 639 ⟨88CZ239⟩

Pfleiderer, G., **5**, 83 ⟨57MI 502-03⟩

Pfleiderer, W., **2**, 203 ⟨89MI 203-02⟩, 409, 410 ⟨84TL5513, 89S497⟩, 780 ⟨64LA(667)113⟩; **3**, 292 ⟨88HCA531⟩; **4**, 409 ⟨72LA(759)84⟩; **5**, 215 ⟨87JHC597⟩

Pfletschinger, J., **2**, 942 ⟨76AG(E)428⟩

Pflieger, D., **1**, 774 ⟨87TL1519⟩

Pflug, J. L., **2**, 17 ⟨77TL363⟩

Pfluger, H. L., **5**, 31, 106 ⟨33JA4079⟩

Pfohl, W., **1**, 665 ⟨60LA(629)1, 60LA(629)210⟩, 666 ⟨60LA(629)207⟩; **2**, 596 ⟨60LA(629)1⟩; **4**, 699 ⟨60LA(629)1⟩

Pfotenhauer, C., **6**, 478 ⟨77ZC411⟩

Pfuentrenreuter, C., **2**, 561 ⟨85N789⟩

Pfützner, L., **4**, 550 ⟨83EGP201449⟩

Pfyffer, J., **4**, 30 ⟨78HCA2482⟩

Pham, D., **4**, 203 ⟨86JOC3811⟩

Pham, H. T., **1**, 777 ⟨81JOC2804⟩

Pham, K. M., **3**, 44, 188 ⟨90JOC2552⟩

Pham, P. Q., **1**, 1092 ⟨83TL1003⟩

Pham, T. N., **2**, 25, 26, 31, 32 ⟨89S614⟩; **5**, 248, 562 ⟨88CC390⟩

Pham, T.-N., **5**, 527 ⟨91TL7411⟩

Pham Ba Chi, **2**, 482, 485, 487, 510 ⟨83ZAAC(501)89⟩

Phan, A., **4**, 353 ⟨89CC89⟩

Phan, T. N., **5**, 553 ⟨86T2025⟩

Phanstiel IV, O., **5**, 328 ⟨92JOC7140⟩

Phan-Tan-Luu, R., **3**, 279 ⟨85BSB897⟩

Phelps, M. E., **2**, 25, 31, 35 ⟨90TL4973⟩, 613 ⟨90JOC3373⟩, 632 ⟨90JOC4560⟩
Philbin, E. M., **3**, 242 ⟨60CI(L)348⟩
Philip, A., **1**, 84 ⟨87JMC1505⟩
Philip, Jr., J. B., **2**, 957 ⟨81JA1847⟩
Philipp, C., **2**, 577 ⟨92CB157⟩
Philippo, C. M. G., **1**, 826 ⟨91JA2762, 93TL3523⟩
Philipps, P., **2**, 984 ⟨91CB2165⟩
Philips, G. B., **1**, 835 ⟨84TL3159⟩
Philips, J. C., **1**, 620, 1074 ⟨71CC22⟩, 904 ⟨65JA1381, 65JA4017⟩; **4**, 777 ⟨75CC107⟩, 779 ⟨72JA1012, 74TL4157⟩
Philips, J. R., **2**, 485, 505, 510, 874 ⟨67CJC675⟩
Philips, R. F., **2**, 564 ⟨71JCS(A)1931⟩
Philips, R. R., **3**, 448, 681 ⟨59OR(10)143⟩
Philipsborn, W. von, **4**, 683 ⟨86OM2413, 90JA7113⟩
Phillion, D. P., **3**, 197 ⟨92JOC2763⟩; **4**, 525 ⟨86JOC1610⟩
Phillipp, W., **1**, 812 ⟨59LA(625)206⟩
Phillips, A. M. M. M., **4**, 570 ⟨91PS(55)41⟩
Phillips, A. P., **1**, 2, 6, 8, 12, 427 ⟨46JA261⟩; **5**, 140 ⟨47JA200⟩
Phillips, B., **4**, 203 ⟨60JOC1699⟩; **5**, 205 ⟨57JA5982⟩
Phillips, B. A., **5**, 685 ⟨B-75MI 517-01⟩; **6**, 417, 418 ⟨73T3309⟩
Phillips, B. T., **3**, 618, 619 ⟨88TL2155⟩
Phillips, C. J., **2**, 343, 763 ⟨76TL233⟩
Phillips, D., **2**, 344 ⟨90JA6696⟩
Phillips, D. D., **3**, 477 ⟨56JA5452⟩; **4**, 836, 837, 838 ⟨68JOC4470⟩
Phillips, D. J., **6**, 563 ⟨72ICA11⟩
Phillips, E. C., **2**, 573 ⟨88OM1877⟩
Phillips, G., **5**, 765, 766 ⟨93JA7584⟩
Phillips, G. T., **1**, 1006 ⟨67JCS(C)2011⟩
Phillips, G. W., **2**, 816 ⟨80JA5866⟩; **4**, 99, 465 ⟨87OS(65)119⟩
Phillips, H., **2**, 286, 732 ⟨28JCS2293⟩, 288, 291 ⟨38JCS341⟩
Phillips, H. K., **2**, 915 ⟨90TL6621⟩
Phillips, I. G., **4**, 547 ⟨88IC4038⟩
Phillips, J. G., **2**, 976 ⟨82JA4708⟩; **5**, 173 ⟨82JA4708⟩; **6**, 472, 473 ⟨82JA4708⟩
Phillips, J. N., **1**, 894 ⟨49JCS1724⟩
Phillips, J. R., **6**, 65, 244 ⟨60JA6228⟩
Phillips, L., **4**, 804 ⟨74JCS(P1)1644⟩
Phillips, L. R., **6**, 491 ⟨89TL7141⟩
Phillips, M. A., **1**, 446 ⟨28JCS3046⟩
Phillips, N. H., **2**, 962 ⟨85JA4101⟩
Phillips, R. B., **1**, 806 ⟨80JA3156⟩
Phillips, R. P., **4**, 681 ⟨76JCS(D)377⟩, 683 ⟨78JCS(D)403⟩
Phillips, W. E., **2**, 168 ⟨71JOC3145⟩
Phillips, W. G., **4**, 71 ⟨72JOC1526⟩; **5**, 245, 478, 490, 491 ⟨72JOC1526⟩, 478, 490 ⟨71JOC3145⟩, 490, 491 ⟨72JOC3818⟩
Phillips, W. V., **1**, 130 ⟨78JOC700⟩; **2**, 420 ⟨78JOC700⟩; **3**, 124 ⟨78JOC700⟩
Phillipson, J. M., **5**, 966 ⟨69BRP1152877⟩
Philoche-Levisalles, M., **5**, 799 ⟨88JCS(D)1591⟩
Philp, D., **4**, 746 ⟨94CC205⟩
Philpott, P. G., **3**, 448 ⟨65JCS7185⟩
Phimmanao, A. S., **1**, 224 ⟨93SL599⟩; **3**, 638 ⟨93SL599⟩
Phipps, J. R., **4**, 892 ⟨76JCS(P1)1241⟩
Phongpradit, T., **3**, 212 ⟨82CL687⟩
Photaki, I., **2**, 273, 275 ⟨62JA3887⟩, 274 ⟨65JA4922⟩, 412 ⟨61CB2644⟩; **5**, 367 ⟨85JCS(P1)2277⟩
Photis, J. M., **1**, 613 ⟨77OS(57)53⟩, 707, 708 ⟨74JA4715, 74JA5801⟩, 708 ⟨76JA4936⟩
Phull, G. S., **6**, 53 ⟨84JCS(P1)455⟩
Phung, H. T., **2**, 494 ⟨80ZAAC(466)179, 81ZAAC(472)75⟩, 496 ⟨79CZ119⟩
Phuong, N. T. M., **1**, 257 ⟨78TL4727⟩
Piana, H., **5**, 954 ⟨91JA2228⟩
Piancatelli, G., **2**, 34 ⟨82SC1127⟩; **3**, 4, 115 ⟨82S245⟩, 67 ⟨81TL1041⟩, 155 ⟨80TL4521⟩
Pianka, M., **4**, 129 ⟨88PS(40)183⟩
Piantadosi, C., **4**, 178, 184 ⟨58JA6613⟩
Piantanida, B., **5**, 652 ⟨61CI(M)1243⟩, 980 ⟨60CI(M)1243⟩
Piantini, U., **4**, 683 ⟨86OM2413, 90JA7113⟩
Pi, R., **1**, 9 ⟨87JOC4299⟩, 880 ⟨86AG(E)349⟩; **2**, 999, 1000 ⟨89OM1688⟩, 1002 ⟨86AG(E)1103⟩
Piasecki, M. L., **4**, 1080 ⟨88MI 424-01⟩
Piatak, D. M., **3**, 294 ⟨73MI 306-01⟩; **5**, 49 ⟨69JOC112, 69JOC116⟩
Piaux, L., **5**, 62 ⟨B-45MI 502-05⟩
Piazza, M. G., **3**, 31 ⟨93JA5821⟩
Piazzesi, A. M., **5**, 142 ⟨79ICA345⟩, 573 ⟨85T5393⟩
Picard, C. W., **5**, 1121 ⟨81EJM175⟩; **6**, 646, 647 ⟨75AF1477⟩
Picard, J.-P., **4**, 933 ⟨72JOM(43)157⟩, 1013 ⟨90JOM(332)13⟩, 1014 ⟨84OM1660⟩; **5**, 416 ⟨72JOM(43)157⟩, 419 ⟨69JOM(19)P33, 79JOC420⟩, 419, 420 ⟨71JOM(26)183⟩, 420 ⟨75JOM(93)51⟩, 420, 794 ⟨90JOM(391)13⟩, 924 ⟨79JA6750⟩; **6**, 124 ⟨71JOM(26)183, 79JOC420⟩, 125 ⟨72JOM(39)C49⟩, 131 ⟨77JOM(141)35⟩, 162, 163 ⟨91JOM(419)C1⟩, 163 ⟨92TL3903⟩, 508 ⟨53CJC896⟩
Picci, M., **3**, 256 ⟨89H(29)349⟩
Picci, N., **3**, 256 ⟨89H(29)349⟩, 620 ⟨93T8423⟩
Piccinelli, P., **3**, 368 ⟨79JCS(P1)1720⟩
Piccinni-Leopardi, C., **3**, 566 ⟨88JOC18⟩
Piccoli, W. A., **2**, 538 ⟨60JA1883⟩
Piccolo, O., **1**, 182 ⟨85JOC3945⟩, 182, 475 ⟨91JOC183⟩, 195, 238 ⟨86T2043⟩; **2**, 9 ⟨89TL4555⟩; **5**, 301 ⟨86T885⟩
Pichat, L., **1**, 536 ⟨82TL63⟩; **2**, 126 ⟨59BSF1544⟩, 232 ⟨59BSF1887⟩, 291 ⟨61T1⟩; **3**, 212 ⟨73S537⟩
Pichl, R., **4**, 570 ⟨87CB789⟩, 571 ⟨92S787⟩, 579 ⟨85ZN(B)352⟩; **5**, 226 ⟨87CB39⟩, 1095 ⟨87CB789⟩; **6**, 702 ⟨92S787⟩, 706 ⟨86AG(E)574⟩
Pichler, E., **2**, 498 ⟨61MI 209-01⟩
Pichler, L., **6**, 642 ⟨80JMC1217⟩
Pichon, C., **2**, 675 ⟨86TL3619⟩; **3**, 113 ⟨86T5627⟩
Pichter, F., **5**, 857 ⟨00CB751⟩
Pick, J. H., **4**, 56 ⟨85JCS(P1)1567⟩
Pick, R., **5**, 8 ⟨88S525⟩
Pickard, A. L., **2**, 597 ⟨67JOM(8)427⟩
Pickard, J., **1**, 1140 ⟨88TL5017⟩
Pickard, P. L., **3**, 405 ⟨61JOC4886⟩, 459 ⟨48JA2012, 50JA876⟩, 459 ⟨51JA42⟩
Pickardt, J., **2**, 565 ⟨84OM69⟩; **4**, 690 ⟨93JOM(459)325⟩
Pickenhagen, W., **1**, 811 ⟨69HCA880⟩, 1082 ⟨71HCA1789⟩
Picker, D., **1**, 1137, 1141 ⟨75JOC124⟩; **2**, 1054 ⟨75JOC124⟩; **5**, 47, 52, 115 ⟨74TL1511⟩
Picker, K., **4**, 841 ⟨77JCS(P1)1723⟩; **5**, 516, 521, 522 ⟨77JCS(P1)1723⟩
Pickering, M. W., **5**, 593 ⟨77JCS(P1)2357⟩
Pickering, R. A., **2**, 104 ⟨83JOC3379⟩
Pickles, G. M., **2**, 653 ⟨74JOM(76)C23, 75CC36⟩
Pickles, W., **1**, 422 ⟨53JCS124⟩
Picotin, G., **2**, 69 ⟨87JOC4796⟩; **4**, 192 ⟨86CB1725⟩; **6**, 729 ⟨87JOM(328)249⟩
Picq, D., **2**, 9, 19 ⟨90TL6527⟩
Pictet, A., **5**, 285 ⟨1887CB3422⟩
Pidacks, C., **1**, 100 ⟨71JOC279⟩
Pidcock, A., **3**, 83 ⟨70CC1703, 75JCS(D)2460⟩, 83 ⟨75JCS(D)2646⟩; **4**, 722, 725 ⟨79JOM(170)95⟩; **5**, 419 ⟨82JOM(225)331⟩, 1140 ⟨77JCS(D)1444⟩; **6**, 381 ⟨92JOM(427)9⟩
Pidvarko, T. I., **6**, 685 ⟨82ZOB1925⟩
Pidvarko, T. V., **6**, 685 ⟨83ZOB1672⟩
Piecara, J. C., **3**, 589 ⟨81JFC(17)265⟩
Piechucki, C., **3**, 642, 643 ⟨74S869⟩
Pieken, W. A., **4**, 53 ⟨90JOC3029⟩
Pielartzik, H., **5**, 122 ⟨85HOU(E5)659⟩, 182 ⟨89HOU(5)633⟩, 692, 699, 700 ⟨85HOU(E5/1)812⟩
Pielichowski, J., **4**, 97 ⟨66MI 403-01⟩, 761, 764 ⟨88PJC483⟩, 793 ⟨88LA595, 89BAP123⟩

Pienta, N. J., **4**, 25 ⟨83JOC2084⟩
Pieper, G., **2**, 347, 348 ⟨58HOU(11)190⟩, 385 ⟨58HOU(11/2)99⟩
Pieper, U., **5**, 782 ⟨92CB2171⟩
Pierce, D., **1**, 84 ⟨79TL1641⟩
Pierce, J. B., **5**, 412 ⟨65JOC2566⟩
Pierce, J. R., **1**, 401 ⟨82JOC798⟩
Pierce, J. S., **4**, 302 ⟨51JA2596⟩
Pierce, J. V., **2**, 214 ⟨54JA1828⟩
Pierce, M. E., **4**, 528 ⟨92MI 411-02⟩, 780 ⟨91JMC2525⟩
Pierce, O. R., **4**, 33 ⟨53JA4091⟩; **6**, 244 ⟨54JA474⟩
Pierce, R. A., **3**, 502 ⟨80JA4970⟩; **4**, 611 ⟨80JA4970⟩
Pierdet, A., **4**, 174 ⟨80T1763⟩
Pierfrancesco, P., **3**, 32 ⟨93CC954⟩
Pierini, A. B., **1**, 474 ⟨87TL4653⟩
Pieronczyk, W., **2**, 829 ⟨79AG655⟩
Piéroni, G., **5**, 135 ⟨90TL4875⟩
Pieroth, M., **2**, 472 ⟨85AG(E)504⟩
Pierpont, C. G., **4**, 682 ⟨75JA7172, 77IC636⟩
Pierre, F., **1**, 815 ⟨87JA5731⟩
Pierre, J. L., **2**, 299 ⟨80T1223⟩, 791 ⟨86SC63⟩
Pierre, S., **1**, 834 ⟨93T3131⟩
Pierron, E., **2**, 100 ⟨65JOC2692⟩
Pierrot, M., **4**, 580 ⟨86CC1208⟩; **5**, 918 ⟨85CC1151⟩
Pierrot, P., **1**, 958 ⟨67CR(C)1492⟩
Piers, E., **1**, 360 ⟨88PAC107⟩, 452 ⟨93JOC11⟩, 469 ⟨82CJC1256⟩, 558 ⟨91TL5219⟩, 615 ⟨87CC1342⟩, 804 ⟨79CC1138, 83CJC3227⟩, 811 ⟨78JOC3630, 83CJC1226⟩, 811, 812 ⟨83CJC1239, 84TL3959⟩, 812 ⟨85TL2735⟩, 831 ⟨89CC1665⟩, 1092 ⟨79JOC3437, 80JOC4263, 84TL3155⟩; **2**, 600 ⟨87CJC78⟩, 917 ⟨86CC626⟩, 947 ⟨87CC1025⟩; **3**, 78 ⟨89JCS(P1)2124⟩, 217 ⟨82CJC1256⟩, 239 ⟨75SC193, 82CJC210⟩, 273 ⟨78CC1033⟩, 696 ⟨86CJC2475⟩, 701 ⟨89JOC1483⟩; **5**, 160 ⟨89T363⟩
Piers, W. E., **2**, 969 ⟨90SL74⟩
Pierson, C., **2**, 20 ⟨71JA438⟩
Pierson, E., **2**, 138 ⟨48JA1450⟩
Pierson, G. O., **5**, 354, 359 ⟨73TL4091⟩
Pierson, R. M., **2**, 175 ⟨58JOC363⟩
Pierson, W., **1**, 548 ⟨60JCS1045⟩
Piet, J., **4**, 993 ⟨87BSF325⟩
Pieter, R., **1**, 140, 141 ⟨74AG(E)479⟩; **2**, 385 ⟨88OSC(6)542⟩; **4**, 392 ⟨74AG(E)479⟩
Pieters, R. J., **2**, 539 ⟨90JOC6082⟩
Pietra, F., **1**, 906 ⟨73CRV293⟩; **3**, 63 ⟨87HCA1400⟩, 374 ⟨75TL2085⟩
Pietra, S., **3**, 399 ⟨64G48⟩, 445 ⟨55G1705⟩; **4**, 416 ⟨76S748⟩
Pietrasanata, Y., **5**, 122 ⟨76MI 503-01⟩
Pietrasanta, F., **1**, 905 ⟨78JCR(S)62⟩
Pietrasanta, Y., **4**, 191 ⟨73T4225⟩, 191 ⟨77T3105⟩; **6**, 7 ⟨87JPS(A)3025, 89JFC(45)215⟩
Pietraszuk, C., **2**, 938 ⟨91JMOC113, 92JMOC307⟩
Piétré, S., **1**, 13, 18, 20 ⟨86JOC3038⟩
Pietroni, B., **1**, 546 ⟨84TL3137⟩
Pietroni, B. R., **2**, 983 ⟨92JOC976⟩
Pietrre, S., **4**, 91, 287, 858, 861 ⟨85T2527⟩
Pietrusiewicz, K. M., **1**, 158 ⟨89TL477⟩, 159 ⟨84JOC1522⟩; **3**, 197, 231 ⟨80TL2287⟩, 220 ⟨88JOC2837⟩; **4**, 120, 804 ⟨89T337⟩, 547, 579 ⟨93AG(E)1735⟩
Pietruszka, J., **1**, 968 ⟨93SL499⟩
Pietrzak, P., **2**, 1008 ⟨86MI 220-1⟩
Pietrzyk, D. J., **6**, 264 ⟨82JFC(20)617⟩
Pietrzykowski, A., **5**, 781 ⟨84ICA85⟩
Pietsch, R., **2**, 498 ⟨61MI 209-01, 64MI 209-01⟩
Piette, J. L., **2**, 279 ⟨75JHC423⟩, 280 ⟨70BSB353, 70BSB367⟩, 280 ⟨70BSB383⟩, 732 ⟨64BSB507⟩; **3**, 256 ⟨76BSF294⟩; **5**, 241 ⟨70BSB383⟩, 1116 ⟨72BSF3559⟩; **5**, 241 ⟨73SA(A)285, 80SA(A)769⟩
Piettre, S., **2**, 1042 ⟨82S1083⟩; **4**, 88, 773, 785 ⟨87T4309⟩, 92 ⟨85T4183⟩
Pifferi, G., **4**, 441 ⟨72JHC827⟩
Piga, E., **6**, 556 ⟨91PS(63)51⟩
Pigarev, S. D., **4**, 629, 638 ⟨85ZOB2396⟩; **5**, 226 ⟨84IZV1897⟩, 432 ⟨82IZV949, 83JOM(248)51, 83ZOR1838, 84MI 509-01⟩; **6**, 131 ⟨83JOM(248)51⟩, 516 ⟨84IZV1897⟩
Pigeon, P., **2**, 19 ⟨88SC1661⟩, 31 ⟨89BSF247⟩
Piggott, H. A., **6**, 439 ⟨45JCS864⟩
Pigott, F., **2**, 822, 841 ⟨79JOC3741⟩
Pigott, H. D., **1**, 236 ⟨74TL3955⟩; **3**, 35 ⟨74TL3955⟩
Pigou, P. E., **1**, 900 ⟨87CC833⟩, 942 ⟨86AJC77⟩
Pihlaja, K., **4**, 213, 213 ⟨70TL4095⟩, 229 ⟨B-80MI 405-01⟩; **5**, 363 ⟨92T531⟩
Pihl, V., **5**, 689 ⟨93JCS(P2)655⟩
Pike, G. A., **4**, 867 ⟨91PS(60)131⟩, 1081 ⟨88JCS(P1)2839⟩
Pike, J. E., **1**, 729 ⟨69JA5364⟩
Pike, P., **1**, 331, 936 ⟨88T6295, 85TL6289⟩
Pike, P. W., **1**, 935 ⟨88OM2220⟩
Pike, R. A., **1**, 624, 782, 783 ⟨53JA3212⟩
Pike, R. M., **6**, 68 ⟨64RTC119⟩
Pike, S., **2**, 1055 ⟨78CC351⟩; **5**, 1000 ⟨78CC351⟩
Pikies, J., **4**, 136 ⟨87LA51⟩
Piklerova, A., **3**, 194 ⟨83CCC3421⟩; **5**, 665 ⟨76CCC3085⟩, 1028 ⟨83CCC3421⟩
Pikna, R. J., **6**, 225, 227 ⟨56USP2755314⟩
Pikul, S., **1**, 374 ⟨89JA5493⟩; **2**, 533 ⟨89JA5493⟩; **3**, 303 ⟨87JA3981⟩
Pilard, S., **1**, 514 ⟨84TL1555⟩
Pilarski, B., **2**, 302 ⟨89OPP135⟩; **4**, 406, 410 ⟨90JCS(P1)541⟩
Pilati, T., **1**, 910 ⟨76JA1999⟩
Pilato, L. A., **4**, 229 ⟨62TL103⟩
Pilette, D., **5**, 941 ⟨92CC1220, 92OM809⟩
Pilgram, K., **3**, 721 ⟨66TL881⟩
Pilgram, K. H., **5**, 1125 ⟨83JHC1533⟩; **6**, 445 ⟨85SC697⟩, 478 ⟨88JHC193⟩, 567 ⟨88JOC38⟩
Pilgrim, F. J., **1**, 257 ⟨48JA3770⟩; **3**, 618 ⟨55JA1701⟩
Pilgrim, W. R., **3**, 194 ⟨80TL931⟩
Pilichowska, P., **2**, 299 ⟨82S922⟩
Pilichowska, S., **3**, 188 ⟨82S922⟩; **4**, 1008 ⟨88PJC165⟩
Pilipovich, D., **5**, 847, 850 ⟨60JA4700⟩; **6**, 283 ⟨78USP4120904⟩, 424 ⟨72IC2531⟩
Pilkus, A. L., **5**, 688 ⟨81JOU1986, 81ZOR2225⟩
Pillai, C. N., **2**, 561 ⟨83JOM(259)1⟩; **5**, 547 ⟨81OPP81⟩, 560 ⟨81OPP71⟩, 1046 ⟨57CJC832⟩
Pillai, T. P., **2**, 385, 386 ⟨85S677⟩
Pillai, V. N. R., **5**, 317 ⟨87MI 507-03⟩
Pillarsky, R., **2**, 609, 614 ⟨31CB1340⟩
Pillay, M. K., **5**, 1145, 1148 ⟨81JOC5425⟩
Pillmann, S., **6**, 380 ⟨87JOM(323)1⟩
Pillot, J.-P., **1**, 963 ⟨81CC1094⟩, 963, 978 ⟨81TL3401⟩; **2**, 916, 923, 940 ⟨75JOM(87)169⟩, 940 ⟨81TL3401⟩; **3**, 140 ⟨79JOC3397⟩, 213 ⟨76TL1871⟩, 230 ⟨81TL3401⟩; **4**, 621 ⟨89JCR(S)146⟩
Pilot, J. F., **2**, 220 ⟨70JA7629⟩, 476 ⟨68JA6726⟩, 831 ⟨71JA5236⟩; **4**, 336 ⟨74PS(4)109⟩; **5**, 247 ⟨70JA7629⟩
Pilote, J., **2**, 355 ⟨73JOC1813⟩
Pilotti, A., **5**, 652 ⟨69ACS818⟩
Piloty, O., **2**, 782 ⟨1897CB1656⟩; **5**, 821 ⟨02CB3101⟩
Pilskaya, V. Yu., **2**, 883 ⟨55IZV472⟩
Pilz, M., **4**, 633 ⟨89AG(E)784⟩, 634 ⟨88AG(E)1370⟩; **5**, 928, 930 ⟨92ZN(B)983⟩, 929 ⟨90AG(E)399⟩; **6**, 184, 389, 398, 714 ⟨89AG(E)781⟩, 201, 397, 715, 716 ⟨90AG(E)1030⟩, 201, 400 ⟨88AG(E)1370⟩, 389, 398, 401, 713 ⟨89AG(E)784⟩, 397, 398 ⟨90AG(E)1032, 90AG(E)399⟩, 712, 713 ⟨90AG(E)401⟩, 714 ⟨90ZN(B)290⟩
Pimentel, G. C., **1**, 923 ⟨67JCP(47)3637⟩
Pimm, A., **2**, 915 ⟨92SL886⟩
Pinazzi, C. P., **6**, 475 ⟨76BSF501⟩
Pinchuk, A. M., **2**, 407 ⟨67ZOB852, 67ZOB2693⟩, 407 ⟨67ZOB856, 75ZOB2394⟩, 407 ⟨89ZOB2135, 89ZOB2136⟩, 407, 408 ⟨77ZOB56, 90ZOB465⟩, 408

⟨78ZOB2686⟩, 428 ⟨90ZOB697⟩; **3**, 494 ⟨90JGU613⟩; **4**, 559 ⟨88ZOB1665⟩; **5**, 369 ⟨88ZOB706⟩, 373 ⟨70ZOB1949⟩, 379 ⟨85ZOB1080⟩; **6**, 14 ⟨80ZOB1897⟩, 242 ⟨93ZOB1240⟩, 242, 682 ⟨92ZOB948⟩, 267 ⟨90ZOB709⟩, 537 ⟨87UKZ395⟩, 619 ⟨73ZOR43⟩, 680, 682 ⟨89ZOB1902⟩, 681 ⟨88ZOB482, 88ZOB1923⟩, 690 ⟨89ZOB2133⟩, 696 ⟨87ZOB2141, 87ZOB2794, 88ZOB1461, 88ZOB2164, 89ZOB959, 89ZOB1904, 90ZOB2814⟩
Pinck, L. A., **3**, 493 ⟨47JA723, 69JA723⟩
Pinder, A. R., **1**, 2, 6, 10, 425, 429 ⟨80S425⟩, 96 ⟨54JCS113⟩; **2**, 117 ⟨51JCS2123⟩; **3**, 94 ⟨80S425⟩, 211 ⟨71OS(51)115⟩
Pindred, H. K., **1**, 45 ⟨54JCS3617⟩
Pindur, U., **3**, 307 ⟨89JHC1563⟩; **4**, 193 ⟨84M587⟩, 193 ⟨84S227⟩, 193 ⟨85LA2472⟩, 277 ⟨91T1925⟩; **6**, 104 ⟨88JHC907⟩, 729 ⟨87CSR75, 89SC2307⟩
Pine, R. D., **2**, 680 ⟨85JOC1212⟩
Pine, S. H., **1**, 145 ⟨90JOC2234⟩, 398, 407 ⟨70OR(18)403⟩, 674 ⟨91COS(5)1115⟩, 766 ⟨93OR(43)1⟩, 767, 769 ⟨80JA3270, 91S165⟩; **2**, 303 ⟨70OR(18)403⟩, 310 ⟨71JOC829, 80JA3270⟩, 680 ⟨85JOC1212, 93OR(43)1⟩, 681 ⟨90OS(69)72⟩
Pineda, N., **4**, 30 ⟨78HCA2482⟩
Pinel, R., **3**, 255 ⟨73BSF3339⟩, 366 ⟨74BSF(2)471⟩
Pines, H., **1**, 294 ⟨74S309⟩, 795 ⟨B-77MI 118-01⟩, 860 ⟨59JA2568⟩
Pineschi, M., **2**, 135 ⟨92SL303⟩
Pinetti, A., **2**, 22 ⟨90JCR160, 92T4579⟩, 24 ⟨89JCR(S)104, 89JCR(S)360⟩, 24 ⟨91G559, 91SC489⟩, 150 ⟨93SC1759⟩
Pinhas, A. R., **5**, 346 ⟨88CC710, 90JOC2943⟩
Pinheiro, S., **4**, 279 ⟨92TA1003⟩
Pinhey, J., **3**, 685 ⟨84AJC1231⟩
Pinhey, J. T., **1**, 490 ⟨74TL3391, 80TL965⟩, 490 ⟨88JCS(P1)2847⟩, 1145 ⟨89JCS(P1)333⟩; **2**, 1102 ⟨91AJC1353⟩; **3**, 542 ⟨72CC451⟩; **4**, 107 ⟨85AJC1505⟩
Piniella, J. F., **4**, 689 ⟨95JOM(489)C65⟩, 763 ⟨94JA2153⟩
Pink, M., **5**, 616 ⟨94TL1977⟩
Pinkas, A. R., **5**, 448 ⟨82OM1730⟩
Pinkerton, F. H., **2**, 813 ⟨75JOC1090⟩
Pinkina, L. N., **1**, 422 ⟨46IZV633⟩; **4**, 802, 805 ⟨60BAU1851⟩
Pinkney, P. S., **1**, 884 ⟨43OSC(2)116⟩
Pinkos, R., **1**, 574 ⟨94AG(E)117⟩
Pinna, G. A., **5**, 657 ⟨78JHC1513⟩
Pinnell, R. P., **2**, 183, 272 ⟨65JOC38⟩
Pinner, A., **4**, 883 ⟨1895CB473⟩; **5**, 727 ⟨1878CB1825⟩, 744 ⟨1878CB4⟩, 748 ⟨1877CB1889⟩, 755 ⟨1890CB2917⟩, 835, 859 ⟨1884CB183⟩; **6**, 70 ⟨1883CB352, 1883CB1643⟩
Pinney, K. G., **1**, 966 ⟨93JOC7180⟩
Pinnick, H. W., **1**, 200, 238 ⟨88JOC1475⟩, 637, 638 ⟨71JA4316⟩; **2**, 186, 273 ⟨79JOC160⟩; **3**, 18, 134 ⟨90OR(38)655⟩, 31, 156 ⟨75JOC3427⟩; **4**, 185 ⟨81JOC2981⟩, 198 ⟨79JOC3727⟩, 199 ⟨78JOC3964⟩, 203 ⟨83JOC1765⟩, 541 ⟨90OR(38)655⟩; **5**, 141 ⟨78JOC371⟩
Pino, P., **1**, 297 ⟨77JA1058⟩, 434 ⟨66JOC321⟩, 999 ⟨66JOC329⟩; **5**, 293 ⟨52JA4496⟩; **6**, 524 ⟨64CI(M)206⟩
Pintér, I., **5**, 1079 ⟨64AG(E)228, 87CAR(166)101⟩
Pinto, B. M., **4**, 238 ⟨91TL4435⟩; **6**, 100, 102 ⟨89H(28)389⟩, 101 ⟨88JOC5668⟩, 101, 102 ⟨86CJC732, 88JOC3766⟩
Pinto, I., **1**, 248 ⟨88CC81⟩
Pinto, P., **6**, 553 ⟨81TL3485⟩
Pinzani, D.,**3**, 34 ⟨92SL803⟩
Pinzelli, R., **4**, 701 ⟨69JOM(16)83⟩
Pioch, D., **3**, 280 ⟨86JOC2128⟩
Pioch, R. P., **4**, 970 ⟨82EUP49618⟩
Piorko, A., **6**, 30 ⟨90TL6831⟩
Piotrowicz, J. B., **2**, 300, 808 ⟨77AG(E)107⟩
Piotrowska, H., **2**, 395 ⟨82T1321⟩; **4**, 428 ⟨79PJC187⟩; **5**, 692 ⟨88BAP139⟩
Piotrowska, K., **1**, 928 ⟨90CJC1698⟩
Piotrowski, A., **2**, 1006 ⟨83TL2043⟩; **3**, 508 ⟨83TL2043⟩; **4**, 387 ⟨82JOC5051⟩, 696 ⟨83TL2043⟩; **5**, 413 ⟨82JOC5051⟩
Piotrowski, A. M., **1**, 680 ⟨88JOC2829⟩; **4**, 698 ⟨88JOC2829⟩, 1061 ⟨85JA7219⟩
Piotrowski, D. W., **1**, 1042 ⟨87JA2224⟩
Piovosi, E., **4**, 170 ⟨89JOC5171⟩
Pipal, J. R., **4**, 673 ⟨71JA5441⟩
Pipe, D. F., **5**, 1073 ⟨85JCS(P1)1471⟩
Piper, J. R., **2**, 214 ⟨67JOC1261⟩, 255 ⟨79JMC631⟩
Piper, J. U., **3**, 247 ⟨73JOC1451⟩, 601 ⟨77JOC4261⟩
Pipko, S. E., **5**, 713 ⟨90JGU849, 90ZOB963⟩
Pippard, D. A., **6**, 465 ⟨87CC410⟩
Pippen, E. L., **3**, 63 ⟨58JOC1580⟩
Pippin, W., **1**, 419, 420, 794 ⟨73T3781⟩
Piquet, A., **3**, 722 ⟨04CB4295⟩
Piras, P., **2**, 386 ⟨81JCS(P2)1449⟩
Piras, P. P., **1**, 975 ⟨92JA4051⟩
Piratla, R. K., **6**, 489 ⟨90AX2374⟩
Pirazzini, G., **5**, 410 ⟨80TL3927⟩, 793, 803 ⟨81JCS(P2)1043⟩
Pires, J., **5**, 330 ⟨91TA1209⟩
Piria, R., **1**, 883 ⟨1856MI 119-01⟩
Pirie, D. K., **3**, 559 ⟨74TL1515⟩
Pirisi, F. M., **2**, 9 ⟨75CC950⟩, 9, 35 ⟨74CC879⟩
Pirkle, W. H., **1**, 966 ⟨78JOC1950, 78JOC2091⟩; **5**, 274, 294 ⟨86JA6054⟩
Piro, O. E., **2**, 1008 ⟨87CJC2000⟩
Piroddi, A. M., **1**, 398 ⟨90JOC5532⟩, 432 ⟨90JOC5386⟩; **4**, 179 ⟨90S313⟩
Pirogov, P. A., **2**, 359 ⟨81ZOB609⟩
Pirozhenko, V. V., **4**, 99 ⟨88JGU2192⟩, 110 ⟨82JOU2240⟩, 1008, 1009 ⟨86ZOB2681⟩; **5**, 655 ⟨88JGU2192, 88ZOB2465⟩, 659 ⟨86ZOB2681⟩, 659, 788, 789 ⟨86JGU2141⟩, 669 ⟨89UKZ1071⟩
Pirrin, A., **4**, 962, 963, 964 ⟨92SL886⟩
Pirrung, M. C., **1**, 196, 410, 820 ⟨86JA6060⟩, 224 ⟨89HCA1301⟩, 382 ⟨81JA82⟩, 874 ⟨80JOC1066⟩; **3**, 707 ⟨86JOC2103⟩
Pisanenko, D. A., **1**, 492 ⟨72ZOR1484⟩
Pisanenko, N. P., **5**, 57 ⟨73S545⟩; **6**, 621 ⟨78JGU1078⟩, 656 ⟨74ZOB1264⟩
Pisareva, I. V., **2**, 915 ⟨77IZV2389⟩
Pisareva, S. A., **4**, 556 ⟨68IZV1417, 68IZV2062⟩
Pisarnitskii, D. A., **2**, 433, 473 ⟨91ZOB1016⟩, 476 ⟨90JGU398⟩; **4**, 458 ⟨91ZOB1478, 92ZOB946⟩, 459, 484 ⟨91ZOB1016⟩
Piscitelli, M., **1**, 1112 ⟨68TL4039⟩
Piscopio, A. D., **1**, 106 ⟨93CC617⟩, 558, 828 ⟨93CC619⟩, 583 ⟨94JOC332⟩; **4**, 355 ⟨91TL855⟩
Piscopo, L., **3**, 306, 307 ⟨93JCS(P1)273⟩
Piskala, A., **1**, 721 ⟨75C341⟩, 730 ⟨70CB2814⟩
Piskov, V. B., **5**, 655 ⟨78JOU758, 78ZOR820⟩
Piskunova, I. P., **6**, 625 ⟨77MI 620-02⟩
Pisutjaroenpong, S., **3**, 62 ⟨85TL3613⟩
Piszkiewicz, L. W., **1**, 85 ⟨67JOC2013⟩
Pitacco, G., **2**, 396 ⟨89T2099⟩, 738, 757, 758 ⟨B-82MI 215-01⟩, 783 ⟨73T3011, 74JCS(P1)660, 74T2741⟩; **5**, 354 ⟨83JHC305⟩, 355 ⟨91G401⟩
Pitchen, P., **2**, 147 ⟨84JA8188⟩, 147 ⟨85PAC1911⟩; **6**, 543, 550 ⟨91T6381⟩
Pitchford, A., **1**, 764 ⟨85TL5093⟩
Pitchumani, K., **2**, 886 ⟨85CJC2285⟩
Piteau, M., **2**, 303 ⟨84JOC2081⟩; **4**, 58 ⟨81EUP40153, 84JOC1700⟩; **6**, 229 ⟨87JA3708⟩, 428 ⟨84JOC2081⟩, 541 ⟨86EUP176412⟩
Piteau, M. D., **4**, 59 ⟨90JOC1847⟩; **6**, 423 ⟨90JOC1847⟩
Piterskikh, I. A., **4**, 311 ⟨83IZV2568⟩
Pitman, I. H., **2**, 338 ⟨69JCS(C)626⟩
Pitochelli, A. R., **2**, 94 ⟨66JA4531⟩
Pitombo, L. R. M., **2**, 732 ⟨59CB40⟩
Pitsinos, E., **4**, 867 ⟨91PS(58)207⟩

Pitt, B. H., **2**, 125 ⟨54JA1082⟩
Pitt, B. M., **4**, 64, 65, 66, 69 ⟨55JA572⟩
Pitt, C., **2**, 490 ⟨91PS(57)1⟩
Pitt, C. G., **2**, 482 ⟨86JOM(308)281⟩, 482, 504 ⟨90POL319⟩, 508, 511 ⟨86OM1266, 87JOM(325)C7⟩; **4**, 597 ⟨91MI 413-01, 91POL319⟩, 597 ⟨91PS(57)1⟩
Pitt, C. P., **2**, 511 ⟨88JOM(354)287⟩
Pitta, I. D. R., **3**, 241 ⟨79JHC821⟩
Pitteroff, W., **4**, 381 ⟨67LA(703)1⟩
Pittman, A. G., **5**, 4 ⟨66JOC2316⟩
Pittman, C. U., **3**, 3 ⟨81JOC1901⟩; **4**, 851 ⟨76S489⟩
Pittman, J. H., **1**, 780 ⟨87TL3901⟩; **2**, 610 ⟨87TL3901⟩
Pittman, Jr., C. U., **1**, 456 ⟨79JOM(178)43⟩, 891 ⟨66APO(4)305, B-73MI 119-05⟩, 893, 897 ⟨65JA2997⟩, 894 ⟨65JA2998⟩, 897, 901, 911 ⟨63JA2991⟩, 902, 912 ⟨64JA1871⟩, 904 ⟨65JA5632⟩, 909 ⟨74JOC378, 65JA2153⟩, 1125 ⟨69CC122⟩; **6**, 43 ⟨72CRV357⟩, 121 ⟨76BCJ1996⟩, 121, 698 ⟨75CC960⟩
Pittman, Jr., C. V., **1**, 427 ⟨93TL3263⟩
Pittman, R. W., **2**, 278, 292 ⟨57JCS799⟩, 289 ⟨50JCS1371⟩; **3**, 382 ⟨50JCS1371⟩, 391 ⟨57JCS799⟩
Pittol, C. A.,**1**, 522 ⟨91JCS(P1)1255⟩
Pitts, Jr., E. H., **3**, 293 ⟨61JOC4151⟩
Pitts, W. D., **5**, 465 ⟨72CC672⟩
Piva, O., **3**, 35, ⟨94T3349⟩, 114 ⟨92TL2459⟩, 275 ⟨88TL2321⟩, 294 ⟨91TL3993⟩
Pivawer, P. M., **3**, 190 ⟨66JOC3152⟩
Pivazyan, V. A., **5**, 1120 ⟨80CA(93)95215⟩
Piven', V. A., **1**, 500 ⟨91ZOR488⟩
Piven, Yu. V., **5**, 861 ⟨66JGU1101⟩
Piveteau, D., **2**, 325 ⟨88T2457⟩; **3**, 422 ⟨88T2457⟩
Piveteau, E., **3**, 181 ⟨91TL2359⟩
Pivnitskii, K. K., **3**, 209 ⟨83JOU259⟩
Pivovarov, M. P., **2**, 453 ⟨78BRP1517876⟩
Piwinski, J. J., **1**, 807 ⟨79JA1611⟩, 840 ⟨82JA7181⟩; **3**, 21 ⟨82JA7181⟩
Pizey, J. S., **2**, 21, 25 ⟨B-74MI 201-01⟩, 27 ⟨B-74MI 201-02⟩, 247, 364 ⟨79PS(7)325⟩, 608 ⟨B-77MI 212-01⟩, 872, 881 ⟨75SC393⟩; **4**, 21 ⟨B-74MI 401-01⟩; **5**, 485 ⟨80PS(8)1⟩
Pizzano, A., **6**, 327, 330 ⟨93IC5569⟩
Pizzo, C. F., **1**, 828 ⟨81JOC3896⟩
Pizzo, F., **2**, 146 ⟨93SC3157⟩; **3**, 65, 232 ⟨87HCA1429⟩
Pizzolato, G., **1**, 91 ⟨77JOC135⟩, 697, 700, 703 ⟨88JOC4282⟩
Pizzorno, M. T., **2**, 675 ⟨72CI(L)425⟩
Pizzotti, M., **5**, 386 ⟨75JOM(88)237, 76JCS(D)1777⟩
Plaas, D., **4**, 679 ⟨83JCS(D)2661⟩
Plabst, D., **5**, 953 ⟨74CB3326⟩
Place, P., **1**, 810 ⟨76TL677⟩, 958 ⟨78TL3801⟩, 974 ⟨81T1359⟩; **2**, 9 ⟨73CC821, 76BSF169⟩
Pla'chik, R. I., **3**, 540 ⟨65DOK(164)357, 68DOK(179)892⟩
Plachky, M., **1**, 650 ⟨76S199⟩
Plackett, J. D., **1**, 822 ⟨68CC186⟩
Plackett, S. J., **5**, 655 ⟨87JCS(P1)1153⟩; **6**, 572 ⟨87JCS(P1)1153⟩
Plagge, H., **6**, 129 ⟨89LA975⟩
Plaia, U., **3**, 680 ⟨87CB2031⟩
Plaisance, C.-B., **5**, 546, 550 ⟨85BSF891⟩
Plamondon, J. E., **1**, 658 ⟨74JA316⟩
Plamondon, L., **1**, 1067 ⟨92JOC5060⟩
Plampin, J. N., **6**, 635 ⟨88S460⟩, 641, 645 ⟨86JOC1882⟩
Planes, J., **5**, 680 ⟨88H(27)1935⟩
Planinsek, Z., **5**, 6 ⟨89TL6095⟩
Plank, J., **3**, 519 ⟨84JOM(264)327⟩, 544 ⟨79JA3133⟩
Plankl, W., **1**, 885 ⟨77JCS(P1)460⟩
Plant, A., **3**, 18, 134 ⟨90SL725⟩
Plant, D., **6**, 540 ⟨55JA2479⟩
Plant, G. P., **2**, 102 ⟨21MI 202-02⟩
Plant, S. G. P., **2**, 128 ⟨48JCS1892⟩; **4**, 309 ⟨48JCS2174⟩
Plantema, O. G., **1**, 747 ⟨78JCS(P1)304⟩
Plaquevent, J.-C., **1**, 180 ⟨88T5495⟩, 573 ⟨93CC116⟩; **4**, 89 ⟨88TL5893⟩; **5**, 283 ⟨87TL2595⟩
Plashkin, V. S., **6**, 42 ⟨73ZOR2013⟩
Plass, H., **2**, 1096 ⟨68JOM(14)21⟩
Plass, W., **4**, 552 ⟨91ZAAC(596)139⟩
Plata, D. J., **1**, 816, 831 ⟨88JOC5189⟩
Platanov, V. E., **6**, 12 ⟨69IZV196⟩
Platau, G. O., **1**, 156 ⟨54JA3972⟩
Plate, A. F., **1**, 795, 801 ⟨76RCR469⟩
Plate, R., **1**, 834 ⟨87JOC370⟩; **3**, 433 ⟨86JOC309⟩
Plath, M. W., **4**, 425 ⟨90S951⟩
Plath, P., **2**, 746 ⟨82S318⟩
Platonova, V. E., **4**, 737 ⟨85JFC(28)99⟩; **5**, 654 ⟨87RCR1125⟩, 712 ⟨88RCR234, 88UK405, 88JFC(40)217⟩, 1070 ⟨89ZOR1523, 89ZOR1689⟩; **6**, 272 ⟨88ZOR1513⟩, 609 ⟨77JFC(9)505, 83JFC(22)439, 86ZOR1297⟩, 616 ⟨84ZOR1197⟩, 619 ⟨88JFC(40)217, 93JPO319⟩
Platoshkin, A. M., **3**, 527 ⟨67DOK(173)1117⟩
Platt, A. E., **6**, 7 ⟨67JCS(C)1150⟩
Platt, J., **1**, 420 ⟨73T3797⟩
Plattner, E., **2**, 177 ⟨74S887⟩
Plattner, J. J., **3**, 494 ⟨70JA3429⟩; **6**, 417 ⟨92EUP486948⟩
Plattner, P. A., **2**, 659 ⟨B-48MI 213-01⟩; **3**, 229 ⟨47HCA905⟩
Platz, K. H., **5**, 206 ⟨70JPR(312)2⟩
Platz, L., **5**, 843 ⟨05CB2986⟩
Platz, M. S., **3**, 565 ⟨92MI 308-03⟩; **4**, 138, 154 ⟨88TL5863⟩
Platzer, N., **2**, 594 ⟨93TL7575⟩; **3**, 304 ⟨77JHC861⟩
Plau, B., **1**, 30 ⟨79CC300, 80JCS(P1)2554⟩; **2**, 11 ⟨79S437⟩
Plaut, A. G., **4**, 527, 528 ⟨90JBC(265)3738⟩
Plaut, H., **3**, 160 ⟨84CC29⟩
Plavac, N., **5**, 418, 924 ⟨82OM994⟩
Plavec, J., **1**, 946 ⟨93T5189, 93T9079⟩
Plazzi, P. V., **5**, 347 ⟨84FES788⟩
Ple, G., **3**, 63 ⟨90SC2983⟩, 65 ⟨90TL3129⟩
Plé, G., **4**, 749 ⟨93S1071⟩; **5**, 315, 316, 318 ⟨93SC2423⟩
Pleass, C. M., **2**, 421 ⟨57JCS2078⟩
Pledger, Jr., H., **4**, 745 ⟨56JA6101⟩
Plemenitas, A., **3**, 89 ⟨84S504⟩
Plénat, F., **1**, 795 ⟨62BSF1325⟩, 905 ⟨78JCR(S)62⟩; **2**, 848 ⟨88S912⟩, 936 ⟨81TL4705⟩; **3**, 195 ⟨B-94MI 304-01⟩, 493, 496 ⟨B-94MI 313-01⟩, 650 ⟨66BSF1315⟩
Plenio, H., **4**, 800 ⟨86CB3150⟩; **5**, 1084 ⟨89ZN(B)94⟩
Plenkiewicz, H., **4**, 736 ⟨84JFC(24)387⟩
Plenkiewicz, J., **5**, 828 ⟨72MI 521-01⟩
Plenkina, O. N., **2**, 949 ⟨78ZOB639⟩
Plentl, A. A., **3**, 531, 532 ⟨41JOC669⟩
Plesch, P. H., **1**, 896 ⟨65JCS6019⟩
Plesch, W., **4**, 311 ⟨84LA1494⟩
Plescia, S., **2**, 770 ⟨71AC(R)391⟩
Plesek, J., **4**, 629 ⟨92CRV175⟩
Pleshkova, A. P., **2**, 356, 357 ⟨91IZV1069⟩, 641 ⟨75DOK(220)146⟩
Pleskov, Y. V., **1**, 248 ⟨87TCC149⟩
Pless, J., **6**, 469 ⟨79HCA398⟩
Plessas, N. R., **2**, 25 ⟨69JOC2163⟩
Plesse, M., **4**, 410 ⟨82JPR832⟩
Plessi, L., **2**, 25, 31 ⟨83S306⟩, 103 ⟨85TL3369⟩
Plesske, K., **1**, 862 ⟨62AG(E)312, 62AG(E)394⟩
Pletcher, D., **3**, 4, 116 ⟨78TL1601⟩
Pletcher, J., **2**, 1009 ⟨87AX(C)313⟩
Pletnev, M. A., **2**, 891 ⟨88JGU1040⟩
Pletnev, S. I., **4**, 799 ⟨91BAU1705⟩; **5**, 22 ⟨91BAU1705⟩, 679 ⟨91IZV1918⟩
Pleurdeau, A., **6**, 475 ⟨76BSF501⟩
Plevey, R. G., **6**, 53 ⟨84JCS(P1)455⟩
Plevyack, J. E., **1**, 494 ⟨79JOC4078⟩, **5**, 157 ⟨77JOC3903⟩
Plieninger, H., **1**, 725 ⟨76LA1475⟩; **2**, 748 ⟨71CB3983⟩; **3**, 261 ⟨70S587⟩, 450 ⟨55CB1956⟩; **5**, 577, 590, 608, 609 ⟨75T1209⟩
Plimmer, J. R., **5**, 62 ⟨60QR292⟩
Plobeck, N. A., **2**, 395 ⟨92TL131⟩
Ploeger, W., **6**, 367 ⟨74GEP2237879⟩

Ploger, W., **4**, 1003, 1010 ⟨71CB2021⟩
Plöger, W., **6**, 158, 159 ⟨72ZAAC(389)119⟩, 449, 452 ⟨71CB969⟩
Plonka, J. H., **1**, 232 ⟨73JA1547⟩
Ploquin, J., **3**, 722 ⟨47BSF901⟩
Plotnikova, G. I., **6**, 83 ⟨60IZV1901⟩
Plotnikova, M. T., **5**, 345 ⟨69ZOR279⟩
Ploux, O., **2**, 229 ⟨87JOC3154⟩
Pluchet, H., **1**, 82 ⟨63APO(1)155⟩
Plug, J. P. M., **3**, 91, 107 ⟨92S1221⟩
Pluijgers, C. W., **6**, 580 ⟨66TL3695⟩
Plumb, J. B., **2**, 849 ⟨66JOC2455⟩; **6**, 61 ⟨65JCS2101⟩
Plumet, J., **1**, 217 ⟨90H(31)1997⟩; **3**, 221 ⟨90TL7669⟩, 414 ⟨85TL4403⟩, 417 ⟨84JCR(S)144⟩; **4**, 304 ⟨84JCR(M)144⟩; **5**, 696 ⟨91OM11⟩
Plumitallo, A., **4**, 972 ⟨88FES103⟩
Plummer, C. W., **4**, 429, 431 ⟨61USP2991315⟩, 429, 431 ⟨67USP3316311⟩; **6**, 145, 150 ⟨67USP3316311⟩, 150 ⟨61USP2991315⟩
Plummer, D. T., **3**, 723 ⟨84JOM(260)347⟩
Plummer, J. S., **1**, 828 ⟨91TL4267⟩, 831 ⟨91JOC353⟩; **5**, 227 ⟨91JOC353⟩
Plummer, M., **3**, 303 ⟨86JA1056⟩
Plummer, T., **3**, 723 ⟨82IC2178⟩
Plunkett, J. J., **1**, 543 ⟨81JOC807⟩
Pluscec, J., **2**, 257 ⟨92JA1521⟩; **3**, 713 ⟨79JOC1218⟩; **5**, 346 ⟨82JOC5160⟩
Plusquelle, D., **2**, 133 ⟨91T5185⟩; **5**, 183 ⟨83TL4195⟩
Pluta, Ch., **2**, 1101 ⟨93JOM(453)C11⟩
Pluygers, C. W., **2**, 273 ⟨73OSC(5)223⟩; **5**, 1026 ⟨55RTC1262⟩
Plyamovatyi, A. Kh., **4**, 132 ⟨91JGU1459⟩
Plzak, K. J., **1**, 829 ⟨93TL1851⟩
Po, S.-Y., **4**, 232 ⟨90TA143⟩
Pocar, D., **1**, 45 ⟨87S665⟩; **3**, 438 ⟨87S719⟩, 467 ⟨63CB802, 65CB2715⟩; **5**, 497, 499 ⟨74JCS(P2)1483⟩, 762 ⟨76TL1839⟩, 762, 816 ⟨78JCS(P1)309⟩
Pocci, M., **3**, 620 ⟨93T8423⟩
Poch, M., **2**, 318 ⟨91TL6935⟩; **3**, 243 ⟨90TL7505⟩
Pochapsky, S. S., **6**, 134 ⟨90MI 604-01⟩
Pochat, F., **3**, 654, 655 ⟨78TL2683⟩; **5**, 675, 682 ⟨84S146⟩
Poché, D., **6**, 416, 430, 486 ⟨88TL5859⟩
Pochini, A., **3**, 60 ⟨75JCS(P1)1527⟩, 97 ⟨78JCS(P1)318⟩
Pöchlauer, P., **2**, 860 ⟨84HCA1238⟩; **5**, 373 ⟨85CC1764⟩
Pocker, Y., **1**, 384 ⟨58JCS403⟩
Podall, H. E., **2**, 626 ⟨58JOC280⟩; **6**, 522, 523 ⟨60JA1325⟩
Poddubnyi, V. G., **5**, 850 ⟨69ZOR1404⟩, 854 ⟨69ZOR1411⟩
Podenko, L. S., **5**, 860 ⟨78KGS991⟩
Podestá, J. C., **1**, 950 ⟨88JOM(340)317⟩; **2**, 653 ⟨75CC36⟩; **3**, 281 ⟨73CI(L)189⟩
Podesva, C. P., **5**, 1125 ⟨62CJC1403⟩
Podewell, C. C., **3**, 468 ⟨74JOC1444⟩
Podkoscielny, W., **6**, 546 ⟨90MI 617-03⟩
Podlaka, J., **5**, 655 ⟨89CCC3245⟩
Podolesov, B. D., **5**, 735 ⟨81MI 518-01⟩
Podovik, A. N., **6**, 538, 539, 584 ⟨88ZOB1489⟩
Podrebarac, E. G., **6**, 653 ⟨63JMC283⟩
Podschadly, O., **1**, 991 ⟨92CB2667⟩
Podvisozkaya, L. S., **6**, 406 ⟨67JOM(7)385⟩
Poë, A. J., **4**, 687 ⟨74JCS(D)1856, 74JCS(D)607⟩
Poehler, T., **4**, 876 ⟨82CC335⟩
Poehler, T. O., **4**, 874 ⟨80CC867⟩; **6**, 591, 594 ⟨80CC866B⟩
Poehlmann, H., **3**, 241 ⟨77CB491⟩
Poel, T.-J., **1**, 209 ⟨93JOC7304⟩
Poels, E. K., **4**, 200, 208 ⟨78JOC3548⟩
Poersch, F., **2**, 432 ⟨67AG(E)355⟩; **5**, 1058 ⟨69AG(E)203⟩
Poeth, T. P., **4**, 726 ⟨71IC522⟩
Pogany, S., **4**, 58, 211 ⟨83JOC5280⟩
Pogonowski, C. S., **1**, 210 ⟨74JOC732⟩, 602, 607 ⟨75JOC542⟩
Pogosyan, A. S., **2**, 446 ⟨80AKZ408⟩; **4**, 914 ⟨80CA186471c⟩
Pohjala, E., **6**, 263 ⟨92JCS(P2)835, 92PS(70)183⟩, 263 ⟨93TL4551⟩
Pohl, D., **4**, 231 ⟨92JOC2217⟩
Pohl, D. G., **2**, 20 ⟨79JOC3728⟩
Pohl, E., **4**, 940, 1014 ⟨92CB2051⟩; **5**, 958 ⟨92CB2051⟩
Pohl, H., **5**, 546 ⟨07CB1303⟩
Pohland, A., **5**, 291 ⟨68JA4706⟩
Pohland, A. E., **1**, 469 ⟨66CRV161⟩; **3**, 238, 266 ⟨64JOC385⟩, 238, 658 ⟨66CRV161⟩; **4**, 970 ⟨66CRV161⟩
Pohland, E., **2**, 811 ⟨26CB2215⟩
Pohland, H. W., **5**, 597, 599 ⟨63CB526⟩
Pohl, S., **2**, 401 ⟨79ZAAC(452)103⟩, 729 ⟨91CB1315⟩; **4**, 88, 288 ⟨91CB1315⟩; **5**, 889 ⟨87AG(E)776⟩, 1085 ⟨90AG(E)90, 91POL1147⟩; **6**, 334, 355, 356 ⟨90CB2325⟩
Pohlmann, S., **4**, 649 ⟨81ZAAC(482)65⟩, 1050 ⟨85JOM(291)277⟩; **6**, 290, 379, 380 ⟨84ZAAC(510)169⟩, 379 ⟨85JOM(291)277, 85ZN(B)1023⟩
Pohloudek-Fabini, R., **5**, 1027 ⟨65AP(298)51⟩, 1047 ⟨68MI 527-01⟩, 1111 ⟨67AP583⟩
Pohlschmidt, A., **3**, 347, 363 ⟨92TL5693⟩
Pohmakor, M., **4**, 395 ⟨76AG(E)437⟩
Pohmakotr, M., **3**, 62 ⟨85TL3613⟩, 210 ⟨84TL4141⟩, 212 ⟨82CL687⟩; **4**, 392 ⟨79CB1420⟩, 962 ⟨93T6483⟩; **5**, 552 ⟨79TL2271⟩
Pohmer, L., **2**, 667 ⟨56CB1334⟩
Poiana, M., **3**, 405 ⟨88SC1501⟩
Poilblanc, R., **4**, 569 ⟨78ICAL294⟩, 1039 ⟨79ICAL67⟩
Poindexter, G. S., **2**, 320 ⟨93SC1329⟩, 942 ⟨87TL1851⟩; **4**, 651 ⟨87TL1851⟩; **5**, 302 ⟨82JOC3787⟩
Pointner, A., **1**, 681, 683 ⟨79CC52⟩; **2**, 677 ⟨79CC52⟩
Poirier, J.-M., **1**, 453 ⟨77JA8356⟩; **2**, 759 ⟨77JA8356, 79JOC3585, 82BSF(2)297, 83JCR(M)2101, 83JCR(S)222, 91S649⟩; **3**, 27 ⟨86OPP79⟩, 76, 262 ⟨83JCR(S)222⟩, 80 ⟨75BSF329⟩; **4**, 15 ⟨73TL4237⟩
Poisel, H., **2**, 338 ⟨75CB2917, 76AG(E)294⟩; **4**, 304 ⟨77CB948⟩; **5**, 315 ⟨75CB2547, 76AG(E)294⟩
Poisson, J., **4**, 176 ⟨87TL3569⟩
Poisson, P., **1**, 69 ⟨85BSF75⟩
Poist, J. E., **1**, 999 ⟨67JOM(8)239⟩
Poje, J. A., **1**, 442 ⟨39JA2418⟩
Poje, M., **4**, 268 ⟨80TL3089⟩
Pojer, P. M., **1**, 36 ⟨68AJC1375⟩; **4**, 222 ⟨78AJC1031⟩, 303 ⟨72AJC1737⟩
Pokatun, V. P., **4**, 561 ⟨83ZOB541, 84URP1067006⟩
Pokorski, U., **2**, 313 ⟨90TL3023⟩
Pokrovskaya, I. E., **2**, 886 ⟨81JGU1056⟩
Pokrovskaya, M., **4**, 222 ⟨91TL7593⟩
Pokrovskaya-Dukhnenko, E. M., **4**, 57 ⟨86ZOR2327⟩
Pola, J., **4**, 352 ⟨73CCC1522⟩; **6**, 410 ⟨81CZP208251, 90CZP264358⟩
Polakowski, J., **2**, 118, 124 ⟨93JMC460⟩
Polanc, S., **5**, 764 ⟨76JHC487⟩, 836 ⟨73TL1677, 74JHC103⟩, 836 ⟨74JOC2143⟩
Poland, J. S., **3**, 483, 484 ⟨70JCS(A)2954⟩; **6**, 672 ⟨70JCS(A)2954⟩
Polanyi, M., **1**, 71 ⟨34MI 103-01⟩
Polaskova, K., **5**, 630 ⟨87CCC2087⟩
Polborn, K., **3**, 606 ⟨90TL4429⟩; **4**, 484 ⟨90TL6999⟩, 710 ⟨89CB1901⟩; **5**, 944 ⟨93JOM(460)C25⟩
Pole, D. L., **3**, 324 ⟨92JA8751⟩
Polechnowski, S. W., **2**, 572 ⟨78JA7565⟩
Polenov, Y. V., **6**, 580 ⟨88MI 618-01⟩
Poleschner, H., **4**, 874 ⟨79ZC192⟩; **6**, 93 ⟨86ZC138⟩, 590, 592, 593 ⟨92JOM(427)213⟩, 630, 731 ⟨84S667⟩
Poletaev, V. A., **4**, 656, 658 ⟨74IZV2861⟩
Poletto, J. F., **1**, 556 ⟨80JOC4702⟩; **2**, 28 ⟨80JOC4702⟩; **3**, 140, 156 ⟨79JOC1438⟩
Polezhaeva, N. A., **3**, 200 ⟨90S515⟩; **4**, 1006 ⟨73IZV1112⟩
Poli, G., **1**, 305 ⟨87HCA2201⟩; **2**, 47 ⟨89TL7385⟩; **3**, 35 ⟨93JOC3165⟩
Poli, M., **1**, 879 ⟨83JOC2957⟩

Poli, R., **2**, 881, 883, 886, 890 ⟨84JCS(D)2365⟩, 890, 891, 893 ⟨88HOU(E13)761⟩; **4**, 562 ⟨89POL1293⟩
Poliakoff, M., **2**, 792 ⟨78ACS(B)625⟩; **3**, 677 ⟨77ACS(B)848⟩
Polianski, M., **4**, 222 ⟨91TL7593⟩
Policastro, P. P., **1**, 520 ⟨84JA2443⟩
Polichnowski, S. W., **4**, 580 ⟨77JA6097, 79JA7282⟩
Polikarpov, Yu. M., **4**, 555, 561 ⟨67IZV591⟩, 556 ⟨67IZV949, 68IZV1417, 68IZV2062⟩, 560, 564 ⟨70IZV1326⟩
Polishchuk, V. R., **4**, 697 ⟨71IZV620⟩; **6**, 62 ⟨72IZV85⟩
Politzer, I., **3**, 61 ⟨73JOC36⟩
Politzer, I. R., **3**, 286 ⟨69JA5886⟩; **4**, 306 ⟨88OSC(6)905⟩
Polivanov, A. N., **4**, 156 ⟨78JGU769⟩
Polizzi, C., **1**, 466 ⟨91JOM(417)289⟩, 1016 ⟨87TL6073⟩
Poljakowa, A. M., **3**, 531, 532 ⟨35CB850⟩
Polk, D., **2**, 746 ⟨82SC939⟩
Polk, M., **6**, 388 ⟨86CB1117⟩
Poll, W., **4**, 607, 656 ⟨91JOM(421)175⟩; **6**, 190, 192, 195, 393 ⟨91JOM(421)175⟩, 604 ⟨92ZN(B)351⟩
Polla, E., **5**, 368 ⟨84TL3701⟩
Pollack, J. M., **6**, 607, 609 ⟨58JCS764⟩
Pollack, M. A., **1**, 414 ⟨38JA1905⟩
Pollack, P. I., **6**, 506 ⟨68USP3539569⟩
Pollack, R. M., **1**, 130, 869 ⟨89T4913⟩, 877 ⟨88JA1981, 89JA6419⟩, 877 ⟨90JOC1019, 91JA1344⟩, 877 ⟨91JA3833⟩, 878 ⟨93JOC7100⟩
Pollack, S. K., **1**, 129, 162 ⟨81JOC1693⟩; **2**, 525, 1005 ⟨81JOC1693⟩
Pollak, A., **2**, 17 ⟨76JOC4002, 77JOC1559⟩
Pollak, I. E., **4**, 317 ⟨67JOC2891⟩, 333 ⟨66JOC3514, 67JOC3101⟩; **5**, 351, 366 ⟨91TL133⟩
Pollak, P. J., **1**, 795 ⟨55JA5166⟩
Pollard, G. E., **4**, 836, 837, 838 ⟨68JOC4470⟩
Pollart, D., **3**, 9, 123 ⟨85TL705⟩
Pollart, K. A., **2**, 461 ⟨63JOC3430⟩, 854 ⟨62JOC4444⟩
Poller, R. C., **2**, 600 ⟨83JOM(255)473⟩, 809 ⟨79COC(3)1114⟩, 815 ⟨79COC(3)1073⟩, 946 ⟨79COC(2)1061⟩, 1100 ⟨B-70MI 223-01⟩; **4**, 156 ⟨72JOM(40)115⟩; **5**, 1017 ⟨65JCS5055⟩
Pollet, P., **3**, 212 ⟨78S142⟩
Pollicino, S., **1**, 821 ⟨78JOC4826, 79JOC4128⟩, 821 ⟨81JOC3315⟩; **2**, 143 ⟨81JOC3315⟩, 156 ⟨88JOC5689⟩
Pollini, G. P., **2**, 181 ⟨74S877⟩, 301 ⟨91S479⟩, 356 ⟨91TL2517⟩, 395 ⟨72S44⟩; **3**, 648 ⟨85S969⟩; **4**, 475 ⟨82S653⟩
Pollock, C. M., **4**, 652, 655 ⟨93JOC7474⟩
Pollock, J. M., **5**, 1016 ⟨63BRP923581⟩
Pollock, R. M., **1**, 797 ⟨B-89MI 118-01⟩
Pollock, T., **2**, 446 ⟨85CB3105⟩
Pollok, T., **1**, 51, 446 ⟨87CB1403⟩; **4**, 548 ⟨84HCA2175⟩, 551, 560, 1023 ⟨85CB3105, 86AG(E)348⟩
Polly, P. W., **2**, 575 ⟨82COMC-I(6)37⟩
Polniaszek, R. P., **1**, 776 ⟨86JOC3023⟩; **2**, 455 ⟨91JOC3137⟩; **5**, 229 ⟨86TL5683⟩
Polo, A., **5**, 801 ⟨86OM1369⟩
Polo, C., **3**, 706 ⟨89S641, 91LA843⟩
Poloński, T., **2**, 342 ⟨74TL2453⟩; **5**, 7, 15 ⟨88S475⟩, 572 ⟨94TA149⟩
Polozov, A. M., **3**, 200 ⟨90S515⟩
Polsky, B., **6**, 489 ⟨93MI 615-03⟩
Polster, R., **1**, 628 ⟨57LA(612)102⟩
Polston, N., **1**, 420 ⟨73T3797⟩
Polston, N. L., **1**, 278, 488, 531, 662, 1103 ⟨68JA6243⟩, 449, 509, 1102 ⟨71JA3395⟩, 509, 1103 ⟨70JA4068⟩; **2**, 900 ⟨68JA6243, 71JA3395⟩
Polt, R. L., **2**, 759, 965 ⟨86JA304⟩; **3**, 406 ⟨82JOC2663⟩; **4**, 445 ⟨85TL695⟩
Polushin, N. N., **2**, 410 ⟨93JOC4606⟩
Polya, J. B., **4**, 424, 425 ⟨51RTC269⟩; **5**, 849 ⟨54JCS3319⟩
Polyachenko, L. K., **5**, 888 ⟨84PS(19)189⟩; **6**, 684 ⟨84ZOB965⟩, 689 ⟨84PS(19)189⟩, 693 ⟨83ZOB693⟩
Polyakov, A. E., **6**, 608 ⟨73ZOR1759⟩
Polyakova, N. Yu., **6**, 145 ⟨86ZOR1393⟩
Polyakov, A. V., **2**, 447 ⟨91MI 208-01⟩; **4**, 674 ⟨91JOM(418)C24⟩
Polyuyan, O. S., **5**, 208 ⟨83ZOR1764⟩
Polzonetti, V., **2**, 396 ⟨92S355⟩
Pombo-Villar, E., **1**, 835 ⟨91TL7233⟩
Pomerantz, M., **5**, 374 ⟨85JOC1712⟩
Pomeranz, C., **3**, 612 ⟨07LA(351)354⟩
Pomeroy, J. H., **3**, 617 ⟨59JA6340⟩
Pommelet, J.-C., **1**, 592 ⟨81AG(E)585⟩, 835 ⟨91JOC970⟩; **3**, 541 ⟨85TL833, 86CC369, 87TL885, 88TL5919⟩; **4**, 71 ⟨81AG(E)585⟩, 900 ⟨91JOC970⟩; **6**, 44 ⟨81AG(E)585⟩
Pommer, H., **1**, 525, 832 ⟨71LA(747)60, 69LA(729)52⟩
Pommerening, H., **2**, 893 ⟨83OM1573⟩
Pommerening, K., **6**, 504 ⟨66CB2937⟩
Pommeret, J., **4**, 183 ⟨71TL231⟩
Pommerville, J., **3**, 531 ⟨88JOC4877⟩
Pommier, A., **1**, 565, 681, 683, 785 ⟨93S441⟩; **3**, 541 ⟨93JCS(P1)1549⟩; **5**, 161 ⟨93S441⟩
Pommier, J.-C., **2**, 813 ⟨80JOM(193)165⟩, 816 ⟨76JOM(120)217, 77JOM(137)23⟩, 816 ⟨80JOM(186)C9, 85JOC3863⟩; **3**, 406 ⟨80JOM(186)C9⟩, 407 ⟨82SC495⟩; **4**, 637, 660 ⟨77JOM(136)7⟩; **5**, 388 ⟨76JOM(107)63⟩
Pomp, C., **4**, 150 ⟨88IC3796⟩
Pompeo, M. P., **4**, 678 ⟨91OM2541⟩
Pomytkin, I. A., **6**, 261 ⟨88IZV466⟩
Ponaras, A. A., **1**, 42, 833 ⟨80JOC4413, 83JOC3866⟩, 833 ⟨80TL4803, 83TL3⟩; **3**, 209 ⟨88JOC1110⟩, 237 ⟨86JOC4751⟩; **6**, 557 ⟨86JOC4741, 87JOC5630, 88JOC1110⟩
Poncet, J., **2**, 749 ⟨86S419⟩; **3**, 180 ⟨86JCS(P1)1895⟩; **5**, 333 ⟨88TL2661, 89JOC617⟩
Poncini, L., **1**, 1026 ⟨84JOC2031⟩; **4**, 192 ⟨81JCR(M)4016⟩, 192 ⟨84JOC2031⟩
Pong, R. G. S., **3**, 543 ⟨73JA248⟩
Pongó, L., **6**, 564 ⟨90JHC1249⟩
Pongor-Csakvari, M., **3**, 310 ⟨77JCS(P1)789⟩
Ponner, C., **2**, 471 ⟨77ZN(B)1244⟩
Ponomarchuk, M. P., **2**, 467 ⟨79JGU34⟩
Ponomarenko, V. A., **4**, 639 ⟨70IZV2280⟩; **6**, 62 ⟨67ZOB1040⟩, 267 ⟨57IZV199⟩
Ponomarev, A. B., **1**, 1010 ⟨84IZV1561, 85DOK(105)630⟩
Ponomarev, D. A., **4**, 343 ⟨87ZOB2300⟩
Ponomarev, S. V., **2**, 471 ⟨78JGU207⟩, 1027 ⟨87MI 221-01⟩, 1093 ⟨71JGU122⟩; **3**, 540 ⟨73AG(E)675, 84ZOB1817⟩; **4**, 619, 628 ⟨70ZOB939⟩, 626 ⟨91ZOB921⟩, 640 ⟨76ZOB594⟩, 814 ⟨89JGU2040⟩, 1056, 1059, 1063 ⟨87ZOB1741⟩, 1057 ⟨84ZOB1817, 85ZOB2801, 78ZOB2147⟩; **5**, 376, 389 ⟨76ZOB594⟩
Ponomarev, V. G., **6**, 219 ⟨92RCR75⟩
Ponomaryov, A. B., **1**, 1012 ⟨84S728⟩; **3**, 290, 305 ⟨85TL4819⟩
Pons, A., **3**, 218 ⟨79BSF381⟩
Pons, J.-M., **1**, 565, 681, 683, 785 ⟨93S441⟩, 687 ⟨88T4295⟩; **3**, 541 ⟨89TL1833, 93JCS(P1)1549⟩; **4**, 765 ⟨89TL1833⟩; **5**, 161 ⟨93S441⟩
Pons, P., **2**, 938, 941 ⟨92JOC4705⟩; **6**, 379 ⟨88JOM(358)31⟩
Ponsford, R. J., **5**, 234 ⟨80CC1085⟩, 235 ⟨80CC429, 80TL5071⟩
Ponsinet, G., **2**, 212 ⟨85TL4093⟩; **3**, 496 ⟨85TL5451⟩; **5**, 688 ⟨87S262⟩
Ponsold, K., **3**, 494 ⟨65CB604⟩; **5**, 984 ⟨60CB519⟩, 1118 ⟨70TL1125⟩
Ponti, F., **2**, 299 ⟨80SC611⟩; **3**, 4, 115 ⟨78S534⟩
Ponticelli, F., **2**, 316 ⟨92JHC321⟩
Ponticello, G. S., **3**, 305 ⟨91TL721⟩
Pontzen, T., **2**, 473 ⟨83AG(E)775⟩
Ponzio, G., **4**, 429 ⟨06JPR494⟩; **5**, 819 ⟨36G116⟩, 819, 837 ⟨31G561⟩, 854 ⟨10G77⟩
Poochaivatanon, P., **4**, 773 ⟨83TL531⟩

Pook, K.-H., **1**, 631 ⟨65CB4022⟩; **6**, 444 ⟨85AF217⟩, 504 ⟨86CB2553⟩
Pook, L. M., **1**, 924 ⟨93TL7819⟩
Pool, C. R., **4**, 137 ⟨76JCS(P1)513⟩, 140 ⟨75JCS(D)2177⟩
Poole, A. J., **3**, 378 ⟨72TL1835⟩
Poole, D. R., **2**, 174, 688 ⟨57JOC536⟩
Poole, V. D., **5**, 639 ⟨50JCS3389⟩
Poon, C.-D., **4**, 1083 ⟨84JCS(P1)1561⟩
Poon, D. J., **4**, 410 ⟨93JOC6949⟩
Poon, Y. C., **5**, 924 ⟨79JA6750, 82JA5667⟩
Poon, Y.-F., **2**, 517 ⟨89JOC5814⟩; **4**, 22 ⟨90JOC5719⟩, 52 ⟨84TL1019⟩
Pooranchand, D., **5**, 731 ⟨87S547⟩
Pope, H. W., **2**, 608, 626 ⟨58JOC1412⟩
Pope, K. R., **4**, 602 ⟨84OM354⟩
Pope, P., **2**, 26, 32 ⟨71T2617⟩
Pope, W. J., **2**, 160 ⟨22JCS1052⟩, 170 ⟨22JCS594⟩; **6**, 411, 412 ⟨20JCS1410⟩
Popelis, J., **4**, 929 ⟨84H(22)987⟩
Popelis, J. J., **6**, 124 ⟨84JOM(265)237⟩
Popenoe, E. A., **4**, 427 ⟨53JA3469⟩
Popic, V. V., **3**, 193 ⟨91S195⟩
Popjak, G., **5**, 99 ⟨52BJ163⟩
Popkova, V. Ya., **2**, 641 ⟨76DOK(229)870, 76DOK(231)864, 77DOK(233)591, 78IZV1193, 78IZV2775, 78IZV493⟩; **5**, 59 ⟨78DOK(239)330⟩, 475 ⟨90BAU1725⟩, 583 ⟨90IZV1898⟩
Pople, J. A., **1**, 909 ⟨78JA4301⟩, 1109 ⟨70JA2141, 71JA808⟩, 1109 ⟨72CPL385⟩, 1112, 1113 ⟨77JOC3004⟩, 1143 ⟨81CPL408⟩; **6**, 386, 401, 404 ⟨76JA5419⟩
Poplowska, M., **5**, 692 ⟨88BAP139⟩
Popoff, I. C., **4**, 465 ⟨63JOC2898⟩, 556 ⟨65IC198, 67USP3332986⟩
Popoka, V. Y., **2**, 168 ⟨82IZV1123⟩
Popov, A. A., **5**, 1070 ⟨88JFC(40)217, 89ZOR1523⟩, 1070 ⟨89ZOR1689⟩; **6**, 619 ⟨88JFC(40)217, 93JPO319⟩
Popov, A. F., **4**, 970 ⟨88ZOR2374⟩
Popov, A. G., **5**, 369 ⟨88ZOB2628⟩
Popov, B. N., **2**, 440 ⟨85URP1145022⟩
Popov, V. I., **2**, 881 ⟨76JGU2477⟩; **6**, 50 ⟨79ZOR1561⟩, 232, 234, 236, 237 ⟨85S667⟩, 233 ⟨82JFC(21)365⟩
Popova, I. Y., **6**, 497 ⟨88ZOB26⟩
Popović, Z., **4**, 697 ⟨84CCA689, 91JOM(405)59, 92CSC1116⟩, 697 ⟨93OM4708⟩; **6**, 207 ⟨84CCA689, 91JOM(411)19⟩
Popovich, T. P., **4**, 794 ⟨80JOU1762⟩; **5**, 657, 659 ⟨79JOU32⟩, 657, 659 ⟨79ZOR38⟩
Popovich, V. I., **2**, 478 ⟨90K668⟩
Popowski, E., **4**, 602, 607, 637, 660 ⟨90ZAAC(583)195⟩, 637 ⟨93JA6025⟩
Pöpp, F., **4**, 592 ⟨49CB152⟩
Popp, F. D., **3**, 637 ⟨80S384⟩; **5**, 558 ⟨85JHC1065⟩
Poppe, M., **3**, 688 ⟨89AG(E)53⟩; **6**, 511 ⟨89AG(E)53⟩
Popplewell, A. L., **2**, 15 ⟨94CC343⟩
Popsavin, M., **2**, 9 ⟨93CCC902⟩
Popsavin, P., **2**, 9 ⟨93CCC902⟩
Porada, M., **5**, 370 ⟨92AP325⟩
Porai-Koshits, B. A., **2**, 779 ⟨64ZOB2094⟩
Porai-Koshits, M. A., **2**, 1003 ⟨79UKZ811⟩; **4**, 717 ⟨80JOM(201)269⟩
Porcher, H., **4**, 179 ⟨79JA2171⟩
Porchia, M., **3**, 442 ⟨87ICA275⟩; **6**, 519 ⟨87JCS(D)2039⟩
Porchina, V. A., **4**, 550, 554 ⟨65ZOB1602⟩
Porciatti, E., **2**, 918, 940 ⟨92SL981⟩
Porco, J. A., **2**, 92 ⟨91COS(5)151⟩; **3**, 240 ⟨89JOC4721⟩
Pore, V. S., **2**, 394 ⟨87SC1421⟩
Porkert, H., **6**, 634 ⟨70CB256⟩
Pornet, J., **1**, 963 ⟨87JOM(319)333⟩, 978 ⟨81TL1327, 81TL3609, 81TL455, 82JOM(236)177, 84TL651, 85OM333, 86T2017, 86T2501, 87TL4689, 88JOM(349)43, 90JOM(396)289⟩; **3**, 60 ⟨88TL4717⟩, 721 ⟨71TL967⟩; **4**, 1047 ⟨82TL4083⟩; **5**, 754 ⟨74BCF994⟩
Poroshin, L. E., **2**, 380 ⟨59ZOB3154⟩
Porritt, C. J., **2**, 416 ⟨70CI(L)499⟩
Porritt, G. M., **2**, 255 ⟨90TL1319⟩
Porritt, W. H., **2**, 733 ⟨27JCS27⟩
Porrmann, H., **5**, 630 ⟨62CB627⟩, 630, 633 ⟨61CB2856⟩
Porsch, P. H., **3**, 706 ⟨73CB3382⟩, 720 ⟨79LA1444⟩
Pörschke, K.-R., **4**, 583 ⟨85CB298⟩
Porshnev, Yu. N., **1**, 916 ⟨77RCR530⟩
Porskamp, P. A. T. W., **2**, 399 ⟨81S368⟩, 400 ⟨83JOC4582⟩; **3**, 368 ⟨78TL807, 78TL811⟩, 370 ⟨83JOC4582⟩; **4**, 365 ⟨84JOC1691⟩, 570, 926 ⟨84JOC263⟩; **5**, 562 ⟨78TL807, 78TL811, 83RTC400⟩, 562, 645 ⟨84JOC263⟩
Porta, F., **5**, 386 ⟨75JOM(88)237, 76JCS(D)1777⟩
Porta, O., **1**, 923 ⟨77JA7960⟩; **2**, 81 ⟨89JOC3872⟩, 305 ⟨90TL2069⟩; **3**, 311 ⟨74JCS(P2)1699⟩
Portal Olea, D., **4**, 858 ⟨88CJC2975⟩
Port, M., **4**, 465, 478 ⟨92TL77⟩
Port, V., **5**, 1064, 1066 ⟨88CB271⟩
Porte, C., **2**, 381 ⟨78BSF(2)602⟩, 381, 383, 384 ⟨75BSF1627⟩
Portella, C., **3**, 237 ⟨91TL83⟩; **4**, 252 ⟨92S965⟩, 353 ⟨92S995⟩
Portell, M., **2**, 498 ⟨63AC(R)1869⟩
Porter, A. E. A., **1**, 945 ⟨80JCS(P1)2657⟩; **2**, 1000 ⟨88JCS(P1)1023⟩; **3**, 373 ⟨78CC83, 79JCS(P1)2624⟩, 713 ⟨88TL3343⟩; **5**, 1054 ⟨80JCS(P1)2657⟩
Porter, B., **3**, 533 ⟨84JOC2052⟩
Porter, C. C., **3**, 616 ⟨44JA1652⟩
Porter, C. W., **4**, 61 ⟨89JMC997⟩
Porter, G., **3**, 366 ⟨72CC632⟩
Porter, J. J., **6**, 45, 69 ⟨57JA5493⟩, 81 ⟨60JA6118⟩
Porter, J. W. G., **5**, 26, 94 ⟨45JCS751⟩
Porter, K. E., **3**, 327 ⟨83JCR(S)262⟩
Porter, L. C., **4**, 586 ⟨87OM1992, 88IC836⟩
Porter, N. A., **1**, 320, 326, 327, 328, 919 ⟨91ACR296⟩, 324 ⟨93TL4489⟩, 326 ⟨89JA8311, 90TL1679, 91JA1791⟩, 326, 327 ⟨91JA7788⟩, 326, 327 ⟨91TL707⟩, 326, 327 ⟨92JA7664⟩, 327, 328 ⟨90JA6740, 91JA7002⟩, 340 ⟨89JA8309⟩, 340, 341 ⟨86JA2787⟩, 838 ⟨90JA1266⟩, 949 ⟨75JOC3614, 84JOC1345⟩; **2**, 676 ⟨88JA6276⟩
Porter, P. K., **4**, 47 ⟨32OSC(1)369⟩; **5**, 305 ⟨41OSC(1)457⟩
Porter, Q. N., **4**, 206 ⟨71JCS2166⟩
Porter, R. A., **2**, 147 ⟨93TA2139⟩; **4**, 262 ⟨93TA2139⟩; **5**, 278 ⟨86CC344⟩
Porter, R. D., **1**, 894 ⟨72JA2044⟩, 895 ⟨71JA464⟩, 914 ⟨72JA2034⟩
Porter, R. S., **2**, 98 ⟨57JA5628⟩; **6**, 231 ⟨57JA5628⟩
Porter, S. J., **4**, 712 ⟨84JCS(D)2553⟩
Porter, T. M., **3**, 639 ⟨87S411⟩
Porter, W. J., **1**, 831 ⟨90TL5285⟩
Portier, P., **3**, 419 ⟨76BSF1222⟩
Portlock, D. E., **1**, 860 ⟨74JOC2006⟩
Portner, M., **6**, 268 ⟨75ZN(B)965⟩
Portnoy, M., **3**, 90 ⟨89CC1816⟩
Portnoy, N. A., **1**, 517 ⟨69JOC520⟩
Portoghese, P. S., **3**, 343 ⟨91JMC1292⟩, 638 ⟨80SC495⟩
Portonovo, P. S., **3**, 437 ⟨93TL6229, 94TL9351⟩
Porwisiak, J., **6**, 10 ⟨93JFC(60)233⟩
Porz, Ch., **6**, 687 ⟨86CB2748⟩
Porzi, G., **2**, 87 ⟨81S793⟩; **3**, 158 ⟨84JOC701⟩; **6**, 469 ⟨82JOC4626⟩
Poschel, P. H., **5**, 306 ⟨87JMC498⟩
Pöschka, K. R., **2**, 575 ⟨76AG(E)621⟩
Posey, I. Y., **1**, 869, 871, 881 ⟨68JOC3477⟩
Poshkus, A. C., **2**, 717 ⟨63JOC2766⟩
Posibaev, V. M., **4**, 455 ⟨78ZOB1008⟩
Posler, J., **3**, 322 ⟨67JA5497⟩
Posner, G. H., **1**, 23 ⟨85JOC2589⟩, 23, 236 ⟨81JA2886⟩, 116, 118 ⟨91COS(3)207⟩, 116, 118, 206, 540, 1005, 1045 ⟨75OR(22)253⟩, 117, 118 ⟨67JA3911⟩, 117, 118, 466 ⟨68JA5615⟩, 119 ⟨73SC281⟩, 120 ⟨73JA7788⟩, 125 ⟨70TL315⟩, 304 ⟨77TL3215⟩, 305 ⟨87ACR72⟩, 311,

539, 540 ⟨72OR(19)1⟩, 444 ⟨85JA1424⟩, 482, 485 ⟨73CC907⟩, 484 ⟨73JOC2747⟩, 778 ⟨89JOC389⟩, 788 ⟨90JOC3967, 91JOC4339, 91JOC6981⟩, 1005 ⟨B-80MI 121-01⟩, 1142 ⟨B-80MI 122-01⟩; **2**, 396 ⟨90T7509⟩, 585 ⟨B-80MI 211-02⟩, 588 ⟨73JA7788⟩, 610 ⟨77JOC3173⟩, 629, 970 ⟨67JA4245⟩, 665 ⟨92T9111⟩, 713 ⟨78JOC4131⟩, 917, 960 ⟨75OR(22)253⟩, 917 ⟨B-80MI 218-01⟩, 1094 ⟨B-80MI 223-01⟩; **3**, 12, 128 ⟨72OR(19)1⟩, 119 ⟨76TL3499, 77S555⟩, 130 ⟨72JA5106, 88OSC(6)248⟩, 165 ⟨77TL3227⟩, 170 ⟨B-83MI 304-02⟩, 170 ⟨B-88MI 304-04⟩, 220 ⟨78TL4205⟩, 286 ⟨70TL4647⟩, 287 ⟨73JA7788, 73TL1815⟩; **4**, 745, 746 ⟨75TL1373⟩, 764 ⟨91JOC6987⟩, 863 ⟨73JOC2747⟩, 958 ⟨78TL3995⟩, 960, 964 ⟨81PAC2307⟩; **5**, 159 ⟨90JOC3967⟩, 1044 ⟨94JMC170⟩

Posner, T., **2**, 193 ⟨05CB646⟩

Posner, T. B., **5**, 720 ⟨79JOM(175)273⟩

Poss, A. J., **2**, 16 ⟨91JOC5962⟩; **3**, 149 ⟨91JOC5962⟩, 200 ⟨87JOC4810⟩

Poss, K. M., **1**, 826 ⟨85JA4359⟩

Poss, M. A., **3**, 60 ⟨85TL2391⟩, 171 ⟨92TL7291⟩; **4**, 649 ⟨85TL2391⟩; **6**, 641 ⟨92TL5933⟩

Possel, O., **3**, 694, 709 ⟨75TL3487, 77TL4229⟩; **4**, 331 ⟨75TL3487⟩, 538 ⟨77TL4229⟩

Pošta, A., **4**, 5 ⟨71CCC1867⟩, 33 ⟨84JFC(26)467⟩; **6**, 4 ⟨72CLY937⟩, 5 ⟨66CCC3584, 67CCC3888⟩, 5 ⟨74CCC1330⟩

Post, E. W., **4**, 706 ⟨78JOM(161)C17⟩

Post, H. W., **2**, 108 ⟨48JOC280⟩, 227 ⟨57JOC145⟩, 262 ⟨51JOC395⟩; **4**, 191 ⟨40JOC244⟩; **5**, 379 ⟨62RTC430⟩; **6**, 68, 77, 81 ⟨B-43MI 603-01⟩, 68, 79 ⟨33JA3851⟩, 296, 297, 308 ⟨48JOC265⟩

Postawka, A., **2**, 1009 ⟨85H(23)1513⟩

Postema, M. H. D., **1**, 244 ⟨93CC429⟩, 942 ⟨93JOC6857⟩

Postich, M. J., **1**, 372 ⟨92JA1329⟩; **4**, 533, 537 ⟨92JOC6354, 92JA1329⟩

Postich, M. P., **3**, 457 ⟨92JA1329⟩

Postma, J. C. W., **4**, 884 ⟨56RTC1377⟩

Postnov, V. N., **4**, 719 ⟨90MI 416-01⟩

Postovoi, S. A., **4**, 820 ⟨82BAU2498⟩

Postovskii, I. Ya., **2**, 361 ⟨60ZOB1431⟩; **5**, 641 ⟨59ZOB1333⟩; **6**, 109 ⟨67KGS1096⟩, 653 ⟨64ZOB2855⟩

Postra, A., **6**, 5 ⟨71CCC1867⟩

Posvic, H., **2**, 384 ⟨74JOC1588⟩

Posyagina, E. Y., **5**, 354 ⟨81ZOR2153⟩

Posyagin, G. S., **5**, 354 ⟨81ZOR2153⟩

Pot, J., **1**, 403, 795 ⟨64RTC1173⟩

Potacek, M., **3**, 361 ⟨88JPR665⟩

Potapov, A. M., **2**, 461 ⟨67ZOB1173⟩, 462 ⟨68ZOB609⟩

Potapov, V. A., **2**, 708 ⟨83IZV230⟩, 1036 ⟨90ZOR2635⟩; **4**, 777 ⟨90ZOR2056⟩

Potekhin, K. A., **2**, 445 ⟨87ZOR2120⟩; **6**, 116 ⟨90ZOR1129⟩

Potemp, S. J., **2**, 357, 688 ⟨46CRV269⟩

Potenza, J. A., **1**, 909 ⟨86JA134⟩; **5**, 379 ⟨88JOC4773⟩

Potier, P., **1**, 332, 935 ⟨85TL5943⟩, 344 ⟨87TL1413⟩, 602 ⟨84CC1298, 85T4347⟩, 935 ⟨88T5479⟩; **2**, 642 ⟨91T9823⟩; **3**, 635 ⟨87T4297⟩; **5**, 212 ⟨73CR(C)1039⟩, 1115 ⟨86T4983⟩; **6**, 546, 567 ⟨85TL5943⟩

Potkin, V. I., **4**, 762 ⟨91ZOB56⟩, 771 ⟨93ZOR56⟩, 797 ⟨91JOU48⟩

Potlock, S. J., **3**, 195, 220 ⟨87S1055⟩

Potoski, J. R., **4**, 979 ⟨66TL2311⟩

Potrzebrowski, M., **2**, 256 ⟨82CC1362⟩

Potschka, V., **2**, 1032 ⟨63N302⟩

Potter, A., **1**, 935 ⟨85JOC2149⟩; **2**, 95 ⟨84JA5261⟩

Pötter, B., **3**, 380 ⟨82IC3147, 84CB3255⟩

Potter, B., **3**, 689 ⟨85JA980⟩

Potter, D. M., **4**, 288 ⟨84JCS(P2)429⟩

Potter, G. J., **2**, 24 ⟨77JCS(P1)226⟩

Potter, G. W. H., **6**, 642 ⟨65JMC446⟩

Potter, J. L., **2**, 127 ⟨65ANC164⟩

Potter, S. E., **1**, 706, 709 ⟨73CC520⟩

Potthast, R., **2**, 853 ⟨68TL1755⟩

Potthoff, B., **3**, 59, 75 ⟨87S1⟩, 71 ⟨77S1⟩

Potts, G. O., **5**, 205 ⟨70JMC948⟩

Potts, Jr., I. W., **1**, 595 ⟨67JOC102⟩

Potts, J. S., **5**, 1029 ⟨65JOC2465⟩

Potts, K. T., **1**, 436 ⟨85JOC1125⟩, 907 ⟨74CRV189⟩; **3**, 267 ⟨83JOC623⟩, 311 ⟨71JOC3368⟩, 357 ⟨73CC833, 76JOC818⟩, 372 ⟨72JA6215, 74JA4268⟩; **4**, 843 ⟨86OS(64)189⟩, 853, 855 ⟨82JOC3027⟩, 980 ⟨83JOC623⟩; **5**, 500 ⟨B-72MI 511-01⟩; **6**, 115 ⟨76JOC813⟩, 631 ⟨84JOC2404⟩

Potvin, P., **4**, 847, 848 ⟨80CJC2780⟩

Pou, R., **3**, 339 ⟨74CR(C)259⟩, 361, 376 ⟨76BSF(2)120⟩; **5**, 634 ⟨76BSF120⟩

Pougny, J. R., **1**, 1009 ⟨89TL4951⟩, 1064 ⟨92JOC651⟩

Pouliquen, J., **3**, 601 ⟨72TL1139⟩; **5**, 44 ⟨86JCE1013⟩

Poulter, C. D., **1**, 794 ⟨70JA7408⟩, 840 ⟨91JA4895⟩, 894 ⟨74JA7591⟩; **2**, 133 ⟨91JA3176⟩; **4**, 553, 562 ⟨87JA5544⟩; **6**, 264 ⟨86JOC4768, 87JA5542, 87JOC1794⟩

Poulton, G. A., **1**, 200, 255 ⟨90SC3347⟩; **5**, 560 ⟨73JCS(P1)1571⟩

Poulton, G., **1**, 498 ⟨80T719⟩

Poupaert, J. H., **1**, 786 ⟨88JOC4859⟩; **5**, 558 ⟨75BSB61⟩, 727 ⟨72S622⟩

Poupart, M.-A., **1**, 810 ⟨93JOC4245⟩, 815 ⟨88TL273⟩; **5**, 174, 277 ⟨83JA7130, 80JA7578⟩

Poupart, M.-A., **5**, 76 ⟨90OS(69)173⟩

Poupat, C., **2**, 642 ⟨91T9823⟩; **5**, 212 ⟨73CR(C)1039⟩

Pourabass, S., **5**, 522 ⟨77JHC1321⟩

Pourahmady, N., **1**, 7, 19 ⟨82JOC2590⟩

Pourcelot, G., **1**, 956 ⟨66BSF3016, 66BSF3024⟩; **2**, 1030, 1031, 1034 ⟨66BSF3016⟩

Pourreau, D. B., **4**, 676 ⟨88OM289⟩, 678 ⟨84JOM(273)333⟩

Pourrias, B., **5**, 759 ⟨72FRP2081556⟩

Pousaz, P., **5**, 12 ⟨90TL75⟩

Poutsma, M. L., **2**, 2, 21 ⟨B-73MI 201-02⟩, 2 ⟨B-69MI 201-01, B-69MI 201-02⟩, 606 ⟨66T2167⟩; **3**, 448 ⟨62JA4887, 62JA4892⟩

Pouzar, V., **4**, 198 ⟨88CLY402⟩

Povarnitsyna, T. N., **3**, 436 ⟨85ZOB2077⟩; **5**, 438 ⟨66JGU1945⟩

Poveda, M. L., **4**, 583 ⟨84OM1438⟩; **5**, 445 ⟨84JA3214⟩, 800 ⟨89JOM(377)157, 90OM583⟩, 948 ⟨94JA791⟩; **6**, 332 ⟨88OM2577⟩

Pover, K. A., **5**, 101 ⟨79CC215, 80JCS(P1)2338⟩

Povey, D. C., **3**, 358 ⟨75CC756⟩

Povlock, T. P., **2**, 36 ⟨67TL4131⟩

Povolotskii, M. I., **2**, 407 ⟨75ZOB2394⟩, 428 ⟨90ZOB697⟩; **5**, 713 ⟨90JGU849, 90ZOB963⟩, 787 ⟨89JGU1496⟩; **6**, 612, 727 ⟨74ZOR449⟩, 655, 685 ⟨91ZOB401⟩, 683 ⟨85ZOB2797, 85ZOB1437⟩, 685, 686 ⟨85ZOB221⟩, 687 ⟨84ZOB715⟩, 689 ⟨84ZOB2800⟩, 690 ⟨89ZOB2133⟩, 692 ⟨92ZOB474⟩, 693 ⟨91ZOB79⟩, 696 ⟨90ZOB2814⟩, 685,686 ⟨90ZOB2238⟩; **6**, 694 ⟨89ZOB2131⟩

Powell, A. K., **4**, 711 ⟨90CB661⟩

Powell, A. R., **5**, 229 ⟨64CI(L)544⟩

Powell, B. F., **2**, 379 ⟨83JHC121⟩

Powell, C., **4**, 300 ⟨63JCED600⟩

Powell, C. E., **2**, 376 ⟨75JOC1213⟩

Powell, D. M., **1**, 133 ⟨91CC1399⟩; **2**, 640 ⟨91CC1399⟩

Powell, D. R., **1**, 637 ⟨84JOC670⟩

Powell, D. W., **1**, 22, 942 ⟨82JA2046⟩, 410, 821 ⟨78JOC4831⟩

Powell, G., **2**, 608 ⟨43JA1430⟩

Powell, H. R., **6**, 400 ⟨88POL2049⟩

Powell, J., **4**, 692 ⟨77JA4707⟩

Powell, J. R., **5**, 1120 ⟨63OSC(4)213⟩

Powell, J. S., **2**, 837 ⟨50JA4291⟩

Powell, J. W., **1**, 631 ⟨59T305⟩

Powell, N. A., **5**, 971 ⟨92JOC7364⟩

Powell, R. L., **6**, 14 ⟨89JFC(45)86⟩, 18 ⟨79CC964⟩

Powell, S. G., **5**, 59 ⟨32OSC(1)168⟩
Powell, W. J., **6**, 69 ⟨21JCS1222⟩
Power, J. M., **2**, 511 ⟨86CC1543⟩; **5**, 889 ⟨88CC753⟩, 890 ⟨88AG(E)838⟩
Power, M. B., **1**, 666 ⟨90OM3086⟩; **3**, 127 ⟨90TL323⟩
Power, P. P., **1**, 847 ⟨85JA2174⟩; **2**, 414, 423 ⟨B-80MI 207-01⟩, 475 ⟨91JA3597⟩, 543 ⟨B-80MI 210-01⟩, 1005 ⟨87JA2541, 91OM1282⟩; **4**, 453 ⟨84IC2550⟩, 597, 598 ⟨76CC623, 80JCS(D)2428⟩, 656 ⟨90OM1720⟩; **6**, 165, 167, 168 ⟨80JCS(D)2428⟩, 167 ⟨84CC1669⟩, 168 ⟨83IC2573⟩, 194, 195 ⟨85JA4337, 88CC1007⟩, 349 ⟨85CC1674⟩
Powers, D. B., **2**, 75, 529, 557 ⟨90JA6339⟩
Powers, J. W., **4**, 161 ⟨57JA6562⟩
Powers, L. D., **2**, 890 ⟨33JA1161⟩
Powers, T. S., **5**, 953, 956 ⟨94JOC6882⟩
Poyser, J. P., **3**, 64 ⟨89JCS(P1)1939⟩
Pozdnev, V. F., **5**, 189 ⟨86ZOB613⟩, 189 ⟨92MI 504-01⟩; **6**, 463 ⟨71ZOR273⟩
Pozhidaev, V. M., **4**, 343, 561 ⟨78ZOB1001⟩
Pozo, M., **6**, 461 ⟨93T10725⟩
Pozsgay, V., **5**, 368 ⟨87TL5091⟩
Pozzi, G., **3**, 643 ⟨90SC965⟩
Praas, H.-W., **4**, 867 ⟨92CB571⟩; **6**, 80, 237 ⟨92CB571⟩
Prabhakar, S., **5**, 212 ⟨82TL1391, 91CC1127⟩, 214, 635 ⟨84S829⟩
Prabhu, V. S., **3**, 300 ⟨84IJC(B)163⟩
Pracejus, H., **6**, 207 ⟨74JPR557⟩
Pradat, C., **2**, 716 ⟨85JOC1327⟩
Pradella, G., **1**, 834 ⟨81JOC2228⟩
Pradère, J.-P., **3**, 414 ⟨86PS(27)327⟩; **4**, 109 ⟨89CJC1125⟩, 328 ⟨91CR(313)517⟩; **5**, 633 ⟨85JCS(P1)1875, 86PS(27)327, 91BSF574⟩, 760 ⟨80JHC767⟩
Pradham, S. K., **1**, 548 ⟨81JOC2622⟩
Pradhan, J., **1**, 327, 328 ⟨91JA7002⟩
Prados, P., **2**, 316 ⟨91JOC452⟩
Praeckel, U., **2**, 501 ⟨81ZN(B)70⟩
Praefcke, K., **1**, 1040 ⟨65CB3311⟩; **2**, 171 ⟨79LA784⟩; **3**, 256 ⟨81JOM(208)35⟩, 349 ⟨75CB673, 78ZN(B)417⟩, 365 ⟨79CB2744, 80JHC1655, 80ZN(B)372⟩, 550 ⟨70CB949⟩; **4**, 232 ⟨80JHC1655⟩, 487 ⟨82PS(13)319⟩; **5**, 238 ⟨77AG(E)318⟩, 241 ⟨79JOM(166)373, 81JOM(205)167⟩
Prager, B., **5**, 858 ⟨05LA(338)360⟩; **6**, 118 ⟨B-18MI 607-01⟩
Prager, J. H., **2**, 94 ⟨65JA230⟩
Prager, R. H., **2**, 348 ⟨83AJC719⟩
Pragrell, M. J., **1**, 445 ⟨65JCS4120⟩
Pragst, F., **5**, 631 ⟨88ZC182⟩
Praill, P. F. G., **2**, 94 ⟨56JCS3933⟩
Prain, H. D., **6**, 647 ⟨77JMC901⟩
Prajapati, D., **3**, 417 ⟨85SC335⟩, 536 ⟨91S1027⟩
Prakasa, A. S. C., **3**, 212 ⟨74S778⟩
Prakash, G. K. S., **1**, 80 ⟨78S5397⟩, 180, 212, 462 ⟨91COS(3)293⟩, 202 ⟨91JOC2089⟩, 208 ⟨90JOC2016, 85MI 111-01⟩, 891 ⟨83AG(E)390⟩, 891, 917 ⟨B-85MI 119-04⟩, 894, 899, 900 ⟨77JOC2666⟩, 899 ⟨81PNA(78)1998⟩, 904 ⟨80JA6640, 82JA1628, 86JA1575⟩, 906 ⟨77JA5683, 80JA683, 82JA7105, 82JOC3903⟩, 906 ⟨83ACR440, 83JOC2146⟩, 918 ⟨89JA746⟩; **2**, 6, 17, 18 ⟨93S693⟩, 11 ⟨81HCA2528, 83HCA1028⟩, 15, 17 ⟨86JFC(33)377⟩, 22 ⟨84JOC2032⟩, 36 ⟨B-91MI 201-01⟩, 144 ⟨90JA3697⟩, 384 ⟨92JOC1585⟩, 397 ⟨92SL327⟩; **3**, 14, 90, 101 ⟨84S228⟩, 18, 134 ⟨80S220⟩, 133 ⟨93SL427⟩, 152 ⟨84JOC2032⟩, 632, 698 ⟨90JOC2016⟩, 707 ⟨90S646⟩; **4**, 45 ⟨B-92MI 402-01⟩, 216 ⟨90S104⟩, 428, 429 ⟨93JOC763⟩, 697, 698 ⟨88JA3231⟩; **5**, 367, 992 ⟨83JOC3358⟩, 1045 ⟨90S646⟩; **6**, 13, 22 ⟨B-92MI 601-01⟩, 13, 22 ⟨B-92MI 601-01⟩, 14 ⟨92MI 601-02, B-92MI 601-03⟩, 726 ⟨89JA8020⟩
Prakash, I., **2**, 405 ⟨90SC2353⟩; **3**, 181 ⟨85SC649⟩; **4**, 184 ⟨86OS(64)138⟩; **5**, 366 ⟨87SC409⟩, 897 ⟨84JA6082⟩
Prakash, N. J., **4**, 73 ⟨89JA1127⟩
Prakash, O., **1**, 109 ⟨89JA6443, 90JA1297⟩; **2**, 44 ⟨86ACR244⟩; **3**, 162 ⟨87JOC150⟩, 165 ⟨93SC1455⟩; **4**, 183 ⟨93JCR(S)430⟩, 184 ⟨86ACR244⟩; **5**, 356 ⟨93JCR(S)244, 84JA6082⟩
Praly, J. P., **4**, 432 ⟨94TL89⟩, 432, 433 ⟨90CC431⟩, 1079 ⟨81CAR(C)1⟩
Pramod, K., **1**, 248 ⟨87AG(E)928⟩
Pranc, P., **5**, 291 ⟨68JA4706⟩
Prangé, T., **1**, 810 ⟨93JOC2349⟩, 1039 ⟨81CC363⟩
Prange, U., **2**, 749 ⟨80GEP2912344⟩; **5**, 27 ⟨78CZ1⟩
Prankprakma, V., **4**, 849 ⟨82S579⟩
Prapas, A. G., **5**, 187 ⟨59JOC388⟩
Prasad, A. S. B., **2**, 316, 325 ⟨92T4623⟩
Prasad, B. V., **6**, 322 ⟨93OM4267⟩
Prasad, C. V. C., **4**, 226 ⟨90JA6263⟩; **5**, 513 ⟨90JA6263⟩
Prasad, G., **1**, 225 ⟨90JOC324⟩, 413 ⟨86JOC1124⟩; **2**, 86 ⟨90JOC324⟩; **3**, 141 ⟨86JOC1124⟩; **6**, 507 ⟨80IJC(B)653⟩
Prasad, H. S., **2**, 882 ⟨71JCS(A)562⟩, 886 ⟨71CJC2529, 72JOM(36)323⟩, 891 ⟨73JOM(50)129⟩; **5**, 1133 ⟨73JOM(50)129, 79JOM(166)365⟩
Prasad, J. S., **1**, 688, 690 ⟨88JA6914⟩, 978 ⟨88TL4253⟩, 1047 ⟨87TL1857⟩; **3**, 273 ⟨87TL1857⟩
Prasad, J. V. N. V., **1**, 487, 488 ⟨88JOC239⟩, 816, 817 ⟨93TL5217⟩; **2**, 74 ⟨90TL1803⟩, 1085 ⟨88JOC1391⟩
Prasad, K., **1**, 1023 ⟨92JOC6344⟩
Prasad, K. M., **2**, 690 ⟨85JOM(291)185⟩
Prasad, K. S., **2**, 373, 376 ⟨72JOC2259⟩
Prasad, P. N., **1**, 986 ⟨93CM357⟩
Prasad, V. A. V., **2**, 869 ⟨74JA7269⟩
Prasanna, S., **4**, 346 ⟨80S72⟩
Prashad, M., **4**, 472 ⟨90TL1759⟩
Prasit, P., **5**, 172 ⟨88JOC4421⟩
Prasitpan, N., **1**, 184 ⟨87JCS(P2)1221⟩
Prati, L., **1**, 208 ⟨86S1032⟩
Prato, M., **4**, 162 ⟨91TL6215⟩, 304 ⟨84G405⟩
Pratt, A. C.,**3**, 430 ⟨86JCS(P1)1691⟩
Pratt, A. J., **2**, 79 ⟨89JCS(P1)1521, 89JCS(P1)1529⟩, 230 ⟨91T5329⟩; **4**, 387 ⟨82CC1115⟩; **5**, 346 ⟨87TL4319⟩
Pratt, B. C., **3**, 614 ⟨54JA5364⟩
Pratt, D. E., **2**, 520 ⟨63JCS1582⟩
Pratt, D. R., **4**, 746 ⟨82JA2223⟩
Pratt, D. V., **1**, 833 ⟨88TL4229⟩; **2**, 642 ⟨87JA5553⟩
Pratt, E. F., **3**, 87 ⟨61JOC2973⟩; **4**, 297 ⟨61JOC4029⟩
Pratt, J. R., **2**, 813 ⟨75JOC1090⟩, 942 ⟨73JOC4271⟩
Pratt, R. N., **4**, 886 ⟨65CC574, 67JCS(C)1569⟩
Pravdić-Sladović, N., **5**, 566, 618 ⟨57CCA319⟩
Pravdina, V. V., **1**, 647 ⟨67ZOR598⟩, 738 ⟨65BAU1434, 65IZV1474⟩
Pream, J. B., **5**, 1121 ⟨81EJM175⟩
Precedo, L., **6**, 255 ⟨91CJC2127⟩
Prechtl, F., **1**, 785 ⟨91JOC5782⟩
Preckel, M., **3**, 266 ⟨73JA7862⟩
Preckel, M. M., **3**, 535 ⟨82JA6779⟩
Precopio, F., **4**, 573 ⟨58USP2835651⟩
Predvoditelev, D. A., **2**, 410 ⟨76ZOB477, 80ZOB1702⟩
Pregaglia, G. F., **1**, 79 ⟨74JOM(70)89⟩; **3**, 534 ⟨68MI 316-01, 68MI 316-03⟩
Pregnolato, M., **2**, 350 ⟨92JOC2726⟩; **3**, 120 ⟨92TL7245⟩; **5**, 588 ⟨93JHC1079⟩
Preiner, G., **3**, 483 ⟨81CB3518⟩; **6**, 163 ⟨87CB1357⟩, 177, 178, 382, 387, 396, 709, 710 ⟨87ZN(B)1062⟩, 177, 380, 382, 392, 673, 709, 710 ⟨81CB3518⟩, 290, 291, 380, 381, 392, 709, 710 ⟨81CB2087⟩, 373, 380, 382, 392, 709, 710 ⟨81CB3505⟩, 382, 709, 710 ⟨87ZN(B)1055⟩, 392, 709, 710 ⟨77AG(E)328⟩, 673 ⟨88CB1407⟩, 709, 710 ⟨88ZN(B)1468⟩
Prejean, G. W., **2**, 435, 435, 1060, 1064 ⟨69JOC2446⟩, 874 ⟨68JOC1681⟩
Prejzner, J., **5**, 1100 ⟨68MI 530-01⟩

Prelog, V., **2**, 71 ⟨53HCA325⟩, 662 ⟨47HCA1883, 47HCA675⟩
Prema, S., **4**, 272, 277 ⟨81S995⟩
Premila, M. S., **4**, 94 ⟨84H(22)585⟩
Prempree, M., **3**, 210 ⟨80TL1169⟩
Prengel, C., **2**, 431 ⟨84IC4170⟩
Prengl, C., **2**, 439 ⟨86PS(26)143⟩
Prentice, J. B., **4**, 563 ⟨68JOM(13)199⟩; **6**, 263 ⟨68JOM(13)199⟩, 263, 518 ⟨67JOC4111⟩
Prenzler, P. D., **5**, 385 ⟨89IC815⟩
Preobrashenski, N. A., **3**, 531, 532 ⟨35CB850⟩
Preobrashenski, W. A., **3**, 531, 532 ⟨35CB850⟩
Preobrazhenskaya, M. N., **2**, 812 ⟨78JOU1101⟩; **5**, 690 ⟨82CHE1302⟩
Preobrdzhenskaya, M. N., **5**, 690 ⟨82KGS1682⟩
Presas, M. J., **4**, 262 ⟨90MI 403-01⟩
Prescott, W., **2**, 614 ⟨78CI(L)56⟩
Presilla-Marquez, J. D., **5**, 1096 ⟨91JCP(95)5612⟩
Press, B. J., **3**, 305 ⟨92JHC247⟩
Press, J. B., **2**, 674 ⟨72TL2677⟩
Pressler, W., **3**, 75, 76
Preston, J., **5**, 279 ⟨65JA2492⟩
Preston, N. W., **1**, 738, 742 ⟨69JCS(C)2464⟩
Preston, P. N., **3**, 580 ⟨70TL1157⟩; **5**, 794, 803 ⟨71MI 520-02⟩; **6**, 443 ⟨89JCS(P1)1727⟩
Preston, S. B., **2**, 317 ⟨89S450⟩
Prestwich, G. D., **4**, 654 ⟨89TL403⟩
Preti, G., **1**, 100 ⟨79TL4985⟩
Pretzer, W. R., **4**, 692 ⟨76JA8072, 78JA2090⟩
Pretztzer, W. R., **2**, 430 ⟨85IC4437⟩
Preugschat, D., **3**, 694, 716 ⟨91AG(E)1644⟩, 703, 722 ⟨88CB1445, 92CC1523⟩, 716 ⟨93AG(E)1456⟩; **4**, 821 ⟨90AG(E)315⟩; **6**, 611 ⟨88CB1445⟩
Preus, M. W., **1**, 85 ⟨83JOC3370⟩
Preuschhof, H., **2**, 374, 378, 380 ⟨68LA(713)74⟩, 374, 380 ⟨70JPR349⟩, 403 ⟨88OSC(6)78⟩, 806 ⟨71JPR642⟩
Preuss, F., **2**, 570 ⟨90ZN(B)191⟩; **4**, 656 ⟨82ZN(B)957⟩
Preut, H., **2**, 238 ⟨90ZN(B)1625⟩, 881 ⟨86AX(C)1875⟩; **5**, 222 ⟨89ZAAC(574)130, 90ZAAC(582)37⟩; **6**, 610, 727 ⟨92ZAAC(617)136⟩, 613 ⟨93ZN(B)19⟩, 615, 728 ⟨87JFC(37)259⟩
Prevost, C., **1**, 127 ⟨64BSF2485⟩, 999 ⟨66BSF971⟩
Prewo, R., **3**, 680 ⟨88HCA1177⟩; **4**, 179 ⟨79JA2171⟩, 683 ⟨86OM2413⟩; **5**, 622 ⟨84HCA534⟩
Pri-Bar, I., **1**, 8 ⟨86JOC734⟩; **3**, 89, 95 ⟨90BSB345⟩, 90, 95 ⟨84JOC4009⟩
Pribor, L. I., **6**, 629 ⟨75ZOR1628⟩
Pribytkova, I. M., **4**, 603, 625, 637, 701 ⟨71JOM(30)C57⟩
Price, A. K., **4**, 160 ⟨68JOC2692⟩
Price, C. C., **1**, 62 ⟨67JOC4126⟩; **2**, 138 ⟨46JOC475⟩, 166 ⟨46JOC469⟩, 675 ⟨61JA1773⟩; **3**, 238 ⟨63OSC(4)185⟩, 326 ⟨62JA2094⟩, 498 ⟨62JA489⟩, 612 ⟨41JA2796⟩, 626 ⟨55OSC(3)174⟩; **4**, 196 ⟨50JA2613, 63OSC(4)558⟩; **5**, 106 ⟨55OSC(3)272⟩, 652 ⟨55OSC(3)649⟩, 748 ⟨59JOC1309⟩
Price, D. T., **3**, 684, 686 ⟨84JOC4595⟩
Price, G. G., **5**, 106 ⟨63CI(L)775⟩
Price, J. A., **3**, 475 ⟨57JOC245⟩
Price, J. D., **1**, 990 ⟨86TL4679⟩
Price, M. E., **3**, 218 ⟨89TL5865⟩
Price, M. F., **1**, 832 ⟨84JOC722⟩
Price, P. J., **6**, 606 ⟨87JCS(P1)1069⟩
Price, S. J., **5**, 655 ⟨87JCS(P1)1153⟩; **6**, 572 ⟨87JCS(P1)1153⟩
Price, T. S., **2**, 122 ⟨09JCS1725⟩, 227 ⟨41JA658⟩
Price, W. A., **3**, 278 ⟨87JOC1972⟩; **4**, 780 ⟨91JMC2525⟩, 856 ⟨87JOC1972⟩
Prichard, W. W., **5**, 103 ⟨55OSC(3)302⟩, 285 ⟨71JMC535⟩
Prickett, K. S., **4**, 734 ⟨83JOC4661⟩
Pridgen, H. S., **1**, 417, 829 ⟨60JOC1815⟩
Pridgeon, L. N., **1**, 471, 473 ⟨82JOC4319⟩; **3**, 311 ⟨84S1048⟩; **6**, 415 ⟨89JOC3231⟩, 558 ⟨83JHC1223⟩
Priebe, E., **5**, 402 ⟨59CB3183⟩
Priebe, W., **4**, 174 ⟨91TL3313⟩
Priepke, H., **1**, 552 ⟨90CB555⟩
Priest, D. C., **2**, 941 ⟨55JA6227⟩
Priest, H. F., **6**, 212 ⟨50IS171⟩
Priester, R. D., **5**, 445 ⟨84JA3214⟩
Priester, W., **4**, 605 ⟨76JA8412⟩; **6**, 206 ⟨76JA8426⟩
Priester, W. de, **6**, 43 ⟨69TL5003⟩
Prieto, C. A., **6**, 689 ⟨88IC784⟩
Prieto, J. A., **1**, 649 ⟨88T3781⟩, 763 ⟨81TL1575⟩; **2**, 420 ⟨88SC253⟩; **4**, 652 ⟨88T3781⟩; **6**, 26 ⟨87SC1047⟩
Prieto, P., **4**, 826 ⟨92SL893⟩
Prietzel, H., **6**, 623 ⟨70GEP1915668⟩
Prigorelov, G. A., **2**, 165 ⟨89IZV197⟩
Prijs, B., **4**, 21 ⟨54HCA90⟩; **5**, 365 ⟨56HCA1253⟩
Prikhodchenko, D. Y., **6**, 635 ⟨92IZV425⟩
Prikhod'ko, Yu. V., **2**, 455 ⟨85JGU2198⟩; **4**, 339 ⟨91ZOB1263⟩
Prikoszovich, W., **2**, 821, 822, 824, 825, 845 ⟨69CB2914⟩, 844, 848 ⟨69CB2922⟩
Prikule, D. E., **1**, 11 ⟨77JOU1033⟩
Prilezhaeva, E. N., **2**, 90 ⟨09CB4811⟩; **4**, 770, 779 ⟨90G235⟩, 854 ⟨82IZV2327⟩, 861 ⟨84IZV194⟩, 866 ⟨84IZV611⟩, 854 ⟨78IZV2654⟩
Prillwitz, P. E., **2**, 614 ⟨84MI 212-01⟩
Priloszowich, W., **2**, 854 ⟨67M1792⟩
Prilutskii, G. M., **2**, 190 ⟨55ZOB1943⟩
Primo, J., **3**, 297 ⟨86H(24)2511⟩
Principe, L. M., **2**, 926 ⟨87JOC1483⟩; **3**, 218 ⟨89JOC5148⟩, 270 ⟨85TL4851⟩
Principe, P. A., **3**, 704 ⟨88JAN454⟩
Pringle, P. G., **4**, 151 ⟨92OM3879⟩, 151, 585 ⟨90JCS(D)1553⟩
Prins, H. J., **2**, 54 ⟨23RTC1050⟩; **6**, 4 ⟨14JPR415, 35RTC249⟩, 4 ⟨35RTC307⟩
Prinzbach, H., **1**, 521 ⟨66AG(E)1039, 68AG(E)67⟩, 522 ⟨66AG1057⟩, 564 ⟨87JA4626⟩, 574 ⟨94AG(E)117⟩, 826 ⟨71HCA1589⟩, 918 ⟨83TL5857, 89JA746⟩; **2**, 173 ⟨87S683⟩; **6**, 48, 729 ⟨66AHC(7)39⟩
Prior, M. J., **4**, 357 ⟨86JCS(P1)683⟩
Prior, S., **5**, 1127 ⟨81AP(314)577, 81MI 530-01⟩
Prisbe, E. J., **1**, 939 ⟨85SC401⟩
Prisbylla, M., **3**, 186, 228 ⟨80TL805⟩
Prischchenko, A. A., **6**, 679 ⟨81DOK(256)1401⟩
Prischenko, A. H., **2**, 476 ⟨78JGU1088⟩
Prischl, G., **4**, 415 ⟨80LA1573⟩
Prishchenko, A. A., **2**, 433 ⟨87URP1294811⟩, 433, 473 ⟨91ZOB1016⟩, 434 ⟨86ZOB2794, 87ZOB884⟩, 451 ⟨84JGU2237, 89JGU2130⟩, 467 ⟨79JGU616⟩, 476 ⟨90JGU398⟩; **4**, 112, 114, 115 ⟨84JGU1354, 84JGU2250⟩, 116 ⟨85JGU1065⟩, 126 ⟨86JGU1258⟩, 458 ⟨91ZOB1478, 92ZOB946⟩, 459 ⟨91URP1618747⟩, 459, 484 ⟨91ZOB1016⟩, 546 ⟨76URP539892, 81ZOB484⟩, 546, 550, 562 ⟨77ZOB775⟩, 547 ⟨76ZOB571⟩, 550 ⟨79ZOB712, 80ZOB949, 80ZOB1881⟩, 551 ⟨80ZOB989⟩, 552 ⟨77ZOB2689⟩, 552, 562 ⟨79ZOB470, 79ZOB1446⟩, 555 ⟨77ZOB2636, 90ZOB1420⟩; **5**, 401 ⟨87JGU1254⟩, 876, 877 ⟨84ZOB1520⟩, 884 ⟨81DOK(256)1401, 81ZOB482⟩; **6**, 117 ⟨86ZOB2427⟩, 117, 490 ⟨88PS(35)329⟩, 265, 679 ⟨81ZOB2630⟩, 281 ⟨84ZOB2504⟩, 313 ⟨88ZOB2168⟩, 314 ⟨87JGU1260, 88ZOB1447, 88ZOB1927⟩, 490 ⟨87JGU1484⟩, 680, 681 ⟨84ZOB1520⟩, 696 ⟨85ZOB1194⟩
Pritchard, G. J., **5**, 438 ⟨90AG(E)282⟩
Pritchard, G. O., **4**, 1080 ⟨88MI 424-01⟩
Pritchard, J. G., **1**, 1144 ⟨60JPS1271⟩; **5**, 305 ⟨70CRV439⟩
Pritchard, M., **1**, 481 ⟨92JCS(P1)3419⟩
Pritchard, R. G., **2**, 197 ⟨92JCS(P1)2371⟩, 881 ⟨87AG(E)264⟩; **4**, 274 ⟨90JCS(P1)773⟩, 281 ⟨92JCS(P1)2371⟩
Pritchenko, A. A., **4**, 914 ⟨87ZOB1406⟩

Pritula, N. A., **2**, 436 ⟨82IZV2390⟩
Pritzkow, H., **1**, 910 ⟨84AG(E)371⟩; **2**, 903 ⟨88AG(E)299, 89CB401, 89ZN(B)1179, 90CB2273⟩; **3**, 375 ⟨89AG(E)221⟩; **4**, 567 ⟨93PS(76)57⟩, 572 ⟨91AG(E)709, 93AG(E)1359⟩, 588 ⟨91OM938⟩, 632 ⟨83JA2582, 89ZN(B)96⟩, 665 ⟨92CB1341⟩, 690 ⟨90AG(E)686⟩, 691 ⟨87AG(E)127, 93CC1459⟩; **5**, 476 ⟨87CB1499⟩, 502 ⟨88CB1435, 90AG(E)60⟩, 652 ⟨89AG(E)221⟩, 886 ⟨89AG(E)740⟩, 888, 893 ⟨93CC673⟩, 892 ⟨92AG(E)316, 93PS(76)21⟩, 918 ⟨93AG(E)731⟩, 919 ⟨94AG(E)456⟩, 927 ⟨92AG(E)1603⟩, 928 ⟨84AG(E)371⟩; **6**, 154 ⟨89AG929⟩, 165, 691, 733 ⟨91AG(E)709⟩, 183 ⟨90AG(E)292⟩, 188, 389, 401 ⟨89CB1881⟩, 253 ⟨89JFC(45)225⟩, 254 ⟨86CB3631⟩, 254, 532 ⟨87CB1499⟩, 255, 275, 304, 306 ⟨91CB2025⟩, 258 ⟨89AG(E)221, 90AG(E)60⟩, 275 ⟨80AG(E)942, 94CB533⟩, 389, 398, 401 ⟨86AG(E)1112⟩, 691 ⟨93CC673, 93CC1585⟩, 691, 733 ⟨93PS(76)21⟩, 691, 734 ⟨89JA6853⟩, 702 ⟨91CB329⟩
Pritzkow, W., **2**, 689 ⟨76JPR1019⟩, 778 ⟨65JPR123⟩
Privett, T. A., **5**, 179 ⟨89LA1233⟩
Privezentseva, N. F., **5**, 822 ⟨60ZOB2409⟩; **6**, 242 ⟨61CRV313⟩, 603 ⟨69JGU1694⟩
Proba, Z. A., **6**, 540 ⟨83TL865⟩
Prober, M., **3**, 567 ⟨56JA2274⟩; **4**, 609 ⟨56JA2274⟩, 731 ⟨53JA968⟩
Probst, A., **4**, 730 ⟨87JFC(37)223⟩; **6**, 3 ⟨87JFC(37)223⟩
Probst, D., **2**, 825 ⟨68LA(715)1⟩
Probst, W. J., **1**, 1076 ⟨72JOC4467⟩, 1143 ⟨72JOC1553⟩
Pröbstl, A., **1**, 367 ⟨84BSB511⟩
Procença, M. F. J. R. P., **3**, 282 ⟨83JCS(P1)1075⟩, 687 ⟨83JCS(P1)1067⟩; **5**, 690 ⟨90JCS(P1)1705⟩, 728 ⟨83JCS(P1)1067⟩
Procter, G., **1**, 215 ⟨81TL1751⟩, 216 ⟨84CC585⟩, 218 ⟨85TL4111⟩, 828 ⟨87TL2041⟩; **2**, 914, 915 ⟨90TL1051⟩, 921 ⟨87TL2041, 88T3953⟩; **4**, 279 ⟨90TL1043⟩; **5**, 326, 330 ⟨89SL32⟩, 341 ⟨81T1283⟩
Procter, S. A., **4**, 886 ⟨65CC574, 65JCS5877⟩
Proctor, G., **5**, 347 ⟨67JCS(C)58⟩
Proctor, G. R., **3**, 235 ⟨72JCS(P1)889⟩, 257 ⟨68JCS(C)2481⟩, 294 ⟨75JCS(P1)622⟩; **3**, 438 ⟨81JCS(P1)2443⟩; **4**, 105 ⟨67JCS(C)58⟩, 105, 300 ⟨67JCS(C)2696, 81JCS(P1)2435⟩, 304, 305 ⟨81JCS(P1)2443⟩; **5**, 696 ⟨81JCS(P1)2435⟩
Proctor, S. A., **4**, 886 ⟨67JCS(C)1569⟩
Prodger, J. C., **1**, 165 ⟨89T7643⟩; **2**, 143 ⟨89T7643⟩; **3**, 20, 29, 31, 137, 157, 160, 169 ⟨89T7643⟩; **4**, 244, 250 ⟨89T7643⟩, 263 ⟨91SL84⟩
Proell, W. A., **2**, 177 ⟨48IEC1129⟩
Proess, G., **4**, 288 ⟨92TL269⟩
Profitt, J. A., **3**, 621 ⟨75JOC127⟩
Prohorova, S. R., **4**, 342 ⟨92IZV196⟩
Proidakov, A. G., **2**, 1093 ⟨81IZV1280⟩
Prokai, B., **1**, 658 ⟨66JOM(5)506⟩; **2**, 1083 ⟨64JOM(1)384⟩; **5**, 779 ⟨66JCS(A)1584⟩; **6**, 61 ⟨66JA1834⟩, 270 ⟨67JOM(7)P20⟩
Prokhorova, V. V., **2**, 440 ⟨63ZOB1919⟩
Prokipcak, J. M., **3**, 618 ⟨71CJC1321⟩; **5**, 689 ⟨76CJC647⟩
Prokof'ev, A. K., **4**, 155 ⟨68IZV270⟩; **6**, 62 ⟨67ZOB1040⟩
Prokofiev, E. P., **2**, 748 ⟨68T4677⟩
Prokopenko, V. P., **5**, 789 ⟨88JGU30⟩
Prokopiou, P. A., **1**, 16 ⟨81JCS(P1)1501⟩
Proksch, E., **3**, 113 ⟨76AG(E)761⟩
Prol, Jr., J., **6**, 415 ⟨89JOC3231⟩
Proll, T., **3**, 355 ⟨79S941⟩
Promonenkov, V. K., **4**, 923 ⟨67ZOB454⟩; **5**, 105 ⟨81MI 502-01⟩, 615 ⟨89ZOR1323⟩
Prons, V. N., **2**, 468 ⟨71JGU475⟩
Pronskii, L. F., **6**, 619 ⟨75ZOR2243⟩, 651 ⟨74ZOR42⟩
Proops, W. R., **1**, 397 ⟨54JA494⟩
Proshutinskii, V. I., **2**, 949 ⟨78ZOB639⟩; **6**, 82 ⟨71ZOR1887⟩, 92 ⟨71ZOR473⟩
Proskurnina, M. V., **2**, 436 ⟨67ZOB2080, 68MI 208-01⟩, 451 ⟨84JGU2237⟩; **3**, 195 ⟨72CA(78)136370⟩; **4**, 344 ⟨72ZOB2418, 73PS(3)55⟩, 552 ⟨76ZOB568⟩, 913 ⟨76ZOB565⟩, 914 ⟨79ZOB1910⟩, 966 ⟨73MI 420-01⟩; **6**, 281 ⟨84ZOB2504⟩
Proskurovskaya, I. V., **1**, 103 ⟨75IZV888⟩
Pross, A., **1**, 129, 162 ⟨81JOC1693⟩; **2**, 525, 1005 ⟨81JOC1693⟩; **4**, 29 ⟨70AJC989⟩
Pross, N., **1**, 1110 ⟨80JOC4309⟩
Prossel, G., **4**, 424 ⟨74LA539⟩
Prosser, F. P., **6**, 33 ⟨58JA4282⟩
Prosser, T., **2**, 25 ⟨63OSC(4)169⟩; **5**, 55 ⟨63OSC(4)169⟩
Prosser, T. J., **5**, 998 ⟨57JA2544⟩
Protasiewicz, J., **1**, 925 ⟨85JOC3220⟩
Protasova, L. D., **2**, 448 ⟨68CA-P106877⟩
Protasova, L. E., **3**, 341 ⟨87KGS1694⟩
Protasov, E. M., **4**, 660 ⟨77ZOB1751⟩
Protopopov, I. S., **2**, 825, 829 ⟨63ZOB3050⟩, 825, 829 ⟨64ZOB1446⟩
Protopopova, M. N., **1**, 353, 354 ⟨94T4519⟩
Protopopova, T. U., **3**, 69 ⟨64JGU109⟩
Protopopova, T. V., **6**, 140 ⟨67ZOR1749⟩
Proudfoot, J. R., **1**, 439, 545 ⟨86TL2187⟩
Prousa, R., **5**, 589, 603, 611 ⟨67JPR294⟩
Prouse, L. J. S., **5**, 802 ⟨89JCS(D)345, 90JCS(D)781⟩
Proust, M., **5**, 561 ⟨81S963⟩
Prout, F. S., **5**, 34 ⟨63OSC(4)93⟩
Prout, K., **5**, 453 ⟨83CC1202, 85JOM(285)213⟩
Pruett, R. L., **2**, 573 ⟨79JOM(172)405⟩; **6**, 46 ⟨50JA3642⟩
Pruitt, J. R., **3**, 198 ⟨91TL2581⟩
Prusse, W., **4**, 987 ⟨81CB2001⟩
Pruszynski, P., **6**, 643 ⟨87CJC626⟩, 646 ⟨91AG(E)1356⟩
Pryce, R. J.,**1**, 522 ⟨91JCS(P1)1255⟩
Prydal, B. S., **5**, 994 ⟨54JCS4018, 57JCS1091⟩
Pryde, C., **1**, 869 ⟨65JOC3212⟩
Pryde, E. H., **2**, 778 ⟨65MI 215-02⟩
Pryor, W. A., **1**, 919 ⟨B-66MI 119-06, B-71MI 119-06⟩; **4**, 220 ⟨72JOC2885⟩
Przemyslaw, R., **6**, 546 ⟨85MI 617-03⟩
Przeslawski, R. M., **1**, 147 ⟨89TL1613⟩
Przheval'skii, N. M., **5**, 639 ⟨93S463⟩, 736 ⟨90ZOR214⟩
Przybilla, K. J., **1**, 232 ⟨89CB347⟩
Przybyla, C. A., **3**, 200 ⟨84TL4741⟩
Przybytek, J. T., **4**, 183 ⟨72TL5133⟩
Psarras, T., **1**, 846 ⟨66JA460⟩
Psaume, B., **2**, 940 ⟨80TL163⟩
Psiorz, M., **2**, 340, 357 ⟨84S217⟩
Psotskaya, G. V., **6**, 610, 727 ⟨73ZOR39⟩
Ptitsyna, O. A., **2**, 884 ⟨53DOK(91)1341⟩
Pu, L., **5**, 944 ⟨87JA7578, 89JA4129⟩, 944 ⟨90JA6232⟩
Puar, M. S., **1**, 91 ⟨81T3615⟩
Pubanz, D., **4**, 695 ⟨93JOM(459)1⟩
Pucci, B., **4**, 191 ⟨77T3105⟩
Puchalski, A. E., **4**, 3, 6 ⟨86JA7739⟩
Puchnarevič, V. B., **2**, 19 ⟨74CCC2616⟩; **4**, 32 ⟨74CCC2616⟩
Puckett, P. M., **1**, 1126 ⟨78JA993⟩
Puckett, R. T., **2**, 219 ⟨61JOC850⟩
Puckette, T. A., **2**, 816 ⟨80JA5866⟩
Puddephatt, R. J., **2**, 589 ⟨82COMC-I(2)765⟩, 1101 ⟨68JCS(C)317, 72JOM(40)143⟩; **4**, 585, 691 ⟨82CC614, 79IC2808⟩, 692 ⟨92OM2224⟩; **5**, 716, 717 ⟨66JOM(5)590⟩; **6**, 247 ⟨90IC2496, 92CM979⟩
Pudjaatmaka, A. H., **1**, 869 ⟨65JA384⟩
Pudova, O. A., **2**, 914 ⟨86ZOR140⟩; **4**, 1044 ⟨88JOM(346)297⟩
Pudova, T. A., **2**, 1033 ⟨84DOK(274)822⟩
Pudovic, A. N., **4**, 915 ⟨68ZOB139⟩
Pudovik, A. N., **2**, 254 ⟨93PS(79)179⟩, 259 ⟨90ZOB476⟩, 266 ⟨81ZOB2140, 83URP956484⟩, 446 ⟨87ZOB296⟩, 454 ⟨60ZOB2348, 69ZOB2418⟩, 461 ⟨88JGU2161⟩, 476 ⟨90JGU397, 91MI 208-03⟩, 820 ⟨84ZOB1917⟩, 845

⟨82ZOB2721⟩, 854 ⟨69ZOB341⟩, 877 ⟨84IZV719⟩, 1074 ⟨89TL577, 89ZOB1979⟩, 1074 ⟨91ZN(B)1578⟩; **3**, 231, 269 ⟨68JGU301⟩; **4**, 106 ⟨90JGU1541⟩, 124, 125, ⟨69JGU577⟩, 124, 125, 128 ⟨71IZV1159⟩, 125, 127, 128 ⟨68JGU2006⟩, 339 ⟨86ZOB711⟩, 342 ⟨71ZOB1964⟩, 343 ⟨72ZOB1861⟩, 343, 561 ⟨78ZOB1001⟩, 345 ⟨81ZOB2140, 83ZOB46⟩, 345 ⟨86ZOB1905, 86ZOB216⟩, 345 ⟨87DOK(292)360⟩, 455 ⟨52DOK(83)865, 63ZOB3353⟩, 455 ⟨78ZOB1008⟩, 455, 481 ⟨90ZOB1718⟩, 463 ⟨53DOK(92)773⟩, 476 ⟨69ZOB1235⟩, 478 ⟨86JGU632, 90JGU397, 93ZOB220⟩, 485 ⟨80ZOB751⟩, 486 ⟨83IZV432⟩, 488 ⟨79ZOB493⟩, 489 ⟨72ZOB1227, 82ZOB2444⟩, 493 ⟨88JGU876, 88JGU1148⟩, 555 ⟨68ZOB2071⟩, 561 ⟨63ZOB1045, 64ZOB3938, 65ZOB354, 66USP3277068⟩, 913 ⟨76ZOB2385⟩; **5**, 221 ⟨89IZV418⟩, 369 ⟨78ZOB2468, 89ZOB2392⟩, 369 ⟨91IZV2103, 92IZV220⟩, 369, 370 ⟨78ZOB695, 88ZOB2424⟩, 373 ⟨73ZOB1451⟩, 398 ⟨88JGU1965⟩, 402 ⟨78JGU414⟩, 405 ⟨77JGU505, 89JGU2388⟩, 406 ⟨85JGU1956, 88JGU1548⟩, 406 ⟨91BAU661⟩, 535 ⟨89JGU1548⟩, 581 ⟨79IZV861, 86ZOB2660⟩, 643 ⟨78ZOB843⟩, 643, 644 ⟨79IZV805, 79ZOB2323⟩, 644 ⟨85ZOB1526, 86ZOB2359⟩, 652 ⟨74JGU2371⟩, 708 ⟨86JGU1096, 86ZOB1245⟩, 709 ⟨84JGU1969, 84ZOB2203⟩, 713 ⟨85JGU1526, 85ZOB1718⟩, 737 ⟨73DOK(211)113, 77JGU916, 85ZOR559⟩, 738 ⟨79JGU2323⟩, 786 ⟨75JGU2512⟩, 1128 ⟨82IZV939⟩, 1131 ⟨91IZV237⟩, 1135 ⟨90ZOB476⟩, 786,1715 ⟨72JGU1637⟩; **6**, 118 ⟨79S81⟩, 121 ⟨92HAC115⟩, 160 ⟨75ZOB1450⟩, 538, 539 ⟨89IZV909⟩, 563 ⟨81ZOB530⟩, 564 ⟨81ZOB28⟩, 581 ⟨88ZOB1468⟩, 582 ⟨76IZV455⟩, 583 ⟨77IZV1177⟩, 657 ⟨90ZOB778⟩, 658 ⟨82IZV1426⟩
Pudovik, A. W., **2**, 254 ⟨87ZOB1429⟩
Pudovik, D. A., **4**, 345 ⟨86ZOB1905⟩; **5**, 406 ⟨85JGU1956, 88JGU1548⟩, 535 ⟨89JGU1548⟩, 737 ⟨85ZOR559⟩
Pudovik, M. A., **2**, 259 ⟨90ZOB476⟩; **4**, 455 ⟨63ZOB3353⟩, 478 ⟨86JGU632, 90JGU397⟩, 478 ⟨93ZOB220⟩; **5**, 369 ⟨78ZOB2468, 89ZOB2392⟩, 369 ⟨91IZV2103, 92IZV220⟩, 369, 370 ⟨78ZOB695⟩, 402 ⟨78JGU414⟩, 1135 ⟨90ZOB476⟩; **6**, 121 ⟨92HAC115⟩, 563 ⟨81ZOB530⟩, 564 ⟨81ZOB28⟩, 657 ⟨90ZOB778⟩
Pudovik, M. H., **2**, 254 ⟨87ZOB1429⟩
Pudovik, O. A., **4**, 563 ⟨70ZOB499⟩
Puertas, S., **5**, 263 ⟨93T4007⟩
Pues, C., **2**, 903 ⟨84ZN(B)1042⟩; **4**, 1048, 1054, 1057 ⟨84ZN(B)19042⟩, 1054, 1063 ⟨84AG(E)313⟩; **5**, 928 ⟨84AG(E)372⟩; **6**, 713 ⟨84ZN(B)1042⟩
Pufahl, D., **4**, 676 ⟨89OM412⟩
Puff, H., **2**, 31 ⟨87JOC5560⟩; **3**, 361, 579 ⟨86CB162⟩; **4**, 979 ⟨86CB162⟩; **5**, 1066 ⟨86CB2114⟩
Pugh, A., **5**, 214 ⟨88JMC499⟩
Pugh, C., **3**, 632 ⟨45JCS535⟩
Pugin, B., **3**, 411 ⟨83JOC6877⟩
Puglia, G., **3**, 97 ⟨78JCS(P1)318⟩
Puglis, J., **1**, 4, 7, 9, 17, 28 ⟨78JOC2259⟩
Puigh, S., **1**, 802 ⟨85JOC915⟩
Pujari, H. K., **4**, 437 ⟨88IJC(B)912⟩
Pukhnarevich, V. B., **4**, 137 ⟨76CCC386⟩
Pulaski, S. P., **1**, 504 ⟨83JOC2270⟩
Puleo, R., **3**, 38, 174, 207 ⟨81JOC2920⟩
Pulido, F. J., **1**, 495 ⟨89T413⟩; **2**, 545 ⟨89S963⟩, 929 ⟨86CC1010, 88TL1825⟩, 929 ⟨91JCS(P1)2811⟩, 929, 930 ⟨89T349, 90CC1030⟩, 930 ⟨92JCS(P1)327⟩; **3**, 203 ⟨89SC275⟩, 356 ⟨88SC1847⟩, 577 ⟨89SC1039⟩, 653 ⟨86TL2027⟩; **5**, 417 ⟨89SC275⟩
Pullaiah, K. C., **1**, 585 ⟨89JOC2217⟩
Pullen, K. M., **3**, 489 ⟨75CC459⟩
Pullin, A. D. E., **6**, 607, 609 ⟨58JCS764⟩
Pullman, B. J., **2**, 486 ⟨63AJC636⟩; **6**, 49 ⟨63AJC722⟩
Pulman, D. A., **1**, 724 ⟨74JCS(P1)2470⟩; **4**, 746 ⟨74JCS(P1)2470⟩
Pulst, M., **3**, 73 ⟨72T5197⟩, 354 ⟨88ZC345⟩, 356 ⟨90ZC247⟩, 361 ⟨88JPR665⟩
Pulver, M. J., **1**, 571 ⟨81OS(60)53⟩, **5**, 174 ⟨81T4059⟩
Pummer, W. J., **4**, 761 ⟨66USP3277068⟩; **6**, 222 ⟨57JA5654⟩
Pump, J., **5**, 375, 378 ⟨62M352⟩, 718 ⟨64CB627⟩
Punja, N., **1**, 990 ⟨67JCS(C)194⟩; **2**, 271 ⟨70JCS(C)2409⟩; **4**, 840 ⟨68JCS(C)3011⟩
Puntambeker, S. V., **5**, 43 ⟨32OSC(1)524⟩
Pupko, L. S., **6**, 653 ⟨57ZOB2134, 65ZOR735⟩, 654 ⟨72ZOR39⟩
Puppe, L., **6**, 465 ⟨93EUP558996, 93EUP559001⟩
Puranik, D. B., **2**, 896 ⟨93JOM(447)197⟩; **4**, 1041 ⟨92OM2743⟩
Purcell, R. H., **3**, 389 ⟨38JCS409⟩
Purcell, W. L., **5**, 386 ⟨79IC937⟩
Purchase, II, C. F., **3**, 188 ⟨91JOC457⟩
Purchase, R., **2**, 388 ⟨72JCS(P1)1692⟩; **5**, 366 ⟨72JCS(P1)1692⟩
Purdie, T., **2**, 89 ⟨03JCS1021⟩
Purdum, W. R., **2**, 445 ⟨75JOC2801⟩
Purdy, A. P., **2**, 482 ⟨86JOM(308)281⟩, 482, 504 ⟨90POL319⟩, 490 ⟨91PS(57)1⟩, 508, 511 ⟨86OM1266, 87JOM(325)C7⟩, 511 ⟨88JOM(354)287⟩; **4**, 597 ⟨91MI 413-01, 91POL319⟩, 597 ⟨91PS(57)1⟩
Puri, L., **5**, 272 ⟨84MI 507-02⟩
Purick, R., **1**, 80 ⟨70JOC3510⟩
Purkayastha, M. L., **1**, 428 ⟨82S876⟩, 485 ⟨85JCS(P1)1289⟩
Purkner, E., **5**, 718 ⟨90S525⟩
Purlee, E. L., **1**, 895 ⟨56JA5807⟩
Purohit, P. C., **1**, 802 ⟨91T8155⟩
Purpura, J. M., **1**, 808 ⟨86JOC183⟩
Purrello, G., **3**, 549 ⟨76JCS(P1)692, 84JCR(S)370⟩; **5**, 734 ⟨82SC865⟩, 734, 760 ⟨84JCS(P1)897⟩
Purrington, S. T., **1**, 210 ⟨82TL2379⟩, 399 ⟨79JOC2348⟩, 780 ⟨87TL3901⟩; **2**, 15, 17 ⟨86CRV(86)997⟩, 16 ⟨86TL2715, 87JOC4307⟩, 16 ⟨90JOC3423⟩, 334 ⟨86CRV997⟩, 610 ⟨87TL3901⟩; **3**, 26, 148 ⟨86TL2715⟩, 224 ⟨82TL2379⟩; **5**, 313 ⟨83JOC761⟩
Purvis, G. D., **1**, 1143 ⟨78JA6930⟩
Puschel, F., **2**, 191 ⟨64CB2903⟩
Pushin, A. N., **2**, 280 ⟨93TL703⟩; **4**, 494 ⟨87DOK(297)1132, 87IZV1680⟩; **5**, 1076 ⟨87IZV1177⟩
Pusset, J., **1**, 63 ⟨76JOM(110)139⟩
Pustinger, J. V., **4**, 554 ⟨64JOC2427⟩; **5**, 1093 ⟨64JOC2427⟩
Pustoslemsek, P., **3**, 399 ⟨76AG(E)704⟩
Putchnarevich, V. B., **1**, 61 ⟨83ZOB806⟩
Puterbaugh, W. H., **1**, 881 ⟨55JOC1531⟩
Putinas, J. M., **3**, 411 ⟨87JA5047⟩
Putman, D., **1**, 707, 993 ⟨90JA4072⟩
Putman, R. E., **4**, 5 ⟨61JA382⟩
Putnik, C. F., **4**, 671 ⟨79JA6933⟩
Putsykin, Yu. G., **3**, 434 ⟨84ZOR988⟩
Putt, S. R., **1**, 594, 616 ⟨85JOC3622⟩; **3**, 190 ⟨78S478⟩
Pütter, E., **5**, 1102 ⟨64CB3012⟩
Pütter, H., **3**, 355 ⟨77CB2506⟩
Pütter, M., **1**, 708 ⟨81AG(E)677⟩
Pütter, R., **2**, 803 ⟨65HOU(10/3)16, 65HOU(10/3)22, 65HOU(10/3)23, 65HOU(10/3)28, 65HOU(10/3)44, 65HOU(10/3)51, 65HOU(10/3)89, 65HOU(10/3)8⟩, 804 ⟨65HOU(10/3)46, 65HOU(10/3)576, 65HOU(10/3)584⟩, 805 ⟨65HOU(10/3)552, 65HOU(10/3)71⟩, 806 ⟨65HOU(10/3)73, 65HOU(10/3)86⟩
Pütter, R., **3**, 722 ⟨66CB2361⟩; **5**, 360 ⟨65HOU(10)67⟩, 1100 ⟨67AG(E)206⟩; **6**, 629 ⟨64CB3022⟩, 632 ⟨64CB3027⟩
Puttmann, W., **1**, 248 ⟨86AG(E)720⟩
Puy, M. V. D., **2**, 614 ⟨88JOC4398⟩
Puza, M., **3**, 490 ⟨71S481⟩
Puzin, Yu. I., **4**, 267 ⟨90ZOR1259⟩; **6**, 302 ⟨86MI 610-01⟩

Pyatakov, N. F., **6**, 144 ⟨59UK484⟩
Pyatin, B. M., **4**, 881 ⟨69MI 420-01⟩
Pye, K., **6**, 604 ⟨91MI 620-01⟩
Pye, P. L., **6**, 720, 721 ⟨76CC644, 77JCS(D)1272⟩, 721 ⟨77JCS(D)1283, 77JCS(D)2160, 77JCS(D)2172, 77MI 622-01, 78JCS(D)826, 78JCS(D)837, 79JCS(D)1929⟩
Pye, W. E., **3**, 311 ⟨82JA536⟩
Pyman, F. L., **5**, 743, 744, 755 ⟨23JCS3359⟩, 757 ⟨27JCS2318⟩
Pyne, S. G., **1**, 21, 23 ⟨86CC1688⟩, 504, 513 ⟨83TL4883⟩, 548 ⟨83TL2821⟩, 1015, 1016 ⟨83TL3291⟩; **2**, 351 ⟨83TL2821⟩, 362 ⟨86JOC81⟩; **3**, 108 ⟨83TL3291⟩, 421 ⟨90JOC1086⟩
Pyszora, H., **2**, 419 ⟨65JCS4256⟩; **5**, 1016 ⟨65JCS4256⟩, 1133 ⟨66MI 530-01⟩, 1144 ⟨66MI 530-02⟩
Pytlewski, T., **4**, 84 ⟨78S883⟩
Pyun, C., **1**, 238 ⟨90TL2713⟩; **2**, 519, 531 ⟨89JA1754⟩, 530 ⟨85JOC5586⟩; **5**, 418, 428 ⟨87TL3261⟩

Q

Qabar, M., **5**, 325 ⟨89JOC5852⟩
Qais, N., **2**, 328 ⟨88TL6949⟩; **3**, 422 ⟨88TL6949⟩
Qasi, A. M., **1**, 97 ⟨78JOC1007⟩
Qi, H., **1**, 324 ⟨93TL4489⟩
Qian, X., **2**, 29 ⟨94T2391⟩
Qiao, K., **4**, 690 ⟨93ZN(B)1558⟩, 715 ⟨94JOM(471)249⟩
Qi, Y., **2**, 454 ⟨86S821⟩; **4**, 467 ⟨86S821⟩
Qimu Liao, **1**, 739 ⟨88JOM(346)181⟩
Qing, W., **2**, 411 ⟨90AG(E)522⟩
Qiu, Z., **6**, 556 ⟨88MI 617-02⟩
Quader, A., **1**, 480 ⟨83TL5137⟩, 970 ⟨83TL1297⟩
Quaedlieg, M., **2**, 175 ⟨55HOU(9)1⟩, 797 ⟨55HOU(9)285⟩
Quagliato, D. A., **3**, 211 ⟨82OM1240⟩; 211, 253 ⟨85JOC1621⟩; **4**, 936 ⟨82OM1243⟩, 942 ⟨85JOC1621⟩
Quail, J. W., **1**, 1112 ⟨82CJC735⟩
Quain, P. L., **5**, 699 ⟨78JOC388⟩
Quallich, G. J., **2**, 66 ⟨93TL4145⟩; **3**, 244 ⟨84JOC392⟩, 310 ⟨92JOC761⟩
Qualls, S., **4**, 214 ⟨91JA8168⟩
Quan, M. Le, **1**, 973 ⟨73JOM(54)153⟩, 1000 ⟨65BSF35, 68BSF1085⟩
Quan, P. M., **4**, 424 ⟨65JOC2769⟩
Quang, L., **4**, 465 ⟨85S62⟩
Quante, J., **5**, 554 ⟨80CB1898⟩, 558 ⟨80AP(313)1029⟩; **6**, 538 ⟨80CB1898⟩
Quarterman, L. A., **2**, 613 ⟨75JOC807⟩
Quashie, S., **6**, 689 ⟨85CC1687, 86OM593⟩
Quasmi, D., **6**, 538 ⟨90SC2769⟩
Quast, H., **1**, 63 ⟨87CB1049⟩; **2**, 176 ⟨74S489⟩, 338 ⟨83CB3931⟩, 377 ⟨81LA943⟩, 401 ⟨81CB774⟩; **3**, 351 ⟨91CB1747⟩, 406, 416 ⟨80LA1814⟩, 561, 582 ⟨79LA83⟩, 562, 563 ⟨79AG(E)788⟩, 564 ⟨89T259⟩; **4**, 122 ⟨82CB3384⟩; **5**, 364 ⟨81CB802⟩, 657 ⟨79LA74⟩, 1073 ⟨75AG(E)428, 85CB526⟩, 1073 ⟨89T259⟩; **6**, 318, 361 ⟨68CB1137⟩
Quast, U., **5**, 1121 ⟨93JMC2004⟩
Quattlebaum, J. M., **6**, 220 ⟨56USP2739989⟩
Quayle, O. R., **3**, 449 ⟨56JA5584⟩
Quayle, P., **1**, 160 ⟨92TL543⟩; **3**, 283 ⟨84JA2469⟩; **4**, 651 ⟨92TL543⟩; **5**, 286 ⟨84JA2469⟩, 934 ⟨91TL7759⟩, 944 ⟨94TL3801, 94TL8883, 95UP 525-01⟩
Quazzotti, S., **3**, 704 ⟨92T7581⟩
Queen, J., **4**, 352 ⟨80S318⟩
Quéguiner, G., **2**, 398 ⟨87BCJ4492⟩, 942 ⟨83T2009⟩; **3**, 107 ⟨68BSF4117⟩, 109 ⟨76BSF789⟩, 310 ⟨81JOM(216)139⟩
Quelet, R., **1**, 181 ⟨62BSF1882⟩
Querci, C., **2**, 329 ⟨89JOC1748⟩
Quevauviller, A., **5**, 751, 753 ⟨56MI 519-02⟩
Qui, Y., **3**, 634, 635 ⟨87CL887⟩
Qui Chi Mir, **5**, 663 ⟨90JFC(48)367, 91IC535⟩
Quici, S., **2**, 9 ⟨75S430⟩, 35 ⟨79JOC3436⟩; **3**, 7 ⟨87JOC2559⟩, 31 ⟨89JOC2970⟩, 613 ⟨79JOC2029⟩
Quick, A. J., **2**, 498 ⟨22JA805⟩
Quick, J., **1**, 876 ⟨78JOC2705⟩
Quick, J. K., **4**, 419 ⟨91TL2683⟩
Quick, M. H., **3**, 723 ⟨82IC2178⟩
Quick, S. J., **3**, 256 ⟨84TL1039⟩
Quiclet-Sire, B., **1**, 828 ⟨93CC960⟩, 944 ⟨92T9261⟩
Quigley, M. A., **5**, 431 ⟨60JA5102⟩
Quignard, F., **1**, 674 ⟨85CC1816⟩
Quiiou, H., **4**, 852, 969 ⟨73BSF581⟩
Quilico, A., **4**, 439 ⟨69OPP5⟩; **5**, 821 ⟨59T24⟩; **6**, 6 ⟨56CI(M)371⟩
Quimby, O. T., **4**, 563 ⟨68JOM(13)199⟩; **6**, 263 ⟨68JOM(13)199⟩, 263, 518 ⟨67JOC4111, 70USP3497313⟩
Quimpère, M., **6**, 134 ⟨92CJC2335⟩
Quin, G. S., **5**, 403 ⟨91JOC4791⟩
Quin, L. D., **2**, 435 ⟨78JOC3508⟩, 440 ⟨82JOC905⟩, 441 ⟨83JOC4466, 84JA7021⟩, 449 ⟨71JOC1285⟩, 450 ⟨86JCS(P2)1029⟩, 453 ⟨82JA1893⟩, 462 ⟨76JOC1578⟩, 833, 834, 840, 841 ⟨63JOC3315⟩, 833, 844 ⟨61JA4124⟩, 838 ⟨62JOC1012⟩; **4**, 424 ⟨65JOC2769⟩; **5**, 403 ⟨91JOC4791⟩
Quin, T., **3**, 58 ⟨78CL859⟩
Quinchon, J., **2**, 465 ⟨60BSF1794⟩
Quincoces Suarez, J., **3**, 357 ⟨91EGP294941⟩
Quincy, D. A., **3**, 4 ⟨83SC817⟩
Quiniou, H., **2**, 744 ⟨74S298⟩; **3**, 73 ⟨72BSF4576⟩, 338 ⟨81PS(10)1⟩, 339 ⟨87BSF517⟩; **5**, 597, 599, 614 ⟨73BSF581⟩, 633 ⟨85JCS(P1)1875, 85PS(25)279, 88T139⟩, 760 ⟨80JHC767, 83PS(15)143⟩
Quinkert, G., **1**, 558 ⟨91TL3357⟩; **3**, 55 ⟨59CB1564⟩, 227 ⟨88HCA1719⟩, 538 ⟨62TL221, 64CB1799⟩, 540 ⟨60JCS1, 65AG(E)211⟩, 540 ⟨77CB3582⟩; **5**, 148 ⟨80AG(E)1029⟩, 174 ⟨87AG(E)362⟩
Quinlou, H., **5**, 760 ⟨84CC157⟩
Quinn, J. F., **5**, 765 ⟨56JA1461⟩, 838 ⟨58USP2851489⟩
Quinn, P. L., **3**, 684 ⟨77JCS(P1)1457⟩
Quinn, S., **5**, 459 ⟨82JA1096⟩
Quintard, J.-P., **1**, 128 ⟨B-87MI 104-05⟩, 452 ⟨B-87MI 110-05⟩, 919 ⟨B-86MI 119-05⟩, 1020 ⟨92TL3647⟩; **2**, 599 ⟨B-87MI 211-07⟩, 1094 ⟨B-87MI 223-05⟩; **3**, 182 ⟨86TL2361⟩; **4**, 300, 533 ⟨84S495⟩, 358 ⟨88CC503⟩, 388 ⟨81JOM(212)C31, 83JOC1559, 87CC29, 92JOM(437)C19⟩, 388, 389 ⟨85BSF787⟩, 533 ⟨86USP4617409, 88JOM(339)267⟩; **5**, 467 ⟨85TL6075⟩, 468 ⟨83JOM(251)175⟩; **6**, 172, 204 ⟨B-87MI 606-01⟩
Quintero, C., **3**, 179 ⟨91CJC2059⟩
Quintero, L., **1**, 935, 938, 941 ⟨89CRV1413⟩; **2**, 140 ⟨89CRV1413⟩
Quintero, M. G., **6**, 425 ⟨90JMC2101⟩
Quintily, U., **4**, 304 ⟨84G405⟩
Quiram, E. R., **2**, 137 ⟨63JOC1952⟩
Quiric, M. G., **1**, 472, 1013 ⟨82T631⟩, 509, 532 ⟨81T2617⟩
Quirion, J.-C., **1**, 1028 ⟨90TL2879⟩
Quirk, R. P., **1**, 67 ⟨76JA5973⟩
Quiroga, M. L., **4**, 325 ⟨92JCS(P1)291⟩
Quiron, J.-C., **2**, 331 ⟨88TL3311⟩
Quis, P., **4**, 792 ⟨63CB1035⟩; **6**, 452, 728 ⟨64CB3162⟩
Quitt, P., **6**, 430 ⟨57HCA604⟩
Qureshi, A. K., **2**, 389, 765 ⟨69JCS(C)1073⟩
Qureshi, M. I., **5**, 360, 361 ⟨67CJC1539⟩

R

Raab, G., **1**, 455 ⟨61JOC2934⟩; **4**, 811, 812 ⟨60AG77, 61JOC2934⟩, 812 ⟨62JA4266⟩
Raab, K., **1**, 517 ⟨84CB3169⟩; **4**, 730 ⟨87JFC(37)223⟩; **6**, 3 ⟨87JFC(37)223⟩, 32 ⟨91EUP461563, 91GEP3937567⟩
Raab, R., **3**, 679 ⟨72CB1258⟩; **5**, 657 ⟨72CB1258⟩
Raab, W., **1**, 560 ⟨79S127⟩; **2**, 739 ⟨80S320⟩, 741 ⟨79S127⟩; **3**, 638, 657 ⟨79S127⟩
Raabe, D., **6**, 29 ⟨87JPR1131⟩
Raabe, E., **2**, 1003 ⟨88CB843⟩; **3**, 61 ⟨89CB585⟩; **4**, 671 ⟨86AG(E)925⟩; **6**, 338 ⟨88OM812⟩
Raabe, G., **2**, 759 ⟨91S649⟩; **3**, 204 ⟨93AG(E)598⟩, 501, 502 ⟨85CRV419⟩; **4**, 425 ⟨90S951⟩, 656 ⟨93JA7215⟩, 720 ⟨90ZAAC(581)41⟩; **5**, 923 ⟨85CRV419⟩, 951 ⟨93CB1849⟩; **6**, 177, 709 ⟨85CRV419⟩, 177 ⟨B-89MI 606-01, B-89MI 622-02⟩
Ra, C. S., **1**, 885 ⟨88JOC4978⟩
Raadt, A. de, **5**, 118 ⟨91TL341⟩
Raaen, V.,**3**, 68 ⟨53OS(33)15⟩
Raap, B., **2**, 872 ⟨80JOM(186)39⟩; **5**, 791 ⟨80JOM(186)39⟩
Raap, R., **1**, 1077 ⟨68CJC1057⟩; **3**, 494 ⟨71CJC2143⟩, 580 ⟨71CJC1792⟩; **4**, 852 ⟨68CJC2251⟩, 903 ⟨68CJC2255⟩; **5**, 730 ⟨71CJC1792⟩, 845 ⟨65CJC126⟩, 1123 ⟨71CJC2139⟩; **6**, 625 ⟨63PCS370⟩
Raasch, M., **5**, 736, 736 ⟨71USP3631071⟩
Raasch, M. S., **1**, 91 ⟨62JOC1406⟩; **2**, 478 ⟨80JOC2151⟩; **3**, 358 ⟨79JOC632, 87JOC1874⟩, 550 ⟨66CC577, 70JOC3470⟩, 587 ⟨69USP3462489, 72JOC1347⟩; **4**, 5 ⟨49JA490⟩, 905 ⟨78JOC2500⟩; **5**, 59 ⟨73OSC(5)393⟩, 253, 736 ⟨72JOC3820⟩, 487, 511, 551, 563 ⟨72JOC1347⟩, 511 ⟨70JOC3470⟩; **6**, 2 ⟨49JOC747⟩, 304, 305, 311, 312 ⟨70JOC3470⟩, 314 ⟨78JOC2500⟩
Rabadeux, J.-C., **6**, 475 ⟨76BSF501⟩
Raban, M., **2**, 357, 358, 361, 767 ⟨89CRV689⟩, 365 ⟨71JA2692, 80JOC1688⟩, 767 ⟨84TL1337, 85JOC2205⟩, 768 ⟨89UP-01⟩; **3**, 436, 437 ⟨89CR689⟩
Rabe, E., **6**, 55 ⟨73LA40, 75LA195⟩
Rabe, J., **1**, 364 ⟨88T3899⟩
Raber, D. J., **1**, 1114 ⟨70JA2538⟩
Rabideau, P. W., **1**, 87 ⟨92OR(42)1⟩, 95 ⟨78JOC4283⟩, 95, 889 ⟨89T1579⟩, 96 ⟨81JOC1593⟩, 97 ⟨80SC627, 84JOC1544⟩, 890 ⟨87TL2481⟩; **2**, 682, 937 ⟨92OR(42)1⟩
Rabier, A., **5**, 955 ⟨94OM4676⟩
Rabiller, C., **5**, 760 ⟨80JHC767⟩
Rabinovich, G. L., **1**, 56 ⟨71MI 102-01⟩
Rabinovitz, M., **5**, 201 ⟨64JA965⟩
Rabinowith, J., **2**, 451 ⟨63HCA2464⟩
Rabinowitz, I., **5**, 748 ⟨26JA732⟩
Rabinowitz, J., **2**, 104 ⟨64HCA1448⟩, 456 ⟨61HCA61⟩
Rabinowitz, J. L., **2**, 35 ⟨75JOC248⟩
Rabinowitz, R., **1**, 942 ⟨57JA5326, 59JA1243⟩; **2**, 127 ⟨59JA1137⟩; **4**, 738 ⟨62JA1312⟩; **6**, 244 ⟨62JA1312⟩
Rabizzoni, A., **6**, 522 ⟨57JA3611⟩
Rabjohn, I., **2**, 86 ⟨76OR(24)261⟩
Rabjohn, N., **1**, 839 ⟨76OR(24)261⟩; **2**, 645 ⟨44JA1272⟩; **3**, 215 ⟨76OR(24)261⟩, 452 ⟨62JOC2765⟩, 567 ⟨61JOC45⟩; **4**, 437 ⟨69JHC187⟩; **5**, 34 ⟨55OSC(3)557⟩, 37 ⟨49OR(5)331, 63OSC(4)441⟩, 275 ⟨63OSC(4)513⟩
Rabold, H., **2**, 441 ⟨56ZAAC(287)208⟩
Rabone, K. L., **2**, 87 ⟨69JCS(C)1605⟩
Raby, C., **6**, 618, 623 ⟨82MI 620-02, 82MI 620-01⟩
Racha, S., **1**, 212 ⟨93TL3539⟩
Racherla, H. S., **2**, 521 ⟨91JOC401⟩
Racherla, U. S., **1**, 898 ⟨89JA7659⟩, 1037 ⟨85JOC1577⟩, 1045 ⟨84TL2411⟩, 1055 ⟨93TL15⟩, 1101 ⟨85TL2187⟩; **2**, 75, 521 ⟨90JA2389⟩, 514 ⟨85TL2187⟩, 520 ⟨86JOC427⟩, 523 ⟨81JOC4313⟩, 525 ⟨83JOC1389, 86JOC895⟩, 535 ⟨87JOC728⟩; **3**, 273 ⟨84TL2411⟩
Racherla, V. S., **2**, 521 ⟨94JOC365⟩
Rachita, M. A., **5**, 342 ⟨93TL6821⟩
Rachlin, S., **6**, 645 ⟨80JMC13⟩
Rachmankulov, D. L., **6**, 555 ⟨91JPR139⟩
Rachoñ, J., **2**, 998 ⟨93JOC546⟩; **3**, 706 ⟨81LA1693, 81LA709⟩, 717, 718 ⟨81LA99⟩; **4**, 1008 ⟨91PS(55)97⟩; **4**, 460 ⟨73ZC254, 75RZC397, 76C187, 76RZC477, 78C253, 78TL1609, 79ZC253⟩, 476 ⟨74ZC152⟩; 816, 818 ⟨93JOC546⟩
Rachwal, B., **1**, 260 ⟨87JCS(P1)805, 89JOC6022⟩; **4**, 296, 300, 310 ⟨87JCS(P1)791⟩, 301, 308 ⟨89JOC6022⟩, 306, 310 ⟨90JCS(P1)1717⟩, 308, 445 ⟨87JCS(P1)811⟩, 309, 445 ⟨87JCS(P1)799⟩, 323 ⟨92S1295⟩
Rachwal, S., **1**, 260 ⟨87JCS(P1)805, 89JOC6022⟩; **2**, 301 ⟨91OPP399⟩, 304 ⟨90CJC456⟩; **4**, 296, 300, 310 ⟨87JCS(P1)791⟩, 301 ⟨91S279⟩, 301, 308 ⟨89JOC6022⟩, 306, 310 ⟨90JCS(P1)1717⟩, 308, 445 ⟨87JCS(P1)811⟩, 309, 445 ⟨87JCS(P1)799⟩, 322 ⟨87JCS(P1)781⟩, 323 ⟨92S1295⟩, 323, 406, 410, 445 ⟨91T2683⟩
Rack, E. P., **2**, 35 ⟨71JA1389⟩
Rackow, S., **3**, 614, 633 ⟨65CB3662⟩; **4**, 889, 972 ⟨66CB2302⟩; **6**, 624 ⟨65CB3662⟩
Raczko, J., **3**, 303 ⟨87JA3981⟩
Radchenko, N. F., **6**, 453 ⟨71ZOR2229⟩
Radchenko, S. I., **2**, 731 ⟨65ZOR2115⟩, 1021, 1025, 1027, 1030, 1036 ⟨89RCR948⟩, 1032 ⟨64ZOB1899, 65ZOR2157⟩; **4**, 786 ⟨74ZOR1986, 77ZOR254⟩, 786 ⟨85ZOR244⟩, 841 ⟨72ZOR1329, 74ZOR2269, 80ZOR720⟩, 872 ⟨74ZOR2456⟩, 905, 910 ⟨77ZOR504⟩
Raddatz, P., **4**, 849 ⟨84LA450⟩; **5**, 174 ⟨84LA450⟩, 174 ⟨84LA462⟩, 174 ⟨84LA474⟩
Raddatz, S., **5**, 1064 ⟨71TL2765⟩, 1064, 1066 ⟨80CB1095⟩; **6**, 620 ⟨80CB1095⟩
Radeglia, R., **6**, 590, 592, 593 ⟨92JOM(427)213⟩
Radel, P. A., **1**, 829 ⟨88JOC1922⟩; **5**, 159 ⟨88JOC1922⟩
Radell, J., **2**, 94 ⟨61JA3985⟩, 772 ⟨56JOC1051⟩
Rademacher, B., **4**, 607, 656 ⟨91JOM(421)175⟩; **6**, 190, 192, 195, 393 ⟨91JOM(421)175⟩, 195 ⟨93ZAAC(619)675, 94JOM(469)135⟩
Rademacher, P., **6**, 689, 690 ⟨87ZAAC(545)7⟩
Räder, H.-J., **1**, 880 ⟨89TL159⟩
Radhakrishna, A. S., **4**, 422 ⟨84JOC4272⟩; **5**, 999 ⟨84JOC4272⟩
Radhakrishnan, T. V., **1**, 548 ⟨81JOC2622⟩
Radhe, R., **1**, 797 ⟨76JA7319⟩
Radics, U., **2**, 310 ⟨93JHC1061⟩
Radies, L., **3**, 239 ⟨75S708⟩
Radimerski, P., **3**, 543 ⟨74HCA2583⟩
Radinov, R. N., **1**, 1097 ⟨88TL5645⟩; **2**, 71 ⟨88TL5645⟩, 919 ⟨92HCA170⟩; **3**, 310 ⟨86S886⟩
Radke, M., **5**, 477, 584, 587, 620, 641 ⟨70LA(739)201⟩, 584, 587, 641 ⟨73LA636⟩; **6**, 444, 501, 577 ⟨79LA1756⟩
Rad'ko, L. V., **5**, 13 ⟨79MI 501-01⟩
Radlick, P., **1**, 418 ⟨67TL4901⟩, 864 ⟨66JA3461⟩; **2**, 320 ⟨74S290⟩
Radner, F., **2**, 631 ⟨88JOC3548⟩
Radom, L., **1**, 96, 890 ⟨80JA3370, 80JA4074, 80JA6430, 81JA284⟩, 129, 162 ⟨81JOC1693⟩, 850 ⟨76AJC1635⟩; **2**, 525, 1005 ⟨81JOC1693⟩, 1022 ⟨82JOC1869, 89JA8297⟩
Radzik, D. M., **1**, 778 ⟨89JOC389⟩
Radziszewski, B., **5**, 709 ⟨1884CB1289⟩
Radziszewski, B. R., **5**, 275 ⟨1885CB355⟩
Radzitzky, P. de, **5**, 551 ⟨55JA5997⟩

Rae, I. D., **3**, 212 ⟨75AJC2669⟩; **4**, 303 ⟨72AJC1737⟩; **5**, 624, 625, 626 ⟨76IJS519⟩, 725 ⟨67CJC1⟩
Raecke, B., **5**, 28 ⟨58AG1⟩, 109 ⟨73OSC(5)813⟩
Raedeker, S., **2**, 569, 570, 571 ⟨93CB2093⟩
Raether, W., **6**, 533 ⟨82EUP62834⟩
Raevskaya, O. E., **4**, 1000, 1001 ⟨72ZOB1727⟩
Raevskii, O. A., **4**, 126, 127 ⟨79IZV2073⟩, 476 ⟨69ZOB1235⟩
Raffelson, H., **5**, 76 ⟨52JA1730⟩, 838 ⟨58USP2851489⟩
Rafferty, M. A., **3**, 210 ⟨87JOC1603⟩
Rafikov, S. R., **4**, 141 ⟨81DOK(261)474⟩
Rafla, F. K., **3**, 357 ⟨59JCS2588, 62JCS1857⟩
Rafta, F. K., **2**, 121 ⟨70JCS(C)1578⟩
Ragan, J. A., **1**, 1079 ⟨89JOC4267⟩
Raggio, M. L., **3**, 643 ⟨76JOC1873⟩
Raggon, J., **1**, 815 ⟨82JOC4815⟩
Raghanan Nair, M. G., **2**, 452 ⟨68JOC579⟩
Raghavan, S., **4**, 232 ⟨93JA1580⟩
Raghuraman, T. S., **3**, 300 ⟨86SC485⟩
Ragimova, Sh. Sh., **6**, 558 ⟨84ZOR2158⟩
Ragnarsson, U., **2**, 299 ⟨87S275, 91ACR285⟩; **5**, 689 ⟨93JCS(P2)655⟩, 977 ⟨88S992⟩
Ragni, A., **4**, 687 ⟨67G1327⟩
Ragonnet, B., **1**, 959 ⟨75BSF1407⟩
Ragué Schleyer, P. von, **2**, 999, 1000 ⟨89OM1688⟩, 999, 1006 ⟨84JA6467⟩; **5**, 436 ⟨87JA2553⟩
Ragulin, L. I., **4**, 833 ⟨71KGS1645⟩; **5**, 502 ⟨71BAU959⟩
Rahaingoson, F., **2**, 28 ⟨92SC1923⟩
Rahim, M., **6**, 517 ⟨93OM2401⟩
Rahm, A., **1**, 919 ⟨B-86MI 119-05⟩; **2**, 599 ⟨B-87MI 211-07⟩, 601 ⟨75JOM(88)79⟩; **5**, 952, 958 ⟨93JA4602⟩, 956 ⟨93OM597⟩; **6**, 172, 204 ⟨B-87MI 606-01⟩
Rahm, M. L., **4**, 29, 40 ⟨66JOC1857⟩
Rahman, A., **3**, 281 ⟨73CI(L)189⟩; **4**, 903 ⟨84S250⟩
Rahman, A. H. A.-E., **5**, 22 ⟨91PS(60)159⟩
Rahman, M., **6**, 563 ⟨91POL823⟩
Rahman, M. A., **6**, 643, 733 ⟨84CB502⟩
Rahman, M. T., **2**, 632 ⟨74JIC1018⟩
Rahman, S., **5**, 944 ⟨94TL3801, 95UP 525-01⟩
Rahman, S. S., **3**, 539 ⟨89CC303⟩
Rahman, Z. A., **6**, 343 ⟨83OM1044⟩
Rahmoune, M., **6**, 156 ⟨89NJC891⟩
Rahn, C. H., **3**, 302 ⟨79JOC3420⟩
Rahrs, E. J., **5**, 140 ⟨55OSC(3)650⟩
Rahtz, D., **3**, 66, 233 ⟨57CB2265⟩
Rai, A. K., **2**, 695 ⟨B-90MI 213-01⟩, 889 ⟨90PS(47)357⟩
Rai, B., **6**, 557 ⟨90OPP128⟩
Rai, K. M. L., **3**, 684 ⟨89S57, 92OPP91⟩
Rai, R. S., **1**, 502 ⟨87JOC3126⟩; **3**, 83 ⟨86JA2608⟩
Raiford, L. C., **2**, 173 ⟨35JA2172⟩
Raikas, T., **5**, 864 ⟨83ACS(A)631⟩
Raimondo, L.,**3**, 431 ⟨89G253⟩
Raina, S., **3**, 151 ⟨92TL6021⟩
Rainer, N. B., **5**, 1003 ⟨65USP3218345⟩
Raines, S., **4**, 557 ⟨64JOC2036⟩
Rainville, D. P., **2**, 527 ⟨76JOC3031⟩
Raisi, A., **6**, 25 ⟨83M813⟩
Raiskuma, I. K., **6**, 555 ⟨92KGS1117⟩
Raison, C. G., **5**, 757 ⟨47JCS160, 49JCS3319⟩
Raistrick, H., **5**, 73 ⟨35BJ(29)606⟩
Raithby, P. R., **3**, 700 ⟨91JOM(403)C15⟩; **4**, 682 ⟨87JOM(319)C51, 88JCS(D)1421⟩, 718 ⟨93OM4545⟩; **6**, 340 ⟨80CC781⟩, 352 ⟨80CC812⟩, 522 ⟨B-93MI 616-01⟩
Raizada, M. S., **5**, 390 ⟨84IJC(A)952⟩
Raja, M., **5**, 801 ⟨82IC1278, 86ICA67⟩
Rajadhyaksha, S. N., **2**, 397 ⟨89JOC5783⟩; **4**, 203 ⟨83JOC1765⟩
Rajadhyaksha, V. J., **1**, 866 ⟨70JA2163⟩
Rajagopal, D., **2**, 999, 1002 ⟨93JOC588⟩; **3**, 162 ⟨93JOC588⟩; **4**, 53, 54, 58 ⟨93JOC588⟩, 153 ⟨92JOM(438)11⟩
Rajagopalan, K., **1**, 810 ⟨92JCS(P1)2727, 93JOC5482⟩, 810 ⟨93SC311⟩, 813 ⟨85TL6137⟩; **3**, 144 ⟨93JOC5482⟩
Rajagopalan, P., **3**, 323 ⟨67TL2689⟩; **4**, 490 ⟨71CC490⟩; **5**, 1073 ⟨65JOC3369⟩
Rajagopalan, S., **1**, 512 ⟨84S113⟩, 1052 ⟨84S111⟩; **2**, 623 ⟨86S480⟩, 922 ⟨84S113⟩
Rajagopalan, T. G., **5**, 1035 ⟨79JCS(P1)2001⟩
Raj, I., **3**, 62 ⟨63JIC114⟩
Rajakumar, P., **2**, 1034, 1035 ⟨89CC154⟩
Rajalakshmi, N., **6**, 335 ⟨92C130⟩
Rajamannar, T., **1**, 961 ⟨86TL3777⟩
Rajamohanan, P. P., **6**, 627 ⟨91S220⟩
Rajamohanan, P. R., **1**, 898 ⟨89JA7659⟩
Rajan, S., **3**, 7, 120 ⟨79JOC138⟩; **4**, 473 ⟨65JA2567⟩
Rajan, S. J., **5**, 282 ⟨80JOC165⟩
Rajanbabu, T. V., **1**, 174, 320, 919 ⟨91ACR139⟩, 176 ⟨88JOC4522⟩, 338 ⟨89JA1759⟩, 587 ⟨90JA6408⟩; **2**, 85 ⟨89JA4525⟩, 681 ⟨86JOC5458⟩, 916 ⟨85JOC3666⟩, 988 ⟨88JA7128⟩; **4**, 170 ⟨86JOC1704⟩; **6**, 557 ⟨87JA609⟩
Raj, P., **2**, 881 ⟨85POL251⟩, 891 ⟨86POL1581, 89POL581⟩; **5**, 1145 ⟨86POL677⟩; **5**, 318, 369 ⟨89POL581⟩, 318, 372 ⟨83IJC(A)120, 84SRI269⟩, 372 ⟨85POL251, 87IJC(A)146⟩
Raj, P.,**5**, 372 ⟨93SRI1011⟩
Rajapaksa, D., **4**, 253 ⟨93TL1463⟩
Rajappa, S., **1**, 194 ⟨91CC372⟩; **3**, 361 ⟨82T1673⟩; **4**, 94 ⟨82IJC(B)1, 82T1673⟩, 796, 901 ⟨81T1453⟩, 901, 904 ⟨90JCS(P1)1217⟩, 901, 904 ⟨93T2101⟩, 973 ⟨77IJC(B)297⟩; **5**, 611 ⟨90TL1327⟩, 616 ⟨92TL7957⟩, 760 ⟨79JCS(P1)1762⟩, 1035 ⟨79JCS(P1)2001⟩; **6**, 627 ⟨91S220⟩, 647 ⟨82PIA441⟩
Raj, T. T., **5**, 690 ⟨90JIC260⟩
Rajca, G., **5**, 1082 ⟨86ZAAC(538)36, 86ZN(B)1223⟩
Rajcoomar, V., **1**, 215 ⟨81TL1751⟩
Rajendra, G., **1**, 832 ⟨89TL2469⟩; **5**, 285 ⟨87TL6257⟩, 285, 322 ⟨87JOC4471⟩, 285, 327, 328 ⟨85TL5385⟩
Rajeswari, S., **4**, 94 ⟨84H(22)585⟩; **5**, 264 ⟨87TL5099⟩
Rajewsky, M. F., **2**, 412 ⟨90AG(E)782⟩
Rajopadhye, M., **3**, 309 ⟨89JOC3264⟩
Raju, B., **3**, 91 ⟨87IJC(B)914⟩
Raju, N., **6**, 71 ⟨83TL5571⟩
Rajviroongit, S., **1**, 787 ⟨94CC889⟩; **2**, 158 ⟨91T8297⟩
Rakhimova, M. R., **2**, 442 ⟨67ZOB2522⟩
Rakhimov, D. N., **5**, 977 ⟨89ZOR885⟩
Rakhit, S., **4**, 323 ⟨79CJC1153⟩
Rakhlin, V. I., **2**, 1091 ⟨86JGU849⟩; **4**, 617 ⟨87IZV1424⟩
Rakhmankulov, D. L., **2**, 265 ⟨87ZOR1981⟩; **4**, 775 ⟨90CA230883w⟩, 1078 ⟨86MI 424-01⟩, 1080 ⟨83ZOB416⟩; **6**, 127 ⟨88ZOB1066⟩; 478 ⟨81ZOR1539, 85ZOR2206⟩, 478 ⟨86JOU270, 87JGU365⟩; **6**, 70 ⟨81MI 603-01⟩, 89 ⟨88URP1421744⟩
Rakhnovich, G. B., **4**, 461 ⟨80IZV1125⟩
Rakiewicz, D. M., **1**, 248 ⟨85JA1488⟩, 347, 1132 ⟨85JA1448, 85T3943⟩
Rakitin, O. A., **6**, 150 ⟨90IZV1620⟩, 150 ⟨90IZV1623⟩
Rakoczi, J., **2**, 788 ⟨69MIP156590⟩
Rákosi, M., **3**, 368 ⟨77LA1529⟩
Rakotomanana, F., **5**, 206 ⟨85T4047⟩
Rakotomanomana, N., **4**, 176 ⟨88CAR(181)246⟩, 225 ⟨88TL4293⟩
Rakotonirina, R., **4**, 255 ⟨85BSF881⟩; **5**, 556 ⟨83TL4203, 85BSF881⟩, 556 ⟨86TL1505, 89T2041⟩, 616 ⟨85T1289⟩
Rakow, D., **5**, 567, 619, 625 ⟨74JHC397⟩
Rakowski, M. C., **1**, 93 ⟨75JA237, 76IC2379⟩
Rakowski DuBois, M., **1**, 93 ⟨78JA2405⟩
Raksha, M. A., **2**, 19 ⟨59JGU2125⟩, 288 ⟨60ZOB4069⟩; **4**, 554 ⟨69ZOB845⟩; **6**, 54 ⟨59JGU2125⟩
Rakshit, D., **4**, 327 ⟨84JCS(P1)2327⟩
Rakshys, J. W., **2**, 836, 839 ⟨68JA5236⟩; **5**, 1129 ⟨68JA5236⟩
Raksis, J. W., **5**, 998 ⟨80TL2223⟩
Ralbovsky, J. L., **2**, 306 ⟨90SC573⟩
Rall, K., **3**, 375 ⟨89AG(E)221, 90CB595⟩; **5**, 502 ⟨90AG(E)60⟩; **6**, 258 ⟨89AG(E)221, 90AG(E)60⟩
Rall, K. B., **4**, 841 ⟨80ZOR1748⟩

Ralli, P., **4**, 46 ⟨89JOC5520⟩, 843 ⟨86OS(64)189⟩, 853, 855 ⟨82JOC3027⟩
Ralston, R. W., **5**, 28 ⟨B-48MI 502-02⟩
Ram, S., **1**, 18, 19 ⟨88TL3741⟩; **6**, 566 ⟨85JHC1269⟩
Ram, V. J., **4**, 983 ⟨92JPR190⟩; **6**, 573 ⟨70MI 618-01⟩
Ramachandran, J., **5**, 81, 546, 549, 550 ⟨83S605⟩, 547 ⟨81OPP81⟩, 560 ⟨81OPP71⟩
Ramachandran, K., **3**, 65, 232 ⟨87HCA1429⟩
Ramachandran, P. K., **3**, 304 ⟨63JOC398⟩
Ramachandran, P. V., **2**, 58 ⟨88JA1549, 89JOC1577, 89JOC4504, 94TA1075⟩, 514, 515, 516, 517, 519, 520, 523, 526, 531 ⟨87MI 210-01⟩, 517 ⟨85JOC5583⟩, 526 ⟨88JA1539, 88JOC2916⟩, 526 ⟨92JOC2379⟩, 526, 531 ⟨85JOC5446⟩, 526, 531 ⟨86JOC3394, 94TL2141⟩
Ramachandran, S., **4**, 743 ⟨62JA995⟩
Ramadan, N., **6**, 349, 585 ⟨85TL1091⟩
Ramadas, K., **6**, 580 ⟨93TL6447⟩
Ramadas, S. R., **5**, 81, 546, 549, 550 ⟨83S605, 81OPP81⟩, 550 ⟨76IJC794⟩, 560 ⟨77JPR169, 81OPP71⟩, 561 ⟨72CC345⟩
Ramage, D. L., **5**, 988 ⟨92OM1242⟩
Ramage, R., **2**, 1004 ⟨75JCS(P1)2513⟩; **4**, 213 ⟨84JCS(P1)1531⟩; **5**, 79 ⟨84JCS(P1)1531⟩, 264 ⟨85JCS(P1)461⟩, 280 ⟨84TL4825⟩; **6**, 425 ⟨92TL385⟩
Ramaiah, M., **1**, 320, 336, 919, 929, 935, 1126 ⟨87T3541⟩, 421 ⟨84S529⟩; **3**, 206, 216 ⟨84S529⟩; **5**, 131 ⟨85JOC4991⟩
Ramaiah, P., **4**, 428, 429 ⟨93JOC763⟩
Ramaiah, T. S., **5**, 355 ⟨93JHC61⟩
Ramakrishnan, K., **6**, 549 ⟨83JA7187⟩
Ramakrishnan, V. T., **2**, 385, 386 ⟨85S677⟩; **5**, 963 ⟨87JCS(P1)2163⟩, 1036 ⟨72IJS139⟩, 1036, 1037 ⟨72JOC1360⟩, 1039 ⟨89OPP228⟩; **6**, 620 ⟨86JOC1719⟩
Ramalingam, K., **2**, 517 ⟨88T5597⟩; **3**, 707 ⟨89OPP511⟩
Ramalingram, K., **3**, 704 ⟨85SC267⟩
Ramamoorthy, P. S., **1**, 327, 328 ⟨93T4841⟩
Ramamurthy, V., **2**, 202 ⟨84JOC615⟩; **3**, 345, 368 ⟨83JOC214⟩, 346 ⟨80CC243⟩, 346 ⟨83JOC4482⟩, 352 ⟨88JOC332, 82JOC127⟩
Ramana, M. M. V., **2**, 791 ⟨88SC2161⟩
Raman, K., **3**, 129 ⟨83JA5935⟩
Raman, K. S. V., **4**, 11 ⟨44JIC112⟩
Ramana Reddy, K. N., **4**, 272, 275 ⟨89SUL79⟩
Ramana Reddy, M. V., **4**, 272, 275 ⟨89SUL79⟩
Ramanathan, H., **1**, 338, 342, 1127 ⟨83TL1871⟩
Ramaprasad, K. R., **5**, 1016 ⟨68JA6247⟩
Rama Rao, A. V., **2**, 22 ⟨93JCS(P1)3171⟩; **3**, 40 ⟨89TL6769⟩, 709 ⟨83SC331⟩; **5**, 276, 298 ⟨74T2899⟩
Ramarijaona, O., **5**, 1092 ⟨89TL177⟩; **6**, 680 ⟨89TL177⟩
Ramasubbu, A., **6**, 7 ⟨87JCS(P1)1515⟩
Ramaswamy, D., **1**, 74 ⟨84JCS(D)1477⟩
Ramaswamy, P. S., **3**, 133 ⟨77CI(L)454⟩
Ramaswamy, S., **1**, 178, 265 ⟨80CJC716⟩, 1023 ⟨92JOC6344⟩; **3**, 636 ⟨83CJC2006⟩
Rambabu, M., **3**, 618 ⟨87IJC(B)407, 89SC1431⟩; **5**, 189 ⟨87IJC(B)407⟩
Rambacher, P., **5**, 183 ⟨68AG(E)465⟩
Rambaud, M., **2**, 998 ⟨84TL835⟩; **3**, 28 ⟨80S644⟩, 152 ⟨84TL835⟩, 152, 154 ⟨81S68⟩, 210 ⟨88S564⟩; **4**, 14, 22 ⟨80S644⟩, 138, 146, 667 ⟨85BSF825⟩, 139 ⟨86BSF470⟩, 146 ⟨84TL835⟩; **5**, 103 ⟨87OS(66)220⟩
Ramberg, L., **1**, 617 ⟨40MI 114-01⟩, 704, 706 ⟨40AK(13A)1⟩
Ramcharitar, S. H., **1**, 340 ⟨92T3413⟩
Ramdani, M., **1**, 834 ⟨92T10315⟩
Ramer, R. M., **1**, 725 ⟨69JCS(C)2728⟩
Ramer, W. H., **4**, 323 ⟨87JCS(P1)775⟩
Ramert, R., **5**, 354 ⟨75LA1264⟩
Ramesh, D., **3**, 7, 120 ⟨79JOC138⟩
Ramesh, M., **1**, 247 ⟨88TL3533⟩, 944 ⟨90JA891⟩; **4**, 240 ⟨90JA891⟩
Ramesh, S., **2**, 325 ⟨88JOC6022⟩; **3**, 231, 303 ⟨85JA5219⟩
Ramesh, V., **3**, 345, 368 ⟨83JOC214⟩
Ramesha, A. R., **2**, 228 ⟨92SC3277⟩; **5**, 252 ⟨94JCS(P1)767⟩
Ramey, K. C., **3**, 560 ⟨72JOC2364⟩; **4**, 698 ⟨65JPC3418⟩
Ramey, P. S., **1**, 1065 ⟨54JA4558⟩
Ramezanian, M., **4**, 993 ⟨90JOC1768⟩
Ramirez, A., **4**, 439 ⟨73T4195⟩
Ramirez, F., **2**, 476 ⟨68JA6726⟩, 831 ⟨71JA5236⟩, 869 ⟨74JA7269⟩; **3**, 451 ⟨52JA4331, 53JA6026⟩, 493 ⟨57JOC41⟩, 495 ⟨62JA1745⟩, 498 ⟨58JOC2036, 65TL201, 66T567⟩; **4**, 27 ⟨68JOC25⟩, 336 ⟨74PS(4)109⟩, 493 ⟨75JA38⟩, 553 ⟨61JA3539⟩, 745 ⟨62JA1745⟩; **5**, 911, 1093 ⟨61JA3539⟩; **6**, 416 ⟨91SA(A)1619⟩, 694 ⟨62JA1745⟩
Ramirez, J. A., **1**, 507 ⟨88IC4429⟩
Ramiz, M. M. M., **4**, 883 ⟨89S775⟩
Ramkumar, D., **2**, 150 ⟨93S1057⟩
Ramler, E. O., **4**, 18 ⟨43JA389⟩
Rammler, D. H., **5**, 185 ⟨63JA1997⟩
Ramnath, N., **3**, 345, 368 ⟨83JOC214⟩
Ramon, D. J., **2**, 553 ⟨92T3585⟩; **3**, 31, 161 ⟨92JOC426, 94T3349, 93T4923⟩
Ramon, F., **4**, 619 ⟨72JOC2662⟩; **6**, 267, 379 ⟨72JOC2662⟩
Ramos, A., **3**, 418 ⟨87OPP181⟩; **5**, 215 ⟨91JCS(P1)223⟩
Ramos, M. N., **5**, 506, 507 ⟨89JCS(F2)1945⟩
Ramoshkene, É. B., **5**, 361 ⟨74ZOR2301⟩
Ramos Tombo, G. M., **4**, 566 ⟨91HCA983⟩
Ramp, F. L., **1**, 78, 421 ⟨52JA4867⟩
Rampal, J. B., **2**, 160 ⟨82JOC552⟩
Rampersaud, A., **4**, 662, 962 ⟨86OM1906⟩
Ramphal, J. Y., **1**, 1134 ⟨89S898⟩
Ramsay, M. V. J., **4**, 229 ⟨92TL4625⟩
Rämsby, S., **5**, 277 ⟨87JOC2033⟩
Ramsden, C. A., **2**, 11 ⟨77S634⟩; **3**, 371, 372 ⟨84CHEC(6)1027⟩; **5**, 1002 ⟨76TL2691⟩; **5**, 641 ⟨68CC499, 74JCS(P1)627⟩, 690 ⟨92JCS(P1)2789⟩; **6**, 611 ⟨79JCS(P1)724⟩
Ramsden, H. E., **2**, 433 ⟨59USP2912465⟩
Ramsden, J. A., **5**, 1097 ⟨92JA5890, 93JA3824⟩
Ramser, H., **2**, 110 ⟨30CB1136⟩
Ramsey, B., **1**, 915 ⟨67JA5259⟩
Ramsey, B. G., **2**, 1010 ⟨81JOC179⟩
Ramuz, H., **4**, 883 ⟨83HCA809⟩
Ramzan Ali, M., **5**, 631 ⟨85IJC(B)977⟩
Rana, K. N., **3**, 612 ⟨40JIC481⟩
Ranade, A. C., **1**, 664 ⟨78CI(L)234⟩; **4**, 698 ⟨70JOM(25)C6⟩; **5**, 110 ⟨76S123⟩
Ranaivonjatovo, H., **2**, 866 ⟨83TL3625⟩; **3**, 501 ⟨94CCR427⟩; **4**, 485 ⟨91JOM(415)327⟩; **6**, 167 ⟨87PS(31)81⟩, 374, 375 ⟨83TL2769, 83TL3625⟩, 709 ⟨94CCR427⟩, 724 ⟨92OM2748⟩
Ranasinghe, M. G., **2**, 217 ⟨88SC227⟩
Rance, M. J., **2**, 777 ⟨75JCS(P1)26⟩
Ranchfuss, T. B., **2**, 824 ⟨77IC2966⟩
Rancourt, J., **1**, 327 ⟨91JA9701, 92JA4912⟩, 327 ⟨93TL5885⟩, 340, 1127 ⟨93TL2091⟩, 831 ⟨90TL5285⟩
Rand, ,C. L., **2**, 926, 931, 976 ⟨83JA6344⟩
Rand, C. L., **1**, 142, 1053 ⟨81JOC5041⟩, 464 ⟨88JA5383⟩, 542 ⟨82TL27⟩, 542, 543 ⟨81JOC4093⟩, 544 ⟨83OM563⟩, 1094 ⟨83JA6344⟩; **2**, 973 ⟨83OM563⟩, 975 ⟨81JOC4093⟩; **4**, 392 ⟨81JOC5041⟩, 809 ⟨88JA5383⟩
Rand, L., **5**, 320 ⟨73OSC(5)663⟩, 998 ⟨60JOC1837, 65JOC48⟩
Randaccio, L., **4**, 585 ⟨74JOM(74)C23⟩, 696 ⟨79JOM(174)121⟩
Randad, R. S., **2**, 75 ⟨89JOC1570, 90T4463⟩, 75, 521 ⟨90JA2389⟩, 75, 521 ⟨90T4457, 90TL455⟩
Randall, C. J., **4**, 604, 647 ⟨84JA6108⟩, 604, 647 ⟨86JA5949⟩
Randall, E. W., **4**, 692 ⟨74JOM(64)271⟩
Randall, J. L., **4**, 45 ⟨84JA4189⟩
Randall, L., **5**, 944 ⟨94TL6717⟩, 955 ⟨95JOM(494)C12⟩
Randall, L. O., **6**, 508 ⟨47JA3150⟩
Randau, G., **6**, 580 ⟨69LA(722)52, 69LA(722)80⟩

Randhawa, H. S., **5**, 506 ⟨77JST25⟩, 506, 507 ⟨74JST123⟩
Randles, D., **1**, 37 ⟨82JOC1047⟩
Rand-Meir, Z., **1**, 108 ⟨61JCS3452⟩; **4**, 33 ⟨61JCS2813⟩
Rando, R. R., **1**, 1135 ⟨75ACR281⟩; **3**, 476 ⟨72JA1629⟩, 476, 532 ⟨70JA6706⟩
Randolph, Jr., C. L., **2**, 418, 419 ⟨49JA3451⟩
Randriamahefa, S., **1**, 186 ⟨85S493⟩; **5**, 38 ⟨85S493⟩
Randrianoelina, B., **1**, 978 ⟨81TL1327, 84TL651⟩, 978 ⟨86T2501, 90JOM(396)289, B-85MI 122-01⟩; **4**, 192 ⟨81JAP(K)56002956⟩
Randria-Raharimanana, C., **3**, 310 ⟨90H(31)427⟩
Rane, A., **4**, 630, 631 ⟨93JA6065⟩
Rane, R. A., **3**, 4 ⟨90S127⟩
Raner, K. D., **1**, 925 ⟨93ACR219⟩; **2**, 21 ⟨90ACR219⟩
Rangaishenvi, M. V., **1**, 443 ⟨89JA384, 91JOC1543⟩, 838 ⟨90TL7115⟩; **2**, 529 ⟨91JOC3286⟩, 530 ⟨86JOC3150⟩, 535 ⟨87JOC728⟩, 740 ⟨89JA384⟩
Rangaishnvi, M. V., **2**, 526 ⟨92JOC2379⟩
Ranganathan, D., **2**, 784, 785 ⟨80JOC1185⟩; **4**, 311 ⟨91JA1042⟩; **5**, 355 ⟨85TL5739⟩
Ranganathan, S., **2**, 129 ⟨91CC934⟩, 784, 785 ⟨80JOC1185⟩; **3**, 708 ⟨88TL1435⟩
Ranganayakulu, K., **1**, 918 ⟨81JA588⟩
Rani, K. S., **2**, 303 ⟨90TL1199⟩
Rani, S. S., **5**, 1106 ⟨90SC2313⟩
Ranise, A., **5**, 597 ⟨83FES101, 85FES120⟩, 597 ⟨91FES1033⟩
Ranjan, A., **2**, 881 ⟨85POL251⟩; **5**, 318, 372 ⟨83IJC(A)120, 84SRI269⟩, 372 ⟨85POL251⟩
Rank, W., **3**, 190 ⟨91TL5353⟩; **4**, 432, 433 ⟨87CJC166⟩
Rankel, L. A., **4**, 673 ⟨78JA802⟩, 674 ⟨78JA5764⟩; **5**, 940 ⟨77IC677⟩
Rankin, D. W. H., **2**, 417 ⟨89JCS(D)779⟩, 622 ⟨89CB2213⟩; **5**, 369, 713 ⟨80JCS(D)1768⟩, 535 ⟨84JCS(D)2301⟩, 538, 541 ⟨76JCS(D)1661⟩, 539 ⟨82JCS(D)1131⟩, 913 ⟨80CB1637⟩, 1081, 1085 ⟨72JCS(D)869⟩; **6**, 165, 167 ⟨82OM1720⟩, 172 ⟨91MI 606-01⟩, 173, 378, 379 ⟨90AG(E)201⟩
Rankin, J. G., **3**, 686 ⟨82TL5453⟩
Rano, B. C., **3**, 61 ⟨87SC155⟩
Rano, T. A., **6**, 488 ⟨92JA8008⟩
Ransley, D. L., **1**, 463 ⟨66JOC3595⟩; **4**, 18 ⟨69JOC2618⟩
Ransom, J. H., **5**, 698 ⟨01CB1615⟩
Ranu, B. C., **1**, 14 ⟨92JCS(P1)1881⟩, 804 ⟨83JOC4453, 84TL2447⟩; **2**, 43 ⟨94T6579⟩, 395 ⟨92T1327⟩, 398 ⟨92T5317⟩; **3**, 18, 133 ⟨88JOC878⟩, 161 ⟨90TL7663⟩, 168 ⟨92JOC7349⟩, 426 ⟨91TL3579⟩; **4**, 258 ⟨92JOC7349⟩
Ranus, Jr., W. J., **1**, 81 ⟨67JOC893⟩; **6**, 642 ⟨68JMC1129⟩
Rao, A. D., **6**, 561 ⟨85IJC(B)580, 85MI 617-04⟩
Rao, A. S., **1**, 842 ⟨83T2323⟩; **2**, 90 ⟨91COS(7)357⟩, 91 ⟨83T2323⟩, 901 ⟨91COS(7)38⟩
Rao, A. T., **4**, 260 ⟨90JCS(P1)3217⟩
Rao, A. V. R., **1**, 1134 ⟨93TL707⟩; **5**, 226 ⟨81CI(L)164⟩
Rao, A. V. S., **5**, 67 ⟨79MI 502-07⟩
Rao, B., **1**, 1124 ⟨71CJC3135, 71TL927⟩
Rao, B. E. V. S., **5**, 215 ⟨87IJC(B)620⟩
Rao, C. B., **2**, 384 ⟨92JOC1585⟩, 784, 785 ⟨80JOC1185⟩, 907 ⟨90JOC46⟩
Rao, C. G., **1**, 905 ⟨78JOC3667, 79JOC133⟩, 905 ⟨79JOC3536, 80JOC2113⟩; **2**, 659 ⟨74CC584⟩, 747 ⟨76CC103⟩; **3**, 13, 82 ⟨79S704⟩; **5**, 226 ⟨82SC177⟩
Rao, Ch. S., **4**, 253 ⟨90T2195⟩
Rao, C. J., **1**, 468, 471 ⟨93T29⟩
Rao, C. K., **6**, 448 ⟨85EUP136147⟩
Rao, C. N. R., **5**, 506, 507 ⟨74JST123⟩, 1046 ⟨63JOC194⟩; **6**, 537 ⟨63CJC1643⟩
Rao, C. P., **1**, 827 ⟨91SC1455⟩; **2**, 48 ⟨89JOC2015⟩
Rao, C. S., **1**, 1066 ⟨89TL5455⟩; **3**, 363 ⟨76IJC(B)999⟩, 618 ⟨87IJC(B)407, 89SC1431⟩; **4**, 253 ⟨92S1075⟩
Rao, D. R., **3**, 344 ⟨71JOC3885⟩
Rao, D. V., **1**, 562 ⟨61JA1705⟩; **2**, 655 ⟨81TL2337⟩
Rao, G. C., **5**, 710, 714 ⟨92CCC540⟩
Rao, G. S., **6**, 218, 219, 220, 226 ⟨58JCS4245⟩
Rao, G. S. K., **3**, 87, 96 ⟨79S144⟩
Rao, G. S. R. S., **1**, 342, 1132 ⟨89TL225⟩
Rao, H. S. P., **3**, 151 ⟨85SC991⟩
Rao, J. M., **3**, 135 ⟨81CC1149⟩, 639 ⟨82IJC(B)1046⟩; **4**, 373 ⟨81CC1149⟩
Rao, K. R., **2**, 216 ⟨91BMC507⟩; **6**, 450 ⟨88S994⟩
Rao, K. S. R. K. M., **5**, 215 ⟨87IJC(B)620⟩
Rao, K. S. S. P., **1**, 943 ⟨93CC1636⟩
Rao, K. V. B., **5**, 690 ⟨86JHC1757⟩
Rao, M. N., **4**, 514 ⟨85TL5001⟩; **6**, 128 ⟨85TL5001⟩
Rao, M. S. C., **1**, 260 ⟨89JCS(P1)2297⟩
Rao, M. V., **2**, 318 ⟨92TA1361⟩; **3**, 18, 134 ⟨91JCS(P1)2056⟩
Rao, N. S., **2**, 440 ⟨82JOC905⟩
Rao, P. B., **5**, 1026 ⟨70IJC759⟩
Rao, P. N., **3**, 219 ⟨65JCS1356⟩, 246 ⟨69JOC2767⟩
Rao, S. A., **1**, 123 ⟨91JA5735⟩, 486 ⟨90JOC5446⟩, 541 ⟨88TL4313⟩; **2**, 69 ⟨90TL7575⟩, 74 ⟨90JA6146⟩, 925 ⟨91JA5735⟩; **4**, 396 ⟨92T2025⟩, 709 ⟨90JOC5446⟩, 772, 963 ⟨91JA5735⟩
Rao, S. J., **3**, 305 ⟨84TL5439⟩
Rao, S. N., **3**, 141 ⟨82CC828⟩; **4**, 100 ⟨90JOC3244⟩
Rao, T., **2**, 36 ⟨85TL2717⟩
Rao, V. B., **1**, 19 ⟨85JA5732⟩; **3**, 121, 294 ⟨83S572⟩; **5**, 277 ⟨85JA5732⟩
Rao, V. J., **3**, 368 ⟨82JOC127⟩
Rao, V. N. M., **6**, 8 ⟨93M 601-01⟩
Rao, V. P., **3**, 352 ⟨88JOC332⟩
Rao, V. S., **2**, 224, 225 ⟨75JOC524⟩, 760 ⟨70TH134⟩; **3**, 639 ⟨82IJC(B)1046⟩; **5**, 219, 574 ⟨85CJC3089⟩
Rao, V. V. R., **1**, 1106 ⟨79JOM(179)C7⟩
Rao, Y. K., **1**, 176 ⟨88TL107⟩; **4**, 277 ⟨84S757⟩
Rapala, R. T., **5**, 115 ⟨55OSC(3)820⟩
Raphael, R. A., **1**, 422 ⟨53JCS2247, 76JCS(P1)410⟩, 504 ⟨61T16⟩, 998 ⟨B-55MI 121-01⟩, 1000 ⟨50JCS115⟩, 1023 ⟨76TL1539⟩, 1064 ⟨60JCS3614, 64JCS1151⟩; **2**, 776 ⟨70JOC1670⟩; **3**, 38 ⟨82TL2407⟩, 273 ⟨84JCS(P1)535⟩; **4**, 427 ⟨55JCS4280⟩
Raphy, G., **6**, 425 ⟨92TL385⟩
Rapkin, A. I., **2**, 170 ⟨88IZV2367⟩; **4**, 104 ⟨86IZV810⟩; **5**, 699, 710 ⟨81IZV2370, 81JFC(18)553⟩, 813 ⟨80IZV2623⟩; **6**, 42 ⟨85IZV659⟩, 56, 282 ⟨76IZV489⟩, 251 ⟨74IZV335⟩
Rapoport, H., **1**, 34 ⟨51JA2718⟩, 259 ⟨71JA2738⟩, 579, 584 ⟨94JOC394⟩, 861 ⟨81JOC2059⟩, 862 ⟨60JA934⟩, 1047 ⟨85JOC3972⟩; **2**, 691 ⟨84JOC1348⟩, 1011 ⟨86JOC5320⟩; **3**, 17, 132 ⟨82JA4446⟩, 40, 184 ⟨87JA236⟩, 58, 494 ⟨71JA5311⟩, 179, 184 ⟨88JA7447⟩, 180, 210 ⟨83JOC2260⟩, 180, 273 ⟨85JOC3972⟩, 180, 284 ⟨84JA1095⟩, 278, 308 ⟨72JOC3618, 81JA6157⟩, 419 ⟨81JOC4914, 86S735⟩, 494 ⟨70JA3429⟩; **4**, 175 ⟨82JOC946⟩; **5**, 54 ⟨62JOC629⟩, 742, 758 ⟨81JOC2455⟩; **6**, 504 ⟨90JOC3699⟩, 644, 647 ⟨77JOC3608⟩, 646 ⟨68JOC552⟩, 651 ⟨73JOC1591, 79JOC4536⟩
Rapoport, S., **5**, 560 ⟨88ZC269⟩
Rappa, A., **1**, 101 ⟨84TL2089⟩
Rapp, E., **3**, 323 ⟨78JOC1071⟩
Rapp, J., **3**, 370 ⟨87JA902⟩
Rapp, K. E., **6**, 46 ⟨50JA3642⟩
Rapp, L. R., **2**, 109 ⟨53JA199⟩
Rapp, R., **4**, 174 ⟨66JOC3032⟩
Rappé, A. K., **1**, 150 ⟨92JOC3819⟩
Rappe, C., **1**, 871 ⟨67JOC3700, 67JOC4127⟩, 871 ⟨68T6287, 68TL2317⟩, 871 ⟨68TL2921⟩; **2**, 28 ⟨88OSC(6)711⟩, 758 ⟨76ACS(B)1000⟩; **4**, 19 ⟨66ACS253⟩, 20 ⟨62ACS2467⟩
Rappoli, B. J., **5**, 461 ⟨87JA5145⟩
Rappoport, Z., **1**, 1110 ⟨71JCS(B)871, 79JA5095, 80JOC4309⟩, 1110, 1112 ⟨73JCS(P2)301, 75JA821⟩, 1110, 1112, 1114 ⟨72JOC1174⟩, 1111 ⟨70TL3233, 79JA1343, 79JA3967⟩, 1112 ⟨71TL4039, 74JA586,

74JA4518, 75JA836⟩, 1112, 1114 ⟨69JA5246⟩, 1122 ⟨81JA2770⟩, 1133 ⟨B-83MI 122-01⟩; **1**, 144 ⟨B-88MI 104-02⟩, 704 ⟨B-88MI 115-01, B-88MI 115-02⟩, 704 ⟨B-88MI 115-03⟩, 707 ⟨B-88MI 115-04⟩, 801 ⟨B-87MI 118-02⟩, 839 ⟨B-88MI 118-01⟩, 869 ⟨B-90MI 119-01⟩, 888 ⟨B-64MI 119-01⟩, 889 ⟨62JCS396, 62JCS1741⟩, 889 ⟨70JCS(B)469⟩, 1109, 1110 ⟨69APO(7)1⟩, 1109, ,1110, 1117 ⟨B-79MI 122-01⟩, 1109 ⟨B-64MI 122-01, 70JA4985⟩, 1110 ⟨71JCS(B)1461⟩; **2**, 142 ⟨B-88MI 203-01⟩, 144 ⟨B-88MI 203-02⟩, 154 ⟨B-88MI 203-03⟩, 159, 196 ⟨B-88MI 203-04⟩, 547 ⟨B-89MI 210-01⟩, 636, 643 ⟨88ACR442⟩, 636 ⟨B-90MI 213-01⟩, 645 ⟨84JA477, 85JA3669⟩, 645 ⟨87JA2112⟩, 646 ⟨85JA1007⟩, 930 ⟨90TL555⟩; **3**, 170 ⟨B-88MI 304-03, B-88MI 304-04⟩, 501 ⟨B-89MI 314-01⟩, 564 ⟨88JOC679⟩, 641, 642, 647, 656, 657, 658, 661, 662, 669 ⟨B-83MI 319-01⟩, 641, 661, 662, 663, 664, 665, 666, 673 ⟨B-83MI 319-02⟩; **4**, 390 ⟨B-88MI 408-01⟩, 759, 764 ⟨87JA6687⟩, 763, 771 ⟨93JOC3421⟩, 894 ⟨82JOC1397⟩, 1076 ⟨69APO(7)1⟩; **5**, 406 ⟨91JOC4241⟩, 652 ⟨B-70MI530-01⟩, 833 ⟨B-70MI 521-02⟩, 1107 ⟨89JOC5967⟩; **6**, 308, 311, 312, 313, 317 ⟨B-86MI 610-02⟩, 308, 311, 312, 313, 317 ⟨B-87MI 610-01⟩, 709 ⟨B-89MI 622-02⟩
Rapson, W. S., **6**, 25 ⟨44JCS74⟩
Raptis, R. G., **4**, 586 ⟨87OM1992, 88IC836⟩
Rasala, D., **1**, 260 ⟨89JCS(P1)225⟩; **2**, 310 ⟨89JCS(P1)225⟩; **4**, 309 ⟨89JCS(P1)225⟩
Rasberger, M., **2**, 771 ⟨69M64⟩
Rascher, W., **5**, 49 ⟨76CB41⟩
Rasetti, V., **3**, 268 ⟨85HCA1235⟩
Rash, F. H., **3**, 546, 547 ⟨68JOC2738⟩; **5**, 511 ⟨68JOC2738⟩
Rasheed, K., **6**, 313 ⟨79JOC267⟩, 553 ⟨81JHC1581⟩
Rashid, M. A., **1**, 497 ⟨92SL721⟩, 1015 ⟨93TL2139⟩
Rashidi, M., **4**, 692 ⟨92OM2224⟩
Raskina, A. D., **6**, 243 ⟨89ZOB2628⟩
Rasmussen, C. R., **6**, 635 ⟨88S460⟩
Rasmussen, J. B., **3**, 360 ⟨81T197⟩
Rasmussen, J. K., **2**, 69 ⟨85TL981⟩; **3**, 47, 192 ⟨74JOC2558⟩, 157 ⟨83TL4075⟩, 615 ⟨78S219⟩, 633 ⟨73S682⟩, 651 ⟨86SC617⟩; **5**, 344 ⟨76CRV389⟩, 348 ⟨73TL2783⟩, 1134 ⟨79S523⟩
Rasmussen, J. R., **1**, 16 ⟨81JOC4843⟩
Rasmussen, M., **1**, 1001 ⟨68AJC2989⟩
Rasmusson, G. H., **1**, 725 ⟨61JOC4278⟩, 882 ⟨65JOC2513⟩
Raso, A. G., **1**, 482, 1027 ⟨92JOC678⟩
Rasoul, H. A. A., **4**, 827 ⟨82JOC2080, 82TL603⟩
Rasshofer, W., **4**, 294, 300, 307, 308 ⟨91HOU(E14a/2)1⟩, 309 ⟨91HOU(E14a/2)170⟩, 704 ⟨91HOU(E14a/2)248⟩; **5**, 1061 ⟨84HOU(4)883⟩
Rassu, G., **1**, 432 ⟨89S28⟩
Rastädter, K., **2**, 403 ⟨56CB685⟩
Rasteikiene, L., **2**, 712 ⟨77RCR548⟩
Rastelli, A., **2**, 354 ⟨90JOC3427⟩
Rastetter, W. H., **2**, 241 ⟨82JOC2785⟩; **5**, 338 ⟨82JOC2785⟩
Rastogi, M. K., **5**, 1132 ⟨87SRI525⟩
Rastogi, R., **2**, 891 ⟨86POL1581⟩; **5**, 372 ⟨84SRI269, 87IJC(A)146⟩, 1145 ⟨86POL677⟩
Rastogi, R. C., **5**, 370 ⟨91CI(L)353⟩; **6**, 627 ⟨91JCR(S)160⟩
Rastogi, S. N., **4**, 982 ⟨66JIC650⟩
Raston, C. L., **1**, 114 ⟨B-87MI 104-03⟩, 452, 453 ⟨B-87MI 110-03⟩; **2**, 557 ⟨B-87MI 211-03⟩, 559 ⟨88CC289⟩, 559, 560 ⟨84CC1702⟩, 565 ⟨82COMC-I(3)559, 82COMC-I(3)635⟩, 690 ⟨76JCS(D)2307⟩, 1003, 1005 ⟨89JCS(D)105, 82CC14⟩, 1005 ⟨84JCS(D)1801, 84JCS(D)321⟩, 1005 ⟨85JCS(D)337⟩; **3**, 545 ⟨84CC220⟩; **4**, 602, 650 ⟨82CC1323⟩, 603, 647 ⟨84JCS(D)311⟩, 603, 647, 657 ⟨82CC1023⟩, 603, 649 ⟨91OM3167⟩, 604 ⟨83CC1419⟩, 604, 646, 647 ⟨82CC14⟩, 604, 646, 647 ⟨84JCS(D)1801⟩, 604, 646, 647 ⟨84JCS(D)321⟩, 604, 647 ⟨88JCS(D)2403⟩, 604, 647, 648 ⟨90JCS(D)1161⟩, 626, 662 ⟨87JCS(D)1647⟩, 646 ⟨89JCS(D)105⟩, 646, 647 ⟨84CC1708, 85JCS(D)337⟩, 647, 657 ⟨80CC1284⟩, 647, 660 ⟨92JCS(D)775⟩, 654 ⟨87OM2110, 88CC652, 88JOC3134⟩, 656 ⟨82JOM(233)C28, 87JCS(D)3085⟩, 660 ⟨84JCS(D)331⟩, 856 ⟨80AJC1073⟩; **6**, 168 ⟨91CC1560⟩, 174, 190 ⟨82CC1323⟩, 190 ⟨84CC1708, 84JCS(D)1801⟩, 190, 195 ⟨83CC1419⟩, 190, 196 ⟨86CC672⟩, 195 ⟨84CC612⟩, 196 ⟨87AG(E)681⟩, 196 ⟨88CC336⟩, 196 ⟨93JCS(D)2653⟩, 393 ⟨88JOC3134⟩
Raston, R. L., **4**, 647, 656 ⟨86JCS(D)603⟩
Rastrup-Andersen, N., **3**, 704 ⟨80T515⟩
Rasuwajev, G., **2**, 408 ⟨29CB598⟩
Ratajczak, E., **1**, 1127 ⟨86JPC402⟩
Ratajczyk, J. D., **1**, 1018 ⟨87JMC254⟩
Ratananakul, P., **1**, 1015 ⟨88TL5565⟩; **3**, 166 ⟨82TL2509⟩, 167 ⟨83TL117⟩; **4**, 69 ⟨83TL117⟩
Ratcliff, M. A., **4**, 300 ⟨71JOC3112⟩
Ratcliffe, A. H., **6**, 416 ⟨75SC47⟩
Ratcliffe, C. T., **2**, 98 ⟨70CC216⟩, 166 ⟨68JA5403⟩; **6**, 231 ⟨71CC784⟩
Ratcliffe, N. M., **4**, 165 ⟨83JCS(P1)1131⟩
Ratcliffe, R. W., **4**, 463 ⟨73TL4645⟩, 847 ⟨82TL3535⟩; **5**, 374 ⟨80TL31⟩; **6**, 128 ⟨80JA6161⟩
Räth, K., **6**, 444 ⟨20CB601⟩
Rath, N. P., **4**, 713 ⟨89OM2077⟩
Rather, B., **5**, 247 ⟨92SUL199⟩
Rather, E. M., **4**, 669 ⟨78JOC2170⟩
Rathke, A., **5**, 1026 ⟨1872CB799⟩
Rathke, B., **2**, 166 ⟨1870CB858⟩, 239 ⟨1873LA(167)212⟩; **6**, 641 ⟨1884CB297⟩
Rathke, J., **2**, 524 ⟨72IC1150⟩; **4**, 142, 143 ⟨72IC1150⟩, 342 ⟨73PS(3)1⟩
Rathke, J. W., **2**, 348, 349 ⟨90JOC3233⟩; **4**, 342 ⟨70JOC2310⟩
Rathke, M. N., **1**, 283 ⟨78JOC1947⟩
Rathke, M. W., **1**, 125 ⟨69JA2147⟩, 125, 280 ⟨68JA818⟩, 126 ⟨69JA2150⟩, 130 ⟨73JA3050⟩, 130, 881 ⟨70JA3222⟩, 162, 163 ⟨73JA2715⟩, 162, 765 ⟨72JA6854⟩, 273 ⟨67JA5708, 67JA5710⟩, 274 ⟨68JA4166⟩, 274, 946 ⟨68JA4165⟩, 280 ⟨68JA1911⟩, 464 ⟨87JOC3697⟩, 467 ⟨77JA4833⟩, 556 ⟨77JOC3961⟩, 830 ⟨73SC67⟩, 873, 874, 876 ⟨80JA3959⟩; **2**, 29, 34 ⟨71TL3995, 68JA5038⟩, 34 ⟨88T2751⟩, 69 ⟨75OR(22)423⟩, 420, 696 ⟨71JA2318⟩, 525 ⟨72JA6854, 80JA3959⟩, 941, 1023 ⟨78JOC376⟩; **3**, 206 ⟨77JOC3961, 81JOC5348⟩, 213 ⟨73TL1297⟩, 417 ⟨80JOC4952⟩, 537 ⟨78JOC376⟩; **4**, 526 ⟨76JOM(122)145⟩, 648 ⟨74TL1403⟩, 664 ⟨72JA6854⟩, 832 ⟨68JA818⟩; **5**, 70 ⟨85JOC4877⟩, 466 ⟨72JOC1732⟩; **6**, 190 ⟨76TL2737⟩, 190 ⟨77JOC2038⟩
Rathke, R. W., **1**, 957 ⟨72TL4249⟩
Rathkie, M. W., **3**, 537 ⟨77JOC2038⟩
Räthlein, K.-H., **4**, 585, 587 ⟨78IS140⟩, 587 ⟨74CB102⟩
Rathore, R., **3**, 88 ⟨84CL2131⟩, 134 ⟨86JOC3063⟩, 190 ⟨93TL1859⟩, 293 ⟨86SC1493⟩, 294 ⟨86JCR(S)458⟩, 375 ⟨91MI 308-01⟩; **6**, 369 ⟨93TL1859⟩
Rathousky, J., **2**, 478 ⟨68AG(E)112⟩, 535 ⟨B-65MI 210-01, B-77MI 210-01⟩
Rathsam, G., **3**, 549 ⟨20HCA853⟩
Rathsburg, H., **6**, 222, 226 ⟨18CB669⟩
Rathum, M. L., **3**, 704 ⟨88JAN454⟩
Ratier, M., **2**, 1034, 1035 ⟨86JOM(304)115⟩; **4**, 386 ⟨87JOM(331)181⟩
Ratka, J. S., **3**, 446 ⟨63JA2784⟩
Ratner, M. A., **6**, 386 ⟨81TCA285⟩
Ratovelomanana, V., **1**, 467, 1009 ⟨81TL315⟩, 468 ⟨84TL6001⟩, 468, 533, 1009 ⟨87TL1649⟩; **4**, 853 ⟨78JCR(S)68, 78SC87⟩; **5**, 554 ⟨75CR(C)1327⟩
Ratovskii, G. V., **2**, 448 ⟨91ZOB2161⟩; **4**, 766 ⟨89ZOB997⟩; **6**, 620 ⟨91ZOB643, 92ZOB306⟩
Ratsino, E. V., **6**, 147 ⟨69ZOR2246⟩, 149 ⟨73ZOR269⟩
Rattenbury, K., **4**, 115, 123, 133 ⟨61JA2299⟩

Ratton, S., **1**, 448 ⟨90TL6179⟩; **2**, 945 ⟨90TL6179⟩
Ratts, K. W., **1**, 399 ⟨68JOC70⟩, 724 ⟨63JA2790⟩, 988 ⟨69JA6112⟩; **2**, 168 ⟨71JOC3145⟩; **3**, 372 ⟨66JOC1185⟩, 497 ⟨62JA854⟩, 499 ⟨63JOC465⟩, 564 ⟨66JOC3907⟩, 591, 592, 603 ⟨69JA6112⟩; **4**, 71 ⟨72JOC1526⟩, 425 ⟨74JOC3745⟩, 728 ⟨70USP3479373⟩, 737 ⟨60JA1260⟩, 737, 738, 745, 751 ⟨62JA854⟩, 738 ⟨65OS(45)33⟩; **5**, 245, 478, 490, 491 ⟨72JOC1526⟩, 478, 490 ⟨71JOC3145⟩, 490, 491 ⟨72JOC3818⟩, 495 ⟨66JOC1689⟩, 860 ⟨66JA4996⟩, 895, 896 ⟨63JOC465⟩, 901 ⟨65JA5603⟩; **6**, 695 ⟨62JA854⟩
Ratunge-Bandarage, P. R. R., **4**, 721 ⟨94OM500, 94OM511⟩
Ratz, H., **2**, 104 ⟨63HOU(6/2)325⟩
Ratz, R., **2**, 259 ⟨66JHC14⟩; **5**, 656 ⟨62JOC2433⟩, 852 ⟨58JOC1931⟩; **6**, 603 ⟨62JOC2589⟩
Ratzkin, H., **3**, 618 ⟨61JOC3356⟩
Rau, A., **1**, 451 ⟨79JA4426⟩; **5**, 953 ⟨92CB2137⟩
Rau, H., **2**, 637 ⟨91AG(E)416⟩
Raubenheimer, H. C., **6**, 540 ⟨92POL893⟩
Raubenheimer, H. G., **5**, 651 ⟨88JOM(355)337⟩, 951 ⟨81AG(E)1055, 85OM275⟩, 951 ⟨94JOM(479)C12⟩; **6**, 333 ⟨84JOM(262)69⟩, 723 ⟨90CC1722, 94AG(E)672⟩
Rauch, E., **3**, 447 ⟨65CB912⟩
Raucher, S., **1**, 816 ⟨86TL1557⟩, 828 ⟨76JA1583, 79TL3057, 80TL4335, 83T3731, 86JOC123⟩; **2**, 280 ⟨77JOC2950, 78JOC4885⟩, 280, 623 ⟨77TL3909⟩, 281 ⟨78JOC4252⟩, 423, 908 ⟨78JOC3794⟩, 1025 ⟨87JOC2332⟩; **3**, 175 ⟨78TL2261⟩; **4**, 202 ⟨86TL1557⟩, 289, 965 ⟨78JOC3794⟩, 375, 400, 964 ⟨78JOC4252⟩; **5**, 159 ⟨76JA1583⟩, 568 ⟨81JOC3558⟩, 570, 610, 618 ⟨80TL4061⟩
Rauchfuss, T. B., **2**, 827 ⟨81IS178, 82IS175⟩; **3**, 363 ⟨86TL3445⟩; **5**, 513 ⟨86TL3445⟩
Rauchschwalbe, G., **1**, 113 ⟨93T10175⟩; **2**, 529 ⟨78JA3258⟩
Raude, E., **4**, 99 ⟨81CB3421⟩
Rauenbusch, C., **1**, 674 ⟨87AG(E)702⟩
Rauhut, M. M., **2**, 428 ⟨59JA1103, 61JOC5138⟩, 450 ⟨61JOC4626⟩, 454 ⟨61JOC4628⟩
Rauk, A., **1**, 918 ⟨81JA588⟩; **2**, 337 ⟨88JA1719⟩; **4**, 391 ⟨65JA5498⟩
Raulet, C., **4**, 739 ⟨73MI 417-01⟩, 739, 746 ⟨70CR(C)1467⟩, 743 ⟨63BSF2147⟩
Raulin, F., **5**, 728 ⟨76BSF667⟩
Raulins, N. R., **1**, 411, 414, 806, 827, 835 ⟨75OR(22)1⟩; **2**, 656 ⟨75OR(22)1⟩; **3**, 21, 22, 141 ⟨75OR(22)1⟩
Rault, I., **2**, 489 ⟨90TL1147⟩
Rausch, D. A., **4**, 31, 34, 44 ⟨63JOC494⟩; **6**, 649 ⟨67JOC1662, 68JOC2522⟩
Rausch, M. D., **2**, 566 ⟨73JA6263, 74JOM(74)85⟩, 963 ⟨86JOM(310)27⟩; **4**, 689, 690 ⟨73IC2396⟩, 706 ⟨60JA76, 62IC414, 70JOC3888, 73JOM(51)1, 78JA6382⟩, 706, 724 ⟨74JOC1420⟩, 707 ⟨57JOC900, 78JOM(153)59⟩, 724 ⟨63JOC3337⟩
Rauschenbach, P., **5**, 361 ⟨77ZN(B)98⟩
Rauscher, H., **2**, 780 ⟨42LA(550)182⟩
Rauser, M. E., **3**, 673 ⟨85JOC5892⟩
Rauss-Godineau, J., **1**, 953 ⟨61BSF2176⟩, 986 ⟨66BSF2885⟩; **2**, 1015 ⟨68BSF193⟩
Raut, S. V., **1**, 1079 ⟨93TL5621⟩
Rautenstrauch, V., **1**, 104 ⟨72HCA2233⟩, 396, 405, 816, 818 ⟨70CC4⟩, 820 ⟨71HCA739⟩; **3**, 225 ⟨84JOC950⟩; **4**, 950 ⟨72HCA594⟩; **5**, 356 ⟨80AG(E)726⟩
Rauwald, W., **6**, 549 ⟨88LA75⟩
Rav, A. V. R., **6**, 572 ⟨74S289⟩
Ravard, A., **1**, 180 ⟨88T5495⟩
Ravasio, N., **6**, 428 ⟨93JOM(451)243⟩
Rave, T. W., **2**, 470 ⟨70CA88061⟩; **3**, 453 ⟨65JA393⟩
Raveendranath, P. C., **1**, 14 ⟨92T9427⟩
Ravenscroft, P. D., **3**, 121, 294 ⟨83S572⟩
Raventos, J., **6**, 16 ⟨57BRP767779⟩
Raverso, G., **3**, 73 ⟨58CB1224⟩
Raverty, W. D., **5**, 96 ⟨82JCS(P1)1763⟩
Ravichandran, R., **2**, 220 ⟨79JOC2654⟩
Ravida, L., **2**, 907 ⟨91JOC1951⟩
Ravikumar, K., **2**, 22 ⟨93JCS(P1)3171⟩
Ravikumar, P. R., **5**, 138 ⟨75JA4144⟩
Ravikumar, V. T., **1**, 813 ⟨85TL6137⟩; **3**, 194 ⟨88T1603⟩
Ravindran, N., **1**, 488 ⟨76JOC1785⟩, 488, 510, 531, 1103 ⟨73JOC1617⟩, 660, 1103 ⟨73JA6456⟩, 661, 1103 ⟨73JA5786⟩, 1094, 1102 ⟨76JA1785⟩, 1102 ⟨76JA1798, 77JA7097⟩; **2**, 41 ⟨73JOC182⟩, 525 ⟨77JOC2533, 77S695⟩, 525 ⟨80JOC384⟩, 525, 900 ⟨76JA1785, 76JA1798, 77IC2938⟩, 606, 628 ⟨73JA5786⟩, 623 ⟨73JA6456⟩, 623, 628 ⟨73JOC1617, 89JOC6068⟩, 900, 901 ⟨89JOC6075⟩; **4**, 631, 635 ⟨76JA1798⟩; **6**, 181 ⟨76JA1798⟩
Ravindranath, K. R., **4**, 979 ⟨79IJC(B)478⟩
Ravindranathan, M., **1**, 294 ⟨86CRV353⟩, 905 ⟨78JOC3667⟩; **4**, 54 ⟨71JCS(C)1772⟩, 58 ⟨71JCS(C)3344⟩; **5**, 275, 295 ⟨82JOC4812⟩
Ravindranathan, R. V., **6**, 504 ⟨62JOC2622⟩
Ravindranathan, T., **1**, 450, 662, 1103 ⟨72JA4013⟩
Ravve, A., **4**, 88 ⟨48JA1244⟩
Rawal, V. H., **1**, 929 ⟨93TL5197⟩; **5**, 367 ⟨94TL4947⟩
Rawalay, S. S., **4**, 420, 424, 449 ⟨64JA1701, 82JCED475⟩
Rawdah, T. N., **1**, 891 ⟨83AG(E)390⟩
Rawlings, T. J., **3**, 356, 361 ⟨72JCS(P1)41⟩; **4**, 850 ⟨72JCS(P1)41⟩
Ray, C., **4**, 170 ⟨87T4433⟩
Ray, D. G., **2**, 148 ⟨90JA5672⟩
Rayadh, A., **3**, 60 ⟨88TL4717⟩
Ray, F. E., **5**, 114 ⟨55OSC(3)420⟩
Ray, III, D. G., **3**, 163 ⟨88JA2987⟩
Ray, J., **1**, 549 ⟨87JOC4875⟩; **5**, 989 ⟨87JOC4875⟩
Ray, N. C., **1**, 14 ⟨92T9427⟩
Ray, N. H., **2**, 165 ⟨61JCS2066⟩, 165 ⟨61JCS2070⟩
Rây, P., **6**, 640 ⟨61CRV313⟩
Ray, P. S., **1**, 1010 ⟨87JOC3997⟩; **5**, 773 ⟨82TL3357⟩
Ray, R., **3**, 35 ⟨82JOC2479⟩; **4**, 362 ⟨80JA7590⟩, 526 ⟨83OM1536⟩
Ray, R. E., **3**, 227 ⟨57JA6308⟩
Ray, S., **1**, 46 ⟨85IJC(B)59⟩; **2**, 474 ⟨86SRI67⟩
Ray, T., **3**, 215 ⟨85JCS(P1)267⟩
Ray, W. J., **1**, 379 ⟨58JA3625, 59JA481⟩
Rayment, I., **6**, 183 ⟨77JCS(D)136⟩
Raymond, A. J., **2**, 94 ⟨61JA3985⟩
Raymond, D. W., **5**, 1140 ⟨79JOM(175)303⟩
Raymond, K. N., **1**, 890 ⟨74JA1348⟩; **4**, 703 ⟨74JA7586⟩
Raymond, M. A., **1**, 726 ⟨63JOC372⟩; **3**, 494 ⟨63JOC217⟩
Raymond, P., **3**, 55 ⟨62CR(255)1127⟩, 463 ⟨66JCS(C)467⟩
Raynaud, G., **5**, 759 ⟨72FRP2081556⟩
Raynaud, P., **5**, 827 ⟨83S220⟩
Rayner, C. M., **2**, 116 ⟨94MI 203-01⟩, 147 ⟨92TL7237⟩, 197 ⟨92JCS(P1)2371⟩, 713, 714 ⟨94MI 214-01⟩; **3**, 102, 103 ⟨92JA3910⟩; **4**, 284 ⟨92JA3910⟩
Rayner, L. S., **2**, 403, 404 ⟨52JCS4191⟩
Raynham, T. M., **1**, 195 ⟨88TL4709⟩
Raynolds, P., **2**, 167 ⟨74JA3146⟩
Raynolds, P. W., **2**, 831 ⟨76HCA695⟩; **5**, 481 ⟨81JOC3911⟩
Raynolds, S., **2**, 422 ⟨60JA472⟩
Raynor, C. M., **4**, 281 ⟨92JCS(P1)2371⟩
Raza, S. M., **4**, 728 ⟨71JCS(C)1836⟩
Razay, H., **4**, 712 ⟨81CC867, 81CC1255⟩, 712 ⟨84JCS(D)1563, 84JCS(D)1581⟩
Razmanova, Z. P., **2**, 882 ⟨53MI 217-01⟩
Razniak, S. L., **5**, 727 ⟨73JOC2242, 83JOC1544⟩
Raznikiewicz, T., **1**, 428 ⟨73OSC(5)149⟩
Razumov, A. I., **2**, 451 ⟨45MI 208-01⟩, 453 ⟨56ZOB2463⟩, 459 ⟨74ZOB2621⟩, 461 ⟨67ZOB1173⟩, 462 ⟨68ZOB609⟩, 465 ⟨60DOK(135)1150⟩, 466 ⟨52IZV894⟩; **3**, 49 ⟨73RCR538⟩; **4**, 340 ⟨71ZOB2575, 72ZOB1647⟩, 728 ⟨71ZOB1873⟩, 884, 913 ⟨77CA121332u⟩, 912 ⟨67ZOB1623⟩; **5**, 403

⟨69JGU2541⟩, 404 ⟨67JGU2606⟩, 707 ⟨68JGU2501, 68ZOB2586⟩; **6**, 314 ⟨67ZOB2137⟩
Razumov, A. M., **5**, 484 ⟨74IZV55⟩
Razumova, N. A., **2**, 454 ⟨68DOK(181)877, 69ZOB2367⟩, 477 ⟨77ZOB1432⟩
Razuvaev, G. A., **2**, 492, 509 ⟨78JOM(162)289⟩, 566 ⟨72SRI33⟩, 569 ⟨77JOM(131)43, 79DOK(245)198⟩, 569 ⟨82JOM(239)335⟩, 832 ⟨69IZV2234⟩, 879 ⟨60JGU3203⟩, 1093 ⟨73JGU626⟩; **4**, 637, 641 ⟨75JOM(96)213, 76IZV1342⟩, 656 ⟨76DOK(231)1138⟩, 719 ⟨77JOM(141)313⟩; **5**, 390 ⟨81ZOB1318⟩, 390, 391 ⟨88IZV859⟩, 1076, 1077 ⟨85IZV2824, 87JOM(327)303⟩, 1081 ⟨87ZOB405⟩, 1082 ⟨88IZV739⟩; **6**, 324, 335 ⟨63BAU1384⟩, 628, 629 ⟨64JGU4149⟩
Razvodovskaya, L. V., **2**, 468, 469, 470 ⟨71ZOB1644⟩; **4**, 131 ⟨68JGU2563⟩
Razzino, P., **1**, 328 ⟨93T6419⟩; **4**, 102 ⟨92T5163⟩
Razzuk, A., **2**, 753 ⟨86JOC5157, 87JOC2619⟩
Re, A., **2**, 9, 19 ⟨79JCS(P1)2248⟩
Read, C. M., **2**, 690 ⟨81JCS(P1)1473⟩
Read, J., **1**, 628 ⟨52JCS153⟩; **4**, 39 ⟨27JCS538⟩
Read, R. R., **5**, 64 ⟨32OSC(1)321⟩
Read, R. W., **4**, 409 ⟨89JOC4771⟩; **5**, 526 ⟨79CC645, 80JCS(P1)2191⟩, 526, 685, 694 ⟨80JCS(P1)2184⟩
Read, T. O., **3**, 497 ⟨61JA1617⟩
Readio, J. D., **5**, 1011 ⟨69JOC4088⟩
Reagan, D. R., **1**, 197 ⟨80JOC4740⟩; **2**, 940 ⟨80JOC4740⟩
Reagen, M. T., **2**, 249 ⟨68JA4096⟩
Re, L., **4**, 205, 210 ⟨79CB148⟩
Reale, A., **1**, 689 ⟨82JCS(P2)77⟩
Reamer, R. A., **1**, 940 ⟨88TL281⟩; **2**, 66 ⟨93TL3243⟩, 528 ⟨91JOC751, 91JOC763⟩; **4**, 247 ⟨89JOC3718⟩; **6**, 446 ⟨91JOC751⟩
Reames, D. C., **1**, 213, 478 ⟨78JOC3800⟩
Reap, J. J., **1**, 598, 601 ⟨74TL1097⟩; **2**, 139 ⟨74TL1097⟩
Rearson, R. L., **2**, 104 ⟨78S452⟩
Réau, R., **3**, 682 ⟨92JA6059⟩
Reau, R., **4**, 589 ⟨93PS(76)49⟩
Réau, R., **5**, 871, 872, 1148 ⟨92JA6059⟩; **6**, 187 ⟨94OM436⟩
Reau, R., **6**, 665, 670, 675 ⟨92JA6059⟩
Rebane, E., **2**, 291 ⟨67ACS652⟩
Rebban, D., **2**, 483 ⟨69CB3955⟩
Rebek, J., **5**, 869 ⟨74TL1391⟩
Rebek, Jr., J., **1**, 326 ⟨91JA5918⟩, 326, 327, 328 ⟨92JA7007⟩; **3**, 299 ⟨84JA1813⟩; **5**, 151 ⟨87JA2426⟩, 166 ⟨84JA1813⟩
Rebelo, R. A., **1**, 489 ⟨92T8117⟩; **5**, 652 ⟨87SC1741⟩
Reber, G., **1**, 51, 446 ⟨87CB1403, 89CB259⟩; **3**, 545 ⟨86AG(E)643⟩; **4**, 548, 1023 ⟨88CB1241⟩, 629 ⟨88ZN(B)727⟩, 647 ⟨86JOM(316)C4⟩; **6**, 383 ⟨89ZN(B)796⟩, 392, 710 ⟨87OM35⟩, 673 ⟨86CC591⟩
Rebertus, R. L., **6**, 261 ⟨67JOC1944, 67USP3355492⟩, 261, 285 ⟨67JOC4045⟩
Rebiere, F., **2**, 152, 198 ⟨91JOC5991⟩, 198 ⟨89TL3659⟩; **3**, 147 ⟨92TL1763⟩; **4**, 261 ⟨90SL643⟩, 706 ⟨90TL3121⟩
Rebolledo, F., **5**, 263 ⟨93T4007⟩, 353 ⟨91S350, 92IJC(B)906⟩, 359 ⟨93S287⟩
Reboul, O., **1**, 833 ⟨74BSF2072⟩; **6**, 41 ⟨74BSF2072⟩
Rebrova, O. A., **1**, 741 ⟨72BAU2056, 72IZV2113⟩; **2**, 447 ⟨85IZV1200⟩
Rebrovic, L., **2**, 1020 ⟨81JOC4324, 84JOC4700⟩
Rebsdat, S., **6**, 104 ⟨68CB41⟩
Recca, A., **1**, 689 ⟨82JCS(P2)77⟩
Recchi, E., **2**, 882 ⟨39MI 217-01⟩
Rechka, J. A., **1**, 677 ⟨88TL2599⟩; **2**, 1000 ⟨88JCS(P1)1023⟩
Reck, R., **4**, 434 ⟨82CB860⟩; **6**, 240, 453, 611 ⟨82CB860⟩
Recknagel, A., **5**, 893 ⟨89AG(E)445⟩
Recktenwald, R., **3**, 562, 563 ⟨79AG(E)788⟩
Record, K. A. F., **5**, 531 ⟨78JCS(P2)1167⟩
Rector, C. H., **3**, 615 ⟨45JOC441⟩
Rector, D. H., **5**, 131 ⟨87JOC3468⟩
Rector, D. L., **5**, 675, 678 ⟨81JMC532⟩
Redda, K., **4**, 949 ⟨80CJC130⟩
Reddie, R. N., **2**, 192 ⟨87SC1129⟩
Redding, R. N., **4**, 650, 656 ⟨93OM338⟩
Reddy, A. V., **1**, 524 ⟨79TL2625⟩; **6**, 550 ⟨81CC756⟩
Reddy, C. K., **2**, 14 ⟨92T8329⟩, 14, 36 ⟨89TL5663⟩, 34 ⟨90TL1919⟩; **5**, 363 ⟨86BCJ1575⟩
Reddy, G. S., **1**, 338 ⟨89JA1759⟩, 471 ⟨84OM630⟩, 669, 766, 767, 769 ⟨78JA3611⟩; **2**, 477 ⟨66IC164⟩, 680 ⟨78JA3611⟩, 681 ⟨86JOC5458⟩; **3**, 508 ⟨78JA3611⟩; **4**, 569 ⟨79JOC1057⟩, 725 ⟨78JA3611⟩
Reddy, G. V., **3**, 437 ⟨94JOC3243⟩; **4**, 516 ⟨90TL1335, 91TL679⟩
Reddy, J. I., **4**, 901, 904 ⟨93T2101⟩
Reddy, J. P., **2**, 350 ⟨93TL3271⟩
Reddy, J. S., **2**, 145 ⟨92CC84⟩
Reddy, K. K., **1**, 1002 ⟨92TL135⟩
Reddy, K. M., **2**, 22 ⟨93JCS(P1)3171⟩
Reddy, K. P., **4**, 657 ⟨89JOM(364)105⟩
Reddy, K. R., **3**, 530 ⟨91JOC6917⟩
Reddy, K. S., **1**, 752 ⟨87TL3075⟩
Reddy, K. V., **3**, 128 ⟨93T965⟩; **5**, 611 ⟨90TL1327⟩, 616 ⟨92TL7957⟩
Reddy, N. K., **1**, 168 ⟨92TL7687⟩, 1002 ⟨91TL3605⟩; **2**, 43 ⟨94SC1019⟩; **3**, 133 ⟨92SC2587⟩
Reddy, N. L., **2**, 546 ⟨92JCS(P1)3277⟩
Reddy, N. P., **3**, 5, 118 ⟨85TL6257⟩
Reddy, P. P., **3**, 617, 662 ⟨88SC2179⟩; **5**, 219 ⟨88SC2179⟩, 363 ⟨86BCJ1575⟩
Reddy, P. S., **3**, 709 ⟨83SC379, 88T7243⟩; **5**, 368 ⟨88SC545⟩
Reddy, P. S.,**6**, 511 ⟨88SC545⟩
Reddy, P. S. N., **3**, 617, 662 ⟨88SC2179⟩; **5**, 219 ⟨88SC2179⟩, 363 ⟨86BCJ1575⟩, 706 ⟨90SC23⟩
Reddy, R. E., **2**, 44 ⟨93PAC633⟩, 420 ⟨92OS(71)146⟩; **3**, 437 ⟨92JOC6387, 93PAC633⟩, 437 ⟨93TL5943⟩
Reddy, R. S., **2**, 145 ⟨92CC84⟩; **4**, 186 ⟨93S1069⟩, 246 ⟨93S67⟩
Reddy, R. T., **2**, 150 ⟨92JA1428⟩, 150 ⟨92JOC7274⟩, 713 ⟨92TA629⟩; **3**, 437 ⟨92JOC5841⟩; **4**, 262 ⟨92JA1428⟩
Reddy, T. I., **4**, 901, 904 ⟨90JCS(P1)1217⟩
Reddy, V. G., **5**, 706 ⟨90SC23⟩
Reddy, V. M., **6**, 561 ⟨85IJC(B)580, 85IJC(B)1298⟩, 561 ⟨85MI 617-04⟩
Reddy, V. P., **1**, 403, 420 ⟨87JA8051⟩
Reded, J., **3**, 433 ⟨83SC649⟩
Redeker, U., **2**, 747 ⟨81TL4263⟩; **5**, 290 ⟨81TL4263⟩
Redel, J., **3**, 55 ⟨62CR(255)1127⟩
Redemann, C. E., **2**, 385 ⟨55OSC(3)244⟩; **3**, 465 ⟨55OSC(3)244⟩
Reden, J., **3**, 87, 99 ⟨84TL2901⟩
Redhouse, A. D., **5**, 932 ⟨65AG(E)1082⟩
Redjal, A., **3**, 642, 648 ⟨73T2437⟩
Redlich, D., **2**, 627 ⟨45HCA628⟩
Redlich, H., **1**, 4, 7 ⟨76CB3598⟩, 5, 7 ⟨81LA1215⟩, 940, 941 ⟨85T4253⟩
Redmond, J. W., **1**, 458 ⟨69NAT1212⟩; **5**, 999 ⟨65TL4039⟩
Redmond, W., **2**, 614 ⟨61JOC5149, 63OS(43)12⟩
Redmond, W. A., **1**, 442 ⟨63JOC568⟩
Redmore, D., **2**, 694 ⟨85SC411⟩, 694, 845, 846 ⟨86JOC179⟩, 847 ⟨91JOC833⟩; **3**, 215 ⟨63JCS2420⟩; **4**, 451 ⟨B-76MI 410-01⟩, 462, 464 ⟨78JOC992⟩, 470 ⟨78JOC996, 88JCR(S)34⟩
Redwane, N., **4**, 812, 820 ⟨82JFC(20)699⟩
Redwood, M. E., **6**, 232 ⟨65CJC1893⟩
Ree, B. R., **1**, 894 ⟨70JA1660⟩
Reed, D., **3**, 459 ⟨87JCS(D)1071⟩
Reed, D. E., **3**, 465 ⟨62OS(41)16⟩; **6**, 176, 381 ⟨85JCS(P2)1687⟩
Reed, J. N., **2**, 305 ⟨83TL3795⟩, 328 ⟨88TL5725⟩; **5**, 199 ⟨92T4863⟩
Reed, Jr., S. F., **4**, 16 ⟨65JOC2195⟩, 104 ⟨68JOC1861⟩
Reed, J. W., **1**, 402 ⟨91COS(5)899⟩, 801 ⟨91COC(5)899⟩, 806, 811 ⟨92OR(41)1⟩

Reed, K. L., **3**, 162 ⟨89SC2595⟩
Reed, L. J., **2**, 127, 214 ⟨55JA416⟩; **5**, 81 ⟨57MI 502-04, 61MI 502-01⟩
Reed, N. V., **1**, 601 ⟨81T263⟩
Reed, R., **5**, 888 ⟨93CC1354⟩; **6**, 691, 704 ⟨93CC1354⟩
Reed, R. W., **5**, 777, 778 ⟨87JOM(331)161⟩
Reed, S. F., **2**, 148 ⟨68JOC3976⟩
Reed, W. L., **4**, 795 ⟨70JOC4236⟩; **6**, 144, 150 ⟨64CRV19⟩
Reeder, R. A., **2**, 133 ⟨76JOC134⟩; **3**, 411 ⟨80JOC1489⟩
Reedy, P. E., **4**, 565 ⟨85JA5272⟩
Reefschlaeger, J., **6**, 512 ⟨87EGP242812⟩
Reen, D. C., **3**, 73 ⟨79JOC1476⟩
Rees, A. H., **3**, 239 ⟨74CJC610⟩
Rees, B., **4**, 677 ⟨81AG(E)887⟩
Rees, C. W., **1**, 811 ⟨81JCS(P1)3225⟩; **2**, 160 ⟨76JCS(P1)2166⟩, 244, 773, 792, 795 ⟨84CC374⟩, 777 ⟨75JCS(P1)26⟩; **3**, 106 ⟨87JCS(P1)931⟩, 196 ⟨92CC285⟩, 372 ⟨87JCS(P1)207, 89JCS(P1)2489⟩; **4**, 461 ⟨90SL415⟩, 539 ⟨90SL413, 91JCS(P1)323⟩, 856 ⟨86JCS(P1)501⟩, 1080 ⟨B-69MI 424-01⟩; **5**, 226 ⟨83JCS(P1)49⟩, 381 ⟨B-88MI 507-01⟩, 655, 657 ⟨87JCS(P1)1389⟩, 742, 850 ⟨84CHEC⟩, 764 ⟨76JCS(P1)2166⟩, 766 ⟨75JCS(P1)1964⟩, 1073 ⟨85JCS(P1)1471⟩; **6**, 141 ⟨69JCS(C)2251⟩, 255 ⟨90JCS(P1)509⟩
Rees, P. M., **6**, 625 ⟨85TL4149⟩
Rees, R. G., **2**, 855 ⟨72JCS(P2)380⟩
Rees, R. J. W., **4**, 206 ⟨50JPP764⟩
Rees, W. M., **5**, 461 ⟨87JA5145⟩
Reese, C. B., **1**, 5, 9 ⟨83S304⟩, 74 ⟨70CC1519, 76T2157⟩, 1082 ⟨81S276⟩; **2**, 127 ⟨90JCS(P1)969⟩, 255 ⟨90TL1319⟩, 1003 ⟨82S419⟩; **3**, 18, 133 ⟨78S919⟩, 25 ⟨81S276⟩, 39, 174 ⟨82JCS(P1)2719⟩, 68 ⟨64T2091⟩, 465 ⟨82S419⟩, 621 ⟨77CC280⟩; **4**, 186 ⟨67JA3366⟩, 190 ⟨85S751⟩, 317 ⟨81JCS(P1)1569⟩, 841 ⟨84CC1028⟩
Reetz, M., **2**, 565, 566 ⟨B-86MI 211-01⟩; **3**, 478 ⟨68LA(716)204⟩; **5**, 844, 866, 874 ⟨68LA(716)204⟩
Reetz, M. T., **1**, 127 ⟨80CC1202⟩, 133 ⟨82AG(E)96, 93TL7395⟩, 186 ⟨81CC237⟩, 189, 200 ⟨84CB322⟩, 202 ⟨83TL2837, 90JOM(382)121⟩, 650 ⟨76S199⟩, 664 ⟨77AG(E)44a, 77AG(E)44b, 80LA171⟩, 767 ⟨B-86MI 116-01⟩, 839 ⟨91TL4481⟩; **2**, 67 ⟨82TCC1, 84AG(E)556, B-86MI 202-02⟩, 70 ⟨93CC328⟩, 348, 355 ⟨91TL4481⟩, 422 ⟨80LA1471⟩, 533 ⟨90TL3863⟩, 543 ⟨90TL6725⟩, 566 ⟨80AG(E)900, 80AG(E)1011⟩, 566 ⟨82TL5259, 86T2931⟩, 568 ⟨86TL5711⟩, 698 ⟨86TL4721⟩, 702 ⟨81TL4691⟩, 945 ⟨78AG(E)278⟩; **3**, 10, 124 ⟨82AG(E)96⟩, 39 ⟨91AG(E)1531⟩, 40 ⟨87AG(E)1141, 93TL7395⟩, 139, 145 ⟨79TL4971⟩, 209 ⟨83CB3702⟩, 613 ⟨81AG(E)1017⟩, 613, 627, 698 ⟨83T961⟩, 616 ⟨88TL3295⟩, 631 ⟨85AG(E)989⟩, 634 ⟨84TL3301⟩; **4**, 202 ⟨84TL511⟩, 614 ⟨77AOC(16)33⟩; **5**, 152 ⟨89AG(E)1706⟩
Reeve, W., **5**, 27, 62 ⟨71S131⟩, 1121 ⟨80CJC2784⟩; **6**, 25 ⟨64JOC1148, 71S131⟩, 25 ⟨80JOC5214⟩, 27 ⟨66CJC575, 68CJC2233⟩, 27 ⟨80CJC485⟩
Reeves, R. L., **3**, 405 ⟨B-70MI 310-02⟩
Reeves, W. P., **5**, 1024 ⟨64JOC3098⟩, 1047 ⟨81SC781⟩
Reformatsky, S. N., **2**, 702 ⟨1887CB1210⟩
Regaila, H. A., **5**, 615 ⟨90JCS(P1)1267⟩
Regan, A. C., **3**, 200 ⟨92TL813⟩
Regan, M. T., **1**, 670 ⟨74JA1518⟩; **5**, 458 ⟨74JA1518⟩
Regel, E., **3**, 311 ⟨77LA145⟩
Regel, W., **5**, 992 ⟨73CB3753⟩
Regeling, H., **3**, 171, 368 ⟨84TL5947⟩; **5**, 562 ⟨85JOC2930⟩
Regen, S.-L., **2**, 4 ⟨88IJC(B)1129⟩, 7 ⟨75JOC1669⟩, 25 ⟨79JOC1961⟩, 25, 31 ⟨77JOC875⟩, 35 ⟨79JOC3436⟩, 132 ⟨81SC409⟩; **3**, 114 ⟨82JOC3331⟩, 613 ⟨79JOC2029⟩, 662 ⟨79JOC4443⟩
Regenye, R. W., **5**, 55 ⟨90OS(69)10⟩
Reger, D. L., **1**, 545 ⟨86JA1940, 88ACR229⟩; **2**, 581 ⟨90OM16, 91OM902⟩, 954 ⟨83OM101⟩, 978 ⟨84OM134⟩; **6**, 717, 727 ⟨78JOM(153)67⟩
Regev, H., **2**, 292 ⟨65JCS7511⟩; **6**, 49 ⟨65JCS7511⟩
Reggel, L., **5**, 25, 108 ⟨51JA436⟩, 1016 ⟨43JA1783⟩
Reggelin, M., **3**, 213 ⟨89TL2919⟩; **4**, 841 ⟨89TL2915⟩
Reginato, G., **1**, 1048 ⟨91TL2169⟩; **2**, 140 ⟨91JOC7323⟩, 541 ⟨93TL873⟩, 918 ⟨91S1201, 93TL3311⟩, 918, 939 ⟨89JOC1473⟩, 918, 940 ⟨92SL981⟩, 935 ⟨92SL883⟩, 1018 ⟨91TL2169⟩; **3**, 34 ⟨92SL803⟩, 336, 340, 346, 363 ⟨91JOC7323⟩, 346, 353 ⟨93TL873⟩, 675 ⟨91TL2169⟩; **4**, 930 ⟨87TL4093⟩, 944 ⟨92SL499, 92SL883⟩, 951 ⟨91TL1899⟩, 1018 ⟨92PAC439⟩; **5**, 416, 423 ⟨87TL4093⟩, 427 ⟨92SL332, 92TL1507⟩, 427, 428 ⟨92SL329⟩, 467 ⟨89JOC2966⟩, 648, 650 ⟨89JOC19⟩
Regis, R. R., **4**, 81 ⟨82TL2539⟩
Regitz, M., **1**, 258 ⟨B-86MI 106-01⟩, 320 ⟨89HOU(E19a)⟩, 395 ⟨69CB3877⟩, 522 ⟨86CB244⟩, 637 ⟨B-86MI 114-04⟩, 919 ⟨89HOU(1)1, 89HOU(2)1⟩, 985 ⟨92SL635⟩; **2**, 404 ⟨65LA(687)214⟩, 449 ⟨91AG(E)207, 91SL433⟩, 478 ⟨82HOU(E1)1, 90MI 208-03⟩, 799 ⟨69LA(728)99⟩, 859, 865 ⟨89AG(E)225⟩, 861 ⟨77LA1309⟩, 1047 ⟨72CB2963⟩, 1048, 1068, 1069, 1070 ⟨74CB2513⟩, 1066 ⟨76CB546⟩, 1068 ⟨77LA1766, 79AG(E)531⟩; **3**, 48 ⟨B-86MI 301-02⟩, 193 ⟨B-86MI 304-05⟩, 467, 473 ⟨70LA(734)70⟩, 471, 472 ⟨67AG786⟩, 471, 472 ⟨67AG(E)733⟩, 471, 532 ⟨67TL739, 72S351⟩, 472 ⟨68CB1263, 68CB2622, 70LA(739)174⟩, 473 ⟨65LA(687)214⟩, 475 ⟨68OS(48)36, 69OPP99⟩, 476 ⟨81AG118, 81AG(E)113⟩, 476, 477 ⟨81BSB615⟩, 476, 490, 579 ⟨68CB2351, 80TL909⟩, 477 ⟨72CB2963, 72CB2975, 72S571, 73LA1505, 74CB2513, 82S149⟩, 478, 482 ⟨79LA1002⟩, 479 ⟨80CB1632⟩, 480 ⟨71LA(748)207, 75AG259, 75AG(E)222, 75CB2040, 80CB3303⟩, 481 ⟨69LA(730)194, 79CB2509, 79S805, 80LA590⟩, 481, 489 ⟨74TL1825⟩, 482 ⟨76CB2039, 78CB3068, 80LA305, 81LA1865, 85T819⟩, 489 ⟨B-86MI 312-01⟩, 498 ⟨67CI324, 67T2701⟩, 579 ⟨79AG(E)320⟩, 606, 609 ⟨93S1247⟩, 688 ⟨88AG(E)1484, 92BSB359, 94JHC663⟩, 688, 689 ⟨86TL1903, 90CRV191⟩, 689 ⟨87JOM(338)329⟩; **4**, 55, 57 ⟨68CB3604⟩, 100 ⟨88AG(E)304⟩, 110 ⟨89AG(E)225, 91CB1207⟩, 115, 567 ⟨89TL3951⟩, 336 ⟨82HOU(E1/E2)⟩, 341 ⟨87JOM(332)C1⟩, 479, 480 ⟨86JOM(306)39, 87AG(E)1257, 87CB1645⟩, 483 ⟨91S1099⟩, 484 ⟨86S31⟩, 489 ⟨69CB2216⟩, 547, 549 ⟨B-90MI 412-01, B-90MI 412-02⟩, 549 ⟨87AG(E)1011, 92AG(E)1055⟩, 549 ⟨B-90MI 412-02, 93CC569, 93PS(76)1⟩, 557 ⟨76CB2039⟩, 558 ⟨78CB705, 81LA1044⟩, 562 ⟨86TL1903⟩, 565 ⟨85ZN(B)1077⟩, 567 ⟨93S285⟩, 889 ⟨83AG(E)553⟩; **5**, 395 ⟨91CB2843, 91S1099⟩, 403, 404 ⟨72CB3357⟩, 548 ⟨82LA1451⟩, 595 ⟨67LA(710)118⟩, 827 ⟨72CB2975⟩, 839 ⟨87S689⟩, 839, 878, 879 ⟨86S31⟩, 839, 880 ⟨85ZN(B)1077⟩, 844 ⟨B-77MI 521-01⟩, 855 ⟨63CB3120⟩, 865, 868, 874 ⟨B-86MI 522-01⟩, 865 ⟨B-78MI 522-01⟩, 868 ⟨B-86MI 522-04⟩, 869 ⟨67AG(E)733, 68CB3734, 69CB2216⟩, 869, 871 ⟨76CB2039, B-78MI 522-02⟩, 869 ⟨B-78MI 522-03⟩, 870 ⟨85T2625, B-86MI 522-02, B-86MI 522-03⟩, 871 ⟨72TL3799, 76CB546, 85T811, 85T825, 86T1815, 89CB565⟩, 871, 874 ⟨71LA(748)207⟩, 872 ⟨73OSC(5)179⟩, 876 ⟨82HOU(E1)27, B-90MI 523-02⟩, 878, 879, 1155 ⟨86JOM(306)39⟩, 883 ⟨88AG(E)272⟩, 883, 1151, 1152, 1154, 1159 ⟨88AG(E)1484⟩, 885 ⟨92AG(E)1055, 93CB1945⟩, 924 ⟨88CB1363⟩, 1152 ⟨92AG(E)758⟩, 1152, 1159 ⟨87JOM(338)329⟩, 1152, 1159 ⟨90CRV191, B-90MI 531-01⟩, 1153 ⟨89PS(29)31⟩, 1154, 1155 ⟨87CB1645, 89MI 531-01⟩, 1155 ⟨87PS(30)479⟩; **6**, 143 ⟨71LA(748)1⟩, 143 ⟨74CB1931⟩, 152 ⟨88AG(E)1484⟩, 153 ⟨89AG(E)1013⟩, 153 ⟨90CRV191⟩, 153 ⟨91AG(E)207⟩, 153 ⟨92AG(E)584⟩, 153 ⟨92AG(E)879⟩, 153 ⟨93JOC4105⟩, 154 ⟨87AG(E)908⟩, 154 ⟨91CC1305⟩, 154 ⟨92AG(E)1055⟩, 512, 582 ⟨91S1099⟩, 623, 660, 668 ⟨79LA1002⟩, 666 ⟨68CB3734, 68TL3171, 72S351⟩, 667, 668, 689 ⟨89S511, 85JOM(290)33⟩, 668, 669 ⟨91AG(E)674⟩, 670 ⟨71LA(748)207⟩, 678, 688, 692, 693 ⟨B-90MI 622-01⟩, 692 ⟨81TCC71, 82HOU(E1)583⟩

Regnier, B., **2**, 36 ⟨79TL1801⟩
Regondi, V., **2**, 134, 143 ⟨83S630⟩, 180 ⟨81S149⟩, 180 ⟨89S957⟩; **3**, 16, 132 ⟨81S51⟩, 283, 309 ⟨87S311⟩; **4**, 249, 250 ⟨81TL1821⟩, 846, 854 ⟨87G227⟩; **6**, 302, 554, 729 ⟨88S22⟩, 472 ⟨80S375, 81S149⟩, 553, 554 ⟨86S894⟩
Regragni, R., **4**, 711 ⟨90OM2234⟩
Regragui, R., **5**, 400 ⟨88NJC547⟩
Regulski, T. W., **1**, 1110 ⟨73JA160⟩
Rehberg, C. E., **5**, 122 ⟨55OSC(3)46⟩
Rehder, W., **3**, 564 ⟨68AG(E)304⟩
Rehiem, A. G. A., **2**, 26 ⟨68CC976⟩
Rehman, M. A., **4**, 105 ⟨67JCS(C)58⟩, 300 ⟨67JCS(C)2696⟩; **5**, 347 ⟨67JCS(C)58⟩
Rehn, D., **5**, 652 ⟨77JCR(S)119⟩
Reho, A., **5**, 282 ⟨83S1013⟩
Rei, M.-H., **1**, 905 ⟨78JOC3667⟩; **2**, 89 ⟨69JA5646⟩
Reibel, W. R. K., **1**, 233 ⟨85CB1230⟩
Reibenspies, J. H., **1**, 573 ⟨92JOC5557⟩, 630 ⟨85JOC5460⟩, 944 ⟨93JA8050⟩; **3**, 29 ⟨84JOC1378⟩
Reich, C. J., **2**, 64 ⟨73TL1505⟩
Reich, H. J., **1**, 108 ⟨78JA4888⟩, 592, 593, 601 ⟨79ACR22⟩, 592, 608 ⟨74JOC428⟩, 592, 713, 714, 716 ⟨79JA6638⟩, 592, 713, 715, 716 ⟨75CC790⟩, 593, 713, 714, 715, 716 ⟨79JA6648⟩, 601, 602, 604 ⟨78JOC1697⟩, 601, 602, 604, 605, 606, 607, 608, 609, 610, 781 ⟨75JA5434⟩, 601, 602, 604, 605, 781 ⟨73JA5813⟩, 601, 602, 610, 713, 714, 715, 781 ⟨75JA3250⟩, 601, 781 ⟨76JOC2503⟩, 603 ⟨75JOC3313, 81JOC3721⟩, 605, 983, 1073 ⟨80JA5967⟩, 606, 610 ⟨74JOC2133⟩, 609 ⟨75JOC2570, 82JA7051⟩, 620 ⟨B-83MI 114-01⟩, 788 ⟨83JA2503⟩, 971 ⟨80JA1423⟩, 971, 985 ⟨86JA7791⟩, 1047 ⟨91JOC2911⟩, 1089 ⟨80JOC5227⟩, 1092 ⟨84JOC3438⟩; **2**, 279 ⟨77JOC1773, 79JA6638⟩, 279 ⟨79JA6648⟩, 280 ⟨79ACR22, B-78MI 204-01⟩, 281 ⟨75CC790, 78JOC1697⟩, 282 ⟨74JOC428, 88OS(67)23⟩, 282, 287 ⟨75JA5434, 75JOC2570⟩, 290 ⟨82TL3319⟩, 368 ⟨75JOC3313, 88JOC2389⟩, 728, 908 ⟨81JOC2775⟩, 938 ⟨93JA6625⟩, 938, 939 ⟨82JA1119, 83T949⟩, 962 ⟨85JA4101⟩; **3**, 56 ⟨82JOC1618⟩, 80 ⟨79JA6638⟩, 135 ⟨87JA5549⟩, 173, 177, 207 ⟨75JA5434⟩, 173, 207 ⟨73JA5813⟩, 175 ⟨B-87MI 304-06⟩, 177 ⟨85T4771⟩, 256 ⟨79TL4433, 81JA3112⟩; **4**, 71, 88, 288 ⟨86JOC2981⟩, 89 ⟨75CJC1922⟩, 91, 92 ⟨78JOC1697, 81JOC3721⟩, 197 ⟨81JA3112⟩, 239, 375 ⟨79JA6638⟩, 285 ⟨75JA5434, 90JA5609⟩, 289, 871, 947, 965 ⟨81JOC2775⟩, 375, 399 ⟨75JOC2570⟩, 398 ⟨77JA263, B-87MI 408-03⟩, 399 ⟨79JA6648, 83MI 408-02⟩, 621 ⟨93JA6625⟩, 701 ⟨88JOC455⟩
Reich, H.-J., **4**, 947 ⟨82JOC1618⟩, 964 ⟨80JOC5227⟩; **5**, 331 ⟨91JOC2911⟩, 334 ⟨87JA5549⟩, 410 ⟨79JA2225⟩, 412 ⟨77JOC1773⟩, 413 ⟨84JOC5282⟩, 423, 427 ⟨82JA1119⟩, 423, 427 ⟨83T949⟩; **6**, 100 ⟨79JA6638⟩, 130 ⟨77JOC1773⟩, 134 ⟨78MI 604-01⟩, 135 ⟨90JA5609, 93AG(E)1469⟩, 136 ⟨75JOC2570⟩, 327, 336 ⟨93AG1489⟩
Reich, I. E., **2**, 938 ⟨90JOC2282⟩
Reich, I. L., **1**, 601, 602, 604, 605, 606, 607, 608, 609, 610, 781 ⟨75JA5434⟩, 601, 602, 604, 605, 781 ⟨73JA5813⟩, 603 ⟨81JOC3721⟩, 606, 610 ⟨74JOC2133⟩, 1092 ⟨84JOC3438⟩; **2**, 282, 287 ⟨75JA5434⟩, 962 ⟨85JA4101⟩; **3**, 173, 177, 207 ⟨75JA5434⟩, 173, 207 ⟨73JA5813⟩; **4**, 91, 92 ⟨81JOC3721⟩, 285 ⟨75JA5434⟩, 621 ⟨93JA6625⟩
Reich, P., **4**, 889, 972 ⟨66CB2302⟩; **6**, 533 ⟨65CB2063⟩
Reichard, A., **6**, 213 ⟨B-72MI 607-02⟩
Reichard, D., **5**, 453 ⟨66IC1177⟩
Reichardt, C., **2**, 1008 ⟨73OSC(5)1135⟩; **3**, 64 ⟨72CB1815⟩, 74, 704 ⟨73AG(E)323⟩, 75, 76 ⟨73TL3979, 84LA649⟩, 356 ⟨91LA111⟩
Reich, S. H., **1**, 826 ⟨88JOC3195⟩
Reiche, A., **5**, 276, 286 ⟨61AG621⟩
Reichel, C. J., **2**, 1008 ⟨80JOC524⟩
Reichel, S., **2**, 873 ⟨67MI 217-01⟩; **6**, 60, 267 ⟨64CB1673⟩, 61 ⟨66CB793⟩
Reichelderfer, R. F., **5**, 919 ⟨74JA2847⟩
Reichelt, H., **3**, 214 ⟨86CB722⟩
Reichelt, W., **2**, 961 ⟨75AG(E)218⟩
Reichenbach, G., **5**, 799 ⟨91G101⟩
Reichenbach, T., **1**, 653, 654 ⟨74TL543⟩; **2**, 607, 617, 622, 628, 922 ⟨74TL543⟩
Reichert, J., **5**, 601 ⟨78JPR313⟩
Reichert, K., **5**, 339, 689 ⟨80LA1075⟩
Reichert, K. H., **6**, 265 ⟨87CZ176⟩
Reichert, W. L., **6**, 430 ⟨73IC769⟩
Reichle, W. R., **2**, 726 ⟨60OS(40)93⟩
Reichle, W. T., **2**, 891 ⟨68JOM(13)529⟩
Reichstein, T., **3**, 612 ⟨32HCA254⟩; **5**, 144 ⟨49HCA647⟩
Reid, D. E., **1**, 1075 ⟨58JOC665⟩
Reid, D. H., **1**, 866 ⟨56CI(L)1504, 58T339⟩, 866, 910 ⟨59JCS2773, 65QR274⟩, 910 ⟨63JCS3295⟩; **3**, 73 ⟨69CC466⟩, 73, 255 ⟨70JCS(C)2412⟩, 73, 376 ⟨68JCS(C)2543⟩, 341 ⟨70JCS(C)145, 73JCS(P1)657⟩, 341, 363 ⟨66CC401⟩, 341, 376 ⟨69JCS(C)913⟩, 356, 377 ⟨79JCS(P1)2340⟩, 357 ⟨68CC863⟩, 371 ⟨75JCS(P1)2097, 76JCS(P1)228⟩, 376, 377 ⟨72JCS(P1)1447, 74JCS(P1)722⟩, 377 ⟨74JCS(P1)242, 75JCS(P1)775, 76JCS(P1)880⟩, 388 ⟨79JCS(P1)2334⟩, 388, 389 ⟨88CC1494⟩, 389, 398 ⟨71JCS(C)3187⟩; **5**, 625 ⟨93S870⟩
Reid, E. B., **4**, 162 ⟨54JCS520⟩
Reid, E. E., **2**, 116, 117, 137, 138 ⟨38JA2452⟩, 118, 135 ⟨21JA880⟩, 119 ⟨32JA1674⟩, 126 ⟨26JA776⟩, 132, 137 ⟨B-60MI 203-01⟩, 132, 213 ⟨B-58MI 203-01⟩, 165 ⟨21JA591⟩, 191 ⟨31JA3407⟩, 192 ⟨B-65MI 203-03⟩, 217 ⟨B-60MI 203-02⟩, 261 ⟨38JA2370⟩; **4**, 287 ⟨26JA520⟩; **5**, 33 ⟨43OSC(2)474⟩, 275, 276 ⟨31JA1879, 37JA401⟩; **6**, 471 ⟨B-62MI 615-01⟩, 560, 561 ⟨B-62MI 617-01⟩
Reid, G., **4**, 584, 585 ⟨82JOM(232)C78, 86JOM(315)255⟩
Reid, J. A., **3**, 171 ⟨92TL7259⟩; **6**, 641 ⟨92TL5933⟩
Reid, J. A. W., **4**, 207 ⟨64JCS1217⟩
Reid, J. C., **3**, 83, 96 ⟨46JA2502⟩
Reid, J. D., **2**, 413 ⟨59JOC1419⟩
Reid, J. G., **3**, 331 ⟨84JA4617⟩
Reid, R. G., **3**, 652 ⟨86CC1230⟩
Reid, R. J., **6**, 225, 227 ⟨56USP2755314⟩
Reid, S. T., **5**, 332 ⟨71CC1609⟩
Reid, W., **3**, 379, 380 ⟨88CB1005⟩; **5**, 1033 ⟨54LA(590)123⟩; **6**, 111 ⟨75LA958⟩, 616, 617 ⟨75CB2290⟩
Reider, P. J., **4**, 210 ⟨79JA2501⟩, 247 ⟨89JOC3718⟩; **5**, 329 ⟨94JOC1231⟩
Reider, W., **1**, 242 ⟨78M609⟩
Reidlinger, A. A., **3**, 674 ⟨61USP3006948⟩
Reier, F. W., **2**, 434 ⟨87GEP3612629⟩; **4**, 577 ⟨85JOM(297)C30⟩, 578 ⟨82JOM(235)287, 83JOM(255)305, 84AG(E)522, 84JOM(269)21⟩; **6**, 706 ⟨81JOM(209)C10⟩
Reif, L., **1**, 336, 346 ⟨83JOC1373⟩, 857 ⟨62JA258⟩
Reif, W., **1**, 525, 832 ⟨71LA(747)60⟩; **3**, 616 ⟨88TL3295⟩
Reifegerste, D., **5**, 324 ⟨85S331⟩
Reiff, H., **1**, 881 ⟨68AG(E)7⟩
Reiff, H. F., **2**, 1076 ⟨69JOM(17)165⟩
Reiff, L. P., **1**, 19 ⟨67TL2329⟩
Reiffen, M., **6**, 630 ⟨77CB37⟩
Reiffenrath, V., **6**, 421, 422 ⟨92GEP4023107⟩
Reifschneider, W., **2**, 273 ⟨73OSC(5)107⟩; **5**, 767 ⟨56JA3825⟩
Reig, J. G., **1**, 821 ⟨84JA4617⟩
Reiger, B., **6**, 421, 422 ⟨92EUP515258⟩
Reiher, U., **5**, 718 ⟨87CB657⟩, 1119 ⟨87CB1597⟩
Reikhsfeld, V. O., **2**, 540 ⟨88RCR447⟩; **4**, 569 ⟨83ZOB1831⟩; **5**, 583 ⟨86ZOB1547⟩; **6**, 537 ⟨86ZOB1547⟩
Reil, S., **1**, 259 ⟨71JA2738⟩
Reilly, J., **1**, 799 ⟨75JA3837⟩; **5**, 841 ⟨57JOC692⟩, 857 ⟨53JA5309⟩; **6**, 641 ⟨53JA4053⟩
Reilly, J. L., **3**, 560 ⟨72JOC2364⟩

Reilly, P. J., **1**, 700, 703 ⟨88JOC1064⟩
Reilly, W. L., **5**, 815 ⟨56JA6032⟩
Reimann, E., **1**, 560 ⟨92M939⟩; **2**, 12 ⟨75LA1252⟩; **4**, 563 ⟨81AP(314)218⟩
Reimann, H., **3**, 255 ⟨60AG415⟩
Reimann-Andersen, S., **5**, 502 ⟨90AG(E)60⟩; **6**, 258 ⟨90AG(E)60⟩, 306 ⟨94CB533⟩
Reimann, R. H., **2**, 439 ⟨76JOC972⟩
Reimann, W., **4**, 661, 701, 1062 ⟨85OM1044⟩, 746, 813, 1069 ⟨86OM1991⟩, 1060, 1062 ⟨85JOM(281)163⟩, 1063, 1065, 1069, 1070 ⟨87JOM(322)151⟩, 1065 ⟨87JOM(322)141⟩; **6**, 209 ⟨85OM1044, 86OM1991⟩
Reimer, G., **4**, 193 ⟨84JCS(P1)1643⟩
Reimer, K. J., **2**, 487 ⟨89CRV713⟩; **5**, 222 ⟨92MI 504-02⟩
Reimlinger, H., **1**, 795, 882 ⟨69S97⟩, 795, 882 ⟨70S405⟩, 956 ⟨68JCS(C)606⟩; **2**, 1049 ⟨70TL3141⟩; **3**, 465, 467 ⟨61AG221, 61CB2547⟩, 471 ⟨60AG33, 60CB2162⟩, 490 ⟨62AG252, 62AG(E)216, 68LA(713)113⟩; **4**, 740 ⟨64CB3503⟩, 740, 755 ⟨62AG153, 64CB339⟩, 829 ⟨71CB2786⟩; **5**, 773 ⟨71CB639⟩; **6**, 607 ⟨78CB1619⟩
Rein, B. M., **2**, 7, 627 ⟨64JA964⟩
Reinach-Hirtzbach, F. de, **2**, 149, 200 ⟨76CJC3012⟩
Rein, F. W. Von, **1**, 539 ⟨71TL3777⟩
Rein, K., **2**, 321 ⟨89JOC3002⟩; **4**, 526 ⟨91COS(3)65⟩, 531 ⟨89JOC3002⟩
Rein, T., **2**, 308 ⟨88JOC3845⟩; **3**, 65 ⟨88ACS569⟩
Reinboldt, H., **2**, 172, 194 ⟨39CB657⟩
Reindl, D., **4**, 580 ⟨90JOM(385)351⟩; **5**, 954 ⟨93JOM(454)133⟩
Reinehe, K. E., **2**, 440 ⟨86EUP194716⟩
Reinehr, D., **1**, 495 ⟨72JOM(34)1⟩
Reineke, L. M., **1**, 241 ⟨71TL3791⟩
Reinemer, P., **4**, 553 ⟨92ZN(B)725⟩
Reiner, L., **2**, 163 ⟨51JA459⟩
Reinert, K., **2**, 1041 ⟨60LA(639)33⟩
Reinertshofer, J., **3**, 466 ⟨52LA(575)174⟩
Reinhard, M. F. D., **2**, 401, 402 ⟨83JOC1⟩
Reinhardt, G., **1**, 1003 ⟨85JA6546⟩
Reinhart, J., **2**, 386 ⟨55JA2453⟩
Reinheckel, H., **2**, 187 ⟨66CB1718⟩, 393 ⟨67M1944, 78ZC214⟩; **3**, 458 ⟨64CB2661⟩; **4**, 659 ⟨76ZC65⟩, 698 ⟨80JOM(201)97⟩; **5**, 595 ⟨64CB2661, 66CB11⟩
Reinhold, D. F., **6**, 25 ⟨86TL3845⟩
Reinhold, G., **1**, 1072 ⟨69ZC229⟩
Reinhoudt, D. N., **1**, 47 ⟨87T3541⟩, 518 ⟨72TL5269, 86JOC2004⟩; **3**, 428 ⟨86RTC103⟩, 434 ⟨90BSF704⟩, 435 ⟨83JOC1816⟩; **5**, 325 ⟨87H(26)437, 88RTC381⟩; **6**, 556 ⟨90CC1432⟩, 562 ⟨92JA10573⟩
Reininger, K., **5**, 166 ⟨70AG(E)457⟩
Reinke, H., **5**, 246 ⟨73JFC(3)383⟩, 1018 ⟨69CB2718⟩; **6**, 432 ⟨67AG(E)705, 70AG(E)466⟩, 432 ⟨73JFC(3)383⟩, 432, 435 ⟨69CB2718⟩
Reinking, P., **2**, 95 ⟨83JOC3126⟩
Reinmuth, O., **1**, 114 ⟨B-54MI 104-01⟩; **2**, 67 ⟨B-54MI 202-01⟩, 557 ⟨B-54MI 211-01⟩, 958 ⟨B-54MI 219-01⟩
Reinshagen, H., **4**, 983 ⟨77TL3619⟩
Reinstein, M., **2**, 1041, 1047 ⟨64AG(E)506⟩; **4**, 885 ⟨64AG571⟩
Reintjes, M., **1**, 273 ⟨65JA4585⟩
Reisacher, H. U., **4**, 559 ⟨88PS(37)241⟩, 1026 ⟨84TL3687⟩; **5**, 884 ⟨88TH17⟩, 884, 1092 ⟨84TL3687⟩
Reisacher, H.-U., **5**, 887 ⟨86AG(E)454⟩
Reis, H., **5**, 27 ⟨68AG(E)329⟩, 142 ⟨68ACS2019⟩
Reisch, J., **1**, 576, 1010 ⟨88LA69, 88LA543⟩; **2**, 687 ⟨90M709⟩; **6**, 109 ⟨90JHC1953⟩
Reischer, R. J., **3**, 379 ⟨73JA4287⟩
Reisdorf, D., **2**, 680 ⟨78MI 213-01⟩
Reisenauer, H. P., **3**, 337 ⟨91CB2609⟩, 358 ⟨83CB273, 92CB265⟩, 504 ⟨85AG(E)1065⟩, 703 ⟨81JOC1046⟩, 710 ⟨92AG(E)1212⟩; **4**, 610 ⟨81AG(E)597⟩; **5**, 1122 ⟨80AG(E)566⟩, 1154 ⟨81AG(E)197⟩; **6**, 713 ⟨85AG(E)1065⟩
Reiser, O., **1**, 194 ⟨93AG(E)547⟩; **2**, 44 ⟨94AG(E)69⟩; **3**, 163 ⟨93JOC3169⟩
Reiser, R., **6**, 469 ⟨92CB2493⟩
Reisfeld, M. J., **2**, 873 ⟨B-86MI 217-01⟩
Reisinger, F., **2**, 1006, 1009 ⟨87JA882, 88JOC4379⟩
Reiske, T., **4**, 602, 607, 637, 660 ⟨90ZAAC(583)195⟩
Reisman, D., **2**, 194, 195 ⟨73TL3563⟩
Reisner, D. B., **3**, 627 ⟨63OSC(4)144⟩
Reiss, J. G., **2**, 570 ⟨73JCS(D)2436⟩
Reiss, W., **1**, 999 ⟨62ZAAC(317)54⟩; **2**, 1076, 1081 ⟨59N321⟩, 1077 ⟨65N59⟩, 1082, 1083 ⟨62ZAAC(317)54⟩
Reissert, A., **5**, 566, 567 ⟨04CB3708⟩, 580, 592 ⟨24CB981⟩
Reissig, H. U., **1**, 801 ⟨B-87MI 118-01⟩, 88TCC73⟩, 955 ⟨92AG(E)1033⟩, 978 ⟨88JOC2450⟩; **2**, 322 ⟨93LA1155⟩, 689 ⟨82CB399, 82CB657⟩; **3**, 170 ⟨82CB3687⟩, 224 ⟨90CB363⟩, 434 ⟨88TL1685⟩; **4**, 108 ⟨88LA1169⟩; **5**, 150 ⟨82JA1735⟩, 164 ⟨88AG(E)268⟩, 166 ⟨88JOC2440⟩, 553 ⟨88LA465⟩, 729 ⟨88LA471⟩, 957 ⟨91CB957⟩
Reist, E. J., **2**, 861 ⟨60JOC666⟩; **4**, 389 ⟨92MI 408-01⟩
Reitel, C., **6**, 448, 452, 454 ⟨80AG(E)201⟩
Reiter, B., **4**, 676 ⟨75JOM(97)245⟩, 677 ⟨81AG(E)887⟩
Reiter, F., **3**, 372 ⟨79CB260⟩, 687 ⟨81LA1025⟩
Reiter, J., **6**, 564 ⟨90JHC1249⟩
Reiter, P. L., **2**, 656 ⟨73AG(E)775⟩
Reith, B. A., **3**, 479 ⟨75T597⟩; **5**, 867, 870 ⟨75T597⟩, 898, 901, 902 ⟨72RTC37, 74JOC2728⟩; **6**, 699 ⟨72RTC37, 74JOC2728⟩
Reithinger, S., **2**, 878 ⟨90TL6331⟩; **4**, 594 ⟨90TL6331⟩; **6**, 374 ⟨90TL6331⟩
Reitinger, S., **3**, 608 ⟨88TL463⟩
Reitsma, B. H., **6**, 79 ⟨88JA4144⟩
Reitz, A. B., **1**, 638, 639, 647, 721, 744 ⟨89CRV863⟩, 722 ⟨89TL1361⟩, 730 ⟨85JA217⟩; **2**, 1003, 1004 ⟨89CRV863⟩; **3**, 48, 195, 493 ⟨89CRV863⟩; **6**, 694 ⟨89CRV863⟩
Reitz, D. B., **1**, 148 ⟨84CRV471⟩; **2**, 302 ⟨84CR471⟩, 1002 ⟨84CRV471⟩; **3**, 182 ⟨81JOC4316⟩, 183 ⟨84CRV471⟩; **4**, 382, 392 ⟨78CRV275, 78JA5428⟩, 529, 535, 537, 1014, 1015, 1016 ⟨84CRV471⟩; **5**, 237 ⟨77TL1839⟩
Reitz, T. J., **2**, 152 ⟨77TL4009⟩
Reitzel, L., **5**, 1137 ⟨88CB1115⟩
Reizig, K., **5**, 400 ⟨86CB1857, 86ZN(B)1096⟩; **6**, 518 ⟨86AG(E)755, 87CB1421⟩, 680, 690 ⟨85AG(E)53, 86CB1857⟩
Reliquet, A., **3**, 363 ⟨91S543⟩; **5**, 606 ⟨82PS(12)325⟩, 633 ⟨85PS(25)279⟩, 760 ⟨80JHC767, 83PS(15)143⟩
Reliquet, F., **3**, 363 ⟨91S543⟩; **5**, 633 ⟨85PS(25)279⟩, 760 ⟨80JHC767, 83PS(15)143⟩
Rellahan, W. L., **1**, 869, 871, 881 ⟨57JA881, 59JOC709⟩
Relles, H. M., **5**, 14 ⟨74JA6469⟩
Relyea, D. I., **2**, 117 ⟨56JA348⟩
Rembarz, G., **4**, 440 ⟨67ZC421⟩, 728 ⟨59JPR150⟩; **6**, 360, 361 ⟨78JPR335⟩
Remers, W. A., **1**, 100 ⟨71JOC279⟩
Remes, M., **2**, 777 ⟨73MI 215-01⟩
Rémion, J., **1**, 592, 593, 713, 716 ⟨76TL1385⟩, 592, 593, 713, 716 ⟨76TL3743⟩
Remion, J., **2**, 758 ⟨79CC1180⟩; **3**, 475 ⟨79CC1180⟩; **4**, 790 ⟨79CC1180⟩; **5**, 2 ⟨79CC1180⟩
Remiszewski, S. W., **2**, 245 ⟨84JOC3243, 86TL1853⟩
Remli, M., **2**, 10 ⟨81JFC(17)127⟩; **6**, 52, 54 ⟨81JFC(18)93⟩
Remmens, G., **2**, 862 ⟨77ZAAC(431)221⟩
Rempel, G. L., **1**, 79, 80 ⟨75S448⟩
Rempfer, H., **5**, 744, 745, 751 ⟨59CB837⟩; **6**, 139, 140 ⟨59CB329⟩
Rempfler, H., **3**, 477, 479 ⟨78CRV363⟩
Remuson, R., **1**, 522 ⟨83TL1129⟩
Remy, D. C., **4**, 97 ⟨79JOC3117⟩; **5**, 847 ⟨62USP3061590⟩
Renaldo, A. F., **1**, 413 ⟨90JA3945⟩, 471 ⟨87JOC422⟩, 809 ⟨83TL3757⟩
Renard, G., **1**, 905 ⟨78JCR(S)62⟩
Renard, S. H., **5**, 565, 579 ⟨49BSF272⟩
Renaud, A., **3**, 309 ⟨84JOC4409⟩

Renaud, J., **1**, 452 ⟨93JOC11⟩
Renaud, P., **4**, 285, 286 ⟨91HCA1305⟩
Renaud, R., **2**, 377 ⟨54CJC545⟩
Renaud, R. N., **4**, 311 ⟨68CJC385⟩; **6**, 20 ⟨75CJC529, 79CJC2617⟩
Renauld, E., **3**, 620 ⟨1895CB1682⟩
Renaut, P., **1**, 287 ⟨82JA6668⟩; **6**, 558 ⟨91S265⟩
Renaut, R., **5**, 146 ⟨86JA284⟩
Ren, W.-Y., **5**, 690 ⟨86JHC1757⟩
Rendell, R. W., **6**, 37 ⟨78T197⟩
Rendle, D. F., **2**, 1004 ⟨75JCS(D)1284⟩
René, L., **2**, 744 ⟨88S138⟩, 749 ⟨86S419⟩, 787, 789 ⟨82S260⟩; **3**, 46 ⟨81S878⟩, 264 ⟨88S138⟩; **4**, 193 ⟨81S878⟩
Renfrew, Jr., W. B., **1**, 881 ⟨43OSC(2)268⟩, 884 ⟨37JA1823⟩; **5**, 998 ⟨37JA121⟩
Renga, J. M., **1**, 234, 821 ⟨78TL519⟩, 410, 821 ⟨78JOC4831⟩, 410, 823 ⟨78TL523⟩, 601, 602, 604, 605, 606, 607, 608, 609, 610, 781 ⟨75JA5434⟩, 601, 602, 604, 605, 781 ⟨73JA5813⟩, 603 ⟨75JOC3313⟩, 606, 610 ⟨74JOC2133⟩; **2**, 282 ⟨88OS(67)23⟩, 282, 287 ⟨75JA5434⟩, 368 ⟨75JOC3313⟩, 683 ⟨84SC69⟩; **3**, 173, 177, 207 ⟨75JA5434⟩, 173, 207 ⟨73JA5813⟩; **4**, 71, 88, 288 ⟨86JOC2981⟩, 285 ⟨75JA5434⟩; **5**, 375 ⟨85TL1175⟩; **6**, 26 ⟨85TL1175⟩
Rengaraju, S., **2**, 659 ⟨74CC584⟩, 747 ⟨76CC103⟩; **4**, 473 ⟨72JOC3304⟩
Renger, B., **1**, 210 ⟨77AG(E)264⟩; **2**, 385 ⟨88OSC(6)542⟩, 388 ⟨78HCA1622⟩; **4**, 529 ⟨78CB2630⟩; **5**, 177 ⟨77AG(E)244⟩
Renhowe, P. A., **2**, 936 ⟨93JA9305⟩
Renier, P., **2**, 216 ⟨90TL5007⟩
Renk, E., **3**, 622, 638 ⟨63CRV489⟩
Renk, E. B., **1**, 664 ⟨79JOC3431⟩
Renk, T., **2**, 1083, 1094 ⟨75ZAAC(418)273⟩
Renk, Th., **5**, 1136 ⟨75ZAAC(418)273⟩
Renkema, K. B., **4**, 650, 656 ⟨93OM338⟩
Renken, T. L., **3**, 335, 369 ⟨82JA3119⟩; **4**, 258 ⟨82JA3119⟩
Renn, K., **4**, 995 ⟨22CB644⟩
Rennecke, A., **5**, 1038 ⟨58CB683⟩
Rennels, R. A., **1**, 1015 ⟨91TL5243⟩
Renner, R., **3**, 621 ⟨87CC1007⟩
Rennert, A. M., **4**, 648 ⟨75CC536⟩
Rennhard, H. H., **1**, 906 ⟨55CI(L)415⟩
Rennick, L. E., **1**, 904 ⟨65JA1381, 65JA4017⟩
Rennie, R. A. C., **3**, 684 ⟨88CC1339⟩; **5**, 652 ⟨76CC240⟩
Renoll, M., **4**, 178 ⟨55OSC(3)502⟩
Renoll, M. W., **4**, 3, 30, 33 ⟨36JA889⟩, 18 ⟨39JA938⟩
Renou, J. P., **5**, 577, 608, 609 ⟨79T1087⟩
Rens, M., **4**, 790 ⟨88OSC(6)282⟩; **5**, 2 ⟨79OS(59)26⟩
Renshaw, R. R., **4**, 135 ⟨25MI 403-01⟩
Rensing, A., **4**, 156, 708 ⟨85CB370⟩; **6**, 203, 204, 210 ⟨80TL2807⟩, 209 ⟨85CB380⟩
Renson, M., **2**, 238 ⟨75T2099⟩, 279 ⟨75JHC423⟩, 280 ⟨70BSB353, 70BSB367⟩, 280 ⟨70BSB383⟩, 288 ⟨68BSB153⟩, 732 ⟨64BSB507⟩; **3**, 74 ⟨73JHC267⟩, 256 ⟨76BSF294⟩; **4**, 54 ⟨61BSB77⟩, 205 ⟨68BSB379⟩, 211 ⟨62BSB379⟩; **5**, 238 ⟨62BSB260⟩, 239 ⟨76JCS(P1)2452⟩, 241 ⟨70BSB383, 73SA(A)285⟩, 334 ⟨89BSB395⟩, 476 ⟨61BSB519⟩, 515, 516 ⟨62BSB563⟩, 515, 516 ⟨64BSB419⟩, 624, 625 ⟨63BSB304⟩, 1051, 1056 ⟨62BSB531⟩, 1116 ⟨72BRP1298049⟩; **6**, 596 ⟨63BSB149⟩, 598 ⟨62BSB541⟩
Renson, M. J., **2**, 632 ⟨85JOC1779⟩; **5**, 334 ⟨92TL3863⟩
Renström, L., **4**, 426 ⟨80T2505⟩
Rentschler, H., **1**, 502 ⟨44HCA1297⟩
Renwick, J. A. A., **4**, 3 ⟨66JOC916⟩
Renzi, G., **3**, 578 ⟨78JCS(P1)1113⟩; **4**, 894 ⟨78JCS(P1)1113⟩
Renzoni, G. E., **1**, 681 ⟨86T1581⟩, 785 ⟨86JA7121⟩
Repenning, K., **1**, 104 ⟨58LA(614)37⟩
Repenning, R., **2**, 1016 ⟨58LA(614)37⟩
Repic, O., **1**, 1023 ⟨92JOC6344⟩
Repina, L. A., **5**, 713 ⟨84JGU1925, 84ZOB2154⟩, 713 ⟨86ZOB1262⟩, 713, 788 ⟨86JGU1113⟩
Repke, D. B., **1**, 1041 ⟨88CC439⟩
Replogle, L. L., **5**, 1109 ⟨68JOC823⟩
Reppe, W., **1**, 530, 890 ⟨48LA(560)1⟩; **2**, 44 ⟨53LA(582)133⟩; **4**, 186, 187 ⟨35USP2000252⟩; **6**, 524 ⟨53LA(582)116⟩
Rìera, J., **4**, 17 ⟨91JOC5445⟩
Rescifina, A., **2**, 668 ⟨91TL5923⟩
Reshetnikov, S. M., **2**, 891 ⟨88JGU1040⟩
Reshetnyak, M. V., **6**, 138 ⟨89T3487⟩
Reske, E., **4**, 192 ⟨70CB643⟩; **6**, 78, 81 ⟨70CB639⟩, 79 ⟨70CB643⟩
Resmini, M., **1**, 812 ⟨91TL6969⟩
Resnati, G., **1**, 143 ⟨85TL2913⟩; **2**, 15 ⟨90TA661⟩, 16 ⟨91JOC4925⟩, 350 ⟨92JOC2726⟩; **3**, 32 ⟨93CC984⟩, 120 ⟨92TL7245⟩, 150 ⟨88CC218⟩; **4**, 170 ⟨89JOC5171⟩
Resnick, P., **3**, 448, 449 ⟨62JA3514⟩
Resnick, P. R., **6**, 18 ⟨61JA1767⟩
Resnick, T. M., **2**, 689 ⟨94JOC6683⟩
Respess, W. L., **1**, 311 ⟨66JOC3128⟩, 426 ⟨69JOM(18)263⟩
Respondek, J., **2**, 447 ⟨83C10⟩; **6**, 705 ⟨83C10⟩
Respondek, M., **5**, 692 ⟨89S256⟩
Ressler, C., **2**, 212, 214 ⟨74JOC253⟩; **3**, 618 ⟨61JOC3356⟩
Rest, A. J., **4**, 344 ⟨69TFS1697⟩
Restelli, A., **2**, 29 ⟨90JOC5400⟩; **5**, 607 ⟨85T4929⟩
Restivo, R. J., **4**, 692 ⟨77IC172⟩
Retey, J., **2**, 11 ⟨82TL1549⟩
Retherford, C., **3**, 201 ⟨90TL1833⟩
Retta, N., **2**, 436 ⟨83JCS(D)905⟩; **6**, 165, 372 ⟨83JCS(D)905⟩, 387 ⟨82JOM(235)265⟩, 393 ⟨80JOM(190)101⟩, 395 ⟨83JOM(249)23⟩
Rettberg, N., **5**, 548, 559 ⟨91TL4679⟩
Rettig, S. J., **1**, 928 ⟨90CJC1698⟩; **2**, 573 ⟨88OM1877⟩, 583 ⟨87CC1110⟩; **4**, 545 ⟨86OM2220⟩, 706 ⟨85OM2196⟩, 1034 ⟨86OM1320⟩
Retzlaff, G., **6**, 494 ⟨79GEP2921130⟩
Reubke, K.-J., **5**, 640 ⟨67AG(E)368⟩, 640, 846 ⟨69CB2117⟩, 641 ⟨68TL5973⟩, 833 ⟨B-70MI 521-03⟩, 852, 853 ⟨70LA(736)166⟩
Reum, M. E., **3**, 113 ⟨93CC954⟩
Reuman, M., **1**, 473, 539 ⟨85T837⟩; **2**, 653 ⟨85T837⟩; **4**, 306 ⟨85T837⟩
Reuss, H., **6**, 383 ⟨84ZAAC(512)93⟩
Reuss, R. H., **2**, 22, 28 ⟨74JOC1785⟩, 8022 ⟨74JOC553⟩; **3**, 26 ⟨74JOC1785⟩, 31, 156 ⟨75JOC3427⟩
Reutenaur, G., **3**, 617 ⟨46MI 318-01⟩
Reuter, D. C., **3**, 204 ⟨89TL4617⟩
Reuter, F., **5**, 114 ⟨35JCS1467⟩
Reuter, G., **2**, 403 ⟨56CB685⟩
Reuter, J. M., **3**, 272 ⟨78JOC2438⟩
Reuter, K., **4**, 360 ⟨79CB936⟩
Reuter, K. A., **5**, 874 ⟨85TL1845⟩
Reuter, K. H., **1**, 1071 ⟨91JOC2045⟩
Reuter, M., **4**, 555 ⟨60USP2937207⟩, 683 ⟨92OM2916⟩
Reuter, S., **4**, 751 ⟨57JA4174⟩
Reuterhäll, A., **5**, 652 ⟨69ACS818⟩
Reuther, I., **6**, 572 ⟨73JPR144⟩
Reuther, K. H., **6**, 545 ⟨60CB3056⟩
Reuther, W., **4**, 826 ⟨82GEP3240287⟩, 890 ⟨93LA207⟩
Reutova, I. O., **1**, 1010 ⟨88ZOR650⟩
Reutov, O. A., **1**, 65, 454 ⟨78T2827⟩, 65, 66 ⟨68PAC79⟩, 647 ⟨67ZOR598⟩, 737 ⟨68JOM(13)263⟩, 738 ⟨65BAU1434, 65IZV1474⟩, 741 ⟨72AOC(10)273, 72IZV2113⟩, 844 ⟨B-78MI 119-04⟩, 845 ⟨B-68MI 119-01⟩, 859 ⟨77MI 119-01, 81JOU381⟩; **2**, 446 ⟨75IZV2845⟩, 447 ⟨85IZV1200⟩, 877 ⟨84DOK(274)606, 84IZV700, 86IZV181⟩, 878 ⟨87JOU1417⟩, 884 ⟨52DOK(87)991, 53DOK(91)1341⟩, 884 ⟨54IZV410, 54ZOB1259⟩; **3**, 328 ⟨68JOM(13)263⟩, 430 ⟨B-83MI 311-01⟩, 496 ⟨89IZV1182⟩; **4**, 152 ⟨89MI 403-03⟩, 587 ⟨64IZV772, 65JOM(4)202, 66DOK(171)111, 66ZOR942, 68ZOR1685⟩, 642 ⟨80IZV2429⟩, 674 ⟨85JOM(297)C25,

86ZOR2457, 91JOM(418)C24, 93JOM(455)C9⟩, 720 ⟨79JOM(182)537⟩, 721 ⟨77DOK(236)371, 79IZV1528⟩, 1033 ⟨80JOM(202)C58⟩, 1067 ⟨78JOM(162)271⟩; **5**, 902 ⟨63DOK(151)856⟩, 919 ⟨65JOM(4)202, 70DOK(195)98⟩, 1142, 1145 ⟨79JOM(177)101⟩
Reutrakal, V., **4**, 773 ⟨83TL531⟩
Reutrakul, V., **2**, 612 ⟨83TL725⟩, 616 ⟨79TL617⟩; **3**, 62 ⟨77TL1377⟩, 153 ⟨77TL1225⟩, 154 ⟨79CL209⟩, 212 ⟨82CL687⟩; **4**, 74, 778 ⟨83TL725⟩, 77 ⟨77TL1225, 79CL209⟩, 77 ⟨79TL617⟩, 742 ⟨83TL527⟩
Reutter, W., **2**, 445 ⟨76GEP1768461⟩
Reuvers, A. J. M., **1**, 115 ⟨70T2683⟩
Reuvers, J. T. A., **1**, 600 ⟨81TL4137⟩; **3**, 63 ⟨81TL4137⟩
Revankar, G. R., **2**, 364 ⟨88JHC1857⟩
Revel, M., **4**, 459 ⟨83TL1975⟩; **5**, 705 ⟨78PS(4)317⟩, 705, 706 ⟨77JHC749⟩
Revelle, L., **2**, 184 ⟨78JA3622⟩
Revelle, L. K., **3**, 337 ⟨77CC287⟩
Revel'skii, I. A., **4**, 561 ⟨87ZOB2294⟩
Revesz, C., **5**, 652 ⟨75JMC581⟩
Revial, G., **1**, 813 ⟨83JA625⟩; **3**, 415 ⟨85JA273⟩; **4**, 386 ⟨83JA625⟩
Revis, A., **4**, 831 ⟨90JOC2972⟩
Revitt, D. M., **2**, 504 ⟨72JCS(D)847⟩
Revol, J. M., **1**, 804 ⟨83JOC4453⟩
Rewcastle, G. W., **2**, 1002 ⟨93AHC(56)155⟩; **5**, 567 ⟨93JMC2459⟩
Rewich, R. T., **2**, 762 ⟨66MI 215-01⟩
Rewicki, D., **1**, 844 ⟨68MI 119-05⟩, 866 ⟨65LA(689)1⟩, 867 ⟨67AG(E)635⟩, 954 ⟨65CB2611⟩
Rey, M., **1**, 398, 399 ⟨83JCS(P1)1009⟩, 812 ⟨82HCA451⟩, 826 ⟨65HCA1985, 71HCA1589⟩; **3**, 237 ⟨87HCA321⟩, 526 ⟨68HCA1466⟩, 527 ⟨70HCA417, 70HCA2159⟩, 704 ⟨80T515⟩; **4**, 768 ⟨68HCA1466⟩, 979 ⟨77HCA1781⟩
Reychler, A., **2**, 180 ⟨1898BSF120⟩
Reye, C., **1**, 490 ⟨93CRV1371⟩; **2**, 68 ⟨89T171⟩, 77 ⟨87JOM(328)C17⟩, 547 ⟨93CRV1371⟩
Reyes, F. D., **5**, 798 ⟨87JOM(327)339⟩
Reynaud, P., **5**, 584, 586, 612 ⟨65BSF3623⟩, 586 ⟨65BSF3628⟩, 611 ⟨66CR(C)788⟩, 615 ⟨93H(36)1979⟩, 743 ⟨76BSF301⟩, 750 ⟨61BCF2067⟩, 753 ⟨64CR4067⟩; **6**, 644 ⟨66BSF73⟩
Reynolds, B. E., **6**, 635 ⟨88S460⟩
Reynolds, C. D., **5**, 263 ⟨93JCS(P1)451⟩
Reynolds, C. H., **6**, 726 ⟨91CC975⟩
Reynolds, D. D., **2**, 121 ⟨62JOC93⟩; **6**, 504 ⟨57JA5717⟩
Reynolds, D. P., **1**, 1024, 1134 ⟨80JCS(P1)852, 79JCS(P1)2954⟩
Reynolds, D. W., **1**, 248 ⟨87ACR65⟩; **4**, 730 ⟨90JOC4448⟩
Reynolds, G. A., **3**, 71 ⟨71JOC600⟩, 370 ⟨75S638⟩; **5**, 62 ⟨63OSC(4)633⟩
Reynolds, K. A., **1**, 991 ⟨93JOC1298⟩
Reynolds, M. A., **2**, 186, 273 ⟨79JOC160⟩
Reynolds, M. E., **1**, 338 ⟨88JA6911⟩
Reynolds, R. D., **4**, 298 ⟨78JOC3838⟩
Reynolds, S. J., **6**, 447 ⟨89TL3229, 91TL259⟩
Reynolds, W. F., **1**, 652 ⟨76JOM(121)293⟩; **2**, 914, 921, 922 ⟨76JOM(121)293⟩; **5**, 924 ⟨79JA6750⟩
Reynoldson, T., **6**, 265, 269 ⟨70JCS(A)31⟩
Reys, M., **3**, 526 ⟨70HCA120⟩
Rezanka, D. S., **1**, 209 ⟨93JOC7304⟩
Rezende, M. C., **5**, 652 ⟨87SC1741, 89SC1181⟩, 1107 ⟨78CC133⟩
Reznicek, D. L., **4**, 524 ⟨69IC275, 88JOM(349)11⟩, 566 ⟨69IC862⟩
Reznik, V. S., **6**, 57 ⟨80IZV2356⟩
Reznikov, V. A., **5**, 669 ⟨91BAU382, 91IZV444⟩
Rezzano, I. N., **3**, 64 ⟨90TL1249⟩
Rhee, C. K., **3**, 161 ⟨88SC1927⟩; **5**, 12 ⟨91TL7165⟩
Rhee, H., **3**, 625 ⟨93TL1399⟩
Rhee, I., **1**, 1035 ⟨69TL4593⟩; **3**, 90 ⟨64BCJ341⟩, 293 ⟨77S776⟩
Rhee, R. P., **2**, 664 ⟨78JOC178, 79JA1628⟩; **4**, 346 ⟨80S72⟩
Rhee, S. G., **2**, 919 ⟨75JOM(86)143⟩; **4**, 1068 ⟨75LA565⟩
Rheinboldt, H., **2**, 134 ⟨32JPR257, 37CB675⟩, 287 ⟨46JA2671⟩, 735 ⟨50BSF245, 55CB1⟩; **5**, 821 ⟨27LA(451)161⟩, 823 ⟨25LA(444)113, 28CB32, 1051 ⟨55HOU(9)951⟩
Rheingold, A., **5**, 944 ⟨87JA7578, 89JA4129⟩, 956 ⟨93OM597⟩
Rheingold, A. L., **1**, 551 ⟨93JA10671⟩; **2**, 489 ⟨86OCS633⟩, 496 ⟨86OM393, 90JA4037⟩, 496 ⟨91OM3764⟩, 500 ⟨90IC798⟩, 504 ⟨93JCS(D)663⟩, 507 ⟨75JOM(102)445, 77JOM(128)155⟩, 507 ⟨82JOM(231)119, 86OCS618⟩, 574, 978 ⟨88JA2575⟩, 945 ⟨93JA1586⟩, 895, 896 ⟨91CB2453⟩, 927, 945 ⟨93JA9846⟩, 1083 ⟨90JA1847⟩; **3**, 545 ⟨88JA4855⟩, 597 ⟨93JA9846⟩; **4**, 134, 135 ⟨75JOM(102)437, 75JOM(102)445⟩, 135 ⟨75SRI199⟩, 584 ⟨89JOM(375)265⟩, 656, 657 ⟨93JA5527⟩, 676 ⟨88IC3248, 88OM289⟩, 683 ⟨90JA7113⟩, 718 ⟨87JOM(334)117⟩; **5**, 375, 376 ⟨90OM1557⟩, 385 ⟨86OM336⟩, 444 ⟨89JA149⟩, 451 ⟨87JA1584⟩, 944 ⟨90JA6232, 95OM2102⟩; **5**, 946 ⟨90JA3220, 92JA6735⟩, 946, 957 ⟨93JA3806⟩, 950 ⟨82OM628⟩, 953, 956 ⟨94JOC6882⟩, 988 ⟨92OM1242⟩, 1096 ⟨88JA4855, 88JA6926⟩; **6**, 243 ⟨93JCS(D)663⟩, 338 ⟨89JA149, 90AX2374⟩, 340 ⟨86IC2472⟩, 344 ⟨86OM2561⟩, 517 ⟨93OM2401⟩, 521 ⟨86JA5355⟩
Rheinwald, G., **1**, 674 ⟨76MAC1349⟩
Rheude, U., **4**, 854 ⟨85CB2852⟩; **5**, 502 ⟨85CB2208⟩; **6**, 254 ⟨83CB1285, 85CB2208⟩
Rhine, W. E., **4**, 1068 ⟨74JA1941⟩
Rhines, S. P., **2**, 1071 ⟨81TL1875⟩
Rhinesmith, H. S., **2**, 30 ⟨43OSC(2)177⟩; **5**, 60 ⟨43OSC(2)177⟩
Rho, A. K., **4**, 203 ⟨91JOC3283⟩
Rhoads, S. J., **1**, 411, 414, 806, 827, 835 ⟨75OR(22)1⟩, 826 ⟨69JA2815⟩, 869 ⟨63T1625, 65JOC3212⟩; **2**, 656 ⟨75OR(22)1⟩; **3**, 21, 22, 141 ⟨75OR(22)1⟩
Rhode, O., **3**, 156 ⟨91CL281⟩
Rhodes, C. A., **3**, 307 ⟨80JCR(S)42⟩
Rhodes, C. J., **1**, 922 ⟨88TL1437⟩
Rhodes, R. A., **3**, 621 ⟨88SC681⟩
Rhodes, S. P., **4**, 631 ⟨69JA4306⟩
Rhyne, L. D., **1**, 472 ⟨79JOC3445⟩
Rhyne, T. C., **5**, 1003 ⟨72USP3666784⟩
Riad, B. Y., **5**, 615 ⟨87SUL105⟩
Riad, Y., **2**, 26 ⟨68CC976⟩
Riahi, A., **3**, 37 ⟨85JOM(280)269⟩, 72 ⟨85TL839⟩
Riahi, M. A., **5**, 760 ⟨84CC157⟩
Riant, O., **1**, 374, 375 ⟨92CRV1007⟩
Riba, M., **6**, 695 ⟨83TL3387⟩
Ribeill, Y., **2**, 444 ⟨88S911⟩; **4**, 847 ⟨87BSF823⟩
Ribeiro, O., **1**, 15 ⟨83JCS(P1)87⟩
Ribi, M., **6**, 441 ⟨73GEP2342860⟩
Riblet, L. A., **5**, 334 ⟨86JMC359⟩
Ribon, V., **5**, 45 ⟨86JCS(P1)521⟩
Ribota, D., **2**, 825 ⟨80JCS(D)55⟩
Ricard, L., **2**, 152, 198 ⟨91JOC5991⟩, 894 ⟨93BSF521⟩; **4**, 545, 565 ⟨93PS(77)254⟩; **6**, 154 ⟨91CC1305⟩, 692 ⟨90OM793⟩
Ricca, A., **5**, 821 ⟨59T24, 67TL327⟩
Ricca, D. J., **1**, 816, 819 ⟨86JOC4090⟩
Ricci, A., **1**, 242 ⟨83TL2311⟩, 667 ⟨87TL2751⟩, 1048 ⟨91TL2169⟩; **2**, 140 ⟨91JOC7323⟩, 265 ⟨83TL2311⟩, 541 ⟨93TL873⟩, 652 ⟨86S633⟩, 907 ⟨91JOC1951⟩, 908 ⟨89S647⟩, 918 ⟨91S1201, 92PAC439, 93TL3311⟩, 918, 939 ⟨89JOC1473⟩, 918, 940 ⟨92SL981⟩, 935 ⟨92SL883⟩, 942 ⟨85S633⟩, 1018 ⟨91TL2169⟩, 1019 ⟨89S461⟩; **3**, 34 ⟨92SL803⟩, 105 ⟨67JCS(C)779⟩, 336, 340, 346, 363 ⟨91JOC7323⟩, 346, 353 ⟨93TL873⟩, 675 ⟨91TL2169⟩; **4**, 358 ⟨90T2999⟩, 929 ⟨89S647⟩, 930 ⟨87TL4093⟩, 944 ⟨92SL499, 92SL883⟩, 951 ⟨91TL1899⟩, 1018 ⟨92PAC439⟩; **5**, 409 ⟨89S647⟩, 410 ⟨80TL3927⟩, 413 ⟨85AG(E)1068⟩, 416, 423 ⟨87TL4093⟩, 417

⟨88JOC3612⟩, 419 ⟨85JOC130⟩, 427 ⟨92SL332, 92TL1507⟩, 427, 428 ⟨92SL329⟩, 467 ⟨89JOC2966⟩, 648, 650 ⟨89JOC19⟩, 793, 803 ⟨81JCS(P2)1043⟩; **6**, 134 ⟨86S633⟩, 325, 335, 520 ⟨86TL5985⟩, 349, 585 ⟨85TL1091⟩, 520, 702 ⟨85AG(E)1068⟩
Ricci, Jr., J. S., **3**, 570 ⟨70JA3489, 71JA2391⟩; **4**, 493 ⟨75JA38⟩
Ricciuti, C., **5**, 205 ⟨55JA4037⟩
Riccoboni, L., **2**, 589 ⟨41CB1089⟩
Rice, C. W., **2**, 590 ⟨75IC2402⟩
Rice, D. A., **2**, 570 ⟨72JCS(D)2313, 73JCS(D)961⟩
Rice, F. O., **2**, 385 ⟨55OSC(3)244⟩; **3**, 389 ⟨34JA2381⟩, 465 ⟨55OSC(3)244⟩
Rice, G. W., **4**, 579 ⟨83OM154⟩; **6**, 707 ⟨83OM154⟩
Rice, J. R., **4**, 614, 656 ⟨74JOM(78)C35⟩
Rice, K., **1**, 364, 365 ⟨90PAC507⟩
Rice, K. C., **2**, 33 ⟨87JOC5466⟩; **3**, 62 ⟨75JHC1325⟩, 242 ⟨84JOC1051⟩; **4**, 107 ⟨89TL911⟩, 179 ⟨88S233⟩; **5**, 1029 ⟨79NAT854⟩
Rice, M. J., **4**, 269 ⟨91TL7743⟩
Rice, R. E., **4**, 151 ⟨88JCS(D)1773⟩
Rice, R. H., **5**, 226 ⟨77JC355⟩
Rich, D. H., **1**, 816, 817 ⟨93TL5217⟩; **2**, 74 ⟨90TL1803⟩; **3**, 39 ⟨B-90MI 301-01⟩, 373 ⟨83JOC1999⟩; **5**, 280 ⟨85JA4342⟩
Rich, E. M., **3**, 474 ⟨25JA1728⟩
Rich, J. D., **2**, 945 ⟨89JA5886, 89OM2609⟩
Richard, C. S., **3**, 427, 430 ⟨87T551⟩
Richard, J. J., **4**, 554, 563 ⟨61JA1722⟩, 557, 575 ⟨63JOC123⟩
Richards, C. J., **3**, 597 ⟨90CC307⟩
Richards, D. H., **2**, 104 ⟨93T7037⟩, 386 ⟨88TL2735⟩
Richards, D. J., **5**, 81, 107 ⟨81S487⟩
Richards, H. J., **5**, 359 ⟨70OPP275⟩
Richards, J. L., **6**, 730 ⟨68T6485⟩
Richards, K. E., **1**, 387 ⟨69T4117⟩
Richards, K. R., **1**, 4, 9 ⟨83AJC2243⟩
Richards, R. L., **2**, 579 ⟨82COMC-I(5)541⟩; **4**, 547 ⟨88JOM(338)C31⟩; **6**, 622 ⟨75JOM(101)C1, 77JCS(D)2015, 79JOM(181)69⟩, 721 ⟨69CC1322, 71JCS(A)21⟩
Richardson, A. C., **6**, 550 ⟨93CAR(242)281⟩, 561 ⟨91CAR(216)271⟩
Richardson, A. P., **2**, 133 ⟨91CC475⟩
Richardson, D. H., **4**, 18 ⟨33JCS496⟩
Richardson, D. P., **1**, 67, 68 ⟨84T2317⟩
Richardson, G., **1**, 700, 703 ⟨86JOC4840⟩; **2**, 7 ⟨75CC622⟩; **5**, 13 ⟨75CC622⟩, 177 ⟨86JOC4840⟩
Richardson, G. D., **4**, 214 ⟨83CC1064⟩, 327 ⟨77TL4519, 80JCS(P1)1407⟩, 327 ⟨81TL1551⟩
Richardson, J., **4**, 575 ⟨82ZAAC(494)55⟩
Richardson, J. F., **6**, 621, 728 ⟨85IC2453⟩
Richardson, Jr., J. W., **6**, 334 ⟨85OM1226⟩
Richardson, K. S., **1**, 107 ⟨B-87MI 104-01⟩, 845 ⟨B-87MI 119-02⟩, 886, 893 ⟨B-87MI 119-05⟩, 891 ⟨B-87MI 119-06⟩, 904 ⟨B-87MI 119-07⟩, 915 ⟨B-87MI 119-09⟩
Richardson, M. L., **6**, 411, 442 ⟨B-93MI 614-01⟩
Richardson, P. F., **2**, 24 ⟨91SL733⟩
Richardson, R. D., **4**, 817 ⟨68JOC286⟩
Richardson, R. L., **5**, 328 ⟨81S121⟩
Richardson, R. S., **5**, 264 ⟨85JCS(P1)461⟩
Richardson, R. W., **1**, 506 ⟨49JCS737⟩; **3**, 54, 231 ⟨49JCS737⟩
Richardson, S., **3**, 60 ⟨85TL2391⟩; **4**, 649 ⟨85TL2391⟩
Richardson, S. G., **1**, 36 ⟨79TL1181⟩
Richardson, S. K., **4**, 856 ⟨85CC618, 90JCS(P1)2967⟩
Richardson, W. H., **1**, 690 ⟨81SC895⟩, 1144, 1145 ⟨59JA4719⟩
Richaud, M. G., **3**, 541 ⟨85TL833⟩
Riche, C., **1**, 626 ⟨90TL1865⟩; **3**, 59 ⟨81HCA1140⟩; **4**, 905, 911 ⟨85CC845⟩
Riche, M. A., **4**, 63 ⟨1855MI 402-01⟩
Richelme, S., **4**, 141 ⟨81JOM(205)311⟩, 521 ⟨72JOM(34)C18, 79JOM(168)43⟩
Richer, J.-C., **2**, 40 ⟨71CJC2586⟩
Richers, C., **2**, 1098 ⟨64CB3317⟩
Richert, S. A., **3**, 113 ⟨90JA1936⟩
Richeson, D. S., **3**, 545 ⟨89JA9057⟩; **4**, 691 ⟨89OM2001⟩
Richey, H. G., **1**, 801 ⟨73TL2545⟩
Richey, J. M., **1**, 904 ⟨65JA4017⟩
Richey, Jr., H. G., **1**, 495 ⟨69TL2313, 71TL3785⟩, 539 ⟨68TL1457, 71TL3777⟩, 801 ⟨76TL575⟩, 847 ⟨B-70MI 119-01⟩, 894, 900 ⟨62JA2016⟩, 897, 901, 911 ⟨62JA1498⟩, 897, 901, 911 ⟨63JA2991⟩, 904 ⟨65JA1381, 65JA4017⟩, 905 ⟨59JOC2044⟩, 912, 913 ⟨65JA4533⟩; **2**, 343, 763 ⟨76TL233⟩, 561 ⟨71TL2183, 81JOC3773⟩
Richman, J. E., **1**, 309 ⟨73TL2595⟩; **4**, 857 ⟨73TL4707⟩; **6**, 365 ⟨74T1769⟩
Richmond, H. H., **4**, 294 ⟨48JA3659⟩
Richmond, R. E., **3**, 206 ⟨83JA575⟩, 533 ⟨84JA4001⟩
Richmond, T. G., **6**, 717, 722 ⟨84OM314⟩
Richtarsky, G., **5**, 407 ⟨78CC946⟩
Richter, A. M., **4**, 870 ⟨83S840, 84EGP211356A1⟩; **6**, 500 ⟨79JPR827⟩, 729 ⟨82JPR669⟩
Richter, B., **3**, 91 ⟨89M765⟩; **6**, 511 ⟨78JOC1544⟩
Richter, D., **4**, 442 ⟨83PAC1799⟩; **5**, 674, 1053 ⟨80ZC97⟩
Richter-Addo, G. B., **4**, 715 ⟨90CJC41⟩
Richter, E., **2**, 906 ⟨01LA(315)26⟩
Richter, F., **4**, 555 ⟨92PS(73)67⟩, 563 ⟨89ZC409⟩; **6**, 213 ⟨B-28MI 607-01, B-41MI 607-01⟩, 213 ⟨B-59MI 607-01⟩
Richter, G. H., **2**, 133 ⟨29JA3131⟩; **4**, 728 ⟨50JOC1285⟩
Richter, H., **4**, 561 ⟨84EGP214610⟩; **5**, 924 ⟨88CB1363⟩, 1040 ⟨40CB843⟩
Richter, J. F. P., **6**, 297 ⟨72MI 610-01⟩
Richter, L., **3**, 706 ⟨93CB427, 93LA427⟩
Richter, M., **6**, 565 ⟨84JPR633⟩
Richter, P., **5**, 374 ⟨60CB2804⟩
Richter, P. H., **6**, 556 ⟨92PHA251⟩
Richter, R., **2**, 25 ⟨83JOC2625⟩; **3**, 550 ⟨79JOC4877⟩, 684 ⟨67JOC2303⟩; **4**, 315 ⟨B-77MI 407-01⟩; **5**, 657 ⟨89ZAAC(576)203⟩, 962, 1017, 1018 ⟨B-77MI 526-01⟩, 971 ⟨74JOC2897⟩; **6**, 143 ⟨68CB3002, 69CB931, 73JOC2614⟩, 364 ⟨68TL5037⟩, 445 ⟨81JOC5226, 86JOC4483⟩, 448 ⟨84JOC3675⟩, 500 ⟨81JOC3011⟩, 602, 608 ⟨B-77MI 620-01⟩, 620 ⟨66CB2454⟩
Richter, R. H., **6**, 509 ⟨79USP4154931⟩
Richter, W., **1**, 415, 827 ⟨71CB3679⟩, 737 ⟨76CB473⟩; **2**, 252 ⟨90S361⟩; **4**, 574 ⟨77CB677⟩, 598 ⟨75CB2649, 76CB473⟩, 598 ⟨77CB1312⟩; **5**, 370 ⟨90S661⟩, 685 ⟨86TL2743, 89CB657⟩, 685 ⟨89CB669⟩, 920, 922 ⟨77CB677, 77CB1312⟩; **6**, 702, 707, 708 ⟨77CB677⟩
Richter, W. J., **2**, 436 ⟨82AG(E)292, 83CB3293⟩, 436 ⟨85CB97⟩, 436, 443, 450 ⟨86JOM(301)289⟩, 443 ⟨84CB2328⟩; **3**, 376 ⟨76JA3615⟩; **5**, 908 ⟨84T3273⟩
Richtmyer, N. K., **4**, 223 ⟨43JA1477⟩, 235 ⟨66CAR(2)461⟩
Richtzenhain, K., **2**, 943 ⟨79CB2829⟩
Rick, E. A., **3**, 661 ⟨71JOM(28)287⟩
Rickard, C. E. F., **3**, 486 ⟨84CC1002⟩; **4**, 582 ⟨91JCS(D)609⟩; **5**, 400 ⟨88JOM(353)355⟩, 941 ⟨92OM809⟩; **6**, 247 ⟨90JOM(388)391, 90JOM(394)615⟩, 722 ⟨88JOM(338)393⟩
Rickards, R. W., **1**, 100 ⟨69AJC1321⟩, 520, 1134 ⟨84JOC1898⟩, 1091 ⟨81TL1437, 82OS(60)81⟩; **3**, 58, 273 ⟨81TL1437⟩
Rickborn, B., **1**, 212 ⟨67JA5850⟩, 217 ⟨73JOC4346⟩, 384 ⟨91COS(3)721⟩, 386 ⟨71JA1693, 91COS(3)733⟩, 854 ⟨61JA3688⟩, 871, 881 ⟨61JA3678⟩; **2**, 59 ⟨70JA6894⟩, 524 ⟨67JA2782⟩
Ricks, M. J., **6**, 28 ⟨83USP4419514⟩
Rico, I., **2**, 19 ⟨86TL1499, 89BSF256⟩; **6**, 229 ⟨81T4209⟩, 233 ⟨81TL323⟩
Rico, J. G., **1**, 688 ⟨89JOC3748⟩, 1089 ⟨87JOC4817⟩
Rico, M., **4**, 314 ⟨69JHC921⟩
Rico, R., **6**, 469 ⟨88CJC2234⟩
Ridd, J. H., **2**, 798 ⟨72CC641, B-76MI 215-02⟩
Riddick, R. D., **2**, 370 ⟨71JCS(A)1125⟩
Riddle, C., **2**, 474 ⟨69JCS(A)2704⟩

Ridenour, M., **1**, 935 ⟨88OM2220⟩
Ridge, R. J., **2**, 230 ⟨82CL921⟩
Ridgway, C.,**1**, 1137 ⟨93JCS(D)2629⟩; **5**, 943 ⟨90OM2621⟩
Ridgway, L. R., **2**, 882 ⟨22JCS104⟩
Ridi, M., **5**, 583 ⟨59G526⟩
Riding, G. H., **4**, 674 ⟨80JOM(202)C49⟩, 706 ⟨84JA2337⟩
Ridley, D., **2**, 421 ⟨65JCS1870, 67JCS(A)56⟩
Ridley, D. D., **1**, 1001 ⟨68AJC2989⟩
Ridley, D. R., **2**, 504 ⟨71MI 209-01⟩, 891 ⟨74IC1252⟩
Ridley, W. P., **2**, 584 ⟨79JA1442⟩
Riebel, A. H., **2**, 782 ⟨60JA1801⟩
Riebel, H. J., **6**, 632 ⟨88GEP(O)3634929⟩
Rieber, M., **1**, 157 ⟨49LA(562)177⟩; **2**, 830 ⟨49LA(562)187⟩; **5**, 1016 ⟨65JCS4256⟩
Rieber, N., **3**, 328 ⟨67AG(E)884⟩, 474, 478, 487 ⟨69CB488⟩, 474, 487 ⟨67AG238, 67AG(E)261⟩, 484 ⟨67AG906, 69LA(730)1⟩; **5**, 844, 871, 874 ⟨69CB488⟩, 871, 874 ⟨67AG(E)261⟩, 873 ⟨67AG(E)884⟩
Riebsomer, J. L., **5**, 62 ⟨55OSC(3)326⟩
Rieche, A., **2**, 96 ⟨58AG251, 58CB1212⟩, 96 ⟨65AG(E)524⟩, 182, 617 ⟨59CB83⟩, 686 ⟨63CB3044⟩, 886 ⟨64LA(678)167⟩, 888 ⟨67CB1544⟩; **3**, 92, 101 ⟨63CB308⟩, 99, 101 ⟨60CB88⟩; **4**, 212 ⟨56CB1254⟩; **6**, 39 ⟨67OS(47)47⟩, 43, 251, 429 ⟨61CB544⟩, 69, 82 ⟨61CB538⟩, 79 ⟨58CB1942⟩, 229 ⟨58CT515⟩
Rieck, H.-P., **2**, 176 ⟨76CZ391⟩; **4**, 282 ⟨76ZN(B)153⟩; **6**, 255 ⟨76CZ391⟩
Rieck, J. N., **2**, 768 ⟨73JOC4339⟩
Ried, V., **3**, 453 ⟨69LA(726)81⟩
Ried, W., **2**, 662 ⟨68LA(717)54⟩, 744 ⟨76S535⟩; **3**, 64 ⟨59LA(622)37⟩, 85 ⟨58AG165⟩, 281 ⟨62LA(653)181⟩, 399 ⟨76AG(E)704⟩, 470 ⟨66LA(691)50⟩, 490 ⟨65LA(681)45⟩, 560 ⟨66LA(700)32⟩; **4**, 307 ⟨65LA(687)187⟩, 738 ⟨64LA(679)51⟩, 901 ⟨86CB1745, 88CB805⟩, 972, 982 ⟨63CB3306⟩; **5**, 482 ⟨64LA(673)124⟩, 586 ⟨60AG268, 66LA(695)217⟩, 597, 599, 614 ⟨64LA(673)132⟩, 599 ⟨72LA(757)147⟩, 658 ⟨86LA1997⟩, 690 ⟨88LA1107⟩, 718 ⟨87CB657⟩, 726 ⟨64LA(676)114⟩, 739 ⟨70CB32⟩, 755 ⟨71CB3329⟩, 855 ⟨52CB470⟩, 856 ⟨64LA(678)132⟩, 857 ⟨53LA(581)44⟩, 858 ⟨53LA(581)23⟩, 1119 ⟨87CB1597⟩; **6**, 27 ⟨56CB2578⟩, 111 ⟨88LA599⟩, 360 ⟨88CZ382⟩, 535 ⟨64LA(674)124⟩, 537 ⟨64LA(590)123⟩, 540 ⟨80CB750⟩, 578 ⟨61LA(646)96⟩, 616 ⟨85CB1371⟩, 620 ⟨76CB1643, 76CB2921, 79CB640⟩, 620, 621 ⟨80CB2583⟩, 620, 621, 623 ⟨80S619, 86LA1997⟩, 621 ⟨87CZ339⟩
Riede, J., **2**, 446 ⟨85CB3105⟩; **3**, 386 ⟨84AG(E)726, 89JOM(364)155⟩, 386 ⟨89JOM(377)105⟩; **4**, 94 ⟨85OM1208⟩, 551, 560, 1023 ⟨84AG(E)247, 85CB3105⟩; **5**, 108, 887 ⟨85JOM(287)⟩, 605, 614, 627 ⟨88CB2095⟩, 1088 ⟨86CB2080⟩; **6**, 164, 371, 373 ⟨88ZN(B)1416⟩, 383 ⟨89ZN(B)796⟩, 392, 710 ⟨84JOM(271)381⟩, 392, 710 ⟨87OM35⟩, 610 ⟨86OM2187, 87OM32⟩
Riedel, H., **5**, 674 ⟨83AP(316)21⟩
Riedel, K., **2**, 802 ⟨68AG(E)138⟩
Riedel, R., **5**, 1077 ⟨93CB2569⟩
Rieder, W., **2**, 160 ⟨75T505⟩, 160, 164 ⟨74TL3319⟩; **5**, 766 ⟨74TL3319⟩
Riediger, W., **4**, 989 ⟨82JPR309⟩
Riediker, M., **1**, 536 ⟨81TL2629, 81TL4655⟩; **2**, 80 ⟨89AG(E)494⟩, 676, 982 ⟨82JA5842⟩
Riedl, B., **5**, 363 ⟨92CC1186⟩
Riedmüller, S., **5**, 516, 520 ⟨74CB915, 75GEP(O)2410157⟩
Riefling, B., **1**, 492 ⟨78JOC1468⟩, 1108 ⟨78JOC131⟩
Riegel, B., **1**, 881 ⟨45JA1273⟩; **3**, 227 ⟨57JA6308⟩
Riegel, R., **5**, 62 ⟨43OSC(2)126⟩
Rieger, B., **2**, 437 ⟨88CB1123⟩
Rieger, D. L., **2**, 532 ⟨92T2127⟩
Rieger, H., **3**, 88 ⟨90CB1357⟩, 539 ⟨88LA943⟩
Rieger, R., **5**, 199 ⟨90S697⟩
Riehl, J. J., **1**, 108, 115 ⟨74TL1451⟩; **3**, 54 ⟨79S507⟩
Riehl, L., **2**, 183 ⟨55AG32⟩
Rieke, R. D., **1**, 115 ⟨90JOC5045⟩, 116 ⟨89JOC3247⟩, 121 ⟨84JOC5280, 87JOC5056⟩, 121 ⟨88JOC4482, 92JA5110, 93JOC2483, 93JOC2492⟩, 123, 467 ⟨91JOC1445⟩, 207 ⟨93TL3063⟩, 472 ⟨79JOC3445⟩, 478 ⟨87JOC5057⟩, 687, 692 ⟨88CRV733⟩; **2**, 74, 585 ⟨92JA5110⟩, 558 ⟨77ACR301⟩, 558, 559 ⟨87JOC3674⟩, 558, 560, 959 ⟨74JA1775⟩, 558, 958 ⟨72JA7178⟩, 561 ⟨89JOC3247⟩, 563, 959 ⟨90JOC5045⟩, 590 ⟨81JOC4323⟩, 591 ⟨85JOC416⟩, 597 ⟨74JOM(67)C64, 74SRI373⟩, 958 ⟨73CC269, 73CC879, 75TCC1, 89SCI1260⟩, 958, 959 ⟨84JOC5280, 88JOC4482⟩, 959 ⟨91JOC1445⟩; **3**, 128 ⟨92JOC2493⟩, 130 ⟨91JOC1445⟩, 130, 286 ⟨89SC1833⟩, 146, 289, 303 ⟨83TL2451, 88JOC4482⟩, 286 ⟨88TL4513, 91JA4672, 93JOC2492⟩, 289 ⟨85JOC1373⟩; **5**, 1139 ⟨79JOC3069⟩
Riekens, R., **6**, 267 ⟨71ZN(B)480⟩
Rieker, A., **1**, 922 ⟨67TL3877⟩
Rieker, C., **6**, 355, 357 ⟨85CC1800⟩
Rieker, M., **2**, 419 ⟨65JCS4256⟩
Rieker, W. F., **1**, 149 ⟨83JA2082⟩, 149 ⟨84JA3270⟩, 355 ⟨84JOC1353⟩, 539 ⟨88JA7178⟩
Riemann, A., **4**, 1049 ⟨85CB2493⟩; **6**, 390 ⟨85CB2493⟩
Riemenschneider, J. L., **1**, 906 ⟨73JA8698⟩
Riemenschneider, K., **1**, 325, 920 ⟨87JOC205⟩
Riemer, M., **4**, 453 ⟨85ZC172⟩; **5**, 883 ⟨83ZC99⟩; **6**, 686 ⟨83ZC99⟩
Riemer, R., **4**, 571 ⟨92S787⟩; **5**, 226 ⟨87S931⟩; **6**, 702 ⟨92S787⟩
Riemschneider, R., **2**, 1012, 1013 ⟨62M952⟩, 1014 ⟨61LA(640)5⟩; **5**, 1043 ⟨59ZN(B)814⟩; **6**, 27 ⟨51M1008⟩
Rienäcker, G., **5**, 560 ⟨88ZC269⟩
Rienäcker, R., **1**, 795 ⟨85AG(E)320⟩; **4**, 694 ⟨69AG(E)677⟩, 698 ⟨77LA1633⟩
Riener, T., **3**, 621 ⟨45JA601⟩
Riener, T. W., **5**, 1043 ⟨45JA1178⟩
Riéra, A., **1**, 217 ⟨89TL6963⟩
Riera, A., **1**, 530 ⟨90JA9388⟩; **2**, 84, 318 ⟨91TL6931, 91TL6935⟩, 1026 ⟨82T1505, 87OS(65)68⟩, 1027 ⟨88CC942⟩; **4**, 763 ⟨94JA2153⟩, 851 ⟨90TL2169⟩
Riera, J., **2**, 1010 ⟨71JA2215⟩; **4**, 764 ⟨86JOC1413⟩; **6**, 30 ⟨91JOC103⟩
Riera, V., **4**, 674 ⟨93JA3786⟩, 678 ⟨92OM370⟩, 683 ⟨79JCS(D)1801⟩; **5**, 799 ⟨88JCS(D)1591⟩; **6**, 327 ⟨92AG81⟩, 328 ⟨91JOM(420)C12⟩, 328 ⟨93OM1394⟩, 329 ⟨94JOM(467)231⟩, 329, 330 ⟨92POL2713, 93OM2888, 94OM1336⟩, 331 ⟨87CC472, 91OM384, 91OM1683⟩
Riesel, L., **2**, 105 ⟨90ZAAC(580)217⟩, 449 ⟨91ZAAC(603)145⟩, 471 ⟨92ZAAC(607)161⟩; **5**, 1131 ⟨92EGP300936A7⟩; **6**, 243 ⟨90ZAAC(588)26⟩, 661, 733 ⟨78JPR389⟩
Rieser, J., **3**, 373 ⟨76LA641⟩
Riesle, L., **2**, 252 ⟨86GEP231075⟩
Riess, J. G., **2**, 10 ⟨72TL847, 74T3793⟩, 483 ⟨72SRI53⟩; **4**, 579 ⟨80JOM(187)331⟩
Riesz, P., **1**, 895 ⟨57JA3724⟩
Rieth, K., **1**, 619 ⟨89TL3131⟩; **4**, 282 ⟨66AG(E)594⟩, 331 ⟨90CB1563⟩; **5**, 489 ⟨90CB1563, 90CB1989⟩, 489, 490 ⟨86TL167⟩
Rieveschl, Jr., G., **5**, 114 ⟨55OSC(3)420⟩
Riezebos, G., **1**, 832 ⟨72JCS(P1)367⟩
Riffel, H., **2**, 1073 ⟨87ZAAC(548)63⟩, 1074 ⟨88ZAAC(567)39⟩
Rifi, M. R., **1**, 865 ⟨63JA3041⟩
Rifqui, M., **6**, 264 ⟨91TL5965⟩
Rigamonti, J., **2**, 614 ⟨61JOC5149⟩
Rigat, L., **3**, 361 ⟨92JCS(P1)351⟩
Rigaudy, J., **2**, 304 ⟨89TL2927, 89TL3977⟩, 305 ⟨90TL4735⟩; **5**, 380 ⟨80TL3367⟩
Rigby, G. W., **2**, 89 ⟨32JA2088⟩; **6**, 2 ⟨49JOC747⟩
Rigby, H. L., **2**, 29 ⟨93SC993⟩
Rigby, J. H., **1**, 145 ⟨93SL829⟩, 218, 221 ⟨87JOC34⟩, 815 ⟨82JOC4815⟩; **2**, 658 ⟨78TL1667⟩; **4**, 25 ⟨89S188⟩; **5**, 325 ⟨89JOC5852⟩, 606 ⟨93TL5397⟩, 990

⟨89JOC224⟩, 993 ⟨84JOC4569, 86JOC1374⟩, 993 ⟨89JOC4019⟩
Rigby, R. B., **2**, 193 ⟨73JCS(P1)676⟩; **6**, 45 ⟨72JCS(P1)155, 72JCS(P1)159⟩, 45 ⟨72JCS(P1)2438⟩, 45, 219 ⟨72JCS(P1)1506⟩, 233 ⟨72JCS(P1)2180⟩
Riggs, R. M., **4**, 227 ⟨92JMC533⟩
Righetti, P. P., **1**, 826 ⟨93T2093⟩; **4**, 986 ⟨79JPR43⟩
Righi, G., **1**, 10 ⟨91JOC6206, 92TL7429⟩
Righi, P., **2**, 391, 392 ⟨90OPP707⟩; **3**, 46, 191 ⟨90OPP707⟩
Righini-Tapie, A., **1**, 144, 237, 483 ⟨83T3283⟩
Rigny, R., **5**, 215 ⟨68CR(C)1303⟩
Rigo, B., **3**, 619 ⟨86TL347⟩, 659 ⟨85SC473⟩; **5**, 223, 374 ⟨88JHC59⟩, 227 ⟨89TL3073, 88JHC49⟩, 718 ⟨86JHC183⟩
Riguera, R., **3**, 412 ⟨82T1569⟩
Rihs, G., **1**, 918 ⟨83TL5857⟩; **2**, 74 ⟨92JA2321⟩; **3**, 268 ⟨85HCA1235⟩; **4**, 1034 ⟨92OM613⟩; **5**, 1124 ⟨85HCA1155⟩; **6**, 64 ⟨88JOM(339)17⟩
Rikers, R., **3**, 105 ⟨88S316⟩
Riks, G., **1**, 564 ⟨87JA4626⟩
Riley, C. M., **3**, 618, 619 ⟨80SC479⟩
Riley, D. P., **2**, 348 ⟨85JOC1563, 88JOC1695⟩
Riley, H. L., **2**, 101 ⟨35JCS901⟩
Riley, P. E., **4**, 679 ⟨80JA1752⟩
Riley, P. I., **2**, 565 ⟨82COMC-I(3)559, 82COMC-I(3)635⟩, 949 ⟨80JOM(202)5⟩; **6**, 194 ⟨79JOM(181)25⟩, 195 ⟨81JCS(D)814⟩
Riley, T. A., **5**, 516, 522 ⟨85JOC1741⟩
Rima, G., **4**, 141 ⟨88SRI317⟩, 141, 372 ⟨84SRI21⟩, 362 ⟨88SRI317⟩, 627 ⟨88JOM(345)39⟩; **5**, 388, 389 ⟨88EJM257⟩
Rim, J. G., **4**, 849 ⟨90JOC5555⟩
Rimmler, G., **5**, 489, 490 ⟨86TL167⟩
Rimoldi, J. M., **6**, 550 ⟨93JOC3798⟩
Rimpler, M., **2**, 214 ⟨78CB3159⟩; **3**, 27 ⟨85CB4288⟩
Rinderknecht, H., **1**, 881 ⟨51JA5770⟩
Rindone, B., **2**, 649 ⟨81T2313⟩
Rinehart, Jr., K. L., **3**, 461 ⟨53JOC1594⟩; **5**, 98 ⟨63OSC(4)444⟩
Ring, H., **3**, 694 ⟨50MI 321-01⟩
Ring, M. A., **6**, 173 ⟨61JA802⟩
Ring, R. N., **3**, 451 ⟨61JOC579⟩
Ringe, K., **1**, 979 ⟨92TL8017⟩
Ringer, J. W., **4**, 701 ⟨88JOC455⟩
Ringold, C., **1**, 314 ⟨88JOC50⟩
Ringold, H. J., **1**, 555 ⟨67JCS(C)1720⟩, 795, 871, 881 ⟨65JA3228⟩, 871 ⟨62JA3402⟩, 871, 881 ⟨63JA1538, 64JA1997, 66JA1332⟩, 871, 881, 882 ⟨62TL669, 57JOC602⟩; **3**, 218 ⟨56JA876⟩, 234 ⟨56JOC1432⟩; **4**, 191 ⟨54JA1359⟩
Ringsdorf, H., **1**, 654 ⟨78CC358⟩; **3**, 499 ⟨67JPS(B)1039⟩
Ringwald, E. L., **4**, 205, 213 ⟨47JA2358⟩
Rini, J., **1**, 140 ⟨85JOC3676⟩; **3**, 350 ⟨85JOC3676⟩
Rinkes, I. J., **2**, 1040 ⟨28RTC960⟩
Rinzema, L. C., **4**, 844, 846 ⟨59RTC354⟩; **6**, 83, 84 ⟨59RTC354⟩
Rio, G., **2**, 209 ⟨69BSF3609⟩
Riobe, D., **4**, 950 ⟨77CR(C)281⟩
Riobe, O., **3**, 243 ⟨71CR(272)1045⟩
Riondel, A., **4**, 211 ⟨87SC1467⟩; **5**, 1105 ⟨88T1619⟩
Riordan, P. D., **3**, 698 ⟨83T2989⟩, 698, 721 ⟨82CC942⟩; **6**, 515 ⟨83T2989⟩
Riordan, R. P., **4**, 465 ⟨63JOC2898⟩
Rios, R., **3**, 310 ⟨93SC985⟩
Rioult, P., **5**, 558 ⟨77RTC254⟩
Rioult, R., **3**, 351 ⟨77BSF688⟩
Ripamonti, A., **2**, 959 ⟨78CC927⟩
Ripka, W. C., **2**, 19 ⟨74OR(21)1⟩; **4**, 8 ⟨74OR(21)1⟩
Ripoll, J.-L., **1**, 772 ⟨93S659⟩, 786 ⟨78T19⟩, 786, 787 ⟨85S121⟩, 787 ⟨79NJC195, 80T2497⟩; **2**, 636 ⟨79NJC195⟩; **3**, 233 ⟨85S121⟩, 335 ⟨90JOC2596⟩, 335 ⟨93JA324, 90TL6183⟩, 359 ⟨91TL2013⟩, 369 ⟨91MI 318-01⟩, 550 ⟨86TL4313⟩, 564 ⟨88T4207⟩, 582 ⟨78TL463, 83CC238⟩, 582 ⟨86JCR(S)188⟩; **5**, 559 ⟨93TL2605⟩, 563 ⟨91CC964⟩
Rippel, H. C., **5**, 896 ⟨89TL5261⟩
Ripperger, H., **2**, 343 ⟨92LA1091⟩; **5**, 574 ⟨87JPR1039⟩
Risaliti, A., **2**, 783 ⟨66TL1821, 66TL6331⟩, 783 ⟨68T1889, 73T3011⟩, 783 ⟨74JCS(P1)660, 74T2741⟩; **4**, 273, 274 ⟨91JOC3530⟩; **5**, 354 ⟨79TL4761, 83JHC305⟩, 354 ⟨85G663⟩
Risbood, P. A., **2**, 30 ⟨78JA4919⟩, 373, 376, 403 ⟨84JOC336⟩
Rise, F., **1**, 546 ⟨87JA3161⟩; **6**, 134 ⟨86JOM(303)189⟩
Risener, R., **2**, 910 ⟨86JOC2381⟩
Rising, A., **2**, 782 ⟨01LA(316)292⟩
Risius, A. C., **1**, 74 ⟨76T2157⟩
Risley, H. A., **5**, 735 ⟨67JOC392⟩
Rissolo, M. A., **3**, 39 ⟨91SC419⟩
Rist, C. E., **6**, 276 ⟨69JOC1642⟩
Rist, G., **1**, 7, 833 ⟨83JOC860⟩; **2**, 444 ⟨87PS(33)41⟩; **4**, 105 ⟨87PS(33)41⟩, 673 ⟨88JOM(342)45⟩
Ristic-Petrovic, D., **4**, 715 ⟨92OM864⟩
Ritchey, W. M., **5**, 389 ⟨69ICA495⟩
Ritchie, C. D., **1**, 869 ⟨67JA1721, 67JA2752⟩, 869 ⟨67JA2960, 68JA2821⟩, 869 ⟨68JA3415, 69JA6749⟩
Ritchie, E., **1**, 36 ⟨68AJC1375⟩, 1001 ⟨68AJC2989⟩
Ritchie, T. J., **3**, 292 ⟨90T2135⟩
Ritieni, A., **2**, 21 ⟨90S212⟩
Ritschel, W., **3**, 590 ⟨80CB811⟩
Ritsema, J. C., **3**, 398 ⟨03JPR72⟩
Rittby, C. M. L., **5**, 1096 ⟨91JCP(95)5609⟩
Rittby, M., **3**, 582 ⟨90JA3779⟩
Ritter, A., **1**, 1045, 1138 ⟨63CB3280⟩; **2**, 262 ⟨61CB1263⟩, 541 ⟨63CB3289⟩; **3**, 273 ⟨63CB3280⟩; **5**, 374 ⟨60CB2804⟩, 376 ⟨61CB1263⟩, 379 ⟨63AG93⟩, 718 ⟨63AG96, 63CB1473⟩, 781 ⟨77JOM(135)183⟩; **6**, 507 ⟨60CB2810⟩
Ritter, A. R., **3**, 698, 707 ⟨83TL655⟩; **5**, 330 ⟨94JOC4602⟩
Ritter, D. M., **2**, 109 ⟨53JA192⟩; **6**, 173 ⟨61JA802⟩
Ritter, G., **5**, 192 ⟨86GEP3511050⟩
Ritter, J. J., **1**, 637 ⟨48JA4045⟩, 658 ⟨68IC1014⟩; **4**, 632, 644 ⟨68JOM(12)269⟩; **5**, 281, 282 ⟨48JA4045⟩
Ritter, J. R., **2**, 308 ⟨48JA4045⟩
Ritter, K., **1**, 433, 480 ⟨93S735⟩, 831 ⟨90TL869⟩; **2**, 689 ⟨93S735⟩; **3**, 285 ⟨85TL1285⟩
Ritter, W., **4**, 437 ⟨66CB1678⟩
Rittersbacher, H., **5**, 1018 ⟨64CB849⟩
Rittle, K. E., **4**, 417 ⟨87JOC2442⟩
Ritz, M., **3**, 470 ⟨66LA(691)50⟩
Ritzel, R., **2**, 99 ⟨86SUL161⟩
Riva, A., **5**, 652 ⟨61CI(M)1243⟩, 980 ⟨60CI(M)1243⟩
Riva, S., **2**, 151 ⟨92TA1063⟩
Rivadeneira, E., **2**, 533 ⟨90TL3863⟩
Rivalle, C., **1**, 42 ⟨76T829⟩
Rivarda, E., **2**, 695, 886 ⟨72JOM(38)91⟩
Rivas-Enterrios, J., **5**, 53 ⟨87JOC689⟩
Rivat, G., **4**, 59 ⟨19CR(169)1143⟩
Rivera, A. D., **4**, 212 ⟨84JOC4958⟩
Rivera, I., **2**, 58, 526 ⟨90TL4677⟩; **4**, 353 ⟨89JOC4051, 90TL4677⟩, 935 ⟨89TL3919⟩
Rivera, M., **4**, 897 ⟨86RTC456⟩
Rivero, B. E., **2**, 1008 ⟨87CJC2000⟩
Rivero, I. A., **2**, 8 ⟨93SC711⟩; **6**, 416 ⟨91SC285⟩
Rivero, R. A., **1**, 573, 581, 584 ⟨92JA9419⟩
Rivers, G. T., **1**, 632 ⟨77JA3414⟩
Rivetti, F., **1**, 318 ⟨78JOC1521⟩; **6**, 465 ⟨93EUP534545⟩
Rivier, H., **5**, 476, 477 ⟨43HCA1722⟩, 566, 618, 735 ⟨37HCA691⟩
Riviere, H., **3**, 83 ⟨80CC432⟩
Riviere, M., **3**, 406 ⟨74S812⟩
Rivière, P., **1**, 766 ⟨80JA5413⟩; **2**, 815 ⟨83RTC65⟩, 946, 948 ⟨82COMC-I(2)399⟩, 947, 949 ⟨73JOM(56)1, 80JOM(202)5⟩; **3**, 457 ⟨87JOM(331)11⟩, 503 ⟨80JA5413⟩; **4**, 360 ⟨92JOM(433)49⟩, 361 ⟨71AG(E)267, 82JOM(232)137⟩, 519 ⟨82COMC-

I(2)365⟩, 521 ⟨72JOM(34)C18, 79JOM(168)43⟩; 432 ⟨90JOM(397)7, 92JOM(433)49⟩
Rivière-Baudet, M., **2**, 815 ⟨83RTC65⟩, 946, 948 ⟨82COMC-I(2)399⟩, 949 ⟨80JOM(202)5⟩; **4**, 519 ⟨82COMC-I(2)365⟩, 521 ⟨79JOM(168)43⟩; **5**, 388 ⟨92MI 507-01⟩, 389 ⟨68BSF4093, 92JOM(423)C5, 92PS(70)75⟩
Rivkin, B. B., **4**, 290 ⟨91HAC307⟩, 290, 291 ⟨88KGS1144⟩
Rivlin, J., **3**, 65 ⟨65TL821⟩
Rix, Th. R., **4**, 884 ⟨54MI 420-01⟩; **6**, 74 ⟨53MI 603-01⟩
Rizk, M., **1**, 816 ⟨70LA(734)106⟩
Rizkalla, N., **5**, 192 ⟨82USP4335058⟩
Rizpolozhenskii, N. I., **2**, 451 ⟨52MI 208-01⟩, 460 ⟨66DOK(168)807⟩, 462 ⟨57MI 208-01, 66IZV493⟩, 463 ⟨67IZV1976⟩
Rizvi, A. Q. A., **2**, 718 ⟨81JOC3467⟩
Rizvi, R., **6**, 553 ⟨81TL3485⟩
Rizvi, S. S. A., **5**, 22 ⟨91MI 507-04⟩
Rizzacasa, M. A., **4**, 856 ⟨87JCS(P1)2017⟩
Rizzardo, E., **2**, 674 ⟨75CC644⟩; **4**, 107 ⟨85AJC1505⟩
Rizzo, V. L., **4**, 442 ⟨70JHC439⟩
Rizzolio, M., **1**, 579 ⟨90T2411⟩
Roach, B. L., **1**, 141, 234, 821 ⟨78JOC1185⟩
Roark, D. N., **4**, 614, 621 ⟨72JA5837⟩
Robak, E. A., **3**, 361, 371 ⟨68CJC1855⟩
Robaugh, D. A., **1**, 922 ⟨86JA3224⟩
Robba, M., **3**, 102 ⟨70BSF1445⟩, 103 ⟨70BSF1838, 77BSF142⟩, 105 ⟨67BSF2495⟩; **5**, 706 ⟨84JHC91⟩
Robb, C. M., **5**, 60 ⟨73OSC(5)635⟩
Robb, M. A., **4**, 610 ⟨93JA3322⟩
Robbe, S., **2**, 589 ⟨92CB2409⟩
Robbins, C. M., **1**, 649 ⟨76CC697⟩; **4**, 357 ⟨76CC697⟩, 519 ⟨81JCS(P1)1934⟩
Robbins, J., **3**, 513 ⟨90JA3875⟩
Robbins, J. D., **2**, 252 ⟨92PS(68)147⟩
Robbins, L. V., **3**, 351 ⟨74JOC2509⟩
Robbins, M. D., **2**, 206 ⟨74JOC2459⟩
Robbins, R. F., **6**, 511 ⟨66CI(L)1266⟩
Röber, K. C., **6**, 207 ⟨74JPR557⟩
Roberge, G., **2**, 667 ⟨81S381⟩
Roberge, J. Y., **1**, 558 ⟨91TL5219⟩
Robers, E., **2**, 780 ⟨50JCS2657⟩
Robert, A., **3**, 650 ⟨86TL5491⟩; **4**, 183 ⟨71TL231⟩, 870 ⟨90CC470⟩; **5**, 25, 56 ⟨86T2275⟩, 652 ⟨87S306⟩; **6**, 30 ⟨91JOC103⟩, 120, 123 ⟨90CC470⟩, 629 ⟨78JOC3732⟩
Robert, D. U., **2**, 10 ⟨72TL847⟩
Robert, J. B., **2**, 852 ⟨76JPC2417⟩
Robert, P., **1**, 1071 ⟨91SL539⟩
Robert, P. C., **1**, 774 ⟨87TL1519⟩
Robert, R. A., **2**, 913, 914 ⟨61JA4385⟩
Roberto, D., **5**, 356 ⟨90OM1245⟩
Roberts, A. T. L., **6**, 394, 395 ⟨86CC908⟩
Roberts, B. G., **6**, 252 ⟨72JOC1458⟩
Roberts, B. P., **1**, 5, 7, 9, 10, 16, 17, 331 ⟨91JCS(P1)103⟩, 5 ⟨B-83MI 101-01⟩, 75 ⟨68JCS(C)1474, 90TL5093⟩, 331, 936 ⟨89CC1387⟩, 657 ⟨72JCS(P2)803⟩, 919 ⟨B-71MI 119-05⟩, 935 ⟨B-83MI 119-01⟩, 937 ⟨B-83MI 119-02⟩, 945 ⟨78JCS(P2)1313⟩; **2**, 96, 98 ⟨76CC1022⟩, 536 ⟨89CC1387⟩
Roberts, B. W., **2**, 219 ⟨61JOC850⟩; **5**, 104 ⟨85OS(64)63⟩
Roberts, C. W., **4**, 22 ⟨73JOC153⟩
Roberts, D. A., **1**, 700 ⟨78JCS(P1)834⟩; **2**, 561 ⟨89TL6753⟩
Roberts, D. D., **1**, 894 ⟨79JOC2860⟩
Roberts, D. G. M., **6**, 317 ⟨92BMC1607⟩
Roberts, D. H., **1**, 936 ⟨86JA5893⟩
Roberts, D. L., **1**, 726 ⟨68JOC3566⟩
Roberts, E., **2**, 1034 ⟨60JA3472⟩
Roberts, E. C., **1**, 884 ⟨68JOC460⟩
Roberts, F. E., **2**, 188, 271, 719 ⟨63JOC593⟩, 528 ⟨91JOC763⟩; **4**, 48 ⟨75TL3979⟩
Roberts, H. L., **2**, 165 ⟨61JCS2066⟩, 165 ⟨61JCS2070⟩
Roberts, J. C., **2**, 533 ⟨89TL7357⟩; **5**, 74 ⟨86TL4537⟩
Roberts, J. D., **1**, 383 ⟨59JA4390⟩, 852 ⟨50JA408⟩, 895 ⟨52JA5943⟩; **2**, 11 ⟨51JA2509⟩, 89 ⟨58JA2584, 59T36⟩, 647 ⟨57JA1458⟩; **3**, 237 ⟨60JA3102⟩, 556 ⟨81OMR(17)180⟩; **4**, 9 ⟨69JA1034, 72JA2020⟩, 38 ⟨70JA7359⟩, 137 ⟨51JA1879⟩, 179 ⟨58JA5837⟩
Roberts, J. L., **2**, 45 ⟨74JCS(P1)1858⟩, 533, 697 ⟨87JOC1347⟩; **5**, 168 ⟨74JCS(P1)1864⟩
Roberts, J. R., **2**, 388 ⟨73JA3228⟩; **4**, 25 ⟨71JA4027⟩
Roberts, J. S., **1**, 232 ⟨73JA1547⟩; **2**, 214 ⟨88USP4721813⟩
Roberts, K. A., **2**, 1020, 1028 ⟨87JA228⟩, 1023, 1024 ⟨86JA7125⟩; **4**, 875, 877 ⟨84JOC1653⟩
Roberts, L. R., **1**, 692 ⟨92TL3899⟩
Roberts, M., **1**, 918 ⟨72JA7063⟩
Roberts, M. R., **1**, 209 ⟨78JOC788⟩; **3**, 37 ⟨81JA724⟩
Roberts, N. K., **2**, 442 ⟨79JA6254⟩
Roberts, P. J., **4**, 582, 583 ⟨74CB3706⟩; **6**, 620 ⟨80JCS(P2)867⟩
Roberts, R. A., **1**, 19 ⟨84JA6690⟩, 816 ⟨87JOC2960⟩
Roberts, R., **2**, 385 ⟨55OSC(3)244⟩; **3**, 465 ⟨55OSC(3)244⟩
Roberts, R. M., **1**, 180, 212 ⟨B-84MI 105-01⟩, 795 ⟨67JA1404⟩, 827 ⟨73JA3348⟩; **5**, 106 ⟨55OSC(3)272⟩, 743 ⟨54JA2411⟩, 746 ⟨49JA3848, 49JOC277⟩; **6**, 74 ⟨58JA1247⟩, 78 ⟨55JA3801⟩
Roberts, R. M. G., **4**, 724 ⟨86JOM(307)231⟩
Roberts, S., **3**, 527 ⟨70HCA417⟩
Roberts, S. J., **2**, 340, 344, 346 ⟨79COC(2)185⟩
Roberts, S. M., **1**, 4, 9 ⟨91JCS(P1)656⟩, 173 ⟨90JCS(P1)2352⟩
Roberts, S. M.,**1**, 522 ⟨91JCS(P1)1255⟩
Roberts, S. M., **1**, 826 ⟨83CC932⟩, 835 ⟨90JCS(P1)1493⟩, 936 ⟨79S471⟩, 1024, 1134 ⟨79JCS(P1)2954, 80JCS(P1)852⟩; **2**, 18 ⟨93JCS(P1)1795⟩, 19 ⟨88JCS(P1)549⟩, 628 ⟨88CC971⟩; **3**, 539 ⟨89CC303⟩; **4**, 650 ⟨91JCS(P1)2276⟩; **5**, 262 ⟨B-89MI 502-01⟩; **6**, 543 ⟨91JCS(P1)787⟩
Roberts, T. G., **1**, 756 ⟨85JCS(P1)1953⟩
Roberts, W., **5**, 106, 298 ⟨65JCS1290⟩
Roberts, W. L., **1**, 399 ⟨79JOC2348⟩
Robertson, A., **3**, 532 ⟨48JCS1674⟩
Robertson, B. E., **2**, 897 ⟨82IC3731⟩
Robertson, F. M., **3**, 686 ⟨85JCS(P1)1517⟩
Robertson, G. B., **4**, 688 ⟨76IC97⟩
Robertson, G. M., **1**, 497 ⟨85T4001⟩, 690 ⟨86TL399, 88JCS(P1)1085⟩; **3**, 422 ⟨89JOC1236⟩
Robertson, G. R., **5**, 94 ⟨32OSC(1)406⟩
Robertson, H. E., **2**, 622 ⟨89CB2213⟩; **6**, 173, 378, 379 ⟨90AG(E)201⟩
Robertson, J., **3**, 143 ⟨88CC1404⟩
Robertson, J. E., **2**, 89 ⟨65JHC152⟩
Robertson, N., **6**, 457 ⟨93JCS(D)1031⟩
Robertson, W. A. H., **3**, 255, 360, 361 ⟨65JCS32⟩, 371 ⟨62JCS5104⟩; **6**, 113 ⟨65JCS32⟩
Robey, R. L., **1**, 1076 ⟨72AG(E)48⟩; **2**, 631, 964 ⟨71JA4845, 73JA982⟩, 982 ⟨71JA6267⟩; **4**, 194 ⟨76JA3037⟩
Robichaud, A. J., **3**, 422 ⟨89JOC1236⟩
Robin, J.-P., **1**, 473 ⟨77TL2015⟩
Robin, M. B., **5**, 685 ⟨B-70MI 517-01⟩
Robin, M. L., **5**, 809 ⟨84JA4266⟩; **6**, 301 ⟨87JFC(37)327⟩, 603 ⟨84JA4266⟩
Robin, P., **5**, 763 ⟨20CR(171)38, 21CR(172)452⟩
Robins, D. J., **2**, 323 ⟨93MI 205-01⟩
Robins, M. J., **1**, 939 ⟨81JA932, 83JA4059⟩, 1010 ⟨81TL421, 83JOC1854⟩, 1010 ⟨90TL3731⟩; **2**, 617, 624 ⟨90TL5633⟩; **4**, 61 ⟨88TL5729⟩, 61, 73, 79 ⟨90JOC4757, 91JOC6878⟩, 62, 63 ⟨93JOC3800⟩, 218 ⟨94JOC544⟩; **6**, 46 ⟨90JOC4757⟩, 533 ⟨81JA932⟩, 535 ⟨83JA4059⟩
Robins, R. K., **2**, 364 ⟨88JHC1857⟩; **4**, 306 ⟨61JA2574⟩; **5**, 516, 522 ⟨85JOC1741⟩, 625 ⟨83JMC445⟩
Robinson, A., **4**, 239 ⟨50JA354⟩
Robinson, B. H., **4**, 721 ⟨94OM500, 94OM511⟩; **6**, 402 ⟨92JOM(437)323⟩
Robinson, B. L., **2**, 18 ⟨83JOC3195⟩

Robinson, C. H., **1**, 13 ⟨54JCS3045⟩; **2**, 24 ⟨59JA2191⟩, 389 ⟨66JOC524⟩
Robinson, E. A., **4**, 281 ⟨81SA(A)819⟩
Robinson, E. D., **1**, 552 ⟨89JOC5854⟩, 818 ⟨88TL3547, 88TL4913⟩, 818 ⟨90JOC227⟩, 957 ⟨90JOC3450⟩; **3**, 222 ⟨90JOC227⟩
Robinson, E. J., **4**, 1029 ⟨91JCS(D)351⟩
Robinson, G., **4**, 688 ⟨64PCS187⟩
Robinson, G. C., **1**, 665 ⟨82COMC-I(7)365⟩; **3**, 302, 303 ⟨66JOC4252⟩
Robinson, G. E., **6**, 625 ⟨80EUP14064⟩
Robinson, G. H., **4**, 563 ⟨88MI 412-01⟩
Robinson, G. M., **4**, 300, 315 ⟨21JCS1470⟩
Robinson, III, J., **6**, 611 ⟨92TL3113⟩
Robinson, J. A., **1**, 614 ⟨88JCS(P1)53, 88TL6357⟩
Robinson, J. E., **2**, 625 ⟨87TL3857⟩
Robinson, L. R., **4**, 1048 ⟨85JA3935⟩
Robinson, M., **2**, 747 ⟨75JCS(P1)1237⟩; **5**, 289 ⟨75JCS(P1)1237⟩
Robinson, M. D., **5**, 593 ⟨77JCS(P1)2357⟩
Robinson, M. R., **5**, 593 ⟨76JCS(P1)33⟩
Robinson, P. D., **1**, 694 ⟨90JOC2128⟩; **4**, 224 ⟨88JOC845⟩; **5**, 540 ⟨88JOC845⟩
Robinson, P. J., **1**, 652 ⟨69JCS(B)652⟩
Robinson, P. L., **2**, 171 ⟨29JCS1048⟩, 830 ⟨85JA5210⟩
Robinson, R., **1**, 1000 ⟨61T18⟩; **2**, 631 ⟨42JCS684⟩; **3**, 474 ⟨28JCS1310⟩; **4**, 318 ⟨B-49MI 407-01⟩; **5**, 559 ⟨67T3785⟩
Robinson, R. A., **1**, 829 ⟨93JA10400⟩
Robinson, R. E., **1**, 51, 52, 101 ⟨55JA3230⟩
Robinson, S. D., **5**, 781 ⟨92JCS(D)3199, 93JCS(D)2827⟩, 781 ⟨93POL563⟩
Robison, D. H., **2**, 402 ⟨68JA7173⟩
Robison, J. L., **4**, 606, 620, 623, 653 ⟨90OM2677⟩, 623 ⟨92OM3464⟩; **6**, 190 ⟨92OM3464⟩, 191 ⟨90OM2677⟩
Robison, M. M., **3**, 233 ⟨63JOC3132⟩
Robl, C., **5**, 1097 ⟨90AG(E)318⟩
Robl, J. A., **3**, 13, 85 ⟨90JCS(P1)757⟩, 433 ⟨85JOC5913⟩; **4**, 182 ⟨87JOC188⟩
Roblin, Jr., R. O., **4**, 181 ⟨33JA3741⟩
Roblin, R. O., **2**, 177 ⟨50JA4890⟩
Robson, E., **2**, 1051 ⟨68JCS(C)1324⟩, 1056 ⟨63AG1033⟩
Robson, J. H., **2**, 386 ⟨55JA2453⟩
Robson, J. M., **4**, 206 ⟨50JPP764⟩
Robson, M. J., **4**, 162 ⟨75AG(E)473⟩
Robson, P., **2**, 145 ⟨69JCS(C)2334⟩, 194 ⟨63JCS3692⟩; **4**, 74, 75 ⟨69JCS(C)2334⟩; **5**, 709 ⟨67JCS(B)1151⟩
Robyr, C., **1**, 195 ⟨88TL4709⟩, 195 ⟨94T415⟩; **2**, 394 ⟨89JOC1233⟩
Roch, J., **2**, 484 ⟨32CB409⟩
Rochas, P., **1**, 212 ⟨48BSF818, 48BSF822⟩, 212 ⟨48BSF825, 48BSF827⟩, 475 ⟨48BSF811⟩
Rochat, W. V., **5**, 246 ⟨69JOC4173⟩
Roche, D., **2**, 713 ⟨92TA629⟩
Roche, E. G., **1**, 1145 ⟨89JCS(P1)333⟩
Roche, J., **5**, 84 ⟨55FOR349⟩
Roche-Dolson, C. A., **2**, 608 ⟨85S669⟩
Rochet, P., **1**, 955, 971 ⟨93SL105⟩
Rochin, C., **2**, 392 ⟨87JOC701⟩
Rochov, E. J., **2**, 872 ⟨60JINC213⟩
Rochow, E. G., **1**, 881 ⟨64JOM(1)476⟩; **2**, 108 ⟨B-73MI 202-03⟩, 444, 824, 882 ⟨57JA5739⟩, 538 ⟨50IS50⟩, 600 ⟨53JA4103, 53JA4105⟩, 696 ⟨63AG(E)617⟩, 699 ⟨64JOM(1)476⟩, 815 ⟨67JOM(9)231⟩; **3**, 440, 456, 457 ⟨67JOM(9)231⟩, 456 ⟨63CB2132⟩, 568 ⟨63AG(E)617⟩; **4**, 140 ⟨60IS37⟩, 602, 652 ⟨55JA907⟩, 607, 638 ⟨54JOC250⟩, 668 ⟨53JOC1739⟩; **5**, 718 ⟨64CB627⟩, 721 ⟨63CB2138⟩
Rochs, J., **2**, 1097 ⟨83JCS(P1)1087⟩
Rock, M. H., **3**, 150 ⟨92SL483⟩; **4**, 949 ⟨92SL483⟩
Rockenbauer, A., **2**, 34 ⟨93T5133⟩; **6**, 546 ⟨90JCS(P1)2009⟩
Röcker, E., **4**, 19, 27, 29 ⟨83OMR(21)64⟩
Rockett, B. W., **4**, 706 ⟨65JA1241, 68JCS(C)656⟩
Rockett, J., **4**, 513 ⟨51JA5130⟩
Rockey, B., **5**, 291 ⟨68JA4706⟩
Rocklage, S., **3**, 517 ⟨80JMOC73⟩
Rocklage, S. M., **3**, 512 ⟨81JA1440⟩
Rockland, L. B., **5**, 84 ⟨47MI 502-01⟩
Rocks, R. R., **4**, 526 ⟨83OM1536⟩
Rockwell, N., **2**, 21 ⟨80TL177⟩
Rod, T., **2**, 858 ⟨73ACS1939⟩
Roddick, D. M., **6**, 519 ⟨85JA4670⟩
Rodé-Gowal, H., **3**, 293 ⟨74HCA2201⟩
Rodeheaver, G. T., **4**, 193 ⟨72TL3595⟩
Rodehorst, R. M., **2**, 957 ⟨81JA1847⟩
Roder, H., **4**, 343 ⟨70CB2984⟩
Röder, T., **4**, 574 ⟨81AG(E)1038, 88CB1509⟩, 574 ⟨91CB199⟩
Roder, T. M., **3**, 73 ⟨53JA1647⟩
Rodes, R., **4**, 1067 ⟨83JCS(P1)1087⟩
Rodewald, G., **2**, 868 ⟨80JOM(190)53⟩
Rodewald, H., **1**, 634 ⟨87TL1985⟩, 705 ⟨84CB966, 86CB1540⟩, 708 ⟨84CB1868⟩
Rodewald, P. G., **1**, 336, 926 ⟨74JA7977⟩
Rodewold, H., **3**, 616 ⟨85LA566⟩
Rodger, M., **2**, 278, 294 ⟨73JCS(D)1716⟩
Rodgers, A. S., **1**, 1109 ⟨69CRV279⟩
Rodgers, J. D., **1**, 821 ⟨88JA4822⟩; **3**, 331 ⟨88JA4822⟩; **6**, 598 ⟨91JMC3187⟩
Rodgers, J. R., **4**, 713 ⟨82JOM(240)299⟩
Rodgers, M., **2**, 1028 ⟨71T2651⟩
Rodgers, S. L., **2**, 30 ⟨78JOC2793⟩
Rodia, J. S., **5**, 751, 757 ⟨53JA1761⟩
Rodier, N., **1**, 526 ⟨87BSF1006⟩
Rodighiero, G., **4**, 427 ⟨51G80⟩
Rodina, L. L., **1**, 393 ⟨67RCR260⟩; **5**, 271, 292 ⟨67RCR260⟩
Rodini, D. J., **1**, 524, 525 ⟨79JA5283, 80JOC2773⟩
Rodionov, E. S., **5**, 722 ⟨72JGU1859⟩, 722, 978 ⟨72ZOB1870⟩, 968, 969 ⟨73ZOB1502⟩, 1015 ⟨71ZOB2340, 72ZOB879, 73ZOB1044⟩
Rodionov, I. L., **4**, 914 ⟨84ZOB416⟩
Rodionova, N. P., **5**, 822 ⟨60ZOB2409⟩
Rodios, N. A., **5**, 355 ⟨91JHC593⟩, 356 ⟨82S824⟩
Rodler, M., **2**, 636 ⟨83JST(97)47⟩; **3**, 582 ⟨84CPL447⟩
Rodríguez, M. A., **1**, 990 ⟨88JA5567⟩; **5**, 1044 ⟨87CC1491⟩, 1108 ⟨90TL7375⟩
Rodríguez, M. S., **1**, 931 ⟨86TL383, 87TL3397⟩, 931 ⟨93TL4865⟩
Rodricks, J. V., **1**, 785 ⟨63JA3524⟩; **6**, 651 ⟨73JOC1591⟩
Rodrigues, H. R., **3**, 90 ⟨79OR(26)1⟩
Rodrigues, R., **2**, 728 ⟨76JOM(114)281⟩; **4**, 88, 90, 348 ⟨76JOM(114)281⟩; **5**, 903 ⟨76JOM(114)281, 77JOM(124)1⟩; **6**, 697 ⟨76JOM(114)281⟩
Rodriguez, A., **3**, 621 ⟨84JOC276⟩
Rodriguez, A. D., **2**, 710 ⟨85T4443⟩
Rodriguez, B., **2**, 325 ⟨88S347⟩
Rodriguez, C. M., **2**, 155 ⟨92T3571⟩
Rodriguez, E., **3**, 173 ⟨78JOC4552⟩
Rodriguez, H., **5**, 759 ⟨82MI 519-01⟩
Rodriguez, H. R., **2**, 554, 653, 907, 908, 942, 965, 997 ⟨79OR(26)1⟩; **4**, 539 ⟨79OR(26)1⟩, 948 ⟨76OR(26)1⟩
Rodriguez, J., **1**, 573 ⟨93CC1038⟩, 797 ⟨89BSF799⟩, 832 ⟨93TL1149⟩, 955 ⟨88CC237, 93JOC5709⟩, 1131 ⟨87TL2009⟩; **2**, 3 ⟨93S1177⟩, 21 ⟨91SL477⟩; **3**, 21 ⟨93S371⟩
Rodriguez, J. B., **6**, 461, 476 ⟨88ZN(B)1038⟩, 477 ⟨89ZN(B)983⟩
Rodriguez, J.-G., **4**, 325 ⟨92JCS(P1)291⟩
Rodriguez, J. H., **3**, 170, 631 ⟨92JOC7235⟩
Rodriguez, M. A., **2**, 31, 35 ⟨87CC1491⟩, 1019 ⟨87S661⟩; **4**, 749, 819, 1064 ⟨88JA5567⟩, 760 ⟨90TL2751⟩, 772 ⟨90TL7375⟩
Rodriguez, M., **2**, 305 ⟨91TL923⟩; **3**, 44 ⟨90TL7319⟩
Rodriguez, M. J., **3**, 412 ⟨82T1569⟩

Rodriguez Amo, J., **4**, 858 ⟨93JCR(S)60⟩
Rodriguez, R., **4**, 249 ⟨93TL1141⟩; **5**, 513 ⟨86CL1251⟩
Rodriguez, R. C., **3**, 170 ⟨89JOC5620⟩
Rodriguez-Luis, F., **1**, 831 ⟨91TL3209⟩
Rodriguez Mieles, L., **1**, 759 ⟨89OS(67)125⟩
Rodriques, K. E., **1**, 1042 ⟨88TL3455⟩; **2**, 343, 351 ⟨88TL3455⟩; **3**, 379 ⟨91TL1275⟩; **5**, 285, 377 ⟨85TL1803⟩
Rodriquez, A., **1**, 821 ⟨84TL3555⟩
Rodriquez, I., **2**, 532 ⟨90JA2767⟩
Roduner, E., **1**, 922 ⟨88TL1437⟩
Roe, A. D., **2**, 432 ⟨87JOM(323)C1⟩
Roe, D. C., **2**, 979 ⟨84OM1223⟩
Roe, E. T., **5**, 275, 276, 277, 278 ⟨49JA2215⟩
Roe, Jr., R., **4**, 768 ⟨68TL1977, 70JOC1515⟩
Roe, R., **3**, 526 ⟨68TL1977⟩, 527 ⟨69JA5679⟩
Roebuck, A., **2**, 45 ⟨55OSC(3)217⟩
Roeda, D., **6**, 414 ⟨77MI 614-01, 78MI 614-03⟩, 414 ⟨81MI 614-01, 83MI 614-01⟩
Roedel, M. J., **3**, 490 ⟨45JA955⟩
Roedig, A., **3**, 247 ⟨58CB320⟩, 475 ⟨64LA(675)47⟩, 590 ⟨80CB811⟩; **4**, 776, 854 ⟨81CB684⟩, 829, 970 ⟨82CB1733⟩, 851 ⟨77CB1484⟩; **6**, 69 ⟨67CB2946⟩, 230 ⟨69CB3127⟩
Roefke, P., **1**, 363 ⟨89LA739⟩; **5**, 343 ⟨89LA739⟩
Roehr, J., **4**, 309, 310 ⟨63LA(664)130⟩; **6**, 48, 82, 89 ⟨61LA(648)21⟩, 111 ⟨61LA(641)21⟩
Roekens, B., **2**, 938, 944 ⟨87TL397⟩; **3**, 424 ⟨87TL397⟩, 636 ⟨86TL5099⟩
Roelen, H. C. P. F., **2**, 467 ⟨92TL2357⟩; **4**, 184 ⟨87RTC545⟩
Roelofs, W. L., **1**, 307 ⟨66JOC646⟩, 503 ⟨83JOC2274⟩
Roelofsen, D. P., **3**, 406 ⟨72RTC605⟩; **4**, 177 ⟨71RTC1141⟩, 177 ⟨72S419⟩, 177 ⟨77RTC44⟩
Roemer, A., **2**, 235 ⟨86GEP3443467, 88GEP3626554⟩
Roemer, D., **6**, 469 ⟨79EUP7498⟩
Roensch, E., **2**, 291, 292 ⟨65ZAAC(338)195⟩
Roesch, L., **4**, 355 ⟨81AG(E)581⟩; **5**, 793 ⟨76ZN(B)281⟩; **6**, 125 ⟨81AG(E)581⟩
Roeschenthaler, G. V., **4**, 337 ⟨78ZN(B)131⟩, 455 ⟨84ZN(B)356⟩
Roesky, H. M., **5**, 389 ⟨78IJ132⟩
Roesky, H. W., **2**, 173 ⟨74CB508⟩, 368 ⟨65CB2429⟩, 489 ⟨89CB1247⟩, 862 ⟨69ZN(B)1250, 70ZAAC(375)140⟩, 862 ⟨77ZAAC(431)221⟩, 872 ⟨89JOM(366)73⟩, 880 ⟨91JOM(402)55⟩; **3**, 370 ⟨90CB1475⟩; **4**, 719 ⟨88ZN(B)1461⟩, 800 ⟨86CB3150⟩; **5**, 606 ⟨85ZN(B)124⟩, 782 ⟨88JOM(345)C1⟩, 893 ⟨89AG(E)445⟩, 995 ⟨74CB508⟩, 1075 ⟨70CB1664⟩, 1084 ⟨89ZN(B)94⟩
Roesrath, U., **1**, 1047 ⟨89TL1073⟩, 1047, 1048 ⟨91LA903⟩; **5**, 513, 549, 560 ⟨89TL1073⟩, 560, 572, 597, 617 ⟨91LA903⟩
Roessler, F., **2**, 546, 917, 977 ⟨84JCS(P1)1805⟩, 917 ⟨80CC276⟩917, 918, 940, 977 ⟨81JCS(P1)2527⟩; **3**, 271 ⟨80CC276⟩; **4**, 1059 ⟨81JCS(P1)2527⟩
Roethling, T., **4**, 106 ⟨79EGP137839⟩
Roffia, S., **3**, 310 ⟨91JOC1822⟩
Rogan, J. B., **5**, 599 ⟨56JA4135⟩
Roger, R., **5**, 685, 687, 690, 726, 742, 769, 849 ⟨61CRV179⟩, 768, 769, 770, 847, 849, 850, 855 ⟨70CRV151⟩
Rogers, B. D., **1**, 826 ⟨87JA1160⟩
Rogers, D., **2**, 384 ⟨74JOC1588⟩; **5**, 944 ⟨83JCS(D)2397⟩
Rogers, D. H., **4**, 239 ⟨94JA4697⟩
Rogers, D. J., **6**, 241 ⟨75JCS(D)2225⟩
Rogers, Jr., F. F., **6**, 506 ⟨66JA1024⟩
Rogers, H. R., **3**, 113 ⟨75JOC3577⟩
Rogers, J., **4**, 19, 27 ⟨81JCS(P1)78⟩
Rogers, J. B., **2**, 170 ⟨67JA5838⟩; **4**, 71 ⟨67JA5838⟩
Rogers, J. H., **2**, 358 ⟨66JOC842, 70JOC279⟩
Rogers, M. A. T., **1**, 626 ⟨55JCS769⟩; **4**, 902 ⟨50JCS3350⟩
Rogers, M. T., **1**, 923 ⟨67JCP(46)3193⟩
Rogers, N. H., **1**, 763 ⟨85JCS(P1)541⟩
Rogers, R. B., **6**, 28 ⟨83USP4419514⟩
Rogers, R. D., **1**, 810 ⟨92HCA1755⟩, 815 ⟨88T3139⟩, 826 ⟨91JOC3841⟩, 1098 ⟨90JA277, 90JA5562⟩; **3**, 519 ⟨84JOM(264)327⟩; **4**, 581 ⟨89JOM(362)117⟩, 656 ⟨78CC140⟩, 703 ⟨93OM2609⟩, 934 ⟨91OM2121⟩, 950, 954 ⟨88T3139⟩; **5**, 445 ⟨84JA3214⟩, 955 ⟨89OM1275⟩, 958 ⟨91OM737, 91OM2121⟩; **6**, 195 ⟨78CC140⟩, 522 ⟨79JA5079⟩
Rogers, R. G., **3**, 626 ⟨55OSC(3)174⟩
Rogers, W. J., **3**, 618 ⟨77S41⟩
Rogerson, C. V., **2**, 1008 ⟨83JOC3043⟩
Rogic, M., **4**, 229 ⟨65JOC855⟩
Rogić, M. M., **1**, 125 ⟨69JA2147, 69JA6854⟩, 125, 280 ⟨68JA818⟩, 126 ⟨69JA2150⟩, 273 ⟨67JA5708⟩, 273 ⟨67JA5710⟩, 274 ⟨68JA4166⟩, 274, 946 ⟨68JA4165, 68JA1911⟩, 464 ⟨69JA6852⟩; **2**, 21, 619 ⟨81JOC4486⟩, 34 ⟨68JA5038, 88T2751⟩; **3**, 192 ⟨79JOC275⟩, 622 ⟨69JA6854⟩; **4**, 832 ⟨68JA818⟩
Rogic, M. R., **2**, 391 ⟨79JOC275⟩
Roginskii, V. A., **1**, 921 ⟨89JOU449⟩
Roglans, A., **2**, 324 ⟨92SC1249⟩
Rogozhin, S. V., **4**, 461 ⟨73IZV955⟩; **6**, 68 ⟨86UK1803, 78IZV468⟩
Rohde, C., **1**, 135, 159 ⟨84JA6467⟩; **2**, 999, 1006 ⟨84JA6467⟩; **4**, 378, 391, 816 ⟨84JA6467⟩
Rohde, G., **3**, 616 ⟨34JPR17⟩
Rohde, M., **4**, 970 ⟨84MI 419-01⟩
Rohde, W., **4**, 718 ⟨91ZN(B)1169⟩
Rohe, D. M. M., **3**, 113 ⟨83HCA177⟩
Rohloff, J. C., **5**, 138 ⟨88JA3672⟩; **6**, 450 ⟨84CC630⟩, 611 ⟨92TL3113⟩
Rohm, K. H., **5**, 1127 ⟨85MI 530-01⟩
Rohmer, M., **1**, 1077 ⟨88CC830⟩
Röhnert, H., **5**, 354 ⟨65ZC302⟩
Röhr, A., **5**, 344, 639 ⟨80TL4247⟩
Rohr-Aehle, R., **4**, 136 ⟨87LA51⟩, 352 ⟨88JOM(354)139⟩, 352, 353 ⟨92CB591⟩; **5**, 224 ⟨88JOM(334)139⟩
Rohr, G., **2**, 7 ⟨77LA806⟩; **5**, 790 ⟨91PS(63)95⟩; **6**, 156 ⟨91PS(63)95⟩
Rohr, W., **2**, 746 ⟨82S318⟩; **5**, 196 ⟨62CB2073⟩
Rohrbaugh, D. K., **4**, 771 ⟨93JFC(63)253⟩; **6**, 528 ⟨92JFC(59)91, 93JFC(60)85⟩
Rohrer, C., **1**, 688 ⟨88JCS(P1)1729⟩
Röhrig, D., **5**, 152 ⟨89AG(E)1706⟩
Rohrig, P., **1**, 248 ⟨88AG(E)411⟩
Röhrl, E., **3**, 655 ⟨74LA1661⟩
Rohwedder, K. H., **4**, 143 ⟨57LA(604)168⟩
Rohwer, H. E., **6**, 173 ⟨78CB3573⟩
Roidot, N., **6**, 11 ⟨91TL7525⟩
Roitman, J. N., **1**, 854 ⟨71JA2225, 71JA2231⟩
Rojas, A., **3**, 638 ⟨93SL599⟩
Rojas, I., **3**, 620 ⟨83SC35⟩
Rojo, T., **2**, 316 ⟨91JOC452⟩
Rokach, J., **1**, 513 ⟨83TL4899⟩, 514 ⟨84TL2179⟩, 940 ⟨93TL8245⟩, 1002, 1134 ⟨83TL5185⟩; **2**, 7 ⟨86JOC789⟩, 135 ⟨83JOC1357⟩; **3**, 62 ⟨83TL4899⟩, 65 ⟨81TL979⟩, 65, 232 ⟨83TL5185, 84TL35⟩, 307 ⟨81TL4901, 83JOC3214⟩; **5**, 172 ⟨88JOC4421⟩, 276 ⟨64JOC1855, 65JOC3361⟩; **6**, 626 ⟨89JA2995⟩
Rokhlin, E. M., **3**, 495 ⟨81RCR280, 82RCR1⟩, 588 ⟨65BAU731, 75JCS(P1)2513⟩; **4**, 697 ⟨79IZV1831⟩, 731 ⟨75JFC(6)227⟩, 736 ⟨76IZV895⟩, 802, 803 ⟨76BAU853, 88BAU1686⟩, 803 ⟨76BAU873⟩, 894 ⟨72IZV1347, 74IZV358⟩; **6**, 694 ⟨82RCR1⟩
Rokicki, A., **6**, 516 ⟨88AOC(28)139⟩, 517 ⟨87IC526⟩
Rol, C., **1**, 290 ⟨92CC59⟩
Roland, D. M., **4**, 255 ⟨79JA4732, 79JA7104⟩; **5**, 557 ⟨79JA4732, 79JA7104⟩; **6**, 134 ⟨79JA4732⟩
Roland, S., **4**, 465, 478 ⟨92TL77⟩
Rolando, C., **3**, 16 ⟨85TL2333⟩
Rold, T. L., **6**, 54 ⟨76JOC1285⟩
Rolefson, G. K., **2**, 103 ⟨32JA977⟩

Rolewicz, H. A., **3**, 565 ⟨62T79⟩; **6**, 150 ⟨62T79⟩
Rolfs, A., **5**, 614 ⟨94CC1437⟩
Roling, P. V., **4**, 706, 724 ⟨74JOC1420⟩
Rolla, F., **1**, 4, 6, 7, 9, 10 ⟨81JOC3909⟩; **2**, 4, 23, 30 ⟨80JOC3527⟩, 9 ⟨75S430⟩, 19 ⟨74S428⟩, 25 ⟨74S37⟩, 32 ⟨78S771⟩; **3**, 3, 114 ⟨79CI(L)213, 79S134⟩; **5**, 29, 106 ⟨82JOC154⟩
Roll, K., **5**, 652 ⟨89AG(E)221⟩
Roll, L. J., **5**, 89, 99 ⟨43OSC(2)382, 43OSC(2)140⟩, 99 ⟨43OSC(2)368⟩
Rollefson, G. K., **6**, 221 ⟨34JA1455⟩
Rollema, H., **2**, 652 ⟨88JMC1080⟩
Röller, H., **1**, 747 ⟨67JA5292⟩
Roller, P. P., **5**, 214 ⟨87JOC4933⟩
Rollin, P., **4**, 176 ⟨77JCS(P1)2513⟩; **5**, 665 ⟨90TL1417⟩
Rol'nik, L. S., **6**, 555 ⟨91JPR139⟩
Rol'nik, L. Z., **6**, 478 ⟨85ZOR2206, 86JOU270⟩, 478 ⟨87JGU365⟩
Roloff, A., **3**, 361, 579 ⟨86CB162⟩; **4**, 979 ⟨86CB162⟩
Roloff, B., **5**, 631 ⟨88ZC182⟩
Roloff, H. R., **2**, 438 ⟨65CB2091⟩
Rom, P., **2**, 776 ⟨29LA(468)117⟩
Roma, G., **3**, 361 ⟨81JHC549⟩; **4**, 891 ⟨71G269, 82FES387⟩
Romakhin, A. S., **2**, 881 ⟨85JGU1332⟩
Roman, C., **2**, 939 ⟨73TL2585⟩; **3**, 214 ⟨73TL2585⟩; **5**, 427 ⟨73TL2585⟩
Roman, D. P., **5**, 1027 ⟨57JA1236⟩
Roman, S. A., **1**, 747 ⟨71TL1821⟩; **4**, 983 ⟨75BEP821282⟩
Roman, T. von, **4**, 949, 950 ⟨84AG(E)366⟩
Roman, V. K., **5**, 1135 ⟨74URP449912, 82S277⟩
Romanelli, A. L., **1**, 928 ⟨88JA3300⟩, 949 ⟨93JOC6851⟩
Romanelli, M. G., **2**, 753 ⟨73OSC(5)552⟩
Romanelli, M. N., **6**, 290 ⟨86JOM(315)C5⟩, 381 ⟨86JCS(P2)1363, 93JOM(451)45⟩
Romanenko, E. A., **5**, 669 ⟨89UKZ1071⟩, 703 ⟨85JOU1166, 85ZOR1281⟩, 1069 ⟨88ZOR986⟩; **6**, 625 ⟨85ZOR1281⟩
Romanenko, L. S., **4**, 940 ⟨91ZOB1893⟩
Romanenko, V., **6**, 660 ⟨91TL2775⟩
Romanenko, V. D., **2**, 406 ⟨79ZOB1908⟩, 407 ⟨80ZOB2621⟩, 408 ⟨81ZOB331⟩, 835, 837 ⟨80S823⟩, 866 ⟨85ZOB2141⟩; **3**, 602 ⟨86JGU221⟩, 689 ⟨92TL2981⟩; **4**, 111 ⟨89T6019⟩, 121, 122 ⟨80JGU802, 80S823⟩; **5**, 876 ⟨89T6019⟩, 888 ⟨84PS(19)189⟩, 1158 ⟨92TL2981⟩; **6**, 655, 685 ⟨91ZOB401⟩, 678 ⟨86ZOB253, B-88MI 622-01, 89T6019⟩, 680 ⟨85ZOB1188⟩, 680, 681, 689 ⟨92IC3493⟩, 684 ⟨84ZOB965⟩, 685 ⟨82ZOB1925, 83ZOB1672, 85ZOB1437⟩, 685, 686 ⟨85ZOB221⟩, 685, 686, 689 ⟨84PS(19)189, 87ZOB901⟩, 689 ⟨84ZOB2800, 87ZOB1433⟩, 690 ⟨89ZOB2133, 92TL2981⟩, 692 ⟨92ZOB474⟩, 692, 693 ⟨83ZOB693, 92MI 622-01⟩, 693 ⟨91ZN(B)1539⟩, 694 ⟨89ZOB2131⟩, 685,686 ⟨90ZOB2238⟩
Romanet, R., **1**, 503 ⟨53CR(236)1044⟩
Romanet, R. F., **1**, 309 ⟨73TL2595⟩; **4**, 855 ⟨73TL4711⟩
Romani, R., **5**, 858 ⟨48G332⟩
Romani, S., **2**, 218 ⟨82ZPC(363)1461⟩
Romaniko, S. V., **5**, 314, 384 ⟨91T7447⟩
Romann, A. J., **1**, 831 ⟨88TL3291⟩
Romano, A. R., **5**, 597 ⟨85FES120⟩
Romano, L. J., **2**, 9 ⟨75JOC1514⟩, 24 ⟨75JA1599⟩
Romano, M. J., **3**, 414 ⟨85CB3652⟩
Romano, U., **1**, 934 ⟨85TL4803⟩; **6**, 465 ⟨93EUP534545⟩
Romanov, C. V., **2**, 845 ⟨82ZOB2721⟩
Romanov, G. V., **2**, 446 ⟨87ZOB296⟩, 454 ⟨69ZOB2418⟩, 877 ⟨84IZV719⟩; **4**, 343, 561 ⟨78ZOB1001⟩, 455 ⟨78ZOB1008⟩, 455, 481 ⟨90ZOB1718⟩; **5**, 1128 ⟨82IZV939⟩; **6**, 658 ⟨82IZV1426⟩
Romanova, I. P., **2**, 894 ⟨83IZV1152⟩; **4**, 342 ⟨83IZV1374⟩, 561 ⟨91ZOB2698⟩
Romanova, T. Y., **1**, 469 ⟨89IZV2623⟩
Romanuk, M., **3**, 430 ⟨88LA1091⟩, 620, 620 ⟨81CCC1682⟩
Romao, C., **2**, 585 ⟨91AG(E)185⟩
Romao, C. C., **2**, 585 ⟨93CB45⟩
Romão, M. J., **1**, 545 ⟨84JOM(265)323⟩; **2**, 435 ⟨85JOM(296)411⟩; **4**, 550 ⟨85JOM(296)411⟩
Romberg, E., **6**, 522 ⟨35ZAAC(221)321⟩
Romberger, M. L., **4**, 652 ⟨88TL25, 93JA10754⟩
Romea, P., **5**, 360 ⟨89JOC3209⟩, 361 ⟨91JOC7038⟩
Romeo, A., **3**, 5 ⟨84S955⟩; **4**, 201 ⟨87S250⟩
Romeo, G., **1**, 1003 ⟨82CC959⟩; **3**, 618 ⟨87S168⟩
Romer, D. R., **5**, 322, 330 ⟨93JOC6083⟩
Romero, A., **5**, 941 ⟨92CC1220⟩
Romero, J. R., **3**, 7, 120 ⟨90SC399⟩
Romero, M., **1**, 23 ⟨51JA1528⟩
Romero-Avila, C., **4**, 858 ⟨88CJC2975⟩
Romesberg, F. E., **1**, 131 ⟨91JA5751, 93TL5213⟩
Romine, J. L., **1**, 434, 815 ⟨87TL31⟩, 558, 575, 584 ⟨90JA9284⟩, 815, 1098 ⟨88JA879⟩; **3**, 206 ⟨87TL31⟩; **6**, 107 ⟨91TL597⟩
Romines, K. R., **1**, 811 ⟨88TL975⟩
Rommel, I., **2**, 1009 ⟨88JOC4379⟩
Rømming, C., **1**, 706, 707 ⟨81ACS(B)263⟩; **2**, 256 ⟨77ACS(B)56, 81ACS(B)239⟩; **3**, 346, 353 ⟨82T993⟩; **5**, 570, 571, 613 ⟨81ACS(B)239⟩, 1117 ⟨89ACS518⟩
Romney-Alexander, T. M., **3**, 661, 662, 669 ⟨87CRV779⟩
Romo, D., **1**, 134 ⟨91T9503⟩, 1067 ⟨92JOC5060⟩
Romo, J., **1**, 23 ⟨51JA1528⟩
Romodin, V. P., **2**, 461 ⟨62ZOB3062⟩
Romussi, G., **5**, 344 ⟨86FES539⟩
Ron, E., **1**, 498 ⟨86JOC3643⟩
Rona, P., **1**, 966 ⟨68JA4733⟩
Rona, R. J., **1**, 648, 650 ⟨75JOC2263⟩, 759, 761 ⟨75JOC2264⟩; **2**, 623, 682, 747 ⟨77JA1993⟩; **4**, 357 ⟨77TL591⟩
Ronald, R. C., **1**, 1136 ⟨79CC124⟩; **2**, 653 ⟨83T2031⟩; **4**, 26 ⟨79SC341⟩, 71 ⟨73TL3831⟩
Ronan, B., **4**, 261 ⟨89JOM(370)43⟩
Ronayne, J., **1**, 155 ⟨73JOM(47)417⟩
Roncaglia, H., **5**, 1126 ⟨85JHC1025⟩
Roncetti, L., **1**, 195, 238 ⟨86T2043⟩
Ronchetti, F., **1**, 1121 ⟨73CI(L)1067⟩
Roncin, J., **1**, 923 ⟨77JA8348⟩
Ronco, A., **1**, 104 ⟨49HCA489⟩, 502, 837 ⟨47HCA1911⟩
Ronco, K., **5**, 365 ⟨56HCA1253⟩
Rondan, N. G., **1**, 135, 159 ⟨84JA6467⟩, 415 ⟨88JA2314⟩; **2**, 41 ⟨84T2257⟩, 999, 1006 ⟨84JA6467⟩; **3**, 313 ⟨80JA1504⟩; **4**, 378, 391, 816 ⟨84JA6467⟩; **6**, 55 ⟨81JA6164⟩
Rondesgveet, C. S., **2**, 190 ⟨54JA3945⟩
Rondestvedt, C. S., **2**, 121 ⟨61JOC3024⟩
Rondestvedt, Jr., C. R., **2**, 112 ⟨55JOC9⟩
Rondestvedt, Jr., C. S., **1**, 447 ⟨55JOC9⟩, 486 ⟨76OR(24)225⟩; **5**, 201 ⟨63OSC(4)766⟩, 723 ⟨77JOC1786⟩
Rondon, N. G., **5**, 866 ⟨81JA6164⟩
Rondot, B., **1**, 940 ⟨93TL8245⟩
Roneo, A., **3**, 62 ⟨47HCA1911⟩
Ronge, F., **3**, 297 ⟨56AG618⟩
Ronman, P., **3**, 418 ⟨78JOC731⟩; **5**, 796, 802 ⟨78JOC731⟩
Rönsch, H., **1**, 2, 10 ⟨65LA(681)196⟩; 1127 ⟨72JPR382⟩
Rönsh, E., **2**, 369 ⟨65ZAAC(338)22⟩
Ronsisvalle, G., **5**, 769 ⟨89MI 519-01⟩
Ronsmans, B., **3**, 602 ⟨85JA2192⟩
Ronzini, G., **1**, 466 ⟨86JOM(312)343⟩
Ronzini, L., **1**, 467 ⟨85JCS(P1)1115⟩, 484 ⟨89TL243⟩, 1065 ⟨74JCS(P1)340⟩, 1112 ⟨72JCS(P1)1136⟩; **2**, 931 ⟨88TL3705⟩; **3**, 130, 285, 306 ⟨87TL2053⟩; **4**, 773 ⟨87S1034⟩
Ronzioni, L., **3**, 285 ⟨84TL4805, 85TL3595⟩
Roof, G. L., **6**, 301, 302 ⟨68JOC3333⟩, 729 ⟨67TL2747⟩
Roof, M. B., **6**, 467 ⟨93TL395⟩
Rooijman, H., **4**, 716 ⟨93OM3187⟩
Rooney, J. J., **1**, 674 ⟨85CC874, 87JA899⟩
Roos, E. C., **3**, 628 ⟨88TL4485⟩

Roos, G. H. P., **3**, 498 ⟨88T4653⟩
Roos, M., **6**, 723 ⟨90CC1722⟩
Roozpeikar, B., **1**, 74 ⟨76T2157⟩
Rop, D., **1**, 879 ⟨83JOC2957⟩
Ropalo, P. P., **5**, 502 ⟨71BAU959⟩
Rope, H., **1**, 469 ⟨16CB25⟩
Roper, T. D., **2**, 528 ⟨92TL6907⟩
Roper, W. A., **3**, 387 ⟨83JOM(244)C53⟩
Roper, W., **5**, 941 ⟨92OM809⟩
Roper, W. R., **3**, 387, 522 ⟨83JA5939⟩, 486 ⟨84CC1002⟩, 520, 522 ⟨86JOM(300)167⟩, 690 ⟨86AOC(25)121⟩; **4**, 150 ⟨79JA503⟩, 150, 722 ⟨83JA5939⟩, 582 ⟨91JCS(D)609⟩, 583 ⟨84JOM(273)C17⟩; **5**, 400 ⟨88JOM(353)355⟩, 460 ⟨68JA2282⟩, 650 ⟨78JOM(159)73⟩, 942 ⟨86AOC(25)121⟩; **6**, 247 ⟨90JOM(388)391, 90JOM(394)615⟩, 247 ⟨90JOM(397)209⟩, 593 ⟨83JOM(244)C53⟩, 716, 717, 718, 722, 727 ⟨88CRV1293⟩, 716, 717, 721 ⟨86AOC(25)121⟩, 717, 718 ⟨80JA1206⟩, 717, 718, 722 ⟨82JOM(234)C9, 82JOM(236)C7⟩, 718 ⟨83CC719, 86JOM(300)167⟩, 718, 722 ⟨82JOM(233)C59⟩, 718, 727 ⟨84JOM(269)C55⟩, 719 ⟨74JOM(80)C35, 76JOM(113)C45, 78JOM(159)73⟩, 719, 721 ⟨74JOM(81)C7⟩, 720, 732 ⟨75JOM(90)C34, 76JOM(107)C37, 75JOM(91)C61⟩, 722 ⟨88JOM(338)393⟩
Ropero, M., **3**, 342 ⟨83PS(17)47⟩
Ropp, G. A.,**3**, 68 ⟨53OS(33)15⟩
Roque, J. P., **4**, 215 ⟨75T809⟩
Roques, B., **3**, 105 ⟨67BSF2495⟩
Roques, C., **2**, 949 ⟨85TL1315⟩
Roques, Ch., **6**, 689 ⟨89TL6869⟩
Roques, J., **6**, 142 ⟨75CR(C)563⟩
Rordan, N. G.,**1**, 323 ⟨86JOC2874⟩
Rorig, K., **2**, 191 ⟨45JA1578⟩
Ros, J., **4**, 689 ⟨95JOM(489)C65⟩, 711 ⟨86JOM(315)C22⟩, 713 ⟨90JOM(395)305⟩; **5**, 455 ⟨90JOM(390)C57⟩
Ros, P., **2**, 1003 ⟨84T765⟩
Ros, R., **3**, 314 ⟨72JOM(46)C29⟩; **6**, 247 ⟨89JCS(D)1149⟩
Rosa, E., **5**, 361 ⟨89JCR(S)260⟩, 361, 362, 711 ⟨85JCR(S)132⟩, 362, 657, 711 ⟨88CC1249, 92JCS(P2)281⟩
Rosa, M. E. N., **4**, 312 ⟨84JCR(S)108⟩, 444 ⟨83JCR(S)108⟩
Rosado, O., **5**, 305, 307 ⟨B-70MI 506-02⟩
Rosamond, J. D., **3**, 560 ⟨74JHC241⟩
Rosan, A., **2**, 577 ⟨77JA8426⟩; **3**, 220 ⟨75JOC3621⟩
Rosario, O., **3**, 135 ⟨80SC813⟩; **4**, 356 ⟨77JOM(132)301⟩, 357 ⟨78JOM(146)C8, 80JOM(198)15⟩
Rosas, N., **1**, 976 ⟨92MI 119-18, 93OM1871⟩
Rosati, O., **2**, 44 ⟨89TL3175⟩; **3**, 129 ⟨91T7403⟩
Rosati, R. L., **3**, 479 ⟨79JOC4697⟩
Rosato, G. C., **1**, 290 ⟨92CC59⟩
Rosay, A., **2**, 577 ⟨76JOM(108)93⟩
Rösch, L., **2**, 473, 474 ⟨74CB854, 70JOM(21)411⟩, 918 ⟨80TL4069⟩; **4**, 355 ⟨83TL45⟩
Rösch, V. L., **2**, 918, 923 ⟨82CZ441⟩
Rösch, W., **3**, 689 ⟨87JOM(338)329⟩; **4**, 479, 480 ⟨86JOM(306)39⟩, 479, 480 ⟨87CB1645⟩, 549 ⟨87AG(E)1011⟩, 565 ⟨85ZN(B)1077⟩; **5**, 839 ⟨87S689⟩, 839, 880 ⟨85ZN(B)1077⟩, 878, 879, 1155 ⟨86JOM(306)39⟩, 1152, 1159 ⟨87JOM(338)329, 87CB1645⟩, 1155 ⟨87PS(30)479⟩
Rosche, F., **4**, 562 ⟨93PS(77)258⟩
Röschenthaler, G. V., **2**, 360 ⟨77ZN(B)721⟩, 440 ⟨84ZN(B)356⟩, 466 ⟨77S580⟩; **4**, 94 ⟨88CZ69⟩, 338 ⟨89CZ320⟩; **6**, 234 ⟨94S145⟩
Roscher, N. M., **2**, 94 ⟨71CC474⟩
Rose, B. G., **3**, 388, 389 ⟨88CC1494⟩
Rose, E., **5**, 955 ⟨94JOM(476)C25⟩
Rose, E. H., **1**, 1073 ⟨78JA4852⟩
Rose, F. L., **3**, 474 ⟨58JCS1134⟩; **6**, 650 ⟨66JCS(C)2031⟩
Rose, J. D., **2**, 393 ⟨47QR358⟩; **3**, 638 ⟨47JCS1500⟩; **5**, 820 ⟨49JCS42⟩
Rose, J. G., **5**, 277, 299 ⟨70TL1791⟩
Rose, J. P., **2**, 1009 ⟨87AX(C)313⟩
Rose, L. M., **2**, 255 ⟨79JMC631⟩
Rose, M. E., **5**, 273, 285, 300 ⟨79T2169⟩
Rosegay, A., **5**, 199 ⟨64CI(L)1867⟩
Rösel, P., **3**, 494 ⟨78CB248⟩
Rose-Munch, F., **5**, 955 ⟨94JOM(476)C25⟩
Rosen, M., **2**, 158 ⟨70JOC905⟩
Rosen, P., **1**, 130, 889 ⟨65JA275⟩; **2**, 696 ⟨65JA275⟩; **3**, 234 ⟨70JOC3045⟩
Rosen, S., **5**, 985 ⟨69JOC1677⟩
Rosen, T., **1**, 505 ⟨90OSC(7)226⟩; **4**, 171 ⟨84JOC3994⟩; **5**, 103 ⟨85OS(64)108⟩
Rosen, W., **1**, 864 ⟨66JA3461⟩
Rosenberg, E., **1**, 450 ⟨71JOM(33)321⟩; **4**, 682 ⟨89JOM(365)163⟩, 683 ⟨80JOM(187)81⟩, 692 ⟨74JOM(64)271⟩
Rosenberg, H., **3**, 49 ⟨69JOC755⟩; **4**, 706 ⟨69TL4011⟩, 724 ⟨57JOC900⟩
Rosenberg, H. E., **2**, 764 ⟨78JCS(P1)1117⟩
Rosenberg, H. M., **1**, 516 ⟨70JOC1237⟩
Rosenberg, I., **4**, 130 ⟨85CCC1507⟩
Rosenberg, J., **1**, 521 ⟨55JA73⟩
Rosenberg, Jr., J. L. von, **1**, 917 ⟨62JA2842⟩
Rosenberg, M., **3**, 61 ⟨80HCA1665⟩
Rosenberg, R. M., **2**, 166 ⟨62IC756⟩
Rosenberg, S. D., **2**, 600 ⟨53JA2507⟩
Rosenberg, S. G., **4**, 563 ⟨92TL3927⟩
Rosenberg, S. H., **2**, 118, 124 ⟨93JMC460⟩; **5**, 236 ⟨91TL6507⟩
Rosenberger, M., **1**, 1134 ⟨80JA5425, 83JA3656⟩; **4**, 166 ⟨72HCA249⟩, 168, 170 ⟨83JA3661⟩
Rosenblatt, D. H., **2**, 336, 339 ⟨67JOC1115⟩
Rosenblitt, D. H., **2**, 192 ⟨74S295⟩
Rosenblum, L., **1**, 657 ⟨55JA5016⟩
Rosenblum, M., **1**, 481 ⟨84JA7264, 88OS(66)95⟩, 519 ⟨82OM397⟩, 862 ⟨52JA2125, 52JA3458⟩; **2**, 577 ⟨76JOM(108)93, 77JA8426⟩; **3**, 138 ⟨80JA5930, 81JOC4103, 84JA7264⟩, 220 ⟨75JOC3621⟩; **4**, 690 ⟨68JA1060⟩
Rosendahl, F., **3**, 696, 700 ⟨73OSC(5)300⟩, 721 ⟨61OS(41)13⟩
Rosendahl, F. K., **3**, 714 ⟨63LA(666)65⟩
Rosenfarb, J., **6**, 507 ⟨79AG(E)503⟩
Rosenfeld, B., **5**, 819 ⟨30LA(484)236⟩
Rosenfeld, J., **1**, 899 ⟨73ACR53⟩, 900 ⟨68JA6882⟩
Rosenfeld, M. N., **2**, 615 ⟨79CC914⟩, 690 ⟨81JCS(P1)1473⟩
Rosenfeld, R. S., **1**, 16 ⟨55JA4367, 57JA5540⟩
Rosenfeld, S. M., **5**, 62 ⟨82OPP249⟩
Rosenfelder, W. J., **1**, 773 ⟨51JCS1048⟩
Rosengart, V. I., **2**, 1033 ⟨91RCR885⟩
Rosenkilde, S., **5**, 626 ⟨85JCS(P1)907⟩
Rosenkrantz, G., **1**, 23 ⟨51JA1528⟩, 881 ⟨57JOC602⟩; **3**, 218 ⟨56JA876⟩, 234 ⟨56JOC644⟩; **5**, 288 ⟨56JOC520⟩
Rosenkranz, R. E., **3**, 245 ⟨63HCA1259⟩
Rosenman, H., **3**, 294 ⟨67TL3665⟩
Rosenmund, K. W., **2**, 884 ⟨21CB438⟩; **3**, 661 ⟨19CB1749⟩; **6**, 431 ⟨28AP(266)277⟩, 484 ⟨34GEP676049⟩
Rosenquist, E. P., **5**, 350 ⟨68IJ587⟩
Rosenquist, N. R., **1**, 981 ⟨76JOC3326⟩; **3**, 542 ⟨75JA6577⟩
Rosenstein, F. U., **1**, 797 ⟨87JOC2586⟩
Rosenstein, I. J., **1**, 326, 327 ⟨91JA7788⟩, 326, 327 ⟨92JA7664⟩
Rosenstock, P. D., **5**, 652 ⟨68JMC534⟩
Rosental, Z., **4**, 205 ⟨89JOC4866⟩
Rosenthal, A. F., **4**, 555 ⟨83JOC4775⟩
Rosenthal, D., **2**, 129 ⟨65JOC3689⟩; **3**, 638 ⟨65JOC3689⟩
Rosenthal, I., **2**, 648 ⟨76TL2809⟩
Rosenthal, M. V., **2**, 260 ⟨80PS(9)107⟩
Rosenthal, R. J., **1**, 355 ⟨84JOC1353⟩; **3**, 679 ⟨84JOC3174⟩

Rosenthal, S., **1**, 284 ⟨87TL4455⟩, 1043 ⟨88S621⟩; **2**, 902 ⟨90T2573⟩, 936 ⟨90CSR147⟩; **4**, 357 ⟨90T2573⟩, 929 ⟨90CSR147⟩; **5**, 304 ⟨88S621⟩, 409, 411, 412, 417 ⟨90CSR147⟩, 410, 423, 424 ⟨90T2573⟩, 423 ⟨86TL5421, 86TL2527⟩
Rosenthal, T. C., **5**, 772, 850, 852, 855 ⟨71JOC1155⟩; **6**, 55 ⟨71JOC1155⟩
Rosenthal, U., **4**, 580 ⟨73AG(E)854, 75JOM393⟩, 580 ⟨78JOM(160)35⟩
Rosenthaler, L., **3**, 628 ⟨08MI 318-01⟩
Rosentsein, I. J., **1**, 326 ⟨91JA1791⟩
Rosenwirth, B., **6**, 489 ⟨88JMC1831⟩
Roser, J., **1**, 999 ⟨86SC983⟩
Roshan-Ali, Y., **3**, 669, 670 ⟨83T3881⟩
Rosini, G., **1**, 37 ⟨83S137, 85S269⟩, 39 ⟨90JOC5159⟩, 834 ⟨81JOC2228⟩; **2**, 245 ⟨81JCS(P1)2322⟩, 318 ⟨87S713⟩, 391, 392 ⟨90OPP707⟩, 391, 783 ⟨88S833⟩, 393, 394 ⟨83S1014, 86S849⟩, 394, 1002 ⟨91COS(2)321⟩, 395 ⟨86S237, 87S711⟩, 396 ⟨90T7531⟩; **3**, 46 ⟨84T3809⟩, 46, 191 ⟨90OPP707⟩, 47, 192 ⟨86S237⟩, 191 ⟨84S607, 88S833⟩, 451 ⟨66JA3865⟩, 480 ⟨75S44⟩, 618 ⟨73JOC1060⟩
Roskamp, E. J., **1**, 400, 410, 820 ⟨86JA6062⟩; **3**, 171, 200 ⟨92TL1131⟩; **5**, 329 ⟨93SL850⟩
Rosling, A., **1**, 482 ⟨90ACS842⟩
Rosmus, P., **3**, 337 ⟨76JA6054⟩, 382 ⟨84CB187⟩, 547 ⟨77CPH453, 77JA1663⟩
Rosnack, K. J., **2**, 344 ⟨90JA6696⟩
Rosner, H., **2**, 511, 512 ⟨80CB3318, 80CB1180⟩
Rosolovskii, V. Ya., **6**, 369 ⟨77IZV1965⟩
Rosowsky, A., **5**, 1118 ⟨66JHC387⟩; **6**, 483, 489 ⟨91MI 615-02⟩, 489, 490 ⟨90BBR(172)288⟩
Ross, A., **3**, 451 ⟨61JOC579⟩; **5**, 831 ⟨62JOC304⟩
Ross, A. M., **1**, 797 ⟨76JA7319⟩, 877 ⟨88JA1981⟩
Ross, Jr., B., **3**, 113 ⟨90JA1936⟩
Ross, B., **2**, 400 ⟨81ZAAC(473)80⟩, 409, 503 ⟨90CB975⟩, 504 ⟨80CB2928⟩
Ross, B. C., **4**, 421 ⟨81CC1269⟩, 834 ⟨83TL3373⟩; **5**, 574 ⟨89TL1997⟩; **6**, 625 ⟨85TL4149⟩, 628, 633 ⟨85TL1105⟩
Ross, D., **5**, 1025 ⟨55LA(594)233⟩
Ross, D. A., **2**, 647 ⟨73CJC2347⟩
Ross, D. L., **5**, 763, 815 ⟨66TL3537⟩, 813, 814, 815 ⟨70JOC3093⟩; **6**, 144 ⟨68JCED437⟩
Ross, D. S., **5**, 812 ⟨72JA8776⟩
Ross, E. J. F., **2**, 421 ⟨66JCS(A)706⟩
Ross, J., **5**, 593, 595 ⟨33JA3672⟩
Ross, J. F., **3**, 686 ⟨82TL5453, 85JCS(P1)1517⟩, 686 ⟨86JCR(S)156⟩; **4**, 232 ⟨81JCS(P1)2991⟩
Ross, L. O., **4**, 170 ⟨67JOC2595⟩
Ross, R. A., **2**, 438 ⟨88JOC3016⟩
Ross, R. J., **1**, 246 ⟨85JOC5200⟩
Ross, S. D., **6**, 143 ⟨87ZPK1829⟩
Ross, S. T., **3**, 236, 494, 499 ⟨62JOC998⟩; **5**, 402 ⟨60ANC1896⟩, 895, 896 ⟨62JOC998⟩
Ross, W. I., **3**, 235 ⟨72JCS(P1)889⟩
Rossano, L. T., **1**, 816, 831 ⟨88JOC5189⟩
Rosscup, R. J., **4**, 18 ⟨60JA6115⟩
Rosseeu, A., **3**, 464 ⟨11LA(381)229⟩
Rossell, O., **2**, 1027 ⟨88CC942⟩
Rosser, R. M., **1**, 212 ⟨93T3007⟩; **4**, 704 ⟨93T3007⟩
Rössert, M., **5**, 731 ⟨81LA65⟩
Rossetti, R., **4**, 687 ⟨67JOM(8)149⟩, 692 ⟨74ICA5, 74JOM(64)271⟩; **6**, 352 ⟨82IC3781, 83CC1613⟩
Rossetto, G., **3**, 442 ⟨87ICA275⟩; **6**, 519 ⟨87JCS(D)2039⟩
Rossetto, O., **5**, 1038 ⟨61JA2729, 69JOC4032⟩
Rossi, C., **2**, 713 ⟨92TA629⟩
Rossi, E., **5**, 762 ⟨87T4785⟩
Rossi, F., **3**, 451 ⟨66JA3865⟩; **5**, 597 ⟨91FES1033⟩
Rossi, G., **5**, 1097 ⟨88CC46⟩
Rossi, J. C., **1**, 940 ⟨93TL8245⟩
Rossi, K., **2**, 556 ⟨89JA765, 91JA5720⟩
Rossi, L., **1**, 10 ⟨92TL7429⟩
Rossi, L. M., **2**, 771 ⟨78S465⟩
Rossi, M., **2**, 147 ⟨86TL6257⟩; **4**, 261 ⟨86TL6257⟩
Rossi, R., **1**, 467 ⟨86TL2529⟩, 472, 1013 ⟨82T631⟩, 506 ⟨79T2039⟩, 509, 532 ⟨81T2617⟩, 514 ⟨77S561⟩, 1006, 1013 ⟨84T2773⟩, 1007 ⟨86TL4351, 88TL2239⟩, 1010 ⟨85T621⟩, 1010, 1013 ⟨84S571⟩, 1035 ⟨85TL523⟩; **2**, 917, 922 ⟨92G65⟩; **4**, 946 ⟨89TL2699⟩
Rossi, R. A., **1**, 474 ⟨73JOC3020, 87TL4653⟩, 483 ⟨74JA112⟩; **2**, 438 ⟨84JOC3387, 84JOC4609⟩, 874 ⟨82JOC77, 84JOM(270)177⟩; **3**, 626 ⟨76JOC3371⟩
Rossi, S., **5**, 652 ⟨61CI(M)1243⟩, 980 ⟨60CI(M)1243⟩
Rossier, J., **4**, 214 ⟨83CC1064⟩
Rossiiskaya, P. A., **4**, 920 ⟨57CA(51)10366g⟩; **6**, 455 ⟨45BAU364, 57IZV48⟩
Rossini, G., **2**, 784 ⟨91COS(2)329⟩; **5**, 419 ⟨85JOC130⟩
Rossinskaya, I. M., **6**, 522 ⟨38BRP479774⟩
Rossiter, B. E., **1**, 305, 306 ⟨92CRV771⟩; **2**, 588 ⟨92CRV771⟩; **3**, 128 ⟨93T965⟩
Rosslein, L., **1**, 178 ⟨88HCA47⟩
Rossler, M., **2**, 428 ⟨78ZAAC(443)42⟩; **5**, 395 ⟨82ZAAC(495)73⟩, 395, 879 ⟨81ZAAC(473)7⟩, 400 ⟨80USP781220⟩
Rössler, P., **6**, 485 ⟨03CB1280⟩
Rossman, D. I., **4**, 762 ⟨93JFC(60)61⟩, 771 ⟨93JFC(63)253⟩; **6**, 528 ⟨92JFC(59)91, 93JFC(60)85⟩
Rossman, G., **3**, 246 ⟨90CB1175⟩
Rossmanith, E., **6**, 647 ⟨78LA1543⟩
Rossmann, M. G., **4**, 770, 779 ⟨65T2899⟩
Rossnecht, H., **2**, 469 ⟨71ZN(B)81⟩
Rössner, H., **5**, 911 ⟨80CB3318⟩
Rossodivita, A., **4**, 205, 210 ⟨79CB148⟩
Rossow, A. G., **3**, 446 ⟨49JA3836⟩
Ross-Petersen, K. J., **2**, 230 ⟨72JOC3550⟩
Rost, W., **3**, 547 ⟨83AG(E)321⟩
Rostovshchikova, T. N., **2**, 23 ⟨83RCR259⟩
Rostovskaya, M. F., **2**, 455 ⟨88JGU693⟩; **4**, 339 ⟨91ZOB909⟩
Roswell, D. G., **6**, 679 ⟨65JCS6875⟩
Rot, I., **6**, 49 ⟨65JCS7515⟩
Rotard, W., **3**, 64 ⟨82LA1216⟩
Rotchford, J., **2**, 328 ⟨88TL5725⟩
Rotello, V., **3**, 419 ⟨89JA371⟩
Rotem, M., **2**, 676 ⟨83OM1689⟩; **4**, 206 ⟨83OM1689⟩
Rotermund, G. W., **2**, 523 ⟨65LA(682)21⟩, 1083, 1084, 1087, 1088, 1097 ⟨68LA(717)21⟩; **6**, 183 ⟨64TL1667⟩
Roth, A., **2**, 809 ⟨69CB3731⟩; **3**, 197 ⟨87TL611⟩, 198 ⟨89JMC1886⟩; **5**, 370 ⟨89JMC1886⟩
Roth, B., **3**, 208 ⟨82TL1751⟩
Roth, D., **3**, 605 ⟨93CB1867⟩
Roth, G., **5**, 942 ⟨95CB373⟩, 954 ⟨93JOM(454)133⟩
Roth, G. A., **5**, 993 ⟨92SC411⟩
Roth, H. J., **4**, 900 ⟨86AP(319)161⟩; **5**, 337 ⟨89AP(322)343⟩
Roth, J. J., **2**, 259 ⟨90TL501⟩
Roth, J. S., **5**, 85 ⟨61JCE217⟩
Roth, K., **1**, 446 ⟨79BCJ2942⟩; **2**, 831 ⟨79AG945⟩; **3**, 603 ⟨77AG(E)877, 81TL1681, 82CB161⟩; **4**, 843 ⟨79CB28⟩; **6**, 75 ⟨86S397⟩
Roth, K.-D., **2**, 301 ⟨93TL2919⟩
Roth, M., **1**, 326, 327 ⟨93TL2445, 92SL442⟩; **2**, 824 ⟨71HCA710⟩; **3**, 268 ⟨71HCA710⟩
Roth, S., **4**, 589 ⟨89OM1192⟩
Roth, W., **1**, 427 ⟨83ZN(B)1697⟩
Roth, W. R., **1**, 400, 799 ⟨69TL2607⟩, 402 ⟨66C229⟩, 402, 794 ⟨70CB426⟩, 404 ⟨63AG(E)688⟩, 411, 806 ⟨63AG(E)115⟩, 412 ⟨62T67, 64TL1009⟩, 795 ⟨65LA(688)28⟩, 801 ⟨82CB2508⟩, 803 ⟨63T715⟩
Roth, Z., **3**, 654 ⟨85JOC3558⟩
Rothe, L., **5**, 567, 584, 618 ⟨71PHA271⟩
Rothenberg, A. S., **1**, 786 ⟨83AG(E)560⟩
Rothenberger, S. D., **1**, 967 ⟨86TL4845⟩
Rothenburg, R., **3**, 328 ⟨1894CB790⟩, 455 ⟨1893CB2053⟩

Rothe-Streit, P., **2**, 74 ⟨92JA2321⟩
Rothfield, M., **5**, 731 ⟨72JCS(P1)2326⟩
Rothgery, E., **5**, 228 ⟨74LA112⟩
Rothman, A. M., **1**, 539 ⟨68TL1457⟩
Rothman, E. S., **3**, 218 ⟨60JOC1966⟩, 281 ⟨70JOC2351⟩, 537 ⟨68MI 316-02⟩; **5**, 3 ⟨62CI(L)1304, 69JOC2486⟩
Rothman, L. A., **1**, 428 ⟨60JOC2203⟩
Rothrock, H. S., **4**, 102, 313 ⟨63JOC1825⟩, 314 ⟨63JOC2082⟩; **5**, 985 ⟨63JOC2082, 64JOC143⟩, 986 ⟨63JOC1825⟩
Rothrock, R. K., **5**, 949 ⟨73JCS(D)1267⟩
Rothschild, A. J., **5**, 873 ⟨77JA6995⟩
Rothstein, E., **3**, 278 ⟨49JCS1946⟩
Rothwell, I. P., **2**, 582 ⟨78CC344⟩; **3**, 512 ⟨86JA1502⟩; **6**, 516, 519 ⟨88CRV1059⟩
Rotondo, E., **4**, 151 ⟨90JOM(396)115⟩
Rotscheidt, K., **3**, 59 ⟨86CB1737⟩
Rotstein, D., **1**, 4, 7, 9, 17, 28 ⟨78JOC2259⟩; **2**, 398 ⟨76JOC3328⟩
Rotstein, D. M., **1**, 223 ⟨93TA1521⟩
Rott, J., **4**, 629 ⟨88ZN(B)727, 89ZN(B)285⟩, 639 ⟨90ZN(B)961⟩
Rotter, H., **1**, 574, 584 ⟨91AG(E)1372⟩
Röttger, D., **4**, 671 ⟨93AG(E)1623⟩
Röttinger, E., **6**, 352 ⟨80CB1654⟩
Rotunno, D., **1**, 484 ⟨89TL243⟩
Rouault, G. F., **5**, 76 ⟨43JA745⟩
Roubineau, A., **5**, 388 ⟨76JOM(107)63⟩
Rouchaud, J., **2**, 155 ⟨80BSF411⟩
Roué, J., **4**, 680 ⟨81CC862⟩
Rouessac, A., **1**, 786 ⟨78T19⟩
Rouessac, F., **1**, 786 ⟨78T19⟩, 881 ⟨61T45, 63BSF1925⟩, 881 ⟨63BSF1930⟩; **5**, 82 ⟨84S876⟩, 199 ⟨91T5481⟩
Rougee, M., **5**, 1097 ⟨77CC648⟩; **6**, 717 ⟨77CC648⟩
Rougny, A., **2**, 102 ⟨76BSF833⟩
Rouillard, L., **2**, 79 ⟨93SL81⟩
Rouiller, L., **3**, 497 ⟨81JCS(P1)1520⟩
Rouilly, M., **5**, 179 ⟨89LA1233⟩
Roulleau, F., **5**, 183 ⟨83TL4195⟩
Roumestant, M.-L., **1**, 495 ⟨78JCR(S)416⟩, 810 ⟨76TL677⟩, 959 ⟨72TL4465, 76S755, 77T2813, 79TL1845⟩, 959, 960 ⟨74BSF1119⟩; **2**, 9 ⟨73CC821, 76BSF169⟩
Roumian, C., **2**, 226 ⟨77JPR1036⟩
Roundhill, D. M., **4**, 550 ⟨91POL1713⟩, 562 ⟨86IC1290⟩; **5**, 385 ⟨86IC9, 86OM336⟩; **6**, 516 ⟨91OM2781⟩
Rouot, B., **6**, 611 ⟨79BSF(2)520⟩, 619 ⟨74TL3765⟩
Rousch, H., **5**, 1127 ⟨87ZC64⟩
Roush, D., **4**, 784 ⟨86USP4621078⟩
Roush, D. M., **1**, 525 ⟨79JA1906, 80JOC2773⟩; **4**, 827 ⟨79JA6023⟩
Roush, W. R., **1**, 221 ⟨83TL1377⟩, 305 ⟨83TL2231⟩, 374, 375 ⟨91COS(5)513⟩, 434 ⟨84JA721⟩, 489 ⟨90TL6509⟩, 698 ⟨85JOC5465⟩, 700, 703 ⟨82TL4879⟩, 749 ⟨84TL2183⟩, 752 ⟨89JOC3009⟩, 788 ⟨93JA2268⟩; **2**, 75 ⟨88JA3979, 90JA6348⟩, 75, 529, 557 ⟨90JA6339⟩, 529 ⟨85JA8186, 86JA294⟩, 529 ⟨90TL7563, 92T1981⟩, 530 ⟨86JA3422⟩, 900, 904 ⟨93JA2268⟩, 976 ⟨84JOC1772, 84JOC4332⟩, 1011 ⟨84JOC3429⟩; **3**, 159 ⟨91COS(2)1⟩, 621 ⟨86TL3353⟩; **4**, 179 ⟨80JOC4283⟩; **5**, 152 ⟨83TL2231⟩, 152 ⟨89JA2984⟩, 155 ⟨84TL2183⟩, 179 ⟨83JOC758⟩, 869 ⟨74TL1391⟩
Roussakis, C., **5**, 296 ⟨85SC1233⟩
Rousseau, D., **2**, 395 ⟨91SL907⟩
Rousseau, G., **3**, 140 ⟨81TL649⟩, 213 ⟨85TL4195⟩, 232 ⟨81TL645⟩, 539 ⟨73CC795⟩; **4**, 831 ⟨87SC1⟩
Roussel, J., **2**, 414 ⟨70BSF192⟩
Rousselet, G., **5**, 750 ⟨93TL6395⟩
Rousselle, J. C., **2**, 767 ⟨85TL4443⟩
Rousset, C. J., **1**, 1036 ⟨91JA1440⟩; **2**, 987 ⟨90CC182⟩, 988 ⟨89TL5105⟩; **4**, 652 ⟨90JOC5406⟩
Roussi, G., **1**, 477 ⟨77JOM(127)371⟩; **2**, 561 ⟨66TL875⟩
Roussi, P., **4**, 610, 611, 621 ⟨85OM581⟩
Roussis, V., **2**, 457 ⟨89JOC627⟩; **3**, 198 ⟨89JOC627⟩, 245 ⟨86JOC2525, 88JOC2011⟩
Routledge, A., **4**, 811 ⟨93T4229⟩
Rouvillois, J., **6**, 522 ⟨28CR(187)564⟩
Rouvinen, J., **5**, 1028 ⟨91JHC1091⟩
Rouwette, P. H. F. M., **3**, 709 ⟨81JOC5159⟩
Roux, C. Le, **3**, 159 ⟨93JOC1835⟩
Roux, J. Le, **1**, 640 ⟨89CC1464⟩; **3**, 197 ⟨90TL2591⟩
Roux, M. C., **3**, 442 ⟨81TL1927⟩
Roux-Schmitt, M. C., **3**, 634 ⟨87S56⟩
Rouzic, A. Le, **4**, 405, 411 ⟨75BSF196⟩
Rovai, D., **6**, 552 ⟨93POL13⟩
Rovnyak, G. C., **4**, 905 ⟨90JMC1510⟩
Rowan, A. J., **5**, 455 ⟨80JCS(D)1845⟩
Rowatt, R. J., **4**, 455 ⟨69IC1336⟩, 566 ⟨69IC862⟩
Rowbottom, C. A., **5**, 317 ⟨90T7247⟩, 687, 698 ⟨93AJC1213⟩
Rowbottom, K. T., **5**, 849 ⟨72JCS(P1)2927⟩
Rowe, B. A., **1**, 490 ⟨80TL965⟩
Rowe, C. D., **2**, 753 ⟨69JOC3430⟩
Rowe, J. E., **2**, 95 ⟨84JA5261⟩; **5**, 675, 703 ⟨84JOC3083⟩, 685, 728 ⟨80S114⟩
Rowe, Jr., C. A., **1**, 797 ⟨68JOC221⟩, 871, 881 ⟨62JA3164⟩, 882 ⟨65JA3244⟩
Rowe, K., **2**, 533 ⟨76JCS(D)2087⟩; **4**, 523 ⟨75JCS(P1)129, 75JCS(P1)138⟩
Rowe, R. A., **2**, 379 ⟨55JA790⟩
Rowland, R. L., **2**, 274 ⟨50JA3595⟩
Rowland, S. P., **2**, 643 ⟨44JA1873, 45JA386⟩, 643, 645 ⟨43JA992⟩, 643, 645, 646 ⟨44JA1109⟩
Rowlands, M., **1**, 765 ⟨89TL5643, 89TL5647⟩; **2**, 525, 905 ⟨89TL5643⟩, 525, 905 ⟨89TL5647⟩; **3**, 127 ⟨89TL5643⟩
Rowles, D. K., **4**, 798 ⟨86JCS(P1)1171⟩
Rowlett, Jr., R. J., **1**, 45 ⟨46JA1813⟩
Rowley, A. G., **2**, 692 ⟨77SC365⟩; **3**, 492, 493 ⟨B-79MI 313-01⟩
Rowley, G. L., **6**, 56 ⟨72JOC152⟩
Rowley, M., **1**, 495 ⟨89T413⟩, 566 ⟨87JCS(P1)2259⟩, 831 ⟨89JA2735⟩, 1099 ⟨88TL4909⟩; **5**, 223 ⟨88TL4909⟩
Rowley, R. J., **4**, 140, 607 ⟨73JOM(57)261⟩, 533, 534 ⟨75JOM(97)159⟩
Rowley, S. P., **4**, 453 ⟨90JA6139⟩
Rowsell, D. G., **2**, 430 ⟨67JA1102⟩, 430, 438 ⟨69T1097⟩, 487, 875 ⟨68JOM(12)133⟩; **4**, 116 ⟨70OMR(2)81⟩, 123, 125 ⟨66JA3572⟩, 728 ⟨67JSP(23)32⟩, 805 ⟨68JOM(12)133⟩
Rowson, G. P., **6**, 611 ⟨79JCS(P1)724⟩
Roy, A. K., **2**, 466, 467 ⟨84IC2062⟩; **4**, 129 ⟨87PS(33)147⟩, 570 ⟨86MM2089⟩, 572 ⟨84IC2063⟩
Roy, C. H., **4**, 546 ⟨66USP3251907⟩, 555, 563 ⟨66BRP1026366, 68JOM(13)199⟩; **6**, 263 ⟨68JOM(13)199⟩
Roy, G., **1**, 414 ⟨50BSF121⟩; **2**, 680 ⟨79CC822⟩; **3**, 11 ⟨79CC822⟩; **4**, 355 ⟨78CC297⟩, 650 ⟨82OM553⟩
Roy, J., **1**, 601 ⟨71JOC2561⟩; **2**, 249 ⟨65CJC3407⟩, 287 ⟨71JOC2561⟩
Roy, M., **3**, 537 ⟨73CC882⟩
Roy, M. A., **1**, 68 ⟨80JCS(P1)1420⟩; **4**, 152, 153 ⟨79S893⟩
Roy, N., **3**, 4, 88, 115 ⟨82S580⟩
Roy, N. K., **4**, 472 ⟨68JA4495, 68JOC3090⟩
Roy, P. J., **1**, 608 ⟨86TL2949⟩; **3**, 228 ⟨86TL2949⟩
Roy, R., **6**, 546 ⟨92MI 617-04⟩
Roy, S., **6**, 492 ⟨92AP(325)267⟩, 514 ⟨90BCJ3658⟩
Roy, S. C., **3**, 206 ⟨82JOC5353⟩
Roy, S. K., **3**, 234 ⟨61JCS1583⟩; **5**, 298 ⟨81S977⟩
Roy, U. V., **1**, 498 ⟨69CC1225⟩
Roy, W., **1**, 5, 7 ⟨81LA1215⟩
Royce, Jr., R. D., **4**, 240 ⟨82JOC693⟩
Royer, G. P., **1**, 74 ⟨79JOC3442⟩
Royer, J., **1**, 104 ⟨94T3139⟩; **2**, 331 ⟨88TL3311⟩; **3**, 232 ⟨72BSF2510⟩
Royer, L. D., **3**, 492 ⟨66JA1587⟩

Royer, R., **2**, 398 ⟨84S1054⟩, 787, 789 ⟨82S260⟩; **3**, 46 ⟨81S878⟩, 304 ⟨55JCS3693, 77JHC861⟩; **4**, 193 ⟨81S878⟩, 798 ⟨87S1020⟩
Royo, P., **2**, 566, 568 ⟨85JOM(295)159⟩, 881 ⟨78JOM(149)315, 78JOM(154)13⟩; **4**, 579 ⟨89OM1604⟩, 656 ⟨89JOM(375)59⟩; **5**, 1137 ⟨87JOM(137)229⟩, 1138 ⟨73SRI237, 78MI 530-02⟩, 1141 ⟨73MI 530-01⟩, 1143 ⟨77JOM(136)309⟩
Rozanel'skaya, N. A., **2**, 825 ⟨74ZOB760⟩
Rozanova, N. S., **6**, 500 ⟨57JGU2590, 57JGU2882⟩
Rozantsev, E. G., **1**, 922 ⟨73RCR1011⟩
Rozdzinski, W., **6**, 438 ⟨82ZN(B)81⟩
Rozell, Jr., J. M., **4**, 603, 637 ⟨85OM2206⟩
Rozema, M. J., **1**, 123 ⟨91JOC5974⟩, 974, 1056 ⟨92JOC4070⟩, 1004 ⟨93JOC2694⟩; **2**, 69, 72, 591 ⟨92JOC1956, 93TL3115⟩, 478 ⟨91JOC5974⟩, 594 ⟨93JOC2694⟩, 998 ⟨89JOC5202⟩; **3**, 303 ⟨89JOC5202⟩, 623 ⟨89JA6474⟩; **4**, 153 ⟨92JOM(438)11⟩, 385 ⟨89JOC5202⟩
Rozen, S., **2**, 15 ⟨76JA3036, 87JOC2769⟩, 15 ⟨87JOC4928, 88JOC2803⟩, 15, 17 ⟨88ACR307⟩, 15 ⟨B-80MI 218-01⟩, 16 ⟨79CC479, 83JOC724⟩, 16 ⟨85JOC3698, 85S665, 85T1111⟩, 17 ⟨86JOC3607⟩, 18 ⟨80TL4543, 85JOC3342⟩, 19 ⟨79TL1823⟩, 94 ⟨90JOC5155, 91JA2648, 93JA1379⟩, 397 ⟨92JOC7342⟩, 613 ⟨81JOC4629, 84JOC806⟩, 626 ⟨88JOC5545⟩, 631 ⟨90JOC3552⟩; **3**, 148 ⟨85T1111⟩, 234 ⟨85JOC3342⟩; **4**, 44 ⟨84JOC806⟩, 52 ⟨92JA7643⟩; **5**, 202 ⟨80JFC(16)19, 80JOC672⟩, 202 ⟨81JOC4629, 84JOC806, 90JOC2621⟩, 203 ⟨85JOC3698, 85S665⟩
Rozenberg, V. I., **5**, 1142, 1145 ⟨79JOM(177)101⟩
Rozenshtein, S. M., **6**, 54 ⟨66ZOB728⟩
Rozhdestvenskaya, N. N., **6**, 145 ⟨73IZV122⟩
Rozhenko, A. B., **5**, 385 ⟨93S117⟩, 685 ⟨90JOU1408, 90ZOR1629⟩
Rozhkov, I. N., **6**, 54 ⟨63IZV1946⟩
Rozinov, V. G., **4**, 766 ⟨89ZOB997⟩, 926 ⟨82ZOB2375⟩; **5**, 656 ⟨81JGU1015, 81ZOB1201⟩, 656 ⟨87JGU1710, 87ZOB1912⟩; **6**, 619 ⟨90ZOB706⟩, 620 ⟨91ZOB643, 92ZOB306⟩
Rozko, A. N., **1**, 612 ⟨89ZOR654⟩
Rozov, L. A., **4**, 49 ⟨93JOC7382⟩
Rua, L., **1**, 635 ⟨75JOC923⟩
Ruan, M.-D., **3**, 260 ⟨91SC1307⟩; **5**, 625 ⟨94SC1761⟩
Ruane, M., **1**, 178 ⟨66AJC521⟩
Ruano, J. L. G., **3**, 170, 631 ⟨92JOC7235⟩; **6**, 567 ⟨85H(23)1181⟩
Ruasch, M. D., **6**, 522 ⟨79JA5079⟩
Ruasse, M.-F., **1**, 318 ⟨75JA1977, 75TL4555⟩, 318 ⟨77JOC2689⟩; **2**, 30 ⟨90ACR87⟩
Ruback, W., **6**, 627 ⟨82LA231⟩
Ruban, A. V., **2**, 406 ⟨79ZOB1908⟩, 408 ⟨81ZOB331⟩; **6**, 655, 685 ⟨91ZOB401⟩, 678 ⟨B-88MI 622-01⟩, 689 ⟨87ZOB1433⟩, 692 ⟨92ZOB474⟩, 692, 693 ⟨83ZOB693, 92MI 622-01⟩, 693 ⟨91ZOB79⟩, 694 ⟨89ZOB2131⟩
Rubenstein, S. M., **1**, 404, 406, 819 ⟨91TL4647⟩; **2**, 88 ⟨91TL4647⟩
Rubina, K., **3**, 430 ⟨89SC3129⟩
Rubin, K., **2**, 573 ⟨86OM2452⟩
Rubin, M., **2**, 28 ⟨53JA3513⟩
Rubin, M. B., **3**, 469 ⟨80JCS(P1)2670⟩; **4**, 162 ⟨86AG(E)999⟩
Rubin, Y., **2**, 1091 ⟨91JA6943⟩; **4**, 746 ⟨91AG(E)698⟩
Rubini, E., **4**, 446 ⟨83JMC129⟩
Rubino, M. R., **4**, 50, 379, 381, 382, 951, 952 ⟨85JOC3255⟩
Rubinskaya, M. I., **4**, 583 ⟨85JOM(288)119⟩
Rubinstein, H., **5**, 1003 ⟨58JOC1107⟩
Rubio, M. B., **1**, 600, 983 ⟨92TL4985⟩
Rubio, O., **4**, 322 ⟨87JCS(P1)781⟩
Rubio, V., **6**, 575 ⟨79CC675⟩
Rubottom, G. M., **2**, 34 ⟨79JOC1731⟩, 665 ⟨77TL611⟩; **3**, 31, 54 ⟨83T861⟩, 131, 284 ⟨83JOC1550⟩, 155 ⟨79JOC1731⟩
Rubtsov, Yu. I., **6**, 250 ⟨93IZV2513⟩
Ruch, E., **3**, 531 ⟨73AG(E)25⟩
Rüchardt, C., **1**, 266 ⟨84TL3441⟩, 321 ⟨86TCC1⟩, 388 ⟨81TL4069⟩, 637, 638 ⟨88C106, 88TL2805⟩, 725 ⟨67CB1144⟩, 922 ⟨78CB72, 87AG(E)770⟩, 924 ⟨87JA2750⟩, 925 ⟨83CB1042, 83TL185⟩, 1120 ⟨80TCC1⟩; **2**, 404 ⟨74TL265⟩; **3**, 621 ⟨87CB1⟩, 696, 707 ⟨90CB635⟩, 699 ⟨91CB2555⟩, 700 ⟨92CB525⟩; **5**, 657 ⟨88C230⟩; **6**, 71 ⟨67AG(E)949⟩, 509 ⟨72LA(765)94⟩
Ruch-Zaske, G., **5**, 357 ⟨88AP(321)25⟩
Rück, K., **2**, 699 ⟨92SL343⟩
Rücker, C., **1**, 178 ⟨86JOM(310)135⟩, 236 ⟨84TL4349⟩, 651, 1052, 1053, 1055, 1061 ⟨82TL719⟩; **2**, 539, 699 ⟨81TL3455⟩; **4**, 1080 ⟨85CB2137⟩; **5**, 375 ⟨81TL3455⟩
Rückert, T., **1**, 398 ⟨93TL297⟩
Ruckle, Jr., R. E., **1**, 635 ⟨86JA7686⟩, 928 ⟨88JA3300⟩; **3**, 193, 472 ⟨86JOC4077⟩; **5**, 869 ⟨86JOC4077⟩
Ruckstuhl, P., **4**, 129, 473 ⟨49HCA1175⟩
Ruda, W. A., **3**, 78 ⟨77IC3124⟩
Rudashevskaya, T. Yu., **4**, 38 ⟨90IZV1802⟩
Rudavskii, V. P., **5**, 683 ⟨65JGU1206, 65ZOB1202⟩, 774 ⟨71MI 519-01⟩
Rudchenko, V. F., **2**, 355, 356, 357 ⟨93CRV725⟩, 356, 357 ⟨79ZV924⟩, 356, 357 ⟨81T4245⟩
Ruddick, J. D., **1**, 507 ⟨71JCS(A)3224⟩; **6**, 247 ⟨70IC2556⟩
Ruden, R. A., **1**, 650 ⟨75SC15⟩, 999, 1066 ⟨73TL1495⟩; **2**, 941 ⟨74JOC3607⟩; **3**, 603 ⟨73TL1495⟩; **4**, 181 ⟨73JOC834⟩, 433 ⟨77JOC4095⟩
Rudenko, E. A., **5**, 656 ⟨91UKZ530⟩
Ruder, J.-P., **4**, 54, 56 ⟨81CJC3055⟩
Ruder, S. M., **1**, 522 ⟨89JA5948⟩; **3**, 50 ⟨92TL861⟩
Ruderfer, I. B., **3**, 484 ⟨75ZOB819⟩
Rudert, R., **5**, 151, 164 ⟨88JA4640⟩
Rudham, R., **3**, 7, 120 ⟨84TL3363⟩
Rudie, C. N., **6**, 401, 402, 404 ⟨79JOM(178)227⟩
Rudinger, J., **5**, 84 ⟨62RCP3, 63PAC335⟩
Rudis, J. A., **5**, 1047 ⟨81SC781⟩
Rudisill, R., **5**, 48, 49, 83, 103, 114 ⟨83CC392⟩
Rudler, H., **5**, 798 ⟨87JOM(327)339, 89OM2253, 90JOM(381)191⟩, 798, 952 ⟨87JOM(333)245⟩, 944 ⟨86CC514, 91CSR503⟩, 953 ⟨92JA8088, 95JOM(494)43⟩; **6**, 716 ⟨91CSR503⟩
Rudler, M., **5**, 798 ⟨90JOM(381)191⟩, 952 ⟨91COS(6)485⟩, 953 ⟨92JA8088⟩; **6**, 716 ⟨91CRV35⟩
Rudner, B., **2**, 383 ⟨57CRV1021⟩
Rudnick, D., **4**, 627 ⟨80TL1405⟩
Rudnicki, K., **6**, 156 ⟨78MIP87868⟩
Rudolph, K., **2**, 1082 ⟨91ZN(B)1319⟩; **4**, 599, 600 ⟨88ZN(B)739⟩
Rudolph, M., **3**, 302 ⟨91LA843⟩; **5**, 22 ⟨92S1145⟩
Rudolph, W., **3**, 619 ⟨34CB1762⟩; **5**, 674, 767 ⟨34CB1762⟩
Rudomino, M. V., **4**, 451, 469 ⟨68PAC235⟩, 552 ⟨91S125⟩
Rudorf, W.-D., **2**, 123 ⟨B-89MI 203-01⟩; **3**, 251 ⟨77S472, 87TL4267⟩, 265 ⟨77JPR545⟩; **4**, 837 ⟨84T381⟩, 900 ⟨77ZC289, 80T1791⟩, 901, 997 ⟨83S926⟩, 969 ⟨79T551⟩, 970 ⟨77JPR545⟩, 983 ⟨78JPR585⟩
Rudy, T., **3**, 323 ⟨53JOC1030⟩
Ruechardt, C., **5**, 29 ⟨82GEP3116474⟩
Ruediger, E. H., **1**, 811, 812 ⟨83CJC1239⟩
Rüegg, G. M., **3**, 149, 152, 154 ⟨91TL1779⟩; **4**, 131 ⟨91SL395⟩
Rüegg, R., **1**, 156 ⟨57HCA1250⟩; **3**, 71 ⟨59HCA844⟩
Rueggeberg, W. H. C., **2**, 259, 260 ⟨46JA1860⟩
Ruehl, W. J., **2**, 695 ⟨74ZAAC(406)52⟩
Ruel, O., **1**, 963 ⟨90TL213⟩, 1089 ⟨83TL4825⟩; **2**, 359 ⟨87T881⟩; **3**, 54, 210 ⟨83TL4825⟩, 55 ⟨83TL4829⟩; **4**, 853 ⟨78JCR(S)68⟩, 938 ⟨83TL4825, 85BSF1250⟩
Ruess, K.-P., **5**, 746 ⟨69CB2640⟩; **6**, 572 ⟨71LA(743)167⟩
Ruest, L., **6**, 134 ⟨92CJC2335⟩
Ruf, H.-R., **1**, 546, 946 ⟨87PAC363⟩
Ruf, W., **2**, 294 ⟨75ZNB642⟩
Ruff, F., **2**, 148 ⟨82JCS(P2)1075⟩, 160 ⟨76T2763⟩

Ruff, I., **2**, 782 ⟨1897CB1656⟩
Ruff, J. K., **2**, 94 ⟨66JA4531⟩, 334, 335, 762 ⟨67CRV665⟩, 418, 421 ⟨61JA2835⟩, 419 ⟨61JA535, 62JOC1020, 66IC1629⟩; **6**, 57 ⟨68JOC1257⟩, 250 ⟨67CC870⟩
Ruff, M., **3**, 255 ⟨60AG415⟩
Ruff, O., **2**, 108 ⟨05CB53⟩; **6**, 212 ⟨36CB299⟩, 217, 220 ⟨31ZAAC(201)245⟩, 409 ⟨34ZAAC(221)154⟩, 441 ⟨36CB684⟩
Ruffini, A. J., **3**, 267 ⟨83JOC623⟩; **4**, 980 ⟨83JOC623⟩
Ruffner, R. J., **1**, 590 ⟨76JOC3218, 80OS(59)202⟩; **2**, 710 ⟨88OSC(6)737⟩
Rufínska, A., **2**, 984 ⟨86AG(E)640⟩; **4**, 659 ⟨85CB4248⟩, 698 ⟨87JOM(333)155⟩
Ruge, B., **5**, 114 ⟨82EUP46194⟩
Rüger, C., **5**, 984 ⟨77JPR841⟩
Rüger, W., **1**, 686 ⟨81TL4937, 84CB277, 86T1693⟩; **3**, 348 ⟨84CB277⟩, 464 ⟨81TL4937⟩
Rugewitz-Blackholm, B., **2**, 385 ⟨87LA583⟩; **4**, 326 ⟨87LA583⟩; **6**, 561 ⟨87LA583⟩
Rugg, P. W., **2**, 32 ⟨76JCS(P1)1446⟩
Ruggeri, R., **3**, 331 ⟨86JOC1556⟩
Ruggeri, R. B., **2**, 330 ⟨88JA8734⟩; **5**, 277 ⟨87JOC5745⟩
Ruh, R. P., **4**, 5, 32 ⟨47JA279⟩; **6**, 217 ⟨56USP2744148⟩, 217, 226 ⟨56USP2745886⟩, 223 ⟨53USP2639302, 53USP2639301, 59USP2871274⟩, 226 ⟨53USP2658086⟩
Ruhemann, S., **5**, 593, 595 ⟨08JCS621⟩
Ruhl, B. L., **2**, 1007 ⟨91JCS(P2)1563⟩
Rühl, T., **4**, 660 ⟨92LA725⟩
Ruhl, W. J., **2**, 478 ⟨76ZAAC(420)74⟩
Rühlmann, K., **2**, 263 ⟨61CB1876⟩, 813 ⟨68CB743⟩; **3**, 483 ⟨68TL4973⟩; **4**, 318 ⟨59JPR285, 59LA(622)94⟩, 515 ⟨68CB743⟩; **5**, 375 ⟨65LA(686)226⟩, 397 ⟨73JOM(49)149⟩, 720 ⟨77JOM(134)21⟩, 873 ⟨68TL4973⟩
Ruhnau, R., **2**, 1009 ⟨59LA(626)60⟩
Ruhoff, J. R., **5**, 33 ⟨43OSC(2)474⟩, 35 ⟨43OSC(2)292⟩, 49 ⟨43OSC(2)315⟩, 276 ⟨37JA401⟩
Ruhs, A., **5**, 919 ⟨77CB805⟩
Ruhter, G.,**1**, 546, 1036 ⟨87JA3486⟩
Rühter, G., **3**, 333, 335, 337 ⟨85TL5265⟩, 334, 349, 365 ⟨90JOC4200⟩, 349, 365 ⟨82AG63⟩, 349, 365 ⟨88CB1159⟩, 349, 365, 550 ⟨83AG(E)55⟩; **5**, 604 ⟨83AG(E)55⟩
Ruidisch, I., **2**, 264 ⟨63AG(E)328⟩
Ruinaard, R., **4**, 859 ⟨92TL6383⟩; **5**, 562 ⟨92TL6383⟩
Ruitenberg, K., **1**, 536 ⟨82TL2797⟩, 968 ⟨84TL3019⟩, 975, 1004, 1009 ⟨81TL1451⟩, 1004, 1006, 1025 ⟨82RTC405⟩, 1017, 1018 ⟨81JOM(217)267⟩, 1058 ⟨81S551⟩; **5**, 554 ⟨80TL1771⟩
Ruiz, C., **6**, 327 ⟨92JCS(D)2307⟩, 327, 330 ⟨93IC5569⟩, 332 ⟨88OM2577⟩
Ruiz, J. A., **3**, 673 ⟨85TL1827⟩
Ruiz, J., **2**, 475 ⟨90AG(E)1150⟩; **5**, 386 ⟨92JCS(D)53⟩, 799 ⟨88JCS(D)1591⟩
Ruiz, J. P., **4**, 552 ⟨86ICA(115)L29⟩
Ruiz, M. A., **4**, 674 ⟨93JA3786⟩, 678 ⟨92OM370⟩; **6**, 403 ⟨88JOM(355)231⟩
Ruiz, M. O., **2**, 36 ⟨84S608, 88S58⟩; **3**, 57 ⟨84S608⟩
Ruiz, N., **5**, 944 ⟨95OM1095⟩
Ruiz, P., **1**, 175 ⟨92TL5261⟩; **4**, 270, 863 ⟨91TL3695⟩; **6**, 550, 557 ⟨92TL5261⟩
Ruiz, V. M., **6**, 547 ⟨80AQ183⟩
Ruiz-Fernández, P., **1**, 175 ⟨93JOC2894⟩; **6**, 473, 544, 550 ⟨93JOC2894⟩
Ruiz-Montes, J., **1**, 194 ⟨86TL4573⟩
Ruiz-Pérez, C., **4**, 719 ⟨86ZN(B)1431⟩
Rukachaisirikul, T., **4**, 1031 ⟨85JHC513⟩
Rukachaisirikul, V., **2**, 612 ⟨83TL725⟩; **4**, 74, 778 ⟨83TL725⟩
Rukasov, A. F., **3**, 434 ⟨84ZOR988⟩
Rukevich, O. S., **5**, 388 ⟨85ZOB1071⟩
Rukhmankulov, D. L., **4**, 1081 ⟨88KGS852⟩
Ruland, A., **4**, 826 ⟨82GEP3240287⟩, 890 ⟨93LA207⟩
Rule, M., **4**, 746 ⟨82JA2223⟩
Rulev, A. Yu., **2**, 741 ⟨89S446⟩
Rulin, F., **2**, 39 ⟨90JA9439⟩
Rumanowski, E. J., **4**, 297 ⟨64JOC3114⟩, 775 ⟨60USP2954316⟩
Rumbaur, G., **5**, 685 ⟨78ZN(B)636⟩
Rumf, P., **4**, 428 ⟨65CR(260)2252⟩
Rumin, R., **1**, 420 ⟨69BSF3665⟩, 795 ⟨72BSF4238⟩
Ruminski, P. G., **5**, 1119, 1121 ⟨86PS(26)1, 87PS(29)219⟩; **6**, 630 ⟨87PS(29)1⟩
Rumler, A., **5**, 660 ⟨84JPR(326)311⟩
Rumyantseva, E. E., **2**, 28 ⟨87IZV2395⟩
Rumyantsev, E. S., **6**, 410 ⟨89BAU1570⟩
Rundel, W., **2**, 706 ⟨68CB2956⟩, 765 ⟨68HOU(10/4)311⟩; **3**, 462 ⟨60CB1541⟩, 463 ⟨62AG469, 62AG(E)403⟩, 465 ⟨60ZN(B)751, 63CB1712⟩, 485 ⟨57CB1302⟩
Rundle, R. E., **4**, 698 ⟨53JCP(21)986⟩
Runge, F., **2**, 597 ⟨51ZAAC(267)39⟩
Runge, J., **5**, 362 ⟨62JPR223⟩
Runge, W., **3**, 556 ⟨75CL51, B-80MI 317-01⟩
Rungwerth, D., **3**, 366 ⟨74ZC17⟩
Runquist, A. W., **3**, 119 ⟨76TL3499⟩
Rupp, H. D., **6**, 236 ⟨71S478⟩, 529 ⟨74S26⟩
Rupp, J., **5**, 212, 213 ⟨85CAR375⟩
Ruppelt, M., **1**, 979 ⟨91CB247⟩
Ruppert, D., **4**, 528 ⟨89EUP315574⟩
Ruppert, H., **4**, 412 ⟨63CB1630⟩
Ruppert, I., **2**, 70 ⟨84TL2195⟩, 399 ⟨75CB2329⟩; **4**, 563 ⟨74ZAAC(406)131⟩, 564 ⟨75CB919, 77AG(E)718⟩; **5**, 876, 1154 ⟨81AG(E)731⟩, 1082 ⟨77AG(E)311, 77TL1987⟩; **6**, 14 ⟨84TL2195⟩, 243 ⟨86JOM(316)41⟩, 603, 615 ⟨80TL4893⟩, 665, 686 ⟨82AG(E)448⟩, 678, 689, 692 ⟨79AG(E)873, 81AG(E)731⟩
Ruppert, K., **6**, 174 ⟨93JOM(462)31⟩
Ruppin, C., **2**, 676 ⟨91SL755⟩; **5**, 386 ⟨88JOM(344)C35⟩
Ruppli, U., **4**, 683 ⟨86OM2413⟩
Rupprecht, G., **3**, 517 ⟨80JMOC73⟩
Rupprecht, G. A., **3**, 510 ⟨78JA5964⟩, 510, 511, 512 ⟨80JA6236, 81JA1440⟩
Rupprich, B., **5**, 375 ⟨65LA(686)226⟩
Ruprecht, R. M., **6**, 483, 489 ⟨91MI 615-02⟩, 489, 490 ⟨90BBR(172)288⟩
Ruprecht, S., **6**, 213 ⟨B-74MI 607-02⟩
Rusalkina, A. M., **4**, 340 ⟨71ZOB2575⟩
Rusche, J., **2**, 379 ⟨77JPR195⟩; **6**, 43 ⟨66CB2625⟩, 298, 300, 312, 321, 646 ⟨64LA(675)142⟩
Ruschenbach, V., **3**, 59 ⟨87HCA445⟩
Rusek, J. J., **5**, 410 ⟨79JA2225⟩
Rusetskaya, G. M., **2**, 436 ⟨62ZOB2848⟩
Rush, W. A., **2**, 377, 403 ⟨65JOC3278⟩
Rusiecki, V., **2**, 326 ⟨88TL4977⟩
Rusler, M., **1**, 253 ⟨89SC2415⟩
Russ, C. R., **2**, 508 ⟨66AG(E)418⟩
Russ, M., **5**, 17, 19 ⟨81S216⟩
Russe, R., **2**, 740 ⟨79TL1361⟩
Russegger, P., **4**, 949 ⟨79TL4273⟩
Russek, A., **2**, 572 ⟨74ZAAC(412)11⟩
Russel, C. R., **1**, 859 ⟨79JA633⟩
Russel, G. A., **2**, 394 ⟨82JOM(225)43⟩
Rüsseler, W., **4**, 671 ⟨86AG(E)925⟩; **6**, 154 ⟨87AG(E)908⟩
Russell, A., **2**, 177 ⟨41JA3530⟩; **3**, 231 ⟨43OS(23)78⟩
Russell, A. T., **1**, 828 ⟨87TL2041⟩; **2**, 921 ⟨87TL2041, 88T3953⟩; **3**, 696, 719 ⟨94CC85⟩, 712, 718, 719 ⟨91SL551⟩
Russell, C. E., **2**, 681 ⟨83JA943⟩; **3**, 214, 293 ⟨83JA943⟩
Russell, C. G., **1**, 21, 25, 437, 438, 943 ⟨80JA4438⟩, 814 ⟨82JOC1632⟩; **3**, 138 ⟨82JOC1632⟩, 141 ⟨82CC828⟩
Russell, C. R., **6**, 276 ⟨69JOC1642⟩
Russell, D. H., **6**, 375 ⟨85JA8211⟩
Russell, D. R., **2**, 575 ⟨82COMC-I(5)1⟩; **4**, 960 ⟨93JCS(P1)67⟩; **5**, 465 ⟨73CC158⟩, 802 ⟨89JCS(D)345, 90JCS(D)781⟩

Russell, G. A., **1**, 67, 948 ⟨84TL5239⟩, 68, 454 ⟨83JA1398⟩, 156 ⟨66JA5663⟩, 925 ⟨B-73MI 119-12, 89ACR1⟩, 1033 ⟨79CC488⟩; **2**, 174 ⟨B-68MI 203-02⟩; **3**, 164 ⟨85JA4175⟩, 247 ⟨69JOC2339⟩; **4**, 587 ⟨86JOC5498⟩, 898, 907, 908, 911 ⟨85JOC3878⟩, 1004 ⟨92JOC6508⟩
Russell, H. F., **3**, 309 ⟨73JOC3324⟩
Russell, J. G., **1**, 870 ⟨67TL29⟩
Russell, J. J., **1**, 345, 946 ⟨90JCS(P1)2691⟩, 346 ⟨86TL2303⟩; **2**, 579 ⟨90JCS(P1)2691⟩; **3**, 7, 120 ⟨84TL3363⟩
Russell, J. R., **2**, 192 ⟨62JOC2853⟩
Russell, L. J., **3**, 131 ⟨82JOC3006⟩
Russell, M. A., **4**, 906 ⟨84CC670, 85JOC2603⟩, 906 ⟨86JOC1012⟩; **5**, 238 ⟨84CC670⟩
Russell, M. J., **1**, 93 ⟨77CC427⟩
Russell, R. R., **1**, 513 ⟨51JA5444⟩
Russell, S., **1**, 614 ⟨88TL6357⟩
Russell, S. W., **1**, 506 ⟨82JCS(P1)545⟩
Russell, T. W., **3**, 621 ⟨74JOC3050⟩
Russell-Maynard, J. K. L., **2**, 25 ⟨85AJC1879⟩
Russo, A., **6**, 250 ⟨93AG(E)905⟩
Russo, C., **2**, 783 ⟨74JCS(P1)660⟩; **5**, 354 ⟨79TL4761, 83JHC305⟩, 354 ⟨85G663⟩, 355 ⟨91G401⟩
Russo, G., **1**, 1121 ⟨73CI(L)1067⟩
Russo, J. M., **3**, 439 ⟨84S479⟩; **6**, 589 ⟨89JCS(P1)1068⟩
Russo, M., **3**, 382 ⟨65JPS(B)581, 66JPS(A)248⟩, 382 ⟨67JCS(B)117⟩
Russo, S., **6**, 488 ⟨89MI 614-04⟩
Russo, T. J., **1**, 130, 869, 871, 881 ⟨68JOC2222⟩
Russo-Rodriguez, S., **1**, 698 ⟨85JOC5465⟩
Rust, F. F., **2**, 116 ⟨42JOC472⟩, 439 ⟨52JA3282⟩
Rust, M. K., **3**, 494 ⟨75JOC3456⟩
Rustamov, M. A., **5**, 63 ⟨84MI 502-01⟩
Rustemeier, K., **2**, 759 ⟨82TH1⟩
Ruston, S., **1**, 615, 616 ⟨79JCS(P1)1290⟩, 615, 654, 696, 697, 698, 700 ⟨78JCS(P1)829⟩, 754 ⟨76JCS(P1)2386⟩
Rutan, J. F., **3**, 32 ⟨90JOC4897⟩
Rutan, P. V., **5**, 599 ⟨47JA1211⟩
Rutenburg, A. M., **5**, 857 ⟨50MI 521-01⟩
Rüter, J., **1**, 395 ⟨69CB3877⟩; **3**, 472 ⟨67TL739, 68CB1263⟩
Ruth, T. J., **2**, 627 ⟨82CC625⟩
Rüther, R., **2**, 887 ⟨86ICA(113)43⟩
Rutherford, C., **2**, 783, 784 ⟨91TL441⟩
Rutherford, K. G., **1**, 442 ⟨63JOC568⟩; **2**, 614 ⟨61JOC5149, 63OS(43)12⟩; **3**, 294 ⟨66JOC2708⟩; **4**, 12 ⟨55JA3278⟩; **5**, 1002 ⟨57JA213⟩
Rutherford, M., **3**, 439 ⟨85CC582⟩
Rutherford, R. J. D., **1**, 794 ⟨69CJC1555⟩
Ruthven, D. M., **2**, 30 ⟨78JA4919⟩
Rutjes, F. P. J. T., **4**, 106 ⟨88TL6975⟩
Rutkowski, A. J., **2**, 521 ⟨57JA5190, 58JA617⟩
Rutledge, M. C., **3**, 439 ⟨87TL5619⟩
Rutledge, P. S., **1**, 1044 ⟨81JCS(P1)40⟩; **2**, 12 ⟨81JCS(P1)2608⟩, 24 ⟨77JCS(D)1674⟩, 34 ⟨78JCS(P1)126⟩, 45 ⟨74JCS(P1)1858, 80OS(59)169⟩, 631 ⟨76JCS(P1)1161⟩; **3**, 8 ⟨78JCS(P1)1483⟩, 155 ⟨78JCS(P1)126⟩; **4**, 27 ⟨81JCS(P1)2608⟩; **5**, 168 ⟨74JCS(P1)1864⟩, 384 ⟨91JOM(420)387⟩, 729 ⟨81JCS(P1)40⟩, 764 ⟨76AJC357⟩, 1044 ⟨79JCS(P1)757⟩, 1107 ⟨84S1017⟩, 1107, 1109 ⟨81JCS(P1)58⟩, 1115 ⟨78S152⟩; **6**, 131 ⟨88JOM(348)317⟩, 314, 565 ⟨81JCS(P1)52⟩
Rutledge, T. F., **1**, 537 ⟨B-69MI 112-01⟩, 549 ⟨B-69MI 112-02⟩, 953, 954 ⟨B-69MI 120-01⟩, 998 ⟨B-68MI 121-01⟩, 1000 ⟨59JOC840⟩
Rutschmann, S. B., **6**, 72 ⟨89AJC1235⟩
Rutt, J. S., **6**, 558 ⟨91IC1776⟩
Rüttimann, A., **1**, 566 ⟨75HCA153⟩, 683 ⟨75HCA1450⟩
Ruveda, M. A., **4**, 338 ⟨72T6013⟩
Ruyu, C., **4**, 488 ⟨93MI 410-01⟩
Ruzicka, L., **2**, 662 ⟨47HCA1883⟩; **5**, 145 ⟨24HCA390⟩, 566 ⟨33HCA1323⟩
Ruzo, L. O., **4**, 743 ⟨82TL5003⟩
Ryabichev, A. G., **6**, 272 ⟨88ZOR1513⟩, 609 ⟨86ZOR1297⟩
Ryabicheva, T. G., **5**, 1070 ⟨89ZOR1523, 89ZOR1689⟩
Ryabinkin, I. I., **2**, 786 ⟨81MI 215-01⟩
Ryan, B. M.,**5**, 900 ⟨94CC805⟩
Ryan, C. M., **4**, 513 ⟨78JOM(153)369⟩
Ryan, C. W., **2**, 1018 ⟨65USP3172912⟩
Ryan, G., **1**, 907 ⟨67JA5048a⟩
Ryan, J. H., **3**, 113 ⟨93CC954⟩
Ryan, J. J., **6**, 417, 418 ⟨73T3309⟩
Ryan, J. W., **5**, 254, 536, 539 ⟨66JOC3439⟩
Ryan, K., **1**, 564 ⟨94JCS(P1)793⟩
Ryan, K. J., **5**, 476, 477 ⟨68JOC1783⟩, 476, 477 ⟨68JOC3727⟩; **6**, 617 ⟨84JOC528⟩
Ryan, K. M., **1**, 135 ⟨87JOC4745⟩
Ryan, M., **6**, 304 ⟨B-83MI 606-01⟩
Ryan, M. D., **2**, 236 ⟨92TL2129⟩; **5**, 1107 ⟨79CJC2145⟩
Ryan, M. T., **4**, 768 ⟨86JFC(31)405⟩
Ryan, R. C., **1**, 456 ⟨79JOM(178)43⟩
Ryan, R. R., **2**, 357, 365 ⟨79JOC3867⟩
Ryan, T. A., **2**, 615 ⟨85JFC(27)263⟩; **6**, 409, 419 ⟨88EUP253527⟩
Ryan, W. T., **5**, 205 ⟨70JMC948⟩
Ryang, M., **1**, 1035 ⟨69TL4593⟩; **3**, 90 ⟨64BCJ341⟩, 293 ⟨69TL5189, 77S776⟩; **5**, 273, 286, 296 ⟨71JOC2721⟩
Ryazantsev, E. N., **4**, 343 ⟨87ZOB2300⟩
Ryazantsev, V. A., **4**, 941 ⟨79ZOB805, 80ZOB722⟩, 948 ⟨79ZOB829, 80ZOB1053⟩
Ryback, G.,**1**, 522 ⟨91JCS(P1)1255⟩
Rybakov, E. A., **5**, 850 ⟨69ZOR1404⟩
Rybczynski, P. J., **2**, 574, 978 ⟨88JA2575⟩
Rybin, A. G., **5**, 1125 ⟨82ZOR1813⟩
Rybin, L. I., **4**, 656 ⟨84ZOB2025⟩; **5**, 432 ⟨83IZV1675⟩
Rybin, L. V., **4**, 583 ⟨85JOM(288)119⟩
Rybinskaya, M. I., **4**, 694, 717 ⟨87JOM(336)187, 87DOK(293)1137⟩
Rychnovsky, S. D., **1**, 342 ⟨83JA3741⟩; **4**, 380 ⟨89TL3011⟩; **5**, 173 ⟨87JA1564⟩, 173 ⟨87JA1565⟩
Rychnovsky, S. P., **2**, 85 ⟨89JOC4982⟩
Ryckman, D. M., **1**, 1099 ⟨88JOC4151, 88JOC5580⟩
Rycroft, D. S., **2**, 638 ⟨81JA1761⟩
Ryder, D. J., **2**, 515 ⟨78TL4715⟩, 516 ⟨79TL4777⟩
Rydon, H. N., **2**, 8 ⟨53JCS2224, 54JCS2281⟩, 8 ⟨88OSC(6)830⟩; **5**, 90 ⟨73OSC(5)586⟩
Ryer, A. I., **2**, 764 ⟨51JA5675⟩
Rylander, P. N., **1**, 2 ⟨B-85MI 101-01⟩, 71, 72, 87, 89, 92 ⟨B-85MI 103-01⟩, 71, 87 ⟨79MI 103-01⟩, 71, 87, 88, 89 ⟨B-67MI 103-01⟩, 454, 455 ⟨B-85MI 110-01⟩, 502 ⟨B-79MI 112-01, B-85MI 112-01⟩; **5**, 307 ⟨58JA6682⟩
Rymareva, T. G., **4**, 561 ⟨87ZOB2138⟩
Rynbrandt, R. H., **1**, 620 ⟨71JA476⟩; **3**, 479 ⟨71JA476⟩; **4**, 69 ⟨71TL3553⟩, 83 ⟨66JA5682⟩, 83, 85, 86 ⟨71JA476⟩
Ryoden, K., **1**, 976 ⟨91TL7683⟩; **5**, 346 ⟨91TL7683⟩
Ryono, D. E., **3**, 200 ⟨84TL4741⟩, 212 ⟨82JMC250⟩
Ryono, L. S., **1**, 468, 472 ⟨81JA6460⟩
Rys, B., **5**, 597, 607 ⟨93M721⟩
Rys, E. G., **6**, 406 ⟨73JGU848⟩
Ryschkewitsch, G. E., **2**, 435 ⟨74IS129⟩; **4**, 534 ⟨76IC1988⟩, 563 ⟨69IC2671⟩
Ryser, G., **1**, 156 ⟨57HCA1250⟩, 956 ⟨59HCA719⟩
Ryska, M., **4**, 894 ⟨71CCC3314⟩
Ryskiewicz, E. F., **3**, 106 ⟨63OSC(4)831⟩
Rytz, G., **1**, 946 ⟨B-83MI 119-05⟩
Ryu, E. K., **3**, 684 ⟨90H(31)1693⟩; **5**, 664 ⟨92JOC6649⟩
Ryu, E. R., **3**, 151 ⟨92CL603⟩
Ryu, I., **1**, 247 ⟨93OM473⟩, 556 ⟨78TL3455⟩, 666 ⟨86JOC2389, 92JOC17⟩, 936 ⟨93SL143⟩, 1073 ⟨91CL1805⟩; **2**, 281 ⟨88S175⟩, 282 ⟨77S874⟩, 542 ⟨88JA1976⟩, 555 ⟨91AG(E)177⟩, 710 ⟨92JA5902⟩, 730 ⟨92JA7591⟩; **3**, 38, 173 ⟨77S874⟩, 90 ⟨90TL6887⟩, 136

⟨90JA455⟩, 143 ⟨88TL6137⟩, 177 ⟨93CL561⟩, 177, 224 ⟨91TL179⟩; **4**, 239, 334, 660 ⟨90OM1355⟩, 334, 909 ⟨90CL2053⟩, 931 ⟨84JA2440⟩, 940 ⟨92JA5902⟩, 940, 947 ⟨91JA9796⟩, 948 ⟨92JA7591⟩; **5**, 172 ⟨94JA5473⟩, 238, 241 ⟨91JA9796⟩, 415 ⟨84JA2440⟩, 437 ⟨90JA455⟩, 438 ⟨90JA7061⟩
Ryu, J., **6**, 495, 497 ⟨89AG(E)452⟩
Ryu, K. E., **3**, 684 ⟨91H(32)477⟩
Ryzhikov, D. V., **2**, 259 ⟨90ZOB476⟩
Rzemykowska, Z., **2**, 767 ⟨79MI 215-01⟩
Rzepa, H., **1**, 905 ⟨77JA377⟩; **3**, 327 ⟨83JCR(S)262⟩
Rzepa, H. S., **1**, 1109 ⟨77JA7432⟩; **2**, 1000 ⟨88JCS(P1)1023⟩; **5**, 212 ⟨91CC1127⟩, 214, 635 ⟨84S829⟩

S

Saá, J. M., **1**, 433 ⟨90JOC991⟩, 482, 1027 ⟨92JOC678⟩
Saak, W., **2**, 729 ⟨91CB1315⟩; **4**, 88, 288 ⟨91CB1315⟩; **5**, 889 ⟨87AG(E)776⟩, 1085 ⟨90AG(E)90, 91POL1147⟩; **6**, 334, 355, 356 ⟨90CB2325⟩
Saal, W., **4**, 545 ⟨76ZN(B)790⟩
Saalfeld, J. C., **3**, 614 ⟨52CB397⟩
Saalfrank, R., **3**, 603 ⟨69AG(E)216⟩; **4**, 802, 825 ⟨69AG(E)216⟩
Saalfrank, R. W., **1**, 966 ⟨92AG(E)224, 93T9775⟩, 988 ⟨73TL3985, 75TL4405⟩; **2**, 831 ⟨80AG740⟩; **3**, 245 ⟨76CB403⟩, 246 ⟨90CB1175⟩, 547 ⟨83AG(E)321⟩, 603 ⟨73CB2601, 77AG(E)877⟩; **4**, 881 ⟨85S1062⟩, 989 ⟨91CB595⟩, 990 ⟨90CB115⟩, 991 ⟨89CB519, 91JHC1863⟩, 991, 993, 996 ⟨85BSB475⟩; **6**, 75 ⟨83CB1463, 83CB3482⟩, 75 ⟨88T5095⟩
Saari, W. S., **2**, 160, 162 ⟨84JMC1634⟩, 613 ⟨81JMC93⟩; **3**, 40 ⟨90S453⟩; **5**, 342 ⟨86JHC1253⟩
Saarinen, H., **5**, 864 ⟨83ACS(A)631⟩
Saavedra, J. E., **2**, 385, 422, 798 ⟨76S462⟩; **3**, 116 ⟨89SC215⟩, 175 ⟨B-87MI 304-05⟩; **4**, 529 ⟨83JOC2388⟩
Saba, A., **1**, 1033 ⟨82SC557⟩; **5**, 343 ⟨88H(27)867⟩
Sabacky, M. J., **1**, 76 ⟨68CC1445, 77JA5946⟩, 922 ⟨67JA2054⟩
Sabahi, M., **2**, 235 ⟨89JOC1092⟩; **3**, 337 ⟨89JOC1092⟩
Sabat, M., **1**, 991 ⟨93AG(E)554⟩; **4**, 906 ⟨87JOC4693⟩, 1019 ⟨88OM2553⟩; **5**, 936 ⟨92OM1759⟩, 955 ⟨88OM2553⟩; **6**, 706 ⟨93AG(E)554⟩
Sabatier, P., **1**, 71 ⟨1897MI 103-01⟩, 87 ⟨04BSF101, 14BSF228⟩
Sabatino, P., **4**, 232 ⟨85TL2131⟩; **6**, 127, 309 ⟨85TL2131⟩
Saberi, S. P., **1**, 987 ⟨91CC1290, 92JCS(P1)259⟩
Sabharwal, A., **3**, 58 ⟨89CI(L)533⟩
Sabherwal, I. H., **4**, 573 ⟨68JCS(A)1105⟩; **6**, 582 ⟨68JCS(A)1105⟩
Sablayrolles, C., **5**, 772 ⟨83EJM413⟩
Sabo, E. F., **3**, 226 ⟨57JA1130⟩
Sabofina, E. Ya., **2**, 872 ⟨91RCR162⟩
Sabol, J. S., **1**, 828 ⟨89TL3377, 89TL6271⟩; **2**, 611 ⟨93TL3057⟩; **4**, 736 ⟨93TL3057⟩
Sabourault, B., **5**, 63 ⟨82TL1259⟩
Sabourin, E. T., **2**, 753 ⟨79JOC1233, 79JOC3671⟩; **3**, 247 ⟨69JOC2339⟩
Sabuco, J.-F., **1**, 156 ⟨91TL4125⟩
Saburi, M., **2**, 986, 988 ⟨89CL761⟩, 987 ⟨89CC852, 90CC182⟩
Saccomano, N., **1**, 885 ⟨91COS(1)173⟩
Saccomano, N. A., **2**, 344 ⟨90JA6696⟩, 571 ⟨91COS(1)173⟩; **3**, 210 ⟨87TL2087⟩; **5**, 152 ⟨87TL2087⟩
Sacconi, L., **6**, 325 ⟨80AG1055, 81JOM(218)81⟩
Sachdeva, R., **5**, 568, 613 ⟨78AG(E)689⟩
Sachdev, H. S., **1**, 608 ⟨80JOC1486⟩, 1082 ⟨75JOC579⟩; **4**, 286 ⟨80JOC1486⟩, 375 ⟨76TL4223⟩
Sachdev, K., **2**, 936 ⟨76TL4041⟩; **4**, 375 ⟨76TL4223⟩, 649 ⟨76TL4041⟩
Sacher, R. E., **2**, 1093 ⟨67SA(A)1169⟩
Sachleben, R. A., **4**, 123 ⟨88IC1787⟩
Sachs, F., **5**, 595 ⟨03CB585⟩, 652 ⟨04CB874⟩
Sachs, W., **6**, 689, 690 ⟨87ZAAC(545)7⟩
Sachs, W. H., **1**, 871 ⟨67JOC3700, 67JOC4127⟩, 871 ⟨68T6287, 68TL2317⟩, 871 ⟨68TL2921⟩
Sachse, B., **2**, 444 ⟨85GEP3332716⟩
Sackmann, H., **5**, 560 ⟨88ZC269⟩
Sacks, C. E., **1**, 464 ⟨75JA7372⟩; **3**, 18, 133 ⟨76S456⟩
Sacripante, G., **3**, 621 ⟨87JOC3659⟩
Saczewski, F., **1**, 260 ⟨91JOC4439⟩; **3**, 710 ⟨91JOC4439⟩; **4**, 309 ⟨91JOC4439⟩, 323 ⟨85JOC1351⟩; **6**, 547 ⟨92JCS(P1)47⟩
Sadanani, N. D., **2**, 466 ⟨85IC3136⟩; **6**, 518 ⟨85IC4449⟩
Sadat-Aalaee, D., **5**, 5 ⟨90JA9651⟩
Sadd, J. S., **4**, 445 ⟨75JCS(P1)1181⟩
Saddler, J. C., **4**, 948 ⟨88JOC3647⟩
Sadeghi, M. M., **2**, 309 ⟨87S190⟩
Sadeghi-Milani, S., **4**, 874 ⟨79JHC1405⟩
Sadek, K. U., **5**, 676, 677 ⟨82JCED103⟩, 771 ⟨82JCS(P1)2663⟩
Sadekov, I. D., **2**, 731, 732 ⟨79RCR343⟩; **4**, 290 ⟨91HAC307⟩, 290, 291 ⟨88KGS1144⟩; **6**, 50 ⟨79ZOR1561⟩
Sadhu, K. M., **1**, 138 ⟨89JA4399⟩; **2**, 529 ⟨83JA2077⟩; **4**, 146 ⟨86TL795⟩, 362 ⟨83JA2077⟩, 389, 664, 665 ⟨89JA4399⟩, 525, 526, 527 ⟨81JA5241⟩, 526, 527 ⟨84OM614, 84OM1284⟩
Sadikov, G. G., **2**, 1003 ⟨79UKZ811⟩; **4**, 717 ⟨80JOM(201)269⟩
Sadler, A. C., **1**, 633 ⟨80JOM(186)155⟩
Sadrai, S., **5**, 639 ⟨89JHC1341⟩
Sadretdinova, F. F., **4**, 463 ⟨88ZOB2456⟩
Sadukzade, S. I., **2**, 1089 ⟨52DOK(85)1297⟩
Sadykhov, I. D., **5**, 63 ⟨84MI 502-01⟩
Sadykov, A. R., **5**, 644 ⟨85ZOB1526⟩, 713 ⟨85JGU1526, 85ZOB1718⟩
Saednya, A., **3**, 618 ⟨83S748, 85S184⟩; **4**, 1014 ⟨90MI 421-01⟩
Saegusa, M., **3**, 253 ⟨86S137⟩
Saegusa, T., **1**, 38, 945 ⟨68JA4182⟩, 86 ⟨86JOC537⟩, 511, 534, 1106 ⟨87JOC1624⟩, 558 ⟨78JOC1011⟩, 571 ⟨88OSC(6)327⟩, 655 ⟨82JA7609⟩, 1137 ⟨74CC380⟩; **2**, 22 ⟨80JOC2022⟩, 240, 359, 768 ⟨70JOC2979⟩, 939 ⟨90JOC2978⟩; **3**, 80 ⟨78BCJ1186⟩, 139 ⟨80JA6381, 86JOC421⟩, 152 ⟨80JOC2022⟩, 206 ⟨78JOC1011⟩, 214 ⟨90JOC2554⟩, 224 ⟨76JOC2073⟩, 225 ⟨79JA489⟩, 418 ⟨84TL3091⟩, 599 ⟨77TL1009, 80SC233⟩, 714, 715 ⟨71T3795⟩, 723 ⟨79JOC2030, 79TL1039⟩, 724 ⟨86JOM(303)301⟩; **4**, 696 ⟨81JOC192⟩; **5**, 146 ⟨74JOC3273⟩, 146, 692 ⟨75S291⟩, 595, 620 ⟨79TL1039⟩, 760 ⟨71T3795⟩, 785, 1130 ⟨70JOC4238⟩, 785, 792 ⟨67JA2240, 68TL935⟩, 793 ⟨84TL5139⟩, 796, 802 ⟨84TL3091⟩, 1037 ⟨70JOC2118⟩, 1071 ⟨70JOC3995⟩, 1072 ⟨75JOC2981⟩; **6**, 127 ⟨85S698⟩, 448 ⟨71JOC858⟩, 464 ⟨68TL831⟩
Saegusa, Y., **5**, 225 ⟨89BCJ539⟩
Saeki, H., **5**, 173 ⟨79BCJ1989⟩
Saeki, M., **5**, 1102 ⟨89JAP3145452⟩
Saeki, T., **4**, 78 ⟨76S697⟩
Saeling, G., **5**, 595 ⟨74JPR534⟩
Saenger, W., **2**, 170 ⟨67JA5838⟩; **4**, 71 ⟨67JA5838⟩
Saerens, E., **6**, 483 ⟨20JCS708⟩
Saeva, F. D., **4**, 993 ⟨90JOC1768⟩
Sáez, J., **5**, 1079 ⟨82SC573⟩
Safa, K. D., **5**, 925 ⟨85JA4338⟩; **6**, 177, 709, 710 ⟨78JOM(157)C50, 80JOM(191)355⟩, 179, 709, 710 ⟨80JOM(186)309⟩
Safadi, M., **5**, 22 ⟨91PS(60)239⟩, 371, 403 ⟨90JCS(P1)3263⟩
Safaev, M. A., **6**, 561 ⟨81UKZ26⟩
Safak, C., **6**, 561 ⟨92AF1453⟩
Safari, F., **5**, 348 ⟨88ZN(B)1495⟩
Safarik, I., **3**, 530 ⟨88JA1025⟩

Safdy, M. E., **2**, 219 ⟨70JOC4148⟩
Saferstein, L., **2**, 674 ⟨66JA1839⟩
Saffhill, R., **4**, 186 ⟨67JA3366⟩
Safieddine, A., **3**, 232 ⟨72BSF2510⟩
Safiev, O. G., **4**, 775 ⟨90CA230883w⟩, 1081 ⟨88KGS852⟩
Safir, S. R., **2**, 882, 883 ⟨41JA575⟩; **4**, 448 ⟨58JA4921⟩; **5**, 1125 ⟨48JOC924⟩; **6**, 650 ⟨68JA6846⟩
Safiulina, O. Z., **4**, 993 ⟨75ZOR452⟩
Safronara, O. A., **5**, 162 ⟨78JGU1249⟩
Safronova, L. P., **4**, 942 ⟨83ZOB1292⟩, 1055 ⟨80ZOB2044⟩; **5**, 226 ⟨84IZV1897⟩; **6**, 516 ⟨84IZV1897⟩
Saft, M. S., **4**, 333 ⟨81JOC765⟩
Sagan, L. S., **2**, 494, 500 ⟨72JOM(39)301⟩
Sagar, A. J. G., **1**, 1053 ⟨76S321⟩
Sage, J.-M., **1**, 815 ⟨82JOC4815⟩
Sagina, E. I., **4**, 555, 1004, 1008 ⟨79ZOB2217⟩, 561 ⟨79ZOB1470⟩, 800 ⟨79JGU1947⟩; **6**, 58 ⟨79ZOB1025⟩, 286 ⟨79ZOB1470⟩
Sagnard, I., **1**, 828 ⟨93CC960⟩
Sagramora, L., **2**, 123 ⟨76S326⟩, 362 ⟨67CC985⟩; **5**, 766 ⟨73JCS(P2)534⟩
Sagyan, A. S., **5**, 385 ⟨88T5507⟩
Saha, B., **3**, 533 ⟨86TL3913⟩; **4**, 170 ⟨87T4433⟩
Saha, C., **4**, 223 ⟨89JA658⟩
Saha, C. R., **3**, 413 ⟨86CB1095⟩; **5**, 630, 733 ⟨85PS(25)63⟩
Saha, J., **6**, 483, 489 ⟨91MI 615-02⟩, 489, 490 ⟨90BBR(172)288⟩
Saha, M., **2**, 43 ⟨94T6579⟩, 765 ⟨67JA1537, 72JOC1690⟩
Saha, N. N., **3**, 239 ⟨57JA4487⟩
Sahai, M., **2**, 14 ⟨90JOC2927⟩; **3**, 123 ⟨90JOC2927⟩, 654 ⟨85JOC3558⟩; **5**, 20 ⟨90JOC3922⟩
Sahai, R., **3**, 234 ⟨67AJC639⟩
Sah, M. M., **6**, 546 ⟨85JIC164⟩
Sah, P. P. T., **6**, 68 ⟨32JA2964⟩, 70 ⟨28JA516, 33MI 603-01⟩
Sahasrabudhe, A. B., **3**, 709 ⟨83SC379⟩
Sahlberg, C., **1**, 480 ⟨83TL5137⟩, 961 ⟨84JOC4120⟩, 970 ⟨83TL1297⟩
Sahli, A., **5**, 206 ⟨92JOC1009⟩
Sahli, M. S., **1**, 779 ⟨67JOC2059⟩
Sahni, M. K., **5**, 652 ⟨77SC57⟩
Sahni, N. S., **6**, 502 ⟨78BCJ1245⟩
Sahoo, P., **1**, 568 ⟨91S827⟩
Sahoo, S. P., **1**, 498 ⟨80T719⟩, 608 ⟨86TL2949⟩; **3**, 228 ⟨86TL2949⟩
Sahraoui-Taleb, S., **1**, 518 ⟨84TL5043⟩; **3**, 216 ⟨84TL5043⟩
Sahu, A.,**3**, 431 ⟨86MI 311-01⟩
Sahu, D. P., **5**, 273, 285 ⟨81JOC1229⟩
Sahyun, M. R. V., **1**, 19, 43, 624, 782 ⟨62JA1734⟩, 881 ⟨65OS(45)89⟩
Sai, M., **5**, 40, 106 ⟨78TL5211⟩
Saičić, R. N., **1**, 231 ⟨90TL4203, 90TL6085⟩, 348 ⟨86TL5893⟩
Said, M. M., **3**, 356 ⟨91SUL213⟩
Said, S. B., **3**, 617 ⟨89S223⟩
Saiga, Y., **6**, 556, 564 ⟨87CPB2840⟩
Saigh, G. S., **3**, 389 ⟨38JCS409⟩
Saigo, K., **1**, 189 ⟨91TL4311, 93BCJ848⟩, 308 ⟨76CL163⟩; **2**, 141 ⟨90CL1097⟩, 816 ⟨83TL511⟩; **3**, 7 ⟨77BCJ2773⟩, 118 ⟨76BCJ1656⟩, 120 ⟨77CL757⟩; **4**, 193 ⟨76CL769⟩; **5**, 125 ⟨77BCJ1863⟩, 344 ⟨90CL905⟩
Saijo, K., **1**, 248 ⟨85TL1333⟩
Saikachi, H., **3**, 710 ⟨83CPB723⟩; **4**, 442 ⟨83CPB723⟩, 538 ⟨79CPB2857⟩; **5**, 367 ⟨77CPB1651⟩, 898 ⟨68YZ1039⟩
Saikali, E., **1**, 812 ⟨91JOC5696⟩
Saiki, H., **3**, 80 ⟨77H(6)1599⟩
Saiki, M., **4**, 206 ⟨88CC560⟩
Sailer, K., **4**, 411 ⟨90AG(E)933⟩
Saillard, J.-Y., **1**, 165 ⟨82JA7549⟩; **4**, 713 ⟨B-85MI 408-01⟩
Sailor, M. J., **6**, 456 ⟨87JA6015⟩
Saimoto, H., **1**, 14, 221 ⟨82TL3597⟩, 422 ⟨81BCJ2747⟩; **2**, 7 ⟨86TL1607⟩; **3**, 116, 161 ⟨84TL3317⟩, 118, 121 ⟨89BCJ519⟩, 244 ⟨83BCJ3088⟩
Sain, B., **3**, 667 ⟨90JOC2545⟩; **5**, 226 ⟨87H(26)777⟩
Sain, E., **3**, 413 ⟨87H(26)777⟩
Saindane, M., **1**, 523 ⟨83TL2473⟩; **2**, 282 ⟨80TL3643⟩; **3**, 174 ⟨80TL3643, 85T4881⟩, 174, 207 ⟨81TL3043, 82JOC1598⟩; **4**, 401 ⟨81JOC4301⟩
Saini, N., **4**, 183 ⟨93JCR(S)430⟩
Saini, S., **4**, 311 ⟨91JA1042⟩; **6**, 270 ⟨91SRI401⟩
Sainsbury, D. M., **1**, 905 ⟨86JOC3736⟩
Sainsbury, M., **4**, 302 ⟨84CHEC(3)995⟩
Sainte, F., **1**, 375 ⟨82JA1428⟩; **4**, 897 ⟨86RTC456⟩; **5**, 718 ⟨82JA1428⟩
Saitho, Y., **1**, 788 ⟨91AG(E)1173⟩
Saito, A., **6**, 317 ⟨93JAP(K)05062690⟩, 444 ⟨83BCJ2969⟩
Saito, G., **4**, 873 ⟨86CL1861⟩, 877 ⟨86CL311⟩; **6**, 555 ⟨92CC1410⟩
Saito, H., **2**, 769 ⟨89BCJ128⟩; **4**, 948 ⟨92TL5701⟩
Saito, I., **1**, 988 ⟨90TL7469⟩, 993 ⟨89TL4995⟩; **2**, 669 ⟨94TL605⟩, 1072 ⟨79JOC865⟩; **3**, 62 ⟨86TL2757⟩, 427 ⟨73BCJ3198⟩
Saito, J., **5**, 971 ⟨75JAP89344⟩
Saito, K., **1**, 904 ⟨80JA7613⟩; **2**, 229 ⟨79BCJ3601⟩; **3**, 357 ⟨90EUP376724⟩, 626 ⟨89OPP354⟩; **4**, 437, 445 ⟨77S647⟩, 892 ⟨76JCS(P1)783⟩; **5**, 344 ⟨91BCJ106⟩, 1112 ⟨88CC1485⟩; **6**, 449 ⟨87SC1887⟩
Saito, M., **2**, 233 ⟨86CL349⟩, 234 ⟨89CL2111⟩, 915, 923 ⟨81JOM(216)149⟩; **3**, 213 ⟨79TL429⟩, 411 ⟨80BCJ3696⟩; **4**, 419 ⟨90BCJ1160⟩, 419, 420 ⟨84CL1599⟩, 438 ⟨88BCJ609⟩
Saito, N., **1**, 1140 ⟨90CL1411⟩; **2**, 1008 ⟨85H(23)2709⟩; **3**, 419 ⟨87S824⟩; **6**, 127, 325, 327 ⟨90CL1411⟩
Saito, R., **5**, 210 ⟨92CL1437⟩
Saito, S., **1**, 303 ⟨94JA4131⟩, 613 ⟨82CL165⟩; **2**, 84 ⟨90TL221, 91TL667⟩, 233 ⟨86CL349⟩, 636 ⟨76CPL399⟩; **3**, 40 ⟨82SC25⟩, 191 ⟨81CL1677, 82CL83⟩; **4**, 240 ⟨92MI 405-01⟩, 766 ⟨92CPB3261⟩
Saito, T., **2**, 69 ⟨93CC811⟩, 226 ⟨92CC600⟩, 244 ⟨78CL591⟩, 464 ⟨82TL1263⟩; **3**, 158 ⟨93JOC2929⟩, 353, 360 ⟨82TL1263, 85BCJ667⟩, 360 ⟨86BCJ3279, 90CC1665⟩, 360 ⟨92BCJ923, 92JCS(P1)2943⟩, 370 ⟨77JOC3922, 84SR33⟩, 370 ⟨92CC411, 92CC600⟩, 559 ⟨76CC516⟩, 581 ⟨76TL1307⟩, 592 ⟨80BCJ2582, 89JCS(P1)2140⟩; **4**, 220 ⟨78BCJ2401⟩, 249 ⟨80S952⟩, 552 ⟨75CL1259, 76JOM(122)113⟩; **5**, 132 ⟨78BCJ2401⟩, 221 ⟨88BCJ3653⟩, 296 ⟨75CI(L)745⟩, 972 ⟨69JOC1631⟩, 1066 ⟨86CL135⟩
Saito, Y., **1**, 66 ⟨77JOM(135)1⟩, 713, 714, 716 ⟨84JOC1096⟩; **4**, 188 ⟨88CL1101⟩, 642, 648, 659 ⟨84JOC1096⟩
Saitoh, K., **3**, 144 ⟨82CL939⟩
Saitoh, N., **2**, 21 ⟨81TL3193⟩
Saitoh, Y., **6**, 469 ⟨78BCJ3549⟩
Saitou, M., **1**, 375 ⟨91TA1305⟩
Saive, R., **4**, 695 ⟨86JOM(308)153, 87JOM(319)9⟩, 717 ⟨86AG(E)165⟩
Saji, T., **1**, 890 ⟨92CC782⟩
Sakaba, H., **5**, 949 ⟨91JA6673⟩
Sakae, M., **5**, 1120 ⟨92CC1379⟩
Sakaeda, T., **1**, 456, 507 ⟨83JA6273⟩
Sakaguchi, M., **2**, 69 ⟨88CC610⟩
Sakaguchi, R., **3**, 156 ⟨81T473⟩
Sakaguchi, S., **1**, 404, 406, 816, 819 ⟨88TL4763⟩, 816 ⟨91TL5137⟩; **3**, 265 ⟨89TL6173⟩
Sakaguchi, T., **2**, 111 ⟨55MI 202-02⟩
Sakaguchi, Y., **1**, 449 ⟨88OM1869⟩; **5**, 70, 73 ⟨80CC327⟩
Sakai, H., **1**, 816 ⟨93CL1397⟩; **3**, 415 ⟨86CPB1050⟩; **4**, 322, 323 ⟨73CPB2257⟩, 405, 418 ⟨69CPB32⟩

Sakai, I., **3**, 125 ⟨82CPB3380⟩; **5**, 873 ⟨82CPB3380⟩
Sakai, K., **1**, 1021 ⟨88CPB4337⟩; **2**, 24 ⟨94TL737⟩, 187 ⟨68BCJ3015⟩, 219, 227 ⟨84CC843⟩, 398 ⟨85ABC2331⟩; **3**, 113 ⟨84TL961, 92TL6331⟩, 216 ⟨81CL55⟩, 475 ⟨73T1983⟩, 666 ⟨70TL3851⟩; **5**, 138 ⟨87CPB1741⟩, 294 ⟨80S887⟩, 991 ⟨71JOC2387⟩, 1078 ⟨85CPB373⟩; **6**, 539 ⟨85JAP62114966⟩
Sakai, M., **1**, 433 ⟨91CL2017⟩, 468 ⟨93BCJ2446⟩, 918 ⟨72JA6867⟩
Sakai, N., **1**, 297 ⟨93JA7033⟩
Sakai, S., **1**, 364 ⟨89BCJ2348⟩; **2**, 279 ⟨90CL927⟩, 399 ⟨75JOC3291⟩, 564 ⟨86CC475⟩; **5**, 262 ⟨70JOM(25)33⟩, 324 ⟨92GSM79⟩, 557 ⟨82CL123⟩, 1024 ⟨74JOC1970, 75BCJ2981⟩, 1031 ⟨78NKK265⟩, 1035 ⟨77BCJ425⟩, 1078 ⟨76CL811, 77S889⟩; **6**, 73 ⟨76CL891⟩, 300 ⟨71JOC1176, 72JOC4198⟩, 313 ⟨74JOM(72)103⟩, 322, 337 ⟨70MI 611-01⟩, 463 ⟨81CL749⟩, 554 ⟨82BCJ1174⟩, 576 ⟨70MI 618-02⟩, 624 ⟨77BCJ3271⟩
Sakai, S.-I., **5**, 1127 ⟨90TL1287⟩
Sakai, T., **1**, 422 ⟨85CL1137⟩, 430 ⟨85S676⟩, 749 ⟨84TL2183⟩; **2**, 14 ⟨89JOC4982⟩, 540 ⟨83MI 210-02, 85MI 210-02⟩; **3**, 32 ⟨82JOC1101⟩, 145 ⟨87BCJ1063⟩, 145 ⟨87BCJ3421⟩, 245 ⟨86JHC1199⟩, 246 ⟨73CL425, 85TL4727⟩; **4**, 773, 781, 783 ⟨85S676⟩; **5**, 155 ⟨84TL2183⟩
Sakai, Y., **4**, 90 ⟨91COS(7)762⟩
Sakaitani, M., **1**, 991 ⟨84CL2005⟩
Sakaizumi, T., **5**, 1103 ⟨90JSP62⟩
Sakaki, J., **5**, 652 ⟨89CPB2952⟩
Sakaki, J.-I., **2**, 39 ⟨90TL7463⟩; **3**, 543 ⟨90CPB94⟩
Sakaki, K., **2**, 235 ⟨74TL1059⟩; **5**, 349 ⟨75TL649⟩
Sakakibara, A., **5**, 1020 ⟨88OM558⟩
Sakakibara, J., **5**, 340 ⟨88CPB877⟩, 346 ⟨91TL7431⟩
Sakakibara, M., **1**, 829 ⟨87TL2597⟩; **3**, 246 ⟨81T709⟩
Sakakibara, T., **1**, 608 ⟨81CC261⟩, 1143, 1144 ⟨68TL503⟩; **2**, 121 ⟨77JOC2180⟩, 784 ⟨82S261⟩, 784, 786 ⟨81CC261, 82BCJ635⟩; **4**, 335 ⟨82S261⟩, 910 ⟨83S920⟩; **5**, 113 ⟨69TL1019⟩, 183, 199 ⟨79CC179⟩
Sakakibara, Y., **1**, 433 ⟨91CL2017⟩; **3**, 662 ⟨75BCJ3298⟩
Sakakibari, Y., **1**, 468 ⟨93BCJ2446⟩
Sakakura, A., **5**, 329 ⟨94TL1247⟩
Sakakura, T., **1**, 1047 ⟨85TL3463⟩; **3**, 145 ⟨85CC1545⟩, 290 ⟨85TL3463⟩; **4**, 235 ⟨84JA2954⟩; **5**, 173 ⟨84JA2954⟩
Sakami, W., **5**, 140 ⟨42JBC(144)203⟩, 140 ⟨54JA2881⟩, 140 ⟨59JA1751⟩
Sakamoto, A., **5**, 172 ⟨87JA7188⟩
Sakamoto, H., **3**, 564 ⟨75BCJ2891⟩
Sakamoto, J., **1**, 571 ⟨91JOC7177⟩
Sakamoto, K., **4**, 613 ⟨88CL1441⟩, 774, 777 ⟨92BCJ210⟩, 774, 779 ⟨93MI 417-01⟩, 900 ⟨85CC279⟩, 1051, 1052 ⟨88CL965⟩; **6**, 383 ⟨88CL1441⟩
Sakamoto, M., **3**, 366 ⟨84JOC1837⟩; **4**, 736 ⟨93JFC(62)201⟩, 979 ⟨74JOC3198⟩; **5**, 246 ⟨91BCJ2720⟩, 570, 613 ⟨92JOC2419⟩, 630 ⟨88JCS(P1)2203⟩; **6**, 114 ⟨86TL1335⟩
Sakamoto, N., **3**, 621 ⟨73CL379⟩
Sakamoto, S., **4**, 563 ⟨88JAP63185993⟩
Sakamoto, T., **1**, 1009 ⟨92S746⟩, 1009, 1010 ⟨78H(9)271, 78CPB3843⟩, 1010 ⟨79CPB193, 81CPB3543, 82CPB1865, 82CPB2410, 82CPB2417, 83S312, 88CPB1890, 88CPB2248⟩, 1034 ⟨88H(27)475⟩; **2**, 677 ⟨84S245⟩, 1027 ⟨92SL502⟩; **3**, 311 ⟨92TL5373⟩, 430 ⟨92JOC3245⟩, 430, 666 ⟨91S750⟩, 431 ⟨93S563⟩, 617 ⟨93SC1061⟩; **4**, 195 ⟨84S245⟩, 196 ⟨86CPB2362⟩; **5**, 116 ⟨79H(13)235⟩, 328 ⟨94JOC929⟩, 332 ⟨90TL243⟩, 333 ⟨94OPP111⟩, 665, 666, 824 ⟨91S750⟩, 665, 672, 824, 825 ⟨92JOC3245⟩, 666 ⟨93SC1061⟩
Sakamoto, Y., **1**, 818 ⟨93TL5923⟩
Sakamura, S., **1**, 789 ⟨87T5245⟩
Sakan, T., **3**, 286 ⟨77CL645⟩
Sakane, K., **3**, 427 ⟨84JAN532⟩; **5**, 864 ⟨B-87MI 521-01⟩
Sakane, S., **3**, 437, 438 ⟨83JA672⟩; **5**, 731 ⟨83JA2831⟩
Sakane, Y., **6**, 461 ⟨94JAP06041019⟩
Sakano, S.,**3**, 437 ⟨83TL943⟩
Sakari, M., **1**, 752 ⟨88H(27)2077⟩
Sakasai, T., **5**, 116 ⟨79H(13)235⟩
Sakashita, N., **6**, 9 ⟨91MI 601-02⟩
Sakata, G., **6**, 229 ⟨92JFC(59)417⟩
Sakata, J., **1**, 884 ⟨83JA1598⟩; **2**, 853 ⟨88BCJ4323⟩, 858 ⟨87TL6649⟩
Sakata, J.-I., **3**, 384, 394 ⟨87TL6649⟩
Sakata, K., **3**, 183 ⟨86BCJ457⟩; **4**, 74 ⟨85BCJ1983⟩
Sakata, S., **3**, 286 ⟨74BCJ1777⟩, 632 ⟨92TA677⟩; **5**, 138 ⟨88JA3670⟩
Sakata, Y., **1**, 243 ⟨81CL627, 82BCJ182⟩, 243 ⟨87BCJ4027⟩, 594, 611, 621 ⟨81CL447⟩; **3**, 668 ⟨86TL2411⟩; **4**, 402 ⟨81CL447⟩
Sakatsume, O., **2**, 106 ⟨91T8717⟩
Sakdarat, S., **4**, 897 ⟨89TL2349⟩, 898 ⟨90JOC5110⟩
Saket, B. M., **3**, 439 ⟨85CC582⟩
Sakito, Y., **1**, 801, 803 ⟨86CL621⟩; **2**, 314 ⟨88TL223⟩; **4**, 408 ⟨78CL1253⟩
Sakla, A. B., **1**, 981 ⟨61JCS2687⟩
Sako, H., **1**, 444 ⟨83JOC2569⟩; **2**, 843 ⟨83JOC2569⟩; **3**, 496 ⟨83JOC2569⟩
Sako, M., **2**, 214 ⟨84H(22)1017⟩; **6**, 111 ⟨84H(22)13⟩
Sako, R., **5**, 57 ⟨82CL1255⟩
Sakodo, H., **3**, 543 ⟨91T5689⟩
Sakoh, H., **2**, 667 ⟨84CL277⟩
Saksena, R. K., **3**, 155 ⟨89CL1241⟩
Sakuma, K., **1**, 131 ⟨93TL5213⟩, 237 ⟨85JCS(P1)1171⟩, 820 ⟨80CL1061⟩
Sakuma, S., **2**, 677 ⟨93BCJ1156⟩
Sakuraba, H., **2**, 24 ⟨83CC947⟩, 57 ⟨89JOC3482⟩
Sakurai, A., **4**, 437, 445 ⟨77S647⟩
Sakurai, H., **1**, 13, 15, 17 ⟨81S32⟩, 253, 486 ⟨87JOC2947⟩, 314 ⟨77JA1673, 78TL3043, 82PAC1, 89SL1⟩, 468 ⟨86TL2881⟩, 759 ⟨73TL4193⟩, 922 ⟨88JPO197⟩, 929 ⟨93JCS(P1)3007⟩; **2**, 11, 36 ⟨79S379, 78CC874⟩, 77 ⟨88JA4599⟩, 78 ⟨89JA6429, 89TL1099⟩, 300 ⟨88CPB2342⟩, 459 ⟨73CL599, 81S811⟩, 537 ⟨66TL5493⟩, 544 ⟨72JOM(36)C13⟩, 546 ⟨71TL1511, 75JOM(92)C46⟩, 547 ⟨89SL1⟩, 847 ⟨78TL2523⟩, 916 ⟨75JA931⟩, 936 ⟨73TL4193⟩, 941 ⟨79S740⟩; **3**, 140 ⟨77JA1673⟩, 140 ⟨79CL245⟩, 186 ⟨82TL547⟩, 213 ⟨79TL399⟩, 502 ⟨91JA3986⟩, 540 ⟨79S740⟩, 640 ⟨82S915⟩; **4**, 140, 141 ⟨78JOM(154)353⟩, 181 ⟨84JOC2808⟩, 395 ⟨67CC889⟩, 612 ⟨74JA5623⟩, 612, 614 ⟨76JA7424⟩, 613 ⟨88CL1441⟩, 616 ⟨68CC930, 86BCJ1509⟩, 618 ⟨88TL6787, 89JA3748, 93CL267⟩, 620, 653 ⟨73TL4193⟩, 640 ⟨82CL1855⟩, 657 ⟨77JAP7973759, 77JOM(142)C49⟩, 689 ⟨72JOM(39)365⟩, 742 ⟨83TL4727⟩, 1003, 1004 ⟨75BCJ2103⟩, 1048 ⟨77JA3879⟩, 1050 ⟨82JA4288⟩, 1051, 1052 ⟨88CL965, 78JOM(162)C43⟩, 1052 ⟨80AG632, 80TL3077⟩; **5**, 378 ⟨66BCJ2263⟩, 419 ⟨79JOM(175)C1⟩; **6**, 158 ⟨75BCJ1030⟩, 172 ⟨B-92MI 606-01⟩, 176, 193 ⟨89JA3748⟩, 186 ⟨89AG(E)55⟩, 190, 194, 393 ⟨73TL4193⟩, 193 ⟨92CL867, 93CL267⟩, 380 ⟨86JOM(309)247⟩, 383 ⟨88CL1441⟩, 465, 481 ⟨75BCJ108⟩, 466 ⟨74TL803⟩
Sakurai, J., **4**, 696 ⟨1881JCS485⟩
Sakurai, K., **1**, 248 ⟨85TL1333⟩; **2**, 55 ⟨86JOC240⟩
Sakurai, M., **5**, 1138 ⟨76BCJ464⟩
Sakurai, R., **4**, 650 ⟨87CPB1663⟩
Sakurai, S., **1**, 76 ⟨56MI 103-01⟩
Sakurai, Y., **1**, 1041 ⟨84TL1083⟩; **2**, 59 ⟨88S995⟩, 65, 528 ⟨87BCJ395⟩, 314 ⟨89JCS(P1)1548⟩
Sakuruba, S., **2**, 54 ⟨90JA5876⟩
Sakuta, K., **1**, 132 ⟨84TL223, 84TL2487⟩; **3**, 663 ⟨88T2413⟩; **5**, 215 ⟨84TL223⟩
Sala, A., **6**, 608 ⟨88SC1171⟩
Sala, M., **5**, 801 ⟨86OM1369⟩

Saladin, E., **2**, 8, 36 ⟨72AG(E)229⟩
Salakhov, M. S., **6**, 412 ⟨72URP345094⟩
Salakhutdinov, R. A., **2**, 448 ⟨76JGU1213⟩, 459 ⟨74ZOB2621⟩; **5**, 403 ⟨71JGU2513⟩
Salam, A., **6**, 558 ⟨87MI 617-02⟩
Salama, M. A., **5**, 590, 619 ⟨87PS(32)51⟩
Salamon, G., **4**, 973, 974 ⟨70CB133, 70CB147⟩
Salamonczyk, I., **3**, 220 ⟨88JOC2837⟩
Salamonov, T. B., **6**, 284 ⟨92MC52⟩
Salamonov, Yu. B., **5**, 680 ⟨90BAU294, 90IZV352⟩
Salanski, P., **4**, 174 ⟨84TL4797⟩; **5**, 344 ⟨85T2441⟩
Salas, M., **3**, 361 ⟨92JCS(P1)313⟩
Salaski, E. J., **3**, 211 ⟨85JA8066⟩
Salaün, J., **1**, 802, 841 ⟨B-87MI 118-04⟩, 975 ⟨92JA4051⟩, 1115 ⟨75JOC1994⟩; **2**, 309 ⟨93TL4193⟩; **3**, 123 ⟨87S37⟩, 539 ⟨83T1567⟩; **4**, 171 ⟨85OS(63)147⟩
Salazar, G. E., **1**, 983 ⟨87JA4338⟩
Salazar, J. A., **1**, 927 ⟨88JCS(P1)3255⟩, 929 ⟨84TL1953⟩; **2**, 13 ⟨86JOC402⟩; **5**, 1118 ⟨90JCS(P1)2417⟩; **6**, 633, 634 ⟨89CC452⟩
Salbaum, H., **1**, 988 ⟨74TL1275⟩; **5**, 43 ⟨74TL1275⟩
Salbeck, G., **2**, 255 ⟨80PS(9)175⟩
Saldaña, M. C., **4**, 314 ⟨69JHC921⟩
Saleh, G., **1**, 608 ⟨79CB355⟩; **2**, 843 ⟨79CB355⟩; **4**, 348 ⟨79CB355⟩; **6**, 123, 697 ⟨79CB355⟩
Saleh, S. A., **3**, 61 ⟨81JOC4817⟩
Salek, A., **4**, 409 ⟨89JOC4771⟩
Salem, G., **1**, 114 ⟨B-87MI 104-03⟩, 452, 453 ⟨B-87MI 110-03⟩; **2**, 442 ⟨82HOU(13/3b)69⟩, 557 ⟨B-87MI 211-03⟩, 559, 560 ⟨84CC1702⟩; **4**, 604, 647 ⟨88JCS(D)2403⟩, 626, 662 ⟨87JCS(D)1647⟩, 654 ⟨88JOC3134⟩; **6**, 393 ⟨88JOC3134⟩
Salem, G. F., **4**, 191 ⟨81S185⟩
Salem, L., **1**, 805 ⟨72JA8917⟩
Salemink, C. A., **1**, 784 ⟨84JCS(P1)1701⟩
Salemnick, G., **2**, 691 ⟨72T3833⟩
Salerno, G., **1**, 496 ⟨86JOM(317)373⟩; **3**, 60 ⟨75JCS(P1)1527⟩
Sales, F., **4**, 884 ⟨79TL3329⟩
Sales, J., **5**, 1137 ⟨84OM996, 92ICA173⟩, 1138 ⟨76SRI217⟩, 1139 ⟨77JOM(128)291⟩
Sales, K. D., **4**, 672 ⟨78JCS(D)1314⟩
Saleske, H., **2**, 436 ⟨77CB1269⟩, 4366 ⟨76JOM(118)C8⟩
Salgado, H. J., **4**, 540 ⟨82JCS(P1)143⟩
Salgudo, V. O. N., **1**, 566 ⟨93CB1481⟩
Saliente, T. A., **3**, 106, 308 ⟨82SC231⟩
Salih, Z. S., **1**, 447 ⟨72JCS(P2)172⟩
Salikhov, S. G., **4**, 478 ⟨69IZV1757⟩
Salim, J. R., **5**, 652 ⟨89SC1181⟩
Salimgareeva, I. M., **4**, 141 ⟨81DOK(261)474⟩
Salimgareeva, N. M., **1**, 61 ⟨81ZOB420⟩
Salimi, N., **5**, 138 ⟨93TA457⟩
Salinaro, R. F., **4**, 746 ⟨82JA2223⟩
Salinger, R., **1**, 846 ⟨66JA460⟩
Salinger, R. M., **4**, 650 ⟨70JOC1308⟩
Salisbury, L., **2**, 403 ⟨61JA747⟩; **4**, 314 ⟨61JA747⟩
Salishchev, V. G., **4**, 913 ⟨74ZOB2067, 76CA192821t⟩
Saljoughian, M., **6**, 25 ⟨83M813⟩
Sal'keeva, L. K., **4**, 921 ⟨90ZOB1940⟩; **5**, 406 ⟨90JGU1940⟩; **6**, 583 ⟨90ZOB563⟩
Sälker, R., **1**, 466 ⟨93CB1447⟩
Sallé, M., **6**, 592 ⟨89CC1520⟩
Sallese, G., **1**, 471, 483 ⟨91JOC261⟩
Salmain, M., **4**, 674 ⟨90OM2993⟩
Salmi, E. J., **4**, 177, 178 ⟨38CB1803⟩
Salmonaa, P., **2**, 638 ⟨74JA4494⟩
Salmond, W. G., **4**, 738, 739 ⟨77TL1239⟩; **6**, 695 ⟨77TL1239⟩
Salmon-Legagneur, F., **5**, 24 ⟨66MI 502-01, B-47MI 502-01⟩, 24 ⟨B-47MI 502-02⟩, 64 ⟨78JCR(S)198⟩
Salomon, F., **6**, 471 ⟨1872JPR(2)477⟩
Salomon, M., **4**, 220 ⟨88TL3773⟩
Salomon, M. F., **2**, 98 ⟨76JOC3983, 79JA4290⟩; **3**, 490 ⟨75JOC756⟩
Salomon, R. G., **1**, 357 ⟨83T485⟩, 729 ⟨75JA6214⟩; **2**, 98 ⟨76JOC3983, 79JA4290⟩; **3**, 141 ⟨86JOC1393, 75JOC756⟩, 620 ⟨86JA1311⟩
Salouki, B., **3**, 527 ⟨77ACS(B)890⟩
Saloutin, V. I., **2**, 227 ⟨88ZOR1088⟩; **4**, 311 ⟨83IZV2568⟩
Salsburg, J. M., **2**, 271, 273 ⟨59JA4810⟩
Salsbury, J. M., **1**, 45 ⟨46JA1813⟩
Salter, I. D., **6**, 340 ⟨82CC51⟩, 343 ⟨82CC773, 83JCS(D)2599⟩, 343 ⟨83JOM(249)273⟩
Saltza, M. H. von, **2**, 214 ⟨54JA1828⟩
Saltzman, H., **5**, 652 ⟨73OSC(5)660⟩
Salunkhe, A. M., **1**, 486 ⟨91SL684⟩; **2**, 306, 318 ⟨91JOC1170⟩
Salunkhe, M. M., **5**, 367 ⟨89IJC(B)186⟩
Saluzzo, C., **1**, 780 ⟨92JOC714⟩; **2**, 280 ⟨90TL2127, 90TL663⟩; **4**, 787 ⟨90TL2127⟩
Salvá, J., **1**, 831 ⟨91TL3209⟩
Salvadori, P. A., **1**, 506 ⟨79T2039⟩
Salvadori, S., **4**, 422 ⟨85JMC769⟩
Salvi, A., **5**, 1123 ⟨68G511⟩
Salvi, J.-P., **3**, 40 ⟨94TL1181⟩
Salvi-Narkhede, M., **6**, 251 ⟨92MI 608-01⟩
Salvino, J. M., **1**, 873 ⟨85TL3931⟩
Salz, U., **5**, 29 ⟨82GEP3116474⟩
Salzberg, P. L., **3**, 612 ⟨41OSC(1)46⟩
Salzburg, H., **1**, 4, 7 ⟨76CB3598⟩; **6**, 437 ⟨83HOU(E4)1177⟩
Salzer, A., **1**, 885 ⟨B-89MI 119-03⟩; **2**, 446 ⟨83HCA1774⟩; **4**, 683 ⟨86OM2413, 90JA7113⟩, 694 ⟨72AG(E)930, 72SRI239⟩, 694 ⟨77JOM(141)339⟩
Salzman, T. N., **2**, 139 ⟨76JA4887⟩
Salzmann, T. N., **1**, 595, 597, 598, 601, 621 ⟨76JA4887⟩, 598, 779 ⟨73JA6840, 75JOC148⟩; **2**, 139, 140 ⟨73JA6840⟩; **3**, 209 ⟨73JA6840⟩; **4**, 197 ⟨75JOC148⟩; **5**, 374 ⟨80TL31⟩; **6**, 128 ⟨80JA6161⟩
Samaan, S., **2**, 408 ⟨78HOU(13/8)220⟩, 414, 1076, 1080 ⟨78HOU(13/8)33⟩, 481 ⟨78HOU((13)/8)1⟩, 872 ⟨78HOU(13)1, 78HOU(13)45⟩, 879 ⟨78HOU(13)465⟩, 881 ⟨75MI 217-02⟩, 884 ⟨78HOU(13)295⟩, 885 ⟨78HOU(13)167, 78HOU(13)376⟩, 890 ⟨78HOU(13)206⟩, 891 ⟨78HOU(13)284⟩, 1063 ⟨79LA43⟩
Samadi, M., **1**, 333, 935 ⟨92T7083⟩
Sam, D. J., **2**, 619 ⟨82JA4680⟩; **5**, 47, 52, 74, 114, 116 ⟨72JA4024⟩; **6**, 28 ⟨B-82MI 601-02⟩
Samama, J.-P., **5**, 584, 586, 612 ⟨65BSF3623⟩, 586 ⟨65BSF3628⟩
Saman, D., **3**, 430 ⟨88LA1091⟩
Samann, S., **2**, 490 ⟨75PS(5)265⟩
Samano, V., **3**, 206 ⟨90SC131⟩
Samarai, L. I., **4**, 314, 329 ⟨74JOU244, 74JOU1565⟩, 436, 441 ⟨76ZOR2103, 77ZOR2449, 76ZOR231⟩, 992 ⟨78ZOR2624⟩; **5**, 226 ⟨81ZOR2452⟩, 669 ⟨89UKZ1071, 90JOU287, 91UKZ202⟩, 669, 672 ⟨90ZOR339⟩, 708 ⟨63JGU551, 63ZOB557, 64JGU1152⟩, 708, 988 ⟨64ZOB1161⟩, 860 ⟨62JGU3689⟩, 967 ⟨69ZOR119⟩, 967, 987 ⟨67AG(E)864⟩, 982 ⟨90UKZ1313, 62ZOB3759⟩, 988 ⟨63ZOB1587, 65ZOB923⟩, 997 ⟨69ZOB1712⟩, 1008 ⟨76ZOR1963, 76ZOR2473⟩, 1032 ⟨73AG(E)842⟩, 1035 ⟨91UKZ431⟩, 1063 ⟨79S738, 81ZOR1420⟩, 1068 ⟨76ZOR2472⟩, 1068, 1069 ⟨80S85, 84ZOR2543⟩, 1069 ⟨88ZOR986⟩, 1075 ⟨87ZOR2385, 87ZOR2390⟩, 1076 ⟨86ZOR1099⟩, 1083 ⟨80ZOB1678⟩; **6**, 438 ⟨68ZOR720⟩, 445 ⟨69CB2972, 80S85⟩, 453 ⟨71ZOR2229⟩, 657 ⟨78ZOB1425⟩
Samat, A., **6**, 111 ⟨72HCA1782⟩, 550 ⟨92JA4111⟩
Samate, D., **4**, 646, 654 ⟨75JOM(93)71⟩, 648 ⟨77JOM(127)281, 78JOM(144)155⟩
Sambale, C., **4**, 712, 716 ⟨82OM1597⟩

Sambeth, J., **2**, 414 ⟨57CB2075⟩
Sambur, V. P., **6**, 232, 233, 234 ⟨75S721⟩
Samejima, S., **5**, 494 ⟨83MI 511-01⟩
Samek, Z., **2**, 540 ⟨80TL4523⟩
Samén, E., **6**, 91 ⟨49AK(1)231, 50ACS397⟩
Sameshima, K.,**5**, 948 ⟨82JCS(D)793⟩
Sami, M., **5**, 358 ⟨75BCJ365⟩, 771 ⟨80JHC877⟩
Samitov, Yu. Yu., **4**, 489 ⟨72ZOB1227⟩, 561 ⟨68ZOB292⟩, 923 ⟨83PS(16)345⟩
Samizu, K., **2**, 85 ⟨89CC1344⟩; **3**, 60 ⟨89CL1283⟩
Sammakia, T., **1**, 199 ⟨86JA3128⟩, 199, 200 ⟨87JA5749⟩
Sammes, M. P., **3**, 363, 686 ⟨91JCS(P1)183⟩, 627, 634 ⟨71JCS(C)2151⟩; **6**, 309 ⟨91TL4329⟩, 562 ⟨85CC1641⟩, 623, 728 ⟨88JCS(P1)899⟩
Sammes, P. G., **1**, 420 ⟨72CC1034⟩, 595 ⟨70CC1683⟩, 654 ⟨86CC620⟩; **2**, 184 ⟨76CRV113⟩, 362, 363 ⟨91JCS(P1)2169⟩, 664 ⟨78CC118, 78CC162⟩; **3**, 39 ⟨82JCS(P1)307⟩, 225 ⟨90JCS(P1)353⟩, 324 ⟨82CC1056, 83CC666⟩, 324 ⟨83JCS(P1)1261, 86JCS(P1)1725⟩, 324 ⟨87JCS(P1)195⟩, 532 ⟨72JCS(P1)2623⟩; **4**, 139 ⟨88JCS(P1)2585⟩, 212 ⟨87JCS(P1)195⟩, 285 ⟨84JCS(P1)21⟩, 900 ⟨89T1801⟩; **5**, 278 ⟨71JCS(C)3540⟩, 325 ⟨72CC1222⟩, 720 ⟨79JCS(P1)2478⟩
Sammour, A., **4**, 105 ⟨71MI 403-01⟩
Samne, S., **5**, 215 ⟨68CR(C)1303⟩
Samoa, T., **4**, 489 ⟨65AG1138⟩
Samodral, R., **1**, 517 ⟨83JOC1262⟩
Samoilova, Z. E., **4**, 886 ⟨69DOK(186)835⟩
Samour, C. M., **3**, 467 ⟨54JA441⟩
Sampathkumar, P. S., **1**, 84 ⟨79TL1641⟩
Samples, M. S., **5**, 379 ⟨87JOM(332)69⟩
Sampson, E. J., **2**, 807 ⟨71JA4009⟩
Sampson, P., **2**, 457 ⟨86JOC4342⟩; **3**, 198 ⟨86JOC4342⟩, 217, 245 ⟨84TL4467, 88TL6897⟩, 245 ⟨86JOC2525⟩
Sams, J. R., **6**, 65 ⟨70IC1682⟩
Samsel, E. G., **1**, 546, 1024 ⟨81JA7520⟩, 946 ⟨86JA4790⟩
Samson, M., **3**, 634, 636 ⟨78SC231⟩
Samsonova, G. A., **4**, 569 ⟨75ZOB2010⟩
Samuel, E., **2**, 566 ⟨73JA6263⟩
Samuel, E. L., **6**, 81 ⟨67JCS(C)30⟩
Samuel, O., **2**, 147 ⟨85PAC1911⟩, 152, 198 ⟨91JOC5991⟩, 413 ⟨87JOC5320⟩; **4**, 261 ⟨89JOM(370)43⟩, 706 ⟨90TL3121⟩
Samuel, P. A., **2**, 355 ⟨73JOC1813⟩
Samuel, P. L., **2**, 403 ⟨68CJC1149⟩
Samuels, S. B., **3**, 138 ⟨80JA5930⟩
Samuelson, A. G., **6**, 544 ⟨88TL827⟩
Samuelson, B., **2**, 7 ⟨84S168⟩
Samuelsson, B., **1**, 587 ⟨90JOC4273⟩; **2**, 7 ⟨79CC978, 80JCS(P1)2866⟩, 7 ⟨82JCS(P1)681, 88JOC6126⟩, 135 ⟨80JA3663⟩, 397 ⟨84JA3682⟩
Samusenko, Yu. V., **4**, 777 ⟨73ZOR69⟩
Samworth, D. J., **4**, 417, 424 ⟨77JCS(P1)1954⟩
Sanadi, N. E., **1**, 729 ⟨75JA6214⟩
Sancassan, F., **5**, 348 ⟨85G265⟩
Sancéau, J.-Y., **2**, 50 ⟨93JOC1991⟩
Sánchez, B., **1**, 175 ⟨92TL5261, 93JOC2894⟩; **2**, 310 ⟨71JOC829⟩; **6**, 473, 544, 550 ⟨93JOC2894⟩, 550, 557 ⟨92TL5261⟩
Sanchez, E. I., **2**, 28 ⟨76S736, 82JOC1588⟩; **5**, 318 ⟨76S736⟩
Sanchez, F., **3**, 68 ⟨77S642⟩
Sánchez, I. H., **1**, 24 ⟨81S55⟩; **3**, 620 ⟨83SC35⟩
Sanchez, J., **5**, 938 ⟨92OM1442⟩
Sanchez, J. F., **2**, 617 ⟨91JHC939⟩
Sanchez, J. P., **3**, 310 ⟨90JHC1527⟩
Sánchez, L., **2**, 573 ⟨88OM1877⟩; **5**, 445 ⟨84JA3214⟩; **6**, 327, 330 ⟨93IC5569⟩, 327 ⟨89IC2120⟩
Sanchez, M., **2**, 411 ⟨65BSF2368⟩; **3**, 689 ⟨92TL2981⟩; **5**, 1158 ⟨92TL2981⟩; **6**, 660 ⟨91TL2775⟩, 690 ⟨92TL2981⟩
Sanchez, M. G., **3**, 141 ⟨82JOC743⟩
Sanchez, M. M., **3**, 414 ⟨85TL4403⟩
Sanchez, R., **2**, 56, 421 ⟨88TL139⟩; **3**, 404, 408 ⟨86JOC1848⟩
Sanchez, R. A., **3**, 638 ⟨73OSC(5)32⟩
Sánchez Marcos, E., **1**, 130 ⟨87JCPB765⟩
Sanchez-Montero, J. M., **1**, 11, 12 ⟨93T8433⟩
Sanchez-Nieva, F., **4**, 14 ⟨48JA2813⟩
Sand, D. M., **3**, 302 ⟨79JOC3420⟩
Sand, J.,**6**, 207 ⟨03LA(329)116⟩
Sanda, F., **3**, 130 ⟨85TL5529⟩
Sandall, P. L., **1**, 784 ⟨94TL2675⟩
Sandberg, R., **1**, 881 ⟨62OS(42)28⟩
Sandborn, L. T., **5**, 48, 49 ⟨32OSC(1)526⟩, 290 ⟨41OSC(1)111⟩
Sandborn, R. E., **2**, 87 ⟨86JOC3762⟩
Sandell, J., **6**, 194, 195 ⟨85JA4337⟩
Sander, J., **4**, 673 ⟨92CB143⟩
Sander, M., **2**, 428 ⟨62CB473⟩, 451, 820, 821 ⟨60CB1220⟩; **5**, 919 ⟨94AG(E)303⟩
Sander, W., **1**, 634 ⟨87JOC2644⟩; **6**, 531 ⟨86SA(A)1281⟩
Sander, W. W., **3**, 382 ⟨85JOC543⟩
Sanders, A., **1**, 936 ⟨73JOC3055⟩
Sanders, D. A., **4**, 698 ⟨71JA1035⟩
Sanders, F., **2**, 214 ⟨54JA1828⟩
Sanders, H. P., **1**, 1082 ⟨81S276⟩; **3**, 25 ⟨81S276⟩, 39, 174 ⟨82JCS(P1)2719⟩
Sanders, J. A., **2**, 691 ⟨74JOC2641⟩
Sanders, M. J., **3**, 686 ⟨79JOC510, 80JOC3753⟩
Sanderson, D. R., **2**, 916, 932 ⟨86TL2801⟩; **3**, 271 ⟨86TL2801⟩
Sanderson, J. R., **4**, 1078 ⟨87JOC3243⟩
Sanderson, R. T., **1**, 889 ⟨64JOC1261⟩
Sanderson, W. R., **2**, 654 ⟨90SL533⟩
Sanderud, K. A., **2**, 290 ⟨81ACS(A)151⟩; **5**, 1117 ⟨81ACS(A)151⟩
Sandham, D. A., **1**, 133 ⟨91CC1399, 92TL6187⟩; **2**, 640 ⟨91CC1399⟩
Sandhu, A., **5**, 210 ⟨89IJC(A)1528⟩
Sandhu, J. S., **2**, 34 ⟨91CI(L)175⟩; **3**, 413 ⟨87H(26)777⟩, 417 ⟨85SC335⟩, 536 ⟨91S1027⟩, 667 ⟨90JOC2545⟩
Sandhu, P. K., **1**, 827 ⟨94CC251⟩
Sandifer, R. M., **4**, 167 ⟨81JOC2260⟩
Sandin, R. B., **2**, 13 ⟨40JA3098⟩; **5**, 205 ⟨56JA3819⟩
Sandler, S. R., **2**, 26 ⟨70JOC3967⟩, 340, 341, 342, 343, 344, 346, 347, 349, 350, 351, 355, 356 ⟨B-89MI 206-03⟩, 773 ⟨B-68MI 215-03⟩, 776 ⟨B-86MI 215-02⟩; **3**, 425 ⟨B-89MI 311-01⟩, 426 ⟨B-72MI 311-01⟩; **4**, 24 ⟨77OS(56)32⟩, 176 ⟨B-89MI 404-01⟩; **5**, 24 ⟨B-68MI 502-01⟩, 182 ⟨B-86MI 504-01⟩, 322 ⟨B-89MI 507-01⟩, 360 ⟨B-86MI 507-01⟩, 685, 687, 690 ⟨B-72MI 517-01⟩, 726 ⟨B-72MI 518-01⟩, 1061, 1078 ⟨B-71MI 528-01⟩; **6**, 68 ⟨B-86MI 603-01⟩, 500 ⟨B-71MI 616-01⟩
Sandman, D., **3**, 398 ⟨77CC177⟩
Sandmann, H., **4**, 545 ⟨71AG(E)940, 73ZN(B)224⟩
Sandmeier, D., **3**, 547, 581 ⟨75AG(E)53⟩, 547, 603, 609 ⟨75AG(E)634⟩, 603, 604 ⟨80CB274⟩; **5**, 343, 730 ⟨80CB912⟩
Sandmeyer, T., **2**, 102 ⟨1888CB857⟩; **3**, 664 ⟨1884CB1633⟩
Sandoval, S., **3**, 135 ⟨80SC813⟩; **4**, 357 ⟨80JOM(198)15⟩
Sandoval-Ramirez, J., **6**, 101, 102 ⟨88JOC3766, 86CJC732⟩
Sandré-Le Craz, A., **1**, 807, 881 ⟨62TL505⟩
Sandri, E., **1**, 821 ⟨78JOC4826, 79JOC4128⟩, 821 ⟨81JOC3315⟩; **2**, 143 ⟨81JOC3315⟩, 156 ⟨88JOC5689⟩
Sandri, J., **2**, 7 ⟨92SC2945⟩
Sandri, J. M., **6**, 51 ⟨59CI(L)1216⟩
Sandri, S., **2**, 87 ⟨81S585⟩; **3**, 4 ⟨76S394⟩, 4, 115 ⟨76JA6737⟩, 57, 89 ⟨76CC190, 76TL3985⟩, 158 ⟨84JOC701⟩; **5**, 1125 ⟨89TL1679⟩; **6**, 469 ⟨82JOC4626⟩

Sandris, C., **5**, 334 ⟨85JHC1635, 87H(25)569⟩
Sandström, J., **1**, 248 ⟨87CC381⟩; **3**, 344 ⟨83JCS(P2)1321⟩, 355 ⟨91JOC1902, 91JOC4439⟩; **4**, 901 ⟨74S433⟩, 902, 979 ⟨78ACS(B)141⟩, 969, 970, 973 ⟨70ACS3102, 75JST(24)373⟩; **5**, 592, 593 ⟨72S34, 74S433⟩, 631, 632 ⟨66ACS689⟩, 846 ⟨63ACS95, 65ACS2395⟩
Sanehi, R., **4**, 584, 587 ⟨85IJC(A)398⟩, 588 ⟨86JOM(303)351⟩
Sanemitsu, Y., **3**, 35, 165 ⟨92JOC1053⟩; **4**, 103 ⟨88S482⟩; **6**, 478 ⟨92JOC1053⟩, 627 ⟨84JOC1703⟩
San Felix, A., **4**, 264 ⟨92RTC163⟩; **5**, 1039 ⟨84S509⟩
San Filippo, J., **2**, 87 ⟨75JOC1678⟩
San Filippo, Jr., J., **1**, 65, 66 ⟨84JA3703⟩, 67 ⟨70JA6611⟩, 118, 119, 469, 473 ⟨69JA4871⟩, 663 ⟨83OM554⟩, 668 ⟨70JA1396⟩; **2**, 600 ⟨78JA4834, 82JA2831, 82OM1496⟩; **3**, 287 ⟨69JA4871⟩
SanFilippo, L. J., **1**, 231, 243, 620, 685, 704 ⟨88T6241⟩; **3**, 392 ⟨85T4843⟩; **4**, 27 ⟨88S547⟩
Sanfilippo, P. J., **1**, 690 ⟨86JA3513⟩; **3**, 305 ⟨92JCS(P1)3401⟩
Sang, C. S., **2**, 327 ⟨88JHC1383⟩
Sangaiah, R., **1**, 11, 19 ⟨87JOC3205⟩
Sanger, A. R., **1**, 456 ⟨82CJC1363⟩; **2**, 414, 423 ⟨B-80MI 207-01⟩, 421 ⟨69CC1064⟩, 437 ⟨83CJC2214⟩, 543 ⟨B-80MI 210-01⟩, 815 ⟨B-80MI 215-02⟩
Sanger, F., **5**, 83 ⟨51MI 502-02⟩, 85 ⟨52MI 502-01⟩
Sanghvi, Y. S., **2**, 127 ⟨90JCS(P1)969⟩
Sangokoya, S. A., **6**, 151 ⟨87PS(38)79⟩, 151, 369 ⟨82CC930, 86IC2699⟩, 370 ⟨85IC2889⟩
Sangstad, J., **1**, 886 ⟨67JA1827⟩
Sanida, C., **5**, 223 ⟨89LA387⟩
Saniere, M., **3**, 211 ⟨85S766⟩
Sanin, A., **4**, 222 ⟨91TL7593⟩
Sanin, P. I., **6**, 565 ⟨86MI 617-03⟩
Sankar, L. G., **3**, 219 ⟨82CJC2528⟩
Sankaran, D. K., **5**, 103 ⟨55OSC(3)623⟩
Sankaran, V., **5**, 90 ⟨72S464⟩
Sankararaman, S., **6**, 147 ⟨86RTC278⟩
Sankar Lal, G., **2**, 700 ⟨88TL4269⟩
Sankov, B. G., **6**, 653 ⟨73IZV819⟩
Sanner, C., **1**, 179 ⟨90TL2149⟩
Sanner, M. A., **3**, 145 ⟨92TL5287⟩
Sanner, R. D., **6**, 246, 247 ⟨89OM1498⟩
Sannicolò, F., **5**, 354 ⟨87JHC773⟩
Sanno, Y., **5**, 74 ⟨73TL1995⟩
Sano, H., **1**, 194 ⟨87CL1237⟩, 290 ⟨86S776⟩, 471 ⟨84CL1225⟩, 615, 698 ⟨81TL2675⟩, 667 ⟨80S1011⟩, 1041 ⟨84TL2367⟩, 1130 ⟨85CL1689⟩; **2**, 30, 34 ⟨86JA6036⟩, 946 ⟨86S776⟩; **3**, 142 ⟨87CL1371⟩, 253 ⟨86S137⟩, 290, 408 ⟨86CL1197⟩, 290, 409 ⟨86BCJ677⟩, 293 ⟨87BCJ767⟩, 625 ⟨84CL1511⟩; **5**, 27 ⟨81YZ265⟩, 331 ⟨92TL3261⟩, 468 ⟨87BCJ3462⟩
Sano, K., **2**, 657 ⟨75TL1695⟩; **5**, 465 ⟨83CL115⟩
Sano, T., **1**, 202 ⟨90JOC5923⟩, 469 ⟨86BCJ1656⟩; **2**, 87, 118, 124 ⟨72JA679⟩; **3**, 184 ⟨69CJC3489⟩; **4**, 90 ⟨91COS(7)762⟩; **5**, 30, 35 ⟨86SC499⟩
Sans, E. A., **4**, 352 ⟨85TL1115⟩
Sansano, J. M., **5**, 1106 ⟨91T5193⟩
Sansone, E. B., **2**, 379 ⟨84JOC3470⟩
Sansoni, M., **6**, 525 ⟨79JA6110⟩
Sansoulet, J., **2**, 31 ⟨89BSF247⟩
Sansoulet, J., **3**, 634 ⟨87S56⟩; **5**, 132 ⟨85S40⟩
Santa, L. E., **2**, 69 ⟨88TL3887⟩
Santa, T., **2**, 579 ⟨93CPB292⟩
Santacroce, C., **6**, 76 ⟨66G375⟩
Santafianos, D., **1**, 596, 780 ⟨87TL221, 91T6813⟩
Santamaria, J., **2**, 304 ⟨89TL2927, 89TL3977⟩, 305 ⟨90TL4735⟩; **4**, 851 ⟨90TL2169⟩
Santaniello, E., **2**, 149 ⟨80S563⟩, 299 ⟨80SC611⟩, 653 ⟨82CC1157⟩; **3**, 3, 114 ⟨86SC43⟩, 4, 115 ⟨78S534⟩, 88 ⟨80SC75, 83S749⟩; **4**, 168 ⟨87JCS(P1)2301⟩
Santarsiero, B. D., **2**, 477 ⟨90IC5081⟩, 969 ⟨87JA203⟩; **3**, 545 ⟨84JA5178, 86JA3318⟩; **4**, 710 ⟨94OM1078⟩; **5**, 441 ⟨86JA3318⟩; **6**, 519 ⟨85JA4670⟩, 622 ⟨90IC5081⟩
Santel, H. J., **6**, 632 ⟨88GEP(O)3634929⟩
Santella, III, J. B., **4**, 780 ⟨91JMC2525⟩
Santelli, M., **1**, 421 ⟨83S429⟩, 422 ⟨79TL4553⟩, 687 ⟨88T4295⟩, 832 ⟨93TL1149⟩, 955 ⟨93JOC5702⟩, 959 ⟨75BSF1407⟩, 978, 1061 ⟨86JOC1199⟩, 1114 ⟨74T243⟩, 1125 ⟨64CR(259)2251⟩, 1126 ⟨68CC718, 69TL3699, 74T227, 74T235⟩, 1131 ⟨⟨87TL2009⟩; **2**, 446 ⟨88S395⟩, 940 ⟨81TL3179⟩; **3**, 195 ⟨91T3347⟩, 212 ⟨79TL4553⟩, 652 ⟨83TL2847⟩; **4**, 1046 ⟨81TL3179⟩
Santelli-Rouvier, C., **1**, 421 ⟨83S429⟩, 1126 ⟨73BSF1800⟩
Santhakumar, V., **1**, 494 ⟨91TL687⟩, 837 ⟨93T5177⟩
Santhanakrishnan, T. S., **1**, 99 ⟨67JA1044⟩, 1023 ⟨68CC634⟩
Santiago, A. N., **2**, 438 ⟨84JOC3387, 84JOC4609⟩, 438 ⟨88JOC3016⟩
Santiago, B., **1**, 449 ⟨90TL5113⟩, 652 ⟨89TL5693⟩; **2**, 922 ⟨90TL5113⟩, 939 ⟨91TL3457⟩; **3**, 621 ⟨93TL5839⟩; **4**, 655 ⟨89TL5693⟩, 954 ⟨92JOC5844⟩
Santiago, C., **3**, 313 ⟨80JA1504⟩
Santiago, M. L., **4**, 329, 392 ⟨80JA7929⟩; **5**, 731 ⟨80JA7929⟩
Santi, M., **6**, 545 ⟨77S873⟩
Santi, R., **1**, 471, 483 ⟨91JOC261⟩, 494 ⟨92JOC3558⟩, 927 ⟨88TL1441⟩
Santiesteban, H., **2**, 282 ⟨81JOC2605⟩
Santillan, A., **1**, 515 ⟨90SC3477⟩
Santilli, A. A., **2**, 377, 379 ⟨60JA2728⟩
Santini, C., **1**, 813 ⟨84JA4038⟩
Santini, C. C., **2**, 440 ⟨83CJC21⟩; **4**, 341 ⟨84JOM(266)285⟩
Santini-Scampucci, C., **4**, 579 ⟨80JOM(187)331⟩
Santini-Scampucci, G., **2**, 570 ⟨73JCS(D)2436⟩
Santo, G. P., **4**, 422 ⟨85JMC769⟩
Santo, K., **3**, 659, 660 ⟨86CPB4620⟩
Santos, A., **5**, 455 ⟨90JOM(390)C57⟩, 1142 ⟨82ICA127⟩
Santos, J. G., **5**, 586 ⟨93JOC4908⟩; **6**, 476, 551 ⟨85PS(25)91⟩, 546, 547 ⟨83JCS(P1)2641⟩
Santos, M. A., **5**, 214, 635 ⟨84S829⟩
Santsyshkina, A., **4**, 717 ⟨80JOM(201)269⟩
Sanyal, G. S., **5**, 864 ⟨83IJC(A)307, 86SRI113⟩
Sanz, D., **3**, 417 ⟨84JCR(S)144⟩; **4**, 304 ⟨84JCR(M)144⟩
Sanz, F., **2**, 855 ⟨72HCA1991⟩; **6**, 661 ⟨72JCS(D)2267⟩
Sanza-Parcio, J.,**4**, 843 ⟨89H(29)1877⟩
Saotome, K., **5**, 57, 59, 89 ⟨63BCJ1264⟩
Sapi, J., **1**, 628 ⟨92S383⟩
Sapino, C., **5**, 500 ⟨B-72MI 511-01⟩
Sapino, Jr., C., **1**, 1004 ⟨90JOC5833⟩
Sappa, E., **4**, 683 ⟨72JOM(39)169⟩, 687 ⟨67JOM(8)149⟩, 688 ⟨93JOM(444)203⟩, 713 ⟨80JOM(197)335, 81JOM(221)93, 91OM1907⟩
Sapp, S. G., **3**, 222 ⟨83SC207⟩
Sapuppo, N., **3**, 616, 621 ⟨88OSC(6)334⟩
Saquet, M., **1**, 1058 ⟨77T2949⟩; **4**, 255 ⟨74CR(C)695, 75BSF657⟩, 256 ⟨77T2949⟩; **5**, 552 ⟨80TL2165⟩, 589 ⟨93PS(80)31⟩, 589, 612 ⟨90JCR(S)106, 91PS(56)283⟩, 729 ⟨83CI(L)391⟩
Sarac, S., **6**, 561 ⟨92AF1453⟩
Saraf, S. D., **3**, 68 ⟨81JPR673⟩
Saraie, T., **3**, 618, 619 ⟨73TL2121⟩, 674 ⟨70JOC675⟩
Saran, M. S., **4**, 821 ⟨73JA1811⟩
Sarandeses, L. A., **1**, 1029 ⟨86TL1523, 88TL1203⟩
Sarapulova, G. I., **6**, 558 ⟨90ZOR1422⟩
Sard, H., **1**, 805 ⟨81T3943⟩, 812 ⟨82JA7670⟩; **2**, 668, 941, 944 ⟨80JOC4810⟩, 928 ⟨83TL23⟩; **3**, 271 ⟨80JOC4810, 83TL23⟩
Sardarian, A. R., **2**, 215 ⟨83SC611⟩; **3**, 4 ⟨86S285⟩
Sardella, D. H., **1**, 870, 871 ⟨69JOC2817⟩
Sardina, F. J., **2**, 44 ⟨92TL4637⟩; **3**, 207 ⟨89JOC4654⟩
Sarel, S., **1**, 803 ⟨78ACR204⟩; **3**, 65 ⟨65TL821⟩; **4**, 681 ⟨71CC1241⟩; **5**, 34 ⟨56JA5416⟩, 344 ⟨76TL451⟩

Sarett, L. H., **1**, 881 ⟨60JA1684⟩; **2**, 1016 ⟨61JA4663⟩
Sargeant, P. B., **2**, 1006 ⟨69JA415⟩
Sarge, S. M., **4**, 352, 353 ⟨92CB591⟩
Sargent, C. R., **6**, 9, 10 ⟨81AHC(28)1⟩
Sargent, G. D., **2**, 271, 273 ⟨59JA5430⟩
Sargent, L. J., **1**, 2, 8 ⟨58JOC1938⟩
Sargent, M. V., **1**, 732 ⟨74JCS(P1)2435⟩; **2**, 654 ⟨74JCS(P1)1353⟩; **3**, 60 ⟨74JCS(P1)37⟩, 91 ⟨74JCS(P1)1353⟩; **4**, 856 ⟨87JCS(P1)2017⟩
Sarges, R., **4**, 420 ⟨74JOC1710⟩
Sargeson, A. M., **4**, 499 ⟨89CC1648⟩; **5**, 385 ⟨73JA5649, 93IC5285⟩, 386 ⟨75CC974, 81JA3559⟩, 386 ⟨84JA625⟩
Sarina, T. V., **3**, 689 ⟨92TL2981⟩; **5**, 1158 ⟨92TL2981⟩; **6**, 655, 685 ⟨91ZOB401⟩, 680 ⟨85ZOB1188, 85ZOB1437⟩, 690 ⟨92TL2981⟩, 685,686 ⟨90ZOB2238⟩
Sarina-Pidvarko, T. V., **6**, 685, 686 ⟨85ZOB221⟩
Sarin, P. S., **6**, 489 ⟨90BBR(171)458⟩
Sarin, R., **3**, 419 ⟨88IJC(B)132⟩
Saris, L. E., **5**, 626, 732 ⟨75CC617⟩
Sarkar, A., **1**, 138 ⟨89JA4399⟩; **2**, 61 ⟨91TL2157⟩; **4**, 389, 664, 665 ⟨89JA4399⟩; **5**, 955 ⟨95OM547⟩
Sarkar, A. K., **1**, 565 ⟨92JCS(P1)3351⟩, 628 ⟨92JCR(S)309⟩, 651, 943 ⟨90CC1575⟩; **2**, 303 ⟨92JCR(S)309⟩
Sarkar, D. C., **3**, 18, 133 ⟨88JOC878⟩, 61 ⟨87SC155⟩
Sarkar, I. M., **2**, 614 ⟨71JOC631⟩
Sarkar, N., **1**, 628 ⟨92JCR(S)309⟩; **2**, 303 ⟨92JCR(S)309⟩
Sarkar, T., **1**, 655 ⟨80JA6885⟩; **2**, 907 ⟨82COMC-I(7)515⟩; **6**, 172 ⟨82COMC-I(7)515⟩
Sarkar, T. K., **1**, 593 ⟨87TL2061, 90T1885⟩; **4**, 401 ⟨89TL3333⟩
Sarker, P. B., **6**, 422 ⟨33JIC537⟩
Sarkozi, P., **6**, 645 ⟨91MIP54636⟩
Sarlo, F. de, **2**, 320 ⟨90TL3351⟩
Sarma, J., **4**, 266 ⟨91T7091⟩
Sarma, J. C., **1**, 944 ⟨88TL6581⟩
Sarma, P.,J. A. R., **4**, 873 ⟨88TL6177⟩
Sarmah, B. K., **2**, 316 ⟨91T8587⟩; **3**, 12, 127 ⟨91T8587⟩
Sarmah, P., **2**, 36 ⟨89T3569⟩
Saroff, A., **2**, 432 ⟨87JOM(323)C1⟩
Saroja, M., **5**, 18 ⟨86SC1423⟩
Sarpeshkar, A. M., **5**, 232 ⟨79TL703⟩
Sarrazin, L., **1**, 199 ⟨93SL769⟩
Sars, C., **2**, 363, 399 ⟨59CB1910⟩
Sarshar, S., **2**, 49 ⟨93JA3828⟩
Sarti, S., **5**, 647 ⟨81CC822⟩; **6**, 127 ⟨81CC822⟩
Sarti-Fantoni, P., **5**, 99 ⟨80JHC621⟩
Sartor, D., **1**, 177, 178 ⟨92JCS(P1)95⟩
Sartorelli, A. C., **6**, 509 ⟨81JMC1089⟩
Sartori, G., **3**, 97 ⟨78JCS(P1)318⟩
Sartori, M., **2**, 882 ⟨39MI 217-01⟩
Sartori, P., **5**, 220, 403 ⟨86ZN(B)859⟩, 221 ⟨83CZ121, 83CZ257, 83CZ363, 85JFC(27)157, 87CZ61⟩; **6**, 232 ⟨69CB2150⟩
Sarussi, S. J., **2**, 753 ⟨81SC925⟩
Saruta, K., **6**, 487 ⟨93CL1401⟩
Saruyama, T., **5**, 461 ⟨76BCJ546⟩
Sarvey, G., **3**, 71 ⟨59HCA844⟩
Sarwar, G., **5**, 679, 703 ⟨90IC4255⟩, 679, 703 ⟨92ICA527⟩; **6**, 260, 615 ⟨89IC3345⟩, 262 ⟨90HAC167, 92ICA527⟩, 603 ⟨90IC571⟩
Sarycheva, T. A., **1**, 1010 ⟨88ZOR650⟩
Sasada, Y., **2**, 582 ⟨85JCS(D)873⟩
Sasajima, K., **4**, 225 ⟨81CAR(95)308⟩
Sasaka, K., **3**, 540 ⟨79S740⟩
Sasaki, H., **1**, 489 ⟨89JA314⟩; **3**, 710 ⟨83CPB723⟩; **4**, 442 ⟨83CPB723⟩; **5**, 189, 196 ⟨87CPB1262⟩, 242 ⟨92PS(67)27⟩, 506, 509 ⟨92JAP(K)4224595, 92MI 512-01⟩; **6**, 465 ⟨92JAP04054156⟩
Sasaki, K., **1**, 433 ⟨91CL2017⟩, 468 ⟨93BCJ2446⟩, 611 ⟨89CL607⟩, 667 ⟨84CL1897⟩, 747 ⟨68BCJ1252⟩; **2**, 292 ⟨75TL649⟩, 323 ⟨90JA2368⟩, 786 ⟨83TL4443⟩, 941 ⟨79S740⟩; **3**, 39, 176 ⟨89CL607⟩, 43 ⟨90JA2368⟩, 430 ⟨90BCJ1843⟩; **4**, 181 ⟨84JOC2808⟩, 238 ⟨91CL415⟩; **5**, 242 ⟨86CL977⟩
Sasaki, M., **3**, 212 ⟨83JOC2503⟩; **5**, 1125 ⟨86JAP1280464⟩
Sasaki, N., **1**, 469 ⟨77TL3369⟩; **2**, 250 ⟨93SL465⟩
Sasaki, O., **1**, 835 ⟨90TL727⟩
Sasaki, S., **3**, 605 ⟨89TL839⟩; **5**, 234 ⟨77CL483⟩, 1092 ⟨89CC1732, 89TL187⟩, 1092 ⟨89TL839⟩
Sasaki, T., **1**, 518 ⟨75JA4413⟩; **2**, 749 ⟨69JCS(C)1086⟩; **3**, 340 ⟨88MI 308-01⟩, 433 ⟨85S279⟩, 696 ⟨74JOC1239⟩, 698 ⟨81JOC5445⟩; **4**, 165 ⟨74H(2)177⟩, 232 ⟨84JCS(P1)2641⟩, 976 ⟨70JCS(C)476, 71JCS(C)196⟩; **5**, 652 ⟨79JAP7946762⟩, 990 ⟨68BCJ1258⟩, 1054 ⟨90MI 524-01⟩, 1118 ⟨81JOC5445⟩
Sasaki, Y., **2**, 326 ⟨93JMC3526⟩; **4**, 552 ⟨75CL1259, 76JOM(122)113⟩, 766 ⟨92CPB3261⟩
Sasakura, C., **2**, 380 ⟨83CPB423⟩
Sasakura, K., **2**, 810, 811 ⟨79SC515⟩; **3**, 282 ⟨78JA4842, 80TL173⟩
Sasao, S., **2**, 318 ⟨90BCJ1252⟩
Sasaoka, M., **2**, 21 ⟨81TL3193, 84CL877⟩
Sasaoka, S., **3**, 151 ⟨91CC86⟩; **4**, 863 ⟨92MI 419-01⟩
Sasatani, H., **1**, 252 ⟨87JOC4133, 89JOC4998⟩
Sasatani, S., **3**, 437 ⟨83TL3255⟩; **5**, 661, 728 ⟨83TL3255⟩
Sas, W., **1**, 68 ⟨84CC862⟩; **2**, 395 ⟨82T1321⟩
Sasaya, J., **5**, 221 ⟨88BCJ3653⟩
Sasazawa, K., **1**, 471 ⟨77CL301⟩
Sashida, H., **2**, 244 ⟨84H(22)1303⟩
Sasho, M., **1**, 1140 ⟨85TL1549, 86TL195⟩, 1140 ⟨87CPB1405, 87TL4583, 88CPB3897⟩; **2**, 665 ⟨81TL4283, 82JOC4376⟩; **5**, 223 ⟨90CPB1836⟩
Sasho, S., **4**, 970 ⟨91EUP413343⟩
Sasin, G. S., **2**, 120, 180 ⟨58JOC1525⟩
Sassaman, M. B., **1**, 208 ⟨90JOC2016⟩; **2**, 384 ⟨92JOC1585⟩; **3**, 632, 698 ⟨90JOC2016⟩; **4**, 216 ⟨90S104⟩
Sass, V. P., **6**, 36 ⟨74ZOR2031⟩, 251 ⟨76ZOR767⟩
Sasse, H. J., **4**, 881 ⟨56CB2060⟩; **6**, 73, 97, 300 ⟨56CB2060⟩
Sasse, K., **2**, 409 ⟨63HOU(12/1)1⟩, 847, 848 ⟨64HOU(12/1)338⟩; **6**, 439 ⟨74GEP2311662⟩, 453 ⟨75GEP2351556⟩, 534 ⟨57GEP1018054⟩, 537 ⟨70LA(735)158⟩
Sasse, W. H. F., **1**, 517 ⟨71AJC2325⟩
Sassenberg, W., **3**, 258 ⟨57LA(602)14⟩; **5**, 822 ⟨55LA(594)185⟩
Sasson, I., **1**, 84 ⟨79TL1641⟩
Sasson, M., **1**, 346 ⟨88TL5209⟩
Sasson, S., **2**, 166 ⟨80JCS(P1)2228⟩
Sasson, Y., **1**, 79, 80 ⟨75S448⟩, 94 ⟨83TL4139⟩, 1065 ⟨86T3569⟩; **2**, 22 ⟨92CC1200⟩, 25, 31 ⟨86CC1250, 84S34⟩, 31 ⟨87TL1223, 90MI 201-01⟩; **3**, 91 ⟨79TL3753, 86S569⟩, 117 ⟨88JOC3553⟩; **4**, 22, 37 ⟨84S34⟩; **5**, 232, 1030 ⟨73JA3440⟩; **6**, 221, 222, 226 ⟨93BSF599⟩
Sastry, K. A. R., **2**, 4, 30 ⟨81JOC3113⟩, 34 ⟨82SC101⟩, 305 ⟨81JOC4296⟩, 608, 632 ⟨82OPP359⟩, 623 ⟨83SC1027⟩, 907 ⟨83JOM(259)269⟩; **5**, 45, 71, 97 ⟨82CC1273⟩
Sastry, K. U., **2**, 34 ⟨82SC101⟩
Sastry, U., **2**, 907 ⟨83JOM(259)269⟩; **5**, 45, 71, 97 ⟨82CC1273⟩
Sastry, V. V. S. K., **5**, 81, 546, 549, 550 ⟨83S605⟩
Satapathi, T. K., **1**, 593 ⟨90T1885⟩; **4**, 401 ⟨89TL3333⟩
Satchell, D. P. N., **3**, 527 ⟨67JCS(B)360⟩
Satchell, J., **5**, 346 ⟨92OPP147⟩
Satchell, R. S., **1**, 896 ⟨63JCS1930⟩
Satcher, Jr., J. H., **6**, 246, 247 ⟨89OM1498⟩
Satek, L. C., **4**, 551 ⟨75IC656, 80ZN(B)832⟩; **6**, 370 ⟨85IC2889⟩
Satgé, J., **1**, 656 ⟨83JOM(243)281, 90CRV283,

90JOM(400)121⟩, 766 ⟨80JA5413, 91OM1771⟩; **2**, 473, 474 ⟨82PS(12)265, 82PS(12)279⟩, 815 ⟨83RTC65⟩, 946, 948, 949 ⟨B-71MI 218-01⟩, 947, 949 ⟨73JOM(56)1, 87JA4411⟩, 1092 ⟨B-90MI 223-02⟩, 1093 ⟨72JOM(34)83⟩; **3**, 457 ⟨87JOM(331)11⟩, 501 ⟨82AOC(21)241, 90CRV283⟩, 501 ⟨94CCR427⟩, 503 ⟨80JA5413, 87JA4411, 91JOM(403)93, 91POL1153, 92OM3176⟩; **4**, 141 ⟨81JOM(205)311⟩, 141 ⟨88SRI317⟩, 338 ⟨72IC2274, 73JOM(47)67⟩, 360 ⟨92JOM(433)49⟩, 361 ⟨71AG(E)267, 82JOM(232)137⟩, 362 ⟨88SRI317⟩, 455, 481, 998 ⟨75HCA1316⟩, 485 ⟨91JOM(415)327⟩, 519 ⟨82COMC-I(2)365⟩, 521 ⟨72JOM(34)C18, 79JOM(168)43⟩, 574 ⟨70CR(C)351, 70JOM(24)633⟩, 581 ⟨90IC1674⟩, 607, 621, 622, 628, 629 ⟨89OM1585⟩, 627 ⟨88JOM(345)39⟩, 628, 629 ⟨85JOM(282)315, 87OM659⟩, 629 ⟨90JOM(395)27, 91JOM(406)123⟩, 640, 646 ⟨90SRI1373⟩, 641 ⟨90JOM(387)65⟩, 647 ⟨92JOM(440)233⟩, 662 ⟨87JA4411⟩, 663 ⟨92OM3176⟩; **5**, 388, 389 ⟨88EJM257, 68BSF4093⟩, 389 ⟨92PS(70)75⟩, 432 ⟨90JOM(397)7⟩, 432 ⟨92JOM(433)49⟩, 720 ⟨75HCA1316⟩, 927 ⟨82AOC(21)241, 90CRV283⟩, **6**, 131 ⟨77JOM(141)35⟩, 167 ⟨87PS(31)81⟩, 186, 188 ⟨91POL1153⟩, 187, 709, 724 ⟨90CRV283⟩, 265, 266 ⟨78BSF(2)361⟩, 373 ⟨82TL4941⟩, 577 ⟨75JOM(88)C35⟩, 582 ⟨79TL3507⟩, 669 ⟨72JOM(34)83⟩, 709 ⟨90JOM(400)121⟩, 709 ⟨94CCR427⟩, 724 ⟨92OM2748⟩

Satici, H., **1**, 225 ⟨90JOC324⟩

Satina, T. Ya., **4**, 915 ⟨78ZOB267⟩

Sato, A., **1**, 795 ⟨70BCJ1120⟩; **5**, 247 ⟨75CL1231⟩

Sato, E., **1**, 829 ⟨88CC1531⟩; **5**, 159 ⟨88CC1531⟩

Sato, F., **1**, 161, 450, 1092, 1105 ⟨83TL1041⟩, 194 ⟨93CC1595⟩, 228 ⟨88TL343⟩, 457, 507 ⟨93CC386⟩, 469 ⟨86AG(E)935⟩, 651 ⟨83CC921⟩, 1049 ⟨91JOC3205⟩, 1107 ⟨85CC1636⟩; **2**, 4, 23 ⟨76JOM(122)C25⟩, 4, 23 ⟨78CL833⟩, 23, 596 ⟨77JOM(142)71⟩, 55 ⟨80TL2171⟩, 566 ⟨81CC1140⟩, 629 ⟨81CC718⟩, 900 ⟨90JOC5324⟩, 921 ⟨88T4073⟩, 923 ⟨90TL6399⟩, 923, 924 ⟨83TL1041⟩, 923, 931 ⟨92SL849⟩, 935 ⟨82TL4589⟩, 936 ⟨83CC921⟩, 969 ⟨83TL1809, 85JOM(285)53⟩; **3**, 82 ⟨81S871⟩, 130, 285 ⟨79TL4303⟩, 286 ⟨91TL6731⟩, 459 ⟨92JOC5056⟩; **4**, 183 ⟨86T4443⟩, 1060 ⟨84CC1130⟩

Sato, H., **4**, 392 ⟨87CPB1413⟩, 942 ⟨90T4553⟩; **5**, 367 ⟨90CPB3184⟩, 404 ⟨67CPB1183⟩, 560 ⟨86BCJ1403⟩

Sato, J., **1**, 387 ⟨92T3749⟩, 826 ⟨90TL377⟩

Sato, K., **1**, 34 ⟨75BCJ1892⟩, 158 ⟨85JA5301⟩, 158 ⟨90JA5244⟩, 408 ⟨92JOC428⟩, 408, 823 ⟨90JA1999⟩, 451 ⟨86CL1473⟩, 730 ⟨70JOC565⟩, 839 ⟨90BCJ1328⟩; **2**, 77 ⟨88JA4599⟩, 78 ⟨89JA6429⟩, 453 ⟨85TL783⟩, 725 ⟨82CL2007⟩; **3**, 28 ⟨91CL1469⟩, 57 ⟨86CL2035⟩, 114 ⟨82TL3085⟩, 118 ⟨85T5645⟩, 238 ⟨80JHC289⟩, 563 ⟨85CC1773⟩, 655 ⟨88S813⟩; **4**, 167 ⟨81JCS(P1)1015⟩, 507 ⟨90JA1999⟩, 570, 575 ⟨85JA5301, 90JA5244⟩, 654, 655 ⟨92CL2047⟩; **5**, 251 ⟨80SC305⟩, 635 ⟨85BCJ995⟩; **6**, 7 ⟨81CL1169, 85T393⟩, 417, 426 ⟨93S103⟩

Sato, K.-I., **2**, 677 ⟨93BCJ1156⟩, 710 ⟨92JA5902⟩; **4**, 940 ⟨92JA5902⟩

Sato, M., **1**, 161, 450, 1092, 1105 ⟨83TL1041⟩, 480, 1027 ⟨84BCJ108⟩, 481, 1028 ⟨83CPB4346⟩, 514, 587, 692 ⟨82CL157⟩, 651 ⟨81TL1609, 83CC921⟩; **2**, 4, 23 ⟨76JOM(122)C25⟩, 4, 23 ⟨78CL833⟩, 16 ⟨92H(33)105⟩, 17 ⟨94CPB459⟩, 23, 596 ⟨77JOM(142)71⟩, 39 ⟨90TL7463⟩, 55 ⟨80TL2171⟩, 394 ⟨82TL2965⟩, 394, 784 ⟨86JOC4368⟩, 528 ⟨90TL231⟩, 566 ⟨81CC1140⟩, 598 ⟨82OM440⟩, 629 ⟨81CC718⟩, 902 ⟨90TL247⟩, 918 ⟨85T3257⟩, 918, 931 ⟨83JA4491⟩, 935 ⟨82TL4589⟩, 936 ⟨83CC921⟩, 1098 ⟨84BJC108⟩; **3**, 82 ⟨81S871⟩, 126 ⟨84BCJ108⟩, 130, 285 ⟨79TL4189⟩, 231 ⟨78CPB3877⟩, 254 ⟨79JOC930⟩, 363 ⟨92JOM(430)105⟩, 543 ⟨84H(22)2563, 85S224⟩, 543 ⟨90CPB94, 91T5689⟩, 543 ⟨92S977⟩; **4**, 260 ⟨92TL5085⟩, 847, 848 ⟨79JOC930⟩; **5**, 463 ⟨73BCJ2820⟩, 652 ⟨89CPB2952⟩; **6**, 121, 698 ⟨79JOC930⟩, 627 ⟨79S554⟩

Sato, M.-A., **3**, 61 ⟨87JOC4135⟩

Sato, N., **1**, 1010 ⟨90JHC503⟩, 1023 ⟨87S139⟩; **2**, 142 ⟨80CL143⟩, 929 ⟨92CC1358⟩; **5**, 173 ⟨94JA5511⟩, 281 ⟨83CL449⟩

Sato, O., **1**, 990 ⟨82JOC1435⟩

Sato, R., **2**, 141 ⟨91S785⟩, 233 ⟨86CL349⟩, 234 ⟨89CL2111⟩; **3**, 336, 363 ⟨91S785⟩; **4**, 419 ⟨90BCJ1160⟩, 419, 420 ⟨84CL1599⟩; **6**, 553 ⟨87BCJ435⟩

Sato, S., **1**, 235 ⟨82CL1871⟩, 590, 693 ⟨72TL737⟩; **2**, 4, 23 ⟨76JOM(122)C25⟩, 23, 596 ⟨77JOM(142)71⟩, 159 ⟨83CC1470⟩, 368 ⟨85S635⟩, 721 ⟨83S1040⟩, 858 ⟨77BCJ3007⟩, 890 ⟨75JOM(96)57⟩, 1003 ⟨79AOC(17)407⟩; **3**, 51, 202 ⟨85TL4229⟩, 77 ⟨89JA2332, 90JA6120⟩, 139 ⟨86CC469, 88JOM(344)71⟩, 263 ⟨84S629⟩, 559 ⟨85CPB2331, 85CPB4691⟩, 634 ⟨85H(23)3069⟩; **4**, 168 ⟨86CC156⟩, 595 ⟨75JOM57⟩, 628, 629 ⟨90OM2061⟩; **5**, 240, 244, 252 ⟨77BCJ3007⟩, 419 ⟨88JOM(344)71⟩, 635 ⟨85BCJ995⟩, 692 ⟨88BCJ3731⟩, 1039 ⟨85H(23)3069⟩; **6**, 246 ⟨70CPB2334⟩, 463 ⟨81CL749⟩, 554 ⟨82BCJ1174⟩

Sato, S. I., **4**, 595 ⟨72JOM(43)333⟩

Sato, T., **1**, 2, 8 ⟨78BCJ2745⟩, 164 ⟨92SL891⟩, 183 ⟨90CL593⟩, 230 ⟨80TL2181⟩, 230, 1107 ⟨82TL3587⟩, 476 ⟨81CL1307, 82TL3583⟩, 476 ⟨83TL5745⟩, 481 ⟨82BCJ3555⟩, 667 ⟨84CL1897, 85TL2205⟩, 667 ⟨90S259⟩, 687 ⟨73CL1041⟩, 881 ⟨76TL1817⟩, 970 ⟨84TL4007⟩, 980 ⟨85CL1457⟩, 1039 ⟨79CL447⟩, 1097 ⟨81CC951⟩; **2**, 25 ⟨84CL1173⟩, 71 ⟨79JA1455⟩, 215 ⟨90TL3591⟩, 521 ⟨85JA4549, 86JA7402⟩, 698 ⟨86JA8279⟩, 758 ⟨83CL1791⟩, 816 ⟨90JA4897⟩, ⟨90CRV215⟩ ⟨90CPB3331⟩; **3**, 35 ⟨88JA5209⟩, 55 ⟨89JOC2779⟩, 82, 102, 107 ⟨83TL1543⟩, 88 ⟨68JOC3277⟩, 146, 212, 288, 296 ⟨81CL1135⟩, 151 ⟨81JOC509⟩, 159 ⟨90JA901⟩, 260 ⟨84CL2055⟩, 286 ⟨77CL645, 82TL5059⟩, 303 ⟨82CL1559⟩, 427 ⟨83CL1537⟩, 483 ⟨81JA5573⟩, 609 ⟨86JOM(311)C63, 90HAC339⟩; **4**, 67 ⟨89JCS(P1)879, 93JOC2360⟩, 142, 521 ⟨91CL97⟩, 155, 700 ⟨92SL891⟩, 187 ⟨84TL5155⟩, 216 ⟨89T1209⟩, 225, 238 ⟨92TL239⟩, 248 ⟨88TL3971⟩, 412 ⟨91CC462⟩, 699 ⟨91T3281⟩, 708 ⟨86TL4339, 93AG(E)105⟩, 834, 933 ⟨78CC478⟩, 892 ⟨73CL111, 76JCS(P1)783⟩, 924 ⟨87JCS(P1)1095⟩, 932 ⟨77JA4181⟩; **5**, 14 ⟨83CL835, 87OS(66)121⟩, 25, 64 ⟨69CPB1564⟩, 131 ⟨82CL1891⟩, 237 ⟨86JA8279⟩, 251 ⟨90TL3591⟩, 301 ⟨85JA4577⟩, 413 ⟨77JA5827, 78TL259⟩, 414 ⟨76TL1591, 77JA4181, 78CC478⟩, 420 ⟨78BCJ2391⟩, 423 ⟨84BCJ2167⟩, 553 ⟨83CL1901⟩, 557 ⟨86TL5405, 87NKK1414⟩, 639, 846 ⟨54BCJ624, 54JPJ821⟩, 925 ⟨81JA5573, 87OM2337⟩; **6**, 125 ⟨76TL1591, 78BCJ2391⟩, 210 ⟨90S259⟩, 555 ⟨92CC1410⟩, 636 ⟨88CPB4755⟩

Sato, Y., **1**, 290 ⟨88JOM(25)348, 88OM739⟩, 291 ⟨90OM1325⟩, 451 ⟨82CC56⟩, 495 ⟨92TL2589, 94T371⟩, 824 ⟨88CC370, 94JCS(P1)393⟩; **2**, 229 ⟨79BCJ3601⟩, 700 ⟨87OM2568⟩, 759 ⟨81JOM(218)309⟩, 945 ⟨76JOC3559⟩, 1045 ⟨78JOC199⟩; **3**, 264 ⟨65TL2261⟩, 340 ⟨88MI 308-01⟩, 660 ⟨82CC56, 83CC170⟩; **4**, 361 ⟨90SC3245⟩, 506 ⟨92JOC6711⟩, 507 ⟨91S996⟩, 508 ⟨85CC1684, 88S988⟩, 508 ⟨91S117⟩, 509 ⟨74JOM(82)21, 75CC640⟩, 509 ⟨76JOM(118)1, 78JOM(153)193⟩, 512 ⟨76JOM(113)115, 90S36⟩, 513 ⟨92JOC5419⟩, 521 ⟨85TL5547, 86CPB3273⟩, 521 ⟨88JOM(339)259, 88JOM(346)1⟩, 522 ⟨91S169⟩, 604, 625, 638 ⟨88JOM(354)155⟩, 609, 638 ⟨89OM1237⟩, 625 ⟨90OM1325⟩, 625, 637 ⟨86OM1197⟩, 625, 637, 1055, 1058 ⟨88OM739⟩, 625, 662 ⟨87OM2568⟩, 626, 662 ⟨87SC1273⟩, 638, 648, 662 ⟨91JOC347⟩, 638, 663 ⟨88JOM(348)25⟩; **5**, 404 ⟨67CPB1183⟩, 431 ⟨90SC3085⟩, 1121 ⟨73CL1225⟩; **6**, 4 ⟨92JAP04164039⟩, 185 ⟨87SC1273⟩, 185 ⟨88JOM(354)155⟩, 185 ⟨90OM1325⟩, 185, 186 ⟨88OM739⟩

Satoh, A., **1**, 831 ⟨90CC1001⟩
Satoh, H., **2**, 626 ⟨88JOC2093⟩, 1036, 1037 ⟨91JCS(P1)2049⟩
Satoh, J. Y., **1**, 992 ⟨91JA9604⟩; **2**, 33 ⟨81S312⟩, 33, 34 ⟨84CL1509⟩, 631 ⟨84BCJ2691⟩; **3**, 155 ⟨81S312⟩
Satoh, K., **2**, 16 ⟨92JCS(P1)221⟩
Satoh, M., **1**, 279 ⟨87S373⟩, 489 ⟨86TL3745, 89JA314⟩, 489, 531 ⟨86CL1329⟩; **2**, 527, 900, 902 ⟨89TL3789⟩, 901 ⟨87S373⟩; **5**, 403, 404 ⟨80JOC4162⟩, 1054 ⟨90MI 527-01⟩
Satoh, N., **3**, 228 ⟨87BCJ3471⟩
Satoh, S., **2**, 141 ⟨91S785⟩, 234 ⟨92TL2517⟩; **4**, 827 ⟨88CC189⟩
Satoh, T., **1**, 485 ⟨88BCJ1231⟩, 598 ⟨92TL7181⟩, 992 ⟨91JA9604⟩; **2**, 85 ⟨88TL2851⟩, 322 ⟨88TL4093⟩, 620 ⟨88BCJ2226⟩; **3**, 36 ⟨85BCJ2849⟩, 36, 44, 162, 167, 168, 183 ⟨92SL455⟩, 153, 154 ⟨92BCJ2800⟩, 156 ⟨81T473⟩, 158 ⟨91BCJ1582⟩, 158 ⟨91JOC4129⟩, 170 ⟨90BCJ1266, 92TL7181⟩, 183 ⟨86BCJ457⟩, 209 ⟨86TL2471, 87BCJ1839⟩, 227 ⟨88CPB1169⟩; **4**, 73, 74, 75, 77 ⟨92SL455⟩, 74 ⟨85BCJ1983⟩, 74, 77 ⟨92TL1483⟩, 76 ⟨88TL313, 93TL2331⟩, 80 ⟨93BCJ2339⟩, 233 ⟨86BCJ2463, 89CPB184⟩, 777 ⟨93BCJ1866⟩; **5**, 272, 283 ⟨92TL1455⟩
Satoh, Y., **1**, 542, 1008 ⟨88TL1811⟩, 1104 ⟨89TL5153⟩; **2**, 902 ⟨83TL735, 85S406⟩, 902 ⟨89TL5153⟩, 902, 927, 939 ⟨88TL1811⟩; **3**, 27 ⟨85S406⟩; **4**, 765 ⟨89TL5153⟩
Satomi, H., **1**, 86 ⟨86JOC537⟩
Satomi, M., **2**, 3 ⟨82S1021⟩; **3**, 152, 154, 155 ⟨82S1021⟩; **5**, 226 ⟨80CC1029⟩
Satomi, N., **2**, 650 ⟨74T3021⟩
Satou, M., **1**, 690 ⟨82BCJ2434⟩
Satsumabayashi, S., **3**, 353 ⟨92MI 303-02⟩
Sattelburger, A. P., **2**, 434 ⟨90IS305⟩
Sattich, W. E., **4**, 148 ⟨89OM1114⟩
Sattur, P. B., **5**, 344 ⟨87H(26)1051⟩; **6**, 450 ⟨88S994⟩, 561 ⟨85IJC(B)580⟩
Satuanarayana, N., **2**, 312 ⟨89SC565⟩
Satyamurthy, N., **2**, 25, 31, 35 ⟨90TL4973⟩, 613 ⟨90JOC3373⟩, 632 ⟨83JOC4394, 90JOC4560⟩
Satyanarayana, G. O. S., **3**, 206 ⟨82JOC5353⟩
Satyanarayana, V., **1**, 827 ⟨91SC1455⟩
Sau, A. C., **2**, 856 ⟨79JA3790, 83IC1771⟩
Saucy, G., **1**, 156 ⟨57HCA1250⟩, 514 ⟨76JOC3497⟩, 979 ⟨67HCA1158⟩, 980 ⟨80JOC2080⟩; **2**, 39 ⟨76JOC3505, 81HCA1158⟩; **3**, 61 ⟨80HCA1665⟩, 229, 233 ⟨67HCA1158⟩; **4**, 166 ⟨72GEP(O)2060217⟩; **5**, 159 ⟨80JOC2080⟩
Sauer, C. W., **4**, 424 ⟨55JA2559⟩
Sauer, D. R., **1**, 362, 555, 558 ⟨92JA7375⟩, 813, 815 ⟨92TL923⟩
Sauer, G., **1**, 24, 144 ⟨91TL1949⟩, 555 ⟨89T6409⟩
Sauer, I., **2**, 493, 505, 886, 890 ⟨84ZN(B)887⟩, 499, 891 ⟨84ZN(B)1668, 87ZN(B)695⟩
Sauer, J., **1**, 786 ⟨80AG(E)779⟩; **2**, 163 ⟨68TL319⟩, 804 ⟨60AG303⟩; **3**, 680 ⟨59JOC892⟩; **5**, 934 ⟨94TL561⟩, 1013 ⟨68TL319⟩
Sauer, J. C., **1**, 1026 ⟨58JA4607⟩; **3**, 78 ⟨62OS(41)813⟩, 79 ⟨63OSC(4)801⟩, 526, 538 ⟨46OR(3)108⟩; **4**, 192 ⟨58JA4607⟩, 192 ⟨63OSC(4)801⟩
Sauer, J. D., **4**, 183 ⟨73TL1599⟩
Sauer, N. N., **1**, 774 ⟨86JOC113⟩
Sauer, R. J., **4**, 56 ⟨73OSC(5)145⟩
Sauer, R. O., **2**, 538 ⟨46JA1161⟩
Sauerbier, M., **5**, 353 ⟨80HOU(4)736⟩
Sauermann, G., **1**, 885 ⟨70CB265⟩
Sauermilch, W., **4**, 237 ⟨51CB648⟩
Sauers, R. F., **2**, 221, 225 ⟨72JOC2367⟩
Sauers, R. R., **1**, 82 ⟨57JA6210⟩, 214 ⟨66JA2257⟩
Sauerwald, M., **1**, 202 ⟨83TL2837, 90JOM(382)121⟩
Saul, G. A., **4**, 920 ⟨60CA(54)18356h⟩
Sauleau, A., **2**, 373 ⟨85MI 207-01⟩
Sauleau, J., **2**, 373 ⟨85MI 207-01⟩
Saulnier, M. G., **4**, 165 ⟨84JOC4518⟩
Saunder, D. H., **6**, 25 ⟨44JCS74⟩
Saunders, B. C., **2**, 101 ⟨50JCS3125⟩, 104 ⟨45JCS380⟩, 409, 410, 412, 413 ⟨49JCS2921⟩, 412 ⟨48JCS699, 48JCS1313⟩, 845, 860 ⟨57JCS2413⟩, 1071, 1072 ⟨60CI(L)47, 63JCS3351⟩; **4**, 76 ⟨48JCS699⟩
Saunders, D. R., **4**, 676 ⟨86JCS(D)657⟩
Saunders, J., **4**, 414 ⟨82CC282⟩
Saunders, J. H., **3**, 216 ⟨55OSC(3)22⟩; **6**, 425 ⟨51JA3796⟩, 443 ⟨50JA1888⟩
Saunders, J. O., **3**, 59 ⟨84JA4558⟩
Saunders, Jr., W. H., **1**, 779, 783 ⟨78JA2802⟩, 927 ⟨86JA2381⟩; **3**, 407 ⟨B-70MI 310-03⟩
Saunders, L. C.,**4**, 866 ⟨84GEP3408757A1⟩
Saunders, M., **1**, 506 ⟨79T929⟩, 891, 906, 918, 919 ⟨91CRV375⟩, 899 ⟨73ACR53⟩, 900 ⟨68JA6882⟩, 906 ⟨80JA6867, 83ACR440, 83JA3571⟩, 912, 914 ⟨58T178⟩; **5**, 85 ⟨58ANY(70)870⟩
Saunders, R. N., **4**, 419 ⟨91TL2683⟩
Saura-Llamas, I., **4**, 586 ⟨87JOM(331)409, 87JOM(333)129⟩, 586 ⟨88OM997⟩
Saurez, E., **5**, 1119 ⟨87JCS(P1)937⟩
Sauriol-Lord, F., **1**, 209 ⟨81JOC2831⟩
Saus, A., **1**, 46 ⟨69M724⟩; **4**, 328 ⟨63AG1050⟩; **5**, 579, 602 ⟨65M1278⟩, 599 ⟨64AG(E)19⟩
Sausen, G. N., **2**, 7622 ⟨67PC-01⟩; **4**, 104 ⟨68JOC2330⟩; **5**, 698, 761, 812, 814, 817, 819 ⟨66JOC3689⟩, 811 ⟨68JOC2330⟩
Saussez, R., **1**, 53 ⟨71T5523⟩; **4**, 112 ⟨71T5523⟩
Sauter, F., **6**, 25, 31 ⟨88M1427⟩, 444 ⟨83JHC1477⟩
Sauter, H., **1**, 571 ⟨88OSC(6)427⟩; **5**, 570, 608, 610 ⟨78BSB525⟩
Sauvaget, F., **6**, 58 ⟨94TL3537⟩
Sauve, D. M., **1**, 51, 52, 101 ⟨55JA3230⟩, 457, 508 ⟨55JA3378⟩; **5**, 27 ⟨57CRV867⟩
Sauvé, G., **2**, 749 ⟨88TL2295, 88TL2299⟩, 750 ⟨90JOC3002⟩; **5**, 219, 574 ⟨85CJC3089⟩, 568, 571 ⟨93T10489⟩, 588 ⟨93BMC619⟩
Sauve, J. P., **3**, 255 ⟨74BSF1196⟩
Sauvêtre, R., **1**, 429 ⟨78S128⟩, 653 ⟨90JOM(394)37⟩, 833 ⟨74BSF2072⟩, 1017 ⟨82TL4329⟩, 1017, 1067 ⟨82TL4325⟩, 1090 ⟨76S761⟩; **3**, 67, 237 ⟨78S128, 78S458, 85TL3999⟩; **4**, 45 ⟨93JFC(63)157⟩, 751 ⟨90JOM(394)37⟩, 763, 771 ⟨75T891⟩, 763, 771 ⟨75T897⟩, 806 ⟨89JOM(367)1⟩, 806, 819 ⟨84JOM(264)155⟩; **5**, 58 ⟨82S297⟩; **6**, 19 ⟨91TL5963⟩, 41 ⟨74BSF2072⟩
Sauveur, F., **3**, 50 ⟨81T1377⟩
Savabi, F., **4**, 874 ⟨77S765⟩
Savaev, I. A., **5**, 361 ⟨77ZOR1144⟩
Savage, G. P., **6**, 365 ⟨90JCR(S)330⟩
Savage, I., **5**, 234 ⟨89CC717⟩
Savage, W. J., **5**, 538, 541 ⟨76JCS(D)1661⟩
Saval, I., **2**, 824 ⟨81JOM(205)161⟩
Savard, M., **6**, 20 ⟨79CJC2617⟩
Savariar, S., **1**, 518 ⟨90OS(68)32⟩; **2**, 1024, 1025 ⟨88TL4917⟩; **3**, 216 ⟨87TL3299⟩; **4**, 187 ⟨88S36⟩
Savary, D. N.-H., **3**, 67 ⟨77HCA1739⟩
Savchenko, T. I., **6**, 272 ⟨88ZOR1513⟩, 609 ⟨77JFC(9)505, 83JFC(22)439⟩, 609 ⟨86ZOR1297⟩, 616 ⟨84ZOR1197⟩
Savchuk, I. V., **2**, 464, 854 ⟨66ZOB500⟩
Savd'eva, N. I., **4**, 915 ⟨75ZOB556⟩
Savéant, J.-M., **1**, 107 ⟨92JA9576⟩; **2**, 578 ⟨90JA1874⟩
Saveant, J. M., **3**, 497 ⟨72BSF3549, 77JOC1242⟩
Savel'eva, I. S., **3**, 365 ⟨59MI 308-01⟩
Savel'ev, V. S., **4**, 861 ⟨78ZOR478⟩
Savel'eve, N. I., **5**, 376 ⟨73ZOB1753⟩, 376, 389 ⟨74ZOB1753⟩
Savelli, G., **3**, 106 ⟨71CC1441⟩
Savelova, V. A., **5**, 1070 ⟨88JCS(P1)3103, 89ZOR1523⟩, 1070 ⟨89ZOR1689⟩; **6**, 619 ⟨88JFC(40)217, 93JPO319⟩

Savelyeva, N. I., **4**, 886 ⟨69JOM(17)241⟩; **5**, 376 ⟨69JOM(17)241⟩
Savenkov, N. F., **2**, 252 ⟨84ZOB2545⟩, 449 ⟨84ZOB2703⟩; **4**, 346 ⟨92ZOB1268⟩
Savich, I. G., **2**, 1032 ⟨64ZOB1899⟩; **4**, 841 ⟨74ZOR2269⟩
Savignac, M., **1**, 469, 494, 1013 ⟨92SL715⟩; **2**, 304 ⟨93SL680⟩
Savignac, P., **2**, 255 ⟨70CR(C)2086⟩, 456 ⟨87TL405⟩, 458 ⟨79T1345⟩, 618 ⟨75S617⟩, 1057 ⟨76TL835⟩; **3**, 49 ⟨83S634, 84JOM(264)9⟩, 49 ⟨87TL405⟩, 50 ⟨81T1377, 85T427⟩, 50 ⟨92SC219⟩, 197 ⟨78S36⟩, 688 ⟨91AG(E)196⟩; **4**, 131 ⟨91SC793⟩, 132, 802 ⟨88JOM(338)295⟩, 340 ⟨73TL173⟩, 347, 773, 778 ⟨76S107⟩, 557 ⟨85TL4435, 86JOM(312)283⟩, 557, 576 ⟨84JOM(264)9, 87TL405⟩, 557, 576, 801 ⟨86JOM(304)283⟩, 570 ⟨86S831⟩, 570, 585, 588 ⟨82S725, 93PS(75)143⟩, 585 ⟨76S766⟩, 588 ⟨93PS(75)99⟩, 740 ⟨75TL609⟩, 740, 741 ⟨75S535⟩, 745 ⟨76S197⟩; **5**, 103 ⟨78SC19⟩, 1152 ⟨91AG(E)196⟩; **6**, 164, 374 ⟨87JOM(323)135⟩, 266 ⟨75S535⟩, 288 ⟨86JOM(304)283, 87SC1559⟩, 289 ⟨88JOM(338)295⟩
Saville, B., **2**, 122 ⟨63JOC3608⟩
Saville, R. W., **3**, 278 ⟨49JCS1946⟩
Savin, V. I., **2**, 379 ⟨85ZOR124⟩
Savina, T. I., **4**, 282 ⟨83TL87⟩, 864 ⟨79ZOR2416⟩; **5**, 314 ⟨80ZOR1330⟩; **6**, 275, 317 ⟨83ZOR79⟩
Savoia, D., **1**, 24 ⟨84JOC1246⟩, 439 ⟨77JCS(P1)123⟩, 688 ⟨85JOM(280)307⟩; **2**, 79 ⟨88JOM(353)177⟩, 590 ⟨83JOC4108⟩, 959 ⟨78CC927, 85PAC1887⟩; **3**, 310 ⟨91JOC1822⟩; **4**, 698, 708 ⟨74JOM(72)C4⟩
Savory, J., **4**, 34, 36 ⟨61JCS3779⟩, 146 ⟨62JCS1993⟩; **6**, 2, 3 ⟨61JCS3779⟩
Savost'yanova, I. A., **4**, 98, 99 ⟨91JGU1875⟩
Savrda, J., **2**, 243 ⟨73S795⟩
Sawa, T., **3**, 717 ⟨87JAN1199, 87JAN1202⟩; **6**, 288 ⟨88JAP63227595⟩
Sawa, Y., **1**, 181 ⟨75CPB1917⟩; **3**, 293 ⟨69TL5189⟩
Sawada, A., **5**, 134 ⟨86CL871⟩
Sawada, H., **1**, 464 ⟨88JA5383⟩, 544 ⟨84JA6105⟩, 1094 ⟨83JA6344⟩; **2**, 926, 931, 973, 976 ⟨84JA6105⟩, 926, 931, 976 ⟨83JA6344⟩, 973, 976 ⟨88JOC913⟩; **3**, 271 ⟨86TL775⟩; **4**, 809 ⟨88JA5383⟩, 1048, 1062 ⟨84JA6105⟩; **5**, 208 ⟨90JFC(46)423⟩; **6**, 12 ⟨90JFC(46)423⟩, 24 ⟨83SUL131⟩
Sawada, K., **3**, 696, 708 ⟨84TL5075⟩
Sawada, M., **1**, 240 ⟨88JA7185⟩; **3**, 668 ⟨86TL2411⟩; **4**, 426 ⟨82TL3181⟩
Sawada, S., **2**, 24 ⟨80BCJ1390⟩, 617 ⟨81BCJ2843⟩
Sawada, T., **2**, 615 ⟨87SC685⟩
Sawahara, K., **5**, 1112 ⟨88CC1485⟩
Sawahata, M., **4**, 736 ⟨83TL4113⟩
Sawai, H., **2**, 309 ⟨93BMC2883⟩; **5**, 138 ⟨84T145⟩; **6**, 505 ⟨72TL4263⟩, 644 ⟨75JOM(94)333⟩
Sawaki, Y., **2**, 348 ⟨B-89MI 206-04⟩; **3**, 645 ⟨86TL4177⟩; **4**, 221 ⟨87CC122⟩; **5**, 205 ⟨67T3327⟩
Sawamoto, H., **5**, 11 ⟨90MI 501-01⟩
Sawamura, M., **1**, 193 ⟨87TL4849⟩; **2**, 311 ⟨90JOC5935⟩; **3**, 707 ⟨87TL4849, 88TL6321⟩
Sawanishi, H., **5**, 693 ⟨84CPB4694⟩
Saward, C. J., **2**, 627 ⟨75TL4539⟩
Sawatzki, J., **5**, 246, 254 ⟨88ZAAC(566)151⟩
Sawaya, H. S., **4**, 745, 746 ⟨75TL1373⟩
Sawden, J. N., **5**, 118 ⟨90TL7223⟩
Sawhney, K. N., **2**, 163 ⟨83JHC899⟩
Sawistowska, M., **1**, 74 ⟨62CI(L)484⟩
Sawitsch, V., **2**, 1017 ⟨1861LA(119)182⟩
Sawitzki, G., **6**, 183 ⟨89JOM(367)19⟩
Sawluk, A., **5**, 554 ⟨86S968⟩
Sawyer, A. K., **2**, 168 ⟨49JOC272⟩
Sawyer, D. T., **2**, 156 ⟨84JA4283⟩, 156 ⟨85JOC1784⟩; **3**, 113 ⟨90JA1936⟩
Sawyer, J. F., **5**, 418, 924, 925 ⟨89OM693⟩, 924 ⟨87OM1246, 88OM2245, 89OM709, 89OM850⟩
Sawyer, J. S., **1**, 138, 139 ⟨84JA3376⟩, 139 ⟨88JA842⟩, 413 ⟨86JOC1124⟩; **2**, 684 ⟨93JOC3229⟩, 999 ⟨88JA842⟩; **3**, 141 ⟨86JOC951⟩; **4**, 379 ⟨84JA3376, 88JA842⟩
Sawyer, T. K., **3**, 613 ⟨78JOC1017⟩
Sax, H., **5**, 1048 ⟨70AP(303)905⟩
Sax, M., **2**, 478 ⟨67JOC2465⟩, 1009 ⟨87AHC(41)275⟩
Sax, N. R., **3**, 233 ⟨57JOC1611⟩
Sax, S. M., **4**, 25 ⟨59JA992⟩
Saxena, A. K., **2**, 881 ⟨85POL251⟩, 891 ⟨86PAC1133⟩; **5**, 318, 372 ⟨83IJC(A)120, 85POL251⟩, 418, 924, 925 ⟨89OM693⟩, 924 ⟨88OM2245, 89OM850, 91OM2758⟩; **6**, 270 ⟨91SRI401⟩, 387 ⟨87JCS(P2)775⟩, 387, 388, 397 ⟨87JCS(P2)779⟩
Saxena, M. P., **5**, 141 ⟨78AG(E)939⟩
Saxene, N., **3**, 293 ⟨86SC1493⟩
Saxton, R. G., **1**, 815 ⟨77T1211⟩
Sayad, Y. A., **1**, 452 ⟨76JOC1184⟩
Sayadyan, S. V., **2**, 437 ⟨90ZOB313⟩, 439 ⟨86AKZ258⟩
Sayal, P. K., **3**, 58 ⟨89CI(L)533⟩
Sayed, G. H., **5**, 201 ⟨82IJC(B)589⟩
Sayer, B. G., **4**, 713 ⟨82OM225, 85OM1123⟩
Sayer, T. S. B., **1**, 130 ⟨78JOC700⟩, 131 ⟨78JOC2153⟩; **2**, 420 ⟨78JOC700⟩; **3**, 124 ⟨78JOC700⟩
Sayigh, A. A. R., **2**, 726 ⟨66JOC2658⟩; **5**, 965 ⟨70USP3492331⟩, 968 ⟨69JOC3200⟩, 971 ⟨63JOC2076, 74JOC2897⟩, 977 ⟨67AG(E)636, 68JOC2887⟩, 980 ⟨64JOC3344⟩, 980, 981, 982, 1013, 1069 ⟨66AG(E)704⟩, 981 ⟨66JOC2658⟩, 982 ⟨62AG(E)595, 65JOC2779⟩, 982, 1070 ⟨63JCS5558⟩, 995 ⟨67JOC4052⟩, 1003 ⟨64AG(E)585⟩, 1029 ⟨65JOC2465⟩, 1061 ⟨B-70MI 528-01⟩, 1065 ⟨78AG761⟩, 1079 ⟨67TL1731⟩; **6**, 143 ⟨68JOC3928⟩, 445 ⟨63JOC1427, 64JOC2401⟩, 652 ⟨68AG(E)291⟩
Saykowski, F., **4**, 545 ⟨78ZN(B)1208, 82ZAAC(486)39⟩, 1025 ⟨84CB1542⟩
Saynovits, M., **3**, 379, 380 ⟨88CB1005⟩
Sayo, N., **1**, 77 ⟨87JA1596, 87JA4129⟩, 405, 816, 817 ⟨84TL565⟩, 405, 816, 817 ⟨85CL1723⟩, 816 ⟨82TL3931⟩, 816, 817 ⟨84CL255⟩; **2**, 252 ⟨79JOC2938⟩; **3**, 22 ⟨82TL3931⟩; **5**, 681 ⟨83BCJ545⟩
Sayrac, T., **3**, 455 ⟨68CB1354⟩
Sayre, L. M., **1**, 68 ⟨79JOC228⟩
Saytzeff, M., **5**, 652 ⟨1877LA(185)191⟩
Sazonova, O. M., **4**, 335 ⟨82IZV161⟩, 854 ⟨78IZV2654⟩; **6**, 53 ⟨81IZV161⟩, 116, 283, 355 ⟨82IZV161⟩
Sazonova, V. A., **4**, 696 ⟨77JOM(137)217⟩, 696, 723 ⟨78JOM(153)115⟩, 706 ⟨70IZV2133, 71DOK(198)590⟩, 723 ⟨72DOK(202)362, 76DOK(226)1092, 80JOM(185)C6, 82JOM(224)C53⟩
Sazonova, Z. Ya., **2**, 463 ⟨69ZOB1543⟩; **5**, 370 ⟨70ZOB2464⟩
Sbarbati, N. E., **5**, 436 ⟨73JA8118⟩
Sbrana, G., **6**, 524 ⟨64CI(M)206⟩
Scafuro, M., **5**, 597 ⟨83FES101⟩
Scaglioni, L., **6**, 695 ⟨84T1523⟩
Scahill, T. A., **1**, 831 ⟨90TL1977⟩; **3**, 302 ⟨90TL1977⟩
Scaiano, J.,**3**, 391 ⟨76CC205⟩
Scaiano, J. C., **1**, 330, 341, 936 ⟨84JA343⟩, 331 ⟨85JA4594⟩, 657 ⟨72JCS(P2)803, 85JA4595⟩; **3**, 314 ⟨76JPC1901, 85JA7204⟩; **5**, 924 ⟨92OM2317⟩
Scaife, C. W., **2**, 784 ⟨47JCS1471⟩
Scamehorn, R. G., **1**, 871 ⟨68JA6749⟩
Scamuzzi, B., **4**, 946 ⟨89TL2699⟩
Scanlan, J. T., **5**, 275, 276, 277, 278 ⟨49JA2215⟩
Scanley, C., **2**, 595 ⟨51JA2854⟩
Scanley, C. S., **6**, 91 ⟨67USP3333007⟩
Scapini, G., **5**, 615 ⟨70MI 514-01⟩
Scarborough, Jr., R. M., **1**, 606 ⟨81JA1501⟩; **3**, 176 ⟨77TL4361⟩
Scarborough, Jr, R. M., **3**, 176 ⟨78TL1649⟩
Scarpati, M. L., **3**, 92 ⟨90SC2565⟩; **5**, 79, 105 ⟨83G421⟩
Scarpati, R., **3**, 221 ⟨83S125⟩, 327 ⟨94JCS(P1)147⟩; **6**, 76

⟨59G1511, 66G375⟩, 106 ⟨59MI 604-01⟩, 107 ⟨63G942, 68G681⟩, 107 ⟨72JHC1087, 74JHC529⟩, 107 ⟨76JHC205⟩
Scarpetti, D., **1**, 618 ⟨90JA8084⟩; **4**, 273 ⟨90JA8084⟩
Scarpone, S., **3**, 338, 345 ⟨79JOC486⟩
Scarponi, U., **4**, 327 ⟨75CC320⟩; **5**, 300 ⟨75CC320⟩
Scartozzi, M., **5**, 351 ⟨92TL5717⟩
Ščedrov, O., **5**, 566, 618 ⟨57CCA319⟩
Scettri, A., **1**, 578 ⟨85S1159⟩; **2**, 34 ⟨82SC1127⟩; **3**, 4, 115 ⟨82S245⟩, 67 ⟨81TL1041⟩, 155 ⟨80TL4521⟩, 222, 303 ⟨93TL7089⟩, 245 ⟨87TL2297⟩
Schaad, D. R., **5**, 384 ⟨92OM2024⟩
Schaad, L., **1**, 393 ⟨54JA35⟩
Schaad, L. H., **6**, 243 ⟨67JA3446⟩
Schaad, L. J., **3**, 337 ⟨91CB2609⟩, 703 ⟨88AG(E)938⟩; **5**, 1020 ⟨90JA4183⟩
Schaaf, T. K., **3**, 494 ⟨70JA397⟩; **5**, 168 ⟨69JA5675, 70JA397⟩
Schaafsma, S. E., **1**, 929 ⟨68RTC1301, 73TL827⟩; **4**, 406 ⟨67RTC417⟩
Schaal, R., **6**, 97, 100 ⟨75JOC2911⟩
Schaap, A., **1**, 1023 ⟨68RTC1249⟩, 1037 ⟨65RTC1200⟩
Schaap, A. P., **2**, 98 ⟨70JA3223⟩; **3**, 319 ⟨83JA663⟩
Schaap, C. E., **4**, 724 ⟨80JOM(186)427⟩
Schaapsma, Y., **2**, 127 ⟨60T76⟩
Schaarschmidt, R., **3**, 398 ⟨39CB1246⟩
Schaart, F. J., **4**, 908 ⟨79RTC258⟩
Schabaker, V., **2**, 403, 404 ⟨65LA(687)236⟩
Schabl, L., **4**, 493 ⟨76LA36⟩
Schach, T., **4**, 100 ⟨88AG(E)304⟩; **5**, 883 ⟨88AG(E)272⟩
Schachner, H., **6**, 362 ⟨79CB1189⟩, 424, 438 ⟨81JFC(18)259⟩
Schacht, H. T., **6**, 402 ⟨90JOM(381)261⟩
Schack, C. J., **6**, 220 ⟨74IC2811⟩, 231 ⟨69JA2902⟩, 424 ⟨72IC2531⟩
Schad, H., **2**, 625 ⟨85HCA813⟩
Schade, C., **1**, 110, 885 ⟨87AOC(27)169⟩; **2**, 907 ⟨86CC260⟩; **6**, 406 ⟨87AOC(27)169⟩, 691 ⟨87CC1399⟩
Schade, C. L., **1**, 777 ⟨81JOC2804⟩
Schade, G., **1**, 812 ⟨59LA(625)206⟩; **3**, 220 ⟨82TL3543⟩
Schädler, H. D., **6**, 608 ⟨83JPR787⟩
Schadow, H., **4**, 550 ⟨83EGP201449⟩, 561 ⟨84EGP214610⟩
Schadt, F. L., **1**, 1114, 1116 ⟨72JA992⟩, 1114, 1116 ⟨76JA7667⟩
Schaefer, A., **2**, 512 ⟨85CB2472⟩
Schaefer, D., **1**, 545 ⟨84JOM(266)203⟩
Schaefer, F. C., **2**, 428 ⟨59JA1103⟩; **4**, 883 ⟨61JOC412⟩; **5**, 749 ⟨61JOC412⟩, 750 ⟨62JOC1255⟩
Schaefer, H., **2**, 778 ⟨65JPR123⟩; **4**, 728 ⟨67CB591⟩, 987 ⟨79CB484⟩
Schaefer, H. J., **5**, 28 ⟨79MI 502-04⟩
Schaefer, III, H. F., **1**, 1143 ⟨78JA1378, 79ACR288⟩, 1143 ⟨81JA1904⟩; **5**, 928 ⟨84CPL521⟩, 1020 ⟨89JCP1031⟩
Schaefer, J. P., **1**, 881, 884 ⟨67OR(15)1⟩; **2**, 7 ⟨65JOC2635⟩, 7, 837 ⟨73OSC(5)249⟩, 627 ⟨67JOC1607⟩, 749 ⟨67OR(15)1, 73OSC(5)142⟩; **5**, 46 ⟨67OR(15)1⟩
Schaefer, W., **4**, 11 ⟨87JOC3713⟩
Schaefer, W. P., **2**, 969 ⟨87JA203⟩; **3**, 508 ⟨81JA7358⟩
Schaeffer, III, H. F., **1**, 1143 ⟨81CPL412⟩; **2**, 364 ⟨76JA7617⟩
Schaeffer, J. R., **5**, 73 ⟨63OSC(4)19⟩, 117 ⟨73JOC1241⟩, 735 ⟨67JOC392⟩
Schaeffer, R., **2**, 524 ⟨72IC1150⟩; **4**, 142, 143 ⟨72IC1150⟩, 142, 524 ⟨65JA488⟩; **6**, 181, 269 ⟨69JCS(A)1882⟩
Schaeffer, W. D., **5**, 964 ⟨62USP3017420⟩
Schaer, B., **3**, 620 ⟨77JOC3956⟩
Schaerer, U., **5**, 447 ⟨66JOM(6)265⟩
Schafer, A., **3**, 108 ⟨83TL3291⟩
Schäfer, A., **4**, 689 ⟨83JOM(256)147⟩; **5**, 889 ⟨87AG(E)776⟩
Schafer, A. I., **1**, 1015, 1016 ⟨83TL3291⟩
Schäfer, B., **1**, 363 ⟨89JOU449⟩; **5**, 343 ⟨89LA739⟩, 894 ⟨93T1053⟩
Schäfer, F., **4**, 772 ⟨12LA(394)325⟩
Schäfer, G., **3**, 356 ⟨91LA111⟩
Schäfer, H., **1**, 396 ⟨63LA(663)22⟩, 396, 397 ⟨68TL2809⟩; **3**, 80 ⟨80JPR149⟩, 473, 479, 480 ⟨69LA(722)45⟩, 479 ⟨65AG379, 65AG(E)358⟩; **4**, 833 ⟨75JPR337⟩, 983, 986 ⟨77JPR149⟩, 995 ⟨80JPR87⟩; **5**, 1155 ⟨77ZAAC(428)222⟩
Schäfer, H.-C., **2**, 408 ⟨82CB185⟩
Schäfer, H. J., **1**, 346 ⟨92SL743⟩, 514 ⟨81CB292⟩, 561 ⟨88CB1151, 93CB1623⟩, 934 ⟨78AG(E)47, 81AG(E)911⟩, 934 ⟨85T4213, 88TL2797⟩, 934 ⟨88TL2801⟩; **2**, 20 ⟨76AG(E)306⟩, 30 ⟨79S964, 79S966, 81TL623⟩; **3**, 116 ⟨81AG(E)104⟩; **4**, 56 ⟨81AG(E)104⟩; **5**, 207 ⟨85T4195, 85T4213, 87AG(E)1253⟩, 208 ⟨86T1285⟩
Schäfer, H. N., **2**, 1009 ⟨92AX(C)912⟩
Schäfer, J., **4**, 408 ⟨58CB598⟩
Schafer, K., **4**, 988 ⟨80AG208⟩
Schäfer, M., **1**, 991 ⟨93AG(E)1315⟩
Schäfer, P., **5**, 657 ⟨79JPR1034⟩
Schäfer, R., **4**, 698 ⟨66TL2315⟩
Schäfer, T., **4**, 611 ⟨89JOM(363)7⟩
Schafer, T., **5**, 1119, 1121 ⟨86PS(26)1, 87PS(29)219⟩; **6**, 493, 495 ⟨80PS(8)301⟩, 630 ⟨87PS(29)1⟩
Schäfer, V., **2**, 903 ⟨88AG(E)299, 89CB401⟩; **3**, 357 ⟨74T4025⟩
Schafer, W., **1**, 690 ⟨84JA5018⟩
Schäfer, W., **1**, 690 ⟨86JA2932⟩; **2**, 163 ⟨85JHC701⟩, 167 ⟨60LA(637)146⟩; **4**, 328 ⟨63AG1050⟩; **5**, 579, 602 ⟨65M1278⟩, 599 ⟨64AG(E)19⟩, 1061, 1078 ⟨81AG855⟩
Schafer, W. R., **2**, 403 ⟨61JA747⟩; **4**, 314 ⟨61JA747⟩
Schaffer, E., **5**, 300 ⟨54CB1294⟩
Schaffer, H., **4**, 291 ⟨85JA675⟩
Schaffner, K., **1**, 96 ⟨82AG(E)820⟩, 422 ⟨B-80MI 109-03⟩, 794 ⟨63HCA678⟩
Schaffrath, R. E., **4**, 316 ⟨59JOC1035⟩
Schagerer, K., **3**, 74, 704 ⟨73AG(E)323⟩
Schakel, M., **1**, 795 ⟨72TL125, 73RTC605⟩; **2**, 555 ⟨94T5933⟩; **4**, 528 ⟨90RTC305⟩
Schalbar, J., **1**, 727, 739 ⟨70MI 116-03⟩
Schalfer, D., **5**, 101 ⟨84JOM(266)313⟩
Schalk, C., **5**, 328 ⟨92TL5055⟩
Schaller, K., **4**, 141 ⟨92JOM(438)45⟩
Schallner, O., **1**, 900 ⟨73AG(E)399⟩
Schallschmidt, H., **4**, 561 ⟨84EGP214610⟩
Schaltegger, H., **4**, 54, 56 ⟨71HCA1037⟩
Schamberg, E., **2**, 381, 382, 383 ⟨20CB1329⟩
Schamp, N., **2**, 3 ⟨87S188⟩, 22 ⟨79OPP115⟩, 312 ⟨88S587⟩, 322 ⟨88JOC4462⟩, 749 ⟨76T2979⟩; **3**, 24 ⟨82TL2853⟩, 35 ⟨84S46⟩, 152 ⟨87S188⟩, 184 ⟨82S765⟩, 219 ⟨81JOC3226⟩, 409 ⟨85AG(E)881, 85TL2709⟩, 409, 563 ⟨82OPP213⟩, 415, 416 ⟨80OPP49, 79OPP115⟩, 416 ⟨82S43, 86JOC3839, 86S192, 88CC825⟩, 417 ⟨82TL789⟩, 418 ⟨83S632, 86TL1707⟩, 563 ⟨78JOC2670⟩, 658 ⟨76T1449, 79S741⟩; **4**, 12 ⟨79SC575⟩, 183 ⟨83CB3631⟩; **6**, 509 ⟨77BSB663⟩
Schamp, N. B., **3**, 27 ⟨88BSB525⟩
Schanbacher, U., **4**, 465 ⟨82AG(E)776⟩
Schank, K., **1**, 704 ⟨B-88MI 115-01⟩; **2**, ⟨B-75MI 215-03⟩, 154 ⟨B-88MI 203-03⟩, 156 ⟨85HOU⟩, 197 ⟨74CB3026⟩, 617 ⟨66CB1414⟩, 742 ⟨76S406⟩, 803 ⟨B-75MI 215-02⟩, 806 ⟨B-75MI 215-04⟩; **3**, 37 ⟨76S406⟩, 80 ⟨79LA1977⟩, 373 ⟨78S678⟩, 500 ⟨89TL6673⟩; **4**, 43, 55 ⟨81CB1951, 81CB1958⟩, 206 ⟨81CB1938⟩, 234 ⟨66CB48⟩, 235 ⟨78CB2859⟩, 236 ⟨77LA1116⟩, 236, 838 ⟨77CB3235⟩, 265 ⟨91PS(58)207⟩, 266 ⟨73CB1107, 82SUL63⟩, 838 ⟨75LA1484, 76S404, 76S406, 76S408, 77LA1676⟩, 838,

1081 ⟨89T6667⟩; **5**, 528 ⟨78ACS(B)780⟩; **6**, 234 ⟨85HOU(E11)1129⟩, 480 ⟨76S408⟩
Schant, R. E., **1**, 556 ⟨80JOC4702⟩
Schantl, J., **4**, 441 ⟨70M568⟩
Schantl, J. G., **4**, 329, 444 ⟨85M1329⟩, 440 ⟨84S315, 85T5525⟩, 448 ⟨85M1051, 87M851⟩
Schantz, Jr., D. L., **2**, 387 ⟨88JOC1467⟩
Schanz, H., **6**, 27 ⟨23CB979⟩
Schapiro, N., **3**, 444 ⟨29CB2133⟩
Scharf, D. J., **4**, 493 ⟨76JOC28⟩; **5**, 372 ⟨74JOC922⟩
Scharf, E., **6**, 220 ⟨59AG524⟩
Scharf, H., **4**, 181 ⟨81LA1105⟩
Scharf, H.-D., **1**, 184 ⟨88S854⟩; **4**, 181, 182 ⟨88S854⟩, 207 ⟨76AG(E)688⟩, 425 ⟨90S951⟩; **5**, 145 ⟨78CB2206⟩, 206 ⟨87S517⟩
Scharf, W., **4**, 579 ⟨77ZN(B)858⟩, 851 ⟨77CB1484⟩; **5**, 1096 ⟨77ZN(B)858⟩; **6**, 706 ⟨77ZN(B)858⟩
Scharnow, H.-G., **2**, 373 ⟨65CB1588⟩
Scharp, J., **3**, 527 ⟨70RTC23⟩
Scharp, M., **1**, 829 ⟨93S615⟩
Schat, G., **3**, 508 ⟨83TL3935⟩; **4**, 608, 626, 669, 696, 700 ⟨85JOM(288)13⟩, 618 ⟨89TL6195⟩, 669, 670 ⟨86JOM(308)1, 87TL6493⟩; **6**, 190, 192, 198 ⟨86TL6123⟩, 399 ⟨89TL6195, 92JA7302⟩
Schatz, B., **1**, 419, 420 ⟨65TL385⟩, 419, 420, 794 ⟨73T3781⟩
Schätzlein, P., **6**, 75 ⟨83CB1463⟩
Schatzman, G. L., **4**, 755, 782 ⟨92JA360⟩
Schaub, B., **2**, 447 ⟨83C10⟩; **4**, 951 ⟨73HCA2166⟩; **5**, 896 ⟨85TL1623⟩; **6**, 705 ⟨83C10, 84TL4097⟩, 705 ⟨85TL1623⟩
Schaub, R. E., **1**, 535 ⟨78JOC3450⟩; **2**, 28 ⟨80JOC4702⟩
Schauble, J. H., **2**, 18 ⟨87S551⟩; **3**, 155 ⟨86S727⟩
Schauder, J. R., **1**, 592 ⟨83TL1657⟩
Schauer, C. K., **4**, 679 ⟨82JA7325⟩
Schäufele, H., **6**, 154 ⟨89AG929⟩
Schaumann, C. W., **2**, 843 ⟨66IS71⟩
Schaumann, E., **1**, 653 ⟨89T3163⟩, 784 ⟨82AG(E)225⟩, 1040 ⟨B-90MI 121-01⟩; **2**, 202 ⟨84JOC615⟩, 938 ⟨91AG(E)1687⟩, 941 ⟨77TL4307, 80TL4251⟩; **3**, 330 ⟨B-89MI 308-01⟩, 333, 335, 337 ⟨85TL5265⟩, 334, 349, 365 ⟨90JOC4200⟩, 334, 351, 369, 550 ⟨82AG(E)225⟩, 349, 365 ⟨82AG63⟩, 349, 365 ⟨88CB1159⟩, 349, 365, 550 ⟨83AG(E)55⟩, 351 ⟨77AG(E)722⟩, 351, 548 ⟨83CB66⟩, 442 ⟨85TL5629⟩, 546 ⟨79CB2698, 85HOU(E11)233⟩, 546, 547 ⟨82CB2755⟩, 546, 547 ⟨79LA1715, 88T1827⟩, 548 ⟨74CB3562, 77TL4307, 79LA1746, 80CB3024, 83CB509, 84AG(E)439⟩, 550 ⟨91CB1485⟩, 552 ⟨79LA1734⟩, 553 ⟨80TL4251⟩; **4**, 259, 855 ⟨90S271⟩, 266 ⟨B-90MI 406-02⟩, 328 ⟨88T1827⟩, 340 ⟨77LA88⟩, 648, 655 ⟨88TL4281⟩, 833 ⟨85HOU(E111)255⟩, 842 ⟨79LA1715, 85HOU(E111)260⟩, 843, 846, 854 ⟨88CB1165⟩, 851 ⟨80S907⟩, 852 ⟨76CB906⟩, 938 ⟨91JOC717⟩, 956 ⟨74CB3562⟩; **5**, 344 ⟨88CB689⟩, 344, 639 ⟨80TL4247⟩, 472 ⟨91COS(6)419⟩, 539 ⟨80JCS(P2)1557⟩, 550, 589 ⟨88T1827⟩, 552 ⟨80S907⟩, 565 ⟨91COS419⟩, 581, 589, 604 ⟨84CB2205, 85TL5269⟩, 589, 593 ⟨79LA1715⟩, 589, 604 ⟨79LA1702⟩, 589, 604, 605, 614 ⟨80CB3024⟩, 589, 614 ⟨77TL4307⟩, 589, 614 ⟨79LA1746, 76CB906⟩, 604 ⟨80CB3010, 83AG(E)55⟩, 604, 623 ⟨79CB1000⟩, 605 ⟨78AG(E)455, 85HOU(E11)251⟩, 626 ⟨80TL4251⟩, 635, 636, 637 ⟨71S111⟩, 636 ⟨71LA(743)154⟩, 638 ⟨74T4147⟩, 639 ⟨88CB701⟩, 729 ⟨74CB3589, 85S172⟩; **6**, 327, 351 ⟨89TL7033⟩, 337 ⟨81LA264⟩, 624, 625, 633 ⟨87CB339⟩, 637 ⟨82CB2516⟩, 647 ⟨78LA1543⟩
Schaverien, C. J., **3**, 516 ⟨86JA2771⟩; **4**, 676 ⟨84JCS(D)2455⟩, 689, 710 ⟨85JCS(D)2483⟩; **6**, 190, 193, 195 ⟨91OM1704⟩, 195 ⟨94OM69⟩
Scheben, J. A., **5**, 16 ⟨B-76MI 501-01⟩, 42 ⟨61JOC307⟩
Schechter, H., **1**, 1078 ⟨79TL2961⟩; **2**, 34 ⟨63OSC(4)543⟩
Schechter, M. S., **2**, 178 ⟨41JA1764⟩
Schechter, W. H., **2**, 110 ⟨61USP2992072⟩
Scheck, D., **1**, 519 ⟨82OM397⟩
Scheck, D. M., **5**, 450 ⟨80JA2723⟩, 951, 958 ⟨77IC3059⟩
Schecker, H.-G., **6**, 506 ⟨79CZ175⟩
Scheeber, W., **5**, 397 ⟨52CB239⟩
Scheel, D., **1**, 486 ⟨69CB2966⟩
Scheemann, H., **2**, 565 ⟨84OM69⟩
Scheer, H., **6**, 383 ⟨73AG(E)654⟩
Scheer, J. C., **1**, 773 ⟨63RTC1123⟩
Scheer, W., **1**, 524 ⟨67JA1753, 69AG(E)604⟩
Scheeren, H. W., **6**, 75 ⟨77JOC3128⟩
Scheeren, J. W., **2**, 29 ⟨80SC821⟩; **3**, 344, 345, 363 ⟨73S149⟩; **4**, 177, 190 ⟨69CC1175⟩, 203 ⟨73S151⟩, 414 ⟨87JOC365⟩, 760, 765 ⟨93HOU(E15/2)1599⟩, 824, 829 ⟨93HOU(E15/2)1674⟩, 825 ⟨79TL2925⟩, 826 ⟨90RTC399⟩, 830 ⟨73RTC11, 74TL1019, 77JOC3128⟩; **5**, 512, 566, 568, 570, 608 ⟨73S149⟩; **6**, 39 ⟨75S527⟩, 39, 40 ⟨71RTC556⟩, 44 ⟨71RTC1123⟩, 55, 107 ⟨69RTC289⟩, 75 ⟨76JCS(P1)1048, 79TL2925⟩, 75 ⟨81RTC13⟩, 77, 78 ⟨78S283⟩, 79 ⟨66RTC793⟩, 79, 95 ⟨69RTC897⟩
Scheerer, B., **4**, 536 ⟨77CB3034⟩
Scheffel, D., **1**, 1114 ⟨77JA3769⟩
Scheffer, A., **6**, 431 ⟨71JCS(C)3766⟩
Scheffler, K., **1**, 922 ⟨67TL3877⟩; **2**, 825 ⟨82PS(13)331⟩; **6**, 302 ⟨72CB3905⟩
Scheffold, R., **1**, 335, 546 ⟨83JA7200⟩, 345, 946 ⟨80JA3642⟩, 546, 946 ⟨87PAC363, 90T3155⟩, 946 ⟨B-83MI 119-05⟩; **2**, 8, 36 ⟨72AG(E)229⟩; **5**, 152 ⟨83JA7200⟩
Schegk, E., **2**, 463 ⟨57MIP1057112⟩
Scheibel, J. J.,**4**, 10 ⟨81JOC3917⟩
Scheiber, D. H., **1**, 1045, 1137 ⟨56JA1675⟩; **3**, 273 ⟨56JA1675⟩
Scheibler, H., **6**, 69 ⟨35CB2151, 58JPR60⟩, 105 ⟨58JPR70⟩
Scheiblich, S., **3**, 442 ⟨85TL5629⟩; **4**, 259, 855 ⟨90S271⟩, 837 ⟨84MI 419-01⟩, 843, 846, 854 ⟨88CB1165⟩; **5**, 589 ⟨85TL5269⟩
Scheibye, S., **3**, 346, 353 ⟨82T993⟩, 346, 363 ⟨78BSB223⟩, 411 ⟨80BSB247⟩; **5**, 484 ⟨80ACS(B)597⟩, 484, 551, 570 ⟨78BSB293⟩, 570, 571, 608, 610, 618 ⟨78BSB229⟩, 570, 571, 610 ⟨80NJC47⟩, 570, 571, 613 ⟨81ACS(B)239⟩, 570, 608, 610 ⟨78BSB299, 78BSB525⟩
Scheide, G. M., **4**, 129 ⟨87PS(33)147⟩, 570 ⟨87OM959⟩; **5**, 887 ⟨87PS(30)463⟩
Scheidges, C., **1**, 558 ⟨89LA2151⟩
Scheidl, F., **2**, 822, 841 ⟨79JOC3741⟩; **3**, 373 ⟨65TL1757⟩
Scheidt, F., **1**, 799 ⟨72CC716⟩
Scheidt, H., **3**, 586, 587 ⟨83CB741⟩
Scheiffele, E., **2**, 793 ⟨65JA4359⟩
Scheigetz, J., **1**, 13, 18, 20 ⟨86JOC3038⟩; **2**, 612 ⟨88CCC2116⟩
Scheinbaum, M. L., **2**, 223 ⟨88OSC(6)1016⟩, 391 ⟨70JOC2785⟩, 764 ⟨70JOC2790⟩; **3**, 541 ⟨64JOC2200⟩
Scheiner, P., **5**, 706 ⟨85JHC1435⟩
Scheinmann, F., **1**, 826 ⟨83CC932⟩, 827 ⟨68HCA1603⟩, 1053 ⟨75CC817, 76S321⟩; **2**, 656 ⟨68HCA1603⟩, 675 ⟨73JCS(P1)2870⟩; **5**, 101 ⟨79CC215, 80JCS(P1)2338⟩
Scheit, K. H., **2**, 105 ⟨67AG(E)362⟩
Scheithauer, S., **1**, 24 ⟨66CB1393⟩; **3**, 364 ⟨68CB3517⟩; **5**, 226 ⟨85HOU(E5/1)615⟩, 476, 477, 484 ⟨65CB829⟩, 476, 477, 510 ⟨63JPR214⟩, 480 ⟨B-79MI 511-02⟩, 513, 592, 593, 620 ⟨68CB3517⟩, 515, 516 ⟨66CB1393⟩, 546, 549 ⟨85HOU(E5)891⟩, 546, 549 ⟨B-79MI 513-02⟩, 586 ⟨72ZC133⟩
Schelhorn, H., **1**, 981 ⟨70TL4315⟩
Schelkun, R. M., **3**, 201 ⟨90TL1833⟩
Schellenbeck, P., **2**, 433, 434, 462, 824, 825 ⟨66CB1134⟩, 434 ⟨67CB692⟩
Scheller, D., **4**, 338 ⟨79ZC417⟩; **5**, 658 ⟨81M(112)959⟩
Scheller, M. E., **1**, 256 ⟨87SC863⟩; **4**, 359 ⟨85HCA44⟩; **5**, 223, 412, 429 ⟨86HCA44⟩, 410, 424 ⟨86HCA1378⟩,

430 ⟨89HCA264⟩; **6**, 127 ⟨86HCA44, 86HCA1378⟩, 128 ⟨84HCA1734⟩
Schellman, J. A., **5**, 84 ⟨61T176⟩
Scheludyakov, V. D., **5**, 968 ⟨69FRP1563380⟩, 978 ⟨75ZOB2007⟩
Schemenaur, J. E., **3**, 424 ⟨90SC2291⟩
Schempp, H., **4**, 979 ⟨65LA(684)103⟩; **5**, 496, 497, 500 ⟨65LA(684)103⟩
Schenck, G. O., **3**, 65 ⟨63LA(668)19⟩
Schenck, H. U., **6**, 138 ⟨68CB3058⟩
Schenek, R. T. E., **2**, 190 ⟨39JA3093⟩
Schenetti, M. L., **2**, 825 ⟨74JCS(P2)1338⟩
Schenk, C., **3**, 432 ⟨79RTC18, 79T147⟩, 432 ⟨80RTC246, 80RTC278⟩
Schenk, D., **5**, 488, 489 ⟨23CB633⟩
Schenk, G. O., **1**, 839 ⟨58LA(618)202⟩
Schenk, H., **1**, 1072 ⟨65AG(E)1080⟩; **3**, 493 ⟨67AG(E)249⟩; **5**, 900 ⟨67AG(E)249⟩, 996 ⟨65CB2954, 66CB1258⟩
Schenk, W. A., **2**, 484 ⟨90JOM(396)C8⟩; **3**, 370 ⟨92IC723⟩; **6**, 353, 354 ⟨88JOM(394)183⟩
Schenone, P., **2**, 204 ⟨67M1518, 83JHC839⟩; **3**, 80 ⟨77JHC249⟩; **5**, 344 ⟨86FES539⟩, 501 ⟨83JHC1549, 86JHC455⟩, 501 ⟨87JHC633⟩, 597 ⟨83FES101, 85FES120⟩, 597 ⟨91FES1033⟩
Schepers, R. G., **5**, 318 ⟨66JA3051⟩
Schepp, H., **1**, 630 ⟨66LA(691)41⟩
Schepp, N. P., **2**, 637 ⟨87JA4000, 90JA4862⟩, 638 ⟨89CC1548⟩, 1022 ⟨87CJC1719⟩; **6**, 646 ⟨91AG(E)1356⟩
Scherbaum, F., **3**, 576 ⟨88AG(E)1542⟩; **6**, 406 ⟨88AG(E)1544⟩
Scherberich, P., **6**, 445 ⟨86JOC3494⟩
Scheren, S. T., **3**, 71 ⟨59HCA844⟩
Scherer, H., **5**, 403, 404 ⟨72CB3357⟩
Scherer, I. G., **6**, 233 ⟨39AG457⟩
Scherer, O., **6**, 234 ⟨38USP2108606⟩, 236 ⟨67GEP1232954⟩
Scherer, O. J., **2**, 360 ⟨74ZN(B)277⟩, 361, 399 ⟨77ZAAC(432)173⟩, 400 ⟨75AG(E)485⟩, 406 ⟨89IS8⟩, 406, 407 ⟨77CB3874⟩, 406, 408 ⟨74AG(E)811, 74JOM(82)C3, 75AG(E)629⟩, 407, 420 ⟨81ZN(B)515⟩, 409, 416 ⟨79ZN(B)888⟩, 415 ⟨64AG(E)702, 68ZAAC(361)296⟩, 415, 416, 489, 503 ⟨66JOM(6)259⟩, 470 ⟨68CB4148⟩, 472 ⟨69ZAAC(370)171⟩; **3**, 485 ⟨65ZN(B)1009⟩; **4**, 547, 549 ⟨B-90MI 407-01, B-90MI 412-02⟩, 585 ⟨81ZN(B)1663⟩; **5**, 777, 780 ⟨68CB2533⟩, 876 ⟨B-90MI 523-02⟩, 1152 ⟨B-90MI 531-01⟩; **6**, 678, 688, 692, 693 ⟨B-90MI 622-01⟩, 692 ⟨85AG(E)924⟩
Scherer, W., **2**, 585 ⟨93CB45⟩
Schering, H. G. Von, **4**, 609 ⟨88JOM(341)109⟩
Scherm, H. P., **4**, 582 ⟨78JOM(160)41⟩
Schermer, D., **2**, 7 ⟨77LA806⟩
Schern, H. P., **4**, 583 ⟨78CB797⟩
Scherowsky, G., **4**, 181 ⟨67LA85⟩; **6**, 314 ⟨74CB3155⟩
Scherr, P. A., **4**, 698 ⟨71JA1035⟩
Schertz, L. D., **4**, 656 ⟨88OM1828⟩
Scherubel, D. M., **4**, 673, 722 ⟨88OM1715, 89OM1199⟩
Scherubel, G. A., **3**, 526 ⟨73JA7447, 74JOC3790⟩
Scherzer, K., **6**, 720 ⟨76JOM(118)C33⟩
Schetty, G., **2**, 248 ⟨47HCA1650⟩
Scheuble, B., **6**, 421, 422 ⟨92EUP499157⟩
Scheuer, G. F., **4**, 566 ⟨77ZAAC(433)157⟩
Scheuer, P. J., **3**, 694, 712 ⟨92ACR433⟩; **4**, 18, 730 ⟨49JA3439⟩, 743, 750 ⟨50JA3952⟩
Scheuermann, H., **6**, 428 ⟨80GEP2920386⟩
Scheuermann, H.-J., **1**, 248 ⟨84TL2329⟩, 713 ⟨86CB1162⟩
Scheunemann, K. H., **3**, 696 ⟨76JPR127⟩; **5**, 226 ⟨85AG(E)180⟩
Scheuplein, S. W., **1**, 816, 817 ⟨90CB917⟩, 1029 ⟨92CB271, 93TL4177⟩; **3**, 496 ⟨91CB1871⟩
Schevchenko, I. V., **6**, 680, 683 ⟨85ZOB264, 87PS(30)433⟩
Schevchenko, V. A., **6**, 696 ⟨90ZOB2814⟩
Scheytt, C., **2**, 442 ⟨86ZN(B)10⟩
Schiavelli, M. D., **1**, 543 ⟨81JOC807⟩, 993 ⟨77JOC1804⟩, 1114 ⟨70JA1082, 71JA6989, 72JA5061, 73JA7916, 73JOC3054, 76JOC681, 77JA3769⟩, 1122 ⟨65JA2295, 68JA1020, 68JA1023⟩
Schick, E., **1**, 46 ⟨43HCA800⟩; **3**, 229 ⟨46HCA704⟩
Schick, H., **1**, 482 ⟨89S477⟩; **3**, 220 ⟨87S1012⟩, 494 ⟨65CB604⟩; **4**, 1004 ⟨84T733⟩; **5**, 161 ⟨93AG(E)1191⟩
Schick, U., **1**, 494 ⟨94SL189⟩
Schick, W., **6**, 165 ⟨84ZAAC(511)108⟩, 704 ⟨84ZAAC(511)95, 84ZAAC(518)14⟩
Schickaneder, H., **4**, 312, 325 ⟨76JFC(7)471⟩
Schickh, G., **2**, 784 ⟨71HOU(10/1)9⟩
Schickh, O. von, **2**, 391, 392, 394 ⟨71HOU(10/1)12⟩
Schickli, C. P., **4**, 102 ⟨91LA655⟩
Schidt, H., **5**, 1155 ⟨85MI 531-01⟩
Schieb, T., **1**, 248 ⟨86AG(E)508⟩
Schiebel, H. M., **5**, 243, 540 ⟨76JCS(P1)564⟩
Schieber, E., **6**, 213 ⟨B-72MI 607-02⟩
Schiebye, S., **2**, 256 ⟨81ACS(B)239⟩
Schieda, O., **3**, 483 ⟨81CB3518⟩; **6**, 177, 380, 382, 392, 673, 709, 710 ⟨81CB3518⟩, 290, 291, 380, 381, 392, 709, 710 ⟨81CB2087⟩, 373, 380, 382, 392, 709, 710 ⟨81CB3505⟩
Schiedewitz, W., **6**, 478 ⟨77ZC411⟩
Schiehser, G., **5**, 334 ⟨89JMC1024⟩
Schiei, M., **5**, 657 ⟨83HCA1618⟩
Schield, J. A., **2**, 753 ⟨74JOC1758⟩
Schiemann, G., **2**, 609, 614 ⟨31CB1340⟩; **5**, 106 ⟨43OSC(2)299⟩
Schiemenz, G. P., **2**, 824 ⟨73LA1480⟩, 824, 827, 828 ⟨69CB1883⟩; **3**, 298 ⟨76LA1514⟩
Schier, A., **2**, 446 ⟨85CB3105⟩, 447 ⟨86JA976⟩, 448 ⟨83S372⟩, 1004 ⟨93AG(E)1746⟩; **4**, 551, 560, 1023 ⟨85CB3105⟩; **6**, 702 ⟨92CB1053⟩
Schierhorn, A., **4**, 969 ⟨79T551⟩
Schierling, P., **6**, 75 ⟨83CB1463, 83CB3482⟩
Schiess, E., **5**, 857 ⟨00CB751⟩
Schiess, M., **3**, 186 ⟨84HCA1593⟩, 707 ⟨88CB507⟩; **5**, 658 ⟨88CB507⟩; **6**, 622 ⟨88CB507⟩
Schiess, P., **1**, 502 ⟨70HCA1713⟩, 506 ⟨70HCA485⟩, 812 ⟨63HCA483⟩; **3**, 528 ⟨77AG(E)469⟩, 539 ⟨88LA943⟩, 543 ⟨74HCA2583⟩
Schiess, P. W., **1**, 1081 ⟨67AG(E)1⟩
Schiesser, C. H., **1**, 336, 337, 338, 547 ⟨85T3925⟩; **6**, 543, 550, 551 ⟨91T121⟩
Schiff, H., **5**, 856 ⟨00CB2741⟩; **6**, 508 ⟨1877LA(189)159⟩
Schiff, R., **3**, 468 ⟨1881CB1375⟩
Schiffer, C., **1**, 564 ⟨86LA1578⟩
Schilchte, K., **2**, 558 ⟨90CB1517⟩
Schildknecht, H., **4**, 441 ⟨68AG(E)293⟩; **5**, 987 ⟨68AG(E)293⟩
Schildknegt, K., **5**, 572 ⟨94JOC1129⟩
Schill, G., **1**, 727 ⟨64CB3293⟩; **3**, 280 ⟨75CB1570⟩; **4**, 177 ⟨79CB3603⟩, 290 ⟨70CB2271⟩
Schilling, B. E. R., **6**, 186, 192, 195 ⟨86JCS(D)1551⟩, 186, 195 ⟨82CC1407⟩
Schilling, G., **1**, 708 ⟨77CB2978⟩
Schilling, M., **4**, 665 ⟨83CB3182⟩
Schilling, W., **5**, 234, 277 ⟨75JA3515⟩
Schillinger, W. J., **1**, 604 ⟨80JOC2247⟩
Schimdt, R. R., **3**, 259 ⟨76AG(E)171⟩
Schimmel, K. F., **5**, 356 ⟨62JOC413⟩
Schimmelschmidt, K., **2**, 456 ⟨56GEP1023033⟩
Schimpf, R., **3**, 475 ⟨73CB1496⟩
Schinazi, R. F., **4**, 214 ⟨94BMC931⟩, 230 ⟨91JA9377⟩, 231 ⟨92JOC2217⟩
Schindlbauer, H., **2**, 428 ⟨59M148⟩, 429, 822, 823 ⟨59M792⟩, 821, 822, 824, 825, 845 ⟨69CB2914⟩, 829 ⟨65CB2431⟩, 832, 833 ⟨65M1021⟩, 844, 848 ⟨69CB2922⟩, 854 ⟨67M1792⟩
Schindler, H., **1**, 786 ⟨87TL183⟩; **4**, 561 ⟨59CB1695⟩

Schindler, N., **4**, 1003, 1010 ⟨71CB2021⟩; **6**, 158, 159 ⟨72ZAAC(389)119⟩, 367 ⟨74GEP2237879⟩, 449, 452 ⟨71CB969⟩
Schindler, O., **5**, 700 ⟨69HCA569⟩
Schings, U., **2**, 473 ⟨83AG(E)775⟩
Schink, K., **3**, 556, 564 ⟨81CL527⟩
Schinkowski, K., **2**, 378 ⟨64CB49⟩
Schinski, W. L., **1**, 232, 234 ⟨71JA676⟩
Schinz, H., **3**, 55 ⟨51HCA728⟩
Schinzer, D., **1**, 664 ⟨77AG(E)44b⟩, 979 ⟨85TL1831, 86CC829, 88S263, 91CB247, 92TL8017⟩, 1012 ⟨92SL766⟩; **2**, 908 ⟨89S179⟩; **3**, 419 ⟨89MI 310-01⟩; **5**, 437 ⟨81TL1881⟩
Schio, L., **1**, 940 ⟨93TL8245⟩
Schionato, A., **1**, 297, 414 ⟨93MI 107-01⟩
Schipoff, M., **4**, 30 ⟨78HCA2482⟩
Schipor, I., **4**, 396 ⟨92T2025⟩
Schippel, O., **4**, 149, 583 ⟨91JOM(417)149⟩
Schipper, E., **2**, 362, 363 ⟨69JOC2397⟩; **4**, 420 ⟨61JOC4480⟩
Schipperijn, A. J., **1**, 1088 ⟨71RTC1110⟩
Schipprak, P., **3**, 251 ⟨54CB841⟩
Schirlin, D., **5**, 328 ⟨92TL5055⟩
Schirmann, J. P., **6**, 430 ⟨87TL6443⟩
Schirmer, U., **6**, 494 ⟨79GEP2921130⟩
Schirp, H., **5**, 109 ⟨73OSC(5)813⟩
Schkeryantz, J. M., **1**, 190 ⟨92JOC2986⟩; **4**, 22 ⟨90JOC5719⟩
Schlachter, S. T., **5**, 155 ⟨88TL6823⟩
Schläfer, L., **1**, 2, 6, 8, 10 ⟨59CB1700⟩
Schlageter, A., **1**, 991 ⟨92CB2667⟩; **4**, 1018 ⟨91OM389⟩
Schlageter, M., **1**, 1067 ⟨91TL4115⟩
Schlageter, M. G., **2**, 91 ⟨83JA7172⟩; **6**, 327 ⟨83JA7719⟩
Schlapkohl, K., **5**, 561 ⟨77JCR(S)144⟩
Schlatmann, J. L. M. A., **1**, 403, 795 ⟨64RTC1173⟩, 419 ⟨61T146⟩; **6**, 643 ⟨69JMC712⟩
Schlatter, M. J., **3**, 620 ⟨55OSC(3)223⟩
Schlecht, M. F., **1**, 227 ⟨82CC1331⟩; **3**, 151 ⟨88TL6707⟩
Schlecker, R., **1**, 137 ⟨78HCA512⟩; **4**, 382 ⟨78HCA512⟩, 943 ⟨77CB2880⟩; **6**, 317, 734 ⟨77CB2880⟩
Schlee, H. G., **6**, 453 ⟨75GEP2351556⟩
Schlegel, H. B., **1**, 879 ⟨75CJC3365⟩; **6**, 691 ⟨84JPC1981⟩
Schlemper, E. O., **2**, 573 ⟨86OM2452⟩
Schlemper, H., **1**, 574 ⟨87AG(E)1167⟩
Schlenk, H., **3**, 302 ⟨79JOC3420⟩; **6**, 27 ⟨52CB901⟩
Schlenk, W., **2**, 555, 556 ⟨17CB262⟩
Schlesinger, A. H., **2**, 127 ⟨54JA585⟩
Schlesinger, H., **2**, 109 ⟨53JA192, 53JA199⟩, 109 ⟨53JA205⟩, 109, 111 ⟨33JA4020⟩, 109, 111 ⟨39JA673⟩
Schlessinger, R. H., **1**, 209 ⟨78JOC788⟩, 309 ⟨73TL2595⟩; **2**, 331 ⟨87TL2083⟩; **3**, 37 ⟨81JA724⟩, 60 ⟨85TL2391⟩, 369 ⟨70TL791⟩, 372 ⟨69JA3953, 69JA6891⟩, 623 ⟨75TL4647⟩; **4**, 224 ⟨78JA1938⟩, 649 ⟨85TL2391⟩, 855 ⟨73TL4711⟩, 857 ⟨73TL4707⟩
Schletter, I., **4**, 168 ⟨73JA7156⟩
Schleu, J., **5**, 942 ⟨95CB373⟩
Schleyer, P. von R., **1**, 9 ⟨87JOC4299⟩, 104 ⟨70JA2542⟩, 110 ⟨85AOC(24)353⟩, 110, 885 ⟨87AOC(27)169⟩, 135, 159 ⟨84JA6467⟩, 380 ⟨64JA2722⟩, 427 ⟨83ZN(B)1697⟩, 552 ⟨89AG(E)1321⟩, 811 ⟨64TL2503⟩, 846 ⟨86TL4411⟩, 847, 853 ⟨66MI 119-07⟩, 852 ⟨83T1141⟩, 880 ⟨86AG(E)349⟩, 891 ⟨B-68MI 119-02, B-70MI 119-02, B-72MI 119-07, B-73MI 119-04, 73TS253, B-76MI 119-01⟩, 892 ⟨69JA1037⟩, 894 ⟨66JA2321, 71JA3927⟩, 900 ⟨64JA4194, 64JA4195⟩, 900 ⟨85JA1421⟩, 905 ⟨79JA522, 88JOC661⟩, 906 ⟨80JA683⟩, 909 ⟨78JA4301⟩, 910 ⟨86JA561, 89JA1147⟩, 915 ⟨B-72MI 119-12⟩, 1088 ⟨85CB1504, 88JA5932⟩, 1109 ⟨69JA5350, 70JA2141⟩, 1110, 1117 ⟨71JA1513⟩, 1111 ⟨72JA3629⟩, 1111, 1114, 1116 ⟨74JA1100⟩, 1112, 1113 ⟨77JOC3004⟩, 1114 ⟨70JA2538, 70JA2540⟩, 1114 ⟨70JA3789⟩, 1114, 1116 ⟨70JA3802, 72JA992, 76JA7658, 76JA7667⟩, 1122 ⟨74JA1110⟩, 1143 ⟨81CPL408, 82CPL493⟩; **2**, 68 ⟨72JA4628⟩, 689 ⟨70AG(E)273, 74JA1100, 76JOC4099⟩; **3**, 497 ⟨86CB1331⟩; **4**, 1072 ⟨B-68MI 424-01⟩, 378, 391, 816 ⟨84JA6467⟩, 549 ⟨94AG(E)2202⟩, 647 ⟨86JOM(316)C4⟩, 667, 668 ⟨82CC1184, 83AG(E)733⟩, 836, 949 ⟨87JOM(332)99⟩; **5**, 433 ⟨92MI 509-01⟩, 927 ⟨84AG(E)825⟩, 928 ⟨85JA2773⟩, 928, 929 ⟨92AG(E)1238, 92AG(E)1384⟩; **6**, 63 ⟨72JA125⟩, 184, 389, 398, 714 ⟨89AG(E)781⟩, 188 ⟨92AG(E)1384⟩, 201, 715 ⟨92AG(E)1238⟩, 205 ⟨83AG(E)733⟩, 386, 401, 404 ⟨76JA5419⟩, 399 ⟨91CC1536⟩, 406 ⟨85AOC(24)354, 87AOC(27)169⟩, 691 ⟨87CC1399⟩, 697, 700 ⟨86CB1331⟩
Schlich, K., **2**, 70 ⟨84TL2195⟩; **6**, 14 ⟨84TL2195⟩
Schlichting, O., **1**, 530, 890 ⟨48LA(560)1⟩
Schliebs, R., **2**, 851 ⟨70ZN111⟩
Schlippes, K., **4**, 492 ⟨72ZC334⟩
Schlodder, R., **3**, 568 ⟨73CB2144⟩
Schlomann, W. W., **4**, 229 ⟨68JOC2133⟩
Schlor, H., **2**, 463 ⟨57MIP1057112⟩
Schlosberg, R. H., **1**, 901 ⟨69JA3261⟩, 914 ⟨72JA2034⟩
Schlossarczyk, H., **1**, 993 ⟨73HCA875⟩
Schlosser, K., **6**, 734 ⟨79JA6282⟩
Schlosser, M., **1**, 113 ⟨64AG(E)287, 64AG(E)362⟩, 113 ⟨76HCA453, 93T10175⟩, 113 ⟨93TL5441⟩, 177 ⟨74AG50⟩, 397 ⟨89T2649⟩, 465 ⟨74AG(E)82⟩, 568 ⟨72HOU(5/1b)180⟩, 570 ⟨88T2855⟩, 579 ⟨90T2411⟩, 639 ⟨68AG(E)650⟩, 721 ⟨65AG(E)689, 75C341⟩, 721, 727, 728, 729, 730 ⟨67LA(708)1⟩, 721, 728, 730, 732, 733 ⟨B-70MI 116-01⟩, 727, 728 ⟨66AG(E)667, 71BSF(2)453⟩, 729, 730 ⟨66AG(E)126, B-76MI 116-01, 70CB2814⟩, 731 ⟨69S75, 71CB2885⟩, 885 ⟨B-73MI 119-03⟩; **2**, 15 ⟨78T3⟩, 19 ⟨88T2875⟩, 447 ⟨83C10⟩, 529 ⟨78JA3258⟩, 551, 557 ⟨B-73MI 211-01⟩, 556 ⟨64AG(E)326, 88PAC1627⟩, 557, 907 ⟨84TL741⟩, 557, 907, 999 ⟨84TL745⟩, 607, 611 ⟨69S75⟩, 612 ⟨69S38, 84HCA284⟩, 615 ⟨79HCA635, 79HCA1194⟩, 618 ⟨61CB1373⟩, 942 ⟨92SL360⟩, 964 ⟨76HCA453, 76JA4674⟩; **3**, 27 ⟨90T4261⟩, 67 ⟨77HCA1739⟩, 445 ⟨62T1023⟩, 494 ⟨64CB3219, 66AG(E)968⟩, 494 ⟨67LA(708)1, 69S38⟩, 495 ⟨90T5213⟩, 497 ⟨61CB1373⟩; **4**, 347 ⟨61CB1373⟩, 378 ⟨84TL745⟩, 755 ⟨69CB1944⟩, 756 ⟨67CB3893⟩, 818 ⟨78JOC1595⟩, 949 ⟨74AG(E)701⟩, 950 ⟨74S888⟩, 951 ⟨73HCA2166⟩, 957 ⟨76HCA13⟩; **5**, 436 ⟨64AG(E)287⟩, 894 ⟨69S75⟩, 894, 896 ⟨85TL1623, 93TL1925⟩, 897 ⟨61CB1373⟩; **6**, 697 ⟨61CB1373⟩, 705 ⟨83C10, 84TL4097, 85TL1623⟩
Schlossman, A., **5**, 371, 403 ⟨90JCS(P1)3263⟩
Schlubach, H. H., **1**, 104 ⟨58LA(614)37⟩; **2**, 1016 ⟨58LA(614)37⟩
Schlude, H., **2**, 680 ⟨75T89⟩
Schluenz, R. W., **2**, 381 ⟨73JOC2412⟩; **5**, 14 ⟨74JA6469⟩
Schlunke, H. P., **4**, 162 ⟨70HCA1598⟩
Schlüter, G., **5**, 655 ⟨88LA833⟩
Schlüter, K., **1**, 922 ⟨79TL929⟩, 981 ⟨82AG(E)924⟩
Schmalstieg, L., **5**, 867 ⟨84AG(E)717⟩
Schmalz, D., **1**, 876 ⟨87AG(E)1212⟩
Schmalz, H.-G., **1**, 305, 306 ⟨91COS(4)199⟩; **5**, 932 ⟨94AG(E)303⟩
Schmann, R. C., **2**, 533, 698 ⟨90T4663⟩
Schmeider, K. R., **3**, 540 ⟨77CB3582⟩
Schmeiss, H., **3**, 88 ⟨39CB440⟩
Schmeisser, M., **6**, 218, 220, 224 ⟨75ZAAC(418)109⟩, 220 ⟨59AG524⟩, 232 ⟨69CB2150⟩
Schmelzer, H.-G., **5**, 652 ⟨68AG(E)856⟩, 654 ⟨68AG942⟩
Schmerling, L., **1**, 294 ⟨B-64MI 107-01⟩, 1003 ⟨B-64MI 121-01⟩; **2**, 650 ⟨70JOC4028⟩; **6**, 221 ⟨51USP2553800⟩
Schmersahl, P.,**1**, 426 ⟨63AP(296)324⟩
Schmerse, G. C., **3**, 550 ⟨91CB1485⟩; **6**, 624, 625, 633 ⟨87CB339⟩
Schmickler, H., **1**, 248 ⟨85AG(E)1046, 86AG(E)508⟩, 248 ⟨88AG(E)411⟩

Schmid, B., **4**, 927 ⟨86CB472⟩, 957 ⟨77TL3583⟩, 958 ⟨79TL4277⟩
Schmid, C. R., **3**, 17, 132, 665 ⟨83JA6732⟩, 88 ⟨84JA3374⟩
Schmid, G., **1**, 988 ⟨75TL4025⟩; **3**, 547, 581 ⟨75AG(E)53, 74AG(E)473⟩, 604 ⟨74AG(E)273⟩, 604, 605 ⟨75TL4025, 80CB3369⟩; **4**, 148 ⟨87OM861⟩; **5**, 155 ⟨79JA259⟩, 343, 730 ⟨80CB912⟩, 433 ⟨68CB2502⟩, 481 ⟨74AG(E)273⟩; **6**, 340 ⟨72JOM(46)149, 76JOM(113)67, 78CB1239, 78CB1603⟩, 341 ⟨76CB3339⟩
Schmid, G. H., **1**, 318 ⟨77JOC2021, B-78MI 122-02⟩, 1110 ⟨73JA160⟩, 1123 ⟨68JA3466, 76CJC3045⟩, 1123 ⟨76CS76, 76JOC2331⟩; **2**, 4, 5 ⟨B-77MI 201-01, B-89MI 201-02⟩, 40 ⟨71JA4907⟩, 280 ⟨77JOC1776, 78T2869⟩
Schmid, H., **1**, 234, 821 ⟨77HCA1361⟩, 403, 795 ⟨70HCA173⟩, 415 ⟨68HCA371, 74T1959⟩, 415 ⟨75HCA1293⟩, 420 ⟨72HCA2233⟩, 734, 795 ⟨72HCA1385, 72HCA1828⟩, 809 ⟨71HCA937, 73CRV293⟩, 814 ⟨79HCA2630⟩, 827 ⟨63HCA1792, 68HCA1603⟩, 835 ⟨73HCA105⟩, 993 ⟨73HCA875⟩; **2**, 656 ⟨63HCA2444, 68HCA1603⟩, 656 ⟨73HCA14⟩, 660 ⟨73HCA75⟩; **3**, 678 ⟨77H(6)143⟩, 679, 680 ⟨74HCA1382⟩
Schmid, J., **2**, 627 ⟨90JHC2151⟩; **4**, 581 ⟨86JOM(306)203, 87CB583⟩
Schmid, K-H., **6**, 267 ⟨78ZAAC(441)125⟩
Schmid, M., **1**, 809 ⟨71HCA937, 73HCA105⟩; **2**, 182 ⟨59HCA1653⟩; **5**, 9 ⟨59HCA1653⟩; **6**, 52 ⟨59HCA1653⟩
Schmid, R., **1**, 234, 821 ⟨77HCA1361⟩
Schmid, W., **2**, 597 ⟨94T749⟩
Schmidbauer, H., **1**, 51, 446 ⟨87CB1403⟩, 158 ⟨79AG(E)782, 89CB259⟩, 737 ⟨76CB473⟩, 740 ⟨68IC168⟩; **2**, 163 ⟨71CB3234⟩, 437 ⟨92ZN(B)640⟩, 444 ⟨83CB3567, 89AG(E)1700⟩, 445 ⟨80ZN(B)990⟩, 446 ⟨85CB3105⟩, 447 ⟨80CB1612, 84ZN(B)618, 86JA976⟩, 448 ⟨83S372⟩, 449 ⟨83ZC249⟩, 470 ⟨67CB1000⟩, 475 ⟨77AG(E)722, 79ZAAC(458)249⟩, 488 ⟨69CB4136, 73CB1226, 73CB3645, 78CB2702⟩, 488, 504 ⟨71ZAAC(386)139⟩, 492, 494 ⟨64CB449, 68ZAAC(363)84⟩, 502 ⟨71CB1842, 68CB1271⟩, 512 ⟨87JOM(325)159⟩, 590 ⟨70AG(E)101⟩, 597 ⟨67CB1129⟩, 622 ⟨89CB2213⟩, 888 ⟨71ZAAC(386)147⟩, 1004 ⟨75CC668⟩; **3**, 496 ⟨68CB595⟩, 500 ⟨78CB2702⟩, 576 ⟨88AG(E)1542⟩, 878 ⟨87CB1281⟩; **4**, 94 ⟨85OM1208⟩, 136, 140 ⟨86ZN(B)1527⟩, 489 ⟨90CB423⟩, 545, 546, 547, 575, 582 ⟨77ZN(B)762⟩, 546, 575 ⟨78GEP2658127⟩, 548 ⟨84HCA2175⟩, 548, 1023 ⟨88CB1241⟩, 550 ⟨86CB2832⟩, 551, 560, 1023 ⟨85CB3105⟩, 551, 574 ⟨68CB3545⟩, 552, 558 ⟨76ZN(B)721⟩, 553 ⟨77CB3501⟩, 554 ⟨84CB3374⟩, 557, 574, 575, 577, 583, 584, 585, 586, 587, 588 ⟨80JOM(200)287⟩, 562 ⟨91CB275⟩, 563 ⟨77CB1748⟩, 564 ⟨79AG(E)782⟩, 565 ⟨84ZN(B)1456⟩, 565, 593 ⟨87CB1281⟩, 569 ⟨69AG(E)372, 69CB83⟩, 569 ⟨70CB97, 70CB3007, 74ZN(B)485⟩, 569, 587 ⟨68CB595⟩, 570 ⟨78CB2696, 87CB789⟩, 570, 571 ⟨70AG(E)737, 70CB3448⟩, 571 ⟨72CB3173⟩, 571, 574, 588 ⟨67CB1032⟩, 573 ⟨74AG(E)540⟩, 574 ⟨77CB677, 80CB1480⟩, 576 ⟨68CB3556, 80CB902⟩, 576, 577 ⟨82ZN(B)950⟩, 577, 586 ⟨77ZAAC(434)145⟩, 577, 587 ⟨85OM1233⟩, 577, 588 ⟨77CB3528⟩, 578 ⟨75ACR62, 83AG(E)907⟩, 579 ⟨77ZN(B)858, 85ZN(B)352⟩, 580, 581, 584, 585 ⟨82CB1379⟩, 581, 582, 583 ⟨83OM257⟩, 582 ⟨73AG(E)853, 74CB3692, 77CB2200, 78JOM(160)41, 80ZN(B)584⟩, 583 ⟨74CB93, 74CB3684, 78CB797, 81AG(E)586⟩, 583, 584, 585, 586, 588 ⟨77CB3517⟩, 583, 585, 586, 587 ⟨77CB3536⟩, 585 ⟨73AG(E)415, 74CB3697⟩, 585 ⟨75JOM(96)133, 75JOM(97)479⟩, 585 ⟨77MI 412-02, 79JOM(182)251⟩, 585, 587 ⟨78IS140⟩, 586 ⟨73AG(E)416, 75ICA(13)79, 77IJ149, 78ZN(B)1325, 81CB441, 84JCS(D)2859⟩, 586, 588 ⟨76AG(E)542⟩, 586, 656 ⟨75CB1321⟩, 587 ⟨74CB102⟩, 587 ⟨75JOM(99)353⟩, 587, 588 ⟨74CB3674, 76AG(E)541⟩, 597 ⟨68IC168⟩, 598 ⟨75CB2649, 76CB473⟩, 598 ⟨77CB1312⟩, 599 ⟨78CB2702⟩, 624 ⟨76AG(E)367⟩, 629 ⟨88ZN(B)727, 89ZN(B)285⟩, 639 ⟨90ZN(B)961⟩, 695 ⟨82AG(E)73⟩, 1023 ⟨84AG(E)247, 86AG(E)348⟩, 1023 ⟨89CB1857⟩, 1052 ⟨87ZN(B)142⟩; **5**, 226 ⟨87CB39⟩, 894 ⟨75ACR62, 78PAC19⟩, 894 ⟨80PAC1057⟩, 894, 1092, 1096 ⟨83AG(E)907⟩, 897 ⟨83CB1955⟩, 908 ⟨81CB2491, 82ZN(B)950, 83CB1386⟩, 908, 1094 ⟨79CB510⟩, 908, 910, 911 ⟨80CB902⟩, 908, 911, 912, 916 ⟨68CB3545⟩, 910, 911 ⟨81CB1428, 84ZN(B)1456⟩, 911, 1093, 1094 ⟨77CB3501⟩, 912 ⟨68AG(E)220, 70CB3007, 70CB3448⟩, 913 ⟨72CB1084, 75CB2851⟩, 913 ⟨77AG(E)639, 80CB1637⟩, 913, 914 ⟨72CB3173⟩, 913, 916, 920, 1093 ⟨67CB1032, 74AG(E)540⟩, 914 ⟨78CB2696, 78JOM(159)137⟩, 914, 1095 ⟨70CB97⟩, 915 ⟨75CB2834, 75CB2842⟩, 920, 922 ⟨77CB677⟩, 921, 1094 ⟨85OM344⟩, 921, 1094 ⟨77CB1312, 87CB1281⟩, 922 ⟨78CB2702⟩, 1092 ⟨81CB3070⟩, 1093 ⟨68CB595, 75JA6281, 77AG(E)417⟩, 1095 ⟨87CB789⟩, 1096 ⟨77ZN(B)858⟩, 1141 ⟨73JA4855, 84JCS(D)2859⟩; **6**, 123, 698 ⟨83CB1955⟩, 173 ⟨89CB2115⟩, 173 ⟨90CB2087⟩, 173, 378 ⟨86ZN(B)1527⟩, 173, 378, 379 ⟨90AG(E)201⟩, 370 ⟨76ZN(B)721⟩, 373 ⟨70CB3448, 92JOM(429)1⟩, 386, 406 ⟨91AG(E)1488⟩, 406 ⟨88AG(E)1544⟩, 700, 703 ⟨70AG(E)77⟩, 700, 703 ⟨71CB150⟩, 701 ⟨83CB1386, 84OM38, 84ZN(B)1456⟩, 702 ⟨67CB1032, 75ACR62, 92CB1053⟩, 702, 707, 708 ⟨77CB677⟩, 703 ⟨74AG(E)540⟩, 705 ⟨83AG(E)907⟩, 706 ⟨77ZN(B)858⟩, 706 ⟨86AG(E)574, 68IC168⟩, 708 ⟨85OM344, 87CB1281⟩
Schmidbaurand, H., **2**, 452 ⟨78ZN(B)1556⟩
Schmidhammer, H., **1**, 432 ⟨84JMC1575⟩
Schmidhauser, J., **4**, 130 ⟨79CC739⟩
Schmidlin, T., **1**, 565 ⟨92JCS(P1)3351⟩
Schmidpeter, A., **2**, 293 ⟨69ZN(B)179⟩, 468 ⟨68CB3883, 72ZAAC(394)171⟩, 468 ⟨72ZN(B)769⟩, 469 ⟨71ZN(B)81⟩, 476 ⟨76CB3581⟩, 870 ⟨76ZN(B)1058⟩; **3**, 456 ⟨71CB1199⟩; **4**, 112 ⟨84CB1695⟩, 545, 558 ⟨93PS(76)13⟩, 550 ⟨86AG(E)253⟩; **5**, 635 ⟨86PS(27)289⟩, 708, 709, 861 ⟨71CB1199⟩, 881 ⟨85MI 523-02⟩, 882 ⟨84AG(E)903, B-90MI 523-01⟩, 907, 908 ⟨92TL471⟩, 907, 908 ⟨93AG(E)1089⟩, 1128 ⟨77AG747, 80AG767⟩; **6**, 156 ⟨83CB549⟩, 625 ⟨71CB1199⟩, 660, 686 ⟨83AG(E)545⟩, 660, 686 ⟨84AG(E)903⟩, 661 ⟨86AG(E)457⟩, 686, 688 ⟨87PS(30)495⟩
Schmidt, A., **1**, 563 ⟨80AG(E)463⟩; **2**, 504 ⟨68CB3976⟩; **4**, 146 ⟨91CB1253⟩; **5**, 776 ⟨86ZAAC(533)55⟩; **6**, 238, 613, 727 ⟨64CB1286⟩, 438 ⟨82ZN(B)81⟩, 624, 730 ⟨66ZAAC(344)113, 72CB3050⟩, 731 ⟨67CB3319⟩, 733 ⟨67CB3725⟩
Schmidt, A. H., **3**, 354 ⟨80S961⟩, 399 ⟨76AG(E)704⟩; **5**, 17, 19 ⟨81S216⟩
Schmidt, B., **2**, 783 ⟨92AG(E)1578⟩; **4**, 99 ⟨89T6113⟩
Schmidt, C. L., **3**, 310 ⟨86JOC2021⟩
Schmidt, D., **1**, 974 ⟨87T2457⟩; **4**, 472 ⟨69T5569⟩
Schmidt, D. G., **2**, 632 ⟨90JOC4560⟩
Schmidt, E., **4**, 489 ⟨65AG1138⟩; **5**, 453 ⟨72JOM(36)C18⟩, 586 ⟨66LA(695)217⟩, 698, 813 ⟨13CB3616⟩, 726 ⟨64LA(676)114⟩, 822 ⟨48BSF597⟩, 1025 ⟨54LA(585)230, 55LA(594)233⟩, 1026 ⟨59LA(621)1⟩; **6**, 318, 361 ⟨68CB1137⟩, 547 ⟨85ZAAC(525)127⟩, 553 ⟨88ZAAC(556)189⟩, 567 ⟨88ZAAC(556)179⟩, 625 ⟨13CB2447⟩, 627 ⟨66CB3268⟩
Schmidt, E. A., **2**, 639 ⟨72JA7832⟩; **6**, 107 ⟨72JA3201⟩
Schmidt, E. W., **2**, 169, 170 ⟨64CB2614⟩, 170 ⟨63CB3050⟩, 373 ⟨B-84MI 207-01⟩
Schmidt, F. C., **6**, 261 ⟨44JOC419⟩
Schmidt, G., **1**, 241 ⟨68AG(E)456⟩, 438 ⟨79TL2317⟩, 1072 ⟨65AG(E)1080⟩; **2**, 1083 ⟨84HOU(13/3c)1⟩; **3**, 4, 115 ⟨79TL399⟩; **6**, 398 ⟨86AG(E)1111⟩
Schmidt, H., **2**, 100 ⟨90ZN(B)557⟩, 467

⟨86ZAAC(535)47⟩, 478 ⟨90IS249⟩, 883 ⟨22LA(429)144⟩, 883, 888 ⟨20LA(421)174⟩; **3**, 116 ⟨81AG(E)104⟩; **4**, 453 ⟨88JOM(355)71⟩, 463, 470 ⟨48CB477⟩, 566 ⟨86ZAAC(535)47⟩, 568 ⟨82ZAAC(488)75, 87PS(30)349⟩, 997, 998 ⟨79ZAAC(459)131⟩, 1017 ⟨75JOM(97)61⟩; **5**, 784, 881 ⟨78JOM(160)47⟩, 879, 880 ⟨85MI 502-01⟩; **6**, 372 ⟨80ZC419⟩, 372, 374 ⟨80ZC153⟩, 372, 688, 689 ⟨81ZAAC(473)85⟩, 502 ⟨79CB727⟩, 658, 687 ⟨81SRI279⟩, 680, 688 ⟨80JOM(192)33, 81ZC357⟩, 687 ⟨85ZAAC(529)216⟩

Schmidt, H. G., **6**, 252 ⟨91CB1353⟩

Schmidt, H.-J., **1**, 181 ⟨69AG496⟩; **4**, 56 ⟨81AG(E)104⟩

Schmidt, H. M., **1**, 999 ⟨67RTC1138⟩

Schmidt, J., **1**, 690 ⟨87TL3091⟩; **4**, 190 ⟨76TL95⟩; **6**, 505 ⟨75CZ245⟩

Schmidt, K., **4**, 548 ⟨90PS(49/50)355⟩

Schmidt, L., **2**, 481, 486, 489, 505, 506, 510 ⟨65ZAAC(339)298⟩

Schmidt, M., **2**, 228 ⟨87AG(E)887⟩, 262 ⟨63CB1019, 63CB1561⟩, 262 ⟨63CB2822⟩, 264 ⟨63AG(E)328⟩, 294 ⟨70JOM(25)17⟩, 415 ⟨64AG(E)702⟩, 415, 416, 489, 503 ⟨66JOM(6)259⟩, 868 ⟨64CB2395⟩, 895 ⟨67JOM(8)105⟩, 1032 ⟨63N302⟩; **3**, 220 ⟨86TL1999, 88S49⟩, 359 ⟨90CB1143, 91JHC573⟩, 371 ⟨68CB277⟩, 431 ⟨86CB2756⟩, 432 ⟨85CB1105⟩, 485 ⟨65ZN(B)1009⟩; **4**, 232 ⟨90CB1143⟩; **5**, 396 ⟨92ZAAC(608)33⟩, 396, 400 ⟨92ZAAC(607)101⟩, 396, 879, 891 ⟨91ZAAC(605)7⟩; **6**, 127, 132, 134 ⟨75ZAAC(418)208⟩, 132 ⟨76CB1239⟩, 701 ⟨87AG(E)79, 87TL2111⟩

Schmidt, M. F., **4**, 712, 716 ⟨82OM1597⟩

Schmidt, O., **5**, 858 ⟨01CB2001⟩

Schmidt, P., **5**, 769, 852 ⟨55HCA1560⟩; **6**, 118 ⟨B-18MI 607-01⟩

Schmidt, R., **4**, 672 ⟨83JA2905⟩, 673 ⟨83AG(E)330⟩, 900 ⟨77ZC289⟩; **6**, 160 ⟨82ZAAC(485)23⟩

Schmidt, R. E., **6**, 324, 684 ⟨84ZAAC(517)75⟩, 582, 684 ⟨85ZAAC(520)139⟩

Schmidt, R. R., **1**, 520 ⟨84TL4357⟩, 831 ⟨90TL4433⟩; **3**, 239 ⟨69LA(723)111⟩, 247 ⟨76CB2928⟩, 267 ⟨87S896⟩; **4**, 174 ⟨87TL1981⟩, 176 ⟨85TL2065⟩, 176, 198, 202 ⟨91COS(6)33⟩, 198 ⟨86AG(E)212⟩, 200 ⟨86ACR250, 89PAC1257⟩, 426 ⟨84TL1769⟩, 852 ⟨65LA(684)37⟩, 901 ⟨62ZN851, 65CB1385⟩, 927 ⟨86CB472⟩, 936, 950, 951 ⟨87CB213⟩, 949 ⟨79TL4273⟩, 950 ⟨92SL429, 92TL8035⟩, 957 ⟨77TL3583⟩, 958 ⟨79TL4277⟩, 1016 ⟨77S869, 78AG(E)204⟩, 1016 ⟨82S748⟩; **5**, 199 ⟨82S958⟩, 634 ⟨69CB269⟩, 692 ⟨86AG213, 86AG(E)212⟩, 692 ⟨87MI 517-01, 91COS(6)33⟩; **6**, 632 ⟨88GEP(O)3634929⟩

Schmidt, S., **2**, 484, 489, 492 ⟨28CB1911⟩

Schmidt, S. P., **2**, 7, 10 ⟨87TL767⟩

Schmidt, T., **1**, 289 ⟨89JA5902, 93TL7183⟩, 582 ⟨91T8155⟩, 1047 ⟨88JA2301⟩; **3**, 229, 273 ⟨88JA2301⟩

Schmidt, U., **1**, 91 ⟨76CB2628⟩, 1045 ⟨71M214⟩; **2**, 214 ⟨59CB1177⟩, 256 ⟨73M503⟩, 338 ⟨75CB2917, 76AG(E)294⟩, 506, 852 ⟨68CB1381⟩; **3**, 179, 184 ⟨64CB1649, 85LA785⟩, 298 ⟨76LA1514⟩, 412 ⟨79CB107⟩, 699 ⟨64AG(E)641⟩; **4**, 465 ⟨82AG(E)776⟩; **5**, 15 ⟨89S832⟩, 81 ⟨65AG(E)846⟩, 166 ⟨70AG(E)457⟩, 262 ⟨78S752⟩, 315 ⟨75CB2547, 76AG(E)294⟩, 351 ⟨92S482⟩, 363 ⟨92CC1186⟩, 612 ⟨65CB3819⟩, 692 ⟨89S256⟩; **6**, 77 ⟨70AG(E)456⟩

Schmidt, V., **4**, 206 ⟨81CB1938⟩

Schmidt, W., **1**, 473 ⟨93S250⟩; **2**, 175 ⟨42CB34⟩; **5**, 199 ⟨55LA(597)95⟩

Schmidt, W. W., **2**, 25, 31 ⟨75JOC134⟩

Schmidtberg, G., **5**, 377 ⟨71CB3831⟩

Schmidtchen, F. P., **5**, 361 ⟨77ZN(B)98⟩, 1029 ⟨90JOC3749⟩

Schmidt-Collerus, J. J., **5**, 361 ⟨71JOC350⟩

Schmidt-Dunker, M., **2**, 847 ⟨B-76MI 216-01⟩

Schmidt-Lukasch, G., **5**, 928 ⟨88ZN(B)801, 90AG(E)398⟩, 928, 930 ⟨88AG(E)960⟩, 929 ⟨92AG(E)1384⟩; **6**, 188 ⟨92AG(E)1384⟩, 713, 714 ⟨88ZN(B)801, 90AG(E)398⟩

Schmidtmann, H., **3**, 614 ⟨1896CB1171⟩

Schmidtpeter, A., **4**, 550 ⟨86ZN(B)444⟩

Schmidt-Samoa, E., **3**, 480 ⟨65AG1138, 65AG(E)1078⟩

Schmidt-Thome, J., **1**, 795 ⟨36CB882⟩

Schmiedel, R., **3**, 354 ⟨74S578⟩, 355 ⟨79S361⟩

Schmiedl, D., **3**, 547 ⟨74ZC92⟩

Schmierer, R., **5**, 148 ⟨81AG(E)207⟩

Schmiesing, R., **1**, 815 ⟨79CC106⟩

Schmiesing, R. J., **4**, 187 ⟨80CC477⟩

Schmit, C., **1**, 518 ⟨84TL5043⟩; **3**, 216 ⟨84TL5043⟩; **4**, 131 ⟨91SL395⟩

Schmitt, C., **4**, 530 ⟨91S637⟩; **6**, 561 ⟨94S983⟩, 561, 563 ⟨91S637⟩

Schmitt, E., **3**, 406, 416 ⟨80LA1814⟩, 561, 582 ⟨79LA83⟩; **5**, 657 ⟨79JPR1034⟩

Schmitt, G., **1**, 534 ⟨75S632⟩

Schmitt, H.-G., **2**, 197 ⟨74CB3026⟩; **4**, 236 ⟨77LA1116⟩

Schmitt, H. G., **4**, 236, 838 ⟨77CB3235⟩

Schmitt, H.-J., **4**, 681 ⟨73ZN(B)508⟩

Schmitt, J., **1**, 82 ⟨63BSF798⟩

Schmitt, J.-L., **1**, 590 ⟨79JA3283⟩

Schmitt, R. J., **2**, 386 ⟨87JOC2292⟩, 1053 ⟨87JOC2294⟩; **4**, 807 ⟨86S132⟩

Schmittel, M., **1**, 499 ⟨93TL8431⟩, 924 ⟨87JA2750⟩; **3**, 700 ⟨92CB525⟩

Schmittheimer, H. F., **3**, 432 ⟨79OPP249⟩

Schmitthenner, H. F., **1**, 791 ⟨80JOC3372⟩

Schmittling, E. A., **2**, 684 ⟨93JOC3229⟩

Schmitz, A., **3**, 40 ⟨87AG(E)1141⟩; **6**, 77, 78 ⟨85S31⟩

Schmitz, C., **1**, 623 ⟨88T2913⟩

Schmitz, E., **2**, 96 ⟨58CB1212⟩, 374 ⟨61CB2166, 62CB680⟩, 374 ⟨64AG(E)333, 68JPR257⟩, 374 ⟨88S529⟩, 377 ⟨62CB2012⟩, 378 ⟨64CB49⟩, 379 ⟨77JPR195, 85JPR445⟩, 387 ⟨69LA(725)1⟩, 403 ⟨63AG(E)157⟩, 686 ⟨70JPR43⟩; **3**, 238 ⟨85TL4911⟩; **4**, 163, 176 ⟨B-67MI 404-01⟩, 183 ⟨58CB410⟩, 212 ⟨56CB1254⟩, 420 ⟨84JPR279⟩, 989 ⟨86TL4281⟩; **5**, 276, 286 ⟨61AG621⟩, 283 ⟨79AHC(24)63⟩, 353 ⟨65TL1857, 67JPR86⟩, 357 ⟨85JPR(327)893⟩, 365 ⟨67CB142⟩; **6**, 70 ⟨68LA(716)207⟩, 79 ⟨58CB1942⟩

Schmitz, R., **4**, 566 ⟨81ZAAC(475)18⟩; **5**, 888 ⟨81ZAAC(475)18⟩; **6**, 164 ⟨81ZAAC(475)18⟩

Schmitz, R. F., **2**, 555 ⟨94T5933⟩, 682 ⟨75TL3429⟩; **4**, 1047, 1058, 1062 ⟨83T3073⟩; **5**, 424 ⟨83T3073⟩, 424, 468 ⟨83CC239⟩; **6**, 130, 133 ⟨83CC239⟩, 130, 133 ⟨83T3073⟩

Schmitz, T., **3**, 485 ⟨89CB595⟩; **6**, 187, 188, 292, 388, 389, 673, 712 ⟨89CB595⟩, 187, 387, 712, 713 ⟨90CB747⟩

Schmitz-Dumont, O., **5**, 823 ⟨25LA(444)113, 28CB32⟩

Schmock, F., **3**, 488 ⟨73JOM(54)23⟩; **5**, 1143 ⟨75CB664⟩; **6**, 655, 733 ⟨79AG(E)693, 80ZAAC(468)165⟩

Schmuck, A., **2**, 873 ⟨89CB803⟩; **6**, 256 ⟨87AG(E)314⟩, 614 ⟨87AG(E)134⟩

Schmuff, N. R., **1**, 238, 613 ⟨80JA5979⟩; **5**, 354 ⟨75CC725⟩

Schmüser, W., **3**, 365 ⟨77JCR(S)244⟩; **5**, 561 ⟨77JCR(S)144⟩

Schmutzler, R., **2**, 10 ⟨82S326⟩, 448 ⟨70JCS(A)2359⟩, 467 ⟨91CB2705, 91ZN(B)258⟩, 471 ⟨78CB890⟩, 475 ⟨82JFC(20)85⟩, 476 ⟨70JCS(A)1049, 76CC643⟩, 476 ⟨92CB801⟩, 477 ⟨66IC164, 67IS63⟩, 821, 822 ⟨92CB2187⟩, 832, 833 ⟨65CB552⟩, 835 ⟨70JCS(C)878⟩, 836 ⟨64IC410⟩, 836, 837 ⟨67IS64, 67IS69⟩, 848 ⟨64JCS4551, 73OSC(5)1005⟩, 855 ⟨72JCS(D)1590, 72JCS(P2)380⟩, 856 ⟨73CC144⟩, 869 ⟨64JA4500, 68IC1327⟩; **4**, 47 ⟨73OSC(5)218⟩, 115 ⟨63MI 403-01⟩, 116 ⟨65JCS5630⟩, 133 ⟨64IC410, 66OS(46)21⟩, 133 ⟨67IS63⟩, 134 ⟨92CB801⟩, 338 ⟨92PS(72)171⟩, 563, 564 ⟨69ZN(B)1081, 81CB1082⟩, 564 ⟨75CC468, 92CB1325⟩,

564 ⟨93PS(75)233⟩; **5**, 407 ⟨70JCS(C)878, 78CC946⟩, 1093 ⟨67SA(A)681⟩
Schnabel, W., **1**, 1120 ⟨80T3229⟩; **6**, 603 ⟨62JOC2589⟩
Schnabel, W. J., **5**, 966, 1013 ⟨69JOC1162⟩
Schnapp, K. A., **3**, 565 ⟨93MI 317-01⟩
Schnatterer, S., **4**, 550 ⟨86CB2832⟩
Schneckenburger, J., **5**, 674 ⟨83AP(316)21⟩
Schneering, H. G. von, **2**, 437 ⟨90ZN(B)1548⟩
Schneider, A. K., **3**, 527 ⟨57JA6261⟩; **6**, 509 ⟨58JA6409⟩
Schneider, B., **2**, 601 ⟨67LA(707)15⟩, 868 ⟨66LA(692)1⟩; **6**, 270 ⟨69JA1954⟩
Schneider, C., **5**, 151, 164 ⟨88JA4640⟩
Schneider, D., **6**, 428 ⟨80GEP2920386⟩
Schneider, D. F., **1**, 733 ⟨68JCS(C)1982⟩; **3**, 231 ⟨87TL559⟩
Schneider, E., **2**, 166, 171, 172 ⟨51CB911⟩, 428 ⟨78ZAAC(443)42⟩
Schneider, G., **6**, 71 ⟨63PCS374⟩, 73 ⟨67CC13, 67T4181⟩, 489 ⟨91JCS(P1)37⟩
Schneider, H., **3**, 534 ⟨23HCA291⟩, 534, 541 ⟨13CB3539⟩
Schneider, H. J., **4**, 846 ⟨61JOC1987⟩
Schneider, J., **1**, 899 ⟨61LA(650)62⟩; **2**, 596 ⟨54LA(589)91, 60LA(629)121⟩; **3**, 130, 286 ⟨78SC59⟩; **5**, 924 ⟨88CB1363⟩; **6**, 74 ⟨66CB1892⟩, 153 ⟨89AG(E)1013⟩, 341 ⟨83M851⟩
Schneider, J. A., **2**, 24 ⟨73MI 201-01⟩, 68 ⟨80TL1035⟩
Schneider, J. J., **4**, 695 ⟨91AG(E)1124⟩
Schneider, M., **1**, 900 ⟨86JOC2969⟩; **3**, 119 ⟨82JOC364⟩, 352 ⟨85RTC9⟩
Schneider, M. P., **1**, 914 ⟨79JA4426⟩; **5**, 118 ⟨90CC648⟩
Schneider, N., **3**, 586 ⟨87LA183⟩
Schneider, N. S., **2**, 43 ⟨94CC463⟩
Schneider, P., **2**, 558, 559 ⟨84TL3305⟩; **3**, 263 ⟨74JPR469⟩
Schneider, P. B., **4**, 352, 353 ⟨92CB591⟩
Schneider, R., **2**, 449 ⟨91AG(E)207⟩; **5**, 339, 689 ⟨80LA1064⟩, 890 ⟨90PS(49)337⟩, 1152 ⟨92AG(E)758⟩; **6**, 153 ⟨91AG(E)207⟩, 153 ⟨92AG(E)758⟩
Schneider, S., **1**, 287 ⟨89AG215⟩; **2**, 7 ⟨83S139⟩
Schneider, U., **2**, 910, 911, 916, 917, 932, 940, 946 ⟨87JOC4868⟩
Schneider, W., **2**, 300, 303 ⟨B-75MI 205-01⟩; **4**, 442 ⟨58LA(615)34⟩; **6**, 208 ⟨63BSF1462⟩, 430 ⟨56USP2755314⟩
Schneider, W. F., **4**, 42 ⟨93JPC1576⟩
Schneider, W. M., **1**, 453 ⟨65CB824⟩
Schneider, W. P., **1**, 729 ⟨69CC785, 69JA5364⟩, 881 ⟨63OSC(4)132⟩, 884 ⟨50JA2395⟩
Schneiderwind, R. G. K., **6**, 488 ⟨83T2207⟩
Schnell, H., **5**, 652 ⟨67GEP1228249⟩, 966 ⟨66GEP1228249⟩
Schnell, H.-W., **1**, 894 ⟨65AG(E)873⟩, 922 ⟨70T5343⟩
Schnell, M., **3**, 590 ⟨92PS(73)185⟩; **4**, 106 ⟨87JPR871⟩; **5**, 790 ⟨91PS(63)95⟩; **6**, 156 ⟨91PS(63)95⟩
Schneller, S. W., **5**, 350 ⟨78JHC439⟩
Schnepf, E., **6**, 195 ⟨92ZAAC(613)67⟩
Schnepf, J. E. O., **6**, 195, 197 ⟨91ZAAC(595)225⟩
Schnering, H.-G. von, **1**, 566 ⟨93CB1481⟩, 966 ⟨92AG(E)224⟩; **2**, 732 ⟨87AG(E)780⟩ 1003 ⟨83M1097⟩; **3**, 553, 554 ⟨91OM3967⟩, 581 ⟨92JOM(423)329⟩; **4**, 991 ⟨91JHC1863⟩, 991, 993, 996 ⟨85BSB475⟩; **6**, 173, 383 ⟨93ZAAC(619)1494⟩, 381, 385, 391 ⟨83USP467715, 88JOM(341)109⟩
Schnider, O., **5**, 276 ⟨61HCA1546⟩
Schnitt, G., **1**, 745 ⟨64JPR143⟩
Schnizer, A. W., **3**, 534 ⟨53JOC1055⟩
Schnöckel, H., **6**, 588 ⟨81ZN(B)1261⟩
Schnoes, H. K., **5**, 185 ⟨87MIP8707604⟩
Schnur, R., **5**, 1064, 1066 ⟨88CB271⟩
Schnur, R. C., **5**, 757 ⟨79JOC3726⟩
Schnurr, O., **2**, 802 ⟨79CB2120⟩
Schnurr, W., **3**, 689 ⟨87JOM(338)329⟩; **4**, 115, 567 ⟨89TL3951⟩; **5**, 1152, 1159 ⟨87JOM(338)329⟩; **6**, 667, 668, 689 ⟨89S511⟩
Schnute, M. E., **2**, 356 ⟨91JOC6738⟩
Schober, P., **4**, 605, 621 ⟨70ZAAC(372)21, 70ZAAC(372)59⟩, 652 ⟨70ZAAC(377)37⟩; **6**, 268 ⟨70ZAAC(372)21⟩
Schöberl, A., **2**, 213 ⟨48CB210⟩, 214 ⟨78CB3159⟩, 214, 215 ⟨58LA(614)66⟩, 216, 225 ⟨55HOU(9)5⟩, 219 ⟨58LA(617)71⟩, 402 ⟨55HOU(9)260⟩; **5**, 84 ⟨48AG308⟩, 1111 ⟨48CB210⟩, 232 ⟨55HOU(9)745⟩
Schobert, R., **1**, 45 ⟨90SL465⟩, 1134 ⟨83AG(E)780⟩; **3**, 220 ⟨88S49⟩
Schock, L. E., **2**, 969 ⟨85JA8103⟩; **6**, 194 ⟨85JA8103⟩
Schödel, H., **6**, 174 ⟨93CB2227⟩
Schodel, S., **3**, 247 ⟨58CB320⟩
Schoeller, W., **5**, 395 ⟨91CB2843⟩
Schoeller, W. W., **4**, 484 ⟨81AG(E)131⟩; **5**, 887 ⟨81AG(E)131⟩; **6**, 168 ⟨81AG(E)131⟩, 168 ⟨83CC1171⟩
Schoellmann, G., **2**, 22 ⟨84JOC1286⟩
Schoemaker, G. C., **5**, 1089 ⟨90JOM(390)C1⟩
Schoemaker, H. J., **4**, 303 ⟨78T179⟩
Schoen, W., **5**, 106 ⟨63OSC(4)532⟩
Schoenberg, A., **3**, 90 ⟨74JA7761⟩; **5**, 143 ⟨74JOC3318⟩, 273, 296, 301, 304 ⟨74JOC3327⟩
Schoenborn, M., **1**, 1143 ⟨73T4199⟩
Schoenen, F. J., **1**, 831 ⟨84JOC4320⟩
Schoenewaldt, E. F., **1**, 135 ⟨87JOC4745⟩; **5**, 972 ⟨66JA3163⟩; **6**, 496 ⟨68JA3254⟩
Schoental, R., **2**, 773 ⟨69NAT765⟩
Schoerner, Ch., **4**, 455 ⟨80TL1845⟩
Schoettel, G., **3**, 514 ⟨87CC129, 89CC1062⟩; **4**, 692 ⟨92OM2224⟩
Schöffling, B., **1**, 1021 ⟨91TL53⟩
Schoffstall, A. M., **1**, 550 ⟨90JOC414⟩; **2**, 105 ⟨66T399⟩; **3**, 322 ⟨90JOC414⟩, 678 ⟨90MI 320-01⟩
Schofield, C. J., **5**, 329 ⟨90CC720⟩, 376 ⟨90T4733⟩, 1106 ⟨90TA881⟩
Schofield, G., **1**, 264 ⟨89CC1852⟩
Schofield, J., **6**, 71 ⟨90JCS(P1)375⟩
Schofield, K., **2**, 790 ⟨B-80MI 215-01⟩
Schohe, R., **3**, 46 ⟨84T2199⟩, 683 ⟨85MI 320-01⟩
Schöler, H., **2**, 835, 837 ⟨77CB2382⟩; **4**, 112 ⟨77CB2382⟩; **5**, 1093 ⟨76AG(E)702⟩
Schöler, H.-F., **4**, 547, 564 ⟨79CB648⟩; **6**, 164, 703 ⟨79CB648, 79CB1068⟩
Scholerl, A., **2**, 214 ⟨57AG713⟩
Schölkens, B., **4**, 528 ⟨89EUP297872⟩
Scholl, P. C., **3**, 120 ⟨80TL4695⟩
Scholl, R., **3**, 618 ⟨18M240⟩; **5**, 819 ⟨1894CB2816⟩, 838 ⟨1890CB3490⟩
Schöller, W. W., **4**, 567 ⟨83CC1171⟩
Schöllhorn, H., **2**, 999, 1000 ⟨89OM1688⟩
Schöllhorn, R., **2**, 384 ⟨66CB3118⟩
Schöllkopf, K., **2**, 1006, 1009 ⟨87JA882⟩
Schöllkopf, U., **1**, 155 ⟨77AG(E)339⟩, 155 ⟨77LA40⟩, 183, 209 ⟨86AG(E)754⟩, 196, 233, 396 ⟨70AG(E)763, 73AG(E)323⟩, 242 ⟨61LA(642)1, 79LA451⟩, 396 ⟨63LA(663)22⟩, 396, 397 ⟨68TL2809⟩, 398, 399 ⟨69TL2619, 69TL3415⟩, 399 ⟨70LA(737)158⟩, 463 ⟨87AG(E)683⟩, 639, 721 ⟨54CB1318⟩, 683 ⟨75AG(E)765⟩, 800, 816 ⟨66LA(698)80⟩, 816 ⟨70LA(734)106⟩; **2**, 551 ⟨70HOU(13/1)1⟩, 802 ⟨68AG(E)138⟩, 909 ⟨77LA1167⟩, 1023 ⟨79LA219⟩; **3**, 194, 704 ⟨71AG(E)491⟩, 328 ⟨67AG(E)884⟩, 473, 479, 480 ⟨69LA(722)45⟩, 474, 478, 487 ⟨69CB488⟩, 474, 484, 485, 486 ⟨74LA1767⟩, 474, 486 ⟨70AG291, 70AG(E)301⟩, 474, 487 ⟨67AG238⟩, 474, 487 ⟨67AG(E)261⟩, 478 ⟨67AG50, 67AG(E)74, 68LA(716)204⟩, 479 ⟨65AG379, 65AG(E)358⟩, 480 ⟨66TL6199, 69AG577, 69AG(E)612⟩, 484 ⟨67AG906, 69LA(730)1, 72LA(761)137⟩, 494 ⟨54CB1318⟩, 543 ⟨75AG(E)765⟩, 580, 582 ⟨74LA1655⟩, 694, 697 ⟨79PAC1347⟩, 694, 697, 707, 713, 717 ⟨77AG(E)339⟩, 694, 706, 714, 717 ⟨74LA44⟩, 696 ⟨76LA183⟩, 696, 702,

704 ⟨77LA40⟩, 697 ⟨68AG(E)805⟩, 700 ⟨81LA103⟩, 700, 720 ⟨70AG(E)458, 80LA28⟩, 704 ⟨71AG(E)331, 76LA2105, 84LA608⟩, 704, 706, 707 ⟨86AG(E)754⟩, 704, 707, 719 ⟨75CB1580⟩, 704, 717 ⟨73LA611⟩, 706 ⟨73CB3382, 77LA1174, 81LA709, 81LA1693, 84S1033, 93CB427, 93LA427⟩, 706, 707 ⟨73LA1571⟩, 706, 714 ⟨75LA533⟩, 706, 714, 717 ⟨84LA600⟩, 708 ⟨73TL629⟩, 709, 713 ⟨72LA(766)130⟩, 711 ⟨75LA1531, 76CB3964⟩, 713 ⟨76LA969⟩, 713, 717 ⟨72LA(766)116⟩, 714, 715 ⟨69AG(E)672, 77LA1167⟩, 717, 718 ⟨81LA99⟩, 719 ⟨85LA1825⟩, 720 ⟨79LA1444⟩, 6971 ⟨B-79MI 321-0⟩; **4**, 381 ⟨64TL1503, 67LA(704)120⟩, 462 ⟨85AG(E)1067⟩, 535 ⟨79AG(E)863, 83PAC1799⟩, 535 ⟨83T2085⟩, 536 ⟨83LA1133, 86S737⟩, 536 ⟨87AG(E)480, 88AG(E)1194⟩, 536 ⟨88AG(E)433⟩, 537 ⟨77AG(E)339⟩, 538 ⟨76TL609, 80LA28⟩; **4**, 49, 86 ⟨63CB2266⟩, 142 ⟨67AG(E)74⟩, 950 ⟨72LA(763)208⟩, 1004 ⟨85LA555⟩, 1005 ⟨74LA44⟩; **5**, 162 ⟨75AG(E)765⟩, 162 ⟨79LA219⟩, 634 ⟨76TL609⟩, 844, 866, 874 ⟨68LA(716)204⟩, 844, 871, 874 ⟨69CB488⟩, 868 ⟨66TL6199⟩, 871, 874 ⟨67AG(E)261⟩, 872 ⟨74LA1767⟩, 873 ⟨67AG(E)884⟩, 873, 874 ⟨70AG(E)301⟩, 873, 874 ⟨72LA(761)137⟩; **6**, 63 ⟨69OS(49)86, 72JA125⟩, 158 ⟨85LA555⟩, 627 ⟨66LA(694)44⟩, 664 ⟨71LA(753)143⟩, 719, 732 ⟨67AG(E)560⟩
Schöllkopf, V., **4**, 908 ⟨72AG(E)311⟩
Schöllkopft, U., **4**, 331, 908 ⟨72LA(766)130⟩, 487 ⟨73TL633⟩
Schölm, R., **5**, 262 ⟨78S752⟩
Scholten, H. P. H., **4**, 304 ⟨80JOC1880⟩
Scholtissek, C., **6**, 444 ⟨56CB2562⟩
Scholtysik, G., **6**, 646, 647 ⟨75ACR62⟩
Scholz, D., **1**, 707 ⟨81CB909⟩; **2**, 139 ⟨83S944⟩; **3**, 130, 286 ⟨78SC59⟩, 164 ⟨84LA259⟩; **4**, 796 ⟨75C512⟩; **6**, 644 ⟨75C514⟩
Scholz, H., **3**, 479 ⟨33CB1012⟩, 566 ⟨36LA(521)95⟩
Scholz, H.-U., **3**, 484 ⟨72LA(761)137⟩; **5**, 873, 874 ⟨72LA(761)137⟩
Scholz, M., **2**, 872 ⟨89JOM(366)73⟩
Scholz, P., **4**, 343 ⟨72ZC178⟩, 492 ⟨72ZC60⟩, 554 ⟨72ZC103⟩
Scholz, R., **6**, 409, 419 ⟨80EGP139936⟩
Scholz, S., **1**, 558 ⟨89LA2151⟩; **5**, 184 ⟨85LA1935⟩
Scholz, T. H., **3**, 305 ⟨91JOC4147⟩
Schomaker, V., **2**, 507 ⟨45JA2014⟩
Schomburg, D., **1**, 519 ⟨86TL5833⟩; **2**, 476 ⟨76CB3581⟩, 821, 822 ⟨92CB2187⟩; **4**, 353 ⟨91JOM(403)29⟩; **5**, 407 ⟨78CC946⟩; **6**, 253 ⟨81JA406⟩
Schön, I., **5**, 277 ⟨74TL1785⟩
Schön, N., **1**, 415 ⟨61LA(641)1⟩; **2**, 788 ⟨61LA(641)1⟩; **3**, 720 ⟨61LA(641)1⟩; **4**, 882, 974 ⟨61LA(641)1⟩; **5**, 852 ⟨61LA(641)1⟩; 104, 105, 313, 314, 318, 321, 730, 731 ⟨61LA(641)1⟩
Schönauer, K., **3**, 11 ⟨83TL573⟩
Schönberg, A., **1**, 1040 ⟨65CB3311⟩; **2**, 809 ⟨69CB2557⟩; **3**, 357 ⟨58JA6312⟩, 362, 365 ⟨68CB716⟩, 498 ⟨40JCS1374⟩, 547 ⟨33CB237⟩, 549 ⟨62CB3077⟩, 550 ⟨70CB949⟩; **4**, 298 ⟨78T1285⟩; **5**, 243 ⟨35CB163⟩, 548 ⟨80CB3605⟩; **6**, 87 ⟨65CB3303⟩, 143 ⟨83CB2068⟩, 253, 254 ⟨33CB567⟩, 304 ⟨67CB767⟩, 598 ⟨87CB1581⟩
Schönberg, H., **5**, 888, 893 ⟨93CC673, 93PS(76)21⟩; **6**, 691 ⟨93CC673⟩, 691, 733 ⟨93PS(76)21⟩
Schönduve, E., **2**, 808 ⟨74AP(307)73⟩
Schonefeld, J., **4**, 16, 25 ⟨71LA(744)42⟩
Schönwälder, K. H., **2**, 1006, 1009 ⟨87JA882⟩; **3**, 246 ⟨82CB2766⟩
Schoolenberg, J., **3**, 621 ⟨69JOC3684⟩
Schoon, D., **4**, 304 ⟨85JOC4231⟩
Schoonraad, J. L., **5**, 317 ⟨90T7247⟩
Schopf, C., **5**, 73 ⟨50CB390⟩
Schöpke, K., **5**, 658 ⟨86LA1997⟩; **6**, 621 ⟨86LA1997⟩
Schöps, R., **4**, 338 ⟨79ZC417⟩
Schore, N. E., **1**, 516, 517, 530, 551 ⟨88CRV1081⟩, 528, 529 ⟨91OR(40)1⟩, 530 ⟨81JOC5436, 85JOC5215⟩; **2**, 439 ⟨87JOM(320)339⟩, 926, 985 ⟨81JOC5357, 91OR(40)1⟩, 985 ⟨81JOC5436, 91COS(5)1037⟩; **3**, 218 ⟨89TL5865, 90JA441⟩; **4**, 357 ⟨81JOC5357⟩, 566, 568 ⟨79JA7410⟩, 566, 568 ⟨81IC3200⟩
Schorigin, P., **1**, 396 ⟨25CB2028⟩, 396, 397 ⟨24CB1634⟩
Schork, R., **4**, 854 ⟨85CB2852⟩; **6**, 531 ⟨85CB1415⟩
Schorp, M., **4**, 259 ⟨92LA1039⟩
Schorpp, K., **3**, 568 ⟨73CB2144⟩
Schosser, M., **4**, 32 ⟨71CB1921⟩
Schossig, J., **1**, 399 ⟨69TL2619, 70LA(737)158⟩
Schössler, W., **6**, 143 ⟨71LA(748)1⟩, 143 ⟨74CB1931⟩
Schott, G., **4**, 569, 570 ⟨74ZAAC(404)204⟩; **6**, 164 ⟨74ZAAC(404)204⟩
Schott, P., **6**, 450 ⟨66CB3103⟩
Schotz, P., **3**, 534 ⟨23HCA291⟩
Schoufs, H. J. M., **1**, 441 ⟨92JOM(441)1⟩
Schoufs, M., **5**, 547, 552 ⟨78S439⟩
Schoufs, M. J. M., **4**, 93 ⟨81PS(11)139⟩
Schouteden, F., **5**, 833 ⟨58MAC246⟩
Schouteden, F. L. M., **5**, 833 ⟨57MAC25⟩
Schouten, A., **3**, 556 ⟨89AX(C)482⟩
Schow, S., **2**, 91 ⟨83JA7172⟩, 407, 862 ⟨75JCS(D)12⟩; **6**, 327 ⟨83JA7719⟩
Schow, S. R., **1**, 85 ⟨83JOC3370⟩, 754, 755 ⟨86JA2662⟩; **5**, 173 ⟨82JA4015⟩
Schowen, R. L., **4**, 229 ⟨69JOC2080⟩
Schrader, G., **2**, 409 ⟨64HOU(12/2)97⟩, 410 ⟨64HOU(12/2)104⟩, 412 ⟨64HOU(12/2)393⟩, 461 ⟨62BEP609076⟩, 463 ⟨57MIP1057112⟩, 464 ⟨54MIP1032247, 58MIP1067011⟩, 464 ⟨61GEP1104520⟩
Schrader, R., **5**, 379 ⟨88JOM(341)95⟩
Schrader, T., **1**, 1042 ⟨86S372⟩; **4**, 489 ⟨90S1153⟩, 1001 ⟨89S97⟩; **5**, 790 ⟨89S97⟩
Schramm, C. H., **2**, 191 ⟨55JA6231⟩
Schramm, S., **2**, 379 ⟨77JPR195⟩, 387 ⟨69LA(725)1⟩, 403 ⟨63AG(E)157⟩; **4**, 420 ⟨84JPR279⟩; **5**, 353 ⟨65TL1857, 67JPR86⟩
Schramm, V., **5**, 1064, 1066 ⟨88CB271⟩
Schrauder, O., **2**, 176 ⟨23HCA249⟩, 182 ⟨23HCA256⟩
Schrauzer, G. N., **1**, 335, 345 ⟨76AG(E)417⟩, 517 ⟨64CB2451⟩; **2**, 570, 573 ⟨82ZN(B)380, 86OM2452⟩, 578 ⟨68IS65⟩; **4**, 1016 ⟨76AG(E)171⟩
Schrecker, A. W., **5**, 194 ⟨54JA5803⟩
Schreiber, F. G., **1**, 1136 ⟨77JCS(P1)90⟩
Schreiber, J., **1**, 1082 ⟨67HCA2101, 71HCA2896⟩, 1082 ⟨72HCA1276, 76OS(55)114⟩; **2**, 380 ⟨88OSC(6)56⟩, 749 ⟨65CR(261)1992⟩; **3**, 26, 147 ⟨72HCA1276⟩, 228 ⟨53HCA482⟩; **5**, 78 ⟨60HCA113⟩, 131 ⟨63AG(E)211⟩, 749 ⟨86HCA1224⟩; **6**, 104 ⟨63AG296, 65HCA1746⟩
Schreiber, K., **1**, 2, 10 ⟨65LA(681)196⟩; **4**, 64, 69, 80 ⟨68JOC1080⟩
Schreiber, K. C., **2**, 168 ⟨61JOC2478⟩; **4**, 68 ⟨61JOC2478⟩
Schreiber, M., **2**, 783 ⟨92AG(E)1578⟩
Schreiber, S. L., **1**, 149 ⟨90JA697⟩, 199 ⟨86JA3128⟩, 199, 200 ⟨87JA5749⟩, 212 ⟨80JA6163⟩, 397 ⟨87TL1043⟩, 697, 698, 703, 1099 ⟨88JA5198⟩, 813 ⟨84JA4038⟩, 831 ⟨89JOC9⟩, 1045 ⟨84TL1757⟩, 1067 ⟨92JOC5060⟩, 1079 ⟨89JOC4267⟩; **3**, 92 ⟨91JOC1692⟩, 240 ⟨89JOC4721⟩; **5**, 625 ⟨88TL4085⟩
Schreier, E., **2**, 223 ⟨74JPR719⟩
Schreiner, E. P., **6**, 550 ⟨88M127⟩
Schreiner, L., **5**, 24 ⟨1879LA1⟩
Schreiner, S., **2**, 882 ⟨60CB1417⟩
Schrempf, G., **6**, 51 ⟨81ZAAC(474)7⟩
Schrensel, J., **2**, 193 ⟨51JOC767⟩
Schrepfer, J., **5**, 831 ⟨70CB82⟩
Schreuder, A. H., **1**, 47 ⟨87T3817⟩
Schreurs, A. M. M., **4**, 660 ⟨86OM1551⟩
Schrielsheim, A., **2**, 192 ⟨65TL2271⟩, 192, 720 ⟨63TL1131⟩

Schrier, D. J., **6**, 558, 564 ⟨93JFC(61)1⟩, 558, 564 ⟨93JMC1802⟩
Schriesheim, A., **1**, 475 ⟨64MI 111-01⟩, 590 ⟨65CI(L)1768⟩, 600, 612 ⟨63JA2739, 64JA1561⟩, 797 ⟨68JOC221⟩, 871, 881 ⟨62JA3164⟩, 882 ⟨65JA3244⟩, 894 ⟨55JA3051⟩
Schriewer, M., **2**, 947, 949 ⟨83JA897⟩
Schriffert, E. J., **6**, 363 ⟨73JOC1080⟩
Schritt, W., **4**, 416, 424 ⟨61JPR72⟩
Schrivastava, T. N., **5**, 341, 342 ⟨83IJC(A)503⟩
Schriver, G. W., **1**, 891 ⟨87MI 119-10⟩, 905 ⟨85MI 119-06⟩
Schrock, R. R., **1**, 663, 668 ⟨76CRV243⟩, 674, 675 ⟨85JA5987, 86JOM(300)249⟩, 675 ⟨87MM1169, 87MM2640⟩, 675 ⟨93JA4413⟩; **2**, 570 ⟨74JA5288, 76JOM(122)209⟩, 570 ⟨80JA6744⟩, 681 ⟨76JA5399, 86OM398⟩, 979 ⟨82JA6808⟩; **3**, 508 ⟨B-86MI 315-01, 91MI 315-01⟩, 510 ⟨78JA3793, 78JA5964⟩, 510, 511 ⟨78JA3359⟩, 510, 511, 512 ⟨74JA6796, 80JA6236⟩, 511 ⟨75JA6577, 78JA2389, 79JA3210, 80JA6744⟩, 512 ⟨79JOM(171)43, 80JA6608⟩, 512 ⟨81JA1440, 86OM2162, 88JA4964, 90IC1104⟩, 513 ⟨85JA5999, 87OM1373⟩, 513 ⟨90JA3875⟩, 514 ⟨78JA6774, 82JA7483, 90JA8378⟩, 515 ⟨90OM2262⟩, 516 ⟨85OM1937, 86JA2771⟩, 517 ⟨80JMOC73⟩, 518 ⟨82OM148⟩, 521 ⟨83OM1505, 87OM893, 88POL1841⟩, 522 ⟨92JA3367⟩, 690 ⟨86ACR342, 88MI 320-01⟩, 691 ⟨89MM2569⟩; **4**, 549 ⟨93AG(E)756⟩, 581 ⟨84OM476⟩; **5**, 918 ⟨80JA6744⟩
Schröck, W., **6**, 445 ⟨82MI 614-03⟩
Schrödel, R., **3**, 107 ⟨56AG438⟩
Schroder, C., **3**, 413 ⟨83JOC1925⟩
Schröder, E., **2**, 392 ⟨72JPR891⟩
Schröder, F.-W., **6**, 515 ⟨71AG(E)339⟩
Schröder, G., **1**, 789 ⟨64CB3140⟩; **2**, 47 ⟨88JA1968⟩; **6**, 469 ⟨92CB2493⟩
Schröder, J., **3**, 597 ⟨90CB2053⟩; **5**, 950, 951 ⟨90CB2053⟩; **6**, 410 ⟨66CB371⟩
Schröder, M., **2**, 44, 45 ⟨80CRV187⟩; **4**, 165 ⟨84JCS(P1)681⟩
Schroder, R., **3**, 694, 706, 714, 717 ⟨74LA44⟩, 706, 714 ⟨75LA533⟩, 709, 713 ⟨72LA(766)130⟩, 713, 717 ⟨69AG(E)672, 72LA(766)116⟩; **4**, 331, 908 ⟨72LA(766)130⟩, 487 ⟨73TL633, 72AG(E)311⟩, 1005 ⟨74LA44⟩
Schrodt, H., **4**, 881 ⟨56CB2060⟩; **6**, 73, 97, 300 ⟨56CB2060⟩
Schroeck, C. W., **1**, 304 ⟨73JA7418⟩, 710 ⟨73JA7424⟩; **2**, 151 ⟨73S485⟩, 362 ⟨71JA5305⟩; **3**, 378 ⟨71JA5303, 73JA7424⟩
Schroeder, D. C., **5**, 1022 ⟨55CRV181⟩; **6**, 569, 575 ⟨55CRV181⟩
Schroeder, F., **4**, 236 ⟨77LA1116⟩, 838 ⟨77LA1676⟩
Schroeder, G., **4**, 731 ⟨85CJC576⟩; **5**, 208 ⟨64BSF542⟩
Schroeder, H., **5**, 656 ⟨62JOC2585⟩, 852 ⟨58JOC1931⟩; **6**, 603 ⟨62JOC2589⟩, 651 ⟨88JPR900⟩
Schroeder, J. P., **4**, 880, 883 ⟨49JA40⟩
Schroeder, N. C., **6**, 334 ⟨85OM1226⟩
Schroeder, W., **3**, 464 ⟨54JOC718⟩
Schroer, R., **2**, 506, 852 ⟨68CB1381⟩
Schroer, T. E., **1**, 870 ⟨68CC1225⟩
Schröer, U., **2**, 478 ⟨76JOM(105)183⟩
Schroeter, G., **3**, 530 ⟨09CB2336, 09CB3356⟩, 530 ⟨16CB2697⟩
Schroeter, L. C., **4**, 237 ⟨B-66MI 405-01⟩
Schroeter, S. H., **5**, 166 ⟨69TL1591⟩
Schroff, L., **2**, 683 ⟨79GEP(O)2807762⟩
Schroll, A. L., **2**, 219, 230 ⟨89JOC244⟩; **6**, 434, 435, 474, 534, 550, 551, 567 ⟨83JOC4750⟩, 435, 480, 497, 548, 560 ⟨86JOC1866⟩, 548 ⟨90JOC1475⟩
Schroll, G., **1**, 457, 508 ⟨55JA3378⟩
Schromm, K., **6**, 73 ⟨65AG(E)873⟩
Schropl, E., **5**, 1047 ⟨68MI 527-01⟩
Schrot, J., **3**, 358 ⟨92BSF161⟩
Schrotch, W., **3**, 547 ⟨74ZC92⟩
Schröter, D., **3**, 683 ⟨85MI 320-01⟩
Schroth, G., **1**, 1001 ⟨75LA2130⟩; **2**, 1087 ⟨75LA2130⟩; **4**, 583 ⟨72IJ293⟩, 604, 647 ⟨87CB471⟩, 659 ⟨85CB4248⟩; **5**, 908 ⟨84T3273⟩
Schroth, W., **1**, 1072 ⟨69ZC229⟩; **2**, 1008 ⟨82AG(E)555⟩; **4**, 793, 970, 971 ⟨82S199⟩, 837, 860 ⟨82S206⟩; **5**, 600, 621 ⟨89S202⟩, 616 ⟨94TL1977⟩; **6**, 312 ⟨73ZC465⟩, 478 ⟨77ZC411⟩, 608 ⟨83JPR787⟩, 616 ⟨77ZC172⟩, 619 ⟨89LA931⟩
Schryver, F. C. de, **2**, 205 ⟨74TL4781⟩
Schubart, I., **2**, 484 ⟨32CB409⟩
Schubart, R., **2**, 357 ⟨85HOU(E11)68⟩; **6**, 236 ⟨85HOU(E11)63⟩
Schuber, F. J., **4**, 391 ⟨72JA8795⟩
Schubert, B., **1**, 1133 ⟨83CB3212⟩; **3**, 40 ⟨86JOC5294⟩
Schubert, D. M., **2**, 428 ⟨84IC4130⟩, 473, 474 ⟨85IC1107⟩
Schubert, F., **2**, 1001, 1002 ⟨93CB1887⟩
Schubert, G., **6**, 546 ⟨84ZAAC(515)182⟩
Schubert, H., **1**, 153 ⟨84T1345⟩, 738 ⟨70JOC2678⟩; **2**, 877 ⟨70JOC2678⟩; **3**, 179 ⟨68JPR12⟩, 476, 477 ⟨81BSB615, 82S149⟩; **4**, 889 ⟨83AG(E)553⟩; **6**, 651 ⟨74JPR434⟩
Schubert, H. H., **6**, 129, 167, 374 ⟨84JOC5087⟩
Schubert, H. J., **3**, 281 ⟨62LA(653)181⟩
Schubert, H. W., **4**, 462, 478 ⟨63LA(665)91⟩
Schubert, I., **4**, 556 ⟨86EGP238612⟩
Schubert, J., **5**, 954 ⟨93CB657⟩
Schubert, R., **4**, 690 ⟨93ZN(B)1558⟩
Schubert, R. M., **2**, 83 ⟨73JA8486⟩
Schubert, U., **1**, 991 ⟨89CB1237⟩; **2**, 447 ⟨80CB1612⟩; **3**, 508 ⟨B-83MI 315-01⟩, 545 ⟨89JOM(373)203⟩, 553 ⟨81AG(E)593, 83AG(E)981⟩, 553 ⟨85CB873⟩, 690 ⟨88MI 320-01⟩; **4**, 149, 583 ⟨88OM1106⟩, 576 ⟨80CB902⟩, 581 ⟨83CB690⟩, 668 ⟨82CC1184⟩, 695 ⟨82AG(E)73, 83JOM(246)141⟩; **5**, 652 ⟨79IS164⟩, 800 ⟨85CB1517⟩, 897 ⟨73CB1062⟩, 908, 910, 911 ⟨80CB902⟩, 936 ⟨77CB2574, 79CB708⟩, 951 ⟨81AG(E)1055⟩, 952 ⟨73JOM(51)C11⟩, 954 ⟨91JA2228, 93CB657⟩, 954 ⟨94JOM(472)C15⟩, 1092 ⟨81CB3070⟩; **6**, 151 ⟨79AG(E)484⟩, 324 ⟨85CB3151⟩, 553 ⟨85ZAAC(530)101⟩, 674, 723 ⟨90JOM(385)221⟩, 716, 719 ⟨89MI 622-03⟩, 716, 719, 722 ⟨84MI 617-04, 77CB2574⟩, 720, 723 ⟨86CB2900, 93JOM(459)55⟩, 723 ⟨86OM173, 88JOM(355)243⟩, 723 ⟨88JOM(358)215, 89JOM(373)203⟩
Schubert, W. M., **1**, 896 ⟨64JA4727, 66JA120⟩, 1110 ⟨70JA207⟩
Schubina, Yu. V., **4**, 793 ⟨84CHE1231⟩
Schucking, G., **3**, 231 ⟨43CB676⟩
Schuda, P. F., **3**, 195, 220 ⟨87S1055⟩, 278 ⟨87JOC1972⟩; **4**, 305 ⟨76JA6715⟩, 856 ⟨87JOC1972⟩; **5**, 199 ⟨82OS(61)147⟩
Schueler, F. W., **2**, 379 ⟨52JA3693⟩
Schuerch, Jr., C., **5**, 34, 43 ⟨48JA2824⟩
Schuetz, R. D., **6**, 425 ⟨67JOC300⟩
Schuetz, S., **6**, 360 ⟨76GEP2460330⟩
Schugar, H. J., **1**, 909 ⟨86JA134⟩
Schugart, K. A., **6**, 334 ⟨84OM1038⟩
Schuhknecht, C., **3**, 373 ⟨78S678⟩
Schuhmacher, R., **4**, 264 ⟨91CB2489⟩
Schuhn, W., **6**, 687 ⟨86TL1661⟩
Schuiji, P. J. W., **3**, 337 ⟨71IJS(B)85⟩; **5**, 554 ⟨69RTC597⟩
Schuiji-Laros, D., **3**, 337 ⟨71IJS(B)85⟩
Schuijl, P. J. W., **1**, 954 ⟨70QRS1⟩; **3**, 548 ⟨68RTC1236⟩, 548, 552 ⟨66RTC889, 68RTC38⟩; **4**, 834 ⟨68RTC929⟩, 843 ⟨69RTC597⟩; **5**, 550, 563 ⟨66RTC889⟩, 552 ⟨68RTC929⟩, 589, 590, 603 ⟨68RTC38⟩, 589, 614 ⟨68RTC1236⟩
Schülde, F., **6**, 302 ⟨73GEP1694210⟩
Schulenberg, J. W., **5**, 303 ⟨65OR(14)1⟩
Schuler, B., **3**, 334 ⟨89TL4111⟩, 334, 349, 365, 369 ⟨90CB177⟩; **4**, 266 ⟨90CB177⟩

Schuler, R. H., **1**, 923 ⟨65JCP(43)2704⟩
Schülke, U., **4**, 563 ⟨89ZAAC(576)272⟩; **6**, 158 ⟨90PS(51/52)153⟩
Schuller, H., **4**, 408 ⟨60AP(293)203⟩
Schulman, E. M., **2**, 22 ⟨77JA6672⟩; **3**, 152 ⟨77JA6672⟩
Schulmann, E. M., **3**, 369 ⟨92JOC2763⟩
Schulte, C., **2**, 890 ⟨1882CB1952⟩
Schulte, G., **1**, 176 ⟨93JA2581⟩, 826 ⟨88TL1669⟩; **3**, 243 ⟨88JA7434⟩, 442 ⟨85JOC4650⟩; **5**, 572 ⟨90JOC831⟩
Schulte, G. K., **3**, 544 ⟨89JA8738⟩
Schulte, P., **6**, 693 ⟨88PS(36)147⟩
Schulte-Elte, K. H., **1**, 883 ⟨87HCA1858, 87HCA1879⟩, 1082 ⟨67HCA2101⟩
Schulte-Frohlinde, D., **4**, 1078 ⟨80JCS(P2)883, 81JCS(P2)143⟩, 1078 ⟨83AG(E)500⟩
Schulten, H., **3**, 547 ⟨33CB237⟩; **6**, 524 ⟨39ZAAC(240)261⟩
Schulten, W., **4**, 141, 154, 608, 616 ⟨80JOM(201)197⟩
Schultess, A. H., **5**, 696 ⟨81HCA1322⟩
Schultheiss, H., **4**, ⟨74M684⟩
Schultz, A. G., **1**, 303 ⟨76JOC4044⟩, 556 ⟨93TL3021⟩, 693 ⟨78JA2140⟩, 802 ⟨84TL1255, 85JOC915⟩; **2**, 1034 ⟨81TL3347⟩; **3**, 88 ⟨87JOC3905⟩, 303 ⟨86JA1056⟩; **4**, 167 ⟨75JOC1371⟩, 212 ⟨77JOC3458⟩, 513 ⟨92JOC6037⟩
Schultz, A. J., **2**, 602 ⟨86JA40⟩; **4**, 656, 658 ⟨86JA40⟩, 673 ⟨78IC2093⟩
Schultz, C. C., **5**, 957 ⟨94JA9363⟩
Schultz, E. M., **1**, 34 ⟨53JA1128⟩
Schultz, F. J., **2**, 787 ⟨60JOC1877⟩
Schultz, F. W., **1**, 543 ⟨85JOC2124⟩
Schultz, H., **4**, 477, 481, 503 ⟨80ZC152⟩
Schultz, H. S., **1**, 108 ⟨55JA5533⟩; **2**, 156 ⟨63JOC1140⟩
Schultz, M., **1**, 556 ⟨90T2371⟩
Schultz, P. G., **1**, 826 ⟨88SCI426⟩; **2**, 820 ⟨82JA766⟩
Schultz, W., **6**, 143 ⟨82CB1721⟩
Schultze, A., **2**, 881 ⟨75MI 217-02⟩
Schultze, G. R., **6**, 44, 47 ⟨59LA(621)8⟩, 234 ⟨56CB1160⟩
Schultze, K. M. L., **6**, 84 ⟨11CB3235⟩
Schultze, W. A., **2**, 166 ⟨38USP2123082⟩
Schulz, B., **4**, 989 ⟨86TL4281⟩
Schulz, D., **4**, 806 ⟨90LA745⟩
Schulz, D. J., **1**, 343, 931 ⟨93TL6787⟩
Schulz, G., **3**, 64 ⟨66CB3057⟩; **4**, 983 ⟨77TL3619⟩; **6**, 460, 462 ⟨B-91MI 615-01⟩, 471 ⟨B-92MI 615-01⟩
Schulz, H., **2**, 903 ⟨90CB2273, 93ZN(B)719⟩; **3**, 85 ⟨59CB530⟩, 371 ⟨68CB277⟩, 493 ⟨61AG27⟩; **4**, 602, 607, 637, 660 ⟨90ZAAC(583)195⟩
Schulz, J., **6**, 153 ⟨92AG(E)879⟩
Schulz, J. G. D., **4**, 993 ⟨76MI 421-02⟩
Schulz, M., **2**, 374 ⟨74JPR347⟩, 659 ⟨72JPR170⟩; **3**, 599 ⟨91CB285⟩; **4**, 149, 583 ⟨91JOM(417)149⟩, 989 ⟨82JPR309⟩; **5**, 357 ⟨74ZC265⟩, 365 ⟨87ZC27⟩, 944 ⟨91CB1121⟩
Schulz, R., **3**, 353 ⟨81AG(E)570⟩, 358 ⟨82JOC1979⟩, 542 ⟨80TL343⟩, 543 ⟨79TL59⟩, 553 ⟨80AG(E)69⟩
Schulz, S., **2**, 529 ⟨80JOM(186)297⟩
Schulz, W., **6**, 444 ⟨75LA2227⟩
Schulze, A., **2**, 490 ⟨75PS(5)265⟩; **6**, 559 ⟨82CB1252, 82CB1259⟩
Schulze, E., **4**, 536 ⟨88AG(E)1194⟩
Schulze, F., **5**, 112 ⟨43OSC(2)425⟩
Schulze, G., **2**, 365 ⟨61GEP1156403⟩, 367, 368 ⟨69LA(729)40⟩
Schulze, J., **1**, 916 ⟨64JA322, 64JA327⟩, 916 ⟨64JA331⟩; **6**, 678 ⟨86PS(28)239⟩
Schulze, L., **5**, 638 ⟨73LA207⟩
Schulze, M., **6**, 477 ⟨93GEP4140446⟩
Schulze, T., **5**, 164, 168 ⟨88JOC4098⟩
Schulze, W., **2**, 381 ⟨66JPR96⟩; **3**, 380 ⟨89ZAAC(578)110⟩; **4**, 893 ⟨67CB3460⟩
Schulze-Steinen, H. J., **5**, 338 ⟨63JCS4918⟩
Schulze-Weisschu, P., **6**, 564 ⟨81AP(314)587⟩
Schumacher, D. P., **2**, 19 ⟨90JOC5291⟩
Schumacher, G., **3**, 363 ⟨67TL2963⟩
Schumacher, H., **3**, 478 ⟨67AG50, 67AG(E)74⟩, 478 ⟨68LA(716)204⟩; **4**, 142 ⟨67AG(E)74⟩; **5**, 844, 866, 874 ⟨68LA(716)204⟩
Schumacher, H. J., **6**, 409 ⟨51MI 614-01⟩, 418 ⟨37CB1080⟩
Schumacher, O., **2**, 440, 442, 446 ⟨62LA(659)49⟩, 447 ⟨61LA(640)79⟩; **3**, 492 ⟨61LA(640)79⟩; **4**, 112 ⟨62LA(659)49⟩
Schumacher, W., **1**, 204, 1023 ⟨78S26⟩, 204, 1023 ⟨81S490⟩
Schumaker, R. R., **4**, 869, 874 ⟨77CC835⟩; **6**, 590 ⟨83CC235⟩
Schumann, D., **3**, 413 ⟨91SL916⟩, 593 ⟨74T2553⟩; **5**, 742 ⟨85HOU(E5)1305⟩, 850 ⟨85HOU(E5)1304⟩
Schumann, E. L., **2**, 103 ⟨60MI 202-03⟩
Schumann, H., **1**, 936 ⟨81AG(E)1043⟩; **2**, 434 ⟨87GEP3612629⟩, 449 ⟨91ZN(B)887⟩, 473, 474 ⟨70JOM(21)411, 74CB854⟩, 478 ⟨77JCR(S)272⟩, 483 ⟨75JOM(87)83⟩, 509 ⟨89JOM(373)307⟩, 564 ⟨86OM1296⟩, 809 ⟨69CB3731⟩, 867 ⟨77JCR(M)3101⟩, 867, 868 ⟨64CB2395, 69CB2900⟩, 868 ⟨80JOM(190)53⟩, 895 ⟨67JOM(8)105⟩, 969 ⟨85JA8103, 85JA8111⟩, 4744 ⟨70JOM(21)351⟩; **4**, 577 ⟨85JOM(297)C30⟩, 578 ⟨76CZ336, 82JOM(235)287⟩, 578 ⟨83JOM(255)305, 84AG(E)522⟩, 578 ⟨84JOM(269)21⟩, 708, 726 ⟨84ZN(B)798⟩; **6**, 194 ⟨85JA8103⟩, 195 ⟨85JA8091⟩, 706 ⟨81JOM(209)C10⟩
Schumann, H. D., **2**, 291 ⟨60ZAAC(305)88⟩
Schumann, P., **6**, 538 ⟨85EGP256705⟩
Schumann, R. L., **2**, 533 ⟨89TL1293⟩
Schümann, U., **2**, 1004 ⟨85CB3529⟩
Schumb, W. C., **6**, 212 ⟨59JA806⟩
Schümmelfeder, B., **6**, 625, 632 ⟨91S753⟩
Schummer, D., **1**, 16, 17 ⟨90SL705⟩
Schunach, W., **6**, 632 ⟨88AF7⟩
Schunack, W., **4**, 970 ⟨82AP(315)680⟩
Schunk, E., **5**, 29, 34, 43 ⟨67BSF2011⟩
Schunn, R. A., **4**, 576, 577 ⟨63JA2394⟩; **5**, 1139, 1140 ⟨76IC208⟩
Schunn, W., **6**, 687 ⟨88USP4754072⟩
Schupak, G. M., **6**, 233 ⟨79ZOR396⟩
Schure, R. M., **6**, 25 ⟨67JOC1217⟩
Schurig, V., **2**, 12 ⟨87OS(66)151⟩; **5**, 60 ⟨87OS(66)151⟩
Schurz, K., **6**, 163 ⟨87CB1357⟩, 673 ⟨86CC591, 88CB1407⟩, 709 ⟨87CB1605⟩, 709, 710 ⟨88ZN(B)1468⟩
Schuster, E., **3**, 413 ⟨91SL916⟩; **4**, 793 ⟨67CB1087⟩
Schuster, G. B., **3**, 565 ⟨92MI 317-01⟩
Schuster, H. F., **1**, 459 ⟨B-84MI 110-02⟩, 954 ⟨B-84MI 120-02⟩
Schuster, I. I., **2**, 949 ⟨86JA6664⟩
Schuster, M., **5**, 263 ⟨93TL1247⟩
Schuster, R. E., **3**, 696 ⟨63JCS4280⟩, 698 ⟨66JOC3473⟩
Schuster, T., **1**, 983 ⟨92JA5998⟩
Schut, J., **6**, 643 ⟨69JMC712⟩
Schütt, W., **5**, 955 ⟨92CB1571⟩
Schutte, L., **4**, 245 ⟨71TL2321⟩
Schuttler, R., **3**, 541 ⟨75CB844⟩
Schutyser, J., **3**, 375 ⟨74JA2964⟩
Schutz, B. A., **6**, 570, 571 ⟨63JOC3140⟩
Schütz, F., **4**, 881 ⟨85S1062⟩
Schütz, J., **5**, 339, 689 ⟨80LA1064⟩, 702 ⟨80LA1055⟩
Schütz, K., **3**, 330 ⟨83HCA801⟩
Schutz, M., **1**, 686 ⟨84CB277⟩
Schütz, M., **3**, 348 ⟨84CB277⟩
Schütz, U., **6**, 197 ⟨93CB2637⟩
Schützenberger, M. D., **6**, 412 ⟨1869BSF198⟩
Schwab, J. M., **1**, 797 ⟨90CR1203⟩, 971 ⟨85JA6046⟩
Schwab, M., **2**, 209 ⟨88CB75⟩; **5**, 340 ⟨86CB1127⟩, 358 ⟨88CB75⟩

Schwab, P., **3**, 521 ⟨92CB2641, 93AG(E)1480⟩
Schwab, P. A., **3**, 536 ⟨64JA4866⟩; **4**, 50 ⟨63JA4004⟩
Schwab, S. T., **2**, 511 ⟨86CC1543⟩, 512 ⟨86JA1703⟩
Schwab, W., **2**, 392 ⟨86S535⟩; **3**, 46 ⟨86S535⟩; **6**, 417 ⟨80LA122⟩
Schwabe, P., **2**, 867, 868 ⟨69CB2900⟩
Schwager, H., **1**, 248 ⟨87CB79⟩; **2**, 984 ⟨86AG(E)640⟩; **4**, 604, 647 ⟨87CB471⟩
Schwager, I., **4**, 112 ⟨76IC1697⟩
Schwalbe, C. H., **2**, 862 ⟨91JCS(P2)2081⟩
Schwall, H., **2**, 799 ⟨69LA(728)99⟩
Schwalm, W. J., **3**, 343, 349 ⟨69CJC3715⟩
Schwam, H., **5**, 972 ⟨66JA3163⟩; **6**, 496 ⟨68JA3254⟩
Schwamborn, M., **2**, 471 ⟨77ZN(B)1244⟩
Schwan, A. L., **2**, 174 ⟨92TL3973⟩
Schwaneberg, H., **3**, 612 ⟨34CB39⟩
Schwark, J. R., **4**, 382 ⟨92T8377⟩
Schwarm, M., **3**, 40 ⟨93JOC3568⟩
Schwarts, J., **1**, 669 ⟨74JOM(81)C37⟩
Schwarttz, A., **2**, 784 ⟨71HOU(10/1)9⟩
Schwartz, A., **1**, 806 ⟨80JOC1172⟩; **3**, 436 ⟨86JOC5024⟩
Schwartz, C. E., **3**, 90, 96 ⟨92TL7499⟩
Schwartz, E., **3**, 331 ⟨86JOC1556⟩
Schwartz, J., **1**, 94 ⟨81JA5253, 81JMOC397⟩, 204, 1023, 1139 ⟨75TL51⟩, 318 ⟨80JA1333⟩, 491, 536 ⟨77TL1303, 80JA7381⟩, 536 ⟨81TL2629, 81TL4655⟩, 536, 1105 ⟨77JA8045⟩, 568 ⟨92TL6783⟩, 669 ⟨74JA8115⟩, 683 ⟨89JA3069⟩, 1049, 1139 ⟨78JA2244, 75JA679⟩, 1105 ⟨78TL4381, 81JA4466⟩, 1105, 1139 ⟨80TL1497, 80JOC3053⟩; **2**, 4 ⟨74JA8115⟩, 4, 568, 969 ⟨76AG(E)333⟩, 365, 607, 624 ⟨86JA2445⟩, 616, 623, 629, 968 ⟨75JA679⟩, 676, 982 ⟨82JA5842⟩, 681 ⟨83JA640, 84JA8300⟩, 969 ⟨78TL4381, 80JA1333⟩, 969, 991, 992 ⟨77JA638⟩, 969, 991, 992 ⟨79JA3521⟩, 969, 992 ⟨77JA8045⟩; **3**, 23 ⟨76JA262⟩, 140 ⟨80JA1333⟩, 147 ⟨80JOC3053⟩, 173 ⟨80TL1497⟩, 509 ⟨80JOM(184)C1, 83JA640⟩; **4**, 579 ⟨80IC3207⟩, 683 ⟨72CC814⟩, 905 ⟨90JMC1510⟩; **5**, 170 ⟨75TL51⟩, 441 ⟨75JA228⟩, 442 ⟨76JA262⟩, 460 ⟨72JA6226⟩, 467 ⟨79JA3521⟩, 917 ⟨80IC3207⟩
Schwartz, J. A., **2**, 271 ⟨86SC565⟩; **3**, 291 ⟨78JOC985⟩
Schwartz, J. L., **1**, 834 ⟨74JOC1575⟩
Schwartz, K.-H., **4**, 889, 972 ⟨66CB2302⟩
Schwartz, M. A., **1**, 174 ⟨65JA3277⟩; **2**, 345 ⟨92TL1689⟩; **3**, 625 ⟨76JOC2502⟩
Schwartz, N., **1**, 168 ⟨63IC1111⟩
Schwartz, N. V., **2**, 221, 226 ⟨69JOC3991⟩; **5**, 431 ⟨60JA5102⟩
Schwartz, R., **5**, 1024 ⟨58JA3332⟩
Schwartz, R. H., **1**, 451, 1105 ⟨79JOC155⟩; **2**, 923 ⟨79JOC155⟩; **4**, 357 ⟨77TL591, 79JOC155⟩
Schwartz, S. B., **1**, 785 ⟨73CC374⟩
Schwartz, S. J., **4**, 934, 949 ⟨76TL2201⟩, 935, 949 ⟨78JOM(156)123⟩
Schwartz, S. T., **2**, 906 ⟨76TL2201⟩
Schwartz, U., **5**, 148 ⟨80AG(E)1029⟩
Schwartze, P., **1**, 755 ⟨86CB2150⟩; **4**, 577, 578, 579, 581 ⟨86CB2150⟩
Schwartzenbach, G., **4**, 129, 473 ⟨49HCA1175⟩
Schwartzentruber, K., **3**, 706 ⟨90JCS(P1)2115⟩
Schwartzman, L. H., **3**, 86 ⟨59JOC627⟩
Schwartzman, S. M., **3**, 618 ⟨75TL277⟩; **5**, 123, 280, 291 ⟨75TL277⟩
Schwarz, A., **4**, 55, 57 ⟨80LA1919⟩; **5**, 547 ⟨11CB1693⟩
Schwarz, B., **4**, 281 ⟨91LA243⟩
Schwarz, G., **5**, 993 ⟨77AG(E)550⟩; **6**, 538 ⟨85DIS(B)840⟩
Schwarz, H., **1**, 95 ⟨87JOC2804⟩, 552 ⟨89AG(E)1321⟩, 1122 ⟨81JA2770⟩; **2**, 1022 ⟨86AG(E)282⟩; **3**, 475 ⟨71LA(747)21⟩; **6**, 239 ⟨70GEP1926659⟩, 240 ⟨76CA(85)142668⟩
Schwarz, H. H., **2**, 391, 392, 394 ⟨71HOU(10/1)12⟩
Schwarz, K.-H., **5**, 161 ⟨93AG(E)1191⟩
Schwarz, M., **1**, 652 ⟨93TL1593⟩, 1000 ⟨72S567⟩; **2**, 10 ⟨75JOC2410⟩
Schwarz, M. W., **2**, 409, 410 ⟨84TL5513⟩
Schwarz, R., **3**, 251 ⟨87TL4267⟩
Schwarz, W., **2**, 1082 ⟨81ZAAC(474)157⟩, 1098 ⟨78JOM(159)373⟩; **5**, 396 ⟨92ZAAC(608)33⟩, 396, 879, 891 ⟨91ZAAC(605)7⟩, 776 ⟨86ZAAC(533)55⟩, 777 ⟨92ZN(B)453⟩, 780 ⟨79JOM(175)33⟩, 880 ⟨92ZAAC(612)72⟩, 1082 ⟨86ZN(B)1223⟩; **6**, 195 ⟨93ZAAC(619)675⟩, 393 ⟨94JOM(469)135⟩, 395 ⟨92AG(E)1364⟩, 438 ⟨82ZN(B)81⟩
Schwarze, W., **4**, 100 ⟨87CZ149⟩
Schwarzenbach, D., **6**, 695 ⟨76CAR(46)9⟩
Schwarzenbach, F., **2**, 74 ⟨92JA2321⟩
Schwarzenbach, G., **4**, 554 ⟨50M202⟩
Schwarzenbach, K., **2**, 806 ⟨59HCA739⟩; **4**, 534, 586 ⟨61LA(650)1⟩
Schwarzer, R., **2**, 482, 510 ⟨67LA(709)248⟩; **6**, 513 ⟨67LA(709)248⟩, 657 ⟨68JOM(13)363⟩
Schwechten, H.-W., **2**, 627 ⟨32CB1605⟩
Schwede, W., **1**, 1070 ⟨93TL5257⟩
Schweig, A., **3**, 353 ⟨81AG(E)570⟩, 358 ⟨82JOC1979⟩, 359 ⟨89JST(198)307⟩, 542 ⟨80TL343⟩, 543 ⟨79TL59⟩, 553 ⟨80AG(E)69⟩
Schweiger, E. J., **1**, 617, 709 ⟨88JA4866⟩
Schweir, J. R., **2**, 516, 523 ⟨78JOC4395⟩
Schweiser, E. E., **3**, 493 ⟨66CC666⟩, 494 ⟨68JOC3082⟩, 496 ⟨64JA2744, 66JOC467⟩, 496 ⟨68JOC336⟩
Schweiter, U., **3**, 71 ⟨59HCA844⟩
Schweitzer, R., **6**, 616, 617 ⟨75CB2290⟩, 620 ⟨76CB1643, 76CB2921⟩
Schweizer, B., **1**, 797 ⟨93HCA418⟩
Schweizer, D., **2**, 788 ⟨71CB3475⟩; **4**, 881 ⟨71CB3475⟩
Schweizer, E., **2**, 445 ⟨68JOC2590⟩
Schweizer, E. E., **1**, 90 ⟨66JOC870⟩, 267 ⟨78JOC4207⟩, 268 ⟨71JOC2379⟩, 625 ⟨60JA4663⟩, 625, 783 ⟨57JA4729⟩, 734 ⟨66JOC2907⟩; **2**, 446 ⟨82JOC2768⟩; **3**, 494 ⟨76JOC1168⟩; **4**, 7, 16, 24 ⟨63OR(13)55⟩; **5**, 862 ⟨94S1057⟩
Schweizer, E. H., **2**, 29 ⟨62CB803⟩
Schweizer, I., **4**, 686 ⟨78ZN(B)911, 79JOM(165)C17⟩
Schweizer, P., **2**, 1086 ⟨67CB2159⟩
Schweizer, W. B., **1**, 873 ⟨81HCA2617, 85JA5403⟩; **2**, 999 ⟨84HCA224⟩; **4**, 302 ⟨83JA5390⟩; **5**, 430 ⟨89HCA264⟩
Schwellnus, K., **1**, 189, 200 ⟨84CB322⟩; **4**, 202 ⟨84TL511⟩
Schwemlein, H. P., **4**, 682 ⟨89AG(E)1296⟩
Schweng, J., **3**, 585, 586 ⟨76M537⟩
Schwengers, D., **4**, 698 ⟨77LA1633⟩
Schwenk, E., **5**, 36 ⟨48JA3091⟩
Schwenker, G., **5**, 760 ⟨69T5437⟩; **6**, 143 ⟨75CB1142⟩
Schwenkkraus, P., **2**, 248 ⟨82TL5389⟩
Schwerdtfeger, J., **1**, 138 ⟨92AG(E)1505⟩
Schwering, J. E., **2**, 160, 162 ⟨84JMC1634⟩; **4**, 279 ⟨90TL6819⟩; **5**, 342 ⟨86JHC1253⟩
Schwerzel, R. E., **2**, 687 ⟨72JOC3096⟩
Schwerzel, T., **3**, 432 ⟨85CB1105⟩
Schwesinger, R., **1**, 574 ⟨87AG(E)1167⟩, 574, 584 ⟨91AG(E)1372⟩
Schwetlick, K., **3**, 366 ⟨74ZC17⟩; **5**, 984 ⟨77JPR841, 79JPR928⟩
Schwickardi, M., **2**, 561 ⟨82AG(E)199⟩
Schwickardi, R., **4**, 720 ⟨90ZAAC(581)41⟩
Schwickerath, W., **1**, 1044 ⟨84LA85⟩; **2**, 1044 ⟨78TL1951, 81LA1844⟩, 1044 ⟨83LA1185⟩; **5**, 304 ⟨81LA1844⟩, 596, 618, 739 ⟨78TL1951⟩; **6**, 620, 651 ⟨82LA2105⟩
Schwier, J. R., **2**, 515 ⟨79JOC465⟩, 523 ⟨82JOC5069⟩
Schwiersch, W., **5**, 854 ⟨61CB1868⟩
Schwille, D., **3**, 239 ⟨69LA(723)111⟩; **5**, 634 ⟨69CB269⟩
Schwindeman, J., **1**, 422 ⟨79TL1995, 80JOC1046⟩; **3**, 211 ⟨79TL1995⟩; **4**, 369, 938, 942 ⟨80JOC1046⟩
Schwinden, M. D., **4**, 1079 ⟨90TL1819⟩
Schwindt, M. A., **1**, 165 ⟨90JA2264⟩

Schwobel, A., **2**, 793 ⟨83SR215⟩
Schwochau, M., **3**, 273 ⟨64CB1649⟩
Schwoebel, A., **5**, 341 ⟨88PS(35)1⟩
Schwökel, A., **2**, 360 ⟨82TL1243⟩
Schwotzer, W., **4**, 675 ⟨83OM1167⟩
Schwyzer, R., **5**, 32 ⟨61HCA1991⟩, 83 ⟨60MI 502-02⟩
Scialdone, M. A., **1**, 668 ⟨90JA6729⟩
Sciavovelli, O., **2**, 740 ⟨91S225⟩; **4**, 476 ⟨91S225⟩
Scilimati, A., **2**, 56 ⟨87T2181, 91S225⟩; **4**, 476 ⟨91S225⟩, 776 ⟨92TL5121⟩; **5**, 179 ⟨88TL5583⟩
Sclare, A. D., **2**, 140 ⟨91JCS(P1)2329⟩; **3**, 333 ⟨91JCS(P1)2281⟩
Scola, D. A., **5**, 298 ⟨60JOC560⟩
Scola, P. M., **2**, 309 ⟨89TL5709⟩; **3**, 417 ⟨89CR1525⟩
Scolastico, C., **2**, 46 ⟨88T491⟩; **3**, 35 ⟨93JOC3165⟩; **4**, 856 ⟨81JCS(P1)1278⟩
Scollar, M. P., **5**, 135 ⟨85JA7072⟩
Scopes, D. I. C., **1**, 218 ⟨85TL97⟩; **4**, 490 ⟨77JOC376⟩; **6**, 113 ⟨76JOC1112⟩
Scorini, S. J., **6**, 537 ⟨80T539⟩
Scorrano, G., **1**, 1123 ⟨65TL4399, 65TL4405⟩; **2**, 240 ⟨94TL301⟩, 1034 ⟨66JCS(B)1188⟩; **4**, 162 ⟨91TL6215⟩, 304 ⟨84G405⟩; **5**, 322 ⟨94JA916⟩
Scott, A. B., **2**, 507 ⟨28JA536⟩
Scott, A. P., **5**, 317 ⟨90T7247⟩
Scott, B., **4**, 803 ⟨90JA3152⟩
Scott, B. A., **4**, 869, 874 ⟨77JA5909⟩; **6**, 590 ⟨77JA5909⟩
Scott, C. B., **1**, 107 ⟨53JA246⟩, 646 ⟨57JOC1118⟩
Scott, C. G., **2**, 39 ⟨81HCA1158⟩
Scott, D. M., **1**, 326 ⟨89JA8311, 90TL1679⟩, 326 ⟨91JA1791⟩
Scott, E. W., **6**, 653 ⟨43JA2390⟩
Scott, F., **1**, 465 ⟨77TL3263⟩, 1092 ⟨78TL3711⟩; **6**, 723 ⟨90CC1722⟩
Scott, F. L., **3**, 681 ⟨72JCS(P2)44⟩, 710 ⟨70TL4079⟩; **5**, 356 ⟨68JOC753⟩, 356, 703, 845 ⟨65TL841⟩, 358 ⟨73JCS(P2)1422⟩, 703, 704 ⟨71CC667⟩, 840 ⟨65TL3801⟩, 841 ⟨57JOC692, 70JCS(C)968⟩, 846 ⟨68TL2463⟩, 857 ⟨53JA5309⟩; **6**, 611 ⟨64CI(L)1757⟩, 623, 627 ⟨72JCS(P2)1050⟩, 641 ⟨53JA4053⟩
Scott, J. W., **2**, 822, 841 ⟨79JOC3741⟩
Scott, K. R., **5**, 332, 333 ⟨91JMC387⟩
Scott, L. B., **4**, 435, 443 ⟨51JA2769⟩
Scott, L. T., **1**, 1003 ⟨85JA6546⟩, 1019 ⟨90JA4054⟩; **2**, 435 ⟨90JA7823⟩; **3**, 232 ⟨73CC882⟩
Scott, M. E., **1**, 1018 ⟨91TL6085⟩; **2**, 915 ⟨91TL5047⟩; **3**, 540 ⟨93TL2071⟩; **4**, 962 ⟨91TL6085⟩, 962, 963 ⟨91TL5047⟩
Scott, M. J., **4**, 603 ⟨65JOM(4)98⟩; **6**, 173 ⟨63JA2243⟩, 173, 378 ⟨64JOC953⟩, 173, 378 ⟨65JOM(4)98⟩
Scott, R., **4**, 295, 407, 412 ⟨73JOC3288⟩
Scott, R. B., **2**, 175 ⟨55JOC1165⟩, 179 ⟨66JOC2357⟩, 273 ⟨54JOC1165⟩
Scott, R. M., **1**, 399 ⟨70CC576⟩; **6**, 7 ⟨87JCS(P1)1515⟩
Scott, R. W., **3**, 618 ⟨44OSC(2)379⟩
Scott, T. L., **3**, 474 ⟨77JOC3757⟩
Scott, W., **2**, 56, 421 ⟨88TL139⟩
Scott, W. J., **1**, 435, 480, 1027 ⟨86JA3033⟩, 465 ⟨93JOC4866⟩, 471 ⟨87CC1755⟩, 480 ⟨88ACR47, 90OS(68)116⟩, 480, 1027 ⟨80TL4313⟩, 494, 1029 ⟨85JOC2302⟩, 584 ⟨84JA4630, 92JA1491⟩, 1029 ⟨84JA7500⟩, 1071 ⟨91JOC2045⟩; **2**, 689 ⟨83TL979, 84JA4630⟩, 689 ⟨85S335⟩, 931 ⟨84JA7500, 86JA3033⟩; **3**, 271 ⟨84JA7500⟩; **4**, 181 ⟨93JOC7274⟩
Scottano, G., **4**, 1075 ⟨87MI 424-01⟩
Scotti, C., **3**, 477 ⟨65CB623⟩
Scotti, F., **2**, 749 ⟨64JOC1800⟩; **3**, 658 ⟨64JOC1800⟩
Scotton, M., **2**, 316 ⟨92JCS(P1)3363⟩
Scouten, C. G., **1**, 510, 1101, 1103 ⟨79JA96⟩, 906 ⟨89MI 119-09⟩; **2**, 41 ⟨73JOC4092⟩, 900, 901, 906 ⟨79JA96⟩; **4**, 630, 631 ⟨79JA96, 82JA531⟩
Scouten, W. H., **4**, 144 ⟨92JOM(431)255⟩
Scouter, C. G., **2**, 517 ⟨76JA5297⟩
Scovill, J. P., **5**, 1056 ⟨81EJM317⟩
Screttas, C., **2**, 434 ⟨62JOC2573⟩
Screttas, C. G., **2**, 551 ⟨78JOC1064⟩, 964 ⟨70C109⟩; **5**, 43, 110 ⟨72CC752⟩
Scriber, R. M., **2**, 805 ⟨64JOC3429⟩
Scribner, R. M., **1**, 447 ⟨55JOC9⟩; **2**, 19 ⟨74OR(21)1⟩, 112 ⟨55JOC9⟩; **4**, 8 ⟨74OR(21)1⟩
Scriven, C. E., **1**, 248 ⟨86JCS(P1)475⟩
Scriven, E. F. V., **2**, 798 ⟨72CC641⟩; **4**, 310 ⟨94H(37)1951⟩
Scrowston, R. M., **3**, 306 ⟨73JCS(P1)1196, 76JCS(P1)749⟩; **5**, 1111 ⟨80JCS(P1)1029⟩, 1112 ⟨71JCS(C)3262⟩, 1123 ⟨92T7703⟩
Scudder, E. D., **3**, 400 ⟨31CB530⟩
Scudder, M. L., **1**, 685, 686 ⟨93JOC5900⟩
Scudder, P. H., **1**, 801 ⟨81JOC506⟩; **4**, 99 ⟨73JA7813⟩
Scullard, P. W., **3**, 663 ⟨68JA4924⟩
Scully, C. N., **2**, 410 ⟨50JA5491⟩
Scully, Jr., F. E., **3**, 407, 411 ⟨78JOC1467, 80JOC1515⟩; **6**, 559 ⟨89JOC2978⟩
Scully, M. F., **2**, 530 ⟨92TL4209⟩; **4**, 526, 527, 528 ⟨92TL4209⟩
Scuotto, J.-F., **3**, 279 ⟨85BSB897⟩
Scuseria, G., **5**, 1020 ⟨90JA4183⟩
Sealfon, S., **1**, 524, 525 ⟨79JA5283⟩
Seapy, D. G., **2**, 26 ⟨82JOC132⟩; **6**, 550 ⟨82JOC132⟩
Searle, C. E., **2**, 118 ⟨50JCS3309⟩
Searle, H. T., **2**, 407 ⟨61JCS1825⟩
Searle, N. E., **3**, 468 ⟨63OSC(4)424⟩
Searle, R. J. G., **1**, 1079 ⟨65JOC1015⟩; **3**, 274 ⟨65JOC1015⟩; **4**, 476 ⟨66JCS(B)789⟩; **5**, 405 ⟨73PS287⟩
Searles, Jr., S., **1**, 417, 829 ⟨49JA1150⟩, 623 ⟨65TL2899⟩
Searles, S., **1**, 785 ⟨84CHEC(7)363⟩, 1122 ⟨44JA686⟩; **3**, 541 ⟨66JA1242⟩
Sears, A. B., **1**, 74 ⟨73JA4346⟩, 795, 801 ⟨72MI 118-01⟩
Sears, C. T., **1**, 73 ⟨68IS101⟩
Sears, D. S., **5**, 1017 ⟨59USP2873171⟩
Sears, Jr., C. T., **4**, 683 ⟨68JOM(11)644⟩; **5**, 460 ⟨68IC27⟩
Seaton, J. A., **2**, 388 ⟨58JA2516⟩
Sebald, A., **2**, 1088 ⟨92CB1597⟩, 1097 ⟨86SA(A)1107⟩, 1100 ⟨88G1⟩; **4**, 1054, 1063 ⟨86JOM(307)157⟩, 1062 ⟨86JOM(303)73⟩
Sebastian, J. F., **4**, 949, 950 ⟨80JOC4959⟩, 949, 950 ⟨82JOC3094⟩
Sebastian, M., **2**, 428 ⟨90ZAAC(589)187⟩
Sebastiani, G. V., **1**, 290 ⟨92CC59⟩
Sebe, E., **6**, 207 ⟨71MI 606-01⟩
Sebenda, J., **5**, 997 ⟨73CZP148911⟩
Sebok, P., **2**, 624 ⟨91TL827⟩
Secane, P. R., **1**, 193, 288 ⟨89JA7487⟩
Secchi, M., **5**, 226 ⟨85TL1857⟩
Secci, D., **1**, 988 ⟨87G625⟩
Secci, M., **1**, 738 ⟨64MI 116-02⟩; **2**, 147 ⟨61TL607⟩; **4**, 231 ⟨76S797⟩
Secco, F., **6**, 341 ⟨81AG215⟩
Sechi, B., **2**, 553 ⟨94SC591⟩
Seck, M., **3**, 632 ⟨89T3731⟩
Seckar, J. A., **3**, 698 ⟨76IC501⟩
Seckinger, K., **6**, 143 ⟨80HCA1958⟩
Seco, M., **2**, 1027 ⟨88CC909⟩; **5**, 1137 ⟨92ICA173⟩
Secomb, R. J., **2**, 690 ⟨76JCS(D)2307⟩
Seconi, G., **1**, 5, 7, 9, 10, 17, 25 ⟨92JOC2427⟩, 318 ⟨79JOM(177)129⟩, 449 ⟨76JCS(P2)925⟩, 936, 945 ⟨90SL219⟩, 1048 ⟨91TL2169⟩; **2**, 536 ⟨90SL257⟩, 541 ⟨90SL219⟩, 907 ⟨91JOC1951⟩, 918 ⟨91S1201, 92PAC323⟩, 918, 940 ⟨92SL981⟩, 1018 ⟨91TL2169⟩, 1019 ⟨89S461⟩; **3**, 675 ⟨91TL2169⟩; **4**, 358 ⟨90T2999⟩, 1018 ⟨92PAC439⟩; **5**, 410 ⟨80TL3927⟩, 413 ⟨85AG(E)1068⟩, 417 ⟨88JOC3612⟩, 467 ⟨89JOC2966⟩, 793, 803 ⟨81JCS(P2)1043⟩; **6**, 325, 335, 520 ⟨86TL5985⟩, 349, 585 ⟨85TL1091⟩, 520, 702 ⟨85AG(E)1068⟩

Secor, H. V., **5**, 168 ⟨68JOC2819⟩

Secrist, III, J. A., **1**, 1134 ⟨78JA4618⟩; **2**, 629 ⟨78JA4618⟩, 675 ⟨79SC157⟩; **4**, 227 ⟨92JMC533⟩; **5**, 173 ⟨78JA4618⟩

Seddighi, M., **2**, 215 ⟨89S378⟩; **3**, 5 ⟨90T6869⟩

Seddon, K. R., **4**, 691 ⟨79IC2808⟩

Sedelmeier, G., **1**, 564 ⟨87JA4626⟩

Seden, T. P., **2**, 163 ⟨69TL477⟩; **5**, 1035 ⟨80SC489⟩

Sedlak, D., **4**, 373 ⟨83CB1479⟩

Sedlak, J., **2**, 27 ⟨63OSC(4)108⟩

Sedlov, A. I., **4**, 344 ⟨74ZOB1029⟩

Sedor, E. A., **5**, 359 ⟨70OPP275⟩

Sedova, N. N., **4**, 696, 723 ⟨78JOM(153)115, 71DOK(198)590⟩, 723 ⟨72DOK(202)362, 76DOK(226)1092, 80JOM(185)C6, 82JOM(224)C53⟩

Sedova, V. F., **6**, 110 ⟨66IZV98⟩

Sedrani, R., **4**, 176 ⟨88PAC49⟩; **5**, 152 ⟨88TL4411⟩

Seebach, D., **1**, 37, 191 ⟨84HCA261⟩, 67 ⟨83JA7407⟩, 111, 1090 ⟨76TL4839⟩, 137 ⟨78HCA512⟩, 140 ⟨74AG(E)202⟩, 140 ⟨77CB1833⟩, 140, 141 ⟨74AG(E)479⟩, 140, 693 ⟨66JOC4097⟩, 148 ⟨75AG(E)15⟩, 156 ⟨76HCA2213, 79CB234, 86CB575⟩, 156, 157 ⟨77TL1161⟩, 157 ⟨76AG(E)505⟩, 210 ⟨65AG(E)1075, 77AG(E)264⟩, 230 ⟨87HCA1326⟩, 311 ⟨87TL3791⟩, 441 ⟨75CB314⟩, 578 ⟨86TL1449⟩, 683 ⟨82HCA1972⟩, 687 ⟨89C39⟩, 716 ⟨77LA846⟩, 873 ⟨81HCA2617, 85CB764, 85HCA1373, 85JA5403⟩, 876 ⟨88AG(E)1624⟩, 1024 ⟨88CB1315⟩, 1026 ⟨87HCA448⟩, 1038, 1039 ⟨87CB1845⟩, 1048 ⟨81AG(E)569⟩, 1090 ⟨78CB2785⟩; **2**, 67 ⟨83AG(E)31⟩, 89 ⟨75AG(E)762⟩, 272 ⟨74AG(E)202⟩, 279 ⟨75CB314⟩, 281 ⟨69AG(E)450, 74AG(E)806⟩, 281 ⟨77LA846⟩, 315 ⟨87TL3103⟩, 356 ⟨85HCA319⟩, 385 ⟨74JMC1225, 88OSC(6)542⟩, 385, 420 ⟨75CB1293⟩, 388 ⟨78HCA1622⟩, 393 ⟨75CB1924, 75CB1961⟩, 394 ⟨81HCA2264, 82HCA1101⟩, 394 ⟨85JA3601, 90OS(69)19⟩, 420 ⟨93OS(72)62⟩, 565, 566 ⟨81HCA1552, B-83MI 211-01⟩, 632 ⟨80CB1304⟩, 699 ⟨85HCA961⟩, 783 ⟨79C1, 92AG(E)1578⟩, 783, 785, 786 ⟨85T4861⟩, 785 ⟨82S1017, 82TL3897⟩, 785 ⟨84HCA261⟩, 909, 910 ⟨76TL4839, 78CB2785⟩, 910, 911, 931 ⟨77CB867⟩, 931, 936 ⟨74AG(E)83, 77CB852⟩, 961 ⟨74CB847⟩, 997, 1002 ⟨85JOM(285)1⟩, 998 ⟨83HCA308⟩, 999 ⟨78HCA3100, 84HCA224, 88HCA299⟩; **3**, 16, 20, 132, 137, 291, 295 ⟨77S357⟩, 19, 135, 291 ⟨79AG(E)239⟩, 35 ⟨73TL5113⟩, 57 ⟨78AG(E)458⟩, 175 ⟨69AG(E)450⟩, 182 ⟨76AG(E)313⟩, 186 ⟨84HCA1593⟩, 191 ⟨79HCA2258, 85JA3601⟩, 192 ⟨85HCA162⟩, 192 ⟨85T4861⟩, 214 ⟨73AG(E)69, 74TL3171⟩, 497 ⟨67AG(E)443, 72CB487⟩, 497 ⟨72CB511⟩, 537 ⟨85JA5396⟩, 707 ⟨88CB507⟩; **4**, 102 ⟨91LA655⟩, 197 ⟨80HCA1960⟩, 197, 244, 1075 ⟨77S357⟩, 213 ⟨84T1313, 86AG(E)178⟩, 213 ⟨87HCA448, 87TL3791⟩, 213 ⟨89HCA690⟩, 238, 284 ⟨77LA846⟩, 244, 1075 ⟨69S17⟩, 245, 250, 1075 ⟨75JOC231⟩, 250 ⟨65AG1134, 71OS(51)39⟩, 250, 391, 392, 957 ⟨66JOC4097⟩, 254 ⟨74TL3171, 77LA811⟩, 254, 618, 853 ⟨77LA830⟩, 289 ⟨84HCA1070⟩, 289, 398, 400 ⟨69AG(E)450⟩, 290, 402 ⟨75CB314⟩, 302 ⟨83JA5390, 85HCA1243, 88HCA224⟩, 335 ⟨85T4861⟩, 378 ⟨76AG(E)438, 80CB1290⟩, 382 ⟨78HCA512⟩, 392 ⟨74AG(E)202, 74AG(E)479, 79CB1420⟩, 394 ⟨74AG(E)806⟩, 395 ⟨75CB2368, 76AG(E)437⟩, 396 ⟨81HCA357⟩, 399, 400 ⟨72CB511⟩, 416 ⟨86AG(E)345⟩, 419, 424 ⟨85HCA135⟩, 511 ⟨74JMC1225⟩, 529 ⟨78CB2630⟩, 530 ⟨72AG(E)301, 75AG(E)15, 80HCA102⟩, 531 ⟨85JOM(285)1⟩, 540 ⟨76AG(E)505⟩, 620 ⟨74AG(E)83⟩, 620, 937, 962 ⟨77CB852⟩, 746 ⟨84CB1877⟩, 816 ⟨83HCA308⟩, 849 ⟨73CB2277, 75S461⟩, 943 ⟨77CB2880⟩, 964 ⟨77CB867⟩, 979 ⟨91CB2897⟩, 1075 ⟨69AG(E)639⟩; **5**, 148 ⟨81AG(E)971⟩, 161 ⟨87HCA1320⟩, 177 ⟨77AG(E)264⟩, 179 ⟨89LA1233⟩, 410 ⟨67JA434⟩, 411 ⟨77CB841⟩, 552 ⟨79TL2271⟩, 566, 584, 586, 590, 597, 608, 609 ⟨76CB1309⟩, 597 ⟨73AG(E)1014, 77CB3733⟩, 606 ⟨76AG(E)313⟩, 657 ⟨83HCA1618⟩, 658 ⟨88CB507⟩; **6**, 79 ⟨87HCA1320⟩, 82, 86, 88, 302, 303, 317 ⟨72CB3280⟩, 82, 87, 88, 303, 317 ⟨67AG(E)442⟩, 86, 92 ⟨84HCA1070⟩, 87, 88, 302, 317, 697 ⟨72CB487⟩, 88, 317 ⟨72CB3892⟩, 92, 307, 317, 697 ⟨72CB511⟩, 93, 94 ⟨69AG(E)450⟩, 98 ⟨69S17⟩, 127, 325 ⟨67JA434⟩, 128, 135, 326, 327 ⟨77CB841⟩, 134 ⟨81HCA357⟩, 135, 327 ⟨75AG37⟩, 175, 190, 192 ⟨74AG(E)83⟩, 175, 190, 334, 351, 390 ⟨77CB852⟩, 204 ⟨79AG(E)785, 80HCA2046⟩, 302 ⟨72CB3905⟩, 317 ⟨67AG(E)443, 78CB3644, 84HCA1083⟩, 317, 734 ⟨77CB2880⟩, 327 ⟨73CB2277⟩, 622 ⟨88CB507⟩

Seebach, E., **4**, 313 ⟨82AP(315)852⟩

Seebachand, D., **4**, 367, 392 ⟨77CB1833⟩

Seebald, S., **5**, 954 ⟨94JOM(472)C15⟩

Seeberger, L., **5**, 842 ⟨1894CB2121⟩

Seebold, U., **6**, 183 ⟨91CB2645⟩

Seeboth, H., **2**, 647 ⟨67AG(E)307⟩

Seefeld, M. A., **2**, 75 ⟨93CC339⟩

Seefelder, M., **1**, 1026 ⟨60AG836⟩; **5**, 577, 608, 609, 610, 728 ⟨63CB2671⟩, 609, 611 ⟨60AG836⟩; **6**, 51 ⟨60AG836⟩, 70 ⟨63CB2671⟩, 313, 321 ⟨64CB1232⟩

Seeger, N. V., **5**, 965 ⟨44USP2362648⟩

Seeger, R., **1**, 502 ⟨70HCA1713⟩; **6**, 386, 401, 404 ⟨76JA5419⟩

Seegmiller, D. W., **1**, 21 ⟨63JOC2347⟩

Seehafer, J., **1**, 1076 ⟨72JOC4466⟩

Seel, F., **2**, 166 ⟨67AG(E)706⟩, 166 ⟨69AG(E)773⟩, 183 ⟨55AG32⟩, 471 ⟨72CB406⟩; **5**, 4 ⟨58CB2553⟩

Seeler, A. K., **4**, 795 ⟨70JOC4236⟩

Seeliger, A., **5**, 1023 ⟨67AG(E)174⟩

Seeliger, W., **4**, 296 ⟨66LA(697)171⟩; **6**, 107 ⟨66LA(698)174⟩

Seelinger, R., **6**, 253 ⟨80AG(E)203⟩

Seelinger, W., **6**, 107 ⟨66AG(E)875⟩

Seely, F. L., **1**, 924 ⟨88JA1631, 93TL7819⟩; **5**, 152 ⟨88JA1631⟩

Seeman, J. I., **1**, 420, 421 ⟨73PAC197⟩; **5**, 453 ⟨86TL623⟩

Seevogel, K., **4**, 583 ⟨72IJ293⟩

Seewald, A., **6**, 502 ⟨79CB727⟩

Seewald, I., **6**, 538 ⟨85DIS(B)840⟩

Seferov, S. O., **5**, 656 ⟨91UKZ530⟩

Seferth, D., **4**, 34, 137 ⟨75JOM(92)7⟩

Seffl, R. J., **4**, 36, 751 ⟨56JA59⟩

Sefton, M. A., **4**, 409 ⟨89JOC4771⟩

Segal, G. A., **6**, 183 ⟨70JA4158⟩

Ségal, G. M., **5**, 168 ⟨60BSF1412⟩

Segal, J. A., **2**, 684 ⟨75JCS(D)1677⟩

Segall, Y., **1**, 80 ⟨76JOC3682⟩, 445 ⟨80PS(8)243⟩; **2**, 210 ⟨B-81MI 203-01⟩, 824 ⟨78JA5130⟩

Segawa, J., **2**, 263, 812 ⟨79TL4311⟩; **3**, 207 ⟨80TL3779⟩; **4**, 188 ⟨82S1089⟩; **5**, 226, 375, 378 ⟨79TL4311, 81S413⟩

Segawa, K., **1**, 196 ⟨87BCJ1525⟩

Segev, D., **1**, 955 ⟨76TL3181⟩

Seggery, P., **5**, 219 ⟨71JCS(C)1917⟩

Segi, M., **1**, 463 ⟨80CC86⟩, 475 ⟨82BCJ167⟩; **2**, 453 ⟨87CL1915⟩, 542 ⟨88JA1976⟩; **3**, 336 ⟨89SC2431⟩, 336, 340, 385 ⟨88JA1976⟩, 336, 340, 385 ⟨93CE53, 88TL6965⟩, 390, 400 ⟨89JA8749⟩, 394 ⟨89TL2095⟩, 553 ⟨92TL7865⟩; **4**, 931 ⟨86CL181⟩; **5**, 429 ⟨86CL177⟩, 516, 520 ⟨92TL7865⟩; **6**, 481 ⟨89CL1009⟩

Segiri, T., **3**, 668 ⟨80CC791⟩

Segmuller, B. E., **1**, 603 ⟨85JA3891⟩

Segner, J., **2**, 677 ⟨77TL4651⟩

Segnitz, A., **2**, 391, 392, 394 ⟨71HOU(10/1)12⟩, 565 ⟨75HOU(13/7)265⟩, 569 ⟨75HOU(13/7)355⟩, 571 ⟨75HOU(13/7)379, 75HOU(13/7)433⟩, 571 ⟨75HOU(13/7)489⟩, 573 ⟨86HOU(13/9a)1⟩, 575 ⟨86HOU(13/9a)175⟩, 579 ⟨84HOU(13/9b)702, 86HOU(13/9a)121, 86HOU(13/9a)525, 86HOU(13/9a)613, 86HOU(13/9a)671⟩, 784 ⟨71HOU(10/1)9⟩; **3**, 275 ⟨73CB39⟩

Segre, A., **2**, 214 ⟨57JA3503⟩; **4**, 301, 309, 979 ⟨68JCS(C)2721⟩
Seguin, M., **6**, 231 ⟨80JFC(15)201⟩, 441 ⟨80JFC(15)37⟩
Seguin, R., **3**, 65 ⟨81TL979⟩
Segupta, S., **1**, 116, 206, 311, 465, 476 ⟨92OR(41)135⟩
Segura, R., **5**, 361 ⟨84JOC3322⟩
Segur, J. B., **5**, 106 ⟨32OSC(1)391⟩
Seher, A., **6**, 69 ⟨54LA(587)226⟩
Sehgal, R., **3**, 249 ⟨75JOC3179⟩
Sehgal, R. K., **1**, 72 ⟨75JOC3073⟩; **2**, 170 ⟨80SC245⟩
Sehgal, S., **5**, 1105 ⟨87IJC154⟩
Sei, R. C., **1**, 1124 ⟨76JA862⟩
Seib, B., **6**, 630 ⟨70AP(303)625⟩
Seibel, W. L., **1**, 1063 ⟨87TL4921⟩
Seiberlich, P., **4**, 1062 ⟨86JOM(303)73⟩
Seibert, H., **2**, 733 ⟨32CB812⟩
Seibt, H., **4**, 559 ⟨75JPR890⟩; **6**, 281 ⟨70JPR475⟩
Seibt, S., **5**, 145 ⟨24HCA390⟩
Seichter, F. S., **2**, 471 ⟨63JA2462⟩
Seichter, W., **1**, 986 ⟨91BCJ659⟩
Seidehamel, R. J., **6**, 508 ⟨73JMC901⟩
Seidel, B., **2**, 396 ⟨91S629⟩; **3**, 424 ⟨92CB2361⟩; **4**, 209 ⟨89TL4223⟩, 455, 462, 478, 481, 499, 500, 501 ⟨78JOM(149)167⟩; **6**, 161 ⟨78JOM(149)167⟩
Seidel, C., **6**, 487 ⟨91ZN(B)339⟩
Seidel, C.-M., **4**, 416 ⟨76AP503⟩
Seidel, G., **6**, 183 ⟨85AG(E)326⟩
Seidel, M., **3**, 452 ⟨62T3⟩, 680 ⟨59JOC892⟩; **5**, 201, 842 ⟨62T3⟩
Seidel, W., **2**, 452 ⟨56ZAAC(288)201⟩; **4**, 685 ⟨93JOM(445)133⟩
Seidel, W. C., **3**, 614 ⟨85MI 318-01⟩
Seidenschwarz, C., **3**, 502 ⟨91AG(E)444⟩; **4**, 606 ⟨92OM1137⟩
Seidl, H., **6**, 289 ⟨68CB2815⟩
Seidl, P., **1**, 918 ⟨72JA6867⟩
Seidl, V., **1**, 1082 ⟨86JOC3214⟩; **2**, 782 ⟨59CCC1158⟩
Seidler, M. D., **2**, 541, 542 ⟨82JOC1354, 81JOC1283⟩; **5**, 386 ⟨89JA3258⟩
Seidler, N., **5**, 880 ⟨92ZAAC(612)72⟩
Seidner, R. T., **1**, 811 ⟨70JA6641⟩
Seido, N., **3**, 127 ⟨92TL635⟩
Seif-El-Nasr, A., **6**, 88 ⟨91CJC415⟩
Seifert, K., **3**, 670 ⟨83TL2841⟩
Seifert, P., **3**, 55 ⟨51HCA728⟩
Seifert, W. K., **1**, 630 ⟨85JOC5460⟩
Seifman, Yu. V., **6**, 26 ⟨92BAU370⟩
Seigel, J. R., **4**, 162 ⟨54JCS520⟩
Seigel, S., **1**, 455, 456 ⟨91COS(8)417⟩
Seil, P., **4**, 440 ⟨90S1048, 90S803⟩
Seiler, M., **4**, 206 ⟨50JPP764⟩
Seiler, M.-E., **1**, 327 ⟨93HCA441⟩
Seiler, P., **1**, 990 ⟨93AG(E)1187⟩; **2**, 997, 1002 ⟨85JOM(285)1⟩; **4**, 531 ⟨85JOM(285)1⟩, 746 ⟨94CC205⟩
Seiler, S. M., **6**, 317 ⟨92BMC1607⟩
Seiler, W., **3**, 318 ⟨78TL4875⟩
Seilz, C., **3**, 40 ⟨86JOC5294⟩
Seip, R., **6**, 186, 195 ⟨82CC1407⟩
Seipp, U., **2**, 387 ⟨74CB745⟩
Seites, P. G., **2**, 11 ⟨78JOC4120⟩
Seitz, B., **3**, 552 ⟨77T449⟩; **5**, 511 ⟨77T449⟩
Seitz, D. E., **1**, 452 ⟨81SC281⟩; **2**, 544 ⟨79SC451⟩, 632 ⟨80JOM(186)C33⟩; **4**, 154 ⟨83SC129⟩, 652, 660 ⟨80TL3451, 81S557⟩, 660 ⟨81SC673⟩
Seitz, E. P., **1**, 401 ⟨76CC846, 78JOC1050⟩
Seitz, F., **5**, 108, 887 ⟨85JOM(287)⟩, 1088 ⟨86CB2080⟩; **6**, 610 ⟨86OM2187⟩
Seitz, G., **3**, 354 ⟨74S578⟩, 355 ⟨78S831, 79S361⟩, 356 ⟨91CB665⟩; **4**, 229 ⟨84AG(E)890⟩, 865 ⟨73AP(306)389⟩; **6**, 173, 267 ⟨58CB22⟩
Seitz, G. H. R., **3**, 355 ⟨76S445, 76S692⟩
Seitz, H., **4**, 502 ⟨86TL2957⟩; **6**, 161 ⟨86TL2957⟩
Seitz, K., **3**, 553 ⟨85JOM(288)C47⟩
Seitz, S. P., **1**, 604 ⟨79JA3704⟩; **2**, 281 ⟨79JA3704⟩; **4**, 217 ⟨84TL3⟩, 224, 225 ⟨83JA2430⟩; **5**, 179 ⟨82JA6818⟩
Seiz, H., **5**, 753 ⟨57CB1837⟩
Sejpka, H., **3**, 607 ⟨86AG(E)1003⟩, 690 ⟨86AG(E)264⟩; **4**, 1047, 1059 ⟨86AG1020⟩; **5**, 1155, 1157 ⟨85TL5507⟩, 1155, 1157 ⟨86TL171⟩, 1160 ⟨86AG(E)264⟩
Šek, B., **5**, 836 ⟨74JOC2143⟩
Sekera, A., **4**, 428 ⟨65CR(260)2252⟩
Sekera, M. H., **1**, 1126 ⟨73CC679⟩
Sekguchi, T., **5**, 635, 636 ⟨71MI 515-01⟩
Sekhar, B. B. V. S., **1**, 943 ⟨93CC1636⟩
Seki, F., **1**, 881 ⟨58JOC841⟩
Seki, H., **5**, 324 ⟨92GSM79⟩, 652 ⟨78JCS(P1)558⟩
Seki, K., **5**, 278, 291, 303 ⟨81S991⟩
Seki, M., **1**, 447 ⟨61CA(55)14340⟩
Seki, N., **5**, 324 ⟨92H(33)121⟩
Seki, T., **2**, 987 ⟨90CC182⟩
Seki, Y., **2**, 912 ⟨83CC1193, 86JOC3890⟩, 912 ⟨87JOC4864⟩
Sekido, E., **2**, 278 ⟨64ANC1768⟩
Sekigawa, S., **2**, 1089 ⟨93T6359⟩
Sekiguch, A., **2**, 266 ⟨83TL4033⟩
Sekiguchi, A., **1**, 63 ⟨85OM584⟩, 656 ⟨85TL4523⟩; **2**, 936 ⟨79JOC413⟩, 947 ⟨85TL4523⟩, 949 ⟨B-92MI 218-01⟩; **3**, 483 ⟨81JA5573⟩; **4**, 398 ⟨79CC1121⟩, 612, 1049 ⟨87OM1857⟩, 613 ⟨88CL1441⟩, 618 ⟨89JA3748, 93CL267⟩; **5**, 234, 242 ⟨82CL885⟩, 413, 430 ⟨79TL871⟩, 873 ⟨77JA6995, 83CL871⟩, 873 ⟨83TL4245⟩, 925 ⟨85TL2337, 86CL2025⟩, 925 ⟨87OM2337⟩, 1020 ⟨87OM2270, 88OM558⟩; **6**, 172 ⟨B-92MI 606-01⟩, 176, 193 ⟨89JA3748⟩, 186 ⟨89AG(E)55⟩, 193 ⟨92CL867⟩, 193 ⟨93CL267⟩, 383 ⟨88CL1441⟩, 672 ⟨87OM1857⟩
Sekiguchi, A. T., **5**, 925 ⟨81JA5573⟩
Sekiguchi, K., **3**, 543 ⟨84H(22)2563, 85S224⟩
Sekiguchi, M., **1**, 1073 ⟨91CL1805⟩; **2**, 108 ⟨79CPB1700⟩; **4**, 334, 909 ⟨90CL2053⟩
Sekiguchi, Y., **1**, 188 ⟨88CC449⟩, 1023 ⟨87S139⟩, 1084 ⟨87CL1905⟩
Sekihachi, J., **1**, 1140 ⟨86TL195, 87CPB1405⟩; **5**, 223 ⟨90CPB1836⟩
Sekine, M., **2**, 239 ⟨89JOC2321⟩; **3**, 293 ⟨78TL363⟩; **4**, 219 ⟨89JOC5998, 91CL121⟩, 339 ⟨77CL485⟩, 918 ⟨79TL2047⟩; **5**, 403 ⟨78CC285, 84JOC2139⟩, 403 ⟨87T3395⟩, 403, 404 ⟨80JOC4162⟩; **6**, 80 ⟨85S408⟩, 118 ⟨78CC285⟩, 119 ⟨78TL363⟩, 125 ⟨79TL4475, 81CC969⟩, 125 ⟨82BCJ224⟩, 582 ⟨79TL3013⟩
Sekine, T., **1**, 925, 946 ⟨93TL1317⟩
Sekine, Y., **3**, 704 ⟨82ABC1803⟩
Sekiya, A., **1**, 94 ⟨81JA5096⟩; **2**, 631 ⟨74BCJ1680⟩; **3**, 661 ⟨75CL277⟩; **5**, 4 ⟨78BCJ1267⟩, 807 ⟨81JOC1277⟩; **6**, 53 ⟨79JA7640, 81IC1⟩, 229, 230 ⟨91CL1421⟩, 235 ⟨82CL1519⟩, 240 ⟨81JFC(17)463⟩, 256 ⟨79IC919, 80IC1330⟩, 441 ⟨79JFC(14)289⟩, 603 ⟨81JOC1277⟩
Sekiya, K., **1**, 122 ⟨87JA8056, 87TL337⟩, 122 ⟨89JA3091⟩; **2**, 69 ⟨87JA8056⟩
Sekiya, M., **1**, 990 ⟨83CPB3306⟩, 1041 ⟨84CC794⟩, 1045 ⟨71CPB391⟩; **2**, 271 ⟨77S795, 78CPB1837⟩; **3**, 59 ⟨72CPB309⟩, 273 ⟨71CPB391⟩, 406 ⟨85CL1371⟩, 465 ⟨76CPB369, 79COC(2)385⟩; **4**, 313 ⟨84CC794⟩, 317 ⟨77CPB2964, 77CPB3385⟩, 322, 323 ⟨73CPB2257⟩, 325 ⟨79CPB1691, 81CPB2496⟩, 405, 418 ⟨69CPB32⟩, 885 ⟨80CPB3098⟩; **5**, 329 ⟨84CJC574, 89CPB1179⟩
Sekiya, M. J., **3**, 713 ⟨85CC1584⟩
Seko, K., **3**, 32 ⟨82JOC1101⟩
Seko, N., **5**, 1072 ⟨93JOC5118⟩
Seko, T., **1**, 603 ⟨88JCS(P1)575⟩; **2**, 285 ⟨88JCS(P1)575, 89JCS(P1)1927⟩; **5**, 248 ⟨85S878⟩, 249 ⟨89JCS(P1)1927⟩
Sekutowski, J. C., **4**, 583, 584, 585, 586, 588 ⟨77CB3517⟩, 586 ⟨77IJ149⟩, 693 ⟨77AG(E)1, 77CB1763⟩
Sela, M., **5**, 84 ⟨58MI 502-04⟩

Selby, T. P., **6**, 630 ⟨84JHC61⟩
Selegue, J. P., **2**, 579 ⟨82COMC-I(4)967⟩; **5**, 1097 ⟨87JA4731⟩
Seleznev, A. V., **5**, 794 ⟨89MI 520-01⟩
Self, C. R., **1**, 226 ⟨81JCS(P1)270⟩
Self, J. M., **6**, 243 ⟨67JA3446⟩
Self, M. F., **4**, 563 ⟨88MI 412-01⟩
Seligman, A. M., **2**, 414 ⟨54JA655⟩; **5**, 857 ⟨50MI 521-01⟩
Seligman, K. L., **5**, 199 ⟨58JA3711⟩
Seligman, R., **5**, 838 ⟨02CB3884⟩
Selikson, S. J., **3**, 134 ⟨83JOC4087⟩
Selim, A., **1**, 134 ⟨83TL3213⟩; **5**, 148 ⟨81AG(E)207⟩
Selim, M. I. B., **4**, 105 ⟨71MI 403-01⟩
Selim, M. R., **3**, 202 ⟨91JCS(P1)1893⟩
Selinger, Z., **4**, 446 ⟨83JMC129⟩
Sell, E., **5**, 1004 ⟨1874CB1228⟩
Sellars, A., **3**, 589 ⟨85JCS(P1)2185⟩
Sellars, P. J., **6**, 79 ⟨80CC207⟩
Selle, B. J., **2**, 24 ⟨84TL607⟩
Sellén, M., **1**, 187 ⟨90JA6615⟩
Sellers, C., **2**, 806 ⟨65JCS6186⟩, 884 ⟨68JCS(C)2317⟩
Sellers, S. F., **1**, 801 ⟨82JA2498⟩; **4**, 7 ⟨82JA2494⟩
Selliah, R. D., **1**, 831 ⟨93JOC6255, 94T1651⟩
Selling, G. W., **2**, 328 ⟨89JOC5574⟩
Selling, H. A., **2**, 797 ⟨76SC129⟩; **4**, 861 ⟨75T2387⟩
Selman, C. M., **1**, 110 ⟨69JOC641⟩
Selmayr, T., **5**, 936 ⟨77CB2574⟩, 949 ⟨77CB2947⟩; **6**, 719 ⟨77CB2574⟩
Selnick, H. G., **3**, 243 ⟨88JA4368⟩, 276 ⟨90JOC202⟩, 305 ⟨91TL721⟩
Selover, J. C., **3**, 242 ⟨54JA4118⟩; **4**, 150 ⟨86JA1455⟩
Selton, B., **2**, 128 ⟨48JCS1892⟩; **4**, 309 ⟨48JCS2174⟩
Seltzman, H. H., **4**, 297, 304, 325 ⟨75JOC2414⟩
Selva, A., **1**, 630 ⟨64G537⟩; **5**, 343 ⟨88H(27)867⟩
Selva, M., **2**, 14 ⟨91S1003⟩
Selvarajan, R., **2**, 1055 ⟨69JA6122⟩; **3**, 566 ⟨70CC889⟩
Selve, C., **5**, 267 ⟨75TL1219⟩
Selwitz, C. M., **2**, 626 ⟨81JOC2169⟩, 753 ⟨79JOC1233, 79JOC3671⟩
Semard, D., **3**, 104, 105, 106 ⟨69BSF2511⟩
Semashko, Z. T., **2**, 469 ⟨78ZOB2624⟩
Semba, F., **4**, 774, 777 ⟨92BCJ210⟩
Sembritzki, F., **5**, 833 ⟨1889CB2958⟩
Semeluk, G. P., **1**, 1109 ⟨70CJC955⟩
Semenenko, N. M., **2**, 420, 494, 495, 695 ⟨74JOM(78)107, 73ZOB1264⟩, 695, 700 ⟨69ZOB1174⟩; **5**, 400 ⟨90JGU1817⟩
Semenii, V. Ya., **5**, 683, 707 ⟨87UKZ397⟩
Semenova, M. P., **4**, 698 ⟨75CA(83)10238⟩
Semenov, D. A., **1**, 383 ⟨59JA4390⟩
Semenov, V. P., **2**, 787 ⟨64ZOB1522⟩; **5**, 702 ⟨88JOU2149, 88JOU2154⟩, 702 ⟨88ZOR2384, 88ZOR2389⟩
Semenov, V. V., **3**, 75 ⟨88CCC1519⟩; **4**, 245 ⟨93T2151⟩; **4**, 995 ⟨91IZV1849⟩; **5**, 1076, 1077 ⟨87JOM(327)303⟩, 1077 ⟨85IZV2824⟩; **6**, 56, 323 ⟨77IZV139⟩, 145 ⟨90TL7379⟩, 261, 281, 323 ⟨78IZV2348⟩, 283 ⟨77IZV2530, 75IZV2621⟩, 323 ⟨76IZV2640, 78IZV1091⟩
Semenyi, V. Ya., **2**, 453 ⟨78BRP1517876⟩
Semerano, G., **2**, 589 ⟨41CB1089⟩
Semigran, M. J., **1**, 688, 690 ⟨76TL3265⟩
Semkow, A., **3**, 337 ⟨82JA312⟩
Semmelhack, M. F., **1**, 111, 585 ⟨93TL4121⟩, 127 ⟨67JA2755, 72OR(19)115⟩, 468, 472 ⟨81JA6460⟩, 474 ⟨81T3957⟩, 515 ⟨75JOC3619⟩, 1095, 1099 ⟨72JA9234⟩; **3**, 7 ⟨83JA4492⟩, 17, 132, 665 ⟨83JA6732⟩, 88 ⟨84JA3374⟩, 127 ⟨77JOC3180⟩, 212 ⟨72JA8637⟩, 291 ⟨83OM467⟩, 528 ⟨82JA747⟩, 625 ⟨82JOM(240)C5, 93TL1399⟩; **4**, 707 ⟨79JA768⟩; **5**, 142 ⟨94JA7455⟩, 170 ⟨84JA1496⟩, 301 ⟨85JA4577⟩, 444, 937 ⟨87OM1839⟩, 935 ⟨83JA4099⟩
Semones, M. A., **3**, 320 ⟨93TL7853⟩
Semonsky, M., **4**, 211 ⟨72MI 403-01⟩
Semple, G., **5**, 574 ⟨89TL1997⟩; **6**, 263, 288 ⟨85JOM(291)145, 86JOM(309)C7⟩
Semsel, A. M., **2**, 428 ⟨61JOC5138⟩
Semyonov, V. V., **4**, 1083 ⟨89IZV356⟩
Sen, A., **2**, 573 ⟨91OM244⟩
Sen, A. B., **6**, 443 ⟨87IJC(B)748⟩
Sen, D. N., **4**, 697 ⟨67IJC210⟩
Sen, V. D., **5**, 1070 ⟨87IZV1829⟩
Senanayake, C. B. W., **2**, 356 ⟨90JA311, 90T4857⟩
Senaratne, K. P. A., **1**, 817, 835 ⟨84TL2303⟩
Senatore, L., **2**, 195 ⟨70JA3035⟩, 688 ⟨73JA2918⟩; **5**, 766 ⟨73JCS(P2)534⟩
Senboku, H., **1**, 931 ⟨92JCS(P1)1837⟩
Sencherey, B. B., **3**, 131 ⟨82JOC3006⟩
Senczuk, L., **2**, 767 ⟨79MI 215-01⟩
Senda, H., **1**, 619 ⟨88CL2009⟩
Senda, K., **4**, 506 ⟨80BCJ789⟩
Sendelbeck, R., **4**, 697 ⟨80JOM(191)7, 83JOM(243)245⟩; **5**, 1141 ⟨83JOM(243)245⟩; **6**, 206, 207 ⟨80JOM(191)7⟩, 206, 207 ⟨83JOM(243)245⟩
Senderens, J. B., **1**, 71 ⟨1897MI 103-01⟩, 87 ⟨04BSF101⟩; **2**, 89 ⟨26CR(182)612⟩
Senderoff, S. R., **3**, 219 ⟨89JOC4702⟩
Senders, J. R., **2**, 1014 ⟨64AG(E)755⟩
Sendlinger, S. C., **2**, 895, 896 ⟨91CB2453⟩
Sendo, Y., **5**, 81 ⟨78GEP2725519⟩
Sendra, J. C., **4**, 636 ⟨70IC1060⟩
Sendurev, M. V., **2**, 448 ⟨90ZOB2394⟩; **4**, 914 ⟨84ZOB457⟩
Sene, A., **4**, 847 ⟨89TL3415⟩; **6**, 121, 316 ⟨89TL3415⟩, 123 ⟨88PS(40)1, 90TL1151⟩
Senet, J.-P., **2**, 303 ⟨84JOC2081⟩; **4**, 211 ⟨87SC1467⟩, 58 ⟨81EUP40153, 90JOC2240⟩, 58, 59 ⟨85JOC3953, 86S627⟩, 59 ⟨88S407, 90JOC1847⟩, 60 ⟨86TL6319, 87S1027⟩, 920, 921 ⟨90JOC5982⟩, 921 ⟨89FP2610926⟩; **5**, 282 ⟨87TL5801⟩, **5**, 994 ⟨88T6079⟩, 1105 ⟨88T1619⟩; **6**, 423 ⟨90JOC1847⟩, 428 ⟨84JOC2081⟩, 429 ⟨90JOC2240, 90JOC5982⟩, 431 ⟨88S407⟩, 541 ⟨86EUP176412⟩, 643, 645 ⟨90T1839⟩
Seng, F., **1**, 268 ⟨62AG(E)116⟩; **4**, 921 ⟨76CA122034u⟩
Sengupta, S., **1**, 86 ⟨89SL64⟩, 486 ⟨93JCS(P1)1943⟩, 845, 859 ⟨88PIA187⟩; **2**, 942 ⟨90TL4267⟩; **3**, 12, 128 ⟨92OPP335⟩; **4**, 383 ⟨87TL1847⟩, 520, 540 ⟨87TL5419⟩, 841, 950 ⟨90JOC5680⟩
Senichev, V. S., **4**, 541 ⟨80ZOB1427⟩
Senio, A., **3**, 335 ⟨93JA324⟩
Senise, Jr., P. P., **4**, 298 ⟨77JOC4217⟩
Senkler, C. A., **1**, 1111, 1114, 1116 ⟨74JA1100⟩; **2**, 689 ⟨74JA1100⟩
Senkler, Jr., G. H., **3**, 376 ⟨74JA5648, 74JA5650⟩, 376 ⟨74JA5651, 75JA2718⟩
Senkus, M., **4**, 212 ⟨46JA734⟩
Senn, J., **4**, 179 ⟨79JA2171⟩
Sennewald, K., **6**, 222, 227 ⟨32CB546⟩, 612 ⟨31LA(489)7⟩
Senning, A., **2**, 204 ⟨65CRV385⟩, 361 ⟨82SUL33⟩, 400 ⟨76CB1069⟩, 478 ⟨86ACS(B)609⟩, 722 ⟨92SR1⟩, 768 ⟨72ACS2877⟩, 809, 892 ⟨65ACS1755⟩; **3**, 355, 356 ⟨84AG(E)729⟩, 370 ⟨90SUL83⟩, 550 ⟨91CB1485⟩, 550, 551 ⟨91JOC509⟩; **4**, 77 ⟨68ACS3256⟩, 282 ⟨73S211⟩, 862 ⟨73TL2389, 74CB2345⟩; **5**, 247 ⟨79AG(E)941⟩, 336 ⟨85T5145⟩, 478, 494 ⟨85SUL177⟩, 478, 494, 582 ⟨87SUL153⟩, 483, 590 ⟨86T739⟩, 528 ⟨78ACS(B)780⟩, 550 ⟨87JOC1156⟩, 592, 593 ⟨72S314⟩, 995 ⟨66AG(E)1047⟩; **6**, 232 ⟨93TL2973⟩, 233 ⟨60ACS2230, 62ACS117⟩, 233 ⟨63ACS2570⟩, 235, 535 ⟨84JOC3854⟩, 238, 540 ⟨72CB2854⟩, 254, 275 ⟨87SUL59⟩, 273, 274, 275, 535, 553, 554 ⟨85T5145⟩, 274 ⟨86T739⟩, 274, 275, 309 ⟨76CB1069⟩, 274, 275, 554 ⟨84SUL85⟩, 301 ⟨91SUL247⟩, 435 ⟨86SUL93⟩, 530 ⟨92SUL275⟩, 533, 536 ⟨85SUL61⟩, 535 ⟨84JA263, 86ACS(B)609⟩, 535 ⟨86SUL203, 91SUL143⟩, 537 ⟨80T539⟩, 553 ⟨71CC314⟩, 555 ⟨86ACS(B)593⟩, 652 ⟨67ACS1293⟩

Sennyey, G., **4**, 58, 59 ⟨85JOC3953, 86S627⟩, 59 ⟨88S407⟩, 60 ⟨87S1027⟩; **5**, 282 ⟨87TL5801⟩; **6**, 322 ⟨81CC780⟩, 431 ⟨88S407⟩, 643, 645 ⟨90T1839⟩
Seno, K., **3**, 36, 172 ⟨79TL3167⟩; **5**, 277, 295 ⟨80TL841⟩
Seno, M., **2**, 35 ⟨81TL1017⟩, 298 ⟨78JOC3345⟩
Senoff, C. V., **5**, 652 ⟨77IS95⟩, 950, 951 ⟨74JOM(70)273⟩
Sen Sharma, D. K., **1**, 905 ⟨85JA3747⟩
Sentenac-Roumanou, H., **5**, 388, 389 ⟨88EJM257⟩
Sentman, R. C., **4**, 827 ⟨82TL603⟩
Sentz, R. C., **1**, 884 ⟨52JA1704⟩
Senyavina, L. B., **6**, 104 ⟨62TL701⟩
Senyushov, L. N., **6**, 36 ⟨74ZOR2031⟩, 251 ⟨76ZOR767⟩
Senzaki, T., **4**, 419, 420 ⟨84CL1599⟩
Seo, J. B., **3**, 82 ⟨90MI 303-01⟩
Seo, Y., **5**, 637 ⟨91S951⟩
Seoane, C., **3**, 427 ⟨83OPP41⟩; **5**, 343 ⟨93TL5627⟩
Seoane, E., **1**, 49 ⟨86T2429⟩
Seoane, G., **2**, 325 ⟨88JOC2094⟩, 326 ⟨87SC1155⟩
Seoane, P. R., **1**, 288 ⟨87JA615⟩
Seong, C. M., **1**, 340 ⟨90JA9401⟩; **3**, 653 ⟨90JA9401⟩
Sepelak, D. J., **4**, 735 ⟨81JOM(205)1⟩
Seper, K. T., **2**, 102 ⟨88JOC2997⟩
Seper, K. W., **3**, 415 ⟨88JOC2991⟩
Sepiol, J. A., **4**, 783 ⟨84JOC1125⟩
Sepiol, J. J., **4**, 783 ⟨84JOC1125⟩
Sepp, D. T., **4**, 49, 51 ⟨67JOC607⟩
Seppelt, K., **2**, 761 ⟨69ZN(B)774⟩, 872 ⟨92AOC(34)207⟩, 873 ⟨89CB803⟩; **3**, 379, 380 ⟨78AG(E)516, 88CB1977⟩, 380 ⟨79AG(E)944, 82IC3147, 83CB645, 84CB3255, 87AG(E)799, 89ZAAC(578)110, 91JA233⟩, 689 ⟨85JA980⟩, 689 ⟨88AG(E)1534⟩; **4**, 868 ⟨88AG(E)1534, 89CB463⟩, 868 ⟨92CB557⟩; **5**, 348, 767 ⟨84HOU(4)883⟩, 1085 ⟨85IC4171⟩; **6**, 228 ⟨77AG325⟩, 238 ⟨79JA347⟩, 256 ⟨87AG(E)314⟩, 275 ⟨80AG(E)942, 88IC2135⟩, 430 ⟨81ZAAC(472)26⟩, 603, 607, 615, 616 ⟨85IC4171⟩, 614 ⟨87AG(E)134⟩
Sera, A., **2**, 579 ⟨92CL377⟩; **3**, 426 ⟨90SL477⟩, 621 ⟨87S631⟩; **5**, 325 ⟨90TL6547⟩
Serafinowska, H. T., **4**, 190 ⟨85S751⟩
Serafinowski, P., **1**, 585 ⟨90S411⟩
Seraglia, R., **2**, 147 ⟨84S325⟩
Serckx-Poncin, B., **1**, 375 ⟨82JA1428⟩; **5**, 718 ⟨82JA1428⟩
Serebis, A. K., **1**, 337 ⟨83AJC545⟩
Serebryakova, E. S., **1**, 1006 ⟨74JOU1990⟩
Serebryakov, E. P., **1**, 516 ⟨79T77⟩; **2**, 13 ⟨65JCS2438⟩
Sereda, S. V., **5**, 703 ⟨87JOU75, 87ZOR86⟩, 1075 ⟨87ZOR2390⟩
Seredenko, V. I., **5**, 778 ⟨71DOK(199)1328⟩
Seredkina, S. G., **2**, 1036, 1037 ⟨90IZV1865⟩; **4**, 765 ⟨88IZV514⟩, 772 ⟨90ZOB227, 91ZOB1084⟩, 787 ⟨90IZV1693, 92ZOB2730⟩, 804 ⟨91JGU983⟩, 872 ⟨91MI 419-01⟩, 926 ⟨82ZOB2375⟩
Serelis, A. K., **2**, 404 ⟨87JOC2911⟩
Serendenko, V. I., **5**, 778 ⟨74BAU1593⟩
Sereno, F.,J., **1**, 576, 1021 ⟨86JOC3098⟩
Sereno, J., **4**, 170 ⟨83JOC5315⟩
Seres, P., **5**, 199 ⟨86JCS(P1)927⟩
Serezhkina, N. V., **4**, 142 ⟨88IZV1946⟩
Serfaty, I. W., **6**, 229 ⟨72JOC2651⟩
Serfontein, W. S., **1**, 21 ⟨63JOC1395⟩
Sergeeva, M. B., **4**, 444 ⟨81IZV1603⟩, 610 ⟨85IZV1917⟩, 656, 658 ⟨74IZV2861, 85ZOB1629⟩
Sergeeva, N. S., **6**, 464 ⟨72IZV1635, 72IZV2733⟩, 464 ⟨73IZV562, 73IZV807⟩
Sergeev, A. P., **4**, 793 ⟨70JOU750⟩
Sergeev, G. B., **2**, 23 ⟨83RCR259⟩; **6**, 147 ⟨80MI 605-01⟩, 148 ⟨73IZV1149⟩
Sergeev, N. M., **5**, 1044 ⟨91ZOR340⟩
Sergeev, V. N., **2**, 172 ⟨80ZOB1059⟩, 471 ⟨78JGU207⟩; **4**, 418 ⟨84JGU635⟩
Sergeichuk, V. V., **3**, 361 ⟨80ZOR443, 80ZOR883⟩
Sergeva, Z. I., **2**, 377 ⟨69ZOR1735⟩
Sergienko, L. F., **6**, 242 ⟨62DOK(142)354⟩
Sergienko, L. M., **4**, 766 ⟨89ZOB997⟩
Serguchev, Y. A., **2**, 18 ⟨90JOU1784⟩
Sergutina, V. P., **2**, 373 ⟨80JOU244⟩
Serianz, A., **2**, 773, 775 ⟨76JOC1751⟩
Seri, T., **4**, 247 ⟨87S164⟩
Serina, L., **5**, 615 ⟨63MI 514-01⟩
Serizawa, H., **1**, 542, 1008 ⟨88TL1811⟩; **2**, 902, 927, 939 ⟨88TL1811⟩
Serizawa, Y., **4**, 216 ⟨85CL1933⟩
Serota, S., **5**, 3 ⟨69JOC2486⟩
Serov, Y. V., **4**, 418 ⟨90JGU2516, 91JGU1072⟩
Serra, A. A., **1**, 2, 6 ⟨85JOC3957⟩
Serra, A. C., **1**, 949 ⟨90T3061⟩
Serrano, K. F., **5**, 208 ⟨83ZOR1764⟩
Serrano, R., **3**, 387, 390 ⟨83AG(E)314⟩; **4**, 656 ⟨89JOM(375)59⟩; **5**, 1097 ⟨85CB340⟩, 1143 ⟨77JOM(136)309⟩
Serranon, R., **2**, 585 ⟨84AG(E)515⟩
Serratosa, F., **1**, 529 ⟨85T5855⟩, 1085 ⟨88S707⟩; **2**, 941 ⟨90JOC395⟩, 1026 ⟨81T1441, 82T1505⟩, 1026 ⟨87OS(65)68, 87T2311⟩, 1027 ⟨88CC909, 90JA7405⟩; **3**, 206 ⟨70CC721⟩, 218 ⟨88T2657⟩, 247, 540 ⟨85S1118, 77TL4437⟩, 540 ⟨90JOC395⟩; **4**, 196 ⟨88S707⟩, 767 ⟨77TL4433⟩, 884 ⟨79TL3329⟩
Serra-Zanetti, F., **3**, 618 ⟨83S740⟩; **5**, 354 ⟨87T4249⟩, 367 ⟨85OPP385⟩
Serros, E., **5**, 386 ⟨70CC100⟩
Servé, M. P., **1**, 516 ⟨70JOC1237⟩
Servens, C., **1**, 128 ⟨75JOM(96)225⟩; **2**, 78 ⟨72JOM(35)C20⟩; **6**, 26, 31 ⟨75JOM(102)423⟩
Servi, S., **1**, 77 ⟨90S1⟩, 78 ⟨88JCS(P1)3061⟩; **2**, 39 ⟨90S1, 94MI 202-02⟩; **3**, 30, 161 ⟨90S1⟩
Service, C. F., **6**, 241 ⟨81JFC(19)91⟩
Servis, K. L., **1**, 905 ⟨87JA7263⟩
Sesana, G., **1**, 923 ⟨77JA7632⟩
Seshadri, R., **1**, 10 ⟨81JOC2596⟩; **3**, 174 ⟨81JOC2596⟩
Seshadri, S., **3**, 91 ⟨73MI 303-01⟩, 300 ⟨84IJC(B)163⟩
Seshadri, T. R., **3**, 101 ⟨56MI 303-01⟩
Seshimoto, O., **3**, 66 ⟨82TL873⟩
Sesny, W. J., **6**, 523 ⟨54JA3831⟩
Sessink, P. J. M., **1**, 786 ⟨87TL357⟩
Sessions, R. B., **2**, 383 ⟨84JCS(P2)411⟩, 383, 384 ⟨82JCS(P1)603⟩
Sessler, J. L., **2**, 394 ⟨91OS(70)68⟩
Sestanj, K., **5**, 567, 630 ⟨89JMC2493⟩
Sestrick, M. R., **1**, 987 ⟨92JA4079⟩
Set, L., **1**, 174, 547, 1131 ⟨84JOC1313, 85CC1205⟩; **3**, 621 ⟨84JOC1313⟩
Seth, S., **1**, 1079 ⟨90SL211, 90SL212⟩, 1079 ⟨93TL5621⟩
Sethi, S. P., **2**, 699 ⟨84TL5953⟩
Setiakusuma, E. D., **5**, 606 ⟨74AP(307)801⟩
Setkina, V. N., **4**, 707 ⟨83JOM(244)C21⟩; **5**, 97 ⟨84JOM(267)309⟩; **6**, 456 ⟨84MI 614-01⟩
Seto, H., **2**, 18 ⟨93CL673⟩, 19 ⟨91CL1185⟩
Setoi, H., **1**, 828 ⟨81TL4097⟩; **2**, 1042 ⟨77BCJ3069⟩
Setsune, J., **2**, 579 ⟨92CL377⟩
Setsune, J.-I., **2**, 960 ⟨75JA6461⟩
Setsuo, H., **3**, 340 ⟨88MI 308-01⟩
Settepani, J. A., **6**, 501, 508 ⟨70JOC843, 73JHC639⟩
Setterquist, R. A., **1**, 637 ⟨58JA4588⟩
Setti, E. L., **2**, 9 ⟨86JOC3217⟩
Settlage, D. H., **4**, 53 ⟨51JA5870⟩
Setton, R., **2**, 959 ⟨B-87MI 219-01⟩; **5**, 122 ⟨74JA8113⟩
Setzer, A., **4**, 860 ⟨85M413⟩
Setzer, W. N., **1**, 110 ⟨85AOC(24)353⟩; **6**, 406 ⟨85AOC(24)354⟩
Seubold, Jr., F. H., **1**, 396 ⟨47JA2916⟩
Seufert, R., **5**, 856, 860 ⟨15JPR1⟩
Seufert, W., **3**, 152, 154 ⟨79CB1670⟩
Seuleiman, A., **1**, 988 ⟨74TL1905⟩
Seuring, B., **1**, 140 ⟨77CB1833⟩, 140, 141 ⟨74AG(E)479⟩,

210 ⟨77AG(E)264⟩; **4**, 367, 392 ⟨77CB1833, 74AG(E)479⟩; **5**, 177 ⟨77AG(E)264⟩
Sevast'yanova, E. I., **4**, 141 ⟨81JOM(205)311⟩
Sevast'yanov, V. N., **6**, 619 ⟨75ZOR2243⟩, 651 ⟨74ZOR42⟩
Sevchenko, A. N., **5**, 208 ⟨59DOK(128)1201⟩
Ševčik, J., **5**, 864 ⟨B-75MI 521-03⟩
Sevenair, J. P., **1**, 667 ⟨68JOC2975⟩
Sevendir, H., **5**, 664 ⟨92SRI851⟩
Seveno, A., **3**, 557, 573 ⟨77TL3353, 77TL3349⟩; **4**, 894 ⟨77TL3349⟩
Severengiz, T., **2**, 465 ⟨83PS(18)73⟩, 499 ⟨85ZN(B)848⟩, 509 ⟨82ZN(B)395⟩; **5**, 244 ⟨85AG(E)1041⟩, 255, 542 ⟨87CC820⟩, 543 ⟨85AG1051⟩
Severin, E. S., **3**, 239 ⟨58JGU3024⟩; **5**, 826 ⟨67IZV1820⟩
Severin, T., **1**, 486 ⟨69CB2966⟩; **2**, 788 ⟨65CB3847, 71CB440⟩, 789 ⟨78CB692⟩, 789, 790 ⟨76CB3541, 78S592⟩, 790 ⟨79S981⟩; **3**, 72, 242 ⟨90CB399⟩, 241 ⟨77CB491⟩, 373 ⟨79CB3007⟩
Sevin, A., **1**, 1025 ⟨65TL1953⟩; **3**, 313 ⟨79JA1095, 79JA1101⟩
Sevost'yanova, V. V., **6**, 653 ⟨79IZV813⟩
Sevrin, M., **1**, 21, 25, 437 ⟨76TL2643⟩, 594 ⟨80TL1877⟩, 623 ⟨78TL187⟩, 713, 1089 ⟨78AG(E)526⟩; **2**, 32 ⟨80CC656⟩, 728 ⟨77TL3835⟩, 908 ⟨78AG(E)526⟩; **3**, 207 ⟨78TL187⟩; **4**, 88, 91, 288, 400 ⟨77AG(E)541⟩, 401, 964 ⟨78AG(E)526⟩
Sewald, N., **1**, 1042 ⟨90S115⟩; **3**, 502 ⟨91AG(E)444⟩; **4**, 606 ⟨92OM1137⟩, 651 ⟨93OM4123⟩
Sewell, M. J., **5**, 767 ⟨71JCS(C)1704⟩
Sexton, A., **2**, 558, 560 ⟨91JOC698⟩
Seyam, A. M., **2**, 602 ⟨81JA6650⟩
Seybold, G., **3**, 550 ⟨74TL555⟩, 552 ⟨75AG(E)248, 77CB1225⟩; **4**, 892 ⟨71AG(E)67⟩; **5**, 480 ⟨75AG(E)703⟩, 511 ⟨77CB1225⟩, 550 ⟨75AG710⟩, 589 ⟨75AG(E)248⟩
Seyden-Penne, J., **1**, 640 ⟨84T5153⟩, 747 ⟨85BSF(2)448⟩, 750, 751 ⟨72T4209, 77TL1137⟩; **3**, 442 ⟨81TL1927⟩, 494 ⟨71CR(272)728⟩, 634 ⟨87S56⟩, 642 ⟨77S126⟩, 642, 648 ⟨73T2437⟩
Seyerl, J. von, **2**, 694 ⟨81CB1407⟩; **3**, 556, 557, 560, 585, 586 ⟨81CB3751⟩
Seyferth, D., **1**, 291 ⟨69JOM(16)503⟩, 455 ⟨61JOC2934⟩, 656 ⟨75JA7162⟩, 1091 ⟨61JA3583⟩; **2**, 209 ⟨75JOM(100)237⟩, 267 ⟨79JOM(164)305⟩, 444, 824, 882 ⟨57JA5884⟩, 447 ⟨59CI(L)849⟩, 483, 487 ⟨58JA1336⟩, 524 ⟨61JA1613⟩, 618 ⟨65JOM(3)337⟩, 909, 910, 911 ⟨64JA883⟩, 936 ⟨B-66MI 218-02⟩, 998 ⟨71JOM(30)151⟩, 1071 ⟨69JOC1483⟩; **3**, 136 ⟨85JOC1985, 86TL1473⟩, 158 ⟨92JOC5620⟩, 293 ⟨68JA540⟩, 323 ⟨74JOM(67)341⟩, 480, 482 ⟨70TL2493⟩, 480, 482 ⟨71JOC1379⟩, 483 ⟨68JA1080, 72JOM(44)279⟩, 489 ⟨71JOC128⟩, 494 ⟨64JOM(2)101, 65JA3467⟩, 494 ⟨66JOM(6)205⟩, 495, 497 ⟨61JA1617, 66JOM(5)267⟩, 498 ⟨63JOC2463⟩, 695 ⟨66TL189⟩, 709, 711 ⟨60JOC809⟩; **4**, 8, 25 ⟨72ACR65⟩, 13 ⟨77JA5317⟩, 14 ⟨77JOM(141)71⟩, 16 ⟨65JA4259⟩, 32 ⟨70JOC1297⟩, 35, 137 ⟨73JOC4031, 73JOM(51)77⟩, 138 ⟨70JOM(23)99⟩, 140 ⟨60IS37⟩, 152 ⟨69JA5027⟩, 152, 155, 156 ⟨75JOM(88)255⟩, 153 ⟨82JIC111⟩, 154, 155, 156 ⟨71JOM(30)151⟩, 155 ⟨75JOM(88)287⟩, 356, 642, 643, 709 ⟨59JA1844⟩, 489 ⟨71JOC1379⟩, 513 ⟨68JA1080⟩, 551, 565 ⟨61JA2055⟩, 551, 570, 588 ⟨64JA1100⟩, 551, 570, 588 ⟨69USP3426021⟩, 551, 576, 588 ⟨63JA642⟩, 563, 740, 741, 801 ⟨73JOM(59)237⟩, 566 ⟨60USP2964550⟩, 566, 569, 597, 598, 599 ⟨58JA1336⟩, 566, 656, 660 ⟨61JOC2604⟩, 569, 570, 574, 587, 588 ⟨59CI(L)849⟩, 569, 570, 574, 587, 588 ⟨61JA1610⟩, 569, 570, 574, 587, 588 ⟨61JA1613⟩, 571 ⟨65JA4156, 78JOM(159)137⟩, 576, 577 ⟨82ZN(B)950⟩, 576, 588 ⟨64JOM(2)1⟩, 602, 652 ⟨55JA907⟩, 605, 636 ⟨70JOM(21)103⟩, 606, 620, 623, 653 ⟨90OM2677⟩, 607, 638 ⟨54JOC250⟩, 620 ⟨91OM551⟩, 623 ⟨71JA3709, 92OM3464⟩, 648 ⟨78JOM(156)299⟩, 683 ⟨92OM3736⟩, 688 ⟨70JA5520⟩, 700 ⟨74SRI515⟩, 706, 709 ⟨62IC227⟩, 735 ⟨79JOM(182)455, 81JOM(205)301⟩, 756 ⟨73JOM(49)117⟩, 806 ⟨77JOM(142)39⟩, 806, 807, 820 ⟨62MI 418-01⟩, 806, 811 ⟨62IC78⟩, 811, 812 ⟨60AG77⟩, 812 ⟨61JOC2934⟩, 812 ⟨62JA4266, 76JOM(104)145⟩, 1015 ⟨89OM1371⟩, 1051, 1052 ⟨84JOM(272)123⟩, 1051, 1052 ⟨84OM1897⟩; **5**, 238 ⟨84OM327⟩, 418 ⟨82JA5534⟩, 437 ⟨83JOC1144, 84TL1651⟩, 596, 719, 729 ⟨84TL5251⟩, 868 ⟨67JA4811, 70TL2493⟩, 868 ⟨71JOC1379⟩, 870 ⟨71JOC128⟩, 896 ⟨66JOM(5)267⟩, 908 ⟨82ZN(B)950⟩, 914 ⟨78JOM(159)137⟩, 1136, 1144 ⟨60JOC809⟩, 1148 ⟨71JOM(29)C25⟩; **6**, 38 ⟨64JA2961, 74JOM(67)341⟩, 51 ⟨74JOC158⟩, 60 ⟨67JA1538, 68JA2944⟩, 61 ⟨66JA1834⟩, 173, 195, 267, 270, 293 ⟨70JOM(24)647⟩, 190 ⟨92OM3464⟩, 190, 381, 392 ⟨91OM551⟩, 191 ⟨90OM2677⟩, 258 ⟨74JOM(71)335⟩, 266 ⟨66JOM(5)185⟩, 266, 288 ⟨73JOM(59)237⟩, 267, 270, 293, 399 ⟨70JOM(23)361⟩, 270 ⟨67JA2790, 67JOM(7)P20⟩, 270 ⟨69JA1954, 71JOM(27)19⟩, 293, 399 ⟨74JA6237⟩, 332 ⟨83OM1696, 87OM283⟩, 341, 401 ⟨73JOM(50)265⟩, 371 ⟨73JOM(59)231⟩, 380, 381, 388, 399 ⟨76JOM(116)257⟩, 380, 391 ⟨77JOM(142)39⟩, 401, 402, 404 ⟨77JA5209⟩, 401, 402, 404 ⟨79JOM(178)227, 80JOM(188)329⟩, 608 ⟨71JOC1786⟩, 670 ⟨71JOC1379⟩, 695 ⟨65JOM(3)337⟩
Seyhan, M., **2**, 125 ⟨49CB594⟩
Seykens, D., **5**, 559 ⟨78TL4839⟩
Seymour, D., **6**, 74 ⟨58JA1247⟩
Seymour, D. E., **4**, 206 ⟨50JPP764⟩
Sgarabotto, P., **5**, 497, 499 ⟨74JCS(P2)1483⟩
Sha, C., **3**, 155 ⟨87JOC3919⟩
Sha, C.-K., **2**, 34 ⟨87JOC3919⟩
Shabana, R., **2**, 856 ⟨93T1271⟩; **3**, 346, 353 ⟨82T993⟩, 358 ⟨86BSF283⟩, 360 ⟨81T197⟩, 411 ⟨80BSB247⟩; **5**, 570, 571, 610 ⟨80NJC47⟩, 581, 611, 643, 644 ⟨85PS(25)297⟩, 599 ⟨86ACS(B)534, 86ACS(B)694⟩, 599 ⟨87ACS(B)157, 83PJC647⟩; **6**, 476 ⟨86CI(L)140⟩
Shabanova, M. P., **2**, 464 ⟨59ZOB1671⟩
Shabica, A. C., **2**, 24 ⟨52JA4891⟩
Shabtai, J., **1**, 795 ⟨65JA4107⟩
Shachat, N., **6**, 509 ⟨63JOC991⟩
Shachidayatov, Ch., **2**, 748 ⟨68T4677⟩
Shackelford, S. A., **2**, 17 ⟨77TL363⟩; **5**, 361, 362 ⟨93TL6669⟩
Shadday, J., **2**, 95 ⟨83JOC3126⟩
Shadle, J. K., **5**, 340 ⟨92T8615⟩
Shadmanov, K. M., **1**, 463 ⟨63CA(59)2732⟩
Shaefer, C. G., **3**, 83 ⟨76S767⟩
Shaefer, S., **2**, 448 ⟨83CSR53⟩
Shäfer, F., **2**, 1034 ⟨12LA(394)325⟩
Shafer, P. R., **3**, 322 ⟨67JA5497⟩
Shaffer, E. T., **1**, 734 ⟨66JOC2907⟩
Shafferman, R., **5**, 228 ⟨58JCS3648⟩
Shafiee, A., **1**, 1077 ⟨70AG(E)464, 71JOC2836⟩, 1077 ⟨73JHC655, 77JHC567⟩; **3**, 553 ⟨73JOC338⟩; **4**, 852 ⟨73JHC11⟩, 870 ⟨80JHC117⟩, 874 ⟨73JOC338, 77S765⟩, 874 ⟨79JHC1405⟩; **5**, 639 ⟨89JHC1341⟩, 1057 ⟨84S512⟩; **6**, 630 ⟨80JHC117⟩, 637 ⟨78JHC473⟩
Shafii, B., **2**, 1083 ⟨90JA1847⟩
Shafiullah, **4**, 268 ⟨91JIC368⟩; **5**, 533 ⟨88JCR(S)376⟩
Shafizadeh, F., **4**, 42 ⟨81CAR284⟩
Shagidullin, R. R., **2**, 501 ⟨73JGU825⟩, 809 ⟨67CA76112w⟩, 886 ⟨73JGU1921⟩, 891 ⟨74JGU1506⟩; **4**, 127 ⟨67JGU1768⟩, 132 ⟨91JGU1459⟩, 478 ⟨69IZV1757⟩, 489 ⟨82ZOB2444⟩; **5**, 679 ⟨91JGU2372, 91ZOB2257⟩; **6**, 564 ⟨83ZOB785⟩
Shagi-Mukhametova, M. M., **2**, 476 ⟨90JGU398⟩
Shagova, E. A., **1**, 657 ⟨68DOK(179)1344⟩, 658 ⟨68IZV548⟩; **6**, 662 ⟨87IZV1139⟩
Shagun, V. A., **4**, 765 ⟨88IZV514⟩
Shagvaleev, F. S., **6**, 659 ⟨87ZOB2467⟩

Shah, A., **4**, 214 ⟨91CC947⟩; **6**, 546 ⟨87JIC194⟩, 547 ⟨85IJC(B)685⟩
Shahak, I., **1**, 684, 685, 686 ⟨74JCS(P1)1794⟩; **3**, 343, 348, 464 ⟨74JCS(P1)1794⟩; **5**, 232, 1030 ⟨73JA3440⟩
Shah, K. K., **5**, 998 ⟨66JIC481⟩
Shahal, L., **4**, 873 ⟨88TL6177, 92CC1048⟩
Shah, M., **5**, 339 ⟨87TL5431⟩
Shah, N. V., **4**, 98 ⟨87H(25)221⟩; **5**, 334 ⟨87H(25)221⟩
Shah, R., **4**, 970, 977, 978 ⟨83JCS(P1)1741⟩
Shah, R. D., **5**, 657 ⟨84JOC546⟩
Shah, R. K., **5**, 690 ⟨82IJC(B)989⟩
Shah, S. K., **1**, 592, 713, 714, 716 ⟨79JA6638⟩, 593, 713, 714, 715, 716 ⟨79JA6648⟩, 601, 602, 610, 713, 714, 715, 781 ⟨75JA3250⟩; **2**, 279 ⟨77JOC1773, 79JA6638⟩, 279 ⟨79JA6648⟩; **3**, 80 ⟨79JA6638⟩, 256 ⟨81JA3112⟩; **4**, 197 ⟨81JA3112⟩, 239, 375 ⟨79JA6638⟩, 399 ⟨77JA263, 79JA6648⟩; **5**, 412 ⟨77JOC1773⟩; **6**, 100 ⟨79JA6638⟩, 130 ⟨77JOC1773⟩, 485 ⟨92JMC3731⟩
Shah, V. G., **1**, 802 ⟨91T8155, 91TL1107⟩
Shah, V. H., **6**, 561 ⟨85MI 617-02, 86MI 617-01⟩, 561 ⟨87MI 617-01⟩
Shahin, H., **1**, 950 ⟨93TL5523⟩
Shahriari-Zavareh, H., **5**, 326 ⟨90JCS(P1)1623⟩
Shaikh, M., **6**, 118 ⟨82MI 604-01⟩
Shaikh, S., **3**, 431, 437, 618 ⟨75S502⟩
Shaikhutdinov, I. T., **4**, 336 ⟨B-84MI 407-02⟩
Shainyan, B. A., **4**, 94 ⟨86ZOR1923⟩, 704 ⟨91ZOR2362⟩, 763, 771 ⟨90SUL109, 93JOC3421⟩, 783 ⟨83ZOR1344, 84ZOR972⟩, 783 ⟨85ZOR965⟩, 861 ⟨93JPO59⟩
Shainyn, B. A., **4**, 861 ⟨92ZOR1711⟩
Shaker, M. R., **4**, 673, 722 ⟨88OM1715, 89OM1199⟩
Shakespeare, W. C., **1**, 499 ⟨90JA6416⟩, 1080 ⟨90JA8578⟩
Shakhaliev, S. M., **5**, 1130 ⟨77IZV249⟩
Shakhidayatov, Kh., **1**, 882 ⟨70RCR859⟩; **5**, 977 ⟨89ZOR885⟩
Shakoor, A., **4**, 659 ⟨85CB4248⟩
Shakya, R. P., **4**, 566 ⟨88SRI163⟩, 566, 573 ⟨85MI 412-01⟩
Shalaby, A. F. A. M., **6**, 114 ⟨64T25⟩
Shalom, E., **3**, 431 ⟨89IJ187⟩
Sham, H.-L., **1**, 25, 630 ⟨87JOC3346⟩, 606 ⟨84CC1345⟩; **3**, 215 ⟨87JOC3346⟩, 700 ⟨79JA1609⟩; **6**, 417 ⟨92EUP486948⟩
Sham, W.-C., **6**, 380 ⟨85JOM(291)25⟩
Shambayati, S., **1**, 149 ⟨90JA697⟩
Shambhu, M. B., **5**, 302 ⟨73TL1627⟩; **6**, 57 ⟨73TL4511⟩
Shamblee, D. A., **2**, 168 ⟨84JMC1379⟩, 322 ⟨88SC205⟩
Shamitin, V. V., **2**, 887 ⟨86JGU287⟩
Shamma, M., **4**, 47 ⟨68JOC3335⟩
Shamout, A. R., **1**, 1038 ⟨82LA1750⟩
Shamsevaleev, F. M., **5**, 384 ⟨90ZOB1997⟩, 581, 619 ⟨92ZOB71⟩, 642, 644 ⟨91RCR1128, 90ZOB464⟩
Shamshinov, S. D., **5**, 317 ⟨92MI 507-02⟩
Shamshoum, E. S., **4**, 675 ⟨83OM1167⟩
Shamsuzzaman, **5**, 533 ⟨88JCR(S)376⟩
Shan, S., **6**, 538 ⟨92MI 617-01⟩
Shanazarov, A. K., **4**, 987 ⟨87KGS1477⟩
Shandu, J. S., **5**, 226 ⟨87H(26)777⟩
Shani, A., **1**, 119, 464 ⟨74JA7101⟩; **3**, 55 ⟨81S137⟩; **4**, 188 ⟨81S137⟩
Shankar, B. B., **1**, 214 ⟨85TL6301⟩, 1089, 1090 ⟨87TL4127⟩; **4**, 961 ⟨85TL6301⟩
Shankar, B. K. R., **1**, 717 ⟨82TL2277⟩
Shankar, C. G., **1**, 555 ⟨83S310⟩
Shankar, C. R., **6**, 561 ⟨85IJC(B)580, 85MI 617-04⟩
Shankaran, K., **1**, 547, 1130 ⟨85TL6001⟩, 583 ⟨94JOC332⟩
Shankaranarayana, M. L., **6**, 594 ⟨70ACS351⟩
Shanker, B. B., **1**, 1090 ⟨87TL4123⟩; **4**, 938 ⟨91JOC3556, 87TL4123⟩
Shanklin, J. R., **1**, 622, 710 ⟨73JA6462, 73JA7424⟩; **3**, 378 ⟨73JA7424⟩
Shanmugam, P., **1**, 810 ⟨93SC311⟩
Shannon, I. J., **1**, 1079 ⟨93TL5621⟩
Shannon, M. L., **4**, 1051, 1052 ⟨84OM1897⟩
Shannon, P. V. R., **1**, 72 ⟨66JCS(C)41⟩, 184 ⟨87JCS(P2)1221⟩
Shantalii, O. A., **1**, 612 ⟨89ZOR654⟩
Shanzer, A., **5**, 1085 ⟨76JFC(7)65⟩
Shao, J. G., **3**, 637 ⟨91MI 318-01⟩
Shao, Y., **3**, 153 ⟨92TL3435⟩; **6**, 25, 26, 31 ⟨92TL3435⟩
Shapelko, N. N., **2**, 478 ⟨90K668⟩; **5**, 516, 521 ⟨75JOU1077⟩
Shapiro, B. L., **1**, 870, 871 ⟨69JOC2817⟩; **2**, 783 ⟨63JA2683⟩; **3**, 446 ⟨63JA2784⟩, 475 ⟨58JOC759⟩
Shapiro, D. G., **2**, 379 ⟨73OPP225⟩
Shapiro, G., **1**, 498 ⟨85TL1429, 85TL989⟩; **2**, 235 ⟨90H(32)2099⟩
Shapiro, H., **1**, 114 ⟨44JA1389⟩; **2**, 1101 ⟨B-68MI 223-02⟩; **5**, 1144 ⟨B-68MI 530-03⟩; **6**, 522, 523 ⟨60JA1325⟩
Shapiro, I., **2**, 109 ⟨59JCP(30)1222, 59JPC1319⟩
Shapiro, M. J., **4**, 423 ⟨80JHC1163⟩
Shapiro, R., **2**, 91 ⟨83JA7172⟩; **6**, 327 ⟨83JA7719⟩
Shapiro, R. H., **1**, 630, 631, 632 ⟨76OR(23)405⟩, 630, 631, 632, 633 ⟨67JA5734⟩, 630, 632 ⟨75TL1811⟩, 630, 633 ⟨78JOC1404⟩, 631 ⟨67JA1442⟩, 631 ⟨67JA471, 74JOC2302⟩, 633 ⟨68JA4762, 68TL345⟩, 633 ⟨78JOC1409, 80JOM(186)155⟩, 871 ⟨64JA2832⟩; **2**, 911, 956 ⟨76OR(23)405⟩
Shapiro, S. H., **5**, 31 ⟨43OSC(2)531⟩; **6**, 69, 74, 80 ⟨42JA2525⟩
Shapiro, S. L., **1**, 91 ⟨61JOC818⟩; **6**, 640 ⟨59JA3728⟩
Shapiro, U. G., **1**, 107 ⟨37JCS1177⟩
Shapley, J. R., **4**, 682 ⟨75JA7172⟩, 683 ⟨88JOM(353)103⟩, 686 ⟨80JOM(201)C31⟩; **6**, 345 ⟨86OM1757, 89JOM(371)257⟩
Shapley, P. A., **5**, 456, 801 ⟨90JA8369⟩
Shaposhnikov, S. I., **5**, 379 ⟨85ZOB1080⟩
Shapov, S. P., **4**, 140 ⟨69DOK(185)311⟩
Shapshinskaya, L. A., **2**, 454 ⟨62MI 208-01⟩
Shapter, J. G., **6**, 247 ⟨92CM979⟩
Sharanin, Y. A., **3**, 656 ⟨85S98⟩; **5**, 615 ⟨89ZOR1323⟩
Share, A. C., **1**, 16 ⟨93TA625⟩
Sharefkin, J. G., **5**, 652 ⟨73OSC(5)660⟩
Sharghi, H., **2**, 104 ⟨84HCA906⟩
Sharif, I., **2**, 15 ⟨92CC595⟩, 16 ⟨88JFC(41)297⟩
Shariffudin, R. S., **6**, 395 ⟨87JCS(D)747⟩
Sharif-Nassirian, F., **1**, 725 ⟨76LA1475⟩
Sharkey, W. H., **1**, 498 ⟨58JA751, 59JA2723⟩; **2**, 166, 170 ⟨65JOC1384⟩, 396 ⟨73OSC(5)445⟩, 1012, 1013 ⟨59JA803⟩; **3**, 330 ⟨65JOC1384⟩, 348 ⟨61JA2589⟩, 348, 349, 351 ⟨65JOC1375⟩, 709, 711 ⟨58JA4151⟩; **4**, 5 ⟨61JA382⟩, 256 ⟨65JOC1384⟩; **5**, 473, 474, 479, 480, 494 ⟨65JOC1375⟩, 474 ⟨61JA2589⟩, 1133 ⟨58JA4151⟩; **6**, 49, 253, 528, 529, 532 ⟨65JOC1375⟩, 528 ⟨61JA2589⟩
Sharma, A. H., **6**, 445 ⟨78JCS(P1)1066⟩
Sharma, A. K., **1**, 402, 799 ⟨80JA3972⟩; **2**, 162 ⟨75JOC2759⟩; **5**, 274, 286 ⟨92S439⟩
Sharma, G. V. M., **1**, 6 ⟨87JOC4784⟩, 174 ⟨90TL4931⟩, 174 ⟨91T519⟩, 1066 ⟨90T7033⟩, 1134 ⟨93TL707⟩; **2**, 85 ⟨90TL4495⟩, 317 ⟨89SC3289⟩; **6**, 550 ⟨90TL4931⟩
Sharma, G. V. R., **1**, 547 ⟨88TL6487⟩
Sharma, H. A., **3**, 701 ⟨92TL1593⟩
Sharma, I., **1**, 46 ⟨85IJC(B)59⟩
Sharma, I. K., **5**, 652 ⟨77SC57⟩
Sharma, J. P., **3**, 4, 115 ⟨90T4417⟩
Sharma, K., **5**, 710, 714 ⟨93CCC2150⟩
Sharma, K. K., **5**, 389 ⟨77JOM(142)165⟩
Sharma, K. S., **2**, 221, 225 ⟨72JOC2367⟩
Sharma, M., **1**, 799 ⟨71CC107, 71CC108⟩; **6**, 636 ⟨87TL2641⟩
Sharma, N., **2**, 249 ⟨87TL2641⟩; **5**, 214 ⟨88IJC(B)1159⟩, 328, 332, 710, 714 ⟨89CCC2738⟩, 710, 714 ⟨93CCC2150⟩
Sharma, N. D., **3**, 404 ⟨85JCS(P1)2123⟩, 405 ⟨86JCS(P1)867⟩; **4**, 262 ⟨92JCS(P1)1105⟩

Sharma, N. K., **1**, 232 ⟨78CJC512⟩, 600 ⟨73JA3420⟩; **2**, 149, 200 ⟨76CJC3012⟩, 200 ⟨73TL2863⟩; **4**, 393 ⟨73JA3420⟩
Sharma, P. K., **4**, 183 ⟨93JCR(S)430⟩
Sharma, R., **4**, 429, 430, 431 ⟨93JOC759⟩
Sharma, R. B., **1**, 905 ⟨85JA3747⟩
Sharma, R. D., **2**, 279 ⟨83JOM(255)61⟩; **4**, 291 ⟨86OM805, 87OM2164⟩; **6**, 101, 102 ⟨86CJC732, 88JOC3766⟩
Sharma, R. P., **3**, 40 ⟨79CC875⟩; **5**, 214 ⟨88IJC(B)1159⟩, 665 ⟨84IJC(B)728⟩, 710, 714 ⟨93CCC2150⟩
Sharma, S., **1**, 289 ⟨93TL7183⟩, 327 ⟨92JA7903⟩, 1106 ⟨86TL6161⟩; **2**, 607 ⟨89JOC5064⟩, 903 ⟨88TL261⟩, 917 ⟨91JOC770⟩; **4**, 639, 656, 663 ⟨93OM3979⟩; **5**, 1027 ⟨78S803, 86SR1⟩; **6**, 443 ⟨87IJC(B)748⟩, 562 ⟨92S919⟩, 577 ⟨75JMC913⟩
Sharma, S. D., **4**, 302 ⟨89S677⟩; **5**, 355 ⟨90JOC2196⟩; **6**, 113 ⟨87S990⟩, 115 ⟨84TL1849⟩
Sharma, S. K., **6**, 561 ⟨82FES205⟩
Sharma, S. N., **1**, 459 ⟨68CJC84⟩
Sharma, T. C., **6**, 502 ⟨78BCJ1245⟩
Sharma, V. L., **2**, 320 ⟨90CI547⟩
Sharp, B. W., **5**, 727 ⟨92TL5117⟩
Sharp, D. B., **5**, 842 ⟨46JA588⟩
Sharp, D. L., **5**, 4 ⟨66JOC2316⟩
Sharp, D. W. A., **2**, 421 ⟨72JCS(D)1766, 73JCS(D)1876⟩; **4**, 692 ⟨74JOM(80)C39⟩; **6**, 46 ⟨72JCS(P1)34⟩
Sharp, G. J., **6**, 72 ⟨89AJC1235⟩
Sharp, J. C., **6**, 25, 31 ⟨67JOC2166⟩
Sharp, J. T., **1**, 881 ⟨67JCS(C)1860⟩; **2**, 777 ⟨75JCS(P1)1065⟩
Sharp, K. G., **4**, 112 ⟨76IC1697⟩
Sharp, M. J., **1**, 473 ⟨87TL5093⟩; **2**, 326 ⟨88JA612⟩; **3**, 421 ⟨88JA612, 88TL901⟩; **5**, 572 ⟨94TL3651⟩, 1108 ⟨88JA612⟩
Sharp, P. R., **3**, 511 ⟨78JA2389⟩, 512 ⟨79JOM(171)43⟩
Sharp, R. L., **4**, 142 ⟨68JA2915⟩
Sharpe, D. A., **4**, 856 ⟨90JCS(P1)2967⟩
Sharpe, J., **1**, 700, 703 ⟨88JOC1064⟩
Sharpe, III, V. V., **6**, 76 ⟨77JOC679⟩
Sharpless, K. B., **1**, 14 ⟨82JOC1378, 85JOC5678⟩, 17, 185 ⟨88JA7538⟩, 206 ⟨74JA5254⟩, 217, 478 ⟨85TL4683⟩, 223 ⟨89JOC1295⟩, 587 ⟨81OS(60)29⟩, 601, 602, 603, 607, 608, 781 ⟨73TL1979⟩, 601, 604, 605, 606, 607 ⟨73JA6137⟩, 601, 604, 781 ⟨75JOC947⟩, 601, 608, 781 ⟨73JA2697⟩, 601, 608, 781 ⟨78JOC1689⟩, 601, 610, 611, 621, 646 ⟨75CS9⟩, 603, 608, 781 ⟨74JOC429⟩, 608 ⟨72JA7154⟩, 692 ⟨72JA6538⟩, 716 ⟨75CS9⟩; **2**, 21 ⟨79JOC4204, 79JOC4208⟩, 46 ⟨94TL2495⟩, 47 ⟨88JA1968, 91TL1003, 94T3639⟩, 48 ⟨89JOC2015, 90TL2990, 90TL3003, 90TL3817⟩, 92JOC2768, 94TL543⟩, 49 ⟨90TL4317, 92JA7570, 92JOC5067, 92TL6407, 92TL6411, 93JOC844⟩, 82 ⟨83PAC589⟩, 84 ⟨88JOC5185⟩, 85 ⟨73JA2697⟩, 86 ⟨77JA5526⟩, 91 ⟨79MI 202-02, 80JA5974⟩, 92 ⟨81JA6237⟩, 281 ⟨78JOC1689⟩, 282, 287 ⟨73JA6137⟩, 286 ⟨73TL1979⟩, 292 ⟨73JA7917, 94CRV2483⟩, 298 ⟨76JA269⟩, 322 ⟨89TL2623⟩, 348 ⟨83JOC3608, 90CL547⟩, 367 ⟨84JOC3861⟩, 721 ⟨87JOC2598⟩; **3**, 5, 117 ⟨76TL2503⟩, 37, 173, 207 ⟨73JA6137⟩, 156 ⟨92JOC5067⟩, 175 ⟨73TL1979⟩, 632 ⟨85JOC1557, 87JOC667⟩, 632 ⟨89JOC1295⟩; **5**, 52 ⟨81JOC3936⟩, 1106 ⟨88JA7538⟩
Sharrah, M. L., **4**, 5 ⟨51JA1329⟩
Sharrard, F., **5**, 633 ⟨85PS(25)279⟩
Sharrocks, D. N., **2**, 533 ⟨76JCS(D)2087⟩
Sharts, C. M., **2**, 15, 18 ⟨74OR(21)125⟩, 335 ⟨68JOC1008, B-69MI 206-01⟩; **5**, 361, 362 ⟨93TL6669⟩, 528 ⟨67JOC3709⟩; **6**, 212 ⟨B-69MI 607-01⟩
Sharutkin, V. V., **2**, 874 ⟨88JGU2050⟩; **4**, 719 ⟨77JOM(141)313⟩
Shasha, B. S., **6**, 276 ⟨69JOC1642, 71CAR(16)145⟩
Shashalova, M. I., **6**, 629 ⟨75ZOR1628⟩
Shashkov, A. S., **1**, 530 ⟨86TL1241⟩; **2**, 811 ⟨76BAU1374⟩; **6**, 65 ⟨76IZV2547⟩
Shat, G., **4**, 657, 663, 701 ⟨84TL5191⟩
Shatenshtein, A. I., **1**, 844 ⟨63APO(1)155⟩
Shaturskii, Y. P., **2**, 847 ⟨74ZOB85⟩
Shatzmiller, S., **2**, 314 ⟨92LA1005⟩; **3**, 431 ⟨89IJ187⟩; **4**, 485 ⟨93LA955⟩; **5**, 701 ⟨92LA997⟩
Shavel, Jr., J., **2**, 692 ⟨67TL859⟩; **3**, 251 ⟨72JHC171⟩; **4**, 221 ⟨72JHC175⟩, 304 ⟨69JHC429⟩, 307 ⟨70JHC1311⟩; **5**, 596 ⟨71JOC1742⟩
Shaver, A., **1**, 692 ⟨84JOM(268)C1⟩; **2**, 468 ⟨71IC2362⟩; **5**, 459 ⟨82JA1096⟩
Shaver, F. W., **5**, 1106 ⟨52JA1323⟩
Shavsha, T. G., **6**, 297 ⟨49DOK(68)515⟩
Shaw, A., **1**, 422 ⟨76JCS(P1)410⟩, 1010 ⟨89JOC1218⟩
Shaw, A. C., **2**, 423 ⟨87OS(66)127⟩
Shaw, B. L., **1**, 513 ⟨54JCS3217⟩; **2**, 435 ⟨83JOM(250)617⟩, 824 ⟨71JCS(A)2976⟩; **3**, 195 ⟨80JCS(D)299⟩; **4**, 548, 1023, 1028 ⟨86CC882⟩, 720 ⟨75JOM(102)511⟩, 1023, 1028 ⟨85CC1635⟩, 1023, 1028 ⟨85CC614⟩, 1023, 1028, 1030 ⟨82CC286⟩, 1028, 1030 ⟨82CC287⟩; **5**, 189 ⟨54JCS1860⟩; **6**, 151 ⟨82CC286⟩
Shaw, C. C., **2**, 627 ⟨90JHC2151⟩
Shaw, D. C., **3**, 306 ⟨76JCS(P1)749⟩
Shaw, E., **5**, 330 ⟨49JA1691⟩
Shaw, Jr., E. H., **4**, 287 ⟨26JA520⟩
Shaw, E. N., **1**, 959 ⟨42JA363⟩
Shaw, G., **2**, 32 ⟨76JCS(P1)1446⟩, 451 ⟨55JCS2039⟩; **3**, 247 ⟨61JCS3254⟩, 620 ⟨59JCS4040⟩; **5**, 1048 ⟨58JCS153⟩
Shaw, G. C., **6**, 2 ⟨62JOC1482⟩
Shaw, G. S., **2**, 1009 ⟨89JA5424⟩
Shaw, I. M., **2**, 25 ⟨76AJC133⟩
Shaw, J. E., **2**, 684 ⟨76JOC732⟩; **3**, 275 ⟨71TL4379⟩, 613 ⟨78JOC1017⟩; **5**, 132 ⟨74JOC1968⟩
Shaw, J. T., **2**, 271, 273 ⟨59JA5430⟩; **5**, 1125 ⟨62JOC3890⟩
Shaw, M. A., **3**, 498 ⟨67JCS(C)2442, 70JCS(C)5⟩; **4**, 563 ⟨70JCS(C)504⟩
Shaw, M. J., **2**, 613 ⟨70JA6498, 71JOC2917⟩
Shaw, N., **2**, 127 ⟨92JCS(P1)255⟩
Shaw, R. A., **2**, 465 ⟨64JCS1020⟩, 843, 845 ⟨76JCS(P1)1739⟩, 864 ⟨B-73MI 216-01⟩
Shaw, R., **1**, 1109 ⟨69CRV279⟩; **2**, 444 ⟨65JCS3354⟩
Shaw, R. S., **2**, 446 ⟨72MI 208-05⟩
Shawali, A. S., **3**, 680 ⟨80JHC833⟩, 681 ⟨87JHC577⟩; **5**, 358 ⟨72T5903, 75BCJ365⟩, 358 ⟨78BCJ512⟩, 675 ⟨90H(31)247⟩, 676 ⟨82H(19)57, 82H(19)319, 84JHC797⟩, 676, 677 ⟨82JHC73, 83JHC719⟩, 703 ⟨87JHC577, 88H(27)695⟩, 704 ⟨80JHC1745, 82H(19)1477⟩, 771 ⟨80JHC877⟩; **6**, 654 ⟨85JHC813⟩
Shawe, T. T., **3**, 182 ⟨91JOC2751⟩
Shawl, E. T.,**3**, 397 ⟨67CC670⟩
Shay, W. R., **2**, 920 ⟨92JA6580⟩
Shchegoleva, T. M., **2**, 886 ⟨81JGU1056⟩
Shchekotikhina, N. A., **6**, 80 ⟨77ZOR943⟩
Shchelokov, V. I., **6**, 104 ⟨64TL47, 65T3537⟩
Shchepin, V. V., **4**, 768 ⟨92ZOR522⟩
Shcherbakov, S. V., **5**, 1024 ⟨85ZOB2100⟩; **6**, 476 ⟨85ZOB2100⟩
Shcherbakov, V. I., **5**, 390 ⟨77ZOB2294, 80ZOB886⟩, 390 ⟨81ZOB1318⟩, 391 ⟨87ZOB1643⟩
Shcherbakova, I. V., **1**, 262 ⟨91JCS(P1)3295⟩; **2**, 1008 ⟨90AHC(50)157, 92CJC2040⟩; **4**, 323 ⟨92S1295⟩
Shcherbinin, V. V., **5**, 722 ⟨72JGU1859⟩, 722, 978 ⟨72ZOB1870⟩
Shchipanov, V. P., **5**, 860 ⟨78KGS991⟩
Shchukareva, T. M., **2**, 448 ⟨76JGU1213⟩
Shchukina, L. V., **2**, 1091 ⟨79JGU536⟩
Shchukina, M. N., **5**, 827 ⟨67ZOR1207⟩
Shchukovskaya, L. L., **3**, 540 ⟨65DOK(164)357, 68DOK(179)892⟩, 540 ⟨71ZOB240⟩
Shchupak, E. A., **4**, 641 ⟨76IZV1342⟩
Shdo, J. G., **4**, 631, 697 ⟨64JOC2742⟩, 735 ⟨67JOC2749⟩, 736 ⟨68JOC2173⟩

Shea, K. J., **1**, 806 ⟨80JA3156⟩, 806, 807 ⟨77JA1499, 92JA2635⟩, 811 ⟨86JA5901⟩, 831 ⟨94TL1317⟩, 928 ⟨73JA6728⟩, 1092 ⟨83TL1003⟩; **5**, 688, 694 ⟨89JOC4335⟩
Shealy, O. L., **5**, 549 ⟨51JA25⟩
Shealy, Y. F., **4**, 282 ⟨85EUP146838⟩; **5**, 199 ⟨69JA3075⟩
Shearer, H. M. M., **3**, 458 ⟨71CC1275⟩, 459 ⟨76CC160⟩; **6**, 183 ⟨77JCS(D)136⟩
Shearer, Jr., N. H., **1**, 45 ⟨46JA1813⟩
Sheats, J. E., **3**, 573 ⟨62JOC4053⟩
Sheats, J. R., **3**, 289 ⟨75JA5448⟩
Shechter, H., **1**, 39 ⟨65JA935⟩, 615, 700 ⟨82TL1963⟩, 616, 654 ⟨84TL1219⟩, 616, 654 ⟨88JOC2688⟩, 630 ⟨67JA5736⟩, 631 ⟨65JA659, 68JA531⟩, 653 ⟨82TL3455⟩, 717 ⟨82TL2277⟩, 869, 871 ⟨69JA2797, 70JA4689⟩, 871 ⟨62JA2905, 69JA530⟩, 1077 ⟨78JA7927⟩; **2**, 27 ⟨73OSC(5)328⟩, 36 ⟨63OSC(4)321, 63OSC(4)323⟩, 674 ⟨72TL2677⟩, 1003 ⟨83JA6096⟩; **3**, 65 ⟨74JOC2939, 78JA7934⟩, 65, 232 ⟨78JA7927⟩, 193, 472 ⟨90TL4987⟩, 465 ⟨65JA935⟩, 502 ⟨76JA7846⟩, 577 ⟨66CC689⟩; **4**, 138 ⟨86JOC3369⟩, 352 ⟨85TL1115⟩, 420, 424, 449 ⟨64JA1701⟩, 428 ⟨61JA3535⟩, 430, 431, 432, 994 ⟨57JA4708⟩, 647 ⟨84TL4645⟩, 709 ⟨61JOC1034⟩, 796 ⟨76JOC2112⟩; **5**, 201 ⟨56JOC1473⟩; **6**, 145 ⟨61JOC51⟩, 150 ⟨61JA3535⟩, 170 ⟨57JA4708⟩
Sheehan, D., **1**, 523 ⟨90JA4965⟩
Sheehan, J. C., **3**, 468 ⟨74JOC1444, 78JOC2203⟩; **4**, 70 ⟨78JOC2203⟩; **5**, 81 ⟨59JA3089⟩, 106 ⟨64JOC2006⟩, 140, 279 ⟨56JA1367, 55JA1067⟩, 279 ⟨56JOC439, 65JA2492⟩, 279, 1062 ⟨61JOC2525⟩, 652 ⟨57OR(9)388⟩
Sheehan, J. J., **1**, 958, 1125 ⟨49JA1964⟩
Sheehan, J. T., **6**, 104 ⟨61JOC4315⟩
Sheenan, J. C., **3**, 722 ⟨64JA1356⟩
Sheeran, P. J., **2**, 1048 ⟨68JOC4406⟩
Sheerdina, N. J., **2**, 305 ⟨38JGU1825⟩
Sheers, E. H., **5**, 1121 ⟨60JOC147⟩
Sheetz, D. P., **5**, 1010 ⟨70JOC47⟩
Sheffer, A. R., **4**, 765 ⟨77ZOR1675⟩
Sheffy, F. K., **3**, 218, 303 ⟨84JA4833⟩
Sheft, I., **2**, 109 ⟨53JA192⟩
Shehata, N., **5**, 615 ⟨90JCS(P1)1267⟩
Shei, J. G., **2**, 147, 155 ⟨82JOC3773⟩
Sheibe, G., **2**, 99 ⟨38CB1873⟩
Sheinker, Y. N., **4**, 987 ⟨87KGS1477⟩; **6**, 104 ⟨62TL701⟩
Shekhani, M. S., **4**, 264 ⟨79JA1057⟩
Shelberg, W. E., **3**, 612 ⟨44JA218⟩, 618, 633 ⟨45JA1745⟩
Sheldon, J. C., **2**, 150 ⟨56JCS2705⟩, 460 ⟨59JA6177⟩
Sheldon, R. A., **1**, 560 ⟨72OR(19)279⟩, 924, 925 ⟨70JA4395⟩; **2**, 12 ⟨72OR(19)279⟩, 90 ⟨80JMOC107, B-81MI 202-01⟩; **4**, 27 ⟨72OR(19)279⟩; **6**, 118 ⟨67JCS(C)1547⟩
Sheldon, R. I., **5**, 655 ⟨85JHC1121⟩
Sheldrake, G. M., **4**, 800 ⟨86CB3150⟩
Sheldrake, G. N., **3**, 333, 340 ⟨84CC1469⟩, 333, 340 ⟨84CC922⟩
Sheldrake, P. W., **1**, 1134 ⟨78JA4618, 78JA4620⟩; **2**, 629 ⟨78JA4618⟩; **5**, 173 ⟨78JA4618⟩, 173, 204 ⟨78JA4620⟩
Sheldrick, G., **6**, 440 ⟨91CB2411⟩
Sheldrick, G. M., **2**, 428, 429, 432 ⟨69JINC3684⟩, 880 ⟨91JOM(402)55⟩, 1000 ⟨85AG(E)573⟩, 1085 ⟨87CB1437⟩; **3**, 370 ⟨90CB1475⟩; **5**, 381 ⟨84CB2112⟩, 606 ⟨85ZN(B)124⟩, 663 ⟨83CB724⟩; **6**, 184 ⟨90CC741⟩, 184 ⟨92CB1807⟩, 196 ⟨93CB2247⟩, 252 ⟨92CC1017⟩, 380 ⟨85ZC309⟩, 604 ⟨86JFC(34)46⟩, 726 ⟨85CB5007⟩
Sheldrick, G. S., **5**, 1086 ⟨92AG(E)437⟩
Sheldrick, W. S., **2**, 476 ⟨76CB3581⟩, 856 ⟨73CC144⟩; **5**, 548 ⟨80CB3605⟩, 1128 ⟨80AG169⟩; **6**, 156 ⟨83CB549⟩
Shellhamer, D. F., **2**, 18 ⟨83JOC3195⟩, 30 ⟨78JOC2793⟩, 606 ⟨80JOC4649⟩
Shellman, R. W., **2**, 374 ⟨56JA3874⟩
Shelly, D. P., **2**, 632 ⟨87JCS(P1)1039⟩
Shelly, K. P., **1**, 769 ⟨88CJC1359⟩
Shelnut, J. G., **3**, 464 ⟨75CC114⟩
Shelton, B. R., **5**, 664, 821 ⟨80JOC3916⟩
Shelton, J. R., **1**, 595 ⟨73IJS205⟩; **2**, 403 ⟨68CJC1149, 71S204⟩
Sheluchenko, V. V., **2**, 476 ⟨68ZOB1906⟩
Sheludyakov, D. O., **5**, 968, 969 ⟨75ZOB2025⟩
Sheludyakov, S. V., **5**, 1076, 1077 ⟨78ZOB1943⟩
Sheludyakov, V. D., **4**, 140 ⟨83JGU109⟩, 140 ⟨88JGU81⟩, 662 ⟨91MI 414-01⟩, 1045 ⟨86ZOB2743⟩; **5**, 722 ⟨72JGU1859⟩, 722, 863 ⟨71DOK(199)103⟩, 722, 978 ⟨72ZOB1870⟩, 968 ⟨69ZOB2598, 71BRP1235156⟩, 968 ⟨79RCR473⟩, 968, 969 ⟨70DOK(190)110⟩, 968, 969 ⟨73ZOB1502⟩, 968, 969 ⟨72ZOB2501, 73ZOB623⟩, 977 ⟨69DOK(181)115⟩, 1000 ⟨73UK1451⟩, 1006, ⟨77ZOB2259⟩, 1014 ⟨68DOK(181)115⟩, 1015 ⟨71ZOB2340, 72ZOB879, 73ZOB1044⟩, 1016 ⟨68ZOB1179⟩, 1075 ⟨85ZOB1308⟩, 1079 ⟨86ZOB1525⟩; **6**, 443 ⟨69ZOB2598, 80ZOB875⟩, 445 ⟨69ZOB220, 77ZOB1063⟩, 626 ⟨88ZOB393, 91ZOB1202⟩
Sheludyakova, O. A., **1**, 1010 ⟨88ZOR650⟩
Sheludyakova, S. V., **5**, 721, 722 ⟨78JGU1313⟩, 721, 722 ⟨78ZOB1429⟩, 1006 ⟨77ZOB2259⟩, 1082 ⟨78ZOB1313⟩; **6**, 444 ⟨80ZOB875⟩, 445 ⟨77ZOB1063⟩
Shelyazhenko, S. V., **5**, 112 ⟨85ZOR2407⟩
Shemanina, V. N., **2**, 288 ⟨59ZOB942, 60ZOB4069⟩; **6**, 49 ⟨59ZOB942⟩
Shemin, D., **1**, 80 ⟨43OSC(2)491⟩; **5**, 66 ⟨43OSC(2)519⟩, 90 ⟨43OSC(2)491⟩, 652 ⟨43OSC(2)11⟩
Shemonaeva, T. G., **5**, 674 ⟨76URP534448⟩
Shemyakin, M. M., **1**, 639 ⟨64AG(E)250⟩, 721 ⟨64PAC271⟩, 722 ⟨63IZV1053⟩, 725 ⟨64IZV2003⟩, 725, 733 ⟨63T149⟩, 727, 733 ⟨67T2709⟩; **3**, 494 ⟨67T2709⟩; **4**, 752 ⟨63T149, 64TL2669⟩; **5**, 100 ⟨64AG(E)250⟩; **6**, 104 ⟨62TL701, 63TL439⟩, 104 ⟨64TL47, 65T3537⟩
Shen, C. C., **4**, 1047 ⟨79TL87⟩
Shen, C. C. C., **1**, 859 ⟨79TL327⟩
Shen, G., **1**, 145 ⟨90JOC2234⟩
Shen, G. S., **1**, 769 ⟨91S165⟩
Shen, G. Y., **1**, 1029 ⟨91JA3873⟩
Shen, J., **1**, 901 ⟨73JA4960⟩
Shen, M., **2**, 1034 ⟨81TL3347⟩; **3**, 347 ⟨72CJC3923⟩
Shen, P. D., **6**, 264 ⟨81JOC4573⟩
Shen, Q., **1**, 178 ⟨93JOC6174⟩
Shen, S., **1**, 909 ⟨86JA134⟩
Shen, T., **1**, 405 ⟨89JA2981⟩, 816 ⟨87TL4993⟩; **2**, 556 ⟨89JA2981⟩; **4**, 380 ⟨89JA2981⟩
Shen, T. Y., **3**, 355 ⟨92MI 308-02⟩
Shen, W., **1**, 326, 328 ⟨90JA6738⟩, 339, 927 ⟨88JA5900⟩, 339, 927 ⟨93JA6051⟩
Shen, Y., **1**, 181 ⟨90JOC3961⟩, 270 ⟨88S321⟩, 646 ⟨82AOC(20)115⟩, 648 ⟨86TL2903, 89SC501⟩, 648 ⟨89SC83⟩, 648, 743 ⟨89JOM(366)87⟩, 987 ⟨91JCS(P1)665⟩, 1079 ⟨81TL5283⟩; **2**, 73 ⟨88JCS(P1)2855, 88TL1395⟩, 487, 488 ⟨89JOM(366)87⟩, 936 ⟨90TL543⟩; **3**, 300 ⟨90JOC3961⟩, 500 ⟨88S321⟩, 644 ⟨89SC83⟩, 649 ⟨91JCS(P1)487⟩; **3**, 79, 274 ⟨85S159⟩, 269, 274 ⟨84S35⟩; **4**, 598 ⟨84TL4425⟩, 798 ⟨93SC1⟩
Shen, Y. C., **4**, 598 ⟨82AOC(20)115⟩
Sheng, H., **1**, 738 ⟨65MI 116-01⟩, 1047 ⟨91S235⟩
Sheng, M. N., **2**, 90 ⟨70JOC1839⟩, 348 ⟨70OS(50)56⟩
Shenhav, H., **1**, 889 ⟨70JCS(B)469⟩
Shenk, Jr., W. J., **2**, 643 ⟨45JA1939, 45JA386⟩, 645 ⟨45JA1937⟩, 645, 646 ⟨44JA1464⟩
Shenkenburger, J.,**1**, 455 ⟨77T457⟩
Shenov, P. K., **2**, 7, 837 ⟨73OSC(5)249⟩, 837 ⟨73OSC(5)142⟩
Shenoy, S. J., **5**, 690 ⟨82IJC(B)989⟩
Shenton, K. E., **1**, 204, 1023, 1139 ⟨75TL51⟩; **5**, 170 ⟨75TL51⟩

Shenvi, A. B., **1**, 1134 ⟨80HCA2426⟩; **4**, 528 ⟨84JBC(259)15106, 85USP4499082⟩
Shen Yan-Chang, **1**, 740 ⟨80MI 116-01⟩, 740, 741 ⟨80MI 116-02⟩
Shepard, A. F., **2**, 687 ⟨67USP3308164, 67USP3308165⟩
Shepard, K. L., **6**, 506 ⟨68USP3575975⟩
Shepeleva, E. S., **4**, 119, 125, 126 ⟨53IZV763⟩
Shephard, K. P., **1**, 993 ⟨79JOC1582⟩
Shepherd, J. P., **2**, 44 ⟨72TL4907⟩
Shepherd, M. K., **3**, 227 ⟨86JCS(P1)961⟩; **4**, 20 ⟨88JCS(P1)961⟩, 21 ⟨86JCS(P1)1495⟩
Shepherd, R. A., **1**, 811 ⟨70JA2158⟩
Shepherd, R. G., **3**, 672 ⟨73JHC899⟩; **5**, 82, 92 ⟨83USP4382143⟩, 595 ⟨76JCS(P1)983⟩
Shepherd, T. M., **3**, 196, 500 ⟨82T1975⟩; **6**, 708 ⟨82T3355⟩
Shepleva, E. S., **4**, 130 ⟨50DOK(75)219, 50IZV39⟩
Sheppard, A. C., **2**, 652 ⟨87TL5115⟩; **3**, 171 ⟨90OR(38)655⟩, 183 ⟨88TL4365⟩, 438 ⟨87TL5115⟩
Sheppard, W. A., **6**, 233 ⟨64JOC898⟩
Sheppard, C., **1**, 595 ⟨77JCS(P1)1574⟩; **2**, 153 ⟨77JCS(P1)1574⟩
Sheppard, G. S., **2**, 702 ⟨90JA866⟩
Sheppard, J. H., **2**, 515 ⟨78TL4715⟩, 516 ⟨79TL4777⟩
Sheppard, L., **4**, 742 ⟨90JCS(P1)1881⟩
Sheppard, N., **5**, 506 ⟨49TFS693⟩
Sheppard, R. C., **5**, 278 ⟨60RCP243⟩
Sheppard, R. N., **1**, 780 ⟨91T6813⟩
Sheppard, W. A., **2**, 15, 18 ⟨74OR(21)125⟩, 335 ⟨B-69MI 206-01⟩, 688 ⟨64JOC898⟩, 716 ⟨64OS(44)39⟩, 836, 839 ⟨68JA5236⟩; **3**, 368 ⟨64JA1891⟩; **4**, 8 ⟨71T945⟩, 9 ⟨73OSC(5)959⟩; **5**, 1129 ⟨68JA5236⟩; **6**, 39, 229, 230 ⟨64JOC1⟩, 212 ⟨B-69MI 607-01⟩, 229 ⟨61JA4860⟩, 229, 230, 424 ⟨64JOC11⟩, 233, 234 ⟨64JOC895⟩, 239, 603 ⟨65JA4338⟩
Sheptienko, O. G., **2**, 477 ⟨77ZOB1432⟩
Sher, P. M., **1**, 330, 341, 923, 936 ⟨86JA303⟩, 548 ⟨88OS(66)75⟩
Sheradsky, T., **2**, 298 ⟨B-82MI 205-01⟩, 691 ⟨72T3833⟩, 692 ⟨67JHC413⟩; **4**, 105 ⟨86JOC3123, 87JOC101⟩; **5**, 351, 366 ⟨85JOC5604, 91TL133⟩
Sherbine, J. P., **4**, 365, 366 ⟨85TL2965⟩, 366, 394, 945, 1060 ⟨84JA3245⟩, 652 ⟨93JA10754⟩
Sherburn, M. S., **1**, 929 ⟨89T7835⟩
Sherchenko, I. V., **2**, 467 ⟨91ZN(B)258⟩
Sherer, O. J., **2**, 503 ⟨69JOM(16)P69⟩
Shereshovets, V. V., **4**, 267 ⟨90ZOR1259⟩
Shergina, N. I., **2**, 1093 ⟨81IZV1280⟩
Sheridan, J. B., **4**, 676 ⟨88IC3248, 88OM289⟩; **5**, 449 ⟨89PAC1723⟩, 451 ⟨87JA1584⟩
Sheridan, R. S., **4**, 763 ⟨87JA5029⟩; **6**, 280 ⟨86JA99⟩
Sherif, S. M., **3**, 669 ⟨86H(24)2023⟩, 669,672 ⟨87H(26)497⟩; **5**, 631 ⟨87S228⟩, 704 ⟨80JHC1745⟩, 771 ⟨80JHC877⟩
Sherman, L. R., **4**, 203 ⟨91JOC3283⟩
Shermergorn, I. M., **4**, 121, 126 ⟨71JGU2228⟩, 124 ⟨75JGU1950⟩, 127 ⟨67IZV1535, 67JGU1768⟩, 127, 128 ⟨68IZV1557⟩, 346 ⟨68IZV1625⟩, 556 ⟨67DOK(177)340⟩
Shermolovich, Y. G., **2**, 239 ⟨90ZOR2083⟩, 438 ⟨90ZOB1243⟩; 399 ⟨77ZOR727⟩; **3**, 374 ⟨78ZOR1659⟩, 385 ⟨91ZOR408⟩; **4**, 110 ⟨82JOU2240⟩, 326 ⟨78ZOR1659⟩; **6**, 621 ⟨73ZOR633, 74ZOR1000⟩
Sherrill, R. G., **6**, 598 ⟨91JMC3187⟩
Sherringham, J. A., **1**, 614 ⟨88TL6357⟩
Sherrington, D. C., **2**, 140 ⟨93CSR199⟩
Sherrod, S. A., **1**, 1114 ⟨69JA2115, 71JA1925⟩
Sherry, J. J., **3**, 275 ⟨71TL4379⟩
Sherstobitov, O. E., **2**, 1033 ⟨91RCR885⟩
Sherstyannikova, L. V., **1**, 61 ⟨83ZOB806⟩; **2**, 1093 ⟨81IZV1280⟩
Sherwin, C. S., **4**, 203 ⟨91JOC3283⟩
Sherwin, P. F., **3**, 335, 369 ⟨82JA3119, 76JA1264⟩; **4**, 258 ⟨82JA3119⟩
Sherwood, L. T., **3**, 612 ⟨44JA545⟩
Sherwood, P., **4**, 712 ⟨88JCS(D)2431⟩
Sherwood, P. W., **5**, 24 ⟨60CI(L)1096⟩
Shestakova, A. K., **4**, 654 ⟨92MI 414-01⟩, 6540 ⟨92MIP173178⟩
Shestakov, A. F., **6**, 341 ⟨90JOM(384)279⟩
Shestopalov, A. M., **3**, 656 ⟨85S98⟩; **5**, 615 ⟨89ZOR1323⟩
Sheth, J. D., **1**, 803, 804 ⟨80JOC5020⟩
Shetta, A., **5**, 676, 677 ⟨82JHC73⟩
Shetta, A.-F., **5**, 676 ⟨82H(19)57⟩
Shetta, A. H., **5**, 704 ⟨82H(19)1477⟩
Shetton, J. R., **2**, 153, 184 ⟨67JA718⟩
Shetty, B. V., **5**, 282 ⟨61JOC3002⟩
Shetty, S., **1**, 1010 ⟨92BMC1619⟩
Sheu, C., **3**, 113 ⟨90JA1936⟩
Sheu, J., **5**, 199 ⟨89TL3295⟩, 346 ⟨92OPP147⟩
Sheu, J.-H., **3**, 232 ⟨89JOC5126⟩
Sheu, J. T., **3**, 634 ⟨76TL3799⟩
Sheu, R. G., **5**, 170 ⟨88JOC2371⟩
Shevchenko, I. V., **4**, 564 ⟨92CB1325, 93PS(75)233⟩; **5**, 401 ⟨89JGU1977⟩; **6**, 612, 623 ⟨81ZOR180⟩, 683 ⟨83ZOB473, 85ZOB2797⟩, 689 ⟨85ZOB1862, 85ZOB2214⟩
Shevchenko, M. B., **6**, 612, 623 ⟨81ZOR180⟩
Shevchenko, M. V., **6**, 612, 727 ⟨79ZOR215⟩, 619 ⟨73ZOR43, 75ZOR71⟩, 619 ⟨78ZOR1841⟩, 658 ⟨73ZOR1815⟩
Shevchenko, S. G., **5**, 924 ⟨89JOM(368)163⟩
Shevchenko, V. I., **2**, 407 ⟨67ZOB852, 67ZOB2693⟩, 407 ⟨67ZOB856⟩; **5**, 373 ⟨70ZOB1949⟩, 683 ⟨66JGU1276, 66ZOB1260⟩, 861 ⟨68ZOB326, 68ZOB850⟩; **6**, 616, 728 ⟨72JGU97⟩, 621 ⟨73ZOR633, 74ZOR1000⟩, 621 ⟨78JGU1078⟩, 656 ⟨74ZOB1264⟩
Shevchenko, Z. A., **1**, 1022 ⟨78JOU1095⟩
Shevchuk, L. I., **5**, 902 ⟨77JGU658⟩
Shevchuk, M. I., **2**, 444 ⟨75ZOB1228⟩; **4**, 570 ⟨77ZOB1715⟩; **5**, 597 ⟨69ZOB1282⟩, 631 ⟨75ZOB2116⟩, 904 ⟨71JGU2019⟩
Shevelev, S. A., **4**, 1083 ⟨89IZV356⟩; **6**, 56, 323 ⟨77IZV139⟩, 144 ⟨66UK1740⟩, 145 ⟨68IZV447⟩, 145 ⟨77IZV136⟩, 145 ⟨90IZV1816⟩, 149 ⟨72IZV1717⟩, 161, 283, 369 ⟨85IZV439⟩, 261, 281, 323 ⟨78IZV2348⟩, 281 ⟨75IZV2750⟩, 281, 348 ⟨73IZV350⟩, 283 ⟨77IZV2530, 77IZV375⟩, 283, 349 ⟨74IZV1350⟩, 323 ⟨75IZV2621, 76IZV2640⟩, 323 ⟨78IZV1091⟩, 348, 349, 653 ⟨75IZV1669⟩, 368 ⟨67IZV2708, 73IZV819, 74ZOR1793⟩, 653 ⟨74MI 621-02⟩
Shevlin, P. B., **1**, 321, 591 ⟨76JA951⟩, 600 ⟨78JA3874⟩; **3**, 65 ⟨79JA1303⟩, 465 ⟨66JA4735⟩
Shevschenko, V. I., **5**, 988 ⟨61ZOB3062⟩
Shevtchenko, V. I., **5**, 57 ⟨73S545⟩
Shewchuk, L. M., **2**, 36 ⟨86H(24)3043⟩
Shhepin, V. V., **4**, 775 ⟨91ZOR1796⟩
Shi, G., **1**, 634 ⟨89JFC(44)161⟩
Shi, L., **1**, 646, 647 ⟨86TL4583, B-90MI 114-01⟩, 647 ⟨87JOC3558, 87TL2159⟩, 647 ⟨88LA377, 89JOC2027⟩, 647, 738 ⟨85TL6447⟩, 647, 741 ⟨87TL2155⟩, 647, 741 ⟨88TL3949⟩; **3**, 60 ⟨87JOC3558⟩, 637 ⟨91JCS(P1)1893⟩
Shi, L.-L., **1**, 270 ⟨91TL3999⟩, 738 ⟨81MI 116-01⟩; **2**, 55 ⟨90TL1185⟩; **3**, 48 ⟨85TL6447, 88TL3949⟩, 48, 65 ⟨86TL4583⟩, 49 ⟨B-90MI 301-02⟩, 196 ⟨88S975⟩, 196, 231 ⟨87TL2155⟩, 196 ⟨B-90MI 304-02⟩, 649 ⟨89SC2639⟩
Shi, X., **3**, 265 ⟨90S162⟩; **4**, 899 ⟨90CB541⟩, 900 ⟨90S162⟩
Shi, X.-X., **2**, 22, 28 ⟨93JOC4596⟩
Shi, Y., **1**, 494 ⟨93JA9421⟩; **5**, 953, 956 ⟨94JOC6882⟩, 958 ⟨94JOC5122⟩
Shi, Y.-J., **2**, 66 ⟨93TL3243⟩
Shia, G. A., **2**, 16 ⟨91JOC5962⟩; **3**, 149 ⟨91JOC5962⟩
Shian, W.-I., **2**, 824, 834, 835 ⟨78JA7304⟩
Shiao, M.-J., **5**, 582 ⟨92S1219⟩, 582, 611 ⟨93JOC4742⟩
Shiau, F.-Y., **4**, 643, 645, 646 ⟨84JOC2565⟩

Shiau, G. T., **5**, 1040 ⟨75MI 527-01⟩
Shiba, S., **5**, 251 ⟨80SC305⟩
Shiba, T., **1**, 882 ⟨87TL5887⟩; **5**, 368 ⟨85BCJ352⟩
Shibafuchi, H., **3**, 684 ⟨86BCJ2827⟩
Shibagaki, M., **1**, 456 ⟨92CL1725⟩; **2**, 55 ⟨88BCJ3283, 88CL1633⟩; **3**, 7 ⟨91BCJ312⟩
Shibahara, J., **2**, 398 ⟨88TL4769⟩
Shibahashi, H., **5**, 510, 551, 583, 620 ⟨82LA1229⟩
Shibanuma, T., **5**, 1025 ⟨77CL573⟩, 1078 ⟨77CL575⟩
Shibasaki, K., **2**, 748 ⟨83JOC3566⟩
Shibasaki, M., **1**, 456 ⟨93S643⟩, 456, 507 ⟨85JOC1147⟩, 495 ⟨85JOC3246, 92TL2589, 94T371⟩, 827, 828 ⟨77TL1625⟩, 1024 ⟨86TCC3⟩, 1081 ⟨82TL4607⟩; **2**, 87 ⟨75TL3183⟩, 912, 945 ⟨91JOC3846⟩, 927, 988 ⟨92JOC3519⟩, 1036, 1037 ⟨91JCS(P1)2049⟩; **3**, 59, 220 ⟨87TL1893⟩, 218, 225 ⟨86JA2090⟩; **4**, 154, 155 ⟨81TL2397⟩, 155 ⟨83CL1303⟩; **5**, 375 ⟨85TL2217⟩
Shibata, F., **2**, 907, 908 ⟨87BCJ777⟩; **4**, 806, 818 ⟨87BSJ777⟩
Shibata, H., **3**, 415 ⟨86CPB1050⟩
Shibata, I., **2**, 701 ⟨93CC1067, 93CL1953⟩
Shibata, J., **4**, 66 ⟨82CL587⟩
Shibata, K., **2**, 22 ⟨91BCJ3749⟩; **4**, 847 ⟨89TL2091⟩
Shibata, N., **4**, 223 ⟨93TL4063⟩, 418 ⟨91TL2375⟩; **5**, 377 ⟨89JCS(P1)1862⟩
Shibata, O., **6**, 501 ⟨72CL971⟩
Shibata, S., **2**, 528 ⟨87JA5551, 87JA7925⟩
Shibata, T., **2**, 48 ⟨90TL3817⟩, 49 ⟨93JOC844⟩, 138 ⟨82JA3225⟩; **3**, 166 ⟨82JA3228⟩, 634 ⟨85H(23)3069⟩; **5**, 138 ⟨84T145⟩, 1039 ⟨85H(23)3069⟩
Shibata, Y., **3**, 63 ⟨90JCS(P1)197⟩; **4**, 736 ⟨86TL3173⟩
Shibato, K., **1**, 229 ⟨84TL1159⟩, 1045 ⟨86S421⟩; **3**, 219 ⟨85CL1145⟩
Shibayama, K., **2**, 857 ⟨84CL603⟩; **3**, 605 ⟨84TL1809, 87PS(30)527⟩, 609 ⟨86JOM(311)C63⟩; **5**, 879 ⟨83CL1653⟩, 1086, 1087 ⟨84TL1809⟩
Shibib, S. M., **5**, 288 ⟨85JCS(P1)831⟩
Shibutani, T., **3**, 310 ⟨86TL3899⟩
Shibuya, H., **1**, 1073 ⟨91CL1805⟩
Shibuya, I., **3**, 371 ⟨79BCJ1235⟩; **5**, 320, 335 ⟨91BCJ3732, 86BCJ2017⟩; **6**, 314 ⟨89NKK63⟩, 575 ⟨90JCS(P1)1218⟩
Shibuya, M., **1**, 577 ⟨94JOC406⟩, 827 ⟨93CC793⟩; **5**, 346 ⟨83H(20)1531⟩; **6**, 563, 565 ⟨82JMC557⟩
Shibuya, S., **1**, 41 ⟨78CC414⟩, 290 ⟨87CL495⟩, 651 ⟨78S713⟩, 1077 ⟨78JOC4366, 78S305⟩; **3**, 309 ⟨81JOC2979⟩, 634 ⟨79H(12)681⟩, 654 ⟨78CPB1874⟩; **4**, 475 ⟨92TA377⟩, 648 ⟨78S746⟩; **5**, 284 ⟨87H(26)359⟩, 570, 613 ⟨87H(26)163⟩
Shibuya, T., **5**, 432 ⟨87CL495⟩
Shido, J. G., **4**, 631, 634 ⟨63JA2684⟩
Shieder, G., **2**, 470 ⟨68CB4148⟩
Shieh, W.-C., **1**, 116 ⟨92TL6435, 93TL5995⟩; **2**, 619 ⟨90JHC1549⟩
Shields, C. J., **3**, 564 ⟨89T259⟩; **5**, 1073 ⟨89T259⟩
Shields, D. J., **5**, 532 ⟨65JCS3312⟩
Shields, S., **5**, 460 ⟨74IC1595⟩
Shields, T. C., **1**, 890 ⟨67JA3365⟩; **2**, 131 ⟨69JA5415⟩
Shiess, P. W., **1**, 89 ⟨63JOC6⟩
Shiga, F., **1**, 1009 ⟨92S746⟩
Shiga, M., **4**, 797 ⟨69BCJ3270⟩
Shigeizumi, S., **2**, 24 ⟨94TL737⟩
Shigematsu, N., **5**, 1106 ⟨72CPB427⟩
Shigemoto, T., **5**, 699 ⟨81BCJ3863⟩
Shigetomi, T., **5**, 981 ⟨73JAP7320534⟩
Shigihara, A., **6**, 135 ⟨92CC870⟩
Shih, C., **4**, 58, 211 ⟨83JOC5280⟩
Shih, C. N., **1**, 806 ⟨74JA3705⟩
Shih, H., **6**, 608 ⟨71JOC1786⟩
Shih, H.-M., **6**, 51 ⟨74JOC158⟩
Shih, J., **2**, 11 ⟨83HCA1028⟩
Shih, J. G., **1**, 900 ⟨85JA2764⟩; **2**, 15 ⟨83JOC3356⟩, 15, 17 ⟨86JFC(33)377⟩, 19 ⟨83S713⟩; **6**, 229 ⟨87JA3708⟩
Shih, M.-J., **6**, 466 ⟨82JA1769⟩
Shih, T. L., **1**, 878 ⟨81JA2127⟩
Shih, Y.-J., **1**, 11 ⟨78JOC2676⟩
Shiihashi, S., **3**, 7 ⟨85JOC1332⟩
Shiina, I., **5**, 236 ⟨90CL1019, 91CL1901⟩
Shiina, K., **2**, 941 ⟨B-69MI 215-04⟩
Shikada, T., **1**, 674 ⟨90AC59⟩
Shikano, E., **2**, 252 ⟨79BCJ1431⟩
Shikhaliev, Sh. M., **4**, 481 ⟨82IZV1611⟩; **5**, 1130 ⟨82IZV1611⟩
Shikita, S., **2**, 447 ⟨88JOC2937⟩
Shilcrat, S. C., **3**, 311 ⟨84S1048⟩
Shi Li-lan, **1**, 738 ⟨83MI 116-01⟩
Shilin, S. V., **6**, 620 ⟨91ZOB643⟩
Shilov, I. V., **2**, 414 ⟨72ZOB1936⟩, 469 ⟨71ZOB2372⟩; **4**, 476 ⟨71ZOB2372⟩
Shilov, S. A., **4**, 914 ⟨84ZOB457⟩
Shim, S. B., **5**, 368 ⟨86TL4749⟩
Shim, S. C., **1**, 239 ⟨85JOC147⟩; **2**, 81 ⟨90TL4765⟩; **3**, 83 ⟨87JOC2114⟩; **4**, 257 ⟨87JOC2114⟩
Shim, J., **3**, 187 ⟨93JOC4662⟩
Shim, Y. S., **6**, 559 ⟨82MI 617-03⟩
Shima, K., **2**, 752 ⟨87JOC753⟩; **4**, 420 ⟨85JOC4227⟩
Shima, T., **1**, 734 ⟨75CL103⟩
Shimabayashi, S., **5**, 125, 126 ⟨82S833⟩
Shimada, E., **5**, 125 ⟨77BCJ1863⟩
Shimada, J.-I., **2**, 702 ⟨83TL3341⟩
Shimada, K., **2**, 543 ⟨89TL6047⟩; **3**, 340 ⟨88MI 308-01⟩; **5**, 516, 520 ⟨92CL1843⟩, 625 ⟨90CL1403⟩, 635 ⟨85BCJ995⟩; **6**, 498 ⟨92CL1389⟩, 597 ⟨90CL1403⟩, 636 ⟨84CPB4893⟩
Shimada, M., **3**, 280, 305 ⟨88BCJ455⟩, 282, 295, 296 ⟨90CL783⟩; **5**, 328 ⟨90SC887⟩
Shimada, S., **5**, 344 ⟨90CL905⟩
Shimada, Y., **2**, 106 ⟨72TL411⟩; **5**, 57 ⟨82CL1255⟩
Shimadzu, H., **1**, 358, 498 ⟨92JA8869⟩
Shimagaki, M., **1**, 590, 591 ⟨88TL659⟩; **2**, 402 ⟨78TL3435⟩
Shimakawa, C., **2**, 684 ⟨88CL899⟩
Shimamaki, T., **2**, 720 ⟨91BCJ702⟩
Shimamura, Y., **2**, 305 ⟨91JOC2920⟩
Shimanska, M., **4**, 929 ⟨84H(22)987⟩
Shimanska, M. V., **4**, 514 ⟨84JOM(268)C29⟩
Shimanskaya, M. V., **5**, 224 ⟨87ZOB686⟩
Shimauchi, Y., **3**, 684 ⟨84CC1513⟩
Shimazaki, M., **1**, 595, 597 ⟨88H(27)261⟩; **2**, 61 ⟨89TL5447⟩; **3**, 160 ⟨87TL5891⟩, 208 ⟨90S677⟩
Shimazaki, N., **1**, 982 ⟨87AG(E)335⟩
Shimazaki, Y., **1**, 1084 ⟨87CL1905⟩
Shimazu, K., **3**, 664 ⟨80CC1253⟩
Shimezawa, H., **1**, 485 ⟨88BCJ1231⟩
Shimizu, B., **6**, 444 ⟨83BCJ2969⟩
Shimizu, H., **1**, 108, 115 ⟨86CPB3599⟩, 230 ⟨80TL2181⟩, 623 ⟨83TL2599⟩, 943 ⟨90JCS(P1)3017⟩; **2**, 10, 20 ⟨85TL4207⟩, 143 ⟨90TL3027⟩, 278, 279, 292 ⟨94JOC1011⟩, 281 ⟨90H(31)889⟩, 286, 288, 292 ⟨94CPB811⟩, 288 ⟨87JOC1397, 92JCR(S)393⟩; **3**, 376 ⟨78TL251, 80JOC2468⟩, 635 ⟨88CPB1698⟩, 637 ⟨93SL121⟩; **4**, 232 ⟨84TL757⟩, 288 ⟨91TL105⟩; **5**, 665 ⟨90CL559⟩, 1122 ⟨85JCS(P1)2333, 85TL1321⟩, 1122 ⟨87JCS(P1)2531⟩
Shimizu, I., **1**, 14 ⟨89JA6280⟩, 193 ⟨85JOC1523⟩, 196 ⟨84TL5157⟩, 196 ⟨85JOM(296)269⟩, 228 ⟨93TL2135⟩, 558 ⟨83TL5635, 86T2971⟩, 558, 561 ⟨84S1009, 87S992⟩, 561 ⟨94T487⟩, 690 ⟨82BCJ2434, 83BCJ192⟩, 690 ⟨83BCJ2023⟩; **2**, 3 ⟨89CL577⟩; **3**, 122 ⟨84TL2791⟩, 124 ⟨85JOC3416⟩, 139 ⟨80TL3199, 83CL1325, 83TL1793, 83TL4713, 84TL5157, 85JOC1523⟩, 206 ⟨82JA5844, 83TL1797, 83TL5635, 83TL5639, 86TL2791⟩, 231 ⟨79JA5070⟩, 233 ⟨85JOC537⟩, 623 ⟨87JOC2988⟩, 645 ⟨86CC118⟩; **5**, 154 ⟨86T2971⟩
Shimizu, K., **2**, 314 ⟨89JCS(P1)1548, 90JCS(P1)1859⟩; **4**, 255 ⟨79JA4732, 79JA7104⟩; **5**, 324 ⟨91H(32)221⟩, 332

⟨80JCS(P1)2122⟩, 557 ⟨79JA4732, 79JA7104⟩; **6**, 134 ⟨79JA4732⟩
Shimizu, M., **1**, 21, 23 ⟨92TL5567⟩, 133 ⟨82JA1025⟩, 283 ⟨76TL1699⟩, 485, 1083 ⟨84JOC1230⟩, 620 ⟨81JOC5246⟩, 650 ⟨89TL967⟩; **2**, 11 ⟨87TL663, 89BCJ2024⟩, 18 ⟨91BCJ2596⟩, 18, 280 ⟨89CC1881, 92JA6566⟩, 610 ⟨89TL967⟩; **3**, 7, 120 ⟨82JOC837⟩, 38, 175 ⟨81BCJ3100⟩, 38, 175 ⟨81BCJ3510⟩, 74 ⟨81TL2183⟩, 87 ⟨90TL2607⟩, 147 ⟨81JOC5246⟩, 155 ⟨77JOC4268⟩, 168 ⟨84CL1531⟩, 175 ⟨78TL1277⟩, 228 ⟨80TL3209⟩; **4**, 137 ⟨89TL967⟩, 400 ⟨80TL3209⟩, 751 ⟨88JFC(41)425⟩; **5**, 404 ⟨92JOC6216⟩
Shimizu, N., **1**, 561 ⟨80CPB282⟩; **2**, 14 ⟨83CL1877⟩, 907, 908 ⟨87BCJ777⟩; **4**, 806, 818 ⟨87BSJ777⟩
Shimizu, R., **2**, 323 ⟨90JA2368⟩; **3**, 43 ⟨90JA2368⟩
Shimizu, S., **4**, 407 ⟨79TL5011, 80H(14)97⟩, 520 ⟨86JOC3897⟩; **6**, 61 ⟨89T637⟩
Shimizu, S.-H., **1**, 238, 246 ⟨89BCJ2913⟩
Shimizu, T., **1**, 569 ⟨93JOC5163⟩; **2**, 240, 359, 768 ⟨70JOC2979⟩, 597 ⟨91JOC2538⟩, 997, 1001 ⟨92JCS(P1)1721⟩, 1089 ⟨93MI 202-02⟩; **3**, 684 ⟨84BCJ2531, 86BCJ2827⟩, 684 ⟨91BCJ318⟩, 685 ⟨87BCJ1948⟩; **4**, 794 ⟨89CL905⟩, 1082 ⟨B-95MI 424-01⟩; **5**, 299 ⟨79CC970⟩, 1071 ⟨70JOC3995⟩
Shimizu, Y., **1**, 471 ⟨77CL301⟩; **3**, 54 ⟨66TL4113⟩, 288 ⟨77CL1423⟩, 310 ⟨87T5281⟩; **4**, 321 ⟨72CI(L)380⟩
Shimmizu, Y., **3**, 140 ⟨77JOM(129)36⟩
Shimoji, K., **1**, 141 ⟨73JA2694⟩, 651 ⟨74BCJ2529, 74JA1620⟩; **4**, 648 ⟨74JA1620⟩, 956, 957 ⟨73JA2694⟩
Shimojo, M., **4**, 62, 63 ⟨90JOC6074⟩
Shimokawa, K., **1**, 1048 ⟨90TL5499⟩
Shimokura, M., **2**, 478 ⟨90CPB673⟩
Shimooka, K., **5**, 223 ⟨90CPB1836⟩
Shimoyama, I., **5**, 464 ⟨89JA8018⟩
Shimp, L. A., **6**, 46, 252 ⟨77IC2974⟩, 205 ⟨73JA1343, 79JA2214⟩, 205 ⟨79JOC2311⟩, 206 ⟨81JA5951⟩
Shimura, Y., **2**, 540 ⟨B-83MI 209-01⟩
Shimuzu, H., **2**, 71 ⟨79JA1455⟩; **3**, 356 ⟨74CL1101⟩
Shimuzu, I., **3**, 206 ⟨84CL1133⟩
Shimuzu, M., **3**, 74 ⟨79TL3461⟩
Shin, C., **1**, 648 ⟨85BCJ2812⟩; **3**, 564 ⟨75BCJ2891⟩; **4**, 433 ⟨78BCJ2614⟩
Shin, C.-G., **5**, 371 ⟨79TL3851, 86BCJ3696⟩
Shin, D.-S., **5**, 368 ⟨88SC545⟩
Shin, H., **3**, 303 ⟨91OM366⟩
Shin, H. H., **2**, 717, 718 ⟨92CL1483⟩; **4**, 74, 75 ⟨93S209⟩
Shin, H. W., **4**, 856 ⟨92T7527⟩
Shin, J. M., **2**, 726 ⟨85TL3821⟩
Shin, K., **2**, 578 ⟨88JA5228⟩
Shin, S., **3**, 3 ⟨78BCJ3016⟩
Shin, W. S., **6**, 123 ⟨92TL5375⟩
Shinada, T., **1**, 333 ⟨93TL6505⟩
Shindarov, L. M., **6**, 502 ⟨73GEP2241471⟩
Shindo, M., **1**, 135 ⟨89CPB1120⟩, 473 ⟨92JA8732⟩; **2**, 502 ⟨67MI 209-02⟩, 502, 889 ⟨68JOM(11)299⟩
Shine, H. J., **2**, 772 ⟨63JOC1232⟩; **3**, 297, 328 ⟨B-67MI 306-01⟩, 374 ⟨75JOC3857⟩; **5**, 208 ⟨59JOC107⟩, 215 ⟨58JOC318⟩
Shine, R. J., **2**, 220 ⟨68QRS189⟩; **6**, 587 ⟨B-73MI 619-01⟩, 594, 596 ⟨69JOC3549, B-73MI 619-06⟩
Shiner, C. S., **2**, 87 ⟨75TL3183⟩, 268 ⟨85JA7167⟩, 759, 965 ⟨86JA304⟩; **3**, 241 ⟨82JA3767⟩; **5**, 437 ⟨88JA957⟩
Shiner, Jr., V. J., **1**, 870 ⟨62JA4824⟩, 892 ⟨53JA2925⟩, 1124 ⟨76JA862⟩; **4**, 139 ⟨89JA7199⟩; **5**, 49 ⟨60JCS1593⟩
Shiner, V. J., **5**, 26 ⟨63JCS2918⟩
Shingaki, T., **2**, 249, 401 ⟨79BCJ1102⟩
Shinhama, K., **2**, 243 ⟨80BCJ2023⟩, 621, 627 ⟨78TL4519⟩, 621, 627 ⟨80BCJ1065⟩, 726 ⟨79TL3307⟩
Shin-Ike, T., **5**, 1036 ⟨91TL3503⟩
Shiniya, S., **1**, 429 ⟨89JFC(44)1637⟩
Shinkai, I., **1**, 940 ⟨88TL281⟩; **2**, 66 ⟨93JOC799⟩
Shinoda, E., **3**, 310 ⟨87T5281⟩
Shinoda, K., **2**, 30, 34 ⟨86JA6036⟩; **3**, 701, 704 ⟨92TL6823⟩
Shinoda, M., **1**, 422 ⟨79JA1599, 81BCJ2747⟩, 1024 ⟨86TL87⟩; **3**, 216 ⟨79JA1599⟩, 224 ⟨78TL771, 80BCJ1010⟩, 244 ⟨83BCJ3088⟩
Shinohara, I., **2**, 123 ⟨87H(26)2583⟩; **6**, 472 ⟨87H(26)2583, 89CPB576⟩
Shinohara, T., **4**, 507 ⟨91EUP488681⟩
Shinokubo, H., **2**, 700 ⟨93TL4985⟩
Shinomoto, R. S., **4**, 656 ⟨86JA7964⟩
Shinozaki, F., **4**, 873 ⟨86CL1861⟩
Shinozaki, H., **5**, 157 ⟨89TL3799⟩
Shinozaki, K., **2**, 252 ⟨79BCJ1431, 83BCJ1187⟩
Shinozaki, M., **1**, 1010 ⟨85CPB1761, 92JMC241⟩
Shinozaki, T., **5**, 223 ⟨87CPB1903⟩; **6**, 447 ⟨86JAN230⟩
Shinozuka, K., **2**, 309 ⟨93BMC2883⟩
Shinsaku, S., **5**, 990 ⟨72MI 526-01⟩
Shintoku, Y., **6**, 425 ⟨79JAP7961121⟩
Shinya, S., **4**, 802 ⟨79JA3689, 80JFC(15)543⟩; **6**, 213, 222, 223, 224, 226, 227, 228 ⟨82JFC(20)89⟩
Shinya, S., **6**, 58, 266 ⟨81JFC(18)197⟩
Shinya, T., **2**, 398 ⟨88BCJ3020⟩
Shio, M., **6**, 511 ⟨70JHC807⟩
Shioara, K., **2**, 596 ⟨94T8983⟩
Shioda, H., **3**, 246, 261 ⟨76CL499⟩
Shiohara, T., **3**, 64 ⟨86TL4497, 88BCJ4353⟩
Shioiri, T., **1**, 185, 200 ⟨80S1007⟩; **2**, 89 ⟨90TL5507⟩, 683 ⟨84CPB3759⟩, 753 ⟨84TL429⟩, 944 ⟨91S1163⟩; **3**, 125 ⟨82CPB119, 82CPB3380⟩, 202 ⟨88S228⟩, 531, 532 ⟨82CPB526⟩, 647 ⟨84TL427⟩, 557,558 ⟨91S1163⟩; **4**, 210 ⟨84TL5409⟩; **5**, 42 ⟨84OS(62)191⟩, 132, 873 ⟨81CPB1475⟩, 235 ⟨74JOC3302⟩, 293 ⟨87CPB2698⟩, 300 ⟨73TL1595⟩, 413, 873 ⟨86TL2005⟩, 775 ⟨83CPB2564, 83CPB3139⟩, 867 ⟨82CPB899⟩, 869 ⟨80TL4461, 84OS(62)187⟩, 873 ⟨80TL4619, 81CPB3249⟩, 873 ⟨81H(15)975, 82CPB119⟩, 873 ⟨82CPB3380, 90OS(68)1⟩, 874 ⟨91S1163⟩, 993 ⟨72JA6203⟩, 1112 ⟨82TL447⟩; **6**, 675 ⟨85H(23)2371⟩
Shiokawa, M., **1**, 590, 591 ⟨88TL659⟩
Shiokawa, T., **1**, 1142 ⟨75MI 122-01⟩, 1142, 1143 ⟨78MI 122-03⟩
Shiokawa, Y., **3**, 350 ⟨90EUP379979⟩; **4**, 296 ⟨70CPB1245⟩
Shiomi, D., **5**, 1092 ⟨89TL5433⟩
Shiomi, Y., **3**, 704 ⟨82ABC1803⟩; **4**, 850 ⟨88CC822⟩
Shiono, H., **1**, 210 ⟨75TL4027⟩; **3**, 57 ⟨74TL3625⟩; **4**, 392 ⟨74TL3625⟩
Shiono, M., **1**, 187 ⟨83S804⟩, 469 ⟨86AG(E)935⟩, 590 ⟨75CL871⟩, 590, 693 ⟨74CL1523⟩; **2**, 524 ⟨72BCJ2244⟩; **5**, 30, 35 ⟨86SC499⟩, 1025 ⟨77CL573⟩, 1078 ⟨77CL575⟩
Shiori, T., **3**, 475, 483 ⟨80TL4461⟩, 613 ⟨80S1007⟩, 616 ⟨79TL4663⟩; **5**, 265, 278, 993 ⟨74T2151⟩
Shiota, F., **1**, 290 ⟨87CL495⟩; **5**, 432 ⟨87CL495⟩
Shiota, S., **6**, 244 ⟨88BCJ3002⟩
Shiota, T., **3**, 432 ⟨84CC874, 87TL2383⟩, 638 ⟨87TL6469⟩
Shiotani, H., **5**, 550 ⟨74JOM(82)223⟩
Shiotani, S., **5**, 671 ⟨84JHC725⟩
Shiozaki, M., **2**, 939 ⟨92JA5867⟩; **5**, 208 ⟨91JOC528⟩
Shipchandler, M. T., **2**, 391 ⟨79S666⟩
Shipitsin, A., **6**, 489 ⟨93MI 615-03⟩
Shipman, M., **1**, 527 ⟨89TL7107⟩
Shipov, A. E., **2**, 454 ⟨59ZOB1450⟩, 462 ⟨61ZOB507, 62ZOB3579⟩, 841 ⟨65ZOB1574⟩
Shipov, A. G., **2**, 209 ⟨84ZOB1842⟩; **4**, 98, 99 ⟨91JGU1875⟩, 100 ⟨89JGU200⟩, 104 ⟨89JGU1625⟩, 140 ⟨89JOM(361)147⟩, 418 ⟨84JGU635⟩, 605 ⟨84ZOB1842⟩; **5**, 374 ⟨88ZOB1093⟩; **6**, 132 ⟨84ZOB1842⟩
Shipp, K. G., **4**, 178, 179 ⟨66JOC853⟩; **6**, 297 ⟨68USP3388147⟩
Shippey, M. A., **1**, 650 ⟨76JA1265⟩
Shipton, M. R., **3**, 178 ⟨92SL491⟩

Shiragami, H., **1**, 205, 1024 ⟨84TL5423⟩; **2**, 922, 923 ⟨88T4009⟩; **4**, 654 ⟨85TL2101⟩
Shirahama, H., **1**, 210 ⟨88CL645⟩, 382 ⟨81TL4313⟩; **3**, 311 ⟨93TL331⟩; **5**, 157 ⟨89TL3799⟩
Shirahata, A., **1**, 314 ⟨78TL3043⟩; **2**, 941 ⟨79S740⟩; **3**, 540 ⟨79S740⟩
Shirai, F., **1**, 816, 817 ⟨84CL255⟩
Shirai, H., **2**, 945 ⟨76JOC3559⟩, 1045 ⟨78JOC199⟩; **4**, 509 ⟨74JOM(82)21, 75CC640⟩, 509 ⟨76JOM(118)1, 78JOM(153)193⟩, 512 ⟨76JOM(113)115⟩; **5**, 25, 64 ⟨69CPB1564⟩
Shirai, K., **1**, 178 ⟨86BCJ3097⟩; **3**, 227 ⟨81CL1341⟩; **4**, 226 ⟨81BCJ817⟩; **5**, 1125 ⟨61JOC4122, 86MI 530-02⟩
Shirai, N., **1**, 824 ⟨88CC370, 94JCS(P1)393⟩; **4**, 506 ⟨92JOC6711⟩, 507 ⟨91S996⟩, 508 ⟨88S988, 91S117⟩, 513 ⟨92JOC5419⟩, 521 ⟨85TL5547⟩
Shirai, S., **1**, 907 ⟨82BCJ3257⟩
Shiraishi, M., **5**, 278 ⟨82JCS(P1)2909⟩
Shiraishi, S., **3**, 683 ⟨68BCJ2954⟩; **5**, 699 ⟨81BCJ3863⟩
Shiraishi, T., **3**, 91 ⟨88CPB974⟩; **4**, 872 ⟨86BCJ2909⟩
Shiraishi, Y., **2**, 311 ⟨92JOC1495⟩; **3**, 150 ⟨87CL1153⟩
Shiraiwa, M., **1**, 1010 ⟨81CPB3543, 83S312⟩; **2**, 677 ⟨84S245⟩; **4**, 195 ⟨84S245⟩
Shirakama, H., **5**, 129 ⟨94JOC5532⟩
Shiralkar, V. P., **4**, 901, 904 ⟨90JCS(P1)1217⟩
Shirasaka, T., **3**, 244 ⟨88JA310⟩
Shirata, E., **5**, 227 ⟨85H(23)549⟩
Shiratani, T., **5**, 329 ⟨93TA1573⟩
Shiratori, S., **4**, 808 ⟨93CC1817⟩
Shirhatti, V., **1**, 821 ⟨84TL3555⟩
Shirin, E., **1**, 445 ⟨80PS(8)243⟩
Shirk, J. S., **3**, 543 ⟨73JA248⟩
Shirley, D. A., **2**, 119, 225 ⟨53JOC1591⟩; **3**, 129, 130, 285, 288, 289 ⟨54OR(8)28⟩; **5**, 98 ⟨54OR(8)28⟩
Shiro, M.,**1**, 431 ⟨88T4095⟩; **2**, 46 ⟨88JA4606⟩; **4**, 808 ⟨88CC1076⟩; **5**, 204 ⟨90JA5677⟩; **6**, 555 ⟨92CC1410⟩
Shirokami, R., **5**, 1025 ⟨89CPB2334⟩
Shirokova, E., **6**, 489 ⟨93MI 615-03⟩
Shirouchi, M., **4**, 202 ⟨93TL1491⟩
Shirshov, A. N., **6**, 57 ⟨80IZV2356⟩
Shiryaev, V. I., **4**, 156 ⟨78JGU769⟩, 588 ⟨77ZOB1664⟩, 660 ⟨72ZOB631, 74ZOB806⟩, 660 ⟨75ZOB2448, 76ZOB1043⟩, 660 ⟨77ZOB1751, 89MI 414-01⟩, 662 ⟨91MI 414-01⟩
Shishido, K., **1**, 577 ⟨94JOC406⟩, 827 ⟨93CC793⟩; **3**, 294 ⟨87CC1360⟩, 620 ⟨86JOC3007⟩; **5**, 179 ⟨85CPB532⟩
Shishido, Y., **2**, 212 ⟨84JA5271⟩; **4**, 1077 ⟨82S1078⟩; **5**, 330 ⟨91CC1237, 91TL4325⟩
Shishikura, J., **4**, 80 ⟨93BCJ2339⟩; **5**, 272, 283 ⟨92TL1455⟩
Shishiyama, Y., **3**, 616, 627, 633 ⟨83T967⟩, 627 ⟨81TL4279, 80TL389⟩
Shishkin, V. E., **2**, 469 ⟨75ZOB2554⟩; **5**, 722 ⟨75JGU935, 75ZOB949⟩
Shishkin, V. E.,**5**, 774 ⟨78ZOB78⟩
Shishkov, I. M., **4**, 561 ⟨87ZOB2294⟩
Shishoo, C. J., **4**, 988 ⟨84TL1291⟩
Shitara, E., **3**, 620 ⟨86JOC3007⟩
Shitara, H., **6**, 115 ⟨91CL1873⟩
Shitara, K., **4**, 521 ⟨88JOM(339)259, 88JOM(346)1⟩
Shitikova, O. V., **1**, 801 ⟨91TL1591⟩
Shitkin, V. M., **6**, 149, 653 ⟨73ZOR896⟩, 368 ⟨72MI 612-01⟩
Shitov, L. N., **6**, 613 ⟨70JGU540⟩
Shitov, O. P., **4**, 524 ⟨74MI 411-01⟩; **6**, 145 ⟨73IZV122⟩, 148 ⟨72IZV490⟩, 148 ⟨73IZV1149⟩, 149 ⟨72IZV2603⟩, 169 ⟨74MI 605-01⟩
Shivanyuk, A. F., **5**, 685 ⟨90JOU1408⟩, 788 ⟨79UKZ624⟩
Shive, B., **1**, 104 ⟨41JA2975⟩
Shive, W., **2**, 494, 506 ⟨47JA559⟩
Shixiang Zhang, **1**, 739 ⟨85S57⟩
Shizuri, Y., **3**, 57 ⟨89TL3693⟩
Shklover, V. E., **5**, 615 ⟨89ZOR1323⟩
Shkrob, A. M., **6**, 104 ⟨62TL701, 63TL439⟩, 104 ⟨64TL47, 65T3537⟩
Shkurak, S. N., **2**, 167 ⟨85ZOK1835⟩
Shloss, Jr., E. L., **5**, 109 ⟨43OSC(2)457⟩
Shlyk, Yu. N., **2**, 292 ⟨69ZOB1804⟩
Shlykova, N. I., **4**, 429 ⟨77IZV2384⟩; **6**, 149 ⟨68ZOR231⟩
Shmaryahu, H., **4**, 993 ⟨92CJC1022⟩
Shmatov, Y. N., **4**, 976 ⟨86ZOR2256⟩
Shoaf, C. J., **2**, 109, 532 ⟨56JA3613⟩; **3**, 86 ⟨59TL9, 64JA1079⟩
Sho, K., **6**, 127 ⟨85S698⟩
Shobana, N., **2**, 319 ⟨90TL3999⟩; **4**, 310 ⟨94H(37)1951⟩
Shockravi, A., **5**, 67 ⟨78JOC2702⟩
Shoda, H., **3**, 159 ⟨93TL6281⟩
Shoda, S., **2**, 8 ⟨77CL383⟩; **5**, 234 ⟨76CL741⟩
Shoeb, A., **4**, 983 ⟨92JPR190⟩
Shoef, N., **2**, 304 ⟨89S825⟩
Shoemaker, B. H., **2**, 177 ⟨48IEC1129⟩
Shoemaker, M. J., **1**, 399 ⟨70JA1769⟩
Shoichiro, O., **5**, 985 ⟨73JAP7300455⟩
Shoji, K. J., **2**, 749 ⟨69JCS(C)1086⟩
Shoji, M., **4**, 650 ⟨87CPB1663⟩
Shoji, T., **3**, 628 ⟨90CL1615⟩; **6**, 636 ⟨84CHEC(3)91⟩
Shokol, V. A., **2**, 169 ⟨70ZOB1692⟩; **4**, 1007, 1009, 1011, 1012, 1013 ⟨89ZOB571⟩, 1007, 1010 ⟨77ZOB75⟩, 1007, 1010, 1013 ⟨83ZOB548⟩; **5**, 644 ⟨60ZOB3006, 61ZOB582⟩, 683 ⟨65JGU1019, 65ZOB1014⟩, 707 ⟨68ZOB1867⟩, 707, 861 ⟨68JGU1815⟩, 708 ⟨63JGU551, 63USP3076843⟩, 787 ⟨73JGU10, 90JGU864⟩, 860 ⟨62ZOB1873, 63JGU2984⟩, 1053 ⟨61ZOB2275⟩; **6**, 160 ⟨77ZOB321⟩, 285 ⟨77ZOB2766⟩, 285, 367, 666 ⟨73ZOB544⟩, 286, 606 ⟨88ZOB1516⟩, 286, 666 ⟨75ZOB1965⟩, 367 ⟨75ZOB2093, 89ZOB571⟩, 622, 634, 658 ⟨71ZOB2155⟩
Sholle, V. D., **1**, 922 ⟨73RCR1011⟩
Sholov, V. A., **5**, 860 ⟨62JGU3689⟩
Shon, Y. S., **2**, 60 ⟨93JOC1941⟩, 317 ⟨93SC3047⟩; **3**, 86 ⟨93JOC1941⟩
Shone, R. L., **3**, 140 ⟨90JA7441⟩
Shong Li, **1**, 737, 738 ⟨82OPP373⟩
Shono, T., **1**, 17 ⟨79TL2157⟩, 432 ⟨79JOC4508⟩, 590 ⟨78TL2807⟩, 615, 698 ⟨78CL69⟩, 1042 ⟨84T811⟩; **2**, 81 ⟨89JOC6001, 90TL1303⟩, 314, 352 ⟨91TL525⟩, 323 ⟨88BCJ3029, 90JA2368⟩, 457 ⟨81TL3249⟩; **3**, 7, 120 ⟨79TL165⟩, 7, 120 ⟨79TL3861, 80TL1867⟩, 17, 132 ⟨83JOC3338⟩, 43 ⟨86TL6083, 90JA2368⟩, 146 ⟨77CL1021⟩, 151 ⟨82TL1609⟩, 212 ⟨83JOC2503⟩, 219 ⟨84JOC300⟩, 261 ⟨82JA6697, 82TL1201⟩, 374 ⟨74TL1071⟩, 426 ⟨83JOC2103⟩, 432 ⟨86JOC549⟩, 497 ⟨68JA2728⟩, 618 ⟨89JOC2249⟩; **4**, 197 ⟨89TL5309⟩, 201 ⟨69JA2803, 87S1099⟩, 210 ⟨86JOC546⟩, 211 ⟨64JOC1371⟩, 300, 311 ⟨82JA5753⟩, 303 ⟨75JA4264⟩; **5**, 132 ⟨86JOC546⟩, 351 ⟨88H(27)1881⟩; **6**, 24 ⟨90TL7181⟩, 25, 26 ⟨81TL871, 82TL1609⟩, 26 ⟨82TL4801⟩, 506 ⟨63BCJ1314⟩
Shook, C., **2**, 553 ⟨90JA6389⟩
Shook, C. A., **4**, 652 ⟨93JA10754⟩
Shook, D. A., **3**, 193 ⟨87SC1709⟩
Shoppee, C. W., **1**, 421 ⟨72JCS(P1)2271⟩; **2**, 25 ⟨46JCS1138, 53JCS1709⟩; **3**, 234 ⟨61JCS1583⟩
Shor, G. S., **5**, 850 ⟨69ZOR1404⟩, 854 ⟨69ZOR1411⟩
Shore, J. T., **5**, 887 ⟨87PS(30)463⟩
Shore, S., **2**, 824, 834, 835 ⟨78JA7304⟩
Shore, S. G., **6**, 345 ⟨84CC392⟩
Shorr, L. M., **2**, 1017 ⟨64IJ121⟩
Short, J. H., **1**, 256 ⟨58JOC1281⟩; **3**, 625 ⟨73T1931⟩; **6**, 641 ⟨63JMC275⟩, 642 ⟨68JMC1129⟩
Short, R. P., **1**, 804 ⟨83JOC4453⟩; **2**, 77, 521 ⟨89JA1892⟩, 533 ⟨89TL7357⟩; **4**, 643, 645 ⟨89JA1892⟩
Short, W. F., **5**, 689, 710 ⟨48JCS1514, 48JCS1618⟩, 749, 757 ⟨46JCS147⟩, 750 ⟨47JCS390⟩, 750, 751 ⟨47JCS1110⟩

Shorter, J., **5**, 49 ⟨50JCS3425⟩
Shortland, A., **4**, 656 ⟨70CC1369, 71CC1079⟩
Shortt, A. B., **2**, 254, 255 ⟨84EUP118109⟩
Shortt, M. F., **4**, 775 ⟨93JCS(P1)1945⟩
Shostakovskii, M. F., **2**, 674 ⟨68RCR907⟩, 1101 ⟨67DOK(175)1079⟩; **6**, 60 ⟨67IZV693⟩, 83 ⟨60IZV1901⟩
Shostakovsky, V. M., **3**, 232 ⟨77AG(E)646⟩
Shota, I., **5**, 1012 ⟨89NKK1881⟩
Shoup, T. M., **1**, 178 ⟨87JOC5484⟩, 449 ⟨89TL6487⟩; **2**, 75 ⟨88JA5578⟩, 901 ⟨88S130⟩; **3**, 223 ⟨88S130⟩; **4**, 814 ⟨93SC2937⟩
Showell, J. S., **2**, 192 ⟨62JOC2853⟩; **5**, 28 ⟨62CHE179⟩, 201 ⟨57JA419⟩
Showler, A. J., **5**, 123 ⟨65JCS4594⟩
Shpanskii, V. A., **5**, 822 ⟨60ZOB2409⟩; **6**, 240 ⟨69ZOB199⟩, 242 ⟨62DOK(142)354⟩, 614 ⟨68JGU37⟩, 614, 627 ⟨69JGU183⟩
Shreeve, J. M., **2**, 166 ⟨68JA5403⟩, 166, 170 ⟨71ICA311⟩, 175 ⟨74IC2710⟩, 367 ⟨84JA7496⟩; **3**, 456 ⟨71MI 312-02⟩, 456, 458, 459 ⟨72IC242⟩; **4**, 52, 59 ⟨72JOC3332⟩, 107 ⟨76IC743⟩, 140 ⟨86IC376⟩, 564 ⟨90IC573, 91JFC(52)1⟩, 803 ⟨90JA3152⟩; **5**, 246 ⟨75MI 505-01, 78JFC(11)327⟩, 248 ⟨76JA6545⟩, 313 ⟨74IC2705⟩, 315 ⟨71IC911⟩, 652 ⟨71CC788⟩, 659 ⟨92IC4917⟩, 675 ⟨93IC287⟩, 679 ⟨90JFC(48)395⟩, 679, 703 ⟨90IC4255⟩, 679, 703 ⟨92ICA527⟩, 684 ⟨92JFC(56)335⟩, 692 ⟨88IC3100⟩, 807 ⟨71IC407, 71IC1635⟩; **6**, 51, 52 ⟨75IC1223⟩, 229, 230 ⟨86IC376⟩, 239 ⟨71IC1635⟩, 240 ⟨69JCS(A)431⟩, 250 ⟨67JA1809, 68IC624⟩, 251, 256, 259, 624 ⟨90JFC(48)395⟩, 253 ⟨73JFC(3)17, 77JA4194⟩, 253 ⟨78IC2173, 79JA5949⟩, 253 ⟨81JA406, 86IC275⟩, 257 ⟨67IC1711⟩, 258 ⟨89JA1773⟩, 258, 278 ⟨89CJC1795⟩, 259 ⟨93IC4802⟩, 259, 614, 624 ⟨92JFC(57)293⟩, 259, 643, 651, 653 ⟨92IC488⟩, 260 ⟨92IC4917⟩, 260, 261 ⟨92IC329⟩, 260, 615 ⟨89IC3345, 75MI 608-01⟩, 262 ⟨90HAC167, 92HCA907⟩, 277 ⟨75IC592⟩, 423 ⟨85JFC(28)219⟩, 432 ⟨76JA6545⟩, 436 ⟨87JFC(36)429⟩, 438 ⟨84CC416⟩, 440 ⟨73IC2890, 74JA1770⟩, 603 ⟨90IC571⟩, 603, 607 ⟨76IC14⟩, 613 ⟨70JA3665⟩
Shreeve, R. W., **6**, 253 ⟨79JA5949⟩
Shrestha-Dawadi, P. B., **3**, 687 ⟨93S426⟩
Shreve, J. M., **2**, 334 ⟨90JA2383⟩, 335 ⟨89MI 206-02⟩
Shriner, R. L., **2**, 103 ⟨33JA766⟩, 126 ⟨33JA1242⟩; **3**, 241 ⟨42JA382⟩, 362 ⟨63JA3989⟩, 459 ⟨55OSC(3)562⟩; **4**, 237 ⟨41JOC888⟩; **5**, 89, 99 ⟨43OSC(2)140, 43OSC(2)382⟩, 99 ⟨43OSC(2)368⟩, 114 ⟨43OSC(2)538⟩, 199 ⟨63OSC(4)242⟩, 742, 751, 755 ⟨44CRV351⟩
Shrivastava, R. D., **2**, 778 ⟨65MI 215-01⟩
Shriver, D. F., **5**, 445, 447 ⟨80JA5093⟩; **6**, 338 ⟨79IC1236, 85JA8136⟩, 338 ⟨85JA8147⟩, 340 ⟨80CC778, 80CC780⟩, 340 ⟨82JA5621⟩, 456 ⟨87JA6015⟩, 521 ⟨84AOC(23)219⟩, 717, 722 ⟨84OM314⟩, 717, 727 ⟨85OM1830⟩
Shropshire, E. Y., **4**, 414 ⟨62JOC3664⟩; **6**, 105, 138, 140 ⟨62JOC3664⟩
Shtamburg, V. G., **2**, 356, 357 ⟨81T4245⟩, 356, 357 ⟨91IZV1069⟩
Shtepanek, A. S., **5**, 373 ⟨69JGU1445⟩, 988 ⟨62ZOB3759⟩; **6**, 537 ⟨87UKZ395⟩
Shterenberg, B. Z., **2**, 1091 ⟨86JGU849⟩; **4**, 617 ⟨87IZV1424⟩
Shu, P., **4**, 850 ⟨83JOC4713⟩, 869 ⟨77CC505⟩; **6**, 591 ⟨77CC505⟩
Shu, S.-G., **4**, 711 ⟨93OM108⟩
Shubert, D. C., **1**, 360 ⟨86JA4683⟩, 491 ⟨83JOC1924⟩
Shubert, U., **2**, 1046 ⟨81JA6142⟩
Shubkin, R. L., **2**, 320 ⟨90TL2991⟩; **5**, 453 ⟨66IC1177⟩
Shuchart, C. E., **4**, 711 ⟨94OM1999⟩
Shudo, K., **1**, 492 ⟨89JOC733, 91JA1364⟩; **2**, 316 ⟨91SL649⟩, 691 ⟨80S461⟩; **3**, 264 ⟨73TL4533⟩, 704 ⟨76CPB832⟩; **6**, 506 ⟨83S471⟩
Shue, Y.-K., **3**, 299 ⟨84JA1813⟩; **5**, 166 ⟨84JA1813⟩
Shuey, C. D., **1**, 538 ⟨79JA1284⟩; **5**, 164 ⟨79JA1284⟩
Shufler, S. L., **6**, 524 ⟨55JA3951⟩
Shuhaibar, K. F., **3**, 92 ⟨92SL77⟩, 92, 99, 101 ⟨93T4015⟩
Shukla, K., **1**, 282 ⟨91JOC3908⟩
Shukla, R. J., **2**, 841 ⟨73ZOB997⟩
Shukyurov, A. Kh., **4**, 140 ⟨88JGU81⟩; **5**, 1075 ⟨85ZOB1308⟩
Shulaev, N. P., **2**, 879 ⟨60JGU3203⟩
Shul'gina, V. M., **4**, 106 ⟨88JOU2003⟩
Shull, D. W., **1**, 590 ⟨76JOC3218, 80OS(59)202⟩, 801 ⟨76TL575⟩
Shulman, F. C., **3**, 465 ⟨68T2881⟩
Shulman, J. I., **1**, 834 ⟨70JA5522⟩; **2**, 618, 630, 909 ⟨70TL447⟩, 709 ⟨70JOC777⟩
Shulman, M. J., **1**, 193 ⟨89JOC2759⟩
Shulman, M. L., **2**, 32 ⟨70TL2517⟩
Shulte-Elte, K. H., **3**, 143 ⟨87HCA1858⟩
Shults, R. H., **1**, 966 ⟨78JOC1389⟩, 970 ⟨76JOC3496, 78JOC1382⟩, 970 ⟨78JOC1385⟩
Shultz, A. G., **2**, 326 ⟨88JOC1380⟩
Shumaker, S. H., **5**, 276, 298 ⟨61JA1983⟩
Shum, P. W., **2**, 942 ⟨90TL6785⟩
Shumate, K. M., **1**, 418 ⟨65JA3996⟩, 419 ⟨66JA1073⟩
Shundo, Y., **2**, 226 ⟨92CC600⟩; **3**, 370 ⟨92CC600⟩
Shurnbura, A. K., **4**, 1009 ⟨89ZOB2492⟩; **5**, 655 ⟨87ZOB475, 86ZOB1262⟩, 655, 788, 790 ⟨87JGU415⟩, 713 ⟨89ZOB2492⟩, 713, 788 ⟨86JGU1113, 89JGU2228⟩
Shusherina, N. P., **4**, 322 ⟨77ZOR1930⟩
Shusterman, A. J., **4**, 580 ⟨79JA7282⟩; **5**, 942 ⟨85OM736⟩
Shustorovich, E. M., **1**, 866 ⟨57DOK(117)1021⟩
Shustov, G. V., **2**, 337 ⟨88JA1719⟩; **3**, 559, 560 ⟨92BAU2039⟩; **5**, 317 ⟨85T575⟩
Shuto, K., **4**, 970 ⟨91EUP413343⟩; **5**, 225 ⟨90JAP280254⟩
Shutske, G. M., **2**, 22, 28 ⟨89JCS(P1)1544⟩
Shutt, F. E., **4**, 213 ⟨84JCS(P1)1531⟩; **5**, 79 ⟨84JCS(P1)1531⟩
Shutt, J. R., **2**, 843 ⟨69JCS(C)2038⟩
Shuttleworth, S. J., **5**, 351, 363 ⟨91T10053⟩; **6**, 515 ⟨91T10053⟩
Shuvalov, V. F., **4**, 1080 ⟨83ZOB416⟩
Shuyama, H., **5**, 205 ⟨86JFC(31)37⟩; **6**, 12 ⟨86JFC(31)37⟩
Shvaika, O. P., **5**, 627 ⟨81ZOB1842⟩, 642 ⟨85ZOB2318⟩, 848 ⟨72UKZ171⟩; **6**, 115 ⟨81ZOB1842⟩
Shvarts, I. S., **6**, 653 ⟨79IZV813⟩
Shvartsberg, M. S., **1**, 1006 ⟨70IZV1144, 71IZV1534⟩, 1006 ⟨72IZV472⟩, 1010 ⟨80IZV726⟩
Shvekhgeimer, G. A., **5**, 687 ⟨91CHE299, 91KGS369⟩; **6**, 144 ⟨59UK484⟩, 145 ⟨71ZOR30⟩, 149 ⟨67ZOB1163⟩, 149, 369 ⟨73RZC1243⟩
Shvets, A. A., **2**, 834, 835, 840 ⟨78ZOB232⟩, 841 ⟨76ZOB1701⟩
Shvetsova-Shilovskaya, K. D., **5**, 966 ⟨66ZOB1707⟩
Shvetsov, Y. S., **6**, 57 ⟨80IZV2356⟩
Shvily, R., **2**, 14 ⟨90JOC2927⟩; **3**, 123 ⟨90JOC2927⟩
Shvo, Y., **2**, 676 ⟨83OM1689⟩; **4**, 206 ⟨83OM1689⟩, 899, 902, 903 ⟨69T4649⟩; **5**, 213 ⟨81JOC2148⟩
Shvyadas, V. K., **4**, 462 ⟨92ZOB1472⟩
Shy, H. C., **4**, 396 ⟨86JOM(315)135⟩, 583 ⟨90JCS(D)2509⟩
Shy, J. C., **2**, 218 ⟨87JOC4620⟩
Shymanska, M. V., **3**, 430 ⟨89SC3129⟩; **4**, 1080 ⟨87CC31⟩; **6**, 124 ⟨84JOM(265)237⟩
Siabalis, N., **5**, 397, 880 ⟨87AG(E)779⟩
Sianesi, D., **2**, 175 ⟨70TL1313⟩; **6**, 42 ⟨66JOC2312⟩, 251 ⟨72GEP2111696⟩
Sibgatullina, F. G., **2**, 469 ⟨77IZV1437⟩
Sibi, M. P., **1**, 464 ⟨84JOC737⟩; **2**, 59 ⟨92SL497⟩, 673 ⟨86JOC271⟩, 942 ⟨83TL4515⟩; **4**, 165 ⟨84JOC4518⟩
Sibilia, J. P., **5**, 998 ⟨70USP3483242⟩, 1001 ⟨69USP3465024⟩

Sibille, S., **3**, 146 ⟨86TL4175⟩; **6**, 22 ⟨92CC53⟩
Sibley, K. L., **6**, 174, 175, 176, 289, 380, 391, 393, 395 ⟨92OM2938⟩
Sibrikov, S. G., **4**, 979 ⟨91ZOR382⟩
Sibtain, F., **1**, 202 ⟨88JOC3306⟩, 437 ⟨83TL5329⟩
Sica, D., **6**, 76 ⟨66G375⟩, 107 ⟨63G942⟩
Sicard, G., **3**, 681 ⟨88JA2663⟩; **5**, 871, 872, 873 ⟨88JA2663⟩, 1131 ⟨88AG289⟩; **6**, 665, 667, 668, 669 ⟨88JA2663, 89JOC4426⟩
Sicé, J., **3**, 105 ⟨53JA3697⟩
Sicher, J., **1**, 389 ⟨65JCS2513⟩
Sicken, M., **1**, 248 ⟨88AG(E)411⟩
Sicken, U., **2**, 811 ⟨77JOM(141)249⟩
Sicker, D., **5**, 324 ⟨85S331⟩; **6**, 486 ⟨89S875⟩
Sicker, U., **4**, 1015 ⟨82JOM(231)191⟩; **6**, 662 ⟨82JOM(231)191⟩
Sicker, V., **5**, 795, 796 ⟨82JOM(231)191⟩
Sickinger, A., **2**, 437 ⟨87JOM(335)59⟩
Sidall, T. L., **2**, 254, 299 ⟨85TL4867⟩
Sidani, A., **2**, 759 ⟨74AG272⟩; **3**, 601 ⟨74AG(E)267⟩
Sidani, A. R., **2**, 758 ⟨80OS(59)26⟩; **4**, 790 ⟨88OSC(6)282⟩; **5**, 2 ⟨79OS(59)26⟩
Siddall, J. B., **1**, 540 ⟨69JA1853⟩; **2**, 976 ⟨80JA6580⟩
Siddall, T. H., **4**, 342 ⟨75JOC2056⟩, 554 ⟨65JCED303⟩
Siddhanta, A. K., **2**, 636, 638 ⟨88ACR135⟩
Siddiq, M., **2**, 621 ⟨86IJC328⟩, 626 ⟨88JOC1799⟩
Siddiqi, S., **3**, 530 ⟨85JOC135⟩
Siddiqui, A. A., **4**, 441, 444 ⟨81IJC1003⟩
Siddiqui, A., **4**, 229 ⟨92TL4625⟩
Siddiqui, A. H., **5**, 355 ⟨93JHC61⟩
Siddiqui, A. U., **5**, 355 ⟨93JHC61⟩
Siddiqui, B. S., **5**, 1125 ⟨82ZN(B)1481⟩
Siddiqui, I. H., **4**, 268 ⟨91JIC368⟩
Siddiqui, M. A., **3**, 34 ⟨83CJC1872⟩
Siddiqui, M. Z., **5**, 785 ⟨85POL1151⟩
Siddiqui, S., **4**, 413 ⟨82ZN(B)512⟩; **5**, 1125 ⟨82ZN(B)1481⟩
Sidduri, A., **2**, 69, 72, 591 ⟨92JOC1956⟩
Sidduri, A. R., **1**, 1004 ⟨93JOC2694⟩; **4**, 815 ⟨92TL3717⟩
Sidebottom, P. J., **3**, 182 ⟨78CC753⟩
Siderius, H., **3**, 709 ⟨72TL2369⟩; **4**, 331, 538 ⟨72TL2367⟩
Siderov, E. O., **5**, 860 ⟨78KGS991⟩
Sidgewick, N. V., **2**, 102 ⟨21MI 202-02⟩; **3**, 694 ⟨30JCS1876⟩
Sidky, M. M., **3**, 357 ⟨58JA6150⟩; **4**, 487 ⟨82PS(13)319⟩
Sidler, D. R., **1**, 208 ⟨88JOC4892⟩, 524 ⟨85JOC2309⟩; **2**, 574, 978 ⟨88JA2575⟩; **3**, 218 ⟨87TL2233, 88JOC4892⟩; **5**, 329 ⟨94JOC1231⟩
Sidler, J. D., **1**, 396, 397 ⟨66JA78⟩
Sidorov, E. O., **5**, 567, 571 ⟨87KGS845⟩
Sidot, C., **1**, 254 ⟨83JOC1919⟩
Sie, E. R. H. B., **6**, 119 ⟨92SL975⟩, 628, 636 ⟨94T3195⟩
Siebald, H., **3**, 487 ⟨92JOM(438)83⟩
Siebeneick, H.-U., **2**, 824, 827, 828 ⟨69CB1883⟩
Siebenthall, F., **5**, 727 ⟨73JOC2242⟩
Sieber, P., **1**, 654 ⟨77HCA2711⟩; **6**, 489 ⟨68HCA622⟩
Sieber, W., **1**, 420 ⟨72HCA3005⟩, 993 ⟨73HCA875⟩; **2**, 878 ⟨83AG(E)1001⟩, 878, 896 ⟨84CB427⟩
Sieber, W. J., **2**, 896 ⟨83OM1266⟩
Siebert, E., **4**, 772 ⟨22CB1014⟩
Siebert, H., **2**, 488 ⟨53ZAAC(273)161⟩; **3**, 618, 619 ⟨1893LA(274)312⟩
Siebert, Von W., **5**, 1136 ⟨75ZAAC(418)273⟩
Siebert, W., **1**, 862 ⟨85AG(E)943⟩, 910 ⟨84AG(E)371⟩; **2**, 294 ⟨72CB454, 75ZNB642⟩, 903 ⟨88AG(E)299, 89CB401⟩, 903 ⟨89ZN(B)1179, 90CB2173⟩, 903 ⟨93ZN(B)719⟩, 1060, 1061 ⟨68JOM(15)69⟩, 1083 ⟨82HOU(13/3b)1⟩, 1083, 1094 ⟨75ZAAC(418)273⟩; **4**, 574 ⟨70ZN(B)314⟩, 631, 635, 635 ⟨90ZN(B)15⟩, 632 ⟨83JA2582, 85AG(E)943⟩, 665 ⟨92CB1341⟩; **5**, 927 ⟨92AG(E)1603⟩, 928 ⟨84AG(E)371⟩; **6**, 183 ⟨90AG(E)292⟩, 188, 389, 401 ⟨89CB1881⟩, 389, 398, 401 ⟨86AG(E)1112⟩
Siebling, W., **1**, 432 ⟨11CB2134⟩
Siebrand, W., **4**, 1078 ⟨90JA4284⟩
Sieburth, S. M., **1**, 841 ⟨81T3967, 81TL2471⟩; **3**, 217 ⟨83JA3348⟩; **2**, 928 ⟨92JOC5279⟩; **3**, 144 ⟨81T3967⟩, 144 ⟨81TL2471⟩
Siedle, A. R., **4**, 850 ⟨75JOC2570⟩; **6**, 120 ⟨75JOC2570⟩
Siedlecka, R., **3**, 7 ⟨90TL2177⟩
Siefken, W., **5**, 636 ⟨54CB825⟩
Siegel, E., **5**, 205 ⟨73OSC(5)904⟩
Siegel, H., **1**, 76 ⟨68AG(E)942⟩, 296 ⟨74CA(81)120222t, 80AG(E)178⟩; **4**, 145 ⟨82TCC55⟩, 746 ⟨84CB1877⟩; **6**, 204 ⟨79AG(E)785, 80HCA2046⟩
Siegel, S., **1**, 72 ⟨66MI 103-01⟩, 74, 82 ⟨75JOC3599⟩
Siegel, W., **2**, 375 ⟨54LA(590)1⟩
Siegemund, G., **2**, 864 ⟨68ZAAC(363)176⟩; **6**, 410, 419, 420 ⟨74GEP2261108⟩, 414, 419, 420 ⟨73AG(E)918⟩
Sieger, G. M., **2**, 111 ⟨61JPS193⟩
Siegfried, B., **1**, 36, 486 ⟨77JOC2431⟩, 442 ⟨77JOC3494⟩; **2**, 621, 627 ⟨77JOC2426⟩; **4**, 19 ⟨76CC433⟩, 19, 27 ⟨76JA1627⟩
Siegfried, R., **1**, 905 ⟨83JA950⟩
Siegmann, K., **1**, 473 ⟨93JA9276⟩
Siegwart, J., **3**, 552 ⟨20HCA833⟩; **5**, 476, 477 ⟨20HCA824⟩
Sieh, D. H., **2**, 387 ⟨80JA3883, 81JA442⟩
Siehl, H.-U., **1**, 1141 ⟨85TL1493⟩
Sieja, J. B., **4**, 59 ⟨71JA2481⟩
Siele, V. I., **4**, 481 ⟨77JHC337⟩
Sielecki, T. M., **1**, 149 ⟨92JA8483⟩
Sieler, J., **5**, 657 ⟨89ZAAC(576)203⟩
Siemeling, U., **4**, 603 ⟨92JOM(429)C14⟩
Siemsen, L., **4**, 176 ⟨88CJC187⟩
Siepak, J., **4**, 470 ⟨73RZC929⟩
Sierakowski, A. F., **1**, 472 ⟨82JCS(P1)2299⟩; **2**, 844 ⟨82JCS(P1)2299⟩
Sierakowski, C., **3**, 710 ⟨92AG(E)1218⟩
Sierra, A., **4**, 779 ⟨74TL4157⟩
Sierra, M., **4**, 545, 565 ⟨93PS(77)254⟩
Sierra, M. A., **5**, 696 ⟨91OM11, 93JOC3886⟩
Sierra, M. L., **2**, 894 ⟨93BSF521⟩
Sies, H., **2**, 235 ⟨86EUP165534, 86GEP3443467⟩
Sieskind, O., **1**, 180 ⟨93TL1197⟩
Sieveking, H. U., **3**, 657 ⟨69AG(E)458⟩
Sieveking, S., **5**, 638 ⟨74T4147⟩, 729 ⟨74CB3589⟩
Sievers, R., **6**, 686 ⟨83CB1873⟩
Sievert, A. C., **6**, 645 ⟨91MIP 601-01⟩
Siew, N. P. Y., **6**, 387 ⟨80JOM(202)157⟩, 445 ⟨78JCS(P1)1066⟩
Siewers, S., **4**, 98 ⟨87CZ247⟩
Siewing, M., **4**, 690 ⟨93JOM(459)325⟩
Sigalov, M. V., **2**, 225 ⟨92ZOR927⟩; **4**, 140 ⟨76IZV2200⟩, 775 ⟨92ZOR1780⟩; **6**, 565 ⟨86ZOR2145⟩
Sigel, G. E., **6**, 349 ⟨85CC1674⟩
Sigg, H. P., **3**, 468, 471 ⟨67HCA1327⟩
Siggia, S., **2**, 181 ⟨48MI 203-01⟩, 478 ⟨57ANC75⟩; **5**, 864 ⟨80ANC967⟩
Sigmund, G., **4**, 983 ⟨77TL3619⟩
Sigmund, S. K., **5**, 363 ⟨87TL285⟩
Signer, M., **2**, 331 ⟨92JA5900⟩; **3**, 189 ⟨92JA5900⟩; **4**, 107 ⟨92JA5900⟩
Sigolaev, Y. F., **1**, 748 ⟨80ZOB76⟩
Sigurdson, E., **2**, 575 ⟨75JOM(99)C19⟩
Sigurdson, E. R., **2**, 602 ⟨77JCS(D)812⟩; **4**, 656 ⟨77JCS(D)812⟩
Sigwalt, P., **6**, 488 ⟨89MI 614-04⟩
Sigwarth, B., **2**, 694 ⟨81CB1407⟩; **6**, 375 ⟨86ZN(B)191⟩
Sih, C. J., **1**, 700, 703 ⟨86JA4603⟩, 1001 ⟨74TL423⟩; **2**, 651 ⟨78CC242⟩, 656 ⟨76TL1637⟩; **5**, 138 ⟨75JA4144⟩, 179 ⟨88TL5583⟩
Sih, J. C., **1**, 1023, 1024 ⟨73TL3899⟩
Siimer, E., **1**, 552 ⟨89CA(111)23052⟩
Siirala-Hansén, K., **5**, 464 ⟨75JA1184⟩
Sik, V., **1**, 522 ⟨91JCS(P1)1255⟩; **4**, 283 ⟨87JCS(D)757⟩

Sikirica, M., **4**, 268 ⟨80TL3089⟩, 697 ⟨84CCA689, 91JOM(405)59⟩; **6**, 207 ⟨82CSC1571, 82JOM(238)327⟩, 207 ⟨83JOM(253)283, 84CCA689⟩, 207 ⟨84JOM(276)1, 86JOM(306)1⟩, 207 ⟨87JOM(319)1, 91JOM(411)19⟩, 405 ⟨74CC646, 78JOM(153)1⟩
Sikkema, D. J., **3**, 574 ⟨76RTC154⟩
Siklosi, M. P., **1**, 401, 800, 883 ⟨78JA2134⟩
Sikoiskij, P., **2**, 561 ⟨82AG(E)199⟩
Sikora, D. J., **6**, 522 ⟨79JA5079⟩
Sikorski, J. A., **1**, 1034 ⟨86JOC4521⟩; **2**, 522, 526 ⟨82JOC863⟩, 522, 526 ⟨82JOC872, 86JOC5264⟩, 1086 ⟨86JOC4521⟩, 5266 ⟨80JOC4540⟩; **3**, 82 ⟨86JOC5264⟩
Sikstrom, R. A., **3**, 363 ⟨64JOC1355⟩
Silaghi-Dimitrescu, L., **2**, 891 ⟨83JOM(252)295⟩
Silberman, L., **3**, 244 ⟨77JOC3846⟩
Silberman, R. G., **1**, 907 ⟨67JA5048b⟩
Silbermann, J., **2**, 600 ⟨78JA4834, 82JA2831⟩
Silberrad, O., **3**, 468 ⟨02JCS598⟩
Silbert, L. S., **2**, 393 ⟨70TL699⟩; **4**, 842 ⟨76T2507⟩; **5**, 28 ⟨62CI(L)1304⟩, 205 ⟨73OSC(5)904⟩, 555 ⟨76T2507⟩, 1109 ⟨78MI 530-01⟩
Siles, S., **2**, 346, 349 ⟨79TL4391⟩
Silhanek, J., **6**, 236, 531 ⟨69CC878⟩
Šilhavý, P., **4**, 17 ⟨83CCC1710⟩; **6**, 7 ⟨80CCC3488, 80CCC3502⟩
Silina, E. B., **4**, 1007, 1009, 1011, 1012, 1013 ⟨89ZOB571⟩; **6**, 367 ⟨89ZOB571⟩
Silkina, N. N., **4**, 660 ⟨89MI 414-01⟩, 662 ⟨91MI 414-01⟩; **6**, 444 ⟨80ZOB875⟩
Silks, III, L. A., **1**, 485 ⟨85JA4679, 86JOC4687⟩; **3**, 217 ⟨85JA4679, 86JOC4687⟩, 707 ⟨91JCS(P1)2495⟩; **5**, 1054 ⟨91JCS(P1)2495⟩; **6**, 595 ⟨91JCS(P1)2495, 91JOC6733⟩
Sille, F., **6**, 52 ⟨78CB921⟩
Sillion, B., **3**, 113 ⟨79TL3⟩
Sillitoe, A., **3**, 685 ⟨83JCS(P1)293⟩
Silva, C., **5**, 347 ⟨84FES788⟩
Silva, S. O. de, **5**, 199 ⟨92T4863⟩
Silveira, A., **2**, 524 ⟨75JOC814⟩
Silveira, C. C., **1**, 999 ⟨86SC283⟩; **2**, 1036, 1038 ⟨86SC283⟩; **4**, 93 ⟨93JOM(460)31⟩
Silveira, Jr., A., **1**, 133 ⟨79TL845⟩, 983 ⟨80JOC3522⟩, 1006 ⟨78JOC358⟩, 1070 ⟨80JOC2526⟩; **2**, 976 ⟨80JOC2526, 80S1034⟩; **4**, 372 ⟨75JOC814⟩
Silver, H. B., **2**, 834, 835 ⟨61CI(L)24⟩
Silver, J., **2**, 690 ⟨80JOM(188)255⟩; **4**, 724 ⟨86JOM(307)231⟩
Silver, J. E., **3**, 345 ⟨91JOC5932⟩
Silver, M. E., **4**, 650, 656 ⟨93OM338⟩
Silver, M. S., **1**, 383 ⟨59JA4390⟩
Silver, S. M., **1**, 443, 649, 759 ⟨73TL3497⟩
Silverman, J., **2**, 1010 ⟨71JPC1246⟩
Silverman, R. B., **3**, 381, 382, 391 ⟨B-73MI 309-01⟩; **4**, 141, 142 ⟨87MI 403-01⟩, 416 ⟨93SC1467⟩
Silverman, S. B., **1**, 20 ⟨81OS(60)108⟩
Silversmith, E. F., **1**, 248 ⟨B-87MI 119-10⟩, 862 ⟨B-87MI 119-03⟩; **6**, 143 ⟨71JOC2262⟩
Silverstein, R., **6**, 506 ⟨66JA1024⟩
Silverstein, R. M., **2**, 1042 ⟨65JOC1027⟩; **3**, 106 ⟨63OSC(4)831⟩, 497 ⟨65JOC1027, 65JOC2543⟩; **4**, 732 ⟨65JOC1027, 65JOC2543⟩, 732 ⟨65TL521⟩; **4**, 8, 732 ⟨64TL1461⟩
Silverstein, R. M., **4**, 9 ⟨73OSC(5)390⟩; **5**, 53 ⟨87JOC689⟩; **6**, 695 ⟨65JOC1027⟩
Silverthorn, W., **2**, 482 ⟨67IS160⟩, 874 ⟨68IC2003⟩
Silverton, J. V., **2**, 412 ⟨85JOC1815⟩
Silverwood, H. A., **2**, 735 ⟨62JOC3401⟩
Silvester, M. J., **6**, 262 ⟨85JCS(P1)1191⟩
Silvestre, J., **4**, 691 ⟨89OM2001⟩
Silvestri, M., **2**, 675 ⟨82TL703⟩
Silvestri, M. G., **1**, 173, 587, 688 ⟨78JOC3249⟩, 635 ⟨85JOC2798⟩
Silvestru, C., **5**, 222 ⟨92JOM(426)49⟩
Silvey, G. A., **6**, 410 ⟨52JA5792⟩
Silvey, G. L., **6**, 491 ⟨84IC4322⟩
Silvi, B., **2**, 1022, 1023 ⟨88JPC37⟩
Silvon, M. P., **3**, 344 ⟨71JOC3885⟩
Sim, G. A., **2**, 1007 ⟨65JCS1097⟩; **6**, 721 ⟨69CC1322⟩
Sim, I. S., **2**, 625 ⟨86JMC213⟩
Simalty, M., **1**, 988 ⟨74TL1905⟩; **4**, 863 ⟨72T3545⟩
Simalty-Siemiatycki, M., **1**, 999 ⟨66BSF971⟩
Simamura, O., **1**, 442 ⟨70T4609⟩; **2**, 1054 ⟨69CC1393, 74BCJ946⟩
Simándi, L. I., **1**, 1034 ⟨B-83MI 121-02⟩; **2**, 726 ⟨93TL6105⟩; **5**, 987 ⟨90AG(E)1147, 93TL2839⟩
Simbera, J., **5**, 62 ⟨85MI 502-02⟩
Simchen, G., **2**, 209 ⟨84LA39⟩, 539 ⟨82S1⟩, 745 ⟨74CB1545⟩, 942 ⟨76AG(E)428⟩, 943 ⟨86T1299⟩; **3**, 307 ⟨85TL5035⟩, 627, 637 ⟨65AG(E)951, 71CB924⟩, 640 ⟨77S636⟩; **4**, 806 ⟨90LA745⟩, 831 ⟨77S867, 83LA816⟩, 834 ⟨77S247⟩, 881 ⟨61AG493, 63CB1350⟩, 979 ⟨71CB792⟩; **5**, 586, 614, 616 ⟨71CB2709⟩, 718 ⟨76S259, 85S403⟩, 718 ⟨87LA687, 90S525⟩, 1144 ⟨79S204⟩; **6**, 43 ⟨82S1⟩, 55 ⟨65CB1078⟩, 68, 81, 95, 104, 138, 317 ⟨85HOU(E5)3⟩, 104 ⟨68CB41⟩, 106 ⟨62AG(E)331⟩, 111, 115 ⟨73CB3725, B-79MI 604-02⟩, 137 ⟨B-74MI 605-02⟩, 138 ⟨68CB3058⟩, 138, 140 ⟨68CB51⟩, 618 ⟨68AG(E)464⟩, 634 ⟨70CB256⟩, 643 ⟨85LA2178⟩
Simchenko, L. I., **2**, 1068 ⟨74JGU1461⟩
Simcheu, G.,**3**, 436 ⟨86LA428⟩
Simcox, J. R., **2**, 192 ⟨74S295⟩
Simeon de Buochberg, M., **5**, 772 ⟨83EJM413⟩
Simeonov, M., **5**, 363 ⟨92T531⟩
Simić, D., **1**, 727 ⟨70LA(739)211⟩; **3**, 494 ⟨70LA(739)211⟩; **5**, 1141 ⟨70LA(739)211⟩
Simig, Gy., **5**, 1118 ⟨74TL2939⟩
Simion, C., **4**, 152 ⟨92RRC393⟩
Simirskaya, G. P., **4**, 1063 ⟨78ZOB2147⟩
Simkin, R. D., **3**, 317 ⟨64TL1847⟩
Simkins, N., **3**, 72 ⟨87TL989⟩
Simkins, N. S., **1**, 943 ⟨90T545⟩
Simm, U., **6**, 630 ⟨89EGP270530⟩
Simmonds, M. S. J., **1**, 596, 780 ⟨87TL221⟩
Simmonds, R. J., **5**, 770 ⟨84JCS(P1)993⟩
Simmonneaux, G., **3**, 699 ⟨78JA4312⟩
Simmons, H. E., **1**, 299 ⟨73OR(20)1⟩; **2**, 591 ⟨73OSC(5)855⟩, 640 ⟨64JOC1876⟩; **3**, 237 ⟨60JA3102⟩; **4**, 164, 183 ⟨60JA2288⟩, 1083 ⟨58JA5323⟩; **5**, 47, 52, 74, 114, 116 ⟨72JA4024⟩, 1119 ⟨65JA2296⟩; **6**, 365 ⟨74T1769⟩
Simmons, Jr., A., **5**, 1047 ⟨81SC781⟩
Simmons, Jr., H. D., **2**, 618 ⟨65JOM(3)337⟩; **4**, 16 ⟨65JA4259⟩; **6**, 695 ⟨65JOM(3)337⟩
Simmons, N. P. C., **5**, 1152 ⟨76CPL(42)460, 78JA446⟩, 1152 ⟨80JSP(82)185⟩, 1153 ⟨79JCS(D)1405⟩; **6**, 679 ⟨76CC513, 78JA446⟩
Simmons, R. G., **4**, 642, 659 ⟨80IC1021⟩
Simmons, T. C., **1**, 942 ⟨56JA6414⟩; **2**, 458 ⟨57JA3570⟩; **6**, 257 ⟨56JA5637⟩
Simms, N., **1**, 481 ⟨88OS(66)95⟩
Simokawa, K., **3**, 150 ⟨87CL1153⟩
Simolike, J. B., **1**, 656 ⟨74JOM(81)41⟩
Simon, A., **2**, 101 ⟨60ZAAC(303)53⟩, 291 ⟨60ZAAC(305)88⟩, 411, 412, 413 ⟨60CB1323⟩; **3**, 61 ⟨81TL625⟩, 380 ⟨79AG(E)944⟩, 689 ⟨85JA980⟩; **5**, 1077 ⟨93CB2569⟩; **6**, 564 ⟨89SUL123⟩
Simon, C., **1**, 103 ⟨49HCA1414⟩; **6**, 238 ⟨80JCS(P1)2254⟩
Simon, C. D., **1**, 254 ⟨88TL2257⟩
Simon, E., **6**, 5 ⟨53JOC328⟩
Simon, E. S., **3**, 408 ⟨86JCS(P1)2243⟩
Simon, F., **2**, 697 ⟨68BSF4265⟩
Simon, H., **4**, 530 ⟨83S61⟩, 913, 918, 925 ⟨84CB2622⟩; **5**, 238 ⟨77AG(E)299⟩, 361 ⟨77ZN(B)98⟩
Simonaitis, R., **2**, 1028 ⟨71T2651⟩
Simon, J., **1**, 474 ⟨88CC203⟩; **6**, 605 ⟨87CB421⟩, 680, 682, 685 ⟨88CB281⟩

Simon, K., **2**, 127 ⟨22CB2423⟩, 1008 ⟨75CSC135⟩; **5**, 220 ⟨89CB1107⟩
Simon, L. J., **2**, 102 ⟨19CR(164)795⟩
Simon, M. S., **2**, 170 ⟨67JA5838⟩; **4**, 71 ⟨67JA5838⟩
Simon, R. L., **6**, 144 ⟨68JCED437⟩, 287 ⟨73S605⟩
Simon, S., **4**, 156 ⟨81JOM(214)191⟩
Simon, W., **5**, 179 ⟨89LA1233⟩
Simonenko, L. S., **6**, 147 ⟨69IZV2309⟩
Simonet, J., **1**, 615 ⟨B-78MI 114-01⟩; **3**, 352 ⟨90BSF427⟩, 652, 654 ⟨83T1551⟩; **4**, 277 ⟨90TL3149⟩; **6**, 302 ⟨74CC634⟩, 648 ⟨91CJC861⟩
Simonetta, M., **1**, 910 ⟨76JA1999, 79AX(B)1846⟩
Simoni, D., **2**, 744 ⟨83S902⟩; **3**, 220 ⟨81S199⟩, 263 ⟨83S902⟩; **4**, 475 ⟨82S653⟩
Simonidesz, V., **2**, 286 ⟨82TL4371⟩; **4**, 240 ⟨82TL4371⟩
Simonis, M.-T., **1**, 774 ⟨87TL1463⟩
Simonneaux, G., **3**, 703 ⟨92ICA73⟩
Simonnin, M.-P., **2**, 436 ⟨85IC533⟩, 1063, 1065, 1066 ⟨67CR(C)995⟩
Simonoff, R., **1**, 11, 12, 18, 19, 21 ⟨53OR(7)263⟩
Simonova, A. A., **5**, 850 ⟨69ZOR1404⟩, 854 ⟨69ZOR1411⟩
Simonov, A. M., **2**, 802 ⟨72KGS1690⟩
Simonov, A. P., **2**, 470 ⟨69JGU2161⟩
Simonovitch, C., **1**, 729 ⟨69JA5364⟩
Simons, G., **2**, 436 ⟨84PS(19)65⟩, 824, 825 ⟨83PS(15)165⟩
Simons, J., **1**, 1143 ⟨78JA6930⟩
Simons, J. B., **6**, 107 ⟨78JOC1459⟩
Simons, J. H., **4**, 2 ⟨37JA1407⟩, 18 ⟨43JA389⟩; **6**, 212 ⟨39JA2962⟩, 212, 217 ⟨40JA3477⟩, 214 ⟨49MI 607-01, 52USP2616927⟩, 215, 216, 221 ⟨39JA435⟩, 216 ⟨50IS37⟩, 218 ⟨46JA968, 49JA2499⟩, 229 ⟨50USP2500388⟩, 419 ⟨46JA1672⟩
Simons, J. K., **4**, 423 ⟨44JA222⟩
Simons, M. J., **3**, 537 ⟨70JCS(D)699⟩
Simonsen, J. L., **4**, 24, 29 ⟨07JCS816⟩
Simonsen, O., **2**, 1002 ⟨81AX(B)344⟩
Simonyan, S. O., **1**, 530 ⟨89S472⟩
Simova, E., **3**, 310 ⟨86S886⟩
Simpkins, N. S., **1**, 24 ⟨B-93MI 101-01⟩, 135, 842 ⟨91TA1⟩, 140, 146 ⟨B-93MI 104-01⟩, 144 ⟨92SL197⟩, 144 ⟨93JCS(P1)3099⟩, 144, 439 ⟨90T6951⟩, 145, 1090 ⟨87TL989⟩, 230 ⟨B-93MI 105-01⟩, 244 ⟨88TL1315, 89TL3865⟩, 482 ⟨87CC207⟩, 612, 617 ⟨B-93MI 114-01⟩, 696 ⟨B-93MI 115-02⟩, 704 ⟨B-93MI 115-01⟩; **2**, 154, 158 ⟨B-93MI 203-01⟩, 280 ⟨88TL6787⟩, 714, 715 ⟨90T6951⟩, 1033 ⟨B-93MI 221-01⟩; **3**, 124 ⟨91TA1⟩, 170 ⟨B-93MI 304-01⟩, 170 ⟨B-93MI 304-02⟩, 171 ⟨B-93MI 304-03⟩, 221 ⟨89TL4309⟩; **4**, 79 ⟨B-93MI 402-02⟩, 187, 826 ⟨89SC21⟩, 286 ⟨88TL6787, 91T323⟩, 390, 391, 395 ⟨B-93MI 408-01⟩, 938, 939, 961 ⟨87TL989⟩, 960 ⟨90T6951⟩
Simpson, H. D., **1**, 898 ⟨71JA6688⟩
Simpson, J., **4**, 721 ⟨94OM500, 94OM511⟩; **6**, 402 ⟨92JOM(437)323⟩
Simpson, J. H., **1**, 468, 1006, 1007 ⟨87JA2138⟩
Simpson, J. M., **1**, 801 ⟨73TL2545⟩; **5**, 360 ⟨79S100⟩
Simpson, M., **2**, 690 ⟨78MI 213-02⟩
Simpson, P., **1**, 999 ⟨69JCS(C)2273⟩; **2**, 1071 ⟨69JCS(C)2273⟩, 1071, 1072 ⟨60CI(L)47⟩, 1071, 1072 ⟨63JCS3351, 71JCS(C)2872⟩
Simpson, P. J., **6**, 80 ⟨88EUP260979⟩
Simpson, R. A., **6**, 651 ⟨73JOC1591, 79JOC4536⟩
Simpson, S. J., **2**, 603 ⟨81IC2991⟩; **4**, 673 ⟨80JCS(D)29⟩
Simpson, T. D., **1**, 898 ⟨67JA3405⟩
Simpson, T. J., **5**, 199 ⟨88TL1595⟩
Simpson, V. G., **2**, 168 ⟨49JOC272⟩
Sims, C. L., **1**, 822 ⟨77JA5453⟩; **5**, 553 ⟨77JA5453⟩
Sims, H., **1**, 628 ⟨73OSC(5)608⟩
Sims, H. J., **3**, 106 ⟨52JA5110⟩
Sims, J., **3**, 327 ⟨73JA7287⟩
Sims, L. B., **1**, 629 ⟨91JA8821⟩, 783 ⟨83JA3714⟩
Sims, L. L., **4**, 23 ⟨57JA6270⟩, 23, 38 ⟨55JA3465⟩
Sims, P., **2**, 651 ⟨53JCS3623⟩
Sims, R. J., **1**, 745 ⟨81JCS(P1)3087⟩
Simson, J. M., **2**, 298 ⟨69JA5107⟩
Sin, N., **1**, 816, 819 ⟨93TL753⟩
Sinai-Zingde, G., **1**, 86 ⟨87TL5287⟩, 804 ⟨87JOC4641⟩; **2**, 326 ⟨87SC1155⟩
Sinaÿ, P., **1**, 68 ⟨84JA8313⟩, 513 ⟨83TL4833⟩; **4**, 172 ⟨83TL4833⟩, 176 ⟨77JCS(P1)2513⟩, 213 ⟨94TL2537⟩, 235 ⟨89CAR(188)81⟩, 238 ⟨93SL522⟩, 950, 952 ⟨86TL6201⟩; **6**, 549 ⟨91T5149⟩
Sinchai, W., **4**, 856 ⟨80AJC1073⟩
Sinclair, I. W., **3**, 294 ⟨75JCS(P1)622⟩
Sinclair, J. A., **1**, 279, 1033 ⟨73JA3080⟩, 1033, 1034 ⟨86JOC4507⟩, 1035, 1103 ⟨76JOC1078⟩, 1050 ⟨77JA954⟩; **2**, 520, 1086 ⟨76JOC1078⟩, 1084 ⟨77JOM(131)163, 86JOC4507⟩; **3**, 147 ⟨77JA955⟩
Sinclair, P. J., **4**, 102 ⟨88JA1547⟩
Sinden, A. W., **2**, 399 ⟨74JINC993⟩
Sindona, G., **2**, 350 ⟨83JHC1207⟩; **3**, 618 ⟨87S168⟩
Sinegouskaya, L. M., **1**, 441 ⟨91SR1⟩; **6**, 565 ⟨90ZOR1364⟩
Sinenko, T. A., **4**, 332 ⟨77JOU2224⟩, 865 ⟨78ZOR1947, 79ZOR1477⟩, 866 ⟨77ZOR508, 78UKZ183⟩; **5**, 595 ⟨73ZOR347⟩; **6**, 275 ⟨76ZOR558⟩, 275, 277, 316 ⟨76ZOR2583⟩
Sineokov, A. P., **4**, 891 ⟨71ZOB482⟩; **6**, 324, 335 ⟨63BAU1384⟩, 493 ⟨68ZOR1661⟩, 628, 629 ⟨64JGU4149⟩
Sinex, F. M., **5**, 83 ⟨54MI 502-03⟩
Singal, K. K., **4**, 422 ⟨85SC829⟩
Singaram, B., **1**, 60 ⟨87JOC4014⟩, 163 ⟨83TL623⟩, 163 ⟨83TL631⟩, 443 ⟨89JA384, 91JOC1543⟩, 660, 764 ⟨83TL635⟩, 831 ⟨89JA3441⟩, 1034 ⟨88JA1529, 91JOC1500⟩; **2**, 59 ⟨94JCS(P2)117⟩, 60 ⟨92TL4533, 93TL257⟩, 60 ⟨94TL1515, 94TL5201⟩, 306, 318 ⟨91JOC1170⟩, 514 ⟨B-86MI 210-01, 87JOC4014⟩, 515 ⟨79JOC465, 84JOC945⟩, 516, 523 ⟨78JOC4395⟩, 519 ⟨93TL7693⟩, 519, 531 ⟨82JOC5069, 89JA1754⟩, 525 ⟨83TL631⟩, 525, 1005 ⟨83TL623⟩, 529 ⟨91JOC3286⟩, 530 ⟨85JOC5586⟩, 531 ⟨85JA4980⟩, 532 ⟨92JOC499, 92JOC2716⟩, 532, 698 ⟨89JA3441⟩, 534 ⟨81JOC2712⟩, 535 ⟨85JA460⟩, 740 ⟨89JA384⟩, 905 ⟨83TL635, 83TL637⟩; **3**, 19 ⟨83TL631⟩, 135 ⟨88JA1529⟩; **4**, 524 ⟨91JOC5691, 93HOU(E15/2)1674⟩, 642, 645, 1077 ⟨83TL637⟩, 664 ⟨83TL623, 83TL631⟩; **6**, 199 ⟨83TL623⟩
Singaram, S., **1**, 200, 238 ⟨88JOC1475⟩, 509, 510 ⟨83TL1433⟩; **2**, 520, 900, 901 ⟨83TL1433⟩; **3**, 3, 114 ⟨83TL1433⟩; **5**, 275 ⟨79JOC4727⟩
Singer, A. W., **5**, 115 ⟨55OSC(3)740⟩
Singer, B., **4**, 321 ⟨89CB2311⟩
Singer, E., **1**, 1040 ⟨65CB3311⟩; **2**, 809 ⟨69CB2557⟩; **4**, 298 ⟨78T1285⟩; **5**, 548 ⟨80CB3605⟩; **6**, 143 ⟨83CB2068⟩, 304 ⟨67CB767⟩, 598 ⟨87CB1581⟩
Singer, F.,**6**, 207 ⟨03LA(329)116⟩
Singer, G., **4**, 232 ⟨87LA451⟩
Singer, L. A., **3**, 557 ⟨67JA598⟩, 584, 586 ⟨74CC962⟩, 584, 586 ⟨74JOC3780⟩, 584, 586 ⟨77JA2622⟩, 596 ⟨69TL5093⟩
Singer, M. I. C., **1**, 54 ⟨67CI(L)787⟩, 647, 648, 737, 738, 743 ⟨72T343⟩, 648 ⟨67CC1042⟩, 737, 738, 743 ⟨72T353⟩, 738 ⟨68CI(L)1277⟩; **2**, 1004 ⟨67CC1042⟩; **3**, 373 ⟨67CI(L)118⟩, 498 ⟨67CI324, 72T353⟩, 500 ⟨71JCS(C)2941, 72T343⟩
Singer, M. S., **4**, 310 ⟨68MI 407-01⟩
Singer, R. D., **1**, 467 ⟨93CRV2117⟩; **2**, 67, 590 ⟨93CRV2117⟩; **3**, 162 ⟨93CRV2117⟩
Singer, S., **1**, 954 ⟨64T2177⟩
Singer, S. J., **5**, 465 ⟨72CC672⟩
Singer, S. P., **1**, 234, 821 ⟨78TL519⟩, 410, 821 ⟨78JOC4831⟩, 601, 610, 611, 621, 646 ⟨75CS9⟩, 619 ⟨78JOC4884⟩, 716 ⟨75CS9⟩

Singh, A., **1**, 21, 25, 437, 438, 943 ⟨80JA4438⟩, 243 ⟨82JOC1641⟩; **2**, 132 ⟨81SC409⟩; **3**, 62 ⟨63JIC114⟩, 543 ⟨76JA3728⟩; **6**, 195 ⟨80IC3637⟩
Singh, A. K., **2**, 690 ⟨85JOM(291)185⟩; **3**, 300 ⟨86SC485⟩; **5**, 1110 ⟨78T1731⟩; **6**, 116 ⟨86MI 604-01⟩
Singh, A. P., **4**, 246 ⟨93S67⟩
Singh, B., **1**, 452 ⟨78JCS(D)657⟩; **2**, 563 ⟨78JCS(D)657⟩; **4**, 422 ⟨85SC829⟩; **5**, 277 ⟨71TL321⟩, 756 ⟨71CC498⟩
Singh, B. P., **1**, 5, 9 ⟨83JOC3667⟩, 915 ⟨84JA3265⟩, 918 ⟨81JA588⟩; **2**, 15 ⟨83JOC3356⟩, 19 ⟨83S713⟩, 36, 142 ⟨83JOC3667⟩; **5**, 39 ⟨83JOC3667⟩; **6**, 229 ⟨87JA3708⟩
Singh, D., **6**, 111 ⟨90IJC(B)180⟩
Singh, G., **1**, 485 ⟨85JCS(P1)1289⟩; **2**, 618 ⟨65JOM(3)337⟩, 936 ⟨B-66MI 218-02⟩; **3**, 495, 497 ⟨66JOM(5)267⟩; **4**, 56 ⟨82JOC2582⟩, 569 ⟨79JOC1057⟩, 571 ⟨65JA4156⟩; **5**, 555 ⟨82S693⟩, 896 ⟨66JOM(5)267⟩; **6**, 266 ⟨66JOM(5)185⟩, 507 ⟨80IJC(B)653⟩, 507, 576 ⟨80S1001⟩, 608 ⟨86JOM(301)15⟩, 695 ⟨65JOM(3)337⟩
Singh, G. B., **4**, 416 ⟨61AP(294)404, 63JIC777⟩
Singh, G. S., **5**, 362 ⟨84BCJ1667⟩
Singh, H., **3**, 165 ⟨87IJC(B)1111⟩, 419 ⟨88IJC(B)132⟩, 634 ⟨79SC569⟩; **4**, 29 ⟨81T1215⟩, 863 ⟨80ACS(A)577⟩; **5**, 597 ⟨79JIC545⟩; **6**, 111 ⟨90IJC(B)180⟩, 559 ⟨82JCS(P1)653⟩, 564 ⟨89IJC(B)439⟩, 643 ⟨69JMC558⟩
Singh, H. B., **2**, 690 ⟨85JOM(291)185⟩
Singh, H. K., **4**, 332, 428, 429 ⟨83JOC332⟩
Singh, J., **1**, 307 ⟨74JA6181⟩, 548 ⟨79JA7107⟩; **2**, 34, 43 ⟨88JOC187⟩, 492 ⟨65JA1213⟩; **3**, 58 ⟨89CI(L)533⟩; **4**, 56 ⟨82JOC2582⟩
Singh, J. D., **2**, 690 ⟨90IJC(A)703⟩; **5**, 210 ⟨87IJC(A)260⟩
Singh, J. N., **2**, 11 ⟨83IJC(B)421⟩
Singh, K., **2**, 656 ⟨79CJC3304⟩; **3**, 419 ⟨88IJC(B)132⟩
Singh, L., **5**, 1105 ⟨87IJC154⟩
Singh, M., **1**, 14 ⟨92JOC1618⟩; **2**, 238 ⟨82IJC(A)305⟩; **3**, 7, 120 ⟨89S403⟩, 637 ⟨82JCS(P1)1303⟩; **5**, 342 ⟨79ICAL99⟩, 1132 ⟨78IJC(A)778⟩
Singh, N., **5**, 331 ⟨73TL2711⟩
Singh, N. K., **5**, 385 ⟨86JMC199⟩
Singh, N. P., **2**, 157 ⟨88SC1061⟩
Singh, O. M., **4**, 253 ⟨92S1075⟩
Singh, O. V., **1**, 556 ⟨92JCS(P1)2565⟩; **2**, 311 ⟨91SL668⟩
Singh, P., **1**, 282 ⟨91JOC3908⟩; **2**, 16 ⟨87JOC4307⟩, 405 ⟨72JA5077⟩; **4**, 29 ⟨81T1215⟩
Singh, P. R., **1**, 68 ⟨83TL1411⟩; **3**, 234 ⟨67AJC639⟩
Singh, R., **1**, 931 ⟨92T3385⟩, 1009 ⟨89JOC4453⟩; **2**, 127 ⟨91JOC7328⟩, 633 ⟨88SC1327⟩, 656 ⟨79CJC3304⟩; **5**, 631 ⟨85IJC(B)977⟩, 685 ⟨93MI 517-01⟩
Singh, R. K., **1**, 521 ⟨79JA7001⟩; **2**, 7 ⟨92CL1505⟩, 665 ⟨77JA5810⟩, 667 ⟨78JOC379⟩; **3**, 244 ⟨78JOC379⟩; **4**, 204 ⟨75JOC3807, 76JOC1668⟩, 204 ⟨81OS(60)66⟩
Singh, R. P., **3**, 4, 115 ⟨79T1789⟩, 294 ⟨64JOC574⟩
Singh, S., **1**, 364, 365 ⟨90PAC507⟩; **2**, 613 ⟨87JA7194⟩; **5**, 210 ⟨89IJC(A)1528⟩
Singh, S. B., **5**, 223 ⟨87CJC2390⟩, 362 ⟨84BCJ1667⟩
Singh, S. K., **1**, 841 ⟨81T3967⟩; **3**, 144 ⟨81T3967⟩
Singh, S. M., **1**, 667 ⟨87CJC1821⟩, 1037 ⟨85JOC1577⟩, 1045 ⟨84TL2411⟩; **2**, 901 ⟨84S920⟩; **3**, 273 ⟨84TL2411⟩; **5**, 229 ⟨84S920⟩
Singh, S. P., **3**, 536 ⟨91S1027⟩; **4**, 183 ⟨72TL5133⟩; **5**, 356 ⟨93JCR(S)244⟩, 1105 ⟨87IJC154⟩; **6**, 573 ⟨80JHC1369⟩
Singh, T., **1**, 538 ⟨50JA494⟩
Singh, V., **3**, 126 ⟨83T345⟩
Singh, V. K., **2**, 528 ⟨87JA7925⟩; **3**, 151 ⟨92TL6021⟩
Singh, W. P., **3**, 708 ⟨88TL1435⟩
Singha, A. S., **5**, 665 ⟨84IJC(B)728⟩
Singhal, G. C., **5**, 821 ⟨58JIC893⟩
Singhal, G. H., **3**, 556, 557 ⟨64JOC34⟩, 561 ⟨69JCED408⟩; **5**, 1078 ⟨67JOC2895⟩
Singhal, K., **2**, 881 ⟨85POL251⟩, 891 ⟨86POL1581⟩; **5**, 372 ⟨84SRI269, 85POL251⟩, 372 ⟨87IJC(A)146, 93SRI1011⟩, 391 ⟨87IJC(A)1046⟩, 1145 ⟨86POL677⟩
Singhawangcha, S., **6**, 448 ⟨86JOC3781⟩
Singla, M., **6**, 650, 655 ⟨80MI 621-01⟩
Singleton, A., **4**, 319 ⟨84JCS(P1)865⟩, 843 ⟨84JCS(P1)859⟩; **5**, 552, 559 ⟨84JCS(P1)865⟩, 558, 559 ⟨84JCS(P1)859⟩
Singleton, D. A., **1**, 470 ⟨93TL3041⟩, 949 ⟨90TL5551⟩; **2**, 43 ⟨92TL1017⟩, 902, 922 ⟨91TL7365⟩, 904 ⟨90JA7423, 92JOC5768⟩
Singleton, E., **1**, 456 ⟨85JMOC213⟩; **3**, 694 ⟨83AOC(22)209⟩
Singleton, H. M., **5**, 272, 293, 304 ⟨38JA540⟩
Singleton, K. A., **5**, 341 ⟨81T1283⟩
Singley, Jr., J. E., **6**, 33, 81, 228 ⟨56JA479⟩
Sinha, A., **3**, 272 ⟨78JOC2438⟩
Sinha, G., **5**, 144 ⟨81IJC(B)411⟩
Sinha, N. D., **1**, 473 ⟨80JA790⟩; **2**, 409, 410 ⟨91JOC4608⟩
Sinha, S. C., **1**, 822 ⟨89TL1575, 90CC418⟩, 822 ⟨90T4887, 94CC789⟩; **2**, 49 ⟨92TL6407⟩
Sinhababu, A. K., **1**, 11, 15 ⟨82SC983⟩, 34 ⟨83SC677⟩; **2**, 398 ⟨83TL227⟩
Sinha-Bagchi, A., **2**, 49 ⟨92TL6407, 92TL6411⟩
Sinharay, A., **4**, 972, 982 ⟨63CB3306⟩
Sinico, A., **3**, 365 ⟨68CC724⟩
Sinisterra, J. V., **1**, 11, 12 ⟨93T8433⟩
Sinitsa, A. D., **2**, 464 ⟨86JGU1567⟩; **4**, 106 ⟨68JGU1276⟩, 304 ⟨69ZOR2181⟩, 487 ⟨69ZOB2192⟩, 927 ⟨85ZOB838⟩, 966 ⟨73ZOB211⟩, 1004 ⟨86ZOB805⟩, 1008, 1009 ⟨86ZOB2681, 89ZOB2492⟩, 1009 ⟨90ZOB229, 90ZOB1187⟩; **5**, 369, 713 ⟨90ZOB1187⟩, 643 ⟨86ZOB1467⟩, 644 ⟨89ZOB1917, 90ZOB1389⟩, 652 ⟨90JGU1165⟩, 655 ⟨87ZOB475⟩, 655, 788, 790 ⟨87JGU415⟩, 657 ⟨78JOU1457, 78ZOR1559⟩, 657 ⟨87JGU1104, 87ZOB1233⟩, 659 ⟨86ZOB2681⟩, 659, 788, 789 ⟨86JGU2372⟩, 660 ⟨87ZOB1514⟩, 660, 788 ⟨87JGU1349⟩, 713 ⟨84ZOB2154, 86ZOB1262, 89ZOB2492, 90JGU849, 90JGU1057, 90ZOB1186, 90ZOB963⟩, 713, 788 ⟨86JGU1113, 89JGU2228, 90JGU1059⟩, 737 ⟨86ZOR1654⟩, 738 ⟨90JGU1389⟩, 776 ⟨85ZOB776⟩, 787 ⟨85JGU402, 89JGU1496⟩, 788 ⟨86JGU1467, 89JGU1129⟩, 789 ⟨88JGU30, 89JGU1123⟩, 863 ⟨80ZOB2413⟩, 965 ⟨70ZOB1933⟩, 1063 ⟨82ZOR668⟩; **6**, 687 ⟨84ZOB715⟩
Sinitsa, D. A., **5**, 713 ⟨84JGU1925⟩
Sink, C. W., **2**, 278 ⟨69JCS(D)1023, 72JCP(57)4434⟩
Sinn, E., **5**, 325 ⟨81IC1823⟩
Sinnema, P.-J., **4**, 656 ⟨93OM184⟩
Sinnema, Y. A., **6**, 110 ⟨57RTC949⟩
Sinnige, H. J. M., **4**, 894 ⟨72JOC1960⟩
Sinnreich, J., **1**, 925 ⟨68T4509⟩; **2**, 157 ⟨72JCS(P1)1543⟩
Sinou, D., **2**, 85 ⟨88JOM(346)C7⟩, 330 ⟨91CC1684⟩; **3**, 632 ⟨88JMOCL23, 91TA178⟩
Sinsky, A. J., **3**, 48 ⟨89JA3924⟩
Sintoni, M., **4**, 405 ⟨86G229⟩
Sinyashin, O. G., **2**, 461 ⟨88JGU2161⟩, 820 ⟨84ZOB1917⟩; **5**, 398 ⟨88JGU1965⟩; **6**, 581 ⟨88ZOB1468⟩, 583 ⟨77IZV1177⟩
Sinyavskii, V. G., **5**, 967 ⟨69ZOR1888⟩
Sioli, G., **2**, 190 ⟨74JCS(P1)771⟩
Sipio, W. J., **1**, 591, 594, 716 ⟨79CC83⟩; **2**, 282 ⟨79S982⟩; **3**, 38, 174 ⟨79S982⟩
Sipos, F., **1**, 389 ⟨65JCS2513⟩
Sipos, G., **1**, 900 ⟨85JA1421⟩
Sipos, W., **1**, 705 ⟨86CB1540⟩, 708 ⟨87CB5⟩
Sipyagin, A. M., **2**, 1003 ⟨86IZV134⟩
Siragusa, J. A., **1**, 548 ⟨60JCS1045⟩
Sirakawa, K., **2**, 228 ⟨70CPB235⟩
Sirat, H. M., **5**, 63 ⟨78CC961⟩
Siray, M., **3**, 607 ⟨84AG(E)619⟩; **5**, 675 ⟨85CB1632⟩, 1058 ⟨84AG(E)970⟩, 1092 ⟨84AG(E)587⟩; **6**, 680, 684 ⟨84TL4447⟩, 680, 687 ⟨84AG(E)579⟩, 680, 687 ⟨84AG(E)970⟩
Sircar, I., **2**, 28 ⟨88TL6835⟩
Sircar, J. C., **2**, 398 ⟨67JOC4134⟩

Sire, J., **6**, 653 ⟨77ACS(B)589⟩
Siriwardane, U., **5**, 690 ⟨88JOC5309⟩
Sirna, A., **2**, 26 ⟨78TL4575⟩; **3**, 206 ⟨77S773⟩
Siroi, T., **2**, 21 ⟨81TL3193⟩
Sironi, A., **5**, 940 ⟨86OM1976⟩; **6**, 345 ⟨84CC1089⟩
Sirotkina, E. E., **4**, 310 ⟨78ZOR1723⟩
Sisak, A., **6**, 517 ⟨85M1103⟩
Sisani, E., **1**, 44 ⟨79CC959⟩
Sishta, C., **2**, 584 ⟨86CC787⟩
Sisido, K., **1**, 884 ⟨64JOC2781⟩; **5**, 29 ⟨59JA5817⟩
Sisko, J., **1**, 106 ⟨93PAC1153⟩, 369 ⟨91JOC3210⟩, 1042 ⟨90JOC393⟩; **2**, 306 ⟨90SC573⟩, 310 ⟨90JOC2563⟩; **3**, 438 ⟨90JOC393, 90JOC3037⟩; **5**, 345 ⟨91JOC3210⟩
Sisler, H. H., **2**, 278 ⟨71JOC1700⟩, 374 ⟨56JA3874⟩, 379 ⟨81JINC3389⟩, 383 ⟨57CRV1021⟩, 384 ⟨56JA1211, 61JOC1819⟩, 387 ⟨62JCS3818, 66IC1835⟩, 387 ⟨68IC698⟩, 435 ⟨74IS129⟩, 471 ⟨69IC1006⟩, 490 ⟨70IC862⟩, 506 ⟨69IC1032⟩, 859 ⟨61JOC611⟩, 892 ⟨67IC1178, 72IC1226⟩; **4**, 569 ⟨63JOC272⟩; **5**, 354 ⟨90IC309⟩
Sisti, A. J., **1**, 115 ⟨72JOC4090⟩; **3**, 407 ⟨74JOC3923⟩
Sisti, M., **1**, 812 ⟨91TL6969⟩; **3**, 59 ⟨88S697⟩, 169 ⟨87S1113⟩; **4**, 462 ⟨92SC107⟩; **5**, 295 ⟨80S385⟩
Sistig, F., **1**, 428 ⟨90S403⟩
Sit, S. Y., **2**, 85 ⟨84JA2949⟩, 625 ⟨81TL3745⟩
Sita, L. R., **2**, 532 ⟨85ACR380⟩
Sitkareva, I. V., **2**, 28 ⟨87IZV2395⟩
Sitkin, A. I., **4**, 993 ⟨68ZOR2116, 75ZOR452⟩, 993, 994 ⟨71ZOR1670⟩
Sitnikova, S. P., **3**, 548 ⟨78ZOB2137⟩; **5**, 540 ⟨75JGU2421⟩
Sittenthaler, W., **6**, 25, 31 ⟨88M1427⟩
Sitzmann, M. E., **2**, 392 ⟨77JOC563⟩; **3**, 662 ⟨73JOC4363⟩; **5**, 366 ⟨82JCED97⟩; **6**, 145 ⟨86JHC81⟩, 145 ⟨89MI 605-01⟩, 145 ⟨90SC3303⟩, 272 ⟨83JEM95, 83USP467715⟩, 308 ⟨83JOC3354⟩, 449 ⟨82JCED97⟩, 449, 492 ⟨85JOC5879⟩
Siu, T.-W., **1**, 709 ⟨78TL4269⟩
Siu, Y. K., **5**, 354 ⟨71CJC1598⟩
Sivadasan, L., **5**, 226 ⟨81CI(L)164⟩
Sivaramakrishnan, R., **1**, 226 ⟨81JCS(P1)270⟩
Sivaram, S., **1**, 294 ⟨86CRV353⟩; **5**, 275, 295 ⟨82JOC4812⟩
Sivasubramanian, S., **6**, 611 ⟨90JOC3045⟩
Sivavec, T. M., **2**, 670 ⟨85TL2159⟩
Sivibi, L., **1**, 556 ⟨90T2371⟩
Sivriev, Kh., **4**, 113, 120 ⟨70JGU255⟩, 555 ⟨70ZOB285⟩
Siwapinyoyos, T., **3**, 210 ⟨80TL1169⟩
Sixma, F. L. J., **1**, 773 ⟨63RTC1123⟩
Sizov, Y. A., **4**, 1023 ⟨72ZOB346⟩; **6**, 614 ⟨76IZV2381⟩
Sizykh, O. L., **4**, 872 ⟨91MI 419-01⟩
Sjöberg, B., **2**, 123 ⟨72ACS279⟩
Sjoberg, B., **4**, 229 ⟨42CB13⟩, 426 ⟨80T2505⟩
Sjöberg, K., **2**, 123 ⟨72ACS279⟩
Sjogren, E. B., **5**, 329 ⟨86TL3119, 86TL4957⟩, 329 ⟨86TL4961⟩
Sjorgen, E. B., **2**, 341 ⟨86TL4961⟩
Skabara, P. J., **6**, 698 ⟨93PS(74)279⟩
Skapski, A. C., **4**, 656 ⟨71CC1079, 71CC1477⟩
Skarnulis, A. J., **2**, 558, 560 ⟨91JOC698⟩
Skarzewski, J., **3**, 7 ⟨90TL2051⟩
Skattebøl, L., **1**, 706, 707 ⟨81ACS(B)263⟩, 808 ⟨65JA4506⟩, 990 ⟨63ACS1683, 84TL2887⟩, 1046 ⟨58JCS1054⟩; **2**, 149 ⟨67JOC3111⟩; **3**, 71 ⟨70JOC3200⟩, 79 ⟨58JCS1054⟩, 471 ⟨60CB2162⟩; **4**, 24 ⟨64JOC2951⟩, 25 ⟨73TL1367⟩
Skattenbol, L.,
Skawinski, W., **6**, 365 ⟨90MI 612-01⟩
Skechter, H., **4**, 37 ⟨91JOC1663⟩
Skell, P. S., **1**, 114 ⟨62JA4660⟩, 232 ⟨73JA1547⟩, 299 ⟨59JA3383⟩, 928 ⟨73JA6728⟩, 1143 ⟨72JA1578, 72JA7862⟩; **2**, 27 ⟨74JA5616⟩, 35 ⟨64JA2956⟩, 961 ⟨73ACR97⟩; **3**, 490 ⟨60JA247⟩; **4**, 24 ⟨56JA5430⟩, 385 ⟨73JOC3937⟩; **5**, 203, 205 ⟨83JA3999⟩, 318 ⟨86JOC4959⟩, 874 ⟨73JOC3937⟩
Skelly, W. G., **5**, 84 ⟨63CRV625⟩
Skelton, B., **4**, 654 ⟨87OM2110⟩
Skelton, B. W., **2**, 1003, 1005 ⟨82CC14, 89JCS(D)105⟩, 1005 ⟨84JCS(D)1801⟩; **4**, 584 ⟨81CC698, 82JOM(228)C71⟩, 604, 646, 647 ⟨82CC14, 84JCS(D)1801⟩, 604, 647, 648 ⟨90JCS(D)1161⟩, 646 ⟨89JCS(D)105⟩, 647, 660 ⟨82JOM(233)C28, 92JCS(D)775⟩, 678 ⟨94JOM(464)191⟩; **6**, 190 ⟨84JCS(D)1801⟩, 589, 593 ⟨88AJC549⟩
Skerlj, R. T., **2**, 917 ⟨86CC626⟩, 947 ⟨87CC1025⟩
Sket, B., **1**, 517 ⟨77JA3504⟩; **2**, 21 ⟨84T2865⟩, 24 ⟨90T2503⟩, 28 ⟨86JOC929, 89SC2481⟩
Skibo, E. B., **2**, 361, 792 ⟨76JOC1333⟩
Skibo, E. S., **2**, 358 ⟨77JOC875⟩
Skiles, R. D., **5**, 1125 ⟨83JHC1533⟩; **6**, 567 ⟨88JOC38⟩
Skinner, A. C., **1**, 115 ⟨63JCS577⟩, 889 ⟨66JCS(C)154⟩
Skinner, C. G., **6**, 570, 571 ⟨70JMC377⟩
Skinner, D. L., **1**, 1063 ⟨67JOC105⟩
Skinner, G. S., **3**, 468 ⟨24JA731⟩
Skinner, H. A., **2**, 491 ⟨52JCS4331⟩
Skirl, R., **4**, 556, 557, 699 ⟨90ZAAC(586)25⟩; **5**, 1074 ⟨89ZAAC(577)165⟩
Skita, A., **3**, 616 ⟨08CB2925⟩
Skiyanova, A. M., **4**, 813 ⟨73JGU679⟩
Skladnev, A. A., **5**, 474, 484 ⟨67BAU324⟩, 645 ⟨63ZOB3293⟩
Sklarz, B., **2**, 389, 765 ⟨69JCS(C)1073⟩; **5**, 49 ⟨67QR3⟩
Skobeleva, S. E., **5**, 390, 391 ⟨88IZV859⟩, 1076, 1077 ⟨87JOM(327)303⟩, 1081 ⟨87ZOB405, 88MI 528-01⟩, 1082 ⟨88IZV739⟩
Skobin, A. S., **2**, 458 ⟨79JGU285, 79JGU1721⟩
Skobun, A. S., **4**, 339 ⟨79ZOB1956, 91ZOB1263⟩
Skoda, L., **4**, 1050 ⟨85JOM(291)277⟩; **6**, 372 ⟨84ZN(B)1500⟩, 379 ⟨85JOM(291)277⟩, 380 ⟨84ZAAC(510)175, 85ZN(B)1023⟩
Skoda-Földes, R., **4**, 510 ⟨91JOM(408)297⟩
Skoglund, M., **3**, 502 ⟨76JA7846⟩
Skoglund, M. J., **1**, 797 ⟨82JOC4758⟩, 797, 882 ⟨82JOC4766⟩
Skold, C. N., **3**, 71 ⟨76SC119⟩, 369 ⟨70TL791⟩
Skoldinov, A. P., **3**, 69 ⟨64JGU109⟩; **6**, 140 ⟨67ZOR1749⟩
Skolik, J., **2**, 1008 ⟨B-86MI 219-01⟩
Skolnick, P., **5**, 1029 ⟨B-79MI 517-01⟩
Skolov, V. B., **5**, 22 ⟨93PS(72)153⟩, 672 ⟨91KFZ33⟩
Skonecka, A., **5**, 597, 607 ⟨93M721⟩
Skoog, I. H., **4**, 864 ⟨77USP4018810⟩; **5**, 228 ⟨64JOC492⟩
Skopenko, V. V., **2**, 1003 ⟨79UKZ811⟩; **5**, 1084 ⟨88USP4762945⟩
Skorcz, J. A., **1**, 881 ⟨62JOC3836, 68OS(48)53⟩; **3**, 626 ⟨73OSC(5)263⟩
Skorianetz, W., **4**, 437 ⟨70HCA251⟩
Skorna, G., **3**, 696, 699, 721, 722 ⟨77AG(E)259⟩; **6**, 462 ⟨77AG(E)259⟩
Skorobogatova, E. V., **5**, 1044 ⟨91ZOR340⟩
Skorobogatova, S. Ya., **4**, 551 ⟨78ZOB757, 80ZOB989⟩
Skotnicki, J. S., **2**, 160 ⟨84JOC2500⟩
Skowronska, A., **5**, 1107, 1112 ⟨83T4175⟩
Skowronski, G., **5**, 597 ⟨76SC387⟩
Skramstad, J., **3**, 620 ⟨70ACS3424⟩
Skripets, V. I., **5**, 674 ⟨76URP534448⟩, 712, 714 ⟨75JOU1719⟩, 712, 714 ⟨75ZOR1724⟩
Skripkin, Yu. V., **4**, 717 ⟨80JOM(201)269⟩
Skrydstrup, I., **6**, 595 ⟨89H(28)269⟩
Skrydstrup, T., **1**, 818 ⟨92TL4563⟩, 1024 ⟨90TL7145⟩; **5**, 832 ⟨87ACS(B)29⟩
Skrynnikova, A. A., **3**, 171 ⟨91JOU1231⟩
Skvorcov, S., **2**, 439 ⟨91ZAAC(594)66⟩; **4**, 550 ⟨91ZAAC(594)66⟩
Skvortsov, N. K., **4**, 569 ⟨83ZOB1831⟩
Skwarski, D., **6**, 546 ⟨84MI 617-01, 86MI 617-04⟩, 546 ⟨88MI 617-01⟩

Sky, A. F., **1**, 922 ⟨88TL6839⟩; **5**, 765, 766 ⟨93JA7584⟩
Skyrzypczynski, Z., **2**, 857 ⟨62MI 216-01⟩
Slabko, M. G., **4**, 339 ⟨91ZOB909⟩
Slack, R., **5**, 769, 847, 848 ⟨56JCS2253⟩, 858 ⟨54JCS2968⟩
Slack, W. E., **3**, 249 ⟨70JA5258, 75JOC3179⟩
Slade, J., **3**, 294 ⟨80JOC2785⟩; **5**, 80 ⟨80JOC2785⟩
Slade, M., **4**, 310 ⟨68MI 407-01⟩
Slade, R. M., **2**, 824 ⟨71JCS(A)2976⟩
Sladkov, A. M., **1**, 1005 ⟨68JOU21⟩; **2**, 1018, 1019 ⟨63IZV2213, 64IZV392, 65ZOR415⟩, 1054 ⟨66JOU1440⟩, 1077 ⟨72IZV529⟩, 1093 ⟨66JGU570⟩
Sladkov, I. B., **6**, 223, 226, 227, 228 ⟨91MI 607-01⟩, 223, 228 ⟨83MI 607-01⟩
Sladky, F., **5**, 1117 ⟨91CB1131⟩; **6**, 355, 356, 357, 358 ⟨89CB2279⟩, 355, 357 ⟨85CC1800⟩, 357, 358 ⟨85JOM(295)C1⟩, 357, 358 ⟨89CB1255⟩, 357, 358 ⟨91CB1131⟩, 357, 358 ⟨91OM2101⟩
Slagel, R. C., **2**, 384 ⟨68JOC1374⟩; **5**, 997 ⟨67CJC2625⟩
Slagle, I. R., **1**, 1127 ⟨84JA4356, 86JPC402⟩
Slagle, J. R., **5**, 208 ⟨59JOC107⟩
Slapak, C., **4**, 646 ⟨81CC179⟩
Slater, M. J., **2**, 926 ⟨87JOC1483⟩; **3**, 218 ⟨89JOC5148⟩
Slater, S. C., **5**, 351 ⟨90JOC6037⟩
Slates, H. L., **3**, 231 ⟨51JA719⟩; **5**, 173 ⟨68T2443⟩
Slaugh, L. H., **2**, 970 ⟨66T1741⟩
Slawin, A. M. Z., **1**, 571 ⟨92JA6330⟩, 987 ⟨91CC1290⟩; **2**, 1000 ⟨88TL6009⟩
Slaylor, M., **4**, 413 ⟨55AJC512⟩
Slaymaker, S. C., **3**, 446 ⟨57JOC204⟩
Slebocka-Tilk, H., **2**, 735 ⟨84JOC3106⟩
Sledeski, A. W., **1**, 133 ⟨92TL1855⟩
Slegir, A. R., **3**, 436 ⟨86JOC5024⟩
Sleigh, J. H., **6**, 450 ⟨83JFC(22)521⟩
Sleiter, G., **4**, 951 ⟨73HCA2166⟩
Slemon, C. E., **4**, 54, 56 ⟨81CJC3055⟩
Slessor, K. N., **3**, 639 ⟨87S411⟩
Sletten, E., **2**, 1008 ⟨66ACS1874⟩
Sletten, J., **3**, 255 ⟨73ACS3881⟩, 354 ⟨70ACS1464, 71ACS3577⟩, 354 ⟨73ACS2517⟩
Sletzinger, M., **1**, 80 ⟨70JOC3273⟩; **4**, 48 ⟨75TL3979, 79S970⟩; **5**, 762, 816 ⟨65JOC259⟩
Sleurich, T., **6**, 630 ⟨89EGP270530⟩
Slimmer, M., **4**, 761 ⟨03CB289⟩
Slinger, C. J., **1**, 16 ⟨81JOC4843⟩
Slish, D. F., **3**, 613 ⟨89JOC4476⟩
Sliskovic, D. R., **4**, 19, 27 ⟨81JCS(P1)78⟩; **5**, 631 ⟨84S946⟩
Sliwa, H., **1**, 808 ⟨77MI 118-02⟩; **3**, 310 ⟨90H(31)427⟩
Sliwa, W., **2**, 1009 ⟨85H(23)1513⟩
Sliwinski, W. F., **6**, 63 ⟨72JA125⟩
Sliwka, H.-R., **1**, 708 ⟨84HCA434⟩
Sloan, A. D. B., **1**, 1000 ⟨61T18⟩
Sloan, C. P., **1**, 547, 1130 ⟨85TL6001⟩, 927 ⟨90JA896⟩
Sloan, K. B., **4**, 54, 58 ⟨83JOC635⟩, 58, 211 ⟨83JOC5280⟩, 437 ⟨69JHC187⟩
Sloan, M. P., **5**, 800 ⟨87JOM(332)371⟩
Sloat, T. K., **6**, 215, 216, 221 ⟨39JA435⟩
Slocombe, R. J., **6**, 425 ⟨51JA3796⟩, 443 ⟨50JA1888⟩
Slocum, D. W., **1**, 844 ⟨B-73MI 119-02⟩; **4**, 706 ⟨65JA1241⟩; **5**, 273, 285 ⟨73JOC1677⟩, 445 ⟨72JOM(44)1⟩
Slomczynska, U., **6**, 435 ⟨84MI 614-02⟩
Slomski, B. A., **4**, 881 ⟨85EUP230127⟩
Slonkin, A., **4**, 343 ⟨86IZV640⟩
Slopianka, M., **3**, 264 ⟨73CB450⟩
Sloss, D. G., **1**, 178 ⟨93JOC6174⟩
Slota, Jr., P. J., **2**, 409 ⟨58JA1107⟩, 431 ⟨60JA2145, 60JA2148⟩
Slotta, K. H., **5**, 1023 ⟨30CB888⟩
Slough, G. A., **2**, 574, 978 ⟨88JA2575⟩; **5**, 342 ⟨93TL6821⟩, 446 ⟨85JA7788⟩
Slougui, N., **3**, 232 ⟨81TL645⟩; **4**, 831 ⟨87SC1⟩
Slovetskii, V. I., **6**, 147 ⟨80MI 605-01⟩, 261, 283 ⟨70IZV2553, 71IZV1073⟩, 283 ⟨72IZV1721⟩, 284, 364, 369 ⟨75IZV2838⟩, 364, 368 ⟨77IZV1563⟩, 650 ⟨75MI 621-01⟩
Slovokhotova, T. A., **1**, 56 ⟨65MI 102-01⟩
Slovokhotov, Yu. L., **4**, 688 ⟨81JOM(206)119⟩
Slowey, P. D., **1**, 990 ⟨84TL2887⟩; **4**, 742 ⟨82TL3795⟩
Sluboski, B. C., **5**, 327 ⟨77JOC3301⟩
Slusarchyk, W. A., **5**, 346 ⟨82JOC179, 84H(21)191⟩
Slusarczuk, G. M. J., **4**, 728 ⟨71JOC37⟩; **6**, 142 ⟨70CC469⟩
Slusarska, E., **2**, 299 ⟨86LA402⟩; **5**, 707 ⟨79S691⟩
Sluski, R. J., **5**, 1012 ⟨73JA3957⟩
Slutskii, V. I., **6**, 562 ⟨81KGS907⟩
Slutsky, J., **1**, 836 ⟨73JOC3658⟩, 838 ⟨72JA2515, 73JA8678⟩
Slyadenskaya, O. S., **5**, 661 ⟨92CHE579, 91KGS1205⟩
Slynko, N. M., **4**, 560 ⟨84DOK(277)371, 85BBA(818)96⟩
Slysarenko, E. I., **5**, 1008 ⟨66ZOB1639⟩
Slyusarenko, T. F., **4**, 140 ⟨83JFC(23)339⟩, 156 ⟨78JGU769⟩, 1045 ⟨86ZOB2743⟩
Smadja, S., **3**, 344 ⟨76RTC172⟩
Smadja, W., **1**, 320, 326, 327, 328, 919 ⟨94SL1⟩, 999 ⟨66BSF971⟩
Smagin, V. M., **5**, 44 ⟨78IZV2094⟩
Smaill, J. B., **1**, 640 ⟨93SL774⟩
Smak, T., **3**, 212 ⟨81TL3351⟩
Small, A., **4**, 296 ⟨91SL485⟩
Small, G. H., **1**, 13 ⟨75JOC3151⟩
Small, L. E., **1**, 883 ⟨78JOC815⟩
Smalla, H., **2**, 100 ⟨62AG(E)97⟩, 793 ⟨62AG(E)89⟩
Smalley, R. K., **3**, 542 ⟨66TL3465, 68JCS(C)2730⟩; **4**, 312 ⟨B-68MI 407-02⟩; **5**, 183, 194, 198 ⟨64JCS755⟩, 688 ⟨90TL123⟩; **6**, 111 ⟨90TL123⟩
Smallridge, A. J., **2**, 928 ⟨87S1032⟩; **5**, 941 ⟨90AOC(30)1⟩
Smart, B. E., **2**, 619 ⟨82JA4680⟩; **3**, 495 ⟨83JOC3531⟩; **4**, 42 ⟨85JA4565⟩, 1076 ⟨84JA5544⟩; **6**, 28 ⟨B-82MI 601-02⟩, 229 ⟨85JA4565⟩
Smart, B. W., **5**, 90 ⟨63OSC(4)55⟩
Smart, C. J., **3**, 244 ⟨89TL1189⟩
Smart, G. N. R., **2**, 798 ⟨48MI 215-01⟩
Smart, J., **2**, 540 ⟨80TL4523⟩
Smearing, T. J., **6**, 573 ⟨68JOC851⟩
Smeaton, E., **4**, 918, 1075 ⟨86TL4635⟩
Smedal, H. S., **3**, 568 ⟨67JOM(8)547, 67ZAAC(354)69⟩, 568 ⟨73CB2144⟩
Smeets, F., **3**, 278 ⟨52BSB694⟩
Smeets, W. J. J., **4**, 656 ⟨93OM3624, 93OM3955⟩, 716 ⟨93OM3187⟩; **6**, 406 ⟨91AOC(32)147⟩
Šmejkal, J., **5**, 280 ⟨66CI(L)1529⟩
Smentowski, F. J., **1**, 378 ⟨57JA5455⟩
Smerat, G., **4**, 173 ⟨85TL6329⟩
Smets, G., **5**, 1014 ⟨73JOC675⟩; **6**, 629 ⟨65T1537⟩
Smid, J., **1**, 859 ⟨66JA307, 66JA318⟩, 859 ⟨69JA4580⟩; **5**, 1007 ⟨91JPS(A)1097⟩
Smid, P., **6**, 549 ⟨92MI 617-05⟩
Smidt, J.,**4**, 894 ⟨72JOC1960⟩
Smiles, S., **2**, 144 ⟨08JCS1833⟩, 172 ⟨08CB4115⟩, 172 ⟨10JCS2585⟩, 688, 688 ⟨36JCS327⟩; **5**, 1111 ⟨25JCS2887⟩; **6**, 83 ⟨26JCS2263⟩
Smiley, R. A., **3**, 613 ⟨60JOC257⟩
Smiley Irelan, J. R., **2**, 1070 ⟨69JOC3349⟩
Smirnov, A. S., **4**, 715 ⟨88MI 416-01⟩
Smirnov, E. A., **2**, 448 ⟨68ZOB1551⟩
Smirnov, K. N., **5**, 807 ⟨67ZOB1413⟩; **6**, 613 ⟨67JGU1343, 69JGU1301⟩
Smirnov, V. N., **5**, 644 ⟨85ZOB1526⟩, 713 ⟨85JGU1526, 85ZOB1718⟩
Smirnov, V. V., **2**, 23 ⟨83RCR259⟩; **5**, 838 ⟨52DOK(83)243⟩
Smirnova, E. I., **5**, 370 ⟨81PS(10)261⟩
Smirnova, N. P., **4**, 1017 ⟨86ZOR582⟩
Smirnova, O. G., **4**, 820 ⟨76JGU2367⟩
Smirnova, T. V., **6**, 475 ⟨85ZOR1670⟩

Smirnova, V. A., **6**, 118 ⟨72MI 602-01⟩
Smischek, M., **4**, 686 ⟨82JOM(226)C59⟩
Smissman, E. E., **2**, 683 ⟨72JOC3944⟩; **6**, 506 ⟨73JOC1652⟩
Smit, D. N., **4**, 682 ⟨88JCS(D)1421⟩
Smit, P. J., **2**, 621, 627 ⟨66RTC857⟩; **3**, 617 ⟨63RTC757⟩
Smit, R., **4**, 414 ⟨87JOC365⟩
Smit, V. A., **1**, 529 ⟨89IZV495⟩
Smit, W. A., **1**, 529 ⟨89TL4021⟩, 530 ⟨86TL1241⟩; **3**, 169 ⟨83TL961⟩
Smit, W., **1**, 530 ⟨89S472⟩
Smith, A., **5**, 750, 751, 753 ⟨73JCS(P1)453⟩
Smith, A. B., **1**, 382 ⟨84JOC4094⟩; **3**, 115 ⟨81S567⟩, 206 ⟨75CC274, 75TL4225⟩, 206 ⟨81JA1996, 83JA575⟩, 223, 275 ⟨86S184⟩, 271 ⟨78TL4661⟩, 303 ⟨84JA2115⟩; **6**, 488 ⟨92JA8008⟩
Smith, III, A. B., **1**, 147 ⟨91Txi⟩, 166 ⟨93TL63⟩, 217 ⟨89TL6963⟩, 516 ⟨82JA5568, 84JA2115, 86S184⟩, 573, 581, 584 ⟨92JA9419⟩, 606 ⟨81JA1501⟩, 754, 755 ⟨86JA2662⟩; **2**, 617 ⟨82JA2659⟩, 664 ⟨85TL4419⟩; **3**, 176 ⟨77TL4361⟩, 176 ⟨78TL1649⟩, 533 ⟨74CC695, 76JA7456⟩, 533 ⟨77JOC3165, 84JA3995⟩, 533 ⟨84JA4001⟩, 538 ⟨69TL4517⟩; **4**, 950 ⟨87JA1269⟩; **5**, 173 ⟨82JA4015⟩, 179 ⟨86JA2451⟩; **6**, 469 ⟨93JOC3703⟩
Smith, Jr., A. C., **2**, 600 ⟨53JA4103, 53JA4105⟩
Smith, A. C. B., **5**, 142 ⟨73JCS(P1)2796⟩, 1120 ⟨87JMC951⟩
Smith, A. E., **4**, 843 ⟨71TL2241⟩
Smith, A. H., **5**, 146 ⟨73JOC1941⟩
Smith, A. J., **5**, 943 ⟨90OM2621⟩
Smith, A. K., **4**, 19, 27 ⟨81JCS(P1)78⟩, 1029 ⟨91JCS(D)351⟩, 1030 ⟨86JOM(309)C56, 87CC826⟩
Smith, A. L., **4**, 617 ⟨61JA3345⟩
Smith, A. S., **3**, 286 ⟨48JOC592⟩
Smith, A. V., **1**, 94 ⟨84JCS(D)1747⟩
Smith, B., **1**, 560 ⟨62ACS(B)523⟩; **2**, 444 ⟨65JCS3354⟩, 683 ⟨56ACS1006⟩; **6**, 78, 79 ⟨56ACS1006⟩
Smith, B. C., **2**, 452 ⟨65JCS5516⟩, 465 ⟨64JCS1020⟩
Smith, B. F., **2**, 147, 155 ⟨82JOC3773⟩; **3**, 210 ⟨73TL5037⟩
Smith, B. H., **3**, 357, 362 ⟨76JOC3784⟩
Smith, B. J., **2**, 1022 ⟨89JA8297⟩
Smith, C., **1**, 407, 408 ⟨69CC293⟩, 407, 823 ⟨80JCS(P1)1436⟩, 807, 821, 822 ⟨68CC186, 69CC99⟩, 835 ⟨90JCS(P1)1493⟩; **2**, 89 ⟨88PS(37)25⟩, 197 ⟨92JCS(P1)2371⟩; **4**, 281 ⟨92JCS(P1)2371⟩; **5**, 330 ⟨92TL273⟩; **6**, 621, 728 ⟨90JMC434⟩
Smith, C. A., **1**, 926 ⟨88JA2210⟩, 1038 ⟨88TL4217⟩, 1065 ⟨88JOC2655⟩; **5**, 1142 ⟨70JCS(A)2675⟩
Smith, C. D., **1**, 777 ⟨66TL6279⟩
Smith, C. G., **5**, 374 ⟨85JOC1712⟩
Smith, C. L., **6**, 174, 175, 176, 289, 380, 391, 393, 395 ⟨92OM2938⟩
Smith, C. P., **2**, 476 ⟨68JA6726⟩; **3**, 498 ⟨65TL201, 66T567⟩; **4**, 27 ⟨68JOC25⟩, 336 ⟨74PS(4)109⟩
Smith, C. S., **2**, 134 ⟨88TL4477, 92T5933⟩; **4**, 249 ⟨92T5933⟩
Smith, C. W., **3**, 530, 531 ⟨63OSC(4)348⟩; **5**, 51 ⟨57JOC1680⟩
Smith, D. B., **1**, 831 ⟨89JOC9⟩, 837 ⟨94TL15⟩, 1079 ⟨89JOC4267⟩
Smith, D. C., **5**, 576 ⟨94JOC348⟩
Smith, D. F., **1**, 883 ⟨59JA1617⟩; **2**, 327 ⟨91SL29⟩
Smith, D. J., **5**, 1144 ⟨88JOM(355)33⟩
Smith, D. J. H., **1**, 232 ⟨78CJC512, 78S579⟩, 232 ⟨81JCS(P1)1846⟩, 297 ⟨83CC1270⟩; **2**, 441 ⟨84JCR(S)32⟩, 860 ⟨72JCS(P1)1184⟩, 864 ⟨79COC(2)1301⟩; **4**, 117 ⟨79COC(2)1233⟩; **5**, 142 ⟨84MIP84/1376A1⟩; **6**, 465 ⟨87CC147⟩
Smith, D. K., **2**, 19 ⟨86CC791⟩, 609 ⟨85TL2233⟩; **5**, 4 ⟨86CC791⟩
Smith, D. L., **4**, 67 ⟨83JOC1096⟩, 377 ⟨79IC2696⟩
Smith, D. O., **1**, 732 ⟨74JCS(P1)2435⟩
Smith, D. R., **1**, 571 ⟨92JA6330⟩
Smith, D. S., **1**, 104 ⟨63OSC(4)377⟩
Smith, D. W., **5**, 334 ⟨86JMC359⟩, 655 ⟨85JHC1121⟩
Smith, E., **5**, 984 ⟨69GEP1910303⟩
Smith, E. L., **5**, 92 ⟨50JPP409⟩
Smith, E. M., **1**, 664 ⟨70TL4355⟩; **2**, 765 ⟨67JA1537, 72JOC1690⟩, 765 ⟨72JOC3584, B-74MI 215-01⟩; **3**, 211, 576 ⟨73JOC2129⟩, 310 ⟨70TL4355, 72JOC3584⟩
Smith, F., **4**, 3 ⟨50JCS2689⟩; **5**, 98 ⟨46MI 502-02⟩; **6**, 9 ⟨52MI 601-01⟩
Smith, F. A., **3**, 107 ⟨51JOC1121⟩
Smith, F. B., **6**, 185 ⟨52JA1418⟩
Smith, F. C., **1**, 1079 ⟨93TL5621⟩
Smith, F. X., **3**, 309 ⟨75JOC2613⟩
Smith, G., **1**, 993 ⟨71JCS(C)1530⟩; **5**, 73 ⟨35BJ(29)606⟩; **6**, 247 ⟨93JOM(453)299⟩
Smith, G. B., **2**, 1099 ⟨60JA3560⟩
Smith, G. B. L., **2**, 764 ⟨51JA5675⟩
Smith, G. C., **4**, 107 ⟨85AJC1505⟩
Smith, G. D., **1**, 826 ⟨73B3492⟩
Smith, G. E. P., **3**, 616 ⟨34JA2095⟩
Smith, Jr., G. E. P., **5**, 589, 595, 620 ⟨49JOC962⟩
Smith, G. G., **1**, 773 ⟨69MI 117-01, 71MI 117-01⟩
Smith, G. H., **2**, 452 ⟨65JCS5516⟩
Smith, G. L., **1**, 266 ⟨67JA3600⟩
Smith, G. M., **2**, 602 ⟨86JA40⟩; **4**, 656, 658 ⟨86JA40, 86JA6805⟩
Smith, G. P., **2**, 358 ⟨49JOC921⟩
Smith, H., **1**, 96 ⟨54JCS113⟩, 98 ⟨51JCS1945⟩, 889 ⟨58QR17⟩
Smith, H. D., **1**, 811 ⟨87TL1853⟩, 812 ⟨91JOC3817⟩; **3**, 193 ⟨87SC983⟩
Smith, H. L., **5**, 999 ⟨75JOC3554⟩
Smith, H. O., **1**, 1109 ⟨71TCA65⟩
Smith, H. Q., **3**, 561 ⟨69JCED408⟩
Smith, Jr., J., **5**, 997 ⟨65JOC1260⟩
Smith, J. A., **1**, 631 ⟨65JA659⟩; **3**, 465 ⟨65JA935⟩
Smith, J. D., **1**, 45 ⟨46JA1813⟩; **2**, 435 ⟨68JCS(A)1909⟩, 436 ⟨83JCS(D)905⟩, 872 ⟨B-75MI 217-01⟩; **6**, 165, 289, 372, 373 ⟨89JOM(366)39⟩, 165, 372 ⟨83JCS(D)905⟩, 173, 394 ⟨93JOM(462)45⟩, 174, 176, 381 ⟨92JAP06033008⟩, 174, 176, 381, 393 ⟨86CC1043⟩, 174, 289, 390 ⟨83CC827⟩, 174, 391 ⟨86CC969⟩, 176, 289, 391 ⟨83CC1390⟩, 176, 394 ⟨91JOM(405)149⟩, 186, 195 ⟨78JOM(153)253⟩, 387 ⟨82JOM(235)265, 84JOM(272)1, 89JCS(D)447⟩, 390 ⟨93JCS(D)3259⟩, 392, 394 ⟨92JOM(437)41⟩, 393 ⟨80JOM(190)101, 89CC273, 94AG(E)1268, 94JOM(469)129, 94JOM(480)199, 94OM753⟩, 393, 394, 396 ⟨88JCS(D)381⟩, 394 ⟨88CC1389, 90CC1471, 95UP 613-01⟩, 394, 395 ⟨86CC908, 87JOM(320)137, 83JOM(249)23⟩, 395 ⟨87JCS(D)747⟩, 396 ⟨84CC870, 84JOM(263)C23, 85CC534, 85CC1380, 87JOM(325)117⟩, 396, 399 ⟨87CC1183, 88JOM(355)33⟩
Smith, J. E., **6**, 444 ⟨88JOC6145⟩
Smith, J. G., **1**, 204, 580, 842 ⟨84S629⟩, 468, 472 ⟨81JA6460⟩; **2**, 87 ⟨86JOC3762⟩; **3**, 9, 122 ⟨84S629⟩, 658 ⟨78S894⟩; **6**, 507 ⟨71JOC2516⟩
Smith, J. H., **2**, 169 ⟨68JCS(C)2713⟩, 824, 834, 835 ⟨78JA7304⟩
Smith, J. N., **2**, 219 ⟨71JOC488⟩
Smith, J. R. L., **2**, 620 ⟨83TL3117⟩; **4**, 445 ⟨75JCS(P1)1181⟩
Smith, K., **1**, 75 ⟨B-88MI 103-01⟩, 449 ⟨B-88MI 110-01⟩, 509 ⟨B-88MI 112-01, 91COS(8)703⟩, 509 ⟨B-88MI 112-02⟩, 657, 658, 946 ⟨79COC(3)791⟩, 765 ⟨89TL5643, 89TL5647⟩, 988 ⟨92JCS(P1)747⟩, 1035 ⟨75CC857⟩, 1035, 1103 ⟨76TL4385⟩; **2**, 4 ⟨81JCR(S)376⟩, 41 ⟨94AG(E)851⟩, 418 ⟨79COC(3)925⟩, 514 ⟨74CSR443⟩, 514, 518, 519, 520, 523, 528, 529, 531, 533 ⟨B-88MI 210-01⟩, 514, 900 ⟨B-85MI 210-01, 91COS(8)703⟩, 525, 900

⟨80CC1169⟩, 533 ⟨76JCS(D)2087⟩, 620 ⟨79COC(3)687, 85S1157, 88TL1319⟩, 900 ⟨79COC(3)689⟩, 900, 903, 904, 906 ⟨B-88MI 218-01⟩, 966 ⟨B-88MI 219-01⟩, 1005 ⟨91PAC403⟩, 1086 ⟨75CC857, 76TL4385⟩; **3**, 19 ⟨79S701⟩, 127 ⟨89TL5643⟩; **4**, 250 ⟨78TL2345⟩, 296 ⟨91SL485⟩, 364 ⟨79S701⟩, 373 ⟨74CC863⟩, 523 ⟨75JCS(P1)129, 75JCS(P1)138⟩, 846 ⟨84TL995⟩; **5**, 234 ⟨77JCS(P1)1672⟩, 381 ⟨B-88MI 507-01⟩, 433 ⟨75CC719⟩, 438 ⟨90AG(E)282⟩; **6**, 172 ⟨79COC(3)687⟩, 172, 198, 201 ⟨91COS(1)487⟩, 172, 198 ⟨B-88MI 606-01⟩, 317 ⟨84TL991⟩
Smith, K. M., **3**, 64 ⟨90TL1365⟩
Smith, K. R., **1**, 795 ⟨72TL125⟩
Smith, Jr., L. O., **5**, 293 ⟨47JA2007⟩
Smith, L. A., **5**, 329 ⟨93SL850⟩
Smith, L., **3**, 527 ⟨70TL2963⟩; **4**, 768 ⟨70JOC1501⟩
Smith, L. H., **2**, 31 ⟨55OSC(3)793⟩
Smith, L. I., **2**, 30 ⟨55OSC(3)350⟩; **3**, 444 ⟨55OSC(3)351⟩; **5**, 76 ⟨43JA745⟩, 103 ⟨55OSC(3)302⟩, 305 ⟨55OSC(3)151⟩
Smith, L. L., **6**, 255 ⟨91CJC2127⟩
Smith, L. R., **2**, 506, 894 ⟨75JOM(84)1⟩, 1047 ⟨62JA1868⟩; **4**, 171 ⟨82JOC3140⟩; **5**, 566, 577, 611 ⟨63JOC3492⟩, 995 ⟨62JOC3742, 63JOC1805⟩, 995 ⟨65JOC4306⟩, 996 ⟨62JOC4361⟩; **6**, 57 ⟨65JOC4303⟩
Smith, M. A. R., **5**, 465 ⟨74CC841⟩
Smith, M. B., **2**, 7, 8 ⟨84JOC431⟩; **5**, 199 ⟨89TL3295⟩, 346 ⟨92OPP147⟩
Smith, M. G., **1**, 831 ⟨88JA854⟩
Smith, M. J., **4**, 470 ⟨80PS(8)67⟩
Smith, M. L., **3**, 696, 719 ⟨94AG(E)1315⟩
Smith, M. R., **2**, 279 ⟨75JOM(88)175⟩, 498, 504 ⟨66JOM(6)17⟩
Smith, N. L., **2**, 384 ⟨61JOC1819⟩, 859 ⟨61JOC611⟩; **4**, 569 ⟨63JOC272⟩
Smith, N. P., **6**, 280 ⟨79JCS(P2)1298⟩
Smith, N. R., **3**, 455 ⟨54JA4931⟩
Smith, N. R. M., **6**, 621, 728 ⟨85IC2453⟩
Smith, P., **3**, 573 ⟨62JOC4053⟩; **6**, 450 ⟨83JFC(22)521⟩
Smith, P. A., **2**, 319 ⟨47OR(4)337⟩
Smith, P. A. S., **1**, 389 ⟨60OR(11)157⟩; **2**, 298 ⟨B-65MI 205-02⟩, 373, 387, 388, 402, 404 ⟨B-83MI 207-01⟩, 376 ⟨58JOC1595⟩, 384 ⟨57JOC358⟩, 753 ⟨69JOC3430⟩, 757 ⟨46OR(3)337, 63OSC(4)819⟩, 764 ⟨64JA1139⟩, 765 ⟨B-66MI 215-03⟩, 798 ⟨B-66MI 215-05⟩; **3**, 407 ⟨B-63MI 310-01⟩, 711 ⟨58JOC1599⟩; **4**, 96 ⟨B-65MI 403-01⟩; **5**, 37 ⟨73OSC(5)612⟩, 620 ⟨61JOC1132⟩, 989 ⟨46OR(3)337, B-63MI 526-01⟩, 989 ⟨B-66MI 526-01⟩, 1000 ⟨B-65MI 527-01⟩, 1040 ⟨60JA3076⟩, 1047 ⟨64JOC2261⟩, 1051 ⟨B-86MI 527-01⟩
Smith, P. H., **4**, 550, 551, 576 ⟨80IC3195⟩; **5**, 322 ⟨94TL5157⟩; **6**, 370 ⟨85IC2889⟩
Smith, P. J., **2**, 599 ⟨82COMC-I(2)519⟩; **3**, 90 ⟨70JCS(C)227⟩; **4**, 86 ⟨74JOC1449⟩
Smith, P. V., **2**, 119, 120 ⟨46JA2103⟩; **6**, 572 ⟨55OSC(3)773⟩
Smith, P. W., **1**, 869 ⟨66JCS(B)241⟩
Smith, R., **1**, 524 ⟨86JOC1159⟩; **2**, 445 ⟨65JCS1539⟩; **3**, 711 ⟨84SC639⟩; **4**, 516, 537 ⟨84SC639⟩, 518 ⟨83CC210, 85T3559⟩; **5**, 728 ⟨83CC210⟩
Smith, R. A., **2**, 3 ⟨75JOC3420⟩, 143 ⟨78OS(58)138⟩, 233 ⟨80JOC5155⟩, 725 ⟨79JOC4135⟩; **5**, 1111 ⟨79S181⟩
Smith, R. A. J., **1**, 207 ⟨90JA4063⟩, 313 ⟨94JA2902⟩; **2**, 589 ⟨90JA4063⟩; **3**, 16, 132 ⟨79SC301⟩, 98 ⟨84S166⟩, 283 ⟨87T1847⟩; **4**, 655 ⟨81AJC181⟩; **6**, 88, 317 ⟨75CC216⟩
Smith, R. D., **2**, 591 ⟨73OSC(5)855⟩; **4**, 28, 39, 40 ⟨73CI(L)331⟩, 1083 ⟨58JA5323⟩
Smith, R. E., **3**, 263 ⟨67JOC3817⟩
Smith, R. F., **1**, 1143 ⟨70JA7606⟩; **2**, 383 ⟨82SC801⟩; **3**, 617 ⟨62JOC4372⟩, 626 ⟨62JOC3836⟩, 666 ⟨66JOC4100⟩; **5**, 772 ⟨77JOC1862⟩, 772, 850, 852, 855 ⟨71JOC1155⟩, 772, 850, 855 ⟨73JOC1344⟩
Smith, R. F., **5**, 774 ⟨83JHC69⟩
Smith, R. F., **6**, 55 ⟨71JOC1155⟩, 573 ⟨68JOC851⟩
Smith, R. G., **1**, 637, 638 ⟨71JA4316⟩, 922 ⟨67JA2054⟩; **6**, 194, 195 ⟨88CC1007⟩
Smith, Jr., R. H., **2**, 387 ⟨88JOC1467⟩, 387, 388 ⟨83S476⟩
Smith, R. J., **2**, 133 ⟨91CC475⟩
Smith, R. L., **1**, 3, 4, 6 ⟨64CJC572⟩, 660 ⟨66JOC3498⟩; **2**, 219, 220 ⟨68JA2677, 69JA7525⟩
Smith, R. S., **1**, 467 ⟨76JOC502⟩; **4**, 698 ⟨82JOM(225)71⟩
Smith, S., **1**, 234 ⟨82CC589⟩; **2**, 224, 225 ⟨75JOC524⟩; **6**, 604 ⟨91MI 620-01⟩
Smith, S. A., **3**, 113 ⟨79CC145⟩
Smith, S. C., **3**, 156 ⟨90SC3359⟩
Smith, S. G., **1**, 148 ⟨83JA2080⟩, 148 ⟨88JA8145⟩; **5**, 476, 477 ⟨61JA4285⟩, 476, 477 ⟨64TL3363⟩, 476, 477 ⟨65JOC2882⟩
Smith, S. J., **1**, 957, 974, 1060 ⟨86JOC2623⟩; **6**, 167 ⟨84CC1669⟩, 169 ⟨87JOM(320)C27⟩
Smith, S. L., **1**, 404 ⟨62JOC2291⟩
Smith, T., **1**, 515 ⟨78CC30⟩; **2**, 561 ⟨78CC30⟩
Smith, T. A., **2**, 194 ⟨63JCS3692⟩
Smith, T. A. K., **1**, 231 ⟨85CC897⟩
Smith, T. C., **3**, 325 ⟨93JOC7635⟩
Smith, T. D., **5**, 1129 ⟨69JCS(A)2190⟩
Smith, T. L., **2**, 616 ⟨85TL279⟩
Smith, T. W., **3**, 269 ⟨78TL2537⟩
Smith, W. B., **2**, 632 ⟨90JOC2543⟩
Smith, W. C., **2**, 476, 835, 836, 881 ⟨60JA6176⟩, 848 ⟨57USP2950306⟩, 861 ⟨57JOC265⟩; **4**, 8 ⟨60JA543⟩; **5**, 5 ⟨60JA543⟩; **6**, 19, 39, 52 ⟨60JA543⟩, 238 ⟨61JA3422⟩
Smith, W. E., **3**, 91, 94 ⟨72JOC3972⟩, 323 ⟨74JOM(67)341⟩; **5**, 344 ⟨85JOC4993⟩; **6**, 38 ⟨74JOM(67)341⟩, 608 ⟨71JOC1786⟩
Smith, W. L., **4**, 687 ⟨90OM2745⟩
Smith, W. N., **1**, 1000 ⟨73JOC3588⟩, 1000, 1001 ⟨74S441⟩; **3**, 273 ⟨73JOC3588⟩
Smith, Jr, W. T., **2**, 378, 386 ⟨68JMC504⟩
Smith, Y., **2**, 94 ⟨86ICA179⟩
Smitheman, K. A., **2**, 832 ⟨61JOC3547⟩
Smithers, R., **1**, 450, 451 ⟨89OR(37)57⟩; **2**, 73, 907 ⟨89OR(37)57⟩; **6**, 107 ⟨72JA3201⟩
Smithers, R. H., **2**, 624 ⟨78JOC2833⟩, 1025, 1026 ⟨85S556⟩; **3**, 67 ⟨83JOC2095⟩, 495 ⟨78JOC2833⟩; **4**, 767 ⟨85S556, 89JOC1479⟩, 818 ⟨83JOC2095⟩
Smithey, W. R., **3**, 274 ⟨51JOC51⟩
Smithies, A. J., **1**, 347 ⟨94TL2413, 94TL2417⟩
Smith-Palmer, T., **2**, 631 ⟨76JCS(P1)1161⟩
Smithwick, E. L., **2**, 219 ⟨67JOC97⟩
Smitrovich, J. H., **2**, 568, 587 ⟨91JOC6494⟩, 926 ⟨92JOC3178⟩
Smits, A., **1**, 1018 ⟨88T6699⟩
Smits, J. F. M., **5**, 478, 491, 492, 562 ⟨77RTC139⟩
Smits, J. M. M., **3**, 433 ⟨86JOC309⟩
Smolarsky, M., **5**, 90 ⟨73JOC457⟩
Smolders, R. R., **2**, 624 ⟨71C59⟩
Smolders, W., **2**, 232 ⟨69BSB191⟩
Smolensky, B., **5**, 631 ⟨64JOC1115⟩
Smolii, O. B., **4**, 99 ⟨88JGU2192⟩, 998 ⟨92ZOB1084⟩; **5**, 655 ⟨86JGU2483, 86ZOB2802⟩, 655 ⟨88JGU2192, 88ZOB2465⟩
Smolikiewicz, A., **1**, 108, 115 ⟨74TL1451⟩
Smolikova, J., **3**, 430 ⟨88LA1091⟩
Smolinski, S., **6**, 304 ⟨82BCJ1106⟩
Smolinsky, G., **1**, 795, 1023 ⟨60JA4918⟩, 862 ⟨60JA934⟩; **3**, 564 ⟨62JOC3557⟩
Smook, M. A., **4**, 44 ⟨50JA4378⟩
Smoot, J., **2**, 217 ⟨74S59⟩
Smorada, R. L., **1**, 485, 1032 ⟨79JOC3444⟩
Smout, G. J., **2**, 597 ⟨90POL2763⟩
Smucker, L. D., **3**, 496 ⟨66JOC467⟩
Smulders, E., **2**, 390, 397, 403, 405 ⟨71CB917⟩

Smutny, E. J., **4**, 843 ⟨71TL2241⟩; **5**, 559 ⟨67T3785⟩
Smysloma, E. I., **4**, 152 ⟨77IZV2417⟩; **5**, 385 ⟨88ZOB62⟩
Smyth, C. P., **6**, 213 ⟨57JA20⟩
Smyth, M. S., **5**, 404 ⟨92TL4137⟩
Smythe, G. A., **3**, 364 ⟨69AJC239⟩
Snaith, R., **3**, 456 ⟨78JCS(D)986⟩, 458 ⟨71CC1275, 72JCS(D)370⟩, 459 ⟨72JCS(D)1501, 87JCS(D)1071⟩, 459 ⟨87JCS(D)2141⟩
Snapper, M. L., **1**, 542 ⟨87TL2221⟩
Snatzke, G., **2**, 103 ⟨69T761⟩
Snavely, C. M., **5**, 972 ⟨76JOC2071⟩
Sneddon, D. W., **4**, 150 ⟨89JCS(D)761⟩
Sneddon, E. A., **6**, 418 ⟨90MI 614-02⟩
Sneddon, L. G., **2**, 902 ⟨87JA5867⟩; **6**, 205 ⟨75CC302⟩
Sneeden, R. P. A., **2**, 571 ⟨B-75MI 211-01⟩; **5**, 27 ⟨B-69MI 502-01⟩; **5**, 932 ⟨B-75MI 525-01⟩
Sneen, R. A., **1**, 871, 879 ⟨56JA6269⟩, 887 ⟨64JA5503⟩
Snegirev, V. F., **4**, 731 ⟨86IZV91⟩, 737 ⟨81JFC(17)441⟩
Snider, B. B., **1**, 7 ⟨87JOC307⟩, 176 ⟨85TL5675⟩, 336, 346, 926 ⟨88JOC2137⟩, 340 ⟨91T8663⟩, 346 ⟨88TL5209, 93JOC6217⟩, 346, 926 ⟨93JOC7640⟩, 347 ⟨90JA2759⟩, 498 ⟨86JOC3643⟩, 499 ⟨83JOC4370⟩, 519, 524 ⟨76JOC3061, 80ACR426⟩, 524 ⟨91COS(5)1⟩, 524, 525 ⟨79JA5283, 77TL2831⟩, 525 ⟨78JOC2161, 79JA1906⟩, 525 ⟨80JOC2773⟩, 583 ⟨92JOC3615⟩, 805 ⟨88JOC1848⟩, 810 ⟨91JOC321⟩, 815, 832 ⟨88JOC4508⟩, 820 ⟨76JA7115⟩, 926 ⟨93TL1701⟩, 929 ⟨92JOC2399, 93JOC7228⟩, 1105 ⟨78JA4624⟩; **2**, 20 ⟨77JA905⟩, 22 ⟨87JOC307⟩, 672 ⟨89JOC38⟩, 924 ⟨78JA4624⟩, 925, 926 ⟨79JOM(179)C37⟩; **3**, 119 ⟨86T2951⟩, 170 ⟨91JOC328⟩, 528 ⟨88JOC4508, 91JOC321⟩, 601 ⟨88CRV793⟩; **4**, 212 ⟨85JOC5167⟩, 827 ⟨79JA6023⟩, 1060 ⟨78JA4624⟩
Snider, D. M., **2**, 478 ⟨79JA5432, 86JA3466⟩
Snider, S. W., **6**, 636 ⟨88JOC3120⟩
Snider, T. E., **3**, 639 ⟨88OSC(6)932⟩
Snieckus, V., **1**, 215, 576 ⟨82TL1343⟩, 429, 1089 ⟨93TL415⟩, 448 ⟨89JOC4372⟩, 464 ⟨84JOC737⟩, 473 ⟨87TL5093, 87TL5097⟩, 524 ⟨78AG(E)476⟩, 547, 1130 ⟨85TL6001⟩, 828 ⟨84JOC2682⟩, 845 ⟨B-92MI 119-01⟩, 927 ⟨90JA896⟩; **2**, 305 ⟨83TL3795⟩, 554, 653, 942 ⟨90CRV879⟩, 627 ⟨84TL483, 89JOC4372⟩, 673 ⟨83TL4515, 86JOC271⟩, 942 ⟨90TL4267⟩, 965 ⟨82ACR306⟩; **3**, 90 ⟨82ACR306, 88BSF67⟩, 90, 98 ⟨89JOC24⟩, 137, 157, 182 ⟨90JOC5680⟩, 296, 298, 300, 305 ⟨90CRV879⟩, 308 ⟨91S1079⟩, 309, 310 ⟨80JA1457⟩, 311 ⟨85JOC5436⟩; **4**, 511 ⟨89TL5841⟩, 513 ⟨89TL5837⟩, 604, 646 ⟨89JOC4372⟩, 768 ⟨93TL411⟩, 841 ⟨93TL415⟩, 841, 950 ⟨90JOC5680⟩; **5**, 199 ⟨92T4863⟩, 289 ⟨89TL5841⟩; **6**, 162 ⟨89TL5841⟩, 556 ⟨92S112⟩
Snitman, D. L., **3**, 48 ⟨78JOC2066⟩
Snoble, K. A. J., **1**, 721, 729 ⟨73JA5778⟩; **3**, 494 ⟨73JA5778⟩
Snoddy, A. O., **2**, 204 ⟨63OSC(4)529⟩; **5**, 73 ⟨63OSC(4)19⟩
Snodin, D. J., **2**, 402 ⟨78JCS(P1)1580⟩
Snoussi, M., **5**, 103 ⟨78SC19⟩
Snow, D. H., **2**, 393 ⟨67JA725⟩
Snow, J. T., **1**, 278, 487, 662 ⟨71JA6309⟩, 278, 531, 662, 1103 ⟨72JA6560⟩, 534 ⟨68JA7139⟩, 658, 963 ⟨70JA1427⟩; **2**, 900 ⟨70JA1427⟩, 949 ⟨85TL1281⟩; **4**, 756 ⟨82S127⟩; **5**, 1136 ⟨72JA6560⟩
Snow, M. R., **4**, 713 ⟨82JOM(240)299⟩
Snow, R. A., **3**, 16 ⟨76SC575⟩
Snowden, R. L., **1**, 635 ⟨90HCA640⟩, 643, 755 ⟨77JCS(P1)550⟩, 883 ⟨87HCA1858, 87HCA1879⟩; **2**, 346, 349 ⟨79TL4391⟩; **3**, 143 ⟨87HCA1858⟩; **4**, 650 ⟨77JCS(P1)550⟩
Snustad, D. C., **3**, 544 ⟨92JA2991⟩
Snyatkova, E. V., **2**, 449 ⟨76JGU428⟩
Snyder, D. C., **4**, 510 ⟨86JOM(301)137, 87JOM(320)163⟩
Snyder, D. M., **1**, 379 ⟨86JA3466⟩
Snyder, E. I., **2**, 7 ⟨72JOC1466⟩, 14 ⟨65JOC3985⟩
Snyder, H., **4**, 20 ⟨58JA1942⟩
Snyder, H. R., **1**, 58 ⟨38JA115⟩, 88 ⟨61JA3160⟩, 257 ⟨48JA3770, 49JA663⟩, 258 ⟨57JA2217⟩, 447 ⟨48JA232⟩; **2**, 30 ⟨43OSC(2)171⟩, 120 ⟨63OSC(4)638⟩, 128, 136 ⟨47JA2675⟩, 129, 136 ⟨47JA2672⟩, 194 ⟨52JA2006⟩, 648 ⟨74JOC3343⟩, 757 ⟨53JA2014⟩, 906 ⟨38JA105⟩; **3**, 106 ⟨52JA5110⟩, 107 ⟨63OSC(4)539⟩, 247 ⟨62JOC3888⟩, 667 ⟨70JOC30, 73JOC481⟩; **4**, 407 ⟨42OS(22)65⟩, 437 ⟨63JOC1144⟩; **5**, 25, 275 ⟨54JA3039⟩, 31 ⟨43OSC(2)531⟩, 51 ⟨43OSC(2)333⟩, 87 ⟨43OSC(2)489⟩, 1109 ⟨65JOC184⟩
Snyder, J. A., **5**, 285 ⟨71JMC535⟩
Snyder, J. K., **6**, 504 ⟨80JOC886⟩
Snyder, J. P., **2**, 217 ⟨70TL3551⟩
Snyder, J. P., **2**, 377, 378, 403 ⟨73TL2859⟩, 390 ⟨75JOC1395⟩, 404 ⟨69TL4643, 71JA3816⟩, 773 ⟨74JA5158⟩; **3**, 351 ⟨90JA3029⟩, 603 ⟨69AG(E)216⟩; **4**, 802, 825 ⟨69AG(E)216⟩
Snyder, L., **1**, 134 ⟨93JOC7507⟩
Snyder, M. P., **2**, 337 ⟨82IC2545⟩
Snyder, R., **1**, 658, 659 ⟨66JOC2777⟩
Snyder, Jr., R. C., **1**, 894 ⟨79JOC2860⟩
Snyder, Sr. R., **4**, 142 ⟨66JOC2773⟩
Snyder, R. D., **1**, 440 ⟨91JA7439⟩; **4**, 812 ⟨91JA7439⟩
Snyder, W. H., **2**, 675 ⟨61JA1773⟩
So, S., **2**, 447 ⟨88JOC2937⟩; **3**, 208, 209, 241 ⟨79JOC3066⟩
Soai, K., **1**, 308 ⟨76BCJ779⟩, 1039 ⟨79CL447, 90JCS(P1)937⟩; **2**, 67 ⟨92CRV833⟩, 69 ⟨93CC811⟩, 71 ⟨79BCJ3371, 79JA1455⟩, 71 ⟨88JOC927, 89JCS(P1)109⟩, 71 ⟨90JOC4832⟩, 269 ⟨84CL251⟩, 331 ⟨92CC1097⟩; **3**, 632 ⟨92TA677⟩
Soares, C. J., **1**, 835 ⟨87TL1031⟩
Sobanov, A. A., **4**, 339 ⟨86ZOB711⟩
Sobchuk, T. I., **4**, 561 ⟨66ZOB296⟩
Sobczak, R. L., **5**, 161 ⟨76JA7874⟩
Sobezak, A., **4**, 476 ⟨74ZC152⟩
Sobieray, D. M., **4**, 253 ⟨81JOC3555⟩; **5**, 355 ⟨91T9599⟩
Sobirov, M. M., **6**, 547 ⟨85KPS844⟩
Sobkowiak, A., **3**, 113 ⟨90JA1936⟩
Sobolewski, A., **6**, 546 ⟨84MI 617-01⟩
Sobolewski, H., **2**, 767 ⟨79MI 215-01⟩; **6**, 546 ⟨84MI 617-01, 86MI 617-04⟩, 546 ⟨88MI 617-01⟩
Soborovskii, L. Z., **2**, 430 ⟨66ZOB484⟩, 440 ⟨61DOK(139)1359, 63ZOB1919⟩; **4**, 561 ⟨64ZOB2897⟩; **6**, 58 ⟨59JGU1115⟩, 366 ⟨67ZOB2513⟩
Sobtsova, N. I., **6**, 149 ⟨67ZOB1163⟩, 149, 369 ⟨73RZC1243⟩
Sochilin, E. G., **4**, 114 ⟨75JGU1668⟩
Sock, O., **5**, 44 ⟨85TL1509, 86GEP3522304⟩, 44 ⟨86JCE1013⟩
Socol, S. M., **4**, 575 ⟨82ZAAC(494)55⟩
Soda, S.-I., **1**, 985 ⟨92CL1357⟩
Sodemann, S., **4**, 715 ⟨94JOM(471)249⟩
Sodeoka, M., **1**, 456 ⟨93S643⟩, 456, 507 ⟨85JOC1147⟩, 495 ⟨85JOC3246, 94T371⟩
Soder, L., **4**, 1074 ⟨59HCA1733⟩
Söderbäck, E., **5**, 1109 ⟨54ACS1851⟩
Söderberg, B. C., **5**, 949 ⟨91OM3951⟩
Soderquist, J. A., **1**, 449 ⟨90TL5113⟩, 652 ⟨89TL5693⟩; **2**, 58 ⟨88T4403⟩, 58, 526 ⟨90TL4677⟩, 516, 517 ⟨81JOC4599⟩, 902 ⟨90TL4545⟩, 902, 922 ⟨89JA4873⟩, 906 ⟨90TL5113, 93JA6065⟩, 936 ⟨88TL2425, 88TL2777⟩, 939 ⟨91TL3457⟩, 946 ⟨80JOC541⟩; **3**, 203 ⟨89JA4873⟩, 240 ⟨83TL2361⟩; **4**, 248 ⟨86TL6305⟩, 353 ⟨89JOC4051, 90TL4677⟩, 353, 643 ⟨78JOM(156)C12⟩, 354 ⟨92JA10078⟩, 356, 643 ⟨80JOC3571⟩, 356, 643, 645 ⟨88T4033⟩, 614 ⟨80TL429⟩, 630, 631 ⟨93JA6065⟩, 630, 631, 633, 644 ⟨89JA4873⟩, 643, 645, 646 ⟨84JOC2565⟩, 655 ⟨89TL5693⟩, 697 ⟨78JOM(159)237⟩, 929 ⟨88TL1899⟩, 929, 930 ⟨90OS(68)25⟩, 929, 930, 949, 951, 952 ⟨82OM830⟩, 929, 930, 950, 951, 952 ⟨80JA1577,

80JOC541⟩, 935 ⟨89TL3919⟩, 951 ⟨83TL2361⟩, 954 ⟨92JOC5844⟩, 1056 ⟨86JOC1330⟩; **5**, 416 ⟨82OM830, 90OS(68)25⟩, 417, 425 ⟨77JOM(131)C1⟩, 422 ⟨92JA10078⟩, 425, 431, 468 ⟨80JA1577⟩, 897 ⟨92TL6915⟩

Sodeyama, T., **3**, 618 ⟨73CL577⟩

Sodhi, A., **5**, 385 ⟨86JMC199⟩

Sodhi, G. S., **6**, 563 ⟨83JIC806⟩

Soeda, Y., **2**, 429 ⟨83BCJ219⟩

Soejima, T., **2**, 81 ⟨89JOC6001⟩

Soelch, R. R., **4**, 160, 755 ⟨89JOC5502⟩

Soenen-Svilarich, M., **3**, 568 ⟨72T4431⟩

Sofer, I., **4**, 1083 ⟨88CPL439⟩

Soffer, M. D., **1**, 103 ⟨52JA4126⟩, 881 ⟨50JA3704⟩; **3**, 451 ⟨55JA1003⟩

Sofia, M. J., **1**, 62 ⟨86JA6826⟩, 1021 ⟨86JMC230⟩; **2**, 540 ⟨86JA6826⟩

Sof'ina, E. N., **2**, 779 ⟨64ZOB2094⟩

Sogadji, K., **3**, 642 ⟨77S126⟩

Sogah, G. D. Y., **5**, 152 ⟨B-87MI 503-01⟩

Sogay, T., **2**, 43 ⟨89CL7573⟩

Sögstad, J., **3**, 695 ⟨74ACS(A)255⟩

Sohani, S. V., **1**, 548 ⟨81JOC2622⟩

Sohar, P., **2**, 528 ⟨91JOC763⟩; **4**, 120 ⟨86ACH145⟩, 273 ⟨89MRC760⟩, 319 ⟨87H(26)2381⟩; **5**, 547 ⟨72JOC1727⟩; **6**, 111 ⟨91JHC481⟩, 489 ⟨91JCS(P1)37⟩, 557 ⟨83T1829⟩, 558 ⟨83JHC1181, 85T1353⟩, 558 ⟨90MRC1045⟩, 562 ⟨91M1047⟩, 564 ⟨90JHC1249⟩

Sohda, T., **1**, 149, 594, 716 ⟨86JOC3108⟩; **4**, 532 ⟨86JOC3108⟩

Sohmiya, H., **2**, 177, 179 ⟨92CL891⟩, 179 ⟨91CL1391⟩

Sohn, D., **6**, 607 ⟨83MI 620-01⟩

Sohn, E.,**3**, 295, 296, 298 ⟨83CB1195⟩

Sohn, E., **5**, 209 ⟨83CB1183⟩

Sohn, J. E., **1**, 874 ⟨80JOC1066⟩

Sohn, W. H., **5**, 161 ⟨72CC472⟩

Soicke, H., **3**, 171 ⟨82LA1096⟩

Soifer, G. B., **4**, 794 ⟨80JOU1762⟩; **5**, 655, 659 ⟨77JOU1471⟩, 655, 659 ⟨77ZOR1594⟩, 657, 659 ⟨79JOU32⟩, 657, 659 ⟨79ZOR38⟩

Soine, T. O., **5**, 305 ⟨63OSC(4)106⟩

Soja, P., **1**, 607 ⟨77JHC681⟩; **2**, 139 ⟨76JOC3587⟩; **4**, 210 ⟨75JOC2962⟩

Sokal'skaya, L. I., **5**, 403, 404 ⟨72JGU1256, 74JGU1685⟩

Sokalskii, M. A., **6**, 613 ⟨72JGU293, 72JGU799⟩

Sokol, P. E., **5**, 116 ⟨73OSC(5)706⟩

Sokol, S., **2**, 122 ⟨50JA1687⟩

Sokolova, N. B., **5**, 516, 521 ⟨75JOU1077⟩

Sokolova, V. M., **1**, 63 ⟨81JOM(214)145⟩; **2**, 266 ⟨81JOM(214)415⟩

Sokolov, L., **6**, 410 ⟨74URP424809⟩

Sokolov, M. P., **3**, 49 ⟨73RCR538⟩; **4**, 132 ⟨91JGU1459⟩, 804 ⟨92JGU1222⟩; **6**, 366, 659 ⟨90ZOB1980⟩, 622 ⟨90ZOB223⟩, 623 ⟨89ZOB714⟩, 634 ⟨86ZOB1427, 90ZOB222⟩, 659 ⟨87ZOB2467, 90ZOB1288⟩, 659 ⟨90ZOB1293, 90ZOB1986⟩, 659 ⟨90ZOB2694⟩, 660 ⟨91ZOB776⟩

Sokolov, S. V., **6**, 36 ⟨74ZOR2031⟩, 410 ⟨74URP424809⟩

Sokolov, V, I., **4**, 721 ⟨77DOK(236)371⟩

Sokolov, V. B., **4**, 494 ⟨87DOK(297)1132, 87IZV1680⟩, 494 ⟨88IZV936⟩; **5**, 652 ⟨88IZV1691⟩, 669 ⟨87BAU204, 87IZV226, 87JGU1478, 87JGU1479, 87ZOB1658, 87ZOB1659, 88BAU1506, 89BAU672, 89IZV745, 89IZV2849, 89KFZ1317, 90BAU415, 90IZV477⟩, 669, 672 ⟨87BAU1000, 87IZV1086, ⟨89BAU2201, 89BAU2612, 89BAU2628, 89IZV2865, 90BAU1007, 90IZV1122, 91PS(60)85⟩, 672 ⟨89IZV2390, 90JGU1718, 90ZOB1923, 91JGU1152, 91ZOB1268, 92BAU236, 92IZV315⟩, 683 ⟨87BAU1314, 87IZV1422⟩, 1076 ⟨87IZV1177⟩; **6**, 430 ⟨89IZV2849⟩

Sokolov, V. I., **4**, 152 ⟨89MI 403-03⟩, 152 ⟨91MI 403-01⟩, 674 ⟨85JOM(297)C25, 86ZOR2457⟩; **4**, 674 ⟨91JOM(418)C24⟩, 720 ⟨79JOM(182)537⟩, 721 ⟨79IZV1528⟩, 1033 ⟨80JOM(202)C58⟩, 1067 ⟨78JOM(162)271⟩

Sokol'skii, G. A., **2**, 183, 249 ⟨75IZV2530⟩, 208 ⟨72IZV2510⟩, 407 ⟨56ZOB3378⟩; **4**, 833 ⟨71KGS1645⟩, 867 ⟨80IZV892⟩; **5**, 386 ⟨74ZOR1428⟩, 502 ⟨71BAU959, 72AG(E)583⟩, 666, 671, 692 ⟨86MI 517-02⟩; **6**, 42 ⟨86ZOR1842⟩, 42, 46 ⟨86ZOR1834⟩, 241 ⟨60DOK(132)602⟩

Sokolyanskaya, L. V., **6**, 562 ⟨81ZOR2269⟩

Sokova, N. B., **4**, 140 ⟨88JGU81⟩

Solaja, B., **4**, 238, 241 ⟨87T4875⟩

Solanki, A. K., **5**, 226 ⟨80IJC(A)69⟩

Solans, X., **4**, 711 ⟨86JOM(314)13⟩, 713 ⟨90JOM(395)305⟩, 858 ⟨93JCR(S)60⟩; **5**, 1137 ⟨84OM996⟩; **6**, 331 ⟨87CC472, 91OM1683⟩

Solcániová, E., **2**, 103 ⟨90OM301⟩

Soldan, F., **4**, 979 ⟨61CB3109⟩; **6**, 138 ⟨61CB3109⟩

Soldatov, B. G., **5**, 105 ⟨80MI 502-01⟩

Soleihavoup, M., **4**, 548 ⟨93AG(E)1167⟩, 589 ⟨93PS(76)49⟩

Soleiman, M., **3**, 579 ⟨88JCS(P2)1071⟩; **6**, 162 ⟨88RTC641⟩

Soliman, A., **5**, 652 ⟨91ACH35⟩

Soliman, F. M., **3**, 356 ⟨91SUL213⟩, 475, 489 ⟨79AG1012⟩, 475, 489 ⟨79AG(E)947⟩

Soliman, G., **3**, 357 ⟨59JCS2588, 62JCS1857⟩

Soll, H., **2**, 347, 348 ⟨58HOU(11)190⟩, 385 ⟨58HOU(11/2)178⟩; **5**, 742, 743, 748, 749 ⟨58HOU(9)39⟩

Söll, M., **2**, 596 ⟨60LA(629)121⟩

Soll, R. M., **2**, 627 ⟨90JHC2151⟩

Solladié, G., **1**, 140 ⟨81S185⟩, 210 ⟨93JOC2181⟩, 269, 647 ⟨83TL4419⟩, 466 ⟨84TL5769⟩, 739, 740 ⟨83SC1193⟩; **2**, 61 ⟨92TL1605⟩, 67 ⟨B-83MI 202-02⟩, 144, 154, 158 ⟨91COS(6)133⟩, 9 ⟨72JA4184⟩; **2**, 997 ⟨81S185⟩; **3**, 170 ⟨91COS(6)133⟩, 170 ⟨93JOC4555⟩; **4**, 229 ⟨93TA1547⟩, 245 ⟨93JOC2181⟩, 270, 863 ⟨91TL3695⟩, 391 ⟨81S185⟩, 959 ⟨84C233, 84JA6097⟩; **6**, 317 ⟨91TL6329⟩

Solladie-Cavallo, A., **2**, 318 ⟨92JOC5831⟩; **3**, 405 ⟨93SC1683⟩, 694 ⟨92TA287⟩, 704 ⟨92TA39⟩; **4**, 540 ⟨87JOM(330)357⟩

Sollenberger, P. Y., **2**, 773 ⟨B-68MI 215-04⟩

Söllenböhmer, F., **4**, 619, 621 ⟨89JA4127⟩

Sollhuber, M. M., **3**, 616 ⟨86TL3285⟩

Solliard, R., **1**, 180 ⟨88T5495⟩

Solly, R. K., **1**, 807 ⟨68MI 118-01⟩

Solodenko, V. A., **4**, 451, 469, 493 ⟨87RCR859⟩, 462 ⟨92ZOB1472⟩

Solodushchenko, G. F., **5**, 683, 707 ⟨87UKZ397⟩

Sologub, L. S., **6**, 609, 619 ⟨83KGS798⟩

Solomakhina, F. Kh., **2**, 882 ⟨57MI 217-01, 60MI 217-01⟩

Soloman, L., **4**, 731 ⟨73JFC(3)63⟩

Solomatina, A. I., **2**, 430 ⟨69JOM(17)201⟩

Solomon, M. F., **1**, 803 ⟨76CC89⟩

Solomon, R. D., **1**, 635 ⟨86JA6276⟩

Solomon, R. G., **1**, 803 ⟨76CC89⟩

Solomon, S., **1**, 808 ⟨65JA4506⟩; **2**, 149 ⟨67JOC3111⟩

Solomonov, B. N., **4**, 922 ⟨92HAC127⟩

Solomons, I. A., **3**, 612 ⟨53JA3430⟩

Solooki, D., **1**, 1010 ⟨90SL427⟩

Soloshonok, I. V., **1**, 109 ⟨88JOU530⟩; **5**, 205 ⟨88ZOR593⟩

Soloshonok, V. A., **2**, 15 ⟨91RCR850⟩; **3**, 588 ⟨86JOU1640⟩; **5**, 385 ⟨93S117⟩; **6**, 9 ⟨94TA937⟩

Soloski, E. J., **2**, 600 ⟨63JOC237⟩

Solouki, B., **3**, 335, 369 ⟨82JA3119⟩, 337 ⟨76JA6054, 77CC287⟩, 342 ⟨92MC67⟩, 380 ⟨79AG(E)944⟩, 382 ⟨84CB187⟩, 547 ⟨77CPH453, 77JA1663⟩; **4**, 258 ⟨82JA3119⟩

Solov'eva, N. P., **4**, 706 ⟨81JOM(209)233⟩, 987 ⟨87KGS1477⟩

Solov'ev, A. V., **2**, 239 ⟨90ZOR2083⟩, 438 ⟨90ZOB1243⟩
Solov'ev, N. A., **6**, 149 ⟨86ZOR2618⟩, 149 ⟨87ZOR2624⟩, 149 ⟨88ZOR644⟩
Solov'ev, V. N., **5**, 370 ⟨88ZOB2424⟩
Solov'yanov, A. A., **1**, 318 ⟨78RCR425⟩, 859 ⟨77MI 119-01, 81JOU381⟩
Soloway, H., **1**, 91 ⟨61JOC818⟩
Soltani-Neshan, A., **2**, 893, 894 ⟨86ZN(B)321⟩; **6**, 168, 192 ⟨83POL291⟩
Soltin, L., **2**, 105 ⟨77S737⟩
Soltzberg, L. J., **2**, 1010 ⟨71JPC1246⟩
Sólyom, S., **1**, 578 ⟨88JPR309⟩, 979 ⟨85TL1831, 86CC829⟩; **5**, 595, 597, 599 ⟨77CB285⟩
Soma, N., **1**, 822 ⟨72TL2923⟩
Somanathan, R., **6**, 416 ⟨91SC285⟩
Somayaji, V., **1**, 14 ⟨83JOC3091⟩; **2**, 608, 632 ⟨82OPP359⟩, 623 ⟨84S919⟩, 623, 628 ⟨89JOC6068⟩, 629 ⟨84JOC4822⟩, 900, 901 ⟨89JOC6075⟩
Somayajulu, G. R., **6**, 221, 223 ⟨79JCED251⟩, 223, 226, 227, 228 ⟨79JMR559⟩
Somberg, H., **2**, 836, 837 ⟨81CB3485⟩
Somei, M., **3**, 106 ⟨92H(34)1295⟩; **5**, 324 ⟨91H(32)221⟩
Somers, P. K., **4**, 226 ⟨90JA6263⟩; **5**, 513 ⟨90JA6263⟩; **6**, 134 ⟨87JA2504⟩
Somers, T. C., **1**, 13 ⟨87JA4424⟩; **3**, 310 ⟨90T6501⟩
Someswara Rao, C., **5**, 189 ⟨87IJC(B)407⟩
Somfai, E., **6**, 645 ⟨91MIP54636⟩
Somfai, P., **1**, 697, 700, 703 ⟨87T4395⟩; **2**, 50 ⟨94TL4631⟩, 533 ⟨89TL7357⟩; **5**, 342 ⟨88T613⟩, 342, 345 ⟨91TL283⟩
Somlo, T., **6**, 501 ⟨72GEP2206366⟩
Somm, F., **1**, 812 ⟨59CCC2371⟩
Sommariva, A., **6**, 412 ⟨28G443⟩
Sommazzi, A., **1**, 933 ⟨92TL1787⟩
Sommelet, M., **1**, 824 ⟨37CR(205)56⟩; **4**, 429 ⟨07BSF394⟩
Sommer, H., **2**, 428, 430 ⟨84CB3400⟩, 435 ⟨85JOM(296)411⟩, 1001 ⟨90CB1853⟩; **4**, 546 ⟨82AG(E)376, 89IS120⟩, 550 ⟨83PS(18)279, 84CB3400⟩, 550 ⟨85JOM(296)411⟩, 829, 970 ⟨82CB1733⟩
Sommer, H. M., **2**, 1094, 1095 ⟨B-87MI 223-01⟩, 1094 ⟨B-87MI 223-02, B-87MI 223-03⟩, 1094 ⟨B-87MI 223-04⟩
Sommer, J., **1**, 462 ⟨85MI 111-01⟩, 891, 917 ⟨B-85MI 119-04⟩, 893 ⟨67JA3576⟩
Sommer, J. M., **2**, 799 ⟨69CC1515⟩
Sommer, K., **2**, 430 ⟨70ZAAC(379)56⟩, 432 ⟨70ZAAC(376)37⟩, 486, 490, 510 ⟨70ZAAC(377)278⟩, 489 ⟨70ZAAC(377)128⟩, 490 ⟨70ZAAC(376)150⟩, 493 ⟨70ZAAC(377)120, 71ZAAC(383)136⟩, 494, 497 ⟨70ZAAC(375)55⟩, 890 ⟨69ZAAC(370)227⟩; **4**, 135 ⟨70ZAAC(377)128⟩, 545, 550, 562 ⟨70ZAAC(376)37⟩
Sommer, K. von, **4**, 592, 593 ⟨69ZAAC(370)31⟩, 592, 593 ⟨70ZAAC(377)120⟩
Sommer, L. H., **1**, 112 ⟨70JOC22⟩, 428 ⟨69JOC638⟩, 450 ⟨54JA1613⟩, 648 ⟨47JA1551⟩, 652 ⟨48JA2869⟩; **2**, 537 ⟨47JA1976⟩, 538 ⟨54JA1030⟩, 812 ⟨79JOM(8)95⟩; **3**, 501 ⟨74JA7105⟩, 502 ⟨79JA2491⟩; **4**, 137 ⟨47JA1976⟩, 513 ⟨51JA5130⟩, 602, 607 ⟨54JA1619⟩, 610, 611 ⟨74JA7105, 75JA7371⟩; **5**, 421 ⟨54JA1613⟩
Sommer, R., **2**, 601 ⟨64LA(705)10, 68JOC802⟩, 868 ⟨66LA(692)1⟩; **3**, 570, 571 ⟨66AG(E)514, 68LA(718)1⟩; **5**, 780 ⟨65LA(688)14⟩
Sommer, S., **3**, 453 ⟨77TL117⟩
Sommer, U., **3**, 239 ⟨69LA(723)111⟩
Sommerfeld, P., **4**, 360, 382 ⟨92SL764⟩
Sommerfeld, T., **2**, 420 ⟨93OS(72)62⟩
Sommerhoff, J., **5**, 246 ⟨73JFC(3)383⟩; **6**, 432 ⟨70AG(E)466, 73JFC(3)383⟩
Sommers, A. H., **1**, 881 ⟨61JOC644⟩
Sommers, P. K., **6**, 134 ⟨90JA3696⟩
Sommerslade, R. H., **2**, 340, 357 ⟨86T2703⟩
Somov, G. V., **1**, 331 ⟨61IZV1099⟩, 925 ⟨61DOK(136)1099, 61IZV2065⟩
Somoza, C., **4**, 100 ⟨88T7007⟩
Somu, S., **5**, 272 ⟨84MI 507-02⟩
Son, N. K., **3**, 255 ⟨73BSF3339⟩, 366 ⟨71CR(C)278, 74BSF(2)471⟩
Son, P., **4**, 979 ⟨67JOC990⟩
Son, P. N., **4**, 971, 979 ⟨65JOC71⟩
Son, T., **6**, 470 ⟨92OM171⟩
Sonada, N., **1**, 556 ⟨78TL3455⟩; **4**, 931 ⟨84JA2440⟩
Sonawame, H. R., **1**, 801 ⟨92TL1645⟩, 802 ⟨91T8259, 91TL1107⟩, 801, 802 ⟨93SL875⟩
Sondej, S., **4**, 9 ⟨86JOC3508⟩
Sondengam, B. L., **1**, 457, 508 ⟨80TL1069⟩; **2**, 310 ⟨77TL3469⟩
Sondheimer, F., **1**, 104 ⟨72JA2128⟩, 268 ⟨58JA4386⟩, 506 ⟨49JCS737⟩, 881 ⟨57JA2906, 58JA5220⟩, 881 ⟨58JA6296⟩, 954 ⟨73TL3181⟩, 990 ⟨73JOC864⟩, 1000 ⟨50JCS115⟩, 1034 ⟨74OS(54)1⟩; **2**, 40 ⟨61JOC630⟩; **3**, 328 ⟨58JA4386⟩; **5**, 288 ⟨56JOC520⟩
Sondhi, S. M., **1**, 11, 19 ⟨82JOC4304⟩
Sone, M., **4**, 979 ⟨73USP3734954⟩
Sonegawa, M., **5**, 138 ⟨88JA3670⟩
Song, C. E., **3**, 628 ⟨94SC103⟩
Song, Y. H., **3**, 6, 118 ⟨84CC762, 86TL2875⟩, 215 ⟨85TL2875⟩
Song, Z., **1**, 148 ⟨88JA8145⟩
Songstad, J., **1**, 737 ⟨64ACS655⟩; **2**, 290 ⟨81ACS(A)151⟩, 858 ⟨73ACS1939⟩; **5**, 1117 ⟨77ACS(A)884, 81ACS(A)151⟩, 1117 ⟨89ACS518⟩
Sönke, H., **2**, 112 ⟨36JPR251⟩; **5**, 227 ⟨37JPR251⟩
Sonmers, A., **2**, 449 ⟨44JA82⟩
Sonnek, G., **4**, 397 ⟨80JOM(194)9⟩, 659 ⟨76ZC65⟩, 698 ⟨80JOM(201)97⟩
Sonnenberg, F. M., **2**, 656 ⟨70JOC3166⟩
Sonnenberg, J., **4**, 24 ⟨62JOC748⟩; **6**, 143 ⟨66JA3825⟩
Sonnenfeld, E., **2**, 103 ⟨1894CB3350⟩
Sonnenschein, H., **2**, 379 ⟨85JPR445⟩; **3**, 238 ⟨85TL4911⟩
Sonnet, P. E., **1**, 592, 636, 640, 644, 650 ⟨80T557⟩, 727 ⟨74JOC3793, 74OPP269⟩; **2**, 10 ⟨75JOC2410, 76SC21⟩, 26 ⟨88OSC(6)424⟩, 26, 33 ⟨76JOC3279⟩; **5**, 138 ⟨87JOC3477⟩
Sonntag, N. O. V., **5**, 278, 304 ⟨53CRV237⟩
Sono, S., **2**, 305 ⟨87T4079⟩
Sonobe, H., **4**, 258 ⟨90T5093⟩
Sonobe, K., **6**, 316 ⟨60JAP6018199⟩
Sonobe, T., **1**, 769 ⟨80JOC2005⟩
Sonoda, A., **1**, 488 ⟨77JA5652⟩, 531, 663 ⟨77CC852⟩
Sonoda, H., **2**, 19 ⟨85CL233⟩
Sonoda, N., **1**, 86 ⟨89BCJ1682⟩, 186, 225, 229 ⟨89JA7938⟩, 212 ⟨84JA6093⟩, 245, 437 ⟨87AG(E)1187⟩, 247 ⟨93OM473⟩, 601 ⟨B-86MI 114-02⟩, 666 ⟨86JOC2389, 92JOC17⟩, 1073 ⟨91CL1805⟩; **2**, 279, 282 ⟨79CL873, 77S874⟩, 281 ⟨88S175⟩, 285 ⟨92JOC111⟩, 311 ⟨87CL1275⟩, 542 ⟨88JA1976⟩, 554 ⟨87AG(E)1187⟩, 555 ⟨91AG(E)177⟩, 710 ⟨92JA5902⟩, 730 ⟨92JA7591⟩, 912 ⟨83CC1193, 86JOC3890⟩, 912 ⟨87JOC4864⟩; **3**, 20 ⟨83JA1370⟩, 38, 173 ⟨77S874⟩, 38, 174 ⟨82TL4813⟩, 90 ⟨90TL6887⟩, 136 ⟨90EUP379979⟩, 143 ⟨88TL6137⟩, 176 ⟨87S1096⟩, 177 ⟨93CL561⟩, 177, 224 ⟨91TL229⟩, 246, 543 ⟨75AG(E)636⟩, 293 ⟨77S776⟩, 336 ⟨89SC2061⟩, 336, 340, 385 ⟨88JA1976, 88TL6965⟩, 394 ⟨89TL2095⟩; **4**, 239, 334, 660 ⟨B-90MI 407-01⟩, 240 ⟨77TL851⟩, 334, 909 ⟨90CL2053⟩, 652 ⟨87AG(E)1187⟩, 940 ⟨92JA5902⟩, 940, 947 ⟨91JA9796, 91JOC5721⟩, 948 ⟨92JA7591⟩; **5**, 172 ⟨94JA5473⟩, 238, 241 ⟨91JA9796⟩, 240, 243, 253 ⟨91JOC3776, 93MI 505-01⟩, 415 ⟨84JA2440⟩, 437 ⟨90JA455⟩, 438 ⟨90JA7061⟩, 510 ⟨91COS(6)461⟩, 624 ⟨91COS(6)476⟩, 625 ⟨85JOC384⟩, 625, 626 ⟨85BCJ1448⟩, 1036 ⟨91TL3503⟩, 1052 ⟨72BCJ2937⟩; **6**, 7 ⟨81AG(E)475⟩, 465 ⟨71TL4885⟩, 465, 481 ⟨75BCJ108⟩, 466 ⟨74TL803⟩, 479 ⟨88TL4767⟩, 479 ⟨89CL1009, 90TL4773⟩, 495, 497 ⟨89AG(E)452⟩, 498 ⟨79S561, 90SC703⟩, 505 ⟨79S735⟩, 510 ⟨71JA6344⟩, 593 ⟨72BCJ2937⟩

Sonoda, T., **1**, 570 ⟨89TL1641⟩, 1112 ⟨76BCJ2560, 76CC612⟩, 1120 ⟨76CC180⟩; **2**, 19 ⟨85CL233⟩, 906 ⟨92TL337, 92TL3779⟩; **4**, 736 ⟨89TL1641, 89TL6379⟩, 736 ⟨91CL961, 92TL337⟩, 736 ⟨92TL3779⟩
Sonogashira, K., **1**, 130 ⟨91COS(2)551⟩, 456 ⟨66BCJ1178⟩, 472, 1008, 1009, 1012, 1137 ⟨75TL4467⟩, 1010, 1012, 1137 ⟨80S627⟩, 1045, 1048 ⟨77S777⟩; **3**, 273 ⟨77S777⟩; **4**, 686 ⟨72CC993⟩; **5**, 304 ⟨77S777⟩
Sonota, A., **1**, 842 ⟨72BCJ1777⟩
Sonveaux, E., **3**, 558 ⟨73JA5417⟩
Sood, A., **5**, 1147 ⟨89MI 530-01⟩; **6**, 492, 515 ⟨91T6915⟩, 514 ⟨90EJM301, 92MI 616-01⟩
Sood, C. K., **6**, 492, 515 ⟨91T6915⟩, 514 ⟨90EJM301⟩
Sood, M., **6**, 608 ⟨86JOM(301)15⟩
Sood, N., **6**, 608 ⟨86JOM(301)15⟩
Sood, R., **1**, 1001 ⟨74TL423⟩
Sood, V. K., **3**, 472 ⟨74S577⟩
Sootome, N., **1**, 667 ⟨84CL1897, 85TL2205⟩; **4**, 699 ⟨91T3281⟩
Sopchik, A. E., **2**, 413 ⟨85JOC1271⟩
Soper, F. G., **2**, 338 ⟨54JCS1750⟩
Soper, Q. F., **2**, 643 ⟨45JA386⟩; **5**, 584, 586 ⟨48JA2837⟩
Sorato, C., **2**, 48 ⟨90TL2990⟩
Sörensen, H., **2**, 1027 ⟨90TL7597⟩
Sörensen, N. A., **5**, 168 ⟨50ACS1080⟩
Sørensen, O. N., **6**, 238, 540 ⟨72CB2854⟩
Sorensen, T. S., **1**, 381 ⟨76ACR257⟩, 895 ⟨78JA5134⟩, 912 ⟨65CJC2744, 65JA5075⟩, 912 ⟨B-70MI 119-04⟩, 918 ⟨81JA588⟩, 919 ⟨90JA8135⟩; **3**, 244 ⟨63JOC687⟩, 530, 531 ⟨86JOC3558⟩; **5**, 447 ⟨90CJC502⟩
Sorenson, C. M., **1**, 633 ⟨80JOM(186)155⟩
Sorenson, E. B., **2**, 181 ⟨32JA3364⟩
Sörenson, H., **1**, 530 ⟨90JA9388⟩
Sorenson, R. J., **4**, 226 ⟨81JHC587⟩
Sorgeloos, D., **3**, 558 ⟨82TL2909⟩; **6**, 652 ⟨81BSB63⟩
Sorgi, K. L., **1**, 703 ⟨84TL2085⟩; **4**, 187 ⟨80CC477⟩
Soria, J. J., **3**, 624 ⟨86JOC1753⟩; **5**, 702 ⟨86TL4701, 91TL2457⟩
Sorkina, T. I., **1**, 421 ⟨60JGU765⟩
Sorm, F., **6**, 57 ⟨58CCC452⟩
Sorochinskii, A. E., **6**, 19 ⟨82ZOR228⟩
Soroka, M., **4**, 460 ⟨74RZC1119, 76RZC661, 78C253, 79PJC2327⟩, 465 ⟨89S547⟩, 467 ⟨84S495⟩, 476 ⟨73TL5201⟩
Sorokina, T. V., **2**, 442 ⟨63ZOB2631⟩
Sorokin, A. Y., **4**, 641 ⟨75ZOB1905⟩
Sorokin, M. S., **4**, 570 ⟨71MI 412-01⟩
Sorokin, N. M., **2**, 420, 494, 495, 695 ⟨73ZOB1264, 74JOM(78)107⟩, 695 ⟨74JOM(78)115⟩
Sorokin, V. D., **2**, 257 ⟨91ZOB1115⟩, 357 ⟨90RCR832⟩
Soroos, H., **1**, 114 ⟨44JA1389⟩; **2**, 494 ⟨42JA392⟩; **6**, 216, 217 ⟨45JA1642⟩
Sorrell, T. N., **3**, 13 ⟨80JOC3449⟩, 83, 94, 102 ⟨78TL2473⟩
Sorrenti, P., **2**, 394 ⟨83S1014⟩; **3**, 46 ⟨84T3809⟩, 191 ⟨84S607⟩
Sosnovsky, G., **1**, 181 ⟨71S203⟩, 181 ⟨72S142⟩, 919 ⟨B-64MI 119-02⟩; **2**, 98 ⟨66CRV529⟩, 194 ⟨56JCS3139⟩, 359 ⟨79S228⟩, 359, 361 ⟨82LA121⟩, 684 ⟨72JOC2267⟩; **3**, 617 ⟨79S722⟩, 618 ⟨78S703, 78ZN(B)1033⟩, 618 ⟨79ZN(B)511⟩; **4**, 206, 207 ⟨60JOC899, 65T871⟩, 220, 226 ⟨62T15⟩, 1080 ⟨B-64MI 424-01⟩; **5**, 334 ⟨79S228⟩
Sosnovsky, G. N., **2**, 96 ⟨58JOC1322⟩
Sostero, S., **3**, 442 ⟨87ICA275⟩; **4**, 151 ⟨92OM3879⟩, 151, 585 ⟨90JCS(D)1553⟩
Sota, K., **6**, 511 ⟨70JHC807⟩
Sotgiu, F., **6**, 321 ⟨72G558⟩
Sotgiu, G., **1**, 10 ⟨91JOC6206⟩
Sotgiv, F., **4**, 347 ⟨84PS(19)61⟩
Sotiropoulos, J., **4**, 854, 855 ⟨84T2951⟩; **5**, 593 ⟨67CR(C)99, 75BSF829⟩, 595 ⟨93PS(78)97⟩
Sotiropoulos, J.-M., **2**, 1068 ⟨92BSF367⟩; **5**, 872 ⟨91AG(E)1154, 92BSF367⟩; **3**, 485 ⟨91AG1174, 91AG(E)1154⟩; **6**, 668, 669, 671 ⟨92BSF367⟩, 669 ⟨91AG(E)1154⟩
Soto, J. L., **3**, 427 ⟨83OPP41⟩; **5**, 606 ⟨85S86, 92H(34)1573⟩
Sotoya, K., **2**, 121 ⟨77JOC2180⟩
Sotoyama, T., **1**, 278 ⟨79BCJ1865⟩
Sottofattori, E., **3**, 361 ⟨81JHC863⟩
Souchi, T., **1**, 76 ⟨84T1245⟩
Soucy, P., **5**, 95 ⟨72CJC2047⟩
Soufiaoui, M., **6**, 186, 188 ⟨91POL1153⟩
Soukup, M., **5**, 345, 346 ⟨88JOC5507⟩
Soula, C., **5**, 749 ⟨87MI 519-01⟩
Soula, G., **5**, 1106 ⟨85JOC3717⟩; **6**, 229, 230 ⟨86BSF925⟩
Soulen, R. L., **3**, 495 ⟨71JOC3386⟩, 649 ⟨73JOC479⟩; **4**, 739 ⟨73JOC479⟩, 740 ⟨76JOC556⟩, 762, 771 ⟨71JOC3386⟩, 783 ⟨84JOC1125⟩
Soulie, B., **5**, 772 ⟨83EJM413⟩
Soulié, J., **1**, 223, 224 ⟨87T2701⟩; **2**, 1083, 1085 ⟨66BSF3846⟩, 1084 ⟨60CR(251)727, 66BSF1981⟩, 517 ⟨74TL3435⟩, 521 ⟨70CR(C)589⟩; **4**, 213 ⟨70BSF332⟩
Souma, Y., **3**, 93 ⟨91CC1313⟩; **5**, 27 ⟨81YZ265⟩
Soundarajan, N., **3**, 324 ⟨87TL1011⟩
Soundararajan, N., **4**, 138, 154 ⟨88TL5863⟩
Soundararajan, R., **2**, 23 ⟨88JOC5664⟩, 900 ⟨90JOC2274⟩; **4**, 631, 635 ⟨90JOC2274⟩; **5**, 318, 320, 390 ⟨83JOM(255)295⟩
Souppe, J., **2**, 56 ⟨84JOC2047⟩; **3**, 7, 119 ⟨84JOC2045⟩
Souquet, F., **2**, 308 ⟨93SC817⟩
Sousa, L. R., **1**, 232 ⟨83JOC494⟩
Sousa Fernandes, M. L. de, **4**, 312 ⟨84JCR(S)108⟩
Sousa Lobo, M. J., **5**, 209 ⟨78TL2171⟩
South, M. S., **5**, 990 ⟨91JHC1003⟩; **6**, 82 ⟨91JHC1013⟩
Southam, R. M., **3**, 463 ⟨66JCS(C)467⟩, 538 ⟨67TL545⟩
Souther, S. K., **2**, 8 ⟨81JOC824⟩
Southern, J. M., **3**, 130 ⟨90JOC3368⟩
Southgate, R., **4**, 419 ⟨91TL2683⟩; **5**, 234 ⟨80CC1085⟩, 235 ⟨80CC429, 80TL5071⟩; **6**, 111 ⟨74JCS(P1)22⟩, 445 ⟨89JAN367⟩
Southon, I. W., **4**, 979 ⟨74JCS(P1)1611⟩
Southwick, P. L., **2**, 643 ⟨44JA1873⟩, 643, 645, 646 ⟨44JA1109⟩, 787 ⟨69JOC3279, 74JOC3351⟩; **5**, 27 ⟨70S628⟩
Soutome, M., **4**, 193 ⟨92JHC1625⟩
Souza, N. J. de, **3**, 433 ⟨83SC649⟩
Souza Barbosa, J. C. de, **3**, 12 ⟨85TL829⟩, 12, 128 ⟨85JOC5761⟩
Souza Gomes, A. de, **6**, 111, 114 ⟨69JHC729⟩
Sowa, F. J., **3**, 238 ⟨36JOC163⟩; **5**, 979 ⟨61USP3013045⟩
Sowa, F. S., **3**, 618 ⟨37JA1202⟩
Sowa, J. R., **1**, 62 ⟨67JOC4126⟩
Sowda, R., **2**, 367 ⟨64JPR128⟩
Sowden, J. C., **5**, 62 ⟨57MI 502-01⟩, 66 ⟨38JA235⟩
Sowell, Sr., J. W., **5**, 106 ⟨85JHC1429⟩
Sowerby, D. B., **2**, 504 ⟨72JCS(D)847⟩; **5**, 222 ⟨92JOM(426)49⟩
Sowerby, J. D., **3**, 458 ⟨71CC1275⟩, 459 ⟨76CC160⟩
Sowerby, R. L., **1**, 591, 693 ⟨72JA4758⟩, 889 ⟨71JA1027⟩
Sowinski, A. F., **2**, 24 ⟨75JA1599⟩
Sowler, J., **2**, 110 ⟨52JCS11⟩
Soyka, R., **5**, 215 ⟨87JHC597⟩
Spaar, R., **1**, 1113 ⟨69TL1439⟩, 1113, 1114 ⟨70HCA2119⟩
Spadaro, A., **1**, 1063 ⟨85CC1571⟩
Spagnolo, P., **3**, 166 ⟨88TL2381⟩, 359 ⟨85CC1441⟩
Spagnuolo, C. J., **4**, 961 ⟨85TL169⟩
Spahic, B., **2**, 612 ⟨80HCA1236⟩; **4**, 734 ⟨83JOC4661⟩, 951 ⟨73HCA2166⟩
Spahn, M., **4**, 562 ⟨91ZAAC(601)65, 93PS(77)258⟩
Spalluto, G., **2**, 301 ⟨91S479⟩, 356 ⟨91TL2517⟩
Spaltenstein, A., **1**, 697, 698, 703 ⟨86TL2095⟩, 697, 698, 703 ⟨87JOC3759⟩
Spande, T. F., **4**, 424 ⟨70JA343⟩
Spang, C., **2**, 489 ⟨89CB1247⟩

Spang, P., **2**, 1003 ⟨88CB843⟩
Spang, W., **1**, 1117 ⟨80CB2015⟩
Spangenberg, B., **4**, 327 ⟨58CB1432⟩
Spangler, C. W., **1**, 400, 402, 403, 794, 795, 799, 800, 838 ⟨76CRV187⟩, 420 ⟨79JCS(P2)810⟩; **3**, 59 ⟨85SC371⟩, 65 ⟨88SC51⟩
Spangler, R. J., **1**, 869 ⟨63T1625⟩
Spanton, S., **4**, 279 ⟨92TL1573⟩
Spanyar, E., **5**, 359 ⟨92SC871⟩
Sparberg, M. S., **2**, 191 ⟨43OSC(2)558⟩
Spargo, P. L., **2**, 2 ⟨94MI 201-01⟩; **3**, 212 ⟨86S944⟩
Sparkes, M. J., **4**, 565 ⟨77MI 412-01, 78BJ(169)239⟩
Sparks, M. A., **2**, 935 ⟨90TA801⟩
Sparrow, J. T., **4**, 218 ⟨70JOC215⟩
Spassov, A., **5**, 770, 852 ⟨62JGU3330, 66CB3734⟩
Spatola, A. F., **1**, 427 ⟨85TL1381⟩; **5**, 558, 573, 587, 611 ⟨84JCS(P1)785⟩
Spatz, S. M., **1**, 881 ⟨45JA2106⟩
Spaude, S., **5**, 769, 848 ⟨63AG(E)217, 64CB3436⟩
Spawn, T. D., **1**, 429 ⟨89JFC(44)1637⟩; **3**, 237 ⟨86BSF876⟩
Spaziante, P. M., **6**, 240 ⟨69JCS(A)431⟩
Speakman, P. R. H., **2**, 145 ⟨69JCS(C)2334⟩; **4**, 74, 75 ⟨69JCS(C)2334⟩; **5**, 208 ⟨78CI(L)579⟩
Spear, K. L., **1**, 141, 234, 821 ⟨78JOC1185⟩
Spear, R. J., **1**, 902 ⟨75JA1539⟩, 904 ⟨74JA5855⟩, 915 ⟨76JA6284⟩, 1122 ⟨75JA1845⟩
Spears, C. P., **2**, 281 ⟨88S133⟩; **5**, 1113 ⟨88S133⟩
Spec, Th. G., **1**, 80 ⟨72RTC1057⟩
Specian, Jr., A. C., **1**, 514 ⟨76JOC3497⟩
Speck, A. L., **4**, 656 ⟨93OM3955⟩
Speck, W., **6**, 382 ⟨81ZAAC(481)60⟩
Speckamp, W. N., **1**, 774 ⟨89JA2588⟩, 978 ⟨83TL1407, 84JOC1149, 84TL3115, 86TL1411, 92T3445⟩; **2**, 324, 326 ⟨87TL3285⟩; **3**, 422 ⟨88JOC3882, 89JA2588⟩, 422 ⟨90T4049⟩; **4**, 98 ⟨75TL4065⟩, 106 ⟨88TL6975⟩, 298 ⟨87TL3285⟩, 303 ⟨71SC103, 75T1437, 78T179⟩, 327 ⟨94JOC3721⟩, 426 ⟨78T2399⟩; **5**, 374 ⟨86TL1411⟩; **6**, 645 ⟨78RTC51⟩
Spector, M. L., **5**, 974 ⟨66JOC596⟩
Speer, H., **4**, 958 ⟨79TL4277⟩
Speer, W., **2**, 720 ⟨24JCS1463⟩
Speers, L., **6**, 36 ⟨71JMC517⟩, 36, 43 ⟨71JMC593⟩, 38, 43 ⟨72JMC604⟩, 38, 43 ⟨72JMC606⟩
Speeter, M. E., **5**, 15 ⟨54JA6208⟩
Speh, P., **5**, 744 ⟨83S35⟩; **6**, 105, 109 ⟨72CB1340⟩, 115 ⟨88LA605⟩, 138 ⟨79S342⟩, 139 ⟨90CZ176⟩, 140 ⟨83S905⟩, 296, 297, 298, 300, 313, 320, 321 ⟨77S73⟩, 643 ⟨84LA108⟩
Spehar, A., **1**, 4, 7, 9, 10, 17, 429 ⟨69JOC3923⟩
Spehr, M., **3**, 395 ⟨91CB423⟩; **6**, 256 ⟨91CB51⟩
Speier, J., **2**, 941 ⟨53JA2821⟩
Speier, J. L., **2**, 536 ⟨57JA974, 79AOC(17)407⟩, 913 ⟨B-79MI 218-01⟩; **4**, 140 ⟨51JA824⟩, 140 ⟨51JA826⟩, 506 ⟨51JA3867⟩
Speigel, G. U., **2**, 428, 430 ⟨87ZN(B)579⟩
Speirs, R. A., **3**, 388, 389 ⟨88CC1494⟩
Spek, A. L., **1**, 127 ⟨82JA6609⟩; **3**, 512 ⟨90RTC446⟩; **4**, 656 ⟨93OM3624⟩, 660 ⟨86OM1551⟩, 671 ⟨93OM2268⟩, 716 ⟨93OM315⟩, 93OM3187⟩, 723 ⟨93OM2227⟩; **5**, 377 ⟨91JA5742⟩, 890 ⟨92OM3844⟩; **6**, 195 ⟨86OM1726⟩, 406 ⟨91AOC(32)147⟩
Spellmeyer, D. C.,**1**, 323 ⟨86JOC2874⟩, 336 ⟨87JOC959⟩
Spencer, A., **1**, 494 ⟨82JOM(233)267, 84JOM(265)323⟩
Spencer, C. F., **1**, 624, 782, 783 ⟨53JA3212⟩
Spencer, G. M., **4**, 138, 369, 372 ⟨92JOM(437)111⟩
Spencer, H. K., **2**, 290 ⟨77JA1470⟩; **4**, 869 ⟨75CC867⟩; **5**, 1117 ⟨77JA1470⟩
Spencer, J. L., **2**, 914, 915 ⟨90TL1051⟩; **4**, 691 ⟨75CC451⟩, 692 ⟨80JCS(D)2182⟩
Spencer, J. T., **3**, 394 ⟨87TL3887⟩
Spencer, T., **1**, 493 ⟨75JOM(99)C8⟩
Spencer, T. A., **1**, 5, 8, 9, 10 ⟨71TL137⟩, 14 ⟨92T9427⟩, 72 ⟨80JOC805⟩; **3**, 253 ⟨73JA250⟩, 322 ⟨67JA5497, 67TL1865⟩; **4**, 274, 277 ⟨92JOC3496⟩
Spensley, C. R. C., **1**, 828, 832 ⟨82TL109⟩
Speranza, G., **1**, 77 ⟨87T4481, 88T1299⟩; **3**, 18, 134 ⟨86SC803⟩, 58 ⟨87T4481⟩
Speranza, M., **1**, 1142 ⟨93CRV2933⟩
Sperati, C. A., **2**, 643 ⟨44JA1873⟩
Sperber, F., **6**, 645 ⟨91MIP54636⟩
Sperber, N., **5**, 36 ⟨48JA3091⟩
Sperk, G., **1**, 432 ⟨84JMC1575⟩
Sperka, J., **2**, 30 ⟨93H(35)701⟩
Sperl, U., **5**, 344 ⟨86AP(319)682⟩
Spero, D. M., **1**, 339, 1130 ⟨86TL2833⟩; **2**, 241 ⟨82JOC2785⟩; **3**, 129 ⟨91T1765⟩; **5**, 338 ⟨82JOC2785⟩
Spero, G. B., **1**, 729 ⟨69JA5364⟩
Speroni, G., **6**, 76 ⟨59G1511⟩
Speroni, S., **3**, 686 ⟨82S719⟩
Spes, H., **3**, 536 ⟨60GEP1081455⟩
Spessard, G. O., **1**, 722 ⟨75JA3512⟩; **3**, 698, 707 ⟨83TL655⟩
Speth, D., **4**, 679 ⟨81AG(E)1049⟩, 686 ⟨82CB878⟩
Speyer, E., **6**, 523 ⟨27CB1424⟩
Speziale, A. J., **1**, 723 ⟨63JA3878⟩, 724 ⟨63JA2790⟩; **2**, 8 ⟨60JA909⟩, 121 ⟨50OS(30)35⟩, 176, 177 ⟨56JA5580⟩, 202 ⟨54JA5011⟩, 1047 ⟨62JA1868⟩; **3**, 266 ⟨64JOC794⟩, 497 ⟨62JA854⟩, 499 ⟨63JOC465⟩, 568 ⟨65JA5068⟩; **4**, 737 ⟨60JA1260⟩, 737, 738, 745, 751 ⟨62JA854⟩, 738 ⟨65OS(45)33⟩, 743, 791 ⟨60JA903, 73OSC(5)387⟩; **5**, 76 ⟨73OSC(5)387⟩, 566, 577, 611 ⟨63JOC3492⟩, 860 ⟨65JA5068⟩, 895, 896 ⟨63JOC465⟩, 901 ⟨65JA5603⟩, 995 ⟨62JOC3742, 63JOC1805⟩, 995 ⟨65JOC4306⟩, 996 ⟨62JOC4361⟩; **6**, 57 ⟨65JOC4303⟩, 509 ⟨62JOC3079⟩, 695 ⟨62JA854⟩
Speziale, J., **2**, 120 ⟨63OSC(4)401⟩; **4**, 728 ⟨70USP3479373⟩
Speziale, V., **1**, 68, 454 ⟨79S891⟩, 531, 662 ⟨82T2355⟩
Spialter, L., **1**, 759 ⟨70JA7424⟩; **2**, 377, 403 ⟨65JOC3278⟩, 772 ⟨56JOC1051⟩, 908 ⟨76JOM(107)229⟩, 941 ⟨55JA6227⟩; **4**, 654 ⟨70JA7424⟩, 930 ⟨76JOM(107)229⟩; **5**, 416 ⟨76JOM(107)229⟩
Spialter, W. A. L., **2**, 298, 320 ⟨B-65MI 205-01⟩
Spicer, L. D., **1**, 18, 19 ⟨88TL3741⟩
Spiegel, ,B. I., **2**, 925, 969, 976 ⟨78JA2254⟩
Spiegel, B. I., **1**, 468, 471, 533, 1020, 1097, 1104 ⟨78JA2254⟩
Spiegel, G. U., **4**, 546 ⟨90CB989⟩
Spiegler, L., **2**, 23 ⟨39JA940⟩
Spielman, M. A., **5**, 599 ⟨46OR(3)83⟩
Spielvogel, B. F., **5**, 1147 ⟨89MI 530-01⟩; **6**, 491 ⟨84IC4322, 86S833⟩, 491 ⟨89IS79⟩, 491, 515 ⟨79JINC1223⟩, 492, 515 ⟨91T6915⟩, 514 ⟨90EJM301⟩, 514 ⟨92MI 616-01, 76JA5702⟩
Spielvogel, B. T., **6**, 514 ⟨89EUP336772⟩
Spiers, J. C., **4**, 584 ⟨79JOM(182)C20⟩, 584, 585 ⟨86JOM(315)255⟩
Spies, H., **6**, 628, 631 ⟨71JPR804⟩
Spiesecke, H., **5**, 506 ⟨56CB1110⟩
Spiess, A. J., **4**, 61 ⟨89JMC997⟩
Spietschka, E., **1**, 395 ⟨55CB934⟩; **3**, 532 ⟨51LA(573)17, 52CB225⟩, 532 ⟨55CB934⟩
Spiewak, J. W., **1**, 620 ⟨71JA476⟩; **3**, 479 ⟨71JA476⟩; **4**, 83, 85, 86 ⟨71JA476⟩
Spillane, L. J., **5**, 103 ⟨55OSC(3)302⟩
Spillane, R. J., **3**, 83, 94, 102 ⟨78TL2473⟩
Spillane, W. J., **2**, 366 ⟨80CRV151⟩, 366, 797 ⟨73IJS469⟩, 367 ⟨82JCR(S)84, 82JCS(P1)677⟩, 367 ⟨83S192, 85S935⟩, 368 ⟨86S1021⟩; **5**, 342 ⟨86S1021⟩
Spille, J., **2**, 1009 ⟨56CB209⟩; **3**, 687 ⟨56CB209⟩; **4**, 881 ⟨56CB2060⟩
Spillett, M. J., **1**, 240 ⟨66CC757⟩
Spilling, C. D., **1**, 156 ⟨88TL5733⟩; **2**, 394 ⟨89JOC1233⟩; **4**, 897 ⟨89TL2349⟩, 898 ⟨89JOC992⟩, 906 ⟨89JOC4723⟩; **5**, 238 ⟨89JOC4723⟩

Spillmann, P., **5**, 566, 590, 606, 609, 610 ⟨72T199⟩
Spillner, C. J., **1**, 894 ⟨74JA7591⟩
Spina, K. P., **2**, 31 ⟨84SC1313⟩
Spindell, D., **1**, 525 ⟨80JOC2550⟩
Spineti, G. G., **1**, 834 ⟨81JOC2228⟩
Spinicelli, L. F., **5**, 772 ⟨77JOC2091⟩
Spinner, M. E., **5**, 142 ⟨73JCS(P1)2796⟩
Spinoza, G. R. de, **2**, 25 ⟨76JCS(P1)104⟩
Spirikhin, L. V., **1**, 491 ⟨89S633⟩
Spitznagel, G. W., **1**, 135, 159 ⟨84JA6467⟩, 846 ⟨86TL4411⟩; **2**, 999, 1006 ⟨84JA6467⟩; **4**, 378, 391, 816 ⟨84JA6467⟩
Spitzner, D., **4**, 648 ⟨88JOC2274⟩; **5**, 62 ⟨87OS(66)37⟩
Spitzner, R., **4**, 793, 970, 971 ⟨82S199⟩, 837, 860 ⟨82S206⟩; **5**, 616 ⟨94TL1977⟩; **6**, 616 ⟨77ZC172⟩, 619 ⟨89LA931⟩
Spivak, A. Y., **1**, 1022 ⟨85IZV1109⟩
Spivak, J. D., **2**, 432 ⟨84JOC2906⟩
Spivey, A. C., **1**, 264 ⟨89CC1852⟩; **5**, 1106 ⟨90TA881⟩
Spliethoff, B., **2**, 558 ⟨90CB1517⟩, 559, 959 ⟨83AG(E)728⟩
Splitter, J. S., **6**, 110 ⟨71JA4075⟩
Spoerri, P. E., **3**, 452 ⟨55JA984⟩
Spohn, R. J., **3**, 293 ⟨68JA540⟩
Spohnholz, R., **3**, 353 ⟨91CB1397⟩
Spönlein, W., **1**, 797 ⟨86S921⟩
Sporman, W., **2**, 94 ⟨36ZAAC(228)341⟩
Spotswood, T. M., **4**, 424, 425 ⟨51RTC269⟩
Sprague, G. S., **2**, 388 ⟨66JA1078⟩
Sprague, J. M., **2**, 120, 125 ⟨36JA1348⟩, 120, 125 ⟨37JA1837⟩, 176, 177 ⟨51JOC621⟩, 177 ⟨37JA2439⟩
Sprague, L. G., **3**, 197 ⟨88JOC1523⟩; **4**, 802 ⟨90JFC(49)75⟩; **6**, 58, 266 ⟨89JOC613, 90JFC(49)75⟩
Sprangers, W. J. J. M., **5**, 438 ⟨74TL3377, 76C199⟩, 438 ⟨76JCS(P2)1895⟩
Spraul, M., **3**, 148 ⟨81CB1858⟩, 148 ⟨83CB2641⟩
Spreafico, F., **1**, 182 ⟨85JOC3945⟩
Sprecher, G. Von, **4**, 123 ⟨90EUP402312⟩
Sprecher, M., **1**, 1111 ⟨69CC1044⟩; **2**, 660 ⟨94JOC1831⟩; **4**, 204 ⟨69T1679⟩; **5**, 191 ⟨69TL4563⟩, 403 ⟨68TL4405⟩
Sprecker, M. A., **3**, 372 ⟨74JA1817⟩
Spreitzer, H., **1**, 564 ⟨86LA1578, 92M587⟩
Sprengeler, E. P., **2**, 373, 374, 379, 381 ⟨59JA2597⟩
Sprengeler, P. A., **1**, 166 ⟨93TL63⟩
Sprenger, G. H., **5**, 246 ⟨75MI 505-01, 78JFC(11)327⟩; **6**, 261 ⟨75MI 608-01⟩, 440 ⟨73IC2890, 74JA1770⟩
Sprenger, W. A., **5**, 1073 ⟨71JOC2142⟩
Spring, F. S., **1**, 383 ⟨56JCS2419⟩; **3**, 229 ⟨54JCS1302⟩
Springal, H. D., **2**, 769 ⟨B-66MI 215-04⟩; **5**, 85 ⟨B-62MI 502-03⟩
Springer, J. P., **1**, 521 ⟨87JA5167⟩, 704, 708 ⟨77JA6946⟩; **3**, 244 ⟨85JA1274⟩; **4**, 417 ⟨87JOC2442⟩; **5**, 705 ⟨81TL2731⟩, 934 ⟨86JA520⟩; **6**, 55, 280, 633 ⟨81JOC5048⟩, 626 ⟨89JA2995⟩
Springfield, J. R., **5**, 723, 827 ⟨71JOC284⟩, 824 ⟨76JOC148⟩
Springhorn, K. F., **3**, 89 ⟨92SC1967⟩
Sprinz, J., **1**, 194 ⟨93TL1769⟩
Sprio, V., **2**, 770 ⟨71AC(R)391⟩
Sproesser, L., **1**, 1113 ⟨78JA7066⟩
Sprügel, W., **5**, 898 ⟨71S431⟩
Sprules, T. J., **1**, 810 ⟨93TL247⟩
Sprung, M. M., **2**, 783 ⟨36JA225⟩; **4**, 294, 295 ⟨40CRV297⟩
Spry, D. O., **5**, 697 ⟨84H(22)2487⟩
Spuch, J. D., **2**, 558, 560 ⟨83ICAL241⟩
Spunta, G., **2**, 316 ⟨90JOC4199⟩, 544 ⟨90PAC595⟩; **3**, 253 ⟨83JOC123⟩, 440, 456 ⟨87TL5369⟩, 456, 457, 458 ⟨90PAC605⟩, 457 ⟨90JOC4199, 91SL243⟩; **5**, 287 ⟨87TL5369⟩, 377 ⟨91SL243⟩
Spurlock, J. J., **4**, 48 ⟨39JOC234⟩
Spurr, P. R., **1**, 564 ⟨87JA4626⟩
Spyronis, S., **2**, 390 ⟨75S445⟩
Spyroudis, S., **2**, 998 ⟨89LA167⟩; **4**, 867 ⟨85JA7178⟩, 1081 ⟨86JOC3453⟩; **5**, 318, 321, 322 ⟨83S207⟩
Squires, R. R., **1**, 846 ⟨88JA607⟩, 866 ⟨86JA5078⟩
Squires, T. G., **2**, 25, 31 ⟨75JOC134⟩, 147, 155 ⟨82JOC3773⟩
Srebnik, M., **1**, 417 ⟨93MI 109-01⟩, 830 ⟨93MI 118-01⟩, 876 ⟨93MI 119-01⟩, 999 ⟨88TL2631⟩; **2**, 58 ⟨89JOC1577⟩, 478 ⟨93CRV763⟩, 514, 515, 516, 517, 519, 520, 523, 526, 531 ⟨87MI 210-01⟩, 516 ⟨87TL3771⟩, 526 ⟨88JOC2916⟩, 527 ⟨88JA6246⟩, 650 ⟨83S1046⟩, 1085 ⟨88JOC1391⟩, 1085; **3**, 135 ⟨87JA5420⟩
Sredojević, S., **3**, 464 ⟨70T1101⟩
Sredy, J., **5**, 567, 630 ⟨89JMC2493⟩
Sreekumar, C., **1**, 139 ⟨80JA1201⟩; **4**, 378, 379 ⟨80JA1201⟩
Sreekumar, K., **5**, 317 ⟨87MI 507-03⟩
Sreekumar, R., **2**, 331 ⟨93TA2095⟩
Sreenivasan, R., **3**, 361 ⟨82T1673⟩; **4**, 94 ⟨82T1673⟩, 973 ⟨77IJC(B)297⟩; **5**, 760 ⟨79JCS(P1)1255⟩, 1035 ⟨79JCS(P1)2001⟩
Sridar, V., **1**, 925 ⟨86JA1708⟩
Sridharan, V., **1**, 494 ⟨91TL687, 94T359⟩, 546 ⟨93TL153, 93TL157⟩; 837 ⟨93T5177⟩, 1015 ⟨88TL5565⟩; **2**, 158 ⟨91T8297⟩, 245 ⟨88T4941⟩; **4**, 311 ⟨85CC1183⟩
Srikrishna, A., **1**, 416 ⟨92JCS(P1)311⟩, 547 ⟨88TL6487⟩, 826 ⟨90T3601, 93JOC7751⟩, 828 ⟨89TL6577⟩, 829 ⟨93JCS(P1)667⟩
Srimal, R. C., **5**, 685 ⟨93MI 517-01⟩
Srimannarayana, C., **1**, 827 ⟨91SC1455⟩
Srimannarayana, G., **1**, 555 ⟨83S310⟩
Srinivasachar, K., **4**, 304 ⟨85JOC4231⟩
Srinivasan, A., **2**, 309 ⟨89JOC1940⟩
Srinivasan, C. V., **3**, 308 ⟨81JOC3760⟩
Srinivasan, K., **5**, 801 ⟨81JA1467⟩
Srinivasan, K. G., **3**, 699 ⟨73CC699⟩
Srinivasan, K. S., **2**, 221, 225 ⟨72JOC2367⟩
Srinivasan, K. V., **2**, 716 ⟨92S825⟩; **3**, 283, 295, 296, 305 ⟨91S322⟩
Srinivasan, N., **6**, 580 ⟨93TL6447⟩
Srinivasan, P. C., **3**, 144 ⟨84S1010⟩
Srinivasan, P. S., **3**, 618 ⟨87IJC(B)407, 89SC1431⟩; **5**, 81, 546, 549, 550 ⟨83S605⟩, 189 ⟨76IJC794, 87IJC(B)407⟩, 560 ⟨77JPR169⟩, 561 ⟨72CC345⟩
Srinivasan, R., **1**, 418 ⟨63JA4045⟩, 418, 419 ⟨62JA4141, 68JA4498⟩
Srinivasen, C., **2**, 886 ⟨85CJC2285⟩
Srinivas Rao, C., **3**, 40 ⟨89TL6769⟩
Srivanavit, C., **4**, 1031 ⟨70JHC1⟩
Srivasatva, T. N., **5**, 210 ⟨87IJC(A)260⟩
Srivastava, A., **2**, 307 ⟨92CC853⟩
Srivastava, D. K., **2**, 434, 485, 872 ⟨90IC3502⟩, 484 ⟨90POL289⟩
Srivastava, G., **2**, 920 ⟨78JOM(152)39⟩; **5**, 389 ⟨73JOM(63)221⟩; **6**, 61 ⟨69JOM(16)33⟩
Srivastava, K. C., **3**, 639 ⟨88OSC(6)932⟩
Srivastava, P. C., **2**, 623 ⟨83SC827⟩, 907 ⟨83JOM(259)269⟩; **5**, 625 ⟨83JMC445⟩
Srivastava, P. K., **5**, 1038 ⟨69JCED110⟩; **6**, 641 ⟨68JMC1268⟩
Srivastava, R. C., **2**, 414, 423 ⟨B-80MI 207-01⟩, 543 ⟨B-80MI 210-01⟩, 815 ⟨B-80MI 211-02⟩; **5**, 28 ⟨73S327⟩, 341, 342 ⟨79ICAL99, 83IJC(A)503⟩, 402 ⟨89JOM(366)281⟩
Srivastava, R. K., **1**, 459 ⟨68CJC84, 70JOC347⟩
Srivastava, R. L., **2**, 434 ⟨85JCR(S)330⟩
Srivastava, R. R., **4**, 187 ⟨89SC901⟩; **5**, 184, 195, 198 ⟨92TL593⟩
Srivastava, S., **1**, 988 ⟨81JA4638⟩; **2**, 690 ⟨90IJC(A)703⟩; **5**, 341 ⟨86IJC(A)57⟩, 734 ⟨86IJC57⟩
Srivastava, S. K., **2**, 690 ⟨90IJC(A)703⟩
Srivastava, T. N., **2**, 238 ⟨82IJC(A)305, 90IJC(A)703⟩; **5**, 342 ⟨79ICAL99⟩, 1016 ⟨62CJC739⟩, 1136 ⟨68JINC1399, 74MI 530-03⟩, 1143, 1144 ⟨72JIC1143⟩
Srivastava, V. K., **5**, 341, 342 ⟨83IJC(A)503⟩

St. John, N. B., **5**, 112 ⟨43OSC(2)45⟩
Staab, H. A., **1**, 233 ⟨85CB1230⟩, 922 ⟨70CB1101⟩, 1143 ⟨72CB2290⟩; **2**, 205 ⟨60CB2902⟩; **3**, 85 ⟨62AG(E)351⟩, 286 ⟨62LA(655)90⟩; **4**, 851 ⟨80AG(E)66⟩; **5**, 1029 ⟨62LA(657)104⟩; **5**, 196 ⟨62CB2073⟩, 278 ⟨62AG(E)351⟩, 477 ⟨64AG(E)132⟩, 973 ⟨61AG657, 61LA(648)72⟩
Staab, H. A., **5**, 973 ⟨62AG(E)361⟩; **6**, 416 ⟨56AG754⟩, 417 ⟨61AG66⟩, 500 ⟨73OSC(5)201⟩, 504 ⟨57LA(609)83⟩, 578 ⟨62LA(657)98⟩
Staats, P. A., **5**, 1129 ⟨60IS84⟩
Stabinsky, Y., **2**, 154 ⟨67CC270⟩
Stäbler, F., **5**, 351 ⟨92S482⟩
Stabrovskaya, L. A., **4**, 489 ⟨72ZOB1227⟩
Stacey, F. R., **2**, 104 ⟨58JA4546⟩
Stacey, F. W., **1**, 919, 948, 950 ⟨63OR(13)150⟩; **2**, 4, 29, 392, 623 ⟨63OR(13)150⟩, 166 ⟨63JA749⟩; **3**, 337 ⟨63JA963⟩; **5**, 474, 475 ⟨63JA749⟩, 808 ⟨63USP3076843⟩; **6**, 46, 233, 234 ⟨61JA840⟩, 625 ⟨62USP3115513⟩
Stacey, G. J., **2**, 101 ⟨50JCS3125⟩, 104 ⟨45JCS380⟩, 409, 410, 412, 413 ⟨49JCS2921⟩, 412 ⟨48JCS699⟩; **4**, 76 ⟨48JCS699⟩
Stacey, M., **2**, 182, 364 ⟨57JCS2574⟩; **4**, 730 ⟨62CI(L)1472⟩; **5**, 193, 199 ⟨54JCS2006⟩; **6**, 9 ⟨52MI 601-01⟩, 20 ⟨60AFC(1)166⟩
Stacey, N. A., **1**, 602 ⟨87CC880⟩
Stach, H., **3**, 190 ⟨86HCA1614⟩
Stach, J., **6**, 565 ⟨81ZAAC(476)7⟩
Stach, K., **1**, 42 ⟨43CB1252⟩
Stachel, H.-D., **2**, 211 ⟨57LA(606)75⟩; **4**, 259 ⟨92LA1039⟩, 890, 891 ⟨60CB1059⟩, 983 ⟨81AP65⟩; **6**, 321 ⟨59AG246⟩
Stachel, M.-D., **4**, 889 ⟨80AP(314)65⟩
Stachel, S. J., **1**, 944 ⟨93TL6209⟩
Stachowski, G., **2**, 689 ⟨76JPR1019⟩
Stachulski, A. V., **3**, 495 ⟨88TL3003⟩; **4**, 738 ⟨88TL3003⟩
Stacino, J.-P., **1**, 439 ⟨86T2475⟩, 439, 700 ⟨82TL3265, 82TL2465⟩
Stack, D. E., **1**, 121 ⟨92JA5110, 93JOC2483⟩, 207 ⟨93TL3063⟩; **2**, 74 ⟨92TL6575⟩, 74, 585 ⟨92JA5110⟩; **3**, 286 ⟨91JA4672⟩
Stack, G. M., **6**, 509 ⟨78JOC922⟩
Stack, J. G., **1**, 326 ⟨91JA5918⟩, 326, 327, 328 ⟨92JA7007⟩
Stackhouse, J., **3**, 376 ⟨74JA5648, 74JA5650⟩, 376 ⟨74JA5651, 75JA2718⟩, 479 ⟨81JOC1891⟩
Stacy, G. W., **4**, 319 ⟨52JA3885, 55JA3869⟩
Stadelbauer, K., **5**, 630 ⟨65CB1556⟩
Stader, C., **4**, 661 ⟨87JOM(321)C1⟩
Stadler, A., **6**, 97 ⟨72HCA75⟩
Stadler, D., **2**, 404 ⟨65LA(687)214⟩; **3**, 473 ⟨65LA(687)214⟩, 477 ⟨68CB2351⟩
Stadler, H., **1**, 832 ⟨91JOC5063⟩
Stadler, M., **4**, 633 ⟨89AG(E)784⟩; **6**, 184, 389, 398, 714 ⟨89AG(E)781⟩, 389, 398, 401, 713 ⟨89AG(E)784⟩
Stadler, P. A., **3**, 291 ⟨72HCA75, 77OS(56)8⟩; **6**, 86 ⟨77OS(56)8⟩, 104 ⟨63HCA2306⟩
Stadlwieser, J., **4**, 49 ⟨85S490⟩
Stadmüller, H.,**5**, 937 ⟨95OM3163⟩
Stadnichuk, M. D., **1**, 748 ⟨80ZOB76⟩; **2**, 1089 ⟨92RCR1091⟩; **4**, 807 ⟨90JGU967⟩, 941 ⟨71ZOB719, 72ZOB870⟩, 941 ⟨79ZOB805, 79ZOB812⟩, 941 ⟨80ZOB722, 84ZOB309⟩, 942 ⟨84ZOB116⟩, 948 ⟨79ZOB829, 80ZOB1053⟩
Stadtmüller, H., **2**, 591 ⟨93JA7027, 93TL7911⟩
Staeger, M. A., **5**, 351 ⟨92SC1⟩
Staendeke, H., **2**, 448 ⟨73AG(E)877⟩
Stafford, S. L., **2**, 904 ⟨60JA6238⟩; **4**, 812 ⟨60JA6232⟩, 813 ⟨60JA6238⟩; **6**, 427, 477 ⟨79S600, 81USP224776⟩
Stafforst, D., **2**, 909 ⟨77LA1167⟩; **3**, 694, 706, 714, 717 ⟨74LA44⟩, 700, 720 ⟨80LA28⟩, 714, 715 ⟨77LA1167⟩; **4**, 538 ⟨80LA28⟩, 1005 ⟨74LA44⟩
Stagé, J., **2**, 814, 815 ⟨B-71MI 215-02⟩
Stahl, C. R., **2**, 478 ⟨57ANC75⟩
Stahl, D., **1**, 1122 ⟨81JA2770⟩
Stahl, I., **1**, 241 ⟨78AG(E)274⟩; **3**, 132 ⟨81S135⟩; **4**, 1073 ⟨85CB1798⟩
Stahl, L., **4**, 678 ⟨92OM3617⟩
Stahl, U., **2**, 318 ⟨91S769⟩
Stahle, H., **6**, 642 ⟨80JMC1217⟩
Stähle, M., **4**, 950 ⟨74S888⟩
Stahly, G. P., **6**, 253 ⟨89JFC(43)53⟩
Stahnecker, P., **3**, 616 ⟨85LA566⟩
Stahnke, M., **1**, 24, 144 ⟨91TL1949⟩
Staib, R. R., **2**, 245 ⟨82T3087⟩; **5**, 326, 330 ⟨82T3087⟩
Stájer, G., **6**, 557 ⟨83T1829⟩, 558 ⟨82H(19)1191, 83JHC1181⟩, 558 ⟨84JHC1373, 85T1353⟩, 558 ⟨90MRC1045⟩
Stakem, G., **1**, 1096 ⟨82JOC1267⟩
Staklis, A., **5**, 999 ⟨65JA1141, 75JOC3554⟩
Staley, D. L., **4**, 584 ⟨89JOM(375)265⟩
Staley, S. W., **1**, 844 ⟨85MI 119-05⟩, 862 ⟨73JA3382⟩, 891 ⟨81MI 119-01⟩, 910 ⟨73JA3384⟩
Stalick, W. M., **1**, 795 ⟨B-77MI 118-01⟩; **2**, 677 ⟨86JOC3577⟩, 1025 ⟨72OPP89, 77OS(57)65⟩, 1025 ⟨88S287⟩
Stalke, D., **2**, 872 ⟨89JOM(366)73⟩; **3**, 370 ⟨90CB1475⟩; **5**, 782 ⟨92CB2171⟩, 893 ⟨89AG(E)445⟩; **6**, 184 ⟨90CC741⟩, 440 ⟨91CB2411⟩
Ställberg-Stenhagen, S., **5**, 13 ⟨47JA2568⟩
Stallman, H., **6**, 524 ⟨43MI 616-01⟩
Stallman, J. B., **1**, 831 ⟨93JA8847⟩
Stallman, J. L., **1**, 831 ⟨93JOC5873⟩
Stam, C. H., **2**, 1003 ⟨84T765⟩; **3**, 599 ⟨93OM3864⟩
Stamatis, N., **5**, 928, 929 ⟨92AG(E)1238⟩; **6**, 201, 715 ⟨92AG(E)1238⟩
Stambouli, A., **3**, 40 ⟨88BSF95⟩
Stamhuis, E. J., **1**, 1122 ⟨61RTC797, 63RTC394⟩
Stamm, A., **2**, 99 ⟨47CB219⟩
Stamm, H., **2**, 319 ⟨87T565⟩, 454, 457 ⟨83CB2936⟩; **4**, 792 ⟨70TL1937⟩; **5**, 345 ⟨86CB2678⟩
Stamm, W., **2**, 485 ⟨64AG(E)66⟩; **5**, 1017 ⟨65JOC693⟩
Stammer, C. H., **3**, 13 ⟨79CC79⟩; **5**, 367 ⟨89JOC145⟩
Stammler, H., **2**, 866 ⟨91CB725⟩
Stammler, H.-G., **4**, 603 ⟨91CB2391⟩; **5**, 893 ⟨91CB989⟩; **6**, 187 ⟨94OM436⟩, 187, 387 ⟨91OM1647, 91OM3838⟩
Stammler, R., **2**, 944 ⟨94SL92⟩
Stamos, I. K., **3**, 635 ⟨85TL477⟩
Stamouli, P., **1**, 546, 946 ⟨87PAC363⟩; **5**, 1041, 1108 ⟨85CC533⟩; **6**, 558 ⟨86HCA1898⟩
Stamper, J. G., **6**, 174, 391 ⟨86CC969⟩, 270 ⟨85CC5⟩, 401 ⟨87T1019⟩
Stamper, W. E., **2**, 608, 626 ⟨58JOC1412⟩
Stampf, E. J., **4**, 813 ⟨76JOM(108)1⟩
Stampfli, U., **3**, 273 ⟨81C336⟩
Stanchen, S., **2**, 70 ⟨93CC328⟩
Stańczyk, W. A., **6**, 290, 391 ⟨87CC1461⟩, 381 ⟨93JCS(P2)59⟩
Stanek, J., **4**, 223 ⟨59CCC64⟩, 223, 225 ⟨63M290⟩
Stanescu, C., **3**, 632 ⟨92TL1431⟩
Stãnescu, L., **1**, 518 ⟨73TL9⟩
Stanescu, M., **3**, 539 ⟨88RRC83⟩
Stanetty, P., **6**, 25, 31 ⟨88M1427⟩, 444 ⟨83JHC1477⟩
Stanforth, J. C., **3**, 145 ⟨85JCS(P1)2667⟩
Stanforth, S. P., **1**, 998 ⟨B-91MI 121-01⟩
Stang, A. F., **6**, 69, 296, 297 ⟨65JOC411⟩
Stang, P. J., **1**, 108 ⟨92AG(E)274⟩, 177, 224, 433, 1021 ⟨82S85⟩, 541, 1018 ⟨87JA7561⟩, 550, 1143 ⟨75JA3854⟩, 981 ⟨75JOC657⟩, 988 ⟨79JA4772, 81JA6437⟩, 988 ⟨87JA782, 88CC301⟩, 988 ⟨88JOC3122, 91T4539⟩, 988, 1143 ⟨82ACR348⟩, 988, 1145 ⟨81JA5429⟩, 1006 ⟨92SL199⟩, 1016 ⟨91JOC3912⟩, 1018 ⟨91JCS(P1)2892⟩, 1070 ⟨74JOC581, 81CC459⟩, 1070 ⟨85S962⟩, 1071, 1143 ⟨74JA4562⟩, 1109, ,1110, 1117 ⟨B-79MI 122-01⟩, 1110,

1117 ⟨71JA1513⟩, 1111 ⟨69JA4600, 74JCS(P2)843⟩, 1111, 1114, 1116 ⟨74JA1100⟩, 1114 ⟨71TL2519, 70JA3802⟩, 1117 ⟨76T37⟩, 1118, 1143 ⟨78CRV383, 75CC17⟩, 1143 ⟨75JA1459, 75JA6478⟩, 1143 ⟨78ACR107⟩, 1145 ⟨80JA6813⟩; **2**, 204 ⟨71CC1614⟩, 280 ⟨93TL703⟩, 445, 1021 ⟨91S1073⟩, 445, 1021 ⟨92JOC1861⟩, 689 ⟨70AG(E)521, 74JA1100⟩, 689 ⟨74OS(54)79, 76JOC4099⟩, 689 ⟨80S283, 82S85⟩, 1008 ⟨83JOC3038⟩, 1012, 1028, 1029 ⟨92AG(E)274⟩, 1019, 1022 ⟨91ACR304⟩, 1020 ⟨88JOC4105, 90AG(E)287⟩, 1020 ⟨90JA6437, 90TL4828⟩, 1020 ⟨91JA5870, 91JOC3912⟩, 1020, 1022, 1028 ⟨85JA1452⟩, 1020, 1028 ⟨87JA228⟩, 1020, 1028, 1033 ⟨91JA4571⟩, 1021 ⟨91JA8997, 92MI 221-01, 92TL1419, 93JA11702⟩, 1023, 1024 ⟨86JA7125⟩, 1028 ⟨86JA7832, 89JA2225⟩, 1029 ⟨88JA3272⟩, 1033 ⟨92SL535⟩, 1036 ⟨93JOC5235⟩; **3**, 273 ⟨79S438, 81S29, 85S962⟩, 496 ⟨89JOC2783, 90JOC5033⟩; **4**, 604, 647 ⟨84JA6108⟩, 604, 647 ⟨86JA5949⟩, 717 ⟨93OM996⟩, 875, 877 ⟨84JOC1653⟩, 985 ⟨84JA6015⟩; **5**, 205, 321 ⟨90TL4821⟩, 944 ⟨92JOM(431)247, 93OM3510⟩, 1108 ⟨92SL535⟩; **6**, 127 ⟨77JOC1667⟩, 129, 167, 374 ⟨84JOC5087⟩, 153 ⟨93JOC4105⟩, 562 ⟨87JOC4117⟩

Stang, P. L., **2**, 1094 ⟨81JA6437⟩

Stange, A., **1**, 16, 938 ⟨81S743⟩, 940 ⟨82TL2019⟩

Stanger, A., **3**, 502 ⟨92OM2326⟩

Stanger, H., **2**, 203 ⟨36CB2753⟩

Stanghellini, P. L., **4**, 692 ⟨74ICA5, 74JOM(64)271⟩; **6**, 352 ⟨82IC3781, 83CC1613⟩

Stangl, H., **3**, 678 ⟨62AG(E)50⟩, 679 ⟨72CB1258⟩; **5**, 657 ⟨72CB1258⟩

Stanhope, B., **1**, 598 ⟨92T33⟩

Staniland, P. A., **1**, 72 ⟨66JCS(C)41⟩

Stanior, U., **6**, 562 ⟨84AP(317)1042⟩

Stankevich, V. K., **3**, 707 ⟨89ZOR1573⟩

Stanko, V. I., **6**, 406 ⟨68BAU414⟩

Stankowiak, A., **2**, 446 ⟨92S577⟩; **5**, 226 ⟨87S931⟩

Stanley, G. C., **2**, 439 ⟨85JA7423⟩

Stanley, G. G., **4**, 545 ⟨89PS(42)97⟩, 550 ⟨85JA7423, 88JA5585⟩, 550 ⟨89IC1872⟩

Stanley, R. L., **6**, 434 ⟨77JOC3686⟩

Stanoeva, E., **2**, 312 ⟨88S587⟩; **4**, 189 ⟨94S427⟩

Stanovnik, B., **3**, 451 ⟨84CHEC(3)1⟩; **5**, 22 ⟨92H(33)657⟩, 609, 626, 836 ⟨81S299⟩, 764 ⟨76JHC487⟩, 824, 834 ⟨78H(9)1327, 82JHC1397⟩, 834 ⟨79JOC1695, 86S807⟩, 835 ⟨82MI 521-01, 87MI 521-02⟩, 835 ⟨B-87MI 521-01⟩, 836 ⟨73TL1677, 74JHC103⟩, 836 ⟨74JOC2143, 81H(16)2173⟩, 836 ⟨82JHC577, 84H(22)1545⟩, 836 ⟨86M(117)221, 87M601⟩, 837 ⟨81H(15)293, 84CS209⟩, 837 ⟨85H(23)2819⟩, 845 ⟨86H(24)1899⟩, 1028 ⟨90JHC407⟩, 1122 ⟨84MI 530-01⟩, 1118, 1122 ⟨83MI 530-02⟩; **6**, 558 ⟨90JHC407, 90JHC643⟩

Stansberry, M., **3**, 145 ⟨92TL5287⟩

Stansbury, Jr., H. A., **5**, 823 ⟨63JOC1673⟩

Stansfield, F., **3**, 419 ⟨86JCS(P1)619⟩, 638 ⟨77JCS(P1)1811⟩; **4**, 321 ⟨77JCS(P1)1811⟩, 899 ⟨52JCS4067⟩; **6**, 114 ⟨74JCS(P1)540⟩

Stansfield, R. F. D., **4**, 692 ⟨80JCS(D)2182⟩

Stanton, J., **1**, 821 ⟨71CC1639⟩; **6**, 88 ⟨91CJC415⟩

Stanton, J. L., **3**, 54 ⟨75JA4018⟩

Stanzione, R. C., **6**, 641, 645 ⟨86JOC1882⟩

Stapleford, K. J. S., **1**, 100 ⟨69AJC1321⟩

Stapleton, I. W., **2**, 220 ⟨62AJC570⟩

Stapp, P. R., **3**, 567 ⟨61JOC45⟩

Staps, R. J. F. M., **4**, 830 ⟨73RTC11⟩

Staral, J. S., **1**, 909 ⟨76JA6290⟩, 914 ⟨78JA6299⟩

Starcher, P. S., **5**, 205 ⟨57JA5982⟩

Starck, D., **6**, 236 ⟨67GEP1232954⟩

Staricco, E. H., **6**, 607 ⟨87JFC(37)29⟩

Starichenko, V. F., **1**, 103 ⟨91ZOR1137⟩

Starikova, Z. A., **2**, 886 ⟨81JGU1056⟩; **5**, 778 ⟨87IZV165⟩

Staring, E. G. J., **5**, 162 ⟨82JA166⟩

Stark, A., **5**, 944 ⟨95CB49⟩

Stark, Jr., C. J., **1**, 24 ⟨82JOC1193⟩, 711 ⟨85PAC763⟩

Stark, H., **2**, 400 ⟨69M959⟩; **5**, 148 ⟨80AG(E)1029⟩

Starke, R., **2**, 360 ⟨77ZN(B)721⟩, 466 ⟨77S580⟩

Starkemann, C., **1**, 305 ⟨87HCA2201⟩; **5**, 347 ⟨90TL4117⟩

Starker, ,B., **4**, 851 ⟨80AG(E)66⟩

Starkey, J. D., **1**, 869 ⟨67JA2751⟩

Starkovsky, N. A., **1**, 89 ⟨64JA118⟩

Starks, C. M., **3**, 613 ⟨73JA3613⟩; **6**, 5, 7 ⟨B-74MI 601-01⟩

Starky, C. M., **2**, 619 ⟨90JHC1549⟩

Starn, R. E., **5**, 706, 845 ⟨55JA4571⟩

Starner, W. E., **3**, 438 ⟨86TL3957⟩

Starobin, Yu. K., **6**, 251 ⟨76ZOR767⟩

Staros, J. V., **5**, 332 ⟨87MI 507-02⟩

Staroverova, I. N., **2**, 471 ⟨83JGU2244⟩

Starr, D., **1**, 91 ⟨43OSC(2)566⟩

Starr, L., **3**, 90 ⟨61JOC2225⟩

Stary, H., **2**, 468 ⟨72ZAAC(394)171⟩

Starý, I., **1**, 194 ⟨89JA4981⟩; **2**, 112 ⟨94T529⟩

Starz, K. A., **4**, 581 ⟨83CB1070, 83CB1209⟩

Starzewski, K-H. A. O., **1**, 737 ⟨76CB473⟩; **4**, 598 ⟨76CB473⟩

Staschewski, D., **1**, 18 ⟨59AG726⟩

Staskun, B., **3**, 86 ⟨65JCS5775⟩

Staszak, M. A., **5**, 328 ⟨94TL6021⟩

Statnishevskii, L. S., **3**, 274 ⟨81JGU1053⟩

Staude, H.-J., **6**, 564 ⟨82AP(315)103⟩

Staudinger, H., **2**, 810, 853, 859 ⟨19HCA635⟩; **3**, 463 ⟨11CB2197, 16CB1897⟩, 463 ⟨16CB1923⟩, 474, 475 ⟨16CB1978⟩, 498 ⟨19HCA619⟩, 526 ⟨B-12MI 316-01⟩, 530 ⟨08LA(356)51⟩, 530, 531 ⟨05CB1735, 09CB4213, 06CB3062⟩, 534 ⟨23HCA291⟩, 534, 541 ⟨08CB2208, 13CB3539⟩, 549 ⟨20HCA853⟩, 552 ⟨20HCA833⟩, 555, 591 ⟨19HCA635⟩, 561 ⟨07LA(356)55⟩, 561, 584, 591 ⟨20CB72⟩, 584 ⟨21HCA887⟩, 586 ⟨21HCA897⟩; **4**, 18 ⟨09CB3966⟩; **5**, 19 ⟨13CB1423⟩, 145 ⟨24HCA390⟩, 286 ⟨17CB1035⟩, 476, 477 ⟨20HCA824⟩, 974 ⟨21HCA861⟩; **6**, 412, 417, 418 ⟨13CB1426⟩

Stauffer, E., **6**, 254 ⟨1890CB3226⟩

Stauffer, P., **5**, 441 ⟨85JA4440⟩

Stauffer, R. D., **1**, 468, 472 ⟨81JA6460⟩, 515 ⟨75JOC3619⟩; **3**, 127 ⟨77JOC3180⟩

Staunton, J., **2**, 664 ⟨79CC1165⟩; **3**, 212 ⟨86S944⟩; **5**, 66, 74 ⟨73OSC(5)8⟩

Stavaux, M., **3**, 339, 356, 357 ⟨80BSF(2)530⟩, 354, 378 ⟨71BSF4429, 80BSF(2)539⟩; **5**, 605, 613 ⟨83BSF(2)277⟩; **6**, 48 ⟨80AHC(27)151⟩, 637, 732 ⟨86NJC51⟩

Stavber, S., **2**, 15 ⟨91JOC7347, 94CC149⟩, 16 ⟨89T6003⟩, 17 ⟨93JCS(P1)2851⟩, 18 ⟨87JOC919⟩, 613 ⟨81CC148⟩, 615 ⟨92CC274⟩; **4**, 45 ⟨93TL4355⟩; **5**, 6 ⟨89TL6095⟩

Stavchansky, S., **5**, 350 ⟨88TL739⟩

Stavholt, K., **5**, 168 ⟨50ACS1080⟩

Stavroskaya, A. V., **6**, 140 ⟨67ZOR1749⟩

Stayner, R. D., **2**, 451 ⟨54USP2686803, 54USP2693482⟩

Stearns, J. F., **6**, 644, 647 ⟨77JOC3608⟩

Stearns, R. S., **1**, 848 ⟨47JA2025⟩

Stebnik, M., **4**, 208 ⟨89SC197⟩

Stec, D., **2**, 943 ⟨93S611⟩

Stec, W., **2**, 441 ⟨71AG(E)494⟩, 452 ⟨70CC1495⟩

Stec, W. J., **2**, 412 ⟨85JOC1815⟩; **3**, 439 ⟨78S521, 82S270⟩; **4**, 466 ⟨78S469⟩; **5**, 694 ⟨91TL4193⟩, 1057 ⟨75S785⟩, 1135 ⟨78S154⟩; **6**, 122 ⟨75S785, 78JCR(S)41⟩

Steckel, T. F., **5**, 1121 ⟨80CJC2784⟩

Steckel, T. P., **3**, 120 ⟨80TL4695⟩

Steckhan, E., **4**, 206 ⟨90H(31)1959⟩, 743 ⟨86JCS(P1)733⟩; **6**, 77, 78 ⟨86T553⟩, 117 ⟨89T1691⟩

Stedehouder, P. L., **6**, 81, 91 ⟨33RTC437⟩, 306 ⟨33RTC923⟩, 306, 307 ⟨33RTC1039⟩

Stedronsky, E. R., **1**, 668 ⟨70JA1426, 72JA5258⟩

Steege, W., **4**, 303 ⟨71SC103⟩

Steegmuller, D., **3**, 280 ⟨88CB117⟩

Steel, C., **4**, 648 ⟨75CC536⟩
Steel, G., **3**, 488 ⟨63NAT903⟩
Steel, P. J., **1**, 185 ⟨90JOC1338⟩; **2**, 404 ⟨87JOC2911⟩, 1008 ⟨92CJC2040⟩; **5**, 597 ⟨90JCS(P1)1869⟩
Steele, B. R., **1**, 1130 ⟨53JCS1199⟩; **4**, 31 ⟨54JCS3747⟩; **6**, 2 ⟨54JCS923⟩
Steele, J. E., **4**, 742 ⟨90JCS(P1)139⟩
Steele, R. B., **1**, 514, 534 ⟨67JA2754, 67JA5085⟩, 680 ⟨66TL6021⟩; **6**, 208 ⟨66TL6021⟩
Steelman, S. L., **2**, 1016 ⟨61JA4663⟩
Steen, K., **1**, 414, 415, 827 ⟨64HCA2425⟩; **4**, 882 ⟨64HCA2425⟩
Steen, R. J., **6**, 332, 351 ⟨83IC3700⟩
Steenhoek, A., **1**, 505 ⟨71RTC961⟩
Steenkamp, J. A., **3**, 24, 146 ⟨92TL2383⟩
Steenken, S., **4**, 1072 ⟨89JA4967⟩; **6**, 734 ⟨79JA6282⟩
Steer, I. A., **2**, 492 ⟨88JCS(D)451⟩
Steer, R. P., **3**, 335 ⟨74CJC3518⟩
Steeves, B. H., **2**, 465, 858 ⟨65JOM(4)320⟩
Stefanchik, M., **3**, 527 ⟨76JOC3303⟩
Stefancich, G., **4**, 856 ⟨82JHC1493⟩
Stefani, A., **1**, 297 ⟨77JA1058⟩
Stefani, H. A., **3**, 177 ⟨87SC443⟩, 177 ⟨91OM1178⟩
Stefanovsky, Y. N., **5**, 607 ⟨94T7193⟩
Steffan, R. J., **2**, 627 ⟨90JHC2151⟩
Steffen, J., **1**, 979 ⟨86CC829⟩
Steffen, U., **5**, 151, 164 ⟨88JA4640⟩
Steffen, W. L., **2**, 442 ⟨82HOU(13/3b)69⟩
Steffgen, F. W., **3**, 295 ⟨61JOC670⟩
Stefinovic, M., **1**, 993 ⟨93SL931⟩
Stegel, F., **2**, 28 ⟨87G83⟩; **4**, 415 ⟨83JOC162⟩
Steger, B., **2**, 437 ⟨87JOM(335)59⟩
Steger, W., **5**, 369, 713 ⟨80JCS(D)1768⟩
Stegk, A., **3**, 540 ⟨77CB3582⟩
Steglich, W., **1**, 537 ⟨71AG(E)653⟩, 985 ⟨75AG(E)58⟩, 1042 ⟨86S372⟩; **2**, 747 ⟨81TL4263⟩; **3**, 680 ⟨71CB3816⟩; **4**, 99 ⟨88T5403⟩, 100, 311 ⟨91TL3163⟩, 102 ⟨87S223⟩, 297 ⟨74CB1488⟩, 297, 327 ⟨66CB1944⟩, 305 ⟨65CB487⟩, 309 ⟨66CB1932⟩, 320 ⟨62LA(658)128⟩, 465 ⟨83MI 410-01⟩, 489 ⟨90S1153⟩, 1001 ⟨89S97⟩; **5**, 124 ⟨85OS(63)183⟩, 290 ⟨81TL4263⟩, 790 ⟨89S97⟩; **6**, 502 ⟨79CB727⟩
Stegmair, C. M., **5**, 955 ⟨92CB1571⟩
Stegman, H. B., **2**, 825 ⟨82PS(13)331⟩
Stegmann, H. B., **1**, 201 ⟨81S125⟩; **6**, 302 ⟨72CB3905⟩
Stehling, L., **3**, 233 ⟨85AG(E)496⟩
Stehouwer, P. M., **1**, 975 ⟨83JOC1103⟩
Steibitz, B., **6**, 512 ⟨87EGP242812⟩
Steichen, D. S., **2**, 687 ⟨79TL4363⟩
Steigel, A., **6**, 72 ⟨91SL160⟩
Steigelmann, O., **2**, 444 ⟨89AG(E)1700⟩; **4**, 1023 ⟨89CB1857⟩; **6**, 173 ⟨89CB2115⟩, 173, 378, 379 ⟨90AG(E)201⟩, 386, 406 ⟨91AG(E)1488⟩
Steiger, A., **5**, 749 ⟨86HCA1224⟩
Steiger, R. E., **5**, 85, 275 ⟨55OSC(3)66⟩, 85, 275 ⟨55OSC(3)88⟩
Steim, J. M., **6**, 489 ⟨90BBR(171)458⟩
Steimann, M., **5**, 943 ⟨92JOM(441)63⟩; **6**, 154 ⟨93AG(E)985⟩
Steimetz, R., **3**, 65 ⟨63LA(668)19⟩
Steimmig, G., **6**, 467 ⟨28GEP516281⟩
Stein, A. R., **5**, 723 ⟨74CJC4050⟩
Stein, F., **4**, 940, 1014 ⟨92CB2051⟩; **5**, 934 ⟨91AG(E)1658⟩, 946 ⟨93SL486⟩, 958 ⟨92CB2051⟩
Stein, G., **5**, 28 ⟨41AG146⟩
Stein, H. H., **2**, 118, 124 ⟨93JMC460⟩
Stein, J. L., **1**, 7 ⟨80TL287⟩
Stein, K., **1**, 402, 794 ⟨70CB426⟩
Stein, M. L., **3**, 469 ⟨65G983⟩
Stein, N. L., **4**, 414 ⟨86JOC2228⟩
Stein, P. M., **1**, 48 ⟨79AG(E)939⟩
Stein, R. A., **2**, 622 ⟨66MI 212-01⟩
Stein, S. E., **1**, 922 ⟨86JA3224⟩
Stein, S. J., **3**, 273 ⟨76SC503⟩
Stein, U., **2**, 1038 ⟨81CB673⟩; **6**, 349 ⟨90JOM(398)65⟩
Steinacker, K. H., **6**, 80 ⟨53CB790, 54CB205⟩
Steinbach, A., **2**, 1009 ⟨88JOC4379⟩
Steinbach, R., **1**, 127 ⟨80CC1202⟩, 186 ⟨81CC237⟩; **2**, 566 ⟨80AG(E)900, 80AG(E)1011⟩
Steinbach, S., **5**, 1131 ⟨92EGP300936A7⟩
Steinberg, D., **5**, 83 ⟨57MI 502-05⟩
Steinberg, H., **1**, 384 ⟨54JA2753⟩, 929 ⟨68RTC1301, 73TL827⟩; **2**, 109, 111, 112 ⟨B-64MI 202-01⟩, 418, 419 ⟨60JA6242, 61JA2122⟩, 531 ⟨B-64MI 210-01⟩; **4**, 49 ⟨81RTC194⟩, 299 ⟨73SC189, 74RTC294⟩, 406 ⟨67RTC417⟩
Steinberg, L., **5**, 321 ⟨65CC449⟩
Steinberger, H., **2**, 468 ⟨74ZN(B)611⟩, 469 ⟨75ZAAC(413)266⟩
Steinbock, H., **5**, 821 ⟨02CB3101⟩
Steinborn, D., **2**, 558 ⟨79JOM(182)313⟩
Steinbruckner, C., **3**, 694 ⟨61CB2802⟩, 695 ⟨59AG386⟩
Steiner, D., **4**, 707 ⟨89JOM(368)67⟩
Steiner, E., **4**, 12 ⟨85T4057⟩, 14 ⟨80HCA1947⟩
Steiner, E. C., **1**, 869 ⟨65JA382, 67JA2751⟩, 869 ⟨67JA3905⟩
Steiner, G., **5**, 347 ⟨78LA635⟩
Steiner, H., **1**, 419 ⟨64JCS3080⟩; **3**, 44 ⟨90HCA405⟩
Steiner, J., **2**, 764 ⟨22CB875⟩
Steiner, R. P., **1**, 811 ⟨70JA2158⟩
Steiner, W., **3**, 616 ⟨34JPR275⟩
Steinert, P., **5**, 944 ⟨95CB49⟩
Steinfeld, A. S., **6**, 608 ⟨73JA2982⟩
Steinfink, H., **4**, 668 ⟨84IC3717⟩
Steingart, C., **5**, 1125 ⟨90TL3681⟩
Steingross, W., **2**, 1016 ⟨66JOM(6)109⟩
Steinhardt, J., **5**, 83 ⟨45MI 502-01⟩
Steinhoff, G., **2**, 947 ⟨88TL5245⟩; **3**, 494 ⟨66AG(E)968⟩
Steininger, E., **2**, 427 ⟨62M230⟩, 429, 822, 823 ⟨59M792⟩, 431, 453 ⟨63CB3184⟩, 451, 461, 838, 851 ⟨62CB2993⟩; **6**, 109 ⟨65AG427⟩
Steinkopf, W., **2**, 484 ⟨32CB409⟩, 484, 489, 492 ⟨28CB1911⟩, 507 ⟨28CB1906⟩; **5**, 819 ⟨11JPR687⟩, 819, 821, 822 ⟨11JPR453⟩, 820 ⟨08JPR198, 10JPR216⟩, 822 ⟨10JPR224⟩, 838 ⟨10JPR209⟩; **6**, 409 ⟨20JPR79⟩
Steinman, H. G., **2**, 884 ⟨46JA1987⟩
Steinman, M., **6**, 550 ⟨86JOC4488⟩
Steinmann, D. H., **5**, 146 ⟨86JA284⟩
Steinmann, M., **4**, 710 ⟨89CB1901⟩
Steinmetz, G. R., **4**, 679 ⟨81JA1278⟩
Steinmetz, M. G., **4**, 1054 ⟨89JOM(378)1⟩
Steinseifer, F., **1**, 54 ⟨85CB1039⟩; **4**, 156, 708 ⟨85CB370⟩, 401, 599, 708, 724 ⟨82CB1810⟩, 596, 600 ⟨80AG(E)723⟩, 596, 600 ⟨85CB1039⟩, 599, 600, 663 ⟨78TL4391⟩
Steinwand, M., **2**, 836 ⟨82CB2181⟩
Steliou, K., **2**, 221 ⟨84JA759⟩, 236 ⟨78JA1222⟩, 411 ⟨78TL3989⟩, 541 ⟨76S721, 82JA3104⟩; **3**, 336, 340, 346, 347, 363, 392 ⟨82JA3104⟩; **5**, 174, 277 ⟨83JA7130, 80JA7578⟩, 338 ⟨78JA1222⟩, 512, 575 ⟨82JA3104⟩; **6**, 577, 578 ⟨78JOC337⟩
Stella, L., **1**, 7 ⟨80TL287⟩; **2**, 323, 326, 337 ⟨83AG(E)337, 81TL61⟩; **3**, 23 ⟨84TL3457⟩; **6**, 52 ⟨79AG(E)615⟩
Stella, S., **6**, 342 ⟨86AG283, 88NJC621⟩
Stelmaszynska, T., **2**, 213 ⟨61MI 203-02⟩
Stelzel, P., **6**, 486 ⟨74HOU(15/2)187⟩
Stelzer, O., **2**, 10 ⟨82S326⟩, 427, 430 ⟨90ZN(B)203⟩, 428, 430 ⟨84CB3400⟩, 428, 430 ⟨87ZN(B)579⟩, 428, 431 ⟨80CB1356⟩, 429 ⟨87CB1707⟩, 435 ⟨85JOM(296)411⟩, 437 ⟨86CB2921⟩, 440 ⟨83IC3624, 84IC3304⟩, 466 ⟨86CB2767⟩, 471 ⟨78CB890⟩, 472 ⟨90JOM(399)C4⟩, 474 ⟨70JOM(21)411⟩, 475 ⟨82JFC(20)85⟩, 867, 868 ⟨69CB2900⟩; **4**, 112, 546 ⟨82AG(E)376, 90CB995⟩, 546 ⟨87CB1707, 89IS120, 90CB989⟩, 550 ⟨83PS(18)279,

84CB3400⟩, 550 ⟨85JOM(296)411, 88ZN(B)31⟩; **6**, 165 ⟨87CB1707⟩
Stemerick, D. M., **1**, 439 ⟨90TL5449⟩, 440 ⟨91JA7439⟩; **2**, 309 ⟨90TL3417⟩; **4**, 61, 79, 780, 782 ⟨90TL5449⟩, 119, 734 ⟨90TL5571⟩, 812 ⟨91JA7439⟩; **6**, 266 ⟨90TL5571⟩, 487 ⟨91JMC569⟩
Stemke, J. E., **1**, 630, 632, 633, 1091, 1094 ⟨78JOC147⟩, 633 ⟨75TL1815⟩, 633, 1094 ⟨76TL2947⟩; **2**, 911 ⟨78JOC147⟩
Stemmle, B., **3**, 616 ⟨71CB3594⟩
Stempfle, W., **4**, 583 ⟨72IJ293⟩
Stenberg, V. I., **4**, 177 ⟨74JOC2815⟩; **5**, 122 ⟨71JOC2548⟩; **6**, 573 ⟨80JHC1369⟩
Stendel, R., **2**, 100 ⟨90ZN(B)557⟩, 460 ⟨83ZAAC(504)147⟩
Stendel, W., **4**, 94 ⟨88GEP3639877A1⟩
Stening, G. B., **6**, 512 ⟨79EUP3008⟩
Stensen, W., **1**, 498 ⟨88ACS(B)259⟩
Stenson, J. P., **6**, 721 ⟨69CC423⟩
Stenstrom, Y., **5**, 937 ⟨86OM178⟩
Stenzel, J., **5**, 1074 ⟨67ZN(B)785⟩, 1076, 1077 ⟨67CB3368⟩
Step, G., **1**, 664 ⟨76JCS(P1)234⟩
Stepanek, A. S., **5**, 988 ⟨61ZOB3062⟩
Stepaniak, R. F., **5**, 802 ⟨76JOM(112)209⟩; **6**, 719 ⟨76JOM(112)227⟩
Stepanian, M., **1**, 769 ⟨88JOC1903⟩
Stepanov, A. A., **2**, 471 ⟨78JGU207⟩
Stepanov, A. S., **6**, 223, 228 ⟨66ZOB1355⟩
Stepanov, B. I., **2**, 825 ⟨69ZOB1544, 74ZOB760⟩
Stepanov, G. S., **2**, 259 ⟨90ZOB476⟩; **5**, 1135 ⟨90ZOB476⟩; **6**, 121 ⟨92HAC115⟩
Stepanova, T. Y., **6**, 658 ⟨82IZV1426⟩
Stepanova, Yu. Z., **6**, 685, 686, 689 ⟨87ZOB901⟩
Stepan'yan, A. E., **6**, 223, 226, 227, 228 ⟨72MI 606-01⟩
Stepanyants, A. U., **6**, 233, 234, 237 ⟨65ZOB1628⟩
Stephan, D., **4**, 768 ⟨89MI 417-01⟩
Stephan, J.-P., **3**, 232 ⟨58BSF1458⟩
Stephan, M., **2**, 437, 444 ⟨90TL6357⟩
Stéphan, M., **4**, 557, 570, 575 ⟨93PS(77)199⟩
Stephan, W., **1**, 664 ⟨77AG(E)44a, 80LA171⟩; **6**, 143 ⟨83CB2068⟩, 598 ⟨87CB1581⟩
Stephen, A., **4**, 728 ⟨66M695⟩
Stephen, A. M., **1**, 41 ⟨55JCS2044⟩
Stephen, H., **3**, 328 ⟨25JOC1874⟩
Stephens, C. R., **3**, 618 ⟨55JA1701⟩
Stephens, C. W., **6**, 241 ⟨81JCS(P1)455⟩
Stephens, F. S., **1**, 53 ⟨88JA4346⟩
Stephens, J. A., **2**, 202 ⟨54JA5011⟩
Stephens, R., **2**, 194 ⟨63JCS3692⟩; **5**, 34, 63, 100 ⟨61JCE217⟩
Stephens, R. D., **1**, 457, 515, 1004, 1005 ⟨64JA4358⟩, 516 ⟨66JA4964⟩, 1004, 1142 ⟨63JOC3313⟩; **3**, 621 ⟨66JA4964⟩
Stephens, R. W., **5**, 1126 ⟨91SC2025⟩
Stephens, T. B., **4**, 66, 67, 68, 69 ⟨69JOC31⟩
Stephens, V. C., **4**, 728 ⟨52JA3868⟩
Stephenson, A., **3**, 547 ⟨33CB237⟩; **6**, 253, 254 ⟨33CB567⟩
Stephenson, B., **2**, 9 ⟨72JA4184⟩
Stephenson, D. S., **2**, 155 ⟨85CB116⟩; **6**, 253, 254, 530 ⟨83CB1623⟩
Stephenson, E. F. M., **2**, 27 ⟨63OSC(4)984⟩
Stephenson, L., **2**, 19 ⟨88JCS(P1)549⟩
Stephenson, R. A., **1**, 1024 ⟨75CC287⟩
Stephenson, T. A., **5**, 229 ⟨64CI(L)193⟩
Stepina, E. M., **4**, 660 ⟨72ZOB631, 75ZOB2448⟩, 660 ⟨76ZOB1043, 77ZOB1751⟩, 662 ⟨91MI 414-01⟩
Steppan, H., **2**, 764 ⟨57LA(606)24⟩; **3**, 470 ⟨58LA(617)20⟩
Steppan, W., **2**, 539 ⟨82S1⟩; **6**, 43 ⟨82S1⟩
Steppel, R. N., **4**, 50 ⟨70JA5664⟩
Sterba, V., **2**, 805 ⟨72CCC3467, 73CCC706⟩; **6**, 558, 565 ⟨84CCC1577⟩
Stercho, Y. P., **2**, 422 ⟨84JOC2438⟩; **3**, 439 ⟨87JOC702⟩
Sterk, H., **1**, 889 ⟨68TL4309⟩; **3**, 527 ⟨68M1958⟩; **5**, 340 ⟨69M959⟩
Sterk, L., **2**, 374 ⟨68JPR257⟩
Sterlin, R. N., **4**, 802, 805 ⟨60BAU1851⟩, 805, 812, 813 ⟨61MI 418-01⟩, 894 ⟨74MI 420-01⟩
Sterlin, S. R., **4**, 152 ⟨73JOM(57)423⟩, 926 ⟨69CA(70)57950j⟩; **6**, 62 ⟨72GEP2101107, 72JOM(34)299⟩, 64 ⟨73JOM(57)423⟩, 537 ⟨81URP804634⟩
Sterling, J., **4**, 176 ⟨87CAR(162)145⟩
Sterling, J. J., **1**, 120 ⟨73JA7788⟩; **2**, 588 ⟨73JA7788⟩; **3**, 287 ⟨73JA7788⟩
Sterling, Jr., J. D., **6**, 220 ⟨B-56MI 614-01⟩, 226 ⟨56USP2729687⟩
Stermitz, F. R., **1**, 628 ⟨60JA4692⟩
Stern, A. G., **5**, 215 ⟨90SC2901⟩
Stern, D., **2**, 237, 238 ⟨92JA9737⟩; **6**, 118 ⟨13LA(396)1⟩
Stern, E. S., **1**, 104 ⟨46JCS396⟩
Stern, E. W., **5**, 974 ⟨66JOC596⟩
Sternbach, B., **2**, 110 ⟨61JA3384⟩
Sternbach, D. D., **5**, 363 ⟨75JOC3450⟩
Sternbach, L. H., **1**, 3, 4, 6 ⟨71CC465⟩; **4**, 412 ⟨74JOC167⟩; **5**, 354 ⟨65JOC1311⟩
Sternberg, E., **1**, 363 ⟨87CB1933⟩; **4**, 649 ⟨85TL301⟩
Sternberg, E. D., **1**, 541 ⟨84JOC1574⟩; **4**, 1061 ⟨84JOC1574⟩
Sternberg, H. W., **2**, 670 ⟨58JA1009⟩; **6**, 524 ⟨55JA3951⟩
Sternberg, J. A., **2**, 398 ⟨88JOC1251⟩; **3**, 190 ⟨88JOC1251⟩
Sternfeld, F., **4**, 236 ⟨86T4333⟩
Sternhell, S., **1**, 490 ⟨74TL3391⟩; **2**, 630 ⟨62JCS470⟩; **3**, 464 ⟨62JCS470⟩; **4**, 29 ⟨70AJC989⟩
Sternson, L. A., **2**, 649 ⟨84JOC4295⟩
Sterzycki, R., **3**, 9, 123 ⟨79S724⟩; **4**, 178, 179 ⟨79S724⟩
Stetin, C., **2**, 816 ⟨85JOC3863⟩; **3**, 407 ⟨82SC495⟩
Stetinova, J., **3**, 257 ⟨75CCC1750⟩
Stetter, H., **2**, 174 ⟨68AG970⟩, 174 ⟨69CB3357⟩, 390, 397, 403, 405 ⟨71CB917⟩; **3**, 292 ⟨76AG(E)639⟩, 621, 633 ⟨88OSC(6)866⟩; **4**, 166 ⟨77LA1807⟩, 192 ⟨70CB643⟩, 408 ⟨51CB834, 58CB598⟩; **5**, 29 ⟨74S36⟩, 48 ⟨55AG769⟩, 701, 827 ⟨69CB1641⟩; **6**, 78, 81 ⟨70CB639⟩, 79 ⟨70CB643⟩, 80 ⟨53CB790, 54CB205⟩, 80 ⟨73S207⟩, 140 ⟨73CB2523⟩
Stetter, W., **3**, 211 ⟨77CB1007⟩
Steudel, H.-P., **5**, 244 ⟨69LA(730)121⟩; **6**, 435, 546 ⟨81LA1244⟩
Steudel, O.-W., **1**, 665, 666 ⟨60LA(629)53⟩; **4**, 699 ⟨60LA(629)53⟩
Steudel, R., **6**, 529, 530 ⟨67AG649⟩, 529, 530 ⟨68ZAAC(361)180⟩
Steuerle, U., **1**, 634 ⟨87TL6159⟩
Stevenart-De Mesmaeker, N., **1**, 1131 ⟨87TL2637, 87TL2887⟩
Stevens, A. E., **1**, 93 ⟨78JA2405⟩, 1143 ⟨83CPL124⟩
Stevens, C., **2**, 22 ⟨92BSB237⟩; **3**, 54 ⟨90BSB41⟩, 416 ⟨93S89⟩; **5**, 697 ⟨90T6753⟩
Stevens, C. L., **2**, 3 ⟨48JA165⟩, 29 ⟨53JOC1112⟩; **3**, 108 ⟨70JOC841⟩, 556, 557 ⟨64JOC34⟩, 558, 560, 561 ⟨54JA4398, 53JA657⟩, 561 ⟨65JOC3718⟩; **4**, 12 ⟨55JA3278⟩, 28, 38 ⟨56JA2264⟩, 183 ⟨58JA4072⟩, 827 ⟨46JA1922⟩; **5**, 770, 849 ⟨65JOC3718⟩, 1078 ⟨67JOC2895⟩; **6**, 70, 80 ⟨46JA1922⟩, 74 ⟨51JA3807⟩
Stevens, C. S., **2**, 312 ⟨91T3407⟩
Stevens, C. V., **1**, 152 ⟨93TL4693⟩
Stevens, D. R., **2**, 684 ⟨55JOC1232⟩
Stevens, E. D., **4**, 995 ⟨89AX(C)1751⟩
Stevens, E. P., **1**, 371 ⟨94TL2641⟩
Stevens, G., **2**, 89 ⟨69JCS(B)749⟩, 402 ⟨78JCS(P1)1580⟩
Stevens, H. C., **5**, 815 ⟨52JA4876⟩
Stevens, H. P., **2**, 120 ⟨02JCS79⟩; **5**, 589, 595, 620 ⟨49JOC962⟩
Stevens, I. D. R., **6**, 280 ⟨79JCS(P2)1298⟩
Stevens, J. D. R., **2**, 374 ⟨88OSC(6)936⟩
Stevens, K. E., **2**, 911 ⟨84CJC2415⟩; **3**, 634 ⟨82CJC825⟩
Stevens, M. F. G., **5**, 770 ⟨84JCS(P1)993⟩

Stevens, N. J., **2**, 112 ⟨68MI 202-02⟩
Stevens, R. C., **4**, 697, 698 ⟨88JA3231⟩
Stevens, R. E., **5**, 117 ⟨73JOC1241⟩, 735 ⟨67JOC392⟩
Stevens, R. R., **2**, 524 ⟨69IC255⟩; **4**, 522 ⟨67TL995⟩
Stevens, R. V., **1**, 91 ⟨68JOC747⟩, 624, 626 ⟨86JA1039⟩, 776 ⟨86JOC3023⟩, 1134 ⟨78CC754⟩; **2**, 91 ⟨83JA7719⟩; **3**, 119 ⟨80JOC2030, 82TL4647⟩, 219 ⟨76TL3799, 90JOC2237⟩; **5**, 278 ⟨86JA1039⟩, 821 ⟨76T1599⟩; **6**, 327 ⟨83JA7719⟩
Stevens, R. W., **2**, 701 ⟨82CL353, 84T1381⟩
Stevens, T. B., **4**, 66 ⟨66CI(L)1555⟩
Stevens, T. E., **2**, 386 ⟨58JOC311⟩, 403 ⟨61JOC2531⟩, 763 ⟨68JOC2664⟩; **3**, 620 ⟨61JOC2531⟩; **4**, 428 ⟨59JOC865⟩, 439 ⟨72JOC922⟩; **5**, 314 ⟨69JA4778⟩; **6**, 55 ⟨67JA182⟩
Stevens, T. S., **1**, 39 ⟨36JCS584⟩, 39, 630, 631 ⟨52JCS4735⟩, 377 ⟨B-73MI 109-01⟩, 397 ⟨60JCS3521⟩, 398 ⟨28JCS3193, 30JCS2107⟩, 398 ⟨32JCS1926, 32JCS55⟩, 398 ⟨63JCS3397⟩, 399 ⟨32JCS69⟩, 407, 628 ⟨32JCS1932⟩, 630 ⟨67JCS(C)1964⟩; **2**, 379 ⟨70JCS(C)1088⟩; **3**, 407 ⟨B-73MI 310-01⟩, 435 ⟨B-73MI 311-01⟩, 446 ⟨61JCS1743⟩, 446, 464 ⟨52JCS4735⟩
Stevens, W., **4**, 83 ⟨48RTC451⟩; **6**, 79 ⟨66RTC793⟩, 79, 95 ⟨69RTC897⟩
Stevens, W. R., **3**, 614 ⟨B-85MI 303-01⟩
Stevenson, D. F. M., **4**, 56 ⟨85JCS(P1)1567⟩
Stevenson, G. W., **3**, 620 ⟨61JBC(236)715⟩
Stevenson, H. B., **4**, 102, 313 ⟨63JOC1825⟩; **5**, 986 ⟨63JOC1825⟩
Stevenson, J. K., **6**, 16 ⟨41JA3478⟩
Stevenson, J. M., **6**, 25 ⟨86TL3845⟩
Stevenson, P., **1**, 468 ⟨85CC971⟩, 1015 ⟨88TL5565⟩; **6**, 7 ⟨87JCS(P1)1515⟩
Stevenson, P. J., **1**, 823 ⟨88TL3993⟩, 828 ⟨91TL4199⟩; **3**, 414 ⟨83S1009⟩
Stevenson, R., **1**, 383 ⟨56JCS2419⟩, 1136 ⟨77JCS(P1)90⟩; **4**, 827 ⟨88H(27)1929⟩
Stevenson, R. W., **1**, 34 ⟨59JA4850⟩
Stevenson, T. A., **5**, 205 ⟨88TL2033⟩
Steward, O. W., **6**, 174, 390, 392 ⟨73JOM(55)209⟩
Stewart, A., **2**, 222 ⟨80CC100⟩
Stewart, A. O., **2**, 351 ⟨89JOC1221⟩; **5**, 214 ⟨88JMC1960⟩
Stewart, A. P., **5**, 403 ⟨70JCS(C)1263⟩
Stewart, Jr., A. T., **4**, 300, 405 ⟨55JA1098⟩
Stewart, A. W., **3**, 535 ⟨07NAT510, 07PCS229⟩
Stewart, C. A., **6**, 165, 167 ⟨82OM1720⟩, 167 ⟨84TL3519⟩, 689 ⟨84JA7015⟩
Stewart, D., **3**, 242 ⟨75CC772⟩; **6**, 118 ⟨77JCS(P1)1898, 77TL1065⟩
Stewart, E. L., **4**, 193 ⟨79JOC4825⟩
Stewart, E. T., **6**, 25 ⟨44JCS74⟩
Stewart, F. H. C., **2**, 26 ⟨54JCS2819⟩, 445 ⟨54JCS2832⟩
Stewart, I. M., **3**, 397 ⟨69AJC891⟩
Stewart, J. A. G., **2**, 813 ⟨78JOM(152)39⟩
Stewart, J. M., **2**, 128, 136 ⟨47JA2675⟩, 129, 136 ⟨47JA2672⟩, 167 ⟨53JA243⟩, 179, 222 ⟨52JA5880⟩
Stewart, K. T., **3**, 280 ⟨89JHC1547⟩
Stewart, L. J., **2**, 34, 628 ⟨87TL4497⟩, 34, 629, 631 ⟨88JOC4477⟩
Stewart, O. J., **4**, 806 ⟨68JOC472⟩, 817 ⟨68JOC286⟩
Stewart, R. A., **1**, 881 ⟨50JA3704⟩
Stewart, R., **1**, 871 ⟨67JCS(B)1173⟩; **5**, 28 ⟨B-64MI 502-01⟩
Stewart, R. C., **5**, 26 ⟨73TL1569⟩
Stewart, R. F., **2**, 657 ⟨78JOC3078⟩
Stewart, S. K., **1**, 489 ⟨93TL3599⟩
Stewart, T. D., **4**, 300 ⟨32JA4172⟩
Stewen, U., **1**, 922 ⟨89JA5845⟩
Steyermark, P. R., **5**, 994 ⟨64USP3155700⟩, 994, 1016 ⟨63JOC586⟩
Steyn, P. S., **4**, 56 ⟨65JCS3075⟩; **6**, 138 ⟨76T2625⟩
Stezowski, J. J., **1**, 1088 ⟨85CB1504⟩; **2**, 1006, 1009 ⟨87JA882⟩, 1008 ⟨85JA2067, 88JOC4379⟩; **4**, 179 ⟨79JA2171⟩
Stgeorgiev, V., **3**, 434 ⟨90HCA169⟩
Stibbard, J. H. A., **3**, 38 ⟨82TL2407⟩
Stibilj, V., **5**, 834 ⟨86S807⟩, 845 ⟨86H(24)1899⟩
Stibr, B., **4**, 629 ⟨92CRV175⟩; **6**, 406 ⟨92CRV225⟩
Štibrányi, L., **5**, 350 ⟨86CCC2167⟩
Stichter, H., **1**, 194 ⟨88TL3579⟩; **2**, 324 ⟨88TL3579⟩
Stick, R. V., **2**, 8 ⟨76JCS(P1)2112⟩, 747 ⟨75JCS(P1)1237⟩; **5**, 289 ⟨75JCS(P1)1237⟩, 516, 522, 526 ⟨76JCS(P1)2112⟩; **6**, 309 ⟨76JCS(P1)2112⟩, 477 ⟨87AJC795⟩, 544 ⟨82AJC1709⟩, 589, 593 ⟨88AJC549⟩, 612 ⟨88AJC563⟩
Stiebitz, B., **6**, 491 ⟨86ZAAC(542)37⟩
Stieglitz, J., **5**, 698 ⟨1896JA751, 01CB1615⟩, 989 ⟨1896MI 526-01⟩; **6**, 452 ⟨1895MI 614-01⟩
Stiehm, T., **1**, 990 ⟨90TL1841⟩
Stier, M., **5**, 327 ⟨89OS(67)69⟩
Stiles, A. R., **2**, 439 ⟨52JA3282⟩, 451 ⟨53USP2659714⟩
Stiles, M., **1**, 385 ⟨59JA1497⟩, 906 ⟨57JOC1243⟩; **5**, 70, 95 ⟨63JA616⟩
Stilke, R., **2**, 943 ⟨74CB3717⟩
Stilkerieg, B., **3**, 365 ⟨77ACS(B)679⟩
Still, I. W. J., **2**, 154 ⟨79SC923⟩, 229 ⟨75CJC455⟩, 713, 714 ⟨91COS(8)403⟩; **3**, 365 ⟨81S295⟩, 368 ⟨92CJC964⟩; **4**, 70 ⟨81JOC4911⟩, 330 ⟨84CJC586⟩, 704 ⟨90CJC1408⟩
Still, J. K., **1**, 1030 ⟨88JA1557⟩
Still, W. C., **1**, 68 ⟨86JA2094⟩, 136 ⟨74JA5561⟩, 138 ⟨78JA1481⟩, 139 ⟨80JA1201⟩, 217 ⟨81JOC5242⟩, 269, 738 ⟨81JA1283⟩, 405, 406, 816, 817, 820 ⟨78JA1927⟩, 548 ⟨79JA7107⟩, 584 ⟨92JA4128⟩, 751, 752 ⟨83TL4405⟩, 813 ⟨77JA4186, 79JA2493⟩, 813 ⟨83JA625⟩; **2**, 40 ⟨86JA2094⟩, 68 ⟨80TL1031, 80TL1035⟩, 546 ⟨76JOC3063⟩, 600 ⟨77JA4836⟩; **3**, 60 ⟨84JA260⟩, 196 ⟨81JA1283⟩, 203, 204 ⟨78TL2659⟩, 217 ⟨77JA4186⟩, 231 ⟨84JA1148⟩; **4**, 165 ⟨84JA1148⟩, 199, 378, 379, 383, 386, 387 ⟨78JA1481⟩, 355 ⟨76JOC3063⟩, 378 ⟨78JA1927, 79TL593⟩, 378, 379 ⟨80JA1201⟩, 386 ⟨83JA625⟩
Stille, C., **5**, 551 ⟨88BBA269⟩
Stille, J. K., **1**, 94 ⟨81JA5096⟩, 228, 479 ⟨89T979⟩, 296 ⟨91COS(4)913⟩, 298 ⟨76JA1806, 76JA1810⟩, 435, 480, 1027 ⟨86JA3033⟩, 468 ⟨87JA813⟩, 468, 1006, 1007 ⟨87JA2138⟩, 468, 1045, 1096 ⟨86AG(E)508⟩, 471 ⟨77ACR434, 87JOC422⟩, 471, 473, 1145 ⟨79JA4992⟩, 480 ⟨90OS(68)116⟩, 482, 1027 ⟨87JA5478⟩, 494, 1029 ⟨85JOC2302⟩, 670 ⟨74JA1518⟩, 1015 ⟨84JA6417, 84JA4630⟩, 1027 ⟨89JA5417⟩, 1029 ⟨84JA7500⟩, 1092 ⟨82JOC3630⟩; **2**, 86 ⟨88JA4039, 89T979⟩, 582, 960 ⟨77ACR434⟩, 583 ⟨91OM1183⟩, 689 ⟨84JA4630⟩, 689, 1092 ⟨86AG(E)508⟩, 931 ⟨84JA7500, 86JA3033⟩, 931 ⟨87JA5478⟩, 932 ⟨87JA813⟩; **3**, 3 ⟨87JA7122, 91COMC-I(4)913⟩, 23, 89, 94, 95, 97, 103 ⟨86JA452⟩, 58, 89 ⟨83JA7175⟩, 89, 288, 296, 300 ⟨86AG(E)508⟩, 211, 288 ⟨83JOC4634⟩, 218, 303 ⟨84JA4833⟩, 243 ⟨90JOC3114⟩, 271 ⟨79JOC1613, 84JA7500⟩, 288, 296 ⟨78JA3636⟩, 290, 295, 296 ⟨88JA1557⟩, 541 ⟨64JA4871⟩; **5**, 24 ⟨72MI 502-02⟩, 142 ⟨75JOC335⟩, 143 ⟨76JA5832⟩, 143 ⟨79TL133⟩, 143, 170 ⟨80JA4193⟩, 290 ⟨80JOC2139⟩, 458 ⟨74JA1518⟩, 463 ⟨74JA4983⟩, 652 ⟨75JOC532⟩, 842 ⟨66JHC155⟩
Stille, J. R., **1**, 769 ⟨83PAC1733⟩, 835 ⟨92JOC7188, 93JOC5095⟩; **2**, 702 ⟨83JA1664⟩; **3**, 508 ⟨83PAC1733⟩; **5**, 572 ⟨94TL1669⟩
Stiller, E. T., **2**, 138 ⟨48JA3667⟩
Stiller, K., **5**, 759 ⟨68TL3277⟩
Stillings, M. R., **5**, 639 ⟨86JMC2280⟩
Stilze, W., **6**, 107 ⟨64GEP1161285⟩
Štimac, A., **5**, 836 ⟨82JHC577, 84H(22)1545⟩
Stimson, R. E., **5**, 445, 447 ⟨80JA5093⟩
Sting, M., **2**, 459 ⟨90S132⟩
Stinn, D. E., **3**, 130, 286 ⟨89SC1833⟩

Stirling, C., **1**, 144 ⟨B-88MI 104-02⟩, 839 ⟨B-88MI 118-01⟩; **3**, 170 ⟨B-88MI 304-04⟩; **4**, 390 ⟨B-88MI 408-01⟩

Stirling, C. J. M., **1**, 145 ⟨93JCS(P2)1617⟩, 145 ⟨93JCS(P2)2163⟩, 186 ⟨92T7383⟩, 612 ⟨75CC940⟩, 704 ⟨B-88MI 115-01, B-88MI 115-02⟩, 704 ⟨B-88MI 115-03⟩, 707 ⟨B-88MI 115-04⟩, 993 ⟨71JCS(C)1530⟩; **2**, 142 ⟨B-88MI 203-01⟩, 144 ⟨88MI 202-01⟩, 154 ⟨B-88MI 203-03⟩, 159 ⟨67CC131⟩, 159, 196 ⟨B-88MI 203-04⟩, 172, 796, 797 ⟨71IJS(B)277⟩, 721 ⟨63JCS5741⟩, 742 ⟨64JCS5863, 66JCS217⟩; **3**, 267 ⟨66JCS(B)1217⟩, 714 ⟨86JA3153, 89CC1231⟩; **4**, 207 ⟨64JCS1217⟩, 218, 235 ⟨62JCS3686⟩, 330 ⟨B-81MI 407-01⟩, 798 ⟨86JCS(P1)1171⟩, 861 ⟨87JCS(P2)1253⟩

Stirling, I., **3**, 72 ⟨77CC808⟩

Stirnweis, V., **2**, 569 ⟨79ZAAC(458)89⟩

Stirton, A. J., **2**, 190 ⟨52JA2526⟩, 190 ⟨63OSC(4)8623⟩

Stivala, S. S., **6**, 504

Stiver, S., **3**, 426 ⟨83CC50⟩

Stobart, S. R., **2**, 439 ⟨83JCS(P1)861⟩; **4**, 141 ⟨74JCS(D)2537⟩, 566 ⟨83JCS(P1)861⟩, 685 ⟨86CC285, 89CC498⟩

Stobie, A., **3**, 8, 121 ⟨81CC1232⟩, 132 ⟨82TL957⟩

Stobie, I., **1**, 782 ⟨82JCS(P1)1845⟩

Stocco, G. C., **5**, 1141 ⟨72MI 530-01⟩

Stocco Gattuso, F., **5**, 1141 ⟨72MI 530-01⟩

Stock, A., **2**, 811 ⟨26CB2215⟩

Stock, A. M., **2**, 192 ⟨60JOC1901⟩

Stock, L. M., **2**, 608 ⟨61JA4605⟩; **5**, 123, 126 ⟨65JOC927⟩; **6**, 504 ⟨80JOC886⟩

Stockel, K., **1**, 1034 ⟨74OS(54)1⟩

Stockel, R. F., **2**, 783 ⟨66JCE144⟩

Stocken, L. A., **2**, 119, 259, 260, 889 ⟨47JCS592⟩

Stocker, J. H., **1**, 6 ⟨68CC934⟩, 88 ⟨62JOC2288⟩; **4**, 177, 179 ⟨55JOC1695⟩

Stöcklin, P., **2**, 720 ⟨24JCS1463⟩

Stockmayer, W. H., **5**, 207 ⟨50JA5426⟩

Stockton, J. D., **3**, 527 ⟨71JOC1486⟩

Stoddard, G. J., **1**, 807 ⟨92JA2635⟩

Stoddart, J. F., **1**, 571 ⟨92JA6330⟩

Stodola, F. H., **5**, 33 ⟨55OSC(3)213⟩

Stoeckli-Evans, H., **5**, 441 ⟨77JCS(D)2297⟩

Stoelwinder, J., **3**, 699, 713, 718 ⟨92JOC2249⟩, 714, 716 ⟨90S568⟩

Stoessel, S. J., **1**, 494, 1029 ⟨85JOC2302⟩

Stoetzel, R., **4**, 647 ⟨84JOM(269)C40⟩, 650 ⟨84JOM(273)C57⟩

Stoffel, P. J., **5**, 1027 ⟨57JA1236⟩; **6**, 509 ⟨62JOC3079⟩

Stoffelsma, J., **4**, 844, 846 ⟨59RTC354⟩; **6**, 83, 84 ⟨59RTC354⟩

Stoffyn, A., **5**, 285 ⟨72CAR(23)251⟩

Stoffyn, P., **5**, 285 ⟨72CAR(23)251⟩

Stofko Jr., J. J., **1**, 411, 811 ⟨73CC319⟩

Stöhr, H., **2**, 769 ⟨53CB1073⟩

Stohrer, W.-D., **2**, 1006, 1009 ⟨87JA882⟩, 1009 ⟨88JOC4379⟩; **6**, 252 ⟨92CC1017⟩

Stoicheva, D. S., **6**, 502 ⟨73GEP1694210⟩

Stojanac, N., **5**, 630 ⟨66C175⟩

Stojanac, Ž., **5**, 566, 618 ⟨57CCA319⟩

Stojda, R. J., **4**, 179 ⟨79JA2171⟩

Stojiljković, A., **3**, 464 ⟨70T1101⟩, 620 ⟨65TL461⟩

Stoka, A. M., **6**, 461, 476 ⟨88ZN(B)1038⟩, 477 ⟨89ZN(B)983⟩

Stokes, C. S., **6**, 32 ⟨52JA1974, 52JA848⟩

Stokes, J. B., **6**, 647 ⟨72IJS(A)105⟩

Stokstad, E. L. R., **2**, 214 ⟨54JA1828⟩

Stöldt, E., **2**, 388 ⟨78AG(E)203⟩

Stoll, A. T., **1**, 464 ⟨88JA5383⟩, 1045, 1047 ⟨83TL5121⟩, 1094 ⟨85TL5671⟩; **3**, 130, 288 ⟨83TL5181⟩, 271 ⟨86TL775⟩; **4**, 809 ⟨88JA5383⟩

Stoll, K., **2**, 471 ⟨83ZAAC(504)47⟩

Stoll, O., **2**, 99 ⟨38CB1873⟩

Stoll, T., **2**, 157 ⟨81GEP2926671⟩

Stolle, A., **1**, 975 ⟨92JA4051⟩; **2**, 309 ⟨93TL4193⟩

Stollé, R., **3**, 328 ⟨04CB3023⟩

Stolle, W. T., **1**, 630 ⟨77TL135⟩; **2**, 678 ⟨79JA249⟩

Stoller, C., **3**, 694 ⟨89OPP451⟩

Stolte, H., **5**, 1056 ⟨1886CB2350⟩

Stoltefuss, J., **6**, 360 ⟨76GEP2460330⟩

Stolyarova, N. E., **5**, 390 ⟨77ZOB2294, 80ZOB886⟩, 390 ⟨81ZOB1318⟩, 391 ⟨87ZOB1643⟩

Stolyarov, V. P., **2**, 165 ⟨89IZV197⟩; **5**, 698 ⟨82IZV937⟩, 698, 762, 813 ⟨78IZV2802⟩, 762 ⟨73IZV2159⟩

Stolz, K., **3**, 560 ⟨86CB3411⟩

Stolz-Dunn, S. K., **1**, 230 ⟨88TL5069, 89TL3487⟩, 230 ⟨90SL341⟩; **3**, 25 ⟨89TL6629⟩

Stolzenberg, H., **3**, 726 ⟨82CC184⟩; **5**, 1089 ⟨82CC184⟩, 1147 ⟨89ZN(B)419⟩; **6**, 675 ⟨82COMC-I(4)523⟩

Stölzer, C., **2**, 411, 412, 413 ⟨60CB1323⟩

Stone, C., **1**, 195 ⟨89TL5883⟩; **2**, 436 ⟨88JOC4425⟩, 992 ⟨86TL1537⟩; **6**, 461 ⟨91HCA465⟩

Stone, D. B., **2**, 420 ⟨67JOC510⟩

Stone, F., **6**, 266 ⟨90JOM(399)189⟩

Stone, F. G. A., **2**, 109 ⟨53JA228⟩, 444, 824, 882 ⟨57JA5884⟩, 481, 482, 509 ⟨54JA386⟩, 872 ⟨82COMC-I(2)⟩, 904 ⟨60JA6218, 60JA6238⟩; **3**, 552 ⟨70JCS(A)944⟩, 690 ⟨93AG(E)1480⟩; **4**, 712 ⟨81CC1255⟩; **5**, 447 ⟨75JOM(94)62⟩, 927 ⟨87JOM(325)185⟩, 937 ⟨82COMC-I⟩, 948, 949 ⟨73JCS(D)975⟩; **6**, 65, 244 ⟨60JA6228⟩, 181, 269 ⟨69JCS(A)1882⟩, 243, 244 ⟨64AOC(1)143⟩, 247 ⟨61JA3593⟩, 265, 269 ⟨70JCS(A)31⟩, 340 ⟨82CC51⟩, 343 ⟨82CC773, 83JCS(D)2599⟩, 343 ⟨83JOM(249)273⟩, 403 ⟨88JOM(355)231⟩, 719 ⟨74JCS(D)1189⟩, 719, 732 ⟨74JCS(D)351⟩, 721 ⟨74JCS(D)357⟩, 721, 732 ⟨75JCS(D)939⟩

Stone, G. B., **1**, 836 ⟨90JOC4614⟩

Stone, G. R., **2**, 1049 ⟨58JOC1389⟩

Stone, H., **2**, 34 ⟨63OSC(4)543⟩, 36 ⟨63OSC(4)321, 63OSC(4)323⟩

Stone, H. W., **1**, 869, 871 ⟨69JA2797⟩

Stone, K. J., **1**, 359 ⟨85JA2495⟩, 527, 944 ⟨85JOC2400⟩

Stone, R. G., **1**, 88 ⟨58JA5270⟩

Stone, W. E., **2**, 632 ⟨83JOC1542⟩

Stonebraker, P. M., **3**, 578 ⟨76JA6036⟩

Stonemark, F. E., **5**, 273, 285 ⟨73JOC1677⟩

Stoner, E. J., **1**, 601, 781, 785 ⟨93JOC322⟩

Stonner, F. W., **5**, 205 ⟨70JMC948⟩

Stoodley, R. J., **2**, 197 ⟨92JCS(P1)2371⟩; **3**, 245 ⟨90JCS(P1)1339⟩; **4**, 274 ⟨90JCS(P1)773⟩, 281 ⟨92JCS(P1)2371⟩, 448 ⟨84CC1289⟩

Stopp, G., **1**, 415 ⟨61LA(641)1⟩; **2**, 788 ⟨61LA(641)1⟩; **3**, 720 ⟨61LA(641)1⟩; **4**, 882, 974 ⟨61LA(641)1⟩; **5**, 852 ⟨61LA(641)1⟩; **6**, 104, 105, 313, 314, 318, 321, 730, 731 ⟨61LA(641)1⟩

Stoppioni, P., **6**, 552 ⟨93POL13⟩

Storace, L., **2**, 778 ⟨80TL2025⟩

Storbeck, O., **2**, 386 ⟨1893CB310⟩

Storch, W., **2**, 467 ⟨85CB2030⟩, 520 ⟨71MI 210-01⟩, 811 ⟨77CB2783⟩; **6**, 405 ⟨81JMR(44)54⟩

Storer, A. C., **5**, 558 ⟨82CJC199⟩

Storhoff, B. N., **2**, 439 ⟨85IC2344, 90JOM(393)143⟩, 824 ⟨81JOM(205)161⟩

Stork, G., **1**, 62 ⟨86JA6826⟩, 84 ⟨83JA1072⟩, 99 ⟨56JA4604⟩, 130 ⟨68JA4462, 68JA4464⟩, 130, 872 ⟨75PAC553⟩, 130, 889 ⟨65JA275⟩, 214 ⟨74JA5268, 74JA5270⟩, 226 ⟨90JA1661⟩, 307 ⟨63JA207, 73JA6152⟩, 307 ⟨74JA6181, 78TL2445⟩, 309 ⟨74JA5272⟩, 320, 322, 1127, 1129, 1130 ⟨82JA2321⟩, 330, 341, 923, 936 ⟨86JA303⟩, 338 ⟨88JA6911⟩, 338, 1127, 1129 ⟨86TL4529⟩, 342 ⟨83JA3741⟩, 392 ⟨60JA4307⟩, 482 ⟨73JOC1775⟩, 525 ⟨76JA6747⟩, 548 ⟨79JA7107⟩, 548, 1127, 1131, 1132 ⟨87JA2829⟩, 548, 1132 ⟨83JA3720⟩, 828 ⟨76JA1583⟩, 919 ⟨88BCJ149⟩,

934 ⟨63JA3419⟩, 1039 ⟨83TL4887⟩, 1129 ⟨85TL5927⟩; **2**, 46 ⟨83TL3951⟩, 539, 696 ⟨68JA4462⟩, 540 ⟨85JA500, 86JA6826⟩, 630 ⟨65JA275, 89TL2173⟩, 738, 758 ⟨63JA207⟩, 759, 965 ⟨86JA304⟩, 921 ⟨71JA2080, 74JA3684⟩, 931 ⟨73JA6152⟩; **3**, 10, 124 ⟨54JA2029⟩, 18 ⟨71JA2080⟩, 20, 55 ⟨78TL5175⟩, 21 ⟨82TL2073⟩, 59 ⟨53JA384, 63JA2178⟩, 129 ⟨88TL2283⟩, 134 ⟨74JA3683⟩, 210 ⟨87TL2087⟩, 241 ⟨82JA3767⟩, 275 ⟨64JA471⟩, 495 ⟨89TL2173⟩, 638 ⟨79TL771⟩; **4**, 175 ⟨83JA3720⟩, 197 ⟨89TL287⟩, 199 ⟨75JA6260⟩, 357 ⟨75PAC553⟩; **5**, 152 ⟨87TL2087⟩, 159 ⟨76JA1583⟩, 173 ⟨82JA4686⟩, 173 ⟨87JA1564⟩, 173 ⟨87JA1565⟩, 896 ⟨89TL2173⟩

Storm, D. L., **3**, 322 ⟨67JA5497, 67TL1865⟩

Storr, A., **2**, 421 ⟨66JCS(A)706, 68JCS(A)2605⟩, 421 ⟨71JCS(A)1237, 72JCS(D)326⟩

Storr, R. C., **2**, 777 ⟨75JCS(P1)26⟩; **3**, 579 ⟨90JCS(P1)485⟩

Story, P. R., **1**, 904, 917 ⟨B-72MI 119-11⟩

Story, P. W., **3**, 326 ⟨65JA737⟩

Stothers, J. B., **1**, 886 ⟨B-80MI 119-03⟩; **5**, 55, 57, 59, 62 ⟨56JA2255⟩

Stotter, P. L., **2**, 3 ⟨72TL4067⟩, 28, 29 ⟨73JOC2576⟩; **3**, 62 ⟨73TL2417⟩

Stotz, D. S., **1**, 130, 869, 871, 881 ⟨68JOC2222⟩

Stötzel, R., **1**, 229 ⟨87CB1695⟩; **4**, 620, 646, 653 ⟨87CB1695⟩

Stout, C. A., **3**, 541 ⟨66JA1242⟩

Stout, T. J., **3**, 92 ⟨91JOC1692⟩, 493 ⟨90JA7001⟩

Stove, E. R., **5**, 9 ⟨63JCS491⟩, 219 ⟨71JCS(C)1917⟩

Stöver, H. D. H., **2**, 560 ⟨87JOC4644⟩

Stovnov, N. M., **2**, 683 ⟨93H(36)13⟩

Stowasser, B., **1**, 982 ⟨86AG(E)466⟩

Stowell, J. C., **1**, 844 ⟨B-79MI 119-01⟩; **2**, 373, 402, 404 ⟨B-75MI 207-01⟩, 377, 403 ⟨67JOC2360⟩, 378, 381 ⟨69JOC2254⟩, 390 ⟨71JOC3055⟩; **3**, 31, 161 ⟨84CRV409, 79S132⟩, 219 ⟨81JOC2428⟩; **5**, 586, 609 ⟨89JOC1212⟩, 665 ⟨80LA791⟩; **6**, 450, 509 ⟨69JOC2254⟩

Stowell, J. K., **4**, 769 ⟨94JA789⟩

Stowell, M. H. B., **4**, 563 ⟨89TL411⟩

Stowers, J. R., **4**, 905, 911 ⟨85CC845⟩

Stoyanovich, F. M., **6**, 83 ⟨60IZV1828⟩, 111 ⟨60MI 604-01, 70BSF2013⟩, 111, 140 ⟨65IZV2179⟩

Strachan, R. G., **5**, 972 ⟨66JA3163⟩

Strachan, W. S., **1**, 383 ⟨56JCS2419⟩

Strack, E., **3**, 612 ⟨34CB39⟩

Stracker, E. C., **1**, 449 ⟨89TL6487⟩; **2**, 630 ⟨91TL3329⟩

Stradi, R., **3**, 467 ⟨77T865⟩; **5**, 762 ⟨76TL1839, 78TL2175⟩, 762 ⟨87T4785⟩, 762, 816 ⟨78JCS(P1)309⟩

Strähle, J., **5**, 781 ⟨88ZN(B)1628⟩; **6**, 341 ⟨80AG411, 81AG94⟩

Straka, R., **5**, 263 ⟨77CCC560⟩

Strametz, H., **2**, 408, 415 ⟨63AG(E)218⟩

Strampach, N., **2**, 570, 573 ⟨82ZN(B)380⟩

Strand, L. B., **2**, 378, 379 ⟨71JMC1052⟩

Strande, P., **4**, 54, 57 ⟨89ACS74⟩

Strandtmann, M. von, **3**, 251 ⟨72JHC171⟩; **4**, 221 ⟨72JHC175⟩, 304 ⟨69JHC429⟩, 307 ⟨70JHC1311⟩; **5**, 596 ⟨71JOC1742⟩

Strang, H., **6**, 632 ⟨88GEP(O)3634929⟩

Strangeways, W. I., **2**, 890 ⟨32JCS2505⟩

Stranghoner, R., **2**, 749 ⟨76S741⟩

Stransky, W., **1**, 229 ⟨76CB3375⟩, 727 ⟨74S798, 76CB1694⟩; **3**, 494 ⟨76CB1694, 78CB248⟩

Strasser, M., **6**, 488 ⟨93ACS125⟩

Strating, J., **2**, 1034 ⟨54RTC565⟩; **3**, 368 ⟨69TL4461, 73JCS(P1)73⟩, 369 ⟨70QRS79⟩, 479 ⟨73TL5207⟩, 527 ⟨70RTC23⟩; **4**, 218, 235 ⟨64TL543⟩, 218, 235 ⟨67TL3057⟩, 234 ⟨64TL547, 67TL3061⟩, 235 ⟨70QRS67⟩, 331 ⟨72RTC209, 88OSC(6)981⟩, 331, 538 ⟨72TL2367⟩, 332 ⟨62RTC966⟩, 770, 776 ⟨53RTC813⟩, 777 ⟨68RTC49⟩; **5**, 51 ⟨72SC79⟩, 88 ⟨72SC423⟩, 488, 489 ⟨64RTC94, 67RTC641⟩, 490 ⟨70RTC687⟩, 562 ⟨67TL3453, 68TL2871⟩, 562 ⟨71RTC614⟩, 867 ⟨73TL5207⟩, 898, 901, 902 ⟨72RTC37, 74JOC2728⟩; **6**, 531 ⟨69TL4461⟩, 699 ⟨72RTC37, 74JOC2728⟩

Stratling, J., **2**, 126, 127 ⟨50RTC638⟩

Stratton, C. D., **2**, 741 ⟨73JMC1113⟩

Straub, H., **2**, 591 ⟨73HOU(13/2b)⟩, 1097 ⟨74HOU(13/2b)26, 74HOU(13/2b)9⟩; **6**, 172, 204, 207 ⟨74HOU(13/2b)1⟩

Straub, J. A., **3**, 48 ⟨91JOC3729⟩

Straub, M., **4**, 609 ⟨88JOM(341)109⟩; **6**, 391 ⟨88JOM(341)109⟩

Straughan, B. P., **5**, 1129 ⟨75JCS(D)2368⟩

Straus, D. A., **3**, 508 ⟨81JA7358⟩

Straus, D., **1**, 769 ⟨83PAC1733⟩; **3**, 508 ⟨83PAC1733⟩

Straus, F., **1**, 898 ⟨21JPR1⟩; **4**, 48 ⟨32LA(493)191⟩, 48, 49, 50 ⟨32LA(498)101⟩, 745 ⟨30CB1868⟩

Strauss, F., **2**, 1016 ⟨30CB1886⟩, 1016, 1017 ⟨30CB1863⟩

Strauss, H., **1**, 906 ⟨58HCA414⟩

Strauss, H. E., **2**, 30 ⟨52JCS2669⟩

Strauss, H. F., **1**, 828 ⟨91JCS(P1)561⟩

Strauss, H. W., **3**, 305 ⟨84JOC2288⟩

Strauss, K., **6**, 565 ⟨84JPR633⟩

Strauss, S., **5**, 567, 619, 625 ⟨74JHC397⟩

Strauss, S. H., **5**, 445, 447 ⟨80JA5093⟩

Strauss, W., **6**, 245 ⟨91JOM(407)1⟩

Strausz, O. P., **1**, 1143 ⟨73TCA183⟩; **3**, 335 ⟨83NJC269⟩, 337 ⟨65CJC717⟩, 353 ⟨80NJC703⟩, 488 ⟨68TL5237⟩, 527 ⟨70JA1766, 70JA1768⟩, 530 ⟨80PAC1623, 88JA1025⟩, 552 ⟨B-79MI 313-03, 82NJC401⟩; **5**, 874 ⟨68TL5237⟩

Strautman, A. K., **2**, 411 ⟨62ZOB3070⟩

Strautmanis, J. R., **4**, 704 ⟨90CJC1408⟩

Straw, T. A., **4**, 133 ⟨91CC234⟩; **6**, 659, 667 ⟨91CB1739⟩

Strawn, K. G., **2**, 397 ⟨68JOC313⟩

Strazewski, P., **1**, 873 ⟨86HCA1041⟩

Strazzolini, P., **2**, 326 ⟨91S447⟩; **3**, 405 ⟨88SC1501⟩

Strebelle, M., **5**, 751 ⟨78BSB717⟩

Strecker, A., **3**, 616 ⟨1850LA(75)27⟩; **5**, 164, 168 ⟨88JOC4098⟩

Strecker, B., **3**, 599 ⟨90AG(E)275, 91CB285⟩

Strecker, W., **2**, 101, 102 ⟨28LA(462)186⟩

Street, D. G., **1**, 869, 871 ⟨67JCS(B)827⟩

Street, L. J., **1**, 196, 410 ⟨93TL4385⟩; **3**, 324 ⟨82CC1056, 83CC666⟩, 324 ⟨83JCS(P1)1261, 86JCS(P1)1725⟩

Street, S. D. A., **1**, 160 ⟨92TL543⟩, 700, 703 ⟨87JCS(P1)2171⟩; **2**, 915 ⟨92SL886⟩; **4**, 651 ⟨92TL543⟩, 953 ⟨87JCS(P1)2189⟩, 962, 963, 964 ⟨92SL886⟩; **5**, 173 ⟨85CC1386⟩

Streeter, D. L., **1**, 886 ⟨89JOC1671⟩

Strege, P. E., **1**, 599 ⟨74JA7165⟩, 698 ⟨76TL3477⟩; **5**, 158 ⟨77JA1649⟩

Strehlow, T., **3**, 353 ⟨91CB1397⟩

Streinz, L., **3**, 620, 620 ⟨81CCC1682⟩

Streith, J., **3**, 536 ⟨83LA1393⟩, 617 ⟨76HCA2786⟩; **5**, 330 ⟨89TL7061, 90TL2139⟩, 330 ⟨91TA1209⟩, 330, 332 ⟨92HCA109⟩

Streith, S., **4**, 12 ⟨85T4057⟩

Streitwieser, Jr., A., **1**, 844, 848 ⟨65MI 119-06⟩, 847 ⟨67JA63, 69JA529⟩, 848 ⟨69JA527⟩, 854, 856, 857, 858 ⟨65MI 119-08⟩, 857 ⟨62JA258⟩, 859 ⟨79TL327⟩, 869 ⟨65JA384⟩, 892 ⟨58JA2326⟩, 910 ⟨B-61MI 119-02⟩; **5**, 436 ⟨87JA2553⟩

Strekowski, L., **1**, 763 ⟨84TL5603⟩; **4**, 245 ⟨93T2151⟩, 411 ⟨90JOC4777⟩

Strelenko, Y. N., **5**, 362 ⟨91MI 507-06⟩

Strelenko, Yu. A., **4**, 658 ⟨85ZOB1629⟩; **5**, 680 ⟨90BAU294, 90IZV352⟩, 680 ⟨91MI 517-01⟩

Strelitz, J. Z., **5**, 595 ⟨62JOC4412⟩

Strelkova, E. N., **4**, 489 ⟨82ZOB2444⟩

Strem, J., **5**, 84 ⟨61T176⟩

Stremler, K. E., **4**, 553, 562 ⟨87JA5544⟩; **6**, 264 ⟨86JOC4768, 87JA5542⟩
Strenkowski, L., **1**, 226 ⟨83S284⟩
Strenli, C. A., **2**, 840, 841, 854 ⟨67T1065⟩
Stres-Bratoš, J., **5**, 764 ⟨76JHC487⟩
Stretch, P. M., **4**, 148 ⟨86ICA(119)177⟩
Streubel, R., **2**, 471, 474 ⟨91CB765⟩; **5**, 955 ⟨94CC2457⟩; **6**, 166 ⟨90CB1245⟩
Strickland, D., **2**, 328 ⟨88TL5725⟩
Strickland, J. B., **3**, 531 ⟨88JOC4877⟩; **4**, 353 ⟨89CC89⟩
Strickland, J. H., **5**, 664 ⟨86SC763⟩
Strickland, R. C., **1**, 722 ⟨72JOC2579⟩
Strickler, P. D., **5**, 277 ⟨63JOC2915⟩
Strickmann, G., **4**, 979 ⟨67CB1661⟩; **5**, 516 ⟨68ZC335⟩
Striegler, H., **2**, 379 ⟨77JPR195⟩
Stright, P. L., **2**, 1030, 1031, 1034 ⟨56JA4783⟩
Strijtveen, B., **2**, 124 ⟨86JOC3664⟩
Strijtveen, H. J. M., **4**, 859 ⟨80RTC45⟩
Stringer, O. D., **2**, 150 ⟨82JA5412⟩, 251 ⟨83TL3191⟩, 287 ⟨83TL1213⟩; **4**, 312 ⟨93OSC(8)546⟩
Stringfellow, Jr., C. R., **6**, 305 ⟨62JOC4068⟩
Stringfellow, G. B., **2**, 879 ⟨91CM96⟩
Stringham, R. S., **2**, 762 ⟨66MI 215-01⟩; **4**, 46 ⟨79JOC2813⟩; **6**, 42 ⟨75JFC(5)25⟩, 250 ⟨79JOC2813⟩
Strobach, D. R., **2**, 610 ⟨71JOC818⟩, 1042 ⟨71JOC1438⟩; **4**, 8, 9 ⟨71JOC818⟩
Strobel, C. W., **3**, 90 ⟨61JOC2225⟩
Stroh, H.-H., **2**, 373 ⟨65CB1588⟩; **3**, 450, 451 ⟨58CB2645, 57CB352⟩, 451 ⟨63CB651, 63CB658⟩
Stroh, J. G., **2**, 344 ⟨90JA6696⟩
Stroh, R., **2**, 334, 337 ⟨62HOU(5)796⟩; **5**, 810 ⟨62HOU(5/3)503⟩, 1100, 1101 ⟨60AG1000⟩; **6**, 28 ⟨62HOU(5/3)735⟩
Strohmeier, W., **1**, 73 ⟨72JOM(42)163⟩
Strokin, Yu. V., **6**, 549 ⟨88ZOR117⟩
Strologo, S., **2**, 703 ⟨85IC654⟩
Ström, P., **5**, 277 ⟨87JOC2033⟩
Stromar, M., **3**, 425 ⟨74MI 311-01⟩
Strommen, D. P., **5**, 1016, 1017 ⟨66JOM(5)383⟩
Stromski, V., **2**, 408 ⟨29CB598⟩
Stronach, M. W., **5**, 953 ⟨93JA1341⟩
Strong, J. G., **3**, 249 ⟨70JOC1709⟩
Strong, J. S., **5**, 580 ⟨56JOC772⟩
Strong, P. M., **3**, 534 ⟨23CB1172⟩
Stropnik, C., **5**, 609, 626, 836 ⟨81S299⟩
Stroud, P. M., **6**, 343 ⟨89CC1029, 89OM1270⟩, 344 ⟨91JCS(D)1017⟩
Strouse, C. E., **2**, 580 ⟨82JA4865, 83JA5804⟩; **4**, 150 ⟨86JA1455⟩, 678 ⟨92OM3617⟩; **5**, 449 ⟨86JA1462⟩, 917 ⟨80ZN(B)1289⟩
Strouse, Ch. E., **6**, 706 ⟨80ZN(B)1289⟩
Strozier, R. W., **3**, 327 ⟨73JA7287⟩
Strube, M., **6**, 563 ⟨84MI 617-03⟩
Strubin, T., **3**, 539 ⟨88LA943⟩
Struchkova, M. I., **1**, 60 ⟨82JOM(226)115⟩
Struchkov, Y. T., **2**, 444 ⟨91ZOB1075⟩, 445 ⟨87ZOR2120⟩; **5**, 385 ⟨88ZOB62⟩, 615 ⟨89ZOR1323⟩; **6**, 116 ⟨88JOC3609, 90ZOR1129⟩, 284 ⟨92MC52⟩, 619 ⟨93JPO319⟩
Struchkov, Yu. T., **4**, 282 ⟨83TL87⟩, 461 ⟨80IZV1125⟩, 549 ⟨82IZV2718, 84IZV2023⟩, 557 ⟨88ZOB1987, 88ZOB1998⟩, 583 ⟨85JOM(288)119⟩, 672 ⟨68CC1365⟩, 674 ⟨91JOM(418)C24⟩, 688 ⟨81JOM(206)119⟩, 694, 717 ⟨87JOM(336)187⟩, 696 ⟨77JOM(137)217⟩, 696, 723 ⟨78JOM(153)115⟩, 721 ⟨79IZV1528⟩; **5**, 703 ⟨87JOU75, 87ZOR86⟩, 1075 ⟨87ZOR2390⟩; **6**, 138 ⟨89T3487⟩, 341 ⟨90JOM(384)279⟩, 562 ⟨81KGS907⟩
Struck, E., **3**, 661 ⟨19CB1749⟩
Struck, R. F., **4**, 554 ⟨61JCS2423⟩
Struhar, S., **6**, 695 ⟨92TL683⟩
Strukov, O. G., **6**, 613 ⟨69JGU1235⟩
Strunin, B. N., **1**, 665 ⟨65JOM(4)349⟩
Strunk, S., **1**, 397 ⟨89T2649⟩; **2**, 557, 907 ⟨84TL741⟩
Strunz, G. M., **3**, 219 ⟨82CJC2528⟩
Struthers, R. J. F., **6**, 207 ⟨27JCS2658⟩
Strutz, G., **2**, 344, 347, 357 ⟨85S807⟩
Strutz, H., **3**, 513 ⟨85JA5999⟩
Struve, W. S., **1**, 393 ⟨42OR(1)38⟩; **3**, 474, 475 ⟨42OR(1)38⟩; **5**, 276 ⟨42OR(1)38⟩
Struzka, V., **1**, 150 ⟨91TL5501⟩
Strychkov, Y. T., **1**, 530 ⟨86TL1241⟩
Strydom, P. J., **6**, 558 ⟨90JOC5230⟩
Stryker, J. M., **1**, 86 ⟨88JA291⟩, 458, 516 ⟨90TL2397⟩, 1039 ⟨83TL4887⟩; **3**, 12 ⟨89TL5677⟩, 12, 128 ⟨88JA291, 88TL3749⟩; **4**, 151 ⟨89OM2973⟩, 585 ⟨88OM2082⟩; **5**, 944 ⟨90JA464⟩, 948 ⟨89OM2075⟩
Strzalko, T., **3**, 634 ⟨87S56⟩
Strzelecka, H., **2**, 412 ⟨57RZC879⟩; **4**, 863 ⟨72T3545⟩
Stuart, A. L., **6**, 689 ⟨84JA7015⟩
Stuart, F., **2**, 781 ⟨55OSC(3)668⟩
Stuart, J. G., **3**, 247 ⟨91H(32)949⟩
Stuart, R. S., **1**, 850 ⟨69CC1067⟩
Stubbe, M., **1**, 310 ⟨82TL3171⟩
Stubbs, C. A., **3**, 119 ⟨82TL4647⟩
Stubbs, C. E., **3**, 274 ⟨67JA5722⟩
Stubbs, J. W., **1**, 1114 ⟨76JOC681⟩
Stubbs, M. E., **3**, 17, 404 ⟨79TL3201, 85JCS(P1)2123⟩
Stubenrauch, G., **5**, 1126 ⟨80MIP-DE283404⟩
Stubentauch, G., **2**, 368 ⟨81AG(E)151⟩
Stuber, F. A., **2**, 655 ⟨81TL2337⟩; **6**, 143 ⟨68JOC3928⟩, 448 ⟨84JOC3675⟩
Stucchi, E., **5**, 427 ⟨92SL332⟩, 427, 428 ⟨92SL329⟩
Stuchkov, Yu. T., **2**, 878 ⟨87JOU1417⟩
Stucki, H., **4**, 307 ⟨66JOC1232⟩
Stückler, P., **3**, 522 ⟨76JOM(120)C6⟩; **5**, 919 ⟨77CB3040⟩; **6**, 722 ⟨77JOM(129)197⟩
Stuckwisch, C. G., **3**, 183 ⟨72JOC318⟩; **4**, 576 ⟨75JOC1173⟩
Stucky, G., **1**, 1026 ⟨87HCA448⟩; **2**, 570 ⟨80JA6744⟩; **3**, 511 ⟨80JA6744⟩; **4**, 213 ⟨86AG(E)178, 87HCA448⟩; **5**, 918 ⟨80JA6744⟩
Stucky, G. D., **4**, 668 ⟨90JA2425⟩, 671 ⟨79JA6933⟩, 709 ⟨69JA2538⟩, 1068 ⟨74JA1941⟩
Stucky, R., **4**, 706 ⟨78JA6382⟩
Studabaker, W. B., **3**, 519 ⟨87CR411⟩; **5**, 937 ⟨85OM943⟩, 942 ⟨87CRV411⟩
Stüdemann, T., **2**, 591 ⟨93JA7027⟩
Studnev, Y. N., **2**, 165 ⟨89IZV197⟩; **4**, 104 ⟨86IZV810⟩, 119 ⟨86JGU2394⟩, 135 ⟨71JGU481⟩, 556 ⟨87ZOB54⟩; **5**, 474, 484 ⟨67BAU324⟩, 645 ⟨63ZOB3293⟩, 698 ⟨82IZV937⟩, 699, 710 ⟨81JFC(18)553, 81IZV2370⟩, 813 ⟨80IZV2623⟩; **6**, 42 ⟨85IZV659⟩, 56, 282 ⟨76IZV489⟩, 251 ⟨74IZV335⟩, 259 ⟨78IZV486⟩
Studt, W. L., **6**, 647 ⟨78AF1435⟩
Stuebe, C., **2**, 253 ⟨56JA976⟩, 462, 842 ⟨55JA3526⟩
Stuehr, J., **1**, 870 ⟨67JA2826⟩
Stüetten, J., **4**, 704 ⟨89LA187⟩
Stufkens, D. J., **5**, 1089 ⟨90JOM(390)C1⟩
Stuhl, L. S., **1**, 93 ⟨76IC2379, 78JA2405⟩
Stuhl, O., **2**, 907 ⟨B-89MI 218-01⟩
Stühler, H., **2**, 488 ⟨73CB1226⟩; **5**, 913 ⟨72CB1084⟩
Stukalo, E. A., **2**, 238 ⟨83ZOR343⟩; **4**, 1011 ⟨80ZOB343⟩; **5**, 349 ⟨77ZOR2055⟩
Stukalo, V. A., **6**, 622, 634, 658 ⟨71ZOB2155⟩
Stull, A., **2**, 279, 292 ⟨27JA2011⟩
Stull, P. D., **3**, 320 ⟨88JA2894, 89JOC817⟩
Stults, B. R., **2**, 421 ⟨76JA4477⟩; **3**, 335 ⟨85JOC3216⟩; **4**, 681 ⟨76JA1774⟩; **5**, 321 ⟨79JOC1447⟩
Stults, J. S., **3**, 331 ⟨86JOC1556, 88JOC2226⟩, 331, 333, 335, 337 ⟨88JA5452⟩
Stumbrevichute, Z. A., **2**, 876, 881 ⟨72IZV213⟩; **4**, 134 ⟨72IZV215⟩, 152 ⟨73JOM(57)423⟩
Stump, Jr., E. C., **4**, 45 ⟨71FCR77⟩; **6**, 5 ⟨64JOC1198⟩
Stumpf, R., **1**, 991 ⟨92CB2667⟩; **2**, 936 ⟨93SL863⟩

Stumpf, T., **3**, 710 ⟨92AG(E)1218⟩
Stupp, S. I., **5**, 314 ⟨90JOC3374⟩
Stur, T., **3**, 370 ⟨92H(34)1295⟩
Sturgeon, M.-E., **1**, 356 ⟨82CJC425⟩
Sturgess, M. A., **4**, 1019 ⟨86TL3811, 88OM2553⟩; **5**, 652 ⟨88T5615⟩, 955 ⟨88OM2553⟩
Sturgis, B. M., **3**, 350 ⟨35JA1126⟩
Sturino, C. F., **4**, 651 ⟨91JOC1944⟩
Sturkovich, R., **4**, 1044 ⟨88JOM(346)297⟩
Sturkovits, R. Y., **2**, 914 ⟨86ZOR140⟩
Sturm, B., **6**, 434, 534, 535 ⟨84ZAAC(508)136⟩, 481 ⟨84ZAAC(513)183⟩, 482 ⟨84ZAAC(514)120⟩, 534 ⟨83ZAAC(502)7, 84ZAAC(509)67⟩, 536 ⟨84ZAAC(509)61⟩, 565 ⟨84ZAAC(512)231⟩
Sturm, E., **5**, 774 ⟨78GEP2758173⟩
Sturm, H., **5**, 657 ⟨72CB1258⟩
Sturm, H. J., **3**, 678 ⟨62AG(E)50⟩, 679 ⟨72CB1258⟩
Sturm, W., **5**, 944 ⟨87OM1424⟩
Sturmer, R., **2**, 43 ⟨93JOC5307⟩, 530 ⟨89CB1783, 90SL759⟩
Stürmer, R., **5**, 173 ⟨86AG(E)1028⟩
Sturts, G., **2**, 457 ⟨79PS(7)101⟩
Sturtz, G., **2**, 457 ⟨85S1048⟩, 1042, 1050, 1073 ⟨65CR(260)1984⟩, 1072 ⟨65CR(261)1019, 66BSF1707⟩; **3**, 78 ⟨88S907⟩, 231 ⟨67BSF2477⟩; **4**, 560 ⟨94PS(88)1⟩, 923 ⟨73CR(C)395⟩
Sturz, G., **4**, 560 ⟨91S661⟩
Stüsser, R., **5**, 488, 489 ⟨23CB633⟩, 652 ⟨23CB1557⟩
Stüssi, R., **1**, 595 ⟨78HCA998⟩; **4**, 256 ⟨78HCA2351⟩; **5**, 247 ⟨76HCA1307⟩
Stutte, B., **6**, 340 ⟨76JOM(113)67, 78CB1239⟩, 340 ⟨78CB1603⟩
Stutten, J., **6**, 32 ⟨89LA187⟩
Stütz, A., **1**, 512 ⟨79TL3145, 84JMC1539⟩, 512 ⟨86JMC112⟩
Stutz, H., **1**, 1124 ⟨74TL2457, 77JA1669⟩
Stütz, P., **3**, 291 ⟨72HCA75, 77OS(56)8⟩; **6**, 86 ⟨77OS(56)8⟩, 97 ⟨72HCA75⟩
Stutz, R. E., **2**, 126 ⟨33JA1242⟩
Stutz, U., **3**, 722, 724 ⟨78AG(E)688⟩
Styan, G. E., **2**, 482, 487, 876 ⟨65JOM(4)151⟩, 487 ⟨66JOM(6)364, 65JOM(3)406⟩; **4**, 137, 141, 153 ⟨66JOM(6)633⟩, 344 ⟨65JOM(4)151⟩
Stylianides, N., **3**, 351 ⟨90JA3029⟩; **5**, 513 ⟨90JA3029⟩
Su, D., **4**, 803 ⟨90JA3152⟩; **6**, 258 ⟨89JA1773⟩, 258, 278 ⟨89CJC1795⟩
Su, D.-B., **6**, 219, 227 ⟨92CC807⟩
Su, H., **2**, 583 ⟨91OM1183⟩
Su, J., **5**, 949 ⟨92JA10665⟩
Su, Q., **1**, 324 ⟨93TL4489⟩
Su, S. R., **3**, 570 ⟨70JOM(21)P21, 74ICA55⟩
Su, T.-M., **6**, 110 ⟨71JA4075⟩
Su, W., **1**, 504, 513 ⟨83TL4883⟩
Su, W.-G., **3**, 222 ⟨87TL5241⟩
Suami, T., **1**, 827 ⟨88TL655⟩, 827, 829 ⟨87JOC1201⟩
Suarato, A., **3**, 284 ⟨84CC530⟩
Suárez, E., **1**, 927 ⟨88JCS(P1)3255⟩, 929 ⟨84TL1953⟩, 931 ⟨86TL383, 87TL3397⟩, 931 ⟨92AG(E)772, 93TL4865⟩; **2**, 13 ⟨86JOC402⟩; **3**, 409 ⟨86S1055⟩; **5**, 1118 ⟨90JCS(P1)2417⟩, 1125 ⟨86MI 530-01⟩; **6**, 633, 634 ⟨89CC452⟩
Suarez, J., **2**, 420 ⟨88SC253⟩
Suarez, J. L., **2**, 18, 610 ⟨91JOC2234⟩
Suarez, J. Q., **4**, 967 ⟨88MI 421-01⟩
Suau, R., **3**, 294 ⟨75H(3)449⟩, 362 ⟨74JCS(P1)2559⟩, 363 ⟨76JA6218⟩; **5**, 634 ⟨74TL91⟩
Suba, L., **5**, 1119, 1121 ⟨87PS(29)219⟩
Suba, L. A., **5**, 1121 ⟨86PS(26)1⟩; **6**, 630 ⟨87PS(29)1⟩
Subbaraj, A., **6**, 620 ⟨86JOC1719⟩
Subba Rao, B. C., **2**, 40 ⟨56JA5694, 57JOC1136⟩, 40 ⟨59JA4108⟩, 126 ⟨56JA2582⟩; **3**, 83 ⟨58JA5377⟩
Subba Rao, G. S. R., **1**, 96 ⟨82JCS(P1)875⟩, 98 ⟨82JCS(P1)1381⟩, 99 ⟨67TL857⟩, 434 ⟨70AJC1641⟩, 797 ⟨68TL3797⟩
Subbarao, H. N., **3**, 4, 115 ⟨79T1789⟩
Subba Rao, N. H., **1**, 809, 813 ⟨67TL227⟩
Subbarow, Y., **5**, 1125 ⟨48JOC924⟩
Subit, M., **3**, 286 ⟨75JOM(99)41⟩
Subluskey, L. A., **2**, 125 ⟨48JA1176⟩, 212 ⟨51JA2647⟩
Subong, A. P., **2**, 390 ⟨89TL6495⟩
Subotkowska, L., **4**, 467 ⟨79S985⟩
Subrahmanyam, C., **1**, 126 ⟨75TL1633⟩; **2**, 623, 628 ⟨89JOC6068⟩, 900, 901 ⟨89JOC6075⟩
Subrahmanyam, G., **1**, 871, 881 ⟨66JA1332⟩
Subrahmanyan, C., **2**, 904 ⟨76JCS(P1)2419, 76JCS(P1)2435⟩
Subrahmanyan, L., **3**, 445 ⟨67AJC339⟩
Subramanian, G., **6**, 322 ⟨93OM4267⟩
Subramanian, L. R., **1**, 177, 224, 433, 1021 ⟨82S85⟩, 433 ⟨84S481⟩, 1109, ,1110, 1117 ⟨79MI 119-05⟩, 1111, 1113 ⟨78LA1894⟩, 1111, 1117 ⟨72AG(E)714⟩, 1111, 1117, 1124 ⟨77JOC174⟩; **2**, 36 ⟨87TL6441⟩, 556 ⟨93HOU(E/19d)713⟩, 689 ⟨74S293, 76JOC4099, 82S85⟩, 1089 ⟨81JOM(204)315⟩; **3**, 92, 101 ⟨90CC1571⟩, 635 ⟨91S1205⟩; **4**, 214 ⟨87S49⟩
Subramanian, R., **1**, 940, 941 ⟨77JCS(P1)1718⟩; **2**, 8 ⟨76JCS(P1)2112⟩, 657 ⟨85JOC5430⟩; **5**, 516, 522, 526 ⟨76JCS(P1)2112⟩; **6**, 309 ⟨76JCS(P1)2112⟩
Subramanian, R. S., **1**, 827 ⟨88TL6797⟩; **2**, 683 ⟨89SC1255⟩
Subramanian, S., **5**, 324 ⟨90SC1733⟩
Subramanyam, C., **2**, 516 ⟨79TL4777⟩, 533 ⟨76JCS(D)2087⟩; **5**, 342 ⟨91BMC733⟩
Suchanck, D., **3**, 10 ⟨63AG(E)683⟩
Suchár, G., **5**, 1054 ⟨72CCC3066⟩
Suchland, B., **5**, 729 ⟨88LA471⟩
Suchomel, H., **2**, 471 ⟨83ZAAC(498)57⟩
Suchy, H., **2**, 112 ⟨68ZAAC(358)44⟩
Suciu, E. N., **2**, 390 ⟨75JOC1395⟩
Suckling, C. W., **6**, 16 ⟨57BRP767779⟩
Sucrow, W., **1**, 39 ⟨88CB1039⟩, 415, 827 ⟨71CB3679⟩; **3**, 264 ⟨73CB450⟩
Suda, K., **3**, 36 ⟨81CC263⟩; **5**, 346 ⟨81S610⟩, 410 ⟨92TL1355⟩
Suda, M., **4**, 731 ⟨80TL2555⟩, 734, 764 ⟨81TL1421⟩, 738 ⟨81TL759⟩; **6**, 233, 234 ⟨81TL1997⟩, 695 ⟨81TL1421⟩
Suda, S., **2**, 453 ⟨87CL1915⟩; **3**, 281, 298 ⟨91CL1059⟩
Suda, T., **5**, 1111 ⟨74CL881⟩
Sudakova, T. M., **4**, 1000, 1001 ⟨72ZOB1727⟩; **5**, 652 ⟨74JGU2371⟩, 786,1715 ⟨72JGU1637⟩
Sudarsanam, V., **5**, 690 ⟨82IJC(B)989⟩; **6**, 647 ⟨82PIA441⟩
Sudau, W., **1**, 940, 941 ⟨85T4253⟩
Suddaby, D., **4**, 206 ⟨50JPP764⟩
Sudhakar, A. R., **1**, 216, 979 ⟨93TL7849⟩
Sudheendranath, C. S., **2**, 751 ⟨82S717⟩
Sudo, K., **2**, 683 ⟨84CPB3759⟩
Sudo, R., **5**, 731 ⟨67BCJ2705⟩
Sudo, T., **3**, 412 ⟨89NKK1748⟩
Sudoh, R., **1**, 608 ⟨81CC261⟩; **2**, 121 ⟨77JOC2180⟩, 397 ⟨71TL2053, 75BCJ3413⟩, 784 ⟨82S261⟩, 784, 786 ⟨81CC261, 82BCJ635⟩; **4**, 335 ⟨82S261⟩; **5**, 183, 199 ⟨79CC179⟩
Sudokova, T. M., **5**, 737 ⟨73DOK(211)113⟩
Sudol, M., **5**, 1057 ⟨75S785⟩; **6**, 122 ⟨75S785⟩
Sudweeks, W. B., **1**, 1037 ⟨75JOC1131⟩
Sue, R. E., **2**, 559 ⟨88CC289⟩
Suehiro, H., **4**, 856 ⟨78CPB685⟩
Suemitsu, R., **3**, 15, 90 ⟨82BCJ1663⟩, 137 ⟨89BCJ2728⟩; **5**, 43, 112 ⟨81NKK1333⟩
Suemmermann, K., **5**, 386 ⟨85JOM(288)237, 85JOM(295)C21⟩
Suemune, H., **1**, 1021 ⟨88CPB4337⟩
Suemune, S., **3**, 113 ⟨84TL961⟩
Suen, Y.-H., **3**, 461 ⟨66BSF1819⟩; **5**, 290 ⟨65BSF1454⟩

Suenaga, H., **2**, 626 ⟨89BCJ3373⟩
Suenaga, K., **5**, 329 ⟨94TL1247⟩
Suenaga, M., **3**, 709 ⟨82TL5335⟩
Suenaga, T., **5**, 157 ⟨89TL6529⟩
Suenobu, K., **3**, 671 ⟨85CL1487⟩
Sueyoshi, S., **5**, 1118, 1124 ⟨82CPB3125⟩
Suffert, J., **1**, 706 ⟨88TL909⟩, 1029 ⟨92CB271, 93TL4177⟩
Suffolk, R. J., **3**, 582 ⟨90JA3779⟩
Sufrin, J. R., **1**, 318 ⟨73TL1513⟩; **4**, 61 ⟨89JMC997⟩, 99 ⟨89SC1479⟩
Suga, H., **3**, 9 ⟨88S394⟩, 27 ⟨90T4261⟩, 200 ⟨87BCJ2463⟩; **4**, 515 ⟨84H(22)1955⟩; **5**, 821 ⟨87BCJ2463⟩
Suga, K., **2**, 308 ⟨69BCJ3606⟩
Suga, S., **1**, 182 ⟨79BCJ2377⟩, 212, 475 ⟨75BCJ960⟩, 463 ⟨80CC86⟩, 463 ⟨82BCJ167, 83BCJ1089⟩; **2**, 71 ⟨88PAC1597, 89JA4028⟩, 542 ⟨88JA1976⟩; **3**, 336 ⟨89SC2061⟩, 336, 340, 385 ⟨88JA1976, 88TL6965⟩, 390, 400 ⟨89JA8749⟩, 394 ⟨89TL2095⟩; **4**, 931 ⟨86CL181⟩; **5**, 429 ⟨86CL177⟩; **6**, 481 ⟨89CL1009⟩
Suga, T., **5**, 47 ⟨65BCJ893⟩
Sugahara, K., **4**, 736 ⟨93JFC(62)201⟩; **5**, 1003 ⟨73BCJ1275⟩
Sugahara, T., **5**, 324 ⟨87CC1471, 90JOC1744⟩
Sugai, K., **1**, 590, 591 ⟨88TL659⟩
Sugai, S., **3**, 419 ⟨85HCA745⟩; **4**, 979 ⟨79CPB2787⟩
Sugai, T., **1**, 9 ⟨92JA10138⟩; **5**, 118 ⟨84MI 502-03⟩, 179 ⟨89CL1775⟩
Sugano, Y., **3**, 631 ⟨88TL6957⟩
Sugasake, K., **5**, 864 ⟨B-87MI 521-01⟩
Sugasawa, S., **1**, 744 ⟨60CPB819⟩
Sugasawa, T., **2**, 810, 811 ⟨79SC515⟩; **3**, 282 ⟨78JA4842, 80TL173⟩, 632 ⟨82CL1401⟩, 638 ⟨79SC553⟩, 664 ⟨90SC15⟩
Sugata, K., **5**, 1134 ⟨88JAP63135393⟩
Sugata, T., **1**, 268 ⟨64T1695⟩
Sugawa, T., **4**, 935 ⟨83CL933⟩
Sugawara, A., **6**, 553 ⟨87BCJ435⟩
Sugawara, R., **1**, 481, 1028 ⟨83CPB4346⟩
Sugawara, S., **6**, 553 ⟨81CPB3158⟩
Sugawara, T., **2**, 794, 795 ⟨92JA1830⟩; **3**, 19 ⟨83SC1149⟩; **4**, 935 ⟨83SC1149⟩; **5**, 640, 642 ⟨80CPB2116⟩; **6**, 82 ⟨74BCJ2457, 75BCJ2496⟩, 99 ⟨74BCJ1496⟩, 304 ⟨82BCJ1106⟩, 558 ⟨93TL3745⟩
Sugaya, T., **3**, 80 ⟨78BCJ1186⟩; **4**, 218 ⟨82CL1555⟩
Suggs, J. W., **1**, 5, 9, 10, 936 ⟨75JOC2554⟩; **2**, 675 ⟨73JOC3224⟩; **3**, 57 ⟨75TL2647⟩, 113 ⟨79JA489⟩; **5**, 459 ⟨78JA640⟩, 800 ⟨79JA489⟩
Sughara, K., **1**, 828 ⟨94CC901⟩
Sughiara, H., **3**, 16, 32 ⟨78S881⟩
Sugi, K. D., **1**, 498 ⟨92JOC1146⟩
Sugi, S., **4**, 398 ⟨92CL819⟩
Sugie, Y., **3**, 251 ⟨87CL939⟩
Sugihara, T., **1**, 596 ⟨92TA853⟩; **2**, 85 ⟨89CC1344⟩, 980 ⟨93T5471⟩
Sugihara, Y., **1**, 17 ⟨79TL2157⟩, 432 ⟨79JOC4508⟩, 581 ⟨93TA1795⟩; **3**, 261 ⟨82JA6697, 82TL1201⟩; **4**, 88, 89 ⟨88CL1317⟩; **5**, 235 ⟨75CJC3693⟩
Sugimoto, A., **2**, 176, 180 ⟨74S876⟩; **4**, 233 ⟨89CPB184⟩
Sugimoto, H., **2**, 156 ⟨84JA4283⟩, 156 ⟨85JOC1784⟩; **4**, 586, 587 ⟨80BCJ3176⟩; **6**, 85 ⟨77JOC1543⟩, 543, 557, 563, 578 ⟨88JOC2263⟩
Sugimoto, I., **6**, 698 ⟨83TL3469⟩
Sugimoto, J., **4**, 737, 743 ⟨91CL1319⟩, 737, 743 ⟨92BCJ1513⟩
Sugimoto, K., **1**, 613, 700, 1073 ⟨86JOC3830⟩; **5**, 642 ⟨73BCJ2921⟩
Sugimoto, N., **5**, 1106 ⟨72CPB427⟩
Sugimoto, S., **4**, 674 ⟨93OM2925⟩
Sugimoto, S. I., **6**, 472 ⟨88H(27)2327, 89CPB576⟩
Sugimoto, T., **2**, 21 ⟨77CI(L)538, 79BCJ255⟩, 21 ⟨88OSC(6)90⟩, 444 ⟨82MI 208-01⟩, 445 ⟨81S53⟩; **3**, 73 ⟨89TL5289⟩, 292 ⟨88HCA531⟩, 326 ⟨94JCS(P1)413⟩; **4**, 250 ⟨81TL3243⟩, 847 ⟨81S53⟩; **6**, 120 ⟨81S53⟩, 124, 698 ⟨84TL4227, 83TL3469⟩
Sugimoto, Y., **1**, 816 ⟨91TL5137⟩
Sugimura, H., **1**, 191, 238 ⟨84CL1505⟩; **2**, 81 ⟨89CL1401⟩; **4**, 959 ⟨80CL1209⟩
Sugimura, N., **4**, 206 ⟨80BCJ1061⟩
Sugimura, T., **1**, 816 ⟨87JA3017⟩, 816, 817 ⟨86JA3841⟩; **2**, 52 ⟨91CC795⟩
Sugimura, Y., **3**, 559 ⟨76CC516⟩, 581 ⟨76TL1307⟩, 634 ⟨85H(23)3069⟩; **4**, 842 ⟨82TL1075⟩; **5**, 1039 ⟨85H(23)3069⟩
Sugino, E., **3**, 309 ⟨81JOC2979⟩
Sugino, K., **5**, 584 ⟨81LA1798⟩, 1118 ⟨57NKZ567, 60JOC1043⟩, 1125 ⟨61JOC4122⟩
Suginome, H., **1**, 228 ⟨82JOM(233)13⟩, 279, 532 ⟨81TL127, 82JOC2117⟩, 532, 1020 ⟨85JA972⟩, 927 ⟨90CC1074, 90CC1076⟩, 929 ⟨87TL3963, 93JCS(P1)3007⟩, 931 ⟨87T3371, 92JCS(P1)1837⟩; **2**, 104 ⟨82JCS(P1)111, 82JCS(P1)1963⟩, 323 ⟨88CL1289⟩, 337 ⟨94JCS(P1)777⟩, 765 ⟨76JCS(P1)2365⟩; **3**, 24 ⟨82JOC2117⟩; **5**, 332 ⟨89JOC5945⟩, 362 ⟨80BCJ3225⟩; **6**, 49 ⟨36BCJ157⟩
Suginome, M., **2**, 916 ⟨91JOC1948⟩; **4**, 1054 ⟨91JOC1948⟩; **5**, 802 ⟨88BCJ3649⟩; **6**, 384, 671 ⟨91JA8899⟩, 550 ⟨89CC1494⟩
Sugisawa, H., **1**, 450 ⟨81JOM(217)43⟩; **4**, 615 ⟨83JOM(250)109⟩, 1047 ⟨78JOM(161)299⟩, 1048 ⟨79JOM(179)377, 80JOM(194)147⟩, 1053 ⟨82OM1473⟩
Sugise, R., **6**, 7 ⟨81AG(E)475⟩
Sugishima, M., **6**, 488 ⟨92JAP05339542⟩
Sugita, J., **5**, 48, 114 ⟨64CPB403⟩
Sugita, K., **1**, 208 ⟨90CL481⟩; **2**, 84 ⟨90TL7467⟩; **3**, 615 ⟨89CL1393⟩, 632 ⟨90CL481⟩
Sugita, M., **2**, 279, 288 ⟨67NKK991⟩
Sugita, N., **1**, 545 ⟨87JOC4859⟩, 611, 716 ⟨85CC1037⟩; **3**, 90 ⟨76BCJ1681⟩, 146 ⟨90JCS(P1)1697⟩; **5**, 241 ⟨87JOM(334)C45⟩, 242 ⟨87JOM(334)C43, 89JOM(359)215⟩
Sugita, T., **1**, 212, 475 ⟨78BCJ174⟩, 609, 983 ⟨92CC46, 93JOC3697⟩; **2**, 631 ⟨82CL1481⟩
Sugita, Y., **2**, 17 ⟨94CPB459⟩, 39 ⟨90TL7463⟩; **3**, 543 ⟨91T5689⟩
Sugiura, M., **1**, 503 ⟨82JOC1221⟩; **2**, 929 ⟨92CC1358⟩, 1045 ⟨78JOC199⟩; **4**, 425 ⟨78JHC949⟩; **5**, 212 ⟨85CPB998⟩, 1127 ⟨78YZ1361, 79H(12)164⟩
Sugiura, S., **1**, 263 ⟨75JOC2009⟩
Sugiura, T., **1**, 193 ⟨85JOC1523⟩, 962 ⟨86CC922, 87S603⟩, 975 ⟨86TL731⟩; **3**, 120 ⟨86JOC155⟩, 139 ⟨85JOC1523⟩, 233 ⟨85JOC537⟩
Sugiura, Y., **1**, 120 ⟨93TL5127⟩, 1038, 1098, 1140 ⟨84TL4233⟩, 1098, 1140 ⟨84JOC3904⟩; **4**, 954 ⟨84TL4233⟩
Sugiya, T., **2**, 445 ⟨91JAP70374395⟩
Sugiyama, H., **4**, 508 ⟨88S988⟩; **5**, 1025 ⟨78YGK1090⟩
Sugiyama, M., **6**, 383 ⟨91JA7790⟩
Sugiyama, N., **2**, 228 ⟨73BCJ2253⟩; **3**, 343 ⟨73BCJ2253⟩
Sugiyama, S., **5**, 1023 ⟨92OPP346⟩
Sugiyama, T., **2**, 631 ⟨81BCJ2847⟩
Sugiyama, Y., **2**, 263 ⟨83TL1273⟩
Sugizaki, M., **3**, 8, 122 ⟨79CC303⟩
Sugo, T., **5**, 864 ⟨86MI 521-02, 87MI 521-03⟩
Suguira, Y., **2**, 69, 564 ⟨84JOC3904⟩
Sugumi, H., **1**, 239 ⟨87CL1691, 87CL1695⟩
Suh, H., **2**, 681 ⟨84TL395⟩
Suh, J. T., **6**, 118 ⟨82MI 604-01⟩
Suhara, H., **1**, 619 ⟨88CL2009⟩
Suhara, Y., **1**, 242 ⟨86BCJ2321⟩; **6**, 314 ⟨89NKK63⟩
Suhr, H., **3**, 465 ⟨63CB1712⟩
Suhr, Y., **4**, 354 ⟨88JOC1569⟩; **5**, 412, 413, 425 ⟨88JOC1569⟩
Sühs, K., **4**, 574 ⟨81AG(E)1038, 88CB1509⟩
Sui, Z., **4**, 978 ⟨91HCA579⟩, 981 ⟨90S959⟩

Suib, S. L., **4**, 671 ⟨79JA6933⟩
Suissa, M. R., **1**, 248 ⟨87CC381⟩
Suits, J. A., **3**, 223, 275 ⟨86S184⟩
Suits, J. Z., **1**, 516 ⟨86S184⟩
Sujeeth, P. K., **4**, 18 ⟨78JOC4367⟩; **6**, 28 ⟨78JOC4367⟩
Suji, J. T., **4**, 14 ⟨64JA4851⟩
Sujino, K., **5**, 700, 701 ⟨92JCS(P1)2127⟩
Sukata, K., **2**, 156 ⟨84BCJ613⟩, 544 ⟨89JOC2015⟩; **3**, 634 ⟨84BCJ613⟩; **5**, 1114 ⟨89JOC2015, 90BCJ825⟩
Sukhai, R. S., **2**, 941 ⟨79RTC55⟩, 1032 ⟨77RTC179⟩; **3**, 548 ⟨77RTC179, 82RTC310⟩, 548, 553 ⟨79RTC55⟩, 552 ⟨77S888⟩; **4**, 93 ⟨81RTC10⟩, 842 ⟨79S455⟩; **5**, 589 ⟨82RTC310⟩, 589, 590, 603, 611, 626 ⟨77S888, 79RTC55⟩
Sukhanovskaya, E. V., **4**, 469 ⟨92ZOB2708⟩
Sukhomlinova, L. I., **1**, 486 ⟨90IZV2665⟩
Sukirthalingam, S., **1**, 1015 ⟨88TL5565⟩
Sukumaran, K. B., **2**, 659 ⟨79JOC4265⟩
Sulay, P. B., **2**, 309 ⟨89AP(322)181⟩; **4**, 981, 982 ⟨87LA1101⟩; **5**, 599 ⟨87LA1101⟩
Suld, G., **3**, 326 ⟨62JA2094⟩
Suleimanova, M. G., **4**, 556 ⟨67IZV949⟩
Suleske, R. T., **3**, 184 ⟨80S329⟩
Sulikowski, G. A., **1**, 573 ⟨92JOC5557⟩; **6**, 488 ⟨92JA8008⟩
Sulikowski, M. M., **1**, 573, 581, 584 ⟨92JA9419⟩
Sulimov, I. G., **4**, 872 ⟨74ZOR2456⟩
Süling, C., **6**, 469 ⟨92CB2493⟩
Sullivan, A. B., **2**, 217 ⟨70TL3547⟩
Sullivan, A. C., **5**, 444 ⟨88JCS(D)53⟩; **6**, 174, 176, 381 ⟨92JAP06033008⟩, 174, 176, 381, 393 ⟨86CC1043⟩, 174, 289, 390 ⟨83CC827⟩, 174, 391 ⟨86CC969⟩, 176, 289, 391 ⟨83CC1390⟩, 393, 394, 396 ⟨88JCS(D)381⟩, 394, 395 ⟨86CC908⟩, 395 ⟨87JCS(D)747⟩, 396 ⟨84CC870, 84JOM(263)C23⟩, 396 ⟨85CC534, 85CC1380⟩
Sullivan, D., **1**, 957 ⟨72TL4249⟩
Sullivan, D. F., **1**, 130 ⟨73JA3050⟩, 556 ⟨77JOC3961⟩, 830 ⟨73SC67⟩; **3**, 206 ⟨77JOC3961⟩, 213 ⟨73TL1297⟩, 537 ⟨77JOC2038⟩; **4**, 648 ⟨74TL1403⟩; **6**, 113 ⟨76JOC1112⟩, 190 ⟨77JOC2038⟩
Sullivan, F. R., **2**, 405 ⟨74CC217⟩
Sullivan, G. R., **2**, 64 ⟨73TL1505⟩
Sullivan, J. M., **5**, 620 ⟨61JOC1132⟩
Sullivan, M. P., **2**, 985 ⟨68IC1298⟩
Sullivan, P. J., **1**, 921 ⟨63JA384⟩
Sullivan, R., **4**, 766 ⟨69JOC1841⟩
Sulmon, P., **2**, 322 ⟨88JOC4462⟩; **3**, 184 ⟨82S765⟩, 409 ⟨85AG(E)881⟩, 416 ⟨86S192⟩, 416, 416 ⟨86JOC3839⟩
Sulsky, R., **2**, 351 ⟨89TL31⟩
Sulsky, R. B., **3**, 623 ⟨75TL4647⟩; **4**, 563 ⟨92USP5103036⟩
Sulston, J. E., **4**, 186 ⟨67JA3366⟩
Sultanbawa, M. U. S., **1**, 681, 682 ⟨76JCS(P1)884, 68TL4569⟩
Sultanov, A. V., **6**, 281 ⟨75IZV2750⟩, 281, 348 ⟨73IZV350⟩, 283 ⟨77IZV375⟩, 283, 349 ⟨74IZV1350⟩, 348, 349, 653 ⟨74MI 621-02, 75IZV1669⟩
Sultanov, R. M., **2**, 561 ⟨83IZV218, 83IZV2053⟩
Sultanova, D. B., **4**, 340 ⟨71ZOB2575, 72ZOB1647⟩
Sultanova, R. B., **4**, 463 ⟨88ZOB2456⟩
Sultanova, R. Kh., **2**, 885 ⟨72JGU1960⟩
Sulzbach, R. A., **5**, 709 ⟨77GEP751220⟩
Sulzbacher, M., **4**, 178 ⟨48JA2827⟩; **5**, 66 ⟨48JA1153⟩
Sulzer, J., **6**, 565 ⟨81CCC1970⟩
Sulzle, D., **6**, 32 ⟨87TL2769⟩
Sum, F.-W., **1**, 482 ⟨79CJC1431, 84OS(62)14⟩; **3**, 217 ⟨79CJC1431⟩; **5**, 103 ⟨84OS(62)14⟩
Sum, P.-E., **4**, 197 ⟨78CJC2700⟩
Sumarokova, T. N., **5**, 389 ⟨76ZOB75⟩
Sumi, K., **1**, 141 ⟨85JOC657⟩, 431 ⟨88T4095⟩, 592, 616, 668 ⟨82TL2205⟩, 600 ⟨82CC281⟩, 1080 ⟨85CC697⟩; **2**, 1020, 1063 ⟨85TL4501⟩; **4**, 235, 236 ⟨86CL1655⟩; **6**, 98, 101, 132, 134 ⟨81TL2005⟩, 101 ⟨85JOC657⟩, 119, 126, 132 ⟨85JOC662⟩
Sumi, S., **3**, 116 ⟨83CL379⟩
Sumida, K., **1**, 569 ⟨93JOC5163⟩
Sumida, S., **1**, 496 ⟨92TL7029⟩
Sumino, K., **6**, 207 ⟨71MI 606-01⟩
Sumita, S., **4**, 766 ⟨92CPB3261⟩
Sumitani, K., **1**, 466, 470 ⟨76BCJ1958⟩, 466, 473 ⟨78OS(58)127⟩, 471 ⟨72JA9268⟩
Sumiya, F., **1**, 824 ⟨88CC370⟩
Sumiya, T., **1**, 471 ⟨84CL1225⟩; **3**, 144 ⟨84BCJ242⟩, 293 ⟨87BCJ767⟩
Summerbell, K., **4**, 47, 48, 49, 51 ⟨35JA2364⟩
Summerford, C., **3**, 458 ⟨68CC250B, 70JCS(A)2010⟩, 459 ⟨70JCS(A)2016⟩
Summerhays, L. R., **1**, 942 ⟨84JOC5206⟩
Sümmermann, K., **5**, 386 ⟨88JOM(344)C35⟩
Summers, A. J. H., **5**, 642 ⟨68CC549⟩, 854 ⟨71JCS(C)409⟩; **6**, 110 ⟨71JCS(C)409⟩
Summers, J., **2**, 471 ⟨69IC1006⟩
Summers, J. B., **1**, 1042 ⟨88TL3455⟩; **2**, 343, 351 ⟨88TL3455⟩; **5**, 214 ⟨88JMC1960⟩, 329 ⟨91JOC5482⟩
Summers, J. C., **2**, 490 ⟨70IC862⟩
Summers, L., **2**, 601 ⟨54CRV101⟩; **4**, 46, 47 ⟨55CRV301⟩
Summers, L. A., **4**, 327 ⟨74AJC1579⟩; **5**, 532 ⟨65JCS3312⟩
Summers, M. F., **5**, 690 ⟨88JOC5309⟩
Summerson, W. H., **2**, 412 ⟨51JA5202⟩
Summerville, R. H., **1**, 1111 ⟨69JA4600, 72JA3629⟩, 1111, 1114, 1116 ⟨70JA3802, 74JA1100⟩, 1122 ⟨74JA1110⟩; **2**, 689 ⟨70AG(E)273, 74JA1100⟩; **4**, 973 ⟨77IJC(B)297⟩
Summy, J. M., **2**, 886, 887 ⟨73JOM(55)143⟩
Sumner, Jr., C. E., **4**, 679 ⟨80JA1752, 82OM1350⟩
Sumoto, I. K., **3**, 267 ⟨73JOC4324⟩
Sumoto, K., **2**, 160, 161 ⟨74S277⟩, 161 ⟨72TL4137⟩, 357, 691 ⟨73JOC1239⟩, 401 ⟨76T431⟩; **3**, 252 ⟨76JCS(P1)1146⟩
Sumpter, W. C., **3**, 453 ⟨48JA1980⟩
Sumrell, G., **1**, 881 ⟨50JOC850⟩; **2**, 35 ⟨64CJC2710⟩
Sun, C., **1**, 871 ⟨67JOC492⟩
Sun, C. H., **5**, 384 ⟨90OM560⟩
Sun, C.-M., **1**, 489 ⟨89JOC868⟩; **2**, 630 ⟨89JOC868⟩
Sun, D.-J., **3**, 227 ⟨89SC1593⟩
Sun, D. K., **6**, 489 ⟨90AX2374⟩
Sun, G., **2**, 456 ⟨87MI 208-01⟩; **3**, 163 ⟨93TL779⟩
Sunagawa, K., **3**, 359 ⟨76TL3815, 79BCJ496⟩, 359 ⟨82BCJ243⟩
Sun, J., **2**, 879 ⟨B-85MI 211-04⟩
Sun, K. K., **4**, 436 ⟨65JOC74⟩
Sun, M. S., **2**, 448 ⟨76SRI115⟩
Sun, R. C., **1**, 16 ⟨93OS(72)48⟩; **2**, 822, 841 ⟨79JOC3741⟩
Sun, S.-H., **4**, 438 ⟨89S307⟩
Sun, W., **5**, 955 ⟨93JA12526⟩
Sun, Y., **2**, 1083 ⟨93JOM(449)27⟩
Sunay, U., **2**, 282 ⟨81JOC2605⟩
Sunazuka, T., **5**, 235 ⟨93JA5302⟩
Sundarababu, G., **1**, 826 ⟨90T3601⟩; **2**, 331 ⟨92JA5900⟩; **3**, 189 ⟨92JA5900⟩; **4**, 107 ⟨92JA5900⟩
Sundaralingam, M., **1**, 909 ⟨63JA3302, 66JA198⟩; **2**, 1003 ⟨87CAR(168)7⟩, 1009 ⟨72JA1668⟩
Sundaram, P. M., **2**, 414 ⟨84JOC1784⟩
Sundar, N. S., **1**, 96 ⟨82JCS(P1)875⟩, 98 ⟨82JCS(P1)1381⟩
Sundararaman, P., **3**, 57 ⟨77JOC813⟩
Sundari, B., **3**, 346 ⟨83JOC4482⟩
Sunday, B. R., **1**, 91 ⟨81T3615⟩
Sundberg, J. E., **1**, 474 ⟨75CPB2620⟩; **3**, 144 ⟨76JOC1702⟩
Sundberg, R. J., **3**, 309 ⟨73JOC3324, 75JOC2613⟩
Sundell, M., **4**, 553 ⟨90SC1865⟩
Sunder, S., **3**, 468 ⟨75JHC1191⟩
Sunderling, C. E., **6**, 261 ⟨44JOC419⟩
Sundermann, M. J., **1**, 991 ⟨93JOC1298⟩
Sundermann, R., **5**, 962 ⟨83HOU(E4)738⟩
Sundermeyer, J., **5**, 918 ⟨93AG(E)731⟩, 918, 919 ⟨92CC1631⟩
Sundermeyer, W., **2**, 155 ⟨85CB116⟩, 209 ⟨88CB75⟩, 761

⟨69ZN(B)774⟩; **3**, 334 ⟨89TL4111⟩, 334, 349, 365, 369 ⟨90CB177⟩, 375 ⟨89AG(E)221, 90CB595⟩, 440, 457 ⟨76CB1491⟩, 615, 628 ⟨73CB587⟩, 628, 632 ⟨73TL1449⟩; **4**, 3 ⟨66AG(E)314⟩, 265 ⟨89CB1757⟩, 266 ⟨90CB177⟩, 854 ⟨85CB2852⟩; **5**, 340 ⟨86CB1127, 86CB3405⟩, 358 ⟨88CB75⟩, 476 ⟨87CB1499⟩, 476, 492 ⟨92CB1889⟩, 478, 492 ⟨91CB1311⟩, 488, 492 ⟨89CB1757⟩, 488, 492 ⟨85TL5505, 90JST171⟩, 492, 494 ⟨88S349⟩, 502 ⟨85CB2208, 88CB1435⟩, 502 ⟨90AG(E)60⟩, 652 ⟨89AG(E)221⟩, 721 ⟨79CB2158⟩, 721, 722 ⟨76CB2542⟩, 994 ⟨76CB1491⟩, 1015 ⟨73CB1752⟩, 1076, 1077 ⟨67CB3368⟩; **6**, 128 ⟨91CB1805⟩, 128, 132 ⟨93CB537⟩, 240, 439 ⟨79CB2158⟩, 253 ⟨80AG(E)203, 84JFC(26)359⟩, 253 ⟨89JFC(45)225⟩, 253, 254, 530 ⟨83CB1623⟩, 254 ⟨83CB1285, 85CB2208⟩, 254 ⟨86CB3631⟩, 254, 530, 532 ⟨88JFC(39)329⟩, 254, 532 ⟨87CB1499⟩, 255, 275, 304, 306 ⟨91CB2025⟩, 258 ⟨89AG(E)221, 90AG(E)60, 94CB533⟩, 362 ⟨79CB1189⟩, 419, 442, 451, 452, 453 ⟨73CB1752⟩, 424, 438 ⟨81JFC(18)259⟩, 436 ⟨85CB4553⟩, 439 ⟨86JFC(34)251⟩, 442 ⟨67AG(E)871⟩, 530, 531 ⟨85CB1415, 86CB269⟩, 531 ⟨86SA(A)1281⟩, 607 ⟨83MI 620-01⟩

Sunder-Plassman, P., **1**, 1023 ⟨68CC634⟩

Sundquist, W. I., **2**, 1009 ⟨89JA1299⟩

Sundt, E., **1**, 1082 ⟨67HCA2101⟩

Sundukova, E. N., **4**, 119 ⟨86JGU501⟩

Sundukova, T. A., **1**, 1010 ⟨80IZV726⟩

Sung, E., **5**, 138 ⟨71MI 503-01⟩

Sung, Z.-Z., **2**, 662 ⟨91CPB1132⟩

Sunggak Kim, **1**, 739, 740 ⟨91TL2913⟩

Sung-Yu, N. K., **5**, 469 ⟨86ICA107⟩

Sunjic, V., **2**, 308, 325 ⟨79S161⟩

Sunjić, V., **5**, 1078 ⟨85S509⟩

Sunkara, P. S., **1**, 440 ⟨91JA7439⟩; **4**, 73 ⟨89JA1127⟩, 812 ⟨91JA7439⟩

Sunkel, C., **6**, 573 ⟨70JMC334⟩

Sünkel, K., **4**, 677, 707, 723 ⟨93JOM(458)181⟩, 707 ⟨88AG(E)939, 88CB799⟩, 707 ⟨89JOM(368)67⟩, 822 ⟨91CB2449⟩

Sunko, G. E., **1**, 255 ⟨58JA4405⟩

Sunthankar, S. V., **3**, 134 ⟨77CI(L)742⟩

Supniewski, J. V., **3**, 612 ⟨41OSC(1)46⟩

Sura, T. P., **2**, 791 ⟨88SC2161⟩

Suratwala, N. S. I., **4**, 28, 39, 40 ⟨73CI(L)331⟩

Surber, B. W., **2**, 1020, 1022, 1028 ⟨85JA1452⟩, 1020, 1028 ⟨87JA228⟩; **3**, 535 ⟨82JA6779⟩

Surendrakumar, S., **2**, 158 ⟨91T8297⟩, 352 ⟨91T4007, 92T6929⟩

Suriano, J. A., **2**, 985 ⟨93JA1154⟩

Suri, S. C., **1**, 637 ⟨84JOC670⟩, 814 ⟨82JOC1632⟩; **2**, 386 ⟨88S743⟩; **3**, 138 ⟨82JOC1632⟩, 141 ⟨82CC828⟩; **4**, 435, 447 ⟨88S743⟩; **5**, 361, 362 ⟨88S743⟩

Suri, S. K., **1**, 741 ⟨80SC457, 80ZN(B)95⟩

Surkov, V. D., **4**, 995 ⟨72ZOR2457⟩; **6**, 145 ⟨74ZOR2003⟩

Surles, J. R., **2**, 252 ⟨86JMC2114⟩

Surmatis, J. D., **3**, 62 ⟨58JOC157⟩, 229 ⟨70HCA974⟩, 233 ⟨57JOC1611⟩

Surrey, A. R., **3**, 618 ⟨45OS(25)63⟩

Surridge, H. H., **4**, 14 ⟨71JOC3324⟩

Surridge, J. H., **1**, 72 ⟨72JOC1182⟩; **2**, 23, 24, 30 ⟨71JOC3324⟩, 24 ⟨71JOC2088⟩, 631 ⟨70JOC3436⟩

Surtees, J. R., **2**, 1099 ⟨65AJC14⟩

Sury, E., **1**, 537 ⟨53HCA1809⟩

Surya Prakash, G. K., **1**, 900 ⟨85JA2764, 88JA8599⟩, 900 ⟨90JA6422⟩, 904 ⟨90JOC6061⟩; **2**, 70 ⟨89JA393⟩, 104 ⟨93S165⟩, 537 ⟨77S677⟩, 650 ⟨91SL39⟩; **4**, 161, 212 ⟨86S513⟩; **6**, 212 ⟨B-92MI 607-02⟩

Suryawanshi, S. N., **4**, 174 ⟨86JOC902⟩, 202 ⟨90SC687⟩, 749, 750 ⟨79IJC(B)500⟩

Surzur, J. M., **1**, 7 ⟨80TL287⟩, 926, 935, 942 ⟨88PAC1659⟩, 949 ⟨92JOC6118⟩; **2**, 337 ⟨81TL61⟩, 1016 ⟨56BSF1615⟩

Süs, O., **3**, 470 ⟨58LA(617)20⟩

Susanne, C., **5**, 955 ⟨94JOM(476)C25⟩

Suschitzky, H., **2**, 614 ⟨65AFC(4)1⟩, 757 ⟨70CI(L)1188⟩, 806 ⟨65JCS6186⟩, 884 ⟨68JCS(C)2317⟩; **3**, 542 ⟨66TL3465, 68JCS(C)2730⟩; **5**, 183, 194, 198 ⟨64JCS755⟩, 593 ⟨76JCS(P1)33, 77JCS(P1)2357⟩, 1111 ⟨70JCS(C)1530⟩; **6**, 28 ⟨89JCS(P1)283⟩, 104 ⟨82JCS(P1)2871⟩

Suslick, K. S., **3**, 163 ⟨92TL7647⟩

Suslova, E. N., **4**, 368 ⟨81JOM(204)13⟩

Süss-Fink, G., **6**, 143 ⟨85CB3959⟩

Sustmann, R., **1**, 786 ⟨80AG(E)779⟩, 892 ⟨69JA1037⟩, 922 ⟨90APO(25)131⟩, 937 ⟨92CB287, 93CB1759⟩, 946 ⟨87CB1197⟩, 1109 ⟨69JA5350⟩; **2**, 243 ⟨93JOC82⟩; **5**, 1 ⟨91COS(6)301⟩, 654 ⟨85HOU(E5/1)628, 85HOU(E5/1)632⟩, 664 ⟨85HOU(E5/1)631⟩

Susuki, I., **6**, 426 ⟨91TL4371⟩

Susuki, T., **6**, 7 ⟨70JOC2982⟩

Suszko, J., **3**, 710 ⟨75JCS(P1)2122⟩

Suszko, P. R., **4**, 706 ⟨84JA2337⟩

Sutcliffe, H., **6**, 604 ⟨64JCS4066⟩

Sutcliffe, L. H., **5**, 49 ⟨55TFS1105⟩

Sutcliffe, R., **2**, 687 ⟨81JCS(P2)1512⟩

Suter, C., **1**, 502 ⟨70HCA1713⟩, 794 ⟨70JA7408⟩

Suter, C. M., **2**, 100 ⟨43OSC(2)112⟩, 101 ⟨38JA538⟩, 102 ⟨43OSC(2)111⟩, 128 ⟨45JA594⟩; **4**, 81 ⟨B-40MI 402-01⟩

Sutharchanadevi, M., **3**, 710 ⟨90JCS(P1)1847⟩

Sutherland, A. G., **1**, 4, 9 ⟨91JCS(P1)656⟩, 618 ⟨89TL3267⟩; **4**, 84 ⟨89TL3267⟩, 650 ⟨91JCS(P1)2276⟩

Sutherland, B. L. S., **4**, 182 ⟨74AJC679⟩

Sutherland, G. B. B. M., **5**, 84 ⟨52MI 502-02⟩

Sutherland, I. O., **1**, 398, 399 ⟨83JCS(P1)1009⟩, 400, 408, 410, 821, 823 ⟨81JCS(P1)1953⟩, 401 ⟨80JCS(P1)1458⟩, 407, 408 ⟨69CC293⟩, 407, 823 ⟨80JCS(P1)1436, 80JCS(P1)1450⟩, 706, 709 ⟨73CC520⟩, 807, 821, 822 ⟨69CC99⟩, 822 ⟨68CC186, 81JCS(P1)1963⟩, 839 ⟨83JCS(P1)1029⟩; **2**, 224 ⟨90JCS(P1)1615⟩, 764 ⟨59JCS2094⟩; **5**, 96 ⟨B-83MI 502-01⟩, 122 ⟨79COC(2)841⟩

Sutherland, J. K., **1**, 108 ⟨77CC321⟩; **2**, 593 ⟨91COS(3)341⟩; **3**, 218 ⟨78CC852⟩, 528 ⟨87CC1728⟩

Sutherland, M. D., **1**, 812 ⟨68CC1229, 70CC892⟩

Sutherland, R. G., **1**, 910 ⟨63JCS3295⟩; **4**, 706 ⟨70CC555⟩; **6**, 30 ⟨90TL6831, 91JOM(419)357⟩

Sutin, N., **5**, 385 ⟨92IC2318⟩

Sutor, D. J., **2**, 1002 ⟨54AX145⟩

Sutowardoyo, K., **2**, 85 ⟨88JOM(346)C7⟩; **3**, 632 ⟨91TA437⟩

Sutrisno, R., **3**, 355 ⟨78S831, 79S361⟩

Sutter, P., **1**, 651 ⟨86JOC558⟩; **5**, 375 ⟨86JOC558⟩

Sutterlin, W., **2**, 110 ⟨31ZAAC(202)1, 31ZAAC(202)22⟩, 111 ⟨35ZAAC(222)92⟩

Sutthivaiyakit, S., **3**, 625 ⟨81TL4107⟩

Sutton, C., **5**, 386 ⟨70CC100⟩

Sutton, Jr., C., **1**, 145 ⟨90JOC2234⟩

Sutton, C. H., **5**, 1147 ⟨91IC1046⟩; **6**, 515 ⟨91IC1046, 92IC4911⟩

Sutton, D., **4**, 713 ⟨82CC371⟩; **6**, 517 ⟨85OM478⟩

Sutton, K. H., **5**, 453 ⟨86CC607, 86TL3787⟩

Sutton, L. E., **3**, 694 ⟨30JCS1876⟩

Sutton, R., **3**, 362 ⟨63JA3989⟩

Sutton, T. M., **3**, 306 ⟨73JCS(P1)1196⟩

Suvalova, E. A., **5**, 644 ⟨89ZOB1917, 90ZOB1389⟩, 660 ⟨87ZOB1514⟩, 660, 788 ⟨87JGU1349⟩, 738 ⟨90JGU1389⟩

Suvorova, D. N., **4**, 719 ⟨77JOM(141)313⟩

Suvorova, I. V., **4**, 941 ⟨84ZOB309⟩, 942 ⟨84ZOB116⟩

Suvorova, S. E., **6**, 653 ⟨74MI 621-01⟩

Suvorov, B. V., **5**, 105 ⟨79MI 502-09⟩

Suvorov, N. N., **5**, 567, 581 ⟨76KFZ41⟩, 1025 ⟨78KGS1497⟩

Suwabe, E. M., **3**, 356 ⟨90JCS(P1)3003⟩

Suwabe, Y., **6**, 471 ⟨93EUP561363⟩
Suwara, Y., **5**, 28 ⟨77MI 502-02⟩
Suya, K., **3**, 139 ⟨86CC469⟩
Suyama, T., **5**, 774 ⟨92NKK1344⟩; **6**, 335 ⟨75YGK136⟩
Suzer, S., **4**, 697, 698 ⟨88JA3231⟩
Suziki, H., **2**, 790 ⟨77S217⟩
Suzki, K., **4**, 325 ⟨81CPB2496⟩
Suzucki, H., **5**, 237, 277 ⟨86CL915⟩, 237, 277 ⟨87CL293⟩
Suzukamo, G., **1**, 801, 803 ⟨86CL621⟩; **2**, 314 ⟨88TL223⟩
Suzuki, A., **1**, 125 ⟨77BCJ2199⟩, 185 ⟨90JOC6356⟩, 198 ⟨79SC809⟩, 198 ⟨82ACR178⟩, 228 ⟨82JOM(233)13⟩, 273 ⟨67JA5708, 67JA5709⟩, 274 ⟨71JA1508⟩, 274, 546 ⟨70JA3503⟩, 276 ⟨75CL523, 76TL255⟩, 276 ⟨78BCJ339⟩, 277 ⟨71JA2792⟩, 277, 977 ⟨73S305⟩, 278 ⟨79BCJ1865⟩, 279 ⟨86TL6369, 87CL25⟩, 279 ⟨87S373, 91SL687⟩, 279, 1033 ⟨73JA3080⟩, 279, 532 ⟨81TL127⟩, 281 ⟨91TL6749⟩, 462 ⟨79CL1003⟩, 465, 473, 489 ⟨91PAC419⟩, 469 ⟨77TL3369⟩, 472, 489, 531 ⟨81JOM(213)C53⟩, 472, 532, 1104 ⟨79CC866⟩, 473, 489 ⟨81SC513⟩, 480, 482, 489 ⟨93JOC2201⟩, 488 ⟨86MI 111-01, 86TL3745⟩, 489 ⟨88CL809, 89JA314⟩, 489 ⟨90OS(68)130, 92BCJ2863⟩, 489 ⟨92SL207, 93CL1429⟩, 489, 531 ⟨81CL879, 86CL1329⟩, 532 ⟨82JOC2117, 85PAC1749⟩, 532, 1020 ⟨79TL3437⟩, 532, 1020 ⟨85JA972⟩, 532, 1094 ⟨80BCJ1471⟩, 532, 1104 ⟨80TL2865⟩, 542, 1008 ⟨88TL1811⟩, 657 ⟨72JA6549⟩, 663 ⟨80BCJ1652⟩, 1009 ⟨90BCJ2124⟩, 1033, 1034 ⟨75TL1961, 86JOC4507⟩, 1050 ⟨92CL695⟩, 1091, 1094 ⟨86CL459⟩, 1104 ⟨89TL5153⟩; **2**, 17 ⟨83CL1135, 84CL1241⟩, 19 ⟨85CL1307, 87CL1675⟩, 34 ⟨71JA1508⟩, 305 ⟨87T4079⟩, 514 ⟨85TCC23⟩, 527, 900, 902 ⟨89TL3789⟩, 528 ⟨90TL231⟩, 615 ⟨87SC685, 89SC865⟩, 623 ⟨78CL665⟩, 623, 628, 902 ⟨83TL731⟩, 624, 902 ⟨83S1005⟩, 690 ⟨91CL459⟩, 901 ⟨87S373⟩, 901, 902 ⟨83TL735, 93JOC5351⟩, 902 ⟨85S406, 89TL5153⟩, 902 ⟨90TL247, 93JA7219⟩, 902, 927, 939 ⟨88TL1811⟩, 904 ⟨74TL2961⟩, 1086 ⟨86JOC4507⟩; **3**, 19 ⟨83SC1149⟩, 24 ⟨82JOC2117⟩, 27 ⟨85S406⟩, 135 ⟨81CL1059, 83SC367⟩, 147 ⟨92CL695⟩, 223 ⟨70JA3503⟩, 228 ⟨87BCJ3471⟩, 233 ⟨89CL1959⟩, 290 ⟨91BCJ1999⟩, 290, 295 ⟨91TL6923⟩, 290, 296, 301, 306, 311 ⟨93TL7595⟩, 290, 305, 311 ⟨92TL4465⟩, 552 ⟨79CL535⟩, 622 ⟨75CL523⟩, 645 ⟨74BCJ2511⟩; **4**, 765 ⟨89TL5153⟩, 815 ⟨93SC2851⟩, 935 ⟨81CL1059, 83CL933⟩, 935 ⟨83SC1149⟩, 935, 949 ⟨80CL591⟩, 954 ⟨89CL1959, 92BCJ2303⟩; **5**, 142 ⟨77BCJ1639⟩, 157 ⟨81CL879⟩; **6**, 634 ⟨83SC367⟩
Suzuki, E., **2**, 108 ⟨93MI 202-01⟩; **3**, 369 ⟨90JST(238)71⟩
Suzuki, F., **2**, 651 ⟨78CC242⟩
Suzuki, H., **1**, 24 ⟨85BCJ2421⟩, 37 ⟨85BCJ1067⟩, 232 ⟨91JCS(P2)1703⟩, 251 ⟨89TL2413⟩, 257 ⟨90JOC408⟩, 469, 1005 ⟨89SC2199⟩, 623 ⟨83TL2599⟩, 648 ⟨89CC1749⟩, 743 ⟨88CL849⟩, 1042 ⟨89CL1079⟩; **2**, 13 ⟨70BCJ3299⟩, 14 ⟨84BCJ303⟩, 71 ⟨89CC1700⟩, 142 ⟨80CL143⟩, 216 ⟨92BCJ626⟩, 235 ⟨86H(24)303⟩, 290 ⟨81NKK472⟩, 293 ⟨85CL679⟩, 311 ⟨92JOC1495⟩, 614 ⟨89CL2213⟩, 630 ⟨88S236⟩, 632 ⟨85CL411⟩, 701 ⟨87CL293⟩, 708, 723 ⟨84CL139⟩, 727 ⟨89CL769⟩, 730, 731, 735 ⟨92S549⟩, 731 ⟨81CL1117⟩, 753 ⟨86CL549⟩, 1001 ⟨83TL5109⟩, 1082 ⟨92BCJ3504⟩; **3**, 117 ⟨85TL2107⟩, 143 ⟨88TL6137⟩, 150 ⟨87CL1153⟩, 165 ⟨89BCJ1358⟩, 196 ⟨93TL8457⟩, 196, 500 ⟨90BCJ950⟩, 210 ⟨89TL3685⟩, 231 ⟨79JA5070⟩, 438 ⟨85CL679, 88CL847⟩, 618 ⟨78S905⟩, 625 ⟨83CL193⟩, 626 ⟨85CL1779⟩, 634, 635 ⟨85BCJ765, 87CL887⟩; **4**, 155 ⟨83CL1303⟩, 247 ⟨91TL2039⟩, 286 ⟨89BCJ1358⟩, 432 ⟨93S1218⟩, 653 ⟨91TL2049⟩, 657 ⟨86JOM(299)C35⟩, 681 ⟨92OM989⟩, 750 ⟨88S236⟩, 774, 777 ⟨92BCJ210⟩, 872 ⟨85S497⟩; **5**, 242, 253 ⟨90JCR(S)56⟩, 281 ⟨83CL449⟩, 289, 302 ⟨91CL1443⟩, 344 ⟨91CL105⟩, 387 ⟨75CL197, 75CL641⟩, 630 ⟨90CL1651⟩, 665 ⟨90CL559⟩, 1055, 1113 ⟨79S705⟩; **6**, 45, 49 ⟨85S497⟩, 150 ⟨88BCJ2927⟩, 482 ⟨89BCJ2117⟩, 498 ⟨85CL1671⟩, 709 ⟨90BCJ950⟩
Suzuki, I., **3**, 440, 456 ⟨89TL4275⟩
Suzuki, J., **1**, 233, 400 ⟨72JOC1721⟩; **3**, 373, 378 ⟨72JOC1721⟩; **5**, 735 ⟨77JAP73023033⟩
Suzuki, K., **1**, 389 ⟨83TL4997⟩, 408, 821 ⟨85JOC2965⟩, 467 ⟨87JA1257⟩, 473 ⟨80TL845⟩, 990 ⟨83CPB3306⟩, 1039 ⟨79CL447, 80CL255⟩, 1039 ⟨84CL405⟩, 1071 ⟨91TL6735⟩, 1098 ⟨85TL861⟩; **2**, 61 ⟨89TL5447⟩, 70 ⟨92TL5965⟩, 71 ⟨79JA1455⟩, 213 ⟨66JA2328, 67JOC727⟩, 271 ⟨77S795, 78CPB1837⟩, 712 ⟨93CL1⟩, 931 ⟨84TL83, 92SL129⟩; **3**, 108 ⟨84CPB4914⟩, 137 ⟨84BCJ613⟩, 142, 146 ⟨83TL4997⟩, 160 ⟨87TL3515⟩, 160 ⟨87TL5891⟩, 276 ⟨86TL373⟩, 281, 298 ⟨92CL1751⟩, 281, 298 ⟨93BCJ3729⟩, 294 ⟨66JA2328⟩; **4**, 219 ⟨79CL1277⟩, 433 ⟨78BCJ2614⟩, 938, 958 ⟨84BCJ1863⟩, 960 ⟨87BCJ1523⟩; **5**, 1138 ⟨76BCJ464⟩; **6**, 229, 230 ⟨92TL4173⟩, 460 ⟨86JAP61118349⟩, 553 ⟨87H(25)123, 87BCJ435⟩
Suzuki, M., **1**, 132 ⟨85JA3348, 88JA4718⟩, 132 ⟨89JOC1785⟩, 339 ⟨90TL2315⟩, 504 ⟨84AG(E)847⟩, 607 ⟨81TL1809⟩, 710 ⟨84JA4547⟩, 800 ⟨90JOC1506⟩, 1056, 1061 ⟨90JOC441⟩, 1140 ⟨84CL1543, 86CPB1531⟩; **2**, 61 ⟨88TL5419⟩, 539 ⟨81T3899⟩; **3**, 144 ⟨82CL939⟩, 293 ⟨87BCJ767⟩, 543 ⟨90CL1705⟩, 615 ⟨81T3899⟩; **4**, 73 ⟨91TL1463⟩, 181, 182 ⟨80TL1357⟩, 192 ⟨80JA3248⟩, 192 ⟨82TL3595⟩, 193 ⟨81TL1809⟩, 413 ⟨76TL1199⟩; **5**, 93 ⟨79CPB1931⟩, 652 ⟨70TL1901⟩; **6**, 124 ⟨80TL1357⟩
Suzuki, N., **1**, 20 ⟨87S736⟩, 158 ⟨85JA5301⟩, 197 ⟨93JA8485⟩, 442 ⟨87JCS(P1)1951⟩, 505, 507 ⟨83BCJ353⟩, 515 ⟨85T2387⟩; **2**, 429 ⟨83BCJ219⟩, 621, 627 ⟨84CC1523⟩, 927 ⟨93CC1042⟩; **3**, 108 ⟨84CPB4914⟩, 208 ⟨90S677⟩, 664 ⟨80CC1253⟩; **4**, 575 ⟨85JA5301⟩
Suzuki, O., **3**, 563 ⟨85CC1773⟩
Suzuki, R., **2**, 936 ⟨B-66MI 218-02⟩
Suzuki, S., **1**, 156 ⟨85SC1123⟩, 186, 225, 229 ⟨89JA7938⟩, 187 ⟨83S804⟩, 212 ⟨84JA6093⟩, 521 ⟨78CL649⟩, 613 ⟨86JOC3834⟩, 892 ⟨58JA2326⟩; **3**, 57 ⟨86BCJ3287⟩, 59 ⟨76CL29⟩; **4**, 183 ⟨86TL69⟩; **5**, 327 ⟨81JGU2125⟩, 385 ⟨86BCJ2203⟩, 419 ⟨80TL1653, 87OM974⟩
Suzuki, T., **1**, 14, 221 ⟨82TL3597⟩, 226 ⟨90JA1661⟩, 290 ⟨88OM739⟩, 473 ⟨86BCJ2044⟩, 620 ⟨88CC1044⟩, 648 ⟨85BCJ2812⟩, 829 ⟨88CC1531⟩, 983 ⟨89TL4271⟩, 1120 ⟨76CC180⟩; **2**, 922, 931 ⟨82TL3387⟩; **3**, 160 ⟨87TL4293⟩, 337 ⟨91JA7782⟩, 353 ⟨89TL4271⟩, 551 ⟨89TL1249⟩; **4**, 186 ⟨92BCJ304⟩, 321 ⟨72CI(L)380⟩, 509 ⟨78JOM(153)193⟩, 625, 637, 1055, 1058 ⟨88OM739⟩, 837 ⟨89TL1249⟩, 1044 ⟨90JOM(391)19, 90JOM(396)299⟩; **5**, 159 ⟨88CC1531⟩, 607 ⟨85H(23)2213, 86H(24)1247⟩; **6**, 185, 186 ⟨88OM739⟩
Suzuki, Y., **2**, 935 ⟨82TL4589⟩; **3**, 9, 123 ⟨89CL901⟩, 97 ⟨83CPB1751⟩, 593 ⟨69TL3407⟩; **4**, 398 ⟨92CL819⟩, 736 ⟨86TL3173⟩; **6**, 632 ⟨85BCJ3379⟩
Suzuki, Z., **2**, 10 ⟨68BCJ1724⟩, 26 ⟨67JOC31⟩; **3**, 498 ⟨68BCJ2815⟩
Suzuli, M., **1**, 34 ⟨73JOC2731⟩
Svanholm, U., **5**, 1054 ⟨66ACS278⟩; **6**, 598 ⟨66ACS278⟩
Svanholt, H., **3**, 553 ⟨79CC99⟩
Svara, J., **2**, 427, 430 ⟨90ZN(B)203⟩, 1073 ⟨87ZAAC(548)63⟩
Sveda, M., **2**, 366 ⟨44JOC89⟩
Svendsen, A., **3**, 246 ⟨75JOC1927⟩
Svendsen, J. S., **1**, 498 ⟨88ACS(B)259⟩; **2**, 48 ⟨89JOC2263⟩
Svendsen, K., **3**, 462 ⟨81ACS(B)179⟩
Svensson, A., **1**, 1072 ⟨83JHC729, 84JOC2018⟩
Svensson, K., **2**, 652 ⟨88JMC1080⟩
Svergun, V. I., **4**, 656, 658 ⟨74IZV2861⟩
Svetkin, Y. V., **5**, 193 ⟨58ZOB1864⟩
Svetlakov, N. V., **2**, 25, 31 ⟨69JOU971⟩
Svetlik, J., **5**, 655 ⟨84CCC1505⟩

Svilarich-Soenen, M., **4**, 800 ⟨72T5149⟩

Sviridov, A. F., **3**, 130 ⟨83TL4355⟩, 211 ⟨83TL4359⟩; **5**, 239 ⟨83TL4355⟩

Sviridova, L. A., **3**, 341 ⟨90KGS1285⟩

Sviridov, E. P., **4**, 344 ⟨74ZOB1712⟩, 999, 1000 ⟨75ZOB12⟩

Svistunova, N. Y., **5**, 385 ⟨93S117⟩

Svoboda, J., **4**, 970 ⟨81CCC1389⟩; **5**, 324 ⟨91CCC1317⟩

Svoboda, J. J., **2**, 1006 ⟨72S307⟩, 1008 ⟨72S306⟩; **5**, 688, 694 ⟨89JOC4335⟩

Swain, C. G., **1**, 107 ⟨53JA246⟩, 883 ⟨61JA1951⟩, 892 ⟨55JA1122⟩; **3**, 459 ⟨47JA2306⟩; **5**, 207 ⟨50JA5426⟩

Swain, C. J., **1**, 338 ⟨92TL3543⟩, 697, 700, 703 ⟨85CC1326, 87JCS(P1)1623⟩; **2**, 351 ⟨91CC1248⟩; **3**, 121, 294 ⟨83S572⟩, 273 ⟨89TL985⟩; **6**, 472, 473 ⟨85CC1326⟩

Swain, M. L., **5**, 1035 ⟨75JCS(P1)922⟩

Swain, M. S., **1**, 892 ⟨55JA2801⟩

Swallen, L. C., **2**, 128, 132 ⟨30JA651⟩; **4**, 51 ⟨30JA651⟩

Swallow, J. C., **4**, 914, 920 ⟨70JOC3583⟩

Swallow, W. H., **1**, 387 ⟨69T4117⟩

Swami, K., **4**, 700 ⟨87OM834, 92JOM(429)311⟩, 701 ⟨83OM909, 84OM1687⟩

Swaminathan, K., **5**, 433 ⟨75CC719⟩

Swaminathan, S., **1**, 813 ⟨85TL6137⟩, 885 ⟨58T88⟩; **2**, 57 ⟨90JOC6328⟩, 526 ⟨91TL6691⟩; **3**, 106 ⟨52JA5110⟩, 144 ⟨84S1010⟩, 216 ⟨71CRV429⟩

Swan, G. A., **1**, 885 ⟨48JCS1408⟩; **2**, 38, 52, 86, 89 ⟨B-65MI 202-01⟩

Swan, J. M., **2**, 122 ⟨63JOC3608⟩, 172, 220 ⟨62JCS2172⟩, 176, 177 ⟨59JCS2969, 62AJC570⟩, 228 ⟨65JCS2901⟩; **4**, 454, 471 ⟨83AJC2095⟩; **5**, 601, 621 ⟨61JCS1194⟩

Swanepoel, A. D., **4**, 228 ⟨90TL2759⟩

Swanick, M. G., **5**, 937 ⟨71JCS(A)794⟩

Swank, D. D., **2**, 831 ⟨71JA5236⟩

Swank, D. W., **2**, 160 ⟨79JOC2632⟩

Swann, B. O., **2**, 982 ⟨71JA7331, 73JA1296⟩

Swann, D. A., **2**, 125 ⟨64T1265⟩; **5**, 238 ⟨70CC1310⟩

Swann, E., **1**, 327, 328 ⟨91JA7002, 90JA6740⟩

Swanson, D. R., **1**, 467 ⟨91TL4453⟩, 536 ⟨87TL917⟩; **2**, 923, 969 ⟨91JOC2590⟩, 927, 986, 988 ⟨89JA3336, 88TL1631⟩, 986, 987 ⟨87CL623⟩, 986, 988 ⟨89CL761⟩, 986, 991 ⟨91OM825⟩; **4**, 652 ⟨90JOC5406⟩, 809 ⟨89JA3336⟩, 1061, 1066 ⟨87TL917⟩

Swanson, E. D., **1**, 530 ⟨86TL1241⟩

Swanson, S., **2**, 612 ⟨88TL3365⟩; **4**, 61, 73, 86 ⟨88TL3365⟩

Swanson, S. B., **2**, 684 ⟨76JOC732⟩

Swapna, V., **3**, 294 ⟨92JOC5841⟩

Swärd, K., **1**, 494, 1029 ⟨85JOC2302⟩

Swaringen, Jr., R. A., **4**, 193 ⟨79JOC4825⟩

Swarts, F., **1**, 426 ⟨20MI 110-01⟩; **6**, 68 ⟨26BSB412⟩, 214, 225 ⟨33BSB102⟩, 216 ⟨03MI 607-01⟩

Swarts, W. A., **6**, 147 ⟨62JOC1455⟩

Swartz, J. E., **2**, 81 ⟨89JA1211⟩

Swartz, T. D., **1**, 905 ⟨86JOC3736⟩

Swartz, W. H., **4**, 513 ⟨92JOC6037⟩

Sweat, F. W., **3**, 88 ⟨67CRV247⟩

Swedo, R. J., **2**, 11 ⟨78JOC2291⟩, 250, 362, 366 ⟨77S798⟩

Sweeney, A., **1**, 803 ⟨71CC1248⟩

Sweeney, J. B., **1**, 264 ⟨89CC1852⟩

Sweeney, J. J., **4**, 656 ⟨79JCS(D)767⟩; **6**, 395 ⟨79JCS(D)767⟩

Sweeney, J. N. A., **4**, 213 ⟨84JCS(P1)1531⟩; **5**, 79 ⟨84JCS(P1)1531⟩

Sweeney, R. F., **4**, 160 ⟨68JOC2692⟩

Sweeney, T. J., **1**, 826 ⟨90T4487⟩

Sweeney, W. M., **4**, 761, 763 ⟨56JA1685⟩

Sweet, L. A., **2**, 260 ⟨48JA1762⟩

Sweeting, O. J., **2**, 176 ⟨63JOC3537⟩

Sweetman, B. J., **4**, 216 ⟨69JOC1799⟩

Sweger, R. W., **1**, 785, 786, 1034 ⟨91COS(5)551⟩

Swenson, D. C., **4**, 746, 748, 760, 817, 821 ⟨93JA5430⟩

Swenson, R., **1**, 690 ⟨87TL3209⟩

Swenson, W., **1**, 1016, 1108 ⟨84JA3551⟩

Swenton, J. S., **1**, 209 ⟨83JOC1455⟩; **2**, 14 ⟨74JOC2300⟩; **4**, 646 ⟨81CC179⟩

Swepston, P. J., **2**, 602 ⟨74JOM(67)61, 84JA6841⟩

Swepston, P. N., **6**, 194 ⟨85JA8103⟩, 195 ⟨85JA8091⟩, 332 ⟨83IC411⟩

Sweptston, P. N., **2**, 969 ⟨85JA8103⟩

Swerczewski, G., **2**, 561 ⟨66TL875⟩

Swering, J. E., **2**, 162 ⟨83TL1011⟩

Swern, D., **1**, 330, 934 ⟨79COC(1)909⟩; **2**, 90 ⟨53OR(7)378⟩, 95, 96 ⟨79COC(4)909⟩, 95, 97, 98 ⟨B-71MI 202-02⟩, 120, 180 ⟨58JOC1525⟩, 160 ⟨74JOC2148⟩, 160 ⟨77JOC593⟩, 162 ⟨74JOC3365⟩, 162 ⟨75JOC2570⟩, 162 ⟨78JOC4537⟩, 163 ⟨71JOC2207⟩, 163 ⟨76S731⟩, 164 ⟨78TL503⟩, 164 ⟨79JOC2510⟩, 164 ⟨84SC915⟩, 192 ⟨62JOC2853⟩, 348 ⟨49CRV34⟩, 478 ⟨71TL1761⟩, 793 ⟨76JOC2207⟩, 1055 ⟨81S165⟩; **3**, 6, 118 ⟨78JOC2480⟩, 6, 118 ⟨78T1651⟩, 6, 88, 118 ⟨81S165⟩; **4**, 205 ⟨75JHC749⟩; **5**, 28 ⟨62CI(L)1304, B-55MI 502-01⟩, 205 ⟨55JA2519, 73OSC(5)904⟩, 209 ⟨81S165⟩, 275, 276, 277, 278 ⟨49JA2215⟩, 350 ⟨71TL1761, 85JOC2224⟩, 985 ⟨67JOC3665, 69JOC1677⟩

Swestock, J., **2**, 321 ⟨93JOC3299⟩

Swetkin, J. W., **4**, 824 ⟨70MI 419-01⟩

Swett, L. R., **1**, 881 ⟨61JOC644⟩; **2**, 684 ⟨86SC1043⟩, 1049 ⟨58JOC1389⟩

Swienty-Busch, J., **1**, 570 ⟨91COS(6)949⟩

Swierczewski, G., **1**, 316 ⟨69TL707⟩, 477 ⟨77JOM(127)371⟩, 1094 ⟨75T2735⟩; **2**, 561 ⟨75CC242⟩

Swigert, J., **2**, 404, 405 ⟨71JA7337⟩

Swincer, A. G., **5**, 941 ⟨83AOC(22)59⟩, 944 ⟨80AJC2605⟩

Swindall, J. J., **4**, 656 ⟨77JCR(S)116⟩; **6**, 394 ⟨77JCR(S)116⟩

Swindell, C. S., **1**, 195 ⟨83S701⟩, 480 ⟨84JOC4894⟩, 482 ⟨79JA2246⟩; **3**, 231 ⟨90JOC3⟩

Swindell, R. F., **3**, 456 ⟨71MI 312-02⟩, 456, 458, 459 ⟨72IC242⟩; **5**, 807 ⟨71IC1635⟩; **6**, 239 ⟨71IC1635⟩

Swingle, M. N., **3**, 128 ⟨93T965⟩

Swingle, N. M., **1**, 305, 306 ⟨92CRV771⟩; **2**, 588 ⟨92CRV771⟩

Swiridoff, W., **4**, 632 ⟨83JA2582⟩

Swisher, J. V., **1**, 620, 1074 ⟨71CC22⟩; **2**, 913, 914 ⟨61JA4381⟩

Swiss, K. A., **2**, 280 ⟨91COS(7)515⟩

Switalski, J., **2**, 172 ⟨71JOC2530⟩

Switzer, C. Y., **1**, 195 ⟨88TL6433⟩

Switzer, F. L., **5**, 367 ⟨89JOC145⟩

Swor, R. A., **3**, 492 ⟨66JA1953⟩

Sworin, M., **3**, 422 ⟨83JOC2685⟩

Swyke, C., **2**, 407, 862 ⟨75JCS(D)12⟩

Sy, A. O., **5**, 998 ⟨80TL2223⟩

Sy, M., **3**, 91, 98 ⟨55BSF1594⟩

Sy, W.-W., **2**, 392 ⟨85TL1193⟩, 631 ⟨89TL3769, 90SC877⟩

Sydnes, L. K., **1**, 356 ⟨82CC1331⟩, 498 ⟨88ACR229⟩

Sygula, A., **5**, 685 ⟨89JCR(S)56⟩

Sykes, P., **3**, 361 ⟨72JCS(P1)2305⟩; **4**, 321 ⟨67JCS(C)1411⟩; **5**, 693, 758, 759 ⟨77JCS(P1)1791⟩; **6**, 314 ⟨71JCS(C)103⟩

Sykes, P. J., **2**, 23 ⟨68JCS(C)416⟩; **3**, 87 ⟨68JCS(C)2915⟩

Sykes, R. B., **3**, 704 ⟨88JAN454⟩

Syldatk, C., **4**, 353 ⟨91JOM(403)29⟩, 506, 513 ⟨91JOM(415)39⟩

Sylvester, L., **3**, 307 ⟨83JHC1565⟩

Symeonides, K., **5**, 485 ⟨80PS(8)1⟩

Symmes, C., **2**, 462 ⟨76JOC1578⟩

Symon, D. A., **4**, 688 ⟨76JOM(120)C13⟩

Symon, J. D., **3**, 73 ⟨69CC466⟩, 73, 255 ⟨70JCS(C)2412⟩, 377 ⟨74JCS(P1)242⟩

Symons, M. C. R., **1**, 338 ⟨92TL3543⟩, 892 ⟨62T333⟩

Synder, J. P., **3**, 603 ⟨73CB2601⟩; **5**, 513 ⟨90JA3029, 90JA3040⟩

Syndikus, D., **2**, 473, 508 ⟨92ZAAC(608)43⟩
Syper, L., **2**, 279 ⟨84S439⟩, 654 ⟨89S167⟩; **3**, 87, 88 ⟨67TL4193⟩, 87, 93, 99 ⟨66TL4493⟩, 89, 96, 98, 99, 108 ⟨84S747⟩; **4**, 288 ⟨84S439⟩; **6**, 92 ⟨84S439⟩
Sypniewski, G. C., **1**, 777 ⟨81JOC2804⟩
Syrkin, A. M., **4**, 775 ⟨90CA230883w⟩
Syrkin, V. G., **4**, 724 ⟨72ZOB2450⟩
Syrova, G. P., **6**, 236 ⟨68ZOB2591⟩
Syschitzky, J. L., **3**, 297 ⟨85JMC1832⟩
Sysoeva, L. P., **3**, 681 ⟨76ZOR1676⟩; **5**, 356 ⟨75ZOR173, 75ZOR348⟩, 652 ⟨76ZOR1557⟩, 860 ⟨74ZOR2200⟩
Syvret, R. G., **2**, 15 ⟨92CC595⟩
Szabo, A., **2**, 274 ⟨64JOC3561⟩
Szabó, A. E., **6**, 558 ⟨82H(19)1191, 84JHC1373⟩, 558 ⟨85T1353, 90MRC1045⟩
Szabo, A. G., **3**, 60 ⟨87JOC4047⟩
Szabó, E. A., **6**, 557 ⟨83T1829⟩, 558 ⟨83JHC1181⟩
Szabo, G. T., **2**, 34 ⟨93T5133⟩
Szabo, J., **4**, 273 ⟨89MRC760⟩, 319 ⟨72ACH363, 87H(26)2381⟩; **6**, 111 ⟨91JHC481⟩
Szabó, J. A., **6**, 558 ⟨85S1149⟩
Szabo, J. L., **2**, 138 ⟨48JA3667⟩
Szabo, L., **5**, 1108 ⟨53BSF771⟩
Šzabó, L., **5**, 567 ⟨93JOC6076⟩
Szabó, Z., **3**, 361 ⟨91BSF976⟩
Szajáni, B., **5**, 1065 ⟨87S520⟩, 1069 ⟨88S349⟩
Szajda, M., **4**, 317, 416 ⟨86S804⟩
Szakacz, S., **5**, 709 ⟨82MI 517-01⟩
Szalontai, G., **4**, 446 ⟨86S817⟩; **5**, 1039 ⟨86S817⟩; **6**, 565 ⟨86S817⟩
Szameitat, J., **4**, 112 ⟨88JOM(344)61⟩, 112, 116, 117 ⟨86ZN(B)974⟩; **6**, 265 ⟨92ZN(B)321⟩, 266 ⟨90JOM(399)189⟩, 678 ⟨86PS(28)239⟩, 679 ⟨87PS(30)401⟩
Szántay, C., **1**, 747 ⟨66JOC1447⟩; **2**, 1015 ⟨92TL683⟩; **5**, 567 ⟨93JOC6076⟩; **6**, 695 ⟨92TL683⟩
Szántay, Cs., **2**, 379 ⟨85JPR445⟩
Szantay, Jr., C., **4**, 901 ⟨89TL4449⟩; **5**, 566, 613 ⟨89T6771, 89TL4449⟩
Szanyi, J., **6**, 489 ⟨91JCS(P1)37⟩
Szarek, W. A., **2**, 9 ⟨87CJC412⟩; **4**, 45 ⟨84CL1751⟩, 190 ⟨78S48⟩
Szargan, R., **6**, 565 ⟨81ZAAC(476)7⟩
Szary, A. C., **4**, 683 ⟨79JCS(D)1801⟩, 849 ⟨73JCS(P1)2272⟩; **6**, 128 ⟨73JCS(P1)2272⟩
Szatmari, E., **2**, 444 ⟨B-84MI 213-01⟩
Szczepaniak, W., **4**, 470 ⟨73RZC929⟩
Szczepanska, Z., **6**, 546 ⟨86MI 617-04⟩
Szczepanski, S. W., **4**, 358 ⟨85JOC5393⟩, 359 ⟨88OR(35)513⟩; **5**, 412 ⟨88OR(35)1⟩, 413, 425 ⟨85JOC5393⟩
Szczygielska-Nowosielska, A., **5**, 277 ⟨80JA7578⟩
Szedlak, I., **6**, 425 ⟨74MIP 614-01⟩
Szefraniec, L. J., **6**, 528 ⟨92JFC(59)91⟩
Szegö, F., **5**, 1079 ⟨64AG(E)228⟩
Szeimies, G., **3**, 467 ⟨65CB1138⟩, 478 ⟨66CB475⟩; **4**, 650 ⟨86T1845⟩
Szeja, W., **4**, 176 ⟨89JCR(S)152⟩; **6**, 562 ⟨88S224⟩
Szen-Gezgin, A., **3**, 477, 479 ⟨78CRV363⟩
Szeverenyi, N. M., **1**, 873, 873 ⟨77JA4954⟩
Szewczyk, J., **4**, 462 ⟨83CJC2425⟩
Szewczyk, J. M., **3**, 437 ⟨93TL6229⟩
Szilagyi, G., **6**, 477 ⟨83MI 622-01⟩
Szilagyi, K., **1**, 578 ⟨88JPR309⟩
Szilágyi, L., **6**, 88 ⟨88T2063⟩
Szilagyi, S., **2**, 154 ⟨79SC923⟩; **6**, 509 ⟨78JOC61⟩
Szirtes, T., **5**, 277 ⟨74TL1785⟩
Szmant, H., **4**, 197 ⟨75TL4543⟩
Szmant, H. H., **1**, 887 ⟨68AG(E)120⟩; **3**, 93 ⟨B-89MI 303-02⟩, 444 ⟨50JA2890⟩; **6**, 409, 411, 412, 415 ⟨B-89MI 614-01⟩
Szmuszkovicz, J., **1**, 307 ⟨63JA207⟩; **2**, 738, 751 ⟨B-63MI 215-01⟩, 738, 758 ⟨63JA79⟩; **3**, 10, 124 ⟨54JA2029⟩, 295 ⟨48JA3352⟩
Sznaidman, M. L., **3**, 618 ⟨93TL1581⟩
Szöcs, L., **6**, 503 ⟨81ACH57⟩
Szonyi, F., **5**, 991 ⟨92JFC(56)85⟩, 1022 ⟨91JFC(53)1⟩; **6**, 608 ⟨92TL2339⟩
Szönyi, S., **2**, 675 ⟨93BSF664⟩
Szpala, A., **4**, 922 ⟨81JCS(P1)1363⟩
Szperl, L., **3**, 382 ⟨32RZC270⟩; **5**, 528, 529 ⟨30RZC652⟩
Szucs, E., **4**, 273 ⟨89MRC760⟩
Szucs, S. S., **1**, 495 ⟨71TL3785⟩
Szulejko, J. E., **1**, 1141 ⟨84JA521⟩
Szulgin, K., **2**, 105 ⟨17MI 202-01⟩
Szura, D. P., **1**, 372 ⟨92JA1329⟩; **3**, 404 ⟨88TL761⟩, 457 ⟨92JA1329⟩; **4**, 537 ⟨88TL761, 92JA1329⟩
Szur, A. J., **6**, 36 ⟨71JMC517⟩, 36, 43 ⟨71JMC593⟩, 38, 43 ⟨72JMC604⟩, 38, 43 ⟨72JMC606⟩
Szurdoki, F., **2**, 374 ⟨88S529⟩
Szwarc, M., **1**, 893 ⟨B-72MI 119-08, B-74MI 119-05⟩
Szymanski, R., **6**, 68 ⟨85MI 603-01⟩
Szymonifka, M. J., **2**, 249 ⟨89TL2869⟩; **6**, 129 ⟨89TL2873⟩

T

Taaffee, T. H., **5**, 282 ⟨80JOC165⟩
Taagaard, M., **2**, 469 ⟨86JCR(S)26⟩
Taal, R., **4**, 177 ⟨77RTC44⟩
Tabaei, S.-M. H., **1**, 427 ⟨93TL3263⟩
Tabakovic, I., **6**, 562 ⟨92JHC383⟩
Tabakovic, K., **6**, 562 ⟨92JHC383⟩
Tabashi, K., **5**, 678 ⟨84CPB4437⟩
Tabata, M., **1**, 1035 ⟨75CC857⟩, 1035, 1103 ⟨76TL4385⟩; **2**, 1086 ⟨75CC857, 76TL4041⟩
Tabata, S. H., **6**, 642 ⟨62JOC2504⟩
Tabbron, H. E., **5**, 943 ⟨82CC215⟩, 944 ⟨83JCS(D)2397⟩
Tabche, S., **2**, 529 ⟨82CB2357⟩
Tabei, K., **2**, 776 ⟨67BCJ1538⟩; **5**, 220 ⟨86JHC1657⟩
Tabei, T., **1**, 404, 406, 816, 820 ⟨86TL4511⟩
Taber, D. F., **1**, 307 ⟨78TL2445⟩, 548 ⟨79JA7107⟩, 635 ⟨86JA7686⟩, 944 ⟨93TL6209⟩; **2**, 54 ⟨91TL4227⟩, 326 ⟨88JOC2968⟩, 988 ⟨88JA7128, 89JA6435⟩; **3**, 61 ⟨81JOC4817⟩, 129 ⟨83JA5935⟩, 193, 472 ⟨86JOC4077⟩, 209 ⟨79JOC450⟩; **5**, 869 ⟨86JOC4077⟩
Taber, T. R., **1**, 873, 876 ⟨82TS1⟩; **2**, 532, 698 ⟨81JA3099⟩; **4**, 832 ⟨81JA3099⟩
Tabernero, I., **4**, 764 ⟨86JOC1413⟩
Tabohashi, T., **4**, 615 ⟨83JOM(250)109, 83OM351⟩
Tabor, D. C., **1**, 74 ⟨83JOC1638⟩
Tabor, T. E., **4**, 50 ⟨67JA6573, 68JOC2934⟩
Taboury, F., **2**, 125 ⟨04CR(138)982⟩
Tabti, B., **5**, 199 ⟨91T5481⟩
Tabuchi, H., **5**, 129 ⟨94JOC4749⟩
Tabuchi, T., **1**, 195 ⟨86TL1195⟩, 680 ⟨86TL3891, 87CL2101⟩, 962 ⟨86TL5237⟩, 975 ⟨87CL2275⟩; **3**, 368 ⟨91JCS(P1)3043⟩; **4**, 900 ⟨93MI 420-01⟩
Tabusa, F., **1**, 1039 ⟨84CL405⟩
Tabushi, E., **2**, 702 ⟨83CPB128⟩
Tabyaoui, B., **3**, 77 ⟨88SC1475⟩
Tacconi, G., **1**, 826 ⟨93T2093⟩; **4**, 986 ⟨79JPR43⟩; **5**, 664 ⟨84G133⟩
Tachdjian, C., **1**, 332, 935 ⟨92T7091⟩, 615, 700, 939 ⟨91TL2703⟩, 700, 939 ⟨92T7109⟩; **5**, 516, 523 ⟨92T7109⟩, 516, 526 ⟨91TL2703⟩; **6**, 541, 550 ⟨91TL2703⟩
Tachibana, K., **4**, 699 ⟨91T3281⟩
Tachibana, Y., **1**, 698, 703 ⟨89BCJ3132⟩; **2**, 853 ⟨88BCJ4323⟩, 858 ⟨87TL6649⟩; **3**, 384, 394 ⟨87TL6649⟩
Tachikawa, M., **4**, 682 ⟨75JA7172⟩
Tachon, C., **4**, 547 ⟨87T1793⟩, 1025 ⟨91TL3687⟩; **5**, 1092 ⟨89TL177⟩; **6**, 264 ⟨87T1793⟩, 680 ⟨89TL177⟩, 688 ⟨90JOC5750⟩
Tack, L. M., **3**, 582 ⟨84CPL447⟩
Tacke, R., **2**, 547 ⟨79TCC1⟩; **4**, 136 ⟨82LA1946, 87LA51⟩, 136, 368 ⟨90JOM(388)57⟩, 141 ⟨92JOM(438)45⟩, 142 ⟨91ZN(B)275⟩, 142, 362 ⟨88JOM(354)147⟩, 352 ⟨88JOM(354)139⟩, 352, 353 ⟨92CB591, 91JOM(403)29⟩, 353 ⟨92JOM(424)273⟩, 506, 513 ⟨91JOM(415)39⟩, 602, 621, 929 ⟨92CB607⟩; **5**, 224 ⟨88JOM(334)139⟩, 411, 416 ⟨90JOM(388)47, 92CB607⟩; **6**, 127 ⟨90JOM(388)47⟩
Tackle, A., **1**, 224 ⟨90T4503⟩
Tada, H., **1**, 829 ⟨88CC1531⟩; **5**, 159 ⟨88CC1531⟩
Tada, M., **1**, 946 ⟨82JOC1775⟩
Tada, N., **2**, 21 ⟨84CL877⟩
Tada, S., **1**, 592, 616, 668 ⟨82TL2205⟩, 600 ⟨82CC281⟩
Tada, S.-I., **4**, 392 ⟨84CPB1829⟩
Tada, Y., **3**, 268 ⟨77H(6)1985⟩
Tadano, K., **1**, 827 ⟨88JOC3868, 88TL655⟩, 827, 829 ⟨87JOC1201⟩, 828 ⟨89JOC1223, 90JOC2108⟩
Taddei, F., **2**, 600 ⟨73JCS(P2)638⟩, 825 ⟨74JCS(P2)1338⟩; **4**, 772 ⟨63CA(59)11313c⟩
Taddei, M., **1**, 242 ⟨83TL2311⟩, 667 ⟨87TL2751⟩; **2**, 265 ⟨83TL2311⟩, 304 ⟨89SC3139⟩, 652 ⟨86S633⟩, 907 ⟨91JOC1951⟩, 917, 938, 977 ⟨85S899⟩, 918, 939 ⟨89JOC1473⟩, 942 ⟨85S633, 85S898⟩, 1019 ⟨89S461⟩; **4**, 358 ⟨90T2999⟩; **5**, 199 ⟨86TL2913⟩; **6**, 134 ⟨86S633⟩
Taddia, R., **2**, 181 ⟨74S877⟩
Tadey, T., **3**, 324 ⟨92JA8751⟩
Tadj, F., **1**, 223 ⟨85TL3121⟩
Tadokoro, T., **1**, 1015 ⟨93TL2139⟩
Tadros, W., **1**, 981 ⟨61JCS2687⟩
Tadros, Z., **3**, 616 ⟨93BSF513⟩
Tae, J. K., **2**, 327 ⟨88JHC1383⟩
Tafel, A., **6**, 127 ⟨90JOM(388)47⟩
Taffer, I. M., **1**, 14 ⟨81JOC5214⟩
Taft, Jr., R. W., **1**, 895 ⟨52JA5372, 53JA1253, 53JA3955, 56JA5807, 75JA2902, 57JA3724⟩, 896 ⟨60JA4729⟩; **2**, 836, 839 ⟨68JA5236⟩, 1006 ⟨65JA2490⟩; **5**, 1129 ⟨68JA5236⟩
Tag, K., **6**, 349 ⟨87IC1488⟩
Taga, T., **1**, 189 ⟨93JOC1207⟩; **5**, 171 ⟨85JA6404⟩; **6**, 116 ⟨86JOC4737, 87NKK1447⟩
Tagaki, S., **3**, 209 ⟨80BCJ169⟩
Tagaki, W., **1**, 731 ⟨74TL2587⟩; **2**, 148 ⟨80CL1063⟩; **3**, 494 ⟨74TL2587⟩; **5**, 1144 ⟨86MI 530-03⟩; **6**, 82 ⟨64T417⟩
Tagami, K., **2**, 713 ⟨85CL805, 87JOC1078⟩
Taganliev, A., **6**, 88 ⟨88MI 603-01, 86JOU270⟩, 478 ⟨85ZOR2206⟩
Tagashira, M., **1**, 677, 1099 ⟨86JA6048⟩
Tagat, J., **1**, 1094 ⟨77TL4145⟩
Tagawa, J., **4**, 346 ⟨79TL2415⟩
Tagawa, J.-I., **3**, 480 ⟨79TL2415⟩
Tagawa, Y., **5**, 215 ⟨86CPB4984⟩
Tagle, G., **5**, 242 ⟨82SC897⟩; **6**, 475 ⟨82SC897⟩
Tagliaferri, E., **5**, 1122 ⟨80AG(E)566⟩
Tagliapietra, S., **3**, 18, 134 ⟨86SC803⟩
Tagliavini, E., **1**, 688 ⟨85JOM(280)307⟩; **2**, 79 ⟨88JOM(353)177⟩, 533, 698 ⟨91JOC5820⟩, 590 ⟨83JOC4108, 90CC1680⟩; **3**, 31 ⟨93JA7001⟩, 159 ⟨92T1299⟩
Tagliavini, G., **2**, 79 ⟨88JOM(341)345⟩
Tagmann, E., **1**, 537 ⟨53HCA1809⟩
Taguchi, H., **1**, 565 ⟨75TL1545⟩, 651 ⟨74BCJ2529, 74JA1620⟩; **2**, 900 ⟨90JOC5324⟩; **3**, 54 ⟨73TL2465⟩, 125 ⟨77BCJ1592⟩; **4**, 19 ⟨74JA3010, 77BCJ1588⟩, 648 ⟨74JA1620⟩; **6**, 25, 31 ⟨74JA3010⟩, 298, 300 ⟨86JAP(K)61289091⟩
Taguchi, K., **3**, 406 ⟨71JOC1570⟩
Taguchi, M., **5**, 726 ⟨85BCJ2721⟩
Taguchi, T., **1**, 197 ⟨92TL1295⟩, 197 ⟨93JA8835⟩, 197, 1057 ⟨92TL3769⟩, 1133 ⟨76CC606⟩; **2**, 19 ⟨82JCS(P1)85⟩, 34 ⟨92TL1299, 93TL2165⟩, 123 ⟨69TL3631⟩, 361 ⟨70TL3411⟩; **3**, 55 ⟨77CL345⟩; **4**, 734 ⟨85CPB5137⟩, 736 ⟨90CPB1104⟩; **5**, 898 ⟨70BCJ2539⟩, 1121 ⟨73CL1225⟩; **6**, 472 ⟨72CPB2348, 73CPB604⟩, 472 ⟨74TL4479, 78CPB3807⟩, 554 ⟨80CPB110⟩
Taguchi, Y., **1**, 242 ⟨86BCJ2321⟩; **5**, 320, 335 ⟨91BCJ3732⟩; **6**, 314 ⟨89NKK63⟩
Taha, A. A., **3**, 712 ⟨85T1931⟩
Tahery, A. P., **2**, 367 ⟨82JCS(P1)677⟩
Tahir, S. H., **1**, 821 ⟨91TL335⟩
Tahiro, S., **3**, 496 ⟨89JOC974⟩
Tai, A., **2**, 52 ⟨91CC795⟩; **5**, 63 ⟨82BCJ2186⟩
Tai, D. F., **3**, 299 ⟨84JA1813⟩; **5**, 152 ⟨89JA2984⟩, 166 ⟨84JA1813⟩
Tai, K. M., **3**, 406 ⟨58JOC535⟩
Tailby, G. R., **4**, 687 ⟨69AJC1143⟩

Tailhan, C., **1**, 231 ⟨89H(28)171⟩, 938 ⟨90TL2565⟩; **3**, 136 ⟨88CC308⟩; **6**, 551 ⟨88CC308, 89H(28)171⟩
Taillades, J., **3**, 616 ⟨93BSF513⟩
Taipale, K. A., **2**, 388 ⟨22MI 207-01⟩
Taira, K., **5**, 402 ⟨84T3215⟩
Tait, B. D., **1**, 648, 649, 759 ⟨87JOC281⟩
Tait, S. J. D., **3**, 4, 116 ⟨78TL1601⟩
Tait, T. A., **5**, 557 ⟨78TL4657⟩
Taits, S. Z., **1**, 103 ⟨75IZV888⟩
Tajana, A., **4**, 852 ⟨74AC(R)305⟩
Tajbakhsh, M., **1**, 777 ⟨94T639⟩
Tajika, M., **1**, 1097 ⟨82JA180⟩; **4**, 1033 ⟨76JA3718, 82JA180⟩
Tajima, K., **4**, 187 ⟨84TL5155⟩; **5**, 693 ⟨84CPB4694⟩
Tajima, S., **5**, 199 ⟨90TL709⟩
Tajima, T., **4**, 899, 979 ⟨84JHC133⟩
Tajima, Y., **5**, 234 ⟨84TL2793⟩, 236 ⟨85TL673⟩
Taka, K., **2**, 74, 572 ⟨89JOC4732⟩
Takabe, K., **2**, 27 ⟨77CL1025⟩; **3**, 57 ⟨82CL1987⟩
Takabuchi, A., **4**, 674 ⟨93OM2925⟩
Takachi, H., **5**, 1110 ⟨92CL509⟩
Takachi, T., **1**, 985 ⟨92CL1357⟩
Takacs, J. M., **1**, 876 ⟨80TL4233, 81TH⟩, 878 ⟨81PAC1109⟩; **3**, 11 ⟨79JA371⟩; **4**, 339 ⟨78JA3467⟩, 917 ⟨79JA371⟩; **5**, 148 ⟨81PAC1109⟩
Takada, A., **5**, 706 ⟨85GSM96⟩
Takada, E. I., **1**, 1050 ⟨92CL695⟩
Takada, N., **3**, 223 ⟨90JA902⟩
Takada, R., **2**, 677 ⟨93BCJ1156⟩
Takada, S., **1**, 400, 799 ⟨69JA4322⟩
Takada, Y., **5**, 601, 602 ⟨76PS75, 72IJS89⟩
Takadate, A., **3**, 627 ⟨79JOC67⟩
Takaegami, Y., **3**, 597 ⟨80JOM(190)C39⟩
Takagaki, H., **1**, 133 ⟨79BCJ1241⟩
Takagaki, T., **1**, 3, 6 ⟨81TL2583⟩
Takagi, A., **3**, 593 ⟨86JHC597⟩
Takagi, K., **1**, 433 ⟨91CL2017⟩, 468 ⟨93BCJ2446⟩, 493 ⟨81TL5231⟩, 1036 ⟨91JA1440⟩; **2**, 403 ⟨72JA7406⟩, 632 ⟨80BCJ3691⟩, 707 ⟨85CL1307⟩, 711 ⟨87CL2221⟩; **3**, 662 ⟨75BCJ3298⟩; **4**, 102 ⟨91JCS(P1)49⟩, 108 ⟨91CPB3120⟩, 658 ⟨89JA3089⟩, 979 ⟨91JHC485⟩; **5**, 1110 ⟨92CL509⟩
Takagi, M., **1**, 493 ⟨92JOM(423)405⟩; **5**, 222 ⟨77BCJ2021⟩
Takagi, S., **1**, 422 ⟨80BCJ169⟩
Takagi, T., **5**, 223 ⟨88OM366⟩, 510, 551, 583, 620 ⟨82LA1229⟩, 551 ⟨77AG(E)786⟩, 609 ⟨77AG(E)787⟩
Takagi, Y., **3**, 632 ⟨89CL1505, 89TL3811⟩
Takagishi, S., **2**, 942 ⟨92SL360⟩
Takahara, J. P., **1**, 195 ⟨88JA4473⟩; **2**, 80 ⟨88JA4473⟩
Takahashi, A., **2**, 713 ⟨87JOC1043⟩; **3**, 59, 220 ⟨87TL1893⟩; **5**, 1125 ⟨86JAP47675⟩
Takahashi, E., **1**, 481, 1028 ⟨83CPB4346⟩
Takahashi, F., **3**, 378 ⟨77TL3633, 79T317⟩
Takahashi, H., **1**, 141 ⟨73JA2694⟩, 141, 834 ⟨73JA2693⟩, 545 ⟨87JOC4859⟩; **2**, 54 ⟨90JA5876⟩, 72 ⟨89TL7095⟩, 313 ⟨92TL235⟩, 381 ⟨81CPB3387⟩, 568 ⟨92T5691⟩; **3**, 21 ⟨73JA2693⟩, 97 ⟨83CPB1751⟩, 223 ⟨90JA902⟩, 470 ⟨73TL4573⟩, 599 ⟨70JOM(23)275⟩, 632 ⟨69BCJ1345⟩; **4**, 231 ⟨92CL2173⟩, 956, 957 ⟨73JA2694⟩; **5**, 241 ⟨87JOM(334)C45⟩, 242 ⟨87JOM(334)C43, 89JOM(361)63⟩, 356 ⟨67JOM(10)511⟩, 552 ⟨73JA5803⟩, 1049 ⟨79CPB1137⟩
Takahashi, I., **1**, 156 ⟨93T537⟩; **2**, 725 ⟨70JAP7321096-R⟩; **4**, 323 ⟨91S1147⟩, 937 ⟨86CL2089⟩
Takahashi, K., **1**, 193 ⟨85JOC1523⟩, 225 ⟨93TL8263⟩, 415, 826 ⟨87TL5879⟩, 456 ⟨92CL1725⟩, 525 ⟨87CL2347⟩, 558 ⟨86T2971⟩, 693 ⟨82CL1697, 90JA4035⟩, 1041 ⟨89JCS(P1)1211⟩, 1079 ⟨82TL343⟩; **2**, 55 ⟨88BCJ3283, 88CL1633⟩, 118, 124, 213, 217 ⟨68TL5907⟩, 444 ⟨88JAP63119491⟩, 445 ⟨91JAP70374395⟩, 659 ⟨92TL4603⟩, 748 ⟨83JOC3566, 84S1043⟩, 1000 ⟨82CL1697⟩; **3**, 7 ⟨91BCJ312⟩, 31, 157 ⟨86TL3665⟩, 139 ⟨85JOC1523⟩, 146 ⟨86CL1597⟩, 255 ⟨81CL805⟩, 417, 658 ⟨78S892⟩, 626 ⟨89OPP354⟩, 665 ⟨77S245⟩, 671 ⟨85CL1487⟩; **4**, 21 ⟨84JHC1157⟩, 205 ⟨76CPB102⟩, 395 ⟨82CL1697⟩, 847 ⟨89TL2091⟩, 854, 857 ⟨82JOC5404⟩, 854, 857 ⟨83CL767, 86JOC326⟩, 857 ⟨86JOC700⟩; **5**, 154 ⟨86T2971⟩, 344 ⟨91BCJ106⟩; **6**, 28 ⟨77JHC881, 78JHC893⟩
Takahashi, M., **1**, 37 ⟨84TL4417⟩, 234, 822 ⟨87CL1569⟩, 1040 ⟨80TL849⟩, 1084 ⟨87CL1905⟩, 1106 ⟨78TL2161⟩; **2**, 4 ⟨78JA290, B-82MI 223-01⟩, 794, 795 ⟨92JA1830⟩; **3**, 117, 161 ⟨84TL4417⟩, 282, 295, 296 ⟨90CL783⟩, 336 ⟨89SC2431⟩, 385 ⟨88TL6965⟩, 635 ⟨85BCJ765⟩; **4**, 346 ⟨87CL1569⟩; **5**, 59 ⟨84EUP115085⟩, 160 ⟨80TL849⟩, 164 ⟨80S403⟩, 657 ⟨88H(27)155⟩; **6**, 413 ⟨88JAP88319205⟩
Takahashi, N., **4**, 856 ⟨86MI 419-01⟩; **5**, 322 ⟨93BCJ327⟩, 557 ⟨82CL123⟩
Takahashi, O., **1**, 217 ⟨91TL1059⟩, 404, 406, 816, 820 ⟨86TL4511⟩, 405, 816, 819 ⟨85CL1729⟩, 816, 819 ⟨87CL69, 91SL629⟩, 1107 ⟨85CC1636⟩; **2**, 85 ⟨90TL7031⟩
Takahashi, S., **1**, 1010, 1012, 1137 ⟨80S627⟩; **2**, 620 ⟨87CL1901⟩; **3**, 64 ⟨86TL4497, 88BCJ4029⟩, 151 ⟨91CC86⟩; **4**, 856 ⟨78CPB685⟩; **5**, 801 ⟨88CC561, 91JCS(D)1531⟩, 801 ⟨92AG(E)851⟩, 1089 ⟨92CL1515⟩; **6**, 68 ⟨65T2059⟩
Takahashi, T., **1**, 34 ⟨75BCJ1892⟩, 166 ⟨94SL121⟩, 193 ⟨84TL5921⟩, 197 ⟨93JA8485⟩, 468 ⟨88OS(66)60⟩, 468, 533, 536 ⟨87JA2393⟩, 473 ⟨88OS(66)67⟩, 491 ⟨87JOM(334)181, 86TL2829⟩, 536 ⟨87TL917⟩, 536, 542 ⟨85MI 112-04⟩, 544 ⟨88S1⟩, 813 ⟨85BCJ146, 87CC1196⟩, 816, 818 ⟨86JOC4315, 87H(25)139⟩, 818 ⟨91JOC3465, 93TL5923⟩, 1006, 1035 ⟨87CC477⟩, 1023 ⟨93SL37⟩, 1040 ⟨80TL849⟩, 1041 ⟨84CC794⟩; **2**, 565, 919, 925, 926, 969, 986 ⟨88S1⟩, 568, 986, 988 ⟨86TL2829⟩, 909, 923, 961 ⟨86JA3402⟩, 927 ⟨93CC1042⟩, 927, 986, 988 ⟨89JA3336⟩, 927, 987, 988 ⟨93TL687⟩, 969 ⟨85MI 219-01⟩, 986, 987 ⟨87CL623⟩, 986, 988 ⟨89CL761, 89CC852⟩, 987 ⟨90CC182⟩, 987, 988 ⟨94TL695⟩, 992 ⟨87JA2393⟩; **3**, 175 ⟨78TL799⟩, 206 ⟨86TL2791⟩213 ⟨83TL3489, 83TL4695⟩, 213 ⟨89JOC4273⟩, 217 ⟨88BCJ1207⟩, 553 ⟨92TL7647⟩; **4**, 313 ⟨84CC794⟩, 761, 771 ⟨76BCJ1931⟩, 808, 1059 ⟨86JA3402⟩, 809 ⟨89JA3336⟩, 950 ⟨81JA5259⟩, 1061 ⟨86TL2829⟩; 1061, 1066 ⟨87TL917⟩; **5**, 160 ⟨80TL849⟩, 516, 520 ⟨92TL7865⟩, 693 ⟨81JOC1769⟩
Takahashi, Y., **1**, 276 ⟨78BCJ339⟩, 462 ⟨79CL1003⟩; **3**, 210, 256 ⟨82CL815, 81CC714⟩, 622 ⟨75CL523⟩, 717 ⟨87JAN1199, 87JAN1202⟩; **6**, 288 ⟨88JAP63227595⟩, 553 ⟨87BCJ435⟩
Takahaski, M., **2**, 793, 794, 795 ⟨89CL2087⟩
Takahata, H., **1**, 209 ⟨88SC1159⟩, 834 ⟨86H(24)3347⟩, 1041 ⟨89JCS(P1)1211⟩; **3**, 180, 284 ⟨84JA1095⟩, 262 ⟨79JHC527⟩; **4**, 902 ⟨82CPB3959, 87JCS(P1)1501⟩; **5**, 284 ⟨87CC1627⟩, 292 ⟨81CPB1063⟩, 597 ⟨88H(27)1953⟩, 597, 599 ⟨84CPB1658⟩, 597, 599 ⟨84SC1257⟩, 597, 599 ⟨87CPB3139⟩, 597, 599, 614 ⟨83S225⟩, 607 ⟨85H(23)2213, 86H(24)1247⟩, 607 ⟨86S1063⟩
Takahori, T., **3**, 162 ⟨93S371⟩
Takai, I., **1**, 608 ⟨81CC261⟩; **2**, 784, 786 ⟨81CC261⟩
Takai, K., **1**, 228 ⟨90JOC1705⟩, 415 ⟨81TL3985⟩, 457 ⟨90TL365, 92JOC1615⟩, 480, 1027 ⟨84BCJ108⟩, 670, 678, 767 ⟨78TL2417⟩, 676 ⟨87JA951, 87TL1443⟩, 677 ⟨86JA7408⟩, 677, 1099 ⟨83TL5281, 86JA6048⟩, 678 ⟨80BCJ1698, 85TL5579⟩, 678, 769 ⟨85TL5581⟩, 766, 770 ⟨87JOC4410⟩, 1027 ⟨80TL2531, 81TL1609⟩, 1038, 1140 ⟨85TL5585⟩; **2**, 73 ⟨88TL5263⟩, 80 ⟨89TL4389⟩, 618, 625, 630 ⟨86JA7408⟩, 681 ⟨87JOC4410⟩, 699 ⟨79CL379⟩, 700 ⟨88TL1065⟩, 927 ⟨90TL365, 90TL369, 91JOC5980, 92JOC1615⟩, 927, 945 ⟨90JOC5310⟩, 927, 988, 989 ⟨90JOC1707⟩, 927, 989 ⟨92BCJ1543, 92CL99⟩, 1098 ⟨84BJC108⟩; **3**, 5 ⟨81TL1605⟩, 5, 117 ⟨80TL4843⟩, 5, 117 ⟨88BCJ3607⟩, 59 ⟨85TL5585⟩, 126

⟨84BCJ108⟩, 161 ⟨87JOC4410⟩, 627 ⟨92BCJ1543⟩; **5**, 344 ⟨91CL1479⟩
Takai, T., **3**, 156 ⟨91CL281⟩, 156 ⟨93CL1047⟩, 158 ⟨91CL1499⟩, 615 ⟨91CL1447⟩
Takai, Y., **2**, 19 ⟨92CL1161⟩; **4**, 426 ⟨82TL2665⟩
Takaishi, N., **5**, 282 ⟨77S632⟩
Takaishi, Y., **1**, 577 ⟨94JOC406⟩, 827 ⟨93CC793⟩
Takajo, T., **4**, 437 ⟨84S259⟩, 437, 438 ⟨84S256⟩, 437, 445 ⟨77S647⟩, 437, 445 ⟨82S1080⟩, 444 ⟨85S100, 81S151⟩; **5**, 606 ⟨85S92⟩
Takakai, W., **2**, 148 ⟨66AOC(4)146⟩
Takakawa, Y., **2**, 543 ⟨89TL6047⟩
Takaki, K., **2**, 312 ⟨90CL203⟩, 478 ⟨80CC1183⟩, 663 ⟨82JOC1200⟩; **3**, 699, 721 ⟨92CL1143⟩; **4**, 943 ⟨87JOC1256⟩, 1046 ⟨87JCS(P1)2269⟩
Takaku, H., **2**, 106 ⟨72TL411, 91T8717⟩, 107 ⟨83CPB2157⟩, 239 ⟨88CL857⟩, 410, 411 ⟨86CL1401⟩
Takaku, M., **3**, 373 ⟨67T4279⟩
Takakuwa, T., **5**, 550 ⟨74JOM(82)223⟩
Takamatsu, M., **4**, 885 ⟨80CPB3098⟩
Takamatsu, N., **1**, 835 ⟨71TL4661⟩
Takamatsu, T., **5**, 284 ⟨87CC1627⟩
Takami, H., **6**, 134 ⟨88BCJ2147⟩
Takami, M., **3**, 427 ⟨73BCJ3198⟩
Takami, N., **1**, 1073 ⟨91CL1805⟩
Takami, Y., **5**, 70 ⟨74CL1035⟩
Takamizawa, A., **2**, 224 ⟨70TL441, 71CPB1022⟩; **5**, 404 ⟨67CPB1183⟩, 1048 ⟨63BCJ1214⟩; **6**, 85 ⟨69CPB1931⟩, 113 ⟨69CPB1924⟩, 114 ⟨71CPB2222⟩
Takamoto, T., **2**, 121 ⟨77JOC2180⟩, 397 ⟨71TL2053, 75BCJ3413⟩
Takamoto, Y., **2**, 630 ⟨88S236⟩; **4**, 750 ⟨88S236⟩
Takamuku, S., **2**, 106 ⟨91PS(55)195⟩, 107 ⟨88BCJ3359⟩
Takamura, N., **1**, 434, 635 ⟨84JA3353⟩, 637 ⟨81TL4163⟩; **3**, 357 ⟨90EUP376724⟩, 468 ⟨71TL4495, 75T227⟩
Takana, H., **2**, 30 ⟨81JOC3312⟩
Takana, K., **5**, 1122 ⟨85TL1321⟩
Takana, N., **2**, 35 ⟨87CL715⟩
Takanagi, H., **5**, 1025 ⟨89CPB2334⟩
Takanami, T., **3**, 496 ⟨85TL2199, 87CPB4960⟩; **5**, 410 ⟨92TL1355⟩
Takane, S., **4**, 1075 ⟨91BCJ2751⟩
Takano, K., **2**, 122 ⟨83YGK776⟩
Takano, M., **1**, 212 ⟨93CPB55⟩; **6**, 636 ⟨93BCJ148⟩
Takano, S., **1**, 5, 7 ⟨84S949⟩, 188 ⟨88CC449⟩, 216, 748 ⟨89TL4845⟩, 220 ⟨93CC614⟩, 234, 822 ⟨87CL1569⟩, 248 ⟨85TL1333⟩, 339 ⟨90TL2315⟩, 581 ⟨93TA1795⟩, 596 ⟨92TA853⟩, 596, 597 ⟨76JA3022⟩, 776 ⟨89H(28)1861, 89H(28)249⟩, 1023 ⟨87S139⟩, 1084 ⟨87CL1905⟩; **2**, 85 ⟨89CC1344⟩, 323 ⟨89H(29)1861⟩; **3**, 60 ⟨89CL1283⟩, 221 ⟨89CC271⟩; **4**, 168 ⟨86CC156⟩, 346 ⟨87CL1569⟩, 412 ⟨91CC462⟩, 827 ⟨88CC189⟩; **5**, 155 ⟨89TL4845⟩, 164 ⟨80S403⟩, 166 ⟨81H(16)381, 81H(16)951⟩, 607 ⟨76JA7084⟩, 607, 616 ⟨82TL881, 76CC776⟩, 616 ⟨82CL529⟩
Takano, Y., **3**, 10, 123 ⟨83JA5703⟩
Takao, N., **5**, 212 ⟨85CPB998⟩
Takao, S., **5**, 277, 295 ⟨80TL841⟩; **6**, 116 ⟨86JOC4737, 87NKK1447⟩
Takaoka, A., **2**, 19 ⟨79BCJ3377⟩; **5**, 6 ⟨79BCJ3377⟩
Takaoka, H., **6**, 150 ⟨88BCJ2927⟩
Takaoka, K., **1**, 37 ⟨85BCJ1067⟩; **6**, 45, 49 ⟨85S497⟩
Takaoka, T., **3**, 617 ⟨83S142⟩
Takaoka, Y., **1**, 1016, 1145 ⟨90CC118⟩, 431 ⟨88T4095⟩, 431, 1068 ⟨88JA6565⟩; **2**, 917, 918 ⟨88JA6565⟩, 1020 ⟨90CC118⟩; **4**, 808 ⟨89TL6701⟩; **5**, 204 ⟨90JA5677⟩
Takarada, M., **6**, 498 ⟨79S597⟩
Takase, I., **1**, 1060 ⟨87CL1519⟩
Takase, K., **2**, 769 ⟨89BCJ128⟩; **3**, 255 ⟨81CL805⟩, 293 ⟨77CL587⟩, 411 ⟨80BCJ3696⟩; **4**, 847 ⟨89TL2091⟩
Takase, M., **3**, 32 ⟨89BCJ3038⟩
Takashi, T., **5**, 1012 ⟨89NKK1881⟩
Takashi, Y., **2**, 488 ⟨67JOM(8)225⟩
Takashina, N., **3**, 498 ⟨62JA489⟩
Takasi, A., **5**, 1012 ⟨89NKK1881⟩
Takasu, K., **6**, 7 ⟨78JOC1734⟩
Takata, M., **2**, 214 ⟨86SUL1⟩; **4**, 398 ⟨79CC1121⟩
Takata, T., **1**, 242 ⟨84TL4011⟩, 983 ⟨86T1989⟩; **2**, 146 ⟨86BCJ1275⟩, 146 ⟨86TL1591⟩, 193 ⟨79TL821⟩, 193 ⟨81T37⟩, 232 ⟨83TL3631⟩, 724 ⟨81BCJ1443⟩; **3**, 351 ⟨83JA6151⟩, 372 ⟨77H(6)1173⟩; **4**, 25 ⟨84JA8174⟩, 181 ⟨90CL2019⟩, 226 ⟨85T2133⟩, 318 ⟨85CL939, 86S139⟩, 704 ⟨91S368⟩, 836 ⟨84TL4011⟩; **6**, 298, 300 ⟨92MM3829⟩
Takata, Y., **3**, 279, 292 ⟨70MI 306-01⟩, 390, 400 ⟨89JA8749⟩, 638 ⟨77JOC2973⟩
Takatani, M., **2**, 82 ⟨83PAC589⟩
Takatoh, K., **5**, 246 ⟨91BCJ2720⟩
Takats, J., **4**, 710 ⟨94OM1078⟩
Takaya, A., **6**, 79 ⟨88JA4144⟩
Takaya, H., **1**, 2 ⟨78JA1765⟩, 76 ⟨84T1245⟩, 77 ⟨86JA7117, 87JA1596⟩, 77 ⟨87JA4129, 87JOC3174⟩, 77 ⟨90ACR345⟩, 297 ⟨93JA7033⟩, 456 ⟨91COS(8)443⟩; **2**, 53 ⟨93JA3318, 93TL2351⟩, 816 ⟨90JA4897⟩, 841 ⟨88OS(67)20⟩; **3**, 127 ⟨92TL635⟩
Takayama, A., **5**, 555 ⟨77JOC3383⟩
Takayama, H., **1**, 620 ⟨88CC1044⟩; **2**, 235 ⟨86H(24)303⟩; **5**, 324 ⟨92H(33)121⟩, 1127 ⟨90TL1287⟩; **6**, 135 ⟨92CC870⟩
Takayama, K., **3**, 543 ⟨90CPB94⟩
Takayama, S., **1**, 671 ⟨65JOC1593⟩
Takayanagi, H., **1**, 635, 1075 ⟨83T2869⟩, 818 ⟨93TL8453⟩; **3**, 646 ⟨94TL1581⟩
Takazawa, O., **3**, 120 ⟨77CL757⟩; **4**, 192, 193 ⟨84BCJ1876⟩
Takebe, M., **1**, 475 ⟨82BCJ167⟩
Takechi, H., **3**, 427 ⟨89S206⟩; **4**, 1077 ⟨82S1078⟩; **5**, 572 ⟨92S778⟩
Takechi, K., **2**, 14 ⟨84BCJ303⟩
Takechi, S., **1**, 173 ⟨90H(31)1189⟩, 1042 ⟨88TL5391⟩
Takeda, A., **1**, 422 ⟨85CL1137⟩, 568 ⟨87BCJ2917⟩, 571 ⟨91JOC7177⟩, 797 ⟨84CL1541⟩, 979 ⟨82JOC4478⟩, 980 ⟨84JOC1204⟩; **2**, 14 ⟨89JOC4982⟩; **3**, 32 ⟨82JOC1101⟩, 145 ⟨87BCJ1063, 87BCJ3421⟩, 245 ⟨86JHC1199⟩, 246 ⟨73CL425, 85TL4727⟩, 469 ⟨68AG(E)950⟩; **5**, 357 ⟨82CL1607⟩; **6**, 561 ⟨87BCJ1807, 88BCJ3205⟩
Takeda, E., **3**, 147 ⟨92CL695⟩
Takeda, H., **1**, 192 ⟨92SL323⟩; **2**, 54 ⟨90JA5876⟩, 54, 305 ⟨89TL363, 89TL367⟩; **3**, 223 ⟨90JA902⟩
Takeda, K., **1**, 238 ⟨88TL4105⟩, 239 ⟨90TL265⟩, 428, 429 ⟨89JOC5308⟩, 813 ⟨64JCS4578, 70JCS(C)2697⟩, 813 ⟨70JCS(C)2704, 74T1525⟩, 1140 ⟨88TL6951⟩; **3**, 61 ⟨87JOC4135⟩, 145 ⟨90TL265, 92TL951⟩; **5**, 226 ⟨80TL1467⟩, 235 ⟨85JOC273⟩, 1025 ⟨89CPB2334⟩, 1049 ⟨79CPB1137⟩, 1111 ⟨90H(30)367⟩; **6**, 28 ⟨78JHC893⟩, 507 ⟨82SC213⟩
Takeda, M., **2**, 25, 32 ⟨72TL4339⟩
Takeda, N., **5**, 234 ⟨84TL2793⟩, 236 ⟨85TL673⟩; **6**, 553 ⟨81CPB3158⟩
Takeda, R., **3**, 38, 175 ⟨81BCJ3510⟩, 74 ⟨79TL3461⟩
Takeda, S., **2**, 28 ⟨85BCJ1599⟩, 293 ⟨85CL679⟩; **3**, 360 ⟨86BCJ3279⟩, 438 ⟨85CL679⟩
Takeda, T., **1**, 236 ⟨85CL1149⟩, 624, 626 ⟨86JA1039⟩, 1053 ⟨87CL1345⟩; **3**, 118 ⟨92CL1185⟩, 121 ⟨92CL423⟩, 137 ⟨84BCJ1863⟩, 168 ⟨86TL3029⟩, 634 ⟨87CL1963⟩, 651 ⟨86TL5109⟩; **4**, 369 ⟨87CL1963⟩, 398 ⟨92CL819⟩, 774 ⟨91TL6563⟩, 938, 958 ⟨84BCJ1863⟩; **5**, 236 ⟨74CL187⟩, 278 ⟨86JA1039⟩
Takeda, Y., **1**, 1107 ⟨85CC1636⟩; **2**, 35 ⟨88S403⟩; **3**, 309 ⟨93H(35)573⟩; **6**, 636 ⟨93AOC(35)211⟩
Takegami, Y., **2**, 631 ⟨82CL1481⟩; **3**, 83 ⟨71BCJ2569⟩, 84 ⟨73TL3535, 75BCJ2490⟩, 84 ⟨75TL1063⟩; **6**, 457 ⟨72CL483⟩
Takegan, Y., **2**, 310 ⟨74TL1879⟩
Takehara, H., **5**, 292 ⟨81CPB1063⟩
Takehara, S., **3**, 628 ⟨90CL1615⟩; **4**, 654, 655 ⟨92CL2047⟩

Takehira, K., **3**, 87 ⟨90TL2607⟩
Takehira, Y., **1**, 969 ⟨75CC174⟩
Takehisa, C., **6**, 544 ⟨92CPB2279⟩
Takei, H., **1**, 133 ⟨79BCJ1241⟩, 238, 483 ⟨79TL3425⟩, 483, 484 ⟨80TL87, 79TL43⟩, 485 ⟨73CL1097⟩, 558 ⟨94T655⟩; **2**, 401 ⟨65BCJ1989⟩; **3**, 55 ⟨77CL345⟩, 166 ⟨78BCJ3008⟩, 224 ⟨88CL921⟩, 286 ⟨73JA4763, 74BCJ1777⟩, 374 ⟨74BCJ909⟩; **4**, 206 ⟨80BCJ1061⟩, 959 ⟨78CL517, 80CL1209⟩; **5**, 179 ⟨80TL4611⟩, 997 ⟨63BCJ95⟩, 1001 ⟨61JOC4381⟩
Takei, R., **4**, 731 ⟨76BCJ2491⟩; **6**, 4 ⟨92JAP04164039⟩, 58, 266 ⟨81JFC(18)197⟩
Takei, S., **3**, 66 ⟨56CB2224⟩
Takei, Y., **1**, 706, 708 ⟨84CL833⟩, 706, 708 ⟨87JOC1703⟩
Takeishi, Y., **5**, 625 ⟨90CL1403⟩; **6**, 597 ⟨90CL1403⟩
Takekuma, S., **1**, 899 ⟨87BCJ3721, 88NKK157⟩
Takemae, M., **5**, 225, 375 ⟨89BCJ2111⟩
Takemasa, T., **5**, 74 ⟨86TL4537⟩
Takemori, A. E., **3**, 343 ⟨91JMC1292⟩
Takemori, T., **1**, 1090 ⟨89JOC1757⟩; **4**, 938 ⟨89JOC1757⟩
Takemoto, D. J., **2**, 743 ⟨91S970⟩
Takemoto, K., **5**, 1065 ⟨81MAC3411⟩, 1078 ⟨85SC171⟩
Takemoto, M., **3**, 292 ⟨83CPB3951⟩
Takemoto, T., **2**, 314 ⟨91BCJ602⟩, 314, 318 ⟨88CL1759⟩
Takemoto, Y., **3**, 59 ⟨88JOC1623⟩
Takemura, S., **1**, 836 ⟨93T7657⟩; **4**, 519 ⟨84H(22)701⟩; **5**, 701 ⟨71CPB1606⟩; **6**, 542 ⟨92TL2543⟩
Takenaka, N. E., **4**, 46 ⟨90JA6715⟩
Takenaka, S., **2**, 684 ⟨88CL899⟩; **4**, 302 ⟨87BCJ4079⟩, 519 ⟨84CL279⟩, 925 ⟨83TL767⟩
Takenaka, Y., **4**, 250 ⟨75S720⟩
Takenish, T., **2**, 192 ⟨67JOC3032⟩
Takeno, N., **1**, 554 ⟨87BCJ4468⟩; **2**, 28 ⟨85BCJ1599⟩
Takenoshita, H., **2**, 43 ⟨89CL7573⟩
Takenouchi, K., **3**, 224 ⟨88CL921⟩
Takeo Saegusa, **5**, 747 ⟨69BCJ3310⟩
Takeshima, T., **3**, 256 ⟨76JCS(P1)1706⟩, 338 ⟨82CL101⟩, 338, 339 ⟨89JCS(P1)1241⟩; **4**, 842 ⟨75JCS(P1)1277⟩, 852 ⟨76JCS(P1)1706, 77JCS(P1)1273⟩; **5**, 547 ⟨69JOC730, 73JCS(P1)1009⟩, 547 ⟨77JCS(P1)1273, 79JCR(S)140⟩, 547 ⟨80S566, 82YGK123⟩, 555 ⟨77JOC3383⟩, 580 ⟨70JOC2438⟩, 652 ⟨78JCS(P1)558⟩, 1112 ⟨80TL635⟩; **6**, 324 ⟨73JCS(P1)1009⟩, 627 ⟨79S554⟩
Takeshita, H., **1**, 690 ⟨86BCJ1109⟩, 690, 806, 807 ⟨88CC354⟩; **5**, 184 ⟨87BCJ4325⟩
Takeshita, K., **2**, 912 ⟨83CC1193, 86JOC3890⟩, 912 ⟨87JOC4864⟩
Takeshita, M., **3**, 213 ⟨84CL1259⟩
Takeshita, T., **2**, 15 ⟨B-80MI 218-01⟩
Taketani, H., **3**, 687 ⟨76BCJ3124⟩
Takeuchi, A., **2**, 61 ⟨88TL5419⟩
Takeuchi, H., **5**, 1066 ⟨87NKK1280⟩
Takeuchi, K., **1**, 867 ⟨90JOC996⟩, 905 ⟨85JOC826⟩, 907 ⟨82BCJ3257⟩; **2**, 323 ⟨90JA2368⟩; **3**, 43 ⟨90JA2368⟩; **5**, 282 ⟨76JCS(P1)2205⟩
Takeuchi, M., **2**, 579 ⟨92CC1728⟩
Takeuchi, N., **2**, 662 ⟨80CC593⟩
Takeuchi, R., **1**, 1010 ⟨90JHC503⟩; **2**, 929 ⟨92CC1358, 93JCS(P1)1031⟩
Takeuchi, S., **1**, 24 ⟨85BCJ2421⟩
Takeuchi, T., **3**, 716, 717 ⟨87JAN1207, 87JAN1199⟩, 717 ⟨87JAN1202⟩; **6**, 288 ⟨88JAP63227595⟩
Takeuchi, Y., **1**, 248 ⟨87JOC5063⟩, 603 ⟨82CC871, 82TL4733⟩; **2**, 130 ⟨60BCJ1743⟩, 282, 284 ⟨82TL1361, 81CL1715⟩, 284 ⟨82CC871, 82CL1733, 84JOC3235⟩, 284, 392, 785 ⟨82TL4733⟩, 368 ⟨85S212, 82CL1109⟩, 1036, 1038 ⟨82CL1244⟩; **3**, 348 ⟨60BCJ1743⟩, 412 ⟨79TL3453⟩, 414 ⟨83BCJ641⟩, 627 ⟨91JA6318⟩, 637 ⟨82CC871, 82TL1361⟩, 669 ⟨72JCS(P1)2054⟩; **4**, 52, 101, 109 ⟨88JCS(P1)1149⟩, 79, 781 ⟨87CPB3959⟩, 101 ⟨87JCS(P1)2203⟩, 102 ⟨91JCS(P1)49⟩, 108 ⟨87JOC5061, 91CPB3120⟩, 108, 780 ⟨89JOC5453⟩, 270, 854, 859, 860 ⟨86SC233⟩, 285 ⟨85TL6205⟩, 376 ⟨84JOC3235⟩, 417 ⟨82CPB1036⟩, 420 ⟨85JOC4227⟩, 516 ⟨86CL1193⟩; **5**, 335 ⟨85S212⟩, 360 ⟨91JA6318⟩, 652 ⟨71TL3595⟩, 1116 ⟨81CL1069⟩; **6**, 43, 47, 54, 57 ⟨86JOC955⟩
Takeura, M., **5**, 1025 ⟨89CPB2334⟩
Takeyama, T., **1**, 426, 427 ⟨85CL1491⟩
Takeyama, Y., **1**, 1131 ⟨89TL3159⟩; **2**, 920 ⟨91BCJ1461⟩; **6**, 21 ⟨89TL3159⟩
Takeyasu, T., **2**, 920, 946 ⟨88JOC3539⟩; **3**, 660 ⟨86TL1841⟩
Takezaki, Y., **3**, 90 ⟨76BCJ1681⟩
Taki, H., **3**, 407 ⟨85CC613⟩; **5**, 171 ⟨87JOC4319⟩
Takido, T., **3**, 357 ⟨92JHC841⟩
Takigawa, T., **1**, 604 ⟨80JOC2247⟩, 836 ⟨80JA7587⟩; **4**, 250 ⟨75S720⟩
Takikawa, Y., **3**, 340 ⟨88MI 308-01⟩; **5**, 516, 520 ⟨92CL1843⟩, 625 ⟨90CL1403⟩, 635 ⟨85BCJ995⟩; **6**, 498 ⟨92CL1389⟩, 597 ⟨90CL1403⟩
Takimoto, F.,**5**, 652 ⟨75BCJ3259⟩
Takimoto, S., **3**, 223, 275 ⟨79CL1021⟩; **5**, 125 ⟨87TL4711⟩, 173 ⟨79CL1019⟩, 173 ⟨79CL1021⟩, 196 ⟨82MI 504-01⟩
Takinami, S., **1**, 1091, 1094 ⟨86CL459⟩; **2**, 623, 628, 902 ⟨83TL731⟩
Takino, Y., **2**, 328 ⟨89H(28)63⟩
Takita, S., **3**, 156 ⟨81T473⟩
Takiyama, N., **1**, 1038, 1098, 1140 ⟨84TL4233⟩; **2**, 69 ⟨85TL4763, 89JA4392⟩; **4**, 954 ⟨84TL4233⟩
Takizawa, M., **3**, 430, 666 ⟨91S750⟩; **5**, 665, 666, 824 ⟨91S750⟩
Takizawa, S., **5**, 635 ⟨85BCJ995⟩
Takizawa, T., **3**, 470, 593 ⟨69TL3403⟩, 540, 543 ⟨72TL1849⟩, 540, 543 ⟨75CPB2933⟩, 582 ⟨71CC587, 69TL3407⟩, 596 ⟨75BCJ2287⟩; **6**, 503 ⟨74YGK936⟩, 505 ⟨72TL4263⟩, 644 ⟨75JOM(94)333⟩
Takken, H. J., **1**, 633, 1094 ⟨76TL2287⟩; **3**, 64 ⟨76RTC308, 77TL2129⟩, 69 ⟨77T2027⟩
Takuku, M., **2**, 1000 ⟨67T3049⟩
Takundwa, C., **1**, 1010, 1012 ⟨81S364⟩
Takusagawa, F., **3**, 410 ⟨92JA5466⟩, 516 ⟨93OM2565⟩; **4**, 673, 722 ⟨89OM1199, 88OM1715⟩
Takuwa, A., **3**, 140 ⟨91CL1013⟩
Talalaeva, T. V., **2**, 882 ⟨53MI 217-01⟩
Talalay, P., **5**, 1044 ⟨94JMC170⟩
Talamas, F. X., **1**, 837 ⟨94TL15⟩
Talamov, V. S., **4**, 110 ⟨82JOU2240⟩
Talanian, R. V., **4**, 129 ⟨89B8270⟩
Talât-Erben, M., **3**, 573 ⟨55JA3710⟩
Talaty, E. R., **4**, 28 ⟨51JIC675⟩
Talbert, J., **2**, 588, 590 ⟨88JOC2390⟩; **3**, 287 ⟨88JOC2390⟩
Talbiersky, J., **3**, 259 ⟨76AG(E)171⟩; **4**, 949 ⟨79TL4273⟩, 1016 ⟨77S869, 78AG(E)204⟩
Talbot, A., **2**, 84 ⟨91SL248⟩
Talbot, J.-M., **3**, 256 ⟨76BSF294⟩
Talbott, R. L., **2**, 98 ⟨68JOC2095⟩; **6**, 251 ⟨71USP3585218⟩, 260, 284 ⟨68USP3387033⟩, 272 ⟨68JOC2095⟩
Taleb, C., **2**, 852 ⟨76JPC2417⟩
Taleb-Sahraoui, S., **3**, 602 ⟨81AG(E)879⟩
Talham, D., **4**, 876 ⟨82CC336⟩
Taljaard, H. C., **4**, 237 ⟨88JCS(P1)2663⟩
Taller, R. A., **3**, 446, 448 ⟨63JA3624⟩
Talley, B. G., **2**, 8 ⟨81JOC824⟩
Talley, J. J., **1**, 904 ⟨80JA1214, 80JA4138⟩, 904 ⟨80JA7613⟩; **3**, 262 ⟨81TL823⟩
Talley, J. T., **3**, 615 ⟨78TL3773⟩
Talley, P., **1**, 1108 ⟨77TL3545⟩; **2**, 607, 617 ⟨77TL3545⟩
Tallman, J., **5**, 1029 ⟨79NAT854⟩
Tallyn, W. H., **3**, 535 ⟨25JA1779⟩
Talmet, M., **2**, 447 ⟨65AJC903⟩
Talow, J. C., **2**, 194 ⟨63JCS3692⟩
Talpas, G. S., **6**, 111 ⟨91JHC481⟩
Talwalkar, P. K., **4**, 979 ⟨79IJC(B)478⟩

Tam, C. C., **1**, 187 ⟨79JOC3428⟩, 503 ⟨82JOC1221⟩, 828 ⟨82JA7174⟩; **4**, 466 ⟨82S188⟩, 849 ⟨82JA7174⟩
Tam, J. N. S., **1**, 104 ⟨68CJC2821⟩
Tam, S. W., **1**, 2, 8 ⟨79JOC641⟩; **3**, 129 ⟨79JOC632⟩
Tam, S. Y., **1**, 1089 ⟨90TL597⟩
Tam, W., **1**, 471 ⟨84OM630⟩; **2**, 580 ⟨79JA1589⟩; **4**, 581 ⟨79JA5440⟩; **6**, 516 ⟨85OM939⟩
Tam, W. W., **3**, 119 ⟨82TL4647⟩
Tamagaki, S., **2**, 235 ⟨74TL1059⟩, 292 ⟨75TL649⟩; **3**, 177 ⟨80BCJ817⟩, 374 ⟨77CL725⟩; **4**, 862 ⟨76H(4)963⟩; **5**, 349 ⟨75TL649⟩; **6**, 303, 307 ⟨72BCJ960⟩
Tamagawa, M., **2**, 380 ⟨83CPB423⟩
Tamagnan, G., **5**, 152 ⟨89CC31⟩
Tamai, K., **1**, 476 ⟨82TL3583⟩
Tamai, S., **1**, 251 ⟨89TL2413⟩; **3**, 210 ⟨89TL3685⟩
Tamai, Y., **6**, 70 ⟨76JAP(K)76108012⟩
Tamaki, A., **1**, 669 ⟨74JA6140⟩; **2**, 590 ⟨73JCS(D)2620, 74JA6140⟩, 590, 960 ⟨73JA6487, 72JOM(40)C81⟩
Tamaki, H., **1**, 362 ⟨90CC1436⟩
Tamaki, K., **1**, 180 ⟨86BCJ3617⟩, 246 ⟨86JCS(P1)1983⟩; **2**, 287 ⟨84CC426⟩; **3**, 176 ⟨86JCS(P1)1983⟩
Tamami, B., **3**, 4 ⟨91SC1387⟩
Tamamura, H., **1**, 189 ⟨93JOC1207⟩; **5**, 327 ⟨81JGU2125⟩
Tamao, K., **1**, 160 ⟨83JOC2120⟩, 160 ⟨84TL1905⟩, 160 ⟨84TL1909⟩, 160 ⟨84TL1913⟩, 451 ⟨75JA5603⟩, 465 ⟨91COS(3)435⟩, 466 ⟨73JOM(50)C12⟩, 466, 470 ⟨76BCJ1958⟩, 466, 473 ⟨78OS(58)127⟩, 467 ⟨76CL1237⟩, 467 ⟨87JA1257⟩, 471 ⟨72JA9268, 75JA4405⟩, 472 ⟨77TL4089, 80TL845⟩, 477 ⟨91COS(3)468⟩, 490 ⟨79TL1137⟩, 490, 515 ⟨79TL1141⟩, 490, 535 ⟨82OM542⟩, 1094 ⟨87JOC4412⟩, 1106 ⟨78TL2161⟩; **2**, 4 ⟨78JA290, 78TL1809⟩, 4 ⟨82OM355, 82OM369⟩, 50 ⟨88JA3712, 88MI 202-01⟩, 51 ⟨90TL7333⟩, 536, 540 ⟨88MI 210-03⟩, 545, 546 ⟨92JA3989⟩, 630 ⟨87JOC1100⟩, 914 ⟨88TL6955⟩, 915 ⟨74JOM(76)105, 74JOM(76)95, 76JOM(114)C19⟩, 920 ⟨89JA6478⟩, 923, 924 ⟨82JOM(226)C9⟩, 926 ⟨89JOC3517⟩, 926, 985 ⟨92SL539⟩, 931 ⟨84TL83⟩, 935 ⟨88T3997⟩, 939 ⟨92JA5867⟩; **3**, 18 ⟨84TL321⟩, 145 ⟨76CL1239⟩, 158 ⟨86TL65⟩; **4**, 140 ⟨71JOM(30)349⟩, 614 ⟨66JOM(5)226⟩, 651 ⟨90TL2925⟩, 651, 655 ⟨84TL1913⟩, 654 ⟨77IJ265, 84TL1905, 84TL4245⟩, 654, 655 ⟨84TL4249⟩, 1033 ⟨76JA3718⟩, 1052 ⟨83JOM(254)13⟩
Tamao, T., **1**, 490 ⟨89TL6051⟩
Tamaoka, M., **3**, 196 ⟨88CPB613⟩
Tamaoka, T., **1**, 1055 ⟨84CL1939⟩; **3**, 634, 651 ⟨84JA7890⟩
Tamaoki, A., **2**, 725 ⟨70JAP7321096-R⟩
Tamara, O., **2**, 331 ⟨92JA5900⟩
Tamaru, M., **6**, 72 ⟨80BCJ3687⟩
Tamaru, Y., **1**, 122, 123, 187 ⟨87JOC4418⟩, 238 ⟨83JOC1293⟩, 807, 834 ⟨80JA2392⟩; **2**, 206 ⟨90JOC1823⟩, 588, 591 ⟨89AG(E)351, 89OS(67)98⟩; **3**, 5, 118 ⟨83JOC1286⟩, 130 ⟨85TL5529⟩, 161 ⟨88JOC1343⟩, 290, 295, 296 ⟨83TL3869⟩; **4**, 230 ⟨84TL5797⟩; **5**, 170 ⟨85TL3207⟩, 170 ⟨87TL325⟩, 595, 614 ⟨81TL3409⟩, 607 ⟨78JA1923, 80JA7806⟩, 607 ⟨81TL3413, 82JA4018⟩, 607 ⟨84TL5793⟩, 607, 614 ⟨81TL4245⟩, 607, 616 ⟨78TL2167⟩, 607, 616 ⟨80MI 514-01⟩
Tamás, J., **2**, 374 ⟨88S529⟩, 379 ⟨85JPR445⟩; **3**, 342 ⟨87IZV2398⟩; **4**, 273 ⟨89MRC760⟩
Tamatsukuri, S., **2**, 409, 467 ⟨86TL199⟩
Tamblyn, W. H., **1**, 353 ⟨82TL2261⟩, 400, 408, 409, 822, 823, 825 ⟨81JOC5094⟩
Tamborski, C., **1**, 426 ⟨64JOC3042, 69JOM(18)263⟩; **2**, 600 ⟨63JOC237⟩, 825 ⟨80JFC(15)239⟩; **4**, 768 ⟨86JFC(31)405⟩; **6**, 11 ⟨81JFC(17)345⟩, 63 ⟨75JFC(5)475⟩
Tambuté, A., **1**, 316 ⟨69TL707⟩, 397 ⟨69TL821⟩
Tamiya, H., **5**, 332 ⟨90TL243⟩
Tamm, C., **1**, 178 ⟨88HCA47⟩, 873 ⟨86HCA1041⟩; **5**, 135, 138 ⟨83HCA2501⟩; **6**, 541 ⟨87HCA138⟩
Tamm, L. A., **2**, 838, 839 ⟨73ZOB2178⟩
Tamm, M., **3**, 711, 723 ⟨91AG(E)196⟩
Tammer, T., **2**, 444 ⟨85GEP3332716⟩
Tamori, S., **4**, 827 ⟨88CC189⟩
Tamoutsidis, E., **4**, 921 ⟨83CB3141, 83ZN(B)726⟩
Tamura, C., **2**, 1003 ⟨79AX(B)2443⟩
Tamura, H., **3**, 290, 408 ⟨86CL1197⟩; **4**, 947 ⟨92JOC5610⟩; **6**, 550 ⟨82YGK352⟩, 556, 566, 652 ⟨91CPB1939⟩, 636 ⟨88CPB4755, 92MI 621-02⟩
Tamura, K., **3**, 374 ⟨77CL725⟩; **5**, 659 ⟨93JOC32⟩
Tamura, M., **1**, 117, 1092 ⟨71JA1483⟩, 117, 119, 465 ⟨71S303⟩, 119 ⟨72JOM(42)205⟩, 491 ⟨72BCJ1120⟩; **2**, 987 ⟨89CC852⟩; **3**, 632 ⟨92SL483⟩; **5**, 236 ⟨86CL1817, 87CL743⟩
Tamura, N., **1**, 942 ⟨83JCS(P1)403⟩; **3**, 156 ⟨81T473⟩
Tamura, O., **3**, 189 ⟨92JA5888⟩; **4**, 107 ⟨92JA5900⟩; **5**, 342 ⟨94HCA554⟩, 377 ⟨89JCS(P1)1862⟩; **6**, 447 ⟨92TL3483⟩
Tamura, R., **1**, 37, 330, 443, 548, 946, 1131 ⟨85T4013⟩, 37, 441 ⟨83JA4017⟩, 37, 945 ⟨81TL1705⟩, 250 ⟨82JA3727, 86JOC4375⟩, 250 ⟨91S423⟩, 251 ⟨89TL2413⟩, 257 ⟨90JOC408⟩, 617 ⟨78TL763⟩, 617, 638 ⟨81CL1139⟩; **2**, 394 ⟨87S258⟩, 394, 784 ⟨86JOC4368⟩; **3**, 169 ⟨90JOC595⟩, 184 ⟨86TL5759⟩, 192 ⟨87S207⟩, 210 ⟨89TL3685⟩, 635, 645 ⟨78TL763⟩; **5**, 665 ⟨90CL559⟩, 935 ⟨83JA4099⟩
Tamura, S., **5**, 1144 ⟨86MI 530-03⟩
Tamura, T., **1**, 697, 700, 703 ⟨90JA7407⟩; **2**, 3 ⟨76CPB820⟩, 521 ⟨86JA7402⟩; **4**, 943 ⟨87JOC1256⟩; **5**, 971 ⟨75JAP89344⟩
Tamura, Y., **1**, 187 ⟨84JA1079⟩, 192 ⟨87TL3497⟩, 195 ⟨93TL7619⟩, 822 ⟨73TL2351⟩, 827, 828 ⟨89CPB2047⟩, 1061, 1062 ⟨90JOC4853⟩, 1140 ⟨85TL1549, 86TL195⟩, 1140 ⟨87CPB1405, 87TL4583⟩, 1140 ⟨88CPB3897⟩; **2**, 69 ⟨87JOC4418⟩, 157 ⟨78CC367⟩, 160, 161 ⟨74S277⟩, 161 ⟨72TL4137⟩, 257 ⟨79TL5015, 83JOC1293⟩, 263, 812 ⟨79TL4311⟩, 289 ⟨79TL2251⟩, 357, 379, 384 ⟨77S1⟩, 357, 691 ⟨73JOC1239⟩, 401 ⟨76T431⟩, 665 ⟨81TL4283, 82JOC4376⟩, 837 ⟨64JOC2572⟩, 1000 ⟨81JCS(P1)212⟩; **3**, 108 ⟨82CPB1257⟩, 115 ⟨79TL1401⟩, 158 ⟨85TL3837⟩, 207 ⟨80TL3779⟩, 249 ⟨79TL1129⟩, 267 ⟨73JOC4324⟩, 380 ⟨74JOC3519, 78CC197⟩, 626 ⟨82CPB3574⟩, 635 ⟨78CPB784, 86JCS(P1)1763⟩, 669 ⟨80JCS(P1)1132⟩; **4**, 67 ⟨81S534⟩, 188 ⟨82S1089⟩, 226 ⟨85T2133⟩, 303 ⟨61JOC2273⟩, 318 ⟨85CL939, 86S139⟩, 828 ⟨93SL273⟩; **5**, 189, 195, 282 ⟨86JOC4150⟩, 189, 199 ⟨84TL6027⟩, 215 ⟨89S334⟩, 223 ⟨90CPB1836⟩, 226, 375, 378 ⟨79TL4311⟩, 294 ⟨80S887, 81S413⟩, 595 ⟨79S887⟩, 1041, 1107 ⟨77TL4417⟩, 1048 ⟨78CI(L)806⟩, 1055, 1114 ⟨79TL2251, 79TL1129⟩, 1107 ⟨81JCS(P1)1577⟩; **6**, 484 ⟨78TL3737⟩
Tamuru, Y., **2**, 69 ⟨88CC610⟩; **3**, 378 ⟨74JCS(P1)1125⟩
Tan, K., **5**, 765, 824 ⟨86JHC1861⟩
Tan, P.-C., **2**, 670 ⟨81JA7677⟩
Tan, R. P. K., **3**, 59 ⟨85SC371⟩
Tan, S.-H., **5**, 723 ⟨74CJC4050⟩
Tan, S. I., **2**, 377, 381, 383 ⟨68JHC397⟩
Tan, S. L., **1**, 357 ⟨85TL4707⟩
Tan, T.-L., **2**, 646 ⟨48JA602⟩
Tan, T. S., **1**, 216 ⟨84CC585⟩; **2**, 921 ⟨88T3953⟩; **4**, 707, 720 ⟨77JOM(141)85⟩
Tanabe, K., **5**, 291 ⟨78BCJ1866⟩; **6**, 38 ⟨58JA3002⟩, 69 ⟨60JA1398⟩
Tanabe, M., **1**, 104 ⟨63JA3522⟩, 430 ⟨73TL447⟩, 832 ⟨80JA862⟩, 871, 881 ⟨69CC1498⟩, 1082 ⟨67TL3739, 67TL3943⟩; **5**, 429 ⟨86CL177⟩; **6**, 111 ⟨84H(22)13⟩
Tanabe, T., **2**, 207 ⟨78BCJ1459⟩, 683 ⟨79JOC638⟩
Tanabe, Y., **1**, 228 ⟨82JOM(233)13⟩; **2**, 539 ⟨94TL8413⟩; **3**, 35, 165 ⟨92JOC1053⟩; **4**, 103 ⟨88S482⟩; **6**, 478 ⟨92JOC1053⟩, 539 ⟨89JAP03020256⟩
Tanaka, A., **1**, 195 ⟨93TL7619⟩
Tanaka, C., **2**, 131 ⟨84TL2675⟩
Tanaka, F., **1**, 134 ⟨91TL7281⟩

Tanaka, H., **1**, 243 ⟨90TL227⟩, 284 ⟨91SL919⟩, 496 ⟨92TL7029⟩, 561 ⟨74JOC2486⟩, 604 ⟨80JOC3537⟩, 836 ⟨80JA7587⟩; **2**, 21 ⟨81TL3193, 84CL877⟩, 243, 359 ⟨78JOC3223⟩, 243, 359 ⟨79JOC1554⟩, 252 ⟨79JOC2938⟩, 281 ⟨89S434⟩, 285 ⟨92JOC111⟩, 312 ⟨88TL3811⟩, 313 ⟨89TL4161, 90TL3023⟩; **3**, 234 ⟨76CL495⟩, 360 ⟨82CL793, 85BCJ667⟩, 361 ⟨89CC505⟩, 366 ⟨78CPB2092⟩, 426 ⟨83CL607⟩; **4**, 197 ⟨80CL617⟩, 240 ⟨92MI 405-01⟩; **5**, 10, 367 ⟨91SL919⟩, 235 ⟨93JA5302⟩, 336 ⟨79JOC1554⟩; **6**, 26, 31 ⟨89JOC444⟩

Tanaka, J., **1**, 232 ⟨89BCJ164⟩; **2**, 27 ⟨77CL1025⟩, 908 ⟨88BCJ3957⟩, 930, 939 ⟨90BCJ51⟩; **4**, 263 ⟨90BCJ466⟩, 367, 936, 938 ⟨88BCJ3957⟩, 367, 946 ⟨90BCJ51⟩, 506, 513 ⟨85BCJ1991⟩

Tanaka, K., **1**, 107 ⟨76CJC1643⟩, 210 ⟨85TL4747⟩, 235 ⟨81CL315, 82CL351⟩, 240 ⟨82BCJ2965, 87CL809⟩, 240 ⟨88JA7185⟩, 444 ⟨80JOM(193)37⟩, 825 ⟨81CL1289⟩, 983 ⟨85TL631⟩; **2**, 156 ⟨81CJC718⟩, 588 ⟨91CC101⟩, 1042 ⟨77BCJ3069⟩; **3**, 309 ⟨80H(14)1939, 88CPB2023⟩, 628 ⟨90JOC181⟩, 629 ⟨91SL563, 92JOC6778⟩, 644 ⟨79S890⟩; **4**, 3 ⟨66BCJ219⟩, 220 ⟨78BCJ2401⟩, 236 ⟨80BCJ3619⟩, 388 ⟨92JA9715⟩, 392 ⟨76S413⟩, 673 ⟨69APO(7)1⟩, 735 ⟨78TL4809⟩, 949 ⟨77CL1379⟩; **5**, 132 ⟨78BCJ2401⟩, 179 ⟨85CPB532⟩, 277 ⟨74TL3995⟩, 664 ⟨84BCJ2184⟩, 681 ⟨87JHC1391⟩, 958 ⟨92JA9715, 93JA9015⟩, 1122 ⟨85JCS(P1)2333, 87JCS(P1)2531⟩; **6**, 628, 730 ⟨72BCJ489⟩

Tanaka, K.-I., **5**, 307 ⟨85CPB5515, 87CPB364⟩; 652 ⟨88CPB3125⟩

Tanaka, M., **1**, 188 ⟨89JA4864⟩, 188 ⟨89JOC4055⟩, 188 ⟨92JOC1024⟩, 227 ⟨89T435⟩, 367 ⟨85TL3011⟩, 469, 1005 ⟨89SC2199⟩, 690 ⟨88JA3578⟩, 813, 814 ⟨80BCJ2958⟩, 1015 ⟨81CC333⟩, 1045, 1138 ⟨78TL2301⟩, 1047 ⟨85TL3463⟩; **2**, 310 ⟨74TL1879⟩, 945 ⟨93OM2065⟩; **3**, 15, 289 ⟨79TL2601⟩, 83 ⟨71BCJ2569⟩, 84 ⟨73TL35[illegible]5⟩, 93 ⟨91CC1551, 92JOC2677⟩, 142 ⟨80BCJ2958⟩, 145 ⟨85CC1545⟩, 238, 270, 273 ⟨78TL2301⟩, 276 ⟨81CC333⟩, 289 ⟨81BCJ637⟩, 290 ⟨85TL3463⟩, 406 ⟨85S967⟩; **4**, 369 ⟨85TL3011⟩, 618 ⟨92JOM(428)1⟩, 674 ⟨93OM2925⟩; **5**, 158 ⟨89JA4864⟩, 226 ⟨82JOM(213)C12⟩, 655 ⟨79JA4245⟩, 801 ⟨79JOM(168)97⟩, 1144 ⟨68JOM(12)241, 80TL2959⟩; **6**, 457 ⟨72CL483⟩, 533 ⟨91JAP03218336⟩

Tanaka, N., **2**, 753 ⟨83CC1322⟩; **4**, 516 ⟨83CC1322⟩

Tanaka, R., **1**, 1136 ⟨75JOC1932⟩; **2**, 1028 ⟨71T2651, 71TL1753⟩, 1055 ⟨83CC329⟩

Tanaka, S., **1**, 192 ⟨93JOC16⟩, 206 ⟨74JA5254⟩, 690, 806, 807 ⟨88CC354⟩, 986 ⟨84CL131⟩; **2**, 176, 180 ⟨74S876⟩, 312 ⟨90CL203⟩, 633 ⟨83JOC3297⟩, 793, 795 ⟨94H(37)693⟩; **3**, 54 ⟨73TL2465⟩, 191 ⟨89CC110⟩, 326 ⟨94JCS(P1)643⟩, 622 ⟨93JOC16⟩, 684 ⟨92NKK420⟩; **5**, 251 ⟨77BCJ220⟩, 325 ⟨82JOC5283⟩, 678 ⟨81H(16)717⟩, 1032, 1056 ⟨77BCJ722⟩, 1036 ⟨77BCJ2785, 78BCJ659⟩; **6**, 465 ⟨93EUP559212⟩, 636 ⟨84CHEC(3)91⟩

Tanaka, T., **1**, 240 ⟨88JA7185⟩, 274 ⟨73TL787⟩, 1018 ⟨91JCS(P1)2892⟩; **2**, 21 ⟨88TL5783⟩, 61 ⟨83TL2653⟩, 409, 467 ⟨86TL199⟩, 626 ⟨89BCJ439⟩, 914 ⟨88TL6955⟩; **3**, 192 ⟨87T813⟩, 228 ⟨80TL3209⟩, 307 ⟨80SC773⟩, 668 ⟨86TL2411⟩; **4**, 279 ⟨91MI 406-01⟩, 400 ⟨80TL3209⟩, 426 ⟨86H(24)25⟩, 743 ⟨94JCS(P1)3⟩, 842 ⟨82TL1075⟩; **5**, 1120 ⟨92CC1379⟩; **6**, 628, 730 ⟨72BCJ489⟩

Tanaka, Y., **1**, 161, 450, 1092, 1105 ⟨83TL1041⟩, 210 ⟨83JOC5133⟩, 806 ⟨80JOC1172⟩; **2**, 24 ⟨83CC947⟩, 923, 924 ⟨83TL1041⟩; **3**, 108 ⟨84CPB4914⟩; **5**, 681 ⟨83BCJ545, 84BCJ539⟩

Tanami, T., **4**, 167 ⟨81JCS(P1)1015⟩

Tandanier, J., **1**, 769 ⟨90CAR(201)185⟩

Tandon, K. N., **6**, 514 ⟨71IC1896⟩

Tandon, P., **5**, 385 ⟨86JMC199⟩

Tandon, S. K., **5**, 1136 ⟨68JINC1399⟩

Tandon, V. K., **2**, 318 ⟨93TL4403⟩

Tandura, S. N., **2**, 1091 ⟨79JGU536⟩; **4**, 660 ⟨89MI 414-01⟩, 662 ⟨91MI 414-01⟩

Tandy, Jr., T. K., **3**, 465 ⟨65JOC4257⟩

Tane, J. P., **2**, 926 ⟨80AG(E)1023⟩; **3**, 270 ⟨80AG(E)1023⟩

Taneka, H., **2**, 81 ⟨89JOC444⟩

Taneko, K., **2**, 71 ⟨89CC1700⟩

Tanemura, K., **4**, 186 ⟨92BCJ304⟩

Tang, C.-C., **2**, 257 ⟨91S454⟩

Tang, D., **5**, 905 ⟨76JOC2102⟩

Tang, D. Y., **3**, 269 ⟨74JOC3076⟩

Tang, D. Y. H., **1**, 321, 923 ⟨74JA2201⟩

Tang, J.-Y., **2**, 253 ⟨89JA2321⟩

Tang, K., **4**, 223 ⟨89JA658⟩

Tang, K.-C., **3**, 197 ⟨83JOC5001⟩

Tang, P.-C., **1**, 551 ⟨81JA7677⟩; **2**, 670 ⟨84JOC2293, 85JA1060⟩; **5**, 934 ⟨84JA434⟩

Tang, P. W., **2**, 713 ⟨78JOC4131⟩; **4**, 958 ⟨78TL3995⟩

Tang, Q., **3**, 307 ⟨90TL995⟩

Tang, R., **1**, 53 ⟨71JA6205⟩; **2**, 430, 438 ⟨69T1097⟩, 462 ⟨73CC273⟩

Tang, R. T., **2**, 652 ⟨73JA7114⟩

Tang, S., **2**, 44 ⟨92TL7823⟩

Tang, S. C., **3**, 339 ⟨73IC2589, 73JA613⟩

Tang, Y., **4**, 444 ⟨83CC721, 84CJC749⟩; **5**, 82 ⟨86MI 502-01⟩, 101 ⟨84TL1797⟩

Tang, Y. S., **2**, 637 ⟨84JA460⟩

Tangari, N., **5**, 307 ⟨75CC71⟩

Tangerman, A., **3**, 370 ⟨77TL259, 79RTC127⟩; **5**, 478, 491, 492, 562 ⟨77RTC139, 75JCS(P2)916⟩; **6**, 536 ⟨75JCS(P2)916⟩

Tangor, B., **4**, 478 ⟨89MI 410-01⟩

Tanguy, G., **4**, 582 ⟨85JOM(280)C31, 89JOM(359)205⟩; **5**, 452 ⟨86CC1754⟩

Tani, H., **1**, 24 ⟨85BCJ2421⟩, 666 ⟨73JOC1130⟩; **2**, 708, 723 ⟨84CL139⟩; **3**, 621 ⟨87S631⟩; **4**, 247 ⟨91TL2039⟩; **5**, 242, 253 ⟨90JCR(S)56⟩, 383 ⟨66MI 507-01⟩, 665 ⟨90CL559⟩

Tani, K., **1**, 243 ⟨87BCJ4027⟩, 457, 507 ⟨93CC386⟩; **2**, 442 ⟨71JA4301, 77JA7876⟩, 816 ⟨90JA4897⟩

Tani, M., **3**, 307 ⟨88H(27)1855⟩, 309 ⟨80H(14)1939, 88CPB2023⟩, 309 ⟨90CPB3261⟩

Tani, S., **1**, 1016, 1145 ⟨90CC118⟩; **2**, 1020 ⟨90CC118⟩

Tanida, H., **5**, 588 ⟨75TL3063⟩

Tanier, S., **4**, 465, 478 ⟨92TL77⟩; **6**, 541 ⟨85TL4941⟩

Tanigawa, H., **2**, 588, 591 ⟨89AG(E)351⟩

Tanigawa, Y., **1**, 193 ⟨82TL5549⟩, 969 ⟨80JOC4536⟩; **2**, 606 ⟨84OS(62)39⟩, 940 ⟨80JOC4536⟩; **3**, 138 ⟨79JA4413⟩

Taniguchi, H., **1**, 462 ⟨86CL357⟩, 657, 658 ⟨79BCJ2942⟩, 985 ⟨92CL1357⟩, 1018 ⟨91JCS(P1)2892⟩, 1112 ⟨76BCJ2560, 76CC612⟩, 1113, 1114 ⟨80CC103⟩, 1120 ⟨76CC180, 79TL1619⟩, 1120 ⟨80T2531⟩; **2**, 89, 683 ⟨88CL1773⟩, 523 ⟨81JOC4313⟩; **3**, 267 ⟨73JOC4324⟩, 380 ⟨74JOC3519⟩; **4**, 21 ⟨78CL73⟩; **5**, 197 ⟨82CC132⟩, 948 ⟨82CC1178⟩; **6**, 552, 554 ⟨86CE311⟩, 552, 554 ⟨B-87MI 619-01⟩

Taniguchi, M., **2**, 313 ⟨90TL3023⟩; **3**, 238 ⟨86TL4759, 86TL4767⟩, 442 ⟨86TL4763⟩; **4**, 418, 419 ⟨81T1487⟩, 944 ⟨91TL2783⟩; **5**, 215 ⟨89S334⟩; **6**, 134 ⟨88BCJ2147⟩, 464 ⟨87BCJ1204⟩

Taniguchi, N., **2**, 16 ⟨92JCS(P1)221⟩

Taniguchi, R., **3**, 280, 296, 298, 305 ⟨80S139⟩

Taniguchi, T., **6**, 488 ⟨92JAP05339226⟩

Taniguchi, Y., **1**, 192 ⟨93JOC1538⟩; **2**, 308, 344, 351 ⟨88TL2973⟩; **3**, 361 ⟨89CC505⟩

Tanikaga, R., **1**, 635 ⟨88BCJ3211⟩; **3**, 16, 32 ⟨78S881⟩, 57 ⟨82JOC5017⟩

Tanikawa, H., **1**, 63 ⟨85OM584⟩; **5**, 873 ⟨83TL4245⟩

Tanimoto, S., **1**, 438 ⟨90CC1284⟩; **2**, 243, 358 ⟨83S716⟩, 247 ⟨86BCJ3897⟩, 391 ⟨88S805⟩, 444 ⟨82MI 208-01⟩, 445 ⟨81S53⟩, 720 ⟨91BCJ702⟩; **3**, 169 ⟨92CL239⟩, 428 ⟨88S805⟩, 429 ⟨88JOC4897⟩, 456 ⟨89JCS(P1)279⟩, 458 ⟨88JCS(P1)945⟩, 498 ⟨64TL1653⟩; **4**, 250 ⟨81TL3243⟩,

416 ⟨75BCJ357⟩, 761, 771 ⟨76BCJ1931⟩, 847 ⟨81S53⟩, 970 ⟨70MI 421-01⟩; **6**, 95 ⟨77MI 603-01⟩, 120 ⟨81S53⟩, 550 ⟨89CC684, 89JOC5603⟩, 552 ⟨82BCJ339⟩, 552, 555 ⟨81TL5195⟩
Tanimura, H., **6**, 80 ⟨85S408⟩
Tanino, K., **3**, 159 ⟨93TL6281⟩, 187 ⟨92JOC7009⟩
Tanio, M., **5**, 1041, 1107 ⟨77TL4417⟩, 1048 ⟨78CI(L)806⟩
Tanis, S. P., **1**, 216 ⟨83JOC4572⟩, 760 ⟨88TL4521⟩, 831 ⟨90TL1977⟩; **3**, 302 ⟨90TL1977⟩
Taniseki, Y., **1**, 195 ⟨93TL7619⟩
Taniuchi, H., **6**, 547 ⟨89BCJ967⟩
Taniyasu, R., **4**, 416 ⟨75BCJ357⟩, 761, 771 ⟨76BCJ1931⟩
Tanji, K., **1**, 1010 ⟨82CPB1865, 88CPB1935⟩
Tanke, R. S., **2**, 915 ⟨90CC1056, 90JA7984⟩; **5**, 336 ⟨86JOC5332⟩; **6**, 369 ⟨90JA7984⟩
Tanko, J., **1**, 801, 802 ⟨89CRV165⟩
Tanko, J. M., **5**, 318 ⟨86JOC4959⟩
Tannenbaum, G. A., **4**, 160, 755 ⟨89JOC5502⟩
Tanner, D., **1**, 697, 700, 703 ⟨87T4395⟩, 754 ⟨84CC349⟩; **2**, 50 ⟨94TL4631⟩, 331 ⟨88TL6493⟩; **5**, 342 ⟨88T613⟩, 342, 345 ⟨91TL283⟩
Tanner, D. D., **1**, 935 ⟨85JOC2149⟩, 945, 946 ⟨81JA1557, 90JOC3321⟩; **2**, 23 ⟨68JOC38⟩, 33, 95 ⟨68JA808, 84JA5261⟩, 398 ⟨88JOC3897⟩
Tanner, E. M., **3**, 267 ⟨65JCS3610⟩, 542 ⟨66TL3465⟩
Tanner, H., **4**, 320 ⟨62LA(658)128⟩
Tannert, A., **1**, 649, 656, 678, 680, 759 ⟨81TL5031⟩
Tanno, M., **5**, 1118, 1124 ⟨82CPB3125⟩
Tanol, M., **5**, 129 ⟨94JA9009⟩
Tanoue, Y., **3**, 101 ⟨89BCJ2736⟩
Tanoury, G. J., **1**, 676 ⟨88JA1636⟩
Tansjö, L., **2**, 417 ⟨59ACS35⟩
Tantasheva, F. R., **4**, 861 ⟨78ZOR478⟩
Tantrarat, T., **1**, 336, 346 ⟨83JOC1373⟩
Tanuma, M., **2**, 478 ⟨84T4553⟩
Tanuma, T., **6**, 4 ⟨92JAP04108765⟩
Tanyeli, C., **3**, 162 ⟨90SC2279⟩
Tanzawa, T., **1**, 568 ⟨92TL6783⟩
Tao, F., **2**, 43 ⟨88TL2967⟩
Tao, N. S., **5**, 586 ⟨72ZC133⟩
Tao, W., **2**, 877 ⟨88MI 217-01⟩
Tao, W. J., **1**, 1009 ⟨90JOC63⟩
Tao, X., **2**, 879 ⟨85MI 217-01⟩
Tao, Y.-T., **1**, 791 ⟨87JOC244⟩
Tao-Pen, C., **4**, 851 ⟨80AG(E)66⟩
Tapia, I., **4**, 232 ⟨90T1783⟩, 266 ⟨89AQ22⟩; **5**, 554 ⟨90T1057⟩
Tapia, R., **6**, 547 ⟨80AQ183⟩
Taplev, C. L., **2**, 439 ⟨90JOM(393)143⟩
Tapper, A., **3**, 485 ⟨89CB595⟩; **6**, 186, 387, 712 ⟨89CB1057⟩, 187, 188, 292, 388, 389, 673, 712 ⟨89CB595⟩, 187, 387, 712, 713 ⟨90CB747⟩, 188, 388, 391, 712 ⟨87CB1069⟩
Taqui-Khan, B., **5**, 55, 89 ⟨62PCS117⟩
Tar, C., **6**, 477 ⟨83MI 622-01⟩
Tara, H., **1**, 454 ⟨74BCJ2663⟩
Taraban, M. B., **4**, 656 ⟨84ZOB2025⟩
Tarakanova, A. V., **1**, 795, 801 ⟨76RCR469⟩
Tarama, K., **1**, 73, 74, 79, 80 ⟨72BCJ2723⟩
Tarantelli, T., **5**, 1053 ⟨62AC(R)79⟩, 1054 ⟨63AC(R)1113⟩
Tarantello, M., **5**, 734 ⟨82SC865⟩
Tarasenko, A. E., **3**, 336 ⟨92IZV160⟩
Tarasenko, A. I., **5**, 334 ⟨81ZOR533⟩
Tarasenko, **3**, 336 ⟨91URP1680699, 92URP1710560⟩
Tarasiuk, B., **6**, 546 ⟨90MI 617-03⟩
Tarasov, A. K., **5**, 989 ⟨89MI 526-01⟩
Tarasova, O. A., **6**, 565 ⟨86ZOR2145, 90ZOR1364⟩
Tarasov, V. A., **1**, 529 ⟨89TL4021⟩, 530 ⟨89S472⟩
Taraz, K., **5**, 246 ⟨85TL4349, 85ZN(B)1563⟩
Tarbell, D. S., **1**, 80 ⟨63OSC(4)136⟩, 637 ⟨46JA502⟩, 827 ⟨44OR(2)1⟩; **2**, 162 ⟨41JA2938⟩; **3**, 449 ⟨48JA1381⟩, 475 ⟨57JOC245⟩; **4**, 20 ⟨54OSC(4)807⟩; **5**, 106 ⟨55OSC(3)267⟩, 232 ⟨63OSC(4)924⟩, 652 ⟨55OSC(3)661⟩; **6**, 83, 84 ⟨53JA1668⟩, 434 ⟨77JOC3686⟩, 489 ⟨88OSC(6)207⟩, 540 ⟨55JA2479⟩, 730 ⟨68T6485⟩
Tarbin, J. A., **2**, 155 ⟨83TL1505⟩
Tarchini, C., **1**, 721 ⟨75C341⟩; **2**, 615 ⟨79HCA635⟩
Tardella, P. A., **1**, 132 ⟨69TL1117⟩; **2**, 699 ⟨69TL1117⟩; **3**, 153 ⟨78S227⟩, 181 ⟨83TL593⟩, 181 ⟨94T3829⟩
Tardivel, R., **2**, 16 ⟨83TL903, 86BSF955⟩, 16 ⟨87TL2359, 89T4431⟩; **3**, 149 ⟨87TL2359⟩
Tarent'eva, S. A., **2**, 476 ⟨90JGU397⟩
Tarhouni, R., **1**, 664 ⟨88JOM(346)C1⟩; **2**, 998 ⟨84TL835⟩; **3**, 152 ⟨84TL835⟩, 152, 154 ⟨81S68⟩; **4**, 138, 146, 667 ⟨85BSF825⟩, 139 ⟨86BSF470⟩, 146 ⟨84TL835⟩, 146 ⟨88JOM(346)C1⟩
Tarkard, M., **5**, 304 ⟨93TL2677⟩
Tarkoy, M., **1**, 797 ⟨93HCA418⟩
Tarli, A., **3**, 540 ⟨93TL2071⟩
Tarli, F., **6**, 42 ⟨66JOC2312⟩
Tarlin, H. I., **1**, 2, 3, 6 ⟨63JA1609⟩
Tarlton, E. J., **5**, 1125 ⟨62CJC1403⟩
Tarnchompoo, B., **1**, 1084 ⟨81CL1241⟩;**4**, 849 ⟨82S579⟩
Tarnow, H., **3**, 699 ⟨69AG(E)20⟩; **5**, 652 ⟨68AG(E)856⟩, 654 ⟨68AG942⟩
Tarnus, C., **5**, 328 ⟨92TL5055⟩
Tárraga, A., **5**, 350 ⟨93H(36)1263⟩, 1066 ⟨90JCS(P1)1727⟩; **6**, 560 ⟨89S923⟩, 563 ⟨88S690⟩
Tarrago, G., **4**, 984 ⟨71BSF283⟩
Tarrant, P., **4**, 45 ⟨71FCR77⟩, 730 ⟨54JA5423⟩, 731 ⟨55JA2783⟩, 806 ⟨68JOC472⟩, 817 ⟨68JOC286⟩; **6**, 5 ⟨55JA768, 64JOC1198⟩, 32 ⟨84JFC(25)69⟩
Tarsi, L., **3**, 310 ⟨91JOC1822⟩
Tartakovskii, V. A., **4**, 429 ⟨77IZV2384⟩, 995 ⟨75IZV2342⟩; **5**, 362 ⟨91IZV1108, 91MI 507-06⟩, 380 ⟨83IZV1413⟩, 680 ⟨91MI 517-01⟩; **6**, 65 ⟨76GEP2451635⟩, 145 ⟨70IZV948⟩, 145 ⟨73IZV122, 81IZV2146⟩, 148 ⟨72IZV490, 73IZV1149⟩, 149 ⟨62IZV272, 67ZOB1163, 68IZV621, 68ZOR231, 72IZV2603⟩, 149, 653 ⟨73ZOR896⟩, 169 ⟨74MIP 614-01⟩, 368 ⟨60IZV505, 68IZV1638⟩, 368 ⟨68IZV2839, 69IZV2304⟩
Tartakovsky, E., **2**, 626 ⟨82CC778⟩
Tartar, H. V., **2**, 186 ⟨38JA544⟩
Taru, H., **5**, 718 ⟨68JPS(B)389, 69JPS(B)17⟩
Tarzia, G., **1**, 871, 881 ⟨69CC1497⟩; **2**, 16 ⟨72CC122⟩
Tasai, H., **2**, 213 ⟨75CL813⟩
Tasaka, A., **6**, 558 ⟨89CPB1249⟩
Tasaka, K., **4**, 27 ⟨68JOC25⟩
Taschner, M. J., **1**, 505 ⟨90OSC(7)226⟩, 664 ⟨78JOC4235⟩; **3**, 65 ⟨78JOC4235⟩; **4**, 171 ⟨84JOC3994⟩, 846 ⟨78JOC4235⟩; **5**, 329 ⟨90TL5297⟩; **6**, 134 ⟨78JOC4235⟩
Tashchi, O. A., **3**, 434 ⟨84ZOR988⟩
Tashchi, V. P., **3**, 434 ⟨84ZOR988⟩
Tashika, H., **3**, 137 ⟨89BCJ2728⟩
Tashiro, J., **1**, 420 ⟨73T3797⟩
Tashiro, M., **1**, 57 ⟨75OPP189⟩, 232 ⟨91JHC289⟩, 426, 428 ⟨87OPP442⟩, 462 ⟨72JA7448⟩; **2**, 794 ⟨76H1243⟩; **3**, 263 ⟨84S663⟩, 309 ⟨90CPB3261⟩, 475 ⟨73T1983⟩; **4**, 21 ⟨84JHC1157⟩, 273 ⟨92JCS(P1)1179⟩, 307 ⟨67TL3769⟩; **5**, 996 ⟨68T2583⟩
Ta-Shma, R., **1**, 1110 ⟨71JCS(B)871, 71JCS(B)1461⟩
Tashma, Z., **5**, 371, 403 ⟨92TL2067⟩; **6**, 581 ⟨82JOC3012, 83JOC3966⟩
Tashtoush, H., **1**, 68, 454 ⟨83JA1398⟩; **3**, 419 ⟨86T4517⟩; **5**, 350 ⟨90SC1811⟩
Tashtoush, H. I., **1**, 937 ⟨92CB287, 93CB1759⟩
Tasi, M., **6**, 517 ⟨85M1103, 85OM1523⟩
Tasker, P. A., **1**, 518 ⟨73JCS(P1)2818⟩
Taskinen, E., **2**, 675 ⟨93T11389⟩
Tasumi, M., **5**, 246 ⟨86CL935⟩
Tatake, J. G., **3**, 134 ⟨77CI(L)742⟩
Tatarenko, A. N., **2**, 882 ⟨51ZOB93⟩
Tatarinova, A. A., **1**, 441 ⟨91SR1⟩; **5**, 1135 ⟨77IZV1839⟩
Tataruch, F., **2**, 1046 ⟨76C387⟩; **6**, 644 ⟨75C514⟩

Tate, B. E., **4**, 888, 889 ⟨45JA202⟩; **6**, 70 ⟨51JA2233⟩
Tate, D. P., **2**, 127, 188 ⟨60JA2872⟩
Tate, H. D., **2**, 101 ⟨51USP2529493⟩
Tatemitsu, H., **3**, 668 ⟨86TL2411⟩
Taticchi, A., **1**, 601 ⟨B-86MI 114-03⟩; **3**, 106 ⟨71CC1441⟩; **4**, 873 ⟨77JCS(P2)775⟩
Tatlow, J. C., **1**, 426 ⟨62JCS1801, 62JCS3253⟩, 1005 ⟨67CC1259, 72JCS(P1)639⟩; **2**, 15 ⟨86JFC(33)227⟩, 182, 364 ⟨57JCS2574⟩; **3**, 589 ⟨85JCS(P1)2185⟩; **4**, 3 ⟨73JFC(3)329⟩, 730 ⟨62CI(L)1472⟩; **5**, 34, 63, 100 ⟨61JCE217⟩, 116 ⟨62JCS186⟩, 193, 199 ⟨54JCS2006⟩; **6**, 9 ⟨52MI 601-01⟩, 20 ⟨60AFC(1)166, 74CC166⟩, 20 ⟨78JCS(P1)202⟩, 43 ⟨75JFC(6)37⟩, 53 ⟨84JCS(P1)455⟩, 450 ⟨83JFC(22)521⟩
Tatsumi, K., **1**, 707 ⟨76T1705⟩; **2**, 30 ⟨77CL1117⟩; **4**, 589 ⟨84JA5920⟩; **5**, 441 ⟨85JA4440⟩, 1097 ⟨80CC509, 81JA3328⟩
Tatsuno, T., **2**, 18, 19 ⟨81CL761, 88CL1877⟩; **3**, 152, 154, 155 ⟨82S1021⟩
Tatsuro, K., **5**, 208 ⟨90JFC(46)423⟩
Tatsuta, K., **3**, 716, 717 ⟨87JAN1199, 87JAN1207⟩, 717 ⟨87JAN1202⟩; **4**, 176, 198, 202, 214 ⟨93CRV1503⟩; **5**, 199 ⟨90TL709⟩; **6**, 288 ⟨88JAP63227595⟩, 550 ⟨79JA6116⟩
Tattershall, B. W., **2**, 360 ⟨64JCS5892⟩; **6**, 236 ⟨67JINC2819⟩, 241 ⟨66JINC1823⟩
Tatzuzaki, Y., **3**, 222 ⟨90JCS(P1)639⟩
Tau, S. I., **1**, 291 ⟨92CC13⟩
Taub, B., **6**, 504 ⟨61JOC5238⟩
Taub, D., **4**, 979 ⟨76TL2405⟩; **5**, 173 ⟨68T2443⟩
Taub, W., **5**, 164 ⟨72T1275⟩
Taube, D. J., **6**, 517 ⟨87IC526⟩
Taube, H., **1**, 94 ⟨88JA7906⟩; **5**, 386 ⟨74CC622⟩, 948 ⟨94JA11584⟩
Tauber, G., **2**, 393 ⟨67M1944⟩
Tauber, H., **2**, 844 ⟨53JA683, 53JA1379⟩
Taubert, R., **5**, 848 ⟨71ZN(B)296⟩
Taufer, I., **1**, 134 ⟨83TL3213⟩
Taunton, J., **3**, 203 ⟨88JOC6158⟩; **4**, 138, 154 ⟨88TL5863⟩
Taurins, A., **3**, 657 ⟨53CJC1211⟩; **4**, 442 ⟨55CJC923⟩; **5**, 749 ⟨61CJC761⟩, 1108 ⟨73CJC1741⟩
Taurke, I., **2**, 108 ⟨05CB1661⟩
Tauster, S. J., **1**, 88 ⟨87ACR389⟩
Tauzher, G., **2**, 583 ⟨86ICA(121)27⟩
Tavakalyan, N. B., **5**, 317 ⟨85T575⟩
Tavakkoli, K., **6**, 173, 394 ⟨93JOM(462)45⟩, 393, 394, 396 ⟨88JCS(D)381, 88CC1389⟩, 394 ⟨(95UP 613-01)⟩
Tavani, C., **1**, 486 ⟨93T235⟩
Tavares, D. F., **4**, 236 ⟨69CJC2875⟩, 410 ⟨78TL4995⟩
Tavares, R. F., **5**, 354 ⟨65JOC1311⟩
Tavecchia, P., **2**, 9 ⟨87S920⟩
Tavernier, D., **4**, 231 ⟨75T327⟩
Tavs, P., **2**, 792 ⟨66AG(E)1048⟩, 845, 847, 848 ⟨70CB2428, 70T5529⟩; **5**, 338 ⟨63JCS4918⟩
Tawamata, T., **1**, 561 ⟨80CPB282⟩
Tawara, K., **3**, 139 ⟨80JA6381⟩
Tawara, Y., **3**, 356 ⟨73JA3043⟩
Tawarayama, Y., **1**, 1098, 1140 ⟨84JOC3904⟩; **2**, 69, 564 ⟨84JOC3904⟩
Tawfik, A. M., **3**, 481 ⟨79S805⟩
Tawfik, M., **3**, 482 ⟨81LA1865⟩
Tawney, P. O., **1**, 311 ⟨41JA2308⟩
Tax, W. J. M., **4**, 203 ⟨73S151⟩
Taya, S., **1**, 404, 405, 816, 818 ⟨81JA6492⟩, 404, 406, 816 ⟨81TL69⟩
Tayaka, H., **6**, 107 ⟨67T45⟩
Tayano, T., **1**, 1104 ⟨89TL5153⟩; **2**, 902 ⟨85S406⟩; **3**, 27 ⟨85S406⟩; **4**, 765 ⟨89TL5153⟩
Taylor, A., **3**, 706 ⟨90JCS(P1)2115⟩, 712 ⟨85T1931⟩
Taylor, A. D., **6**, 380 ⟨90JOM(398)59⟩
Taylor, A. W., **1**, 422 ⟨76JCS(P1)410⟩; **3**, 559 ⟨77GEP(O)2728884⟩, 559, 581 ⟨77TL3831⟩
Taylor, B. F., **5**, 446 ⟨84CC420⟩
Taylor, B. J., **3**, 5, 115 ⟨79S815⟩
Taylor, C. W., **6**, 241 ⟨62JOC1064⟩
Taylor, D. A., **3**, 69 ⟨80T2125⟩
Taylor, D. P., **5**, 334 ⟨86JMC359⟩
Taylor, D. R., **1**, 498, 518 ⟨71CC1144, 74JCS(P1)1209⟩, 518 ⟨74JCS(P1)1844⟩; **2**, 1013 ⟨68TL3909⟩; **4**, 116 ⟨78JFC(11)441⟩, 117 ⟨77JFC(10)27⟩, 736 ⟨70TL5215⟩
Taylor, E. A., **1**, 575 ⟨73JOC26⟩
Taylor, E. C., **1**, 47 ⟨62JA3744⟩, 81 ⟨65JA1984⟩, 491 ⟨76OS(55)48⟩, 860 ⟨74JA8095⟩, 1010 ⟨87JOC3997⟩, 1076 ⟨72AG(E)48⟩; **2**, 160 ⟨82JOC552⟩, 160 ⟨84JOC2500⟩, 598 ⟨68JOM(15)500⟩, 621 ⟨70JOC528⟩, 626 ⟨69TL1623, 72JOC88⟩, 631 ⟨69TL2427, 71JA4841⟩, 631, 964 ⟨71JA4845⟩, 652 ⟨70JA3520⟩, 738, 751 ⟨B-63MI 215-01⟩, 776 ⟨70JOC1670⟩, 982 ⟨71JA6267, 73JA982⟩, 1021, 1025, 1027 ⟨B-60MI 221-01⟩; **3**, 146 ⟨77TL1827⟩, 303 ⟨67T897⟩, 527 ⟨72OS(52)36⟩, 664 ⟨84TL5473⟩; **4**, 191 ⟨77S467⟩, 194 ⟨73JA3635⟩, 194 ⟨76JA3037⟩, 194 ⟨76JA6750⟩, 194 ⟨77TL1827⟩, 194 ⟨82COMC-I(7)465⟩, 306 ⟨61JA2574⟩, 349 ⟨86JOC1537⟩, 422 ⟨84JOC113⟩; **5**, 8 ⟨88OSC(6)549⟩, 28 ⟨60JOC2245⟩, 183 ⟨68JA2422⟩, 355 ⟨91T9599⟩, 580 ⟨60JA2656⟩, 743, 746, 753 ⟨63JOC1108⟩, 746 ⟨77JOC1523, 80JOC2485⟩, 769, 849 ⟨72JOC3958⟩, 1135 ⟨78JOC2280⟩; **6**, 121 ⟨91JOC1816⟩, 504 ⟨62JOC2622⟩
Taylor, E. J., **1**, 635 ⟨76JA2275⟩
Taylor, E. R., **2**, 29 ⟨41OSC(1)115⟩; **5**, 60 ⟨32OSC(1)115⟩, 108 ⟨43OSC(2)588⟩, 115 ⟨43OSC(2)135⟩
Taylor, G., **3**, 414 ⟨83JOC5348⟩; **6**, 18 ⟨79CC964⟩
Taylor, G. A., **4**, 886 ⟨65CC574, 65JCS5877⟩, 886 ⟨67JCS(C)1569⟩; **5**, 325 ⟨87JCS(P1)1635⟩
Taylor, G. E., **4**, 674 ⟨80JOM(202)C49, 82JCS(D)1783⟩, 679 ⟨82JCS(D)1297⟩
Taylor, G. F., **5**, 288 ⟨76TL3089⟩, 991 ⟨78JOC2164⟩
Taylor, G. J., **1**, 1042 ⟨88T5451⟩
Taylor, G. W., **6**, 508 ⟨53JA1120⟩
Taylor, H. M., **3**, 616 ⟨73OSC(5)437⟩, 657 ⟨60JA1786⟩
Taylor, H. T., **1**, 422 ⟨53JCS124, 59JCS4017⟩
Taylor, H. W., **2**, 374 ⟨76JOC3805⟩
Taylor, J. A., **1**, 1006 ⟨67JCS(C)578, 69JCS(C)2173⟩, 1018 ⟨68TL2919, 71JCS(C)186⟩
Taylor, J. B., **2**, 163 ⟨80CSR477⟩; **3**, 39 ⟨82JCS(P1)307⟩; **4**, 330 ⟨75CSR189, 80CSR477⟩
Taylor, J. M., **4**, 802 ⟨93JOC1531⟩
Taylor, J. W., **1**, 562 ⟨61JA1705⟩
Taylor, K. G., **2**, 404, 405 ⟨71JA7337⟩; **4**, 224 ⟨89TL2179⟩; **6**, 107 ⟨78JOC1459⟩
Taylor, M. C., **2**, 94 ⟨25JA395⟩
Taylor, M. V., **1**, 595 ⟨70CC1683⟩
Taylor, M. W., **4**, 114, 802 ⟨69JA1929, 69JA1934⟩
Taylor, N. F., **1**, 2, 6 ⟨60JCS298⟩
Taylor, N. J., **1**, 448 ⟨89JOC4372⟩; **2**, 627 ⟨89JOC4372⟩; **4**, 604, 646 ⟨89JOC4372⟩, 711 ⟨90OM2234⟩, 713 ⟨91OM1907⟩; **6**, 195 ⟨80IC3637⟩, 626 ⟨83CJC1213⟩
Taylor, P., **5**, 943 ⟨90OM2621⟩, 944 ⟨94TL8883⟩; **6**, 49, 50 ⟨75JCS(D)488⟩
Taylor, P. C., **2**, 241, 792 ⟨88CC950, 94JCS(P1)1935⟩
Taylor, P. G., **1**, 447, 450 ⟨B-89MI 110-01⟩; **2**, 943 ⟨B-89MI 218-03⟩; **4**, 653 ⟨84TL2705⟩
Taylor, R., **1**, 772, 773, 774, 775, 777 ⟨B-79MI 117-02⟩, 773 ⟨72JCS(P2)165⟩, 774 ⟨76JCS(P2)280⟩, 774, 778 ⟨75JCS(P2)1025, 82JCS(P2)1175⟩, 777 ⟨82JCS(P2)1179, 83JCS(P2)291, 86JCS(P2)1255, 88JCS(P2)177, 88JCS(P2)737, 91JCS(P2)1703⟩; **2**, 43 ⟨94CC463⟩; **6**, 475 ⟨83JCS(P2)291⟩
Taylor, R. B., **3**, 638 ⟨44JA725, 45JA1071⟩
Taylor, R. J., **1**, 233, 235 ⟨88TL6005⟩, 400 ⟨90T6501⟩; **2**, 1000 ⟨88TL6009⟩; **3**, 614 ⟨88OM1761⟩; **4**, 171 ⟨82JOC3140⟩
Taylor, R. J. K., **1**, 130, 304 ⟨85S364⟩, 541 ⟨85TL3285⟩, 618 ⟨89TL3267⟩, 619 ⟨89T455⟩, 708 ⟨88CC454⟩; **2**, 232 ⟨80S910⟩; **3**, 12, 128 ⟨85S364⟩, 64 ⟨91S320⟩, 65 ⟨85CC782, 89JCS(P1)683⟩, 256 ⟨84TL1039⟩, 642 ⟨77S566⟩; **4**, 84 ⟨89TL3267⟩

Taylor, R. T., **1**, 74 ⟨91OR(40)91⟩, 1094 ⟨77TL159⟩; **2**, 911, 946 ⟨77TL159⟩; **3**, 156 ⟨89JOC4576⟩; **5**, 30 ⟨89OS(68)198⟩, 205 ⟨88TL2033⟩
Taylor, S. A., **4**, 799 ⟨88JCS(D)2781⟩
Taylor, S. J., **2**, 398 ⟨89OS(68)8⟩; **4**, 327 ⟨93OSC(8)550⟩, 897 ⟨89TL2349⟩, 898 ⟨89JOC992⟩, 906 ⟨87JOC4693, 90OR(38)655⟩
Taylor, S. K., **1**, 209 ⟨93JOC7304⟩, 475 ⟨81JOC2709⟩
Taylor, S. L., **1**, 582 ⟨90OSC(7)506⟩
Taylor, S. R., **1**, 178 ⟨93JOC4469⟩; **3**, 124 ⟨93JOC4469⟩
Taylor, S. T., **2**, 942 ⟨83JOC4156⟩
Taylor, T. E., **5**, 932 ⟨84JA1576⟩
Taylor, W. C., **1**, 36 ⟨68AJC1375⟩, 688 ⟨90AJC1439⟩, 1001 ⟨68AJC2989⟩
Taylor, W. G., **4**, 739 ⟨80S554⟩, 740 ⟨81JOC4290⟩; **6**, 695 ⟨80S554⟩
Taylor, W. H., **4**, 317 ⟨33JA4588⟩
Tazima, H., **3**, 704 ⟨76CPB832⟩
Tazoe, M., **3**, 704 ⟨82ABC1803⟩
Tazuma, J. J., **5**, 965 ⟨59USP2884435⟩
Tchakirian, A., **6**, 221, 222, 227 ⟨37CR(204)1927⟩
Tchelitcheff, S., **3**, 685 ⟨63BSF140⟩
Tcherniac, J., **3**, 619 ⟨07CB2709⟩
Tchernoff, G., **5**, 145 ⟨59MI 503-01⟩
Tchoubar, B., **1**, 55 ⟨63BSF1428⟩; **3**, 494 ⟨71CR(272)728⟩
Teager, D. S., **3**, 28 ⟨93JOC5493⟩
Tea-Gokou, C., **5**, 760 ⟨80JHC767⟩
Teague, H. J., **3**, 357 ⟨67JOC3140⟩
Teah, I., **4**, 655 ⟨81AJC181⟩
Teasdale, A., **1**, 494 ⟨91TL687⟩
Teasdale, A.,**1**, 546 ⟨93TL153⟩
Tebbe, F. N., **1**, 669 ⟨71JA3793⟩, 669, 766, 767, 769 ⟨78JA3611⟩; **2**, 680 ⟨78JA3611⟩, 979 ⟨80JA6149⟩; **3**, 508 ⟨78JA3611⟩; **4**, 672 ⟨71JA3793⟩, 725 ⟨78JA3611⟩
Tebbe, K. F., **2**, 473 ⟨83AG(E)775⟩; **4**, 1025 ⟨84CB1542⟩
Tebbe, R. F., **6**, 220 ⟨47JA947⟩
Tebben, P., **1**, 138 ⟨90AG(E)1422⟩; **3**, 213 ⟨89TL2919⟩; **4**, 841 ⟨89TL2915⟩
Tebboth, J. A., **1**, 862 ⟨52JCS632⟩
Tebby, J. C., **1**, 51 ⟨83JCS(P2)1923⟩; **2**, 810, 892 ⟨72JCS(P1)2793⟩; **3**, 498 ⟨61JCS2126, 67JCS(C)2442⟩, 498 ⟨69JCS(C)1100, 70JCS(C)5⟩, 498 ⟨89JCS(P1)2425⟩; **4**, 558 ⟨86CC871, 92JCS(P1)479⟩, 563 ⟨70JCS(C)504⟩, 922 ⟨81JCS(P1)1363⟩
Tech, B., **4**, 970 ⟨84MI 419-01⟩
Teche, A., **4**, 757 ⟨81EUP30887⟩
Téchy, B., **4**, 790, 791 ⟨77NJC369⟩
Techy, P., **4**, 790 ⟨80TL223⟩
Tecilla, P., **5**, 11 ⟨90JA9586⟩
Tedder, J. M., **1**, 815 ⟨66CC775⟩, 860 ⟨B-65MI 119-03⟩, 919, 925 ⟨82AG(E)401⟩, 919 ⟨B-79MI 119-04, 82T313⟩, 935 ⟨73JCS(P2)1570⟩; **2**, 1051 ⟨68JCS(C)1324⟩, 1056 ⟨63AG1033⟩; **3**, 212 ⟨53JCS3628⟩; **4**, 30 ⟨71JCS(B)1723⟩, 1077 ⟨B-79MI 424-01⟩; **5**, 110 ⟨77JFC(10)437⟩
Tedeschi, R. J., **1**, 503, 506 ⟨62JOC2398⟩, 503, 506 ⟨62JOC4323, 67ANY(145)91⟩
Tedga, N., **2**, 759 ⟨83JAP58113195, 83JCR(S)222⟩; **3**, 76, 262 ⟨83JCR(S)222⟩
Tedjo, E. M., **1**, 600 ⟨81TL4137⟩
Teeter, H. M., **2**, 94 ⟨63OSC(4)125⟩, 778 ⟨65MI 215-02⟩
Tefertiller, B. A., **1**, 871, 881 ⟨68JOC935⟩
Tefteller, A. T., **3**, 304 ⟨63JOC398⟩
Tegeler, J. J., **2**, 350 ⟨76TL4037⟩
Tegeler, M., **2**, 893, 894 ⟨86ZN(B)321⟩
Tegg, D., **6**, 69, 296, 297 ⟨65JOC411⟩
Teichman, S., **4**, 109 ⟨89T2957⟩
Teichmann, G., **6**, 268 ⟨60ZAAC(303)85⟩
Teichmann, H., **2**, 1099 ⟨63CB1462⟩; **3**, 361, 558 ⟨77ZC93⟩; **4**, 106 ⟨87JPR871⟩
Teichmann, K., **4**, 417, 423 ⟨58CB2432⟩, 442 ⟨56CB1134⟩
Teichmann, S., **1**, 1069 ⟨89CB377⟩
Teital, S., **3**, 294 ⟨71JA4321⟩
Teixeira-Dias, J. J. C., **5**, 506, 507 ⟨89JCS(F2)1945⟩, 506, 508 ⟨90JST67, 87JST381⟩
Tejel, C., **6**, 627 ⟨83S375⟩
Tejero, T., **3**, 43 ⟨92TL4221⟩; **5**, 1026 ⟨89BSB289⟩; **6**, 563 ⟨91S147⟩, 605 ⟨83S375, 84S520⟩
Tejima, T., **5**, 129 ⟨94JOC4749⟩
Teleha, C. A., **2**, 206 ⟨93TL3667⟩
Teles, J. H., **3**, 686 ⟨87AG(E)155⟩, 703 ⟨88AG(E)938⟩
Telford, R. P., **1**, 74, 441 ⟨77JCS(P1)443⟩; **3**, 13, 83, 93 ⟨80JCS(P1)27⟩
Teller, E., **4**, 53, 56 ⟨67CB2120⟩
Teller, R., **3**, 66 ⟨68LA(715)106⟩
Teller, W., **5**, 1050 ⟨72TL1513⟩
Tellez, M. R., **6**, 589 ⟨89JCS(P1)1068⟩
Tellier, F., **6**, 19 ⟨91TL5963⟩
Telschow, J. E., **3**, 156 ⟨78JOC188⟩
Teltz, H. P., **4**, 215 ⟨59LA(620)1⟩
Temciuc, M., **5**, 595 ⟨93S483⟩; **6**, 566 ⟨93S483⟩
Temin, S. C., **2**, 748 ⟨57JOC1714⟩
Tempel, N., **2**, 1082 ⟨81ZAAC(474)157⟩
Tempesti, E., **2**, 190 ⟨74JCS(P1)771⟩
Temple, Jr., C., **3**, 357, 362 ⟨76JOC3784⟩; **6**, 109 ⟨65JOC829⟩
Temple, Jr., D. L., **5**, 334 ⟨86JMC359⟩
Temple, J. S., **1**, 536 ⟨80JA7381, 81TL2629⟩
Temple, P., **1**, 591 ⟨74JCS(P1)433⟩
Temple, R. D., **5**, 51, 75 ⟨70JOC1275⟩
Temple, S., **2**, 176 ⟨68JOC344⟩
Templeton, D. H., **1**, 890 ⟨74JA1348⟩
Templeton, G. F., **5**, 179 ⟨89LA1233⟩
Templeton, J. L., **4**, 675 ⟨83JA2599⟩; **5**, 940 ⟨85JA4474⟩, 952, 953 ⟨85OM2102⟩, 1097 ⟨91JA2324⟩
Temu, C. T., **3**, 116 ⟨89SC215⟩
Tenaglia, A., **2**, 85 ⟨88TL4851⟩
TenBrink, R., **1**, 879 ⟨83JOC2957⟩
Tencer, M., **3**, 722 ⟨15CB953⟩
Tener, G. M., **2**, 105, 106 ⟨59CI(L)542⟩, 105, 106 ⟨61JA159⟩
Tenesheva, N. Kh., **2**, 883 ⟨83JGU1832⟩
Tenforde, T., **5**, 1063 ⟨72JOC3372⟩
Teng, K., **1**, 1135 ⟨82JOC2549⟩; **3**, 273 ⟨82JOC2549⟩
Tengi, J. P., **5**, 346 ⟨85JOC3462⟩
Tenma, H., **6**, 529 ⟨85JAP62113712, 86JAP62176910⟩, 533 ⟨85JAP62120358, 90JAP04066566⟩, 533 ⟨90JAP04128263, 90JAP05032616⟩
Tennant, G., **2**, 342, 345 ⟨79COC(2)504⟩; **3**, 404 ⟨79COC(2)385⟩; **3**, 431 ⟨79COC(2)500⟩, 663 ⟨79COC(2)528⟩; **5**, 1022 ⟨79COC(2)513⟩
Tennent, H. G., **2**, 575 ⟨72JA2512⟩
Teo, C. C., **3**, 438 ⟨84TL5951⟩
Teodoradze, G. A., **2**, 594 ⟨75JOM(88)1⟩
Teodori, E., **4**, 232 ⟨87JMC1934⟩
Teodorovic, A. V., **2**, 526 ⟨91TL6691, 92JOC2379⟩
Teofili, E., **6**, 420 ⟨26G847⟩
Teoh, S. C., **6**, 563 ⟨78JINC451⟩
Teoh, S.-G., **2**, 1008 ⟨92POL2351⟩
Terabe, S., **4**, 752 ⟨67TL1123⟩
Terada, A., **3**, 101 ⟨89BCJ2736⟩, 374 ⟨70CPB490⟩
Terada, H., **5**, 678 ⟨81H(16)717⟩; **6**, 563, 565 ⟨82JMC557⟩
Terada, M., **1**, 374 ⟨91TA643⟩; **2**, 80 ⟨89JA1940, 90JA3949⟩, 80 ⟨92JA6566, 93CC327⟩; **3**, 615 ⟨92CL1169⟩
Terada, S., **5**, 171 ⟨85JA6404⟩
Terada, T., **2**, 19 ⟨82JCS(P1)85⟩; **3**, 7, 30 ⟨86S774⟩; **4**, 837 ⟨85CPB2256⟩
Teraji, T., **1**, 729 ⟨71JCS(C)3252⟩; **3**, 427 ⟨84JAN532⟩
Terakawa, T., **1**, 1042 ⟨89CL1079⟩
Teramoto, K., **3**, 252 ⟨76JCS(P1)1146⟩
Teramoto, S., **6**, 552, 554 ⟨86CE311⟩, 552, 554 ⟨B-87MI 619-01⟩
Teramura, D. H., **1**, 828 ⟨81JOC1485⟩; **2**, 70 ⟨87JA7838⟩
Teramura, K., **3**, 684 ⟨84BCJ2531, 86BCJ2827⟩, 684 ⟨91BCJ318⟩, 685 ⟨87BCJ1948⟩

Teranaka, T., **1**, 590, 591 ⟨88TL659⟩
Teranishi, A., **3**, 557 ⟨80JOC5385⟩
Teranishi, A. Y., **1**, 601, 604, 605, 606, 607 ⟨73JA6137⟩; **2**, 282, 287 ⟨73JA6137⟩, 702 ⟨73JA3310⟩; **3**, 37, 173, 207 ⟨73JA6137⟩
Teranishi, S., **1**, 692 ⟨78JOC2477⟩, 1004, 1107 ⟨79JOC55, 74TL1067⟩; **2**, 91 ⟨79JA159⟩, 616 ⟨74TL1067, 76CL449⟩; **3**, 117 ⟨83TL5009⟩, 121 ⟨76CL231⟩, 620 ⟨85CL1339⟩; **5**, 143 ⟨68TL633⟩
Terao, K., **1**, 602 ⟨84JOC3796, 85JCS(P1)373⟩; **2**, 282 ⟨84JOC3796, 85JCS(P1)373⟩; **3**, 37, 174 ⟨84JOC3796⟩; **5**, 285 ⟨86JCS(P1)1837, 86JOC1724⟩, 285 ⟨87JOC2018⟩
Terao, S., **5**, 278 ⟨82JCS(P1)2909⟩
Terao, Y., **1**, 367 ⟨85TL3011, 86TL4039⟩, 368 ⟨88H(27)981⟩, 1045 ⟨71CPB391⟩; **3**, 273 ⟨71CPB391⟩, 367 ⟨87CL1851⟩, 465 ⟨76CPB369, 79CPB682⟩; **4**, 317 ⟨77CPB2964⟩, 325 ⟨81CPB2496⟩, 369 ⟨85TL3011, 86H(24)1571⟩, 369 ⟨87CPB1734⟩, 937 ⟨86CL2089⟩; **6**, 129 ⟨86H(24)1571⟩, 129, 282 ⟨87CPB1734⟩
Teraoka, T., **6**, 261 ⟨92JAP04110383⟩
Terapane, J. F., **1**, 777 ⟨66TL6279⟩
Terasawa, I., **3**, 634 ⟨73OSC(5)266⟩
Terasawa, M., **1**, 455 ⟨79MI 110-01⟩; **2**, 616 ⟨76CL449⟩
Terasawa, S., **2**, 683 ⟨84CPB3759⟩
Terashima, K., **3**, 659 ⟨76CPB3011⟩; **5**, 249, 563, 640 ⟨88BCJ3931⟩
Terashima, M., **1**, 91 ⟨68JOC747⟩; **3**, 309 ⟨93H(35)573⟩; **5**, 301 ⟨82CC741⟩
Terashima, S., **1**, 33, 34 ⟨81T2797⟩, 455 ⟨79MI 103-01⟩, 1029 ⟨91TL3405⟩, 1140 ⟨84CL1543, 86CPB1531⟩; **3**, 32 ⟨89BCJ3038⟩, 61 ⟨72TL1815⟩; **6**, 447 ⟨92TL3483⟩
Terashita, Z.-i., **6**, 558 ⟨89CPB1249⟩
Terauchi, M., **1**, 235 ⟨81CL315, 82CL351⟩
Terauchi, Y., **3**, 352, 353 ⟨88CL717⟩
Ter Borg, A. P., **1**, 402, 795 ⟨69RTC266⟩, 404 ⟨65RTC241⟩; 794 ⟨63RTC741⟩
Tercel, M., **5**, 567 ⟨93JMC2459⟩
Terent'ev, A. P., **2**, 478 ⟨61ZOB217⟩
Terent'eva, S. A., **4**, 478 ⟨86JGU632, 90JGU397⟩, 478 ⟨93ZOB220⟩
Terentev, P. B., **5**, 116 ⟨68JIC1109⟩
Tereshchenko, G. F., **5**, 270, 283 ⟨71RCR835⟩
Ter-Gabrielyan, E. G., **3**, 588 ⟨75BAU1274⟩; **4**, 894 ⟨75IZV1274⟩, 979 ⟨86IZV2344⟩; **6**, 22 ⟨90IZV491⟩
Terlouw, J. K., **2**, 1022 ⟨86AG(E)282⟩; **3**, 686 ⟨91TL1487⟩
Ter Mer, E., **4**, 432 ⟨1876LA(181)1⟩
Termine, E. J., **3**, 133 ⟨82SC15⟩
Termont, D., **1**, 356 ⟨78TL2057⟩
Termura, K., **6**, 51 ⟨76JOC3794⟩
Ternansky, R. J., **1**, 364, 365 ⟨90PAC1597⟩; **3**, 639 ⟨90TL2805⟩
Ternay, A. L., **2**, 478 ⟨67JOC2465, 69JOC940⟩
Ternay, Jr., A. L., **2**, 1000 ⟨83JHC399, 84AX(C)842⟩
Ternikova, T. V., **4**, 995 ⟨75IZV2342⟩
Ternovskoi, L. A., **2**, 18 ⟨87JOU230⟩
Terpinski, J., **2**, 19 ⟨84JOC3216⟩; **4**, 830 ⟨87JA3811⟩; **5**, 4 ⟨84JOC3216⟩; **6**, 280 ⟨83JA6513⟩, 280, 614 ⟨86TL419⟩
Terpko, M., **5**, 157 ⟨77JOC3903⟩
Terpstra, J. W., **2**, 104 ⟨83JOC3379⟩, 619 ⟨87JOC1619⟩
Terrel, R., **2**, 738, 758 ⟨63JA79⟩
Terrell, R., **1**, 307 ⟨63JA207⟩; **3**, 10, 124 ⟨54JA2029⟩
Terrell, R. C., **6**, 36 ⟨71JMC517⟩, 36, 43 ⟨71JMC593⟩, 38, 43 ⟨72JMC604⟩, 38, 43 ⟨72JMC606⟩, 44, 45, 46 ⟨65JOC4011⟩, 74 ⟨73USP3745220⟩
Terrence, K. M., **2**, 915 ⟨90TL6621⟩
Terrett, N. K., **1**, 244 ⟨88TL1315, 89TL3865⟩, 654 ⟨83TL4151, 83TL4153⟩, 815 ⟨84TL5103⟩, 943 ⟨90T545⟩
Terrier, F., **6**, 97, 100 ⟨75JOC2911⟩
Terry, M. R.,**5**, 946 ⟨90JA3220, 92JA6735⟩
Terry, P. H., **1**, 1076 ⟨66JOC2867⟩; **6**, 655 ⟨67JMC118⟩
Ter-Sarkisyan, G. S., **2**, 811 ⟨76IZV1823⟩
Tertakovskii, V. A., **4**, 524 ⟨74MI 411-01⟩
Tertov, B. A., **4**, 539 ⟨70MI 411-01⟩
Terui, T., **5**, 276 ⟨69JCS(C)874⟩
Terunuma, D., **4**, 142, 521 ⟨91CL97⟩, 506 ⟨80BCJ789⟩, 610, 617, 622 ⟨87BCJ2263⟩
Ter Wiel, J., **4**, 237 ⟨70TL935⟩
Terzis, A., **5**, 366 ⟨90JHC1741⟩
Tesar, V., **4**, 317 ⟨58LA(619)169⟩
Tesařik, K., **4**, 5 ⟨71CCC1867⟩; **6**, 5 ⟨71CCC1867⟩
Tesch, A. G., **4**, 576 ⟨59JA5519⟩
Teschner, M., **3**, 35 ⟨73TL5113⟩
Teshirogi, T., **6**, 417 ⟨83S908⟩
Tesky, F.-M., **2**, 401 ⟨79ZAAC(452)103⟩
Teslya, I. A., **5**, 778 ⟨87IZV165⟩
Tesno, S. L., **4**, 140 ⟨85SRI321⟩; **5**, 224 ⟨85SRI321⟩
Tesorero, J. M., **3**, 170 ⟨89JOC5620⟩
Tesse, M. W., **4**, 473 ⟨84AJC205⟩
Tesser, G. I., **2**, 409 ⟨84TL1307⟩; **5**, 1028 ⟨76MI 527-01⟩
Tessier, C. A., **1**, 1010 ⟨90SL427⟩
Tessier, J., **1**, 751 ⟨86MI 116-01⟩; **4**, 757 ⟨81EUP30887⟩
Tessier-Youngs, C., **4**, 652 ⟨86IS95⟩
Testa, E., **2**, 103 ⟨62HCA358⟩, 182, 248 ⟨62HCA717⟩; **5**, 566 ⟨68FES642⟩
Testaferri, L., **1**, 484 ⟨83T2289, 91G59⟩, 782 ⟨93JOC1349⟩; **2**, 278, 729 ⟨83JOC4289⟩, 709 ⟨83JOC4795⟩, 711 ⟨83S751⟩, 1036, 1037 ⟨93SL211⟩; **4**, 92 ⟨91JOC4529⟩; **5**, 238 ⟨86T4577⟩
Tester, R. W., **1**, 400 ⟨92JOC3479⟩
Testoni, G., **4**, 440 ⟨79S380⟩
Tetel'baum, B. I., **4**, 923 ⟨67ZOB454⟩; **6**, 54 ⟨66ZOB728⟩
Tetenbaum, M. T., **2**, 391 ⟨79JOC275⟩; **3**, 192 ⟨79JOC275⟩
Tetenyi, P., **2**, 4 ⟨88IJC(B)1129⟩
Tetreault-Ryan, L., **2**, 201 ⟨78TL2353⟩
Tetsch, C., **2**, 134 ⟨37CB675⟩
Tetsuo, O., **1**, 243 ⟨80SC595⟩
Tetzlaff, C., **4**, 1080 ⟨88JOC1806⟩
Tetzlaff, G., **2**, 745 ⟨69JPR162⟩
Teuben, J. H., **1**, 670 ⟨87CC849⟩; **2**, 569 ⟨76JOM(110)327, 87OM2354⟩, 981 ⟨87OM2053⟩; **3**, 510 ⟨89JA5977⟩; **4**, 656 ⟨93OM184, 93OM1936⟩, 671 ⟨93OM2268⟩, 703 ⟨93OM2609⟩; **5**, 797 ⟨91OM2665⟩; **6**, 194 ⟨87OM1509⟩, 195 ⟨86OM1726⟩, 195 ⟨88CC962⟩
Teuber, H. J., **3**, 231 ⟨70TL4069⟩
Teuber, L., **4**, 842 ⟨88ACS629⟩
Teulade, J. C., **5**, 360 ⟨90CPB2352⟩
Teulade, M. P., **2**, 456 ⟨87TL405⟩; **3**, 49 ⟨87TL405⟩; **4**, 132, 802 ⟨88JOM(338)295⟩, 557 ⟨85TL4435, 86JOM(312)283⟩, 557, 576 ⟨87TL405⟩, 557, 576, 801 ⟨86JOM(304)283⟩, 570 ⟨86S831⟩; **6**, 164, 374 ⟨87JOM(323)135⟩, 288 ⟨86JOM(304)283, 87SC1559⟩, 289 ⟨88JOM(338)295⟩
Teutsch, G., **6**, 511, 605 ⟨86T2677⟩
Teva, D., **2**, 617 ⟨91JHC939⟩
Tewari, R. S., **1**, 268, 738, 739 ⟨76JOM(108)175⟩, 733 ⟨76IJC(B)419⟩, 738 ⟨73JOM(60)247, 75JOM(85)173⟩, 738 ⟨75JOM(96)237, 76JOM(112)279⟩, 738 ⟨78IJC(B)623, 78S616⟩, 738 ⟨79IJC(B)637⟩, 740, 742 ⟨77TL3843⟩, 740, 742 ⟨79IJC(B)859⟩, 741 ⟨80SC457, 80ZN(B)95⟩; **2**, 841 ⟨73ZOB997⟩; **3**, 254 ⟨78SC315⟩, 500 ⟨78IJC(B)623, 79IJC(B)637⟩; **5**, 678, 679 ⟨81JCED418⟩
Tews, H., **1**, 925 ⟨93SL286⟩
Texier, F., **3**, 567 ⟨75CR(C)51⟩; **6**, 592 ⟨89CC1494⟩
Texier-Boullet, F., **2**, 394 ⟨86TL493⟩, 457 ⟨86TL3515⟩; **3**, 191 ⟨86TL493⟩, 406 ⟨85S679⟩, 684 ⟨94JCR(S)116⟩; **4**, 338 ⟨86TL3515⟩
Teyssie, Ph., **1**, 549 ⟨78TL1239⟩
Tezuka, M., **2**, 927 ⟨90TL369, 91JOC5980⟩
Tezuka, T., **1**, 690 ⟨83BCJ192⟩, 795 ⟨70BCJ1120⟩; **5**, 357 ⟨87H(26)2081, 90TL7633⟩
Thacher, A. F., **5**, 103 ⟨73OSC(5)509⟩
Thain, E. M., **2**, 127 ⟨52JCS800⟩
Thaisrivongs, S., **1**, 67 ⟨83JA7407⟩, 218, 831 ⟨79CJC1743,

83JA1988⟩, 831 ⟨80JA1155, 81JA2446⟩; **2**, 900 ⟨77JOC3243⟩
Thakar, K. A., **3**, 91 ⟨77IJC(B)1056⟩
Thaker, V. B., **1**, 548 ⟨81JOC2622⟩
Thakkar, D. C., **2**, 376 ⟨73OPP219⟩, 380, 381, 382 ⟨73OPP233⟩
Thakker, D. R., **6**, 491 ⟨89TL7141⟩
Thakore, A. N., **2**, 26, 32 ⟨71T2617⟩
Thal, A. F., **3**, 612 ⟨41OSC(1)107⟩; **5**, 34 ⟨32OSC(1)436⟩
Thaler, W. A., **6**, 434 ⟨68CC527⟩
Thaller, V., **3**, 712 ⟨85T1931⟩
Tham, F. S., **2**, 1006 ⟨93HOU(E/19d)713⟩
Thamer, D., **4**, 313 ⟨73S302⟩, 327 ⟨73S303⟩; **5**, 94 ⟨76AP(308)954⟩, 122 ⟨75AP(308/12)954⟩, 361 ⟨76S241⟩
Thames, S. F., **2**, 813 ⟨75JOC1090⟩, 942 ⟨73JOC4271⟩
Thamm, R., **2**, 429 ⟨82ZN(B)965⟩; **4**, 453, 454, 481 ⟨73JPR526⟩; **6**, 490 ⟨81ZN(B)910⟩
Thamnusan, P., **2**, 616 ⟨79TL617⟩; **4**, 77 ⟨79TL617⟩
Thanassi, J. W., **6**, 38 ⟨60JOC2009⟩
Thangaraj, K., **3**, 144 ⟨84S1010⟩
Thang, S. H., **2**, 321 ⟨93SC2355⟩
Thankarajan, N., **4**, 697 ⟨67IJC210⟩
Thaoubane, S. A., **2**, 473, 474 ⟨84PS(20)81⟩
Thatte, S. D., **1**, 1076 ⟨66JOC2867⟩
Thavonekham, B., **1**, 177, 178 ⟨89JOC5831⟩
Thayer, F. K., **5**, 98 ⟨32OSC(1)398⟩
Thayer, J. S., **2**, 1076 ⟨74JOM(76)265⟩; **3**, 698 ⟨76IC501⟩; **5**, 1016, 1017 ⟨66JOM(5)383⟩
Thé, K. I., **2**, 476 ⟨77JA7379⟩, **4**, 437 ⟨72CC841, 73CJC2448⟩; **6**, 365 ⟨73CJC2448⟩
Thea, S., **2**, 168 ⟨88TL2865⟩, 173, 717 ⟨87TL5193⟩, 194 ⟨85JOC5592⟩; **4**, 907 ⟨74JCS(P1)2357⟩; **5**, 336 ⟨90TA141⟩
Thebtaranonth, C., **1**, 787 ⟨94CC889⟩, 919 ⟨90T1385⟩; **3**, 210 ⟨80TL1169⟩; **4**, 849 ⟨82S579⟩
Thebtaranonth, Y., **1**, 474 ⟨81T3957, 94CC889⟩, 823 ⟨81JCS(P1)1963, 90T1385⟩, 1084 ⟨81CL1241⟩; **4**, 849 ⟨82S579⟩
Theibault, H., **2**, 16 ⟨87TL2359⟩
Theilacker, W., **2**, 374 ⟨B-64MI 207-01⟩; **5**, 199 ⟨55LA(597)95⟩
Theim, K.-W., **2**, 618 ⟨72CB1683⟩
Theis, M., **4**, 667, 668 ⟨83AG(E)733⟩; **6**, 172, 204, 205 ⟨83AG(E)733, 87TCC(138)1⟩, 205 ⟨86OS378⟩, 405 ⟨84AG(E)995⟩
Theis, R. J., **1**, 1075 ⟨66JOC624⟩
Theis, W., **2**, 1003 ⟨91JOC1453⟩; **5**, 870 ⟨85T2625⟩
Theisen, C. Th., **2**, 889, 890 ⟨68JOM(11)627⟩
Theisen, D., **5**, 701, 827 ⟨69CB1641⟩
Thenappan, A., **4**, 118 ⟨90JOC2311⟩; **5**, 894 ⟨90JOC2311⟩
Thenn, W., **3**, 680 ⟨73AG(E)155⟩; **4**, 311, 493 ⟨73AG(E)502⟩
Theobald, D. W., **2**, 675 ⟨64JCS1029⟩
Theobald, F., **1**, 526 ⟨87BSF1006⟩
Theobald, J. M., **5**, 122 ⟨63JCS1927⟩
Theobald, P. G., **1**, 439 ⟨87TL6565⟩
Theocharis, A. B., **5**, 366 ⟨90JHC1741⟩
Theodorakis, E. A., **1**, 777 ⟨94T639⟩, 944 ⟨92JA5904, 93JA8050⟩; **2**, 721 ⟨91T9167⟩; **5**, 1110 ⟨92T2613⟩
Theodoridis, G., **4**, 843 ⟨86OS(64)189⟩, 853, 855 ⟨82JOC3027⟩
Theodoropoulos, S., **3**, 106 ⟨63JOC3052⟩, 498 ⟨68T2241⟩
Theopold, K. H., **2**, 895, 896 ⟨91CB2453⟩; **4**, 687 ⟨81JA2489, 82OM219⟩, 691 ⟨89OM2001⟩
Theopold, K. M., **2**, 571 ⟨90ACR263⟩
Theriault, N., **2**, 107 ⟨77JA1277⟩
Thérien, M., **3**, 6, 119 ⟨81S394⟩; **4**, 247 ⟨88TL6729⟩, 248 ⟨88TL6733⟩
Thétaz, C., **5**, 1122 ⟨80AG(E)566⟩
Theunisse, A. W. G., **1**, 834 ⟨87JOC370⟩
Theuns, H. G., **1**, 784 ⟨84JCS(P1)1701⟩
Thévenot, G., **2**, 906 ⟨01LA(315)19⟩
Thewalt, U., **2**, 878, 896 ⟨86JOM(317)175⟩, 963 ⟨86JOM(308)289⟩, 999, 1000 ⟨89OM1688⟩; **6**, 268 ⟨90CB945⟩, 338 ⟨90ZN(B)447⟩
Thewissen, D. H. M. W., **1**, 72, 88 ⟨90RTC87⟩
Thewlis, B. H., **4**, 13, 23 ⟨51JCS1145⟩; **6**, 28 ⟨51JCS1145⟩
Thiam, M., **2**, 381 ⟨90TL1429⟩; **3**, 41 ⟨90TL1429, 92BSF161⟩
Thibaudeau, C., **1**, 946 ⟨93T9079⟩
Thibault, J., **3**, 293 ⟨69BSF915⟩
Thibault-Starzyk, F., **4**, 257 ⟨92SC1359⟩; **5**, 554 ⟨93SL148⟩; **6**, 121, 325 ⟨92SC1359⟩
Thibaut, P., **5**, 239 ⟨76JCS(P1)2452⟩
Thie, I., **1**, 48 ⟨79AG(E)939⟩
Thiebault, A., **1**, 474 ⟨88CC203⟩
Thiebault, H., **2**, 16 ⟨83TL903, 86BSF955⟩; **3**, 149 ⟨87TL2359⟩
Thiéffry, A., **2**, 684 ⟨83JOC441⟩; **3**, 8, 121 ⟨81TL2885⟩
Thiel, F., **3**, 220 ⟨87S1012⟩
Thiel, M., **2**, 167 ⟨60LA(637)146⟩, 233 ⟨59LA(627)195⟩; **4**, 317 ⟨58LA(619)169⟩, 318 ⟨59LA(622)83, 59LA(622)94⟩
Thiel, W., **5**, 554 ⟨77ZC366, 89JPR243⟩, 559 ⟨86JPR(328)812⟩, 561 ⟨77ZC92⟩, 565 ⟨88SR1⟩, 601 ⟨88ZC233⟩
Thiel, Y., **2**, 556 ⟨91JA5720⟩
Thiele, A., **4**, 1004 ⟨85LA555⟩; **6**, 158 ⟨85LA555⟩
Thiele, E., **4**, 728 ⟨00LA353⟩
Thiele, G., **1**, 574, 584 ⟨91AG(E)1372⟩; **2**, 507 ⟨78AG(E)135⟩
Thiele, J., **2**, 386, 388, 389, 403 ⟨10LA(376)239⟩; **4**, 203 ⟨60LA355⟩
Thiele, K.-H., **2**, 566 ⟨68ZAAC(356)195, 71JOM(32)335⟩, 569 ⟨72PAC575⟩, 569, 570 ⟨77ZAAC(431)88⟩, 572 ⟨72ZAAC(394)293, 74ZAAC(412)11⟩, 586 ⟨68JOM(12)225⟩
Thiele, M., **4**, 545, 558 ⟨93PS(76)13⟩
Thielepape, E., **5**, 723 ⟨35CB751⟩
Thielmann, M., **1**, 519 ⟨86TL5833⟩
Thiem, J., **4**, 225 ⟨93CAR(249)197⟩, 912 ⟨75MI 420-01⟩; **5**, 141 ⟨80HOU(6/1a)874⟩
Thiemann, T., **4**, 309 ⟨91SL87⟩
Thieme, E., **5**, 720 ⟨77JOM(134)21⟩
Thieme, H. K., **3**, 564 ⟨70CB1982⟩; **4**, 894 ⟨70CB1982⟩
Thieme, P. C., **5**, 1044 ⟨78LA227⟩
Thien, W. E., **2**, 388, 389 ⟨69JOC2997⟩
Thienen, N. von, **4**, 109 ⟨89T2819⟩
Thier, W., **1**, 74 ⟨61TL353⟩; **6**, 107 ⟨66AG(E)848⟩
Thierry, J., **1**, 344 ⟨87TL1413⟩, 602 ⟨85T4347⟩, 935 ⟨84CC1287, 88T5479⟩; **3**, 635 ⟨87T4297⟩; **5**, 1115 ⟨86T4983⟩
Thiers, M., **2**, 291 ⟨61T1⟩
Thies, R. W., **1**, 401 ⟨76CC846, 78JOC1050⟩, 401 ⟨82JOC798⟩, 811 ⟨83CJC1226⟩
Thieser, R., **1**, 1065 ⟨86T3569⟩
Thiessen, G. W., **5**, 28 ⟨60RCP243⟩
Thijs, L., **1**, 925 ⟨79JOC4123⟩, 1023 ⟨89TL2153⟩; **2**, 743 ⟨83S628⟩; **3**, 37 ⟨83S628⟩, 368 ⟨69TL4461, 73JCS(P1)73⟩; **5**, 229 ⟨92RCT211⟩, 489 ⟨67RTC641⟩, 490 ⟨70RTC687⟩, 562 ⟨68TL2871, 71RTC614⟩; **6**, 316 ⟨73TL3589⟩, 531 ⟨69TL4461⟩
Thil, L., **1**, 108, 115 ⟨74TL1451⟩
Thimm, K., **3**, 365 ⟨75TL537, 77JCR(S)244⟩
Thind, S. S., **2**, 11 ⟨80JCS(P1)1890⟩; **5**, 1107 ⟨80JCS(P1)865⟩
Thiollet, G., **4**, 569 ⟨78ICAL294⟩, 1039 ⟨79ICAL67, 80AX(B)1344⟩
Thiruvengadam, T. K., **6**, 620 ⟨86JOC1719⟩
Thistlethwaite, G. H., **2**, 1101 ⟨72JOM(40)143⟩
Thivolle-Cazat, J., **4**, 722 ⟨91OM4005⟩
Thiys, L., **2**, 209 ⟨77JCS(P1)1469⟩
Thom, C., **2**, 251 ⟨92S582⟩; **5**, 343 ⟨92S582⟩
Thom, E., **3**, 92 ⟨79JHC799⟩, 283 ⟨63JOC1379⟩; **4**, 168, 170 ⟨83JA3661⟩
Thoma, G., **1**, 327 ⟨91TL6307, 92JA4436⟩, 327

⟨93JA8585⟩, 937 ⟨89TL2907, 91HCA1135⟩, 937 ⟨91HCA1143⟩; **4**, 284 ⟨92JA4436⟩
Thoma, R. J., **4**, 566 ⟨88OM572⟩, 567 ⟨85IC1993⟩, 573 ⟨87TL6121⟩; **5**, 893 ⟨85IC1993⟩; **6**, 129 ⟨87TL6121⟩, 689 ⟨88IC784⟩, 692 ⟨84OM1132⟩
Thoman, C. J., **3**, 613 ⟨89JOC4476⟩
Thomann, M., **4**, 632 ⟨93AG(E)295⟩, 635 ⟨93OM2423⟩
Thomas, A., **3**, 588 ⟨74CC45⟩; **4**, 704 ⟨88T1667⟩; **5**, 293 ⟨57JCS3032⟩, 731 ⟨88JCS(P1)169, 88T1667⟩
Thomas, A. F., **1**, 816 ⟨74HCA2084⟩, 840 ⟨69JA3281⟩; **3**, 275 ⟨86TL505⟩
Thomas, A. H., **5**, 985 ⟨66FRP1440591⟩
Thomas, A. J., **2**, 22 ⟨87OS(66)194⟩
Thomas, A. M., **2**, 219 ⟨71JOC488⟩, 219, 220 ⟨69JA7525⟩
Thomas, A. P., **4**, 1046 ⟨87JCS(P1)2269⟩
Thomas, B. R. J., **6**, 9 ⟨52JCS3490⟩
Thomas, B. S., **2**, 421 ⟨72JCS(D)326⟩
Thomas, C. B., **2**, 388 ⟨72JCS(P1)1692⟩; **5**, 366 ⟨72JCS(P1)1692⟩
Thomas, C. C., **1**, 278, 487, 662, 1103 ⟨76JOC3947⟩
Thomas, C. H., **4**, 761 ⟨55JA3886⟩
Thomas, C. J., **2**, 484 ⟨92SRI461, 93POL89⟩
Thomas, C. L., **5**, 549 ⟨28JA3106⟩
Thomas, C. R., **5**, 277 ⟨54JCS1188⟩
Thomas, C. W., **3**, 532 ⟨66T209⟩
Thomas, D., **2**, 386 ⟨75AP(308)954⟩; **3**, 92 ⟨92SL77⟩; **4**, 829 ⟨75JOC2552⟩; **6**, 107 ⟨73TL1807⟩, 558 ⟨91S265⟩
Thomas, D. G., **3**, 439, 440, 456 ⟨83JOC289⟩
Thomas, E., **2**, 394 ⟨82HCA1101⟩
Thomas, E. G., **2**, 166, 173, 174 ⟨61JOC1996⟩
Thomas, E. G.,**5**, 745 ⟨61JOC2563⟩
Thomas, E. J., **1**, 38 ⟨79CC345⟩, 235, 246 ⟨80TL395⟩, 410, 822 ⟨82JCS(P1)2757⟩, 591 ⟨74JCS(P1)433⟩, 595, 601, 624 ⟨91COS(6)1011⟩, 601 ⟨81T263⟩, 602 ⟨87CC880⟩, 603 ⟨91JCS(P1)429⟩, 822 ⟨80TL399⟩, 945 ⟨81CC901⟩, 1127 ⟨89CC480⟩; **2**, 79 ⟨88TL2479, 89JCS(P1)1521⟩, 79 ⟨89JCS(P1)1529, 90TL6239⟩, 184, 196 ⟨91COS(6)1011⟩; **4**, 57 ⟨93JCS(P1)1493⟩, 90 ⟨80TL395, 82JCS(P1)2757⟩, 387 ⟨82CC1115⟩, 519 ⟨81JCS(P1)1934⟩; **5**, 63 ⟨78CC961⟩, 234 ⟨89CC717⟩; **6**, 473, 550 ⟨87CC1228⟩
Thomas, E. W., **2**, 942 ⟨85S945⟩; **6**, 445 ⟨89JMC228⟩
Thomas, F. L., **5**, 142 ⟨73MI 503-02⟩
Thomas, G., **1**, 502 ⟨87JOC3126⟩; **2**, 4333 ⟨61CB2244⟩; **3**, 83 ⟨86JA2608⟩
Thomas, G. G., **5**, 30 ⟨52JCS1607⟩
Thomas, G. H., **2**, 34 ⟨55JA3826⟩
Thomas, H. G., **6**, 77, 78 ⟨85S31⟩
Thomas, I. M., **2**, 112, 421 ⟨61CJC1386⟩, 409, 420, 421 ⟨62CJC449⟩, 409, 421 ⟨62CJC1355⟩
Thomas, J. A., **1**, 382 ⟨81JOC3376⟩
Thomas, J., **4**, 1050, 1052 ⟨84JOM(271)107⟩; **6**, 178, 290 ⟨84JOM(271)107⟩
Thomas, J. K., **3**, 514 ⟨90JA8378⟩
Thomas, J. R., **1**, 163 ⟨70JOM(24)263⟩; **4**, 633 ⟨70JOM(24)263⟩; **6**, 181, 182 ⟨70JOM(24)263⟩
Thomas, K., **4**, 416, 423 ⟨75LA2318⟩; **5**, 606 ⟨75LA2318⟩
Thomas, K. M., **6**, 186 ⟨76CC261⟩
Thomas, M. D. O., **4**, 692 ⟨80JCS(D)2182⟩
Thomas, M. G., **4**, 692 ⟨76JA8072, 78JA2090⟩
Thomas, M. J., **4**, 3, 6 ⟨86JA7739⟩
Thomas, M. T., **2**, 229 ⟨75CJC455⟩
Thomas, O. H., **1**, 51, 52, 101 ⟨55JA3230⟩
Thomas, P. J., **1**, 612 ⟨75CC940⟩; **2**, 654 ⟨91SC2335⟩; **4**, 1079 ⟨90TL1819⟩
Thomas, R., **1**, 638, 639, 744, 745 ⟨74CRV87⟩; **3**, 655 ⟨56CB1263⟩
Thomas, R. C., **1**, 278, 487, 662, 1103 ⟨76JOC3947⟩; **3**, 244 ⟨77JOC3846⟩
Thomas, R. C. L. J., **2**, 214 ⟨55JA5446⟩
Thomas, R. D., **3**, 250 ⟨83CC485⟩; **4**, 602, 650 ⟨89JOM(359)285⟩, 620 ⟨89JOC1784⟩; **6**, 190 ⟨89JOC1784⟩
Thomas, R. J., **1**, 835 ⟨90JCS(P1)1493⟩
Thomas, R. M., **1**, 1140 ⟨88TL5017⟩
Thomas, S., **6**, 721 ⟨79JCS(D)1929⟩
Thomas, S. E., **1**, 987 ⟨91CC1290, 92JCS(P1)259⟩; **2**, 535 ⟨B-91MI 210-01⟩; **3**, 597 ⟨90CC307⟩; **5**, 425 ⟨92T7629⟩
Thomasco, L. M., **1**, 174 ⟨85JOC546⟩; **5**, 155 ⟨88TL6823⟩
Thomas-Miller, M. E., **4**, 673, 722 ⟨89OM1199⟩
Thometzek, P., **4**, 582 ⟨85CB261⟩
Thomi, S., **5**, 342 ⟨89TL6009⟩
Thommen, W., **1**, 403, 734 ⟨75HCA1016⟩; **3**, 494 ⟨75HCA1016⟩
Thompsen, P. A., **2**, 530 ⟨83JOC5398⟩
Thompson, A. F., **1**, 959 ⟨42JA363⟩
Thompson, A. M., **5**, 567 ⟨93JMC2459⟩, 613, 615 ⟨93JCS(P1)1835⟩
Thompson, A. S., **1**, 450 ⟨86JA3516⟩, 751 ⟨88JA2248⟩, 754, 755 ⟨86JA2662⟩; **2**, 528 ⟨93JOC2880⟩; **4**, 199 ⟨88JA2248⟩; **5**, 173 ⟨82JA4015⟩
Thompson, B., **1**, 417, 829 ⟨60JOC1815⟩
Thompson, B. B., **4**, 299 ⟨68JPS715⟩
Thompson, C. M., **1**, 165, 879 ⟨91T4223⟩
Thompson, D., **1**, 6 ⟨60CI(L)405⟩
Thompson, D. F., **1**, 881 ⟨54JOC1490⟩
Thompson, D. J., **1**, 212 ⟨B-91MI 105-02⟩
Thompson, D. M., **3**, 136 ⟨90JCS(P1)2721⟩; **5**, 457 ⟨90JCS(P1)2721⟩
Thompson, D. T., **1**, 668 ⟨68JCS(A)190⟩
Thompson, D. W., **1**, 543 ⟨81JOC807, 85JOC2124⟩
Thompson, F. M., **2**, 379 ⟨74JMC948⟩
Thompson, H. W., **1**, 72 ⟨73JA6379⟩
Thompson, J., **1**, 566, 681, 682 ⟨81JOC3359⟩, 652 ⟨69JCS(B)852⟩; **3**, 468 ⟨06CB1379⟩, 476 ⟨06CB1383⟩
Thompson, J. G., **3**, 493 ⟨66CC666⟩, 494 ⟨68JOC3082⟩, 496 ⟨68JOC336⟩
Thompson, J. J., **5**, 1110 ⟨78T541⟩
Thompson, J. L., **1**, 729 ⟨69CC303⟩; **3**, 494 ⟨69CC303⟩
Thompson, J. M., **5**, 464 ⟨75JOC2667⟩
Thompson, J. M. C., **4**, 569 ⟨59BRP815231, 59USP2889349⟩
Thompson, J. W., **5**, 1003 ⟨72USP3666784⟩; **6**, 10 ⟨78IJ129⟩
Thompson, K. L., **4**, 697 ⟨78JOM(159)237⟩
Thompson, L., **2**, 769 ⟨21MI 215-01⟩
Thompson, M., **4**, 200 ⟨91T1329⟩
Thompson, M. D., **2**, 133 ⟨76JOC134⟩; **6**, 546 ⟨87JHC581⟩
Thompson, M. E., **1**, 147 ⟨84JOC1700⟩; **2**, 969 ⟨87JA203⟩; **4**, 394 ⟨84JOC1700⟩
Thompson, M. J., **1**, 848 ⟨61JCS73⟩; **2**, 791 ⟨89T191⟩
Thompson, M. R., **6**, 403 ⟨79JA2768⟩
Thompson, M. W., **3**, 170 ⟨88JOC906⟩
Thompson, N., **2**, 147, 156 ⟨87S1015⟩
Thompson, N. T., **3**, 17, 404 ⟨79TL3201, 85JCS(P1)2123⟩
Thompson, P. G., **2**, 94 ⟨65JA230⟩, 99 ⟨67JA4316⟩; **6**, 250 ⟨67JA1811⟩
Thompson, Q. E., **2**, 99 ⟨65JOC2703⟩, 100 ⟨65JOC2692⟩, 202 ⟨54JA5011⟩, 361 ⟨70QRS245⟩, 452, 694 ⟨61JA845⟩; **3**, 526 ⟨49JA2242⟩; **5**, 195 ⟨49JA2242⟩
Thompson, R. B., **5**, 103 ⟨43OSC(2)10⟩
Thompson, R. C., **1**, 810 ⟨93JOC4952⟩; **2**, 427 ⟨48JA2295⟩
Thompson, S. K., **6**, 487 ⟨93MIP9305026⟩
Thompson, T. M., **4**, 200 ⟨93SL111⟩
Thompson, W., **1**, 529 ⟨88JOM(354)233⟩
Thompson, W. J., **2**, 108 ⟨90JA6936⟩; **4**, 97 ⟨88EUP264183⟩, 279 ⟨90TL6819⟩; **6**, 417 ⟨92TL2781⟩
Thoms, H., **1**, 432 ⟨11CB2134⟩
Thomsen, I., **5**, 570, 575, 608, 610, 618 ⟨77BSB321⟩, 600, 621 ⟨75ACS(B)538⟩, 700 ⟨93ACS1004⟩; **6**, 608, 611 ⟨89LA985⟩
Thomsen, T., **4**, 433 ⟨84ACS113⟩
Thomson, B. J., **5**, 944 ⟨80AJC2605⟩
Thomson, G. A., **2**, 642 ⟨92CC1256⟩; **3**, 371 ⟨90JCS(P1)2881⟩
Thomson, J., **4**, 308, 445 ⟨87JCS(P1)811⟩

Thomson, R. H., **1**, 922 ⟨B-68MI 119-04⟩
Thomson, R. I., **1**, 25 ⟨86CC451⟩
Thomson, S. K., **4**, 658 ⟨88POL1953⟩
Thomson, T., **1**, 398 ⟨32JCS55⟩, 399 ⟨32JCS69⟩, 407, 628 ⟨32JCS1932⟩
Thomson, W., **4**, 58 ⟨93JCS(P1)2303⟩
Thoransch, P., **2**, 472 ⟨78PS(4)137⟩
Thorat, M. T., **5**, 367 ⟨89IJC(B)186⟩
Thorausch, P., **2**, 431 ⟨77PS(3)203⟩
Thoraval, J. Y., **6**, 60 ⟨90T3859⟩, 165 ⟨90TL3859⟩
Thoren, S., **3**, 62 ⟨73JOC1380⟩
Thorgeirsson, S. S., **5**, 214 ⟨87JOC4933⟩
Thorimbert, S., **4**, 465, 478 ⟨92TL77⟩
Thorkilsen, B., **5**, 1101 ⟨64ACS825⟩
Thorn, D. L., **2**, 926, 927, 988 ⟨87JA2788⟩, 979 ⟨84OM1223⟩; **4**, 692 ⟨78JA2090⟩
Thorn, G. D., **6**, 560, 561 ⟨B-62MI 617-02⟩
Thorne, A. J., **2**, 852 ⟨82CC691⟩; **4**, 660 ⟨82JOM(233)C28⟩; **6**, 186 ⟨84CC480⟩, 186, 192, 195 ⟨86JAP(K)61289091⟩, 186, 195 ⟨82CC1407⟩
Thornton, D. A., **5**, 823 ⟨61MAC62⟩
Thornton, D. E., **3**, 527 ⟨70JA1768⟩
Thornton, D. M., **4**, 560 ⟨88JOM(346)341⟩; **6**, 123 ⟨88PS(40)1⟩, 264 ⟨88JOM(340)93⟩
Thornton, Jr., S. D., **3**, 461 ⟨53JOC1594⟩
Thornton, S. R., **2**, 307, 345, 346, 347, 349 ⟨93CC169⟩
Thornton-Pett, M., **1**, 837 ⟨93T5177⟩; **2**, 352 ⟨91T4007⟩; **4**, 548 ⟨92JCS(D)2353⟩, 719 ⟨80JCS(D)2315⟩, 1023, 1028 ⟨85CC614, 85CC1635⟩; **4 6**, 338 ⟨83CC1557⟩, 339 ⟨81CC189, 81CC2496⟩
Thorpe, F. G., **1**, 493 ⟨75JOM(99)C8⟩; **2**, 653 ⟨71CC1475, 74JOM(76)C23⟩, 653 ⟨75CC36⟩
Thorpe, R. S., **2**, 88 ⟨41JA1118⟩
Thorsen, M., **5**, 558, 573, 587 ⟨83T3429⟩, 558, 573, 587, 611 ⟨84JCS(P1)785⟩, 573 ⟨81TL3635⟩
Thorsett, E. D., **3**, 468 ⟨82TL1875⟩; **4**, 409 ⟨78JOC4276⟩; **5**, 36 ⟨81JOC5351⟩
Thorstad, W. T., **1**, 879 ⟨89JOC773⟩
Thorwart, W., **1**, 1143 ⟨72CB2290⟩
Thottathil, J. K., **3**, 200 ⟨84TL4741⟩
Thrasher, J. S., **3**, 703 ⟨89AG(E)1256⟩; **5**, 348, 767 ⟨84IC2217⟩, 675 ⟨90JFC(48)407⟩, 809 ⟨84JA4266⟩, 1085 ⟨82IC1616, 85IC4171⟩; **6**, 440 ⟨86IS10⟩, 558, 565 ⟨84CB1707⟩, 603 ⟨84JA4266, 88IC570⟩, 603, 607, 615, 616 ⟨85IC4171⟩, 616, 651 ⟨82JFC(19)411⟩, 617 ⟨83JFC(23)593⟩
Threadgill, M. D., **1**, 442 ⟨86JCS(P1)873⟩; **6**, 493, 494 ⟨89JOC2940⟩
Threlfall, T. L., **1**, 428 ⟨73OSC(5)346⟩
Thrush, A. M., **5**, 184 ⟨53JCS2117⟩
Thu, T. T. M., **2**, 612 ⟨80HCA1236⟩
Thuenker, W., **4**, 181 ⟨81LA1105⟩
Thuillier, A., **1**, 787 ⟨80T2497⟩, 1058 ⟨77T2949⟩; **2**, 617 ⟨75BSF2493⟩; **3**, 562 ⟨88BSF883, 91JCS(P1)3389⟩, 562 ⟨91PS(59)141⟩, 582 ⟨78TL463⟩; **4**, 255 ⟨80CR(C)183⟩, 256 ⟨77T2949, 78TL2717⟩; **5**, 546 ⟨78TL3847⟩, 550 ⟨85PS(23)253⟩, 551 ⟨84SUL147⟩, 551, 729, 731 ⟨84T1573⟩, 552 ⟨80TL2165⟩, 553, 559, 729 ⟨78TL2715⟩, 557, 559 ⟨82TL4087⟩, 558, 559 ⟨80TL2421⟩, 561 ⟨79JOC2807⟩, 589, 612 ⟨90JCR(S)106⟩, 589, 612 ⟨91PS(56)283⟩, 729 ⟨83CI(L)391⟩, 731 ⟨91JCS(P1)3389, 91PS(59)141⟩
Thümmler, C., **2**, 997, 999 ⟨92CB2265⟩
Thurkauf, A., **2**, 33 ⟨87JOC5466⟩; **4**, 179 ⟨88S233⟩
Thurman, E. N., **1**, 128 ⟨29JA1491⟩
Thurman, J. C., **1**, 772 ⟨65JCS6202⟩; **5**, 767 ⟨71JCS(C)1704⟩
Thurman, N., **2**, 594 ⟨48OSC(1)519⟩
Thurmann, U., **6**, 65 ⟨94JOM(465)153⟩, 245 ⟨92JOM(433)63, 92JOM(434)159⟩
Thurston, D. E., **2**, 304 ⟨90TL6903⟩
Thurston, J., **1**, 949 ⟨87TL4011⟩
Thust, U., **4**, 570 ⟨87EGP251134, 87EGP251136⟩
Thuy, V. M., **3**, 89 ⟨89BSB221, 93BSB299⟩, 219 ⟨87JHC497⟩; **4**, 181 ⟨80BSB759⟩
Thweatt, J. G., **1**, 355 ⟨64JOC940⟩; **2**, 751 ⟨63JCS1462⟩
Thyagarajan, B. S., **1**, 515 ⟨90SC3477⟩; **2**, 214 ⟨58CRV439⟩
Thyagarajan, G., **2**, 469 ⟨81CRV415⟩; **3**, 196 ⟨81CRV415⟩; **4**, 340 ⟨81CRV415⟩
Thym, S., **6**, 438 ⟨78GEP2706683⟩
Thys, L., **5**, 562 ⟨67TL3453⟩
Tice, C. M., **1**, 813 ⟨81JOC9⟩
Tice, D. S., **4**, 658 ⟨85OM1308⟩
Tichelaar, G. R., **5**, 34 ⟨63OSC(4)93⟩
Tichenor, G. J. W., **2**, 710 ⟨72JOC2391⟩
Tichniouin, M., **2**, 373 ⟨85MI 207-01⟩
Tichy, M., **1**, 389 ⟨65JCS2513⟩
Ticozzi, C., **1**, 738 ⟨75G109⟩; **4**, 584 ⟨76JOM(111)361⟩
Tidbury, R., **3**, 38 ⟨82TL2407⟩
Tideswell, J., **1**, 754 ⟨78JCS(P1)590⟩
Tidwell, T. T., **2**, 928, 941 ⟨93JA10097⟩, 930 ⟨90JOC3634⟩, 1022 ⟨87CJC1719⟩; **3**, 6, 118 ⟨90OR(39)297⟩, 6, 118 ⟨90S857⟩; **4**, 222 ⟨90OR(39)297⟩
Tiecco, M., **1**, 484 ⟨83T2289, 85JOC3828⟩, 484 ⟨91G59⟩, 782 ⟨93JOC1349⟩; **2**, 10, 658 ⟨88S749⟩, 278, 729 ⟨83JOC4289⟩, 709 ⟨83JOC4795⟩, 711 ⟨83S751⟩; **4**, 92 ⟨91JOC4529⟩; **5**, 238 ⟨86T4577⟩
Tieckelmann, H., **2**, 105 ⟨66T399⟩, 262 ⟨51JOC395⟩, 381 ⟨75JOC1070⟩; **6**, 296, 297, 308 ⟨48JOC265⟩
Tiefensee, K., **5**, 8 ⟨88S525⟩
Tiekink, E. R., **4**, 711 ⟨91JOM(407)391⟩, 660, 700 ⟨93OM2788⟩
Tieman, C. H., **4**, 983 ⟨75BEP821282⟩
Tiemann, F., **3**, 616 ⟨1880CB382⟩; **5**, 825 ⟨1884CB1685, 1885CB727⟩, 825 ⟨1891CB3453⟩, 830 ⟨1886CB1668⟩, 833 ⟨1884CB126, 1889CB1936⟩, 833 ⟨1891CB3420⟩; **6**, 69 ⟨1882CB2685⟩
Tien, J. M., **2**, 374 ⟨55JA1843⟩
Tien, N. B., **3**, 91, 98 ⟨55BSF1594⟩
Tien, R. Y., **1**, 83 ⟨73JOM(61)167⟩
Tiensripojamarn, A., **3**, 154 ⟨79CL209⟩; **4**, 77 ⟨79CL209⟩
Tierney, F. J.,**4**, 866 ⟨84GEP3408757A1⟩
Tiers, G. V. D., **2**, 181, 182 ⟨63JOC1244⟩
Tietze, E., **1**, 840 ⟨26LA(449)81⟩
Tietze, L., **1**, 282 ⟨94SL511⟩
Tietze, L.-F., **2**, 412 ⟨90AG(E)782⟩, 658 ⟨82S683⟩, 1008 ⟨76AG(E)239⟩; **3**, 71 ⟨82TL1147⟩, 75, 265 ⟨89CB83⟩, 246 ⟨88S274⟩; **4**, 182 ⟨88LA559⟩, 195 ⟨88S274⟩, 950 ⟨87LA311⟩
Tiffany, B. D., **1**, 881 ⟨51JA4158⟩
Tiffin, P. D., **3**, 89 ⟨92SC1967⟩
Tigchelaar, M., **1**, 544 ⟨81RTC337⟩, 967 ⟨82JOM(240)329⟩
Tigchelaar-Lutjeboer, H. D. A., **6**, 74 ⟨60RTC888⟩
Tignol, A. P., **2**, 360 ⟨80EUP10477⟩
Tihomirov, S., **4**, 230 ⟨88JOC5179⟩
Tijerina, T., **2**, 680 ⟨85JOC1212⟩
Tijhuis, M. W., **4**, 264 ⟨81JOC3273⟩, 304 ⟨80JOC1880⟩; **5**, 14 ⟨83OS(61)1⟩
Tikhomirov, D. A., **4**, 793 ⟨84CHE1231⟩; **5**, 661 ⟨91KGS1205, 92CHE579⟩
Tikhonov, A. Ya., **2**, 340 ⟨86S704⟩; **3**, 427 ⟨86S704⟩; **5**, 212 ⟨85ZOR196⟩
Tikhonova, L. I., **2**, 464 ⟨60IZV651⟩
Tikhonova, N. A., **6**, 80 ⟨77ZOR943⟩
Tikkanen, W. R., **4**, 657 ⟨84OM825⟩
Tikwe, L., **1**, 686 ⟨86T1693⟩
Tilak, B. D., **3**, 305 ⟨84TL5439⟩, 370 ⟨66T7⟩, 370, 376 ⟨74TL3911⟩
Tilhard, H. J., **4**, 565 ⟨80TL2803⟩
Tilichenko, M. N., **2**, 455 ⟨85JGU2198⟩, 458 ⟨79JGU1721, 79JGU285⟩; **4**, 339 ⟨79ZOB1956⟩
Tilichenko, V. N., **2**, 458 ⟨79JGU1714⟩
Tilkunova, N. A., **6**, 240, 615 ⟨80JFC(15)169⟩, 616 ⟨77IZV2379⟩, 643 ⟨78MI 612-01⟩

Tilles, H., **2**, 137 ⟨59JA714⟩
Tillett, J. G., **2**, 362, 364 ⟨B-90MI 206-03⟩; **5**, 340 ⟨B-90MI 507-01⟩
Tilley, B. P., **2**, 1086 ⟨67CB2159⟩; **5**, 779 ⟨66JCS(A)1584⟩, 1038 ⟨64JCS2108⟩
Tilley, J. N., **5**, 971 ⟨63JOC2076⟩, 980 ⟨64JOC3344⟩; **6**, 445 ⟨64JOC2401⟩
Tilley, J. W., **1**, 1029 ⟨88JOC386, 89JMC1820⟩; **5**, 284 ⟨85JOC2220⟩
Tilley, T. D., **3**, 598 ⟨87JA2049⟩; **4**, 353 ⟨88JA313⟩, 355 ⟨87JA3318⟩; **5**, 422 ⟨87JA2049, 88JA313⟩, 422, 431 ⟨92OM2198⟩, 444 ⟨89JA149⟩; **6**, 338 ⟨89JA149⟩, 521 ⟨85JA4084, 85JA6409⟩, 521 ⟨86JA5355, 87JA2049⟩
Tillotson, E. W., **2**, 260 ⟨48JA1762⟩
Tillyer, R. D., **3**, 78 ⟨89JCS(P1)2124⟩
Tilney-Bassett, J. F., **4**, 692 ⟨59JA4757, 61JCS577⟩
Tils, L., **4**, 69 ⟨64CB179⟩
Tilstam, U., **3**, 175 ⟨89TL2665⟩; **4**, 88, 89, 93 ⟨89TL2665⟩
Timar, E., **3**, 634 ⟨78OPP211⟩
Timberlake, J. W., **1**, 690 ⟨79JOC4729⟩, 900 ⟨86JOC2969⟩; **2**, 373, 402, 404 ⟨B-75MI 207-01⟩, 403 ⟨72S632⟩; **3**, 369 ⟨84CHEC(7)449⟩; **4**, 441 ⟨80T1753⟩; **6**, 509 ⟨78JOC61⟩
Timergaliev, R. G., **6**, 659 ⟨90ZOB2694⟩
Timm, U., **1**, 1077 ⟨79CB3728⟩; **5**, 511 ⟨78JHC697, 79JHC1295⟩, 511 ⟨80CB2519⟩
Timmer, K., **1**, 72, 88 ⟨90RTC87⟩
Timmermans, G. J., **1**, 522 ⟨71S139⟩
Timmermans, P. B. M. W. M., **4**, 780 ⟨91JMC2525⟩; **6**, 645 ⟨78RTC51⟩
Timmers, D. A.,**5**, 937 ⟨91JA927⟩
Timmins, G., **1**, 905 ⟨83CJC2403⟩
Timmler, H., **2**, 258 ⟨71CB3740⟩; **3**, 448 ⟨57CB1060⟩
Timmons, C. J., **3**, 215 ⟨63JCS2420⟩
Timms, P. L., **2**, 961 ⟨72MI 219-01⟩
Timofeeva, A. Y., **1**, 500 ⟨87ZOR2545⟩
Timofeeva, G. N., **6**, 233, 234 ⟨84ZOR115⟩
Timofeev, A. M., **4**, 123 ⟨86DOK(286)19⟩
Timofeev, S. V., **4**, 121, 122 ⟨88JGU465⟩
Timofeev, V. E., **4**, 340 ⟨72ZOB479⟩
Timofeev, V. S., **4**, 1045 ⟨86ZOB2743⟩
Timokhina, L. V., **3**, 341 ⟨87KGS1694⟩, 356 ⟨78ZOR666, 79ZOR82⟩
Timokhin, B. V., **2**, 434 ⟨78URP635102⟩, 448 ⟨91ZOB2161⟩, 458 ⟨83JGU1059⟩
Timony, P. E., **1**, 658 ⟨73JOM(60)19⟩; **2**, 7, 621 ⟨88JOC2858, 90JOC3415⟩
Timpanaro, P. L., **1**, 1114 ⟨73JOC3054⟩
Timpe, H.-J., **3**, 77 ⟨86CCC2167⟩; **5**, 350 ⟨86CCC2167⟩, 773 ⟨70JPR601⟩, 856 ⟨70JPR610⟩; **6**, 555 ⟨91JPR139⟩
Timpe, K., **6**, 546, 554, 555, 567 ⟨93H(35)77⟩
Tin, K.-C., **1**, 142 ⟨74CC990⟩; **2**, 151 ⟨74CJC761⟩, 202, 207 ⟨70CJC845⟩; **4**, 74, 75, 76 ⟨70TL4643⟩, 391 ⟨74CJC761⟩
Tinant, B., **1**, 995 ⟨93TL1779⟩; **5**, 655 ⟨89CCC3245⟩; **6**, 653 ⟨90T3897⟩
Tindall, J. B., **6**, 241 ⟨44USP2365981⟩
Tinder, J. R., **5**, 942 ⟨93JOM(445)143⟩
Tinembart, O., **1**, 546, 946 ⟨87PAC363, 90T3155⟩
Tinga, M. A. G. M., **6**, 190, 192, 198 ⟨86TL6123⟩
Ting, J.-S., **1**, 119 ⟨73SC281⟩
Ting, P. C., **1**, 1015 ⟨86JOC2230⟩; **2**, 351 ⟨88JA1633⟩
Ting, P. L., **3**, 527 ⟨74JOC763⟩
Ting, W. Y., **1**, 738 ⟨65MI 116-01⟩
Ting, Y., **6**, 556 ⟨90CC1432⟩
Tingoli, M., **1**, 480 ⟨84JOC4894⟩, 484 ⟨83T2289, 85JOC3828⟩, 484 ⟨91G59⟩, 782 ⟨93JOC1349⟩; **2**, 278, 729 ⟨83JOC4289⟩, 709 ⟨83JOC4795⟩, 711 ⟨83S751⟩, 1036, 1037 ⟨93SL211⟩; **4**, 92 ⟨91JOC4529⟩; **5**, 238 ⟨86T4577⟩
Tinkelberg, A., **1**, 774 ⟨72RTC3⟩
Tinker, J. H., **4**, 45 ⟨93MI 402-01⟩
Tinker, J. M., **2**, 23 ⟨39JA940⟩
Tinkler, R. B., **4**, 86 ⟨69JCS(C)652⟩; **5**, 767 ⟨70JCS(C)1429⟩; **6**, 652 ⟨70JCS(C)1429⟩
Tinoco, Jr., I., **5**, 85 ⟨61T134⟩
Tinti, O., **1**, 208 ⟨86S1032⟩
Tintillier, P., **2**, 398 ⟨87BCJ4492⟩
Tinucci, L., **5**, 301 ⟨86T885⟩
Tinyakova, E. I., **2**, 566 ⟨77JOM(142)145⟩
Tiollais, R., **3**, 405 ⟨47BSF716⟩
Tipping, A. E., **2**, 193 ⟨73JCS(P1)676⟩, 339 ⟨68JCS(C)796⟩, 383 ⟨71JCS(C)2404⟩; **3**, 483 ⟨73JCS(D)483⟩, 600 ⟨70CC456, 73JCS(P1)1561⟩; **4**, 136 ⟨76JCS(D)694⟩, 137 ⟨76JCS(P1)513⟩, 140 ⟨75JCS(D)2177⟩, 607, 622 ⟨75JCS(D)1434⟩, 617, 623, 656, 661 ⟨75JCS(D)1832⟩, 654, 658 ⟨75JCS(D)1837⟩, 794 ⟨73JCS(P1)1066⟩, 979 ⟨68JCS(C)796⟩; **5**, 681 ⟨83JFC(23)293, 93JFC(65)157⟩, 681, 703 ⟨93JFC(65)101⟩; **6**, 45 ⟨72JCS(P1)130, 72JCS(P1)159⟩, 45 ⟨72JCS(P1)2438⟩, 45, 219 ⟨72JCS(P1)1506⟩, 46, 47 ⟨76JCS(P1)1178⟩, 54 ⟨71JCS(C)3833⟩, 191 ⟨74JCS(P1)1706, 75JCS(D)1434⟩, 233 ⟨72JCS(P1)2180⟩, 239 ⟨65JCS6141, 67JCS(C)1241, 68JCS(C)796, 80JCS(P1)1544⟩, 241 ⟨75JCS(D)2225, 81JFC(17)591⟩, 241 ⟨81JFC(17)85, 81JFC(19)91⟩, 603 ⟨65JCS5774⟩
Tirado-Rives, J., **1**, 690 ⟨84JOC1627⟩
Tirel, M. D., **5**, 1125 ⟨84JCS(P1)147⟩
Tiriliomis, A., **3**, 386 ⟨88JOM(358)229⟩, 386, 397 ⟨87CC559⟩; **5**, 605, 614, 627 ⟨87CC559⟩, 605, 614, 627 ⟨87JOM(332)105, 88CB2095⟩
Tiripicchio, A., **4**, 678 ⟨92OM370⟩, 683 ⟨80JOM(187)81⟩, 713 ⟨80JOM(197)335, 81JOM(221)93⟩, 713 ⟨91OM1907⟩, 718 ⟨83POL77⟩
Tiripicchio-Camellini, M., **4**, 678 ⟨92OM370⟩, 713 ⟨80JOM(197)335, 91OM1907⟩
Tirouflet, J., **2**, 566 ⟨80CC214⟩; **3**, 83 ⟨85JOM(292)225⟩
Tirpak, R. E., **5**, 70 ⟨85JOC4877⟩
Tischler, A. N., **2**, 379 ⟨74JMC948⟩; **4**, 196 ⟨86TL1653⟩
Tishaninova, A. A., **4**, 335 ⟨82IZV161⟩; **6**, 116, 283, 355 ⟨82IZV161⟩, 134 ⟨81IZV161⟩, 283, 349, 355 ⟨89IZV2106⟩
Tishchenko, G. N., **2**, 478 ⟨90K668⟩
Tishchenko, N. P., **5**, 238 ⟨77ZOR100⟩
Tishler, M., **1**, 503 ⟨82JOC1221⟩; **2**, 138 ⟨48JA1450⟩; **3**, 231 ⟨51JA719⟩; 451 ⟨84CHEC(3)1⟩; **4**, 466 ⟨82S188⟩; **5**, 22 ⟨92H(33)657⟩, 609, 626, 836 ⟨81S299⟩, 764 ⟨76JHC487⟩, 824, 834 ⟨78H(9)1327, 82JHC1397⟩, 834 ⟨79JOC1695, 86S807⟩, 835 ⟨83H(20)1591⟩, 836 ⟨73TL1677, 74JHC103, 74JOC2143, 81H(16)2173, 82JHC577, 84H(22)1545, 86M(117)221, 87M601⟩, 837 ⟨81H(15)293, 84CS209, 85H(23)2819⟩, 845 ⟨86H(24)1899⟩, 1118, 1122 ⟨83MI 530-02, 84MI 530-01⟩, 1028 ⟨90JHC407⟩; **6**, 254 ⟨35JA217⟩, 558 ⟨90JHC407, 90JHC643⟩
Tisnes, P., **2**, 244 ⟨84S1037⟩; **5**, 337 ⟨84S1037⟩; **6**, 375 ⟨84JCR(S)274⟩
Tison, J., **3**, 255 ⟨73BSF3339⟩
Tissot, I., **5**, 330 ⟨91TA1209⟩
Titherley, A. W., **5**, 758 ⟨11JCS1493, 11JCS1507⟩
Titov, A. I., **2**, 887 ⟨53MI 217-02⟩; **5**, 838 ⟨52DOK(83)243⟩
Titov, A. J., **2**, 188 ⟨64DOK(157)139⟩
Titov, Y. A., **1**, 391, 884 ⟨70RCR732⟩
Titova, I. E., **6**, 144 ⟨78MI 605-01⟩
Titova, K. V., **6**, 369 ⟨77IZV1965⟩
Titterington, D. R., **6**, 54 ⟨76JOC1285⟩
Tittle, B., **6**, 7 ⟨67JCS(C)1150⟩
Tittmann, R., **6**, 581 ⟨87ZN(B)77⟩
Titus, G. R., **3**, 267 ⟨83JOC623⟩; **4**, 980 ⟨83JOC623⟩
Titus, P. E., **5**, 689 ⟨76CJC647⟩
Titzmann, R., **5**, 821 ⟨52CB344⟩
Tius, M. A., **1**, 221, 478 ⟨83JOC4131⟩, 556 ⟨92JA5959⟩, 580 ⟨88SC1905⟩, 1002 ⟨91TL3605⟩; **2**, 593 ⟨80JA1742⟩, 611 ⟨92SC1461, 93SL207⟩; **3**, 11 ⟨80JA1742, 80TL3535⟩, 55, 59 ⟨86TL2571⟩, 57 ⟨88SC1905⟩; **4**, 187 ⟨88S36⟩

Tiwari, K. N., **4**, 227 ⟨92JMC533⟩
Tjin A-Lim, D. S., **4**, 468 ⟨84TL4259⟩
Tkachenko, E. N., **5**, 373 ⟨69JGU1445⟩
Tkachenko, S. E., **4**, 119 ⟨86JGU1430, 90JGU1351⟩, 121, 122 ⟨88JGU465⟩
Tkachenko, T. K., **5**, 603 ⟨89IZV1319⟩
Tkachev, A. S., **5**, 1006 ⟨75ZOB477, 77ZOB2259⟩; **6**, 445 ⟨77ZOB1063⟩
Tkachev, A. V., **3**, 638 ⟨93SL580⟩
Tkachev, V. V., **5**, 1070 ⟨87IZV1829⟩
Tkatchenko, I., **1**, 228, 296, 297 ⟨82COMC-I(8)101⟩
Tkeshelashvili, R. Sh., **5**, 720 ⟨70IZV1157, 72IJS319⟩
Toba, T., **2**, 348 ⟨90CL547⟩
Tobe, Y., **1**, 561, 569 ⟨93S460⟩, 811 ⟨81CC786⟩; **3**, 122 ⟨85CL1437⟩
Tobey, S. W., **1**, 907 ⟨64JA1459⟩; **2**, 1006 ⟨64JA1459⟩; **4**, 4, 22, 23, 30 ⟨66JA2481⟩
Tobias, B., **1**, 1014, 1029 ⟨90JA9330⟩
Tobias, M. A., **3**, 249 ⟨70JOC1709⟩
Tobias, R. S., **2**, 590 ⟨75IC2402⟩; **5**, 1141 ⟨72MI 530-01⟩
Tobie, W. C., **5**, 89 ⟨32OSC(1)23⟩
Tobinaga, S., **2**, 662 ⟨80CC593⟩; **3**, 363 ⟨90H(30)905, 91COS(7)151⟩, 365 ⟨89CPB1999⟩
Tobita, E., **6**, 14 ⟨81CL1719⟩
Tobita, H., **4**, 640 ⟨82CL1855⟩, 1050 ⟨82JA4288⟩, 1052 ⟨80AG632, 80TL3077⟩
Tobitsuka, J., **6**, 629 ⟨73CPB62⟩
Tobler, E., **5**, 388 ⟨62ZN(B)135⟩
Tobolla, U., **2**, 836, 837 ⟨81CB3485⟩
Tocher, D. A., **4**, 152 ⟨85CC1278⟩; **5**, 781 ⟨92JCS(D)3199⟩
Tochesche, R., **5**, 652 ⟨47AG65⟩
Tochtermann, W., **3**, 231 ⟨89CB1653⟩
Tocker, S., **6**, 513 ⟨77GEP2708024⟩
Toda, F., **1**, 969 ⟨75CC174⟩, 983 ⟨85TL631⟩; **3**, 352, 360 ⟨87CL1299⟩, 539 ⟨76BSJ2645⟩, 621 ⟨76BCJ2643⟩
Toda, M., **1**, 19 ⟨72BCJ264⟩
Toda, S., **1**, 348 ⟨88JA1288⟩; **4**, 259 ⟨90JCS(P1)2035⟩
Toda, T., **1**, 982 ⟨87AG(E)335⟩; **2**, 136 ⟨92CL1655⟩, 318 ⟨93CL1787⟩; **4**, 420 ⟨71BCJ3445⟩; **5**, 226 ⟨87H(25)79⟩
Todd, A., **2**, 764 ⟨59JCS2094⟩; **5**, 712 ⟨55CI(L)1087⟩
Todd, A. R., **2**, 105 ⟨45JCS382⟩, 108 ⟨53JCS2257⟩, 412 ⟨45JCS660, 47JCS674⟩, 414 ⟨58JCS3039⟩; **5**, 712 ⟨56JCS1231⟩, 975 ⟨45JCS660⟩; **6**, 221 ⟨47JCS674⟩
Todd, C. W., **3**, 449 ⟨48JA1381⟩
Todd, D., **1**, 11, 18, 19, 41, 887 ⟨48OR(4)378⟩; **5**, 115 ⟨73OSC(5)617⟩
Todd, L. J., **4**, 142, 524 ⟨65JA488⟩, 152 ⟨69JA5027⟩, 681 ⟨75JOM(99)281⟩; **5**, 1147 ⟨89CC900, 91IC1046⟩; **6**, 38 ⟨64JA2961⟩, 169 ⟨82COMC-I(1)543⟩, 515 ⟨89CC900, 91IC1046⟩, 515 ⟨92IC4911⟩
Todd, R. S., **2**, 346 ⟨91TL6789⟩
Todd, S. L., **5**, 835 ⟨92JOC4648⟩
Toder, B. H., **1**, 797 ⟨82JOC163⟩; **3**, 206 ⟨75TL4225, 81JA1996⟩, 434 ⟨90HCA169⟩, 533 ⟨76JA7456, 84JA3995⟩, 533 ⟨84JA4001⟩
Todesco, P. E., **2**, 145 ⟨60G383⟩, 148 ⟨71IJS(A)130⟩, 626 ⟨71JCS(C)3652⟩; **4**, 405 ⟨86G229⟩, 646, 653 ⟨88CC807⟩
Todo, E., **3**, 539 ⟨76BSJ2645⟩
Todo, W., **3**, 539 ⟨76BSJ2645⟩
Todres, Z. V., **1**, 890 ⟨89JOM(371)279⟩
Todt, E., **2**, 597 ⟨63ZAAC(321)120⟩
Todt, K., **1**, 876 ⟨66AG(E)899⟩
Toennies, G., **5**, 140 ⟨42JBC(144)203⟩, 140 ⟨54JA2881⟩, 140 ⟨59JA1751⟩
Toepel, T., **1**, 530, 890 ⟨48LA(560)1⟩
Toepffer, H., **1**, 21 ⟨28CB1750⟩
Toepfl, W., **4**, 843, 852, 899 ⟨62CB2861⟩; **6**, 619 ⟨80CP1076132⟩
Toeplitz, B., **4**, 227 ⟨78JOC79⟩
Toerien, J. G., **5**, 951 ⟨94JOM(479)C12⟩; **6**, 723 ⟨94AG(E)672⟩
Toft, J., **5**, 640 ⟨61ACS1109⟩
Toft, P., **2**, 675 ⟨70CC111⟩
Togni, A., **4**, 1034 ⟨90JOC1649, 92OM613⟩
Togo, A., **1**, 1053 ⟨87CL1345⟩
Togo, H., **1**, 934 ⟨91CL1691, 91TL6559⟩, 934 ⟨93JCS(P1)2417⟩; **2**, 708 ⟨83BCJ3802, 84BCJ232⟩; **4**, 266 ⟨91T7091⟩, 327 ⟨85T5507⟩, 432 ⟨93TL5097⟩; **5**, 700, 701 ⟨92JCS(P1)2127⟩
Toh, H. T., **2**, 15 ⟨76JA3034, B-80MI 218-01⟩, 16 ⟨72CC122⟩, 335 ⟨74JCS(P1)732⟩; **5**, 313, 314, 316 ⟨74JCS(P1)732⟩; **6**, 231 ⟨74JCS(P1)732⟩
Tohda, Y., **1**, 472, 1008, 1009, 1012, 1137 ⟨75TL4467⟩, 1045, 1048 ⟨77S777⟩; **3**, 273 ⟨77S777⟩, 428 ⟨88BCJ461⟩; **5**, 304 ⟨77S777⟩
Tohjo, T., **4**, 418 ⟨91TL2375⟩
Tohma, T., **2**, 51 ⟨90TL7333⟩
Tohnishi, M., **4**, 900 ⟨82CL1933⟩; **5**, 630 ⟨88JCS(P1)2203⟩
Toia, R. F., **4**, 271 ⟨92MI 406-02⟩
Tojo, E., **1**, 626 ⟨88H(27)2367⟩
Tojo, G., **2**, 103 ⟨92SC73⟩; **3**, 106 ⟨87JCS(P1)931⟩
Tojo, M., **6**, 467 ⟨88JAP63238043⟩
Tokareva, E. V., **2**, 379 ⟨85ZOR124⟩
Tokbolatov, R. B., **2**, 392 ⟨90JOU583⟩
Töke, L., **1**, 747 ⟨66JOC1447⟩; **2**, 34 ⟨93T5133⟩; **5**, 1065 ⟨87S520⟩, 1069 ⟨88S397⟩; **6**, 503 ⟨81ACH57⟩
Toki, T., **1**, 212 ⟨84JA6093⟩
Tokida, A., **3**, 66 ⟨82TL873⟩
Tokitoh, N., **1**, 242 ⟨89JOC4660⟩, 983 ⟨87TL1803, 89TL4271⟩; **2**, 22 ⟨91BCJ3749⟩, 949 ⟨91TL6877⟩; **3**, 351 ⟨87TL4833, 89TL2955⟩, 351, 551 ⟨90TL3571⟩, 353 ⟨89JOC4660, 89TL4271⟩, 353 ⟨89TL4825⟩, 366 ⟨89TL1249, 90TL7641⟩; **4**, 311 ⟨87BCJ3291⟩, 653 ⟨91TL2049⟩, 837 ⟨89TL1249⟩; **6**, 129 ⟨87CL2177⟩
Tokiyama, H., **2**, 626 ⟨87CL627⟩
Tokizawa, M., **5**, 1001 ⟨61JOC4381⟩
Tokles, M., **2**, 588 ⟨87JA2040⟩
Tokoroyama, T., **1**, 180 ⟨87CC358⟩; **3**, 84 ⟨76S526⟩, 226 ⟨90JCS(P1)1745⟩; **4**, 1034 ⟨91CC1390⟩; **5**, 68 ⟨74S443⟩
Tokuda, M., **1**, 276 ⟨75CL523, 78BCJ339⟩, 552 ⟨91SL651⟩, 1133 ⟨76CC606⟩; **2**, 323 ⟨88CL1289⟩, 337 ⟨94JCS(P1)777⟩; **3**, 622 ⟨75CL523⟩
Tokumaru, K., **3**, 666 ⟨70CC1601⟩
Tokumasu, H., **4**, 696 ⟨73NKK381⟩
Tokumasu, S., **4**, 925 ⟨84BCJ2127⟩; **6**, 122 ⟨85CL1099⟩
Tokumitsu, T., **4**, 796 ⟨85JOC1547⟩; **5**, 597, 599 ⟨86BCJ3871⟩
Tokunaga, H., **2**, 853 ⟨72BCJ506⟩
Tokunaga, T., **2**, 185 ⟨88SC2083⟩
Tokunaga, Y., **3**, 220 ⟨90JOC4497⟩, 352, 360 ⟨87CL1299⟩; **4**, 92 ⟨93TL1311⟩; **6**, 256 ⟨93TL1311⟩
Tokura, N., **2**, 797 ⟨73BCJ1254⟩; **5**, 282 ⟨76JCS(P1)2205⟩; **6**, 428 ⟨73TL2287⟩
Tokutake, N., **5**, 280 ⟨85CL123⟩
Tolberg, W. E., **2**, 762 ⟨66MI 215-01⟩
Tolbert, L. M., **1**, 357 ⟨82JA1742⟩, 499 ⟨90JA6416⟩
Tolbert, T. L., **3**, 405 ⟨61JOC4886⟩
Tolcher, P., **2**, 101 ⟨55MI 202-01⟩
Tolchinskii, S. E., **4**, 885 ⟨80ZOB984⟩, 976 ⟨79ZOR1824, 81ZOR1169⟩; **5**, 691 ⟨91CHE424, 91KGS532⟩
Toldy, I., **1**, 578 ⟨88JPR309⟩
Toldy, L., **1**, 85 ⟨79MIP15154⟩
Tolgyesi, W. S., **1**, 893, 899 ⟨63JA1328, 64JA1360⟩, 914 ⟨62JOC3441⟩; **2**, 13 ⟨B-64MI 201-01⟩; **3**, 282 ⟨63JA1328⟩; **4**, 41 ⟨B-64MI 402-01⟩
Tollari, S., **1**, 601 ⟨84JCS(P1)1593⟩
Tolls, E., **4**, 546 ⟨78ZN(B)691⟩
Tolman, C. A., **3**, 614 ⟨85MI 318-01⟩; **4**, 551 ⟨75IC656⟩
Tolman, G. L., **2**, 693 ⟨83OS(61)116⟩
Tolman, V., **4**, 3, 32 ⟨77CCC2537⟩
Tolochko, A. F., **5**, 631 ⟨75ZOB2116⟩
Tolschinskii, S. E., **4**, 885 ⟨91ZOB1192⟩
Tolstaya, T. P., **1**, 486 ⟨90IZV2665⟩; **2**, 838, 839 ⟨57DOK(117)996⟩, 876 ⟨58DOK(122)614⟩; **4**, 726 ⟨73DOK(209)1113⟩
Tolstikov, G. A., **1**, 469 ⟨89IZV2623⟩, 491 ⟨89S633⟩, 801

⟨91TL4591⟩, 1022 ⟨85IZV1109⟩; **2**, 595 ⟨83ZOB2519⟩; **3**, 8 ⟨89S940⟩; **4**, 267 ⟨90ZOR1259⟩, 669 ⟨89IZV2562⟩, 698 ⟨88IZV1448⟩, 1068 ⟨88ZOB1567⟩; **5**, 603 ⟨89IZV1319⟩
Toma, S., **3**, 567 ⟨71TL3367⟩
Tomada, S., **2**, 284, 392, 785 ⟨82CL1109, 82TL4733⟩
Tomalia, D. A., **5**, 1010 ⟨67JHC178, 70JOC47⟩, 1033 ⟨66JHC384⟩, 1073 ⟨71JOC2142⟩
Toman, J. J., **6**, 386 ⟨81TCA285⟩
Tomari, M., **5**, 1078 ⟨70T5731⟩
Tomarovskaya, M. G., **3**, 707 ⟨89ZOR1573⟩
Tomás, A., **2**, 299 ⟨81JCS(P1)574⟩
Tomas, M., **1**, 524 ⟨84JOC3314⟩; **2**, 920 ⟨87CC1195⟩; **3**, 679, 680 ⟨86JA6739⟩; **4**, 421 ⟨89TL4573⟩, 513 ⟨84JOC3314⟩, 517 ⟨83TL4303⟩, 692 ⟨88OM2279⟩, 692 ⟨89JCS(D)169, 93JA4145⟩; **5**, 728 ⟨85JOC4415; **6**, 571 ⟨89S228⟩, 575 ⟨79CC675⟩
Tomaschewski, G., **5**, 358, 707 ⟨88ZC24⟩, 1062 ⟨69TL3191⟩
Tomasi, R. A., **1**, 649, 763 ⟨62JOC3647⟩; **3**, 603 ⟨62JOC3647⟩
Tomasić, S., **5**, 566, 618 ⟨57CCA319⟩
Tomasic, V., **2**, 478 ⟨61CI(L)469⟩
Tomasik, P., **2**, 802 ⟨70RZC1131⟩
Tomasik, W., **3**, 156 ⟨81TL607⟩
Tomasini, C., **2**, 23 ⟨94CC735⟩; **3**, 158 ⟨84JOC701⟩; **4**, 99 ⟨92JOC1082⟩; **5**, 1125 ⟨89TL1679⟩
Tomasz, M., **3**, 275 ⟨64JA471⟩
Tomatis, R., **4**, 422 ⟨85JMC769⟩
Tom, T. K., **2**, 270 ⟨76JMC1423⟩
Tombo, G. M., **1**, 1039 ⟨90SL547⟩
Tomboulian, P., **1**, 881 ⟨59JOC229⟩
Tombret, F., **3**, 62 ⟨81JOC3741⟩
Tom Dieck, H., **1**, 519 ⟨83AG(E)778⟩; **4**, 298 ⟨84CB694⟩, 580 ⟨72JOM(42)C35⟩
Tomelini, S. A., **6**, 25 ⟨86TL3845⟩
Tomer, K., **1**, 633 ⟨68JA4762⟩
Tometzki, G. B., **1**, 196 ⟨87JA2176⟩
Tomi, K., **3**, 84 ⟨75TL1063⟩
Tomida, I., **3**, 66 ⟨56CB2224⟩
Tomie, Y., **4**, 673 ⟨69BCJ545⟩
Tomiguchi, A., **4**, 902 ⟨82CPB3959⟩
Tomii, S., **6**, 544 ⟨92CPB2279⟩
Tomimatsu, M., **2**, 197 ⟨80CPB134⟩
Tominaga, K., **2**, 537 ⟨66TL5493⟩
Tominaga, M., **1**, 503, 1052 ⟨75T1846⟩
Tominaga, Y., **1**, 253, 486 ⟨87JOC2947⟩; **2**, 77 ⟨87CC1517, 89JOC3254⟩, 78 ⟨90JOC2415⟩, 140 ⟨92TL85⟩, 300 ⟨88CPB2342⟩, 1003 ⟨79AX(B)2443⟩; **3**, 361, 362 ⟨91JHC1245⟩; **4**, 231 ⟨87CC1442⟩, 368 ⟨92TL85⟩, 650 ⟨87CPB1663⟩, 842 ⟨75CPB2390⟩, 844 ⟨73CPB2770, 75CPB2749⟩, 844 ⟨91JHC1245⟩, 900 ⟨88JHC959, 90JOC5308⟩, 901 ⟨87JHC519⟩, 979 ⟨73USP3734954⟩, 983, 986 ⟨85JHC937⟩; **5**, 559 ⟨79H(12)485⟩, 560 ⟨75YZ980⟩; **6**, 634 ⟨91TL5987⟩
Tomino, I., **2**, 64 ⟨84JA6709⟩
Tomioka, H., **1**, 14, 221 ⟨82TL3597⟩, 442 ⟨87JCS(P1)1623⟩; **2**, 619 ⟨85TL1651⟩, 922, 931 ⟨82TL3387⟩; **3**, 5 ⟨81TL1605⟩, 5, 117 ⟨80TL4843⟩, 118 ⟨82TL539⟩, 158, 161 ⟨86BCJ105⟩, 542 ⟨75JA6577⟩; **4**, 343 ⟨70JOC597⟩, 415 ⟨84TL1023⟩; **5**, 404 ⟨92JOC6216⟩; **6**, 670 ⟨89MI 621-01⟩
Tomioka, I., **1**, 909 ⟨80TL947⟩
Tomioka, K., **1**, 55 ⟨76TL57, 76TL61⟩, 135 ⟨89CPB1120⟩, 199 ⟨89TL851⟩, 473 ⟨92JA8732⟩; **2**, 50 ⟨89CC1921, 90TL1741⟩; **3**, 49, 60 ⟨78JOC3788⟩, 291 ⟨85TL3027, 85TL3031⟩, 639 ⟨84TL1599⟩; **6**, 128, 327 ⟨85TL3031⟩
Tomioka, N., **1**, 143 ⟨86BCJ2801⟩
Tomioka, T., **6**, 575 ⟨75S384⟩
Tomisawa, H., **5**, 199 ⟨69TL2465⟩
Tomisawa, S., **4**, 74 ⟨71S89⟩
Tomita, I., **3**, 707, 708 ⟨89TL1257⟩
Tomita, K., **1**, 816 ⟨91TL5137⟩; **2**, 16 ⟨86TL4465, 90JA8563⟩, 381 ⟨81CPB3387⟩; **3**, 149 ⟨90JA8563⟩, 380 ⟨78CC197⟩; **4**, 368 ⟨92TL85⟩, 979 ⟨79CPB2787⟩
Tomita, S., **3**, 60 ⟨89CL1283⟩; **4**, 346 ⟨87CL1569⟩
Tomita, S.-I., **1**, 234, 822 ⟨87CL1569⟩
Tomita, T., **1**, 836 ⟨86CL501⟩
Tomita, Y., **3**, 137 ⟨79CL321⟩
Tomizawa, G., **2**, 16 ⟨90JA8563⟩; **3**, 149 ⟨90JA8563⟩; **4**, 62, 63 ⟨86BCJ3625⟩
Tomizawa, N., **3**, 357 ⟨92JHC841⟩
Tomlinson, G. D., **1**, 280 ⟨89TL2013⟩
Tommes, P., **6**, 489 ⟨93PS(83)65⟩
Tomoda, S., **1**, 603 ⟨82CC871, 82TL4733⟩; **2**, 280 ⟨89CL1235⟩, 281 ⟨88CC1283⟩, 282, 284 ⟨82TL1361, 81CL1715⟩, 284 ⟨82CC871, 82CL1733⟩, 284 ⟨84JOC3235⟩, 368 ⟨85S212⟩, 1036, 1038 ⟨82CL1244⟩; **3**, 412 ⟨79TL3453⟩, 414 ⟨83BCJ641⟩, 528 ⟨82JA747⟩, 637 ⟨82CC871, 82TL1361⟩; **4**, 376 ⟨84JOC3235⟩, 516 ⟨86CL1193⟩; **5**, 335 ⟨85S212⟩, 1116 ⟨81CL1069⟩
Tomoeda, M., **2**, 131 ⟨62JCS1967⟩
Tomoi, M., **2**, 35 ⟨79BCJ1653⟩, 614 ⟨90CL769⟩
Tomooka, K., **1**, 405 ⟨92TL5795⟩, 552 ⟨91JOC5973⟩, 810 ⟨93T1025⟩, 816 ⟨91CL43⟩, 818 ⟨93SL527⟩; **2**, 931 ⟨92SL129⟩; **4**, 379 ⟨92TL5795⟩
Tomoskozi, I., **1**, 726 ⟨61TL579⟩; **3**, 239 ⟨75S708⟩
Tomoto, A., **3**, 635 ⟨88CPB1698⟩
Tompkins, J., **2**, 15 ⟨90TL937⟩
Tompsett, A. M., **4**, 309 ⟨48JCS2174⟩
Tone, H., **5**, 173, 199 ⟨87TL4569⟩
Toneganva, F., **2**, 252 ⟨79BCJ1431⟩
Tonellato, U., **1**, 318 ⟨78JOC1521⟩, 1112 ⟨68CC1363, 68CC1676⟩, 1112 ⟨69CC1520, 71JCS(B)1569⟩, 1112 ⟨71JCS(B)374, 71JCS(B)381⟩, 1112 ⟨72JCS(P2)2026⟩; **3**, 365 ⟨66CC586, 70JCS(B)325⟩; **5**, 1038 ⟨69JOC4032⟩
Tonesdale, L. K., **2**, 298 ⟨76JA269⟩
Tong, L. K. J., **3**, 418 ⟨B-70MI 310-04⟩
Tong, M., **2**, 8 ⟨86JOC2637⟩
Tong, W., **4**, 276 ⟨91T3431⟩
Tong, W.-H., **1**, 115 ⟨68JOC2203⟩
Tong, Y. C., **2**, 149 ⟨78JOC2717⟩
Tongpenyai, N., **3**, 625 ⟨81TL4107⟩
Toniolo, C., **5**, 573 ⟨85T5393⟩
Toniolo, L., **3**, 92 ⟨80JOM(194)221⟩; **5**, 142 ⟨79IC2928⟩
Tonker, T. L., **5**, 940 ⟨85JA4474⟩, 952, 953 ⟨85OM2102⟩
Tonn, C. E., **2**, 84 ⟨88TL2701⟩
Tonnard, F., **2**, 489 ⟨90TL1147⟩; **4**, 328 ⟨91CR(313)517⟩; **6**, 156 ⟨89NJC891⟩
Tonne, P., **3**, 473, 479, 480 ⟨69LA(722)45⟩
Tonnel, J., **3**, 616 ⟨93BSF513⟩
Tonneson, G. L., **2**, 632 ⟨80JOM(186)C33⟩
Tonomura, Y., **1**, 595, 597 ⟨88H(27)261⟩
Tonooka, K., **3**, 618, 697, 722 ⟨61JOC2202⟩; **6**, 74 ⟨61JOC2202⟩
Tonosaki, F., **4**, 852 ⟨77JCS(P1)1273⟩; **5**, 547 ⟨77JCS(P1)1273⟩
Tonosaki, M., **6**, 465 ⟨93EUP528498⟩
Tontapanish, N., **4**, 894 ⟨71JOC1685⟩
Toofan, J., **3**, 5, 116 ⟨86T719⟩
Toogood, J. B., **1**, 506 ⟨49JCS1827, 49JCS2028⟩; **3**, 273 ⟨49JCS1827⟩
Toole, A. J., **5**, 956 ⟨94ICA215⟩
Toomey, J. E., **1**, 87 ⟨84AHC(37)167⟩
Toone, E. J., **3**, 159 ⟨92JOC426⟩
Toong, Y. C., **2**, 229 ⟨78CC205⟩
Toops, D., **1**, 1098 ⟨90JA277⟩
Toorongian, S. A., **5**, 203 ⟨88JMC362⟩
Top, A. W. H., **6**, 77, 78 ⟨78S283⟩
Top, S., **1**, 202 ⟨87JOM(336)143⟩; **2**, 309 ⟨81JOC78⟩
Topet, L., **6**, 635 ⟨89JOC1185⟩
Topfl, W., **3**, 247 ⟨62LA(659)90⟩, 547 ⟨62CB2861, 62CB2871⟩; **4**, 900, 970 ⟨62CB2871⟩
Topiol, S., **6**, 386 ⟨81TCA285⟩
Topol, A., **5**, 1107 ⟨89JOC5967⟩
Topolski, M., **2**, 998 ⟨93JOC546⟩; **4**, 816, 818 ⟨93JOC546⟩, 1008 ⟨91PS(55)97⟩

Toppet, S., **3**, 238 ⟨73T4241⟩, 558 ⟨82TL2909⟩; **5**, 604 ⟨84BSB405⟩, 622 ⟨91JCS(P1)607, 93JCS(P1)1719⟩, 638 ⟨80JOC4366⟩, 661 ⟨92S705⟩, 1123 ⟨81JOC4478⟩; **6**, 652 ⟨81BSB63⟩
Toppett, S. M., **2**, 23 ⟨93T3193⟩
Topping, A. E., **4**, 34 ⟨76JCS(P1)2349⟩
Toprek, S., **5**, 1079 ⟨93OM1023⟩
Topsom, R. D., **1**, 258 ⟨60JOC463⟩; **5**, 725 ⟨65JCS5230⟩
Tor, Y., **2**, 613 ⟨81JOC4629, 84JOC806⟩; **4**, 44 ⟨84JOC806⟩; **5**, 202 ⟨81JOC4629, 84JOC806⟩
Torabi, H., **2**, 381 ⟨73JMC736⟩
Torck, B., **5**, 710 ⟨66BSF1376⟩
Tordeux, M., **4**, 187, 208 ⟨83SC629⟩; **5**, 4 ⟨82SC513⟩; **6**, 11 ⟨87CC1701⟩, 233, 234 ⟨84CC793⟩, 233, 234 ⟨85JOC4047⟩, 233, 234 ⟨91CC993⟩, 423 ⟨82SC513⟩
Tordo, P., **4**, 485 ⟨91TL2125⟩
Toreki, R., **3**, 522 ⟨92JA3367⟩
Toren, P. E., **6**, 42 ⟨70JOC2054⟩, 261, 285 ⟨67JOC4045⟩
Torgacheva, N. A., **2**, 361 ⟨75ZOB1005⟩
Torgomyan, A. M., **2**, 446 ⟨80AKZ408⟩; **4**, 914 ⟨80CA186471c, 80CA6622a⟩, 914 ⟨83CA(98)4614z, 86CA(104)19636z⟩
Torgov, I. V., **1**, 79 ⟨56ZOB1175⟩
Tori, K., **1**, 813 ⟨70JCS(C)2697⟩; **6**, 111 ⟨79CC396⟩
Tori, S., **2**, 312 ⟨88TL3811⟩
Torii, K., **1**, 238 ⟨88TL4105⟩, 239 ⟨90TL265⟩
Torii, S., **1**, 496 ⟨92TL7029⟩, 497 ⟨92SL721⟩, 561 ⟨74JOC2486⟩, 946 ⟨85JOC5875⟩, 1015 ⟨92TL3503, 93TL2139⟩; **2**, 21 ⟨81TL2291, 81TL3193⟩, 21 ⟨84CL877⟩, 28 ⟨80JOC2731⟩, 30 ⟨81JOC3312⟩, 81 ⟨89JOC444⟩, 83 ⟨90JOC1548⟩, 243, 359 ⟨78JOC3223⟩, 243, 359 ⟨79JOC1554⟩, 252 ⟨79JOC2938⟩, 313 ⟨89TL4161, 90TL3023⟩; **3**, 7, 120 ⟨91BCJ796⟩, 7, 31 ⟨90JOC462⟩, 46 ⟨86JOC254, 86JOC155⟩, 129 ⟨92CL1895⟩, 152, 154, 155 ⟨80JOC2731⟩, 174 ⟨80TL1863⟩, 234 ⟨76CL495⟩, 426 ⟨83CL607⟩, 469 ⟨68BCJ2707⟩; **4**, 181 ⟨83CL1349⟩, 197 ⟨80CL617⟩; **5**, 336 ⟨79JOC1554⟩; **6**, 26, 31 ⟨89JOC444⟩
Torisawa, Y., **1**, 1081 ⟨82TL4607⟩; **4**, 154, 155 ⟨81TL2397⟩, 155 ⟨83CL1303⟩
Toriumi, K., **3**, 5, 117 ⟨80TL4843⟩
Tork, D. S., **4**, 181 ⟨93JOC7274⟩
Tornare, J.-M., **1**, 519 ⟨84JOC2510⟩
Tornetta, B., **5**, 615 ⟨70MI 514-01⟩
Tornroos, K. W., **4**, 89 ⟨90JOM(391)165⟩
Torocheshnikov, V. N., **4**, 603, 625, 637, 701 ⟨71JOM(30)C57⟩
Torode, J. S., **6**, 461 ⟨93S1109⟩
Torok, E., **4**, 307 ⟨65LA(687)187⟩
Toromanoff, E.,**1**, 503 ⟨65JOC1752⟩
Torossian, G. O., **5**, 1106 ⟨90ZN(B)1091⟩
Tor-Poghassian, G., **2**, 470 ⟨69M311⟩
Torra, N., **5**, 360, 361 ⟨89TL863⟩
Torralba, A. F., **2**, 456 ⟨57JOC972⟩
Torr, R. S., **1**, 643 ⟨83JCS(P1)1169⟩, 753 ⟨83JCS(P1)1173⟩
Torras, J., **1**, 527 ⟨85TL6393, 87TL4745⟩; **3**, 218 ⟨85TL6385, 87TL4745⟩
Torre, G., **2**, 147 ⟨68JCS(C)1317⟩, 337 ⟨83JOC2640⟩; **3**, 406 ⟨70S141⟩
Torre, M., **3**, 492 ⟨82G25⟩
Torregrosa, J. L., **1**, 531, 662 ⟨82T2355⟩
Torreilles, E., **3**, 4, 116 ⟨86TL1775⟩; **5**, 373 ⟨91S382⟩
Torrence, P. F., **6**, 264 ⟨91BMC357⟩
Torres, E., **1**, 981 ⟨88TL1355⟩; **4**, 1048, 1058, 1059, 1060, 1062 ⟨86OM2274⟩, 1060 ⟨88TL1355⟩
Torres, F. F., **6**, 589 ⟨89JCS(P1)1068⟩
Torres, J.-M., **5**, 570, 618 ⟨81JCR(S)58⟩
Torres, L., **4**, 290 ⟨90JOM(381)69⟩
Torres, L. E., **3**, 130, 202 ⟨85JOC5260⟩, 225 ⟨86TL2223⟩; **4**, 831 ⟨88JOC633⟩
Torres, M., **3**, 335 ⟨83NJC269⟩, 353 ⟨80NJC703⟩, 530 ⟨80PAC1623, 88JA1025⟩, 552 ⟨79NJC149, 82NJC401⟩
Torres, R. O., **5**, 320, 321 ⟨84AQ49⟩
Torres, S., **1**, 1073, 1085 ⟨90SC2733⟩
Torres, T., **2**, 163 ⟨85JHC701⟩
Torssell, K. B. G., **2**, 345 ⟨B-88MI 206-01⟩, 352 ⟨B-88MI 206-02⟩, 354 ⟨B-88MI 206-03⟩, 356 ⟨B-88MI 206-04⟩, 487, 488 ⟨53ACS1293⟩; **3**, 57 ⟨81ACS(B)643⟩, 427, 430, 433 ⟨B-88MI 311-01⟩, 683 ⟨84T2985⟩, 683, 684 ⟨B-88MI 320-02⟩; **4**, 540 ⟨88MI 411-01⟩; **5**, 652 ⟨69ACS818⟩, 700 ⟨93ACS1004⟩; **6**, 149 ⟨67ACS(C)1392⟩
Torssell, T. B. G., **6**, 608, 611 ⟨89LA985⟩
Tortelli, V.,**6**, 250 ⟨92JFC(58)143⟩
Tortora, J. A., **5**, 315 ⟨78JOC3713⟩
Tortorella, V., **5**, 307 ⟨75CC71⟩
Tortoreto, P., **4**, 99 ⟨92JOC1082⟩
Toru, T., **1**, 243 ⟨88JA4815, 90TL6669⟩, 333 ⟨93JA464⟩, 594, 668 ⟨92JOC3145⟩, 603 ⟨88JCS(P1)575⟩, 946 ⟨93JA10464⟩; **2**, 285 ⟨88JCS(P1)575, 89JCS(P1)1927⟩; **5**, 173 ⟨75JA2287⟩, 248 ⟨85S878⟩, 248, 249 ⟨87NKK1442, 88BCJ2205⟩, 249 ⟨89JCS(P1)1927⟩, 985 ⟨73JAP7300455⟩
Toryashinova, D. D., **4**, 140 ⟨91IZV1039⟩
Toscano, P. C., **4**, 821 ⟨89CC1159⟩
Toscano, V. G., **2**, 454 ⟨61CB1317⟩, 834, 835 ⟨61CB2122⟩, 857 ⟨68CB3070⟩; **4**, 575, 576, 577 ⟨61CB1987⟩, 576 ⟨61CB1323, 62CB581⟩
Toschner, M. J., **5**, 103 ⟨85OS(64)108⟩
Toseland, P., **6**, 598 ⟨56JA4458⟩
Toselli, M., **5**, 992 ⟨92JCS(P1)1101⟩
Toshida, Y., **1**, 635, 1075 ⟨83T3279⟩
Toshima, K., **1**, 956 ⟨93CC1406⟩; **4**, 176, 198, 202, 214 ⟨93CRV1503⟩
Toshimitsu, A., **1**, 602 ⟨84JOC3796, 85JCS(P1)373⟩, 602, 781 ⟨80TL5037⟩, 603 ⟨81JOC4727, 83JOC5246⟩, 604 ⟨82TL2105⟩, 611 ⟨82TL1177, 83JOM(250)203⟩; **2**, 281 ⟨80CC412, 85T5301⟩, 282 ⟨78JCS(P1)1273, 79JCS(P1)1206, 80JOC1953, 81JOC3021, 82TL2105, 84JOC3796, 85JCS(P1)373⟩, 284 ⟨81JOC4727⟩, 286 ⟨81CC546⟩, 287 ⟨83CC1501⟩, 597 ⟨76BCJ2762⟩, 617 ⟨81BCJ2843⟩; **3**, 37, 174 ⟨84JOC3796⟩, 169 ⟨92CL239⟩, 174 ⟨82TL2105⟩; **5**, 282 ⟨81CC546⟩, 282, 300 ⟨81JOC4727⟩, 285 ⟨86JCS(P1)1837, 86JOC1724⟩, 285 ⟨87JOC2018⟩, 1051 ⟨B-87MI 527-02⟩, 1053 ⟨78BCJ1818⟩, 1113, 1115 ⟨75BCJ1925⟩, 1113 ⟨B-87MI 530-01, 78JCS(P1)1273⟩
Toshimitsu, Y., **5**, 329 ⟨93TA1573⟩
Toshiro, I., **3**, 357 ⟨91EUP462531⟩
Toske, S. G., **5**, 351 ⟨90JOC6037⟩
Tost, W., **3**, 246 ⟨89CB1179⟩
Tosto, S. S., **2**, 504 ⟨77SRI195⟩
Tosunyan, A. I., **1**, 965 ⟨63IZV499⟩
Totani, T., **2**, 1085 ⟨65IC657⟩
Toth, B., **1**, 700, 703 ⟨88JOC1064⟩
Toth, B. R., **1**, 905 ⟨86JOC3736⟩
Toth, G., **2**, 34 ⟨93T5133⟩; **5**, 199 ⟨86JCS(P1)927⟩, 1079 ⟨87CAR(166)101⟩
Toth, J. E., **2**, 726 ⟨93JOC3469⟩
Toth, K., **2**, 440, 441, 443 ⟨80JOC3691⟩
Totino, F., **1**, 983 ⟨80JOC3513⟩
Totleban, M. J., **1**, 347 ⟨90SL773, 92JA6050⟩, 347, 937 ⟨92SL943⟩; **2**, 565 ⟨90SL773, 92SL943⟩, 587 ⟨92JOC1740⟩
Toto, S. D., **2**, 7 ⟨87JOC4999⟩, 441, 444 ⟨91PS(56)27⟩
Totoki, T., **3**, 90 ⟨84JOM(270)283⟩
Touaux, B., **3**, 684 ⟨94JCR(S)116⟩
Toube, T. P., **1**, 725 ⟨73JCS(P1)1416⟩; **3**, 57 ⟨75JCS(P1)1457⟩
Toubro, N. H., **3**, 687 ⟨78JCS(P1)1445⟩
Touchard, D., **5**, 316, 318 ⟨74JOC3136⟩, 941 ⟨89AOC(29)163⟩
Touet, J., **2**, 64 ⟨92TA841⟩
Touillaux, R., **4**, 790 ⟨80TL223⟩
Toullec, J., **1**, 869 ⟨82APO(18)1, B-90MI 119-04⟩, 871 ⟨69CC292⟩

Toupance, G., **5**, 728 ⟨76BSF667⟩
Toupet, L., **1**, 622 ⟨93SL839⟩; **2**, 47 ⟨89AG(E)755⟩, 324 ⟨87TL1757⟩; **3**, 414 ⟨86OS(64)164⟩; **4**, 109 ⟨89CJC1125⟩, 249 ⟨93TL1141⟩, 989 ⟨91CB595⟩, 993 ⟨87BSF325⟩; **5**, 633 ⟨85JCS(P1)1875, 86PS(27)327⟩, 633 ⟨88T139, 91BSF574⟩
Tour, J. M., **1**, 456 ⟨90JOC3452⟩, 464 ⟨88JA5383⟩, 496 ⟨88JA5231⟩, 515 ⟨92JOC4786⟩, 536 ⟨85JA2568⟩, 680 ⟨89TL3927⟩, 1070 ⟨85OS(64)44⟩, 1094 ⟨83JA6344⟩; **2**, 86 ⟨89JOC484⟩, 926, 931, 976 ⟨83JA6344⟩, 927, 986, 988 ⟨89JA3336⟩, 931 ⟨85JA8289⟩, 973, 976 ⟨88JOC913⟩; **3**, 270 ⟨85JA2568, 89JA3336⟩; **4**, 809 ⟨88JA5383, 89JA3336⟩, 1061 ⟨85JA2568⟩
Tourbah, H., **3**, 62 ⟨86TL4983⟩
Toure, S. A., **5**, 633 ⟨91BSF574⟩
Tourneau, M. E. Le, **4**, 61, 62, 73, 79, 780 ⟨85JA735⟩
Tournilhac, F., **1**, 474 ⟨88CC203⟩
Touzin, A. M., **2**, 759 ⟨77TL3589⟩; **4**, 465, 478 ⟨92TL77⟩
Tovstenko, V. I., **2**, 835, 837 ⟨80S823⟩; **4**, 121, 122 ⟨80JGU802, 80S823⟩
Towae, F., **2**, 683 ⟨79GEP(O)2807762⟩
Towers, C., **4**, 1034, 1037, 1042 ⟨79JCS(D)1552, 73CC258⟩, 1042 ⟨79JCS(D)814⟩
Towle, P. H., **1**, 624 ⟨49JA3929⟩; **2**, 756 ⟨49JA3929⟩
Towne, T. B., **5**, 944 ⟨93JOC6952⟩
Townsend, C. A., **1**, 539 ⟨88JCS(P1)839⟩
Townsend, J. M., **1**, 5, 8, 9, 10 ⟨71TL137⟩; **2**, 822, 841 ⟨79JOC3741⟩
Townsend, L. B., **3**, 362 ⟨78JOC4910⟩, 633 ⟨83S308⟩; **5**, 275 ⟨86JOC1065⟩, 619 ⟨70JHC863⟩, 1063, 1083 ⟨86JOC1277⟩; **6**, 566 ⟨85JHC1269⟩
Towson, J. C., **3**, 171 ⟨90OS(69)158⟩
Toy, A., **5**, 951 ⟨89JOC4481⟩
Toy, A. D. F., **2**, 108 ⟨48JA3882⟩, 411 ⟨59IS69⟩, 412, 413 ⟨42JA1337, 42JA1553⟩, 413 ⟨59IS73⟩, 847 ⟨48JA186⟩; **4**, 115, 123, 133 ⟨61JA2299⟩, 473 ⟨65JA4757⟩
Toy, M. S., **4**, 46 ⟨79JOC2813⟩; **6**, 42 ⟨75JFC(5)25⟩, 250 ⟨79JOC2813⟩
Toyama, T., **1**, 233, 400 ⟨72JOC1721⟩; **3**, 373, 378 ⟨72JOC1721⟩
Toye, J., **2**, 748 ⟨75JA2276⟩, 758 ⟨80OS(59)26⟩; **4**, 790 ⟨88OSC(6)282⟩; **5**, 2 ⟨79OS(59)26⟩
Toyoda, J., **3**, 324 ⟨86TL4383, 87CL2135⟩
Toyoda, T., **2**, 810, 811 ⟨79SC515⟩; **3**, 282 ⟨78JA4842, 80TL173⟩, 638 ⟨79SC553⟩; **5**, 173, 235 ⟨82JA5523⟩
Toyofuku, M., **3**, 139 ⟨84CL1227⟩
Toyooka, K., **6**, 563, 565 ⟨82JMC557⟩
Toyo'oka, T., **2**, 945 ⟨76JOC3559⟩; **4**, 509 ⟨75CC640⟩
Toyoshima, K., **1**, 459 ⟨80JOC1604⟩
Toyota, A., **2**, 17 ⟨94CPB459⟩
Toyota, K., **2**, 857, 865 ⟨93TL1043⟩; **3**, 605 ⟨84TL1809, 85PS(25)237⟩, 605 ⟨87PS(30)527, 90TL2311⟩, 607 ⟨90CL827⟩, 609 ⟨86JOM(311)C63, 90HAC339⟩, 689 ⟨92CL1053⟩; **4**, 559 ⟨90CL643, 91CL2213⟩; **5**, 879 ⟨83CL1653⟩, 884 ⟨88TL333⟩, 1086 ⟨90BCJ2736⟩, 1086, 1087 ⟨84TL1809⟩, 1086, 1087 ⟨88POL2213⟩, 1092 ⟨84CC689⟩
Tozune, S., **1**, 233, 400 ⟨72JOC1721⟩; **3**, 373, 378 ⟨72JOC1721⟩
Traas, P. C., **1**, 633, 1094 ⟨76TL2287⟩; **3**, 64 ⟨76RTC308, 77TL2129⟩, 69 ⟨77T2027⟩
Trabalka, C., **6**, 223 ⟨81MIP68355⟩
Trabelsi, M., **2**, 216 ⟨90TL5007⟩; **4**, 289 ⟨92SL638⟩
Traber, R. P., **2**, 850 ⟨78JOC1867, 79JOC4918⟩
Trabold, P., **5**, 192 ⟨89GEP3811344⟩
Trace, R. L., **5**, 938 ⟨92OM1442⟩
Trach, F., **1**, 946 ⟨93CB1187⟩
Trach, S. S., **3**, 374 ⟨78ZOR1659⟩; **4**, 326 ⟨78ZOR1659⟩; **5**, 810 ⟨75ZOR1981⟩
Trachevskii, V., **5**, 709 ⟨87MI 517-02⟩
Traenckner, H. J., **4**, 381 ⟨67LA(704)120⟩
Träger, P., **3**, 483 ⟨68TL4973⟩; **5**, 873 ⟨68TL4973⟩
Trahanovsky, W. S., **1**, 655 ⟨88JA6579⟩, 774 ⟨86JOC113⟩, 787 ⟨81JA6691, 86JOC4208⟩; **3**, 535 ⟨82JA6779⟩, 543 ⟨73JA5412⟩
Trainor, G. L., **5**, 446 ⟨85JA7788⟩; **6**, 229, 230 ⟨85MI 607-01⟩
Trakarnpruk, W., **4**, 681 ⟨94OM2423⟩
Trambarulo, R., **3**, 694 ⟨50MI 321-01⟩
Tramonanto, A.,**1**, 532, 1070 ⟨84JA462⟩
Tramontano, A., **1**, 657 ⟨78JOM(156)203⟩; **2**, 517 ⟨79JA2352⟩; **3**, 676 ⟨84JA462⟩
Tramonti, M., **2**, 147 ⟨61TL607⟩
Tramontini, M., **1**, 318 ⟨73S744⟩, 635 ⟨90T1791⟩, 1041 ⟨73S703⟩; **2**, 316 ⟨73S703⟩; **3**, 184 ⟨90T1791⟩; **4**, 96 ⟨90T1791⟩, 315, 322 ⟨73S703⟩
Trampisch, H., **2**, 252 ⟨83ZN(B)365⟩
Tran, H. W., **2**, 299 ⟨83JCS(P1)1131⟩
Tranchepain, I., **1**, 486, 1032 ⟨89JOC5324⟩
Tranchier, J.-P., **2**, 381 ⟨92JOC4563⟩; **3**, 42 ⟨92JOC4563⟩
Tranne, A., **4**, 304 ⟨85JOC4231⟩
Trapani, G., **5**, 282 ⟨83S1013⟩
Trapp, H., **1**, 1144 ⟨63ZN(B)1125⟩; **3**, 495 ⟨66CB689⟩; **4**, 145, 817 ⟨63ZN(B)1125⟩, 758 ⟨66CB670⟩; **6**, 62 ⟨65TL969⟩
Trapp, O. D., **3**, 573 ⟨59JA4878, 60JA5394⟩
Trathnigg, B., **6**, 73 ⟨82MI 603-01⟩
Trauth, L. J., **5**, 370 ⟨70AF790⟩
Trautman, R. J., **6**, 517 ⟨85JA2355⟩
Trautner, K., **3**, 672 ⟨62CB281⟩
Trautwein, W.-P., **6**, 143 ⟨69CB931⟩
Trave, R., **3**, 489 ⟨77JOC1352⟩
Trave, S., **3**, 3, 114 ⟨86SC43⟩
Traverso, G., **3**, 255 ⟨58G1226⟩, 398 ⟨57AC(R)1244⟩
Traverso, O., **4**, 151, 585 ⟨90JCS(D)1553⟩
Travis, J. D., **6**, 143 ⟨68JOC3931⟩
Traylor, T. G., **1**, 666, 671 ⟨71JA6278⟩; **2**, 1007 ⟨66JA4934⟩
Traynard, J.-C., **2**, 1071 ⟨69CR(268)358⟩
Traynelis, V. J., **1**, 881 ⟨64JOC221⟩; **2**, 768 ⟨73JOC4339⟩; **4**, 28, 38 ⟨56JA2264⟩
Traynham, J. G., **1**, 113, 114 ⟨50JA849⟩, 869, 871 ⟨69JA2797⟩, 890 ⟨86JCE930⟩; **2**, 420 ⟨67JOC510⟩
Traynor, S. G., **1**, 420 ⟨75TL3555, 78T2783⟩, 797 ⟨80JOC900⟩
Treadgold, R. C., **2**, 540 ⟨87JCS(P1)1107⟩
Treadwell, J., **6**, 36 ⟨71JHC557⟩, 36, 43 ⟨71JMC593⟩
Trebbien, C., **3**, 639 ⟨83ACS(B)639⟩
Trebellas, J. C., **1**, 413, 809 ⟨66JOM(6)412⟩
Treber, J., **2**, 558 ⟨90CB1517⟩
Trecker, D. J., **2**, 1566 ⟨63JA3992⟩; **5**, 307 ⟨69JA3992⟩
Trede, A., **2**, 100 ⟨62AG(E)97⟩, 793 ⟨62AG(E)89⟩
Trefonas, L. M., **4**, 458 ⟨72JHC1295⟩; **6**, 509 ⟨78JOC922⟩
Tréfouel, T., **2**, 398 ⟨87BCJ4492⟩
Trehan, A., **1**, 752 ⟨88TL419⟩; **3**, 162 ⟨93TL7335⟩; **4**, 192 ⟨93TL7335⟩
Trehan, I. R., **5**, 274, 286 ⟨92S439⟩
Trehan, S., **3**, 162 ⟨93TL7335⟩; **4**, 192 ⟨93TL7335⟩
Trehearne, T. E., **2**, 439 ⟨90JOM(393)143⟩
Treiber, A. J. H., **4**, 16 ⟨65JA4259⟩
Treiber, H. J., **6**, 140 ⟨63CB1505⟩
Treiber, K. D., **5**, 944 ⟨93JOC6952⟩
Treibs, A., **3**, 306, 308 ⟨70LA(733)27⟩; **5**, 549 ⟨69LA(723)129⟩, 561 ⟨70LA(737)179⟩, 638 ⟨73LA207⟩
Treibs, W., **5**, 362 ⟨62JPR223⟩
Treichel, P., **5**, 453 ⟨66IC1177⟩
Treichel, P. M., **2**, 573 ⟨82COMC-I(4)1⟩; **5**, 802 ⟨73JOM(61)415⟩; **6**, 243, 244 ⟨64AOC(1)143⟩, 247 ⟨61JA3593⟩, 517 ⟨71JA5424⟩, 721 ⟨69CC423⟩
Treier, K., **3**, 386 ⟨89CC667, 90JOM(384)305⟩
Treiger, L. M., **1**, 56 ⟨71MI 102-01⟩
Trejo, A., **1**, 837 ⟨94TL15⟩
Tremaine, J. F., **1**, 862 ⟨52JCS632⟩
Trembovla, V. N., **6**, 456 ⟨84MI 607-01⟩
Tremeer, E. J., **2**, 540 ⟨87JCS(P1)1107⟩
Tremelling, M., **1**, 104 ⟨71JA1544⟩

Tremmel, S., **4**, 312, 325 ⟨76JFC(7)471⟩
Tremmling, M. J., **2**, 622 ⟨78JOC3076⟩
Tremper, H. S., **1**, 13 ⟨69JOC4, 71JOC758⟩
Trenbeath, S., **2**, 651 ⟨78CC242⟩, 656 ⟨76TL1637⟩
Trend, J. E., **1**, 601, 602, 604 ⟨78JOC1697⟩, 601, 781 ⟨76JOC2503⟩; **2**, 281 ⟨78JOC1697⟩; **4**, 89 ⟨75CJC1922⟩, 92 ⟨78JOC1697⟩
Trenholm, J. E., **6**, 255 ⟨83CI(L)205⟩
Trenkle, A., **2**, 431 ⟨79ZN(B)642⟩
Trenner, N. R., **3**, 231 ⟨51JA719⟩
Trent, J. O., **2**, 1004 ⟨91JCS(P2)1077⟩; **4**, 802 ⟨93JOC1531⟩
Trentmann, B., **1**, 500 ⟨91CB2335⟩
Trenwith, A. B., **4**, 5 ⟨53JCS2083⟩
Trepka, R. D., **3**, 645 ⟨62JOC422⟩
Treppendahl, S.,**5**, 652 ⟨77T3137⟩
Treptow, B., **1**, 1001 ⟨93S558⟩
Treptow, W., **2**, 689 ⟨80S283⟩
Trethewey, A. N., **2**, 289, 292 ⟨88JCS(P1)1913⟩; **3**, 384 ⟨86CC1152⟩, 636 ⟨88JCS(P1)1913⟩; **4**, 290 ⟨88JCS(P1)1913⟩; **5**, 336 ⟨88JCS(P1)1913⟩, 336, 1113 ⟨86CC1152⟩
Tretter, J. R., **4**, 420 ⟨74JOC1710⟩
Trevoy, L. W., **1**, 3, 4, 7, 8, 10 ⟨49JA1675⟩
Triantaphylides, C., **5**, 226 ⟨85TL1857⟩
Tricarico, K. A., **2**, 118, 124 ⟨93JMC460⟩
Trickes, G., **5**, 1003, 1013 ⟨77AG(E)555⟩
Triem, H., **4**, 328 ⟨63AG1050⟩; **5**, 599 ⟨64AG(E)19⟩
Trifan, D., **1**, 381 ⟨49JA2953⟩, 381, 905 ⟨52JA1154, 52JA1147⟩
Triffeneau, M., **5**, 62 ⟨42MI 502-02⟩
Trifunovich, I. D., **1**, 582 ⟨90CC84⟩
Trigo, G. G., **3**, 616 ⟨86TL3285⟩
Trimarco, P., **2**, 771 ⟨78S465⟩
Trimble, L. A., **2**, 375 ⟨86JA6395b⟩
Trimitsis, G. B., **1**, 879 ⟨83JOC2957⟩; **5**, 43 ⟨73JOC1491⟩
Trimitsis, J. B., **1**, 318 ⟨75JA7193, 76JA1997⟩
Trimmer, R. W., **3**, 577 ⟨76JOC13⟩
Trinchera, C., **3**, 445 ⟨55G1705⟩
Trindle, C., **1**, 649 ⟨73JOC2664⟩; **2**, 936 ⟨73JOC2664⟩
Tring, P. C., **1**, 175 ⟨88JA1633⟩
Trinks, U. P., **1**, 224 ⟨89HCA1301⟩
Trinquier, G., **3**, 497 ⟨79JA7169⟩, 502 ⟨81T2875⟩, 596 ⟨89AG(E)621⟩; **4**, 567, 1082 ⟨89AG(E)621⟩; **6**, 125 ⟨89AG617⟩, 694 ⟨79JA7169⟩
Tripathi, D. N., **4**, 773 ⟨93TL7645⟩
Tripathi, V. K., **3**, 267 ⟨79JCS(P1)36⟩; **4**, 213 ⟨79JCS(P1)36⟩
Tripathy, P. B., **1**, 138 ⟨89JA4399⟩; **4**, 144 ⟨90S200⟩, 389, 664, 665 ⟨89JA4399⟩, 814 ⟨74JOM(69)53⟩; **6**, 182 ⟨70JOM(21)P6⟩, 182, 396 ⟨74JOM(69)53⟩
Trippet, S., **3**, 328 ⟨71JCS(C)1114⟩, 492 ⟨60CI933⟩, 493, 494 ⟨65JCS7140, 62JCS2337⟩
Trippet, S. J., **3**, 494 ⟨66JCS67⟩
Trippetta, L., **5**, 320, 321 ⟨84AQ49⟩
Trippett, S., **1**, 104 ⟨64QR406⟩, 638 ⟨63QR406⟩, 721 ⟨56CI(L)80⟩, 733 ⟨66JCS(C)1090⟩, 738, 739, 741 ⟨71JCS(C)1114⟩, 747 ⟨61CI(L)990⟩, 832 ⟨81JCS(P1)2127⟩, 1079 ⟨59JCS3874, 62JCS2333⟩, 1079 ⟨64JCS543⟩; **2**, 442 ⟨61JCS4263⟩, 843 ⟨69JCS(C)2038⟩, 849 ⟨82TL2603⟩, 860 ⟨72JCS(P1)1184⟩; **3**, 494, 642 ⟨61JCS1266⟩, 591 ⟨59JCS3874⟩, 618 ⟨79TL1725⟩, 674 ⟨62JCS2333, 64JCS543⟩; **4**, 337 ⟨61JCS2813⟩; **5**, 403 ⟨70JCS(C)1263⟩, 786 ⟨B-70MI 520-01⟩, 900 ⟨68JCS(C)1612⟩, 920 ⟨71JCS(C)1114⟩; **6**, 118 ⟨67JCS(C)1547⟩
Trischler, F. D., **5**, 967 ⟨67PP1149⟩
Trischmann, H., **4**, 183 ⟨61CB2258⟩, 436 ⟨67LA(706)107⟩
Trishin, Ju. G., **2**, 1074 ⟨89TL577, 91ZOB611⟩
Trishin, Yu. G., **2**, 467 ⟨79JGU616⟩; **4**, 493 ⟨88JGU1148⟩
Trisler, J. C., **5**, 1124 ⟨75TL3125⟩
Tristam, E. W., **3**, 426 ⟨59JA2151⟩
Tristram, E., **3**, 54 ⟨57JA456⟩; **4**, 213 ⟨57JA456⟩
Tritschler, W., **4**, 193 ⟨71S312⟩; **6**, 105 ⟨71S312, 72S32⟩, 105 ⟨72S418, 73S423⟩, 105 ⟨75S272⟩
Trivedi, B., **6**, 561 ⟨89JIC60⟩
Trivedi, G. K., **5**, 324 ⟨90SC1733⟩
Trivedi, J. P., **6**, 537 ⟨63CCC2047⟩
Trivedi, N., **6**, 229 ⟨87JA3708⟩
Trivelli, G., **3**, 612 ⟨32HCA254⟩
Trnka, T., **1**, 4, 10 ⟨84CAR(130)103⟩
Troepol'skaya, T. V., **3**, 448 ⟨63BAU408⟩
Troffkin, H. J., **2**, 388 ⟨66JA1078⟩
Trofimenko, S., **3**, 102 ⟨64JOC3046⟩, 565 ⟨63JOC217⟩; **4**, 437 ⟨70JA5118⟩, 764 ⟨78JOC43⟩; **5**, 779 ⟨67JA7014⟩; **6**, 362 ⟨70JA5118⟩
Trofimov, B. A., **2**, 674 ⟨68RCR907⟩, 708 ⟨83IZV230⟩, 722 ⟨92SR1⟩; **3**, 430 ⟨85ZOR766⟩; **5**, 1025 ⟨88ZOR1771⟩; **6**, 562 ⟨81ZOR2269⟩, 565 ⟨86ZOR2145, 87ZC24⟩
Tröger, J., **5**, 858 ⟨21JPR1⟩
Trogler, W. C., **4**, 336 ⟨90JOC2644⟩; **6**, 516 ⟨87OM2451⟩
Trogu, E. F., **4**, 687 ⟨67G1327⟩
Troise, C. A., **1**, 806 ⟨85JA5739⟩
Troisi, L., **4**, 188 ⟨89TL257⟩
Troitskaya, L. L., **4**, 720 ⟨79JOM(182)537⟩, 721 ⟨77DOK(236)371, 79IZV1528⟩, 1033 ⟨80JOM(202)C58⟩
Troll, C., **1**, 991 ⟨92CB2667⟩; **3**, 390 ⟨91JOM(415)211⟩; **5**, 954 ⟨93JOM(454)133⟩
Trombini, C., **1**, 24 ⟨84JOC1246⟩, 688 ⟨85JOM(280)307⟩; **2**, 79 ⟨88JOM(353)177⟩, 590 ⟨83JOC4108⟩, 698 ⟨90CC1680⟩, 959 ⟨78CC927, 85PAC1887⟩; **3**, 31 ⟨93JA7001⟩, 159 ⟨92T1299⟩
Tromboni, C., **1**, 439 ⟨77JCS(P1)123⟩
Trometer, J. D., **1**, 226, 227 ⟨89T391⟩, 226, 227, 751 ⟨88JOC4274⟩; **2**, 86 ⟨88JOC4274⟩
Tromifov, B. A., **6**, 565 ⟨90ZOR1364⟩
Trompen, W. P., **5**, 662 ⟨71RTC463⟩, 662, 809, 810 ⟨71RTC458, 70JCS(C)875⟩; **6**, 284 ⟨72RTC331⟩, 609 ⟨74TL1687, 74TL1691⟩, 621 ⟨70JCS(C)875⟩
Tronchet, J., **3**, 247 ⟨70HCA1489⟩
Tronchet, J. M., **5**, 211 ⟨87MI 504-02⟩, 212, 213 ⟨85CAR375⟩
Tronchet, J. M. J., **1**, 725 ⟨79HCA2091⟩; **2**, 341, 344 ⟨90TL531⟩; **3**, 247 ⟨70HCA1489⟩; **5**, 155 ⟨79HCA2091⟩, 214 ⟨85HCA1893⟩, 895, 896 ⟨79HCA1401⟩; **6**, 695 ⟨76CAR(46)9⟩
Trong-Anh, N., **3**, 614 ⟨69CC112⟩
Tronich, W., **1**, 740 ⟨68IC168⟩; **3**, 323 ⟨74JOM(67)341⟩, 496 ⟨68CB595⟩; **4**, 551, 574 ⟨68CB3545⟩, 569, 587 ⟨68CB595⟩, 571, 574, 588 ⟨67CB1032⟩, 576 ⟨68CB3556⟩, 597 ⟨68IC168⟩; **5**, 24, 105 ⟨83GEP3148422⟩, 908, 911, 912, 916 ⟨68CB3545⟩, 912 ⟨68AG(E)220⟩, 913, 916, 920, 1093 ⟨67CB1032, 68CB595⟩; **6**, 38 ⟨74JOM(67)341⟩, 51 ⟨74JOC158⟩, 608 ⟨71JOC1786⟩, 702 ⟨67CB1032⟩, 708 ⟨68IC168⟩
Tronov, B. V., **1**, 107 ⟨26MI 104-01⟩; **2**, 631 ⟨60MI 212-01⟩
Troost, J. J., **3**, 618 ⟨82TL1505⟩, 622 ⟨76TL255⟩
Trope, A. F., **3**, 165 ⟨80TL4301⟩
Tropitzsch, R., **1**, 620 ⟨74TL3987⟩; **2**, 188 ⟨74TL3987⟩
Tropper, F. D., **6**, 546 ⟨92MI 617-04⟩
Troschutz, R., **4**, 900 ⟨86AP(319)161⟩
Trost, B. M., **1**, 130, 871, 872, 881 ⟨65JOC2502⟩, 142, 599 ⟨75JOC2014⟩, 146 ⟨92SL27⟩, 192 ⟨80ACR385⟩, 192 ⟨93TL7513⟩, 192, 212 ⟨82COMC-I(8)799⟩, 193, 287, 358, 360, 361, 362 ⟨86AG(E)1⟩, 193, 288 ⟨89JA7487, 89TL1495⟩, 194 ⟨80TL2591⟩, 194, 288, 289 ⟨89JA6482⟩, 194, 288, 546 ⟨87JA3483⟩, 194, 289 ⟨90JA408⟩, 196 ⟨84JA6837, 87JA2176, 87T4817⟩, 232, 234 ⟨71JA676⟩, 233 ⟨B-75MI 105-01⟩, 237 ⟨88JA5216⟩, 237, 484, 1031 ⟨86JA1098⟩, 238 ⟨90JOC1127⟩, 238, 613 ⟨80JA5979⟩, 239, 484 ⟨84JA7260⟩, 248 ⟨86TCC3⟩, 287 ⟨79JA6429, 82JA6668, 88JA1602, 89AG215⟩, 287, 361 ⟨91JA9007⟩, 287, 789 ⟨86TL1445⟩, 288 ⟨86JA6051,

87JA615, 88JA3687⟩, 289 ⟨82JA3733, 82JA6879⟩, 289 ⟨89H(28)321, 89JA5902, 91JA7363, 93TL7183⟩, 289, 362 ⟨91JA7350, 92JOC686⟩, 307 ⟨66JOC646⟩, 316 ⟨92JA7903⟩, 361 ⟨92S151⟩, 390 ⟨73JA3068, 92JA5432⟩, 398 ⟨B-75MI 109-01⟩, 436, 596, 598, 599, 778 ⟨78CRV363⟩, 438 ⟨81TL3463⟩, 439 ⟨83JA5075⟩, 457 ⟨89TL4657⟩, 494 ⟨93JA9421⟩, 496 ⟨88JA5231⟩, 496, 1051 ⟨90JA2816⟩, 522 ⟨83TL1129⟩, 523 ⟨89TL7157⟩, 525 ⟨85JA1781, 85TL4887, 87TL1611⟩, 538 ⟨79JA1284⟩, 546 ⟨87JA3161⟩, 546, 1036 ⟨87JA3486⟩, 564 ⟨71TL2603⟩, 595, 597, 598, 601, 621 ⟨76JA4887⟩, 595, 598 ⟨93JOC1579⟩, 596, 597, 778, 779 ⟨78ACR453, 75TL4197⟩, 598 ⟨76JA5017, 77JA8116⟩, 598, 779 ⟨73JA6840⟩, 599 ⟨74JA7165⟩, 613 ⟨86JOC3896⟩, 615, 698, 702 ⟨77JOC2036⟩, 632 ⟨85JA1293⟩, 675 ⟨91JA1850, 93S824⟩, 676 ⟨88JA1636⟩, 696 ⟨88BCJ107⟩, 698 ⟨76TL3477⟩, 747 ⟨67JA5292⟩, 779 ⟨75JOC148⟩, 801 ⟨81JOC506⟩, 821 ⟨71CC1639, 73JA962, 73JOC3438⟩, 871, 872, 881 ⟨65JOC1341⟩, 882 ⟨65JOC2513⟩, 1031 ⟨84JA4263⟩, 1047 ⟨88JA2301⟩, 1050 ⟨93TL2271⟩; **2**, 86 ⟨89JOC484⟩, 138 ⟨82JA3225⟩, 139 ⟨76JA4887⟩, 139, 140 ⟨73JA6840⟩, 139, 140 ⟨75PAC563⟩, 155 ⟨81TL1287⟩, 212 ⟨80JOC1839⟩, 540 ⟨81TL4999, 83JOC3252⟩, 555 ⟨84OS(62)58⟩, 658 ⟨78TL1667⟩, 696 ⟨65JOC2502⟩, 709 ⟨83JA5075⟩, 724 ⟨77JA4405⟩, 753 ⟨81JA2483⟩, 912, 945 ⟨83TL4895⟩, 927 ⟨91MI 218-02, 93TL4735⟩, 928 ⟨89JOC2271⟩, 938 ⟨86JOC3435⟩, 980 ⟨88JA7255⟩; **3**, 9, 122 ⟨91TL3039⟩, 54 ⟨75JA4018⟩, 89, 100 ⟨81JOC4617⟩, 117 ⟨84TL173⟩, 138 ⟨79JA4413⟩, 139 ⟨77T2615, 80TL2591⟩, 164 ⟨78CRV363⟩, 165 ⟨80JA3548⟩, 166 ⟨82JA3228⟩, 170 ⟨93JOC1579⟩, 209 ⟨73JA6840, 78CB363⟩, 209 ⟨83JOC3252⟩, 210 ⟨79JOC3448⟩, 219 ⟨75JA6911⟩, 220 ⟨82TL1047⟩, 229, 273 ⟨88JA2301⟩, 242 ⟨76JA5017⟩, 249 ⟨78JOC1031⟩, 266 ⟨73JA7862⟩, 283 ⟨84JA2469⟩, 374 ⟨B-75MI 308-01⟩, 470 ⟨70JA2591⟩, 492 ⟨91COS⟩, 536 ⟨66JA1587⟩, 537 ⟨65JOC2502⟩, 620 ⟨87TL1611⟩, 636 ⟨91JA1044⟩; **4**, 64, 71 ⟨74JOC2648⟩, 99 ⟨73JA7813⟩, 162 ⟨74ACR85⟩, 172 ⟨75JA7182⟩, 197 ⟨75JOC148⟩, 206 ⟨92AG(E)1335⟩, 220 ⟨77JA4405, 78JA7103, 86T3323⟩, 244, 257 ⟨78CRV363⟩, 250 ⟨82TL1047⟩, 266 ⟨90CAR(202)1⟩, 274 ⟨81TL1287⟩, 280 ⟨91TL2193, 92TL1831, 92TL717⟩, 329 ⟨81JA2483⟩, 329, 392 ⟨80JA7929⟩; **5**, 146 ⟨86JA284⟩, 158 ⟨77JA1649⟩, 158 ⟨89AG(E)1173⟩, 164 ⟨79JA1284⟩, 286 ⟨84JA2469⟩, 731 ⟨80JA7929⟩

Trost, M. K., **1**, 675 ⟨91JA1850⟩

Trostmann, U., **3**, 318 ⟨79TL4049, 81CB2979⟩, 318 ⟨81TL4953, 85CB4035⟩

Trostyanskaya, I. G., **4**, 820 ⟨76JGU2367⟩, 915 ⟨78ZOB267⟩

Trotman-Dickenson, A. F., **2**, 872 ⟨B-75MI 215-01⟩

Trötsch-Schaller, I., **4**, 567 ⟨87TL2693⟩

Trotta, F., **2**, 14 ⟨91S1003⟩

Trott, P., **6**, 16 ⟨47JA1820⟩

Trotter, J., **2**, 361 ⟨73CJC2504⟩; **4**, 656 ⟨91OM2857, 93OM4572⟩

Trotter, J. W., **3**, 527 ⟨77JOC4157⟩

Troupel, M., **3**, 620 ⟨93T5091⟩; **5**, 44 ⟨85TL1509, 86GEP3522304, 86JCE1013⟩

Troupet, L., **2**, 1003 ⟨91AX(C)328⟩

Trouve, B., **4**, 386 ⟨87JOM(331)181⟩

Trova, M., **2**, 1024, 1025 ⟨88TL4917⟩

Trova, M. P., **1**, 134 ⟨93T4147⟩, 708 ⟨86TL1895, 88JA8197⟩, 776 ⟨87TL2795, 90JA228⟩

Troxell, H. A., **5**, 298 ⟨55HOU(9)951⟩

Troyansky, E. I., **2**, 365 ⟨85T4279⟩

Trozzolo, A. M., **3**, 314 ⟨70JA1402⟩

Trska, P., **6**, 5 ⟨74CCC1330⟩

Trubitsin, A. E., **6**, 55, 282 ⟨87ZOR1657⟩

Truby, F. K., **2**, 213 ⟨65CI(L)680⟩

Truce, W. E., **1**, 379 ⟨58JA3625, 59JA481⟩, 379 ⟨70JOC1828, 70OR(18)99⟩, 379 ⟨86JA3466⟩, 470 ⟨74JOC3174⟩, 485, 1032 ⟨79JOC3444⟩; **2**, 127 ⟨62JA1621⟩, 127, 188 ⟨60JA2872⟩, 157 ⟨70JOC4217⟩, 182 ⟨74JOC238⟩, 182 ⟨74JOC245⟩, 188, 271, 719 ⟨63JOC593⟩, 203 ⟨54JA5357⟩, 204 ⟨69JOC3097⟩, 207 ⟨70JOC1226⟩, 478 ⟨86JA3466⟩, 710 ⟨72JOC2391, 70OR(18)99⟩, 742 ⟨66JOC3543, 70JOC3275⟩, 1030, 1034 ⟨56JA2760, 60JA3799⟩, 1034 ⟨66JOC3545, 71JOC1727⟩; **3**, 92 ⟨57OR(9)37⟩; **4**, 64, 65, 80 ⟨52JA3594, 74JOC1449⟩, 281 ⟨63JOC1420⟩, 332 ⟨69JOC3104⟩, 391, 393 ⟨54JA3577⟩, 770 ⟨58JA1916, 58JA6450⟩, 770, 779 ⟨65T2899⟩, 971, 979 ⟨65JOC71, 67JOC990⟩; **5**, 496 ⟨63JA3231⟩, 670, 819 ⟨66CJC297⟩; **6**, 16, 44, 45, 234 ⟨52JA3594⟩

Truces, W. E., **2**, 478 ⟨79JA5432⟩

Truchet, F., **2**, 900 ⟨87TL4169⟩; **5**, 199 ⟨87TL4169⟩

Truchet, R., **2**, 1016 ⟨31AC(P)309⟩; **5**, 24 ⟨B-48MI 502-01⟩, 28 ⟨42MI 502-01⟩

Trudell, M. L., **3**, 62 ⟨89JA7504⟩, 206 ⟨84JOC1196⟩

Truebik, M. M., **4**, 656 ⟨70JOM(25)C36⟩

Truelock, M. M., **4**, 656 ⟨71SAP7004922, 72CC613⟩; **6**, 396 ⟨71SAP7004922⟩

Truesdale, L. K., **2**, 541 ⟨77JA5009⟩; **3**, 474 ⟨76JOC3335⟩, 615 ⟨73CC55, 74JOC914⟩, 615, 616 ⟨73TL4929⟩; **4**, 216 ⟨77JA5009⟩; **5**, 1134 ⟨74JOC914⟩

Truesdell, J. W., **4**, 187 ⟨88S36⟩

Trujillo, H., **2**, 439 ⟨85IC2344⟩

Trukhina, T. V., **5**, 390 ⟨81ZOB1318⟩

Trullinger, D. P., **1**, 980 ⟨80JOC2080⟩; **5**, 159 ⟨80JOC2080⟩

Trumbull, E. R., **1**, 78, 421 ⟨52JA4867⟩, 624, 627, 628, 637, 640, 782, 783 ⟨60OR(11)317⟩; **2**, 346 ⟨60OR(11)317⟩

Trummer, I., **2**, 746 ⟨82ZN(B)105⟩; **3**, 264 ⟨81S225⟩

Trumper, P. K., **4**, 204 ⟨81JA5618⟩

Trumtel, M., **3**, 240 ⟨86S337⟩; **4**, 235 ⟨89CAR(188)81⟩

Truong, V., **5**, 946 ⟨93TL227⟩

Truscheit, E., **1**, 734 ⟨62LA(658)65⟩

Truscott, R. J. W., **2**, 906 ⟨90JOC4581⟩

Trusenev, A. G., **4**, 345 ⟨86ZOB216⟩; **5**, 369 ⟨91IZV2103⟩, 405 ⟨89JGU2388⟩

Truskanova, T. D., **5**, 666, 671, 692 ⟨86MI 517-02⟩

Truter, M. R., **5**, 725 ⟨60JCS997⟩

Trybulski, E. J., **1**, 1134 ⟨78JA4618, 78JA4620⟩; **2**, 629 ⟨78JA4618⟩; **4**, 20 ⟨76TL4577⟩; **5**, 173 ⟨78JA4618⟩, 173, 204 ⟨78JA4620⟩

Trzeciak, A.,**5**, 279, 281 ⟨89TL1927⟩

Trzupek, L. S., **5**, 436 ⟨73JA8118⟩

Tsai, C.-Y., **1**, 489 ⟨89JOC868⟩; **2**, 630 ⟨89JOC868⟩; **3**, 227 ⟨88JOC5⟩; **4**, 526 ⟨83OM1536⟩

Tsai, D. J.-S., **1**, 405, 816, 817, 818 ⟨84JOC1842⟩; 651, 762 ⟨81TL2751⟩, 764 ⟨83OM236⟩; 816 ⟨85JA3915⟩; **2**, 34 ⟨85JA3915⟩, 529, 530, 935 ⟨81TL2751⟩; **4**, 642 ⟨83OM236⟩

Tsai, I. L., **1**, 627 ⟨88H(27)751⟩

Tsai, J.-H., **1**, 991 ⟨93OM3971⟩

Tsai, L., **5**, 275 ⟨57JA2530⟩

Tsai, M., **1**, 1139 ⟨76JOC1669⟩

Tsai, M.-R., **4**, 387 ⟨82JOC5051⟩; **5**, 413 ⟨82JOC5051⟩

Tsai, M.-Y., **1**, 218, 221 ⟨77JOC394⟩

Tsai, T. Y. R., **1**, 498 ⟨80T419⟩

Tsai, Y., **1**, 1060 ⟨85JA7233⟩

Tsai, Y.-M., **1**, 341 ⟨84JA8209⟩; **2**, 936 ⟨92JOC7010⟩; **3**, 659 ⟨81JOC3671⟩; **4**, 358 ⟨85JOC5393⟩, 359 ⟨88OR(35)513⟩, 661 ⟨92TL7895⟩, 930 ⟨92JOC7010⟩, 936 ⟨83TL4387⟩, 937 ⟨84JA8209⟩; **5**, 410, 411 ⟨92JOC7010⟩, 411 ⟨91TL3515⟩, 412 ⟨88OR(35)1⟩, 413, 425 ⟨85JOC5393⟩

Tsanaktsidis, J., **1**, 900 ⟨87CC833⟩; **2**, 13 ⟨89TL6967⟩

Tsang, R., **1**, 828 ⟨89JOC5350⟩

Tsang, T., **2**, 328 ⟨88TL6715⟩; **3**, 419 ⟨88TL6715⟩

Tsang, W., **1**, 921 ⟨85JA2872⟩

Tsao, C. H., **1**, 445 ⟨86JOM(312)53⟩

Tsao, C.-H., **1**, 53 ⟨85JOC4329⟩, 444 ⟨85JCR(S)18⟩; **2**, 432 ⟨85JCR(S)18, 85JOC4329⟩, 432 ⟨86JOM(312)53⟩

Tsao, J.-H., **1**, 654 ⟨78CC358⟩
Tsareva, L. A., **4**, 891 ⟨71ZOB482⟩; **6**, 425 ⟨71JOU1943⟩
Tsarkova, I. A., **4**, 469 ⟨92ZOB2708⟩
Tsatsas, G., **5**, 751, 753 ⟨56MI 519-01⟩, 751, 753 ⟨56MI 519-02⟩
Tsay, S.-C., **3**, 13, 85 ⟨90JCS(P1)757⟩, 426 ⟨90T7413⟩, 620 ⟨91JCS(P1)1493⟩; **5**, 637, 829 ⟨90T7413⟩
Tsay, Y. H., **2**, 575 ⟨76AG(E)621⟩; **3**, 597 ⟨86CB3141⟩; **4**, 582, 583 ⟨74CB3706⟩; **4**, 659 ⟨85CB4248⟩, 693 ⟨77CB1763⟩; **5**, 917 ⟨85OM1310⟩, 951 ⟨81AG(E)1055⟩
Tschamber, T., **5**, 330 ⟨90TL2139, 91TA1209⟩
Tschampel, M., **4**, 690 ⟨93ZN(B)1558⟩
Tschanber, T., **3**, 536 ⟨83LA1393⟩
Tscherniak, J., **5**, 838 ⟨1875CB114⟩
Tscherniak, M., **6**, 610 ⟨1878BSF(2)185⟩
Tschitschibabin, A. E., **3**, 363 ⟨29CB3048⟩; **6**, 77 ⟨05CB561⟩
Tschoetsch, C., **4**, 870 ⟨84EGP211356A1⟩
Tschoetsch, C. K. J., **4**, 870 ⟨83S840⟩
Tschugaeff, L., **2**, 279 ⟨09CB49⟩
Tse, A., **3**, 182 ⟨81JOC4316⟩
Tse, C.-L., **1**, 16, 940 ⟨93TL2733⟩, 777 ⟨94T639⟩; **6**, 550 ⟨93TL2733⟩
Tse, C.-W., **1**, 801, 802 ⟨89CRV165⟩
Tse, H. L. A., **1**, 1092 ⟨84TL3155⟩; **4**, 229 ⟨92TL4625⟩; **6**, 134 ⟨90JA4552⟩
Tse, J. S., **6**, 247 ⟨90IC2496⟩
Tse, R. L., **3**, 66, 70 ⟨56JOC638⟩; **4**, 743 ⟨56JOC638⟩
Tselinskii, I. V., **2**, 386 ⟨78ZOR240⟩; **4**, 418 ⟨90JGU2516, 91JGU1072⟩; **5**, 361 ⟨77ZOR1144⟩; **6**, 55 ⟨73IZV1424⟩, 55, 282 ⟨87ZOR1657⟩, 145 ⟨86ZOR1393⟩, 149 ⟨87ZOR2624⟩
Tseng, C. C., **1**, 187 ⟨86JOC2884⟩
Tseng, C. K., **6**, 656 ⟨74OMR(6)494⟩
Tseng, C.-P., **2**, 160 ⟨82JOC552⟩
Tseng, S. S., **2**, 405 ⟨73JA1677⟩
Tsentsiper, I. L., **6**, 145 ⟨80MI 605-02⟩
Tsetlina, E. O., **2**, 1091 ⟨79JGU536⟩; **3**, 356 ⟨78ZOR666⟩; **4**, 813 ⟨73JGU679⟩
Tsetlina, O. E., **3**, 356 ⟨79ZOR82⟩
Tsetlin, J. S., **3**, 363 ⟨72IJS(A)0216⟩
Tsetlin, Ya. S., **3**, 356 ⟨78ZOR666, 79ZOR82⟩, 360 ⟨75ZOR1945⟩, 361 ⟨80ZOR1842⟩
Tsichiya, T.,**3**, 439 ⟨76BCJ550⟩
Tsiliopoulos, E., **2**, 16, 336 ⟨88JCS(P1)2805⟩, 334 ⟨91JFC(52)389⟩; **4**, 108 ⟨86JFC(32)461⟩; **5**, 22 ⟨91JFC(52)389⟩
Tsil'ko, A. E., **5**, 691 ⟨88JOU1049, 88ZOR1165⟩
Tsipis, C. A., **2**, 35 ⟨85CC1404⟩
Tsipouras, A., **5**, 274, 294 ⟨86JA6054⟩
Tsivunin, V. S., **2**, 461 ⟨65URP173230, 66ZOB1436⟩; **4**, 120 ⟨71JGU2183⟩, 340 ⟨70ZOB2560⟩; **5**, 403 ⟨71JGU2513⟩
Tsi-yu, K., **2**, 320 ⟨90SC3563⟩; **3**, 24 ⟨91S539⟩
Tso, C. C., **2**, 360, 361 ⟨71CC1413⟩; **5**, 457 ⟨87JA5844⟩
Tso, H.-H., **1**, 790 ⟨86JCS(P1)1039, 87CC934⟩, 791 ⟨84CC1323, 87JOC244⟩; **5**, 140 ⟨79OPP93⟩
Tso, H.-W., **3**, 624 ⟨93CC669⟩
Tsolomitis, A., **5**, 334 ⟨85JHC1635, 87H(25)569⟩
Tsotadze, M. V., **5**, 969 ⟨75ZOB2185⟩
Tsotinis, A., **2**, 929, 938 ⟨88TL1833, 86TL2761⟩; **5**, 199 ⟨86TL2761⟩
Tsou, H., **4**, 193 ⟨81JOC2557⟩
Tsou, T. T., **1**, 1099 ⟨79JA7547⟩
Tsou, U.-P. E., **1**, 776 ⟨87JA553⟩
Tsoukalas, S. N., **6**, 238, 239, 240, 242 ⟨58JA3604⟩
Tsubaki, A., **5**, 199 ⟨90JOC1679⟩
Tsubaki, K., **1**, 122, 123, 187 ⟨87JOC4418⟩; **2**, 69 ⟨87JOC4418⟩
Tsubaki, Y., **2**, 312 ⟨90CL203⟩
Tsubata, K., **1**, 17 ⟨79TL2157⟩, 432 ⟨79JOC4508⟩; **2**, 457 ⟨81TL3249⟩; **3**, 219 ⟨84JOC300⟩, 261 ⟨82JA6697, 82TL1201⟩, 618 ⟨89JOC2249⟩
Tsuboi, A., **4**, 118 ⟨85JCR(S)38⟩
Tsuboi, M., **3**, 575 ⟨87JSP(123)476⟩
Tsuboi, S., **1**, 568 ⟨87BCJ2917⟩, 571 ⟨91JOC7177⟩, 797 ⟨84CL1541⟩, 979 ⟨82JOC4478⟩, 980 ⟨84JOC1204⟩; **2**, 14 ⟨89JOC4982⟩, 39 ⟨94TL4569⟩; **3**, 246 ⟨73CL425⟩; **6**, 561 ⟨87BCJ1807, 88BCJ3205⟩
Tsubota, T., **1**, 156 ⟨93T537⟩
Tsuboyama, K., **1**, 238 ⟨88TL4105⟩; **5**, 235 ⟨85JOC273⟩, 1025 ⟨89CPB2334⟩
Tsubuki, M., **1**, 558 ⟨93TA21⟩; **3**, 206 ⟨91JCS(P1)954⟩, 222 ⟨90JCS(P1)1733, 90JCS(P1)639⟩, 303 ⟨90JCS(P1)2677⟩
Tsuchida, T., **3**, 716, 717 ⟨87JAN1207⟩
Tsuchihashi, G., **1**, 389 ⟨83TL4997⟩, 408 ⟨80JA2125⟩; **2**, 712 ⟨68TL5071⟩; **3**, 61 ⟨78TL1305⟩, 276 ⟨86TL373⟩; **4**, 74 ⟨70BCJ2271, 71S89⟩, 74, 75 ⟨69TL5259⟩, 74, 77 ⟨71BCJ1726, 70S588⟩, 197 ⟨71TL3151⟩, 221 ⟨74JA4280⟩, 233 ⟨70CC1689⟩, 326, 856, 857 ⟨74JA1960⟩, 326, 856, 857 ⟨78TL375⟩, 856 ⟨72TL1383, 79BCJ2013, 80S736⟩, 856, 857 ⟨74CL659, 79S880⟩; **5**, 552 ⟨80S736⟩
Tsuchihashi, G.-I., **1**, 210 ⟨72TL4605⟩, 1098 ⟨85TL861⟩; **2**, 200, 222 ⟨69TL2743⟩, 202 ⟨82TL5427⟩, 931 ⟨92SL129⟩; **3**, 142, 146 ⟨83TL4997⟩, 160 ⟨87TL3515⟩, 160 ⟨87TL5891⟩; **4**, 233 ⟨72BCJ2023⟩
Tsuchihashi, T., **2**, 398 ⟨87CL191⟩
Tsuchija, H., **2**, 402 ⟨78TL3435⟩
Tsuchisa, T., **2**, 244 ⟨84H(22)1303⟩
Tsuchiya, K., **3**, 494 ⟨70BRP1202353⟩
Tsuchiya, M., **3**, 282 ⟨89BCJ3371⟩
Tsuchiya, R., **2**, 859 ⟨83CL441⟩; **4**, 763 ⟨80MI 417-01⟩
Tsuchiya, R. T., **5**, 118 ⟨B-88MI 502-01⟩
Tsuchiya, T., **1**, 561 ⟨91H(32)2089⟩; **2**, 13 ⟨88TL6287⟩; **4**, 808 ⟨93CC1309, 93CC1817⟩; **5**, 215 ⟨86TL3239⟩, 320, 335 ⟨91BCJ3732⟩, 693 ⟨84CPB4694⟩; **6**, 575 ⟨90JCS(P1)1218⟩
Tsuchiya, Y., **3**, 615 ⟨91CL537⟩
Tsuda, A., **1**, 827 ⟨93CC793⟩
Tsuda, E., **1**, 729 ⟨71JCS(C)3252⟩
Tsuda, H., **2**, 797 ⟨79CL149⟩
Tsuda, K., **2**, 35 ⟨85CC400, 88S403⟩, 540 ⟨B-85MI 208-01⟩, 660 ⟨63JOC783⟩
Tsuda, M., **6**, 533 ⟨91JAP03218336⟩
Tsuda, T., **1**, 86 ⟨86JOC537⟩, 511, 534, 1106 ⟨87JOC1624⟩, 1137 ⟨74CC380⟩; **2**, 228 ⟨70CPB235⟩, 939 ⟨90JOC2978⟩; **3**, 139 ⟨80JA6381⟩, 139 ⟨86JOC421⟩, 214 ⟨90JOC2554⟩; **4**, 696 ⟨81JOC192⟩; **6**, 448 ⟨71JOC858⟩, 464 ⟨68TL831⟩
Tsuda, Y., **1**, 45 ⟨81CPB3489⟩; **3**, 88 ⟨78CL1397⟩; **4**, 90 ⟨85CPB1745, 91CPB1365⟩; **6**, 541 ⟨86CPB430⟩
Tsuetkov, E. N., **2**, 821, 824, 847 ⟨69ZOB841⟩
Tsuge, A., **4**, 273 ⟨92JCS(P1)1179⟩
Tsuge, O., **1**, 372 ⟨87JOC2523⟩, 552 ⟨89AHC(45)231⟩; **2**, 249 ⟨70BCJ3543, 71BCJ2750⟩, 667 ⟨84CL277⟩, 793 ⟨71MI 215-01⟩, 908 ⟨88BCJ3957⟩, 930, 939 ⟨90BCJ51⟩; **3**, 189 ⟨B-89MI 304-03⟩, 200 ⟨87BCJ2463⟩, 269 ⟨89BCJ860⟩, 372 ⟨77H(6)1173⟩, 407, 586 ⟨84JOC2688⟩, 409 ⟨86BCJ1809⟩, 475 ⟨73T1983⟩, 683 ⟨90H(30)719⟩; **4**, 263 ⟨90BCJ466⟩, 302 ⟨87BCJ4079⟩, 307 ⟨67TL3769⟩, 367, 936, 938 ⟨88BCJ3957⟩, 367, 946 ⟨90BCJ51⟩, 422 ⟨79JOC4543⟩, 506, 513 ⟨85BCJ1991⟩, 512, 518 ⟨86BCJ2537⟩, 515 ⟨84H(22)1955⟩, 516 ⟨83CL1131⟩, 517 ⟨84JOC2688⟩, 517, 518 ⟨85H(23)2489⟩, 519 ⟨84CL279⟩, 979 ⟨86JOC1997⟩; **5**, 631 ⟨75H(3)1081⟩, 729 ⟨87JOC2523⟩, 821 ⟨87BCJ2463⟩, 996 ⟨68T2583⟩, 1022 ⟨B-77MI 527-02⟩, 1031 ⟨84JOC2688⟩
Tsuge, S., **3**, 82, 102, 107 ⟨83TL1543⟩
Tsugoshi, T., **5**, 189, 195, 282 ⟨86JOC4150⟩
Tsui, S. K., **1**, 91, 92 ⟨79CJC1977⟩
Tsuji, A., **3**, 32 ⟨82JOC1101⟩
Tsuji, J., **1**, 14 ⟨89JA6280⟩, 130, 889 ⟨65JA275⟩, 192 ⟨82PAC197⟩, 192 ⟨B-80MI 105-01⟩, 193 ⟨84TL5921⟩, 193 ⟨85JOC1523⟩, 196 ⟨84TL5157⟩, 196

⟨85JOM(296)269⟩, 228 ⟨88TL343⟩, 248 ⟨86T4361⟩, 493 ⟨84T2699⟩, 521 ⟨78CL649⟩, 558 ⟨83TL5635, 86T2971⟩, 558, 561 ⟨87S992, 84S1009⟩, 635, 1075 ⟨83T3279⟩, 674 ⟨80TL2955⟩, 816, 818 ⟨86JOC4315, 87H(25)139⟩, 826 ⟨90TL4041⟩, 962 ⟨86CC922, 87S603⟩, 962 ⟨93TL2161⟩, 975 ⟨86TL731, 93JA5865⟩, 975 ⟨93JOM(451)15⟩, 976 ⟨91JOM(417)305, 91SL697⟩, 976 ⟨91TL3397, 91TL7683⟩, 976, 1025 ⟨90TL7179⟩, 1040 ⟨80TL849⟩; **2**, 45 ⟨90JOC766⟩, 616 ⟨78SC103⟩, 622, 627 ⟨66TL4713⟩, 696 ⟨65JA275⟩; **3**, 57, 215 ⟨86TL1805⟩, 58 ⟨77S721, 78CL859⟩, 113 ⟨84S369, 87TL3683⟩, 114 ⟨82TL3085⟩, 118 ⟨85T5645⟩, 122 ⟨84TL2791⟩, 124 ⟨85JOC3416⟩, 139 ⟨80TL3199, 83CL1325, 83TL1793, 83TL4713, 84TL5157⟩, 139 ⟨85JOC1523⟩, 175 ⟨78TL799⟩, 206 ⟨82JA5844, 83TL1797⟩, 206 ⟨83TL5635, 83TL5639, 86TL2791⟩, 213 ⟨83TL3489, 83TL4695, 89JOC4273⟩, 231 ⟨79JA5070⟩, 233 ⟨85JOC537⟩, 431 ⟨87TL4557⟩, 470 ⟨73TL4573⟩, 599 ⟨70JOM(23)275⟩, 623 ⟨87JOC2988⟩, 645 ⟨86CC118⟩; **4**, 193 ⟨84S369⟩, 942 ⟨90T4553⟩, 950 ⟨81JA5259⟩; **5**, 142 ⟨63TL1437⟩, 154 ⟨86T2971⟩, 158 ⟨64TL605, 77BCJ553⟩, 160 ⟨80TL849⟩, 281 ⟨83CL449⟩, 346 ⟨91TL7683⟩, 356 ⟨67JOM(10)511⟩, 419 ⟨80TL1653, 87OM974⟩, 458 ⟨66JA3452⟩, 459 ⟨69S1⟩, 974 ⟨66CC828⟩; **6**, 7 ⟨70JOC2982, 81CL1169⟩, 7 ⟨85T393⟩, 461 ⟨90TL5629, 91JOM(417)305⟩

Tsuji, K., **5**, 558 ⟨84JCS(P1)85⟩

Tsuji, M., **1**, 250 ⟨86JOC4375⟩, 698, 703 ⟨89BCJ3132⟩

Tsuji, S., **5**, 281 ⟨81S287⟩

Tsuji, T., **1**, 564 ⟨91JOC997⟩, 800 ⟨90JOC1506⟩, 842 ⟨72BCJ1777⟩, 1071 ⟨79CL481⟩; **2**, 16 ⟨82TL1165⟩, 88 ⟨83JA2908⟩, 344 ⟨78CL1057⟩, 402 ⟨71JA1992⟩; **3**, 149 ⟨82TL1165⟩, 206 ⟨84CL1133⟩, 620 ⟨63CPB296⟩; **5**, 81 ⟨78GEP2725519⟩; **6**, 641 ⟨64CPB946⟩

Tsuji, Y., **1**, 192 ⟨93JOC16⟩, 196 ⟨85CL1477⟩, 196 ⟨89JOM(369)C51⟩; **2**, 753 ⟨84BCJ2440, 84BCJ2440⟩; **3**, 290 ⟨88JOM(345)397⟩, 405, 426 ⟨90CL635⟩, 622 ⟨93JOC16⟩; **5**, 286 ⟨86JOM(309)333⟩

Tsujiguchi, Y., **1**, 975 ⟨93JA5865⟩

Tsujihara, K., **1**, 621 ⟨71T4921, 71T6101⟩; **6**, 504 ⟨82CPB534⟩

Tsujii, Y., **6**, 9 ⟨80GEP3008081, 91MI 601-02⟩

Tsujikawa, T., **2**, 228 ⟨70CPB235⟩

Tsujimoto, A., **1**, 694 ⟨90JOC2128⟩

Tsujimoto, K., **4**, 850 ⟨85CC1803⟩

Tsujimoto, M., **2**, 650 ⟨89CL1267⟩

Tsujino, Y., **2**, 757 ⟨77BCJ1513⟩

Tsujita, H., **4**, 699 ⟨91T3281⟩

Tsuk, R., **6**, 25 ⟨80JOC5214⟩

Tsukagoshi, F., **4**, 213 ⟨78JOC992⟩

Tsukamoto, A., **3**, 85, 100 ⟨64JA1089⟩

Tsukamoto, G., **5**, 370 ⟨90JMC2192⟩

Tsukamoto, M., **1**, 831 ⟨89JA2735⟩

Tsukamoto, T., **4**, 736 ⟨92SL977⟩

Tsukanaka, M., **3**, 224 ⟨74JA3713, 80BCJ1010⟩

Tsukanaka, T., **1**, 505, 507 ⟨83BCJ353⟩, 515 ⟨85T2387⟩

Tsukano, K., **3**, 357 ⟨92JHC841⟩

Tsukazaki, M., **1**, 429, 1089 ⟨93TL415⟩, 448 ⟨89CPB36⟩; **4**, 768 ⟨93TL411⟩, 841 ⟨93TL415⟩

Tsukervanik, I. P., **1**, 212 ⟨60MI 105-01⟩, 475 ⟨58DOK31, 60DOK(131)27⟩

Tsukida, K., **1**, 613 ⟨93JCS(P1)987⟩; **3**, 63 ⟨90JCS(P1)197⟩

Tsukiyama, K., **1**, 94 ⟨82CL603⟩

Tsukiyama, T., **1**, 1026 ⟨93SL413⟩

Tsukube, H., **2**, 7, 27 ⟨93JOC4382⟩

Tsukui, T., **4**, 657 ⟨86JOM(299)C35⟩

Tsumaki, H., **3**, 417 ⟨82TL3073⟩; **5**, 234, 242 ⟨82CL885⟩, 323 ⟨83SC1053⟩

Tsumori, Y., **4**, 326 ⟨74JOC3412⟩

Tsumuraya, T., **1**, 656 ⟨85TL4523⟩; **2**, 949 ⟨91AG(E)902⟩; **4**, 628, 629 ⟨90OM2061, 88OM1882⟩, 629 ⟨89OM2286⟩

Tsumuruya, T., **2**, 947 ⟨85TL4523, 86TL3251⟩

Tsunekawa, M., **3**, 380 ⟨74JOC3519⟩

Tsunemi, H., **4**, 607, 654 ⟨67JOM(10)111⟩, 616 ⟨69CC207⟩

Tsuneoka, K., **5**, 567 ⟨69CPB550⟩

Tsung, O., **2**, 796 ⟨71BCJ1896⟩

Tsuno, Y., **2**, 14 ⟨83CL1877⟩, 907, 908 ⟨87BCJ777⟩; **4**, 806, 818 ⟨87BSJ777⟩; **5**, 318, 998 ⟨71BCJ1632, 71BCJ1639⟩, 998 ⟨71BCJ1644⟩

Tsunoda, K., **3**, 101 ⟨89BCJ545⟩

Tsunoda, T., **1**, 215 ⟨84TL5781⟩, 835 ⟨90TL727⟩; **2**, 39 ⟨90JA9439⟩, 73 ⟨89CL1647⟩; **4**, 181, 182 ⟨80TL1357⟩, 528 ⟨91TL1975⟩; **6**, 124 ⟨80TL1357⟩

Tsunoi, S., **1**, 936 ⟨93SL143⟩; **5**, 172 ⟨94JA5473⟩

Tsuritani, K., **5**, 292 ⟨81CPB1063⟩

Tsuruda, T., **3**, 31, 157 ⟨86TL3665⟩

Tsurugi, J., **2**, 211 ⟨59JOC807, 70BCJ1890⟩, 211 ⟨72IJS199⟩, 211, 722 ⟨71JOC3677, B-68MI 203-03⟩, 688 ⟨72CL441⟩, 723 ⟨61JOC2483⟩; **5**, 247 ⟨70BCJ1890, 71JOC3677⟩

Tsurumi, H., **6**, 623 ⟨76CPB26⟩

Tsuruoka, M., **1**, 483 ⟨86H(24)3337⟩; **2**, 271 ⟨86H(24)3019⟩

Tsuruoka, T., **3**, 357 ⟨90JAN1150⟩

Tsuruta, T., **4**, 660 ⟨90BCJ3036⟩

Tsusaka, M., **1**, 333 ⟨93JA464⟩, 946 ⟨93JA10464⟩

Tsushima, T., **2**, 16 ⟨82TL1165⟩; **3**, 149 ⟨82TL1165⟩; **4**, 8 ⟨85TL2445⟩; **5**, 588 ⟨75TL3063⟩

Tsutsui, H., **5**, 177 ⟨84TL2163⟩

Tsutsui, M., **2**, 602 ⟨76ACR217⟩; **4**, 719, 726 ⟨75IC78⟩; **5**, 327 ⟨81JHC555⟩

Tsutsui, T., **1**, 990 ⟨82JOC1435⟩

Tsutsumi, H., **5**, 243 ⟨89ZN(B)1050⟩

Tsutsumi, K., **1**, 943 ⟨90JCS(P1)3017⟩; **2**, 143 ⟨90TL3027⟩, 288 ⟨87JOC1397, 92JCR(S)393⟩

Tsutsumi, S., **1**, 1035 ⟨69TL4593⟩, 1143, 1144 ⟨68TL503⟩; **3**, 90 ⟨64BCJ341⟩, 293 ⟨69TL5189⟩; **5**, 273, 286, 296 ⟨71JOC2721⟩, 1052 ⟨72BCJ2937⟩; **6**, 465 ⟨71TL4885⟩, 510 ⟨71JA6344⟩, 593 ⟨72BCJ2937⟩

Tsutumi, Y., **4**, 651 ⟨90TL2925⟩

Tsuyama, K., **4**, 854, 857 ⟨83CL767⟩

Tsuzuki, ,D., **1**, 831 ⟨80JA6891⟩

Tsuzuki, K., **6**, 529 ⟨85JAP62113712, 86JAP62176910, 88JAP01257116, 89JAP03037110, 89JAP03060417, 89JAP03060418, 89JAP03197310, 91JAP03060419⟩, 533 ⟨85JAP61229860, 85JAP61229861, 85JAP62120358, 86GEP3616009, 87JAP01075462, 87JAP01102058, 88JAP01301654, 88JAP02009858, 90JAP03220173, 90JAP04066566, 90JAP04128263, 90JAP05032616, 91JAP03193759, 93JAP05058990⟩, 537, 538 ⟨85JAP61233662⟩, 537, 538 ⟨85JAP61233663⟩

Tsuzuki, Y., **4**, 661 ⟨91SL911, 92JCS(P1)2813⟩

Tsvetkov, E. N., **2**, 430, 437 ⟨86S198⟩, 450 ⟨60DOK(135)323, 63IZV1227⟩, 451 ⟨57DOK(117)817⟩, 678 ⟨84URP1035028⟩, 838, 839, 851 ⟨62ZOB3351⟩, 848 ⟨68ZOB2285⟩; **4**, 113, 120 ⟨70JGU255⟩, 118, 132, 133 ⟨69JGU1490⟩, 119 ⟨86JGU1430, 86JGU501, 90JGU1351⟩, 119, 120, 121 ⟨80IZV491⟩, 120 ⟨62DOK(143)211⟩, 121, 122 ⟨88JGU465⟩, 124, 126, 128 ⟨89JGU1778⟩, 124, 128 ⟨92JGU39⟩, 126 ⟨89IZV850⟩, 134 ⟨90TL5361⟩, 339 ⟨73ZOB534, 84PS(21)59⟩, 546, 553 ⟨86S198⟩, 552 ⟨91S125⟩, 555 ⟨70ZOB285⟩

Tsvunin, V. S., **4**, 911 ⟨68ZOB1523⟩

Tsyban, A. V., **1**, 1030 ⟨80S904⟩

Tsymbal, I. F., **5**, 775 ⟨84ZOR1784⟩

Tsypina, G. M., **2**, 460, 851 ⟨62ZOB4017⟩

Tsypyshev, O. Y., **1**, 469 ⟨89IZV2623⟩

Tsyzin, B. M., **2**, 879 ⟨60JGU3203⟩

Tu, C., **2**, 86 ⟨93SL145⟩

Tu, C. J., **2**, 28 ⟨87JOC5624⟩

Tu, J., **5**, 706 ⟨85JHC1435⟩

Tu, S., **1**, 25, 630 ⟨87JOC3346⟩; **3**, 215 ⟨87JOC3346⟩

Tubergen, M. W., **2**, 4, 23, 30 ⟨90JA7431⟩, 4, 23, 30 ⟨93JA3071⟩

Tubéry, F., **1**, 833 ⟨87TL6461⟩, 834 ⟨90TL523⟩; **3**, 355 ⟨87TL6461⟩

Tubis, M., **4**, 420, 424, 449 ⟨64JA1701⟩
Tuček, E., **5**, 1054 ⟨67CB1367⟩; **6**, 596 ⟨67CB1459⟩, 596, 598 ⟨67CB1373⟩
Tuck, B., **5**, 401 ⟨89JCS(P1)1319⟩; **6**, 117 ⟨89JCS(P1)1319⟩
Tuck, D. G., **2**, 594 ⟨79PAC2005⟩, 597 ⟨82COMC-I(1)683⟩; **4**, 153, 588 ⟨91OM2159⟩, 153, 698 ⟨86OM525⟩
Tuck, R. D., **2**, 1008 ⟨92CJC2040⟩
Tucker, B., **2**, 25 ⟨83JOC2625⟩, 726 ⟨66JOC2658⟩; **5**, 968 ⟨69JOC3200⟩, 977 ⟨67AG(E)636, 68JOC2887⟩, 981 ⟨66JOC2658⟩, 995 ⟨67JOC4052⟩, 1079 ⟨67TL1731⟩; **6**, 143 ⟨68JOC3928⟩, 445 ⟨81JOC5226⟩, 448 ⟨84JOC3675⟩, 511 ⟨78JOC1544⟩, 652 ⟨68AG(E)291⟩
Tucker, B. W., **6**, 509 ⟨79USP4154931⟩
Tucker, C. E., **1**, 486 ⟨90JOC5446⟩, 991 ⟨91JA9888, 94OM94⟩
Tucker, C. E., **2**, 69 ⟨90TL7575⟩, 73 ⟨92JOC6384⟩, 591 ⟨93JA7027⟩, 900 ⟨92JOC3482⟩, 1096 ⟨93TL4623⟩; **3**, 614, 646, 665, 670, 675 ⟨93TL4623⟩; **4**, 153 ⟨92JOM(438)11⟩, 709 ⟨90JOC5446, 93S530⟩, 723 ⟨94OM94⟩
Tucker, J. N., **5**, 332 ⟨71CC1609⟩
Tucker, J. R., **3**, 519 ⟨83JA258⟩; **4**, 260 ⟨76JOC3975⟩; **5**, 937 ⟨83JA258⟩; **6**, 128 ⟨76JOC3975⟩
Tucker, P. M., **6**, 181, 269 ⟨69JCS(A)1882⟩
Tucker, S. H., **5**, 293 ⟨46JCS454⟩
Tucker, T. J., **4**, 279 ⟨90TL6819⟩
Tucker, W. P., **3**, 357 ⟨67JOC3140⟩; **4**, 223 ⟨62JA4789⟩; **6**, 301, 302 ⟨68JOC3333⟩, 729 ⟨67TL2747⟩
Tückmantel, W., **2**, 903, 919 ⟨86TL2007⟩, 918 ⟨85TL4629, 86CB1581⟩, 918, 923 ⟨86T4427⟩; **3**, 725 ⟨91SL654⟩; **5**, 1052 ⟨91SL654⟩
Tueting, D., **1**, 228, 479 ⟨89T979⟩; **2**, 86 ⟨88JA4039, 89T979⟩, 439 ⟨87JOM(320)339⟩; **3**, 211, 288 ⟨83JOC4634⟩
Tufariello, J. J., **2**, 350 ⟨76TL4037, 79JA7114⟩, 524 ⟨66JA4757⟩; **3**, 433 ⟨80JA373⟩; **4**, 206 ⟨66TL6145⟩, 832 ⟨67JA6804⟩
Tukada, H., **3**, 519 ⟨85JA2924⟩, 520 ⟨82JA3761⟩; **6**, 594 ⟨86BCJ1741⟩
Tuladhar, S. M., **1**, 900 ⟨90JA3228⟩; **2**, 93, 138 ⟨87TL523⟩
Tulecki, J., **6**, 546 ⟨85MI 617-03⟩
Tuleen, D. L., **4**, 63, 69 ⟨67JOC204⟩, 66 ⟨66CI(L)1555, 69JHC115⟩, 66, 67, 68, 69 ⟨69JOC31⟩
Tulegenov, S., **6**, 88 ⟨88MI 601-01⟩
Tuley, W. F., **5**, 34 ⟨32OSC(1)289⟩, 115 ⟨55OSC(3)822⟩
Tulich, L., **2**, 4, 34 ⟨81JOC1821⟩
Tulimowski, Z., **2**, 461 ⟨62RZC1781⟩
Tulip, T. H., **2**, 979 ⟨84OM1223⟩
Tullar, B. F., **2**, 27 ⟨63OSC(4)921⟩
Tull, R., **5**, 579, 580 ⟨69AG(E)278⟩
Tull, R. J., **6**, 506 ⟨68USP3539569⟩
Tuller, F. N., **1**, 871 ⟨69JOC3070⟩
Tulley, A., **5**, 570 ⟨64JCS1190⟩
Tulloch, I. F., **5**, 1120 ⟨87JMC951⟩
Tullock, C. W., **2**, 19 ⟨74OR(21)1⟩, 169 ⟨60JOC2016⟩, 855 ⟨60JOC2015⟩; **3**, 419 ⟨62JA2337⟩; **4**, 8 ⟨74OR(21)1⟩, 130 ⟨60JOC2016⟩; **5**, 675, 975 ⟨64JA357⟩, 808 ⟨60JOC2016⟩, 808, 989 ⟨62JA2337⟩, 1003 ⟨62JA4275⟩; **6**, 52, 53, 238, 259, 409, 423, 438 ⟨62JA4275⟩, 236 ⟨60JOC2016⟩, 241 ⟨64USP3121112⟩, 242 ⟨62JA2337⟩, 408, 410 ⟨60IS155⟩
Tully, C. R., **4**, 647 ⟨79JOM(172)11⟩
Tuloup, R., **4**, 109 ⟨89CJC1125⟩
Tumanski, B. L., **6**, 537 ⟨81URP804634⟩
Tumas, W., **1**, 675 ⟨89SCI907⟩, 883 ⟨88JA2714⟩
Tumey, M. L., **3**, 502 ⟨76JA7844⟩
Tumkevicius, S., **5**, 652 ⟨88KGS1559⟩, 690 ⟨87CHE907, 87KGS1131, 88CHE1297⟩
Tuncay, A., **1**, 318 ⟨75JA7193, 76JA1997⟩; **3**, 163 ⟨92TL7647⟩; **5**, 43 ⟨73JOC1491⟩
Tuncay, C. I., **3**, 163 ⟨92TL7647⟩
Tüncher, W., **1**, 443 ⟨77S339⟩; **2**, 403 ⟨77S339⟩; **5**, 366 ⟨77S339⟩
Tundo, P., **2**, 14 ⟨91S1003⟩, 25, 31 ⟨87JCS(P1)2157⟩, 35 ⟨76S516, 79S952⟩, 35 ⟨79TL5055⟩; **6**, 50 ⟨74JHC943⟩
Tune, D. J., **4**, 693 ⟨77CB1763⟩
Tunemoto, D., **1**, 144 ⟨75TL1007⟩, 159 ⟨74AG(E)407⟩, 299, 354 ⟨77TL109⟩; **2**, 1000 ⟨67T4279⟩; **3**, 56 ⟨75TL1007⟩, 373 ⟨67T4279⟩; **4**, 347 ⟨74JCS(P1)1279⟩
Tung, C. C., **3**, 88 ⟨50JA2992⟩; **4**, 102 ⟨87JHC945⟩; **6**, 445 ⟨87JHC945⟩
Tung, R. D., **5**, 280 ⟨85JA4342⟩
Tunggal, B. D., **5**, 403, 404 ⟨72CB3357⟩
Tun-kyi, A., **2**, 229 ⟨79HCA2507⟩
Tunstad, L. M., **2**, 89 ⟨88PS(37)25⟩
Tuong, H. B., **1**, 721 ⟨75C341⟩
Tuong, T. D., **1**, 1038 ⟨81JOM(205)4⟩
Tupchienko, S. K., **5**, 776 ⟨85ZOB776⟩
Tuppy, H., **5**, 652 ⟨59N35⟩
Turaiello, J. J.,**3**, 431 ⟨B-84MI 311-01⟩
Turbeville, M. J., **5**, 949 ⟨91OM3951⟩
Turco, A., **5**, 1138 ⟨84JOM(276)431⟩
Turconi, M., **6**, 425 ⟨90JMC2101⟩
Turcotte, J. G., **6**, 489 ⟨90AX2374⟩
Tureček, F., **2**, 636 ⟨88JA7984⟩
Turkel, R. M., **4**, 152 ⟨69JA5027⟩
Turkel'taub, G. N., **6**, 626 ⟨91ZOB1202⟩
Turley, P. C., **4**, 162 ⟨74FCF73⟩
Turley, R., **2**, 381, 382 ⟨73ZN(B)829, 73CB1052⟩, 387 ⟨71AG(E)813⟩
Turnbloom, E. W., **2**, 475 ⟨71JA4065⟩
Turnbull, J. H., **2**, 125 ⟨64T1265⟩, 413 ⟨58JCS1963⟩; **5**, 238 ⟨70CC1310⟩
Turnbull, K., **6**, 453 ⟨93TL239⟩
Turnbull, M. M., **2**, 577 ⟨91JOM(419)245⟩; **6**, 120 ⟨78S309⟩
Turnbull, Jr., S. P., **3**, 311 ⟨91S897⟩
Turner, A. B., **1**, 555 ⟨67JCS(C)1720⟩; **3**, 226 ⟨67JCS1720⟩
Turner, A. F., **4**, 56, 57 ⟨64JCS766⟩
Turner, A. T., **2**, 19 ⟨86JCR(S)300⟩
Turner, C., **5**, 1108 ⟨53JCS959⟩
Turner, E. E., **1**, 117 ⟨20MI 104-01⟩; **2**, 875 ⟨23JCS2489⟩
Turner, E. T., **2**, 528 ⟨91JOC751⟩
Turner, G. K., **4**, 693 ⟨75CC615, 77JOM(129)429⟩
Turner, H. A., **5**, 748 ⟨53MI 519-01⟩, 751, 753 ⟨49JCS1308⟩
Turner, H. S., **1**, 442 ⟨42JCS748, 43JCS86⟩
Turner, H. W., **2**, 603 ⟨81IC2991⟩; **3**, 512 ⟨80JA6608⟩
Turner, I., **4**, 413 ⟨93TL6329⟩
Turner, J. O., **1**, 894 ⟨65JA3000⟩, 912 ⟨65JA2153⟩, 912, 913 ⟨65JA4533⟩
Turner, J. V., **1**, 823 ⟨73JOC2915⟩; **2**, 699 ⟨84TL5099⟩
Turner, K. A., **3**, 113 ⟨93CC954⟩
Turner, K., **6**, 721 ⟨72CC851, 74JCS(D)1591⟩, 721 ⟨75CC929, 78JCS(D)348⟩
Turner, L., **3**, 86 ⟨56JCS1686⟩
Turner, N. J., **2**, 38 ⟨94MI 202-02⟩; **5**, 118 ⟨90TL7223⟩, 262 ⟨89MI 506-03⟩
Turner, R. A., **5**, 363, 364 ⟨47JA875⟩
Turner, R. J., **5**, 591 ⟨61JOC3002⟩
Turner, R. M., **4**, 554 ⟨90TL5381⟩
Turner, R. W., **2**, 163 ⟨69TL477⟩, 667 ⟨83CC466⟩; **5**, 292 ⟨77S31, 77S33⟩, 1118, 1124 ⟨75S332⟩
Turner, S., **5**, 142 ⟨73JCS(P1)2796⟩
Turner, S. E., **4**, 834 ⟨83TL3373⟩
Turner, W., **5**, 559 ⟨67T3785⟩
Turner, W. E. S., **5**, 24 ⟨05JCS747⟩
Turner, W. V., **6**, 231, 232 ⟨85AG48⟩
Turner Jones, E. T., **2**, 528 ⟨91JOC763⟩
Turney, T. W., **4**, 688 ⟨76IC90, 76IC97⟩, 689 ⟨76IC107⟩
Turocy, G., **5**, 367 ⟨89JOC145⟩
Turos, E., **1**, 831 ⟨89JA8231⟩; **2**, 73 ⟨91TL3001⟩; **5**, 329 ⟨82SC989⟩
Turov, A. V., **5**, 1069 ⟨88ZOR986⟩

Turovskii, A. A., **5**, 530 ⟨82MI 512-01⟩
Turovskii, N. A., **5**, 530 ⟨82MI 512-01⟩
Turowsky, L., **6**, 275 ⟨88IC2135⟩
Turpin, J., **4**, 586 ⟨87JOM(331)409, 87JOM(333)129⟩
Turro, N. J., **3**, 145 ⟨86TL4671⟩, 538 ⟨67TL545⟩, 679 ⟨84JOC3174⟩, 679, 680 ⟨86JA6739⟩; **4**, 162 ⟨66JA3672⟩, 406 ⟨67TL3085⟩; **5**, 728 ⟨85JOC4415⟩
Tursch, B., **2**, 130 ⟨62JOC1041⟩
Tuseev, A. P., **2**, 410 ⟨80ZOB1702⟩
Tustin, G. J., **5**, 425 ⟨92T7629⟩
Tuszynski, W. J., **1**, 905 ⟨82TL3119⟩
Tuthill, P. A., **1**, 606 ⟨84CC1345⟩
Tutkunkardes, S., **2**, 173 ⟨74CB508⟩; **5**, 995 ⟨74CB508⟩
Tuzhikov, O. I., **2**, 427 ⟨91ZOB1021⟩, 428 ⟨90ZOB697⟩, 440 ⟨85URP1145022⟩
Tuzikov, A. B., **3**, 214 ⟨88JOM(348)123⟩; **4**, 642 ⟨80IZV2429⟩
Tuzun, C., **5**, 664, 821 ⟨89OPP517⟩
Tverdokhlebov, V. P., **6**, 145 ⟨86ZOR1393⟩
Tvorogov, A. N., **2**, 700 ⟨68JOM(14)229⟩
Twanmoh, L.-M., **4**, 420 ⟨72JHC1145⟩
Twelves, R. R., **4**, 32 ⟨57JOC730⟩; **6**, 2 ⟨63USP3076041⟩
Twigg, M. V., **1**, 212 ⟨B-91MI 105-02⟩
Twiss, D. F., **2**, 122 ⟨09JCS1725⟩, 227 ⟨41JA658⟩, 237 ⟨14JCS1672⟩
Twiss, P., **2**, 1003, 1005 ⟨89JCS(D)105, 84JCS(D)321⟩; **4**, 604, 646, 647 ⟨84JCS(D)321⟩, 604, 647 ⟨88JCS(D)2403⟩, 604, 647, 648 ⟨90JCS(D)1161⟩, 646 ⟨89JCS(D)105⟩, 660 ⟨84JCS(D)331⟩
Twitchin, B., **2**, 699 ⟨84TL5099⟩
Twong, H. B., **2**, 447 ⟨83C10⟩
Tyagi, B. S., **2**, 660 ⟨62JOC1430⟩
Tyczkowski, E. A., **4**, 3 ⟨58JA6533⟩
Tye, F. L., **1**, 895 ⟨52JCS2167⟩, 895, 896 ⟨52JCS2172⟩
Tyers, K. G., **4**, 713 ⟨80IC693, 82CC371⟩
Tyka, R., **2**, 454 ⟨88SC425⟩; **4**, 461 ⟨78PJC321, 78PJC959⟩, 461, 463 ⟨77S239⟩, 465 ⟨88SC425⟩, 467 ⟨78PJC1949⟩, 487 ⟨77S571⟩
Tykwinski, R., **2**, 445, 1021 ⟨92JOC1861, 92TL1419⟩, 1021 ⟨93JA11702⟩, 1033 ⟨93T3043⟩, 1036 ⟨93JOC5235⟩
Tyler, C. J., **5**, 62 ⟨B-52MI 502-04⟩
Tyler, D. C., **4**, 712 ⟨84JCS(D)2553⟩
Tyler, J. W., **5**, 223 ⟨85JCS(P1)1927⟩
Tyler, L. J., **2**, 538 ⟨54JA1030⟩
Tyler, P. C., **4**, 235 ⟨77CAR(58)397⟩
Tyler, S. J., **4**, 689 ⟨89JOM(361)231⟩
Tyman, J. H. P., **5**, 110 ⟨73TL4839, 79JCS(P1)2079⟩, 110 ⟨80JCS(P1)1658⟩
Tyndall, G. S., **4**, 42 ⟨93JPC1576⟩
Tyrala, A., **1**, 613, 622 ⟨87S1142⟩
Tyree, Jr., S. Y., **2**, 460 ⟨59JA6177⟩
Tyrlik, S., **1**, 687, 688 ⟨73BSF2147⟩
Tyrra, W., **2**, 483, 873 ⟨87JOM(334)323⟩; **5**, 205, 321 ⟨89JFC(45)401⟩; **6**, 49 ⟨94AG(E)323⟩, 220 ⟨89JFC(45)401⟩, 220, 246 ⟨91CJC327⟩, 243 ⟨87JOM(334)323⟩, 243 ⟨91JOM(417)C47⟩, 245 ⟨91JOM(407)1, 89JOM(368)131⟩
Tyrrell, A. W. R., **5**, 334 ⟨85JMC1661⟩
Tyrrell, E., **1**, 199 ⟨93SL769⟩
Tyrrell, N. D., **1**, 38 ⟨79CC345⟩, 945 ⟨81CC901⟩
Tyryshkin, N. J., **4**, 922 ⟨92HAC127⟩
Tyssee, D. A., **2**, 807 ⟨73JA8066⟩
Tytko, S. F., **6**, 87 ⟨87TL4153⟩
Tyukhteneva, Z. I., **5**, 350 ⟨92ZOR568⟩
Tyuleneva, V. V., **3**, 495 ⟨81RCR280, 82RCR1⟩; **4**, 970 ⟨71DOK(201)1359⟩; **6**, 694 ⟨82RCR1⟩
Tyurikov, V. A., **4**, 312 ⟨73TL1889⟩; **6**, 469 ⟨82JGU131⟩
Tyurin, A. Yu., **5**, 680 ⟨91MI 517-01⟩
Tyurin, V. S., **6**, 12 ⟨69IZV196⟩
Tzai, L., **4**, 973 ⟨86CB2208⟩
Tzeng, D., **4**, 795 ⟨91JOC537⟩, 994 ⟨85JOC2736⟩; **5**, 967 ⟨91JOC3043⟩
Tzeng, Y.-L., **2**, 936 ⟨90JA9356⟩
Tzirenstchikow, T., **6**, 444 ⟨89JHC231⟩
Tzodikov, N. R., **4**, 948 ⟨76JOC2312, 76JOC2874⟩
Tzschach, A., **2**, 253, 260 ⟨82TL1345⟩, 414, 482, 503 ⟨64ZAAC(326)280⟩, 427, 437 ⟨59CB1118⟩, 431 ⟨59CB704, 60CB1852⟩, 440 ⟨84ZC365⟩, 482, 489 ⟨65ZAAC(336)36⟩, 482, 510 ⟨67LA(709)248⟩, 485, 510 ⟨65CB977⟩, 486, 487, 510 ⟨70ZC118⟩, 486, 506, 510 ⟨61AG26⟩, 507 ⟨71JPR259⟩, 872 ⟨80JOM(186)39⟩, 876 ⟨71JPR254⟩, 890 ⟨70JOM(21)231⟩; **4**, 317, 499, 500 ⟨81ZC403⟩, 451, 452, 455, 481 ⟨84ZC365⟩, 454, 495, 496, 501, 502 ⟨74JOM(81)187⟩, 455 ⟨80JOM(193)307⟩, 455, 462, 478, 481, 499, 500, 501 ⟨78JOM(149)167⟩, 495, 496 ⟨70JOM(21)131, 76ZC64⟩, 495, 497 ⟨73EGP99803⟩, 496, 498, 499 ⟨73JOM(60)95⟩, 497, 498, 499, 500 ⟨85ZC369⟩, 551, 570, 588 ⟨79JOM(182)465⟩, 570 ⟨87EGP251134⟩, 570 ⟨87EGP251136, 80EGP142887⟩, 700 ⟨84JOM(277)335⟩; **5**, 390 ⟨79JOM(182)465⟩, 791 ⟨78ZC452, 80JOM(186)39⟩, 791, 892 ⟨79JOM(166)175⟩; **6**, 119 ⟨83TL5481⟩, 161 ⟨78JOM(149)167, 78ZC452⟩, 513 ⟨67LA(709)248⟩, 657 ⟨68JOM(13)363⟩
Tzschach, V. A., **4**, 483, 499, 500 ⟨74JPR851⟩, 495, 496, 497 ⟨73JPR65⟩; **5**, 790 ⟨76JPR409⟩

U

Uang, B.-J., **1**, 343 ⟨82JOC2231⟩; **3**, 62 ⟨90S1033⟩; **4**, 230 ⟨90TL257⟩, 232 ⟨90TA143⟩
Ubbels, H. J. C., **1**, 127 ⟨82JA6609, 82OM1366⟩
Ubertini, F. M., **4**, 47, 48 ⟨50JOC715⟩
Ubukata, M., **5**, 342 ⟨86TL3907⟩
Ucan, H. I., **5**, 664 ⟨90SRI437⟩
Uccella, N., **2**, 350 ⟨83JHC1207⟩; **3**, 618 ⟨87S168⟩
Ucciardi, T., **6**, 38, 43 ⟨72JMC604⟩, 44, 45, 46 ⟨65JOC4011⟩
Ucciardi, T. R., **6**, 36 ⟨71JHC557⟩
Ucciardi, T. U., **6**, 36, 43 ⟨71JMC593⟩
Uchibori, Y., **2**, 18 ⟨93CL673⟩, 19 ⟨91CL1185⟩
Uchida, A., **1**, 538 ⟨68T4285⟩, 1035 ⟨72JOC3749⟩; **3**, 253 ⟨68T4285⟩; **4**, 863 ⟨92MI 419-01⟩
Uchida, I., **1**, 217 ⟨86TL2015⟩
Uchida, K., **1**, 161, 1094 ⟨76JOC2941⟩, 161, 450, 1106 ⟨76JOC2215⟩, 207, 1100 ⟨77T1945⟩, 511 ⟨77TL3641, 77T2987⟩; **2**, 323 ⟨88BCJ3029⟩, 922 ⟨76JOC2215, 76JOC2941⟩, 922 ⟨77T2987⟩; **3**, 219 ⟨84JOC300⟩
Uchida, M., **1**, 600 ⟨72TL649⟩; **3**, 137 ⟨89BCJ2728⟩
Uchida, N., **6**, 554 ⟨82BCJ1174⟩
Uchida, S., **2**, 518 ⟨91TL1479⟩
Uchida, T., **1**, 490, 535 ⟨82OM542⟩; **2**, 753 ⟨87CE551⟩, 986, 988 ⟨89CL761⟩; **5**, 362 ⟨80BCJ3225⟩
Uchida, Y., **1**, 140 ⟨B-88MI 104-01⟩, 704 ⟨B-88MI 115-02⟩; **2**, 986, 988 ⟨89CL761⟩, 987 ⟨89CC852, 90CC182⟩; **3**, 170 ⟨B-88MI 304-03⟩; **4**, 390 ⟨B-88MI 408-01⟩; **5**, 143 ⟨75BCJ2075⟩, 334, 336 ⟨81CC510⟩, 463 ⟨73BCJ2820⟩, 666 ⟨84BCJ2011⟩; **6**, 456 ⟨74NKK732⟩
Uchikawa, M., **1**, 404, 406, 816, 819 ⟨86TL4581⟩, 816 ⟨86TL4577⟩
Uchimaru, T., **4**, 235 ⟨84JA2954⟩; **5**, 173 ⟨84JA2954⟩
Uchimaru, Y., **2**, 945 ⟨93OM2065⟩; **4**, 618 ⟨92JOM(428)1⟩
Uchimoto, T., **2**, 532 ⟨77CL153⟩
Uchimura, J., **2**, 41 ⟨89CC1429⟩
Uchino, M., **2**, 271 ⟨77S795, 78CPB1837⟩
Uchino, N., **1**, 468 ⟨93BCJ2446⟩
Uchiro, H., **5**, 236 ⟨89CL1001, 90CL1019⟩
Uchiyama, A., **6**, 533, 540 ⟨62JOC4509⟩
Uchiyama, D., **1**, 1009 ⟨92S746⟩
Uchiyama, H., **1**, 651 ⟨83CC921⟩; **2**, 936 ⟨83CC921⟩; **4**, 21 ⟨78CL73⟩
Uchiyama, K., **2**, 321 ⟨90CC1385⟩, 715 ⟨84S323⟩
Uchiyama, M., **1**, 411 ⟨65JPC1278⟩
Uchiyama, T., **1**, 1004, 1107 ⟨79JOC55⟩; **2**, 616 ⟨76CL449⟩
Uda, H., **1**, 209 ⟨74CL385⟩; **2**, 713 ⟨85CL805, 87JOC1043⟩; **3**, 54 ⟨90JCS(P1)1901⟩, 210, 256 ⟨82CL815⟩, 250, 256 ⟨79CL785⟩; **4**, 956, 957 ⟨79CL785⟩
Uda, M., **6**, 563, 565 ⟨82JMC557⟩
Udachin, Y. M., **3**, 361 ⟨80ZOR443, 80ZOR883⟩, 361, 376 ⟨78ZOR2459⟩
Udayakumar, B. S., **4**, 1054 ⟨89JOM(378)1⟩
Udding, J. H., **4**, 327 ⟨94JOC3721⟩
Uddrich, P., **6**, 304 ⟨65LA(689)179⟩
Ude, H., **1**, 599 ⟨76SC357⟩
Udenfriend, S., **2**, 650 ⟨54JBC(208)731⟩
Ueamura, S., **2**, 597 ⟨76BCJ2762⟩, 598 ⟨B-87MI 211-06⟩
Uebel, R., **5**, 1074 ⟨89ZAAC(568)35⟩
Ueda, C., **2**, 219, 227 ⟨84CC843⟩
Ueda, H., **3**, 706 ⟨85ABC1761⟩; **4**, 368 ⟨92TL85⟩, 686 ⟨77BCJ2250⟩; **5**, 138 ⟨87ABC1345⟩
Ueda, I., **5**, 208 ⟨91CL61⟩
Ueda, K., **1**, 1137 ⟨74CC380⟩; **2**, 657 ⟨87H(26)65⟩; **3**, 151 ⟨81JOC509⟩; **6**, 229, 230 ⟨91CL1421⟩, 636 ⟨84CHEC(3)91⟩
Ueda, M., **1**, 141 ⟨85JOC657⟩, 826 ⟨90TL4041⟩; **2**, 715 ⟨84S323⟩; **3**, 28 ⟨91CL1469⟩; **4**, 235, 236 ⟨86CL1655⟩; **5**, 126, 280, 300 ⟨85JOC760⟩, 278, 291, 303 ⟨81S991, 82S933⟩, 303 ⟨92BCJ1636⟩; **6**, 98, 101, 132, 134 ⟨81TL2005⟩, 101 ⟨85JOC657⟩, 119, 126, 132 ⟨85JOC662⟩, 417 ⟨83S908⟩
Ueda, N., **4**, 232 ⟨84TL757⟩
Ueda, S., **6**, 4 ⟨92EUP473105⟩
Ueda, T., **1**, 456, 507 ⟨83JA6273⟩; **5**, 340 ⟨88CPB877⟩, 1054 ⟨90MI 527-01⟩; **6**, 46, 47 ⟨66BCJ2191⟩, 641 ⟨64CPB946⟩
Ueda, Y., **3**, 626 ⟨82CPB3574⟩; **4**, 76 ⟨88TL313⟩; **5**, 193 ⟨86CZ377⟩, 733 ⟨93TL7933⟩, 1106 ⟨63MI 530-01⟩
Uedelhoven, W., **5**, 919 ⟨78CB3283⟩; **6**, 720, 722 ⟨76JOM(113)C31⟩
Uehara, A., **2**, 859 ⟨83CL441⟩; **4**, 763 ⟨80MI 417-01⟩
Uehara, K., **3**, 263 ⟨84S663⟩
Uehara, S., **5**, 1065 ⟨79BCJ533⟩
Uehling, D. E., **1**, 1079 ⟨89JOC4267⟩
Ueki, M., **2**, 252 ⟨79BCJ1431, 83BCJ1187⟩; **5**, 221 ⟨88BCJ3653⟩
Uem, K., **5**, 54 ⟨76CZ3⟩
Uematsu, T., **1**, 165 ⟨85T2751⟩
Uemura, J., **5**, 733 ⟨93TL7933⟩
Uemura, M., **1**, 195 ⟨89JOC469⟩, 202 ⟨85TL767, 86JOM(299)119⟩, 202 ⟨86S386, 86TL2479⟩, 202 ⟨86TL967⟩, 202, 203 ⟨86JOC2859, B-91MI 105-01⟩; **5**, 652 ⟨90TL3603⟩
Uemura, N., **4**, 67 ⟨93JOC2360⟩
Uemura, S., **1**, 66, 454 ⟨80JCS(P1)1098⟩, 84 ⟨76BCJ1437⟩, 180 ⟨86BCJ3617⟩, 246 ⟨86JCS(P1)1983⟩, 437, 545 ⟨82TL1181⟩, 452 ⟨B-87MI 110-04⟩, 454 ⟨74BCJ2663, 81JCS(P1)991⟩, 493 ⟨80BCJ553, 87JOC4859⟩, 602 ⟨84JOC3796, 85JCS(P1)373⟩, 602, 781 ⟨80TL5037⟩, 603 ⟨81JOC4727, 83JOC5246⟩, 604 ⟨82TL2105⟩, 609, 983 ⟨92CC46⟩, 611 ⟨82TL1177, 83JOM(250)203, 85TL895, 85TL921, 90CE261⟩, 611, 620, 716 ⟨83TL4347⟩, 611, 716 ⟨83JA2748, 85CC1037⟩, 620 ⟨85JCS(P1)471⟩, 983 ⟨93JOC3697⟩; **2**, 24 ⟨74BCJ143, 74BCJ692, 74BCJ3121, 80BCJ1390⟩, 144, 723, 733, 734 ⟨91COS(7)757⟩, 281 ⟨80CC412⟩, 282 ⟨78JCS(P1)1273, 79JCS(P1)1206, 80JOC1953, 81JOC3021, 82TL2105, 84JOC3796, 85JCS(P1)373⟩, 284 ⟨81JOC4727⟩, 286 ⟨81CC546⟩, 287 ⟨83CC1501, 83JA2748⟩, 287 ⟨84CC426⟩, 287, 288 ⟨85JCS(P1)471, 86JCS(P1)1983⟩, 617 ⟨77JCS(P1)676, 81BCJ2843⟩, 617, 629 ⟨75CC925⟩, 624, 629 ⟨79CL1357⟩, 631 ⟨74BCJ147⟩, 633 ⟨83JOC3297⟩; **3**, 37, 174 ⟨84JOC3796⟩, 146 ⟨90JCS(P1)1697⟩, 174 ⟨82TL2105⟩, 176 ⟨86JCS(P1)1983⟩, 289 ⟨93JOM(443)253⟩, 290 ⟨92CC453⟩, 664 ⟨72T3025⟩; **4**, 330 ⟨91COS(7)762⟩, 1068 ⟨81JCS(P1)991⟩; **5**, 241 ⟨87JOM(334)C45⟩, 242 ⟨87JOM(334)C43, 89JOM(361)63⟩, 251 ⟨77BCJ220⟩, 282 ⟨81CC546⟩, 282, 300 ⟨81JOC4727⟩, 285 ⟨86JCS(P1)1837, 86JOC1724, 87JOC2018⟩, 1032, 1056 ⟨77BCJ722⟩, 1036 ⟨77BCJ2785⟩, 1042 ⟨75BCJ3205, 79BCJ3611⟩, 1051 ⟨B-87MI 527-02⟩, 1053 ⟨78BCJ1818⟩, 1105 ⟨74BCJ2745⟩, 1113 ⟨80BCJ553⟩, 1113, 1115 ⟨75BCJ1925, B-87MI 530-01⟩, 1115 ⟨78JCS(P1)1273⟩
Ueng, S., **3**, 707, 714 ⟨79JOC2042⟩
Uenishi, J., **1**, 14 ⟨82CC1292⟩, 489 ⟨87JA4756⟩, 677, 1038, 1099 ⟨86JA5644⟩; **2**, 7, 27 ⟨93JOC4382⟩; **4**, 227 ⟨91CC1421⟩
Uenishi, J. I., **4**, 67 ⟨81S534⟩

Uenishi, K., **1**, 782 ⟨94LA41⟩
Ueno, K., **2**, 266 ⟨83TL4033⟩, 1008 ⟨85H(23)2709⟩; **3**, 113 ⟨84TL961⟩, 409 ⟨86BCJ1809⟩; **4**, 25 ⟨84JA8174⟩, 1075 ⟨91BCJ2751⟩
Ueno, M., **1**, 5, 8 ⟨86S570⟩; **3**, 129 ⟨86S570⟩; **4**, 1069 ⟨93CL1881⟩
Ueno, R., **6**, 558 ⟨93TL3745⟩
Ueno, T., **1**, 243 ⟨88JA4815⟩
Ueno, Y., **1**, 243 ⟨88JA4815, 90TL6669⟩, 333 ⟨93JA464⟩, 594, 668 ⟨92JOC3145⟩, 603 ⟨88JCS(P1)575⟩, 615, 698 ⟨81TL2675⟩, 667 ⟨80S1011⟩, 936 ⟨82JA5564⟩, 942 ⟨82TL443⟩, 946 ⟨93JA10464⟩, 1127, 1130, 1131 ⟨84CL1437⟩, 1130 ⟨82TL2575⟩; **2**, 131 ⟨84TL2675⟩, 156 ⟨80BCJ3615⟩, 188 ⟨84CL2125⟩, 210 ⟨86TL215⟩, 256 ⟨81S547⟩, 285 ⟨88JCS(P1)575, 89JCS(P1)1927⟩, 287 ⟨77CL835⟩, 362 ⟨77TL2413⟩, 369 ⟨92JCS(P1)2247⟩, 946 ⟨86S776⟩; **3**, 118 ⟨76TL4597⟩, 119 ⟨83S815⟩, 634 ⟨76TL2967⟩, 636, 637 ⟨77CL835⟩; **4**, 187 ⟨86JCS(P1)1351⟩, 847 ⟨78CC74⟩, 850 ⟨76JA7440⟩; **5**, 249 ⟨88BCJ2205, 89JCS(P1)1927⟩, 560 ⟨77TL4607⟩; **6**, 309 ⟨72BCJ1797⟩, 561 ⟨88BCJ2203⟩, 731 ⟨79S182⟩
Uenyama, K., **6**, 21 ⟨91TL7425⟩
Uerdingen, W., **6**, 80 ⟨73S207⟩
Uesaka, N., **2**, 927, 988 ⟨92JOC3519⟩
Ueshima, T., **3**, 120 ⟨83CC479⟩
Uesugi, T., **3**, 609 ⟨93JOM(461)81⟩
Ueta, K., **1**, 779 ⟨78TL4903⟩
Ueyama, M., **6**, 111 ⟨79CC485⟩
Ueyama, N., **2**, 227 ⟨88CPB791⟩
Uff, B. C., **1**, 74, 82, 84 ⟨67JCS(C)1120⟩, 859 ⟨84CHEC(2)315⟩; **5**, 558 ⟨85JHC1065⟩
Uggeri, F., **1**, 391 ⟨84AG(E)413, 85S505⟩; **2**, 29 ⟨86AG(E)259, 87JOC3018⟩; **5**, 28 ⟨84AG(E)413⟩
Ugghetto-Monfrin, J., **6**, 550 ⟨88CAR(181)253⟩
Ugi, I., **1**, 38 ⟨61CB1157⟩; **3**, 35 ⟨87S159⟩, 60 ⟨79S266⟩, 328 ⟨58AG702⟩, 531 ⟨73AG(E)25⟩, 694 ⟨61CB2802, 82AG(E)810, 91COS(4)1083⟩, 694, 696, 699, 700, 721, 722, 724, 725 ⟨65AG(E)472⟩, 694, 699, 721, 722, 724, 725 ⟨B-71MI 321-01⟩, 695 ⟨59AG386⟩, 696 ⟨61CB2814, 82AG(E)374, 82JCR(S)79⟩, 696, 699, 721, 722 ⟨77AG(E)259⟩, 696, 699, 721, 722 ⟨85S400⟩, 696, 700 ⟨73OSC(5)300⟩, 696, 708 ⟨89S718⟩, 696, 721, 722 ⟨72AG(E)929⟩, 696, 721, 722, 723 ⟨60CB239⟩, 714 ⟨63LA(666)65⟩, 721 ⟨61OS(41)13, 64CA(60)563, 72AG(E)530⟩; **4**, 20 ⟨79LA278⟩, 313 ⟨95T755⟩, 493 ⟨75JA38, 76LA36⟩; **5**, 370 ⟨90S661⟩, 652 ⟨77JCR(S)119⟩, 654, 655, 657 ⟨62CB126⟩, 795 ⟨B-71MI 520-01⟩, 809 ⟨69AG572⟩; **6**, 417 ⟨82JCR(S)79⟩, 422 ⟨75S598⟩, 425, 487 ⟨78AG(E)361⟩, 462 ⟨77AG(E)259⟩, 485, 487 ⟨79LA278⟩, 485 ⟨B-71MI 615-01, 76AG(E)681⟩, 488 ⟨83T2207, 85ACS(B)761, 93ACS125⟩, 607 ⟨69AG(E)606⟩
Ugi, I. K., **4**, 313 ⟨93T9495⟩
Ugo, R., **1**, 79 ⟨74JOM(70)89⟩; **3**, 467 ⟨63CB802⟩
Ugolini, A., **1**, 615, 698, 703 ⟨86JA2776⟩; **4**, 171 ⟨83JOC4427⟩; **6**, 122 ⟨82TL913⟩
Ugrak, B. I., **1**, 500 ⟨91ZOR488⟩; **4**, 995 ⟨91IZV1849⟩; **5**, 314, 384 ⟨91T7447⟩
Uguagliati, P., **3**, 723 ⟨84IC961⟩
Uguen, D., **1**, 24 ⟨76BSF513⟩, 231 ⟨90TL357⟩, 483 ⟨87BSF487⟩; **2**, 678 ⟨91SL505⟩; **3**, 20, 31 ⟨91SL501⟩, 172 ⟨89TL7389⟩; **4**, 200 ⟨91SL501⟩, 201 ⟨91SL503⟩
Uh, D. S., **6**, 340 ⟨93IC5883⟩
Uhing, E., **4**, 115, 123, 133 ⟨61JA2299⟩
Uhing, E. H., **4**, 117 ⟨67USP3314900⟩, 473 ⟨65JA4757⟩
Uhl, G., **2**, 478 ⟨90IS249⟩; **5**, 395 ⟨82ZAAC(495)73⟩, 1086 ⟨83PS(18)7⟩, 1086, 1087, 1092 ⟨83JA7194⟩; **6**, 324, 684 ⟨84ZAAC(517)75⟩, 582, 684 ⟨85ZAAC(520)120⟩, 582, 684 ⟨85ZAAC(520)139⟩, 584, 680, 684 ⟨84ZAAC(518)21, 84ZAAC(517)89⟩
Uhl, W., **2**, 478 ⟨90IS249⟩; **3**, 688 ⟨81ZN(B)16⟩; **4**, 549 ⟨81ZAAC(479)41⟩, 659, 698 ⟨90POL277, 91JOM(415)181⟩, 698, 708 ⟨91CB1511⟩; **5**, 395, 879 ⟨80TH158, 81ZAAC(473)7⟩, 882 ⟨81ZAAC(479)41⟩, 1086 ⟨83PS(18)7⟩, 1154 ⟨81ZN(B)16⟩; **6**, 195 ⟨89JOM(364)289, 89JOM(368)139, 90POL277, 92CB1547, 92ZAAC(613)67⟩, 195, 197 ⟨89ZAAC(579)75⟩, 195, 197 ⟨91ZAAC(595)225⟩, 197 ⟨93CB2637⟩, 391, 394 ⟨91AG(E)324⟩, 395 ⟨92AG(E)1364⟩, 684 ⟨80ZAAC(463)144⟩
Uhlemann, E., **3**, 364 ⟨65AG(E)154⟩, 397 ⟨74ZC288, 75ZC66⟩; **5**, 633 ⟨81ZC271⟩
Uhlenbrauck, H., **1**, 1045, 1138 ⟨63CB3280⟩; **3**, 273 ⟨63CB3280⟩
Uhlenbrock, W., **6**, 608 ⟨71CB3989⟩
Uhlenbroeck, J. H., **2**, 176, 180, 271 ⟨57RTC129⟩
Uhlich, U., **2**, 240 ⟨59CB1048⟩
Uhlig, E., **4**, 473 ⟨65ZC109⟩
Uhlig, F., **4**, 570 ⟨87EGP251134⟩
Uhlig, G., **4**, 570 ⟨87EGP251134⟩
Uhlig, W., **2**, 253, 260 ⟨82TL1345⟩, 1089 ⟨91ZAAC(603)109⟩; **4**, 570 ⟨87EGP251134, 87EGP251136⟩
Uhlinger, R. H., **2**, 491, 495 ⟨19IEC105⟩
Uhm, ,S., **2**, 958 ⟨73CC269⟩
Uhm, S. J., **2**, 590 ⟨81JOC4323⟩
Uhrhammer, R., **4**, 656 ⟨93JA8493⟩
Ujjainwalla, M., **5**, 1107 ⟨79CJC2145⟩
Ujszaszy, K., **3**, 342 ⟨87IZV2398⟩
Ujvary, I., **4**, 120 ⟨86ACH145⟩
Ukachukwu, V. C., **4**, 818 ⟨85JOC2195⟩
Ukader, G. S., **2**, 461 ⟨66ZOB1436⟩
Ukaji, Y., **1**, 178 ⟨88BCJ571⟩, 183 ⟨92BCJ3487⟩, 352 ⟨92CL61⟩; **3**, 123 ⟨89CL1623⟩; **5**, 148 ⟨86CL81⟩
Ukawa, K., **6**, 558 ⟨89CPB1249⟩
Ukhin, L. Y., **1**, 1005 ⟨68JOU21⟩; **2**, 1018, 1019 ⟨63IZV2213⟩, 1018, 1019 ⟨64IZV392⟩, 1018, 1019 ⟨65ZOR415⟩
Ukida, M., **2**, 243, 359 ⟨78JOC3223⟩, 243, 359 ⟨79JOC1554⟩; **5**, 336 ⟨79JOC1554⟩
Ukita, T., **1**, 440 ⟨83CL1457⟩, 440, 616, 668 ⟨83CC619⟩, 931 ⟨87CL133⟩; **2**, 46 ⟨88JA4606⟩; **3**, 143 ⟨84CC1007⟩; **4**, 392 ⟨84CPB1829⟩
Ukwueze, A. C., **5**, 1140 ⟨86ICAL17⟩
Ulan, J. G., **1**, 502 ⟨87JOC3126⟩
Ulbight, T. A., **2**, 364 ⟨73JA5831⟩
Ulbrich, U., **4**, 352, 353 ⟨92CB591⟩
Ulibarri, G., **1**, 818 ⟨92TL4563⟩
Ulibarri, T. A., **3**, 517 ⟨93JA8167⟩
Ulich, L. H., **4**, 53 ⟨21JA660⟩; **5**, 10 ⟨20JA599⟩
Ulivi, P., **4**, 944 ⟨92SL499, 92SL883⟩
Ullah, H., **3**, 361 ⟨72JCS(P1)2305⟩
Ullas, G. V., **4**, 988 ⟨84TL1291⟩
Ullenius, C., **1**, 311 ⟨77ACS(B)667⟩, 812 ⟨72JA5910⟩; **2**, 587 ⟨82TL3823⟩; **5**, 151 ⟨90T46⟩
Ullerich, K., **2**, 693 ⟨52N353⟩
Ullman, E. F., **2**, 405 ⟨72JA5077, 73JA1677⟩; **3**, 317 ⟨62JA1315, 64JA3814⟩
Ullman, J., **3**, 492 ⟨89S685⟩
Ullmann, H., **2**, 726 ⟨61CB1067⟩
Ullmann, J., **2**, 445, 1021 ⟨91S1073⟩
Ullrich, F.-W., **3**, 59 ⟨86CB1737⟩, 59, 75 ⟨87S1⟩, 71 ⟨77S1⟩
Ullrich, J. W., **3**, 420 ⟨83JA6160⟩
Ullrich, W., **5**, 955 ⟨92CB1571⟩
Ullyot, G. E., **4**, 420, 421, 426 ⟨55JA745⟩
Ulm, K., **4**, 730 ⟨87JFC(37)223⟩; **6**, 32 ⟨88MI 601-01⟩
Ulmand, K., **6**, 3 ⟨87JFC(37)223⟩
Ulmschneider, D., **1**, 58 ⟨55AG710, 57CB2733⟩
Ulpts, R., **2**, 387 ⟨23CB679⟩
Ulrich, B., **4**, 694 ⟨77JOM(141)339⟩
Ulrich, H., **2**, 726 ⟨66JOC2658⟩; **3**, 550 ⟨79JOC4877⟩, 681 ⟨B-79MI 320-01⟩; **4**, 315 ⟨B-77MI 407-01⟩; **5**, 654 ⟨B-68MI 517-01⟩, 656 ⟨62JOC2585⟩, 664, 665, 669 ⟨B-68MI 517-02⟩, 675, 676, 678, 679, 682 ⟨B-68MI 517-03⟩, 699 ⟨B-68MI 517-04⟩, 800 ⟨B-68MI 521-01⟩, 962, 1017,

1018 ⟨B-77MI 526-01⟩, 965 ⟨70USP3492331⟩, 968 ⟨69JOC3200⟩, 971 ⟨74JOC2897⟩, 977 ⟨67AG(E)636, 68JOC2887⟩, 980 ⟨64JOC3344⟩, 980, 981, 982, 1013, 1069 ⟨66AG(E)704⟩, 981 ⟨66JOC2658⟩, 982 ⟨62AG(E)595, 65JOC2779⟩, 982, 1070 ⟨63JCS5558⟩, 995 ⟨67JOC4052⟩, 1003 ⟨64AG(E)585⟩, 1018 ⟨65CRV369⟩, 1029 ⟨65JOC2465⟩, 1061 ⟨B-70MI 528-01⟩, 1065 ⟨78AG761⟩, 1079 ⟨67TL1731⟩; **6**, 143 ⟨68JOC3928, 73JOC2614⟩, 445 ⟨63JOC1427, 64JOC2401, 81JOC5226⟩, 500 ⟨81JOC3011⟩, 509 ⟨79USP4154931⟩, 511 ⟨78JOC1544⟩, 512 ⟨68USP3413382⟩, 602, 606, 614, 619, 620, 621, 622 ⟨B-68MI 620-01⟩, 602, 608 ⟨B-77MI 620-01⟩, 603 ⟨62JOC2589⟩, 652 ⟨68AG(E)291⟩
Ulrich, J., **6**, 666 ⟨75MI 621-02, 75OMS(10)1021⟩
Ulrich, M., **5**, 838, 855 ⟨14CB2938⟩
Ulrich, P., **4**, 199 ⟨76TL809⟩
Ulrich, S. E., **1**, 113 ⟨50JA5127⟩
Ulrich, T. A., **3**, 494 ⟨68JOC3082⟩, 496 ⟨68JOC336⟩
Ultee, A. J., **3**, 615 ⟨06CB1856⟩
Ultee, W., **1**, 624, 779, 783 ⟨78JA3927⟩
Ultee, W. J., **2**, 150 ⟨76CC496⟩
Uma, V., **2**, 249 ⟨65CJC3407⟩
Umani-Ronchi, A., **1**, 24 ⟨84JOC1246⟩, 439 ⟨73CC351, 77JCS(P1)123⟩, 688 ⟨85JOM(280)307⟩; **2**, 79 ⟨88JOM(353)177⟩, 590 ⟨83JOC4108⟩, 698 ⟨90CC1680⟩, 959 ⟨78CC927⟩; **3**, 20 ⟨78JOC1598⟩, 31 ⟨93JA7001⟩, 159 ⟨92T1299⟩, 621 ⟨76S596⟩, 621, 625 ⟨75JCS(P1)1273⟩, 146 ⟨71T6109⟩, 698, 708 ⟨74JOM(72)C4⟩; **5**, 832 ⟨67G1664⟩
Umano, S., **2**, 55 ⟨88JOC3752⟩
Umbach, W., **5**, 138 ⟨71MI 503-01⟩
Umbrasas, B. N., **3**, 212 ⟨75AJC2669⟩
Umbreit, M. A., **1**, 587 ⟨81OS(60)29⟩, 692 ⟨72JA6538⟩; **2**, 86 ⟨77JA5526⟩
Umbricht, G., **1**, 194, 353 ⟨91HCA232⟩
Umeda, A., **2**, 286, 288, 292 ⟨94CPB811⟩
Umeda, I., **3**, 621 ⟨72JOC1542⟩
Umeda, K., **5**, 981 ⟨73JAP7301653, 73JAP7320534⟩
Umeda, Y., **2**, 14 ⟨80BCJ770⟩
Umei, Y., **4**, 943 ⟨87JCS(P1)589⟩
Umelanczuk, J., **2**, 124 ⟨87S1009⟩
Umemomto, T., **6**, 22, 233, 234 ⟨90TL3579⟩
Umemoto, T., **1**, 108 ⟨87BCJ3307, 91BCJ2008⟩, 109 ⟨87BCJ3823⟩; **2**, 16 ⟨90JA8563, 90OS(69)129⟩; **3**, 149 ⟨90JA8563⟩, 151 ⟨90TL3579⟩; **4**, 62, 63 ⟨86BCJ3625⟩; **5**, 205 ⟨86JFC(31)37, 87BCJ3307⟩, 360 ⟨83JFC(22)91⟩; **6**, 12 ⟨84TL81, 86JFC(31)37⟩, 13, 237 ⟨93JA2156⟩, 233, 234 ⟨82TL3929⟩, 233, 234 ⟨86BCJ447⟩, 235 ⟨82CL1519⟩
Umemotu, T., **2**, 16 ⟨86TL4465⟩
Umemura, K., **1**, 241 ⟨89JOC2374⟩
Umemura, M., **3**, 634 ⟨93T1831⟩
Umen, M. J., **1**, 275 ⟨72JA5495⟩, 1093 ⟨75JOC1191⟩, 1142 ⟨73JOC3893⟩; **2**, 587 ⟨73JOC3893⟩
Umeno, M., **1**, 471 ⟨91CL2073⟩; **2**, 18 ⟨93CL673⟩, 19 ⟨91CL1185⟩
Umetani, T., **5**, 74 ⟨73TL1995⟩
Umezawa, B., **6**, 449 ⟨87SC1887⟩
Umezawa, H., **1**, 690 ⟨83BCJ2023⟩; **3**, 716, 717 ⟨87JAN1199, 87JAN1207⟩, 717 ⟨87JAN1202⟩
Umezawa, I., **1**, 212 ⟨93CPB55⟩
Umezawa, J., **1**, 217 ⟨91TL1059⟩
Umezawa, S., **6**, 49 ⟨36BCJ157⟩
Umezawa, Y., **3**, 40 ⟨82TL525⟩
Umezu, Y., **2**, 440, 829 ⟨91TA667⟩
Umhoefer, S. G., **3**, 617 ⟨79S722⟩
Umino, H., **4**, 657 ⟨77JAP7973759, 77JOM(142)C49⟩
Umino, K., **5**, 635, 636 ⟨71MI 515-01⟩
Umino, N., **6**, 504 ⟨82CPB534⟩
Umpleby, J. D., **2**, 561 ⟨75CC242⟩, 562 ⟨72TL2285⟩
Undavia, N. K., **6**, 561 ⟨89JIC60⟩
Undeheim, K., **3**, 271 ⟨87ACS(B)536⟩
Underhill, A. E., **4**, 850 ⟨87S837⟩
Underhill, M., **4**, 682, 683 ⟨75JCS(D)1614⟩; **6**, 721, 732 ⟨75JCS(D)939⟩
Underiner, T. L., **1**, 187 ⟨87JOC897⟩, 187 ⟨89JOC3239⟩; **4**, 716 ⟨92ICA557⟩; **5**, 949 ⟨91JA6673⟩
Underwood, A. L., **1**, 103 ⟨65JOC1154⟩
Underwood, G. R., **5**, 315 ⟨84JOC5225⟩
Underwood, W. G. E., **1**, 595 ⟨70CC1683⟩; **5**, 278 ⟨71JCS(C)3540⟩
Undheim, K., **3**, 247 ⟨79ACS(B)669⟩, 303 ⟨90ZC247⟩, 357 ⟨69ACS1975⟩, 370 ⟨75JCS(P1)2099⟩; **4**, 202 ⟨89ACS706⟩, 232 ⟨88ACS(B)515⟩; **5**, 324 ⟨93ACS1141⟩; **6**, 134 ⟨86JOM(303)189⟩
Unelius, R., **2**, 1011 ⟨85SC463⟩
Uneme, H., **1**, 240 ⟨82BCJ2965⟩; **4**, 900 ⟨93MI 420-01⟩
Uneyama, K., **1**, 580 ⟨92T10265⟩, 1047 ⟨91TL1459⟩; **2**, 21 ⟨81TL2291⟩, 30 ⟨81JOC3312⟩; **3**, 174 ⟨80TL1863⟩; **4**, 92 ⟨93TL1311⟩, 284 ⟨91TL375⟩; **5**, 733 ⟨93TL7933⟩; **6**, 20 ⟨89TL109⟩, 21 ⟨91TL375⟩, 256 ⟨93TL1311⟩
Ung, C. S., **1**, 86 ⟨89SL64⟩
Ungari, F., **2**, 890, 891, 893 ⟨88HOU(E13)761⟩
Ungaro, R., **3**, 60 ⟨75JCS(P1)1527⟩, 97 ⟨78JCS(P1)318⟩
Ungefug, G. A., **4**, 22 ⟨73JOC153⟩
Unger, P., **5**, 1029 ⟨89JHC1735⟩
Unger, S. H., **3**, 307 ⟨85JMC1037⟩
Ungnade, H., **2**, 393 ⟨63OSC(4)724⟩
Ungnade, H. E., **2**, 647 ⟨55OSC(3)130⟩; **4**, 440 ⟨57JOC1662⟩, 994 ⟨66JOC369⟩; **5**, 825 ⟨58JOC1517⟩
Ungváry, F., **2**, 430 ⟨67ZN(B)1238⟩; **5**, 456 ⟨84CC824⟩; **6**, 517 ⟨85M1103⟩
Unkovskii, B. V., **6**, 566 ⟨91KGS416⟩
Unni, A. K. V., **4**, 706 ⟨70CC555⟩
Unni, M. K., **1**, 661 ⟨67JA4531⟩; **2**, 623 ⟨67JA4531⟩, 623, 628 ⟨89JOC6068⟩
Unno, K., **1**, 829 ⟨88CC1531⟩; **5**, 159 ⟨88CC1531⟩
Unno, M., **4**, 608 ⟨87CL2293⟩; **5**, 635 ⟨87H(25)183⟩
Uno, F., **1**, 631 ⟨68JA800⟩
Uno, H., **1**, 1042 ⟨89CL1079⟩; **2**, 205 ⟨82CPB332⟩, 311 ⟨92JOC1504⟩; **3**, 150 ⟨87CL1153⟩; **4**, 774, 777 ⟨92BCJ210⟩, 774, 779 ⟨93MI 417-01⟩
Uno, K., **3**, 683 ⟨68BCJ2954⟩; **5**, 970 ⟨65JOC1158, 66JOC142⟩
Uno, M., **3**, 151 ⟨91CC86⟩
Uno, T., **3**, 66 ⟨81JOC3175⟩, 487 ⟨86OM356⟩
Uno, Y., **2**, 944 ⟨91S1163⟩; **3**, 557,558 ⟨91S1163⟩; **5**, 874 ⟨91S1163⟩
Untch, K. G., **1**, 179 ⟨81JOC2985⟩, 458 ⟨72JA9256⟩, 517 ⟨79JOC248, 79JOC253⟩, 808 ⟨65JA4501⟩; **3**, 613 ⟨81JOC2985⟩; **4**, 187, 199 ⟨72JA7827⟩
Untereiner, G. L., **2**, 377, 403 ⟨65JOC3278⟩
Unterhals, B., **5**, 122 ⟨75AP(308/12)954⟩
Unterhalt, B., **2**, 385 ⟨90HOU(E16a/2)1147⟩, 386 ⟨75AP(308)954⟩; **3**, 710 ⟨81AP(314)459⟩; **4**, 69 ⟨64CB179⟩, 313 ⟨73S302, 82AP(315)852⟩, 327 ⟨73S303⟩, 447 ⟨85S973⟩; **5**, 94 ⟨76AP(308)954⟩, 361 ⟨76S241⟩
Untersteller, E., **4**, 213 ⟨94TL2537⟩
Unverferth, K., **5**, 984 ⟨77JPR841, 79JPR928⟩
Unwalla, R. J., **4**, 138, 139 ⟨92OM859⟩
Uohama, M., **3**, 351 ⟨79JOC2244⟩
Uosaki, Y., **5**, 199 ⟨86TL4609⟩
Uotani, K., **6**, 529 ⟨86JAP62176910⟩
Uotani, T., **6**, 529 ⟨85JAP62113712, 88JAP01257116, 89JAP03037110, 89JAP03060417, 89JAP03060418, 89JAP03197310, 91JAP03060419⟩, 533 ⟨85JAP61229860, 85JAP61229861, 85JAP62120358, 86GEP3616009, 88JAP01301654, 88JAP02009858, 90JAP03220173, 91JAP03193759⟩, 537, 538 ⟨85JAP61233662, 85JAP61233663⟩
Uozumi, Y., **5**, 301 ⟨86CC841⟩
Upham, R. A., **2**, 219 ⟨70JOC4148⟩; **6**, 508 ⟨65CJC1798⟩
Uphoff, J., **1**, 500 ⟨87AG(E)357⟩
Upton, C. J., **1**, 928 ⟨72JA301, 76JOC523⟩; **4**, 394 ⟨75JOC3052⟩; **6**, 555 ⟨75JOC3052⟩

Upton, R. M., **5**, 720 ⟨79JCS(P1)2478⟩
Ura, K., **2**, 855 ⟨74JAP7401534⟩
Ura, T., **3**, 117 ⟨88JOC3587, 88SC869⟩
Urabe, H., **1**, 194 ⟨93CC1595⟩, 471 ⟨82JA6831⟩, 1127 ⟨86TL1355⟩; **2**, 537 ⟨84JOC1140⟩; **3**, 7, 120 ⟨82JOC837⟩, 10, 123 ⟨83JA5703⟩, 74 ⟨81TL2183⟩, 144 ⟨82JA6831⟩, 459 ⟨92JOC5056⟩; **4**, 552 ⟨76JOM(122)113⟩; **5**, 12, 96 ⟨87OS(66)87⟩
Urabe, K., **3**, 279 ⟨92CL1987⟩
Urabe, Y., **4**, 13 ⟨75TL997⟩
Uramoto, M., **1**, 72, 81 ⟨71BCJ166⟩
Urano, N., **2**, 159 ⟨92JCS(P1)1497⟩
Urano, Y., **4**, 354 ⟨88JA4826⟩
Urasaki, I., **2**, 650 ⟨74T3021⟩; **6**, 501 ⟨68MI 606-01⟩
Urata, H., **5**, 194 ⟨87JAP6289642, 88TL4435⟩; **6**, 22 ⟨91TL91⟩
Urata, Y., **3**, 684 ⟨94JCS(P1)413⟩; **6**, 32 ⟨62NKZ936⟩
Uray, G., **4**, 96 ⟨75ZN(B)245⟩
Uray, S., **4**, 55, 57 ⟨80LA1919⟩
Urayama, S., **4**, 604, 625, 638 ⟨88JOM(354)155⟩; **6**, 185 ⟨88JOM(354)155⟩
Urayama, T., **5**, 296 ⟨83JOC4058⟩
Urbach, H., **1**, 325, 326 ⟨83TL5343⟩
Urbain, E., **4**, 59 ⟨19CR(169)1143⟩; **6**, 412, 413 ⟨19CR(169)17⟩
Urban, E., **3**, 91 ⟨89M765⟩
Urban, E. J., **6**, 143 ⟨71JOC2262⟩
Urban, F. J., **1**, 806 ⟨70CC165⟩
Urbaniak, W., **4**, 566 ⟨88SRI695⟩, 568 ⟨88MIP4121⟩
Urbanski, T., **4**, 302 ⟨63AHC(2)311, 78AHC(23)1⟩
Urbanski, T. S., **4**, 428 ⟨79PJC187⟩
Urbansky, M., **5**, 567 ⟨93SC829⟩
Urbas, L., **2**, 784, 785, 787 ⟨B-70MI 215-02⟩; **4**, 540 ⟨70MI 411-02⟩
Urbech, H.-J., **4**, 528 ⟨89EUP315574⟩
Urbigkit, M. J., **1**, 869 ⟨63T1625⟩
Urch, C. J., **3**, 640 ⟨85JOM(285)173⟩; **4**, 365 ⟨89JCS(P1)115⟩
Urech, E., **1**, 537 ⟨53HCA1809⟩
Urech, F., **3**, 615 ⟨1872LA(164)255⟩
Urenovitch, J. V., **3**, 711 ⟨62JCS548⟩
Ureta, C., **5**, 586 ⟨93JOC4908⟩
Uriel, S., **6**, 563 ⟨91S147⟩
Urinovich, E. M., **6**, 118 ⟨72MI 602-01⟩
Urleb, U., **5**, 836 ⟨87M601⟩, 837 ⟨85H(23)2819⟩, 845 ⟨86H(24)1899⟩, 1028 ⟨90JHC407, 90JHC433⟩; **6**, 558 ⟨90JHC407, 90JHC643⟩
Urogdi, L., **1**, 260 ⟨89JCS(P1)225⟩, 260, 261 ⟨89S323⟩, 262, 1041 ⟨89TL3303⟩; **2**, 302 ⟨89OPP135⟩, 306 ⟨89TL3303⟩, 310 ⟨89JCS(P1)225⟩; **3**, 710 ⟨90JCS(P1)1847⟩; **4**, 309 ⟨89JCS(P1)225⟩, 329 ⟨90S565⟩, 406, 410 ⟨90JCS(P1)541⟩, 417 ⟨89CC337⟩; **5**, 1031, 1067 ⟨90S565⟩
Urpí, F., **2**, 317 ⟨87TL5941⟩, 341 ⟨90T587⟩; **5**, 234 ⟨86TL4623⟩, 360 ⟨89JOC3209⟩, 360, 361 ⟨84JOC3322, 89TL863⟩
Urquhart, G. G., **2**, 120 ⟨55OSC(4)363⟩
Urquhart, P. W., **1**, 669 ⟨74JOM(81)C37⟩
Urrios, P., **3**, 182 ⟨88S314⟩
Urry, G., **2**, 5411 ⟨68IC1253⟩; **4**, 622, 624 ⟨70IC2372⟩, 636 ⟨70IC1060⟩
Urry, W. H., **3**, 466 ⟨64JA1815⟩; **6**, 5 ⟨47JA1100⟩
Uruma, T., **1**, 938 ⟨89CC1265⟩; **5**, 513 ⟨89CC1265⟩; **6**, 550 ⟨89CC1265⟩
Urushadze, M. V., **4**, 894 ⟨72IZV1347⟩
Uryak, B. I., **2**, 334, 365 ⟨91T7447⟩
Uryukina, I. G., **2**, 802 ⟨72KGS1690⟩
Usami, T., **4**, 513 ⟨92JOC5419⟩
Usami, Y., **1**, 241 ⟨87CC449⟩; **3**, 161 ⟨92S751⟩; **4**, 235, 236 ⟨86CL1655⟩
Uschak, E. A., **3**, 90, 97 ⟨77JOC3308⟩
Uschmann, J., **6**, 299, 305, 362 ⟨86JOC370⟩
Uschold, R. E., **1**, 869 ⟨67JA1721, 67JA2752⟩, 869 ⟨67JA2960, 68JA2821⟩, 869 ⟨68JA3415⟩
Usgaonkar, R. N., **5**, 184 ⟨66JIC615⟩
Ushakov, M. V., **4**, 658 ⟨75IZV1600⟩
Ushakov, S. N., **4**, 137 ⟨37JGU2495⟩
Ushio, H., **2**, 71 ⟨89CC1700⟩
Ushio, J., **5**, 932 ⟨84JA5892⟩, 933 ⟨83JA426⟩
Ushio, Y., **1**, 155 ⟨90S401⟩; **2**, 60 ⟨88BCJ2684⟩, 312 ⟨90S401⟩
Usifer, D., **1**, 436 ⟨85JOC1125⟩
Usifoh, C. O., **6**, 109 ⟨90JHC1953⟩
Usković, M. R., **1**, 91 ⟨77JOC135⟩, 576, 1021 ⟨86JOC3098⟩, 697, 700, 703 ⟨88JOC4282⟩, 749, 828 ⟨87JOC1372⟩, 832 ⟨74HCA771⟩; **2**, 342, 347 ⟨85T3455⟩; **4**, 170 ⟨83JOC5315⟩
Uskokovic, W. R., **2**, 519 ⟨73JA7171⟩
Usman, N., **6**, 58 ⟨94TL3227⟩, 540 ⟨83TL865⟩
Usmanova, L. N., **2**, 254 ⟨87ZOB1429⟩
Usón, A., **4**, 585 ⟨88JCS(D)341⟩, 586 ⟨83ICA63, 88SRI69⟩
Usón, R., **4**, 584 ⟨85JOM(290)125⟩, 585 ⟨88JCS(D)341⟩, 586 ⟨83ICA63, 88SRI69⟩, 692 ⟨88OM2279, 89JCS(D)169, 93JA4145⟩, 1138 ⟨73SRI237, 84ICA215⟩, 1141 ⟨73MI 530-01⟩
Usov, I. A., **2**, 167 ⟨64ZOB1948⟩
Usov, V. A., **3**, 341 ⟨87KGS1694⟩, 354 ⟨86MI 308-01⟩, 356 ⟨78ZOR666, 79ZOR82⟩, 360 ⟨75ZOR1945⟩, 361 ⟨80ZOR1842⟩, 363 ⟨72IJS(A)0216⟩
Usova, T. L., **3**, 341 ⟨87KGS1694⟩
Ustenko, S. N., **2**, 476 ⟨91JGU464⟩; **4**, 116, 133 ⟨90JGU1536⟩, 570 ⟨91MI 412-02⟩
Ustunol, A., **5**, 82 ⟨83JOC4076⟩
Ustynyih, Y. A., **4**, 603, 625, 637, 701 ⟨71JOM(30)C57⟩
Ustynyuk, T. K., **1**, 391, 884 ⟨70RCR732⟩
Ustynyuk, Y. A., **4**, 627 ⟨80IZV2668⟩; **6**, 387, 395 ⟨79BAU2222⟩
Usui, M., **5**, 125 ⟨77BCJ1863⟩
Usui, S., **1**, 818 ⟨93TL8453⟩
Usuki, M., **5**, 1055, 1113 ⟨79S705⟩
Usuki, Y., **2**, 280 ⟨89CL1235⟩
Usyshkin, O. G., **6**, 261, 283 ⟨70IZV2553⟩
Utagawa, A., **3**, 217 ⟨88BCJ1207⟩
Utaka, M., **1**, 571 ⟨91JOC7177⟩; **2**, 14 ⟨89JOC4989⟩, 39 ⟨94TL4569⟩; **3**, 32 ⟨82JOC1101⟩, 145 ⟨87BCJ1063⟩, 145 ⟨87BCJ3421⟩; **5**, 357 ⟨82CL1607⟩; **6**, 561 ⟨88BCJ3205⟩
Utamapanya, S., **1**, 1084 ⟨81CL1241⟩
Utebaev, U., **2**, 876 ⟨75IZV897⟩; **3**, 588 ⟨75JCS(P1)2513⟩; **4**, 731 ⟨75JFC(6)227⟩, 736 ⟨76IZV895⟩, 803 ⟨76BAU853, 76BAU873⟩
Utermohlen Jr., W. P., **3**, 632 ⟨45JA1505⟩
Utikal, H.-P., **3**, 556 ⟨75CJC1333⟩
Utimoto, J., **1**, 761 ⟨77TL1807⟩
Utimoto, K., **1**, 160, 1094 ⟨77TL1805⟩, 160, 451 ⟨84TL1999⟩, 161, 1094 ⟨76JOC2941⟩, 161, 450, 1106 ⟨76JOC2215⟩, 205, 1024 ⟨84TL5423⟩, 207 ⟨74T3037⟩, 207, 1100 ⟨77T1945⟩, 208 ⟨87JOC1013, 90TL6209⟩, 228 ⟨90JOC1705⟩, 264 ⟨90TL6379⟩, 274 ⟨73TL787⟩, 330, 925, 946 ⟨89BCJ143⟩, 333 ⟨88TL1041, 89TL3155⟩, 451 ⟨75TL2825⟩, 457 ⟨90TL365, 92JOC1615⟩, 480 ⟨87CL2203⟩, 485 ⟨74BCJ503⟩, 491 ⟨87CL5⟩, 511 ⟨77TL3641⟩, 650, 761 ⟨76JOC2940⟩, 676 ⟨87JA951, 87TL1443⟩, 677 ⟨86JA7408⟩, 677, 1099 ⟨86JA6048⟩, 766, 770 ⟨87JOC4410⟩, 801 ⟨88TL1543⟩, 884 ⟨64JOC2781⟩, 925, 938, 946, 1131 ⟨88TL6127⟩, 939, 946, 1130 ⟨88TL6125⟩, 1033 ⟨73TL1847, 74T2159⟩, 1043 ⟨88TL1785⟩, 1045, 1138 ⟨78TL2301⟩, 1094 ⟨77T2987, 88BCJ2693⟩, 1130, 1132 ⟨87JA2547⟩, 1131 ⟨89TL3159⟩; **2**, 73 ⟨88TL5263⟩, 74, 572 ⟨89JOC4732⟩, 77 ⟨91TL1069⟩, 80 ⟨89TL4389⟩, 86 ⟨90JOC1705⟩, 319 ⟨90TL6379⟩, 618, 625, 630 ⟨86JA7408⟩, 681 ⟨87JOC4410⟩, 700 ⟨88TL1065, 93TL4985⟩, 710 ⟨87CL1647⟩, 903, 919 ⟨86TL2007⟩, 913 ⟨93BCJ2348, 93BCJ2356⟩, 918, 919, 939 ⟨88T4277⟩, 918, 923 ⟨86T4427⟩, 920 ⟨91BCJ1461⟩, 922 ⟨76JOC2215,

76JOC2941, 77T2987⟩, 922, 923 ⟨87JA2547, 88T4009⟩, 924 ⟨77TL1805, 79JOM(177)145⟩, 927 ⟨90TL365, 90TL369⟩, 927 ⟨91JOC5980, 92JOC1615⟩, 927, 945 ⟨90JOC5310⟩, 927, 988, 989 ⟨90JOC1707⟩, 927, 989 ⟨92BCJ1543, 92CL99⟩, 930, 946 ⟨87BCJ3468, 88BCJ2693⟩, 946 ⟨87TL3709⟩, 972 ⟨84TL1999⟩, 983 ⟨86JA2753⟩, 1088 ⟨76TL3969⟩; **3**, 5, 117 ⟨88BCJ3607⟩, 6, 118 ⟨93CL571⟩, 12, 128 ⟨91BCJ2585⟩, 30 ⟨92CL2173⟩, 118, 121 ⟨89BCJ519⟩, 161 ⟨87JOC4410⟩, 238, 270, 273 ⟨78TL2301⟩, 289 ⟨75TL4239⟩, 410 ⟨91JOC5812⟩, 615 ⟨91CL1447⟩, 616, 627, 633 ⟨83T967⟩, 627 ⟨81TL4279, 92BCJ1543⟩, 632 ⟨90TL6209⟩, 632, 698 ⟨80TL389, 87JOC1013⟩, 638 ⟨90TL6379⟩; **4**, 188 ⟨89BCJ2050⟩, 231 ⟨90JA8189, 92CL2173⟩, 639 ⟨88BCJ2693⟩, 654 ⟨85TL2101⟩, 737, 743 ⟨91CL1319, 92BCJ1513⟩, 808, 811, 813 ⟨89TL3155⟩, 811 ⟨92TL7031⟩, 944 ⟨91TL2783⟩, 1045 ⟨88T4277⟩, 1060 ⟨84TL1999⟩, 1061, 1062 ⟨79JOM(177)145⟩; **5**, 344 ⟨91CL1479⟩; **6**, 21 ⟨89TL3159⟩, 60, 62 ⟨93T3227⟩, 550 ⟨88TL6125⟩

Utjes-LeGall, M., **5**, 64 ⟨78JCR(S)198⟩

Utke, A. R., **1**, 889 ⟨64JOC1261⟩

Utkina, N. K., **6**, 138 ⟨89T3487⟩

Utley, J. H. P., **1**, 6 ⟨70CC738⟩; **3**, 497 ⟨81JCS(P1)1520⟩

Utne, T., **4**, 213 ⟨57JA456⟩; **6**, 25 ⟨86TL3845⟩

Utpal, R. C., **3**, 233 ⟨81T4035⟩

Utrapiromsuk, N., **5**, 322 ⟨84JOC106⟩

Utvary, K., **2**, 387 ⟨66IC1835, 68IC698⟩, 468 ⟨69JINC271⟩; **5**, 1078 ⟨68M2498⟩

Uvarov, V. I., **6**, 261 ⟨88IZV466, 88IZV2639⟩

Uwe, J., **6**, 502 ⟨83GEP3135111⟩

Uyama, H., **4**, 210 ⟨86JOC546⟩; **5**, 132 ⟨86JOC546⟩; **6**, 24 ⟨90TL7181⟩

Uyeda, R. T., **1**, 797 ⟨64JA5466⟩

Uyeda, S., **4**, 78 ⟨76S697⟩

Uyehara, T., **1**, 188 ⟨90AG(E)609⟩; **2**, 379 ⟨79JA2076⟩; **3**, 440, 456 ⟨89TL4275⟩; **6**, 426 ⟨91TL4371⟩

Uyeo, S., **3**, 357 ⟨91CPB672⟩, 634 ⟨73SC265⟩

Uyzbaev, K. M., **4**, 343 ⟨79MI 407-01⟩

Uzan, A. T., **5**, 698, 762, 813 ⟨78IZV2802⟩

Uzan, R., **2**, 7 ⟨91TL3977⟩

Uziel, J., **4**, 465, 478 ⟨92TL77⟩

Uzlova, L. A., **5**, 394, 404 ⟨80RCR843⟩

Uznanski, B., **5**, 1135 ⟨78S154⟩

Uzuki, N., **2**, 474 ⟨85JA3501⟩

Uzun, A. T., **5**, 762 ⟨73IJS389⟩; **6**, 259 ⟨78IZV486⟩

Uzzell, P. S., **3**, 419 ⟨55CI(L)1039⟩

V

Vaalburg, W., **5**, 88 ⟨72SC423⟩
Vacca, A., **1**, 507 ⟨88IC4429⟩
Vacca, J. P., **3**, 207 ⟨80JA3964⟩, 209 ⟨82TL2623⟩; **5**, 164 ⟨80JA3964⟩
Vaccaro, H. A., **1**, 573, 581, 584 ⟨92JA9419⟩
Vaccaro, W., **1**, 1042 ⟨86TL4241⟩
Vacher, B., **5**, 727 ⟨92TL5117⟩, 728 ⟨88T3501⟩; **6**, 550 ⟨92JA4111⟩
Vadde, R., **5**, 272 ⟨84MI 507-02⟩
Vadecard, J., **1**, 573 ⟨93CC116⟩
Vadnere, M. K., **4**, 141, 142 ⟨87MI 403-01⟩
Vaerman, J. L., **5**, 670, 671 ⟨87BSB51⟩
Vågberg, J.-O., **1**, 193 ⟨89TL617⟩
Vaglio, G. A., **4**, 682 ⟨75G939, 75ICA155⟩
Vahldiek, M., **2**, 428 ⟨90ZAAC(589)187⟩
Vahrenhorst, A., **1**, 271 ⟨83CB473⟩, 647, 667, 717 ⟨78TL4399⟩; **4**, 565, 596, 600 ⟨83CB473⟩, 599, 600, 663 ⟨78TL4391⟩
Vahrenkamp, H., **2**, 431 ⟨79ZN(B)642⟩, 511 ⟨80CB3517⟩; **4**, 683 ⟨93CB373, 93CB2017⟩, 685 ⟨93AG(E)1048⟩, 711 ⟨90CB661⟩; **6**, 352 ⟨80CB1654⟩, 402 ⟨90JOM(381)261⟩, 518 ⟨86ZN(B)283⟩
Vaid, B. K., **2**, 405 ⟨90SC2353⟩; **3**, 194 ⟨88T1603⟩; **4**, 206 ⟨89TL3019⟩
Vaid, R. K., **1**, 74 ⟨87SC703⟩, 109 ⟨89JA6443, 90JA1297⟩, 1076 ⟨87CC711⟩; **2**, 405 ⟨90SC2353⟩, 1020 ⟨90SL365⟩; **3**, 162 ⟨87JOC150⟩, 194 ⟨88T1603⟩; **4**, 206 ⟨89TL3019⟩, 1081 ⟨90S431⟩
Vaidya, O. C., **4**, 337, 344 ⟨73CJC1136⟩
Vaidya, V. M., **3**, 370 ⟨66T7⟩
Vaidyanathaswamy, R., **1**, 459 ⟨67JOC4143, 71TL2075⟩, 459 ⟨75S194⟩; **2**, 463 ⟨81IJC(B)932⟩; **5**, 371 ⟨89CI(L)389⟩, 464 ⟨80JOC2270⟩
Vail, D. K., **4**, 32 ⟨48JA1550⟩
Vail, P. D., **4**, 333 ⟨81JOC765⟩
Vail, S. L., **4**, 458, 477, 481 ⟨74JHC407⟩
Vainer, L. M., **4**, 560 ⟨84DOK(277)371, 87IZV860⟩
Vainiotalo, P., **5**, 1028 ⟨91JHC1091⟩; **6**, 263 ⟨92JCS(P2)835, 92PS(70)183⟩
Vairamani, M., **1**, 927 ⟨86JA2381⟩
Vaissermann, J., **4**, 674 ⟨90OM2993, 94OM2244⟩; **5**, 798 ⟨90JOM(381)191⟩, 953 ⟨92JA8088, 95JOM(494)43⟩
Vajda, E., **6**, 564 ⟨89SUL123⟩
Vakatar, V. V., **3**, 134 ⟨77CI(L)742⟩
Vakhomchik, L. P., **4**, 805, 812, 813 ⟨61MI 418-01⟩
Vakhrin, M. I., **6**, 656 ⟨⟨78ZOB811⟩
Vakoulis, F. D., **2**, 35 ⟨85CC1404⟩
Vakul'skaya, T. I., **6**, 131 ⟨87IZV2872, 88MI 604-01⟩
Valada, J. C. R., **3**, 640 ⟨67PNA(265)939⟩
Valade, J., **1**, 132 ⟨67CR(C)1204⟩; **2**, 600 ⟨71JOM(33)C4⟩, 601 ⟨68BSF3385, 71JOM(33)C7⟩, 701 ⟨67BSF1928, 68JOM(11)97⟩
Valange, P., **2**, 1014 ⟨64AG(E)747, 64AG(E)755⟩
Valasinas, A., **1**, 259 ⟨71JA2738⟩; **3**, 106 ⟨74JOC2872⟩
Valcarcel, M. V., **5**, 711 ⟨85SC643⟩
Valdeolmillos, A. M., **6**, 556 ⟨84S172⟩
Valderrama, J. A., **6**, 546, 547 ⟨80AOC(18)1, 83JCS(P1)2641⟩
Valdes, J. A., **3**, 427 ⟨83OPP41⟩
Valdes, V., **3**, 206 ⟨90SC131⟩
Valdez, C., **2**, 487, 875 ⟨68JOM(12)133⟩; **4**, 805 ⟨68JOM(12)133⟩
Valente, L. F., **1**, 467 ⟨80JA3298⟩, 1106 ⟨80S1034⟩; **2**, 629, 976 ⟨80JA3298⟩, 919, 925 ⟨78JOM(156)C20, 80S1034⟩
Valenti, E., **2**, 941 ⟨90JOC395⟩, 1026 ⟨87T2311⟩, 1027 ⟨90JA7405⟩; **3**, 243 ⟨90TL7319⟩
Valentin, E., **2**, 396 ⟨89T2099⟩, 738, 757, 758 ⟨B-82MI 215-01⟩, 783 ⟨66TL1821, 66TL6331⟩, 783 ⟨68T1889, 73T3011⟩, 783 ⟨74T2741⟩; **5**, 354 ⟨83JHC305⟩
Valentine, D., **2**, 442, 443 ⟨B-84MI 208-01⟩
Valentine, Jr., D., **2**, 440, 441, 443 ⟨80JOC3691⟩
Valentine, J. D., **2**, 822, 841 ⟨79JOC3741⟩
Valentine, J. S., **2**, 87 ⟨75JOC1678⟩
Valentino, D. S., **3**, 153 ⟨78S139⟩
Valenty, S. J., **4**, 385 ⟨73JOC3937⟩; **5**, 874 ⟨73JOC3937⟩
Valetdinov, R. K., **2**, 440 ⟨85URP1145022⟩, 442 ⟨63ZOB2631, 67ZOB2522⟩; **4**, 336 ⟨84MI 407-03⟩, 342 ⟨78ZOB2653, 79DOK(247)609⟩
Valette, G., **1**, 212 ⟨75S256⟩
Valeur, M., **2**, 494 ⟨27CR(185)956⟩, 494, 505, 506 ⟨27CR(185)779⟩
Valkanas, G., **4**, 30 ⟨63HCA1818⟩
Valkanus, G., **1**, 838 ⟨59JCS2720⟩
Valkó, K., **2**, 374 ⟨88S529⟩
Valkov, P. G., **2**, 358 ⟨79ZOB841⟩
Valle, G., **3**, 5, 117 ⟨87JOC5467⟩, 268 ⟨84JOC4696⟩, 382 ⟨65JCS807, 67JCS(B)117⟩, 706 ⟨89S641⟩; **4**, 249 ⟨90BSF734⟩, 628, 701 ⟨88OM210⟩, 861 ⟨88MI 419-02⟩; **6**, 556 ⟨91PS(63)51⟩
Valle, L., **6**, 120 ⟨86T1963⟩
Vallée, Y., **1**, 772 ⟨93S659⟩; **2**, 140 ⟨90JOC2596⟩, 234 ⟨93CC806⟩; **3**, 335 ⟨90JOC2596, 93JA324⟩, 337 ⟨88TL5899⟩, 351 ⟨90TL6183⟩, 369 ⟨91NJC533⟩, 550 ⟨86TL4313⟩; **4**, 870 ⟨92TL6131⟩; **5**, 552 ⟨83TL3617, 85T2691⟩, 559 ⟨93TL2605⟩, 563 ⟨91CC964, 92TL6131⟩
Valli, V. L. K., **3**, 59 ⟨90TL7521⟩, 116, 158, 161 ⟨90TL5785⟩
Valls, J., **3**, 206 ⟨70CC721⟩
Valmas, M. D., **4**, 327 ⟨84JCS(P1)2327⟩
Valnot, J.-Y., **1**, 180 ⟨88T5495⟩, **2**, 22 ⟨91SL807⟩, 28 ⟨73BSF1465⟩
Valoti, E., **1**, 182 ⟨85JOC3945⟩, 182, 475 ⟨91JOC183⟩; **5**, 301 ⟨86T885⟩
Valpey, R. S., **1**, 653 ⟨84JOC1317⟩
Valueva, Z. P., **4**, 724 ⟨62IZV1683⟩; **6**, 457 ⟨87IZV1671⟩
Valvano, N. L., **5**, 330 ⟨91JOC3750⟩
Valverde, S., **1**, 187 ⟨90JOC2294⟩, 579 ⟨92JOC1613⟩
Vamamoto, H.,**3**, 437 ⟨83TL943⟩
Vaman Rao, M., **1**, 330 ⟨88JOC1432⟩
van Acker, Z., **4**, 231 ⟨75T327⟩
Van, D. Le, **2**, 293, 499, 500 ⟨79ZN(B)1171⟩, 894 ⟨81ZN(B)48⟩; **4**, 112 ⟨89ZN(B)175⟩, 112, 116 ⟨90ZN(B)148, 91ZN(B)978⟩; **5**, 786, 1087 ⟨91ZN(B)978⟩; **6**, 49 ⟨90JOM(386)321⟩, 242, 683 ⟨93ZN(B)58⟩, 258, 679 ⟨90ZN(B)148⟩, 258, 680, 683 ⟨86ZN(B)149⟩, 264 ⟨85PS(21)349⟩, 265 ⟨90ZN(B)299, 92ZN(B)321⟩, 265, 266, 679, 680, 691 ⟨84AG(E)710⟩, 266 ⟨90JOM(399)189⟩, 678 ⟨86PS(28)239⟩, 679 ⟨87PS(30)401⟩, 684 ⟨93PS(76)265⟩, 685 ⟨90CB2317⟩, 700 ⟨89NJC363⟩
Van, T. T., **2**, 938, 944 ⟨87TL397⟩; **3**, 424 ⟨87TL397⟩
Vanag, E. V., **5**, 823 ⟨60ZOB506⟩
Vanagisawa, A., **2**, 74 ⟨89JOC5198⟩
Van Allan, J. A., **3**, 71 ⟨71JOC600⟩, 616 ⟨55OSC(3)275⟩; **4**, 47 ⟨49JOC754⟩, 179 ⟨63OSC(4)21⟩; **5**, 62 ⟨63OSC(4)633⟩, 302 ⟨55OSC(3)765⟩

van Asch, A., **2**, 245, 250 ⟨81TL583⟩
van Asselt, A., **3**, 511 ⟨86JA5347⟩
van Asten, J. J. A., **2**, 150 ⟨76CC496⟩
Van Audenhove, M., **1**, 356 ⟨78TL2057⟩
van Auwers, K., **3**, 612 ⟨23CB1172⟩
van Baar, B. L. M., **6**, 610 ⟨93JOC6930⟩
van Bekkum, H., **1**, 522 ⟨71S139⟩, 80 ⟨72RTC1057⟩, 96 ⟨72RTC1069⟩, 115 ⟨70T2683⟩; **3**, 406 ⟨72RTC605⟩; **4**, 177 ⟨71RTC1141⟩, 177 ⟨72S419⟩, 177 ⟨77LA624⟩
van Benthem, R. A. T. M., **2**, 311 ⟨90TA163⟩
van Bergen, T. J., **2**, 168 ⟨74JA5508⟩, 761 ⟨74JA5512⟩
van Boeckel, C. A. A., **2**, 675 ⟨81S305⟩
Van Boom, J. H., **1**, 185 ⟨91S347⟩, 648 ⟨89TL229⟩, 955, 1125 ⟨65RTC31⟩; **2**, 106 ⟨91TL6021⟩, 409 ⟨84TL1253⟩, 467 ⟨92TL2357⟩, 469 ⟨86JCR(S)26⟩, 675 ⟨81S305⟩, 1032 ⟨66RTC580, 66RTC952⟩, 1032 ⟨68RTC97⟩; **3**, 61 ⟨65RTC31⟩; **4**, 176, 199, 352 ⟨90TL2197⟩, 184 ⟨87RTC545⟩, 186 ⟨73S169⟩, 220 ⟨91T1547⟩; **6**, 317 ⟨91S347⟩, 549 ⟨92MI 617-05⟩
VanBrocklin, H. F., **6**, 134 ⟨90MI 604-01⟩
Van Broeck, D., **6**, 619 ⟨79AG(E)333⟩
van Camp, A., **3**, 558, 560 ⟨80TL3081⟩
Van Campen, Jr., M. G., **1**, 447 ⟨38JA121⟩; **2**, 112 ⟨38JA121⟩
Van-Catledge, F. A., **4**, 1076 ⟨84JA5544⟩
Vance, R. L., **1**, 415 ⟨88JA2314⟩
Vanchikov, A. N., **1**, 486 ⟨90IZV2665⟩
van Coppenolle, J., **2**, 238 ⟨75T2099⟩, 279 ⟨75JHC423⟩
Van Daalen, J. J., **3**, 540 ⟨61RTC810⟩
van Dauren, A. M. G., **6**, 549 ⟨92MI 617-05⟩
Van de Castle, J. F., **3**, 87 ⟨61JOC2973⟩
Vande Griend, L., **2**, 476 ⟨77JA7379, 83PS(18)89⟩
van de Heisteeg, B. J. J., **4**, 657, 663, 701 ⟨84TL5191⟩, 669, 670 ⟨86JOM(308)1, 87TL6493⟩; **6**, 190, 192, 198 ⟨86TL6123⟩
van de Kamp, F.-P., **5**, 1040 ⟨56CB133⟩
Van De Mark, M. R., **3**, 120 ⟨80TL4695⟩
Van Den Berghe, J., **5**, 599 ⟨48JA2425⟩
Van den Bril, M., **3**, 687 ⟨80BCB433⟩
Van den Broek, L. A. G. M., **4**, 365 ⟨84JOC1691⟩
van den Broek, P. J., **5**, 1035 ⟨75JCS(P1)922⟩
Vandenbulcke-Coyette, B., **4**, 798 ⟨81HCA1208⟩; **5**, 670, 671 ⟨87BSB51⟩
Van Den Elst, H., **4**, 220 ⟨91T1547⟩
Van den Elzen, R., **1**, 142 ⟨74CJC761⟩, 823 ⟨72JA9261⟩; **2**, 151 ⟨74CJC761⟩, 200 ⟨74JA935⟩; **3**, 87 ⟨74JA935⟩; **4**, 391 ⟨74CJC761⟩
Vanden Eynde, J. J., **5**, 773 ⟨87T4185⟩
van den Hark, T. E. M., **1**, 890 ⟨74AX(B)833⟩
van den Hende, J. H., **2**, 1030 ⟨56RTC1459⟩; **3**, 389 ⟨66JA5045⟩
Vandenheste, T., **1**, 1098 ⟨90JA5562⟩
van den Hoek, W. J.,**4**, 894 ⟨72JOC1960⟩
van den Marel, G. A., **2**, 467 ⟨92TL2357⟩
Vandensavel, J. M., **5**, 1014 ⟨73JOC675⟩
Vanderah, D. J., **4**, 298 ⟨90JOC1459⟩
van der Baan, J. L., **1**, 194 ⟨88TL3579⟩, 194 ⟨93TL3309⟩, 415, 826 ⟨86TL6267⟩; **2**, 324 ⟨88TL3579⟩, 925 ⟨87TL2889⟩; **4**, 1061 ⟨87TL2889⟩
van der Burg, W. J., **5**, 769, 852 ⟨55RTC257⟩
van der Does, T., **3**, 710 ⟨88AG(E)936⟩
van der Donk, W. A., **2**, 527 ⟨91JA6139, 92JA9350⟩
van der Drift, J. A. M., **3**, 432 ⟨80RTC278⟩
Vanderesse, R., **1**, 430, 431 ⟨81JOC1270⟩, 437 ⟨88TL2963⟩, 472 ⟨89JOC4840⟩
Van der Eycken, E., **1**, 19 ⟨87TL4759⟩; **3**, 532 ⟨84SC163⟩
Van der Gen, A., **1**, 248 ⟨83JOM(256)C1⟩, 873 ⟨78TL491⟩; **2**, 10 ⟨79RTC371⟩, 34 ⟨79TL2817⟩, 36 ⟨83SC387⟩, 311 ⟨90TA163⟩, 316 ⟨91RTC25⟩, 330 ⟨90T1653⟩, 722 ⟨69RTC519⟩, 739 ⟨79TL2433, 80TL2671⟩, 739 ⟨84RTC305⟩; **3**, 11 ⟨79TL2433, 80TL2671⟩, 21 ⟨78TL491⟩, 27, 35 ⟨79TL2817⟩, 182 ⟨81TL2799⟩, 628 ⟨88TL4485, 90SUL83⟩, 628 ⟨91SC1387⟩, 631 ⟨93T1063⟩; **4**, 200, 208 ⟨78JOC3548⟩, 210 ⟨76TL1725⟩, 468 ⟨84TL4259⟩, 489 ⟨90JOC4063, 90JOC4069⟩, 829 ⟨83TL1303⟩, 64 ⟨77TL885⟩, 199 ⟨79RTC371⟩; **6**, 48, 120 ⟨77TL885⟩, 116 ⟨83TL1303⟩
Vanderhaar, R. W., **4**, 754 ⟨71JFC(1)381⟩
Vanderhaeghe, H., **2**, 167 ⟨76JOC2561⟩
van der Heide, T. A. J., **1**, 194 ⟨93TL3309⟩; **2**, 521 ⟨93TL4655⟩
van der Heiden, R., **4**, 792 ⟨87S76⟩
van der Helm, D., **1**, 146 ⟨90JOC3840⟩, 637 ⟨84JOC670⟩
Van Der Heyden, A., **2**, 756 ⟨66RTC56⟩
van der Horn, D. E., **2**, 532, 697 ⟨79TL1665⟩
Vander Jagt, D. L., **1**, 905 ⟨85JOC1165⟩
Van der Kerk, G. J. M., **2**, 273 ⟨73OSC(5)223⟩, 601 ⟨56CI(L)352⟩, 1096 ⟨78JOM(153)C37, 79JOM(174)129⟩, 1100 ⟨64RTC295⟩; **4**, 699 ⟨70JOM(23)471⟩, 724 ⟨77JOM(127)391⟩; **5**, 1026 ⟨55RTC1262⟩; **6**, 209 ⟨70JOM(23)471⟩, 405 ⟨65RTC43⟩, 560 ⟨51RTC917, 51RTC949⟩
van der Klein, P. A. M., **1**, 185 ⟨91S347⟩; **6**, 317 ⟨91S347⟩
van der Knaap, T. A., **1**, 445 ⟨83T4225⟩; **2**, 441 ⟨83T4225⟩, 1003 ⟨84T765⟩; **4**, 340 ⟨83T3189⟩, 483 ⟨84T991⟩
Van der Kooi, J. P., **2**, 241, 358 ⟨70TL4983⟩
van der Leij, M., **1**, 704 ⟨80RTC39⟩; **3**, 368 ⟨78TL807, 78TL3383⟩, 368 ⟨78TL811⟩; **4**, 270, 272 ⟨80RTC39⟩, 859 ⟨80RTC45⟩, 936 ⟨78TL3383⟩; **5**, 562 ⟨78TL807, 83RTC400⟩
van der Louw, J., **1**, 194 ⟨88TL3579⟩; **2**, 324 ⟨88TL3579⟩, 925 ⟨87TL2889, 92T6087⟩, 925 ⟨92T6105⟩; **4**, 1061 ⟨87TL2889⟩
van der Marel, G. A., **1**, 185 ⟨91S347⟩, 648 ⟨89TL229⟩; **2**, 106 ⟨91TL6021⟩, 409 ⟨84TL1253⟩; **4**, 176, 199, 352 ⟨90TL2197⟩, 184 ⟨87RTC545⟩, 220 ⟨91T1547⟩; **6**, 317 ⟨91S347⟩, 549 ⟨92MI 617-05⟩
van der Merwe, J. P., **1**, 733 ⟨68JCS(C)1982⟩
van der Merwe, T. L., **4**, 190 ⟨91H(32)1445⟩
van der Plas, H. C., **3**, 419 ⟨84TL3763⟩
van der Plas, M. C., **4**, 893 ⟨77RTC68⟩
Vanderpool, S., **1**, 801 ⟨73JA8096⟩
Van der Puy, M., **2**, 16 ⟨91JOC5962⟩, 614 ⟨82JFC(21)385⟩; **3**, 149 ⟨91JOC5948⟩; **4**, 391 ⟨76JOC1883⟩; **6**, 7 ⟨93JFC(61)133⟩
van der Schaaf, P. A., **4**, 656 ⟨93OM3624, 93OM3955⟩
Van der Schaaf, P., **4**, 704 ⟨91NJC301⟩
VanderSchueren, D., **3**, 12 ⟨92JOC5761⟩
Vanderslice, C. W., **1**, 4, 7, 9, 10, 17, 429 ⟨69JOC3923⟩
van der Sluis, P., **3**, 512 ⟨90RTC446⟩
Vanderstede, E., **5**, 622 ⟨91JCS(P1)607⟩
Van der Steen, F. H., **5**, 22 ⟨91T7503⟩, 377 ⟨91JA5742⟩
Vanderstichele, P. L., **4**, 761 ⟨20JCS691⟩
Vander Stouw, G. G., **3**, 465 ⟨65JA935⟩
Van Der Veek, A. P. M., **6**, 39, 40 ⟨71RTC556⟩, 43 ⟨69TL5003⟩, 46, 48 ⟨72RTC349⟩
VanDerveer, D., **1**, 886 ⟨76JA7835, 89JOC1671⟩; **4**, 165 ⟨89JOC91⟩
van der Veken, B. J., **4**, 130 ⟨85JCP1517⟩
van der Welle, R. A., **1**, 1048 ⟨73RTC667⟩; **3**, 675 ⟨73RTC667⟩
Vander Werf, C. A., **1**, 42 ⟨56JOC918⟩, 268 ⟨62JA677⟩, 881 ⟨55OSC(3)44⟩; **2**, 436, 442, 443 ⟨59JA248, 64JA2378⟩, 842 ⟨59JA3805⟩, 848 ⟨56JA3063⟩; **5**, 203 ⟨64JOC2993⟩
Vander Zanden, J. M., **4**, 206 ⟨55RTC1429⟩
Vander Zwan, M. C., **1**, 1143 ⟨74JOC761⟩; **2**, 679 ⟨73JOC2910⟩

van de Steen, ,M., **4**, 156 ⟨81JOM(214)191⟩
Vandevelde, O., **2**, 938, 944 ⟨87TL397⟩; **3**, 424 ⟨87TL397⟩
van de Ven, L. J. M., **1**, 403, 799 ⟨70CC1168⟩
Vandewalle, J. J. M., **5**, 773 ⟨71CB639⟩
Vandewalle, M., **1**, 19 ⟨87TL4759⟩, 356 ⟨78TL2057⟩; **3**, 206 ⟨81T2085⟩, 532 ⟨84SC163⟩, 634, 636 ⟨78SC231⟩; **6**, 128 ⟨91TL2821⟩
Van Dine, G. W., **1**, 894 ⟨66JA2321⟩; **6**, 63 ⟨72JA125⟩
van Dolah, R. W., **2**, 103 ⟨55CRV485⟩
Van Doorn, J. A., **2**, 432 ⟨92RTC170⟩, 439 ⟨86EUP184872⟩
van Doorn, R., **4**, 50 ⟨79JA7617⟩
Van Dorp, A. D., **3**, 62 ⟨48RTC973⟩
van Dorp, D. A., **2**, 676 ⟨48RTC973⟩
Van Dort, M., **6**, 606 ⟨87JMC1241⟩
van Drunen, J. A. A., **3**, 364 ⟨75RTC1⟩
van Duijneveldt, F. B., **1**, 1114 ⟨78JOC763⟩
Van Duuren, B. L., **4**, 47 ⟨72CEN55, 72CEN62⟩
Van Duyne, G., **1**, 690 ⟨84JA5018, 86JA2932⟩
Van Duyne, G. D., **3**, 545 ⟨89JA9057⟩
van Duzer, J. H., **3**, 419 ⟨89JA371⟩
Van Dyke, C. H., **4**, 360 ⟨67IC1751, 67IC1989⟩, 638 ⟨70MI 414-01⟩, 638, 641 ⟨72IC408⟩; **6**, 173 ⟨67IC1751⟩
Van Dyke, J. W., **3**, 247 ⟨62JOC3888⟩
Vandyukova, I. I., **4**, 132 ⟨91JGU1459⟩
Vane, F. M., **3**, 446 ⟨63JA2784⟩
Van Eenam, D. N., **1**, 399 ⟨67JOC1155⟩
van Eenoo, M., **2**, 36 ⟨79TL1801⟩
van Eijk, P. J. S. S., **3**, 428 ⟨86RTC103⟩
van Eikema Hommes, N. J. R., **4**, 618 ⟨88TL5237, 89TL6195⟩, 668, 699 ⟨87RTC514⟩, 669 ⟨88RTC393⟩; **6**, 399 ⟨89TL6195⟩
van Elburg, P. A., **1**, 348 ⟨89TL2501⟩; **5**, 325 ⟨87H(26)437, 88RTC381⟩
van Elburg, P., **2**, 36 ⟨83SC387⟩
Vanelle, P., **1**, 156 ⟨91TL4125⟩
Van Emster, K., **1**, 380 ⟨22CB2500⟩
Van Ende, D., **1**, 21, 25, 437 ⟨76TL2643⟩, 712, 713 ⟨89H(28)1203⟩; **2**, 278, 294 ⟨85T4793⟩, 281 ⟨75AG(E)700⟩, 728, 936 ⟨79TL485⟩; **4**, 238, 239, 283, 375 ⟨75TL1613⟩, 287 ⟨75AG(E)700⟩, 287, 288 ⟨85T4793, 79TL485⟩; **6**, 94 ⟨79JOM(177)1⟩, 136 ⟨78JOM(149)C10⟩
Van Engen, D., **4**, 232 ⟨89JA6881⟩; **6**, 604 ⟨78TL1391⟩
van Erkelens, P. A. E., **5**, 1124 ⟨78S775⟩
van Es, T., **3**, 86 ⟨65JCS5775⟩; **4**, 228 ⟨76CAR(46)237, 79CAR(70)217⟩, 238, 287 ⟨77CAR(59)351⟩
Vanetten, R. L., **5**, 1127 ⟨85MI 530-01⟩
Vangedal, S., **3**, 704 ⟨80T515⟩; **4**, 743, 748 ⟨60T149⟩
van Gennep, H. E., **3**, 708 ⟨73TL627⟩
Vangheluwe, P., **5**, 604 ⟨84BSB405⟩
van Gheluwe, P., **6**, 296, 297 ⟨74TL4409⟩
Van Ginkel, C. H. D., **3**, 548 ⟨67RTC907, 68RTC1236⟩; **5**, 589, 614 ⟨68RTC1236⟩
van Gösseln, H. J.,**6**, 566 ⟨92AP(325)173⟩
van Guldener, D. B., **3**, 574 ⟨76RTC154⟩
van Ham, G., **2**, 211 ⟨58LA(617)62⟩
Van Hamme, M. H. J., **6**, 695 ⟨77JFC(10)131⟩
van Hamme, M. J., **4**, 753 ⟨77JFC(10)131, 79JFC(13)407, 79JOM(169)123⟩; **6**, 695 ⟨79JOM(169)123⟩
van Heerden, F. R., **4**, 190 ⟨91H(32)1445⟩
Van Heerden, P. S., **3**, 24, 146 ⟨92TL2383⟩
van Hemert, A. W., **3**, 709, 720 ⟨85RTC50⟩; **4**, 328 ⟨85RTC177⟩; **6**, 122 ⟨85RTC177⟩
Vanhessche, K., **6**, 128 ⟨91TL2821⟩
van Heyningen, E., **6**, 560 ⟨65JMC174⟩
Van Heyningen, E. M., **5**, 81 ⟨69JA5674⟩
Van Hijfte, L., **1**, 944 ⟨85JOC3940⟩
Van Hoeven, H., **6**, 570, 645 ⟨75JMC90⟩
Van Hook, J. O., **1**, 1069 ⟨54JA1700⟩; **4**, 195 ⟨49JA2741⟩, 195, 195 ⟨49JA2736⟩
Van Horn, D. E., **1**, 468 ⟨77JA3168⟩, 468, 471, 533, 1020, 1097, 1104 ⟨78JA2254⟩, 468, 533, 536 ⟨87JA2393⟩, 468, 536, 1104 ⟨78TL1027⟩, 468, 543 ⟨78JA2252⟩, 534 ⟨76TL1927⟩, 542, 543 ⟨81JOC4093⟩, 544 ⟨83OM563⟩; **2**, 532, 698 ⟨79TL2229⟩, 629, 975, 976 ⟨79S501⟩, 919, 925 ⟨78JOM(156)C20, 925, 969, 976 ⟨78JA2254⟩, 925, 974, 975 ⟨78JA2252⟩, 925, 974, 975 ⟨85JA6639⟩, 969 ⟨77JA3168, 78TL1027⟩, 973 ⟨83OM563⟩, 974, 975 ⟨81JOC4093, 86OCS467⟩, 992 ⟨87JA2393⟩; **4**, 1066 ⟨85JA6639⟩
Vanhorn, W. F., **2**, 217 ⟨70TL3551⟩
Van Horn, W. F., **2**, 221 ⟨70JA7224⟩
Vanhorne, J. L., **2**, 172 ⟨70JOC3267⟩
Vanhove, A., **6**, 480 ⟨90MI 614-02⟩
Van Huelsen, E., **4**, 382 ⟨85CB2822⟩
Van Hulsen, E., **4**, 384 ⟨85TL411⟩
van Hummel, C. J., **4**, 1074, 1075 ⟨84JA3785⟩
van Hummel, G. J., **1**, 518 ⟨86JOC2004⟩; **6**, 92 ⟨84JA3784⟩
Vanier, N. R., **1**, 142, 155 ⟨75JA7006⟩, 831 ⟨79CJC1743⟩; **4**, 391 ⟨76JOC1883⟩
van Kamp, H., **4**, 748 ⟨65RTC904⟩
Vankar, P., **3**, 134 ⟨86JOC3063⟩
Vankar, Y. D., **1**, 5, 8, 9 ⟨79S59⟩, 5, 8, 9 ⟨80JOC3531⟩; **2**, 6, 11, 17, 392, 615 ⟨79JOC3872⟩, 10 ⟨91TL1081⟩, 34 ⟨84TL233⟩, 36 ⟨85TL2717⟩, 785 ⟨85SC1253⟩; **3**, 151, 153 ⟨79TL3653⟩, 155 ⟨84TL233⟩, 155 ⟨89CL1241⟩, 618 ⟨78S702⟩, 618, 662 ⟨79S227⟩, 620 ⟨79S36⟩; **4**, 32 ⟨79JOC3842⟩; **5**, 4 ⟨79JOC3872⟩, 125, 189, 196, 297, 303 ⟨82S506⟩, 321 ⟨84TL233⟩; **6**, 439 ⟨79JOC3872⟩
Vankateswaran, R. V., **3**, 528 ⟨88CC1421⟩
van Koeveringe, J. A., **1**, 957 ⟨76RTC80⟩
van Koten, G., **1**, 127 ⟨82JA6609, 82OM1366⟩, 1018 ⟨88T6699⟩; **2**, 585, 589 ⟨82COMC-I(2)709⟩, 940 ⟨93JOC5849⟩; **3**, 512 ⟨90RTC446⟩; **4**, 656 ⟨93OM3624, 93OM3955⟩, 660 ⟨86OM1551⟩, 695 ⟨75JOM(84)117, 75JOM(84)129⟩, 696 ⟨75JOM(102)551⟩, 704 ⟨91NJC301⟩, 708 ⟨73JOM(55)419⟩, 723 ⟨93OM2227⟩, 724 ⟨80JOM(186)427⟩; **5**, 22 ⟨91T7503⟩, 377 ⟨91JA5742⟩
van Leusen, A. M., **1**, 785 ⟨59RTC551⟩; **2**, 156 ⟨79S733⟩; **3**, 194, 709 ⟨77TL4233⟩, 214 ⟨84TL2585⟩, 479 ⟨73TL5207, 75T597⟩, 621 ⟨73TL1357, 77JOC3114⟩, 694 ⟨72SC281⟩, 694, 709 ⟨77TL4229⟩, 694, 709 ⟨80S325⟩, 694, 709 ⟨85RTC50⟩, 694, 709 ⟨B-87MI 321-01⟩, 696, 709 ⟨88OSC(6)987⟩, 696, 711 ⟨86SC865⟩, 699, 713, 718 ⟨92JOC2249⟩, 702, 713 ⟨91RTC402⟩, 706 ⟨91S531⟩, 708 ⟨73TL627⟩, 709 ⟨72TL2369, 75TL3487⟩, 709 ⟨81JOC5159⟩, 709, 720 ⟨85RTC177⟩, 712 ⟨95TL2109⟩, 714, 716 ⟨90S568⟩, 714, 720 ⟨82RTC202⟩, 718 ⟨91RHA270⟩; **4**, 218, 235 ⟨64TL543⟩, 218, 235 ⟨67TL3057, 67TL3061⟩, 235 ⟨70QRS67⟩, 328 ⟨85RTC177⟩, 331 ⟨75TL3487, 81JOC2069⟩, 331 ⟨84RTC41, 88OSC(6)981⟩, 331 ⟨88OSC(6)987⟩, 331, 538 ⟨72TL2367⟩, 332 ⟨62RTC966⟩, 517 ⟨86SC865⟩, 538 ⟨77TL4229, 80S325⟩, 908 ⟨79RTC258, 80MI 420-01⟩, 908 ⟨82RTC202, 84TL2581⟩; **5**, 566, 618 ⟨76TL143⟩, 867 ⟨73TL5207⟩, 867, 870 ⟨75T597⟩, 898, 901, 902 ⟨72RTC37, 74JOC2728⟩; **6**, 122 ⟨85RTC177⟩, 605, 624 ⟨81JHC1127⟩, 699 ⟨72RTC37, 74JOC2728⟩
van Leusen, D., **1**, 400, 823 ⟨84JOC1917⟩; **3**, 194, 709 ⟨77TL4233⟩, 479 ⟨73TL5207, 75T597⟩, 621 ⟨77JOC3114⟩, 694, 709 ⟨80S325⟩, 702, 713 ⟨91RTC402⟩, 706 ⟨81JOC5159, 91S531⟩, 712 ⟨95TL2109⟩, 718 ⟨91RTC393⟩; **4**, 331 ⟨84RTC41⟩, 538 ⟨80S325⟩, 908 ⟨79RTC258, 84TL2581⟩; **5**, 867 ⟨73TL5207⟩, 867, 870 ⟨75T597⟩
van Lombard, A.,**5**, 951 ⟨85OM275⟩
van Look, G., **2**, 539 ⟨B-88MI 210-04⟩
van Loon, J.-D., **1**, 990 ⟨93AG(E)1187⟩; **6**, 556 ⟨90CC1432⟩

Vanltier, M., **2**, 317, 469 ⟨92BSF71⟩
van Mechelen, J. B., **6**, 190, 193, 195 ⟨91OM1704⟩
Van Meerssche, M., **3**, 46 ⟨78BSB693⟩, 416, 416 ⟨86JOC3839⟩, 566 ⟨88JOC18⟩, 602 ⟨82JA2920⟩; **4**, 700 ⟨84JOM(277)335⟩; **5**, 481 ⟨75AG(E)53⟩, 601 ⟨85T693⟩; **6**, 74 ⟨78BSB693⟩, 105 ⟨83TL1035⟩
Van Meervelt, L., **5**, 604 ⟨84BSB405⟩
van Melick, J. E. W., **4**, 177, 190 ⟨69CC1175⟩; **6**, 44 ⟨71RTC1123⟩
VanMiddlesworth, F., **1**, 700, 703 ⟨86JA4603⟩
Van Mier, G. P. M., **5**, 377 ⟨91JA5742⟩
van Mourik, G. L., **2**, 423 ⟨76TL2565⟩
Vannes, P., **3**, 529 ⟨81JA4616⟩
van Niel, M. B., **1**, 165 ⟨89T7643⟩; **2**, 749 ⟨88JCS(P1)269⟩; **3**, 20, 29, 31, 137, 157, 160, 169 ⟨89T7643⟩, 43, 182 ⟨87CC775⟩; **4**, 244, 250 ⟨89T7643⟩, 849 ⟨88JCS(P1)269⟩, 1076 ⟨85CC742⟩
van Nispen, S. P. J. M., **4**, 264 ⟨81JOC3273⟩, 331 ⟨81JOC2069⟩
Vannoorenberghe, Y., **2**, 64 ⟨88TL3235⟩
van Ornum, J. V., **4**, 894 ⟨71JOC1685⟩
Vanotti, E., **3**, 284 ⟨84CC530⟩
Van Peppen, J. F., **1**, 73 ⟨70CC495, 70CC497⟩, 73 ⟨70CC571⟩
van Rantwijk, F., **1**, 80 ⟨72RTC1057⟩
Van Reijendam, J. W., **1**, 731 ⟨70T1291⟩; **3**, 494 ⟨70T1291⟩; **5**, 991 ⟨73S413⟩
van Rens, E. M. M., **6**, 316 ⟨73TL3589⟩
Van Rheenen, V., **1**, 993 ⟨79JOC1582⟩
Van Rheenen, V. H., **4**, 1003, 1004, 1011 ⟨85USP4501695⟩
Van Rijn, P. E., **1**, 981 ⟨82JOM(233)C25⟩, 1037 ⟨81S459⟩; **2**, 939 ⟨82JOM(232)C1⟩, 940 ⟨82JOM(233)C25⟩, 1096 ⟨78JOM(153)C37, 79JOM(174)129⟩
van Rogy, P., **1**, 585 ⟨89JOC2217⟩
van Rooyen, P. H., **4**, 720 ⟨93AG(E)710⟩; **5**, 651 ⟨88JOM(355)337⟩; **6**, 723 ⟨94AG(E)672⟩
van Rozendaal, H. L. M., **5**, 562 ⟨85JOC2930⟩
van Schaik, H. P., **5**, 890 ⟨92OM3844, 93PS(77)252⟩
Van Schaik, T. A. M., **4**, 829 ⟨83TL1303⟩; **6**, 116 ⟨83TL1303⟩
Van Sickle, D. E., **1**, 857 ⟨62JA258⟩
van Soolingen, J., **1**, 95, 889 ⟨90SC2165⟩
Van Strien, R. E., **1**, 648 ⟨47JA1551⟩
van Sweiten, A. P.,**5**, 438 ⟨74TL3377, 76C199⟩
van Tamelen, E. E., **1**, 13, 18, 21 ⟨74JA5290⟩, 21, 23, 435 ⟨62OR(12)356⟩, 49, 51, 157 ⟨70JA2139⟩, 82 ⟨65CC71⟩, 174 ⟨65JA3277⟩, 216 ⟨72JA8225⟩, 438 ⟨66OR(12)356⟩; **3**, 59 ⟨53JA384⟩, 232 ⟨56JA4405⟩, 493 ⟨70JA2139⟩; **5**, 166 ⟨50JA1357⟩, 166 ⟨58JA3079⟩
van't Hoff, J. H., **6**, 216, 221 ⟨1877CB678⟩
van Tilborg, M. W. E. M., **4**, 50 ⟨79JA7617⟩
van Tilborg, W. J. M., **4**, 299 ⟨73SC189, 74RTC294⟩, 406 ⟨67RTC417⟩
van Veen, R., **2**, 523 ⟨73JOM(47)33⟩
van Vliet, L. A., **4**, 859 ⟨92TL6383⟩
van Vuuren, G., **2**, 244 ⟨84CC1144⟩
VanWinkle, J. L., **2**, 462 ⟨53USP2642461⟩
Van Woerden, H. F., **2**, 100, 101 ⟨63CRV557⟩
Van Zanten, B., **6**, 414 ⟨77MI 614-01, 78MI 614-03⟩
van Zomeren, J. A. J., **3**, 708 ⟨78JOC1972⟩
Van Zwieten, P. A., **6**, 645 ⟨78RTC51⟩
Van Zyl, C. M., **2**, 611, 630, 916 ⟨86JOC3561⟩, 916, 932 ⟨86TL2801⟩; **3**, 271 ⟨86TL2801⟩
VanZyl, G., **5**, 166 ⟨50JA1357⟩
Van Zyl, W. E., **6**, 540 ⟨92POL893⟩
van Zyl, W. J., **5**, 573 ⟨90SC2235⟩
Vaquer, J., **2**, 902 ⟨90TL4545⟩, 906 ⟨93JA6065⟩; **4**, 630, 631 ⟨93JA6065⟩
Vaquero, J. J., **2**, 249 ⟨92TL3677⟩, 301 ⟨91SC535⟩; **3**, 171 ⟨92TL3677⟩, 306 ⟨93JOC5862⟩
Vara, M. C., **1**, 657 ⟨68JCS(B)1044⟩
Varache, M., **6**, 194 ⟨72JOM(40)C53⟩
Varadarajan, A., **1**, 18, 564 ⟨84JA6702⟩, 917 ⟨83JOC1431⟩
Váradi, G., **4**, 688 ⟨81JOM(206)119⟩
Varakin, G. S., **4**, 35 ⟨83ZOR1625⟩
Vara Prasad, J. V. N., **2**, 42 ⟨88BCJ93, 88JOC2911⟩, 517 ⟨85JOC1582, 85JOC5583⟩, 519, 531 ⟨86JA2049⟩, 519, 531 ⟨87JOC310⟩, 561 ⟨83JOM(259)1⟩, 901 ⟨88JOC239⟩
Vara-Prasad, W. V. N., **2**, 519 ⟨86JOC4296, 86JOC4526⟩
Varaprath, S., **1**, 493 ⟨84JA5274⟩
Varasi, M., **3**, 684, 686 ⟨93JCS(P1)2151⟩
Varela, A. P. G. V. de, **2**, 749 ⟨68T2481⟩
Varella, E. A., **5**, 765 ⟨83CCC596⟩
Varenne, J., **2**, 410 ⟨84TL4511⟩
Varetti, E. L., **6**, 230 ⟨67MI 607-02⟩
Varga, I., **4**, 319 ⟨72ACH363⟩
Varga, S. L., **4**, 97 ⟨88EUP264183⟩
Vargas, L. A., **4**, 555 ⟨83JOC4775⟩
Vargha, L., **2**, 177 ⟨44ZPC(281)156⟩
Varia, T., **2**, 381 ⟨84JHC1709⟩
Varick, T. R., **3**, 621 ⟨90TL2975⟩
Varie, D. L., **3**, 331 ⟨82JA1445, 86JOC1556⟩
Varkey, T. E., **2**, 162 ⟨74JOC3365⟩
Varkony, H., **2**, 86 ⟨88OSC(6)43⟩
Varkony, T. H., **2**, 86 ⟨75JOC2141⟩
Varlet, J. M., **3**, 50 ⟨81T1377⟩
Varley, J. H., **2**, 291 ⟨76CC61⟩
Varley, J. S., **6**, 312, 436 ⟨84JFC(24)485⟩
Varma, K. S., **4**, 850 ⟨87S837⟩
Varma, M., **3**, 426 ⟨85SC1325, 85TL6013⟩, 426 ⟨86SC91⟩, 714 ⟨86JA3153⟩; **4**, 307 ⟨82JCS(P1)2771⟩
Varma, P. S., **4**, 11 ⟨44JIC112⟩
Varma, R., **5**, 1016 ⟨68JA6247⟩
Varma, R. S., **2**, 341 ⟨85JOC133, 87JHC767⟩, 393, 398, 784 ⟨87OPP283⟩, 398 ⟨84SC1093, 85SC151⟩, 398 ⟨90T7443⟩; **3**, 165 ⟨85CL243, 85SC443⟩, 426 ⟨85JOC133, 85SC1325⟩, 426 ⟨85TL6013, 86SC91⟩; **4**, 307 ⟨82JCS(P1)1193, 82JCS(P1)2771⟩
Varma, V., **1**, 546, 1024 ⟨81JA7520⟩, 1023, 1024 ⟨76CC907⟩, 1023, 1024 ⟨78JOC353⟩; **5**, 170 ⟨76CC907⟩
Varnaeva, G. N., **2**, 391 ⟨89RCR470⟩
Varney, M. D., **1**, 450 ⟨84JA3668⟩, 831 ⟨86JOC635⟩
Varsanyi, G., **2**, 608, 619 ⟨57JCS1823⟩
Varsel, C., **5**, 168 ⟨68JOC2819⟩
Varshavskii, S. L., **2**, 448 ⟨68CA-P106877⟩, 449 ⟨68DOK(182)337⟩
Vartanyan, S. A., **1**, 965 ⟨63IZV499⟩; **5**, 690 ⟨77KFZ38, 90CHE1170⟩, 690 ⟨90KGS1403⟩
Varvoglis, A., **2**, 390 ⟨75S445⟩, 619, 626, 631 ⟨82JCR(S)150⟩, 998 ⟨89LA167, 90AX(C)1300⟩; **3**, 500 ⟨87JCR(S)374, 88S913⟩; **4**, 866 ⟨89JCS(P1)379⟩, 867 ⟨85JA7178⟩, 1081 ⟨88JCS(P1)2839⟩; **5**, 205 ⟨86JCS(P1)795⟩, 318, 321, 322 ⟨83S207⟩; **6**, 254 ⟨88JCR(S)306⟩, 278, 699, 708 ⟨88S913⟩, 562 ⟨87JCS(P1)2759⟩
Varvounis, G., **6**, 28 ⟨89JCS(P1)283⟩
Vasella, A., **1**, 321, 946 ⟨83HCA2210⟩; **2**, 292 ⟨73JA7917⟩, 397 ⟨83HCA789⟩; **3**, 428 ⟨85HCA2254⟩; **4**, 190 ⟨93HCA211⟩, 226 ⟨93HCA995⟩, 229 ⟨93HCA1779⟩, 485, 489 ⟨82HCA1953, 85HCA1730⟩, 912 ⟨91HCA451⟩; **5**, 212 ⟨85HCA1730⟩; **6**, 549 ⟨93HCA995⟩
Vasileff, R. T., **5**, 81 ⟨69JA5674⟩
Vasil'eva, A. S., **2**, 171 ⟨59ZOB3792⟩; **6**, 229 ⟨58JGU2539⟩, 233, 234, 238 ⟨59ZOB3786⟩, 236 ⟨59ZOB3784⟩, 236, 528, 533, 534 ⟨59ZOB3792⟩, 440 ⟨57ZOB2243⟩
Vasil'eva, G. A., **2**, 566 ⟨72SRI33⟩
Vasil'eva, I. A., **2**, 458 ⟨79JGU1714⟩

Vasil'eva, L. L., **5**, 162 ⟨78JGU1249⟩
Vasil'eva, M. A., **6**, 562 ⟨84ZOR2114⟩
Vasil'eva, M. N., **6**, 613 ⟨71ZOR2267⟩
Vasileva, N. V., **1**, 103 ⟨91ZOR1137⟩
Vasil'eva, T. P., **2**, 173 ⟨81IZV159⟩, 211 ⟨81IZV1850⟩, 227 ⟨80IZV2111⟩
Vasil'eva, T. T., **6**, 5 ⟨87IZV808⟩
Vasileva, V., **4**, 339 ⟨84PS(21)59⟩; **6**, 502 ⟨73GEP2131789⟩
Vasilev, G., **6**, 558 ⟨89MI 617-01⟩
Vasil'ev, L. S., **2**, 533 ⟨61DOK(139)385⟩
Vasil'ev, N. V., **5**, 666, 671, 692 ⟨86MI 517-02⟩
Vasil'ev, V. V., **2**, 399 ⟨77ZOR727⟩; **3**, 374 ⟨78ZOR1659⟩; **4**, 326 ⟨78ZOR1659⟩, 573 ⟨77URP403314, 78URP351438⟩, 574 ⟨72ZOB293⟩
Vasilevski, S. F., **1**, 1010 ⟨80IZV726⟩
Vasilevsky, V. I., **2**, 617 ⟨79S533⟩
Vasilvizky, A. E., **3**, 232 ⟨77AG(E)646⟩
Vasilyeva, A. S., **5**, 476 ⟨57JGU2301⟩
Vasilyeva, E. I., **5**, 38, 89 ⟨57JGU2481⟩
Vasilyeva, T. T., **5**, 38, 89 ⟨57JGU2481⟩, 55, 60, 89 ⟨62T69⟩
Vasilyev, L. S., **1**, 60 ⟨82JOM(226)115⟩
Vasisht, S.-K., **6**, 193, 399 ⟨91AG(E)93⟩, 608 ⟨86JOM(301)15⟩
Vasishtha, S. C., **5**, 716 ⟨67JCS(C)1309⟩
Vaska, L., **1**, 73 ⟨62JA679⟩
Vass, A., **4**, 446 ⟨86S817⟩; **5**, 1039 ⟨86S817⟩; **6**, 565 ⟨86S817⟩
Vasser, M., **1**, 727 ⟨77JOC2783⟩; **3**, 494 ⟨77JOC2783⟩
Vassilatos, S. N., **4**, 539 ⟨83T2023⟩
Vassilios, V. P., **3**, 429 ⟨91OM3967⟩
Vasta, J. A., **5**, 1017 ⟨54JOC1300⟩
Vastine, F. D., **5**, 460 ⟨68JA2282⟩
Vasudevan, A., **5**, 999 ⟨88JOC5158⟩
Vasvari-Debreczy, L., **3**, 310 ⟨77CZ133⟩; **4**, 422 ⟨81T4337⟩
Vasyanina, M. A., **2**, 854 ⟨69ZOB341⟩; **4**, 125, 127, 128 ⟨68JGU2006, 78JGU1813⟩, 129 ⟨78JGU1168⟩, 555 ⟨68ZOB2071⟩; **5**, 534 ⟨73JGU2635⟩; **6**, 538, 539 ⟨89IZV909⟩, 538, 539, 584 ⟨88ZOB1489⟩
Vasyukov, S. E., **6**, 54 ⟨66ZOB728⟩
Vatakencherry, P. A., **1**, 389 ⟨64JA478⟩
Vatele, J. M., **1**, 955, 971 ⟨93SL105⟩; **2**, 9 ⟨81TL3579⟩, 314 ⟨90TL2277⟩, 678 ⟨84TL5997⟩, 930 ⟨91TL1187⟩; **3**, 180 ⟨90TL2277⟩; **4**, 764 ⟨85CAR(136)177⟩, 766 ⟨83TL1239, 86T4443⟩
Vatelino, M. G., **2**, 740 ⟨94TL219⟩
Vatsuro, K. V., **1**, 870 ⟨67TL4525⟩
Vaughan, D., **5**, 226 ⟨83JCS(P1)49⟩
Vaughan, D. J., **3**, 459 ⟨50JA876⟩
Vaughan, J., **1**, 258 ⟨60JOC463⟩
Vaughan, J. F. S., **1**, 570 ⟨94JCS(P1)3115⟩
Vaughan, J. G., **3**, 171 ⟨93CC1434⟩
Vaughan, Jr., J. R., **4**, 420 ⟨60JA2731⟩; **5**, 193 ⟨51JA5553⟩
Vaughan, K., **4**, 444 ⟨83CC721, 84CJC749⟩
Vaughan, L. G., **2**, 440 ⟨68JOC3088⟩, 909, 910, 911 ⟨64JA883⟩
Vaughan, R. J., **3**, 476 ⟨68JA4088⟩
Vaughan, W. E., **2**, 116 ⟨42JOC472⟩, 439 ⟨52JA3282⟩
Vaughan, W. R., **1**, 393 ⟨59OS(39)37⟩; **2**, 27 ⟨62JA769⟩, 278 ⟨55JOC1019⟩; **5**, 146 ⟨59OS(39)37⟩
Vaughan-Williams, G. F., **1**, 212 ⟨93T3007⟩; **4**, 704 ⟨93T3007⟩
Vaughn, G. D., **4**, 150 ⟨86JA1455⟩; **5**, 449 ⟨86JA1462⟩
Vaughn, H., **4**, 338 ⟨71JOC3843⟩; **6**, 118 ⟨71JOC3843⟩, 263 ⟨71JOC1835⟩
Vaughn, T. H., **1**, 1000 ⟨37JOC1⟩; **2**, 1018 ⟨32JA787, 33JCS741⟩, 1018, 1019 ⟨33JA2150⟩
Vaultier, M., **1**, 514 ⟨84TL1555⟩; **2**, 182 ⟨92SL539⟩, 318 ⟨87T1799⟩, 321 ⟨93SL519⟩, 324 ⟨87TL1757⟩, 900 ⟨87TL4169⟩; **4**, 329, 392 ⟨80JA7929⟩; **5**, 199 ⟨87TL4169⟩, 494 ⟨75JOC2990⟩, 731 ⟨80JA7929⟩
Vaupel, A., **2**, 591 ⟨93TL7911⟩
Vauzeilles, B., **4**, 238 ⟨93SL522⟩
Vavaprasad, J. V. N., **3**, 276 ⟨88JOC1391⟩
Vavasourn, G. R., **2**, 456 ⟨53CJC976⟩
Vavayannis, A., **6**, 241 ⟨81JFC(17)331⟩
Vaver, V. A., **1**, 722 ⟨63IZV1053⟩; **2**, 1084 ⟨57IZV812⟩; **4**, 752 ⟨64TL2669⟩
Vaverde, S., **1**, 942 ⟨92CC613⟩
Vavilova, A. N., **6**, 562 ⟨81ZOR2269⟩
Vawter, E. J., **1**, 1015 ⟨90TL4393⟩; **2**, 702 ⟨86JOC3993⟩, 980 ⟨90TL4393⟩; **3**, 128 ⟨86JOC3993⟩
Vaya, J., **2**, 166 ⟨80JCS(P1)2228⟩
Vaz, A. D. N., **2**, 22 ⟨84JOC1286⟩
Vazeux, M., **1**, 1058 ⟨77JOM(140)257⟩; **3**, 343 ⟨78RTC121⟩, 351 ⟨77BSF688⟩; **4**, 233 ⟨86JOC2276⟩
Vázquez, J., **2**, 347 ⟨93TL141⟩
Vázquez de Miguel, A., **4**, 687 ⟨83JCS(D)1441⟩
Včelák, J., **2**, 19 ⟨74CCC2616⟩; **4**, 32 ⟨74CCC2616⟩, 137 ⟨76CCC386⟩
Vdovenko, S. I., **6**, 609, 619 ⟨83KGS798⟩
Vdovin, V. M., **4**, 444 ⟨81IZV1603⟩, 610 ⟨85IZV1917⟩, 622 ⟨69DOK(189)334⟩, 656, 658 ⟨74IZV2861, 75IZV1600⟩, 658 ⟨85ZOB1629⟩
Veal, H. S., **2**, 164 ⟨78TL503⟩
Veale, C. A., **4**, 226 ⟨90JA6263⟩; **5**, 513 ⟨90JA6263⟩; **6**, 134 ⟨87JA2504⟩
Veazey, R. L., **6**, 76 ⟨77JOC679⟩
Veber, D. F., **2**, 223 ⟨68TL3057⟩; **4**, 417 ⟨87JOC2442⟩; **5**, 972 ⟨66JA3163⟩; **6**, 496 ⟨68JA3254⟩
Vecchiani, S., **2**, 30 ⟨86JOC4224⟩
Vecera, M., **6**, 565 ⟨81CCC1970⟩
Veciana, A. de, **3**, 361 ⟨92JCS(P1)351⟩
Veda, S., **3**, 412 ⟨80BCJ281⟩
Vedejs, E., **1**, 18 ⟨75OR(22)401⟩, 22, 942 ⟨82JA2046⟩, 141, 234, 821 ⟨78JOC1185⟩, 234, 410, 822 ⟨84ACR358⟩, 234, 821 ⟨78TL519⟩, 321 ⟨B-88MI 108-01⟩, 408, 524, 821, 823 ⟨79JA6452⟩, 410, 821 ⟨75JA6878, 78JOC4831⟩, 410, 823 ⟨78TL523⟩, 470, 736 ⟨75TL1359⟩, 471 ⟨92JA6556⟩, 491 ⟨74TL3207⟩, 524, 1140 ⟨89TL2627⟩, 617, 704 ⟨82T2857⟩, 619 ⟨78JOC4884⟩, 630 ⟨77TL135⟩, 640, 641 ⟨71JA4070⟩, 640, 641 ⟨73JA822⟩, 721 ⟨89JA5861⟩, 721, 729 ⟨73JA5778⟩, 736 ⟨73JOC3625⟩, 811 ⟨70JA2158⟩, 821 ⟨83JA127, 84JA4617, 87JA5878, 88JA4822⟩, 823 ⟨83JOC4773⟩, 834 ⟨84JOC1840⟩; **2**, 213 ⟨79JOC3230⟩, 324 ⟨86CRV941⟩, 445 ⟨79JA6452⟩, 678 ⟨79JA249⟩; **3**, 16 ⟨75TL2017⟩, 55 ⟨71JOC366⟩, 156 ⟨78JOC188⟩, 331 ⟨82JA1445, 83JA1683, 84JA4617, 86JOC1556⟩, 331 ⟨88JA4822, 88JOC2226⟩, 331, 332 ⟨86JA2985⟩, 331, 333, 335, 337 ⟨88JA5452⟩, 331, 335 ⟨88JOC2220⟩, 334 ⟨80JOC2601⟩, 351 ⟨84ACR358⟩, 494 ⟨73JA5778, 75TL1359⟩, 496 ⟨79JA6452⟩; **4**, 222, 365, 393 ⟨75TL2017⟩, 232 ⟨86JOC117⟩, 517 ⟨79JA6452, 80JA7993⟩, 517 ⟨83JOC4773⟩; **5**, 331 ⟨92TL3261⟩, 554 ⟨82JOC4384⟩
Vedejs, F., **5**, 177 ⟨87JOC3937⟩
Vederas, J. C., **1**, 230 ⟨87JA4649⟩; **2**, 375 ⟨86JA6395b⟩; **3**, 189 ⟨72JA9128⟩; **5**, 85 ⟨88JA2237, 91OS(70)1⟩, 1078 ⟨89JA3973⟩
Vedernikov, A. N., **4**, 922 ⟨92HAC127⟩
Vedjes, E., **3**, 496 ⟨86CRV941⟩; **5**, 554 ⟨80JOC2601⟩
Veenboer, J. T., **6**, 414 ⟨78MI 614-02⟩
Veeneman, G. H., **4**, 220 ⟨91T1547⟩
Veenstra, G. E., **4**, 272 ⟨78T1585⟩, 859 ⟨76RTC202⟩; **5**, 478, 491, 492, 562 ⟨77RTC139⟩
Veenstra, S. J., **2**, 521 ⟨85JA4549⟩
Veeraraghavan, S., **3**, 637 ⟨80S384⟩

Vega, J. A., **2**, 249 ⟨92TL3677⟩; **3**, 171 ⟨92TL3677⟩
Vega, J. C., **5**, 242 ⟨82SC897⟩, 513 ⟨86CL1251⟩; **6**, 475 ⟨82SC897⟩, 476, 551 ⟨85PS(25)91⟩, 479 ⟨90S825⟩, 534 ⟨86S760⟩, 546, 547 ⟨80AOC(18)1, 83JCS(P1)2641⟩, 558 ⟨85S975⟩
Vega, S., **5**, 655 ⟨87S628⟩
Vegas, A., **5**, 386 ⟨92JCS(D)53⟩
Vegh, D., **4**, 764 ⟨88MI 417-01⟩, 970 ⟨88ZOR2374⟩
Veinberg, G., **6**, 558 ⟨85KGS339⟩
Veit, A., **1**, 326 ⟨91JA1791⟩, 326, 327 ⟨93HCA441, 93TL2445⟩, 946 ⟨90SL166⟩; **2**, 490, 507 ⟨82AG(E)375⟩, 490, 507 ⟨85JOM(290)307⟩
Veit, J. R., **2**, 564 ⟨90JOC3565⟩
Veits, Yu. A., **2**, 471 ⟨83JGU2244⟩, 694 ⟨69ZOB2479⟩; **5**, 395, 396, 399 ⟨89JGU1817⟩, 395, 396, 399, 400 ⟨88JGU201⟩, 395, 396, 399, 400 ⟨88JGU2372⟩, 396, 399 ⟨87DOK114⟩, 398 ⟨91JGU114⟩, 398, 400 ⟨90JGU1810⟩, 399 ⟨86JGU636, 91JGU108⟩, 400 ⟨90JGU1817, 90JGU1822⟩
Vekemans, J., **4**, 833 ⟨83T2493⟩
Velarde, E., **1**, 2, 4, 6 ⟨71LA(748)123⟩; **3**, 235 ⟨73JA6655⟩
Vel'der, Y. L., **1**, 491 ⟨89S633⟩
Velezheva, V. S., **5**, 1025 ⟨78KGS1497⟩
Velić, M., **3**, 458 ⟨67AG818, 67AG(E)803⟩
Vellekoop, A. S., **1**, 313 ⟨94JA2902⟩
Velleman, K. D., **2**, 471 ⟨72CB406⟩
Velluz, L., **1**, 403 ⟨49BSF501⟩, 419 ⟨55BSF1341⟩; **3**, 226 ⟨57BSF1289⟩; **5**, 62 ⟨45BSF17⟩
Velsvik, M., **3**, 255 ⟨73ACS3881⟩
Veltheer, J. E., **4**, 656 ⟨93OM4572⟩
Veltmann, V., **6**, 695 ⟨77TL399⟩
Venanzi, L. M., **2**, 588 ⟨92TL5857⟩; **3**, 411 ⟨83JOC6877⟩; **4**, 186 ⟨93TL5269⟩, 566 ⟨91HCA983⟩
Vendley, R., **2**, 1014 ⟨60JOC1626⟩
Vene, J., **5**, 28 ⟨45BSF506⟩
Venegas, M. G., **2**, 403 ⟨78JOC2921⟩
Venemalm, L., **3**, 309 ⟨87TL3741⟩
Venerable, J. T., **4**, 825 ⟨50JA1661⟩; **6**, 69 ⟨50JA1661⟩
Veneziani, G., **3**, 682 ⟨92JA6059⟩; **5**, 871, 872, 1148 ⟨92JA6059⟩; **6**, 665, 670, 675 ⟨92JA6059⟩
Vengizhemovskii, V. A., **2**, 256 ⟨80ZOB2459, 87ZOB1078⟩
Venien, F., **4**, 405, 411 ⟨75BSF196⟩
Venier, C. G., **2**, 147, 155 ⟨82JOC3773⟩; **3**, 368 ⟨78TL807⟩; **4**, 73 ⟨74OPP77⟩, 74, 75, 76, 78 ⟨73CC319, 73JOC17⟩, 742 ⟨74JOC501⟩; **5**, 562 ⟨78TL807⟩, 867 ⟨75JA3238⟩
Venit, J., **2**, 543, 811 ⟨81TL1787⟩
Venkatachalam, T., **2**, 8 ⟨93OPP249⟩
Venkataramani, P. S., **3**, 267 ⟨79JCS(P1)36⟩; **4**, 213 ⟨79JCS(P1)36⟩
Venkateswarlu, A., **2**, 539 ⟨72JA6190⟩; **5**, 223, 374 ⟨72JA6190⟩
Venkateswarlu, R., **4**, 230 ⟨85JCS(P1)587⟩
Venkatramanan, M. K., **3**, 640 ⟨87JOC2427⟩; **4**, 520 ⟨89JOC644⟩
Venkov, A. P., **3**, 187 ⟨80TL809⟩; **4**, 488 ⟨91SC1951, 92S263⟩
Vennstra, G. E., **2**, 197 ⟨75S519⟩; **4**, 235 ⟨75S519⟩
Veno, Y., **1**, 290 ⟨86S776⟩
Venter, E. M. M., **1**, 828 ⟨91JCS(P1)561⟩
Venton, D. L., **2**, 374 ⟨81JOC5413⟩
Venturello, C., **2**, 51 ⟨89S295⟩; **3**, 117, 161 ⟨91JOC5924⟩
Venturello, P., **2**, 25, 31 ⟨87JCS(P1)2157⟩, 35 ⟨79S952⟩; **3**, 653 ⟨89JCS(P1)105⟩; **6**, 126 ⟨92T2501⟩
Venturini, A., **4**, 610 ⟨93JA3322⟩; **6**, 691 ⟨86JA5395⟩
Venturini, M., **6**, 341 ⟨81AG215⟩
Venugopal, D., **1**, 282 ⟨91JOC3908⟩
Venugopalan, B., **4**, 923 ⟨81TL191⟩
Venuti, M. C., **1**, 755 ⟨81JOC459⟩; **2**, 326 ⟨88S985⟩, 445 ⟨81USP4296042⟩; **3**, 310 ⟨90T6623⟩
Vepachedu, S. R., **1**, 174 ⟨90TL4931⟩, 174 ⟨91T519⟩; **6**, 550 ⟨90TL4931⟩
Vepsäläinen, J., **5**, 1028 ⟨91JHC1003⟩; **6**, 263 ⟨92JCS(P2)835, 92PS(70)183⟩, 263 ⟨93TL4551⟩
Veracini, C. A., **3**, 374 ⟨75TL2085⟩
Verani, G., **6**, 474 ⟨80T1451⟩, 595 ⟨80JHC571⟩
Verardo, G., **2**, 326 ⟨91S447⟩
Verbanc, J. J., **5**, 962 ⟨57CRV47⟩
Verbeek, J., **2**, 942 ⟨84CC257⟩
Verbeek, W., **4**, 3 ⟨66AG(E)314⟩; **6**, 442 ⟨67AG(E)871⟩
Verbeke, M., **5**, 622 ⟨93JCS(P1)1719⟩
Verber, D. F., **4**, 296 ⟨88OSC(6)5⟩
Verbicky, Jr., J. W., **2**, 622 ⟨82TL371⟩
Verboom, W., **1**, 47 ⟨87T3541⟩; **2**, 939 ⟨82JOM(232)C1⟩; **3**, 434 ⟨90BSF704⟩; **4**, 93 ⟨81RTC10⟩; **5**, 1105 ⟨78S577⟩; **6**, 556 ⟨90CC1432⟩, 562 ⟨92JA10573⟩
Verbruggen, A., **5**, 1123 ⟨79TL49, 81JOC4478⟩
Verbruggen, C., **4**, 461 ⟨94S23⟩
Verbruggen, R., **3**, 46 ⟨78BSB693⟩; **6**, 74 ⟨78BSB693⟩
Verček, B., **5**, 764 ⟨76JHC487⟩, 834 ⟨78H(9)1327, 79JOC1695⟩, 836 ⟨73TL1677, 74JHC103⟩, 836 ⟨74JOC2143, 82JHC577⟩, 837 ⟨84CS209⟩, 1118, 1122 ⟨83MI 530-02⟩
Verdaguer, X., **2**, 84, 318 ⟨91TL6931⟩; **4**, 763 ⟨94JA2153⟩
Verdet, L., **3**, 113 ⟨79TL3⟩
Verdun, D. L., **3**, 375 ⟨69CJC4509, 71JA6304⟩
Verenchikov, S. P., **1**, 926 ⟨72JOU2515⟩
Verenikin, O. V., **2**, 170 ⟨88IZV2367⟩; **4**, 104 ⟨86IZV810⟩; **5**, 710 ⟨81IZV2370⟩; **6**, 42 ⟨85IZV659⟩
Vereshchagina, N. N., **6**, 109 ⟨67KGS1096⟩
Vereshchagin, L. I., **1**, 1006 ⟨74JOU1990⟩, 1045 ⟨67ZOR219⟩; **3**, 267 ⟨84JGU28⟩; **4**, 106 ⟨85JOU806, 88JOU2003⟩
Veretenov, A. L., **1**, 529 ⟨89IZV495⟩
Verfürth, U., **4**, 273 ⟨90JCS(P1)2919⟩
Verge, J. P., **5**, 593 ⟨76JCS(P1)33⟩
Vergne, F., **1**, 224 ⟨93SL599⟩
Vergne, G., **4**, 60 ⟨86TL6319⟩
Verhe, R., **1**, 115 ⟨B-83MI 104-02⟩; **2**, 2, 3, 16, 21, 28 ⟨B-88MI 201-01⟩, 312 ⟨88S587⟩, 749 ⟨76T3063⟩; **3**, 24 ⟨82TL2853⟩, 26, 32, 35, 44, 50 ⟨B-88MI 301-01⟩, 35 ⟨84S46⟩, 148, 152, 155 ⟨B-88MI 304-01⟩, 162, 165 ⟨B-88MI 304-02⟩, 178, 194 ⟨B-88MI 304-05⟩, 184 ⟨82S765⟩, 219 ⟨81JOC3226⟩, 409 ⟨85TL2709⟩, 409, 563 ⟨82OPP213⟩, 415, 416 ⟨80OPP49, 82S43⟩, 416 ⟨86S192, B-88MI 310-02⟩, 417 ⟨82TL789⟩, 418 ⟨83S632⟩, 563 ⟨78JOC2670⟩, 658 ⟨76T3063, 79S741⟩; **4**, 12 ⟨79SC575⟩, 183 ⟨83CB3631⟩; **6**, 509 ⟨77BSB663⟩
Verhelst, W. F., **1**, 1122 ⟨74JA6692⟩
Verheyden, J. P. H., **2**, 8 ⟨70JOC2319⟩
Verhoeven, T. R., **1**, 192, 212 ⟨82COMC-I(8)799⟩, 698 ⟨76TL3477⟩; **2**, 66 ⟨93TL3243⟩, 91 ⟨79MI 202-02⟩
Verhulst, J., **3**, 278 ⟨52BSB694⟩
Verkade, J. G., **4**, 342 ⟨70JOC2310, 73PS(3)1⟩, 575 ⟨82ZAAC(494)55⟩, 917 ⟨B-81MI 420-02⟩; **6**, 79 ⟨88JA4144⟩
Verke, R., **3**, 416 ⟨88CC825⟩
Verkode, J. G., **4**, 342 ⟨66JA1140⟩
Verkruijsse, H. D., **1**, 141 ⟨78CC596⟩, 880 ⟨86AG(E)465⟩, 954, 957 ⟨B-84MI 120-01⟩, 1025 ⟨79S292⟩, 1037 ⟨81S459⟩, 1045, 1046 ⟨88JOM(338)289⟩; **2**, 621 ⟨81TL2495, 86CC260⟩, 907 ⟨86RTC66⟩, 907, 908, 909, 939, 942, 946 ⟨B-87MI 218-01⟩, 939 ⟨82JOM(232)C1⟩; **3**, 273 ⟨88JOM(338)289⟩; **4**, 1017 ⟨87MI 421-01⟩; **4**, 836, 949 ⟨87JOM(332)99⟩
VerKruijsse, H. D., **4**, 843 ⟨85S958⟩
Verkruijsse, M., **2**, 551 ⟨B-87MI 211-02⟩
Verkuijlen, E., **1**, 674 ⟨74CC793, 77CC198⟩
Verlander, M. S., **5**, 972 ⟨76MI 526-01⟩
Verlhac, J.-B., **3**, 182 ⟨86TL2361⟩, 303 ⟨91OM366⟩; **4**, 358 ⟨88CC503, 90JOM(391)283⟩, 533 ⟨88JOM(339)267⟩;

5, 415 ⟨92JOM(437)C13⟩, 467 ⟨85TL6075⟩; 6, 449 ⟨91OM366⟩
Verloop, A., 1, 403 ⟨57RTC689⟩
Verma, P. K., 6, 608 ⟨91ZAAC(605)131⟩
Verma, R. D., 2, 335 ⟨89MI 206-02⟩; 3, 162 ⟨93TL7335⟩; 4, 192 ⟨93TL7335⟩; 5, 210 ⟨89IJC(A)1528⟩
Verma, R. P., 5, 384 ⟨76JIC1230⟩
Verma, S., 3, 604 ⟨82ACH35⟩; 6, 264 ⟨80IJC(B)610⟩
Verma, V. K., 5, 631 ⟨85IJC(B)977⟩
Vermeer, P., 1, 466 ⟨85JOC3042⟩, 496 ⟨76TL947⟩, 514 ⟨79S430⟩, 536 ⟨82TL2797⟩, 541 ⟨77TL2023⟩, 544 ⟨79TL3327, 81JOM(206)257⟩, 544 ⟨81RTC337⟩, 966 ⟨89JOC3726⟩, 967 ⟨79S390, 82JOM(240)329⟩, 967 ⟨85JOC5143⟩, 968 ⟨84TL3019⟩, 975 ⟨83JOC1103, 84TL5571⟩, 975, 1004, 1009 ⟨81TL1451⟩, 989 ⟨75TL2923⟩, 999 ⟨74RTC240⟩, 1004, 1006, 1025 ⟨82RTC405⟩, 1017, 1018 ⟨81JOM(217)267, 82RTC97⟩, 1058 ⟨81S551⟩, 1094 ⟨77TL1823⟩; 2, 589 ⟨79JOM(172)377⟩, 616, 624 ⟨77RTC168⟩, 924, 925 ⟨77TL1823⟩, 940 ⟨79S390⟩; 3, 548 ⟨74RTC26⟩, 642 ⟨77S784, 78S454⟩, 642 ⟨79S430⟩; 4, 245 ⟨74RTC242⟩, 704 ⟨92S432⟩, 809 ⟨77TL1823⟩, 843 ⟨73RTC1067⟩, 843 ⟨74RTC184⟩, 940 ⟨84JOM(276)317⟩, 960, 963 ⟨75RTC14, 86JOM(302)1⟩, 963 ⟨75TL2923⟩; 5, 510 ⟨74RTC51⟩, 547, 552 ⟨78S439⟩, 548, 550, 559, 560 ⟨79S432⟩, 550 ⟨73RTC601, 75RTC83⟩, 554 ⟨80TL1771⟩
Vermeeren, H. P. W., 2, 150 ⟨76CC496⟩
Vermehren, J., 4, 799 ⟨89CB331⟩
Vermes, J.-P., 3, 43 ⟨78CJC119⟩
Vermeyen, C., 4, 917 ⟨84TL2883⟩
Vermylen, L., 6, 105 ⟨83TL1035⟩
Vernekar, S. P., 5, 990 ⟨92MI 526-01⟩
Verner, E. J., 1, 398, 405 ⟨92JA375⟩
Vernice, G. G., 4, 49 ⟨93JOC7382⟩
Vernière, C., 1, 958 ⟨81TL103⟩, 974 ⟨81T1359⟩
Vernon, J. M., 4, 54 ⟨75JCS(P1)2048⟩, 303 ⟨90JCS(P1)83⟩
Vernon, P. G., 1, 942 ⟨91JCS(P1)1433⟩
Vernon, R. H., 2, 288 ⟨20JCS889⟩
Veronese, A. C., 3, 268 ⟨84CC1593, 84JOC4696⟩, 268 ⟨88JCR(S)246⟩
Veronkov, M. G., 4, 32 ⟨74CCC2616⟩
Verpeaux, J.-N., 1, 144, 237, 483 ⟨83T3283⟩, 439 ⟨86T2475⟩, 474 ⟨88CC203⟩, 485 ⟨85BSF762, 85BSF772⟩, 485, 700 ⟨82TL2457, 82TL2469⟩, 700 ⟨82TL2465⟩; 2, 623 ⟨89JOM(379)201⟩, 628 ⟨90BSF275⟩, 678 ⟨91SL505⟩; 4, 138 ⟨90SL769⟩, 147 ⟨92SL133⟩, 742, 746 ⟨89JOM(379)201⟩
Verrado, G., 3, 405 ⟨88SC1501⟩
Verrall, K. A., 2, 892 ⟨70JPR751⟩
Verruijsse, H. D., 3, 58 ⟨86RTC66⟩
Vershchurnski, I. V., 2, 448 ⟨72ZOB1285⟩
Vershinina, G. E., 4, 913 ⟨76ZOB2385⟩
Versleijen, J. P. G., 3, 712 ⟨95TL2109⟩
Verzele, M., 2, 177 ⟨65BSB21⟩
Veschambre, H., 3, 16, 132 ⟨90TL2599⟩
Veselovskii, V. V., 1, 60 ⟨82JOM(226)115⟩
Vesely, J. A., 2, 650 ⟨70JOC4028⟩
Vesley, G. F., 5, 122 ⟨71JOC2548⟩
Vesper, H. G., 6, 221 ⟨34JA1455⟩
Vesper, J., 4, 545 ⟨71AG(E)940, 73ZN(B)224⟩
Vesprozvannyi, M. A., 6, 261, 283 ⟨70IZV2553⟩
Vessal, B., 2, 215 ⟨83SC611⟩
Vessiere, R., 1, 706, 707 ⟨79CJC1958⟩, 1124 ⟨72TL2983⟩; 2, 249 ⟨85BSF463⟩; 4, 779 ⟨77CJC3190⟩
Vessiéres, A., 4, 718 ⟨93OM4545⟩
Vester, A., 6, 197 ⟨93CB2637⟩
Vetesnik, P., 4, 764 ⟨88MI 417-01⟩
Vetter, H., 2, 44 ⟨53LA(582)133⟩
Vetter, H.-J., 2, 408, 415 ⟨63AG(E)218⟩, 410 ⟨61CB1505⟩, 414, 415, 891 ⟨64ZAAC(330)233, 64ZAAC(331)35⟩
Vetter, W. M., 2, 573 ⟨91OM244⟩
Vevera, E., 1, 98 ⟨56CB2105⟩
Vey, M., 5, 640 ⟨92S637⟩
Veyrières, A., 4, 235 ⟨89CAR(188)81⟩
Veyron, B., 2, 9, 19 ⟨90TL6527⟩
Veysoglu, T., 2, 152 ⟨77TL4009⟩
Viala, J., 1, 580 ⟨86TL299⟩; 2, 446 ⟨88S395⟩; 3, 195 ⟨91T3347⟩
Vial, C., 3, 494 ⟨77HCA1161⟩
Vialle, J., 2, 130 ⟨62BSF2126, 67BSF1213⟩, 229 ⟨83T3839⟩; 3, 339 ⟨72CR(C)623, 75BSF(2)1435⟩, 343 ⟨76TL4297, 77T3089⟩, 343 ⟨78T2289, 80JOC2517⟩, 343, 351 ⟨79TL2677⟩, 344 ⟨73BSF(2)1973⟩, 352 ⟨72BSF3138⟩, 376 ⟨74CR(C)793⟩, 378 ⟨79BSF(2)199, 79BSF(2)205⟩; 5, 552 ⟨83TL3617⟩, 553 ⟨86T2025⟩; 6, 309 ⟨67BSF2239⟩
Vial-Reveillon, F., 6, 7 ⟨90JFC(47)95⟩
Viana, L. H., 5, 241 ⟨92SC1619⟩
Viani, F., 1, 143 ⟨85TL2913⟩
Viau, R., 1, 143 ⟨71JA3077⟩
Viaud, M. C., 5, 665 ⟨90TL1417⟩
Vibet, A., 3, 343 ⟨78T2289⟩
Vibuljan, P., 3, 231 ⟨80CC486⟩
Vice, S. F., 2, 767 ⟨82TL821⟩
Vicens, J., 1, 603 ⟨79TL2697⟩; 2, 281, 284 ⟨79TL2697⟩
Vicens, J. J., 3, 375 ⟨80JA2460⟩
Vicens, M., 2, 444 ⟨91T403⟩
Vicente, C., 5, 386 ⟨92JCS(D)53⟩
Vicente, H., 3, 64 ⟨90TL1365⟩
Vicentini, G., 2, 732 ⟨59CB40⟩
Vicioso, M. M., 6, 124 ⟨91T3171⟩
Vick, S. C., 4, 700 ⟨74SRI515⟩, 1051, 1052 ⟨84JOM(272)123⟩, 1051, 1052 ⟨84OM1897⟩
Vicker, N., 5, 587, 593 ⟨93JMC1604⟩
Vickers, D., 3, 128 ⟨93T965⟩
Vickersveen, L., 2, 773 ⟨88ACS(B)354⟩
Vickery, E. H., 1, 7, 19 ⟨82JOC2590⟩
Vicković, I., 2, 1000 ⟨84AX(C)842⟩; 4, 268 ⟨80TL3089⟩; 6, 692 ⟨84OM1132⟩
Vickrey, T. M., 4, 1016 ⟨76AG(E)171⟩; 6, 221, 223 ⟨79JCED251⟩, 223, 226, 227, 228 ⟨79JMR559⟩
Victor, R., 4, 681 ⟨71CC1241⟩
Vida, J. A., 2, 25 ⟨70TL3447⟩
Vidal, A., 3, 585 ⟨90CC829, 91JOC4008⟩; 5, 622, 1067 ⟨89TL2847⟩, 1066 ⟨89T4263, 90CC7⟩, 1067, 1079 ⟨90T1063⟩
Vidal, J., 1, 613 ⟨86TL3733⟩
Vidal, M., 1, 84 ⟨87BSF462⟩; 2, 444 ⟨91T403⟩
Vidal, Y., 2, 650 ⟨86BSF625⟩
Vidari, G., 4, 246 ⟨90TL5815⟩
Videau, B., 3, 263 ⟨81S880⟩
Vidyapati, T. J., 4, 357 ⟨85JOM(280)31⟩
Vidyasagar, V., 4, 428 ⟨87JOC4781⟩
Viehe, H. G., 1, 571 ⟨85T3331⟩, 592 ⟨81AG(E)585⟩, 919, 922 ⟨85ACR148⟩, 919, 922 ⟨88PAC1635⟩, 956 ⟨68JCS(C)228⟩, 982 ⟨73AG(E)566⟩, 995 ⟨93TL1779⟩, 998 ⟨B-69MI 121-01⟩, 999 ⟨62CB319⟩, 999, 1016 ⟨B-69MI 121-05⟩, 1043 ⟨59CB1950⟩; 2, 746 ⟨68T4217⟩, 758 ⟨69AG468⟩, 1011 ⟨77HOU(5/2a)110⟩, 1011, 1013, 1014, 1017, 1018, 1019 ⟨B-69MI 221-01⟩, 1012, 1013, 1062 ⟨62CB319⟩, 1013, 1047 ⟨70CB1198⟩, 1014 ⟨64AG(E)747, 64AG(E)755, 65AG(E)746⟩, 1014, 1015 ⟨59CB1270⟩, 1014, 1015, 1016 ⟨59CB1950, 70CB1216⟩, 1016 ⟨62CB3064⟩, 1017 ⟨66USP3249630⟩, 1039 ⟨B-69MI 222-01⟩, 1041, 1047 ⟨64AG(E)506⟩, 1042 ⟨70CB564, 82S1083⟩, 1046 ⟨76C387, 77JCR(S)105⟩, 1046, 1049 ⟨66AG(E)584⟩, 1047 ⟨70CB1209, 70TL3141⟩, 1051, 1052 ⟨69AG(E)273, 75C515⟩, 1052 ⟨75C516⟩, 1076 ⟨62GEP1126388⟩; 3, 46 ⟨78BSB693⟩,

409 ⟨89SC2825⟩, 419 ⟨B-76MI 310-01⟩, 473 ⟨80AG754, 80AG(E)716⟩, 601 ⟨B-76MI 317-01⟩; **4**, 71 ⟨81AG(E)585⟩, 88, 773, 785 ⟨87T4309⟩, 91, 287, 858, 861 ⟨85T2527⟩, 92 ⟨85T4183⟩, 764 ⟨69MI 417-01⟩, 794 ⟨78BSB391⟩, 796 ⟨75C512⟩, 798 ⟨78AG(E)936, 81HCA1208⟩, 817 ⟨70CB1198⟩, 885 ⟨64AG571⟩, 891, 904, 905 ⟨69AG(E)454⟩, 904, 905 ⟨69T3447⟩, 966 ⟨B-76MI 403-01⟩, 976, 977 ⟨67AG(E)767⟩, 976, 980 ⟨B-69MI 421-01, B-69MI 421-02⟩, 979 ⟨68JCS(C)1726, 70JCS(C)881⟩; **5**, 132 ⟨85TL4439⟩, 481 ⟨75AG710⟩, 577, 617 ⟨85T3331⟩, 601 ⟨85S79, 85T693⟩, 601 ⟨90TL6339⟩, 601, 612 ⟨85S77⟩, 670, 671 ⟨87BSB51⟩; **6**, 44 ⟨81AG(E)585⟩, 52 ⟨79AG(E)615⟩, 57 ⟨68T4217⟩, 74 ⟨78BSB693⟩, 105 ⟨83TL1035⟩, 238 ⟨73AG837⟩, 506 ⟨72AG993⟩, 609, 610, 614, 619, 620 ⟨B-76MI 620-01⟩, 610, 727 ⟨89SC2825⟩, 619 ⟨78BSB391⟩, 619 ⟨79AG(E)333⟩, 619, 728 ⟨71AG(E)573, 75C209⟩, 644 ⟨75C514⟩, 653 ⟨90T3897⟩, 727, 728, 729 ⟨B-76MI 623-01⟩

Vieillefosse, R., **2**, 182 ⟨47BSF351⟩

Vieira, M. A. M. S. A., **1**, 949 ⟨90T3061⟩

Vieler, R., **1**, 331, 342 ⟨90S411⟩

Vierhapper, F. W., **1**, 242 ⟨78M609⟩; **2**, 160 ⟨76CC1003⟩; **5**, 338 ⟨79JOC2863⟩

Vierling, K., **6**, 467 ⟨39GEP740366⟩

Vieroth, C., **2**, 421 ⟨69ZAAC(364)316⟩

Vieta, R. S., **1**, 808 ⟨80JOC4793⟩

Vietmeyer, N. D., **1**, 420, 421 ⟨73PAC197⟩, 633 ⟨68JA4762⟩

Viets, D., **6**, 252 ⟨91CB1353, 92CC1017⟩, 253 ⟨92CB535⟩

Vig, A. K., **1**, 745 ⟨75JIC538⟩

Vig, O. P., **1**, 745 ⟨75JIC538⟩; **3**, 62 ⟨63JIC114⟩

Vigalok, I. V., **2**, 336 ⟨83JOU1203⟩, 459 ⟨74ZOB1261⟩

Vigevani, A., **4**, 441 ⟨72JHC827⟩

Vigevani, E., **3**, 467 ⟨65CB2715⟩

Viggiano, M., **4**, 453 ⟨84IC2550⟩

Vignaux, P., **3**, 159 ⟨93JOC1835⟩

Vigneron, J. P., **1**, 503, 505 ⟨80TL1739⟩

Vij, A., **3**, 162 ⟨93TL7291⟩; **4**, 192 ⟨93TL7335⟩

Vijayakumaran Nair, G., **3**, 303 ⟨71T3323⟩

Vijayalakshmi, S., **4**, 272, 275 ⟨89SUL79⟩

Vijh, A. K., **5**, 28 ⟨67CRV623⟩

Vijn, R. J., **3**, 422 ⟨88JOC3882⟩

Vikane, O., **2**, 889 ⟨73ACS1337, 73ACS2135⟩

Vikhreva, L. A., **2**, 15 ⟨91RCR885⟩, 1033 ⟨83IZV1662, 84DOK(274)822⟩

Vikić-Topić, D., **4**, 697 ⟨93OM4708⟩

Viktorova, E. A., **4**, 775 ⟨88ZOR443, 90MI 417-01⟩

Viktorov, N. A., **4**, 142 ⟨85JGU1079⟩, 142 ⟨91MI 403-02⟩, 142 ⟨92MI 403-02⟩, 1058 ⟨86ZOB1535⟩

Vilaplana, J., **5**, 680 ⟨88H(27)1935⟩

Vilaplana, M. J., **2**, 121 ⟨85TL469⟩; **5**, 1107 ⟨82S472⟩

Vilar, E. T., **2**, 9 ⟨91S353⟩, 36 ⟨87TL6441⟩; **3**, 92, 101 ⟨90CC1571⟩

Vilarrasa, J., **2**, 317 ⟨87TL5941⟩, 341 ⟨90T587⟩; **5**, 234 ⟨86TL4623⟩, 353, 354 ⟨87TL341⟩, 360 ⟨89JOC3209⟩, 360, 361 ⟨84JOC3322, 89TL863⟩, 361 ⟨91JOC7038⟩

Vil'chevskaya, V. D., **2**, 982 ⟨54IZV1008⟩

Vílchez, J. E., **2**, 394 ⟨91TL3225⟩

Vil'davskaya, A. I., **4**, 841 ⟨80ZOR1748⟩

Vile, S., **4**, 200 ⟨91T1329⟩

Viljoen, J. C., **5**, 951 ⟨85OM275⟩

Vilkas, M., **1**, 546 ⟨62CR(254)1817⟩; **6**, 538 ⟨90SC2769⟩

Vill, J. J., **1**, 268 ⟨62JA3944⟩

Villa, A. C., **5**, 946, 959 ⟨94ICA233⟩

Villa, L., **2**, 53 ⟨91TL4381⟩

Villa, M., **2**, 29 ⟨89T4243⟩; **3**, 294 ⟨88TL5561⟩

Villa, R., **2**, 46 ⟨88T491⟩

Villacata, G. M., **2**, 588 ⟨82JA5284⟩

Villacorta, G. M., **3**, 287 ⟨82JA5824⟩

Villadsen, B., **5**, 241 ⟨79JOM(166)373⟩

Villalgordo, J. M., **5**, 611 ⟨93T7215⟩

Villalobos, A., **1**, 700, 703 ⟨89JOC12⟩

Villalobos, A. C., **2**, 36 ⟨88S58⟩

Villamana, J., **3**, 638 ⟨81S375⟩

Villani, F. J.,**1**, 503 ⟨54JA4446⟩

Villani, Jr., F. J., **1**, 1006 ⟨78JOC358⟩; **6**, 635 ⟨88S460⟩

Villani, R., **4**, 12 ⟨88JA5533⟩

Villarica, R. M., **3**, 322 ⟨67JA5497⟩

Ville, G., **2**, 7 ⟨75BSF607⟩; **4**, 173 ⟨89JOC610⟩

Villemin, D., **1**, 674 ⟨80TL1715⟩, 934 ⟨80CC732⟩; **2**, 36 ⟨90SC41⟩; **3**, 193 ⟨89SC2567⟩, 251 ⟨91S301⟩; **4**, 246 ⟨89SC31⟩, 257 ⟨92SC1359⟩, 281 ⟨90SC925⟩, 842 ⟨85CC870⟩; **5**, 199 ⟨93SC419⟩, 554 ⟨93SL148⟩, 615 ⟨93SC2259⟩; **6**, 58 ⟨94TL3537⟩, 82 ⟨89SC31⟩, 121, 325 ⟨92SC1359⟩

Villeras, J., **2**, 530 ⟨93TL6899⟩

Villieras, J., **1**, 164 ⟨77BSF765⟩, 191 ⟨87TL5521⟩, 238, 483 ⟨77JOM(142)1⟩, 465 ⟨77TL3263⟩, 466 ⟨77JOM(128)1⟩, 477 ⟨77JCR(M)2101⟩, 491, 1108 ⟨74JOM(77)269⟩, 540 ⟨74BSF1656, 75JOM(96)471⟩, 664 ⟨88JOM(346)C1⟩, 833 ⟨74BSF2072⟩, 1017, 1018 ⟨75TL1465, 80T1215⟩, 1067 ⟨74BSF(2)1731, 75S458⟩, 1067 ⟨79S502⟩, 1090 ⟨76S761⟩, 1108 ⟨74JOM(77)281⟩; **2**, 420 ⟨75JOM(97)325⟩, 624 ⟨74S803⟩, 629 ⟨74JOM(77)269⟩, 676 ⟨77BSF693⟩, 679 ⟨75TL3833⟩, 682 ⟨75JOM(93)415⟩, 998 ⟨84TL835⟩, 1015 ⟨75S458⟩; **3**, 28 ⟨80S644⟩, 125 ⟨79S968⟩, 152 ⟨84TL835⟩, 152, 154 ⟨81S68⟩, 210 ⟨88S417⟩; **4**, 14, 22 ⟨75BSF1797, 80S644⟩, 138, 146, 667 ⟨85BSF825⟩, 139 ⟨86BSF470⟩, 146 ⟨84TL835⟩, 146 ⟨88JOM(346)C1⟩, 147 ⟨71MI 403-03⟩, 738, 739, 745 ⟨71TL1035⟩, 740, 741 ⟨75S458⟩, 763, 771 ⟨75T891⟩, 763, 771 ⟨75T897⟩, 963 ⟨77BSF693⟩, 1016, 1017 ⟨76AG417⟩; **5**, 103 ⟨87OS(66)220⟩, 296 ⟨85SC1233⟩, 296, 297 ⟨87TL59⟩, 606 ⟨82PS(12)325⟩; **6**, 244 ⟨67BSF1520, 71MI 607-01⟩, 41 ⟨74BSF2072⟩

Villiers, J., **3**, 236 ⟨78S29⟩

Villiger, V., **1**, 897 ⟨02CB3031⟩

Villiras, J., **2**, 186 ⟨69JOM(121)123⟩

Vilsmaier, E., **1**, 620 ⟨74TL3987⟩; **2**, 160 ⟨75T89⟩, 188 ⟨74TL3987⟩, 188 ⟨75S55⟩, 195 ⟨71S590⟩; **4**, 47, 48 ⟨69LA(728)12⟩, 1080 ⟨88JOC1806⟩; **5**, 339, 689 ⟨80LA1064⟩, 339, 689 ⟨76S46, 80LA1075⟩, 702 ⟨80LA1055⟩, 898 ⟨71S431⟩

Vin, D. Le, **3**, 689 ⟨90CB2317⟩

Vinader, M. V., **5**, 1066 ⟨89CB307⟩

Vinander, M. V., **5**, 1066 ⟨88JOC4654⟩

Vinayak, R. S., **1**, 1010 ⟨90TL3731⟩

Vincens, H., **1**, 180 ⟨88T5495⟩

Vincens, M., **1**, 84 ⟨87BSF462⟩

Vincent, B. F., **2**, 341, 764 ⟨65JOC2880⟩, 763 ⟨65JOC2877⟩

Vincent, B. R., **1**, 702 ⟨88CJC2860⟩; **2**, 232 ⟨91JOC904⟩

Vincent, J. E., **1**, 1045 ⟨78TL937⟩; **3**, 273 ⟨78TL937⟩

Vincent, M., **1**, 826 ⟨79CC584⟩

Vincent, P., **1**, 536 ⟨82TL63⟩; **2**, 968 ⟨82TL63⟩

Vincente, J., **4**, 584 ⟨89JOM(364)407⟩, 586 ⟨87JOM(331)409, 87JOM(333)129⟩, 586 ⟨88OM997, 91JCS(D)2579⟩

Vincenzi, C., **3**, 64 ⟨67G397⟩

Vinczer, P., **2**, 1015 ⟨92TL683⟩; **6**, 695 ⟨92TL683⟩

Vines, S. M., **2**, 151, 688 ⟨76JOC3987⟩, 363 ⟨77TL963⟩

Viney, M. M., **1**, 456 ⟨85JMOC213⟩

Vineyard, B. D., **1**, 76 ⟨77JA5946⟩; **4**, 315 ⟨66JCED620⟩

Vinick, F. J., **3**, 633 ⟨78TL4221⟩

Vinigradov, A. S., **6**, 240, 615 ⟨80JFC(15)169⟩

Vinkler, E., **4**, 319 ⟨72ACH363⟩

Vinod, T. K., **3**, 271 ⟨88TL885⟩

Vinograd, L. K., **5**, 567, 581 ⟨76KFZ41⟩

Vinogradoff, A. P., **6**, 636 ⟨88JOC3120⟩

Vinogradov, A. S., **5**, 1076 ⟨85IZV700⟩; **6**, 615 ⟨83IZV694, 83MI 620-02⟩, 615 ⟨85IZV700⟩, 616 ⟨81IZV2632⟩

Vinogradov, M. G., **1**, 926 ⟨72JOU2515⟩; **3**, 214 ⟨88JOM(348)123⟩; **5**, 168 ⟨73CC693⟩
Vinogradov, V. M., **6**, 145 ⟨90IZV1816⟩
Vinogradova, T. K., **4**, 998 ⟨92ZOB1084⟩, 999 ⟨84ZOB288⟩; **5**, 655, 656 ⟨84JGU255⟩, 655, 656 ⟨84ZOB288⟩
Vinogradova, V. N., **4**, 669, 684, 722 ⟨73IZV2796⟩, 685 ⟨72IZV1600⟩
Vinogradova, V. S., **4**, 339 ⟨76IZV1416⟩, 1006 ⟨73IZV1112⟩
Vinokur, E., **1**, 651 ⟨70TL1137⟩; **4**, 646 ⟨70TL1137⟩; **6**, 192 ⟨70TL1137⟩
Vinokurova, G. M., **2**, 439 ⟨59DOK(127)1217⟩
Vinokurova, N. G., **5**, 162 ⟨78JGU1249⟩
Vinson, W. A., **4**, 734 ⟨83JOC4661⟩
Vio, L., **5**, 769 ⟨89MI 519-01⟩
Viola, A., **1**, 412 ⟨65JA1150⟩, 412, 810 ⟨67JA3462, 70JOC856⟩
Viola, H., **3**, 364 ⟨68CB3517⟩; **4**, 256 ⟨88ZC269⟩; **5**, 476, 477 ⟨75ZC348⟩, 513, 592, 593, 620 ⟨68CB3517⟩, 550 ⟨76ZC355⟩, 554 ⟨77ZC366, 78ZC90⟩, 561 ⟨77ZC92⟩, 601 ⟨78JPR313, 88ZC233⟩; **6**, 536 ⟨79ZC289⟩
Violeau, B.,**3**, 429 ⟨91BSF730⟩
Viols, H., **5**, 360 ⟨90CPB2352⟩
Vipond, P. W., **2**, 169 ⟨68JCS(C)2713⟩
Viragh, C., **5**, 1126 ⟨76JCS(P2)524⟩
Viret, J., **2**, 374 ⟨87T891⟩
Virgil, S., **2**, 50 ⟨89JA9243⟩
Virgil, S. C., **3**, 517 ⟨93JA8167⟩
Virgili, A., **1**, 209 ⟨87TL1881⟩
Virgili, P., **4**, 100 ⟨86TL3199⟩
Virin, L. I., **6**, 497 ⟨88ZOB26⟩
Virkari, S. D., **1**, 707 ⟨75IJC251⟩
Virtanen, A. I., **5**, 83 ⟨55AG381⟩
Virtanen, P. O. I., **1**, 229 ⟨73ACS3944⟩
Vis, J. H., **2**, 485, 505, 510, 874 ⟨67CJC675⟩
Viscontini, M., **2**, 227 ⟨54HCA375⟩
Visentin, G., **1**, 182 ⟨85JOC3945⟩
Vishinskaya, L. I., **2**, 569 ⟨77JOM(131)43⟩
Vishnevskii, F. N., **5**, 1015 ⟨72ZOB879⟩
Vishnevskii, O. V., **6**, 438 ⟨68ZOR720⟩, 453 ⟨71ZOR2229⟩
Vishnivetskaya, L. A., **5**, 669 ⟨91BAU382, 91IZV444⟩
Vishnuvajjala, B., **3**, 625 ⟨76JOC2502⟩
Vishnyakova, T. P., **6**, 621 ⟨81MI 620-02⟩, 646 ⟨81MI 621-01⟩
Vishwakarma, J. N., **4**, 704 ⟨88T1667⟩; **5**, 731 ⟨88JCS(P1)169, 88T1667⟩
Vishwakarma, L. C., **4**, 312 ⟨93OSC(8)546⟩; **5**, 106 ⟨85JHC1429⟩
Vismara, E., **1**, 934 ⟨85TL4803, 89H(28)489⟩, 934 ⟨90JHC79, 91JOC2866⟩; **3**, 311 ⟨89H(28)489, 91JOC2866⟩
Visnich, M., **1**, 226 ⟨83S284⟩
Visnick, M., **1**, 763 ⟨84TL5603⟩
Viso, A., **1**, 600, 983 ⟨92TL4985⟩
Visser, F., **2**, 1003 ⟨84T765⟩
Visser, G. W. M., **2**, 613 ⟨86JOC1886⟩
Visser, J., **6**, 218 ⟨79MI 607-01⟩
Visser, M. S., **1**, 316 ⟨94JA3123⟩
Visser, P., **4**, 483 ⟨84T991⟩
Visser, R., **3**, 428 ⟨86RTC103⟩
Visser, R. G., **1**, 981 ⟨81RTC34, 81TL2827⟩, 1037 ⟨81S459⟩; **5**, 427 ⟨81TL2827⟩, 605, 614 ⟨81RTC414⟩
Visser, T., **2**, 693 ⟨59JBC(234)1⟩
Vissers, A., **4**, 1025 ⟨84CB1542⟩
Viswanathan, N., **4**, 718 ⟨87JOM(334)117⟩, 979 ⟨79IJC(B)478⟩
Vit, J., **3**, 84 ⟨70MI 303-01⟩
752⟩
Vita Finzi, P., **5**, 1004 ⟨65TL4645⟩
Vitale, A. C., **1**, 115 ⟨72JOC4090⟩
Vitali, D., **2**, 893 ⟨84IC1504⟩
Vitali, F., **5**, 347 ⟨84FES788⟩
Vitarius, J., **1**, 255 ⟨90CC1532⟩; **2**, 302 ⟨90CC1532⟩
Vitcha, J. F., **6**, 38, 43 ⟨72JMC604⟩, 44, 45, 46 ⟨65JOC4011⟩
Vitek, J., **6**, 410 ⟨81CZP208251, 90CZP264358⟩
Viterbo, R., **2**, 214 ⟨57JA3503⟩
Viteva, L. Z., **5**, 607 ⟨94T7193⟩
Viti, S. M., **1**, 14 ⟨82JOC1378⟩; **2**, 82 ⟨83PAC589⟩
Vitkovskii, V. Yu., **4**, 140 ⟨91IZV1039⟩, 783 ⟨85ZOR965⟩, 861 ⟨92ZOR1711⟩; **5**, 924 ⟨89JOM(368)163⟩; **6**, 620 ⟨91ZOB643⟩
Vitt, J., **5**, 766 ⟨59CB2563⟩; **6**, 651 ⟨59CB2563⟩
Vitt, S. V., **5**, 385 ⟨85JA4252⟩, 652 ⟨88JCS(P1)305⟩
Vittal, J. J., **4**, 692 ⟨92OM2224⟩
Vittorelli, P., **1**, 415 ⟨68HCA1457, 75HCA1293⟩
Viturro, H. R., **2**, 1008 ⟨87CJC2000⟩
Vivarelli, P., **2**, 825 ⟨74JCS(P2)1338⟩
Vizgert, R. V., **5**, 1108 ⟨58MI 530-01⟩
Vladimirova, I. I., **2**, 136 ⟨90KGS1477⟩
Vladimirova, I. L., **5**, 774 ⟨78ZOB2465⟩
Vladuchick, W. C., **1**, 598 ⟨77JA8116⟩; **3**, 165 ⟨80JA3548⟩
Vladzimirs'ka, O. V., **3**, 344, 354 ⟨65MI 308-01⟩
Vlahof, R., **3**, 59 ⟨73TL2309⟩
Vlahov, I., **3**, 157, 161 ⟨91T3753⟩
Vlasenko, S. D., **4**, 140 ⟨83JGU109⟩
Vlasova, N. N., **5**, 1076, 1077 ⟨85ZOB621⟩
Vlassa, M., **5**, 735 ⟨77RRC1111⟩; **6**, 558 ⟨85JPR1007⟩
Vlattas, I., **4**, 956, 957 ⟨76JA2008⟩
Vlazov, V. M., **2**, 1064 ⟨75JGU1832⟩
Vleggaar, R., **6**, 138 ⟨76T2625⟩
Vliet, E. B., **1**, 772 ⟨43OSC(2)416⟩; **5**, 33 ⟨43OSC(2)416⟩
Vliet, L. A. V., **5**, 562 ⟨92TL6383⟩
Vll'yasov, A., **4**, 342 ⟨78ZOB2653⟩
Vo, N. H., **1**, 826 ⟨91JA2762⟩, 929 ⟨93JOC7228⟩
Voaden, M., **1**, 160 ⟨92TL543⟩; **4**, 651 ⟨92TL543⟩
Vob, E., **4**, 195 ⟨88S274⟩
Vocelle, D., **2**, 337 ⟨72CJC1167⟩, 675 ⟨64CJC212⟩
Vodicka, L., **2**, 617 ⟨88CCC110⟩
Vodolazskaya, V. M., **3**, 540 ⟨73ZOB2088⟩
Vodopivec, B., **5**, 1045 ⟨90TL7229⟩
Vodop'yanov, V. G., **5**, 716 ⟨76JGU714, 76ZOB714⟩, 1081 ⟨82ZOB921⟩
Voeffray, R., **4**, 489 ⟨82HCA1953⟩
Voek, M. V., **5**, 1083 ⟨80ZOB1678⟩
Voelker, C., **6**, 155 ⟨82OM1114⟩
Voelter, W., **2**, 9 ⟨87LA617⟩; **6**, 422 ⟨80CC1265, 83LA248⟩, 487 ⟨93ZN(B)1000⟩
Voevodskaya, T. I., **5**, 385 ⟨88ZOB62⟩
Vofsi, D., **2**, 157 ⟨64JCS4962⟩; **4**, 82 ⟨64JCS4962⟩; **6**, 7 ⟨63JCS1887⟩, 116 ⟨86MI 604-01⟩
Vogel, A. I., **1**, 860 ⟨60JCS4454⟩; **2**, 19 ⟨63OSC(4)525⟩, 25 ⟨43JCS636⟩; **6**, 412 ⟨B-56MI 614-01⟩
Vogel, D., **6**, 556 ⟨92PHA251⟩
Vogel, D. E., **1**, 832 ⟨85JOC4664⟩; **3**, 141 ⟨85JOC4664⟩; **5**, 105 ⟨87OS(66)29⟩
Vogel, E., **1**, 248 ⟨85AG(E)1046, 86AG(E)720⟩, 248 ⟨87AG(E)928, 88AG(E)411⟩, 419, 421 ⟨65TL391⟩, 419, 812 ⟨63AG(E)1⟩, 811 ⟨58LA(615)1⟩, 811 ⟨61LA(644)172, 63AG(E)739⟩, 875 ⟨79JA6120⟩, 910 ⟨65AG(E)354, 66AG(E)420⟩, 910 ⟨68HCA225, 70TL1941⟩; **2**, 532, 698 ⟨79JA6120⟩, 532, 698 ⟨81JA3099⟩; **3**, 61 ⟨83AG(E)410⟩; **4**, 832 ⟨81JA3099⟩; **5**, 867 ⟨84AG(E)717⟩
Vogel, G. E., **2**, 547 ⟨61JA996⟩
Vogel, H. A., **5**, 1102 ⟨68USP3733349⟩
Vogel, H. H., **5**, 28, 286 ⟨70S99⟩
Vogel, J., **6**, 606, 607, 618 ⟨69JPR15⟩, 610 ⟨86OM2187⟩
Vogel, K., **2**, 1008 ⟨73OSC(5)1135⟩
Vogel, M., **4**, 724 ⟨57JOC900⟩

Vogel, M. K., **2**, 21 ⟨80TL441⟩
Vogel, P., **1**, 519 ⟨84JOC2510⟩, 891 ⟨B-85MI 119-03⟩, 899 ⟨73ACR53⟩; **2**, 616 ⟨89HCA1008⟩; **3**, 60 ⟨74JCS(P1)37⟩, 163 ⟨89CC1634⟩; **5**, 584 ⟨93SL801⟩
Vogel, P. W., **2**, 255 ⟨65JA4447⟩, 258 ⟨56JA4447⟩, 845 ⟨55JA1864⟩
Vogelbacher, U., **4**, 479, 480 ⟨86JOM(306)39⟩; **5**, 878, 879, 1155 ⟨86JOM(306)39, 87PS(30)479⟩
Vogelbacher, U. J., **3**, 582 ⟨88JA1337⟩; **4**, 100 ⟨88AG(E)304⟩; **5**, 883 ⟨88AG(E)272⟩
Vogelfanger, E., **1**, 905 ⟨65JA376⟩
Voges, M., **5**, 867 ⟨89CB1147⟩
Vogiazoglou, D., **1**, 464 ⟨87JOC3697⟩
Vogl, O., **6**, 33 ⟨89JA8510⟩
Voglozin, A., **5**, 633 ⟨91BSF574⟩
Vogt, C. E., **1**, 494 ⟨88JA2655⟩
Vogt, H., **6**, 243 ⟨90ZAAC(588)26⟩
Vogt, H. J., **6**, 379 ⟨76ZAAC(419)2⟩
Vogt, P., **1**, 1067 ⟨91TL4115⟩
Vogt, P. F., **4**, 236 ⟨69CJC2875⟩
Vogt, R., **2**, 467 ⟨91CB2705⟩
Vogt, R. R., **1**, 1000 ⟨37JOC1⟩; **2**, 606, 623 ⟨36JA1806⟩, 1016, 1018 ⟨37JA1307⟩; **4**, 16 ⟨39JA1460⟩, 24 ⟨36JA1806⟩
Vogt, W., **2**, 749 ⟨80GEP2912344⟩
Vögtle, F., **1**, 232 ⟨75AG(E)559, 89CB1113⟩, 232 ⟨89CB347⟩, 814 ⟨86AG(E)1026⟩; **2**, 31 ⟨87JOC5560⟩, 132 ⟨78CB2445⟩; **3**, 90 ⟨86AG(E)1026⟩
Vohra, R., **3**, 212 ⟨91CC568⟩
Voigt, B., **5**, 755, 758 ⟨78PHA568⟩
Voigt, D., **4**, 569 ⟨78ICAL294⟩
Voigt, K., **1**, 494 ⟨94SL189⟩
Voisin, D., **5**, 191, 196 ⟨85TL1503⟩
Voitenko, A. D., **5**, 224 ⟨87ZOB686⟩
Volante, R. P., **1**, 940 ⟨88TL281⟩; **2**, 118 ⟨81TL3119⟩
Volatron, F., **5**, 932 ⟨86JA2173⟩
Volbach, W., **2**, 70 ⟨84TL2195⟩; **6**, 14 ⟨84TL2195⟩
Volchkov, N. V., **2**, 615 ⟨89IZV1935⟩
Volford, J., **6**, 449 ⟨81MI 614-02⟩
Volger, H. C., **1**, 518 ⟨72TL5269⟩; **2**, 1030 ⟨56RTC1459⟩
Volgin, Yu. V., **4**, 696 ⟨77JOM(137)217⟩, 723 ⟨76DOK(226)1092⟩
Volina, E. A., **1**, 63 ⟨81JOM(214)145⟩
Volk, H., **4**, 609 ⟨88JOM(341)109⟩; **6**, 178, 190, 191, 385, 391 ⟨83ZAAC(497)21⟩, 381, 385, 391 ⟨83ZAAC(497)119⟩
Volk, M., **6**, 391 ⟨88JOM(341)109⟩
Vol'kenshtein, Y. B., **2**, 13 ⟨77RCR891⟩
Völker, H., **2**, 432 ⟨61CB392⟩
Volkholz, M., **2**, 471 ⟨78CB890⟩
Völkl, A., **3**, 680 ⟨87CB2031⟩; **4**, 1017 ⟨82ZN(B)1044⟩
Volkman, R. A., **3**, 436 ⟨91COS(2)994⟩, 440 ⟨91COS(2)934⟩
Volkmann, R. A., **1**, 119 ⟨83JOC1767⟩, 204 ⟨83JA5946⟩, 1041 ⟨83TL4503⟩, 1042 ⟨91COS(1)355⟩; **2**, 344 ⟨90JA6696⟩, 345 ⟨91COS(1)355⟩; **6**, 553 ⟨92JOC4352⟩
Volkonskii, A. Yu., **4**, 697 ⟨79IZV1831⟩
Volkov, A. N., **6**, 562 ⟨81ZOR2269⟩
Volkov, N. D., **6**, 236 ⟨79S972⟩
Volkova, A. N., **3**, 674 ⟨77RCR374⟩
Volkova, L. I., **5**, 718 ⟨88BAU378, 88IZV460⟩
Volkova, V. N., **5**, 1128 ⟨82IZV939⟩
Volkulich, P. M., **3**, 271 ⟨78TL4661⟩
Voll, U., **2**, 1004 ⟨74CB1949⟩
Vollhardt, J., **1**, 146 ⟨88TL1529⟩; **2**, 997, 1000 ⟨85AG(E)859, 91JA4002⟩, 1000 ⟨86AG(E)653, 86AG(E)939⟩, 1000 ⟨88TL1259⟩; **4**, 936, 938 ⟨85AG(E)696⟩
Vollhardt, K. P. C., **1**, 64 ⟨77AG(E)399, 80JA5253⟩, 95 ⟨87JOC2804⟩, 447, 448, 530 ⟨84AG(E)539⟩, 448 ⟨77JA5483, 83CC502, 85JA1379, 87AG(E)1035⟩, 451 ⟨86JA1359⟩, 541 ⟨84JOC1574⟩, 991 ⟨85IJ131⟩, 1009 ⟨85TL709, 86AG(E)268⟩, 1064 ⟨76JA2667⟩; **2**, 627 ⟨75TL4539⟩, 632 ⟨76CC833⟩, 926 ⟨80AG(E)1023⟩, 944 ⟨85JA5670, 77JA5483⟩, 944, 984 ⟨84AG(E)539⟩; **3**, 270 ⟨80AG(E)1023⟩; **4**, 690 ⟨83JA4846⟩, 1061 ⟨84JOC1574⟩; **5**, 386 ⟨84JOC4786⟩; **6**, 402 ⟨81JOM(217)105, 86OM394⟩, 403 ⟨79JA2768, 83JA1384⟩
Vollmar, A., **4**, 203 ⟨86JOC3811⟩
Vollmer, H., **2**, 749 ⟨80GEP2936210⟩
Vollmer, S. H., **5**, 469 ⟨80JA5393⟩; **6**, 516 ⟨81JA2206⟩
Vollmerhaus, R., **6**, 93 ⟨92CC1539⟩
Vollrath, R., **6**, 360, 361 ⟨70CB766, 70ZC30⟩
Vollrath, W., **5**, 1052 ⟨75CB3071⟩
Volmar, C., **2**, 489 ⟨22BSF545⟩
Volnina, E. A., **2**, 266 ⟨81JOM(214)415⟩
Volodarskii, L. B., **2**, 340 ⟨86S704⟩; **3**, 427 ⟨86S704⟩; **5**, 212 ⟨85ZOR196⟩, 669 ⟨91BAU382, 91IZV444⟩
Vologdina, L. P., **4**, 106 ⟨88JOU2003⟩
Voloshchuk, V. G., **2**, 287, 732 ⟨68ZOB2509⟩; **6**, 49 ⟨77ZOR2008⟩, 236 ⟨68ZOB2591⟩, 237 ⟨68ZOB2509⟩
Volpe, A. A., **4**, 29 ⟨76S313⟩
Volpin, M. E., **1**, 281 ⟨80CC748⟩
Vol'Pin, M. E., **1**, 493 ⟨74JOM(72)163⟩
Volpp, G. P., **5**, 580 ⟨70JOC2102⟩, 832 ⟨70CI(L)238⟩
Volpp, W., **3**, 215 ⟨88CB1795⟩
Volringer, H., **6**, 221, 222, 227 ⟨37CR(204)1927⟩
Volterra, E., **6**, 6 ⟨56CI(M)371⟩
Volves, J., **6**, 5 ⟨74CCC1330⟩
Volynskaya, E. M., **5**, 904 ⟨71JGU2019⟩
Volz, H., **1**, 894 ⟨63AG(E)622, 65AG(E)873⟩, 922 ⟨70T5343⟩; **2**, 1006 ⟨81LA1415⟩, 1008 ⟨75LA128⟩
Volz, W., **3**, 369 ⟨90CB2065⟩
Volz-de Lecea, M., **2**, 1008 ⟨75LA128⟩
Von Banholzer, R.,**6**, 444 ⟨85AF217⟩
von Brachel, H.,**1**, 524 ⟨62LA(651)141⟩
Vonderheid, C., **1**, 560 ⟨79S127⟩; **2**, 741 ⟨79S127⟩, 748 ⟨75S512⟩; **3**, 638, 657 ⟨79S127⟩
Von Eggers Doering, W.,**1**, 806 ⟨85JA5739⟩
Vong, A. K. K., **1**, 231, 949 ⟨90TL823⟩
VonGeldern, T. W., **1**, 248 ⟨87H(25)75⟩
Vonk, G. P., **2**, 401, 402 ⟨83JOC1⟩
von Schnering, H. G.,**6**, 178 ⟨81ZAAC(475)87⟩, 704 ⟨84ZAAC(511)95⟩
Vonwiller, S. C., **1**, 486 ⟨90AJC1375⟩
Voorhoeve, R. J. H., **2**, 538 ⟨B-67MI 210-03⟩
Vo-Quang, L., **3**, 8 ⟨89S64⟩, 72 ⟨83TL5209⟩, 230 ⟨80T1331⟩; **4**, 487 ⟨91SC1847⟩
Vo-Quang, Y., **3**, 8 ⟨89S64⟩, 72 ⟨83TL5209⟩, 230 ⟨80T1331⟩; **4**, 465 ⟨85S62⟩, 487 ⟨91SC1847⟩
Vora, M. M., **3**, 308 ⟨81H(16)399⟩
Vorbrüggen, H., **3**, 220 ⟨86S41⟩, 424 ⟨90T3849⟩, 671 ⟨83S316⟩; **4**, 191 ⟨63MI 404-02⟩; **5**, 131 ⟨63AG(E)211⟩
Vordermaier, G., **5**, 397 ⟨77AG(E)269, 78ZN(B)1457⟩, 403 ⟨79CB1456, 79CB763⟩, 403 ⟨79CB793⟩
Vorländer, D., **2**, 194, 239 ⟨19CB413⟩; **3**, 452 ⟨23CB1136, 24CB1108⟩; **5**, 833 ⟨1891CB803⟩
Vornberger, W., **4**, 571 ⟨70AG(E)737, 72CB3173⟩; **5**, 913 ⟨72CB1084⟩, 913, 914 ⟨72CB3173⟩
Vorob'eva, E. M., **5**, 389 ⟨76ZOB75⟩
Vorob'eva, V. V., **1**, 45 ⟨70KGS495⟩
Voronko , V. M. G., **1**, 63 ⟨81JOM(214)145⟩
Voronkov, A. N., **6**, 54 ⟨79IZV1911⟩
Voronkov, M., **2**, 266 ⟨81JOM(214)415⟩
Voronkov, M. G., **1**, 61 ⟨83ZOB806⟩; **2**, 19 ⟨74CCC2616⟩, 338, 365 ⟨89PS(45)35⟩, 547 ⟨66PAC35, 79TCC77⟩, 741 ⟨89S446⟩, 1091 ⟨79JGU536, 86JGU849⟩, 1093 ⟨81IZV1280⟩; **3**, 341 ⟨87KGS1694⟩, 356 ⟨78ZOR666, 79ZOR82⟩, 360 ⟨75ZOR1945⟩, 361 ⟨80ZOR1842⟩, 363 ⟨72IJS(A)0216⟩, 430 ⟨85ZOR766⟩, 548 ⟨78ZOB2137⟩; **4**, 137 ⟨76CCC386⟩, 140 ⟨75IZV1576, 76IZV2200,

76JGU2076, 89JOM(361)147, 91IZV1039⟩, 152 ⟨88JOM(341)225⟩, 312 ⟨84ZOR1502⟩, 342 ⟨79DOK(247)609⟩, 368 ⟨81JOM(204)13⟩, 507 ⟨87JOM(326)159⟩, 569 ⟨75ZOB2010⟩, 617 ⟨87IZV1424⟩, 765 ⟨88IZV514⟩, 786 ⟨85ZOR2467, 87ZOR60, 88ZOR509, 89ZOR1470, 89ZOR1773⟩, 787 ⟨90ZOR978⟩, 812 ⟨79JGU625⟩, 813 ⟨73JGU679⟩, 926 ⟨82ZOB2375⟩; **5**, 376 ⟨86ZOB1196, 87ZOB956⟩, 540 ⟨75JGU2421⟩, 718 ⟨84JGU426, 84ZOB477, 88BAU378, 88IZV460⟩, 924 ⟨88IZV1947, 89JOM(368)163⟩, 1025 ⟨88ZOR1771⟩, 1076, 1077 ⟨85ZOB621⟩, 1135 ⟨74URP449912, 77IZV1839, 82S277⟩, 1142 ⟨76DOK(230)853, 88JOM(341)225⟩; **6**, 131 ⟨87IZV2872, 88MI 604-01⟩

Voronova, N. P., **2**, 436 ⟨68MI 208-01⟩

Vorontsov, V. M., **4**, 966 ⟨80IZV1147⟩

Vorontsova, L. G., **6**, 662 ⟨82IZV2370⟩

Voropaeva, T. I., **5**, 742, 751 ⟨72ZOR1246⟩

Vorozhtsov, Jr., N. N., **6**, 12 ⟨69IZV196⟩

Vors, J.-P., **2**, 446 ⟨79S538⟩

Vorspohl, K., **1**, 924 ⟨89T951⟩; **3**, 92 ⟨89T951⟩; **5**, 924 ⟨87OM1246, 87OM2128⟩

Vos, B. A., **4**, 280 ⟨92TL717⟩

Vos, G. J. M., **1**, 518 ⟨86JOC2004⟩

Vose, C. E., **3**, 618 ⟨44OSC(2)379⟩

Vose, C. W., **1**, 15 ⟨83JCS(P1)87⟩

Vosejpka, P. C., **4**, 581 ⟨90JA3713⟩, 680 ⟨88OM934⟩; **5**, 450 ⟨90JA4083⟩

Voskuil, W., **1**, 999 ⟨62RTC993⟩; **2**, 434 ⟨63RTC302⟩, 1059, 1079, 1081 ⟨64RTC1301⟩, 1061 ⟨62RTC993⟩, 1066 ⟨69RTC71⟩

Vosooghi, M., **4**, 870 ⟨80JHC117⟩; **6**, 630 ⟨80JHC117⟩

Voss, B., **3**, 61 ⟨83AG(E)410⟩

Voss, E., **3**, 75, 265 ⟨89CB83⟩, 246 ⟨88S274, 89CB1179⟩

Voss, G., **6**, 650 ⟨73ZC58⟩

Voss, H., **5**, 244 ⟨85AG(E)1041⟩, 543 ⟨85AG1051⟩

Voss, J., **3**, 330 ⟨83HCA801⟩, 345, 346 ⟨80CB2255⟩, 345, 349 ⟨82JCR(S)314, 83LA1694⟩, 348 ⟨77AG(E)725, 77JCR(S)146⟩, 349, 365 ⟨75TL537, 81LA10⟩, 365 ⟨77JCR(S)244⟩, 369 ⟨90CB2065⟩; **5**, 233, 510, 512, 513 ⟨91COS(6)435⟩, 472 ⟨77SST141, 79SST139⟩, 480 ⟨B-79MI 511-01⟩, 485 ⟨80AG(E)563, 81PS(10)213⟩, 510, 516, 523 ⟨77AG743, 79JCR(S)160⟩, 546 ⟨B-79MI 513-01⟩, 550 ⟨89PS(43)129⟩, 554 ⟨78S849, 86S968⟩, 561 ⟨77JCR(S)144⟩, 565 ⟨77SST159, 79SST156⟩, 565 ⟨B-70MI 514-01⟩, 618, 619 ⟨75LA275⟩, 620, 623, 624 ⟨62LA(660)60⟩, 623 ⟨77AG(E)725⟩, 629, 638, 639, 640 ⟨B-70MI 515-01⟩, 645, 739 ⟨75LA1808⟩, 725, 726, 733 ⟨B-70MI 518-01⟩; **6**, 96 ⟨89S684⟩, 579 ⟨71LA(746)92⟩

Voss, K.-U., **1**, 466 ⟨93CB1447⟩

Voss, P., **2**, 689 ⟨73LA20⟩

Voss, V. F., **1**, 596 ⟨91SL873⟩

Voss, W., **2**, 100 ⟨31LA(485)258⟩

Vossius, D., **3**, 407 ⟨58CB12⟩

Vostokov, I. A., **5**, 968, 969 ⟨73ZOB623⟩, 968, 969 ⟨72ZOB2501, 75ZOB2025⟩, 978 ⟨75ZOB2007⟩, 1085 ⟨89MI 528-01⟩

Vostretsova, N. L., **6**, 160 ⟨75ZOB1450⟩

Vostrikov, N. S., **3**, 8 ⟨89S940⟩

Vostrikova, O. S., **1**, 316 ⟨85JOM(285)43⟩; **2**, 561 ⟨83IZV218, 83IZV2053⟩, 561 ⟨85JOM(285)43⟩

Vostrowsky, O., **1**, 229 ⟨76CB3375⟩, 620 ⟨74TL3987⟩, 727 ⟨74TL207, 76CB1694⟩; **2**, 188 ⟨74TL3987⟩, 997 ⟨83TCC85⟩; **3**, 494 ⟨78CB248⟩, 494 ⟨76CB1694⟩

Voswinckel, H., **5**, 859 ⟨03CB2484⟩

Voticky, Z., **5**, 655 ⟨84CCC1505⟩

Votral, R. J., **3**, 496 ⟨66JOC467⟩

Vött, V., **1**, 104 ⟨67TL3871⟩, 1123 ⟨68TL4617, 70LA(733)5⟩

Vottero, P. J. A., **4**, 224 ⟨93CAR(248)377⟩

Vougioukas, A. E., **1**, 208 ⟨87TL5513⟩; **3**, 615, 632 ⟨87TL5513⟩; **4**, 566 ⟨88SRI163⟩, 566, 573 ⟨85MI 412-01⟩

Vovk, M. V., **2**, 256 ⟨87ZOB1078⟩; **4**, 446 ⟨92ZOR2119⟩; **5**, 669 ⟨89UKZ1071, 91UKZ202⟩, 982 ⟨90UKZ1313⟩, 1035 ⟨91UKZ431⟩, 1044 ⟨91UKZ89⟩, 1063 ⟨79S738, 81ZOR1420⟩, 1068, 1069 ⟨89ZOR759⟩, 1071 ⟨87ZOR2023⟩, 1076 ⟨86ZOR1099⟩; **6**, 539 ⟨92ZOR607⟩

Vovsi, B. A., **3**, 526 ⟨69RCR487⟩

Vowinkel, E., **2**, 683 ⟨66CB1479⟩; **3**, 618, 662 ⟨74CB1221⟩

Voyle, M., **2**, 684 ⟨88JOM(347)365⟩

Voznesenskaya, A. K. L., **2**, 477 ⟨77ZOB1432⟩

Vozza, J. F., **1**, 861 ⟨62JOC3856⟩

Vrba, L., **4**, 211 ⟨72MI 404-01⟩

Vreeke, M., **1**, 107 ⟨90JA3554⟩

Vrencur, D. J., **2**, 204 ⟨69JOC3097⟩, 207 ⟨70JOC1226⟩

Vrielink, J. J., **2**, 682 ⟨75TL3429⟩

Vries, G. de, **2**, 273 ⟨73OSC(5)223⟩; **4**, 206 ⟨55RTC1429⟩; **5**, 1026 ⟨55RTC1262⟩

Vries, J. G. de, **4**, 234 ⟨80JOC4126⟩

Vries, L. de, **1**, 912 ⟨61JA2392⟩; **3**, 595 ⟨71JOC3442⟩, 608 ⟨93BCJ3729⟩

Vries, M. J. de, **4**, 56 ⟨65JCS3075⟩

Vrieze, K., **1**, 73 ⟨68IS101⟩; **2**, 399 ⟨74JOM(74)289⟩; **4**, 716 ⟨92ICA689, 93OM315⟩; **5**, 781 ⟨77JOM(135)183, 77MI 519-02⟩

Vrijland, M. S. A., **2**, 244 ⟨74TL837⟩; **5**, 809 ⟨70TL3763⟩, 1112 ⟨77OS88, 88OSC(4)727⟩

Vrkoc, J., **4**, 223 ⟨59CCC64⟩

Vrobel, K., **1**, 993 ⟨92AG(E)90⟩; **5**, 1041 ⟨92AG(E)90⟩

Vroegop, P. J., **1**, 420 ⟨73T1393⟩

Vu Anh Dang, **6**, 421 ⟨90JOC1851⟩, 423 ⟨90JOC1847⟩, 541 ⟨86EUP176412⟩

Vuazankin, N. S., **6**, 131 ⟨88MI 604-01⟩

Vuitel, L., **2**, 746 ⟨76S543⟩

Vukicevic, R., **2**, 286 ⟨87TL4343⟩

Vukojevic, N., **2**, 9 ⟨93CCC902⟩

Vukov, R., **1**, 400, 799 ⟨69JA4322⟩

Vulakh, E. L., **5**, 13 ⟨79MI 501-01⟩

Vunnam, R. R., **2**, 220 ⟨75JOC250⟩

Vuorinen, E., **2**, 1000 ⟨91T8611⟩

Vuper, M., **4**, 614, 616 ⟨81JA6788⟩

Vutthikongsirigool, W., **4**, 422 ⟨81T4337⟩

Vvedenskii, V. Y., **5**, 376 ⟨86ZOB1196, 87ZOB956⟩

Vyas, D. M., **3**, 635 ⟨75JCS(P1)180⟩; **4**, 271 ⟨75JCS(P1)180⟩; **6**, 611 ⟨84TL487⟩

Vyazankina, N. S., **6**, 131 ⟨83JOM(248)51⟩, 516 ⟨84IZV1897⟩

Vyazankin, N. S., **1**, 986 ⟨87IZV695⟩; **2**, 1093 ⟨73JGU626⟩; **3**, 483 ⟨77JOM(142)155⟩, 533 ⟨75IZV199, 76ZOB930⟩; **4**, 628 ⟨76IZV1887⟩, 629, 638 ⟨85ZOB2396⟩, 637, 641 ⟨75JOM(96)213, 80IZV1451⟩, 641 ⟨76IZV1342⟩, 656 ⟨84ZOB2025⟩, 660 ⟨72JOM(46)51⟩, 766 ⟨83IZV2153⟩, 933 ⟨81IZV654⟩, 940 ⟨77IZV596⟩, 942 ⟨83ZOB1292⟩, 1055 ⟨80ZOB2044⟩; **5**, 226 ⟨84IZV1897⟩, 391, 718 ⟨87ZOB1199⟩, 432 ⟨82IZV949, 83IZV1675, 83ZOR1838⟩, 718 ⟨87JGU1073⟩, 719 ⟨83JGU200, 83JGU1851, 83ZOB229⟩, 719, 720 ⟨83BAU1373, 83IZV1515, 83ZOB2051⟩, 874 ⟨80RCR357⟩, 924 ⟨88IZV1947, 89JOM(368)163⟩, 1137 ⟨69ZOB2365⟩; **6**, 62 ⟨72IZV85, 72JOM(34)299⟩, 131 ⟨87IZV2872⟩, 185 ⟨77JOM(142)155⟩

Vyazankina, O. A., **4**, 629, 638 ⟨85ZOB2396⟩, 656 ⟨84ZOB2025⟩; **5**, 226 ⟨84IZV1897⟩, 391, 718 ⟨87ZOB1199⟩, 432 ⟨82IZV949, 83IZV1675, 83JOM(248)51, 83ZOR1838, 87JGU1073⟩, 719 ⟨83JGU200, 83JGU1851, 83ZOB229⟩, 719, 720 ⟨83BAU1373, 83IZV1515, 83ZOB2051⟩; **6**, 131 ⟨83JOM(248)51, 84IZV1897⟩

Vycudiliak, W., **2**, 162 ⟨68TL3607⟩

Vydzhak, R. N., **4**, 998 ⟨92ZOB1423⟩
Vyglazov, O. G., **1**, 626 ⟨88MI 114-01⟩
Vyplel, H., **2**, 15, 17 ⟨85C305⟩
Vyshenskaya, L. I., **2**, 566 ⟨72SRI33⟩, 569 ⟨82JOM(239)335⟩, 569 ⟨79DOK(245)198⟩
Vyshinskii, N. N., **2**, 569 ⟨82JOM(239)335⟩
Vysotskii, M. F., **4**, 339 ⟨91ZOB1263⟩
Vysotskii, V. I., **2**, 455 ⟨85JGU2198, 88JGU693⟩, 458 ⟨79JGU1714, 79JGU1721⟩, 458 ⟨79JGU285⟩; **4**, 339 ⟨79ZOB1956, 91ZOB1315⟩, 339 ⟨91ZOB909⟩
V'yunov, K. A., **2**, 5 ⟨81RCR151⟩
Vyvyan, J. R., **6**, 719 ⟨93OM2806⟩

W

Waack, R., **1**, 850 ⟨66JA1272⟩
Waalkes, T. P., **4**, 30 ⟨46JA496⟩
Waard, E. R. de, **1**, 318 ⟨77TL4528⟩, 619, 707 ⟨81T417⟩, 708 ⟨77TL1173, 80H(14)1739⟩
Waas, J. R., **3**, 157 ⟨92TL3717⟩; **4**, 815 ⟨92TL3717⟩
Wacher, V. J., **4**, 271 ⟨92MI 406-02⟩
Wachholz, G., **1**, 1069 ⟨89CB377⟩
Wachter, M., **5**, 11 ⟨91S18⟩
Wachter, W. A., **4**, 703 ⟨74JA7586⟩
Wächterhäuser, G., **1**, 1145 ⟨64AG(E)651⟩
Wächtershäuser, G., **3**, 466 ⟨67LA(707)44⟩
Wacker, C.-D., **1**, 981 ⟨85AG(E)851⟩
Wackerle, L., **3**, 80 ⟨80JOC4059⟩; **6**, 415 ⟨80JOC4059⟩, 422 ⟨75S598⟩
Wada, A., **2**, 665 ⟨81TL4283, 82JOC4376⟩; **6**, 32 ⟨91EUP450584⟩
Wada, E., **2**, 667 ⟨84CL277⟩; **3**, 269 ⟨89BCJ860⟩
Wada, F., **1**, 486 ⟨81JOC4885, 81T31⟩, 486 ⟨83JOC1333⟩, 1033 ⟨87JCS(P1)1511⟩; **2**, 932 ⟨84TL5789⟩; **3**, 90 ⟨84JOM(270)283⟩, 290 ⟨82CL35⟩; **5**, 197, 198 ⟨80TL2877⟩, 198 ⟨81JOC4413⟩, 220, 222 ⟨84TL5789⟩; **6**, 70 ⟨76JAP7630034⟩
Wada, H., **1**, 568 ⟨87BCJ2917⟩; **6**, 460 ⟨86JAP61118349⟩, 465 ⟨92JAP04356446⟩
Wada, K., **1**, 690 ⟨86BCJ2209⟩; **5**, 344 ⟨91BCJ106⟩
Wada, M., **1**, 179 ⟨72TL1279⟩, 1041 ⟨84TL1083⟩; **2**, 87, 118, 124 ⟨72JA679⟩, 325 ⟨87H(26)929⟩, 695 ⟨88TL4719⟩; **3**, 55 ⟨77CL345⟩, 186 ⟨84TL5405⟩, 561 ⟨71BCJ1362⟩; **4**, 118 ⟨85JCR(S)38⟩, 955 ⟨75CC899, 78JOM(159)417⟩, 955 ⟨80JOM(201)477⟩; **5**, 287 ⟨87TL3377⟩, 735 ⟨71BCJ1362⟩, 940 ⟨78JOM(159)417⟩, 948 ⟨82CC1178, 82JCS(D)793⟩, 1137 ⟨76IC2346, 83JCS(D)1841⟩; **6**, 517 ⟨76IC2346⟩
Wada, T., **2**, 7 ⟨92CL1505⟩; **3**, 60 ⟨83TL1171⟩; **4**, 806, 811 ⟨62IC78⟩
Wada, Y., **4**, 67 ⟨89JCS(P1)879⟩; **5**, 143 ⟨75BCJ2075⟩
Waddell, S. T., **1**, 928 ⟨86TL1553⟩
Waddey, W. E., **5**, 25 ⟨55OSC(3)560⟩
Waddington, T. C., **1**, 897 ⟨62JCS600⟩
Wade, A. M., **6**, 107 ⟨71TL4519⟩
Wade, A. R., **4**, 193 ⟨84JCS(P1)1643⟩
Wade, K., **1**, 885 ⟨B-67MI 119-02⟩; **2**, 815 ⟨76JCS(D)1⟩; **3**, 341, 376 ⟨69JCS(C)913⟩, 456 ⟨78JCS(D)986⟩, 457 ⟨64JCS1649, 76JCS(D)1⟩, 458 ⟨60JCS2614, 65JCS2662, 65JCS5083, 67JCS(A)1222, 68CC250B, 70JCS(A)2010, 71CC1275, 72JCS(D)370⟩, 459 ⟨70JCS(A)2016, 72JCS(D)1501, 76CC160, 87JCS(D)1071, 87JCS(D)2141⟩, 461 ⟨68JCS(A)57⟩; **5**, 383 ⟨68JCS(A)2535⟩
Wade, L. E., **1**, 355 ⟨77JA4417⟩
Wade, Jr., L. G., **4**, 220 ⟨78TL731⟩, 235 ⟨79TL689⟩; **5**, 259 ⟨B-84MI 506-01⟩
Wade, M. J., **5**, 624, 625, 626 ⟨76IJS519⟩
Wade, P. A., **1**, 200, 238 ⟨88JOC1475⟩, 368 ⟨91COS(4)1111⟩; **2**, 393, 395 ⟨73JOC1418⟩, 394 ⟨82JOC365⟩; **3**, 45, 160, 189, 678 ⟨91COS(4)1111⟩, 314 ⟨91COS(4)1159⟩, 684, 686 ⟨84JOC4595⟩; **4**, 108 ⟨91JA8807⟩, 333 ⟨81JOC765⟩; **5**, 1145, 1148 ⟨81JOC5425⟩; **6**, 611 ⟨90JOC3045⟩
Wade, T. M.,**4**, 10 ⟨80JOC5333⟩
Wade, T. N., **2**, 10 ⟨78TL4507⟩
Wadepohl, H., **4**, 632 ⟨83JA2582⟩, 690 ⟨90AG(E)686⟩, 691 ⟨87AG(E)127, 93CC1459⟩
Wadgaonkar, P. P., **3**, 426 ⟨90SC2453⟩; **5**, 367 ⟨89IJC(B)186⟩, 990 ⟨92MI 526-01⟩
Wadi, M., **3**, 357 ⟨90EUP376724⟩
Wadia, M. S., **3**, 133 ⟨77CI(L)454⟩
Waditschatka, R., **6**, 25, 31 ⟨88M1427⟩
Wadley, E. F., **1**, 912, 914 ⟨58T178⟩
Wadman, S. N., **1**, 481 ⟨89JA2363, 92JCS(P1)3419, 92JCS(P1)3431⟩
Wadsworth, A. H., **1**, 1024, 1134 ⟨80JCS(P1)852, 79JCS(P1)2954⟩; **3**, 254 ⟨82TL561⟩
Wadsworth, D. H., **1**, 635 ⟨73OS(53)13⟩; **2**, 463 ⟨58JA3945⟩
Wadsworth, Jr., W. S., **1**, 638, 639, 640 ⟨77OR(25)73⟩, 639, 987 ⟨61JA1733⟩; **2**, 1003 ⟨77OR(25)73⟩; **3**, 195 ⟨77OR(25)73⟩, 639 ⟨61JA1733⟩, 587 ⟨62JA1316, 64JOC2816⟩; **5**, 103 ⟨73OSC(5)547⟩, 975 ⟨62JA1316, 67JOC1279⟩, 975, 1031, 1069 ⟨64JOC2816⟩
Waegell, B., **1**, 633 ⟨75T1035⟩, 797 ⟨89BSF799⟩, 929 ⟨B-79MI 119-03⟩; **2**, 85 ⟨88TL4851⟩; **5**, 316 ⟨78JOC3750⟩
Waespe-Šarčevič, N., **2**, 900 ⟨79JOC3374⟩; **5**, 135, 138 ⟨83HCA2501⟩
Wagatsuma, N., **4**, 325 ⟨78GEP2705994⟩
Wagenaar, A., **3**, 368 ⟨73JCS(P1)73⟩; **6**, 316 ⟨73TL3589⟩
Wagenhofer, H., **3**, 678 ⟨62AG848⟩
Wagenknecht, J. H., **1**, 103 ⟨81MI 103-01⟩; **3**, 83 ⟨72JOC1513⟩; **4**, 850 ⟨74CC751⟩
Wager, J. S., **4**, 850 ⟨75JOC2577⟩, 972, 988 ⟨71JOC1613⟩; **5**, 745, 753 ⟨71JOC1613⟩; **6**, 120 ⟨75JOC2577⟩
Waggoner, K. M., **2**, 475 ⟨91JA3597⟩
Wagh, S. B., **5**, 131 ⟨83T1475⟩
Wagland, A. M., **1**, 7 ⟨87TL5339⟩
Wagler, T. R., **5**, 385 ⟨89JOC1584⟩
Waglund, T., **5**, 132 ⟨87JOC3777⟩
Wagman, A. S., **6**, 58, 542 ⟨92TL1839⟩
Wagner, A., **1**, 504, 623 ⟨88CB2239⟩; **2**, 10 ⟨89TL557⟩, 216, 225 ⟨55HOU(9)5⟩, 402 ⟨55HOU(9)260⟩, 724 ⟨60CB2736⟩, 1018 ⟨90TL3141⟩; **5**, 199 ⟨82S958⟩, 232 ⟨55HOU(9)745⟩
Wagner, B., **5**, 936 ⟨92JOM(426)343⟩, 955 ⟨91CB2431⟩
Wagner, B. J., **2**, 247 ⟨90JOC4156⟩
Wagner, B. O., **1**, 665 ⟨75JOM(85)1⟩
Wagner, C., **4**, 651 ⟨93OM4135⟩
Wagner, D., **2**, 190 ⟨68TL4479⟩, 809 ⟨60AG209⟩
Wagner, F., **4**, 506, 513 ⟨91JOM(415)39⟩, 563 ⟨77CB1748⟩; **5**, 747 ⟨80LA344⟩
Wagner, F. C., **2**, 191 ⟨31JA3407⟩
Wagner, G., **1**, 380 ⟨1899CB2307⟩; **2**, 573 ⟨58LA(618)24⟩; **4**, 239 ⟨64AP(297)461⟩; **5**, 567, 584, 618 ⟨71PHA271⟩, 587 ⟨68ZC376⟩, 622 ⟨85ZC364⟩, 755, 758 ⟨78PHA568⟩; **6**, 174, 289, 291, 392, 709, 710 ⟨86CB1455⟩, 177, 178, 382, 387, 396, 709, 710 ⟨87ZN(B)1062⟩, 178, 387, 392, 709 ⟨83AG(E)1005⟩, 178, 387, 673, 710 ⟨86CB1467⟩, 382, 709, 710 ⟨87ZN(B)1055⟩, 383 ⟨89ZAAC(579)75⟩, 392, 710 ⟨84JOM(271)381⟩, 392, 710 ⟨85AG(E)229, 87OM35⟩, 710 ⟨87OM32⟩
Wagner, H., **4**, 570 ⟨87EGP251134, 87EGP251136⟩, 998 ⟨82AG559⟩
Wagner, H.-P., **4**, 883 ⟨83HCA809⟩
Wagner, I., **2**, 729 ⟨91CB1315⟩; **4**, 88, 288 ⟨91CB1315⟩; **6**, 334, 355, 356 ⟨90CB2325⟩, 355, 356 ⟨88CB2109, 93ZAAC(619)1693⟩
Wagner, J., **3**, 163 ⟨89CC1634⟩; **6**, 195 ⟨92ZAAC(613)67⟩
Wagner, K., **3**, 490 ⟨65LA(681)45⟩; **4**, 701 ⟨91CB503⟩;

5, 599 ⟨72LA(757)147⟩, 658 ⟨78S38⟩, 1061, 1078 ⟨81AG855⟩; **6**, 209 ⟨91CB503⟩, 297, 299 ⟨72S599⟩
Wagner, K. P., **5**, 802 ⟨73JOM(61)415⟩
Wagner, M., **4**, 632 ⟨93AG(E)295⟩
Wagner, O., **2**, 859, 865 ⟨89AG(E)225⟩; **4**, 110 ⟨89AG(E)225, 91CB1207⟩, 480 ⟨87AG(E)1257⟩; **5**, 924 ⟨88CB1363⟩, 1155 ⟨87PS(30)479⟩; **6**, 153 ⟨89AG(E)1013⟩
Wagner, P., **4**, 305 ⟨69CB2651⟩, 1079 ⟨91MI 424-02⟩
Wagner, P. H., **3**, 247 ⟨76AG(E)382⟩
Wagner, P. J., **1**, 841 ⟨B-80MI 118-03⟩; **3**, 541 ⟨66JA1242⟩; **4**, 3, 6 ⟨86JA7739⟩
Wagner, R., **1**, 815 ⟨82JOC3190⟩; **4**, 628, 639 ⟨86TL4015⟩
Wagner, R. B., **1**, 393 ⟨50JA2884⟩; **3**, 526 ⟨B-53MI 316-01⟩
Wagner, R. I., **2**, 427, 429 ⟨53JA3869⟩, 430 ⟨67JA1102⟩, 453, 837, 845 ⟨67JOC2172⟩, 474 ⟨53JA3872⟩, 821 ⟨62JOC2192⟩; **4**, 545 ⟨63USP3086053, 63USP3086056⟩
Wagner, R. M., **2**, 790 ⟨72AG(E)315⟩; **3**, 231 ⟨75LA874⟩
Wagner, S., **5**, 720 ⟨77JOM(134)21⟩; **6**, 709, 710 ⟨91CB1981⟩
Wagner, U., **2**, 366 ⟨72LA(762)93⟩
Wagner, W. J., **1**, 632 ⟨64JOC2788⟩; **2**, 16 ⟨91JOC5962⟩; **3**, 149 ⟨91JOC5962⟩
Wagner, W. M., **4**, 16 ⟨59PCS229, 62RTC925⟩; **6**, 25 ⟨59PCS229⟩
Wagner-Jauregg, T., **2**, 412 ⟨51JA5202⟩; **5**, 201 ⟨57HCA852⟩; **6**, 629, 630 ⟨58HCA377⟩
Wagstaff, K. E., **1**, 895 ⟨78JA5134⟩
Wahab, A.-M. A. A., **4**, 838, 1081 ⟨89T6667⟩
Wahl, A., **6**, 500 ⟨34BSF244⟩
Wahl, G. H., **3**, 532 ⟨65CI(L)425⟩
Wahl, H., **6**, 653 ⟨62MI 621-01⟩
Wahl, R., **6**, 104 ⟨68CB41⟩
Wahle, B., **2**, 944 ⟨83CB2564⟩
Wahr, N., **3**, 521 ⟨92CB2641⟩
Wahren, R., **3**, 621 ⟨78JA1119⟩
Wai, J. S. M., **1**, 615 ⟨87CC1342⟩
Wai, K.-F., **3**, 363, 686 ⟨91JCS(P1)183⟩
Waid, P. P., **2**, 13 ⟨90SC1011⟩
Waid, R. D., **2**, 439 ⟨84IC778⟩
Waidhas, M., **1**, 445 ⟨86JCE930⟩; **2**, 490 ⟨86JCS(D)997⟩
Waigh, R. D., **1**, 436 ⟨86SC779⟩
Waight, E. S., **1**, 838 ⟨59JCS2720⟩; **5**, 735 ⟨74JCS(P1)1499⟩
Wailes, P. C., **2**, 968 ⟨71JOM(27)373⟩; **3**, 108 ⟨53JCS1740⟩
Waissermann, J., **4**, 722 ⟨91OM4005⟩
Wai Tan, W. H. L., **2**, 466 ⟨90T6715⟩
Waite, A. A. C., **1**, 51 ⟨83JCS(P2)1923⟩
Waite, N. E., **3**, 498 ⟨69JCS(C)1100⟩
Waites, J. A., **5**, 244 ⟨85JOC4164⟩
Waitkins, G. R., **5**, 28 ⟨45CRV235⟩
Waits, H. P., **1**, 924 ⟨64JA1911⟩
Wajima, M., **5**, 1125 ⟨86JAP47675⟩
Wakabayashi, N., **1**, 561 ⟨61JA927⟩, 832 ⟨69TL3253⟩
Wakabayashi, S., **2**, 7, 27 ⟨93JOC4382⟩; **3**, 634, 651 ⟨84JA7890, 86TL5109⟩; **4**, 220 ⟨80TL2557⟩, 227 ⟨91CC1421⟩
Wakabayashi, T., **2**, 300 ⟨80JOC357⟩; **6**, 636 ⟨84CHEC(3)91⟩
Wakabayashi, Y., **3**, 616, 627, 633 ⟨83T967⟩, 627 ⟨81TL4279⟩
Wakamatsa, T., **1**, 102 ⟨80H(14)1441⟩
Wakamatsu, K., **1**, 710 ⟨84JA4547⟩; **2**, 106 ⟨85BJC2425⟩, 710 ⟨87CL1647⟩, 903, 919 ⟨86TL2007⟩, 918 ⟨85TL4629⟩, 918, 923 ⟨86T4427⟩, 946 ⟨87TL3709⟩
Wakamatsu, M., **5**, 549 ⟨74JOM(78)405⟩
Wakamatsu, T., **1**, 813 ⟨86TL219⟩; **3**, 217 ⟨86TL155⟩
Wakamiya, T., **1**, 882 ⟨87TL5887⟩; **6**, 487 ⟨93CL1401⟩
Wakamutsu, K., **3**, 289 ⟨85BCJ2425⟩
Wakasa, M., **1**, 449 ⟨88OM1869⟩
Wakasa, N., **1**, 195, 477 ⟨85TL3259⟩
Wakasugi, M., **1**, 493 ⟨80BCJ553⟩; **5**, 1113 ⟨80BCJ553⟩
Wakatsuki, Y., **1**, 992 ⟨91JA9604, 93BCJ987⟩; **2**, 938 ⟨91CC703⟩, 940 ⟨93BCJ987⟩; **4**, 714 ⟨83OM726⟩
Wakefield, B. J., **1**, 3, 6 ⟨63PCS219⟩, 110 ⟨B-74MI 104-01⟩, 663 ⟨B-74MI 114-01⟩, 664 ⟨82COMC-I(7)85⟩, 665 ⟨66MI 114-01⟩, 826 ⟨83CC932⟩, 885 ⟨66MI 119-08, B-74MI 119-01, B-88MI 119-02⟩; **2**, 67, 68 ⟨B-88MI 202-02⟩, 551 ⟨B-74MI 211-01, 82COMC-I(7)1, B-88MI 211-01⟩, 653 ⟨B-74MI 213-01⟩, 821 ⟨B-74MI 233-01⟩, 907, 908, 909, 910, 942 ⟨B-88MI 218-02⟩; **3**, 539 ⟨89CC303⟩; **4**, 843 ⟨90S991⟩; **5**, 664 ⟨81JCR(S)129⟩; **6**, 28 ⟨89JCS(P1)283⟩
Wakefield, E., **4**, 860 ⟨92T7551⟩
Wakeshima, I., **6**, 469 ⟨78BCJ3549⟩
Wakharkar, R. D., **6**, 557 ⟨90OPP128⟩
Waki, Y., **1**, 907 ⟨82BCJ3257⟩
Wakita, A., **1**, 235 ⟨84JOC1985⟩
Wakita, H., **3**, 355 ⟨89MI 308-02⟩
Wakita, S., **3**, 303 ⟨79TL4577⟩
Wakselman, C., **1**, 522 ⟨75TL363⟩; **2**, 10 ⟨79JOC3406⟩, 19 ⟨79TL353⟩; **3**, 67 ⟨85S754⟩; **4**, 187, 208 ⟨83SC629⟩, 731 ⟨87JOC290⟩, 736 ⟨92JOC5530⟩; **5**, 4 ⟨82SC513⟩, 481, 487, 582 ⟨87JFC(35)523⟩; **6**, 11 ⟨87CC1701⟩, 63 ⟨77JOC565⟩, 229 ⟨81T4209⟩, 233 ⟨81TL323⟩, 233, 234 ⟨84CC793, 85JOC4047, 91CC993⟩, 423 ⟨82SC513⟩
Wakselman, G., **3**, 237 ⟨89JOC5640⟩
Wakselman, M., **6**, 477 ⟨78T3105⟩
Wakuda, K., **5**, 557 ⟨82CL123⟩
Walace, E. M., **4**, 138, 355 ⟨92JOC386⟩
Walaschewski, E. G., **2**, 481, 482, 486, 491, 492, 498 ⟨53JCS1552⟩, 486, 491 ⟨53CB272⟩; **6**, 64 ⟨53JCS3607⟩
Walaton, D. R. M., **2**, 1089 ⟨64JOM(2)97⟩
Walborsky, E. C., **6**, 403 ⟨83JA1384⟩
Walborsky, H. M., **1**, 21, 24 ⟨50JA4296⟩, 38 ⟨78JOC2396⟩, 155 ⟨55JA6026, 74JA3711, 77JA2631⟩, 268 ⟨64T1695⟩, 491 ⟨76JA3732⟩, 851 ⟨62JA2465, 64JA3283⟩, 871 ⟨70JA2445, 70JA3697⟩, 885 ⟨62JA4838⟩; **2**, 268 ⟨74JOC600⟩, 554 ⟨64JA3283⟩, 653 ⟨70JA2042⟩, 942, 949 ⟨78JOC731⟩, 998 ⟨93JOC546⟩; **3**, 218, 276, 408 ⟨82JOC52⟩, 418 ⟨78JOC731, 81JOC5405⟩, 418, 697 ⟨69JA7778, 74JOC600⟩, 526 ⟨52JA4962⟩, 694 ⟨B-83MI 321-01⟩, 696, 699, 721, 722 ⟨72JOC187⟩, 696, 700 ⟨88OSC(6)751⟩, 697 ⟨70JA6675, 71OS(51)31, 71TL4965, 74JOC604, 74JOC611, 78JOC734⟩; **4**, 33, 35, 37 ⟨91JOC2193⟩, 816, 818 ⟨93JOC546⟩; **5**, 68, 420, 792, 796, 797 ⟨74JOC600⟩, 88 ⟨74JOC604⟩, 792, 796, 797 ⟨70JA6675⟩, 796 ⟨69JA7778, 82JOC52⟩, 796, 797 ⟨71OS(51)31⟩, 796, 802 ⟨78JOC731⟩
Wälchli, R., **5**, 272, 283 ⟨82HCA2299⟩, 272, 284 ⟨85HCA484⟩
Walczak, K., **4**, 706 ⟨78JA6382⟩
Walczak, M., **4**, 706 ⟨78JA6382⟩
Waldbach, T. A., **4**, 720 ⟨93AG(E)710⟩
Walden, P., **1**, 898 ⟨02CB2018⟩
Walder, L., **1**, 345, 946 ⟨80JA3642⟩, 546, 946 ⟨87PAC363, 87OM1606⟩, 946 ⟨B-83MI 119-05⟩
Waldi, J., **6**, 128 ⟨91CB1805⟩
Waldman, M. C., **2**, 487 ⟨70CJC1885⟩; **6**, 65 ⟨70IC1682⟩
Waldmann, H., **1**, 370, 371 ⟨94AG(E)683⟩; **4**, 176, 180 ⟨91COS(6)631⟩
Waldmüller, D., **6**, 72 ⟨91SL160⟩
Waldner, A., **1**, 624, 626 ⟨86JA1039⟩; **2**, 245 ⟨89TL3061⟩; **5**, 278 ⟨86JA1039⟩
Waldon, P. T., **5**, 813 ⟨1897JA129⟩
Waldron, R. W., **4**, 557 ⟨72IC2340⟩
Walgraeve, L., **2**, 29 ⟨69S39⟩
Wali, A., **1**, 499 ⟨90JCS(P1)2185⟩
Walia, A. S., **4**, 195 ⟨76JOC3765⟩
Walia, J. S., **4**, 195 ⟨76JOC3765⟩

Walia, M., **3**, 162 ⟨93TL7335⟩; **4**, 192 ⟨93TL7335⟩
Walinsky, S. W., **1**, 801 ⟨72JA8934⟩, 821 ⟨72JA8932⟩; **4**, 1074 ⟨72JA8932⟩
Walker, A. J., **1**, 140, 143 ⟨92TA961⟩; **2**, 153 ⟨92TA961⟩; **5**, 840, 860 ⟨25JCS975⟩, 841 ⟨25JCS1687⟩
Walker, A. M., **3**, 14 ⟨81S303⟩
Walker, B. J., **1**, 744, 745 ⟨B-79MI 116-01⟩, 1069 ⟨83JCS(P1)1689⟩; **2**, 105 ⟨B-72MI 202-01⟩; **3**, 493, 494 ⟨65JCS7140⟩; **4**, 451, 493 ⟨B-68MI 410-01⟩
Walker, C. V., **5**, 573 ⟨83T1075, 87TL2171⟩
Walker, D., **3**, 54 ⟨67CRV153⟩
Walker, D. F., **2**, 260 ⟨48JA1762⟩
Walker, D. G., **1**, 523 ⟨81JOC2065⟩; **2**, 622 ⟨91JOC4066⟩; **3**, 221, 227, 275 ⟨84JA5585⟩; **5**, 772 ⟨77JOC1862⟩
Walker, D. M., **1**, 747 ⟨61CI(L)990⟩, 1079 ⟨59JCS3874⟩; **3**, 51, 200 ⟨88JOC4069⟩, 492 ⟨60CI933⟩, 494, 642 ⟨61JCS1266⟩, 591 ⟨59JCS3874⟩
Walker, E. R. H., **1**, 382 ⟨72JCS(P1)1274⟩; **2**, 362, 363 ⟨91JCS(P1)2169⟩
Walker, F. J., **1**, 14 ⟨85JOC5678⟩; **2**, 82 ⟨83PAC589⟩
Walker, G. L., **2**, 288 ⟨67JOC2580⟩
Walker, G. N., **1**, 19 ⟨58JOC133⟩; **2**, 381 ⟨61JOC2740⟩; **5**, 184 ⟨57JA3508⟩
Walker, J., **5**, 24 ⟨B-51MI 502-01⟩
Walker, J. A., **1**, 278, 487, 662, 1103 ⟨76JOC3947⟩, 822 ⟨72CC354⟩, 831 ⟨73CC117⟩; **2**, 478 ⟨84JOC4733⟩; **5**, 553 ⟨72CC354⟩; **6**, 627 ⟨72CC354⟩
Walker, J. C., **1**, 134 ⟨85CC209, 86CC495⟩; **5**, 453 ⟨84TL2709, 85CC209, 86CC495, 86CC609, 86TL3787⟩
Walker, J. D., **4**, 562 ⟨89POL1293⟩
Walker, K. A. M., **1**, 82 ⟨66JCS(C)1894, 71AJC513⟩, 85 ⟨67TL3457⟩, 223 ⟨93TA1521⟩; **2**, 135 ⟨77TL4475⟩
Walker, L. E., **3**, 617 ⟨62JOC4372⟩, 626 ⟨62JOC3836⟩
Walker, L. G., **5**, 222 ⟨60CJC472⟩
Walker, M. A., **1**, 738, 739, 741 ⟨71JCS(C)1114⟩; **3**, 328 ⟨71JCS(C)1114⟩; **5**, 920 ⟨71JCS(C)1114⟩
Walker, M. P., **4**, 98 ⟨85CC686⟩
Walker, N., **6**, 42, 231 ⟨75JA13⟩
Walker, R. T., **2**, 625 ⟨86JMC213⟩
Walker, S., **4**, 232 ⟨89JA6881⟩
Wälki, M., **2**, 188 ⟨70JCS(C)270⟩
Walkinshaw, M. D., **3**, 686 ⟨86JCR(S)156⟩
Walkowiak, U., **4**, 343 ⟨80PJC233⟩
Walkup, R. D., **5**, 557 ⟨85T5089⟩
Walkup, R. E., **3**, 107 ⟨85TL2155⟩
Wall, A., **1**, 619, 707, 949 ⟨86JA4568⟩, 707 ⟨83JA6165⟩; **2**, 195, 196 ⟨85JA1783⟩; **4**, 82, 83, 85 ⟨86JA4568⟩, 138, 393 ⟨87JOC809⟩, 368 ⟨85TL1425⟩
Wall, D. K., **5**, 872 ⟨68JCS(C)2201⟩
Wall, G. D., **3**, 537 ⟨74CC513⟩
Wall, G. M., **5**, 247 ⟨91T6127⟩
Wall, L. A., **4**, 761 ⟨66USP3277068⟩; **6**, 222 ⟨57JA5654⟩
Wall, M., **2**, 129 ⟨65JOC3689⟩; **3**, 638 ⟨65JOC3689⟩
Wall, M. E., **2**, 274 ⟨65JOC33⟩; **3**, 218 ⟨60JOC1966⟩
Wallace, D. J., **1**, 831 ⟨91TL7601⟩; **3**, 208, 209, 241 ⟨79JOC3066⟩
Wallace, E. M., **1**, 648 ⟨91SL764⟩; **3**, 203 ⟨92JOC386⟩, 713 ⟨93CC1760⟩
Wallace, J. G., **5**, 46 ⟨B-62MI 502-01⟩
Wallace, K. C., **3**, 512 ⟨86OM2162, 88JA4964⟩, 512 ⟨90IC1104⟩
Wallace, P., **1**, 643 ⟨84TL357, 92JCS(P1)3407⟩, 643, 757 ⟨88JCS(P1)2971⟩
Wallace, R. G., **5**, 8 ⟨90S1143⟩
Wallace, R. W., **2**, 355 ⟨73JOC1813⟩
Wallace, T. J., **1**, 590 ⟨65CI(L)1768⟩, 600, 612 ⟨64JA1561, 63JA2739⟩; **2**, 192 ⟨65TL2271⟩, 192, 720 ⟨63TL1131⟩, 213 ⟨66JOC1217, B-66MI 203-01⟩, 214, 215 ⟨66JOC3071⟩
Wallace, T. W., **1**, 472 ⟨82JCS(P1)2299⟩; **2**, 844 ⟨82JCS(P1)2299⟩
Wallace, W. E., **2**, 645 ⟨44JA1274⟩
Wallach, D. P., **2**, 103 ⟨60MI 202-03⟩
Wallach, J., **1**, 975 ⟨93TL3129⟩
Wallach, O., **2**, 1014 ⟨1880LA(203)83⟩; **5**, 652 ⟨1877LA(184)79⟩
Wallach, P., **5**, 380 ⟨80TL3367⟩
Wallbank, P. J., **1**, 489 ⟨92T8117⟩
Wallbillich, G., **3**, 452 ⟨62T3⟩, 680 ⟨59JOC892⟩; **5**, 201, 842 ⟨62T3⟩
Wallbridge, M. G. H., **5**, 1142 ⟨70JCS(A)2675⟩
Wallenfels, K., **3**, 611, 627 ⟨B-70MI 318-01⟩, 641, 642, 661 ⟨B-70MI 319-01⟩
Waller, C. W., **1**, 34 ⟨58JA1654⟩
Waller, F. D., **1**, 627 ⟨67JCS(B)509⟩
Walley, A. R., **2**, 161 ⟨80JCS(P2)154⟩
Walley, D. R., **1**, 568 ⟨83TL1475⟩; **3**, 494 ⟨82SC469⟩; **4**, 846 ⟨84TL5729⟩
Wallian, F., **3**, 303 ⟨91OM366⟩; **6**, 449 ⟨91OM366⟩
Wallin, A., **5**, 689 ⟨93JCS(P2)655⟩
Wallin, A. P., **4**, 392 ⟨87JA5403⟩
Walling, C., **1**, 595, 779 ⟨64JOC2699⟩, 905 ⟨83ACR448⟩, 919 ⟨B-57MI 119-01⟩, 919, 923, 924 ⟨85T3887⟩, 924, 925 ⟨65JA3413⟩, 942 ⟨57JA5326, 59JA1243⟩; **2**, 4, 116, 117, 439 ⟨40CRV351⟩, 95 ⟨62JOC2976⟩, 96 ⟨53JA4372⟩, 104 ⟨58JA4546⟩, 127 ⟨59JA1137⟩, 338 ⟨73OSC(5)184⟩, 429 ⟨48USP2437795, 48USP2437797⟩; **4**, 23 ⟨40CRV351⟩; **6**, 5 ⟨63OR(13)91⟩
Walling, C. S., **1**, 919, 924, 948 ⟨63OR(13)91⟩
Wallingford, V. H., **5**, 591 ⟨61JOC3002⟩
Wallington, T. J., **4**, 42 ⟨93JPC1576⟩
Wallis, C. J., **2**, 680 ⟨79JCS(P2)3099⟩; **3**, 11 ⟨79JCS(P1)3099⟩; **4**, 343 ⟨77CC314⟩, 949 ⟨79JCS(P1)3099⟩
Wallis, E. S., **1**, 113 ⟨50JA4842⟩, 394 ⟨41JA1674⟩; **2**, 320, 756 ⟨46OR(3)267⟩; **5**, 998 ⟨46OR(3)267⟩, 1000 ⟨41JA1674⟩
Wallis, J. M., **2**, 512 ⟨87JOM(325)159⟩
Wallis, R. C., **5**, 944 ⟨80AJC2605⟩
Wallis, T. G., **1**, 704, 708 ⟨77JA2815, 77JA6946⟩
Walliser, B., **3**, 44 ⟨90HCA405⟩
Wallquist, O., **1**, 527, 944 ⟨85JOC2400⟩
Wallwork, S. C., **5**, 781 ⟨86JCS(D)1359⟩
Walpole, A. L., **3**, 474 ⟨58JCS1134⟩
Walraven, H. G. M., **3**, 310 ⟨76RTC220⟩
Walsh, C. T., **1**, 520 ⟨84JA2443⟩; **4**, 528 ⟨89B3541⟩
Walsh, D. A., **2**, 168 ⟨84JMC1379⟩
Walsh, Jr., E. J., **1**, 810, 815 ⟨68JA4729⟩; **3**, 83 ⟨66JA571⟩
Walsh, E. N., **2**, 411 ⟨59IS69⟩, 413 ⟨59IS73⟩, 833, 837 ⟨55JA929⟩
Walsh, J. J., **5**, 264 ⟨94TL4865⟩
Walsh, R. A., **1**, 104 ⟨63JA3522⟩, 1144 ⟨70JOC1086⟩
Walsh, R., **1**, 402, 794, 795 ⟨69CRV103⟩, 1069 ⟨89CB377⟩, 1109 ⟨69CRV279⟩
Walsh, R. J. A., **5**, 587, 593 ⟨93JMC1604⟩
Walshaw, K. B., **2**, 675 ⟨64JCS1029⟩
Walshe, N. D. A., **1**, 700, 703 ⟨86JOC4840⟩; **5**, 177 ⟨86JOC4840⟩
Walsingham, R. W., **6**, 515 ⟨69CC462⟩
Walson, D. R. M., **4**, 1048 ⟨70JOM(24)529⟩
Walter, A., **2**, 439 ⟨91ZAAC(592)121⟩
Walter, B., **2**, 460 ⟨83ZAAC(504)147⟩
Walter, D., **1**, 396, 397 ⟨68TL2809⟩
Walter, D. S., **1**, 347 ⟨94TL2413, 94TL2417⟩; **5**, 639 ⟨86JMC2280⟩, 1120 ⟨87JMC951⟩
Walter, H., **2**, 780 ⟨64LA(667)113⟩
Walter, K., **2**, 380 ⟨67ZC151⟩
Walter, L. A., **4**, 69 ⟨45JA655, 45JA657⟩
Walter, M., **5**, 276 ⟨61HCA1546⟩, 633 ⟨81ZAAC(480)21⟩
Walter, R., **1**, 472 ⟨90AG(E)977⟩, 601 ⟨71JOC2561⟩; **2**, 287 ⟨71JOC2561⟩; **3**, 422 ⟨91COS(3)744⟩; **5**, 851 ⟨08JPR489, 1898JPR223⟩, 859 ⟨1896JPR475⟩

Walter, R. W., **2**, 217 ⟨80MI 203-02, 80MI 203-03⟩
Walter, S. R., **4**, 24 ⟨83JA3411⟩
Walter, W., **2**, 814 ⟨77AG(E)535⟩; **3**, 355 ⟨79S941⟩, 413 ⟨86CB1095⟩, 548 ⟨74CB3562⟩; **4**, 898 ⟨73LA443⟩, 903 ⟨75LA19⟩, 956 ⟨74CB3562⟩; **5**, 472 ⟨77SST141, 79SST139⟩, 477, 584, 587, 620, 641 ⟨70LA(739)201⟩, 506 ⟨77JST25⟩, 565 ⟨77SST159, 79SST156, B-70MI 514-01⟩, 565, 579 ⟨66AG(E)447⟩, 579, 580, 623, 624 ⟨60CB1511⟩, 579, 581, 592, 601 ⟨66LA(698)131⟩, 584, 587, 641 ⟨73LA636⟩, 590, 609 ⟨63LA(669)66⟩, 590, 609, 623 ⟨68LA(712)58⟩, 591, 601, 623, 624 ⟨62LA(660)74⟩, 592, 609 ⟨71LA(753)187⟩, 592, 623 ⟨65LA(681)64⟩, 606 ⟨69CB3786⟩, 608 ⟨69LA(725)234, 70LA(733)195, 71LA(755)145⟩, 618, 619 ⟨75LA275⟩, 620, 623, 624 ⟨62LA(660)60⟩, 623 ⟨65AG(E)1095⟩, 623, 624 ⟨60LA(633)35, 68LA(712)46, 71LA(752)115, 76LA1584⟩, 623, 735 ⟨69LA(727)50⟩, 624, 625, 732 ⟨76CB956⟩, 624, 697 ⟨75LA305⟩, 629, 638, 639, 640 ⟨B-70MI 515-01⟩, 630, 733 ⟨85PS(25)63⟩, 631 ⟨73LA476⟩, 635, 636, 637 ⟨71S111⟩, 636 ⟨61LA(643)29, 71LA(743)154⟩, 638 ⟨74T4147⟩, 638, 735 ⟨69LA(727)22⟩, 640 ⟨67AG(E)368⟩, 640, 846 ⟨69CB2117⟩, 641 ⟨68TL5973⟩, 645 ⟨75AG(E)427⟩, 645, 739 ⟨75LA1808⟩, 725, 726, 733 ⟨B-70MI 518-01⟩, 729 ⟨74CB3589⟩, 735, 736 ⟨71LA(753)169, 73LA462⟩, 736 ⟨74LA776⟩, 746 ⟨69CB2640⟩, 770, 771, 839, 850 ⟨72LA(758)162⟩, 833 ⟨B-70MI 521-03⟩, 852, 853 ⟨70LA(736)166⟩; **6**, 238, 240, 540 ⟨72LA(755)145⟩, 444, 501, 577 ⟨79LA1756⟩, 533 ⟨65LA(681)64, 67AG(E)281⟩, 537, 539, 540 ⟨70LA(733)195⟩, 556 ⟨85PS(21)291⟩, 571 ⟨79LA263⟩, 572 ⟨71LA(743)167⟩, 580 ⟨69LA(722)52, 69LA(722)80⟩, 627 ⟨82LA231⟩
Walter, W. F., **1**, 21, 23, 24, 435, 438 ⟨62CRV347⟩
Walters, M. A., **1**, 372 ⟨86JA2769⟩, 835 ⟨91TL179⟩; **3**, 49, 198 ⟨89TL4787⟩, 623 ⟨91TL179, 93TL1453⟩
Walters, M. E., **1**, 432 ⟨78JOC4797⟩
Walters, P. M., **2**, 29 ⟨42JA1963⟩; **6**, 74, 80 ⟨42JA1963⟩
Walters, T. R., **2**, 349 ⟨88JOC5856⟩, 390 ⟨88S808⟩; **4**, 108, 109 ⟨91JOC316⟩
Walther, B., **4**, 338 ⟨79ZC417⟩; **6**, 645, 666 ⟨73ZC192⟩
Walther, D., **2**, 561 ⟨85N789⟩
Walther, E., **2**, 532 ⟨90JA2767⟩
Walther, G., **5**, 196 ⟨62CB2073⟩, 1029 ⟨62LA(657)104⟩; **6**, 578 ⟨62LA(657)98⟩
Walti, M., **2**, 214 ⟨73E389⟩
Walton, D. R. M., **1**, 64 ⟨74JCS(P2)596⟩, 447 ⟨72JCS(P2)172⟩, 452 ⟨73JCS(P1)137⟩, 455 ⟨80JOM(201)233⟩, 999, 1000 ⟨64JOM(1)420, 65JOM(4)217⟩, 1000, 1045, 1138 ⟨72JOM(37)45⟩, 1017, 1018 ⟨72TL5209⟩, 1122 ⟨65JCS384⟩; **2**, 43 ⟨94CC463⟩, 132 ⟨70JOM(25)389⟩, 176 ⟨68JCS(C)1265⟩, 535 ⟨B-84MI 210-02⟩, 608 ⟨72CC840⟩, 753 ⟨74CC256⟩, 769 ⟨75S738⟩, 941, 1032 ⟨76CC1008⟩, 1055 ⟨78CC351⟩; **3**, 270 ⟨72JOM(37)45⟩, 287 ⟨71MI 306-01⟩, 579 ⟨89JCS(P2)1841⟩, 664 ⟨72JOM(46)267⟩; **4**, 568 ⟨87JSP(121)304⟩; **5**, 410 ⟨80TL3927⟩, 511, 589 ⟨76CC1008⟩, 1000 ⟨78CC351⟩, 1130 ⟨87JSP304⟩
Walton, D. R. M., **6**, 60, 191, 289 ⟨71JOM(29)389⟩, 174, 289, 380, 387, 390 ⟨70JOM(24)529⟩, 177, 709, 710 ⟨78JOM(157)C50⟩, 179, 709, 710 ⟨80JOM(186)309⟩, 349, 585 ⟨85TL1091⟩
Walton, D. R. N., **3**, 548 ⟨76CC1008, 77JOM(127)C1⟩
Walton, E. D., **4**, 552, 563 ⟨80IC1982⟩; **6**, 151 ⟨80PS(9)123⟩
Walton, J. C., **1**, 919 ⟨B-79MI 119-04, 88JCS(P2)1989⟩, 919, 923 ⟨92CSR105⟩, 923 ⟨90CC1589⟩; **4**, 1077 ⟨B-74MI 424-02⟩
Walts, A. E., **1**, 434 ⟨84JA721⟩; **2**, 530 ⟨86JA3422⟩, 1011 ⟨84JOC3429⟩
Walvogel, E., **6**, 546, 552 ⟨92HCA907⟩
Walz, G., **4**, 282 ⟨66AG(E)594⟩
Walz, K., **3**, 709 ⟨62AG(E)334, 65LA(686)92⟩
Walz, L., **2**, 437 ⟨90ZN(B)1548⟩
Walz, P., **1**, 189, 200 ⟨84CB322⟩; **3**, 139, 145 ⟨79TL4971⟩
Wambsgans, A., **2**, 401, 402 ⟨79JOC2278, 79JOC2061⟩; **3**, 439 ⟨87JOC702⟩
Wamhoff, H., **2**, 746 ⟨75S426⟩; **4**, 972 ⟨73LA573⟩; **5**, 690 ⟨85LA1910⟩, 849 ⟨72LA(755)101⟩, 1066 ⟨86CB2114⟩
Wamprecht, C., **3**, 417 ⟨86LA132⟩, 586 ⟨87LA183⟩
Wan, C.-N., **1**, 229 ⟨75JOC2963⟩; **2**, 816 ⟨75JOC2963⟩
Wan, E., **6**, 183 ⟨74JCS(D)665⟩
Wan, P., **4**, 1073 ⟨86CJC1949, 88JA4336⟩
Wanat, R. A., **1**, 153 ⟨85JA2078⟩
Wancowicz, D. J., **6**, 429 ⟨78JOC752⟩
Wander, M., **5**, 910 ⟨79CB1093⟩
Wang, A. B., **1**, 78 ⟨37JCS1136⟩
Wang, B.-C., **2**, 1009 ⟨87AHC(41)275⟩
Wang, B. S. L., **1**, 436 ⟨88JOC4466⟩
Wang, C. H., **2**, 403 ⟨55JA2457⟩; **5**, 108 ⟨81MI 502-05⟩, 117 ⟨85JOC1992⟩, 140 ⟨79OPP93⟩, 183, 195 ⟨90JCR(S)84⟩
Wang, C.-J., **1**, 489 ⟨89JOC868⟩
Wang, C. J., **2**, 630 ⟨89JOC868⟩
Wang, C.-L. J., **1**, 209, 651 ⟨75CC537⟩; **2**, 19 ⟨85OR(34)319⟩, 206 ⟨93TL3667⟩; **4**, 8 ⟨85OR(34)319⟩; **6**, 10 ⟨85OR(34)319⟩
Wang, D., **1**, 161 ⟨91TL2879⟩, 216 ⟨84CC1273⟩, 1091 ⟨88CC1280⟩; **2**, 78 ⟨89JOC5768⟩; **4**, 506 ⟨89TL3041⟩
Wang, D. G., **3**, 474 ⟨24JA2551⟩
Wang, E., **1**, 1041 ⟨89JCS(P1)1211⟩
Wang, E. C., **1**, 209 ⟨88SC1159⟩
Wang, E.-C., **5**, 597, 599 ⟨87CPB3139⟩
Wang, G., **2**, 70 ⟨87JA7838⟩; **4**, 343 ⟨92S258⟩, 467 ⟨90S256⟩, 470, 488 ⟨91S490⟩
Wang, G-Z., **2**, 312 ⟨92CC980⟩
Wang, G. Z., **3**, 119, 294 ⟨92CC337⟩
Wang, H., **6**, 692 ⟨89JOM(368)C29⟩
Wang, H. B., **6**, 251 ⟨92MI 608-01⟩
Wang, J., **1**, 120 ⟨93TL1111⟩; **5**, 252 ⟨90MI 505-01⟩
Wang, J. B., **1**, 120 ⟨93TL2911, 93TL7557⟩
Wang, J. T., **4**, 360 ⟨67IC1989⟩
Wang, J.-X., **5**, 183, 195 ⟨90JCR(S)84⟩, 253 ⟨90JCR(S)230⟩
Wang, K. K., **1**, 455, 1033 ⟨84JOC5175⟩, 497, 1057 ⟨93JOC1651⟩, 977 ⟨83JOC5376⟩, 1034 ⟨86JOC4521⟩, 1059 ⟨85JOC2193⟩; **2**, 904 ⟨86JOC767⟩, 904, 939 ⟨92JOC794⟩, 930 ⟨86JOC745⟩, 939, 940 ⟨83JOC5376⟩, 1086 ⟨86JOC4521⟩; **3**, 135 ⟨84JOC5175⟩; **4**, 630, 631 ⟨82JA531⟩
Wang, K.-Y., **1**, 922 ⟨89JOC5648⟩
Wang, M., **2**, 585 ⟨93CB45⟩
Wang, M.-D., **1**, 242 ⟨89JOC20⟩
Wang, M.-L., **2**, 922, 929 ⟨91JOC2468⟩
Wang, N., **2**, 388 ⟨89JOC1070⟩, 393 ⟨88SC21⟩
Wang, N. C., **6**, 648 ⟨71CJC2315⟩
Wang, P. A., **1**, 96 ⟨93JA2205⟩, 890 ⟨90JA1280⟩
Wang, P.-C., **2**, 683 ⟨84SC69⟩; **5**, 375 ⟨85TL1175⟩; **6**, 26 ⟨85TL1175⟩
Wang, P. J., **1**, 809 ⟨76JOC2048⟩
Wang, P. M. C., **4**, 61 ⟨83JFC(22)557⟩
Wang, Q., **2**, 6, 17, 18 ⟨93S693, 92CC1133⟩, 144 ⟨90JA3697⟩, 573 ⟨86OM2452⟩, 650 ⟨91SL39⟩, 879 ⟨85MI 217-01⟩; **4**, 109 ⟨92S710⟩; **5**, 992 ⟨90JOC4282⟩
Wang, Q.-Z., **5**, 600, 621 ⟨91OPP435⟩
Wang, R., **1**, 1048 ⟨93TL5911⟩
Wang, R.-J., **1**, 986 ⟨91BCJ659⟩; **2**, 436 ⟨90POL2901⟩
Wang, R.-T., **3**, 675 ⟨93TL5911⟩
Wang, S., **1**, 826 ⟨93TL3523⟩; **3**, 531 ⟨93CB297⟩; **4**, 587 ⟨88OM2415⟩
Wang, S.-L., **3**, 355 ⟨91JOC4919⟩; **4**, 230 ⟨90TL257⟩; **6**, 334 ⟨85OM1226⟩
Wang, S. L. B., **5**, 949 ⟨92JA10665⟩

Wang, S. S., **4**, 413 ⟨59JA3789⟩; **6**, 558 ⟨90JOC5230⟩
Wang, S.-Z., **1**, 192 ⟨92SL323⟩; **4**, 259 ⟨90JCS(P1)2035⟩
Wang, T., **1**, 1067 ⟨92JA10653⟩; **2**, 936 ⟨90TL543⟩; **3**, 649 ⟨91JCS(P1)487⟩
Wang, T. C., **1**, 870 ⟨69JA1146⟩
Wang, T.-C., **4**, 899 ⟨68JA3781⟩
Wang, T.-C. L., **6**, 345 ⟨84CC392⟩
Wang, T.-F., **1**, 807 ⟨80TL2125, 81TL1179⟩, 807 ⟨84JA718⟩; **3**, 61 ⟨81TL1179⟩
Wang, T. S., **1**, 881 ⟨61JOC644⟩
Wang, T.-Z., **1**, 613 ⟨87JA3730⟩
Wang, T. Z., **1**, 826 ⟨93TL3523⟩
Wang, W., **1**, 316 ⟨87JA2442⟩, 646, 647 ⟨90JOM(400)121, 89JOC2027⟩
Wang, W.-B., **1**, 270 ⟨91TL3999⟩; **2**, 55 ⟨90TL1185⟩; **3**, 49 ⟨B-90MI 301-02⟩, 196 ⟨B-90MI 304-02⟩; **5**, 329 ⟨93SL850⟩
Wang, W.-L., **3**, 158 ⟨92JOC5620⟩; **5**, 437 ⟨83JOC1144, 84TL1651⟩
Wang, W.-N., **2**, 1009 ⟨87T1273⟩
Wang, X., **1**, 584 ⟨92JA4128⟩, 1057 ⟨92JOC1242, 92JOC3387⟩; **3**, 156 ⟨89TL6497⟩, 424 ⟨92CB2361⟩; **4**, 188 ⟨89TL6497⟩; **5**, 199 ⟨92T4863⟩, 227 ⟨90TL1245, 90TL2259⟩
Wang, Y., **1**, 134 ⟨93T4147⟩, 622 ⟨93SL839⟩, 647 ⟨86TL4583, 89JOC2027⟩, 944 ⟨93TL6209⟩, 963 ⟨90TL213⟩, 978 ⟨87JOC1370⟩, 1099 ⟨87JOC4825, 88JOC4151, 88JOC5580⟩; **3**, 369 ⟨92SL909⟩; **4**, 396 ⟨86JOM(315)135⟩, 583, 585 ⟨86JOM(309)225⟩, 716 ⟨92ICA557⟩
Wang, Y. F., **3**, 45 ⟨77JA944⟩, 48 ⟨89JA3924, 93TL403⟩, 48, 65 ⟨86TL4583⟩, 631 ⟨89TL1917⟩
Wang, Y. G., **3**, 410 ⟨92JA5466⟩
Wang, Y.-P., **2**, 870 ⟨81JA1785⟩
Wang, Z., **1**, 281 ⟨91TL4619⟩; **2**, 301 ⟨89OM1093⟩; **3**, 62 ⟨90TL5579⟩, 68, 69 ⟨88CB999⟩, 629 ⟨93TL3971⟩; **4**, 105 ⟨90JHC139⟩; **5**, 625 ⟨88TL4085⟩; **6**, 216 ⟨86MI 607-01⟩
Wang, Z.-M., **2**, 48 ⟨92JOC2768, 94TL543⟩, 49 ⟨92TL6407, 92TL6411⟩; **3**, 222 ⟨91TL1467⟩
Wang, Z. Y., **1**, 135 ⟨93TA1957⟩
Wani, M. C., **5**, 131 ⟨87JOC3468⟩
Wanka, U., **5**, 911 ⟨82JOM(236)C1⟩
Wanless, G. G., **4**, 15, 24, 29 ⟨65JA3151⟩
Wanmaker, W. L., **5**, 749 ⟨51RTC638⟩
Wannagat, U., **1**, 991 ⟨84JOM(264)135, 84JOM(271)181⟩; **2**, 503 ⟨69JOM(19)367⟩, 547 ⟨79TCC1⟩, 697 ⟨63AG(E)47⟩; **3**, 456 ⟨63CB2132⟩; **5**, 375, 378 ⟨62M352⟩, 379 ⟨88JOM(341)95⟩, 721 ⟨63CB2138⟩
Wannamaker, M. W., **2**, 306, 927 ⟨89JA4486⟩, 928 ⟨87JOC4760⟩, 987 ⟨89JA776⟩
Wann, R. E., **1**, 623 ⟨65TL2899⟩; **6**, 552, 553 ⟨61JOC4047⟩
Wanzke, W., **5**, 561 ⟨87CB429⟩; **6**, 235, 253, 275, 726 ⟨87CB429⟩
Wanzlick, H.-W., **2**, 742 ⟨66CB1580⟩; **4**, 404, 405, 408, 410 ⟨53CB1463⟩; **6**, 55 ⟨73LA40, 75LA195⟩
Warburton, W. K., **2**, 708, 847 ⟨57CRV1011⟩
Ward, A. D., **2**, 24 ⟨84TL1197⟩; **5**, 714, 723 ⟨69AJC161⟩
Ward, A. S., **4**, 658 ⟨85OM1308⟩
Ward, B. F., **2**, 220 ⟨70JOC1118⟩
Ward, C. F., **2**, 29 ⟨22JCS1161⟩
Ward, D. D., **4**, 42 ⟨81CAR284⟩
Ward, D. E., **1**, 231 ⟨91TL843⟩; **3**, 161 ⟨88SC1927⟩; **5**, 12 ⟨91TL7165⟩
Ward, D. G., **2**, 76 ⟨88OM2289⟩
Ward, D. L., **1**, 232 ⟨83JOC494⟩
Ward, E. L. M., **5**, 944 ⟨94TL3801, 94TL8883⟩
Ward, H. P., **2**, 385 ⟨55OSC(3)244⟩; **3**, 465 ⟨55OSC(3)223⟩
Ward, H. R., **1**, 809 ⟨69JA522⟩
Ward, J. E. H., **4**, 134 ⟨72JA5702⟩; **5**, 950, 951 ⟨74JOM(70)273⟩
Ward, J. S., **3**, 212 ⟨90JHC1709⟩
Ward Jr., L. F., **4**, 836, 837, 838 ⟨68JOC4470⟩
Ward, M. A., **2**, 446 ⟨72MI 208-05⟩; **5**, 867 ⟨75JA3238⟩
Ward, M. D., **1**, 94 ⟨81JA5253, 81JMOC397⟩; **3**, 28 ⟨93JOC5482⟩
Ward, M. L., **2**, 645, 646 ⟨46JA389⟩
Ward, N. D., **1**, 74, 82, 84 ⟨67JCS(C)1120⟩
Ward, P., **6**, 445 ⟨78JCS(P1)1066⟩
Ward, R. S., **3**, 498 ⟨67JCS(C)2442, 69JCS(C)1100⟩, 526 ⟨B-80MI 316-01⟩, 530 ⟨75JCS(P1)1600⟩; **4**, 230 ⟨85JCS(P1)587⟩, 563 ⟨70JCS(C)504⟩, 840, 961 ⟨90JCS(P1)2775⟩
Ward, T. R., **4**, 566 ⟨91HCA983⟩
Ward, W. H., **5**, 83 ⟨49MI 502-01⟩
Wardell, J. L., **1**, 110, 111, 166 ⟨B-87MI 104-02⟩, 452, 453 ⟨B-87MI 110-02⟩; **2**, 134 ⟨B-74MI 203-02⟩, 551 ⟨82COMC-I(1)43⟩, 551, 556 ⟨B-87MI 211-01⟩, 555 ⟨B-85MI 211-01⟩, 590 ⟨82COMC-I(2)863⟩, 707 ⟨B-74MI 214-01⟩, 872 ⟨82COMC-I(2)⟩, 879 ⟨80JOM(188)345⟩; **4**, 138, 369, 372 ⟨92JOM(437)111⟩, 154 ⟨74JOM(78)395, 75JCS(D)1786⟩, 344 ⟨82COMC-I(2)681⟩, 377, 390, 398 ⟨B-87MI 408-02⟩, 527 ⟨87MI 411-01⟩; **6**, 707 ⟨82COMC-I(2)681⟩
Wardle, R. B., **5**, 315 ⟨82JOC4605⟩
Wardleworth, M., **3**, 66 ⟨83CC74⟩; **5**, 1035 ⟨83CC74⟩
Wardman, P., **2**, 160, 162 ⟨84JMC1634⟩
Ware, J., **3**, 584, 586 ⟨77JA2622⟩
Ware, W. R., **3**, 363 ⟨76JA6218⟩
Wargotz, B., **2**, 175 ⟨58JOC363⟩
Wariishi, K., **4**, 188 ⟨88CL1101⟩
Waring, A. J., **1**, 306 ⟨79COC(1)1043⟩
Waring, A. M., **3**, 666 ⟨66JOC4100⟩
Waring, P., **4**, 855 ⟨82AJC785⟩
Wark, J. C., **3**, 474 ⟨24JA2551⟩
Warkentin, J., **1**, 339 ⟨93JOC2115⟩, 871 ⟨68JA4629, 68JOC1301⟩, 1082 ⟨78CJC308⟩; **2**, 373, 376, 403 ⟨84JOC336⟩; **3**, 323 ⟨83CJC619, 89CJC1753⟩, 324 ⟨92JA8751⟩, 573 ⟨60JA5394⟩; **4**, 314 ⟨69CJC3983, 69JOC3230⟩; **5**, 707 ⟨84JOC3466⟩
Warkentin, J. D., **6**, 313 ⟨79JOC267⟩, 553 ⟨81JHC1581⟩
Warmerdam, E. G. J. C., **3**, 631 ⟨93T1063⟩
Warmus, J. S., **1**, 150 ⟨91TL5501⟩, 150 ⟨91TL5505⟩, 150 ⟨91TL5509⟩
Warnan, M., **2**, 472 ⟨78CB3838⟩
Warnant, J., **3**, 234 ⟨61BSF559⟩
Warne, P. J.,**4**, 866 ⟨84GEP3408757A1⟩
Warnecke, H.-U., **3**, 71 ⟨73LA2078⟩
Warner, C. D., **1**, 885 ⟨75S395⟩
Warner, C. M., **5**, 431 ⟨60JA5102⟩
Warner, D. T., **2**, 366 ⟨58JOC1133⟩
Warner, G. H., **6**, 229, 255 ⟨56JA6070⟩
Warner, J. C., **2**, 385 ⟨84CC451, 84TL2619⟩
Warner, J. S., **6**, 596 ⟨63JOC1642⟩
Warner, M., **6**, 56 ⟨72JOC152⟩
Warner, P., **1**, 918 ⟨72JA6867⟩; **4**, 57 ⟨76JA6752⟩; **5**, 453 ⟨84CC956, 84TL2709, 85JOM(285)213, 85TL2125, 86TL623⟩
Warner, R. C., **5**, 83 ⟨46MI 502-01⟩
Warnhoff, E. W., **2**, 21 ⟨63OSC(4)162⟩; **3**, 208 ⟨63OSC(4)162⟩
Warning, K., **6**, 620 ⟨73CB2093, 74CB698⟩
Warning, U., **2**, 453 ⟨76CB805⟩
Warnock, J., **3**, 4, 115 ⟨78JOC2618⟩
Warnock, W. J., **2**, 352 ⟨92T6929⟩
Warolin, C., **4**, 508 ⟨76CR(C)473⟩
Warren, C. D., **1**, 882 ⟨64TL1191, 68JCS(C)1903⟩
Warren, L., **4**, 664 ⟨93T2988⟩
Warren, S., **1**, 599 ⟨92JCS(P1)2303⟩, 641 ⟨81CC100⟩, 641, 642 ⟨83TL5293⟩, 641, 642, 757 ⟨83TL111⟩, 641, 643 ⟨88JCS(P1)1799⟩, 641, 753, 756 ⟨85JCS(P1)2307⟩, 641, 755 ⟨77JCS(P1)1452, 84JCS(P1)243⟩, 643

⟨83JCS(P1)1169, 84TL357, 85TL505, 86TL2265, 91JCS(P1)1485, 92JCS(P1)2155, 92JCS(P1)3407⟩, 643, 755 ⟨77JCS(P1)550⟩, 643, 757 ⟨88JCS(P1)2971⟩, 753 ⟨83JCS(P1)1173, 83TL3931⟩, 754 ⟨85TL2119, 85TL5709⟩, 756 ⟨83TL593, 86TL645, 89TL601⟩, 757 ⟨87JCS(P1)2569, 89TL4581⟩, 836 ⟨93JCS(P1)1275, 93JCS(P1)2913⟩; **2**, 680 ⟨79JCS(P2)3099⟩; **3**, 11 ⟨79JCS(P1)3099⟩, 35 ⟨77JCS(P1)1131, 77JCS(P1)2263, 91JCS(P1)451⟩, 166 ⟨79JCS(P1)1074⟩, 197 ⟨87JCS(P1)2569, 88JCS(P1)1799⟩; **4**, 218 ⟨92JCS(P1)2303⟩, 250 ⟨78TL2345⟩, 343 ⟨77CC314⟩, 346, 347 ⟨77JCS(P1)2263⟩, 650 ⟨77JCS(P1)550⟩, 923 ⟨82TL4167⟩, 923, 924 ⟨87JCS(P1)967⟩, 949 ⟨79JCS(P1)3099⟩; **6**, 121 ⟨77JCS(P1)2263, 85JCS(P1)2585⟩
Warren, S. E., **1**, 498 ⟨82JA3676⟩
Warren, S. G., **2**, 807 ⟨B-67MI 215-01⟩; **5**, 786 ⟨B-67MI 520-01⟩
Warrener, R. N., **1**, 784 ⟨75TL1335⟩; **5**, 1048 ⟨58JCS153⟩
Warrier, U., **1**, 1140 ⟨88TL2923⟩
Warshaw, J. A., **3**, 434 ⟨89JOC1736⟩
Warshawsky, A., **2**, 304 ⟨89S825⟩
Warshawsky, A. M., **3**, 234 ⟨88SC1323⟩
Wartanessian, S., **6**, 178 ⟨81ZAAC(475)87⟩, 290 ⟨79JOM(178)11⟩
Wartenberg, F. H., **1**, 339, 341 ⟨93TL5251⟩
Warthen, D., **1**, 1001 ⟨68JMC371, 68JMC373⟩
Wartski, L., **3**, 442 ⟨81TL1927⟩
Warwel, S., **1**, 534 ⟨75S632⟩, 674 ⟨87AG(E)702⟩
Warwick, G. P., **3**, 104 ⟨53JCS1742⟩
Warwick, P. J., **1**, 1010 ⟨89JOC1218⟩
Warzynski, R. J., **3**, 218 ⟨53JA3268⟩
Wasburn, R. M., **2**, 906 ⟨63OSC(4)68⟩
Waseda, T., **1**, 1046 ⟨83CL35⟩
Waser, J., **2**, 507 ⟨45JA2014⟩
Wasgindt, M., **3**, 716 ⟨93AG(E)650⟩
Wash, G., **1**, 104 ⟨41JA2975⟩
Washburn, L. C., **2**, 626 ⟨76S545⟩
Washburn, R. M., **2**, 832 ⟨61JOC3547⟩, 863 ⟨61JA4466⟩
Washburne, S. S., **1**, 656 ⟨74JOM(81)41⟩; **2**, 416, 417 ⟨69MI 207-03⟩; **4**, 623 ⟨71JA3709⟩; **5**, 992 ⟨72JOC1738, 73JOC2982⟩; **6**, 60 ⟨67JA1538⟩, 179, 709, 710 ⟨80JOM(186)309⟩
Washecheck, D. M., **6**, 525 ⟨79JA6110⟩
Washington, J., **4**, 710 ⟨94OM1078⟩
Washioka, Y., **5**, 296 ⟨83JOC4058⟩
Washnock, Jr., J., **5**, 595, 617 ⟨39JA105⟩
Wasielewski, C., **4**, 460 ⟨73ZC254, 75RZC397, 76C187, 76RZC477, 78PJC1315, 78TL1609, 79ZC253⟩, 476 ⟨74ZC152⟩
Wasley, J., **3**, 307 ⟨83JHC1565⟩
Wasmund, L. M., **5**, 575 ⟨89JCR(S)155⟩
Wasmus, T., **3**, 616 ⟨06CB1181⟩
Wasmuth, D., **4**, 197 ⟨80HCA1960⟩; **5**, 148 ⟨81AG(E)971⟩
Wassef, W. N., **6**, 25 ⟨89JCS(P2)251⟩, 25, 31 ⟨83CC283, 84JCS(P2)1247, 86JCS(P2)471⟩
Wassenaar, S., **1**, 685 ⟨72JOC4045⟩; **3**, 368 ⟨72JOC4045⟩
Wasser, D. J., **6**, 253 ⟨86JFC(32)89⟩
Wasserman, D., **4**, 46 ⟨89JOC5520⟩
Wasserman, H. H., **1**, 1021 ⟨86TL4913⟩, 1134 ⟨85TL2237⟩; **2**, 326 ⟨88TL4977⟩, 693 ⟨64JOC1817⟩; **3**, 247 ⟨73JOC1451⟩, 267 ⟨89JOC6012⟩, 419 ⟨89JA371⟩; **4**, 162 ⟨74FCF73, 80JOC2874⟩, 299, 406 ⟨70TL1729⟩, 827 ⟨58T321⟩; **5**, 174 ⟨81T4059⟩, 759 ⟨68TL3277⟩
Wasserman, H. J., **2**, 979 ⟨82JA6808⟩; **4**, 581 ⟨82IC3913, 84OM476⟩; **6**, 343 ⟨83OM1179⟩
Wassiliadou-Micheli, N., **5**, 140 ⟨70CB37⟩
Wassmundt, F. W., **5**, 294 ⟨67JA7131⟩
Wasson, R. C., **5**, 362, 367 ⟨87JCS(P1)2511⟩
Wasylishen, R. E., **4**, 323 ⟨79JHC415⟩
Waszkuc, W., **3**, 158 ⟨84S1025⟩
Waszkuc, W. T., **4**, 343 ⟨77PS(3)345⟩
Wat, E. K. W., **5**, 332 ⟨63JA827⟩
Watabe, K.-I., **1**, 257 ⟨90JOC408⟩
Watabe, T., **2**, 457 ⟨79BCJ466⟩; **4**, 854 ⟨79BCJ466⟩
Watabe, Y., **5**, 183, 199 ⟨79CC179⟩
Watanabe, A., **3**, 665 ⟨84CL773⟩; **4**, 432, 433 ⟨87T693⟩
Watanabe, F., **1**, 428, 429 ⟨89JOC5308⟩; **2**, 631 ⟨89BCJ1349⟩
Watanabe, H., **1**, 57 ⟨75OPP189⟩, 161, 450, 1092, 1105 ⟨83TL1041⟩, 165 ⟨85T3663⟩, 165 ⟨86T295⟩, 1036 ⟨85JOC565⟩, 1047 ⟨91TL1459⟩; **2**, 543 ⟨89TL6047⟩, 546 ⟨77CC534⟩, 629 ⟨81CC718⟩, 915 ⟨80JOM(186)51⟩, 915, 923 ⟨81JOM(216)149⟩, 923, 924 ⟨83TL1041⟩, 1085 ⟨65IC657⟩; **3**, 597 ⟨80JOM(190)C39⟩; **4**, 422 ⟨79JOC4543⟩; **5**, 184 ⟨87BCJ4325⟩, 344 ⟨91BCJ106⟩, 385 ⟨86BCJ2203⟩, 652 ⟨88CPB3354⟩, 659 ⟨93JOC32⟩, 733 ⟨93TL7933⟩; **6**, 465 ⟨92JAP04356446, 93JAP05255197⟩
Watanabe, I., **2**, 401 ⟨65BCJ1989⟩; **3**, 628 ⟨86BCJ893⟩
Watanabe, J., **4**, 854, 857 ⟨82JOC5404⟩; **5**, 410 ⟨92TL1355⟩
Watanabe, J.-I., **4**, 271 ⟨81TL4499⟩
Watanabe, K. A., **3**, 275 ⟨82CPB63⟩; **4**, 893 ⟨80H(14)619, 82CPB63⟩
Watanabe, K., **1**, 48 ⟨59BCJ861⟩, 442 ⟨87JCS(P1)1951⟩, 1036 ⟨85JOC565⟩; **2**, 14 ⟨83CL1877⟩, 150 ⟨77TL1345⟩, 300 ⟨80JOC357⟩, 657 ⟨75AG(E)820⟩, 1000 ⟨67T3049⟩; **3**, 91 ⟨88CPB974⟩, 373 ⟨67T4279⟩, 626 ⟨85CL1779⟩, 708 ⟨85MI 321-01⟩; **4**, 696 ⟨81JOC192⟩; **5**, 118 ⟨81MI 502-05⟩; **6**, 561 ⟨87BCJ1807⟩
Watanabe, M., **1**, 102 ⟨80CRV99⟩, 178 ⟨86BCJ3097⟩, 405 ⟨92TL5795⟩, 448 ⟨89CPB36⟩, 505 ⟨84JOC747⟩, 592 ⟨75JA2553⟩, 636 ⟨72TL3827, 72TL4659⟩, 822 ⟨90CC418, 90T4887⟩, 936 ⟨82JA5564⟩; **2**, 71 ⟨88JOC927, 89JCS(P1)109⟩, 244, 795 ⟨80JOC1691⟩, 631 ⟨88CL795⟩, 694 ⟨90CPB2637⟩; **3**, 260 ⟨84CL2055⟩, 309, 310 ⟨80JA1457, 87T5281⟩; **4**, 166 ⟨92JOC742⟩, 187 ⟨86JCS(P1)1351⟩, 226 ⟨81BCJ817⟩, 379 ⟨92TL5795⟩, 421 ⟨75CL607⟩, 936 ⟨90CPB902⟩; **5**, 206 ⟨88CPB178⟩, 1120 ⟨92CC1379⟩; **6**, 554 ⟨80CPB110⟩, 608 ⟨72TL3827⟩
Watanabe, N., **1**, 178 ⟨86H(24)3149⟩, 241 ⟨89JOC2374⟩, 603 ⟨83JOC5246⟩; **2**, 1000 ⟨90CPB329⟩; **3**, 634 ⟨87H(26)913⟩; **4**, 265, 853, 854 ⟨90CPB3242⟩; **5**, 1042 ⟨75BCJ3205, 79BCJ3611⟩, 1053 ⟨78BCJ1818⟩, 1066 ⟨87NKK1280⟩, 1105 ⟨74BCJ2745⟩
Watanabe, O., **3**, 335 ⟨91BCJ1389⟩
Watanabe, S., **1**, 47 ⟨87S846⟩, 495 ⟨92TL2589⟩; **4**, 213 ⟨78JOC1248⟩, 736 ⟨89JOC5630, 93JFC(62)201⟩; **5**, 570, 613 ⟨92JOC2419⟩, 630 ⟨88JCS(P1)2203⟩
Watanabe, T., **1**, 72, 81 ⟨71BCJ166⟩, 489 ⟨92SL207⟩; **2**, 537 ⟨66TL5493⟩; **3**, 168 ⟨90CPB3314⟩, 293 ⟨77S776⟩, 307 ⟨88H(27)1855⟩, 373 ⟨88S178, 89CL973⟩, 431 ⟨83TL1049⟩, 667 ⟨79CL1427⟩; **4**, 211 ⟨72CPB2123⟩; **5**, 225 ⟨89BCJ539⟩
Watanabe, W. H., **1**, 414 ⟨57JA2828⟩; **4**, 306 ⟨57JA2825, 57JA2833⟩
Watanabe, Y., **1**, 1036 ⟨85JOC565⟩
Watanabe, Y., **1**, 1036 ⟨86BCJ3201⟩, 1071 ⟨79CL481⟩; **1**, 196 ⟨85CL1477, 89JOM(369)C51⟩, 243 ⟨90TL6669⟩, 333 ⟨93JA464⟩, 493 ⟨92JOM(423)405⟩, 518 ⟨94AG(E)580⟩, 590 ⟨75CL871⟩, 590, 693 ⟨74CL1523⟩, 594, 668 ⟨92JOC3145⟩, 821 ⟨87JA5878⟩, 946 ⟨93JA10464⟩, 1010 ⟨92JMC241⟩; **2**, 8 ⟨77CL383⟩, 26 ⟨77CL1013⟩, 106 ⟨92TL1313⟩, 145 ⟨81S204⟩, 210 ⟨86TL215⟩, 310 ⟨74TL1879⟩, 344 ⟨78CL1057⟩, 676 ⟨86TL2125⟩, 753 ⟨84BCJ2440⟩; **3**, 83 ⟨71BCJ2569⟩, 84 ⟨73TL3535, 75BCJ2490, 75TL1063⟩, 87 ⟨90TL2607⟩, 162 ⟨87JOC2230⟩, 290 ⟨88JOM(345)397⟩, 597 ⟨80JOM(190)C39⟩, 634 ⟨81S204⟩, 696 ⟨77CL697⟩; **5**, 226 ⟨81CL1367, 81CL285⟩, 234 ⟨76CL741⟩, 286

⟨86JOM(309)333⟩, 322, 330 ⟨85JOC1818, 89JOC4088⟩, 979 ⟨77CL1345⟩; **6**, 339 ⟨89JOM(368)199⟩, 457 ⟨72CL483⟩, 561 ⟨88BCJ2203⟩
Watanbe, S., **2**, 308 ⟨69BCJ3606⟩
Watanuki, J., **6**, 461 ⟨94JAP06041019⟩
Watarai, H., **5**, 254 ⟨81S370⟩
Watari, F., **3**, 335 ⟨91BCJ1389⟩, 369 ⟨90JST(238)71⟩
Waterfeld, A., **5**, 662, 663 ⟨82CC839⟩, 662, 809 ⟨83CB724, 85CB4459⟩; **6**, 251 ⟨91JFC(55)313⟩, 252 ⟨83JFC(23)325, 90IC4588, 91CB1353, 92CC1017⟩, 253 ⟨86IC275, 92CB535⟩, 253, 275, 726, 729 ⟨90CB1635⟩, 726 ⟨85CB4997, 85CB5007, 94CB597⟩
Waterhouse, A., **1**, 966 ⟨78JOC1389⟩, 970 ⟨76JOC3496, 78JOC1382, 78JOC1385⟩
Waterhouse, I., **1**, 590, 591, 615, 700 ⟨77TL4223⟩, 616, 696, 700 ⟨80JCS(P1)1045⟩, 698, 1073 ⟨78TL2625, 79JCS(P1)2429⟩
Waterman, H., **6**, 218 ⟨50USP2531372⟩
Waterman, H. I., **2**, 416, 417 ⟨52MI 207-01, 53MI 207-01, 54MI 207-01, 53MI 207-03⟩, 417 ⟨53MI 207-04⟩, 417, 421 ⟨53MI 207-02⟩
Waterous, J. M., **4**, 713 ⟨82CC371⟩
Waters, J. H., **6**, 144, 145, 369 ⟨63T(S)177⟩
Waters, J. M., **3**, 387, 522 ⟨83JA5939⟩; **4**, 150, 722 ⟨83JA5939⟩; **6**, 718, 727 ⟨84JOM(269)C55⟩
Waters, K. L., **5**, 62 ⟨47CRV585⟩
Waters, R. M., **1**, 652 ⟨93TL1593⟩, 832 ⟨69TL3253⟩, 1000 ⟨72S567⟩
Waters, W. A., **1**, 919 ⟨B-46MI 119-01⟩; **2**, 379, 388 ⟨66JCS(C)813⟩, 626 ⟨50JCS573⟩, 631 ⟨50JCS3694⟩, 782 ⟨64PCS230⟩, 889 ⟨50JCS18⟩; **5**, 28 ⟨B-64MI 502-02⟩
Waters, W. L., **1**, 1125 ⟨68JA6741⟩
Waterson, D., **1**, 836 ⟨93JCS(P1)1275⟩; **3**, 52, 203 ⟨84CC28⟩
Waterston, D., **2**, 546 ⟨92JCS(P1)3277⟩
Wates, P. J., **5**, 214 ⟨88JMC499⟩
Watkin, D., **1**, 1140 ⟨88TL5017⟩
Watkins, C. L., **2**, 434, 485, 872 ⟨90IC3502⟩, 484 ⟨92SRI461, 93POL89⟩, 507, 894 ⟨87IC1638⟩, 893 ⟨88ICA(150)51⟩
Watkins, P. M., **2**, 370 ⟨71JCS(A)1125⟩
Watkins, R. J., **1**, 983 ⟨87JA4338⟩
Watkins, S. F., **5**, 465 ⟨72CC672⟩
Watkins, T. I., **2**, 176, 177 ⟨51JCS3292⟩
Watnick, C., **4**, 429, 430, 431 ⟨93TL6677⟩
Watson, A. A., **2**, 218 ⟨64JCS2100⟩; **5**, 826 ⟨77JOC1610⟩
Watson, ,B. T., **2**, 987 ⟨89JA776⟩
Watson, B. T., **2**, 306, 927 ⟨89JA4486⟩, 969 ⟨87TL3895⟩, 986, 987 ⟨86JA7411, 87JA2544⟩
Watson, D. G., **2**, 889 ⟨70JAP7321096-R⟩
Watson, H. A., **3**, 234 ⟨80JOC1109⟩
Watson, H. G., **2**, 409, 410, 412, 413 ⟨49JCS2921⟩
Watson, H. R., **4**, 545 ⟨62JCS1490⟩
Watson, J. M., **1**, 827 ⟨73JA3348⟩
Watson, J. S., **2**, 383, 388 ⟨56JCS3677⟩
Watson, J. V., **2**, 43 ⟨92TL1017⟩
Watson, K. G., **4**, 232 ⟨73AJC2491⟩; **5**, 605 ⟨72TL4191⟩
Watson, K. M., **5**, 768, 769, 772, 773 ⟨B-75MI 519-03⟩, 830, 847 ⟨B-75MI 521-02⟩
Watson, P. L., **4**, 578 ⟨83CC276⟩, 670, 702 ⟨83JA6491⟩
Watson, R. A., **3**, 128 ⟨86TL1437⟩
Watson, R. M., **2**, 455 ⟨51JA4101, 51JA5466⟩
Watson, S. P., **2**, 595 ⟨78USP4118409⟩
Watson, T. W., **2**, 638 ⟨81JA1761⟩
Watson, W. D., **2**, 620 ⟨85JOC2145⟩
Watson, W. H., **2**, 150 ⟨82JA5412⟩, 1000 ⟨83JHC399, 84AX(C)842⟩; **3**, 345 ⟨91JOC5932⟩; **4**, 429, 430, 431 ⟨93JOC759⟩, 566 ⟨88OM572⟩; **6**, 317 ⟨93T3035⟩, 350, 694 ⟨84CC698⟩, 692 ⟨84OM1132⟩
Watt, A. N., **5**, 132 ⟨68JOC2143⟩
Watt, C. I. F., **1**, 883 ⟨88JCS(P2)983, 89CC161⟩
Watt, D. S., **1**, 155 ⟨84OR(31)1⟩, 630 ⟨85JOC5460⟩, 789 ⟨72TL4651⟩; **2**, 631 ⟨85JOC5384, 85TL2043⟩; **3**, 29 ⟨84JOC1378⟩, 48 ⟨78JOC2066⟩, 80 ⟨74JOC2799, 74SC127⟩, 134 ⟨83JOC4087⟩, 162 ⟨90SC2279⟩, 280 ⟨84JOC4226⟩, 292 ⟨84OR47⟩, 615 ⟨93JOC159⟩, 620 ⟨84OR1⟩, 621 ⟨75JOC127⟩, 627, 634, 635, 636, 639, 651 ⟨76JOC2846⟩, 643 ⟨76JOC1873, 77JA182⟩; **4**, 94 ⟨84JOC1084⟩, 365, 938 ⟨81JOC5182, 83JOC34⟩, 938 ⟨83JOC383⟩, 938, 939 ⟨83JA619⟩; **5**, 272, 283, 284 ⟨77JA182⟩, 306 ⟨86TL919⟩; **6**, 119 ⟨77S181⟩
Watt, G. W., **2**, 427 ⟨48JA2295⟩; **4**, 697 ⟨64JINC1531⟩
Watt, J. A. C., **4**, 610 ⟨66JCS(C)1137⟩
Watta, G., **2**, 135 ⟨25G973⟩
Wattanasin, S., **1**, 767, 769 ⟨89SC2659⟩; **3**, 131, 273, 284 ⟨84TL811⟩; **6**, 127 ⟨79TL1827, 80JCS(P1)2678⟩
Watterson, S. H., **1**, 980 ⟨93JA3776⟩
Watthey, J. W. H., **1**, 421 ⟨65TL377⟩
Watts, C. T., **2**, 374 ⟨88OSC(6)936⟩
Watts, J. C., **5**, 285 ⟨71JMC535⟩
Watts, J. D., **3**, 582 ⟨90JA3779⟩; **6**, 270 ⟨85CC5⟩, 401 ⟨87T1019⟩
Watts, L., **4**, 724 ⟨65JA3254⟩
Watts, M. E., **2**, 160, 162 ⟨84JMC1634⟩
Watts, O., **5**, 453 ⟨84JOM207⟩
Watts, P. C., **5**, 70 ⟨74JOC1676⟩
Watts, Jr., P. H., **2**, 1004 ⟨73JA6158⟩
Watts, R. D., **3**, 246 ⟨81JOC4047⟩
Watts, R. O'B., **4**, 137 ⟨76JCS(P1)513⟩
Watts, W. E., **1**, 377 ⟨B-73MI 109-01⟩, 380 ⟨64JA2722⟩, 528 ⟨73JCS(P1)975, 73JCS(P1)977⟩, 811 ⟨64TL2503⟩, 900 ⟨64JA4195⟩, 1032 ⟨63JCS2990⟩; **2**, 985 ⟨73JCS(P1)977⟩; **3**, 242 ⟨75CC772⟩, 407 ⟨B-73MI 310-01⟩, 435 ⟨B-73MI 311-01⟩, 492 ⟨63JCS2990⟩; **4**, 688 ⟨76JOM(120)C13⟩
Waugh, F., **1**, 1000, 1045, 1138 ⟨72JOM(37)45⟩; **3**, 270 ⟨72JOM(37)45⟩
Wawrzenczyk, C., **4**, 654 ⟨89TL403⟩
Wawzonek, S., **1**, 103 ⟨42JA2365⟩; **2**, 95 ⟨60JOC621⟩, 421 ⟨55JA1796⟩; **5**, 554 ⟨74JOC511⟩; **6**, 304 ⟨83MI 610-01⟩
Wax, G., **2**, 825 ⟨82PS(13)331⟩
Way, G. M., **6**, 183 ⟨73CC532, 75JCS(D)148⟩
Wayaku, M., **3**, 121 ⟨76CL231⟩
Wayda, A., **1**, 205 ⟨86AG(E)760⟩
Wayda, A. L., **1**, 890 ⟨81JOC3925⟩; **2**, 564 ⟨81CC292⟩; **5**, 440 ⟨81CC706⟩
Waykole, L., **1**, 576 ⟨86JA3739⟩, 634 ⟨87JA3174⟩
Wayne, W. J., **2**, 224 ⟨57JA6483⟩
Wayner, D. D. M., **1**, 331 ⟨85JA4594⟩, 936 ⟨85JA4595⟩, 946 ⟨90JOC3321⟩; **2**, 141, 265 ⟨91TL2853⟩
Wayte, D. E., **3**, 328 ⟨51OS(31)25⟩
Wazeer, M. I. M., **5**, 201 ⟨89T5979⟩
Wazer, J. R., **2**, 493, 495, 496 ⟨70JA3969⟩, 507 ⟨69JA4986⟩
Wazneh, L., **3**, 337 ⟨88TL5899⟩, 337, 351 ⟨91T3753⟩
Wear, R. L., **6**, 241 ⟨62JOC1064⟩
Weatherbee, C., **4**, 301 ⟨64JOC407⟩
Weatherford, III, W. D., **2**, 95 ⟨83JOC3126⟩
Weaver, B. M., **2**, 381 ⟨61JOC2740⟩
Weaver, C., **1**, 447 ⟨48JA232⟩; **2**, 162 ⟨41JA2938⟩
Weaver, E. R., **5**, 346 ⟨82JOC5160, 84H(21)191⟩
Weaver, S. L., **1**, 498 ⟨90JOC711⟩
Weaver, T. D., **3**, 322 ⟨67JA5497⟩
Weaver, W. I., **4**, 423 ⟨44JA222⟩
Weaver, W. M., **4**, 161 ⟨57JA6562⟩
Weavers, R. T., **1**, 1131 ⟨88TL1085⟩; **4**, 655 ⟨81AJC181⟩, 811 ⟨93T4229⟩
Webb, A., **6**, 394 ⟨95UP 613-01⟩
Webb, A. D., **6**, 173, 394 ⟨93JOM(462)45⟩, 390 ⟨93JCS(D)3259⟩
Webb, C. N., **5**, 652 ⟨41OSC(1)82⟩

Webb, G., **4**, 736 ⟨70TL5215⟩
Webb, G. G., **1**, 906 ⟨90JA8991⟩
Webb, J. L., **2**, 653 ⟨70JA2042⟩; **4**, 30 ⟨73JOC2091⟩
Webb, M., **3**, 271 ⟨86TL775⟩; **4**, 656, 663, 709, 724 ⟨81JCS(D)1593⟩; **5**, 334 ⟨89JMC1024⟩
Webb, M. B., **1**, 536 ⟨86JOC4080⟩; **3**, 270 ⟨86JOC4080⟩
Webb, M. W., **2**, 95 ⟨83JOC3126⟩
Webb, R., **5**, 326 ⟨79TL1185⟩
Webb, II, R. E., **5**, 326 ⟨81JA3173⟩
Webb, R. F., **5**, 712 ⟨55CI(L)1087, 56JCS1231⟩, 823 ⟨61MAC62⟩
Webb, R. I. I., **5**, 700, 701 ⟨90MI 517-01⟩
Webb, R. L., **1**, 852 ⟨50JA408⟩; **6**, 251, 625 ⟨82JHC1205⟩, 632 ⟨87JHC275⟩
Webb, R. M., **2**, 127 ⟨65MI 203-01⟩
Webb, II, R. R., **1**, 243 ⟨83TL1357⟩, 575 ⟨85JA1421⟩, 961 ⟨82TL3051⟩
Webb, S. B., **4**, 804 ⟨74JCS(P1)1644⟩
Webb, T. H., **5**, 155 ⟨88TL6823⟩
Webb, T. R., **2**, 440 ⟨88ICA(146)111⟩
Webb, V. J., **4**, 662, 962 ⟨86OM1906⟩
Webber, A., **3**, 497 ⟨81JCS(P1)1520⟩
Webber, G. M., **2**, 660 ⟨64JA742⟩
Webber, J. A., **5**, 81 ⟨69JA5674⟩
Webber, S., **1**, 1001 ⟨84CC350⟩
Webber, S. E., **2**, 623, 624 ⟨84JA5734, 86S453⟩
Weber, A., **4**, 236 ⟨77LA1116⟩, 838 ⟨73CB1107, 75LA1484⟩
Weber, B., **1**, 194, 353 ⟨91HCA232⟩; **2**, 112 ⟨94T6117⟩, 1068 ⟨79AG(E)531⟩; **3**, 478, 482 ⟨79LA1002, 80LA305⟩; **5**, 876 ⟨82HOU(E1)27⟩; **6**, 143 ⟨71LA(746)92⟩, 623, 660, 668 ⟨79LA1002⟩
Weber, C., **4**, 709 ⟨87JOM(321)257⟩; **5**, 322, 971 ⟨70S589⟩
Weber, D., **5**, 1050 ⟨64TL799⟩, 1052 ⟨68CB3475⟩
Weber, E., **1**, 986 ⟨91BCJ659⟩
Weber, E. J., **2**, 539 ⟨87JOC165⟩
Weber, F. G., **2**, 24 ⟨69T4283⟩
Weber, G., **2**, 39 ⟨81HCA1158⟩, 104 ⟨64HCA1647⟩, 451 ⟨63HCA2464⟩; **4**, 432 ⟨80ZC437⟩; **5**, 655 ⟨88TL3475⟩, 657, 1053 ⟨88JPR241⟩
Weber, H., **1**, 886 ⟨78CB2480⟩; **2**, 1009 ⟨85CB4259, 87AHC(41)275⟩; **5**, 357 ⟨88AP(321)25, 88AP(321)551⟩; **6**, 580 ⟨81JPR41⟩
Weber, H.-J., **4**, 48, 49, 50 ⟨32LA(498)101⟩
Weber, H. P., **1**, 835 ⟨91TL7233⟩
Weber, J., **1**, 1109 ⟨76JA875, 76JCP(64)4159⟩; **3**, 119 ⟨82JOC364⟩, 233 ⟨57JOC1611⟩; **4**, 737 ⟨92TL4537⟩
Weber, J. F., **6**, 151 ⟨90MI 605-01⟩
Weber, J. P., **1**, 910 ⟨68HCA225⟩
Weber, J. V., **4**, 827 ⟨88H(27)1929⟩
Weber, K. A., **1**, 950 ⟨91JOC1309⟩
Weber, K., **5**, 918 ⟨93AG(E)731⟩, 918, 919 ⟨92CC1631⟩
Weber, L., **2**, 866 ⟨91CB725⟩; **3**, 379 ⟨83AG(E)516⟩; **4**, 551 ⟨84ZN(B)1518⟩, 580 ⟨77JOM(131)49, 84CB1103, 84CB3340, 85CB3570, 85CB541, 85ZN(B)968⟩, 580, 581 ⟨90CB739⟩, 1029 ⟨85CB3560⟩; **5**, 400 ⟨86CB1857, 86JOM(311)269, 86ZN(B)1096⟩, 641 ⟨88JCS(P2)821⟩, 893 ⟨91CB989⟩, 894 ⟨B-82MI 523-01⟩, 1092 ⟨90CB733⟩; **6**, 152 ⟨84ZN(B)1518⟩, 518 ⟨86AG(E)755, 87CB1421⟩, 680, 690 ⟨85AG(E)53, 86CB1857⟩, 705 ⟨B-82MI 622-01⟩
Weber, M., **6**, 630 ⟨89EGP270530⟩
Weber, P., **2**, 266 ⟨81JOM(210)163⟩, 751 ⟨89S117⟩
Weber, R., **3**, 74 ⟨73JHC267⟩; **5**, 239 ⟨76JCS(P1)2452⟩, 953 ⟨91CB1635⟩
Weber, R. O., **6**, 570, 571 ⟨66PHA23⟩
Weber, S., **2**, 102 ⟨02MI 202-01⟩; **5**, 173 ⟨68T2443⟩
Weber, T., **1**, 182 ⟨88CB421⟩, 182, 475 ⟨87AG(E)142⟩, 1042 ⟨87JA2224⟩; **6**, 45 ⟨89S463⟩
Weber, U., **2**, 512 ⟨84JOM(260)281, 85ZN(B)1430⟩; **5**, 879, 880 ⟨85MI 502-01⟩, 1155 ⟨85MI 531-01⟩; **6**, 375 ⟨86ZN(B)191⟩
Weber, W., **2**, 439 ⟨85CB2365⟩; **4**, 443 ⟨93JA9347⟩, 1027, 1030 ⟨85CB2365⟩
Weber, W.-D., **5**, 148 ⟨80AG(E)1029⟩
Weber, W. P., **1**, 229 ⟨85JOC2782⟩, 450, 451 ⟨73JA3416⟩, 490, 1106 ⟨71TL4701⟩, 648 ⟨B-83MI 114-02⟩, 656 ⟨84OM1340⟩; **2**, 44 ⟨72TL4907⟩, 133 ⟨B-77MI 203-02⟩; **3**, 632, 633 ⟨82JOC2873⟩, 695, 696, 700 ⟨88OSC(6)232⟩, 695, 721 ⟨72TL1637, 72AG(E)530⟩; **4**, 357 ⟨B-83MI 408-01⟩, 570 ⟨87JPS(A)1967⟩; **6**, 172, 189, 190 ⟨B-83MI 606-01⟩
Weberndörfer, V., **5**, 842 ⟨67CB71⟩; **6**, 304, 305 ⟨61E566⟩
Webster, B. R., **2**, 615 ⟨85JFC(27)263⟩
Webster, C. J., **1**, 794 ⟨69CJC1555⟩
Webster, D., **4**, 565 ⟨77MI 412-01, 78BJ(169)239⟩
Webster, D. E.,**1**, 447 ⟨60JCS179⟩
Webster, E. R., **1**, 905 ⟨59LA(623)217⟩
Webster, F. X., **5**, 53 ⟨87JOC689⟩
Webster, G. L., **5**, 751, 757 ⟨53JA1761⟩
Webster, J. A., **2**, 536 ⟨57JA974⟩
Webster, J., **4**, 675 ⟨86CC542⟩
Webster, M., **6**, 343 ⟨89CC1029, 89OM1270⟩, 344 ⟨91JCS(D)1017⟩
Webster, O. W., **1**, 861 ⟨67JOC39⟩, 862 ⟨65JA1820, 66JA3046, 66JA4055⟩; **2**, 22 ⟨92CC1200⟩; **5**, 152 ⟨B-87MI 503-01⟩; **6**, 221, 222, 226 ⟨93BSF599⟩
Webster, R. G., **3**, 341 ⟨73JCS(P1)657⟩, 371 ⟨75JCS(P1)2097⟩, 376, 377 ⟨72JCS(P1)1447, 74JCS(P1)722⟩, 377 ⟨75JCS(P1)775⟩, 388 ⟨79JCS(P1)2334⟩
Webster, S. T., **3**, 284 ⟨57JA393⟩
Webster, W., **5**, 744, 750, 754 ⟨47JCS738⟩
Wechsberg, M., **6**, 231 ⟨69JA4432⟩
Wechter, W. J., **1**, 108 ⟨54JA6040⟩
Weck, G. de, **5**, 932 ⟨88JA8413⟩, 953 ⟨88JA2122⟩
Weclas, L., **3**, 365 ⟨85JOC32⟩
Wedding, D. L., **2**, 381 ⟨73JMC736⟩
Wedegaertner, D. A., **2**, 256 ⟨87JA1227⟩
Wedekind, B., **5**, 765, 766 ⟨57CB1638⟩
Wedekind, E., **5**, 488, 489 ⟨23CB633⟩, 652 ⟨23CB1557⟩
Wedemann, P., **1**, 527 ⟨85AG(E)316⟩
Wedler, M., **4**, 719 ⟨88ZN(B)1461⟩; **5**, 782 ⟨88JOM(345)C1, 90JOM(338)21, 92CB2171⟩
Wedmid, Y., **3**, 302 ⟨79JOC3420⟩
Wee, A. G. H., **1**, 688, 690 ⟨88JA6914⟩
Weeden, B. C. L., **5**, 625 ⟨52QR380⟩
Weedon, A. C., **1**, 356 ⟨82CC1331⟩, 688 ⟨82TL5485, 85CC1282⟩, 690 ⟨87JOC2905⟩, 797 ⟨83S952⟩; **3**, 335, 346, 351 ⟨78NJC331⟩, 347 ⟨78TL4621⟩, 358 ⟨79JOC1977⟩, 552 ⟨82NJC401⟩
Weedon, B. C., **3**, 267 ⟨46JCS45⟩
Weedon, B. C. L., **1**, 6 ⟨70CC738⟩, 503 ⟨53JCS3294⟩, 506 ⟨49JCS1827, 49JCS2028⟩, 725 ⟨73JCS(P1)1416⟩, 725, 733 ⟨68JCS(C)1997, 68JCS(C)1984⟩, 871, 934 ⟨63MI 119-04⟩, 882 ⟨60T246⟩; **2**, 445 ⟨75BRP1391806⟩; **3**, 57 ⟨75JCS(P1)1457⟩, 62 ⟨46JCS937⟩, 64 ⟨51JCS2687⟩, 273 ⟨49JCS1827⟩; **5**, 28 ⟨B-60MI 502-02⟩, 820 ⟨49JCS42⟩
Weeks, P. D., **1**, 491 ⟨74TL3207⟩; **3**, 234 ⟨80JOC1109⟩
Weerasooriya, U., **1**, 495 ⟨83JOC5251⟩, 1078 ⟨82JOC1837⟩; **2**, 680, 739 ⟨83JOC448⟩; **3**, 11 ⟨83JOC448⟩
Weerasuria, K. D. V., **1**, 53 ⟨88JA4346⟩
Weeratunga, G., **5**, 1125 ⟨90TL3681⟩
Weese, K. J., **2**, 1005 ⟨87JA2541⟩
Weetman, D. G., **3**, 235 ⟨69JOC34⟩
Weferling, N., **2**, 427, 430 ⟨90ZN(B)203⟩, 455 ⟨87ZAAC(548)55⟩; **4**, 112, 546 ⟨90CB995⟩
Wege, D., **1**, 774 ⟨79AJC1819⟩
Wege, P. M., **1**, 1093 ⟨76JOC3144⟩
Wegemund, B., **3**, 538 ⟨62TL221⟩
Wegener, P., **1**, 428 ⟨90S403⟩
Wegener, W., **4**, 343 ⟨72ZC178⟩, 492 ⟨72ZC334, 72ZC60⟩, 554 ⟨72ZC103⟩

Wegler, R., **2**, 365 ⟨57GEP1101407⟩; **3**, 527 ⟨68CB1120⟩; **4**, 901, 904 ⟨68CB1131⟩, 920 ⟨59CA(53)10040f⟩; **6**, 492 ⟨B-70MI 615-01, B-77MI 615-01, B-82MI 615-01⟩
Wegmann, B., **5**, 692 ⟨87MI 517-01⟩
Wegner, E., **2**, 374 ⟨B-64MI 207-01⟩
Wegner, G., **5**, 32 ⟨85TL6051⟩
Wegner, P., **1**, 1134 ⟨84T2537⟩; **2**, 437 ⟨87JOM(335)59⟩
Wegrzyn, M., **4**, 905, 911 ⟨85CC845⟩
Weh, S.-J., **3**, 227 ⟨89SC1593⟩
Wehinger, E., **1**, 1143 ⟨72CB2290⟩
Wehlacz, J. T., **2**, 977 ⟨75JA7171⟩
Wehl, J., **5**, 599 ⟨64AG(E)705⟩
Wehle, D., **1**, 248 ⟨84TL2329⟩, 713 ⟨86CB1162⟩
Wehman, A. T., **4**, 688 ⟨70JA5520⟩
Wehmeyer, R. H., **3**, 286 ⟨88TL4513⟩
Wehmeyer, R. M., **1**, 121 ⟨87JOC5056⟩, 123, 467 ⟨91JOC1445⟩, 478 ⟨87JOC5057⟩; **2**, 959 ⟨91JOC1445⟩; **3**, 130 ⟨91JOC1445⟩, 130, 286 ⟨89SC1833⟩
Wehner, G., **1**, 627 ⟨79LA1925⟩; **3**, 292 ⟨80CB302⟩, 292, 295, 303, 311 ⟨79CB2045⟩; **5**, 1134 ⟨79S522⟩
Wehner, J. M., **1**, 200 ⟨91TL7195, 93H(35)349⟩
Wehrli, A., **4**, 197 ⟨80HCA1960⟩
Wehrli, F. W., **1**, 800 ⟨82JOC4786⟩
Wehrli, H., **1**, 794 ⟨63HCA678⟩
Wehrli, P. A., **3**, 620 ⟨77JOC3956⟩
Wehrli, P., **3**, 659 ⟨69AG(E)343⟩
Wehrli, S., **1**, 800 ⟨82JOC4786⟩
Wehrmann, R., **4**, 634 ⟨85AG(E)788⟩, 1082 ⟨84AG(E)826⟩; **5**, 927, 928 ⟨84AG(E)369, 85AG(E)788⟩, 928 ⟨84AG(E)372, 84AG(E)826⟩; **6**, 201 ⟨85AG(E)788⟩
Wehrmeister, H. L., **2**, 357, 688 ⟨46CRV269⟩; **4**, 47, 50, 51 ⟨49JA4007⟩
Wehrsig, A., **2**, 543 ⟨90TL6725⟩
Wei, C. C., **5**, 345, 346 ⟨85JOC3462, 88JOC5507⟩
Wei, C.-C., **6**, 543 ⟨88S226⟩
Wei, C. H., **6**, 352 ⟨84IC2973⟩, 524 ⟨66JA1821⟩
Wei, J., **1**, 939 ⟨93JOC5028⟩; **2**, 652 ⟨87TL5115⟩, 700 ⟨88TL4269⟩; **3**, 438 ⟨87TL5115⟩
Wei, K.-T. H., **2**, 1001 ⟨74JA4099⟩
Wei, L., **3**, 109 ⟨91JOC4296⟩
Wei, S.-Y., **1**, 552 ⟨91JOC5973⟩, 816 ⟨91CL43⟩
Wei, S. Y., **1**, 810 ⟨93T1025⟩
Wei, Y., **2**, 973, 976 ⟨88JOC880⟩
Wei, Z. Y., **2**, 78 ⟨89JOC5768⟩
Weiberth, F. J., **3**, 131, 405 ⟨87JOC3901⟩
Weibo Wang, **1**, 269, 738 ⟨88TL5295⟩
Weichert, A., **3**, 250 ⟨91JOC4098⟩
Weichert, R., **3**, 227 ⟨82S34⟩
Weichert, U., **6**, 549 ⟨88LA75⟩
Weichmann, H., **2**, 431, 454, 822, 854 ⟨68CB2197⟩; **4**, 551, 570, 588 ⟨79JOM(182)465⟩, 555 ⟨92PS(73)67⟩, 563 ⟨89ZC409⟩, 570 ⟨80EGP142887, 87EGP251136⟩; **5**, 390 ⟨79JOM(182)465⟩
Weichmann, J., **3**, 387, 390 ⟨83AG(E)314⟩
Weichselbaum, T. E., **5**, 90 ⟨43OSC(2)384⟩
Weickert, P., **4**, 580, 582 ⟨83ZN(B)711⟩; **5**, 448 ⟨82CB1488⟩
Weidemann, P., **5**, 755 ⟨71CB3329⟩
Weidenbruch, M., **2**, 812 ⟨78ZN(B)1465⟩; **3**, 581 ⟨92JOM(423)329⟩; **5**, 889 ⟨87AG(E)776⟩, 1085 ⟨90AG(E)90, 91POL1147⟩, 1134 ⟨78ZN(B)1465⟩
Weidenhammer, K., **3**, 544 ⟨79JA3133⟩
Weidenhaupt, H.-J., **3**, 233 ⟨87CB23⟩
Weidinger, H., **1**, 1026 ⟨60AG836⟩; **5**, 577, 608, 609, 610, 728 ⟨63CB2671⟩, 609, 611 ⟨60AG836⟩; **6**, 51 ⟨60AG836⟩, 70 ⟨63CB2671⟩, 313, 321 ⟨64CB1232⟩
Weidlein, J., **2**, 488 ⟨69CB4136, 73CB1226⟩, 488, 504 ⟨⟨71ZAAC(386)139⟩, 509, 511 ⟨89ZAAC(569)62⟩, 888 ⟨71ZAAC(386)147⟩, 1082 ⟨81ZAAC(474)157⟩, 1098 ⟨78JOM(159)373⟩; **5**, 780 ⟨78ZN(B)24, 79JOM(175)33⟩, 1082 ⟨86ZAAC(538)36, 86ZN(B)1223⟩, 1084 ⟨89JOM(359)1⟩; **6**, 393 ⟨94JOM(469)135⟩
Weidmann, B., **2**, 67 ⟨83AG(E)31⟩, 565, 566 ⟨81HCA1552, B-83MI 211-01⟩; **4**, 396 ⟨81HCA357⟩; **6**, 134 ⟨81HCA357⟩
Weidmann, E., **1**, 725 ⟨76LA1475⟩
Weidmann, H., **1**, 688 ⟨87S1071, 88JCS(P1)1729⟩; **2**, 678 ⟨86CC1149⟩; **4**, 655 ⟨91JOM(414)295⟩
Weidmann, T., **5**, 955 ⟨91CB2431⟩
Weidner-Wells, M. A., **5**, 325 ⟨89JOC5746⟩
Weier, R. M., **1**, 1008 ⟨87JMC193⟩
Weigand, W., **3**, 707 ⟨88CB507⟩; **5**, 388 ⟨92ZN(B)1165⟩, 658 ⟨88CB507⟩, 1088, 1089 ⟨88JOM(347)101⟩; **6**, 622 ⟨88CB507⟩
Weigel, F. C., **6**, 492, 515 ⟨81JINC457⟩
Weigel, H., **1**, 18 ⟨64CI(L)1689⟩
Weigel, L. O., **3**, 494 ⟨75JOC3456⟩; **4**, 649 ⟨78JA2916⟩
Weigele, M., **1**, 755 ⟨88TL2401⟩; **2**, 749 ⟨70USP3542848⟩; **5**, 315 ⟨78JOC3713⟩, 346 ⟨85JOC3462⟩
Weigert, F., **4**, 38 ⟨70JA7359⟩; **5**, 232 ⟨03CB1007⟩
Weigert, F. J., **4**, 730 ⟨93JFC(65)67⟩; **5**, 984 ⟨73JOC1316⟩; **6**, 12 ⟨93JFC(61)1⟩, 645 ⟨91MIP 601-01⟩
Weiglt, C., **3**, 145 ⟨92TL5287⟩
Weigmann, R., **2**, 1053 ⟨87AG(E)923⟩
Weigmann, R. H., **2**, 1054, 1058 ⟨89TL6147⟩
Weigold, H., **2**, 968 ⟨71JOM(27)373⟩
Weihe, G. R., **1**, 85 ⟨83JOC3370⟩
Weii, E., **2**, 1004 ⟨85CB3529⟩
Weik, C., **4**, 485 ⟨93LA955⟩
Weil, J. K., **2**, 190 ⟨52JA2526, 63OSC(4)8623⟩
Weil, L., **2**, 1012, 1013 ⟨62M952⟩
Weiland, J., **6**, 314 ⟨74CB3155⟩
Weiler, H., **5**, 1011 ⟨74TL2445⟩
Weiler, H.-U., **6**, 237 ⟨85CB943⟩
Weiler, L., **1**, 216 ⟨83CJC214, 86CJC584⟩, 482 ⟨79CJC1431, 84OS(62)14⟩, 769 ⟨88CJC1359⟩, 942 ⟨90CJC1636⟩, 1124 ⟨71CJC3135, 71TL927⟩; **3**, 217 ⟨79CJC1431⟩, 578 ⟨76JA6036⟩; **4**, 197 ⟨78CJC2700⟩; **5**, 103 ⟨84OS(62)14⟩
Weill, P., **5**, 105 ⟨B-45MI 502-02⟩
Weill-Raynal, J., **2**, 393, 394, 396 ⟨B-73MI 207-01⟩
Weimar, Jr, W. R., **2**, 32 ⟨69CI(L)200⟩
Weimaster, J. F., **1**, 797 ⟨76JA7319⟩
Weimer, D. F., **3**, 198 ⟨86TL4265, 87JOC4185, 90JOC2842⟩
Weimer, N., **3**, 634 ⟨37JA744⟩
Weinand, R., **3**, 553 ⟨85CC1145⟩
Weinberg, A. E., **1**, 1118 ⟨56JA4654⟩
Weinberg, D. S., **2**, 7 ⟨65JOC2635⟩
Weinberg, E. L., **2**, 566 ⟨84JOM(267)265⟩
Weinberg, J. S., **4**, 54 ⟨79JOC4722⟩
Weinberg, K. G., **2**, 834, 835 ⟨75JOC3586⟩
Weinberg, N. L., **6**, 107 ⟨66JOC4054⟩
Weinberger, B., **3**, 696, 710 ⟨80AG(E)480⟩; **4**, 582 ⟨85JOM(280)C31, 89JOM(359)205⟩
Weinberger, M. A., **2**, 108 ⟨67CJC495⟩
Weiner, L. M., **4**, 560 ⟨85BBA(818)96⟩
Weiner, M. A., **1**, 1091 ⟨61JA3583⟩; **2**, 887 ⟨67JOC3707⟩
Weiner, M. L., **5**, 745, 754 ⟨60JOC2245⟩
Weiner, N., **5**, 35 ⟨43OSC(2)376⟩
Weingand, C., **2**, 468 ⟨72ZAAC(394)171⟩
Weingart, F., **3**, 628 ⟨91TL2605⟩
Weingarten, H., **3**, 406, 407 ⟨67JOC3246⟩, 601 ⟨70JOC3970⟩; **4**, 405 ⟨66JOC4041, 67JOC213⟩, 972 ⟨66JA850, 68JOC1506⟩, 972, 988 ⟨71JOC1613⟩, 975 ⟨67JOC3293, 68T2767⟩; **5**, 652 ⟨70CJC983⟩, 745, 753 ⟨71JOC1613⟩; **6**, 139 ⟨68T2767⟩, 140 ⟨71JOC2885⟩, 141 ⟨66JA850, 66JOC2874, 68JOC1246⟩, 361 ⟨66JA2885⟩
Weingarten, H. I., **6**, 361 ⟨70USP3551418⟩
Weingarten, M. D., **3**, 320 ⟨93TL7853⟩

Weinges, K., **1**, 705 ⟨84CB966, 86CB1540⟩, 708 ⟨74CB1915, 76CB793, 77CB2978, 84CB1868, 87CB5, 87LA679⟩, 782 ⟨93LA1133⟩; **3**, 616 ⟨71CB3594, 80CB710, 80LA212, 85LA566⟩
Weinheimer, A. J., **4**, 183 ⟨58JA4072⟩
Weinhold, E., **5**, 151, 164 ⟨88JA4640⟩
Weinhouse, S., **2**, 448, 489 ⟨49JOC429⟩
Weinig, P., **2**, 69, 564 ⟨90JOC4990⟩
Weining, P., **2**, 568 ⟨86TL5711⟩
Weinkam, R. J., **2**, 377 ⟨83JOC4116⟩
Weinkauff, D. J., **1**, 76 ⟨77JA5946⟩
Weinman, S., **2**, 626 ⟨82CC778⟩
Weinmann, W., **5**, 927 ⟨92AG(E)1603⟩
Weinmayr, V., **6**, 2 ⟨63USP3076041⟩
Weinmayr, W., **2**, 122 ⟨66JOC957⟩
Weinreb, S. M., **1**, 369 ⟨91JOC3210⟩, 408 ⟨86T2979⟩, 791 ⟨80JOC3372⟩, 937 ⟨88T4671⟩, 1042 ⟨90JOC393⟩, 1047 ⟨81TL3815⟩; **2**, 244, 245 ⟨84JA7867, 82T3087⟩, 245 ⟨83JA4499, 84JOC3243, 86TL1853, 88TL4233⟩, 249 ⟨86TL2099⟩, 306 ⟨90SC573⟩, 309 ⟨89TL5709⟩, 310 ⟨90JOC2563⟩, 366, 793 ⟨88ACR313, 91COS(5)401⟩; **3**, 13, 85, 273, 285 ⟨81TL3815⟩, 294 ⟨93JOC4745⟩, 311 ⟨82JA536⟩, 417 ⟨89CR1525⟩, 437 ⟨92BSB381⟩, 438 ⟨90JOC3037, 90JOC393⟩, 618 ⟨79TL4907⟩; **4**, 12 ⟨88JA5533⟩, 209 ⟨71JA746⟩, 346 ⟨80S72⟩, 370 ⟨86T2979⟩; **5**, 30 ⟨82TL2347⟩, 235 ⟨77JOC3960⟩, 273, 296 ⟨76JOC725⟩, 277 ⟨77TL4171⟩, 326, 330 ⟨82T3087⟩, 329 ⟨82SC989⟩, 345 ⟨91JOC3210⟩; **6**, 23, 25 ⟨90JOC1281⟩
Weinreg, S. M., **2**, 245 ⟨84JA7861⟩
Weinreich, G. H., **2**, 190, 208 ⟨65BSF787⟩
Weinschneider, S., **3**, 345, 349 ⟨83LA1694⟩
Weinshenker, N. M., **1**, 936 ⟨75JOC1966⟩; **3**, 494 ⟨70JA397⟩; **5**, 168 ⟨69JA5675, 70JA397⟩
Weinstein, A. H., **2**, 175 ⟨58JOC363⟩
Weinstein, B., **1**, 811 ⟨69TL3459⟩
Weinstein, E. A., **2**, 373, 378, 381 ⟨64JMC493⟩
Weinstein, G. N., **3**, 339 ⟨73IC2589, 73JA613⟩
Weinstein, R. M., **5**, 418 ⟨82JA5534⟩, 437 ⟨83JOC1144⟩
Weinstein, S., **1**, 889 ⟨62JCS1741⟩
Weinstein, S. Y., **2**, 88 ⟨64JA5326⟩
Weinstock, J., **1**, 627 ⟨76OS(55)3⟩; **3**, 60 ⟨88SC1913⟩; **4**, 102 ⟨85JHC957⟩; **5**, 991 ⟨61JOC3511⟩
Weinstock, L. M., **2**, 342, 346 ⟨88JOC1768⟩; **3**, 193 ⟨81SC947⟩, 696, 699, 704 ⟨88OSC(6)620⟩; **4**, 48 ⟨75TL3979, 79S970⟩, 302 ⟨84TL4337⟩; **5**, 579, 580 ⟨69AG(E)278⟩; **6**, 25 ⟨86TL3845, 87JOC944⟩
Weintraub, L., **5**, 852 ⟨68JOC1679⟩
Weintritt, V., **1**, 331, 342 ⟨90S411⟩
Weir, J. R., **1**, 32, 457, 507, 1009 ⟨80JOC4926⟩
Weirich, R., **1**, 472 ⟨90AG(E)977⟩
Weis, C., **5**, 816 ⟨73GEP2242772⟩
Weis, C. D., **1**, 521, 1084 ⟨62JOC3520⟩, 571 ⟨88OSC(6)427⟩, 637, 638 ⟨90JOC5801⟩, 651 ⟨86JOC558⟩; **3**, 108 ⟨76HCA211⟩; **5**, 375 ⟨86JOC558⟩; **6**, 612, 616 ⟨69HCA812⟩
Weis, G., **2**, 399 ⟨56AG678⟩
Weis, K., **6**, 71 ⟨56CB314⟩
Weisblat, D. I., **2**, 392 ⟨49JA3079⟩
Weischedel, F., **5**, 856 ⟨71LA(750)12⟩
Weischel, C., **2**, 171 ⟨79LA784⟩
Weise, A., **5**, 1004 ⟨67AG(E)168⟩
Weisel, C. A., **3**, 638 ⟨45JA1071⟩
Weisel, C., **3**, 638 ⟨44JA725⟩
Weisel, K. H., **4**, 796 ⟨77AP(310)30⟩
Weisenberger, F., **4**, 352, 353 ⟨92CB591⟩
Weisenfeld, R. B., **4**, 845 ⟨79JOC4744⟩, 956 ⟨79JOC3601⟩; **5**, 101 ⟨79JOC3601⟩
Weisgerber, C. A., **5**, 64 ⟨55OSC(3)114⟩
Wei-Shan, Z., **3**, 220 ⟨89CC612⟩
Weisheit, R., **2**, 460 ⟨83ZAAC(504)147⟩
Weiske, T., **2**, 1022 ⟨86AG(E)282⟩
Weisman, G. R., **2**, 382 ⟨73TL2321, 76JA5269⟩, 843, 845 ⟨79JOC2226⟩; **6**, 140 ⟨81TL4365⟩
Weisman, R. B., **1**, 924 ⟨86JA7935⟩
Weismiller, M. C., **1**, 147 ⟨89TL1613, 91T173⟩; **3**, 171 ⟨90OS(69)158⟩
Weisner, K., **3**, 59 ⟨73TL2309⟩
Weiss, B., **3**, 275 ⟨89HCA117⟩; **4**, 991, 993, 996 ⟨85BSB475⟩
Weiss, D., **3**, 353 ⟨92S751⟩
Weiss, D. S., **1**, 841 ⟨77TL763⟩
Weiss, E., **1**, 104 ⟨67AG(E)801⟩, 158 ⟨79AG(E)782⟩, 885 ⟨70CB265, 77CB717⟩, 1133 ⟨83CB3212⟩; **2**, 551 ⟨93AG(E)1501⟩, 1022, 1023 ⟨63HCA1121⟩, 1096 ⟨68JOM(14)21⟩; **4**, 564 ⟨79AG(E)782⟩, 577, 587 ⟨85OM1233⟩, 681 ⟨62CB1155, 76JOM(108)213⟩; **5**, 592 ⟨29HCA554⟩, 1097 ⟨81JOM(213)451, 84JOM(263)67⟩
Weiss, E. R., **5**, 597 ⟨68JOC2357⟩
Weiss, F., **1**, 810 ⟨68FRP1519701⟩
Weiss, F. T., **4**, 765 ⟨40JA1849⟩
Weiss, G., **2**, 365 ⟨61GEP1156403⟩, 367, 368 ⟨69LA(729)40⟩
Weiss, H., **5**, 770, 771, 839, 850 ⟨72LA(758)162⟩, 852, 853 ⟨70LA(736)166⟩
Weiss, H. D., **2**, 463 ⟨58JA3945⟩; **5**, 122 ⟨57JA4759⟩
Weiss, H. G., **2**, 109 ⟨59JCP(30)1222, 59JPC1319⟩
Weiss, I., **6**, 253 ⟨92CB535⟩
Weiss, J., **4**, 632 ⟨83JA2582⟩; **6**, 622 ⟨69AG(E)455⟩
Weiss, J. V., **2**, 476 ⟨76CC643⟩
Weiss, K., **3**, 690 ⟨88MI 320-01⟩; **5**, 949 ⟨74CB3554, 76CB1120, 76CB1128⟩, 953 ⟨73CB1277, 76CB1868⟩; **6**, 716, 719, 722 ⟨84MI 617-04⟩
Weiss, K. T., **6**, 111 ⟨84H(22)281⟩
Weiss, L. B., **1**, 573 ⟨78JA6483⟩
Weiss, M., **2**, 25, 31 ⟨86CC1250⟩
Weiss, M. J., **1**, 100 ⟨71JOC279⟩, 534 ⟨72TL4083⟩, 556 ⟨80JOC4702⟩; **2**, 28 ⟨80JOC4702⟩, 421 ⟨49JOC310⟩; **3**, 140, 156 ⟨79JOC1438⟩
Weiss, R., **2**, 734 ⟨12CB1835⟩, 768 ⟨24CB1077⟩, 1006 ⟨74AG(E)1⟩, 1008 ⟨80S227⟩, 1009 ⟨92AOC(34)207⟩, 1046 ⟨81JA6142⟩; **3**, 256 ⟨12CB1835⟩; **4**, 870 ⟨70TL481⟩, 1034 ⟨77JA3537⟩, 1037 ⟨78JA5748⟩; **5**, 1097 ⟨78AG(E)781⟩; **6**, 314, 317 ⟨70TL481⟩
Weiss, U., **2**, 122 ⟨50JA1687⟩
Weiss, W., **2**, 439 ⟨85CB2365⟩; **6**, 434 ⟨64GEP1224720⟩, 529 ⟨65GEP1219455⟩
Weissbach, K., **2**, 172 ⟨30CB2856⟩
Weissbach, O., **6**, 213 ⟨B-72MI 607-02⟩
Weissberger, A., **3**, 468 ⟨32CB265⟩; **5**, 652 ⟨55OSC(3)108⟩
Weissenfels, M., **3**, 356 ⟨90ZC247⟩
Weissensteiner, W., **2**, 949 ⟨86JA6664⟩
Weissermel, K., **1**, 1023 ⟨56CB853⟩; **4**, 187 ⟨63CB77⟩; **6**, 215, 220, 224 ⟨B-93MI 607-02⟩
Weissflog, E., **3**, 382 ⟨84CB187⟩; **6**, 127, 132, 134 ⟨75ZAAC(418)208⟩, 132 ⟨76CB1239⟩
Weisshuhn, C. M., **5**, 551 ⟨79CB1650⟩
Weisshuhn, M. C., **3**, 324 ⟨75AG(E)422⟩; **5**, 560 ⟨76CB740⟩
Weissler, M., **4**, 326 ⟨87LA583⟩
Weissman, B.-A., **1**, 1126 ⟨73CC679⟩
Weissman, E. C., **5**, 798, 799, 800 ⟨79JOM(172)251⟩
Weissman, P. M., **2**, 126 ⟨66JA1458⟩
Weissman, S. A., **2**, 57 ⟨90JOC6328⟩, 446 ⟨87TL603⟩
Weissman, S. I., **2**, 1010 ⟨64JA2538⟩
Weisz, A., **5**, 1078 ⟨68M2498⟩
Weisz, I., **6**, 503 ⟨78IZV220⟩
Weith, W., **3**, 698 ⟨1873CB210⟩; **5**, 1036 ⟨1873CB210⟩
Weitkamp, H., **2**, 848 ⟨70T5529⟩
Weitl, F. L., **6**, 431 ⟨67JOC2633⟩
Weitz, H. M., **2**, 391, 392 ⟨80S261⟩; **3**, 190 ⟨79LA612⟩, 190 ⟨80S261⟩

Weiyu, D., **6**, 708 ⟨84TL4425⟩
Weizmann, A., **1**, 89 ⟨49JA4154⟩
Weizmann, C., **5**, 66 ⟨48JA1153⟩
Weizmann, Ch., **2**, 656 ⟨39JCS398⟩
Welbourn, A. P., **5**, 639 ⟨86JMC2280⟩
Welch, A. J., **3**, 684 ⟨91JCS(P2)249⟩; **5**, 838 ⟨91JCS(P2)249⟩, 952 ⟨80JA7811⟩
Welch, A. J. E., **6**, 439 ⟨45JCS864⟩
Welch, C. A., **2**, 1016, 1033 ⟨63JA1648⟩, 1062 ⟨62JA2020⟩
Welch, D. E., **4**, 551, 570, 588 ⟨64JA1100⟩, 551, 576, 588 ⟨63JA642⟩, 576, 588 ⟨64JOM(2)1⟩, 812 ⟨62JA4266⟩
Welch, J., **2**, 6 ⟨74S653⟩, 11 ⟨74S652⟩, 12 ⟨74S896⟩, 22 ⟨92BSB237⟩; **3**, 3, 114 ⟨77S419⟩, 8, 121 ⟨78JA5396⟩; **5**, 6 ⟨78JA5396⟩; **6**, 423 ⟨74S654⟩
Welch, J. T., **1**, 415, 827, 828 ⟨85JOC5909, 91TL4267⟩, 828, 834 ⟨87JA6716⟩, 830 ⟨93TL2251⟩, 831 ⟨91JOC353⟩, 876 ⟨B-88MI 119-01⟩; **2**, 6, 11, 17, 392, 615 ⟨79JOC3872, 87T3123⟩, 102 ⟨88JOC2997⟩, 325 ⟨91SL573⟩; **3**, 415 ⟨88JOC2991⟩; **4**, 32 ⟨79JOC3842⟩, 137 ⟨90JOC4782⟩; **5**, 4 ⟨79JOC3872⟩, 227 ⟨91JOC353⟩; **6**, 9 ⟨B-91MI 601-03⟩, 439 ⟨79JOC3872⟩
Welch, M. J., **2**, 9 ⟨87JOC658⟩, 13 ⟨86CJC138⟩, 18 ⟨86JFC(31)99⟩; **5**, 796 ⟨82MI 520-01⟩; **6**, 134 ⟨90MI 604-01⟩
Welch, S. C., **1**, 432 ⟨78JOC4797⟩, 434 ⟨85JOC2668⟩, 693 ⟨81JOC4072⟩; **3**, 208 ⟨72TL1853⟩, 220 ⟨87JOC1440⟩
Welch, W. M., **3**, 559 ⟨74TL1515⟩
Welch, Z. D., **6**, 220, 225 ⟨47CI(L)427⟩
Welcman, N., **2**, 292 ⟨63JCS1268, 65JCS7511⟩, 499, 500 ⟨62JCS2529⟩; **6**, 49 ⟨62JCS2290, 63JCS1268, 65JCS7511, 65JCS7515⟩, 256, 536, 537 ⟨76ZAAC(427)114⟩, 276 ⟨74CZ511⟩
Welinder, H., **5**, 26 ⟨71JA5821⟩
Welke, S., **4**, 673 ⟨86AG(E)909⟩
Welker, M. E., **1**, 134 ⟨85TL3075⟩
Well, M., **2**, 476 ⟨92CB801⟩; **4**, 134 ⟨92CB801⟩, 338 ⟨92PS(72)171⟩
Wellard, N. K., **1**, 1023 ⟨76TL1539⟩
Weller, F., **5**, 776 ⟨82ZAAC(486)61⟩, 900 ⟨93OM4056⟩; **6**, 207 ⟨75ZAAC(415)233⟩
Weller, H. N., **2**, 91 ⟨83JA7172⟩; **3**, 119 ⟨80JOC2030⟩; **6**, 327 ⟨83JA7719⟩
Weller, M. E., **2**, 577 ⟨92CRV97⟩
Weller, T., **2**, 783 ⟨79C1⟩
Wellington, C. A., **1**, 801 ⟨62JPC1671⟩
Wells, D. V., **2**, 845 ⟨73JOC2703⟩; **3**, 363 ⟨70T817⟩
Wells, G. J., **1**, 248 ⟨82TL263⟩, 284 ⟨84JA6442⟩, 690 ⟨84JOC3610, 84JOC3618, 84JOC3641⟩; **3**, 220, 639 ⟨84JOC3604⟩; **4**, 780 ⟨91JMC2525⟩
Wells, J. S., **3**, 704 ⟨88JAN454⟩
Wells, P. B., **1**, 80 ⟨64MI 103-01⟩, 505 ⟨64MI 112-01⟩
Wells, P. R., **4**, 660, 663 ⟨85OM821⟩; **6**, 208, 405 ⟨84JOM(266)37⟩
Wells, R. J., **3**, 92 ⟨78AJC1533⟩
Wells, R. L., **2**, 482 ⟨86JOM(308)281⟩, 482, 504 ⟨90POL319⟩, 490 ⟨91PS(57)1⟩, 508, 511 ⟨86OM1266, 87JOM(325)C7⟩, 511 ⟨88JOM(354)287⟩; **4**, 597 ⟨91MI 413-01, 91POL319⟩, 597 ⟨91PS(57)1⟩
Welmaker, G. S., **5**, 467 ⟨92SL537⟩
Welsh, D. A., **2**, 787 ⟨69JOC3279⟩
Welstead, Jr., W. J., **3**, 451 ⟨64JA732⟩
Welter, A., **2**, 235 ⟨86EUP165534, 88GEP3626554⟩
Welter, W., **3**, 481 ⟨79CB2509⟩, 481, 489 ⟨74TL1825, 78CB3068⟩; **5**, 871 ⟨72TL3799⟩
Welton, A. F., **1**, 1029 ⟨89JMC1820⟩
Welton, D. E., **5**, 1125 ⟨47JA365⟩
Welvart, Z., **1**, 55 ⟨54MI 102-01, 54MI 102-02⟩, 55 ⟨63BSF1428⟩
Welzel, J., **6**, 49 ⟨90JOM(386)321⟩
Welzel, P., **1**, 687 ⟨83MI 115-01⟩, 690 ⟨87TL3091⟩; **3**, 157, 161 ⟨91T3753⟩; **5**, 223 ⟨87T2045⟩
Wember, K., **5**, 762, 815 ⟨53CB400⟩
Wempen, I., **2**, 212 ⟨61JOC3401⟩
Wemple, J., **2**, 148 ⟨77S791⟩; **4**, 61, 63, 73 ⟨77S791⟩; **5**, 232 ⟨79TL703⟩, 237 ⟨75TL3255, 76TL2577⟩
Wen, X., **1**, 647, 741 ⟨87TL2155⟩, 647, 741 ⟨88TL3949⟩; **3**, 48 ⟨88TL3949⟩, 196, 231 ⟨87TL2131⟩
Wen, Y.-S., **6**, 348 ⟨91OM2485⟩
Wendel, A., **2**, 235 ⟨86EUP165534, 86GEP3443467⟩
Wendel, K., **2**, 205 ⟨60CB2902⟩; **6**, 500 ⟨73OSC(5)201⟩
Wendelborn, D. F., **1**, 601, 602, 604 ⟨78JOC1697⟩; **2**, 281 ⟨78JOC1697⟩; **4**, 92 ⟨78JOC1697⟩
Wender, I., **2**, 539 ⟨58JOC50⟩, 670 ⟨58JA1009⟩; **6**, 524 ⟨55JA3951⟩
Wender, P. A., **1**, 120 ⟨88JA2218⟩, 267 ⟨71JA4318⟩, 364, 365 ⟨90PAC1597⟩, 523 ⟨89JA6432⟩, 542 ⟨87TL2221⟩, 706 ⟨88TL909⟩, 799 ⟨92JA5878⟩, 811 ⟨76JOC3490⟩, 841 ⟨81T3967, 81TL2471⟩, 889 ⟨78JA292⟩; **2**, 22 ⟨85JA7771⟩, 988 ⟨90JA4956⟩; **3**, 144 ⟨81T3967⟩, 144 ⟨81TL2471⟩, 217 ⟨83JA3348⟩, 270 ⟨90TL3691⟩, 325 ⟨89JA8954, 89JA8957⟩, 616, 621 ⟨77JOC2001⟩, 616, 621 ⟨88OSC(6)334⟩
Wenderoth, B., **1**, 127 ⟨80CC1202⟩; **2**, 566 ⟨82TL5259⟩
Wendisch, D., **4**, 170 ⟨87LA607⟩
Wendler, N. L., **1**, 79 ⟨83JOC1991⟩; **3**, 231 ⟨51JA719⟩; **4**, 130 ⟨69TL4647⟩; **5**, 173 ⟨68T2443⟩
Wendschuh, P. H., **1**, 420, 421 ⟨73PAC197⟩
Wendt, G., **2**, 274 ⟨58JOC1449⟩
Wendt, H., **2**, 403 ⟨78AG(E)370⟩, 403, 420 ⟨78AG(E)202⟩
Wendt, M. W., **5**, 199 ⟨58JA3711⟩
Weng, N. S., **1**, 924 ⟨B-91MI 119-03⟩
Weng, W., **5**, 1097 ⟨92JA5890, 93JA3824⟩
Weng, W.-W., **1**, 1127 ⟨93JOC5574⟩
Wengel, J., **3**, 48 ⟨91JOC3591⟩
Wengenroth, K. J., **1**, 1026 ⟨86JOC2607⟩; **3**, 208 ⟨86S870⟩
Wenger, R., **1**, 794 ⟨63HCA678⟩
Wengjuan Xiao, **1**, 739 ⟨87S370⟩
Wenglowsky, S., **2**, 671 ⟨91JA9873⟩
Wengrovius, J., **3**, 517 ⟨80JMOC73⟩
Wengrovius, J. H., **3**, 518 ⟨82OM148⟩
Wenis, E., **3**, 107 ⟨51JOC1121⟩; **4**, 424 ⟨57JOC984⟩; **5**, 10 ⟨61JOC1514⟩
Wenk, P., **1**, 1119 ⟨76TL4191, 76TL4195⟩
Wenker, H., **5**, 976, 977 ⟨36JA2608⟩
Wenkert, D., **5**, 48, 49, 83, 103, 114 ⟨83CC392⟩
Wenkert, E., **1**, 91 ⟨68JOC747⟩, 195 ⟨83S701⟩, 436 ⟨85JOC1125⟩, 436, 438, 484 ⟨82CC840⟩, 480 ⟨84JOC4894⟩, 482 ⟨79JA2246, 79CC637, 83T2289, 84CC617⟩, 484 ⟨85JOC3828⟩, 487 ⟨88CC975⟩, 808 ⟨77MI 118-02⟩, 970 ⟨85JOC1122⟩, 1078 ⟨72JA8084, 72SC331⟩; **2**, 562 ⟨72TL2285⟩; **3**, 65, 232 ⟨87HCA1429⟩, 209, 241 ⟨83JA2030⟩, 224 ⟨78JA1267⟩, 226 ⟨76JA3645⟩, 274 ⟨72SC331⟩, 533 ⟨84JOC2052⟩; **4**, 179 ⟨77SC409⟩; **5**, 31 ⟨58JA217⟩, 325 ⟨88SC1331⟩
Wenner, W., **5**, 275 ⟨63OSC(4)760⟩
Wennerbeck, I., **4**, 969, 970, 973 ⟨70ACS3102, 75JST(24)373⟩
Wenschuh, E., **2**, 99 ⟨86SUL161⟩, 201 ⟨76S829⟩, 258 ⟨93PS(79)131⟩, 363 ⟨70JPR129⟩
Wensing, M., **4**, 139, 141, 146, 156 ⟨85CB391⟩
Went, M. J., **4**, 712 ⟨81CC867⟩
Wentrup, C., **1**, 835 ⟨91JOC970⟩, 1076 ⟨78AG(E)609⟩; **2**, 799 ⟨70HCA1637⟩, 1040 ⟨80AG(E)720⟩; **3**, 366 ⟨84AG(E)800⟩, 541 ⟨86CC369⟩, 552 ⟨83AG(E)543⟩, 578 ⟨80AG(E)564⟩, 579 ⟨80AG(E)720⟩, 582 ⟨88JA1337⟩, 584 ⟨92CC1571, 92CC487, 92CC488⟩, 680 ⟨86AG(E)85⟩, 680, 682 ⟨94AG(E)527⟩, 681 ⟨85AG(E)56⟩, 686 ⟨79AG(E)467, 91CRV363, 91TL1487⟩, 703 ⟨81JOC1046⟩, 710 ⟨81JOC1045⟩, 722, 724 ⟨78AG(E)688⟩; **4**, 894 ⟨92CC487⟩, 900 ⟨91JOC970⟩; **5**, 698 ⟨92CC1571⟩, 1086, 1087, 1092 ⟨83JA7194⟩, 1101 ⟨66ACS2107⟩, 1104 ⟨66ACS2123⟩,

1122 ⟨72HCA565, 80AG(E)566⟩, 1123 ⟨81JOC1045⟩; **6**, 593 ⟨66ACS2091⟩
Wenzel, H. W., **6**, 687 ⟨88USP4754072⟩
Wenzel, T. T., **1**, 890 ⟨84CC1654⟩
Wenzel, W., **6**, 230 ⟨69CB3127⟩
Wenzke, H. H., **2**, 1019 ⟨34JA1106⟩
Wepplo, P., **3**, 634 ⟨89SC1533⟩; **5**, 233 ⟨89SC1533⟩
Wepplo, P. J., **4**, 855 ⟨73TL4711⟩, 857 ⟨73TL4707⟩
Wepster, B. M., **1**, 115 ⟨70T2683⟩
Werbel, L. M., **2**, 496, 498 ⟨57JOC452⟩
Werchan, H., **6**, 650 ⟨73ZC58⟩
Werchan, H. G., **6**, 498 ⟨82TL2833⟩
Werle, P., **2**, 890 ⟨71CB3782⟩
Werley, R. T., **1**, 816, 818 ⟨91JOC1185⟩
Wermann, K., **6**, 538 ⟨85DIS(B)840⟩
Wermuth, C.-G., **2**, 304 ⟨89SC3139⟩
Wermuth, C. G., **6**, 619 ⟨74TL3765⟩
Wernberg, A. A., **4**, 93 ⟨85JA6298⟩
Werner, A., **2**, 103 ⟨1893CB1567, 1894CB3350⟩; **3**, 683 ⟨1894CB2193⟩, 722 ⟨04CB4295⟩; **5**, 821 ⟨1894CB2846⟩, 821, 834 ⟨1894CB2193⟩, 826 ⟨1896CB1161, 1899CB1975⟩
Werner, E. A., **3**, 466 ⟨19JCS1093⟩; **5**, 1035 ⟨1891JCS396⟩
Werner, E., **5**, 897, 898 ⟨66M1797⟩
Werner, F., **3**, 80 ⟨79LA1977⟩
Werner, G., **1**, 790 ⟨73LA437, 74LA1190⟩
Werner, H., **1**, 862 ⟨77AG(E)1⟩, 991 ⟨93AG(E)1315⟩; **2**, 437 ⟨90ZN(B)1548⟩; **3**, 375, 387 ⟨84AG(E)58⟩, 387, 391 ⟨83AG(E)316⟩, 521 ⟨92CB2641, 93AG(E)1480⟩, 553 ⟨81AG(E)593, 83AG(E)981, 85CB873, 85CC1145⟩, 553, 554 ⟨91OM3967⟩, 599 ⟨90AG(E)275, 91CB285⟩; **4**, 148 ⟨85JOM(289)141⟩, 149 ⟨87JOM(336)413⟩, 149, 583 ⟨88OM1106, 91JOM(417)149⟩, 582 ⟨81JOM(219)C29, 82JOM(236)C65, 85CB261, 85CB3032, 85JOM(281)317⟩, 693 ⟨77CB1763, 79CC814, 79JOM(179)421, 79JOM(179)439, 80CB1072⟩, 694 ⟨72AG(E)930, 72SRI239, 77AG(E)1, 77JOM(141)339⟩; **5**, 458 ⟨79CB834⟩, 652 ⟨69JOM(16)29⟩, 800 ⟨85CB1517⟩, 944 ⟨91CB1121, 95CB49⟩, 953 ⟨71JOM(28)359⟩; **6**, 324 ⟨85CB3151⟩, 335 ⟨86JOM(311)63⟩, 335, 336 ⟨85ZN(B)1351⟩, 352 ⟨79AG663, 80CB1654⟩
Werner, J., **5**, 692 ⟨89S256⟩
Werner, J. A., **1**, 196, 410, 820 ⟨86JA6060⟩
Werner, K. von, **4**, 730 ⟨87JFC(37)223⟩; **6**, 3 ⟨87JFC(37)223⟩
Werner, N. D., **3**, 696 ⟨63JCS4280⟩, 698 ⟨66JOC3473⟩
Werner, N. W., **2**, 757 ⟨73OSC(5)273⟩
Werner, P., **4**, 649 ⟨81ZAAC(482)65⟩
Werner, R. L., **5**, 1054, 1057 ⟨65TL3003⟩
Werner, S., **3**, 394 ⟨90AG(E)609⟩; **4**, 695 ⟨91AG(E)1124⟩, 725 ⟨90JA9620, 91OM3791⟩, 725 ⟨92CB1953, 92JA8531⟩, 725 ⟨92OM3517, 93OM4979⟩
Wernick, W., **1**, 623 ⟨1898CB1553⟩
Wernicke, U., **6**, 649 ⟨65ZN(B)1165⟩
Wershofen, S., **1**, 184 ⟨88S854⟩; **4**, 181, 182 ⟨88S854⟩
Werst, G., **3**, 450 ⟨55CB1956⟩
Werstiuk, N. H., **1**, 631 ⟨72JA7081⟩, 868 ⟨83T206⟩, 905 ⟨83CJC2403⟩; **3**, 21, 137 ⟨83T205⟩, 168 ⟨B-87MI 304-02⟩, 344, 345 ⟨92CJC974⟩
Werstmijze, H., **4**, 940 ⟨84JOM(276)317⟩
Wert, K., **6**, 632 ⟨87JHC275⟩
Wertheim, B., **2**, 216 ⟨69BBR(37)593⟩
Werthemann, L., **1**, 414, 416, 827 ⟨70JA741⟩; **5**, 159 ⟨70JA741⟩
Wesche, H., **2**, 1076 ⟨23LA(432)39⟩
Wesdorp, J. C., **4**, 843 ⟨74RTC184⟩
Wesemann, J., **2**, 821, 822 ⟨92CB2187⟩
Wesolek, M., **3**, 515 ⟨82CC514⟩
Wess, J., **6**, 444 ⟨89AF539⟩
Wesseler, E. P., **4**, 743 ⟨72JOC683⟩
Wessells, F., **5**, 276 ⟨66JOC3829⟩
Wessels, P. L., **6**, 138 ⟨76T2625⟩
Wessely, H.-J., **2**, 507 ⟨82ZAAC(486)70⟩, 508, 509 ⟨80ZAAC(462)113⟩; **4**, 549 ⟨81ZAAC(479)41⟩; **5**, 882 ⟨81ZAAC(479)41⟩, 892 ⟨81ZN(B)1080⟩, 1086 ⟨83PS(18)7⟩, 1086, 1087, 1092 ⟨83JA7194⟩
Wessely, H.-J., **6**, 582, 684 ⟨85USP4558160⟩
Wessjohann, L., **3**, 44, 187 ⟨89S359⟩
Wessling, M., **2**, 318 ⟨91CB2303⟩
West, A. A., **6**, 563 ⟨88JCS(D)2363⟩
West, B. O., **2**, 483 ⟨62AJC710⟩, 486 ⟨56JCS3631, 63AJC636⟩, 507 ⟨70AJC1559⟩; **6**, 49 ⟨63AJC722⟩
West, C. T., **5**, 282 ⟨74JOC2740⟩
West, F. G., **1**, 400 ⟨92JOC3479, 93JA1177⟩, 823 ⟨83JOC4773⟩; **2**, 324 ⟨86CRV941⟩; **3**, 496 ⟨86CRV941⟩; **4**, 517 ⟨83JOC4773⟩
West, J. A., **2**, 738, 751 ⟨63JCE194⟩
West, J. P., **6**, 221 ⟨51USP2553800⟩
West, J. R., **5**, 730 ⟨73JOC3951⟩
West, K. F., **3**, 275 ⟨87JOC2537⟩
West, P. J., **3**, 366 ⟨72CC632, 73JCS(P1)1580⟩; **5**, 560 ⟨73JCS(P1)1571⟩
West, P. R., **4**, 314 ⟨69JOC3230⟩
West, R., **1**, 907 ⟨64JA1459⟩; **2**, 131 ⟨72JA4784⟩, 1006 ⟨64JA1459⟩, 1058 ⟨71JA1714⟩; **3**, 502 ⟨81JA1845, 83JA6176⟩, 502 ⟨84JA525⟩, 565, 711 ⟨70JOM(25)385⟩, 568 ⟨71JA1714⟩; **4**, 4, 22, 23, 30 ⟨66JA2481⟩, 358, 371 ⟨74JA3227⟩, 359, 362, 371 ⟨74JA3214⟩, 368 ⟨72JA4784⟩, 368, 371 ⟨74JA3222⟩, 516 ⟨70JOM(25)385⟩, 604, 648 ⟨71JA1714⟩, 605 ⟨76JA8412⟩, 668 ⟨53JOC1739⟩, 1058 ⟨84AG(E)454⟩; **5**, 1019 ⟨88JA7228⟩, 1020 ⟨87OM2617⟩; **6**, 125 ⟨91OM3466⟩, 162 ⟨70JOM(25)385⟩, 185, 206 ⟨65JA3788, 73JA3324⟩, 206 ⟨69JA6156, 74MI 606-01⟩, 206 ⟨76JA8426⟩
West, S. D., **2**, 439 ⟨90JOM(393)143⟩
West, W., **2**, 539 ⟨82S1⟩; **4**, 834 ⟨77S247⟩; **6**, 43 ⟨82S1⟩
Westcott, N. D., **2**, 16 ⟨72CC122⟩
Westcott, S. A., **2**, 43 ⟨93CJC930⟩, 527 ⟨92JA9350⟩
Westdorp, I., **1**, 688 ⟨82JOC248⟩
Westeppe, U., **2**, 558 ⟨90CB1517⟩
Westera, G., **6**, 414 ⟨81MI 614-01⟩
Westerback, S. K., **4**, 473 ⟨65JA2567⟩
Westerhaus, A., **4**, 112 ⟨83CA(98)16507e⟩; **5**, 887 ⟨80AG(E)556⟩, 1154 ⟨81AG(E)197, 81TL2159⟩; **6**, 288 ⟨83CB114⟩, 288, 693 ⟨82TL2017⟩, 293, 693 ⟨82AG(E)80⟩, 688 ⟨81TL2159⟩, 688, 689 ⟨81TL4957⟩
Westerhausen, M., **4**, 568 ⟨87PS(30)349⟩, 607, 656 ⟨91JOM(421)175⟩; **5**, 396 ⟨92ZAAC(608)33⟩, 396, 400 ⟨92ZAAC(607)101⟩, 396, 879, 891 ⟨91ZAAC(605)7⟩, 777 ⟨92ZN(B)453⟩, 879, 880 ⟨85MI 502-01, 92ZAAC(612)72⟩, 1155 ⟨85MI 531-01⟩; **6**, 190, 192, 195, 393 ⟨91JOM(421)175⟩, 195 ⟨93ZAAC(619)675, 94JOM(469)135⟩
Westerhuis, E., **6**, 301 ⟨52RTC1071⟩
Westerman, P. W., **1**, 58, 841 ⟨72JA7859⟩, 895 ⟨74JA3548⟩, 904 ⟨74JA5855⟩
Westermann, H., **2**, 472 ⟨91MI 208-02⟩
Westermann, J., **1**, 127 ⟨80CC1202⟩, 186 ⟨81CC237⟩; **2**, 566 ⟨80AG(E)1011, 80AG(E)900⟩
Westfelt, L., **3**, 293 ⟨81JOC5414⟩
Westheimer, F. H., **1**, 883 ⟨63PCS253⟩; **2**, 694 ⟨58JA5541⟩; **3**, 406 ⟨71JOC1570⟩, 479 ⟨81JOC1891⟩; **4**, 189 ⟨60CI(L)656⟩
Westheimer, F. J., **3**, 476 ⟨68JA4088⟩, 482 ⟨76JA7327⟩
Westlake, D. J., **5**, 452 ⟨71JCS(A)367⟩
Westlake, H. E., **2**, 227 ⟨42JA149⟩
Westlake, P. J., **1**, 931 ⟨92T4027⟩
Westlake, S., **2**, 843, 846 ⟨82T1511⟩, 862 ⟨82T3073⟩
Westland, R. D., **2**, 129 ⟨72JMC968⟩; **3**, 468 ⟨54JA2884, 54JA2887⟩

Westling, M., **1**, 524 ⟨86JOC1159⟩; **3**, 707 ⟨87JA590, 92JA4089⟩, 707 ⟨92T2113⟩
Westman, T. L., **5**, 144 ⟨63JOC3512⟩
Westman, T. R., **1**, 777 ⟨81JOC2804⟩
Westmijze, H., **1**, 496 ⟨76TL947⟩, 514 ⟨79S430⟩, 536 ⟨82TL2797⟩, 541 ⟨77TL2023⟩, 544 ⟨79TL3327, 81JOM(206)257⟩, 967 ⟨79S390⟩, 975 ⟨83JOC1103⟩, 989 ⟨75TL2923⟩, 1004, 1006, 1025 ⟨82RTC405⟩, 1058 ⟨81S551⟩, 1094 ⟨77TL1823⟩; **2**, 589 ⟨79JOM(172)377⟩, 616, 624 ⟨77RTC168⟩, 924, 925 ⟨77TL1823⟩, 940 ⟨79S390⟩; **3**, 642 ⟨77S784, 78S454⟩, 642 ⟨79S430⟩; **4**, 704 ⟨92S432⟩, 809 ⟨77TL1823⟩, 963 ⟨75TL2923⟩; **5**, 548, 550, 559, 560 ⟨79S432⟩, 554 ⟨80TL1771⟩
Westmoreland, D. L., **2**, 511 ⟨88JOM(341)C1, 88POL77⟩, 511 ⟨90CM221⟩, 893 ⟨90AX(C)774⟩
Weston, A. F., **2**, 214, 215 ⟨73TL655⟩, 358 ⟨72JOM(44)37⟩
Weston, A. W., **3**, 105 ⟨50JA1422, 63OSC(4)915⟩
Weston, J. B., **2**, 625 ⟨87TL3857⟩
Westphal, B., **4**, 690 ⟨93JOM(459)325⟩
Westphal, G., **2**, 726 ⟨69ZC305⟩
Westphal, O., **2**, 373, 378, 380, 382 ⟨41CB759⟩, 378, 380, 383 ⟨41CB1365⟩, 388 ⟨41CB774⟩
Westrum, L. J., **1**, 134 ⟨93TL7701⟩
Westwell, A. D., **2**, 147 ⟨92TL7237⟩
Westwood, D., **2**, 147 ⟨90SL457⟩, 749 ⟨88JCS(P1)269⟩; **3**, 43, 182 ⟨87CC775⟩; **4**, 261 ⟨90SL457⟩, 263 ⟨89JCS(P1)2441⟩, 849 ⟨88JCS(P1)269⟩
Westwood, N. P. C., **5**, 1152 ⟨78JA446⟩, 1153 ⟨79JCS(D)1405⟩; **6**, 679 ⟨78JA446⟩
Westwood, R., **5**, 1123 ⟨92T7703⟩; **6**, 111 ⟨90TL123⟩
Westwood, S., **5**, 177 ⟨87JOC3937⟩
Westwood, S. W., **1**, 245, 1127 ⟨88JA2565⟩
Wetlaufer, D. B., **5**, 85 ⟨62MI 502-01⟩
Wettach, R. H., **2**, 1020 ⟨81JOC4324⟩
Wetter, A., **2**, 806 ⟨04CB629⟩
Wetter, H., **1**, 827, 828 ⟨79HCA1922⟩; **2**, 262, 537, 539 ⟨85TL5515⟩, 537, 538, 539 ⟨85TL5511⟩; **5**, 224, 375 ⟨85TL5515⟩
Wetterholm, A., **6**, 144, 150 ⟨63T(S)155⟩
Wetterich, F., **1**, 327 ⟨92TL2673, 93JA8585⟩, 327 ⟨93TL5885⟩, 923 ⟨93JOC1083⟩
Wettermark, U. G., **2**, 466, 467 ⟨84IC2062⟩; **4**, 129 ⟨87PS(33)147⟩, 570 ⟨87OM959⟩, 572 ⟨84IC2063⟩; **5**, 887 ⟨87PS(30)463⟩
Wettling, T., **2**, 449 ⟨91AG(E)207, 91SL433⟩; **4**, 549 ⟨93PS(76)1⟩; **5**, 1152 ⟨92AG(E)758⟩; **6**, 153 ⟨89AG(E)1013⟩, 153 ⟨91AG(E)207⟩, 153 ⟨92AG(E)758⟩
Wettstein, A., **2**, 95 ⟨61E475⟩
Wetzel, C. R., **5**, 711 ⟨65JOC3724⟩
Wetzel, D. M., **1**, 97 ⟨84JOC1544⟩
Wetzel, J. M., **4**, 182 ⟨85JOC3946⟩, 182 ⟨87JOC188⟩
Weuter, W., **6**, 243 ⟨89JOM(378)125⟩
Wevers, J. H., **2**, 360 ⟨75CC413⟩
Wewers, D., **4**, 551 ⟨84ZN(B)1518⟩, 580 ⟨84CB1103, 85CB3570⟩, 580 ⟨85CB541, 85ZN(B)968⟩, 1029 ⟨85CB3560⟩; **6**, 152 ⟨84ZN(B)1518⟩
Wexler, R. R., **4**, 780 ⟨91JMC2525⟩
Weyenberg, D. R., **4**, 604 ⟨66JOC2036⟩, 610 ⟨65JOC2618⟩
Weyerstahl, P., **3**, 349 ⟨75CB673⟩; **4**, 35 ⟨71TL3869⟩, 36 ⟨73TL611⟩, 759, 764 ⟨87JA6687⟩; **5**, 184 ⟨85LA1935⟩
Weygand, F., **1**, 393 ⟨60AG535⟩, 395 ⟨59CB528⟩; **2**, 270 ⟨62ZN(B)807⟩; **3**, 531, 532 ⟨59CB528⟩; **4**, 69 ⟨55ZN296, 60CB2340⟩, 297, 327 ⟨66CB1944⟩, 305 ⟨65CB487⟩, 309 ⟨66CB1932⟩, 320 ⟨62LA(658)128⟩; **5**, 185, 190 ⟨69ZN(B)314⟩, 271, 292 ⟨59CB528⟩, 304 ⟨55CB301⟩
Weyler, W., **1**, 571 ⟨88OSC(6)210⟩; **3**, 539 ⟨76OS(55)32⟩
Weyler, Jr., W., **3**, 546 ⟨70JA4132, 71JA2812⟩
Weymuth, C., **1**, 546, 946 ⟨87PAC363⟩
Weyna, P. L., **6**, 76 ⟨59JA2579⟩
Whale, T. W., **2**, 250 ⟨78JOC4659⟩
Whalen, D. L., **1**, 797 ⟨76JA7319⟩, 877 ⟨88JA1981, 91JA1344⟩, 878 ⟨93JOC7100⟩
Whalen, R., **2**, 20 ⟨71JA438⟩
Whaley, A. M., **2**, 25 ⟨41OSC(1)142⟩; **6**, 16 ⟨41JA3478⟩
Whalley, M., **5**, 199 ⟨54JCS3722⟩
Whalley, T. W., **2**, 873 ⟨B-86MI 217-01⟩
Whalley, W., **1**, 412, 813 ⟨70TL509⟩
Whalley, W. B., **1**, 197 ⟨76JCS(P1)1384⟩; **4**, 420, 421, 426 ⟨55JA745⟩; **6**, 217 ⟨47CI(L)427⟩
Whalon, M. R., **5**, 135 ⟨90JA2849⟩
Whang, E. E., **2**, 1009 ⟨89JA1299⟩
Whangbo, M. H., **5**, 1097 ⟨80CC509⟩
Wharton, P. S., **1**, 812 ⟨68JOC4258, 73JOC4117⟩; **4**, 827 ⟨58T321⟩
Wheatland, D. A., **4**, 557 ⟨72IC2340⟩
Wheaton, G. A., **4**, 735 ⟨76TL895, 83JOC917⟩, 737 ⟨78JOC2643⟩; **6**, 224 ⟨78JOC2643⟩
Wheeler, A. S., **5**, 549 ⟨28JA3106⟩
Wheeler, D. M. S., **1**, 1038, 1140 ⟨91TL1707⟩
Wheeler, G. L., **2**, 1004 ⟨73JA6158⟩
Wheeler, H. S., **5**, 813 ⟨1897JA129⟩
Wheeler, K. J., **5**, 288 ⟨85JCS(P1)831⟩
Wheeler, K. W., **3**, 616 ⟨46JA847⟩
Wheeler, M., **2**, 655 ⟨81TL3451⟩
Wheeler, M. M., **1**, 1038, 1140 ⟨91TL1707⟩
Wheeler, O. H., **1**, 731 ⟨65JOC1473⟩; **3**, 494 ⟨65JOC1473⟩; **5**, 305, 307 ⟨B-70MI 506-02⟩
Wheeler, R. A., **3**, 545 ⟨89JA9057⟩
Wheeler, T. N., **1**, 549 ⟨87JOC4875⟩; **5**, 989 ⟨87JOC4875⟩; **6**, 695 ⟨84JOC706⟩
Wheeler, T. S., **3**, 241 ⟨55JCS862⟩
Whelan, J., **6**, 472, 473 ⟨86IJ250⟩
Wheland, G. W., **1**, 379, 380 ⟨B-60MI 109-01⟩, 848 ⟨47JA2025⟩, 848, 869 ⟨32JA1212⟩, 870 ⟨B-60MI 119-04⟩
Wheland, R. C., **2**, 16 ⟨85JOC3698⟩; **4**, 869 ⟨76JA3916⟩; **5**, 203 ⟨85JOC3698⟩
Whetstone, R. R., **4**, 836, 837, 838 ⟨68JOC4470⟩
Whimp, P. O., **4**, 688 ⟨76IC97⟩
Whipple, G. H., **6**, 212 ⟨58GEP1040011⟩
Whipple, W. L., **1**, 1047 ⟨91JOC2911⟩; **3**, 131, 305 ⟨91JOC2911⟩; **5**, 331 ⟨91JOC2911⟩
Whistler, R. L., **2**, 7 ⟨78CAR(61)511⟩, 129 ⟨69JOC97⟩; **4**, 238 ⟨82H(19)1719⟩
Whitaker, C. R., **4**, 626, 662 ⟨87JCS(D)1647⟩
Whitaker, K. E., **3**, 468 ⟨71JCS(C)1384⟩, 667 ⟨70JOC30⟩
Whitby, R. J., **1**, 481 ⟨92JCS(P1)3419⟩; **3**, 324 ⟨87JCS(P1)195⟩, 597 ⟨92CC7⟩; **4**, 212 ⟨87JCS(P1)195⟩
Whitcher, W. J., **4**, 761, 765 ⟨42JA2635⟩
Whitcombe, G. P., **3**, 5, 117 ⟨87CC1625⟩
Whitcombe, M. J., **1**, 41 ⟨80TL4031⟩
White, A. C., **2**, 346 ⟨91TL6789⟩
White, A. D., **3**, 5, 117 ⟨87CC1625⟩
White, A. H., **2**, 690 ⟨76JCS(D)2307⟩, 1003, 1005 ⟨89JCS(D)105, 82CC14⟩, 1005 ⟨84JCS(D)1801, 84JCS(D)321, 85JCR(S)330⟩; **3**, 545 ⟨84CC220⟩; **4**, 584 ⟨81CC698, 82JOM(228)C71⟩, 602, 650 ⟨82CC1323⟩, 603, 647 ⟨84JCS(D)311⟩, 604 ⟨83CC1419⟩, 604, 646, 647 ⟨82CC14, 84JCS(D)1801, 84JCS(D)321⟩, 604, 647 ⟨88JCS(D)2403⟩, 604, 647, 648 ⟨90JCS(D)1161⟩, 626, 662 ⟨87JCS(D)1647⟩, 646 ⟨89JCS(D)105⟩, 646, 647 ⟨84CC1708, 85JCS(D)337⟩
White, A. H., **4**, 647, 656 ⟨86JCS(D)603⟩, 647, 660 ⟨92JCS(D)775⟩, 654 ⟨87OM2110, 87JCS(D)3085⟩, 660 ⟨82JOM(233)C28, 84JCS(D)331⟩, 678 ⟨94JOM(464)191⟩, 856 ⟨80AJC1073⟩; **5**, 1127 ⟨85JA3910⟩; **6**, 168 ⟨91CC1560⟩, 174, 190 ⟨82CC1323⟩, 190 ⟨84CC1708, 84JCS(D)1801⟩, 190, 195 ⟨83CC1419⟩, 190, 196 ⟨86CC672⟩, 195 ⟨84CC612⟩, 196 ⟨87AG(E)681, 88CC336, 93JCS(D)2653⟩, 589, 593 ⟨88AJC549⟩

White, A. M., **1**, 894 ⟨72JA2044⟩, 894, 899 ⟨68JA1884, 69JA5801⟩, 906 ⟨69JA3958, 70JA4627⟩, 909 ⟨69JA3667⟩; **2**, 95, 96 ⟨53JCS1541, 54JCS2200⟩, 1007 ⟨69JA5801⟩
White, A. W., **1**, 120 ⟨88JA2218⟩; **2**, 715 ⟨84S215⟩
White, C. A., **5**, 946, 957 ⟨93JA3806⟩
White, C., **1**, 93 ⟨77CC427, 78JCS(D)617⟩; **6**, 247 ⟨91CC165⟩, 517 ⟨93OM2401⟩
White, C. M., **6**, 529, 535 ⟨74IC1778⟩
White, C. R. H., **4**, 970, 977, 978 ⟨80EUP10396⟩
White, D. A., **2**, 395 ⟨73TL3597⟩
White, D. H., **1**, 690 ⟨84JA5018, 86JA2932⟩; **2**, 13 ⟨83JOC4158, 86CJC138⟩
White, D. K., **6**, 445 ⟨78JOC4530⟩
White, D. M., **5**, 375, 378, 718 ⟨66JA3390⟩; **6**, 143 ⟨66JA3825⟩
White, D. R., **4**, 179 ⟨79JA2171⟩
White, E. H., **1**, 28, 32, 35 ⟨B-68MI 102-01⟩, 624 ⟨B-68MI 114-01⟩, 627 ⟨B-68MI 114-02⟩, 637 ⟨B-68MI 114-03⟩; **2**, 176 ⟨87JOC2162⟩; **3**, 471 ⟨64JOC2070⟩; **5**, 36, 108 ⟨55JA6011⟩, 361 ⟨61JA1191, 66JOC3038⟩
White, E. W., **4**, 116 ⟨78JFC(11)441⟩, 117 ⟨77JFC(10)27⟩
White, F. H., **1**, 74 ⟨83JOC1638⟩, 149 ⟨88JA4778⟩
White, F. L., **6**, 70 ⟨35JA2480⟩
White, G. R., **6**, 647 ⟨77JMC901⟩
White, J., **3**, 307 ⟨77JOC4248⟩; **6**, 364 ⟨90S151⟩
White, J. B., **1**, 815 ⟨82JOC3190⟩
White, J. D., **1**, 13 ⟨87JA4424⟩, 216 ⟨93JA2970⟩, 357 ⟨68JA6171⟩, 382 ⟨81JOC3376⟩, 457, 508 ⟨78JOC4555⟩, 573 ⟨92JA9673⟩, 582 ⟨92JOC2270⟩, 829 ⟨89JOC736⟩, 1140 ⟨88TL2923⟩; **3**, 496 ⟨92TL577⟩; **5**, 177 ⟨77JA646⟩, 351 ⟨90JOC6037⟩
White, J. E., **2**, 172 ⟨73PNA(70)328⟩
White, J. F., **1**, 668, 669 ⟨73JA4451, 76JA6521⟩; **4**, 194 ⟨76JA6750⟩
White, K. S., **1**, 832 ⟨89JA8878⟩, 993 ⟨87JOC4031⟩
White, L. L., **1**, 205 ⟨83SC237⟩
White, L. R., **4**, 912 ⟨81MI 420-01⟩
White, M. R., **1**, 988, 1145 ⟨81JA5429⟩
White, P., **1**, 927 ⟨90JA6842⟩
White, P. S., **5**, 1097 ⟨91JA2324⟩
White, R. A., **5**, 1035 ⟨54CI(L)785, 56JCS659⟩, 1035 ⟨63OSC(4)700⟩
White, R. E., **5**, 317 ⟨75JA1180⟩
White, R. W., **1**, 107 ⟨56JA3484⟩
White, S., **1**, 152 ⟨81JA3081, 81JA3088⟩; **3**, 124 ⟨81JA3081⟩
White, T. G., **5**, 1121 ⟨81EJM175⟩; **6**, 646, 647 ⟨75ACR62⟩
White, W. A., **3**, 406, 407 ⟨67JOC3246⟩; **4**, 405 ⟨66JOC4041, 67JOC213⟩, 972 ⟨66JA850⟩; **6**, 141 ⟨66JA850, 66JOC2874⟩, 361 ⟨66JA2885, 70USP3551418⟩
White, W. N., **1**, 99 ⟨56JA4604⟩, 383 ⟨59JA4390⟩, 826 ⟨70JOC3585⟩
Whitear, B., **1**, 862 ⟨61PCS117⟩
Whitear, B. R. D., **1**, 537 ⟨67JCS(C)1986⟩
Whitehead, B. J., **4**, 558 ⟨92JCS(P1)479⟩
Whitehead, C. W., **5**, 584, 586 ⟨48JA2837⟩; **6**, 104 ⟨53JA671⟩, 504 ⟨63OSC(4)515⟩
Whitehead, J. F., **1**, 173 ⟨90JCS(P1)2352⟩; **2**, 612 ⟨88TL3365⟩; **4**, 61, 73, 86 ⟨88TL3365⟩
Whitehouse, M. W., **5**, 62 ⟨58MI 502-02⟩
Whitehouse, R. D., **1**, 601 ⟨70CC86⟩
Whitehurst, D. D., **3**, 541 ⟨64JA4871⟩
Whitehurst, J. S., **1**, 826 ⟨90JCS(P1)11⟩; **4**, 14 ⟨90JCS(P1)3317⟩
Whiteker, G. T., **3**, 356 ⟨91TL5019⟩
Whiteley, C. G., **2**, 522 ⟨85JOC1969⟩
Whiteley, M., **5**, 944 ⟨94JCS(D)1671⟩
Whiteley, R. H., **4**, 685, 706, 719 ⟨79JOM(181)425⟩
Whiter, P. F., **1**, 965 ⟨66JCS(C)1223⟩
Whitesell, J. K., **1**, 839 ⟨91JA3526⟩; **2**, 152, 721 ⟨91JOC4552⟩, 245 ⟨85CC1449, 90JA7653⟩, 245, 331 ⟨91JA3526⟩, 751 ⟨83S517⟩, 976 ⟨83JOC1556⟩; **3**, 10, 124, 404 ⟨83S517⟩, 407 ⟨77JOC377⟩
Whitesell, M. A., **2**, 751 ⟨83S517⟩; **3**, 10, 124, 404 ⟨83S517, 77JOC377⟩; **5**, 789 ⟨83TL1679⟩
Whitesides, G. M., **1**, 67 ⟨70JA6611, 74JA870⟩, 118, 119, 469, 473 ⟨69JA4871⟩, 119, 312 ⟨74JOC400⟩, 311 ⟨66JOC3128⟩, 465, 1090 ⟨76OS(55)103⟩, 491, 1105 ⟨71JA1379⟩, 668 ⟨70JA1426, 72JA5258⟩, 668, 669 ⟨73JA4451, 76JA6521⟩, 86OM1473⟩, 700, 703 ⟨87JOC2838⟩, 1010 ⟨88JOC2489⟩; **2**, 124 ⟨93JOC633⟩, 127 ⟨91JOC7328⟩, 589 ⟨74JA2806⟩, 596 ⟨79JOC3603⟩, 597 ⟨94JOC3714⟩, 958 ⟨80JA217, 80JA226, 80JA231, 80JA239, 80JA2493⟩; **3**, 48 ⟨91JOC3849⟩, 113 ⟨75JOC3577⟩, 159 ⟨93JOC1835⟩, 287 ⟨69JA4871⟩; **5**, 138 ⟨84JA7250⟩, 436 ⟨73JA8118⟩, 441 ⟨76JA6529⟩
Whitesides, T. H., **5**, 652 ⟨71CC1514⟩
Whitfield, G. F., **2**, 160 ⟨74JOC2148⟩, 162 ⟨74JOC3365⟩, 478 ⟨71TL1761⟩; **5**, 350 ⟨71TL1761⟩
Whitfield, Jr., L. L., **6**, 558 ⟨85S423⟩
Whitfield, R. E., **1**, 881 ⟨66JOC989⟩
Whitham, G. H., **1**, 97 ⟨78JCS(P1)1479⟩, 231 ⟨85CC897⟩, 591 ⟨74JCS(P1)433, 88JCS(P1)813, 88JCS(P1)817⟩, 640, 641, 753 ⟨74CC142⟩, 649 ⟨76CC697⟩, 756 ⟨85JCS(P1)1953⟩, 797 ⟨66JCS(B)164⟩; **2**, 195 ⟨81CC741⟩, 195 ⟨81CC742⟩; **3**, 66 ⟨77JCS(P1)1346⟩; **4**, 139 ⟨88T4087⟩, 319 ⟨84JCS(P1)865⟩, 357 ⟨76CC697, 86JCS(P1)683⟩, 519 ⟨81JCS(P1)1934⟩, 843 ⟨84JCS(P1)859⟩; **5**, 431 ⟨85JCS(P1)409⟩, 552, 559 ⟨84JCS(P1)865⟩, 558, 559 ⟨84JCS(P1)859⟩, 709 ⟨64JCS1189⟩, 1125 ⟨82CC335⟩; **6**, 559 ⟨88JCS(P1)813⟩
Whiting, A., **1**, 489 ⟨93TL3599⟩; **2**, 530 ⟨93T177⟩; **3**, 204 ⟨93T177⟩; **4**, 448 ⟨84CC1289⟩
Whiting, D. A., **2**, 660 ⟨91COS(3)803⟩; **3**, 317 ⟨90JCS(P1)153⟩, 318 ⟨87TL3155, 90JCS(P1)1193⟩; **5**, 327 ⟨93T9137⟩
Whiting, J., **2**, 366 ⟨72CJC3502⟩
Whiting, M. C., **1**, 80 ⟨54JCS3197⟩, 513 ⟨54JCS1854, 54JCS3217⟩, 631 ⟨59T305⟩, 862 ⟨52JA2125, 52JA3458⟩, 959 ⟨60JCS341⟩, 1035 ⟨52JCS2014, 54JCS147⟩, 1046 ⟨58JCS1054⟩, 1066 ⟨53OS(33)68⟩, 1069 ⟨52JCS2010⟩, 1120 ⟨67JCS(B)991⟩; **2**, 25, 31 ⟨84TL5949⟩, 402 ⟨78JCS(P1)1580⟩; **3**, 63 ⟨56JCS4082⟩, 79 ⟨58JCS1054⟩, 444 ⟨70OS(50)3⟩, 463 ⟨66JCS(C)467⟩; **5**, 74, 102, 103 ⟨58JCS1313⟩, 102 ⟨56CI(L)548⟩, 106 ⟨63CI(L)775⟩, 106, 298 ⟨65JCS1290⟩, 168 ⟨59JCS1313⟩, 189 ⟨54JCS1860⟩
Whitla, W. A., **2**, 476 ⟨77JA7379⟩
Whitlock, B. J., **5**, 70 ⟨74JOC3144⟩
Whitlock, H. W., **1**, 266 ⟨67JA3600⟩; **4**, 421 ⟨76JOC3050⟩
Whitlock, Jr., H. W., **1**, 872 ⟨69JOC1962⟩, 1142 ⟨89JA3640⟩; **5**, 70 ⟨74JOC3144⟩
Whitman, G. H., **1**, 871 ⟨68JCS(C)338⟩
Whitman, G. M., **3**, 614 ⟨54JA5364⟩
Whitman, M. D., **2**, 340 ⟨90JOC1981⟩
Whitman, P. J., **5**, 971 ⟨74JOC2897⟩
Whitmarsh, C. K., **4**, 606, 654 ⟨91OM1336⟩
Whitmire, K., **6**, 340 ⟨80CC778, 80CC780⟩
Whitmire, K. H., **2**, 880 ⟨91JOM(402)55⟩; **6**, 247 ⟨90IC3252⟩, 340 ⟨82JA5621, 86IC2472⟩
Whitmore, F. C., **1**, 128 ⟨29JA1491⟩, 301 ⟨41JA758⟩, 450 ⟨54JA1613⟩, 648 ⟨47JA1551⟩, 652 ⟨48JA2869⟩; **2**, 88 ⟨41JA1118⟩, 187 ⟨48OSC(1)492⟩, 393 ⟨41OSC(1)401⟩, 537 ⟨47JA1976⟩, 560 ⟨32JA3714⟩, 594 ⟨48OSC(1)519⟩; **3**, 612 ⟨33JA4161⟩, 638 ⟨44JA725, 45JA1071⟩; **4**, 137 ⟨47JA1976⟩; **5**, 115 ⟨32OSC(1)159⟩, 421 ⟨54JA1372⟩
Whitmore, M. C., **2**, 393 ⟨41OSC(1)401⟩
Whitmore, W. G., **2**, 1000 ⟨69AX(B)73⟩
Whitney, C. C., **1**, 278, 487, 661, 1103 ⟨67JA3652⟩, 278,

487, 662 ⟨71JA6309⟩, 278, 488, 531, 662, 1103 ⟨68JA6243⟩, 278, 531, 662, 1103 ⟨72JA6560⟩, 534 ⟨68JA7139⟩; **2**, 518 ⟨66JOC4178⟩, 629 ⟨67JA2753⟩, 900 ⟨67JA3652, 68JA6243⟩; **5**, 1136 ⟨72JA6560⟩
Whitney, R. A., **1**, 67 ⟨86JOC2024⟩
Whitney, S. E., **1**, 120 ⟨86T2873⟩; **2**, 588 ⟨86T2873⟩
Whittaker, D., **1**, 891 ⟨81MI 119-02⟩; **2**, 89 ⟨69JCS(B)350⟩
Whittaker, G., **4**, 737 ⟨78CC475⟩; **6**, 28 ⟨82EUP65358⟩
Whittaker, M., **4**, 200 ⟨93SL111⟩
Whittaker, M. R., **1**, 209 ⟨93JOC7304⟩
Whittaker, N., **1**, 726 ⟨61MI 116-01⟩, 747 ⟨63JCS1461⟩
Whittaker, S. M., **6**, 380, 381 ⟨93TH61301⟩
Whittall, N., **4**, 419 ⟨91TL2683⟩
Whitten, C. E., **1**, 120 ⟨73JA7788⟩; **2**, 588 ⟨73JA7788⟩; **3**, 130 ⟨72JA5106, 88OSC(6)248⟩, 286 ⟨70TL4647⟩, 287 ⟨73JA7788, 73TL1815⟩
Whitten, J. P., **3**, 673 ⟨86JOC3228⟩; **5**, 1127 ⟨85TL6273⟩; **6**, 10, 84 ⟨86TL4861⟩
Whittern, D., **1**, 769 ⟨90CAR(201)185⟩
Whittle, B. J. R., **5**, 342 ⟨89JCS(P1)1815⟩
Whittle, C. P., **3**, 445 ⟨67AJC339⟩
Whittle, E. L., **3**, 468 ⟨54JA2884, 54JA2887⟩
Whittle, R. R., **2**, 244, 245 ⟨84JA7867, 84JA7861⟩, 245 ⟨84JOC3243⟩; **3**, 45 ⟨84JA5598⟩; **4**, 678 ⟨84JOM(273)333⟩, 706 ⟨84JA2337⟩
Whittlesey, B. R., **2**, 512 ⟨86JA1703, 87IC4065⟩, 866 ⟨83CC881⟩, 866, 893 ⟨84IC2582⟩; **6**, 167 ⟨84TL3519⟩, 168 ⟨84IC2582⟩
Whittlesley, B. R., **6**, 169 ⟨83CC881⟩
Whitworth, S. M., **1**, 919 ⟨90JA8135⟩
Whörle, C., **1**, 499 ⟨93TL8431⟩
Whybrow, D., **3**, 232 ⟨81JCS(P1)3147⟩
Wiatkowski, G. K., **1**, 752 ⟨88JA5653⟩
Wiatschka, A., **5**, 586, 609 ⟨82M(113)593⟩
Wiaux-Zamar, C., **3**, 259 ⟨76AG(E)371⟩, 529 ⟨81JA4616⟩; **4**, 1016, 1017 ⟨76AG417⟩
Wibberley, D. G., **4**, 413 ⟨71JCS(C)780⟩
Wiberg, E., **1**, 665 ⟨52CB593⟩; **2**, 110 ⟨31ZAAC(202)1, 31ZAAC(202)22, 53ZN608⟩, 111 ⟨35ZAAC(222)92⟩, 487 ⟨56ZN(B)750⟩, 879 ⟨56ZN(B)751, 56ZN(B)753⟩
Wiberg, K. B., **1**, 809 ⟨85JOC3393⟩, 841 ⟨73TL2727⟩, 894 ⟨68JA63⟩, 919, 923 ⟨89CRV975⟩, 928 ⟨86TL1553⟩; **2**, 556 ⟨91JA5720⟩; **5**, 275 ⟨55JA2519⟩, 709 ⟨53JA3961⟩
Wiberg, N., **3**, 483 ⟨81CB3518⟩, 501 ⟨84JOM(273)141⟩; **4**, 699 ⟨72JOM(36)267, 72JOM(36)277⟩, 975, 985 ⟨68AG(E)766⟩; **5**, 1076, 1077, 1083 ⟨78ZN(B)575⟩; **6**, 163 ⟨87CB1357⟩, 172 ⟨91MI 606-02⟩, 174, 289, 291, 392, 709, 710 ⟨86CB1455⟩, 177, 178, 382, 387, 396, 709, 710 ⟨87ZN(B)1062⟩, 177, 178, 392, 709, 710 ⟨86JOM(315)9⟩, 177, 180, 187, 193, 382, 709, 711 ⟨84JOM(273)141⟩, 177, 380, 382, 392, 673, 709, 710 ⟨81CB3518⟩, 178, 387, 392, 709 ⟨83AG(E)1005⟩, 178, 387, 673, 710 ⟨86CB1467⟩, 186, 188, 292, 388, 397, 711 ⟨86CB2966⟩, 188, 388, 397, 673, 711 ⟨86CB2980⟩, 188, 673, 711 ⟨87CB1203⟩, 193, 399 ⟨91AG(E)93⟩, 290, 291 ⟨87CB653⟩, 290, 291, 380, 381, 392, 709, 710 ⟨81CB2087⟩, 290, 291, 393, 709, 710 ⟨89CB409⟩, 373, 380, 382, 392, 709, 710 ⟨81CB3505⟩, 382, 709, 710 ⟨87ZN(B)1055⟩, 383 ⟨89ZN(B)796⟩, 392, 709, 710 ⟨77AG(E)328⟩, 392, 710 ⟨84JOM(271)381, 87OM35⟩, 608 ⟨71CB3989, 76ZN(B)1317⟩, 617 ⟨77ZN(B)1003⟩, 673 ⟨86CC591, 87CB1213, 88CB1407⟩, 709 ⟨87CB1605⟩, 709, 710 ⟨88ZN(B)1468, 91CB1981, 85AG(E)229⟩, 710 ⟨87OM32⟩
Wicha, J., **1**, 210 ⟨93TA399⟩, 697, 698, 700, 703 ⟨85TL5597⟩; **3**, 294 ⟨73TL3635⟩; **4**, 393 ⟨91CC297⟩; **5**, 49 ⟨69JOC116⟩
Wichmann, B., **2**, 569, 570, 571 ⟨93CB2093⟩
Wichrowska, N., **4**, 715 ⟨90CJC41⟩
Wichterle, O., **2**, 782 ⟨59CCC1158⟩
Wick, A., **1**, 566 ⟨75HCA153⟩, 683 ⟨75HCA1450⟩
Wick, A. E., **1**, 414, 415, 827 ⟨64HCA2425⟩, 415 ⟨69HCA1030⟩; **4**, 882 ⟨64HCA2425⟩
Wick, A. K., **5**, 298 ⟨60HCA1473⟩
Wickberg, B., **4**, 299 ⟨77JCS(P1)684⟩
Wickenkamp, R., **2**, 910, 911, 916, 917, 932, 940, 946 ⟨87JOC4868⟩, 916, 930 ⟨85CC354⟩; **6**, 395, 399, 401, 405 ⟨85JOM(291)179⟩
Wickham, G., **1**, 67, 69 ⟨81JOC563⟩
Wickham, P. P., **1**, 1071 ⟨91JOC2045⟩
Wickramaratne, M., **1**, 828 ⟨90SC423⟩
Wickramasinghe, A., **2**, 687 ⟨90M709⟩
Wickramasinghe, J. A. F., **1**, 871 ⟨68JCS(C)338⟩
Wicks, G. E., **1**, 510 ⟨85JA3626⟩; **2**, 397 ⟨88JOC5353⟩
Wicks, Z. W., **2**, 33 ⟨46JA2402⟩
Widdowson, D. A., **1**, 473 ⟨86T2111⟩; **2**, 195, 688 ⟨70CC1466⟩, 613 ⟨90SL761⟩, 615 ⟨79CC914⟩, 711 ⟨91T8621⟩; **3**, 652, 654 ⟨73CC781⟩
Wideburg, N., **1**, 769 ⟨90CAR(201)185⟩
Widener, R. K., **2**, 525 ⟨87JA931⟩; **4**, 645, 1057 ⟨87JA931⟩
Widera, R., **5**, 657 ⟨84T405⟩
Widera, R. P., **3**, 695, 696, 700 ⟨88OSC(6)232⟩
Widiger, G. N., **1**, 503 ⟨73JOC3652⟩; **3**, 90 ⟨70JCS(C)227⟩
Widlanski, T. S., **4**, 769 ⟨94JA789⟩
Widler, L., **1**, 683 ⟨82HCA1972⟩; **4**, 396 ⟨81HCA357⟩; **6**, 134 ⟨81HCA357⟩
Widmer, E., **1**, 77 ⟨76HCA1832⟩, 1067 ⟨91TL4115⟩
Widmer, U., **1**, 809 ⟨73HCA75⟩; **2**, 660 ⟨73HCA75⟩; **5**, 126 ⟨83S135⟩
Wieber, G. M., **5**, 375 ⟨89JOC4649⟩, 953 ⟨90JA6255⟩
Wieber, M., **2**, 262 ⟨63CB1019, 63CB1561⟩, 262 ⟨63CB2822⟩, 476 ⟨68TL5333⟩, 477 ⟨74CB639⟩, 481 ⟨B-77MI 209-01, B-81MI 209-01, B-82MI 209-01, B-83MI 209-01, B-86MI 209-01⟩, 489 ⟨83ZAAC(505)134⟩, 493, 505, 886, 890 ⟨84ZN(B)887⟩, 496 ⟨78ZAAC(439)139⟩, 496, 501 ⟨76ZAAC(423)40⟩, 496, 695 ⟨75ZAAC(418)167⟩, 499, 891 ⟨84ZN(B)1668⟩, 500 ⟨73CB2738, 74ZAAC(408)261, 79ZAAC(448)89⟩, 695 ⟨75ZAAC(418)279⟩, 837 ⟨70M776⟩, 872 ⟨B-77MI 217-01⟩, 891 ⟨87ZN(B)695⟩, 1080 ⟨B-82MI 223-01⟩, 1082 ⟨B-81MI 223-01, B-81MI 223-02⟩; **4**, 114, 128 ⟨73CB2733⟩, 135 ⟨73CB2742⟩, 595 ⟨92ZAAC(607)157, 92ZAAC(607)164⟩, 599, 600 ⟨88ZN(B)739⟩; **6**, 547 ⟨85ZAAC(525)127⟩, 553 ⟨88PS(35)93, 88ZAAC(556)189⟩, 567 ⟨88ZAAC(556)179⟩
Wiechel, K.-H., **2**, 824 ⟨63LA(665)1⟩
Wiechell, H., **1**, 1144 ⟨1894LA(279)337⟩
Wiechert, R., **1**, 5, 9 ⟨83LA705⟩, 24, 144 ⟨91TL1949⟩, 555 ⟨89T6409⟩, 558 ⟨89JOC5831⟩, 1070 ⟨93TL5257⟩; **2**, 38 ⟨83AG(E)406⟩, 103 ⟨69T761⟩, 1018 ⟨84AG(E)539⟩; **3**, 64 ⟨66CB3057⟩; **6**, 25 ⟨58CB2664⟩
Wieckowski, T., **2**, 1007 ⟨91JCS(P2)1563⟩
Wieczorek, C., **5**, 928 ⟨90AG(E)398⟩; **6**, 713, 714 ⟨90AG(E)398⟩
Wieczorek, J., **4**, 13 ⟨70JOC508⟩
Wieczorek, J. J., **2**, 660 ⟨64JA742⟩
Wieczorek, J. S., **2**, 847, 848, 850 ⟨79S452⟩; **4**, 463 ⟨80JPR213⟩, 464 ⟨77S625⟩
Wieczorek, M., **2**, 184 ⟨89AG(E)97⟩
Wieczorek, M. W., **6**, 116 ⟨88JOC3609⟩, 124 ⟨91TL4189⟩
Wieczorkowski, J., **2**, 187, 199 ⟨60JCS1665⟩, 412 ⟨57RZC879⟩
Wiedenfeld, D., **1**, 927 ⟨91JA8977⟩
Wiedenmann, R., **5**, 605, 616, 634 ⟨68CB1428⟩
Wiederhold, M., **4**, 679 ⟨94JOM(476)101⟩
Wiedmann, H., **3**, 258 ⟨70M157⟩
Wiegand, N. H., **3**, 564 ⟨86JOC3176⟩
Wiegel, K., **4**, 607, 621 ⟨76ZAAC(419)157⟩
Wiegel, L. O., **3**, 83 ⟨76S767⟩
Wieghardt, K., **4**, 150 ⟨88IC3796⟩
Wiegman, A. M., **2**, 1030 ⟨59RTC252⟩
Wiegräbbe, W., **2**, 863 ⟨68CB1414⟩

Wiegrebe, L., **3**, 694 ⟨30CB1650⟩
Wiegrebe, W., **5**, 1127 ⟨81AP(314)577, 81MI 530-01⟩
Wieiler, M., **2**, 385 ⟨87LA583⟩
Wieland, D., **5**, 203 ⟨88JMC362⟩
Wieland, D. M., **1**, 217, 226 ⟨71JA3047, 73JOC4263⟩
Wieland, G., **6**, 643 ⟨85LA2178⟩
Wieland, H., **2**, 381, 382, 383 ⟨20CB1329⟩, 388 ⟨12LA(392)133⟩; **3**, 464 ⟨11LA(381)229⟩; **5**, 819 ⟨30LA(484)236⟩, 838 ⟨06CB1480⟩, 867 ⟨84AG(E)717⟩
Wieland, J. H. S., **3**, 493 ⟨61MI 313-01⟩
Wieland, M., **2**, 1076 ⟨23LA(432)39⟩
Wieland, P., **1**, 1077 ⟨70HCA171⟩, 1082 ⟨67HCA2101, 67HCA2108⟩; **2**, 12 ⟨76HCA1027⟩, 95 ⟨61E475⟩
Wieland, T., **2**, 7 ⟨77LA806⟩, 570 ⟨90ZN(B)191⟩; **5**, 83 ⟨57MI 502-03⟩, 84 ⟨51AG7, 54AG507⟩, 84 ⟨57AG362, 63AG(E)358⟩, 1038 ⟨58CB683⟩
Wieleseck, R., **2**, 1017 ⟨73OSC(5)921⟩
Wielstra, Y., **2**, 981 ⟨87OM2053⟩
Wiemer, D. F., **2**, 457 ⟨89JOC627, 91JOC5556⟩; **3**, 14, 82, 100, 102 ⟨84JOC2279⟩, 50, 198 ⟨91JOC5556⟩, 198 ⟨86JOC4342, 89JOC627⟩, 245 ⟨84TL4467, 88JOC2011⟩
Wiemers, D. M., **2**, 444 ⟨85JFC(27)85⟩; **4**, 117 ⟨85JFC(27)85⟩; **6**, 15, 246 ⟨85JA5014, 86JA832⟩
Wienand, A., **5**, 957 ⟨91CB957⟩
Wiener, D. F., **2**, 457 ⟨86JOC4342⟩
Wiener, M., **5**, 936 ⟨79CB708⟩
Wienreb, S. M., **2**, 354, 369 ⟨B-87MI 206-01⟩
Wierda, D. A., **2**, 71 ⟨90JOC784⟩; **3**, 521 ⟨88CC1489⟩
Wierenga, W., **2**, 805 ⟨72JOC1597⟩
Wierengo, C. J., **3**, 558 ⟨74JHC633⟩
Wiernik, M., **2**, 489 ⟨15CB1473⟩
Wierschke, S. G., **1**, 149 ⟨90JA697⟩
Wiersig, J. R., **2**, 900 ⟨79JOC3374⟩
Wiersum, U. E., **1**, 774 ⟨88MI 117-01⟩, 776, 778 ⟨89JOC5811⟩, 789 ⟨72CC347, 81MI 117-01⟩
Wiesboeck, R. A., **2**, 762 ⟨66IC1629⟩; **6**, 57 ⟨68JOC1257⟩
Wieschollek, R., **4**, 673 ⟨92CB143, 93CB79⟩, 673 ⟨93CB89⟩
Wiese, H. C., **2**, 18 ⟨83JOC3195⟩
Wiesenberger, E., **4**, 761 ⟨52M1⟩
Wiesenberger, F., **4**, 142 ⟨91ZN(B)275⟩
Wiesenfeld, R. B., **3**, 233 ⟨81T4035⟩
Wieser, K., **2**, 778 ⟨75AG(E)70⟩
Wieser, M., **4**, 710 ⟨93AG(E)923⟩
Wiesner, K., **1**, 498 ⟨80T419⟩; **3**, 293 ⟨72CJC1925⟩
Wiesner, M., **4**, 225 ⟨93CAR(249)197⟩
Wiessler, M., **2**, 385 ⟨84LA1468⟩; **3**, 233 ⟨69TL4987⟩; **4**, 311 ⟨84LA1494⟩, 312 ⟨84LA1468⟩; **6**, 561 ⟨87LA583⟩, 562 ⟨84AP(317)1042⟩
Wietzke, M., **1**, 686 ⟨84CB277⟩; **3**, 348 ⟨84CB277⟩
Wife, R. L., **2**, 439 ⟨86EUP184872⟩
Wigert, H., **4**, 416 ⟨61CB2209⟩
Wigfield, Y., **4**, 391 ⟨72JA8795⟩
Wiggen, U., **4**, 54, 57 ⟨89ACS74⟩
Wigger, N., **2**, 124 ⟨72HCA481⟩
Wiggins, J. M., **4**, 305 ⟨90JOC2999⟩
Wiggins, L. F., **1**, 21 ⟨50JCS3000⟩
Wightman, R. H., **4**, 472 ⟨90TL1759⟩
Wihler, H.-D., **2**, 7 ⟨76CB3446⟩; **5**, 1093 ⟨76AG(E)702⟩, 1095 ⟨78CB2054⟩
Wiimann, W., **2**, 388 ⟨73CB3097⟩
Wijata, A., **5**, 774 ⟨90MI 519-01⟩
Wijers, H., **2**, 278, 279 ⟨63RTC68⟩
Wijers, H. E., **1**, 956 ⟨63RTC1040⟩; **2**, 722 ⟨69RTC519⟩, 1032 ⟨62RTC583, 64RTC208⟩; **3**, 548 ⟨67RTC907, 68RTC1236⟩; **5**, 589, 614 ⟨68RTC1236⟩
Wijers, H. W., **1**, 954 ⟨70QRS1⟩
Wijnberg, J. B. P. A., **1**, 584 ⟨94JOC374⟩; **3**, 122 ⟨85JOC2650⟩; **4**, 303 ⟨75T1437, 78T179⟩, 426 ⟨78T2399⟩
Wijngaarden, B. H. von, **1**, 505 ⟨71RTC961⟩
Wijsman, A., **4**, 64 ⟨77TL885⟩; **6**, 48 ⟨79JOC1847⟩, 48, 120 ⟨77TL885⟩
Wikholm, D. M., **5**, 820 ⟨50JA4697⟩
Wikholm, R. J., **2**, 876 ⟨75JOM(85)179⟩
Wikström, H., **2**, 652 ⟨88JMC1080⟩
Wilberg, E., **2**, 110 ⟨30CB1136⟩
Wilbert, G., **5**, 990, 991 ⟨67CI(L)1406⟩
Wilbrandt, D., **2**, 435 ⟨91PS(61)261⟩
Wilbur, D. J., **2**, 387 ⟨80JA3883⟩
Wilbur, D. S., **2**, 632 ⟨83JOC1542⟩, 685 ⟨82JOC359⟩
Wilburn, J. C., **2**, 466 ⟨79IC347⟩; **4**, 571 ⟨77CC308, 79IC347⟩
Wilby, A. H., **1**, 74, 455, 457 ⟨85CRV129⟩; **2**, 379 ⟨82T419⟩
Wilcott, III, M. R., **1**, 600 ⟨78JA3874⟩
Wilcox, B. B., **2**, 498 ⟨06MI 209-01⟩
Wilcox, Jr., C. F., **1**, 905 ⟨82TL3119, 84TL3403⟩
Wilcox, C. S., **1**, 174 ⟨85JOC546⟩, 174, 341 ⟨86JA3102⟩, 218, 831 ⟨83JA1988, 79CJC1743⟩, 831 ⟨80JA1155⟩, 873 ⟨91COS(2)99⟩; **2**, 681 ⟨84TL395⟩, 699 ⟨84TL699⟩; **4**, 831 ⟨84TL699⟩; **5**, 155 ⟨88TL6823⟩
Wilcox, G., **2**, 1050 ⟨80TL3241⟩
Wilcsek, R. J., **4**, 1054 ⟨78JOC950⟩; **6**, 182, 183, 199, 292 ⟨73JA5096⟩, 182, 199 ⟨78JOC950⟩, 183, 200, 292, 386, 400 ⟨73JOM(57)231⟩
Wilczok, U., **2**, 558 ⟨90CB1517⟩
Wild, F., **2**, 412 ⟨48JCS699⟩; **4**, 76 ⟨48JCS699⟩
Wild, H., **1**, 1098 ⟨89TL647⟩; **3**, 422 ⟨89TL647⟩
Wild, H.-J., **1**, 422 ⟨76JCS(P1)410⟩
Wild, J., **4**, 465 ⟨82AG(E)776⟩
Wild, J. D., **2**, 874 ⟨68ICA379⟩
Wild, S. B., **1**, 53 ⟨88JA4346⟩; **2**, 438 ⟨86IC3396, 90PAC1139⟩, 438 ⟨92JCS(D)1751, 94CC475⟩, 442 ⟨79JA6254, 82HOU(13/3b)69⟩, 487 ⟨90CC1169⟩
Wildbredt, D.-A., **4**, 484 ⟨81AG(E)131⟩, 558 ⟨80CB1549⟩, 567 ⟨83CC1171⟩; **5**, 887 ⟨81AG(E)131⟩; **6**, 168 ⟨81AG(E)131, 83CC1171⟩, 694 ⟨81CC72⟩
Wilde, H., **3**, 328 ⟨69JPR604⟩; **4**, 432 ⟨80ZC437⟩, 880 ⟨85JPR297⟩; **5**, 324 ⟨85S331⟩
Wilde, M. P. von, **1**, 71 ⟨1874CB352⟩
Wilde, R. G., **3**, 331 ⟨86JOC1556⟩, 331, 332 ⟨86JA2985⟩, 331, 333, 335, 337 ⟨88JA5452⟩, 331, 335 ⟨88JOC2220⟩; **4**, 232 ⟨86JOC117⟩
Wildeman, J., **2**, 156 ⟨79S733⟩; **3**, 709, 720 ⟨85RTC177⟩, 714, 720 ⟨82RTC202⟩; **4**, 328 ⟨85RTC177⟩, 331 ⟨81JOC2069⟩, 908 ⟨82RTC202⟩; **5**, 566, 618 ⟨76TL143⟩; **6**, 122 ⟨85RTC177⟩
Wilder, R., **6**, 486 ⟨92JOC2755⟩
Wilder, R. G., **6**, 69 ⟨55JA6644⟩
Wilder Smith, A. E., **1**, 871 ⟨60HCA1915⟩
Wilding, I. G. E., **2**, 409, 410, 412, 413 ⟨49JCS2921⟩, 412 ⟨48JCS699⟩; **4**, 76 ⟨48JCS699⟩
Wilds, A. L., **1**, 394 ⟨48JOC763⟩
Wildschut, G. A., **6**, 88 ⟨67M1043⟩
Wildsmith, E., **2**, 299 ⟨84S941⟩; **3**, 188 ⟨84S941⟩
Wilemon, G. M., **3**, 3 ⟨81JOC1901⟩
Wilen, S. H., **2**, 532 ⟨B-82MI 209-01⟩
Wiles, L. A., **5**, 334 ⟨64JCS5679⟩
Wiley, D. W., **2**, 640 ⟨64JOC1876⟩; **4**, 164, 183 ⟨60JA2288⟩
Wiley, G. A., **2**, 7, 627 ⟨64JA964⟩
Wiley, M. R., **1**, 173, 321, 334 ⟨85T4079⟩; **6**, 541, 550, 557 ⟨85T4079⟩
Wiley, P. F., **5**, 754 ⟨49JA3746⟩
Wiley, R. H., **2**, 676 ⟨55OSC(3)853⟩; **3**, 446 ⟨57JOC204, 59JOC1925⟩, 455 ⟨54JA4931⟩; **5**, 25 ⟨55OSC(3)560⟩, 264 ⟨63OSC(4)5⟩, 298 ⟨50JOC800⟩
Wilhelm, E., **2**, 1004 ⟨77CB3168⟩; **3**, 603 ⟨85CB1720⟩
Wilhelm, H., **6**, 581 ⟨92PS(73)81⟩
Wilhelm, R. S., **1**, 118, 465 ⟨81JA7672⟩, 119 ⟨82JA4696⟩, 204, 207 ⟨82JA2305⟩, 217, 478 ⟨84JOC3928⟩; **2**, 1095

⟨81JA7672⟩; **3**, 128 ⟨84T5005⟩, 325 ⟨89JA8954, 89JA8957⟩
Wilhelms, R. E., **3**, 499 ⟨67JPS(B)1039⟩
Wilk, B. K., **3**, 613 ⟨93SC1775⟩; **6**, 280 ⟨89TL2473⟩, 280, 632, 731 ⟨87TL1969⟩
Wilke, C. A., **6**, 208, 405 ⟨75OMS(10)18⟩
Wilke, G., **1**, 60 ⟨83CB1336⟩, 248 ⟨87CB79⟩, 511, 534, 542 ⟨60LA(629)222⟩, 665 ⟨60LA(629)172⟩, 812 ⟨69LA(727)194⟩; **2**, 96, 97 ⟨62LA(652)7⟩, 967 ⟨56CB444⟩, 967, 974 ⟨60LA(629)222⟩, 984 ⟨86AG(E)508⟩; **3**, 233 ⟨85AG(E)496⟩, 397 ⟨74ZC288, 75ZC66⟩; **4**, 583 ⟨72IJ293, 85CB298⟩, 604, 647 ⟨87CB471⟩, 671 ⟨66AG(E)898⟩, 676 ⟨92ICA741⟩, 698 ⟨60LA(629)222⟩; **6**, 208 ⟨60LA(629)222, 63BSF1462⟩
Wilke, M., **1**, 686 ⟨84CB277⟩; **3**, 348 ⟨84CB277⟩
Wilke, S., **5**, 366 ⟨88JA1638⟩
Wilken, P. H., **3**, 453 ⟨48JA1980⟩
Wilkening, D., **5**, 146 ⟨84SC227⟩
Wilkening, D. W., **2**, 309 ⟨89JOC1940⟩
Wilkenloh, J., **4**, 16, 25 ⟨90CB583⟩, 1080 ⟨84JCR(S)396, 87TL5489⟩
Wilkes, B., **1**, 54 ⟨88ZAAC(560)141⟩; **6**, 237 ⟨89ZAAC(576)225, 92ZAAC(608)69⟩
Wilkes, M. C., **3**, 535 ⟨82JA6779⟩
Wilkie, C. A., **4**, 619 ⟨72JOC2662⟩; **6**, 267, 379 ⟨72JOC2662⟩
Wilkins, C., **2**, 45 ⟨73S156⟩
Wilkins, J. D., **2**, 570 ⟨72JCS(D)2313, 73JCS(D)961⟩; **5**, 781 ⟨74JOM(80)349⟩, 797 ⟨74JOM(67)269⟩
Wilkins, J. M., **1**, 1093 ⟨75JOC1460⟩
Wilkins, M. A., **6**, 492, 515 ⟨81JINC457⟩
Wilkins, R. F., **2**, 316 ⟨88TL2377, 90TL4229⟩; **3**, 186 ⟨90SL619⟩
Wilkinson, D. L., **4**, 489 ⟨90CB423⟩, 1023 ⟨89CB1857⟩; **5**, 1097 ⟨88AG(E)583⟩
Wilkinson, G., **1**, 73 ⟨65CC131, 65CI(L)560⟩, 73, 74 ⟨68JCS(A)2665⟩, 73, 79, 80 ⟨67JCS(A)1574⟩, 85 ⟨67JCS(C)270⟩, 507 ⟨71JCS(A)3224⟩, 862 ⟨52JA2125, 59MI 119-04⟩; **2**, 478 ⟨67JCS(A)1347, 68JCS1133⟩, 574 ⟨76JCS(D)2204⟩, 575 ⟨75JOM(99)C19⟩, 592 ⟨79IS253⟩, 602 ⟨77JCS(D)812⟩, 825 ⟨80JCS(D)55⟩; **4**, 591 ⟨82COMC-I(1)⟩, 656 ⟨70CC1369, 71CC1079, 71CC1477, 77JCS(D)812, 82POL31⟩, 656, 677 ⟨76JCS(D)2204⟩, 672 ⟨78JCS(D)446⟩, 678 ⟨91CC181⟩, 680 ⟨79JA4128, 80JCS(D)1771⟩, 687 ⟨64PCS401, 65IC1270⟩, 719 ⟨80JCS(D)2315⟩, 750 ⟨79IS262⟩, 966 ⟨76AG(E)609⟩; **5**, 229 ⟨64CI(L)193⟩, 368 ⟨B-80MI 507-01⟩, 444 ⟨88JCS(D)53⟩, 455 ⟨63JCS4096⟩, 461 ⟨71JCS(A)850⟩, 464 ⟨67JCS(A)865⟩, 797, 801 ⟨81JCS(D)2088⟩, 798 ⟨92JCS(D)19⟩, 937 ⟨82COMC-I⟩; **6**, 338 ⟨83CC1557⟩, 339 ⟨81CC189, 81CC2496⟩, 457 ⟨68IS99⟩, 523 ⟨63JCS1133⟩
Wilkinson, J. A., **1**, 916 ⟨92PNA(89)915⟩; **2**, 15 ⟨92CRV505⟩
Wilkinson, J. F., **2**, 882 ⟨22JCS91⟩
Wilkinson, J. R., **4**, 681 ⟨75JOM(99)281⟩
Wilkinson, L. W., **2**, 519 ⟨93TL7693⟩
Wilkinson, R. W., **1**, 34 ⟨58JA1654⟩
Wilkinson, S. G., **2**, 40 ⟨79COC(1)594⟩
Will, B., **1**, 520 ⟨87CB531⟩
Will, W., **6**, 463 ⟨80S485⟩
Willans, J. L., **2**, 840 ⟨57CI(L)235⟩
Willard, A. K., **1**, 131, 416, 829, 874, 876 ⟨76JA2868⟩, 416, 417, 829, 830, 876 ⟨75TL3975⟩, 490, 1106 ⟨71TL4701⟩; **2**, 420, 696 ⟨76JA2868⟩; **5**, 227 ⟨75TL3975, 76JA2868⟩
Willard, C., **3**, 106 ⟨63OSC(4)831⟩
Willard, G. F., **1**, 636 ⟨78JOC4750⟩
Willard, J. E., **2**, 33 ⟨60JA3816⟩; **6**, 221, 226 ⟨42JA1342⟩
Willard, P. G., **2**, 551 ⟨91COS(1)1⟩
Willcott III, M. R., **1**, 74 ⟨73JA4346⟩, 795, 801 ⟨72MI 118-01⟩
Willeford, B. R., **4**, 726 ⟨71HCA1651⟩
Willem, R., **4**, 700 ⟨84JOM(277)335⟩
Willemart, A., **2**, 1059 ⟨60CR(250)866⟩, 1084 ⟨60CR(251)727⟩
Willems, J. F., **6**, 575 ⟨63FCF554⟩
Willemsens, L. C., **4**, 699 ⟨70JOM(23)471⟩; **6**, 209 ⟨70JOM(23)471⟩, 405 ⟨65RTC43⟩
Willer, R. L., **1**, 241 ⟨76JOC1052⟩; **3**, 429 ⟨85H(50)5123⟩; **4**, 434, 435, 440 ⟨84JOC5147⟩; **5**, 215 ⟨90SC2901⟩, 338 ⟨79JOC2863⟩
Willershausen, P., **5**, 928, 929 ⟨92AG(E)1238⟩, 928, 930 ⟨92ZN(B)983, 92AG(E)1384⟩; **6**, 188 ⟨92AG(E)1384⟩, 201, 397, 715, 716 ⟨90AG(E)1030⟩, 201, 715 ⟨92AG(E)1238⟩, 389, 714 ⟨91AG(E)594⟩
Willett, R. D., **4**, 803 ⟨90JA3152⟩; **5**, 1063 ⟨89JOC1048⟩
Willette, G. L., **3**, 73 ⟨60JOC53⟩
Willette, R. E., **4**, 54, 57 ⟨73JOC3878⟩
Willetts, A. J., **2**, 18 ⟨93JCS(P1)1795⟩
Willey, F. G., **1**, 1075 ⟨64AG(E)138⟩
Willey, P. R., **4**, 366, 394, 945, 1060 ⟨84JA3245⟩
Willfang, G., **4**, 181 ⟨41CB145⟩
Willgerodt, **5**, 276 ⟨09JPR192⟩
Willhalm, A., **5**, 882 ⟨84AG(E)903⟩; **6**, 660, 686 ⟨83AG(E)545⟩, 660, 686 ⟨84AG(E)903⟩, 661 ⟨86AG(E)457⟩, 686, 688 ⟨87PS(30)495⟩
Willhalm, B., **1**, 403, 734 ⟨75HCA1016⟩; **3**, 494 ⟨75HCA1016⟩
Williams, A., **2**, 194 ⟨85JOC5592⟩, 847 ⟨76JCS(P2)1444⟩; **4**, 94 ⟨86JCS(P1)1681⟩, 130, 131 ⟨89JCS(P1)563⟩; **5**, 1061 ⟨81CRV589⟩
Williams, A. C., **3**, 707 ⟨92JA4089⟩
Williams, A. D., **1**, 23 ⟨87JOC2317⟩
Williams, C. C., **6**, 86 ⟨72JOC2757⟩
Williams, C. R., **5**, 336 ⟨94JOC806⟩
Williams, C. S., **2**, 421 ⟨68JCS(C)1704⟩
Williams, D., **2**, 381 ⟨73JMC736⟩
Williams, D. H., **1**, 630 ⟨65CB3236⟩, 871 ⟨64JA2832⟩; **3**, 498 ⟨69JCS(C)1100⟩; **4**, 563 ⟨70JCS(C)504⟩
Williams, D. J., **1**, 410, 822 ⟨82JCS(P1)2757⟩, 571 ⟨92JA6330⟩, 987 ⟨91CC1290⟩; **2**, 949 ⟨82JA6136⟩, 1000 ⟨88JCS(P1)1023, 88TL6009⟩; **3**, 308 ⟨84JOC3503⟩; **4**, 90 ⟨82JCS(P1)2757⟩, 448 ⟨84CC1289⟩, 678 ⟨91CC181⟩, 799 ⟨88JCS(D)2781⟩; **5**, 387, 388 ⟨86JCS(D)813, 575 ⟨91CC116⟩; **6**, 445 ⟨89JMC228⟩, 473, 550 ⟨87CC1228⟩
Williams, D. R., **1**, 152 ⟨81JA3081⟩, 152 ⟨81JA3088⟩, 535, 1091 ⟨77TL3847⟩, 697, 700, 703 ⟨86JOC3916⟩; **2**, 85 ⟨84JA2949⟩, 282 ⟨80TL4417⟩, 350 ⟨93TL3271⟩, 354 ⟨89TL1327⟩, 625 ⟨81TL3745⟩, 816 ⟨78JOC3245⟩, 976 ⟨82JA4708⟩; **3**, 38 ⟨80TL4417⟩, 71 ⟨87S908⟩, 124 ⟨81JA3081⟩, 244 ⟨81TL3565⟩, 312 ⟨93TL7705⟩, 407 ⟨76JA3032⟩; **5**, 173 ⟨82JA4708⟩; **6**, 134 ⟨90JA4552⟩, 472, 473 ⟨82JA4708⟩
Williams, E., **1**, 604 ⟨80JOC3537⟩; **5**, 239 ⟨78JOC1283⟩
Williams, E. A., **2**, 163 ⟨86JOC829⟩
Williams, E. B., **2**, 219 ⟨71JOC488⟩
Williams, F., **1**, 445 ⟨74JA5032⟩
Williams, F. D., **3**, 382, 389 ⟨68CC459⟩
Williams, Jr., F. T., **1**, 869, 871 ⟨70JA4689⟩
Williams, G., **2**, 810, 811 ⟨B-61MI 215-02⟩
Williams, G. D., **4**, 718 ⟨87JOM(334)117⟩; **5**, 937 ⟨91JA927⟩
Williams, G. J., **4**, 174 ⟨66JOC3032⟩
Williams, G. M., **1**, 1045 ⟨92JOC5818⟩; **2**, 27 ⟨90TL7509⟩
Williams, H., **3**, 498 ⟨67JCS(C)2442⟩
Williams, H. B., **3**, 363 ⟨70T817⟩
Williams, H. J., **1**, 1082, 1083 ⟨77JOC2380⟩; **4**, 171 ⟨82JOC3140⟩
Williams, H. R., **2**, 95 ⟨54JA2984, 54JA2987⟩
Williams, H. W. R., **2**, 612 ⟨88CJC1479⟩
Williams, I. D., **2**, 1083 ⟨93JOM(449)27⟩; **4**, 689, 710 ⟨85JCS(D)2483⟩

Williams, J. C., **2**, 1060 ⟨76MI 222-01⟩; **5**, 351 ⟨86JCS(P1)1655⟩
Williams, J. E., **1**, 1109 ⟨69JA5350⟩
Williams, Jr., J. E., **1**, 892 ⟨69JA1037⟩, 1109 ⟨70JA2141⟩
Williams, J. H., **1**, 78 ⟨37JCS1136⟩; **2**, 889 ⟨50JCS18⟩; **4**, 555 ⟨47JCS1465⟩
Williams, J. K., **1**, 498 ⟨58JA751, 59JA2723⟩
Williams, J. M., **2**, 602 ⟨86JA40⟩; **3**, 131 ⟨90SC2683⟩; **4**, 235, 236 ⟨86TL4355⟩, 656, 658 ⟨86JA40⟩, 673 ⟨78IC2093⟩; **6**, 87 ⟨87TL2769⟩
Williams, Jr., J. M., **1**, 620 ⟨68JA429, 68JA435⟩; **4**, 83, 84 ⟨68JA435⟩
Williams, J. M. J., **1**, 192, 194 ⟨92TA1089⟩, 194 ⟨93TL3149⟩
Williams, J. O.,**1**, 522 ⟨91JCS(P1)1255⟩
Williams, J. P., **4**, 22 ⟨90JOC5719⟩; **5**, 799 ⟨89OM1523⟩
Williams, J. R., **1**, 78 ⟨83JOC3162⟩; **4**, 71, 78 ⟨79JOC1177⟩; **5**, 490 ⟨79JOC1177⟩
Williams, J. W., **3**, 86 ⟨55OSC(3)626⟩, 86, 100 ⟨39JA2248⟩, 535 ⟨40JOC122⟩, 536, 541 ⟨35JA774⟩
Williams, K. C., **2**, 602 ⟨67JOC4062⟩; **4**, 699 ⟨70JOM(22)141, 70JOM(23)465⟩
Williams, L., **1**, 163 ⟨83TL623⟩, 163 ⟨83TL627⟩; **2**, 525, 1005 ⟨83TL623⟩, 525, 1005 ⟨83TL627⟩, 622 ⟨82TL371⟩; **4**, 664 ⟨83TL623⟩; **6**, 199 ⟨83TL623⟩
Williams, Jr., L. P., **1**, 378, 886 ⟨61JA412, 61JA2537⟩
Williams, L. R., **4**, 88 ⟨48JA1244⟩
Williams, L. T. D., **2**, 458 ⟨54BRP810930⟩
Williams, M. W., **5**, 122 ⟨63JCS1927⟩
Williams, N. E., **2**, 257 ⟨80TL4959⟩
Williams, N. R., **1**, 432 ⟨55JCS522⟩; **2**, 694 ⟨55JCS522⟩
Williams, P. D., **3**, 325 ⟨89JA8954, 89JA8957⟩
Williams, P. H., **3**, 244 ⟨87CC756⟩; **4**, 213 ⟨59JOC284⟩, 1076 ⟨85CC742⟩; **5**, 709 ⟨61JOC651, 61JOC659⟩
Williams, P. J., **1**, 747, 748 ⟨69AJC1737⟩, 747, 748 ⟨69AJC2145⟩; **5**, 376 ⟨83JOC4111⟩
Williams, P. L., **3**, 362 ⟨48JCS1181⟩
Williams, R. E., **1**, 315 ⟨80JA4973⟩; **3**, 138 ⟨80JA4973⟩; **4**, 629 ⟨92CRV175⟩
Williams, R. G., **1**, 421 ⟨73JA3932⟩
Williams, R. H., **2**, 450 ⟨52JA5418⟩, 451 ⟨60USP2957931⟩, 454 ⟨55JA3411⟩
Williams, R. J., **5**, 62 ⟨B-43MI 502-01⟩
Williams, R. M., **1**, 166 ⟨B-89MI 104-02⟩, 488, 983, 1033, 1034 ⟨91T343⟩, 1004, 1135 ⟨88JA2501⟩, 1042 ⟨90JOC4657⟩, 1042, 1135 ⟨88T5425⟩; **2**, 901, 904 ⟨91T343⟩; **3**, 285, 295 ⟨87JOC2615⟩; **4**, 102 ⟨88JA1547, 90JOC4657⟩, 302 ⟨B-89MI 407-01⟩, 815 ⟨91SM2535⟩; **5**, 84 ⟨B-89MI 502-01⟩
Williams, R. O., **1**, 807, 809 ⟨60TL29⟩, 826 ⟨68CC587⟩; **4**, 165 ⟨80JCS(P1)2535⟩
Williams, R. V., **1**, 146 ⟨90JOC3840⟩, 233 ⟨82JA2560⟩, 284 ⟨87TL4455⟩, 654 ⟨83JOC4976⟩, 1043 ⟨88S621⟩; **4**, 936 ⟨89CC1872⟩; **5**, 304 ⟨88S621⟩
Williams, S. B., **3**, 5, 294 ⟨79JOC2955⟩, 310 ⟨86JOC2021⟩
Williams, T. H., **2**, 822, 841 ⟨79JOC3741⟩; **4**, 170 ⟨83JOC5315⟩
Williams, T. M., **3**, 46 ⟨85TL6269⟩
Williamsen, P., **1**, 983 ⟨80JOC3513⟩
Williamson, A. E., **3**, 612 ⟨1854JPR(61)60⟩
Williamson, A. W., **6**, 68 ⟨1854LA(92)346, 1854PRS135⟩
Williamson, B. L., **1**, 1006 ⟨92SL199⟩, 1016 ⟨91JOC3912⟩; **2**, 1020 ⟨91JA5870, 91JOC3912⟩; **5**, 1108 ⟨92SL535⟩
Williamson, D. H., **1**, 73, 79, 456 ⟨76OR(24)1⟩, 803 ⟨79JCS(P2)962⟩
Williamson, K., **2**, 845 ⟨64JCS1240⟩
Williamson, M., **3**, 544 ⟨85JA6715⟩
Williamson, W., **2**, 88 ⟨1852JCS106, 1852JCS229⟩
Williams-Smith, D. L., **1**, 232 ⟨73JA1547⟩
Williard, P. G., **1**, 624, 626 ⟨86JA1039⟩, 873 ⟨85JA3345, 85TL3931⟩, 873 ⟨86JA462, 87JA5539⟩, 885 ⟨91COS(1)1⟩; **2**, 997, 1003 ⟨89OM2308⟩; **4**, 173 ⟨84TL5009⟩; **5**, 278 ⟨86JA1039⟩, 705 ⟨81TL2731⟩, 958 ⟨93JA9015⟩; **6**, 172, 189, 190 ⟨91COS(1)1⟩
Willis, A. C., **2**, 438 ⟨92JCS(D)1751, 94CC475⟩, 487 ⟨90CC1169⟩; **4**, 499 ⟨89CC1648⟩; **6**, 102 ⟨86CJC732⟩
Willis, B. J., **1**, 240 ⟨81JOC5027⟩, 684, 685, 686 ⟨72JCS(P1)305⟩, 800 ⟨82JOC4786⟩
Willis, C., **2**, 596 ⟨79JOC3603⟩
Willis, C. J., **1**, 157 ⟨85HOU(V/2c)127⟩; **4**, 7 ⟨60JA1888, 60JA5298⟩; **6**, 232 ⟨65CJC1893⟩, 244 ⟨60JA5298, 62JA898⟩, 245 ⟨60JA1888⟩
Willis, C. L., **1**, 177 ⟨93CC1111⟩; **3**, 62 ⟨87TL2175⟩
Willis, C. R., **1**, 5, 7, 9, 10, 16, 17, 331 ⟨91JCS(P1)103, 90TL5093⟩, 331, 936 ⟨89CC1387⟩; **2**, 536 ⟨89CC1387⟩
Willis, F., **5**, 828, 832 ⟨68CJC1873⟩
Willis, P. A., **1**, 1130 ⟨88CC283⟩
Willis, R. G., **1**, 504 ⟨61T16⟩
Willis, R. R., **4**, 683 ⟨92OM154⟩
Willis, W. W., **3**, 56 ⟨82JOC1618⟩
Willis, Jr., W. W., **1**, 605, 983, 1073 ⟨80JA5967⟩, 1089 ⟨80JOC5227⟩; **2**, 290 ⟨82TL3319⟩, 728, 908 ⟨81JOC2775⟩; **3**, 177 ⟨85T4771⟩; **4**, 289, 871, 947, 965 ⟨81JOC2775⟩, 947 ⟨82JOC1618⟩, 964 ⟨80JOC5227⟩
Willison, D., **1**, 529 ⟨88JOM(354)233⟩; **2**, 103 ⟨90OM301⟩
Willitzer, M., **4**, 893 ⟨67CB3460⟩
Willmes, A., **3**, 592 ⟨82LA80⟩
Willms, G., **2**, 249 ⟨85S66⟩
Willms, L., **3**, 171 ⟨85S66⟩
Willner, H., **6**, 588 ⟨81ZN(B)1261⟩
Willoughby, C. A., **2**, 330 ⟨92JA7562⟩
Wills, K. D., **1**, 100 ⟨67JOC2794⟩
Wills, M., **2**, 362 ⟨93TA2159⟩; **3**, 29 ⟨89TL2971⟩; **5**, 340 ⟨91JCS(P1)3383⟩, 340, 347 ⟨93JCS(P1)1581, 90TL4117⟩
Willscher, S., **5**, 365 ⟨87ZC27⟩
Willson, C. G., **1**, 608 ⟨80JOC1486⟩; **4**, 286 ⟨80JOC1486⟩
Willson, J. M., **5**, 1125 ⟨90TL3681⟩
Willson, T., **1**, 654, 713, 716 ⟨87CC106⟩
Willy, W. E., **2**, 31 ⟨76BCJ1989⟩
Wilms, H., **4**, 920 ⟨59CA(53)10040f⟩; **5**, 199 ⟨50LA(568)1⟩
Wils, E. R. J., **4**, 177 ⟨71RTC1141⟩
Wilshire, C., **3**, 352 ⟨75CC557, 79JCS(P1)1166⟩, 628 ⟨88AJC1697⟩
Wilshire, J. F. K., **6**, 580 ⟨88AJC995⟩
Wilsmore, N. T. M., **3**, 534, 535 ⟨07JCS1938, 07NAT510⟩, 535 ⟨07PCS229⟩
Wilson, A. C., **4**, 490 ⟨82TL3835⟩
Wilson, B. M., **2**, 101 ⟨50JCS2088⟩
Wilson, C. J., **5**, 465 ⟨74JCS(D)421⟩
Wilson, C. L., **1**, 853 ⟨36JCS623⟩; **5**, 1110 ⟨80T1839⟩
Wilson, Jr., C. O., **2**, 453, 837, 845 ⟨67JOC2172⟩
Wilson, C. V., **2**, 12, 627 ⟨57OR(9)332⟩, 45 ⟨57OR(9)350⟩; **4**, 27 ⟨57OR(9)332⟩; **5**, 184 ⟨55OSC(3)28⟩
Wilson, D. A., **2**, 274 ⟨75JCS(P1)1925⟩; **3**, 433 ⟨80JCR(S)366⟩
Wilson, D. K. T., **3**, 368 ⟨92CJC964⟩
Wilson, D. M., **2**, 236 ⟨91JOC3520⟩
Wilson, E. R., **6**, 151 ⟨84BRP2123829⟩
Wilson, G., **1**, 1137, 1141 ⟨75JOC124⟩; **2**, 1054 ⟨75JOC124⟩
Wilson, G. E., **4**, 65 ⟨73JOC2160⟩, 229 ⟨68JOC2133⟩
Wilson, Jr., G. E., **4**, 64, 71, 80 ⟨70JOC3002⟩; **5**, 237 ⟨80JOC2766⟩
Wilson, G. L., **2**, 470 ⟨69CJC4281⟩
Wilson, J., **4**, 979 ⟨71JHC241⟩, 1077 ⟨86TL5033⟩
Wilson, J. A. Z., **1**, 218, 221 ⟨87JOC34⟩
Wilson, J. D., **2**, 342, 354 ⟨91TL4431⟩; **4**, 850 ⟨74CC751, 75JOC2577⟩, 972, 988 ⟨71JOC1613⟩; **5**, 652 ⟨70CJC983⟩, 745, 753 ⟨71JOC1613⟩; **6**, 120 ⟨75JOC2577⟩, 141 ⟨68JOC1246⟩, 661, 733 ⟨72JOC2730⟩
Wilson, Jr., J. H., **2**, 461 ⟨66USP3261743⟩

Wilson, J. M., **1**, 883 ⟨89CC161⟩
Wilson, III, J. R., **1**, 45 ⟨46JA1813⟩
Wilson, J. S., **1**, 939 ⟨81JA932, 83JA4059⟩; **6**, 533 ⟨81JA932⟩, 535 ⟨83JA4059⟩
Wilson, J. W., **1**, 163 ⟨80JOM(186)297⟩, 163 ⟨83TL623⟩, 163 ⟨83TL627⟩, 163 ⟨83TL631⟩, 163 ⟨85OM1690⟩, 660, 764 ⟨83TL635⟩, 1124 ⟨69JA3238⟩; **2**, 525 ⟨80JOM(186)297, 83TL631⟩, 525, 1005 ⟨83TL623⟩, 525, 1005 ⟨83TL627⟩, 905 ⟨83TL635, 83TL637, 85OM1690⟩; **3**, 19 ⟨83TL631⟩; **4**, 420, 421, 426 ⟨55JA745⟩, 630, 632, 634 ⟨82JCR(S)132⟩, 642, 645, 1077 ⟨83TL637⟩, 664 ⟨83TL623, 83TL631⟩, 664 ⟨93T2988⟩, 665 ⟨93T2979⟩, 1057 ⟨81JOM(209)1⟩; **5**, 106 ⟨55OSC(3)267⟩; **6**, 199 ⟨82JCR(S)132⟩, 199 ⟨83TL623⟩, 201 ⟨85OM1690⟩
Wilson, K. D., **1**, 240 ⟨87JOC218⟩, 363 ⟨86JOC4627⟩
Wilson, K. E., **1**, 722 ⟨75JA3512⟩
Wilson, K. J., **5**, 953, 956 ⟨94JOC6882⟩
Wilson, L. A., **2**, 476 ⟨68TL85⟩
Wilson, L. J., **2**, 688 ⟨92JOC1948⟩; **3**, 164 ⟨92JOC1948⟩; **4**, 230 ⟨91JA9377⟩; **5**, 334 ⟨92JOC1948⟩
Wilson, M. A., **1**, 631 ⟨69T3107⟩
Wilson, M. E., **5**, 441 ⟨76JA6529⟩
Wilson, M. W., **6**, 558, 564 ⟨93JMC1090⟩
Wilson, N., **4**, 935, 949 ⟨78JOM(156)123⟩
Wilson, R., **4**, 650 ⟨91JCS(P1)2276⟩
Wilson, R. C., **5**, 108 ⟨66JCS(C)840⟩
Wilson, R. D., **6**, 220 ⟨74IC2811⟩, 283 ⟨78USP4120904⟩, 424 ⟨72IC2531⟩
Wilson, R. M., **1**, 253 ⟨81JOC3293⟩
Wilson, S. E., **1**, 514 ⟨76TL15⟩
Wilson, S. R., **1**, 400, 401, 803, 805 ⟨93OR(43)93⟩, 401, 414, 800 ⟨77TL2559⟩, 804 ⟨80JA6351, 78CC479⟩, 832 ⟨84JOC722⟩; **2**, 74 ⟨89JOC3087⟩, 998 ⟨91JA723⟩, 1004 ⟨91JA1468⟩; **3**, 22 ⟨77TL2559⟩, 349 ⟨82JOC3319⟩, 679 ⟨84TL405⟩; **4**, 152 ⟨92JA2592⟩, 680 ⟨93CC284, 94OM2309⟩, 696 ⟨93JOM(449)203⟩; **5**, 413 ⟨82JOC747⟩; **6**, 345 ⟨89JOM(371)257⟩
Wilson, T., **2**, 78 ⟨88JA984⟩; **3**, 240 ⟨90TL3433⟩
Wilson, T. M., **2**, 78 ⟨88JA984⟩
Wilson, V. P., **2**, 4, 23, 30 ⟨90JA7431⟩, 4, 23, 30 ⟨93JA3071⟩
Wilson, W., **3**, 452 ⟨52JCS1321⟩
Wilson, W. C., **5**, 74 ⟨32OSC(1)276⟩
Wilson, W. F., **5**, 798 ⟨88JCS(D)27⟩
Wilt, J. W., **1**, 81 ⟨67JOC893⟩, 337 ⟨85T3979⟩, 382 ⟨84JOC2937⟩, 632 ⟨64JOC2788⟩, 925, 927 ⟨B-73MI 119-13⟩, 934 ⟨64JOC921⟩; **3**, 642 ⟨71JOC2026⟩; **4**, 28 ⟨63OSC(4)254⟩
Wimmer, E., **1**, 415 ⟨88JA2314⟩
Wimmer, J., **1**, 1004, 1005 ⟨83JCR(M)167⟩
Wimmer, P., **4**, 229 ⟨91HOU(E14a/1)785⟩
Wimmer, T., **4**, 562 ⟨91CB275⟩
Wimmer, Z., **3**, 430 ⟨88LA1091⟩
Winans, R. E., **2**, 613 ⟨81JA1964⟩
Winberg, H. E., **5**, 548 ⟨65JA2776⟩; **6**, 104 ⟨66USP3239519⟩, 105 ⟨64USP3121084⟩
Winchester, W. R., **2**, 997, 1003 ⟨89OM2308⟩; **4**, 603, 648, 649 ⟨90JA1382⟩, 603, 649 ⟨90OM1314⟩, 648 ⟨90JA2582⟩
Winckelmann, I. B., **3**, 66 ⟨81T2375⟩
Winders, J. A., **1**, 522 ⟨91JCS(P1)1255⟩
Windholz, T. B., **2**, 1016 ⟨61JA4663⟩
Windridge, G. C., **5**, 279 ⟨71JA6318⟩
Wineburg, J. P., **4**, 205 ⟨75JHC749⟩
Wineman, R. J., **4**, 12 ⟨58JA6233⟩; **5**, 1093 ⟨62JA4349⟩
Winfield, J. M., **2**, 421 ⟨72JCS(D)1766, 73JCS(D)1876⟩
Wingard, A. K., **1**, 778 ⟨89JOC389⟩
Wingard Jr., R. E., **1**, 707, 708 ⟨74JA5801⟩
Wingbermühle, D., **2**, 569, 570, 571 ⟨93CB2093⟩, 577 ⟨92CB157⟩; **4**, 673 ⟨86AG(E)909, 93CB79⟩
Wingen, R., **2**, 249 ⟨85RTC153⟩; **3**, 171 ⟨85S66⟩
Wingert, H., **1**, 522 ⟨86CB244⟩; **2**, 1028 ⟨89JA2225⟩, 1029 ⟨88JA3272⟩
Wingfield, J. N., **3**, 570 ⟨69CC967⟩, 570, 571 ⟨72JCS(D)787⟩
Winiarski, J., **3**, 634 ⟨89LA203⟩, 634, 638 ⟨87BSB(96)303⟩
Winitz, M., **5**, 84 ⟨57ANY(69)94⟩
Wink, D. J., **1**, 890 ⟨81JOC3925⟩
Winkel, C., **5**, 15 ⟨89RTC51⟩
Winkelman, D. V., **2**, 216 ⟨69TL3901⟩
Winkelmann, E., **6**, 533 ⟨82EUP62834⟩
Winkelmann, E. H., **2**, 27 ⟨59AG349⟩
Winkelmann, H., **4**, 454 ⟨74SRI191⟩, 454, 481 ⟨76ZAAC(424)97⟩
Winkelmann, H. D., **3**, 709 ⟨64AG(E)647⟩; **5**, 852 ⟨66CB850⟩
Winkelmuller, W., **5**, 106 ⟨43OSC(2)299⟩
Winkhaus, V., **3**, 605, 607 ⟨86CB2466, 84AG(E)619⟩, 609 ⟨86JOM(307)93⟩; **4**, 1026 ⟨86CB2466⟩; **5**, 1092 ⟨84AG(E)587⟩; **6**, 680, 687 ⟨84AG(E)579⟩
Winkle, M. R., **2**, 653 ⟨83T2031⟩
Winkler, E., **5**, 1088 ⟨74JOM(65)51⟩
Winkler, F. W., **3**, 615 ⟨1832LA(4)246⟩
Winkler, H., **2**, 461 ⟨64TL175⟩, 886 ⟨64TL3271⟩; **3**, 492 ⟨65LA(685)1⟩
Winkler, J., **1**, 479 ⟨76TL4405⟩
Winkler, J. D., **1**, 925 ⟨86JA1708⟩; **3**, 241 ⟨82JA3767⟩
Winkler, T., **1**, 415 ⟨68HCA1457⟩; **4**, 12 ⟨85T4057⟩, 239 ⟨89TL6311⟩, 922 ⟨80HCA402⟩; **5**, 355 ⟨86HCA333⟩, 570, 608, 610 ⟨78BSB525⟩, 1124 ⟨85HCA1155⟩; **6**, 160 ⟨90PS(54)197⟩
Winn, L. S., **1**, 1009 ⟨85TL709⟩
Winnewisser, M., **3**, 703 ⟨81JOC1046⟩
Winnik, W., **5**, 593 ⟨93JHC887⟩
Winnitz, M., **5**, 83 ⟨B-61MI 502-01⟩
Wins, A., **2**, 624 ⟨71C59⟩
Winstead, M. B., **4**, 38 ⟨70JA7359⟩
Winstein, S., **1**, 6 ⟨60CI(L)405⟩, 381 ⟨49JA2953⟩, 381, 905 ⟨52JA1154⟩, 396 ⟨47JA2916⟩, 421 ⟨65TL377⟩, 868 ⟨65JA914, 65JA915⟩, 886 ⟨68JA6562⟩, 904, 917 ⟨B-72MI 119-10⟩, 905 ⟨52JA1147, 65JA376, 65JA3267⟩, 917 ⟨66JA2047, 69QR141⟩, 918 ⟨72JA6867⟩, 1114, 1116 ⟨48JA846, 51JA2700, 57JA4146⟩; **2**, 598 ⟨64TL3393⟩; **4**, 24 ⟨62JOC748⟩; **6**, 71 ⟨43JA613⟩, 74 ⟨58JA1247⟩
Winston, A., **6**, 25, 31 ⟨67JOC2166⟩
Wintel, T., **3**, 706 ⟨81LA709, 84S1033⟩
Winter, B. M., **3**, 482 ⟨82JOC1284⟩; **6**, 118, 119, 636, 666 ⟨82JOC1284⟩
Winter, C. H., **4**, 724 ⟨92OM3169⟩
Winter, E., **4**, 203 ⟨60LA355⟩, 728 ⟨00LA353⟩
Winter, G., **6**, 44 ⟨58LA(616)1⟩, 44, 47 ⟨59LA(621)8⟩, 234 ⟨56CB1160⟩, 479 ⟨73AJC755⟩
Winter, H. W., **1**, 1076 ⟨78AG(E)609⟩; **2**, 1040 ⟨80AG(E)720⟩; **3**, 366 ⟨84AG(E)800⟩, 578 ⟨80AG(E)564⟩, 579 ⟨80AG(E)720⟩, 582 ⟨88JA1337⟩, 703 ⟨81JOC1046⟩, 710 ⟨81JOC1045⟩; **5**, 1122 ⟨80AG(E)566⟩, 1123 ⟨81JOC1045⟩
Winter, J., **6**, 302 ⟨73GEP1694210⟩
Winter, M., **1**, 178 ⟨63HCA1792⟩, 1082 ⟨67HCA2101⟩; **3**, 272 ⟨75CB237⟩
Winter, M. J., **1**, 517, 530 ⟨B-85MI 112-03⟩; **4**, 580 ⟨91JCS(D)1117⟩, 674 ⟨80JOM(202)C49⟩; **5**, 446 ⟨84CC420, 88JOM(342)C5⟩, 943 ⟨82CC215, B-90MI 523-02⟩, 944 ⟨83JCS(D)2397, 94TL6717⟩, 955 ⟨95JOM(494)C12⟩
Winter, R., **4**, 868 ⟨88IC4329, 89IC3766⟩; **6**, 252 ⟨90IC4588⟩
Winter, R. A. E., **6**, 555 ⟨65JA934⟩
Winter, S. R., **5**, 452 ⟨72JA1788⟩
Winter, T. J., **2**, 672 ⟨87TL6109⟩
Winter, W., **3**, 64 ⟨76ZN(B)1654⟩, 549 ⟨82CB3653⟩; **4**, 581 ⟨83CB1070⟩

Winter-Extra, S., **3**, 334, 349, 365 ⟨90JOC4200⟩; **4**, 259, 855 ⟨90S271⟩
Winterfeld, K., **2**, 176, 177 ⟨62AP(295)615⟩; **4**, 408 ⟨60AP(293)203⟩, 416 ⟨61AP(294)404⟩
Winterfeldt, A., **6**, 32 ⟨92GEP4025227⟩
Winterfeldt, E., **1**, 519 ⟨86TL5833⟩; **2**, 60 ⟨75S617⟩; **3**, 13 ⟨75S617⟩, 71, 72 ⟨64CB1959⟩, 593 ⟨74T2553⟩; **4**, 849 ⟨84LA450⟩; **5**, 174 ⟨84LA450⟩, 174 ⟨84LA462⟩, 174 ⟨84LA474⟩
Winterle, J. S., **5**, 830 ⟨81JMC1124⟩
Winter-Mihaly, E., **5**, 212, 213 ⟨85CAR375⟩
Winternitz, F., **6**, 506 ⟨82JMC178⟩
Winters, L. J., **2**, 386, 798 ⟨65JOC2471⟩; **4**, 994 ⟨63T49⟩
Wintner, C., **1**, 1082 ⟨76OS(55)114⟩; **2**, 380 ⟨88OSC(6)56⟩; **4**, 996 ⟨69TL2689⟩
Winton, P. M., **2**, 1064 ⟨84CC200⟩
Wintzer, A., **2**, 982 ⟨58AG166⟩
Winzenberg, K. N., **1**, 754, 755 ⟨86JA2662⟩; **5**, 173 ⟨82JA4015⟩, 1127 ⟨79CI(L)319, 85JA3910⟩
Wiorogorski, W., **3**, 382 ⟨32RZC270⟩
Wipf, P., **1**, 207 ⟨93AG(E)1068⟩, 414, 876 ⟨91COS(5)827⟩, 417, 795, 830 ⟨91JOC3572⟩, 417, 830, 876 ⟨91JOC650⟩; **2**, 568, 587 ⟨91JOC6494⟩, 586, 702, 910, 919 ⟨93S537, 92JOC1740⟩, 587 ⟨92SL718, 93JOC3455⟩, 656 ⟨91COS(5)827⟩, 926 ⟨92JOC3178⟩, 975 ⟨93AG(E)1068⟩; **3**, 140 ⟨90JOC1425⟩; **4**, 831 ⟨91JOC650⟩; **5**, 575 ⟨87HCA1001⟩
Wipf, R., **3**, 680 ⟨88HCA1177⟩
Wipke, W. T., **1**, 794 ⟨64PAC539⟩; **4**, 524 ⟨93HOU(E15/2)1674⟩
Wippel, H. G., **1**, 639, 752, 753 ⟨58CB61⟩, 639, 752, 753 ⟨59CB2499⟩; **2**, 454, 844 ⟨58CB64⟩; **4**, 575, 576, 577 ⟨59CB2499⟩, 576 ⟨58CB61⟩
Wippermann, T., **4**, 603 ⟨84CB1885⟩
Wippich, B., **2**, 21 ⟨92LA559⟩
Wiriyachitra, P., **2**, 652 ⟨77JOC2274⟩
Wirkner, C., **2**, 467 ⟨86ZAAC(535)47⟩; **4**, 566 ⟨86ZAAC(535)47⟩, 568 ⟨82ZAAC(488)75⟩; **6**, 372 ⟨80ZC419⟩, 372, 374 ⟨80ZC153⟩, 372, 688, 689 ⟨81ZAAC(473)85⟩, 658, 687 ⟨81SRI279⟩, 680, 688 ⟨81ZC357⟩
Wirkus, M., **5**, 284 ⟨85JOC2220⟩
Wirtel, A. F., **6**, 504 ⟨26JA1736⟩
Wirth, D., **2**, 489 ⟨83ZAAC(505)134⟩
Wirth, D. D., **1**, 248 ⟨87ACR65⟩
Wirth, H., **2**, 847 ⟨56CB688⟩, 881 ⟨56N59⟩
Wirth, H. O., **2**, 631 ⟨60LA(634)84⟩; **5**, 1145 ⟨68SAP6903273⟩
Wirth, R. P., **4**, 220 ⟨78TL731⟩
Wirth, T., **5**, 657 ⟨88C230⟩
Wirth, U., **4**, 990 ⟨90CB115⟩, 991 ⟨89CB519, 91JHC1863⟩
Wirz, B., **5**, 97 ⟨83MI 502-01⟩
Wirz, J., **2**, 637 ⟨87JA4000⟩, 638 ⟨86HCA1331⟩, 1022 ⟨89AG(E)181, 89JA2355⟩; **3**, 366 ⟨72CC632⟩; **6**, 646 ⟨91AG(E)1356⟩
Wisansky, W. A., **2**, 753 ⟨55OSC(3)307⟩
Wischnewskij, O. W., **6**, 445 ⟨69CB2972⟩
Wise, D. S., **3**, 633 ⟨83S308⟩; **6**, 566 ⟨85JHC1269⟩
Wise, P. H., **2**, 627 ⟨41JA2847⟩
Wise, R. M., **3**, 613 ⟨56JA450⟩; **5**, 25, 35 ⟨56JA450⟩
Wiseman, G. E., **1**, 437 ⟨54JA1706⟩
Wishart, N., **1**, 968 ⟨93SL219, 93SL499⟩; **2**, 512 ⟨90JOM(399)C21⟩, 591 ⟨92JOC3397⟩
Wishnia, A., **5**, 85 ⟨58ANY(70)870⟩
Wisian-Neilson, P., **2**, 861, 862 ⟨77IC1460⟩; **4**, 129 ⟨87PS(33)147⟩, 566 ⟨82IC3568, 83IC575⟩, 570 ⟨86MM2089, 87OM959⟩, 572 ⟨84IC2063⟩; **6**, 165, 167, 193, 688 ⟨85OM339⟩, 350, 694 ⟨84CC698⟩, 491 ⟨84IC4322⟩, 491, 515 ⟨79JINC1223⟩, 492, 515 ⟨81JINC457⟩
Wiskott, E., **2**, 802 ⟨68AG(E)138⟩; **6**, 627 ⟨66LA(694)44⟩
Wisniewski, V., **2**, 36 ⟨88SC119⟩
Wisniewski, W., **4**, 120, 804 ⟨89T337⟩
Wisotsky, M. J., **1**, 894, 900 ⟨62JA2016⟩, 897, 901, 911 ⟨62JA1498⟩
Wisowaty, J. C., **4**, 193 ⟨79JOC4825⟩
Wisser, G. V., **2**, 613 ⟨84CC655⟩
Wissing, E., **4**, 656 ⟨93OM3624⟩
Wissinger, J. E., **3**, 217 ⟨84JA3869⟩
Wissmann, H., **5**, 264 ⟨80AG(E)133⟩
Wissner, A., **3**, 157, 166 ⟨79JOC4617⟩; **4**, 832 ⟨78TL2749⟩; **5**, 15 ⟨78JOC3972⟩
Wisted, E. C., **1**, 778 ⟨87JOC3688⟩
Wistuba, E., **1**, 388 ⟨81TL4069⟩; **6**, 488 ⟨92GEP4219418⟩
Witczak, Z. J., **3**, 699, 708 ⟨86TL155⟩; **4**, 238 ⟨82H(19)1719⟩; **5**, 1053, 1057 ⟨85T4781⟩
Witek, S., **5**, 671 ⟨90JPR731⟩
With, J. de, **2**, 570 ⟨90OM2207⟩
Withers, G. P., **2**, 623, 682, 747 ⟨77JA1993⟩; **5**, 652 ⟨79JOC2945⟩
Withers Jr., H. P., **6**, 404 ⟨80JOM(188)329⟩
Witiak, D. T., **2**, 129 ⟨70JOC4209⟩
Witiak, J. L., **1**, 522 ⟨72JA6229⟩
Witkop, B., **4**, 424 ⟨70JA343⟩; **6**, 108 ⟨52CB949, 52JA3855⟩
Witnauer, L. P., **2**, 190 ⟨52JA2526⟩
Witold, W. E., **1**, 745 ⟨69RZC95⟩
Witschel, M., **3**, 605 ⟨93CB2151⟩
Witt, K. E., **1**, 688, 690 ⟨78JOC3609⟩
Witte, C., **2**, 387 ⟨23CB679⟩
Witte, H.,**3**, 680 ⟨68LA(713)1⟩
Witte, H., **4**, 522 ⟨65LA(687)1⟩, 523 ⟨69LA(722)21⟩
Witte, J., **5**, 611, 710 ⟨58CB972, 58CB1129⟩
Witte, J. F., **4**, 563 ⟨89TL411⟩
Wittenberg, D., **2**, 553 ⟨58JOC2044⟩; **5**, 417 ⟨58JA4529⟩
Wittenberger, S., **3**, 331 ⟨86JOC1556⟩
Wittenberger, S. J., **1**, 821 ⟨88JA4822⟩; **3**, 331 ⟨88JA4822⟩; **6**, 417 ⟨92EUP486948⟩
Wittenbrook, L. S., **2**, 619 ⟨67JA4487⟩; **4**, 64, 69, 80 ⟨68JOC1080⟩, 69 ⟨67JA4483⟩, 82 ⟨67JA4487⟩, 85 ⟨73OSC(5)231⟩; **5**, 1124 ⟨75JHC37⟩
Wittig, G. A., **1**, 104 ⟨68LA(711)46⟩
Wittig, G., **1**, 104 ⟨68LA(711)65⟩, 157 ⟨49LA(562)177⟩, 271, 647, 738, 743 ⟨60JA563⟩, 396 ⟨47LA(557)205⟩, 396, 397 ⟨42LA(550)260⟩, 399 ⟨48LA(560)116⟩, 445 ⟨64CB769⟩, 446 ⟨52LA(577)26, 52LA(578)136⟩, 519 ⟨62AG(E)415⟩, 628 ⟨57LA(612)102⟩, 638 ⟨53LA(580)44⟩, 639, 721 ⟨54CB1318⟩, 640, 721, 733 ⟨55CB1654⟩, 727 ⟨58LA(619)10⟩, 881 ⟨64CB3548, 68AG(E)7⟩, 881 ⟨70OS(50)66⟩, 986 ⟨63CB1535⟩, 1084 ⟨63CB342⟩; **2**, 88 ⟨71BSF1921⟩, 420 ⟨71LA(746)149⟩, 487, 488 ⟨53ACS1293⟩, 618 ⟨61CB1373⟩, 667 ⟨56CB1334⟩, 680 ⟨62CB2514⟩, 810 ⟨62AG(E)53⟩, 824 ⟨71LA(751)17⟩, 830 ⟨49LA(562)187, 64CB741⟩, 873 ⟨52LA(578)136⟩, 873, 879 ⟨52LA(577)26⟩, 873, 881 ⟨64CB789⟩; **3**, 10 ⟨63AG(E)683⟩, 80 ⟨73OSC(5)571⟩, 218 ⟨68LA(712)79⟩, 445 ⟨62T1023⟩, 493 ⟨53LA(580)44, 58LA(619)10⟩, 494 ⟨54CB1318, 55CB1654⟩, 497 ⟨61CB1373⟩, 500 ⟨53LA(560)57⟩; **4**, 340 ⟨62CB2514⟩, 347 ⟨61CB1373⟩, 385 ⟨67LA(702)24⟩, 534, 586 ⟨61LA(650)1⟩; **5**, 48 ⟨48AG169⟩, 897 ⟨61CB1373⟩; **6**, 697 ⟨61CB1373, 62CB2514⟩
Wittig, U., **1**, 363 ⟨87CB1933⟩
Wittle, A. J., **3**, 234 ⟨81TL3031⟩
Wittman, H., **2**, 1003 ⟨82M1451⟩
Wittman, M. D., **1**, 643 ⟨87JOC4303⟩, 816 ⟨90TL4301⟩, 816, 819 ⟨88JOC4631⟩
Wittmann, H., **2**, 400 ⟨69M959, 83M1097⟩, 1003 ⟨83M783⟩; **5**, 340 ⟨69M959⟩
Wittmann, R., **3**, 697, 721 ⟨61AG219⟩
Wittwer, M., **2**, 219 ⟨22CB1424⟩; **6**, 467 ⟨28GEP516281⟩
Wittwer, W., **1**, 1001 ⟨93S558⟩

Witty, M. J., **1**, 596 ⟨91SL873⟩
Witucki, E. F., **4**, 444 ⟨82JCED94⟩; **6**, 56 ⟨72JOC152⟩
Wityak, J., **1**, 1019 ⟨91SC977⟩; **4**, 528 ⟨92MI 411-02⟩
Witz, M., **2**, 613 ⟨87JA7194⟩; **6**, 253 ⟨84JFC(26)359, 89JFC(45)225⟩, 439 ⟨86JFC(34)251⟩
Witzel, H., **6**, 104 ⟨88JHC907⟩, 729 ⟨87CSR75⟩
Witzel, T., **1**, 172, 325, 330 ⟨84AG(E)69⟩; **3**, 621 ⟨84AG(E)69⟩
Witzeman, J. S., **1**, 811 ⟨86JA5901⟩
Witzgall, K., **3**, 497 ⟨86CB1331⟩; **5**, 413 ⟨85AG(E)1068⟩; **6**, 520, 702 ⟨85AG(E)229⟩, 697, 700 ⟨86CB1331⟩
Wiygyl, F. M., **6**, 592 ⟨83CC294⟩
Wizinger, R., **4**, 1074 ⟨59HCA1733⟩
Wladislaw, B., **4**, 197 ⟨62CI(L)1868⟩, 197 ⟨62JOC281⟩
Wlassics, I. D., **5**, 551 ⟨88BBA269⟩
Wlodarczak, G., **5**, 1160 ⟨94MI 506-04⟩
Wlodecki, B., **3**, 234 ⟨80JOC1109⟩
Wlostowska, J., **3**, 465 ⟨77BSB739, 78SC569⟩; **6**, 55, 280, 633 ⟨81JOC5048⟩, 280, 632 ⟨82JOC4177⟩
Wlostowski, M., **3**, 613 ⟨93SC2323⟩; **4**, 830 ⟨87JA3811⟩
Wnuk, S., **4**, 306 ⟨83S654, 88S111⟩
Wnuk, S. F., **4**, 61 ⟨88TL5729⟩, 61, 73, 79 ⟨90JOC4757⟩, 61, 73, 79 ⟨91JOC6878⟩, 62, 63 ⟨93JOC3800⟩, 218 ⟨94JOC544⟩; **6**, 46 ⟨90JOC4757⟩
Wnuk, T. A., **2**, 306 ⟨76JA5678⟩
Wodzki, W., **2**, 436 ⟨84JOM(269)107⟩, 437, 443 ⟨79TL1069⟩
Woell, J. B., **1**, 297 ⟨83CC1270⟩; **5**, 142 ⟨84MIP84/1376A1⟩
Woerner, F. P., **4**, 829 ⟨71CB2786⟩
Woerns, K. P., **4**, 617 ⟨84ZAAC(512)103⟩
Woerpel, K. A., **1**, 353 ⟨91JA726⟩
Woessner, W. D., **6**, 86 ⟨72JOC2757⟩, 88 ⟨76CL43⟩
Wohl, A., **5**, 856 ⟨00CB2741⟩
Wohl, R. A., **2**, 679 ⟨74S38⟩; **5**, 346 ⟨73HCA1826⟩
Wöhler, F., **5**, 19 ⟨1832MI 501-01⟩
Wohlers, K., **5**, 623, 624 ⟨71LA(752)115⟩
Wohlleben, A., **4**, 563 ⟨77CB1748⟩
Wohlleben-Hammer, A., **4**, 586 ⟨78ZN(B)1325⟩; **5**, 908, 1094 ⟨77AG(E)417, 79CB510⟩
Wojahn, H., **5**, 735 ⟨52AP(285)375⟩
Wojcicki, A., **3**, 570 ⟨70JOM(21)P21, 74ICA55⟩; **4**, 267 ⟨73JA6962⟩, 683 ⟨92OM154⟩, 711 ⟨94OM1999⟩; **5**, 451 ⟨90JOM(383)C1⟩, 452 ⟨66IC889⟩, 943 ⟨93JA9986⟩
Wojciechowski, J., **6**, 156 ⟨78MIP87868⟩
Wojciechowski, K., **3**, 625 ⟨88LA203⟩; **5**, 746 ⟨84TL3651⟩
Wojciechowski, P. S., **4**, 869 ⟨84USP4465845A⟩, 874 ⟨81CC669⟩
Wojcik, B. H., **5**, 76 ⟨43OSC(2)260⟩
Wojnowich, L., **6**, 515 ⟨76JA5702⟩
Wojtkowski, P., **4**, 832 ⟨67JA6804⟩
Wojtkowski, P. W., **2**, 524 ⟨70TL215⟩, 532 ⟨71JOC1790⟩; **5**, 575 ⟨75JA5628⟩
Wojtowicz, H., **4**, 124 ⟨87S477⟩, 1083 ⟨90JOC5420⟩
Wokulat, J., **2**, 415 ⟨68ZAAC(361)296⟩
Wolber, E. K. A., **3**, 700 ⟨92CB525⟩
Wolber, G. J., **5**, 237 ⟨76TL2577⟩
Wolbring, W., **5**, 858 ⟨05JPR366⟩
Wolcott, R. A., **4**, 1042 ⟨78ICAL331⟩
Wolczanski, P. T., **3**, 545 ⟨89JA9057⟩
Wold, S., **1**, 905 ⟨80JCS(P2)1447⟩
Woldring, M. G., **5**, 88 ⟨72SC423⟩
Wolf, A., **4**, 691 ⟨93CC1459⟩
Wolf, A. D., **6**, 439 ⟨81USP4255353⟩
Wolf, A. P., **1**, 268 ⟨62JA676⟩, 896 ⟨60JA4729⟩; **2**, 614 ⟨83CC108⟩; **3**, 465 ⟨66JA4735⟩, 492 ⟨62JA676⟩
Wolf, D., **5**, 769, 848 ⟨64CB3436, 63AG(E)217⟩
Wolf, E., **4**, 555 ⟨60USP2920094⟩
Wolf, F., **6**, 254 ⟨64ZC457⟩
Wolf, G., **1**, 255 ⟨58JA4405⟩; **4**, 603 ⟨91CB2391⟩
Wolf, G. C., **2**, 182 ⟨74JOC238⟩, 1034 ⟨71JOC1727⟩
Wolf, H., **2**, 383, 384 ⟨59CB1842⟩, 1046 ⟨81JA6142⟩; **4**, 94 ⟨88GEP3639877A1⟩; **5**, 49 ⟨76CB41⟩; **6**, 523 ⟨27CB1424⟩
Wolf, H. R., **3**, 319 ⟨79HCA1645⟩
Wolf, J., **1**, 991 ⟨93AG(E)1315⟩; **2**, 866 ⟨83TL3625⟩; **3**, 521 ⟨92CB2641, 93AG(E)1480⟩, 553 ⟨83AG(E)981⟩; **4**, 149 ⟨87JOM(336)413⟩, 149, 583 ⟨91JOM(417)149⟩; **5**, 944 ⟨95CB49⟩
Wolf, J. F., **1**, 895 ⟨75JA2902⟩
Wolf, J.-G., **6**, 374, 375 ⟨83TL2769, 83TL3625⟩, 375 ⟨84JAP59193897⟩
Wolf, J. R., **1**, 675 ⟨93JA4413⟩
Wolf, P., **5**, 282 ⟨87TL5801⟩
Wolf, P. F., **2**, 650 ⟨69JOC3441⟩
Wolf, P. R., **4**, 59 ⟨90JOC1847⟩; **6**, 423 ⟨90JOC1847⟩
Wolf, R., **2**, 411 ⟨65BSF2368⟩, 693 ⟨58MI 213-01⟩; **4**, 478 ⟨81PS(11)87⟩, 743 ⟨86JCS(P1)733⟩; **5**, 370 ⟨86BSF101⟩, 1158 ⟨92TL2981⟩; **6**, 690 ⟨92TL2981⟩
Wolfarth, E. F., **1**, 826 ⟨70JOC3585⟩
Wolf, U., **1**, 986 ⟨79LA1388⟩, 995 ⟨78TL4263⟩
Wolf, V., **2**, 624 ⟨70TL1629⟩, 1018 ⟨60LA(638)33⟩; **4**, 976 ⟨60LA(638)33⟩
Wolf, W., **4**, 598 ⟨75CB2649⟩; **5**, 321 ⟨65CC449⟩, 913 ⟨75CB2851⟩, 915 ⟨75CB2834, 75CB2842⟩
Wolf, W. A., **3**, 193, 472, 473 ⟨68JOC3610⟩
Wolf, W. H. de, **4**, 1083 ⟨89JA8491⟩
Wolfbeis, O., **2**, 787 ⟨77CB2480⟩
Wolfbeis, O. S., **2**, 746 ⟨82ZN(B)105⟩; **3**, 264 ⟨81S225⟩
Wolfe, J. F., **1**, 860 ⟨74JOC2006⟩; **5**, 24 ⟨B-79MI 502-01, B-91MI 502-01⟩, 122 ⟨B-92MI 503-01⟩, 342 ⟨92JOC5328⟩
Wolfe, J. W., **5**, 203, 320 ⟨70JOC2056⟩
Wolfe, S., **1**, 562 ⟨79S117⟩, 849 ⟨72ACR102⟩, 879 ⟨75CJC3365⟩; **2**, 23, 26 ⟨66CJC2339⟩; **4**, 390 ⟨B-85MI 408-01⟩, 391 ⟨65JA5498⟩, 417 ⟨72CJC2902⟩, 1075 ⟨85MI 424-01⟩; **5**, 9 ⟨63JA643, 68CJC2549, 67JA5287⟩, 347 ⟨83JA7755⟩, 813 ⟨71CJC1384⟩
Wölfel, G., **1**, 1041 ⟨84AG(E)53, 87S848⟩; **4**, 313 ⟨84AG(E)53⟩
Wolfers, E., **6**, 465 ⟨93GEP4134688⟩
Wolff, A., **4**, 611 ⟨93CB575⟩, 651 ⟨93OM4123⟩
Wolff, C., **3**, 231 ⟨89CB1653⟩
Wolff, H., **4**, 211 ⟨48JA2624⟩, 580 ⟨79CB2709⟩; **5**, 283, 1001 ⟨46OR(3)307⟩
Wolff, H.-M., **5**, 776 ⟨80CB1394, 82ZAAC(486)61⟩
Wolff, I. A., **1**, 955 ⟨65JOC2983⟩
Wolff, L., **1**, 393 ⟨12LA(394)23⟩; **3**, 475, 530 ⟨12LA(394)23, 02LA(325)129⟩, 530 ⟨04LA(333)1⟩
Wolff, M. E., **1**, 927 ⟨63CRV55⟩; **2**, 412 ⟨57JA1970⟩
Wolff, R., **3**, 689 ⟨92TL2981⟩; **4**, 417 ⟨73MI 409-01⟩
Wolff, S., **1**, 19 ⟨85JA5732⟩, 596 ⟨86JA3385⟩; **2**, 166 ⟨87GEP3543567⟩; **3**, 241 ⟨85TL703⟩, 413 ⟨83JOC1925⟩; **4**, 99 ⟨89T6113⟩; **5**, 277 ⟨85JA5732⟩
Wolffenstein, R., **1**, 623 ⟨1898CB1553⟩
Wolff-Jensen, J., **5**, 1101 ⟨69ACS1567⟩
Wolfgardt, P., **2**, 421 ⟨76ZN(B)697⟩
Wolfgrubber, M., **3**, 545 ⟨88JOM(355)267⟩
Wolfgruber, M., **2**, 878 ⟨83AG(E)1001⟩, 878, 896 ⟨84CB427⟩, 878, 896 ⟨86JOM(317)175, 83OM1266⟩
Wolfhugel, J., **3**, 65 ⟨73TL1635⟩
Wolfindale, B. A., **2**, 615 ⟨85JFC(27)263⟩
Wolfinger, M. D., **4**, 68, 71, 81, 83, 84 ⟨74JOC2516⟩, 80 ⟨74JOC2521⟩
Wolford, L. T., **4**, 447 ⟨88JCS(P1)2137⟩; **5**, 963 ⟨88JCS(P1)2137⟩, 964, 1119 ⟨89JCR(S)331⟩; **6**, 653 ⟨91CL569⟩
Wolf-Pflugmann, M., **6**, 444 ⟨89AF539⟩
Wolfram, E., **2**, 288, 289, 292 ⟨67ZAAC(352)167⟩
Wolfram, J., **5**, 282 ⟨82JOC3787, 82JOC3881⟩
Wolfram, P., **2**, 434 ⟨87GEP3612629⟩
Wolfrom, M. L., **1**, 24 ⟨49JA2870⟩, 45 ⟨43JA1516⟩; **2**, 171 ⟨63JOC2992⟩; **5**, 49 ⟨56JA2489⟩

Wolfsberger, W., **2**, 433 ⟨91ZN(B)1578, 92ZN(B)155⟩, 437 ⟨88ZN(B)295, 91CZ7⟩, 439 ⟨90CZ353⟩, 440 ⟨85CZ317⟩, 448 ⟨83CZ77⟩, 468 ⟨89CZ349⟩, 470 ⟨67CB1000, 74MI 208-02⟩, 875 ⟨85CZ53⟩; **4**, 566, 573 ⟨89CZ349⟩
Wolin, M. S., **4**, 519, 520 ⟨75JOC437⟩, 519, 520 ⟨75JOM(99)385⟩
Wolinski, L., **2**, 262 ⟨51JOC395⟩
Wolinsky, J., **1**, 563 ⟨68JA113⟩, 1144 ⟨65JA1142⟩; **2**, 21 ⟨80TL177, 81TL5019⟩; **4**, 742 ⟨66CC567, 70JOC1211⟩
Wolinsky, J. L., **2**, 622, 625 ⟨65JOC2208⟩
Wolkoff, P., **3**, 681, 687 ⟨75CJC1333⟩; **5**, 641 ⟨74CJC879, 75CJC3211⟩; **6**, 111, 114, 115 ⟨76ACS(B)837⟩
Wollenberg, E., **5**, 357 ⟨88AP(321)25, 88AP(321)551⟩
Wollenberg, R. H., **1**, 535 ⟨76TL4705⟩, 535, 1091, 1093 ⟨75JOC2265⟩, 1048, 1081, 1091 ⟨74JA5581, 77JA7365⟩; **2**, 676 ⟨77JA7365⟩, 970, 971 ⟨74JA5582, 75JOC2265⟩; **3**, 60 ⟨77JA7365⟩, 63 ⟨78TL717⟩; **5**, 234 ⟨78TL3523⟩
Wollensak, J. C., **1**, 1000 ⟨58JA6626⟩
Wollenwerber, M., **1**, 574 ⟨94AG(E)117⟩
Woller, P., **5**, 1066 ⟨86CB2114⟩
Wollmann, K., **6**, 158, 159 ⟨72ZAAC(389)119⟩
Wollmann, T. A., **2**, 698 ⟨86JA8279⟩; **5**, 237 ⟨86JA8279⟩
Wollmans, T. A., **2**, 521 ⟨86JA7402⟩
Wollny, B., **6**, 96 ⟨89S684⟩
Wollowitz, S., **1**, 601, 602, 604 ⟨78JOC1697⟩, 609 ⟨82JA7051⟩, 788 ⟨83JA2503⟩; **2**, 281 ⟨78JOC1697⟩, 290 ⟨82TL3319⟩; **4**, 92 ⟨78JOC1697⟩
Wollweber, H. J., **3**, 722, 724 ⟨78AG(E)688⟩
Wolman, Y., **5**, 1061, 1078 ⟨B-80MI 528-01⟩
Wolmershäuser, G., **2**, 360 ⟨74ZN(B)277⟩, 361, 399 ⟨77ZAAC(432)173⟩, 400 ⟨75AG(E)485⟩
Wolochowicz, I., **1**, 687, 688 ⟨73BSF2147⟩
Wolosinski, H. T., **4**, 18, 730 ⟨49JA3439⟩, 743, 750 ⟨50JA3952⟩
Wolstenholme, G. E. W., **5**, 84 ⟨B-58MI 502-01⟩, 85 ⟨B-53MI 502-01⟩
Wolter, A., **3**, 468 ⟨30CB702⟩
Wolterman, A., **4**, 536 ⟨77CB3034⟩
Woltermann, A., **1**, 647, 667, 717 ⟨78TL4399⟩, 667 ⟨77AG(E)862⟩; **4**, 401 ⟨77AG(E)710⟩, 600 ⟨82CB645⟩
Wolters, D., **6**, 518 ⟨86ZN(B)283⟩
Wolters, E., **3**, 256 ⟨62CB1237⟩; **4**, 207 ⟨76AG(E)688⟩; **5**, 1113 ⟨62CB1237⟩; **6**, 465 ⟨93EUP558996, 93EUP559001⟩
Wolters, E. T., **2**, 219 ⟨71JOC488⟩
Wolters, J., **1**, 248 ⟨83JOM(256)C1⟩
Womack, G. W., **6**, 332 ⟨83OM1696⟩
Wong, A., **5**, 451, 937 ⟨83OM1852⟩
Wong, C. F., **1**, 986 ⟨92CPL243⟩
Wong, C.-H., **1**, 6 ⟨92JA741⟩, 9 ⟨92JA10138⟩; **2**, 38 ⟨B-94MI 202-01⟩; **3**, 32 ⟨90JOC4897⟩; **3**, 48 ⟨89JA3924, 93TL403⟩, 159 ⟨91S499⟩, 159 ⟨92JA741⟩, 631 ⟨89TL1917⟩; **5**, 263 ⟨93TL1247⟩
Wong, C. K., **1**, 21, 25, 437, 438, 943 ⟨80JA4438⟩, 942 ⟨78CC41⟩; **2**, 286 ⟨77CC725, 78CC441⟩
Wong, C. M., **2**, 214 ⟨73SC317⟩, 656 ⟨79CJC3304⟩; **3**, 16, 132 ⟨72S561⟩, 618 ⟨75SC299, 75SC423⟩; **4**, 220 ⟨73CC224⟩; **5**, 191 ⟨73SC63⟩
Wong, C. S., **1**, 22, 23, 437 ⟨85JOC5413⟩
Wong, D. H., **3**, 6, 119 ⟨81S394⟩
Wong, F., **4**, 168, 170 ⟨83JA3661⟩
Wong, F. S., **4**, 713 ⟨82JOM(240)299⟩
Wong, G. B., **5**, 107 ⟨72SC389⟩
Wong, G. S. K., **1**, 342 ⟨85JOC5620⟩, 1127 ⟨85TL957⟩; **3**, 358 ⟨79JOC1977⟩, 621 ⟨84JOC276⟩, 640 ⟨87JOC2427⟩; **4**, 512 ⟨87JOC2427⟩, 520 ⟨89JOC644⟩
Wong, H. C., **6**, 550 ⟨89SC547⟩, 554 ⟨88SC1531⟩
Wong, H. N. C., **1**, 801, 802 ⟨89CRV165⟩; **3**, 103 ⟨92T1039⟩
Wong, J. Y., **1**, 936 ⟨75JOC1966⟩
Wong, L. L., **5**, 366 ⟨80CJC1316⟩; **6**, 626 ⟨83CJC1213⟩
Wong, L. T. L., **5**, 291 ⟨69JOC2766⟩
Wong, M.-S., **2**, 152, 721 ⟨91JOC4552⟩; **6**, 550 ⟨89SC547⟩
Wong, M. Y., **4**, 19, 27 ⟨81JCS(P1)78⟩
Wong, M. Y. H., **4**, 253 ⟨77TL1617, 78JA3548⟩
Wong, P. C., **1**, 1127 ⟨81JOC2327⟩; **4**, 780 ⟨91JMC2525⟩
Wong, P. K., **5**, 142 ⟨75JOC335⟩, 143 ⟨76JA5832⟩, 652 ⟨75JOC532⟩
Wong, P. L., **3**, 421 ⟨93JA1434⟩; **5**, 126 ⟨93JA11434⟩
Wong, R. Y., **3**, 208 ⟨72TL1853⟩
Wong, S., **1**, 1000 ⟨86TL5445⟩
Wong, S. F., **6**, 720, 732 ⟨76JOM(105)C19⟩
Wong, S. M., **1**, 858 ⟨71JA2235⟩
Wong, S. S., **2**, 999 ⟨88HCA299⟩
Wong, T. C., **5**, 226 ⟨83JCS(P2)1471⟩, 521 ⟨81JA7055⟩, 521, 523 ⟨82JCS(P2)473⟩
Wong, V. K., **5**, 242, 252 ⟨78CC838⟩, 242, 252 ⟨79T2329⟩; **6**, 475 ⟨78CC838, 79T2329⟩
Wong, W., **5**, 465 ⟨72CC672⟩
Wong, W.-K., **2**, 436 ⟨90POL2901⟩, 580 ⟨79JA1589⟩; **4**, 578 ⟨90POL875⟩, 581 ⟨79JA5440⟩, 656 ⟨82POL31⟩; **6**, 338 ⟨83CC1557⟩, 339 ⟨81CC189, 81CC2496⟩
Wong Chi Man, M., **4**, 606 ⟨93JOM(451)C1⟩, 607, 608, 654 ⟨93TL2111⟩
Wong-ng, W., **4**, 487 ⟨82PS(13)319⟩
Wong-Ng, W., **5**, 923, 924 ⟨85AX(C)1632, 82JA5667⟩
Wonnacott, A., **1**, 757 ⟨85T3825⟩; **4**, 236 ⟨86T4333⟩
Woo, C.-M., **5**, 288 ⟨81JHC691⟩
Woo, H. G., **4**, 353 ⟨88JA313⟩; **5**, 422 ⟨88JA313⟩, 422, 431 ⟨92OM2198⟩
Woo, S. L., **1**, 1024, 1139 ⟨76TL1769⟩
Wood, A., **2**, 591 ⟨92JOC3397⟩; **3**, 288 ⟨89CC644⟩
Wood, C. D., **3**, 510 ⟨78JA3793, 78JA5964⟩, 511 ⟨79JA3210⟩
Wood, D. C., **5**, 1139 ⟨70JCS(A)1847⟩
Wood, G. L., **2**, 474 ⟨88CB871⟩
Wood, G. P., **3**, 174 ⟨80JOC80⟩; **5**, 728, 732 ⟨80JOC80⟩
Wood, H. B., **4**, 913 ⟨92JOC178⟩; **6**, 118, 666 ⟨92JOC178⟩
Wood Jr., H. B., **4**, 420 ⟨72JHC1145⟩
Wood, J. D., **1**, 91, 92 ⟨79CJC1977⟩
Wood, J. F., **6**, 409, 419, 422 ⟨48JCS2183⟩
Wood, J. H., **3**, 88 ⟨50JA2992⟩
Wood, J. L., **3**, 618 ⟨79TL4907⟩; **5**, 1111 ⟨53JA4655⟩; **6**, 488 ⟨92JA8008⟩
Wood, J. M., **2**, 584 ⟨79JA1442⟩
Wood, M. L., **5**, 8 ⟨90S1143⟩
Wood, N. F., **4**, 111 ⟨70JCS(C)744⟩; **5**, 31 ⟨64TL2969⟩
Wood, R. D., **5**, 285 ⟨82JA3233⟩
Wood, S. D., **5**, 263 ⟨93JCS(P1)451⟩
Wood, S. G., **1**, 1010 ⟨90TL3731⟩; **5**, 516, 522 ⟨85JOC1741⟩
Wood, T. A., **1**, 173 ⟨91CC690⟩
Woodard, R., **3**, 704 ⟨85SC267⟩
Woodard, R. A., **1**, 66, 69, 454 ⟨82JOC3707⟩
Woodard, S. S., **2**, 82 ⟨83PAC589⟩
Woodbridge, D. T., **2**, 287, 291 ⟨62JCS2089⟩; **6**, 594 ⟨61JCS2922⟩
Woodbury, R. P., **1**, 283 ⟨78JOC1947⟩, 556 ⟨77JOC3961⟩; **2**, 941, 1023 ⟨78JOC376⟩; **3**, 537 ⟨77JOC2038, 78JOC376⟩; **6**, 190 ⟨77JOC2038⟩
Woodcock, D., **4**, 442, 443 ⟨49JCS1638⟩
Woodcock, D. C., **4**, 44 ⟨94MIP9322265⟩
Woodcock, D. J., **1**, 28, 32, 35 ⟨B-68MI 102-01⟩, 624 ⟨B-68MI 114-01⟩, 627 ⟨B-68MI 114-02⟩, 637 ⟨B-68MI 114-03⟩; **2**, 1051 ⟨68JCS(C)1324⟩
Woodcock, G. J., **2**, 409, 410, 412, 413 ⟨49JCS2921⟩
Woodcock, M., **2**, 160, 162 ⟨84JMC1634⟩
Wooden, G. P., **4**, 59 ⟨88S407⟩, 60 ⟨86TL6319⟩; **6**, 431 ⟨88S407⟩
Woodfuff, R. A., **4**, 34, 137 ⟨75JOM(92)7⟩
Woodgate, P. D., **1**, 1044 ⟨81JCS(P1)40⟩; **2**, 12

⟨81JCS(P1)2608⟩, 24 ⟨77JCS(D)1674⟩, 34 ⟨78JCS(P1)126⟩, 631 ⟨76JCS(P1)1161⟩; **3**, 8 ⟨78JCS(P1)1483⟩, 155 ⟨78JCS(P1)126⟩; **4**, 27 ⟨81JCS(P1)2608⟩; **5**, 384 ⟨91JOM(420)387⟩, 729 ⟨81JCS(P1)40⟩, 1044 ⟨79JCS(P1)757⟩, 1107 ⟨84S1017⟩, 1107, 1109 ⟨81JCS(P1)58⟩, 1115 ⟨78S152⟩; **6**, 131 ⟨88JOM(348)317⟩, 314, 565 ⟨81JCS(P1)52⟩
Woodgate, S. D., **5**, 1107 ⟨84S1017⟩
Woodman, D. J., **3**, 577 ⟨66JA3169, 68JA1371⟩, 577 ⟨69JOC3451⟩, 578 ⟨76JA6036, 77H(7)247⟩; **4**, 894 ⟨71JOC1685⟩, 979 ⟨69JOC3451⟩
Woodruff, M., **3**, 342 ⟨82AJC1747⟩
Woodruff, R. A., **4**, 14 ⟨77JOM(141)71⟩, 137 ⟨73JOC4031⟩; **6**, 258 ⟨74JOM(71)335⟩
Woods, G. F., **1**, 853 ⟨40JA2933⟩; **4**, 186 ⟨47JA2246⟩
Woods, J. M., **2**, 390 ⟨88S808⟩; **4**, 108, 109 ⟨91JOC316⟩
Woods, L. A., **1**, 117 ⟨52JOC1630⟩; **2**, 129 ⟨45JA1844⟩, 274 ⟨48JA550⟩
Woods, M., **4**, 236 ⟨93TL5649⟩
Woods, R. J., **3**, 64 ⟨51JCS2687⟩
Woods, T. L., **1**, 787 ⟨81JA6691⟩
Woods, W. G., **2**, 904 ⟨66JOC2766⟩
Woodside, A. B., **6**, 264 ⟨86JOC4768⟩
Woodstock, H., **2**, 456 ⟨45USP2471472⟩
Woodstock, W. H., **2**, 833, 837 ⟨55JA929⟩
Woodward, C. E., **2**, 1022 ⟨82JOC1869⟩
Woodward, F. N., **2**, 129 ⟨48JCS1892⟩
Woodward, G. E., **5**, 109 ⟨32OSC(1)408⟩, 115 ⟨32OSC(1)159⟩
Woodward, K. N., **3**, 262 ⟨76JCS(P1)904⟩
Woodward, P., **4**, 679 ⟨84JCS(D)2293⟩, 680 ⟨81CC862, 83JCS(D)2099⟩, 681 ⟨76JCS(D)377⟩, 686 ⟨82JCS(D)1309⟩, 692 ⟨80JCS(D)2182⟩, 712 ⟨81CC867⟩; **6**, 340 ⟨82CC51⟩, 343 ⟨82CC773, 83JCS(D)2599⟩
Woodward, P. R., **3**, 197 ⟨87TL345⟩; **5**, 277 ⟨87TL3019⟩
Woodward, R. A., **4**, 206 ⟨82JOC3707⟩
Woodward, R. B., **1**, 349, 400, 417, 418, 419, 422 ⟨69AG(E)781⟩, 400 ⟨65JA2511⟩, 417, 418, 419, 422 ⟨65JA395⟩, 417, 422 ⟨B-70MI 109-01⟩, 538 ⟨50JA494⟩, 794 ⟨B-70MI 118-01⟩, 808 ⟨59T19⟩, 862 ⟨52JA2125, 52JA3458⟩; **2**, 223 ⟨88OSC(6)1016⟩, 626 ⟨40JA1625⟩; **3**, 247 ⟨46JA2229⟩, 331, 341 ⟨60JA3800, 90T7599⟩, 577 ⟨61JA1007, 61JA1010⟩, 577 ⟨66JA3169, 66T(S)415, 68JA1371⟩; **4**, 225 ⟨81JA3215⟩, 996 ⟨69TL2689⟩; **5**, 164 ⟨58T1⟩, 304 ⟨73PAC145⟩, 905 ⟨78JA8214⟩; **6**, 105 ⟨68PAC519⟩
Woodward, R. W., **2**, 517 ⟨88T5597⟩; **3**, 13 ⟨84T3387⟩
Woodward, S., **4**, 580 ⟨91JCS(D)1117⟩; **5**, 446 ⟨88JOM(342)C5⟩
Woodward, S. S., **2**, 92 ⟨81JA6237⟩
Woodworth, R. C., **1**, 299 ⟨59JA3383⟩
Woody, R. W., **5**, 85 ⟨61T134⟩
Wooldridge, K. R., **3**, 275 ⟨56JCS3070⟩
Woolery, D. O., **4**, 444 ⟨83JOC611⟩
Woolf, C., **4**, 767 ⟨57USP2803665, 57USP2803666⟩; **5**, 1102 ⟨71USP3761501⟩; **6**, 16 ⟨64FRP1369784⟩, 224, 226 ⟨59USP2885450⟩
Woolf, L., **6**, 409 ⟨71GEP2105907⟩
Woolford, R. G., **5**, 55, 57, 59, 62 ⟨56JA2255⟩
Woolhouse, A. D., **1**, 784 ⟨93JHC873⟩
Woollins, J. D., **5**, 575 ⟨91CC116⟩
Woolsey, N. F., **2**, 13 ⟨65JCS2438⟩
Woosley, M. H., **1**, 808 ⟨66JA5846⟩
Wooster, C. B., **1**, 95 ⟨37JA596⟩
Wooster, N. F., **1**, 929 ⟨87CC1238, 88TL955⟩
Wootton, G., **5**, 362, 367 ⟨87JCS(P1)2511⟩
Woplin, J. R., **2**, 855 ⟨72JCS(D)1590⟩
Worakun, T., **1**, 468 ⟨85CC971⟩, 1015 ⟨88TL5565⟩
Worden, L. R., **3**, 82 ⟨63JOC2918⟩, 90 ⟨70JCS(C)227⟩; **5**, 76 ⟨73OSC(5)251⟩
Wordie, J. D., **5**, 108 ⟨50JOC617⟩
Worley, J. W., **4**, 255, 394 ⟨72JA597⟩
Wormack, G. B., **6**, 332 ⟨87OM283⟩
Worman, J. J., **3**, 347 ⟨72CJC3923⟩
Worms, K.-H., **2**, 847 ⟨B-76MI 216-01⟩; **4**, 561 ⟨79ZAAC(457)203⟩; **6**, 118 ⟨73ZAAC(399)1⟩, 158 ⟨79ZAAC(457)214⟩, 158, 159 ⟨72ZAAC(389)119⟩, 159 ⟨79ZAAC(457)209⟩
Wormsbacher, D., **3**, 553 ⟨82CB1332, 84JOM(270)93⟩, 553 ⟨84JOM(272)C40⟩
Wormser, H. C., **4**, 427 ⟨75JHC595⟩
Wormser, U., **4**, 446 ⟨83JMC129⟩
Wörns, K.-P., **6**, 383 ⟨84ZAAC(512)93⟩
Worrall, D. E., **5**, 593 ⟨24JA2834⟩, 593, 595 ⟨18JA415⟩, 593, 595 ⟨20JA1055⟩, 593, 595 ⟨28JA1456, 25JA2974⟩, 595 ⟨37JA1486, 39JA104⟩, 595, 617 ⟨39JA105⟩
Worrall, I. J., **2**, 597 ⟨74JOM(81)329⟩
Worrall, R., **5**, 193, 199 ⟨54JCS2006⟩
Worrell, D. E., **2**, 888 ⟨30JA2046⟩
Worrell, M., **4**, 870 ⟨92TL6131⟩; **5**, 563 ⟨92TL6131⟩
Wörsching, A., **6**, 383 ⟨84ZAAC(512)131, 84ZAAC(512)93⟩
Worsley, M., **3**, 616 ⟨70CJC1881⟩
Worstell, J. H., **2**, 824 ⟨81JOM(205)161⟩
Wortel, Th. M., **4**, 177 ⟨77RTC44⟩
Worth, B. R., **1**, 941 ⟨78HCA690⟩; **3**, 92 ⟨78AJC1533⟩
Worth, D. F., **2**, 373, 378, 381 ⟨64JMC493⟩
Worthing, C. R., **6**, 642 ⟨65JMC446⟩
Worthington, P. A., **4**, 165 ⟨83JCS(P1)1131⟩
Wos, J. A., **1**, 831 ⟨93JA8847⟩
Woshurne, S. S., **1**, 291 ⟨69JOM(16)503⟩
Woticky, W., **5**, 855 ⟨57LA(605)191⟩
Wotiz, J. H., **1**, 954 ⟨B-69MI 120-02⟩, 1040 ⟨54JOC1580, 56JA1221⟩; **2**, 420, 422 ⟨73JOC489⟩, 1014 ⟨60JOC1626⟩
Woulfe, S. R., **2**, 241 ⟨85TL3891⟩; **5**, 334 ⟨85TL3891⟩, 334, 336 ⟨86JOC3133, 86TL3095⟩
Wouters, J. M. A., **3**, 599 ⟨93OM3864⟩
Wovkulich, P. M., **1**, 606 ⟨81JA1501⟩; **2**, 342, 347 ⟨85T3455⟩, 1003 ⟨91COS(1)843⟩
Woyrsch, O.-F., **3**, 467 ⟨53CB76⟩
Wozniak, J., **3**, 368 ⟨89T3741⟩
Wozniak, L., **2**, 108, 452 ⟨85TL4965⟩
Wozny, J. C., **4**, 30 ⟨78JFC(11)527⟩
Wrackmeyer, B., **2**, 903 ⟨77ZN(B)1400, 81ZN(B)1265⟩, 903 ⟨85CC1199, 86CC1170⟩, 903 ⟨86CC403, 89T1119⟩, 1048 ⟨78JOM(145)17⟩, 1083 ⟨84HOU(13/3c)1⟩, 1088 ⟨91AG(E)1370, 92CB1597⟩, 1097 ⟨86SA(A)1107⟩, 1102 ⟨90JOM(399)1⟩; **4**, 661 ⟨86JOM(310)151, 87JOM(321)C1⟩, 665 ⟨83CB3182, 84OM1, 85CC1199, 91ZN(B)1207, 93CB1361⟩, 701 ⟨91CB503⟩, 1054, 1063 ⟨86JOM(307)157⟩, 1062 ⟨86JOM(303)73⟩, 1068 ⟨81JOM(205)1, 88JOM(338)195⟩; **6**, 183 ⟨85AG(E)326⟩, 209 ⟨91CB503⟩, 209, 401, 405 ⟨91SA(A)849⟩, 390 ⟨93JCS(D)3259, 81JMR(44)54⟩
Wragg, R. T., **2**, 478 ⟨64JCS5482⟩
Wratten, S. J., **3**, 699 ⟨77JA7367⟩; **6**, 604 ⟨78TL1391, 78TL1395⟩, 604 ⟨77JA7367⟩
Wray, V., **2**, 428, 431 ⟨80CB1356⟩
Wreford, S. S., **4**, 678 ⟨83JA1679⟩
Wren, C. M., **1**, 777 ⟨72JCS(P2)1744, 72JCS(P2)2359⟩, 777 ⟨72JCS(P2)926⟩
Wren, D. L., **1**, 967 ⟨91JOC1083⟩
Wrench, E., **3**, 295 ⟨56JCS4943⟩
Wriede, U., **3**, 349, 365 ⟨82AG63⟩, 349, 365, 550 ⟨83AG(E)55⟩; **4**, 843, 846, 854 ⟨88CB1165⟩, 851 ⟨80S907⟩; **5**, 552 ⟨80S907⟩, 581, 589, 604 ⟨84CB2205, 83AG(E)55⟩
Wrigglesworth, R., **2**, 52 ⟨94CC1775⟩; **6**, 632 ⟨89JOC1062⟩
Wright, A., **2**, 131 ⟨72JA4784⟩; **4**, 358, 371 ⟨74JA3227⟩,

359, 362, 371 ⟨74JA3214⟩, 368 ⟨72JA4784⟩, 368, 371 ⟨74JA3222⟩
Wright, A. H., 3, 387, 522 ⟨83JA5939⟩; 4, 150, 722 ⟨83JA5939⟩, 582 ⟨91JCS(D)609⟩, 583 ⟨84JOM(273)C17⟩; 6, 718 ⟨82JOM(236)C7⟩, 718, 722 ⟨82JOM(233)C59⟩
Wright, A. N., 3, 488 ⟨63NAT903⟩
Wright, B., 3, 194 ⟨80TL931⟩; 6, 37 ⟨78T197⟩
Wright, B. T., 1, 340, 341 ⟨86JA2787⟩
Wright, C. D., 5, 762, 811, 817 ⟨73JOC1075⟩, 808 ⟨73USP3751312⟩, 811, 815 ⟨67JOC3859⟩, 814 ⟨73JOC1065⟩, 815, 817, 818 ⟨73USP3726903⟩; 6, 279 ⟨72USP3707555⟩, 279, 281 ⟨73JOC1065⟩, 284, 362 ⟨72USP3689560⟩, 284, 362, 613, 650 ⟨73JOC1075⟩, 284, 616, 649 ⟨67JOC3859⟩, 558, 564 ⟨93JMC1802⟩
Wright, C. M., 4, 428 ⟨63T(S)3⟩; 6, 151 ⟨75USP3883377⟩
Wright, D., 5, 985 ⟨69BRP1142991⟩
Wright, D. J., 5, 664 ⟨81JCR(S)129⟩
Wright, Jr., D. P., 5, 1126 ⟨89USP4857550⟩
Wright, D. R., 1, 783 ⟨83JA3714⟩
Wright, G., 3, 700 ⟨62CJC863⟩
Wright, G. C., 5, 849 ⟨70JPS105⟩
Wright, G. E., 4, 129 ⟨89B8270⟩
Wright, G. F., 1, 1043 ⟨31JA4192⟩; 2, 798 ⟨48MI 215-01⟩; 3, 466 ⟨47JA3028⟩; 4, 294 ⟨48JA3659⟩, 434, 440 ⟨49MI 409-01⟩; 5, 293 ⟨53JOC432⟩; 6, 655 ⟨57CJC527⟩
Wright, G. J., 4, 182 ⟨70T1311⟩
Wright, I. G., 5, 316 ⟨79USP771219⟩
Wright, J., 1, 558, 575, 584 ⟨90JA9284⟩, 816 ⟨87JOC2960⟩
Wright, J. C., 5, 998 ⟨68MI 526-01⟩
Wright, J. G., 5, 369, 713 ⟨80JCS(D)1768⟩
Wright, J. J., 1, 91 ⟨81T3615⟩
Wright, J. M., 3, 573 ⟨60JA5386⟩
Wright, K. J., 6, 440 ⟨73IC2890⟩
Wright, L. J., 6, 717, 718 ⟨80JA1206⟩, 722 ⟨88JOM(338)393⟩
Wright, M. E., 1, 808 ⟨56JA2547⟩, 1015 ⟨84JA6417⟩; 3, 33 ⟨93JA2059⟩
Wright, N. D., 1, 518 ⟨73JCS(P1)2818⟩
Wright, O. L., 3, 295 ⟨53JOC244⟩
Wright, P. B., 2, 409, 410 ⟨91JOC4608⟩
Wright, P. M., 5, 1035 ⟨83CB3413⟩
Wright, P. W., 1, 754 ⟨78JCS(P1)590⟩
Wright, S. H. B., 2, 783 ⟨88S379⟩, 906 ⟨88S923⟩; 4, 509 ⟨91JCS(P1)1091⟩; 5, 8 ⟨89S598⟩; 6, 443 ⟨89JCS(P1)1727⟩
Wright, S. W., 1, 504 ⟨90JOC1670⟩, 601 ⟨87TL4917⟩; 5, 334, 336 ⟨92TL153⟩, 636 ⟨84TL2639⟩
Wright, T. C., 2, 512 ⟨87IC4065⟩
Wright, Jr., W. B., 5, 302 ⟨62JOC1042⟩; 6, 416 ⟨65JHC41⟩, 508 ⟨66JMC852⟩
Wrighton, M. S., 4, 674 ⟨77ICA85⟩, ⟨78JOM(159)137⟩ ⟨78JOM(157)41⟩
Wrigley, T. I., 6, 642 ⟨65JMC446⟩
Wrobel, D., 1, 991 ⟨84JOM(264)135, 84JOM(271)181⟩
Wrobel, J., 5, 567, 630 ⟨89JMC2493⟩
Wrobel, J. E., 1, 204, 478, 1023 ⟨84JA3693⟩, 451 ⟨83JOC3761⟩
Wrobel, J. T., 3, 634 ⟨87S452⟩
Wrobel, L., 6, 365 ⟨90JCR(S)330⟩
Wrobel, R. R., 3, 134 ⟨83JOC4087⟩
Wrobel, Z., 1, 404, 406, 819 ⟨91TL4647⟩
Wrobleski, D. A., 2, 827 ⟨81IS178, 82IS175⟩
Wroblewski, A. E., 6, 79 ⟨88JA4144⟩
Wroblowsky, H. S.-J., 2, 340, 342, 350, 353, 355, 356 ⟨90HOU(E16)1⟩
Wu, A., 1, 693 ⟨89JOC1375⟩; 5, 992 ⟨90JOC4282⟩
Wu, A-h., 2, 941 ⟨89S568⟩; 3, 495 ⟨90S885⟩; 5, 8 ⟨89S566⟩, 10 ⟨91S1177⟩
Wu, C., 6, 216 ⟨86MI 607-01⟩
Wu, C.-H. S., 3, 573 ⟨60JA5386, 60JA5394⟩
Wu, C.-M., 3, 62 ⟨90S1033⟩
Wu, D. K., 4, 332 ⟨69JOC3104⟩
Wu, F.-J., 4, 583, 585 ⟨86JOM(309)225⟩; 6, 518 ⟨85IC4449, 87JA7764⟩, 518 ⟨90JOM(383)295⟩
Wu, G., 2, 972 ⟨90TL493⟩; 4, 526 ⟨76JOM(122)145⟩, 658 ⟨89JA3089⟩, 716 ⟨93OM1213⟩; 5, 464 ⟨89JA8018⟩
Wu, G.-P., 2, 257 ⟨91S454⟩
Wu, G.-Y., 4, 900 ⟨91SC1213⟩
Wu, H., 1, 623 ⟨87TL801, 88JOC4862⟩; 3, 176, 647 ⟨88JOC4862⟩; 4, 660, 700 ⟨93OM2788⟩
Wu, H.-J., 2, 668 ⟨94TL729⟩; 3, 221 ⟨91CC770⟩
Wu, I.-Y., 1, 991 ⟨93OM3971⟩
Wu, J., 2, 301 ⟨91OPP399⟩, 304 ⟨90CJC456⟩
Wu, J.-C., 1, 244 ⟨91T2237⟩; 4, 276 ⟨90T2587⟩
Wu, M. T., 4, 979 ⟨76TL2405⟩; 5, 199 ⟨65CI(L)648⟩
Wu, R., 1, 680 ⟨89TL3927⟩
Wu, S., 3, 150 ⟨92JOC3807⟩; 6, 38 ⟨89JFC(44)433⟩
Wu, S.-W., 2, 751 ⟨92JOC3807⟩; 6, 15 ⟨89JCS(P1)2385⟩, 232, 234 ⟨89CC705⟩
Wu, T.-C., 1, 115 ⟨90JOC5045⟩, 121 ⟨93JOC2483, 93JOC2492⟩, 478 ⟨87JOC5057⟩, 1099 ⟨87JOC4819, 87JOC4823⟩; 2, 563, 959 ⟨90JOC5045⟩; 3, 128 ⟨92JOC2493⟩, 130, 286 ⟨89SC1833⟩, 286 ⟨93JOC2492⟩; 4, 354 ⟨54JA2502⟩
Wu, T.-T., 3, 311 ⟨82JA536⟩
Wu, W., 4, 677 ⟨92OM1480⟩, 955 ⟨91JA9406, 93OM3431⟩
Wu, W.-X., 1, 324 ⟨93TL4489⟩, 326, 327 ⟨91JA7788⟩, 326, 327 ⟨91TL707⟩, 326, 327 ⟨92JA7664⟩; 6, 604 ⟨90JOC1503⟩
Wu, X.-M., 3, 113 ⟨92TL6331⟩; 4, 154 ⟨85JCS(P1)1185⟩, 387 ⟨83TL4257⟩
Wu, Y., 1, 647 ⟨86TL4583⟩; 3, 48, 65 ⟨86TL4583⟩; 5, 329 ⟨93TL7371⟩; 6, 641 ⟨92JOC2497, 93TL3389⟩
Wu, Y.-C., 1, 784 ⟨85H(23)3085⟩
Wu, Y.-D., 1, 816, 817 ⟨90JOC1421⟩; 2, 41 ⟨84T2257⟩, 58, 521 ⟨86JA7404⟩
Wu, Y.-J., 1, 690 ⟨88TL4369⟩
Wu, Z., 4, 199 ⟨88TL6549⟩
Wu, Z.-P., 2, 639 ⟨87PAC1577⟩
Wu, Z.-Z., 3, 427 ⟨85JA3338⟩
Wucherer, E. J., 6, 525 ⟨79JA6110⟩
Wucherpfenning, W., 2, 245 ⟨67AG109⟩, 796 ⟨66TL1671, 67AG(E)149⟩, 796 ⟨B-68MI 215-05⟩
Wudl, F., 2, 187 ⟨67JA4099⟩, 478 ⟨84JOC4733⟩; 4, 78 ⟨67JA4099⟩, 291 ⟨85JA675⟩, 874 ⟨80CC866⟩, 876 ⟨82JA1154⟩, 877 ⟨81TL4199⟩; 6, 591, 594 ⟨80CC866A⟩, 628, 730 ⟨80CC866⟩, 631 ⟨81JOM(217)329⟩
Wuehsthoff, M. T., 2, 59 ⟨70JA6894⟩
Wuerthwein, E.-U., 3, 417 ⟨85CB3089⟩, 422 ⟨86CB857⟩
Wuerzer, B., 6, 494 ⟨79GEP2921130⟩
Wuest, H., 1, 450 ⟨77TL4305⟩, 651, 736 ⟨71HCA1767⟩, 823 ⟨74JA7573⟩, 826 ⟨89HCA996⟩, 983 ⟨79JOC4116⟩, 1131 ⟨79JOC546⟩; 3, 60 ⟨69JOC1122⟩, 260 ⟨71JOC609⟩, 493 ⟨71HCA1767⟩
Wuest, J. D., 1, 24 ⟨85JA6121⟩; 2, 900 ⟨77JOC3243⟩; 3, 527 ⟨77JOC2111⟩
Wuest, M., 1, 635 ⟨90HCA640⟩
Wuhrmann, J. C., 5, 220, 221 ⟨81CB2272⟩
Wuis, P. J., 1, 898 ⟨37RTC535⟩
Wülfert, P., 6, 720, 723 ⟨93JOM(459)55⟩
Wulff, G., 6, 71 ⟨71CAR(19)139⟩
Wulff, J., 3, 600 ⟨69CB1833⟩; 5, 1068 ⟨69CB1848⟩
Wulff, W., 1, 474 ⟨81T3957⟩; 3, 625 ⟨82JOM(240)C5⟩
Wulff, W. D., 1, 551 ⟨81JA7677, 93JA10671⟩; 2, 607 ⟨86JOC277⟩, 670 ⟨81JA7677, 84JOC2293⟩, 670 ⟨85JA1060⟩, 670, 979 ⟨91COS(5)1065⟩, 671 ⟨84JA7565, 91JA9873⟩; 3, 270 ⟨85JA503⟩, 680 ⟨90OM2867⟩; 4, 613 ⟨79JOM(168)23⟩, 614 ⟨78JA6236⟩, 955 ⟨85JA503, 87JOC3263⟩; 5, 696 ⟨88JA307⟩, 934 ⟨84JA434, 86JA520⟩, 934 ⟨89JOC3249⟩, 949 ⟨92JA10665⟩, 952,

958 ⟨93JA4602⟩, 953, 956 ⟨94JOC6882⟩, 955, 956 ⟨87JOC3263⟩, 955, 959 ⟨91COS(5)1065, 85JA503⟩, 956 ⟨89TL4061, 93OM597⟩, 956 ⟨94ICA215⟩, 958 ⟨94JOC5122⟩
Wulff, W. P., **1**, 551 ⟨87JOM(334)9⟩
Wulfman, C. E., **1**, 447 ⟨55JOC9⟩; **2**, 112 ⟨55JOC9⟩
Wuller, S. P., **6**, 357, 358 ⟨93JA545⟩
Wults, A. E., **2**, 529 ⟨85JA8186⟩
Wunder, D., **3**, 232 ⟨62LA(654)131⟩
Wunderlich, D., **1**, 860 ⟨59JA2568⟩
Wunderlich, H., **2**, 861 ⟨91ZAAC(606)119⟩, 1009 ⟨85CB4259⟩
Wunderlich, K., **6**, 73 ⟨60LA(632)38⟩
Wünsch, E., **2**, 218 ⟨82ZPC(363)1461⟩; **5**, 279 ⟨66CB110⟩; **6**, 486 ⟨74HOU(15/1)46⟩
Wunsch, M., **2**, 880 ⟨87AG(E)229⟩; **4**, 1039 ⟨87AG(E)229⟩, 1041 ⟨86AG(E)571⟩
Wursthorn, K. R., **4**, 735 ⟨79JOM(182)455, 81JOM(205)301⟩
Würthwein, E.-U., **2**, 1053 ⟨87AG(E)923⟩; **3**, 459 ⟨93CB957⟩, 567, 580, 585, 586 ⟨90CB2339⟩; **4**, 321 ⟨89CB2311⟩; **5**, 694 ⟨85CB3089⟩; 625 ⟨86CB3236, 87CB1271⟩, 730 ⟨92CB2487⟩
Wurtz, C., **2**, 647 ⟨1867LA(144)121⟩
Würzer, B., **6**, 632 ⟨86GEP(O)3504453⟩
Wust, H. H., **3**, 719 ⟨85LA1825⟩
Wüster, H., **1**, 990 ⟨85CC1812⟩
Wustrow, D. J., **4**, 208 ⟨85TL5411⟩
Wuts, P. G. M., **2**, 25 ⟨93SC2199⟩, 326 ⟨88JOC5989⟩, 529 ⟨82JOC2498⟩, 530 ⟨83JOC5398⟩, 539 ⟨B-91MI 210-02⟩, 658, 683 ⟨B-91MI 213-01⟩; **3**, 9, 16 ⟨B-91MI 301-01⟩, 123, 132 ⟨B-91MI 304-02⟩; **4**, 176, 180, 184, 185, 186, 198 ⟨B-91MI 404-01⟩, 176 ⟨B-91MI 404-02⟩, 203 ⟨77JOC1794, B-91MI 404-03⟩, 209 ⟨B-91MI 404-04⟩; **5**, 22 ⟨B-91MI 507-01⟩, 223 ⟨B-91MI 504-01⟩, 260 ⟨B-91MI 506-02⟩; **6**, 487 ⟨B-91MI 615-03⟩
Wuttke, F., **4**, 353 ⟨91JOM(403)29, 92JOM(424)273⟩; **6**, 127 ⟨90JOM(388)47⟩
Wuyts, H., **3**, 618 ⟨32BSB196⟩
Wyant, K., **2**, 625 ⟨91JOC6199⟩
Wyatt, B. K., **3**, 458 ⟨68CC250B⟩, 459 ⟨70JCS(A)2016⟩; **5**, 383 ⟨68JCS(A)2535⟩
Wychuyse, F., **4**, 12 ⟨79SC575⟩
Wyckoff, J. C., **1**, 925 ⟨71JA2065⟩; **2**, 20 ⟨71JA438⟩
Wyckoff, K. K., **3**, 102 ⟨68JHC95⟩
Wydra, R. L., **4**, 411 ⟨90JOC4777⟩
Wykypiel, W., **2**, 1002 ⟨78HCA3100⟩; **4**, 529 ⟨78CB2630⟩; **5**, 24, 105 ⟨83GEP3148422⟩
Wyler, R., **4**, 912 ⟨91HCA451⟩
Wylie, B. B., **5**, 215 ⟨65JPS1373⟩
Wylie, C. M., **3**, 627, 634 ⟨71JCS(C)2151⟩
Wylie, R. D., **1**, 357 ⟨80HCA1198⟩
Wyman, B. M., **2**, 35 ⟨64CJC2710⟩
Wyman, D. P., **4**, 13 ⟨64JOC1956⟩
Wyman, L., **5**, 460 ⟨74IC1595⟩
Wymann, W. E., **2**, 683 ⟨88SC1379⟩
Wynberg, H., **1**, 690 ⟨77JA602⟩, 970 ⟨88CC748⟩; **2**, 738, 751 ⟨63MI 208-01⟩, 1021, 1025, 1027 ⟨60MI 217-01⟩; **3**, 91 ⟨82OR(28)1⟩, 186 ⟨94SC103⟩, 250 ⟨69RTC994⟩, 527 ⟨70RTC23⟩; **4**, 17 ⟨82OR(28)1⟩; **5**, 28 ⟨60JOC2245⟩, 51 ⟨72SC79⟩, 88 ⟨72SC423⟩, 162 ⟨82JA166⟩, 411 ⟨71TL3445⟩
Wyne, K. J., **2**, 288 ⟨69JA1649⟩, 292 ⟨70IC106⟩
Wynn, W. K., **4**, 3 ⟨66JOC916⟩
Wyrick, S. D., **6**, 514 ⟨92MI 616-01⟩
Wysocki, Jr., R. J., **1**, 1130 ⟨89JOC1238⟩
Wysong, R. D., **2**, 626, 685 ⟨67JOC2358⟩
Wyss, C., **1**, 326, 327 ⟨93TL2445⟩
Wystrach, V. P., **2**, 428 ⟨59JA1103, 61JOC5138⟩, 440, 446 ⟨61JA168⟩, 835, 836 ⟨65IC764⟩, 837 ⟨65IC765⟩; **3**, 449 ⟨48JA1381⟩
Wythes, M. J., **1**, 968 ⟨93SL219, 93SL499⟩; **2**, 591 ⟨92JOC3397⟩; **3**, 130, 634 ⟨90SL735⟩, 288 ⟨89CC644⟩
Wyvratt, J. M., **6**, 25 ⟨87JOC944⟩
Wyvratt, M. J., **1**, 785 ⟨70JA1763⟩

X

Xavier, L. C., **2**, 528 ⟨91JOC751, 91JOC763⟩; **6**, 446 ⟨91JOC751⟩
Xi, Z., **1**, 244 ⟨91T2237⟩; **2**, 987, 988 ⟨94TL695⟩; **4**, 276 ⟨91T3431⟩
Xia, H., **3**, 539 ⟨91JOC6094⟩
Xia, J., **2**, 843 ⟨84S781, 86S691⟩
Xia, W. J., **1**, 647, 741 ⟨87TL2155⟩, **3**, 196, 231 ⟨87TL2155⟩
Xia, Y., **3**, 220 ⟨87JOC1375⟩
Xiang, J., **1**, 417, 795, 830 ⟨91JOC3572⟩
Xiang, Y., **1**, 987 ⟨91JCS(P1)665⟩
Xiang, Y. B., **1**, 374 ⟨89JA5493⟩; **2**, 533 ⟨89JA5493⟩
Xiao, C., **3**, 23 ⟨88TL6697⟩; **4**, 651, 724 ⟨88TL6697⟩
Xiao, M., **4**, 652 ⟨93JA10754⟩
Xiao, W., **1**, 646, 647 ⟨90JOM(400)121⟩
Xiao, W. J., **3**, 49 ⟨B-90MI 301-02⟩, 196 ⟨B-90MI 304-02⟩
Xiao, X., **3**, 436 ⟨B-94MI 304-02⟩
Xiaolan, L., **4**, 488 ⟨93MI 410-01⟩
Xie, G., **2**, 43 ⟨88TL2967⟩
Xie, L., **1**, 262 ⟨93JOC4376⟩, 623 ⟨87TL801, 88JOC4862⟩; **3**, 176, 647 ⟨88JOC4862⟩; **6**, 558 ⟨93OPP83⟩
Xie, L.H., **4**, 446 ⟨93S45⟩
Xie, Q., **1**, 890 ⟨92JA3978⟩
Xie, Y., **5**, 395 ⟨86JST(139)255⟩
Xie, Z., **1**, 1021 ⟨88CPB4337⟩
Xie, Z.-M., **4**, 566, 572 ⟨83OM921⟩; **6**, 165, 167, 193, 688 ⟨85OM339⟩, 350, 694 ⟨84CC698⟩
Xin, H. Y., **2**, 753 ⟨83JOC4397⟩; **3**, 626 ⟨83JOC4397⟩
Xin, Y., **3**, 269, 274 ⟨84S35⟩
Xin, Y. C., **4**, 213 ⟨94TL2537⟩
Xing, Y., **1**, 841 ⟨88JA3601⟩
Xingya, L., **1**, 367 ⟨83TL4181⟩; **3**, 367 ⟨81JA7032, 83TL4181⟩
Xin Yuan-Kong, **1**, 740, 741 ⟨80MI 116-02⟩, 740 ⟨B-80MI 110-01⟩
Xiong, H., **1**, 115 ⟨90JOC5045⟩, 116 ⟨89JOC3247⟩; **2**, 561 ⟨89JOC3247⟩
Xiong, Y., **1**, 900 ⟨90JA3225⟩
Xu, C., **1**, 599 ⟨87SC1839⟩
Xu, D., **2**, 48 ⟨92JOC2768⟩, 49 ⟨92JA7570⟩, 670 ⟨91JA7418⟩; **3**, 544 ⟨93JOC159⟩; **5**, 946 ⟨93TL227⟩
Xu, D. Q., **2**, 46 ⟨94TL2495⟩
Xu, H.-T., **6**, 219 ⟨89MI 607-02⟩
Xu, J.-Z., **6**, 461 ⟨91HCA465⟩
Xu, L., **2**, 43 ⟨88TL2967⟩
Xu, L. X., **4**, 737 ⟨92TL4537⟩
Xu, R. X., **3**, 307 ⟨81TL4899⟩
Xu, S. L., **2**, 668 ⟨91TL5923⟩, 671 ⟨92JA5881⟩; **3**, 321 ⟨93JOC4646⟩, 539 ⟨91JOC6094⟩
Xu, W., **4**, 513 ⟨92JOC6037⟩; **5**, 82 ⟨86MI 502-01⟩, 101 ⟨84TL1797⟩
Xu, W. H., **5**, 71 ⟨85CC1478⟩
Xu, X., **6**, 194, 195 ⟨85JA4337⟩
Xu, Y., **1**, 196 ⟨87JOC974⟩, 196 ⟨88JOC4419⟩, 634 ⟨89JFC(44)161⟩; **2**, 843 ⟨84S781, 86S691⟩, 846 ⟨84S778, 86S240⟩; **5**, 956 ⟨94ICA215⟩; **6**, 461 ⟨88JOC4419⟩
Xu, Y.-C., **1**, 551 ⟨93JA10671⟩; **4**, 955 ⟨87JOC3263⟩; **5**, 955, 956 ⟨87JOC3263⟩
Xu, Y. Y., **1**, 738 ⟨B-81MI 105-01⟩; **4**, 735 ⟨92TL1221, 93JCS(P1)795⟩
Xu, Z., **1**, 316 ⟨91JA5079, 91JA8950⟩, 316 ⟨92JA6692, 93JA6614⟩; **2**, 561 ⟨91JA5079, 91JA8950⟩, 562 ⟨92JA6692⟩
Xu, Z.-Q., **2**, 16 ⟨91CC179⟩; **3**, 149 ⟨91CC179⟩, 639 ⟨92JCS(P1)313⟩; **4**, 704 ⟨92JCS(P1)313⟩, 763 ⟨89JFC(42)69⟩
Xue, C.-B., **2**, 133 ⟨92TL1435⟩
Xueqing Wen, **1**, 739 ⟨87S370⟩
Xun, X., **2**, 331 ⟨89SC1423⟩; **3**, 415 ⟨89SC1423⟩
Xuong, N. D., **3**, 91, 98 ⟨55BSF1594⟩, 280 ⟨61JOC2401⟩

Y

Yabe, A., **3**, 351 ⟨89TL2955⟩
Yabe, M., **5**, 210 ⟨92CL1437⟩
Yabe, Y. T., **3**, 356 ⟨90JCS(P1)3003⟩
Yablokova, N. V., **4**, 637, 641 ⟨75JOM(96)213⟩
Yablokov, V. A., **4**, 637, 641 ⟨75JOM(96)213⟩
Yablonski, Z., **5**, 595 ⟨86KGS1405⟩
Yablunky, H. L., **2**, 882 ⟨41JA207⟩
Yabuki, Y., **2**, 1088 ⟨76TL3969⟩
Yachida, K., **4**, 750 ⟨67BCJ1275⟩
Yadagiri, P., **5**, 368 ⟨88SC545⟩; **6**, 511 ⟨88SC545⟩
Yadau, V. K., **2**, 456 ⟨90SC239⟩
Yadav, J. S., **1**, 1002 ⟨92TL135⟩, 1066 ⟨89TL5455, 90T7033⟩; **2**, 85 ⟨90TL4495⟩; **3**, 40 ⟨89TL6769⟩, 709 ⟨83SC331, 83SC379⟩, 709 ⟨88T7243, 90TL6217⟩
Yadav, L. D. S., **2**, 256 ⟨81S547⟩; **6**, 562 ⟨92S919⟩, 564 ⟨89IJC(B)439⟩
Yadav, V. G., **6**, 647 ⟨82PIA441⟩
Yadav, V. K., **2**, 141 ⟨88TL897⟩
Yadhav, P. K., **2**, 516 ⟨81JOC5047⟩
Yadla, R., **3**, 639 ⟨82IJC(B)1046⟩
Yaeger, D. B., **2**, 606 ⟨80JOC4649⟩
Yagi, H., **2**, 659 ⟨76JOC977⟩; **3**, 430 ⟨90BCJ1843⟩
Yagi, T., **6**, 469 ⟨94CL495⟩
Yagihara, M., **5**, 242 ⟨92PS(67)27⟩, 506, 509 ⟨92JAP(K)4224595⟩, 506, 509 ⟨92MI 512-01⟩
Yagihara, T., **1**, 233, 400 ⟨72JOC1721⟩, 820 ⟨71JOC1732⟩; **3**, 373, 378 ⟨72JOC1721⟩; **5**, 546 ⟨72MI 513-01⟩, 550 ⟨72T3203⟩
Yagii, T., **5**, 359 ⟨90TL5327⟩
Yagodzinskaya, E., **5**, 595 ⟨86KGS1405⟩
Yagodzinski, T., **5**, 595 ⟨86KGS1405⟩
Yagudeev, T. A., **2**, 1076 ⟨82IZV933⟩
Yagupol'skii, L. B., **6**, 232 ⟨52ZOB2216⟩
Yagupol'skii, L. M., **1**, 109 ⟨88JOU530⟩; **2**, 100 ⟨88JOU1473⟩, 287, 732 ⟨68ZOB2509⟩, 476, 614 ⟨89JGU469⟩, 848 ⟨59ZOB3766⟩; **4**, 123 ⟨60JGU3986⟩, 771 ⟨90ZOR740⟩, 777 ⟨73ZOR69⟩, 793 ⟨77JOU1918⟩, 817 ⟨87JOU246⟩, 826 ⟨65ZOB1639⟩, 861 ⟨68ZOB1503⟩, 992 ⟨76UKZ204⟩; **5**,, 58 ⟨85ZOR769⟩, 112 ⟨85ZOR2407⟩, 204 ⟨77S574⟩, 205 ⟨88ZOR593⟩, 566, 567, 610, 657 ⟨76ZOR2213, 76JOU2148⟩; **6**, 14 ⟨93RHA232⟩, 19 ⟨B-89MI 601-02⟩, 232, 234, 236, 237 ⟨85S667⟩, 233 ⟨79ZOR1245, 79ZOR396⟩, 233 ⟨82JFC(21)365⟩, 233, 234 ⟨84ZOR115⟩, 233, 234, 237 ⟨65ZOB1628, 67ZOB1770⟩, 236 ⟨68ZOB2591, 79S972⟩, 237 ⟨68ZOB2509⟩, 238 ⟨66ZOB1309, 71ZOR2000⟩, 501 ⟨82JOU19582220⟩; **6**, 37 ⟨83JFC(22)105⟩, 38, 39 ⟨70ZOR144, 70ZOR2498⟩, 39 ⟨75ZOR1672, 76ZOR1798⟩, 39, 43 ⟨76ZOR1287⟩, 49 ⟨77ZOR2008⟩, 50 ⟨79ZOR1561⟩, 52 ⟨76ZOR2213⟩, 61 ⟨68ZOB2813⟩, 69 ⟨71ZOR2161⟩, 82, 302 ⟨69ZOB1755⟩, 229 ⟨55DOK(105)100, 64ZOB1979⟩, 232 ⟨54JGU885⟩, 232, 233, 234 ⟨75S721⟩
Yagupol'skii, Y. L., **3**, 588 ⟨86JOU1640⟩; **4**, 992 ⟨76UKZ204⟩; **5**, 566, 567, 610, 657 ⟨76ZOR2213⟩; **6**, 52 ⟨76ZOR2213⟩, 235 ⟨90S1151⟩, 275, 317 ⟨83ZOR79⟩
Yagupol'skii, Y. T., **2**, 15 ⟨91RCR1050⟩
Yagupol'skii, Yu. L., **4**,, 282 ⟨83TL87⟩, 864 ⟨79ZOR2416⟩; **5**, 314 ⟨80ZOR1330, 76JOU2148⟩, 657 ⟨81JOU1527, 81ZOR1720⟩; **6**, 234 ⟨94S145⟩
Yagupsky, G., **4**, 656 ⟨70CC1369⟩; **5**, 461 ⟨71JCS(A)850⟩
Yahata, N., **4**, 854, 857 ⟨83CL767, 86JOC508⟩, 858 ⟨90TL4621⟩
Yahner, J. A., **5**, 1111 ⟨78JOC1604⟩
Yahya, A., **4**, 21 ⟨89JCS(P1)2009⟩
Yajima, H., **4**, 218 ⟨89CPB526⟩
Yajima, T., **1**, 195 ⟨93TL7619⟩
Yakei, H., **1**, 191, 238 ⟨84CL1505⟩
Yakhontov, L. N., **1**, 45 ⟨70KGS495⟩
Yakiminskaya, N. Sh., **4**, 342 ⟨78ZOB2653⟩
Yakimovich, S. I., **1**, 225 ⟨72JOU2530⟩
Yakobson, G. G., **2**, 793 ⟨85RCR1706⟩; **4**, 737 ⟨85JFC(28)99⟩; **6**, 12 ⟨69IZV196⟩, 609 ⟨77JFC(9)505⟩
Yakovlev, I. P., **5**, 381, 382 ⟨70DOK(194)595⟩, 381 ⟨71ZOB2501⟩, 778 ⟨74BAU1593⟩
Yakovlev, V. N., **3**, 585, 586, 587 ⟨80JGU45⟩
Yakovleva, G. Y., **4**, 563 ⟨92TL3927⟩
Yakoyama, Y., **2**, 518 ⟨91TL1479⟩
Yakshin, V. V., **2**, 439 ⟨68IZV190⟩; **5**, 395, 397, 402 ⟨68T2995⟩, 403, 404 ⟨74JGU1685, 72JGU1256⟩
Yaktin, H. K., **3**, 102 ⟨68JHC95⟩
Yakubovich, A. Ya., **2**, 470 ⟨69JGU2161⟩; **2**, 498 ⟨50DOK303, 52ZOB1528⟩; **4**, 114 ⟨52JGU1575⟩, 135 ⟨41JGU41⟩, 135, 136 ⟨52JGU1569⟩, 140 ⟨52ZOB1783⟩, 154, 155 ⟨52JGU1827⟩, 555, 569 ⟨58ZOB728⟩, 793 ⟨70JOU750⟩; **5**, 822 ⟨60ZOB2409⟩; **6**, 258, 259 ⟨65ZOB1412⟩
Yakubovich, A. Y., **2**, 166, 170 ⟨65ZOB1412⟩; **5**, 1018 ⟨64ZOB3609⟩; **6**, 54 ⟨66ZOB728⟩, 242 ⟨61DOK(141)357, 62DOK(142)354, 68ZOB709⟩
Yakupov, I. S., **4**, 1078 ⟨86MI 424-01⟩
Yakura, T., **1**, 362 ⟨90CC1436⟩; **3**, 158 ⟨85TL3837⟩
Yakutin, V. I., **6**, 54 ⟨66ZOB728⟩, 613 ⟨72JGU799⟩
Yale, H. L., **1**, 446 ⟨51JA2880⟩; **2**, 757 ⟨43CRV209⟩, 886 ⟨51JA4470⟩, 889 ⟨51JA2880⟩; **5**, 222 ⟨51JA4470⟩, 354 ⟨53JA1933⟩, 652 ⟨71JOC3238⟩, 864 ⟨43CRV209⟩; **6**, 104 ⟨61JOC4315⟩
Yalovskaya, A. I., **2**, 7 ⟨89JOU587, 89JOU642⟩
Yalpani, M., **1**, 60 ⟨83CB1336⟩, 1077 ⟨70AG(E)464, 71JOC2836⟩; **3**, 553 ⟨73JOC338, 74JOC3906⟩; **4**, 874 ⟨73JOC338, 74JOC3906⟩; **5**, 381 ⟨93CB565⟩, 516, 519, 522 ⟨76JOC729⟩, 589, 626 ⟨77S328⟩
Yamabe, K., **2**, 1055 ⟨83CC329⟩; **5**, 607 ⟨86S1063⟩
Yamabe, M., **4**, 117 ⟨81CC1173⟩, 767 ⟨80BP2029827⟩
Yamabe, S., **2**, 931 ⟨90TL2917⟩, 931, 939 ⟨92JOC5610⟩; **4**, 947 ⟨91JCS(P1)1555, 92JOC5610⟩
Yamabe, T., **6**, 555 ⟨92CC1410⟩
Yamada, A., **4**, 517, 538, 1015 ⟨82BCJ1163⟩, 538 ⟨81AG(E)126⟩; **6**, 162 ⟨82BCJ1163⟩
Yamada, E., **2**, 627 ⟨69OPP221⟩
Yamada, H., **1**, 16 ⟨87JOC4878⟩, 166 ⟨94SL1⟩, 228 ⟨88TL343⟩, 827 ⟨88TL655⟩, 827, 829 ⟨87JOC1201⟩, 828 ⟨89JOC1223⟩; **2**, 121 ⟨81CL1457⟩; **3**, 117 ⟨88JOC3587, 88SC869⟩, 426 ⟨90SL477⟩; **4**, 950 ⟨81JA5259⟩; **5**, 118 ⟨90PAC1441⟩, 179 ⟨89CL1775⟩
Yamada, J., **1**, 1026 ⟨86JA7116⟩; **2**, 70 ⟨90JA6118⟩; **4**, 390 ⟨90JA6118⟩
Yamada, J.-T., **2**, 85 ⟨89TL4255⟩
Yamada, K., **1**, 33 ⟨78CC1089⟩, 216 ⟨90JCS(P1)1226⟩, 238, 246 ⟨89BCJ2913⟩, 532, 1020 ⟨79TL3437⟩, 532, 1020 ⟨85JA972⟩, 582 ⟨92BCJ2366⟩, 710 ⟨84JA4547⟩, 938 ⟨89CC1265, 91JCS(P1)1173⟩, 1029 ⟨91TL3405⟩, 1034 ⟨75TL1961⟩, 1090 ⟨89JOC1757⟩; **2**, 4 ⟨82OM369⟩; **3**, 113 ⟨85CL1385⟩, 191 ⟨89CC110⟩, 417, 658 ⟨78S892⟩, 684 ⟨B-92MI 318-01⟩; **4**, 206 ⟨88CC560, 92BCJ2366⟩, 208 ⟨89CL221⟩, 222 ⟨76TL65⟩, 938 ⟨89JOC1757⟩, 960 ⟨87AR(C)197⟩; **5**, 199 ⟨86TL4609⟩, 325 ⟨82JOC5283, 87JOC2327⟩, 329 ⟨94TL1247⟩, 513 ⟨89CC1265, 91JCS(P1)1173, 91JOC2849⟩; **6**, 550 ⟨89CC1265, 91JOC2849⟩
Yamada, M., **1**, 45 ⟨81CPB3489⟩, 409, 821, 822 ⟨72JOC3791⟩, 697, 700, 703 ⟨86JOC3916⟩, 710

⟨84JA4547⟩; **2**, 64 ⟨84JA6709⟩, 118, 124 ⟨67BCJ2380⟩, 121 ⟨77JOC2180⟩, 429 ⟨83BCJ219⟩, 478 ⟨80CC1183⟩, 663 ⟨82JOC1200⟩; **3**, 57, 215 ⟨86TL1805⟩; **5**, 183, 199 ⟨79CC179⟩; **6**, 465 ⟨93JAP05310644⟩
Yamada, N., **1**, 192 ⟨93JOC16⟩; **2**, 850 ⟨82BCJ909⟩; **3**, 62 ⟨82CL1997⟩, 622 ⟨93JCS(P1)2151⟩; **6**, 465 ⟨93EUP528498⟩
Yamada, S., **1**, 102 ⟨69TL3101⟩, 305 ⟨92TL7521⟩, 929 ⟨87TL3963⟩, 931 ⟨87T3371⟩, 1024 ⟨86TCC3⟩; **2**, 235 ⟨86H(24)303⟩; **3**, 468 ⟨71TL4495, 75T227⟩; **4**, 837 ⟨85CPB2256⟩; **5**, 223 ⟨85JCS(P1)2361⟩, 235 ⟨74JOC3302⟩, 265, 278, 993 ⟨74T2151⟩, 300 ⟨73TL1595⟩, 329, 330 ⟨85CC1418⟩, 652 ⟨71TL3595⟩, 745 ⟨69TL699, 72JA6203⟩
Yamada, S.-I., **1**, 55 ⟨69TL3105, 76TL57⟩, 55 ⟨76TL61⟩; **5**, 354 ⟨75TL2701⟩, 746 ⟨78CPB2246⟩, 869 ⟨84OS(62)187⟩
Yamada, T., **1**, 372 ⟨87JOC2523⟩, 1039 ⟨84CL405⟩; **2**, 701 ⟨87CL293⟩; **3**, 57 ⟨82CL1987⟩, 139 ⟨80TL3199⟩, 156 ⟨91CL145⟩; **4**, 220 ⟨78BCJ2401⟩, 517, 518 ⟨85H(23)2489⟩; **5**, 132 ⟨78BCJ2401⟩, 237, 277 ⟨86CL915⟩, 237, 277 ⟨87CL293⟩, 279 ⟨88CC419⟩, 729 ⟨87JOC2523⟩
Yamada, W., **1**, 145 ⟨90JOC2234⟩
Yamada, Y., **1**, 238 ⟨83JOC1293⟩, 243 ⟨88JA4815⟩, 582 ⟨89CPB2882⟩; **2**, 92 ⟨81JA6237⟩, 257 ⟨79TL5015, 83JOC1293⟩, 323 ⟨88CL1289⟩, 855 ⟨68TL5015⟩; **3**, 5, 118 ⟨83JOC1286⟩, 115 ⟨79TL1401⟩, 226 ⟨84TL3621⟩, 290, 295, 296 ⟨83TL3869⟩, 627, 634 ⟨90S768⟩, 659 ⟨69AG(E)343⟩; **4**, 21 ⟨78CL73⟩, 201 ⟨87S1099⟩, 485 ⟨88TL663⟩, 654 ⟨87MI 414-01⟩; **5**, 199 ⟨86JAN1257⟩, 248, 249 ⟨87NKK1442, 88BCJ2205⟩, 250 ⟨87S371⟩; **6**, 143 ⟨81JOC1571⟩
Yamagami, K., **3**, 698 ⟨66JOC4170⟩
Yamagata, H., **5**, 403, 404 ⟨80JOC4162⟩; **6**, 582 ⟨79TL3013⟩
Yamagata, N., **1**, 492 ⟨91JA1364⟩
Yamagichi, T., **1**, 103 ⟨68BCJ2921⟩
Yamagishi, K., **2**, 19 ⟨87CL1675⟩
Yamago, S., **1**, 360 ⟨89JA7285, 90JOC5553⟩, 360 ⟨92JA8707⟩; **4**, 830 ⟨89JA7285⟩
Yamaguchi, A., **2**, 35 ⟨87CL715⟩
Yamaguchi, H., **1**, 243 ⟨81CL627, 82BCJ182⟩, 594, 611, 621 ⟨81CL447⟩, 611, 621 ⟨83S1006⟩, 1069 ⟨81S879⟩, 1080 ⟨85CC697⟩; **2**, 289 ⟨80SC595⟩, 293 ⟨81CL447⟩, 464 ⟨82TL1263⟩, 1020, 1063 ⟨85TL4501⟩; **3**, 353, 360 ⟨82TL1263, 80JOC927⟩; **4**, 402 ⟨81CL447⟩; **5**, 325 ⟨90TL6547⟩, 1113 ⟨80SC595⟩, 1117 ⟨82SC131⟩
Yamaguchi, I., **5**, 1103 ⟨90JSP62⟩
Yamaguchi, J., **3**, 118 ⟨92CL1185⟩, 121 ⟨92CL423⟩
Yamaguchi, K., **1**, 305 ⟨92TL7521⟩, 707 ⟨76T1705⟩, 988 ⟨90TL7469⟩; **2**, 175 ⟨75CL307⟩, 360 ⟨75CL991⟩; **3**, 440 ⟨82S461⟩, 627 ⟨91JA6318⟩; **5**, 235 ⟨85JOC273⟩, 360 ⟨91JA6318⟩, 368 ⟨85MAC1803⟩
Yamaguchi, M., **1**, 14 ⟨87TL4437⟩, 180 ⟨86TL3403⟩, 195 ⟨86TL1195⟩, 204, 1023 ⟨83TL391⟩, 204, 1024 ⟨83TL5121⟩, 223, 1023 ⟨84CC202⟩, 229 ⟨84TL1159, 84TL4549⟩, 404, 406, 816, 819 ⟨86TL4581⟩, 404, 406, 816, 819 ⟨88TL4763⟩, 404, 406, 819 ⟨92BCJ1841⟩, 458, 516 ⟨91TL5559⟩, 680 ⟨86TL3891, 87CL2101⟩, 816 ⟨86TL4577, 87TL803, 91TL5137⟩, 954 ⟨86TL4599⟩, 962 ⟨86TL5237⟩, 975 ⟨87CL2275⟩, 1024 ⟨84T4261⟩, 1038 ⟨91JOC4091⟩, 1045 ⟨81CL1415, 83TL5131, 86S421⟩, 1046 ⟨83CL35⟩, 1047 ⟨83TL1719⟩, 1050 ⟨92CL2479⟩, 1051 ⟨93JA3362⟩, 1134 ⟨77CL959⟩; **2**, 517 ⟨80CL993⟩, 662 ⟨86TL2401⟩, 776 ⟨67BCJ1538⟩, 1100 ⟨92CL2479⟩; **3**, 3, 89, 114 ⟨81BCJ2221⟩, 219 ⟨85CL1145⟩, 273 ⟨83TL5131⟩; **4**, 219 ⟨79CL1277⟩, 1069 ⟨93CL1881⟩; **5**, 125 ⟨87TL4711⟩, 173 ⟨78JOM(159)137, 79BCJ1989, 79CL1021, 81CL1415⟩, 411 ⟨85TL2675⟩; **6**, 126 ⟨85TL2675⟩
Yamaguchi, R., **1**, 1004 ⟨82CL1485⟩, 1042 ⟨83TL1801, 87JOC2094⟩, 1043 ⟨88TL1785⟩, 1060 ⟨87CL1519⟩; **2**, 783 ⟨82CL1505⟩; **4**, 810 ⟨82SC1027⟩; **5**, 146 ⟨87CC83⟩, 460 ⟨84JOC4460⟩
Yamaguchi, S., **1**, 818 ⟨93TL8453⟩; **2**, 64 ⟨73JOC1870⟩; **3**, 668 ⟨86TL2411⟩; **4**, 852 ⟨77JCS(P1)1273⟩; **5**, 547 ⟨77JCS(P1)1273⟩
Yamaguchi, T., **1**, 1010 ⟨92JMC241⟩; **2**, 626 ⟨89BCJ3373⟩; **3**, 353 ⟨92MI 308-01⟩, 357 ⟨90CRV879⟩; **4**, 321 ⟨72CI(L)380⟩, 426 ⟨86H(24)25⟩, 432, 433 ⟨88S106⟩, 902 ⟨67BCJ2641⟩; **5**, 199 ⟨90JOC1679⟩; **6**, 105, 106 ⟨67BCJ2641⟩, 113 ⟨65BCJ2107, 66BCJ2005⟩, 316 ⟨60JAP6018199⟩, 554 ⟨80CPB110⟩
Yamaguchi, Y., **1**, 574 ⟨89TL3335⟩; **2**, 81 ⟨89JOC6001⟩; **3**, 703, 714 ⟨89TL3335⟩
Yamagushi, S., **2**, 939 ⟨92JA5867⟩
Yamaichi, A., **1**, 308 ⟨79CC100, 82H(18)187⟩, 1085 ⟨81TL4171⟩; **6**, 128 ⟨79CC100⟩
Yamaichi, S., **4**, 419 ⟨90BCJ1160⟩
Yamaji, K., **1**, 822 ⟨94CC789⟩
Yamaji, T., **1**, 161, 450, 1092, 1105 ⟨83TL1041⟩; **2**, 923, 924 ⟨83TL1041⟩
Yamakado, Y., **1**, 651 ⟨81JA5568⟩; **4**, 1059, 1060 ⟨81JA5568⟩
Yamakawa, K., **1**, 598 ⟨92TL7181⟩; **2**, 85 ⟨88TL2851⟩, 322 ⟨88TL4093⟩; **3**, 36 ⟨85BCJ2849⟩, 36, 44, 162, 167, 168, 183 ⟨92SL455⟩, 153, 154 ⟨92BCJ2800⟩, 156 ⟨81T473⟩, 158 ⟨91BCJ1582⟩, 158 ⟨91JOC4129⟩, 170 ⟨90BCJ1266⟩, 170 ⟨92TL7181⟩, 183 ⟨86BCJ457⟩, 209 ⟨86TL2471, 87BCJ1839⟩, 227 ⟨88CPB1169⟩; **4**, 73, 74, 75, 77 ⟨92SL455⟩, 74 ⟨85BCJ1983⟩, 74, 77 ⟨92TL1483⟩, 76 ⟨88TL313, 93TL2331⟩, 80 ⟨93BCJ2339⟩, 233 ⟨86BCJ2463, 89CPB184⟩, 777 ⟨93BCJ1866⟩; **5**, 166 ⟨87TL655⟩, 272, 283 ⟨92TL1455⟩
Yamakawa, Y., **2**, 676 ⟨86TL2125⟩; **3**, 162 ⟨87JOC2230⟩
Yamaki, T., **5**, 292 ⟨81CPB1063⟩
Yamamoto, A., **1**, 194 ⟨89TL375⟩, 228 ⟨93TL2135⟩, 415, 826 ⟨89SC2109⟩; **2**, 301 ⟨90TL1743⟩, 331 ⟨89JA6301⟩; **4**, 607, 646 ⟨86JPS(A)1839⟩; **5**, 226 ⟨81BCJ1720⟩, 384, 387 ⟨80BCJ1299⟩, 385 ⟨78BCJ2909⟩, 461 ⟨76BCJ546⟩, 462 ⟨81BCJ2161⟩, 465 ⟨83CL115⟩; **6**, 470 ⟨92OM171⟩
Yamamoto, E., **1**, 1010 ⟨81CPB3543⟩
Yamamoto, F., **5**, 327 ⟨81JGU2125⟩
Yamamoto, G., **1**, 258, 259 ⟨87CL1163⟩; **5**, 1052 ⟨72BCJ2937⟩; **6**, 593 ⟨72BCJ2937⟩
Yamamoto, H., **1**, 112 ⟨91JA5893⟩, 116 ⟨91JA8955⟩, 141 ⟨73JA2694⟩, 141, 834 ⟨73JA2693⟩, 184 ⟨85JA3343⟩, 187 ⟨91SL63⟩, 206 ⟨74JA5254⟩, 303 ⟨94JA4131⟩, 387 ⟨92T3749⟩, 415, 826 ⟨88JA7922, 90JA316⟩, 490, 535 ⟨82OM542⟩, 565 ⟨75TL1545⟩, 651 ⟨74BCJ2529, 74JA1620⟩, 651 ⟨81JA5568, 89TL1265⟩, 825, 826 ⟨90JA7791⟩, 826 ⟨90TL377⟩, 830 ⟨94T3099⟩, 899 ⟨87BCJ3721, 88NKK157⟩, 927 ⟨88TL2215⟩, 971, 1054 ⟨91COS(2)81⟩, 973 ⟨82CL1029⟩, 1041 ⟨84TL2367⟩, 1044, 1059 ⟨82TL3395⟩, 1055 ⟨82JA7667, 86JA483⟩, 1055 ⟨86TL1175⟩, 1056, 1057 ⟨84BCJ2768⟩, 1106 ⟨85AG(E)668⟩, 1130 ⟨85CL1689⟩; **2**, 4 ⟨78JA290⟩, 30, 34 ⟨86JA6036⟩, 73, 563 ⟨91JA8955⟩, 74 ⟨89JOC5198⟩, 420 ⟨75TL1859⟩, 431 ⟨91BCJ2398⟩, 528 ⟨93JA11490, 92JOC6386⟩, 596 ⟨94T8983⟩, 618, 630, 909 ⟨70TL447⟩, 630 ⟨88S236⟩, 679 ⟨86TL1363, 90JOC5814⟩, 769 ⟨89BCJ128⟩, 905 ⟨86JA483⟩, 936 ⟨86BCJ657⟩; **3**, 9, 122 ⟨91SL491⟩, 21 ⟨73JA2693⟩, 26 ⟨93JA3943⟩, 34 ⟨91T6983⟩, 54 ⟨73TL2465⟩, 62 ⟨78TL3051⟩, 62, 72, 271 ⟨78TL3047⟩, 125 ⟨77BCJ1592⟩, 127 ⟨88BCJ2241⟩, 159 ⟨90JA7422⟩, 184 ⟨91COS(6)171⟩, 227 ⟨81CL1341⟩, 244 ⟨88JA310⟩, 265 ⟨90S195⟩, 269, 437 ⟨83JA6312⟩, 407, 438 ⟨82TL3395⟩, 435 ⟨91COS(6)786, 83JA2831⟩, 437 ⟨83TL3255⟩, 437, 438 ⟨83JA672⟩, 639 ⟨84BCJ2768⟩, 700 ⟨79JA1609⟩; **4**, 19 ⟨74JA3010, 77BCJ1588⟩, 122 ⟨92CL407⟩, 174 ⟨85JOC5444, 90JOC5814⟩, 208 ⟨92SL565⟩, 648 ⟨74JA1620⟩, 750 ⟨88S236⟩, 889, 900, 970, 971, 972 ⟨90S195⟩, 956, 957 ⟨73JA2694⟩, 1059, 1060 ⟨81JA5568⟩, 1061 ⟨84BCJ2768⟩; **5**, 30 ⟨88OS(67)76⟩, 421 ⟨90JA7791, 91SL793⟩, 421, 431

⟨91TA647⟩, 552 ⟨73JA5803⟩, 661, 728 ⟨83TL3255⟩, 731 ⟨83JA2831⟩, 777 ⟨81JA4186⟩; **6**, 25, 31 ⟨74JA3010⟩, 150 ⟨88BCJ2927⟩, 558 ⟨89CPB1249⟩

Yamamoto, I., **1**, 430 ⟨85S676⟩; **2**, 447 ⟨88JOC2937⟩; **3**, 557 ⟨80JOC5377⟩, 557, 593 ⟨81JCS(P1)2727⟩; **4**, 339 ⟨77CL485⟩, 365 ⟨84JCS(P1)435⟩, 773, 781, 783 ⟨85S676⟩; **5**, 303 ⟨74JOC3924⟩

Yamamoto, J., **4**, 394 ⟨75JOC3052⟩; **6**, 555 ⟨75JOC3052⟩

Yamamoto, K., **1**, 192 ⟨92SL323⟩, 248 ⟨88BCJ1281⟩, 305 ⟨92TL7521⟩, 521 ⟨78CL649⟩, 635, 1075 ⟨83T2869⟩, 690 ⟨83JA7171, 86BCJ2209⟩, 690 ⟨88JA3578⟩, 788 ⟨91AG(E)1173⟩, 990 ⟨82JOC1435⟩; **2**, 306 ⟨80JOC2739⟩, 537 ⟨90SL259⟩, 915 ⟨75JOM(86)C27⟩; **3**, 58 ⟨77S721, 78CL859⟩, 83 ⟨71BCJ2569⟩, 549 ⟨71T1781⟩; **4**, 892 ⟨73CL111, 76JCS(P1)783⟩, 1032, 1033 ⟨74TL4405, 80BCJ1138⟩, 1048 ⟨79JOM(179)377⟩; **5**, 225, 375 ⟨89BCJ2111⟩, 419 ⟨80TL1653, 87OM974⟩; **6**, 14 ⟨80JCS(P1)2755⟩, 22, 246 ⟨79TL4071⟩, 457 ⟨72CL483⟩

Yamamoto, K.-i., **3**, 684 ⟨84CC1513⟩

Yamamoto, M., **1**, 216 ⟨90JCS(P1)1226⟩, 582 ⟨92BCJ2366⟩, 767, 769 ⟨88BCJ467⟩, 938 ⟨89CC1265, 91JCS(P1)1173⟩, 983 ⟨85TL631⟩, 1090 ⟨89JOC1757⟩; **2**, 434 ⟨82CL977⟩, 941 ⟨90JOC1108⟩; **3**, 191 ⟨89CC110⟩, 374 ⟨75CC289, 76JCS(P1)1688, 92NKK420⟩; **4**, 206 ⟨88CC560, 92BCJ2366⟩, 208 ⟨89CL221⟩, 938 ⟨89JOC1757⟩, 960 ⟨87BCJ1523⟩; **5**, 215 ⟨89S334⟩, 325 ⟨82JOC5283, 87JOC2327⟩, 513 ⟨89CC1265, 91JCS(P1)1173, 91JOC2849⟩; **6**, 550 ⟨89CC1265, 91JOC2849⟩

Yamamoto, S., **1**, 119 ⟨80JA2318⟩, 312 ⟨82JOC119⟩, 430 ⟨85S676⟩; **2**, 84, 445 ⟨88BCJ4476⟩, 88 ⟨83JA2908⟩, 524 ⟨72BCJ2244⟩, 702 ⟨82JOC119⟩, 853 ⟨88BCJ4029⟩; **3**, 118 ⟨92CL1169⟩; **4**, 773, 781, 783 ⟨85S497⟩; **5**, 364 ⟨85H(23)677⟩

Yamamoto, T., **2**, 588, 591 ⟨89AG(E)351⟩, 829 ⟨92CL547⟩; **3**, 38, 174 ⟨82TL4813⟩, 338 ⟨82CL101, 85CC1299⟩, 338, 339 ⟨89JCS(P1)1241⟩, 356 ⟨84JHC1445, 90JCS(P1)3003⟩; **4**, 852 ⟨77JCS(P1)1273⟩; **5**, 226 ⟨81BCJ1720⟩, 384, 387 ⟨80BCJ1299⟩, 461 ⟨76BCJ546⟩, 462 ⟨81BCJ2161⟩, 465 ⟨83CL115⟩, 547 ⟨77JCS(P1)1273⟩, 555 ⟨77JOC3383⟩, 1023 ⟨92OPP346⟩, 1024 ⟨91JAP(K)3193760⟩; **6**, 316 ⟨60JAP6018199⟩

Yamamoto, W., **4**, 943 ⟨87JCS(P1)589⟩

Yamamoto, Y., **1**, 119 ⟨80JA2318⟩, 188 ⟨89JA4864, 89JOC4055, 90AG(E)609, 92JOC1024⟩, 188, 227 ⟨92SL769⟩, 189 ⟨93JOC1207⟩, 189 ⟨93T9479, 89T435⟩, 257 ⟨90JOC408⟩, 305 ⟨87CC1572⟩, 312 ⟨82JOC119⟩, 312, 478, 540 ⟨86AG(E)947⟩, 450, 510 ⟨78CC702⟩, 455 ⟨79MI 110-01⟩, 488 ⟨77JA5652⟩, 531 ⟨75JA5606⟩, 531, 663 ⟨77CC852⟩, 540 ⟨79JOC1744⟩, 713, 714, 716 ⟨84JOC1096⟩, 839 ⟨73CC848⟩, 1001 ⟨80CL669⟩, 1026 ⟨86JA7116⟩, 1038, 1039 ⟨86CC102⟩, 1059 ⟨84CC1004⟩; **2**, 54 ⟨88JA4475⟩, 67 ⟨82H(18)357⟩, 70 ⟨90JA6118⟩, 78 ⟨84T2239, 87ACR243⟩, 79 ⟨92JOC7003⟩, 85 ⟨89TL4255⟩, 250 ⟨93SL465⟩, 268 ⟨74TL793, 75BCJ3682⟩, 268 ⟨75JOC3644, 75TL2689⟩, 355 ⟨76JOC300⟩, 517 ⟨81JA1969, 81JA3229⟩, 524 ⟨72CC71, 72S699⟩, 566 ⟨81TL2895⟩, 586, 702 ⟨86AG(E)947, 82JOC119⟩, 703 ⟨80TL4607⟩, 938 ⟨91CC703⟩, 997 ⟨91COS(2)55⟩, 1004 ⟨75CC668⟩; **3**, 5, 118 ⟨83JOC1286⟩, 31 ⟨93CRV2207⟩, 115 ⟨79TL1401⟩, 117 ⟨85TL2107⟩, 311 ⟨82CPB2003, 82H(19)41⟩, 350 ⟨83CPB1929⟩, 407 ⟨85AG(E)668⟩, 440, 456 ⟨89TL4275⟩, 694 ⟨80CCR193⟩, 706 ⟨85ABC1761⟩; **4**, 142, 143 ⟨71JA2796⟩, 252 ⟨75TL4547⟩, 307 ⟨90CC1304⟩, 390 ⟨90JA6118, 92JOC2981⟩, 395, 397 ⟨87BCJ1189⟩, 528 ⟨91TL1975⟩, 574, 583, 585, 586, 587, 588, 589 ⟨84BCJ43⟩, 574, 588, 589 ⟨82BCJ3025⟩, 577, 587 ⟨83BCJ1772⟩, 585 ⟨75JOM(96)133, 75JOM(97)479⟩, 586 ⟨79BCJ2560, 80CL311⟩, 586, 587 ⟨80BCJ3176⟩, 598 ⟨77CB1312⟩, 614 ⟨58JOC292⟩, 642, 648, 659 ⟨84JOC1096⟩, 1057 ⟨86JOC886⟩; **5**, 158 ⟨89JA4864⟩, 172 ⟨87JA7188⟩, 356 ⟨76S750⟩, 795, 796 ⟨75BCJ3682⟩, 795, 796 ⟨75TL2689⟩, 801 ⟨74IC438, 91JCS(D)1531⟩, 922 ⟨77CB1312⟩; **6**, 61 ⟨71JA2796⟩, 426 ⟨91TL4371⟩, 465 ⟨94EUP581240⟩, 604 ⟨B-91MI 621-01⟩, 719 ⟨75BCJ3691⟩

Yamamura, K., **1**, 252 ⟨86TL3025⟩; **3**, 642 ⟨77TL4429⟩

Yamamura, M., **1**, 466, 1097 ⟨79JOC2408⟩

Yamamura, S., **1**, 19 ⟨72BCJ264⟩; **3**, 57 ⟨89TL3693⟩

Yamamura, Y., **3**, 437, 438 ⟨83JA672⟩

Yamamuro, A., **2**, 521 ⟨86JA7402⟩; **3**, 667 ⟨79CL1427⟩

Yamana, M., **4**, 735 ⟨83TL507, 83TL5657⟩

Yamana, Y., **5**, 240, 243, 253 ⟨91JOC3776⟩

Yamanaka, H., **1**, 993 ⟨89TL4995⟩, 1009 ⟨92S746⟩, 1009, 1010 ⟨78H(9)271, 78CPB3843⟩, 1010 ⟨79CPB193, 81CPB3543⟩, 1010 ⟨82CPB1865, 82CPB2410⟩, 1010 ⟨82CPB2417, 83S312⟩, 1010 ⟨88CPB1890, 88CPB2248⟩, 1034 ⟨88H(27)475⟩; **2**, 677 ⟨84S245⟩, 1027 ⟨92SL502⟩; **3**, 311 ⟨92TL5373⟩; **4**, 195 ⟨84S245⟩, 752 ⟨67TL1123, 68BCJ756⟩; **5**, 116 ⟨79H(13)235⟩; **6**, 51 ⟨76JOC3794⟩

Yamanaka, I., **6**, 469 ⟨94CL495⟩

Yamanaka, S., **2**, 582 ⟨85JCS(D)873⟩

Yamanaka, T., **1**, 561 ⟨91H(32)2089⟩; **2**, 30 ⟨81JOC3312⟩

Yamane, K., **4**, 774, 779 ⟨93MI 417-01⟩

Yamanishi, T., **2**, 563 ⟨87JOM(332)247⟩

Yamanka, H., **4**, 196 ⟨86CPB2362⟩

Yamano, Y., **1**, 613 ⟨93JCS(P1)987⟩

Yamanoi, K., **1**, 882 ⟨87TL5887⟩

Yamanoi, T., **2**, 269 ⟨84CL251⟩

Yamanouchi, T., **4**, 925 ⟨83TL767, 84BCJ2127⟩

Yamanoue, M., **2**, 81 ⟨89JOC444⟩; **6**, 26, 31 ⟨89JOC444⟩

Yamaoka, H., **5**, 134 ⟨77JAP(K)77/102237⟩

Yamaoka, K., **5**, 85 ⟨61T134⟩

Yamaoki, H., **1**, 976 ⟨91JOM(417)305, 91TL3397⟩, 976, 1025 ⟨90TL7179⟩; **6**, 461 ⟨91JOM(417)305⟩

Yamasaki, H., **2**, 631 ⟨87CL2109, 88BCJ600⟩

Yamasaki, K., **2**, 264 ⟨54MI 203-01⟩

Yamasaki, N., **5**, 236 ⟨86CL1013⟩

Yamasaki, T., **1**, 828 ⟨91TL4267⟩; **2**, 321 ⟨90CC1385⟩; **5**, 351 ⟨88H(27)1881⟩, 353 ⟨87S183⟩

Yamasaki, Y., **1**, 186, 225, 229 ⟨89JA7938⟩, 434 ⟨87JOC300⟩; **2**, 853 ⟨87CL357⟩

Yamashia, M., **2**, 438 ⟨85S291⟩

Yamashina, N., **1**, 489 ⟨88CL809⟩

Yamashiro, D., **5**, 185 ⟨87MI 504-01⟩

Yamashiro, R., **3**, 227 ⟨81CL1341⟩

Yamashita, A., **1**, 474 ⟨81T3957⟩, 1085 ⟨75JA891, 76CC959⟩; **3**, 127 ⟨77JOC3180⟩; **5**, 951 ⟨89JOC4481⟩

Yamashita, H., **2**, 85, 135 ⟨88BCJ1213⟩, 429 ⟨83BCJ219⟩, 913 ⟨93OM988⟩; **3**, 465 ⟨79CPB682⟩; **5**, 1066 ⟨86CL135⟩

Yamashita, J., **1**, 483 ⟨86CL407⟩, 676, 678 ⟨84BCJ2335⟩

Yamashita, K., **1**, 726 ⟨72TL2521⟩

Yamashita, M., **2**, 6 ⟨84BCJ2675⟩, 310 ⟨74TL1879, 83BCJ219⟩, 457 ⟨77S179⟩, 459 ⟨90BCJ2728⟩; **3**, 15, 90 ⟨82BCJ1663⟩, 61 ⟨78TL1305⟩, 84 ⟨73TL3535, 75BCJ2490⟩, 84 ⟨75TL1063⟩, 137 ⟨89BCJ2728⟩; **4**, 122 ⟨92CL407⟩, 854 ⟨79BCJ466⟩, 856 ⟨75S385⟩, 924 ⟨86BCJ3293⟩; **5**, 43, 112 ⟨81NKK1333⟩, 403, 404 ⟨80BCJ1625⟩, 403, 404 ⟨90BCJ2728⟩, 1118 ⟨57NKZ567⟩; **6**, 119 ⟨86BCJ3293⟩, 539 ⟨85JAP62114966⟩

Yamashita, N., **2**, 457 ⟨79BCJ466⟩

Yamashita, O., **2**, 949 ⟨89JA1888⟩

Yamashita, S., **1**, 115 ⟨73JOC4268⟩, 115, 452 ⟨73T1069⟩; **2**, 563 ⟨73T1069⟩; **4**, 669 ⟨76BCJ1177⟩; **6**, 26, 31 ⟨89JOC444⟩, 488 ⟨92JAP06033008⟩

Yamashita, T., **1**, 486 ⟨87CL1775⟩, 1079 ⟨82TL343⟩; **2**, 752 ⟨87JOC753⟩; **3**, 122 ⟨85CL1437⟩

Yamashita, Y., **1**, 119, 471, 1092 ⟨83JOC1912⟩; **2**, 468 ⟨92OPP209⟩, 793, 795 ⟨94H(37)693⟩; **3**, 213 ⟨89JOC4273⟩, 307 ⟨83TL3455⟩, 639 ⟨85TL5835⟩

Yamashito, S., **2**, 81 ⟨89JOC444⟩

Yamata, T., **2**, 320 ⟨91JOC5263⟩
Yamataka, H., **1**, 1124 ⟨73TL4767⟩
Yamataka, K., **2**, 244 ⟨72JOC3810⟩
Yam, T. M., **1**, 436 ⟨88JOC4466⟩
Yamato, H., **1**, 408, 409, 410, 820, 821, 822 ⟨72JA3870⟩, 820 ⟨71JOC1732⟩
Yamato, M., **2**, 328 ⟨88TL6949⟩; **3**, 422 ⟨88TL6949⟩
Yamato, T., **1**, 202 ⟨91JOC2089⟩; **6**, 229 ⟨87JA3708⟩
Yamauchi, H.,**3**, 426 ⟨90SL365⟩
Yamauchi, K., **2**, 683 ⟨79JOC638⟩
Yamauchi, M., **3**, 168 ⟨90CPB3314⟩, 373 ⟨88S178, 89CL973⟩
Yamauchi, T., **1**, 180 ⟨86BCJ3617⟩, 246 ⟨86JCS(P1)1983⟩; **2**, 287 ⟨84CC426⟩; **3**, 176 ⟨86JCS(P1)1983⟩
Yamauchi, Y., **5**, 70 ⟨82NKK811⟩
Yamaura, M., **1**, 648 ⟨85BCJ2812⟩
Yamaura, Y., **2**, 326 ⟨93JMC3526⟩; **4**, 816 ⟨88JOC3089⟩
Yamawaki, A., **3**, 245 ⟨86JHC1199⟩
Yamawaki, J., **3**, 116 ⟨83CL379⟩; **5**, 285 ⟨81CL1143⟩
Yamawaki, K., **3**, 117 ⟨88JOC3587, 88SC869⟩
Yamawaki, N., **2**, 385 ⟨88CPB459⟩; **3**, 428 ⟨88BCJ461⟩
Yamaya, M., **1**, 207, 1100 ⟨77T1945⟩, 511 ⟨77TL3641⟩
Yamazaki, A., **2**, 192 ⟨67JOC3032⟩; **3**, 151 ⟨82TL1609⟩; **6**, 25, 26 ⟨82TL1609⟩
Yamazaki, H., **1**, 14 ⟨89JA6280⟩, 992 ⟨76CC841, 91JA9604⟩, 992 ⟨93BCJ987⟩; **2**, 313 ⟨93CPB287⟩; **3**, 90 ⟨90TL6887⟩; **4**, 711 ⟨72JOM(35)367⟩, 714 ⟨83OM726⟩; **5**, 356 ⟨76S750⟩, 801 ⟨74IC438, 91JCS(D)1531⟩; **6**, 719 ⟨75BCJ3691⟩
Yamazaki, N., **2**, 299 ⟨87JOC1956⟩; **3**, 440 ⟨82S461⟩; **5**, 368 ⟨85LA2371⟩; **6**, 575 ⟨75S384⟩
Yamazaki, S., **2**, 931 ⟨90TL2917⟩, 931, 939 ⟨92JOC5610⟩; **3**, 407 ⟨92CL823⟩; **4**, 947 ⟨91JCS(P1)1555, 92JOC5610⟩
Yamazaki, T., **1**, 209 ⟨88SC1159⟩, 834 ⟨86H(24)3347⟩, 1041 ⟨89JCS(P1)1211⟩; **2**, 19 ⟨81CL761⟩, 318 ⟨93CL1787⟩, 325 ⟨91SL573⟩; **3**, 262 ⟨79JHC527⟩; **4**, 736 ⟨89JOC5630⟩, 837 ⟨85CPB2256⟩, 902 ⟨82CPB3959, 87JCS(P1)1501⟩; **5**, 57, 59, 89 ⟨63BCJ1264⟩, 199 ⟨87CPB3453⟩, 284 ⟨87CC1627⟩, 597 ⟨88H(27)1953⟩, 597, 599 ⟨84CPB1658, 84SC1257, 87CPB3139⟩, 597, 599, 614 ⟨83S225⟩, 607 ⟨85H(23)2213, 86H(24)1247⟩, 607 ⟨86S1063⟩, 693 ⟨81JOC1769⟩
Yamazaki, Y., **4**, 937 ⟨85JFC(30)357, 86BSF937⟩; **5**, 652 ⟨90TL3895⟩
Yamazuki, T., **3**, 659 ⟨76CPB3011⟩
Yambushev, F. D., **2**, 881 ⟨89JGU2500⟩, 883 ⟨83JGU1832⟩, 885, 890 ⟨87JGU2496⟩, 886 ⟨73JGU1123⟩, 889 ⟨89JGU2254⟩, 891 ⟨91JGU405⟩; **5**, 369, 373 ⟨88ZOB829⟩; **6**, 564 ⟨89ZOB2520⟩
Yamochi, H., **4**, 873 ⟨86CL1861⟩; **6**, 555 ⟨92CC1410⟩
Yamomoto, Y., **2**, 73 ⟨93CRV2207⟩
Yamoto, M., **4**, 417 ⟨82CPB1036⟩
Yan, B. F. Y., **6**, 481, 482 ⟨91JCS(P2)501⟩
Yan, C.-F., **1**, 521 ⟨79JA7001⟩; **3**, 38 ⟨78SC211⟩
Yan, S., **3**, 406 ⟨82OR(28)37⟩
Yan, T. H., **1**, 248 ⟨82TL263⟩, 690 ⟨84JOC3610⟩; **3**, 220, 639 ⟨84JOC3604⟩; **4**, 946 ⟨90TL673⟩
Yan, X., **6**, 556 ⟨88MI 617-02⟩
Yanada, K., **1**, 568 ⟨90CC730⟩
Yanada, R., **1**, 568 ⟨90CC730⟩
Yanagawa, M., **6**, 561 ⟨88BCJ2203⟩
Yanagi, A., **3**, 311 ⟨82CPB2003, 82H(19)41⟩
Yanagi, K., **1**, 165 ⟨85T2751⟩, 165 ⟨85T3663⟩, 187 ⟨84JA1079⟩; **2**, 331 ⟨89JA5886⟩
Yanagi, R., **3**, 162 ⟨92CL2039⟩; **4**, 213 ⟨78JOC992⟩
Yanagi, S., **5**, 303 ⟨74JOC3924⟩
Yanagi, T., **1**, 473, 489 ⟨81SC513⟩, 613, 700, 1073 ⟨84JA3670⟩; **2**, 78 ⟨90JOC2415⟩; **3**, 552 ⟨79CL535⟩
Yanagi, Y., **2**, 786 ⟨83TL4443⟩; **4**, 900 ⟨90JOC5308⟩
Yanagida, M., **5**, 157 ⟨89TL3799⟩
Yanagida, N., **5**, 179 ⟨80TL4611⟩
Yanagida, S., **2**, 19 ⟨81BCJ1151⟩; **4**, 52 ⟨77TL2893, 81BCJ1151⟩, 72 ⟨75JOC3540⟩; **5**, 6 ⟨77TL2337, 77TL2893⟩, 967 ⟨72JOC3030⟩, 1003 ⟨71JOC3542, 73BCJ1275⟩; **6**, 445 ⟨71BCJ2182⟩, 618 ⟨73BCJ303⟩
Yanagida, T., **3**, 593 ⟨69TL3407⟩
Yanagihara, N., **1**, 208 ⟨87JOC1013⟩; **2**, 983 ⟨86JA2753⟩; **3**, 632, 698 ⟨87JOC1013⟩
Yanagisawa, A., **1**, 112 ⟨91JA5893⟩, 116 ⟨91JA8955⟩, 132 ⟨85JA3348, 88JA4718⟩, 187 ⟨91SL251⟩; **2**, 73, 563 ⟨91JA8955, 92JOC6386⟩; **4**, 192 ⟨82TL3595⟩
Yanagisawa, I., **2**, 3 ⟨76CPB820⟩
Yanagisawa, K., **1**, 466, 1097 ⟨79JOC2408⟩
Yanagita, M., **1**, 881 ⟨58JOC841⟩
Yanagiya, M., **1**, 210 ⟨88CL645⟩; **5**, 9 ⟨85T3625⟩
Yanai, S., **6**, 116 ⟨86MI 604-01⟩
Yanai, T., **1**, 250, 251, 486 ⟨86CC1285⟩
Yanaka, H., **2**, 313, 544 ⟨91JCS(P1)1341⟩
Yanami, T., **2**, 196 ⟨82JA5781⟩
Yanase, G., **6**, 4 ⟨92JAP04164039⟩
Yanase, M., **1**, 776 ⟨89H(28)249⟩
Yanbikov, Ya. M., **1**, 421 ⟨43IZV389⟩
Yancey, J. A., **1**, 895 ⟨52JA5943⟩
Yanchang, S., **6**, 707 ⟨82AOC(20)115⟩
Yanchang, Sh., **6**, 708 ⟨84TL4425⟩
Yanchang Shen, **1**, 739 ⟨85S57, 88JOM(346)181⟩, 742 ⟨89JOM(375)45⟩
Yandovskii, V. N., **2**, 786 ⟨81MI 215-01⟩
Yanemitsu, O., **3**, 264 ⟨65TL2261⟩
Yañez, R., **4**, 711 ⟨86JOM(315)C22⟩, 713 ⟨90JOM(395)305⟩
Yang, B., **1**, 194, 288, 289 ⟨89JA6482⟩; **4**, 798 ⟨93SC1⟩
Yang, B. V., **4**, 60 ⟨93SL195⟩
Yang, C., **1**, 900 ⟨90JA3225⟩
Yang, C.-C., **2**, 133 ⟨91JA3177⟩; **5**, 386 ⟨91IC1862⟩
Yang, C.-T., **4**, 661 ⟨92TL7895⟩
Yang, D. C., **2**, 607 ⟨86JOC277⟩, 671 ⟨84JA7565⟩; **5**, 696 ⟨88JA307⟩
Yang, D.-S., **6**, 247 ⟨90IC2496⟩
Yang, D. S. C., **2**, 1009 ⟨87AX(C)313⟩
Yang, D. T. C., **1**, 41 ⟨77S124⟩, 41, 635 ⟨76JOC574, B-80MI 119-06⟩
Yang, F. A., **4**, 949, 950 ⟨82JOC3094⟩
Yang, F., **1**, 566, 681, 682 ⟨81JOC3359⟩
Yang, G., **3**, 436 ⟨94MI 311-01⟩
Yang, G. K., **4**, 582 ⟨91OM369⟩
Yang, H.-L., **1**, 282 ⟨89TL6829⟩
Yang, I. W., **2**, 632 ⟨79TL2503⟩
Yang, J., **1**, 646, 647 ⟨B-90MI 114-01, 86TL4583⟩, 647 ⟨87JOC3558, 87TL2159⟩, 647 ⟨88LA377⟩, 647, 738 ⟨85TL6447⟩, 647, 741 ⟨87TL2155⟩, 647, 741 ⟨88TL3949⟩; **2**, 245 ⟨86TL1853⟩; **3**, 60 ⟨87JOC3558⟩; **5**, 938 ⟨92OM1442, 94OM971⟩
Yang, J. H., **1**, 585 ⟨90TL3829⟩; **3**, 48 ⟨85TL6447, 88TL3949⟩, 48, 65 ⟨86TL4583⟩, 49 ⟨B-90MI 301-02⟩, 196, 231 ⟨87TL2155⟩, 196 ⟨B-90MI 304-02⟩
Yang, J. T., **5**, 85 ⟨61T143⟩
Yang, K., **1**, 436 ⟨90CC819⟩, 583 ⟨92JOC3615⟩
Yang, K.-W., **5**, 277, 299 ⟨70TL1791⟩
Yang, L., **6**, 216 ⟨86MI 607-01⟩
Yang, M.-H., **1**, 101 ⟨83SC1103⟩
Yang, N. C., **1**, 811 ⟨72JA9228⟩; **2**, 89, 652 ⟨59JA4230⟩
Yang, P. H., **2**, 595 ⟨86JOM(307)273⟩
Yang, R., **5**, 315 ⟨78JOC3713⟩
Yang, R.-Y., **2**, 311 ⟨93S481⟩, 998 ⟨92CC1487⟩
Yang, S., **1**, 4, 7, 9 ⟨83AG(E)562⟩
Yang, S. H., **2**, 632 ⟨87JOC691⟩, 684 ⟨89CC1363⟩
Yang, T., **2**, 877 ⟨88MI 217-01⟩; **3**, 436 ⟨B-94MI 304-02⟩
Yang, T.-K., **1**, 145 ⟨92T8963⟩; **2**, 359, 363 ⟨94JOC914⟩; **3**, 439, 440, 456 ⟨83JOC289⟩
Yang, Y., **1**, 489 ⟨93ACS221⟩; **3**, 303 ⟨92CC656⟩
Yang, Y. L., **2**, 446 ⟨84TL5903⟩
Yang, Y.-P., **4**, 545 ⟨86OM2220⟩; **6**, 465 ⟨77IZV363⟩
Yang, Z., **1**, 483, 494, 1029 ⟨86TL1171⟩; **3**, 103 ⟨92T1039⟩; **4**, 706 ⟨93JOM(463)163⟩; **5**, 596

⟨91JOC2143⟩; **6**, 63 ⟨87JFC(37)171⟩, 143 ⟨90S666⟩, 170 ⟨91JOC2143⟩
Yang, Z.-Y., **1**, 433 ⟨86CC1452⟩; **4**, 139 ⟨89JFC(45)435⟩, 816, 820 ⟨94T2993⟩; **6**, 9, 14, 22, 235, 266 ⟨92T189⟩, 58 ⟨91TL1019⟩
Yang-Chung, G., **4**, 198 ⟨72JA8929⟩
Yangida, N., **4**, 206 ⟨80BCJ1061⟩
Yanilkina, R. L., **4**, 106 ⟨90JGU2208⟩
Yankelevich, S., **1**, 790 ⟨67TL4945⟩
Yankep, E., **2**, 303, 304 ⟨89TL7383⟩
Yan-Kit, C., **6**, 481, 482 ⟨91JCS(P2)501⟩
Yanko, W., **3**, 638 ⟨44JA725⟩
Yanko, W. H., **4**, 205, 213 ⟨47JA2358⟩
Yankov, V. V., **4**, 660 ⟨74ZOB806, 75ZOB2448⟩
Yankovskaya, V. L., **6**, 613 ⟨88KFZ143⟩
Yankowsky, A. W., **2**, 822 ⟨77JOC1236⟩, 824, 825, 826 ⟨77PS(3)191⟩
Yannakopoulou, K., **1**, 260 ⟨89JCS(P1)225⟩, 260, 261 ⟨89S747⟩, 261, 1041 ⟨89S31⟩; **2**, 306 ⟨89S31⟩, 310 ⟨89JCS(P1)225⟩; **3**, 710 ⟨89TL6657⟩; **4**, 309 ⟨89JCS(P1)225⟩, 323 ⟨92S1295⟩
Yannoni, C. S., **1**, 906 ⟨82BSB422, 82JA907⟩, 906 ⟨90JA8991⟩, 914 ⟨79JA4770⟩; **4**, 668 ⟨84IC3717⟩
Yannoni, N. F., **2**, 1010 ⟨71JPC1246⟩
Yano, A., **5**, 547 ⟨79JCR(S)140⟩
Yano, K., **3**, 140 ⟨91CL1181⟩
Yano, S., **1**, 1140 ⟨88TL6951⟩
Yano, S.-G., **3**, 61 ⟨87JOC4135⟩
Yano, T., **1**, 532, 1094 ⟨80BCJ1471⟩, 532, 1104 ⟨80TL2865⟩; **3**, 176 ⟨80JA2456⟩; **6**, 111 ⟨79CC485⟩, 467 ⟨94EUP583789⟩
Yano, Y., **1**, 731 ⟨74TL2587⟩; **3**, 494 ⟨74TL2587⟩
Yanovskaya, L. A., **1**, 747 ⟨64IZV1341, 65IZV1504⟩, 882 ⟨70RCR859⟩, 888 ⟨84RCR744⟩; **2**, 748 ⟨68T4677⟩; **5**, 24 ⟨B-69MI 502-02⟩
Yanovskii, A. I., **2**, 447 ⟨91MI 208-01⟩; **4**, 694, 717 ⟨87JOM(336)187⟩
Yanovsky, A. I., **4**, 674 ⟨91JOM(418)C24⟩; **6**, 341 ⟨90JOM(384)279⟩
Yanuck, M. D., **1**, 530 ⟨85JOC5215⟩, 835 ⟨85JA443⟩
Yany, F., **3**, 562, 563 ⟨79AG(E)788⟩
Yao, A. N., **1**, 399 ⟨68JOC70⟩; **3**, 372 ⟨66JOC1185⟩; **5**, 495 ⟨66JOC1689⟩
Yao, C. F., **4**, 1004 ⟨92JOC6508⟩
Yao, G., **2**, 320 ⟨93JOC2086⟩
Yao, G.W., **4**, 446, 449 ⟨93JOC2086⟩
Yao, H., **4**, 651 ⟨90TL2925⟩
Yao, H. C., **3**, 448, 449 ⟨62JA3514⟩
Yao, K., **3**, 130, 285, 295 ⟨86TL929⟩
Yao, T., **3**, 364, 378 ⟨76BCJ3128⟩; **4**, 834 ⟨76BCJ3128⟩
Yao, Z., **2**, 118, 124 ⟨93JMC460⟩; **3**, 418 ⟨86TL1707⟩
Yaouanc, J.-J., **2**, 344, 347, 357 ⟨85S807⟩; **4**, 582 ⟨89JOM(359)205⟩
Yaozeng, H., **1**, 269, 738 ⟨88TL5295⟩, 737, 738 ⟨82OPP373⟩, 739 ⟨85S57, 87S207⟩, 743 ⟨90TL5897⟩, 766 ⟨93T6821⟩; **6**, 707 ⟨82AOC(20)115⟩, 708 ⟨84TL4425⟩
Yaozhong, J., **2**, 331 ⟨89SC1423⟩; **3**, 415 ⟨89SC1423⟩
Yap, C. H. M., **5**, 325 ⟨87JCS(P1)1635⟩
Y, Pietrasanta, **6**, 7 ⟨76T2295⟩
Yarbro, C. L., **4**, 178, 184 ⟨58JA6613⟩
Yarbrough, K. N., **3**, 363 ⟨70T817⟩
Yaremenko, V. V., **4**, 573 ⟨84ZOB59⟩
Yaremko, B., **1**, 1029 ⟨89JMC1820⟩
Yarham, A. C., **4**, 19, 27 ⟨81JCS(P1)78⟩
Yarkevich, A. N., **4**, 119 ⟨86JGU1430, 86JGU501⟩, 119 ⟨90JGU1351⟩, 121, 122 ⟨88JGU465⟩
Yarkova, É. G., **4**, 493 ⟨88JGU876⟩
Yarnell, T. M., **1**, 1122, 1138 ⟨78TL2549⟩
Yarosh, A. V., **5**, 1025 ⟨78KGS1497⟩
Yarosh, O. G., **2**, 1091 ⟨79JGU536⟩; **4**, 940 ⟨91ZOB1893⟩, 942 ⟨90ZOB526⟩; **6**, 60 ⟨67IZV693⟩, 565 ⟨90ZOB602⟩
Yaroslavsky, C., **2**, 626 ⟨74TL3395⟩
Yarovenko, N. N., **2**, 19 ⟨59JGU2125⟩, 171 ⟨59ZOB3792⟩, 288 ⟨59ZOB942, 60ZOB4069⟩; **5**, 476 ⟨57JGU2301⟩; **6**, 49 ⟨59ZOB942⟩, 54 ⟨59JGU2125⟩, 229 ⟨58ZOB2506⟩, 233, 234, 238 ⟨59ZOB3786⟩, 236 ⟨59ZOB3784⟩, 236, 528, 533, 534 ⟨59ZOB3792⟩, 239 ⟨59ZOB2157⟩, 440 ⟨57ZOB2243⟩, 613 ⟨60JGU4029⟩
Yarovenko, V. N., **6**, 653 ⟨79IZV813⟩
Yarrow, P. I. W., **6**, 195 ⟨81JCS(D)814⟩
Ya Ryzhiova, T., **4**, 455, 481 ⟨90ZOB1718⟩
Yaser, H. K., **1**, 839 ⟨91JA3526⟩; **2**, 245, 331 ⟨91JA3526⟩
Yashina, O. G., **1**, 1045 ⟨67ZOR219⟩
Yashiro, T., **5**, 25, 64 ⟨69CPB1564⟩
Yashkina, L. V., **6**, 323 ⟨82IZV460⟩
Yashunsky, D. V., **3**, 130 ⟨83TL4355⟩, 211 ⟨83TL4359⟩; **5**, 239 ⟨83TL4355⟩
Ya. Skorobogatova, S., **4**, 547 ⟨76ZOB571⟩
Yaslak, S., **2**, 837 ⟨83CC857⟩
Yaslovitskii, A. V., **5**, 967 ⟨69URP241428⟩
Yassin, A. A., **5**, 676 ⟨84JHC797⟩
Yastenko, S. N., **4**, 570 ⟨90ZOB695⟩
Yastrebov, V. V., **3**, 76 ⟨71ZOR2120⟩
Yastrebova, G. E., **3**, 231, 269 ⟨68JGU301⟩; **4**, 561 ⟨68ZOB292⟩, 563 ⟨70ZOB499⟩
Yasuda, H., **2**, 263 ⟨83TL1273⟩, 616 ⟨78SC103⟩; **3**, 627, 634 ⟨90S768⟩, 669 ⟨80JCS(P1)1132⟩; **4**, 188 ⟨82S1089⟩, 603, 649 ⟨83OM21⟩; **5**, 158 ⟨77BCJ553⟩, 189, 195, 282 ⟨86JOC4150⟩, 189, 199 ⟨84TL6027⟩, 383 ⟨66MI 507-01⟩, 718 ⟨68JPS(B)389, 69JPS(B)17⟩, 1107 ⟨81JCS(P1)1577⟩
Yasuda, K., **5**, 1032 ⟨74JOC957⟩, 1144 ⟨69JAP47016293⟩
Yasuda, M., **2**, 702 ⟨83CPB128⟩, 752 ⟨87JOC753⟩
Yasuda, N., **1**, 38, 945 ⟨68JA4182⟩, 133 ⟨79BCJ1241⟩; **5**, 785, 1130 ⟨70JOC4238⟩
Yasuda, S., **3**, 543 ⟨91T5689⟩
Yasuda, T., **2**, 21 ⟨81TL2291⟩, 30 ⟨81JOC3312⟩
Yasufuku, K., **4**, 711 ⟨72JOM(35)367⟩, 714 ⟨83OM726⟩
Yasuhara, A., **1**, 1009 ⟨92S746⟩
Yasuhara, T., **6**, 4 ⟨91EUP421322, 92EUP473105⟩, 510 ⟨71JA6344⟩
Yasuhisa, T., **3**, 357 ⟨91EUP462531⟩
Yasui, E., **1**, 186, 225, 229 ⟨89JA7938⟩; **5**, 249, 563, 640 ⟨88BCJ3931⟩
Yasui, H., **2**, 35 ⟨85CC400⟩
Yasui, K., **1**, 195 ⟨93TL7619⟩
Yasui, S., **1**, 431 ⟨85JOC3283⟩
Yasui, T., **4**, 506 ⟨92JOC6711⟩; **5**, 27 ⟨81MI 502-01⟩
Yasuike, S., **4**, 808 ⟨93CC1309, 93CC1817⟩
Yasukawa, T., **2**, 566 ⟨67JOM(10)P9⟩
Yasumoto, M., **5**, 320, 335 ⟨91BCJ3732⟩; **6**, 575 ⟨90JCS(P1)1218⟩
Yasumura, M., **4**, 943 ⟨87JOC1256⟩
Yasumuro, N., **1**, 595, 597 ⟨88H(27)261⟩
Yasunami, M., **2**, 769 ⟨89BCJ128⟩; **3**, 293 ⟨77CL587⟩, 689 ⟨92CL1053⟩
Yasuo, H., **5**, 93 ⟨79CPB1931⟩
Yasuoka, N., **2**, 244 ⟨72JOC3615⟩; **4**, 686 ⟨77BCJ2250⟩
Yasuzawa, M., **4**, 774, 779 ⟨93MI 417-01⟩
Yatagai, H., **1**, 119 ⟨80JA2318⟩, 312 ⟨82JOC119⟩, 450, 510 ⟨78CC702⟩, 488 ⟨77JA5652⟩, 510, 531, 532, 663, 1104 ⟨80BCJ1670⟩, 531 ⟨75JA5606⟩, 531, 663 ⟨77CC852⟩, 532 ⟨80JOC1640⟩, 540 ⟨79JOC1744⟩, 713, 714, 716 ⟨84JOC1096⟩, 1001 ⟨80CL669⟩; **2**, 517 ⟨81JA1969, 81JA3229⟩, 702 ⟨82JOC119⟩; **4**, 642, 648, 659 ⟨84JOC1096⟩
Yatagi, H., **2**, 78 ⟨84T2239⟩
Yates, B. F., **5**, 1127 ⟨84AJC367⟩
Yates, J. B., **1**, 173, 321, 334 ⟨85T4079⟩, 173, 334 ⟨82JA5829⟩; **5**, 326 ⟨79TL4627⟩; **6**, 541, 550, 557 ⟨85T4079⟩
Yates, J. E., **2**, 112 ⟨71USP3600417⟩
Yates, K., **1**, 318 ⟨77JOC2021⟩, 1109 ⟨71JCP(55)3835⟩, 1110 ⟨73JA160⟩, 1123 ⟨76CJC2839, 76JOC2241⟩; **2**, 176 ⟨75JA1492⟩

Yates, M. J., **1**, 97 ⟨80SC627⟩
Yates, P., **1**, 807, 809 ⟨61T13⟩; **2**, 229 ⟨78CC205⟩; **3**, 256 ⟨71CJC1477⟩, 426 ⟨83CC50⟩, 464, 470 ⟨61JOC2617⟩, 475 ⟨58JOC759⟩, 547 ⟨71CJC1456⟩, 634 ⟨82CJC825⟩; **4**, 852 ⟨71CJC1477⟩; **6**, 88 ⟨91CJC415⟩
Yates, S. W., **2**, 95 ⟨83JOC3126⟩; **3**, 131 ⟨82JOC3006⟩
Yatsenko, R. D., **4**, 802, 805 ⟨60BAU1851⟩
Yatsishin, A. A., **4**, 971 ⟨79ZOR1381⟩, 980, 987 ⟨77ZOR954⟩, 990 ⟨78ZOB803⟩
Yatziv, S., **5**, 1074 ⟨88JOC1744⟩
Yau, C., **3**, 124 ⟨78JOC700⟩
Yau, C.-C., **1**, 130 ⟨78JOC700⟩, 131 ⟨78JOC2153⟩; **2**, 420 ⟨78JOC700⟩
Yau, E. K., **1**, 1074 ⟨90JOC3147⟩; **2**, 253 ⟨90TL1953⟩
Yaunwei, C., **2**, 331 ⟨89SC1423⟩
Yavashev, L., **5**, 98 ⟨76MI 502-01⟩
Yavorskii, B. M., **6**, 456 ⟨84MI 607-01⟩
Yawney, D. B. W., **4**, 688 ⟨67AJC77⟩
Yazawa, H., **5**, 277 ⟨74TL3995⟩
Yazawa, N., **2**, 614 ⟨89CL2213⟩
Yazawa, T., **4**, 696 ⟨81JOC192⟩
Yazdanbakhsch, M., **6**, 235, 254, 303 ⟨77CB916⟩, 252, 274 ⟨74CZ109⟩, 274, 275, 307, 534 ⟨76CB1976⟩
Yazdekhasti, H., **4**, 429, 430, 431 ⟨93TL6677⟩
Yazdikarimy, B., **5**, 639 ⟨89JHC1341⟩
Yde, B., **5**, 118 ⟨B-88MI 502-01⟩, 551 ⟨84T2663⟩, 558, 573, 587 ⟨83T3429⟩
Ye, T., **3**, 508 ⟨94CR1091⟩
Ye, T.-G., **4**, 129 ⟨89B8270⟩
Ye, W., **2**, 456 ⟨87MI 208-01⟩
Yeakey, E. L., **4**, 1078 ⟨87JOC3243⟩
Yeannin, Y., **4**, 720 ⟨90IC1817⟩
Yeap, G.-Y., **2**, 1008 ⟨92POL2351⟩
Yeates, C., **1**, 700, 703 ⟨87JCS(P1)2171⟩, 1094 ⟨83TL3905, 84CC151⟩; **4**, 953 ⟨83TL3905, 87JCS(P1)2189⟩; **5**, 173 ⟨85CC1386⟩
Yeates, C. L., **1**, 481 ⟨92JCS(P1)3419⟩
Yeats, P., **3**, 532 ⟨69T2121⟩
Yeaw, J. S., **5**, 63 ⟨32OSC(1)321⟩
Yechezkel, T., **1**, 360 ⟨93JOC6716⟩
Yee, K. C., **1**, 869, 871 ⟨70JA5926⟩, 869, 871 ⟨70JA5933⟩, 869, 871 ⟨70JA5939⟩
Yee, N. K. N., **1**, 24 ⟨92JOC4598⟩
Yee, V. C., **2**, 573 ⟨88OM1877⟩; **4**, 656 ⟨91OM2857, 93OM4572⟩
Yefsah, R., **3**, 123 ⟨87S37⟩; **5**, 798 ⟨89OM2253, 90JOM(381)191⟩, 952 ⟨91COS(6)485⟩; **6**, 716 ⟨91CRV35⟩
Yeh, A., **5**, 386 ⟨91IC1862⟩
Yeh, C.-L., **1**, 727 ⟨77TL4257⟩
Yeh, H. J. C., **2**, 659 ⟨76JOC977⟩
Yeh, M. C. P., **1**, 123 ⟨88TL2395⟩, 291 ⟨92CC13⟩, 1017 ⟨89TL4799⟩; **2**, 478 ⟨90OS(70)195⟩, 588, 590 ⟨88JOC2390, 92OS(70)195⟩, 998 ⟨89JOC5202⟩; **3**, 23 ⟨88TL6697⟩, 287 ⟨88JOC2390⟩, 303 ⟨89JOC5202⟩, 623 ⟨89JA6474⟩; **4**, 385 ⟨89JOC5202⟩, 651, 724 ⟨88TL6697⟩
Yeh, M.-K., **1**, 154 ⟨76JOC439⟩
Yeh, S.-K., **4**, 583 ⟨90JCS(D)2509⟩
Yeh, W.-Y., **4**, 683 ⟨88JOM(353)103, 93JOM(461)207⟩; **6**, 345 ⟨86OM1757, 89JOM(371)257⟩
Yek, M. C. P., **2**, 69 ⟨88TL3887⟩
Yellin, W., **2**, 164 ⟨67JA2435⟩
Yellowlees, L. J., **6**, 457 ⟨93JCS(D)1031⟩
Yelm, K. E., **1**, 788 ⟨83JA2503⟩, 1092 ⟨84JOC3438⟩; **2**, 938 ⟨93JA6625⟩; **4**, 621 ⟨93JA6625⟩
Yen, C., **1**, 905 ⟨80JCR(S)152⟩
Yen, C.-F., **3**, 232 ⟨89JOC5126⟩
Yen, H.-K., **3**, 684, 686 ⟨84JOC4595⟩
Yen, K. Y., **6**, 538, 539 ⟨83JOC1449⟩
Yen, T. F., **2**, 175 ⟨58JOC363⟩
Yeola, S., **4**, 230 ⟨91JA9377⟩
Yeon, S. H., **4**, 609 ⟨93OM2360⟩
Yeowell, D. A., **4**, 193 ⟨79JOC4825⟩
Yeroushalmi, S., **1**, 917 ⟨83JOC1431⟩
Yeske, P. E., **1**, 106 ⟨93PAC1153⟩, 470 ⟨92JOC1161⟩; **4**, 275 ⟨91JOC6386, 92JOC298⟩
Yeung, B. W. A., **3**, 696 ⟨86CJC2475⟩; **5**, 157 ⟨89CC1160⟩
Yeung, E. S., **6**, 79 ⟨88JA4144⟩
Yeung, K.-S., **1**, 640 ⟨93SL774⟩
Yeung, L. L., **1**, 991 ⟨87CC981, 90JOC1874⟩, 1023 ⟨92SL291⟩; **3**, 365 ⟨90JOC1874⟩
Yeung Lam Ko, Y. Y. C., **4**, 479, 480 ⟨84CC1634⟩, 566 ⟨87PS(30)523⟩; **5**, 785, 792 ⟨83CJC1105⟩; **6**, 60 ⟨90T3859⟩, 117 ⟨84CC1634⟩
Yevich, J. P., **5**, 334 ⟨86JMC359⟩
Yi, C. S., **3**, 308 ⟨81H(16)399⟩; **5**, 946, 957 ⟨93JA3806⟩
Yi, K. Y., **1**, 939 ⟨86JOC2613⟩; **3**, 618, 619, 697, 721 ⟨86TL1925⟩; **5**, 1030, 1078 ⟨85TL1661⟩, 1030, 1078 ⟨86JOC2613, 86TL1925⟩
Yi, L., **1**, 272, 743 ⟨90CC85⟩
Yi, M., **5**, 212 ⟨91CC1127⟩
Yi, Q., **3**, 626 ⟨85CL1779⟩
Yiannikouros, G., **4**, 226 ⟨90JA6263⟩; **5**, 513 ⟨90JA6263⟩
Yiannios, C. N., **2**, 214 ⟨63JOC3246⟩
Yijima, C., **3**, 36 ⟨81CC263⟩; **5**, 346 ⟨81S610⟩
Yi Liao, **1**, 743 ⟨90TL5897⟩
Yilmaz, I., **3**, 411 ⟨79TL2109⟩
Yim, G. K. W., **6**, 577 ⟨80JMC1261⟩
Yimenu, T., **1**, 650 ⟨80JA6894⟩
Yin, B.-Z., **2**, 769 ⟨89BCJ128⟩
Yin, J., **4**, 677 ⟨91OM1278, 92OM1480⟩, 678 ⟨91OM2541⟩, 955 ⟨91JA9406, 93OM3431⟩; **5**, 938 ⟨92OM1442, 94OM971⟩
Yin, T., **1**, 681 ⟨86T1581⟩
Yin, T.-K., **1**, 785 ⟨86JA7121⟩
Ying, F.-H., **2**, 668 ⟨94TL729⟩
Yinglin, H., **2**, 308 ⟨90S122, 90S615⟩; **3**, 179 ⟨90S615⟩
Yip, Y. C., **1**, 801, 802 ⟨89CRV165⟩, 991 ⟨87CC981, 90JOC1874⟩; **3**, 365 ⟨90JOC1874⟩
Yizun, J., **4**, 21 ⟨89JCS(P1)2009⟩
Ylagan, L., **2**, 458 ⟨88SC285⟩
Yoakim, C., **2**, 525 ⟨88PAC1705⟩, 539 ⟨84TL4717⟩; **4**, 217 ⟨90CJC897⟩
Yocklovich, S. G., **2**, 184 ⟨78TL97⟩, 718 ⟨81JOC3467⟩
Yoda, H., **1**, 210 ⟨85TL4747⟩
Yoder, C. H., **2**, 417, 418 ⟨69IC279⟩; **4**, 140 ⟨85SRI321⟩, 513 ⟨78JOM(153)369⟩; **5**, 224 ⟨85SRI321⟩, 375, 376 ⟨90OM1557⟩, 379 ⟨74JA4283, 87JOM(332)69⟩, 720 ⟨72JA5285, 78JOM(161)313⟩
Yoder, C. M. S., **6**, 299, 300 ⟨67JHC166⟩
Yoffe, S. T., **1**, 870 ⟨67TL4525⟩
Yogev, A., **1**, 869, 870 ⟨67JOC2162⟩; **4**, 1083 ⟨88CPL439⟩
Yogo, T., **1**, 198 ⟨79SC809⟩, 481 ⟨77CL103⟩; **3**, 19 ⟨83SC1149⟩, 135 ⟨81CL1059⟩; **4**, 935 ⟨81CL1059, 83CL933⟩, 935 ⟨83SC1149⟩, 935, 949 ⟨80CL591⟩
Yohannes, D., **2**, 655 ⟨87JOC5283⟩
Yokayama, M., **2**, 69, 564 ⟨84JOC3904⟩
Yokoe, M., **6**, 445 ⟨71BCJ2182⟩
Yokoi, M., **2**, 264 ⟨54MI 203-01⟩; **5**, 560 ⟨86BCJ1403⟩
Yokokoji, O., **6**, 29 ⟨92JAP04149143⟩
Yokomatsu, T., **1**, 41 ⟨78CC414⟩, 1077 ⟨78JOC4366, 78S305⟩; **2**, 459 ⟨92TA377⟩; **4**, 475 ⟨92TA377⟩; **5**, 284 ⟨87H(26)359⟩, 570, 613 ⟨87H(26)359⟩
Yokomichi, I., **6**, 9 ⟨80GEP3008081⟩
Yokommatsu, T., **3**, 654 ⟨78CPB1874⟩
Yokoo, T., **3**, 6, 118 ⟨93CL571⟩
Yokota, H., **5**, 692 ⟨88BCJ3731⟩
Yokota, K., **1**, 477 ⟨81CC313⟩; **2**, 619 ⟨80CL1423⟩; **5**, 578, 608, 618 ⟨75JOM(92)139⟩
Yokota, M., **1**, 943 ⟨90JCS(P1)3017⟩; **2**, 442 ⟨77JA7876⟩
Yokota, Y., **2**, 837 ⟨64JOC2572⟩, 859 ⟨67BCJ147⟩; **3**, 698, 721 ⟨65BCJ858⟩
Yokoyama, B., **1**, 815 ⟨88CC1281⟩

Yokoyama, H., **1**, 206, 289, 1092 ⟨82TL1267⟩; **2**, 285 ⟨92JOC111⟩; **4**, 947 ⟨91JOC5721⟩

Yokoyama, K., **1**, 363, 364 ⟨78JA1791⟩; **2**, 28 ⟨88OSC(6)520, 92JOC111⟩; **3**, 225 ⟨72JA1772⟩; **4**, 947 ⟨91JOC5721⟩

Yokoyama, M., **1**, 47 ⟨87S846⟩, 202 ⟨82BCJ643⟩, 515 ⟨87JOC5695⟩, 934 ⟨91CL1691, 91TL6559⟩, 934 ⟨93JCS(P1)2417⟩, 1098, 1140 ⟨84JOC3904, 82CC1042⟩; **2**, 36 ⟨83S460⟩; **3**, 88 ⟨83TL2399⟩, 617 ⟨83S142⟩, 618 ⟨82S591⟩, 655 ⟨88S813⟩; **4**, 432 ⟨93TL5097⟩, 842 ⟨70JOC283⟩, 852 ⟨70BCJ2938⟩, 872 ⟨86BCJ2909⟩, 900 ⟨82CL1933, 85CC279⟩, 988 ⟨82JCS(P1)1059⟩; **5**, 286 ⟨82BCJ643⟩, 546 ⟨84S797⟩, 547 ⟨69JOC730⟩, 551, 560, 574, 575 ⟨84S827⟩, 558 ⟨84JCS(P1)85⟩, 580 ⟨70JOC2438, 71JOC2009⟩, 700, 701 ⟨92JCS(P1)2127⟩, 1112 ⟨80TL635⟩; **6**, 564 ⟨81JCS(P1)2499, 81S908⟩, 627, 730 ⟨85S891⟩

Yokoyama, S., **1**, 698, 703 ⟨89BCJ3132⟩

Yokoyama, Y., **1**, 458, 516 ⟨91TL5559⟩, 608 ⟨77JA5210⟩; **2**, 84 ⟨90TL221⟩, 380 ⟨83CPB423⟩, 629 ⟨79JA4749⟩; **3**, 307 ⟨88H(27)1855⟩, 309 ⟨80H(14)1939, 88CPB2023⟩, 309 ⟨90CPB3261⟩, 644, 656 ⟨77JA5210⟩; **5**, 235 ⟨74JOC3302⟩, 239 ⟨78JOC1283⟩; **6**, 79 ⟨80MM252⟩

Yokozawa, T., **4**, 831 ⟨84TL3987, 84TL3991⟩

Yoldikov, V. N., **2**, 32 ⟨70TL2517⟩

Yon, G. H., **1**, 86 ⟨86TL2409⟩

Yonashiro, M., **5**, 28, 46, 148 ⟨82S521⟩

Yoneda, F., **1**, 188 ⟨85CC1452⟩; **2**, 213 ⟨66JA2328, 67JOC727⟩; **3**, 294 ⟨66JA2328⟩

Yoneda, N., **1**, 462 ⟨79CL1003⟩, 1009 ⟨90BCJ2124⟩; **2**, 6, 15, 18 ⟨91T5329⟩, 17 ⟨83CL1135, 84CL1241⟩, 19 ⟨85CL1693, 87CL1675⟩, 615 ⟨87SC685, 89SC865⟩, 690 ⟨91CL459⟩; **3**, 374 ⟨74BCJ909⟩; **5**, 93 ⟨79CPB1931⟩

Yoneda, R., **1**, 836 ⟨91COS(8)895, 93T7657⟩, 993 ⟨94TL1255⟩; **3**, 647 ⟨84TL427⟩, 659, 660 ⟨86CPB4620⟩; **5**, 347 ⟨86CPB4545⟩, 1111, 1116 ⟨87TL4189⟩; **6**, 542 ⟨91CPB1659, 92T9433⟩, 542 ⟨92TL2543⟩, 544 ⟨92CPB2279⟩

Yoneda, S., **4**, 345 ⟨78JOC1980, 79BCJ3342⟩; **5**, 70, 73 ⟨83CC1487⟩, 1039 ⟨62MI 527-01⟩; **6**, 316 ⟨78JOC595⟩

Yoneda, Sh., **6**, 698 ⟨83TL3469⟩

Yoneda, T., **1**, 243 ⟨90TL6669⟩

Yoneda, Y., **3**, 670 ⟨86CPB4545⟩

Yonekawa, Y., **5**, 557 ⟨86TL5405⟩

Yonekura, M., **5**, 285 ⟨83JA7345⟩

Yonemaru, S., **1**, 429 ⟨92SL833⟩

Yonemitsu, O., **2**, 90 ⟨88TL4139⟩, 717 ⟨86S852⟩; **3**, 310 ⟨76H(4)1859⟩; **4**, 201 ⟨82TL889⟩; **5**, 173, 199 ⟨87TL4569⟩

Yonemoto, K., **5**, 320, 335 ⟨91BCJ3732, 86BCJ2017⟩

Yonemura, H., **2**, 22 ⟨87BCJ809⟩

Yoneyama, Y., **1**, 690 ⟨86BCJ2209⟩

Yoneyoshi, Y., **2**, 314 ⟨88TL223⟩

Yonezawa, K., **2**, 546 ⟨71TL1511⟩; **5**, 146 ⟨74JOC3273⟩

Yonezawa, T., **1**, 1109 ⟨68JA1239⟩; **3**, 487 ⟨86OM356⟩; **5**, 932 ⟨84JA5892⟩, 933 ⟨83JA426⟩; **6**, 115 ⟨69BCJ2323⟩

Yonezawa, Y., **3**, 564 ⟨75BCJ2891⟩; **4**, 433 ⟨78BCJ2614⟩; **5**, 371 ⟨79TL3851, 86BCJ3696⟩

Yong Gil Kim, **1**, 739, 740 ⟨91TL2913⟩

Yonovich, M., **3**, 91 ⟨79TL3753⟩

Yonovich-Weiss, M., **2**, 31 ⟨84S34⟩; **4**, 22, 37 ⟨84S34⟩

Yoo, B. K., **3**, 6, 118 ⟨84CC762⟩

Yoo, B. R., **3**, 502 ⟨91OM1178⟩

Yoo, C. S., **2**, 1009 ⟨87AHC(41)275⟩

Yoo, S., **2**, 629 ⟨78JA4618⟩; **5**, 173 ⟨78JA4618⟩, 173 ⟨79JA7131⟩, 173, 204 ⟨78JA4620⟩

Yoo, S.-E., **1**, 307 ⟨88TL2195⟩, 1134 ⟨78JA4618, 78JA4620⟩; **2**, 324 ⟨88TL2195⟩; **4**, 780 ⟨91JMC2525⟩

Yoo, W.-J., **1**, 15, 16 ⟨93JCS(P1)9⟩

Yoon, D. C., **2**, 150 ⟨88TL6453⟩

Yoon, I.-N., **5**, 648 ⟨91MI 516-01⟩

Yoon, J. S. U. I. N., **6**, 128 ⟨91MI 604-01⟩

Yoon, K. B., **2**, 31 ⟨87CC1013, 89JOC3028⟩

Yoon, M. S., **3**, 82 ⟨88H(27)1595, 88MI 303-01⟩, 82 ⟨90MI 303-01⟩, 82, 99 ⟨87TL6231⟩, 93 ⟨89TL3677⟩

Yoon, N. M., **1**, 14 ⟨68CC1549, 68JA2927⟩, 86 ⟨88SC89⟩; **2**, 60 ⟨66JA1464, 93JOC1941⟩, 83 ⟨68JA2686⟩, 126 ⟨66JA1458⟩, 268 ⟨84JA8001⟩, 317 ⟨93SC3047⟩, 515 ⟨77JOM(135)C10⟩, 516 ⟨77JA5514⟩, 523 ⟨82JOC5069⟩; **3**, 13, 82 ⟨87JOC5400, 72JOC2942⟩, 82 ⟨84JA8001⟩, 83, 107 ⟨93SC1775⟩, 84 ⟨91MI 303-01⟩, 85 ⟨70JA7161⟩, 86 ⟨93JOC1941⟩, 127 ⟨88SC89⟩

Yoon, S. K., **1**, 618, 707, 709 ⟨91JA9682⟩

Yoon, U.-C., **4**, 510 ⟨87JA4421⟩

Yoon, Y. B., **4**, 849 ⟨90JOC5555⟩

Yorifuji, T., **3**, 130, 285 ⟨80S877⟩

Yorke, S. C., **3**, 297 ⟨84JCS(P1)1339⟩

Yorozu, K., **3**, 156 ⟨93CL1579⟩, 409 ⟨86BCJ1809⟩

Yoshiba, H., **2**, 457 ⟨79BCJ466⟩

Yoshida, A., **5**, 234 ⟨84TL2793⟩, 236 ⟨85TL673⟩; **6**, 553 ⟨81CPB3158⟩

Yoshida, E., **4**, 248 ⟨88TL3971⟩; **5**, 616 ⟨76CC776⟩, 631, 632 ⟨85JOC1681⟩

Yoshida, H., **2**, 457 ⟨77S179⟩; **3**, 90 ⟨76BCJ1681⟩, 332, 364 ⟨75BCJ2907⟩, 353 ⟨81S36⟩, 364, 378 ⟨76BCJ3128⟩, 493 ⟨77S626⟩, 687 ⟨76BCJ3124⟩; **4**, 834 ⟨76BCJ3128⟩, 847 ⟨76S552⟩, 854 ⟨79BCJ466⟩; **5**, 226 ⟨85BCJ1903⟩, 403, 404 ⟨80BCJ1625⟩, 728, 729 ⟨77S626⟩, 775 ⟨79BCJ1541⟩, 1020 ⟨88OM558⟩; **6**, 87, 99 ⟨76S551⟩, 383 ⟨91JA7790⟩

Yoshida, J., **1**, 472 ⟨77TL4089⟩, 490, 515 ⟨79TL1141⟩, 490, 535 ⟨82OM542⟩, 651 ⟨87TL211⟩; **2**, 4 ⟨78JA290, 78TL1809⟩, 4 ⟨82OM355, 82OM369⟩, 950 ⟨92SL843⟩; **3**, 703, 709 ⟨94CC549⟩; **4**, 197 ⟨91T615⟩, 197 ⟨92JOC1321⟩, 654 ⟨77IJ265⟩, 937 ⟨93T2011⟩; **5**, 343 ⟨92JOC4877⟩, 417, 421, 423 ⟨91JOC1307⟩

Yoshida, J.-I., **1**, 1106 ⟨78TL2161⟩; **2**, 912, 945 ⟨83TL4895⟩; **3**, 32 ⟨92JOC1321⟩, 36, 165 ⟨88CC1468⟩; **4**, 361 ⟨92SL843⟩; **6**, 124 ⟨90JA1962⟩, 129 ⟨91T615⟩, 131 ⟨92SL843⟩, 133 ⟨89TL219⟩

Yoshida, K., **3**, 20 ⟨89TL6371⟩, 265 ⟨90JCS(P1)1891⟩, 310 ⟨81JA6990, 87CPB4700⟩; **4**, 326 ⟨74JOC3412⟩; **6**, 620 ⟨86JOC1719⟩

Yoshida, M., **1**, 442 ⟨70T4609⟩; **2**, 136 ⟨92CL1655⟩, 227 ⟨91CC167⟩, 478 ⟨90CPB673⟩, 620 ⟨88BCJ2226, 88CL2017⟩, 1054 ⟨69CC1393, 74BCJ946⟩; **4**, 56 ⟨89CC1559⟩, 563 ⟨88JAP63185993⟩; **5**, 208 ⟨90JFC(46)423⟩, 226 ⟨87H(25)79⟩, 1125 ⟨86JAP1280464⟩; **6**, 12 ⟨90JFC(46)423⟩, 21 ⟨89CC1559⟩

Yoshida, N., **3**, 8, 122 ⟨81TL2295⟩; **4**, 223 ⟨93TL4063⟩, 418 ⟨91TL2375⟩; **5**, 657 ⟨88H(27)155⟩

Yoshida, R., **4**, 900 ⟨90JOC5308⟩

Yoshida, S., **2**, 11, 36 ⟨79S379⟩, 650 ⟨89CL1267⟩; **3**, 363 ⟨91COS(7)119⟩, 618 ⟨82S591⟩, 662 ⟨83JOM(243)95⟩; **5**, 769 ⟨82CPB4521⟩

Yoshida, T., **1**, 453 ⟨81JA4985⟩, 456, 507 ⟨83JA6273⟩, 488, 983, 1033, 1034 ⟨91T343⟩, 511, 534, 1106 ⟨87JOC1624⟩, 532, 1100, 1103 ⟨73CC606⟩, 532, 662 ⟨73CC874, 74JOC2321⟩, 544 ⟨83OM563⟩, 613 ⟨82CL165⟩, 669 ⟨84TL3407⟩, 991 ⟨81JA1276⟩, 999 ⟨74CC762⟩; **2**, 524 ⟨75JOC814⟩, 596, 970 ⟨80TL1501⟩, 631 ⟨77JCR(S)215⟩, 798 ⟨B-76MI 215-02⟩, 901, 904 ⟨91T343⟩, 925, 974, 975 ⟨85JA6639⟩, 969 ⟨74CC762, 84TL3407⟩, 973 ⟨83OM563⟩, 976 ⟨81JA4985⟩, 976, 978 ⟨81JA1276⟩; **3**, 117 ⟨88JOC3587, 88SC869⟩, 211 ⟨81JA1277⟩, 223 ⟨73CC606⟩, 289 ⟨75JOC1675⟩, 357 ⟨90JAN1150, 91CPB672⟩; **4**, 247 ⟨86JOC1427⟩, 372 ⟨75JOC814⟩, 750, 1066 ⟨81JA1276⟩, 815 ⟨91T343, 85JA6639⟩; **5**, 208 ⟨90JFC(46)423⟩, 463, 800, 801 ⟨69JA7196⟩; **6**, 12 ⟨90JFC(46)423⟩, 232 ⟨79TL3865⟩, 524 ⟨56JAP10924⟩

Yoshida, Y., **2**, 614 ⟨89CL2213, 89TL7199⟩, 614 ⟨90CL769⟩; **3**, 625 ⟨83CL193⟩

Yoshida, Z., **1**, 187 ⟨84JA1079⟩, 192 ⟨87TL3497⟩, 238 ⟨83JOC1293⟩, 807, 834 ⟨80JA2392⟩, 907 ⟨73TCC47⟩; **2**, 69 ⟨87JOC4418, 88CC610⟩, 157 ⟨78CC367⟩, 579 ⟨92CC1718⟩, 588, 591 ⟨89AG(E)351, 89OS(67)98⟩; **3**, 5, 118 ⟨83JOC1286⟩, 115 ⟨79TL1401⟩, 130 ⟨85TL5529⟩, 290, 295, 296 ⟨83TL3869⟩, 356 ⟨73JA3043⟩, 367 ⟨77AG(E)866⟩; **4**, 74, 80 ⟨68JA4496⟩, 230 ⟨84TL5797⟩, 345 ⟨78JOC1980, 79BCJ3342⟩; **5**, 170 ⟨85TL3207⟩, 170 ⟨87TL325⟩, 595, 614 ⟨81TL3409⟩, 607 ⟨78JA1923, 80JA7806⟩, 607 ⟨81TL3413, 82JA4018⟩, 607 ⟨84TL5793⟩, 607, 614 ⟨81TL4245⟩, 607, 616 ⟨78TL2167⟩; **6**, 302 ⟨81JAP(K)56125326⟩, 316 ⟨78JOC595⟩

Yoshida, Z.-i., **1**, 122, 123, 187 ⟨87JOC4418⟩; **2**, 206 ⟨90JOC1823⟩, 257 ⟨79TL5015, 83JOC1293⟩ 960 ⟨75JA6461⟩; **3**, 161 ⟨88JOC1343⟩, 399 ⟨75CC359⟩; **6**, 124, 698 ⟨84TL4227⟩, 698 ⟨83TL3469⟩

Yoshifuji, M., **1**, 491, 536 ⟨77TL1303⟩; **2**, 851 ⟨75BCJ546⟩, 854 ⟨71CC1186⟩, 857 ⟨84CL603⟩, 857, 865 ⟨93TL1043⟩; **3**, 480 ⟨79TL2415, 84TL1809, 85PS(25)237, 87PS(30)527, 89TL839, 90TL2311⟩, 607 ⟨90CL827, 91CC124⟩, 609 ⟨86JOM(311)C63, 90HAC339⟩, 609 ⟨93JOM(461)81⟩, 634 ⟨79CL723⟩, 689 ⟨92CL1053⟩; **4**, 346 ⟨79TL2415⟩, 559 ⟨90CL643, 91CL2213, 93PS(74)373⟩; **5**, 877 ⟨88CL1733⟩, 879 ⟨83CL1653⟩, 884 ⟨88TL333⟩, 1086 ⟨90BCJ2736⟩, 1086, 1087 ⟨84TL1809⟩, 1086, 1087 ⟨88POL2213⟩, 1092 ⟨84CC689, 89CC1732⟩, 1092 ⟨89TL187, 89TL5433⟩, 1092 ⟨89TL839⟩; **6**, 680, 690 ⟨88CL1733⟩

Yoshifuji, S., **5**, 307 ⟨85CPB5515, 87CPB364⟩, 652 ⟨88CPB3125⟩

Yoshiga, T., **3**, 30, 33 ⟨94T2821⟩

Yoshigi, M., **5**, 189, 195, 282 ⟨86JOC4150⟩, 189, 199 ⟨84TL6027⟩

Yoshihara, K., **1**, 813 ⟨83JA625⟩, 1142 ⟨75MI 122-01⟩, 1142, 1143 ⟨78MI 122-03⟩; **4**, 386 ⟨83JA625⟩

Yoshihara, M., **2**, 8, 25 ⟨80S746⟩; **5**, 694 ⟨87NKK1304⟩

Yoshihara, M. K., **2**, 312 ⟨89BCJ3845⟩

Yoshihiro, I., **3**, 437, 438 ⟨83JA672⟩

Yoshii, E., **1**, 1140 ⟨88TL6951⟩; **2**, 455, 846 ⟨79S110⟩, 459 ⟨78TL3741⟩; **3**, 61 ⟨87JOC4135⟩; **4**, 831 ⟨74CPB2767⟩; **5**, 301 ⟨87JA3402⟩

Yoshiie, S., **4**, 892 ⟨76JCS(P1)783⟩

Yoshiizumi, K., **5**, 370 ⟨90JMC2192⟩

Yoshika, H., **2**, 39 ⟨70B1626⟩

Yoshikane, M., **6**, 87, 99 ⟨76S551⟩

Yoshikawa, K., **1**, 469 ⟨86BCJ1656⟩; **5**, 30, 35 ⟨86SC499⟩

Yoshikawa, M., **3**, 361 ⟨89CC505⟩

Yoshikawa, S., **2**, 757 ⟨77BCJ1513⟩

Yoshikoshi, A., **1**, 822 ⟨89TL1575, 90CC418⟩, 822 ⟨90CL613, 90T4503⟩, 822 ⟨91TL6159, 94CC789⟩, 1136 ⟨75JOC1932⟩; **2**, 783 ⟨82CL1505, 85ACR284⟩; **3**, 39, 175 ⟨80S664⟩, 160 ⟨87TL4293⟩, 192 ⟨89CC841⟩, 410 ⟨90CL239⟩, 435 ⟨90T7569⟩, 637 ⟨82JCS(P1)1303⟩; **4**, 906 ⟨78CC362⟩

Yoshimatsu, M., **2**, 278, 279, 292 ⟨94JOC1011⟩, 286, 288, 292 ⟨94CPB811⟩; **3**, 637 ⟨93SL121⟩; **4**, 288 ⟨91TL105⟩

Yoshimi, M., **5**, 516, 519, 522, 523, 626 ⟨90BCJ835⟩

Yoshimine, M., **1**, 1109 ⟨76JCP(64)4159⟩

Yoshimoto, M., **1**, 822 ⟨72TL2923⟩

Yoshimoto, Y., **5**, 294 ⟨80S887⟩

Yoshimura, H., **3**, 607 ⟨90CL827, 91CC124⟩

Yoshimura, I., **3**, 207 ⟨81CPB2733⟩; **4**, 326, 856, 857 ⟨78TL375⟩

Yoshimura, J., **1**, 648 ⟨85BCJ2812⟩; **3**, 564 ⟨75BCJ2891⟩; **4**, 433 ⟨78BCJ2614⟩; **5**, 371 ⟨79TL3809⟩; **6**, 72 ⟨80BCJ3687, 81CL375⟩

Yoshimura, K., **3**, 280, 305 ⟨88BCJ455⟩

Yoshimura, M., **2**, 60 ⟨88BCJ2684⟩; **6**, 428 ⟨73TL2287⟩

Yoshimura, N., **1**, 155 ⟨90S401⟩

Yoshimura, S., **4**, 856 ⟨78CPB685⟩

Yoshimura, T., **3**, 378 ⟨77TL3633, 79T317⟩, 634, 635 ⟨83BCJ257⟩; **4**, 223 ⟨83BCJ257, 83BCJ266⟩, 227 ⟨83BCJ270⟩

Yoshinaga, E., **2**, 855 ⟨67JAP6703120⟩

Yoshinaga, K., **4**, 226 ⟨82S312⟩

Yoshinaga, M., **4**, 872 ⟨85S497⟩; **6**, 45, 49 ⟨85S497⟩

Yoshinaga, Y., **5**, 329 ⟨84CL369⟩

Yoshinari, T., **2**, 904 ⟨74TL2961⟩

Yoshino, H., **2**, 124, 125 ⟨77CL437⟩

Yoshino, K., **5**, 370 ⟨90JMC2192⟩, 1072 ⟨93JOC5118⟩

Yoshino, S., **6**, 539 ⟨89JAP03020256⟩

Yoshino, T., **1**, 1049 ⟨91JOC3205⟩; **3**, 709 ⟨82TL5335⟩

Yoshio, S., **1**, 818 ⟨93TL8453⟩

Yoshio, Y., **1**, 246 ⟨78CL1007⟩

Yoshioka, H., **1**, 238 ⟨83JOC1293⟩, 650 ⟨89TL967⟩; **2**, 10, 20 ⟨85TL4207⟩, 11 ⟨87TL663, 89BCJ2024⟩, 18 ⟨93CL673⟩, 18, 280 ⟨89CC1881⟩, 19 ⟨91CL1185⟩, 257 ⟨79TL5015, 83JOC1293⟩, 305 ⟨81JOC4296⟩, 855 ⟨68TL5015⟩; **4**, 101, 105 ⟨87JFC(35)677⟩, 137 ⟨89TL967⟩, 751 ⟨88JFC(41)425⟩; **5**, 747 ⟨69BCJ3310⟩

Yoshioka, K., **1**, 942 ⟨83JCS(P1)403⟩

Yoshioka, M., **1**, 208 ⟨70JCS2365⟩; **2**, 72 ⟨89TL1657, 89TL7095⟩, 228 ⟨73BCJ2253⟩, 568 ⟨92T5691⟩; **3**, 343 ⟨73BCJ2253⟩, 614, 616, 633 ⟨77OR(25)255⟩, 616 ⟨72JA4654⟩, 632 ⟨70JCS(C)2365⟩, 634 ⟨72JA4635, 72JA4644⟩; **5**, 81 ⟨78GEP2725519⟩, 345 ⟨84JA1133⟩, 1142 ⟨72JA4635, 72OS5290⟩

Yoshioka, N., **5**, 541 ⟨79NKK1224⟩

Yoshioka, T., **2**, 749 ⟨69JCS(C)1086⟩; **3**, 684 ⟨84CC1513⟩, 716, 717 ⟨87JAN1207⟩; **4**, 201 ⟨82TL889⟩

Yoshioka, Y., **1**, 707 ⟨76T1705⟩; **5**, 990 ⟨68BCJ1258⟩, 993 ⟨83JAP8052⟩

Yoshitake, J., **3**, 58 ⟨78CL859⟩

Yoshitake, M., **1**, 216 ⟨90JCS(P1)1226⟩

Yoshitome, T., **1**, 1004 ⟨82CL1485⟩

Yoshitomi, S., **2**, 620 ⟨87CL1901⟩, 626 ⟨88JOC2093⟩, 710 ⟨91BCJ3485⟩; **5**, 252 ⟨81S637⟩

Yoshiura, H., **4**, 694 ⟨69AG(E)677⟩

Yoshiura, K., **5**, 296 ⟨83JOC4058⟩

Yoshiyasu, K., **5**, 985 ⟨73JAP7300455⟩

Yoshizaki, T., **2**, 1085 ⟨65IC657⟩

Yoshizawa, J., **1**, 438 ⟨93JOC6541⟩

Yoshizawa-Kumagaye, K., **6**, 487 ⟨93CL1401⟩

Yoshizumi, K., **2**, 927, 989 ⟨92BCJ1543⟩; **3**, 627 ⟨92BCJ1543⟩; **5**, 344 ⟨91CL1479⟩

Yoshizumi, T., **2**, 53 ⟨93JA3318⟩

Yoshokoshi, A., **1**, 228 ⟨91JOC6483⟩

Yosshimatsu, M., **2**, 281 ⟨90H(31)889⟩

You, M.-L., **1**, 24 ⟨85TL4495⟩

You, Q.-D., **5**, 600, 621 ⟨91OPP435⟩

You, X.-Z., **4**, 438 ⟨88SC2289, 89S307⟩

You, Z., **1**, 816 ⟨93TL2597⟩

Youds, P., **2**, 19 ⟨88JCS(P1)549⟩

Youn, I. K., **1**, 86 ⟨86TL2409⟩

Youn, J.-H., **2**, 173 ⟨86TL1493⟩, 173, 717 ⟨87S72⟩; **3**, 35 ⟨87S159⟩

Youn, K., **2**, 66 ⟨88TL4453⟩

Younathan, E. S., **2**, 84 ⟨91TL5273⟩

Yound, D. E., **6**, 240 ⟨70CC395⟩

Yound, G. B., **4**, 658 ⟨88POL1953, 89POL57⟩

Younes, H. A., **4**, 303 ⟨90G677⟩

Younes, M. I., **5**, 690 ⟨90S704⟩

Young, A. E., **1**, 851 ⟨64JA3283⟩, 909 ⟨64JA734⟩; **2**, 554 ⟨64JA3283⟩

Young, II, A. R., **2**, 421 ⟨64JA5359⟩

Young, C. A., **2**, 606, 623 ⟨36JA1806⟩; **4**, 24 ⟨36JA1806⟩

Young, C. H., **1**, 503 ⟨51JA1066⟩

Young, C. I., **2**, 638 ⟨74JA4494⟩

Young, C. W., **3**, 459 ⟨51JA42⟩

Young, D., **2**, 210 ⟨88OM1196⟩

Young, D. E., **2**, 94 ⟨69JA1310⟩; **5**, 1102

⟨71USP3761501⟩; **6**, 42, 43 ⟨70JOC3730⟩, 428 ⟨69TL723⟩
Young, D. M., **1**, 97 ⟨84JOC1544⟩
Young, D. P., **5**, 1000 ⟨41JCS263⟩
Young, D. S., **4**, 3 ⟨40JA1171⟩
Young, D. V., **1**, 88 ⟨61JA3160⟩
Young, D. W., **1**, 394 ⟨83JCS(P1)2387⟩, 486 ⟨93TL4667⟩; **3**, 4, 115 ⟨79S401⟩
Young, E. H. P., **5**, 1036 ⟨50JCS1582⟩
Young, F., **4**, 86 ⟨74JOC1449⟩
Young, G., **5**, 652 ⟨58MI 504-01⟩
Young, G. B., **4**, 654, 656 ⟨89JOM(361)123⟩
Young, G. F., **1**, 73 ⟨65CI(L)560⟩
Young, G. H., **4**, 683 ⟨92OM154⟩
Young, G. T., **2**, 223 ⟨64JCS3832⟩; **4**, 218 ⟨64JCS3832⟩; **5**, 84 ⟨B-62MI 502-02⟩, 122 ⟨63JCS1927⟩
Young, H. A., **2**, 213 ⟨37JA811⟩
Young, I. G., **1**, 826 ⟨73B3492⟩
Young, J., **3**, 155 ⟨87JOC3919⟩
Young, J. A., **4**, 1076 ⟨67FCR359⟩; **5**, 361 ⟨71JOC350⟩, 989 ⟨59JA1587⟩; **6**, 238, 239, 240, 242 ⟨58JA3604⟩, 253 ⟨59JA574⟩, 257 ⟨56JA5637, 58JA1889⟩, 440 ⟨58JOC1576⟩
Young, J. C., **1**, 490 ⟨93CRV1371⟩; **2**, 547 ⟨93CRV1371⟩; **4**, 32 ⟨60JCS4503⟩; **5**, 421 ⟨77CC73, 81JOM(205)31⟩
Young, J. F., **1**, 73 ⟨65CC131⟩
Young, J.-J., **2**, 34 ⟨87JOC3919⟩
Young, J. R., **1**, 128 ⟨68JOC1675⟩; **6**, 508 ⟨73JMC901⟩
Young, Jr., V. G., **2**, 939 ⟨93JA3794⟩; **6**, 681, 690 ⟨91JA9379⟩
Young, M. G., **5**, 346 ⟨82JOC179⟩
Young, M. W., **1**, 601, 602, 603, 607, 608, 781 ⟨73TL1979⟩, 601, 604, 781 ⟨75JOC947⟩, 601, 610, 611, 621, 646 ⟨75CS9⟩, 716 ⟨75CS9⟩; **2**, 286 ⟨73TL1979⟩; **3**, 175 ⟨73TL1979⟩
Young, P. R., **5**, 214 ⟨88JMC1960⟩
Young, R. G., **4**, 539 ⟨90SL413, 91JCS(P1)323⟩
Young, R. J., **3**, 428 ⟨59JA5251⟩
Young, R. N., **1**, 532, 662 ⟨86TL539⟩, 894 ⟨79MI 119-05⟩; **2**, 309 ⟨74JCS(P1)2101⟩; **3**, 65 ⟨81TL979⟩; **4**, 247 ⟨88TL6729⟩, 248 ⟨88TL6733⟩
Young, S. D., **1**, 637 ⟨81TL4163⟩, 799 ⟨80JA4841⟩
Young, T. E., **4**, 907 ⟨80JOC933⟩
Young, W. B., **1**, 812 ⟨91JOC5696⟩
Young, W. G., **1**, 837 ⟨B-64MI 118-02⟩
Young, W. R., **1**, 847 ⟨69JA529⟩, 848 ⟨69JA527⟩
Youngdahl, K., **3**, 83 ⟨84OM1601⟩
Youngs, W. J., **1**, 456, 507 ⟨83JA6273⟩, 1010 ⟨90SL427⟩; **2**, 432 ⟨83OM979⟩; **3**, 620 ⟨86JA1311⟩
Yousaf, T. I., **2**, 385 ⟨90CC1598⟩; **3**, 406 ⟨87T5171⟩
Yousif, M. M., **4**, 422 ⟨79JOC4543⟩
Yousif, N. M., **3**, 358 ⟨86BSF283⟩; **5**, 551 ⟨84T2663⟩, 581, 643 ⟨89T4599⟩, 590, 619 ⟨87PS(32)51⟩
Youssef, A. A., **1**, 851 ⟨62JA2465⟩; **2**, 26 ⟨68CC976⟩
Youssef, A. K., **1**, 800 ⟨72JOC2601, 73JOC487⟩
Youssef, B., **6**, 7 ⟨87JPS(A)3025⟩
Youssefi-Tabrizi, M., **4**, 90, 784 ⟨87S169⟩, 577 ⟨86JOM(316)13⟩; **6**, 136 ⟨87S169⟩
Youssefyeh, R. D., **1**, 925 ⟨65CC7⟩; **2**, 26, 33 ⟨63CI(L)609⟩
Yovell, J., **5**, 344 ⟨76TL451⟩
Ypenburg, J. W., **6**, 302 ⟨72RTC1117⟩
Yu, C.-C., **2**, 344 ⟨78JOC2289⟩
Yu, C. H., **2**, 79, 533 ⟨90JA878⟩
Yu, C. M., **1**, 977, 1055 ⟨90JA878⟩; **2**, 76 ⟨89JA5495⟩
Yu, D., **4**, 109 ⟨89T2957⟩
Yu, H., **5**, 466 ⟨72JOC1732⟩
Yu, J., **1**, 179, 183 ⟨93JOC5892⟩, 183 ⟨92JOC3757⟩, 183 ⟨92TL6723⟩
Yu, L.-C., **1**, 408, 821 ⟨85JOC2965⟩; **3**, 721, 722 ⟨90CA(112)118475⟩; **6**, 134 ⟨84JOC605, 85JOC3266⟩
Yu, S., **3**, 123 ⟨86SC1357⟩; **4**, 42 ⟨81JOC571⟩
Yu, S. H., **2**, 13 ⟨74S560⟩, 553, 555 ⟨92JOC3680⟩; **4**, 42 ⟨75JA2293⟩
Yu, S.-L., **3**, 337 ⟨78JA7436⟩; **6**, 603, 607 ⟨76IC14⟩
Yu, T.-H., **3**, 62 ⟨90S1033⟩
Yu, V.,**4**, 833 ⟨70IZV2140⟩
Yu, Y., **3**, 229 ⟨88TL1045, 89JOC1105⟩; **4**, 716 ⟨93OM1213⟩
Yu, Z., **2**, 562 ⟨93JA6614⟩
Yuan, C., **1**, 907, 909 ⟨58JA5991⟩; **2**, 454 ⟨86S821⟩, 456 ⟨87MI 208-01⟩, 848 ⟨90S140⟩; **3**, 477 ⟨58JA5991⟩; **4**, 343 ⟨92S258⟩, 467 ⟨86S821, 90S256⟩, 470, 488 ⟨91S490⟩, 470, 488 ⟨92S1124⟩, 472 ⟨93S955⟩, 801 ⟨91S854⟩; **6**, 118 ⟨92PS(69)75⟩
Yuan, C.-S., **4**, 218 ⟨94JOC544⟩
Yuan, J.-J., **1**, 444 ⟨85JCR(S)18⟩; **2**, 432 ⟨85JCR(S)18⟩
Yuan, K., **1**, 465 ⟨93JOC4866⟩
Yuan, W., **5**, 263 ⟨93TL1247⟩
Yuan, Z., **2**, 1083 ⟨93JOM(449)27⟩
Yuanwwei, C., **3**, 415 ⟨89SC1423⟩
Yuanyao Xu, **1**, 737, 738 ⟨82OPP373⟩
Yuasa, K., **1**, 276 ⟨78BCJ339⟩
Yuasa, Y., **3**, 634 ⟨79H(12)681⟩, 634, 651 ⟨84JA7890⟩
Yu Belavin, I., **4**, 418 ⟨84JGU635⟩
Yudina, K. S., **4**, 555, 561 ⟨67IS63⟩, 556 ⟨66IZV1954⟩
Yudin, B. F., **6**, 223 ⟨59MI 607-03⟩
Yue, C., **1**, 104 ⟨94T3139⟩
Yue, S., **3**, 9 ⟨92SC1217⟩
Yuen, P.-W., **2**, 50 ⟨⟨89JA9243⟩; **4**, 1083 ⟨84JCS(P1)1561⟩
Yufit, D. S., **4**, 282 ⟨83TL87⟩; **6**, 619 ⟨93JPO319⟩
Yuhara, M., **1**, 228 ⟨88TL343⟩, 558, 561 ⟨87S992⟩, 962 ⟨86CC922⟩; **3**, 623 ⟨87JOC2988⟩
Yuhura, M., **3**, 645 ⟨86CC118⟩
Yu Jin, G., **4**, 456, 482 ⟨81TL229, 81TL1105⟩
Yukawa, T., **1**, 946 ⟨85JOC5875⟩
Yukawa, Y., **1**, 636 ⟨74CL165⟩, 1124 ⟨73TL4767⟩; **5**, 318, 998 ⟨71BCJ1632, 71BCJ1639⟩, 998 ⟨71BCJ1644⟩
Yukhno, Yu. M., **2**, 469 ⟨75ZOB2554⟩; **5**, 722 ⟨75JGU935, 75ZOB949⟩
Yukimasa, H., **5**, 332, 333 ⟨92CPB1177⟩
Yukizaki, H., **1**, 476 ⟨83TL5745⟩
Yu. Kopylova, L., **4**, 559 ⟨89ZOB1686⟩
Yuldasheva, E. K., **6**, 78 ⟨50DOK(70)231⟩
Yum, E. K., **1**, 151 ⟨93JOC823⟩
Yun, K.-Y., **3**, 75, 76 ⟨84LA649⟩
Yung, R. T., **2**, 388 ⟨89JOC1070⟩
Yung, T. W. K., **6**, 309 ⟨91TL4329⟩
Yung, Y. M., **3**, 712, 714, 718 ⟨81TL3759⟩
Yunis, M., **3**, 361, 579 ⟨86CB162⟩, 552 ⟨85CB851⟩; **4**, 979 ⟨86CB162⟩; **5**, 616 ⟨85CB851⟩
Yura, T., **3**, 164 ⟨86CL1809⟩; **5**, 557 ⟨88CL1021, 88CL1025⟩
Yura, Y., **1**, 1041 ⟨63CPB1049⟩; **3**, 220, 496 ⟨74TL4223⟩
Yurchenko, A. G., **2**, 428 ⟨88ZOB33⟩; **4**, 337 ⟨86JGC425⟩
Yurchenko, R. I., **2**, 427 ⟨91ZOB1021, 88ZOB33⟩, 428 ⟨90ZOB697⟩
Yurchenko, R. Yu., **4**, 337 ⟨86JGC425⟩
Yurechko, V. V., **6**, 621 ⟨81MI 620-02⟩, 646 ⟨81MI 621-01⟩
Yur'ev, V. P., **1**, 61 ⟨81ZOB420⟩; **4**, 141 ⟨81DOK(261)474⟩
Yur'eva, E. M., **2**, 238 ⟨83ZOR343⟩; **4**, 1011 ⟨80ZOB343⟩
Yus, M., **1**, 136 ⟨93T1327⟩, 145 ⟨89JOC1491⟩, 237 ⟨89TL3837⟩, 332, 343, 948 ⟨88CRV487⟩, 486, 1031 ⟨89TL173⟩, 681 ⟨86CC1665, 84JCS(P1)721⟩, 1089, 1090 ⟨87TL6709, 88CC1135⟩, 1090 ⟨88TL4859, 89JCR(S)200, 89JCS(P1)1387, 90CC1521, 90CC587, 91CC1001⟩; **2**, 24, 31 ⟨86JCR(S)274, 85CC1422⟩, 85 ⟨89JOC1491⟩, 392 ⟨90TL4199⟩, 420, 553 ⟨93TL1649⟩, 551, 553 ⟨91CC398, 88TL2859⟩, 553 ⟨92T4593, 92TL5597, 94T5775⟩, 936 ⟨88S234⟩; **3**, 27, 155 ⟨86S678⟩, 31, 161 ⟨92JOC750⟩, , 31, 161 ⟨93T4923, 94T3437⟩, 58 ⟨85JCS(P1)447⟩, 152 ⟨91JCS(P1)2890⟩,

152, 154 ⟨90S1003⟩, 256 ⟨88JCS(P1)1029, 88JOC4708⟩, 429 ⟨85CC126, 85T699⟩, 638 ⟨81S375⟩; **4**, 741, 747, 756, 758 ⟨89JCS(P1)691⟩, 808 ⟨87JCS(P1)1017⟩, 938, 961 ⟨87TL6709⟩, 938, 961 ⟨88JOC4708⟩; **5**, 71 ⟨83JCS(P1)3019⟩, 284 ⟨81CC1178, 81CC670⟩
Yusa, K., **3**, 282 ⟨89BCJ3371⟩
Yushmanova, T. I., **5**, 718 ⟨88IZV460⟩
Yu Sizov, A., **6**, 236 ⟨92RCR517⟩
Yusuda, S., **4**, 970 ⟨70MI 421-01⟩
Yusuf, H. M., **2**, 405 ⟨69JCS(C)397⟩
Yuuki, M., **4**, 23, 27 ⟨67BCJ594⟩
Yuvchenko, A. P., **2**, 1098 ⟨85JGU1691, 86JGU1812⟩; **5**, 1110 ⟨90MI 530-01⟩
Yuzhakova, G. A., **6**, 656 ⟨78ZOB811, 81ZOB880⟩
Yver, B., **5**, 22 ⟨91JHC1511⟩

Z

Zabik, M. J., **6**, 425 ⟨67JOC300⟩
Zabirov, N. G., **5**, 370 ⟨88ZOB2424⟩, 384 ⟨90ZOB1997⟩, 581 ⟨79IZV861, 86ZOB2660⟩, 581, 619 ⟨92ZOB71⟩, 642, 644 ⟨91RCR1128⟩, 643 ⟨78ZOB843⟩, 643, 644 ⟨79IZV805, 90ZOB464⟩
Zablocka, M., **1**, 158 ⟨89TL477⟩, 159 ⟨84JOC1522⟩; **4**, 120, 804 ⟨89T337⟩, 547, 579 ⟨93AG(E)1735⟩
Zablocki, J., **2**, 931 ⟨88TL6541⟩; **3**, 419, 420 ⟨88TL6541⟩
Zabolotskikh, A. V., **2**, 615 ⟨89IZV1935⟩
Zaboronski, L. M., **2**, 166, 170 ⟨71ICA311⟩; **5**, 807 ⟨71IC1635, 71IC407⟩; **6**, 239 ⟨71IC1635⟩, 613 ⟨70JA3665⟩
Zabotina, E. Ya., **4**, 549 ⟨85ZOB1471⟩; **6**, 678, 691 ⟨91RCR162⟩
Zabrodskaya, I. S., **5**, 597 ⟨69ZOB1282⟩
Zaburdyaevla, S. N., **2**, 886 ⟨89MI 217-02⟩
Zaby, G., **4**, 776, 854 ⟨81CB684⟩, 851 ⟨77CB1484⟩
Zacharias, D. E., **5**, 350 ⟨85JOC2224⟩
Zacharie, B., **1**, 24 ⟨85JA6121⟩; **2**, 749 ⟨88TL2299⟩, 750 ⟨90JOC3002⟩; **4**, 229 ⟨92TL4625⟩; **5**, 568, 571 ⟨93T10489⟩, 588 ⟨93BMC619⟩
Zachystalová, D., **4**, 226 ⟨61CCC2206⟩
Zadel, G., **3**, 131, 284, 303 ⟨92AG(E)1035⟩
Zador, M., **2**, 40 ⟨71CJC2586⟩
Zadornaya, A. B., **4**, 559 ⟨85ZOB1865⟩
Zaev, E. E., **1**, 1006 ⟨72HOU(5/1b)180⟩
Zagidullin, R. N., **6**, 565 ⟨87ZPK1829⟩
Zagnibida, D. M., **5**, 774 ⟨71MI 519-01⟩
Zagurskaya, L. M., **5**, 397 ⟨75BAU816⟩
Zaher, H., **5**, 371, 403 ⟨92TL2067⟩
Zahler, R., **1**, 184 ⟨80S389⟩, 767, 769 ⟨80JA3270⟩; **2**, 680 ⟨80JA3270⟩; **5**, 370 ⟨83JA3743⟩
Zahler, R. E., **1**, 474, 483 ⟨51CRV273⟩
Zahman, A., **3**, 357, 377 ⟨75JCS(P1)1335⟩
Zahn, P., **4**, 699 ⟨72JOM(36)267⟩
Zahn, T., **2**, 512 ⟨84AG(E)812, 85CB2472⟩, 585 ⟨84AG(E)515⟩; **4**, 549 ⟨87AG(E)236⟩, 709 ⟨87JOM(321)257⟩
Zahneisen, T., **4**, 138 ⟨90SL769⟩
Zahnow, E. W., **3**, 638, 665 ⟨69JA4181⟩
Zahr, S., **3**, 60 ⟨79S266⟩
Zahra, J. P., **1**, 1131 ⟨87TL2009⟩
Zahradník, R., **1**, 866 ⟨64CCC1932⟩
Zaidi, N. A., **1**, 929 ⟨89TL3343, 92T6883⟩; **6**, 544, 557 ⟨92T6883⟩
Zaidlewicz, M., **2**, 42 ⟨88JOC2911⟩, 75, 521 ⟨90JA2389⟩, 525 ⟨76JA4917⟩, 557 ⟨87S645, 91JOM(409)103⟩, 900 ⟨82COMC-I(7)143, 82COMC-I(7)161⟩; **4**, 653 ⟨87S645⟩
Zaikin, V. G., **2**, 266 ⟨81JOM(214)415⟩; **4**, 798 ⟨81IZV1603⟩
Zaiko, I. A., **4**, 57 ⟨86ZOR2327⟩
Zaim, Ö., **3**, 209 ⟨88JOC1110⟩, 237 ⟨86JOC4751⟩; **6**, 557 ⟨86JOC4741, 87JOC5630⟩, 557 ⟨88JOC1110⟩
Zaitseva, E. L., **2**, 470 ⟨69JGU2161⟩
Zaitseva, G. S., **4**, 616 ⟨79ZOB2785⟩, 637 ⟨76MI 414-01⟩, 827 ⟨84ZOB1925⟩; **5**, 162 ⟨78JGU1249⟩, 702 ⟨91JGU1268, 91ZOB1389⟩, 723 ⟨89JGU2055, 89ZOB2293⟩
Zajac, Jr., W. W., **2**, 390 ⟨88S808, 89TL6495⟩, 392 ⟨83S545⟩, 349 ⟨88JOC5856⟩, 785 ⟨82TL135⟩; **3**, 190 ⟨81JOC5151⟩, 190 ⟨82JOC1176⟩, 190 ⟨83S545⟩, 445 ⟨62JOC3716⟩; **4**, 108, 109 ⟨91JOC316⟩, 178 ⟨70JOC3375⟩, 199 ⟨72JOC521⟩
Zajacek, J. G., **2**, 90 ⟨70JOC1839⟩, 348 ⟨70OS(50)56⟩
Zajc, B., **2**, 15 ⟨86BCJ1659⟩, 16 ⟨82JOC573, 89T6003⟩
Zajdel, W. J., **1**, 148 ⟨84CRV471⟩; **2**, 302 ⟨84CR471⟩, 1002 ⟨84CRV471⟩; **3**, 183 ⟨84CRV471⟩; **4**, 529, 535, 537, 1014, 1015, 1016 ⟨84CRV471⟩, 530 ⟨84JA1010⟩
Zak, H., **1**, 91 ⟨76CB2628⟩
Zak, Z., **2**, 370 ⟨77CCC1466⟩
Zakaraya, D., **2**, 39 ⟨94TL4985⟩
Zakharkin, L. I., **1**, 462 ⟨87BAU327⟩, 665 ⟨65JOM(4)349⟩; **2**, 434 ⟨58MI 208-01⟩, 598 ⟨64JOM(2)281⟩, 915 ⟨77IZV2389⟩, 1015 ⟨58IZV846⟩, 1093 ⟨67JGU515⟩, 1096 ⟨65JOM(3)257⟩, 1099 ⟨63IZV1146, 64IZV1989⟩, 1099 ⟨65JGU1677⟩; **3**, 82 ⟨64ZOB1029⟩, 84 ⟨62TL619, 63TL2087⟩, 85 ⟨69T5555⟩, 531 ⟨71T1317⟩; **4**, 143 ⟨82IZV1425, 85IZV809⟩, 698 ⟨75CA(83)10238⟩; **5**, 38, 89 ⟨57JGU2481⟩, 44 ⟨78IZV2094⟩; **6**, 208 ⟨68IZV910⟩, 208, 210 ⟨83IZV636⟩, 221 ⟨52MI 607-01⟩, 406 ⟨67JOM(7)385, 73JGU848⟩, 523 ⟨58IZV100⟩
Zakharova, E. S., **5**, 388 ⟨85ZOB1071⟩
Zakharova, R. P., **5**, 390 ⟨81ZOB1318⟩, 1085 ⟨89MI 528-01⟩
Zakharov, L. N., **6**, 562 ⟨81KGS907⟩
Zakharov, V. A., **6**, 145 ⟨77ZOR1559⟩
Zakrewski, J., **1**, 554 ⟨84TL1279⟩; **2**, 221 ⟨81TL3659⟩, 338 ⟨88SC2135⟩; **5**, 387 ⟨89JOM(359)215⟩
Zalar, F. V., **1**, 885 ⟨67JA946⟩
Zalesova, N. N., **2**, 255 ⟨87ZOB786⟩
Zalesov, V. S., **6**, 145 ⟨76KFZ35⟩
Zalesskaya, I. M., **5**, 112 ⟨85ZOR2407⟩
Zaleta, M., **3**, 164 ⟨85JA4175⟩
Zaleta, M. A., **2**, 805 ⟨72JOC1597⟩
Zalinyan, S. A., **2**, 447 ⟨86AKZ457⟩
Zalkin, V. G., **1**, 63 ⟨81JOM(214)145⟩
Zalkow, L. H., **3**, 634 ⟨77TL1117⟩
Zaller, F., **5**, 1025 ⟨54LA(585)230, 55LA(594)233⟩
Zally, W. J., **3**, 190 ⟨79S295⟩; **4**, 412 ⟨74JOC167⟩
Zal'tsman, I. S., **2**, 407 ⟨89ZOB2135⟩; **3**, 494 ⟨90JGU613⟩; **4**, 559 ⟨88ZOB1665⟩; **5**, 369 ⟨88ZOB706⟩
Zalukaev, L., **5**, 823 ⟨60ZOB506⟩
Zálupsky, P., **5**, 665 ⟨76CCC3085⟩
Zaluski, M.-C., **3**, 102 ⟨70BSF1445⟩, 103 ⟨70BSF1838⟩
Zalut, C., **2**, 712 ⟨60JA5359⟩
Zaluzec, E. J., **5**, 934 ⟨89JOC3249⟩
Zamaletdinova, G. U., **4**, 921 ⟨84ZOB1324⟩; **6**, 582 ⟨76IZV455⟩
Zamaletdinova, R. R., **4**, 463 ⟨88ZOB2456⟩
Zamanskii, A. A., **4**, 894 ⟨74MI 420-01⟩
Zambito, A. J., **5**, 90 ⟨73OSC(5)373⟩
Zamboni, R., **1**, 513 ⟨83TL4899⟩, 514 ⟨84TL2179⟩, 532, 662 ⟨86TL539⟩; **2**, 137 ⟨89JOC1230⟩; **3**, 62 ⟨83TL4899⟩, 207 ⟨80JA2097⟩, 219 ⟨81JA3460⟩
Zamkanei, M., **5**, 592, 630 ⟨77TL2139⟩; **6**, 111 ⟨79CB1873⟩
Zamore, M., **5**, 273, 296 ⟨76JOC725⟩
Zamponi, A., **3**, 35, 35 ⟨94T3349⟩
Zamyatina, V. A., **2**, 430 ⟨69JOM(17)201⟩
Zamyshlyaeva, L. I., **1**, 56 ⟨65MI 102-01⟩
Zanathy, L., **3**, 342 ⟨92MC67⟩
Zanazzi, P. F., **6**, 341 ⟨87AF1003⟩
Zandbergen, P., **3**, 628 ⟨91SC1387⟩
Zander, F., **3**, 579 ⟨81CB536⟩
Zander, R., **3**, 547 ⟨81LA1073⟩
Zanella, A. W., **5**, 386 ⟨75IC42⟩
Zanella, P., **3**, 442 ⟨87ICA275⟩; **6**, 519 ⟨87JCS(D)2039⟩

Zanello, P., **1**, 507 ⟨88IC4429⟩
Zang, S.-S., **5**, 253 ⟨90JCR(S)230⟩
Zanger, M., **2**, 35 ⟨75JOC248⟩, 311 ⟨91S176⟩, 436, 442, 443 ⟨64JA2378⟩; **3**, 264 ⟨91S176⟩
Zangrando, E., **4**, 585 ⟨74JOM(74)C23⟩
Zani, P., **2**, 938 ⟨91TL2971⟩; **3**, 369 ⟨86CC964, 90JOC3744⟩; **4**, 232 ⟨85TL2131, 91TL2971⟩, 371, 944 ⟨89JCS(P1)2083⟩, 859 ⟨92TL6383⟩; **5**, 562 ⟨92TL6383⟩, 647, 648 ⟨86JCS(P1)381, 88CC365⟩, 648 ⟨88JCS(P1)1499, 90JOC3744⟩, 648, 650 ⟨89JCS(P1)2083⟩; **6**, 127 ⟨87JCS(P1)2647, 91TL2971⟩, 127 ⟨91TL6801⟩, 127, 128 ⟨90H(31)47⟩, 127, 309 ⟨85TL2131⟩
Zanirato, P., **3**, 359 ⟨85CC1441⟩; **4**, 515 ⟨91JCS(P1)2789⟩; **5**, 647 ⟨81CC822⟩, 992 ⟨92JCS(P1)1101⟩; **6**, 127 ⟨81CC822⟩
Zanirato, V., **2**, 356 ⟨91TL2517⟩; **3**, 648 ⟨85S969⟩
Zank, G. A., **3**, 363 ⟨86TL3445⟩; **5**, 513 ⟨86TL3445⟩
Zanke, D., **5**, 1062 ⟨69TL3191⟩
Zanobini, F., **2**, 438 ⟨84JCS(D)575⟩
Zanon, P., **5**, 1078 ⟨85S509⟩
Zanotti, V., **1**, 37 ⟨83S137⟩; **5**, 1088 ⟨91AG(E)847⟩; **6**, 332 ⟨86G101⟩
Zanotti-Gerosa, A., **2**, 553 ⟨92CC1623⟩; **5**, 943 ⟨92CC1623⟩, 953, 958 ⟨95JOM(486)279⟩, 956 ⟨94TL6377⟩
Zanotto, L., **6**, 247 ⟨90OM1449⟩
Zaoral, M., **5**, 263 ⟨77CCC560⟩
Zapadinskii, B. I., **5**, 24 ⟨73RCR939⟩
Zapata, A., **1**, 552 ⟨89JOC5854⟩; **4**, 154 ⟨83SC129⟩, 652, 660 ⟨80TL3451, 81S557, 81SC673, 93JOM(448)69⟩; **6**, 624 ⟨94TL2365⟩
Zapevalov, A. Y., **6**, 42 ⟨73ZOR2013⟩
Zappelli, P., **4**, 205, 210 ⟨79CB148⟩
Zaragozá, R. J., **1**, 1082 ⟨92JOC50⟩
Zarantonello, P., **3**, 456 ⟨91TL2967⟩
Zarate, E. A., **3**, 620 ⟨86JA1311⟩
Zardaloo, F. S., **2**, 135 ⟨91T9861⟩
Zard, S. M., **2**, 627 ⟨87T4321⟩
Zard, S. Z., **1**, 231 ⟨89H(28)171⟩, 348 ⟨88CC285⟩, 931 ⟨91JA1055⟩, 935 ⟨87T2733, 91TL199⟩, 935, 938 ⟨86PAC675, 90TL2565⟩, 939, 946 ⟨90JA2034⟩, 1076 ⟨91TL5321⟩; **2**, 226 ⟨88T1153⟩, 364 ⟨86TL1153⟩, 785 ⟨82CC551, 83BSF(2)61⟩; **3**, 136 ⟨88CC308⟩, 150 ⟨92TL1285⟩, 368 ⟨89T3741⟩, 405 ⟨84CC337, 84TL3707⟩, 408 ⟨86JCS(P1)2243⟩, 427, 430 ⟨87T551⟩, 713 ⟨88TL3343⟩; **4**, 266 ⟨91T7091⟩, 269 ⟨92TL4913⟩, 327 ⟨85T5507⟩; **5**, 1115 ⟨86T4983⟩; **6**, 547 ⟨89TL4367, 90JA2034⟩, 550 ⟨86T2329⟩, 551 ⟨88CC308, 89H(28)171⟩, 551 ⟨92JA7909⟩
Zarecki, A., **3**, 294 ⟨73TL3635⟩
Zaret, E. H., **1**, 181 ⟨71S203⟩, 181 ⟨72S142⟩; **2**, 684 ⟨72JOC2267⟩
Zaretskaya, I. I., **1**, 421 ⟨57JGU693, 60JGU765⟩, 421, 422 ⟨44IZV65⟩
Zarges, W., **2**, 999, 1005 ⟨91CB543⟩, 1003 ⟨91CB861⟩; **3**, 566 ⟨89AG(E)1392⟩; **4**, 646 ⟨91CB543⟩
Zaripov, Sh. I., **3**, 336 ⟨79ZOB1084⟩
Zarkadis, A. K., **4**, 141, 154, 608, 616 ⟨80JOM(201)197⟩
Zaro, J., **2**, 355 ⟨73JOC1813⟩
Zarrinmayeh, H., **2**, 104 ⟨84HCA906⟩
Zartman, W., **1**, 881 ⟨43OSC(2)606⟩
Zartner, H., **5**, 643 ⟨74ZC56⟩, 1031 ⟨75ZC56⟩
Zarva, T. V., **1**, 1045 ⟨67ZOR219⟩
Zaslona, A. T., **2**, 447 ⟨81JCS(P1)3059⟩; **4**, 923 ⟨82TL4167⟩, 923, 924 ⟨87JCS(P1)967⟩
Zasorina, V. A., **6**, 537 ⟨87UKZ395⟩
Zatorski, A., **1**, 821 ⟨78JOC2518⟩; **2**, 713 ⟨78JOC473⟩; **4**, 346 ⟨79JOC2967⟩, 347 ⟨73S669⟩, 835, 847, 848 ⟨78T3081⟩, 847, 848 ⟨76TL2731⟩; **6**, 48 ⟨77JPR17⟩, 119 ⟨76TL477, 78JOC2518⟩, 119 ⟨79JOC2907⟩
Zaubitzer, T., **4**, 648 ⟨88JOC2274⟩
Zaugg, H. E., **1**, 881 ⟨60JA2895, 61JOC644⟩, 1042 ⟨70S49, 84S85⟩; **2**, 1049 ⟨58JOC1389⟩; **4**, 96, 98 ⟨70S49⟩, 96, 98, 99, 308, 416, 417, 424, 425, 427 ⟨65OR(14)52⟩; **5**, 115 ⟨55OSC(3)820⟩, 161 ⟨50JA2998⟩; **6**, 104 ⟨64JOC2769⟩
Zavadskii, K. S., **4**, 548 ⟨87ZOB2606⟩
Zavadsku, K. S., **5**, 1092 ⟨87JGU2350⟩
Zavalishina, A. I., **5**, 370 ⟨81PS(10)261⟩
Zavgorodini, V. S., **1**, 492 ⟨72ZOR1484⟩; **2**, 1071 ⟨67JGU898⟩, 1080 ⟨62ZOB1349⟩, 1090 ⟨87JGU2482⟩, 1101 ⟨62DOK(143)855⟩; **4**, 222 ⟨91TL7593⟩, 766 ⟨89ZOB2145⟩, 807, 811, 813 ⟨91JGU1295⟩, 810 ⟨91JGU1306⟩, 955 ⟨81ZOB1857⟩
Zavist, A. F., **2**, 176 ⟨48JA3526⟩
Zavlin, P. M., **2**, 469 ⟨67ZOB2787⟩
Zavozin, A. G., **2**, 28 ⟨87IZV2395⟩
Zav'yalov, S. I., **2**, 22 ⟨87IZV2395, 92BAU356⟩
Zav'yalov, V. I., **5**, 691 ⟨92JOU1399⟩, 930 ⟨92ZOR1750⟩
Zawadzki, S., **2**, 308 ⟨92SC1929⟩, 317 ⟨87S487⟩, 373 ⟨87S485⟩; **5**, 273, 285 ⟨79S527, 79S549⟩, 652 ⟨88PS(40)263⟩, 774 ⟨90MI 519-01⟩
Zawaszi, T., **5**, 347 ⟨86FES676⟩
Zawistowski, E. A., **6**, 694 ⟨71JFC(1)347⟩
Zawoiski, S., **1**, 1029 ⟨88JOC386, 89JMC1820⟩
Zaworotko, M. J., **6**, 187 ⟨81JOM(212)C4⟩, 195 ⟨81JAP(K)56125326⟩
Zayed, A. A., **4**, 882 ⟨83MI 420-01⟩
Zayed, E. M., **4**, 984 ⟨77G555⟩
Zayed, M. F., **4**, 487 ⟨82PS(13)319⟩
Zbikowski, I., **2**, 675 ⟨64CJC212⟩
Zbiral, E., **1**, 179 ⟨76HCA2100⟩, 821 ⟨78T317⟩; **2**, 771 ⟨69M64⟩; **3**, 11 ⟨83TL573⟩, 270 ⟨79LA492, 85M77⟩, 270 ⟨91S357⟩, 478 ⟨75T1817⟩, 496 ⟨B-79MI 313-04⟩, 585, 586 ⟨76M537⟩, 613 ⟨76HCA2100⟩; **4**, 339 ⟨91CB175, 91LA229⟩, 561 ⟨79LA492⟩, 916 ⟨86TL5829⟩; **5**, 897, 898 ⟨66M1797, 69LA(721)121⟩, 898, 903 ⟨66M180⟩, 1068 ⟨86M(117)1437⟩, 1107 ⟨76HCA2100⟩; **6**, 550 ⟨88M127⟩
Zbirovsky, M., **6**, 236 ⟨66FRP1437908⟩, 236, 531 ⟨69CC878⟩
Zbrial, E., **3**, 696, 708 ⟨88LA765⟩
Zdanovich, V. I., **1**, 76 ⟨67T793⟩
Zderic, S. A., **1**, 657 ⟨78JOM(156)203⟩; **2**, 517 ⟨79JA2352⟩
Zdero, C., **1**, 736 ⟨73CB3779⟩, 1142 ⟨67CB1193⟩; **3**, 78 ⟨73CB2755⟩
Zdorova, S. N., **2**, 448 ⟨76JGU572⟩; **4**, 566 ⟨76ZOB575⟩, 566, 568, 576 ⟨76ZOB1218⟩, 569 ⟨72ZOB112⟩
Zdravkovski, Z., **3**, 4 ⟨89SC1317⟩
Zdrojewski, T., **6**, 280 ⟨90JPO694⟩
Zdunek, L. Z., **4**, 42 ⟨82JCS(P2)881⟩
Zdunneck, P., **2**, 566 ⟨71JOM(32)335⟩
Zeavin, J. M., **5**, 198 ⟨32JA3738⟩
Zebovitz, T. C., **5**, 157 ⟨77JOC3903⟩
Zecchi, G., **1**, 524 ⟨77JHC947⟩; **3**, 677, 678, 680 ⟨91S181⟩; **4**, 440 ⟨79S380⟩; **5**, 371 ⟨86S772⟩, 676, 677 ⟨81JCS(P1)2245, 84JCS(P1)2535⟩, 677 ⟨89T7427, 92OPP352⟩, 677 ⟨93JHC559⟩, 681 ⟨86JCR(S)210⟩, 862 ⟨86JCS(P1)2177⟩, 905 ⟨84T2165⟩
Zecchini, G. P., **3**, 31 ⟨82T1827⟩
Zech, J., **6**, 173 ⟨90CB2087⟩
Zecher, N., **5**, 46 ⟨62AG(E)626⟩
Zecher, W., **5**, 652 ⟨68AG(E)856⟩, 654 ⟨68AG942⟩; **6**, 239 ⟨63GEP1222917⟩
Zechmeisten, L., **2**, 776 ⟨29LA(468)117⟩
Zedda, A.,**6**, 250 ⟨92JFC(58)143⟩
Zee-Cheng, R. K. Y., **6**, 504 ⟨79JMC28⟩
Zeeck, A., **1**, 166, 183 ⟨93TL1917⟩
Zeegers, H. J., **2**, 199 ⟨81JOC5408⟩; **3**, 634 ⟨81JOC5408⟩
Zeeh, B., **2**, 340, 343, 344, 346, 350, 353, 354, 355, 356 ⟨71HOU(10/1)1091⟩, 382, 383 ⟨70CB2052⟩; **5**, 566 ⟨75LA1984⟩

Zefirov, N. S., **1**, 500 ⟨87ZOR2545, 91ZOR488⟩; **2**, 232 ⟨B-92MI 203-01⟩, 280 ⟨93TL703⟩, 445 ⟨87ZOR2120⟩, 1020 ⟨90TL4828⟩, 1021 ⟨92TL1419⟩; **3**, 374 ⟨78ZOR1659⟩, 429 ⟨68RCR543⟩; **4**, 326 ⟨78ZOR1659⟩, 800 ⟨79JOU2207⟩, 970 ⟨79ZOR1099⟩; **5**, 205, 321 ⟨90TL4821⟩, 316 ⟨77ZOB968⟩, 810 ⟨75ZOR1981⟩, 1044 ⟨91ZOR340⟩, 1107 ⟨92JOC5270⟩; **6**, 368 ⟨68IZV1638⟩

Zeggaf, C., **3**, 240 ⟨86S337⟩

Zegrocka, O., **5**, 344 ⟨91CAR283⟩

Zehavi, U., **3**, 365 ⟨77JOC2821⟩

Zehdner, M., **1**, 326, 327 ⟨93HCA441, 93TL2445⟩, 327 ⟨92SL442⟩, 327, 950 ⟨92JOC3994⟩; **2**, 72 ⟨90AG(E)205⟩, 913 ⟨92JOC3994⟩; **5**, 330 ⟨91TA1209⟩

Zeid, I., **5**, 661 ⟨92T8271⟩; **6**, 618, 621 ⟨92T8271⟩

Zeiffman, Yu. V., **3**, 560, 561, 588 ⟨73BAU1740, 65BAU731⟩, 588 ⟨75BAU1274, 75JFC(6)227, 76BAU573⟩; **4**, 297, 304 ⟨65IZV2046, 66IZV1108⟩, 731 ⟨75JFC(6)227⟩, 820 ⟨82BAU2498⟩, 833 ⟨70IZV2140⟩, 894 ⟨75IZV1274⟩, 970 ⟨71DOK(201)1359⟩; **5**, 475 ⟨77BAU206⟩

Zeigler, F. E., **1**, 807 ⟨80TL2125, 81TL1179⟩, 807 ⟨84JA718⟩

Zeil, W., **2**, 1016 ⟨66JOM(6)109⟩

Zeile, J. V., **5**, 450 ⟨76JA2365⟩, 801 ⟨78IC2369⟩

Zeiler, A. G., **5**, 965 ⟨73CRV75⟩; **6**, 460 ⟨73CRV75⟩

Zeilstra, J. J., **3**, 426 ⟨74S49⟩; **4**, 331 ⟨74RTC11⟩, 333 ⟨74JOC3215⟩

Zeise, W. C., **2**, 119 ⟨1834LA(11)1⟩

Zeiss, H.-J., **2**, 74, 529, 530 ⟨81JOC1309⟩, 529 ⟨80AG(E)218, 80JOM(187)321⟩, 529 ⟨82CB2357⟩

Zeiss, W., **2**, 468 ⟨72ZN(B)769⟩; **3**, 456 ⟨71CB1199⟩; **5**, 708, 709, 861 ⟨71CB1199⟩; **6**, 625 ⟨71CB1199⟩

Zeitler, L., **3**, 251 ⟨54CB841⟩

Zeitz, H.-G., **1**, 326 ⟨89JA8311, 91JA631⟩, 327 ⟨91SL423, 91SL425⟩, 327 ⟨93TL5885⟩

Zeitz, J. R., **2**, 119, 225 ⟨53JOC1591⟩

Zeitz, Jr., J. R., **1**, 665 ⟨82COMC-I(7)365⟩

Zel, R., **4**, 1056, 1059, 1063 ⟨87ZOB1741⟩

Zelawski, Z. S., **1**, 884 ⟨67JOC2257⟩

Zelck, U., **4**, 728 ⟨59JPR150⟩

Zeldin, L., **4**, 430, 431, 432, 994 ⟨57JA4708⟩; **6**, 170 ⟨57JA4708⟩

Zelenina, N. S., **6**, 82 ⟨71ZOR1887⟩

Zelenin, G. E., **6**, 242 ⟨62DOK(142)354⟩

Zelenin, K. N., **2**, 373 ⟨80JOU244⟩

Zelenka, M., **2**, 363, 399 ⟨59CB1910⟩

Zelenova, L. M., **2**, 441 ⟨80IZV734⟩

Zelesko, M. J., **2**, 631 ⟨69TL2427, 71JA4841⟩

Zelinski, R. P., **6**, 145 ⟨50JA5329⟩

Zell, R., **1**, 77 ⟨76HCA1832⟩

Zelle, R. E., **2**, 196 ⟨86JA5908⟩; **3**, 243 ⟨88JA4368⟩; **5**, 282 ⟨84JA6414⟩

Zeller, E., **1**, 368 ⟨88H(27)1575⟩; **2**, 328 ⟨88H(27)1575⟩

Zeller, E. M., **4**, 603 ⟨91CB1575⟩

Zeller, J. R., **2**, 401, 402 ⟨83JOC1⟩

Zeller, K.-P., **1**, 393 ⟨75AG(E)32⟩, 916 ⟨85HOU(V/2c)127⟩; **2**, 591 ⟨73HOU(13/2b)⟩, 1097 ⟨74HOU(13/2b)26, 74HOU(13/2b)9⟩; **3**, 551 ⟨71T5953, 77AG(E)835⟩; **6**, 172, 204, 207 ⟨74HOU(13/2b)1⟩

Zeller, P., **1**, 156 ⟨57HCA1250⟩, 956 ⟨59HCA719⟩

Zeller, W. E., **1**, 510 ⟨84TL1655⟩

Zellers, E. T., **1**, 650 ⟨80JA6894⟩

Zellner, K., **6**, 164, 371, 373 ⟨88ZN(B)1416⟩, 247 ⟨91CC466⟩, 374 ⟨91OM2884⟩, 394 ⟨90CC1621⟩

Zelnitski, A. P., **5**, 208 ⟨83ZOR1764⟩

Zeltner, J., **5**, 566, 618, 735 ⟨37HCA691⟩

Zembayashi, M., **1**, 466, 470 ⟨76BCJ1958⟩, 467 ⟨76CL1237⟩, 472 ⟨77TL4089⟩; **3**, 145 ⟨76CL1239⟩

Zembower, D. E., **4**, 419, 427 ⟨93JMC305⟩

Zemlicka, J., **3**, 40 ⟨76JOC3317⟩, 68, 69 ⟨59CCC2378, 59CCC2385⟩; **6**, 80 ⟨64CI(L)581⟩

Zemlyanskii, N. N., **4**, 627 ⟨80IZV2668⟩; **6**, 187, 710 ⟨83IZV959⟩, 387, 395 ⟨79BAU2222⟩

Zemplen, G., **5**, 641 ⟨53CB697⟩

Zemskov, S., **2**, 15 ⟨B-89MI 201-03⟩, 334 ⟨B-89MI 206-01⟩; **6**, 16 ⟨B-89MI 601-01⟩, 212 ⟨B-89MI 607-01⟩

Zen, S., **2**, 394 ⟨88OSC(6)503⟩, 398 ⟨88OSC(6)797⟩; **3**, 684 ⟨78CPB3254⟩

Zenbayashi, M., **5**, 225 ⟨90JAP280254⟩

Zeng, B., **1**, 877 ⟨91JA3833⟩

Zeng, B. F., **1**, 877 ⟨89JA6419⟩

Zeng, Z., **1**, 1038 ⟨92JA9279⟩

Zenitz, B. L., **2**, 609 ⟨46JOC444⟩

Zenki, S.-I., **3**, 431 ⟨83TL1049⟩, 432 ⟨84CC874, 84H(21)483⟩

Zenneck, U., **1**, 910 ⟨84AG(E)371⟩; **4**, 549 ⟨93PS(77)5⟩, 632 ⟨83JA2582⟩; **5**, 928 ⟨84AG(E)371⟩; **6**, 154 ⟨89AG929⟩

Zenner, F., **6**, 560 ⟨90ZC90⟩

Zenner, K. F., **5**, 964 ⟨64GEP1205087⟩

Zentz, S., **4**, 579 ⟨83OM154⟩; **6**, 707 ⟨83OM154⟩

Zepeda, L. G., **4**, 245 ⟨91SC1369⟩

Zerbetto, F., **4**, 1078 ⟨90JA4284⟩

Zerby, G. A., **1**, 166 ⟨79TL3375⟩; **4**, 236 ⟨79TL3375⟩; **6**, 126, 135 ⟨79TL3375⟩

Zercher, C., **2**, 1010 ⟨87JOC3062⟩

Zerner, B., **3**, 577 ⟨86JA5543⟩; **4**, 892, 979 ⟨86JA5543⟩

Zervas, L., **2**, 273, 275 ⟨62JA3887⟩, 274 ⟨65JA4922⟩, 358 ⟨64JA3660⟩, 412 ⟨55JA5351, 61CB2644⟩, 414 ⟨55JA5354⟩; **6**, 487 ⟨32CB1192⟩

Zeuner, S., **3**, 386 ⟨84AG(E)726, 85JOM(289)C21⟩, 386 ⟨89JOM(377)105⟩, 397 ⟨83ZN(B)1365⟩, 400 ⟨83JOM(252)C63⟩; **4**, 581 ⟨87CB583⟩

Zey, R. L., **5**, 315 ⟨61JA4469⟩, 698, 813 ⟨66JOC3708⟩

Zezza, C. A., **5**, 199 ⟨89TL3295⟩

Zgierski, M. Z., **4**, 1078 ⟨90JA4284⟩

Zgliczynski, J. M., **2**, 213 ⟨61MI 203-02⟩

Zhabinskiy, V. N., **2**, 924, 925 ⟨90TL4937⟩; **4**, 809 ⟨90TL4937⟩

Zhadnova, M. P., **3**, 420 ⟨82RCR469⟩

Zhai, D., **1**, 1004, 1135 ⟨88JA2501⟩; **4**, 102 ⟨88JA1547⟩

Zhai, H.-X., **6**, 219 ⟨89MI 607-02⟩

Zhai, W., **1**, 1004, 1135 ⟨88JA2501⟩, 1042, 1135 ⟨88T5425⟩; **3**, 285, 295 ⟨87JOC2615⟩

Zhang, B., **4**, 652 ⟨93JA10754⟩

Zhang, B. L., **3**, 526 ⟨81T4189⟩

Zhang, C., **1**, 688 ⟨91JOC6447⟩; **4**, 661 ⟨92AG(E)232⟩; **5**, 426 ⟨92TL6751⟩; **6**, 30 ⟨90TL6831, 91JOM(419)357⟩

Zhang, C. H., **2**, 931 ⟨92JOC3270⟩

Zhang, C.-P., **6**, 24 ⟨91SC2189⟩

Zhang, D., **2**, 678 ⟨91SL505⟩; **3**, 20, 31 ⟨91SL501⟩; **4**, 200 ⟨91SL501⟩, 201 ⟨91SL503⟩

Zhang, G.-Z., **2**, 257 ⟨91S454⟩

Zhang, H., **2**, 512 ⟨87IC4065⟩; **3**, 545 ⟨85JA3728⟩; **5**, 439 ⟨86JA1722⟩

Zhang, H. C., **2**, 61 ⟨94TL4891⟩

Zhang, H.-M., **6**, 689 ⟨84JA7015⟩

Zhang, H. X., **2**, 915 ⟨90JOC1857⟩; **4**, 813 ⟨90JOC1857⟩; **6**, 480 ⟨93MI 615-01⟩

Zhang, H.-Z., **1**, 439 ⟨89S42⟩

Zhang, J., **1**, 326, 328 ⟨90JA6738⟩, 647 ⟨87TL2159⟩; **2**, 565 ⟨87JOM(328)81⟩, 846 ⟨84S778⟩

Zhang, L., **1**, 647, 743 ⟨90JCR(S)388⟩

Zhang, L.-J., **1**, 272 ⟨92JOC774⟩; **2**, 73 ⟨91TL6579⟩, 487 ⟨92JOC774⟩

Zhang, N., **5**, 142 ⟨94JA7455⟩

Zhang, P., **3**, 310 ⟨93S1247⟩

Zhang, Q., **1**, 346, 926 ⟨93JOC7640⟩

Zhang, R., **2**, 436, 824, 825 ⟨87S938⟩; **3**, 639 ⟨87S938⟩

Zhang, S., **5**, 924 ⟨92OM2317⟩

Zhang, S.-W., **1**, 493 ⟨92JOM(423)405⟩
Zhang, T., **6**, 580 ⟨90MI 618-01⟩
Zhang, W., **1**, 340 ⟨92TL7307⟩, 340, 919, 928, 929, 931 ⟨93CRV2091⟩
Zhang, X., **2**, 573 ⟨86OM2452⟩; **3**, 119 ⟨88JCS(P1)2423⟩; **5**, 894, 896 ⟨93TL1925⟩
Zhang, X. D., **6**, 467 ⟨89OPP771⟩
Zhang, X.-H., **2**, 49 ⟨92TL6411⟩
Zhang, X.-L., **2**, 48 ⟨92JOC2768⟩
Zhang, X.-M., **1**, 921 ⟨93ACR510, 93JOC6067⟩
Zhang, X.-Y., **2**, 53 ⟨93JA3318⟩
Zhang, Y., **1**, 14 ⟨86TL4343⟩, 536 ⟨86JOC4080⟩, 957, 973 ⟨93TL449⟩, 1070, 1094 ⟨87TL5793⟩; **2**, 30 ⟨87TL5793⟩, 55 ⟨88TL4113⟩, 877 ⟨88MI 217-01⟩, 980 ⟨89JA3454, 90JA8590⟩; **3**, 270 ⟨86JOC4080⟩; **5**, 253 ⟨93SC1403⟩, 272, 293 ⟨87TL3815⟩, 464 ⟨89JA8018⟩, 1044 ⟨94JMC170⟩
Zhang, Y.-Z., **1**, 473 ⟨86T2111⟩
Zhang, Z., **1**, 599 ⟨87SC1839⟩; **5**, 937 ⟨92OM622⟩
Zhang, Z.-D., **1**, 946 ⟨90T3155⟩; **5**, 184 ⟨86JOC3744⟩, 184 ⟨86TL4937⟩, 184, 198 ⟨86TL4933⟩
Zhang, Z.-H., **2**, 73 ⟨91TL6579⟩
Zhang, Z. J., **3**, 323 ⟨91JOC3271⟩
Zhang-Lin, Z., **1**, 766 ⟨93T6821⟩
Zhao, D., **2**, 928, 941 ⟨93JA10097⟩
Zhao, K., **1**, 226 ⟨90JA1661⟩; **2**, 456 ⟨93T363⟩; **3**, 495 ⟨89TL2173⟩; **4**, 197 ⟨89TL287⟩; **5**, 252 ⟨90MI 505-01⟩, 896 ⟨89TL2173⟩
Zhao, L., **1**, 109 ⟨89JA6443, 90JA1297⟩; **5**, 946, 958 ⟨95TL3699⟩, 953 ⟨94CL777⟩
Zhao, M., **2**, 320 ⟨91JOC5263⟩
Zhao, S.-H., **2**, 147 ⟨85PAC1911⟩
Zhao, X., **1**, 262 ⟨91JCS(P1)3295⟩; **2**, 320 ⟨93JOC2086⟩
Zhao, X. H., **4**, 446, 449 ⟨93JOC2086⟩
Zhao, X.-R., **5**, 625 ⟨94SC1761⟩
Zharkov, V. V., **6**, 366 ⟨81ZOB1035⟩
Zharov, A. N., **4**, 654 ⟨92MI 414-01⟩, 6540 ⟨92MIP173178⟩
Zhdankin, V. V., **1**, 988 ⟨91T4539⟩; **2**, 445, 1021 ⟨92JOC1861⟩, 1020 ⟨90JA6437⟩, 1020 ⟨90TL4828⟩, 1020 ⟨91JA5870⟩, 1020, 1028, 1033 ⟨91JA4571, 91JA8997⟩, 1021 ⟨93JA11702⟩, 1029, 1030 ⟨93TL1461⟩; **5**, 205, 321 ⟨90TL4821⟩
Zhdanov, Yu. A., **5**, 394, 404 ⟨80RCR843⟩
Zhebarov, O. Zh., **1**, 61 ⟨81ZOB420⟩; **4**, 141 ⟨81DOK(261)474⟩
Zheltonogova, E. A., **4**, 100 ⟨89JGU200⟩
Zheng, D., **6**, 565 ⟨86MI 617-05⟩
Zheng, J., **1**, 1079 ⟨81TL5283⟩; **6**, 565 ⟨86MI 617-05⟩
Zheng, P., **4**, 716 ⟨93OM1213⟩
Zheng, W., **1**, 199 ⟨93TL2581⟩
Zheng, Y. Y., **5**, 679, 703 ⟨92ICA527⟩; **6**, 256, 261 ⟨84JOC3590⟩, 256, 603 ⟨83JOC4844⟩, 259, 614, 624 ⟨92JFC(57)293⟩, 259, 643, 651, 653 ⟨92IC488⟩, 262 ⟨92ICA527⟩
Zheng, Z.-I., **1**, 176, 940 ⟨90JOC1416⟩, 836, 941 ⟨87TL5973⟩; **6**, 543 ⟨87TL5973⟩
Zhengxiang, G., **6**, 708 ⟨84TL4425⟩
Zhi, L., **3**, 321 ⟨89TL1491, 90JA3100⟩, 321 ⟨92JA593⟩; **4**, 280 ⟨92TL1831⟩
Zhichkin, P. E., **4**, 766 ⟨91ZOB1014⟩
Zhidkova, A. M., **6**, 104 ⟨77UK685⟩
Zhidkova, L. A., **4**, 490 ⟨81ZOB757⟩
Zhi-Wei, G., **5**, 179 ⟨88TL5583⟩
Zhmurova, I. N., **2**, 407 ⟨64ZOB1441, 64ZOB3055⟩, 407 ⟨65ZOB344, 67ZOB896⟩; **5**, 373, 862 ⟨81T437⟩
Zholondkovskaya, T. N., **5**, 864 ⟨69MI 521-01⟩
Zhong, G.-F., **1**, 113 ⟨93TL5441⟩
Zhong, H. M., **1**, 929 ⟨93TL5197⟩; **5**, 367 ⟨94TL4947⟩
Zhong, Q., **3**, 637 ⟨91MI 318-01⟩
Zhou, B., **1**, 196 ⟨87JOC974⟩, 196 ⟨88JOC4419⟩; **6**, 461 ⟨88JOC4419⟩
Zhou, G., **5**, 1007 ⟨91JPS(A)1097⟩
Zhou, H., **4**, 472 ⟨93S955⟩; **6**, 209, 401, 405 ⟨91SA(A)849⟩
Zhou, H.-Y., **5**, 600, 621 ⟨91OPP435⟩
Zhou, J. Y., **1**, 1029 ⟨91JMC2452⟩
Zhou, L.-L., **4**, 826 ⟨91TL1539⟩
Zhou, P., **1**, 147 ⟨90TL1653⟩, 147 ⟨93JOC4890⟩; **2**, 251 ⟨90TL1653⟩; **3**, 437 ⟨94JOC3243⟩
Zhou, Q., **3**, 128 ⟨93TL7725⟩
Zhou, Q.-S., **5**, 226 ⟨82SC709⟩
Zhou, T., **1**, 196 ⟨93TL4497⟩
Zhou, W.-S., **3**, 222 ⟨91TL1467⟩
Zhou, X., **2**, 231 ⟨94TL8833⟩; **5**, 253 ⟨93SC1403⟩
Zhou, X.-J., **2**, 304 ⟨89SC1347⟩, 730 ⟨92S281⟩, 756 ⟨69SC1347⟩; **5**, 625 ⟨94SC1761⟩
Zhou, X.-X., **6**, 488 ⟨85ACS(B)761⟩
Zhou, Y., **2**, 436, 824, 825 ⟨87S938⟩; **3**, 639 ⟨87S938⟩
Zhou, Y. F., **1**, 482 ⟨82SC795⟩; **3**, 209 ⟨82SC795⟩
Zhou, Z., **3**, 637 ⟨91JCS(P1)1931⟩
Zhou, Z. H., **3**, 699, 721 ⟨92CL1143⟩
Zhou, Z.-Y., **6**, 195 ⟨93JOM(443)C39⟩
Zhu, J., **1**, 786 ⟨94TL2787⟩
Zhu, J. L., **1**, 585 ⟨90TL3829⟩
Zhu, L., **1**, 123, 467 ⟨91JOC1445⟩; **2**, 959 ⟨91JOC1445⟩; **3**, 130 ⟨91JOC1445⟩
Zhu, S., **2**, 161 ⟨92TL6503⟩; **4**, 864 ⟨92MI 419-02⟩, 1083 ⟨84MI 424-03⟩; **6**, 699 ⟨93JFC(60)175⟩
Zhu, S.-Z., **2**, 162 ⟨91CC732⟩; **4**, 281, 1081 ⟨90CC1459⟩; **6**, 254 ⟨90CC1459⟩
Zhu, X., **6**, 538 ⟨92MI 617-01⟩
Zhuang, X., **3**, 119 ⟨87BCJ4143⟩
Zhuangyu, Z., **2**, 320 ⟨90SC3563⟩; **3**, 24 ⟨91S539⟩
Zhukhlistova, N. E., **2**, 478 ⟨90K668⟩
Zhukova, S. V., **4**, 312 ⟨73TL1889⟩
Zhukova, Yu. N., **4**, 472 ⟨79IZV2118⟩
Zhukovskaya, N. A., **2**, 1098 ⟨85JGU1691, 86JGU1812⟩; **5**, 1110 ⟨90MI 530-01⟩
Zhukovskii, Yu. G., **2**, 1033 ⟨91RCR885⟩
Zhun', V. I., **4**, 140 ⟨83JGU109⟩
Zhuravkova, L. G., **4**, 926 ⟨69CA(70)57950j⟩
Zhuravleva, L. P., **4**, 556 ⟨67IZV949⟩
Zhut-skii, P. V., **2**, 476 ⟨90JGU398⟩; **4**, 459 ⟨91URP1618747⟩
Zhuzhlikova, S. T., **2**, 878 ⟨87JOU1417⟩; **5**, 902 ⟨63DOK(151)856⟩
Ziadulewicz, E., **3**, 72 ⟨89JCS(P1)1793⟩
Ziani-Chérif, C., **1**, 210 ⟨93JOC2181⟩; **4**, 245 ⟨93JOC2181⟩
Zibarev, A. V., **2**, 793 ⟨85RCR1706⟩
Zibuck, R., **5**, 179 ⟨86JA2451⟩, 179 ⟨89LA1233⟩
Ziche, W., **4**, 651 ⟨92JOM(426)1, 93OM4123⟩; **5**, 925 ⟨92JA4910⟩, 926 ⟨92OM2494⟩
Zidani, A., **2**, 317, 469 ⟨92BSF71⟩
Zieba, A., **3**, 362 ⟨92H(34)247⟩
Zieba, M., **1**, 995 ⟨90PJC123⟩
Zieba-Zoltowska, B., **6**, 546 ⟨90MI 617-03⟩
Ziegenbein, W., **1**, 453 ⟨65CB824⟩, 1002, 1021, 1026, 1037, 1038, 1046, 1047 ⟨B-69MI 121-03⟩, 1063 ⟨60AG391⟩, 1133 ⟨B-69MI 122-01⟩; **2**, 26 ⟨62CB2976⟩; **3**, 69 ⟨60CB2743⟩; **5**, 289 ⟨57CB2291⟩
Ziegenhagen, B., **1**, 686 ⟨84CB277⟩; **3**, 348 ⟨84CB277⟩
Zieger, H. E., **1**, 108 ⟨76JA2580⟩, 110 ⟨82JOC3521⟩, 112 ⟨79JA2207⟩; **6**, 425 ⟨82JOC3521, 86JOC1180⟩
Ziegler, B., **4**, 672, 673 ⟨90ZN(B)221⟩; **6**, 389, 714 ⟨91AG(E)594⟩
Ziegler, C., **2**, 176, 177 ⟨51JOC621⟩
Ziegler, C. B., **3**, 714, 718 ⟨84CC133⟩, 714, 718 ⟨85CC816⟩; **5**, 157 ⟨77JOC3903⟩
Ziegler, E., **2**, 400 ⟨69M959⟩, 746 ⟨82ZN(B)105⟩, 1003 ⟨82M1451, 83M1097⟩, 1003 ⟨83M783⟩; **3**, 264 ⟨81S225⟩, 348 ⟨75ZN(B)760⟩, 527 ⟨68M1958⟩, 542 ⟨64M1053⟩; **4**, 96 ⟨75ZN(B)245⟩, 411 ⟨68M990⟩, 761

⟨52M1⟩; **5**, 340 ⟨69M959⟩, 556 ⟨77LA1751⟩, 1107 ⟨48M316⟩
Ziegler, F., **5**, 558, 587, 593 ⟨73S297⟩
Ziegler, F. E., **1**, 161 ⟨84TL131⟩, 176 ⟨93JA2581⟩, 176, 940 ⟨90JOC1416⟩, 187 ⟨79JOC3428⟩, 267 ⟨71JA4318⟩, 414, 825, 836 ⟨88CRV1423⟩, 473 ⟨80JA790⟩, 690 ⟨82JOC5229⟩, 807 ⟨79JA1611⟩, 826 ⟨88TL1669⟩, 828 ⟨82JA7174⟩, 836, 941 ⟨87TL5973⟩, 840 ⟨82JA7181, 91COC(5)875⟩, 929, 935 ⟨93TL4505⟩; **2**, 923 ⟨84TL131⟩; **3**, 21 ⟨82JA7181⟩, 61 ⟨81TL1179⟩, 233 ⟨81T4035⟩, 291 ⟨78JOC985⟩, 309 ⟨69JA2342⟩, 616, 621 ⟨77JOC2001⟩, 616, 621 ⟨88OSC(6)334⟩; **4**, 842 ⟨78JOC3065⟩, 849 ⟨82JA7174⟩; **6**, 543 ⟨87TL5973⟩
Ziegler, G. R., **2**, 1016, 1033 ⟨63JA1648⟩, 1017 ⟨73OSC(5)921⟩, 1062 ⟨62JA2020⟩
Ziegler, H. J., **2**, 29 ⟨69S39⟩
Ziegler, J., **3**, 118 ⟨76S811⟩
Ziegler, J. B., **2**, 24 ⟨52JA4891⟩, 128, 136 ⟨47JA2675⟩, 129, 136 ⟨47JA2672⟩
Ziegler, J.-C., **5**, 141 ⟨76S811⟩
Ziegler, Jr., C. B., **1**, 494, 1096 ⟨78JOC2941⟩, 988 ⟨84CC152, 87JOC443⟩, 988 ⟨88TL411, 90JOC2983⟩; **5**, 157 ⟨78JOC2952⟩
Ziegler, K., **1**, 663 ⟨50LA(567)179⟩, 665 ⟨55AG541, 60LA(629)172⟩, 665 ⟨60LA(629)1, 60LA(629)210⟩, 665, 666 ⟨60LA(629)53, 60LA(629)14⟩, 881 ⟨32LA(495)84⟩, 901 ⟨23LA(434)34⟩; **2**, 43, 112 ⟨60LA(629)241⟩, 60 ⟨55AG424, 55AG425⟩, 555 ⟨50LA(567)195⟩, 595 ⟨70HOU(13/4)9⟩, 596 ⟨54LA(589)91, 60LA(629)121⟩, 596 ⟨60LA(629)1⟩, 967 ⟨70HOU(13/4)23⟩, 1041 ⟨60LA(639)33⟩, 1098 ⟨70HOU(13/4)154⟩; **4**, 668 ⟨55ZAAC(282)345⟩, 699 ⟨57LA(608)1, 60LA(629)14⟩, 699 ⟨60LA(629)1, 60LA(629)53⟩, 709 ⟨57LA(605)93⟩; **5**, 199 ⟨50LA(568)1⟩; **6**, 204, 208 ⟨70HOU(13/4)1⟩, 206 ⟨55ZAAC(282)345⟩
Ziegler, M. L., **2**, 512 ⟨84AG(E)812, 85CB2472⟩, 585 ⟨84AG(E)515⟩, 865 ⟨88AG(E)709⟩; **3**, 387, 390 ⟨83AG(E)314⟩, 401 ⟨84CC686⟩, 519 ⟨84JOM(264)327⟩, 544 ⟨79JA3133⟩, 607 ⟨86AG(E)1003⟩; **4**, 549 ⟨87AG(E)236, 88AG(E)389⟩, 565 ⟨88AG(E)709⟩, 580 ⟨88ZN(B)1293⟩, 581 ⟨89JOM(371)329⟩, 679 ⟨81AG(E)1049⟩, 681 ⟨73ZN(B)508⟩, 686 ⟨82CB878, 82JOM(226)C59⟩, 686 ⟨83JOM(258)81⟩, 709 ⟨87JOM(321)257⟩, 1047, 1059 ⟨86ACR250⟩; **6**, 154 ⟨88AG(E)389⟩, 154 ⟨88AG(E)709⟩, 622 ⟨69AG(E)455⟩
Ziegler, T., **2**, 316 ⟨91LA269⟩, 321 ⟨90S575⟩, 665 ⟨87CB1347⟩; **3**, 33 ⟨92TL5157⟩, 530, 531 ⟨86JOC3558⟩, 628 ⟨87AG(E)458, 90S575⟩, 628 ⟨90TL1249, 91TL2605⟩, 631 ⟨91LA47⟩; **5**, 135 ⟨91LA47⟩
Ziegler, W. M., **4**, 83, 84 ⟨40JA2596⟩
Ziehn, K.-D., **3**, 618 ⟨71CB2025⟩, 618, 619 ⟨71CB1030⟩; **4**, 552 ⟨71CB2250⟩; **5**, 1063 ⟨71CB1335⟩; **6**, 620 ⟨73CB2093⟩
Ziehn, K. D., **6**, 620 ⟨74CB698⟩
Zieliński, W., **5**, 656 ⟨80PJC745⟩; **6**, 621 ⟨93H(36)1521⟩
Zielke, R., **5**, 584, 589 ⟨75S47⟩
Ziembinski, R., **6**, 186, 387, 712 ⟨89CB1057⟩, 187, 387, 712, 713 ⟨90CB747⟩
Ziemnicka, B. T., **5**, 374 ⟨85JOC1712⟩
Zienty, F. B., **3**, 498 ⟨64JOC3721⟩
Zierold, G., **5**, 1004 ⟨1874CB1228⟩
Žigić-Mamuzić, L., **1**, 91 ⟨67T215⟩
Zigman, A. R., **1**, 81 ⟨67JOC893⟩
Zikán, V., **4**, 211 ⟨72MI 404-01⟩
Zikra, N., **1**, 1093 ⟨82JOC3805⟩
Zil'berman, E. N., **2**, 308 ⟨60RCR331⟩; **3**, 282 ⟨62RCR615⟩
Zilka, A., **2**, 190 ⟨68TL4479⟩
Zilkha, A., **2**, 910, 931 ⟨63CJC2977⟩; **4**, 807 ⟨63CJC2977⟩; **5**, 199 ⟨57JCS4397⟩, 318 ⟨72JOM(35)303⟩
Ziller, J. W., **3**, 517 ⟨93JA8130, 93JA8167⟩, 520, 522 ⟨92JA3974⟩, 521 ⟨83OM1505⟩; **4**, 683 ⟨88JOM(353)103⟩, 703 ⟨93OM2618⟩, 704 ⟨93OM3998⟩, 726 ⟨93JOM(462)141⟩, 883 ⟨91MI 420-01⟩; **6**, 345 ⟨86OM1757⟩
Zilliken, F., **5**, 62 ⟨58MI 502-02⟩
Zima, G., **2**, 280 ⟨78TL4977⟩, 282 ⟨80TL3643⟩, 542 ⟨78TL5087⟩; **3**, 38, 174, 207 ⟨81JOC2920⟩, 39, 175 ⟨78TL5091⟩, 174 ⟨80TL3643⟩, 174 ⟨85T4881⟩, 256 ⟨79SC697⟩; **4**, 239 ⟨78TL5091⟩, 401 ⟨80JOC2736⟩
Ziman, S. D., **6**, 650 ⟨76JOC3253⟩
Zimin, M., **5**, 644 ⟨85ZOB1526⟩
Zimin, M. G., **2**, 266 ⟨81ZOB2140, 83URP956484⟩; **4**, 339 ⟨86ZOB711⟩, 342 ⟨71ZOB1964⟩, 345 ⟨81ZOB2140, 83ZOB46⟩, 485 ⟨80ZOB751⟩; **5**, 581 ⟨79IZV861, 86ZOB2660⟩, 643 ⟨78ZOB843⟩, 643, 644 ⟨79IZV805, 79ZOB2323⟩, 644 ⟨86ZOB2359⟩, 713 ⟨85JGU1526, 85ZOB1718⟩, 737 ⟨77JGU916⟩, 738 ⟨79JGU2323⟩; **6**, 160 ⟨75ZOB1450⟩
Zimin, V. M., **4**, 554 ⟨67ZOB2055⟩, 561 ⟨87ZOB2294⟩; **5**, 1131 ⟨67MI 530-01⟩
Zimmer, A. J., **5**, 105 ⟨B-61MI 502-03⟩
Zimmer, H., **1**, 729, 730 ⟨B-76MI 110-01⟩; **2**, 304 ⟨70JOC2826⟩, 379 ⟨55JA790⟩, 618 ⟨83JOC2782⟩; **5**, 370 ⟨70AF790⟩
Zimmer, M., **2**, 379 ⟨55JA790⟩
Zimmer, R., **1**, 458, 955 ⟨93S165⟩; **2**, 322 ⟨93LA1155⟩, 939 ⟨93S165⟩
Zimmer-Gasser, B., **1**, 158 ⟨79AG(E)782⟩; **2**, 447 ⟨80CB1612⟩; **4**, 564 ⟨79AG(E)782⟩, 576 ⟨80CB902⟩; **5**, 908, 910, 911 ⟨80CB902⟩, 913 ⟨77AG(E)639, 80CB1637⟩; **6**, 151 ⟨79AG(E)484⟩
Zimmerli, A., **3**, 463 ⟨10JCS2156⟩
Zimmerman, D. M., **5**, 1066, 1073 ⟨70TL3453⟩
Zimmerman, F., **3**, 269 ⟨63LA(662)178⟩
Zimmerman, G., **1**, 1045 ⟨71M214⟩
Zimmerman, H. E., **1**, 8 ⟨59JA3634, 59JA3644⟩, 96 ⟨93JA2205⟩, 378 ⟨57JA5455⟩, 378, 886 ⟨61JA1196⟩, 635 ⟨86JA6276⟩, 794 ⟨71ACR272⟩, 802 ⟨77JA2342⟩, 802 ⟨78JOC1493⟩, 871 ⟨68JA6091, B-63MI 119-01⟩, 890 ⟨90JA1280⟩; **3**, 317 ⟨64TL1847⟩
Zimmerman, I., **3**, 450 ⟨59JA2515⟩
Zimmerman, J. E., **5**, 277 ⟨64JA1839⟩, 279 ⟨67JA7151⟩
Zimmerman, R., **4**, 566 ⟨83TL3591, 85CB814⟩; **6**, 166 ⟨85CB814⟩
Zimmerman, W. T., **1**, 632 ⟨77JA3414⟩
Zimmermann, D. C., **2**, 942 ⟨85S945⟩; **6**, 445 ⟨89JMC228⟩
Zimmermann, F., **3**, 269, 530 ⟨62AG32⟩
Zimmermann, G., **1**, 925 ⟨89JOC3571⟩
Zimmermann, H., **1**, 634 ⟨87JOC2644⟩; **2**, 430 ⟨65AG172⟩; **4**, 990 ⟨90CB115⟩
Zimmermann, H.-J., **2**, 698 ⟨75LA2201⟩
Zimmermann, J. P., **6**, 430 ⟨57HCA604⟩
Zimmermann, K., **2**, 440 ⟨77ZAAC(436)20⟩
Zimmermann, M., **1**, 731 ⟨69S75, 71CB2885⟩; **2**, 607, 611 ⟨69S75⟩; **5**, 894 ⟨69S75⟩
Zimmermann, R., **1**, 52 ⟨B-79MI 102-01⟩; **3**, 602 ⟨82HOU(E1)759⟩, 363; **4**, 571 ⟨92CB2081, 92S787⟩; **5**, 894 ⟨82HOU(E1)616⟩; **6**, 694, 695, 697, 698 ⟨82HOU(E1)616⟩, 694, 697, 700, 701, 702 ⟨91COS(6)171⟩, 701 ⟨B-72MI 622-01, 92S787⟩
Zimmermann, R. G., **1**, 466 ⟨84TL5769⟩
Zimmermann, T., **3**, 366 ⟨88JPR35⟩; **4**, 852 ⟨84CHEC(6)890⟩
Zimmermann, W., **2**, 597 ⟨51ZAAC(267)39⟩
Zimmerschied, W. J., **4**, 6 ⟨47JA281⟩
Zimmicki, J., **1**, 745 ⟨69RZC95⟩
Zimont, S. L., **2**, 439 ⟨68IZV190⟩; **5**, 395, 397, 402 ⟨68T2995⟩
Zincke, T., **2**, 688 ⟨12LA(391)55⟩
Zinczuk, J., **5**, 1126 ⟨85JHC1025⟩
Zindel, J., **1**, 166, 183 ⟨93TL1917⟩

Zingaro, R. A., **2**, 260 ⟨80PS(9)107, 84GEP3229565⟩, 261 ⟨80JCS(P1)2287⟩, 289 ⟨79JMC572⟩, 414, 504 ⟨68JOM(12)377⟩, 461 ⟨61JOC5205⟩, 465, 858 ⟨65JOM(4)320⟩, 483, 502, 890 ⟨64IC580⟩, 494, 500 ⟨72JOM(39)301⟩, 497 ⟨75JOM(91)295⟩, 497, 502 ⟨66IC187, 70JOM(22)637⟩, 498, 1080 ⟨71JINC3177⟩, 498, 504 ⟨66JOM(6)17⟩, 499 ⟨71JA5677⟩, 500, 502 ⟨63JCS(C)798, 64JOM(1)369⟩, 872 ⟨83JOM(257)67⟩; **6**, 596 ⟨53JOC292, 63OSC(4)359⟩

Zinic, M., **3**, 425 ⟨74MI 311-01⟩

Zink, J. I., **5**, 1092 ⟨78ACR289⟩, 1094 ⟨73JA7510⟩

Zinke, P., **5**, 727 ⟨83JOC1544⟩

Zinkeisen, W., **5**, 833 ⟨1889CB2946⟩

Zinn, A., **5**, 780 ⟨91ZAAC(596)47⟩, 781 ⟨90ZN(B)1273⟩

Zinner, G., **2**, 211 ⟨54LA(585)142⟩, 340, 357 ⟨84S217⟩; **4**, 405 ⟨65CB4036, 67CB2515⟩, 435 ⟨66CB2686, 73AP(306)161⟩, 437 ⟨66CB1678⟩, 492 ⟨85AG(E)979⟩; **5**, 211, 322 ⟨86AP(319)1037⟩, 212 ⟨85AP(318)977, 85CZ389⟩, 213 ⟨63AP(296)420, 63AP(296)57⟩, 698 ⟨77CZ154⟩, 837 ⟨64AG440⟩, 904, 906 ⟨85AG(E)979⟩, 1089 ⟨87JOM(331)193⟩; **6**, 360 ⟨73CB2315⟩, 360, 361 ⟨70CB766, 70ZC30⟩, 505 ⟨74AP(307)7, 75CZ245⟩, 506 ⟨79CZ175⟩, 570, 571 ⟨61AP765, 66PHA23⟩, 622 ⟨85AP(318)1057, 85AP(318)992⟩, 650 ⟨72CB1709⟩

Zinner, H., **4**, 327 ⟨58CB1432⟩, 416 ⟨61CB2209⟩, 416, 424 ⟨61JPR72⟩, 728 ⟨59JPR150⟩

Zinov'ev, Yu. M., **2**, 448 ⟨68ZOB1551⟩

Zinsmeister, K., **2**, 394 ⟨85BCJ1069⟩

Ziolkowski, F., **1**, 777 ⟨73AJC1259⟩

Zipkin, R., **1**, 754 ⟨84CC349⟩

Zipory, E. S., **5**, 406 ⟨91JOC4241⟩

Zippel, M., **1**, 681 ⟨81CC891⟩

Zipse, H., **1**, 326, 327 ⟨93TL2445⟩

Zirkle, C. L., **5**, 92 ⟨83EUP66456⟩

Zirzow, K.-H., **6**, 661 ⟨86AG(E)457⟩

Zissis, E., **4**, 235 ⟨66CAR(2)461⟩

Zivadinovitsch, R., **5**, 830 ⟨36CB2352⟩; **6**, 625 ⟨36CB2358⟩

Zizelman, P. M., **5**, 948 ⟨89OM2075⟩

Zizvashvili, J., **2**, 626 ⟨82CC778⟩

Zlatoidsky, P., **3**, 40 ⟨94HCA575⟩

Zlobin, V. A., **5**, 989 ⟨89MI 526-01⟩

Zlotin, S. G., **2**, 391 ⟨89RCR470⟩

Zlotskii, S. S., **4**, 1078 ⟨86MI 424-01⟩, 1080 ⟨83ZOB416⟩; **6**, 478 ⟨81ZOR1539, 85ZOR2206, 86JOU270, 87JGU365⟩

Zlotskij, S. S., **6**, 555 ⟨91JPR139⟩

Žmitek, J., **2**, 322 ⟨91OPP721⟩; **5**, 836 ⟨81H(16)2173⟩; **6**, 632 ⟨85OPP256, 91OPP721⟩

Zoda, M., **3**, 625 ⟨76JOC2502⟩

Zoeckler, M. T., **1**, 401 ⟨81JA7661⟩

Zoellner, E. A., **2**, 560 ⟨29JA1576⟩; **5**, 43 ⟨32OSC(1)524⟩

Zolk, R., **3**, 553 ⟨83AG(E)981⟩; **4**, 582 ⟨85CB261⟩

Zoll, E. C., **5**, 999 ⟨73JMC1⟩

Zoller, U., **1**, 707 ⟨B-88MI 115-04⟩; **2**, 186 ⟨B-90MI 203-07⟩; **3**, 365 ⟨85JOC1107⟩; **4**, 100, 309, 327 ⟨75T863⟩

Zollinger, H., **2**, 182 ⟨59HCA1653⟩; **3**, 281 ⟨59HCA1659⟩; **5**, 9 ⟨59G2428⟩; **6**, 52 ⟨59HCA1653⟩, 57 ⟨59HCA1659⟩

Zollinger, J. L., **5**, 762, 811, 817 ⟨73JOC1075⟩, 808 ⟨73USP3751312⟩, 814 ⟨73JOC1065⟩, 815, 817, 818 ⟨73USP3726903⟩; **6**, 279, 281 ⟨73JOC1065⟩, 284, 362, 613, 650 ⟨73JOC1075⟩

Zollinger, J. L. M., **6**, 279 ⟨72USP3707555⟩, 284, 362 ⟨72USP3689560⟩

Zollo, P. H. A., **1**, 513 ⟨83TL4833⟩; **4**, 172 ⟨83TL4833⟩

Zolopa, A. R., **2**, 524 ⟨79JA248⟩

Zolotareva, A. S., **4**, 626 ⟨91ZOB921⟩

Zolotarev, A. S., **4**, 814 ⟨89JGU2040⟩

Zolotov, A. N., **2**, 239 ⟨86ZOB898⟩

Zolotov, M. A., **2**, 410 ⟨80ZOB1702⟩

Zoltewicz, J. A., **2**, 298 ⟨78AHC(22)71⟩; **4**, 413 ⟨73JOC1949, 76JOC1303⟩; **5**, 580 ⟨60JA2656⟩

Zomaya, I. I., **4**, 205 ⟨76TL3739⟩, 824 ⟨79TL391⟩

Zomisch, G., **4**, 900 ⟨78ZC345⟩

Zon, G., **1**, 53 ⟨71JA6205⟩; **2**, 440 ⟨69JA2788⟩, 440, 441 ⟨69JA7027, 69JA7012⟩, 441 ⟨69JA7023⟩, 441, 443 ⟨71FCF61⟩

Zon, J., **4**, 462 ⟨81PJC643⟩, 491 ⟨84S661⟩, 492 ⟨79PJC541⟩, 1005 ⟨81S324⟩

Zondler, H., **4**, 409 ⟨72LA(759)84⟩

Zonis, S. A., **3**, 275 ⟨50JGU1262⟩

Zonnebelt, S., **2**, 167 ⟨74JA3146⟩

Zonnerbelt, S. M., **5**, 282 ⟨74JOC2740⟩

Zontova, V. N., **6**, 635 ⟨90ZOB798⟩

Zook, H. D., **1**, 112 ⟨53JA3975⟩, 130 ⟨71JOC1112⟩, 130, 869, 871, 881 ⟨68JOC2222⟩, 859, 881 ⟨60JA1386⟩, 869, 871, 881 ⟨59JOC709⟩, 869, 871, 881 ⟨68JOC3477, 57JA881⟩, 883 ⟨59JA1617⟩; **3**, 526 ⟨B-53MI 316-01⟩

Zope, U. R., **4**, 429, 430, 431 ⟨93JOC759⟩

Zora, J. A., **6**, 387 ⟨89JCS(D)447⟩

Zoran, A., **1**, 94 ⟨83TL4139⟩

Zorc, B., **3**, 118 ⟨86JOC5482⟩

Zordan, M., **2**, 79 ⟨88JOM(341)345⟩

Zoretic, P. A., **1**, 607 ⟨77JHC681⟩; **2**, 139 ⟨76JOC3587⟩; **3**, 309 ⟨69JA2342⟩; **4**, 210 ⟨75JOC2962⟩

Zorin, V. V., **4**, 775 ⟨90CA230883w⟩, 1080 ⟨83ZOB416⟩, 1081 ⟨88KGS852⟩; **6**, 127 ⟨88ZOB1066⟩, 478 ⟨81ZOR1539⟩

Zorn, H., **2**, 427 ⟨62M230⟩, 829 ⟨65CB2431⟩

Zorn, L., **5**, 151, 164 ⟨88JA4640⟩

Zosel, K., **1**, 665 ⟨60LA(629)1⟩; **2**, 43, 112 ⟨60LA(629)241⟩, 596 ⟨60LA(629)121, 60LA(629)1⟩; **4**, 699 ⟨60LA(629)1⟩

Zoss, A. O., **3**, 453 ⟨41JA1151⟩; **5**, 304 ⟨41JA1151⟩

Zotov, Y. L., **2**, 430 ⟨90ZOB1795⟩; **5**, 774 ⟨78ZOB78⟩

Zoubek, G., **2**, 507 ⟨78AG(E)135⟩

Zoukas, T., **4**, 259 ⟨92LA1039⟩

Zountsas, J., **1**, 1077 ⟨86T1711⟩

Zoutberg, M. C., **3**, 599 ⟨93OM3864⟩

Zrimšek, Z., **5**, 609, 626, 836 ⟨81S299⟩

Zschage, O., **2**, 997, 999 ⟨89AG(E)69, 91AG(E)321⟩; **4**, 382 ⟨92T8377⟩

Zschiesche, R., **3**, 224 ⟨90CB363⟩, 434 ⟨88TL1685⟩; **4**, 108 ⟨88LA1169⟩

Zschunke, A., **2**, 439 ⟨87ZAAC(553)136, 91ZAAC(594)66⟩, 478 ⟨91ZAAC(600)195⟩; **4**, 455 ⟨74SRI453, 74ZC243⟩, 482 ⟨78PS(4)59⟩, 550 ⟨91ZAAC(594)66⟩

Zsély, M., **6**, 88 ⟨88T2063⟩

Zshuke, A., **2**, 253, 260 ⟨82TL1345⟩

Zsindely, J., **1**, 809 ⟨73HCA75⟩; **2**, 660 ⟨73HCA75⟩

Zsolnai, L., **1**, 956 ⟨93CB1077⟩; **2**, 512 ⟨84JOM(260)281, 85ZN(B)1430⟩, 875, 896 ⟨87JOM(336)337⟩, 1079 ⟨90ZN(B)1529⟩; **3**, 556, 557, 560, 586 ⟨84T893⟩, 602 ⟨82CB172⟩; **4**, 714 ⟨93JOM(447)C1⟩; **5**, 1146 ⟨81CB3655⟩, 1146, 1149 ⟨84CB1161⟩; **6**, 375 ⟨86ZN(B)191⟩, 643 ⟨84CB1161⟩

Zuanic, M., **2**, 384 ⟨92JOC1585⟩

Zubanova, N. I., **6**, 283, 349, 355 ⟨89IZV2106⟩

Zub, Y. L., **2**, 1003 ⟨79UKZ811⟩

Zubek, A., **2**, 373 ⟨68ZC41⟩, 377 ⟨72ZC129⟩, 380 ⟨65CB3377, 69ZC105⟩

Zuberi, S., **2**, 613 ⟨87JA7194⟩; **6**, 230 ⟨92JOC4749⟩, 301 ⟨87JFC(37)327⟩

Zubiani, G., **1**, 659, 764 ⟨66TL4315⟩, 679 ⟨70T1281⟩, 680 ⟨67TL1581⟩; **2**, 1006 ⟨70T1281⟩; **4**, 146 ⟨71T6109⟩, 664 ⟨65TL3429⟩, 669 ⟨70T1281⟩; **6**, 198 ⟨65TL3429⟩, 198 ⟨66TL4315⟩

Zubieta, J. A., **4**, 699, 700 ⟨83OM106, 87OM834⟩, 701 ⟨84OM1687⟩; **6**, 552 ⟨70JA3342, 72JA1563⟩

Zubieta, J., **2**, 195, 196 ⟨85JA1783⟩; **4**, 223 ⟨89JA658⟩, 700 ⟨82OM404⟩; **6**, 349 ⟨87IC1488⟩

Zuboff, P., **2**, 87 ⟨01CB3246⟩

Zubrick, J. W., **2**, 393 ⟨75TL71⟩; **3**, 613 ⟨75TL71⟩; **5**, 1134 ⟨75TL71⟩
Zuccarello, G., **1**, 617, 709 ⟨88JA4866⟩, 704, 707, 709 ⟨88JA7247⟩
Zucchini, U., **4**, 725 ⟨71JOM(26)357⟩
Zucco, C., **2**, 636, 638 ⟨88ACR135, 81JA1761⟩; **5**, 652 ⟨87SC1741⟩
Zuckermann, J. J., **1**, 450 ⟨71JOM(33)321⟩; **2**, 538 ⟨64MI 210-02⟩, 859 ⟨71IC1028⟩; **4**, 726 ⟨71HCA1651⟩; **6**, 299, 300 ⟨67JHC166⟩
Zuckermann, R. N., **2**, 606 ⟨85JOC1599⟩
Zuech, E. A., **2**, 942 ⟨57JA4560⟩
Zueva, G. Ya., **4**, 142 ⟨88IZV1946⟩
Zuffanti, S., **2**, 191 ⟨40JA1044⟩
Zugaza-Bilbao, A., **5**, 126, 267, 291 ⟨80S547⟩
Zugravescu, I., **1**, 398 ⟨B-76MI 109-01⟩; **3**, 678 ⟨B-76MI 320-01⟩
Zuikova, A. N., **4**, 342 ⟨78ZOB2653⟩
Zukel, J. W., **2**, 101 ⟨51USP2529493⟩
Zulanf, P., **2**, 999 ⟨93CB1873⟩
Zumach, G., **2**, 270 ⟨62ZN(B)807⟩; **3**, 699 ⟨69AG(E)20⟩; **5**, 989 ⟨B-71MI 526-03⟩, 1036 ⟨70AG(E)54⟩; **6**, 435 ⟨70AG(E)54⟩, 449 ⟨70S542⟩, 453 ⟨68FRP1517379⟩, 602 ⟨67AG(E)649⟩, 602, 604, 605, 606 ⟨B-71MI 620-01⟩
Zündorf, W., **1**, 1145 ⟨67T565⟩
Zupan, M., **1**, 517 ⟨77JA3504⟩; **2**, 15 ⟨86BCJ1659, 91JOC7347⟩, 16 ⟨82JOC573, 89T6003⟩, 17 ⟨76JOC4002, 77JOC1559⟩, 17 ⟨93JCS(P1)2851⟩, 18 ⟨87BCJ3083, 87JOC919⟩, 21 ⟨84T2865⟩, 24 ⟨90T2503⟩, 28 ⟨86JOC929⟩, 613 ⟨81CC148⟩, 615 ⟨92CC274⟩; **3**, 4 ⟨89SC1317⟩; **4**, 45 ⟨93TL4355⟩, 61 ⟨76JFC(8)305⟩, 751 ⟨87JFC(37)313⟩; **5**, 6 ⟨89TL6095⟩; **6**, 10 ⟨90TL3357⟩
Zupet, P., **2**, 24 ⟨90T2503⟩
Zuraw, P. J., **1**, 949 ⟨84JOC1345⟩
Zurawinski, R., **2**, 156, 721 ⟨88T6687⟩
Zurawski, B., **1**, 1109 ⟨73CPL309⟩
Zurc, J., **4**, 554 ⟨50M202⟩
Zurcher, C., **3**, 280 ⟨75CB1570⟩
Zürcher, S. C., **1**, 1000 ⟨75S423⟩
Zurer, P. St. J., **3**, 465 ⟨81JOC4685⟩
Zurmühlen, F., **2**, 449 ⟨91AG(E)207⟩; **4**, 341 ⟨87JOM(332)C1⟩, 565 ⟨85ZN(B)1077⟩; **5**, 839, 880 ⟨85ZN(B)1077⟩; **6**, 153 ⟨91AG(E)207⟩
Zürn, V., **6**, 438 ⟨82ZN(B)81⟩
Zutter, U., **1**, 624, 626 ⟨86JA1039⟩; **5**, 278 ⟨86JA1039⟩
Zutterman, F., **1**, 248 ⟨82TL4385⟩
Zvereva, M. A., **2**, 462 ⟨57MI 208-01⟩
Zverina, V., **2**, 777 ⟨73MI 215-01⟩
Zvezdina, E. A., **3**, 420 ⟨82RCR469⟩
Zvezdkina, L. I., **2**, 436 ⟨82IZV2390⟩
Zvonok, A. M., **3**, 274 ⟨81JGU1053⟩
Zwainz, J. G., **3**, 348 ⟨75ZN(B)760⟩
Zwane, I., **2**, 522 ⟨85JOC1969⟩
Zwanenburg, B., **1**, 582 ⟨91TL3739⟩, 704 ⟨80RTC39⟩, 786 ⟨87TL357, 94TL2787⟩, 1023 ⟨89TL2153⟩, 209 ⟨77JCS(P1)1469⟩, 399 ⟨81S368⟩, 743 ⟨83S628⟩; **2**, 197 ⟨75S519⟩, 400 ⟨83JOC4582, 85HOU(E11)943⟩; **3**, 171, 368 ⟨84TL5947⟩, 365 ⟨81S295, 69TL4461⟩, 368 ⟨73JCS(P1)73, 73S617⟩, 368 ⟨78TL3383, 78TL807⟩, 368 ⟨78TL811, 84CC502⟩, 369 ⟨70QRS79, 82RTC1⟩, 369 ⟨89PS(43)1⟩, 370 ⟨77TL259, 79RTC127⟩, 370 ⟨83JOC4582⟩, 37 ⟨83S628⟩; **4**, 234 ⟨64TL547⟩, 235 ⟨75S519⟩, 237 ⟨70TL935⟩, 270, 272 ⟨80RTC39, 78T1585⟩, 305 ⟨67JA4456⟩, 365 ⟨84JOC1691⟩, 570, 926 ⟨84JOC263⟩, 859 ⟨76RTC202, 80RTC45⟩, 859 ⟨92TL6383⟩, 936 ⟨78TL3383⟩; **5**, 228 ⟨83T2995⟩, 229 ⟨85T6063, 92RCT211⟩, 472, 488 ⟨85HOU(E11/1)911⟩, 472, 488, 562 ⟨82RTC1⟩, 472, 495 ⟨85HOU(E11/2)1326⟩, 478, 491, 492, 562 ⟨77RTC139⟩, 489 ⟨B-67MI 520-01⟩, 490 ⟨70RTC687, 67TL3453⟩, 562 ⟨68TL2871, 71RTC614⟩, 562 ⟨75JCS(P2)916, 78TL807⟩, 562 ⟨78TL811, 83RTC400⟩, 562 ⟨85JOC2930, 89(43)PS1⟩, 562 ⟨92TL6383⟩, 562, 645 ⟨84JOC263⟩, 734, 736 ⟨84RTC342⟩; **6**, 316 ⟨73TL3589⟩, 536 ⟨75JCS(P2)916⟩
Zwangenburg, B., **6**, 531 ⟨69TL4461⟩
Zwaschka, F., **5**, 1128 ⟨77AG747, 80AG767⟩
Zweep, S. D., **2**, 1007 ⟨86JA1692⟩
Zweifel, G., **1**, 163 ⟨73S37⟩, 178 ⟨87JOC5484⟩, 207, 465, 511, 512, 533, 534, 542, 543, 1106, 1139 ⟨84OR(32)375⟩, 278, 487, 661, 1103 ⟨67JA3652⟩, 278, 487, 662 ⟨71JA6309⟩, 278, 488, 531, 662, 1103 ⟨68JA6243⟩, 278, 531, 662, 1103 ⟨72JA6560⟩, 449 ⟨74S339, 89TL6487⟩, 449, 509 ⟨59JA1512⟩, 449, 509, 1102 ⟨71JA3395⟩, 449, 977 ⟨74JA5620⟩, 451 ⟨81S999⟩, 465 ⟨76S625⟩, 487 ⟨75S376⟩, 487, 1100, 1103 ⟨67JA5086⟩, 488, 1103 ⟨78JOM(156)159⟩, 491, 534 ⟨70JA6678⟩, 509 ⟨67JA291, 71JA1308⟩, 509, 1100, 1102 ⟨61JA3834⟩, 509, 1103 ⟨70JA4068⟩, 511 ⟨78JOC2739⟩, 512 ⟨83JA1383, 84S113⟩, 514 ⟨77S52⟩, 514, 534 ⟨67JA5085⟩, 533 ⟨76S816, 67JA2754⟩, 534 ⟨68JA7139, 74S658⟩, 657, 658 ⟨67JA567, 74JA316⟩, 658, 963 ⟨70JA1427⟩, 680 ⟨66TL6021⟩, 977 ⟨81JOC829, 82JOC3364⟩, 977, 1056 ⟨78JA5561⟩, 1019 ⟨83S128⟩, 1048 ⟨80S150⟩, 1052 ⟨84S111⟩, 1052, 1055, 1056, 1058 ⟨83S883, 84JOC4565⟩; **2**, 40 ⟨60JA3222, 63OR(13)1⟩, 41, 900 ⟨67JA291⟩, 42, 57 ⟨64JA393⟩, 42, 57, 519 ⟨64JA397⟩, 75 ⟨88JA5578⟩, 268, 522, 526, 900 ⟨80JA5919⟩, 514 ⟨60JA4708⟩, 516, 517 ⟨63JA2072⟩, 518 ⟨61JA1241, 62JA190, 63JA2066, 66JOC4178⟩, 523 ⟨67JA561⟩, 616 ⟨79JA5101⟩, 629 ⟨67JA2753⟩, 900 ⟨61JA3834, 67JA3652⟩, 900 ⟨68JA6243, 70JA1427. 71JA3395, 73S67, 67JA5086, 77JA5192, 88S130⟩, 902 ⟨81S288⟩, 904 ⟨75S376⟩, 910, 919, 922, 968 ⟨84OR375⟩, 922 ⟨80S803, 84S113⟩, 922, 929 ⟨91JOC2468⟩, 922, 931 ⟨78JOC2739, 81S999⟩, 967 ⟨79COC(3)1013⟩, 968 ⟨B-78MI 219-01⟩, 1005 ⟨66TL2535, 73S37⟩; **3**, 223 ⟨88S130⟩, 675 ⟨80S150⟩; **4**, 631 ⟨61JA3834, 67JA291⟩, 631 ⟨73S37⟩, 664, 665 ⟨66TL2535⟩, 756 ⟨82S127⟩, 758, 816 ⟨79JA5101⟩, 808, 1059 ⟨78JOC2739⟩, 808, 1062, 1068 ⟨84OR(32)375⟩, 808, 811 ⟨81JOC1292⟩, 814 ⟨67JA5086, 77JA5192, 93SC2937⟩, 1056 ⟨77JA3184⟩, 1061 ⟨84JOC4565⟩; **5**, 52 ⟨67JA291⟩, 417 ⟨77JA3184⟩, 417, 426 ⟨81S288, 81JA6217⟩, 1136 ⟨72JA6560⟩; **6**, 198 ⟨73S37⟩, 208 ⟨66TL6021, 71CC1593⟩
Zweig, A., **1**, 378, 886 ⟨61JA1196⟩
Zweitering, L., **1**, 88 ⟨63MI 103-01⟩
Zwettler, R., **4**, 725 ⟨90JA9620⟩
Zwick, W., **1**, 948 ⟨82CB2526⟩
Zwierak, A., **2**, 107 ⟨71T3163⟩
Zwierkak, A., **2**, 308 ⟨92SC1929⟩
Zwierzak, A., **2**, 255 ⟨71ZN(B)386⟩, 299 ⟨86LA402⟩, 300, 808 ⟨77AG(E)107⟩, 317 ⟨87S487⟩, 373 ⟨78TL4731, 87S485⟩, 377 ⟨81S537⟩, 412 ⟨70T3527, 74TL3383⟩, 412 ⟨76T1649⟩, 414 ⟨67T2243⟩, 808 ⟨75S507, 76AG(E)302⟩; **3**, 188 ⟨82S922⟩; **5**, 273, 285 ⟨79S527, 79S546⟩, 300 ⟨81S1005⟩, 371, 373, 707 ⟨93HAC11, 79S691⟩, 1031 ⟨89S300⟩
Zwiezak, A., **2**, 299 ⟨82S922⟩
Zwikker, J. W., **3**, 708 ⟨78JOC1972⟩
Zwilgmeyer, F., **5**, 62 ⟨43OSC(2)126⟩
Zwinger, O., **5**, 851 ⟨1898JPR223⟩
Zwolinski, B. J., **6**, 221, 223 ⟨79JCED251⟩, 223, 226, 227, 228 ⟨79JMR559⟩
Zyablikova, T. A., **2**, 879 ⟨88JGU1173⟩, 894 ⟨83IZV1152⟩; **4**, 124 ⟨75JGU1950, 79DOK(244)610⟩, 342 ⟨78ZOB2653⟩, 341 ⟨80IZV1626⟩, 343 ⟨79IZV1863⟩, 476 ⟨79IZV2783⟩, 556 ⟨67DOK(177)340⟩, 804 ⟨92JGU1222⟩, 911 ⟨68ZOB1523⟩; **6**, 659 ⟨90ZOB1293⟩
Zyabrev, V. S., **5**, 655 ⟨89JOU2329⟩, 655, 1053 ⟨89ZOR2597⟩

Zyat'kov, I. P., **5**, 208 ⟨59DOK(128)1201⟩
Zybill, C., **4**, 118 ⟨87OM2489⟩, 554 ⟨84CB3374⟩; **5**, 897 ⟨83CB1955⟩, 1097 ⟨88AG(E)583⟩; **6**, 123, 698 ⟨83CB1955⟩
Zydowsky, T. M., **3**, 156 ⟨89JOC4576⟩
Zygmunt, J., **4**, 130 ⟨85T4979⟩, 343 ⟨80PJC233⟩, 472, 473 ⟨81PJC713, 78PJC2271⟩, 473 ⟨81PJC411⟩
Zyk, N. V., **2**, 232 ⟨92MI 203-02⟩
Zykova, L. V., **5**, 631 ⟨75ZOB2116⟩
Zykova, T. N., **5**, 567, 581 ⟨76KFZ41⟩
Zykova, T. V., **2**, 459 ⟨74ZOB2621⟩, 1076 ⟨88JGU2050⟩; **4**, 120 ⟨71JGU2183⟩; **5**, 403 ⟨71JGU2513⟩; **6**, 659 ⟨87ZOB2467⟩
Zyl, C. M. V., **3**, 271 ⟨86JOC3561⟩
Zymalkowski, F., **3**, 280 ⟨71AP(304)543⟩

Cumulative Subject Index

PHILIP ASLETT
Marlborough, Wiltshire, UK

Every effort has been made to index as comprehensively as possible, and to standardize the terms used in the index in line with the IUPAC Recommendations. In view of the diverse nature of the terminology employed by the different authors, the reader is advised to search for related entries under the appropriate headings.

The index entries are presented in letter-by-letter alphabetical sequence. Compounds are normally indexed under the parent compound name, with the substituent component separated by a comma of inversion. An entry with a prefix/locant is filed after the same entry without any attachments, and in alphanumerical sequence. For example, 'dienes', '1,3-dienes', and '1-bromo-1,4-dienes' will be filed as:

Dienes
1,3-Dienes
1,4-Dienes, 1-bromo-

The Index is arranged in set-out style, with a maximum of three levels of heading. Location references refer to volume number (in bold) and page number (separated by a comma); major coverage of a subject is indicated by bold, elided page numbers: for example,

Halides, alkynyl, synthesis of, **2**, **1234–55**
  properties of, **5**, 345

See cross-references direct the user to the preferred term; for example,

Mercaptans *See* Thiols

See also cross-references provide the user with guideposts to terms of related interest, from the broader term to the narrower term, and appear at the end of the main heading to which they refer, e.g.

Thiones
  See also
    Thioketone

A

Abscisic acid, synthesis of, **1**, 726
Absorption agents, for the synthesis of phosphaethynes, **5**, 1152
1,5-Abstractions, and radical generation, **1**, 927
(±)Acarenoic acid, synthesis of, **1**, 568
Acetal, acetamide, reactions, with macrocyclic triamines, **6**, 140
Acetal, diazoacetaldehyde dimethyl, synthesis of, **3**, 47
Acetal, *N,N*-dimethylacetamide diethyl, reactions, with enamines, **4**, 987
Acetal, *N,N*-dimethylacetamide dimethyl (DMADMA), use in the synthesis of *N*-heteroarylacetamide oximes, **5**, 836
Acetal, *N,N*-dimethylformamide, reactions, with ethyl isocyanoacetate, **3**, 720
Acetal, dimethylformamide diethyl
  reactions
    with *N*-chloroamides, **6**, 634
    with phenyl isothiocyanate, **6**, 630
  as starting material, for 1-dimethylamino-2-nitroethylene, **2**, 788
Acetal, *N,N*-dimethylformamide diethyl, reactions, with tertiary ethynyl alcohols, **2**, 740
Acetal, dimethylformamide dimethyl
  reactions
    with acylhydrazines, **5**, 773
    with α-amino acids, **5**, 746
    with dinitromethane, **4**, 995
    with hydroxyacids, **1**, 565
    with nitroimidazoles, **3**, 673
    with thiobenzamide, **5**, 760
Acetal, *N,N*-dimethylformamide dimethyl (DMFDMA)
  use in the synthesis of enamines, **2**, 740
  use in the synthesis of oxyimidic amides, **5**, 835
Acetal, dimethylformamide dineopentyl, use in the synthesis of dienes, **1**, 566
Acetal, ketene diethyl, reactions, with chloromethanesulfonyl chloride, **5**, 496
Acetaldehyde
  lithium enolates of, use in the synthesis of carbon-bridged dicobalt complexes, **4**, 687
  mercuration of, **6**, 207
  reactions
    with allylphosphine oxides, **1**, 755
    with arsonium ylides, **1**, 738
    with caesium(IV) ions, **5**, 49
    with dienes, **2**, 140
    with α-methoxyallylphosphine oxide, **4**, 917
    with methoxycarbonylethylidene-triphenylphosphorane, **1**, 725
    with monoisopinocampheylborane, **2**, 531
  as starting material
    for (±)-alanine, **5**, 85
    for enamines, **3**, 49
    for ketenes, **3**, 535
  synthesis of, via mercury-catalyzed hydration, **2**, 982
Acetaldehyde, α,α-bis(benzeneseleno)-, reactions, with morpholinobenzeneselenamide, **6**, 93
Acetaldehyde, bromo-
  reactions
    with thiols, **3**, 35
    with triphenyl arsine, **3**, 48
Acetaldehyde, *t*-butoxy-, isolation of, **2**, 686
Acetaldehyde, 2-carboxy-4,5-dimethoxyphenyl-, from, 2-hydroxy-5,6-dimethoxyindane-1-one, **5**, 73
Acetaldehyde, chloro-, reactions, with potassium *O*-ethyl dithiocarbonate, **6**, 546
Acetaldehyde, dichloro-
  reactions
    with *o,p*-dinitrophenyl hydrazine, **3**, 451
    with phenylhydrazine, **3**, 451
Acetaldehyde, diphenyl-, synthesis of, **1**, 271, 646, 743
Acetaldehyde, phenyl-, synthesis of, **1**, 738
Acetaldehyde, phenylcyano-, reactions, with dialkylformamides, **2**, 749
Acetaldehyde, phenylselenyl-, use in the synthesis of α-selenyl ketones, **3**, 174
Acetaldehyde, trichloro-
  for the oxidation of secondary alcohols, **3**, 119
  reactions
    with arenes, **6**, 27
    with Grignard reagents, **6**, 27
    with silanes, **6**, 27
    with thiols, **4**, 215
  as starting material
    for 2,2-dichlorovinyl chloroformate, **6**, 429
    for trichloromethylcarbinols, **6**, 27
Acetaldehyde, trifluoro-, reactions, with hydrogen sulfide, **4**, 215
Acetaldehyde, trimercurated, synthesis of, **6**, 207
Acetaldehyde, triphenyl arsonium, synthesis of, **3**, 48
Acetaldehyde, tris(chloromercurio)-, synthesis of, **6**, 207
Acetaldehyde, tris(mercurio)-, oxonium nitrate of, **6**, 207
Acetaldehyde enolates, chloro-, synthesis of, **3**, 27
Acetaldehyde enolates, fluoro-, synthesis of, **3**, 27
Acetaldehyde oxime, 2-nitro-, as starting material, for glyoxylic acidoxime oxyimidic chlorides, **5**, 822
Acetaldehydes, 2-(diphenylphosphinoyl)-, reactions, with saturated aldehydes, **2**, 81
Acetaldehydes, α-sulfonyl-, reactions, with aniline, **2**, 742
Acetaldimine, α,α-bis(trimethylsilyl)-*t*-butyl, synthesis of, **4**, 604
Acetaldoximes, *O*-perfluoroalkyl imino-, synthesis of, **5**, 711
Acetalization
  of aldehydes, **4**, 178
  of ketones, **4**, 178
S,N-Acetal *S*-oxides, synthesis of, **4**, 330
Acetals
  alkylation of, **2**, 89
  bromination of, **2**, 29
  carbolithiation of, **2**, 555
  α-chlorination of, **2**, 22
  deprotection of, **4**, 175
  deprotonation of, **4**, 961
  electrochemical oxidation of, for ortho esters, **6**, 77
  from
    acetals, **4**, 184
    O,S-acetals, **4**, 196
    α-acyloxy ethers, **4**, 200
    alcohols, **4**, 186
    aldehydes, **4**, 177
    alkenes, **4**, 193
    alkynes, **4**, 193
    carbonyl ylides, **3**, 319
    β-chlorovinyl carbonyls, **4**, 196
    cyclic hemiacetals, **4**, 201
    dialkylaminoethoxycarbenium tetrafluoroborates, **6**, 320
    1,2-dicarbonyls, **4**, 183
    enol acetates, **4**, 188
    enol ethers, **4**, 186

ethers, **4**, 201
ethyl *N*,*N*-dimethylcarbamate, **6**, 313
geminal dihalides, **4**, 189
α-halo ethers, **4**, 198
halogenated ketones, **4**, 183
hemiacetals, **4**, 200
ketones, **4**, 177
nitrobenzaldehydes, **4**, 183
ortho esters, **4**, 192
silyl enol ethers, **4**, 188
silyl ketones, **6**, 124
α-substituted ethers, **4**, 198
α-sulfonyl ethers, **4**, 200
thioacetals, **4**, 196
vinyl epoxides, **4**, 183
halogenation of, **6**, 250
hydrolysis of, **3**, 123
irradiation of, **4**, 1079
reactions
with acetyl chloride, **4**, 175
with acid chloride, **4**, 48
with 1-acylhydrazines, **5**, 363
with alcohols, **4**, 184
with alkynes, **1**, 1025
with alkynyltributylstannanes, **1**, 1026
with alkynyltrimethylsilanes, **1**, 1026
with allenyltrialkylstannanes, **1**, 1053
with amides, **4**, 310
with amines, **2**, 749
with anhydrides, **4**, 211
with bis(dimethylaluminum) selenide, **3**, 385; **5**, 520
with *t*-butoxy radicals, **4**, 1078
with *t*-butyl isocyanide, **3**, 658
with dimethylcopperlithium, **4**, 388
with β-hydroxycarboxylic acids, **4**, 213
with isocyanates, **4**, 426
with Lewis acids, **2**, 1007
with lithium metal, **1**, 397
with pyrazole, **4**, 437
with silyl enol ethers, **3**, 220
with sulfinic acids, **4**, 236
with thiols, **4**, 249
with 2,2,2-trichloroethanol, **4**, 184
with *S*-trimethylsilyl enethiolate, **5**, 557
reduction of, **1**, 21; **2**, 89
as starting materials
for acetals, **4**, 184
for acyl alkyl acetals, **4**, 211
for aldehydes, **3**, 9
for alkenyl ethers, **2**, 678
for amidines, **5**, 743
for amines, **2**, 310
for α-aminoalkylideneamino enamines, **4**, 987
for α-bromo ethers, **4**, 50
for cations, **4**, 1072
for α-chloro ethers, **4**, 48
for dithioacetals, **4**, 245
for enamines, **2**, 739
for formamidines, **5**, 746
for hemiacetals, **4**, 175
for ketones, **3**, 123
for monoselenoacetals, **4**, 238
for monotelluroacetals, **4**, 238
for monothioacetals, **4**, 215
for ortho esters, **6**, 77
for α-phosphoryl enamines, **4**, 1004
for propargyl ethers, **1**, 1026
for thioaldehydes, **3**, 340
for thioketones, **3**, 347
for toluenes, **1**, 21
stereoselective synthesis of, **4**, 182
substitution of, **1**, 479
synthesis of, **4**, 176
via acetal exchange, **4**, 190
via acetal interchange, **4**, 185
via conjugate addition of alcohols, **4**, 195
via electrochemical methods, **4**, 197
via hetero Diels–Alder reaction, **4**, 202
via palladium(II)-catalyzed oxidation of terminal alkenes, **4**, 193
use in the synthesis of hemiaminals, **4**, 309
use in the synthesis of β-hydroxy ketones, **3**, 159
Acetals, acetaldehyde, synthesis of, **4**, 185
Acetals, acrolein, reactions, with vinyl halides, **1**, 1096
Acetals, acyl alkyl
from
acetals, **4**, 211
alcohols, **4**, 209
aldehydes, **4**, 212
carboxylic acids, **4**, 209
enol esters, **4**, 209
enol ethers, **4**, 208
ethers, **4**, 207
hemiacetals, **4**, 209
ketones, **4**, 212
α,β-unsaturated ketones, **4**, 213
synthesis of, **4**, 207
via Baeyer–Villiger oxidation, **4**, 212
Acetals, acylketene, hydrolysis of, **3**, 159
Acetals, aldehyde
synthesis of, **3**, 159
transacetalization of, **4**, 249
Acetals, aldehyde dimethyl, as starting materials, for thioaldehydes, **3**, 336
Acetals, alkenylketene, reactions, with methyl propiolate, **1**, 521
Acetals, *N*-alkyl-*N*,*O*-ketene, rearrangement of, **1**, 835
Acetals, alkyl silyl
from, silylketene acetals, **4**, 202
synthesis of, **4**, 202
Acetals, alkynic, synthesis of, **4**, 192
Acetals, allylic
reactions, with Grignard reagents, **2**, 679
as starting materials, for 2-cyanoethers, **2**, 89
Acetals, amide
esterification of, **5**, 131
reactions
with allylic alcohols, **1**, 827
with nitromethane, **2**, 788
with prop-2-yn-1-ols, **1**, 979
as starting materials, for ketene aminals, **4**, 975
use in the synthesis of carboxylic esters, **5**, 126
Acetals, amide dialkyl, reactions, with thionoamides, **5**, 633
Acetals, amide diethyl, reactions, with *t*-butyl isocyanoacetate, **3**, 717
Acetals, amino, synthesis of, via reductive amination, **3**, 44
Acetals, α-amino
from
diamines, **3**, 41
diethoxy acetonitriles, **3**, 40
glyoxal, **3**, 40
hydrazino acetals, **3**, 41
organometallics, **3**, 40
Acetals, arylacetaldehyde, from, styrenes, **4**, 197
Acetals, β-azido, synthesis of, **3**, 48
Acetals, *N*-(benzenesulfinyl), synthesis of, **4**, 325
Acetals, benzophenone, synthesis of, **4**, 197, 198
Acetals, 4,6-*O*-benzylidene, cleavage of, **4**, 217
Acetals, bis(methoxyethyl), synthesis of, **4**, 829
Acetals, bis(2,2,2-trichloroethyl), synthesis of, **4**, 184
Acetals, bromo
cyclization of, **1**, 547
as starting materials, for β-nitro aldehyde acetals, **3**, 46
Acetals, (*E*)-bromo, as starting materials, for (*Z*)-2-bromoalkenals, **3**, 68
Acetals, α-bromo, as starting materials, for ketene diethyl acetals, **4**, 826

Acetals, carboxylic amide, as starting materials, for phosphaalkenes, **5**, 883
Acetals, chiral, reactions, with trimethylsilyl cyanide, **3**, 631
Acetals, β-chloroketene, silylation of, **4**, 806
Acetals, α-cyano, synthesis of, **3**, 627
Acetals, cyanoketene, synthesis of, **4**, 762
Acetals, cyclic, from, methyl acetals, **4**, 184
Acetals, cyclopropane, reactions, with *N*-tosylimines, **5**, 344
Acetals, dialkoxyphosphinylketene, use in Claisen rearrangement, **1**, 831
Acetals, dialkyl, reactions, with allylchromiums, **2**, 73
Acetals, dialkylformamide, reactions, with ketones, **3**, 266
Acetals, *N*,*N*-dialkylformamide
  reactions
    with alkynes, **1**, 1025
    with 1-alkynes, **1**, 1026
Acetals, diarylketene, synthesis of, **4**, 826
Acetals, diaryloxy, synthesis of, **4**, 199
Acetals, dichloroketene-, reactions, with sulfonamyl chlorides, **6**, 106
Acetals, dicyanoketene
  from, tetracyanoethylene, **4**, 829
  reactions, with 1,1-disubstituted hydrazines, **5**, 845
Acetals, dihydrofuran, synthesis of, via sulfoxide elimination, **1**, 596
Acetals, dimethyl, synthesis of, **4**, 197
Acetals, dimethylformamide dimethyl, reactions, with arylhydrazines, **5**, 770
Acetals, dipent-4-enyl, for the protection of diols, **4**, 184
Acetals, ethylene, synthesis of, **4**, 190
Acetals, ethyl glyoxylate, synthesis of, **4**, 189
Acetals, formaldehyde, synthesis of, **4**, 185, 190, 198
Acetals, formamide
  reactions
    with amines, **5**, 743, 746
    with 2-amino-3-cyano-5-methylpyridazine, **5**, 746
    with hydroxylamine, **5**, 764
    with thioamides, **5**, 760
    with thioureas, **5**, 760
Acetals, α-functionalized, from, enol ethers, **4**, 187
Acetals, halo
  dehydrohalogenation of, **2**, 1025
  reactions, with tri-*n*-butyltin alkynides, **1**, 1003
Acetals, 2-halo, as starting materials, for α-substituted esters, **1**, 391
Acetals, α-halo
  hydrogen halide elimination from, **4**, 826
  as intermediates, for ortho esters, **6**, 74
  as starting materials
    for ketene acetals, **4**, 826
    for ortho esters, **6**, 68, 69
Acetals, hydrazino
  as starting materials, for α-amino acetals, **3**, 41
  synthesis of, **3**, 41
Acetals, α-hydroxy, synthesis of, **4**, 183, 187
Acetals, α-iodo, as starting materials, for ketene acetals, **4**, 826
Acetals, ketene
  amination of, **3**, 265
  boryloxy compounds of, synthesis of, **4**, 832
  from
    1-alkoxyalkynes, **4**, 827
    carboxylic acids, **4**, 824
    carboxylic esters, **4**, 824
    α,α-difluoro ethers, **4**, 826
    gem-dihaloalkenes, **4**, 829
    α-haloacetals, **4**, 826
    α-halo ortho esters, **4**, 827
    α-iodoacetals, **4**, 826
    ketenes, **4**, 829
    ortho esters, **4**, 825
    phosphonium salts, **4**, 825
    α-tosyloxyacetals, **4**, 826
    triacylmethanes, **4**, 824
    trihalomethanes, **4**, 826
    α,β-unsaturated acetals, **4**, 830
    α,β-unsaturated esters, **4**, 827
  hydrolysis of, **2**, 638
  oxidation of, **1**, 558
  reactions
    with amines, **4**, 887, 972
    with ammonia, **4**, 972, 983
    with azides, **6**, 107
    with dibromo carbene, **4**, 162
    with diethylamine, **4**, 972
    with hydroxamic vinyl esters, **4**, 833
    with isocyanates, **6**, 106
    with ketenes, **3**, 246
  selenocyanation of, **2**, 284
  as starting materials
    for 2-haloalkyl esters, **5**, 139
    for α-hydrazino enamines, **4**, 983
    for hydrazonates, **5**, 845
    for imino esters, **5**, 691
    for ketene aminals, **4**, 971
    for ketene hemiaminals, **4**, 887
    for β-lactams, **6**, 106
    for ortho esters, **6**, 74
  synthesis of, **4**, 806, 824
    via [3+2] cycloaddition reactions, **4**, 830
    via Horner–Wittig reactions, **4**, 828
Acetals, *S*,*N*-ketene
  thio-Claisen rearrangements of, **5**, 616
  use in the synthesis of thionoamides, **5**, 5707
Acetals, ketene bis(phenylthio), synthesis of, **5**, 100
Acetals, ketene diethyl
  from, α-bromo acetals, **4**, 826
  reactions, with *N*-ethylaniline, **4**, 887
Acetals, ketene disilyl, from, carboxylic acids, **4**, 832
Acetals, ketene methyl trialkylsilyl, as silylating agents, for carboxylic acids, **5**, 226
Acetals, ketene *S*-methyl-*S*-(trimethylsilyl), flash vacuum pyrolysis of, **3**, 550
Acetals, ketene silyl
  from
    α-bromo esters, **4**, 831
    ester enolates, **4**, 831
    ethyl bromoacetate, **4**, 831
  reactions
    with bis(dimethylaluminum) selenide, **3**, 553
    with dibromodifluoromethane, **4**, 736
  regioselectivity of, **4**, 831
  stereochemistry of, **4**, 831
  synthesis of, **4**, 831
Acetals, α-keto
  cleavage of, **4**, 161, 212
  from, dihalo ketones, **4**, 189
Acetals, methoxybenzylidene, synthesis of, **4**, 201
Acetals, methyl
  as starting materials, for cyclic acetals, **4**, 184
  synthesis of, **4**, 197
Acetals, methylene, from, diols, **4**, 190
Acetals, methylenecyclopropanone, thermolysis of, **1**, 360
Acetals, α-nitro, from, β-phenylthionitroalkenes, **2**, 393
Acetals, β-nitro, synthesis of, **3**, 46
Acetals, nitro aldehyde, synthesis of, **3**, 46
Acetals, β-nitro aldehyde, from, bromo acetals, **3**, 46
Acetals, α-nitroso, production of, **2**, 391
Acetals, α-phenyl selenenyl, synthesis of, **4**, 51
Acetals, silylketene
  for the alkylation of *O*-alkyloximes, **5**, 329
  conjugate addition of, **1**, 304
  as intermediates, **1**, 830
  Ireland–Claisen rearrangement of, **5**, 227
  reactions
    with aryl triflates, **1**, 483

with *N*-silylimines, **3**, 440
3,3 sigmatropic rearrangement of, **1**, 829
as starting materials, for alkyl silyl acetals, **4**, 202
synthesis of, **5**, 227
use in Michael reactions, **1**, 308
Acetals, (*E*)-silylketene, synthesis of, **1**, 417
Acetals, (*Z*)-silylketene, synthesis of, **1**, 417; **2**, 699
Acetals, stannyl
reactions
with acetyl chloride, **4**, 388
with alkenylaluminums, **4**, 388
with alkynylaluminums, **4**, 388
Acetals, symmetrical
from
aldehydes, **4**, 190
ketones, **4**, 190
synthesis of, **4**, 190
Acetals, tetramethylurea diisopropyl, synthesis of, **4**, 974
Acetals, thioacylketene, from, 1,2-dithioles, **3**, 361
Acetals, α-(tosylamino) aldehyde, synthesis of, **2**, 323
Acetals, α-tosyloxy, as starting materials, for ketene acetals, **4**, 826
Acetals, tributylstannyl, as starting materials, for α-alkoxystannanes, **4**, 388
Acetals, trimethylenemethane
cycloaddition of, **1**, 360
reactions, with alkenes, **1**, 360
Acetals, trimethylsilylketene, reactions, with sulfur dioxide, **2**, 209
Acetals, α,β-unsaturated
isomerization of, **4**, 830
reactions, with dimethyl sulfide, **4**, 201
as starting materials, for ketene acetals, **4**, 830
Acetals, unsymmetrical
from, enol ethers, **4**, 186, 188
synthesis of, **4**, 198
Acetals, urea, as starting materials, for ketene aminals, **4**, 974
Acetals, vinylketene
rearrangement of, **1**, 829
synthesis of, **1**, 827
N,N-Acetals, ketene
reactions, with chlorosulfene, **5**, 497
use in the synthesis of chlorosulfene, **5**, 496
N,O-Acetals, ketene
reactions, with amines, **4**, 972
as starting materials, for α-phosphoryl enamines, **4**, 1003
synthesis of, **1**, 827; **4**, 972, 974
N,O-Acetals, *N*-silylketene, from, *N*-phenylimidates, **1**, 828
N,S-Acetals, *N*-acylated ketene, as starting materials, for α-amido enamines, **4**, 981
N,S-Acetals, ketene
from, amides, **2**, 1049
as starting materials
for α-aminoalkylideneamino enamines, **4**, 988
for α-azido enamines, **4**, 990
for α-hydrazino enamines, **4**, 983
for α-phosphoryl enamines, **4**, 1003
for tertiary ynamines, **2**, 1049
for thioketenes, **3**, 550
N,X-Acetals, ketene, reactions, with dialkyl phosphites, **4**, 1003
O,As-Acetals, from, ketones, **4**, 344
O,N-Acetals
from
aldehydes, **4**, 311
alkenes, **4**, 306, 312
amides, **4**, 303
amino acids, **4**, 301
amino alcohols, **4**, 301
2-aminomethylbenzimidazoles, **4**, 296
hydrazones, **4**, 313
hydroxyamides, **4**, 301
imides, **4**, 303
imines, **4**, 304, 312
iminium salts, **4**, 304
ketones, **4**, 311
nitrones, **4**, 312
oximes, **4**, 313
trichloroacetaldehyde imines, **4**, 304
synthesis of, **4**, 293
See also Hemiaminals
O,N-Acetals, *N*-acyl, synthesis of, **4**, 308
O,N-Acetals, *N*-nitroso, synthesis of, **4**, 312
O,O-Acetals
reactions, with tris(phenylseleno)borane, **4**, 238
as starting materials, for monothioacetals, **4**, 216
O,O-Acetals, ketene, reactions, with hydrazines, **5**, 706
O,S-Acetals
as homologation reagents, **4**, 370
oxidation of, **4**, 235
as starting materials, for acetals, **4**, 196
synthesis of, via *S*-alkylation, **4**, 225
O,S-Acetals, ketene
as starting materials, for thioketenes, **3**, 550
synthesis of, **6**, 336
O,S-Acetals, α,β-unsaturated, as starting materials, for ketene monothioacetals, **4**, 836
P-Acetals, synthesis of, **4**, 340
Se,N-Acetals
from
aldehydes, **4**, 333
alkenes, **4**, 335
diazoalkanes, **4**, 335
ketones, **4**, 333
selenocarbonyls, **4**, 333
synthesis of, **4**, 333
Se,P-Acetals
from
phosphates, **4**, 348
phosphine oxides, **4**, 348
phosphines, **4**, 347
synthesis of, **4**, 347
S,N-Acetals
from
acid chlorides, **4**, 328
aldehydes, **4**, 315, 328
imines, **4**, 319, 328
iminium salts, **4**, 319
ketones, **4**, 315, 328
sulfones, **4**, 332
sulfonic acids, **4**, 331
thiocarbonyls, **4**, 319, 328
reactions, with lead tetracetate, **5**, 731
synthesis of, **4**, 315
See also Hemithioaminals
S,N-Acetals, acyclic, synthesis of, **4**, 318
S,N-Acetals, *N*-acyl
from
α-chloroalkylamides, **4**, 327
chloromethyl phenyl sulfide, **4**, 329
synthesis of, **4**, 325
S,N-Acetals, ketene
from
alkynes, **4**, 905
bis(1,1-alkylthio)alkenes, **4**, 899
1-chloro enamines, **4**, 904
ketene *S,S*-acetals, **3**, 265
nitriles, **4**, 905
thioamides, **4**, 899
α,β-unsaturated thioamides, **4**, 901
reactions, with isothiocyanates, **5**, 597
synthesis of, **4**, 898
S,P-Acetals
chemistry of, **4**, 345
from
aldehydes, **4**, 345

alkenes, **4**, 346
diazoalkanes, **4**, 346
ketones, **4**, 345
phosphorus ylides, **4**, 346
thiocarbonyls, **4**, 345
S,S-Acetals, ketene, as starting materials, for ketene *S,N*-acetals, **3**, 265
S,X-Acetals, ketene, as starting materials, for thioketenes, **3**, 550
Te,N-Acetals
from, iminium salts, **4**, 334
synthesis of, **4**, 333
Te,P-Acetals
from, phosphorus ylides, **4**, 348
synthesis of, **4**, 347
Acetamidate, trichloro-, as starting material, for alkyl bromides, **2**, 32
Acetamide
reactions
with bromine, **6**, 506
with diphenylantimony chloride, **5**, 369
with trimethylsilyl chloride, **5**, 379
Acetamide, *N*-acetyl-, reactions, with dimethylbis(pentafluorophenyl)silane, **5**, 719
Acetamide, *N*-acetyltrifluoro-, reactions, with chlorophosphite, **4**, 1009
Acetamide, *N*-alkylidene-, reactions, with phosphorocyanatidites, **4**, 493
Acetamide, *O,N*-bis(*t*-butyldimethylsilyl)-, synthesis of, **5**, 719
Acetamide, *N,N*-bis(difluorophosphino)-, synthesis of, **5**, 713
Acetamide, bis(trimethylsilyl)-, reactions, with bromo difluorophosphine, **5**, 713
Acetamide, *N,O*-bis(trimethylsilyl)- (BSA)
reactions
with *N*-bromosuccinimide, **5**, 378
with *S,S*-diethyl dithiochlorophosphite, **5**, 369
with phosphorus bromodifluoride, **5**, 369
with phosphorus trichloride, **5**, 369
with tetrahalotellurium, **5**, 349
as starting material, for *N*-silyl-*N*-phosphoamides, **5**, 369
synthesis of, **5**, 379
Acetamide, *N,O*-bistrimethylsilyltrifluoro-, as silylating agent, **5**, 226
Acetamide, *N*-bromo-
for the electrophilic bromination of arenes, **2**, 626
reactions, with dialkyl sulfides, **2**, 159
Acetamide, 2-bromo-2-cyano-, use in the synthesis of dialkyl phosphates, **2**, 106
Acetamide, chloro-, as starting material, for chloroacetonitrile, **3**, 627
Acetamide, *N*-chloro-, reactions, with tristrimethylsilylphosphine, **4**, 483
Acetamide, cyano-
deprotonation of, **4**, 872
reactions, with 2-acylcyclohexanones, **3**, 672
thionation of, **5**, 574
use in the synthesis of heterocyclic aromatic nitriles, **3**, 669
Acetamide, α-diazo-, from, methyl diazoacetate, **3**, 475
Acetamide, diazocyano-, as starting material, for thiadiazoles, **5**, 571
Acetamide, *N,N*-dichlorotrifluoro-, synthesis of, **5**, 317
Acetamide, 2,2-dicyano-*N*-phenyl-, reactions, with secondary α-chloro enamines, **4**, 792
Acetamide, dimethyl-, use in the synthesis of aryl sulfides, **2**, 711
Acetamide, *N,N*-dimethyl-, reactions, with ethyl chloroformate, **4**, 882
Acetamide, *N,N*-dimethylbis(trimethylgermyl)-, synthesis of, **4**, 625, 628
Acetamide, *N*-(2,6-dimethylphenyl)-*N*-isothiocyanatomethylchloro-, reactions, with sodium hydrogen sulfide, **6**, 565
Acetamide, *N,N*-dimethyl-2-trimethylsilyl-, deprotonation of, **1**, 283
Acetamide, *N*-ethyl-, oxidation of, **4**, 303
Acetamide, fluorodinitro-, synthesis of, **6**, 57
Acetamide, *N*-(hydroxymethyl)-, synthesis of, **4**, 296
Acetamide, 4-methoxyphenyl-, from, 4-methoxybenzoyl chloride, **5**, 276
Acetamide, *N*-methyl-*N*-(t-butyldimethylsilyl)trifluoro- (MTBSTFA), as silylating agents, **2**, 263
Acetamide, *N*-methyl-*N*-nitroso-, reactions, with potassium hydroxide, **3**, 467
Acetamide, *aci*-nitro-, reactions, with thionyl chloride, **5**, 819
Acetamide, *N*-phenyl-, synthesis of, **5**, 254
Acetamide, α-phenyl-, reactions, with [methoxy(tosyloxy)iodo]-benzene, **5**, 321
Acetamide, trichloro-, reactions, with diphenyldimethoxyselenium, **5**, 340
Acetamide, *N*-(2,2,2-trichloro-1-hydroxyethyl)trichloro-, from, imidoyl chlorides, **5**, 657
Acetamide, trifluoro-
alkylation of, **2**, 299
Hofmann rearrangement, with sodium hypobromite, **6**, 218
reactions, with diaryltellurium oxide, **5**, 349
Acetamide, tris(trimethylgermyl)-, synthesis of, **6**, 185
Acetamide *O*-methyl oxime, *N*-heteroaryl-*N*-methyl-, synthesis of, **5**, 836
Acetamide *O*-methyl oximes, *N*-heteroaryl-, methylation of, **5**, 836
Acetamide oximes, *N*-heteroaryl-, synthesis of, **5**, 836
Acetamides
from, acetic anhydrides, **5**, 277
reactions
with hypochlorites, **3**, 668
with malonaldehydes, **3**, 75
with *N*-sulfonyl isocyanates, **5**, 767
with trimethylchlorosilane, **5**, 719
as starting materials, for α-diazo penicillins, **3**, 471
Acetamides, *N*-acyloxyethyl-*N*-cyano-, from, 4,5-dihydro-2-methyloxazole, **5**, 1123
Acetamides, *N*-acylthioethyl-*N*-cyano-, from, 4,5-dihydro-2-methylthiazole, **5**, 1123
Acetamides, *N*-alkylhalogen
reactions
with boron trihalides, **5**, 660
with dimethylboron bromide, **5**, 660
Acetamides, *N*-alkylisocyano-, synthesis of, **3**, 706
Acetamides, allenyltrichloro-, synthesis of, **1**, 993
Acetamides, α-allyloxy-, 2,3 sigmatropic rearrangement of, **1**, 405
Acetamides, *N*-arylhalogen
reactions
with boron trihalides, **5**, 660
with dimethylboron bromide, **5**, 660
Acetamides, 2-aryl-2-isocyano-, synthesis of, **3**, 706
Acetamides, *N*-aryltrifluoro-, reactions, with bromodimethylborane, **5**, 717
Acetamides, α-bromo-, synthesis of, **4**, 99
Acetamides, cyano-, pyrolysis of, **5**, 997
Acetamides, dialkylboro-, reactions, with aminoboranes, **5**, 382
Acetamides, diaryl-, reactions, with phosphorus pentachloride, **5**, 656
Acetamides, dichloro-, reactions, with trialkylphosphines, **2**, 1047
Acetamides, dicyano-, as starting materials, for ketene aminals, **4**, 977
Acetamides, *N,N*-disubstituted, reactions, with oxalyl chloride, **5**, 996
Acetamides, nitroso-, reactions, with barium oxide, **3**, 471
Acetamides, *N*-nitroso-
as starting materials, for diazopropyne, **3**, 490
thermal decomposition of, **1**, 637

Acetamides, *O*-silylvinyl-, rearrangement of, **5**, 376
Acetamides, *N*-(1,2,2,2-tetrachloroethyl)-, from, imidoyl chlorides, **5**, 657
Acetamides, trichloro-
reactions, with tributylphosphine, **4**, 791
as starting materials, for α,β-trichloroenamines, **2**, 1047
Acetamides, α,α,α-trichloro-, reactions, with sulfur, **5**, 602
Acetamides, trifluoro-
reactions, with alkylidene phosphoranes, **2**, 751
as starting materials, for isothiocyanates, **5**, 1030
Acetamidine, reactions, with tripropylborane, **5**, 778
Acetamidine, *N*,*N*′-dimethyl-
reactions
with trialkylaluminum, **5**, 780
with trialkylgallium, **5**, 780
with trialkylindium, **5**, 780
Acetamidine, thiobenzoyl-, synthesis of, **5**, 760
Acetamidine, *N*,*N*,*N*′-trifluoro-, synthesis of, **5**, 762
Acetamidine hydrochlorides
reactions
with benzoyl chloride, **5**, 758
with potassium chloride, **5**, 763
Acetamidines, reactions, with dimethyl sulfoxide, **6**, 55
Acetamidines, *N*′-acyl-*N*,*N*-dimethyl-, synthesis of, **5**, 759
Acetamidines, 2,2-dichloro-2-(dialkoxyphosphinyl)-, synthesis of, **6**, 58
Acetamidines, *N*,*N*′-disubstituted, synthesis of, **5**, 753
Acetamidomalonate, diethyl
as starting material, for diethyl aminomalonate, **5**, 90
synthesis of, **5**, 90
Acetamidrazone, N^1,N^1-dimethyl-, alkylation of, **5**, 772
Acetamidrazone, phenyl-, synthesis of, **5**, 769, 848
Acetamidrazone, *N*-thionoacetyl amino-, synthesis of, **5**, 640
Acetamidrazones, synthesis of, **5**, 773
Acetanilide, 2-chloroaceto-, reactions, with arenediazonium salts, **5**, 843
Acetanilide, *N*-cyano-, synthesis of, **5**, 1123
Acetanilide, dichloro-, use in the synthesis of formazans, **5**, 859
Acetanilide, α-nitro-, synthesis of, **3**, 685
Acetanilide, *N*-nitroso-
as starting material
for arylazo-functionalized ylides, **5**, 905
for phenyl radicals, **6**, 323
Acetanilides
acylation of, **3**, 299
reactions, with amides, **5**, 754
as starting materials, for sulfenyl chlorides, **5**, 491
Acetanilides, *N*-chloro-, rearrangement of, **5**, 315
Acetanilides, α-mercapto-
chlorination of, **5**, 245
as starting materials, for α-chloroalkanesulfenyl chlorides, **4**, 71
Acetate
rearrangement of, **1**, 923
as starting material, for hydroxymethylsilanes, **4**, 352
Acetate, acetyl chloro-, synthesis of, **5**, 193
Acetate, acetyl trifluoro-, synthesis of, **5**, 193
Acetate, alkyl, reactions, with octacarbonyldicobalt, **3**, 20
Acetate, ammonium
reactions
with aldehydes, **5**, 600
with ketones, **5**, 600
with salicyl aldehyde, **4**, 438
with thioimidic esters, **5**, 757
Acetate, anhydrous mercuric, reactions, with acetonitrile, **6**, 207
Acetate, benzenetellurinyl, synthesis of, **5**, 210
Acetate, benzoyl diazo-, use in the synthesis of diazonium salts, **2**, 799
Acetate, *t*-butyl aceto-, reactions, with tosyl azide, **3**, 475
Acetate, *t*-butyl azido-
reactions, with thiols, **2**, 768
as starting material, for glycine *t*-butyl ester, **5**, 90
Acetate, *t*-butyl bis(trimethylsilyl), as starting material, for bistrimethylsilylketene, **3**, 537
Acetate, *t*-butyl diazo-, synthesis of, **3**, 471, 475
Acetate, *t*-butyl diazoaceto-, reactions, with sodium methoxide, **3**, 475
Acetate, *t*-butyl diazonitro-
reactions, with trifluoroacetic acid, **3**, 480
synthesis of, **3**, 480
Acetate, *t*-butyl (diphenylphosphinyl)isocyano-, use in the synthesis of steroids, **3**, 718
Acetate, *t*-butyl isocyano-
reactions, with amide diethyl acetals, **3**, 717
synthesis of, **3**, 704
Acetate, *t*-butyl trimethylgermyl, deprotonation of, **4**, 663
Acetate, carboxymethyl dithiophenyl, as thioacylating reagent, **5**, 586
Acetate, chiral, synthesis of, **1**, 458
Acetate, chlorodifluoro-, reactions, with tributylphosphine, **6**, 695
Acetate, chromyl, for the oxidation of semicarbazones, **4**, 441
Acetate, crassin, synthesis of, **1**, 1002
Acetate, 6-cyano-2-picolyl-*N*-oxide, oxidation of, **4**, 206
Acetate, cyclohexyl trifluoro-, pyrolysis of, **1**, 774
Acetate, cyclopentanone enol, reactions, with tetrafluorohydrazine, **5**, 314
Acetate, dimethylaminomethyl, synthesis of, **4**, 305
Acetate, 4-(dimethylamino) phenylmercuric, reactions, with potassium cyanide, **5**, 1141
Acetate, dimethyltin bis(ethyl) diazo-, synthesis of, **3**, 484
Acetate, 7(*E*),9(*Z*)-dodecadien-1-yl, synthesis of, **1**, 253
Acetate, ethyl
from, α-fluoroalkyl esters, **4**, 52
reactions
with benzaldehyde, **5**, 104
with phenylselenenyl chloride, **1**, 606
as starting material
for ketenes, **3**, 535
for triethyl trithioorthoacetate, **6**, 84
use in the synthesis of alkylmanganese(II) compounds, **2**, 574
Acetate, ethyl aceto-
acidity of, **1**, 853
alkylation of, **1**, 165
dichlorination of, **4**, 12
double alkylation of, **1**, 107
reactions, with isothiocyanates, **5**, 593
as starting material, for acetoacetic acid, **5**, 62
Acetate, ethyl bromo-
reactions, with tris(cyclopentyl)borane, **4**, 832
as starting material, for ketene silyl acetals, **4**, 831
Acetate, ethyl bromocyano-, reactions, with *N*,*N*-dimethyl-*N*′-phenylthiourea, **6**, 559
Acetate, ethyl bromodiazo-, synthesis of, **3**, 478
Acetate, ethyl 2-chloro-, reactions, with arenediazonium salts, **5**, 843
Acetate, ethyl α-chlorobenzoyl, reactions, with baker's yeast, **1**, 8
Acetate, ethyl chlorodiazo-, synthesis of, **3**, 478
Acetate, ethyl chloro(hydroxyimino)-, as starting material, for ethoxycarbonylformonitrile oxide, **3**, 684
Acetate, ethyl α-chlorophenyl, hydrolysis of, **5**, 55
Acetate, ethyl cyano-
reactions
with acyl isothiocyanates, **5**, 631
with carbonyl sulfide, **4**, 833
with cinnamaldehyde, **3**, 366
with dimethyl acetylenedicarboxylate, **1**, 537
with ethanethiol, **4**, 905
with isobutylene oxide, **4**, 883
with isothiocyanates, **5**, 593
Acetate, ethyl cyanodiazo-, synthesis of, **3**, 477

Acetate, ethyl cyclohexylidene, from, cyclohexanone, **5**, 103
Acetate, ethyl diazo-
  from, glycine ethyl ester hydrochloride, **3**, 468
  metallation of, **3**, 484, 486
  reactions
    with aldehydes, **3**, 474
    with benzenetellurenyl bromide, **4**, 93
    with bis(dimethylamino)dimethylstannane, **3**, 484
    with dihaloplatinum complexes, **4**, 151
    with dimethylaminodimethylarsane, **3**, 482
    with dinitrogen pentoxide, **3**, 479
    with Grignard reagents, **3**, 486
    with mercury bis[bis(trimethylsilyl)amide], **3**, 488
    with methylmagnesium bromide, **3**, 486
    with 4-nitrobenzenediazonium chloride, **5**, 843
    with *N*-phenylselenophthalimide, **4**, 335
    with potassium selenocyanates, **5**, 1056
    with trialkynylboranes, **1**, 1033
  silylation of, **5**, 873
  as starting material
    for dihydro-1,2,4,5-tetrazine-3,6-dicarboxylic acid, **5**, 93
    for *β*-keto esters, **1**, 389
  synthesis of, **3**, 471, 473
Acetate, ethyl diazodimethylarsyl, synthesis of, **3**, 482
Acetate, ethyl diazonitro-, reactions, with 1,1-dimethylcyclohexane-3,5-dione, **3**, 473
Acetate, ethyl diazopentamethyldisilanyl, synthesis of, **3**, 483
Acetate, ethyl diazotrimethylsilyl, reactions, with aldehydes, **3**, 474
Acetate, ethyl dibromofluoro-, reactions, with sodium azide, **6**, 54
Acetate, ethyl dibromonitro-, reactions, with sodium azide, **6**, 57
Acetate, ethyl dicyano-, reactions, with hydrogen chloride, **4**, 792
Acetate, ethyl(dimethylantimony)diazo-, synthesis of, **3**, 482
Acetate, ethyl(dimethylbismuth)diazo-, synthesis of, **3**, 482
Acetate, ethyl(dimethylsulfuranylidene), reactions, with trichloroacetonitrile, **5**, 494
Acetate, ethyl fluoro(phenoxy)-, bromination of, **6**, 43
Acetate, ethyl iododiazo-, synthesis of, **3**, 478
Acetate, ethyl isocyano-
  reactions
    with amines, **3**, 706
    with *N*,*N*-dimethylformamide acetal, **3**, 720
  as starting material, for 1-isocyanocyclopropanecarboxylates, **3**, 704
Acetate, ethyl, lithium salt, trimethylgermylation of, **6**, 185
Acetate, ethyl nitroso(triphenylphosphoranylidene), synthesis of, **5**, 904
Acetate, (ethylthio)fluoro-, bromination of, **6**, 47
Acetate, ethyl trifluoro-, as starting material, for trifluoromethyl ketones, **3**, 150
Acetate, ethyl (trimethylgermyl)
  reactions, with carboxylic compounds, **1**, 291
  synthesis of, **6**, 185
Acetate, ethyl (trimethylsilyl), reactions, with Grignard reagents, **1**, 650
Acetate, ethyl (triphenylphosphoranylidene), reactions, with acid chlorides, **5**, 103
Acetate, ethyl tris(trimethylgermyl), synthesis of, **6**, 185
Acetate, (±)-fauronol, synthesis of, **3**, 325
Acetate, fluorocarbonylsulfenic trifluoro-, synthesis of, **5**, 246
Acetate, 2-fluoroheptyl, hydrolysis of, **5**, 250
Acetate, fluoro(phenylthio), bromination of, **6**, 47
Acetate, (6*E*,11*Z*)-hexadeca-6,11-dien-1-yl, synthesis of, **1**, 455
Acetate, iodomagnesium ethyl diazo-, synthesis of, **3**, 486
Acetate, isoprenyl, as dehydrating agent, in the synthesis of cyclic anhydrides, **5**, 199
Acetate, isopropenyl
  reactions
    with alcohols, **4**, 188
    with carbonyls, **2**, 675
    with malonic acids, **4**, 204
  use in *trans*-esterification reactions, **5**, 135
Acetate, isopropyl
  reactions, with isopropoxymethanetriamines, **4**, 973
  as starting material, for acetyl iodide, **5**, 20
  synthesis of, **5**, 142
Acetate, mercuric
  reactions
    with acetic anhydride, **6**, 207
    with longifolene, **4**, 748
    with methylenediphosphines, **6**, 371
    with methylenediphosphonates, **6**, 371
    with 1-propenylmercury acetate, **4**, 697
  as reagent, in complete mercuration of cyclopentadiene, **4**, 697
Acetate, *α*-metallated ethyl isocyano-
  reactions
    with aldehydes, **3**, 717
    with ketones, **3**, 717
Acetate, methanesulfinyl, synthesis of, **5**, 209
Acetate, methoxymethyl, from, dimethoxymethane, **4**, 211
Acetate, methyl benzoyl diazo-, use in the synthesis of diazonium salts, **2**, 799
Acetate, methyl bromodifluoro-
  as starting material, for iodotrifluoromethane, **6**, 219
  thermal decomposition of, **6**, 227
Acetate, methyl 2-chloro-2-cyclopropylidene
  reactions
    with diazomethane, **4**, 109
    with diphenyldiazomethane, **4**, 109
Acetate, methyl chlorodifluoro-
  decomposition of, **4**, 737
  as starting material
    for difluorocarbene, **6**, 224
    for iodotrifluoromethane, **6**, 219
  thermal decomposition of, **6**, 227
Acetate, methyl cyano-
  reactions
    with hydroxo–platinum complexes, **3**, 568
    with magnesium methoxide, **3**, 653
  as starting material, for ketene dithioacetals, **4**, 843
Acetate, methyl diazo-, as starting material, for *α*-diazo acetamide, **3**, 475
Acetate, methyl dichlorofluoro-, reactions, with sodium hydride, **4**, 752
Acetate, methyl *α*-dichlorophosphine-*α*-cyano-, synthesis of, **3**, 639
Acetate, methyl dithiophenyl, reactions, with amines, **5**, 586
Acetate, methyl fluorosulfonyldifluoro-
  reactions, with copper(I) iodide, **6**, 15
  trifluoromethylthiolation with, **6**, 232
Acetate, methyl (*Z*)-*β*-ionylidene, synthesis of, **1**, 748
Acetate, methyl nitro-, synthesis of, **2**, 398
Acetate, methyl perfluoroadamantane, as starting material, for trifluoromethyl esters, **6**, 230
Acetate, methyl phenyldiazo-, as starting material, for ketenimines, **3**, 596
Acetate, methyl trichloro-, as alkylating agent, **2**, 683
Acetate, *p*-nitrophenyl diazo-, use in the synthesis of *α*-diazo amides, **3**, 476
Acetate, norbornyl, synthesis of, **1**, 381
Acetate, 9-oxonon-1-yl, reactions, with (*E*)-pent-2-entyltriphenylphosphonium bromide, **1**, 734
Acetate, phenyl
  reactions
    with phosphorus pentachloride, **4**, 766
    with phosphorus trichloride, **4**, 742
  rearrangement of, **2**, 656
Acetate, (+)-8-phenylmenthyl isocyano-, synthesis of, **3**, 704
Acetate, phenyl selenyl trifluoro-, use in the synthesis of nitroalkenes, **2**, 786

Acetate, pregnenolone, reactions, with trioctynylboron, **1**, 1050
Acetate, pyrrolidinium, reactions, with 4-*aci*-nitrocrotonic aldehydes, **2**, 789
Acetate, thallic trifluoro-, for the synthesis of aryl iodides, **2**, 631
Acetate, thallous, for the *o*-iodination of phenols, **2**, 631
Acetate, trichlorovinyl dichloro-, from, trichloroacetyl chloride, **4**, 768
Acetate, trifluoromethyl, synthesis of, **6**, 230
Acetate, triisopropylsilylphenyl, as starting material, for acid bromides, **5**, 17
Acetate, trimethyllead, reactions, with 2-lithio-1,3-oxathiane, **6**, 133
Acetate, trimethyl orthodimethoxy, synthesis of, **6**, 74
Acetate, trimethyl orthodiphenyl, synthesis of, **6**, 74
Acetate, trimethyl orthophenyl, elimination of methanol from, **4**, 825
Acetate, trimethylsilyl, as reagent, for the synthesis of acetoxyiodobenzene triflate, **5**, 205
Acetate, trimethylsilyl trichloro-, reactions, with aldehydes, **6**, 26
Acetate, trimethylsilyl trifluoro-, elimination of, for vinyl iodides, **2**, 628
Acetate, trimethylsilyl (trimethylsilyloxy), reactions, with carbonyls, **4**, 213
Acetate, tris(trimethylsilylethynl)methyl, reactions, with triethylaluminum, **1**, 970
Acetate, vinyl
  hydroformylation of, **1**, 297
  reactions
    with alcohols, **4**, 188
    with phosphoric acid, **2**, 693
    with phosphorus pentachloride, **4**, 766
  use in *trans*-esterification reactions, **5**, 135
Acetate anions, elimination of, **1**, 582
Acetate anions, cyano-, for the displacement of alcohols, **1**, 179
Acetate elimination, transition metal-catalyzed, **1**, 582
Acetates
  oxyselenylation of, **3**, 175
  pyrolysis of, **1**, 773
  synthesis of, **1**, 698
Acetates, aceto-, reactions, with glutarates, **2**, 662
Acetates, alkenyl
  demercuration of, **1**, 67
  from, alkenyl halides, **2**, 682
  reactions, with methyllithium, **2**, 696
  as starting materials, for diols, **1**, 67
Acetates, alkyl, pyrolyses of, **1**, 773
Acetates, alkyl cyano-
  reactions, with aryl halides, **3**, 625
  use in the synthesis of nitriles, **3**, 620
Acetates, alkyl dicyano-, use in the synthesis of nitriles, **3**, 621
Acetates, alkynylmercuric, synthesis of, **2**, 1097
Acetates, allyl, carbonylation of, **1**, 192
Acetates, allylic
  as allylating reagents, **2**, 74
  cyanation of, **1**, 192
  metal-catalyzed rearrangement of, **1**, 838
  as nucleophilic allylating reagents, **2**, 80
  oxidative cleavage of, **3**, 162
  reactions
    with alkenes, **1**, 195
    with alkynes, **1**, 195
    with allenyltrialkylstannanes, **1**, 1053
    with hydroxylamines, **2**, 344
    with samarium(II) iodide, **1**, 195
  3,3 sigmatropic rearrangement of, **1**, 836
  as starting materials, for amines, **2**, 308
  substitution of, **1**, 476
  umpolung, **1**, 195
Acetates, allylic aceto-, rearrangement of, **1**, 832
Acetates, arylmethyl, oxidation of, **4**, 206
Acetates, azido fluoro-, synthesis of, **4**, 109
Acetates, benzylic, reactions, with silyl enol ethers, **1**, 202
Acetates, bis(cyano)enol, synthesis of, **5**, 359
Acetates, bistrifluoro-, reactions, with trimethylsilyl cyanide, **3**, 701
Acetates, 1-bromocyclopropyl, synthesis of, **4**, 57
Acetates, 2-butenyl, asymmetric allylic amination of, **2**, 301
Acetates, chloro-, pyrolysis of, **1**, 774
Acetates, α-chloroaceto-, reactions, with isoamyl nitrite, **5**, 824
Acetates, cycloalkyl, as starting materials, for alkenes, **1**, 773
Acetates, diazo-
  decomposition of, **1**, 825
  photolysis of, **4**, 385
  reactions
    with dimethylaminotrimethylgermane, **3**, 483
    with triethylmethoxystannane, **3**, 484
    with α,β-unsaturated acyl chlorides, **3**, 475
Acetates, α-diazo-
  reactions
    with phenyllithium, **1**, 486
    with vinyl chloroboranes, **1**, 486
  synthesis of, **3**, 470
    via Bamford–Stevens reactions, **3**, 470
Acetates, dicyano-, as starting materials, for ketene aminals, **4**, 977
Acetates, dienyl, regioselectivity of, **1**, 192
Acetates, enol
  allylation of, **3**, 139
  bromination of, **3**, 54
  as intermediates, **1**, 657
  nitration of, **3**, 190
  oxidation of, **1**, 558
  reactions
    with alkoxydialkylstibines, **2**, 494
    with alkoxystannanes, **2**, 701
    with aminodialkylstibines, **2**, 494
    with aminostilbines, **2**, 695
    with arylsulfonyl peroxides, **3**, 163
    with Selectfluor reagents, **2**, 16
    with selenium compounds, **1**, 606
    with tributyltin methoxide, **1**, 558
  as starting materials
    for acetals, **4**, 188
    for α-iodo ketones, **3**, 155
    for nitro compounds, **2**, 392
    for α-selenocarbonyls, **2**, 282
  synthesis of, **4**, 915
Acetates, enol trifluoro-
  reactions
    with alkoxydialkylstibines, **2**, 494
    with alkoxystilbines, **2**, 695
    with aminodialkylstibines, **2**, 494
Acetates, ethyl bromodiazo-, synthesis of, **5**, 844
Acetates, ethyl chlorodiazo-, synthesis of, **5**, 844
Acetates, ethyl diazo-, synthesis of, **3**, 483
Acetates, ethyl iododiazo-, synthesis of, **5**, 844
Acetates, fluorinated enol, synthesis of, **4**, 768
Acetates, α-fluoroalkyl, synthesis of, **4**, 52
Acetates, D-glucal, reactions, with alkynyltrimethylsilanes, **1**, 1026
Acetates, halo-, reactions, with cuprates, **1**, 465
Acetates, iododifluoro-, reactions, with trimethyl(vinyl)silanes, **4**, 139
Acetates, ketene silyl
  reactions
    with bromotrichloromethane, **4**, 743
    with tetrachloromethane, **4**, 743
Acetates, phosphonofluoro-, synthesis of, **4**, 131
Acetates, propargyl, reactions, with trialkynylalanes, **1**, 1025
Acetates, propargyl aceto-, pyrolysis of, **3**, 233
Acetates, propargylic

reactions
with acetic acid, **4**, 206
with organocuprates, **1**, 960
with pentynoates, **1**, 975
with terminal alkynes, **1**, 1025
silylcupration of, **2**, 940
as starting materials, for α-acetoxy ketones, **3**, 162
Acetates, thiol, cleavage of, **2**, 230
Acetates, trifluoro-
hydrolysis of, **2**, 86
use in elimination reactions, **1**, 981
Acetates, α-unsubstituted bromo-, reactions, with nitriles, **3**, 461
Acetates, vinyl
reactions
with arenesulfenyl halides, **4**, 56
with benzeneselenenyl fluoride, **4**, 52
with bromine, **4**, 57
as starting materials
for α-bromoalkyl esters, **4**, 57
for cyanohydrins, **3**, 627
for α-fluoroalkyl esters, **4**, 51
synthesis of, **3**, 537
o-Acetates, reactions, with alkynylmagnesium halides, **1**, 1026
O,N-Acetates, ketene, synthesis of, **4**, 881
Acetate salts, tetraalkylammonium, for the acetolysis of alkyl tosylates, **2**, 87
Acethydroxamoyl bromide, trifluoro-, synthesis of, **5**, 664
Acethydroxamoyl chloride, synthesis of, **5**, 666
Acetic acid
n-amyl esters of, hydrolysis of, **5**, 30
electrochemical fluorination of, **6**, 214
reactions
with aliphatic triazenes, **1**, 1119
with alkenylboranes, **1**, 1102
with bis(methylseleno)alkynes, **4**, 841
with bis(1-naphthyl)bismuthinate, **5**, 630
with chlorodiphenylphosphine, **6**, 118
with chlorosulfuric acid, **4**, 282
with diazo hydroxides, **3**, 462
with dichloro(chloromethyl)[trimethylsilyl]-methylsilane, **5**, 224
with diphenylantimony(III) chloride, **5**, 222
with diphenylmercury, **1**, 65
with hydroiodic acid, **1**, 432
with norbornyl brosylate, **1**, 381
with organoboranes, **1**, 58
with platinum oxide, **1**, 88
with propargylic acetates, **4**, 206
with 1-selenoalkynes, **4**, 841
with silirenes, **4**, 1051
with 1-telluroalkynes, **4**, 841
with triethylborane, **5**, 227
with *O*-trimethylsilyl thiocarboxylates, **5**, 539
as reagent, for the synthesis of α-hydroxylamine alkyl phosphonates, **5**, 789
for the reduction of amines, **1**, 34
as starting material, for ketenes, **3**, 535
synthesis of, **5**, 49
Acetic acid, aceto-, from, ethyl acetoacetate, **5**, 62
Acetic acid, *p*-aminophenyl-, synthesis of, **5**, 33
Acetic acid, benzoyl-, synthesis of, **5**, 70
Acetic acid, 2,2-bis(trifluoromethylthio)-, oxidation and dehydration of, **4**, 867
Acetic acid, bromo(hydroxyimino)-, hydrolysis of 320032.440
Acetic acid, 2-carboxyphenyl-2-oxo-, synthesis of, **5**, 74
Acetic acid, chloro-, reactions, with ketene, **5**, 193
Acetic acid, chlorodifluoro-, electrolysis of, **6**, 225
Acetic acid, α-chlorophenyl-
as starting material, for Grignard reagents, **4**, 147
synthesis of, **5**, 55
Acetic acid, chloro-, sodium salt, as starting material, for *S*-carboxylmethyl esters, **5**, 550
Acetic acid, cyano-
reactions
with cyclohexanone, **3**, 644
with ethyl orthoformate, **2**, 749
as starting material, for β-ketonitriles, **3**, 633
synthesis of, **5**, 95
Acetic acid, diazo-, reactions, with α-chloroenamines, **3**, 475
Acetic acid, dichloro-
from
chloral hydrate, **4**, 14; **5**, 59
trichloroacetic acid, **4**, 13
Acetic acid, di-(*p*-chlorophenyl)-, synthesis of, **5**, 38
Acetic acid, diethyl-, from, 3-pentanone, **5**, 85
Acetic acid, 6,11-dihydro[*b,e*]thiepin-11-on-3-yl, carbonation of, **5**, 82
Acetic acid, dithiophenyl-, synthesis of, **5**, 546
Acetic acid, N^1-fluoroimino-N^2,N^2-difluoroaminodifluoro-, synthesis of, **5**, 812
Acetic acid, α-mesityl-, synthesis of, **5**, 34
Acetic acid, 2-methoxy-2-trifluoromethyl-2-phenyl-, synthesis of, **5**, 12
Acetic acid, *p*-nitrophenyl-, from, *p*-nitrobenzyl cyanide, **5**, 94
Acetic acid, peroxytrifluoro-
for the oxidation of arenes, **2**, 650
as oxidizing reagent, **2**, 650
reactions, with dienes, **4**, 46
Acetic acid, phenyl-
silylation of, **5**, 224
as starting material, for 1,1'-dibromobenzyl sulfone, **4**, 83
synthesis of, **5**, 627
Acetic acid, tetrazolyl-, reactions, with 2,2-dimethylpropanoyl chloride, **5**, 193
Acetic acid, 2-thiohydroxy-, reactions, with ortho esters, **6**, 95
Acetic acid, thiol
reactions
with alkenes, **2**, 117
with allyl chlorides, **2**, 117
use in the synthesis of disulfides, **2**, 227
Acetic acid, trichloro-
decomposition of, **6**, 25, 30
reactions, with ortho esters, **4**, 1081
as starting material, for dichloroacetic acid, **4**, 13
use in the synthesis of acetals, **4**, 195
Acetic acid, trifluoro- (TFA)
bromination of, **6**, 218
as catalyst, for dienone–phenol rearrangement, **2**, 660
electrochemical fluorination of, **6**, 214
electrolysis of, **6**, 20
reactions
with alkylidene-1,3-dithianes, **4**, 252
with 2-(alkylsulfinyl)benzamides, **5**, 336
with amines, **5**, 659
with arenes, **1**, 475
with azo compounds, **1**, 154
with *t*-butyl diazonitroacetate, **3**, 480
with *t*-butyl isocyanide dichloride complexes, **5**, 963
with dimethyl sulfoxide, **5**, 209
with dithiols, **6**, 84
with enamines, **2**, 759
with ortho esters, **4**, 1081
with silanes, **1**, 281
as starting material
for cobaltacyclopentadiene complexes, **4**, 689
for trifluoromethyl radicals, **6**, 11
use in the cleavage of carbon–silicon bonds, **1**, 61
use in degermylation reactions, **1**, 449
use in dehydration reactions, **1**, 578
use in modified Gabriel syntheses, **2**, 299
use in the synthesis of alkanoic acids, **5**, 81

use in the synthesis of amines, **2**, 300
use in the synthesis of amino acids, **5**, 85
use in the synthesis of benzaldehydes, **3**, 91
use in the synthesis of *N*-fluoroimines, **5**, 807
Acetic acid, tris(acetoxymercurio)-, relationship with mercuretin, **6**, 207
Acetic acid amide, *N*-cyano phenoxy-, metal salts from, **5**, 723
1-Acetic acid anhydrides, adamantane-, synthesis of, **5**, 191
Acetic acid chloride, 2-chloro-2,2-dicyano-*N*-chloroimino-, synthesis of, **5**, 810
Acetic acid dianions, phenoxy-, reactions, with 9-borabicyclo[3.3.1]nonyls, **1**, 185
Acetic acid esters, 2-chloro-, reactions, with diazonium halides, **5**, 676
Acetic acid ethyl ester, benzoylfluoro-, synthesis of, **4**, 52
Acetic acids, as starting materials, for enamines, **2**, 742
Acetic acids, α-alkoxy-, synthesis of, **5**, 66
Acetic acids, alkyl-, synthesis of, **5**, 38
Acetic acids, α-alkyl-α-sulfonyl-, synthesis of, **4**, 81
Acetic acids, aryl-, as starting materials, for 2-chloro-2-tosylazostyrenes, **4**, 799
Acetic acids, α-aryl-α-sulfonyl-, synthesis of, **4**, 81
Acetic acids, α-bromo-, synthesis of, **5**, 56
Acetic acids, α-hydroxyaryl-, synthesis of, **5**, 67
Acetic acids, α-substituted α-hydroxyphenyl-, synthesis of, **5**, 80
Acetic acids, tertiary alkyl-, synthesis of, **5**, 38
Acetic acids, trigermyl-, synthesis of, **6**, 185
Acetic anhydride
for the acylation of oximes, **5**, 215
as dehydrating agent, for the synthesis of anhydrides, **5**, 184
hydrolysis of, **5**, 37
as oxidizing agent, for the synthesis of bisacetoxyiodobenzene, **5**, 204
reactions
with acid chlorides, **5**, 187
with aryl aldehyde, **5**, 98
with benzenetellurinic anhydride, **5**, 210
with chlorodiethylphosphine, **4**, 911
with dialkyl phosphites, **4**, 921
with 3,6-diphenyl-1,2,4,5-tetraaza-3,6-diborane, **5**, 383
with 1-ethoxyisochroman, **4**, 212
with formic acid, **6**, 85
with hexafluoroacetone imine, **4**, 304
with mercuric acetate, **6**, 207
with methoxyamine, **5**, 332
with 1-methoxy-2-benzopyran-4-one, **4**, 212
with 3-methylbenzenethiol, **6**, 84
with nitric acid, **6**, 365
with pyridine, **2**, 747
with sulfenamides, **2**, 361
with *sym*-diarylthioureas, **5**, 1035
with tellurium tetrachloride, **4**, 291
with thallium(III) oxide, **5**, 229
with ylides, **4**, 993
as starting material, for acetyl iodide, **5**, 19
synthesis of, **5**, 192
use in the synthesis of 1-aminoalkylphosphonic acids, **4**, 470
Acetic anhydride, α-alkyl-, reactions, with aryl aldehyde, **5**, 98
Acetic anhydride, diphenyl-, reactions, with diphenylketene, **4**, 829
Acetic anhydride, trifluoro- (TFAA)
for alcohol activation, **2**, 8
bromination of, **6**, 218
as catalyst, for dienone–phenol rearrangement, **2**, 660
as dehydrating agent, in the synthesis of cyclic anhydrides, **5**, 199
reactions
with diphenylphosphinic chloride, **5**, 220
with heptamethylsilazane, **5**, 377
with isocyanic acid, **5**, 995
with steroidal dienones, **2**, 660
use in the synthesis of *N*-fluoroimines, **5**, 807
use in the synthesis of mixed anhydrides, **5**, 193
Acetic anhydrides, as starting materials, for acetamides, **5**, 277
Acetic anhydrides, aminopropanoic, synthesis of, **5**, 194
Acetic esters, nitro-, reactions, with alkyl orthoformates, **2**, 787
Acetic nitrate, trifluoro-, synthesis of, **2**, 791
Acetimidate, *O*,*N*-bis(trimethylsilyl)trifluoro-, synthesis of, **5**, 721
Acetimidate, *O*-*t*-butyltrichloro-, reactions, with alcohols, **2**, 90
Acetimidate, *p*-methoxybenzyl trichloro-, use in the synthesis of benzyl ethers, **2**, 90
Acetimidate, propargylic trichloro-, thermolysis of, **1**, 993
Acetimidates, *t*-butyl-, reactions, with *t*-butyl hydroperoxide, **2**, 98
Acetimidates, diazo, synthesis of, **3**, 477
Acetimidates, tributylstannyl *N*-cyano-, synthesis of, **5**, 716
Acetimidoyl chloride, *N*-(α-chlorobenzyl)trichloro-, synthesis of, **5**, 659
Acetimidoyl chloride, *N*-chlorotrichloro-, synthesis of, **5**, 810
Acetimidoyl chloride, *N*-(dichlorophosphono)chloro-, synthesis of, **5**, 683
Acetimidoyl chloride, *N*-(2-hydroxyethyl)ethoxycarbonyl-, synthesis of, **5**, 661
Acetimidoyl chloride, *N*-(pentachloroethyl)trichloro-, synthesis of, **5**, 659
Acetimidoyl chloride, *N*-phenyltrifluoro-, synthesis of, **5**, 657
Acetimidoyl chloride, 1,2,2-trichlorocyano-, synthesis of, **5**, 832
Acetimidoyl chloride, *N*-2,2-trichlorocyano-, synthesis of, **5**, 809
Acetimidoyl chloride, *N*-(trichloroethenyl)trichloro-, photochlorination of, **5**, 659
Acetimidoyl chlorides, *N*-alkyltrifluoro-, synthesis of, **5**, 659
Acetimidoyl chlorides, *N*-aryltrifluoro-, synthesis of, **5**, 659
Acetimidoyl chlorides, 2,2,2-trihalo-, as starting materials, for *gem*-phosphorylamino phosphorylalkenes, **4**, 1008
Acetimidoyl fluoride, *N*-cyclohexyl-, synthesis of, **5**, 658
Acetimidoyl fluoride, perfluoroamino-, synthesis of, **5**, 663
Acetimidoyl iodide, *N*-phenyl-, synthesis of, **5**, 661
Acetoin, periodate oxidation of, **5**, 49
Acetonates, acetyl-, alkylation of, **2**, 577
Acetone
bromination of, **1**, 853
chlorination of, **1**, 853
electrochemical fluorination of, **6**, 214
iodination of, **1**, 853
mercuration of, **6**, 208
pyrolysis of, **3**, 535
reactions
with arsines, **2**, 1079
with carbenes, **3**, 314
with carbenoids, **4**, 146
with lithium chloride, **2**, 25
with methyl chloroformate, **5**, 630
with nitro compounds, **3**, 427
with potassium hypoiodite, **6**, 216
with red selenium, **6**, 93
with sodium hypobromite, **6**, 215
with 1-thioglycerol, **4**, 229
as starting material
for α-aminoisobutyric acid, **5**, 85
for ketenes, **3**, 535
Acetone, aceto-, reactions, with ammonia, **3**, 263

Acetone, acetyl-
acidity of, **1**, 853
reactions
with acyl isothiocyanates, **5**, 631
with hydrogen selenide, **3**, 397
with tellurium(IV) chloride, **2**, 690
Acetone, benzylidene-, reduction of, **1**, 86
Acetone, chloro-
reactions
with alkynes, **1**, 1035
with arenediazonium salts, **5**, 843
with diazonium halides, **5**, 676
as starting material, for oxyimidic chlorides, **5**, 823
Acetone, α-chloroacetyl-, reactions, with malononitrile, **3**, 672
Acetone, 1-chloro-1-(4-nitrophenyl)-, reactions, with diazonium halides, **5**, 676
Acetone, chloropentafluoro-
reactions, with perfluoro-2,3-dialkyloxaziridines, **5**, 825
use in the synthesis of perfluoroenol phosphate, **2**, 640
Acetone, 1-chloro-1-(4-toluenesulfonyl)-, reactions, with diazonium halides, **5**, 676
Acetone, diazo-, reactions, with 4-nitrobenzenediazonium chloride, **5**, 843
Acetone, 1,3-dibromo-, nitrosation of, **5**, 667
Acetone, α,α'-dibromo-, reactions, with iron nonacarbonyl, **1**, 364
Acetone, 3,3-dibromo-1,1,1-trifluoro-, reactions, with sodium acetate, **4**, 161
Acetone, 1,3-dichloro-, reactions, with caesium dithiolates, **2**, 133
Acetone, dichlorotetrafluoro-, reactions, with ethanol, **4**, 164
Acetone, 1,3-dichlorotetrafluoro-, use in the synthesis of oxyimidic halides, **5**, 825
Acetone, 1,3-dichloro-1,1,3,3-tetrafluoro-, reactions, with arenesulfinamides, **4**, 311
Acetone, geranyl-
reactions, with 1,4-bis(triphenylphosphonio)butane dibromide, **1**, 721
reduction of, **6**, 461
Acetone, hexachloro-
as chlorinating agent, for aryl chlorides, **2**, 621
for the chlorination of enamines, **3**, 152
decomposition of, **6**, 31
as starting material, for trichloromethyl anions, **6**, 31
Acetone, hexafluoro-
chloroethynylation of, **2**, 1016
reactions
with ammonia, **4**, 297
with arsines, **4**, 344
with *N*,*N'*-bis(tributylsilyl)carbodiimides, **5**, 1083
with *N*,*N'*-bis(trimethylsilyl)carbodiimides, **5**, 1083
with hydrogen sulfide, **4**, 215
with methylarsine, **2**, 482
with phosphines, **4**, 337
with thionoamides, **5**, 633
synthesis of, **6**, 16
Acetone, nitro-
reactions, with imines, **3**, 258
as starting material, for 2-chloro-2-nitro enamines, **4**, 796
Acetone, perfluoro-, reactions, with potassium fluoride, **5**, 987
Acetone, stannyl-, reactions, with trimethylbromogermane, **2**, 700
Acetone, thioacetyl-, synthesis of, **3**, 344
Acetone, *p*-tolyl-, synthesis of, **1**, 738
Acetone, trichlorotrifluoro-, as reagent, for the synthesis of anhydrides, **5**, 196
Acetone, triethylgermyldiazo-, reactions, with bis(triethylgermyl)mercury, **6**, 185
Acetone, 1,1,1-trifluoro-
reactions, with 2-chloroethanol, **4**, 164
as starting material
for α-bromo alcohols, **4**, 42
for Michael acceptors, **4**, 766
Acetone, trifluoroacetoxymercuricperfluoro-, synthesis of, **6**, 64
Acetone, 1,1,1-tris(triethylgermyl)-, synthesis of, **6**, 185
Acetone enol, perfluoro-, *See* Propen-2-ols, perfluoro-
Acetoneimine, *N*-thionoacyl perfluoro-, synthesis of, **5**, 633
Acetone oxime, reactions, with benzyl chloride, **2**, 103
Acetonitrile
condensation of, **3**, 654
electrochemical fluorination of, **6**, 214
reactions
with amines, **5**, 750
with anhydrous mercuric acetate, **6**, 207
with *t*-butyllithium, **2**, 1058
with cyanotrimethylsilane, **2**, 305
with cyclohexanone, **3**, 644
with α-methylene lactones, **3**, 652
with thiobenzamide, **5**, 760
as solvent, for the synthesis of ketene aminals, **4**, 971
as starting material
for dichlorocerium anions, **2**, 69
for *N*-fluoroimino fluorides, **5**, 807
trimethylgermylation of, **6**, 185
use in reduction reactions, **1**, 74
use in the synthesis of alkanethiols, **2**, 121
use in Wittig reactions, **4**, 738
Acetonitrile, aryl-
from
α-cyanomethyl phosphonate, **3**, 626
methylthioacetonitrile, **3**, 626
Acetonitrile, α-aryl-α-thiomethyl-, synthesis of, **3**, 626
Acetonitrile, bromo-
reactions
with arylzinc chlorides, **3**, 625
with tributylantimony, **3**, 639
as starting material
for cyanomethyl(aryl) sulfones, **3**, 635
for β-hydroxynitriles, **3**, 632
Acetonitrile, *t*-butyl-, reactions, with sodium amide, **3**, 567
Acetonitrile, (*R*)-(+)-α-[*t*-butyldimethylsilyl)oxyl]phenyl-, reactions, with methylmagnesium iodide, **2**, 316
Acetonitrile, *o*-carboxyphenyl-, synthesis of, **3**, 626
Acetonitrile, chloro-
from, chloroacetamide, **3**, 627
reactions
with dibutyltellurium, **3**, 637
with selenocyanate, **3**, 636
with sodium *N*,*N*-dimethyl dithiocarbamate, **6**, 561
with sodium trithiocarbonate, **3**, 634
with triphenylphosphine, **3**, 639
as starting material, for α-thionitriles, **3**, 634
Acetonitrile, chlorodiphenyl-
reactions
with triethyl phosphite, **3**, 568
with triphenylphosphine, **5**, 860
Acetonitrile, diazo-
reactions, with trialkylboranes, **3**, 622
synthesis of, **3**, 477
Acetonitrile, diazophenyl-, synthesis of, **3**, 477, 478
Acetonitrile, dibromo-, as starting material, for cyano thioformaldehyde, **3**, 334
Acetonitrile, dichloro-
anions derived from, **1**, 126
reactions, with trialkylboranes, **3**, 622, 627
Acetonitrile, diphenyl-
reactions, with ethyl propiolate, **1**, 537
synthesis of, **5**, 860
Acetonitrile, ferrocenyl-
reactions, with methylmagnesium chloride, **3**, 567
synthesis of, **3**, 626
Acetonitrile, fluorodinitro-, hydrolysis of, **6**, 57
Acetonitrile, α-lithiotrimethylgermyl-, stability of, **4**, 662

Acetonitrile, α-(*N*-methylanilino)-, as starting material, for α-cyano enamines, **2**, 749
Acetonitrile, α-(*N*-methylanilino)-α-(trimethylsilyl)-, reactions, with formaldehyde, **2**, 748
Acetonitrile, methyleneamino-, use in the synthesis of glycine, **5**, 85
Acetonitrile, monolithio-, reactions, with trimethylsilyl chloride, **4**, 604
Acetonitrile, nitro-, as starting material, for 3-amino-2-nitroacrylonitrile, **2**, 787
Acetonitrile, *o*-nitroaryl-, use in the displacement of alcohols, **1**, 200
Acetonitrile, phenyl-
  reactions
    with aldehyde, **3**, 625
    with amines, **5**, 751
Acetonitrile, α-(phenylsulfinyl)-, reactions, with aldehydes, **3**, 651
Acetonitrile, *N*,*N*,2,2-tetrafluoroamino-, use in the synthesis of imino esters, **5**, 692
Acetonitrile, thioformyl-, synthesis of, **3**, 333
Acetonitrile, trialkoxy-, as starting material, for orthocarbonates, **6**, 296
Acetonitrile, trichloro-
  chlorination of, **5**, 810
  reactions
    with alcohols, **5**, 691
    with alkoxides, **6**, 296
    with amines, **5**, 749
    with chlorine, **5**, 662
    with ethyl(dimethylsulfuranylidene)acetate, **5**, 494
    with hydroxylamine, **5**, 822
    with neopentyl alcohols, **2**, 7
    with phenylsubstituted benzotrichlorides, **5**, 658
    with propargylic alcohols, **4**, 1014
    with trialkylmethoxyplumbanes, **5**, 717
    with trialkyltin alkoxides, **5**, 716
  as starting material, for orthocarbonates, **6**, 296
  use in the synthesis of haloimidic fluorides, **5**, 809
Acetonitrile, triethoxy-, reactions, with pyrrolidine, **6**, 140
Acetonitrile, trifluoro-
  reactions
    with alcohols, **5**, 691
    with *N*-chloroperfluoroalkylhydrazine, **5**, 679
    with hydrogen sulfide, **5**, 546, 580
    with phenylmagnesium bromide, **3**, 627
  as starting material, for hydrazonyl halides, **5**, 679
  synthesis of, **3**, 627
  use in the synthesis of haloimidic fluorides, **5**, 809
Acetonitrile, trifluoromethylimino trimethylsilyloxy-, synthesis of, **5**, 721
Acetonitrile, trimethylgermyl-, reactions, with bromotrimethylgermane, **6**, 185
Acetonitrile, trimethylsilyl-
  deprotonation of, **4**, 648
  oxasilylation of, **4**, 358
  silylation of, **3**, 660
  as starting material, for β-silyloxynitriles, **3**, 632
  synthesis of, **4**, 604
Acetonitrile, trinitro-
  in cycloaddition reactions, **6**, 149
  synthesis of, **6**, 150
3-Acetonitrile, indole-, synthesis of, **1**, 256
Acetonitrile anions, bis(trimethylgermyl)-, synthesis of, **6**, 185
Acetonitrile oxide, reactions, with *N*-trimethylsilyldiethylamine, **5**, 765
Acetonitrile oxides, triphenyl-, isolation of, **3**, 683
Acetonitriles
  reactions
    with *N*-bromosuccinimide, **6**, 143
    with silyloxiranes, **4**, 519
  synthesis of, **6**, 143
  use in the synthesis of α,β-alkenic nitriles, **3**, 644
Acetonitriles, amino-, lithiation–germylation of, **4**, 521
Acetonitriles, aryl-, reactions, with phosgene, **5**, 1002
Acetonitriles, α-aryl-, synthesis of, **3**, 626
Acetonitriles, 2-(arylsulfinyl)-, use in the synthesis of nitriles, **3**, 651
Acetonitriles, chloro-, reactions, with aldehydes, **3**, 644
Acetonitriles, *N*-chloroamino-, synthesis of, **5**, 658
Acetonitriles, 2-cyano-3-indole-, synthesis of, **3**, 670
Acetonitriles, 2-dialkylamino-2-(trimethylsilyl)-, reactions, with Grignard reagents, **2**, 305
Acetonitriles, diethoxy-, as starting materials, for α-amino acetals, **3**, 40
Acetonitriles, halo-, as starting materials, for thionitriles, **3**, 634
Acetonitriles, heteroaryl-, as starting materials, for formazans, **5**, 858
Acetonitriles, (*E*)-2-hydroxyimino-2-aryl-, synthesis of, **3**, 430
Acetonitriles, selenamido-
  reactions
    with 2-chloro-3-cyanopyridines, **3**, 656
    with 1,3-diketones, **3**, 656
Acetonitriles, α-silyl-
  enolate of, synthesis of, **3**, 639
  for the opening of epoxides, **1**, 209
Acetonitriles, sulfonyl-, as starting materials, for ketenimines, **3**, 566
Acetonitriles, trialkoxy-
  reactions, with Grignard reagents, **6**, 78
  as starting materials
    for ortho esters, **6**, 77
    for trialkyl orthobenzoate esters, **6**, 78
Acetonitriles, triethoxy-, reactions, with Grignard reagents, **6**, 78
Acetonitriles, trihalogenated, reactions, with bromine, **5**, 662
α-Acetonitriles, *N*-unsubstituted α-alkylated pyrrole-, synthesis of, **1**, 256
Acetonitrolic acid, synthesis of, **5**, 838
Acetophenone
  reactions
    with amyl nitrite, **5**, 823
    with chlorosulfene, **5**, 499
    with *N*-chlorosulfonamides, **5**, 345
    with potassium carbonate, **5**, 70
  as starting material
    for aryl bromides, **2**, 626
    for α-phenylpropionic acid, **5**, 85
  synthesis of, **5**, 10, 437
Acetophenone, benzoyldiazo-, use in the synthesis diazonium salts, **2**, 799
Acetophenone, benzylidene-, synthesis of, **1**, 737
Acetophenone, ω-chloro-, reactions, with arenediazonium salts, **5**, 843
Acetophenone, diazo-, reactions, with silver oxide, **3**, 530
Acetophenone, dichloro-, hydrolysis of, **5**, 66
Acetophenone, 2-(dimethylsulfuranylidene)-
  reactions
    with bromine, **5**, 495
    with iodine, **5**, 495
Acetophenone, 2,2-dimorpholino-, reactions, with hydrochloric acid, **4**, 309
Acetophenone, *p*-fluoro-, reduction of, **1**, 426
Acetophenone, 4-methoxy-, reduction of, **1**, 98
Acetophenone, 4-methoxy-3-nitro-
  as starting material
    for α-ketothioamides, **5**, 485
    for α-ketothioesters, **5**, 485
Acetophenone, *p*-methyl-, as starting material, for terephthalic acid, **5**, 114
Acetophenone, *p*-nitro-, reactions, with arsonium ylides, **1**, 738
Acetophenone, ω-nitro-, reactions, with acetyl chloride, **5**, 822

Acetophenone, p,α,α-tribromo-, hydrolysis of, **5**, 66
Acetophenone, 2,2,2-trichloro-, reactions, with sulfonamides, **5**, 345
Acetophenone, trifluoro-, as starting material, for vinyl fluorides, **2**, 611
Acetophenone oxime, α-chloro-, reactions, with sodium *O*-methyl dithiocarbonate, **6**, 546
Acetophenones
  cyclization in, **4**, 1079
  reactions
    with dimorpholine disulfide, **5**, 599
    with ethylidenetriphenylphosphoranes, **1**, 730
    with isothioureas, **4**, 980
    with morpholine, **5**, 599
    with potassium thiocyanate, **3**, 165
    with sulfur dichloride, **5**, 602
  reduction of, **1**, 96
  as starting materials, for aminals, **4**, 406
  synthesis of, **1**, 1112; **3**, 285
  use in the synthesis of α-keto oxyimidic thioesters, **5**, 830
Acetophenones, alkynyl-, as starting materials, for naphthols, **2**, 672
Acetophenones, azido-, reactions, with benzaldehydes, **3**, 258
Acetophenones, α-thiocyanato-, synthesis of, **3**, 165
Acetophenones, α-thio-α-sulfinyl-, from, esters, **4**, 857
Acetophenones, trifluoro-, synthesis of, **4**, 732
Acetophenones, α,α,α-trifluoro-, synthesis of, **5**, 109
Acetoxime, reactions, with perfluorinated nitriles, **5**, 711
Acetoximidic fluoride, difluorochloro-, from, 1,2,2-trifluoro-1,2-dichloro-1-nitroethane, **5**, 822
Acetoxy anhydrides, synthesis of, **5**, 194
α-Acetoxy bromides
  reactions, with aldehydes, **1**, 677
  as starting materials, for alkenes, **1**, 677
Acetoxy groups, reactions, with phosphorus pentachloride, **4**, 766
Acetylacetone, 3-chloro-, reactions, with arenediazonium salts, **5**, 843
Acetylacetone, tetrabutylammonium-, reactions, with carbon disulfide, **5**, 547
Acetylacetone, 3-(trinitromethylmercury)-, synthesis of, **6**, 368
Acetylation
  of alcohols, **5**, 129
  reagents for, **3**, 282
Acetyl azide, trichloro-, Curtius rearrangement of, **6**, 606
Acetyl bromide
  reactions
    with bis triethylphosphine complexes, **5**, 460
    with nitroalkenes, **5**, 820
    with triethyltin isocyanate, **5**, 1006
  synthesis of, **5**, 17
Acetyl bromide, trichloro-, thermal decomposition of, **6**, 221
Acetyl bromide, trifluoro-
  for the synthesis of aryl bromides, **2**, 626
  thermal decomposition of, **6**, 218
Acetyl chloride
  as dehydrating agent, in the synthesis of cyclic anhydrides, **5**, 199
  electrochemical fluorination of, **6**, 214
  reactions
    with acetals, **4**, 175
    with arsenic(III) oxide, **4**, 592
    with benzyl 6-aminohexanoate, **5**, 214
    with bis(trimethylsilyl) telluride, **5**, 244
    with diisopropylcarbodiimide, **5**, 1147
    with enecarbamates, **3**, 261
    with ferrocene, **1**, 862
    with fluoroarenes, **1**, 448
    with heptasulfur imide, **5**, 338
    with imines, **4**, 100
    with keto diacids, **4**, 204
    with lithiumsilaneselenolate, **5**, 541
    with lithium silanethiolate, **5**, 538
    with ω-nitroacetophenone, **5**, 822
    with oxaplatinacyclobutane complexes, **4**, 151
    with palladiumphosphine complexes, **5**, 463
    with phenylbis(diethyldiamino)methane, **4**, 455
    with *N*-(1-piperidinylmethyl)benzamide, **4**, 99
    with secondary enaminones, **3**, 413
    with stannyl acetals, **4**, 388
    with triaryl orthoformates, **6**, 44
    with triphenylsilyllithium, **5**, 417
    with uranium triscyclopentadienyl, **5**, 469
    with zinc chloride, **3**, 9
  as starting material
    for acetyl fluoride, **5**, 4
    for cyclopentadienylmolybdenum acyl compounds, **5**, 445
  synthesis of, **5**, 745
  thermal decomposition of, **6**, 215
Acetyl chloride, chloro-
  reactions
    with 7-carboxymethylheptanoic acid, **5**, 193
    with imines, **4**, 100
    with tri(2-chloroethyl) phosphate, **4**, 768
Acetyl chloride, 2-chloro-2-(chlorosulfenoimino)-, synthesis of, **5**, 675
Acetyl chloride, chloronitro-, hydrolysis of, **5**, 822
Acetyl chloride, di(1-adamantyl)-, synthesis of, **5**, 8
Acetyl chloride, diazo-
  reactions, with 1,4-diazabicyclo[2.2.2]octane, **3**, 489
  synthesis of, **3**, 475
Acetyl chloride, diphenyl-, reactions, with triethyl phosphite, **4**, 920
Acetyl chloride, fluoro-, reactions, with dimethylphosgeniminium chloride, **4**, 794
Acetyl chloride, menthoxy-, synthesis of, **5**, 10
Acetyl chloride, phenyl-, synthesis of, **5**, 14
Acetyl chloride, tosyl hydrazono-, reactions, with (*E*)-but-2-en-1-ol, **3**, 470
Acetyl chloride, trichloro-
  reactions
    with hydrogen sulfide, **5**, 506
    with organozincs, **3**, 152
  as starting material
    for trichlorothioacetic acid, **5**, 506
    for trichlorovinyl dichloroacetate, **4**, 768
  thermal decomposition of, **6**, 215
Acetyl chloride, trifluoro-
  reactions, with amines, **5**, 721
  thermal decomposition of, **6**, 219
Acetyl chlorides, tribromo-, as starting materials, for zinc enolates, **4**, 768
Acetyl chlorides, trichloro-, as starting materials, for zinc enolates, **4**, 768
Acetyl complexes, x-ray crystal structure of, **5**, 443
Acetyl cyanide
  reactions
    with dialkyltin diisopropoxides, **5**, 1144
    with trialkyltin isopropoxides, **5**, 1144
Acetylene *See* Ethyne
Acetylenedicarboxylic acid *See* 2-Butynedioic acid
Acetylenes *See* Alkynes
Acetylenides
  reactions
    with cyanogen chloride, **3**, 675
    with phenyl cyanate, **3**, 675
    with *p*-toluenesulfonyl cyanide, **3**, 675
Acetyl fluoride, from, acetyl chloride, **5**, 4
Acetyl fluoride, *N*-fluorimino-, synthesis of, **5**, 812
Acetyl fluoride, N^1-fluoroimidoyl-N^2,N^2-difluoroaminodifluoro-, hydrolysis of, **5**, 812
Acetyl fluoride, trichloro-, synthesis of, **5**, 5
Acetyl fluoride, trifluoro-, thermal decomposition of, **6**, 219

2-Acetyl hydrazines, 1,2-diaryl-, synthesis of, **5**, 356
Acetyl hypofluorites, as fluorinating agent, **6**, 264
3-Acetylindole, 1-methyl-, recovery of, **3**, 670
Acetyl iodide
  from
    acetic anhydride, **5**, 19
    isopropyl acetate, **5**, 20
Acetyl iodide, heptadecafluoro-, reactions, with carbon dioxide, **5**, 59
Acetyl iodide, trichloro-, thermal decomposition of, **6**, 221
Acetyl isocyanate, trifluoro-, fluorination of, **6**, 439
Acetylium group transfer, mechanism, **5**, 37
Acetyl nitrate, trifluoro-
  for the oxidation of hydroxyarenes, **2**, 651
  synthesis of, **2**, 651
Acetyl peroxide, 4-chlorobenzoyl, synthesis of, **5**, 208
Acetyl trifluoromethyl peroxide, trifluoro-, synthesis of, **6**, 231
Acid amides, as starting materials, for alkylammonium tosylates, **2**, 320
Acid analogue, with two chalcogens other than oxygen, **5**, 545
Acid anhydrides
  as acylating reagents, for the synthesis of aromatic ketones, **3**, 280
  for the acylation of amines, **5**, 264
  for the acylation of tellurolate anions, **5**, 241
  as dehydrating agents, **3**, 684
  reactions
    with bis(dimethylamino)methane, **4**, 309
    with carboxylic acid chlorides, **5**, 538
    with ethyl *N*-benzoyl glycinate, **3**, 182
    with halophosphite, **5**, 403
    with pyridinium poly(hydrogen fluoride), **5**, 4
    with titanocene methylidene complexes, **2**, 702
    with *N*-trimethylsilylphosphineimide, **2**, 860
  reduction of, **2**, 54
  as starting materials
    for acid iodides, **5**, 21
    for bismercurated carboxylic acids, **4**, 697
    for ketones, **3**, 130
Acid bromides
  from
    acid chlorides, **5**, 17
    aldehydes, **5**, 18
    carboxylic acids, **5**, 17
    esters, **5**, 17
    trialkylsilyl esters, **5**, 17
    triisopropylsilyl *p*-nitrobenzoate, **5**, 17
    triisopropylsilyl phenylacetate, **5**, 17
  synthesis of, **5**, 17
  use in the synthesis of amides, **5**, 264
Acid bromides, α-bromo, synthesis of, via Hell–Vollhard–Zelinsky reaction, **5**, 264
Acid catalysis
  in enolization reactions, **1**, 555
  for the synthesis of acetals, **4**, 186
Acid-catalyzed hydrolysis, of ortho esters, **6**, 70
Acid-catalyzed migrations, of hydrogen, **1**, 795
Acid chloride, α-keto, synthesis of, **5**, 15
Acid chloride, α,β-unsaturated, reactions, with thiovinylsilane, **3**, 253
Acid chlorides
  as acylating reagents, **5**, 322
  alcoholysis of, mechanism, **5**, 130
  dehalogenation of, **1**, 933
  as dehydrating agents, **3**, 684
  from
    acylcobalts, **5**, 456
    aldehydes, **5**, 15
    alkyl chlorides, **5**, 16
    carboxylic acid esters, **5**, 14
    carboxylic acids, **5**, 7
    1-chloro-*N*,*N*-2-trimethylpropenylamine, **5**, 1
    p-nitrobenzoic acid, **5**, 13
    penicillins, **5**, 9
  in Heck reactions, **1**, 494
  hydrolysis of, **5**, 195
  imination of, **3**, 404
  peptide coupling reactions with, **5**, 9
  reaction rates, **5**, 130
  reactions
    with acetals, **4**, 48
    with acetic anhydride, **5**, 187
    with alkenes, **3**, 211
    with β-alkoxyvinylstannanes, **3**, 245
    with 1-alkynes, **1**, 1045
    with alkynylcoppers, **1**, 1045
    with alkynyl cuprates, **3**, 273
    with alkynylsilanes, **1**, 1045
    with alkynyltins, **1**, 1045
    with alkynylzinc halides, **1**, 1045
    with alkynylzincs, **1**, 1045
    with allenic organomercury compounds, **1**, 1060
    with allylic mercury(II) iodides, **3**, 140
    with bis(trialkylstannyl)carbodiimides, **5**, 389
    with bis(trimethylstannyl)carbodiimide, **5**, 1121
    with carboxylic acids, **5**, 184
    with chromium pentacarbonyl dianion, **5**, 444
    with cyanomethylenetriphenylphosphorane, **3**, 674
    with cyclopentadienylrhodium compound, **5**, 458
    with cyclopentadienylsamarium complexes, **5**, 439
    with dialkylbenzenes, **3**, 296
    with dialkylcuprates, **3**, 286
    with diazoalkane, **3**, 192
    with diazomethane, **1**, 393; **3**, 474
    with dichloromethane, **5**, 129
    with dienes, **3**, 230
    with *S*-(*N*,*N*-diethylthiocarbamoyl)sulfenamide, **5**, 335
    with dimercaptoalkanes, **4**, 1073
    with enamines, **3**, 261
    with ethyl α-methyl thioacetate, **3**, 166
    with ethyl (triphenylphosphoranylidene)acetate, **5**, 103
    with formyl phosphoranes, **3**, 79
    with Grignard reagents, **3**, 129
    with hexabutylditin, **5**, 467
    with hexamethyldisilazane, **5**, 376, 379
    with hexamethylditin, **5**, 467
    with hexaphenylcarbodiphosphorane, **2**, 1063
    with hydrazines, **5**, 351
    with hydroperoxides, **5**, 206
    with 2-hydroxythiophenol, **4**, 1073
    with Lawesson's reagent, **5**, 590, 619
    with lithium alkynolates, **2**, 1030
    with magnesiopyrroles, **3**, 308
    with mesitylmagnesium bromide, **2**, 646
    with monoacyl peroxides, **5**, 206
    with morpholinosulfur trifluoride, **5**, 5
    with *N*-nitromethylamine, **5**, 361
    with organocuprates, **3**, 130
    with organolithiums, **3**, 212
    with organomanganese halides, **3**, 130
    with organometallics, **3**, 198
    with organozinc halides, **3**, 130
    with phenyl trimethylsilyl telluride, **5**, 242
    with phosphinic acids, **6**, 118
    with phosphites, **6**, 118
    with phosphorus(III) esters, **5**, 870
    with phosphosilylimines, **5**, 373
    with potassium fluoride, **5**, 4
    with potassium thiocyanate, **5**, 1047
    with propargylicsilanes, **3**, 230
    with pyridines, **5**, 129
    with pyridinium poly(hydrogen fluoride), **5**, 4
    with rhenium pentacarbonyls, **5**, 449
    with rhodium compounds, **5**, 458

with silylalkynes, **3**, 270
with *N*-silylformamidines, **5**, 759
with *O*-silyl selenon esters, **5**, 244
with *O*-silyl thion esters, **5**, 243
with silyltinalkenes, **3**, 271
with sulfimines, **5**, 339
with tetrakis(triphenylphosphine)platinum, **5**, 464
with thiopyridyl salt, **1**, 231
with titanocene methylidene complexes, **2**, 702
with transition metal complexes, **5**, 441
with trialkyl stannyllithiums, **5**, 467
with trialkyltin alkynides, **1**, 1135
with triethylamine, **3**, 528
with trifluoromethanesulfonic acid, **5**, 209
with trimesitylgermylamine, **5**, 389
with trimethylsilyl azide, **5**, 992
with trimethylsilyl bromide, **5**, 17
with trimethylsilyl cyanide, **3**, 627
with triphenylbismuth, **5**, 402
with triphenylphosphine, **5**, 402
with tristrimethylsilylphosphorus acid, **5**, 404
with vinylsilanes, **3**, 211
with vinyltins, **3**, 240
with zinc fluoride, **5**, 4
reduction of, **2**, 59
as starting materials
for *S*,*N*-acetals, **4**, 328
for acid bromides, **5**, 17
for acid fluorides, **5**, 4
for acid halides, **5**, 3
for acid iodides, **5**, 19
for acylmanganese pentacarbonyl complexes, **5**, 446
for acyl peroxides, **5**, 205
for alkynylphosphonium oxides, **2**, 1067
for anhydrides, **5**, 189
for carbodiimides, **5**, 1063
for ketene dithioacetals, **4**, 846
for ketenes, **3**, 527
for ketones, **3**, 129
for nitriles, **3**, 618
for silyloxymethylenephosphanes, **5**, 878
for thioketenes, **3**, 545
for titanocene acyl compounds, **5**, 441
for *O*-trimethylsilyl thiocarboxylates, **5**, 539
for α,β-unsaturated ketones, **3**, 261
synthesis of, **5**, 7, 128, 264
with benzoyl chloride, **5**, 15
with cyanuric chloride, **5**, 15
with oxalyl chloride, **5**, 10
with phosphorus compounds, **5**, 13
with phthaloyl chloride, **5**, 15
reagents for, **5**, 1
summary of methods, **5**, 16
with thionyl chloride, **5**, 7
for the synthesis of diacyl peroxides, **5**, 206
thiolation–elimination reactions of, **4**, 846
use in the synthesis of α-chloroalkyl esters, **4**, 53
use in the synthesis of hydroxamates, **5**, 328
use in the synthesis of α,β-unsaturated ketones, **3**, 229
Acid chlorides, α-alkoxy
decarbonylation of, **4**, 49
as starting materials, for α-chloro ethers, **4**, 49
Acid chlorides, alkyl, reactions, with arylzincs, **3**, 288
Acid chlorides, aromatic
synthesis of
with sulfuryl chloride, **5**, 15
via Friedel–Crafts reaction with phosgene, **5**, 15
Acid chlorides, α-aryloxy
decarbonylation of, **4**, 49
as starting materials, for α-chloro ethers, **4**, 49
Acid chlorides, carboxylic, reactions, with potassium *O*-alkyl dithiocarbonates, **6**, 546
Acid chlorides, ^{14}C-labeled, synthesis of, **5**, 13
Acid chlorides, α-halo, reactions, with 1,3-diones, **3**, 246
Acid chlorides, heteroaromatic, use in the synthesis of aromatic ketones, **3**, 288
Acid chlorides, α-keto, synthesis of, **5**, 14
Acid chlorides, β-keto, synthesis of, theoretical, **4**, 768
Acid chlorides, α,α,γ-trichloro, hydrolysis of, **5**, 57
Acid chlorides, α,β-unsaturated
reactions, with alkenes, **3**, 213
reduction of, **3**, 61
as starting materials
for alkenylketenes, **2**, 668
for α,β-unsaturated ketones, **3**, 212
Acid compounds
reactions
with alkoxymethyl anions, **3**, 162
with amines, **5**, 295
with dimethylsulfoxonium methylide, **3**, 171
with phosphonate carbanions, **3**, 196
with sulfonyl-stabilized carbanions, **3**, 170
with sulfoxide-stabilized anions, **3**, 170
Acid dichlorides, phosphonic, alcoholysis of, **4**, 131
Acid diesters, phosphonothioic
reactions
with ammonia, **4**, 469
with carbonyls, **4**, 469
Acid esters, alkyl chloromethylphosphinic, synthesis of, **4**, 126
Acid esters, aromatic, synthesis of, **5**, 560
Acid esters, 2-aryl acetic, oxidation of, **4**, 197
Acid esters, aryl chloromethylphosphinic, synthesis of, **4**, 126
Acid esters, *N*,*N*-bis(diphenylarsinomethyl) amino, synthesis of, via Mannich reactions, **4**, 500
Acid esters, β-bromo-α-isocyanoacrylic, synthesis of, **3**, 716
Acid esters, α-chlorophosphonic, synthesis of, **4**, 131
Acid esters, diboronic, as reagents, for the synthesis of carbon-bridged dimercury complexes, **4**, 696
Acid esters, α-halophosphinic, synthesis of, **4**, 125
Acid esters, α-halophosphonic, synthesis of, **4**, 131
Acid esters, heteroaromatic, synthesis of, **5**, 560
Acid esters, hydrazonic
from
aromatic hydrazonoyl halides, **5**, 844
α-arylhydrazono nitroalkanes, **5**, 703
carboxylic ortho esters, **5**, 705
hydrazines, **5**, 704
hydrazonoyl halides, **5**, 703
imino esters, **5**, 704
selenoesters, **5**, 704
as starting materials, for carboxylic acid hydrazides, **5**, 704
synthesis of, **5**, 703
See also Hydrazonates
Acid esters, hydroxamic, *See also* Imidates, *N*-alkoxy-
Acid esters, imidic, from, thioimidates, **5**, 694
Acid esters, lactic, reduction of, **3**, 32
Acid esters, oxetanylimidic, synthesis of, **5**, 692
Acid esters, phosphinic, reactions, with nucleophilic halogen, **4**, 125
Acid esters, sulfonic
synthesis of, **4**, 86
use in the synthesis of α,β-unsaturated ketones, **3**, 209
Acid esters, vinyl hydroxamic, synthesis of, **4**, 833
Acid fluorides
from
acid chlorides, **5**, 4
alcohols, **5**, 6
aldehydes, **5**, 6
carboxylic acids, **5**, 5
lack of reactivity with neutral oxygen nucleophiles, **5**, 5
synthesis of, **5**, 3
with cyanuric fluoride, **5**, 5
with selenium tetrafluoride, **5**, 6
Acid halides

as catalysts, for Friedel–Crafts acylation, **1**, 462
dehydrohalogenation of, with tertiary amines, **3**, 526
as electrophiles, for dithiocarboxylic enethiolates, **4**, 842
from
acid chlorides, **5**, 3
carboxylic acid esters, **5**, 3
carboxylic acids, **5**, 1
1-halo-*N*,*N*-2-trimethylpropenylamines, **5**, 1
reactions
with alkynes, **3**, 238
with *N*-aminomethylamines, **4**, 99
with bisphosphine complexes of iridium, **5**, 460
with chlorophosphazines, **3**, 619
with organotins, **3**, 288
with phosphoranes, **3**, 274
as starting materials, for ketenes, **3**, 526
synthesis of, **5**, **1–21**
thermally unstable, synthesis of, **5**, 2
Acid halides, alkynic, reactions, with carbon nucleophiles, **3**, 275
Acid halides, α-halo, as starting materials, for α-phosphorylated enol phosphates, **4**, 920
Acid halides, inorganic, use in the synthesis of alkyl halides, **2**, 6
Acid halides, α,β-unsaturated, as starting materials, for ketenes, **3**, 527
Acidic cation-exchange resins, as catalysts, for Friedel–Crafts acylation, **1**, 462
Acidic cleavage, of *S*-benzyl thioethers, **5**, 475
Acidic hydrocarbons, metallation of, **6**, 205
Acidic hydrolysis, of nitriles, **5**, 269
Acid iodides
from
acid anhydrides, **5**, 21
acid chlorides, **5**, 19
carboxylic acids, **5**, 19
esters, **5**, 20
synthesis of, **5**, 19
Acids
as activating groups for alkenes, **1**, 300
as catalysts, in Cope rearrangements, **1**, 808
from, pyridyl sulfides, **1**, 333
halogenation of, **2**, 3
reactions
with alcohols, **1**, 576
with alkenes, **1**, 895
with [bis(trifluoroacetoxy)iodo]benzene, **1**, 934
with [bis(trifluoroacetoxy)iodo]-pentafluorobenzene, **1**, 934
with isocyanide dichlorides, **5**, 1004
reduction of, **3**, 64
reductive halogenation of, **2**, 14
as starting materials
for *N*-alkylalkenoamides, **5**, 295
for *N*-alkylamides, **5**, 276
for *N*-alkyl arylamides, **5**, 299
for *N*-arylamides, **5**, 290
for *N*-arylarylamides, **5**, 302
for α-bromo ketones, **3**, 154
for α-chloro ketones, **3**, 152
for α-fluoro ketones, **3**, 150
for nitriles, **3**, 617
for primary amides, **5**, 274
for primary arylamides, **5**, 298
for α-silyl ketones, **3**, 202
for the synthesis of diacyl peroxides, **5**, 208
Acids, allenic, from, propargyl glycolates, **1**, 980
Acids, aromatic
from
anhydrides, **5**, 109
aromatic aldehydes, **5**, 114
ketones, **5**, 114
reactions
with aromatic amines, **5**, 303
with lithium dihydrogen phosphide, **5**, 879
as starting materials, for amides, **5**, 298
synthesis of, **5**, 549
via carbonation of organometallics, **5**, 110
via hydrolysis, **5**, 105
via oxidation, **5**, 113
Acids, ω-fluoro-substituted, synthesis of, **5**, 55
Acids, α-fluoro-α,β-unsaturated, synthesis of, **5**, 57
Acids, halo, reactions, with ammonia, **5**, 88
Acids, β-quaternary, from, tertiary alcohols, **1**, 186
Acids, α,β-unsaturated
conjugate addition to, **1**, 312
from, esters, **5**, 98
reactions
with *N*-acyl sulfonamides, **5**, 342
with carbon nucleophiles, **3**, 229
synthesis of, **5**, 97
reviews, **5**, 97
via carbonation of organometallics, **5**, 100
via hydrolysis, **5**, 98
via oxidation, **5**, 101
via Ramberg–Bäcklund reactions, **1**, 619
Acids, β,γ-unsaturated, synthesis of, **5**, 98
Aconitic acid, from, citric acid, **5**, 98
Acorenone, synthesis of, **1**, 599
Acridine, reactions, with dimethylketene, **4**, 886
Acridines, 9-chloro-
reactions
with potassium thiocyanate, **5**, 1040
with silver thiocyanate, **5**, 1040
Acridine-9-thiones, synthesis of, **3**, 362
Acridinone, oxidation of, **2**, 687
Acridones, as starting materials, for selenoacridones, **3**, 398
Acrolein
acyclic monothioacetals of, **4**, 218
frontier molecular orbital coefficient of, **1**, 301
reactions
with benzophenone imine, **3**, 44
with bis(cyclooctene)chlororhodium and 1,2-bis(diphenylphosphino)ethane, **6**, 456
with nitrites, **3**, 46
with thiophenol, **3**, 36
with trialkylboranes, **1**, 273
reduction of, **1**, 85
sulfuration of, **3**, 337
unsymmetrical acetals of, **4**, 199
use in the synthesis of ketones, **3**, 119
Acrolein, 3-ethoxy-2-alkyl-, reactions, with enamines, **2**, 759
Acroleine, β-silyl-, reactions, with phenylthio(diphenylphosphinoxido)-methyllithium, **2**, 939
Acroleins, alkoxy-
aminolysis of, **3**, 75
reactions, with thionyl chloride, **3**, 69
Acroleins, 3-alkoxy-, synthesis of, **3**, 72
Acroleins, α-alkoxy-, synthesis of, **3**, 71
Acroleins, β-alkoxy-, synthesis of, **3**, 71
Acroleins, (*Z*)-β-amino-, synthesis of, **3**, 76
Acroleins, α-amino-β-hydroxy-, synthesis of, **3**, 74
Acroleins, α,β-bis(silyl)-, synthesis of, **2**, 938
Acroleins, β-chloro-, synthesis of, **3**, 69
Acroleins, diamino-, synthesis of, **3**, 74
Acroleins, α,β-diamino-, hydrolysis of, **3**, 74
Acroleins, enamino-, synthesis of, **3**, 75
Acroleins, furyl-, reduction of, **1**, 78
Acroleins, β-halo, stability of, **3**, 68
Acrylaldehyde, (*Z*)-β-iodo, reactions, with primary amines, **2**, 313
Acrylamide, *N*-diethyl-, reactions, with sodium benzene sulfinate, **3**, 657
Acrylamides
reactions

with azomethine ylides, **1**, 370
with *t*-butyl radicals, **1**, 326
with Lawesson's reagent, **5**, 571
thionation of, attempted, **5**, 568
Acrylamides, 3-amino-, as starting materials, for iminoketenimines, **3**, 558
Acrylamides, methyl-, radical additions to, **1**, 326
Acrylamides, β'-sulfonyl-, reactions, with alkenes, **1**, 613
Acrylate, β-chloro-, reactions, with ketone enolates, **1**, 470
Acrylate, α-cyano-, reactions, with enol ethers, **4**, 827
Acrylate, ethyl
reaction, with *t*-butanesulfenic acids, **2**, 153
reactions
with benzenethiol, **2**, 138
with chiral radicals, **1**, 328
with ethanethiol, **2**, 138
with methanethiol, **2**, 138
Acrylate, ethyl β-bromo-, alkylation of, **1**, 469
Acrylate, ethyl α-(hydroxymethyl)-, from, paraformaldehyde, **5**, 103
Acrylate, methyl, reactions, with methyl (2-hydroxyethyl)arsine, **2**, 487
Acrylate, methyl (*E*)-β-aziridino-β-bromo-, synthesis of, **4**, 793
Acrylate, methyl 3-benzyloxyamino-2-cyano-3-(2-hydroxyethoxy)-, synthesis of, **4**, 896
Acrylate, methyl α-chloro-, reactions, with morphilinocyclohexene, **2**, 757
Acrylate, methyl trifluoro-, reaction rate of, **6**, 41
Acrylates
reactions
with enamines, **5**, 152
with enol esters, **5**, 152
with organoboranes, **1**, 276
Acrylates, β-alkoxy-, reactions, with alcohols, **4**, 195
Acrylates, β-amino-, reactions, with propargylic aldehydes, **3**, 66
Acrylates, β-azido-, photolysis of, for the synthesis of ketenimines, **3**, 564
Acrylates, 3,3-dichloro-, synthesis of, **4**, 743
Acrylates, (*Z*)-α,β-disubstituted, synthesis of, **1**, 591
Acrylates, β-substituted ethyl α-isocyano-
reactions, with dimethylsulfoxonium methylide, **3**, 704
synthesis of, **3**, 717
Acrylate salts, (*E*)-3-bromo-3-lithio-, synthesis of, **4**, 818
Acrylic acid, reactions, with methyl (2-hydroxyethyl)arsine, **2**, 487
Acrylic acid, α-cyano-β-phenyl-, synthesis of, **5**, 95
Acrylic acid, dichloro-, pyrolysis of, **2**, 1014
Acrylic acid, β,β-dimethyl-, from, 4-methyl-3-penten-2-one, **5**, 103
Acrylic acid, furyl-, anhydrides of, synthesis of, **5**, 189
Acrylic acid, perfluoromethyl-, reactions, with alcohols, **4**, 894
Acrylic acid, (*E*)-3-[(*E*)-3-phenyl-2-propenoxy]-, synthesis of, **5**, 105
Acrylic acid nitriles, *N*-benzoylamino-, as starting materials, for 3-(*N*-benzoylamino)-3-substituted pyruvic acids, **5**, 94
Acrylic acids
reactions, with sodium azide, **5**, 989
reduction of, **1**, 81
Acrylic acids, phosphinico-, from, phosphinic chlorides, **4**, 124
Acrylic acids, 2-trifluoromethyl-3-alkoxy-3-fluoro-, as intermediates, in the synthesis of ketene hemiaminals, **4**, 894
Acrylic acid salts, trichloro-, pyrolysis of, **2**, 1014
Acrylimidoyl chlorides, 3,3-bis(dimethylamino)-*N*-(phenylglyoxyloyl)-, synthesis of, **5**, 658
Acrylonitrile
chlorination of, **3**, 674
reactions
with aldehydes, **3**, 411, 651
with aryl iodides, **3**, 647
with chloroform, **6**, 6
with hydroxyalkylphosphorus compounds, **4**, 342
with iron(III) chloride, **5**, 770
with β-ketothionoamides, **5**, 607
with phenyl azide, **3**, 478
with phosphine, **2**, 428
with phosphines, **3**, 498
with tetrachloromethane, **6**, 6
with trimethylsilyl dialkyl phosphite, **3**, 640
Acrylonitrile, α-alkoxy-, reactions, with Grignard reagents, **3**, 243
Acrylonitrile, 3-amino-2-nitro-, from, nitroacetonitrile, **2**, 787
Acrylonitrile, 3-benzylamino-3-ethoxy-, synthesis of, **4**, 896
Acrylonitrile, β,β-dichloro-, reactions, with alcohols, **4**, 762
Acrylonitrile, 1,2,2-trichloro-, reactions, with sodium methoxide, **4**, 762
Acrylonitriles
as Michael acceptors, **3**, 621
reactions, with sodium toluene-*p*-sulfinates, **4**, 783
as starting materials, for β-cyanoethylhydrazines, **2**, 376
synthesis of, **3**, 622
thermal dimerization of, **1**, 354
Acrylonitriles, β-amino-, isomerization to imidoyl phosphinates, **5**, 786
Acrylonitriles, 2-aryl-, reactions, with triphenylphosphine, **3**, 649
Acrylonitriles, 2-aryl-3,3-dichloro-, synthesis of, **3**, 649
Acrylonitriles, aryl-3,3-dichloro-, synthesis of, **3**, 649
Acrylonitriles, 3,3-(bisdimethylamino)-, reactions, with phenylglyoxyloyl chlorides, **5**, 658
Acrylonitriles, α-chloro-, reactions, with radicals, **3**, 627
Acrylonitriles, dichloro-, reactions, with trifluoromethylthiocopper, **4**, 771
Acrylonitriles, β-ethynyl-, synthesis of, **3**, 648
Acrylonitriles, 2-(1-hydroxyalkyl)-, synthesis of, **3**, 651
Acrylonitriles, β-silyl-, synthesis of, **2**, 920
Acrylonitriles, (*Z*)-β-silyloxy-, synthesis of, **3**, 653
Acrylonitriles, β-substituted-α-[(trimethylsilyl)oxy]-, synthesis of, **3**, 651
Acrylonitriles, trichloro-, pyrolysis of, **3**, 674
Acryloyl azide, β-(5-nitro-2-furyl)-, Curtius rearrangement of, **5**, 990
Acryloyl chloride, reactions, with urea, **6**, 506
Acryloyl chlorides, synthesis of, **4**, 125
Acryloyl fluoride, pentafluoromethyl-, synthesis of, **4**, 736
Acryolyl chloride, from, vinyl chloride, **5**, 16
Actinides
alkylation of, **2**, 602
synthesis of, **5**, 651
Actinides, acyl, synthesis of, **5**, 468
Actinides, alkyl
stability of, **2**, 602
synthesis of, **2**, 602
Actinides, imidoyl-, synthesis of, **5**, 803
Actinium chloride, reactions, with oxaplatina(IV) cyclobutanes, **4**, 151
Activation, of metals, methods, **2**, 958
Active metals, synthesis of, via treatment of metal salts with alkali metals, **2**, 958
Acyclate complexes, synthesis of, **5**, 949
Acyclic anhydrides, synthesis of, **5**, 184
Acyl alkyl compounds, synthesis of, **5**, 450
Acyl allyl complexes, from, quinoline aldehydes, **5**, 459
Acylals, 2-(phenylsulfinyl)-, from, phenylvinyl sulfoxides, **4**, 207
Acylamines
reactions
with alkylnitramines, **4**, 447
with phosgene, **6**, 506
See also Amides

Acyl anion equivalents
  addition of, **1**, 308
  early classes of, **1**, 559
  examples, **1**, 165
  imidoylzincs as, **5**, 803
  lithiated trithioorthoformates, **6**, 317
  synthesis of, **6**, 317
  use in the synthesis of acetals, **4**, 197
  use in the synthesis of aromatic ketones, **3**, 291
Acyl anions
  reactions
    with aldehydes, **3**, 157
    with alkyl isothiocyanates, **5**, 596
    with ketones, **3**, 157
  as starting materials
    for α-amino ketones, **3**, 182
    for ketones, **3**, 136
  synthesis of, via Shapiro reactions, **1**, 634
  use in the synthesis of α-hydroxycarbonyls, **3**, 157
Acyl anions, chiral, as reagents, **4**, 269
Acylaromatics, haloform oxidation of, **5**, 114
Acylate, methyl, reactions, with cyclohexyl radicals, **1**, 331
Acylate complexes
  alkylation of, **5**, 934
  protonation of, **5**, 948
  reactions, with benzenediazonium
      tetrafluoroborate, **5**, 936
  synthesis of, **5**, 933
Acylating agents
  reactions
    with acyl phosphonates, **4**, 920
    with *vic*-diphosphine, **5**, 400
    with nucleophilic phosphorus compounds, **5**, 394
    with phosphorous acids, **5**, 402
  for the synthesis of aromatic ketones, **3**, 283
  for the synthesis of diacetates, **4**, 204
Acylating reagents
  reactions
    with amino acids, **3**, 180
    with α-nitrogen anions, **3**, 182
Acylation
  of alkoxyamines, **5**, 327, 332
  of alkynes, **3**, 273
  of allylsilanes, **3**, 140
  of amidines, **5**, 758
  of amines, **5**, 260
    categories, **5**, 261
  of anilines, **5**, 290
  of benzenamine, **5**, 302
  of Grignard reagents, **3**, 285
  of hydrazine, **5**, 350
  of hydroxamates, **5**, 332
  of hydroxylamines, **5**, 211, 322
  of organocoppers, **3**, 286
  of organolithiums, **3**, 284
  of organometallics, **3**, 289
  of organotins, **3**, 287
  of organozincs, **3**, 288
  of α-silylimines, **4**, 514
  for the substitution of silicon functions, **1**, 284
  of sulfonamides, **5**, 342
  for the synthesis of alkenyl ethers, **2**, 674
  for the synthesis of enamides, **5**, 288
  for the synthesis of tricoordinate phosphorus
      functions, **4**, 481
  of thioamines, **5**, 335
  of vinyl ethers, **3**, 245
Acylation, electrophilic, of organometallics, **3**, 284
C-Acylation
  of ketene mercaptals, **3**, 253
  of nitro compounds, **3**, 190
N-Acylation
  of imidates, **5**, 694
  for the synthesis of *N*-alkylamides, **5**, 277
  use of acyl halides for, **5**, 278
O-Acylation
  of phenolates, **2**, 683
  of phenols, **2**, 683
Acyl chloride
  reactions
    with methoxymethylene triphenylphosphorane, **5**, 897
    with percarboxylic acid, **5**, 207
Acyl chloride, allenyl, reactions, with phosphorus
      ylides, **1**, 987
Acyl chlorides
  for the acylation of oximes, **5**, 215
  electrochemical fluorination of, **6**, 39
  reactions
    with alkyl bis(trimethylsilyl)silyllithium, **5**, 418
    with aluminum trichloride, **3**, 278
    with aryl cyanates, **6**, 616
    with benzyltetrabutylstiborane, **1**, 272
    with complex borohydrides, **3**, 102
    with cyclopropanes, **1**, 285
    with 1,3,5-hexatriazines, **4**, 99
    with hydrazones, **5**, 362
    with hydridoiron tetracarbonyl anions, **3**, 102
    with *N*-hydroxyformamides, **6**, 506
    with hypervalent silicon hydrides, **3**, 107
    with lithium methyl tris(trimethylsilyl)aluminum, **5**, 418
    with lithium phosphide dimethoxyethane
      complex, **5**, 1155
    with lithium tetrakis(trimethylsilyl)aluminum, **5**, 418
    with mercury(II) bis(trifluoromethanethiolate), **6**, 235
    with metal carboxylates, **5**, 198
    with methyl bis(trimethylsilyl)silyllithium, **5**, 418
    with methylcyclopropane, **1**, 285
    with pentaorganostiboranes, **1**, 272
    with phosphorus pentasulfide, **5**, 484
    with potassium acetate, **5**, 182
    with primary phosphines, **5**, 396
    with propargylmercury iodide, **1**, 974
    with selenols, **5**, 238
    with silver cyanate, **5**, 994
    with silver dimesylamine, **5**, 348
    with silyl cuprates, **5**, 417
    with silyl metallics, **5**, 417
    with sodium acetate, **5**, 182
    with sodium azide, **2**, 757
    with sodium carboxylates, **5**, 183
    with sodium diethyldihydroaluminate, **3**, 107
    with sodium tri(*t*-butoxy)aluminum hydride, **3**, 98
    with sulfur-substituted phosphide, **5**, 398
    with thiocyanates, **6**, 618
    with thiols, **5**, 233
    with *N*-trimethylsilylphosphineimide, **2**, 860
    with *O*-trimethylsilyl thiocarboxylates, **5**, 539
    with triphenyl germyllithium, **5**, 432
    with triphenylphospineimides, **5**, 373
    with tris(trimethylsilyl)phosphine, **5**, 1154
  reduction of, **2**, 60; **3**, 98
  reductive silylation of, **5**, 419
  as starting materials
    for acyl isoselenocyanates, **5**, 1056
    for anhydrides, **5**, 182
    for symmetrical alkanoic anhydrides, **5**, 185
    for symmetrical heteroaromatic anhydrides, **5**, 194
    for thioacyl chlorides, **5**, 484
    for trithioortho esters, **6**, 84
    for unsymmetrical alkanoic anhydrides, **5**, 192
  use in the synthesis of alkylmanganese(I)
      compounds, **2**, 573
  use in the synthesis of allenes, **1**, 986
Acyl chlorides, acetoxy-
  reactions, with alkyl hydroperoxides, **6**, 71
  solvolysis of, for acyloxy ortho esters, **6**, 71

Acyl chlorides, α-chloro-α-(chlorosulfenyl)-, synthesis of, **4**, 71
Acyl chlorides, (*N*-chloroformylamino)-, as intermediates, **5**, 970
Acyl chlorides, dichloroalkoxy-, synthesis of, **6**, 39
Acyl chlorides, 2-isocyanato-, synthesis of, **5**, 970
Acyl chlorides, ω-isothiocyanato-, synthesis of, **5**, 1031
Acyl chlorides, α,β-unsaturated
  reactions, with diazo acetates, **3**, 475
  reductive silylation of, **5**, 416
Acyl cleavage, for the synthesis of carbanions, **1**, 883
Acyl complexes
  acylation of, **5**, 937
  dehydration of, **1**, 1145
  stability of, **5**, 464
  as starting materials, for alkoxycarbene complexes, **5**, 937
  synthesis of, **5**, 469, 937
  x-ray structure of, **5**, 469
Acyl compounds
  from, aziridines, **5**, 446
  reactions, with alkoxyamines, **5**, 329
  reduction of, **2**, 379
Acyl compounds, α-hydroxy, synthesis of, **5**, 449
N-Acyl compounds, tautomerism of, **4**, 898
O-Acyl compounds, synthesis of, **1**, 935
Acyl fluorides
  from, carboxylic acids, **5**, 234
  reactions
    with perfluoromethanamines, **5**, 313
    with *N*-silylhydrazines, **5**, 353
    with tetrafluorohydrazine, **6**, 52
  use in the synthesis of amides, **5**, 264
Acyl fluorides, α-fluorosulfinyl-, synthesis of, **2**, 175
Acyl fluorides, perfluoro-, as starting materials, for α-fluoroalkyl esters, **4**, 52
Acyl functions, imino-
  linked to any heteroatom other than halogen, chalcogen or nitrogen, synthesis of, **5**, **805–864**
  linked to any other heteroatom, synthesis of, **5**, **783–804**
Acyl halides
  for the acylation of amines, **5**, 264
  ammonolysis of, **5**, 295
  decarbonylation of, **1**, 667
  hydrolysis of, **5**, 26
  reactions
    with [(alkoxycarbonyl)methylidene] phosphoranes, **1**, 986
    with arylstannanes, **3**, 287
    with benzyltriethylammonium tetrathiomolybdate, **5**, 252
    with organothiomethylenephosphoranes, **5**, 898
    with phosphoranes, **1**, 986
    with silver salts, **1**, 1137
    with silyl isothiocyanates, **5**, 1048
    with sodium carboxylates, **5**, 183
    with sulfonamides, **5**, 342
    with tetraalkylammonium pentacarbonylacylmetalates, **5**, 949
    with trimethylsilyl selenides, **5**, 238
    with water, **5**, 26
  as starting materials
    for acyl isothiocyanates, **5**, 1047
    for alkanethiols, **2**, 130
    for *N*-alkylamides, **5**, 278
    for alkynoic amides, **5**, 304
    for aromatic ketones, **3**, 278
    for primary amides, **5**, 274
    for selenocarboxylic esters, **5**, 238
    for tellurocarboxylic esters, **5**, 241
    for thiocarboxylic esters, **5**, 233
  synthesis of, **5**, 26
  use in *N*-acylation reactions, **5**, 278
  use in the synthesis of α-diazo ketones, **3**, 474
Acyl halides, 2-halo-, as starting materials, for ketenes, **3**, 530
Acyl halides, α-halo-, dehydrohalogenation of, with tertiary amines, **3**, 526
Acyl halides, imino-, synthesis of, **5**, **653–723**
Acyl iodides, reactions, with silver cyanide, **3**, 704
O-Acylisoamides, *C*-substituted, synthesis of, **3**, 687
Acylotropy, definition of, **5**, 406
Acyloxy compounds, synthesis of, **5**, **181–229**
Acyloxy halides, α- and β-, solvolysis of, for ortho esters, **6**, 71
Acyloxy iodides, synthesis of, **5**, 204
Acyl radicals
  cyclization of, **1**, 548
  from
    acylcobalt salophens, **5**, 457
    seleno ester, **1**, 933
  reactions, with pyridine, **3**, 311
  synthesis of, **1**, 933
  use in the synthesis of aromatic ketones, **3**, 292
  use in the synthesis of ketones, **3**, 136
2-Adamantal, oxidation of, **5**, 51
Adamantane
  chlorination of, **2**, 20
  fluorination of, **2**, 15
  reactions, with thionyl chloride, **2**, 174
Adamantane, 1-amino-, synthesis of, **2**, 298
Adamantane, bromo-, reactions, with trinitromethane salts, **6**, 145
Adamantane, 1-bromo-
  reactions, with thioureas, **6**, 635
  reactivity of, **1**, 893
  synthesis of, **1**, 116
Adamantane, 1-chloro-, as starting material, for 1-isocyanoadamantane, **3**, 698
Adamantane, α-chloronitroso-, reactions, with Grignard reagents, **3**, 432
Adamantane, 1-(*N*,*N*-difluoroamino)-, synthesis of, **2**, 335
Adamantane, *N*,*N*-difluoroamino-, synthesis of, **2**, 335
Adamantane, 1,2-diiodo-, reactions, with *n*-butyllithium, **1**, 661
Adamantane, 1,3-diisocyanato-, synthesis of, **5**, 993
Adamantane, 1-isocyano-, from, 1-chloroadamantane, **3**, 698
Adamantane bromide, 1-phenylphosphonium-, reduction of, **1**, 445
Adamantanes, from, trimethylsilyl isothiocyanate, **5**, 1045
Adamantanes, 1,3-disubstituted halo-, reactions, with phosphine-aluminum chloride complexes, **2**, 427
Adamantanethione, from, adamantanone, **3**, 343
Adamantanone
  reductive amination of, **1**, 131
  as starting material, for adamantanethione, **3**, 343
Adamantenes, as intermediate, **1**, 661
Adamantyl anhydrides, synthesis of, **5**, 191
Adamantyl azides, 3-homo-, reactivity of, **5**, 1066
Adamantyl cations, synthesis of, **1**, 900
1-Adamantyl chloride, reactions, with *N*-alkyl-*N*-trimethylsilyl cyanamides, **5**, 1118
Addition
  of allenyl carbanions, **1**, 1054
  for the formation of carbon–carbon bonds, **1**, **461–500**
  for the formation of carbon–hydrogen bonds, **1**, **425–460**
  of propargyl carbanions, **1**, 1054
  for the synthesis of alkynes, **1**, 1037
  for the synthesis of carbon–carbon bonds, **1**, **501–551**
  1,2- vs. 1,4-, **1**, 301
Addition, acid-catalyzed, for the synthesis of tertiary phosphines, **2**, 439
Addition, base-catalyzed
  for the synthesis of primary phosphines, **2**, 428
  for the synthesis of tertiary phosphines, **2**, 439
Addition, conjugate, of trialkylborane to α,β-unsaturated ketones, **2**, 698

Addition, electrocyclic
regioselectivity of, **1**, 350
stereoselectivity of, **1**, 351
for the synthesis of alkenes, **1**, 516
to carbon–carbon multiple bonds, **1**, 349
use of term, **1**, 349
Addition, electrophilic, for the synthesis of α-carbocations of alkyne-bridged dimolybdenum complexes, **4**, 674
Addition, nucleophilic, side reactions, **2**, 1012
Addition, palladium-catalyzed
of alkynes, **1**, 546
of disilanes, **2**, 915
Addition, thermal [2+2], of alkenes, **1**, 354
1,4-Addition
of alkynes, **1**, 1048
of allenyl metals, **1**, 1061
of propargyl metals, **1**, 1061
to carbon–carbon bonds, **1**, 300
1,6-Addition, of allylsilanes, **1**, 314
[2+2] Addition, metal-catalyzed, of alkenes, **1**, 357
[2+2] Addition, photochemical, of alkenes, **1**, 355
endo Addition, mechanism, **1**, 351
exo Addition, mechanism, **1**, 351
Addition–elimination
for the desulfonization of vinyl sulfones, **1**, 440
for the synthesis of selenium compounds, **4**, 785
of vinyl halides, **1**, 469
Addition–elimination, nucleophilic, for the synthesis of vinyl cations, **1**, 1110
1,2-Addition–elimination
of allenyl carbanions, **1**, 1060
of propargyl carbanions, **1**, 1060
for the synthesis of alkynes, **1**, 1044
Addition–elimination process, side reactions, **2**, 1012
Addition–elimination reaction, for the synthesis of alkoxycarbene complexes, **5**, 942
Addition reaction
for the synthesis of aryl sulfides, **2**, 712
for the synthesis of hydrazonamides, **5**, 847
for the synthesis of primary phosphines, **2**, 428
for the synthesis of secondary phosphines, **2**, 430
α-Addition reactions, for the synthesis of isocyanides, **3**, 694
Adenines, as starting materials, for purines, **1**, 36
Adipamide, *meso*-2,5-dithiol-*N*,*N*,*N'*,*N'*-tetramethyl- (*meso*-DTA), for the reductive cleavage of disulfides, **2**, 127
Adipate, diethyl α,δ-dibromo-, as starting material, for muconic acid, **5**, 103
Adipates, bis(silyl), synthesis of, **5**, 227
Adipic acid
from, cyclohexene, **5**, 52
synthesis of, **5**, 46
Adipic acid, α-chloro, synthesis of, **5**, 55
Adipic acid, *meno*-2,5-diphenyl-, synthesis of, **5**, 42
AD-mixes *See cis*-Dihydroxylation-mixes, asymmetric
Adociane, 7,20-diisocyano-, synthesis of, **3**, 701
Adpoc fluoride, from, 2-(1-adamantyl)propan-2-ol, **6**, 422
Aerial oxidation, of thiols, **2**, 213
Aerobic fungi, for the deprotection of arylamines, **2**, 304
Aerocyanidin, synthesis of, **3**, 704
A-frame platinum complexes, reactions, with diazomethane, **4**, 691
Agelasidine, synthesis of, **2**, 159
Agelasidine A
from, farnesol, **6**, 472
synthesis of, **6**, 472
Agelaspongin, dibromo-, occurrence, in natural products, **6**, 138
Agelnopsis aperta, venom of, for the synthesis of polyamines, **2**, 344
Air, reactions, with platinum, **5**, 48
Alanates
reactions
with cyanogen, **1**, 534
with iodine, **1**, 961
synthesis of, **1**, 534
Alanates, alkynyl-, reactions, with oxiranes, **1**, 1024
Alanates, allenyl-
alkylation of, **1**, 1052
carbonation of, **1**, 1058
Alane, chlorodialkyl-, reactions, with organolithium, **2**, 990
Alane, diisobutyl-, for the hydroalumination of alkynes, **2**, 968
Alane complexes, reactions, with bis(trifluorovinyl)mercury, **4**, 815
Alanes
reactions
with alkenes, **2**, 43
with cyanogen, **1**, 534
with enamines, **1**, 443
with enones, **1**, 1049
reduction of, **2**, 60
as starting materials, for dialkylaminoboranes, **2**, 418
Alanes, alkenyl-
alkylation of, **1**, 1106
arylation of, **1**, 471
degree of aggregation of, **2**, 953
dimerization of, **1**, 491
as intermediates, **1**, 533
protonation of, **1**, 542
reactions
with *n*-butyllithium, **1**, 543
with electrophiles, **2**, 976
with epoxides, **1**, 1106
with organic halides, **1**, 542
transmetallation of, **2**, 926
use in organic synthesis, **2**, 968
Alanes, (*E*)-alkenyl-, synthesis of, **1**, 468
Alanes, (*E*)-1-alkenyl-, synthesis of, via hydroalumination, **2**, 967
Alanes, (*Z*)-1-alkenyl-, synthesis of, **2**, 974
Alanes, alkenyldichloro-, synthesis of, via transmetallation, **2**, 991
Alanes, alkenyldiisobutyl-, reactions, with alkynyl halides, **1**, 1020
Alanes, alkynyl-
bishydroalumination of, **6**, 208
conjugate addition to enones, **3**, 147
from, alkynyllithiums, **2**, 1098
reactions
with alkylidenemalonates, **1**, 1050
with dialkylaluminum hydrides, **6**, 208
with enones, **1**, 1049
with nitroalkenes, **1**, 1049
with tertiary alkyl halides, **1**, 1003
synthesis of, **2**, 1098; **6**, 208
Alanes, alkynyldialkyl-
as alkynylating reagents, **1**, 1021
bishydroalumination of, **6**, 208
Alanes, alkynyldiethyl-
reactions
with aldehydes, **1**, 1037
with episulfonium tetrafluoroborates, **1**, 1031
with ketones, **1**, 1037
Alanes, alkynyldimethyl-, reactions, with 2-(6-carbomethoxyhexyl)-4-hydroxy-2-cyclopentenone, **1**, 1049
Alanes, allyl-, reactions, with alkynes, **1**, 543
Alanes, allyldialkyl-, as reagents, for zirconium-catalyzed carboalumination, **2**, 974
Alanes, benzyl-
reactions, with alkynes, **1**, 543
as reagents, for zirconium-catalyzed carboalumination, **2**, 974
Alanes, 1,1-bis(boryl)-
from
1,1-diborylalkenes, **4**, 634

vinylboranes, **4**, 631
synthesis of, **4**, 629
via changing the groups on the central methylene, **4**, 634
via changing the ligands on boron, **4**, 634
via hydroboration of alkynes, **4**, 630
via hydroboration of vinylboranes, **4**, 631
via quenching a carbanion with a boryl electrophile, **4**, 632
Alanes, dialkylalkynyl-, synthesis of, **2**, 1098
Alanes, diethylalkynyl-
reactions
with alkenyl phosphates, **1**, 1027
with enones, **1**, 1048
Alanes, mixed, synthesis of, **2**, 1099
Alanes, α-silylalkenyl-, synthesis of, **1**, 1106
Alanes, α-silylmethyl-, synthesis of, **4**, 659
Alanes, β-stannylalkenyl-, synthesis of, **1**, 1106
Alanes, trialkyl-
reactions
with aldehydes, **1**, 1037
with ethyne, **2**, 974
with ketones, **1**, 1037
stability of, **2**, 1098
Alanes, trialkynyl-
explosive decomposition of, **2**, 1099
reactions, with propargyl acetates, **1**, 1025
Alanes, vinyl-
alkylation of, **1**, 476
as intermediates, in the synthesis of vinylsilanes, **2**, 918
reactions, with alkyl sulfones, **1**, 483
synthesis of, **1**, 128, 468
use in the substitution of vinyl halides, **1**, 468
Alanes, vinylic, reactions, with vinylic halides, **1**, 1106
Alanine, *N*-acetylphenyl-, synthesis of, **1**, 76
Alanine, (*S*)-3-(3,4-dihydroxyphenyl)- ((*S*)-DOPA), synthesis of, **1**, 76
Alanine, phenyl-, synthesis of, **1**, 80
Alanine, L-phenyl-, reactions, with triphosgene, **6**, 486
Alanine, *S*-phenyl-, synthesis of, **1**, 76
Alanine, (±)-β-phenyl-, synthesis of, **5**, 87
Alanine, β-phenyl-, synthesis of, **5**, 90
(±)-Alanine, from, acetaldehyde, **5**, 85
2-Alanine, *N*-(benzoyloxycarbonyl)-β-(pyrazol-1-yl)-, synthesis of, **5**, 85
D-Alanine, as starting material, for *trans*-2,5-dimethylpyrrolidine, **2**, 331
L-Alanine
as starting material
for (2*S*,5*S*)-2,5-dimethylpyrrolidine, **2**, 325
for *trans*-2,5-dimethylpyrrolidine, **2**, 331
β-Alanine, synthesis of, **5**, 85, 86
(±)-Albene, synthesis of, **1**, 287
Alcohol, reactions, with diarylphosphinous halides, **2**, 840
Alcohol, allyl, reactions, with hydrogen cyanide, **3**, 622
Alcohol, benzyl
hydrogenation of, **1**, 88
reactions, with *n* -butyllithium, **4**, 383
synthesis of, **1**, 398
Alcohol, *t*-butyl
reactions
with chromium trioxide, **5**, 47
with tetraalkylvanadium, **2**, 569
as starting material, for *t*-butyl fluoroformate, **6**, 422
Alcohol, 2-chloroethyl, reactions, with alkali metal methylarsenide, **2**, 482
Alcohol, cinnamyl, use in the synthesis of (*E*)-3-[(*E*)-3-phenyl-2-propenoxy]acrylic acid, **5**, 105
Alcohol, darvon, use in the reduction of ketones, **2**, 64
Alcohol, fenchyl, protonation of, **1**, 381
Alcohol, hydroxybenzyl, reactions, with chlorobenzene, **2**, 684
Alcohol, α-methylallyl, silylation of, **1**, 287
Alcohol, neopentyl, as starting material, for cyanates, **5**, 1102
Alcohol, 4-nitrobenzyl, reactions, with tetrakis(pyridino)cobalt(II) dichromate, **3**, 99
Alcohol, propargyl, reactions, with Grignard reagents, **4**, 1065
Alcohol, 2,2,2-trichloro-*t*-butyl, reactions, with triphosgene, **6**, 487
Alcohol, *p*-(triphenylsilyl)benzyl, as starting material, for *p*-(triphenylsilyl)benzoic acid, **5**, 114
Alcoholates
reactions
with formamimidinium chlorides, **6**, 632
with thiophosphonic acid dihalides, **2**, 463
with transition metal carbyne complexes, **5**, 942
as starting materials, for chlorothioformates, **6**, 533
Alcohol complexes, propargyl, protonation of, **4**, 688
Alcohols
acetylation of, **5**, 129
for the alkylation of amides, **5**, 284
benzoylation of, **5**, 129
catalyzed exchange reactions, with magnesium chloride, **6**, 79
dehydration of, **1**, 575, 692
deoxygenation of, **1**, 16, 939
displacement of, **1**, 177
via cyanoacetate anions, **1**, 179
elimination of, **1**, 579
esterification of, **5**, 128
exchange reactions
with orthoformate esters, **6**, 78
with trialkyl ortho esters, **6**, 78
Friedel–Crafts alkylation of, **1**, 475
from
alkenes, **2**, 40
allylic halides, **2**, 87
amines, **1**, 383; **2**, 88
carbonyls, **2**, 52
cyclohexene oxide, **1**, 388
diethyl α,α-bis(methylthio)phosphonate, **6**, 121
epoxides, **1**, 842
racemates, **4**, 390
O-methylation of, **2**, 89
Michael addition to α,β-unsaturated nitriles, **3**, 632
O-nitration of, **2**, 104
optically active, retention of configuration, **6**, 425
oxidation of, **3**, 57, 64; **5**, 46, 141
chromium reagents for, **3**, 115
protection of, **2**, 539
protection of as acetals, **4**, 186, 198
protonation of, **2**, 5
reactions
with acetals, **4**, 184
with acetyl hypoiodite, **2**, 95
with acids, **1**, 576
with *O*-acylated hydroxylamines, **5**, 125
with acylimines, **4**, 304
with aldehydes, **2**, 678; **4**, 177
with alkali cyanides, **5**, 1103
with alkanesulfenyl chlorides, **2**, 194
with alkenes, **2**, 89
with β-alkoxy acrylates, **4**, 195
with *S*-alkyl alkyl chlorodithiophosphonates, **2**, 464
with *S*-alkyl alkyl chlorothiophosphonates, **2**, 464
with alkyl(alkylthio)chlorophosphines, **2**, 461
with alkyl(chloro)alkoxyarsines, **2**, 496
with alkyldihaloarsines, **2**, 497
with alkyl metals, **2**, 112
with *Se*-alkyl selenocarboxylates, **5**, 519
with alkynes, **4**, 194
with alkynylselenolate anions, **5**, 519
with alkynyl selenols, **5**, 519
with allylidene complexes, **5**, 941
with aminoarsines, **2**, 494
with aminomethylenephosphines, **6**, 685

with 3-amino-propynals, **3**, 76
with areneselenenyl halides, **2**, 734
with arylarsenic anhydrides, **2**, 885
with arylarsenic imines, **2**, 885
with arylhydrazines, **3**, 451
with arylmethylenenitrile, **3**, 625
with arylmethylnitriles, **3**, 625
with aryl selenocyanates, **2**, 734
with arylsulfinamides, **2**, 200
with bis(diethylamino)chlorophosphite, **6**, 234
with 1,1-bis(diethylamino)ethylene, **4**, 888
with boranes, **2**, 528
with boron trihalides, **2**, 110
with α-bromo-β-selenenyl ethers, **4**, 51
with *N*-(*t*-butoxycarbonyl)phosphoramidate, **2**, 299
with *N*-butoxycarbonyl *p*-toluenesulfonamide, **2**, 309
with *O*-*t*-butyltrichloroacetimidate, **2**, 90
with caesium fluoroxy sulfate, **5**, 6
with carbodiimides, **6**, 633
with carbonyl ylides, **3**, 319
with carboxylic acids, **5**, 122
with chlorine, **2**, 94
with chloroformate esters, **6**, 461
with 3-chloro-1-methoxypropene, **4**, 199
with chlorophosphines, **2**, 450
with chlorothionoformates, **6**, 541
with cyanamide, **6**, 731
with cyanamides, **6**, 632
with cyanogen halides, **5**, 1101
with 1,1-dialkoxycyclopropanes, **6**, 76
with dialkylhaloarsines, **2**, 494
with dialkylhalobismuthines, **2**, 494
with dialkylhalostibines, **2**, 494
with diaminoarsines, **2**, 496
with diazoalkanes, **4**, 198
with α-diazo alkyl phosphonates, **4**, 913
with dibenzoyl peroxide, **4**, 220
with diborane, **2**, 109
with β,β-dichloroacrylonitrile, **4**, 762
with dichlorocarbenes, **4**, 767
with dichloroiminium salts, **6**, 729
with dichloro(methyl)arsine, **2**, 496
with di(disulfonamido) disulfides, **2**, 100
with diethylaminosulfur trifluoride, **4**, 97
with diethyl azodicarboxylate, **2**, 299
with diethyl phosphochloridodithionite, **2**, 106
with difluorocarbene, **2**, 89
with difluorophosphonium ylides, **2**, 476
with dimethylmethyleneammonium chloride, **4**, 305
with 2-diphenylacetoxytetrahydrofuranyls, **4**, 200
with diphenyldiazomethane, **2**, 89; **4**, 198
with diphosgene, **6**, 426
with disilanes, **2**, 539
with disulfur dichloride, **2**, 99, 100
with enol ethers, **4**, 186
with α-epoxy ketones, **3**, 242
with ethoxy(thiocarbonyl) isocyanate, **6**, 559
with ethylmagnesium halides, **2**, 112
with *N*-fluoroimines, **6**, 279
with α-haloalkanesulfonyl halides, **4**, 86
with haloalkenes, **6**, 41
with α-haloalkyl carboxylates, **4**, 209
with α-halo ethers, **4**, 198
with α-halo sulfides, **4**, 218
with hydridosilanes, **2**, 539
with hydriodic acid, **2**, 86
with hydrogen halides, **2**, 6
with hydrogen peroxide, **2**, 95
with hydrogen sulfide, **2**, 118
with hydroxy furanones, **4**, 211
with hypochlorous acid, **2**, 94
with imidate ester hydrochlorides, **6**, 70
with imidoyl chlorides, **4**, 312; **5**, 685
with *N*-iodosuccinimide, **2**, 95
with isocyanates, **6**, 484, 622
with isonitriles, **5**, 692
with isopropenyl acetate, **4**, 188
with isothiocyanates, **6**, 558
with isothiocyanatophosphazenes, **6**, 558
with Jones reagent, **1**, 957
with ketones, **2**, 678; **4**, 177
with metal alkoxides, **2**, 87
with metal chlorides, **2**, 112
with metals, **2**, 112
with methanesulfonyl chloride, **4**, 780
with *N*-methyl *p*-toluenesulfonamide, **2**, 309
with mono-(2-cyanoethyl)phosphate, **2**, 105
with nitrile oxides, **5**, 699, 826
with nitriles, **4**, 883; **5**, 691, 742
with α-nitronitriles, **5**, 692
with nitrous acid, **2**, 104
with organophosphorus selenocyanates, **5**, 1055
with oxathiolium salts, **6**, 97
with perfluoromethylacrylic acid, **4**, 894
with perfluoro-2-phosphapropene, **6**, 683
with *N*-phenylsulfenylimides, **2**, 135
with phosgene, **6**, 425, 460
with phosphines, **2**, 445
with phosphinic halides, **2**, 454
with phosphonic dichlorides, **2**, 456
with phosphonous acids, **2**, 456
with phosphonous chlorides, **2**, 838
with phosphorous acid, **2**, 106
with phosphorous trihalides, **2**, 104
with phosphorus pentasulfide, **2**, 119, 258
with potassium alkynyl selenolates, **5**, 519
with potassium nitrate, **2**, 104
with propargylic esters, **3**, 72
with pyrophosphoryl chloride, **2**, 105
with selenium dioxide, **3**, 109
with selenophosphonic dihalides, **2**, 466
with silicon disulfide, **2**, 108
with silirenes, **4**, 1051
with silyl enol ethers, **4**, 188
with sodium borohydride, **2**, 532
with sulfamoyl chlorides, **2**, 367
with sulfinyl chlorides, **2**, 196
with *N*-sulfinylsulfonamides, **2**, 100
with *N*-sulfinyl-*p*-toluenesulfonamide, **2**, 202
with sulfonimidoyl chlorides, **2**, 365
with sulfonyl chlorides, **2**, 202
with sulfur chloride, **2**, 101
with sulfuric acid, **1**, 894
with tetrafluoroformamidine, **5**, 814
with thiiranes, **2**, 128
with thiocarbamoyl chlorides, **6**, 557
with thiocarbonyl chloride fluoride, **6**, 533
with thiocarbonyldiimidazole, **6**, 543
with *N*,*N*′-thiocarbonyldiimidazole, **6**, 557
with thioketenes, **5**, 511
with thionyl chloride, **2**, 101; **4**, 97
with thiophosgene, **6**, 533
with thiophosphinic chlorides, **2**, 462; **4**, 128
with toluene sulfonic acid, **1**, 578
with tosyl cyanide, **2**, 721
with tributylphosphine, **2**, 135
with trichloroacetonitrile, **5**, 691
with trichloromethanesulfenyl chloride, **2**, 194
with trifluoroacetonitrile, **5**, 691
with trimethylaluminum, **2**, 596
with triphenylphosphine, **2**, 299
with tris(difluoroamino)methyl isocyanate, **6**, 363
with uranium hexafluoride, **5**, 6
with urea, **6**, 484
with vinyl acetate, **4**, 188
with vinyl azides, **4**, 894

with vinylphosphine oxides, **2**, 453
with vinyl sulfide, **3**, 634
with vinylsulfonyl chlorides, **2**, 721
with ynamines, **4**, 885
as reagents, for the synthesis of carbon-bridged dialuminum compounds, **4**, 699
reduction of, **1**, 11, 17
saponification of, **5**, 32
selective oxidation of, **3**, 161
as solvents, for the synthesis of nitriles, **3**, 613
as starting materials
for acetals, **4**, 186
for acid fluorides, **5**, 6
for acyl alkyl acetals, **4**, 209
for aldehydes, **3**, 3
for alkanesulfinic esters, **2**, 202
for alkanethiols, **2**, 118
for alkenes, **1**, 383
for alkyl bromides, **2**, 31
for alkyl chlorides, **2**, 8, 25
for alkyl fluorides, **2**, 19
for alkyl halides, **2**, 5
for alkyl iodides, **2**, 35
for alkyl selenocyanates, **5**, 1114
for alkyl xanthates, **2**, 118
for α-chloro ethers, **4**, 47
for chloroformate esters, **6**, 424
for α-chloro ketones, **3**, 151
for chlorothioformates, **6**, 533
for cyanates, **5**, 1101
for *O*,*S*-dialkyl dithiocarbonates, **6**, 549
for dialkyl sulfides, **2**, 135
for dithiocarbonates, **6**, 550
for ethers, **6**, 36
for α-fluoroaliphatic ketones, **3**, 148
for fluoroformate esters, **6**, 422
for fluorohydrins, **2**, 15
for isoselenocyanates, **5**, 1053
for isothiocyanates, **5**, 1041
for isothiouronium salts, **2**, 120
for ketones, **3**, 114
for nitriles, **3**, 613
for pyridyl ketones, **3**, 311
for radicals, **1**, 172
for α-silyl aldehydes, **3**, 51
for α-silyl ketones, **3**, 202
for thiocyanates, **5**, 1107
for thiophosphates, **2**, 258
for trihalomethyl sulfides, **6**, 234
for α,β-unsaturated aldehydes, **3**, 57
sulfonation of, **2**, 5
synthesis of, **1**, 204; **2**, 38
via alkylative addition, **2**, 67
via biotransformations, **2**, 38
via displacement of amines, **2**, 88
via hydroboration, **2**, 40
via oxidation, **2**, 86
via rearrangement reactions, **2**, 88
tetrahydropyranylation of, **4**, 186
type I, synthesis of, **1**, 1055
type II homopropargylic, synthesis of, **1**, 1055
type III homopropargylic, synthesis of, **1**, 1056
type IV homopropargylic, synthesis of, **1**, 1056
type V homopropargylic, synthesis of, **1**, 1057
type VI homopropargylic, synthesis of, **1**, 1057
umpolung, **1**, 195
use in the disposal of phosgene, **6**, 411
use in the synthesis of thionoamides, **5**, 581
Alcohols, acyclic alkynic, reactions, with silylaluminums, **2**, 86
Alcohols, acyclic allylic, epoxidation of, **2**, 90
Alcohols, aliphatic, reactions, with 1-alkoxyalkynes, **4**, 828
Alcohols, *syn*-1,3-alkoxy, synthesis of, **2**, 61
Alcohols, alkoxyamino, reactions, with dialkylaminodichlorophosphines, **2**, 410
Alcohols, alkyl
displacement of, **1**, 177
phenylation of, **2**, 684
Alcohols, alkynic
cyclization of, **5**, 944
hydration of, **3**, 244
oxidation of, **3**, 275
reactions
with molybdenum complexes, **5**, 944
with transition metals, **5**, 944
Alcohols, α-alkynic
rearrangement of, **3**, 59
as starting materials, for α,β-unsaturated aldehydes, **3**, 59
Alcohols, β-alkynic, isomerization of, **1**, 956
Alcohols, alkynyl homopropargylic, as starting materials, for allenynones, **1**, 957
Alcohols, allenic
reactions, with thiophosgene, **6**, 436
synthesis of, **1**, 609
Alcohols, α-allenic
from, propargylic bromides, **1**, 967
synthesis of, **1**, 956
Alcohols, β-allenic, from, pent-3-yn-1,5-diols, **1**, 961
Alcohols, allenyl, synthesis of, **1**, 971
Alcohols, α-allenyl
from
but-2-yn-1,4-diols, **1**, 961
α-halo-α'-hydroxy alkynic compounds, **1**, 958
synthesis of, **1**, 966
Alcohols, allyl
alkylation of, **1**, 476
allylmagnesiation of, **2**, 561
from, α-halo epoxides, **1**, 664
reactions, with enol ethers, **4**, 188
synthesis of, **1**, 162, 664; **2**, 927
Alcohols, (*E*)-allyl, synthesis of, **1**, 592
Alcohols, allylic
acid-catalyzed rearrangement of, **1**, 837
asymmetric epoxidation of, rate of reaction, **2**, 92
as byproducts, **3**, 69
displacement of, **1**, 186
Eschenmoser rearrangement of, **1**, 828
from
allylic selenoxides, **1**, 608
chalcone, **4**, 269
α,β-unsaturated carbonyls, **2**, 57
hydroboration of, **2**, 42
isomerization of, **2**, 639
for the synthesis of enols, **2**, 639
Johnson rearrangement of, **1**, 828
kinetic resolution of, **2**, 44, 92
natural products, **1**, 786
oxidation of, **3**, 215, 229; **5**, 154
oxyselenylation of, **3**, 175
as precursors of aldehydes, **3**, 8
reactions
with alkyl vinyl ethers, **1**, 414
with amide acetals, **1**, 827
with aminosulfenyl chlorides, **2**, 363
with *t*-butyl hydroperoxide, **2**, 91
with 1,2-diketones, **2**, 73
with Lewis acids, **1**, 364
with lithium aluminum hydride, **1**, 429
with nickel, **1**, 195
with ortho esters, **1**, 416; **5**, 159
with osmium tetroxide, **2**, 46
with pyridinium chlorochromate, **3**, 78
with tetramethylethylenediamine, **1**, 316
with titanium, **1**, 173
with titanium tetrachloride, **1**, 364
with ynamines, **1**, 828

reactivity of, **1**, 316
rearrangement of, **1**, 831
reduction of, **1**, 77, 84, 440
Sharpless epoxidation of, **2**, 318
as starting materials
for alkenes, **1**, 12
for allyl cations, **1**, 364
for allylic chlorides, **2**, 8
for allylic halides, **2**, 7
for allylic sulfinamides, **2**, 247
for chlorides, **2**, 25
for dienoic esters, **1**, 788
for ketones, **3**, 122
for γ-lactones, **5**, 170
synthesis of, **1**, 513, 531, 677; **1**, 716, 779; **2**, 70
from selenoxide pyrolysis, **1**, 781
via elimination reactions, **1**, 608
via oxidation of alkenes, **2**, 86
via oxidation of allylic tellurides, **1**, 611
via ring-opening of vinyl epoxides, **2**, 86
via [2,3] sulfoxide-sulfenate rearrangement, **1**, 839
triethylboron *ate* complex of, **1**, 133
use in Claisen rearrangement, **1**, 415
Alcohols, (*E*)-*homo*-allylic, synthesis of, **1**, 757
Alcohols, allylic iodo, synthesis of, **1**, 961
Alcohols, allylic tertiary, oxidation of, **3**, 57
Alcohols, allylic thioenol, hydrolysis of, **3**, 54
Alcohols, amido
as intermediates, in the synthesis of methylenebisamides, **4**, 424
reactions, with nitriles, **4**, 425
Alcohols, amino
reactions
with boronic acids, **2**, 528
with diphosgene, **5**, 972
with phosphonous compounds, **2**, 443
with ynamines, **5**, 691
as starting materials
for *O*,*N*-acetals, **4**, 301
for amino aldehydes, **3**, 40
for carboxylic esters, **5**, 140
synthesis of, **2**, 305
Alcohols, 1,2-amino
from, 2-silyloxynitriles, **2**, 317
oxidation of, **3**, 184, 260
reactions, with boranes, **2**, 64
synthesis of, **2**, 311, 514; **3**, 39
Alcohols, 1,3-amino, synthesis of, **2**, 43
Alcohols, 2-amino
semipinacol rearrangement of, **1**, 389
synthesis of, **2**, 43
Alcohols, 3-amino, synthesis of, **2**, 320
Alcohols, *erythro*-β-amino, synthesis of, **2**, 331
Alcohols, *threo*-1,2-amino, synthesis of, **2**, 70
Alcohols, α-amino, synthesis of, **2**, 310
Alcohols, β-amino, synthesis of, **2**, 318, 321, 519
Alcohols, ω-amino, reactions, with isothiocyanates, **6**, 570
Alcohols, 1-aryl-1-alkylethyl, reactions, with *N*-benzyloxyamine, **2**, 351
Alcohols, 1,2-azido, synthesis of, **2**, 84
Alcohols, α-azido, synthesis of, **4**, 313
Alcohols, azo, as starting materials, for acylhydrazines, **5**, 357
Alcohols, benzyl
carbonylation of, **1**, 202
deprotonation of, **4**, 387
reactions, with xenon difluoride, **4**, 45
as starting materials, for ethers, **2**, 89
synthesis of, **1**, 162
Alcohols, benzylic
allylation of, **1**, 200
displacement of, **1**, 200
oxidation of, **2**, 43; **3**, 87
reactions
with manganese dioxide, **3**, 98
with titanium, **1**, 173
reduction of, **1**, 12
as starting materials, for chlorides, **2**, 25
use in the synthesis of dialkylhydrogenphosphates, **2**, 106
Alcohols, 2-(benzylideneamino), *O*-allylation of, **2**, 312
Alcohols, 1,2-*syn*-*N*-benzyloxyamino, synthesis of, **2**, 350
Alcohols, bishomoallylic, synthesis of, **1**, 481
Alcohols, α-bromo
from
aldehydes, **4**, 42
levoglucosenone, **4**, 42
1,1,1-trifluoroacetone, **4**, 42
1,1,1-trifluoropentane-2,4-dione, **4**, 42
Alcohols, butenolide, as starting materials, for α-chloroalkyl esters, **4**, 54
Alcohols, chiral, as starting materials, for chiral titanium compounds, **2**, 568
Alcohols, chiral amino, use in the synthesis of amines, **2**, 314
Alcohols, α-chloro, from, 2-hydroxy 1,3-diketones, **4**, 43
Alcohols, β-chloro
reactions, with cyanide, **3**, 632
synthesis of, **3**, 167
Alcohols, cyclopropyl
elimination of, **1**, 576
redox reactions of, **1**, 929
Alcohols, diallyl, hydrosilylation of, **2**, 51
Alcohols, 2,3-difluoroallylic, as starting materials, for α-fluoro epoxides, **4**, 45
Alcohols, 2,2-difluorohomoallylic, synthesis of, **2**, 77
Alcohols, dinitrofluoro, dimerization of, **6**, 57
Alcohols, α,α-diphosphono, synthesis of, **6**, 118
Alcohols, 3,3-disubstituted allylic, synthesis of, **2**, 81
Alcohols, 2-(2-dithianyl), synthesis of, **2**, 66
Alcohols, diyne, reduction of, **1**, 504
Alcohols, epoxy
reductive ring-cleavage of, **2**, 84
ring-opening of, **2**, 84, 318
Alcohols, epoxy propargylic, reactions, with methylcuprate, **1**, 966
Alcohols, erythro, synthesis of, **4**, 343
Alcohols, α-fluoro, stability of, **4**, 42
Alcohols, fluoroallenic, synthesis of, **1**, 958
Alcohols, halo, reactions, with cuprates, **1**, 465
Alcohols, 3-halo, synthesis of, **2**, 70
Alcohols, α-halo
stability of, **4**, 41
synthesis of, **4**, 41
Alcohols, homoallylic
asymmetric epoxidation of, **2**, 92
from
aldehydes, **2**, 73
B-allyldiazaborolane, **2**, 76
3-hydroxy sulfoxides, **2**, 85
oxidation of, **3**, 218
ozonolysis of, **3**, 31
reactions, with potassium permanganate, **4**, 168
1,3 sigmatropic rearrangement of, **1**, 401
synthesis of, **1**, 230, 388, 404, 478; **1**, 543, 633, **793–842**, 1107; **2**, 80
Alcohols, homopropargyl, synthesis of, **1**, 1054
Alcohols, homopropargylic, as starting materials, for α-allenic ketones, **1**, 957
Alcohols, hydroxylamino, reactions, with hydroxylamino ketones, **3**, 189
Alcohols, α-iodo, from, 4,4-diiodo-1,1-dimethyl-1,4-dihydroquinolinium cation, **4**, 43
Alcohols, (*Z*)-3-iodoallyl, synthesis of, **2**, 629
Alcohols, isocyano, synthesis of, via the opening of epoxides, **1**, 209
Alcohols, γ-isocyano, synthesis of, **3**, 704
Alcohols, α-keto, cleavage of, **5**, 114

Alcohols, mercapto
  reactions
    with aldehydes, **4**, 229
    with ketones, **4**, 229
Alcohols, methoxy, synthesis of, **4**, 343
Alcohols, 2-methylbenzyl, synthesis of, **2**, 43
Alcohols, neopentyl, reactions, with trichloroacetonitrile, **2**, 7
Alcohols, nitrilo, from, 2-(1-(dimethylamino)ethyl)-3-(2-hydroxyethyl)-1*H*-indole, **1**, 253
Alcohols, nitro
  optically active, synthesis of, **2**, 398
  reactions, with ethyl chloroformate, **6**, 477
  synthesis of, **2**, 394; **4**, 906
Alcohols, 1,2-nitro, synthesis of, **3**, 191
Alcohols, 1,3-nitro, synthesis of, **3**, 191
Alcohols, 2-nitro
  dehydration of, **2**, 784
  oxidation of, **3**, 46
Alcohols, β-nitro
  from, silyl nitronates, **2**, 394
  as intermediates, **2**, 785
Alcohols, ω-nitro, production of, **2**, 396
Alcohols, 2-nitroso, production of, **2**, 390
Alcohols, 3-oxyallylic, oxidation of, **3**, 247
Alcohols, phosphinoyl, synthesis of, **1**, 641
Alcohols, phthalide, as starting materials, for α-chloroalkyl esters, **4**, 54
Alcohols, 2-polyhalomethyl, synthesis of, **2**, 81
Alcohols, primary
  from, carboxylic acids, **2**, 59
  oxidation of, for aldehydes, **3**, 3
  reactions, with dimethylamine hydrochloride, **2**, 309
  as starting materials, for primary amines, **2**, 309
  synthesis of, **1**, 316
Alcohols, propargyl
  anti-hydroalumination of, **2**, 973
  hydrozirconation of, **2**, 969
  isomerization of, **1**, 1085
  reactions
    with acyl isothiocyanates, **6**, 629
    with allylzinc bromide, **1**, 543
    with Grignard reagents, **1**, 539
    with lithium aluminum hydride, **1**, 513
  as starting materials
    for *gem*-halophosphorus allenes, **4**, 802
    for 1-oxa-2-alumina-3-cyclopentene, **2**, 970
    for vinyl iodides, **2**, 629
  synthesis of, **1**, 162, 1037, 1073; **2**, 1061
  tetrahydropyranyl ether of, metallation of, **1**, 1134
Alcohols, propargylic
  displacement of, **1**, 197
  ethyne–vinylidene rearrangement of, **5**, 941
  hydrogenation of, **1**, 506
  isomerization of, **1**, 197
  oxidation of, **3**, 78
  reactions
    with amines, **3**, 76
    with copper compounds, **1**, 197
    with hydridoaluminates, **2**, 970
    with pentacarbonyl complexes, **5**, 954
    with sodium bis(methoxyethoxy)aluminum hydride, **4**, 948
    with sulfenyl chloride, **1**, 993
    with trichloroacetonitrile, **4**, 1014
  rearrangement of, **3**, 216
  as starting materials
    for α-silyl enamines, **4**, 1014
    for thioketones, **3**, 352
    for trienynes, **1**, 962
Alcohols, prop-2-ynyl, aminomercuration of, **3**, 417
Alcohols, secondary, oxidation of, for ketones, **3**, 114
Alcohols, secondary benzylic, oxidation of, **3**, 294
Alcohols, selenyl, oxidation of, **3**, 174
Alcohols, 2-silyl, synthesis of, **2**, 58
Alcohols, α-silyl, oxidation of, **5**, 412
Alcohols, silyl allenyl, synthesis of, **2**, 86
Alcohols, α-silyl allylic, Simmons–Smith cyclopropanation of, **5**, 429
Alcohols, β-silyl allylic, oxidation of, **3**, 272
Alcohols, stannyl, stability of, **4**, 386
Alcohols, β-stannyl, synthesis of, **1**, 664
Alcohols, stannylated, use in the synthesis of arenesulfenates, **2**, 720
Alcohols, terminal, as starting materials, for selenides, **1**, 604
Alcohols, terpene, synthesis of, **1**, 839
Alcohols, tertiary
  as byproducts, **3**, 284
  hydrogenation of, **1**, 12
  reactions, with hypobromous acid, **2**, 95
  reduction of, **1**, 13, 16
  as starting materials
    for carbamates, **6**, 417
    for carbon-centered radicals, **1**, 939
    for hydroperoxides, **2**, 95
    for β-quaternary acids, **1**, 186
  synthesis of, via oxidation of hydrocarbons, **2**, 86
  for the synthesis of radicals, **1**, 940
  trifluoroacetylation of, **1**, 582
Alcohols, tertiary benzylic, acid-catalyzed dehydration of, **1**, 577
Alcohols, tertiary ethynyl, reactions, with *N,N*-dimethylformamide diethyl acetal, **2**, 740
Alcohols, tricyclopropyl, rearrangement of, **1**, 576
Alcohols, 2-trifluoromethyl, synthesis of, **2**, 70
Alcohols, β-trimethylsilyl, synthesis of, **1**, 678
Alcohols, γ-trimethylsilyl, oxidative fragmentation of, **1**, 655
Alcohols, 1-trimethylsilylpropargyl, as starting materials, for trimethylsilyloxyallenes, **1**, 963
Alcohols, trisubstituted allylic, synthesis of, **2**, 68, 970
Alcohols, unsaturated
  oxidation of, **5**, 101
  reduction of, **1**, 986
  synthesis of, **1**, 548
    via elimination reactions, **1**, 643
Aldazines
  reactions
    with ethyl phosphonites, **4**, 460
    with iodine compounds, **4**, 198
  as starting materials
    for α-aminophosphonic acids, **4**, 460
    for α-hydrazinophosphonic acids, **4**, 460
Aldazines, aryl, reactions, with dialkyl phosphites, **4**, 460
Aldehyde, aryl
  reactions
    with acetic anhydride, **5**, 98
    with α-alkylacetic anhydride, **5**, 98
    with hexamethylphosphorus triamide, **2**, 740
Aldehyde, α-chloro, reactions, with methylhydrazine, **2**, 771
Aldehyde, 3-indole, synthesis of, **3**, 107
Aldehyde, β-nitro, elimination of, **3**, 57
Aldehyde, salicyl
  as electrophile, for dithiocarboxylic enethiolates, **4**, 842
  reactions
    with ammonium acetate, **4**, 438
    with bisiminoalkanes, **4**, 438
    with 4-dimethylamino benzaldehyde, **4**, 438
    with enamines, **4**, 307
    with *S*-imidoyl sulfenamides, **5**, 736
    with *N*-styrylmorpholine, **4**, 307
    with ylides, **1**, 734
Aldehyde acetals, α-amino, synthesis of, **2**, 331
Aldehyde acids, synthesis of, reviews, **5**, 62
Aldehyde dehydrogenase (AldDH), use in the synthesis of L-α-hydroxy acids, **5**, 117
Aldehyde 1,1-diols, α-keto, synthesis of, **4**, 161

Aldehyde enolates, synthesis of, **2**, 701
Aldehydes
acetalization of, **4**, 178
as activating groups for alkenes, **1**, 300
alkenation of, **2**, 680; **3**, 642
alkylation of, **3**, 10
allenylation of, **2**, 79
arylation of, **3**, 24
asymmetric addition of, **3**, 628
asymmetric addition of hydrogen cyanide to, **3**, 629
bearing an oxygen function, synthesis of, **3**, 28
bearing a metalloid function, synthesis of, **3**, 51
bearing a nitrogen function, synthesis of, **3**, 39
bearing a selenium function, synthesis of, **3**, 37
bearing a sulfur function, synthesis of, **3**, 34
bearing a tellurium function, synthesis of, **3**, 37
boron-Wittig reaction of, **1**, 988
bromination of, **1**, 853; **3**, 27
α-bromination of, **2**, 28
bromoacylation of, **4**, 56
α-chlorination of, **2**, 21
chloroacylation of, **4**, 53
chloromethylenation of, for vinyl chlorides, **2**, 618
cyanoselenylation of, **3**, 644
decarbonylation of, **5**, 459
α,α-dichlorination of, **4**, 12
dithioacetalization of, **4**, 247
elimination of, **1**, 606
enantioselective alkylation of, **2**, 71
enantioselective propargylation of, **1**, 1055
formyl group, deprotonation of, **5**, 437
from
acetals, **3**, 9
acylpyrroles, **3**, 106
alcohols, **3**, 3
aldehydes, **3**, 10
alkanes, **3**, 2
alkenes, **3**, 2
alkyl aryl sulfides, **3**, 16
alkyl halides, **3**, 3
alkynes, **3**, 3
allylic alcohols, **3**, 8
amides, **1**, 34
amines, **3**, 17
carboxylic acids, **3**, 13
diols, **3**, 7
enol esters, **3**, 9
enol ethers, **3**, 9
epoxides, **1**, 387; **3**, 9
esters, **3**, 8
ethers, **3**, 8
formyl anion equivalents, **3**, 20
hydrazones, **3**, 18
ketones, **3**, 10
N-methoxy-*N*-methylamides, **3**, 13
nitroalkanes, **3**, 18
organoboranes, **1**, 273; **3**, 19
organosilanes, **3**, 18
oximes, **3**, 18
pyridyl sulfides, **1**, 333
silyl ethers, **3**, 8
sulfur, **3**, 15
thioaldehydes, **1**, 834
(*E*)-vinylboranes, **1**, 163
vinylsilanes, **3**, 18
zirconium acyl compounds, **5**, 442
α-functionality of, **3**, 12
C-germylation of, **4**, 360
haloalkylation of, **4**, 47
halogenation of, **3**, 54
α-halogenation of, **3**, 26
homologation of, **3**, 10; **5**, 873
hydrogenation of, **2**, 52
imination of, **3**, 404
iodination of, **3**, 27
α-iodination of, **2**, 34
mercuration of, **6**, 207
metallation of, **3**, 158
methylenation of, **1**, 769
β- and more remotely unsaturated, synthesis of, **3**, 21
niobium-promoted reactions, with alkynes, **2**, 988
one-carbon homologation of, **5**, 417
oxidation of, **1**, 1077; **3**, 54; **5**, 48, 276, 368
α-oxidation of, **3**, 31
oximes of, synthesis of, **3**, 425
Peterson alkenation of, **2**, 950
phenyl selenylation of, **3**, 37
propargylation of, **2**, 79
reactions
with α-acetoxy bromides, **1**, 677
with acrylonitrile, **3**, 411, 651
with acyl anions, **3**, 157
with alcohols, **2**, 678; **4**, 177
with alkenylaluminums, **1**, 534
with alkenylboranes, **1**, 1103
with alkenylchromiums, **1**, 1099
with 3-alkenylpropargylic boranes, **1**, 977
with alkenylzirconocene chlorides, **2**, 70
with 5-alkoxyallylstannanes, **2**, 79
with alkoxyamines, **3**, 434
with α-alkoxyleads, **4**, 390
with 2-alkoxyorganoplumbanes, **2**, 70
with alkyl diphenylphosphine oxides, **1**, 641
with alkyl halides, **6**, 58
with *N*-alkylhydrazines, **3**, 446
with alkylidene ylides, **1**, 727
with alkylthiotrimethylsilanes, **2**, 541
with alkylzincs, **2**, 69
with alkylzirconocene chlorides, **2**, 70
with 1-alkynes, **1**, 1038
with alkynezirconocenes, **2**, 987
with *B*-1-alkynyl-9-borabicyclo[3.3.1]nonanes, **1**, 1037
with alkynylchromiums, **1**, 1038
with alkynyldiethylalanes, **1**, 1037
with 5-alkyn-1-yllithiums, **2**, 68
with alkynyllithium triisopropoxides, **1**, 1039
with alkynylmagnesiums, **1**, 1039
with alkynyl metals, **1**, 1037
with alkynyltins, **1**, 1038
with alkynyltitanium triisopropoxides, **1**, 1038, 1039
with alkynylzincs, **1**, 1039
with 9-allenyl-9-borabicyclo[3.3.1]nonyl, **1**, 1055
with allenylboronic esters, **1**, 1055
with allenylmagnesiums, **1**, 1055
with allenyl metals, **1**, 1054
with allenylsilanes, **1**, 1056, 1057
with allenylstannanes, **1**, 1057
with allenyltitaniums, **1**, 1056
with allenylzinc chlorides, **1**, 1055
with allenylzirconiums, **1**, 1057
with allylboranes, **1**, 281; **2**, 75
with allyl organometallics, **3**, 31
with allyltrialkylstannanes, **2**, 78
with allyltrichlorosilanes, **2**, 78
with aluminum compounds, **1**, 1055
with amidrazones, **5**, 772
with amines, **2**, 739; **4**, 294, 405
with (ω-aminoalkyl)phenylphosphines, **4**, 453
with aminosilanes, **4**, 509
with 2-aminothiophenol, **4**, 318
with ammonia, **3**, 617; **4**, 294
with ammoniopropanedinitrile toluene-*p*-sulfonate, **4**, 883
with ammonium acetate, **5**, 600
with ammonium polysulfide, **5**, 639
with arenesulfonylmethyl isocyanides, **4**, 908

with arsonium ylides, **1**, 646
with arylgermyllithiums, **4**, 360
with arylhydrazines, **3**, 446
with arylthiotrimethylsilanes, **2**, 541
with benzotriazole, **4**, 296
with benzyloxynitromethane, **4**, 897
with benzynezirconocenes, **2**, 987
with bisdiphenylthiophosphinoylmethane, **4**, 1022
with bis-α-hydroxyalkylphosphines, **4**, 341
with bis(methylene)sulfurane, **4**, 231
with bismuthonium ylides, **1**, 743
with bis(trifluoroethyl) phosphonates, **1**, 751
with bis(trimethylsilyl)bromomethyllithium, **4**, 805
with bis(trimethylsilyl)methyllithiums, **4**, 936
with 1,3-bis(trimethylsilyl)propenyl anions, **1**, 651
with bis(trimethylsilyl) selenide, **2**, 542
with boranes, **2**, 57
with boron enolates, **3**, 159
with boron trichloride, **2**, 110
with boroxazolidones, **5**, 228
with (bromochloromethyl)phenylmercury, **2**, 618
with bromomalonic esters, **1**, 742
with bromonitromethane, **4**, 798
with *N*-bromosuccinimide, **5**, 18
with 1,2-butadienylmagnesium halides, **1**, 1055
with caesium fluoroxy sulfate, **5**, 6
with chloroacetonitriles, **3**, 644
with 2-chloroallylic chromium, **2**, 80
with 2-(chlorodifluoromethyl)allylsilanes, **2**, 77
with α-chloro-α-isocyanomethyl phosphonates, **3**, 717
with chloromethyl phenyl sulfoxide, **4**, 776
with chloromethylphosphinates, **4**, 126
with chlorophosphines, **4**, 119, 121
with Clayfen, **4**, 431
with cyanide, **5**, 85
with cyanomethylzinc, **3**, 632
with cyclopentadienyltantalum complexes, **4**, 355
with *N*,*N*-dialkylhydrazines, **3**, 446
with 3,3-dialkyl-substituted allenyllithiums, **1**, 1056
with diallenyltin dibromide, **1**, 1055
with diazo ketones, **1**, 1078
with (diazomethyl)diphenylphosphine oxides, **2**, 1066
with diazomethyls, **3**, 474
with diethyl benzenesulfonylmethyl phosphonate, **4**, 925
with diethyl ethoxycarbonylmethyl phosphonate, **1**, 745
with diethyl(isocyanomethyl) phosphonate, **3**, 714
with diethyl malonate, **4**, 559
with diethyl succinate, **5**, 98
with di(ethylthio)phosphine, **2**, 461
with dihydroxy acetone phosphate, **3**, 159
with dimetalloalkynides, **1**, 1037
with dimethoxyvinyllithiums, **3**, 71
with dimethyl *n*-alkyl phosphonothioates, **1**, 646
with dimethyl(diazomethyl) phosphonate, **1**, 1078
with dimethyl diazomethyl phosphonate, **3**, 11
with dimethyl 2-oxoheptyl phosphonate, **1**, 747
with dimethylphenylsilyllithium, **4**, 355
with diphosphonites, **4**, 564
with 1,3-dipoles, **4**, 183
with 1,3-disubstituted allenyltitaniums, **1**, 1057
with disulfonate esters, **4**, 864
with (*Z*)-enolates, **1**, 878
with epoxides, **4**, 181
with ethoxycarbonylmethylides, **1**, 724
with ethyl diazoacetate, **3**, 474
with ethyl diazotrimethylsilyl acetate, **3**, 474
with fluorenylidenetriphenylphosphorane, **1**, 723
with fluorodinitromethanides, **6**, 57
with fluoromethylsulfoximine, **4**, 86
with formyl phosphonium ylides, **3**, 65
with formylphosphoranes, **3**, 60, 65
with 3-furanones, **3**, 241
with germyllithiums, **5**, 431
with Grignard reagents, **2**, 67; **5**, 468
with halonitromethanes, **4**, 798
with halophosphines, **4**, 123
with hexaorganosilylgermanes, **5**, 431
with homoenolate anions, **3**, 161
with hydrazines, **3**, 444
with hydrogen peroxide, **2**, 96
with hydrogen selenide, **3**, 382
with hydrogen sulfide, **2**, 223
with (*E*)-4-hydroxybut-2-enyl triphenylarsonium salts, **1**, 270
with β-hydroxycarboxylic acids, **4**, 213
with β-hydroxyethyltriphenylarsonium bromide, **1**, 269
with hydroxylamine, **3**, 425
with hydroxylamines, **4**, 435
with (3-hydroxypropyl)phenylphosphines, **4**, 337
with iminothiocarbonyl ylides, **2**, 140
with iodoarsonium iodides, **3**, 649
with iodo ketones, **2**, 74
with iodophosphoranes, **3**, 274
with isocyanomethyl phosphonate diethyl ester, **3**, 717
with isothiocyanatotrimethylsilane, **2**, 544
with α-ketophosphoranes, **3**, 220
with α-lithiated amines, **4**, 528
with lithiated silylmethyl methyl ethers, **2**, 680
with lithiated triselenoorthoformates, **6**, 94
with lithiated trithioorthoformates, **6**, 88
with lithiobromomethyl phenyl sulfoxide, **3**, 154
with lithiochloromethyl phenyl sulfoxide, **3**, 153
with α-lithio phosphonamides, **1**, 644
with 2-lithio thiazole, **3**, 29
with lithium aluminum amides, **3**, 405
with lithium amides, **4**, 299
with lithium propynoate, **5**, 154
with lithium tetramethylpiperidide, **5**, 437
with malonic acid, **5**, 98
with mercaptoalcohols, **4**, 229
with α-metallated ethyl isocyanoacetate, **3**, 717
with metallated indoles, **3**, 309
with α-metallated isocyanides, **3**, 713
with metallated pyrroles, **3**, 308
with methoxy allenylaluminums, **1**, 973
with 2-methoxybenzyltriphenylphosphonium bromide, **1**, 732
with 2-methyl-1-aza-1-cycloalkenes, **3**, 413
with 3-methylbenzothiazolium bromide, **3**, 157
with (*R*)-*N*-methylphenylglycinol, **2**, 313
with 2-monosubstituted allylic chromiums, **2**, 73
with nitro compounds, **1**, 252
with nitro enamines, **2**, 789
with nitro(phenylseleno)methane, **4**, 910
with nitro(phenylthio)methane, **4**, 906
with organoantimony alkoxides, **4**, 198
with organoantimony compounds, **2**, 73
with organoboranes, **2**, 531
with organocoppers, **2**, 69
with ortho esters, **4**, 190
with orthoformates, **4**, 190
with (3-oxoalkyl)phenylphosphines, **4**, 454
with phenyl acetonitrile, **3**, 625
with (*R*)-phenylglycinol, **2**, 313
with (phenylseleno)trimethylsilane, **2**, 542
with α-(phenylsulfinyl)acetonitrile, **3**, 651
with phenyl trimethylsilyl selenide, **4**, 239
with phosgene, **4**, 58; **6**, 428
with phosphine oxides, **1**, 757; **3**, 182
with phosphines, **2**, 446; **5**, 1130
with phosphonate carbanions, **1**, 750
with phosphonium ylides, **1**, 754
with phosphoramidites, **4**, 475
with phosphorous acid dialkyl esters, **2**, 457
with phosphorus(III) chlorides, **2**, 454

with phosphorus(III) compounds, **5**, 404
with phosphorus(III) esters, **2**, 454
with α-phosphorus-substituted ketones, **3**, 269
with phosphorus trichloride, **2**, 456; **4**, 130
with piperidines, **2**, 310
with polyhalomethanes, **2**, 81
with potassium *t*-butoxide, **1**, 43
with potassium hydroxide, **1**, 42
with 1,2-propadienyl tributylstannane, **3**, 30
with 1,3-propanedithiol, **5**, 410
with propargylboranes, **1**, 977
with propargyl metals, **1**, 1054
with propargyl trimethylsilanes, **1**, 978
with α-pyridinium salts, **3**, 258
with selenium(II) compounds, **3**, 37
with semicarbazide, **3**, 449
with silicon compounds, **1**, 1038
with silirenes, **4**, 1052
with α-silylcarbanions, **1**, 763
with α-silyl dithioesters, **5**, 559
with silyl enol ethers, **2**, 66; **3**, 159, 220
with silyl ethers, **4**, 181
with sodium benzenesulfonamide, **5**, 345
with sodium bisulfite, **4**, 237
with sodium dithionite, **4**, 234
with 1-substituted allenylboranes, **1**, 1056
with *N*-substituted hydroxylamines, **2**, 344
with α-substituted propargyl bromides, **1**, 1056
with α-sulfinyl carbanions, **4**, 77
with *N*-sulfinyl sulfonamides, **1**, 1042
with *N*-sulfinylsulfonamides, **4**, 311
with *N*-sulfinyl-*p*-toluenesulfonamide, **2**, 306
with sulfonamides, **4**, 423
with sulfonyl carbanions, **1**, 614
with telluronium salts, **1**, 623
with telluronium ylides, **1**, 623
with tetraalkylleads, **4**, 390
with tetrabromomethane, **4**, 745
with tetraiodomethane, **4**, 749
with tetrakis(triphenylphosphine)platinum, **5**, 466
with tetraoxyhydridophosphoranes, **2**, 476
with thiobenzamide, **5**, 606
with α-thiol carboxylic acids, **4**, 230
with thiols, **4**, 247
with 2-thiophenylalkylcyanocuprates, **2**, 69
with thiosemicarbazide, **3**, 449
with titanium complexes, **1**, 991
with tosylhydrazine, **3**, 446
with trialkylalanes, **1**, 1037
with trialkylallylsilanes, **2**, 78
with trialkyl orthoacetates, **6**, 80
with trialkylsilylethynyllithiums, **1**, 1039
with triborylmethide anions, **6**, 182
with tributylallylstannanes, **2**, 79
with tributylstannylmagnesium chloride, **4**, 386
with tributylstibonium ylides, **1**, 647, 743
with tributyl(tribromomethyl)stannane, **6**, 31
with tributyl(trichloromethyl)stannane, **6**, 26
with trichloromethyl anions, **6**, 25
with trichloromethyltrimethylsilane, **6**, 26
with trifluoromethylethyl phosphites, **2**, 476
with trimethylgermyldimethylphenylsilane, **4**, 361
with trimethylsilyl azide, **3**, 665; **5**, 377
with trimethylsilyl chloro sulfonate, **2**, 209
with trimethylsilyl cyanide, **3**, 646
with trimethylsilyldiazomethane, **1**, 1078
with (trimethylsilyl)dichloromethyllithium, **4**, 742
with *S*-trimethylsilyl enethiolate, **5**, 557
with trimethylsilyl sulfides, **4**, 216
with 2-trimethylsilyl thiazole, **3**, 29
with *O*-trimethylsilyl thiocarboxylates, **5**, 540
with α-(trimethylsilyl)tosylmethyl isocyanide, **3**, 720
with trimethylsilyl trichloroacetate, **6**, 26
with triphenylarsonium salts, **1**, 647
with triphenylphosphine, **4**, 745, 749
with tris(dimethylamino)phosphine, **2**, 469
with tris(trimethylsilyl)ketenimines, **3**, 660
with uranium hexafluoride, **5**, 6
with vinyltitaniums, **2**, 70
with Wittig reagents, **2**, 728
with ylides, **1**, 730
with zinc alkylcyanocuprates, **2**, 69
with 1-zinc-1-zirconium alkenes, **1**, 991
reduction of, **1**, 11, 18; **4**, 699
reductive amination of, **3**, 188
reductive halogenation of, **2**, 14
regioselective propargylation of, **1**, 1055
samarium-mediated methylenation of, **1**, 680
selenation of, **3**, 385
silyl anion addition to, **4**, 354
C-stannylation of, **4**, 386
as starting materials
for acetals, **4**, 177
for *O,N*-acetals, **4**, 311
for *Se,N*-acetals, **4**, 333
for *S,N*-acetals, **4**, 315, 328
for *S,P*-acetals, **4**, 345
for acid bromides, **5**, 18
for acid chlorides, **5**, 15
for acid fluorides, **5**, 6
for acyl alkyl acetals, **4**, 212
for acylsilanes, **5**, 410
for *N*-acylsulfonamides, **5**, 345
for aldehydes, **3**, 10
for alkanethiols, **2**, 130
for alkenes, **1**, 593, 630
for alkyl aldehydes, **3**, 21
for alkyl aldehydes with aryl or hetaryl substituents, **3**, 24
for alkynylphosphine oxides, **2**, 1067
for aminals, **4**, 405
for amines, **2**, 310
for α-aminoalkyl phosphonates, **4**, 466
for benzal chlorides, **6**, 28
for benzonitriles, **3**, 662
for α-bromo alcohols, **4**, 42
for bromo dithiacetals, **4**, 245
for α-bromostannanes, **4**, 155
for α-bromo sulfides, **4**, 71
for carbamoyl azides, **6**, 511
for α-chloro ethers, **4**, 47
for α-chloro sulfides, **4**, 69
for cyanohydrins, **3**, 615
for *gem*-diazidoalkanes, **4**, 432
for dibromoalkanes, **4**, 25
for dichloroalkanes, **4**, 18
for difluoroalkanes, **4**, 8
for diselenoacetals, **4**, 287
for dithioacetals, **2**, 541; **4**, 245
for epoxides, **1**, 268
for *gem*-formamido phosphorylalkenes, **4**, 1004
for geminal diacetates, **4**, 203
for hemiacetals, **4**, 164
for hemiaminals, **4**, 294
for hemidithioacetals, **4**, 245
for hemithioaminals, **4**, 324
for homoallylic alcohols, **2**, 73
for α-iodostannanes, **4**, 155
for isocyanates, **5**, 986
for isothiocyanates, **5**, 1042
for ketene monothioacetals, **4**, 834
for ketones, **3**, 123, 127
for monothioacetals, **4**, 217
for nitriles, **3**, 617, 621
for 1,3,4-oxathiazoles, **3**, 686
for phosphorus compounds, **4**, 336

for α-phosphoryl enamines, **4**, 1002
for α-silyl aldehydes, **3**, 51
for *N*-silylamides, **5**, 377
for 1-silyl-1,3-dienes, **2**, 936
for silylenol ethers, **4**, 1049
for α-silyl ketones, **3**, 202
for silyloxy ketones, **3**, 163
for *N*-substituted imines, **3**, 405
for sulfides, **2**, 139
for symmetrical acetals, **4**, 190
for thiazolidines, **4**, 318
for thiiranes, **2**, 140
for thioaldehydes, **3**, 336, 340, 341
for vinyl bromides, **2**, 624
for vinyl chlorides, **2**, 618
for vinyl fluorides, **2**, 611
for vinyl halides, **2**, 607
for vinyl iodides, **2**, 630
in Strecker reaction, **3**, 616
sulfenylation of, **3**, 35
synthesis of, **3**, **1–52**
factors affecting, **1**, 873
use of umpolung methods for, **3**, 19
via aluminum catalysts, **1**, 387
via hydroformylation, **1**, 296
synthesis of functions, **3**, **313–327**
tantalum-promoted reactions, with alkynes, **2**, 988
thiaenolization of, **2**, 709
use in the synthesis of 1,3-azarsa compounds, **4**, 495
use in the synthesis of 1,3-benzazarsolines, **4**, 496
via rearrangement, **1**, 386
Aldehydes, α-acyloxy, synthesis of, **3**, 31
Aldehydes, aliphatic
reactions
with 2-[(alkoxycarbonyl)amino] aldehydes, **2**, 81
with diethyl ethoxycarbonylalkyl phosphonates, **1**, 747
as starting materials
for ketene dithioacetal *S,S*-dioxides, **4**, 858
for α-methoxyaliphatic acids, **5**, 67
Aldehydes, alkenyl, hydrostannylation of, **3**, 78
Aldehydes, alkoxy
anti addition, **3**, 29
syn addition, **3**, 29
Aldehydes, α-alkoxy
synthesis of
via ozonolysis of allylic ethers, **3**, 31
via Pummerer rearrangement, **3**, 32
Aldehydes, β-alkoxy, as starting materials, for β-alkoxyimides, **3**, 32
Aldehydes, 2-[(alkoxycarbonyl)amino], reactions, with aliphatic aldehydes, **2**, 81
Aldehydes, alkyl
alkynyl-substituted, synthesis of, **3**, 25
with aryl or hetaryl substituents, synthesis of, **3**, 24
from
aldehydes, **3**, 21
formylation reactions, **3**, 23
with more than one double bond, synthesis of, **3**, 24
with one double bond, synthesis of, **3**, 21
reactions
with sulfenyl chlorides, **3**, 35
with trimethylsilyl isothiocyanate, **5**, 1042
with triphenylphosphine, **2**, 446
synthesis of, **3**, **1–52**
Claisen rearrangement for, **3**, 21
via oxy-Cope rearrangement, **3**, 22
via Wittig oxy-Cope rearrangement, **3**, 22
Aldehydes, α-alkyl selenyl, synthesis of, **3**, 37
Aldehydes, α-(alkylthio), reactions, with amines, **3**, 418
Aldehydes, alkynyl, reduction of, **1**, 506
Aldehydes, β-allenic, synthesis of, **1**, 979
Aldehydes, allenyl, from, β-chlorovinyl aldehydes, **1**, 981
Aldehydes, allenylstannane, cyclization of, **1**, 1057
Aldehydes, allylate, reactions, with η3-allylmolybdenum complexes, **2**, 76
Aldehydes, amino
from
α-amino acids, **3**, 39
amino alcohols, **3**, 40
optically active, synthesis of, **3**, 40
synthesis of, **3**, 39
via oxidative cleavage of alkenes, **3**, 40
via oxidative cleavage of 1,2-diols, **3**, 40
Aldehydes, 2-amino, allylation of, **2**, 74
Aldehydes, α-amino
from
α,β-epoxy sulfoxides, **3**, 44
α-halo aldehydes, **3**, 43
hydration of, **3**, 39
reactions, with organometallics, **3**, 39
synthesis of, **1**, 261
via 1,2 amino migration, **3**, 43
via formylation of nitrones, **3**, 43
Aldehydes, β-amino
synthesis of
via conjugate addition of amines to enols, **3**, 44
via [1+3] cycloaddition reactions, **3**, 44
via hydroboration of propargylic amines, **3**, 44
via Mannich reactions, **3**, 45
Aldehydes, aromatic
reactions, with formamide oxime, **5**, 837
as starting materials
for aromatic acids, **5**, 114
for dithioic acids, **5**, 549
Aldehydes, aryl
Baeyer–Villiger oxidation of, **2**, 654
from, aryl halides, **3**, 90
reactions
with hexamethylphosphorus triamide, **4**, 475
with 2-hydroxyethyltriphenylarsonium bromide, **1**, 738
with sodium hydrogen selenide, **2**, 735
with tri-*n*-butylstibonium benzylide, **1**, 743
with triphenylstibonium tetraphenylcyclopentadienylide, **1**, 647
as starting materials, for *N*-alkylarylamides, **5**, 302
synthesis of, **3**, **81–109**
methods, **3**, 82
Aldehydes, α-aryl selenyl, synthesis of, **3**, 37
Aldehydes, *o*-azido, reactions, with hydrogen sulfide, **3**, 342
Aldehydes, α-azido, from, α,β-epoxy sulfones, **3**, 48
Aldehydes, β-azido, synthesis of, **3**, 48
Aldehydes, benzylic, reduction of, **1**, 11, 18
Aldehydes, bromo
as starting materials, for α-thiol aldehydes, **3**, 34
synthesis of, **3**, 67
Aldehydes, α-bromo, synthesis of, **3**, 26, 27
Aldehydes, α-t-butyldimethylsilyl, synthesis of, **3**, 51
Aldehydes, carbohydrate, reduction of, **1**, 18
Aldehydes, chiral 2-aryl, synthesis of, **3**, 25
Aldehydes, α-chloro, synthesis of, **3**, 26
Aldehydes, β-chloro, condensation of, **3**, 416
Aldehydes, β-chlorovinyl
reactions, with ethylene glycols, **4**, 196
as starting materials, for allenyl aldehydes, **1**, 981
Aldehydes, ω-cyano, from, cycloalkenes, **3**, 634
Aldehydes, diazo, synthesis of, **3**, 47
Aldehydes, α-diazo, synthesis of, via diazo group transfer, **3**, 473
Aldehydes, 2,3-dibromo, as starting materials, for alkynals, **3**, 79
Aldehydes, dichloro, reactions, with 1,3-diketones, **3**, 219
Aldehydes, α,α-dichloro, synthesis of, **3**, 27
Aldehydes, α,α-difluoro, synthesis of, **3**, 27
Aldehydes, α,α-dihalo, synthesis of, **3**, 27
Aldehydes, β,β-diselenyl, synthesis of, **3**, 39
Aldehydes, enamino, synthesis of, **3**, 76

Aldehydes, α-fluoro, synthesis of, **3**, 26, 27
Aldehydes, ω-functionalized, from, 2-(2-bromoethyl)-1,3-dioxolane, **1**, 119
Aldehydes, furanyl, reactions, with organometallics, **3**, 303
Aldehydes, halo
from
enamines, **3**, 26
enolsilanes, **3**, 26
Aldehydes, α-halo
elimination of, **3**, 54
reactions, with trialkyl phosphites, **3**, 50
reductive halogenation of, **3**, 12
as starting materials
for α-amino aldehydes, **3**, 43
for α-oxygenated aldehydes, **3**, 32
for α-sulfenyl aldehydes, **3**, 35
synthesis of, **3**, 26; **4**, 132
Aldehydes, haloalkyl, synthesis of, **3**, 26
Aldehydes, 3-halogenated α,β-unsaturated, synthesis of, **3**, 68
Aldehydes, halogenated α,β-unsaturated, synthesis of, **3**, 66
Aldehydes, *gem*-halo phosphonamido, synthesis of, **4**, 804
Aldehydes, *O*-hetaryl, synthesis of, **3**, 102
Aldehydes, heteroaryl, synthesis of, **3**, **81–109**
Aldehydes, *N*-heterocyclic, synthesis of, **3**, 106
Aldehydes, *O*-heterocyclic, synthesis of, **3**, 102
Aldehydes, heterocyclic aryl, synthesis of, **3**, 102
Aldehydes, homochiral trialkylsilyl, synthesis of, **3**, 51
Aldehydes, hydroxy
from
acyl dithioacetals, **3**, 30
cyanohydrins, **3**, 30
epoxy sulfones, **3**, 29
keto aldehydes, **3**, 31
stability of, **3**, 31
synthesis of, **3**, 28
Aldehydes, α-hydroxy
from, 1,2-diols, **3**, 30
synthesis of, **3**, 28, 42; **4**, 231
via ozonolysis, **3**, 30
Aldehydes, β-hydroxy
from, formyl anions, **3**, 31
synthesis of, via ozonolysis of homoallylic alcohols, **3**, 31
Aldehydes, hydroxyaryl, oxidation of, via alkaline peroxide, **2**, 655
Aldehydes, hydroxylamino, synthesis of, **3**, 46
Aldehydes, γ-hydroxy-α,β-unsaturated, synthesis of, **1**, 580
Aldehydes, imino, synthesis of, **3**, 411
Aldehydes, α-iodo, synthesis of, **3**, 27
Aldehydes, isothiocyanato, synthesis of, **5**, 1033
Aldehydes, keto
reactions
with diethylzinc, **2**, 72
with 3-methoxyphenylmagnesium bromide, **4**, 166
reduction of, **3**, 161
as starting materials, for hydroxy aldehydes, **3**, 31
Aldehydes, 1,3-keto
selenylation of, **3**, 174
synthesis of, **4**, 250
Aldehydes, α-keto, reactions, with thiols, **4**, 215
Aldehydes, β-keto
amination of, **3**, 263
reactions, with phenyl selenyl chloride, **3**, 38
Aldehydes, α-mercapto
dimerization of, **3**, 34
from, bromo aldehydes, **3**, 34
Aldehydes, α-methyl, reactions, with enethiolates, **5**, 557
Aldehydes, α-methylene
from, siloxycyclopropanes, **1**, 663
synthesis of, **3**, 60
Aldehydes, α-nitroso, synthesis of, **3**, 47
Aldehydes, α-oxo, synthesis of, **4**, 162
Aldehydes, α-oxygenated
from
α-halo aldehydes, **3**, 32
α-halo aldimines, **3**, 32
Aldehydes, β-oxygenated, from, epoxy silyl ethers, **3**, 160
Aldehydes, pentenyl, intramolecular reductive cyclization reactions of, **2**, 81
Aldehydes, polyaromatic, synthesis of, **3**, 99
Aldehydes, polycyclic aryl, from, naphthaldehydes, **3**, 99
Aldehydes, polyhydroxy, synthesis of, **3**, 32
Aldehydes, polyoxygenated, synthesis of, **3**, 29, 32
Aldehydes, potassium enolates, *C*-alkylation of, **3**, 21
Aldehydes, primary, reactions, with tin(II) amides, **2**, 814
Aldehydes, propargylic
reactions, with β-aminoacrylates, **3**, 66
synthesis of, **3**, 78
Aldehydes, quinoline
reactions
with chlorotris(triphenylphosphine)rhodium, **5**, 459
with quinuclidones, **3**, 258
as starting materials, for acyl allyl complexes, **5**, 459
Aldehydes, saturated, reactions, with 2-(diphenylphosphinoyl)acetaldehydes, **2**, 81
Aldehydes, saturated unsubstituted, synthesis of, **3**, 2
Aldehydes, selenium heterocyclic, synthesis of, **3**, 106
Aldehydes, α-selenyl
from, alkenes, **3**, 38
synthesis of, **3**, 37
Aldehydes, β-selenyl, synthesis of, **3**, 39
Aldehydes, silicon-functionalized, synthesis of, **3**, 51
Aldehydes, silyl, synthesis of, **3**, 78
Aldehydes, α-silyl, synthesis of, **3**, 51
Aldehydes, β-silyl, synthesis of, **3**, 52
Aldehydes, silyloxy, from, trialkyl silyl ethers, **3**, 33
Aldehydes, α-silyloxy
from, cyanohydrins, **3**, 33
synthesis of
via the oxidation of silyl enol ethers, **3**, 33
via silylformylation, **3**, 33
Aldehydes, β-silyloxy, synthesis of, via Lewis acid-catalyzed rearrangement, **3**, 33
Aldehydes, β-silyl-α,β-unsaturated, synthesis of, **2**, 928
Aldehydes, α-sulfenyl
alkylation of, **3**, 35
from
alkenylsilanes, **3**, 36
alkenyl sulfides, **3**, 36
α,β-epoxy sulfoxides, **3**, 36
α-halo acids, **3**, 35
α-halo aldehydes, **3**, 35
synthesis of, via sulfenylation of aldehydes, **3**, 35
Aldehydes, β-sulfenyl, synthesis of, via sulfenylation of homoenolate anions, **3**, 36
Aldehydes, α-sulfonyl, synthesis of, **3**, 37
Aldehydes, tellurium heterocyclic, synthesis of, **3**, 106
Aldehydes, thiadiazole, as starting materials, for triazole thionoamides, **5**, 622
Aldehydes, β-thioketo, synthesis of, **3**, 347
Aldehydes, thiophene, reactions, with indole-3-ones, **3**, 257
Aldehydes, 2-thio α,β-unsaturated, synthesis of, **3**, 72
Aldehydes, 3-thio α,β-unsaturated, synthesis of, **3**, 72
Aldehydes, β-thioxo, as starting materials, for enamino thioketones, **3**, 356
Aldehydes, β-trialkylsilyl, synthesis of, **3**, 52
Aldehydes, β-trichlorostannyl, dehydrostannation of, **1**, 663
Aldehydes, 6-trimethylsilyloxy, synthesis of, **3**, 34
Aldehydes, β-trimethylsilyl-α,β-unsaturated, synthesis of, **3**, 77
Aldehydes, α-trisubstituted, synthesis of, **1**, 385
Aldehydes, unhindered aromatic, as starting materials, for allenes, **6**, 706
Aldehydes, unsaturated, synthesis of, **1**, 412
Aldehydes, (*E*)-α,β-unsaturated, synthesis of, **1**, 647
Aldehydes, α,β-unsaturated

acetalization of, **4**, 181
[1,2] addition, tendency for, **1**, 302
alkylation of, **3**, 213
conjugate addition to, **1**, 273, 313
from
alcohols, **3**, 57
α-alkynic alcohols, **3**, 59
bismethylthio allyl anions, **3**, 54
carbonyls, **3**, 68
cyclopropanes, **3**, 65, 66
epoxides, **3**, 62
α-epoxy aldehydes, **3**, 57
ketones, **3**, 54
halogenation of, **3**, 67
nitrogen-substituted, synthesis of, **3**, 74
α-nitrogen-substituted, synthesis of, **3**, 74
β-nitrogen-substituted, synthesis of, **3**, 75
oxygen-substituted, synthesis of, **3**, 71
reactions
with amines, **3**, 44
with benzylidenetriphenylphosphorane, **1**, 733
with cobalt(II) chloride, **3**, 214
with germanium compounds, **1**, 290
with α-ketophosphoranes, **3**, 220
with nucleophiles, **1**, 303
with organolithiums, **1**, 310
with selenates, **3**, 39
with α-silylanions, **1**, 651
with thiols, **2**, 138
with trimethylsilylamines, **2**, 741
with ylides, **1**, 734
selenium-substituted, synthesis of, **3**, 74
as starting materials, for tertiary α-cyano enamines, **2**, 748
in the Strecker reaction, **3**, 617
synthesis of, **3**, **53–79**
via aldol condensation, **3**, 59
via Diels–Alder reaction, **3**, 61
via elimination reactions, **3**, 54
via isomerization, **3**, 61
via Nef reactions, **3**, 57
via reduction, **3**, 61
via Wittig reaction, **3**, 60, 65
Aldehydes, β,γ-unsaturated
from, 1,3-dienes, **5**, 442
synthesis of, **3**, 23
Aldehydes, γ,δ-unsaturated
from
alkyl vinyl ethers, **1**, 415
titanium acyl compounds, **5**, 441
synthesis of, **1**, 141; **3**, 23
Aldehydes, δ,ε-unsaturated, synthesis of, **3**, 24
Aldehydes, ω-unsaturated, synthesis of, **1**, 841
4-Aldehydes, pyrrolidone, reactions, with arenesulfonylmethyl isocyanides, **4**, 908
endo-Aldehydes, reactions, with hexylidenetriphenylphosphorane, **1**, 729
Aldehydosugars, reactions, with ethoxycarbonylmethylidenetriphenylphosphorane, **1**, 725
Aldimine, lithio-, synthesis of, **5**, 792
Aldimines
as intermediates, in the synthesis of acylsilanes, **5**, 419
α-monochlorination of, **3**, 416
reactions
with alkyl nitrates, **2**, 788
with alkynyllithiums, **1**, 1041
with allenyl metals, **1**, 1059
with bis(*p*-methylbenzyl) hydrogen phosphite, **4**, 461
with (chloromethyl)phosphonic diesters, **4**, 473
with dialkyl phosphites, **4**, 462
with dialkyl phosphoriodites, **4**, 462
with diethyl phosphite, **4**, 461
with germanium hydrides, **4**, 521
with Grignard reagents, **2**, 312
with *N*-halosuccinimides, **3**, 416
with organolithiums, **2**, 312
with (3-oxoalkyl)phenylphosphines, **4**, 454
with phosphorus trichloride, **4**, 488
with propargyl metals, **1**, 1059
reductive dimerization of, **2**, 311
as starting materials
for amines, **1**, 1059
for α-aminophosphonic acids, **4**, 461
synthesis of, **3**, 404
via Michael addition, **3**, 406
Aldimines, *N*-alkyl-, hydrolysis of, **3**, 86
Aldimines, aryl-
from
isocyanides, **3**, 411
N-nitrosoamines, **3**, 411
oxaziridines, **3**, 411
synthesis of, **3**, 411
Aldimines, *N*-carbon-substituted, synthesis of, **3**, 406
Aldimines, α-chloro-, as starting materials, for α-cyano enamines, **2**, 749
Aldimines, α-halo-, as starting materials, for α-oxygenated aldehydes, **3**, 32
Aldimines, α-lithio-, synthesis of, **3**, 276
Aldimines, metallated
as intermediates, **3**, 697
synthesis of, **3**, 418
Aldimines, metallic, carbonation of, **5**, 68
Aldimines, metallo-
reactions
with aminohaloboranes, **5**, 795
with chlorotrialkylsilane, **5**, 792
with dialkylhaloboranes, **5**, 795
as starting materials, for imidoylsilicons, **5**, 792
Aldimines, 3-methyl-2-aminopyridyl-, as starting materials, for imidoylrhodiums, **5**, 800
Aldimines, *N*-silyl-, synthesis of, **2**, 544; **3**, 456
Aldimines, stannyl-, synthesis of, **3**, 418
Aldimines, *N*-stannyl-, synthesis of, **3**, 457
Aldimines, *N*-sulfonyl-, reactions, with alkynyllithiums, **1**, 1042
Aldimines, *N*-trimethyl-, synthesis of, **3**, 440
Aldimines, *N*-trimethylsilyl-, synthesis of, **3**, 439, 456
Aldimines, α,β-unsaturated, synthesis of, **2**, 312
N-H Aldimines
from, oxaziridines, **3**, 404
synthesis of, **3**, 404
Aldimine salts, reactions, with diethyl phosphite, **4**, 464
Aldoamides, acetyl-, as starting materials, for acetylaldonic acids, **5**, 66
Aldol addition
for the synthesis of β-lactones, **5**, 161
to carbonyls, **2**, 715
Aldol condensation
for the synthesis of acylsilanes, **5**, 426
for the synthesis of 3(2*H*)-furanones, **3**, 245
for the synthesis of 2-thio α,β-unsaturatedketones, **3**, 247
for the synthesis of α,β-unsaturated aldehydes, **3**, 59, 66
for the synthesis of α,β-unsaturated ketones, **3**, 219, 230
Aldol reaction
of carbene complexes, **5**, 956
enzyme-catalyzed, **3**, 159
nitro compounds in, **2**, 394
reversibility of, **1**, 883
for the substitution of silicon functions, **1**, 283
for the synthesis of β-hydroxy ketones, **3**, 159
for the synthesis of β-telluro ketones, **3**, 176
for the synthesis of thionoamides, **5**, 607
See also Mukaiyama aldol reaction
Aldol reaction, asymmetric, for the synthesis of thiocarboxylic esters, **5**, 236, 237
Aldol reaction, diastereoselective, for the synthesis of chiral boron enolates, **2**, 698

Aldol reaction, enantioselective, for the synthesis of chiral boron enolates, **2**, 698
Aldol reaction, nitro-
for the synthesis of *gem*-halo nitroalkenes, **4**, 798
See also Henry reaction
Aldols, reactions, with sulfone ortho esters, **5**, 166
Aldonic acids, acetyl-, from, acetylaldoamides, **5**, 66
Aldonitrones
reactions
with cyanotrimethylsilane, **3**, 417
with Grignard reagents, **2**, 345
with organometallics, **2**, 345
Aldoses, reactions, with arylhydrazines, **3**, 450
2-Aldoxime, thiophene-, reactions, with chlorine, **3**, 683
Aldoximes
C-alkylation of, **3**, 428
C-arylation of, **3**, 428
Beckmann rearrangement of, **5**, 275
chlorination of, **5**, 765, 821
from, α,β-unsaturated nitroalkenes, **3**, 426
halogenation of, **5**, 664
metallation of, **3**, 427
oxidation of, **3**, 684
reactions
with *N*-bromosuccinimide, **5**, 821
with *t*-butyl hydroperoxide, **3**, 684
with *N*-chlorosuccinimide, **5**, 821
with isocyanates, **3**, 434
as starting materials
for nitrile oxides, **3**, 683
for oxyimidic halides, **5**, 821
for primary amides, **5**, 298
synthesis of, **5**, 215
Aldoximes, *O*-methyl
deoximation of, **3**, 430
synthesis of, **3**, 430
Aldoximes, pyridine, *O*-alkylation of, **3**, 430
(*Z*)-Aldoximes, alkylation of, **3**, 434
α-Aldoximes, hydrolysis of, **5**, 65
β-Aldoximes, hydrolysis of, **5**, 65
Algae, use in the synthesis of ketones, **1**, 677
Alicyclic substrates, aromatization of, **2**, 658
Aliphatic acids, α-methoxy-, from, aliphatic aldehydes, **5**, 67
Aliquat 336, for the reduction of benzene, **1**, 94
Alkadienes, lithio-, synthesis of, **1**, 634
Alkali amide bases, reactions, with aliphatic nitriles, **5**, 750
Alkali arsenes
reactions, with ammonium salts, **4**, 499
use in the synthesis of aminomethylarsines, **4**, 501
Alkali cyanates, reactions, with alkyl halides, **5**, 964
Alkali cyanides
reactions
with alcohols, **5**, 1103
with phenols, **5**, 1103
Alkali earth metals, in liquid ammonia, for the reduction of isonitriles, **1**, 38
Alkali earth metals, trihalomethyl, synthesis of, **6**, 244
Alkali hydroxides, reactions, with ammonium salts, **1**, 627
Alkali metal alkylarsenides
synthesis of, **2**, 510
use in the synthesis of transition metal complexes, **2**, 512
Alkali metal alkynylides, reactions, with brominating reagents, **2**, 1018
Alkali metal amide bases, development of, **1**, 131
Alkali metal amides
reactions, with alkynyl chlorides, **2**, 1041
use in elimination reactions, **1**, 1065
Alkali metal arsenites, reactions, with alkyl halides, **2**, 497
Alkali metal cyanates
reactions
with alkyl halides, **5**, 964
with amines, **6**, 504
with diamines, **6**, 508
Alkali metal cyanides
reactions
with chlorotrimethylsilane, **5**, 1134
with tellurium, **2**, 290
Alkali metal dialkylstibides
reactions
with 1,2-dibromoethane, **2**, 505
with 1,2-dichloroethane, **2**, 505
Alkali metal diorganoarsenides
reactions
with alkenyl halides, **2**, 874
with aryl halides, **2**, 874
Alkali metal diorganostibnides
reactions
with alkenyl halides, **2**, 874
with aryl halides, **2**, 874
Alkali metal diphenylphosphides, synthesis of, **2**, 826
Alkali metal fluorides, use in the synthesis of trifluoromethyls, **6**, 18
Alkali metal halides, reactions, with sulfonate esters, **2**, 9
Alkali metal hydrides, reactions, with dialkylaluminum chlorides, **2**, 596
Alkali metal hydroxides, use in the synthesis of alkynes, **1**, 1064
Alkali metal methylarsenide, reactions, with 2-chloroethyl alcohol, **2**, 482
Alkali metal phosphides, synthesis of, **2**, 475
Alkali metals
in liquid ammonia
for the reduction of isonitriles, **1**, 38
use in the cleavage of nitriles, **1**, 55
for the migration of double bonds, **1**, 795
organic derivatives, reactivity of, **1**, 112
reactions
with arsenic dihalides, **2**, 893
with bis(trimethylsilyl)mercury, **4**, 354
with chlorosilanes, **2**, 545
with dialkylhaloarsines, **2**, 510
with dialkylhalostibines, **2**, 510
with dimethyliodoarsine, **2**, 505
with iodoform, **6**, 32
with liquid ammonia, **1**, 55
with metal salts, **2**, 958
with tertiary phosphines, **2**, 432, 437
synthesis of, **2**, 894; **6**, 369
and transmetallation, **2**, 992
use in the cleavage of carbon–phosphorus bonds, **1**, 53
Alkali metals, trihalomethyl, synthesis of, **6**, 244
Alkali metals, trimethylsilyl, synthesis of, **4**, 354
Alkali metal salts
reactions
with carboxylic acid chlorides, **5**, 937
with dihalomethanes, **4**, 147
synthesis of, **5**, 937
Alkali metal selenocyanates
reactions
with alkyl halides, **5**, 1113
with carbonimidoyl dichlorides, **5**, 1056
Alkali metal stibides, as starting materials, for arylstibines, **2**, 895
Alkali metal tellurocyanates, synthesis of, **2**, 290
Alkali metal tetrakis(trifluoromethyl)cadmates, synthesis of, **6**, 246
Alkali metal thiocyanates, reactions, with halides, **5**, 1038
Alkaline aqueous bases, use in the reduction of carbon–mercury bonds, **1**, 68
Alkaline earth metal thiocyanates, solid-supported, studies, **5**, 1039
Alkaline hydrolysis, of nitriles, **5**, 269
Alkaline metals, and transmetallation, **2**, 992
Alkali phosphides, reactions, with α,ω-dihaloalkanes, **2**, 429
Alkalis, reactions, with phosphines, **2**, 475
Alkaloid (±)-crinines, synthesis of, **1**, 266

Alkaloids
reactions, with diazomethane, **1**, 627
structural determination of, **1**, 627
synthesis of, **1**, 791
Alkaminium salts, *N*-alkyl-*N*-sulfinyl-, reactions, with sulfoxides, **2**, 360
Alkanals, *ω*-*H*-perfluoro-, reactions, with trimethylselenophosphate, **3**, 385
Alkanamine, *N*-trityl, reactions, with bis(trimethylsilyl)phosphonites, **4**, 461
Alkane, bis(trimethylsilyl)amino-, reactions, with phosgene, **6**, 507
Alkane, diazo-
reactions
with acid chloride, **3**, 192
with cobalt, **4**, 686
with rhodium, **4**, 686
as starting material, for carbon-bridged dicobalt complexes, **4**, 685
use in the synthesis of methylene-bridged manganese complexes, **4**, 676
Alkane, *N*,*N*-difluoroamino-, synthesis of, **6**, 52
Alkanedioic acid dichlorides, reactions, with trimethylsiloxyalkyl isocyanates, **5**, 1006
Alkanedioic acids, synthesis of, reviews, **5**, 24
Alkane halides, reactions, with phosphorothiolates, **2**, 120
Alkane–metal complexes, diazo-, synthesis of, **6**, 673
Alkanephosphonates, α-diazo-, synthesis of, **5**, 867
Alkanephosphonates, diisopropyl dichloro-, as starting materials, for (dichloroalkyl)phosphines, **5**, 1152
Alkanes
bromination of, **4**, 19
chlorination of, **4**, 11
cleavage of, **1**, 901
dehydrogenation of, **1**, 554
deprotonation of, **2**, 554
in Friedel–Craft alkylations, **1**, 492
from
alkyl halides, **1**, 1
semicarbazones, **1**, 42
halogenation of, **2**, 2; **4**, 2
iodination of, **4**, 28
photochemical iodination of, **2**, 33
reactions
with alkenes, **1**, 294
with cyanogen azide, **5**, 1119
with haloorganosilanes, **6**, 173
with isocyanates, **5**, 386
with lead tetra(trifluoroacetate), **2**, 86
with phenylphosphinic acid esters, **2**, 845
with sulfur, **2**, 144
with sulfur dioxide, **2**, 191
with sulfur trioxide, **2**, 176
as starting materials
for aldehydes, **3**, 2
for alkyl bromides, **2**, 27
for alkyl chlorides, **2**, 20
for alkyl fluorides, **2**, 15
for alkyl halides, **2**, 2
for alkyl hydroperoxides, **2**, 96
for alkyl iodides, **2**, 33
for amines, **2**, 306
for dibromoalkanes, **4**, 19
for dichloroalkanes, **4**, 11
for difluoroalkanes, **4**, 2
for diiodoalkanes, **4**, 28
for *gem*-dinitroalkenes, **4**, 993
for ketones, **3**, 112
for trialkyl phosphates, **2**, 108
synthesis of
via Raney nickel reduction, **1**, 21
via reduction of alkenes, **1**, 76
via reduction of ketones, **1**, 19
Alkanes, acylaminoimino-, synthesis of, **4**, 449
Alkanes, acylaminomethylazo-, synthesis of, **4**, 448
Alkanes, acylaminomethylimino-, synthesis of, **4**, 448
Alkanes, acyloxyiodo-, synthesis of, **5**, 204
Alkanes, α-aluminosilyl-
from
organolithiums, **4**, 659
organomagnesiums, **4**, 659
synthesis of, **4**, 659
Alkanes, β-amino dichloro-, as starting materials, for ketene aminals, **4**, 979
Alkanes, aryl-
from, aromatic ketones, **1**, 19
reactions, with maleic anhydride, **5**, 201
Alkanes, aryl diazo-
as starting materials, for diarylethylenes, **1**, 717
synthesis of, **3**, 465
via Forster reaction of oximes, **3**, 463
Alkanes, α-arylhydrazono nitro-, as starting materials, for hydrazonic acid esters, **5**, 703
Alkanes, 1-azido 1-methoxy-, synthesis of, **4**, 313
Alkanes, azo-
as starting materials, for carbon-centered radicals, **1**, 944
synthesis of, **4**, 440
Alkanes, 1-(benzotriazol-1-yl)-1-chloro-, displacement of, **4**, 308
Alkanes, α,ω-bis(bisdialkylaminomethylphosphine)-, synthesis of, **4**, 452
Alkanes, 1,1-bis(boryl)-
from
alkynes, **4**, 630
1-alkynes, **2**, 906
See also Alkanes, *gem*-diboryl-
Alkanes, bis(carboxyalkylamino)-, pyrolysis of, **6**, 507
Alkanes, 1,1-bis(dialkoxyboryl)-, synthesis of, **6**, 199
Alkanes, α,ω-bis(dialkylarseno)-, synthesis of, **2**, 486
Alkanes, α,ω-bis(diazo)-, from, dinitroso compounds, **3**, 467
Alkanes, 1,1-bis(difluoroamino)-, dehydrofluorination of, **5**, 815
Alkanes, 1,1-bisgermyl-
from
carbon suboxide, **4**, 628
digermenes, **4**, 627
germenes, **4**, 627
ketenes, **4**, 628
synthesis of, **4**, 625
via changing the groups attached to the central methylene, **4**, 628
via changing the ligands on germanium, **4**, 629
via oxidative addition, **4**, 628
via quenching a carbanion with a germyl electrophile, **4**, 625
via quenching a germyl anion with a carbon electrophile, **4**, 626
via rearrangements, **4**, 628
Alkanes, bis(halosulfonyloxy)fluoro-, synthesis of, **6**, 42
Alkanes, bis(heteroaryl)-, synthesis of, **1**, 262
Alkanes, bisheterocyclic-, uses of, **1**, 262
Alkanes, bisimino-, reactions, with salicyl aldehyde, **4**, 438
Alkanes, bismesitylboryl(trimethylsilyl)-, reactions, with fluoride ions, **1**, 764
Alkanes, bis(primary phosphine)-, synthesis of, **2**, 428
Alkanes, 1,1-bis(silyl)-
from
bis(silyl)ketenes, **4**, 619
di-Grignard reagents, **4**, 608
disilanes, **4**, 614
disilenes, **4**, 609
1,1-disilylated epoxides, **4**, 619
1,2-disilylethanes, **4**, 616
α-lithiosilylalkanes, **4**, 602
α-magnesiosilylalkanes, **4**, 606
α-metallosilylalkanes, **4**, 602

silenes, **4**, 609
1,1,1-tris(trimethylsilyl)alkanes, **4**, 619
as starting materials, for α-lithio compounds, **4**, 653
synthesis of, **4**, 602
via changing the groups attached to the central methylene, **4**, 620
via changing the ligands on silicon, **4**, 621
via Diels–Alder reactions, **4**, 611
via hydrosilylation, **4**, 617
via quenching of carbanions with silyl electrophiles, **4**, 602
via quenching silyl anions with carbon electrophiles, **4**, 609
via rearrangements, **4**, 614
Alkanes, 1,2-bis(silyl)-, from, vinylsilanes, **4**, 617
Alkanes, 1,1-bis(silyl)-1-lithio-, synthesis of, **4**, 618
Alkanes, α,α'-bis(trichlorophosphazo)-α,α,β,β,α'α'β'β'-octachloro-, hydrolysis of, **5**, 57
Alkanes, α-borylgermyl-, synthesis of, **4**, 646
Alkanes, α-borylsilyl-
from
1,3-disilacyclobutanes, **4**, 644
vinylsilanes, **4**, 643
synthesis of, **4**, 642
via changing the groups attached to the central methylene, **4**, 645
via changing the groups attached to the metalloids, **4**, 645
via hydroboration of vinylsilanes, **4**, 643
via quenching a carbanion with a boryl electrophile, **4**, 642
via quenching a carbanion with a silyl electrophile, **4**, 642
via rearrangements, **4**, 642
Alkanes, boryl(stannyl)-, reactions, with alkyllithiums, **6**, 201
Alkanes, bromo-
from, carboxylic acids, **4**, 1077
reactions
with *O*-ethyl triphenyltin dithiocarbonate, **6**, 551
with mercury(II) nitrate, **2**, 103
with prenylpotassiums, **1**, 113
reduction of, **1**, 8
chemoselectivity in, **1**, 9
as starting materials, for nitro compounds, **2**, 393
Alkanes, bromochloro-
from
bromochlorocarbene, **4**, 38
carboxylic acids, **4**, 38
haloalkanes, **4**, 37
haloalkenes, **4**, 38
ketones, **4**, 38
synthesis of, **4**, 37
Alkanes, bromofluoro-
from
bromofluorocarbene, **4**, 35
carboxylic acids, **4**, 35
haloalkanes, **4**, 33
haloalkenes, **4**, 34
as starting materials, for α-fluoro aldehydes, **3**, 27
synthesis of, **4**, 33
Alkanes, bromo iodo-, synthesis of, **4**, 40
Alkanes, chloro-
from, carboxylic acids, **4**, 1077
reactions, with chloroalkenes, **6**, 4
reduction of, **1**, 6, 97
Alkanes, α-chloroazo-, synthesis of, **4**, 109
Alkanes, chlorofluoro-
from
alkenes, **4**, 30
carboxylic acids, **4**, 33
chlorofluoroalkenes, **4**, 32
chlorofluorocarbene, **4**, 32
haloalkanes, **4**, 30
imines, **4**, 32
synthesis of, **4**, 30
Alkanes, *N*-chlorofluoro-, synthesis of, **6**, 51
Alkanes, chloroiodo-
from
carboxylic acids, **4**, 40
haloalkanes, **4**, 39
haloalkenes, **4**, 39
ketones, **4**, 40
synthesis of, **4**, 39
Alkanes, *gem*-chloronitroso-, reactions, with iodine, **4**, 40
Alkanes, 1,1-diacetoxy-, synthesis of, **4**, 203
Alkanes, dialkoxydialkylamino-, as intermediates, in the synthesis of ketene hemiaminals, **4**, 881
Alkanes, N^1,N^2-dialkylidene 1,1-diamino-, reactions, with dialkyl phosphites, **4**, 461
Alkanes, 1,1-dialumino-, synthesis of, **6**, 208
Alkanes, α,ω-diamino-, reactions, with bishydrazidoyl halides, **5**, 681
Alkanes, diaryl-, synthesis of, **1**, 475
Alkanes, diaryl diazo-, reactions, with cobalt isocyanide complexes, **3**, 599
Alkanes, 1,1-diaryl-2-halo-, metallation of, **1**, 378
Alkanes, *gem*-diazido-
from
aldehydes, **4**, 432
gem-dihaloalkanes, **4**, 432
ketones, **4**, 432
trimethylsilyl azide, **4**, 432
synthesis of, **4**, 432
Alkanes, diazo-
acylation of, **3**, 474
antimony-substituted, synthesis of, **3**, 482; **6**, 669
arsenic-substituted, synthesis of, **3**, 482; **6**, 669
bismuth-substituted, synthesis of, **3**, 482; **6**, 669
boron-substituted, synthesis of, **3**, 485; **6**, 672
cadmium-substituted, synthesis of, **3**, 488
as carbene sources, **4**, 669
elimination of nitrogen from, **1**, 637
from
N-alkyl-*N*'-nitro-*N*-nitrosoguanidines, **3**, 466
N-alkyl-*N*-nitrosocarboxamides, **3**, 467
N-alkyl-*N*-nitroso-*p*-toluenesulfonamides, **3**, 467
amides, **3**, 465
N-nitrosoureas, **3**, 466
urethanes, **3**, 465
germanium-substituted, synthesis of, **3**, 482; **6**, 669, 672
halogen-substituted, synthesis of, **3**, 478
lanthanide-substituted, synthesis of, **3**, 487
lead-substituted, synthesis of, **3**, 482; **6**, 671, 675
lithium salts
synthesis of, **6**, 670, 672
use in the synthesis of α-diazophosphonothioic diamides, **6**, 667
lithium-substituted, synthesis of, **3**, 485
magnesium-substituted, synthesis of, **3**, 486
mercury-substituted, synthesis of, **3**, 488
metallation of, **3**, 488
nitrogen-substituted, synthesis of, **3**, 479
phosphorus-substituted, synthesis of, **3**, 480; **6**, 667
reactions
with alcohols, **4**, 198
with alkyl hydrogen selenites, **2**, 101
with alkylmercury ethoxides, **3**, 488
with alkyl nitroso compounds, **2**, 765
with alkylsulfinic acids, **2**, 100
with alkynes, **1**, 549
with arylmercury ethoxides, **3**, 488
with aryl nitroso compounds, **2**, 765
with bromochlorocarbene, **4**, 759
with carbenes, **4**, 755, 759
with chlorophosphines, **4**, 114

with cycloalkenones, **3**, 473
with 1-cyclopropenylphosphine oxides, **3**, 481
with 1-cyclopropenyl phosphonates, **3**, 481
with electrophiles, **5**, 871
with haloalkenes, **4**, 109
with halo metal complexes, **4**, 147
with iodine, **4**, 29
with ketones, **1**, 389
with monomeric carbonyls, **4**, 686
with phosphaalkenes, **4**, 484
with selenones, **4**, 333
with sulfonyl chlorides, **3**, 479
with sulfur dioxide, **2**, 207
with sulfur monoxide, **3**, 368
with thiiriene-1,1-dioxides, **3**, 479
with thioketones, **4**, 328
silicon-substituted, synthesis of, **3**, 482; **6**, 672
silver salts
synthesis of, **6**, 670
use in the synthesis of phosphadiazoles, **6**, 667
silver-substituted, synthesis of, **3**, 487
sodium-substituted, synthesis of, **3**, 485
as starting materials
for *Se,N*-acetals, **4**, 335
for *S,P*-acetals, **4**, 346
for carbenes, **1**, 408
for carbenoids, **6**, 673
for carbon-bridged dimetals, **4**, 709
for carbon-bridged dirhodium complexes, **4**, 686
for α-diazo ketones, **3**, 474
for difluoroalkanes, **4**, 10
for nitrile imides, **3**, 682
for phosphaalkenes, **5**, 880
structure of, **2**, 1003
sulfur-substituted, synthesis of, **3**, 478
synthesis of, **1**, 637; **3**, 465; **5**, 865
for the synthesis of phosphorus-containing heterocycles, **4**, 488
thallium-substituted, synthesis of, **3**, 485
tin-substituted, synthesis of, **3**, 482; **6**, 670, 675
transition metal-substituted, synthesis of, **3**, 486
zinc-substituted, synthesis of, **3**, 488
Alkanes, *gem*-diazo-, synthesis of, **4**, 436
Alkanes, 1-diazo-1-halo-, synthesis of, **5**, 844
Alkanes, diazo nitro-, synthesis of, **4**, 439
Alkanes, *gem*-diboryl-
reactions, with organolithiums, **4**, 665
as starting materials, for α-lithioborylalkanes, **4**, 664
See also Alkanes, 1,1-bisboryl-
Alkanes, 1,1-diboryl-1-metallo-, synthesis of, **4**, 633
Alkanes, 1,2-diboryl-1-silyl-, synthesis of, via borylboration, **4**, 644
Alkanes, 1,1-diboryl-substituted, from, boron halides, **6**, 181
Alkanes, dibromo-
from
aldehydes, **4**, 25
alkanes, **4**, 19
alkenes, **4**, 23
alkynes, **4**, 24
amines, **4**, 27
carboxylic acids, **4**, 27
dibromocarbene, **4**, 24
dihaloalkanes, **4**, 22
hydrazones, **4**, 27
imines, **4**, 27
ketones, **4**, 25
trihaloalkanes, **4**, 23
reactions, with mercuric fluoride, **4**, 4
synthesis of, **4**, 19
Alkanes, 1,2-dibromo-
elimination of, for vinyl bromides, **2**, 622
as starting materials, for alkynes, **1**, 1065
Alkanes, *gem*-dibromo-, as starting materials, for ketene diselenoacetals, **4**, 875
Alkanes, α,ω-dibromo-, reactions, with vinyllithiums, **1**, 464
Alkanes, dichloro-
dehydrohalogenation of, **1**, 1066
from
aldehydes, **4**, 18
alkanes, **4**, 11
alkenes, **4**, 14
alkynes, **4**, 15
amines, **4**, 19
dichlorocarbene, **4**, 16
dihaloalkanes, **4**, 13
imines, **4**, 19
ketones, **4**, 18
trihaloalkanes, **4**, 13
reactions, with lithium dialkylarsenides, **2**, 486
as starting materials, for α-chlorophosphonic dichlorides, **4**, 130
synthesis of, **4**, 11
Alkanes, 1,1-dichloro-, dehydrohalogenation of, **1**, 1067
Alkanes, α,ω-dichloro-
reactions
with alkylarsenides, **2**, 486
with primary phosphines, **2**, 433
Alkanes, dichloroalkoxy-, synthesis of, **6**, 41
Alkanes, dichloroamino-, synthesis of, **6**, 52
Alkanes, 1-(*N,N*-dichloroamino)-2-chloro-, synthesis of, **2**, 338
Alkanes, dichloronitroso-
reactions
with carboxylic acid chlorides, **5**, 669
with trimethylchlorosilane, **5**, 669
as starting materials, for hydroxamoyl halides, **5**, 669
Alkanes, α,ω-dicyano-, reactions, with phosphorus pentachloride, **5**, 57
Alkanes, diethylamino-, synthesis of, **6**, 54
Alkanes, difluoro-
from
aldehydes, **4**, 8
alkanes, **4**, 2
alkenes, **4**, 5
alkynes, **4**, 6
diazoalkanes, **4**, 10
difluorocarbene, **4**, 7
dihaloalkanes, **4**, 3
imines, **4**, 10
ketones, **4**, 8
trihaloalkanes, **4**, 5
metallation of, **6**, 63
synthesis of, **4**, 2
Alkanes, difluoroamino-, synthesis of, **6**, 51
Alkanes, *gem*-difluoroamino-
safety precautions, **4**, 439
synthesis of, **4**, 439
Alkanes, difluoroiodo-, metallation of, **6**, 63
Alkanes, dihalo-
reactions
with alkyllithiums, **2**, 998; **6**, 62
with antimony pentafluoride, **2**, 1006
with dimetallic alkyl arsenides, **2**, 486
with trimethyltinlithium, **4**, 699
as starting materials
for dibromoalkanes, **4**, 22
for dichloroalkanes, **4**, 13
for difluoroalkanes, **4**, 3
synthesis of, **4**, **1–40**
Alkanes, 1,1-dihalo-
Arbuzov reactions of, **4**, 555
deprotonation of, **1**, 164
reactions
with 1,2-dimetallodiphosphines, **4**, 545
with metallated cyclopolyphosphines, **4**, 545
with metal phosphides, **4**, 545
with phosphines, **4**, 546

with phosphorus nucleophiles, **4**, 553
with platinumphosphine complexes, **4**, 584
as starting materials
for alkynes, **1**, 1064
for bis(phosphonium salts), **4**, 553
Alkanes, 1,2-dihalo-, as starting materials, for alkynes, **1**, 1064
Alkanes, *gem*-dihalo-
reactions, with lithium, **4**, 667
as starting materials
for α-chlorophosphonic dihalides, **4**, 130
for *gem*-diazidoalkanes, **4**, 432
Alkanes, α,ω-dihalo-
dialkylation of, **1**, 141
reactions
with alkali phosphides, **2**, 429
with metallocycles, **1**, 116
for the synthesis of spirocyclic compounds, **1**, 148
Alkanes, ω-dihalo-, as starting materials, for cyclic alkoxycarbene complexes, **5**, 943
Alkanes, dihaloamino-, from, amines, **6**, 51
Alkanes, *N,N*-dihaloamino-, synthesis of, **6**, 53
Alkanes, dihalonitro-, synthesis of, **6**, 54
Alkanes, *gem*-dihydrazino-, synthesis of, **4**, 436
Alkanes, *gem*-dihydroxylamino-, synthesis of, **4**, 435
Alkanes, *gem*-diimino-
reactions, with aralkyl ketones, **4**, 445
synthesis of, **4**, 437
Alkanes, diiodo-
from
alkanes, **4**, 28
alkynes, **4**, 29
diiodocarbene, **4**, 29
haloalkanes, **4**, 28
imines, **4**, 29
as starting materials, for carbon-bridged dicobalt complexes, **4**, 686
synthesis of, **4**, 28
Alkanes, α,ω-diiodo-
reactions
with diynes, **1**, 111
with trimethylsilylalkyne, **4**, 810
Alkanes, α,ω-diiodoperfluoro-, as starting materials, for diisocyanates, **5**, 967
Alkanes, *gem*-diisocyanato-
reactions
with phosphorus ylides, **4**, 992
with triphenylphosphine imides, **4**, 441
as starting materials, for *gem*-alkylideneamino isocyanatoalkenes, **4**, 992
Alkanes, 1,1-dilithio-, synthesis of, **4**, 667
Alkanes, 1,1-dimetallo-
reactions
with chlorophosphine oxides, **4**, 556
with chlorophosphines, **4**, 546
synthesis of, **2**, 1006
Alkanes, *gem*-dimetallo-, synthesis of, theoretical, **2**, 1005
Alkanes, dinitro-
reactions, with halogens, **6**, 55
synthesis of, **6**, 55
Alkanes, *gem*-dinitro-
from
halo nitroalkanes, **4**, 432
nitroalkenes, **4**, 431
1,1,1-trinitroethane, **4**, 431
synthesis of, **4**, 428
Alkanes, α,ω-dinitro-, synthesis of, via Kolbe electrolysis, **2**, 396
Alkanes, dinitrohalo-, interconversion of, **6**, 57
Alkanes, *gem*-dinitrosamino-, oxidation of, **4**, 440
Alkanes, *gem*-dinitroso-, synthesis of, **4**, 428
Alkanes, *gem*-diphosphine-, synthesis of, **2**, 428
Alkanes, dithiol-
reactions
with acid chlorides, **4**, 1073
with ethanoic anhydrides, **4**, 1074
Alkanes, fluoro-
chlorination of, **4**, 30
epoxidation of, **6**, 42
reduction of, **1**, 6
Alkanes, fluorodinitro-, synthesis of, **6**, 56
Alkanes, fluorohalo-, synthesis of, **4**, 30
Alkanes, fluoroiodo-
from
carboxylic acids, **4**, 37
fluoroiodocarbene, **4**, 36
haloalkanes, **4**, 36
haloalkenes, **4**, 36
reactions
with dimethyl disulfide, **6**, 45
with methyl sulfide, **6**, 45
synthesis of, **4**, 36
Alkanes, fluoronitro-, synthesis of, **6**, 54
Alkanes, α-galliosilyl-, synthesis of, **4**, 659
Alkanes, geminal diamino-, synthesis of, **4**, 404
Alkanes, geminal dihalo-, as starting materials, for carbenes, **1**, 299
Alkanes, geminal disilyl-, desilylation of, **1**, 160
Alkanes, geminal trialumino-, synthesis of, **6**, 208
Alkanes, geminal trilithio-, synthesis of, methods, **6**, 204
Alkanes, 1-germanyl 1-silyl-, as starting materials, for 1-germanyl 1-silylalkenes, **4**, 1058
Alkanes, 1-germyl-1-silyl-
from
disilenes, **4**, 639
silenes, **4**, 639
synthesis of, **4**, 636, 639
Alkanes, α-germylsilyl-
synthesis of, **4**, 636
via changing the groups attached to the metalloids, **4**, 640
via changing the groups attached to the methylene, **4**, 640
via hydrogermylation, **4**, 639
via quenching a carbanion with a germyl electrophile, **4**, 636
via quenching a carbanion with a silyl electrophile, **4**, 637
via quenching a metalloid anion with a carbon electrophile, **4**, 638
via rearrangements, **4**, 639
Alkanes, halo-
for the alkylation of lithium compounds, **1**, 111
coupling reactions, **1**, 117
reactions
with alkenes, **6**, 4
with alkylaluminums, **1**, 128
with alkyl hydroperoxides, **2**, 98
with allyltins, **1**, 128
with arylmagnesium bromides, **1**, 117
with bis(*p*-tolyl)mercury, **1**, 128
with α,α'-dimetallated intermediates, **1**, 146
with dithiocarbamates, **6**, 561
with fluoroalkenes, **6**, 4
with lithiumtrialkylmanganates, **1**, 125
with organocadmiums, **1**, 127
with organomercury compounds, **1**, 128
with organometallics, **1**, 109
with α-sulfonyl Grignard reagents, **1**, 144
with zinc, **1**, 122
with zinc alkyl compounds, **1**, 127
as starting materials
for bromochloroalkanes, **4**, 37
for bromofluoroalkanes, **4**, 33
for chlorofluoroalkanes, **4**, 30
for chloroiodoalkanes, **4**, 39

for diiodoalkanes, **4**, 28
for fluoroiodoalkanes, **4**, 36
for ortho amides, **6**, 140
Alkanes, α-halo azo-
reactions, with amines, **4**, 444
synthesis of, **4**, 109
Alkanes, (halomethyl)silyl-
reactions
with tin, **4**, 660
with tin dichloride, **4**, 660
Alkanes, halonitro-, as starting materials, for *gem*-dinitroalkanes, **4**, 432
Alkanes, α-halonitro-, as starting materials, for α-nitro sulfones, **4**, 332
Alkanes, hydrazino-, synthesis of, **4**, 440
Alkanes, β-hydroxyaryl-, as intermediates, in the synthesis of diarylalkanes, **1**, 475
Alkanes, α-hydroxy bromofluoro-, synthesis of, **4**, 33
Alkanes, hypervalent halo-, reduction of, **1**, 11
Alkanes, β-iminophosphoryldiazo-, synthesis of, **3**, 481
Alkanes, iodo-
alkylation of, **6**, 128
oxidative acylation of, **5**, 204
radical substitution of, **1**, 10
reactions
with tetra-*n*-butyl *N*-thioformyl dithiocarbamate, **6**, 561
with trichloromethyllithium, **6**, 23
reduction of, **1**, 10
chemoselectivity in, **1**, 10
Alkanes, iododinitro-, synthesis of, **6**, 56
Alkanes, iodofluoro-
as starting materials
for α,α-dichloro sulfides, **6**, 45
for lithium difluoroalkylsulfinates, **6**, 45
Alkanes, iodoperfluoro-
from, perfluorocarboxylic acids, **6**, 32
reactions
with tin(II) chloride, **6**, 65
with tin(II) fluoride, **6**, 65
synthesis of, **6**, 32
Alkanes, isocyanato-, synthesis of, **4**, 441
Alkanes, *gem*-isocyanato phosphoryl-, as starting materials, for *gem*-phosphorodiamidoyl ureidoalkenes, **4**, 1011
Alkanes, α-keto monofluoro-, bromination of, **4**, 33
Alkanes, 1-lithio-1-aryl-1-silyl-, reactions, with tin halides, **4**, 660
Alkanes, α-lithioboryl-
from, *gem*-diborylalkanes, **4**, 664
synthesis of, via deprotonation, **4**, 664
Alkanes, lithiodifluoro-, synthesis of, **6**, 63
Alkanes, α-lithiogermyl-
synthesis of, **4**, 626
via deprotonation, **4**, 662
via halogen–lithium exchange, **4**, 662
Alkanes, 1-lithio-1-silyl-, synthesis of, via deprotonation, **4**, 646
Alkanes, α-lithiosilyl-
as starting materials
for 1,1-bis(silyl)alkanes, **4**, 602
for 1,3-disilacyclobutanes, **4**, 605
synthesis of, **4**, 646
via addition of an organolithium to a vinylsilane, **4**, 651
via deprotonation, **4**, 646
via halogen–lithium exchange, **4**, 652
via sulfur–lithium exchange, **4**, 652
via transmetallation, **4**, 652
transmetallation of, **4**, 660
Alkanes, α-magnesiosilyl-
as starting materials, for 1,1-bis(silyl)alkanes, **4**, 606
synthesis of, **4**, 653, 654
via addition of a Grignard reagent to a vinylsilane, **4**, 654
via halogen–magnesium exchange, **4**, 653
via rearrangements, **4**, 655
via transmetallation, **4**, 655
transmetallation of, **4**, 660
Alkanes, α-metallated nitro-, synthesis of, **4**, 540
Alkanes, 1-metallo-1,1-bisgermyl-, synthesis of, **4**, 628
Alkanes, 1-metallo-1,1-bis(silyl)-, synthesis of, **4**, 620
Alkanes, α-metalloboryl-
reactions, with silyl electrophiles, **4**, 642
synthesis of, **4**, 665
Alkanes, α-metallodiazo-, as starting materials, for alkylidyne complexes, **3**, 691
Alkanes, α-metallogermyl-
as intermediates, for the synthesis of α-germylsilylalkanes, **4**, 637
reactions, with silyl electrophiles, **4**, 636
synthesis of, **4**, 662
Alkanes, 1-metallo-1-silyl-, synthesis of, **4**, 646
Alkanes, α-metallosilyl-
reactions
with boryl electrophiles, **4**, 642
with electrophilic complexes, **4**, 656
with germyl electrophiles, **4**, 636
as starting materials, for 1,1-bis(silyl)alkanes, **4**, 602
transmetallation of, **4**, 660
Alkanes, nitramino-, synthesis of, **4**, 440
Alkanes, nitro-
as acyl anion equivalents, **1**, 165
of α-anions, **4**, 540
cycloaddition of diazomethane to, **4**, 440
deprotonation of, **1**, 156
double deprotonation of, **4**, 540
from
alkyl nitrates, **4**, 428
ketoximes, **2**, 397
nitroalkenes, **2**, 398
nitrosoalkanes, **2**, 398
halogenation of, **5**, 665
Henry reaction of, **4**, 540
as Michael donors, **2**, 394
Nef reaction of, **4**, 540
nitration of, **4**, 428
reactions
with allyl ethers, **2**, 395
with arenediazonium salts, **3**, 449; **4**, 439
with benzeneselenyl chloride, **6**, 116
with *n*-butyllithium, **2**, 784
with carbonyls, **2**, 784
with enals, **3**, 47
with hexamethyldisilathiane, **5**, 637
with silver nitrate, **4**, 428
with sodium nitrite, **4**, 428
with tetranitromethane, **4**, 429
reduction of, **1**, 36
as starting materials
for aldehydes, **3**, 18
for carbon-centered radicals, **1**, 945
for *gem*-dinitroalkenes, **4**, 994
for ketones, **3**, 134
for 2-(trialkylsilyloxy)oxime *O*-trialkylsilyl ethers, **3**, 436
substitution of, **1**, 251
synthesis of, **4**, 439
Alkanes, 1-nitro-, reactions, with triethyl orthoformate, **2**, 789
Alkanes, nitronitroso-, synthesis of, **4**, 439
Alkanes, nitroso-, as starting materials, for nitroalkanes, **2**, 398
Alkanes, perfluoro-, reactions, with copper, **6**, 63
Alkanes, perfluorobromonitro-, reduction of, **5**, 822
Alkanes, perfluorochloronitro-, reduction of, **5**, 822
Alkanes, perhalo-, reactions, with cadmium, **6**, 64
Alkanes, perhalofluoro-, reactions, with ylides, **4**, 118
Alkanes, 1-(phenylthio)-2-(phenylseleno)-, synthesis of, **2**, 285

Alkanes, 1-phenylthio-1-silyl-
  reactions, with *N*-chlorosuccinimide, **4**, 943
  reductive lithiation of, **4**, 652
Alkanes, phosphinodiazo-
  from, phosphaalkynes, **5**, 1159
  as intermediates, in the synthesis of phosphaalkynes, **5**, 1159
Alkanes, phosphoryldiazo-, synthesis of, **3**, 482
Alkanes, α-plumbylsilyl-, synthesis of, **4**, 660
Alkanes, polycyclic, reactions, with dimeric rhodium compounds, **5**, 460
Alkanes, polyhalo-
  reactions, with lithium vapor, **6**, 204
  sulfurization of, **5**, 479
  synthesis of, **1**, 168
  use in the synthesis of alkyl halides, **2**, 7
Alkanes, polylithiated, synthesis of, **6**, 205
Alkanes, α-potassiosilyl-, synthesis of, **4**, 653
Alkanes, primary halo-, reactions, with vinylic lithium compounds, **1**, 111
Alkanes, primary iodo-, reactions, with lithium dialkylcuprates, **1**, 120
Alkanes, primary nitro-
  as starting materials
    for α-nitrocarboxylic acids, **5**, 95
    for *N*-thionoacylhydroxylamines, **5**, 637
Alkanes, β-propargylic nitro-, synthesis of, **1**, 1062
Alkanes, quaternary nitro-, as starting materials, for alkenes, **1**, 637
Alkanes, α-selenosilyl-, transmetallation of, **4**, 652
Alkanes, α-sodiosilyl-, synthesis of, **4**, 653
Alkanes, α-stannylsilyl-
  from, 1-sila-1-stanna-1-alkenes, **4**, 661
  synthesis of, **4**, 660
    via quenching a stannyl anion with a carbon electrophile, **4**, 660
    via transmetallation of α-metallostannylalkanes, **4**, 660
    via transmetallation reactions, **4**, 660
Alkanes, tertiary, synthesis of, **1**, 154
Alkanes, tertiary chloro-, alkylation of, **1**, 128
Alkanes, tertiary halo-, alkylations of, **1**, 141
Alkanes, tertiary nitro-, as radical precursors, **1**, 330
Alkanes, tertiary perfluoroiodo-, synthesis of, safety precautions, **6**, 3
Alkanes, 1,1,2,2-tetrahalo-, dehalogenation of, **1**, 1069
Alkanes, α-thalliosilyl-, synthesis of, **4**, 659
Alkanes, α-thiocyanatoaryl-
  from
    thiocyanogen, **5**, 1110
    thiocyanogen halides, **5**, 1110
Alkanes, thiophosphoryl diazo-, synthesis of, **3**, 480
Alkanes, 1,1,1-trialumino-, synthesis of, **6**, 208
Alkanes, 1,1,1-triamino-
  characteristics of, **6**, 138
  synthesis of, **6**, 137
Alkanes, 1,1,1-triaryl-2-halo-, metallation of, **1**, 378
Alkanes, 1,1,1-trichloro-, reductive silylation of, **6**, 173
Alkanes, trichloromethyl-, synthesis of, **6**, 23
Alkanes, β-(trichloromethyl)nitro-, synthesis of, **6**, 24
Alkanes, trifluoro-, as starting materials, for tertiary ynamines, **2**, 1042
Alkanes, 1,1,1-trigermyl-, synthesis of, **6**, 185
Alkanes, trihalo-
  as starting materials
    for dibromoalkanes, **4**, 23
    for dichloroalkanes, **4**, 13
    for difluoroalkanes, **4**, 5
Alkanes, 1,1,1-trihalo-
  reactions, with sulfur, **5**, 480
  as starting materials
    for ortho esters, **6**, 68
    for triselenoortho esters, **6**, 92
    for trithioortho esters 603079.635
Alkanes, 1,1,1-trilithio-, synthesis of, **6**, 204
Alkanes, 1,1,1-trimetallated, with different metals attached to carbon, **6**, 209
Alkanes, α-trimethylsilyldiazo-, synthesis of, **5**, 873
Alkanes, trimethylsilyl-substituted, reactions, with iodosobenzene, **2**, 1020
Alkanes, trinitro-, synthesis of, **6**, 348
Alkanes, 1,1,1-trinitro-
  characteristics of, **6**, 143
  synthesis of, **6**, 143
    via addition of intact trinitromethyl groups, **6**, 144
Alkanes, 1,1,2-trisboryl-, synthesis of, **4**, 632
Alkanes, 1,1,1-tris(dihaloboryl)-, synthesis of, **6**, 181
Alkanes, 1,1,1-trisilyl-, synthesis of, methods, **6**, 172
Alkanes, tris(phenylthio)-, reactions, with dimethyl(methylthio) sulfonium fluoroborate, **3**, 283
Alkanes, tristannyl-
  from, 1,1-distannyl-2,2-diorganyl-1-alkenes, **6**, 209
  stability of, **4**, 1069
Alkanes, 1,1,1-tristannyl-, synthesis of, **6**, 209
Alkanes, 1,1,1-tris(trimethylsilyl)-, as starting materials, for 1,1-bis(silyl)alkanes, **4**, 619
Alkanes, unactivated
  bromination of, **2**, 27
  chlorination of, **2**, 20
Alkanes, unsaturated diazo-, synthesis of, **3**, 488
Alkanes, α,β-unsaturated diazo-
  from
    N-nitrosoureas, **3**, 489
    N-nitrosourethanes, **3**, 489
  synthesis of, **3**, 489
Alkanes, α,β-unsaturated nitro-, as starting materials, for oximes, **3**, 426
Alkanes, unsymmetrical, oxidation of, **1**, 554
Alkanes, vinyl diazo-, use in the synthesis of α,β-unsaturated diazoalkanes, **3**, 489
Alkanesulfonic esters, perfluoro-, synthesis of, **1**, 433
Alkanethiols
  chlorination of, **2**, 166
  from
    acyl halides, **2**, 130
    alcohols, **2**, 118
    aldehydes, **2**, 130
    alkenes, **2**, 116
    alkyl halides, **2**, 119
    carboxylic acids, **2**, 130
    ketones, **2**, 130
    phosphorothiolates, **2**, 120
    sulfonyl halides, **2**, 125
    thioacids, **2**, 130
    thiocyanates, **2**, 127
    trithiocarbonates, **2**, 123
    xanthates, **2**, 123
  Lewis acid-catalyzed elimination of, **4**, 844
  oxidation of, **2**, 186
  reactions
    with alkyl iodides, **2**, 134
    with aminoarsines, **2**, 500
    with arenesulfonylalkynes, **4**, 861
    with azides, **2**, 240
    with boron dichlorides, **2**, 268
    with boron trichlorides, **2**, 268
    with chlorosilanes, **2**, 262
    with chlorotrialkylsilanes, **2**, 262
    with dialkylhaloarsines, **2**, 499
    with diaryl selenides, **2**, 237
    with ethylmagnesium bromide, **2**, 273
    with hexamethyldisilazane, **2**, 263
    with hydrosilanes, **2**, 264
    with selenoaldehydes, **2**, 234
    with thionitrates, **2**, 220
    with thiophosphoryl tribromide, **2**, 252
    with trichlorophospholines, **2**, 253

silylation of, **2**, 263
as starting materials
for disulfides, **2**, 215
for thiosilanes, **2**, 262
storage as disulfides, **2**, 126
synthesis of, **2**, 116
via heterocyclic ring-cleavage, **2**, 129
t-Alkanethiols, synthesis of, **2**, 125
Alkane transition metal complexes, diazo-, synthesis of, **6**, 675
Alkanimines, α-chloro-, as starting materials, for α-phosphoryl enamines, **4**, 1004
1-Alkanimines, *N*-bromoperhalo-, from, perhalogenated nitriles, **5**, 808
Alkanoamides, *N*-alkenyl-, synthesis of, **5**, 288
Alkanoamides, *N*-alkyl-
from
esters, **5**, 277
nitriles, **5**, 281
synthesis of, **5**, 276
reagents for, **5**, 277
Alkanoamides, *N*-aryl-, synthesis of, **5**, 290, 291
Alkanoamides, *N*-heteroaryl, synthesis of, **5**, 290
Alkanoamides, *N*-unsubstituted, synthesis of, **5**, 274
Alkanoates, ω-isocyanato-, synthesis of, **5**, 998
Alkanoic acid, ^{13}C-labeled oxygen-substituted, synthesis of, **5**, 71
Alkanoic acids
anionic synthesis of, **5**, 76
reactions, with *N*,*N*-dimethylmethoxymethaneiminium methyl sulfate, **3**, 76
as starting materials, for amides, **5**, 274
synthesis of, **5**, 28
reviews, **5**, 28, 62
Alkanoic acids, 6-amino-, lactamization of, **5**, 276
Alkanoic acids, ω-amino-
from, ω-amino-substituted vinyl 1,1-dichlorides, **5**, 88
synthesis of, **5**, 86
Alkanoic acids, aryl-, synthesis of, **1**, 181
Alkanoic acids, α-aryl-, synthesis of, reviews, **5**, 28
Alkanoic acids, (*S*)-2-chloro-, synthesis of, **5**, 60
Alkanoic acids, cyano-, synthesis of, **5**, 95
Alkanoic acids, halo-
synthesis of, **5**, 54
via carbonation of organometallics, **5**, 58
via oxidation, **5**, 59
Alkanoic acids, ω-halo-, for the alkylation of sodium alkynides, **1**, 1001
Alkanoic acids, heteroatom-substituted
synthesis of, **5**, 96
via carbonation of organometallics, **5**, 97
via hydrolysis, **5**, 96
Alkanoic acids, nitrogen-substituted
synthesis of, **5**, 90
via carbonation of organometallics, **5**, 95
via hydrolysis, **5**, 92
via oxidation, **5**, 95
Alkanoic acids, oxygen-substituted
synthesis of, **5**, 62
via carbonation of organometallics, **5**, 68
via hydrolysis, **5**, 62
Alkanoic acids, β-(substituted benzoyl), synthesis of, **5**, 63
Alkanoic acids, sulfur-substituted
synthesis of, **5**, 81
via carbonation of organometallics, **5**, 82
via hydrolysis, **5**, 81
via oxidation, **5**, 83
Alkanoic acids, α,α,γ-trichloro-, synthesis of, **5**, 57
Alkanoic acids, unsubstituted
synthesis of, **5**, 29
via carbonation of organometallics, **5**, 42
via oxidation, **5**, 46
Alkanoic acid trimethylsilyl esters, 2-isothiocyanato-, synthesis of, **5**, 1030
Alkanoic anhydrides, symmetrical
from
acyl chlorides, **5**, 185
carboxylic acids, **5**, 185
metal carboxylates, **5**, 185
synthesis of, **5**, 185
Alkanoic anhydrides, unsymmetrical
from
acyl chlorides, **5**, 192
alkenes, **5**, 193
carboxylic acids, **5**, 193
imines, **5**, 193
ketenes, **5**, 193
metal carboxylates, **5**, 192
synthesis of, **5**, 192
Alkanoic diacids, α,α,β,β,α'α'β'β'-octachloro-, synthesis of, **5**, 57
Alkanols
reactions
with acetyl hypoiodite, **2**, 95
with boric acid, **2**, 111
with chlorosulfonic acid, **2**, 101
with phosphoric trichloride, **2**, 105
with phosphorous, **2**, 108
with phosphorous acid, **2**, 104
with phosphorous oxychloride, **2**, 107
with phosphorous trihalides, **2**, 104
with sulfuric acid, **2**, 101
with sulfur trioxide, **2**, 101
with sulfuryl chloride, **2**, 102
as starting materials
for aliphatic orthocarbonates, **6**, 296
for carbenium ions, **6**, 296
Alkanols, 2-amino-, reactions, with vinyl ethers, **4**, 306
Alkanols, dinitro-, as starting materials, for *gem*-dinitroalkenes, **4**, 994
Alkanols, ω-halo-, for the alkylation of sodium alkynides, **1**, 1001
Alkanols, 1-trifluoromethyl-, synthesis of, **2**, 85
1-Alkanols, 1,1-bis(trimethylsilyl)-, synthesis of, **5**, 420
n-Alkan-1-ols, synthesis of, **2**, 43
Alkanones, α-diazo-α-(trialkylsilyl)-, synthesis of, **3**, 483
Alkanoyl chlorides, reactions, with trimethylsilyl azide, **5**, 991
Alkanoyl chlorides, isocyanato-
reactions
with trimethylsiloxyalkyl isocyanates, **5**, 1006
with trimethylsiloxy isocyanates, **5**, 1007
Alkanoyl chlorides, tertiary, reactions, with trimethylsilyl azide, **5**, 992
Alkenals, 2-bromo-, synthesis of, **3**, 67
Alkenals, (*Z*)-2-bromo-, from, (*E*)-bromoacetals, **3**, 68
Alkenals, β,β-dichloro-, synthesis of, **3**, 70
Alkenals, α-halo-, synthesis of, via Wittig reaction, **3**, 68
Alkenals, α-sulfonyl-, synthesis of, **3**, 72
2-Alkenals, 2-chloro-, synthesis of, **3**, 67
2-Alkenals, 3-chloro-, reactions, with sulfur nucleophiles, **3**, 72
2-Alkenals, 2-fluoro-, synthesis of, **3**, 66
2-Alkenals, 2-halo-, reactions, with secondary aliphatic amines, **2**, 741
2-Alkenals, 2-iodo-, synthesis of, **3**, 67
2-Alkenals, 3-silyl-, synthesis of, **3**, 77
Alkenamine compounds, use of term, **4**, 898
Alkenation
of aldehydes, **3**, 642
of ketones, **3**, 642
for the substitution of phosphorus functions, **1**, 268
for the synthesis of α,β-unsaturated esters, **5**, 155
Alkenation, boron-mediated, for the synthesis of alkenes, **1**, 764
Alkenation, germanium-mediated, for the synthesis of alkenes, **1**, 765

Alkenation, silicon-mediated, **1**, 758
Alkenation, tellurium-mediated, for the synthesis of alkenes, **1**, 766
Alkene, 1,1-bis(bromomagnesio)-, from, alkynes, **4**, 1065
Alkene, 1,2-bis(silyl)-
from, alkynylsilanes, **2**, 923
as starting material, for 2-silyl-1-alkene, **2**, 923
Alkene, 1,1-bis(trichlorostannyl)-, synthesis of, **4**, 1069
Alkene, 1-boryl-1-silyl-
deborination of, **2**, 922
use in the synthesis of alkenylsilanes, **2**, 922
Alkene, 1-bromo-, synthesis of, **2**, 622
Alkene, bromochloro-, reactions, with tributylzinclithium, **1**, 431
Alkene, 1-chloro-, synthesis of, **2**, 616
Alkene, *gem*-chloro nitro-, synthesis of, **4**, 796
Alkene, 1,1-diplumbyl-, synthesis of, **4**, 1069
Alkene, fluorobromo-, dehydrohalogenation of, **2**, 1014
Alkene, 1-magnesio 1-silyl-, from, alkynes, **4**, 1060
Alkene, nitroso-, synthesis of, **2**, 778
Alkene, perfluoro-, reactions, with *t*-butylamine, **3**, 589
Alkene, tetraamino-, reactions, with dialkyl phosphite, **6**, 156
Alkene, 1-trimethylsilyl 1-trialkylstannyl-, synthesis of, **4**, 1062
1-Alkene, 2-silyl-, from, 1,2-bis(silyl)alkene, **2**, 923
π-Alkene complexes, reactions, with dialkyl cuprates, **2**, 580
Alkene exchange, *gem*-dibromo-, kinetic vs. thermodynamic control of, **4**, 818
Alkene inversion, mechanism, **1**, 595
Alkene ligands, isomerization of, **2**, 577
Alkene metathesis, for the synthesis of alkenes, **1**, 674
2-Alkenephosphonates, 2-cyano-
as intermediates, **3**, 646
synthesis of, **3**, 659
Alkene polymerization, mechanism, **1**, 662
Alkenes
acetalization of, **4**, 193
acid-catalyzed hydration of, **2**, 40
acylation of, **1**, 422; **3**, 212
additions, diastereoselectivity of, **1**, 327
additions of haloalkanes to, **6**, 4
alkoxymercuration of, **4**, 193
alkoxyselenenylation of, **2**, 282
alkylation of, **1**, 315
for the alkylation of amides, **5**, 285
for the allylation of aldehydes, **2**, 80
allylic amination of, **2**, 307
β-amidoselenenylation of, **2**, 284
amination of, **2**, 306
arylation of, **1**, 494
bromination of, **5**, 813
electrophilic, **2**, 30
bromofluorination of, **2**, 18
bromoiodination of, **2**, 31
as byproducts, in nitrile synthesis, **3**, 612
carbocupration of, **2**, 589
carbomagnesiation of, **2**, 562
chlorobromination of, **2**, 24
chloroiodination of, **2**, 24
cis-addition of chlorine to, **2**, 24
cyclopropanation of, **1**, 304; **4**, 152
deprotonation of, **1**, 1089
1,2-dihydroxylation of, **2**, 51
dimerization of, **1**, 294
[1+3] dipolar addition of nitrones to, **2**, 354
double silylation of, **4**, 618
electrochemical addition of haloalkenes to, **6**, 7
electrochemical bromination of, **2**, 30
electrocyclic additions to, **1**, 349
electron deficient, use in the synthesis of dialkyl sulfides, **2**, 138
electrophilic addition to, **1**, 294
electrophilic bromination of, **2**, 30
electrophilic selenenylation of, **2**, 284
ethylmagnesiation of, **2**, 561
formylation of, **3**, 58
in Friedel–Craft alkylations, **1**, 492
from
α-acetoxy bromides, **1**, 677
alcohols, **1**, 383
aldehydes, **1**, 593, 630
alkenes, **1**, 383, 568
β-alkoxy selenides, **1**, 716
alkyl selenoxides, **2**, 286
alkyl-*p*-toluenesulfonylhydrazones, **1**, 39
alkyl vinyl ethers, **1**, 415
alkynes, **1**, 455, 502
allenes, **1**, 458
allylic alcohols, **1**, 12
allyl sulfides, **1**, 236
aminodiols, **1**, 513
anthracene adducts, **1**, 785
arenesulfonylhydrazones, **1**, 630
arsonium ylides, **1**, 738; **3**, 48
aziridinium ions, **1**, 629
bis(sulfones), **4**, 275
β-bromo esters, **1**, 568
β-bromo sulfides, **1**, 591
β-chloro sulfides, **1**, 591
cycloalkyl acetates, **1**, 773
cyclopentadienes, **1**, 785
dialkylalkenylboranes, **1**, 487
disilanes, **4**, 1054
enamines, **1**, 443
enol ethers, **1**, 434
episulfides, **1**, 592
epoxides, **1**, 593
β-halo selenides, **1**, 594
β-halosilanes, **1**, 652
α-halo sulfones, **1**, 704
hydrazones, **1**, 635
β-hydroxyacids, **1**, 565, 681
β-hydroxy arsonium salts, **1**, 646
3-hydroxycarboxylic acids, **1**, 785
β-hydroxyphosphine oxides, **1**, 641
β-hydroxyphosphonic acid bisamides, **1**, 644
β-hydroxy selenides, **1**, 592, 716
β-hydroxy sulfides, **1**, 590
β-hydroxy sulfoxides, **1**, 600
β-hydroxy thiols, **1**, 591
ketones, **1**, 593, 630
β-lactones, **1**, 681
metallacyclobutanes, **1**, 666
metallated alkenes, **1**, 453
endo-2-methyl-2-norborneol, **1**, 576
organocobalt complexes, **1**, 335
oxathiolan-5-ones, **1**, 686
phosphonium betaine, **1**, 639
primary tellurides, **1**, 594
pyridinium cations, **1**, 629
quaternary nitroalkanes, **1**, 637
secondary amines, **1**, 636
1,3,4-selenadiazolines, **1**, 684, 686
selenides, **1**, 712
α-seleno ketones, **1**, 593
selenonium ylides, **1**, 712, 716
selenoxides, **1**, 716
stibonium ylides, **1**, 743
β-substituted germanes, **1**, 656
sulfinamides, **1**, 621, 695
sulfones, **1**, 617; **4**, 840
sulfoxides, **1**, 693
sulfoximines, **1**, 710
telluroxides, **1**, 716
1,3,4-thiadiazolines, **1**, 684

tosylhydrazones, **1**, 635
twofold extrusion reaction, **3**, 392
vinylboranes, **1**, 449
vinyl carbon–nitrogen bonds, **1**, 443
vinyl ethers, **1**, 480
vinylgermanes, **1**, 451
vinyl halides, **1**, 429
vinylphosphines, **1**, 446
vinyl selenides, **1**, 438
vinylsilanes, **1**, 450
vinyl sulfides, **1**, 438
vinyl sulfones, **1**, 439
vinyl sulfoxides, **1**, 439
vinyl tellurides, **1**, 438
xanthate esters, **1**, 700
functionalization of, **1**, 608
geometry of, **1**, 112
geometry inversion, **1**, 753
halochlorination of, **2**, 23
halofluorination of, **2**, 17
halogenation of, **2**, 3, 5
haloselenenylation of, **2**, 280
homologation of, **1**, 619
hydration of, **1**, 896; **2**, 40
hydroacylation of, for the synthesis of ketones, **3**, 113
hydroalumination of, **2**, 4, 970
hydroboration of, **1**, 75, 84, 1034, **2**, 4, 514, 525, 592, 970
hydrobromination of, **2**, 29
hydrochlorination of, **2**, 23
hydrofluorination of, **2**, 17
hydroformylation of, **2**, 576; **3**, 2; **5**, 442
hydrogenation of, **1**, 72, 281
hydrohalogenation of, **2**, 4
hydroiodination of, **2**, 34
hydrometallation of, **1**, 458; **2**, 583
hydrosilylation of, **2**, 4, 50, 536
cis-hydroxylation of, **2**, 45
hydroxyselenenylation of, **2**, 281
hydrozirconation of, **2**, 4
industrial production of, **1**, 674
intermolecular, **1**, 294
intramolecular cyclopropanation of, **1**, 354
iodination of, **2**, 35
iodine-catalyzed isomerization of, **1**, 797
iodofluorination of, **2**, 18
iridium-catalyzed carbonylation of, **5**, 416
isomerization of, **1**, 640, 650, 666, 882
contrathermodynamic, **1**, 655
lithiation of, **2**, 965
long-chain, synthesis of, **1**, 662
mercuration of, **2**, 40; **4**, 1067
metal-catalyzed [2+2] additions of, **1**, 357
metallation of, **2**, 965
via deprotonation with bases, **2**, 964
metathesis of, **1**, 674, 675
nitration of, **4**, 797
nitromercuration of, **1**, 668
β-nitroselenenylation of, **2**, 284
non-redox metal-hydrogen exchange of, **2**, 963
oxidation of, **2**, 44, 86, 96
oxidative carbon–hydrogen activation of, **2**, 962
oxidative cleavage of, **3**, 40, 113
oxo reaction, for aldehydes, **3**, 2
oxymercuration of, **1**, 343; **5**, 142
oxymetallation of, **6**, 209
oxypallidation of, **2**, 983
oxyselenocyanation of, **5**, 1115
oxyselenylation of, **3**, 38
ozonization of, **3**, 219
perfluoroalkylation of, **1**, 1130
photochemical [2+2] additions of, **1**, 355
PO-activated, synthesis of, **1**, 744
polymerization of, **6**, 4
protonation of, **1**, 911
radical additions to, **6**, 4
reactions
with acetylthiosulfenyl chloride, **5**, 248
with acid chlorides, **3**, 211
with acids, **1**, 895
with alanes, **2**, 43
with alcohols, **2**, 89
with alkanes, **1**, 294
with alkenes, **1**, 294, 332, 354
with alkenyl anions, **1**, 309
with alkyl anions, **1**, 309
with *N*-alkylhydroxylamines, **2**, 346
with alkyllithiums, **2**, 555
with alkylnitroso compounds, **2**, 352, 354
with alkyl radicals, **1**, 323
with alkyl sulfenic acids, **2**, 153
with alkylsulfenyl halides, **2**, 715
with alkynezirconocenes, **2**, 987
with alkynyl anions, **1**, 309
with allenic esters, **1**, 498
with allyl anions, **1**, 363
with allyl cations, **1**, 363
with allylic acetates, **1**, 195
with aluminum hydride, **2**, 596
with aminoiminoboranes, **5**, 381
with arenesulfenyl halides, **2**, 715
with arsenic trichloride, **4**, 597
with aryl acyl radicals, **3**, 292
with aryl anions, **1**, 309
with arylsulfenyl chlorides, **2**, 712
with 2-azaallyl anions, **1**, 371
with azides, **4**, 515
with azomethine ylides, **1**, 368
with benzeneselenenyl bromide, **2**, 280
with benzeneselenenyl chloride, **4**, 91
with benzeneselenyl bromide, **2**, 730
with benzenesulfenyl chloride, **4**, 106
with benzynezirconocenes, **2**, 987
with biscyclopentadienyl zirconium compounds, **5**, 441
with bis(fluoroxy)difluoromethane, **6**, 250
with bis(trifluoromethyl) trioxide, **6**, 42
with borane, **1**, 75; **2**, 514
with bromine compounds, **1**, 707
with bromomethanesulfonyl bromide, **4**, 83
with *t*-butyl hydroperoxide, **2**, 90
with carbenes, **1**, 298, 352, 988
with carbon monoxide, **5**, 192
with carbon nucleophiles, **1**, 300
with carboxylic acids, **3**, 212
with catecholborane, **2**, 42
with cationic complexes, **2**, 580
with chiral auxiliaries, **1**, 357
with chlorine, **2**, 23
with chloromethanesulfonyl bromide, **4**, 82
with chloromethanesulfonyl chloride, **4**, 82
with 5-chloro-2*H*-1,4-oxazine-2-ones, **5**, 661
with cobalt(I) complexes, **2**, 579
with cobalt tetracarbonyl, **5**, 456
with copper(II) halides, **5**, 1115
with cumyl hydroperoxide, **2**, 90
with cuprous halides, **3**, 698
with cyanamide, **5**, 1080, 1118
with cyanogen chloride, **6**, 606
with cyclic radicals, **1**, 325
with cyclopentadienyliron complexes, **2**, 577
with dialkylthiophosphinous acids, **2**, 462
with diboranes, **2**, 40, 514, 521
with dicyclopentadienyllanthanide hydrides, **2**, 43
with diethyl tetrasulfide, **2**, 137
with diethyl trichloromethyl phosphonate, **6**, 58
with diisobutylgallium hydride, **2**, 597
with dimethylaminosulfenyl chloride, **2**, 359

with diphenyl diselenide, **2**, 284
with electrophiles, **1**, 295
with electrophilic radicals, **1**, 331
with ethyl carbamate, **5**, 979
with ethyl *N*-chlorocarbamate, **5**, 979
with fluorodinitromethanides, **6**, 57
with fluorotrichloromethane, **6**, 4
with gallium hydride dihalide, **2**, 597
with Grignard reagents, **1**, 316
with α-halovinyl selenides, **4**, 92
with heterodienes, **1**, 375
with hydride complexes, **2**, 576
with hydridosilanes, **2**, 536
with hydrogen chloride, **2**, 23
with hydrogen cyanide, **3**, 614
with hydrogen iodide, **2**, 34
with hydrogen peroxide, **2**, 96
with hydrogen sulfide, **2**, 116, 137
with hydroxylamines, **2**, 343
with hypophosphite esters, **2**, 451
with hypophosphorous acid, **2**, 451
with ikarugamycin, **1**, 434
with imidoseleniums, **2**, 298
with iminophosphines, **2**, 468
with iodine, **2**, 44; **5**, 1107
with iodine(I) thiocyanate, **5**, 1044
with iodine isocyanate, **3**, 703; **5**, 984
with iron(III) porphyrin complexes, **2**, 579
with isocyanates, **5**, 386
with isocyanic acid, **5**, 985
with isothiocyanic acid, **2**, 307; **5**, 1043
with isoxazoline *N*-oxides, **2**, 356
with 2-isoxazolines, **3**, 683
with ketene dithioacetals, **1**, 358
with lithium allyl compounds, **1**, 363
with lithium aluminum hydride, **2**, 596
with manganese acetate, **1**, 336
with mercury salts, **2**, 593
with metal carbenoids, **1**, 299
with metal hydride bonds, **1**, 296
with metallacyclopropanes, **2**, 984
with metallacyclopropenes, **2**, 984
with
[methoxy(phenylthio)methyl]trimethylsilane, **4**, 370
with methylborane, **2**, 516
with methylene carbenes, **1**, 1144
with methylenecyclopropanes, **1**, 362
with 2-methyl-2-phenylselenomalonitrile, **4**, 284
with methyl propiolate, **1**, 524
with molybdenum carbene complexes, **3**, 514
with monoalkylboranes, **2**, 516
with nitrenes, **4**, 515
with nitric acid, **4**, 797
with nitrile oxides, **3**, 45
with nitrogen nucleophiles, **3**, 187
with nitrones, **2**, 352; **3**, 45
with nitronic esters, **2**, 355
with organoselenium electrophiles, **2**, 286
with osmium tetroxide, **2**, 50
with ozone, **2**, 98
with peracids, **2**, 90
with 5-phenyl-1,2,4-dithiazole-3-thione, **5**, 634
with phenylselenenyl thiocyanate, **5**, 1107
with (phenylselenothioperoxy)benzoate, **2**, 285
with phenylselenyl cyanide, **3**, 637
with phosphine oxides, **2**, 450
with phosphines, **2**, 438, 440; **6**, 58
with phosphonous acids, **2**, 454
with phosphorous, **2**, 104
with potassium *t*-butoxide, **1**, 882
with potassium selenocyanate, **5**, 1115
with rhodium(III) hydride complexes, **2**, 583
with selenenyl bromides, **2**, 284
with selenenyl chlorides, **2**, 44
with seleno radicals, **2**, 285
with selenosulfonates, **2**, 285
with *N*-substituted-*N*-chlorocyanamides, **5**, 1120
with sulfenes, **5**, 497
with sulfenylating compounds, **2**, 221
with β'-sulfonyl acryl amides, **1**, 613
with sulfonyl isocyanates, **5**, 344
with sulfur, **2**, 131, 137
with sulfur dioxide, **2**, 192
with sulfur monochloride, **2**, 221
with sulfur trioxide, **2**, 206, 208; **6**, 42
with tellurium(IV) chloride, **1**, 595
with tetraalkylammonium chloroiodate salts, **2**, 23
with tetracyanoethylene, **1**, 355
with tetrafluorohydrazine, **4**, 104
with tetrahydrofuran, **1**, 365
with thallium(III) acetate, **5**, 1115
with thallium(III) nitrate, **4**, 194
with thallium(III) salts, **2**, 598
with thallium sulfate, **2**, 40
with thexylborane, **2**, 516, 518
with thexylchloroborane, **2**, 522
with thiocarbonyl ylides, **1**, 367
with thiocyanogen, **5**, 1109
with thiolacetic acid, **2**, 117
with thiolcarboxylic acids, **2**, 117, 124
with thiols, **2**, 137, 712
with thionitroso compounds, **2**, 244
with thionyl chloride, **2**, 188
with tin hydrides, **2**, 601
with tributyltin hydride, **1**, 172
with trichloroamine, **2**, 338
with trichlorogermane, **6**, 62
with trifluoromethyl hypochlorite, **6**, 229
with trifluoromethyl hypofluorite, **6**, 229
with trifluoromethyl radicals, **6**, 20
with triisobutylaluminum, **2**, 596
with trimethylenemethane acetals, **1**, 360
with trimethylsilylphosphine, **2**, 428
with tritylsulfenyl chloride, **2**, 221
with α,β-unsaturated acid chlorides, **3**, 213
with α,β-unsaturated acyl cyanides, **3**, 652
with zirconocene hydrochloride, **3**, 614
reduction of, **1**, 71
remotely-substituted, reduction of, **1**, 84
ring-opening of, **2**, 45
selective displacement of, **1**, 657
selectivity of, **1**, 504
selenocyanation of, **2**, 284
selenosulfonation of, **2**, 285
selenylation of, **3**, 174
as starting materials
for acetals, **4**, 193
for *O*,*N*-acetals, **4**, 306, 312
for *Se*,*N*-acetals, **4**, 335
for *S*,*P*-acetals, **4**, 346
for alcohols, **2**, 40
for aldehydes, **3**, 2
for alkanethiols, **2**, 116
for alkenes, **1**, 383, 568
for alkylboranes, **2**, 57
for alkyl bromides, **2**, 29
for alkyl chlorides, **2**, 23
for alkyl cuprates, **2**, 587
for alkyl fluorides, **2**, 17
for alkyl halides, **2**, 3
for alkyl iodides, **2**, 34
for alkyliron complexes, **2**, 577
for alkyl metals, **2**, 592
for alkynes, **1**, 1063
for amines, **2**, 306
for anhydrides, **5**, 192

for α-bromo ketones, **3**, 153
for chlorofluoroalkanes, **4**, 30
for α-chloro ketones, **3**, 151
for dialkyl sulfides, **2**, 137
for 1,2-dibromides, **1**, 568
for dibromoalkanes, **4**, 23
for dichloroalkanes, **4**, 14
for α,α-dichloro sulfides, **6**, 46
for difluoroalkanes, **4**, 5
for *N,N*-difluoroamines, **2**, 335
for 1,2-dihalides, **1**, 568
for *gem*-dinitroalkenes, **4**, 993
for diols, **2**, 50
for 1,2-diols, **2**, 44
for episulfonium tetrafluoroborates, **1**, 1031
for α-fluoro esters, **4**, 52
for 2-haloalkenylphosphorus compounds, **4**, 802
for hemithioaminals, **4**, 326
for hydrazines, **2**, 376
for β-hydroxynitriles, **3**, 632
for β-hydroxy selenides, **2**, 281
for α-iodo ketones, **3**, 155
for isocyanates, **5**, 984
for 2-isothiazolines, **3**, 686
for isothiocyanates, **5**, 1043
for ketones, **1**, 296; **3**, 113
for nitro compounds, **2**, 392
for β-nitrogen-substituted selenides, **2**, 284
for β-nitro selenides, **2**, 785
for nitroso compounds, **2**, 391
for oxyimidic halides, **5**, 821
for phosphonium ylides, **3**, 498
for 2-pyrazolines, **3**, 452
for selenoxides, **1**, 712
for α-selenyl aldehydes, **3**, 38
for silasilenes, **5**, 926
for silenes, **5**, 926
for α-sulfenyl ketones, **3**, 165
for thioacyl chlorides, **5**, 480
for thiocyanates, **5**, 1110
for α,β-unsaturated ketones, **3**, 211
for unsymmetrical alkanoic anhydrides, **5**, 193
for vicinal dichlorides, **2**, 24
for vinyl fluorides, **2**, 610
for ylides, **3**, 498
stereochemistry of, **1**, 72, 722
stereoselective synthesis of, **1**, 641
stereospecific, synthesis of, **1**, 111
strained, synthesis of, **1**, 618
substituted, synthesis of, **1**, 480
substitution of, organopalladium-mediated, **1**, 493
sulfonation of, **2**, 208
synthesis of, **1**, 63
electronic factors, **1**, 774
stereoselective, **1**, 760
via boron-mediated alkenation, **1**, 764
via cyclic enol ethers, **2**, 88
via cyclobutane pyrolysis, **1**, 784
via Diels–Alder reaction, **1**, 519
via electrocyclic addition, **1**, 516
via elimination reactions, **1**, **589–671**
via elimination of sulfides, **1**, 590
via free radical addition, **1**, 546
via germanium-mediated alkenation, **1**, 765
via Horner reaction, **1**, 752
via hydroalumination, **1**, 511, 533
via hydroboration, **1**, 509
via hydrosilylation, **1**, 535
via hydrostannylation, **1**, 535
via hydrozirconation, **1**, 536
via ionic additions, **1**, 531
via Julia reactions, **1**, 696
via oxidative coupling of hydrocarbons, **1**, 674
via Peterson reaction, **1**, 758
via pyrolysis of silacyclobutanes, **4**, 610
via Ramberg–Bäcklund reaction, **1**, 617, 704
via rearrangements, **1**, 381
via reductive coupling of carbonyls, **1**, 687
via reductive elimination of β-alkoxy sulfides, **1**, 693
via reductive elimination of β-hydroxy sulfides, **1**, 693
via retro-Diels–Alder reaction, **1**, 785
via selenal intermediates, **3**, 385
via selenoxide pyrolysis, **1**, 780
via selones, **3**, 391
via tellurals, **3**, 389
via tellurium-mediated alkenation, **1**, 766
via thermal Cope rearrangement, **1**, 807
via Wittig reaction, **1**, 727
via Wittig reactions, **3**, 48
telomerization of, **6**, 5
thermal [2+2] additions of, **1**, 354
use in the synthesis of dithiocarbonates, **6**, 551
use in the synthesis of hemiacetals, **4**, 167
use in the synthesis of quaternary phosphonium salts, **2**, 445
Alkenes, *gem*-acylhydrazino azido-, from, *gem*-diazidoalkenes, **4**, 991
Alkenes, aliphatic, synthesis of, **1**, 761
Alkenes, alkenyl-substituted, reduction of, **1**, 79
Alkenes, *gem*-alkoxycarbonylamino phosphoryl-, synthesis of, **4**, 1006
Alkenes, alkoxydialkylsilyl-, synthesis of, **1**, 490
Alkenes, 2-alkoxy 1,1-dichloro-, synthesis of, **4**, 738
Alkenes, 2-alkoxy 1,1-disulfonyl-, reactions, with amines, **4**, 866
Alkenes, 1-alkoxy-1-halo-, synthesis of, **4**, 912
Alkenes, alkoxy nitro-, synthesis of, **4**, 898
Alkenes, 1-alkylarsino-2-chloro-, synthesis of, **2**, 876
Alkenes, 2-alkyl 1,1-distannyl-, synthesis of, **4**, 1069
Alkenes, *gem*-alkylideneaminophosphoryl-
from
N-acylamides, **4**, 1009
gem-arylideneamino chloroalkenes, **4**, 1009
imidoyl chlorides, **4**, 1009
synthesis of, **4**, 1009
Alkenes, alkyl-substituted, reduction of, **1**, 78
Alkenes, 1-alkylthio 1-metallo-, halogenation of, **4**, 774
Alkenes, alkynyl-substituted, reduction of, **1**, 79
Alkenes, 1-alumino 1-silyl-, synthesis of, **4**, 1062
Alkenes, amido-, cyclization of, **2**, 323
Alkenes, *gem*-amido carboxy-, as starting materials, for *gem*-amido phosphonioalkenes, **4**, 998
Alkenes, *gem*-amidodiphenylphosphinoyl-, from, *N*-tetrachloroethylamides, **4**, 1000
Alkenes, *gem*-amidophosphinoyl-
reactions, with phosphorus pentachloride, **4**, 1001
as starting materials, for *gem*-chloroarylideneamino phosphinoylalkenes, **4**, 1001
Alkenes, *gem*-amidophosphonoyl-
from
acylaminophosphinic esters, **4**, 1001
N-tetrachloroethylamides, **4**, 1002
synthesis of, **4**, 1001
Alkenes, *gem*-amidophosphoramidoyl-, from, azalactones, **4**, 1010
Alkenes, *gem*-amidophosphoryl-
from
amides, **4**, 1005
gem-amido phosphorylalkenes, **4**, 1006
azalactones, **4**, 1005
α-oxo phosphonates, **4**, 1005
N-tetrachloroethylamides, **4**, 1006
N-trichloroethylidene amides, **4**, 1006
as starting materials
for *gem*-amido phosphorylalkenes, **4**, 1006
for *gem*-arylchloromethyleneamino dichlorophosphorylalkenes, **4**, 1012

synthesis of, **4**, 1005
Alkenes, *β*-amino-, synthesis of, **1**, 659
Alkenes, *gem*-aminobenzylideneamino phosphorodiamidoyl-, from, *gem*-chlorobenzylideneamino phosphorylalkenes, **4**, 1012
Alkenes, *gem*-aminoborylamino silyl-
from
amino haloboranes, **4**, 1015
carbiminosilanes, **4**, 1015
Alkenes, *gem*-aminochloromethyleneamino chloro-, as starting materials, for α-diaminomethyleneamino enamines, **4**, 988
Alkenes, *gem*-arsino fluoro-, synthesis of, **4**, 805
Alkenes, aryl-, synthesis of, **1**, 471, 1097
Alkenes, 2-arylamino-1,1-dinitro-, synthesis of, **4**, 994
Alkenes, arylazo-, synthesis of, **4**, 799
Alkenes, *gem*-arylazo nitro-, from, nitroformaldehyde arylhydrazone, **4**, 995
Alkenes, *gem*-arylchloromethyleneamino dichlorophosphoryl-, from, *gem*-amido phosphorylalkenes, **4**, 1012
Alkenes, 2-aryl 1,1-distannyl-, synthesis of, **4**, 1069
Alkenes, *gem*-arylideneamino chloro-, as starting materials, for *gem*-alkylideneamino phosphorylalkenes, **4**, 1009
Alkenes, aryl nitro-, as starting materials, for *gem*-dinitroalkenes, **4**, 993
Alkenes, aryl-substituted
reactions, with sulfuric acid, **1**, 895
reduction of, **1**, 79
Alkenes, 1-arylthio 1-metallo-, halogenation of, **4**, 774
Alkenes, azido-, cyclization of, **2**, 326
Alkenes, α-azido-, vapor phase pyrolysis of, for the synthesis of ketenimines, **3**, 564
Alkenes, *gem*-azido halo-, synthesis of, **4**, 799
Alkenes, 1-azido-1-phenyl-, as starting materials, for benzonitrile ylides, **3**, 679
Alkenes, *gem*-azido phosphazido-, as intermediates, **4**, 993
Alkenes, azo-
characterization of, **2**, 770
as starting materials, for hydrazones, **6**, 604
synthesis of, **2**, 769; **5**, 367
via 1,4-elimination, **2**, 771
via elimination reactions, **2**, 769
Alkenes, 1-benzenesulfonyl-1-trimethylsilyl-, Michael addition of, **6**, 130
Alkenes, benzyloxynitro-, synthesis of, **4**, 897
Alkenes, bicyclic, synthesis of, **1**, 420
Alkenes, 2,2-bis(acetoxymercurio)-, synthesis of, **4**, 1067
Alkenes, bis(1,1-alkylthio)-
reactions
with amines, **4**, 899
with sulfodiimines, **4**, 899
with sulfoximines, **4**, 899
as starting materials, for ketene *S*,*N*-acetals, **4**, 899
Alkenes, (*Z*)-1,2-bis(dichloroboryl)-, synthesis of, **2**, 903
Alkenes, 1,2-bismetallo-, reactions, with electrophiles, **2**, 917
Alkenes, 1,1-bis(seleno)-, as starting materials, for α-lithiovinyl selenides, **2**, 911
Alkenes, 1,1-bis(silyl)-
from, alkynes, **4**, 1044
reactions, with organolithiums, **4**, 618
Alkenes, 1,2-bis(silyl)-, from, alkynylsilanes, **4**, 1044
Alkenes, 1,1-bis(trimethylstannyl)-, transmetallation of, **4**, 1065
Alkenes, 1-boro 1-germanyl-, synthesis of, **4**, 1058
Alkenes, 1-boronyl 1-silyl-, from, alkynes, **4**, 1056
Alkenes, 1-boro 1-stannyl-, from, alkynes, **4**, 1063
Alkenes, 1-boro-2-(trimethylstannyl)-, synthesis of, **4**, 1068
Alkenes, bromo-
fluorination of, **4**, 34
from
1,2-dibromides, **1**, 574
vinyl bromides, **1**, 430
reactions
with bromine, **4**, 23
with hydrogen bromide, **4**, 23
reactivity of, **1**, 467
synthesis of, **1**, 574
Alkenes, 1-bromo-, synthesis of, **2**, 623, 624
Alkenes, 2-bromo-, synthesis of, **2**, 623
Alkenes, bromochloro-, synthesis of, **4**, 756, 759
Alkenes, 1-bromo-1-chloro-
reactions, with lithium trialkyl zincates, **1**, 431
synthesis of, **4**, 758
Alkenes, 1-bromo-2-chloro-, synthesis of, **2**, 624
Alkenes, bromofluoro-
reactivity of, **6**, 46
synthesis of, **1**, 653
Alkenes, 1-bromo-2-fluoro-, synthesis of, **2**, 624
Alkenes, *gem*-bromo germyl-, synthesis of, **4**, 811
Alkenes, bromo iodo-, synthesis of, **4**, 759
Alkenes, *gem*-bromo lithio-
synthesis of, **4**, 818
See also Lithium compounds, α-bromoalkenyl-
Alkenes, α-bromo nitro-, synthesis of, **4**, 795
Alkenes, *gem*-bromo silyl-
synthesis of, **4**, 805
silylation methods, **4**, 806
Alkenes, chloro-
epoxidation of, **4**, 50
reactions
with arylamines, **4**, 987
with chlorine, **4**, 14
with chloroalkanes, **6**, 4
with chlorosilanes, **4**, 806
with fluorine, **4**, 31
as starting materials
for α-chloro epoxides, **4**, 50
for ketenimines, **3**, 590
synthesis of, **1**, 610
Alkenes, 1-chloro-
hydroalumination of, for vinyl iodides, **2**, 629
synthesis of, **2**, 616
Alkenes, ω-chloro-, reactions, with azides, **3**, 419
Alkenes, chloroalkoxy-, chlorination of, **6**, 41
Alkenes, *gem*-chloroarylideneamino phosphinoyl-, from, *gem*-amido phosphinoylalkenes, **4**, 1001
Alkenes, *gem*-chlorobenzylideneamino phosphoryl-, as starting materials, for *gem*-aminobenzylideneamino phosphorodiamidoylalkenes, **4**, 1012
Alkenes, 1-chloro-2-bromo-, synthesis of, **2**, 624
Alkenes, (*Z*)-1-chloro 1-bromo-, synthesis of, **4**, 756
Alkenes, 1-chloro-1-dialkylamino-
reactions
with phenols, **4**, 889
with silver carboxylates, **4**, 889
with sodium alkoxides, **4**, 889
Alkenes, chlorofluoro-
reactions, with thiols, **6**, 47
reactivity of, **6**, 46
as starting materials, for chlorofluoroalkanes, **4**, 32
Alkenes, 1-chloro-1-fluoro-, synthesis of, **4**, 753, 755
Alkenes, chloro halo-, synthesis of, **4**, 756
Alkenes, *gem*-chloro manganio-, as starting materials, for α-manganio enamines, **4**, 1019
Alkenes, *gem*-chloro metallo-, synthesis of, **4**, 821
Alkenes, α-chloro nitro-, synthesis of, **4**, 795
Alkenes, *gem*-chloro nitroso-, from, α-chloro oximes, **4**, 798
Alkenes, 1-chloro 1-phenylsulfinyl-, synthesis of, **4**, 777
Alkenes, *gem*-chloro silyl-, synthesis of, silylation methods, **4**, 806
Alkenes, (*E*)-*gem*-chloro stannyl-, synthesis of, **4**, 812
Alkenes, conjugated
reactions, with hydrogen cyanide, **3**, 615
synthesis of, **1**, 762
Alkenes, cyano-, from, α-cyanocarboxylates, **1**, 561

Alkenes, cyclic nitro-, from, β-nitro ketones, **2**, 785
Alkenes, 1,1-dialkoxy-
[2+2] cycloaddition reactions, with carbonyls, **6**, 75
[3+2] cycloaddition reactions, 1,3-dipoles, **6**, 76
as intermediates, for ortho esters, **6**, 74
reactions, with epoxides, **6**, 75
as starting materials
for aminoxy ortho esters, **6**, 74
for cyclic ortho esters, **6**, 75
for mixed ortho esters, **6**, 74
for ortho esters, **6**, 74
thermal [4+2] cycloaddition reactions of, for cyclic ortho esters, **6**, 75
Alkenes, 1,1-dialumino-, from, alkynes, **4**, 1068
Alkenes, *vic*-diamino-, halogenation of, **4**, 793
Alkenes, *gem*-diaryl-, as starting materials, for *gem*-dinitroalkenes, **4**, 993
Alkenes, 2,2-diaryl-1,1-dinitro-, synthesis of, **4**, 993
Alkenes, *gem*-diarylphosphinoylamino phosphoryl-, synthesis of, **4**, 1008
Alkenes, *gem*-diazido-
from, *gem*-dichloroalkenes, **4**, 993
as starting materials
for *gem*-acylhydrazino azidoalkenes, **4**, 991
for α-azido enamines, **4**, 989
for *gem*-diphosphazidoalkenes, **4**, 996
for *gem*-diphosphiminoalkenes, **4**, 996
Alkenes, *gem*-diazo-
from
diazonium salts, **4**, 992
3-formyl-1,5-diphenylformazan, **4**, 992
γ-ketocarboxylic acids, **4**, 992
Alkenes, 1,1-diboro-
from, alkynes, **4**, 1054
synthesis of, **4**, 1054
Alkenes, *gem*-diboro-, as starting materials, for *gem*-bromo boronalkenes, **4**, 814
Alkenes, 1,1-diboryl-
hydrogenation of, **4**, 634
as starting materials
for 1,1-bisborylalkanes, **4**, 634
for boriranylideneboranes, **6**, 713
Alkenes, dibromo-
reactions, with alkynyltins, **1**, 1015
synthesis of, **4**, 745
Alkenes, 1,1-dibromo-
dehydrohalogenation of, **1**, 1067
lithium–halogen exchange of, **1**, 1066
reactions, with butyllithium, **4**, 757
as starting materials
for 1,1-diiodoalkenes, **4**, 750
for 1,1-distannylalkenes, **4**, 1069
Alkenes, *gem*-dibromo-, lithium–bromine exchange of, **4**, 793
Alkenes, (*E*)-1,2-dibromo silyl-, synthesis of, **4**, 807
Alkenes, dichloro-, reactions, with tetraethylammonium hydrosulfide, **4**, 851
Alkenes, 1,1-dichloro-
dehydrohalogenation of, **1**, 1067
lithium–halogen exchange of, **1**, 1066
reactions
with alkynes, **1**, 1009
with 1,2-dipotassium-1,2-di-*t*-butyldiphosphide, **4**, 1025
as starting materials
for alkynyl chlorides, **2**, 1015
for 1-phosphacumulenes, **3**, 604
synthesis of, **4**, 737
Alkenes, 1,2-dichloro-
from, alkynes, **2**, 617
selective substitution of, **1**, 467
Alkenes, (*E*)-1,2-dichloro-, reactions, with alkynes, **1**, 1009
Alkenes, *gem*-dichloro-
as starting materials
for alkynyl chlorides, **2**, 1015
for *gem*-diazidoalkenes, **4**, 993
Alkenes, (*Z*)-1,2-dichloro-, reactions, with alkynes, **1**, 1009
Alkenes, 1,1-dichloro-2-fluoro-, reactions, with *n*-butyllithium, **1**, 1069
Alkenes, 1,1-di(chloromercurio)-, synthesis of, **4**, 1067
Alkenes, *gem*-dichlorophosphoryl isocyanato-
reactions, with antimony trifluoride, **4**, 1012
as starting materials
for *gem*-difluorophosphoryl isocyanatoalkenes, **4**, 1012
for *gem*-dihalophosphoryl isocyanatoalkenes, **4**, 1013
Alkenes, 1,1-dicyano-, use in the synthesis of ketenimines, **3**, 570
Alkenes, *gem*-dicyano-, as starting materials, for ketene aminals, **4**, 979
Alkenes, 2,2-dicyano bis(1,1-alkylthio)-, reactions, with aziridine, **4**, 899
Alkenes, difluoro-
reactions
with acetyl hypobromite, **4**, 52
with thiols, **6**, 46
as starting materials, for monofluoroalkenes, **1**, 429
synthesis of, **4**, 730
via elimination reactions, **4**, 730
via ylide–carbene reactions, **4**, 735
Alkenes, 1,1-difluoro-
oxidation of, **6**, 410
ozonization of, **6**, 410
as starting materials
for 1-fluoroalkynes, **2**, 1042
for hydroxamoyl halides, **5**, 666
for imidoyl fluorides, **3**, 588
Alkenes, *gem*-difluoro-
hydrolysis of, **5**, 99
as starting materials, for *gem*-diisothioureidoalkenes, **4**, 992
Alkenes, *gem*-difluorophosphoryl isocyanato-, from, *gem*-dichlorophosphoryl isocyanatoalkenes, **4**, 1012
Alkenes, *gem*-difluorophosphoryl ureido-, from, *gem*-isocyanato phosphorylalkenes, **4**, 1012
Alkenes, 1,1-digermanyl-, from, alkynes, **4**, 1055
Alkenes, 1,1-digermanyl 2-chloro-, synthesis of, **4**, 1055
Alkenes, dihalo-, selective substitution of, **1**, 467
Alkenes, 1,1-dihalo-
reactions
with disulfides, **6**, 46
with sodium alkoxides, **6**, 74
with thiols, **6**, 46
with trifluoromethyl hypochlorite, **6**, 43
as starting materials, for ortho esters, **6**, 74
synthesis of, **4**, 730
Alkenes, 1,2-dihalo-
dehalogenation of, **1**, 1069
synthesis of, **2**, 624
Alkenes, *gem*-dihalo-
amination of, **4**, 793
dehalogenation of, **1**, 990
dimerization of, **1**, 990
reactions, with diphenyltrimethylsilylphosphine, **4**, 997
as starting materials
for ketene acetals, **4**, 829
for ketene dithioacetals, **4**, 851
for α-phosphino enamines, **4**, 997
substitution by alkoxides, **4**, 829
See also Vinyl halides
Alkenes, *gem*-dihalophosphoryl isocyanato-
from
gem-dichlorophosphoryl isocyanatoalkenes, **4**, 1013
dihaloethyl isocyanates, **4**, 1012
haloalkyl isocyanates, **4**, 1012
tetrachloroethyl isocyanates, **4**, 1013
synthesis of, **4**, 1012

Alkenes, *gem*-dihydrazino-
from
azobenzenes, **4**, 985
bis(dialkylamino)-1-methoxymethane, **4**, 985
tricyanoquinodimethanes, **4**, 985
synthesis of, **4**, 985
Alkenes, 1,1-diiodo-
from, 1,1-dibromoalkenes, **4**, 750
synthesis of, **4**, 749
Alkenes, 1,2-diiodo-, synthesis of, **2**, 629
Alkenes, *gem*-diiodo-, synthesis of, **4**, 813
Alkenes, *gem*-diisothioureido-
from
gem-difluoroalkenes, **4**, 992
vinylidene difluorides, **4**, 992
Alkenes, 1,1-dilithio-, synthesis of, **4**, 1064
Alkenes, 1-dimesitylboronyl 1-trimethylsilyl-, synthesis of, **4**, 1057
Alkenes, 1,1-dimetallo-
characteristics of, **2**, 952
configurational stability of, **2**, 909
Alkenes, *gem*-dimethoxazonyl-, from, bismethoxazonylmethane, **4**, 996
Alkenes, dinitro-
reactions
with nitric acid, **6**, 150
with trifluoroacetic anhydride, **6**, 150
Alkenes, *gem*-dinitro-
from
alkanes, **4**, 993
alkenes, **4**, 993
aryl nitroalkenes, **4**, 993
gem-diarylalkenes, **4**, 993
diaryl carbinols, **4**, 993
1,1-diarylethanes, **4**, 993
diazo compounds, **4**, 995
dinitro alkanols, **4**, 994
dinitromethane, **4**, 994
naphthyl ethanols, **4**, 993
nitroalkanes, **4**, 994
nitroalkenes, **4**, 995
synthesis of, **4**, 993
via nitration reactions, **4**, 993
Alkenes, *gem*-diphenylphosphinoyl ureido-, from, *gem*-isocyanato phosphinoylalkenes, **4**, 1001
Alkenes, *gem*-diphosphazido-, from, *gem*-diazidoalkenes, **4**, 996
Alkenes, *gem*-diphosphimino-, from, *gem*-diazidoalkenes, **4**, 996
Alkenes, 1,1-diphosphorus-substituted, addition to, **4**, 560
Alkenes, 1,1-diplumbyl-
from, 1,1-distannylalkenes, **4**, 1069
as starting materials
for 1-germanyl 1-plumbylalkenes, **4**, 1063
for 1-plumbylvinyllithiums, **4**, 1065
transmetallation of, **4**, 1063
Alkenes, 1,1-disilyl-
from
alkynesilane, **4**, 1048
haloalkenes, **4**, 1048
Alkenes, 1,1-distannyl-
from
1,1-dibromoalkenes, **4**, 1069
1-stannylalkynes, **4**, 1068
as starting materials
for 1,1-diplumbylalkenes, **4**, 1069
for 1-germanyl 1-stannylalkenes, **4**, 1063
for 1-plumbyl 1-stannylalkenes, **4**, 1069
transmetallation of, **4**, 1063, 1069
Alkenes, *gem*-distannyl-, halogenation of, **4**, 813
Alkenes, disubstituted, synthesis of, **1**, 468; **5**, 155
Alkenes, 1,1-disubstituted
osmylation of, **2**, 46
reactions
with amyl nitrite, **2**, 786
with butyl nitrite, **2**, 786
synthesis of, **1**, 542
Alkenes, 1,2-disubstituted, synthesis of, **1**, 278; **2**, 875
Alkenes, 2,2-disubstituted-1-nitro-
[4+2] cycloadditions of, **2**, 784
synthesis of, **2**, 784, 785
Alkenes, 1,1-disulfonyl-
from, phenyliodinium ylides, **4**, 866
See also Dithioacetal tetroxides, ketene
See also Dithioacetal *S,S,S',S'*-tetroxides, ketene
Alkenes, *gem*-disulfoximido-, from, ketene dithioacetals, **4**, 997
Alkenes, *gem*-diureido-, from, ketene aminals, **4**, 982
Alkenes, electron-rich, use in the synthesis of carbene complexes, **6**, 721
Alkenes, fluorinated, reactions, with sodium hypobromite, **4**, 45
Alkenes, fluoro-
addition of haloalkanes to, **6**, 4
addition of halogens to, **6**, 2
bromination of, **4**, 34
from
α-fluoro ylides, **5**, 894
vinyl chlorides, **1**, 430
vinyl fluorides, **1**, 429
nucleophilic vinylic substitution of, **1**, 466
reactions
with alkyllithium, **4**, 785
with bromine trifluoride, **6**, 2
with chlorine, **4**, 31
with chlorosilanes, **4**, 140
with hydrazine hydrate, **5**, 839
with iodine bromide, **6**, 2
with iodine chloride, **6**, 2
with iodine pentafluoride, **6**, 2
with phosphines, **4**, 111
with sodium nitrite, **6**, 54
with tetrafluorodiphosphine, **4**, 114
with trialkylstannane hydrides, **4**, 153
as starting materials
for α-fluoroalkyl esters, **4**, 51
for hydrazonoyl halides, **5**, 839
synthesis of, **1**, 446, 650
via dehydroiodination reactions, **1**, 574
Alkenes, 1-fluoro-, synthesis of, **2**, 612
Alkenes, 2-fluoro-, synthesis of, **2**, 611
Alkenes, fluorobromo-, synthesis of, **4**, 751
Alkenes, fluorohalo-
from, 1-halo 1-metalloalkenes, **4**, 754
isomer ratio of, **4**, 752
synthesis of, **4**, 750
Alkenes, *gem*-fluoro nitro-, synthesis of, **4**, 795
Alkenes, *gem*-fluoro phenylazo-, synthesis of, **4**, 799
Alkenes, 1-fluoro 1-phenylsulfinyl-, synthesis of, **4**, 778
Alkenes, *gem*-fluoro potassio-, synthesis of, **4**, 819
Alkenes, *gem*-fluoro silyl-
desilylation of, **4**, 819
synthesis of, **4**, 806
Alkenes, fluorostannyl-, reactions, with iodine, **1**, 455
Alkenes, *gem*-formamido phosphoryl-
from
aldehydes, **4**, 1004
oxazolines, **4**, 1005
tetraethyl formamidomethylenediphosphonate, **4**, 1004
synthesis of, **4**, 1004
Alkenes, functionalized
deprotonation of, **2**, 909
silylation of, **2**, 909
synthesis of, **1**, 762
Alkenes, geminal dibromo-, as starting materials, for 1,1-distannyl-2,2-diorganyl-1-alkenes, **6**, 209

Alkenes, 1-germanyl 1-plumbyl-, from, 1,1-diplumbylalkenes, **4**, 1063
Alkenes, 1-germanyl 1-silyl-, from, 1-germanyl 1-silylalkanes, **4**, 1058
Alkenes, 1-germanyl 1-stannyl-
from
1,1-distannylalkenes, **4**, 1063
1-germanylalkynes, **4**, 1063
Alkenes, halo-
coupling with organoboranes, **1**, 488
reactions
with alcohols, **6**, 41
with diazoalkanes, **4**, 109
with germane hydrides, **4**, 141
with hydrogen halides, **4**, 2
with hypohalites, **6**, 42
with mercuric fluoride, **6**, 64
with oximes, **6**, 42
with tetrafluorohydrazine, **5**, 811
with trifluoromethanesulfonyl chloride, **6**, 46
as starting materials
for bromochloroalkanes, **4**, 38
for bromofluoroalkanes, **4**, 34
for chloroiodoalkanes, **4**, 39
for α,α-dihalo ethers, **6**, 41
for 1,1-disilylalkenes, **4**, 1048
for fluoroiodoalkanes, **4**, 36
for ketenimines, **3**, 588
Alkenes, α-halo acyloxy-, synthesis of, **4**, 768
Alkenes, haloalkoxy-, as starting materials, for α,α-dihalo ethers, **6**, 41
Alkenes, 1-halo-1,2-dialkyl-, synthesis of, **1**, 660
Alkenes, 1-halo-1-dialkylamino-, as starting materials, for ketene hemiaminals, **4**, 889
Alkenes, halogenated, as starting materials, for haloimidic halides, **5**, 811
Alkenes, 1-halo-1-lithio-, synthesis of, **1**, 990
Alkenes, *gem*-halo lithio-, synthesis of, **4**, 816
Alkenes, 1-halo 1-metallo-, as starting materials, for fluorohaloalkenes, **4**, 754
Alkenes, *gem*-halo nitro-
synthesis of, **4**, 795
via condensation between aromatic aldehydes and halonitromethanes, **4**, 798
via halogenation, **4**, 795
via nitration, **4**, 797
Alkenes, *gem*-halo nitroso-, synthesis of, **4**, 798
Alkenes, 1-halo 1-silyl-, lithiation of, **4**, 1059
Alkenes, 1-halo-1-silyl-
synthesis of, **2**, 922, 925
via carboalumination of alkynylsilanes, **2**, 926
Alkenes, *gem*-halo stannyl-, synthesis of, **4**, 812
Alkenes, 1-halo 1-sulfinyl-, synthesis of, **4**, 776
Alkenes, 1-halo 1-thioalkyl-, synthesis of, **4**, 773
Alkenes, heteroalkyl-substituted, reduction of, **1**, 81
Alkenes, heteroaryl-substituted, reduction of, **1**, 79
Alkenes, hetero-substituted, synthesis of, **1**, 763
Alkenes, hindered, synthesis of, **1**, 686, 689
Alkenes, hydroxy-, cyclization of, **2**, 93
Alkenes, *gem*-iminomethylamino lithio-, from, eneamidines, **4**, 1016
Alkenes, *gem*-imino phosphimino-
from
α-bromo esters, **4**, 996
1,3-oxazines, **4**, 996
Alkenes, iodo-
from
α-iodo ylides, **5**, 896
vinyl iodides, **1**, 431
synthesis of, **1**, 113, 333
Alkenes, 1-iodo-, synthesis of, **2**, 628
Alkenes, 2-iodo-, synthesis of, **2**, 628
Alkenes, (*Z*)-1-iodo-, synthesis of, **2**, 630
Alkenes, 1-iodo 1-alkoxy-, as intermediates, **4**, 765
Alkenes, 1-iodo-2-alkyl-, synthesis of, **2**, 629
Alkenes, 1-iodo-2-bromo-, synthesis of, **2**, 629
Alkenes, 1-iodo-2-chloro-, synthesis of, **2**, 629
Alkenes, 1-iodo-2-halo-, synthesis of, **2**, 629
Alkenes, *gem*-iodo lithio-, synthesis of, **4**, 819
Alkenes, (*E*)-1-iodo-2-methyl-, synthesis of, **2**, 629
Alkenes, β-iodonitro-, from, alkylalkynes, **2**, 1052
Alkenes, *gem*-iodo silyl-, synthesis of, **4**, 808, 811
Alkenes, 1-iodo 1-stannyl-, as starting materials, for 1-magnesio 1-stannylalkenes, **4**, 1066
Alkenes, 1-iodo 1-sulfonyl-, synthesis of, **4**, 781
Alkenes, *gem*-iodo trialkylgermyl-, synthesis of, **4**, 812
Alkenes, *gem*-isocyanato phosphinoyl-, as starting materials, for *gem*-diphenylphosphinoyl ureidoalkenes, **4**, 1001
Alkenes, *gem*-isocyanato phosphonoyl-, as starting materials, for *gem*-phosphonoyl ureidoalkenes, **4**, 1002
Alkenes, *gem*-isocyanato phosphoryl-
from
1-chlorovinyl isocyanate, **4**, 1010
1,2-dibromoethyl isocyanate, **4**, 1009
tetrachloroethyl isocyanates, **4**, 1010
as starting materials
for *gem*-difluorophosphoryl ureidoalkenes, **4**, 1012
for *gem*-phosphoramidoyl ureidoalkenes, **4**, 1011
for *gem*-phosphorodiamidoyl ureidoalkenes, **4**, 1011
for *gem*-phosphoryl ureidoalkenes, **4**, 1007
synthesis of, **4**, 1009
Alkenes, *gem*-isocyano lithio-, from, vinylidene isocyanates, **4**, 1015
Alkenes, α-isocyano silyl-, from, vinylidene isocyanides, **4**, 1014
Alkenes, *gem*-isothiocyanato silyl-, from, isocyanides, **4**, 1014
Alkenes, α-lithiated, from, sulfones, **4**, 960
Alkenes, lithio-, reactions, with cyanocuprates, **4**, 953
Alkenes, 1-lithio-1-diphenylmethylsilyl-, as starting materials, for 1-trimethylgermanyl-1-diphenylmethylsilylalkenes, **4**, 1058
Alkenes, 1-lithio 1-plumbyl-
as starting materials, for 1-silyl 1-plumbylalkenes, **4**, 1063
transmetallation of, **4**, 1063
Alkenes, 1-lithio 1-silyl-
from, 1-phenylthio 1-silylalkenes, **4**, 1059
stability of, **4**, 1059
as starting materials
for 1-magnesio 1-silylalkenes, **4**, 1060
for 1-silyl 1-stannylalkenes, **4**, 1062
synthesis of, via metal–halogen exchange, **4**, 1059
use in the synthesis of 1-cupro-1-silylalkenes, **4**, 1061
Alkenes, 1-lithio 1-stannyl-, as starting materials, for 1-magnesio 1-stannylalkenes, **4**, 1066
Alkenes, 1-lithio 1-trimethylsilyl-
reactions, with dimesitylfluoroboranes, **4**, 1057
stability of, **4**, 1059
Alkenes, macrocyclic, synthesis of, **1**, 674, 690, 709
Alkenes, 1-magnesio 1-silyl-
from, 1-lithio 1-silylalkenes, **4**, 1060
as starting materials, for 1-silyl 1-stannylalkenes, **4**, 1062
synthesis of, **4**, 1060
Alkenes, 1-magnesio 1-stannyl-
from
1-iodo 1-stannylalkenes, **4**, 1066
1-lithio 1-stannylalkenes, **4**, 1066
Alkenes, 1-magnesio 1-zinco-, synthesis of, **4**, 1066
Alkenes, metallated, as starting materials, for alkenes, **1**, 453
Alkenes, *gem*-metallo silyl-
configurational instability of, **4**, 808
stability of, **4**, 809
Alkenes, *gem*-methoxycarbonylamino diphenylphosphinoyl-
from

carbamates, **4**, 1000
ethyldiphenylphosphinite, **4**, 1000
Alkenes, 1-methoxy-1-germyl-, hydrolysis of, **2**, 950
Alkenes, mixed dihalo-, from, fluorenyldiazomethane, **4**, 755
Alkenes, mixed 1,1-dihalo-, from, diazofluorene, **4**, 755
Alkenes, monobromo-, metallation of, **4**, 818
Alkenes, monofluoro-, from, difluoroalkenes, **1**, 429
Alkenes, monosubstituted *trans*-difluoro-, synthesis of, **4**, 819
Alkenes, nitro-
cycloaddition of, **2**, 396
[4+2] cycloaddition of, **2**, 356
dihalogenation of, **4**, 795
electroreduction of, **3**, 426
in Friedel–Craft alkylation, **1**, 492
from, ketones, **2**, 785
as Michael acceptors, **2**, 393
in natural product syntheses, **1**, 310
polyhalogenation of, **4**, 797
reactions
with acetyl bromide, **5**, 820
with alkoxides, **3**, 46
with alkynylalanes, **1**, 1049
with 1-aminoalkynes, **3**, 435
with benzene, **5**, 820
with butanoyl bromide, **5**, 820
with enolate equivalents, **3**, 192
with ethyl trithiocarbonate, **1**, 443
with lithium enolates, **3**, 435
with oxygen anions, **3**, 162
with propanoyl bromide, **5**, 820
with thiolate anions, **3**, 165
with trichloromethyl anions, **6**, 24
with zinc borohydride, **3**, 426
reduction of, **1**, 443; **2**, 341, 398; **3**, 426
as starting materials
for α-alkoxy ketones, **3**, 162
for *gem*-dinitroalkanes, **4**, 431
for *gem*-dinitroalkenes, **4**, 995
for nitroalkanes, **2**, 398
for β-nitronitriles, **3**, 638
for oxyimidic chlorides, **5**, 819
for α-phosphoryl enamines, **4**, 1004
for γ,δ-unsaturated ketones, **3**, 143
synthesis of, **1**, 668; **2**, 783; **4**, 335, 906
via dehydration reactions, **1**, 578
via Henry reaction, **2**, 784
uses of, **2**, 783
in [4+2] cycloaddition reactions, **2**, 784
Alkenes, 1-nitro-
reactions, with copper–zinc organometallics, **2**, 785
as starting materials, for primary amines, **2**, 318
synthesis of, **2**, 784
Alkenes, α-nitro-, reactions, with stannylallenes, **1**, 1062
Alkenes, nitrogen-substituted, reduction of, **1**, 82
Alkenes, nitroso-
cycloaddition of, **6**, 127
reactions
with trithiocarbonates, **6**, 309
with ylides, **2**, 779
synthesis of, **2**, 778; **4**, 798
Alkenes, oxo nitro-, reduction of, **3**, 47
Alkenes, perfluorinated, as starting materials, for hydrazonyl halides, **5**, 679
Alkenes, perfluoro-
additions to, **6**, 262
reactions
with halogen monofluorosulfates, **6**, 42
with nucleophiles, **4**, 1076
Alkenes, perhalo-, vapor phase pyrolysis of, **6**, 604
Alkenes, 1-phenyl-, in cyclization reactions, **1**, 316
Alkenes, 1-phenylthio-1,2-bis(trimethylsilyl)-, synthesis of, **2**, 930
Alkenes, β-phenylthionitro-, as starting materials, for α-nitro acetals, **2**, 393
Alkenes, 1-phenylthio 1-silyl-, as starting materials, for 1-lithio 1-silylalkenes, **4**, 1059
Alkenes, (*Z*)-1-phenylthio-1-trialkylsilyl-, synthesis of, **4**, 942
Alkenes, 1-phenylthio 1-trimethylsilyl-, lithiation of, **4**, 1060
Alkenes, *gem*-phosphonio ureido-, from, *N*-tetrachloroethyl ureas, **4**, 999
Alkenes, *gem*-phosphonoyl ureido-, from, *gem*-isocyanato phosphonoylalkenes, **4**, 1002
Alkenes, *gem*-phosphoramidoyl ureido-, from, *gem*-isocyanato phosphorylalkenes, **4**, 1011
Alkenes, *gem*-phosphorodiamidoyl ureido-
from
gem-isocyanato phosphorylalkanes, **4**, 1011
gem-isocyanato phosphorylalkenes, **4**, 1011
Alkenes, *gem*-phosphorylamino phosphoryl-
from
benzoylhydrazones, **4**, 1008
halo tricyanomethanes, **4**, 1008
2,2,2-trihalo acetimidoyl chlorides, **4**, 1008
synthesis of, **4**, 1008
Alkenes, *gem*-phosphoryl ureido-
from
gem-isocyanato phosphorylalkenes, **4**, 1007
N-tetrachloroethylureas, **4**, 1007
synthesis of, **4**, 1007
Alkenes, plumbyl-, synthesis of, **4**, 1063
Alkenes, 1-plumbyl 1-stannyl-
from, 1,1-distannylalkenes, **4**, 1069
as starting materials, for 1-plumbylvinyllithiums, **4**, 1065
Alkenes, polycyclic, synthesis of, via Shapiro reactions, **1**, 634
Alkenes, polyfluoro-
reactions
with alkoxides, **4**, 761
with lithium dialkylamides, **4**, 793
with phosphites, **4**, 803
Alkenes, polyhalo-
displacement of, **4**, 770
reactions
with oxygen bases, **2**, 1012
with sulfur, **5**, 472
sulfurization of, **5**, 472, 479
Alkenes, primary amino-, cyclization of, **3**, 411
Alkenes, 1-seleno 1-nitro-, synthesis of, **4**, 910
Alkenes, siloxy-, photooxygenation of, **5**, 227
Alkenes, silyl-, synthesis of, **4**, 735, 805
Alkenes, 1-silyl-1-alumino-
alkylation of, **2**, 922
dealumination of, **2**, 922
as intermediates, in the synthesis of alkenylsilanes, **2**, 922
transmetallation of, **2**, 922
Alkenes, 1-silyl-1-phosphorus substituted, addition to, **4**, 572
Alkenes, 1-silyl 1-plumbyl-, from, 1-lithio 1-plumbylalkenes, **4**, 1063
Alkenes, 1-silyl 1-stannyl-
from
1-lithio 1-silylalkenes, **4**, 1062
1-magnesio 1-silylalkenes, **4**, 1062
synthesis of, **4**, 1062
transmetallation of, **4**, 1060
Alkenes, (*Z*)-1-silyl-2-stannyl-
palladium-catalyzed coupling of, **2**, 916
synthesis of, via stannylation of alkynes, **2**, 916
Alkenes, silyltin-, reactions, with acid chlorides, **3**, 271
Alkenes, 1-silyl 1-zinco-, synthesis of, **4**, 1061
Alkenes, stannyl-
dihydroxylation of, **2**, 46
reactions, with cyanocuprates, **4**, 953
Alkenes, β-sulfinyl nitro-, reactions, with organolithiums, **2**, 784

Alkenes, sulfonyl-
as starting materials
for 1-sulfonylvinyllithiums, **4**, 961
for α-sulfonyl vinyllithiums, **1**, 145
Alkenes, symmetrical, synthesis of, **1**, 689
Alkenes, terminal
acetalization of, **4**, 193
from, 4-methoxyphenyl telluroxides, **1**, 611
Markovnikov reactions of, **3**, 174
reactions, with *N,N*-dibromophosphoramidate, **2**, 307
as starting materials, for α-hydroxymethyl ketones, **3**, 158
synthesis of, **1**, 676
via selenoxide elimination, **1**, 604
via sulfoxide pyrolysis, **1**, 779
Alkenes, *gem*-thioalkyl nitromethyl-, as starting materials, for ketene aminals, **4**, 979
Alkenes, 2-thiocyanato-, synthesis of, **5**, 1108
Alkenes, (*E*)-tributylstannyl-, synthesis of, **4**, 962
Alkenes, trichloro-, dehalogenation of, **1**, 1069
Alkenes, 1-trichlorogermanyl 1-silyl-, synthesis of, **4**, 1058
Alkenes, trichlorosilyl-, as intermediates, **1**, 1106
Alkenes, 1-trichlorosilyl-1-germanyl-, synthesis of, **4**, 1058
Alkenes, trifluoro-
reactions, with sulfur, **5**, 472
as starting materials, for thioacyl fluorides, **5**, 472
Alkenes, 1,1,2-trifluoro-3-hydroxy-, rearrangement of, **3**, 67
Alkenes, trihalo-, reactions, with alkyllithiums, **6**, 62
Alkenes, trihalo-substituted, synthesis of 221019.1235
Alkenes, 1-trimethylgermanyl-1-diphenylmethylsilyl-, from, 1-lithio-1-diphenylmethylsilylalkenes, **4**, 1058
Alkenes, 1-trimethylsilyl-2-bromo-, synthesis of, **2**, 929
Alkenes, 1-trimethylsilyl 1-stannyl-, synthesis of, **4**, 1062
Alkenes, 1-trimethylsilyl 1-titano-, synthesis of, **4**, 1061
Alkenes, trisilyl-, synthesis of, **4**, 1051
Alkenes, 1,1,2-trisilyl-, from, 1-silylalkynes, **4**, 1045
Alkenes, trisubstituted
hydroboration of, **2**, 43
selective synthesis of, **1**, 489
synthesis of, **1**, 464, 468, 484, 487, 700, 757; **4**, 838
Alkenes, trisulfonyl-, synthesis of, **4**, 861
Alkenes, α,β-unsaturated nitro-
reactions, with chromium(II) chloride, **3**, 426
as starting materials
for aldoximes, **3**, 426
for ketoximes, **3**, 426
Alkenes, α,β-unsaturated-1-nitro-, synthesis of, **2**, 784
Alkenes, unsymmetrical, synthesis of, **1**, 690
1-Alkenes
dehydrogenation of, **1**, 1063
reactions
with phosphorus trihalides, **2**, 448
with trichloromethylsulfonyl chloride, **6**, 24
synthesis of, **1**, 508
1-Alkenes, 1,1-bis(dialkylboryl)-, from, dialky (1-alkynyl)boranes, **6**, 183
1-Alkenes, 1,2-bis(phenylseleno)-, [2,3] sigmatropic rearrangement of, **1**, 1073
1-Alkenes, 1,2-bis(silyl)-, synthesis of, **2**, 923
1-Alkenes, 1-boryl-2-stannyl-, from, 1-stannyl-1-alkynes, **2**, 903
1-Alkenes, *trans* 1-*t*-butyl-, synthesis of, **1**, 644
1-Alkenes, 1,1-dichloro-, reactions, with difluoroamine, **4**, 439
1-Alkenes, *cis*-1,2-dichloro-, dehydrohalogenation of, for alkynyl chlorides, **2**, 1012
1-Alkenes, 1,1-difluoro-2-aryl-, synthesis of, **2**, 906
1-Alkenes, *anti*-3-(diphenylamino)-2-hydroxy-, synthesis of, **2**, 75
1-Alkenes, 1,1-distannyl-, synthesis of, **6**, 209
1-Alkenes, 1,1-distannyl-2,2-diorganyl-
from, geminal dibromoalkenes, **6**, 209
as starting materials, for tristannylalkanes, **6**, 209
1-Alkenes, 2-halo-, synthesis of, **2**, 902
Alk-1-enes, (*Z*)-1-halo-, synthesis of, **1**, 449
1-Alkenes, 1-iodo-, as starting materials, for vinyllithiums, **1**, 1090
1-Alkenes, (*E*)-2-iodo-1-tosyl-, reactions, with amines, **2**, 743
1-Alkenes, 2-metallo 1-silyl-, synthesis of, **4**, 1045
1-Alkenes, perfluoro-, reactions, with phenylhydrazine, **4**, 799
1-Alkenes, 1-silyl-, from, 1-alkynes, **2**, 915
1-Alkenes, 1-silyl-2-stannyl-3-hydroxy-, as starting materials, for allenylsilanes, **2**, 940
1-Alkenes, 1,1,2-trifluoro-3-hydroxy-, synthesis of, **3**, 67
1-Alkenes, 1-trimethylsilyl-, synthesis of, **2**, 913
Alk-2-enes, 1,4-dihydroxy-, synthesis of, **2**, 85
2-(*E*)-Alkenes
from
nona-1,2-diene, **1**, 459
octa-1,2-diene, **1**, 459
cis-Alkenes
asymmetric hydroboration of, **2**, 519
synthesis of, **1**, 247, 1130
(*E*)-Alkenes
from
(*E*)-alkenyldialkylboranes, **1**, 659
1,1-((*Z*)-1-bromoalkenyl)dialkylboranes, **1**, 487
synthesis of, **1**, 485, 677
via Wadsworth–Emmons reaction, **1**, 745
via Wittig reaction, **1**, 722
via Wittig–Schlosser reaction, **1**, 729
n-Alkenes, reactions, with *Candida*, **5**, 117
trans-Alkenes, synthesis of, **1**, 1130
(*Z*)-Alkenes
synthesis of, **1**, 565, 683
via Wadsworth–Emmons reaction, **1**, 750
via Wittig reaction, **1**, 677, 724
α-Alkenes, reactions, with carbon monoxide, **5**, 142
Alkenoamides, *N*-alkenyl-, synthesis of, **5**, 297
Alkenoamides, *N*-alkyl-
from
acids, **5**, 295
nitriles, **5**, 296
synthesis of, **5**, 295
via carbonylation, **5**, 296
2-Alkenoate esters, ω-halo-, reactions, with primary amines, **2**, 323
Alkenoic acids, as starting materials, for amides, **5**, 295
Alkenoic acids, α,β-unsaturated, synthesis of, **5**, 100
Alk-4-enoic esters, 2-hydroxy-3-methyl-4-trimethylsilyl-, synthesis of, **2**, 70
Alkenols
from, iodoalkynes, **5**, 229
as starting materials, for 1,3-diols, **2**, 50
synthesis of, **1**, 756
Alkenols, dialkoxy-, synthesis of, **2**, 79
2-Alken-1-ols, (*Z*)-2-iodo-, synthesis of, **2**, 970
2-Alken-1-ols, (*Z*)-3-iodo-, synthesis of, **2**, 970
(*E*)-Alkenols, synthesis of, **1**, 513
(*Z*)-Alkenols, synthesis of, **1**, 506
Alkenones, reactions, with cuprates, **1**, 120
Alkenones, β-germyl-, synthesis of, **2**, 947
Alkenones, α-methyl-, synthesis of, **3**, 224
Alkenyl anions, reactions, with alkenes, **1**, 309
Alkenyl bromides
alkylation of, **1**, 466
dehydrobromination of, **1**, 1064
palladium-catalyzed reaction of, with monoalkyl phosphonites, **2**, 845
palladium-catalyzed reaction, for tertiary phosphine oxides, **2**, 843
reactions
with cadmium, **4**, 821
with lithium enolates, **1**, 467
with zinc, **4**, 821
as starting materials, for alkenylboronic esters, **2**, 904

Alkenyl carbanions, stability of, **1**, 1088
Alkenyl–chalcogen leaving groups, substitution of, **1**, 484
Alkenyl chlorides
alkylation of, **1**, 466
as intermediates, **1**, 1066
reactions, with alkynes, **1**, 1012
Alkenyl compounds, α-chloro-, stability of, **4**, 820
Alkenyl compounds, 1-halo-, synthesis of, **4**, 814
Alkenyl fluorides, alkylation of, **1**, 466
Alkenyl halides
carbonylation of, **1**, 1015
dehalogenation of, **1**, 5
palladium catalysts with, **2**, 983
palladium-catalyzed alkynylation of, **1**, 1008
reactions
with alkali metal diorganoarsenides, **2**, 874
with alkali metal diorganostibnides, **2**, 874
with alkynyltins, **1**, 1007, 1015
with aryl organometallics, **1**, 471
with carbenoids, **1**, 1067
with copper(I) alkynides, **1**, 1005
with metallated alkynes, **1**, 1006
with sodium toluene-*p*-sulfinates, **4**, 781
with terminal alkynes, **1**, 1006, 1008
reactivity of, **1**, 1007, 1009
as starting materials
for alkenyl acetates, **2**, 682
for alkynes, **1**, 1065
for transition metals, **4**, 821
substitution of, **1**, 1004
synthesis of, **1**, 1064
Alkenyl halides, chiral, alkylation of, **1**, 466
1-Alkenyl halides, reactions, with trialkylboranes, **1**, 279
cis-Alkenyl halides, synthesis of, **1**, 1103
Alkenyl iodides
alkene isomerization of, **1**, 1015
reactions
with alkynyl Grignard reagents, **1**, 1135
with hexynyl metals, **1**, 1015
with trimethylstannyl(trimethylsilyl)ethyne, **1**, 1007
as starting materials, for alkenylboronic esters, **2**, 904
Alkenyl nucleophiles, for the substitution of aryl halides, **1**, 471
(*Z*)-7-Alkenyl-1-ols, synthesis of, **1**, 660
2-Alken-4-ynals, synthesis of, **1**, 1077
Alkenynes, synthesis of, **1**, 1082
Alk-3-en-1-ynes, 3-bromo, as starting materials, for butatrienes, **1**, 967
1-Alken-4-yn-3-ones, synthesis of, **1**, 1030
Alkoxide anions
reactions
with alkynyl halides, **2**, 1028
with phosphorylated chloroalkynes, **2**, 1028
Alkoxide bases, use in the synthesis of imidoylphosphorus compounds, **5**, 786
Alkoxide catalysis, for the synthesis of alkenes, **1**, 451
Alkoxide ions, use in the synthesis of ketene acetals, **4**, 829
Alkoxide ligands, use in the synthesis of dialkoxydialkylvanadiums, **2**, 569
Alkoxides
alkylation of, **2**, 88
anionic oxy-Cope rearrangement of, **1**, 810
cleavage of, **1**, 883
reactions
with alkyl chlorosulfites, **2**, 102
with *O*-alkyllactonium tetrafluorborate salts, **6**, 73
with alkylthiochloroalkynes, **4**, 765
with arenesulfenyl halides, **2**, 720
with *N*-arenesulfonylimidoyl chlorides, **5**, 703
with *N*-arylcarbonimidic dichlorides, **6**, 616
with aziridines, **5**, 693
with carbonimidic dichlorides, **6**, 616, 624
with α-chloroalkyl isocyanates, **4**, 314
with chloromethylpentamethyldisilane, **4**, 614
with 1,2-dichloroalkyne, **4**, 764
with *N*-(dichlorophosphinoyl)imidoyl chlorides, **5**, 707
with *N*-(dichlorophosphinthioyl)imidoyl chlorides, **5**, 707
with difluorocarbene, **6**, 37
with dihalocarbenes, **6**, 37
with 1,1-dinitro-2,2-diphenylethylene, **4**, 898
with dioxenium salts, **6**, 73
with dioxolenium, **6**, 73
with dithiocarbenium salts, **6**, 98
with *N*-fluorinated imidoyl fluorides, **5**, 698
with geminal dihalides, **4**, 189
with α-haloboranes, **4**, 362
with α-haloboronates, **4**, 362
with α-halo ketones, **4**, 183
with hydrazonoyl chlorides, **5**, 703
with hydroxamoyl chlorides, **5**, 699
with *N*-hydroxyimidoyl bromides, **5**, 699
with methanes, **1**, 759
with nitroalkenes, **3**, 46
with *N*-perfluoroalkylcarbonimidic difluorides, **6**, 614
with *N*,*N*-perfluoroformamidines, **5**, 813
with 2-phenylbenzo-1,3-dioxolium tetrafluoroborate, **6**, 73
with phosphoramidates, **2**, 108
with polyfluoroalkenes, **4**, 761
with seleno ester ketones, **1**, 246
with sulfamoyl chlorides, **2**, 366
with thiophosgene, **6**, 540
with tosylhydrazones, **1**, 1077
with trichloroacetonitrile, **6**, 296
with trihalomethanes, **4**, 826
as starting materials, for difluorocarbene, **4**, 1083
synthesis of, **2**, 112
transmetallation of, **2**, 112
use in the synthesis of alkynes, **1**, 1064
use in the synthesis of aminocarbene complexes, **5**, 953
use in the synthesis of heterocuprates, **2**, 588
Alkoxides, allylic
isomerization of, **1**, 133
use in fragmentation reactions, **1**, 1082
Alkoxides, chiral, nucleophilic displacement of, **2**, 600
Alkoxides, lanthanide
as catalysts
for the oxidation of alcohols, **3**, 119
for the synthesis of aldehydes, **3**, 7
for the synthesis of cyanohydrins, **3**, 616
Alkoxides, lanthanoid(III), as catalysts, for the synthesis of β-oxygenated nitriles, **3**, 632
Alkoxides, γ-lithio, from, oxetanes, **5**, 943
Alkoxides, perfluoro, as starting materials, for α-fluoroalkyl esters, **4**, 52
Alkoxy anion, α-fluoro-, tris(dimethylamino)sulfonium salt of, **4**, 42
Alkoxy arene–transition metal complexes, substitution of, **1**, 483
α-Alkoxy carbanions
rearrangements of, **1**, 396
synthesis of, **1**, 138; **4**, 377
Alkoxycarbonylation
for the synthesis of carboxylic esters, **5**, 142
for the synthesis of α,β-unsaturated esters, **5**, 157
Alkoxydiazenium ions, reactions, with carboxylates, **5**, 366
Alkoxyethynes, germanium-substituted, synthesis of, **2**, 1027
Alkoxy groups
elimination of, **3**, 209, 226
use in the synthesis of allyl cations, **1**, 364
β-Alkoxy groups, elimination of, **3**, 62
Alkoxy ligands, in substitution reactions, **1**, 119
Alkoxymercuration
of alkynes, **2**, 676
of enol ethers, **4**, 188

Alkoxymethyl anions, reactions, with acid compounds, **3**, 162
N-Alkoxymethyls, cleavage of, **4**, 102
Alkoxy radicals, synthesis of, **1**, 929
Alkoxy radicals, β-trialkylstannyl, fragmentation of, **3**, 143
Alkoxyselenenylation, of alkenes, **2**, 282
Alkylamines, primary, reactions, with 2,3,5,6-tetraphenylpyrylium cations, **1**, 31
Alkyl anions
reactions, with alkenes, **1**, 309
reactivity of, **1**, 847
Alkylating agents
reactions
with acyliron tricarbonyl complexes, **5**, 451
with thioamides, **5**, 726
with thiolactams, **5**, 726
Alkylating reagents
reactions
with enolates, **3**, 179
with tertiary amines, **2**, 760
Alkylation
of aldehydes, **3**, 10
via enolates, **3**, 10
via nitrogen compounds, **3**, 10
of alkylarsenides, **2**, 485
of alkylarsines, **2**, 482
of alkylstibides, **2**, 485
of amides, **5**, 284, 289
of anionic cobalt compounds, **2**, 575
of arsines, **2**, 482
of carbene complexes, **5**, 955
of carboxylate anions, **5**, 132
of carboxylic acids and esters, **4**, 824
chiral, **1**, 180
competing *O*- and *C*-, **1**, 130
of *N,N'*-dimethoxyurea, **5**, 331
of enolates, **1**, 129
of haloethynylides, **2**, 1012
of hydrazines, process problems, **2**, 376
of hydroxyimides, **5**, 333
of ketones, **3**, 123
via enolates, **3**, 124
via nitrogen compounds, **3**, 124
via silyl enol ethers, **3**, 124
of metal enolates, **3**, 124
remote, of carbene complexes, **5**, 956
for the substitution of silicon functions, **1**, 281
for the synthesis of *N*-alkylhydroxylamines, **2**, 341
for the synthesis of amidines, **5**, 751, 754
for the synthesis of *N,N*-dialkylhydroxylamines, **2**, 344
for the synthesis of formyl phosphonates, **3**, 50
for the synthesis of osmium(II) complexes, **2**, 582
for the synthesis of ruthenium(II) complexes, **2**, 582
for the synthesis of tetraalkyltins, **2**, 600
for the synthesis of thioimidic esters, **5**, 726
for the synthesis of tricoordinate phosphorus functions, **4**, 481
of tertiary arsines, **2**, 487
of tertiary stibines, **2**, 487
of thioimidic esters, **5**, 730
of α,β-unsaturated aldehydes, **3**, 213
Alkylation, asymmetric, examples, **1**, 134
Alkylation, enantioselective, of carbonyls, **2**, 71
Alkylation, intramolecular, for the synthesis of cyclic monothioacetals, **4**, 231
Alkylation, reductive, for the synthesis of amides, **5**, 273
C-Alkylation
of aldoximes, **3**, 428
of enolates, **1**, 130
for the synthesis of iminium salts, **2**, 757
N-Alkylation
of amides, **5**, 292, 296
of amidines, **5**, 754
of arylamides, **5**, 300
for the synthesis of *N*-alkenylarylamides, **5**, 302
for the synthesis of amines, **2**, 300
for the synthesis of iminium salts, **2**, 757
O-Alkylation
of enolates, **1**, 130
of phenolates, **2**, 683
of phenols, **2**, 683
for the synthesis of alkenyl ethers, **2**, 674
of trinitromethanes, **6**, 145
S-Alkylation, for the synthesis of *O,S*-acetals, **4**, 225
α-Alkylation, methods for, **1**, 129
α-Alkylation, asymmetric, for the synthesis of enolates, **1**, 135
B-Alkyl-9-borabicyclo[3.3.1]nonyls, synthesis of, **1**, 654
Alkyl bromides
dehalogenation of, **1**, 937
Friedel–Crafts acylation with, **1**, 462
from
alcohols, **2**, 31
alkanes, **2**, 27
alkenes, **2**, 29
alkyl chlorides, **2**, 31
alkyl fluorides, **2**, 31
alkyl halides, **2**, 31
amines, **2**, 33
ethers, **2**, 32
selenides, **2**, 32
silyl ethers, **2**, 10
tetrahydropyranyl ethers, **2**, 10
trichloroacetamidate, **2**, 32
hydrolysis of, **2**, 87
reactions
with allenyllithiums, **1**, 464
with aluminum, **2**, 595
with 2-butadienyl Grignard reagents, **1**, 1092
with carboxylates, **5**, 132
with cuprates, **1**, 465
with dibutyltin oxide, **2**, 89
with hydrogen peroxide, **2**, 96
with lithium diphenylcuprate, **1**, 119
with potassium carbonate, **6**, 462
with potassium superoxide, **2**, 98
with sulfur, **2**, 226
with zinc, **6**, 64
reduction of, **1**, 2
as starting materials
for alkyl chlorides, **2**, 25
for alkyl fluorides, **2**, 18
for alkyl iodides, **2**, 35
for diethyl alkylphosphoramidates, **5**, 1031
substitution of, **1**, 464; **2**, 25
synthesis of, **2**, 27
via bromoalkylation of arenes, **2**, 33
via bromodecarboxylation of carboxylic acids, **2**, 33
via bromohalogenation of alkenes, **2**, 30
via hydrobromination of alkenes, **2**, 29
use in the synthesis of dialkyl carbonates, **6**, 463
Alkyl bromides, ω-amino-, reactions, with chlorodiethoxyphosphine, **2**, 457
Alkyl bromides, chiral, as starting materials, for palladium acyl compounds, **5**, 463
Alkyl carbanions
stability of, **1**, 846
structure of, **1**, 847
synthesis of, **1**, 845
Alkyl carbocations
stability of, **1**, 893
synthesis of
problems, **1**, 900
via bond scission, **1**, 899
Alkyl cations, α-(alkylthio)-, synthesis of, **2**, 1008
Alkylceriums, indenyl-, synthesis of, **2**, 565

Alkyl chains, introduction of a substituent into, **1**, 129
Alkyl–chalcogen leaving groups, substitution of, **1**, 483
Alkyl chlorides
catalytic carbonylation of, **5**, 16
Friedel–Crafts acylation with, **1**, 462
from
alcohols, **2**, 8, 25
alkanes, **2**, 20
alkenes, **2**, 23
alkyl bromides, **2**, 25
alkyl chloroformates, **2**, 25
alkyl halides, **2**, 25
amines, **2**, 27
arylalkyl ethers, **2**, 26
dialkyl ethers, **2**, 26
imidate ester hydrochlorides, **6**, 70
trialkylboranes, **2**, 23
reactions
with cyanogen chloride, **5**, 962
with 1-(2-hydroxyethyl)-4,6-diphenylpyridine-2-thione, **2**, 121
with isocyanic acid, **5**, 964
with Lewis acid halides, **1**, 900
with silicon, **2**, 538
with sodium, **2**, 599
with sodium hydrogen sulfide, **2**, 119
with sodium iodide, **2**, 35; **3**, 612
with zinc, **6**, 64
reactivity of, **3**, 612
as starting materials
for acid chlorides, **5**, 16
for alkyl bromides, **2**, 31
for alkyl fluorides, **2**, 18
for aminohydrazines, **2**, 373
substitution of, **1**, 464
synthesis of, **2**, 20
via chloroalkylation of arenes, **2**, 27
via chlorodecarboxylation of carboxylic acids, **2**, 27
via halochlorination of alkenes, **2**, 23
via hydrochlorination of alkenes, **2**, 23
Alkyl chlorides, *β*-hydroxy-, carbonation of, **5**, 71
Alkyl chlorides, *α*-keto, reactions, with potassium fluoride, **4**, 3
Alkyl chlorides, tertiary, reduction of, **1**, 4
t-Alkyl chlorides
reactions
with dialkyl cyanamides, **5**, 1146
with trichloroferrate complexes, **5**, 1146
with trimethylsilyl azide, **2**, 317
Alkyl complexes
dimerization of, **5**, 465
reactions, with hydrogen chloride, **5**, 949
as starting materials, for ruthenium carbene complexes, **5**, 938
Alkyl *σ*-complexes, as starting materials, for aminocarbene complexes, **5**, 955
Alkyl compounds
from, alkylarylsulfonyls, **1**, 39
microbial oxidative metabolism of, **5**, 117
reactions
with carbon monoxide, **5**, 442
with phosphines, **5**, 452
stability of, **2**, 566
synthesis of, **2**, 566; **5**, 1093
N-Alkyl compounds, synthesis of, **2**, 389
Alkyl fluorides
Friedel–Crafts acylation with, **1**, 462
from
alcohols, **2**, 19
alkanes, **2**, 15
alkenes, **2**, 17
alkyl bromides, **2**, 18
alkyl chlorides, **2**, 18
alkyl halides, **2**, 18
alkyl iodides, **2**, 18
amines, **2**, 20
silyl ethers, **2**, 10
reduction of, **1**, 2, 3
as starting materials
for alkyl bromides, **2**, 31
for alkyl iodides, **2**, 35
for Grignard reagents, **2**, 558
substitution of, **2**, 25
synthesis of, **2**, 15
via fluorine–halogen addition to alkenes, **2**, 17
via fluorodecarboxylation, **2**, 20
Alkylgermyl halides, reactions, with lithium phosphides, **2**, 867
Alkyl groups
activation of, via a heterocyclic group, **1**, 861
cleavage of, **1**, 58
migratory aptitudes of, **1**, 384
reactivity of, **1**, 859
sigmatropic rearrangements of, **1**, 799
Alkyl halides
for the alkylation of amides, **5**, 284
displacement of halogen atoms in, **4**, 761
displacement of halogens from, **1**, 107
electrochemical reduction of, **1**, 946
as electrophiles, for dithiocarboxylic enethiolates, **4**, 842
from
alcohols, **2**, 5
alkanes, **2**, 2
alkenes, **2**, 3
alkyl halides, **2**, 5
amines, **2**, 11
ethers, **2**, 10
geminal dihalo compounds, **2**, 14
hydrazines, **2**, 12
pyridinium salts, **2**, 11
tetrahydropyranyl ethers, **2**, 10
thiohydroxamate esters, **2**, 13
tosylhydrazines, **2**, 12
vinyl halides, **2**, 14
xanthates, **2**, 8
homologation of, **2**, 15
Michaelis–Arbuzov reaction of, for tertiary phosphine oxides, **2**, 842
optically active, synthesis of, **2**, 6
for the oxidative addition of pentakis (*t*-butylisonitrile)iron, **5**, 799
photoelectrochemical reduction of, **1**, 946
radical cyclization of, **1**, 341
reactions
with *N*-acyl-*N'*-triflylhydrazine, **5**, 363
with aldehydes, **6**, 58
with alkali cyanates, **5**, 964
with alkali metal arsenites, **2**, 497
with alkali metal cyanates, **5**, 964
with alkali metal selenocyanates, **5**, 1113
with alkali selenides, **2**, 278
with *N*-(alkylamino)benzotriazoles, **2**, 302
with alkylantimony anions, **2**, 485
with alkylarsenic anions, **2**, 485
with alkyldialkoxydiphosphines, **2**, 453
with alkylplatinum(II) complexes, **2**, 584
with *α*-alkylthioalkyl phosphonates, **4**, 347
with alkynes, **4**, 781
with allenyllithium, **1**, 971
with allyllithiums, **1**, 112
with antimony, **2**, 486
with arene sulfinate salts, **2**, 715
with arsenic, **2**, 486
with arsines, **2**, 487; **3**, 196
with arylcoppers, **1**, 121
with 2-(arylimino)imidazolines, **6**, 642

with aryl isocyanides, **3**, 158
with barium metal, **2**, 563
with benzyl *N*-cyanocarbamate, **5**, 1121
with benzyllithiums, **1**, 112
with *O*-benzyloxylamine, **2**, 342
with beryllium, **2**, 563
with calcium metal, **2**, 563
with carbanions, **1**, 110, 610
with carbodiimides, **5**, 1146
with chromium(II) salts, **2**, 571
with cobalt complexes, **1**, 335, 345
with Collman's reagent, **2**, 575
with copper, **2**, 585
with cyanamide, **2**, 308; **5**, 1118
with cyanide salts, **3**, 612
with cyanogen chloride, **5**, 962
with 4-cyano-3-oxo-tetrahydrothiophene, **3**, 648
with cyclopentadienylruthenium complexes, **2**, 582
with dialkoxyhalophosphines, **2**, 458
with dialkyl arylphosphonites, **2**, 845
with *S*,*S*-dialkyl dithiocarbonates, **2**, 134, 143
with *N*,*N*-dialkylhydroxylamines, **2**, 349
with dilithiums, **1**, 1001
with diorganocuprates, **1**, 118
with diorganophosphides, **2**, 437
with 1,1-diselenolates, **4**, 875
with disodium disulfide, **2**, 120
with enamines, **2**, 757
with enolate anions, **1**, 8475
with Grignard reagents, **1**, 114
with hydrogen sulfide, **2**, 119
with *N*-hydroxy thioureas, **5**, 831
with indium–magnesium alloy, **2**, 597
with iridium(I) halide complexes, **2**, 583
with isocyanic acid, **5**, 964
with Lewis acids, **1**, 106
with α-lithiated amines, **4**, 528
with lithiated triselenoorthoformates, **6**, 94
with lithiated trithioorthoformates, **6**, 88
with lithium, **2**, 551
with lithium diphenylphosphide, **1**, 52
with mercury(II) selenocyanates, **5**, 1053
with metal(0) complexes, **2**, 582
with metal diarylphosphides, **2**, 826
with metal fulminates, **3**, 683
with metal nitrocyanamides, **5**, 963
with metal phosphides, **2**, 427
with metal sulfides, **2**, 132
with metal thiocyanates, **5**, 1105
with methylphosphine, **2**, 429
with 1-methylpyridine-2-thione, **2**, 121
with monoalkali phosphides, **2**, 429
with nitrile carbanions, **3**, 620
with nitronate metal salts, **4**, 540
with nucleophiles, **4**, 761
with osmium(0) complexes, **2**, 582
with 1,3,2-oxaazaphospholane, **2**, 861
with 1,3,2-oxazaphospholans, **2**, 455
with palladium, **2**, 582
with palladium carbonyl complexes, **5**, 463
with phenacylides, **2**, 877
with phosphonium iodide, **2**, 427
with platinum, **2**, 582
with potassium, **2**, 556
with potassium selenocyanates, **5**, 1053
with rhodium(I) halide complexes, **2**, 583
with ruthenium(0) complexes, **2**, 582
with samarium(II) iodide, **1**, 937
with samarium(II) salts, **2**, 565
with selenoamides, **5**, 732
with silver cyanate, **5**, 965
with silver nitrate, **2**, 103
with silver nitrocyanamide, **5**, 963
with sodium, **2**, 556
with sodium alkynides, **1**, 1000
with sodium disulfite, **2**, 176
with sodium–iron tetracarbonyl complexes, **5**, 451
with sodium–lead alloys, **2**, 601
with sodium telluride, **2**, 279
with sodium thiophosphate, **2**, 255
with sodium trithiocarbonate, **2**, 123
with stannyl anions, **2**, 600
with sulfinate anions, **4**, 235
with sulfites, **2**, 191
with tertiary alkylstibines, **2**, 487
with tertiary phosphines, **2**, 444
with tertiary thioamides, **4**, 899
with tetrachloromethane, **6**, 24
with thioamides, **3**, 638
with thiophenolates, **2**, 710
with thiosaccharin, **2**, 121
with thiourea, **2**, 120
with tin dichloride, **2**, 600
with tin oxide, **2**, 600
with trimethylsilyl cyanide, **3**, 613
with *O*-trimethylsilyl thiocarboxylates, **5**, 540
with Vaska complex, **2**, 583
with Wilkinson's complex, **2**, 583
with ylides, **2**, 446; **4**, 558
reactivity of, **1**, 2, 107, 462
reduction of
catalysts for, **1**, 2
chemoselectivity in, **1**, 4
to haloalkanes, **1**, 1
reviews, **2**, 2
as starting materials
for aldehydes, **3**, 3
for alkanethiols, **2**, 119
for alkyl bromides, **2**, 31
for alkyl chlorides, **2**, 25
for alkyl fluorides, **2**, 18
for alkyl halides, **2**, 5
for alkyl iodides, **2**, 35
for alkyllithiums, **2**, 551
for alkylmagnesiums, **2**, 558
for alkyl xanthates, **2**, 123
for amides, **5**, 272
for amines, **2**, 307
for dialkyl sulfides, **2**, 132
for α-haloalkenylsilicons, **4**, 810
for isocyanates, **5**, 962
for ketene dithioacetal *S*,*S*-dioxides, **4**, 857
for ketones, **3**, 114
for nitriles, **3**, 612
for nitro compounds, **2**, 393
for ortho amides, **6**, 140
for phosphonium salts, **3**, 492
for primary amines, **2**, 298
for thiophosphines, **2**, 254
substitution of, **1**, 462, 1000
via Friedel–Crafts acylation, **1**, 462
via organometallic carbanions, **1**, 463
synthesis of, **2**, **1–36**
general methods, **2**, 2
via haloalkylation of arenes, **2**, 13
via halodecarboxylation, **2**, 12
via halogen-halogen addition to alkenes, **2**, 5
via hydrohalogenation of alkenes, **2**, 4
via sulfonates, **2**, 9
trichloromethylation of, **6**, 24
ultrasonication of, **3**, 14
use in the synthesis of thioimidic esters, **5**, 726
Alkyl halides, alkoxy-, reactions, with thiourea, **2**, 128
Alkyl halides, fluoro-, reactions, with dialkylphosphites, **4**, 131

Alkyl halides, α-metallo-, synthesis of, **2**, 997
Alkyl halides, phenoxy-, as starting materials, for monoalkylhydrazines, **2**, 373
Alkyl halides, 2-silyl-, reactions, with tris(diethylamino)sulfonium difluorotrimethyl silicates, **2**, 70
Alkyl halides, tertiary
  reactions
    with alkynylalanes, **1**, 1003
    with alkynylsilanes, **1**, 1003
t-Alkyl halides
  reactions
    with dialkyl cyanamide antimony(V) chloride complexes, **5**, 1146
    with enolates, **1**, 133
Alkylidenation
  of esters, **1**, 770; **2**, 681
  of lactones, **1**, 770; **2**, 681
Alkylidene, use of term, **3**, 507
Alkylidene carbenoids
  of lithium, chemistry of, **4**, 816
  synthesis of, **4**, 816
μ_2-Alkylidene clusters, synthesis of, **6**, 347
Alkylidene complexes, as starting materials, for alkylidyne complexes, **3**, 690
Alkylidene halides
  displacement of halogen atoms in, **4**, 761
  reactions, with nucleophiles, **4**, 761
Alkylidenes, from, diazo compounds, **1**, 298
Alkylidenes, diazo
  synthesis of, **3**, 488
  as transient intermediates, **3**, 488
n-Alkylides, sulfonium, reactions, with alkyllithiums, **1**, 623
Alkylidyne–carbene complexes
  synthesis of
    via β-addition to unsaturated ligands, **3**, 691
    via formal abstraction of oxide from acyl or carbamoyl ligands, **3**, 691
    via metathesis of metal–metal triple bonds with alkynes, **3**, 691
    via metathesis with nitriles, **3**, 691
Alkylidyne complexes
  chemistry of, **3**, 690
  from
    alkylidene complexes, **3**, 690
    α-metallodiazoalkanes, **3**, 691
    non-alkylidyne precursors, **3**, 690
  synthesis of
    via ligand substitution reactions, **3**, 692
    via modification of alkylidyne-metal complexes, **3**, 691
    via modification of the substituent attached to the carbyne carbon, **3**, 692
    via oxidation and reduction reactions, **3**, 692
Alkylidyne complexes, bis-, synthesis of, **6**, 338
Alkylidyne compounds, acyl, synthesis of, **5**, 446
Alkylidyne–metal complexes
  synthesis of, **3**, 690
    via abstraction of an α-hydrogen atom, **3**, 690
    via rearrangement of carbene complexes, **3**, 690
Alkylidyne–transition metal complexes, synthesis of, **3**, 690
Alkyl iodides
  acylation of, **3**, 289
  dehalogenation of, **1**, 937
  Friedel–Crafts acylation with, **1**, 462
  from
    alcohols, **2**, 35
    alkanes, **2**, 33
    alkenes, **2**, 34
    alkyl bromides, **2**, 35
    alkyl fluorides, **2**, 35
    alkyl halides, **2**, 35
    amines, **2**, 36
    epoxides, **2**, 36
    ethers, **2**, 36
    nitro compounds, **2**, 36
    tetrahydropyranyl ethers, **2**, 10
  hydrolysis of, **2**, 87
  oxidation of, **5**, 204
  photolysis of, **1**, 334
  reactions
    with alkanethiols, **2**, 134
    with *N*-alkylthiohydroxamic acids, **3**, 435
    with allenyllithiums, **1**, 464
    with aluminum, **2**, 595
    with calcium, **1**, 115
    with carboxylates, **5**, 132
    with cerium, **2**, 564
    with copper(I) salts, **2**, 72
    with cuprates, **1**, 465
    with cyclopentadienylcobalt complexes, **2**, 579
    with diethylzinc, **2**, 591
    with ethyl xanthate potassium salts, **6**, 478
    with europium, **2**, 564
    with indium, **2**, 597
    with lanthanide metals, **2**, 564
    with osmium salts, **5**, 455
    with peracids, **1**, 108
    with phosphorus, **2**, 452
    with samarium, **2**, 564
    with silver cyanide, **3**, 694
    with silver perchlorate, **2**, 94
    with silver selenate, **2**, 102
    with silver sulfate, **2**, 102
    with sodium nitrocyanamide, **5**, 964, 1119
    with sodium tetraphenylborate, **3**, 290
    with triaryltin hydrides, **1**, 330
    with 2,4,6-tripyrrolidinobenzene, **2**, 1009
    with vinyl cuprates, **1**, 465
    with ytterbium, **2**, 564
    with zinc, **6**, 64
    with zinc-copper couples, **2**, 591
  reduction of, **1**, 2
  sonication of, **2**, 591
  as starting materials
    for alkyl fluorides, **2**, 18
    for alkyllithiums, **2**, 554
    for organocoppers, **3**, 287
    for trialkylaluminums, **2**, 595
  substitution of, **1**, 464
  synthesis of, **2**, 8, 33
    via Finkelstein reactions, **1**, 107
    via hydroiodination of alkenes, **2**, 34
    via iodine-iodine addition to alkenes, **2**, 35
    via iodoalkylation of arenes, **2**, 36
    via iododecarboxylation of carboxylic acids, **2**, 36
  use in the synthesis of tertiary phosphines, **2**, 433
Alkyl iodides, perfluoro-
  reactions
    with carbon dioxide, **5**, 58
    with diphenyl diselenides, **4**, 284
    with phosphorus pentasulfide, **3**, 348; **5**, 474
Alkyl iodides, β-(*N*-tosyl)amino, substitution of, **1**, 120
Alkyl ligands, loss of, **1**, 119
Alkylmercuration, for the synthesis of radicals, **1**, 948
Alkyl moieties, 1,1-bis(silyl), synthesis of, **4**, 624
Alkyl radicals
  from
    alkylaluminums, **2**, 595
    alkylcobalt salophens, **1**, 667
    diacyl peroxides, **1**, 934
    trialkylboranes, **1**, 333
  oxidation of, **1**, 3217
  reaction rates of, **1**, 920
  reactions
    with alkenes, **1**, 323
    with alkynes, **1**, 333

with quinones, **1**, 275
with samarium diiodide, **1**, 347
stability of, **1**, 321
structure of, **1**, 922
synthesis of, **1**, 330
via homolysis of organocobalt complexes, **1**, 335
via photolysis, **1**, 948
via thermolysis, **1**, 948
Alkyl radicals, tertiary, synthesis of, **1**, 330
Alkylselenyl halides, as starting materials, for alkane selenenamides, **2**, 292
Alkylthiols
reactions
with iminoboranes, **5**, 863
with ketenimines, **5**, 730
with vinyl phosphonates, **3**, 166
Alkyl transfer, selectivity reversal, **4**, 810
Alkynals
from
alkynes, **3**, 79
2,3-dibromoaldehydes, **3**, 79
reactions, with bis(silyloxy)dienes, **1**, 520
synthesis of, **1**, 1082
2-Alkynamines, *N*,*N*-dialkyl-, synthesis of, **1**, 261
Alkyne, bis(trimethylsilyl)-, reactions, with nitryl fluoride, **4**, 807
Alkyne, bromophenyl-, reactions, with pentylmagnesium bromide, **1**, 1017
Alkyne, chloro-, reactions, with arenesulfenyl chloride, **4**, 772
Alkyne, diacetoxy-, reactions, with caesium fluoride, **4**, 768
Alkyne, dichloro-
addition of amines to, **4**, 792
reactions
with benzeneselenenyl bromides, **4**, 787
with benzeneselenenyl chlorides, **4**, 787
with benzeneselenol, **4**, 786
with nitrogen dioxide, **4**, 797
with nitrogen nucleophiles, **4**, 792
with sodium arenethiolates, **4**, 772
as starting material, for ketene dithioacetals, **4**, 851
synthesis of, **4**, 976
Alkyne, 1,2-dichloro-, reactions, with alkoxides, **4**, 764
Alkyne, dicyano-, synthesis of, **1**, 1083
Alkyne, diiodo-
reactions
with iodine, **4**, 797
with nitrous acid, **4**, 797
Alkyne, di(trimethylsilyl)-, reactions, with boracyclopropane, **4**, 1082
Alkyne, ethoxy-
reactions
with allylamines, **4**, 884
with α-pyridylaminodialkylboranes, **4**, 884
Alkyne, 1-ethoxy-, reactions, with *B*-iodo-9-borabicyclo[3.3.1]nonane, **4**, 765
Alkyne, fluoro-
from, 1,1-difluoroethylene, **1**, 1017
reactions, with thiols, **4**, 772
Alkyne, hexachlorophenyl-, reactions, with sodium methoxide, **4**, 764
Alkyne, phenoxy-
reactions
with bromine, **4**, 765
with iodine, **4**, 765
Alkyne, phenyl-
oxidative coupling of, **1**, 1035
reactions, with alkyl *N*,*N*-dichlorocarbamates, **4**, 794
Alkyne, phenyl(trimethylsilyl)-, mercury(II)-catalyzed addition of iodine to, **4**, 808
Alkyne, sulfonyl-, reactions, with *N*-*t*-butylhydroxylamine, **2**, 691
Alkyne, trimethylsilyl-
reactions
with allyl bromide, **4**, 810
with α,ω-diiodoalkanes, **4**, 810
with organic halides, **1**, 1137
Alkyne-bridged complexes, from, tungsten–tungsten-bonded complexes, **4**, 675
Alkyne chlorides, (alkylthio)-, reactions, with phosphorus pentachloride, **4**, 804
Alkyne complexes, alkali aluminum, synthesis of, **2**, 1099
Alkynecyclopentadienylnickel–cyclopentadienylmolybdenum complexes, synthesis of, **4**, 713
Alkyne exchange, for the synthesis of alkyne-bridged dicobalt complexes, **4**, 687
Alkyne hydrosilylation, reversal in the stereochemistry of, **2**, 915
Alkyne insertion
for the synthesis of carbon-bridged ruthenium complexes, **4**, 683
for the synthesis of carbon-bridged trinuclear iron complexes, **4**, 683
for the synthesis of ferracyclopentadiene complexes, **4**, 683
(*E*)-Alkynenones, synthesis of, **3**, 228
Alkynes
acetoxymercuration of, **1**, 454
acylation of, **3**, 273
addition of germyl compounds to, **2**, 946
addition reactions of, **1**, 455
1,4 addition reactions of, **1**, 1048
additions to, **1**, 174
alkoxycarbonylation of, **5**, 157
alkoxymercuration of, **2**, 676
alkylation of, **1**, 178, 1000, 1001
allylmetallation of, **3**, 218
aminosilylation of, **2**, 920
bishydroboration of, **4**, 629
bisilylation of, **2**, 915
bromination of, **2**, 1018
carboboration of, **1**, 542; **2**, 902
carbocupration of, **2**, 976
carbogermylation of, **2**, 946
syn-carbolithiation of, **2**, 972
carbomagnesiation of, **1**, 453
carbometallation of, **1**, 538, 544; **2**, 919, 971
carbonylation of, **3**, 276
carbopalladation of, **2**, 979
carbosilylation of, **2**, 919
catalytic hydrogenation of, **1**, 506
catalytic reactions, with metal carbene, **2**, 981
cleavage of, **6**, 402
cocyclotrimerization of, **2**, 984
Cope rearrangement of, **1**, 808
cyanation of, **3**, 675
cyanogermylation of, **2**, 946
cyclic oligomerization of, **2**, 984
[2+2] cycloaddition reaction of, **1**, 516
cycloaddition reactions of, **1**, 790
in cycloaddition reactions, with α,β-unsaturated esters, **4**, 827
cyclooligomerization of, **1**, 530
cyclotrimerization of, **1**, 530; **5**, 461
dehydroboration of, **6**, 181
deprotonation of, **5**, 589
diboronation of, **4**, 1054
dihydroboration of, **4**, 631
dihydrosilylation of, **4**, 617
dimercuration of, **4**, 1067
dimerization of, **1**, 546
dimetallic compounds of, **2**, 1077
double germylation of, **2**, 946
double silylation of, **2**, 915
electrochemical reduction of, **1**, 1133
as electrophiles, for dithiocarboxylic enethiolates, **4**, 842

ene reactions of, **1**, 518
fluorination of, **4**, 6
formylation of, **3**, 70, 79
Friedel–Crafts alkylation of, **1**, 1025
from
  alkenes, **1**, 1063
  alkenyl halides, **1**, 1065
  alkenylsilanes, **1**, 1080
  alkenyl triflates, **1**, 1070
  5,5-alkyl aryl-3-nitroso-2-oxazolidones, **1**, 1076
  1-alkynyl halides, **1**, 998
  amides, **1**, 1066
  5-aminoisoxazoles, **1**, 1076
  1-amino-1,2,3-triazoles, **1**, 1075
  5-aryl-3-nitroso-2-oxazolidones, **1**, 1076
  (*E*)-1,2-dianisylvinyl bromide, **1**, 1120
  5,5-diaryl-3-nitroso-2-oxazolidones, **1**, 1076
  diazo compounds, **1**, 1077
  1,2-dibromoalkanes, **1**, 1065
  1,1-dihaloalkanes, **1**, 1064
  1,2-dihaloalkanes, **1**, 1064
  4,4-dihalo-5-pyrazolones, **1**, 1076
  1,2-dihydrazones, **1**, 635
  enol benzoxazolium salts, **1**, 1071
  enol compounds, **1**, 1069
  enol sulfonates, **1**, 1081
  3-haloselenophenes, **1**, 1072
  3-halotellurophenes, **1**, 1072
  hydrazones, **1**, 1075
  5-isoxazolones, **1**, 1076
  ketene–anthracene cycloaddition adducts, **1**, 1084
  monoalkylthexyborinates, **1**, 1034
  2-nitroso-5-isoxazolones, **1**, 1076
  1,2,3-selenodiazoles, **1**, 1077
  tetrazoles, **1**, 1078
  1,2,3-thiadiazoles, **1**, 1077
  1-tosylamino-1,2,3-triazoles, **1**, 1075
  vinylic selenoxides, **1**, 605
  ylides, **1**, 1079
germylcupration of, **2**, 946
haloboration of, **1**, 1104
heterogeneous catalytic hydrogenation of, **1**, 502
hydration of, **3**, 200
hydroalumination of, **1**, 128, 468, 511; **4**, 1068
  with diisobutylalane, **2**, 968
hydroboration
    of, **1**, 449, 509, 531, 1100; **2**, 527, 900; **3**, 3, 645
  for the synthesis of ketones, **3**, 114
hydrogenation of, **1**, 455
hydrogermylation of, **2**, 946
hydrometallation of, **1**, 531; **3**, 140
anti-hydrometallation of, **2**, 970
syn-hydrometallation of, **2**, 965, 969
hydrosilylation of, **1**, 1106
hydrostannation of, **4**, 701
hydrostannylation of, **1**, 330
hydrozirconation of, **1**, 453, 515
intramolecular Diels–Alder cycloaddition of, **1**, 523
intramolecular oxymercuration of, **2**, 982
isomerization of, **1**, 954
Mannich reaction of, **1**, 1041
metal-catalyzed homocoupling of, **1**, 992
metallation of, **2**, 917
  with heavy metals, **2**, 982
metalloboration of, **2**, 903
metallo-ene reactions with, **1**, 526
metallogermylation of, **2**, 946
metallometallation of, **2**, 916
methylalumination of, **2**, 976
monometallation of, **1**, 1133
nickel-promoted reaction of, with carbon dioxide, **2**, 988
as non-transferable ligands, **2**, 587
for the opening of epoxides, **1**, 204
oxidative coupling of, **2**, 952, 983
oxidative metallation of, **2**, 983, 989
palladium-catalyzed additions of, **1**, 546
palladium-catalyzed heterocoupling reactions of, **1**, 1036
palladium-catalyzed homocoupling reactions of, **1**, 1036
palladium-catalyzed reactions, **1**, 1027
perfluoroalkylation of, **1**, 1130
phenyselenation of, **2**, 1037
photohydration of, for the synthesis of enols, **2**, 637
photolysis of, **4**, 679
radical addition of hydrogermanes, **2**, 946
reactions
  with acetals, **1**, 1025
  with acid halides, **3**, 238
  with alcohols, **4**, 194
  with alkenezirconocenes, **2**, 986
  with alkenylarsines, **2**, 875
  with alkenyl chlorides, **1**, 1012
  with alkenylketenes, **2**, 668
  with alkenylmagnesium bromides, **1**, 1105
  with alkenylstibines, **2**, 875
  with alkenyl sulfones, **1**, 545
  with alkenyl triflates, **1**, 1029
  with alkylallenes, **1**, 498
  with alkylcopper complexes, **5**, 100
  with alkyl halides, **4**, 781
  with alkyl iridium complexes, **5**, 461
  with alkylmanganese pentacarbonyl complexes, **3**, 218
  with alkyl radicals, **1**, 333
  with alkylsulfonic acid, **2**, 689
  with alkynezirconocenes, **2**, 987
  with allyl alanes, **1**, 543
  with allylic acetates, **1**, 195
  with allylic halides, **1**, 1050
  with allylic sulfides, **1**, 235
  with aluminum hydrides, **1**, 533
  with 2-aminoalkylarsines, **4**, 498
  with arenediazonium salts, **1**, 1033
  with aryl dihalides, **1**, 1009
  with aryl iodides, **1**, 546
  with aryl polyhalides, **1**, 1009
  with aryl triflates, **1**, 1029
  with benzeneselenenyl chloride, **4**, 92
  with benzyl alanes, **1**, 543
  with (benzylideneacetone)tricarbonyliron, **4**, 681
  with benzynezirconocenes, **2**, 987
  with 9-borabicyclo[3.3.1]nonyl, **1**, 1101
  with borenes, **5**, 928
  with boron compounds, **2**, 900
  with bromoallenes, **1**, 1137
  with bromomethanesulfonyl bromide, **4**, 84
  with 1,3-butadienes, **1**, 1036
  with *n*-butyllithium, **5**, 101
  with carbene complexes, **2**, 979
  with carbenes, **1**, 549
  with carbenoids, **1**, 549
  with carbon dioxide, **1**, 545
  with carbon–tellurium bonds, **2**, 730
  with carbonyl ylides, **1**, 524
  with carboxylic acids, **4**, 206
  with catecholborane, **1**, 1101
  with chloroacetone, **1**, 1035
  with chloroborane, **1**, 531
  with chlorodialkylaluminums, **1**, 542
  with chlorohydridozirconocene, **2**, 968
  with chloromethanesulfonyl bromide, **4**, 82
  with chromium carbene complexes, **1**, 551
  with copper(I) salts, **2**, 1094
  with copper cyanide, **3**, 675
  with cyanogen azide, **3**, 476
  with dialkylaluminum hydrides, **6**, 208
  with dialkylboranes, **1**, 278; **2**, 748
  with *N,N*-dialkylformamide acetals, **1**, 1025

with dialkyllithium cuprates, **1**, 1107
with dialkyl sulfates, **1**, 1021
with *N,N*-dialkylthiocarbamoyl chlorides, **1**, 1048
with dialkylzincs, **1**, 544
with diaryl diselenides, **5**, 241
with diaryl disulfides, **5**, 238
with diazoalkanes, **1**, 549
with diboron tetrachloride, **6**, 181
with dibromoborane dimethyl sulfide, **1**, 1102
with 1,1-dichloroalkenes, **1**, 1009
with (*E*)-1,2-dichloroalkenes, **1**, 1009
with (*Z*)-1,2-dichloroalkenes, **1**, 1009
with dichloroketene, **3**, 237
with dicobalt octacarbonyl, **1**, 198
with dienes, **1**, 519
with 1,1-dihalocarbenes, **1**, 550
with diisoamylborane, **1**, 1100
with diisobutylaluminum hydride, **1**, 533
with dimesitylborane, **1**, 510
with dimethylarsine, **2**, 1078
with dimethylphenylsilane, **3**, 77
with diorganoarsines, **2**, 875
with diorganostibines, **2**, 875
with diorganylgermylenes, **2**, 946
with diphenylcarbene, **1**, 549
with diselenides, **4**, 871
with disulfides, **4**, 940
with dodecacarbonyldicobalt, **4**, 687
with dodecacarbonyltriosmium, **4**, 682
with electrophiles, **1**, 1122
with enoates, **1**, 1051
with enols, **1**, 525
with enones, **1**, 1051
with ethene, **1**, 528
with ethylmagnesium bromide, **1**, 1134
with Fischer carbene complexes, **1**, 551
with germanium-bridged cobalt complexes, **4**, 687
with Grignard reagents, **1**, 539
with halogens, **1**, 1123
with heteroaryl dihalides, **1**, 1009
with heteroaryl polyhalides, **1**, 1009
with hexacarbonylbis(η^5-cyclopentadienyl)dimolybdenum, **4**, 674
with hexakis(*t*-butoxy)ditungsten, **3**, 691
with hydrogen, **1**, 455
with hydrogen cyanide, **1**, 545; **3**, 660
with hydrogen sulfide, **3**, 337
with hydrosilanes, **2**, 913
with iminium ions, **3**, 421
with indane-1,2,3-trione, **1**, 980
with iodine isocyanate, **5**, 985
with iron compounds, **1**, 545
with isocyanates, **5**, 297, 386
with isocyanide dichlorides, **1**, 1047
with isothiocyanic acid, **5**, 1108
with isoxazoles, **3**, 683
with keteniminium salts, **3**, 216
with lithium divinyl cuprates, **3**, 222
with lithium trimethylsilylalkylamide, **2**, 306
with maleoylcobalt complexes, **5**, 457
with mercuric chloride, **1**, 1108
with mercuric halides, **2**, 1097
with mercury acetate, **2**, 982
with mercury compounds, **4**, 687
with metal alanates, **2**, 1099
with metal carbene, **2**, 979
with metal carbyne, **2**, 979
with metal complexes, **4**, 712
with metal imine complexes, **4**, 1017
with metallacyclopropanes, **2**, 984
with metallacyclopropenes, **2**, 984
with metal–metal double-bonded cobalt complexes, **4**, 689
with methyl compounds, **5**, 466
with methyl 2,3-pentadienoate, **1**, 1051
with methyl vinyl ketones, **1**, 1051
with monocyclopentadienyltitanium(IV) metal imido complexes, **3**, 410
with nickel complexes, **1**, 545
with niobium(III) chloride, **2**, 672
with nitrile oxides, **3**, 189
with nitrile ylides, **1**, 524
with norbornadiene, **1**, 530
with norbornene, **1**, 529
with nucleophilic radicals, **1**, 324
with octacarbonyldicobalt, **4**, 687
with octakis(phosphorus trifluoride)dirhodium, **4**, 688
with organic halides, **1**, 1137
with organoalanes, **2**, 974
with organoaluminums, **1**, 542
with organoarsenic halides, **2**, 875
with organoborons, **1**, 542
with organocoppers, **1**, 539; **2**, 976
with organocuprates, **3**, 642; **4**, 772
with organolithiums, **1**, 539
with organometallics, **1**, 538
with organometals, **2**, 875, 972
with organotin hydroxides, **2**, 1100
with organotin oxides, **2**, 1100
with organozincs, **1**, 543
with orthocarbonates, **1**, 1025
with ortho esters, **1**, 1025; **4**, 192
with oxazoles, **1**, 522
with oxetanes, **1**, 1024
with oxiranes, **1**, 1022
with Oxone, **3**, 151
with palladium salts, **1**, 1107
with pentacarbonylcyclopentadienylcobaltnickel, **4**, 713
with 5-phenyl-1,2,4-dithiazole-3-thione, **5**, 634
with *N*-phenylselenophthalimide, **4**, 787
with phosphines, **2**, 438
with phosphorus nucleophiles, **4**, 561
with phosphorus pentachloride, **4**, 772
with platinum(II) salts, **5**, 939
with potassium thiocyanate, **5**, 1044
with propargylic halides, **1**, 1002
with propargylzinc bromides, **1**, 544
with rhodium complexes, **4**, 689
with selenenyl halides, **1**, 1123
with selenium compounds, **1**, 545
with seleniumphenyl *p*-tolueneselenosulfonate, **4**, 947
with selenols, **2**, 728
with siliranes, **4**, 1052
with silirenes, **4**, 1051, 1052
with *β*-silylvinyl cuprates, **3**, 271
with sodium hydrazide, **3**, 447
with stannyl radicals, **1**, 548
with sulfenic acids, **2**, 184
with sulfenyl halides, **1**, 1123
with sulfonate esters, **1**, 1021
with sulfonyl isocyanates, **5**, 344
with sulfur, **5**, 603
with sulfur compounds, **1**, 545
with Tebbe reagent, **2**, 979
with tellurium compounds, **1**, 545
with tertiary phosphines, **2**, 446
with tetraalkyltelluriums, **1**, 247
with tetramethyldisilene, **4**, 1052
with thallium acetate, **4**, 1068
with thiocarbamoyl halides, **5**, 597
with thiocyanic acid, **5**, 1044
with thiols, **2**, 710; **4**, 939
with tin compounds, **4**, 962
with titanacyclobutane, **2**, 979
with trialkylaluminums, **1**, 542

with trialkylboranes, **1**, 542
with trimethylaluminum, **1**, 1106
with trimethylenemethane, **1**, 527
with trimethylsilyl cyanide, **3**, 671
with tris(trimethylsilyl)aluminum, **2**, 918
with tungsten complexes, **4**, 1028
reactivity of, **1**, 525
reduction of, **1**, 74, 456, 502
regiochemistry of, **2**, 913
selenosulfonation of, **2**, 285
silylalumination of, **2**, 918; **4**, 1045
silylcupration of, **2**, 917; **4**, 1045, 1059
silylformylation of, **3**, 77
silylmagnesiation of, **2**, 918
silylmanganesation of, **4**, 1045
silylmetallation of, **2**, 917
silyltitanation of, **4**, 1045
silylzincation of, **2**, 918
solvomercuration of, **1**, 66
stannylation of, **1**, 342; **3**, 271
stannylgermylation of, **2**, 947
stannylsilylation of, **2**, 916
as starting materials
for acetals, **4**, 193
for aldehydes, **3**, 3
for alkenes, **1**, 455, 502
for alkenylaluminums, **1**, 207
for alkenyl ethers, **2**, 676
for alkenylgermanes, **2**, 946
for alkenyl metals, **2**, 876
for alkenylsilanes, **2**, 913
for alkynals, **3**, 79
for alkynylarsines, **2**, 1077
for allenes, **4**, 859
for allylic amines, **2**, 307
for 1,1-bisborylalkanes, **4**, 630
for 1,1-bis(bromomagnesio)alkene, **4**, 1065
for 1,1-bis(silyl)alkenes, **4**, 1044
for 1,1-bis(trimethylstannyl)allene, **4**, 1068
for 1-boronyl 1-silylalkenes, **4**, 1056
for 1-boro 1-stannylalkenes, **4**, 1063
for α-chloro ketones, **3**, 151
for cyclic alkoxycarbene complexes, **5**, 944
for 1,1-dialuminoalkenes, **4**, 1068
for 1,1-diboranes, **2**, 41
for 1,1-diboroalkenes, **4**, 1054
for dibromoalkanes, **4**, 24
for dichloroalkanes, **4**, 15
for 1,2-dichloroalkenes, **2**, 617
for difluoroalkanes, **4**, 6
for 1,1-digermanylalkenes, **4**, 1055
for 1,1-digermanylallenes, **4**, 1055
for diiodoalkanes, **4**, 29
for dithianes, **4**, 258
for 1,1-dizincalkene, **4**, 1067
for 1,3-enynes, **1**, 1036
for fulvenes, **1**, 1014
for α-halo enols, **4**, 764
for isothiocyanates, **5**, 1043
for ketene *S,N*-acetals, **4**, 905
for ketene dithioacetal *S*-oxides, **4**, 859
for ketones, **3**, 114
for magnesiocuprates, **1**, 479
for 1-magnesio 1-silylalkene, **4**, 1060
for phosphonium ylides, **3**, 498
for silacyclopropenes, **4**, 1048
for 1-silyl 1-zincoallenes, **4**, 1061
for β-sulfonyl enamines, **2**, 743
for thiocyanates, **5**, 1110
for transition metals, **4**, 822
for α,β-unsaturated ketones, **3**, 222, 238, 244, 252
for vinylboranes, **1**, 1100
for vinyl bromides, **1**, 658; **2**, 623
for vinyl cations, **1**, 1122
for vinyl chlorides, **2**, 616
for vinyl fluorides, **2**, 611
for vinylgermanes, **1**, 452
for vinyl halides, **2**, 606
for vinyl iodides, **1**, 658; **2**, 628
for vinyl radicals, **1**, 1130
for vinyl selenides, **2**, 728
for vinylsilanes, **2**, 913
for vinyl sulfides, **2**, 710
for vinyl tellurides, **2**, 730
for ylides, **3**, 498
stereochemistry of, **2**, 913
stereo-random hydrometallation of, **2**, 970
synthesis of, **1**, **997–1085**, **1**, 1119, 1120
via 1,2 addition–elimination reactions, **1**, 1044
via alkynic nucleophilic substitution, **2**, 1011
via carbon–carbon bond formation, **1**, 1000
via carbon–carbon triple bonds, **1**, 1062
via carbon–hydrogen bond formation, **1**, 998
via dehalogenation, **1**, 1069
via dehydrohalogenation, **1**, 1063
via elimination reactions, **1**, 1062
via fragmentation reactions, **1**, 1081
via isomerization, **1**, 1084
via pericyclic reactions, **1**, 1083
via transition metal-catalyzed cross-coupling reactions, **1**, 1006
thionation of, **5**, 572
titanation of, **4**, 1066
transition metal catalysis of, **2**, 913
transition metal-catalyzed hydrogermylation, **2**, 946
with transition metals
heterometallation of, **2**, 983
oxymetallation of, **2**, 983
uses of, **1**, 998
use in the synthesis of carbon-bridged diiron complexes, **4**, 681
zirconation of, **4**, 1061, 1066
Alkynes, activated, reactions, with stabilized carbanions, **1**, 537
Alkynes, alkoxy-
as activating agents, for the synthesis of anhydrides, **5**, 189
Birch reduction of, **2**, 676
as dehydrating agent, in the synthesis of cyclic anhydrides, **5**, 199
reactions
with carboxylic acids, **5**, 195
with dialkylhalophosphines, **4**, 914
with ketenes, **3**, 247
with organocoppers, **2**, 676
with phosphorus compounds, **4**, 913
with phosphorus trichloride, **4**, 765
reduction of, **3**, 62
as starting materials, for ketenes, **3**, 540
synthesis of, **2**, 1025
Alkynes, 1-alkoxy-
reactions
with aliphatic alcohols, **4**, 828
with hindered phenols, **4**, 827
as starting materials
for 1-haloenol ethers, **4**, 765
for ketene acetals, **4**, 827
for ortho esters, **6**, 74
Alkynes, alkyl-
reactions, with iodopyrimidines, **1**, 1009
as starting materials, for β-iodonitroalkenes, **2**, 1052
Alkynes, alkyl alkoxy-, as starting materials, for alkyl (trimethylsilyl)ketenes, **3**, 540
Alkynes, alkylphenyl-
protodemercuration of, **1**, 454
reactions, with *t*-butyl cations, **1**, 1122
Alkynes, alkylthiochloro-

reactions
with alkoxides, **4**, 765
with selenols, **4**, 787
Alkynes, amino-
hydrogenolysis of, **1**, 506
reactions
with allylzinc bromide, **1**, 544
with diethylzincs, **2**, 1096
with diphenylzincs, **2**, 1096
for the synthesis of manganacyclopentadiene complexes, **4**, 677
Alkynes, 1-amino-, reactions, with nitroalkenes, **3**, 435
Alkynes, arenesulfonyl-, reactions, with alkanethiols, **4**, 861
Alkynes, aryl-
from, aryl halides, **1**, 472
Mannich reactions of, **1**, 1041
reactions
with 2-bromo-1,1-dichloroethane, **2**, 307
with iodopyrimidines, **1**, 1009
with trimethylsilyl cyanide, **3**, 660
as starting materials, for vinyl cations, **2**, 637
Alkynes, azo-
reactions, with cyclohexane, **2**, 1054
stability of, **2**, 1054
synthesis of, **2**, 1054
Alkynes, 1-(benzenesulfonyl)-, reactions, with 3-phenylglycidol, **4**, 840
Alkynes, bis(arenesulfonyl)-, reactions, with methanol, **4**, 840
Alkynes, bis(methylseleno)-, reactions, with acetic acid, **4**, 841
Alkynes, bis(phosphonio)-, synthesis of, **2**, 446
Alkynes, bis(tributylstannyl)-, reactions, with diphenylantimony halides, **2**, 1081
Alkynes, 1-boro-, hydroboration of, **4**, 1054
Alkynes, bromo-
reactions
with crotylzinc bromide, **1**, 544
with monoalkynes, **1**, 1142
with trialkylaluminums, **1**, 1141
as starting materials, for ketene aminals, **4**, 976
synthesis of, **3**, 676
Alkynes, 1-bromo-
as alkylating agents, **1**, 532
hydroboration of, **4**, 814
reactions
with dicyclohexylborane, **1**, 1100
with Grignard reagents, **1**, 999
with thexylborane, **1**, 1100
Alkynes, ω-bromo 1-silyl-, carbamagnesiation of, **4**, 1060
Alkynes, chloro-
reactions
with amines, **4**, 976
with lithium aluminum hydride, **4**, 816
with trialkylaluminum hydrides, **4**, 816
as starting materials
for ketene aminals, **4**, 976
for ketenimines, **3**, 590
synthesis of, **2**, 1012; **4**, 976
Alkynes, 1-chloro-
reactions
with bis(pyridine)iodine(1) tetrafluoroborate, **4**, 760
with disubstituted boranes, **4**, 756
with thexylborane, **1**, 1100
synthesis of, **2**, 616
Alkynes, β-chlorocyano-, synthesis of, **3**, 674
Alkynes, α-cyano-, synthesis of, **3**, 674
Alkynes, dialkenyl-, synthesis of, **1**, 1012
Alkynes, dialkoxy-, synthesis of, **2**, 1026
Alkynes, dialkyl-
dissolving metal reductions of, **1**, 508
reduction of, **1**, 457
Alkynes, diallenyl-, synthesis of, **1**, 1012
Alkynes, diaryl-
as starting materials, for cyanopyrroles, **3**, 671
synthesis of, **1**, 1004, 1012
Alkynes, diaryl-substituted
from, α,α-dichlorodibenzyl sulfides, **1**, 1074
synthesis of, **1**, 1074
Alkynes, diazo-, synthesis of, **3**, 490
Alkynes, dichloro-, as intermediates, **4**, 761
Alkynes, dimethoxyphosphoryl chloro-, reactions, with triphenylphosphine, **2**, 1062
Alkynes, diphosphonyl-, reactions, with alkylamines, **5**, 789
Alkynes, disilyl-, reactions, with dicarbonylcyclopentadienylcobalt, **4**, 689
Alkynes, disubstituted, synthesis of, **2**, 1076
Alkynes, divinyl-, bromination of, **1**, 995
Alkynes, ethoxy-
reactions
with iodotrimethylsilane, **2**, 941
with secondary amines, **4**, 884
as starting materials
for ketene aminals, **4**, 979
for silylketenes, **2**, 941
Alkynes, 1-ethoxy-, hydroboration of, **2**, 901
Alkynes, fluoro-
stability of, **4**, 976
synthesis of, **2**, 1012
Alkynes, 1-fluoro-, from, 1,1-difluoroalkenes, **2**, 1042
Alkynes, fluoro-, alkyl-substituted, synthesis of, **2**, 1014
Alkynes, β-functionalized, synthesis of, **1**, 1054
Alkynes, 1-germanyl-
hydrostannation of, **4**, 1063
as starting materials, for 1-germanyl 1-stannylalkenes, **4**, 1063
Alkynes, germyl-
hydroalumination of, **4**, 812
hydroboration of, **4**, 812
Alkynes, germyl alkoxy-, thermolysis of, **3**, 540
Alkynes, *C*-glycosylated, synthesis of, **1**, 1026
Alkynes, halo-
coupling with organoboranes, **1**, 488
hydroboration of, **3**, 275; **4**, 814
reactions
with bromoborane dimethyl sulfide complexes, **4**, 815
with chromium(II) chloride, **1**, 1140
with isopinocampheylboranes, **4**, 815
with selenyls, **4**, 786
with tertiary lithium enolates, **1**, 1108
with zinc, **2**, 1096
as starting materials
for amidines, **4**, 976
for α-haloalkenyl metals, **4**, 822
for α-halo enols, **4**, 764
for ketene aminals, **4**, 976
for ketenimines, **3**, 588
Alkynes, 1-halo-
aluminum hydride reduction of, **4**, 816
reactions
with alkenylaluminums, **1**, 533
with diisobutylaluminum hydride, **2**, 967
with thexylborane, **1**, 531
Alkynes, heteroaryl-, synthesis of, **1**, 1009
Alkynes, heteroatom-substituted, carbocupration of, **2**, 977
Alkynes, heterosubstituted, carbocupration of, **2**, 977
Alkynes, hydroxy-
acidity of, **2**, 1022
synthesis of, **2**, 1022
See also Ynols
Alkynes, 3-hydroxy-, synthesis of, **1**, 1066; **2**, 85
Alkynes, β-hydroxy-, synthesis of, **1**, 1022
Alkynes, γ-hydroxy-, synthesis of, **1**, 1024
Alkynes, hydroxymethyl-, reduction of, **1**, 961
Alkynes, internal, hydrozirconation of, **2**, 969
Alkynes, iodo-

reactions, with amines, **4**, 976
as starting materials, for alkenols, **5**, 229
synthesis of, **2**, 1019
Alkynes, 1-iodo-
diimide reduction of, **1**, 458
hydroboration of, **4**, 814
reactions
with bis(pyridine)iodine(1) tetrafluoroborate, **4**, 749
with *N*-bromosuccinimide, **4**, 760
with dicyclohexylborane, **1**, 1100
with Grignard reagents, **1**, 999
with *N*-iodosuccinimide, **4**, 760
Alkynes, metallated
reactions
with alkenyl halides, **1**, 1006
with alkenyl phosphates, **1**, 1027
with alkenyl sulfonates, **1**, 1027
with allenyl halides, **1**, 1006
with aryl halides, **1**, 1006
with aryl phosphates, **1**, 1027
with aryl sulfonates, **1**, 1027
Alkynes, metallo-, as starting materials, for α-haloalkenyl metals, **4**, 822
Alkynes, methyl-, as starting materials, for propargylic lithium alanates, **1**, 973
Alkynes, nitro-
reactions, with *N*-chloroguanidines, **4**, 796
synthesis of
oxidation of nitrosoalkynes, **2**, 1052
via dehydrobromination of alkenes, **2**, 1052
Alkynes, α-nitro-, synthesis of, **1**, 1033
Alkynes, nitroso-
from, trialkylstannylalkynes, **2**, 1052
reactivity of, **2**, 1052
synthesis of, **2**, 1051
Alkynes, perfluoroalkyl-, reactions, with iodobenzenes, **1**, 1009
Alkynes, perlithio-, synthesis of, **6**, 205
Alkynes, peroxy-, as starting materials, for alkynylmercury compounds, **2**, 1098
Alkynes, α-phenoxy-, reactions, with boranes, **1**, 198
Alkynes, phenyl-, synthesis of, **1**, 483
Alkynes, phenyliodonium-, reactions, with sodium thiocyanate, **5**, 1108
Alkynes, 1-(phenylsulfonyl)-, from, β-keto sulfones, **2**, 1035
Alkynes, phosphorylated chloro-, reactions, with alkoxide anions, **2**, 1028
Alkynes, platinum-catalyzed, hydrosilation of, **2**, 969
Alkynes, push-pull, use of term, **2**, 1049
Alkynes, selenyl-substituted, reactions, with halogens, **4**, 786
Alkynes, siloxy-, synthesis of, **2**, 1024
Alkynes, silyl-
dihydroboration of, **4**, 632
double hydroboration of, **4**, 644
hydroboration of, **3**, 203
hydroformylation of, **3**, 77
reactions
with acid chlorides, **3**, 270
with alkyl Grignard reagents, **1**, 1105
with aminostannanes, **2**, 1092
with 9-borabicyclo[3.3.1]nonane, **3**, 203
with diboranes, **4**, 1054
with metal carbene complexes, **2**, 927
with organomagnesiocuprates, **1**, 1094
as starting materials
for boryl enol ethers, **5**, 423
for vinylsilanes, **2**, 921
Alkynes, 1-silyl-
cyclizations of, **2**, 945
hydroboration of, **4**, 1056
hydromagnesiation of, **4**, 1060
hydrosilylation of, **4**, 1044
reactions
with butyllithium, **4**, 1058
with trialkoxyboranes, **4**, 1057
as starting materials
for 1-titano 1-silyl allene, **4**, 1061
for 1-trialkylgermanyl 1-silyl allenes, **4**, 1058
for 1,1,2-trisilylalkenes, **4**, 1045
Alkynes, silyl alkoxy-, thermolysis of, **3**, 540
Alkynes, silyl(alkylsilyl)-, hydrosilylation of, **4**, 1054
Alkynes, stannyl-
reactions
with alkylcopper(I) compounds, **1**, 968
with diboranes, **4**, 1054
Alkynes, 1-stannyl-
hydroboration of, **4**, 1063
hydrostannation of, **4**, 1068, 1069
as starting materials
for 2-boro borienes, **4**, 1054
for 1,1-distannylalkenes, **4**, 1068
Alkynes, substituted, as starting materials, for ketene hemiaminals, **4**, 884
Alkynes, substituted bromo-, synthesis of, **2**, 1018
Alkynes, 1-sulfonyl-, as starting materials, for ketene monothioacetal *S,S*-dioxides, **4**, 840
Alkynes, symmetrical distannyl-, synthesis of, **2**, 1100
Alkynes, terminal
bishydroboration of, **4**, 697
carboalumination of, **2**, 975
carbonylation of, **1**, 1040
condensation of, **2**, 1081
for the synthesis of alkynylarsenics, **2**, 1078
dihydroboration of, **1**, 163
from
methyl ketones, **1**, 1070
trimethylsilylethyne, **1**, 1012
haloboration of, for α-bromo aldehydes, **3**, 27
hydroalumination of, **1**, 1020
hydrostannation of, **2**, 971
iodination of, **2**, 1018
oxidative homocoupling reactions of, **1**, 1034
reactions
with alkenyl halides, **1**, 1006, 1008
with alkenyl sulfonates, **1**, 1028
with alkenyl triflates, **1**, 1029
with alkylcuprates, **1**, 541
with allenyl bromides, **1**, 1005
with allenyl halides, **1**, 1008
with allyl halides, **1**, 121
with allylzinc bromide, **1**, 543
with aminostannanes, **2**, 1100
with aryl fluoroalkane sulfonates, **1**, 1029
with aryl halides, **1**, 1006, 1008
with aryl sulfonates, **1**, 1028
with bis(triethyltin)oxide, **1**, 1006
with brominating reagents, **2**, 1017
with carbonates, **1**, 1025
with chlorinating reagents, **2**, 1016
with copper(I) halides, **1**, 1004
with copper(II) acetate, **1**, 1034
with diisobutylaluminum hydride, **1**, 511
with heteroaryl triflates, **1**, 1029
with [hydroxy(tosyloxy)iodo]benzene, **2**, 1020
with propargyl halides, **1**, 121
with propargylic acetates, **1**, 1025
with silylcopper(I) compounds, **2**, 917
with toluene sulfonates, **1**, 1025
with *N*-trialkylstannylamines, **1**, 1006
with trialkylstannyltrialkylsilanes, **2**, 916
as starting materials
for α-acyloxymethyl ketones, **3**, 162
for dichloromethyl ketones, **4**, 16
for lithium alkynides, **1**, 111
for ruthenium complexes, **4**, 683
synthesis of, via protecting groups, **1**, 1005

use in the synthesis of carbon-bridged dimanganese complexes, **4**, 678
Alkynes, terminally substituted, with electron-donating groups, hydroalumination, **2**, 968
Alkynes, thiocyanato-, synthesis of, **5**, 1108
Alkynes, tin-substituted, reactions, with (dicyano)iodonium triflate, **2**, 1021
Alkynes, trialkylsilyl-substituted nitro-, synthesis of, **2**, 1053
Alkynes, trialkylstannyl-, as starting materials, for nitrosoalkynes, **2**, 1052
Alkynes, 1-trimethylsilyl-
hydroalumination of, **4**, 1062
hydrobromination of, **2**, 929
Alkynes, trimethylsilyl-substituted, reactions, with diphenyl diselenidecopper(I) triflate, **2**, 1037
Alkynes, triphenylphosphonium chloro-, synthesis of, **2**, 1062
Alkynes, unactivated, reactions, with organocuprates, **1**, 540
1-Alkynes
acidity of, **1**, 1133
from
1-alkynyl selenides, **1**, 999
1-alkynyl sulfoxides, **1**, 999
1,1-dihalides, **1**, 1065
1,2-dihalides, **1**, 1065
lithium alkynide, **1**, 1034
isomerization of, to 2-alkynes, **1**, 1084
reactions
with acid chlorides, **1**, 1045
with aldehydes, **1**, 1038
with alkyl chlorothionoformates, **1**, 1047
with alkynyltins, **1**, 1038
with aluminum hydride, **4**, 698
with aminochloroboranes, **2**, 1085
with butyllithium, **1**, 1134
with *N*,*N*-dialkylcarbamoyl chlorides, **1**, 1048
with *N*,*N*-dialkylformamide acetals, **1**, 1026
with (*Z*)-1,2-dichloroethene, **1**, 1065
with dimethylzinc, **2**, 1096
with diorganozincs, **2**, 1096
with *N*,*N*-disubstituted aminoalkyl ethers, **1**, 1041
with ethylmagnesium bromide, **2**, 1076
with heterocyclic halides, **1**, 1137
with hexaalkyldisilanes, **2**, 1090
with imidoyl chlorides, **1**, 1047
with imines, **1**, 1040
with ketones, **1**, 1038
with organogermaniums, **2**, 1093
with propargylic halides, **1**, 1004
with tribromoborane, **2**, 902
with triorganosilanes, **2**, 1090
as starting materials
for alkenylboranes, **2**, 900
for 1,1-bis(boryl)alkanes, **2**, 906
for carboxylic acids, **1**, 1103
for 2-halo-1-alkenylboranes, **2**, 902
for 1-silyl-1-alkenes, **2**, 915
use in the synthesis of alkynylalanes, **2**, 1099
1-Alkynes, ω-carboxy-, monoalkylation of, **1**, 1001
1-Alkynes, 1-chloro-, reactions, with lithium aluminum hydride, **4**, 758
1-Alkynes, *N*,*N*-disubstituted-1-amino-, synthesis of, **2**, 1039
1-Alkynes, 1-halo-
hydroboration of, **2**, 901
reactions, with *n*-hexyldibromoborane, **2**, 902
1-Alkynes, ω-hydroxy-, monoalkylation of, **1**, 1001
1-Alkynes, 1-silyl-
hydroboration of, **2**, 901
Pauson–Khand [2+2+1] cycloaddition of, **2**, 926
as starting materials, for allenylsilanes, **2**, 940
1-Alkynes, 1-stannyl-
hydrostannation of, **6**, 209
reactions
with trialkylboranes, **2**, 903
with trimethyltin hydride, **6**, 209
as starting materials, for 1-boryl-2-stannyl-1-alkenes, **2**, 903
1-Alkynes, 1-trialkylsilyl-, reactions, with Grignard reagents, **1**, 160
1-Alkynes, 1-trimethylsilyl-
hydromagnesiation of, **1**, 161
as starting materials, for 1,1-bis(trimethylsilyl) allenes, **4**, 1047
2-Alkynes
isomerization of, **1**, 1085
synthesis of, **1**, 1063, 1084
2-Alkynes, 1-amino-, synthesis of, **1**, 1041
2-Alkynes, 1-bromo-, reduction of, **1**, 958
2-Alkynes, 1-chloro-, hydroboration of, **1**, 963
3-Alkynes, synthesis of, **1**, 1084
Alkynethiols, *S*-trimethylsilyl-, reactions, with imines, **5**, 614
Alkyne–vinylidene rearrangement
mechanism, **5**, 941
for the synthesis of aminocarbene complexes, **5**, 954
Alkyne Zipper reaction, mechanism, **1**, 1085
Alkynic carbanions, synthesis of, **1**, 1132
Alkynic carbocations, synthesis of, **1**, 1141
Alkynic compounds
homologation of, **1**, 995
isomerization of, **1**, 954
reactions
with diboron tetrachloride, **2**, 903
with di-*t*-butyl diboron dichloride, **2**, 903
with secondary amines, **2**, 751
with tetrafluorohydrazine, **2**, 762
Alkynic compounds, α-halo-α'-hydroxy, as starting materials, for α-allenyl alcohols, **1**, 958
Alkynic halides
dehalogenation of, **1**, 958
reduction of, **1**, 959
α-Alkynic halides, hydrodehalogenation of, **1**, 958
Alkynic ions, synthesis of, **1**, 1132
Alkynic radicals, synthesis of, **1**, 1132, 1142
Alkynide anions
reactions
with carbamoyl chlorides, **5**, 304
with isocyanates, **5**, 304
stabilization of, **1**, 1133
Alkynide complexes, alkylation of, **5**, 946
Alkynide σ-complexes, use in the synthesis of carbene complexes, **5**, 940
Alkynide ions
reactions, with dialkyl phosphoro halides, **2**, 1072
synthesis of, **1**, 1133
Alkynides
reactions
with allylic halides, **1**, 1001
with dialkylaluminum chlorides, **6**, 208
with propargylic halides, **1**, 1001
use in the synthesis of alkynylantimony compounds, **2**, 1081
Alkynides, 1,3-dilithio, reactions, with carbon dioxide, **5**, 101
Alkynides, dimetallo, reactions, with aldehydes, **1**, 1037
Alkynides, lithio
reactions, with bromoalkenylborinates, **3**, 275
stability of, **6**, 461
synthesis of, **3**, 273
Alkynides, metallated, use in alkylation reactions, **1**, 1133
Alkynides, silyl, acylation of, **3**, 273
Alkynides, tri-*n*-butyltin, reactions, with halo acetals, **1**, 1003
Alkynides, trimethylsilyl, use in 1,4 addition reactions, **1**, 1049
Alkynoates, synthesis of, **1**, 1078
2-Alkynoates, ethyl, synthesis of, **1**, 1079
Alkynoic acids

intramolecular oxypalladation of, **2**, 983
reduction of, **1**, 503
as starting materials, for amides, **5**, 304
synthesis of, **1**, 1076
2-Alkynoic acids, synthesis of, **1**, 1040, 1047, 1072
3-Alkynoic acids, synthesis of, **1**, 1058
Alkynolate anions, from, α-halo ketones, **2**, 1023
Alkynols
carbonylation of, **1**, 546
from, 1,2-halo ethers, **1**, 1066
hypothetical, compounds of, **2**, 1021
intramolecular alkoxymercuration of, **2**, 676
isomerization of, **1**, 1085
synthesis of, **1**, 1056; **2**, 1048
3-Alkyn-1-ols, reactions, with trimethylaluminum, **1**, 543
ω-Alkyn-1-ols, synthesis of, **1**, 1085
Alkynones
amination of, **3**, 267
from
amides, **1**, 1046
1,3-β-diketones, **3**, 274
α,β-epoxy ketones, **1**, 1082
lactones, **1**, 1045
reactions, with cuprates, **3**, 271
synthesis of, **1**, 1045, 1078, 1082; **3**, 273
Alkynones, (2-hydroxyaryl)-, reactions, with silver nitrate, **3**, 242
γ-Alkynones, from, α,β-enones, **1**, 1139
Alkyn-2-on-1-yl chlorides, reactions, with lithium trialkylgermanium, **4**, 1055
Alkynoxides, synthesis of, **5**, 162
Alkynyl acids, reactions, with phenylhydrazine, **3**, 453
Alkynyl anions, reactions, with alkenes, **1**, 309
Alkynylation
intramolecular, for the synthesis of alkynes, **1**, 1012
radical induced, for the synthesis of alkynylgermanes, **2**, 1093
1-Alkynylation, of diorganoboron halides, **2**, 1083
Alkynyl bromides
from, alkali metal alkynylides, **2**, 1018
reactions
with alkenylboranes, **1**, 1020
with alkylmagnesium halides, **1**, 1017
with γ-alkynylcarboxylic acids, **1**, 1020
with lithium selenoates, **2**, 1037
with lithium telluroates, **2**, 1037
with methylene(triphenyl)phosphorane, **1**, 1016
with sodium diethyl phosphite, **2**, 1072
with trialkylaluminums, **1**, 1019
stability of, **2**, 1012
synthesis of, **2**, 1012, 1017
Alkynyl carbanions
addition reactions of, **1**, 1036
alkylation of, **1**, 1000, 1001
substitution of, **1**, 1000
Alkynyl cations, detection of, **1**, 1141
Alkynyl chlorides
from
cis-1,2-dichloro-1-alkenes, **2**, 1012
metal hypochlorites, **2**, 1016
sodium chloroethynylide, **2**, 1015
terminal alkynes, **2**, 1016
reactions
with alkali metal amides, **2**, 1041
with enolates, **1**, 1016
with Grignard reagents, **1**, 1017
with imines, **2**, 1053
with lithium amides, **2**, 1041
reactivity of, **1**, 1018
stability of, **2**, 1012
as starting materials, for alkynylphosphines, **2**, 1060
synthesis of, **2**, 1012, 1014
Alkynyl chlorides, 3-hydroxy-, synthesis of, **2**, 1015
Alkynyl compounds
stability of, **2**, 1076
as starting materials, for alkynylantimony compounds, **2**, 1082
synthesis of, **2**, **1075–1102**
toxicity of, **2**, 1076
Alkynyl fluorides
reactions, with organometallics, **1**, 1017
as reagents in organic syntheses, **2**, 1014
stability of, **2**, 1012
synthesis of, **2**, 1012, 1013
Alkynyl groups, conjugate addition of, **1**, 311
Alkynyl halides
alkenylation of, **1**, 1021
nickel-catalyzed reaction with trialkylalanes, **1**, 1019
nucleophilic attack of, on azide ions, **2**, 1055
properties of, **2**, 1012
reactions
with alkenyldiisobutylalanes, **1**, 1020
with alkenyltins, **1**, 1020
with alkoxide anions, **2**, 1028
with δ-alkynic malonate enolates, **1**, 1020
with carbon nucleophiles, **1**, 1016
with enolates, **1**, 1016
with Grignard reagents, **1**, 1017
with lithium diorganocuprates, **1**, 1018
with organocopper(I) compounds, **1**, 1017
with organolithiums, **1**, 1017
with organomagnesium halides, **1**, 1018
with phosphines, **2**, 1062
with sodium diethyl phosphonate, **2**, 850
with tertiary amines, **2**, 1042
with thiocyanate ions, **2**, 1056
with thiolates, **2**, 1033
with trialkyl phosphites, **2**, 1071
reduction of, **1**, 998
as starting materials
for alkynyl phosphonates, **2**, 1071
for alkynylphosphonium salts, **2**, 1062
for ammonium salts of ynamines, **2**, 1042
for tertiary ynamines, **2**, 1041
substitution reactions of, **1**, 1016
synthesis of, **2**, **1011–1038**
elimination methods, **2**, 1012
general methods, **2**, 1011
theoretical, **2**, 1054
1-Alkynyl halides
reactions, with copper hydrides, **1**, 999
as starting materials, for alkynes, **1**, 998
Alkynyl–heteroatom functions, reduction of, **1**, 998
Alkynylides
chlorination of, **2**, 1016
reactions
with arylsulfonyl chlorides, **2**, 1016
with halogens, **2**, 1012
with thiolating reagents, **2**, 1032
Alkynyl iodides
reactions
with 3-buten-2-one, **1**, 1021
with chromium(II) chloride, **1**, 1038
with methyl-2-propenoate, **1**, 1021
stability of, **2**, 1012
synthesis of, **2**, 1012, 1018
use in cyclization reactions, **1**, 1038
Alkynyl ligands
from, 3-methoxy-3-methyl-1-butyne, **1**, 119
in substitution reactions, **1**, 119
Alkynyl–silicon bonds, cleavage of, deprotection conditions, **1**, 999
Alkyoxycarbonylpalladation, intramolecular, **2**, 980
Allenamines, α-lithio, synthesis of, **4**, 1015
Allenamines, α-potassio, from, ynamines, **4**, 1017
Allene, reactions, with bis(phenyldimethylsilyl)cuprate, **2**, 929

Allene, 1,1-bis(trimethylstannyl)-, from, alkynes, **4**, 1068
Allene, cyclopropyl-, reduction of, **1**, 459
Allene, 1,3-dilithiated bis(trimethylsilyl)-, as starting material, for 4,4-bis(trimethylsilyl)-1-phosphabutatriene, **3**, 607
Allene, 3,3-dimethyl-1-trimethylsilyl-, reactions, with α,β-unsaturated acyl cyanides, **1**, 1061
Allene, 1,1-diphenyl-, synthesis of, **1**, 760
Allene, 1-(diphenylphosphino)-1,3-diphenyl-, synthesis of, **1**, 446
Allene, 1-methyl-1-(trimethylsilyl)-, reactions, with 5-ethoxy-2-pyrrolidinone, **1**, 1060
Allene, 3-methyl-1-(triphenylstannyl)-, reactions, with 2-cyclohexenone, **1**, 1061
Allene, tetra-*t*-butyl-, from, 1,1,3,3-tetra-*t*-butylprop-2-en-1-ol, **1**, 981
Allene, 1-titano 1-silyl-, from, 1-silylalkynes, **4**, 1061
Allene, trimethylsilyl-
  reactions
    with 2-cyclohexenone, **1**, 1061
    with α,β-unsaturated acyl cyanides, **1**, 1061
Allene, triphenyl-, from, 1,1,3-triphenylpropyne, **1**, 957
Allene complexes, use in the synthesis of α-carbocations of alkyne-bridged dimolybdenum complexes, **4**, 674
Allene 1,3-dicarboxylic acids, synthesis of, **5**, 101
Allene equivalents, synthesis of, **1**, 566
Allene metallation, use in the synthesis of α-ketoacylsilanes, **5**, 423
Allene oxide, *t*-butyl-, synthesis of, **1**, 652
Allene oxides
  as intermediates, in the synthesis of α-fluoro ketones, **3**, 148
  as starting materials, for β-keto phosphonates, **3**, 200
Allene oxides, vinyl-, as starting materials, for pentadienyl cations, **1**, 423
Allenes
  addition reactions of, **1**, 458
  Cope rearrangement of, **1**, 808
  cycloaddition of, **6**, 124
  cycloaddition reactions, **1**, 497
  deprotonation of, **1**, 111
  Diels–Alder cycloadditions of, **1**, 498
  as electrophiles, for dithiocarboxylic enethiolates, **4**, 842
  ene reactions of, **1**, 518
  free radical additions to, **1**, 497
  from
    β-alkoxyalkenylboranes, **1**, 655
    alkynes, **4**, 859
    allylic mesylates, **1**, 983
    boron compounds, **1**, 976
    1-bromo-4-ene-2-ynes, **1**, 958
    carbonyls, **1**, 986
    β-chloroalkenylboranes, **1**, 655
    β-hydroxy selenides, **1**, 440
    α-lactones, **1**, 785
    phosphorus ylides, **1**, 986
    propargyl halides, **1**, 958
    propargylic carbamates, **1**, 966
    propargylic compounds, **1**, 966
    (*Z*)-selenoxides, **1**, 1073
    unhindered aromatic aldehydes, **6**, 706
    vinylic selenoxides, **1**, 605
    vinylphosphines, **1**, 446
  hydrogermylation of, **2**, 946
  as intermediates, in the synthesis of α-ketoacylsilanes, **5**, 423
  nucleophilic addition to, **1**, 495
  reactions
    with allyl metals, **1**, 496
    with arylsulfenyl halides, **1**, 1125
    with carbenes, **1**, 499
    with electrophiles, **1**, 1125
    with hydrogen, **1**, 458
    with iodine isocyanate, **5**, 985
    with iodomethane, **4**, 1024
    with isocyanates, **5**, 386
    with ketenes, **1**, 498; **3**, 216
    with methanol, **4**, 736
    with sulfonyl isocyanates, **5**, 344
    with tetrafluorohydrazine, **2**, 762
    with trimethylsilyl cyanide, **3**, 660
  reduction of, **1**, 80
  selenosulfonation of, **2**, 285
  semi hydrogenation of, **1**, 458
  silylcupration of, **2**, 929
  silyllithiation of, **2**, 929
  silylmagnesiation of, **2**, 929
  silylstannylation of, **2**, 930
  solvolysis of, **1**, 1126
  as starting materials
    for alkenes, **1**, 458
    for alkenylsilanes, **2**, 929
    for allenylsilanes, **2**, 939
    for pheromones, **1**, 966
    for vinyl cations, **1**, 1125
    for vinyl radicals, **1**, 1130
    for vinylsilanes, **2**, 929
  substituted, synthesis of, **1**, 464
  synthesis of, **1**, 197, **953–995**, **4**, 1045
    via allenyl metal intermediates, **1**, 975
    via elimination of β-hydroxysilanes, **1**, 651
    via hydroboration, **1**, 963
    via propargylic rearrangement, **1**, 955
    via sulfoxide elimination, **1**, 600
Allenes, acyl-, synthesis of, **1**, 983
Allenes, alkenyl-
  from
    enynes, **1**, 955
    propargylic carbonates, **1**, 976
Allenes, α,ω-alkenyl-, cyclization of, **1**, 496
Allenes, alkoxy-
  as starting materials, for allenylsilanes, **2**, 939
  substitution of, **1**, 496
Allenes, 1-alkoxy-, reactions, with cuprates, **1**, 496
Allenes, alkyl-
  from, methyl alkyl cuprates, **1**, 966
  reactions, with alkynes, **1**, 498
Allenes, 1-alkyl-3-bromo-1-pentafluorophenyl-, synthesis of, **1**, 987
Allenes, alkynyl-, from, 1,4-diynes, **1**, 954
Allenes, aminomethyl-, synthesis of, **1**, 960
Allenes, aryloxymethyl-, synthesis of, **1**, 961
Allenes, bicyclic, synthesis of, **1**, 990
Allenes, 1,1-bis(silyl)-, from, pentamethyl (3'-trimethylsilyloxy-1-alkynyl)disilanes, **2**, 940
Allenes, 1,1-bissilyl-, synthesis of, **1**, 983
Allenes, 1,1-(bis)silylvinyl-, synthesis of, **4**, 1046
Allenes, bis(trifluoromethyl)amino-, reactions, with trifluoronitrosomethane, **3**, 600
Allenes, 1,1-bis(trimethylsilyl)-, from, 1-trimethylsilyl-1-alkynes, **4**, 1047
Allenes, bromo-
  reactions, with alkynes, **1**, 1137
  reactivity of, **1**, 466
  synthesis of, **1**, 983
Allenes, 3-bromo-, synthesis of, **4**, 794
Allenes, chiral, synthesis of, **1**, 609, 667
Allenes, chloro-, reactivity of, **1**, 466
Allenes, cyano-
  reactions, with germylenes, **2**, 947
  synthesis of, **3**, 649
Allenes, cyclic, synthesis of, **1**, 990
Allenes, 1,2-cyclic, reduction of, **1**, 459
Allenes, 1-dialkoxyboronyl silyl-, synthesis of, **4**, 1057
Allenes, 1,3-dialkyl-, synthesis of, **1**, 966
Allenes, dialkylphosphinomethyl-, synthesis of, **1**, 960

Allenes, 1,3-diaryl-, synthesis of, **6**, 706
Allenes, diazo-, as intermediates, **3**, 489, 490
Allenes, difluoro-, synthesis of, **1**, 634
Allenes, 1,1-digermanyl-, from, alkynes, **4**, 1055
Allenes, 1,1-dilithio-, synthesis of, **4**, 1064
Allenes, 1,1-dimagnesio-, synthesis of, **4**, 1065
Allenes, 1,1-disilyl-
as starting materials
for alkyne disilanes, **4**, 1046
for 1-silylbutynes, **4**, 1046
Allenes, 1,3-disubstituted, synthesis of, **1**, 962
Allenes, exocyclic, synthesis of, **1**, 978
Allenes, α-halo-, solvolysis of, **1**, 1114
Allenes, α-hydroxy-
from
alkynyl cyclic carbonates, **1**, 962
3-phenylselenobuta-1,3-dienes, **1**, 985
synthesis of, **1**, 962
Allenes, hydroxymethyl-, synthesis of, **1**, 961
Allenes, isothiocyanate-substituted, synthesis of, via [3,3] sigmatropic rearrangements, **5**, 1041
Allenes, γ-keto-, synthesis of, **1**, 982
Allenes, lithiated trimethylsilyl-, as starting materials, for 1-phosphabutatrienes, **3**, 607
Allenes, lithio-
from, 1,3,3-tris(trimethylsilyl)propyne, **4**, 1047
synthesis of, **1**, 111
Allenes, γ-lithio-γ-alkoxy-, as starting materials, for allenic compounds, **1**, 971
Allenes, methoxy-, α-sulfination of, **4**, 837
Allenes, γ-methoxy-γ-alkyl-, as starting materials, for allenic compounds, **1**, 971
Allenes, methoxy hydroxy-, synthesis of, **1**, 973
Allenes, γ-methoxy-γ-lithio-, reactions, with tetrahydrofuranons, **1**, 971
Allenes, methyl-, reactions, with iodine isocyanate, **5**, 985
Allenes, 1-phenyl-3-alkyl-, synthesis of, **1**, 966
Allenes, silyl-
reactions
with 3-butyn-2-one, **1**, 527
with triphenylgermane, **4**, 639
as starting materials, for allenic compounds, **1**, 971
synthesis of, **1**, 970
Allenes, silylated, from, trimethylsilyl propargylic sulfinates, **1**, 967
Allenes, 1-silyl 1-zinco-, from, alkynes, **4**, 1061
Allenes, stannyl-
reactions, with α-nitroalkenes, **1**, 1062
synthesis of, **1**, 968
Allenes, terminal, synthesis of, **1**, 760
Allenes, 1,1,3,3-(tetraalkoxycarbonyl)-, synthesis of, **1**, 986
Allenes, 1,1,3,3-tetrakis(dialkylamino)-, from, 1,1,3,3-tetrakis(dialkylamino) allylic chlorides, **1**, 982
Allenes, tetraphenyl-, synthesis of, **1**, 763
Allenes, tetraphosphine, cleavage of, **1**, 51
Allenes, 1-trialkylgermanyl 1-silyl-, from, 1-silylalkynes, **4**, 1058
Allenes, trimethylsilyl-
rearrangement of, **6**, 124
as starting materials, for β-lactams, **6**, 128
Allenes, trimethylsilyloxy-
from, 1-trimethylsilylpropargyl alcohols, **1**, 963
synthesis of, **1**, 985
Allenes, vinyl-
as starting materials, for hydrindane, **1**, 499
synthesis of, **1**, 975, 983
Allenes, vinylic, synthesis of, **1**, 970
β-Allenes, isomerization of, **3**, 229
Allenic compounds
from
γ-lithio-γ-alkoxyallenes, **1**, 971
γ-methoxy-γ-alkylallenes, **1**, 971
silylallenes, **1**, 971
Allenols, synthesis of, **1**, 963, 971
α-Allenols, synthesis of, **1**, 967
Allenyl anions, synthesis of, **1**, 111
Allenyl bromides
reactions
with terminal alkynes, **1**, 1005
with trimethylsilylbutadiyne, **1**, 1006
Allenyl carbanions
1,2 addition–elimination reactions of, **1**, 1060
addition reactions of, **1**, 1054
substitution of, **1**, 1051
Allenyl halides
alkylation of, **1**, 468
reactions
with copper(I) alkynides, **1**, 1005
with cyanocuprates, **1**, 1015
with Grignard reagents, **1**, 466
with heterocuprates, **1**, 466
with homocuprates, **1**, 466
with lithium dialkylcuprates, **1**, 1015
with metallated alkynes, **1**, 1006
with organocoppers, **1**, 1015
with terminal alkynes, **1**, 1008
reactivity of, **1**, 1009
1,3-substituted reactions of, **1**, 1015
substitution of, **1**, 1004
Allenyl halides, chiral, alkylation of, **1**, 466
Allenyl iodides, alkylation of, **1**, 466
Allenyl radicals
cyclization of, **1**, 341
synthesis of, **1**, 330
use in the synthesis of cyclopentanes, **1**, 339
Allenylthiols, reactions, with dodecacarbonyl triirons, **4**, 683
Allenynes
from, propargylic halides, **1**, 975
synthesis of, **1**, 472, 1004, 1025, 1137
Allenynones, from, alkynyl homopropargylic alcohols, **1**, 957
Allophanoyl chlorides, reactions, with isocyanates, **5**, 980
Allophanoyl chlorides, 2,4-dialkyl-, synthesis of, **5**, 980
Allophanoyl chlorides, tolylene-, as intermediates, **5**, 966
Alloys
for the activation of metals, **2**, 958
oxidative metallation with, **2**, 959
All-(*E*)-trienes, synthesis of, **1**, 737
Allyl anions
from, vinylboranes, **1**, 162
reactions, with alkenes, **1**, 363
as starting materials, for cyclopentanes, **1**, 363
Allylation
of radicals, **1**, 173
of saturated aldehydes, **3**, 21
of saturated ketones, **3**, 139
for the synthesis of allylsilanes, **2**, 77
Allylation, copper(I)-catalyzed, of organozincs, **1**, 122
Allylation, nucleophilic, of carbonyls, **2**, 73
Allyl bromide
reactions
with allylcoppers, **1**, 121
with copper(I) salts, **1**, 1004
with rhenium vinyls, **5**, 450
with trialkylstibines, **2**, 73
with trimethylsilylalkyne, **4**, 810
as starting material, for 1-bromo-1,4-dienes, **2**, 624
Allyl bromides
reactions
with 1,1-dimethylethanethiolate, **4**, 373
with ethanethiolate, **4**, 373
Allyl carbanions, review, **1**, 107
Allyl cations
from, allylic alcohols, **1**, 364
reactions
with alkenes, **1**, 363

n, **1**, 191
olybdenum, **1**, 191
ckel, **1**, 191
lladium, **1**, 191
with tungsten, **1**, 191
synthesis of, **1**, 1125
Allyl chloride, 2-lithio, reactions, with electrophiles, **2**, 74
Allyl chlorides
for the alkylation of hydrazines, **2**, 378
deprotonation of, **4**, 138
reactions
with stannyl furans, **3**, 303
with thiolacetic acid, **2**, 117
η^1,η^3-Allyl complexes, protonation of, **4**, 680
π-Allyl compounds
synthesis of, **1**, 106
via the opening of vinyl epoxides, **1**, 228
Allyl groups
for the opening of epoxides, **1**, 207
for the protection of anomeric hydroxyl groups, **4**, 176
Allyl halides
reactions
with acylstannanes, **3**, 142
with *N*-(alkylamino)benzotriazoles, **2**, 302
with carbonyls, **2**, 73
with cyanomethylcopper, **3**, 623
with Grignard reagents, **1**, 119
with iodomethylzinc iodide, **3**, 623
with iron pentacarbonyl, **5**, 452
with nickel carbonyl, **1**, 526
with organoaluminums, **1**, 973
with organomercury compounds, **1**, 128
with organometallics, **1**, 119
with organozincs, **1**, 123
with potassium nitronates, **1**, 156
with sodium nitronates, **1**, 156
with terminal alkynes, **1**, 121
reactivity of, **3**, 21
as starting materials, for γ,δ-unsaturated nitriles, **3**, 623
Allylic anions, reactions, with carbonyls, **4**, 938
Allylic bromides
metallation of, **2**, 558
reactions, with triphenylarsine, **1**, 269
reduction of, **1**, 8, 84
synthesis of, **2**, 3
Allylic carbanions, stability of, **1**, 847
Allylic carbocations
reactions, with phenyl groups, **1**, 901
synthesis of, **1**, 897
Allylic cations
stability of, **1**, 901
synthesis of, **1**, 902
Allylic chlorides
from, allylic alcohols, **2**, 8
reactions, with barium, **1**, 116
reduction of, **1**, 84
synthesis of, **2**, 26
Allylic chlorides, 1,1,3,3-tetrakis(dialkylamino)-, as starting materials, for 1,1,3,3-tetrakis(dialkylamino)allenes, **1**, 982
Allylic complexes, [3+2] cycloadditions of, **2**, 577
Allylic compounds, deoxygenation of, **1**, 940
Allylic fluorides, synthesis of, **1**, 780
Allylic halides
as alkylating agents, **5**, 555
alkylation of, **1**, 464
formylation of, **3**, 23
from, allylic alcohols, **2**, 7
metallation of, **2**, 559
oxidation of, **3**, 57
reactions
with alkynes, **1**, 1050
with alkynides, **1**, 1001
with allenylmagnesium halides, **1**, 1052
with chromium salts, **2**, 571
with propargylic organoboranes, **1**, 977
with sodium alkynides, **1**, 1001
with transition metals, **1**, 1004
reactivity of, **1**, 2
as starting materials
for alcohols, **2**, 87
for amines, **2**, 308
thermal rearrangement of, **1**, 838
π-Allylic intermediates, use in the displacement of allylic alcohols, **1**, 191
Allylic iodides
oxidation of, **3**, 57
reactions, with indium(I) chloride, **2**, 597
Allylic rearrangement
for the synthesis of fluorohaloalkenes, **4**, 755
in the synthesis of nitriles, **3**, 612
in the synthesis of radicals, **1**, 925
Allylidene complexes, reactions, with alcohols, **5**, 941
Allyl iodide
reactions
with lithium (ethyl diazoacetate), **3**, 474
with silver cyanide, **3**, 693
Allyl iodides, reactions, with peracids, **2**, 88
Allylmagnesiation, of allyl alcohols, **2**, 561
Allylmanganese chloride, reactions, with 1-bromoundecane, **1**, 125
Allylmetallation–iodination, for the synthesis of α-haloalkenylsilicons, **4**, 809
Allyl radicals
cyclization of, **1**, 338
synthesis of, **1**, 330
via intramolecular 1,5-hydrogen transfer, **1**, 339
Allylthiols, double deprotonation of, **4**, 367
Allylzincation, of alkynylsilanes, **2**, 925
Alpine-Borane
reactions
with α,β-alkynic ketones, **2**, 57
with ketones, **2**, 57
Alumina
as catalyst, for the Strecker reaction, **3**, 616
reactions
with 1-chloro-2,3,4,5-tetramethylarsole, **2**, 893
with cyano esters, **3**, 659
as reagent, for the synthesis of *N*-alkylamides, **5**, 276
Alumina–metal catalysts, for the dealkylation of alkylbenzenes, **1**, 56
Aluminate, reactions, with epichlorohydrin, **1**, 224
Aluminate, chloro-, synthesis of, **1**, 914
Aluminates, iodination of, **4**, 122
Aluminates, aldimino-, synthesis of, **3**, 458
Aluminates, alkenyl-
for the monotransmetallation of zirconocene dichloride, **2**, 991
synthesis of, **1**, 479
Aluminates, alkynyl-, synthesis of, **2**, 1099
Aluminates, amido-, from, amides, **5**, 383
Aluminates, hydrido-, reactions, with propargylic alcohols, **2**, 970
Aluminates, hydridoalkyl-, synthesis of, **2**, 1099
Aluminates, imido-, from, imides, **5**, 384
Aluminates, imino-, synthesis of, **3**, 458
Aluminates, (methylnitrilium)trimethyl-, synthesis of, **5**, 1147
Aluminates, mixed alkylalkynyl-, synthesis of, **2**, 1099
Aluminides, synthesis of, **1**, 72
Aluminum
activation of, **2**, 595
for the displacement of allylic sulfones, **1**, 237
elimination of, **1**, 662
reactions
with alkyl bromides, **2**, 595

with alkyl iodides, **2**, 595
with antimony trichloride, **2**, 55
with dibromomethane, **4**, 698
with dichloromethane, **1**, 680; **4**, 698
with organomercury compounds, **4**, 698
with trimethylsilylpropargyl bromide, **1**, 971
Aluminum, (2,6-di-*t*-butoxy-4-methylphenoxy)dimethyl-, use in the synthesis of ketones, **3**, 127
Aluminum, diethylhexynyl-, reactions, with 3,4-epoxycyclopentene, **1**, 1024
Aluminum, diisobutyl- (DIBAL), reactions, with isonitriles, **5**, 803
Aluminum, (ethoxy)ethyl(iodomethyl)-, synthesis of, **4**, 153
Aluminum, (iodomethyl)diethyl-, synthesis of, **4**, 153
Aluminum, isobutyl-, elimination of, **1**, 662
Aluminum, trialkyl-
reactions
with *N*,*N'*-dimethylacetamidine, **5**, 780
with ketones, **2**, 699
Aluminum, triethyl-
reactions
with alkenylboranes, **1**, 1030
with benzyl halides, **1**, 128
with isothiocyanates, **5**, 595
with tris(trimethylsilylethynl)methyl acetate, **1**, 970
stereocontrol of, **1**, 17210
use in the synthesis of tris(cyclooctatetraene)dititanium, **4**, 671
Aluminum, trihexynyl-, reactions, with secondary alkyl sulfonates, **1**, 1021
Aluminum, triisobutyl-
as catalyst, for the synthesis of cyclic alkenyl ethers, **2**, 679
reactions, with alkenes, **2**, 596
as reducing agent, **1**, 663
as starting material, for diisobutylaluminum hydride, **2**, 596
use in the synthesis of hemiacetals, **4**, 174
Aluminum, trimethyl-
photolysis of, **4**, 698
reactions
with alcohols, **2**, 596
with alkynes, **1**, 1106
with 3-alkyn-1-ols, **1**, 543
with alkynyllithiums, **2**, 1099
with butyllithium, **1**, 220
with 2-deuterio-1-phenylethyne, **2**, 975
with dimethylberyllium, **4**, 709
with dimethylmagnesium, **4**, 709
with hydrazines, **5**, 350
with methylisonitrile, **5**, 1147
with *N*-phenylbenzamide, **5**, 718
with sulfonamides, **5**, 342
with tungsten tetrachloride, **2**, 572
as reagent, for zirconium-catalyzed carboalumination, **2**, 974
Aluminum, trineopentyl-, elimination of, **1**, 663
Aluminum, tripropyl-, reactions, with propene, **1**, 662
Aluminum, tris(thiophenyl)-, reactions, with esters, **5**, 415
Aluminum, tris(trimethylsilyl)-
reactions
with alkynes, **2**, 918
with methyl benzoate, **6**, 125
with *S*-2-pyridyl esters, **5**, 418
Aluminum acyl compounds, synthesis of, **5**, 466
Aluminum aldimines, synthesis of, **3**, 458
Aluminum alkoxides
as oxidants, for the synthesis of aldehydes, **3**, 7
synthesis of, **2**, 112
use in Meerwein–Ponndorf–Verley reduction, **2**, 55
Aluminum alkoxides, dialkyl-, synthesis of, **2**, 596
Aluminum alkyls
for the alkylation of arsenic oxide, **2**, 485
as catalysts, for the synthesis of cyanohydrins, **3**, 616
reactions, with zirconium alkyls, **4**, 725
Aluminum alkyne, reactions, with cobalt complexes, **1**, 199
Aluminum alkynides
selectivity of, **1**, 1139
synthesis of, **1**, 1139
Aluminum alkynides, dimethyl-, reactions, with diol oxiranes, **1**, 1024
Aluminum alkynides, methoxymethyl-, reactions, with oxiranes, **1**, 1024
Aluminum amalgam
for the reduction of sulfoxides, **1**, 23
for the synthesis of hydroxylamines, **2**, 764
use in the synthesis of α-aminoalkyl phosphonates, **4**, 472
Aluminum amides
use in the synthesis of *N*-alkylalkanoamides, **5**, 277
use in the synthesis of amides, **5**, 262, 263
use in the synthesis of amidines, **5**, 749
Aluminum anions, α-chloroalkenyl-, synthesis of, **4**, 816
Aluminum azide, diethyl-, use in the synthesis of acyl azides, **5**, 367
Aluminum-based compounds, for the reduction of acids to aldehydes, **3**, 82
Aluminum benzeneselenolate, diisobutyl-, use in the synthesis of monoselenoacetals, **4**, 238
Aluminum bis(4-bromo-2,6-di-*t*-butyl) phenoxide, methyl- (MAPH)
reactions, with 1,3,5-trioxane, **3**, 159
use in the synthesis of β-silyloxy aldehydes, **3**, 34
Aluminum bromide, reactions, with diisobutylaluminum hydride, **1**, 13
Aluminum bromide adducts, of molybdenum acyl compounds, **5**, 445
Aluminum chloride
as catalyst
for dealkylation, **1**, 57
in the synthesis of amidines, **5**, 749
reactions
with alkyllithiums, **4**, 698
with benzene, **1**, 57
with benzimide, **5**, 384
with chloromethylpentamethyldisilane, **4**, 614
with cyclopentadienylzirconium complexes, **5**, 466
with α,α-dichloroamides, **1**, 571
with α-hydrazononitriles, **3**, 671
with lithium bis(trimethylsilyl)methanide, **6**, 196
with perfluoro iodo ethers, **4**, 768
with phthalimide, **5**, 384
with succinimide, **5**, 384
with tertiary amides, **5**, 578
with trialkylaluminums, **2**, 596
with tris(trimethylsilylmethyl)lithium, **6**, 395
use in amination reactions, **2**, 306
use in the synthesis of amidines, **5**, 743
use in the synthesis of aromatic ketones, **3**, 278
use in the synthesis of α-chloro ethers, **4**, 49
Aluminum chloride, alkynylethyl-, use in the synthesis of alkynes, **1**, 1031
Aluminum chloride, dialkyl-, as reagents, **2**, 699
Aluminum chloride, diethyl-, reactions, with alkynyllithium, **1**, 314
Aluminum chlorides, alkyl-, reactions, with alkynylsilanes, **2**, 925
Aluminum chlorides, dialkyl-
reactions
with alkali metal hydrides, **2**, 596
with alkyllithiums, **2**, 595
with alkynides, **6**, 208
synthesis of, **2**, 596
Aluminum complexation, for the synthesis of hydroxamates, **5**, 329
Aluminum complexes, allylic
condensation with aldehydes, **1**, 714
synthesis of, **1**, 714

Aluminum cyanide, diethyl-
as cyanide source, for the synthesis of cyanohydrins, **3**, 616
reactions
with cyclopropylketones, **3**, 634
with epoxides, **3**, 632
as starting material, for γ-ketonitriles, **3**, 634
synthesis of, **5**, 1143
use in the opening of epoxides, **1**, 223
Aluminum cyanides, dialkyl-, synthesis of, **2**, 596; **5**, 1142
Aluminum dialkoxides, methyl-, synthesis of, **2**, 596
Aluminum dichlorides, acyl-, synthesis of, **5**, 466
Aluminum dichlorides, alkyl-, from, aluminum sesquichlorides, **2**, 595
Aluminum dihydrides, bis(dialkylamino)-, for the reduction of benzoic acids, **3**, 82
Aluminum enolates
alkylation of, **1**, 534
for the opening of epoxides, **1**, 209
synthesis of, **2**, 699
Aluminum etherate, tris(trimethylsilyl)-, use in the synthesis of α-hydroxysilanes, **4**, 355
Aluminum ethoxide, reactions, with tetraethylhypophosphate, **2**, 107
Aluminum halides
for the hydrolysis of aromatic esters, **5**, 106
reactions, with tungsten complexes, **3**, 515
use in the synthesis of alkanoic acids, **5**, 40
Aluminum halides, alkyl-, reactions, with alkyllithiums, **4**, 698
Aluminum halides, dialkyl-
reactions
with carbon dioxide, **5**, 44
with lithium alkynides, **1**, 1139
Aluminum heterocycles, synthesis of, **2**, 816
Aluminum hydride
reactions
with alkenes, **2**, 596
with 1-alkynes, **4**, 698
with ethene, **2**, 112
synthesis of, **2**, 596
Aluminum hydride, bis(diisopropylamino)-, for the conjugate reduction of enones, **3**, 127
Aluminum hydride, bis(*N*-methylpiperazinyl)-
reactions, with 3-furoic acid, **3**, 102
for the reduction of benzoic acid, **3**, 82
Aluminum hydride, bis(4-methyl-1-piperazinyl)- (BMPA), for the reduction of aromatic esters, **3**, 84
Aluminum hydride, diethyl-
reactions
with hydrogen cyanide, **5**, 1143
with phenylcarbyne complexes, **3**, 522
use in the synthesis of tris(cyclooctatetraene)dititanium, **4**, 671
Aluminum hydride, diisobutyl- (DIBAL-H)
for the conjugate reduction of enones, **3**, 127
from, triisobutylaluminum, **2**, 596
reactions
with alkynes, **1**, 533
with 1-alkynyl epoxides, **2**, 83
with alkynylstannanes, **2**, 967
with aluminum bromide, **1**, 13
with 1-haloalkynes, **2**, 967
with nitriles, **3**, 458
with terminal alkynes, **1**, 511
with 1-trimethylgermanyl-2-phenylethyne, **4**, 1063
as reagent
for hydroalumination, **2**, 967
for the reduction of esters to aldehydes, **3**, 13
for the reduction of allenes, **1**, 459
for the reduction of amides, **3**, 85
for the reduction of aromatic esters, **3**, 84
for the reduction of benzylic sulfones, **1**, 24
for the reduction of glycidols, **1**, 14
for the reduction of ketene dithioacetals, **4**, 253
for the reduction of ketones, **1**, 20
for the reduction of oxiranes, **1**, 14
for the reduction of thiol esters, **3**, 85
use in hydroalumination reactions, **1**, 511
use in the reduction of α-amino acids, **3**, 39
use in the reduction of carbonyls, **2**, 60
use in the synthesis of aryl aldehydes, **3**, 82
use in the synthesis of hemiacetals, **4**, 169
use in the synthesis of pyridine carboxaldehydes, **3**, 107
Aluminum hydride, (*N*-methylpiperazinyl)-, use in the synthesis of aldehydes, **3**, 13
Aluminum hydride, triethoxy-, use in the synthesis of 3-formylpyridines, **3**, 107
Aluminum hydrides
for the conjugate reduction of enones, **3**, 127
reactions, with alkynes, **1**, 533
for the reduction of amides, **3**, 85
for the reduction of amidines, **1**, 45
for the reduction of hydrazones, **1**, 43
use in the reduction of amines, **1**, 33
use in the reduction of carbonyls, **2**, 60
use in the reduction of organomercury compounds, **1**, 66
use in the synthesis of aldehydes, **1**, 34
Aluminum hydrides, dialkyl-
reactions
with alkynes, **6**, 208
with alkynylalanes, **6**, 208
as reagents, for hydroalumination, **2**, 967
as starting materials, for trialkylaluminums, **2**, 596
synthesis of, **2**, 596
Aluminum hydrides, trialkyl-, reactions, with chloroalkynes, **4**, 816
Aluminum hydride–triethylamine, reactions, with cobaltacyclopentadiene complexes, **4**, 726
Aluminum imines, *N*-diisobutyl-
reactions
with Grignard reagents, **2**, 316
with lithium compounds, **2**, 316
Aluminum iodide, diethyl-
reactions
with bis(perfluoroisopropyl)mercury, **3**, 395
with diazomethane, **4**, 153
as starting material, for tellurocarbonyl difluoride, **6**, 588
Aluminum isopropoxide, for the reduction of alcohols, **2**, 55
Aluminum–lanthanide compounds, synthesis of, **4**, 726
Aluminum–magnesium alloys, for the synthesis of trialkylaluminums, **2**, 595
Aluminum–mercury amalgams
for the reduction of α,β-unsaturated sulfones, **1**, 439
use in the reduction of acyl ylides, **1**, 51
Aluminum methanesulfonate, diisobutyl-, reactions, with carbamates, **4**, 383
Aluminum methylselenolate, dimethyl-, reactions, with esters, **5**, 239
Aluminum–nickel(II) chloride, use in the synthesis of amines, **2**, 316
Aluminum oxide
as catalyst, in the synthesis of β-amino ketones, **3**, 187
as support, **1**, 48
Aluminum phenyltellurolate, diisobutyl-, use in the synthesis of β-telluro ketones, **3**, 39
Aluminums
in Claisen rearrangements, **1**, 415
reactions
with aldehydes, **1**, 1055
with ketones, **1**, 1055
with 1-metalloalkylphosphines, **4**, 587
with orthoformates, **4**, 192
with phosphorus ylides, **4**, 587
with trimethyl orthoacetate, **4**, 192
selectivity of, **1**, 222
synthesis of, **6**, 208

Aluminums, acyl-, synthesis of, **5**, 466
Aluminums, alkenyl-
from, alkynes, **1**, 207
reactions
with aldehydes, **1**, 534
with aryl halides, **1**, 533
with 1-haloalkynes, **1**, 533
with ketones, **1**, 534
with methyllithium, **1**, 534
with stannyl acetals, **4**, 388
with vinyl halides, **1**, 533
synthesis of, **1**, 1106
Aluminums, α-alkoxy-, synthesis of, **4**, 383
Aluminums, alkyl-
as Lewis acids, **2**, 595
reactions
with haloalkanes, **1**, 128
with osmium complexes, **2**, 581
with palladium bis(acetylacetonate, **2**, 581
as starting materials, for alkyl radicals, **2**, 595
synthesis of, **2**, 595
via transmetallation, **2**, 595
Aluminums, alkynyl-
alkylation of, **1**, 1003
reactions
with oxiranes, **1**, 1023
with secondary alkyl sulfonates, **1**, 1021
with stannyl acetals, **4**, 388
synthesis of, **1**, 314; **2**, 1098
Aluminums, allenyl-, reactions, with allenylmagnesium halides, **1**, 1055
Aluminums, aryl-, reactions, with aryl iodides, **3**, 290
Aluminums, β-branched alkyl-, elimination of, **1**, 662
Aluminums, chlorodialkyl-, reactions, with alkynes, **1**, 542
Aluminums, dialkyl-, reactions, with trimethylsilyl cyanide, **5**, 1142
Aluminums, dimethylalkynyl-, reactions, with β-propiolactones, **1**, 1024
Aluminums, formimidoyl-, synthesis of, **5**, 803
Aluminums, α-halovinyl-, synthesis of, **4**, 815
Aluminums, imidoyl-, synthesis of, **5**, 803
Aluminums, methoxy allenyl-, reactions, with aldehydes, **1**, 973
Aluminums, monosilyl-, for the synthesis of vinylsilanes, **2**, 918
Aluminums, phenyl-, elimination of benzene from, **2**, 816
Aluminums, silyl-, reactions, with acyclic alkynic alcohols, **2**, 86
Aluminums, trialkyl-
dimeric structure of, **4**, 698
from
alkyl iodides, **2**, 595
dialkylaluminum hydrides, **2**, 596
reactions
with alkynes, **1**, 542
with alkynyl bromides, **1**, 1019
with aluminum chloride, **2**, 596
with amides, **5**, 383
with aryl triflates, **1**, 482
with azides, **2**, 776
with bromoalkynes, **1**, 1141
with carbonyls, **1**, 663
with hydrogen cyanide, **5**, 1142
with manganese(II) salts, **6**, 523
with nitriles, **3**, 458
with phosphorus trichloride, **2**, 434
with tetrachloromethane, **2**, 596
with zinc chloride, **2**, 596
synthesis of, **2**, 595; **4**, 699
via carboalumination, **2**, 595
via hydroalumination, **2**, 595
Aluminums, triaryl-, dimeric structure of, **4**, 698
Aluminums, trihalomethyl-, synthesis of, **6**, 245
Aluminums, tris(silylalkyl)-, reactions, with sulfur trioxide, **4**, 397
Aluminums, vinyl-, synthesis of, **1**, 453
Aluminum selenide, reactions, with thioic acid *O*-alkyl esters, **5**, 563
Aluminum sesquichlorides, as starting materials, for alkylaluminum dichlorides, **2**, 595
Aluminum sesquichlorides, alkyl-, synthesis of, **2**, 595
Aluminum sulfide, ethyl-
use in the synthesis of thionoalkanoamides, **5**, 610
use in the synthesis of thionoamides, **5**, 578
Aluminum sulfides, reactions, with esters, **5**, 235
Aluminum telluride, as starting material, for alkanetellurols, **2**, 278
Aluminum tetramethylpiperidide, dialkyl-, reactions, with carbonyls, **2**, 699
Aluminum thioalkoxides, dialkyl-, synthesis of, **2**, 596
Aluminum thiocyanates, dialkyl-, synthesis of, **2**, 596
Aluminum thiolates, dialkyl-, use in the synthesis of thioimidic esters, **5**, 731
Aluminum thiophenoxide, diethyl-, use in the synthesis of monothioacetals, **4**, 216
Aluminum tribromide
reactions, with triethylsilane, **1**, 61
as starting material, for carbonic bromide chloride, **6**, 420
use in the synthesis of 1,1-bis(silyl)alkanes, **4**, 609
Aluminum trichloride
as catalyst, for Friedel–Crafts acylation, **1**, 463
as chlorinating agents, for aryl chlorides, **2**, 620
reactions
with acyl chlorides, **3**, 278
with alkylfurans, **3**, 302
with alkyllithiums, **2**, 595
with alkynylsodiums, **2**, 1098
with arylsulfonamides, **5**, 674
with dialkylbenzenes, **3**, 296
with ethers, **6**, 39
with lithium alkynides, **1**, 1003
with magnesium compounds, **2**, 595
with α-silylmethyl Grignard reagents, **4**, 659
with sodium borohydride, **2**, 59
with trichloramine, **2**, 298
with (trichloromethyl) benzene, **3**, 296
with vinylalkoxymethyl ethers, **2**, 89
with ylides, **4**, 397
as starting material, for phosgene, **6**, 413
for the synthesis of aryl iodides, **2**, 631
transmetallation reaction, with alkenylzirconocene chloride, **2**, 991
use in Diels–Alder reactions, **1**, 373
use in the synthesis of carbon-bridged dialuminums, **4**, 698
Aluminum triiodide
as starting material, for selenocarbonyl difluoride, **6**, 588
for the synthesis of aryl iodides, **2**, 631
Aluminum trimethylamine complex, tris(trifluorovinyl)-, synthesis of, **4**, 815
Aluminum tris(2,6-triphenylphenoxide) (ATPH), reactions, with carbonyl groups, **1**, 303
Aluminum–yttrium complexes, synthesis of, **4**, 726
Alyl bromide, reactions, with alkynylphosphines, **2**, 1063
Alylic bromides, reactions, with indium, **2**, 597
Alynes, reactions, with *n*-butyl *t*-butyl telluride, **4**, 947
Amalgams
for the activation of metals, **2**, 958
oxidative metallation with, **2**, 959
Amaryllidaceae alkaloids, synthesis of, **3**, 422
Amberlite IR-100, use in the synthesis of alkanoic acids, **5**, 30
Amberlite IR-120, use in the synthesis of carboxylic acids, **5**, 95
Amberlyst-15
as catalyst, for the synthesis of ketones, **3**, 133
reactions, with amides, **5**, 36
use in the synthesis of aldehydes, **3**, 9, 18

use in the synthesis of ketones, **3**, 123
Amberlyst A-26 resin, iodide form of, use in the synthesis of chloroiodoalkanes, **4**, 39
Ambiphilic radicals
definition of, **1**, 921
intermolecular addition reactions, **1**, 325
Amidases, for the hydrolysis of amides to carboxylic acids, **5**, 118
Amidate anions, α,β-unsaturated, reactions, with organolithiums, **4**, 651
Amidates, allylic amino-, [2,3] rearrangement of, **1**, 839
Amide, *N*-benzyl-, reactions, with potassium superoxide, **2**, 649
Amide, *N*,*O*-bis(trimethylsilyl)-, reactions, with diorganohaloboranes, **5**, 383
Amide, *N*-chloromethyl-, reactions, with ethyl diphenylphosphinite, **4**, 487
Amide, diisopropyl-, use in the synthesis of lithium compounds, **4**, 600
Amide, ethyl oxalic acid oxyimidic, reactions, with nitrosyl chloride, **5**, 825
Amide, hydroxyethyl-, pyrolysis of, **6**, 142
Amide, *N*-propylthiobenzoyl-, reactions, with nitrogen tetroxide, **5**, 361
Amide, silyl-, use in the synthesis of hydroxamic acids, **5**, 325
Amide, tertiary, reactions, with phosphorus pentachloride, **4**, 790
Amide acetals
from
amides, **6**, 107
diaminoalkoxymethanes, **6**, 140
α,α-difluoroamines, **6**, 105
N,*N*-dimethylformamide dimethyl acetal, **6**, 104
iminium chlorides, **6**, 105
isocyanates, **6**, 107
nitrates, **6**, 105
ortho esters, **6**, 104
4-oxa-β-lactam, **6**, 107
tetrakis(dimethylamino)ethene, **6**, 105
ynamines, **6**, 106
reactions
with amides, **5**, 759
with amines, **5**, 746
with arylphosphines, **6**, 686
with phosphites, **6**, 116
with thiols, **6**, 111
as starting materials
for ortho esters, **6**, 80
for thioamide acetals, **6**, 111
structure of, **6**, 104
synthesis of
via transacetalation of other amide acetals, **6**, 104
via transamidation, **6**, 104
Amide acetals, bicyclic, synthesis of, **6**, 107
Amide anions
reactions
with carbon disulfide, **5**, 232
with diazonium salts, **2**, 776
Amide bond formation, rate of, **5**, 260
Amide chlorides
reactions, with lithium amides, **2**, 1045
synthesis of, **2**, 1045
Amide complexes, synthesis of, **5**, 385
Amide complexes, oxyimidic, synthesis of, **5**, 864
Amide dithioacetal, from, dithioortho esters, **6**, 111
Amide monothioacetals
from, 1,3,4-thiadiazolines, **6**, 111
synthesis of, **6**, 111
Amide oximes, nitrosation of, **5**, 667
Amides
as activating groups for alkenes, **1**, 300
alkenation of, **5**, 957
alkylation of, **4**, 881; **5**, 284, 289
intramolecular, **5**, 285
N-alkylation of, **5**, 292, 296
O-alkylation of, **3**, 419
amination of, **5**, 354
asymmetric *cis*-dihydroxylation of, **2**, 49
bromination of, **5**, 318
as chiral auxiliaries, **1**, 305
chlorination of, **5**, 315; **6**, 607
α-chlorination of, **2**, 23
dehydration of, **3**, 619, 674
deprotonation of, **1**, 148; **5**, 384
elimination of, **1**, 637
fluorination of, **2**, 335; **5**, 313
from
alkanoic acids, **5**, 274
alkenoic acids, **5**, 295
alkyl halides, **5**, 272
alkynoic acids, **5**, 304
amides, **5**, 273
aromatic acids, **5**, 298
bis(chloromethyl)phosphinic chlorides, **4**, 127
cinnamic acids, **5**, 613
α-diazo ketones, **1**, 395
enzymes, **5**, 270
heteroaromatic acids, **5**, 298
ketenimines, **2**, 1040
ketones, **5**, 276
β-lactams, **4**, 825
nitriles, **5**, 268
oxaziridines, **5**, 272
pyridyl sulfides, **1**, 333
halogenation of, **2**, 3
hydrolysis of, **5**, 26, 108
intramolecular alkylation of, reagents for, **5**, 285
α-iodination of, **2**, 34
Mannich aminomethylation of, **4**, 322
metallation of, **2**, 302; **4**, 530; **5**, 385
methylenation of, **1**, 766
for the migration of double bonds, **1**, 795
nitrosation of, **5**, 360
oxidation of, **5**, 324
oxidative rearrangement of, **5**, 999
phosphorolation of, **5**, 368
phosphorylation of, **5**, 370
pyrolysis of, **1**, 782
reactions
with acetals, **4**, 310
with acetanilides, **5**, 754
with acetyl hypobromite, **5**, 320
with alkoxyamines, **5**, 329
with alkylboranes, **5**, 381
with allenyllithium, **1**, 971
with Amberlyst-15, **5**, 36
with amide acetals, **5**, 759
with aminols, **4**, 427
with benzanilide, **5**, 754
with benzotriazole, **4**, 449
with bis(trifluoroacetoxy)iodobenzene, **5**, 999
with boron compounds, **5**, 381
with bromine, **2**, 756
with *N*-bromosuccinimide, **4**, 138
with *n*-butyl nitrite, **5**, 35
with caesium fluoroxy sulfate, **2**, 334
with chlorine, **2**, 756
with chloromethylgermanes, **4**, 521
with chlorosilanes, **2**, 811
with diazonium salts, **5**, 365
with dibromoisocyanuric acid, **5**, 318
with dibutyl iodomethylboronates, **4**, 525
with dimethylformamide, **5**, 999
with formaldehyde, **4**, 416
with haloboranes, **5**, 381
with halogens, **5**, 657

with hydrazines, **5**, 772
with hypobromite, **5**, 997
with iodine, **5**, 320
with isocyanates, **5**, 754
with lead tetraacetate, **5**, 997
with lithium aluminum hydride, **1**, 34
with methyl trichloropyruvate, **3**, 417
with nitrosonium tetrafluoroborate, **5**, 36
with organoarsenic dihalides, **2**, 892
with oxalyl chloride, **5**, 576, 743, 995
with phenylselenium chloride, **5**, 335
with phosgene, **2**, 746; **4**, 971; **5**, 995, 996
with phosphorus oxychloride, **5**, 576
with phosphorus pentachloride, **5**, 743
with phosphorus pentahalides, **5**, 655
with phosphorus pentaselenide, **5**, 626
with phosphorus pentasulfide, **4**, 424; **5**, 613
with phosphorus trihalides, **5**, 655
with pyridines, **3**, 310
with silyl halides, **6**, 162
with silylmethyl halides, **4**, 513
with sodium, **5**, 723
with sodium nitrite, **5**, 36
with *N*-sulfinyloxazolidinones, **2**, 363
with *N*-sulfinyl sulfonamides, **5**, 767
with sulfur transfer reagents, **5**, 333
with sulfuryl chloride, **5**, 657
with tetrahalotins, **5**, 389
with tetrakis(dimethylamino)titanium, **4**, 972
with thionyl chloride, **5**, 657
with trialkylaluminums, **5**, 383
with triethylgermyllithium, **6**, 515
with triethyloxonium tetrafluoroborate, **5**, 751
with trimethylchlorosilane, **5**, 718
with triphenylarsine dibromide, **2**, 809
with tris(alkylthio)boranes, **6**, 113
reduction of, **2**, 54; **3**, 85
for secondary phosphines, **2**, 822
reduction to aldehydes, **3**, 13
silylation of, **5**, 374, 379
as starting materials
for *O,N*-acetals, **4**, 303
for *N*-acyliminophosphoranes, **5**, 373
for *N*-acyltellurium imines, **5**, 349
for *N*-acyl triazenes, **5**, 365
for aldehydes, **1**, 34
for aliphatic amidines, **5**, 751
for alkynes, **1**, 1066
for alkynones, **1**, 1046
for amide acetals, **6**, 107
for amides, **5**, 273
for amidines, **5**, 743
for amidoaluminates, **5**, 383
for *gem*-amido phosphorylalkenes, **4**, 1005
for amines, **2**, 319
for aminobis(phosphonates), **6**, 158
for aromatic amidines, **5**, 757
for benzonitriles, **3**, 666
for bis(trimethylsilyl)aminomethanes, **6**, 162
for *N*-boroamides, **5**, 381
for *N*-bromoamides, **5**, 318
for carboxylic acids, **5**, 108
for *N*-chloroamides, **5**, 315
for cyanamides, **5**, 1125
for diazoalkanes, **3**, 465
for *N,N*-dibromoamides, **5**, 318
for *N,N*-dichloroamides, **5**, 317
for *N,N*-difluoroamines, **2**, 335
for dithyreanitrile, **4**, 245
for *N*-fluoroamides, **5**, 313
for hydrazonamides, **5**, 851
for imidoyl halides, **5**, 654
for *N*-iodo(I)amides, **5**, 320
for *N*-iodo(III)amides, **5**, 321
for isocyanates, **5**, 995
for isothiocyanates, **5**, 1030
for ketene *N,S*-acetals, **2**, 1049
for ketene aminals, **4**, 971, 972
for ketene hemiaminals, **4**, 881
for keteniminium salts, **3**, 601
for ketones, **3**, 131
for metal amides, **5**, 388
for monothioacetals, **6**, 111
for nitriles, **3**, 619
for *N*-nitroamides, **5**, 361
for *N*-nitrosoamides, **5**, 360
for oxyimidic chlorides, **5**, 834
for phosphaalkenes, **5**, 882
for *N*-phosphinoamides, **5**, 368
for sulfinamides, **5**, 340
for thioamides, **2**, 1049; **5**, 333
for thioimidates, **4**, 517
for thiolactams, **5**, 333
for thionoamides, **5**, 566
for transition metal amides, **5**, 384
for ynamines, **2**, 1050
sulfonylation of, **5**, 345
synthesis of, **3**, 530; **5**, **257–307**
general methods, **5**, 258
via rearrangements, **5**, 270
for the synthesis of radicals, **1**, 925
tautomeric forms of, **5**, 685
thionation of, **5**, 570, 610, 613
types of, **5**, 259
uses of, **5**, 258
Vilsmeier reaction of, **3**, 76
See also Acylamines
Amides, acetic, as starting materials, for ketenes, **3**, 535
Amides, *N*-acyl-
as starting materials, for *gem*-alkylideneamino phosphorylalkenes, **4**, 1009
synthesis of, **5**, 305
Amides, *N*-acyloxy-, methylation of, **5**, 700
Amides, alkano-, as starting materials, for thionoamides, **5**, 610
Amides, *N*-alkenylaryl-, synthesis of, **5**, 302
Amides, *N*-alkoxy-
alkylation of, **5**, 700
as starting materials
for *N*-alkoxyimidoyl bromides, **5**, 666
for *N*-alkoxyimidoyl halides, **5**, 824
See also Hydroxamic acids
Amides, ω-alkoxycarbonyl alkyl-, reactions, with boranedimethyl sulfide, **2**, 326
Amides, *N*-alkoxymethyl-, as starting materials, for 1-aminoalkylphosphonium salts, **4**, 492
Amides, *N*-alkyl-
from
acids, **5**, 276
active esters, **5**, 277
acyl halides, **5**, 278
anhydrides, **5**, 277
azides, **5**, 278
esters, **5**, 277
nitriles, **5**, 281
synthesis of
via alkylation of amides, **5**, 284
via Beckmann rearrangement, **5**, 282
via carbonylation reactions, **5**, 285
via coupling reagents, **5**, 278
via oxaziridine–amide rearrangement, **5**, 283
via rearrangement reactions, **5**, 282
via Schmidt rearrangement, **5**, 283
Amides, *N*-alkyl-*N'*-alkoxy-, synthesis of, **5**, 700
Amides, *N*-alkylaryl-
from

acids, **5**, 299
aryl aldehydes, **5**, 302
synthesis of, **5**, 299
via carbonylation reactions, **5**, 301
via hydrolysis of nitriles, **5**, 300
Amides, *N*-alkyl-*N*-chloromethyl-, synthesis of, **4**, 99
Amides, *N*-alkyl-*N*-nitroso-, elimination of nitrogen from, **5**, 108
Amides, *N*-alkylthiomethyl-, reactions, with electrophiles, **4**, 100
Amides, *N*-alkyltosyl-, anodic oxidation of, **2**, 323
Amides, alkynic, synthesis of, **2**, 1044
Amides, alkynoic
from
acyl halides, **5**, 304
esters, **5**, 304
Amides, β-allenic, synthesis of, **1**, 979
Amides, allyl-, as starting materials, for enamides, **5**, 290
Amides, amino-
condensation of, **4**, 417
reactions
with Lawesson's reagent, **5**, 571
with thionyl chloride, **5**, 340
Amides, *N*-amino-, synthesis of, **5**, 354
Amides, α-aminooxyimidic, from, α-aminonitriles, **5**, 833
Amides, β-amino-α,β-unsaturated, as starting materials, for α-lithio enamines, **4**, 1016
Amides, *N*-(1-arenesulfonyl-2-chloroethenyl)-, reactions, with phosphorus pentachloride, **5**, 656
Amides, aromatic oxyimidic, synthesis of, **5**, 835
Amides, arsonic
from, arsonic esters, **2**, 891
synthesis of, via amide exchange, **2**, 891
Amides, aryl-
N-alkylation of, **5**, 300
use in the synthesis of *N*-substituted phenyl ketones, **3**, 299
Amides, *N*-aryl-
from
acids, **5**, 290
isocyanates, **5**, 293
oxidation of, **5**, 325
synthesis of
via carbonylation, **5**, 293
via coupling reagents, **5**, 291
via oxaziridine rearrangement, **5**, 292
via oxidation reactions, **5**, 293
via rearrangement reactions, **5**, 292
via Semmler–Wolff aromatization, **5**, 293
via Smiles rearrangement, **5**, 292
via Wolff rearrangement, **5**, 292
Amides, *N*-arylalkeno-, synthesis of, **5**, 297
Amides, *N*-arylaryl-
from, acids, **5**, 302
synthesis of, **5**, 302
reagents for, **5**, 302
via coupling reactions, **5**, 303
via rearrangement reactions, **5**, 303
Amides, *N*-(1-aryldiphenylphosphonyl-2-dichloroethenyl)-, reactions, with phosphorus pentachloride, **5**, 656
Amides, benzylic, reactions, with thionyl chloride, **2**, 27
Amides, 2,2'-binaphthyl-, use in the synthesis of δ-lactones, **5**, 172
Amides, α,α-bis(silyl)-, synthesis of, **4**, 604
Amides, *N*,*N*-bis(sulfonyl)-
hydrolysis of, **5**, 348
synthesis of, **5**, 348
Amides, bis(trialkylsilyl)-
lactamization of, **5**, 377
reactions, with iodine, **5**, 377
Amides, *N*,*N*-bis(tributylstannyl)-, synthesis of, **5**, 391
Amides, bis(trimethylsilyl)-, reactions, with phosgene, **5**, 997
Amides, *N*-boro-
from
amides, **5**, 381
aminoiminoboranes, **5**, 381
N-silylamides, **5**, 381
synthesis of, **5**, 380
via acylation of aminoborane, **5**, 382
Amides, bromo-, reactions, with diaryltellurium, **5**, 341
Amides, *N*-bromo-
from, amides, **5**, 318
reactions, with dimethyl sulfide, **5**, 338
as starting materials, for *N*-iodo(I)amides, **5**, 320
Amides, α-bromo-
from, methyl *N*-acetyl-α-hydroxyglycinate, **4**, 99
as starting materials, for *C*-bromoketenimines, **3**, 557
Amides, *N*-*t*-butyl, as starting materials, for nitriles, **3**, 619
Amides, carboxylic
reactions, with alkynyl metals, **1**, 1046
as starting materials, for *C*-aminophosphaalkenes, **5**, 880
synthesis of, **1**, 1043
via Wolff reactions, **1**, 394
via Wolff rearrangement, **1**, 395
Amides, chiral, from, hexatriazines, **4**, 99
Amides, *N*-chloro-
from
amides, **5**, 315
imines, **5**, 315
reactions
with arylselenium chloride, **5**, 341
with dimethylformamide diethyl acetal, **6**, 634
with dimethyl sulfide, **5**, 338, 689
with selenium, **5**, 341
with sulfur, **5**, 349
as starting materials, for α,β-unsaturated amino acids, **5**, 315
synthesis of, **5**, 315
reagents for, **5**, 315
Amides, α-chloro-
reactions, with phosphorus pentachloride, **3**, 561
as starting materials, for *N*-trimethylsilyloxymethyl lactams, **4**, 100
synthesis of, **2**, 23; **4**, 98
Amides, *N*-α-chloroalkyl-, characteristics of, **4**, 96
Amides, α-chloroalkyl-, as starting materials, for *N*-acyl *S*,*N*-acetals, **4**, 327
Amides, *N*-chloromethyl-, synthesis of, **4**, 99
Amides, α-cyanamido-, synthesis of, **5**, 1118
Amides, cyano-
reactions
with Grignard reagents, **5**, 743
with organolithiums, **5**, 743
Amides, *N*-cyano-, synthesis of, **5**, 1120
Amides, cyclic, synthesis of, methods, **5**, 260
Amides, dialkyl-
deprotonation of, **1**, 876
as starting materials, for enolates, **1**, 131
Amides, *N*,*N*-dialkyl-, reactions, with alkynyllithiums, **1**, 1046
Amides, dialkylisobutyryl-, as starting materials, for haloenamines, **2**, 8
Amides, α-diazo-, from, α-diazo esters, **3**, 475
Amides, *N*,*N*-dibromo-
from, amides, **5**, 318
as starting materials, for *N*,*N*-diiodoamides, **5**, 321
Amides, dichloro-, as starting materials, for transition metal amides, **5**, 387
Amides, *N*,*N*-dichloro-
from
amides, **5**, 317
perfluoroacyl iminosulfur difluorides, **5**, 317
handling precautions, **5**, 317
reactions, with *S*,*S*'-dialkyl disulfides, **5**, 349
Amides, α,α-dichloro-, reactions, with aluminum chloride, **1**, 571
Amides, *N*,*N*-difluoro-, from, α-diketones, **5**, 314

Amides, *N,N*-diiodo-, from, *N,N*-dibromoamides, **5**, 321
Amides, dimethyl-, reactions, with triethylgermyllithium, **6**, 131
Amides, *N,N*-dimethyl-
reduction of, **3**, 85
synthesis of, **5**, 280
Amides, disilyl-
from
disilylamines, **5**, 379
N-haloamides, **5**, 379
synthesis of, **5**, 379
via silylation of amides, **5**, 379
Amides, *N,N*-disubstituted carboxylic, reduction of, **3**, 85
Amides, *N*-fluoro-
from
amides, **5**, 313
fluoroamines, **5**, 314
perfluoromethanimine, **5**, 313
O-silyl enol ethers, **5**, 313
as intermediates, in the synthesis of *N,N*-difluoroamines, **2**, 335
synthesis of, reagents for, **5**, 313
Amides, *N*-formyl-, synthesis of, **5**, 760
Amides, *N*-germyl-
from, trialkylgermaniumacetamides, **5**, 390
synthesis of, **5**, 389
Amides, glucosamine, oxidative cleavage of, **3**, 40
Amides, *N*-halo-
from, sulfinamides, **5**, 341
reactions, with enol ethers, **3**, 43
as starting materials
for *N*-acyl sulfinimidoyl chlorides, **5**, 349
for *N*-acyl sulfonium salts, **5**, 338
for disilylamides, **5**, 379
for thiolactams, **5**, 335
synthesis of, **5**, 313
Amides, α-halo-
from, α-hydroxy compounds, **4**, 98
synthesis of, **4**, 98
via cleavage of *gem*-diheteroatomic compounds, **4**, 99
via halogenation, **4**, 99
Amides, *N*-haloalkyl-, synthesis of, **4**, 99
Amides, *N*-(haloalkyl)-, synthesis of, **4**, 308
Amides, *N*-α-haloalkyl-, reactions, with nucleophiles, **4**, 96
Amides, halomagnesium dialkyl-, synthesis of, **2**, 697
Amides, *N*-hetarylalkeno-, synthesis of, **5**, 297
Amides, *N*-hetero-, as starting materials, for acylhydrazines, **5**, 353
Amides, *C*-heteroaryl-substituted oxyimidic, from, heterocyclic nitriles, **5**, 834
Amides, *N*-heteroaryl-substituted oxyimidic, synthesis of, **5**, 80667
Amides, *N*-hetero-*N*-chloro-, from, hydroxamates, **5**, 317
Amides, *N*-hetero-*N*-silyl-, synthesis of, **5**, 380
Amides, *N*-heterosubstituted, synthesis of, **5**, **309–391**
Amides, hydroxy-
as starting materials, for *O,N*-acetals, **4**, 301
use in the synthesis of 1,3-oxazines, **4**, 302
Amides, *N*-hydroxy-, reactions, with thionyl chloride, **5**, 666
Amides, α-hydroxy-
reactions
with chlorophosphines, **4**, 465
with tetraphenyl bisphosphines, **4**, 465
as starting materials
for alkylaminomethyl phosphinic acids, **4**, 465
for α-aminoalkylphosphorus compounds, **4**, 465
Amides, β-hydroxy-, alkylation of, **3**, 210
Amides, *N*-(hydroxyalkyl)-, synthesis of, **4**, 324
Amides, hydroxymethyl-
reactions, with phosphorus trichloride, **4**, 469
synthesis of, **4**, 417
Amides, *N*-hydroxymethyl-, as starting materials, for 1-aminoalkylphosphonium salts, **4**, 492
Amides, β-hydroxyphosphinothioic
elimination of, **1**, 643
reactions, with methyl iodide, **1**, 643
Amides, γ-hydroxy-α,β-unsaturated, synthesis of, **2**, 49
Amides, *N*-iodo-, reactions, with dialkyldithiocarbamates, **5**, 335
Amides, *N*-iodo(I)
from
amides, **5**, 320
N-bromoamides, **5**, 320
stability of, **5**, 320
Amides, *N*-iodo(III)
from
amides, **5**, 321
N-silylamides, **5**, 321
Amides, 4-keto-, synthesis of, **1**, 1045
Amides, α-keto-, in Shapiro reactions, **1**, 634
Amides, β-keto-
dehydrogenation of, **1**, 556
use in the synthesis of 1,3-azarsa compounds, **4**, 495
Amides, 2-keto-*N*-methoxy-*N*-methyl-, as starting materials, for diols, **2**, 59
Amides, α'-lithioallyl-, alkylation of, **1**, 148
Amides, α-magnesio-α-silyl-, synthesis of, **4**, 655
Amides, metallo-, use in the synthesis of amides, **5**, 262
Amides, methacrylic, as starting materials, for α-silyl enamines, **4**, 1014
Amides, *N*-methoxy-
reactions, with tetrachloromethane, **3**, 666
synthesis of, **5**, 327
Amides, *O*-methoxy-, use in the synthesis of *N*-alkoxyimidoyl halides, **5**, 824
Amides, α-methoxy-, reactions, with trimethylsilyl cyanide, **3**, 638
Amides, *N*-methoxy-*N*-methyl-
as acylating agents, **3**, 285
reactions, with alkynyllithiums, **1**, 1047
reactivity of, **1**, 1047
as starting materials
for aldehydes, **3**, 13
for ketones, **3**, 131
Amides, *N*-methylene-, reactions, with trialkyloxonium salts, **3**, 422
Amides, *N*-methyl-*N*-nitro-, synthesis of, **5**, 361
Amides, *N*-methylsilyl-, synthesis of, **5**, 377
Amides, *N*-methylthiomethyl-, synthesis of, **4**, 327
Amides, monoalkyl-, reactions, with dimethylmalonyl chlorides, **4**, 881
Amides, monosilyl-, as starting materials, for *N*-silyl-*N*-phosphoamides, **5**, 369
Amides, monosilylated, as equilibrium mixtures of *O*-and *N*-silylated compounds, **5**, 720
Amides, 2-nitro-, reactions, with Lawesson's reagent, **5**, 603
Amides, *N*-nitro-
from
amides, **5**, 361
imidoyl chlorides, **5**, 362
nitroamines, **5**, 361
silylamides, **5**, 362
Amides, *N*-nitroso-
from
amides, **5**, 360
imidoyl chlorides, **5**, 361
nitrosoamines, **5**, 360
thioamides, **5**, 361
reduction of, **5**, 353
as starting materials, for *N*-nitrosoketene hemiaminals, **4**, 896
synthesis of, **5**, 360, 711
thionation of, **5**, 571
Amides, oxygen-substituted, synthesis of, **5**, 322
Amides, oxyimidic
from

aliphatic nitro compounds, **5**, 838
amidines, **5**, 835
formamidoxime, **5**, 837
nitrile oxide oximes, **5**, 837
nitriles, **5**, 833
nitrolic acids, **5**, 838
nitrosolic acids, **5**, 838
oxyamide oximes, **5**, 838
oxyimidic esters, **5**, 835
oxyimidic halides, **5**, 834
nitrosative deamination of, **5**, 824
as starting materials, for oxyimidic chlorides, **5**, 825
synthesis of, **5**, 833, 838
See also Amidoximes
See also Oxamidines
Amides, perfluoroalkanesulfinic, reactions, with chlorosulfonyl isocyanate, **5**, 995
Amides, phenoxy-, as starting materials, for hydroxy anilides, **5**, 292
Amides, *N*-[(*S*)-1-phenylethyl]-*N*-chloromethyl-, synthesis of, **4**, 99
Amides, phosphenimidous, reactions, with 2-lithium-2-trimethylsilyl-1,3-dithiane, **6**, 684
Amides, phosphinic, synthesis of, **4**, 127
Amides, phosphinothioic, use in the synthesis of alkenes, **1**, 643
Amides, *N*-phosphoryl-, synthesis of, via phosphorylation of amides, **5**, 370
Amides, piperidine, synthesis of, **1**, 979
Amides, polyoxyimidic, synthesis of, **5**, 833
Amides, primary
t-butoxycarbonylation of, **5**, 329
from
acids, **5**, 274
acyl halides, **5**, 274
aldoximes, **5**, 298
anhydrides, **5**, 274
carboxylic acids, **5**, 275
esters, **5**, 274
nitriles, **5**, 275
Hofmann degradation of, **4**, 422
reactions
with 1-(hydroxymethyl)benzotriazole, **4**, 449
with phosphorus pentachloride, **5**, 373
as starting materials, for nitriles, **5**, 569
synthesis of
via oxidation of aldehydes, **5**, 276
via photochemical radical reaction, **5**, 276
via rearrangement reactions, **5**, 275
Amides, primary alkeno-, synthesis of, **5**, 295
Amides, primary aryl-
from, acids, **5**, 298
synthesis of
via hydrolysis of nitriles, **5**, 298
via rearrangement reactions, **5**, 298
Amides, *N*-propargyl-, azo-Claisen rearrangement of, **3**, 624
Amides, secondary
reactions
with carbonic difluoride, **6**, 438
with dibromotriphenylphosphorane, **3**, 557
with phosphorus pentachloride, **3**, 560
with phosphorus pentoxide, **3**, 556
with tetrakis(dimethylamino)titanium, **5**, 753
as starting materials
for imidate chloroformates, **6**, 429
for ketenimines, **3**, 556, 557
Amides, silyl-
reactions
with boron compounds, **5**, 381
with chlorophosphites, **5**, 369
for the silylation of alkanethiols, **2**, 263
as starting materials
for *N*-chloroimidates, **5**, 317
for *N*-nitroamides, **5**, 362
for *N*-phosphinoamides, **5**, 369
for sulfinamides, **5**, 341
use in the synthesis of enolates, **1**, 874
Amides, *N*-silyl-
from
aldehydes, **5**, 377
azides, **5**, 377
carbodiimides, **5**, 377
N-silylimines, **5**, 377
reactions
with iodonium sulfonates, **5**, 321
with silyloxiranes, **4**, 513
with trifluoromethyliodine(III) difluoride, **5**, 321
with trihaloboranes, **5**, 381
as starting materials
for *N*-acyltellurium imines, **5**, 349
for *N*-boroamides, **5**, 381
for *N*-iodo(III)amides, **5**, 321
for thioamides, **5**, 335
for thiolactams, **5**, 335
synthesis of, **5**, 374
via acylation of *N*-silylamines, **5**, 376
via lactamization of bis(trialkylsilyl)amides, **5**, 377
via rearrangement of amido organosilanes, **5**, 376
via silylation of amides, **5**, 374
via transsilylation, **5**, 376
Amides, α-silyl-, metallation of, **4**, 648
Amides, silylated, reactions, with ketenes, **4**, 886
Amides, *N*-silylated, as starting materials, for arylimines, **3**, 412
Amides, silylboro-, synthesis of, **5**, 383
Amides, *N*-sulfonylcarboxy-
reactions, with amines, **5**, 767
as starting materials, for *N*-imidoylsulfonamides, **5**, 767
Amides, tertiary
acylation of, **4**, 881
reactions
with aluminum chloride, **5**, 578
with diethylthiocarbamoyl chloride, **5**, 578
with hydroboranes, **5**, 381
as starting materials, for α-phosphoryl enamines, **4**, 1003
synthesis of, **5**, 292
thionation of, **5**, 575, 617
Amides, *N*-tetrachloroethyl-
reactions
with diethyl phenylphosphinite, **4**, 1002
with ethyl diphenylphosphinite, **4**, 1000
as starting materials
for *gem*-amido diphenylphosphinoylalkenes, **4**, 1000
for *gem*-amido phosphonioalkenes, **4**, 999
for *gem*-amido phosphonoylalkenes, **4**, 1002
for *gem*-amido phosphorylalkenes, **4**, 1006
Amides, β-thioalkoxy-, reactions, with sulfuryl chloride, **2**, 168
Amides, *N*-(thiocarbamoylthio)-, synthesis of, **5**, 335
Amides, thiocarboxy-
reactions, with hydrazine, **5**, 769
as starting materials, for amidrazones, **5**, 769
See also Thioamides
Amides, γ-thiocyanatophosphoryl-, synthesis of, **5**, 1112
Amides, thiol, condensation of, **4**, 325
Amides, β-thiol, as starting materials, for thiazines, **4**, 325
Amides, γ-thiol, reactions, with paraformaldehyde, **4**, 325
Amides, thiono-, synthesis of, reagents for, **5**, 566, 618
Amides, thionoalkano-, synthesis of, **5**, 610
Amides, thionoalkeno-, synthesis of, **5**, 613
Amides, thionoalkynylo-, synthesis of, **5**, 617
Amides, thionoaryl-, synthesis of, **5**, 618
Amides, thionohetaryl-, synthesis of, **5**, 618
Amides, thiophosphinic, synthesis of, **4**, 128
Amides, *N*(trialkylstannyl)-, synthesis of, **5**, 390
Amides, *N*(trialkyltin)-, synthesis of, **5**, 390

Amides, *N*-trichloroethyl-, as starting materials, for *gem*-amido phosphonioalkenes, **4**, 999
Amides, *N*-trichloroethylidene-, as starting materials, for *gem*-amido phosphorylalkenes, **4**, 1006
Amides, trihalomethyl-, synthesis of, **6**, 239
Amides, *N*-trihalomethyl-, synthesis of, **6**, 239
Amides, trimethylsilyl-, reactions, with arylselenium trichloride, **5**, 341
Amides, *N*-(trimethylsilyl)-, synthesis of, **5**, 379
Amides, *N*-trimethylsilylacyl-, reactions, with cyanogen chloride, **5**, 1082
Amides, α,β-unsaturated
  reactions, with organocuprates, **1**, 313
  synthesis of, **1**, 607
Amides, β,γ-unsaturated, thionation of, **5**, 570
Amides, γ,δ-unsaturated, synthesis of, **1**, 827
Amides, γ,δ-unsaturated carboxylic, synthesis of, **1**, 415
Amides, α,β-unsaturated tertiary, thionation of, **5**, 613
Amides, *N*-unsubstituted, synthesis of, methods, **5**, 259
Amides, *N*-unsubstituted aryl-, synthesis of, **5**, 298
Amides, ynamine-, synthesis of, **2**, 1044
α-Amides, β-hydroxylated, stereoselectivity, in intermolecular radical additions, **1**, 327
Amide salts, as starting materials, for ortho amides, **6**, 138
Amidc salts, alkyl-, as starting materials, for thiocarbamates, **6**, 494
Amide substitution reactions, examples, **5**, 385
Amide thioacetals
  from
    alkoxyiminium cations, **6**, 111
    chloroiminium salts, **6**, 111
    α-chlorothioacetals, **6**, 111
  reactions, with aryl isocyanates, **6**, 115
  synthesis of, **6**, 111
Amidine–antimony complexes, synthesis of, **5**, 776
Amidine–arsenic complexes, synthesis of, **5**, 776
Amidine hydrochloride, synthesis of, **5**, 748
Amidine hydrochloride, *N*-thioacyl-, hydrolysis of, **5**, 630
Amidine–lanthanide metal complexes, synthesis of, **5**, 782
Amidine metal comlexes, *N*-silylated, synthesis of, **5**, 780
Amidines
  acylation of, **5**, 758
  alkylation of, **1**, 573
  N-alkylation of, **5**, 743, 754
  bromination of, **5**, 763
  chlorination of, **5**, 762
  from
    acetals, **5**, 743
    amides, **5**, 743
    benzonitriles, **5**, 756
    formanilide, **5**, 575
    haloalkynes, **4**, 976
    4-methylbenzonitrile, **5**, 755
    1-naphthonitriles, **5**, 758
    nitriles, **5**, 742
    ortho esters, **5**, 743
    thioamides, **5**, 743
    thioesters, **5**, 743
    trimethylsilylmethyl isocyanide, **4**, 518
    vinylidene dihalides, **4**, 970
    ynamines, **4**, 977
  halogenation of, **5**, 866
  lithiation of, **4**, 511
  monoalkylation of, **5**, 754
  mono *N*-halogenation of, **5**, 816
  reactions
    with alkoxy hypohalides, **5**, 816
    with alkylsulfonyl chlorides, **5**, 767
    with arylsulfonyl chlorides, **5**, 767
    with *s*-butyllithium, **4**, 107
    with chlorobis(diethoxyphosphonyl)methyl isocyanate, **6**, 367
    with cyanamide, **5**, 1121
    with dialkylthioborinates, **5**, 778
    with hexaethyl phosphite, **5**, 863
    with hydrogen sulfide, **5**, 611
    with isothiocyanates, **5**, 760
    with sulfonium salts, **5**, 766
    with thiochloroformates, **5**, 760
    with trialkylboranes, **5**, 778
  reduction of, **1**, 45
    for the synthesis of aminals, **4**, 413
  silylation of, **4**, 511
  as starting materials
    for *N*-chloroamidines, **5**, 816
    for diazirines, **6**, 280
    for *N*-haloamidines, **5**, 815
    for hydrazonamides, **5**, 850
    for imidic acid esters, **5**, 694
    for iminocyanamides, **5**, 1125
    for isocyanates, **5**, 987
    for ketene aminals, **4**, 979
    for oxyimidic amides, **5**, 835
  sulfenation of, **5**, 765
  sulfonation of, **5**, 767
  synthesis of, **1**, 1044; **5**, **741–782**
    general methods, **5**, 742
    reviews, **5**, 742
    via alkylation, **5**, 751
  as tautomers of ketene aminals, **4**, 969
  thioacylation of, **5**, 760
  thiophosphorylation of, **5**, 774
  use in the synthesis of alkanoic acids, **5**, 31
Amidines, acyl-
  as starting materials, for *N*-acylaminals, **4**, 420
  synthesis of, **5**, 758
Amidines, *N*-acyl-, synthesis of, **5**, 758
Amidines, N^2-acyl-N^1,N^1-dimethyl-, reactions, with hydrogen sulfide, **5**, 631
Amidines, aliphatic
  from
    amides, **5**, 751
    nitriles, **5**, 748
    ortho esters, **5**, 753
    thioamides, **5**, 753
    thioimidic esters, **5**, 753
  synthesis of, **5**, 748
Amidines, *N*-alkyl-
  phosgenation of, **5**, 967
  reactions, with phosgene, **5**, 987
  sulfenation of, **5**, 766
Amidines, alkynyl-, synthesis of, **5**, 755
Amidines, amino-
  acetylation of, **5**, 758
  benzoylation of, **5**, 758
Amidines, aromatic
  from
    amides, **5**, 757
    nitriles, **5**, 755
    thioamides, **5**, 757
    thioimidic esters, **5**, 757
  synthesis of, **5**, 755
Amidines, aryl-
  from
    aryl cyanides, **5**, 755
    1-naphthonitrile, **5**, 755
Amidines, *N*-aryl-
  alkylation of, **5**, 744, 754
  reactions, with phosgene, **5**, 987
  as starting materials, for *N*-chloro-*N'*-arylamidines, **5**, 816
  synthesis of, **5**, 742, 751
Amidines, N^2-[bis(dimethylamino)phosphinoyl]-, synthesis of, **5**, 775
Amidines, *N*-boryl-, synthesis of, **5**, 778
Amidines, *N*-bromo-, synthesis of, **5**, 763
Amidines, *N*-chloro-

from
amidines, **5**, 816
N-chlorocyanoformimide chloride, **5**, 816
reactions, with dialkyl sulfides, **5**, 766
synthesis of, **5**, 762
Amidines, *N*-chloroaryl-, reactions, with triphenylphosphine, **5**, 861
Amidines, *N*-chloro-*N'*-aryl-, from, *N*-arylamidines, **5**, 816
Amidines, *N*-cyano-, synthesis of, **5**, 1121
Amidines, *N*-(dialkylboryl)-, synthesis of, **5**, 778
Amidines, *N*-(diarylboryl)-, synthesis of, **5**, 778
Amidines, α-diazo, synthesis of, **3**, 477
Amidines, *N*,*N'*-dibenzoyl-, synthesis of, **5**, 759
Amidines, *N*,*N'*-diboryl-, synthesis of, **5**, 779
Amidines, *N*,*N'*-dichloro-, from, *N*,*N'*-dichlorooxalimide chloride, **5**, 816
Amidines, N^1,N^1-diethyl-N^2-tetraethyldiamidophosphito-, synthesis of, **5**, 863
Amidines, disilyl-, synthesis of, **5**, 777
Amidines, *N*,*N*-disubstituted, reactions, with phenylhydrazine, **5**, 851
Amidines, *N*-fluoro-
from, *N*-fluoroimines, **5**, 817
safety precautions, **5**, 761
synthesis of, **5**, 761, 762
Amidines, *N*-halo-
from
amidines, **5**, 815
N-haloimidic halides, **5**, 816
synthesis of, **5**, 761, 815
Amidines, *N*-iodo-, synthesis of, **5**, 763
Amidines, monobenzoyl-, synthesis of, **5**, 759
Amidines, monochloro-, synthesis of, **5**, 762
Amidines, monosilyl-, synthesis of, **5**, 777
Amidines, perfluoro-
from
perfluoroamidines, **5**, 818
tris(difluoroamino)methyls, **5**, 818
reactions, with methanol, **5**, 814
as starting materials, for perfluoroamidines, **5**, 818
synthesis of, **5**, 817
Amidines, *N*,*N*-perfluoro-, as starting materials, for *N*-haloimidic esters, **5**, 813
Amidines, *N*-phosphoryl-, synthesis of, **5**, 774
Amidines, primary, synthesis of, **5**, 748
Amidines, selenoacyl-, synthesis of, **5**, 761
Amidines, silyl-, reactions, with boron halogen compounds, **5**, 778
Amidines, *N*-silyl-, synthesis of, **5**, 777
Amidines, *N*-substituted, synthesis of, **5**, **741–782**
Amidines, *N*-thioacyl-, synthesis of, **5**, 760
Amidines, thiobenzoyl-, synthesis of, **5**, 760
Amidines, *N*-thionoacyl-, synthesis of, **5**, 633
Amidines, *N*-(trialkylboryl)-, synthesis of, **5**, 778
Amidines, *N*,*N*,*N'*-trichloro-, synthesis of, **5**, 763
Amidines, 2,2,2-trichloroacetic-, reactions, with trialkyl phosphites, **6**, 58
Amidines, trifluoro-, reactions, with radicals, **5**, 818
Amidines, *N*,*N*,*N'*-trifluoro-, synthesis of, **5**, 815
Amidines, trimethylsilyl-, reactions, with diisopropylphosphorus chloride, **5**, 775
Amidines, trisilyl-, synthesis of, **5**, 778
Amidines, ynamine, synthesis of, **2**, 1046
(*Z*)-Amidines, synthesis of, **5**, 761
Amidine–tantalum complexes, synthesis of, **5**, 781
Amidine–transition metal complexes, synthesis of, **5**, 780
Amidinium chlorides, as starting materials, for sulfonylguanidines, **6**, 651
Amidinium salts
reactions
with diethyl phosphite, **6**, 156
with diphenylarsane, **6**, 161
reduction of, for the synthesis of aminals, **4**, 413
as starting materials
for *N*-alkenyliminocarbonyls, **6**, 644
for *N*-alkyliminocarbonyls, **6**, 642
for *N*-aryliminocarbonyls, **6**, 644
for ortho amides, **6**, 138
for *N*-unsubstituted iminocarbonyls, **6**, 641
synthesis of, **5**, 749
Amidinium salts, chloro-, as starting materials, for *N*,*N*-dialkyliminium compounds, **6**, 660
Amidinium salts, phosphorylated, synthesis of, **6**, 661
Amidinophosphonates, synthesis of, **6**, 657
Amidophosphonates, alkylidene-, synthesis of, **6**, 657
β-Amidoselenenylation, of alkenes, **2**, 284
Amidoxime, phenyl-
acetylation of, **5**, 220
acylation of, **5**, 220
Amidoximes
acylation of, **5**, 220
See also Amides, oxyimidic
See also Hydroxylamines, *N*-imidoyl-
Amidoximes, acyl-, synthesis of, **5**, 219
Amidoximes, aryl-, reactions, with antimony pentachloride, **5**, 776
Amidoximes, cyclic, ring-expansion of, **6**, 510
Amidrazone, N^3-benzoyl-, synthesis of, **5**, 774
Amidrazones
N-acylation of, **5**, 773
N-alkylation of, **5**, 772
from
hydrazanoyl halides, **5**, 771
nitriles, **5**, 768
thiocarboxyamides, **5**, 769
nomenclature, **5**, 768
oxidation of, **6**, 660
reactions, with aldehydes, **5**, 772
reduction of, **5**, 855
synthesis of, **5**, 768; **6**, 659
See also Hydrazidines
Amidrazones, *N*-acyl-, synthesis of, **5**, 773
Amidrazones, N^1-acyl-, synthesis of, **5**, 773
Amidrazones, N^3-acyl-, synthesis of, **5**, 774
Amidrazones, N^1-formyl-, synthesis of, **5**, 773
Amidrazones, *N*-substituted
from
amines, **5**, 771
ammonia, **5**, 771
hydrazines, **5**, 769
Amidyl radicals, reactions of, **1**, 950
Aminal, silyl, reactions, with silylated trinitromethane, **6**, 147
Aminals
from
acetophenones, **4**, 406
aldehydes, **4**, 405
amines, **4**, 405
cyclopropanone, **4**, 406
diamines, **4**, 407
dihalo compounds, **4**, 411
dihalomethylbenzenes, **4**, 411
fluorocarbons, **4**, 411
formamidinium salts, **4**, 414
geminal dihalo compounds, **4**, 410
imidazolines, **4**, 414
imines, **4**, 411
iminium salts, **4**, 411
ketones, **4**, 405
selenoaldehydes, **4**, 415
terephthalaldehyde, **4**, 406
reactions
with organolithiums, **3**, 42
with sulfur, **5**, 599
solvolysis of, **4**, 309
stability of, **4**, 404

as starting materials
for α-amidoalkyl sulfones, **4**, 332
for β-amino ketones, **1**, 255
for formamidines, **5**, 746
for ketene aminals, **4**, 975
for stannanes, **4**, 533
synthesis of, **4**, 404
use in the synthesis of *S,N*-acetals, **4**, 322
use in the synthesis of trisaminomethylphosphines, **4**, 459
Aminals, acyclic, synthesis of, **4**, 405
Aminals, *N*-acyl
from
acylamidines, **4**, 420
α-aminonitriles, **4**, 418
synthesis of, **4**, 416
Aminals, β-acylketene, synthesis of, **4**, 979
Aminals, *N*-acyl-*N'*-sulfonyl alkylidene, synthesis of, **4**, 427
Aminals, *gem*-arylazo nitroketene, synthesis of, **4**, 995
Aminals, asymmetrical, synthesis of, **4**, 405
Aminals, *N*-benzoyl, reactions, with 3-methylthiophenol, **4**, 323
Aminals, cyanoketene
from, cyano ynamines, **4**, 976
synthesis of, **4**, 972
Aminals, cyclic, synthesis of, **4**, 407
Aminals, β,β-diacylketene, reactions, with methanol, **4**, 979
Aminals, diaza heterocyclic ketene, synthesis of, **4**, 973
Aminals, dicyanoketene
reactions, with trichloromethyldimethylamine, **4**, 987
synthesis of, **4**, 977
Aminals, ketene
N-acylation of, **4**, 975, 980
deacylation of, **4**, 979
from
alkoxy tris(dimethylamino)methanes, **4**, 973
amide acetals, **4**, 975
amides, **4**, 971, 972
amidines, **4**, 979
aminals, **4**, 975
β-amino dichloroalkanes, **4**, 979
aminopyridines, **4**, 979
2-amino-2*H*-pyrido[1,2-*a*]pyrimidines, **4**, 979
anhydrides, **4**, 972
azines, **4**, 979
benzodiazepines, **4**, 979
1,2-benzodithiole-3-thione, **4**, 979
bis(dialkylamino)malononitriles, **4**, 975
bis(formamidinium) ethers, **4**, 975
bromoalkynes, **4**, 976
carbodiimides, **4**, 979
carboxylic acids, **4**, 972
chloroalkynes, **4**, 976
chloro enamines, **4**, 971
cyanamides, **4**, 975
diaziridines, **4**, 979
dichlorocarbenes, **4**, 979
dicyanoacetamides, **4**, 977
dicyanoacetates, **4**, 977
gem-dicyanoalkenes, **4**, 979
dicyanothioacetates, **4**, 977
1,3-dimethyl-2-ethoxy-4,5-dihydroimidazolium tetrafluoroborate, **4**, 975
1,2-dithiolimines, **4**, 979
enamines, **4**, 979
esters, **4**, 972
ethoxyalkynes, **4**, 979
haloalkynes, **4**, 976
imidazoles, **4**, 973
imidazolines, **4**, 979
imino esters, **4**, 972
5-indolones, **4**, 979
isothioureas, **4**, 973
isoxazoles, **4**, 979
isoxazolium salts, **4**, 979
ketene acetals, **4**, 971
ketene aminals, **4**, 979
ketene dithioacetals, **4**, 969
nitriles, **4**, 979
ortho esters, **4**, 975, 979
oxazoles, **4**, 979
pyrimidines, **4**, 978
tetraamines, **4**, 979
tetrakis(dimethylamino)methane, **4**, 975
tetramethylchloroformamidinium chloride, **4**, 974
N,N,N',N'-tetramethylethoxyformamidinium fluoroborate, **4**, 975
gem-thioalkyl nitromethylalkenes, **4**, 979
thioesters, **4**, 979
thioformamidinium salts, **4**, 973
thioureas, **4**, 973
trichlorocyclopropenes, **4**, 979
tricyanomethanide, **4**, 977
trihalomethanes, **4**, 979
urea acetals, **4**, 974
vinylidene azides, **4**, 979
vinylidene diazoles, **4**, 979
vinylidene dihalides, **4**, 970
ynamines, **4**, 976
oxidation of, **1**, 558
reactions
with arenediazonium ions, **3**, 260
with dibromotriphenylphosphorane, **4**, 990
stability of, **4**, 969
as starting materials
for α-amido enamines, **4**, 980
for α-aminoalkylideneamino enamines, **4**, 987
for *gem*-diureidoalkenes, **4**, 982
for ketene aminals, **4**, 979
for ketene hemiaminals, **4**, 888
for α-phosphimino enamines, **4**, 990
for α-phosphoryl enamines, **4**, 1003
for α-ureido enamines, **4**, 981
β substituents, introduction of, **4**, 979
synthesis of, **4**, 969, 980
Aminals, methylene
from, formaldehyde, **4**, 410
synthesis of, **4**, 405
Aminals, monoacyl
from, β-lactams, **4**, 419
synthesis of, **4**, 417
Aminals, morpholine, synthesis of, **4**, 406
Aminals, nitroketene, synthesis of, **4**, 974
Aminals, *N*-sulfonyl
from, arenesulfonamides, **4**, 423
synthesis of, **4**, 423
Amination
of organometallics, **2**, 305
reagents for, **2**, 298
for the synthesis of arylamines, **2**, 752
Amination, electrophilic, for the synthesis of *gem*-chloro carbamates, **4**, 794
Amination, reductive
of carbonyls, **2**, 310
for the synthesis of perhydroazepines, **2**, 327
O-Amination, of phenol, **2**, 691
S-Amination, of thioamides, **5**, 736
Amine, allyl-, reactions, with metalloalkoxycarbenes, **5**, 798
Amine, allyl(2-chloroethyl)-, reactions, with monopotassium phosphide, **2**, 427
Amine, 2-aminobenzyl-, reactions, with carbon disulfide, **6**, 575
Amine, benzhydryl-, reactions, with 1-bromo-3-chloropropane, **2**, 322
Amine, benzylnitro-, synthesis of, **2**, 386
Amine, *N*-benzyloxy-
reactions, with 1-aryl-1-alkylethyl alcohols, **2**, 351

use in the synthesis of *N*-alkoxy primary amines, **2**, 351
Amine, benzyloxy-(1-aryl-1-alkylethyl)-, synthesis of, **2**, 351
Amine, *O*-benzyloxyl-
reactions, with alkyl halides, **2**, 342
synthesis of, **2**, 351
Amine, bis(benzotriazol-1-ylmethyl)-, reactions, with Grignard reagents, **1**, 260
Amine, bis(bistrifluoromethylarsino)-, synthesis of, **2**, 503
Amine, bis(bromocarbonyl)methyl-, synthesis of, **6**, 453
Amine, bis(di-*t*-butylarsino)-, synthesis of, **2**, 502
Amine, bis(diphenylboryl)-, synthesis of, **2**, 811
Amine, bis(1-hydroxycyclopropyl)-, synthesis of, **4**, 299
Amine, bis(methanethio)trimethylsilylbenzylidenemethyl-, synthesis of, **6**, 326
Amine, bis(trimethylsilyl)-, reactions, with hydrazine sulfate, **6**, 626
Amine, *N*-bromo-*N*,*N*-bis(trifluoromethyl)-, synthesis of, **6**, 240
Amine, *N*-bromo-*t*-butyl-, for the *o*-bromination of phenols, **2**, 626
Amine, *t*-butyl-, reactions, with sulfur dichloride, **2**, 361
Amine, *t*-butyldifluoro-, reactions, with boron trifluoride, **2**, 762
Amine, chloro-, reactions, with trialkylarsines, **2**, 504
Amine, *N*-chloro-*N*,*N*-bis(trifluoromethyl)-, synthesis of, **6**, 240
Amine, *N*-chlorodialkyl-
for the *o*-chlorination of phenol, **2**, 620
reactions, with phosphonous acid methyl esters, **2**, 469
Amine, *N*-chlorodiethyl-, reactions, with lithium phenylalkynide, **4**, 976
Amine, 1-chloro-2-methyl-*N*,*N*-tetramethylene propenyl-, use in the synthesis of allenes, **1**, 970
Amine, *N*-chloro-*N*-perfluoropropyl-(2-chlorotetrafluoroethyl)-, synthesis of, **5**, 659
Amine, *N*-chloro-*N*-perfluoropropyl perfluoroethyl-, synthesis of, **5**, 659
Amine, *β*-cyanoethyl-, hydrolysis of, **5**, 85
Amine, cyclohexyl-, reactions, with phosphorus acid, **4**, 470
Amine, dibenzyl-, reactions, with 2,3-dihydro-6-methoxy-2-thioxo-4*H*-1,3,5-thiadiazin-4-one, **6**, 559
Amine, *N*,*N*-dichloroperhaloalkyl-, thermolysis of, **5**, 807
Amine, difluoro-
addition to aldehydes and ketones, **4**, 439
reactions
with aziridines, **1**, 636
with 1,1-dichloro-1-alkenes, **4**, 439
with methylvinyl ketone, **4**, 439
with primary amines, **1**, 35
with tetramethyl-1,3-cyclobutanedione, **5**, 314
Amine, diisopropyl-, reactions, with trifluoromethylphosphane, **3**, 689
Amine, dimethoxy-, reactions, with cyclohexanone, **5**, 331
Amine, dimethyl-
reactions
with diphenyl carbonate, **6**, 502
with tetrakis(*t*-butylisocyanide)rhodium salt, **6**, 664
Amine, *N*,*N*-dimethyl-, chlorination of, **6**, 238
Amine, dimethylethyl-, reactions, with silaneimine, **6**, 163
Amine, dimethylfluoro-, reactions, with hydrogen chloride, **2**, 762
Amine, *N*,*N*-dimethyl-2-phenylethynyl-, synthesis of, **2**, 1051
Amine, dimethyl(tribromomethyl)-, synthesis of, **6**, 238
Amine, *N*,*N*-dimethyltrimethylsilyl-, reactions, with diphenyldithiophosphonic anhydride, **2**, 862
Amine, diphenylmethylene-, reactions, with phenyl isothiocyanate, **6**, 571
Amine, diphenylvinyl-, reactions, with *i*-amyl nitrite, **2**, 800
Amine, 2,2-diphenylvinyl-, reactions, with nitrosyl chloride, **1**, 1118
Amine, *N*-fluoro-*N*,*N*-bis(trifluoromethyl)-, synthesis of, **6**, 240
Amine, *N*-(formylaminomethyl)-, from, *N*-(formylaminomethyl)benzotriazole, **4**, 446
Amine, *N*-formylethanol-, reactions, with benzenesulfonyl chloride, **3**, 714, 715
Amine, *N*-iodobis(ditrifluoromethyl)-, reactions, with hexafluoropropene, **6**, 54
Amine, *N*-iodo-*N*,*N*-bis(trifluoromethyl)-, synthesis of, **6**, 240
Amine, methoxy-
reactions
with acetic anhydride, **5**, 332
with alkyllithiums, **2**, 305
as starting material, for primary amines, **2**, 305
Amine, *N*-methylene-t-butyl-, addition of amides to, **4**, 418
Amine, *N*-methyl(*o*-hydroxybenzylidene)-, reactions, with dichlorophenylphosphine, **4**, 478
Amine, *α*-naphthyl-, reactions, with nitrosyl sulfuric acid, **2**, 780
Amine, *N*-nitroso(methyl)vinyl-, as starting material, for *N*-nitroso(1-methoxy-2-chloroethyl) methylamine, **4**, 896
Amine, (*S*)-2-octyl-, synthesis of, **2**, 299
Amine, *N*-tributylstannyldiphenylmethylene-, reactions, with methyl isothiocyanate, **6**, 571
Amine, trichloro-, reactions, with alkenes, **2**, 338
Amine, trichloromethyl-, synthesis of, **6**, 238
Amine, *N*-(2,4,6-trichlorophenyl)-*N*,*N*-bis(trichloromethyl)-, synthesis of, **6**, 238
Amine, triethylgermylphenyl-, reactions, with ketenes, **5**, 389
Amine, trifluoromethyl-, synthesis of, **6**, 238
Amine, trimesitylgermyl-, reactions, with acid chlorides, **5**, 389
Amine, trimethyl-, electrochemical fluorination of, **6**, 214
Amine, *N*-trimethylsilyl-, reactions, with phosgene, **6**, 443
Amine, tris(2-aminoethyl)-, reactions, with 5-hydroxy-1-naphthalenesulfonic acid, **2**, 309
Amine, tris(chloromethyl)-
from, hexamethylenetetramine, **4**, 307
reactions, with hydrazines, **4**, 437
silylation of, **4**, 511
Amine, tris(phenylthiomethyl)-, synthesis of, **4**, 317
Amine, tris(trifluoromethyl)-, synthesis of, **6**, 214, 239
Amine, *N*,*N*,*O*-tris(trimethylsilyl)-, use in the synthesis of hydroxamic acids, **5**, 323
Amine, vinyldipropargyl-, reactions, with triethylsilane, **2**, 920
Amine–borane complexes, reactions, with cyclodextrins, **2**, 57
Amine functions, *α*-metallated, synthesis of, **4**, 526
Amine hydrobromides, alkyl-, reactions, with silver tricyanomethanide, **4**, 977
Amine hydrochloride, reactions, with phosgene, **5**, 965
Amine hydrochlorides, reactions, with *N*-dichloromethylenesulfonamides, **5**, 1071
Amine hydrochlorides, allylfurfuryl-, as starting materials, for 1,2-dihydroisoindolines, **2**, 329
Amine oxide, trifluoro-, for the fluorination of amines, **2**, 334
Amine oxides
elimination of, **1**, 623
from, 2-amino-3-phenylbutane, **1**, 624
for the oxidation of benzyl halides, **3**, 89
[2,3] sigmatropic rearrangement of, **1**, 839
as starting materials, for hydroxylamines, **1**, 839
Amine oxides, tertiary, reactions, with trialkylboranes, **2**, 112
Amine *N*-oxides
Cope rearrangement of, **2**, 346
reactions, with sulfur dioxide, **2**, 357
rearrangement of, **2**, 355
synthesis of, **2**, 347
via retro-Cope rearrangements, **2**, 349
via tertiary amines, **2**, 347
Amine *N*-oxides, salts, synthesis of, **2**, 350

Amine *N*-oxides, *N*,*N*,*N*-trialkyl-, alkylation of, **2**, 350
Amine *N*-oxides, *N*,*N*,*N*-trialkyl-, salts, **2**, 356
Amine protection, use of *t*-butyl fluoroformate for, **6**, 422
Amines
activation of, **2**, 88
acylation of, **5**, 260, 359
N-arylthiomethylation of, **4**, 316
asymmetric synthesis of, **2**, 330
autoxidation of, **2**, 348
carbolithiation of, **2**, 555
chlorination of, **2**, 337; **6**, 607
cleavage of, **2**, 553
α-cyanation of, **2**, 305
cyclopalladation of, **4**, 720
dealkylation of, **2**, 302
dehydrogenation of, **3**, 620
demethylation of, **2**, 302
diazotization of, **1**, 1032; **2**, 802; **3**, 462, 468
elimination of, **1**, 635
via deamination in Mannich bases, **1**, 635
enantioselective synthesis of, **2**, 345
from
acetals, **2**, 310
aldehydes, **2**, 310
aldimines, **1**, 1059
alkanes, **2**, 306
alkenes, **2**, 306
alkyl halides, **2**, 307
allylic acetates, **2**, 308
allylic halides, **2**, 308
amides, **2**, 319
amines, **2**, 300
azides, **2**, 317
aziridines, **2**, 319
carboxylic acids, **2**, 319
epoxides, **2**, 318
heterocycles, **2**, 320
hydrazines, **2**, 315
hydroxyls, **2**, 308
imines, **2**, 311; **3**, 404
iminium salts, **2**, 316
ketals, **2**, 310
ketones, **2**, 310
nitriles, **2**, 316
nitro compounds, **2**, 318
oximes, **2**, 314
α,β-unsaturated carbonyls, **2**, 314
halogenation of, **2**, 334
N-halo(trihalomethyl)-, synthesis of, **6**, 240
intramolecular alkylation of, **2**, 300
iodination of, **2**, 339
nitrosation of, **2**, 779
oxidation of, **1**, 624; **2**, 782; **3**, 407, 620; **5**, 324
for imines, **3**, 17
oxidative acylation of, **5**, 213
protection of, **2**, 543
reactions
with acetals, **2**, 749
with acetonitrile, **5**, 750
with acid compounds, **5**, 295
with *N*-acylcarbamidic acid esters, **6**, 506
with *N*-acyl oxaziranes, **5**, 353
with acylphosphorus compounds, **5**, 403
with aldehydes, **2**, 739; **4**, 294, 405
with alkali metal cyanates, **6**, 504
with alkanesulfonyl chlorides, **4**, 331
with alkoxycarbene complexes, **5**, 953
with alkoxycarbonyl isothiocyanates, **6**, 572
with 2-alkoxy 1,1-disulfonylalkenes, **4**, 866
with β-alkoxy-α,β-unsaturated ketones, **3**, 264
with *N*-alkylanilines, **4**, 316
with alkylantimony halides, **2**, 502
with alkylarsenic halides, **2**, 502
with alkylbismuth halides, **2**, 502
with *O*-alkyl chlorothioformates, **6**, 556
with alkylfluoro sulfates, **2**, 366
with alkyl nitrates, **2**, 798
with alkyl pyruvates, **2**, 745
with alkylsulfinyl chlorides, **2**, 362
with α-(alkylthio)-aldehydes, **3**, 418
with alkylthiosulfines, **5**, 587
with alkynic ethers, **4**, 884
with alkynic sulfones, **2**, 742
with alkynyl alkyl sulfides, **5**, 590
with alkynyl ketones, **3**, 267
with alkynyl thioethers, **5**, 611
with amide acetals, **5**, 746
with aminoboranes, **2**, 358
with ammonia, **2**, 298
with arenesulfenates, **2**, 358
with aromatic diazonium salts, **2**, 776
with aryl alkyl ketones, **5**, 599
with *O*-aryl chlorothioformates, **6**, 556
with arylfluoro sulfates, **2**, 366
with *N*-arylimidoyl dichlorides, **5**, 1070
with aryl isonitriles, **5**, 747
with arylphosphinic halides, **2**, 860
with arylphosphoryl chlorides, **2**, 861
with arylselenenyl halides, **2**, 369
with aryl thioketones, **3**, 361
with benzaldehyde, **2**, 335
with benzonitrile, **5**, 750
with benzotriazole, **4**, 406
with benzoyl isocyanate, **5**, 721
with bis(alkoxycarbonyl)diazines, **6**, 506
with bis(1,1-alkylthio)alkenes, **4**, 899
with bis(chloromethyl) ether, **4**, 411
with bis(dialkoxyphosphoryl) disulfides, **2**, 361
with bis-*N*-(hydroxyalkyl)ureas, **4**, 416
with 2,2-bis[methylsulfonyl]vinyl ethyl ether, **2**, 743
with bis(trichloromethyl)pentathiodiperoxy carbonate, **5**, 1029
with boron trichloride, **2**, 533
with 2-(1-bromoalkyl)oxiranes, **2**, 318
with *t*-butyl isocyanide, **6**, 644
with carbamic acid chlorides, **6**, 501
with carbamic acid esters, **6**, 501
with carbon dioxide, **5**, 975
with carbon diselenide, **5**, 1055; **6**, 594
with carbon disulfide, **5**, 1023; **6**, 560
with carbonic acid esters, **6**, 502
with carbonimidic dichlorides, **6**, 619, 645
with carbonimidic difluorides, **6**, 615
with carbon monoxide, **5**, 974
with carbon oxide sulfide, **6**, 503
with carbonyl chloride, **5**, 965
with *N*,*N*'-carbonyldiimidazole, **5**, 972
with carbonyls, **3**, 407, 416
with carbyne complexes, **5**, 953
with catechol sulfate, **2**, 366
with α-chloroalkanesulfenyl chlorides, **4**, 72
with α-chloroalkyl alkyl carbonates, **4**, 60
with chloroalkylbenzotriazoles, **4**, 445
with chloroalkynes, **4**, 976
with chloroformamidinium salts, **6**, 641, 646
with 2-chloro-1-methoxy-3-nitropyridinium perchlorate, **5**, 674
with 2-(chloromethyl)benzimidazole, **4**, 411
with chloromethylphosphonic acid, **4**, 473
with chlorophosphines, **4**, 116
with chlorosilanes, **2**, 811
with chlorosulfonic acid, **2**, 366
with chlorosulfonyl azide, **2**, 368
with α-chloro sulfoxides, **2**, 322
with *S*-chlorothiocarbonyl chloride, **6**, 501
with 4-chloro-5-tosyl-1,2,3-dithiazole, **6**, 651

with 1-chlorovinyliminophosphorane, **5**, 861
with cyanamidinium hexachloroantimonates, **6**, 643
with cyanates, **6**, 632
with cyanides, **6**, 505
with 2-cyanobenzaldehydes, **4**, 419
with *N*-cyano-*N'*,*S*-dimethylisothiourea, **6**, 647
with cyanogen bromide, **4**, 413; **5**, 1125; **6**, 632
with cyanogen chloride, **6**, 632
with cyanoguanidines, **5**, 1121
with cyclohexane-1,3-diones, **3**, 263
with *N*,*N*-dialkylaminosilanes, **2**, 358
with dialkylhalophosphines, **2**, 466
with dialkyl phosphites, **5**, 975
with dialkyl phosphorochloridates, **5**, 975
with diarylsulfonyl peroxides, **2**, 357
with dibenzyl peroxide, **2**, 340, 351
with diborane, **2**, 418
with dichlorine heptoxide, **2**, 339
with dichlorocarbene, **3**, 695
with 2,2-(dichloro)dihydro-4,6-pyrimidinediones, **6**, 360
with 2,2-dichloro-4,5-imidazolidinediones, **6**, 360
with dichloroiminium salts, **6**, 619
with dichlorophosphoranes, **2**, 468
with 4,6-dichloro-5-substituted pyrimidines, **4**, 978
with dichlorovinyl ketones, **4**, 970
with dihalogens, **2**, 334
with 1,3-diketones, **2**, 744
with α-diketones, **3**, 260
with dimethyl chlorothiophosphate, **5**, 614
with dimethylcyanodithioimidocarbonate, **6**, 647
with dimethylformamide, **5**, 745
with dimethyl sulfate, **2**, 760
with dinitrogen tetroxide, **2**, 88
with 1,2-diones, **3**, 418
with diphenyl carbonate, **5**, 976
with diphenylphosphinyl peroxide, **2**, 344, 357
with 4,6-diphenylthieno[3,4-*d*]-1,3-dioxol-2-one 5,5-dioxide, **6**, 502
with diphosgene, **5**, 965, 972
with di-2-pyridyl thiocarbonate, **5**, 1030
with disulfides, **2**, 358
with 1,3-dithiane-2-thione, **6**, 305
with 1,2,4-dithiazolium cations, **6**, 114
with dithioacids, **5**, 614
with dithionomalonic esters, **4**, 895
with enamines, **4**, 971
with 1,3-enynes, **3**, 413
with 1,2-epoxy-3-ols, **2**, 318
with 1,2-epoxypenten-4-ol, **2**, 318
with ester aminals, **5**, 747
with ethoxy(thiocarbonyl) isocyanate, **6**, 559
with ethyl chlorothioformate, **6**, 492
with ethyl isocyanoacetate, **3**, 706
with ethyl orthoacetate, **5**, 753
with ethyl thionoformate, **5**, 586
with ethynyl ketones, **6**, 565
with fluorinating agents, **2**, 334
with 1-fluoro enol ethers, **4**, 894
with fluoroiminonitriles, **5**, 761
with formaldehyde, **1**, 255; **4**, 295
with formamide acetals, **5**, 743, 746
with formyl phosphonates, **3**, 49
with geminal dihalo compounds, **4**, 410
with germylmethyl halides, **4**, 519
with glyoxal, **4**, 298
with guanidines, **2**, 749
with α-haloazoalkanes, **4**, 444
with halomethylsilanes, **4**, 506
with hexachloro-1,3,5-trithiane, **6**, 254
with hydrazonate esters, **5**, 772
with 2-hydroxyphenyl sulfamate esters, **2**, 367
with hypochlorite ions, **2**, 761
with hypochlorous acid, **2**, 761
with imidic esters, **5**, 748
with imidic ester salts, **5**, 742
with imidoyl chlorides, **5**, 743
with *S*-imidoyl sulfenates, **5**, 736
with imines, **4**, 411
with iminium salts, **4**, 412; **6**, 156
with iodoalkynes, **4**, 976
with (*E*)-2-iodo-1-tosyl-1-alkenes, **2**, 743
with isocyanates, **6**, 504
with isonitriles, **5**, 747
with isothiocyanates, **6**, 569
with ketene acetals, **4**, 887, 972
with ketene *N*,*O*-acetals, **4**, 972
with ketene dithioacetals, **4**, 969
with ketenimines, **5**, 753
with ketones, **4**, 294, 405
with β-keto sulfones, **2**, 743
with manganese pentacarbonyls, **5**, 447
with 2-mercaptobenzothiazole, **2**, 131, 358
with mercury(II) chloride, **5**, 1055
with mesityl 2-pyridyl ketone, **2**, 302
with 1-methoxycarbonyl-3-methyl-(η^3-allyl)-tetracarbonyliron tetrafluoroborate, **2**, 331
with methyl cyanodithioformate, **5**, 587
with 2-methylcyclopentane-1,3-dione, **3**, 263
with methyl dithiophenylacetate, **5**, 586
with methyl isothiocyanate, **6**, 647
with *S*-methylisothiouronium salts, **6**, 641
with 2-methyl-1-naphthonitrile, **5**, 756
with *S*-methyl-*N*-nitroisothioureas, **6**, 654
with methyl selenonic esters, **2**, 370
with mineral acids, **2**, 759
with nitrilamines, **6**, 659
with nitriles, **2**, 749
with *N*-nitrosocarbamidic acid esters, **6**, 506
with nitrosyl chloride, **3**, 462
with nitrous acid, **2**, 798; **3**, 462
with organic acids, **2**, 759
with organolead halides, **2**, 816
with organotin halides, **2**, 815
with 1,3,4-oxadiazolium salts, **6**, 109
with oxalyl chloride, **5**, 973
with palladiumalkene complexes, **5**, 464
with pentafluoroguanidine, **6**, 650
with pentafluoroguanidines, **5**, 817
with perfluoro-2-azapropene, **5**, 1076
with phenylacetonitrile, **5**, 751
with *N*-phenylcarbonimidic dichloride, **6**, 619
with phenylglyoxal, **4**, 300
with phenyl isoselenocyanate, **5**, 1056
with phosgene, **5**, 965; **6**, 500
with phosphaalkenes, **4**, 116; **6**, 683
with phosphineboranes, **2**, 444
with phosphinic anhydrides, **4**, 127
with phosphinic chlorides, **4**, 127
with phosphonate anions, **5**, 603
with phosphonate esters, **6**, 512
with phosphonic dichlorides, **4**, 133
with phosphorus ylides, **5**, 603
with *O*-phosphorylisouronium salts, **6**, 641
with *o*-phthalaldehyde, **3**, 671
with phthalimides, **2**, 358
with potassium salts, **4**, 977
with propargylic alcohols, **3**, 76
with 2-pyridylthiolsulfonates, **2**, 364
with pyrimidinediones, **6**, 360
with pyrylium salts, **1**, 30
with selenium tetrafluoride, **2**, 369
with selenoalkynes, **4**, 910
with selenoalkynyl carboxylates, **5**, 626
with selenocarbonyl difluoride, **6**, 49
with selenoketenes, **5**, 626

with selenothiocyanates, **2**, 368
with selenothioesters, **5**, 626
with silyloxiranes, **4**, 506
with sodium methoxide, **2**, 760
with sodium nitrite, **2**, 803
with sulfamoyl chlorides, **2**, 367
with sulfenes, **4**, 282
with sulfenyl halides, **2**, 358, 767
with sulfenyl thiocyanates, **2**, 360
with sulfides, **2**, 160
with sulfinic acids, **2**, 363
with sulfonic acids, **2**, 364
with *N*-sulfonylcarboxyamides, **5**, 767
with sulfonyl chlorides, **2**, 364, 797
with sulfonyl halides, **2**, 248
with sulfonyls, **2**, 364
with sulfur dichloride, **2**, 361
with sulfur tetrafluoride, **6**, 51
with sulfuryl chloride, **2**, 367, 368, 797
with tetraethyl orthocarbonate, **6**, 625
with tetrafluoroethene hexamer, **3**, 589
with tetrasulfones, **4**, 866
with thiazolium cations, **6**, 114
with *N*-thioacylbenzimino chlorides, **5**, 761
with thioacyl bromides, **5**, 582
with thioacyl chlorides, **5**, 583
with thioamides, **5**, 753
with *N*-thiobenzoyl azoles, **5**, 587
with *N*-thiobenzoylbenzoxazolones, **5**, 587
with *N*-thiobenzoylimidazoles, **5**, 587
with *N*,*N'*-thiocarbonyl diimidazole, **5**, 1029
with 1,1'-(thiocarbonyldioxy)dibenzotriazole, **5**, 1030
with 1,1'-thiocarbonyl-2,2'-pyridone, **5**, 1030
with thiocarbonyls, **5**, 1022, 1028
with thiocyanates, **6**, 635
with thioimidic esters, **5**, 742, 753, 757
with thioisocoumarins, **5**, 586
with thioketenes, **5**, 589
with thioketenoids, **5**, 589
with α-thiol ketones, **4**, 318
with thiols, **2**, 240, 358
with thionyl chloride, **5**, 968
with thiophosgene, **5**, 1022, 1026; **6**, 576
with thiophosphinic chlorides, **4**, 128
with thiosulfonates, **2**, 358
with *N*(trialkylstannyl)imides, **5**, 390
with triarylphosphine dihalides, **2**, 859
with 1,3,5-triazine, **5**, 747
with trichloroacetonitrile, **5**, 749
with trifluoroacetic acid, **5**, 659
with trifluoroacetyl chloride, **5**, 721
with trifluoronitrosomethane, **6**, 242
with 2,4,6-trimethylbenzonitrile, **5**, 756
with triphenylphosphine dibromide, **5**, 975
with triphenylphosphine dihalides, **2**, 860
with triphenylpyrylium nitrate, **2**, 104
with triphosgene, **5**, 965
with tris(dimethylamino)borane, **2**, 533
with α,β-unsaturated aldehydes, **3**, 44
with α,β-unsaturated nitriles, **3**, 638
with urea, **6**, 502
with Vilsmeier salts, **5**, 744
with vinylidene dihalides, **4**, 970
with vinylphosphine oxides, **2**, 453
with ynamines, **4**, 976
with ynediamines, **4**, 976
and reductive deaminations, **1**, 32
reductive methylation of, **2**, 300
selenoacylations of, **5**, 626
silylmethylation of, **4**, 507
as starting materials
for alcohols, **1**, 383; **2**, 88
for aldehydes, **3**, 17
for alkyl bromides, **2**, 33
for alkyl chlorides, **2**, 27
for alkyl fluorides, **2**, 20
for alkyl halides, **2**, 11
for alkyl iodides, **2**, 36
for *N*-alkylphosphorimidic acid triamides, **2**, 407
for aminals, **4**, 405
for amines, **2**, 300
for α-aminoalkylphosphorus compounds, **4**, 466
for *N*-aminoguanidines, **6**, 653
for β-aminonitriles, **3**, 638
for arylamines, **2**, 756
for aryl isocyanides, **3**, 721
for azo compounds, **2**, 403
for azoxy compounds, **2**, 405
for bis(dialkylamino)boranes, **2**, 418
for bis(dialkylamino)haloarsines, **2**, 414
for bis(dialkylamino)phosphoryl halides, **2**, 413
for bis(dialkylamino)silanes, **2**, 416
for *N*-chloroamines, **2**, 337
for cyanamides, **5**, 1125
for cyclic imines, **3**, 411
for dialkylaminoalkyliminophosphines, **2**, 406
for dialkylaminoboranes, **2**, 418
for (dialkylamino)dihaloarsines, **2**, 414
for *N*,*N*-dialkylaminophosphinic dihalides, **2**, 409
for dialkylaminosilanes, **2**, 416
for dibromoalkanes, **4**, 27
for dichloroalkanes, **4**, 19
for *N*,*N*-dichloroalkylamines, **2**, 338
for *N*,*N*-difluoroalkylamines, **2**, 335
for dihaloaminoalkanes, **6**, 51
for, α-aminonitriles, **3**, 638
for α-haloimines, **3**, 415
for hexaalkylphosphoric triamides, **2**, 413
for hydrazines, **2**, 374, 379, 382
for hydroxamic acids, **5**, 324
for hydroxylamines, **2**, 340
for imidoyl halides, **5**, 659
for *N*-imidoylhydroxylamines, **5**, 765
for imidoylrheniums, **5**, 801
for imidoyl-tricoordinated phosphorus compounds, **5**, 786
for imino esters, **5**, 691
for isocyanates, **5**, 965; **6**, 443
for isocyanides, **3**, 693
for isoselenocyanates, **5**, 1055
for isothiocyanates, **5**, 1022, 1030
for ketenimines, **3**, 588, 590
for ketones, **3**, 132
for lithium dithiocarbamates, **5**, 1024
for nitriles, **3**, 620
for *N*-nitroamines, **2**, 385
for nitro compounds, **2**, 397
for *N*-nitrosoamines, **2**, 385
for nitroso compounds, **2**, 389
for *N*-perchlorylamines, **2**, 339
for phosphoramidates, **2**, 412
for quaternary hydrazinium salts, **2**, 384
for selenals, **3**, 388
for selenoureas, **6**, 597
for *N*-substituted amidrazones, **5**, 771
for sulfamic acids, **2**, 366
for sulfilimines, **2**, 401
for sulfinylamines, **2**, 399
for sulfonimidamides, **2**, 402
for sulfoximines, **2**, 402
for sulfurdiimines, **2**, 399
for sulfurimino dihalides, **2**, 400
for *N*,*N*,*N'*,*N'*-tetraalkylphosphorodiamidates, **2**, 413
for tetrakis(dialkylamino)germanes, **2**, 418
for tetrakis(dialkylamino)silanes, **2**, 417
for thiocarbamoyl chlorides, **6**, 537

for thiocyanates, **5**, 1107
for tris(dialkylamino)arsines, **2**, 415
for tris(dialkylamino)boranes, **2**, 419
for tris(dialkylamino)silanes, **2**, 417
synthesis of, **1**, 1041; **2**, 1039
reviews, **2**, 298
via allylation of imines, **2**, 313
via amination of organometallics, **2**, 305
via dealkylation, **2**, 302
via Gabriel synthesis, **2**, 299
via reduction, **2**, 312
thioacylation of, **5**, 582, 609, 611
N-thioalkylation of, **4**, 324
use in the disposal of phosgene, **6**, 411
use in dissolving metal reductions, **1**, 101
use in the reduction of oxazaborolidines, **2**, 66
via 1,2 Stevens rearrangement, **1**, 399
Amines, *N*-acyl-, synthesis of, **5**, 331
Amines, acyl alkoxy-, synthesis of, **2**, 352
Amines, *N*-acyl-*N*-nitroso-, cleavage of, **3**, 465
Amines, *N*-acyloxy-
hydrolysis of, **2**, 347
synthesis of, **2**, 344
use of term, **2**, 350
Amines, acyl-protected, synthesis of, methods, **5**, 260
Amines, *N*-acyl-*N*-sulfinyl-
reactions
with dienes, **5**, 340
with ketenes, **5**, 340
as starting materials, for sulfinamides, **5**, 340
Amines, acylvinyl nitro-, reactions, with bromine, **4**, 106
Amines, alkenyl-
reactions, with *N*-bromosuccinimide, **2**, 339
selective *N*-chlorination of, **2**, 337
Amines, 4-alkenyl-, cyclization of, **2**, 323
Amines, alkenyl *N*-alkoxy-, cyclization of, **2**, 353
Amines, alkoxy-
acylation of, **5**, 327, 332
reactions
with acyl compounds, **5**, 329
with aldehydes, **3**, 434
with amides, **5**, 329
with carbonyl diimidazole, **5**, 327
with cyclic anhydrides, **5**, 333
with dicyclohexylcarbodiimide, **5**, 327
with esters, **5**, 329
with lactams, **5**, 329
with lactones, **5**, 329
with 1,3-oxazolidinine-2-thiones, **5**, 329
with thiazolidine-2-thiones, **5**, 329
Amines, *N*-alkoxy-
N-alkylation of, **2**, 353
chlorination of, **2**, 357
reactions, with alkanesulfinyl chlorides, **2**, 363
Amines, *N*-(α-alkoxyalkyl)-, synthesis of, **4**, 299, 309
Amines, *N*-alkoxyl-, use of term, **2**, 350
Amines, *N*-alkoxyl secondary, from, nitroxides, **2**, 355
Amines, *N*-alkoxy primary, synthesis of, **2**, 350
Amines, *N*-alkoxy secondary
alkylation of, **2**, 350
synthesis of, **2**, 353
Amines, alkyl-
bromination of, **2**, 339
carbonylation of, with palladium, **6**, 448
fluorination of, **2**, 335
reactions
with 5-(arylimino)-4-chloro-5*H*-1,2,3-dithiazoles, **5**, 736
with arylselenenyl bromides, **2**, 368
with benzeneselenenyl chloride, **2**, 368
with calcium, **1**, 508
with 1,4-dibromobutane-2,3-dione, **6**, 564
with dihalocarbene complexes, **6**, 722
with diphosphonylalkynes, **5**, 789
with ketenimines, **4**, 977
with lithium, **1**, 508
with sulfamides, **2**, 367
with sulfur dichloride, **2**, 360
with urea, **6**, 502
as starting materials
for cyanamides, **5**, 1125
for isothiocyanates, **5**, 1024
sulfamation of, **2**, 366
synthesis of, **2**, 306, **297–332**
Amines, α-alkyl-, synthesis of, **3**, 407
Amines, alkylselenyl-, as starting materials, for alkane selenenamides, **2**, 292
Amines, alkynyl-, intramolecular aminopalladation of, **3**, 410
Amines, 2-alkynyl-, as starting materials, for *N*,*N*-dialkyl allenyl amines, **1**, 956
Amines, α-alkynyl-, synthesis of, **1**, 1003
Amines, α-allenic tertiary, synthesis of, **1**, 973
Amines, allenyl-, from, 1-dialkylamino-4-methoxy-4-methylpent-2-ynes, **1**, 981
Amines, α-allenyl-, synthesis of, **1**, 967
Amines, allyl-
isomerization of, **2**, 816
reactions
with ethoxyalkyne, **4**, 884
with nitrones, **2**, 345
synthesis of, **5**, 271
Amines, (*Z*)-allyl-, synthesis of, **1**, 1107
Amines, allylic
alkylation of, **2**, 322
allylmagnesiation of, **2**, 561
displacement of, via vinyl cuprates, **1**, 486
from
alkynes, **2**, 307
bis(sulfones), **4**, 279
palladium(0)-catalyzed dealkylation of, **2**, 304
reactions
with *t*-butyllithium, **1**, 1090
with catecholborane, **2**, 43
with diazo compounds, **1**, 823
Amines, *N*-amino α-halo-, synthesis of, **4**, 105
Amines, aminomethylhydroxyl-, synthesis of, **4**, 442
Amines, *N*-(arenesulfonyloxy)-, synthesis of, **2**, 357
Amines, aromatic
nitration of, **2**, 791
oxidation of, **2**, 772, 775
reactions
with aromatic acids, **5**, 303
with chloromethylsilanes, **4**, 507
with thionyl chloride, **2**, 793
as starting materials, for azodioxy compounds, **2**, 775
use in the reduction of carbon–mercury bonds, **1**, 68
Amines, aryl-, reactions, with isocyanates, **3**, 684
Amines, 4-aryl-1,3-butadiynyl)-, synthesis of, **2**, 307
Amines, *N*-arylhydroxyl-, synthesis of, **2**, 764
Amines, arylsulfinyl-, as starting materials, for arenediazonium salts, **2**, 800
Amines, *N*-arylsulfinyl-, as starting materials, for isothiocyanates, **5**, 1032
Amines, *N*-aryltrifluoro-, synthesis of, **6**, 602
Amines, *N*-azido-, synthesis of, **2**, 388
Amines, *N*-[1-(benzotriazol-1-y-1)alkyl]-, synthesis of, **4**, 309
Amines, 1-(benzotriazol-1-yl)alkyl-, synthesis of, **4**, 323
Amines, *N*-benzoyloxy-, synthesis of, **2**, 340
Amines, benzylic
reactions, with nitrosonium ions, **3**, 665
synthesis of, **2**, 317
Amines, benzylidenealkyl-, reactions, with fluorine, **5**, 659
Amines, *N*-benzylidineallyl-, reactions, with trimethylsilyloxyphosphorus(III) compounds, **4**, 461
Amines, *N*-(benzyloxy)-, reactions, with hydrogen peroxide, **3**, 435

Amines, *N*-benzyloxy-, synthesis of, via Meisenheimer rearrangement, **2**, 355
Amines, *O*-benzyloxy-, hydrogenolysis of, **2**, 340
Amines, bis(chlorocarbonyl)-, synthesis of, **6**, 449
Amines, bis(halocarbonyl)-, synthesis of, **6**, 437
Amines, 2,2-bis[methylsulfonyl]vinyl-, synthesis of, **2**, 743
Amines, *N*,*N*-bis(silyl)-
  reactions, with carbonyls, **3**, 406
  as starting materials, for Schiff bases, **3**, 406
Amines, bis(trifluoromethyl)-
  from
    trichloromethyl isocyanide dichloride, **6**, 238
    trifluoromethyl isocyanide difluoride, **6**, 238
Amines, bis(trifluoromethyl)alkyl-, from, metal bis(trifluoromethyl)amides, **6**, 238
Amines, bis(trimethylsilyl)-
  reactions, with *O*-aryl chlorothioformates, **6**, 556
  use in the synthesis of carbodiimides, **5**, 1070
Amines, α-branched, synthesis of, **2**, 312
Amines, *N*-bromo-, synthesis of, **2**, 761
Amines, *N*-bromoadamatyl-, as starting materials, for azo compounds, **2**, 403
Amines, *N*-bromoalkyl-, synthesis of, **2**, 339
Amines, β-bromoalkylbis(trifluoromethyl)-, synthesis of, **6**, 239
Amines, bulky, reactions, with vanadium oxytrichloride, **2**, 570
Amines, *N*-*t*-butoxy carbonyl allylic, synthesis of, **4**, 307
Amines, *t*-butyldimethylsilyl-, reactions, with *t*-butylmagnesium chloride, **5**, 376
Amines, carbinol-, as starting materials, for α-aminoalkylphosphorus compounds, **4**, 465
Amines, chiral
  reactions, with cyclohexane-1,2-diones, **3**, 260
  as starting materials, for chiral chloroformates, **6**, 426
Amines, chiral primary, from, *N*-trimethylsilyl benzaldehyde imine, **2**, 544
Amines, *N*-chlorinated perfluoroalkyl-, as starting materials, for hydrazonyl halides, **5**, 679
Amines, chloro-
  for the chlorination of amines, **2**, 337
  reactions
    with disulfides, **2**, 360
    with selenenyl halides, **2**, 369
Amines, *N*-chloro-
  for the chlorination of alkanes, **2**, 20
  from, amines, **2**, 337
  photolysis of, **4**, 414
  reactions
    with phosphorous(III) compounds, **2**, 860
    with potassium superoxide, **3**, 407
    with sulfur dioxide, **2**, 368
  as starting materials
    for aminyl radicals, **2**, 337
    for α-chloroamines, **4**, 98
    for nitrosoamines, **2**, 385
    for sulfenamides, **2**, 359
  synthesis of, **2**, 761
Amines, α-chloro-
  from, *N*-chloroamines, **4**, 98
  reactions, with trinitromethane salts, **6**, 145
  synthesis of, **4**, 97
Amines, β-chloro-
  elimination of, **1**, 636
  synthesis of, **2**, 312
Amines, *N*-chloro-*N*-alkoxyl-
  as starting materials, for *N*,*N*-dialkoxylamines, **2**, 356
  synthesis of, **2**, 357
Amines, *N*-chloroalkyl-
  from, trimethylsilylamines, **2**, 338
  synthesis of, **2**, 337
Amines, α-chloroalkyl-, use in the synthesis of α-aminosilanes, **4**, 510
Amines, *N*-chlorodialkyl-, reactions, with isocyanides, **5**, 1146
Amines, α-chloronitroso-, synthesis of, **2**, 385
Amines, *N*-chloro-*N*-perchloryl-, synthesis of, **2**, 339
Amines, *N*-chloro primary, reactions, with azomethines, **2**, 377
Amines, α-chloro tertiary
  reactions
    with lithium dialkyl(aryl)phosphines, **4**, 455
    with lithium diphenylphosphide, **4**, 455
Amines, cyano-, reduction of, for the synthesis of aminals, **4**, 413
Amines, *N*-cyano-, reduction of, for the synthesis of aminals, **4**, 413
Amines, cyanomethylene-
  hydrogenation of, **5**, 1003
  as starting materials, for isocyanates, **5**, 1003
Amines, cyclic
  reactions, with phenyl thionoacetate, **5**, 586
  synthesis of, **2**, 322
Amines, cyclopentene *N*-methoxy-, synthesis of, **2**, 351
Amines, cyclopropylidene-, synthesis of, **3**, 416
Amines, (2-cyclopropylidineethyl)-, synthesis of, **2**, 309
Amines, *N*,*N*-dialkoxyl-
  from
    N-chloro-*N*-alkoxylamines, **2**, 356
    N-substituted-*N*-alkoxylamines, **2**, 356
  synthesis of, **2**, 355
  transesterification of, **2**, 356
  use of term, **2**, 350
Amines, dialkyl-
  reactions
    with dichlorocarbene, **6**, 107
    with dichloromethylenephosphoramidic compounds, **6**, 655
    with ethylphosphonous acid ethyl ester, **2**, 469
    with α-methylene ketones, **3**, 187
  as starting materials, for *N*-fluoroamines, **2**, 335
  synthesis of, **2**, 300
Amines, *N*,*O*-dialkyl-, from, α-chloronitro compounds, **2**, 353
Amines, α,α-dialkyl-, synthesis of, **3**, 407
Amines, *N*,*N*-dialkyl allenyl-, from, 2-alkynyl amines, **1**, 956
Amines, *N*,*N*-dialkyl[1-(trimethylsilyl)alkyl]-, synthesis of, **2**, 305
Amines, *N*,*N*-dichloro-
  nitration of, **2**, 798
  as starting materials, for sulfinylamines, **2**, 399
  synthesis of, **6**, 53
Amines, *N*,*N*-dichloroalkyl-
  from, amines, **2**, 338
  reactions, with nitroso compounds, **2**, 775
  synthesis of, **2**, 338
Amines, α,α-dichloroalkyl-, synthesis of, **6**, 51
Amines, difluoro-, as starting materials, for fluorosulfites, **2**, 763
Amines, *N*,*N*-difluoro-
  from
    alkenes, **2**, 335
    amides, **2**, 335
    sulfamides, **2**, 335
    ureas, **2**, 335
    urethanes, **2**, 335
Amines, α,α-difluoro-, as starting materials, for amide acetals, **6**, 105
Amines, *N*,*N*-difluoroalkyl-
  from, amines, **2**, 335
  synthesis of, **2**, 335
Amines, *N*,*N*-difluoro α-halo-, synthesis of, **4**, 104
Amines, *N*,*N*-dihalo-, synthesis of, **6**, 230
Amines, α,α-dihalo-, from, haloiminium salts, **6**, 53
Amines, *N*,*N*-dihalo-1,1-difluoro-, as intermediates, **5**, 809
Amines, di-*N*-methylated aromatic, synthesis of, **2**, 753

Amines, (*N,N'*-diperchloryl-, synthesis of, **2**, 339
Amines, diphenylphosphinyloxy-
  hydrolysis of, **2**, 344, 347
  synthesis of, **2**, 340
Amines, *N*-(diphenylphosphinyloxy)-, synthesis of, **2**, 357
Amines, disilyl-
  as starting materials, for disilylamides, **5**, 379
  synthesis of, **2**, 417
Amines, α-disubstituted primary, synthesis of, **2**, 321
Amines, *N,N*-disulfenyl-, nomenclature, **2**, 767
Amines, *N*-ethoxymethyl dialkyl-, reactions, with α-oxo-phosphonic acid diethyl ester, **4**, 458
Amines, fluoro-
  as starting materials, for *N*-fluoroamides, **5**, 314
  synthesis of, **4**, 97
Amines, *N*-fluoro-
  for the fluorination of amines, **2**, 334
  from
    dialkylamines, **2**, 335
    imines, **2**, 335
Amines, α-fluoro-, from, dicyanodifluoroethenes, **4**, 104
Amines, *N*-fluoroalkyl-, synthesis of, **2**, 334
Amines, *N*-fluorochloro-, synthesis of, **6**, 53
Amines, fluoromethyl dialkyl-, characteristics of, **4**, 96
Amines, *N*-halo-
  reactions
    with Grignard reagents, **2**, 305
    with sulfides, **2**, 160
  synthesis of, **2**, 761
Amines, α-halo-
  covalent character of, **4**, 96
  in equilibrium, with iminium salts, **4**, 96
  from
    enamines, **4**, 100
    imines, **4**, 100
    methylenediamines, **4**, 97
  reactions
    with electrophiles, **4**, 100
    with trialkyl(aryl)phosphites, **4**, 464
  synthesis of, **4**, 97
    reviews, **4**, 96
Amines, δ-halo-, as starting materials, for pyrrolidines, **2**, 323
Amines, ε-halo-, cyclization of, **2**, 326
Amines, α-halo alkenyl sulfinyl-, synthesis of, **4**, 800
Amines, *N*-haloalkyl-, synthesis of, **2**, 334
Amines, *N*-halo(trihalomethyl)-, synthesis of, **6**, 240
Amines, heterocyclic
  reactions, with trisformamidomethane, **5**, 836
  as starting materials, for thioformyl compounds, **5**, 836
  use in the synthesis of oxyimidic amides, **5**, 835
Amines, heterocyclic tertiary, cleavage of, **1**, 625
Amines, homoallylic
  from, *N*-allyl aminomethylstannanes, **1**, 406
  1,3 sigmatropic rearrangements of, **1**, 401
  synthesis of, **1**, 82, 407, 408; **2**, 322
Amines, homopropargyl-, synthesis of, **1**, 1059
Amines, 2-hydroperoxy-, synthesis of, **2**, 96
Amines, 2-hydroxy-, dehydration of, **3**, 413
Amines, α-hydroxy-, reactions, with phosphorus, **4**, 466
Amines, β-hydroxy-, synthesis of, **2**, 320; **5**, 999
Amines, 3-hydroxy alkyl-, synthesis of, **4**, 383
Amines, *N*-hydroxymethyl dialkyl-
  reactions
    with α,ω-bisphosphinoalkanes, **4**, 452
    with phosphorus, **4**, 452
Amines, *N*-hydroxymethyl-*N*-nitro-, synthesis of, **4**, 311
Amines, α-hydroxy-*N*-sulfonyl-, oxidation of, **5**, 347
Amines, *N*-iodoalkyl-, synthesis of, **2**, 339
Amines, (*Z*)-γ-iodo allylic, synthesis of, **2**, 313
Amines, keto-, rearrangement of, **5**, 272
Amines, α-lithiated
  reactions
    with aldehydes, **4**, 528
    with alkyl halides, **4**, 528
  synthesis of, **4**, 527
Amines, lithioallyl-, synthesis of, **2**, 1001
Amines, lithium bis(trimethylsilyl)-, use in the synthesis of carbodiimides, **5**, 1070
Amines, metallated, acylation of, **5**, 389
Amines, α-methoxy-, as starting materials, for iminium ions, **3**, 421
Amines, 1-methoxylated cyclic *N*-formyl-, as starting materials, for *N*-acyl enamines, **2**, 748
Amines, *N*-methyldialkyl-, deprotonation of, **1**, 148
Amines, methylene-, as starting materials, for *N*-(phosphinomethyl)arylamines, **4**, 453
Amines, monocyclic, as starting materials, for *gem*-dinitramines, **4**, 435
Amines, nitro-
  from, alkyl halides, **2**, 386
  Mannich reaction for, **2**, 394
  Michael addition of, **2**, 386
  as starting materials
    for hydrazines, **2**, 379
    for *N*-nitroamides, **5**, 361
Amines, *N*-nitro-
  from
    amines, **2**, 385
    β-Methylaminopropionitrile, **2**, 386
    nitrimines, **2**, 386
  reactions, with formaldehyde, **4**, 311
  synthesis of, **2**, 385
Amines, β-nitro-, synthesis of, **2**, 394
Amines, *N*-nitro α-halo-, synthesis of, **4**, 106
Amines, nitroso-
  from, *N*-chloroamines, **2**, 385
  as starting materials, for azo compounds, **2**, 403
Amines, *N*-nitroso-
  cleavage of, **3**, 465, 471
  from, amines, **2**, 385
  lithiation of, **4**, 511
  oxidation of, **2**, 386
  silylation of, **4**, 511
  as starting materials
    for aryl aldimines, **3**, 411
    for hydrazines, **2**, 379, 381
Amines, nitroso acyl-, as starting materials, for diazonium salts, **2**, 805
Amines, nitrovinyl-
  from, 1-chloro-2-nitroethylenes, **2**, 787
  synthesis of, **2**, 787
Amines, *N*-oxy α-halo-, synthesis of, **4**, 104
Amines, (*Z,E*)-penta-2,4-dienyl-, synthesis of, **2**, 303
Amines, *N*-perchloryl-
  chlorination of, **2**, 339
  from, amines, **2**, 339
  handling precautions, **2**, 339
  synthesis of, **2**, 339
Amines, perfluoro-
  bromination of, **4**, 794
  reductive defluorination of, **5**, 808
Amines, perfluoroalkylated, synthesis of, **2**, 311
Amines, perhalo-, as starting materials, for carbonimidic dichlorides, **6**, 609
Amines, *N*-phosphinyl-, cleavage of, **3**, 439
Amines, *N*-phosphinyloxy-, synthesis of, **2**, 357
Amines, *N*-phosphoryl α-halo-, synthesis of, **4**, 106
Amines, poly(isocyanatoethyl)-, synthesis of, **5**, 1003
Amines, primary
  α-alkylation of, **2**, 302
  diazotization of, **1**, 258; **5**, 868
  from
    alkyl halides, **2**, 298
    methoxyamine, **2**, 305
    1-nitroalkenes, **2**, 318

potassium 2,2,6,6-tetramethyl-2,6-disilapiperidide, **2**, 300
primary alcohols, **2**, 309
trimethylsilylimines, **3**, 439
kinetic resolution of, via enantioselective *N*-acetylation, **2**, 331
phosgenation of, **5**, 965
reactions
with arenesulfinic esters, **2**, 361
with carbon disulfide, **6**, 574
with diazonium salts, **2**, 777
with α,α'-dibromo ketones, **3**, 417
with difluoroamine, **1**, 35
with 3,4-dihydro-4,4-dimethyl-2*H*-pyran-2-one, **2**, 320
with dithionooxamides, **5**, 591
with ω-halo-2-alkenoate esters, **2**, 323
with hydroxylamine-*O*-sulfonic acid, **1**, 35
with (*Z*)-β-iodo acryl aldehyde, **2**, 313
with iodomethaneboronates, **4**, 525
with mesityl oxide, **3**, 465
with organoarsenic dihalides, **2**, 892
with phenyl glyoxal, **3**, 417
with phosgene, **6**, 443
with pyrylium ions, **3**, 420
with rhena-β-diketones, **5**, 801
with sulfates, **2**, 322
with sulfenyl chlorides, **2**, 361
with sulfur halides, **2**, 796
with tetrakis(hydroxymethyl)phosphonium salts, **4**, 456
with trimethylsilyl chloride, **2**, 300
with triphenylphosphine dibromide, **5**, 1065
as starting materials
for bisphosphinic acids, **4**, 470
for carbamoyl chlorides, **6**, 443
for dihydrazidines, **5**, 854
for *N*-substituted imines, **3**, 405
synthesis of, via Ritter reaction, **2**, 308
Amines, primary alkyl-, reactions, with dimethylalkylboranes, **2**, 300
Amines, primary allyl-, from, 4-amino-1-azadienes, **2**, 322
Amines, primary aromatic
oxidation of, **2**, 772
as starting materials, for nitramines, **2**, 798
Amines, primary aryl-, oxidation of, **2**, 772
Amines, primary *N*-fluoroalkyl-, synthesis of, **2**, 334
Amines, primary α-lithio-, synthesis of, **4**, 529
Amines, primary trihalomethyl-, synthesis of, **6**, 238, 238
Amines, primary vinyl-, nitrosation of, **2**, 800
Amines, propargyl-
isomerization of, for the synthesis of tertiary ynamines, **2**, 1050
oxidative deamination of, **3**, 78
reactions, with lithium aluminum hydride, **1**, 514
synthesis of, **1**, 1041; **2**, 301, 1049
via Beckmann rearrangements, **1**, 1041
Amines, propargylic, hydroboration of, **3**, 44
Amines, quaternary, rearrangement of, **1**, 399
Amines, secondary
alkylating properties of, **1**, 258
alkylation of, **2**, 302
from
arenediazonium salts, **2**, 305
1-(1-(arylamino)alkyl)benzotriazoles, **1**, 260
tertiary amines, **6**, 443
N-phenylation of, **2**, 300
reactions
with alkanoyl isothiocyanates, **6**, 572
with alkynic compounds, **2**, 751
with 2-(benzeneseleninyl)vinyl, **4**, 874
with bis(acyl) tellurides, **5**, 242
with 1-bromoalkyl isothiocyanates, **6**, 571
with dialkyl chlorophosphites, **5**, 788
with dibenzoyl peroxide, **2**, 344
with 1,1-difluorothiols, **5**, 474
with *N*,*N*-dimethylthionoformamide, **5**, 591
with 1,3,2-dioxophospholes, **6**, 572
with ethoxyalkynes, **4**, 884
with formamidinium salts, **6**, 138
with *exo*-7-halobicyclo[3.2.0]hept-2-en-6-ones, **2**, 301
with *N*-methyl-3,4-dinitropyrrole, **4**, 415
with nitroethene, **2**, 301
with 2-nitrothiophene, **4**, 907
with α-oxoacids, **5**, 603
with peroxides, **2**, 343
with phosgene, **5**, 971; **6**, 443, 500
with phosphaalkenes, **2**, 467
with secondary arsines, **4**, 500
with silicon(IV) isothiocyanate, **6**, 572
with thiophosgene, **6**, 537
with trichloromethyl isocyanide, **6**, 605
with tris(dimethylaminomethyl)phosphines, **4**, 456
with tris(hydroxymethyl)phosphines, **4**, 456
with trisulfanes, **5**, 485
as starting materials
for alkenes, **1**, 636
for *N*-chlorocarbonyls, **6**, 443
for cyclic aminomethylphosphines, **4**, 453
for *N*,*N*-disubstituted carbamoyl bromides, **6**, 452
for α-hydroxylaminonitriles, **3**, 638
for nitrones, **3**, 432
for tertiary amines, **2**, 298
synthesis of
via dealkylation of tertiary amines, **2**, 303
via reductive amination, **2**, 311
use in the substitution of nitrogen functions, **1**, 258
Amines, secondary aliphatic, reactions, with 2-halo-2-alkenals, **2**, 741
Amines, secondary *N*-fluoroalkyl-, synthesis of, **2**, 334
Amines, secondary isocyano-, synthesis of, **3**, 710
Amines, secondary trihalomethyl-, synthesis of, **6**, 238
Amines, α-selenenyloxyalkyl-, synthesis of, **4**, 311
Amines, silyl-
reactions
with cyanogen chloride, **5**, 1082
with diaminotellurenamide tetrafluorides, **2**, 370
with ketenes, **4**, 886
with tellurium hexafluoride, **2**, 370
as starting materials, for isocyanates, **6**, 443
synthesis of, **2**, 812
Amines, *N*-silyl-
acylation of, **5**, 376
as nucleophiles, **4**, 417
Amines, α-silyl-, synthesis of, **4**, 511
Amines, silylated
reactions, with Meldrum's acid, **5**, 227
for the synthesis of aminals, **4**, 406
Amines, *N*-silylated, as starting materials, for trialkylsilyl imidates, **5**, 721
Amines, *N*-silylated chloro-, as oxidation reagents, for the synthesis of methylenephosphines, **6**, 693
Amines, *N*-(silylmethyl)-, desilylation of, **1**, 285
Amines, silylmethyldialkyl-, reactions, with benzyne, **2**, 945
Amines, *N*-silyloxy-, use of term, **2**, 350
Amines, α-silyloxyalkyl-, synthesis of, **4**, 311
Amines, silyl primary, reactions, with phosgene, **5**, 968
Amines, stannyl-, reactions, with phenyl isothiocyanate, **6**, 571
Amines, (stannylethynyl)-, reactions, with aryl isothiocyanates, **5**, 739
Amines, *N*-substituted-*N*-alkoxyl-, as starting materials, for *N*,*N*-dialkoxylamines, **2**, 356
Amines, α-substituted allyl-, synthesis of, via palladium(0)-catalyzed azidation, **2**, 309
Amines, γ-substituted allylic, synthesis of, **1**, 1090
Amines, *N*-substituted α-halo-, synthesis of, **4**, 98
Amines, α-substituted tertiary, synthesis of, **2**, 310

Amines, sulfato-, synthesis of, **2**, 357
Amines, sulfinyl-
 from
  amines, **2**, 399
  N,N-dichloroamines, **2**, 399
  N-substituted hexamethyldisilazanes, **2**, 399
  N-sulfinylsulfonamides, **2**, 399
  sulfurdiimines, **2**, 399
 as starting materials
  for diazonium salts, **2**, 800
  for sulfurdiimines, **2**, 399
  for sulfurimino dichlorides, **2**, 400
  for thiocarbonyl *S*-imides, **3**, 370
 synthesis of, **2**, 399; **4**, 110
Amines, *N*-sulfinyl-
 reactions
  with Grignard reagents, **2**, 363
  with ynamines, **5**, 623
 as starting materials, for isocyanides, **3**, 695
Amines, *N*-sulfinylaryl-, as starting materials, for aryl isocyanides, **3**, 721
Amines, *N*-sulfinyloxy-, from, hydroxylamines, **2**, 357
Amines, sulfito-, synthesis of, **2**, 357
Amines, sulfonyl-
 from
  sulfamyl chlorides, **2**, 401
  sulfonylureas, **2**, 401
 synthesis of, **2**, 401
Amines, *N*-sulfonyl-, synthesis of, **2**, 793; **3**, 438
Amines, *N*-sulfonyloxyl-, synthesis of, **2**, 357
Amines, tertiary
 bromination of, **2**, 339
 dealkylation of, **2**, 303
 N-dealkylation of, with α-chloroethyl chloroformate, **6**, 428
 demethylation of, **2**, 303
 from, secondary amines, **2**, 298
 metallation of, **4**, 527
 oxidation of, **2**, 347; **4**, 493
 reactions
  with alkylating reagents, **2**, 760
  with alkynyl halides, **2**, 1042
  with arenesulfenyl chlorides, **2**, 360
  with cyanogen bromide, **2**, 33
  with cyanogen halides, **5**, 1126
  with dialkylboron triflates, **2**, 532
  with ethyl chloroformate, **2**, 27
  with fluorine, **2**, 336
  with nitrous acid, **2**, 779
  with phosgene, **3**, 696; **6**, 443, 500
  with phosphines, **2**, 474
  with propiolate esters, **1**, 835
  with *p*-toluenesulfonic acid, **1**, 256
 reactivity of, **1**, 1064
 [1,5] sigmatropic rearrangements of, **1**, 838
 silylmethylation of, **4**, 507
 as starting materials
  for *N*-chloroammonium salts, **2**, 338
  for cyanamides, **5**, 1126
  for secondary amines, **6**, 443
 use in alkylation reactions, **1**, 256
 use in the substitution of nitrogen functions, **1**, 255
Amines, tertiary alkyl-
 oxidation of, **2**, 347
 synthesis of, via Ritter reaction, **2**, 309
Amines, tertiary allenyl-, synthesis of, **1**, 978
Amines, tertiary α-halo-, synthesis of, **4**, 97
Amines, tertiary propargyl-
 reduction of, **1**, 512
 synthesis of, **2**, 301, 306
Amines, tertiary trihalomethyl-, synthesis of, **6**, 238
Amines, *N*-thio-α-halo-, synthesis of, **4**, 104
Amines, *N*-thionitroso-
 from
  hydrazines, **2**, 385
  sulfinylhydrazines, **2**, 385
 synthesis of, **2**, 385, 792
Amines, *N*-thionyl-, synthesis of, **2**, 399
Amines, thiosulfinyl-, synthesis of, **2**, 399
Amines, *n*-tosyl-, oxidation of, **3**, 43
Amines, (*N*-tosyl)propargyl-, as starting materials, for *N*-tosyl-β-lactams, **5**, 346
Amines, trialkyl-
 reactions, with silver fluoride, **1**, 370
 as starting materials, for dialkyl arylamines, **2**, 301
Amines, *N*-trialkylstannyl-, reactions, with terminal alkynes, **1**, 1006
Amines, trichloromethyl-
 from, carbonimidic dichlorides, **6**, 238
 reactions, with triethyl phosphite, **6**, 366
Amines, trichlorovinyl-, synthesis of, **4**, 791
Amines, trifluoro-, as intermediates, for the synthesis of carbonimidic difluorides, **6**, 602
Amines, trifluoromethyl-
 dehydrofluorination of, **6**, 602
 from
  carbonimidic difluorides, **6**, 238, 239
  N,N-disubstituted formamides, **6**, 440
 hydrolysis of, **6**, 440
 as starting materials
  for carbamoyl fluorides, **6**, 440
  for carbonimidic difluorides, **6**, 602
Amines, *N*-trifluorovinyl secondary
 from, potassium perfluoro aminopropanoates, **4**, 794
 synthesis of, via elimination of carbon dioxide, **4**, 794
Amines, trihalomethyl-, synthesis of, **6**, 238
Amines, trimethylsilyl-
 bromination of, **2**, 339
 cleavage of, **2**, 761
 reactions
  with benzeneselenenyl chloride, **2**, 368
  with benzoyl-*t*-butylcarbodiimide, **5**, 377
  with isothiocyanates, **6**, 571
  with ketene, **5**, 376
  with thiophosgene, **6**, 537
  with α,β-unsaturated aldehydes, **2**, 741
 as starting materials, for *N*-chloroalkylamines, **2**, 338
Amines, *N*-trimethylsilyl-
 reactions, with selenium oxide, **2**, 370
 as starting materials, for isothiocyanates, **5**, 1030
Amines, tris(diethoxyphosphinyl)methyl-, synthesis of, **6**, 366
Amines, trityl *N*-halo-, Steigliz rearrangement of, **3**, 407
Amines, unsaturated, as intermediates, **2**, 306
Amines, β,γ-unsaturated, from, phenyl hydroxylamine, **2**, 307
Amines, β,γ-unsaturated primary, synthesis of, **2**, 306
Amines, unsymmetrical secondary, synthesis of, **2**, 300
Amines, vinyl-
 uses of, **2**, 738
 See also Enamines
Amines, vinylbisdifluoro-, stability of, **2**, 762
Amines, vinyldifluoro-, synthesis of, **2**, 762
Amines, vinylidene cyclic, from, amino propargyl trimethylsilanes, **1**, 978
Amine salts
 reactions
  with cyanamide, **6**, 642
  with sodium hypochlorite, **2**, 337
Amine salts, *N*-alkoxy-*N,N,N*-trialkyl-, synthesis of, **2**, 350
Amine salts, aromatic, synthesis of, **2**, 759
Amine–sulfene complexes, synthesis of, **5**, 502
Amine transfer agents, reactions, with hydroxylamine, **2**, 691
Aminimides, trialkyl-, pyrolysis of, **5**, 997
Amino acid anhydrides, synthesis of, **5**, 189
Amino acid anilides, synthesis of, **5**, 293

Amino acid azides, Curtius rearrangement of, **4**, 426, 446
Amino acid chlorides
from, α-amino acids, **5**, 15
use in peptide coupling, **5**, 15
α-Amino acid chlorides, reduction of, **3**, 39
α-Amino acid chlorides, *N*-acyl, reactions, with Grignard reagents, **3**, 180
Amino acid ester hydrochlorides, reactions, with phosgene, **5**, 971
Amino acid ester isocyanates, synthesis of, **5**, 971
Amino acid esters, reactions, with *N*-thioacylbenzimidazolones, **5**, 587
Amino acid esters, 4-dimethyl-, synthesis of, **5**, 560
α-Amino acid esters
N-chloroamines of, synthesis of, **2**, 338
reactions, with isoamylnitrile, **3**, 468
as starting materials, for β-keto esters, **1**, 34
Amino acid fluorides, [(9-fluorenylmethyl)oxy]carbonyl, synthesis of, **5**, 5
Amino acid hydrochlorides, synthesis of, **5**, 964
Amino acids
optically active, synthesis of, **1**, 166
oxidation of, **4**, 311
reactions
with acylating reagents, **3**, 180
with bis(trichloromethyl) carbonate, **6**, 430
with butyldiphenylborinates, **5**, 228
with dibutyltin oxide, **5**, 277
with diphosgene, **5**, 972
with *o*-nitrobenzenesulfenyl chloride, **2**, 358
with organometallics, **3**, 180
with sodium tetraphenylborate, **5**, 228
with trialkylboranes, **5**, 228
as sources of chiral auxiliaries, **1**, 134
as starting materials
for *O*,*N*-acetals, **4**, 301
for aminocarbene complexes, **5**, 953
for nitriles, **3**, 620
for oxazolinones, **1**, 134
for peptides, **5**, 259
structure of, reviews, **5**, 84
synthesis of, **1**, 76; **3**, 288; **5**, 83
reviews, **5**, 83
via carbonation of organometallics, **5**, 88
via hydrolysis, **5**, 85
for the synthesis of α-aminoalkylphosphines, **4**, 455
thioacylation of, **5**, 587
uses of, reviews, **5**, 84
Amino acids, *N*-acyl, reactions, with phosgene, **1**, 985
Amino acids, *N*-acyloxycarbonyl, synthesis of, **5**, 89
Amino acids, allenic, synthesis of, **1**, 968
Amino acids, *N*-benzoyloxycarbonyl-, synthesis of, **5**, 89
Amino acids, *N*-benzyloxycarbonyl-
reactions
with 1-chloro-*N*,*N*-2-trimethyl-1-propen-1-amine, **5**, 15
with *N*-(dimethylaminopropyl)-*N*′-ethylcarbodiimide hydrochloride, **5**, 185
Amino acids, *N*-butoxycarbonyl-, reactions, with triphosgene, **6**, 486
Amino acids, *t*-butoxycarbonyl-, reactions, with dicyclohexylcarbodiimide, **5**, 185
Amino acids, *N*-*t*-butyloxycarbonyl-, reactions, with *N*-(dimethylaminopropyl)-*N*′-ethylcarbodiimide hydrochloride, **5**, 185
Amino acids, cyclic, as catalysts, for the synthesis of asymmetric cyanohydrins, **3**, 628
Amino acids, α-fluoro-, synthesis of, **4**, 101
Amino acids, fluoroallenyl-, synthesis of, **1**, 958
Amino acids, protected
as starting materials, for thionoesters, **5**, 586
thionation of, **5**, 568
Amino acids, α-substituted, synthesis of, **2**, 140
Amino acids, *N*-tosyl-, oxidative rearrangement of, **5**, 346
Amino acids, α,β-unsaturated, from, *N*-chloroamides, **5**, 315
Amino acids, β,γ-unsaturated, synthesis of, **1**, 1135
2-Amino acids, diazotization of, **2**, 88
D-Amino acids, oxidation of, **5**, 118
α-Amino acids
decarbonylation of, **3**, 419
diazotization of, **2**, 11
from
acylimidazoles, **3**, 40
acylpyrazoles, **3**, 40
2-aminoketene dithioacetal monosulfoxides, **4**, 857
thiol esters, **3**, 40
phosphonic acid compounds of, synthesis of, **4**, 466
reactions
with chloramine-T, **3**, 620
with copper, **5**, 385
with dimethylformamide dimethyl acetal, **5**, 746
with dithiocarbamic esters, **6**, 575
with nickel, **5**, 385
with palladium, **5**, 385
with phosgene, **5**, 970
reduction of, **1**, 55; **3**, 39
as starting materials
for amino acid chlorides, **5**, 15
for amino aldehydes, **3**, 39
for formamidines, **5**, 746
for α-hydrazinoacids, **2**, 374
for nitriles, **3**, 620
synthesis of, **1**, 166, 280; **4**, 102; **5**, 88
trifluoroacetylation of, **4**, 320
α-Amino acids, *N*-acyl, β-elimination of, **5**, 315
α-Amino acids, *N*-carboxy anhydrides, from, bis(trichloromethyl) carbonate, **6**, 416
α-Amino acids, *N*-formyl, hydrolysis of, **5**, 88
α-Amino acids, *N*-hydroxy, synthesis of, **2**, 340
β-Amino acids, chemistry of, **1**, 120
ω-Amino acids, *N*-protected, synthesis of, **5**, 993
Amino acid synthon, chiral, reactions, with triflates, **1**, 180
Amino acid trimethylsilyl esters, as starting materials, for isothiocyanates, **5**, 1027
Amino alcohols, as starting materials, for thiocarbamic acid *O*-esters, **6**, 557
1,2-Amino alcohols, reactions, with carbon diselenide, **6**, 595
1-Amino-1-alkanols, synthesis of, **4**, 294
α-Amino alkoxides, synthesis of, **4**, 299
α-Aminoalkyl phosphonites, from, ethyl phenyl phosphonite, **4**, 460
α-Aminoalkyl radicals, synthesis of, **2**, 1010
α-Amino anions, synthesis of, **4**, 527
Aminobis(phosphonates)
from
amides, **6**, 158
nitriles, **6**, 159
phosphorus trihalides, **6**, 158
Aminobis(trichloromethyl), oxidation of, **6**, 242
α-Amino carbanions
asymmetric alkylation of, **1**, 149
synthesis of, **1**, 151
Amino compounds
from, carbonyls, **1**, 881
protection of, **6**, 486
as starting materials, for *S*,*S*-dialkyl iminodithiocarbonates, **6**, 627
synthesis of, **4**, 495; **5**, 949
Amino compounds, benzyloxymethyl-, cleavage of, **4**, 103
Amino compounds, diisopropyl-, synthesis of, **3**, 689
Amino compounds, *N*-ethoxycarbonyl, synthesis of, **3**, 181
α-Amino compounds
diazotization of, **5**, 868
synthesis of, **3**, 178
1-Amino-4,6-diphenyl-2-pyridone, pyrolysis of, **5**, 1002
α-Amino fatty acid esters, hydrolysis of, **5**, 24
Aminols

reactions
with amides, **4**, 427
with sulfonamides, **4**, 427
Aminomercuration, for the synthesis of (*Z*)-γ-amino enamines, **2**, 740
Amino oxides, benzylidencyanalkyl-, as starting materials, for imino esters, **5**, 698
Aminoselenation, for the synthesis of pyrrolidines, **2**, 325
Aminosilylation
of alkynes, **2**, 920
of ethynedicarboxylate, **2**, 920
Aminosilyl compounds, *N*-α-chloroalkyl, synthesis of, **4**, 107
Amino sugars, acylation of, **5**, 221
Aminosulfenium salts, acyl, oxidation of, **5**, 350
β-Amino sulfone, synthesis of, **1**, 145
Aminosulfuration, for the synthesis of thionoamides, **5**, 599, 621
Aminyl radicals
cyclization of, **1**, 950
from, *N*-chloroamines, **2**, 337
Ammeline, fluorination of, **5**, 815
Amminium radicals, in radical addition reactions, **1**, 950
Ammonia
reactions
with acetoacetone, **3**, 263
with aldehydes, **3**, 617; **4**, 294
with alkoxycyclopentenones, **3**, 264
with alkylantimony halides, **2**, 502
with alkylarsenic halides, **2**, 502
with alkylbismuth halides, **2**, 502
with amines, **2**, 298
with areneseleninyl chlorides, **2**, 736
with benzaldehydes, **4**, 437
with bromomethanesulfonyl chloride, **4**, 86
with *n*-butyl chloroformate, **6**, 483
with carboxylic acids, **5**, 274
with chloral, **4**, 297
with chlorodi(*t*-butyl)arsine, **2**, 502
with *N*-cyanoimido carbonate, **2**, 322
with 1,5-dialdehydes, **2**, 330
with diazo ketones, **5**, 276
with di(chloromethyl)phosphinic acid, **4**, 473
with diethyl α-bromovinyl phosphonate, **4**, 473
with dithionomalonic esters, **4**, 895
with *N*-ethyl-*N*-phenylcarbamic acid chloride, **6**, 501
with (*Z*)-fluoroimines, **5**, 761
with fluoroiminonitriles, **5**, 761
with formaldehyde, **4**, 407
with haloacids, **5**, 88
with hexafluoroacetone, **4**, 297
with hydrazonate esters, **5**, 772
with hydrogen isocyanate, **6**, 500
with iodomethanesulfonyl chloride, **4**, 86
with isothiocyanates, **6**, 569
with ketene acetals, **4**, 972, 983
with ketones, **3**, 405
with lithium alkynides, **1**, 1000
with nitriles, **5**, 780
with 1-nitroaldehydehydrazones, **5**, 854
with nitrobenzene, **2**, 806
with 3-nitro-2,5-dihydrofuran, **2**, 307
with oxiranes, **1**, 1022
with oxyimidic chlorides, **5**, 765
with pentafluoroguanidines, **5**, 817
with phosphonothioic acid diesters, **4**, 469
with phthalaldehyde, **4**, 298
with pyrylium ions, **3**, 420
with sodium, **1**, 508
with sodium amide, **1**, 1065
with sulfonyl chlorides, **2**, 726
with sulfonyl halides, **2**, 248
with *N*-thioacylbenzimino chlorides, **5**, 761
with thioamides, **5**, 753
with thioimidic esters, **5**, 753
with (trifluoroguanylazo)difluoromethyleneimine, **6**, 363
with Vilsmeier salts, **6**, 641
with ytterbium, **1**, 508
as starting material
for *N*-imidoylhydroxylamines, **5**, 765
for *N*-substituted amidrazones, **5**, 771
use in the synthesis of arylamines, **2**, 752
use in the synthesis of primary amides, **5**, 275
Ammonia, trimethyl-, reactions, with Simmons–Smith reagents, **4**, 534
Ammonias, aldehyde, dehydration of, **4**, 407
Ammonium, triethylbenzyl-, as phase-transfer catalyst, **5**, 4
Ammonium azide, tetrabutyl-, reactions, with carboxylic acid chlorides, **5**, 991
Ammonium betaine, (*E*)-(carboxyvinyl)trimethyl-, as intermediate, **5**, 105
Ammonium bisulfate, tetra-*n*-butyl-, as phase-transfer catalyst, **5**, 998
Ammonium bromide, cetyltrimethyl-, as catalyst, for the synthesis of cyanates, **5**, 1104
Ammonium bromide, dibenzyldimethyl-, rearrangement of, **1**, 399
Ammonium bromide, tetrabutyl-, use in the synthesis of isothiocyanates, **5**, 1031
Ammonium bromide, tetra-*n*-butyl-, reactions, with sodium hydrogen carbonate, **5**, 183, 195
Ammonium bromide, tetraethyl-, as catalyst, for the synthesis of acetals, **4**, 181
Ammonium chloride
reactions
with dimethyltrichlorophosphorane, **2**, 468
with diphenylphosphinous chloride, **2**, 864
with methyltetrachlorophosphorane, **2**, 468
with zinc, **2**, 764
Ammonium chloride, azidochloromethylenedimethyl-, use in the synthesis of α-diazo ketones, **3**, 473
Ammonium chloride, benzyltriethyl-
as catalyst, for acid chlorides, **5**, 8
as phase-transfer reagent, **1**, 1013; **4**, 792
use in the chlorobromination of alkenes, **2**, 24
Ammonium chloride, *t*-butylchloromethylenedimethyl-, from, *N*,*N*-dimethyl pivalamide, **5**, 526
Ammonium chloride, (*Z*)-4-chloro-2-butenyl-, synthesis of, **2**, 325
Ammonium chloride, *N*-chlorotrimethyl-, stability of, **2**, 338
Ammonium chloride, *N*,*N*-dimethylchloromethylene-, use in the synthesis of acid chlorides, **5**, 14
Ammonium chloride, dimethylmethylene-, reactions, with alcohols, **4**, 305
Ammonium chloride, *N*,*N*-dimethylmethylene-, reactions, with lithium butanetellurolate, **4**, 334
Ammonium chloride, methyltrioctyl-
as catalyst
for the oxidation of dithioacetals, **4**, 266
for the synthesis of dialkyl carbonates, **6**, 462
Ammonium chloride, stearyltrimethyl-, as phase-transfer catalyst, **6**, 425
Ammonium chloride, tetra-*n*-butyl-, reactions, with sodium hydroxide, **5**, 183
Ammonium chloride, tricaprylmethyl-, use in the synthesis of purple benzene, **5**, 47
Ammonium chloride, triethylbenzyl-
as catalyst, for the synthesis of carbodiimides, **5**, 1065
as starting material, for alkyl fluoroformates, **6**, 423
Ammonium chlorides, alkoxymethylene-, synthesis of, **5**, 526
Ammonium chlorides, *N*-chlorotrialkyl-, stability of, **2**, 338
Ammonium chlorides, diaryloxymethyltriethyl-, from, chlorodiaryloxymethanes, **6**, 105
Ammonium chlorides, quaternary, reactions, with methyltetrachlorophosphorane, **2**, 477

Ammonium chloroiodate salts, tetraalkyl-, reactions, with alkenes, **2**, 23
Ammonium compounds, *N*-fluoro-, detection of, **2**, 762
δ-Ammonium compounds, elimination of, **3**, 228
Ammonium dichlorobromate, tetrabutyl-, for the chlorobromination of alkenes, **2**, 24
Ammonium dichloroiodate, benzyltrimethyl-
  for the *para* iodination of aryl ethers, **2**, 631
  as reagent, for the α-chlorination of carbonyls, **2**, 21
  for the synthesis of aryl iodides, **2**, 631
Ammonium fluoride, reactions, with phosphonium ylides, **2**, 475
Ammonium fluoride, benzyltrimethyl-, reactions, with enol ether, **1**, 133
Ammonium fluoride, tetra-*n*-butyl- (TBAF)
  as catalyst, for the synthesis of aminals, **4**, 406
  for the desilylation of vinylsilanes, **1**, 451
  for the displacement of trifluoromethyl from silicon, **6**, 14
  as elimination agent, for the synthesis of 1-adamantylphosphaethyne, **5**, 1155
  reactions, with (trimethylsilyl)trifluoromethane, **2**, 70
  use in the synthesis of fluorobromoalkenes, **4**, 751
  use in the synthesis of *gem*-isothiocyanato silylalkenes, **4**, 1015
Ammonium fluoride hydrofluoride, electrolysis of, **6**, 410
Ammonium fluorides, tetraalkyl-, for the displacement of halides, **2**, 19
Ammonium halides, quaternary
  reactions
    with phenyllithium, **1**, 628
    with potassium amide, **1**, 628
Ammonium halides, tetraalkyl-, cleavage of, **1**, 34
Ammonium halides, tetrabutyl-, reactions, with sulfonate esters, **2**, 9
Ammonium halides, triethyl-, carbonylation of, **1**, 254
Ammonium halide salts, reactions, with Grignard reagents, **1**, 253
Ammonium hydrochloride, trimethyl-, reactions, with thiopyrophosphates, **2**, 257
Ammonium hydrogen fluoride, reactions, with phosgene, **6**, 409
Ammonium hydroxide
  reactions
    with mercury(II) oxide, **5**, 1026
    with 4-(trifluoromethyl)imidazoles, **3**, 673
Ammonium iodate, tetrabutyl-, for the oxidation of disulfides, **2**, 232
Ammonium iodide, dibenzyltrimethyl-, as starting material, for Sommelet–Hauser products, **1**, 399
Ammonium iodide, quaternary, synthesis of, **1**, 627
Ammonium iodide, tetrabutyl-, for the synthesis of aryl iodides, **2**, 631
Ammonium iodide, tetraethyl-, use in the synthesis of alkanethiols, **2**, 121
Ammonium iodides, allyl trialkyl-, substitution of, **1**, 486
Ammonium iodides, aryl trialkyl-, substitution of, **1**, 486
Ammonium iodides, silyl-, fragmentation of, **1**, 655
Ammonium ions, quaternary, synthesis of, **5**, 1127
Ammonium nitrate, use in the synthesis of disulfides, **2**, 222
Ammonium nitrate, ceric (CAN), reactions, with organotins, **4**, 252
Ammonium nitrate, ceric triethyl- (CTAN), reactions, with tetrahydrofuran, **4**, 201
Ammonium perbromides, tetraalkyl-, use in the synthesis of alkyl bromides, **2**, 28
Ammonium permanganate, cetyltrimethyl-, reactions, with arylhydrazones, **3**, 134
Ammonium perruthenate, tetrapropyl- (TPAP), for the oxidation of sulfides, **2**, 155
Ammonium salt, dicyclohexyl-, reactions, with thionyl chloride, **5**, 9
Ammonium salts
  as alkylating reagents, **2**, 299
  S-alkylation of, **6**, 479
  dehydration of, **3**, 617
  deprotonation of, **1**, 407
  reactions
    with alkali arsenes, **4**, 499
    with alkali hydroxides, **1**, 627
  synthesis of, **2**, 760; **4**, 455
Ammonium salts, acyl-, as intermediates, for the synthesis of ketenes, **3**, 526
Ammonium salts, alkyl-, synthesis of, **4**, 977
Ammonium salts, aryl-, synthesis of, **4**, 977
Ammonium salts, bisquaternary, pyrolysis of, **1**, 1075
Ammonium salts, *N*-chloro-
  from, tertiary amines, **2**, 338
  for the *para* chlorination of arenes, **2**, 620
Ammonium salts, cyano-, synthesis of, **5**, 1126
Ammonium salts, *N*-cyano-*N*,*N*,*N*-trialkyl-, synthesis of, **5**, 1120
Ammonium salts, cyclobutylidene-, synthesis of, **2**, 758
Ammonium salts, dialkoxymethyl-, reactions, with malonates, **4**, 193
Ammonium salts, *N*-fluoro-, synthesis of, **2**, 336
Ammonium salts, *N*-oxo-, use in the synthesis of hydroxy aldehydes, **3**, 31
Ammonium salts, quaternary
  condensation of, **1**, 255
  deprotonation of, **1**, 823
  dequaternization of, **2**, 303
  of dithioic acids, **5**, 546
  elimination of, **1**, 627
  from
    2-aryl-3-(*N*,*N*-dimethylamino)-1-propenes, **1**, 253
    diisopropylethylamine, **1**, 254
  reactions
    with diphenylphosphine, **4**, 483
    with enolate anions, **1**, 253
    with tertiary phosphines, **2**, 446
    with tributylphosphine, **2**, 447
  reduction of, **1**, 33
  synthesis of, **4**, 480; **5**, 497
  use in the substitution of nitrogen functions, **1**, 252
Ammonium salts, *N*-silyl-, deprotonation of, **4**, 512
Ammonium salts, β-silyltetraalkyl-, elimination of, **1**, 629
Ammonium salts, tetraalkyl-
  pyrolysis of, **4**, 674
  synthesis of, **2**, 1000
  use in electroreduction of carbonyls, **2**, 81
Ammonium salts, tetraethyl-, as coupling reagents, **6**, 469
Ammonium salts, tetramethyl-, protonation of, **5**, 933
Ammonium salts, triethyl-, phosphorylation of, **5**, 535
Ammonium salts, trimethylsilyl-, desilylation of, **1**, 823
Ammonium salts, (trimethylsilylethyl)-, as starting materials, for ylides, **4**, 512
Ammonium tosylates, alkyl-
  from, acid amides, **2**, 320
  synthesis of, **5**, 999
Ammonium triacetoxyborohydride, tetramethyl-, use in the reduction of ketones, **2**, 62
Ammonium triacetoxyborohydride, tetramethyl- (TABH), use in the synthesis of *N*-alkoxy primary amines, **2**, 350
Ammonium tribromide, benzyltrimethyl-
  for the synthesis of aryl bromides, **2**, 626
  use in the synthesis of dialkyl sulfoxides, **2**, 149
Ammonium tribromide, tetraalkyl-, use in the synthesis of alkyl bromides, **2**, 30
Ammonium tribromide, tetrabutyl-, for the *para* bromination of phenols, **2**, 626
Ammonium trinitromethanide, reactions, with nitryl fluoride, **6**, 364
Ammonolysis
  of bis(trifluoromethyl)chloroarsine, **2**, 503
  as side reaction, in the synthesis of amidines, **5**, 749

11(20)-Amphilectene, 8,15-diisocyano-, synthesis of, **3**, 701
Amphotericin, synthesis of, **1**, 750
Amyl nitrite, reactions, with acetophenone, **5**, 823
i-Amyl nitrite, reactions, with diphenylvinylamine, **2**, 800
Andersen procedure
  for the synthesis of dialkyl sulfoxides, **2**, 151
  for the synthesis of sulfoxides, **2**, 173
Anderson synthesis, for the synthesis of chiral vinyl sulfoxides, **2**, 713
Andrena wilkella, as a source of spiroketal pheromones, **1**, 165
Androstane, 17-keto-, homologation of, **3**, 643
5-Androstene-17β-carboxylic acid, 3β-acetoxy-, methyl anhydride of, hydrolysis of, **5**, 66
5-Androstene-3,17-dione, acidity of, **1**, 877
Androst-5-en-17-one, 3β-hydroxy-, reactions, with nitromethane, **2**, 785
Anesthetics
  bromochlorotrifluoroethane, **6**, 16
  halothane, **6**, 8
  isoflurane, **4**, 49
  synthesis of, **6**, 36
Angeli's salt, reactions, with pyrrolidine, **3**, 453
Angelic acid
  as starting material, for tigloyl chloride, **5**, 12
  synthesis of, reviews, **5**, 98
Angeloyl chloride, synthesis of, **5**, 13
Anhydride, acetic
  reactions
    with quinazoline-3-oxide, **3**, 722
    with silver acetate, **3**, 537
  as reagent, for the synthesis of nonclassical thiophenes, **3**, 372
Anhydride, benzeneseleninic
  as reagent
    for aldehyde regeneration, **3**, 18
    for oxime and hydrazone cleavage, **3**, 134
Anhydride, bis(trifluoroacetic), synthesis of, **3**, 534
Anhydride, citraconic, reactions, with *t*-butyl radicals, **1**, 327
Anhydride, formic acetic, reactions, with oximes, **3**, 712
Anhydride, isobutyric, pyrolysis of, **3**, 534
Anhydride, 1,8-naphthalic, reactions, with potassium hydroxide, **5**, 109
Anhydride, propanoic, use in the synthesis of tris(cyclopentadienyl)dinickel salts, **4**, 694
Anhydride, triflic, reactions, with triphenylphosphine oxide, **3**, 662
Anhydride, trifluoroacetic (TFAA)
  for the dehydration of amides, **3**, 619
  reactions
    with glyoxal α-oximes, **3**, 673
    with ketene dithioacetals, **3**, 253
Anhydride, *p*-trifluoromethylbenzoic, use in the synthesis of aromatic ketones, **3**, 281
Anhydride, triphenylaceticbenzoic, reactions, with lithium metal, **3**, 84
Anhydrides
  as acylating reagents, **5**, 322
    for the synthesis of aromatic ketones, **3**, 280
  from
    acid chlorides, **5**, 189
    acyl chlorides, **5**, 182
    alkenes, **5**, 192
    anhydrides, **5**, 193
    carboxylic acids, **5**, 184
    metal carboxylates, **5**, 182
    phosphorodiamidic chlorides, **5**, 188
    trimethylsilyl esters, **5**, 189, 196
  hydrolysis of, **5**, 26
  olfactory properties of, **5**, 183
  oxidation of, **5**, 205
  phosgenation of, **5**, 995
  reaction rates, **5**, 130
  reactions
    with acetals, **4**, 211
    with bis(trialkylstannyl)carbodiimides, **5**, 389
    with bis(trimethylstannyl)pentafluoroaniline, **5**, 389
    with esters, **6**, 39
    with formate esters, **6**, 39
    with hydrazones, **5**, 362
    with isocyanates, **5**, 305
    with pyridylzincs, **3**, 311
    with sulfonamides, **5**, 342
    with sulfur tetrafluoride, **6**, 39
    with triphenylphosphineimides, **5**, 373
  reduction of, **3**, 84
  as starting materials
    for *N*-alkylamides, **5**, 277
    for anhydrides, **5**, 193
    for aromatic acids, **5**, 109
    for isocyanates, **5**, 995
    for ketene aminals, **4**, 972
    for primary amides, **5**, 274
    for trithioortho esters, **6**, 84
  synthesis of, **5**, 182
    in situ, **3**, 280
    polymer assisted, **5**, 184
  use in *N*-acylation reactions, **5**, 277
Anhydrides, acetic, as starting materials, for ketenes, **3**, 534
Anhydrides, acetic nitronic, synthesis of, **3**, 435
Anhydrides, alkyl aryl, synthesis of, **5**, 199
Anhydrides, aromatic
  from
    arenediazonium tetrafluoroborates, **5**, 197
    arenes, **5**, 197
Anhydrides, aryl alkyl
  pyrolysis of, **5**, 197
  synthesis of, **5**, 197
Anhydrides, carboxylic
  reactions
    with alkynyl metals, **1**, 1045
    with lithium alkynyltrifluoroborates, **1**, 1045
Anhydrides, crotonic, as starting materials, for vinylketenes, **3**, 535
Anhydrides, cyclic
  for the acylation of amines, **5**, 305
  from
    cyclic anhydrides, **5**, 199
    dicarboxylic acids, **5**, 199
  reactions
    with alkoxyamines, **5**, 333
    with bis(triphenylphosphine)nickel dicarbonyl, **1**, 564
  as starting materials, for cyclic anhydrides, **5**, 199
  synthesis of, **5**, 184, 196
Anhydrides, diphenylacetic, synthesis of, **3**, 534
Anhydrides, ethanoic, reactions, with dithiolalkanes, **4**, 1074
Anhydrides, halo, synthesis of, **5**, 191
Anhydrides, heteroaromatic
  from, carboxylic acids, **5**, 198
  synthesis of, **5**, 197
Anhydrides, heterocyclic
  from, heterodienes, **5**, 201
  synthesis of, **5**, 196
Anhydrides, malonic acid
  as starting materials
    for dialkylketenes, **3**, 541
    for ketenes, **3**, 534
Anhydrides, mixed
  from
    carboxamides, **5**, 716
    carboximidic acids, **5**, 709
    ethyl chloroformate, **5**, 264
    isonitriles, **5**, 714
    nitrobenzoic acids, **5**, 714
  reactions
    with sodium azide, **5**, 991

with sodium borohydride, **2**, 305
as starting materials, for ketenes, **3**, 534
for the synthesis of *N*-acylamides, **5**, 306
use in the synthesis of amides, **5**, 264
use in the synthesis of carboxylic esters, **5**, 123
use in the synthesis of ketones, **3**, 296
Anhydrides, phosphinic, reactions, with amines, **4**, 127
Anhydrides, symmetrical aromatic
from
carboxylic acids, **5**, 195
metal carboxylates, **5**, 194
synthesis of, **5**, 194
Anhydrides, symmetrical heteroaromatic
from
acyl chlorides, **5**, 194
carboxylic acids, **5**, 195
synthesis of, **5**, 194
Anhydrides, unsymmetrical aromatic
from
arenediazonium tetrafluoroborates, **5**, 198
carboxylic acids, **5**, 198
synthesis of, **5**, 197
polymer assisted, **5**, 198
Anhydronucleosides, x-ray structural analysis of, **6**, 322
Anhydrosugars
ring-opening of, **2**, 1099
synthesis of, **4**, 228
trans diaxial openings in, **1**, 218
Anilide, thionyl, reactions, with chlorotrimethylsilane, **4**, 370
Anilides, hydroxy, from, phenoxy amides, **5**, 292
Aniline
phenylselenomethylation of, **4**, 333
reactions
with acetylsulfenyl chloride, **5**, 254
with azodicarboxylate, **2**, 777
with 2-benzoylbenzoic acid, **4**, 299
with chloroformamidinium salts, **6**, 645
with epoxides, **2**, 318
with maleic anhydride, **5**, 95
with perfluoroacrylic ester, **4**, 894
with perfluoromethacrylic esters, **3**, 588
with phenoxyfluoroxonium (fluorosulfonyl) methylide, **4**, 282
with *O*-phosphorylisouronium salts, **6**, 645
with α-sulfonylacetaldehydes, **2**, 742
with *N*,*N*'-thiocarbonylbis (3,5-dimethyl)pyrazole, **6**, 578
with titanium tetrachloride, **4**, 796
with *p*-toluenesulfinic acid, **2**, 797
with vinylidene dihalides, **4**, 970
with vinylsilanes, **4**, 507
as starting material, for hydrazones, **3**, 446
Aniline, 4-aminobenzyl-, reactions, with phosgene, **5**, 971
Aniline, *N*-benzyl-, reactions, with Caro's acid, **2**, 782
Aniline, benzylidene-
reactions
with phosphorus trichloride, **5**, 785
with triphenylarsonium benzylide, **1**, 741
Aniline, benzylidine-, reactions, with 4-hydroxycoumarin, **4**, 304
Aniline, *N*,*N*-bis(*s*-butyl)-, synthesis of, **2**, 301
Aniline, *N*,*N*-bisfluorocarbonyl-, from, phenyl isocyanate, **6**, 438
Aniline, bis(trimethylstannyl)pentafluoro-, reactions, with anhydrides, **5**, 389
Aniline, bromo-*N*,*N*-bis(trimethylsilyl)-, synthesis of, **2**, 813
Aniline, *N*-*s*-butyl-, alkylation of, **2**, 301
Aniline, 4-chloro-, reactions, with (*Z*)-fluoroimines, **5**, 761
Aniline, *N*-chloro-, synthesis of, **2**, 761
Aniline, *p*-chloro-, reactions, with *N*-acylketenimines, **4**, 980, 981
Aniline, 5-chloro-2-nitro-, reactions, with 1,3-thiazolidine-2-thione 6: 295–31830
Aniline, *N*,*N*-dichloro-, synthesis of, **2**, 761
Aniline, dimethyl-, reactions, with methyl benzenesulfonate, **2**, 760
Aniline, 2,6-dimethyl-, reactions, with cyclic ketones, **5**, 292
Aniline, *N*,*N*-dimethyl-
chlorination of, **6**, 607
metallation of, **1**, 148
reactions
with fluoroborate salts, **1**, 109
with phosphorus trichloride, **2**, 829
synthesis of, **1**, 50
Aniline, *N*-ethyl-, reactions, with ketene diethyl acetals, **4**, 887
Aniline, *N*-ethyl-*N*-methyl-, oxidation of, **4**, 311
Aniline, fluoro-, reduction of, **1**, 426
Aniline, *N*-methyl-
N-methoxymethylation of, **4**, 300
reactions
with alkoxydichloromethyl chlorocarbonyl disulfides, **6**, 497
with *N*-thiobenzoylimidazole, **5**, 587
with *N*-thiobenzoyl 1,2,4-triazole, **5**, 587
with triethyl orthoformate, **6**, 140
use in the synthesis of aminals, **4**, 412
Aniline, 4-nitro-
reactions
with bisdiethylthiocarbamoyl disulfide, **5**, 1029
with bisdiethylthiocarbamoyl sulfide, **5**, 1029
Aniline, *p*-nitro-
reactions
with difluorodinitromethane, **6**, 260
with triphenylarsine oxide, **2**, 810
Aniline, (nitrophenylamino)-, reactions, with *O*-phenyl chlorothioformate, **6**, 556
Aniline, *p*-nitroso-*N*,*N*-dimenthyl-, reactions, with 4,6-diphenylpyridinium-2-carboxylates, **3**, 433
Aniline, *p*-nitroso-*N*,*N*-dimethyl-, synthesis of, **2**, 779
Aniline, pentafluoro-, as starting material, for pentafluorophenylcarbonimidic dichloride, **6**, 609
Aniline, *N*-(phenylselenomethyl)-, synthesis of, problems, **4**, 333
Aniline, *N*-pivaloyl-, reactions, with carbon monoxide, **5**, 438
Aniline, *N*-sulfinyl-, reactions, with Grignard reagents, **2**, 797
Aniline, 2,4,6-trichloro-
as starting material
for dibromodinitromethane, **6**, 260
for dichlorodinitromethane, **6**, 260
Aniline, *N*-(trimethylsilyl)-, reactions, with imino chlorides, **5**, 777
Anilines
acylation of, reagents for, **5**, 290
condensation of, **2**, 326
cyanation of, **3**, 664
diazotization of, for aryl halides, **2**, 609
protection of, **2**, 543
reactions
with *S*-alkylisothiouronium salts, **6**, 644
with cyanogen bromide, **5**, 1125
with 2,2-dichloro-1,3-benzodioxole, **6**, 646
with dichlorocarbenes, **6**, 140
with sulfinyl chlorides, **2**, 796
with sulfonic acid, **6**, 645
reduction of, **1**, 87, 89
as starting materials
for aryl bromides, **2**, 626, 627
for aryl chlorides, **2**, 621
for aryl cyanamides, **5**, 1125
for aryl fluorides, **2**, 614
for aryl iodides, **2**, 632
for *N*-aryl *S*-methyl dithiocarbamates, **6**, 563
use in the synthesis of *N*-arylamides, **5**, 292

Anilines, acyl-, synthesis of, **3**, 300
Anilines, *N*-alkyl-
oxidation of, **4**, 311
reactions
with amines, **4**, 316
with cyanogen bromide, **5**, 1126
Anilines, *N*-alkyl-*N*-allyl-, 3-aza-Cope rearrangement of, **1**, 835
Anilines, 2-alkyl-substituted, reduction of, **1**, 99
Anilines, benzylidene-, reactions, with thiols, **4**, 319
Anilines, *N*-benzylidene-, synthesis of, **3**, 412
Anilines, *N,N*-bis(arylthiomethyl)-, synthesis of, **4**, 316
Anilines, bromo-, reactions, with *n*-butyllithium, **2**, 813
Anilines, *C*-chlorobenzylidene-, as starting materials, for imidoylsilanes, **5**, 793
Anilines, *o*-cyano-, synthesis of, **3**, 664
Anilines, *N*-cyanomethyl-*N*-methyl-, synthesis of, **2**, 305
Anilines, diisopropyl-, synthesis of, **1**, 1071
Anilines, *N,N*-dimethyl-
oxidation of, **2**, 305
reactions
with iodosylbenzene, **4**, 443
with trimethylsilyl azide, **4**, 443
as starting materials, for azidomethyls, **4**, 443
Anilines, *N*-fluorocarbonyl-*N*-trifluoromethyl-, synthesis of, **6**, 438
Anilines, *N*-(methoxymethyl)-, as starting materials, for imines, **4**, 412
Anilines, methyl-*N*-acetyl-, as starting materials, for *N*-acetylamino benzaldehydes, **3**, 99
Anilines, *p*-nitro-, synthesis of, **2**, 791
Anilines, secondary, oxidation of, **3**, 412
Anilines, *o*-silyl-*N*-dialkyl-, synthesis of, **2**, 945
Anilines, thiocyanato-, phosgenation of, **5**, 966
Anilines, tris(phosphoryl)methyl-
from, *N*-phenyldichloroimine, **6**, 367
synthesis of, **6**, 367
Aniline salts, reactions, with tetrahydro-2*H*-1,3-oxazin-2-one, **2**, 320
Anion centres
flanked by a halogen and an oxygen, **4**, 1076
flanked by oxygen and phosphorus, **4**, 1075
flanked by two halogens, **4**, 1076
flanked by two phosphorus atoms, **4**, 1076
flanked by two sulfur atoms, **4**, 1075
Anion exchange resins, for the deprotection of amines, **2**, 304
Anionic cleavage, use of term, **1**, 883
Anionic complexes, reactions, with iodo epoxides, **5**, 955
Anionotropic transformation, of carbodiimides, **5**, 1080
Anions
flanked by a boron and sulfur, **4**, 1077
flanked by silicon, **4**, 1077
from, α-alkoxy alkylphosphorus compounds, **4**, 917
reactions
with electrophiles, **4**, 917
with iron complexes, **5**, 951
with trialkylboranes, **1**, 125
synthesis of, **1**, **843–951**, **4**, 1075
via *N*-deprotonation of amides, **5**, 723
use in the synthesis of carboxylic acids, **5**, 76
Anions, allylic, reactions, with carbonyls, **2**, 73
Anions, boron-stabilized, alkylation of, **1**, 178
Anions, dimesitylboron-stabilized, for the opening of epoxides, **1**, 212
Anions, diphenylphosphinoyl-stabilized, acylation of, **1**, 756
Anions, dipole-stabilized, alkylation of, **1**, 148
Anions, nucleophilic, reactions, with epoxides, **5**, 946
Anions, organometallic, reactions, with ferrocenoyl chloride, **4**, 722
Anions, selenium-stabilized, synthesis of, **1**, 713
Anions, silicon-stabilized, from, methoxy bis(trimethylsilyl)methane, **5**, 417
Anions, α-silylalkyl, synthesis of, **2**, 1005
Anions, α-sulfenyl, as starting materials, for dialkyl sulfides, **2**, 143
Anions, sulfoxide-stabilized, reactions, with acid compounds, **3**, 170
Anions, sulfur-stabilized
chemistry of, **4**, 390
for the opening of epoxides, **1**, 210
Anions, tellurium-stabilized, alkylation of, **1**, 716
Anisaldehyde, reactions, with 4-nitrobenzylidenetriphenylphosphorane, **1**, 731
p-Anisate, methyl, hydrolysis of, **5**, 106
Anisole
para iodination of, for the synthesis of aryl iodides, **2**, 631
reactions
with potassium cyanate, **5**, 986
with sulfuryl chloride, **4**, 47
reduction of, **1**, 98
use in the synthesis of alkoxyaryl ketones, **3**, 298
Anisole, *p*-chloro-
reduction of, **1**, 428
synthesis of, **4**, 47
Anisole, *p*-iodo-, synthesis of, **2**, 631
Anisole, *o*-lithiated, acylation of, **3**, 298
Anisole complexes, reactions, with butyllithium, **4**, 707
Anisoles, substitution of, **3**, 664
Anisoles, *o*-cyano-, synthesis of, **3**, 664
Anisoles, *p*-cyano-, synthesis of, **3**, 664
(-)Anisomycin, synthesis of, via xanthate pyrolysis, **1**, 776
β-Anisyl-α,β-unsaturated acids, oxidative decarboxylation of, **1**, 1121
Annulation
of acyclic fragments, for the synthesis of phenols, **2**, 662
for the synthesis of γ-lactones, **5**, 170
[3+3] Annulation, for the synthesis of phenols, **2**, 662
[4+2] Annulation
of alkenylketenes, **2**, 668
for the synthesis of phenols, **2**, 663
cis,cis,cis,trans-[9]Annulene anions, synthesis of, **1**, 886
Annulenes
metal reduction of, **1**, 890
synthesis of, **1**, 1012
Antarafacial addition, mechanism, **1**, 349
Antheraea polyphemus, sex pheromones, **1**, 455
Antheridiogen-A_n, synthesis of, **1**, 801, 804
Anthracene
formylation of, **3**, 101
Friedel–Crafts acylation of, **3**, 301
from, tetranitromethane, **6**, 147
reactions
with magnesium, **2**, 561, 959
with thioacyl halides, **5**, 492
reduction of, **1**, 101
use in the synthesis of alkylmagnesiums, **2**, 559
use in the synthesis of perhalogenated aliphatic halosulfines, **5**, 492
Anthracene, 9-benzoyl-, reactions, with sodium cyanide, **3**, 667
Anthracene, dicyano-, use in the synthesis of carbonyl ylides, **3**, 317
Anthracene, 9,10-dicyano-, synthesis of, **3**, 667
Anthracene, 9-formyl-, synthesis of, **3**, 101
Anthracene, 9-phenyl-, synthesis of, **3**, 284
Anthracene adducts
pyrolysis of, **1**, 785
as starting materials
for alkenes, **1**, 785
for silenes, **5**, 926
Anthracenes
irradiation of, for the synthesis of ketenes, **3**, 538
retro-Diels–Alder cleavage of, **1**, 787
synthesis of, **4**, 258
Anthracenes, 9-*t*-alkyl-, synthesis of, **1**, 265
Anthracines, 9,10-dimethyl-, retro-Diels Alder reaction of, **5**, 326

Anthracyclinones, synthesis of, **3**, 539
Anthranil, 3-methyl-, oxidation of, **2**, 782
Anthranilamides, reactions, with ethyl propiolate, **4**, 422
Anthranilic acid
  diazotization of, with alkyl nitrites, **6**, 99
  reactions, with phosgene, **5**, 969
Anthroic acid, 5-formyl-4-phenyl-, synthesis of, **5**, 115
Anthrone
  reactions
    with *t*-alkyl-magnesium halides, **1**, 265
    with 2-cyanopropyl radicals, **1**, 265
    with dimethylazobisisobutyrate, **1**, 266
Antibiotics
  A-32390A, synthesis of, **3**, 719
  antitumor macrolide, **1**, 1134
  bicyclomycin, synthesis of, **1**, 654
  O′,*O′*-dimethylxanthocillin, **3**, 715
  elaiomycin, **2**, 773
  frustulosin, synthesis of, **1**, 1136
  (+)-hitachimycin, synthesis of, **6**, 488
  (-)-ptilocaulin, synthesis of, **1**, 434
  (+)-streptazolin, synthesis of, **2**, 331
  synthesis of, **1**, 702; **3**, 717; **4**, 205
  tunicamycins, synthesis of, **4**, 239
Antibodies, catalytic monoclonal, synthesis of, **1**, 826
Anti-elimination, under acidic conditions, **1**, 649
Antihypertensive agents, pyrrolidines as, **1**, 256
Anti-inflammatory drugs, ibuprofen, **1**, 298; **5**, 118
Antimalarial drugs, synthesis of, **1**, 45
Anti-Markovnikov addition
  of hydrogen chloride, via hydroboration, **2**, 23
  of hydrogen halides, **2**, 4
  mechanism, **1**, 948
  for the synthesis of alkanethiols, **2**, 117
  in the synthesis of alkyl bromides, **2**, 29
  for the synthesis of tertiary phosphines, **2**, 439
  to alkenes, **3**, 614
  via hydroalumination, **2**, 34
  via hydroboration, **2**, 34
  via hydrozirconation, **2**, 34
Antimonane, 1-chloro-, synthesis of, **2**, 491
Antimonate, chlorodiphenylcarbenium hexachloro-, synthesis of, **2**, 1006
Antimonate, (4-chlorophenyl)-(2,6-dimethylphenyl)trifluoromethylsulfonium hexafluoro-, as trifluoromethylation agent, **6**, 233
Antimonate, 2,2-dichloroethylene diazonium hexachloro-, synthesis of, **2**, 800
Antimonate, diphenylboryl hexachloro-, use in the synthesis of aromatic ketones, **3**, 281
Antimonate, nitrosyl hexachloro-, reactions, with isocyanates, **2**, 800
Antimonate, triethyloxonium hexachloro-, reactions, with α-diazo esters, **2**, 799
Antimonate, trityl tetrachloro-, reactions, with ortho esters, **4**, 1074
Antimonates, alkyl carbocation hexafluoro-, stability of, **1**, 899
Antimonates, alkyl carbocationic hexafluoro-, synthesis of, **1**, 899
Antimonates, *N*-benzoyl-*C*-(methylthio)nitrilium hexachloro-, synthesis of, **5**, 1146
Antimonates, cyanamidinium hexachloro-, synthesis of, **5**, 1146
Antimonates, nitrilium hexachloro-, synthesis of, **3**, 687
Antimonates, oxocarbonium hexafluoro-, use in the synthesis of aromatic ketones, **3**, 282
Antimonate salts, hexachloro-
  synthesis of, **3**, 687; **6**, 610, 729
  use in the synthesis of arylarsonic acids, **2**, 884
Antimonate salts, hexafluoro-, as starting materials, for aryl fluorides, **2**, 614
Antimonium bromide, tributylcyanomethyl-, synthesis of, **3**, 639
Antimony
  analogues of ketenes, theoretical, **3**, 602
  α-haloalkenyl compounds of, synthesis of, **4**, 805
  heteroaromatic compounds of, **2**, 879
  reactions, with alkyl halides, **2**, 486
  as starting material
    for phenylstibines, **2**, 872
    for vinylstibines, **2**, 872
  synthesis of functions, **4**, **591–600**
    with no halogen, chalcogen or nitrogen, **4**, **1021–1042**
Antimony, dialkanethiotrimethyl-, synthesis of, **2**, 502
Antimony, diethoxytrimethyl-
  reactions
    with thioacetic acid, **2**, 502
    with thiobenzoic acid, **2**, 502
Antimony, doubly-bonded, synthesis of functions, **6**, **677–724**
Antimony, pentaaryl-, reactions, with carboxylic acids, **2**, 888
Antimony, pentakis[(trimethylsilyl)methyl]-, synthesis of, **3**, 500
Antimony, pentaphenyl-, reactions, with chlorine, **1**, 446
Antimony, triaryl-, reactions, with carboxylic acids, **2**, 888
Antimony, tributyl-, reactions, with bromoacetonitrile, **3**, 639
Antimony, triethynyl-, synthesis of, **2**, 1081
Antimony, trimethylbis(trimethylsiloxy)-, reactions, with hydrogen chloride, **2**, 492
Antimony, triphenyl-
  reactions
    with *N*-chloro-*N*,*N′*-dimethylbenzamidine, **5**, 776
    with diazo compounds, **1**, 54
  as starting material, for diphenylantimony(III) acetate, **5**, 222
Antimony, tri(phenylethynyl)-
  reactions, with dicobaltoctacarbonyl, **2**, 1082
  synthesis of, **2**, 1081
Antimony, tris(trifluorovinyl)-, synthesis of, **4**, 805
Antimony(I), vinyl-, synthesis of, **2**, 879
Antimony(III), chlorodicyano-, reactions, with sodium cyclopentadienide, **5**, 1132
Antimony(III) acetate, diphenyl-
  from, triphenylantimony, **5**, 222
  synthesis of, **5**, 222
Antimony(III) amides, reactions, with carbon disulfide, **2**, 889
Antimony(III) chloride, as starting material, for dichlorodifluoromethane, **6**, 225
Antimony(III) chloride, diphenyl-, reactions, with acetic acid, **5**, 222
Antimony(III) compounds, synthesis of, **5**, 222
O-Antimony(III) enolates, synthesis of, **2**, 695
Antimony(III) enolates, of 1,3-dicarbonyls, **2**, 695
Antimony(III) fluoride
  for the synthesis of tertiary trifluoromethylamines, **6**, 238
  for the synthesis of trifluoromethylsilane, **6**, 243
  for the synthesis of trifluoromethyl sulfides, **6**, 232
Antimony(III) halides
  oxidative hydrolysis of, **2**, 498
  reactions, with organometallics, **2**, 872
Antimony(III) phenolates, synthesis of, **2**, 695
Antimony(III) salt, diphenyl-, reactions, with sodium enolates, **2**, 695
Antimony(V), as catalyst, for the synthesis of trihalides, **6**, 8
Antimony(V) acid compounds, alkyl-, synthesis of, **2**, 498
Antimony(V) analogues, bis(acyloxy)triaryl-, synthesis of, **5**, 222
Antimony(V) azide, tetrachloro-, reactions, with tetrachloromethane, **6**, 733
Antimony(V) chloride
  as catalyst, **6**, 217
  reactions
    with cyanogen bromide, **6**, 613

with cyanogen iodide, **6**, 613
Antimony(V) chloride complexes, dialkyl cyanamide, reactions, with *t*-alkyl halides, **5**, 1146
Antimony(V) enolates, synthesis of, **2**, 695
Antimony(V) fluoride
  Friedel–Crafts (trifluoromethyl)thiothioacylations with, **6**, 235
  reactions, with epoxides, **3**, 122
Antimony(V) halides, synthesis of, **2**, 492
Antimony–amidine complexes, synthesis of, **5**, 776
Antimony anions, alkyl-, reactions, with alkyl halides, **2**, 485
Antimony–antimony bonds, cleavage of, **2**, 491
Antimony–benzamidine complexes, synthesis of, **5**, 776
Antimony bromides, *t*-butyl-, for the formation of antimony–antimony bonds, **2**, 509
Antimony–carbon bonds, cleavage of, **2**, 490
Antimony chloride, diphenyl-
  reactions
    with acetamide, **5**, 369
    with succinimide, **5**, 369
Antimony chlorides, for liquid phase reactions, **6**, 220
Antimony compounds
  alkali metals of, synthesis of, **2**, 509
  main group metal compounds of, synthesis of, **2**, 510
  synthesis of, **4**, 598, 1042
  transition metals of, synthesis of, **2**, 511
Antimony compounds, acyloxy-, synthesis of, **5**, 222
Antimony compounds, alkyl-, synthesis of, **2**, **479–512**
Antimony compounds, alkynyl-
  from
    alkynyl compounds, **2**, 1082
    alkynyllithiums, **2**, 1080
  methylation of, **2**, 1082
  safety precautions, **2**, 1076
  synthesis of, **2**, 1076, 1080
Antimony compounds, α-aminoalkyl-, synthesis of, **4**, 504
Antimony compounds, cyano-, synthesis of, **5**, 1132
Antimony compounds, α-halo-, synthesis of, **4**, 136
Antimony compounds, imidoyl-, synthesis of, **5**, 792
Antimony compounds, pentaalkyl-, synthesis of, problems, **2**, 488
Antimony compounds, tetracoordinated, synthesis of, **6**, 707
Antimony compounds, triaryl-, reactions, with 1,3-diazocarbonyls, **3**, 196
Antimony compounds, triethynyl-, stability of, **2**, 1076
Antimony dibromide, trialkyl-, reactions, with alkyllithium, **2**, 488
Antimony dibromide, tris[(trimethylsilyl)methyl]-, trimethylsilylmethylation of, **3**, 500
Antimony dibromides, tricyclohexyl-, reactions, with sodium sulfide, **2**, 502
Antimony dibromides, triethylcyclohexyl-, reactions, with sodium sulfide, **2**, 502
Antimony dichloride, phenyl-
  reactions, with 1,1-di-*n*-butyl-2,5-dimethylstannole, **4**, 1041
  use in the synthesis of stiboles, **4**, 1039
Antimony dichloride, triethyl-, reactions, with diethylmagnesium, **2**, 488
Antimony dichloride, trimethyl-
  reactions, with thiols, **2**, 502
  as starting material, for trimethylantimony pseudohalides, **2**, 504
  synthesis of, **2**, 492
Antimony dichloride, tris(trifluoromethyl)-, synthesis of, **2**, 492
Antimony dihalides, alkyl-, reactions, with sodium alkoxides, **2**, 496
Antimony dihalides, trialkyl-
  reactions, with organometallics, **2**, 488
  synthesis of, **2**, 492
  thermal decomposition of, **2**, 489
Antimony diisocyanates, trialkyl-, synthesis of, **5**, 1017
Antimony enolates, synthesis of, **2**, 694
Antimony fluorides, reactions, with hydrogen fluoride, **6**, 16
Antimony functions
  elimination of, **1**, 1079
  substitution of, **1**, 271, 1032
  synthesis of, **4**, 595
Antimony functions, acyl-, synthesis of, **5**, **393–407**
Antimony functions, aryl-, synthesis of, **2**, **871–897**
Antimony functions, dicoordinated, synthesis of, **5**, 892
Antimony functions, imidoyl-, synthesis of, **5**, 784
Antimony functions, tetracoordinated, synthesis of, **5**, 922
Antimony functions, tricoordinated, synthesis of, **5**, 894
Antimony functions, vinyl-, synthesis of, **2**, **871–897**
Antimony groups, elimination of, **1**, 647
Antimony halide, dimethyl-, nucleophilic metallation of, **2**, 512
Antimony halide catalysts, mixed, use in the synthesis of aryl sulfides, **2**, 712
Antimony halides
  reactions
    with imides, **5**, 372
    with metal alkynides, **2**, 1080
    with organometallics, **2**, 483
  synthesis of, **2**, 880
Antimony halides, alkyl-
  reactions
    with amines, **2**, 502
    with ammonia, **2**, 502
    with metal amides, **2**, 502
    with organometallics, **2**, 483
    with transition metal complex anions, **2**, 511
  reduction of, **2**, 481
Antimony halides, dialkyl-, synthesis of, **2**, 489
Antimony halides, diphenyl-, reactions, with bis(tributylstannyl)alkynes, **2**, 1081
Antimony hydrides, synthesis of, **1**, 446
Antimony hydrides, dialkyl-, synthesis of, **2**, 481
Antimony imides, synthesis of, **5**, 369
Antimony isocyanates, synthesis of, **5**, 1016
Antimony oxide, for the synthesis of tertiary alkylstibines, **2**, 485
Antimony oxide, triphenyl-, synthesis of, **1**, 54
Antimony oxides, trialkyl-
  reactions
    with carboxylic acids, **5**, 222
    with isocyanic acid, **5**, 1017
Antimony oxides, triaryl-, reactions, with carboxylic acids, **5**, 222
Antimony pentachloride
  reactions
    with arylamidoximes, **5**, 776
    with ethyne, **2**, 881
    with 9-fluorenecarbaldehyde tosylhydrazone, **2**, 800
    with nitriles, **5**, 389
  use in the synthesis of dichloroalkanes, **4**, 18
Antimony pentafluoride
  reactions
    with dihaloalkanes, **2**, 1006
    with hydrogen fluoride, **6**, 38
    with tetrahalides, **6**, 726
  use in the synthesis of difluoroalkanes, **4**, 4
Antimony phenolates, synthesis of, **2**, 695
Antimony pseudohalides, trimethyl-, from, trimethylantimony dichloride, **2**, 504
Antimony sulfides, alkynyl-, synthesis of, problems, **2**, 1082
Antimonytriamide, hexamethyl-, reactions, with 2-butynedioic acid, **2**, 876
Antimony trichloride
  reactions
    with aluminum, **2**, 55
    with thiols, **2**, 261
    with trimethylsilylmethylmagnesium chloride, **4**, 598
    with zinc, **2**, 55

use in the synthesis of α-fluoro sulfides, **4**, 61
Antimony trichloride, diethyl-, hydrolysis of, **2**, 498
Antimony trichlorides
as catalysts, **6**, 229
reactions
with bis(trimethylsilyl)methane, **6**, 168
with *t*-butylmagnesium chloride, **2**, 484
with lithium tetraethylaluminate, **2**, 483
Antimony trifluoride
as fluorinating reagent, **6**, 222
reactions
with bromochloro ethers, **6**, 38
with dichloroalkyl sulfides, **6**, 45
with dichloro ethers, **6**, 38
with *gem*-dichlorophosphoryl isocyanatoalkenes, **4**, 1012
with dichloro sulfides, **6**, 47
with phosgene, **6**, 409
with silicon tetraisocyanate, **5**, 1016
with sulfonyl chlorides, **2**, 183
as starting material, for carbonic chloride fluoride, **6**, 419
use in fluorination reactions, **6**, 9
use in the synthesis of chlorofluoroalkanes, **4**, 30
use in the synthesis of difluoroalkanes, **4**, 4
Antimony trihalides
and guanidines, **6**, 656
reactions
with dialkylhalostibines, **2**, 490
with Grignard reagents, **2**, 483
with organometallics, **2**, 488
with tertiary alkylstibines, **2**, 490
Antimony trihalides, dialkyl-, thermolysis of, **2**, 489
Antimony ylides
from, stibines, **3**, 500
synthesis of, **3**, 500, **491–500**, **5**, 922
Antiradiation agents, 2-aminoethanethiols, **2**, 237
Anti selectivity, in radical addition reactions, **1**, 326
Apatite, hydroxy-, use in the synthesis of alcohols, **2**, 82
Appel reaction
for the synthesis of carboxylic esters, **5**, 131
for the synthesis of phosphoramidates, **2**, 413
Appel reagents, in the synthesis of iminocarbonyl chlorides, **6**, 617
Aprotic solvents, use in the synthesis of carboxylic esters, **5**, 129
β-L-Arabinopyranoside, methyl 2-*O*-methyl-, reactions, with Viehe's salt, **6**, 544
β-L-Arabinopyranoside, methyl 2-*O*-methyl-3,4-*O*-thiocarbonyl-, synthesis of, **6**, 544
Arachidonic acid, synthesis of, **1**, 1016
Arachidonic acid, 5,6-dehydro-, synthesis of, **1**, 1091
Aratani ligands, development of, **1**, 353
Arbuzov reaction
of alkynyl halides, with trialkyl phosphites, **2**, 1070
of aryl halides, with trialkyl phosphites, **2**, 848
of *N*-bromoimides, **5**, 372
of carbyne complexes, **5**, 952
of 1,1-dihaloalkanes, **4**, 555
of 1-haloalkylphosphorus compounds, **4**, 555
of 1-haloalkylsilanes, **4**, 569
on iodotrifluoroethene, **4**, 802
of phosphinate acetals, **6**, 117
of phosphite with phosgene, **6**, 455
for the synthesis of *gem*-alkylideneamino phosphorylalkenes, **4**, 1009
for the synthesis of *gem*-amido phosphorylalkenes, **4**, 1006
for the synthesis of α-aminoalkylphosphorus compounds, **4**, 464
for the synthesis of α-aminophosphorus compounds, **4**, 487
for the synthesis of carbon–phosphorus bonds, **4**, 340
for the synthesis of α-cyanophosphonates, **3**, 639
for the synthesis of formyl phosphonates, **3**, 50
for the synthesis of β-keto phosphonates, **3**, 196
for the synthesis of phosphinecarboxylates, **6**, 490
for the synthesis of phosphines, **4**, 913
for the synthesis of α-phosphonoacylsilane, **5**, 426
for the synthesis of phosphoramides, **5**, 371
for the synthesis of *N*-phosphorylamidines, **5**, 774
for the synthesis of polyphosphines, **2**, 472
for the synthesis of α,β-unsaturated ketones, **3**, 270
for the synthesis of α-unsubstituted phosphonates, **5**, 869
of trialkyl phosphites, **6**, 120
use in the synthesis of acylphosphorus compounds, **5**, 402
use in the synthesis of phosphinic acids, **2**, 453
See also Michaelis–Arbuzov reaction
Arbuzov rearrangement
of alkoxybis(dialkylamino)phosphines, **2**, 469
of 1,3,2-dioxaphosphacycloheptane, **2**, 453
of 1,3,2-oxazaphospholane, **2**, 453
Arene, thionitroso-, as intermediate, **2**, 793
π-Arene complexes, reactions, with methyl iodide, **2**, 582
Arene dichlorides, iodo-, as starting materials, for α-chlorophosphonium salts, **5**, 897
Arene oxides, synthesis of, **1**, 789
Arenes
for the activation of metals, **2**, 958
alkylation of, **1**, 462, 486
bromination of, for aryl bromides, **2**, 625
bromoalkylation of, **2**, 33
bromomethylation of, **2**, 13
^{77}Br-radiolabeled, synthesis of, **1**, 449
carboxylation of, **5**, 197
chlorination of, *ortho*:*para* ratio, **2**, 620
chloroalkylation of, **2**, 27
chloromethylation of, **2**, 14
dissolving metal reduction of, **1**, 95
electrochemical reduction of, **1**, 103
electrophilic hydroxylation of, **2**, 649
electrophilic oxidation of, **2**, 689
electrophilic substitution, **5**, 201
electrophilic sulfonylation of, **2**, 708
fluorination of, for aryl fluorides, **2**, 613
formylation of, **3**, 91
Friedel–Crafts sulfonylation of, **2**, 715
from
arylarsines, **1**, 445
aryl bismenes, **1**, 445
arylboranes, **1**, 447
aryl carbon–phosphorus bonds, **1**, 444
N-aryl-2,4-diphenyl-5,6-dihydrobenzo[*h*]quinolinium fluorides, **1**, 31
arylgermanes, **1**, 448
aryl halides, **1**, 425
arylsilanes, **1**, 447
arylstannanes, **1**, 452
arylstibenes, **1**, 445
aryl trimethylstannanes, **1**, 452
metallated arenes, **1**, 452
1-methoxy-2-oxazolinylarenes, **1**, 483
phenols, **1**, 432
haloalkylation of, **2**, 13
heterogeneous hydrogenation of, **1**, 87
homogeneous hydrogenation of, **1**, 93
hydride reduction of, **1**, 102
hydroxylation of, **2**, 39
iodination of, for aryl iodides, **2**, 631
iodoalkylation of, **2**, 36
iodomethylation of, **2**, 13
^{131}I-radiolabeled, synthesis of, **1**, 449
lithiation of, **2**, 965
metal-hydrogen exchange of, **2**, 956
metallation of, **2**, 965
via deprotonation with bases, **2**, 964
via electrophilic substitution, **2**, 963
meta photocycloaddition of, **1**, 364

non-redox metal-hydrogen exchange of, **2**, 963
ortho-lithiation of, **2**, 952
oxidative carbon–hydrogen activation of, **2**, 962
reactions
with carbon monoxide, **3**, 93
with chlorosulfonyl isocyanates, **5**, 1005
with α-cyano sulfoxides, **3**, 635
with electrophilic sulfur, **2**, 712
with epoxides, **2**, 85
with formyl cations, **3**, 101
with formyl fluoride, **3**, 93
with hexamethylenetetramine, **3**, 91
with lithium, **2**, 551
with [methoxy(phenylthio)methyl]trimethylsilane, **4**, 370
with organometals, **2**, 875
with selenium oxychloride, **2**, 729
with selenium tetrachloride, **2**, 729
with sulfamoyl chlorides, **2**, 365
with trichloroacetaldehyde, **6**, 27
with tris(phenylthio)methane, **3**, 98
reduction of, **1**, 87
as starting materials
for arenechromium tricarbonyl complexes, **1**, 202
for arenesulfinic acids, **2**, 718
for aromatic anhydrides, **5**, 197
for aryl chlorides, **2**, 619
for benzaldehydes, **3**, 97
for diaryl selenides, **2**, 729
for isocyanates, **5**, 984
for the synthesis of active metals, **2**, 959
trichloromethylation of, **6**, 29
Arenes, acyl-
Baeyer–Villiger oxidation of, **2**, 654
reactions, with methylmagnesium bromide, **2**, 655
as starting materials, for phenols, **2**, 654
synthesis of, **3**, 290
Arenes, acyloxyiodo-, synthesis of, **5**, 204
Arenes, (arenesulfonylimino)iodo-, as starting materials, for nitrenoids, **2**, 726
Arenes, bis(acyloxy)iodo-
from, bis(trifluoroacetoxy)iodobenzene, **5**, 205
synthesis of, **5**, 205
Arenes, 1,2-bis(cyanomethyl)-
use in the synthesis of benzazepines, **1**, 255
use in the synthesis of indenes, **1**, 255
Arenes, bis(halo)iodo-, acyloxylation of, **5**, 205
Arenes, bis(trifluoroacetoxyiodo)-, synthesis of, **5**, 204
Arenes, bromo-
from, aryl bromides, **1**, 428
reduction of, **1**, 428
Arenes, chloro-
from, aryl chlorides, **1**, 427
reduction of, **1**, 427
Arenes, *o*-di(formamido)-, as starting materials, for *o*-diisocyanoarenes, **3**, 723
Arenes, dihalo-
selective formylation of, **3**, 95
selective substitution of, **1**, 471
Arenes, *o*-diisocyano-, from, *o*-di(formamido)arenes, **3**, 723
Arenes, dimethyl-, dimetallation of, **5**, 43
Arenes, fluoro-
from, aryl fluorides, **1**, 426
reactions, with acetyl chloride, **1**, 448
as starting materials, for 2-fluorobenzaldehydes, **3**, 96
Arenes, halo-
coupling with organoboranes, **1**, 488
formylation of, **3**, 96
as starting materials
for alkyl aryl sulfides, **2**, 711
for benzaldehydes, **3**, 95
Arenes, iodo-
from, aryl iodides, **1**, 429
oxidative acylation of, **5**, 204
reactions
with 9-alkyl-9-borabicyclo[3.3.1]nonyls, **3**, 290
with arylboronic acids, **3**, 290
with diaryl cuprates, **1**, 473
with naphthyl boronic acids, **3**, 301
Arenes, iodoso-, acyloxylation of, **5**, 205
Arenes, metallated
oxidation of, **2**, 652
reactions
with aryl selenocyanates, **2**, 729
with diaryl diselenides, **2**, 729
as starting materials, for arenes, **1**, 452
Arenes, 1-methoxy-2-oxazolinyl-, as starting materials, for arenes, **1**, 483
Arenes, nitro-
acylation of, **3**, 299
formylation of, **3**, 99
from, aryl carbon–nitrogen bonds, **1**, 441
nucleophilic substitution of, **3**, 99
reactions
with Grignard reagents, **2**, 318
with phenylthiomethyl isocyanide, **3**, 702
with sodium oximate, **2**, 648
as starting materials, for phenols, **2**, 648
Arenes, nitroso-, reactions, with phenacylpyridinium bromide, **2**, 765
Arenes, polyhalo-, selective substitution of, **1**, 471
Arenes, stannylated, reactions, with chlorosulfonyl isocyanate, **2**, 726
Arenes, tetrakis(trifluoroacetoxy)iodo-, synthesis of, **5**, 205
Arenes, thionitroso-
from
N-(arylaminothio)phthalimides, **2**, 792
N,*N*-thiobis(arylamines), **2**, 792
Arenes, trichloromethyl-
reactions
with 1-acylhydrazines, **5**, 363
with hydroxylamines, **3**, 429
reactivity of, **6**, 27
synthesis of, **6**, 27
Arenes, trifluoromethyl-, synthesis of, **6**, 9
Arenes, trihydroxy-, synthesis of, **2**, 664
Arenes, trimethylgermylfluoro-, degermylation of, **1**, 449
Arenes, trinitro-, synthesis of, **6**, 147
Arenethiols
chlorination of, **2**, 716
from
arylamines, **2**, 707
aryl halides, **2**, 706
phenols, **2**, 707
oxidation of, **2**, 719
oxygen oxidation of, **2**, 720
reactions
with aryl bromides, **2**, 711
with disulfur dichloride, **2**, 723
with sulfuryl chloride, **2**, 717
as reducing reagents, **2**, 718
as starting materials
for arenesulfinic acids, **2**, 719
for arylsulfanes, **2**, 722
synthesis of, **2**, 706
via Herz reaction, **2**, 708
via reduction, **2**, 708
use in the synthesis of arenesulfinyl chlorides, **2**, 717
Arenonium ions
stability of, **1**, 914
stable, NMR spectra of, **1**, 914
synthesis of, **1**, 912, 914
Argentates, cyano-, synthesis of, **2**, 589
Argentates, dialkyl-, synthesis of, **2**, 589
Argentates, onium dicyano-, use in the synthesis of isocyanides, **3**, 695

Arndt–Eistert homologation
  for the synthesis of carboxylic esters, **5**, 144
  for the synthesis of primary amides, **5**, 276
Arndt–Eistert reaction, use in the synthesis of α-diazosilicons, **5**, 873
Arndt–Eistert synthesis, mechanism, **1**, 394
Aromatic acids
  reactions
    with 9-borabicyclo[3.3.1]nonane, **3**, 82
    with metal hydrides, **3**, 82
Aromatic halides
  acetylation of, **3**, 293
  cyanation of, **3**, 661
Aromatic halides, heterocyclic, displacement of, **3**, 669
Aromatic iodides, as starting materials, for arylhydrazines, **2**, 769
Aromatic *ipso*-desilylation, mechanism, **1**, 64
Aromatics
  reactions
    with chlorothioformamidinium salts, **5**, 731
    with ethoxycarbonyl isothiocyanate, **5**, 620
    with thiocarbamoyl chlorides, **5**, 593
  reductive silylation of, **2**, 945
  thiocarbamoylation of, **5**, 592
Aromatics, phenylazoethynyl-, synthesis of, **1**, 1137
Aromatization, of alicyclic substrates, **2**, 658
Aromatization, reductive, of dienones, **2**, 660
Aroyl chlorides
  reactions
    with alkylzincs, **3**, 288
    with benzylic halides, **3**, 289
    with sodium telluride, **5**, 244
    with toluene, **3**, 295
    with trimethylsilyl azide, **5**, 992
  as starting materials, for aryl metals, **2**, 957
Aroyl complexes, synthesis of, **5**, 451
Aroyl fluorides, synthesis of, **5**, 4
Aroyl halides
  reactions, with arylselenols, **5**, 238
  as starting materials, for ketenes, **3**, 527
Arsaalkenes
  dimerization of, **5**, 891
  as intermediates, in the synthesis of tetraarsacubane, **6**, 155
  reactions, with cobalt(II) chloride, **6**, 155
  stability of, **6**, 678
  as starting materials, for arsaalkynes, **3**, 689
  synthesis of, **5**, 890
  See also Arsanes, methylene-
Arsaalkynes
  from, arsaalkenes, **3**, 689
  as reaction intermediates, **6**, 155
  synthesis of, **5**, 1160
1-Arsaallenes, analogues of ketenes, **3**, 602
Arsabenzenes
  reactions
    with chromium, **4**, 1041
    with propargylic ketones, **3**, 269
$1\lambda^3$-Arsabutatriene, synthesis of, **3**, 609
1-Arsabutatrienes, dimerization of, **6**, 374
Arsacumulenes, as starting materials, for dialkyl-1,3-diarsetanes, **2**, 878
1-Arsacumulenes, synthesis of, **3**, 602, 609
Arsacymantrenes
  Friedel-Crafts acetylation of, **4**, 1042
  synthesis of, **4**, 1042
Arsacymantrenes, 2,5-dimethyl-, synthesis of, **4**, 1042
Arsaethyne, *t*-butyl-, synthesis of, theoretical, **5**, 1160
Arsaethyne, 2,4,6-tris(*t*-butyl)phenyl-, synthesis of, **5**, 1160
Arsaferrocene, 2,5-dimethyl-, synthesis of, **4**, 1039
1,1'-Arsaferrocene, 2,2',5,5'-tetradeuterio-, from, 1,1'-diarsaferrocene, **4**, 1039
Arsaferrocenes, synthesis of, **4**, 1039
Arsane, aryl-, reactions, with dichloroarsane, **6**, 168
Arsane, bis(dimethylamino)methyldiphenyl-, synthesis of, **6**, 161
Arsane, bis(trimethylsilyl)methyldichloro-, synthesis of, **6**, 374
Arsane, chlorodiphenyl-, reactions, with methylene(triphenyl)phosphorane, **5**, 911
Arsane, diazomethyldimethyl-, synthesis of, **3**, 482
Arsane, dichloro-, reactions, with arylarsane, **6**, 168
Arsane, dimesitoyl-, synthesis of, **5**, 891
Arsane, dimethylaminodimethyl-, reactions, with ethyl diazoacetate, **3**, 482
Arsane, diphenyl-, reactions, with amidinium salts, **6**, 161
Arsane, diphenylvinyl-, oxidation of, **2**, 877
Arsane oxide, diphenylvinyl-, synthesis of, **2**, 877
Arsane oxide, triphenyl-, condensation of, with dimethylphenacylsulfonium tetraphenylborate, **5**, 920
Arsanes, synthesis of, **2**, 896
Arsanes, acyloxy-, synthesis of, **5**, 222
Arsanes, acylsilyl-, rearrangement of, **5**, 891
Arsanes, alkylbis(trimethylsilyl)-, reactions, with pivaloyl chloride, **5**, 891
Arsanes, aminomethylene-, synthesis of, **5**, 891
Arsanes, arylbis(trimethylsilyl)-, reactions, with pivaloyl chloride, **5**, 891
Arsanes, chloro-, metallation of, **6**, 169
Arsanes, diazomethyl-, synthesis of, **3**, 482
Arsanes, dichloro-, with compounds having acidic methylene groups, **6**, 708
Arsanes, disilyl-, as starting materials, for *N*-aryl pivaloylimide chlorides, **5**, 891
Arsanes, metallo-, reactions, with ylides, **5**, 911
Arsanes, methylene-
  synthesis of, **5**, 890
  See also Arsaalkenes
Arsanes, pivaloyl(silyl)-, synthesis of, **5**, 891
Arsanes, silanyloxy-, reactions, with carboxylic acid chlorides, **5**, 222
Arsanes, silyl-, reactions, with dimethylformamide, **5**, 892
Arsanes, silylmethylene-, synthesis of, **5**, 892
Arsanyl iodide, synthesis of, **5**, 250
Arsaoxane, ethyl-
  large scale production of, **2**, 495
  as starting material, for dichloro(ethyl)arsine, **2**, 491
Arsaoxane, methyl-
  reactions, with sulfur, **2**, 500
  as starting material, for dichloro(methyl)arsine, **2**, 491
  synthesis of, **2**, 495
Arsaoxanes, alkyl-
  reactions, with alkylarsines, **2**, 507
  as starting materials, with alkylarsonous acids, **2**, 493
  synthesis of, **2**, 495
Arsaphosphetes, dihydro-, synthesis of, **2**, 894
Arsapropyne, synthesis of, **5**, 1160
Arsasilenes, reactions, with benzophenone, **5**, 892
Arsenane, 1-methyl-, synthesis of, **2**, 486
Arsenane, 2-naphthylmethylene triphenyl-, synthesis of, **1**, 268
Arsenate, dimethyl(selenocyanato)sulfonium hexafluoro-, synthesis of, **2**, 238
Arsenic
  analogues of ketenes, **3**, 602
  cyclic compounds, synthesis of, **4**, 594
  α-haloalkenyl compounds of, synthesis of, **4**, 805
  heteroaromatic compounds of, **2**, 879
  reactions, with alkyl halides, **2**, 486
  as starting material
    for phenylarsines, **2**, 872
    for vinylarsines, **2**, 872
  synthesis of functions, **4**, **591–600**
    with no halogen, chalcogen or nitrogen, **4**, **1021–1042**
Arsenic, *t*-butyldichloro-, reactions, with di-Grignard reagents, **2**, 484

Arsenic, doubly-bonded, synthesis of functions, **6**, **677–724**
Arsenic, triphenyl-, reactions, with *N*-chloro-
N,N'-dimethylbenzamidine, **5**, 776
Arsenic, tris(trifluorovinyl)-, synthesis of, **4**, 805
Arsenic(III) amides, reactions, with carbon disulfide, **2**, 889
Arsenic(III) amides, alkenyl-, from, chloroarsenic amides, **2**, 891
Arsenic(III) amides, aryl-, from, chloroarsenic amides, **2**, 891
Arsenic(III) fluoride, trifluoromethylation with, **6**, 243
Arsenic(III) halides, reactions, with organometallics, **2**, 872
Arsenic(III) nitriles, synthesis of, **5**, 1132
Arsenic(III) oxide
reactions
with acetyl chloride, **4**, 592
with potassium acetate, **2**, 494
Arsenic(III) phenolates, synthesis of, **2**, 695
Arsenic(V) halides, synthesis of, **2**, 492
Arsenic acids, alkynyl-, from, aminoarsenic chlorides, **2**, 1080
Arsenicals, alkynyl-, synthesis of, **2**, 1076
Arsenic amides
from, ethyldiarylarsinates, **5**, 370
as starting materials
for thioarsenic esters, **2**, 889
for thioarsinic esters, **2**, 889
Arsenic amides, chloro-
reactions, with organometallics, **2**, 891
as starting materials
for alkenylarsenic(III) amides, **2**, 891
for arylarsenic(III) amides, **2**, 891
Arsenic–amidine complexes, synthesis of, **5**, 776
Arsenic anhydrides, aryl-, reactions, with alcohols, **2**, 885
Arsenic anions, alkyl-, reactions, with alkyl halides, **2**, 485
Arsenic–arsenic bonds, cleavage of, **2**, 482, 491
Arsenic–carbon bonds, cleavage of, **2**, 490
Arsenic chloride, dimethyl-, use in the synthesis of alkynylarsines, **2**, 1078
Arsenic chloride, tetraphenyl-, reactions, with sodium succinimide, **5**, 372
Arsenic chlorides, amino-, as starting materials, for alkynylarsenic acids, **2**, 1080
Arsenic chlorides, bis(trimethylsilylmethyl)-substituted alkyl-, synthesis of, **2**, 489
Arsenic compounds
alkali metals of, synthesis of, **2**, 509
main group metal compounds of, synthesis of, **2**, 510
synthesis of, **4**, 597, 599, 1042; **5**, 1092
transition metals of, synthesis of, **2**, 511
Arsenic compounds, acyloxy-, synthesis of, **5**, 221
Arsenic compounds, alkyl-, synthesis of, **2**, **479–512**
Arsenic compounds, alkynyl-
from, alkynylarsines, **2**, 1079
reviews, **2**, 1076
safety precautions, **2**, 1076
synthesis of, **2**, 1076
Arsenic compounds, 1,3-aminoalkyl-
synthesis of
via cyclocondensation reactions, **4**, 495
via Mannich reactions, **4**, 499
Arsenic compounds, α-aminoalkyl-
synthesis of
via alkylation reactions, **4**, 501
via cycloaddition reactions, **4**, 502
Arsenic compounds, cyano-, synthesis of, **5**, 1132
Arsenic compounds, α-halo-, synthesis of, **4**, 134
Arsenic compounds, imidoyl-
from
imidoyl chlorides, **5**, 791
imidoyl ethers, **5**, 790
nitrile ylides, **5**, 791
synthesis of, **5**, 790
Arsenic compounds, tetracoordinated, synthesis of, **6**, 707
Arsenic compounds, triethynyl-, stability of, **2**, 1076
Arsenic-containing heterocycles, synthesis of, **4**, 502
Arsenic dichloride, trimethyl-, reactions, with methyllithium, **2**, 488
Arsenic dichloride, triphenyl-
reactions
with hydrogen selenide, **2**, 890
with phenylnitromethane, **5**, 921
Arsenic dichloride, tris(trifluoromethyl)-, synthesis of, **2**, 492
Arsenic dichlorides, alkyl-, reactions, with malonic esters, **4**, 594
Arsenic dichlorides, aryl-, reactions, with malonic esters, **4**, 594
Arsenic difluoride, tributyl-, synthesis of, **2**, 492
Arsenic difluoride, trimethyl-, synthesis of, **2**, 492
Arsenic dihalides, reactions, with alkali metals, **2**, 893
Arsenic dihalides, alkyl-, reactions, with sodium alkoxides, **2**, 496
Arsenic dihalides, aryl-, as starting materials, for cyclopolyarsines, **2**, 893
Arsenic dihalides, trialkyl-
ammonolysis of, **2**, 504
synthesis of, **2**, 492
Arsenic diisocyanates, triphenyl-, synthesis of, **5**, 1017
Arsenic enolates, synthesis of, **2**, 694
Arsenic functions
elimination of, **1**, 1079
substitution of, **1**, 268, 1032
synthesis of, **4**, 592
Arsenic functions, acyl-, synthesis of, **5**, **393–407**
Arsenic functions, aryl-, synthesis of, **2**, **871–897**
Arsenic functions, dicoordinated, synthesis of, **5**, 890; **6**, 678
Arsenic functions, imidoyl-, synthesis of, **5**, 784
Arsenic functions, tetracoordinated, synthesis of, **5**, 920
Arsenic functions, tricoordinated, synthesis of, **5**, 894
Arsenic functions, vinyl-, synthesis of, **2**, **871–897**
Arsenic–gallium bonds, synthesis of, **2**, 511
Arsenic groups, elimination of, **1**, 646
Arsenic halide, dimethyl-, nucleophilic metallation of, **2**, 512
Arsenic halides
reactions
with British antilewisite, **2**, 889
with metal alkynides, **2**, 1076
with organometallics, **2**, 483
as starting materials
for thioarsenates, **2**, 260
for thioarsenic esters, **2**, 889
for thioarsinic esters, **2**, 889
synthesis of, **2**, 880
Arsenic halides, alkyl-
reactions
with amines, **2**, 502
with ammonia, **2**, 502
with bis(ethylthio)lead, **2**, 499
with bis(trifluoromethylthio)mercury, **2**, 499
with metal amides, **2**, 502
with organometallics, **2**, 483
with transition metal complex anions, **2**, 511
reduction of, **2**, 481
as starting materials, for alkylarsines, **2**, 481
Arsenic halides, amino-
reactions
with Grignard reagents, **2**, 504
with organometallics, **2**, 503
Arsenic halides, perfluoroalkyl-, reduction of, **2**, 481
Arsenic imines, aryl-, reactions, with alcohols, **2**, 885
Arsenic–indium bonds, synthesis of, **2**, 511
Arsenic iodide, dimethyl-, reactions, with sugar thioacetates, **2**, 260
Arsenic isocyanates, synthesis of, **5**, 1016
Arsenic β-keto ylides, alkylation of, **2**, 878
Arsenic–nitrogen bonds, cleavage of, **2**, 491
Arsenic oxide, phenyl-, reactions, with dibromomethane, **4**, 592

Arsenic oxide, tertiary, reactions, with methylene compounds, **5**, 920
Arsenic oxides
alkylation of, **2**, 485
as starting materials
for thioarsenic esters, **2**, 889
for thioarsinic esters, **2**, 889
Arsenic–oxygen bonds, cleavage of, **2**, 491
Arsenic phenolates, synthesis of, **2**, 695
Arsenic trichloride
reactions
with alkenes, **4**, 597
with bis(1,2-oxarsolanyl) oxide, **2**, 497
with *N*-chloro-*N*,*N'*-dimethyl benzamidine, **5**, 776
with ethanedithiol, **2**, 260
with ethyne, **2**, 881
with organosilicons, **2**, 489
with tribenzylarsine, **2**, 490
with trimethylsilylmethylmagnesium chloride, **4**, 597
Arsenic trichloride, dicyclohexyl-, synthesis of, **2**, 492
Arsenic trichloride, dimethyl-, synthesis of, **2**, 492
Arsenic trichlorides, reactions, with bis(trimethylsilyl)methane, **6**, 168
Arsenic trifluoride, as starting material, for carbonic chloride fluoride, **6**, 419
Arsenic trihalides
reactions
with Grignard reagents, **2**, 483
with organometallics, **2**, 488
Arsenic trihalides, dialkyl-, synthesis of, **2**, 492
Arsenic triiodide, reactions, with bis(trifluoromethyl)mercury, **2**, 483
Arsenic trioxide, synthesis of, **2**, 1080
Arsenic ylides
from, arsines, **3**, 500
as starting materials, for ketenylidene λ^5-arsorane, **3**, 609
synthesis of, **3**, **491–500**, 500
Arsenide, diethyl, reactions, with tris(dimethylamino)titanic bromide, **2**, 512
Arsenide anions, reactions, with protic acids, **1**, 445
Arsenides
alkylation of, **2**, 482; **4**, 134
[2+1] cyclocondensation of, **2**, 508
from, arsines, **2**, 510
reactions, with chlorosilanes, **2**, 508
use in the synthesis of arsenic–gallium bonds, **2**, 511
Arsenides, alkyl
alkylation of, **2**, 485
reactions, with α,ω-dichloroalkanes, **2**, 486
synthesis of, reviews, **2**, 509
Arsenides, dialkyl, synthesis of, **2**, 510
Arsenides, dimetallic alkyl, reactions, with dihaloalkanes, **2**, 486
Arsenides, methyl-, synthesis of, **2**, 510
Arsenins, synthesis of, via hetero-Diels–Alder reaction, **2**, 879
Arsenite esters, prop-2-ynyl, thermal isomerization of, **2**, 879
N-Arseno compounds, synthesis of, **2**, 470
Arsenoketenes, synthesis of, **2**, 896
Arsenous(III) chloride, reactions, with (methylseleno)trimethylsilane, **2**, 293
Arsenous acids, reactions, with mercuric oxide, **1**, 446
Arsenous chloride, use in the synthesis of α-acyl enamines, **2**, 745
Arsenoxides
reactions, with thiols, **2**, 260
as starting materials, for thioarsenates, **2**, 260
Arsinate, aryl, synthesis of, **2**, 695
Arsinates, ethyldiaryl, as starting materials, for arsenic amides, **5**, 370
Arsine, amino-, reactions, with phenols, **2**, 695
Arsine, aminobistrifluoromethyl-, synthesis of, **2**, 503
Arsine, aminodi(*t*-butyl)-, synthesis of, **2**, 502
Arsine, azidodimethyl-, synthesis of, **2**, 504
Arsine, benzyldichloro-, synthesis of, **2**, 490
Arsine, bis(diethylamino)chloro-
reactions
with Grignard reagents, **2**, 498
with trimethylsilylmethylmagnesium, **4**, 597
Arsine, bis(dimethylamino)methyl-, reactions, with ethylenedithiol, **2**, 501
Arsine, bis(dimethylamino)trifluoromethyl-, synthesis of, **2**, 504
Arsine, bis(trifluoromethyl)chloro-
ammonolysis of, **2**, 503
synthesis of, **2**, 492
Arsine, bis(trifluoromethyl)fluoro-, synthesis of, **2**, 492
Arsine, bis(trifluoromethyl)iodo-
oxidation of, **2**, 498
reactions, with silver sulfide, **2**, 499
Arsine, bistrifluoromethyl(trifluoromethylthio)-, synthesis of, **2**, 499
Arsine, bis(trimethylsilylmethyl)chloro-, synthesis of, **4**, 597
Arsine, bromo(benzyl)methyl-, synthesis of, **2**, 490
Arsine, butyl-, synthesis of, **2**, 482
Arsine, *t*-butyldichloro-
reactions, with dilithium *t*-butylarsenide, **2**, 507
use in the synthesis of arsenic–transition metal bonds, **2**, 512
Arsine, *t*-butyl(dimethylamino)phenyl-, enantioselective synthesis of, **2**, 891
Arsine, chloro-, reactions, with *N*-metallated aminophosphine, **2**, 472
Arsine, chlorodi(*t*-butyl)-, reactions, with ammonia, **2**, 502
Arsine, chlorodimethyl-
hydrolysis of, **2**, 493
reactions
with cyclopentadienyllithium, **2**, 484
with methanethiol, **2**, 499
Arsine, chloro(dimethyl)-
reactions, with trimethylsilanol, **2**, 494
synthesis of, **2**, 492
Arsine, chlorodiphenyl-, reactions, with bis(arsines), **6**, 155
Arsine, cyclopentadienyl(dimethyl)-
from, cyclopentadienyltrimethylsilane, **2**, 484
synthesis of, **2**, 484
Arsine, diacetoxy(propyl)-, synthesis of, **2**, 496
Arsine, dialkylethynyl-, reactions, with hydrogen chloride, **2**, 876
Arsine, (diazotrimethylgermylmethyl)dimethyl-, synthesis of, **6**, 670
Arsine, dibenzylmethyl-, reactions, with bromine, **2**, 490
Arsine, dibromo(methyl)-, reactions, with potassium carbonate, **2**, 495
Arsine, di-*n*-butylethynyl-, reactions, with tributyltin hydrides, **2**, 876
Arsine, dichloro-, dehydrohalogenation of, for the synthesis of arsoranes, **4**, 598
Arsine, dichloro(diethylamino)-
alkylation of, **2**, 503
as starting material, for alkylarsinic acids, **2**, 498
Arsine, dichloro(ethyl)-, from, ethylarsaoxane, **2**, 491
Arsine, dichloro(3-hydroxypropyl)-, as starting material, for 2-chloro-1,2-oxarsolane, **2**, 494
Arsine, dichloro(methyl)-
from, methylarsaoxane, **2**, 491
large-scale production of, **2**, 491
reactions
with alcohols, **2**, 496
with *o*-phenylenebis(lithiophenylphosphine), **2**, 508
Arsine, dichloro(propyl)-, reactions, with silver acetate, **2**, 496
Arsine, dichloro(trifluorovinyl)-, synthesis of, **4**, 805
Arsine, dichloro(trimethylsilylmethyl)-, synthesis of, **4**, 597
Arsine, dicyclohexyl-
reactions, with phenyllithium, **2**, 510

as starting material, for lithium dicyclohexylarsenide, **2**, 485
Arsine, difluoro(methyl)-, synthesis of, **2**, 491
Arsine, difluoromethylene-, from, stannylarsines, **6**, 691
Arsine, diiodomethyl-, synthesis of, **2**, 491
Arsine, dilithiophenyl-, reactions, with 1,4-dichlorobuta-1,3-diene, **4**, 1039
Arsine, dimethyl-
alkylation of, **2**, 487
from, methylarsine, **2**, 482
reactions
with alkynes, **2**, 1078
with 2,3-dichlorotetrafluoropropene, **4**, 805
with hexafluoropropene, **6**, 60
with tetrakistrifluorodiphosphine, **2**, 508
as starting material, for tetramethyldiarsine, **2**, 506
synthesis of, **2**, 482
Arsine, dimethylaminodiethyl-, reactions, with hydrogen chloride, **2**, 492
Arsine, dimethylaminodimethyl-
reactions
with carboxylic acids, **5**, 222
with cyclopentadiene, **2**, 484
Arsine, dimethylchloro-, reactions, with silver trifluoroacetate, **5**, 222
Arsine, dimethyl(dimethylamino)-
reactions
with diols, **2**, 494
with polyethylene glycols, **2**, 494
as starting material, for alkylarsines, **2**, 504
Arsine, dimethyl(ethylthio)-, synthesis of, **2**, 499
Arsine, dimethylfluoro-, synthesis of, **2**, 491
Arsine, dimethyliodo-
reactions
with alkali metals, **2**, 505
with silver azide, **2**, 504
with sodium sulfide, **2**, 499
synthesis of, **2**, 491
Arsine, dimethyl(methanethio)-, synthesis of, **2**, 499
Arsine, diphenyl-, metathesis of, **2**, 893
Arsine, diphenylchloro-
reactions
with stibinomethyllithium, **4**, 596
with tributylstannylmethyllithium, **4**, 600
Arsine, diphenylethyl-, reduction of, **1**, 445
Arsine, *R*-ethylmethyl-*n*-propyl-, synthesis of, **2**, 487
Arsine, 1-ethynyl-, reactions, with sodium carbonate, **5**, 1160
Arsine, imidoyl-, hydrochloride salt of, **5**, 791
Arsine, iodomethyldiphenyl-, as starting material, for diphenylarsinomethyllithium, **4**, 599
Arsine, methyl-
reactions
with hexafluoroacetone, **2**, 482
with hydrogen sulfide, **2**, 500
as starting material, for dimethylarsine, **2**, 482
Arsine, methyl(2-hydroxyethyl)-
reactions
with acrylic acid, **2**, 487
with methyl acrylate, **2**, 487
synthesis of, **2**, 482
Arsine, methylphenylchloro-, substitution of, for the synthesis of acyloxyarsenic compounds, **5**, 221
Arsine, methylphenyltrifluoroacetoxy-, synthesis of, **5**, 222
Arsine, phenyl-, reactions, with *t*-butyldipropynylphosphine, **4**, 1030
Arsine, phenyl-2-aminoethyl-, reactions, with acetimino methyl ester hydrochloride, **5**, 790
Arsine, phenylbis(trimethylsilyl)-, reactions, with pivaloylimidoyl chloride, **5**, 791
Arsine, poly-*t*-butyl-, synthesis of, **2**, 507
Arsine, polytrifluoromethyl-, as starting material, for trifluoromethylarsine, **2**, 482
Arsine, secondary methyl-, reactions, with diboranes, **2**, 509
Arsine, tribenzyl-, reactions, with arsenic trichloride, **2**, 490
Arsine, triethyl-, reactions, with organyl bromides, **2**, 1079
Arsine, trifluoromethyl-, from, polytrifluoromethylarsine, **2**, 482
Arsine, trimethyl-
from, tris(dimethylamino)arsine, **2**, 484
reactions
with chloromethyltrimethylsilane, **4**, 597
with xenon difluoride, **2**, 492
Arsine, triphenyl-
elimination of, **1**, 647
reactions
with allylic bromides, **1**, 269
with bromo acetaldehyde, **3**, 48
with 2-bromomethylnaphthalene, **1**, 268
with chloramine-T, **2**, 809
with sodium *N*-chlorotoluene-*p*-sulfonamide, **2**, 892
reduction of, **1**, 445
Arsine, tris(dimethylamino)-
reactions, with trifluoromethyl iodide, **2**, 504
as starting material, for trimethylarsine, **2**, 484
use in the synthesis of aminals, **4**, 405
Arsine, tris(trifluoromethyl)-, synthesis of, **2**, 483, 486
Arsine, tris(trimethylsilyl)-
reactions, with methyllithium, **5**, 1160
as starting material, for lithium bis(trimethylsilyl)arsine complex, **2**, 509
Arsine, tris(trimethylsilylmethyl)-, synthesis of, **4**, 597
Arsine anions, phenylethyl-, synthesis of, **1**, 445
Arsine chalcogenides, bisdialkyl-, synthesis of, **2**, 499
Arsine dibromide, triphenyl-, reactions, with amides, **2**, 809
Arsine dichlorides, chloroalkenyl-, synthesis of, **2**, 881
Arsine dihalides, alkyl-, synthesis of, **2**, 489
Arsine halides, alkyl-
from, alkylarsonous acids, **2**, 491
reactions, with halogens, **2**, 492
synthesis of, **2**, 488
Arsine oxide, elimination of, **1**, 646
Arsine oxide, bisdiethyl-, reactions, with tetraethyldiarsine, **2**, 494
Arsine oxide, bisdimethyl-, synthesis of, **2**, 493, 494
Arsine oxide, 2-hydroxyethyldiphenyl-, synthesis of, **1**, 738
Arsine oxide, methyldiphenyl-, elimination of, **1**, 646
Arsine oxide, tributyl-, elimination of, **1**, 647
Arsine oxide, tri-*n*-butyl-, synthesis of, **2**, 497
Arsine oxide, triphenyl-
reactions, with *p*-nitroaniline, **2**, 810
as starting material
for arsinimines, **2**, 892
for arsonium ylides, **6**, 708
Arsine oxides
reactions, with 1,3-dicarbonyls, **3**, 48, 196
as starting materials, for arsinous esters, **2**, 888
synthesis of, **2**, 888
Arsine oxides, alkyl-, synthesis of, **2**, 497
Arsine oxides, bisdialkyl-
from, alkylarsinous acids, **2**, 493
synthesis of, **2**, 493
Arsine oxides, trialkyl-, synthesis of, **2**, 497
Arsine oxides, triaryl-
from, arenediazonium salts, **2**, 885
reactions, with isocyanates, **2**, 809
synthesis of, via oxidation of triarylarsine sulfides, **2**, 888
Arsines
alkylation of, **2**, 482, 509, 512
from
alkylarsinic acids, **2**, 482
alkylarsonic acids, **2**, 482
bis(diphenylarsino)methane, **4**, 593
halogenation of, **2**, 881
methylation of, **2**, 1079
oxidation of, **2**, 495

reactions
with acetone, **2**, 1079
with alkyl halides, **2**, 487; **3**, 196
with 2-diazo 1,3-dicarbonyls, **3**, 196
with formaldehyde, **2**, 1079
with hexafluoroacetone, **4**, 344
with organolithiums, **2**, 510
as starting materials
for arsenic ylides, **3**, 500
for arsenides, **2**, 510
synthesis of, **2**, 872; **4**, 597
Arsines, alkenyl-, reactions, with alkynes, **2**, 875
Arsines, alkoxy-
cleavage of, **2**, 491
oxidation of, **2**, 498
synthesis of, **2**, 494
Arsines, alkoxy(dialkyl)-, from, dialkyl(diethylamino)arsines, **2**, 494
Arsines, alkoxydialkyl-, synthesis of, **2**, 494
Arsines, alkyl-
alkylation of, **2**, 482
from
alkylarsenic halides, **2**, 481
dimethyl(dimethylamino)arsine, **2**, 504
metallation of, **2**, 510
oxidation of, **2**, 495
reactions
with alkylarsaoxanes, **2**, 507
with alkyldihaloarsines, **2**, 507
with imidoyl chlorides, **5**, 791
with pentamethylcyclopentaarsine, **2**, 507
with phenyl isocyanate, **6**, 513
with sulfur, **2**, 499
as starting materials
for diarsines, **2**, 506
for dilithium alkyl arsenide, **2**, 510
synthesis of, **2**, 481
Arsines, alkyl(alkoxy)halo-, synthesis of, **2**, 497
Arsines, alkylbis(trimethylsilyl)-, synthesis of, **2**, 509
Arsines, alkyl(chloro)alkoxy-, reactions, with alcohols, **2**, 496
Arsines, alkyldialkoxy-
from, α,ω-diols, **2**, 496
synthesis of, **2**, 496
Arsines, alkyldi(alkylthio)-, synthesis of, **2**, 501
Arsines, alkyldihalo-
hydrolysis of, **2**, 495
reactions
with alcohols, **2**, 497
with alkylarsines, **2**, 507
with selenols, **2**, 500
with thiols, **2**, 500
reduction of, **2**, 507
synthesis of, **2**, 489
via redistribution reactions, **2**, 490
use in the synthesis of arsenic–transition metal bonds, **2**, 512
Arsines, alkylfluoro-, synthesis of, **2**, 491
Arsines, alkylhalo-
reactions, with metal azides, **2**, 504
synthesis of, **2**, 489
Arsines, alkyliminochloro-, synthesis of, **2**, 408
Arsines, alkyliminofluoro-, synthesis of, **2**, 408
Arsines, alkynyl-
from
alkynes, **2**, 1077
arsinic acids, **2**, 1078
magnesium compounds of, reactions, **2**, 1079
organometallic complexes of, **2**, 1079
oxidation of, **2**, 1080
as starting materials, for alkynylarsenic compounds, **2**, 1079
Arsines, amidino-, synthesis of, **6**, 657
Arsines, amino-
amino-exchange reaction of, **2**, 504
cleavage of, **2**, 491
hydrolysis of, **2**, 493
reactions
with alcohols, **2**, 494
with alkaneselenols, **2**, 500
with alkanethiols, **2**, 500
as starting materials
for diarsine oxides, **2**, 493
for organoarsenic compounds, **2**, 504
for tertiary alkylarsines, **2**, 484
for thioarsenates, **2**, 260
synthesis of, **2**, 502
Arsines, aminoalkyl-, reactions, with carbonyls, **4**, 495
Arsines, 1-aminoalkyl-, synthesis of, **4**, 499
Arsines, 2-aminoalkyl-
reactions
with alkynes, **4**, 498
with imino ester hydrochlorides, **5**, 790
as starting materials
for 1,3-azarsolidines, **4**, 498
for 1,3-benzazarsolines, **4**, 498
Arsines, α-(*N*-arenesulfonamido)benzyl-, synthesis of, **4**, 503
Arsines, aryl-
reactions, with phenyl isocyanate, **6**, 513
as starting materials, for arenes, **1**, 445
Arsines, azido-, synthesis of, **2**, 504
Arsines, bicyclic, synthesis of, via dipolar cycloaddition reactions, **4**, 502
Arsines, bis(dialkylamino)halo-
from
amines, **2**, 414
(dialkylamino)haloarsines, **2**, 415
tris(dialkylamino)arsines, **2**, 415
synthesis of, **2**, 414
Arsines, bis(halomethyl)-, synthesis of, **4**, 135
Arsines, bistrifluoromethyl-
from, tetrakistrifluoromethyldiarsine, **2**, 482
synthesis of, **2**, 510
Arsines, α-bromo-, synthesis of, **4**, 135
Arsines, (bromomethyl)-, synthesis of, **4**, 135
Arsines, carbamoyl-, synthesis of, **6**, 513
Arsines, chloro-
from, bis(trimethylsilyl)methyllithium, **4**, 597
reactions
with Grignard reagents, **4**, 597
with potassium cyanide, **5**, 1132
with trimethylsilylmethyllithium, **4**, 597
synthesis of, **4**, 592
Arsines, α-chloro-, synthesis of, **4**, 134
Arsines, chloro(dialkyl)-, oxidation of, **2**, 498
Arsines, (chloromethyl)-, synthesis of, **4**, 134
Arsines, *N*-cyclohexylcarbamoyl-, synthesis of, **6**, 513
Arsines, dialkoxy(methyl)-, synthesis of, **2**, 496
Arsines, dialkyl-, condensation of, **2**, 506
Arsines, dialkyl(alkanethio)-, synthesis of, **2**, 500
Arsines, α-dialkylaminoalkyl-, synthesis of, via Mannich reactions, **4**, 500
Arsines, (dialkylamino)dihalo-
from
amines, **2**, 414
isobutylamine, **2**, 414
tris(dialkylamino)arsines, **2**, 414
synthesis of, **2**, 414
Arsines, (dialkylamino)halo-, as starting materials, for bis(dialkylamino)haloarsines, **2**, 415
Arsines, dialkyl(diethylamino)-
reactions, with thiols, **2**, 500
as starting materials, for alkoxy(dialkyl)arsines, **2**, 494
Arsines, dialkyl(dimethylamino)-, synthesis of, **2**, 503
Arsines, dialkylhalo-
condensation of, **2**, 506

hydrolysis of, **2**, 493
reactions
with alcohols, **2**, 494
with alkali metals, **2**, 510
with alkaneselenol, **2**, 500
with alkaneselnols, **2**, 499
with alkanethiols, **2**, 499
with bis(alkylseleno)mercury, **2**, 500
with halogens, **2**, 492
with metal sulfides, **2**, 499
reduction of, **2**, 505
synthesis of, **2**, 489
via redistribution reactions, **2**, 490
Arsines, dialkyl(organochalcogeno)-, synthesis of, **2**, 499
Arsines, dialkyl(organoseleno)-, synthesis of, **2**, 499
Arsines, dialkyl(organosulfuro)-, synthesis of, **2**, 499
Arsines, diamino-
reactions
with alcohols, **2**, 496
with thiols, **2**, 501
Arsines, diazomethyl-, synthesis of, **6**, 671
Arsines, difluorohalomethyl-, synthesis of, **6**, 243
Arsines, dihalo(alkyl)-, hydrolysis of, **2**, 495
Arsines, dihalo(halomethyl)-, synthesis of, **4**, 135
Arsines, dimethylacyloxy-, synthesis of, **5**, 222
Arsines, α-dimethylaminomethyl-, synthesis of, **4**, 501
Arsines, α-fluoro-, synthesis of, **4**, 134
Arsines, fluorodi-, synthesis of, **6**, 60
Arsines, heterocyclic, synthesis of, **2**, 486
Arsines, imidoyl silyl-, synthesis of, **5**, 791
Arsines, α-iodo-, synthesis of, **4**, 135
Arsines, (iodomethyl)diphenyl-, synthesis of, **4**, 135
Arsines, lithio diaryl-, reactions, with alkyl triflates, **3**, 196
Arsines, methyl-, oxidation of, **2**, 495
Arsines, methylene-, characteristics of, **6**, 678
Arsines, methylene-, *C*-amino-substituted, stabilization of, **6**, 691
Arsines, methylene-, *C*,*C*-dihetero-substituted, synthesis of, **6**, 691
Arsines, phenyl-, from, arsenic, **2**, 872
Arsines, *N*-phenylcarbamoyl-, synthesis of, **6**, 513
Arsines, polyalkyl-
from, primary alkylarsines, **2**, 507
synthesis of, **2**, 506
Arsines, primary, synthesis of, **4**, 495
Arsines, primary alkyl-
oxidation of, **2**, 498, 507
reactions, with dibenzylmercury, **2**, 507
as starting materials, for polyalkylarsines, **2**, 507
Arsines, primary alkyldialkoxy-, oxidation of, **2**, 498
Arsines, primary alkyldihalo-, oxidation of, **2**, 498
Arsines, primary aryl-, as starting materials, for cyclopolyarsines, **2**, 894
Arsines, propynyl-, synthesis of, **2**, 1078
Arsines, secondary
alkylation of, **2**, 487
reactions
with *N*-arenesulfonylbenzaldimines, **4**, 503
with bis(dimethylamino)methane, **4**, 501
with secondary amines, **4**, 500
as starting materials, for tertiary alkylarsines, **2**, 487
use in the synthesis of aminomethylarsines, **4**, 501
Arsines, secondary alkyl-
reactions
with cobalt complexes, **2**, 512
with iridium complexes, **2**, 512
use in the synthesis of arsenic–gallium bonds, **2**, 511
use in the synthesis of arsenic–indium bonds, **2**, 511
Arsines, secondary aminoalkyl-, synthesis of, **4**, 495
Arsines, secondary *O*-aminophenyl-, synthesis of, via cyclocondensation reactions, **4**, 496
Arsines, silyl-, use in the synthesis of aminomethylarsines, **4**, 501
Arsines, stannyl-, as starting materials, for difluoromethylenearsine, **6**, 691
Arsines, tertiary
alkylation of, **2**, 487
quaternization of, **5**, 1094
reactions
with azides, **2**, 809
with halogens, **2**, 492
with stibines, **2**, 492
synthesis of, **2**, 1078
Arsines, tertiary alkyl-
from
aminoarsines, **2**, 484
aminobismuthines, **2**, 484
aminostibines, **2**, 484
2-chloro-1,3,2-tetrahydrodioxarsolane, **2**, 485
secondary arsines, **2**, 487
oxidation of, **2**, 497
oxidative fluorination of, **2**, 492
reactions, with chalcogens, **2**, 502
synthesis of, **2**, 483
Arsines, trialkyl-
reactions
with alkyl dihalides, **4**, 135
with chloroamine, **2**, 504
with chloromethyltrimethylsilane, **4**, 597
with selenium, **2**, 502
with sulfur, **2**, 502
Arsines, triaryl-
cleavage of, **2**, 510
photochemical synthesis of, **2**, 874
reactions, with cyanogen halides, **5**, 1132
Arsines, tris(dialkylamino)-
from
amines, **2**, 415
stannanes, **2**, 415
as starting materials
for bis(dialkylamino)haloarsines, **2**, 415
for (dialkylamino)dihaloarsines, **2**, 414
synthesis of, **2**, 415
Arsines, tris(3,3-dimethyl-1-butynyl)-, synthesis of, **2**, 1076
Arsines, unsymmetrical tertiary alkyl-, synthesis of, **2**, 483
Arsines, vinyl-, from, arsenic, **2**, 872
Arsine selenide, triphenyl-, synthesis of, **2**, 890
Arsine sulfide, bisdimethyl-
reactions, with sodium, **2**, 510
synthesis of, **2**, 499
Arsine sulfide, ethyl(methoxycarbonylthio)phenyl-, synthesis of, **6**, 476
Arsine sulfides, methyl-, synthesis of, **2**, 500
Arsinic acid, butylpropyl-, synthesis of, **2**, 498
Arsinic acid, dimethyl-
reactions, with phosphorus trichloride, **2**, 492
reduction of, **2**, 482, 506
Arsinic acids
reductive halogenation of, **2**, 883
as starting materials, for alkynylarsines, **2**, 1078
Arsinic acids, alkyl-
from
alkali alkylarsonites, **2**, 498
dichloro(diethylamino)arsine, **2**, 498
reduction of, **2**, 482
for the synthesis of alkylarsine halides, **2**, 491
as starting materials, for arsines, **2**, 482
synthesis of, **2**, 497
use in the synthesis of diarsines, **2**, 506
Arsinic acids, bistrifluoromethyl-, synthesis of, **2**, 498
Arsinic acids, diaryl-, synthesis of, **2**, 885
Arsinic acids, α-halo-, synthesis of, **4**, 135
Arsinic dichlorides, alkenyl-, synthesis of, **2**, 883
Arsinic dichlorides, aryl-, synthesis of, **2**, 883
Arsinidenes, chloro-, reactions, with carbon nucleophiles, **2**, 896

Arsinimine, *N-p*-nitrophenyltriphenyl-, synthesis of, **2**, 810
Arsinimine, *N*-pentafluorophenyl-, synthesis of, **2**, 810
Arsinimine, triphenyl-*N-p*-tolylsulfonyl-, synthesis of, **2**, 810
Arsinimines
  detection of, **2**, 810
  from, triphenylarsine oxide, **2**, 892
  synthesis of, **2**, 809
Arsinimines, organostannylphenylamino-, synthesis of, **2**, 809
3-Arsinoacroleins, synthesis of, **2**, 879
Arsinogallanes, synthesis of, **2**, 511
Arsinooximes, properties of, **3**, 435
Arsinosilanes, alkyl-, synthesis of, **2**, 508
Arsinous acid, esters, synthesis of, **2**, 494
Arsinous acids, alkyl-
  as starting materials, for bisdialkylarsine oxides, **2**, 493
  synthesis of, **2**, 493
Arsinous amide, *N*-acetyl-As,As-diphenyl-, synthesis of, **5**, 369
Arsolane, 2-chloro-1,3,2-tetrahydrodiox-, as starting material, for tertiary alkylarsines, **2**, 485
Arsolane, 1-methyl-, synthesis of, **2**, 486
Arsole, 1-chloro-2,3,4,5-tetramethyl-, reactions, with alumina, **2**, 893
Arsole, 2,5-dimethyl-1-phenyl-
  reactions
    with decacarbonyldimanganese, **4**, 1042
    with potassium, **4**, 1039
Arsole, 1-phenyl-, synthesis of, **4**, 1039
Arsole complexes, tetraphenyl-, synthesis of, **4**, 1042
3-Arsolenes, synthesis of, **2**, 484
Arsoles
  manganese complexes of, **4**, 1042
  reactions, with iron(II) chloride, **4**, 1039
  rhenium complexes of, **4**, 1042
Arsoles, 1-(1-chloroalkyl)-, synthesis of, **4**, 134
Arsolyl anions, reactions, with cyclopentadienyldicarbonyliron iodide, **4**, 1039
Arsonane, dibromotris(trimethylsilylmethyl)-, disproportionation of, **4**, 135
Arsonane, pentamethyl-, synthesis of, **2**, 488
Arsonanes, alkyl-, synthesis of, **2**, 488
Arsonate, sodium methyl-, reduction of, **2**, 491
Arsonic acid, butyl-, reduction of, **2**, 482
Arsonic acid, (chloromethyl)-, reduction of, **4**, 134
Arsonic acid, 2-ethylhexyl-, synthesis of, **2**, 498
Arsonic acid, methyl-
  reduction of, **2**, 507
  synthesis of, **2**, 498
Arsonic acid, *O*-nitrophenyl-, reduction of, **4**, 496
Arsonic acids
  reduction of, **2**, 495
  reductive halogenation of, **2**, 883
  as starting materials, for arsonic esters, **2**, 886
Arsonic acids, alkyl-
  reduction of, **2**, 482, 495, 507
    for the synthesis of alkylarsine halides, **2**, 491
  as starting materials, for arsines, **2**, 482
  synthesis of, **2**, 497
Arsonic acids, aryl-
  from, aryldiazenium salts, **2**, 884
  as starting materials, for cyclopolyarsines, **2**, 893
Arsonic acids, α-halo-, synthesis of, **4**, 135
Arsonic esters
  from, arsonic acids, **2**, 886
  as starting materials, for arsonic amides, **2**, 891
Arsonites, alkali alkyl-, as starting materials, for alkylarsinic acids, **2**, 498
Arsonium bis(benzenesulfonyl)methylide, triphenyl-, stability of, **6**, 708
Arsonium bromide, reactions, with iodine, **3**, 649
Arsonium bromide, benzyltriphenyl-, reactions, with sodium hydride, **1**, 740
Arsonium bromide, *p*-chlorobenzyltriphenyl-, reactions, with sodium hydride, **1**, 740
Arsonium bromide, 2-hydroxyethyltriphenyl-, reactions, with aryl aldehydes, **1**, 738
Arsonium bromide, β-hydroxyethyltriphenyl-, reactions, with aldehydes, **1**, 269
Arsonium bromide, methyltriphenyl-, reactions, with sodium amide, **2**, 1004
Arsonium bromide, (2-naphthyl)methyl triphenyl-, synthesis of, **1**, 268
Arsonium bromide salts, vinyl-, synthesis of, via β-elimination of halides, **2**, 877
Arsonium halides, aminotrialkyl-, reactions, with potassium amide, **2**, 504
Arsonium iodides, iodo-, reactions, with aldehydes, **3**, 649
Arsonium iodides, (iodomethyl)trimethyl-, synthesis of, **4**, 135
Arsonium salts
  for the condensation of diphenylselenium oxide, **5**, 920
  from, methylenearsorane, **4**, 598
  as intermediates, **1**, 742
  reactions, with butyllithium, **1**, 738
  synthesis of, **2**, 487; **3**, 48; **4**, 598
  use in the synthesis of nitriles, **3**, 649
Arsonium salts, alkenyl-, from, arsorane ylides, **2**, 877
Arsonium salts, alkyl-, synthesis of, **2**, 487
Arsonium salts, alkynyl-, reduction of, **1**, 999
Arsonium salts, allyltriphenyl-, deprotonation of, **1**, 740
Arsonium salts, dialkylarsinomethyl-, deprotonation of, **5**, 921
Arsonium salts, α-halo-, synthesis of, **4**, 135
Arsonium salts, β-hydroxy-
  reactions, with sodium hydride, **1**, 646
  as starting materials
    for alkenes, **1**, 646
    for epoxides, **1**, 646
Arsonium salts, (*E*)-4-hydroxybut-2-enyl triphenyl-, reactions, with aldehydes, **1**, 270
Arsonium salts, tetramethyl-, synthesis of, **2**, 488
Arsonium salts, trimethylsilylmethyl-, deprotonation of, **5**, 921
Arsonium salts, triphenyl-, reactions, with aldehydes, **1**, 647
Arsonium salts, tris(dialkylamino)-, synthesis of, **2**, 415
Arsoniumtrimethylsilylmethylide, triphenyl-, synthesis of, **6**, 708
Arsonous acid esters, alkyl-, synthesis of, **2**, 496
Arsonous acid halides, alkyl-, synthesis of, **2**, 497
Arsonous acids, alkyl-
  as starting materials
    for alkylarsaoxanes, **2**, 493
    for alkylarsine halides, **2**, 491
  synthesis of, **2**, 495
Arsorane, *p*-bromobenzoylmethylene triphenyl-, reactions, with phenylsulfine, **5**, 920
Arsorane, carbethoxymethylene triphenyl-, reactions, with phenylsulfene, **5**, 920
Arsorane, diarsenylmethylene-, synthesis of, **6**, 708
Arsorane, fluorenylidenetriphenyl-, reactions, with *p*-dimethylaminobenzaldehyde, **1**, 646
Arsorane, α-germylmethyl trimethyl-, disproportionation of, **5**, 922
Arsorane, ketenylidene λ^5-, synthesis of, **3**, 609
Arsorane, methylene-, as starting material, for arsonium salts, **4**, 598
Arsorane, methylenetriphenyl-, synthesis of, **2**, 1004
Arsorane, pentamethyl-, cleavage of, **2**, 488
Arsorane, 3-(trimethyllsilyl)-2-propynylidenetriphenyl-, reactions, with α,β-unsaturated ketones, **1**, 270
Arsoranes
  [2+2]-photodimerization of, **4**, 594
  synthesis of, **2**, 872; **4**, 597
Arsoranes, bis(arenesulfonyl)methylene-, from, diazo compounds, **6**, 708

Arsoranes, bis(arenesulfonyl)methylenetriaryl-, synthesis of, **6**, 707
Arsoranes, bis(methylene)-, synthesis of, problems, **6**, 693
Arsoranes, iminotrialkyl-, synthesis of, **2**, 504
Arsoranes, methylene-, synthesis of, **5**, 920
Artificial sweeteners, sulfamate esters as, **2**, 367
Aryl acyl radicals, reactions, with alkenes, **3**, 292
Arylamine
reactions
with potassium nitrite, **1**, 442
with sodium nitrite, **1**, 442
Arylamines
deprotection of, **2**, 304
N,*N*-dialkylation of, **2**, 304
from
amines, **2**, 756
aryl halides, **2**, 753
azides, **2**, 753
carboxylic acids, **2**, 756
ethers, **2**, 756
hydrides, **2**, 752
nitro compounds, **2**, 752
phenols, **2**, 753
oxidation of, **2**, 651
reactions
with alkyl nitrites, **1**, 36
with chloroalkenes, **4**, 987
with diethylthiocarbamoyl chloride, **5**, 1029
with dihalocarbene complexes, **6**, 722
with dimethyl sulfoxide, **2**, 793
with 1-naphthonitrile, **5**, 756
with thiophosgene, **5**, 1027
with Vilsmeier salts, **5**, 745
as starting materials
for areneselenols, **2**, 727
for arenethiols, **2**, 707
for diaryl diselenides, **2**, 734
for diaryl selenides, **2**, 729
for diaryl tellurides, **2**, 730
synthesis of, **2**, 311, 752
via reduction of aryl nitro groups, **1**, 441
Arylamines, *N*(alkoxymethyl)-, synthesis of, **4**, 300
Arylamines, *N*-allyl-, synthesis of, **2**, 318
Arylamines, *N*,*N*-bis(benzotriazolylmethyl)-, reactions, with Grignard reagents, **2**, 304
Arylamines, *N*-(bis(diphenylphosphinyl)methylene)-, synthesis of, **6**, 665, 666
Arylamines, *N*-butoxycarbonyl-, as starting materials, for 1,2,3,4-tetrahydroquinolines, **2**, 328
Arylamines, dialkyl-, from, trialkylamines, **2**, 301
Arylamines, *N*,*N*-dialkyl-, synthesis of, **2**, 304
Arylamines, *N*,*N*-disilyl-, synthesis of, **2**, 814
Arylamines, *N*-methoxymethyl-, reactions, with trialkyl phosphites, **4**, 467
Arylamines, *N*-(phosphinomethyl)-
from, methyleneamines, **4**, 453
synthesis of, **4**, 453
Arylamines, primary, reactions, with 2,3,5,6-tetraphenylpyrylium cations, **1**, 31
Arylamines, salts, synthesis of, **2**, 757
N-Arylamines, *N*-(alkoxymethyl)-, reactions, with secondary phosphines, **4**, 453
Aryl anions, reactions, with alkenes, **1**, 309
Arylation
of saturated aldehydes, **3**, 24
of saturated ketones, **3**, 144
of thioamides, **5**, 727
C-Arylation, of aldoximes, **3**, 428
Arylazo compounds, of bis(arylsulfonyl)chloromethanes, **6**, 277
Aryl bromides
alkylation of, **1**, 466
displacement of, **2**, 874
from
arenediazonium salts, **2**, 627
aryl metallics, **2**, 627
palladium-catalyzed reaction, for tertiary phosphine oxides, **2**, 843
reactions, with arenethiols, **2**, 711
reduction of, **1**, 428
as starting materials
for aryl chlorides, **2**, 621
for aryl iodides, **2**, 632
for bromoarenes, **1**, 428
synthesis of
via electrophilic aromatic substitution, **2**, 625
via halogen exchange, **2**, 627
via Sandmeyer reaction, **2**, 609
Aryl chalcogen leaving groups, substitution of, **1**, 484
Aryl chlorides
alkylation of, **1**, 466
from
arenediazonium salts, **2**, 621
aryl metallics, **2**, 620
arylsulfonyl chlorides, **2**, 622
benzoyl chlorides, **2**, 622
reactions
with silicon, **2**, 538
with sodium amide, **1**, 1066
reduction of, **1**, 427
as starting materials
for aryl iodides, **2**, 632
for chloroarenes, **1**, 427
substitution of, **1**, 470
synthesis of, **2**, 609
via electrophilic aromatic substitution, **2**, 619
via halogen exchange, **2**, 621
via Sandmeyer reaction, **2**, 609
Aryl complexes, dimerization of, **5**, 465
Aryl σ-complexes, synthesis of, **2**, 1009
Aryl compounds
reactions, with sulfonylisocyanates, **5**, 344
synthesis of, **6**, 9, 150
N-Aryl compounds, synthesis of, **2**, 777
Aryl dihalides, reactions, with alkynes, **1**, 1009
Aryl fluorides
alkylation of, **1**, 466
from
anilines, **2**, 614
arenediazonium salts, **2**, 614
aryl metallics, **2**, 613
reduction of, **1**, 426
as starting materials, for fluoroarenes, **1**, 426
substitution of, **1**, 470
synthesis of, **2**, 609
via chloride displacement, **2**, 614
via electrophilic aromatic substitution, **2**, 612
via halogen exchange, **2**, 614
via Schiemann reaction, **2**, 609
Aryl groups
and aryl metals, **2**, 952
migratory aptitudes of, **1**, 384
sigmatropic rearrangements of, **1**, 799
Aryl halides
alkylation of, **1**, 468
Arbuzov reaction of, with trialkyl phosphites, **2**, 848
aromatic nucleophilic displacement of, **2**, 683
carbonylation of, **1**, 1015; **3**, 90
conversion to benzylic nitriles, **3**, 625
electrolysis of, **3**, 620
electron deficient, use in the synthesis of aryl diselenides, **2**, 734
from
aryl metallics, **2**, 608
diazotization of anilines, **2**, 609
hydrogenolysis of, **3**, 94

lithiation of, **2**, 735
metal-halogen exchange of, **2**, 961
nucleophilic aromatic substitution of, **1**, 473
oxidative metallation of, **2**, 957
using activated metals, **2**, 959
palladium catalysts with, **2**, 983
radical cyclization of, **1**, 341
reactions
with alkali metal diorganoarsenides, **2**, 874
with alkali metal diorganostibnides, **2**, 874
with alkenylaluminums, **1**, 533
with alkyl cyanoacetates, **3**, 625
with arene sulfinate salts, **2**, 715
with aryllithium, **2**, 717
with aryl tellurates, **2**, 731
with cobalt complexes, **1**, 345
with copper(I) cyanide, **3**, 661
with *N*-copper compounds, **2**, 753
with cuprous alkynides, **1**, 472
with diselenide dianions, **2**, 734
with 2-ethyldifluorosilyl thiophene, **3**, 305
with guanidines, **6**, 642
with *N*-hydroxyphthalimide, **2**, 692
with lithium diphenylphosphide, **1**, 52
with lithium methaneselenate, **2**, 729
with metal cyanides, **3**, 661
with metal diarylphosphides, **2**, 826
with metallated alkynes, **1**, 1006
with metal sulfides, **2**, 711
with nitrile enolates, **3**, 626
with organofluorosilanes, **3**, 290
with palladium(II) salts, **3**, 662
with potassium phthalimide, **2**, 753
with sodium cyanide, **3**, 661
with terminal alkynes, **1**, 1006, 1008
with tetraorganostannanes, **1**, 471
with *N*-tin compounds, **2**, 753
with vinylsilanes, **1**, 490
reduction of, **1**, 425
as starting materials
for alkyl aryl selenides, **2**, 729
for arenes, **1**, 425
for areneselenols, **2**, 727
for arenethiols, **2**, 706
for aryl aldehydes, **3**, 90
for arylalkynes, **1**, 472
for arylamines, **2**, 753
for α-arylnitriles, **3**, 626
for diaryl selenides, **2**, 729
for diaryl tellurides, **2**, 730
substitution of, **1**, 470, 1004
synthesis of, **2**, **605–633**
via electrophilic substitution, **2**, 608
via halide exchange, **2**, 609
with transition metal complexes
oxidative addition of, **2**, 960
oxidative displacement of, **2**, 960
trifluoromethylation of, **6**, 14
and Wurtz coupling, **2**, 958
Aryl heterocycles, deprotonation of, **3**, 721
Aryl iodide
reactions
with alkylstannanes, **1**, 471
with allyltributyltin, **1**, 471
Aryl iodides
carbonylation of, **1**, 1015
from
anilines, **2**, 632
arenediazonium salts, **2**, 632
aryl metallics, **2**, 632
hydrolysis of, **2**, 647
reactions
with acrylonitrile, **3**, 647
with alkenylzincs, **1**, 1097
with alkynes, **1**, 546
with arylaluminums, **3**, 290
with carbon monoxide, **3**, 89
with copper(I) alkynides, **1**, 1004
with hydrides, **1**, 429
with sodium–iron complexes, **5**, 452
with sodium telluride, **2**, 735
with stannylthiols, **2**, 711
with thiourea, **2**, 707
with trimethylsilyl cyanide, **3**, 661
with zinc compounds, **3**, 290
reactivity of, **1**, 1009
reduction of, **1**, 429
as starting materials
for aryl chlorides, **2**, 621
for iodoarenes, **1**, 429
synthesis of, **2**, 609; **5**, 1132
via electrophilic aromatic substitution, **2**, 630
via halogen exchange, **2**, 632
Arylmethyl carbocations, stabilization of, **1**, 894
Aryl migration, **2**, 874
Aryl nucleophiles
conjugate addition of, **3**, 24
to α,β-unsaturated ketones, **3**, 145
for the substitution of aryl halides, **1**, 472
Aryloic acids, 2-mercapto-, flash vacuum pyrolysis of, **3**, 552
Aryl polyhalides, reactions, with alkynes, **1**, 1009
Aryl radicals
cyclization of, **1**, 338, 341, 346, 547
from, decomposition of diazonium salts, **2**, 647
reactions
with propargyloxy groups, **3**, 104
with samarium diiodide, **1**, 347
synthesis of, **1**, 330; **2**, 351
Arylthiolation, of ylides, **5**, 898
β-Aryl-α,β-unsaturated acids, synthesis of, **5**, 98
Arynes
as starting materials, for phosphonium ylides, **3**, 498
synthesis of, **1**, 1027, 1066, 1080
via fragmentation reactions, **1**, 1081
Ascorbic acid, synthesis of, reviews, **5**, 98
Aspartame, synthesis of, **5**, 283
Aspartates, 3-hydroxy-, synthesis of, **2**, 44
DL-Aspartic acid, synthesis of, **5**, 90
DL-Aspartic acid, glycyl-, synthesis of, **5**, 36
Aspergillus niger, use in the dihydroxylation of alkenes, **2**, 51
Aspidosperma alkaloids, synthesis of, **3**, 422
Astatine, characteristics of, **1**, 107
Asymmetric addition, of aldehydes, **3**, 628
Asymmetric compounds, synthesis of, **2**, 485
Asymmetric induction, for the synthesis of α,β-unsaturated ketones, **3**, 244
Ate complexes, synthesis of, **1**, 278
Ate complexes, tetraalkyl, synthesis of, **2**, 596
Atherton–Todd procedure, for the synthesis of phosphoramidates, **2**, 412
Atom transfer conditions, in radical macrocyclizations, **1**, 340
Atom transfer method
for intermolecular radical addition, **1**, 334
in tandem cyclizations, **1**, 348
Atom transfer reaction, for radical cyclizations, **1**, 344
Atrolactamide, hydrolysis of, **5**, 65
Atrolactic acid, synthesis of, **2**, 71; **5**, 63, 65
Aurate, tetraethylammonium tetrabromo-, use in the oxidation of dithioacetals, **4**, 260
Aurates, dialkyl
alkylation of, **2**, 590
synthesis of, **2**, 590
Aurates, tetraalkyl, synthesis of, **2**, 590
Aureomycin, reactions, with boric acid, **2**, 111
Aurones

from, β-hydroxy chalcones, **3**, 242
synthesis of, **3**, 241
Autoxidation
of amines, **2**, 348
for the synthesis of alkyl hydroperoxide, **2**, 96
Avermectins, synthesis of, **1**, 596, 780
Axial dissymmetry
of ketenimines, **3**, 556, 564
of phosphaallenes, **3**, 605
Axisonitrile-4, synthesis of, **3**, 716
Axulenes, hydro-, synthesis of, **1**, 812
Azaalkanes, fluoro-
carbonylation of, **5**, 989
as starting materials, for fluoroalkyl isocyanates, **5**, 989
Azaalkenes, perfluoro-
epoxidation of, **6**, 256
hydrolysis of, **5**, 988
2-Azaallene ions, synthesis of, **3**, 422
1-Azaallenes, ketenimines described as, **3**, 556
2-Azaallenium salts, 1-alkoxy-, synthesis of, **3**, 422
Azaallyl anions
additions of, **1**, 371
from, imines, **3**, 404
intramolecular addition of, **1**, 372
reactions, with styrene, **1**, 350
Azaallyl anions, non-stabilized, synthesis of, **1**, 372
2-Azaallyl anions
α-alkylation of, **2**, 302
reactions, with alkenes, **1**, 371
Azaallyl carbanions, from, 1,2-diarylethane-1,2-diamines, **3**, 408
Azaarenes, ethynyl-, reactions, with sodium methoxide, **4**, 195
1,3-Azaarsolines
from, *N*-(2-chloroethyl)-benzimidoyl chloride, **5**, 791
synthesis of, **5**, 790
7-Azabenzotriazole, 1-hydroxy-, use in coupling reactions, **5**, 279
1-Azabicyclo[3.1.0]hexanes
synthesis of, **2**, 324
thermolysis of, **2**, 325
5-Azabicyclo[2.2.2]oct-5-en-3-ones, 6-chloro-2-oxa-, synthesis of, **5**, 661
1,2-Azaborolidines, from, enaminodialkylboranes, **2**, 811
1-Azabutadienes, 4-amino-, as starting materials, for pyrimidine-2(1*H*)-thiones, **6**, 575
1-Azabuta-1,3-dienes, reactions, with vinyl ethers, **4**, 305
1-Aza-1,3-butadienes, synthesis of, **3**, 413
2-Azabutadienes
cycloaddition reactions of, **4**, 421
reactions, with methyl propiolate, **1**, 522
synthesis of, **1**, 375
2-Azabutadienes, (*E*,*E*)-4-amino-1-aryl-3-cyano-4-methoxy-, synthesis of, **4**, 883
2-Aza-1,3-butadienes
reactions, with dialkyl azodicarboxylates, **3**, 414
synthesis of, **3**, 414
via base-catalyzed isomerization, **3**, 414
use in Diels–Alder reactions, **3**, 414
2-Azabuta-1,3-dienes, 1-aryl-, electrocyclization of, **3**, 414
1-Aza-2-butenes, *trans*-1-silyl-, synthesis of, **2**, 813
α-Azacarbanions, synthesis of, **2**, 1001
Azacarbaphosphaboretanes, synthesis of, **4**, 484
Azacrown compounds, as starting materials, for α-aminoalkylphosphines, **4**, 453
1-Azacumulenes, synthesis of, **3**, 584, 592, 597
Azacycloalkanes, alkynyl-, synthesis of, **1**, 1047
Azacycloalkanes, α-alkynyl-, synthesis of, **1**, 1041
1-Aza-1-cycloalkenes, 2-methyl-
reactions
with aldehydes, **3**, 413
with ketones, **3**, 413
2-Azacyclobutanes, 1,3-disila-, synthesis of, **4**, 606
Azacyclohexanone, 3-difluoroamino-2-(*N*-fluoro)-, synthesis of, **5**, 314
1-Azadienes, 4-amino-, as starting materials, for primary allylamines, **2**, 322
1-Aza-1,3-dienes, 1-alkyl-4-aryl-, reactions, with triethoxysilane, **2**, 813
1-Aza-1,3-dienes, 4-amino-, as starting materials, for 1,3-diamines, **2**, 312
1-Aza-1,3-dienes, *N*-[bis(trimethylsilyl)methyl]-, thermally induced cyclization of, **3**, 409
Aza-1,3-dienes, synthesis of, **3**, 413; **4**, 518
2-Aza-1,3-dienes, synthesis of, **3**, 414
2-Aza-1,3-dienes, 1-aryl-, synthesis of, **3**, 414
2-Aza-1,3-dienes, 3-ethoxycarbonyl-, synthesis of, **3**, 414
Azadi-*p*-methanes, rearrangement of, **1**, 800
Azadioxabicyclo[3.3.0]octane, synthesis of, **6**, 105
Azadiphosphetanes, synthesis of, **4**, 563
1,2,4-Azadiphosphetines, from, iminophosphanes, **5**, 885
Azadiphosphiranes, synthesis of, **2**, 471
Azadirachtin, synthesis of, **1**, 780
Azaenolates, synthesis of, **1**, 147
1-Azaenolates, α-trimethylsilylation of, **3**, 416
1-Azafulvene, 6-dimethylamino-, reactions, with alkyllithiums, **1**, 256
Azagermines, synthesis of, **4**, 521
3-Azagermoles, synthesis of, **2**, 949
Azalactones
as starting materials
for *gem*-amido phosphoramidoylalkenes, **4**, 1010
for *gem*-amido phosphorylalkenes, **4**, 1005
synthesis of, **4**, 1002, 1006
Azamines, synthesis of, **2**, 405
5-Azanonamdienes, 8-methyl-, synthesis of, **3**, 536
1-Azapenems, synthesis of, **4**, 421
1-Azapentane, perfluoro-, as starting material, for perfluoro-1-azapentene-1; **5**, 808
1-Azapentene-1, perfluoro-, from, perfluoro-1-azapentane, **5**, 808
Azaphosphaallenes, as intermediates, in the synthesis of aminophosphaethyne, **5**, 1157
Azaphosphabutadiyne, synthesis of, **5**, 1159
Azaphosphahexatriyne, synthesis of, **5**, 1153
1,3-Azaphosphanes
from, diallylaminomethylphosphines, **4**, 458
synthesis of, **4**, 473
1,3-Azaphosphepane, synthesis of, **4**, 458
1,2-Azaphosphetanes, synthesis of, **4**, 548
1,3-Azaphosphetanes, synthesis of, via heteroatom exchange, **4**, 457
Azaphosphetidenes, synthesis of, **2**, 469
1,3-Azaphosphetidine, synthesis of, **6**, 687
1,3λ^3-Azaphosphinine, synthesis of, **4**, 484
Azaphospholanes, as intermediates, in the synthesis of alkyl halides, **2**, 8
1,2-Azaphospholanes, synthesis of, **4**, 567
1,3-Azaphospholanes, synthesis of, **4**, 453
1,3-Azaphospholan-5-one, 1-amino-, from, carboxyalkylphosphines, **4**, 482
Azaphospholenes, synthesis of, **2**, 449
1,3-Azaphospholines
from
diazaphospholes, **5**, 785
dilithio phenylphosphine, **5**, 784
nitrile ylides, **5**, 785
2-phosphinoethylamines, **5**, 784
vinylisonitrile, **5**, 784
Azaphosphoridines, synthesis of, via cycloaddition reactions, **4**, 493
1,3-Azaphosphorines, tetrahydro-, synthesis of, **4**, 454
Azapines, diformylation of, **3**, 76
2-Azapropene, perfluoro-
reactions
with amines, **5**, 1076

with α-aminoalkyl phosphonates, **5**, 1076
as starting material, for trifluoromethyl isocyanate, **5**, 988
2-Azapropenylamines, trifluoroalkyl-, synthesis of, **5**, 1076
1,3-Azarsa compounds, synthesis of, **4**, 495
1,3-Azarsinanes
synthesis of, via cyclocondensation reactions, **4**, 495
use in the synthesis of α-aminoalkylarsenic compounds, **4**, 501
Azarsolanes, synthesis of, via alkylation, **4**, 496
1,3-Azarsolidines
from, 2-aminoalkylarsines, **4**, 498
synthesis of, via cyclocondensation reactions, **4**, 495
Azasilaboratedines, from, methyleneboranes, **6**, 187
Azasilines, synthesis of, **4**, 509
Azasilolines
from, *m*-chlorophenylsilanes, **4**, 508
synthesis of, **4**, 507
4-Azaspiro[2,4]hept-4-enes, 7-oxo-5-phenyl-6-oxo-, hydrolysis of, **5**, 87
1-Aza-8-stannabicyclo[3.2.0]octa-2,4,6-trienes, synthesis of, **6**, 196
Azasulfonium chlorides
for the condensation of ylides, **5**, 900
reactions, with phosphonium ylides, **5**, 898
Azasulfonium salts, synthesis of, **2**, 360
Azasulfonium salts, methoxy, synthesis of, **2**, 360
Azathiabutadienes, synthesis of, **6**, 571
3-Aza-1-thiabutadienes, 2-amino-, structure of, **6**, 571
2-Aza-1-thiacephems, desulfurization of, **4**, 421
3-Azathiopyrylium salts
reactions
with 2,6-di-*t*-butyl phenol, **5**, 634
with 1,3,3-trimethyl-2-methyleneindoline, **5**, 634
Azelaric acid, from, ricinolenic acid, **5**, 52
Azeotropic distillation, use in the synthesis of carboxylic esters, **5**, 122
Azepine, *N*-(azidocarbonyl)-, synthesis of, **6**, 511
Azepines, perhydro-
alkylation of, **1**, 151
synthesis of, via reductive amination, **2**, 327
Azepinone, 4-perhydro-, synthesis of, **2**, 324
Azetes, as starting materials, for phosphaalkenes, **5**, 883
Azetes, *N*-hydroxy-1,2-dihydro-, ring-opening of, **3**, 428
Azetidine
reactions
with bis(isocyanide) complexes, **6**, 663
with coordinated isocyanides, **6**, 663
with hydrido alkyl isocyanide complexes, **6**, 663
Azetidine, *N*-chlorocarbonyl-, synthesis of, **6**, 450
Azetidine, *N*-hydroxy-, oxidation of, **5**, 325
Azetidine, 2-imino-, hydrolysis of, **5**, 348
Azetidine, *N*-tosyl-, acetoxylation of, **2**, 323
Azetidine, 1,3,3-trinitro-, synthesis of, **4**, 429
2-Azetidine, 2-ethoxy-1-methanesulfonyl-4-trichloromethyl-3-trimethylsilyl-, ring-opening of, **5**, 702
Azetidine-2-one, 4-acetoxy-, synthesis of, **4**, 306
Azetidines
cleavage of, **2**, 320
from, β-chloroimines, **3**, 416
ring-opening of, **4**, 891
synthesis of, **2**, 322
Azetidines, amino-, synthesis of, **4**, 414
Azetidines, 2-methylene-, from, *N*-aryl β-chloro ketimines, **2**, 322
Azetidines, 2-substituted, synthesis of, **2**, 323
Azetidines, 2-*N*-(*p*-toluenesulfonyl)imino-, as starting materials, for β-amino-*N*-acyl sulfonamides, **5**, 348
Azetidinium salts, synthesis of, **2**, 323
Azetidinones
as starting materials, for trithiocarbonates, **6**, 553
synthesis of, **5**, 286
See also β-Lactams
Azetidinones, 4-acetoxy-, condensation of, **5**, 236
Azetidinones, 2-(hydroxyamino)-, synthesis of, **4**, 442
Azetidinones, 4-(*N*-imidazolo)-, synthesis of, **4**, 417
Azetidinones, 4-silyl-, synthesis of, **4**, 514
Azetidinones, *N*-substituted, as starting materials, for nitrogen-functionalized ylides, **5**, 905
Azetidin-2-ones, 4-acetoxy-
as starting materials
for 2-oxoazetidin-4-yl phosphinates, **4**, 487
for 2-oxoazetidin-4-yl phosphonates, **4**, 487
2-Azetidinones, 4-thioxo-, synthesis of, **5**, 639
Azetidino[3,2-*b*]pyridine, as intermediate, for the synthesis of ketenimines, **3**, 584
Azetidinylidene complexes, imino-, synthesis of, **5**, 954
Azetidin-4-yl phosphinates, 2-oxo-, from, 4-acetoxyazetidin-2-ones, **4**, 487
Azetine, 2,3-dichloroperfluoro-*N*-methyl-, pyrolysis of, **6**, 613
Azetines, reactions, with diphenylketene, **4**, 829
Azibenzil, as starting material, for diazophenylmethane, **3**, 475
Azide, ammonium, for the nucleophilic displacement of halides, **4**, 432
Azide, *p*-carboxybenzenesulfonyl
as diazo group transfer reagent, **3**, 473
reactions, with arylsulfonylmethylenetriphenylphosphorane, **3**, 479
Azide, 4-dimethylaminobenzenesulfonyl, reactions, with β-*H*-alkynyl ethers, **3**, 477
Azide, methanesulfonyl, reactions, with β-diketones, **3**, 472
Azide, *p*-nitrophenyl, thermolysis of, **2**, 810
Azide, pentafluorophenyl, thermolysis of, **2**, 810
Azide, phenyl
reactions
with acrylonitrile, **3**, 478
with mesityl(diphenylmethylene)phosphines, **4**, 483
as starting material, for α-aminoalkylphosphorus compounds, **4**, 490
Azide, phenylthiomethyl, reactions, with magnesium, **2**, 753
Azide, phosphoryl, reactions, with magnesium, **2**, 753
Azide, tetramethyl guanidinium, for the nucleophilic displacement of chloride, **4**, 447
Azide, 4-toluenesulfonyl, decomposition of, **2**, 161
Azide, *p*-toluenesulfonyl, reactions, with 1,2,3,4-tetraphenylcyclopentadiene, **3**, 490
Azide, *p*-tolyl, reactions, with phosphine, **6**, 151
Azide, tosyl
as diazo group transfer reagent, **3**, 473
as diazo transfer reagent, **5**, 869
reactions
with alkyl 1-formylalkane sulfonates, **5**, 867
with benzene, **4**, 483
with bis(benzotriazoyl)methane, **6**, 364
with *t*-butyl acetoacetate, **3**, 475
with dithioacetals, **4**, 860
with formyl enamines, **3**, 473
with magnesium, **2**, 753
with potassium nitronates, **4**, 331
with sodium phenylalkynide, **2**, 1055
with trimethylsilylmethane, **3**, 483
Azide, tributylstannyl, use in the synthesis of 1,2-azido alcohols, **2**, 84
Azide, trimethylsilyl
reactions
with acid chlorides, **5**, 992
with aldehydes, **3**, 665; **5**, 377
with alkanoyl chlorides, **5**, 991
with *t*-alkyl chlorides, **2**, 317
with aroyl chlorides, **5**, 992
with *N*,*N*-dimethylanilines, **4**, 443
with 1,2-dimethyl-1,2-cyclopropanedicarboxylic anhydride, **5**, 992
with enol ethers, **1**, 263
with sugar lactones, **4**, 432

with tertiary alkanoyl chlorides, **5**, 992
with thenoyl chloride, **5**, 992
with trialkylboranes, **2**, 307
with vinyltriethylsilane, **2**, 813
as starting material
for *gem*-diazidoalkanes, **4**, 432
for phospheneimides, **2**, 859
use in the synthesis of acyl azides, **5**, 367
Azide, trimethylsilylmethyl (TMSMA)
reactions, with magnesium, **2**, 753
reduction of, **3**, 440
stability of, **4**, 516
as starting material, for isothiocyanates, **5**, 1031
synthesis of, **4**, 516
Azide anions
reactions
with hexafluoropropene, **4**, 799
with thiazolium cations, **6**, 113
Azide ions
nucleophilic attack of, **2**, 1055
reactions, with carbon disulfide, **5**, 1046
Azides
acylation of, **5**, 367
as aminyl radicals, **1**, 932
as diazo transfer reagents, **3**, 472
reactions
with acylphosphoranes, **3**, 476
with alkanethiols, **2**, 240
with alkenes, **4**, 515
with alkynyl ethers, **3**, 477
with aminomethyl halide, **4**, 444
with ω-chloroalkenes, **3**, 419
with cyclohexyldichloroborane, **2**, 318
with dichlorocarbene, **6**, 608
with 1,3-diketones, **3**, 194
with enals, **3**, 48
with enamines, **3**, 467
with Grignard reagents, **2**, 776
with ketene acetals, **6**, 107
with ortho esters, **6**, 108
with phosphines, **2**, 1068
with sulfides, **2**, 161
with tertiary arsines, **2**, 809
with tetrachloromethane, **2**, 773
with trialkylaluminums, **2**, 776
with trimethyl phosphite, **2**, 317
with triphenylphosphine, **2**, 324
reduction of, **1**, 5; **5**, 90
as starting materials
for *N*-alkylamides, **5**, 278
for *N*-alkylphosphorimidates, **2**, 407
for *N*-alkylphosphorimidic acid triamides, **2**, 407
for amines, **2**, 317
for arylamines, **2**, 753
for isocyanates, **5**, 983
for *N*-silylamides, **5**, 377
for sulfenamides, **2**, 359
for triazenes, **2**, 387
synthesis of, **4**, 897; **5**, 278
use in the substitution of nitrogen functions, **1**, 263
Azides, acyl
for the acylation of amines, **5**, 265
Curtius rearrangement of, **2**, 757
from, α-thioalkanoyl chlorides, **4**, 328
reactions
with hexabutyldistannane, **5**, 391
with hydrazines, **5**, 351
with metal complexes, **5**, 386
with phosphines, **5**, 373
as starting materials
for *N*-acylated α-ureido enamines, **4**, 982
for *N*-acyl iminophosphoranes, **5**, 373
for isocyanates, **5**, 989
for transition metal amides, **5**, 386
synthesis of, **5**, 367, 991
via acylation of azides, **5**, 367
via oxidation of acylhydrazines, **5**, 367
via oxidation of aldehydes, **5**, 368
via ring-opening of *N*-aminobenzotriazinone, **5**, 368
use in peptide coupling reactions, **5**, 278
Azides, acylaminomethyl, synthesis of, **4**, 447
Azides, 1-adamantyl, reactivity of, **5**, 1066
Azides, 2-adamantyl, reactivity of, **5**, 1066
Azides, alkali, reactions, with sulfamoyl halides, **2**, 368
Azides, 1-alkoxyalkyl, synthesis of, **4**, 313
Azides, alkyl
reactions
with hydrazine, **2**, 317
with phosphines, **5**, 373
as starting materials, for imines, **3**, 407
Azides, *t*-alkyl, reactions, with triethyl phosphite, **2**, 317
Azides, aminomethyl, synthesis of, **4**, 443
Azides, arenesulfinyl, use in the synthesis of arene thiosulfinate esters, **2**, 724
Azides, aryl
cycloaddition of, **2**, 771
photochemical decomposition of, **3**, 565
reactions
with bis[2-pyridylbis(trimethylsilyl) methyl]stannylene, **6**, 196
with carbon monoxide, **5**, 984
with carbonyl complexes, **5**, 983
with potassium cyanide, **5**, 1124
as starting materials, for imines, **3**, 407
Azides, 1-arylalkyl, reduction of, **2**, 317
Azides, carbamoyl, synthesis of, **6**, 511
Azides, cycloalkyl, as starting materials, for imines, **3**, 407
Azides, 2,2-dicyanovinyl, as starting materials, for secondary α-chloro enamines, **4**, 792
Azides, α-halo, synthesis of, **4**, 109
Azides, 2-hydroxy, reactions, with triphenylphosphine, **2**, 860
Azides, β-hydroxy-α-silyl, from, silyl oxiranes, **4**, 516
Azides, keto, as starting materials, for cations, **4**, 1072
Azides, nitroacyl, Curtius rearrangement of, **4**, 440
Azides, organic
reactions
with isocyanides, **5**, 1071
with organyldichloroboranes, **2**, 317
with phosphorous compounds, **5**, 1065
Azides, perfluorosulfonyl, photolysis of, **2**, 161
Azides, phenylpropioyl, synthesis of, **2**, 1055
Azides, polymer-supported, use in the synthesis of acyl azides, **5**, 367
Azides, silyl
reactions
with carboxylic acid anhydrides, **5**, 992
with silaethene, **6**, 673
synthesis of, **4**, 516
Azides, silylmethyl, reduction of, **4**, 512
Azides, sulfamoyl, synthesis of, **2**, 368
Azides, sulfonyl
reactions
with β-alkynyl ethers, **3**, 477
with ethoxyethyne, **5**, 827
as starting materials, for sulfonamides, **2**, 365
Azides, thienoyl, reactions, with tetrachloromethane, **5**, 990
Azides, thionoacyl, synthesis of, **5**, 634
Azides, thiophosphoryl, synthesis of, **3**, 481
Azides, tosyl, as starting materials, for bis(phosphinyl)iminocarbonyls, **6**, 666
Azides, α-(trimethylsiloxy)alkyl, synthesis of, **5**, 377
Azides, 1-trimethylsilyloxyalkyl, synthesis of, **4**, 313
Azides, vinyl
reactions, with alcohols, **4**, 894
as starting materials, for ketenimines, **3**, 564

synthesis of, **2**, 1055
Azides, vinylidene, as starting materials, for ketene aminals, **4**, 979
Azidinium salts, reactions, with (acylmethyl)triphenylphosphonium salts, **3**, 481
α-Azido acids, from, α,α-dichlorooxiranes, **6**, 42
2-Azido-1,3-dithiole, as starting material, for tetrathioorthocarbonates, **6**, 305
Azidoformate, *t*-butyl, synthesis of, **6**, 489
Azidomethyls, from, *N*,*N*-dimethylanilines, **4**, 443
Azine, acetone, as starting material, for 2-pyrazoline hydrochloride salts, **3**, 452
Azine, ketenimine-, synthesis of, **3**, 586
Azines
cyclization of, **3**, 452
cycloaddition reactions of, **1**, 790
from
hydrazine, **3**, 444
hydrazones, **3**, 444
nitriles, **3**, 445
halogenation of, **5**, 677
reactions
with Grignard reagents, **2**, 377
with hydrazine, **3**, 444
reduction of, **2**, 377
as starting materials
for acylhydrazones, **5**, 364
for α-aminoalkylorganophosphorus compounds, **4**, 460
for hydrazines, **2**, 377
for ketene aminals, **4**, 979
for ylides, **3**, 497
synthesis of, **6**, 611
Azines, cyclic, synthesis of, **3**, 451, 453
Azines, phosphine, as starting materials, for phosphonium ylides, **3**, 497
Azines, tetrabromo-, synthesis of, **6**, 611
2*H*-Azirenes
as starting materials, for nitrile ylides, **3**, 679
synthesis of, **3**, 678
2*H*-Azirenes, 3-phenyl, from, benzonitrile ylides, **3**, 679
Aziridimes, 2,2-dialkoxy-, as starting materials, for ortho esters, **6**, 77
Aziridine
reactions
with benzonitrile oxides, **5**, 765
with 2,2-dicyano bis(1,1-alkylthio)alkenes, **4**, 899
with methyl bromopropiolate, **4**, 793
with oxyimidic chlorides, **5**, 765
with thiols, **4**, 899
with thiophosgene, **5**, 1033
Aziridine, 1-benzyl-2,3,3-trifluoro-2-(trifluoromethyl)-, synthesis of, **6**, 54
Aziridine, 1-phenylthiocarbonyl-, thermal rearrangement of, **5**, 1010
Aziridine, α-stannyl-, synthesis of, **4**, 534
Aziridine, *N*-tosyl-, cleavage of, **1**, 264
Aziridine, 1,2,3-triphenyl-, synthesis of, **1**, 741
1-Aziridinecarboimidoyl chloride, *N*-aryl-, synthesis of, **5**, 1073
Aziridine-*N*-carbonyl fluorides
from, trifluormethyl hypofluorite, **6**, 441
rearrangement of, **6**, 441
Aziridine hydrazones, amino-, as starting materials, for diazo compounds, **1**, 1077
Aziridines
N-chlorination of, **2**, 337
cleavage of, **2**, 553
elimination of, **1**, 636
from
acetoxy quinazolones, **4**, 515
6*H*-1,2-oxazines, **2**, 322
1,3-insertion of, into manganese–hydride bonds, **5**, 448
nitrosation of, **1**, 636
reactions
with alkoxides, **5**, 693
with difluoroamine, **1**, 636
with diorganocopperlithiums, **2**, 319
with hydride complexes, **5**, 446
with isocyanates, **6**, 510
with lithium heptynide, **1**, 1032
with manganese pentacarbonyls, **5**, 447
with phosgene, **5**, 1010
with phosphorous acids, **2**, 457
with prop-2-ynyl triphenylphosphonium bromide, **1**, 267
rearrangement of, **6**, 571
ring-opening of, **1**, 263; **2**, 129, 319
as starting materials
for acyl compounds, **5**, 446
for amines, **2**, 319
for azomethine ylides, **1**, 368
for 1,2-diamines, **2**, 319
for isocyanates, **5**, 1010
for ketenimines, **3**, 574
for phosphonium ylides, **3**, 498
for thiocyanates, **5**, 1107
synthesis of, **2**, 322; **3**, 32
Aziridines, acryl-, reactions, with lithium dibutylcopper, **1**, 263
Aziridines, 2-acyl-, as starting materials, for ketenimines, **3**, 595
Aziridines, *N*-acyl-, rearrangement of, **1**, 266
Aziridines, 2-alkoxy-, from, 1-azirines, **4**, 305
Aziridines, *N*-alkyl-
deamination of, **1**, 636
reactions, with *m*-chloroperbenzoic acid, **1**, 636
Aziridines, amino-, reactions, with alkyl isocyanides, **6**, 337
Aziridines, *N*-amino-
oxidation of, **1**, 636
synthesis of, **2**, 380
use in the reduction of hydrazones, **1**, 43
Aziridines, *cis*-*N*-aryl-, synthesis of, **2**, 322
Aziridines, *N*,*N'*-bis-protected 2-(aminomethyl)-, as starting materials, for 3-substituted-1,2-diaminopropanes, **2**, 309
Aziridines, carbethoxy-, synthesis of, **4**, 514
Aziridines, *N*-chloro-, nitrogen inversion barrier in, **2**, 337
Aziridines, *N*-cyano-, synthesis of, **5**, 1119
Aziridines, 1,3-diaryl-2,2-dibromo-, reactions, with triethylamine, **3**, 574
Aziridines, 2,2-dihalo-, synthesis of, **6**, 51
Aziridines, 2,2-dimethyl-, ring-opening of, **2**, 319
Aziridines, α-hydroxy-, reactions, with thionyl chloride, **2**, 364
Aziridines, lithiated, ring-opening of, **1**, 371
Aziridines, *N*-methoxy-, synthesis of, **6**, 114
Aziridines, silyl-, synthesis of, **4**, 514
Aziridines, *C*-silyl-, synthesis of, **4**, 514
Aziridines, *N*-silylmethyl-, synthesis of, **4**, 515
Aziridines, *N*-sulfonyl-
reactions
with diethyl malonate, **5**, 345
with sodium ethylmethylphosphinate, **2**, 454
ring-opening of, **5**, 345
synthesis of, **2**, 365
Aziridines, *N*-thioacyl-, ring-expansion of, **6**, 571
Aziridines, *N*-tosyl-
reactions
with cuprates, **1**, 486
with trimethylsilyl cyanide, **3**, 638
ring-opening of, **2**, 319
as starting materials
for *N*-tosyl β-aminonitriles, **2**, 319
for β-*N*-tosylaminonitriles, **3**, 638
Aziridines, 1,3,3-triaryl-2,2-dichloro-, as starting materials, for ketenimines, **3**, 574

Aziridines, triethylsilyl-*N*-trimethylsilyl-, synthesis of, **2**, 813
Aziridines, *N*-unsubstituted, from, cyclic sulfates, **2**, 322
Aziridines, vinyl-
  rearrangement of, **1**, 267
  ring-opening of, **2**, 326
Aziridines, *N*-vinyl-, rearrangement of, **1**, 266
2,3-Aziridines, reactions, with phenyl isothiocyanate, **6**, 571
Aziridinium ions, as starting materials, for alkenes, **1**, 629
Aziridinones, reactions, with cyanamides, **5**, 1118
n-Aziridinylurethane, ethyl-, synthesis of, **6**, 483
2*H*-Azirine, 2,2-bis(trifluoromethyl)-3-methoxy-, thermal rearrangement of, **5**, 1011
Azirines
  from, ylides, **3**, 600
  photochemical ring-opening of, for nitrile ylides, **5**, 785
  reactions, with diphenylketene, **4**, 829
1-Azirines, as starting materials, for 2-alkoxyaziridines, **4**, 305
2*H*-Azirines, 2,2-dialkoxy-, as starting materials, for ortho esters, **6**, 77
Azlactones
  of α-benzoylaminocinnamic acid, **5**, 87
  as intermediates, **5**, 656
  as starting materials, for ketenimines, **3**, 578
Azobenzenes
  reactions, with 2-methyl-1-propenyl triflate, **4**, 985
  as starting materials, for *gem*-dihydrazinoalkenes, **4**, 985
1-Azobicyclo[2.2.2]octane, *N*-chloro-, stability of, **2**, 338
2,2'-Azobisisobutyronitrile (AIBN)
  as mercurating agent, **6**, 368
  as radical initiator, **1**, 330
    for the synthesis of diethylformimidoylphosphines, **5**, 785
  thermal decomposition of, **1**, 265; **3**, 573
  use in the synthesis of alkyltins, **2**, 601
  use in the synthesis of cyanophosphines, **5**, 1130
  use in the synthesis of formyl thiophenes, **3**, 105
  use in the synthesis of radicals, **1**, 924
Azocarbonamides, *N*-chloro, fragmentation of, **2**, 806
Azo compounds
  from
    amines, **2**, 403
    arylhydrazones of chloral, **4**, 799
    azoxy compounds, **2**, 403
    N-bromoadamatylamines, **2**, 403
    diazenium salts, **2**, 403
    hydrazines, **2**, 403, 772
    hydrazones, **2**, 403
    hydroxylamine-*O*-sulfonic acid, **2**, 403
    ketazines, **2**, 404
    nitrosoamines, **2**, 403
    sulfamides, **2**, 403
  metallation of, **4**, 536
  reactions
    with dichlorocarbene, **6**, 608
    with trifluoroacetic acid, **1**, 154
  reduction of, **2**, 377
  as starting materials
    for azoxy compounds, **2**, 404
    for formazans, **5**, 858
    for hydrazines, **2**, 377, 382, 383
    for hydrazones, **3**, 448
    for nitroso compounds, **2**, 390
  synthesis of, **2**, 402, 769
  use in the synthesis of radicals, **1**, 924
Azo compounds, aminomethyl-, synthesis of, **4**, 444
Azo compounds, aromatic, synthesis of, **2**, 772
Azo compounds, aryl-, reactions, with vinylstannanes, **1**, 486
Azo compounds, α,α'-dicarbonyl-, reactions, with dinitromethanes, **6**, 284
Azo compounds, α,β-unsaturated, from, α-halohydrazones, **2**, 771
N-Azo compounds, synthesis of, **2**, 470
Azo decoupling, for the synthesis of diazonium salts, **2**, 804
Azo-*N*,*N*'-dioxides, as starting materials, for azoxy compounds, **2**, 404
N,*N*'-Azodioxides, synthesis of, **2**, 389, 405
Azodioxy compounds, from, aromatic amines, **2**, 775
Azoles, *N*-(chloroalkyl)-, synthesis of, **4**, 322
Azoles, *N*-(α-chloroalkyl)-, use in the synthesis of *O*,*N*-acetals, **4**, 308
Azoles, *N*-(hydroxyalkyl)-, acylation of, **4**, 310
Azoles, 1-(isocyanomethyl)-, synthesis of, **3**, 710
Azoles, *N*-(phenylthioalkyl)-, oxidation of, **4**, 323
Azoles, *N*-thioacyl-, for the thioacylation of amines, **5**, 584
Azoles, thiobenzoyl-, for the thioacylation of amines, **5**, 620
Azoles, *N*-thiobenzoyl-, reactions, with amines, **5**, 587
Azomethines, *N*-*t*-alkyl-, reactions, with ketene, **4**, 421
Azoxy compounds
  from
    amines, **2**, 405
    aromatic nitro compounds, **2**, 775
    aromatic nitroso compounds, **2**, 775
    azo compounds, **2**, 404
    azo-*N*,*N*'-dioxides, **2**, 404
    diamines, **2**, 404
    hydrazines, **2**, 404
    hydroxylamine, **2**, 405
    nitroso compounds, **2**, 405
    nitroso dimers, **2**, 404
  reduction of, **2**, 403
  as starting materials
    for azo compounds, **2**, 403
    for nitroso dimers, **2**, 405
  synthesis of, **2**, 404, 773
  uses of, **2**, 773
Azoxy compounds, aromatic, synthesis of, **2**, 773
Azoxy compounds, α,β-unsaturated, synthesis of, **2**, 773
AZT (pharmaceutical) phosphonates, in HIV-research, **6**, 489
Azulene
  chemistry of, **1**, 916
  reactions, with tetraethyl orthocarbonate, **1**, 898
Azulene, 1-nitrotrimethyl-, protonation of, **1**, 917
Azulenes, reactions, with oxirane, **1**, 212
Azulenes, nitro-, synthesis of, **2**, 790
Azulenes, silyl-, from, tropylium ions, **2**, 944
Azulenes, tellurocyanato-, synthesis of, **5**, 1116
Azulenyl carbocations, synthesis of, **1**, 916

B

Bacteria, use in the synthesis of carboxylic acids, **5**, 117
Bacteriostatic agents, synthesis of, **5**, 1047
Baeyer–Villiger oxidation
of acylarenes, **2**, 654
of α-alkoxy ketones, **4**, 212
of β-silyl ketones, **1**, 650
for the synthesis of acyl alkyl acetals, **4**, 212
for the synthesis of dialkyl carbonates, **6**, 466
for the synthesis of lactones, **5**, 161
for the synthesis of γ-lactones, **5**, 171
Baker's yeast
for carbonyl reduction, **2**, 39
for the cleavage of hydrazones, for aldehydes, **3**, 18
for the cleavage of oximes, for aldehydes, **3**, 18
control of enantioselection, **2**, 40
for the enantioselective reduction of alkenes, **1**, 77
for the oxidation of thiols, **2**, 216
reactions, with ethyl α-chlorobenzoyl acetate, **1**, 8
as reagent, for the hydrolysis of oximes and hydrazones, **3**, 134
for the reduction of β-oxo dithioesters, **5**, 557
for the resolution of β-hydroxynitriles, **3**, 632
use in reductions, **3**, 30
use in the synthesis of amines, **2**, 314
use in the synthesis of β-hydroxy ketones, **3**, 161
Bakkenolide A, synthesis of, **1**, 822; **5**, 553
Baldwin's rules
and cycloalkanone formation, **3**, 124
in the opening of epoxides, **1**, 213
for ring-closure reactions, **1**, 304
Balz–Schiemann reaction
for the synthesis of aryl halides, **2**, 609
See also Schiemann reaction
Bamberger reaction, for the synthesis of guanidines, **6**, 653
Bamberger rearrangement, of *N*-arylhydroxylamines, **2**, 649
Bamford–Stevens reaction
aprotic conditions, **1**, 630
of α-keto tosylhydrazones, **3**, 470
mechanism, **1**, 630, 631
regioselectivity of, **1**, 632
and Shapiro reaction, compared, **1**, 631
stereoselectivity of, **1**, 633
for the synthesis of diazomethylcarboranes, **5**, 873
for the synthesis of α-diazophosphorus compounds, **5**, 870
for the synthesis of α-diazo phosphoryls, **3**, 480
for the synthesis of α-diazosilyls, **5**, 873
for the synthesis of silicon-substituted diazoalkanes, **3**, 483
of tosylhydrazones, **3**, 464, 489
use of tosylhydrazones in, **3**, 446
Barbier reaction
mechanism, **2**, 73, 551
modification of, **2**, 422
samarium diiodide mediated, **2**, 565
for the synthesis of alkyllithiums, **2**, 551
for the synthesis of alkylzincs, **2**, 591
for the synthesis of α-fluorovinyltin, **4**, 812
use in the synthesis of alkylchromium(III) compounds, **2**, 572
Barbituric acid acetals, synthesis of, **6**, 106
Barium
reactions, with allylic chlorides, **1**, 116
use in Barbier reactions, **2**, 73
Barium(II) iodide, reactions, with lithium biphenylide, **1**, 116
Barium ferrate
for the oxidation of alcohols, **3**, 117
for the oxidation of hydroxymethyl furans, **3**, 102
Barium hydrazide, use in the synthesis of hydrazonamides, **5**, 848
Barium hydroxide
for the hydrolysis of aromatic esters, **5**, 106
use in Horner–Wadsworth–Emmons reaction, **1**, 640
use in the synthesis of amino acids, **5**, 85
Barium iodide, reactions, with lithium biphenylide, **2**, 563
Barium manganate, use in the synthesis of polyaromatic aldehydes, **3**, 100
Barium manganate(VII), for alcohol oxidation, **3**, 116
Barium metal, reactions, with alkyl halides, **2**, 563
Barium oxide, reactions, with nitrosoacetamides, **3**, 471
Barium permanganate, use in the synthesis of disulfides, **2**, 215
Bariums, alkyl-, synthesis of, **2**, 563
Bariums, allyl-
alkylation of, **1**, 116
structure of, **1**, 116
synthesis of, **1**, 116; **2**, 563
Barium sulfate, reactions, with palladium, **1**, 503
Barium triselenocarbonate
alkylation of, **6**, 592
as starting material, for triselenocarbonic acid, **6**, 590
Barium trithiocarbonate, as starting material, for trithiocarbonic acid salts, **6**, 552
Barton esters
as starting materials, for thiocyanates, **5**, 1110
use in the synthesis of alkenes, **1**, 443
Barton–McCombie reaction, mechanism, **1**, 16
Barton method
for intermolecular radical addition, **1**, 332
for radical cyclization reactions, **1**, 344
for tandem cyclizations, **1**, 348
Bart reaction
use in the synthesis of arylarsonic acids, **2**, 884
use in the synthesis of diarylarsinic acids, **2**, 885
Bases
as catalysts, for the synthesis of cyanohydrins, **3**, 615
choice of, for the synthesis of ylides, **3**, 493
for nitrile deprotonation, **3**, 620
reactions, with α-halooximes, **2**, 778
for the synthesis of amides, **5**, 273
Bases, tertiary, use in the synthesis of chloroformate esters, **6**, 425
Baudisch reaction, mechanism, **2**, 779
Béchamp reaction, use in the synthesis of arylarsonic acids, **2**, 884
Beckmann fragmentation
of ketoximes, **3**, 663
for the synthesis of enamides, **2**, 747
Beckmann rearrangement
of aldoximes, **5**, 275
of arylthiohydroxamic acid anhydrides, **5**, 832
of *O*-chlorooxalates, **3**, 687
of ketoxime *O*-chloroformates, **3**, 687
of oxime carbonates, **1**, 1041
of oxime sulfonates, **1**, 1044, 1059; **3**, 407; **5**, 731
of sulfonimines, **3**, 437
for the synthesis of *N*-alkylamides, **5**, 282
for the synthesis of amides, **5**, 270
for the synthesis of amidines, **5**, 755
for the synthesis of ω-aminonitriles, **3**, 638
for the synthesis of *N*-arylamides, **5**, 292
for the synthesis of aryl isocyanides, **3**, 722
for the synthesis of cyclic ureas, **6**, 510
for the synthesis of enamides, **5**, 288

for the synthesis of imidoyl chlorides, **5**, 656
for the synthesis of ketenimines, **3**, 563
for the synthesis of phosphoramidates, **5**, 371
for the synthesis of primary amides, **5**, 275
for the synthesis of primary arylamides, **5**, 298
Belleau's reagent
characteristics of, **5**, 573
reactions
with glycines, **5**, 574
with peptides, **5**, 574
for the thionation of peptides, **5**, 574
Benezenamine, acylation of, **5**, 302
Benkeser reduction
mechanism, **1**, 101
for the synthesis of calcium enolates, **2**, 697
Bente salts, use in the synthesis of trisulfides, **2**, 233
Bentonitic earth, as catalyst, for the synthesis of dithioacetals, **4**, 246
Benzal chlorides
from, aldehydes, **6**, 28
reactions, with isobutyl hypochlorite, **6**, 28
Benzaldazines, chlorination of, **5**, 679
Benzaldehyde
industrial synthesis of, **3**, 93
reactions
with *S*-acetyldiphenylthiophosphinous acid, **4**, 345
with allylic ylides, **1**, 721
with amines, **2**, 335
with (*o*-aminobenzyl)phenylphosphine oxide, **4**, 471
with benzylphosphonium salts, **1**, 733
with bislithio compounds, **4**, 938
with *N*-chlorosuccinimide, **5**, 15
with crotonaldehyde dithioacetals, **4**, 849
with α-cyclocitral, **3**, 66
with diazomethylenephosphorane, **2**, 1068
with dibromophosphoranes, **4**, 745
with (diethoxymethyl)phosphine oxide, **4**, 828
with diethylzinc, **2**, 71
with dimethyl methylene bis(sulfonylacetate), **4**, 277
with diphenyl cyanomethyl phosphonate, **1**, 751
with dithiopropanoic acid *S*-ethyl ester, **5**, 557
with ethoxycarbonylmethylides, **1**, 722
with ethyl acetate, **5**, 104
with isothiocyanates, **4**, 1014
with 2-lithio-1,1,3,3,5,5-hexamethyl-1,3,5-trisilacyclohexane, **4**, 623
with α-methoxyallylphosphine oxide, **4**, 917
with (methoxymethylene)bisphosphonic esters, **4**, 916
with methylene(triphenyl)phosphorane, **1**, 639
with *N*-phenylphosphoramidites, **4**, 476
with thiophenol, **4**, 216
with trifluoroethyl phosphonate, **1**, 752
with triphenylarsonium benzoyl ylide, **1**, 737
with tris(*p*-substituted phenyl)arsonium ylides, **1**, 739
as starting material
for benzoic acid, **5**, 114
for 1,1-dichlorostyrene, **4**, 738
for ethylmethylhydrazine, **2**, 379
synthesis of, **3**, 84, 93; **5**, 10, 439
Benzaldehyde, 4-*t*-butyl-, from, *t*-butyl benzene, **3**, 94
Benzaldehyde, 2,5-dimethoxy-, reactions, with lithio diethyl(isocyanomethyl) phosphonate, **3**, 716
Benzaldehyde, 4-dimethylamino-, reactions, with salicyl aldehyde, **4**, 438
Benzaldehyde, *p*-(dimethylamino)-
reactions
with disulfonyl fluoride, **4**, 864
with fluorenylidenetriphenylarsorane, **1**, 646
with triphenylarsonium fluorenylide, **1**, 738
Benzaldehyde, 4-fluoro-
reactions
with morpholine, **4**, 298
with piperidine, **4**, 298
Benzaldehyde, *m*-(4-fluorophenyl)-, as starting material, for *m*-(4-fluorophenyl)benzoic acid, **5**, 114
Benzaldehyde, 2-hydroxy-, reactions, with *N*-alkylbenzothiazolium cation, **6**, 111
Benzaldehyde, 3-methoxy-, reactions, with *S*-diamines, **4**, 408
Benzaldehyde, 4-methyl-, reactions, with ozone, **5**, 208
Benzaldehyde, *o*-nitro-, reactions, with potassium cyanide, **2**, 783
Benzaldehyde, *p*-nitro-
reactions
with benzoyl ylides, **1**, 741
with phenacylidenetriphenylphosphorane, **1**, 724
with triarylarsonium benzoyl ylides, **1**, 740
with triphenylarsonium benzylide, **1**, 738
synthesis of, **3**, 99
Benzaldehyde, 2,3,6-trichloro-, as starting material, for 2,3,6-trichlorobenzoic acid, **5**, 114
Benzaldehyde dimethyl acetals, electrochemical oxidation of, **6**, 77
Benzaldehyde hydrazones, heterocyclic, as starting materials, for formazans, **5**, 857
Benzaldehyde imine, *N*-trimethylsilyl-, as starting material, for chiral primary amines, **2**, 544
Benzaldehyde oxime, *p*-methoxy-, reactions, with 1-trichlorophosphazopentachloroethane, **5**, 683
Benzaldehydes
aminosulfuration of, **5**, 600, 621
fluorine-substituted, synthesis of, **3**, 95
from
arenes, **3**, 97
aryl Grignard reagents, **3**, 90
aryllithiums, **3**, 90
aryl organometallics, **3**, 89
arylpalladiums, **3**, 89
haloarenes, **3**, 95
nitrogen-substituted, synthesis of, **3**, 98
oxygen-substituted, synthesis of, **3**, 96
reactions
with *N*-acetyl indole-3-one, **3**, 257
with alkyldichlorophosphines, **4**, 123
with ammonia, **4**, 437
with arsonium ylides, **1**, 738
with aryldichlorophosphines, **4**, 123
with azidoacetophenones, **3**, 258
with morpholine, **4**, 406
with phosphoroamidothioates, **4**, 467
with α-sulfonamido ketones, **3**, 257
with 2-thioxoindole, **5**, 615
with ylides, **1**, 268
reduction of, **1**, 18
substituted, synthesis of, **3**, 93
sulfur-substituted, synthesis of, **3**, 98
synthesis of
via, alkene hydroformylation process, **1**, 52
via oxidation of toluenes, **3**, 86
via Sonn and Müller procedure, **3**, 85
use in the synthesis of 1,3-benzazarsolines, **4**, 496
use in the synthesis of ketene dithioacetal *S*-oxides, **4**, 856
Benzaldehydes, *N*-acetylamino-, from, methyl-*N*-acetylanilines, **3**, 99
Benzaldehydes, alkyl-, synthesis of, **3**, 93
Benzaldehydes, amino-, synthesis of, **3**, 99
Benzaldehydes, 4-amino-, synthesis of, **3**, 99
Benzaldehydes, 2-cyano-, reactions, with amines, **4**, 419
Benzaldehydes, 2-fluoro-, from, fluoroarenes, **3**, 96
Benzaldehydes, halo-, synthesis of, **3**, 94
Benzaldehydes, 2-hydroxy-, synthesis of, **3**, 87
Benzaldehydes, *o*-hydroxy-
Knoevenagel condensation of, **3**, 671
synthesis of, **3**, 91
Benzaldehydes, methyl-, synthesis of, **3**, 93
Benzaldehydes, 2-methyl-, synthesis of, **3**, 93

Benzaldehydes, nitro-
as starting materials, for acetals, **4**, 183
synthesis of, **3**, 98
Benzaldehydes, *o*-*S*-substituted, synthesis of, **3**, 98
Benzaldehydes, *p*-substituted
reactions, with 2-methyl-2-nitropropane, **3**, 433
synthesis of, **3**, 94
Benzaldimine, *N*-*t*-butyl-, oxidation of, **4**, 312
Benzaldimine, *N*-(phenylsulfenyl)-, reactions, with ester enolates, **5**, 335
Benzaldimine, *N*-(trimethylsilyl)-, reactions, with alkyllithiums, **2**, 313
Benzaldimines
oxidation of, **2**, 775
reactions, with diisopropyl peroxido carbonate, **3**, 666
Benzaldimines, *N*-arenesulfonyl-, reactions, with secondary arsines, **4**, 503
Benzaldoximes, *O*-alkyl-4-nitro-, reactions, with sodium hydride, **3**, 666
Benzamide
reactions, with oxyimidic chlorides, **5**, 765
thionation of, **5**, 575
Benzamide, *o*-amino-, for the condensation of *O*-ethyl selenocarboxylates, **5**, 522
Benzamide, *N*-chloro-*N*-methyl-, hydrolysis of, **5**, 709
Benzamide, *N*,*N*-dichloro-, reactions, with tris(*N*-ethyl-*N*-phenylamido) phosphite, **5**, 379
Benzamide, *N*-(2,2-dichloroethenyl)-, from, imidoyl chlorides, **5**, 657
Benzamide, *N*-(diethylaminomethyl)-, reactions, with benzylthiol, **4**, 322
Benzamide, *N*,*N*-dimethyl-
reduction of, **3**, 85
silylation of, **6**, 162
as starting material, for bis(silyl)methylamines, **4**, 511
Benzamide, hydro-
reactions, with dialkyl phosphites, **4**, 460
as starting material, for α-aminoalkylorganophosphorus compounds, **4**, 460
Benzamide, iodo-, oxidation of, **5**, 321
Benzamide, 2-mercapto-, reactions, with triethyl orthoformate, **6**, 111
Benzamide, *N*-phenyl-, reactions, with trimethylaluminum, **5**, 718
Benzamide, *N*-(piperidinomethyl)-, reactions, with benzylthiol, **4**, 322
Benzamide, *N*-(1-piperidinylmethyl)-, reactions, with acetyl chloride, **4**, 99
Benzamide, *N*-(trimethylsilyl)-, reactions, with sulfur dichloride, **5**, 335
Benzamides
as starting materials
for benzimidoyl chlorides, **5**, 657
for α,ω-bromides, **2**, 33
for naphthols, **2**, 673
Benzamides, 2-(alkylsulfinyl)-
reactions
with thionyl chloride, **5**, 336
with trifluoroacetic acid, **5**, 336
Benzamides, 2-amino-, reactions, with 1,3-bis(methoxycarbonyl)-*S*-methylisothiourea, **6**, 364
Benzamides, *N*-arenesulfonyl-, as starting materials, for *N*-arenesulfonyl benzimidoyl chlorides, **5**, 674
Benzamides, *o*-azido-, as starting materials, for iminophosphoranes, **5**, 1066
Benzamides, *N*-[α-(benzylideneamino)benzyl]-, as starting materials, for *N*,*N*'-alkylidene bisamides, **4**, 424
Benzamides, *N*,*N*-diethyl-, *o*-metallation of, **5**, 199
Benzamides, 2-hydroxy-, reactions, with Lawesson's reagent, **5**, 571
Benzamides, *N*-phosphoryl-, synthesis of, **5**, 776
Benzamides, *N*-(α-dialkylaminobenzyl)-, displacement of, **4**, 322
Benzamidine
condensation of, **4**, 988
reactions
with benzenesulfenyl chloride, **5**, 766
with 3,5-dichlorobenzenesulfenyl chloride, **5**, 766
with iodomethanesulfonyl chloride, **4**, 86
with toluenesulfenyl chloride, **5**, 766
Benzamidine, *N*,*N*'-bis(trimethylsilyl)-, synthesis of, **5**, 777
Benzamidine, bis(trimethylsilyl)-*N*-methyl-, reactions, with thiophosgene, **5**, 1051
Benzamidine, *N*-bromo-*N*'-methyl-, synthesis of, **5**, 763
Benzamidine, *N*-chloro-*N*,*N*'-dimethyl-
reactions
with arsenic trichloride, **5**, 776
with triphenylantimony, **5**, 776
with triphenylarsenic, **5**, 776
Benzamidine, *N*-iodo-, synthesis of, **5**, 763
Benzamidine, *N*-(iodomethanesulfonyl)-, synthesis of, **4**, 86
Benzamidine, *N*-iodo-4-methoxy-, synthesis of, **5**, 763
Benzamidine, *N*-methyl-
acylation of, **5**, 759
bromination of, **5**, 763
Benzamidine, *N*-methyl-*N*'-trimethylsilyl-, synthesis of, **5**, 777
Benzamidine, *N*-phenyl-, methylation of, **5**, 755
Benzamidine, *N*-phenyl 2-methyl-, synthesis of, **5**, 756
Benzamidine, *N*,*N*,*N*'-trimethylsilyl-, reactions, with thallium(III) chloride, **5**, 780
Benzamidine, tris(trimethylsilyl)-, reactions, with copper(I) chloride, **5**, 781
Benzamidine, *N*,*N*,*N*'-tris(trimethylsilyl)-, synthesis of, **5**, 778
Benzamidine hydrochloride, reactions, with benzoyl chloride, **5**, 758
Benzamidines
reactions, with sodium hydroxide, **5**, 763
synthesis of, **5**, 758
Benzamidines, arsoranyl-, synthesis of, **5**, 776
Benzamidines, *N*-chloro-
reactions
with phosphorus trichloride, **5**, 775
with tribenzyl phosphite, **5**, 774
with triphenylphosphine, **5**, 775
Benzamidines, *N*-methoxy-, synthesis of, **5**, 699
Benzamidines, *N*-phosphoryl-, synthesis of, **5**, 774
Benzamidines, thiocarbonyl-, synthesis of, **5**, 761
Benzamidoyl chlorides, *N*-(dichloromethylenedichlorophosphoryl)-, synthesis of, **5**, 656
Benzamidrazone, synthesis of, **5**, 769
Benzamidrazone, N^1,N^3-diphenyl-, reactions, with benzoyl chloride, **5**, 774
Benzanilide, reactions, with amides, **5**, 754
Benzannulation, transition metal-meditated, for the synthesis of phenols, **2**, 669
[3+2+1] Benzannulation, for the synthesis of phenols, **2**, 669
Benzatriazoles, 1-substituted-*N*-formylamino-, synthesis of, **3**, 710
1,3-Benzazarsolines
from, 2-aminoalkylarsines, **4**, 498
synthesis of
via alkylation reactions, **4**, 497
via cyclocondensation reactions, **4**, 496
Benzazepines, synthesis of, **1**, 282
Benzazepines, tetrahydro-, synthesis of, **1**, 149
Benzazepines, *N*-tosyl-, halogenation of, **4**, 105
Benzazocine, reactions, with sodium amide, **2**, 742
Benzazonine, reactions, with sodium amide, **2**, 742
Benzenamine, *N*-methylene-2,6-diethyl-, reactions, with phosgene, **4**, 102
Benzene

acylation of, **3**, 295
alkylation of, **1**, 475
chlorination of, **6**, 215
electrochemical reduction of, **1**, 103
Friedel–Crafts reaction
with phosphorus tribromide, **2**, 832
with phosphorus trichloride, **2**, 832
methylation of, **1**, 914
phenylethylation of, **1**, 916
reactions
with aluminum chloride, **1**, 57
with bis(4-toluyl)methyl selenocyanates, **5**, 1054
with carbonic diazide, **6**, 511
with ethene, **1**, 364
with hexanitroethane, **4**, 994
with nitroalkenes, **5**, 820
with potassium permanganate, **5**, 47
with tosyl azide, **4**, 483
with triphenyltin hydride, **1**, 174
reduction of, **1**, 87, 93, 889
reduction rates, **1**, 103
as solvent, in the synthesis of anhydrides, **5**, 183
as starting material
for benzoic anhydride, **5**, 197
for 1,2-*cis*-dihydroxycyclohex-3,5-diene, **2**, 39
for phenyl ketones, **3**, 295
synthesis of, **1**, 56
use in the substitution of boron functions, **1**, 277
Benzene, allyl-
as starting material, for (2-fluoroallyl)benzene, **2**, 610
synthesis of, **1**, 788
Benzene, bi(acetoxy)iodo-, synthesis of, **5**, 204
Benzene, 1,4-bis[bis(trimethylsilyl)methyl]-, metallation of, **6**, 190
Benzene, 1,2-bis(dimethylsilyl)-, reactions, with styrene, **4**, 618
Benzene, *p*-bis(iodosyl)-, reactions, with bis(trimethylsilyl)ethyne, **2**, 1021
Benzene, bis(silyl)-, desilylation of, **2**, 943
Benzene, bis(sulfonylethynyl)-, synthesis of, **2**, 1036
Benzene, bis(trichloroacetoxy)iodo-, synthesis of, **5**, 205
Benzene, [bis(trifluoroacetoxy)iodo]pentafluoro-, reactions, with acids, **1**, 934
Benzene, bis(trifluoroacetoxy)iodo- (PIFA)
reactions
with acids, **1**, 934
with amides, **5**, 999
with *p*-diethynylbenzene, **2**, 1019
as starting material, for bis(acyloxy)iodoarenes, **5**, 205
Benzene, *o*-(bistrimethylsilyl)-, reactions, with hexachlorides, **4**, 290
Benzene, *p*-bromofluoro-, reduction of, **1**, 428
Benzene, *p*-bromoiodo-, reduction of, **1**, 429
Benzene, *t*-butyl-, as starting material, for 4-*t*-butylbenzaldehyde, **3**, 94
Benzene, chloro-
reactions
with hydroxybenzyl alcohol, **2**, 684
with isopropylmagnesium chloride, **1**, 470
with triethylsilane, **1**, 428
Benzene, 2-cyanoazo-, from, 3-amino-2-phenyl indazole, **2**, 773
Benzene, decafluoroethyl-, as starting material, for octafluorostyrene, **4**, 730
Benzene, 1,2-diamino-
reactions
with 1,3-dithietanes, **4**, 415
with *N,N'*-thiocarbonylbis (3,5-dimethyl)pyrazole, **6**, 578
Benzene, 1,2-dibromo-, monoalkynylation of, **1**, 1009
Benzene, 1,3-dichloro-, metallation of, **2**, 942
Benzene, (dichloroiodo)-
for the α-chlorination of ethers, **4**, 47
reactions, with ethyl phenyl sulfoxide, **4**, 75
Benzene, (dichloro)iodo-, reactions, with lithium phenylethynylide, **2**, 1019
Benzene, dichloromethyl-, synthesis of, **4**, 11
Benzene, 3-(β,β-dichlorovinyl)-4-methylchloro-, oxidation of, **5**, 116
Benzene, *o,o'*-dicyanoazoxy-, synthesis of, **2**, 775
Benzene, *p*-diethynyl-, reactions, with (bistrifluoroacetoxy)iodobenzene, **2**, 1019
Benzene, dihydropropyl-, synthesis of, **1**, 97
Benzene, 1,2-dihydroxy-, cleavage of, **5**, 54
Benzene, 1,2-diiodo-, monoalkynylation of, **1**, 1009
Benzene, 1,3-dimethoxy-, reactions, with phosphorus trichloride, **2**, 829
Benzene, 1,4-dimethyl-, reduction of, **1**, 93
Benzene, 2,4-dimethyltrichloromethyl-, synthesis of, **6**, 29
Benzene, 1,4-dinitro-
reactions
with potassium superoxide, **2**, 648
with sodium borohydride, **2**, 773
Benzene, 4,4'-dinitroazo-, synthesis of, **2**, 773
Benzene, (diphenylthio)methyl-, synthesis of, **4**, 255
Benzene, ethyl-, synthesis of, **1**, 96
Benzene, ethylpropyl-, manufacture of, **1**, 492
Benzene, fluoro-
as starting material, for organocalciums, **1**, 116
synthesis of, **1**, 428; **2**, 613
Benzene, (2-fluoroallyl)-, from, allylbenzene, **2**, 610
Benzene, hexadeuterio-, use in Birch reductions, **1**, 95
Benzene, hexaethynyl-, synthesis of, **1**, 1009
Benzene, hexafluoro-
reactions, with bromotrifluoromethane, **6**, 12
reduction of, **1**, 426
as starting material, for α-fluoro epoxides, **4**, 45
Benzene, hexahydroxy-, from, dihydroxyethyne, **2**, 1023
Benzene, hexakis(trimethylgermyl)-, synthesis of, **2**, 949
Benzene, hexamethyl-, hexaalkylation of, **1**, 164
Benzene, [hydroxy(tosyloxy)iodo]-, reactions, with terminal alkynes, **2**, 1020
Benzene, *p*-iodoethyl-, synthesis of, **2**, 631
Benzene, iodoso-
reactions
with trimethylsilyl-substituted alkanes, **2**, 1020
with ylides, **5**, 897
use in the sulfhaloform reaction, **2**, 179
use in the synthesis of dialkyl sulfoxides, **2**, 148
Benzene, iodosyl-
oxidative azidation of anilines, **4**, 443
reactions
with alkynylsilanes, **4**, 808
with *N,N*-dimethylanilines, **4**, 443
Benzene, (+)-1-isobutyl-4-(1'-methyloctyl-, oxidation of, **5**, 118
Benzene, isopropyl-
manufacture of, **1**, 492
synthesis of, **1**, 470
Benzene, 2-lithio(dimethylaminomethyl)-, reactions, with silver bromide, **4**, 708
Benzene, methoxy-, reduction of, **1**, 93, 94
Benzene, 3-methoxy-1,2-bis(cyanomethyl)-, from, 3-methoxy-1-(dimethylaminomethyl)benzene, **1**, 255
Benzene, 3-methoxy-1-(dimethylaminomethyl)-, as starting material, for 3-methoxy-1,2-bis(cyanomethyl)benzene, **1**, 255
Benzene, 3-methoxy-2-hydroxymethyl-1-(dimethylaminomethyl)-, displacement of, **1**, 255
Benzene, [methoxy(tosyloxy)iodo]-, reactions, with α-phenylacetamide, **5**, 321
Benzene, nitro-
acetylation of, **3**, 293
acylation of, **3**, 300
carbonylation of, **5**, 984
Friedel–Crafts alkylation of, **1**, 181

iodination of, **2**, 631
for the oxidation of aryl organometallics, **2**, 652
reactions, with ammonia, **2**, 806
reduction of, **1**, 94; **2**, 781
Benzene, *p*-nitrofluoro-, synthesis of, **2**, 614
Benzene, nitroso-
reactions, with hydroxylamine, **2**, 806
synthesis of, **2**, 781
Benzene, pentafluoro-
reduction of, **1**, 426
synthesis of, **1**, 426
Benzene, *n*-pentyl-, synthesis of, **1**, 110
Benzene, propyl-, synthesis of, **1**, 97
Benzene, 1,2,4,5-tetrafluoro-, synthesis of, **1**, 426
Benzene, 1,2,4,5-tetramethyl-, reduction of, **1**, 101
Benzene, 2,3,4,5-tetramethyltrichloromethyl-, synthesis of, **6**, 29
Benzene, 2,3,4,6-tetramethyltrichloromethyl-, synthesis of, **6**, 29
Benzene, 2-thionitrosonitro-, synthesis of, **2**, 792
Benzene, 1,3,5-tri[bis(trimethylsilyl)methyl]-, from, 1,3,5-tri(dichloromethyl)benzene, **4**, 608
Benzene, 3-(β,β,β-trichloroethyl)-4-methylchloro-, oxidation of, **5**, 116
Benzene, (trichloromethyl)-, reactions, with aluminum trichloride, **3**, 296
Benzene, trichloromethyl pentafluoro-, hydrolysis of, **5**, 99
Benzene, 1,3,5-tri(dichloromethyl)-, as starting material, for 1,3,5-tri[bis(trimethylsilyl)methyl]benzene, **4**, 608
Benzene, trihalomethyl-, sulfhydrolysis of, **5**, 549
Benzene, 1,2,3-trimethoxy-, demethoxylation of, **1**, 432
Benzene, 1,3,5-trimethoxy-, reactions, with phosphorus trichloride, **2**, 829
Benzene, 2,4,5-trimethyltrichloromethyl-, synthesis of, **6**, 29
Benzene, 1,3,5-trinitro-, trichloromethylation of, **6**, 30
Benzene, 2,4,6-tripyrrolidino-, reactions, with alkyl iodides, **2**, 1009
Benzene, 1,3,5-tris(*t*-butyl)-, synthesis of, **2**, 626
Benzene bis(trifluoroacetate), iodo-, for the synthesis of aryl iodides, **2**, 631
Benzenecarboxamide, 2-amino-, reactions, with triethyl orthoacetate, **6**, 109
Benzene complexes, chloro-, reactions, with butyllithium, **4**, 707
Benzene complexes, methyl-, synthesis of, **1**, 914
Benzene complexes, ruthena-, synthesis of, **4**, 680
Benzene diacetate, iodo-
use in dehydrogenation reactions, **5**, 366
use in the synthesis of dialkyl sulfoxides, **2**, 148
Benzenediazonium chloride, reactions, with bis(methanesulfonyl)methane, **6**, 114
Benzenediazonium chloride, 4-nitro-
reactions
with diazoacetone, **5**, 843
with diazomethane, **5**, 843
with ethyl diazoacetate, **5**, 843
Benzenediazonium hydroxide, reactions, with γ-ketocarboxylic acids, **4**, 992
Benzenediazonium ions
reactions
with aryl dinitromethanes, **3**, 449
with diaryl nitromethanes, **3**, 449
Benzene diazonium nitrate, synthesis of, **2**, 1057
Benzene diazonium nitrate, dimethylamino-, synthesis of, **2**, 1057
Benzenediazonium salts
reactions
with D-glucose osazone, **5**, 857
with nitromethane, **3**, 449
with phenylhydrazones, **5**, 857
use in the synthesis of hydrazonyl halides, **5**, 677
Benzene-1,3-dicarboxylic acid, 2-iodo-, reactions, with peracetic acid, **5**, 204
Benzene dichloride, iodo-
for the chlorination of alkanes, **2**, 20
for the chlorination of alkenes, **2**, 23
for the oxidation of dithiolanes, **4**, 260
reactions, with ylides, **5**, 895
use in the synthesis of alkyl chlorides, **2**, 25
use in the synthesis of dialkyl sulfoxides, **2**, 148
Benzene-1,2-diol, transesterification reactions, with trialkyl orthobenzoates, **6**, 79
Benzene-1,2-diselenolate, dilithium, reactions, with dibromomethyl methyl ether, **6**, 100
Benzenedisulfonimide, *N*-chloro-, for the synthesis of aryl chlorides, **2**, 619
Benzenedisulfonimide, *N*-fluoro-*O*-, for the synthesis of aryl fluorides, **2**, 613
Benzenedithiol, 4-methyl-1,2-, reactions, with isothiocyanates, **2**, 307
Benzenemethanethiol
reactions
with bisisocyanates, **6**, 121
with methoxycarbonylsulfenyl chloride, **6**, 478
Benzene monohydroperoxides, 2,4,6-trialkyl-, synthesis of, **2**, 687
Benzene–osmium complexes, hydrogenation of, **1**, 94
Benzene oxide, hexafluoro-, synthesis of, **4**, 46
Benzenephosphonous dichlorides, alkyl-, synthesis of, **2**, 832
Benzene rings
aromaticity of, **2**, 646
thiocyanation of, **5**, 1110
Benzeneruthenium, hexamethyl-, reduction of, **1**, 94
Benzenes
reactions
with ethoxycarbonyl isothiocyanate, **5**, 592
with hydroxylamine, **2**, 764
as starting materials, for carboxylic acids, **5**, 115
synthesis of, **1**, 530
Benzenes, activated, alkylation of, **1**, 95
Benzenes, alkyl-
dealkylation of, **1**, 56
iodination of, **2**, 631
nitration of, **2**, 790
reactivity of, **3**, 94
trifluoroacetylation of, **3**, 295
Benzenes, alkyl-substituted, reduction of, **1**, 89
Benzenes, (arylsulfonyliminoiodo)-
carbonylation of, **5**, 987
as starting materials, for isocyanates, **5**, 987
Benzenes, azo-
reactions
with chromium carbene, **5**, 356
with cobalt octacarbonyl, **5**, 356
synthesis of, **2**, 773
Benzenes, bis(acyloxy)iodo-, synthesis of, **5**, 205
Benzenes, bromo-
reactions, with 1-heptyne, **1**, 1009
reduction of, **1**, 428
Benzenes, deuterio-, from, aryltins, **1**, 452
Benzenes, dialkyl-
reactions
with acid chlorides, **3**, 296
with aluminum trichloride, **3**, 296
Benzenes, diamino-, reactions, with *N*-cyanocarbodithioimidates, **5**, 1124
Benzenes, 1,2-dibromo-3-nitro-, alkynylation of, **1**, 1009
Benzenes, dihalo-, acylation of, **3**, 296
Benzenes, dihalomethyl-
as starting materials, for aminals, **4**, 411
sulfhydrolysis of, **5**, 549
Benzenes, diisopropenyl-, reactions, with isocyanic acid, **5**, 985
Benzenes, dimethyl-
reactions, with ceric ammonium nitrate, **3**, 93
reduction of, **1**, 103

Benzenes, fluoro-, synthesis of, **5**, 110
Benzenes, formarylazo-, synthesis of, **5**, 857
Benzenes, halo-
acylation of, **3**, 296
benzoylation of, **3**, 296
reduction of, **1**, 97
Benzenes, haloalkyl-, acylation of, **3**, 296
Benzenes, halofluoro-, substitution of, **1**, 474
Benzenes, hetero-
from, dihydrostannins, **2**, 879
stability of, **2**, 878
Benzenes, iodo-
methoxy-substituted, Stille-type formylation of, **3**, 97
reactions
with perfluoroalkylalkynes, **1**, 1009
with 3,3,3-trifluoro-1-pentynylzinc chloride, **1**, 1009
Benzenes, isopropenyl-, reactions, with isocyanic acid, **5**, 985
Benzenes, 2-lithio-1-(dimethylaminomethyl)-, reactions, with copper(I) bromide, **4**, 695
Benzenes, monoalkyl-, use in the synthesis of monoalkyl phenyl ketones, **3**, 295
Benzenes, nitro-, reduction of, **1**, 89
Benzenes, pentahaloethyl-, loss of chlorine from, **4**, 730
Benzenes, polyalkyl-
acylation of, **3**, 296
oxidation of, **3**, 93
Benzenes, polychlorinated, reactions, with fluorotrichloromethane, **6**, 30
Benzenes, polyhydroxy, synthesis of, **2**, 662
Benzenes, substituted
metallation of, **2**, 964
reactions, with acyl isothiocyanates, **5**, 631
as starting materials, for diols, **2**, 39
Benzeneselenamide, morpholino-
reactions, with α,α-bis(benzeneseleno)acetaldehyde, **6**, 93
as selenating reagent, **6**, 93
Benzeneselenenamides, dialkyl-, as starting materials, for α-selenocarbonyls, **2**, 282
Benzeneselenenic acid
synthesis of, **2**, 368
use in the synthesis of selenides, **2**, 281
Benzeneselenenyl bromide
hydrolysis of, **2**, 733
reactions
with alkenes, **2**, 280
with alkenyllithiums, **4**, 785
with α-diazocycloalkanones, **4**, 90
with (triphenylstannyl)methyllithium, **4**, 401
as starting material, for α-chloro selenides, **4**, 90
Benzeneselenenyl bromides, reactions, with dichloroalkyne, **4**, 787
Benzeneselenenyl chloride
reactions
with alkenes, **4**, 91
with alkyl amines, **2**, 368
with alkynes, **4**, 92
with diazo compounds, **4**, 90
with α-diazocycloalkanones, **4**, 90
with ketones, **4**, 90
with silver fluoride, **4**, 52
with trimethylsilyl amines, **2**, 368
with vinyl chloride, **4**, 91
use in the synthesis of isoselenocyanates, **5**, 1056
use in the synthesis of selenides, **2**, 281
Benzeneselenenyl chlorides, reactions, with dichloroalkyne, **4**, 787
α-Benzeneselenenyl enones, synthesis of, **4**, 90
Benzeneselenenyl fluoride, reactions, with vinyl acetates, **4**, 52
Benzeneselenic acid, use in the synthesis of α-selenosilanes, **4**, 375
Benzeneseleninic acid
as reagent, **2**, 388
synthesis of, **2**, 733
Benzeneseleninic acid, 2-nitro-, synthesis of, **2**, 733
Benzeneseleninic anhydride
reactions, with *O*-cholestannyl selenobenzoate, **5**, 523
use in the synthesis of acyldiazines, **5**, 366
Benzeneseleninyl chloride, in selenoxide elimination reactions, **1**, 609
Benzeneselenol
reactions
with bis(trifluoromethyl)thioketene, **5**, 563
with dichloroalkyne, **4**, 786
with *cis*-1,2-dichloroethene, **4**, 786
with dimethoxymethane, **4**, 238
with formic acid, **6**, 92
with tetrachloroethene, **4**, 786
with trichloroethylene, **4**, 786
use in the synthesis of selenocarboxylic esters, **5**, 239
use in the synthesis of β-selenyl aldehydes, **3**, 39
Benzeneselenolate, sodium
reactions, with dichloromethyl methyl ethers, **6**, 100
synthesis of, **6**, 92
Benzeneselenolate ions, reactions, with aryl α-iodo ketones, **1**, 10
Benzeneselenonic acid, synthesis of, **2**, 734
Benzeneselenyl bromide
reactions
with alkenes, **2**, 730
with (1-bromovinyl)trimethylsilane, **4**, 947
with phosphonate anions, **4**, 929
Benzeneselenyl chloride, reactions, with nitroalkanes, **6**, 116
Benzenesulfenyl acetate, 2,4-dinitro-, synthesis of, **5**, 208
Benzenesulfenyl chloride
reactions
with alkenes, **4**, 106
with benzamidine, **5**, 766
with cyclopropenes, **4**, 775
with α-diazo ketones, **4**, 69
with 1,3-diketone dimedones, **4**, 70
with phosphorus ylides, **5**, 898
with ylides, **6**, 323
Benzenesulfenyl chloride, 3,5-dichloro-, reactions, with benzamidine, **5**, 766
Benzenesulfenyl chloride, 2,4-dinitro-
reactions
with benzimidazoles, **2**, 767
with 3-benzyl-4-methylcarbostyril, **2**, 768
with diarylamines, **2**, 768
with 8-ethyl-4-methylcarbostyril, **2**, 768
with piperidinium tellurocarboxylates, **5**, 250
Benzenesulfenyl chloride, nitro-, reactions, with indoles, **2**, 767
Benzenesulfenyl chloride, *o*-nitro-, reactions, with amino acids, **2**, 358
Benzenesulfenyl chloride, pentachloro-, reactions, with sodium trithiocarbonate, **6**, 535
Benzenesulfenyl chlorides, reactions, with phosphorus-substituted ethynes, **4**, 926
Benzenesulfenyl chlorides, 2,4-dinitro-, reactions, with sodium acetate, **5**, 208
Benzenesulfinamide, *N*-cyclohexyl-, use in the synthesis of polyaromatic aldehydes, **3**, 100
Benzenesulfinamides, *N*-(hexafluoroisopropylidene)-
reactions
with ethanethiol, **4**, 325
with thiophenol, **4**, 325
Benzene sulfinate, methyl, reactions, with chlorotrimethylsilane, **4**, 370
Benzenesulfinic acid
reactions
with dihydropyran, **4**, 236
with 2-methoxytetrahydropyran, **4**, 236
Benzenesulfinyl chloride, reactions, with thionoamides, **5**, 638

Benzenesulfinyl fluoride, synthesis of, **2**, 716
Benzenesulfonamide
  reactions
    with carboxylic ortho esters, **5**, 701
    with formaldehyde, **4**, 423
    with perfluoroisobutene, **3**, 588
Benzenesulfonamide, *p*-amino-, phosgenation of, **5**, 997
Benzenesulfonamide, *N*,*N*-dibromo-
  reactions
    with cyclohexene, **5**, 701
    with formamide, **5**, 701
Benzenesulfonamide, *N*,*N*-dichloro-, use in the synthesis of *N*-acyl benzenesulfonamides, **5**, 345
Benzenesulfonamide, *N*-tosyl-, synthesis of, **5**, 344
Benzenesulfonamide, *N*-(trifluoromethyl)-*N*-nitroso-, as trifluoromethylation agent, **6**, 233
Benzenesulfonamides, *N*-acyl-, synthesis of, **5**, 345
Benzenesulfonate, methyl-, reactions, with dimethylaniline, **2**, 760
Benzenesulfonates, *o*-nitro-, reactions, with potassium thiocyanate, **5**, 1108
Benzenesulfonic acid, reactions, with sodium hydroxide, **2**, 647
Benzenesulfonimidates, alkyl *N*-alkyl-, synthesis of, **2**, 401
Benzenesulfonimidates, aryl *N*-alkyl-, synthesis of, **2**, 401
Benzenesulfonyl azide, 4-carboxy-, reactions, with alkyl 1-formylalkane sulfonates, **5**, 867
N-Benzenesulfonyl carboximidates, synthesis of, **5**, 701
Benzenesulfonyl chloride
  for the dehydration of formamides, **3**, 696
  reactions
    with γ-benzil monoxime, **3**, 722
    with *N*-formylethanolamine, **3**, 714, 715
    with nitro enamines, **4**, 907
    with thiocarboxylic acids, **5**, 252
    with thionoamides, **5**, 638
  as reagent, in the synthesis of β-lactones, **5**, 161
  use in the synthesis of β-lactones, **1**, 566
N-Benzenesulfonylformimidate, (2-bromocyclohexyl), synthesis of, **5**, 701
N^2-Benzenesulfonyl hydrazidoyl chlorides, *N*-methylation of, **5**, 681
Benzenesulfonylhydrazone, cycloheptanone 2,4,6-triisopropyl-, reactions, with *s*-butyllithium, **2**, 785
N-Benzenesulfonylimide, selenocarbonyl, reactions, with thiols, **2**, 235
N-Benzenesulfonylimines, α,β-unsaturated, use in Diels–Alder reactions, **2**, 329
Benzenesulfonyl isocyanates, synthesis of, **5**, 996
Benzenesulfonylnitrene, synthesis of, **2**, 794
Benzenesulfonyl peroxide, nitro-, for the *para* chlorination of arenes, **2**, 620
N-Benzenesulfonyl-1,2,3,4-tetrahydropyridines, synthesis of, **2**, 329
Benzenetellurenic acid, reactions, with thiobenzoic acid, **5**, 249
Benzenetellurenyl bromide, reactions, with ethyl diazoacetate, **4**, 93
Benzenetellurenyl iodide, reactions, with *O*-silyl selenocarboxylates, **5**, 250
Benzenetellurinic acid anhydride, *p*-methoxy-, use in the synthesis of carbodiimides, **5**, 1078
Benzenetellurinic anhydride, reactions, with acetic anhydride, **5**, 210
Benzenethiocarbonic acid chloride, pentafluoro-, synthesis of, **6**, 536
Benzenethiol
  reactions
    with *N*-chlorolactams, **5**, 335
    with cinnamaldehyde, **2**, 138
    with diselenoortho esters, **6**, 102
    with ethyl acrylate, **2**, 138
    with hexenylmercury chloride, **1**, 68
    with phenyl isothiocyanate, **6**, 565
    with tertiary alkylbismuthines, **2**, 501
Benzenethiol, 2-amino-, reactions, with silylethynes, **4**, 939
Benzenethiol, 2-hydroxy-, reactions, with ortho esters, **6**, 95
Benzenethiol, 3-methyl-, reactions, with acetic anhydride, **6**, 84
Benzenethiolate ions, reactions, with aryl α-iodo ketones, **1**, 10
Benzenethiols
  reactions
    with chloroformamidinium salts, **6**, 732
    with di-*t*-butyl dicarbonate, **6**, 477
  as starting materials
    for chlorothiolformate esters, **6**, 434
    for fluorothiolformate esters, **6**, 432
Benzenethiomethane, bis(trimethylsilyl)-, synthesis of, **6**, 351
Benzene thiosulfinate, phenyl, synthesis of, **4**, 232
Benzenethiosulfonic acid *S*-dichlorofluoromethyl ester, thermal decomposition of, **6**, 532
Benzenetriflate, acetoxyiodo-, synthesis of, **5**, 205
Benzenoid compounds, reduction of, **1**, 33
Benzenonium ions, synthesis of, **1**, 914
Benzenonium ions, pentamethyl-, synthesis of, **1**, 914
Benzhydrazidoyl chlorides, oxidation of, **5**, 681
Benzhydroxamic acid, use in the synthesis of *O*-alkylhydroxylamines, **2**, 103
Benzhydroxamic chloride, pyrolysis of, **5**, 1001
Benzhydroxamic chloride, 3-chloro-4-methoxy-, pyrolysis of, **5**, 1001
Benzhydroxamoyl chlorides, synthesis of, **5**, 664
Benzhydryl chloride, reactions, with dimsyl anion, **1**, 142
Benzhydrylthiols, synthesis of, **2**, 123
Benzil, reactions, with triamines, **2**, 321
Benzil dianils, as intermediates, **6**, 575
Benzilic acid, from, benzoin, **5**, 74
Benzimidate, *O*,*N*-bis(trimethylsilyl)-, synthesis of, **5**, 721
Benzimidate, ethyl *N*-chloro-, reactions, with aryl dialkylphosphines, **5**, 708
Benzimidate, ethyl *N*-(dimethyloxyphosphoryl)-, reactions, with chloral, **5**, 709
Benzimidate, ethyl *N*-(methoxyvinyloxyphosphoryl)-, synthesis of, **5**, 709
Benzimidate ester hydrochlorides, as starting materials, for orthobenzoate esters, **6**, 70
Benzimidates
  methoxyaminolysis of, **5**, 699
  pyrolysis of, **1**, 777
Benzimidates, *N*-benzyl methyl-, metallation of, **5**, 696
Benzimidates, *N*-(methoxycarbonyl)methyl-, metallation of, **5**, 696
Benzimidazole
  reactions
    with 1-chloroethyl chloroformate, **4**, 60
    with phosgene, **6**, 444
    with trimethylsilyl cyanide, **4**, 426
Benzimidazole, 2-(chloromethyl)-, reactions, with amines, **4**, 411
Benzimidazole, 2-cyano-1,3-dibenzoyl-2,3-dihydro-, synthesis of, **4**, 426
Benzimidazole, 1,5-dimethyl-, synthesis of, **2**, 783
Benzimidazole, 1,3-dimethyl-2-phenyl-
  for the conjugate reduction of enones, **3**, 128
  for the reduction of α-halo aldehydes, **3**, 12
  use for the reduction to ketones, **3**, 129
  use in the synthesis of deuterated aldehydes, **3**, 13
Benzimidazole, 2-trichloromethyl-, reactions, with thiols, **6**, 81
1*H*-Benzimidazole, 1-(3',5'-*O*-(1,1,3,3-tetraisopropyldisiloxane-1,3-diyl)-β-D-ribofuranosyl)-, reactions, with *p*-tolyl chlorothionoformate, **6**, 541
Benzimidazole-2-carbamates, synthesis of, **6**, 363
Benzimidazoles

reactions, with 2,4-dinitrobenzenesulfenyl chloride, **2**, 767
ring-opening of, **5**, 1033
synthesis of, **4**, 323; **5**, 571
Benzimidazoles, *N*-alkyl-, thiocyanation of, **5**, 1108
Benzimidazoles, 2-aminomethyl-, as starting materials, for *O*,*N*-acetals, **4**, 296
Benzimidazoles, 2-cyanamido-, synthesis of, **5**, 1124
Benzimidazoles, 2-cyano-, synthesis of, **3**, 672
Benzimidazoles, dihydro-, for the reduction of α-bromocarbonyls, **1**, 10
Benzimidazole-2-thiol, reactions, with *N*,*N*-diethylthiocarbamoyl chloride, **6**, 566
Benzimidazoline, synthesis of, **4**, 415
Benzimidazoline, 1,3-dimethyl-, as starting material, for telluroureas, **6**, 599
Benzimidazoline, 1,3-dimethyl-2-phenyl-, use in the synthesis of acyl aldehydes, **3**, 83
Benzimidazoline-2-thione, synthesis of, **6**, 578
Benzimidazolones, phosgenation of, **5**, 1013
Benzimidazolones, methyl-, as intermediates, **5**, 966
Benzimidazolones, *N*-thioacyl-, reactions, with amino acid esters, **5**, 587
Benzimide, reactions, with aluminum chloride, **5**, 384
Benzimidoyl bromide, *N*-methyl-, synthesis of, **5**, 657
Benzimidoyl chloride, *N*-(chlorocarbonyl)-, synthesis of, **5**, 658
Benzimidoyl chloride, *N*-(2-chloroethyl)-
reactions, with dilithio anion of phenylarsine, **5**, 791
as starting material, for 1,3-azaarsolines, **5**, 791
synthesis of, **5**, 658
Benzimidoyl chloride, *N*-[chloro-(iminophenyl)methyl]-, synthesis of, **5**, 657
Benzimidoyl chloride, *N*-(1-chloroisopropyl)-, synthesis of, **5**, 658
Benzimidoyl chloride, *N*-ethoxycarbonyl-, from, *N*-benzoylurethane, **5**, 656
Benzimidoyl chloride, *N*-methyl-, reactions, with dialkyl phosphonates, **5**, 713
Benzimidoyl chloride, *N*-(pentachloroethyl)-, synthesis of, **5**, 659
Benzimidoyl chloride, *N*-phenyl-
reactions
with arylhydrazines, **5**, 850
with 1,1-dimethylhydrazine, **5**, 850
with 1,2-dimethylhydrazine, **5**, 770
with lithio diethylphosphine, **5**, 784
with potassium diphenylphosphine, **5**, 784
as starting material, for *N*-phenylbenzimidoyl triphenylgermanium, **5**, 794
Benzimidoyl chloride, *N*-trichlorovinyl-, synthesis of, **4**, 794
Benzimidoyl chloride, *N*-(2,2,2-trifluoroethyl)-, synthesis of, **5**, 659
Benzimidoyl chlorides
from, benzamides, **5**, 657
reactions
with cyclopentadienyl tricarbonyl tungstonate, **5**, 798
with *N*-hydroxypyrimidines, **5**, 711
with sulfimides, **5**, 766
with trimethylstannyllithium, **5**, 803
with triphenylstannyllithium, **5**, 803
as starting materials
for *N*-(1-chloro-2,2-dihaloethenyl)benzimidoyl chlorides, **5**, 659
for *N*-substituted benzimidoyl isothiocyanates, **5**, 1051
synthesis of, **5**, 655
Benzimidoyl chlorides, *N*-(aminothiocarbonyl)-, synthesis of, **5**, 657
Benzimidoyl chlorides, *N*-arenesulfonyl-
from, *N*-arenesulfonyl benzamides, **5**, 674
synthesis of, **5**, 674
Benzimidoyl chlorides, *N*-aryl-
reactions
with potassium *O*-alkyl dithiocarbonates, **6**, 559
with silver nitrate, **5**, 711
Benzimidoyl chlorides, aryliminochloromethyl-, synthesis of, **5**, 657
Benzimidoyl chlorides, *N*-chloroalkyl-, synthesis of, **5**, 657
Benzimidoyl chlorides, *N*-(1-chloro-2,2-dihaloethenyl)-, from, benzimidoyl chlorides, **5**, 659
Benzimidoyl chlorides, *N*-chlorosulfonyloxy-, synthesis of, **5**, 666
Benzimidoyl chlorides, *N*-(dialkylaminomethylio)-, synthesis of, **5**, 660
Benzimidoyl chlorides, *N*-(1,1-dichloro-2,2,2-trihaloethyl)-, synthesis of, **5**, 659
Benzimidoyl chlorides, *N*-methyl-, synthesis of, **5**, 655
Benzimidoyl chlorides, *N*-pentachloroethyl-, synthesis of, **5**, 658
Benzimidoyl chlorides, *N*-[selenocarbamoyl]-, synthesis of, **5**, 655
Benzimidoyl chlorides, *N*-(1,1,2,2-tetrachloro-2-cyanoethyl)-, synthesis of, **5**, 658
Benzimidoyl chlorides, *N*-[thiocarbamoyl]-, synthesis of, **5**, 655
Benzimidoyl fluoride, *N*-cyclohexyl-, synthesis of, **5**, 658
Benzimidoyl fluoride, *N*-(2-pentyl)-, synthesis of, **5**, 658
Benzimidoyl fluoride, *N*-(3-pentyl)-, synthesis of, **5**, 658
Benzimidoyl fluorides, *N*-alkyl-, synthesis of, **5**, 659
Benzimino chlorides, *N*-thioacyl-
reactions
with amines, **5**, 761
with ammonia, **5**, 761
synthesis of, **5**, 760
1,2-Benziodoxol-3(1H)-one, triacetoxy-1,1,1-dihydro-, synthesis of, **5**, 204
1,2-Benziodoxol-3(1H)-one 1-oxide, 1-hydroxy-, synthesis of, **5**, 204
1,2-Benzisoselenazol-3(2H)-one, 2-aryl-, ring-opening of, **2**, 235
Benzisothiazoles, nomenclature, **2**, 767
Benzisothiazol-3-ones, synthesis of, **5**, 336
Benz[*k*]fluoroanthene, 7,12-dicyano-, reactions, with potassium hydroxide, **5**, 108
Benzoate, benzyl, synthesis of, **5**, 433
Benzoate, *t*-butyl, cleavage of, **5**, 106
Benzoate, *t*-butyl peroxy-, reactions, with trithiane, **6**, 99
Benzoate, ethyl, as starting material, for *O*-ethyl selenobenzoate, **5**, 520
Benzoate, ethyl 3,5-dichloro-4-hydroxy-, hydrolysis of, **5**, 106
Benzoate, ethyl *p*-fluoro-, hydrolysis of, **5**, 106
Benzoate, ethyl 2-ketohexahydro-, hydrolysis of, **5**, 31
Benzoate, methyl
hydrolysis of, **5**, 107
reactions
with potassium tetrafluorocobaltate, **4**, 3
with tris(trimethylsilyl)aluminum, **6**, 125
Benzoate, methyl 3,5-dimethoxy-, reactions, with thallium(III) nitrate, **3**, 433
Benzoate, methyl *m*-nitro-, hydrolysis of, **5**, 106
Benzoate, palladium(II), synthesis of, **5**, 229
Benzoate, (phenylselenothioperoxy)-, reactions, with alkenes, **2**, 285
Benzoate, triisopropylsilyl *p*-nitro-, as starting material, for acid bromides, **5**, 17
Benzoate, vinyl, reactions, with trifluoromethanesulfonyl chloride, **4**, 56
Benzoate, zinc, reactions, with diazomethane, **4**, 385
Benzoate esters
cleavage of, **5**, 107
reactions, with alkanethiotrimethylsilanes, **6**, 84
as starting materials, for trialkyl trithioorthobenzoate esters, **6**, 84
Benzoate groups, displacement of, **1**, 179
Benzoates
hydrolysis of, **5**, 105

pyrolysis of, **1**, 774
reactions, with sodium naphthalenide, **1**, 590
as starting materials, for α-chloro sulfides, **4**, 69
synthesis of, **1**, 700
Benzoates, alkyl
reactions
with *s*-butyllithium, **4**, 382
with trimethylsilyl chloride, **6**, 124
reductive silylation of, **5**, 419
Benzoates, cyclopropylcarbinyl 3,5-dinitro-, solvolysis of, **1**, 894
Benzoates, diol, synthesis of, **2**, 84
Benzoates, thiol, pyrolysis of, **5**, 475
Benzo-3-azagermole, 1,1-dimethyl-2-t-butyl-, synthesis of, **2**, 949
Benzo-1,3-azaphosphanes, synthesis of, **4**, 454
Benzo-1,3-azaphosphepanes, synthesis of, **4**, 454
Benzo-1,3-azaphospholanes, synthesis of, **4**, 454
Benzobarrelene, reactions, with difluorocarbene, **4**, 8
Benzobicyclo[3.3.1]nonanes, synthesis of, **4**, 342
2,3-Benzobicycloocta-2,7 diene, *N-n*-propyl-1,8-di-*t*-butyl-5,6-di-oxo-, as starting material, for *n*-propyl isocyanide, **3**, 699
Benzobiphenylenes, synthesis of, **1**, 255
2-Benzoboroles, reactions, with lithium 2,2,6,6-tetramethylpiperidide, **4**, 642
Benzochlorophospholene, synthesis of, **2**, 449
Benzo[*c*]phenanthrene, 1-carboxy-, synthesis of, **5**, 110
Benzo[*c*]phenanthrenes, carboxy-, from, methylbenzophenanthrenes, **5**, 115
Benzocyclobutane, from, acylsilanes, **1**, 238
Benzocyclobutenediones, as starting materials, for bisketene dienophiles, **3**, 539
Benzocyclobutenes, as starting materials, for *o*-quinodimethanes, **1**, 418
Benzocyclobutenes, bis(trimethylsilyl)-, cobalt-catalyzed synthesis of, **2**, 944
Benzocyclobutenones, from, Weinreb amides, **3**, 285
Benzocycloheptatrienes, deprotonation of, **1**, **843–951**
Benzocycloheptatrienyl anion, synthesis of, **1**, **843–951**
Benzocyclopentadienes, zircona-, as intermediates, in the synthesis of 2-trimethylsilylbenzothiophenes, **2**, 945
Benzocyclopropene, reactions, with sulfonyl isocyanates, **5**, 344
Benzodiazepines, as starting materials, for ketene aminals, **4**, 979
1,2,3-Benzodioxaborole, 2-chloro-, reactions, with lithium (ethyl diazoacetate), **3**, 485
1,3,2-Benzodioxaborole, 2-*t*-butylamino-, reactions, with phenyl isothiocyanate, **5**, 1038
(1,2,3-Benzodioxaborol-2-yl)diazoacetate, ethyl-, synthesis of, **3**, 485
Benzo-1,3,2-dioxaphospholes, reactions, with Grignard reagents, **4**, 126
Benzodioxaphosphorinanes
ethanolysis of, **4**, 126
synthesis of, **4**, 128
2*H*,4*H*-1,3-Benzodioxin, 4-oxo-, synthesis of, **4**, 213
Benzodioxins, 2-acyl, synthesis of, **3**, 240
1,3-Benzodioxole, 2,2-bis(phenylthio)-, from, 2,2-dichloro-1,3-benzodioxole, **6**, 312
1,3-Benzodioxole, 2,2-dichloro-
reactions
with aniline, **6**, 646
with morpholine, **6**, 321
with piperidine, **6**, 321
as starting material, for 2,2-bis(phenylthio)-1,3-benzodioxole, **6**, 312
Benzo-1,3-dioxole, 2,2-dichloro-, synthesis of, **6**, 251
1,3-Benzodioxoles, 2-alkyl-, electrochemical oxidation of, **6**, 77
1,3-Benzodioxoles, 2-aryl-, electrochemical oxidation of, **6**, 77
Benzo-1,3-dioxolium tetrafluoroborate, 2-phenyl-, reactions, with alkoxides, **6**, 73
Benzodioxyphospholenes, reactions, with catechol, **2**, 477
Benzodiphosphaarsole, synthesis of, **2**, 508
1,3,2-Benzodiphosphaarsoles, from, *o*-phenylenebis(phenylphosphine), **2**, 865
Benzo-1,3-diselena-2-selone, from, 1,2,3-benzoselenadiazole, **6**, 590
Benzo-1,3-diselenoles, 2-phosphoryl, Horner–Wittig reaction of, **4**, 875
Benzo-1,3-ditellurole, synthesis of, **4**, 290
Benzoditelluroles, synthesis of, **4**, 290
Benzodithiepines, as starting materials, for dithioacetals, **4**, 250
Benzodithiepins, stepwise alkylation of, **1**, 165
1,4-Benzodithiins, formylation of, **3**, 73
Benzo-1,3-dithiolane *S,S'*-dioxides, Mannich reaction of, **4**, 863
1,3-Benzodithiole, 2-alkoxy-, as starting material, for 2-ethylthio-1,3-benzodithiole, **6**, 82
1,3-Benzodithiole, 2-methoxy-, synthesis of, **6**, 99
1,3-Benzodithiole, 2-methyl-, oxidation of, **4**, 262
1,3-Benzodithiole, 2-thioxo-, reactions, with lead(IV) acetate, **6**, 309
Benzodithioles
as starting materials, for dithioacetals, **4**, 250
synthesis of, **2**, 213
Benzodithioles, 2-alkoxy-
from, anthranilic acid and alkyl nitrites, **6**, 99
reactions, with ethanethiol, **6**, 82
1,3-Benzodithiole tetraoxide, use in the synthesis of bis(sulfones), **4**, 274
1,2-Benzodithiole-3-thione, as starting material, for ketene aminals, **4**, 979
Benzo-1,3-dithiole-2-thione, reactions, with triethyl phosphite 6: 295–31845
Benzo-1,3-dithiole-2-thiones, as starting materials, for thiosulfonate esters, **4**, 862
Benzo-1,3-dithiole-5-thiones, as starting materials, for thiosulfonate esters, **4**, 862
Benzodithiolium salts, use in the synthesis of phosphonates, **6**, 120
1,3-Benzodithiolylium, reactions, with ketones, **3**, 251
Benzoferracyclopentadiene complexes, from, tricarbonylbenzocyclobutadieneiron, **4**, 681
11*H*-Benzo[*a*]fluoren-11-one, 5-bromo methyl-, reactions, with silver nitrate, **3**, 100
Benzofuran, 3-formyl-, synthesis of, **3**, 104
Benzofuran-3-acetic acids, reactions, with pyridine *N*-oxide, **3**, 103
Benzofurans
cleavage of, **3**, 304
formylation of, **3**, 103
from, ketenes, **3**, 529
hydrogenation of, **1**, 91
lithiation of, **3**, 103
ozonolysis of, **3**, 87
reduction of, **1**, 100
synthesis of, **1**, 1120; **4**, 167
Benzofurans, 3-*t*-butyldimethylsiloxy-, photooxygenation of, **5**, 227
Benzofurans, 2,3-dihydro-, synthesis of, **1**, 91
Benzofurans, 4,7-dihydro-, synthesis of, **1**, 100
Benzofurans, 2,3-disubstituted, acylation of, **3**, 304
Benzofurans, 3-trimethylsiloxy-, photooxygenation of, **5**, 227
Benzofurans, 2-trimethylsilyl-, *ipso*-substitution of, **3**, 304
Benzohydrazonamide, *N*-methyl-*N*-phenyl-, synthesis of, **5**, 848
Benzohydrazonoyl chloride, *N*-phenyl-, dehydrochlorination of, **3**, 680
Benzohydroximoyl chloride, reactions, with sodium *O,O*-diethyl phosphorodithioate, **5**, 644

Benzoic acid
from
benzaldehyde, **5**, 114
toluene, **5**, 116
reduction of, **1**, 97, 101
synthesis of, **5**, 110
use in Mitsunobu reactions, **2**, 87
Benzoic acid, 2-amino-, reactions, with cyclohexyl isocyanate, **6**, 504
Benzoic acid, 2-benzoyl-, reactions, with aniline, **4**, 299
Benzoic acid, *o*-chloro-, from, *o*-chlorotoluene, **5**, 115
Benzoic acid, *m*-chloro- (MCPBA), for the oxidation of dithioacetals, **4**, 258, 267, 271
Benzoic acid, 5-chloro-2-methyl-, synthesis of, **5**, 116
Benzoic acid, 3,5-dibromo-4-hydroxy-, as starting material, for nitroso compounds, **2**, 781
Benzoic acid, 3,4-dichloro-, synthesis of, **5**, 116
Benzoic acid, 2,5-dinitro-, from, 2-nitroso-5-nitrotoluene, **5**, 115
Benzoic acid, *p*-fluoro-
reduction of, **1**, 426
synthesis of, **2**, 614
Benzoic acid, *m*-(4-fluorophenyl)-, from, *m*-(4-fluorophenyl)benzaldehyde, **5**, 114
Benzoic acid, 2-formyl-, synthesis of, **3**, 84
Benzoic acid, 2-hydroxy-3-methyl-, as starting material, for 2-hydroxyisophthalic acid, **5**, 115
Benzoic acid, *m*-iodo-, synthesis of, **2**, 631
Benzoic acid, 2-iodosyl-, use in the synthesis of dialkyl sulfoxides, **2**, 148
Benzoic acid, *m*-iodosyl-, as co-oxidant, with bis(triphenylphosphine)ruthenium(II) chloride, **3**, 5
Benzoic acid, *o*-mercapto-, as starting material, for ketenes, **3**, 536
Benzoic acid, 5-methyl-2,3-methylenedioxy-, synthesis of, **5**, 110
Benzoic acid, 4-nitro-, reactions, with *t*-butyldimethylsilyl cyanide, **5**, 226
Benzoic acid, *o*-nitro-, from, *o*-nitromandelonitrile, **2**, 783
Benzoic acid, *p*-nitro-
from, *p*-nitrotoluene, **5**, 115
as starting material, for acid chlorides, **5**, 13
Benzoic acid, 2-nitro-5-thiocyanato-, as starting material, for thiocyanates, **5**, 1111
Benzoic acid, pentachloro-, synthesis of, **5**, 112
Benzoic acid, 2,3,4-trichloro-, from, 2,3,4-trichlorotoluene, **5**, 115
Benzoic acid, 2,3,6-trichloro-, from, 2,3,6-trichlorobenzaldehyde, **5**, 114
Benzoic acid, 2-(β,β,β-trichloroethyl)-4-chloro-, synthesis of, **5**, 116
Benzoic acid, *p*-(4,4,4-trifluoro-*n*-butyl)-, synthesis of, **5**, 112
Benzoic acid, 2,4,6-trimethyl-, from, methyl esters, **5**, 107
Benzoic acid, 2,4,6-trinitro-, from, 2,4,6-trinitrotoluene, **5**, 115
Benzoic acid, *p*-(triphenylsilyl)-
from
p-(triphenylsilyl)benzyl alcohol, **5**, 114
triphenyl-(*p*-tolyl)silane, **5**, 115
Benzoic acid anhydride, 4-bromobenzoyl-, stability of, **5**, 198
Benzoic acid anhydrides, chloro-, synthesis of, **5**, 197
Benzoic acid esters
reactions
with lithium bis(trimethylsilylamide)amide, **5**, 721
with sodium bis(trimethylsilyl)amide, **5**, 721
Benzoic acid *O*-ethyl oxyimidic amide, as starting material, for oxyimidic chlorides, **5**, 825
Benzoic acids
hydrolysis of, **5**, 106
reactions
with benzoyl chloride, **5**, 198
with diethylaminosulfur, **6**, 10
with hypervalent silicon compounds, **3**, 82
with imidoyl chlorides, **5**, 714
with sulfur tetrafluoride, **6**, 10
reduction of, **1**, 21; **3**, 82, 96, 98
as starting materials
for aryl bromides, **2**, 627
for aryl iodides, **2**, 633
synthesis of, **5**, 105
Benzoic acids, 2-alkyl-3-cyano-, synthesis of, **3**, 665
Benzoic acids, halo-, reduction of, **3**, 94
Benzoic acids, nitro-
reduction of, **3**, 99
as starting materials, for mixed anhydrides, **5**, 714
Benzoic anhydride
from, benzene, **5**, 197
synthesis of, **5**, 196, 198
Benzoic anhydride, 4-bromo-, synthesis of, **5**, 198
Benzoic anhydrides, acetic, hydrolysis of, **5**, 37
Benzoimidazolium tetrafluoroborate, 2-chloro-1,3-dimethyl-, as starting material, for tetraaminomethane, **6**, 361
Benzoin, as starting material, for benzilic acid, **5**, 74
Benzoisocyanides, dentate, synthesis of, **3**, 724
Benzoisothiazoles, use in the synthesis of α,β-unsaturated aldehydes, **3**, 61
Benzonitrile
hydrolysis of, **5**, 108
reactions
with amines, **5**, 750
with calcium bis(trimethylsilyl)amide, **5**, 777
with dialkylzincs, **3**, 461
with diisopropyl dithiophosphate, **5**, 737
with dimethylammonium *N*,*N*-dimethyl dithiocarbamate, **5**, 582
with lithium methylamide, **5**, 777
with methanol, **5**, 515
with *N*-methyl-*N*-phenylhydrazine, **5**, 848
Benzonitrile, *p*-amino-, as starting material, for *p*-aminotoluene, **1**, 48
Benzonitrile, *p*-bromo-, reactions, with hydroxylamine, **5**, 763
Benzonitrile, 2-isocyano-, synthesis of, **3**, 722
Benzonitrile, 4-methyl-
reactions, with potassium 9-*sec*-amyl-9-boratabicyclo[3.3.1]nonane, **3**, 93
as starting material, for amidines, **5**, 755
Benzonitrile, 4-nitro-, reactions, with diphenylphosphinous acid, **5**, 786
Benzonitrile, 2,4,6-trimethoxy-, synthesis of, **5**, 832
Benzonitrile, 2,4,6-trimethyl-, reactions, with amines, **5**, 756
Benzonitrile oxide
[1 + 3] dipolar cycloaddition of, **6**, 117, 126
[1 + 3] dipolar cycloadditions to carbon–sulfur or carbon–nitrogen double bonds, **6**, 111
reactions, with dihydrooxazoles, **6**, 107
synthesis of, **3**, 683
Benzonitrile oxide, 2,5-dichloro-, reactions, with hydrosulfide ions, **5**, 832
Benzonitrile oxide, 2,3,5,6-tetramethyl-, reactions, with hydrosulfide ions, **5**, 832
Benzonitrile oxides, reactions, with aziridine, **5**, 765
Benzonitriles
from
aldehydes, **3**, 662
amides, **3**, 666
carbonyls, **3**, 665
2,4,6-trinitrotoluene, **3**, 662
reactions
with *O*,*O*-dialkyldithiophosphoric acids, **5**, 643
with lithium bis(trimethylsilyl)amide, **5**, 777, 778
as starting materials, for amidines, **5**, 756
synthesis of, **3**, 664
Benzonitriles, 4-alkoxy-, synthesis of, **3**, 666

Benzonitriles, halo-, reduction of, **3**, 94
Benzonitrile *N*-sulfide, reactions, with maleic anhydride, **2**, 240
Benzonorbornadiene, reactions, with thiocyanogen, **5**, 1044
1,4-Benzooxathiins, formylation of, **3**, 73
Benzo-1,3,2-oxazaphospholes, synthesis of, **4**, 134
1,4,2-Benzooxazaphosphorinanes, hydrolysis of, **4**, 124
Benzopentathiepin, use in the synthesis of polysulfides, **2**, 234
Benzo[*c*]phenanthrenes, methyl-, as starting materials, for carboxybenzo[*c*]phenanthrenes, **5**, 115
Benzophenenone, reactions, with lithiated dithioacetals, **6**, 98
Benzophenone
  methylenation of, **1**, 646
  reactions
    with arsasilenes, **5**, 892
    with lithium 1,1-bis(trimethylsilyl) alkoxides, **4**, 932
    with lithium silyl arsenides, **5**, 892
    with phosphoranes, **1**, 763
    with stibonium ylides, **1**, 647
    with triphenylmethylstibonium salts, **1**, 271
    with triphenylstibonium methylylide, **1**, 743
  as starting material, for β-phenylcinnamic acid, **5**, 103
  synthesis of, **5**, 402
Benzophenone, 3-cyano-4-hydroxy-, synthesis of, **3**, 664
Benzophenone, 4-nitro-, cyanation of, **3**, 664
Benzophenone, *p*-nitro-, synthesis of, **1**, 723
Benzophenone dichloride, reactions, with thiohydroxamic acids, **5**, 1046
Benzophenone imine
  reactions, with oxalyl bromide, **6**, 453
  as starting material, for carbamoyl fluorides, **6**, 438
Benzophenoneimine, *N*-acetyl, reactions, with diethyl trimethylsilylphosphine, **5**, 720
Benzophenone imine, *N*-(trimethylsilyl)-, for the storage of silaethenes, **6**, 709
Benzophenone oxime, acylation of, **5**, 219
Benzophenone oxime esters, use in the synthesis of alkyl halides, **2**, 13
Benzophenone phenylimine, reactions, with phenyl isocyanate, **6**, 504
Benzophenones
  from, diarylmethanes, **3**, 293
  reactions
    with aryliminomagnesiums, **3**, 412
    with lithium naphthalene, **1**, 136
  reduction of, **2**, 64
  synthesis of, **3**, 279; **4**, 189
Benzophenones, fluoro-, synthesis of, **1**, 448
Benzophosphoranes, 3-amino-, from, phosphorus amides, **4**, 478
Benzopinacolates, bis(trialkylstannyl)-, use in the synthesis of radicals, **1**, 924
Benzopyran, 8-hydrazino-5-chloro-3,4-dihydro-2,6-dimethyl-, reactions, with allyl chlorodithioformates, **6**, 567
2-Benzopyran-4-one, 1-methoxy-, reactions, with acetic anhydride, **4**, 212
Benzopyranones, reactions, with *N*-bromosuccinimide, **4**, 174
Benzopyrano[1,3]oxazines, synthesis of, **4**, 304
Benzopyrans, synthesis of, **3**, 671
Benzopyrans, 2-amino-, synthesis of, **4**, 307
Benzopyrans, arylidene-, from, benzothiopyrones, **1**, 741
2*H*-1-Benzopyrans, 3-hydroxylamino-, synthesis of, **2**, 341
Benzo[*h*]quinolinium fluorides, *N*-aryl-2,4-diphenyl-5,6-dihydro-, as starting materials, for arenes, **1**, 31
1,2-Benzoquinone, reactions, with 1-morpholinocyclohexene, **4**, 307
1,4-Benzoquinone, reactions, with dialkyl acyl phosphoramidites, **5**, 373
1,4-Benzoquinone, 2,5-diazo-3,6-di-*t*-butyl-, use in the synthesis of *t*-butylcyanoketene, **5**, 349
1,4-Benzoquinone, 2,3-dichloro-5,6-dicyano- (DDQ)
  availability of, **1**, 554
  as oxidant, for the synthesis of enones, **1**, 556
  for the oxidation of toluenes, **3**, 87
  reactions
    with cycloheptenones, **1**, 554
    with methoxybenzyl ethers, **4**, 201
    with triazulenylmethanes, **1**, 899
  as reagent, for the synthesis of phenols, **2**, 659
  for the ring-opening of epoxides, **2**, 84
  use in elimination reactions, **1**, 648
  use in the synthesis of polyaromatic aldehydes, **3**, 100
o-Benzoquinone, reduction of, **2**, 695
o-Benzoquinone, tetrachloro-, reactions, with phosphines, **2**, 1074
o-Benzoquinone, 3,4,5,6-tetrachloro-, reactions, with cyclic trithiocarbonates, **6**, 312
p-Benzoquinone, as electrophile, for dithiocarboxylic enethiolates, **4**, 842
Benzoquinones, synthesis of, **3**, 539
1,4-Benzoquinones, 2,5-diazo-3,6-dihexynyl-, pyrolysis of, **3**, 653
Benzoquinuclidine, reactions, with benzoyl peroxide, **2**, 765
1,2,3-Benzoselenadiazole, as starting material, for benzo-1,3-diselena-2-selone, **6**, 590
Benzoselenamide, morpholino-, reactions, with enals, **3**, 74
Benzoselenane, synthesis of, **3**, 383
Benzo-1,2,3-selenodiazole, as starting material, for selenoketenes, **3**, 553
Benzostabase compounds, use in the protection of amines, **2**, 543
Benzo-substituted compounds, synthesis of, **4**, 1074
Benzosulfonamide, *N*-(bisalkylthiolmethylene)-, chlorination of, **5**, 1071
Benzotellurole, bromination of, **4**, 291
Benzotellurophenedione, from, *o*-phthaloyl chloride, **5**, 244
1,2,3-Benzothiadiazole, 7-amino-6-substituted, diazotization of, **2**, 803
Benzothiazaphoranes, synthesis of, **2**, 856
Benzothiazine, dihydro-, synthesis of, **6**, 111
Benzothiazines, reactions, with potassium dimsylate, **1**, 622
4*H*-1,3-Benzothiazin-4-one, 2-thioxo-6-nitro-2,3-dihydro-, synthesis of, **6**, 566
Benzothiazole, 2-amino-, reactions, with oxalyl chloride, **5**, 974
Benzothiazole, 2-hydroxy-, reactions, with copper(I) alkynides, **1**, 1025
Benzothiazole, 2-mercapto-
  reactions, with amines, **2**, 131, 358
  as trapping agent, **1**, 598
1,3-Benzothiazole, 2,2-diethoxy-3-methyl-2,3-dihydro-, synthesis of, **6**, 317
Benzothiazoles
  metallation of, **5**, 595
  reactions, with allenylmagnesium bromide, **1**, 1061
2-Benzothiazolesulfenamide, transamination of, **2**, 359
Benzothiazolines, *N*-substituted, synthesis of, **4**, 318
Benzothiazolin-2-ones, 3-(2-hydroxyethyl)-, reactions, with potassium hydroxide, **6**, 545
Benzothiazolin-2-ones, 3-(3-hydroxypropyl)-, reactions, with potassium hydroxide, **6**, 545
Benzothiazolin-2-thione, 3-chloromethyl-, reactions, with sodium *O*-alkyl dithiocarbonates, **6**, 545
Benzothiazolium bromide, 3-methyl-, reactions, with aldehydes, **3**, 157
Benzothiazolium cation, *N*-alkyl-, reactions, with 2-hydroxybenzaldehyde, **6**, 111
Benzothiazolium cations, cyclization of, **6**, 111
Benzothiazolium salts, use in the synthesis of alkyl halides, **2**, 8
Benzothiazolium salts, fluoro-, use in the synthesis of alkyl iodide, **2**, 36
Benzothiazolium salts, 3-substituted, reduction of, **4**, 321

1,3-Benzothiazolium tetrafluroborate, 2-chloro-3-methyl-, reactions, with resorcinol, **6**, 317
1,3-Benzothiazol-2-one, 3-methyl-, reactions, with triethyloxonium tetrafluoroborate, **6**, 317
2(3*H*)-Benzothiazolones, 3-chloromethyl-, synthesis of, via Friedel–Crafts reactions, **4**, 103
Benzothiazolo[3,2-a][1,3,5]triazine-1,3-diones, 2-(2-benzothiazoyl)-, synthesis of, **5**, 974
Benzothiazolyl sulfones, alkyl, reductive cleavage of, **2**, 188
Benzothieno[2,3-*c*]pyridines, (*R*)-1,2,3,4-tetrahydro-, from, benzothiophenes, **2**, 326
2-Benzo[*b*]thienyl isocyanate, synthesis of, **5**, 992
Benzothiete, thermolysis of, **3**, 359; **4**, 232
Benzothiophene
  Friedel–Crafts acylation of, **3**, 306
  reactions, with Grignard reagents, **1**, 484
  Vilsmeier formylation of, **3**, 105
Benzothiophene, 3-iodo-, reactions, with arylboronic acids, **3**, 306
Benzothiophenes
  desulfurization of, **1**, 436
  as starting materials, for (*R*)-1,2,3,4-tetrahydrobenzothieno[2,3-*c*]pyridines, **2**, 326
Benzothiophenes, 3-aroyl-, synthesis of, **3**, 306
Benzothiophenes, formyl-, synthesis of, **3**, 105
Benzothiophenes, 2-trimethylsilyl-, synthesis of, **2**, 945
1-Benzothiopyrano[3,4-d][1,2,3]-triazoles, synthesis of, **2**, 771
Benzothiopyrones, as starting materials, for arylidenebenzopyrans, **1**, 741
Benzotriazinone, *N*-amino-, ring-opening of, **5**, 368
Benzotriazole
  aminoalkylation of, **4**, 309, 323
  Mannich-type reaction of, **4**, 445
  reactions
    with aldehydes, **4**, 296
    with amides, **4**, 449
    with amines, **4**, 406
    with 2,5-dimethoxytetrahydrofuran, **4**, 309
    with formamides, **3**, 709
    with 9-vinylcarbazole, **4**, 323
  for the stabilization of α-anions, **6**, 170
  use in the synthesis of amino aldehydes, **3**, 42
  use in the synthesis of vinyl sulfides, **2**, 710
Benzotriazole, allenyl-, synthesis of, **1**, 956
Benzotriazole, 1-azidomethyl-, as starting material, for *N*-triphenylphosphorylidene-1-(benzotriazol-1-yl) methylamine, **1**, 262
Benzotriazole, bis(trimethylsilyl)methyl-, synthesis of, **6**, 162
Benzotriazole, 1-chloro-, use in the synthesis of dialkyl sulfoxides, **2**, 148
Benzotriazole, *N*-chloro-, reactions, with bisallyl selenides, **2**, 236
Benzotriazole, 1-(chloromethyl)-, reactions, with thiols, **4**, 322
Benzotriazole, *N*-(formylaminomethyl)-, as starting material, for *N*-(formylaminomethyl)amine, **4**, 446
Benzotriazole, 1-hydroxy-
  active esters of, use in the synthesis of amides, **5**, 264
  as co-reagent, in coupling reactions, **5**, 266
Benzotriazole, hydroxy (HOBT), use in coupling reactions, **5**, 279
Benzotriazole, 1-(hydroxymethyl)-
  as intermediate, for the synthesis of hydroxamic acids, **4**, 449
  reactions, with primary amides, **4**, 449
Benzotriazole, 1-(phenylthiomethyl)-, lithiation of, **4**, 322
Benzotriazole, propargyl-, reactions, with sodium hydroxide, **1**, 956
Benzotriazole, *N*-thiobenzoyl-, as thioacylating reagent, **5**, 641
Benzotriazole, 1-trimethylsilyloxy-, reactions, with thiophosgene, **5**, 1030
1*H*-Benzotriazole, reactions, with *S*-chlorothiocarbonyl chloride, **6**, 501
1*H*-Benzotriazole, 1-(1*H*-benzotriazolocarbonylthio)-, synthesis of, **6**, 501
Benzotriazole anions, 1-(carbazol-9-ylmethyl)-, reactions, with isothiocyanates, **5**, 596
Benzotriazoles
  reactions, with *n*-butyllithium, **3**, 29
  stability of, **5**, 1030
  synthesis of, **4**, 449
  use in the substitution of nitrogen functions, **1**, 260
  use in the synthesis of α-hydroxy ketones, **3**, 157
Benzotriazoles, 1-(1-alkoxyalkyl)-, synthesis of, **4**, 300
Benzotriazoles, *N*-(alkylamino)-
  reactions
    with alkyl halides, **2**, 302
    with allyl halides, **2**, 302
    with lithium ester enolates, **2**, 319
Benzotriazoles, *N*-[1-(alkylthio)alkyl]-, synthesis of, **4**, 318
Benzotriazoles, aminomethyl-
  reactions, with lithium enolates, **3**, 186
  synthesis of, **4**, 445
Benzotriazoles, 1-(1-(arylamino)alkyl)-, as starting materials, for secondary amines, **1**, 260
Benzotriazoles, *N*-[1-(arylthio)alkyl]-, synthesis of, **4**, 318
Benzotriazoles, chloroalkyl-, reactions, with amines, **4**, 445
Benzotriazoles, 1-(dialkylaminoalkyl)-, reactions, with alkynyllithium, **1**, 261
Benzotriazoles, 1-(1-dialkylaminoalkyl)-, use in the synthesis of alkynes, **1**, 1041
Benzotriazoles, 1-hydroxyalkyl-, synthesis of, **4**, 296
Benzotriazoles, 1-(thioalkyl)-, synthesis of, **4**, 322
Benzotriazoles, 1-vinyl-, flash vacuum pyrolysis of, **3**, 579
1-(Benzotriazol-1-yl)methylamine, *N*-triphenylphosphorylidene-, reactions, with Grignard reagents, **2**, 306
Benzotrichloride, 4-chloro-, use in the synthesis of azonia allene salts, **5**, 661
Benzotrichloride, 4-nitro-, synthesis of, **6**, 28
Benzotrichloride, 2,4,6-trimethyl-, reactions, with hydroxylamine, **5**, 666
Benzotrichlorides
  from, 4-nitrotoluene, **6**, 28
  hydrolysis of, **5**, 27
  as starting materials
    for benzotrifluoride, **6**, 9
    for hydrazonic acid esters, **5**, 703
  synthesis of, **6**, 28
Benzotrichlorides, methyl-, synthesis of, **6**, 28
Benzotrichlorides, phenylsubstituted
  reactions
    with dichloromalodinitrile, **5**, 658
    with trichloroacetonitrile, **5**, 658
Benzotrifluoride, from, benzotrichlorides, **6**, 9
Benzotrifluoride, pentafluoro-, copyrolysis of, with chlorodifluoromethane, **4**, 737
1,2,3-Benzotriphosphole, from, *o*-phenylenebis(phenylphosphine), **2**, 865
Benzotropylium ions, synthesis of, **1**, 906
Benzotropylium ions, tri-, synthesis of, **1**, 906
1,3-Benzoxadiole-2-thione, synthesis of, **6**, 543
1,3-Benzoxamines, 2-substituted-*N*-benzyl-4,4,7α-trimethyl-*trans*-octahydro-, ring-opening of, **2**, 306
Benzoxaphospholes, reactions, with tetraoxaspirononanes, **4**, 478
Benzoxarsole, cycloaddition of, **6**, 119
4*H*-3,1-Benzoxathiines, synthesis of, **4**, 232
1,3-Benzoxathiole, synthesis of, **4**, 231
1,3-Benzoxathiolium ions, 2-substituted-, use in the synthesis of ketoindoles, **3**, 309
Benzoxazaphosphorinanes, synthesis of, **4**, 128
1,3-Benzoxazines, synthesis of, **4**, 301
3,1-Benzoxazines, synthesis of, **4**, 306

1,3-Benzoxazin-4-one, 2-amino-, reactions, with sodium cyanamide, **5**, 1120
1,3-Benzoxazin-4-ones, synthesis of, **4**, 302
2*H*-Benzoxazin-3-ones, 2-hydroxy-, synthesis of, **4**, 175
Benzoxazole, 2-chloro-, reactions, with phenol, **1**, 432
Benzoxazoles
  reactions, with allenylmagnesium bromide, **1**, 1061
  ring-opening of, **5**, 1033
  as starting materials, for aryl isocyanides, **3**, 723
1,2-Benzoxazoles, oxidation of, **4**, 206
Benzoxazolium salts, use in the synthesis of alkyl halides, **2**, 8
Benzoxazolium salts, enol
  elimination of, **1**, 1071
  as starting materials, for alkynes, **1**, 1071
Benzoxazolone, synthesis of, **5**, 969
Benzoxazolone, *N*-silyl-substituted, thermal rearrangement of, **5**, 1012
Benzoxazolones, *N*-thiobenzoyl-, reactions, with amines, **5**, 587
1-Benzoxepines, 4,5-epoxy-2,3,4,5-tetrahydro-, ring-opening of, **2**, 318
Benzoyl, reactions, with diphenyl-*N*-propylphosphinous amide, **2**, 835
Benzoylation
  of alcohols, **5**, 129
  for the synthesis of anthryl ketones, **3**, 301
Benzoyl azide, *o*-amino-, synthesis of, **5**, 368
Benzoyl bromide
  reactions, with silver cyanide, **3**, 703
  reduction of, **3**, 83
  synthesis of, **5**, 17, 18
Benzoyl bromides, as starting materials, for aryl bromides, **2**, 627
Benzoyl chloride
  halo-substituted, reduction of, **3**, 94
  reactions
    with acetamidine hydrochlorides, **5**, 758
    with benzamidine hydrochloride, **5**, 758
    with benzoic acids, **5**, 198
    with N^1,N^3-diphenylbenzamidrazone, **5**, 774
    with hexamethylphosphorus triamide, **5**, 407
    with hexyloxydimethylsilanethiol, **5**, 538
    with hydrogen selenide, **5**, 233
    with hydrogen sulfide, **5**, 528
    with 4-methyl quinoline, **1**, 861
    with nickel compounds, **5**, 463
    with nitromethylquinoline, **5**, 823
    with phospholes, **4**, 341
    with potassium iodide, **5**, 19
    with secondary enaminones, **3**, 413
    with sodium hydrogen telluride, **5**, 253
    with *N*-tosyltriphenylbismuthimines, **5**, 344
    with triphenylsilylpotassium, **5**, 417
  as starting material
    for benzoyl isothiocyanate, **5**, 1047
    for selenobenzoic acid, **5**, 508
  use in the synthesis of acid chlorides, **5**, 15
  use in the synthesis of carboxylic esters, **5**, 130
Benzoyl chloride, 4-bromo-, reactions, with tributylstannane, **3**, 95
Benzoyl chloride, isocyanato-, synthesis of, **5**, 970
Benzoyl chloride, 2-isocyanato-, synthesis of, **5**, 995
Benzoyl chloride, 4-methoxy-, as starting material, for 4-methoxyphenylacetamide, **5**, 276
Benzoyl chloride, *p*-methoxy-, as starting material, for *p*-methoxybenzoyl fluoride, **5**, 4
Benzoyl chloride, 4-methyl-
  reactions
    with cadmium-moderated borohydride, **3**, 93
    with sodium tri(*t*-butoxy)aluminum hydride, **3**, 93
Benzoyl chloride, *p*-nitro-, reactions, with hydridoiron tetracarbonyl anions, **3**, 98
Benzoyl chloride, 2,4,6-trichloro-
  reactions, with 4-dimethylaminopyridine, **5**, 173
  use in the synthesis of anhydrides, **5**, 199
Benzoyl chloride, 2,4,6-tris(*t*-butyl)-, reactions, with lithium bis(trimethylsilyl)arsine, **5**, 1160
Benzoyl chlorides
  reactions, with arylmagnesiums, **3**, 285
  as starting materials, for aryl chlorides, **2**, 622
Benzoyl chlorides, methyl-, chlorination of, **4**, 11
Benzoyl compounds, synthesis of, **5**, 444
N-Benzoyl compounds, from, methyl 2-methoxyhippurate, **4**, 327
Se-Benzoyl compounds, synthesis of, **4**, 239
Benzoyl fluoride
  reactions, with fluorodiphenylphosphine, **5**, 407
  synthesis of, **5**, 4, 5, 6
Benzoyl fluoride, 2-bromo-, synthesis of, **5**, 6
Benzoyl fluoride, *p*-methoxy-, from, *p*-methoxybenzoyl chloride, **5**, 4
Benzoyl halides
  reactions, with hexaalkylphosphorus triamides, **5**, 402
  reduction of, **3**, 83
Benzoylhydrazide, N^2-phenyl-, thionation of, **5**, 639
Benzoyl iodide, synthesis of, **5**, 19
Benzoyl nitrates, as reagents, for the synthesis of nitrate esters, **2**, 104
2-Benzoyloxy acids, synthesis of, **2**, 84
α-Benzoyloxy compounds, from, 1,4-thiazin-3-ones, **4**, 226
Benzoyl peroxide, bis-3,4-dichloro-, thermal decomposition of, **5**, 116
Benzoyl peroxide, bis-4-methyl-, synthesis of, **5**, 208
Benzoyl peroxide, di-*t*-butyl- (DTBP), as radical initiator, **1**, 331
Benzoylphosphonate, use in the synthesis of tris(phosphonate)methanes, **6**, 154
Benzthiazoles, allyloxy-, reactions, with copper nucleophiles, **1**, 187
Benzthiazoloneimine, 3-methyl-, reactions, with vinylidene difluorides, **4**, 992
Benzyl alcohol, reactions, with urea, **6**, 484
Benzylamine
  reactions
    with alkoxyenones, **3**, 265
    with 4-arylthiazolium perchlorate, **3**, 339
    with cyanoacetic thionoesters, **4**, 896
    with (*Z*)-methyl-β-bromo-α-isocyanocinnamate, **3**, 720
    with pinanone, **2**, 331
Benzylamine, *N*-benzylidene-, acylation of, **4**, 325
Benzylamine, bis(trimethylsilylmethyl)-, as starting material, for azomethine ylides, **1**, 285
Benzylamine, cyclopalladated *N*,*N*-dimethyl-, reactions, with carbyne complexes, **4**, 716
Benzylamine, *N*-diphenylmethylene-, reactions, with Schiff bases, **3**, 408
Benzylamine, α-methyl-
  kinetic resolution of, **2**, 330
  reactions, with methyl crotonate, **2**, 315
Benzylamine, *N*-trifluoroethylidenyl-, as starting material, for imidoyl phosphonates, **5**, 789
Benzylamines
  alkyl substituted, synthesis of, **1**, 43
  oxidation of, **3**, 89; **4**, 449
  reactions, with glyoxal, **4**, 298
  synthesis of, **1**, 57
Benzylamines, α-cyano-, synthesis of, **2**, 311
Benzylamines, primary, oxidation of, **3**, 412
Benzylamines, (*R*)-α-substituted, synthesis of, **2**, 331; **3**, 415
Benzylamines, (*S*)-α-substituted, synthesis of, **2**, 331; **3**, 415
Benzyl anions
  reactions, with *S*-methyl trithiocarbonate, **5**, 550
  stability of, **1**, 850
Benzylation, of saturated ketones, **3**, 145
Benzyl azide, reactions, with hexafluoropropene, **6**, 54
Benzyl bromide

reactions
with potassium hydrogen carbonate, **6**, 462
with tetramethylbismuthine, **2**, 491
with trialkylstibines, **1**, 272
reactivity of, **1**, 873
Benzyl bromide, 4-nitro-, reactions, with dimethyl selenoxide, **3**, 99
Benzyl bromides
chlorination of, **4**, 37
reactions
with alkenylboranes, **1**, 1104
with dimethyl selenoxide, **3**, 98
with mercury(I) nitrate, **3**, 89
Benzyl carbanions
review, **1**, 107
synthesis of, **1**, 138
Benzyl carbinols, synthesis of, **2**, 655
Benzyl chloride
reactions
with acetone oxime, **2**, 103
with *n*-butyllithium, **1**, 110
Benzyl chloride, 2,4-diisocyanato-, reactions, with trialkyl phosphites, **5**, 1009
Benzyl chloride, 4-isocyanato-, synthesis of, **5**, 971
Benzyl chloride, *p*-nitro-, alkylation of, **5**, 140
Benzyl chlorides
for the alkylation of hydrazines, **2**, 378
reduction of, **1**, 6
as starting materials, for perchlorates, **1**, 904
use in the synthesis of benzyl ethers, **2**, 90
N-Benzylephedrine, from, ephedrine, **2**, 300
Benzyl fluorides, reactions, with benzylic organolithiums, **1**, 108
Benzyl halides
oxidation of, **3**, 88
reactions
with hexamethylenetetramine, **3**, 88
with 2-nitropropane, **3**, 88
with nitrosylruthenium complexes, **2**, 582
with potassium tellurocyanate, **5**, 1117
with potassium thiocyanate, **5**, 1039
with triethylaluminum, **1**, 128
Benzyl halides, nitro-, reactions, with 2-nitropropane anions, **1**, 156
Benzyl hexyl carbonate, synthesis of, **6**, 462
Benzylic bromides
synthesis of, **2**, 3
use in the synthesis of benzyl ethers, **2**, 90
Benzylic chlorides, synthesis of, **2**, 26, 27
Benzylic fluorides, reduction of, **1**, 2, 3, 6
Benzylic halides
alkylation of, **1**, 493
chain elongation of, **5**, 143
metallation of, **2**, 559
reactions, with aroyl chlorides, **3**, 289
reactivity of, **1**, 2
Benzylic iodides, synthesis of, **2**, 36
Benzylic systems, exchange reactions of, **1**, 857
Benzylide, tri-*n*-butylstibonium, reactions, with aryl aldehydes, **1**, 743
Benzylide, triphenylarsonium
reactions
with benzylideneaniline, **1**, 741
with *p*-nitrobenzaldehyde, **1**, 738
Benzylides, synthesis of, **3**, 493
Benzylimine, *N*-methyl-, reactions, with spirophosphoranes, **4**, 478
Benzyl radicals, synthesis of, **1**, 290; **4**, 1072
Benzyl sulfone, 1,1'-dibromo-, from, phenylacetic acid, **4**, 83
N-Benzylsulfonyl acid phenylhydrazides, reactions, with ferric chloride, **5**, 87
Benzyl sulfoxides, alkyl, deprotonation of, **1**, 143
Benzyl sulfoxides, aryl, reactions, with lithium *N*-isopropylcyclohexylamide, **1**, 142
Benzyl systems
deuterium exchange, rates of, **1**, 857
racemization of, **1**, 857
Benzylthiol
reactions
with *N*-(diethylaminomethyl)benzamide, **4**, 322
with formaldehyde, **4**, 215
with *N*-(piperidinomethyl)benzamide, **4**, 322
Benzylthiols
carbonylation of, **1**, 239
double deprotonation of, **4**, 367
Benzylthiols, germyl-, from, benzyl germyl thioethers, **4**, 371
Benzylthiols, α-silyl-, from, benzyl silyl thioethers, **4**, 368
Benzyne
reactions
with carbon disulfide, **6**, 99
with ethyl sulfides, **2**, 713
with 5-phenyl-1,2,4-dithiazole-3-thione, **5**, 634
with phenyl isothiocyanate, **5**, 596
with silylmethyldialkylamines, **2**, 945
use in the synthesis of nitriles, **3**, 626
Benzyne intermediates, carbometallation of, **2**, 974
Benzynes
as intermediates, in the synthesis of phenols, **2**, 647
reactions, with α-aminosilanes, **4**, 508
reactivity of, **1**, 539
as starting materials, for sulfur ylides, **1**, 821
Benzynezirconocenes
from, arylzirconocenes, **2**, 986
reactions
with aldehydes, **2**, 987
with alkenes, **2**, 987
with alkynes, **2**, 987
with ketones, **2**, 987
with nitriles, **2**, 987
Bergmann cycloaromatization, of allenes, **1**, 497
Beryllium, reactions, with alkyl halides, **2**, 563
Beryllium, diethyl-, use in the synthesis of alkylberylliums, **2**, 563
Beryllium, dimethyl-
reactions
with hydrogen cyanide, **5**, 1137
with trimethylaluminum, **4**, 709
Beryllium chloride
reactions
with *N*-lithioketimines, **3**, 459
with phosphorus ylides, **4**, 577
with ylides, **4**, 395
Beryllium cyanide, methyl-, synthesis of, **5**, 1137
Beryllium cyanides, alkyl-, synthesis of, **5**, 1137
Beryllium halides, reactions, with dialkylberylliums, **2**, 563
Beryllium halides, alkyl-, synthesis of, **2**, 563
Beryllium powder, use in the synthesis of alkylberylliums, **2**, 563
Berylliums, acyl-, synthesis of, **5**, 438
Berylliums, alkyl-, synthesis of, **2**, 563
Berylliums, dialkyl-
from
dialkylmercury compounds, **2**, 563
trialkylborons, **2**, 563
reactions, with beryllium halides, **2**, 563
synthesis of, via transmetallation reactions, **2**, 563
Berylliums, diaryl-, synthesis of, **1**, 452
Betaine
reactions, with 1-(1-propenyl)piperidine, **6**, 314
for the synthesis of vinyl fluorides, **2**, 612
Betaine–cation complexes, in Wittig reactions, **1**, 727
Betaine–lithium complexes, reactions, with organolithiums, **1**, 729
Betaines
acylation of, **5**, 140
Brook rearrangement of, **2**, 936
chlorination of, for the synthesis of vinyl chlorides, **2**, 618

from, epoxides, **1**, 640
as intermediates, **4**, 732
vs. oxaphosphetanes, **1**, 639
in Wittig reactions, **1**, 721
protonation of, **2**, 445
synthesis of, **2**, 476; **4**, 342; **5**, 1068
Betaines, nitrilium
classes of, **3**, 677
synthesis of, **3**, 677
in situ, **3**, 678
Betaines, oxaborole, X-ray structure of, **4**, 936
Betaines, triazanium, synthesis of, **2**, 387
Betaine salts, phospho, as intermediates, **4**, 752
Betylates, as starting materials, for halides, **2**, 9
Biacetyl
reactions, with
diethylaminofluoromethylphosphine, **2**, 476
as starting material, for ketenes, **3**, 535
3,3'-(1,1'-Biadamantyl) dications, synthesis of, **1**, 900
Bianthrone, reactions, with malononitrile, **3**, 668
Biaryls
asymmetric synthesis of, **1**, 473
from
aryllithiumbis(trifluoroacetate), **1**, 491
arylsilanes, **1**, 489
aryl triflates, **1**, 482
diaryl selenides, **1**, 484
phenoxide anions, **1**, 474
synthesis of, **1**, 472
Bibenzyl, synthesis of, **1**, 122
Bicyclic salts, opening of, **1**, 241
Bicyclization, diene, zirconium-promoted, **2**, 988
Bicyclization, diyne
carbonylative, iron-promoted, **2**, 985
zirconium-promoted, **2**, 988
Bicyclization, enyne
carbonylative, tungsten-promoted, **2**, 985
zirconium-promoted, **2**, 988
Bicycloalkenes, vinyl-, Cope rearrangement of, **1**, 814
Bicyclo[2.2.1]alkenes, carbonylation of, **5**, 192
Bicyclo[4.n.0]alkenones, synthesis of, **1**, 498
Bicyclobutane, as byproduct, in the synthesis of cyclobutene, **1**, 418
Bicyclobutanebis(ketenimine), synthesis of, **3**, 593
[1.1.0]Bicyclobutanes, synthesis of, **1**, 184
Bicyclobutylidene, perfluoro-, reactions, with metal fluorides, **6**, 18
Bicyclo[6.2.0]decene, from, 1,3-cyclodecadiene, **1**, 418
Bicyclo[3.1.0]diazahexenes, isomerization of, **3**, 481
Bicyclo[5.4.1]dodecapentaenyl anion, paramagnetic ring currents in, **1**, 910
Bicyclo[5.4.1]dodecapentene, synthesis of, **1**, 910
Bicyclo[3.2.0]heptadiene, synthesis of, **1**, 404
Bicyclo[2.2.1]heptane, via rearrangements, **1**, 381
Bicyclo[3.2.0]heptane, rearrangement of, **1**, 400
exo-Bicyclo[3.2.0]heptane, rearrangement of, **1**, 805
Bicyclo[3.2.0]heptanones, hydroxy-, rearrangement of, to lactones, **3**, 539
Bicycloheptaphosphines, synthesis of, **2**, 472
Bicyclo[2.2.1]heptene, 7,7-dimethoxy-, from, 7,7-dimethoxy-1,2,3,4-tetrachlorobicyclo[2.2.1]hept-2-ene, **1**, 430
Bicyclo[2.2.1]hept-2-ene, 7,7-dimethoxy-1,2,3,4-tetrachloro-, as starting material, for 7,7-dimethoxybicyclo[2.2.1]heptene, **1**, 430
Bicyclo[3.2.0]heptene, as starting material, for (-)-$\Delta^{9(12)}$-capnellene, **1**, 802
Bicyclo[2.2.1]heptenes, Cope rearrangement of, **1**, 814
Bicyclo[3.2.0]hept-2-enes, thermal rearrangement of, **1**, 799
Bicyclo[3.3.1]heptenes, 7-vinyl-, Cope rearrangement of, **1**, 815
Bicyclo[2.2.1]heptene systems, reactions, with isothiocyanic acid, **5**, 1043
Bicyclo[2.2.1]hept-5-en-2-one, reactions, with tri-*n*-butylstannyllithium, **4**, 952
Bicyclo[3.2.0]hept-2-en-6-ones, *exo*-7-halo-, reactions, with secondary amines, **2**, 301
Bicyclo[2,2,1]hexene, rearrangement of, **1**, 400
Bicyclo[2.1.1]hex-2-enes, thermal rearrangement of, **1**, 799
Bicyclo[2.2.0]hexenones, synthesis of, **4**, 755
Bicyclo[2.2.0]hex-5-en-2-ones, as starting materials, for hydrates, **4**, 160
Bicyclo[4,3,0]nonadiene, from, cyclononatriene, **1**, 421
Bicyclo[6.1.0]nona-2,4,6-triene, as starting material, for cyclononatetraenide anion, **1**, 863
Bicyclo[2.2.1]non-1-ene 1,2-dicarboxylates, dipolar addition to, **1**, 287
Bicyclo[3.2.1]octa-2,6-diene, deprotonation of, **1**, 865
Bicyclo[2,2,2]octane, synthesis of, **1**, 382
Bicyclo[2.2.2]octane, 1,4-dihydroxy-, as starting material, for cyanates, **5**, 1102
Bicyclo[2.2.2]octane, 2-exomethylene-6-vinyl-, oxy-Cope rearrangements of, **1**, 815
Bicyclo[2.2.2]octane, tetrasila-, synthesis of, **6**, 178
Bicyclo[3,2,1]octane
from, bicyclo[4,2,0]octane, **1**, 382
synthesis of, **1**, 382
Bicyclo[3,3,0]octane, synthesis of, **1**, 382
Bicyclo[4,2,0]octane, as starting material, for bicyclo[3,2,1]octane, **1**, 382
Bicyclo[2.2.2]octane-2-carboxylic acid, 1,3-dimethyl-5-oxo-, synthesis of, **5**, 62
Bicyclo[3.3.0]octane ring systems, synthesis of, **1**, 316
Bicyclooctanes, phosphasilyl-, synthesis of, **2**, 473
Bicyclo[3.3.0]octanols, from, allyl pentenyl ketones, **2**, 81
Bicyclo[4,2,0]octene, from, 1,3-cyclooctadiene, **1**, 418, 419
Bicyclo[2.2.2]octenes, Cope rearrangement of, **1**, 814
Bicyclo[2.2.2]oct-2-enes, 5-(phenylsulfonyl)-, alkylation of, **1**, 146
in-Bicyclo[4.4.4]-1-tetradecyl cations, synthesis of, **1**, 919
Bidentate carbanions, stability of, **2**, 1004
Bidentate complexes, oxidative addition to, **2**, 584
Bifunctional electrophiles, from, vinyl isocyanates, **6**, 448
Bifunctional intermediates, for heterocyclic synthesis, **6**, 445
Biguanide, fluorination of, **5**, 815
Bimetallic complexes, alkylidene, displacement of, **5**, 1088
BINAL-H, synthesis of, **2**, 63
BINAP, for the reduction of alkenes, **1**, 76
2,2'-Binaphthol, for the alkylation of crotonic esters, **1**, 134
(1,1'-Bi-2,2'-naphthol (BINOL), reactions, with lithium aluminum hydride, **2**, 63
Biocatalysts, immobilized, **5**, 117
Biomethylation, mechanism, **2**, 584
Biotin, synthesis of, **1**, 143
Biotransformation, for the synthesis of carboxylic acids, **5**, 117
Biphenyl, 3-chloro-4-cyano-, reactions, with sodium hydroxide, **5**, 108
Biphenyl, 2,2'-dilithio-
synthesis of, **4**, 668
X-ray structure of, **4**, 668
Biphenylenephosphorane, triphenyl-, synthesis of, **2**, 830
Biphosphine disulfides, reactions, with sodium sulfide, **2**, 462
Biphosphine disulfides, alkyl-, desulfurization of, **2**, 431
Biphosphines, organosubstituted, reduction of, **2**, 431
Biphosphines, tetraalkyl-, cleavage of, **2**, 437
Biphosphine sulfide, tetramethyl-, reactions, with bromine, **2**, 1070
Biphosphine sulfides, organosubstituted, reduction of, **2**, 431
Bipyrazoles, ring-opening of, **4**, 989
Bipyridine, oxidative addition to, **2**, 584
Bipyridine–molybdenum complexes, alkylation of, **2**, 573
Bipyridine–tungsten complexes, alkylation of, **2**, 573
Bipyridyl catalysts, in the selective alkylation of aldehydes, **2**, 72
Biradicals, use in the synthesis of radicals, **4**, 1077

1,4-Biradicals, synthesis of, via abstraction of a γ-hydrogen, **4**, 1079
Birch reduction
of alkoxyalkynes, **2**, 676
of aryl alkyl ethers, **2**, 682
and benzylic alcohols, **1**, 13
carbanion formation during, **1**, 889
mechanism, **1**, 95, 890
of phenols, **3**, 218
for the synthesis of sulfenyl selenides, **2**, 238
(*Z*)-α-Bisabolene, synthesis of, **1**, 641, 757
Bisacetate, phenyliodo(III), use in the synthesis of acylhydrazines, **5**, 356
Bis(acyl) complexes, synthesis of, problems, **5**, 441
Bisacyl complexes, anionic, synthesis of, **5**, 449
Bis(acyl) halides, reactions, with thionoamides, **5**, 630
Bis(acyloxy) compounds, from, carboxylic acids, **4**, 205
Bisalkene complexes, synthesis of, **5**, 450
Bis(alkoxythiocarbonyl) compounds, *trans*-esterification of, **5**, 511
Bisalkylidyne complexes, from, cycloalkenes, **3**, 691
Bisallenes
deprotonation of, **1**, 995
synthesis of, **1**, 990
Bisallenyl compounds, from, diynes, **1**, 954
Bis(η^3-allyl) complexes, reactions, with cyclohexadiene, **4**, 694
Bisamides, synthesis of, **4**, 418
Bisamides, alkylidene, synthesis of, **4**, 424
Bisamides, *N*,*N'*-alkylidene
from, *N*-[α-(benzylideneamino)benzyl]benzamides, **4**, 424
synthesis of, via condensation of amides with carbonyls, **4**, 424
Bisamides, N^1,N^2-alkylidene, synthesis of, **4**, 423
Bisanilides, synthesis of, **4**, 282
Bisarsenane oxide, synthesis of, **2**, 493
Bisarsenic acid, synthesis of, **4**, 592
Bisarsenic oxide, dimeric, as starting material, for bisarsinomethanes, **4**, 592
Bisarsenioalkenes, stability of, **4**, 1038
Bis(arsines)
reactions, with chlorodiphenylarsine, **6**, 155
synthesis of, **2**, 1077
Bisarsinomethanes
from
bis(dichloroarsino)methane, **4**, 593
dimeric bisarsenic oxide, **4**, 592
synthesis of, **4**, 592
Bisarsolane oxide, synthesis of, **2**, 493
Bisazide, synthesis of, **6**, 110
Bisaziridines, use in asymmetric *cis*-dihydroxylation reactions, **2**, 50
Bis(azo compounds), synthesis of, **6**, 260
Bis(azoles), thiocarbonyl-, as starting materials, for thionocarbonates, **6**, 543
Bis(benzenethio)methane
metallation of, **6**, 86
reactions
with *n*-butyl lithium, **6**, 86
with dimethyl disulfide, **6**, 86
Bisbenzotriazole, 1,1'-carbonyl-, synthesis of, **6**, 501
Bisbenzotriazole, *N*,*N'*-(carbonyldioxy)-, as starting material, for carbamates, **6**, 417
Bisbenzyl compounds, reactions, with carbon monoxide, **5**, 442
Bis(4-benzylidene-1,4-dihydropyridine), 1,1'-carbonyl-, use in the synthesis of dialkyl carbonates, **6**, 463
1,2-Bis(bromomethyl) compounds, synthesis of, **4**, 143
Bis(calcium iodide), methylene-, synthesis of, **4**, 669
Biscarbamates, N^1,N^2-alkylidene, synthesis of, **4**, 427
2,2-Biscarbamates, synthesis of, **6**, 364
1,4-Bis(chloromethylbenzene), reactions, with sulfur, **5**, 561
Bis(chlorophosphonium salts), hydrolysis of, **4**, 554
Bis(dialkylamide) compounds, reactions, with carbon monoxide, **6**, 516
Bis(diazo) compounds, from, 1,6-diarylhexane-1,3,4,6-tetraones, **3**, 469
Bis(α-diazo β-oxophosphonates), synthesis of, **3**, 482
Bis(dichloroarsino)
reactions, with 1,2-bis(*N*-methylamino)ethane, **4**, 593
use in the synthesis of tetraalkyls, **4**, 593
Bis(dichlorophosphine), methylene-, reactions of, **4**, 550, 562
Bis(dimethylarsines), synthesis of, **4**, 134
Bis(dimethylglyoximato) compounds, use in the synthesis of radicals, **1**, 946
Bis(diphenylphosphide), carbonyl, synthesis of, **6**, 517
Bis(diphenylphosphine), methylene-, reactions of, **4**, 550, 563
Bisdiphenylphosphinoethene–chromium complexes, structure of, **4**, 1030
Bisdithioacetals, from, dicarbonyls, **4**, 245
Bis(dithiocarbamates), synthesis of, **6**, 565
Bis(dithioorthocarbonates), cyclic, from, aryl *N*,*N*-dimethyldithiocarbamates, **6**, 313
Bis(ethanethio)methane, as starting material, for tetraethyl tetrathioorthocarbonate, **6**, 86
Bisformazans
from, arylhydrazones, **5**, 856
synthesis of, **5**, 857
Bisgermanium oxides, reactions, with thiourea, **5**, 1082
Bishemithioaminals, from, terephthalaldehyde, **4**, 317
1,4-Bis(hexafluoroisopropanol), phenyl-, reactions, with cyanogen bromide, **5**, 1102
Bishomoaromatic cations, synthesis of, **1**, 918
Bishomocyclopentadienyl anions
stability of, **1**, 865
synthesis of, **1**, 865
Bishydrazidoyl chlorides, chlorine exchange, **5**, 681
Bishydrazidoyl halides, reactions, with α,ω-diaminoalkanes, **5**, 681
Bishydrazine, methylene-, synthesis of, **4**, 437
Bishydrazoles, reactions, with tetrasulfurtetranide, **2**, 794
Bishydrazones
as starting materials, for tetraazathiapentalenes, **3**, 378
synthesis of, **5**, 522
Bishydrazonoyl halides
reactions
with thiophenolate, **5**, 681
with tributyltin hydride, **5**, 681
Bishydroboration, for the synthesis of carbon-bridged dimercury complexes, **4**, 697
Bis(hydroxamic acids), thiooxalic, from, cyanogen *N*,*N*-dioxide, **5**, 828
Bishydroxylamines, reactions, with ketones, **4**, 436
Bisilylation, of alkynes, **2**, 915
Bisimidates, synthesis of, **5**, 685
Bis(imidazole), oxomethylene-, reactions, with methanesulfonic acid, **3**, 696
Bis(imidazole), thiocarbonyl-, synthesis of, **6**, 578
Bis(imidazole), *N*,*N'*-thiocarbonyl-
as starting material, for thioureas, **6**, 577
synthesis of, **6**, 577
Bisimidoyl chlorides, from, *N*,*N'*-bis(trifluoroacetyl)hydrazine, **5**, 676
Bisimines, from, diphenylmethanimine, **4**, 438
Bisimino complexes, dinuclear, reactions, with ethyne, **4**, 683
Bisiodonium salts, bis(alkynyl), as starting materials, for bis(alkynyl) dithiocyanates, **2**, 1034
Bis(isocyanide) complexes, reactions, with azetidine, **6**, 663
Bisisonitriles, alkylidene-, chemistry of, **4**, 538
Bisketenes, silyl-substituted, synthesis of, **2**, 941
Bis(ketenimine), synthesis of, **3**, 557, 561, 570, 573; **3**, 593
Bislithio compounds, reactions, with benzaldehyde, **4**, 938
Bismaferrocenes, synthesis of, **4**, 1038, 1039
Bis(magnesium bromide), methylene-
reactions

with metallocene dichlorides, **4**, 670
with titanocene dichloride, **4**, 670
synthesis of, **4**, 669
Bismenes, aryl, as starting materials, for arenes, **1**, 445
Bismercuration, of carboxylic acids, **4**, 697
Bis(mercuric iodide), methylene-, reactions, with boron tribromide, **4**, 632
Bis(mercury halides)
as starting materials, for methylenedilithium, **4**, 668
transmetallation of, **4**, 668
Bismetallics
reactions of, **2**, 594
synthesis of, **1**, 679
Bis(methane sulfonate), oxomethylenebis-($3H^+$-imidazolium)-, synthesis of, **3**, 696
2,2-Bis-(4-methoxybenzoyloxy) compounds, from, 1,3-dicarbonyls, **4**, 206
Bismethyl compounds, reactions, with carbon monoxide, **5**, 442
Bis(*N*-methyl dithiocarbazates), synthesis of, **6**, 563
Bismoles
reactions, with iron(II) chloride, **4**, 1039
synthesis of, **4**, 1038
Bismolybdenumtetracarbonyl complexes, synthesis of, **4**, 1023
Bismolyl anions
reactions, with iron(II) chloride, **4**, 1041
synthesis of, **4**, 1039
Bis(monothiocarbonates), oxidation of, **6**, 479
Bismuth
analogues of ketenes, theoretical, **3**, 602
α-haloalkenyl compounds of, synthesis of, **4**, 805
heteroaromatic compounds of, **2**, 879
reactions, with arenediazonium tetrafluoroborates, **2**, 876
synthesis of functions, **4**, **591–600**
with no halogen, chalcogen or nitrogen, **4**, **1021–1042**
Bismuth, diphenylchloro-, synthesis of, **5**, 402
Bismuth, doubly-bonded, synthesis of functions, **6**, **677–724**
Bismuth, triaryl-, reactions, with carboxylic acids, **2**, 888
Bismuth, tribromo-, reactions, with trimethylphosphine, **2**, 473
Bismuth, triphenyl-, reactions, with acid chloride, **5**, 402
Bismuth, tris(trifluorovinyl)-, synthesis of, **4**, 805
Bismuth(III) compounds, synthesis of, via transmetallation, **2**, 882
Bismuth(III) halides
reactions
with organometallics, **2**, 872
with sodium phenolates, **2**, 695
Bismuth(III) halides, alkyl-, synthesis of, via redistribution reactions, **2**, 490
Bismuth(V) compounds, synthesis of, **5**, 222
Bismuth(V) cyanides, synthesis of, **5**, 1133
Bismuth(V) dicyanides, synthesis of, **5**, 1133
Bismuth(V) halides, reactions, with sodium phenolates, **2**, 695
Bismuth(V) halides, triaryl-, reactions, with potassium cyanide, **5**, 1133
Bismuthate, tris(ethylthio)-, synthesis of, **2**, 261
Bismuth–bismuth bonds, cleavage of, **2**, 491
Bismuth bistrifluoroacetate, triphenyl-, reactions, with enols, **2**, 675
Bismuth bromides, alkylation of, **2**, 512
Bismuth carbonate, reactions, with carboxylic acids, **5**, 222
Bismuth carbonate, triphenyl-
for the oxidation of alcohols, **3**, 118
use in the synthesis of diphenyldiazomethane, **3**, 464
Bismuth–carbon bonds, cleavage of, **2**, 490
Bismuth compounds
alkali metal derivatives of, synthesis of, **2**, 509
main group metal derivatives of, synthesis of, **2**, 510
synthesis of, **4**, 599, 600
transition metal derivatives of, synthesis of, **2**, 511
Bismuth compounds, acyloxy-, synthesis of, **5**, 222
Bismuth compounds, alkyl-, synthesis of, **2**, **479–512**
Bismuth compounds, alkynyl-
safety precautions, **2**, 1076
synthesis of, **2**, 1076, 1082
Bismuth compounds, aryl-
carbonylation of, **3**, 290
stability of, **2**, 1082
Bismuth compounds, cyano-, synthesis of, **5**, 1132
Bismuth compounds, dialkynyl-, purification of, **2**, 1082
Bismuth compounds, α-halo-, synthesis of, **4**, 136
Bismuth compounds, imidoyl-, synthesis of, **5**, 792
Bismuth compounds, monoalkynyl-, synthesis of, **2**, 1082
Bismuth compounds, pentaaryl-, synthesis of, reviews, **2**, 872
Bismuth compounds, tetracoordinated, synthesis of, **6**, 707
Bismuth compounds, trialkynyl-, polymerization of, **2**, 1082
Bismuth compounds, triaryl-
reactions
with cyanogen iodide, **5**, 1132
with 1,3-diazocarbonyls, **3**, 196
Bismuth compounds, triethynyl-, stability of, **2**, 1076
Bismuth diacetate, triphenyl-, for the phenylation of alkyl alcohols, **2**, 684
Bismuth dichloride, reactions, with β-diketones, **3**, 196
Bismuth dihalides, alkyl-, reactions, with sodium alkoxides, **2**, 496
Bismuth diiodide, phenyl-, use in the synthesis of bismoles, **4**, 1039
Bismuth enolates, synthesis of, **2**, 694
Bismuth functions
elimination of, **1**, 1079
substitution of, **1**, 1032
synthesis of, **4**, 595
Bismuth functions, acyl-, synthesis of, **5**, **393–407**
Bismuth functions, aryl-, synthesis of, **2**, **871–897**
Bismuth functions, dicoordinated, synthesis of, **5**, 892
Bismuth functions, imidoyl-, synthesis of, **5**, 784
Bismuth functions, tetracoordinated, synthesis of, **5**, 922
Bismuth functions, tricoordinated, synthesis of, **5**, 894
Bismuth functions, vinyl-, synthesis of, **2**, **871–897**
Bismuth groups, elimination of, **1**, 647
Bismuth halide, dimethyl-, nucleophilic metallation of, **2**, 512
Bismuth halides
reactions, with organometallics, **2**, 483
synthesis of, **2**, 880
Bismuth halides, alkyl-
reactions
with amines, **2**, 502
with ammonia, **2**, 502
with metal amides, **2**, 502
with organometallics, **2**, 483
with transition metal complex anions, **2**, 511
reduction of, **2**, 481
synthesis of, **2**, 492
Bismuth halides, diaryl-, reactions, with sodium alkynes, **2**, 1082
Bismuthimines, *N*-tosyltriphenyl-, reactions, with benzoyl chloride, **5**, 344
Bismuthinate, bis(1-naphthyl)-, reactions, with acetic acid, **5**, 630
Bismuthine, azidodimethyl-, synthesis of, **2**, 504
Bismuthine, chlorodimethyl-, reduction of, **2**, 481
Bismuthine, dibromoethyl-, synthesis of, **2**, 489
Bismuthine, dibromo(methyl)-
reactions, with dilithium *o*-phenylene diselenide, **2**, 501
synthesis of, **2**, 489
Bismuthine, dichloro(methyl)-, reduction of, **2**, 481
Bismuthine, diethoxy(methyl)-, transesterification of, **2**, 496
Bismuthine, dimethyl-, synthesis of, **2**, 481
Bismuthine, dimethyltrimethylsiloxy-, synthesis of, **2**, 494
Bismuthine, dimethyl(trimethylsilylmethyl)-, synthesis of, **4**, 599

Bismuthine, methyl-, synthesis of, **2**, 481
Bismuthine, tetramethyl-
  reactions
    with benzyl bromide, **2**, 491
    with hydrogen chloride, **2**, 491
    with iodine, **2**, 491
Bismuthine, triethyl-
  reactions, with tris(pentafluorophenyl)germane, **2**, 509
  synthesis of, **2**, 483
Bismuthine, trimethyl-, reactions, with hydrazoic acid, **2**, 504
Bismuthine, tris(dimethylamino)-, synthesis of, **2**, 416
Bismuthine, tris(trifluoromethyl)-, synthesis of, **6**, 243
Bismuthine oxides, bisdialkyl-, synthesis of, **2**, 493
Bismuthines
  halogenation of, **2**, 881
  as starting materials, for bismuth ylides, **3**, 500
  synthesis of, **4**, 599
Bismuthines, alkoxy-, synthesis of, **2**, 494
Bismuthines, alkyl-
  reactions, with sodium, **2**, 510
  synthesis of, **2**, 481
Bismuthines, alkyldialkoxy-, synthesis of, **2**, 496
Bismuthines, alkyldihalo-
  reactions
    with selenols, **2**, 500
    with thiols, **2**, 500
  synthesis of, **2**, 490
Bismuthines, alkylhalo-, synthesis of, **2**, 489
Bismuthines, amino-
  reactions, with cyclopentadiene, **2**, 485
  as starting materials, for tertiary alkylarsines, **2**, 484
  synthesis of, **2**, 502
Bismuthines, bromo-, reduction of, **2**, 481
Bismuthines, bromo(dialkyl)-, synthesis of, **2**, 490
Bismuthines, chloro(dialkyl)-, synthesis of, **2**, 490
Bismuthines, dialkylhalo-
  reactions, with alcohols, **2**, 494
  synthesis of, **2**, 490
Bismuthines, tertiary alkyl-
  reactions
    with benzenethiol, **2**, 501
    with bismuth trihalides, **2**, 490
  synthesis of, **2**, 483
Bismuthines, trialkyl-
  cleavage of, **2**, 490
  as starting materials, for dibismuthines, **2**, 505
Bismuthines, unsymmetrical tertiary alkyl-, synthesis of, **2**, 483
Bismuth iodide, diphenyl-, as starting material, for tetraphenyldibismuthine, **2**, 893
Bismuth nitrate, reactions, with ethanethiol, **2**, 261
Bismuth nitriles, diaryl-, synthesis of, **5**, 1132
Bismuthonium, synthesis of, **3**, 500
Bismuth phenolates, synthesis of, **2**, 695
Bismuth trichloride, reactions, with trimethylsilylmethylmagnesium chloride, **4**, 599
Bismuth trichlorides, reactions, with lithium tetraethylaluminate, **2**, 483
Bismuth trihalides
  and guanidines, **6**, 656
  reactions
    with Grignard reagents, **2**, 483
    with organometallics, **2**, 488
    with tertiary alkylbismuthines, **2**, 490
Bismuth ylides
  from, bismuthines, **3**, 500
  synthesis of, **3**, 196, 500, **491–500**
Bismutines, methyldiphenyl-, synthesis of, **1**, 54
Bis(nitrile oxides), synthesis of, **3**, 685
Bis(1,2-oxarsolanyl) oxide, reactions, with arsenic trichloride, **2**, 497
Bisoxazoline ligands, for copper catalysts, **1**, 353
Bisoximes, as starting materials, for diazapentalenes, **3**, 378
Bis(oxo) compounds, synthesis of, **2**, 573
Bisperoxides, synthesis of, **2**, 96
Bisphenol A
  reactions
    with phosgene, **6**, 470
    with triphosgene, **6**, 471
Bis(phenylarsenic acid), methylene-
  from, bis(diphenylarsino)methane, **4**, 592
  synthesis of, **4**, 592
Bisphosphiminoalkenes, as starting materials, for α-phosphimino enamines, **4**, 991
Bisphosphineimides, synthesis of, **2**, 864
Bisphosphines
  reactions
    with diphenylphosphinyl chloride, **6**, 151
    with hydrogen chloride, **5**, 398
    with organodihalides, **2**, 444
  for the reduction of alkenes, **1**, 76
  as starting materials, for tris(phosphino)methanes, **6**, 151
  synthesis of, **2**, 431; **4**, 111, 546
Bis(phosphines), alkynyl-, as starting materials, for alkynylphosphonium salts, **2**, 1063
Bisphosphines, methylene-, reactions, with chlorophosphines, **6**, 151
Bisphosphines, tetrachloro-, from, dichloromethane, **2**, 448
Bisphosphines, tetraphenyl-, reactions, with α-hydroxyamides, **4**, 465
Bisphosphinic acids, from, primary amines, **4**, 470
α,ω-Bisphosphinoalkanes, reactions, with *N*-hydroxymethyl dialkylamines, **4**, 452
Bisphosphinoalkenes
  reactions, with organometallics, **4**, 1029
  synthesis of, **4**, 1028
  transition metal complexes of, synthesis of, **4**, 1028
Bisphosphinoalkenes, acyclic
  nomenclature, **4**, 1022
  synthesis of, **4**, 1022
Bisphosphinoalkenes, cyclic, synthesis of, **4**, 1022
1,1-Bisphosphinoalkenes, reactions, with transition metals, **4**, 1028
Bisphosphinomethanes, as starting materials, for phosphinomethylphosphonium salts, **5**, 907
Bis(phosphinyl dichloride), alkynyl-, synthesis of, **2**, 1065
Bis(phosphonates)
  from, isopropyl tetraester, **6**, 263
  synthesis of, **4**, 553; **6**, 123, 263
Bis(phosphonates), carbonyl-, synthesis of, **6**, 518
Bis(phosphonates), (diazomethylene)-, synthesis of, **6**, 666
Bis(phosphonates), methoxymethylene, halogenation of, **6**, 281
Bis(phosphonates), methylene, synthesis of, **4**, 555
Bis(phosphonic acid), alkylidene-, metal salts of, uses, **6**, 287
Bis(phosphonic acids), synthesis of, **6**, 287
Bis(phosphonic acids), halogenated methylene-, uses of, **6**, 287
Bis(phosphonic acids), methylene-, halogenation of, **6**, 263
Bis(phosphonium) disalts, bis(diazo), synthesis of, **3**, 482
Bis(phosphonium salts)
  deprotonation of, **5**, 910
  from
    1,1-dihaloalkanes, **4**, 553
    phthaloyl dichloride, **2**, 1063
  synthesis of, **2**, 447; **4**, 545; **6**, 370
Bis(phosphonium salts), diazo, synthesis of, **3**, 482
Bis(phosphonium salts), diphosphirane, synthesis of, **4**, 550
Bisphosphonium ylides, condensation of, **4**, 801
Bis(phosphonous diamides), (diazomethylene)-, synthesis of, **6**, 665
Bispiperazine, use in asymmetric *cis*-dihydroxylation reactions, **2**, 50
Bisproline ligands, use in the synthesis of organometallics, **2**, 71
Bis(rhenium) compounds, synthesis of, **2**, 583

Bis(rhodium) complexes, for the reduction of benzene, **1**, 93
Bis[rhodium-(1,5-hexadiene)chloride], for the reduction of benzene, **1**, 94
Bis(rhodiumnorbornadiene) chloride complexes, immobilization of, **1**, 94
Bis(secondary phosphines), synthesis of, **2**, 429, 430
1,2-Bis(selenides), from, dialkyl diselenides, **2**, 286
Bisselenoesters, synthesis of, **4**, 288
Bissilanes, (2,6-xylylimino)-, from, (2,6-xylylimino)(trimethylsilyl)methyllithium, **6**, 672
Bis(silylalkynes), reactions, with nitronium tetrafluoroborate, **2**, 1053
Bis(silyl) anions, synthesis of, **4**, 618
α,α-Bis(silyl)carbanions
  as starting materials, for alkenylsilanes, **2**, 936
  synthesis of, via Peterson alkenation, **2**, 936
Bis(silyldiazomethyl) compounds, photolysis of, **6**, 383
Bissulfonamides
  from, *N*-alkyl arylsulfonamides, **4**, 428
  synthesis of, **4**, 282, 427
Bissulfonamides, N^1,N^2-alkylidene, synthesis of, **4**, 427
gem-Bissulfonates, synthesis of, **4**, 214
Bis(sulfones)
  electrochemical oxidation of, **4**, 277
  from
    alkylidene disulfones, **4**, 274
    bis(phenylsulfonyl)methane, **4**, 274
    dithioacetals, **4**, 272
    ketene dithioacetal tetraoxides, **4**, 277
    methylene disulfones, **4**, 274
    sulfonyl fluorides, **4**, 280
  reactions, with dihalides, **4**, 275
  as starting materials
    for alkenes, **4**, 275
    for allylic amines, **4**, 279
  synthesis of, **4**, 272
Bis(sulfones), hydroxy, Mitsunobu coupling of, **1**, 183
Bis(sulfones), hydroxycyclopropyl, synthesis of, **4**, 277
Bis(sulfones), oxaziridine, synthesis of, **4**, 273
Bis(sulfones), quinadyl, synthesis of, **4**, 281
Bissulfonic acids, synthesis of, **4**, 281
Bis(sulfonylacetate), dimethyl methylene-, reactions, with benzaldehyde, **4**, 277
Bis(sulfonylacetic acid), methylene-, decarboxylation of, **4**, 272
Bissulfoxides
  from
    dithioacetals, **4**, 267
    sulfines, **4**, 270
Bissulfoxides, methylene, as starting materials, for bissulfoxides, **4**, 269
Bis(tertiary arsines), as ligands, **2**, 897
Bis(*O*-thioacetate), dimethyltin
  NMR spectrum of, **5**, 541
  synthesis of, **5**, 541
Bisthioalkenes, desulfurization of, **1**, 438
Bisthioamides, *N,N'*-alkylidene, synthesis of, **4**, 424
Bisthioamines, amino, synthesis of, **2**, 361
Bis(*O*-thiobenzoate), dimethyltin, synthesis of, **5**, 541
Bisthiols, reactions, with ketones, **4**, 347
Bisthionoamides, synthesis of, **5**, 580
Bis(thiophosphoryl chloride), methylene-, reduction of, **4**, 553
Bis(tosylhydrazones), ring-opening of, **3**, 668
Bis(trialkyltin) oxides, reactions, with thiourea, **5**, 1082
Bis(-1,2,4-triazole), cyclopentadieneto thiocarbonyl-, cycloaddition of, **6**, 114
Bis(-1,2,4-triazole), 1,1'-thiocarbonyl-, use in the synthesis of thionocarbonates, **6**, 543
Bis(tributyltin) compounds, synthesis of, **4**, 702
Bis(tributyltin) oxide
  reactions, with *N*-phenyl-*N'*-diphenylmethyleneaminothiourea, **6**, 571
  as reagent, for the synthesis of alcohols, **2**, 87
Bis(trichloromethyl) compounds, synthesis of, **6**, 30
Bis(triethyltin)oxide, reactions, with terminal alkynes, **1**, 1006
Bis(trifluoroacetate), phenyl iodosyl [PIFA], for the oxidation of amides, **4**, 422
Bis(trifluoromethyl) compounds, synthesis of, **6**, 30
1,1-Bis(trimethylsilyl) compounds, synthesis of, **4**, 605
Bis(trimethylsilyl)methyls, from, lanthanide complexes, **6**, 194
Bis(trimethylstannane), methylene-, synthesis of, **4**, 699
Bisulfite, in Strecker reaction, **3**, 616
Bisulfite addition compounds, of aldehydes and ketones, **4**, 237
Bisulfite adducts, of carbonyls, for the synthesis of cyanohydrins, **3**, 615
N,N'-Bisurethane-protected carboxamidines, as starting materials, for guanidines, **6**, 641
Bisylide, synthesis of, **3**, 498
Bisylides, mercury-bridged, from, mercury acetate ylide adducts, **5**, 919
1,1-Biszincs, alkyl-, synthesis of, via metallo-Claisen reactions, **2**, 594
Blaise reaction, mechanism, **3**, 461
Blocking groups, for the synthesis of amines, **2**, 300
Bohme procedure, for the synthesis of iminium salts, **3**, 185
Bond dissociation energy
  hydrocarbons of, **1**, 925
  for the measurement of radical stabilization, **1**, 921
σ-Bonded compounds, synthesis of, **4**, 1032
Bond scission
  for the synthesis of vinyl cations, **1**, 1111
  for the synthesis of vinyl radicals, **1**, 1127, 1129
π-Bond stabilization, of vinyl cations, **1**, 1113
Bond strengths, of carbon–halogen, **1**, 107
Bongkrekic acid, synthesis of, **3**, 676
BOP coupling reagents
  for the synthesis of alkynoic amides, **5**, 304
  for the synthesis of amides, **5**, 267
9-Borabicyclononane, from, 1,5-cyclooctadiene, **2**, 516
9-Borabicyclo[3.3.1.]nonane, *n*-hexyl-, reactions, with *N-t*-butylimino pentylidinlithium, **5**, 795
9-Borabicyclo[3.3.1]nonane, *B*-iodo-, reactions, with 1-ethoxyalkyne, **4**, 765
9-Borabicyclo[3.3.1]nonane, *B*-isopropyl-, bromination of, **4**, 143
9-Borabicyclo[3.3.1]nonane, *B*-methyl- (*B*-methyl-9-BBN)
  alkylation of, **1**, 162
  deprotonation of, **1**, 765
9-Borabicyclo[3.3.1]nonane, *B*-3-pinanyl-, synthesis of, **2**, 57
9-Borabicyclo[3.3.1]nonane, potassium 9-thexyloxy-, synthesis of, **2**, 534
9-Borabicyclo[3.3.1]nonane (9-BBN-H)
  as hydroborating reagent, **4**, 631
  for the hydroboration of alkynes, **1**, 510
  reactions
    with aromatic acids, **3**, 82
    with silylalkynes, **3**, 203
  use in the synthesis of trialkylboranes, **2**, 517
9-Borabicyclo[3.3.1]nonanes, acyloxy-, reduction of, **3**, 82
9-Borabicyclo[3.3.1]nonanes, 9-(1-alkynyl)-, reactions, with isocyanates, **1**, 1043
9-Borabicyclo[3.3.1]nonanes, *B*-1-alkynyl-
  reactions
    with aldehydes, **1**, 1037
    with ketones, **1**, 1037
9-Borabicyclo[3.3.1]nonyl, 9-alkyl-, reactions, with bromopyridine, **3**, 311
9-Borabicyclo[3.3.1]nonyl, *B*-1-alkynyl-, use in 1,4-addition reactions, **1**, 1050
9-Borabicyclo[3.3.1]nonyl, 9-allenyl-
  reactions
    with aldehydes, **1**, 1055

with ketones, **1**, 1055
9-Borabicyclo[3.3.1]nonyl, *B*-bromo-, bromination of, **4**, 143
9-Borabicyclo[3.3.1]nonyl, *β*-isopinocampheyl-, synthesis of, **2**, 517
9-Borabicyclo[3.3.1]nonyl, 9-(*ω*-stannylalkyl)-, as organoborons, **1**, 279
9-Borabicyclo[3.3.1]nonyl (9-BBN)
reactions
with alkynes, **1**, 1101
with diacetone glucose, **2**, 57
with dithiocarboxylic acids, **5**, 589
with enamines, **1**, 443
with organocuprates, **2**, 522
synthesis of, **2**, 41
for the thionation of amides, **5**, 578
use in the synthesis of carbodiimides, **5**, 1085
use in the synthesis of thionoalkanoamides, **5**, 610
9-Borabicyclo[3.3.1]nonyl (9-BBN) ligand, use in the synthesis of organoborons, **1**, 126
9-Borabicyclo[3.3.1]nonyls, reactions, with phenoxy acetic acid dianions, **1**, 185
9-Borabicyclo[3.3.1]nonyls, alkyl-, deprotonation of, **4**, 664
9-Borabicyclo[3.3.1]nonyls, 9-alkyl-, reactions, with iodoarenes, **3**, 290
Boraborenes
from
borenes, **5**, 928
dichlorides, **5**, 928
Boracycloalkanes, synthesis of, **6**, 183
1-Boracyclohepta-2,6-diene, 1-phenyl-, reactions, with metal carbonyls, **4**, 695
Boracyclopropane, reactions, with di(trimethylsilyl)alkyne, **4**, 1082
Boradiphosphines, synthesis of, **2**, 474
Boraethenes, synthesis of, via [2+4] cycloreversion processes, **3**, 504
3-Borahomoadamantane, reactions, with methylene(triphenyl)phosphorane, **2**, 447
Boranate, *α*-amido-, synthesis of, **4**, 526
Borane
reactions
with alkenes, **1**, 75; **2**, 514
with catechol, **2**, 528
with diazomethylenephosphorane, **5**, 872
with dithianyl myrtanyls, **2**, 42
with (-)-norephedrine, **2**, 314
with 2-pinene, **2**, 41
with *α*-pinene, **2**, 514
with silyl enol ethers, **1**, 434
for the reduction of oxiranes, **1**, 14
as starting material, for alkoxyboranes, **2**, 109
use in the synthesis of hydroxylamines, **2**, 763
Borane, acetoxy-, as hydroborating reagent, **2**, 43
Borane, acetoxydiethyl-, synthesis of, **5**, 227
Borane, allyl dimesityl-, *γ*-alkylation of, **1**, 163
Borane, aminomethyldimethyl-, synthesis of, **4**, 524
Borane, benzoyldibutyl-, synthesis of, **5**, 433
Borane, benzyl dimesityl-, deprotonation of, **1**, 163
Borane, benzyloxydibutyl-, synthesis of, **5**, 433
Borane, bis(diarylphosphino)phenyl-, synthesis of, **2**, 867
Borane, bis(diisopropylamino)chloro-, reactions, with diazomethane, **3**, 485
Borane, bis(dimethylamino)chloro-, reactions, with lithium diethylarsenide, **2**, 509
Borane, bis(ethylthio)phenyl-, reactions, with isocyanates, **2**, 267
Borane, bis(stannyl)methylene-, detection of, **6**, 713
Borane, bistrifluoromethylarsino-, synthesis of, **2**, 509
Borane, bis[(1-trimethylsilyl)isobutyl]-
hydrolysis of, **4**, 645
methanolysis of, **4**, 645
Borane, [boryl(silyl)methylene]-, synthesis of, **6**, 714
Borane, [boryl(stannyl)methylene]-, synthesis of, **6**, 714
Borane, bromodimethyl-, reactions, with *N*-aryltrifluoroacetamides, **5**, 717
Borane, (1-bromoethyl)diethyl-, synthesis of, **4**, 143
Borane, *n*-butoxydichloro-, as starting material, for borinates, **2**, 419
Borane, *t*-butylperoxydi-(*o*-nitrophenoxy)-, synthesis of, **2**, 99
Borane, chloro-, reactions, with alkynes, **1**, 531
Borane, chloroamino-, reactions, with alkynylsodiums, **2**, 1085
Borane, (*α*-chlorobenzyl)diphenyl-, synthesis of, **4**, 142
Borane, chlorobis(diethylamino)-, synthesis of, **2**, 418
Borane, chlorobis(diisopropylamino)-, reactions, with triisopropylsilyldiazomethane, **5**, 1148
Borane, chlorobis(dimethylamino)-, reactions, with crotylpotassium, **2**, 529
Borane, 4-chlorobutoxydichloro-, synthesis of, **2**, 110
Borane, *β*-chlorodiiso-2-ethyl apopinocampheyl-
synthesis of, **2**, 526
use in the reduction of ketones, **2**, 58
Borane, chlorodiisopinocampheyl-
synthesis of, **2**, 526
use in asymmetric carbonyl reduction, **2**, 58
use in the synthesis of alkoxyboranes, **2**, 531
Borane, chlorodimethoxy-, as starting material, for trimethylborate, **2**, 111
Borane, chloromethyldimethyl-, synthesis of, **4**, 523
Borane, chloro(2,2,6,6-tetramethylpiperidino)-(tris(trimethylsilyl)methyl-, synthesis of, **6**, 388
Borane, chlorothexyl-, as reagent, for the synthesis of *α*-haloalkenylborons, **4**, 815
Borane, cyano-, reactions, with triethyl phosphite, **6**, 514
Borane, *N*-cyclohexyl-*N*-2-butenylaminodichloro-, synthesis of, **2**, 811
Borane, *N*-cyclohexyl-*N*-1-cyclohexenylaminodichloro-, synthesis of, **2**, 810
Borane, cyclohexyldichloro-, reactions, with azides, **2**, 318
Borane, dialkylarsino-, synthesis of, **2**, 509
Borane, dibromo-, synthesis of, **2**, 525
Borane, dibromomethoxy-, stability of, **2**, 110
Borane, dibromophenyl-, reactions, with *N*-trimethylsilylthiobenzanilide, **5**, 739
Borane, di(*t*-butylperoxy)hydroxy-, synthesis of, **2**, 99
Borane, di-*t*-butylperoxy-*o*-nitrophenoxy-, synthesis of, **2**, 99
Borane, dibutyl(trifluoromethyl)-, synthesis of, **6**, 243
Borane, dichloro-
synthesis of, **2**, 525
use in the synthesis of 1,1-bisborylalkanes, **4**, 631
Borane, dichloro(diisopropylamino)-, reactions, with sodium–potassium alloy, **6**, 184
Borane, dichloropent-4-enoxy-, synthesis of, **2**, 110
Borane, (dichloro)vinyl-, reactions, with tetrachlorodiborane, **4**, 632
Borane, dichloro(2-vinyloxy)ethoxy-, synthesis of, **2**, 110
Borane, dicyclohexyl-
reactions
with 1-bromoalkynes, **1**, 1100
with 1-hexyne, **4**, 630
with 1-iodoalkynes, **1**, 1100
synthesis of, **2**, 516
use in the synthesis of alkenes, **1**, 509
use in the synthesis of trialkylboranes, **2**, 517
Borane, diethyl-
reactions, with propargyl chloride, **4**, 631
synthesis of, via equilibration, **2**, 523
use in the synthesis of dialkylzincs, **2**, 592
Borane, difluorovinyl-, hydroboration of, **4**, 632
Borane, di-*n*-hexyl imidoyl-, synthesis of, **5**, 795
Borane, diiodo-, synthesis of, **2**, 525
Borane, diisoamyl-, reactions, with alkynes, **1**, 1100
Borane, [1*S*]-di-2-isocaranyl-, synthesis of, **2**, 42
Borane, diisopinocampheyl-

reactions
with *cis*-but-2-ene, **2**, 42
with tetramethylethylenediamine [1,2-bis(dimethylamino)ethane], **2**, 515, 523
synthesis of, **2**, 42, 515
use in the reduction of carbonyls, **2**, 57
use in the synthesis of alkoxyboranes, **2**, 532
Borane, dimesityl-
hydroboration of, **3**, 3
reactions, with alkynes, **1**, 510
synthesis of, **2**, 520
Borane, dimesityl methyl-, deprotonation of, **1**, 163
Borane, dimethoxy-, synthesis of, **2**, 109, 109
Borane, dimethoxy(trimethylsilylmethyl)-, synthesis of, **4**, 642
Borane, dimethyl-, use in the synthesis of primary amines, **2**, 300
Borane, dimethylaminobis(trifluoromethyl)-, ene reaction of, **6**, 243
Borane, dimethylisocyanato-, synthesis of, **5**, 1015
Borane, di-*exo*-norbornyl-, use in displacement reactions, **1**, 655
Borane, diphenylphosphinoethynyldimesityl-, synthesis of, **2**, 1083
Borane, disiamyl-
hydroboration of, **2**, 40
for the reduction of amides, **3**, 85
synthesis of, **2**, 516
use in the reduction of carbonyls, **2**, 57
use in the synthesis of alkenes, **1**, 509
use in the synthesis of trialkylboranes, **2**, 517
Borane, dithexyl-, as reagent, for the synthesis of α-haloalkenylborons, **4**, 815
Borane, eapine-, synthesis of, **2**, 57
Borane, (ferrocenylethynyl)dimesityl-, synthesis of, **2**, 1083
Borane, fluorenylidene-, reactions, with *t*-butyl isocyanide, **3**, 595
Borane, 9-fluorenylidene(tetramethylpiperidino)-, use in the synthesis of 1,2-oxaboretane-4-thione, **5**, 534
Borane, (fluoromethyl)difluoro-, synthesis of, **4**, 143
Borane, halo-, reactions, with lithium enolates, **2**, 698
Borane, 1-hexenyldichloro-, synthesis of, **2**, 900
Borane, *n*-hexyldibromo-, reactions, with 1-halo-1-alkynes, **2**, 902
Borane, menthoxydihydroxy-, from, trimenthoxyborane, **2**, 111
Borane, mesityl-, use in the synthesis of alcohols, **2**, 41
Borane, methoxy dialkyl-, reactions, with alkynyllithiums, **2**, 1084
Borane, methyl-, reactions, with alkenes, **2**, 516
Borane, monobromo-, synthesis of, **2**, 525
Borane, monochloro-
reactivity of, **2**, 41
synthesis of, **2**, 525
Borane, monoisopinocampheyl-
for the hydroboration of alkenes, **2**, 516
reactions
with acetaldehyde, **2**, 531
with 1-heterocycloalkenes, **2**, 42
synthesis of, **2**, 515, 523
use in asymmetric hydroborations, **2**, 42
Borane, oxybis(diacetoxy)-, reactions, with *t*-butyl hydroperoxide, **2**, 99
Borane, *N*-silylamido-, synthesis of, **5**, 383
Borane, *N*-silyliminoxy-, synthesis of, **5**, 383
Borane, thexyl-
hydroboration of, **2**, 41
reactions
with alkenes, **2**, 516, 518
with 1-bromoalkynes, **1**, 1100
with 1-chloroalkynes, **1**, 1100
with dienes, **2**, 518
with 1-haloalkynes, **1**, 531
with α-pinene, **2**, 515
with vinyl allyl ethers, **2**, 41
use in the reduction of carbonyls, **2**, 57
use in the synthesis of aryl aldehydes, **3**, 82
use in the synthesis of α-haloalkenylborons, **4**, 815
Borane, thexylchloro-
as hydroborating agent, **2**, 526
reactions, with alkenes, **2**, 522
synthesis of, **2**, 526
Borane, thexyldialkyl-, use in the synthesis of ketones, **3**, 134
Borane, tribromo-, reactions, with 1-alkynes, **2**, 902
Borane, tri-*n*-butoxy-, synthesis of, **2**, 111
Borane, tributyl-, reactions, with phenyl isocyanide, **4**, 522
Borane, tricyclohexylphosphinecarboxy-, synthesis of, **6**, 492
Borane, triethoxy-, reactions, with diethylzinc, **2**, 520
Borane, triethyl-
as catalyst, for the synthesis of α-haloalkenylsilicons, **4**, 811
as radical initiator, **1**, 330, 333
reactions
with acetic acid, **5**, 227
with (*E*)-α-iodoalkenylsilanes, **4**, 811
with octahydroxycyclobutane, **1**, 60
with tetraalkynyl stannanes, **2**, 1088
use in the synthesis of vinyl sulfides, **2**, 710
Borane, triethylphosphitecarboxy-, from, triethylphosphitecyanoborane, **6**, 492
Borane, triethylphosphitecyano-, as starting material, for triethylphosphitecarboxyborane, **6**, 492
Borane, triethylphosphite-*N*-ethylcarbamoyl-, synthesis of, **6**, 515
Borane, tri-*n*-hexyl-, cleavage of, **1**, 58
Borane, triisobutyl-, reactions, with 1-methoxyvinyllithium, **4**, 935
Borane, triisopropyl-, reactions, with butanethiol, **2**, 268
Borane, trimenthoxy-, as starting material, for menthoxydihydroxyborane, **2**, 111
Borane, trimethoxy-
reactions
with propargylmagnesium bromide, **2**, 904
with sodium hydride, **2**, 109
Borane, trimethyl-
pyrolysis of, **6**, 183
reactions, with chlorine, **4**, 142
Borane, trimethylaminecarboxy-
reactions
with methanol, **6**, 491
with methyl chloroformate, **6**, 491
with phenylaniline methyl ester, **6**, 514
synthesis of, **6**, 491
Borane, trimethylamine imidazolecarbonyl-, reactions, with sodium ethoxide, **6**, 492
Borane, 1,2,2-trimethylpropyl-, synthesis of, **2**, 515
Borane, trioctyl-, use in the synthesis of dialkylboranes, **2**, 523
Borane, triphenyl-
reactions
with alk-3-en-1-ynylsodium, **2**, 1086
with silyloxylalkyl isonitriles, **5**, 1147
Borane, triphenylphosphinecarboxy-, synthesis of, **6**, 492
Borane, tripropyl-, reactions, with acetamidine, **5**, 778
Borane, tris(*s*-butylamino)-, as starting material, for tris(dialkylamino)boranes, **2**, 419
Borane, tris(cyclopentyl)-, reactions, with ethyl bromoacetate, **4**, 832
Borane, tris(diethylamino)-, reactions, with *t*-butanol, **2**, 419
Borane, tris(difluorosulfoximidoyl)-, use in the synthesis of tetrakis(difluorosulfoximidoyl)methanes, **6**, 362
Borane, tris(dimethylamino)-, reactions, with amines, **2**, 533
Borane, tris(ethylthio)-, reactions, with carboxylic acids, **5**, 234
Borane, tris(phenylseleno)-, reactions, with *O*,*O*-acetals, **4**, 238

Borane, tris(trifluoromethyl)-
acylation of, **5**, 382
synthesis of, **6**, 243
Borane adducts, triethylphosphitecyano-, synthesis of, **6**, 514
Boraneamidine, reactions, with phenylisonitrile, **6**, 662
Borane–amine complexes, reactions, with butyllithium, **2**, 60
Boranecarboximidamides, *N'*-alkyl-, synthesis of, **6**, 662
Borane complexes, reactivity of, **2**, 514
Boranediylborinanes, as starting materials, for borinanes, **6**, 188
Boranediylboriranes, reactions, with stannylenes, **6**, 202
2-Boranediyl-1,3-diboretanes, reactions of, **6**, 184
Borane elimination
for contrathermodynamic alkene isomerization, **1**, 655
regioselectivity of, **1**, 657
for the synthesis of alkenes, **1**, 657
trapping in, **1**, 654
Borane hydrobromides, alkylthioimino-, synthesis of, **5**, 863
Borane method, for intermolecular radical addition, **1**, 333
Borane protecting groups, removal from phosphorus, **1**, 158
Borane quinuclidine, benzylcyano-, reactions, with triethyloxonium tetrafluoroborate, **5**, 1147
Boranes
addition of aminotin and alkoxytins to, **4**, 701
handling problems, **4**, 523
reactions
with alcohols, **2**, 528
with aldehydes, **2**, 57
with 1,2-amino alcohols, **2**, 64
with *t*-butyllithium, **4**, 524
with carboxylic acids, **2**, 57
with isocyanides, **6**, 637
with isonitriles, **5**, 1147
with organometallics, **2**, 529
with oxamidines, **5**, 779
with α-phenoxyalkynes, **1**, 198
with terminal enynes, **2**, 900
for the reduction of carbonyls, **2**, 56
regeneration of, **2**, 531
stability of, **2**, 1086
stepwise dialkylation of, **1**, 163
synthesis of, **5**, 432
use in allylation reactions, **2**, 74
use in the synthesis of alcohols, **2**, 40
use in the synthesis of aryl aldehydes, **3**, 82
Boranes, acyl-
as intermediates, in the carbonylation of organoboranes, **5**, 432
synthesis of, **5**, 432
Boranes, acyloxy-
synthesis of, **5**, 227
use in the synthesis of ethers, **4**, 208
Boranes, alkenyl-
coupling of, **1**, 489
elimination of, **1**, 1080
β-elimination of, **1**, 1033
from
alkenyl metals, **2**, 904
1-alkynes, **2**, 900
alkynylborates, **2**, 904
boron-stabilized carbanions, **2**, 905
hydrolysis of, **1**, 510
mercuration of, **1**, 1095
oxidation of, **1**, 1103
protonolysis of, **1**, 509
reactions
with acetic acid, **1**, 1102
with aldehydes, **1**, 1103
with alkynyl bromides, **1**, 1020
with benzyl bromides, **1**, 1104
with halogens, **1**, 1103
with iodine, **1**, 531
with triethylaluminum, **1**, 1030
with α,β-unsaturated ketones, **1**, 1103
with vinyl halides, **1**, 1097
replacement of, **1**, 532
as starting materials, for alkenylcopper(I) compounds, **1**, 1018
synthesis of, **1**, 531, 654
via boron-Wittig type reaction, **2**, 905
uses of, **1**, 1102
use in the synthesis of ketones, **3**, 135
use in the synthesis of unsaturated ketones, **3**, 140
Boranes, (*E*)-alkenyl-
from, 1-bromo 1-alkenylboranes, **2**, 905
synthesis of, via hydroboration of alkynes, **1**, 1033
Boranes, alkenylacetoxy-, synthesis of, **5**, 229
Boranes, alkenylcatechol-, synthesis of, **1**, 1101
Boranes, alkenyldialkyl-
cleavage of, **1**, 449
deboronobromination of, **1**, 658
as intermediates, **3**, 645
reactions
with bromine, **1**, 658
with iodine, **1**, 658
with palladium acetate, **1**, 660
with potassium cyanide, **2**, 748
Boranes, (*E*)-alkenyldialkyl-, as starting materials, for (*E*)-alkenes, **1**, 659
Boranes, alkenyldibromo-
hydrolysis of, **1**, 487
reactions, with methylcopper, **1**, 488
synthesis of, **1**, 1102
Boranes, (*E*)-1-alkenyldibromo-, synthesis of, **2**, 900
Boranes, alkenyldimesityl-, reactions, with organometallics, **6**, 201
Boranes, 3-alkenylpropargylic, reactions, with aldehydes, **1**, 977
Boranes, alkoxy-
from, borane, **2**, 109
hydrolysis of, **2**, 56
reactions
with lithium aluminum hydride, **2**, 535
with organometallics, **2**, 520, 529
with potassium hydride, **2**, 534
synthesis of, **2**, 109
use in the synthesis of ketones, **3**, 135
Boranes, α-alkoxy-, synthesis of, **4**, 362
Boranes, β-alkoxyalkenyl-, as starting materials, for allenes, **1**, 655
Boranes, 2-alkoxyalkyl-, synthesis of, **1**, 82
Boranes, (*Z*)-3-alkoxyallyl-, from, pinene, **2**, 75
Boranes, alkoxychloro-, disproportionation of, **2**, 111
Boranes, alkoxydichloro-, stability of, **2**, 110
Boranes, alkoxydihalo-, synthesis of, **2**, 110
Boranes, alkoxydihydroxy-, synthesis of, **2**, 111
Boranes, alkoxyhalo-
from, boron trihalides, **2**, 110
as starting materials, for borates, **2**, 111
synthesis of, **2**, 110
Boranes, alkoxyhydroxy-, synthesis of, **2**, 111
Boranes, alkyl-
coupling of, **1**, 489
deprotonation of, **4**, 664
elimination of, **1**, 657
from, alkenes, **2**, 57
oxidation of, **2**, 531
reactions
with amides, **5**, 381
with hydride donors, **2**, 534
with thiols, **2**, 533
reactivity of, **1**, 75
as starting materials, for thioboranes, **2**, 268
synthesis of, **2**, 514
use in the reduction of carbonyls, **2**, 57
use in the substitution of boron, **1**, 489

use in the synthesis of alcohols, **2**, 40
Boranes, alkylaminecarboxy-, esterification of, **6**, 491
Boranes, alkylamino-, synthesis of, **2**, 533
Boranes, alkyl arylketimino-
from
dialkylthioboric acid esters, **2**, 811
ketimines, **2**, 811
Boranes, alkylbis(alkylperoxy)-, as intermediates, **2**, 97
Boranes, alkylchloromethoxy-, as intermediates, **2**, 530
Boranes, alkyldialkoxy-, reactions, with dichloromethyllithium, **2**, 529
Boranes, alkyldichloro-
pyrolysis of, **6**, 183
as starting materials, for alkyl hydroperoxides, **2**, 97
synthesis of, **2**, 526
Boranes, alkyldihalo-, use in the synthesis of alkenes, **1**, 487
Boranes, alkyldimesityl-, α-metallation of, **4**, 664
Boranes, alkylhalo-
synthesis of, **2**, 525
via hydroboration of alkenes, **2**, 525
Boranes, alkyl(α-haloalkyl)-, synthesis of, **4**, 142
Boranes, alkyloxy-
synthesis of, **2**, 527
via hydroboration of alkenes, **2**, 527
Boranes, alkyl(phenylalkynyl)alkoxy-, stability of, **2**, 1086
Boranes, alkylphosphinecarboxy-, synthesis of, **6**, 492
Boranes, (alkylseleno)alkyliodo-, synthesis of, **2**, 294
Boranes, alkylseleno(diiodo)-, synthesis of, **2**, 294
Boranes, alkynyl-
conjugate addition to enones, **3**, 147
detection of, **2**, 1085
isomerization of, **1**, 1085
stability of, **2**, 1083
Boranes, 1-alkynyl-, reduction of, **1**, 999
Boranes, alkynylalkoxy-, synthesis of, **2**, 1085
Boranes, alkynylamino-, synthesis of, **2**, 1085
Boranes, alkynyl(aminocarboxy)-, synthesis of, **2**, 1086
Boranes, alkynylaminochloro-, synthesis of, **2**, 1085
Boranes, alkynyl(bisamino)-, direct carboxylation of, **2**, 1086
Boranes, alkynyldialkyl-, synthesis of, **2**, 1084
Boranes, alkynyldiisopropoxy-, reactions, with enones, **1**, 1050
Boranes, 1-alkynyldiisopropoxy-, synthesis of, **2**, 1086
Boranes, alkynylfluoroisopropoxy-, synthesis of, **1**, 1050
Boranes, allenyl-
cleavage of, **1**, 449
hydrolysis of, **1**, 976
Boranes, allydialkoxy-, synthesis of, **2**, 527
Boranes, allyl-
reactions
with aldehydes, **1**, 281; **2**, 75
with carbonyls, **2**, 74
with ethoxyethyne, **1**, 1030
with ketones, **1**, 281
synthesis of, **2**, 529
Boranes, allyldialkoxy-, synthesis of, **2**, 529
Boranes, allyldialkyl-, synthesis of, **2**, 521
Boranes, amido-, synthesis of, **5**, 380, 660
Boranes, β-amido-, elimination of, **1**, 657
Boranes, amino-
acylation of, **5**, 382
bromine addition to, **6**, 292
reactions
with amines, **2**, 358
with dialkylboroacetamides, **5**, 382
synthesis of, **2**, 810
use in the reduction of carbonyls, **2**, 58
Boranes, α-amino-, synthesis of, **4**, 522
Boranes, aminochloro-, reactions, with 1-alkynes, **2**, 1085
Boranes, aminohalo-
reactions
with dichloromethane, **6**, 243
with metalloaldimines, **5**, 795
as starting materials, for *gem*-aminoborylamino silylalkenes, **4**, 1015
Boranes, aminoimino-
reactions
with alkenes, **5**, 381
with ketones, **5**, 381
as starting materials, for *N*-boroamides, **5**, 381
Boranes, amino(methylene)-
from, 9-fluorenyltetramethylpiperidinoboron halides, **3**, 504
synthesis of, **6**, 712
Boranes, apopinocampheyl-*t*-butylchloro-, use in the reduction of carbonyls, **2**, 58
Boranes, aryl-
coupling of, **1**, 489
cyclization of, **2**, 523
reactions, with α-bromoketones, **1**, 464
reduction of, **1**, 447
as starting materials, for arenes, **1**, 447
use in the synthesis of biaryls, **1**, 473
Boranes, aryl(α-bromoalkyl)(*n*-butoxy)-, synthesis of, **4**, 144
Boranes, aryl(α-haloalkyl)-, synthesis of, **4**, 142
Boranes, asymmetric chloro-, for the synthesis of 2-silyl alcohols, **2**, 58
Boranes, bis(acyloxy)-, synthesis of, **5**, 228
Boranes, bis(dialkylamino)-
from, amines, **2**, 418
synthesis of, **2**, 418
Boranes, boriranylidene-, from, 1,1-diborylalkenes, **6**, 713
Boranes, *B*-boryl(1,3-diboretane-2-ylidene)-, synthesis of, **6**, 714
Boranes, bromo-, use in the synthesis of alkyl bromides, **2**, 32, 33
Boranes, α-bromo-, as starting materials, for hydroxy dialkyls, **2**, 524
Boranes, 1-bromo 1-alkenyl-, as starting materials, for (*E*)-alkenylboranes, **2**, 905
Boranes, (*Z*)-bromoalkenyl-, reactions, with alkynylzincs, **1**, 1008
Boranes, 1,1((*Z*)-1-bromoalkenyl)dialkyl-, as starting materials, for (*E*)-alkenes, **1**, 487
Boranes, 2-bromo-1-alkenyldibromo-, as starting materials, for 2-silyl-1,3-butadienes, **2**, 939
Boranes, α-bromoalkyl-
decomposition of, **6**, 61
as intermediates, **4**, 142
Boranes, (α-bromoalkyl)imino-, synthesis of, **4**, 145
Boranes, caranyl-, reactivity of, **2**, 42
Boranes, caranyl allyl-, as reagents, **2**, 75
Boranes, carbamoyl-, synthesis of, **6**, 514
Boranes, carbimino-, synthesis of, **5**, 660
Boranes, carboxy-, synthesis of, **6**, 491
Boranes, chiral dialkyl-, use in the synthesis of alkylboranes, **2**, 519
Boranes, chloro-
reactions
with alkyllithiums, **2**, 526
with bisstannyldiazomethane, **3**, 682
with organolithiums, **2**, 520
reduction of, **2**, 522
use in the reduction of carbonyls, **2**, 58
use in the synthesis of α-diazo borons, **5**, 872
Boranes, β-chloroalkenyl-, as starting materials, for allenes, **1**, 655
Boranes, α-chloroalkyl-
dehalogenative reduction of, **2**, 529
displacement of, **2**, 530
Boranes, (α-chloroalkyl)-, oxidation of, **4**, 142
Boranes, β-chloroalkyl-, elimination of, **1**, 655
Boranes, (α-chloroalkyl)dioxy-, synthesis of, **4**, 144
Boranes, α-chloroallyldialkoxy-, synthesis of, **2**, 530
Boranes, *cis*-2-chlorocyclohexyl-, stability of, **1**, 655
Boranes, 3-chloromethoxyethoxy chloro-, synthesis of, **2**, 110

Boranes, (chloromethyl)dimethyl-, synthesis of, **4**, 142
Boranes, chloro-*o*-nitrophenoxy-, reactions, with *t*-butyl hydroperoxide, **2**, 99
Boranes, cuprous methyltrialkyl-, reactions, with 1-benzoyl-2-vinylcyclopropane, **1**, 275
Boranes, cyano-
 ethylation of, with triethyloxonium tetrafluoroborate, **5**, 795
 as starting materials, for imidoylboranes, **5**, 795
Boranes, cyclic acyloxy-, synthesis of, **5**, 229
Boranes, dialkenylchloro-
 alcoholysis of, **1**, 1102
 as intermediates, **1**, 531
 reactions
  with dienes, **1**, 1094
  with methylcopper, **1**, 488
 synthesis of, **1**, 1102
Boranes, dialkoxy-
 from
  dialkoxyhaloboranes, **2**, 109
  sodium borohydride, **2**, 109
  sodium trimethoxyborohydride, **2**, 109
  trialkylboroxines, **2**, 110
 synthesis of, **2**, 109
 use in the synthesis of amines, **2**, 330
Boranes, dialkoxyhalo-, as starting materials, for dialkoxyboranes, **2**, 109
Boranes, dialkoxyhydroxy-, synthesis of, **2**, 111
Boranes, dialky(1-alkynyl)-, as starting materials, for 1,1-bis(dialkylboryl)-1-alkenes, **6**, 183
Boranes, dialkyl-
 reactions
  with alkylnitriles, **3**, 457
  with alkynes, **1**, 278; **2**, 748
  with arylnitriles, **3**, 457
  with α-diazo esters, **2**, 533
  with enones, **2**, 533
  with hydrochloric acid, **2**, 526
 synthesis of
  via hydroboration, **2**, 515
  via redistribution of organoboranes, **2**, 522
Boranes, dialkylalkenyl-
 protonolysis of, **1**, 509
 reactions, with lithium alkynide, **1**, 488
 as starting materials, for alkenes, **1**, 487
Boranes, dialkylallyl-, stability of, **2**, 517
Boranes, dialkylamino-
 from
  alanes, **2**, 418
  amines, **2**, 418
  stannanes, **2**, 418
 reactions, with organometallics, **2**, 520
 synthesis of, **2**, 418
Boranes, (dialkylamino)dichloro-, as starting materials, for tris(trifluoromethyl)borane, **6**, 243
Boranes, dialkylchloro-
 reactions
  with acyl carbanions, **2**, 267
  with triphenylphosphonium ylides, **5**, 916
 reduction of, **2**, 526
 as starting materials
  for α-boryl ylides, **5**, 916
  for trialkylboranes, **2**, 526
Boranes, dialkylhalo-, reactions, with metalloaldimines, **5**, 795
Boranes, (dialkylthio)alkyl-, synthesis of, **2**, 533
Boranes, dialkynyl-, synthesis of, **2**, 1083
Boranes, diazoalkyl-, synthesis of, **3**, 485
Boranes, (diborylmethylene)-, synthesis of, **6**, 714, 716
Boranes, α,β-dibromo-, synthesis of, **4**, 145
Boranes, dibutyl(trifluoromethyl)-, synthesis of, **6**, 243
Boranes, dichloroalkyl-
 reactions, with phosphorus ylides, **5**, 916
 as starting materials, for α-boryl ylides, **5**, 916
Boranes, α,α-dichloroalkyl-, synthesis of, **6**, 61
Boranes, dichloro-2-chloroalkoxy-, synthesis of, **2**, 110
Boranes, dichloro ω-chloroalkylalkoxy-, synthesis of, **2**, 110
Boranes, dienyl-
 from, terminal enynes, **2**, 900
 use in the synthesis of 2-(2-but-1,4-dienyl)alkanols, **2**, 75
Boranes, difluoro(trifluoromethyl)-, synthesis of, **6**, 243
Boranes, dihalo-, regioselectivity of hydroborations with, **2**, 525
Boranes, α,α-dihaloalkyl-, reactions, with nucleophiles, **6**, 61
Boranes, diisopinocampheyl-, synthesis of, **2**, 521
Boranes, dimesitylfluoro-, reactions, with 1-lithio 1-trimethylsilylalkenes, **4**, 1057
Boranes, dimethylalkyl-
 reactions
  with chloramine, **2**, 306
  with primary alkylamines, **2**, 300
Boranes, (dimethylamino)organo halo-, reactions, with silver cyanide, **2**, 811
Boranes, dimethylarsino-, synthesis of, **2**, 509
Boranes, diorganohalo-, reactions, with *N,O*-bis(trimethylsilyl)amide, **5**, 383
Boranes, diphenoxyallyl-, for the allylation of aldehydes, **2**, 76
Boranes, *B*-[(*E*)-3-(diphenylamino)prop-2-enyl]diisopinocampheyl-, as allylating reagents, **2**, 75
Boranes, disubstituted, reactions, with 1-chloroalkynes, **4**, 756
Boranes, [2-(1,3-dithianyl)]myrtanyl-, synthesis of, **2**, 42
Boranes, enaminodialkyl-, as starting materials, for 1,2-azaborolidines, **2**, 811
Boranes, fluoro-, synthesis of, **6**, 388
Boranes, halo-
 as hydroborating agents, **2**, 525
 for the hydroboration of alkynes, **1**, 510
 reactions
  with amides, **5**, 381
  with *N*-lithioimines, **3**, 458
  with lithium phosphides, **2**, 474
  with organometallics, **2**, 520, 526
  with *N*-trialkylsilylimines, **3**, 458
 use in hydroboration, **2**, 525
 use in the synthesis of alcohols, **2**, 41
 use in the synthesis of carbodiimides, **5**, 1085
Boranes, 2-halo-, stability of, **1**, 84
Boranes, α-halo-, reactions, with alkoxides, **4**, 362
Boranes, 2-halo-1-alkenyl-, from, 1-alkynes, **2**, 902
Boranes, (1-halo-1-alkenyl)dialkyl-, reactions, with lead(IV) acctatc, **1**, 660
Boranes, β-haloalkyl-, elimination of, **1**, 655
Boranes, (α-haloalkyl)bis(alkoxy)-
 synthesis of, **4**, 144
 use in asymmetric synthesis, **4**, 144
Boranes, (α-haloalkyl)oxy-, synthesis of, **4**, 144
Boranes, haloallyl-, synthesis of, via redistribution, **2**, 526
Boranes, 1-halodienyl-, rearrangement of, **3**, 223
Boranes, halo(α-haloalkyl)-, synthesis of, **4**, 143
Boranes, halo(methylene)-, synthesis of, **6**, 713
Boranes, *trans*-α-halovinyl-, synthesis of, **1**, 1100
Boranes, *trans*-2-heteroaryl cycloalkenyl isopinocampheyl-, oxidation of, **2**, 42
Boranes, homochiral allydialkoxy-, from, dialkyl tartrates, **2**, 529
Boranes, hydro-, reactions, with tertiary amides, **5**, 381
Boranes, imidoyl-
 from, cyanoboranes, **5**, 795
 synthesis of, **5**, 795
Boranes, imino-
 reactions
  with alkylthiols, **5**, 863
  with diphenyldiazomethane, **4**, 145
 synthesis of, **4**, 522

Boranes, (2-isocarenyl)-, synthesis of, **2**, 521
Boranes, isocyanato-, synthesis of, **5**, 1015
Boranes, isopinocampheyl-, reactions, with haloalkynes, **4**, 815
Boranes, 2-isopropylapopinocampheyl-, as reagent for asymmetric hydroboration, **2**, 42
Boranes, isothiocyanato-substituted imino-, synthesis of, **5**, 864
Boranes, α-lithio-, synthesis of, **4**, 664
Boranes, α-metallo-, synthesis of, **2**, 1005
Boranes, methylene-
  dimerization of, **6**, 712
  reactions, with alkyllithiums, **6**, 201
  reactivity of, **6**, 712
  rearrangement of, **6**, 187
  as starting materials
    for azasilaboratedines, **6**, 187
    for polyborylmethides, **6**, 201
  synthesis of, **3**, 501; **6**, 712
  use in the synthesis of borylmethide salts, **6**, 201
Boranes, methylene-, *C*,*C*-dihetero-substituted, synthesis of, **6**, 712
Boranes, methylene-, *C*,*C*-disilyl-substituted
  synthesis of, **6**, 712
  use in the synthesis of metalloid functions, **6**, 187
Boranes, monoacyloxy-, synthesis of, **5**, 228
Boranes, monoalkoxy-, synthesis of, **2**, 109
Boranes, monoalkoxyfluoro-, synthesis of, **2**, 110
Boranes, monoalkyl-
  as hydroboration reagent, **2**, 42
  reactions
    with alkenes, **2**, 516
    with nitriles, **3**, 457
  synthesis of, **1**, 75; **2**, 40
    via hydroboration, **2**, 515
Boranes, monoalkynyldialkoxy-, synthesis of, **2**, 1085
Boranes, monochlorodialkoxy-, synthesis of, **2**, 110
Boranes, monohalo-, selectivity of, **2**, 525
Boranes, α-nitro-, synthesis of, **4**, 524
Boranes, organoperoxy-, synthesis of, **2**, 99
Boranes, organyl-1-alkynylalkoxy-, synthesis of, **2**, 1085
Boranes, organylbis(diisopropylamino)-, reactions, with hydrogen chloride, **2**, 321
Boranes, organyldichloro-, reactions, with organic azides, **2**, 317
Boranes, pentaalkyl-1,5-dicarba-*closo*-penta-, synthesis of, **6**, 183
Boranes, peroxy-, from, alkylborates, **2**, 99
Boranes, phosphine-
  from, phosphine oxides, **1**, 158
  synthesis of, **1**, 157
Boranes, 1-phosphinoalkyl-, synthesis of, **4**, 569
Boranes, phosphinodiaryl-, synthesis of, **2**, 867
Boranes, propargyl-, reactions, with aldehydes, **1**, 977
Boranes, α-pyridylaminodialkyl-, reactions, with ethoxyalkyne, **4**, 884
Boranes, 3-silylpropenyl-, use in the synthesis of *anti*-alk-1-ene-3,4-diols, **2**, 75
Boranes, 2-stannyl-1-alkenyl-, from, 1-alkynylborates, **2**, 904
Boranes, β-substituted alkyl-, elimination of, **1**, 655
Boranes, 1-substituted allenyl-
  reactions
    with aldehydes, **1**, 1056
    with ketones, **1**, 1056
Boranes, thexylalkylalkenyl-, synthesis of, **1**, 1100
Boranes, thexyldialkenyl-, synthesis of, **1**, 1100
Boranes, thexyl-1-halo-1-alkenyl-, synthesis of, **1**, 1100
Boranes, (*Z*)-2-thio-1-alkenyl-, synthesis of, via palladium-catalyzed thioboration of 1-alkynes, **2**, 902
Boranes, thioalkyl-, as electrophilic compounds, **2**, 520
Boranes, α-thioalkyl-
  from
    diphenyl dithioacetals, **4**, 372
    methyl phenyl thioethers, **4**, 372
  synthesis of, **4**, 372
Boranes, trialkenyl-, cleavage of, **1**, 449
Boranes, trialkoxy-
  reactions
    with allyl metals, **2**, 529
    with aryl Grignard reagents, **2**, 906
    with boron trifluoride, **2**, 110
    with diborane, **2**, 109
    with 1-silylalkynes, **4**, 1057
    with sodium hydride, **2**, 109
  reduction of, **2**, 109
Boranes, trialkyl-
  bromination of, **2**, 524
  carbonylation of, with carbon monoxide, **3**, 19
  as catalysts, in Pinner reactions, **5**, 778
  electrolysis of, **3**, 622
  from, dialkylchloroboranes, **2**, 526
  oxidation of, **2**, 112
  reactions
    with acrolein, **1**, 273
    with 1-acyl-2-vinylcyclopropanes, **1**, 275
    with alkenyl Grignard reagents, **1**, 1100
    with 1-alkenyl halides, **1**, 279
    with alkynes, **1**, 542
    with α-alkynic ketones, **3**, 223
    with amidines, **5**, 778
    with amino acids, **5**, 228
    with anions, **1**, 125
    with α-bromo ester enolates, **4**, 832
    with 1,3-butadiene monoepoxide, **1**, 277
    with carbanions, **1**, 278
    with carbon monoxide, **4**, 363
    with crotonaldehydes, **1**, 274
    with diazoacetonitrile, **3**, 622
    with dichloroacetonitrile, **3**, 622, 627
    with 2,2-difluoro-1-lithio-1-vinyltosylate, **2**, 906
    with 1,2-dimethoxy-1-lithioethene, **4**, 935
    with diphenyl disulfide, **2**, 712
    with diphenylphosphinous chloride, **2**, 436
    with dithioacetals, **4**, 372
    with 3,4-epoxybut-1-ynes, **1**, 277, 977
    with ferric selenocyanate, **5**, 1114
    with isocyanides, **4**, 522
    with 5-lithio-2,3-dihydrofurans, **4**, 935
    with mercury acetate, **2**, 592
    with mercury alkoxides, **2**, 592
    with 1-methoxyvinyllithium, **4**, 934
    with nitrogen ylides, **4**, 522
    with oxygen, **2**, 97
    with quinones, **1**, 275
    with radicals, **1**, 546
    with 1-stannyl-1-alkynes, **2**, 903
    with tertiary amine oxides, **2**, 112
    with α-thiocarbanions, **4**, 372
    with [*N*-(*p*-toluenesulfonyl)imino]phenyliodinane, **2**, 311
    with 2,2,2-trifluoroethyl tosylate, **4**, 735
    with trimethylsilyl azide, **2**, 307
    with trimethylsilylmethyllithium, **4**, 642
    with tris(phenylthio)methane anions, **4**, 373
    with vinyllithiums, **1**, 278
    with vinyl sulfones, **1**, 485
  rearrangement of, **2**, 40
  stability of, **1**, 58
  as starting materials
    for alkyl chlorides, **2**, 23
    for alkylnitriles, **3**, 622
    for alkyl radicals, **1**, 333
    for α-borio enamines, **4**, 1015
    for cuprous methyltrialkylborates, **1**, 276
  synthesis of, **1**, 75, 164

via hydroboration, **2**, 517
use in the substitution of boron functions, **1**, 272
use in the synthesis of *N*-borylimines, **3**, 458
use in the synthesis of ketones, **3**, 135
Boranes, tri(alkylperoxy)-, synthesis of, **2**, 99
Boranes, trialkynyl-
reactions, with ethyl diazoacetate, **1**, 1033
synthesis of, **2**, 1083
Boranes, triaryl-, reactions, with nitrogen ylides, **4**, 522
Boranes, triarylsilyl-, reactions, with 1,2-bis(trimethylsilyl)ethyne, **4**, 1058
Boranes, trifluoromethyl-, synthesis of, **6**, 243
Boranes, trihalo-
reactions
with 1-silyl 1-alkyn-2-ol alkyl ethers, **4**, 1057
with *N*-silylamides, **5**, 381
Boranes, trihydridic, reactions, with carbonyls, **2**, 111
Boranes, trimethylamine-, as starting materials, for thioboranes, **2**, 269
Boranes, (trimethylsilyl)ethyl-, synthesis of, **4**, 356
Boranes, tris(alkylseleno)-
from
alkaneselenolates, **2**, 294
alkaneselenols, **2**, 294
Boranes, tris(alkylthio)-, reactions, with amides, **6**, 113
Boranes, tris(dialkylamino)-
from
amines, **2**, 419
stannanes, **2**, 419
tris(dialkylamino)boranes, **2**, 419
synthesis of, **2**, 419
Boranes, tris(trifluoromethyl)-, synthesis of, **6**, 243
Boranes, trisubstituted alkyl-, reactions, with lithiodichloromethoxymethane, **2**, 530
Boranes, trivinyl-, oxidation of, **5**, 417
Boranes, vinyl-
arylation of, **1**, 472
boroboration of, **4**, 632
from, alkynes, **1**, 1100
hydroboration of, **3**, 3; **4**, 629
as intermediates, in the synthesis of 1,4-dicarbonyls, **1**, 126
reactions
with alkyllithiums, **6**, 201
with bromine, **4**, 145
with electrophiles, **4**, 144
with vinyl halides, **1**, 279
reduction of, **1**, 449
as starting materials
for alkenes, **1**, 449
for allyl anions, **1**, 162
for 1,1-bisborylalkanes, **4**, 631
for cycloalk-3-enylboranes, **2**, 43
for β,γ-unsaturated nitriles, **3**, 623
for vinyl bromides, **2**, 623
synthesis of, **2**, 527; **4**, 935
for the synthesis of vinyl iodides, **2**, 628
use in the synthesis of α,β-unsaturated ketones, **3**, 228
Boranes, (*E*)-vinyl-
as starting materials, for aldehydes, **1**, 163
synthesis of, **1**, 278
Boranes, vinylchloro-, reactions, with α-diazoacetates, **1**, 486
Boranes, vinyldialkoxy-, reactions, with organolithiums, **1**, 487
Boranes, vinylic dialkoxy-, reactions, with dichloromethane, **2**, 530
Boranes, vinyloxy-
from, carbonyls, **2**, 532
See also Boron enol ethers
Boranyl, diethyl-, reactions, with dinitromethane salts, **6**, 169
Boraphosphines, synthesis of, **2**, 474
Borataalkyne, *C*-boryl-, reactions, with *t*-butyllithium, **6**, 201
Borataalkynes
reactions, with 1,1-diaryl-2,2-difluorodiborane, **6**, 714
synthesis of, **6**, 714
Borataallenes, synthesis of, **6**, 712
2-Borataallenes
reactions, with electrophiles, **6**, 716
synthesis of, **6**, 715
Boratabenzene complexes, use in the synthesis of multidecker borole complexes of ruthenium, **4**, 695
Borate, arenediazonium tetrafluoro-, photolysis of, for aryl fluorides, **2**, 614
Borate, 2-azido-3-ethylbenzthiazolium tetrafluoro-, diazo group transfer from, **3**, 477
Borate, benzenediazonium tetrafluoro-
reactions
with acylate complexes, **5**, 936
with potassium triphenylcyanoborate, **5**, 1148
Borate, 1,3-benzodithiolium tetrafluoro-, use in the synthesis of aldehydes, **3**, 15
Borate, bis(pyridine)iodine(I) tetrafluoro-
reactions
with alkynyl sulfides, **4**, 772
with 1-chloroalkynes, **4**, 760
with 1-iodoalkynes, **4**, 749
Borate, 2-chloro-1-ethylbenzoxazolium tetrafluoro-, reactions, with methyl thiocarbamates, **5**, 979
Borate, 2-chloro-3-ethylbenzoxazolium tetrafluoro-, use in the synthesis of isocyanides, **3**, 696
Borate, cupric tetrafluoro-, reactions, with cyclopropyl silyl ethers, **1**, 929
Borate, diethoxycarbenium tetrafluoro-, reactions, with ketones, **4**, 193
Borate, 1,3-dimethyl-2-ethoxy-4,5-dihydroimidazolium tetrafluoro-, as starting material, for ketene aminals, **4**, 975
Borate, dimethyl(methylthio) sulfonium fluoro- (DMTSF), reactions, with tris(phenylthio)alkanes, **3**, 283
Borate, dimethylphenacylsulfonium tetraphenyl-, for the condensation of triphenylarsane oxide, **5**, 920
Borate, 1,3-dithienium fluoro-, synthesis of, **4**, 1074
Borate, ethyltriphenylarsonium fluoro-, deprotonation of, **1**, 269
Borate, ethynyl(phenyl)iodonium tetrafluoro-
from, bis(trimethylsilyl)ethyne, **2**, 1020
reactions, with enolate anions, **1**, 1145
Borate, *N*-fluoro-*N*-methylisopropylidenammonium tetrafluoro-, synthesis of, **2**, 762
Borate, (hydroxymethyl)triphenylphosphonium tetrafluoro-, reactions, with diethylaminosulfur trifluoride, **4**, 117
Borate, 1-methyl-2,5-diphenyl-6-nitro-1,4-dithiinium tetrafluoro-, ring-opening of, **4**, 907
Borate, 4-nitrobenzenediazonium tetrafluoro-, reactions, with methylsulfonyldibenzoylmethane, **4**, 332
Borate, *p*-nitrobenzenediazonium tetrafluoro-, use in the synthesis of formazans, **5**, 858
Borate, *N*-nitrocollidinium tetrafluoro-, use in the synthesis of nitrate esters, **2**, 104
Borate, nitronium tetrafluoro-
as oxidant, for the synthesis of ketones from methyl ethers, **3**, 121
reactions, with bis(silylalkynes), **2**, 1053
use in the synthesis of aldehydes, **3**, 8
Borate, nitrosium tetrafluoro-, reactions, with allylsilanes, **3**, 160
Borate, nitrosonium fluoro-, as reagent, for the hydrolysis of amides, **5**, 108
Borate, nitrosonium tetrafluoro-
as oxidant, for the synthesis of ketones from silyl ethers, **3**, 121
reactions, with amides, **5**, 36
for the synthesis of aryl fluorides, **2**, 615
Borate, phenylselenodimethylsulfonium tetrafluoro-, synthesis of, **2**, 238

Borate, *N*-sulfinyldimethylammonium tetrafluoro-, reactions, with methyl dithiobenzoate, **5**, 731
Borate, *N,N,N',N'*-tetramethylethoxyformamidinium fluoro-, as starting material, for ketene aminals, **4**, 975
Borate, trialkyl-, reactions, with α-thiocarbanions, **4**, 372
Borate, triethyloxonium tetrafluoro-
  for the ethylation of cyanoboranes, **5**, 795
  reactions
    with amides, **5**, 751
    with benzylcyanoborane quinuclidine, **5**, 1147
    with ethyl *N,N*-dimethylcarbamate, **6**, 313
    with hydroxamic acids, **5**, 701
    with iron alkyls, **5**, 942
    with thioamide *S*-oxides, **5**, 734
  as reagent, for the *N*-ethylation of cyanoboranes, **5**, 795
  use in the synthesis of amidines, **5**, 751
Borate, triisopropyl-, reactions, with 1-methoxyvinyllithium, **4**, 934
Borate, trimethyl-
  from, chlorodimethoxyborane, **2**, 111
  reactions
    with 1,1-dichloroethane, **4**, 813
    with phenylthiomethyllithium, **4**, 372
    with trimethylsilylmethylmagnesium chloride, **4**, 642
  for transmetallation reactions, **2**, 652
Borate, trimethyloxonium tetrafluoro-
  reactions
    with dialkyl cyanamides, **5**, 1120
    with thioacylhydrazines, **5**, 358
    with triphenylstibine, **1**, 271
Borate, trityl fluoro-
  reactions
    with 1,3-dithiane, **4**, 1074
    with silacyclohexanes, **1**, 656
Borate, trityl tetrafluoro-
  as oxidant, for the synthesis of ketones from silyl ethers, **3**, 121
  as reagent
    for the synthesis of ketones from alcohols, **3**, 120
    for the synthesis of phenols, **2**, 659
  use in the synthesis of aldehydes, **3**, 8
Borate nitrilium salts, tetrafluoro-, synthesis of, **5**, 795
Borates
  elimination of, **1**, 659
  from
    alkoxyhaloboranes, **2**, 111
    borohydrides, **2**, 1087
    boron trihalides, **2**, 1086
    carbohydrates, **2**, 111
  reactions, with alkynyl metals, **2**, 1085
  synthesis of, **2**, 111, 1005, 1083
Borates, alkenyliodonium tetrafluoro-, elimination of, **1**, 1068
Borates, alkenyltrialkyl-, cleavage of, **1**, 450
Borates, alkyl-
  ring-contraction of, **2**, 524
  as starting materials, for peroxyboranes, **2**, 99
  synthesis of, **2**, 43
Borates, *N*-alkyl-5-phenylisoxazolium fluoro-, as starting materials, for α-azido enamines, **4**, 990
Borates, alkynyl-
  reactions, with electrophiles, **2**, 904
  reduction of, **1**, 999
  as starting materials
    for alkenylboranes, **2**, 904
    for 1,3-butadiynes, **1**, 1035
    for 1,4-dicarbonyls, **1**, 126
Borates, 1-alkynyl-
  metal-substituted, synthesis of, **2**, 1088
  as starting materials, for 2-stannyl-1-alkenylboranes, **2**, 904
Borates, alkynyl(phenyl)iodonium tetrafluoro-, synthesis of, **2**, 1020
Borates, (aminoalkyl)alkynyl-, quaternization of, **2**, 1088
Borates, arenediazonium tetrafluoro-
  reactions
    with bismuth, **2**, 876
    with dienes, **5**, 1044
    with phosphorus trichloride, **2**, 844
    with sodium carboxylates, **5**, 198
    with vinylgermanes, **1**, 452
  as starting materials
    for aromatic anhydrides, **5**, 197
    for unsymmetrical aromatic anhydrides, **5**, 198
  use in the synthesis of aryl triflates, **2**, 689
Borates, chloro-, reactions, with alkynylsodium, **2**, 1085
Borates, 5-chloroisoxazolium tetrafluoro-, as starting materials, for 2-substituted 3-isoxazoline-5-thiones, **5**, 532
Borates, cuprous methyltrialkyl-, from, trialkylboranes, **1**, 276
Borates, cyclohexadienyl fluoro-, synthesis of, **1**, 914
Borates, dialkoxycarbenium tetrafluoro-
  reactions
    with sodium hydrosulfide, **5**, 512
    with sodium sulfide, **5**, 512
Borates, α-diazoalkyl-, synthesis of, **3**, 485
Borates, diazonitrilium-, synthesis of, **5**, 1148
Borates, episulfonium tetrafluoro-
  from, alkenes, **1**, 1031
  reactions
    with alkynyldiethylalanes, **1**, 1031
    with lithium dialkynyldiethylaluminates, **1**, 1031
Borates, imidoyl-, synthesis of, **5**, 795
Borates, imino-, synthesis of, **3**, 458
Borates, nitrilium-
  synthesis of, **5**, 1147
    via alkylation of cyanoborates, **5**, 1147
Borates, peroxy-, from, alkyl hydroperoxides, **2**, 99
Borates, phosphonium tetrafluoro-, synthesis of, **2**, 1062
Borates, silyl-substituted ethynyl-, synthesis of, **2**, 1088
Borates, tetraaryl-, in cross-coupling reactions, **1**, 489
Borates, trialkyl-, synthesis of, **2**, 40, 111
Borates, trialkyloxonium tetrafluoro-, reactions, with *N*-alkyl lactams, **5**, 576
Borates, vinyliodonium tetrafluoro-, as vinyl cation equivalents, **1**, 431
Borates, zwitterionic, synthesis of, **2**, 1088
Borate salt, arenediazonium tetrafluoro-, as starting material, for aryl fluorides, **2**, 614
Borate salts, alkyl-, synthesis of, **2**, 1085
Borate salts, amino oxosulfoxonium tetrafluoro-, synthesis of, **2**, 363
Borate salts, cyano-
  as starting materials, for oxazaborolines, **4**, 523
  synthesis of, **4**, 523
Borate salts, diazonium tetrafluoro-, reduction of, **1**, 442
Borate salts, fluoro-
  reactions, with *N,N*-dimethylaniline, **1**, 109
  synthesis of, **1**, 127
Borate salts, quaternary, synthesis of, **4**, 642
Borate salts, tetrafluoro-
  reactions, with methoxide, **4**, 892
  synthesis of, **4**, 1073; **5**, 1146
Borations, double, of 1-alkynes, **2**, 903
Borazine, discovery of, **2**, 811
Borazine, 2,4,6-trichloro-, reactions, with alkynyl Grignard reagents, **2**, 1085
Borazines, alkynyl 2,4,6-trichloro-, synthesis of, **2**, 1085
Borenes
  from
    borenes, **5**, 928
    diboriranes, **5**, 928
    dichlorides, **5**, 928
  sp^2-hybridized, **5**, 930
  reactions

with alkynes, **5**, 928
with 4-*t*-butylpyridine, **5**, 929
with nucleophiles, **5**, 929
with phenols, **5**, 930
as starting materials
for boraborenes, **5**, 928
for borenes, **5**, 928
x-ray structure of, **5**, 929
Borenes, dilithio-, oxidation of, **5**, 929
Borenes, mesityl-bridged, synthesis of, **5**, 929
Borenyl anions, synthesis of, **5**, 928
Borepane, 3,6-dimethyl-
as dihydroborating reagent, **4**, 631
synthesis of, **2**, 516
Borepanes, *trans*-alkenyl-, elimination of, **1**, 660
Boretes, 1,2-dihydro-, thermal cycloreversion of, **6**, 713
Boric acid
reactions
with alkanols, **2**, 111
with aureomycin, **2**, 111
with diols, **2**, 111
with mannitol, **2**, 111
Boric acid, fluoro-
for the desilylation of vinylsilanes, **1**, 450
reactions, with nickelocene, **4**, 694
use in the synthesis of carbon-bridged diruthenium complexes, **4**, 680
Boric acid, tetrafluoro-, use in the synthesis of carbon-bridged diruthenium complexes, **4**, 680
Boric acid, trialkyl esters, synthesis of, **2**, 112
Borides, synthesis of, **1**, 72
Borienes, 2-boro, from, 1-stannylalkynes, **4**, 1054
Borinane, reactions, with 1-iodo-1-butyne, **5**, 229
Borinane, 3,5-dimethyl-, synthesis of, **2**, 516
Borinanes, from, boranediylborinanes, **6**, 188
Borinanes, alkenyl-, synthesis of, **5**, 229
Borinanes, *B*-alkyl-, as starting materials, for carbon-centered radicals, **1**, 946
Borinates
disproportionation of, **2**, 111
from, *n*-butoxydichloroborane, **2**, 419
Borinates, bromoalkenyl-, reactions, with lithioalkynides, **3**, 275
Borinates, butyldiphenyl-, reactions, with amino acids, **5**, 228
Borinates, enol-
hydrolysis of, **1**, 273
α-selenylation of, **3**, 173
Borinates, monoalkylthexy-, as starting materials, for alkynes, **1**, 1034
Borinic acids, dearylation of, **1**, 447
Borinic acids, dialkyl-, synthesis of, **1**, 1102
Borinic esters, vinyl-, reactions, with arylmagnesiums, **2**, 907
Borinic esters, vinylaryl-, synthesis of, **2**, 907
Boriranes, synthesis of, **5**, 929
Borirenes
from, ethynes, **4**, 1058
synthesis of, **4**, 1063
Bornane skeleton, via rearrangements, **1**, 381
2-Bornanones, as starting materials, for chloro nitroso compounds, **4**, 107
Bornene, phenyldimethylsilyl-, synthesis of, **2**, 908
exo-Borneol, 10-mercapto-, as starting material, for 1,3-oxathianes, **4**, 231
Bornyl chloride, from, α-pinene, **1**, 380
Boroboration, of vinylboranes, **4**, 632
Borocanes, synthesis of, **5**, 229
Borohydride
reactions, with tertiary phosphines, **2**, 474
use in the reduction of azides, **2**, 317
Borohydride, cadmium-moderated, reactions, with 4-methylbenzoyl chloride, **3**, 93
Borohydride, trimethoxy-, reactions, with hydrogen chloride, **2**, 109
Borohydride anions, oligomeric, synthesis of, **4**, 524
Borohydrides
for the reduction of amidines, **1**, 45
as starting materials, for borates, **2**, 1087
use in the reduction of amines, **1**, 33
use in the reduction of carbonyls, **2**, 59
use in the synthesis of amines, **2**, 300
Borohydrides, amino-, for the reduction of alkylcyclohexanones, **2**, 60
Borohydrides, α-amino-, synthesis of, **4**, 522
Borohydrides, chiral, synthesis of, **2**, 57
Borohydrides, complex, reactions, with acyl chlorides, **3**, 102
Borohydrides, cyano-, zinc modified, **1**, 39
Borohydrides, ethoxy-, pyrolysis of, **2**, 109
Borohydrides, trialkyl-
carbonylation of, **4**, 364
use in the reduction of carbonyls, **2**, 60
use in the reductive cleavage of epoxides, **2**, 83
Borolane, β-methoxy-2-(trimethylsilyl)-, synthesis of, **4**, 643
Borolane, C2-symmetric, use in asymmetric reduction, **2**, 58
Borolanes, *B*-allyl-, use in the allylation of aliphatic aldehydes, **2**, 76
Borolanes, 2-allyl-1,3,2-dioxa-, synthesis of, **2**, 529
Borolanes, 2,5-dimethyl-, synthesis of, **2**, 521
Borole complexes, triple-decker, synthesis of, **4**, 695
4-Borolene, 1-(diisopropylamino)-2,3-di(trimethylsilyl)-, synthesis of, **4**, 643
3-Borolenes, as starting materials, for (*E*)-alkene-1,3-diols, **2**, 75
Boroles, synthesis of, **4**, 1063
Boroles, 2-boro-, synthesis of, **4**, 1054
Boroles, dihydro-, reactions, with metal carbonyls, **4**, 695
Boron
reactions
with carbon monoxide, **5**, 1020
with oxygen, **2**, 56
1,3 sigmatropic rearrangement of, **1**, 838
stabilization by, for the synthesis of carbenes, **4**, 1082
substitution of, **1**, 487
synthesis of functions, **5**, **409–433**
bearing borons and metals, **6**, 198
bearing a germanium and a boron, **4**, 646
bearing a silicon and a boron, **4**, 642
bearing three borons, **6**, 180
bearing two borons, **4**, 629
metal functions, **6**, 292
two silicon functions, **6**, 292
transmetallation reactions of, **2**, 560
use in the transposition of ketones, **3**, 126
Boron, bis(1,2-dimethylpropyl)-, use in the synthesis of alkynes, **1**, 1034
Boron, bis(trimethylstannylethynyl)-, reactions, with triethyl boron, **4**, 1063
Boron, doubly-bonded
synthesis of, **6**, 712
synthesis of functions, **6**, **677–724**
Boron, thexylmonomethoxy-, use in the synthesis of alkynes, **1**, 1034
Boron, trialkyl-, reactions, with 2-lithiothiophene, **5**, 82
Boron, triethyl-
reactions
with bis(trimethylstannylethynyl) boron, **4**, 1063
with carbenes, **6**, 691
with 1-(trimethylstannyl)ethyne, **4**, 1068
with trimethylstannylethynyl boron, **4**, 1063
use in the synthesis of enolates, **1**, 133
Boron, trimethylstannylethynyl-, reactions, with triethyl boron, **4**, 1063
Boron, tri α-naphthyl-, reactions, with methanol, **1**, 447
Boron, trioctynyl-, reactions, with pregnenolone acetate, **1**, 1050
Boron alkenes, *gem*-bromo, from, *gem*-diboroalkenes, **4**, 814
Boron alkoxides, synthesis of, **2**, 112

Boron amides, synthesis of, **5**, 380
N-Boron amides, *N*-hetero-, synthesis of, **5**, 383
Boronate, α-bromo-, reactions, with sodium methoxide, **4**, 363
Boronate, dibutyl α-styryl-, hydroboration of, **4**, 631
Boronate, dibutyl β-styryl-, as starting material, for tetrabutyl 2-phenylethylidenebisboronate, **4**, 631
Boronate, diisopropyl (1-methoxyvinyl)-
  reactions, with pinanediol, **4**, 934
  synthesis of, **4**, 934
Boronate, pinacol(1-chloroethyl)-, reactions, with trimethylstannyllithium, **4**, 665
Boronate, pinacol(1-iodoethyl)-, reactions, with *t*-butyllithium, **4**, 664
Boronate, pinacol[1-(trimethylstannyl)ethyl]-, synthesis of, **4**, 665
Boronate, (*S*)-pinanediol (1-methoxyvinyl)-, synthesis of, **4**, 934
Boronate, propanediol benzyl-, deprotonation of, **4**, 664
Boronate, tetrabutyl 2-phenylethylidenebis-, from, dibutyl β-styrylboronate, **4**, 631
Boronate, tetrabutyl 2-phenylethylidene-1,1-bis-, reactions, with ethylene glycol, **4**, 634
Boronate complexes
  synthesis of, **1**, 126
  use in the synthesis of aromatic ketones, **3**, 289
Boronate esters, α-(phenylthio)alkyl-, lithiation of, **6**, 135
1-Boronate esters, (*Z*)-7-alkenyl-, oxidation of, **1**, 660
β-Boronate esters, synthesis of, **2**, 530
Boronates
  from, tartaric acid, **2**, 75
  reactions, with lithium benzyl oxide, **4**, 362
  synthesis of, **1**, 838; **4**, 524, 1056
Boronates, alkenylpinacol-, from, trimethylsilylpinacolborylmethyllithium, **2**, 905
Boronates, 1-alkynyl-, reduction of, **1**, 999
Boronates, allenic, synthesis of, via hydroboration of an enyne, **2**, 900
Boronates, allyl-
  displacement of halides from, **4**, 363
  reactions
    with α-ketocarboxylic acids, **1**, 281
    with ketones, **1**, 281
Boronates, aminomethyl-, as chymotrypsin inhibitors, **4**, 526
Boronates, aryl-, in cross-coupling reactions, **1**, 489
Boronates, azidomethyl-, synthesis of, **4**, 526
Boronates, bis(2-trifluoromethylethyl) alkyl-, use in the synthesis of 1,3,2-oxazaborolidines, **2**, 66
Boronates, chiral chloro-, synthesis of, **4**, 362
Boronates, α-chloro-
  substitution of, with sodium thiophenolate, **4**, 373
  synthesis of, **4**, 362, 526
Boronates, dibutyl iodomethyl-, reactions, with amides, **4**, 525
Boronates, (*E*)-enol dialkyl-, Claisen rearrangement of, **1**, 831
Boronates, α-halo-
  reactions, with alkoxides, **4**, 362
  substitution of, with sulfur nucleophiles, **4**, 373
Boronates, α-iodo-, synthesis of, **4**, 525
Boronates, α-metalloalkyl-, synthesis of, **2**, 1005
Boronates, α-methoxy-, synthesis of, **4**, 363
Boronates, α-phenoxy-, synthesis of, **4**, 363
Boronates, α-phenylthiomethyl-, synthesis of, **4**, 372
Boronates, 1-phosphonioalkyl-, synthesis of, **4**, 574
Boronates, vinyl-
  arylation of, **1**, 472
  halogenation of, **4**, 814
  reactions
    with diborane, **4**, 631
    with vinyl halides, **1**, 279
Boron azides, photolytic decomposition of, **2**, 811
Boron bromide, dimethyl-
  reactions
    with *N*-alkylhalogenacetamides, **5**, 660
    with *N*-arylhalogenacetamides, **5**, 660
    with *O*-glycosides, **4**, 225
  use in the synthesis of monothioacetals, **4**, 217
Boron bromide, diphenyl-, use in the synthesis of monothioacetals, **4**, 217
Boron chloride, bis(dimethylamino)-, reactions, with dilithiomethane, **4**, 632
Boron chloride, di-*n*-butyl-
  reactions
    with silver cyanide, **5**, 1136
    with trimethylsilyl cyanide, **5**, 1136
Boron chloride, dimethoxy-, reactions, with chloroform, **6**, 181, 206
Boron chlorides, dialkyl-, reactions, with carboxylic acids, **5**, 228
Boron chlorides, diaryl-, reactions, with carboxylic acids, **5**, 228
Boron chlorides, di(cyclohexyl)-, reactions, with triethylamine, **2**, 532
Boron compounds
  as coupling reagents, for the synthesis of amides, **5**, 268
  for the opening of epoxides, **1**, 207
  of phosphorus functions, synthesis of, **5**, 889
  reactions
    with alkynes, **2**, 900
    with amides, **5**, 381
    with phosphorus ylides, **4**, 574
    with silylamides, **5**, 381
  singly-bonded, synthesis of, **5**, 650
  as starting materials, for allenes, **1**, 976
  synthesis of, **2**, 900; **4**, 1031; **5**, 872, 927
  use in Ireland–Claisen rearrangements, **1**, 417
  uses of, **1**, 162
Boron compounds, acyl-, synthesis of, **5**, 432
Boron compounds, alkenyl-, synthesis of, **1**, 1100; **2**, 900
Boron compounds, alkyl-, synthesis of, **2**, 514, **513–547**
Boron compounds, alkynyl-
  from, ethynylmagnesium bromides, **2**, 1083
  reactions, with enones, **1**, 1050
  synthesis of, **2**, 1083
  for the synthesis of alkynes, **1**, 1002
Boron compounds, allenyl-, for the synthesis of alcohols, **1**, 1055
Boron compounds, aminomethyl-, chemistry of, **4**, 522
Boron compounds, aryl-, synthesis of, **2**, 906, **899–950**
Boron compounds, (1*S*)-(1-chlorobutyl)-, synthesis of, **4**, 144
Boron compounds, cyano-, synthesis of, **5**, 1136
Boron compounds, α-diazo-, synthesis of, **5**, 872
Boron compounds, diazomethyl-, synthesis of, **5**, 872
Boron compounds, α-halo-, synthesis of, **4**, 142
Boron compounds, 1-halo-1-alkenyl-, synthesis of, **2**, 901
Boron compounds, α-haloalkenyl-
  synthesis of, **4**, 813
    via halogenation reactions, **4**, 813
    via hydroboration of haloalkynes, **4**, 814
    via ligand exchange, **4**, 813
Boron compounds, imidoyl-, synthesis of, **5**, 795
Boron compounds, silyl-, addition to alkynes, **2**, 919
Boron compounds, trialkyl-, as starting materials, for dialkylberylliums, **2**, 563
Boron compounds, trialkynyl-, reactions, with methyl vinyl ketone, **1**, 1050
Boron compounds, tricoordinated, synthesis of, **5**, 795
Boron compounds, vinyl-, synthesis of, **2**, **899–950**
Boron cyanide, dimethyl-, synthesis of, **5**, 1136
Boron cyanides, dialkyl-, from, dialkylboron halides, **5**, 1136
Boron cyanides, di-*n*-alkyl-, synthesis of, **5**, 1136
Boron dichlorides, reactions, with alkanethiols, **2**, 268
Boron enolate methodology, concept of, **1**, 133
Boron enolates
  from

carbonyls, **2**, 698
α-diazo ketones, **2**, 697
α,β-unsaturated ketones, **2**, 698
geometry of, **2**, 698
hydrolysis of, **2**, 524
Mannich reaction of, **2**, 302
of nitriles, **3**, 632
reactions
with aldehydes, **3**, 159
with cobalt alkynes, **1**, 199
with *N,N,N',N'*-tetramethyldiaminoethane, **1**, 255
with *N,N,N',N'*-tetramethyldiaminomethane, **2**, 302
synthesis of, **1**, 126, 333; **2**, 697
Boron enolates, chiral, synthesis of, **2**, 698
Boron enol ethers
synthesis of, **2**, 532
use in the synthesis of alkoxyboranes, **2**, 532
Boron esters, cyclic, reactions, with 1,3,5-triphenyl-1,3,5-triazine, **4**, 457
Boron fluoride, dimesityl-, alkynylation of, **2**, 1083
Boron functions
doubly-bonded, synthesis of, **3**, 504
elimination of, **1**, 656, 1080
substitution of, **1**, 272, 1033
Boron halides
reactions
with alkyl metals, **2**, 520
with metal alkynides, **2**, 1083
with perfluorovinyltins, **4**, 813
as starting materials
for 1,1-diboryl-substituted alkanes, **6**, 181
for tris(dihaloboryl)methanes, **6**, 181
Boron halides, dialkyl-, as starting materials, for dialkylboron cyanides, **5**, 1136
Boron halides, 9-fluorenyltetramethylpiperidino-, as starting materials, for amino(methylene)boranes, **3**, 504
Boron–halogen compounds, reactions, with silylamidines, **5**, 778
Boron hydride, for the reduction of epoxides, **2**, 83
Boron hydrides, (α-haloalkyl)-, synthesis of, **4**, 142
Boronic acid, allenyl-, synthesis of, **2**, 904
Boronic acid, aminomethyl-, synthesis of, **4**, 525
Boronic acid, 1-chloro-2-methylpropyl-, synthesis of, **4**, 144
Boronic acid, phenyl-
deboronation of, **1**, 447
reactions, with β-hydroxyacids, **5**, 229
Boronic acids
deboronation of, **1**, 447
from, (iodomethyl)boronic esters, **4**, 144
reactions
with amino alcohols, **2**, 528
with (*S*)-(-)-2-(di-β-naphthylhydroxymethyl)pyrrolidine, **2**, 65
with iodopyridine, **3**, 311
synthesis of, **4**, 525
use in the synthesis of biaryls, **1**, 473
use in the synthesis of monoalkyl phenyl ketones, **3**, 296
Boronic acids, alkenyl-, synthesis of, via hydrolysis of alkenylboronic esters, **2**, 904
Boronic acids, (*E*)-alkenyl-, iodination of, **1**, 658
Boronic acids, alkyl-, use in the substitution of boron, **1**, 489
Boronic acids, α-amino-, synthesis of, **4**, 522
Boronic acids, aryl-
protodeboronation of, **1**, 447
reactions
with iodoarenes, **3**, 290
with 3-iodo benzothiophene, **3**, 306
as starting materials, for aryl iodides, **2**, 632
synthesis of, **2**, 906
Boronic acids, aryl-, *N*-methyldiethanolamine esters, use in the synthesis of aryl fluorides, **2**, 613
Boronic acids, α-halo-, synthesis of, **4**, 144
Boronic acids, naphthyl-, reactions, with iodoarenes, **3**, 301
Boronic esters
reactions, with 1-alkynyllithiums, **2**, 1085
synthesis of, **2**, 531
Boronic esters, alkenyl-
deboronobromination of, **1**, 657
from
alkenyl bromides, **2**, 904
alkenyl iodides, **2**, 904
alkenyllithiums, **2**, 904
alkenylmagnesiums, **2**, 904
synthesis of, via esterification of alkenylboronic acids, **2**, 904
Boronic esters, allenyl-, synthesis of, **2**, 904
Boronic esters, cyclic, synthesis of, **6**, 181
Boronic esters, 2-(1,3-dienyl)-, synthesis of, **2**, 906
Boronic esters, α-phenylthioalkyl-, reactions, with electrophiles, **6**, 124
Boron iodide, dimethyl-, reactions, with iodocyanogen, **5**, 1136
Boron isocyanates, synthesis of, **5**, 1015
Boron methylidenide ions, synthesis of, **2**, 1005
Boron monochlorides, reactions, with thiols, **2**, 267
Boron–nitrogen bonds, synthesis of compounds with, **2**, 810
Boron selenide, as intermediate, for selones, **3**, 392
Boron sulfide
as reagent, for the synthesis of thioketones, **3**, 361
as thionating reagent, **5**, 575
Boron tetrafluoride, reactions, with carbon, **6**, 387
Boron tribromide
reactions
with methylene bis(mercuric iodide), **4**, 632
with pentacarbonyl(dimethylamino-[methoxy(phenyl)methyleneamino]-carbene)chromium(0), **5**, 660
as starting material, for carbonic bromide chloride, **6**, 420
transhalogenation of, with trifluoromethylfluorothiol formate, **6**, 436
use in the synthesis of alkyl bromides, **2**, 32
use in the synthesis of dibromoalkanes, **4**, 26
Boron trichloride
reactions
with aldehydes, **2**, 110
with amines, **2**, 533
with bis(tricyclohexyltin) sulfide, **5**, 575
with diols, **2**, 110
with epoxides, **2**, 110
with α-fluorinated nitriles, **5**, 684
with imines, **4**, 107
with silylgermanes, **4**, 639
with triethyltin hydride, **2**, 526
with tris(dimethoxyboryl)methane, **6**, 181
use in the synthesis of alkanoic acids, **5**, 40
use in the synthesis of 2,4,6-trimethylbenzoic acid, **5**, 107
Boron trichlorides
reactions, with alkanethiols, **2**, 268
use in the synthesis of alkyl chlorides, **2**, 23
Boron triflates, dialkyl-
as borylating agents, **2**, 698
reactions, with tertiary amines, **2**, 532
Boron trifluoride
as catalyst, for the synthesis of aromatic ketones, **3**, 280
reactions
with *t*-butyldifluoroamine, **2**, 762
with diazomethylenephosphorane, **5**, 872
with ethanol, **2**, 110
with Lewis bases, **2**, 1084
with mercuric oxide, **4**, 195
with mesitylmagnesium bromide, **2**, 520
with methanol, **2**, 110
with monosubstituted ureas, **5**, 979
with silenes, **6**, 387
with trialkoxyboranes, **2**, 110
with triethylsilane, **1**, 13

use for carbocation formation, **1**, 386
use in the synthesis of aromatic ketones, **3**, 278
use in the synthesis of borates, **2**, 111
use in the synthesis of furanyl ketones, **3**, 302
use in the synthesis of phenolic ketones, **3**, 296
Boron trifluoride complexes, as alkylation reagents, **1**, 188
Boron trifluoride diethyletherate, reactions, with lithium methoxy alkynyldialkylborinates, **2**, 1084
Boron trifluoride etherate
as catalyst
for the addition of lithium compounds to epoxides, **1**, 204
for the alcoholysis of nitriles, **4**, 883
for the condensation of thiols, **4**, 245
for the synthesis of acetals, **4**, 181
in [2,3] Wittig rearrangements, **1**, 234
reactions
with alkynyllithiums, **1**, 1023, 1045
with alkynylsilanes, **4**, 808
with bis(diisopropylamino)chlorodiazomethylene phosphorane, **3**, 485
with bis(trimethylsilyl)selenide, **5**, 625
with cyanophosphates, **3**, 647
with α-diazo-β-hydroxycarbonyls, **1**, 1078
with α-diazophosphonates, **6**, 124
with ethanethiols, **2**, 304
with hydrogen chloride, **4**, 885
with *threo*-β-hydroxysilanes, **1**, 761
with monoorganocoppers, **1**, 312
with orthocarbonates, **6**, 729
with β-patchouline oxide, **1**, 842
with 1-trifluoroacyl 2,2-dimethylhydrazine, **5**, 383
for the ring-opening of aziridines, **1**, 263
use in the synthesis of amines, **2**, 319
use in the synthesis of ketene *S,N*-acetals, **4**, 905
use in the synthesis of monoalkylboranes, **2**, 515
Boron trifluoride etherate, phenyllithium-, as starting material, for hydroxytosylates, **1**, 223
Boron trihalides
reactions
with alcohols, **2**, 110
with *N*-alkyl halogenacetamides, **5**, 660
with alkylhydroperoxides, **2**, 99
with *N*-aryl halogenacetamides, **5**, 660
with arylmercury compounds, **2**, 906
for the selective deprotection of methyl ethers, **4**, 175
as starting materials
for alkoxyhaloboranes, **2**, 110
for borates, **2**, 1086
Boron triiodide, for the cleavage of ethers, **2**, 36
Boron tris(benzene selenide)
reactions
with triethyl orthoformate, **6**, 92
with trimethyl orthoacetate, **6**, 92
Boron triselenides, exchange reactions, with ortho esters, **6**, 92
Boron tris(methane selenide), reactions, with trimethyl orthoacetate, **6**, 92
Boron trisulfide
reactions, with phosphine oxides, **2**, 462
as reagent, for the synthesis of thiocarboxylates, **5**, 512
Boron-Wittig reaction
of aldehydes, **1**, 988
mechanism, **1**, 657
Boroxazolidines, use in the synthesis of alcohols, **2**, 65
Boroxazolidones, reactions, with aldehydes, **5**, 228
Boroxines, reduction of, **2**, 110
Boroxines, trialkyl-, as starting materials, for dialkoxyboranes, **2**, 110
α-Borylalkenyl compounds, synthesis of, **2**, 1005
Borylamidine complexes, dialkyl-, synthesis of, **5**, 778
Borylamidines, dialkyl-
as cyclic and open chain structures, **5**, 778
synthesis of, **5**, 778
α-Boryl anions, from, geminal diboranes, **1**, 163
Borylboration, for the synthesis of α-borylsilylalkanes, **4**, 644
C-Borylboriranides, thermal isomerization of, **6**, 715
α-Boryl carbanions
stability of, **1**, 162
synthesis of, **1**, 164; **2**, 1005
Boryl compounds, synthesis of, **5**, 927
Boryl compounds, bis(diisopropylamino)-, reactions, with methyllithium, **3**, 682
Boryl compounds, dialkyl-
as intermediates, **4**, 364
reactions, with phenylisonitrile, **6**, 662
2-Boryl-1,3-diboretanides, thermal isomerization of, **6**, 715
Boryl electrophiles, reactions, with α-metallosilylalkanes, **4**, 642
Borylmethide salts, synthesis of, **6**, 198, 199
Boryloxy compounds
from
α-diazo esters, **4**, 832
α-sulfuranylidene esters, **4**, 832
of ketene acetals, synthesis of, **4**, 832
Botrydianes, synthesis of, **1**, 676
Bouveault–Blanc reaction, mechanism, **1**, 101; **2**, 54
Bouveault reaction, mechanism, **2**, 320
Brady's reagent, reactions, with α-halocycloalkanones, **3**, 451
Brassinolide, 2,3 Wittig rearrangements of, **1**, 818
Bredt's rule, **1**, 853
Brefeldin A, synthesis of, **5**, 146
Brefeldin C, synthesis of, **1**, 748
Brefeldin ring systems, synthesis of, **1**, 949
Brefeldins, synthesis of, **1**, 1099
endo-Brevicomin, synthesis of, **3**, 321
exo-Brevicomin, synthesis of, **1**, 165; **2**, 529; **3**, 321
(±)-Breynolide, synthesis of, **1**, 581
Bridgehead, halogen exchange at, **2**, 25
Bridgehead halides, from, sulfonates, **2**, 9
Bridgehead iodides, synthesis of, **4**, 107
Bridgehead radicals
reviews, **1**, 919
structure of, **1**, 923
Bridging carbon atoms, in silver and copper complexes, **4**, 695
British antilewisite, reactions, with arsenic halides, **2**, 889
Bromide
reactions, with diazo compounds, **4**, 90
transhalogenation of, with dimethylcarbamoyl chloride, **6**, 452
Bromide azide
reactions, with vinylsilanes, **4**, 514
as starting material, for *gem*-diazides, **4**, 433
Bromide chloride, thiocarbonyl, synthesis of, **6**, 532
Bromide fluoride, thiocarbonyl, synthesis of, **6**, 532
Bromides
alkylation of, **1**, 465
precursors for radical cyclization, **1**, 341
as radical precursors, **1**, 330
reactions
with silver (ethyl diazoacetate), **3**, 474
with silver nitrate, **1**, 893
as starting materials, for cuprates, **3**, 286
use in the synthesis of disulfides, **2**, 217
α,ω-Bromides, from, benzamides, **2**, 33
Brominating agents, for the synthesis of *N*-bromoamides, choice of, **5**, 318
Brominating reagents, reactions, with alkali metal alkynylides, **2**, 1018
Bromination
of aldehydes, **3**, 27
of alkanes, **4**, 19
of bis(trifluoromethyl)dimethyllead, **6**, 245
of chlorodifluoromethane, **6**, 223

electrochemical, **2**, 28
of hydroxamoyl halides, **5**, 671
of ketones, **4**, 20
of mercury(II) bis(trifluoromethyl)amide, **6**, 240
of methyl isocyanate, **6**, 240
of potassium trichloroacetate, **6**, 221
of silver bromochlorofluoroacetate, **6**, 227
of silver chlorodifluoroacetate, **6**, 223
of silver dichlorofluoroacetate, **6**, 225
of silver trifluoroacetate, **6**, 218
for the synthesis of phenols, **2**, 658
for the synthesis of α,β-unsaturated ketones, **3**, 208
of tetrabromomethane, **6**, 222
of tetrachloromethane, **6**, 221
of trichloromethane, **6**, 221
of trifluoroacetic acid, **6**, 218
of trifluoroacetic acid anhydride, **6**, 218
of trifluoromethane, **6**, 218
vapor phase
of dichlorofluoromethane, **6**, 227
of difluoromethane, **6**, 226
of vinylgermanes, **4**, 811
Bromination, allylic, for the synthesis of alkyl bromides, **2**, 27
Bromination, benzylic, for the synthesis of alkyl bromides, **2**, 27
Bromination, *N*-bromosuccinimide, mechanism, **2**, 339
Bromination, *meta*-, for the synthesis of aryl bromides, **2**, 626
Bromination, *para*-, for the synthesis of aryl bromides, **2**, 626
o-Bromination, for the synthesis of aryl bromides, **2**, 626
Bromination–dehydrobromination, stereochemistry of, **4**, 779
Bromine
for the electrophilic bromination of arenes, **2**, 625
for the oxidation of alcohols to ketones, **3**, 118
reactions
with acetamide, **6**, 506
with acylvinyl nitroamines, **4**, 106
with alkenyldialkylboranes, **1**, 658
with alkyl thiocyanates, **6**, 623
with alkynyl sulfides, **4**, 772
with amides, **2**, 756
with 3-arylhydrazono-2,4-pentanediones, **5**, 679
with benzaldehyde hydrazones, **5**, 841
with benzaldehyde tetrazolyl-5-hydrazone, **5**, 841
with 1,2-bis(hydroxymethyl)dicarbaboranes, **4**, 143
with bromoalkenes, **4**, 23
with α-bromocarboxylic acids, **4**, 27
with carbon monosulfide, **6**, 529
with cyanogen fluoride, **5**, 809
with 1-cyano-2-tosyl compounds, **4**, 105
with dibenzylmethylarsine, **2**, 490
with 1-diethylamino-2-phthalimidostyrenes, **4**, 793
with α,α-dihalogenated nitriles, **5**, 662
with dimethyl benzenesulfamido malonate, **4**, 105
with 2-(dimethylsulfuranylidene)acetophenone, **5**, 495
with diphosphines, **2**, 449
with *N*,*N*-disubstituted thiobenzamides, **5**, 734
with ethyl-2-arylhydrazono-3-oxobutanoates, **5**, 679
with fluorovinylsilanes, **1**, 653
with fumaric acid, **5**, 60
with α-keto acids, **4**, 28
with metallated imides, **5**, 318
with methylselenobut-3-en-1-yne, **4**, 786
with methylthiomethyl ethers, **4**, 216
with organoaluminums, **4**, 758
with perfluoronitriles, **5**, 808
with phenoxyalkyne, **4**, 765
with phenyliodine(III) bis(phthalimide), **5**, 318
with 1-phenyl-2-phospholene 1-oxide, **4**, 122
with 1-(phenylseleno)pent-1-yne, **4**, 786
with phosphaalkenes, **6**, 689
with phosphorus, **2**, 28
with potassium thiocyanate, **5**, 1107
with pyrazolines, **4**, 105
with silver acetate, **5**, 203
with silver acrylate, **4**, 27
with silver cyanate, **5**, 1015
with silver ethyl diazoacetate, **3**, 478
with silver α-fluoro carboxylate, **4**, 35
with sodium dinitromethanide, **6**, 261
with sulfides, **4**, 71
with tetrakis(trimethylsilylmethyl)germane, **4**, 641
with tetramethylbiphosphine sulfide, **2**, 1070
with 1,1,3,3-tetramethyl-1,3-disilacyclobutane, **4**, 622
with tetraphenylgermane, **1**, 449
with thallium(I) salts, **4**, 27
with thioformamides, **6**, 540
with 1,2,3-triazolo[1,5-*a*]pyridine, **4**, 27
with triethoxyphosphonium bis(phenylsulfonyl)methylide, **6**, 278
with trifluorosilyl(vinyl)silanes, **4**, 138
with trihalogenated acetonitriles, **5**, 662
with trimethylsilyl(vinyl)silanes, **4**, 138
with triphenyl phosphite, **4**, 25
with tris(trimethylsilylmethyl), **4**, 597
with vinyl acetates, **4**, 57
with vinylboranes, **4**, 145
with vinyl-1,2-dicarbadodecaboranes, **4**, 143
with vinylgermanes, **1**, 452
with vinylsilanes, **4**, 138, 751
for the synthesis of aryl bromides, **2**, 627
for the synthesis of vinyl bromides, **2**, 624
use in the synthesis of carboxylic esters, **5**, 141
use in the synthesis of disulfides, **2**, 212, 222
Bromine, bisquinuclidine, alcohol oxidations with, **3**, 119
Bromine chloride, as reagent, for the synthesis of *N*-bromoimides, **5**, 318
Bromine compounds
of phosphorus functions, synthesis of, **5**, 877, 895
reactions, with alkenes, **1**, 707
Bromine monofluoride, for the synthesis of aryl bromides, **2**, 626
Bromine trifluoride
reactions
with fluoroalkenes, **6**, 2
with tetrabromomethane, **6**, 217, 218, 222, 226
with tetrachloromethane, **6**, 212, 217, 220, 225, 226
with tetraiodomethane, **6**, 212, 217, 219
with trichlorofluoromethane, **6**, 225
as starting material, for carbonic bromide fluoride, **6**, 420
use in the synthesis of α-fluoro ethers, **4**, 45
Bromoalkylation, of arenes, **2**, 33
Bromoaromatic acids, annulation of, **3**, 284
Bromochlorination, of alkenes, **2**, 24
Bromo compounds, reactions, with sodium azide, **4**, 109
(*R*)-(-)-α-Bromo compounds, from, (*S*)-(-)-methyl phenylvinylphosphine oxide, **4**, 804
Bromodealkylation, for the cleavage of ring systems, **5**, 1126
Bromodecarboxylation
of carboxylic acids, **2**, 33
for the synthesis of dibromoalkanes, **4**, 28
Bromodenitrogenation, of N^2-benzoylarginine ethyl ester, **5**, 1122
Bromodifluoromethyl group, as starting material, for difluoromethyl group, **4**, 5
Bromofluorination, of alkenes, **2**, 18
Bromoform
reactions
with carbosilanes, **6**, 178
with dibromothiolane 1,1-dioxide, **4**, 748
with lithium diisopropylamide, **4**, 746
reduction of, **4**, 23
as starting material
for triaryl trithioortho esters, **6**, 81
for triselenoorthoformates, **6**, 92

use in the synthesis of dibromocarbene, **4**, 24
See also Methane, tribromo-
Bromoiodination, of alkenes, **2**, 31
Bromomethylation, of arenes, **2**, 13
Bromonium ions, use in the synthesis of alkyl bromides, **2**, 30
1,2-(Bromosilyl) compounds, synthesis of, **4**, 1051
Brönsted acids
as catalysts
for the synthesis of dithioacetals, **4**, 246
for the synthesis of isocyanates, **5**, 1004
protonating power of, **1**, 897
reactions
with *S*-benzyl thioethers, **5**, 481
with iodide, **1**, 5
with Lewis acids, **2**, 1007
for the synthesis of organometal(III) halides, **2**, 883
Brönsted–Lewis superacids, as catalysts, for Friedel–Crafts acylation, **1**, 462
Brook rearrangement
of α-aminosilanes, **4**, 506
of betaines, **2**, 936
for the synthesis of alcohols, **1**, 283
See also Reverse Brook rearrangement
Brook rearrangement, retro-, of silylpropargyl silyl ethers, **2**, 938
Brook-type migration, of silicon, from carbon to oxygen, **5**, 793
Brosylate, norbornyl, reactions, with acetic acid, **1**, 381
Brosylates, reactivity of, **1**, 1115
endo Brosylates, synthesis of, **1**, 380
exo Brosylates, rearrangement of, for *endo* brosylates, **1**, 380
Brucinium iodide, *N*-methyl-, reactions, with lithium tribiphenylenephosphate, **2**, 869
Bucherer reaction, use in the synthesis of phenols, **2**, 647
Buckminsterfullerene (C_{60}), reversible electron transfer with, **1**, 890
Bulk metals, characteristics of, **2**, 958
Bullvalene, synthesis of, **1**, 789
(±)-β-Bulnesene, synthesis of, **3**, 325
(±)-β-Bulnesol, synthesis of, **3**, 325
t-Bumeoc fluoride, synthesis of, **6**, 422
Bunte salts
reactions, with sodium hydrogen telluride, **2**, 122
as starting materials
for thioaldehydes, **3**, 333, 340
for thioketones, **3**, 349, 365
for thiols, **2**, 122
synthesis of, **2**, 122; **5**, 601
use in the synthesis of alkanethiols, **2**, 122
use in the synthesis of dialkyl sulfides, **2**, 140
use in thiolysis reactions, **2**, 220
Burgess' reagent
for the dehydration of amides, **3**, 619
mechanism, **1**, 575
use in dehydration reactions, **1**, 575
Butadiene
hydromagnesiation of, **2**, 561
reactions
with diphenylethyne, **1**, 519
with ethyne, **1**, 519
with phosphorus trichloride, **2**, 449
with tetranitromethane, **6**, 149
with ynamines, **2**, 740
telomerization of, **2**, 561
Butadiene, 1,4-bis(trimethylsilyl)-, transition metal-catalyzed hydrosilylation of, **2**, 939
Butadiene, 2,3-diborata-, synthesis of, **6**, 716
Butadiene, 2,3-dimethyl-
magnesium compounds of, for the synthesis of trinuclear iridiacyclopentadiene complexes, **4**, 690
reactions
with (dichloromethyl)phosphine, **4**, 111
with dimethyl dithionooxalate, **4**, 229
Butadiene, 1,4-diphenyl-, reactions, with 1,3-diphosphaindene, **4**, 548
Butadiene, (*E*,*E*)-1,4-diphenyl-, reactions, with osmium tetroxide, **2**, 47
Butadiene, (*E*,*E*)-disilyltetrachloro-, synthesis of, **4**, 806
Butadiene, hexachloro-
fluorination of, **6**, 16
reactions, with 1,3-propanedithiol, **4**, 770
Butadiene, hexafluoro-, as starting material, for hexafluoro-2-butyne, **6**, 18
Butadiene, perfluoro-, reactions, with trifluoronitrosomethane, **6**, 54
Butadiene, (*Z*,*Z*)-1,2,3,4-tetrachloro-, silylation of, **4**, 806
Buta-1,2-diene
reactions
with deuterium, **1**, 459
with hydrogen chloride, **1**, 1125
1,2-Butadiene, 1,1,4,4-tetraphenyl-, synthesis of, **1**, 960
1,3-Butadiene, *trans,trans*:*trans,cis*:*cis,trans*-4-aryl-1-chloro-1,2,3,4-tetrafluoro-, use in the synthesis of carboxylic acids, **5**, 58
1,3-Butadiene, 1,1,-bis(methylthio)-, synthesis of, **4**, 852
1,3-Butadiene, 2,3-bis(phenylsulfonyl)-, as starting material, for allenyl sulfones, **1**, 980
1,3-Butadiene, 1,4-bis(trimethylsilyl)-, silylation of, **4**, 1045
1,3-Butadiene, 1,3-bis(trimethylsilyloxy)-1-methoxy-, reactions, with 1,3-dicarbonyls, **2**, 662
Buta-1,3-diene, 1,4-dichloro-, reactions, with dilithiophenylarsine, **4**, 1039
Buta-1,3-diene, 2,3-dichloro-, reactions, with tributylphosphine, **2**, 446
1,3-Butadiene, 2,3-dilithio-, reactions, with bis(chlorodimethylsilyl)methane, **4**, 621
1,3-Butadiene, (*E*)-1-methoxy-, lithiation of, **2**, 946
1,3-Butadiene, 1-methoxy-3-trimethylsilyloxy-
reactions
with dienophiles, **2**, 665
with Eschenmoser's salt, **4**, 305
with 3-nitrocycloalkanones, **2**, 665
with 3-phenylsulfinylcycloalkenones, **2**, 665
1,3-Butadiene, perchloro-, reactions, with arenethiolates, **4**, 851
1,3-Butadiene, 1*H*-perchloro-, nitration of, **4**, 797
Buta-1,3-diene, perfluoro 2,3-dimethyl-
from
cyclobutanes, **4**, 730
perfluoro 2,3-dimethylbutanes, **4**, 730
1,3-Butadiene, 1-silyl-2-aminomethyl-, synthesis of, **2**, 939
1,3-Butadiene, 1,2,3,4-tetrakis(diphenylphosphino)-, detection of, **4**, 1023
1,3-Butadiene, 1-(trimethylsilyl)-, reactions, with borane dimethyl sulfide, **4**, 643
Buta-1,3-diene, 2-trimethylsilyl-, synthesis of, from 3-sulfolene pyrolysis, **1**, 791
Butadienes
from, vinyl nucleophiles, **1**, 496
as starting materials, for 1,4-dithiocyanato-2-butenes, **5**, 1108
synthesis of, **1**, 145, 173; **4**, 892, 1028
from 3-sulfolene pyrolysis, **1**, 790
Butadienes, (bis)dialkylamino-, synthesis of, **2**, 1050
Butadienes, 1,1-bis(toluene-*p*-sulfinyl)-, synthesis of, **4**, 863
Butadienes, 4,4-bis(trimethylsilyl) 1-phenyl-, synthesis of, **4**, 1049
Butadienes, 1,1-disubstituted, synthesis of, from 3-sulfolene pyrolysis, **1**, 790
Butadienes, 1,4-disubstituted, synthesis of, from 3-sulfolene pyrolysis, **1**, 790
Butadienes, 2-silyloxy-, reactions, with methyl propiolate, **1**, 520
Butadienes, 2-substituted, synthesis of, from 3-sulfolene pyrolysis, **1**, 791
Butadienes, 1-thiocyanato-, synthesis of, **5**, 1108

1,2-Butadienes, 1,1,3,4-tetrasilyl-, synthesis of, **4**, 1045
1,3-Butadienes
from, dinitrotetraalkylethanes, **1**, 638
reactions, with alkynes, **1**, 1036
reduction of, **1**, 79
1,3-Butadienes, 2-alkylazo-substituted, synthesis of, **2**, 770
1,3-Butadienes, 2-cyano-, synthesis of, **3**, 646, 659
1,3-Butadienes, 2,3-dialkyl-, synthesis of, **1**, 495
1,3-Butadienes, (*E,E*)-1,4-diaryl-, from, styryl sulfonyl chlorides, **1**, 485
Buta-1,3-dienes, 1,2-dichloro-, as starting materials, for tris(arylthio)butatrienes, **1**, 982
Buta-1,3-dienes, 2,3-diiodo-, reduction of, **1**, 986
1,3-Butadienes, isothiocyanato-, synthesis of, via [3,3] sigmatropic rearrangements, **5**, 1041
1,3-Butadienes, 2-silyl-, from, 2-bromo-1-alkenyldibromoboranes, **2**, 939
1,3-Butadienes, 1,1,4,4-tetraalkoxy-, from, 1,1,4,4-tetraalkoxy 2-butynes, **4**, 830
Buta-1,3-dienes, 1-thiocyanato-, reactions, with dienophiles, **5**, 1041
2,3-Butadienoic acid, reduction of, **1**, 80
(*E*)-Butadienyl carbonates, synthesis of, **6**, 461
O-Butadienyl carbonates, from, α,β-unsaturated aldehydes, **6**, 461
2-Butadienyl Grignard reagents, reactions, with alkyl bromides, **1**, 1092
Butadiyne, as starting material, for alkynyl chlorides, **2**, 1016
Butadiyne, bis(trimethylsilyl)-, zirconium-catalyzed *trans*-carboalumination of, **4**, 809
Butadiyne, 1,4-diphenyl-, synthesis of, **2**, 1051
Butadiyne, trimethylsilyl-, reactions, with allenyl bromides, **1**, 1006
1,3-Butadiyne, from, 1,4-dichloro-2-butyne, **1**, 1069
Butadiynes, synthesis of, **1**, 1035
Butadiynes, mono(iodonium)-, synthesis of, **2**, 1021
1,3-Butadiynes, from, alkynylborates, **1**, 1035
1,3-Butadiynes, 1,4-dialkyl-, synthesis of, **1**, 1035
Butanal, reactions, with mercury(II) chloride, **4**, 697
Butanamidine hydrochloride, heptafluoro-, reactions, with potassium chloride, **5**, 763
Butandinitrile, 2,3-diphenyl-, synthesis of, **3**, 626
Butane, 2-amino-3-phenyl-, as starting material, for amine oxides, **1**, 624
Butane, 1,4-bis(6-carboxypyridyl-3), synthesis of, **5**, 116
Butane, (+)-2-bromo-, reactions, with *n*-butyllithium, **1**, 112
Butane, 2-bromo-, reactions, with benzylmagnesium bromide, **1**, 114
Butane, 2-chloro-, reactions, with diethyl ether, **2**, 558
Butane, diazoiso-, reactions, with cyanic acid, **5**, 1104
Butane, 1,4-dibromo-
reactions, with 1-methylphosphorane, **2**, 444
as starting material, for 2,7-octanedione, **1**, 141
Butane, 3,4-dibromo-, synthesis of, **4**, 23
Butane, 1,4-dichloro-
reactions
with disodium methylarsenide, **2**, 486
with lithium di-*t*-butylstibide dioxane complex, **2**, 506
Butane, 1,4-dilithio-, use in the synthesis of nickel complexes, **2**, 577
Butane, 2,2-dimethyl-, synthesis of, **1**, 127
Butane, 2,3-*O*-isopropylidene-2,3-dihydroxy-1,4-bis(diphenylphosphino)- (DIOP), as chiral catalyst, **3**, 614
Butane, 2-methyl-, reactions, with 3-methylpentane, **5**, 1020
Butane, 2-methyl-1-phenyl-, synthesis of, **1**, 114
Butane, 2,2,3,3-tetramethyl-, synthesis of, **1**, 114
Butane, 1,2,4-trihydroxy-, reactions, with ortho esters, **6**, 79
Butane, 2,3,3-trimethyl-2-chloro-, reactions, with methylmagnesium chloride, **1**, 114
Butane, 1,3,3-tris(difluoramino)-, synthesis of, **4**, 439
Butane dibromide, 1,4-bis(triphenylphosphonio)-, reactions, with geranylacetone, **1**, 721
Butane-1,4-diols, as starting materials, for γ-lactones, **5**, 171
1,3-Butanedione, 2-chloro-1-(2-furyl)-, reactions, with diazonium halides, **5**, 676
1,3-Butanedione, 2-chloro-1-phenyl-, reactions, with diazonium halides, **5**, 676
Butane-2,3-dione, 1,4-dibromo-, reactions, with alkylamines, **6**, 564
Butanes, diphosphadistannyl-, synthesis of, **2**, 474
Butanes, perfluoro 2,3-dimethyl-, as starting materials, for perfluoro 2,3-dimethylbuta-1,3-diene, **4**, 730
t-Butane sulfenate, *O*-alkyl, synthesis of, **2**, 195
t-Butane sulfenate, *O*-ethyl-, synthesis of, **2**, 194
t-Butanesulfenic acids
reaction, with ethyl acrylate, **2**, 153
synthesis of, **2**, 185
t-Butanesulfenyl chloride, synthesis of, **2**, 172
t-Butanesulfenyl iodide, reactions, with sodium ethoxide, **2**, 194
Butanesulfinic acid, perfluoro-, reactions, with polyenes, **2**, 156
t-Butanesulfinic acid, reactions, with thionyl chloride, **2**, 173
Butanesulfinyl chloride, perfluoro-, synthesis of, **2**, 172
β-*t*-Butane sulfinyl propionate, ethyl-, synthesis of, **2**, 153
Butanesulfonate, trimethylgermyl perfluoro-, reactions, with disulfene, **6**, 132
Butanesulfonates, trialkylsilyl perfluoro-, reactions, with disulfene, **6**, 128
Butanesulfonyl chloride, reduction of, **2**, 125
Butanethiol
reactions
with cuprous oxide, **2**, 273
with di-*t*-butyl dicarbonate, **6**, 477
with dichlorophosphorothiolates, **2**, 252
with hexamethyldisilathiane, **2**, 263
with thexyldimethylsilyl chloride, **2**, 262
with triisopropylborane, **2**, 268
with Viehe's salt, **6**, 589, 593
synthesis of, **2**, 135, 137
t-Butanethiol
reactions
with oxaziridines, **2**, 185
with sulfates, **2**, 134
Butanethiosilane, synthesis of, **2**, 263
Butanimidate, 1,1,5-trihydroperfluoroamyl-*N*-chloroperfluoro-, synthesis of, **5**, 813
Butanoamidine, (*E*)-*N,N,N'*-trifluoro-, synthesis of, **5**, 762
Butanoamidine, *N,N,N'*-trifluoro-, synthesis of, **5**, 762
Butanoamidine, (*Z*)-*N,N,N'*-trifluoro-, acid-catalyzed isomerization of, **5**, 762
Butanoamidine hydrochlorides, reactions, with potassium chloride, **5**, 763
Butanoate, ethyl sodio-2-cyano-, reactions, with 2-(nitromethyl)cyclohexenone, **1**, 251
Butanoates, 3-chloro-, synthesis of, **1**, 182
Butanoates, 2,2-dihydroxy-3-oxo-, synthesis of, **4**, 162
Butanoates, ethyl-2-arylhydrazono-3-oxo-, reactions, with bromine, **5**, 679
Butanoates, 3-sulfonyloxy-, optically active, reactions, **1**, 182
Butanoic acid, 4-amino-, reactions, with hydrogen chloride, **5**, 970
Butanoic acid, 3-hydroxy-, reactions, with α,α-dichlorophthalide, **6**, 69
Butanoic acids, 3-alkyl-4-amino-, synthesis of, **2**, 315
Butanol, 2-*t*-butyl-3,3-dimethyl-, oxidation of, **5**, 47
1-Butanol, carbonylation of, **6**, 464
Butan-1-ol, (*S*)-2-amino-3-methyl-1,1-diphenyl-, use in the reduction of ketones, **2**, 64

1-Butanol, 2-fluoro-2-nitro-, reactions, with sodium hypobromite, **6**, 55
t-Butanol
  reactions
    with isocyanates, **4**, 1007
    with tris(diethylamino)borane, **2**, 419
t-Butanol, perfluoro-, reactions, with chlorine trifluoride, **2**, 98
Butanolides, α,β-diacetoxy-, reduction of, **1**, 16
2-Butanols, *threo* 3-triphenylstannyl-, from, *cis*-2,3-butene oxide, **1**, 664
Butanone, as starting material, for ketenes, **3**, 535
2-Butanone, synthesis of, **1**, 387
Butan-2-one, 3,3-dibromo-, synthesis of, **4**, 24
Butanones, reactions, with thionyl chloride, **3**, 249
Butan-2-oxime complexes, synthesis of, **1**, 63
Butanoyl bromide, reactions, with nitroalkenes, **5**, 820
Butanoyl chloride, 2-(1-adamantyl)-3,3-dimethyl-, synthesis of, **5**, 8
Butatriene, 1,1-bis(dimethylamino)-4,4-bis(methoxycarbonyl)-, synthesis of, **1**, 995
Butatriene, 1,1-bis(dimethylamino)-4,4-dicyano-, synthesis of, **1**, 995
Butatriene, 1,4-diphenyl-, synthesis of, **1**, 988
Butatriene, 1,1,4,4-tetrakis(trimethylsilyl)-, reactions, with α,α-disilyl ketones, **4**, 619
Butatriene, tetraphenyl-, reduction of, **1**, 960
Butatrienes
  from, 3-bromo alk-3-en-1-ynes, **1**, 967
  synthesis of, **1**, 983, 988; **4**, 1045
Butatrienes, alkoxy-, synthesis of, **1**, 990
Butatrienes, alkynyl-, synthesis of, **1**, 990
Butatrienes, fluorinated, synthesis of, **1**, 990
Butatrienes, α-silyl α-silyloxy-, synthesis of, **1**, 981
Butatrienes, symmetrical, synthesis of, **1**, 988
Butatrienes, tetrasubstituted, synthesis of, **1**, 986
Butatrienes, tetra(2-thienyl)-, synthesis of, **1**, 990
Butatrienes, tris(arylthio)-, from, 1,2-dichlorobuta-1,3-dienes, **1**, 982
1,2,3-Butatrienes, synthesis of, **1**, 981; **4**, 821
2-(2-But-1,4-dienyl)alkanols, synthesis of, **2**, 75
But-2-enal, 3-chloro-2-methyl-, reactions, with sodium sulfide, **3**, 72
But-3-enal, 2,2-dimethyl-, reactions, with ethylidenetriphenylphosphorane, **1**, 729
But-3-enal, 2,2-dimethyl-3-phenyl-, irradiation of, **1**, 800
Butene, perfluoroiso-
  reactions, with trimethyltin methoxide, **4**, 762
  synthesis of, **4**, 737
Butene, 2-phospha-1-perfluoro-, from, stannylphosphine, **6**, 679
Butene, polychloro-, reactions, with sodium phenylmethanethiolate, **4**, 770
But-1-ene, synthesis of, **1**, 459
1-Butene, 4-bromo-
  alkylation of, **1**, 168
  as starting material, for cyclic acylsilanes, **5**, 411
But-1-ene, 4,4-dichloro-, synthesis of, **4**, 14
2-Butene, metallation of, **2**, 557
But-2-ene, (*Z*)-2-bromo-, reactions, with lithium, **1**, 1088
But-2-ene, 1,4-dibromo-, reactions, with dialkyl ketones, **1**, 982
But-2-ene, *trans*-2,3-dibromo-, reactions, with trifluoroperacetic acid, **4**, 24
2-Butene, 1,4-dichloro-, as starting material, for 1,3-butenyne, **1**, 1069
2-Butene, (*Z*)-1,4-dichloro-, reactions, with hexamethylenetetramine, **2**, 325
2-Butene, 1,4-dilithio-, synthesis of, **6**, 206
But-2-ene, 2,3-dimethyl-
  reactions
    with chlorodimethylsilane, **2**, 538
    with difluorocarbene, **4**, 7
  See also Borane, thexyl-
2-Butene, (*e*)-1,4-di(trimethylsilyl)-, dimetallation of, **4**, 603
2-Butene, hexakis(trimethylsilyl)-, from, diethyl maleate, **6**, 174
But-2-ene, 2-trimethylsilyl-, reactions, with glycolates, **2**, 70
2-Butene, (*z*)-1,1,4,4-tetra(trimethylsilyl)-, synthesis of, **4**, 603
But-3-ene, 1,2,2,3,4,4-hexafluoro-1,1-diiodo-, as starting material, for 2,2,3,4,4-pentafluorothiobut-3-enoyl fluoride, **5**, 473
cis-But-2-ene, reactions, with diisopinocampheylborane, **2**, 42
(*E*)-But-2-ene, synthesis of, **1**, 459
(*Z*)-2-Butene, synthesis of, **1**, 63
(*Z*)-But-2-ene, synthesis of, **1**, 459
(*E*)-Butene-1,4-diol, synthesis of, **1**, 513
2,3-Butene-1,4-diols, Claisen rearrangement of, **5**, 166
3,4-Butene-1,2-diols, 1,2-erythro-, synthesis of, **2**, 73
2-Butenenitrile, 3-methyl-, synthesis of, **3**, 643
3-Butenenitriles, 2-diethylphosphonoxy-2-methyl-, synthesis of, **3**, 659
cis-2,3-Butene oxide, as starting material, for *threo* 3-triphenylstannyl-2-butanols, **1**, 664
But-1-enes, synthesis of, **1**, 773
But-2-enes, synthesis of, **1**, 773
2-Butenes, 1,4-dithiocyanato-, from, butadienes, **5**, 1108
But-2-enes, phenyl-, synthesis of, **1**, 730
But-3-enes, 1-alkoxy-, from, methyl allyl ethers, **1**, 816
3-Butenes, 2-isothiocyanato-1-aryl-, synthesis of, **5**, 1044
3-Butenoate, methyl 2-methyl-2-(3-trimethylsilylpropyl)-, saponification of, **5**, 96
3-Butenoate, methyl perchloro-, photochlorination of, **6**, 230
But-2-enoic acid, (*Z*)-2-methyl-, synthesis of, **1**, 1088
2-Butenoic acid, (*Z*)-2-trimethylsilyl-, as starting material, for trimethylsilylvinylketene, **2**, 941
But-3-enoic acid, 2,2-dichloro-, synthesis of, **4**, 13
(*E*)-But-2-en-1-ol
  reactions, with tosylhydrazonoacetyl chloride, **3**, 470
  synthesis of, **1**, 516
Butenolides
  bromination of, **4**, 56
  reactions
    with furans, **5**, 171
    with zinc acetate, **5**, 79
  synthesis of, **1**, 599, 763; **5**, 166
Butenolides, bromo, synthesis of, **4**, 57
But-2-en-1-ols, 4-alkyl-, synthesis of, **1**, 277
3-Buten-2-one, reactions, with alkynyl iodides, **1**, 1021
3-Buten-2-one, 4-phenyl-, as starting material, for cinnamic acid, **5**, 103
1-Butenyl complexes, 1-amino-3-oxo-, synthesis of, **4**, 1018
3-Butenyl radicals, cyclization of, **1**, 340
Butenyne, perchloro-, as starting material, for ketenimines, **3**, 590
1,3-Butenyne, from, 1,4-dichloro-2-butene, **1**, 1069
1-Buten-3-yne, synthesis of, **4**, 1045
But-1-en-3-ynes, 4-aryl-, synthesis of, **1**, 1115
1-Buten-3-ynes, 4-aryl-1,1-dichloro-, reactions, with lithium dialkylamides, **2**, 307
But-3-1-ol, reactions, with arylmethoxycarbene complexes, **5**, 942
t-Butoxide
  for the cleavage of allylphosphonium bromide, **1**, 49
  reactions, with iodine, **2**, 95
  use in the cleavage of carbon–silicon bonds, **1**, 62
t-Butoxides, as starting materials, for trialkylated lanthanides, **2**, 565
t-Butoxy chloride, reactions, with morpholinones, **4**, 102
t-Butoxy groups, [1,3]-sigmatropic migration of, **5**, 925
t-Butoxy radicals, reactions, with acetals, **4**, 1078

Butylamine, reactions, with 1,3-oxazin-6-ones, **4**, 981
t-Butylamine
reactions
with perfluoroalkene, **3**, 589
with perfluoro-2-phosphapropene, **5**, 1087
with rhenium complexes, **3**, 522
Butyl bromide, synthesis of, **5**, 17
t-Butyl bromide
reactivity of, **1**, 893
as starting material, for *t*-butyl thiocyanate, **5**, 1105
use in the synthesis of α-bromo ketones, **3**, 154
t-Butyl cations, reactions, with alkylphenylalkynes, **1**, 1122
t-Butyl chloride
hydrolysis of, **1**, 891
reactions
with ammonium thiocyanates, **5**, 1038
with diethylcadmium, **1**, 127
with diethylzinc, **1**, 127
t-Butyl chloride, diphenyl-, for the silylation of amides, **5**, 375
t-Butyldimethylsilyl chloride, reactions, with sodium salt, **5**, 542
t-Butyldimethylsilyl (TBDMS)
use in the synthesis of acyloxysilanes, **5**, 222
use in the synthesis of *N*-silylamides, **5**, 374
t-Butyldiphenylsilyl, use in the synthesis of acyloxysilanes, **5**, 222
s-Butyl iodide, reactions, with silver cyanate, **5**, 1104
t-Butyl isocyanide, reactions, with amines, **6**, 644
Butyllithium, reactions, with 1,8-bis(trimethylsilymethyl)naphthalene, **4**, 603
n-Butyllithium–pentamethyldiethylenetriamine, reactions, with tetramethylsilane, **4**, 602
n-Butyllithium–potassium *t*-butoxide
as metallating agent, **4**, 650
See also Superbase
n-Butyllithium–tetramethylethylenediamine, reactions, with ferrocenes, **4**, 1034
Butyllithium–tetramethylethylenediamine[1,2-bis(dimethylamino)ethane], reactions, with 2,3-dihydro-1,1-dimethyl-1-silaphenalene, **4**, 603
t-Butyl radicals
reactions
with acryl amides, **1**, 326
with citraconic anhydride, **1**, 327
with diethyl citraconate, **1**, 327
with methyl propiolate, **1**, 546
t-Butylthiol, reactions, with imines, **4**, 325
n-Butylthiomethylene group, use in the synthesis of α,β-unsaturated ketones, **3**, 253
t-Butylthiyl radicals, reactions, with thioamide, **4**, 1077
But-2-yn-1,4-diols, as starting materials, for α-allenyl alcohols, **1**, 961
2-Butyn-4-diols, rearrangement of, **3**, 216
1-Butyne, reactions, with *n*-butyllithium, **4**, 605
1-Butyne, 4-bromo-1-(trimethylsilyl)-, as starting material, for 2-methylcyclobutenylmetals, **2**, 976
1-Butyne, 1-iodo-, reactions, with borinane, **5**, 229
1-Butyne, 3-methoxy-3-methyl-, as starting material, for alkynyl ligands, **1**, 119
1-Butyne, 1-phenyl-, dilithiation of, **4**, 668
1-Butyne, 1-trimethylsilyl-, lithiation of, **4**, 1061
1-Butyne, 1,3,3-tris(trimethylsilyl)-, synthesis of, **4**, 605
2-Butyne
photo-carbonylation of, **2**, 670
reactions, with *n*-butyllithium, **4**, 605
But-2-yne
as starting material, for dihydrido diiridium carbon-bridged complexes, **4**, 689
use in the synthesis of chromacyclopentadienes, **4**, 676
2-Butyne, 1,4-dichloro-, as starting material, for 1,3-butadiyne, **1**, 1069
2-Butyne, hexachloro-, reactions, with lithium, **6**, 173
2-Butyne, hexafluoro-
from
hexafluorobutadiene, **6**, 18
hexafluorocyclobutene, **6**, 18
reactions
with cumulenes, **4**, 613
with silver fluoride, **4**, 696
with tetrafluorohydrazine, **2**, 762
But-2-yne, hexafluoro-, reactions, with tetraorganodiarsines, **2**, 876
2-Butyne, hexafluoro-
as starting material, for cyclopropropenylketenimines, **3**, 593
synthesis of, **6**, 16
2-Butyne, 1,1,1,3,3,3-hexafluoro-, as starting material, for 2,3,7,8-tetrakis(trifluoromethyl)-1,4,6,9-tetrathiaspiro[4.4]nona-2,7-diene, **6**, 306
2-Butyne, hexa(trimethylsilyl)-, synthesis of, **6**, 173
2-Butynedioic acid
from, α,β-dibromosuccinic acid, **5**, 103
reactions, with hexamethylantimonytriamide, **2**, 876
2-Butynedioic acid chloride, synthesis of, **5**, 14
Butynes, reactions, with potassium *t*-butoxide, **4**, 1028
Butynes, 1-silyl-, from, 1,1-disilyl allenes, **4**, 1046
Butynes, 1-(trimethylsilyl)-, alkylation of, **1**, 1052
1-Butynes, 3-bromo 1-silyl, reactions, with silyl cuprates, **4**, 1046
But-2-ynes, 1,4-dialkoxy-, reduction of, **1**, 961
But-2-ynes, 4-dialkylamino-1-methoxy-, reduction of, **1**, 961
2-Butynes, 1,4-dihalo-, reductive dehalogenation of, **1**, 986
2-Butynes, 1,4-dihydroxy-, reduction of, **1**, 986
2-Butynes, 1,1,4,4-tetraalkoxy, as starting materials, for 1,1,4,4-tetraalkoxy 1,3-butadienes, **4**, 830
But-2-ynes, 1-trimethylsilyloxy-4-alkoxy-, as starting materials, for butatrienyl ethers, **1**, 981
2-Butynoate, methyl 4-chloro-, from, propargyl chloride, **5**, 103
3-Butynoate, triethyl ortho-, from, tetraalkyl orthocarbonate, **6**, 77
2-Butynoic acid, polysilylation of, **4**, 1047
But-1-yn-4-ol, reactions, with platinum salts, **5**, 944
2-Butyn-1-ol, reduction of, **1**, 516
3-Butyn-2-ol, 2-methyl-
arylation of, **1**, 1013
reactions, with carbon dioxide, **6**, 463
But-3-yn-2-one, reactions, with dienes, **1**, 519
3-Butyn-2-one, reactions, with silylallenes, **1**, 527
2-Butynyl cations, 1,1-diphenyl-, nuclear magnetic resonance analysis of, **1**, 904
Butyramide, *N*,*N*-dimethyliso-, as by-product, in acid halide synthesis, **5**, 2
Butyramidine hydrochloride, heptafluoro-, reactions, with fluorine, **5**, 815
Butyrate, dimethylazobisiso-, reactions, with anthrone, **1**, 266
Butyrate, ethyl α-bromo-, reactions, with magnesium, **1**, 117
Butyrate, *O*-phenyl selenoiso-, synthesis of, problems, **5**, 516
Butyrates, ethyl 3-oxo-4-alkoxy-, oxidation of, **4**, 212
Butyric acid, γ-amino-, synthesis of, **5**, 85
Butyric acid, α-aminoiso-, from, acetone, **5**, 85
Butyric acid, γ-bromo-, synthesis of, **5**, 57
Butyric acid, 4-(*N*,*N*-difluoroamino)-, synthesis of, **2**, 335
Butyric acid, α-hydroxymethyl-, (-) and (+) enantiomers of, **5**, 118
Butyric acid, α-methyl-, use in the synthesis of α-hydroxymethylbutyric acid, **5**, 118
Butyric acid, γ-phenoxy-
cleavage of, **5**, 57
synthesis of, **5**, 63

Butyric acid chloride, 4-difluoroamino-, use in the synthesis of isocyanates, **5**, 989
Butyric acid esters, 2-arylhydrazino-3-oxo-, as starting materials, for hydrazonyl halides, **5**, 678
Butyric acid ethyl esters, 2-phenylhydrazino-3-oxo-, bromination of, **5**, 678
4-Butyrolactim, *O*-methyl-, reactions, with phosgene, **5**, 998
γ-Butyrolactols, β-aryl-, synthesis of, **4**, 168
Butyrolactone, reactions, with epoxides, **6**, 72
4-Butyrolactone, 2-acetyl-, as starting material, for 5-phenyl selenyl 2-pentanone, **3**, 176
γ-Butyrolactones, γ-methylene-, epoxidation of, **4**, 209
Butyronitrile, 4-chloro-, from, 3-chloropropyl bromide, **3**, 612
Butyronitrile, perfluoro-, reactions, with 2,2,3,3,4,4,5,5-octafluoropentyl hypochlorite, **5**, 699
Butyronitrile, γ-phthalimido-, hydrolysis of, **5**, 85

C

Cacodyl oxide, *See* Arsine oxide, bisdimethyl-
Cadet's liquid, synthesis of, **2**, 494
Cadet reaction
  for the synthesis of diarsines, **2**, 506
  for the synthesis of tetraalkyldiarsine oxides, **2**, 494
Cadiot–Chodkiewicz reaction
  mechanism, **1**, 1018, 1142
  palladium-catalyzed version of, **1**, 1019
Cadmium
  reactions
    with alkenyl bromides, **4**, 821
    with dihalodifluoromethanes, **6**, 15
    with perhaloalkanes, **6**, 64
  as trifluoromethylation reagent, **6**, 220
  use in Wittig reactions, **4**, 734
Cadmium, bis(diazomethyl)-, synthesis of, **3**, 488
Cadmium, bis(perfluoroalkyl)-, transmetallation of, **6**, 65
Cadmium, bis(perfluoroisopropyl)-, synthesis of, **4**, 152
Cadmium, bis(trifluoromethyl)-
  reactions
    with dimethyl telluride, **6**, 49
    with phosphorus triiodide, **2**, 434
  as starting material, for fluorocarbene complexes, **6**, 718
  synthesis of, **6**, 246
  trifluoromethylation with, **6**, 245
  use in the synthesis of difluorocarbene, **4**, 7
Cadmium, dialkyl-, reactions, with trichlorophosphine, **2**, 447
Cadmium, diazomethyl-, synthesis of, **3**, 488
Cadmium, diethyl-, reactions, with *t*-butyl chloride, **1**, 127
Cadmium, metallic, reactions, with phosphonates, **6**, 266
Cadmium, perfluoroisopropyl-, metathesis of, **4**, 151
Cadmium(II) chloride, reactions, with diazomethyllithium, **3**, 488
Cadmium bis(ethyl diazoacetate), synthesis of, **3**, 488
Cadmium carbonate, as reagent, for dithiane hydrolysis, **5**, 410
Cadmium chloride
  reduction of, **2**, 590
  use in the synthesis of borenes, **5**, 928
Cadmium chloride, alkyl-, structure of, **6**, 394
Cadmium compounds
  alkylation of, **1**, 465
  reactions, with phosphorus ylides, **4**, 586
  as starting materials, for copper compounds, **6**, 15
  for the substitution of *β*-chloroenones, **1**, 469
Cadmium compounds, alkyl-
  electrochemical synthesis of, **2**, 594
  synthesis of, **2**, 590
    via transmetallation, **2**, 592
Cadmium compounds, bis(perfluoroalkyl)-, transmetallation of, **6**, 65
Cadmium compounds, dialkyl-
  reactions
    with allyliron complexes, **1**, 291
    with trihalo(perfluoroalkyl)germanium compounds, **6**, 62
Cadmium compounds, imidoyl-, synthesis of, **5**, 802
Cadmium compounds, perhalodialkyl-, synthesis of, **6**, 64
Cadmium compounds, trihalomethyl-, synthesis of, **6**, 245
Cadmium compounds, vinyl-, reactions, with copper(I) bromide, **4**, 821
Caesium, trimethylsilylmethyl-, use in deprotonation reactions, **2**, 557
Caesium, triphenylsilyl-, synthesis of, **2**, 545
Caesium, tri-*o*-tolylsilyl-, synthesis of, **2**, 545
Caesium(IV) ions
  reactions
    with acetaldehyde, **5**, 49
    with formaldehyde, **5**, 49
Caesium carbonate
  use in the synthesis of dialkyl sulfides, **2**, 133
  use in the synthesis of macrolactones, **5**, 177
Caesium compounds, synthesis of, **2**, 557
Caesium compounds, alkyl-, synthesis of, **2**, 557
Caesium dithiolates, reactions, with 1,3-dichloroacetone, **2**, 133
Caesium fluoride
  as catalyst, for the synthesis of difluorodioxirane, **6**, 250
  for the desilylation of arylsilanes, **1**, 448
  reactions
    with bisfluoroformyl peroxide, **6**, 424
    with chlorodiphenyl[tris(trimethylsilyl)-methyl]germane, **6**, 710
    with diacetoxyalkyne, **4**, 768
    with difluoroaminocarbonyl fluoride, **6**, 257
    with difluorocyanamide, **6**, 261
    with perfluoroalkylnitriles, **3**, 419
    with perfluoro-2,5-diazahexa-2,4-diene, **6**, 262
    with tris(pentafluoroethyl)phosphine oxide, **2**, 476
  use in the substitution of silicon functions, **1**, 284
Caesium fluoroxy sulfate
  for the fluorination of sulfonamides, **2**, 365
  reactions
    with aldehydes, **5**, 6
    with amides, **2**, 334
  as reagent, for aryl fluorides, **2**, 613
  for the synthesis of aryl fluorides, **2**, 613
Caesium salt, 1,12-bis(hydroxymethyl)decahydrododecaborate, reactions, with hydrogen iodide, **4**, 143
Caesium selenocarboxylates, from, *O*-trimethylsilyl selenocarboxylates, **5**, 542
Caesium thiolcarboxylates, use in the synthesis of thiols, **2**, 124
Calciferol, synthesis of, **1**, 1029
Calcium
  reactions
    with alkylamines, **1**, 508
    with alkyl iodides, **1**, 115
    with 1,3-dienes, **2**, 563
    with diiodomethane, **4**, 669
Calcium, activated, synthesis of, **2**, 563
Calcium(II) halides, reactions, with dimethyl sulfoxide, **2**, 9
Calcium bis(trimethylsilyl)amide, reactions, with benzonitrile, **5**, 777
Calcium borohydride, use in the reduction of carbonyls, **2**, 60
Calcium bromide, reactions, with lithium biphenylide, **2**, 563
Calcium carbonate
  as trapping agent, **1**, 598
  use in elimination reactions, **1**, 644
  use in sulfoxide elimination, **1**, 596
  use in the synthesis of isothiocyanates, **5**, 1028
Calcium chloride, reactions, with sodium hydrogen sulfide, **2**, 764
Calcium compounds, alkyl-, synthesis of, **2**, 563
Calcium compounds, allyl-, homocoupling of, **1**, 116
Calcium compounds, benzyl-, synthesis of, **1**, 115
Calcium cyanamide
  reactions
    with sulfonic acids, **5**, 1119
    with trialkylammoniumtriazines, **5**, 1120

Calcium cyanide, as starting material, for
γ-ketonitriles, **3**, 634
Calcium enolates, synthesis of, **2**, 697
Calcium fluoride
reactions, with phosgene, **6**, 409
as starting material, for carbonic chloride fluoride, **6**, 419
Calcium halides, reduction of, **1**, 115
Calcium hypochlorite
alcohol oxidations with, **3**, 118
for the chlorination of amides, **5**, 317
for the synthesis of aryl chlorides, **2**, 619
Calcium metal, reactions, with alkyl halides, **2**, 563
Calcium metallocycles, for the synthesis of
carbocycles, **1**, 116
Calcium vapor, condensation of, **2**, 563
Calicheamicin
action of, **2**, 233
ring-closure of, **1**, 1012
synthesis of, **1**, 1009; **2**, 233
Calixarenes, use in the synthesis of ethers, **2**, 89
Callitrisaldehyde, as starting material, for callitrisic
acid, **5**, 51
Callitrisic acid, from, callitrisaldehyde, **5**, 51
Camphane, 2-chloro-2-nitroso-, rearrangement of, **3**, 434
(-)-(1*S*,4*R*)-Camphanic acid, synthesis of, **5**, 66
Camphene
reactions
with isothiocyanic acid, **5**, 1043
with *n*-pentylsodium, **1**, 113
Camphenilol, 2-*p*-anisyl-, use in the synthesis of nonclassical
carbocations, **1**, 905
(+)-Camphenilone, reactions, with potassium
t-butoxide, **1**, 868
Camphor, 3-diazo-, synthesis of, **3**, 468
Camphor-derived auxiliaries, use in the synthesis of
enolates, **1**, 134
Camphor-3-glyoxylic acid, reactions, with nitrous
acid, **5**, 325
Camphoric anhydride, (-)-(1*R*,3*R*)-3-chloro-, hydrolysis
of, **5**, 66
Camphorsulfonic acid, use in the synthesis of
γ-lactones, **5**, 171
Camphor-10-sulfonyl chloride, as starting material, for
halosulfines, **5**, 488
Camphor sultams, as chiral auxiliaries, in radical
reactions, **1**, 326, 328
Camptothecins, synthesis of, **6**, 487
Canaline, synthesis of, **2**, 103
Canavanine, synthesis of, **2**, 103
Candida, reactions, with *n*-alkenes, **5**, 117
Candida cylindracea, for the hydrolysis of octyl (*R*)-(+)-2-
chloropropionate, **5**, 118
Candida rugosa, use in the synthesis of carboxylic
acids, **5**, 118
Cannabinol, Δ-6-tetrahydro-, synthesis of, **3**, 539
Cannivonine, dihydro-, synthesis of, **1**, 455
Cannizzaro reaction
of carbonyls, **6**, 25
for the synthesis of 2-furoic acid, **5**, 74
Canonical forms
of ketenimines, **3**, 556, 565
of phosphaallenes, **3**, 603
of phosphacumulenes, **3**, 603
(-)-$\Delta^{9(12)}$-Capnellene
from, bicyclo[3.2.0]heptene, **1**, 802
synthesis of, **1**, 558
$\Delta^{9(12)}$-Capnellene, synthesis of, **1**, 1132
Caprohydroxamic acid hydrochloride, ε-amino-,
phosgenation of, **5**, 971
Caproic acid, ε-benzoylamino-, bromination of, **5**, 57
Caproic acid, ε-benzoylamino-α-bromo-, synthesis of, **5**, 57
Caproic acid, α-bromo-, synthesis of, **5**, 60
n-Caproic acid, α-bromination of, **5**, 60
Caprolactam
α,α-dichlorination of, **4**, 12
N-ethylthiomethylation of, **4**, 324
thionation of, **5**, 574
ε-Caprolactam, reactions, with hydroxylamine
hydrochloride, **5**, 971
Caprothionolactone, reactions, with organolithiums, **4**, 226
Capto-dative methylenes, aminosulfuration of, **5**, 601
Captodative radicals
reviews, **1**, 919
use of term, **1**, 922
cis-Carane, synthesis of, **1**, 72
2-Carbaldehyde, pyrrole-, reactions, with piperidine, **4**, 406
α-Carbalkoxyenethiols, deprotonation of, **1**, 140
Carbalumination, of phenylethyne, **4**, 1068
Carbamate, benzyl *N*-cyano-, reactions, with alkyl
halides, **5**, 1121
Carbamate, *N*,*O*-bis(trimethylsilyl), acylation of, **5**, 1007
Carbamate, *t*-butyl[1-(*t*-butoxycarbonyl)-3-oxo-4-
pentenyl]-, use in the synthesis of carboxylic
acids, **5**, 92
Carbamate, *Se*-*n*-butyl *N*,*N*-diethyselenol-, synthesis
of, **6**, 497
Carbamate, dimethylammonium *N*,*N*'-dimethyl, use in the
synthesis of thionoamides, **5**, 600
Carbamate, ethyl, reactions, with alkenes, **5**, 979
Carbamate, ethyl *N*-chloro-
reactions
with alkenes, **5**, 979
with dialkyl sulfides, **2**, 160
Carbamate, ethyl *N*,*N*-dimethyl, reactions, with
triethyloxonium tetrafluoroborate, **6**, 313
Carbamate, ethyl diphenylmethylene, reactions, with
hydroxylamines, **4**, 448
Carbamate, ethyl *N*-methyl, as starting material, for
dimethyl *N*-methyliminodicarbonate, **6**, 625
Carbamate, 4-(3-methylthioureido)phenyl *N*-
methoxycarbonyl-*N*-methyl, reactions, with dimethyl
sulfate, **6**, 335
Carbamate, trimethylsilyl *N*-trimethylsilyl-*N*-
trimethylsiloxy-, thermolysis of, **5**, 1079
Carbamate esters, reactions, with phosphorus
pentachloride, **6**, 617
Carbamates
O-alkylation of, **6**, 625
deprotonation of, **1**, 138; **3**, 188
from
bis(2-pyridyl) carbonate, **6**, 417
N,*N*'-carbonyldiimidazole, **6**, 416
N,*N*'-(carbonyldioxy)bisbenzotriazole, **6**, 417
chloroformates, **6**, 421
N,*N*'-disuccinimidyl carbonate, **6**, 417
lithium compounds, **4**, 934
4-nitrophenyl chloroformate, **6**, 417
tertiary alcohols, **6**, 417
metallation of, **4**, 530
as protective groups, **6**, 486
pyrolysis of, **1**, 774, 777
reactions
with *s*-butyllithium, **1**, 138; **4**, 382
with diisobutylaluminum methanesulfonate, **4**, 383
with formimidoyl chlorides, **5**, 977
with lithium tetramethylpiperidide, **1**, 1089
with oxalyl chloride, **5**, 977
with phosgene, **5**, 977
with phosphorus halides, **5**, 977
with phosphorus pentachloride, **5**, 976
with phosphorus pentoxide, **5**, 977
with tris(dimethylamino)titanium chloride, **4**, 383
as starting materials
for α-aminoalkylphosphorus compounds, **4**, 466
for iminocarbonates, **6**, 625
for isocyanates, **5**, 976

for *gem*-methoxycarbonylamino diphenylphosphinoylalkenes, **4**, 1000
synthesis of, via Hofmann rearrangement, **5**, 999
thermal decomposition of, **5**, 975, 1005
thermolysis of, **5**, 977
use in modified Gabriel syntheses, **2**, 299
Carbamates, *N*-acyl alkylidene, synthesis of, **4**, 427
Carbamates, alkenyl, Fritsch–Buttenberg–Wiechell rearrangement of, **1**, 1071
Carbamates, alkyl
diazotization of, **6**, 423
as starting materials, for fluoroformate esters, **6**, 423
Carbamates, *O*-alkyl
deprotonation of, **4**, 382
lithiation of, **4**, 382
silylation of, **4**, 360
Carbamates, alkyl *N*-cyano-, alkylation of, **5**, 1119
Carbamates, alkyl *N*,*N*-dichloro-, reactions, with phenylalkyne, **4**, 794
Carbamates, *O*-alkyl (isothiocyanatocarbonyl), synthesis of, **6**, 559
Carbamates, alkyl phosphoryl, thermal decomposition of, **5**, 1017
Carbamates, *O*-alkyl-*N*-(trifluoromethyl), from, isocyanates, **6**, 239
Carbamates, aryl *N*-(benzyloxymethyl), hydrogenolysis of, **4**, 310
Carbamates, benzyl
cleavage of, **2**, 304
as starting materials, for α-aminoalkylphosphorus compounds, **4**, 467
Carbamates, benzyl *N*-alkyl-*N*-cyano-, synthesis of, **5**, 1121
Carbamates, benzylic, alkylation of, **4**, 382
Carbamates, α-bromo
synthesis of, **4**, 102
use in the synthesis of α-amino acids, **4**, 102
Carbamates, chiral, deprotonation of, **4**, 382
Carbamates, *gem*-chloro, synthesis of, **4**, 794
Carbamates, α-chloro-, synthesis of, **4**, 102
Carbamates, α-chloroalkyl, synthesis of, **4**, 60
Carbamates, ω-chloroalkyl, synthesis of, **6**, 445
Carbamates, *N*-cyanothiol, synthesis of, **5**, 1121
Carbamates, cyclic, from, diethyl carbonate, **6**, 417
Carbamates, dialkyl *N*-cyano-, synthesis of, **5**, 1119
Carbamates, enol
α-metallation of, **4**, 841
oxidation of, **1**, 558
α-selenylation of, **4**, 841
Carbamates, enol allyl, dehydrogenation of, **1**, 558
Carbamates, ethyl
reduction of, **4**, 303
as starting materials, for α-aminoalkylphosphorus compounds, **4**, 467
synthesis of, **6**, 363
Carbamates, ethynyl, synthesis of, **2**, 1029
Carbamates, α-halo
synthesis of, **4**, 101
via halogenation, **4**, 102
Carbamates, 4-hydroxy-1,2-alkadienyl, from, propargylic carbamates, **1**, 974
Carbamates, lithiated, synthesis of, **4**, 530
Carbamates, α-lithiodifluorovinyl, reactions, with electrophiles, **1**, 1089
Carbamates, *O*-(1-lithiovinyl), reactions, with 1,2-dibromotetrafluoroethane, **4**, 768
Carbamates, methyl
reactions, with sodium hydrogen telluride, **2**, 304
reduction of, **2**, 304
as starting materials, for α-aminoalkylphosphorus compounds, **4**, 467
use in the synthesis of α-aminoalkylphosphines, **4**, 456
Carbamates, methyl *N*-(α-benzotriazol-1-ylalkyl), reactions, with 2-methylthiophene, **1**, 262
Carbamates, nitraminomethyl, synthesis of, **4**, 447
Carbamates, nitroso-, as starting materials, for diazo compounds, **5**, 867
Carbamates, organostannyl-*t*-butyl, as starting materials, for α-aminoalkylcyanocuprates, **2**, 315
Carbamates, phenyl, alkylation of, **4**, 382
Carbamates, *N*-phenyl, as starting materials, for diisocyanates, **5**, 984
Carbamates, propargylic
as starting materials
for allenes, **1**, 966
for 4-hydroxy-1,2-alkadienyl carbamates, **1**, 974
Carbamates, *N*-silyl, thermolysis of, **5**, 978
Carbamates, *O*-silyl, decomposition of, **5**, 978
Carbamates, *O*-silylated, reactions, with fluorophosphoranes, **2**, 476
Carbamates, 1-sulfonylalkyl, dehydrochlorination of, **4**, 840
Carbamates, sulfonylallyl, alkylation of, **3**, 213
Carbamates, 1-sulfonylallyl, deprotonation of, **4**, 841
Carbamates, 1-sulfonylvinyl, synthesis of, **4**, 840
Carbamates, tertiary, synthesis of, **1**, 138
Carbamates, *N*-thioaroyl, synthesis of, **5**, 593
Carbamates, *N*-thio(hetero)aroyl, as intermediates, in the synthesis of thionoamides, **5**, 592
Carbamates, vinyl, lithiated compounds of, use in the synthesis of ketones, **3**, 137
Carbamate transfer reagents, *N*-[^{3}H]methyl *N*-nitroso-, synthesis of, **6**, 483
Carbametallation, mechanism, **1**, 315
Carbamic acid benzyl ester, synthesis of, **6**, 484
Carbamic acid chloride, *N*-ethyl-*N*-phenyl-, reactions, with ammonia, **6**, 501
Carbamic acid chlorides, reactions, with amines, **6**, 501
Carbamic acid compounds, as starting materials, for isocyanates, **5**, 976
Carbamic acid compounds, *N*-silylated, decomposition of, **5**, 977
Carbamic acid esters
reactions, with amines, **6**, 501
as starting materials, for *N*-chlorocarbonyls, **6**, 449
Carbamic acid esters, halocarbonyl-, synthesis of, **6**, 437
Carbamic amides, halocarbonyl-, synthesis of, **6**, 437
Carbamic halides
nomenclature of, **6**, 437
See also Carbamoyl halides
Carbamide chloride, *N*-chlorobenzyl-, synthesis of, **5**, 658
Carbamidic acid chloride, reactions, with hydroxylamine, **6**, 505
Carbamidic acid chlorides, *N*-(α-chloroalkyliden), from, acyl isocyanates, **5**, 658
Carbamidic acid esters, *N*-acyl-, reactions, with amines, **6**, 506
Carbamidic acid esters, *N*-nitroso-, reactions, with amines, **6**, 506
Carbamimidic acids, synthesis of, **6**, 631
Carbamimidic azides, sulfonyl-, reactions, with triphenylphosphane, **6**, 651
Carbamimidic chlorides
from
carbodiimides, **6**, 621
carbonimidic dichlorides, **6**, 619
cyanamides, **6**, 620
dialkylcyanamides, **6**, 620
dichloroiminium salts, **6**, 619
diisopropylcarbodiimide, **6**, 621
thioureas, **6**, 619
ureas, **6**, 620
synthesis of, **6**, 619
Carbamimidic chlorides, *N*,*N*-dimethyl-, synthesis of, **6**, 619
Carbamimidic chlorides, *N*-sulfonyl-, synthesis of, **6**, 619
Carbamimidic esters *See* Isoureas
Carbamimidic fluorides, synthesis of, **6**, 615
Carbamimidothioic acids, synthesis of, **6**, 634

Carbamimidothioic esters *See* Isothioureas
Carbaminoyl chlorides, *N*-silyl-, as intermediates, **5**, 968
Carbamoyl azides
from, aldehydes, **6**, 511
synthesis of, **6**, 511
Carbamoyl azides, *N*-stannyl-, synthesis of, **5**, 993
Carbamoyl bromide, dimethyl-, synthesis of, **6**, 452
Carbamoyl bromides
from, isocyanates, **6**, 452
synthesis of, **6**, 452
Carbamoyl bromides, α-bromoalkyl-, from, α,β-unsaturated isocyanates, **6**, 452
Carbamoyl bromides, *N*,*N*-disubstituted
from
carbamoyl chlorides, **6**, 452
N,*N*-disubstituted formamides, **6**, 452
secondary amines, **6**, 452
synthesis of, **6**, 452
Carbamoyl chloride
dehydrohalogenation of, **5**, 965
reactions, with sodium azide, **6**, 511
synthesis of, **6**, 443
use in the synthesis of *N*-alkyl arylamides, **5**, 301
Carbamoyl chloride, bis(trifluoromethyl)-, synthesis of, **6**, 438
Carbamoyl chloride, *N*-*t*-butyl-*N*-alkyl, pyrolysis of, **5**, 971
Carbamoyl chloride, chloromethyl-, as intermediate, in the synthesis of isocyanates, **5**, 986
Carbamoyl chloride, *N*-chloromethyl-*N*-phenyl-, synthesis of, **5**, 971
Carbamoyl chloride, dimethyl-
fluorination of, **6**, 438, 440
transhalogenation of, with bromide, **6**, 452
Carbamoyl chloride, *N*-phenyl-*N*-4-isocyanatobenzyl-, synthesis of, **5**, 971
Carbamoyl chlorides
carcinogenic properties of, **6**, 442
from
diphosgene, **6**, 446
formamides, **6**, 449
phosgene, **6**, 443
primary amines, **6**, 443
triphosgene, **6**, 446
reactions
with alkynide anions, **5**, 304
with dithiocarbonates, **6**, 493
with monosubstituted formamides, **5**, 744
with thiols, **6**, 493
stability of, **6**, 442
as starting materials
for *N*,*N*-disubstituted carbamoyl bromides, **6**, 452
for thiocarbamates, **6**, 493
synthesis of, **4**, 102; **6**, 442
thermal cleavage of, **5**, 977
Carbamoyl chlorides, *N*-carbonyl-substituted, synthesis of, **5**, 1007
Carbamoyl chlorides, chloromethyl-, synthesis of, **4**, 102
Carbamoyl chlorides, dialkyl-
as starting materials, for *S*-2-chloroalkyl *N*,*N*-dialkyl thiocarbamates, **6**, 493
use in the synthesis of isocyanates, **5**, 1006
Carbamoyl chlorides, *N*,*N*-dialkyl-
reactions
with 1-alkynes, **1**, 1048
with alkynyl metals, **1**, 1048
Carbamoyl chlorides, disubstituted, synthesis of, **5**, 971
Carbamoyl chlorides, *N*,*N*-disubstituted
stability of, **6**, 443
synthesis of, **6**, 443, 448
Carbamoyl chlorides, *N*-monosubstituted
as starting materials, for isocyanates, **6**, 443
synthesis of, **6**, 448
Carbamoyl chlorides, tolyleneamino-, as intermediates, **5**, 966
η^2-Carbamoyl complexes, synthesis of, **6**, 516
Carbamoyl compounds, α-halo-, synthesis of, **4**, 101, 102
Carbamoyl disulfides, synthesis of, **6**, 480
Carbamoyl fluoride, difluoromethyl-, synthesis of, **6**, 53
Carbamoyl fluoride, diphenyl-, from, carbonic chloride fluoride, **6**, 437
Carbamoyl fluoride, *n*-propyl-*N*-fluoro-, synthesis of, **6**, 439
Carbamoyl fluorides
aromatic substitution of, **6**, 439
from
alkyl isocyanates, **6**, 439
N-aryldifluoroazomethines, **6**, 438
aryl isocyanates, **6**, 439
benzophenone imine, **6**, 438
carbamoyl fluorides, **6**, 439
carbonic difluoride, **6**, 438
carbonic fluoride halides, **6**, 437
N-carboxymethyl-*N*-trimethylsilylhydrazine, **6**, 438
cyanogen chloride, **6**, 441
fluorination, **6**, 439
hydrogen fluoride, **6**, 439
trifluoromethylamines, **6**, 440
as intermediates, for fungicides and insecticides, **6**, 437
nitration of, **6**, 439
stability of, **6**, 437
as starting materials, for carbamoyl fluorides, **6**, 439
synthesis of, **6**, 437
Carbamoyl fluorides, *N*,*N*-disubstituted, uses of, **6**, 437
Carbamoyl fluorides, *N*-monosubstituted, synthesis of, **6**, 439
Carbamoyl halides
nomenclature of, **6**, 437
synthesis of, via halogen–fluorine exchange, **6**, 438
Carbamoyl iodides, synthesis of, **6**, 454
Carbamoyl iodides, α-iodoalkyl-, from, α,β-unsaturated isocyanates, **6**, 454
Carbamoyl–metal complexes, synthesis of, **6**, 517
Carbamoyl phosphonates, synthesis of, **6**, 513
Carbamyl halides
nomenclature of, **6**, 437
See also Carbamoyl halides
Carbanionic compounds, reactions, with phosphonium salts, **3**, 494
Carbanion lone pair pyramidal inversion, **1**, 846
Carbanion reaction, for the synthesis of ketene *S*,*N*-acetals, **4**, 901
Carbanions
adjacent to arsene, synthesis of, **2**, 1003
adjacent to chalcogen, synthesis of, **2**, 999
adjacent to halogen, synthesis of, **2**, 997
adjacent to metalloids, synthesis of, **2**, 1004
adjacent to metals, synthesis of, **2**, 1005
adjacent to nitrogen, synthesis of, **2**, 1001
adjacent to phosphorus, synthesis of, **2**, 1003
formation of, **1**, 398
from
organometallics, **1**, 885
unsaturated acidic hydrocarbons, **1**, 866
general literature survey, **1**, 844
with no stabilizing group, synthesis of, **1**, 109
with one stabilizing group, synthesis of, **1**, 129
quenching of, for the synthesis of 1,1-bis(silyl)alkanes, **4**, 602
reactions
with alkoxythiocarbonyls, **5**, 510
with *O*-alkyl chlorothioformates, **5**, 513
with *O*-alkyl dithiocarbonates, **5**, 513
with alkyl halides, **1**, 110, 610
with bis(methylthio)methanimines, **4**, 901
with dicarbonyls, **4**, 166
with diiron salts, **4**, 680
with electrophiles, **1**, 971; **3**, 620
with enones, **3**, 188

with ethyl iodide, **1**, 858
with 2-isocyanoacrylic esters, **3**, 706
with lactones, **4**, 171
with thioaldehydes, **1**, 1040
with thioketones, **1**, 1040
with tin, **6**, 292
with transition metals, **1**, 845
with trialkylboranes, **1**, 278
rearrangement of, **1**, 886
and SE1 mechanism, **1**, 854
selenylation of, **4**, 285; **6**, 123
solvation of, **1**, 859
sources of, for the synthesis of nitriles, **3**, 621
stability of, **1**, 846
stabilization of, **1**, 301, 867
via conjugation, **1**, 847, 848
sulfenylation of, **6**, 122, 123
synthesis of, **1**, 315, 378, 844
choice of bases and solvents for, **1**, 880
with double bond rearrangement, **1**, 882
reviews, **1**, 1088, 1132
via acyl cleavage, **1**, 883
via addition to carbon–carbon double bonds, **1**, 888
via cleavage of alkoxides, **1**, 883
via cleavage of ketones, **1**, 884
via decarboxylation, **1**, 883
via Michael addition, **1**, 888
via proton loss, **1**, 848
via scission of carbon–carbon bonds, **1**, 883
via scission of carbon–nitrogen bonds, **1**, 887
via scission of carbon–oxygen and carbon–halogen bonds, **1**, 886
thiocarbamoylation of, **4**, 901
with two and three stabilizing groups, synthesis of, **1**, 165
types of, **1**, 845
use in the synthesis of alkenes, **2**, 1003
Carbanions, acyl, reactions, with dialkylchloroboranes, **2**, 267
Carbanions, α-(acylamino), synthesis of, **2**, 1002
Carbanions, α-alkoxy, synthesis of, **2**, 999
Carbanions, α-amino, synthesis of, via nitrosation of secondary amines, **2**, 302
Carbanions, anti-aromatic monocyclic, synthesis of, **1**, 865
Carbanions, aromatic, synthesis of, **1**, 886
Carbanions, aromatic monocyclic, synthesis of, **1**, 861
Carbanions, base-catalyzed, kinetics, **1**, 853
Carbanions, benzylic, stability of, **1**, 847
Carbanions, boron-stabilized
for the methylation of ketones, **1**, 765
as starting materials, for alkenylboranes, **2**, 905
synthesis of, **1**, 162
use in the synthesis of alkenes, **1**, 764
Carbanions, bridgehead, synthesis of, **1**, 847
Carbanions, colored, use in analytical procedures, **1**, 847
Carbanions, dicoordinate, synthesis of, **1**, 1088
Carbanions, dipole-stabilized
reviews, **4**, 382
synthesis of, **1**, 152
use in the synthesis of α-amino compounds, **4**, 529
Carbanions, formamidine-stabilized, use in the synthesis of α-amino ketones, **3**, 182
Carbanions, 2-halo, synthesis of, **2**, 70
Carbanions, α-halo, synthesis of, **2**, 997
Carbanions, α-heteroatom-stabilized, synthesis of, **2**, 997
Carbanions, homoaromatic
synthesis of, **1**, 886
via proton loss, **1**, 865
Carbanions, lithiated, silylation of, **6**, 289
Carbanions, nitrile-stabilized, synthesis of, **1**, 155
Carbanions, α-nitro
synthesis of, **2**, 1002
x-ray crystal structure of, **2**, 1002
See also Nitrosonate anion
Carbanions, nitrogen-stabilized
reviews, **4**, 526
synthesis of, **1**, 147
Carbanions, α-oxygenated, chemistry of, **1**, 135
Carbanions, oxygen-stabilized, synthesis of, **1**, 135
Carbanions, α-phenylsulfonyl, as starting materials, for vinyl bromides, **2**, 623
Carbanions, phosphonate-stabilized, **1**, 744
Carbanions, phosphorus-stabilized
sulfenylation of, **6**, 121
synthesis of, **1**, 157
via conjugate addition, **1**, 159
Carbanions, phosphoryl-stabilized
synthesis of, **2**, 1003
for the synthesis of alkenes, **1**, 752
Carbanions, selenium(II)-stabilized, alkylation of, **1**, 712
Carbanions, selenium(IV)-stabilized, alkylation of, **1**, 712
Carbanions, selenium-stabilized
reactions, with halosilanes, **4**, 374
silylation of, **4**, 374
as starting materials, for β-hydroxy selenides, **2**, 281
synthesis of, **1**, 164; **4**, 398
Carbanions, α-selenoxyl, synthesis of, **2**, 1001
Carbanions, α-selenyl, synthesis of, **2**, 1001
Carbanions, silicon-stabilized, synthesis of, **1**, 159, 759
Carbanions, α-silyl, synthesis of, **2**, 1004, 1005
Carbanions, stabilized
reactions, with activated alkynes, **1**, 537
synthesis of, **2**, 997
Carbanions, α-sulfenyl, synthesis of, **2**, 999
Carbanions, α-sulfinyl, synthesis of, **2**, 1000
Carbanions, sulfinyl-stabilized, as nontransferable cuprate ligands, **4**, 396
Carbanions, α-sulfonimidoyl, x-ray crystal structure of, **2**, 1000
Carbanions, α-sulfonyl
as starting materials, for dialkyl sulfones, **2**, 158
synthesis of, **2**, 1000
Carbanions, sulfonyl-stabilized, reactions, with acid compounds, **3**, 170
Carbanions, sulfoxide-stabilized, use in the synthesis of biotin, **1**, 143
Carbanions, sulfur(II)-stabilized, alkylation of, **1**, 693
Carbanions, sulfur(IV)-stabilized, alkylation of, **1**, 693
Carbanions, sulfur(VI)-stabilized, alkylation of, **1**, 696
Carbanions, sulfur-stabilized
silylation of, **4**, 365
as starting materials, for dialkyl sulfoxides, **2**, 153
synthesis of, **1**, 139
Carbanions, tricoordinate, synthesis of, **1**, 844
α-Carbanions, synthesis of, **1**, 246
6a-Carbaprostaglandin I_2, synthesis of, **1**, 710
Carbatitanation, of 1-trimethylsilyl-2-phenylethyne, **4**, 1061
Carbazate, *t*-butyl, reactions, with α-chloro esters, **4**, 448
Carbazate, ethyl
Mannich-type reaction of, **4**, 442
synthesis of, **2**, 374
Carbazates
diazotization of, **6**, 489
from, hydrazines, **4**, 60
reactions, with phosgene, **6**, 444
synthesis of, **6**, 488
Carbazole, as phase-transfer catalyst, **4**, 792
Carbazole, *N*-acetyl-, reactions, with phosphorus pentachloride, **4**, 791
Carbazole, *N*-(1,2-dichlorovinyl)-, synthesis of, **4**, 791
Carbazole, *N*-nitroso-3-nitro-, as nitroso-group transfer reagent, **2**, 798
Carbazole, *N*-(phenylthiomethyl)-, synthesis of, **4**, 323
Carbazole, 9-vinyl-, reactions, with benzotriazole, **4**, 323
Carbazoles, 9-(1-alkoxyalkyl)-, synthesis of, **4**, 310
Carbene, bromochloro-
reactions, with diazoalkanes, **4**, 759

reactivity of, **4**, 1080
as starting material, for bromochloroalkanes, **4**, 38
synthesis of, **4**, 38
Carbene, bromofluoro-
as starting material, for bromofluoroalkanes, **4**, 35
synthesis of, **4**, 35
Carbene, chlorofluoro-
from
dichlorofluoromethane, **4**, 1083
trichlorofluoromethane, **4**, 1082
reactions, with triphenylphosphonium fluorenylide ylide, **4**, 753
as starting material, for chlorofluoroalkanes, **4**, 32
synthesis of, **4**, 32
Carbene, dibromo-
from
cyclopropanes, **4**, 1083
phenyl tribromomethylmercury, **4**, 748
reactions, with ketene acetals, **4**, 162
reactivity of, **4**, 1080
as starting material, for dibromoalkanes, **4**, 24
synthesis of, **4**, 24
photochemical methods, **4**, 1083
Carbene, dichloro-
addition to dihydrofurans, **6**, 124
from
cyclopropanes, **4**, 1083
ketenes, **4**, 1083
tetrachloromethane, **4**, 1083
for the homologation of phosphaethenes and 1-phosphaallenes, **3**, 607
identification of, **4**, 1082
insertion into germanium–carbon bonds, **1**, 291
as intermediate, for the synthesis of carbonimidic dichlorides, **6**, 609
reactions
with allyl benzyl sulfide, **4**, 775
with amines, **3**, 695
with azides, **6**, 608
with azo compounds, **6**, 608
with cyclohexyl isocyanide, **3**, 596
with dialkylamines, **6**, 107
with diaryldiazomethanes, **4**, 740
with iron(II) tetraphenylporphyrin, **3**, 700
with 2-methylpyridine, **3**, 722
with silanes, **6**, 60
with sulfides, **4**, 775
with thioacetals, **4**, 775
reactivity of, **4**, 1080
as starting material
for carbonimidic dichlorides, **6**, 608
for dichloroalkanes, **4**, 16
for dichloromethylenetriphenylphosphine, **4**, 737
synthesis of, **4**, 16
in situ, **5**, 877
use in semipinacol rearrangements, **1**, 390
Carbene, difluoro-
from
alkoxides, **4**, 1083
bromodifluoromethylphosphonium salts, **4**, 7
(bromodifluoromethyl)triphenylphosphonium bromide, **4**, 737
chlorodifluoromethane, **4**, 8; **6**, 38
methyl chlorodifluoroacetate, **6**, 224
polytetrafluoroethylene, **6**, 12
trifluoromethylphenylmercury, **4**, 1083
reactions
with alcohols, **2**, 89
with alkoxides, **6**, 37
with benzobarrelene, **4**, 8
with cyclopropenes, **4**, 737
with 2,3-dimethylbut-2-ene, **4**, 7
with enols, **6**, 229
with sodium diethyl phosphonate, **6**, 58
with thiols, **6**, 45
reactivity of, **4**, 1080
sources of, **4**, 7
as starting material
for aryl trifluormethyl sulfones, **6**, 235
for difluoroalkanes, **4**, 7
for difluorohalomethylarsines, **6**, 243
for phosphonium ylides, **3**, 495
synthesis of, **4**, 1083; **6**, 243, 244
photochemical methods, **4**, 1083
for the synthesis of bromodifluoroiodomethane, **6**, 224
Carbene, diiodo-
as starting material, for diiodoalkanes, **4**, 29
synthesis of, **4**, 29
Carbene, dimethoxy-, synthesis of, **4**, 1080
Carbene, dinitro-, from, ylides, **4**, 1083
Carbene, diphenyl-
reactions
with alkynes, **1**, 549
with thiobenzoyliminodithiocarbonates, **6**, 314
Carbene, fluoroiodo-
as starting material, for fluoroiodoalkanes, **4**, 36
synthesis of, **4**, 36
Carbene, fluoro(trifluoromethyl)-, reactions, with phosphines, **4**, 111
Carbene, isopropylidene-
reactions, with *trans*-diaryldiphosphines, **4**, 1025
synthesis of, **4**, 985
Carbene, silyl-, rearrangement of, **6**, 383
Carbene adducts, diphenyl-, from, tungsten carbenes, **3**, 517
Carbene complexes
aldol reactions of, **5**, 956
alkylation of, **5**, 955
conjugated reactions of, **5**, 958
deprotonation of, **6**, 674
Fischer synthesis of, **6**, 719
from
carbyne complexes, **6**, 722
isocyanide complexes, **6**, 721
lithium acyl metallates, **6**, 719
metal carbonyls, **6**, 719
metallacyclobutanes, **1**, 666
functionalization of, **5**, 955
reactions
with alkynes, **2**, 979
with *n*-butyllithium, **5**, 946
with *N,N*-dimethylformamide, **6**, 723
with Michael acceptors, **5**, 958
with nucleophiles, **2**, 580
as starting materials
for cyclic alkoxycarbene complexes, **5**, 946
for oxacyclopentylidene complexes, **5**, 946
for thioether complexes, **5**, 951
synthesis of, **3**, 507
See also Transition metal carbene complexes
Carbene complexes, acyloxy-
stability of, **5**, 949
as starting materials, for aminocarbene complexes, **5**, 953
synthesis of, **5**, 949
Carbene complexes, alkoxy-
alkylation of, **5**, 955
under phase-transfer conditions, **5**, 955
characteristics of, **5**, 933
from, acyl complexes, **5**, 937
reactions
with amines, **5**, 953
with lithiodithianes, **5**, 951
with thiols, **5**, 951
stability of, **5**, 933
as starting materials, for aminocarbene complexes, **5**, 953
synthesis of, **5**, 949; **6**, 719
via addition–elimination reactions, **5**, 942

via Fischer reaction, **5**, 933
via Semmelhack reaction, **5**, 937
transacylation of, **5**, 951
Carbene complexes, alkoxy acyclic mononuclear, synthesis of, **5**, 933
Carbene complexes, alkoxy(silyl)-
as starting materials, for amino(silyl)carbene complexes, **6**, 723
synthesis of, **6**, 723
Carbene complexes, alkylthio(silyl)-, synthesis of, **6**, 723
Carbene complexes, amino-
characteristics of, **5**, 952
from
acyloxycarbene complexes, **5**, 953
alkoxycarbene complexes, **5**, 953
alkyl σ-complexes, **5**, 955
amino acids, **5**, 953
carboxylic acid amides, **5**, 953
carbyne complexes, **5**, 953
reactions, with enolates, **5**, 958
structure of, **5**, 932
synthesis of, **5**, 949, 952
via cycloaddition, **5**, 954
via ethyne–vinylidene rearrangements, **5**, 954
Carbene complexes, amino(silyl)-, from, alkoxy(silyl)carbene complexes, **6**, 723
Carbene complexes, arylmethoxy-, reactions, with but-3-1-ol, **5**, 942
Carbene complexes, carbohydrato-, synthesis of, **5**, 942
Carbene complexes, cyclic alkoxy-
from
alkynes, **5**, 944
carbene complexes, **5**, 946
ω-dihaloalkanes, **5**, 943
ω-halocarboxylic acid chlorides, **5**, 943
σ-vinyl complexes, **5**, 948
synthesis of, **5**, 943
Carbene complexes, cyclopropenyl-, synthesis of, **3**, 513
Carbene complexes, dichloro-
modification of, for the synthesis of dihalocarbene complexes, **6**, 718
synthesis of, **5**, 1097
Carbene complexes, difluoro-, synthesis of, **6**, 717
Carbene complexes, dihalo-
from
metal trihalomethyls, **6**, 717
trihalomethyl complexes, **6**, 717
reactions
with alkylamines, **6**, 722
with arylamines, **6**, 722
synthesis of, **6**, 717
Carbene complexes, *C,C*-dihetero-substituted, synthesis of, **6**, 716
Carbene complexes, diiodo-, as intermediates, **5**, 1097
Carbene complexes, dimethyl-, synthesis of, **3**, 518
Carbene complexes, *C,C*-dinitrogen-substituted, synthesis of, **6**, 721
Carbene complexes, fluoro-, from, bis(trifluoromethyl)cadmium, **6**, 718
Carbene complexes, hydridoalkyl-, synthesis of, **6**, 663
Carbene complexes, hydroxy-
stability of, **5**, 948
synthesis of, **5**, 948
Carbene complexes, β-hydroxy-, dehydration of, **5**, 957
Carbene complexes, oxacycloalkylidene-, as starting materials, for 2-acyl-4-alkoxyphenols, **2**, 670
Carbene complexes, phenoxy-, synthesis of, **5**, 936
Carbene complexes, phenyl-, synthesis of, **3**, 514
Carbene complexes, phosphinoiridium-, rearrangement of, **4**, 583
Carbene complexes, silicon-substituted, synthesis of, **6**, 723
Carbene complexes, silyloxy-, from, chlorosilanes, **6**, 719
Carbene complexes, unsaturated, cycloaddition reactions of, **5**, 959
Carbene complexes, α,β-unsaturated
as Michael acceptors, **5**, 958
synthesis of, **5**, 941, 957
Carbene complexes, α,β-unsaturated alkoxy-, reactions, with enolate anions, **5**, 958
Carbene complexes, vinyl-, stability of, **3**, 509
Carbene cycloadditions, stereochemistry of, **1**, 353
Carbene hydride complexes, from, tantalum alkyl complexes, **3**, 512
Carbene moieties, addition to double bonds, **1**, 988
Carbene–rhenium complexes, reactions, with zirconocene chlorohydride, **4**, 716
Carbenes
alkylation–acylation of, **5**, 946
dimerization of, **4**, 830
electrochemical methods, synthesis of, **4**, 1083
flanked by one sulfur and one oxygen, **4**, 1081
flanked by two hetero atoms, **4**, 1080
flanked by two nitrogens, **4**, 1080
flanked by two oxygens, **4**, 1080
flanked by two sulfurs, **4**, 1081
from
arylmercuric compounds, **3**, 321
diazoalkanes, **1**, 408
diazo compounds, **1**, 196; **4**, 1082
α-diazo esters, **1**, 549
1-ethynylvinyl triflates, **1**, 988
geminal dihaloalkanes, **1**, 299
selenoureas, **6**, 143
insertion into carbon–germanium bonds, **1**, 291
as intermediates
in the synthesis of cyclopropanes, **1**, 298
in the synthesis of dithioacetals, **4**, 250
for the synthesis of tetrathioorthocarbonates, **6**, 306
iron complexes of, **6**, 727
as Michael acceptors, **5**, 958
as Michael donors, **5**, 958
molybdenum complexes of, **6**, 727
osmium complexes of, **6**, 727
reactions
with acetone, **3**, 314
with alkenes, **1**, 298, 352, 988
with alkynes, **1**, 549
with allenes, **1**, 499
with allyl methyl sulfide, **1**, 409
with allyl phenyl sulfide, **1**, 409
with carbonyls, **1**, 366
with diazoalkanes, **4**, 755, 759
with *trans*-diphosphene, **6**, 264
with enynes, **1**, 549
with germylenes, **3**, 503
with heteroatoms, **1**, 400
with phosphines, **3**, 497
with silanes, **4**, 137
with thiocarbonyl groups, **1**, 367
with thiocarbonyls, **4**, 862
with triethylboron, **6**, 691
reduction of, **5**, 457
ruthenium complexes of, **6**, 727
sources of, **4**, 709
as starting materials
for carbonyl ylides, **3**, 319
for furans, **3**, 321
for ketenimines, **3**, 596
for methylenephosphonium ions, **6**, 691
for nitrile ylide, **3**, 679
for oxygen heterocycles, **3**, 319, 320
for phosphonium ylides, **3**, 497
for sulfur ylides, **1**, 821
for ylides, **3**, 500
synthesis of, **1**, **1087–1145**, 822; **4**, **1071–1083**, 1080
irradiative methods, **4**, 1083
via α-elimination, **4**, 1081

transition metal complexes of, **6**, 727
use in the synthesis of cyclopropanes, **6**, 351
use in the synthesis of ketene aminals, **4**, 975
use of term, **3**, 507
Carbenes, acyclic, synthesis of, **6**, 663
Carbenes, alkoxy(dialkylamino)-, synthesis of, **6**, 719
Carbenes, alkyl-, use in the synthesis of phenols, **2**, 670
Carbenes, alkylidene-, *in situ* generated, **1**, 495
Carbenes, carbalkoxy-, reactions, with carbonyls, **3**, 319
Carbenes, chloro-
reactions
with phosphaalkenes, **4**, 115
with phosphaalkynes, **4**, 110
Carbenes, dialkoxy-, synthesis of, **3**, 323
Carbenes, dichloro-
from
sodium trichloroacetate, **6**, 25
trichloromethyl anions, **6**, 23
reactions
with alcohols, **4**, 767
with anilines, **6**, 140
with pyrazole, **6**, 141
as starting materials, for ketene aminals, **4**, 979
Carbenes, dichloromethylene-, synthesis of, **1**, 1143
Carbenes, difluoro-
reactions, with disilanes, **6**, 268
synthesis of, **6**, 261, 266
Carbenes, dihalo-
reactions
with alkoxides, **6**, 37
with silyl enol ethers, **3**, 67, 235
synthesis of, **4**, 1082
Carbenes, 1,1-dihalo-, reactions, with alkynes, **1**, 550
Carbenes, diphosphorus-substituted, addition of, **4**, 557
Carbenes, dithiolium-, dimerization of, **4**, 851
Carbenes, fluoro-, reactions, with cyclopentadienes, **2**, 615
Carbenes, α-germyl-, as starting materials, for
germenes, **3**, 503
Carbenes, halo-
from, diazirines, **6**, 280
reactions
with diphosphenes, **6**, 688
with imines, **6**, 51
Carbenes, α-keto-
as intermediates, in the synthesis of ketenes, **3**, 530
reactions, with furans, **3**, 65, 232
Carbenes, metallated
synthesis of, **4**, 1083
See also Carbenoids
Carbenes, metalloalkoxy-
reactions, with allylamine, **5**, 798
as starting materials
for imidoylchromiums, **5**, 798
for imidoyltungstens, **5**, 798
Carbenes, methylene-
reactions
with alkenes, **1**, 1144
with trialkylsilanes, **1**, 1144
synthesis of, **1**, 1143
Carbenes, nonheteroatom-stabilized, synthesis of, **3**, 508
Carbenes, phosphino(silyl)-
addition of, **4**, 567
intramolecular insertion of, **4**, 567
Carbenes, silyl-, as starting materials, for silenes, **3**, 502
Carbenes, α-silyl-, as intermediates, for the synthesis of
silenes, **5**, 925
Carbenes, silyloxy-, rearrangement of, for the synthesis of
acylsilanes, **5**, 421
Carbenes, α-stannyl-, insertion into *C-S* bonds, **4**, 398
Carbenes, triplet, use in the synthesis of
cyclopropanes, **1**, 298
Carbenes, vinyl-
cyclization of, **3**, 322
as intermediates, **1**, 550
reactions
with dienes, **1**, 811
with diphenyl diselenide, **4**, 875
with diphenyl ditelluride, **4**, 877
Carbenes, vinylidene-, synthesis of, **1**, 1143
Carbene transfer
classes of, **1**, 352
mechanism, **1**, 352
Carbenium cations, tris(methylthio)-, as starting materials,
for tetrathioorthocarbonates, **6**, 301
Carbenium hexachloroantimonate, tris(azido)-, synthesis
of, **6**, 733
Carbenium ions
definition of, **1**, 890
from, alkanols, **6**, 296
reactions
with alkyl hydroperoxides, **2**, 98
with hydrogen peroxide, **2**, 98
stability of, **1**, 560
stable salts of, synthesis of, **6**, 729
synthesis of, **6**, 726
Carbenium ions, α-amino-, synthesis of, **2**, 1008
Carbenium ions, imino-, synthesis of, **2**, 1009
Carbenium ions, oxo-, synthesis of, **2**, 1007
Carbenium ions, trialkoxy-
reactions
with lithium dialkylphosphides, **6**, 314
with sodium dialkyl phosphites, **6**, 313
use in the synthesis of orthocarbonates, **6**, 296
Carbenium salts, as starting materials, for
phosphaalkenes, **5**, 883
Carbenium salts, dioxo-
reactions
with sodium hydrosulfide, **6**, 96
with sodium sulfide, **6**, 96
with thiols, **6**, 96
as starting materials
for monothioortho esters, **6**, 96
for ortho esters, **6**, 73
Carbenium salts, trialkoxy-, as starting materials, for
orthocarbonates, **6**, 296
Carbenium salts, tris(trichlorophosphoranediylamino)-,
synthesis of, **6**, 655
Carbenium tetrafluoroborate, triethoxy-, synthesis of, **6**, 729
Carbenium tetrafluoroborate, trimethoxy-, synthesis
of, **6**, 729
Carbenium tetrafluoroborates, dialkylaminoethoxy-, as
starting materials, for acetals, **6**, 320
Carbenium tetrafluoroborates, trialkoxy-, as starting
materials, for tetraalkyl orthocarbonates, **6**, 300
Carbenoid method, in the synthesis of phosphonium
ylides, **6**, 697
Carbenoids
definition of, **4**, 741
from
diazoalkanes, **6**, 673
1,1-dihalocyclopropanes, **1**, 164
reactions
with acetone, **4**, 146
with alkenyl halides, **1**, 1067
with alkynes, **1**, 549
with carbonyls, **1**, 366
with electrophiles, **4**, 817
with iminophosphines, **6**, 693
with methylenephosphines, **6**, 693
synthesis of, **2**, 997; **4**, 145, 1083
transmetallation of, **4**, 820
use of term, **1**, 352; **4**, 816
See also Carbenes, metallated
See also α-Halo carbanions
Carbenoids, alkoxy-, synthesis of, **6**, 63
Carbiimide, dichloro-

as starting material
for pentaalkylguanidines, **6**, 643
for tetraalkylguanidines, **6**, 643
for trialkylguanidines, **6**, 643
Carbiimides, dichloro-, as starting materials, for *N*-aryliminocarbonyls, **6**, 645
Carbiimides, *N*-(2,4-dichlorophenyl)dichloro-, reactions, with ethylenediamine, **6**, 645
Carbiimides, dihalo-, as starting materials, for *N*-alkyliminocarbonyls, **6**, 643
Carbimidate, dimethyl, synthesis of, **5**, 698
Carbinamines, from, *N*-borylimines, **2**, 316
Carbinamines, secondary, synthesis of, **2**, 317
Carbinamines, tertiary, synthesis of, **2**, 312, 316
Carbinol, bis(triethylgermyl)-, from, triethylgermyllithium, **6**, 131
Carbinol, methyldiphenyl-, synthesis of, **1**, 136
Carbinols
oxidation of, **6**, 519
synthesis of, **4**, 513
Carbinols, acetylenyl-, as starting materials, for acylsilanes, **5**, 413
Carbinols, diaryl-, as starting materials, for *gem*-dinitroalkenes, **4**, 993
Carbinols, 1,1-divinyl-, oxidation of, **1**, 422
Carbinols, ethoxyethynyl-, from, lithium diethylamide, **2**, 1025
Carbinols, ethynyl-, as starting materials, for alkynyl chlorides, **2**, 1016
Carbinols, silyl-, synthesis of, **4**, 358
Carbinols, stannyl-, synthesis of, **4**, 533
Carbinols, tertiary, synthesis of, **4**, 935
Carbinols, tribromomethyl-, from, tribromomethyllithium, **6**, 31
Carbinols, trichloromethyl-
from, trichloroacetaldehyde, **6**, 27
hydrolysis of, **5**, 66
oxidation of, **6**, 26
reactions, with cyanamide, **5**, 1121
synthesis of, **6**, 26
Carbinols, α-trichloromethyl-, synthesis of, **6**, 25
Carbinols, trihalo-
from
bromotrichloromethane, **6**, 26
tetrachloromethane, **6**, 26
Carbinols, vinyl alkynyl-, oxy-Cope rearrangement of, **3**, 276
N-Carboalkyloxyamino acid phenylhydrazides, reactions, with ferric chloride, **5**, 87
Carboalumination
of alkynylsilanes, **2**, 925
vs. hydroalumination, **2**, 974
mechanism, **1**, 453; **2**, 596
regiochemistry of, **2**, 980
of terminal alkynes, catalyzed by zirconocene dichloride, **2**, 975
use in the synthesis of trialkylaluminums, **2**, 595
zirconium-catalyzed, **2**, 974
and carbocupration compared, **2**, 977
of homopropargyls, **2**, 977
of propargyls, **2**, 977
trans-Carboalumination, zirconium-catalyzed, of bis(trimethylsilyl)butadiyne, **4**, 809
Carboboration, of alkynes, **2**, 902
Carbocation, formation of, **1**, 380
Carbocationic salts, alkyl, synthesis of, **1**, 899
Carbocationic salts, triphenylmethyl, synthesis of, **1**, 898
Carbocation mechanism, use of term, **1**, 891
Carbocations
adjacent to chalcogen, synthesis of, **2**, 1007
adjacent to halogen, synthesis of, **2**, 1006
adjacent to nitrogen, synthesis of, **2**, 1008
bearing one heteroatom, synthesis of, **2**, 1006
bearing three heteroatom functions, synthesis of, **6**, 726
definition of, **1**, 890
dicoordinate, synthesis of, **1**, 1109
formed in superacid media, **1**, 893
from, radicals, **1**, 336
general literature survey, **1**, 891
as intermediates, **1**, 893
in the synthesis of alkenes, **1**, 631
nomenclature, **1**, 890
nonclassical, synthesis of, **1**, 904
nonclassical model, criticisms, **1**, 906
optimum geometry, **1**, 892
stability of, **1**, 894
stabilization, via hyperconjugation, **1**, 892
structural stability of, **1**, 891
structure of, **1**, 892
synthesis of, **1**, 890; **5**, 38
reviews, **1**, 1109
solvent effects, **1**, 892
via addition to carbon–carbon bonds, **1**, 911
via bond scission, **1**, 891, 897
via electrophilic addition to alkenes, **1**, 294
via electrophilic addition to carbon–carbon bonds, **1**, 895
via protonation of alkenes, **1**, 911
via protonation of dienes, **1**, 911
See also Carbonium ions
Carbocations, acyl, synthesis of, **2**, 1008
Carbocations, α-alkoxy, synthesis of, **2**, 1007
Carbocations, α-amino, synthesis of, **2**, 1008
Carbocations, aryl-substituted, synthesis of, via bond scission, **1**, 897
Carbocations, benzylic, synthesis of, **1**, 894
Carbocations, α-halo, synthesis of, **2**, 1006
Carbocations, α-heteroatom-stabilized, synthesis of, **2**, 1006
Carbocations, homoaromatic, synthesis of, **1**, 917
Carbocations, μ-hydrido-bridged, synthesis of, **1**, 918
Carbocations, α-hydroxy, synthesis of, **2**, 1007
Carbocations, α-selenyl, synthesis of, **2**, 1008
Carbocations, α-sulfenyl, synthesis of, **2**, 1008
Carbocations, tricoordinate, synthesis of, **1**, 890
Carbocations, trithio-substituted, fluorination of, **6**, 275
α-Carbocations
stability of, **4**, 674
stabilization of, in alkyne-bridged dicobalt complexes, **4**, 688
Carbocupration
of alkynes, **1**, 123; **2**, 976
of alkynylsilanes, **2**, 924
of heteroatom-substituted alkynes, **2**, 977
of heterosubstituted alkynes, **2**, 977
and hydroalumination compared, **2**, 974
regiochemistry of, **2**, 980
of silylethyne, **2**, 924
and zirconium-catalyzed carboalumination compared, **2**, 977
Carbocycles
synthesis of, **1**, 116, 123; **2**, 593
via intramolecular Michael addition, **1**, 307
via radical cyclization, **1**, 341
Carbodesilylation, mechanism, **1**, 448
Carbodiarsane, synthesis of, **5**, 1094
Carbodiimide, benzoyl-*t*-butyl-
reactions
with trimethylgermanium dimethylamine, **5**, 390
with trimethylsilylamines, **5**, 377
with trimethyltin dimethylamine, **5**, 390
Carbodiimide, bis(germanyl)-, synthesis of, **5**, 1081
Carbodiimide, bis(pentafluorosulfanyl)-, from, bis(pentafluorosulfanyl)urea, **5**, 1085
Carbodiimide, bis(silyl)-, as *O*-silylation agents, **5**, 536
Carbodiimide, bis(trialkylstannyl)-, as starting material, for metal amides, **5**, 389
Carbodiimide, bis(tributylstannyl)-, acylation of, **5**, 716

Carbodiimide, bis(trifluoromethyl)-, synthesis of, **5**, 1080
Carbodiimide, *N*,*N*-bis(trifluoromethyl)-, hydrolysis of, **6**, 239
Carbodiimide, bis(trimethyl)-, reactions, with 2-chloro-1,3,2-diazaborolidines, **5**, 1085
Carbodiimide, bis(trimethylsilyl)-
reactions, with carboxylic acid chlorides, **5**, 721
as starting material, for carbodiimides, **5**, 1074
synthesis of, **5**, 1076, 1081
Carbodiimide, bis(trimethylstannyl)-, reactions, with acid chlorides, **5**, 1121
Carbodiimide, bis(triphenylstannyl)-, reactions, with 1,1,3-trisubstituted thioureas, **5**, 1121
Carbodiimide, *N*-cyano-*N*'-phenyl-, synthesis of, **5**, 1070
Carbodiimide, diborated, from, *t*-butyliminotetramethylpiperidinoborane, **5**, 1085
Carbodiimide, dicyclohexyl- (dcc)
as coupling reagent, for the synthesis of *N*-alkylamides, **5**, 279
as dehydrating agent, for the synthesis of acylated oximes, **5**, 215
reactions
with alkoxyamines, **5**, 327
with *t*-butoxycarbonyl amino acids, **5**, 185
with carboxylic acids, **3**, 475
with methanols, **6**, 519
with peracids, **5**, 208
with thiocarboxylic acids, **5**, 243
for the synthesis of diacyl peroxides, **5**, 208
use in the synthesis of anhydrides, **5**, 185
use in the synthesis of carboxylic esters, **5**, 124
use in the synthesis of isothiocyanates, **5**, 1023
use in the synthesis of thiocarboxylic esters, **5**, 234
Carbodiimide, difluorophosphoryltrimethylsilyl-, synthesis of, **5**, 1074
Carbodiimide, diisopropyl-
reactions, with acetyl chloride, **5**, 1147
as starting material, for carbamimidic chlorides, **6**, 621
Carbodiimide, *N*,*N*'-diisopropyl-, reactions, with trimethylsilyl cyanide, **5**, 777
Carbodiimide, diphenyl-
reactions
with diethyl phosphite, **6**, 366
with phenyloxirane, **6**, 510
as starting material, for ketenimines, **3**, 593
Carbodiimide, *N*,*N*'-diphenyl-
reactions
with bis(trimethylsilyl)mercury, **6**, 661
with phosphite triesters, **6**, 657
as starting material, for amidinophosphines, **6**, 658
Carbodiimide hydrochloride, *N*-(dimethylaminopropyl)-*N*'-ethyl-
reactions
with *N*-benzyloxycarbonyl amino acids, **5**, 185
with *N*-*t*-butyloxycarbonyl amino acids, **5**, 185
with methyl succinate, **5**, 185
Carbodiimides
for alcohol activation, **2**, 8, 32, 36
as coupling reagents, for the synthesis of *N*-alkylamides, **5**, 279
as dehydrating agents, for the synthesis of anhydrides, **5**, 185
from
acid chlorides, **5**, 1063
bis(trimethylsilyl)carbodiimide, **5**, 1074
N,*N*'-disubstituted ureas, **5**, 1079
heterocycles, **5**, 1073
imidoyl dichlorides, **5**, 1070
isocyanates, **5**, 1061, 1065
isothiocyanates, **5**, 1062, 1065
metal oxides, **5**, 1062
phosphorous compounds, **5**, 1063
N-sulfonyl-*N*'-substituted thioureas, **5**, 1064
thioureas, **5**, 1062
ureas, **5**, 1062
hydrosilylation of, **5**, 759, 777
hydrostannylation of, **5**, 780
as isomers of cyanamides, **5**, 1062
phenylboration of, **5**, 779
polymerizable, synthesis of, **5**, 1065
reactions
with *N*-acylketenimines, **4**, 982
with alcohols, **6**, 633
with alkyl halides, **5**, 1146
with alkynyl metals, **1**, 1044
with bis(fluorosulfonyl)methane, **4**, 866
with carboxylic acids, **6**, 506
with cyanuric chloride, **6**, 622
with Grignard reagents, **5**, 743
with hydrogen chloride, **6**, 728
with organolithiums, **5**, 743
with organometallics, **5**, 754
with organoniobiums, **5**, 781
with organotantalums, **5**, 781
with oxiranes, **6**, 510
with phenyllithium, **5**, 777
with phosgene, **6**, 621
with phosphorus pentachloride, **6**, 622
with stannylated ynamines, **2**, 1044
with thiols, **6**, 635
with trichlorotriazine, **6**, 622
reduction of, **5**, 745
stability of, **5**, 1062
as starting materials
for *N*-alkyliminocarbonyls, **6**, 643
for arylguanidines, **6**, 645
for *N*-aryliminocarbonyls, **6**, 645
for carbamimidic chlorides, **6**, 621
for *N*-chlorocarbonyls, **6**, 445
for cyanoguanidines, **6**, 648
for *N*-cyanoiminocarbonyls, **6**, 648
for hydrazonamides, **5**, 849
for isocyanides, **3**, 695
for isothioureas, **6**, 635
for isoureas, **6**, 633
for ketene aminals, **4**, 979
for metal amides, **5**, 390
for selenoureas, **6**, 596
for *N*-silylamides, **5**, 377
for sulfonylguanidines, **6**, 652
synthesis of, **5**, 1061; **6**, 358
catalysts for, **5**, 1079
uses of, **5**, 1061
water soluble, synthesis of, **5**, 1062, 1065, 1078
Carbodiimides, acyl-, synthesis of, problems, **5**, 1073
Carbodiimides, *N*-(1-acyloxyalkyl)-N^1-aryl-, synthesis of, **5**, 1075
Carbodiimides, *N*'-acyl *N*-silyl-, synthesis of, **5**, 1082
Carbodiimides, N^1-alkyl-N^2-aryl-, synthesis of, **5**, 1070
Carbodiimides, *N*-alkyl-*N*'-nitro-, as intermediates, **5**, 963
Carbodiimides, *N*-alkyl-*N*'-silyl-, from, alkylcyanamides, **5**, 1082
Carbodiimides, ammonio-, use in the synthesis of anhydrides, **5**, 185
Carbodiimides, *N*,*N*'-bis(1-alkenyl)-, synthesis of, **5**, 1078
Carbodiimides, *N*,*N*'-bis(1-chloroalkyl)-, synthesis of, **5**, 1063
Carbodiimides, *N*,*N*'-bis(diphenylphosphine)-, stability of, **5**, 1078
Carbodiimides, bisorganophosphoryl-, synthesis of, **5**, 1074
Carbodiimides, bis(trialkylstannyl)-
reactions
with acid chlorides, **5**, 389
with anhydrides, **5**, 389
Carbodiimides, *N*,*N*'-bis(tributylsilyl)-, reactions, with hexafluoroacetone, **5**, 1083

Carbodiimides, bis(trimethylsiloxy)-, synthesis of, **5**, 1079
Carbodiimides, *N*,*N'*-bis(trimethylsilyl)-, reactions, with hexafluoroacetone, **5**, 1083
Carbodiimides, carbamoyl-, synthesis of, **5**, 1064
Carbodiimides, 1-chloroalkyl-, reactions, with carboxylic acids, **5**, 1075
Carbodiimides, α-chloroalkyl-, from, α-chloroalkyl isocyanates, **5**, 1068
Carbodiimides, conjugated unsymmetrical, synthesis of, **5**, 1066
Carbodiimides, *N*-cyano *N'*-alkoxyphosphoryl-, synthesis of, **5**, 1074
Carbodiimides, cyclic silyl-, synthesis of, **5**, 1077
Carbodiimides, *N*,*N'*-dialkyl-
  as starting materials
    for dialkylguanidines, **6**, 643
    for tetraalkylguanidines, **6**, 643
    for trialkylguanidines, **6**, 643
Carbodiimides, *N*-(dialkylamino)-, reactivity of, **5**, 1063
Carbodiimides, *N'*-(1,2-diarylethenyl)-, synthesis of, **5**, 1066
Carbodiimides, dimeric, synthesis of, **5**, 1082
Carbodiimides, *N*-(dimethylamino)*N'*-phenyl-, characterization of, **5**, 1063
Carbodiimides, heteroaryl-, synthesis of, **5**, 1068
Carbodiimides, imidoyl-, synthesis of, **5**, 1064
Carbodiimides, isocyanatomethyl-, synthesis of, **4**, 441
Carbodiimides, β-keto-, synthesis of, **3**, 194
Carbodiimides, oligomeric, synthesis of, **5**, 1076
Carbodiimides, organic, hydrolysis of, **5**, 1081
Carbodiimides, perfluoroalkyl-
  anionotropic transformations of, **5**, 1080
  synthesis of, **5**, 1069
Carbodiimides, phenyl-, stability of, **5**, 1083
Carbodiimides, phenylphosphonoyl-*N'*-triorganometalloyl-, synthesis of, **5**, 1084
Carbodiimides, phosphinic, synthesis of, **5**, 1062
Carbodiimides, phosphonic, synthesis of, **5**, 1062
Carbodiimides, phosphonyl-*N'*-alkyl-, synthesis of, **5**, 1062
Carbodiimides, polymeric, synthesis of, **5**, 1082
Carbodiimides, polymeric diborylated, synthesis of, **5**, 1085
Carbodiimides, *N*-silylated, reactions, with α-chloro alkyl isocyanates, **4**, 441
Carbodiimides, *N*-sulfonyl-, synthesis of, **5**, 1064
Carbodiimides, symmetrical, synthesis of, **5**, 1076
Carbodiimides, thiocarbamoyl-, synthesis of, **5**, 1064
Carbodiimides, *N'*-trifluoromethyl-*N*-alkyl-, synthesis of, **5**, 1076
Carbodiimides, *N*-(trihalomethyl)-, synthesis of, **6**, 239
Carbodiimides, unsymmetrical, synthesis of, **5**, 1062
Carbodiphosphines
  synthesis of, **5**, 1091
    via halogen–metal exchange, **5**, 1092
Carbodiphosphorane, dimethyltetraphenyl-, reactions, with chlorotrimethylsilane, **6**, 370
Carbodiphosphorane, hexamethyl-
  characteristics of, **5**, 1094
  synthesis of, **5**, 1093
Carbodiphosphorane, hexaphenyl-
  reactions, with acid chlorides, **2**, 1063
  as starting material, for 1-phosphacumulenes, **3**, 604
Carbodiphosphoranes
  from, phosphonium salts, **5**, 1094
  with heteroatoms attached to phosphoranes, synthesis of, **5**, 1095
  phosphorus–arsenic analogues, synthesis of, **5**, 1094
  structure of, **5**, 1092
  synthesis of, **5**, 1092; **6**, 703
  uses of, **5**, 1092
Carbodiphosphoranes, hexaphenyl-
  from, methylenebis(triphenylphosphonium) bromide, **5**, 1093
  synthesis of, **5**, 1093
Carbodithioimidates, *N*-cyano-, reactions, with diaminobenzenes, **5**, 1124
ω-Carboethoxyacyl chlorides, reactions, with alkylzinc chloride, **5**, 76
Carboethoxy group, migratory aptitude of, **2**, 660
Carbogermylation, of alkynes, **2**, 946
Carbohydrates
  cyclization in, **4**, 1079
  dehydration of, **4**, 858
  as hemiacetals, **4**, 176
  as starting materials
    for borates, **2**, 111
    for carboxylic esters, **5**, 140
  synthesis of, **2**, 597; **6**, 309
    via enzymatic aldol reactions, **3**, 159
Carbohydrates, 1-halo-2-acyloxy, conversion to ortho esters, **6**, 71
Carbohydrates, phosphorylated, synthesis of, **2**, 106
Carboimidates, *N*-cyano-, reactions, with nucleophiles, **5**, 1124
β-Carboline, 3,4-dihydro-, α-addition to, **2**, 313
β-Carboline-3-carboxylic acid, 1,2,3,4-tetrahydro-, reactions, with carbon disulfide, **6**, 564
β-Carboline-3-carboxylic acid, 1,2,3,4-tetrahydro-2-(methoxythiocarbonyl)-, synthesis of, **6**, 556
β-Carbolines
  lithiation of, **4**, 532
  synthesis of, **3**, 433
β-Carbolines, 1-substituted-1,2,3,4-tetrahydro-, synthesis of, **2**, 313
Carbolithiation
  of alkynes, cyclic, **2**, 972
  of benzyne intermediates, **2**, 974
  mechanism, **2**, 555
  for the synthesis of alkyllithiums, **2**, 555
syn-Carbolithiation, of alkynes, **2**, 972
Carbomagnesiation
  of alkynes, **2**, 973
  of alkynylsilanes, **2**, 924
  of benzyne intermediates, **2**, 974
  catalyzed by copper iodide, **2**, 973
  mechanism, **1**, 316
  for the synthesis of alkylmagnesiums, **2**, 561
Carbomagnesiation, enantioselective, for the synthesis of alkylmagnesiums, **2**, 562
Carbomagnesiation, intramolecular, for the synthesis of alkylmagnesiums, **2**, 562
Carbometallation
  of alkynes, **1**, 538, 544; **2**, 919, 971
  of alkynylsilanes, **2**, 924; **4**, 809
  of allenes, **1**, 495
  for aryl metals, **2**, 9560
  characteristics of, **2**, 956
  of ethynylsilanes, **4**, 809
  of Grignard reagents, **2**, 973
  of homopropargyls, **2**, 975
  and *syn*-hydrometallation compared, **2**, 972
  mechanism, **2**, 971
  for organometallics, **2**, 952
  with organometals containing groups 1 and 2 metals, **2**, 972
  with organotransition metals, **2**, 978
    catalytic, **2**, 979
    stoichiometric, **2**, 978
  problems with, **2**, 967
  of propargyls, **2**, 975
  reactions, **2**, 984
  for the synthesis of alkylcobalt(I) compounds, **2**, 576
  for the synthesis of alkylcoppers, **2**, 589
  for the synthesis of alkyl metal(II) compounds, **2**, 577
  for vinyl metals, **2**, 956
Carbometallation, intramolecular, of alkylzincs, **2**, 592
anti-Carbometallation, with organotransition metals, **2**, 981
syn-Carbometallation, requirements for, **2**, 978
Carbomethylation, for the synthesis of carboxylic acids, **5**, 27

Carbon
  as catalyst, for the synthesis of phosgene, **6**, 412
  halomethanes as a source of, **5**, 1097
  migration of, **1**, 799
  [1,7] migrations of, **1**, 800
  planar geometry at, **6**, 404
  reactions, with boron tetrafluoride, **6**, 387
Carbon, six-coordinated, structure of, **6**, 406
sp Carbon
  synthesis of oxygen-based functional groups attached to, **2**, 1021
  synthesis of selenium-based functional groups attached to, **2**, 1036
  synthesis of sulfur-based functional groups attached to, **2**, 1030
  synthesis of tellurium-based functional groups attached to, **2**, 1036
Carbon–aluminum bonds, cleavage of, **1**, 69
Carbonamidic dichlorides, *N*-(*β*-chloroalkyl)-, synthesis of, **6**, 606
Carbon–antimony bonds, reduction of, **1**, 54, 444
Carbon–arsenic bonds
  cleavage of, **1**, 268
  reduction of, **1**, 53, 444
Carbonate, acetophenone oxime ethyl, Beckmann rearrangement of, **5**, 661
Carbonate, bis(2-ethylhexyl), synthesis of, **6**, 463
Carbonate, bis(2-pyridyl), as starting material, for carbamates, **6**, 417
Carbonate, bis(succinimido), reactions, with 1,2-diaminoethane, **6**, 507
Carbonate, bis(trichloromethyl)
  from, dimethyl carbonate, **6**, 416
  as phosgene substitute, **6**, 462
  reactions, with amino acids, **6**, 430
  as reagent
    for the synthesis of aldehydes, **3**, 6
    for the synthesis of ketones, **3**, 118
  safety precautions, **6**, 416
  synthesis of, **6**, 230, 416, 462
  use as phosgene alternative, **6**, 416
  use in the synthesis of *N*-carboxy anhydrides of α-amino acids, **6**, 416
  See also Triphosgene
Carbonate, bis(trichloromethyl)pentathiodiperoxy, reactions, with amines, **5**, 1029
Carbonate, *N*-cyanoimido, reactions, with ammonia, **2**, 322
Carbonate, *Se*-decyl *O*-methyl, synthesis of, **6**, 482
Carbonate, *Te*-decyl *O*-methyl, synthesis of, **6**, 482
Carbonate, dialkyl, as alkylating agent, **2**, 683
Carbonate, diallyl
  synthesis of, **6**, 463
  use in allylation reactions, **1**, 193
Carbonate, dibutyl, synthesis of, **6**, 464
Carbonate, di-*s*-butyl, synthesis of, **6**, 463
Carbonate, di-*t*-butyl peroxy-, photolysis of, **2**, 98
Carbonate, diethyl
  industrial synthesis of, **6**, 460
  reactions
    with ethylenediamine, **6**, 484
    with methylhydrazine, **6**, 489
  as starting material, for cyclic carbamates, **6**, 417
  synthesis of, **6**, 465
    via ozonolysis, **6**, 469
    via transesterification, **6**, 467
Carbonate, diethyl *N*-chloroimino, as starting material, for phosphoryl dichloride isocyanate, **5**, 988
Carbonate, diisopropyl peroxido, reactions, with benzaldimines, **3**, 666
Carbonate, dimethyl
  chlorination of, **6**, 462
  from, tetramethoxyethylene, **6**, 469
  industrial synthesis of, **6**, 465
  as octane enhancer, **6**, 467
  photochlorination of, **6**, 230
  as starting material, for bis(trichloromethyl) carbonate, **6**, 416
  synthesis of, **6**, 415, 464, 465
    via transesterification, **6**, 467
Carbonate, dimethyl cyanodithioimido, reactions, with amines, **6**, 647
Carbonate, dimethyl *N*-cyclohexylimino, from, carbonimidic dichloride, **6**, 624
Carbonate, dipentyl, synthesis of, **6**, 463
Carbonate, diphenyl
  reactions
    with amines, **5**, 976
    with dimethylamine, **6**, 502
  transesterification of, **6**, 470
Carbonate, dipropyl, synthesis of, **6**, 464
Carbonate, ethylene, as dehydrating agent, for the synthesis of acetals, **4**, 178
Carbonate, 1-iodo-7-octene-2,4-diol 2,4-cyclic, synthesis of, **6**, 469
Carbonate, isopropenyl, aminolysis of, **6**, 484
Carbonate, methyl, chlorination of, **4**, 59
Carbonate, 4-nitrophenylene, synthesis of, **6**, 467
Carbonate, 7-octene-2,4-diol-2,4-cyclic, synthesis of, **6**, 469
Carbonate, propargyl, synthesis of, **6**, 461
Carbonate, 1,2,2,2-tetrachloroethyl *t*-butyl, synthesis of, **4**, 59
Carbonate, trichloromethyl, as *in situ* source of phosgene, **4**, 58
Carbonate complexes, monoalkyl, synthesis of, **6**, 470
Carbonate esters, ^{14}C-labeled, synthesis of, **6**, 462
Carbonates
  chlorination of, for chloroformates, **6**, 426
  iodocyclization of, **6**, 468
  pyrolysis of, **1**, 774, 777
  reactions, with terminal alkynes, **1**, 1025
  as starting materials
    for dialkyl thiocarbonates, **6**, 474
    for ketones, **3**, 122
  synthesis of
    via electrochemistry, **6**, 469
    via ozonolysis, **6**, 469
  use as phosgene alternatives, **6**, 417
Carbonates, acyl
  as acylating reagents, **5**, 235
  synthesis of, **6**, 469
Carbonates, alkenyl, synthesis of, **2**, 675
Carbonates, 2-alkyl, as starting materials, for 1,2-dien-4-ynes, **6**, 461
Carbonates, *O*-alkyl
  lithiation of, **4**, 382
  silylation of, **4**, 360
Carbonates, alkyl allyl
  from, allyl chloroformate, **6**, 461
  as starting materials, for ketones, **3**, 122
Carbonates, alkynyl cyclic, as starting materials, for α-hydroxyallenes, **1**, 962
Carbonates, allylic
  as allylating reagents, **2**, 74
  as nucleophilic allylating reagents, **2**, 80
  reactions, with palladium(0) complexes, **6**, 470
Carbonates, 4-amino-2-alkynyl, as starting materials, for allenic *β*-lactams, **1**, 976
Carbonates, α-chloroalkyl, from, chloromethyl chloroformates, **4**, 59
Carbonates, chloroalkyl alkyl, synthesis of, **4**, 59
Carbonates, α-chloroalkyl alkyl, reactions, with amines, **4**, 60
Carbonates, 1-chloroethyl, as starting materials, for fluoroformate esters, **6**, 423
Carbonates, chloroimino, reactions, with phosphorus(III) halides, **5**, 988

Carbonates, cyclic
  as starting materials, for urethanes, **6**, 484
  synthesis of, **6**, 467
Carbonates, 1,2-cyclic, synthesis of, **6**, 468
Carbonates, 4,5-cyclic, synthesis of, **6**, 468
Carbonates, dialkyl
  as activating reagents, for the synthesis of anhydrides, **5**, 189
  from
    carbon oxides, **6**, 463
    formates, **6**, 466
    inorganic carbonates, **6**, 462
    ketones, **6**, 466
    phosgene, **6**, 460
    urea compounds, **6**, 463
  reactions
    with diamines, **6**, 507
    with silica, **2**, 108
  as starting materials, for urethanes, **6**, 484
  synthesis of, via iodolactonization, **6**, 468
Carbonates, diaryl, reactions, with ureas, **5**, 981
Carbonates, enol, synthesis of, **6**, 462
Carbonates, α-haloalkyl alkyl, synthesis of, **4**, 59
Carbonates, α-haloalkyl aryl, synthesis of, **4**, 59
Carbonates, α-haloalkyl phenyl, synthesis of, **4**, 59
Carbonates, imino-
  from
    carbamates, **6**, 625
    carbonimidic difluorides, **6**, 624
    carbonimidic halides, **6**, 624
    cyanates, **6**, 624
    cyanogen halides, **6**, 624
    dichlorobis(aryloxy)methanes, **6**, 624
    isothiocyanates, **6**, 624
    orthocarbonates, **6**, 624
  synthesis of, **6**, 623
    via *O*-alkylation of alkoxycarbonyl functions, **6**, 625
    via functional group interconversion, **6**, 625
  See also Carbonimidic acids
Carbonates, inorganic, as starting materials, for dialkyl carbonates, **6**, 462
Carbonates, α-methylene cyclic, synthesis of, **6**, 463
Carbonates, monoalkyl, cyclization of, **6**, 464
Carbonates, oxime
  Beckmann rearrangements of, **1**, 1041
  rearrangement of, **5**, 689
Carbonates, *o*-phenylene, as starting materials, for cyclopentadienylketene, **3**, 543
Carbonates, phosphoryl *O*-alkyl, synthesis of, **6**, 489
Carbonates, propargylic
  as starting materials
    for alkenylallenes, **1**, 976
    for allenic esters, **1**, 975
Carbonates, tetraalkyl, as starting materials, for trialkyl orthobenzoates, **6**, 77
Carbonates, 1,2,2,2-tetrachloroethyl, as starting materials, for fluoroformates, **6**, 423
o-Carbonates, reactions, with alkynylmagnesium halides, **1**, 1026
Carbonate salts, as reagents, for the synthesis of alcohols, **2**, 87
Carbon atoms, pentacoordinate, in dimeric bis(cyclopentadienyl) metal–methyl complexes of lanthanides, **4**, 702
Carbon–bismuth bonds, reduction of, **1**, 54, 444
Carbon–boron bonds
  cleavage of, **1**, 58
  reduction of, **1**, 57, 447
  synthesis of, **4**, 629
Carbon–boron bond scission, for the synthesis of carbon-centered radicals, **1**, 946
Carbon-bridged complexes, synthesis of, **4**, 669
Carbon–bromine bonds, reduction of, **1**, 8
Carbon–carbon bonds
  cleavage of, **1**, 56
  nucleophilic addition to, **1**, 888
  synthesis of
    use of alkyl metals in, **2**, 550
    via addition reactions, **1**, **293–317**, **461–500**, **501–551**
    via elimination reactions, **1**, **553–587**, **589–671**
    via the opening of epoxides, **1**, 204
    via radical addition, **1**, 320
    via radical reactions, **1**, 331
    via sigmatropic rearrangements, **1**, 979
    via substitution, **1**, **171–247**, **249–292**, **461–500**
    via substitution of halogen, **1**, **105–169**
  for the synthesis of alkynes, **1**, 1000
Carbon–carbon bond scission, for the synthesis of radicals, **1**, 928
Carbon–carbon double bonds
  reactions
    with germyl radical, **1**, 950
    with group 14-centered radicals, **1**, 950
    with heteroatom radicals, **1**, 948
    with nitrogen-centered radicals, **1**, 950
    with oxygen-centered radicals, **1**, 948
    with phosphoranesulfenyl bromides, **2**, 256
    with silyl radical, **1**, 950
    with stannyl radicals, **1**, 950
    with sulfonyl radicals, **1**, 949
    with sulfur-centered radicals, **1**, 949
  reduction of, **1**, 71, 75
  synthesis of
    via condensation reactions, **1**, **719–770**
    via pericyclic processes, **1**, **771–791**
  for the synthesis of radicals, **1**, 948
  for the synthesis of thiophosphines, **2**, 255
Carbon–carbon multiple bonds
  addition to carbon radicals, **1**, **319–375**
  electrocyclic additions to, **1**, 349
  reactions
    with carbon-centered radicals, **1**, 320
    with carbon electrophiles, **1**, **293–317**
    with carbon nucleophiles, **1**, **293–317**
Carbon–carbon triple bonds
  reduction of, **1**, 958
  for the synthesis of alkynes, **1**, 1062
Carbon-centered radicals
  bearing three heteroatom functions, synthesis of, **6**, 734
  from
    B-alkylborinanes, **1**, 946
    alkylcobalt(III) compounds, **1**, 946
    azoalkanes, **1**, 944
    epoxides, **1**, 942
    isocyanides, **1**, 945
    isoselenocyanates, **1**, 945
    isothiocyanates, **1**, 945
    methyl xanthates, **1**, 940
    nitroalkanes, **1**, 945
    organoboranes, **1**, 946
    phenylselenyls, **1**, 942
    tertiary alcohols, **1**, 939
  reaction mechanisms, **1**, 320
  reaction rates, **1**, 923
  reactions
    with carbon–carbon multiple bonds, **1**, 320
    with ketones, **1**, 931
  stability of, **1**, 921
  structure of, **1**, 321, 921
  synthesis of, via carbon–halogen bond scission, **1**, 935
Carbon–cobalt bond scission, for the synthesis of carbon-centered radicals, **1**, 946
Carbon compounds
  N,*N*-dimethylaminomethylation of, **2**, 302
  as nucleophiles, **1**, 193

Carbon(η^5-cyclopentadienyl)phosphinerhodium, reactions, with tetracarbonylethyneosmium, **4**, 710
Carbon dioxide
  reactions
    with alkynes, **1**, 545
    with allenyllithium, **1**, 971
    with amines, **5**, 975
    with dialkylaluminum halides, **5**, 44
    with 1,3-dilithioalkynides, **5**, 101
    with dimethoxymagnesium, **5**, 95
    with (diphenylphosphino)ethynylmagnesium bromide, **2**, 1061
    with heptadecafluoroacetyl iodide, **5**, 59
    with iminophosphoranes, **5**, 974
    with lithiated trithioorthoformates, **6**, 88
    with lithium trialkylalkynylborates, **5**, 101
    with 2-methyl-3-butyn-2-ol, **6**, 463
    with organoaluminums, **1**, 973
    with organolithiums, **3**, 284
    with perfluoroalkyl iodides, **5**, 58
    with phenylhydrazine, **5**, 68
    with phosphinimines, **5**, 1079
    with phosphoramidate anions, **5**, 974, 975
    with phosphoranes, **1**, 987
    with propargylic organoboranes, **1**, 977
    with sodium alcoholates, **6**, 463
    with titanium(III) alkyls, **1**, 667
    with vinyllithiums, **5**, 101
Carbon dioxide, ^{14}C-labeled, reactions, with Grignard reagents, **5**, 71
Carbon diselenide
  electrochemical reduction of, **4**, 875
  reactions
    with amines, **5**, 1055; **6**, 594, 597
    with 1,2-amino alcohols, **6**, 595
    with ethynecarboxylate esters, **4**, 874
    with hydrazines, **6**, 594
    with selenolate, **6**, 591
    with sodium alkynide, **6**, 590
    with trialkylphosphines, **6**, 596
  as starting material
    for organometallic selenocarbonates, **6**, 593
    for triselenocarbonates, **6**, 590
Carbon disulfide
  bromination of, **6**, 215
  chlorination of, **2**, 166; **6**, 214, 236
  fluorination of, **6**, 252
  from, thioureas, **6**, 574
  reactions
    with allenyllithiums, **1**, 1058
    with amide anions, **5**, 232
    with amines, **5**, 1023; **6**, 560, 563
    with 2-aminobenzylamine, **6**, 575
    with 5-amino-1,2,4-triazoles, **6**, 564
    with antimony(III) amides, **2**, 889
    with arenediazonium chlorides, **6**, 536
    with arsenic(III) amides, **2**, 889
    with arylhydrazines, **6**, 560
    with aryllithium, **5**, 560
    with azide ions, **5**, 1046
    with benzyne, **6**, 99
    with chlorine, **6**, 255
    with cobalt carbene complexes, **6**, 324
    with cobalt compounds, **6**, 352
    with copper(I) complexes, **6**, 544
    with dialkoxydibutylstannanes, **6**, 300
    with dialkylcyanamides, **6**, 575
    with α,ω-diamines, **6**, 575
    with 1,8-diaminooctane, **6**, 575
    with diethoxytin, **6**, 469
    with dimethylammonium *N*,*N*′-dimethyl dithiocarbamate, **5**, 609
    with dimethylketene, **4**, 836
    with enolates, **3**, 251
    with fluorine, **6**, 46
    with formamide, **6**, 560
    with Grignard reagents, **5**, 546
    with hydrazides, **6**, 558
    with α-hydrazino enamines, **4**, 986
    with hydroxylamines, **6**, 575
    with indoline, **6**, 561
    with lithiated trithioorthoformates, **6**, 88
    with lithium alkynides, **5**, 549
    with lithium enolates, **6**, 550
    with metal sulfides, **6**, 552
    with methylbis(trimethylsilyl)phosphine, **6**, 324
    with 2-methylcyclopentanone, **6**, 324
    with methylhydrazine, **6**, 560
    with 3-methyl thiophenol, **5**, 1033
    with nickel complexes, **6**, 332
    with nitro compounds, **3**, 643
    with *m*-phenylenediamine, **6**, 565
    with phenyliodinium ylides, **4**, 866
    with phosphoranes, **4**, 843
    with phosphorus–nitrogen bonds, **6**, 564
    with phosphorus ylides, **3**, 550
    with pipecolic acid, **6**, 564
    with primary amines, **6**, 574
    with *n*-propylamine, **5**, 1024
    with siladiazoles, **6**, 576
    with sodium cyanide, **5**, 589
    with sulfonamides, **6**, 627
    with tertiary phosphines, **3**, 549
    with tetraaminoethylenes, **5**, 548
    with tetrabutylammoniumacetylacetone, **5**, 547
    with 1,2,3,4-tetrahydro-β-carboline-3-carboxylic acid, **6**, 564
    with 1,2,3,4-tetrahydroisoquinoline-3-carboxylic acid, **6**, 564
    with tetrakis[(phenylsodiophosphino)methyl]-methane, **5**, 1058
    with thallium(I) alkoxides, **6**, 300
    with thiols, **6**, 552
    with α-thioxocarbene, **6**, 305
    with trialkylphosphines, **6**, 121
    with triethylamine, **5**, 590
    with triphenyltelluronium alkoxides, **6**, 547
  as reagent, for the synthesis of thioketones, **3**, 348, 363
  as starting material
    for *O*,*S*-dialkyl dithiocarbonates, **6**, 549
    for α,α-dichloro sulfides, **6**, 46
    for dithiocarbamate, **6**, 563
    for thioketenes, **3**, 549
  use in the synthesis of oximes, **3**, 427
Carbon electrophiles
  reactions
    with carbon–carbon multiple bonds, **1**, **293–317**
    with silyl anions, **2**, 536
Carbon–fluorine bonds
  substitutions, **1**, 108
  synthesis of, **2**, 15
Carbon functions, elimination of, **1**, 559, 1063
Carbon functions, dicoordinate, with heteroatoms other than chalcogen or nitrogen, **5**, **1091–1098**
Carbon–germanium bonds
  cleavage of, **1**, 64
    photochemical methods, **1**, 64
  reduction of, **1**, 447
  for the synthesis of alkynylgermanes, **2**, 1094
Carbon–halogen bond cleavage, homolytic, **2**, 572
Carbon–halogen bonds
  homolysis of, **1**, 935
  polarity reversal of, **1**, 106
  as precursors, for organometallics, **1**, 106
  reduction of, **1**, 425
  strength of, **1**, 107

transformations, **1**, 105
Carbon–halogen bond scission, for the synthesis of carbon-centered radicals, **1**, 935
Carbon–heteroatom bonds, hydrogenolysis of, **1**, 72
Carbon hydrides
as hydrogen atom donors, **1**, 331
use in reduction reactions, **2**, 55
Carbon–hydrogen activation
of ligand of metal catalysts, **2**, 963
oxidative, **2**, 962
Carbon–hydrogen activation, intermolecular, for the synthesis of vinyl metals and aryl metals, **2**, 962
Carbon–hydrogen activation, intramolecular, for the synthesis of vinyl metals and aryl metals, **2**, 963
Carbon–hydrogen activation, oxidative
of alkenes and arenes, **2**, 962
for aryl metals, **2**, 956
for organometallics, **2**, 952
for vinyl metals, **2**, 956
Carbon–hydrogen bonds
acidity of, **1**, 848
rates of proton removal from, **1**, 871
synthesis of
via addition, **1**, **425–460**
via addition to carbon–carbon multiple bonds, **1**, **71–103**
via substitution, **1**, **1–26**, **27–70**, **425–460**
for the synthesis of alkynes, **1**, 998
Carbon–hydrogen bond scission, for the synthesis of radicals, **1**, 925
Carbon–hydrogen–carbon bonds
three-center two electron, **1**, 918
transanular three-center, **1**, 918
Carbonic acid dichloride *See* Carbonic dichloride
Carbonic acid diesters, as starting materials, for tetraaryl orthocarbonates, **6**, 300
Carbonic acid esters
reactions, with amines, **6**, 502
synthesis of, **6**, 462
Carbonic acid halides, reactions, with silylphosphines, **6**, 686
Carbonic acid ortho amides, as starting materials, for acylguanidines, **6**, 647
Carbonic acids, as starting materials, for trialkylsilyl imidates, **5**, 721
Carbonic anhydrides, carboxylic ethyl, use in the synthesis of benzaldehydes, **3**, 84
Carbonic bromide chloride
from
aluminum tribromide, **6**, 420
boron tribromide, **6**, 420
phosgene, **6**, 420
synthesis of, **6**, 420
Carbonic bromide fluoride
from
bromine trifluoride, **6**, 420
carbon monoxide, **6**, 420
sulfur trioxide, **6**, 420
tribromofluoromethane, **6**, 420
reactions, with thiols, **6**, 432
synthesis of, **6**, 410, 420
Carbonic bromide iodide, synthesis of, theoretical, **6**, 421
Carbonic chloride fluoride
as byproduct, **6**, 409
from
anhydrous hydrogen fluoride, **6**, 419
antimony trifluoride, **6**, 419
arsenic trifluoride, **6**, 419
calcium fluoride, **6**, 419
carbonic dichloride, **6**, 419
carbon monoxide, **6**, 419
chlorine monofluoride, **6**, 419
fluorotrichloromethane, **6**, 419
nitrogen trifluoride, **6**, 419
phosgene, **6**, 419
silicon tetrafluoride, **6**, 419
sulfur trioxide, **6**, 419
thionyl fluoride, **6**, 419
reactions
with oxetane, **6**, 424
with oxirane, **6**, 424
with tetrahydrofuran, **6**, 424
with thiols, **6**, 432
as starting material
for diphenylcarbamoyl fluoride, **6**, 437
for fluoroformate esters, **6**, 422
synthesis of, **6**, 409, 418
Carbonic chloride iodide, synthesis of, theoretical, **6**, 421
Carbonic diazide, reactions, with benzene, **6**, 511
Carbonic dibromide
from
oxalyl bromide, **6**, 417
tetrabromomethane, **6**, 417
safety precautions, **6**, 417
synthesis of, **6**, 417
See also Phosgene, bromo-
Carbonic dichloride
detection of, **6**, 411
exposure to, first aid, **6**, 411
handling of, **6**, 411
industrial production of, **6**, 411
properties of, **6**, 411
safety precautions, **6**, 411, 412
as starting material
for carbonic chloride fluoride, **6**, 419
for carbonic difluoride, **6**, 409
toxicity of, **6**, 411
use of term, **6**, 408, 411
See also Phosgene
Carbonic difluoride
as byproduct, in the synthesis of carbonic bromide fluoride, **6**, 410
electrochemical synthesis of, **6**, 410
from
carbonic dichloride, **6**, 409
carbon monoxide, **6**, 409
carbonyl sulfide, **6**, 410
dibromodifluoromethane, **6**, 410
perfluoropropene, **6**, 410
tetrafluoroethylene, **6**, 410
trichloromethyl chloroformate, **6**, 409
metathesis with phosgene, **6**, 419
properties of, **6**, 408
reactions
with oxirane, **6**, 424
with piperidine, **6**, 438
with secondary amides, **6**, 438
as starting material
for fluoroformate esters, **6**, 422
for fluoromethyl fluoroformate, **6**, 423
synthesis of, **6**, 408
industrial processes, **6**, 410
oxidative methods, **6**, 410
safety precautions, **6**, 408
See also Carbonyl difluoride
Carbonic dihalides
as starting materials, for fluorothiolformate esters, **6**, 432
use of term, **6**, 408
Carbonic diiodide
properties of, theoretical, **6**, 418
synthesis of, **6**, 418
Carbonic fluoride halides
as starting materials
for carbamoyl fluorides, **6**, 437
for fluoroformate esters, **6**, 422
Carbonic fluoride iodide
from

carbon monoxide, **6**, 420
iodine pentafluoride, **6**, 420
synthesis of, **6**, 420
Carbonimidic acids
synthesis of, **6**, 623
See also Carbonates, imino-
Carbonimidic bromide chlorides, synthesis of, **6**, 613
Carbonimidic bromide fluorides, synthesis of, **6**, 613
Carbonimidic chloride fluorides, synthesis of, **6**, 612
Carbonimidic chloride iodides, synthesis of, **6**, 613
Carbonimidic dibromide, *N*-4-bromophenyl-, synthesis of, **6**, 610
Carbonimidic dibromide, *N*-phenyl-, synthesis of, **6**, 610
Carbonimidic dibromide, *N*-trifluoromethyl-, synthesis of, **6**, 611
Carbonimidic dibromides
from, isocyanides, **6**, 610
synthesis of, **6**, 610
tautomerization of, **6**, 611
Carbonimidic dichloride, as starting material, for dimethyl *N*-cyclohexyliminocarbonate, **6**, 624
Carbonimidic dichloride, *N*-chloro-, reactions, with sulfur, **6**, 609
Carbonimidic dichloride, *N*-hydroxy-, synthesis of, **6**, 608
Carbonimidic dichloride, pentafluorophenyl-
from, pentafluoroaniline, **6**, 609
production of 620002.184
Carbonimidic dichloride, *N*-phenyl-
reactions
with amines, **6**, 619
with trialkyl phosphites, **6**, 367
synthesis of, **6**, 609
Carbonimidic dichloride, trichloromethyl-
use in the synthesis of bis(trifluoromethyl)amines, **6**, 239
use in the synthesis of orthocarbonates, **6**, 296
See also Isocyanide dichloride, trichloromethyl-
Carbonimidic dichlorides
alkylation of, **6**, 727
N-alkylation of, **6**, 610
fluorination of, **6**, 602
from
butyl isocyanate, **6**, 606
carbonimidic dichlorides, **6**, 609
cyanogen chloride, **6**, 606
dichlorocarbene, **6**, 608
formanilides, **6**, 606
isocyanates, **6**, 606
isocyanides, **6**, 605
isothiocyanates, **6**, 604
methyl isocyanate, **6**, 606
perhaloamines, **6**, 609
tetrachloromethane, **6**, 608
interconversion of, **6**, 609
as natural products, **6**, 604
protonation of, **6**, 727
reactions
with alkoxides, **6**, 616, 624
with amines, **6**, 619, 645
with phosphines, **6**, 686
with phosphorus compounds, **6**, 665
with sulfur nucleophiles, **6**, 617
with thiols, **6**, 626
as starting materials
for bis(phosphinyl)iminocarbonyls, **6**, 665, 666
for carbamimidic chlorides, **6**, 619
for carbonimidic dichlorides, **6**, 609
for carbonimidic difluorides, **6**, 602
for guanidines, **6**, 619
for iminocarbonyl chlorides, **6**, 617
for triazinones, **6**, 619
for trichloromethylamines, **6**, 238
synthesis of, **6**, 604
chlorination methods, **6**, 607
tautomerization of, **6**, 606
See also Isocyanide dichlorides
Carbonimidic dichlorides, acyl-, synthesis of, **6**, 609
Carbonimidic dichlorides, *N*-alkyl-, reactions, with potassium iodide, **6**, 612
Carbonimidic dichlorides, *N*-aryl-
displacement of chloride from, **6**, 619
from, formanilides, **6**, 606
reactions
with alkoxides, **6**, 616
with hydrogen fluoride, **6**, 602
synthesis of, **6**, 606
Carbonimidic dichlorides, perchloroaryl-, synthesis of, **6**, 609
Carbonimidic dichlorides, perfluoroaryl-, synthesis of, **6**, 609
Carbonimidic diesters *See* Carbonates, imino-
Carbonimidic difluoride, *N*-bromo-, from, cyanogen fluoride, **6**, 603
Carbonimidic difluoride, *N*-ethyl-, synthesis of, **6**, 603
Carbonimidic difluoride, *N*-fluoro-, addition reactions of, **6**, 256
Carbonimidic difluoride, *N*-pentachlorophenyl-, synthesis of, **6**, 602
Carbonimidic difluoride, *N*-trifluoromethyl-, synthesis of, **6**, 604
Carbonimidic difluorides
dimerization of, **6**, 602, 615
from
aryl isothiocyanates, **6**, 603
carbonimidic dichlorides, **6**, 602
isocyanides, **6**, 603
isothiocyanates, **6**, 603
nitriles, **6**, 603
tetrafluoroethylene, **6**, 604
trifluoromethylamines, **6**, 602
reactions, with amines, **6**, 615
as starting materials
for iminocarbonates, **6**, 624
for trifluoromethylamines, **6**, 238, 239
synthesis of, **6**, 602
via dehalogenation methods, **6**, 603
Carbonimidic difluorides, *N*-halo-, synthesis of, **6**, 603
Carbonimidic difluorides, perfluoroalkyl-, from, oxazetidines, **6**, 604
Carbonimidic difluorides, *N*-perfluoroalkyl-, reactions, with alkoxides, **6**, 614
Carbonimidic dihalides, use of term, **6**, 602
Carbonimidic diiodides
dissociation of, **6**, 612
synthesis of, problems, **6**, 612
Carbonimidic esters, as starting materials, for isoureas, **6**, 631
Carbonimidic halides
as starting materials
for iminocarbonates, **6**, 624
for iminodithiocarbonates, **6**, 626
for iminothiocarbonates, **6**, 628
for isothioureas, **6**, 635
for isoureas, **6**, 631
Carbonimidic thioesters, as starting materials, for isoureas, **6**, 631
Carbonimidodithioic acids
synthesis of, **6**, 626
See also Dithiocarbonates, imino
Carbonimidodithioic esters *See* Dithiocarbonates, imino
Carbonimidothioic acids, synthesis of, **6**, 628
Carbonimidothioic esters *See* Thiocarbonates, imino
P,*P*'-(Carbonimidoyl)bis(phosphonic diamide), synthesis of, **6**, 668
Carbonimidoyl dichlorides
reactions
with alkali metal selenocyanates, **5**, 1056
with ammonium sulfide, **5**, 1032

with bis(trimethylsilyl) sulfide, **5**, 1032
as starting materials
for isoselenocyanates, **5**, 1056
for isothiocyanates, **5**, 1032
Carbonimidoyl dichlorides, pentafluorophenyl-, use in the synthesis of carbodiimides, **5**, 1070
Carbonimidoyl dichlorides *See also* Isocyanide dichlorides
Carbonitrile, 2-pyridine-, reactions, with hydrazine, **5**, 769
Carbonitrile, 4-pyridine-, as starting material, for 4-pyridinecarboxyamidrazone, **5**, 769
3-Carbonitrile, pyridine-, reactions, with hydroxylamine, **5**, 763
Carbonitrile oxides, benzenesulfonyl-
as intermediates, **5**, 1148
synthesis of, **5**, 1145
Carbonitriles, 1-cyclobutane-, synthesis of, **3**, 645
Carbonitriles, 3-indole-, debenzylation of, **1**, 47
Carbonium ions
conformations of, **1**, 183
definition of, **1**, 380, 890
Friedel–Crafts alkylations of, **1**, 199
reactions
with 1,1-dichloroethylene, **5**, 26
with Grignard reagents, **1**, 202
rearrangement of, **1**, 185
use of term, **1**, 890
validity of, **1**, 905
See also Carbocations
Carbonium ions, oxo-, crystallographic studies of, **2**, 1008
Carbon–mercury bonds
homolysis of, **1**, 948
reduction of, **1**, 65
stability of, **1**, 66
synthesis of, **2**, 593
Carbon–mercury bond scission, for the synthesis of carbon-centered radicals, **1**, 948
Carbon–metal bonds
nature of, **1**, 885
protonolysis of, **1**, 69
reduction of, **1**, 65, 452
reviews, **6**, 172
Carbon–metal bonds, aryl, reduction of, **1**, 452
Carbon–metal bonds, vinyl, reduction of, **1**, 453
Carbon–metal σ bonds, protonolysis of, **1**, 65
Carbon monosulfide
reactions
with bromine, **6**, 529
with chlorine, **6**, 529
with diazo compounds, **3**, 550
with sulfenyl chlorides, **6**, 535
with thiol, **5**, 550
as reagent, for the synthesis of thioketones, **3**, 356
as starting material
for chlorodithioformates, **6**, 535
for thiocarbonyl dibromide, **6**, 529
for thiocarbonyl diiodide, **6**, 530
for thioketenes, **3**, 550
for thiophosgene, **6**, 529
Carbon monoxide
electrolysis of, **6**, 469
elimination of, **1**, 983
fluorination of, **6**, 409
insertion into silylmethyllithiums, **4**, 931
insertion reactions, **6**, 516
as ligand, for transition metals, **1**, 192
oxidative amination of, **6**, 505
reactions
with alkenes, **5**, 192
with α-alkenes, **5**, 142
with alkyl compounds, **5**, 442
with alkyllithiums, **5**, 238
with allylic complexes of iridium, **5**, 461
with amines, **5**, 974
with arenes, **3**, 93
with arylalkoxychromium carbene complexes, **1**, 551
with aryl azides, **5**, 984
with aryl iodides, **3**, 89
with bisbenzyl compounds, **5**, 442
with biscyclopentadienyl bisalkyl hafnium complexes, **5**, 443
with bis(dialkylamide) compounds, **6**, 516
with bismethyl compounds, **5**, 442
with boron, **5**, 1020
with *n*-butyllithium, **5**, 437
with *t*-butyllithium, **5**, 729
with cobalt, **6**, 524
with cyclic titanium alkyl compounds, **5**, 441
with cyclopentadienyliron complexes, **5**, 453
with cyclopentadienylthorium complexes, **5**, 468
with cyclopentadienylvanadium complexes, **5**, 444
with cyclopentadienylzirconium complexes, **6**, 521
with dialkyl complexes, **5**, 462
with dialkylosmiums, **5**, 455
with diamines, **6**, 510
with diaryl diselenides, **5**, 241
with diaryl ditellurides, **5**, 242
with dibenzyl disulfides, **5**, 237
with dicobalt octacarbonyl, **1**, 202
with dimethylsilylene, **5**, 1019
with diorganosilylenes, **5**, 1019
with esters, **5**, 192
with ethylnickel complexes, **5**, 461
with fluorine, **6**, 409
with Grignard reagents, **5**, 438
with hexafluoroazomethane, **5**, 989
with homochiral alkylirons, **5**, 453
with iridium(I) carbonyl azides, **5**, 983
with iridium complexes, **4**, 931
with iridium halides, **6**, 525
with iridium–nitrogen bonds, **6**, 516
with iron complexes, **6**, 518
with lithium *t*-butylamide, **6**, 515
with manganese alkyl compounds, **5**, 446
with metal–phosphido bonds, **6**, 519
with methyl complexes, **5**, 463
with methylimidoosmium complexes, **5**, 456
with methylmanganese pentacarbonyl, **5**, 446
with methylnickel complexes, **5**, 461
with monoalkyltitanocenes, **5**, 441
with nitro compounds, **5**, 984
with nitrogen nucleophile anions, **6**, 517
with organolithiums, **5**, 436
with phenyllithium, **5**, 436
with [2-phenyl-1-(silyl)vinyl]lithium, **5**, 438
with *N*-pivaloylaniline, **5**, 438
with Raney cobalt, **6**, 524
with rhenium heptoxide, **6**, 523
with rhodium, **6**, 525
with rhodium(I) carbonyl azides, **5**, 983
with ruthenium(III) acetylacetone, **6**, 524
with ruthenium–amide bonds, **6**, 516
with silylenes, **5**, 1020
with silylmethyllithiums, **4**, 931
with [1-(silyl)vinyl]lithium, **5**, 438
with silylzirconiums, **5**, 422
with tetraalkylstannanes, **3**, 15
with titanium, **6**, 522
with transition metals, **1**, 212; **6**, 521
with trialkylboranes, **4**, 363
with trifluoromethyl hypofluorite, **6**, 424
with zirconabicycles, **3**, 270
as reagent, for the synthesis of aldehydes, **3**, 15
as starting material
for carbonic bromide fluoride, **6**, 420
for carbonic chloride fluoride, **6**, 419
for carbonic difluoride, **6**, 409

for carbonic fluoride iodide, **6**, 420
for chlorothiolformate esters, **6**, 434
for phosgene, **6**, 412
for thiocarboxylic esters, **5**, 237
in the synthesis of aldehydes, **3**, 19
use in the synthesis of *N*-alkylamides, **5**, 286
use in the synthesis of amides, **2**, 318
use in the synthesis of anhydrides, **5**, 194
use in the synthesis of aryl ketones, **3**, 289
use in the synthesis of decacarbonylmanganese, **6**, 523
use in the synthesis of dialkyl carbonates, **6**, 464
use in the synthesis of isocyanates, **5**, 974
Carbon monoxide, ^{13}C-labeled, reactions, with organoboranes, **5**, 45
Carbon monoxide insertion
for the synthesis of acyl hydrazines, **5**, 356
for the synthesis of *N*-acylsulfonamides, **5**, 346
for the synthesis of lactone complexes, **4**, 688
Carbon–nitrile bonds, cleavage of, **1**, 55
Carbon–nitrogen bonds
β-fragmentation of, **1**, 946
reactions, with phosphines, **2**, 440
reduction of, **1**, 441
reduction to carbon–hydrogen bonds, **1**, 28
synthesis of
for the synthesis of *N*-alkoxy primary amines, **2**, 351
for the synthesis of *N*-alkoxy secondary amines, **2**, 353
Carbon–nitrogen bonds, aryl
reduction of, **1**, 441
as starting materials, for nitroarenes, **1**, 441
Carbon–nitrogen bonds, vinyl
reduction of, **1**, 443
as starting materials, for alkenes, **1**, 443
Carbon–nitrogen bond scission, for the synthesis of carbon-centered radicals, **1**, 944
Carbon–nitrogen double bonds
reduction of, **1**, 38
for the synthesis of isocyanates, **5**, 987
Carbon–nitrogen single bonds, reduction of, **1**, 28
Carbon–nitrogen triple bonds, reactivity of, **3**, 641
^{13}Carbon–NMR chemical shifts, of cationic carbon atoms, **1**, 894
Carbon nucleophiles
conjugate addition to enones, **3**, 128
for the displacement of oxygen leaving groups, **1**, 204
reactions
with alkenes, **1**, 300
with alkynic acid halides, **3**, 275
with alkynyl halides, **1**, 1016
with β-azido enol ether, **1**, 263
with bromotrimethylgermane, **4**, 361
with carbon–carbon multiple bonds, **1**, **293–317**
with chloroarsinidenes, **2**, 896
with imines, **2**, 313
with *S*-methyl (diisopropoxyphosphonyl)dithioformate, **6**, 121
with phenyl isothiocyanate, **5**, 729
with styrene oxide, **1**, 204
with thioimidoyl chlorides, **5**, 728
with α,β-unsaturated acids, **3**, 229
Carbon nucleophiles, organometallic-, reactions, with chlorosilanes, **6**, 172
Carbonochloridate, α-chloromethyl, for the dealkylation of tertiary amines, **2**, 303
Carbonochloridates, order of cleavage, **2**, 303
Carbonochloridoimidic acid, synthesis of, **6**, 616
Carbonochloridoimidothioic acid, synthesis of, **6**, 617
Carbonofluoridoimidic acid, synthesis of, **6**, 614
Carbonohalidic acid, use of term, **6**, 421
Carbonohalidothioic acid *S*-esters, nomenclature, **6**, 431
Carbonothioic acid salts, as starting materials, for thiocarbonates, **6**, 479
Carbon oxides
elimination of, **1**, 560
as starting materials, for dialkyl carbonates, **6**, 463
Carbon oxide sulfide
reactions
with amines, **6**, 503
with diamines, **6**, 509
with 1,3-diaminopropane, **6**, 509
Carbon–oxygen bonds, reduction of, **1**, 11
Carbon–oxygen bond scission, for the synthesis of carbon-centered radicals, **1**, 938
Carbon–oxygen double bonds, reduction of, **1**, 432
Carbon oxysulfide
synthesis of, **6**, 471
use in the synthesis of *S*,*S'*-dialkyl dithiocarbonates, **6**, 471
Carbon–phosphorus bonds
cleavage of, **1**, 49, 268
reduction of, **1**, 49, 444
synthesis of, via Arbuzov rearrangement, **4**, 340
Carbon–phosphorus bonds, aryl
reduction of, **1**, 444
as starting materials, for arenes, **1**, 444
Carbon–phosphorus bonds, vinyl, reduction of, **1**, 446
Carbon radicals
addition to carbon–carbon multiple bonds, **1**, **319–375**
synthesis of, via homolysis, **1**, 935
Carbon–ruthenium bonds, homolysis of, **2**, 583
Carbons, halo-
commercial availability of, **1**, 105
reactions, with lithium vapor, **6**, 204
Carbons, pentacoordinate, in carbon-bridged dimanganese complexes, **4**, 677
Carbons, perchloro-, reactions, with lithium vapor, **6**, 204
Carbons, quaternary, synthesis of, **1**, 150, 176
Carbons, tetracoordinate aryl, in dimeric dimesityliron complexes, **4**, 685
Carbon selenide sulfide, reactions, with dimethyl ethynedicarboxylate, **4**, 869
Carbon–selenium bonds, reduction of, **1**, 21, 435
Carbon–selenium bond scission, for the synthesis of carbon-centered radicals, **1**, 942
Carbon–silicon bonds, reduction of, **1**, 60, 447
Carbon suboxide
addition of aminotin and alkoxytins to, **4**, 701
reactions
with chlorodiphenylphosphine, **4**, 550
with germanium alkoxides, **1**, 65
as starting material, for 1,1-bisgermylalkanes, **4**, 628
Wittig reaction of, **1**, 988
Carbon sulfide, reactions, with Grignard reagents, **5**, 232
Carbon–sulfur(VI) bonds, use in the synthesis of sulfonyl halides, **2**, 175
Carbon–sulfur bonds
cleavage of, **1**, 143; **2**, 127
reduction of, **1**, 21, 435
Carbon–sulfur bond scission, for the synthesis of carbon-centered radicals, **1**, 942
Carbon–sulfur double bonds, properties of, **3**, 330
Carbon surfactants, perfluoro-, synthesis of, **6**, 5
Carbon systems, [2,3] rearrangement of, **1**, 806
Carbon–tellurium bonds
reactions, with alkynes, **2**, 730
reduction of, **1**, 21, 435, 437
Carbon–tellurium bond scission, for the synthesis of carbon-centered radicals, **1**, 943
Carbon tetrabromide *See* Methane, tetrabromo-
Carbon tetrachloride *See* Methane, tetrachloro-
Carbon tetrafluoride *See* Methane, tetrafluoro-
Carbon tetrahalide, reactions, with triphenylphosphines, **2**, 7
Carbon tetraiodide *See* Methane, tetraiodo-
Carbon–tin bonds
cleavage of, **4**, 700
reactions, with electrophiles, **1**, 70
Carbonyl, *t*-butoxy- (*t*-BOC)

for the deprotonation of organometallics, **1**, 149
use in the synthesis of α-aminocarbanions, **1**, 151
Carbonyl acid, *N,N*-dimethylamidetriethylgermyl-, synthesis of, **6**, 516
Carbonyl acid chlorides
reactions
with diphenyldiazidosilane, **5**, 992
with stannyl azides, **5**, 992
Carbonyl anions, acyl tetra-, synthesis of, **5**, 452
Carbonylation
of aryl halides, **3**, 90
of bisarylchromiums, **5**, 444
cobalt-catalyzed, **1**, 242
of ethylmagnesium bromide, **5**, 439
of iron alkyl complexes, **5**, 452
for the synthesis of *N*-alkylalkenoamides, **5**, 296
for the synthesis of *N*-alkylamides, **5**, 285
for the synthesis of *N*-alkylarylamides, **5**, 301
for the synthesis of α,β-alkynoic esters, **5**, 160
for the synthesis of *N*-arylamides, **5**, 293
for the synthesis of *N*-arylarylamides, **5**, 304
for the synthesis of γ-lactones, **5**, 170
for the synthesis of α-methylene lactams, **5**, 296
for the synthesis of thiocarboxylic esters, **5**, 238
for the synthesis of α,β-unsaturated ketones, **3**, 218, 233
of terminal alkynes, **1**, 1040
Carbonylation, iridium-catalyzed, of alkenes, **5**, 416
Carbonylations, palladium-catalyzed, **1**, 228
Carbonylative cyclization
of 1-iodo-1-silyl-1,4-dienes, **2**, 931
of malonic esters, **5**, 464
Carbonyl azides, synthesis of, **5**, 989
Carbonyl azides, hetero-, thermolysis of, **5**, 990
Carbonyl bonds, polarization of, **3**, 330
Carbonyl bromide *See* Carbonic dibromide
Carbonyl bromide fluoride *See* Carbonic bromide fluoride
Carbonyl bromides, imino-
nitrogen-substituted, synthesis of, **6**, 623
sulfur-substituted, synthesis of, **6**, 623
Carbonyl bromofluoride *See* Carbonic bromide fluoride
Carbonyl chalcogenides, synthesis of, **6**, **459–498**
Carbonyl chloride, reactions, with amines, **5**, 965
Carbonyl chloride *See* Carbonic dichloride
Carbonyl chloride, *N*-(chlorosulfinyl)imino-, reactions, with diarylamine, **5**, 1122
Carbonyl chloride, *N*-phosphorylimino-, reactions, with diarylamine, **5**, 1122
Carbonyl chloride fluoride *See* Carbonic chloride fluoride
Carbonyl chloride isocyanate, reactions, with ammonium thiocyanate, **6**, 559
Carbonyl chlorides
reactions
with isocyanic acid, **5**, 994
with metal cyanates, **5**, 994
Carbonyl chlorides, imino-
from
carbonimidic dichlorides, **6**, 617
isocyanides, **6**, 617
isothiocyanates, **6**, 617
methyl thiocyanate, **6**, 618
thiocyanates, **6**, 618
metal-substituted, synthesis of, **6**, 622
nitrogen-substituted, synthesis of, **6**, 619
with one other heteroatom, **6**, 616
oxygen-substituted, synthesis of, **6**, 616
phosphorus-substituted, synthesis of, **6**, 622
selenium-substituted, synthesis of, **6**, 618
sulfur-substituted, synthesis of, **6**, 617
1-Carbonyl chlorides, 2-imidazolidinone-, synthesis of, **5**, 1012
N-Carbonyl chlorides
from, isocyanates, **6**, 448
transhalogenation of, with fluorides, **6**, 438
Carbonyl chlorofluoride *See* Carbonic chloride fluoride
Carbonyl complexes, reactions, with aryl azides, **5**, 983
Carbonyl complexes, halo-, synthesis of, **6**, 455
Carbonyl complexes, iodo-
from
iron pentacarbonyl, **6**, 455
tricarbonyl(cyclohexadienyl)iron cation, **6**, 455
Carbonyl dibromide *See* Carbonic dibromide
Carbonyl dichloride, *See* Carbonic dichloride
Carbonyl difluoride
chlorination of, **6**, 217
reactions
with dimethylformamide, **6**, 52
with oxygen difluoride, **2**, 99
with trifluoromethyl hypofluorite, **2**, 98
See also Carbonic difluoride
Carbonyl dihalides, use of term, **6**, 408
Carbonyl disulfides, alkoxydichloromethyl chloro-, reactions, with *N*-methylaniline, **6**, 497
Carbonyl fluoride
reactions
with chlorine trifluoride, **6**, 231
with cyclobutanones, **4**, 59
with nitriles, **5**, 1003
with potassium cyanate, **5**, 1015
use in the synthesis of trifluoromethyls, **6**, 10
Carbonyl fluoride, difluoroamino-
reactions
with caesium fluoride, **6**, 257
with potassium fluoride, **6**, 257
Carbonyl fluorides, *N,N*-disubstituted, synthesis of, **6**, 438
Carbonyl fluorides, imino-
nitrogen-substituted, synthesis of, **6**, 615
with one other heteroatom, **6**, 614
oxygen-substituted, synthesis of, **6**, 614
phosphorus-substituted, synthesis of, **6**, 616
sulfur-substituted, synthesis of, **6**, 614
N-Carbonyl fluorides, synthesis of, **6**, 438
Carbonyl group, chloro-, migration, oxygen to nitrogen, **6**, 446
Carbonyl group, imino-
synthesis of functions
any element other than a halogen or chalcogen, **6**, **639–675**
at least one chalcogen, **6**, 623
at least one halogen, **6**, **601–637**
Carbonyl groups
and acylphosphorus functions, chemical behavior of, **5**, 393
for the condensation of halides, **1**, 676
directed reduction of, **2**, 62
reactions, **1**, 129
with aluminum tris(2,6-triphenylphenoxide), **1**, 303
with Grignard reagents, **2**, 67
with organometallics, **2**, 67
α-selenylation of, **3**, 173
synthesis of functions, containing two heteroatoms other than a halogen or chalcogen, **6**, **499–526**
via nucleophilic rearrangements, **1**, 379
Carbonyl halides
with one halogen and one other heteroatom function, synthesis of functions, **6**, 421
with two dissimilar halogens, synthesis of functions, **6**, 418
with two similar halogens, synthesis of functions, **6**, 408
use of term, **6**, 408
Carbonyl halides, imino-
with one halogen and one other heteroatom, **6**, 614
with two dissimilar halogen functions, **6**, 612
with two similar halogens, **6**, 602
Carbonyl iodides, imino-
nitrogen-substituted, synthesis of, **6**, 623
sulfur-substituted, synthesis of, **6**, 623
Carbonyl metals, halo-, reactions, with carbonyl metallates, **6**, 525

Carbonyl methylenation, mechanism, **1**, 666
Carbonyl oxides
  reactions, with dicarbonyls, **3**, 326
  synthesis of, **3**, 326
Carbonyl reagents, α-mercurio-, reactivity of, **1**, 128
Carbonyl reduction, stereocontrol in, **2**, 60
Carbonyls
  aldol condensation of, **6**, 25
  alkylation of, **1**, 599
  asymmetric reduction of, **2**, 63
  α-bromination of, **2**, 28
  Cannizzaro reaction of, **6**, 25
  cathodic reduction of, **2**, 80
  α-chlorination of, **2**, 21
  condensation of, **1**, 1075
  deamination of, **1**, 635
  deprotonation of, **2**, 699
    for the synthesis of metal enolates, **2**, 696
  diazo group transfer to, **3**, 471
  dissolving metal reduction of, **2**, 54
  enantioselective alkylation of, **2**, 71
  enantioselective hydrosilylation of, **2**, 64
  enolization of, **1**, 1037; **2**, 645
  fluorination of, **2**, 15
  halogenation of, **2**, 3
  homologation of, **1**, 766; **5**, 284, 897
  hydration of, **4**, 160
  α-iodination of, **2**, 33
  McMurry coupling of, **1**, 687
  methylenation of, **1**, 678, 767
  nitrosation of, **3**, 469
  nucleophilic allylation of, **2**, 73
  reactions
    with alkenylzirconiums, **1**, 1104
    with alkoxyethynylides, **2**, 1027
    with alkylcobalts, **2**, 577
    with alkylirons, **2**, 577
    with alkylzincs, **2**, 71
    with alkynyl metals, **1**, 1037
    with allenylmagnesium halides, **1**, 1055
    with allylboranes, **2**, 74
    with allyl halides, **2**, 73
    with allylic anions, **2**, 73; **4**, 938
    with amines, **3**, 407, 416
    with aminoalkylarsines, **4**, 495
    with aminostannanes, **2**, 701
    with arsonium ylides, **1**, 646, 737, 740, 742; **3**, 48, 57
    with *N*,*N*-bis(silyl)amines, **3**, 406
    with bis(trimethylsilylmethyl) sulfide, **2**, 139
    with 2-bromo esters, **2**, 69
    with carbalkoxycarbenes, **3**, 319
    with carbenes, **1**, 366
    with carbenoids, **1**, 366
    with α-chloro sulfones, **4**, 236
    with cyanide ions, **3**, 667
    with cyano(methylthio)methyltrimethylsilane, **3**, 654
    with cyclopropyl anions, **1**, 284
    with dialkylaluminum tetramethylpiperidide, **2**, 699
    with dialkyl phosphites, **4**, 469, 559
    with dialkylthiophosphinous acids, **2**, 462
    with diethyl 1-fluoro-1-(phenylsulfonyl)methyl
      phosphonate, **4**, 780
    with diphosphinates, **4**, 563
    with 3,3-disubstituted allenyllithiums, **1**, 1056
    with ethylenediamine complexes, **1**, 1037
    with glyoxalates, **2**, 81
    with Grignard reagents, **1**, 662
    with α-halophosphoranes, **3**, 649
    with homoenolate anions, **3**, 31
    with hydrogen peroxide, **2**, 96
    with iminophosphoranes, **4**, 518
    with isocyanic acid, **5**, 986
    with isopropenyl acetate, **2**, 675
    with α-lithio sulfinamides, **1**, 695
    with lithium bis(trialkylsilyl)amides, **3**, 456
    with metallacyclopropanes, **2**, 984
    with metallacyclopropenes, **2**, 984
    with metal phosphides, **6**, 686
    with metals, **1**, 1057
    with *N*-methyl-*N*,*O*-
      bis(trimethylsilyl)hydroxylamine, **3**, 433
    with methylthioacetonitrile, **3**, 654
    with monolithium alkynide, **1**, 1037
    with nitroalkanes, **2**, 784
    with nitrogen-substituted ketones, **3**, 257
    with organoaluminums, **1**, 971
    with organoantimony compounds, **1**, 272
    with organolanthanides, **2**, 69
    with organometallics, **1**, 692
    with organotellurium ylides, **3**, 647
    with phosphines, **4**, 548; **6**, 686
    with phosphonate carbanions, **1**, 744
    with phosphonothioic acid diesters, **4**, 469
    with phosphoramidates, **5**, 1069
    with primary phosphines, **2**, 450
    with propargylic organoboranes, **1**, 977
    with selenium compounds, **1**, 605
    with selenonium ylides, **1**, 720
    with silaethenes, **6**, 178
    with silylallyl anions, **1**, 651
    with α-silyl carbanions, **1**, 649, 758, 761
    with sodium bis(trialkylsilyl)amides, **3**, 456
    with sulfonium ylides, **1**, 720
    with sulfur tetrafluoride, **6**, 39
    with telluronium ylides, **1**, 720
    with thiazole, **3**, 42
    with thioethynylide, **2**, 1031
    with tin(II) triflate, **2**, 701
    with trialkylaluminums, **1**, 663
    with trichloromethyllithium, **6**, 25
    with trifluorovinyllithium, **3**, 67
    with trihydridic boranes, **2**, 111
    with α-trimethylsilyl esters, **1**, 283
    with *N*-(trimethylsilyl)phosphimines, **3**, 440
    with trimethylsilyl (trimethylsilyloxy) acetate, **4**, 213
    with urea, **6**, 508
    with vinylidene phosphoranes, **1**, 986
    with water, **4**, 160
    with Wittig–Horner type reagents, **2**, 713
    with zirconium compounds, **2**, 70
  reduction of, **2**, 55, 531
  reductive amination of, **2**, 310
  reductive halogenation of, **2**, 14
  as starting materials
    for alcohols, **2**, 52
    for alkenylsilanes, **2**, 936
    α-alkoxystannanes, **4**, 386
    for allenes, **1**, 986
    for amino compounds, **1**, 881
    for benzonitriles, **3**, 665
    for boron enolates, **2**, 698
    for *N*-carbon-substituted imines, **3**, 405
    for cyclic monothioacetals, **4**, 229
    for dialkyl sulfides, **2**, 139
    for diazo compounds, **3**, 464
    for α,α-dihalo ethers, **6**, 39
    for enamines, **2**, 739
    for 2-haloalkenylphosphorus compounds, **4**, 801
    for α-haloalkenylsilicons, **4**, 805
    for halo enals, **3**, 68
    for hydroxycarboxylates, **5**, 164
    for methylenes, **1**, 767
    for monoselenoacetals, **4**, 238
    for monotelluroacetals, **4**, 238
    for monothioacetals, **4**, 215
    for oximes, **3**, 425

for oxyimidic halides, **5**, 823
for oxyimidic thioesters, **5**, 830
for silylmethyl imines, **4**, 517
for sulfenimines, **3**, 436
for tertiary phosphines, **2**, 440
for thiophosphates, **2**, 258
for α,β-unsaturated aldehydes, **3**, 68
for α,β-unsaturated ketones, **3**, 268
for vinyl bromides, **2**, 624
for vinyl chlorides, **2**, 617
for vinyl fluorides, **2**, 611
for vinyl iodides, **2**, 630
for vinyloxyboranes, **2**, 532
synthesis of, **6**, 500
thionation of, **5**, 513
use in the synthesis of vinyl cations, **1**, 1111
via pinacol rearrangement, **1**, 384
via semipinacol rearrangement, **1**, 388
Carbonyls, β-acetoxy-α,β-unsaturated, substitution of, **1**, 482
Carbonyls, *N*-acylimino-
from
cyanamides, **6**, 646
guanidines, **6**, 646
S-methylisothiourea, **6**, 646
synthesis of, **6**, 646
Carbonyls, *N*-alkenylimino-
from
amidinium salts, **6**, 644
guanidines, **6**, 644
synthesis of, **6**, 644
Carbonyls, β-alkoxy-α,β-unsaturated, substitution of, **1**, 482
Carbonyls, β-alkylated-α,β-unsaturated, synthesis of, **1**, 467
Carbonyls, *N*-alkylimino-
from
alkyl isocyanides, **6**, 671, 674
amidinium salts, **6**, 642
carbodiimides, **6**, 643
cyanamides, **6**, 643
dihalocarbiimides, **6**, 643
ureas, **6**, 642
synthesis of, **6**, 642
Carbonyls, alkynic, cyclization reactions of, **2**, 81
Carbonyls, *N*-alkynylimino-, synthesis of, **6**, 646
Carbonyls, α-amino-, diazotization of, **3**, 468
Carbonyls, β-amino-, use in the Prelog synthesis of phenols, **2**, 662
Carbonyls, *N*-aminoimino-, synthesis of, **6**, 653, 672, 674
Carbonyls, α-aryl-, synthesis of, via nickelphosphine-catalyzed coupling, **1**, 465
Carbonyls, α-arylated, synthesis of, **1**, 483
Carbonyls, β-arylated, synthesis of, **1**, 483
Carbonyls, *N*-arylimino-
from
amidinium salts, **6**, 644
aryl isocyanides, **6**, 671, 674
carbodiimides, **6**, 645
cyanamides, **6**, 645
dichlorocarbiimides, **6**, 645
ureas, **6**, 644
synthesis of, **6**, 644
Carbonyls, bis(phosphino)imino-, synthesis of, **6**, 665
Carbonyls, bis(phosphinyl)imino-, synthesis of, **6**, 665
Carbonyls, *N*-bromo-, synthesis of, **6**, 452
Carbonyls, α-bromo-
alkylation of, **2**, 758
chlorination of, **4**, 37
reduction of, **1**, 8, 10
Carbonyls, α-bromo-β-silyl-, bromination of, **1**, 653
Carbonyls, *N*-chalcogenoimino-, synthesis of, **6**, 650
Carbonyls, *N*-chloro-
from
carbamic acid esters, **6**, 449
phosgene, **6**, 443
reactivity of, **6**, 442
synthesis of, **6**, 442
Carbonyls, α-chloro, bromination of, **4**, 37
Carbonyls, α-chloro-
reactions, with lithium trimethylsilylacetonitrile, **3**, 650
reduction of, **1**, 6
with iodide, **1**, 7
Carbonyls, β-chloro-, dehydrochlorination of, **1**, 571
Carbonyls, α-chloroalkyl-*N*-chloro-, synthesis of, **6**, 445
Carbonyls, β-chlorovinyl, as starting materials, for acetals, **4**, 196
Carbonyls, *N*-cyanoimino-
from
carbodiimides, **6**, 648
cyanamides, **6**, 648
thioureas, **6**, 647
Carbonyls, β-dialkylamino-, synthesis of, **2**, 316
Carbonyls, diazo-
reactions, with thiophosgene, **6**, 535
x-ray crystal structure of, **2**, 1003
Carbonyls, 1,3-diazo-
reactions
with triarylantimony compounds, **3**, 196
with triarylbismuth compounds, **3**, 196
Carbonyls, α-diazo-
from
N-alkyl-*N*-nitrosocarboxamides, **3**, 471
α-diazocarbonyls, **3**, 475
α-diazo-β-dicarbonyls, **3**, 475
reactions, with phosphaalkynes, **4**, 479
as starting materials
for α-diazocarbonyls, **3**, 475
for diazonium salts, **2**, 799
for ketenes, **3**, 530
for silyl ynol ethers, **2**, 1024
substitution of, **3**, 474
synthesis of, **3**, 468, 469
Carbonyls, α-diazoalkyl-, synthesis of, **5**, 870
Carbonyls, diazohalogenmethyl-, synthesis of, **3**, 478
Carbonyls, α-diazo-β-hydroxy-, reactions, with boron trifluoride etherate, **1**, 1078
Carbonyls, diazomethyl-, reactions, with acyl isocyanates, **3**, 475
Carbonyls, α,α-dichloro-β-imino-, as starting materials, for α,α-dichloroketimines, **3**, 409
Carbonyls, enolizable, as starting materials, for α-selenocarbonyls, **2**, 282
Carbonyls, α-fluoro-
from, enolsilanes, **2**, 16
reduction of, **1**, 6
Carbonyls, α-germyl-α-diazo-, as starting materials, for ketenes, **3**, 533
Carbonyls, halo-, substitution reactions of, **1**, 467
Carbonyls, α-halo-
reactions
with dithiocarbamate anions, **6**, 561
with hydrazines, **3**, 451
with zinc metal, **2**, 702
reduction of, **1**, 2, 5
methods, **1**, 2
Carbonyls, *N*-haloimino-, synthesis of, **6**, 648, 672
Carbonyls, β-halo-β-nitro-α,β-unsaturated, synthesis of, **4**, 795
Carbonyls, *gem*-halophosphono α,β-unsaturated, synthesis of, **4**, 803
Carbonyls, α-hydroimino-, synthesis of, **3**, 428
Carbonyls, hydroxy-
cyclization of, **4**, 164
as starting materials, for cyclic hemiacetals, **4**, 164
synthesis of, **4**, 167
Carbonyls, α-hydroxy-
reduction of, **1**, 13

synthesis of, **3**, 157
Carbonyls, imino-
boron-substituted, synthesis of, **6**, 634
containing two metal functions, **6**, 675
with one group 14 function and one group 15 function, **6**, 668
with one metal function and one group 15 function, **6**, 670
with one metalloid function and one metal function, **6**, 674
with one nitrogen and one group 15 function, synthesis of, **6**, 656
with one nitrogen and one metal function, **6**, 663
with one nitrogen and one metalloid function, **6**, 661
phosphorus-substituted, synthesis of, **6**, 634, 636
selenium-substituted, synthesis of, **6**, 628, 630
silicon-substituted, synthesis of, **6**, 634
sulfur-substituted, synthesis of, **6**, 636
tellurium-substituted, synthesis of, **6**, 630
tin-substituted, synthesis of, **6**, 634
with two group 15 functions, **6**, 665
with two metalloid functions, **6**, 671
with two nitrogen functions, synthesis of, **6**, 640
N-unsubstituted
from, amidinium salts, **6**, 641
from, thioureas, **6**, 641
from, urea, **6**, 641
structure of, **6**, 671
synthesis of, **6**, 640
use in Bamford–Stevens reactions, **3**, 470
Carbonyls, α-iodo-, synthesis of, **2**, 34
Carbonyls, mercapto-, cyclization of, **4**, 227
Carbonyls, α-methoxycarbonylsulfenylated, synthesis of, **6**, 478
Carbonyls, monomeric, reactions, with diazoalkanes, **4**, 686
Carbonyls, perchloro-, atom transfer cyclization of, **1**, 346
Carbonyls, β-phosphoryloxy-α,β-unsaturated, substitution of, **1**, 482
Carbonyls, α-silyl-, metallation of, **4**, 648
Carbonyls, β-silyl-, elimination of, **1**, 653
Carbonyls, α-silyl-α-diazo-, as starting materials, for ketenes, **3**, 533
Carbonyls, *N*-silylimino-, **6**, 673
Carbonyls, β-sulfonyl-, elimination of, **1**, 613
Carbonyls, β-thioalkyl-α,β-unsubstituted, substitution of, **1**, 485
Carbonyls, thiol, cyclization of, **4**, 227
Carbonyls, (*E*)-α,β-unsaturated, synthesis of, **1**, 747
Carbonyls, α,β-unsaturated
acetalization of, **4**, 179
conjugate addition to, **2**, 702
reactions
with allenyllithiums, **1**, 1061
with diethyl phosphorocyanidate, **3**, 659
with 1,3-dithianes, **1**, 308
with Grignard reagents, **1**, 310
with hexamethyl compounds, **2**, 565
with hydrogen cyanide, **3**, 633
with isothiocyanic acid, **5**, 1107
with nitronate anions, **3**, 191
with organocuprates, **1**, 540
with phosphines, **4**, 339
with propargyllithiums, **1**, 1061
with trimethylsilyl cyanide, **3**, 633
reduction of, **1**, 85
as starting materials
for allylic alcohols, **2**, 57
for amines, **2**, 314
for carboxylic acids, **5**, 51
for cyanohydrin, **3**, 633
for γ-ketonitriles, **3**, 633
synthesis of
from selenoxide pyrolysis, **1**, 781
via elimination, **1**, 599
via retro-Diels–Alder reaction, **1**, 786
via selenoxide elimination, **1**, 605
via sulfoxide pyrolysis, **1**, 779
thermal [4+2] cycloaddition reactions of, for cyclic ortho esters, **6**, 75
use in the Prelog synthesis of phenols, **2**, 662
Carbonyls, γ,δ-unsaturated, synthesis of, **1**, 414, 825, 1094
Carbonyls, δ,ε-unsaturated, synthesis of, **1**, 412, 413, 809
Carbonyls, α,β-unsaturated β-chloro-, reactions, with allylidenetriphenylphosphorane, **1**, 736
Carbonyl selenide, as intermediate, in the synthesis of dialkyl carbonates, **6**, 465
Carbonyl sulfide
fluorination of, **6**, 410
as starting material, for carbonic difluoride, **6**, 410
Carbonyl–thiocarbonyl rearrangements, for the synthesis of 1,2-oxaphosphol-3-ene-5-thione, **5**, 534
Carbopalladation
of alkynes, **2**, 979
products of, **2**, 980
regiochemistry of, **2**, 980
Carbophilic addition, to allylic Grignard reagents, **4**, 255
o-Carborane, *C*-methyl, lithiation, **1**, 168
Carboranes
chemistry of, **6**, 169
reactions of, **6**, 406
Carboranes, diazomethyl-, synthesis of, **5**, 873
o-Carboranes, mono alkylation of, **1**, 168
o-Carboranes, 1,2-di-*C*-alkylated-, synthesis of, **1**, 168
Carbosilanes
from
dimercuracycles, **4**, 624
silacyclobutanes, **4**, 610
tetramethylsilanes, **4**, 617
trisilylmethyllithiums, **6**, 385
linear, synthesis of, **6**, 267
metallation of, **6**, 190
reactions
with bromoform, **6**, 178
with butyllithium, **6**, 178
rearrangement of, **6**, 385
silylation of, **6**, 290
as starting materials, for tetrasilylmethanes, **6**, 385
synthesis of, **6**, 382, 383
Carbosilanes, cyclic, synthesis of, **6**, 268
Carbosilanes, fluoro-, synthesis of, **6**, 268
Carbosilanes, tricyclic, synthesis of, **6**, 178
Carbosilylation, of alkynes, **2**, 919
Carbostyril, 3-benzyl-4-methyl-, reactions, with 2,4-dinitrobenzenesulfenyl chloride, **2**, 768
Carbostyril, 8-ethyl-4-methyl-, reactions, with 2,4-dinitrobenzenesulfenyl chloride, **2**, 768
Carbothioates, *S*-ethyl, reactions, with 1-alkynyltrimethylsilanes, **1**, 1045
Carboxaldehyde, 9-anthracene-, as starting material, for 9-anthracenecarboxylic acid, **5**, 114
2-Carboxaldehyde, pyridine, reactions, with sulfur dioxide, **4**, 237
3-Carboxaldehyde, furan, synthesis of, **3**, 102
Carboxaldehydes, benzofuran, synthesis of, **3**, 102
Carboxaldehydes, benzothiophene, synthesis of, **3**, 104
Carboxaldehydes, furan, synthesis of, **3**, 102
Carboxaldehydes, indole, synthesis of, **3**, 106
Carboxaldehydes, pyridine, synthesis of, **3**, 107
Carboxaldehydes, pyrrole, synthesis of, **3**, 106
Carboxaldehydes, quinoline, synthesis of, **3**, 107
Carboxaldehydes, thiophene, synthesis of, **3**, 104
Carboxamide, *N*-cyclohexyl-*t*-butyldiphenylsilyl-, synthesis of, **6**, 515
Carboxamideamides, *N*-halo-, reactions, with *N*-trimethylsilylsuccinimide, **5**, 814
Carboxamides
from, imidate ester hydrochlorides, **6**, 70
reactions

with chlorocarbonylsulfenyl chloride, **3**, 686
with phosphorus pentaselenide, **5**, 624
silylation of, **5**, 718
N-stannylation of, **2**, 815
as starting materials, for mixed anhydrides, **5**, 716
Carboxamides, *N*-alkyl-*N*-nitroso-
as starting materials
for diazoalkanes, **3**, 467
for α-diazocarbonyls, **3**, 471
Carboxamides, (*N*-alkyl-*N*-nitrosoaminomethyl)-, cleavage of, **3**, 465
Carboxamides, azo-, reactions, with nitroform, **6**, 364
Carboxamides, chlorosulfonyl-, as intermediates, **3**, 669
Carboxamides, *N*,*N*-dialkyl-, reactions, with triethyl germyllithium, **5**, 432
Carboxamides, *N*-iodo-, synthesis of, **5**, 321
N-Carboxamides, imidazole-
synthesis of, **5**, 973
use in the synthesis of isocyanates, **5**, 973
Carboximidamides, borane, synthesis of, **6**, 662
Carboximidic acid esters, synthesis of, **5**, 685
Carboximidic acid esters, *N*-toluolsulfonyl-, synthesis of, **5**, 702
Carboximidic acids, as starting materials, for mixed anhydrides, **5**, 709
Carboximidoyl chlorides, adamantane-, synthesis of, **5**, 657
Carboximidoyl chlorides, *N*-benzenesulfonyl, synthesis of, **5**, 674
Carboximidoyl chlorides, *N*-1-chloroalkenylalkane, synthesis of, **5**, 659
Carboximidoyl chlorides, cyclohexane, synthesis of, **5**, 655
Carboximidoyl fluorides, *N*-fluorinated, synthesis of, **5**, 663
N-Carboxyamide acetals, synthesis of, **6**, 107
Carboxyamidrazone, 4-pyridine-, from, 4-pyridinecarbonitrile, **5**, 769
N-Carboxy α-amino acid anhydrides, synthesis of, **6**, 486
N-Carboxy anhydrides (NCA's)
synthesis of, **5**, 972
use in the synthesis of peptides, **5**, 972
use in the synthesis of poly(α-amino acids), **5**, 972
N-Carboxy compounds, synthesis of, **6**, 105
Carboxyl acids, aromatic, esterification of, **5**, 122
Carboxylate, *t*-butyl 2(*S*)-*N*-tosylaziridine, reactions, with Grignard reagents, **1**, 263
Carboxylate, ethyl, reactions, with methoxymethylene triphenylphosphorane, **5**, 897
Carboxylate, ethyl 7-chloro-4-hydroxy-3-quinoline, hydrolysis of, **5**, 106
4-Carboxylate, ethyl 1-hydroxymethylpyrazole-, use in the synthesis of α-fluoroamines, **4**, 105
C-Carboxylate, ketenimine-, synthesis of, **3**, 578
Carboxylate anions
alkylation of, **5**, 132
reactions, with ethylene oxide, **5**, 138
Carboxylate esters, alkynyl, research background, **2**, 1022
Carboxylates
reactions
with alkoxydiazenium ions, **5**, 366
with alkyl bromides, **5**, 132
with alkyl iodides, **5**, 132
as starting materials, for γ-lactones, **5**, 168
Carboxylates, alkynyl
from, alkynyl(phenyl)iodonium carboxylates, **2**, 1029
synthesis of, **2**, 1030
Carboxylates, alkynyl(phenyl)iodonium, as starting materials, for alkynyl carboxylates, **2**, 1029
Carboxylates, allyl α-cyano-, as starting materials, for α,β-unsaturated nitriles, **3**, 645
Carboxylates, α-cyano-, as starting materials, for cyanoalkenes, **1**, 561
Carboxylates, cyclohexyl azo-, methylation of, **1**, 267
Carboxylates, ethyl allenyl, from, [(ethoxycarbonyl)methylidene]phosphoranes, **1**, 986
Carboxylates, α-haloalkyl
reactions
with alcohols, **4**, 209
with zinc, **2**, 999
Carboxylates, hydroxy-, from, carbonyls, **5**, 164
Carboxylates, γ-hydroxy-
cyclization of, **5**, 166
dehydration of, **5**, 164
Carboxylates, iminovinyl, synthesis of, **3**, 413
Carboxylates, ketene, synthesis of, **3**, 545
Carboxylates, methyl dialkyl cyanoazo-, in electrophilic substitution reactions, **1**, 267
Carboxylates, nitro-, reactions, with 1,5-diazabicyclo[5.4.0]undec-5-ene, **4**, 108
Carboxylates, phosphonium, as intermediates, **4**, 571
Carboxylates, selenoalkynyl, reactions, with amines, **5**, 626
Carboxylates, α-substituted α-amino-, hydrolysis of, **1**, 280
Carboxylates, [*N*-(sulfonylimino)benzyl]-, synthesis of, **5**, 714
Carboxylates, thiol, use in the synthesis of thiols, **2**, 124
2-Carboxylates, benzene-1-diazonium-, thermolysis of, **5**, 885
2-Carboxylates, 4,6-diphenylpyridinium-, reactions, with *p*-nitroso-*N*,*N*-dimenthylaniline, **3**, 433
4-Carboxylates, methyl 2-alkylthiazolidine-, synthesis of, **4**, 318
Carboxylate salts
conversion to acid chloride, **5**, 9
synthesis of, **1**, 654
Carboxyl compounds
from, Grignard reagents, **2**, 1079
as starting materials, for dialkyl sulfides, **2**, 140
Carboxyl group activation (CGA)
mechanism, **5**, 122
for the synthesis of β-lactones, **5**, 161
for the synthesis of macrolactones, **5**, 173
Carboxylic acid, 2-adamantane-, synthesis of, **5**, 51
Carboxylic acid, 9-anthracene-, from, 9-anthracenecarboxaldehyde, **5**, 114
Carboxylic acid, deuterium-labeled, synthesis of, **5**, 47
Carboxylic acid, 2'-formyl-2-biphenyl-, synthesis of, **5**, 115
Carboxylic acid, 4-hexadecylamino-2-thiophene-, synthesis of, **5**, 82
Carboxylic acid, *threo*-β-hydroxy-, use in the synthesis of β-lactones, **5**, 161
Carboxylic acid, isoindole-, reactions, with oxalyl chloride, **5**, 11
Carboxylic acid, isoxazole-, conversion to acid chloride, **5**, 10
Carboxylic acid, peroxy-, reactions, with diethyl ketals, **6**, 466
Carboxylic acid, 4-phenanthrene-, reactions, with sodium azide, **5**, 1001
Carboxylic acid, *cis*-4-(phosphonoxy)-2-piperidine-, synthesis of, **3**, 422
1-Carboxylic acid, adamantane-, reactions, with *t*-butyldimethylsilyl cyanide, **5**, 226
2-Carboxylic acid, fluorenone-, synthesis of, **5**, 114
2-Carboxylic acid, thiophene-, reduction of, **1**, 100, 103; **3**, 104
3-Carboxylic acid, 3-dibromomethyl-4'-fluorobiphenyl-, synthesis of, **5**, 116
3-Carboxylic acid, piperidine-, synthesis of, **1**, 91
4-Carboxylic acid, 3-chlorobiphenyl-, synthesis of, **5**, 108
4-Carboxylic acid, 4'-fluorobiphenyl-, synthesis of, **5**, 112
Carboxylic acid acyloxyamides, as starting materials, for oxyimidic esters, **5**, 827
Carboxylic acid alkoxyamides
alkylation of, **5**, 827
as starting materials, for oxyimidic esters, **5**, 827
Carboxylic acid amides
alkylation of, **5**, 688, 690
reactions

with dialkoxychlorophosphine, **5**, 713
with dialkoxy diphenylsulfuranes, **5**, 689
with oxalyl chloride, **5**, 654
with phosgene, **5**, 654
with thiophosgene, **5**, 654
as starting materials
for aminocarbene complexes, **5**, 953
for trialkylsilyl imidates, **5**, 718
synthesis of, **2**, 1040
Carboxylic acid amides, alkali salts, synthesis of, **5**, 723
Carboxylic acid amides, *N*-alkoxy-
alkylation of, **5**, 700
reactions
with phosphorus pentachloride, **5**, 665
with thionyl chloride, **5**, 665
Carboxylic acid amides, *N*-phosphorylated, as starting materials, for *N*-phosphorylated imidoyl halides, **5**, 683
Carboxylic acid amides, tertiary, as starting materials, for (*Z*)-enolates, **1**, 876
Carboxylic acid amides, *N*-toluolsulfonyl-, synthesis of, **5**, 702
Carboxylic acid analogues, synthesis of, **5**, 506
Carboxylic acid anhydrides
as acylating reagents, **3**, 281, 475
reactions
with isocyanatosilanes, **5**, 1006
with silyl azides, **5**, 992
as starting materials, for ketenes, **3**, 534
Carboxylic acid anhydrides, benzimidic, synthesis of, **5**, 714
Carboxylic acid anhydrides, perfluoro-, reactions, with phosphinic acids, **5**, 221
1-Carboxylic acid anhydrides, adamantane-, synthesis of, **5**, 191
Carboxylic acid chlorides
reactions
with acid anhydrides, **5**, 538
with alkali metal salts, **5**, 937
with alkynyl metals, **1**, 1045
with bis(trimethylsilyl)carbodiimide, **5**, 721
with bis(trimethylsilyl) sulfide, **5**, 538
with *N*-chlorocarbonyl isocyanate, **5**, 1008
with dichloronitrosoalkanes, **5**, 669
with dithiocarbamidate, **5**, 537
with isocyanic acid, **5**, 994
with 2-methylthio-4-phenyl-1-triphenylphosphoranylidenaminoimidazole, **5**, 680
with nitrile oxides, **5**, 669
with phosphorus ylides, **1**, 1079
with silanyloxyarsanes, **5**, 222
with sodium azide, **5**, 989
with succinic anhydride, **5**, 538
with tetrabutylammonium azide, **5**, 991
as starting materials, for hydroxamoyl halides, **5**, 669
Carboxylic acid chlorides, aryl-, reactions, with arylcarboxylic acids, **5**, 194
Carboxylic acid chlorides, *ω*-halo-, as starting materials, for cyclic alkoxycarbene complexes, **5**, 943
Carboxylic acid chlorides, *ω*-isocyanatoalkane-, synthesis of, **5**, 970
Carboxylic acid enolates, condensation of, **1**, 681
Carboxylic acid esters
reduction of, **3**, 39
as starting materials
for acid chlorides, **5**, 14
for acid halides, **5**, 3
4-Carboxylic acid esters, thiazolidine-, as starting materials, for *α*,*β*-unsaturated imines, **3**, 412
Carboxylic acid hydrazides
from, hydrazonic acid esters, **5**, 704
reactions, with nitrous acid, **5**, 989
as starting materials, for hydrazonyl halides, **5**, 675
Carboxylic acid lactate esters, *α*-amino-*α*-cyano-, as starting materials, for thionoamides, **5**, 580
Carboxylic acids
as acylating reagents, for the synthesis of aromatic ketones, **3**, 280
for the acylation of amines, **5**, 262
aliphatic, synthesis of, **1**, 391
alkenation of, **2**, 680
alkylation of, **4**, 205, 824
bismercuration of, **4**, 697
α-bromination of, **2**, 28
bromodecarboxylation of, **2**, 33
α-chlorination of, **2**, 21
chlorodecarboxylation of, **2**, 27
cleavage of, **1**, 70
containing an alkene functionality, conversion of, **5**, 8
conversion to potassium salt, **5**, 12
decarboxylation of, **1**, 560, 772, 933; **4**, 1077
as desilylating reagents, **1**, 448
dianions, lactonization of, **1**, 209
electrophilic addition–elimination of, **1**, 1110
elimination of, **1**, 582
esterification of, **5**, 873
fluorodecarboxylation of, **2**, 20
from
1-alkynes, **1**, 1103
amides, **5**, 108
benzenes, **5**, 115
1,1-dibromovinyl groups, **5**, 38
1,1-difluorovinyl groups, **5**, 38
1,1-dihalovinyls, **5**, 26
ethyl esters, **5**, 92
2-halo ketones, **1**, 391
naphthalenes, **5**, 116
nitriles, **5**, 25
pyridines, **5**, 116
α,*β*-unsaturated carbonyls, **5**, 51
zirconium acyl compounds, **5**, 442
halodecarboxylation of, **2**, 12
homologation of, **1**, 394
α-iodination of, **2**, 34
iododecarboxylation of, **2**, 36
lactonization of, **5**, 168
mercuration of, **4**, 697
non-Kolbé electrolysis of, **1**, 561
phosphorus-containing, reviews, **5**, 96
photochemistry of, **5**, 24
as precursors, for unsaturated ketones, **3**, 140
reactions
with acetylurea, **5**, 262
with acid chlorides, **5**, 184
with alcohols, **5**, 122
with alkenes, **3**, 212
with alkoxyalkynes, **5**, 195
with 2-alkyl-2-hydroxymethyl-1,3-dihydroxypropanes, **6**, 71
with alkyllithiums, **3**, 131
with alkynes, **4**, 206
with ammonia, **5**, 274
with arylsilanes, **1**, 448
with bismuth carbonate, **5**, 222
with *N*,*N*-bis(2-oxo-3-oxazolidinyl)phosphordiamidic chloride, **5**, 188
with boranes, **2**, 57
with carbodiimides, **6**, 506
with 2,2'-carbonylbis(3,5-dioxo-4-methyl-1,2,4-oxadiazolidine, **5**, 333
with 1-chloroalkyl carbodiimides, **5**, 1075
with *α*-chloro enamines, **2**, 758
with chlorosulfonyl isocyanate, **3**, 618, 666
with cyanuric fluoride, **5**, 5
with dialkylboron chlorides, **5**, 228
with dialkylboryl triflates, **4**, 832
with diarylboron chlorides, **5**, 228
with dibenzoylethene, **5**, 190

with di-*t*-butyl peroxide, **1**, 925
with dicyclohexylcarbodiimide, **3**, 475
with dihydropyran, **4**, 208
with diiodosilane, **5**, 20
with dimethylaminodimethylarsine, **5**, 222
with diphenylphosphinic chloride, **5**, 220
with diphenyl phosphorazidate, **5**, 993
with dithiaborinanes, **3**, 83
with enol ethers, **4**, 208
with ethane hexamercarbide, **6**, 405
with ethyl enol ethers, **4**, 208
with formamide, **5**, 262
with Grignard reagents, **3**, 285
with α-haloalkyl ethers, **4**, 209
with hexafluoropropene, **5**, 6
with hexamethyldisilazane, **5**, 224, 376
with hydrazines, **5**, 351
with hydrazoic acid, **5**, 1001
with hydrogen peroxide, **2**, 90
with hydroxylamines, **5**, 322
with *N*-hydroxy-2-thiopyridone, **4**, 266
with imidoyl chlorides, **5**, 716
with isocyanates, **5**, 293
with 1-methoxyethyne, **4**, 827
with organolithiums, **3**, 284
with organosilanes, **5**, 225
with 1,1'-oxalyldiimidazole, **5**, 196
with pentaarylantimony, **2**, 888
with phenyl(bromodichloromethyl)mercury, **6**, 38
with phenyl dichlorophosphate, **5**, 239
with phosphinic chlorides, **5**, 234
with phosphorus trichloride bisphosphonic acids, **6**, 118
with selenium tetrafluoride, **5**, 5
with silyl dichlorides, **5**, 224
with sulfonamides, **5**, 342
with sulfur trioxide, **2**, 176
with thionyl chloride, **5**, 184
with trialkylantimony oxides, **5**, 222
with triarylantimony, **2**, 888
with triarylantimony oxides, **5**, 222
with triarylbismuth, **2**, 888
with triarylstibines, **2**, 888
with trimethylsilyl chloride, **5**, 223
with triphenylphosphine dibromide, **5**, 17
with tris(ethylthio)borane, **5**, 234
with urea, **5**, 262
with vinyl compounds, **3**, 210
with vinyllithiums, **3**, 210
reduction of, **1**, 20, 77; **3**, 82, 96
reduction to aldehydes, **3**, 13
reductive silylation of, **6**, 174
silylation of, **5**, 222, 226
choice of catalyst for, **5**, 224
as starting materials
for acid bromides, **5**, 17
for acid chlorides, **5**, 7
for acid fluorides, **5**, 5
for acid halides, **5**, 1
for acid iodides, **5**, 19
for acyl alkyl acetals, **4**, 209
for acyldiazines, **5**, 366
for acyl fluorides, **5**, 234
for acyloxysilanes, **5**, 223
for aldehydes, **3**, 13
for alkanethiols, **2**, 130
for amines, **2**, 319
for anhydrides, **5**, 184
for arylamines, **2**, 756
for aryl bromides, **2**, 627
for aryl-substituted ketones, **3**, 146
for bis(acyloxy) compounds, **4**, 205
for bis(dialkylboryloxy)ethenes, **4**, 832
for bromoalkanes, **4**, 1077
for bromochloroalkanes, **4**, 38
for bromofluoroalkanes, **4**, 35
for chloroalkanes, **4**, 1077
for chlorofluoroalkanes, **4**, 33
for chloroiodoalkanes, **4**, 40
for dibromoalkanes, **4**, 27
for dithioacetals, **4**, 257
for fluoroiodoalkanes, **4**, 37
for 2-haloalkyl esters, **5**, 139
for heteroaromatic anhydrides, **5**, 198
for isocyanates, **5**, 989
for ketene acetals, **4**, 824
for ketene aminals, **4**, 972
for ketene disilyl acetals, **4**, 832
for ketene dithioacetals, **4**, 845
for ketenes, **3**, 535, 536
for ketones, **3**, 129, 131
for nitriles, **3**, 617
for ortho esters, **6**, 71
for primary alcohols, **2**, 59
for primary amides, **5**, 275
for selenocarboxylic esters, **5**, 239
for symmetrical alkanoic anhydrides, **5**, 185
for symmetrical aromatic anhydrides, **5**, 195
for symmetrical heteroaromatic anhydrides, **5**, 195
for thioamides, **5**, 618
for thiocarboxylic esters, **5**, 234
for thiohydroxamate esters, **1**, 332
for thioketenes, **3**, 547
for thionoamides, **5**, 570
for trialkylsilyl imidates, **5**, 721
for triselenoortho esters, **6**, 92
for trithioortho esters, **6**, 84, 84
for unsymmetrical alkanoic anhydrides, **5**, 193
for unsymmetrical aromatic anhydrides, **5**, 198
synthesis of, **1**, 274; **3**, 530; **5**, **23–120**
general methods, **5**, 24
reviews, **5**, 24
via biotransformations, **5**, 117
via carbonation of organometallics, **5**, 27
via hydrocarboxylation, **1**, 296
via hydrolysis, **5**, 24
via oxidation, **5**, 28
via Wolff reactions, **1**, 394
use in decarboxylation reactions, **1**, 934
use in the synthesis of allenes, **1**, 986
X-ray crystal structure of, **2**, 1004
Carboxylic acids, acyclic, synthesis of, **5**, 51
Carboxylic acids, α-alkoximino-, from, α-keto-thiolic acid esters, **3**, 427
Carboxylic acids, alkoxyquinoline-, reactions, with thionyl chloride, **5**, 9
Carboxylic acids, *N*-alkyl-*N*-hydroxy-α-amino-, synthesis of, **5**, 92
Carboxylic acids, γ-alkynyl-, reactions, with alkynyl bromides, **1**, 1020
Carboxylic acids, allenic, synthesis of, **1**, 971
Carboxylic acids, aromatic, synthesis of, reviews, **5**, 105
Carboxylic acids, aryl-, reactions, with arylcarboxylic acid chlorides, **5**, 194
Carboxylic acids, azetidinone-, silylation of, **5**, 223
Carboxylic acids, 2,2-bis(ethylthio)-, as starting materials, for α-ketoacids, **1**, 265
Carboxylic acids, bismercurated, from, acid anhydride, **4**, 697
Carboxylic acids, α-bromo-
iodo decarboxylation of, **4**, 40
reactions
with bromine, **4**, 27
with *N*,*N*-dialkyl dithiocarbamates, **6**, 561
Carboxylic acids, β-bromo-, as starting materials, for β-lactones, **5**, 161
Carboxylic acids, α-chloro-

reactions, with lead tetraacetate, **4**, 40
synthesis of, **2**, 12; **5**, 55
Carboxylic acids, β-chloro-, synthesis of, **5**, 55
Carboxylic acids, α-chloro-α,β- ethylenic, synthesis of, **5**, 103
Carboxylic acids, ^{14}C-labeled, synthesis of, reviews, **5**, 24
Carboxylic acids, cyclopropyl-, synthesis of, **1**, 284
Carboxylic acids, dibasic, synthesis of, reviews, **5**, 24
Carboxylic acids, diene-, synthesis of, **5**, 101
Carboxylic acids, difluoroaminoperfluoro-, decarboxylation of, **5**, 811
Carboxylic acids, fluorenone-, reactions, with thionyl chloride, **5**, 246
Carboxylic acids, α-fluoro-, synthesis of, **2**, 11
Carboxylic acids, ω-fluoro-substituted, synthesis of, **5**, 55
Carboxylic acids, halo-, reactions, with metal phosphides, **2**, 430
Carboxylic acids, α-halo-, synthesis of, **2**, 11
Carboxylic acids, ω-halo-, synthesis of, via microorganism oxidation of alkyl or alkenyl halides, **5**, 117
Carboxylic acids, 2-haloaryl-, decarboxylative fragmentation of, **1**, 1081
Carboxylic acids, 3-hydroxy-, as starting materials, for alkenes, **1**, 785
Carboxylic acids, α-hydroxy-
reactions, with pivaldehyde, **4**, 213
synthesis of, **1**, 281
Carboxylic acids, β-hydroxy-
dehydrative cyclization of, **5**, 161
reactions
with acetals, **4**, 213
with aldehydes, **4**, 213
synthesis of, **5**, 71
transesterification reactions, with trialkyl ortho esters, **6**, 79
Carboxylic acids, 2-hydroxy-4-keto-, synthesis of, **5**, 79
Carboxylic acids, α-keto-, reactions, with allylboronates, **1**, 281
Carboxylic acids, γ-keto-
reactions, with benzenediazonium hydroxide, **4**, 992
as starting materials, for *gem*-diazoalkenes, **4**, 992
Carboxylic acids, 3-methylene-, synthesis of, **1**, 481
Carboxylic acids, monobasic, synthesis of, reviews, **5**, 24
Carboxylic acids, nitrato-, synthesis of, **5**, 95
Carboxylic acids, α-nitro-
α-fluorination of, **4**, 108
from, primary nitroalkanes, **5**, 95
Carboxylic acids, nitrogen-substituted, synthesis of, reviews, **5**, 90
Carboxylic acids, ω-oximino-, synthesis of, **3**, 428
Carboxylic acids, 3-oxo-
elimination of, **1**, 1072
as starting materials, for α-fluoroalkyl esters, **4**, 52
Carboxylic acids, peptide, reactions, with dimethylphosphinothioyl chloride, **5**, 221
Carboxylic acids, perfluoro-
as starting materials, for iodoperfluoroalkanes, **6**, 32
synthesis of, **5**, 58
Carboxylic acids, peroxy-, as reagents, for the synthesis of bis(sulfones), **4**, 273
Carboxylic acids, pyridyl-, use in the synthesis of aryl pyridyl ketones, **3**, 310
Carboxylic acids, silyl-, reactions, with trimethylsilyl isocyanate, **5**, 1006
Carboxylic acids, α-silyl-, deprotonation of, **4**, 648
Carboxylic acids, sterically hindered, conversion to acid chlorides, **5**, 8
Carboxylic acids, α-sulfonyl-
decarboxylation of, **4**, 84
as starting materials, for α-halo sulfones, **4**, 81
Carboxylic acids, ω-thiocyanato-, from, 1,1-hydroxy hydroperoxides, **5**, 1110
Carboxylic acids, thiol, reactions, with alkenes, **2**, 117, 124
Carboxylic acids, α-thiol
reactions
with aldehydes, **4**, 230
with ketones, **4**, 230
with sulfenyl chlorides, **2**, 126
Carboxylic acids, unsaturated
synthesis of, **1**, 393
via elimination reactions, **1**, 642
Carboxylic acids, α,β-unsaturated
conjugate addition to, **1**, 276
reactions, with alkyl(alkylthio)chlorophosphines, **2**, 463
reduction of, **1**, 74
as starting materials, for vinyl bromides, **2**, 625
synthesis of, **1**, 534
Carboxylic acids, β,γ-unsaturated, iodolactonization of, **5**, 161
Carboxylic acids, γ,δ-unsaturated, synthesis of, **1**, 417
1-Carboxylic acids, naphthalene-, reduction of, **1**, 97
2-Carboxylic acids, arenediazonium-, decarboxylative fragmentation of, **1**, 1081
2-Carboxylic acids, *cis*-aziridine-, synthesis of, **3**, 437
5-Carboxylic acids, 2,7-nonadiene-, synthesis of, **1**, 831
9-Carboxylic acids, homochroman-, synthesis of, **5**, 110
Carboxylic acid silyl esters, isocyanato-, synthesis of, **5**, 992
Carboxylic acid trimethylsilyl esters, α-isocyanato-, synthesis of, **5**, 1012
Carboxylic anhydride analogues, synthesis of, **5**, 528
Carboxylic anhydrides, perfluoro-, reactions, with isocyanic acid, **5**, 995
Carboxylic chlorides, reactions, with phosphorus(III) compounds, **5**, 403
Carboxylic compounds, reactions, with ethyl (trimethylgermyl) acetate, **1**, 291
Carboxylic ester analogues, synthesis of, **5**, 509
Carboxylic esters
dealkylation of, **5**, 32
from
amino alcohols, **5**, 140
carbohydrates, **5**, 140
ketones, **5**, 141
halogenation of, **2**, 3
reactions
with lithium diisopropylamide, **6**, 550
with triphenyl germyllithium, **5**, 432
reductive silylation of, **5**, 420
as starting materials
for methylenephosphanes, **5**, 880
for ortho esters, **6**, 71
for trithioortho esters, **6**, 84
synthesis of, **5**, **121–179**
via Baeyer–Villiger oxidation, **5**, 141
via *trans*-esterification, **5**, 134
via esterification reactions, **5**, 122
via one carbon chain elongation, **5**, 142
via oxidation reactions, **5**, 141
via ozonolytic cleavage of double bonds, **5**, 141
Carboxylic esters, aromatic, from, halides, **5**, 143
Carboxylic esters, dialkylphosphinosubstituted, synthesis of, **2**, 433
β-Carboxythiophosphinates, *S*-alkyl, synthesis of, **2**, 463
Carbozincation
of alkynes, **2**, 973
catalyzed by zirconocene diiodide, **2**, 973
for the synthesis of alkylzincs, **2**, 592
Carbyne clusters, as starting materials, for metal clusters, **6**, 402
Carbyne complexes
Arbuzov reaction of, **5**, 952
deprotonation of, **1**, 1145
protonation of, **3**, 516
reactions
with amines, **5**, 953
with cyclopalladated *N*,*N*-dimethylbenzylamine, **4**, 716
with methyllithium, **3**, 518

with trimethylphosphane, **5**, 919
as starting materials
for alkyne-bridged tungsten complexes, **4**, 675
for aminocarbene complexes, **5**, 953
for carbene complexes, **6**, 722
for phosphinocarbene complexes, **5**, 952
for tungsten carbene complexes, **3**, 516
synthesis of, **6**, 338
Carbyne complexes, amino-, synthesis of, **3**, 691
Carbyne complexes, cationic, reactions, with nitrite anions, **5**, 449
Carbyne complexes, cobalt-bridged, metal substitution of, **6**, 402
Carbyne complexes, *trans*-halo-, synthesis of, **3**, 690
Carbyne complexes, phenyl-, reactions, with diethylaluminum hydride, **3**, 522
Carbyne complexes, silyl-substituted, synthesis of, **6**, 404
Carbyne ligands
synthesis of, **6**, 401
in the synthesis of monochalcogenomethanes, **6**, 336
Carbynes, as starting materials, for rhenium acyl compounds, **5**, 451
Carbynes, alkyl-, use in the synthesis of phenols, **2**, 670
Carbynes, μ_3-methoxy-, synthesis of, **6**, 347
Carcinogens
bis(chloromethyl) ether, **4**, 47
carbamoyl chlorides, **6**, 442
elaiomycin, **2**, 773
N-ethyl-*N*-phenylcarbamic acid chloride, **6**, 501
nitrosoamines as, **2**, 379, 385
Car-3-ene, hydrogenation of, **1**, 72
Carnanion, stabilization of, via α-silyl groups, **6**, 392
Carnidazole, x-ray irradiation of, **6**, 335
Caro's acid
reactions, with *N*-benzylaniline, **2**, 782
use in the synthesis of carboxylic esters, **5**, 141
use in the synthesis of nitroso compounds, **2**, 782
β,γ-Carotene, synthesis of, **1**, 748
Carpyrinic acid, synthesis of, reviews, **5**, 90
Carroll reaction, for the synthesis of nitriles, **3**, 623
Carroll rearrangement, mechanism, **1**, 831
Carvone
reduction of, **1**, 79
synthesis of, **1**, 598
α-Caryophyllene, synthesis of, **1**, 382
Castanospermine, dipropionylation of, **5**, 135
Castro–Stephens reaction, for the synthesis of alkynes, **1**, 1004
Catalysts
for the alkylation of alcohols, **1**, 180
choice of
for the substitution of aryl halides, **1**, 470
for the synthesis of alkynes, **1**, 1051
for Friedel–Crafts acylation, **1**, 462
for hydrogenation reactions, **1**, 71
for metathesis reactions, **1**, 674
for the synthesis of alkanethiols, **2**, 117
for the synthesis of alkenes, **1**, 502
for the synthesis of asymmetric cyanohydrins, **3**, 628
for the synthesis of dithioacetals, **4**, 246
Catalysts, chiral, for the alkylation of carbonyls, **2**, 71
Catalytic cycles, transmetallation in, **2**, 992
Catechol
reactions
with benzodioxyphospholenes, **2**, 477
with borane, **2**, 528
with dimethylamino bicyclic phosphoranes, **2**, 477
use in the synthesis of thionocarbonates, **6**, 543
Catechol, 4-nitro-
reactions
with phosgene, **6**, 467
with triphosgene, **6**, 467
Catecholborane
for the conjugate reduction of enones, **3**, 127
for the hydroboration of alkynes, **1**, 510
reactions
with alkenes, **2**, 42
with alkynes, **1**, 1101
with 1,3-dienes, **2**, 527
with dithiocarboxylic acids, **5**, 589
with tosylhydrazones, **1**, 635
as reducing agent, for the synthesis of *p*-toluenesulfonylhydrazones, **1**, 41
for the reduction of ketene dithioacetals, **4**, 253
synthesis of, **2**, 528
use in the reduction of hydroxy ketones, **2**, 63
use in the synthesis of alkyloxyboranes, **2**, 527
Catechol carbonic acid diethylamide, synthesis of, **6**, 484
Catechols
methylenation of, **4**, 190
synthesis of, **2**, 655
Catechol sulfate, reactions, with amines, **2**, 366
Cathepsin G, inhibition by α-aminoboronic acids, **4**, 526
Cationic carbene complexes, synthesis of, **6**, 732
Cationic centres
flanked by an oxygen and a sulfur, **4**, 1073
flanked by a nitrogen and an oxygen, **4**, 1072
flanked by two oxygens, **4**, 1072
flanked by two selenium atoms, **4**, 1075
flanked by two sulfur atoms, **4**, 1073
Cationic complexes
from, trifluoromethyl complexes, **6**, 727
reactions, with alkenes, **2**, 580
reactivity of, **2**, 582
synthesis of, **2**, 579
Cation radical oxidants, for the oxidation of substituted cyclopentanones, **1**, 556
Cations
bearing chalcogen and metal and nitrogen functions, synthesis of, **6**, 732
bearing chalcogen and nitrogen functions, synthesis of, **6**, 729
bearing halogen and chalcogen functions, synthesis of, **6**, 726
bearing halogen and other elements, synthesis of, **6**, 727
bearing nitrogen and other element functions, synthesis of, **6**, 733
bearing phosphorus and silicon functions, synthesis of, **6**, 733
bearing three chalcogen functions, synthesis of, **6**, 729
bearing three halogens, synthesis of, **6**, 726
bearing three nitrogen functions, synthesis of, **6**, 732
from
acetals, **4**, 1072
keto azides, **4**, 1072
stabilization of, **1**, 909
synthesis of, **1**, **843–951**; **4**, 1072
group abstraction methods, **4**, 1074
photochemical methods, **4**, 1075
Cations, α-alkoxyalkyl, synthesis of, **2**, 1007
Cations, α-arylvinyl, synthesis of, **1**, 1122
Cations, *β*-bromovinyl, structure of, **1**, 1109
Cations, diarylchloromethyl, crystal structure of, **2**, 1007
Cations, α-haloalkyl, synthesis of, **2**, 1006
Cations, α-hydroxyalkyl, synthesis of, **2**, 1007
Cations, phosphaallylic, reactions, with chloromethylene immonium salts, **5**, 882
Cations, tertiary cycloalkyl
in super acid media, **1**, 895
synthesis of, **1**, 895
Cations, *β*-thiovinyl, stability of, **1**, 1109
Cations, tricoordinated stabilized, synthesis of, **6**, **725–734**
Cecropia, juvenile hormones, synthesis of, **1**, 747
Cefazolin, synthesis of, **5**, 193
Cellosolve, methyl, reactions, with hydrogen chloride, **5**, 82
Cembranoid diterpenes, synthesis of, **1**, 818

Cephalosporine, azido-, reactions, with ethyl vinyl ether, **5**, 697
Cephalosporins
from, penicillin sulfoxides, **2**, 184
synthesis of, **1**, 1131; **6**, 485, 488
thionation of, **5**, 575
Cephalosporins, 3-desacetyl-, reactions, with benzyl isocyanate, **6**, 485
Cephems, diazo-, synthesis of, **3**, 479
Cephems, 2α-methoxy-, synthesis of, **4**, 227
Ceric ammonium nitrate (CAN)
as oxidant, for the synthesis of ketones from ethers, **3**, 121
for the oxidation of toluenes, **3**, 86
reactions
with cyclohexanone, **5**, 95
with cyclopentanone, **5**, 95
with dimethyl benzenes, **3**, 93
with norbornanone, **5**, 95
as reagent, for dithiane hydrolysis, **5**, 410
use in nitration reactions, **2**, 791
use in the synthesis of aldehydes, **3**, 8
use in the synthesis of carboxylic acids, **5**, 107
Ceric chloride, reactions, with organolithiums, **4**, 954
Cerium
oxidation states of, **2**, 565
reactions, with alkyl iodides, **2**, 564
Cerium(II) chloride, for the mediation of the reaction of organometallics with carbonyls, **2**, 69
Cerium(III) chloride, use in the reduction of carbonyls, **2**, 59
Cerium(IV) compounds, alkyl-, synthesis of, **2**, 565
Cerium alkynides, synthesis of, **1**, 1140
Cerium catalysts, use in the synthesis of Grignard reagents, **1**, 482
Cerium chloride
reactions
with alkyllithiums, **2**, 313
with lithium compounds, **1**, 1140
with organolithiums, **1**, 1098
Cerium chlorides, use in the synthesis of phosphineboranes, **1**, 158
Cerium complexes, cyclopentadienyl-, synthesis of, **2**, 565
Cerium compounds
in Peterson alkenation, **1**, 649
with two carbon-bridges, **4**, 703
Cerium compounds, alkenyl-
reactions, with ketones, **1**, 1098
synthesis of, **1**, 1098
Cerium compounds, β-silylvinyl-, reactions, with ketones, **1**, 1098
Cerium compounds, trimethylsilylethynyl-, uses of, **1**, 1140
Cerium compounds, α-trimethylsilylvinyl-, reactions, with ketones, **1**, 1098
Cerium dichlorides, alkynyl-, reactions, with ketones, **1**, 1038
Cerium iodide
reactions
with lithium compounds, **1**, 1140
with organolithiums, **1**, 1098
Ceriums, silylvinyl-, as reagents, in the synthesis of alkenylsilanes, **2**, 931
Cerium trichloride, use in methylenation reaction, **1**, 759
Chain reaction
fragmentation method, **1**, 334
initiation of, **1**, 330
mediation of
via germanium hydrides, **1**, 331
via silicon hydrides, **1**, 331
via tin hydrides, **1**, 330
in radical reactions, **1**, 329
Chalcogen, synthesis of functions, **4**, **41–93**
Chalcogen compounds
of arsenic functions, synthesis of, **5**, 891
of phosphorus functions, synthesis of, **5**, 878
Chalcogenides, as catalysts, for Friedel–Crafts acylation, **1**, 462
Chalcogenides, alkenyl, synthesis of functions, **2**, **635–703**
Chalcogenides, alkyl, synthesis of, **2**, **37–112**, **113–275**, **277–295**
Chalcogenides, alkynyl, synthesis of, **2**, 1021, **1011–1038**
Chalcogenides, aryl
synthesis of, **2**, **705–736**
synthesis of functions, **2**, **635–703**
Chalcogenides, vinyl, synthesis of, **2**, **705–736**
N-Chalcogenides, synthesis of, **2**, 469
Chalcogens
and group 15 elements, synthesis of functions, **4**, **293–349**
at least one other than oxygen, synthesis of functions, **5**, **1021–1059**
reactions
with tertiary alkylarsines, **2**, 502
with tertiary alkylstibines, **2**, 502
with tetraalkyldibismuthines, **2**, 499
with tetraalkyldistibines, **2**, 499
substitution of, **1**, 483
synthesis of functions, **6**, **103–136**
with any group other than a halogen or a chalcogen, **4**, **879–965**
containing one chalcogen, **6**, **319–358**
containing a silicon, germanium, boron or metal, **4**, **351–402**
containing three or four chalcogens, **6**, **295–318**
containing two chalcogens, **4**, **823–877**; **5**, **545–563**; **6**, **319–358**
containing two chalcogens other than oxygen, **4**, **243–291**
containing two halogens, **6**, 250
with no halogens, **6**, **67–102**
and one halogen, **6**, 272
synthesis of mixed functions, **6**, 95
Chalcone, as starting material, for allylic alcohols, **4**, 269
Chalcones, β-hydroxy, as starting materials, for aurones, **3**, 242
Chapman rearrangement
of acylphosphines, **5**, 396
for the synthesis of acylhydrazines, **5**, 358
for the synthesis of *N*-arylarylamides, **5**, 303
Charcoal, animal-derived, chlorination of, **6**, 215
Charge transfer complexes, thermolysis of, **4**, 870
Chelates, synthesis of, **5**, 879
Chelation control
in the alkylation of ketones, **2**, 67
in carbonyl alkylation, **2**, 68
in the reduction of carbonyl groups, **2**, 61
Chelation transition state model, and facial diastereoselectivity, **2**, 345
Chelidonic acid, from, diethyl 2,4,6-trioxo-1,7-heptandioate, **5**, 62
Chemical warfare agents, Lewisite, **2**, 881
Chemoselectivity, problems of, **1**, 241
Chinchona alkaloids, use in asymmetric *cis*-dihydroxylation reactions, **2**, 48
Chiral auxiliaries
in asymmetric syntheses, **1**, 305
covalent inclusion of, **1**, 134
from, hydrazones, **1**, 153
reactions, with alkenes, **1**, 357
for the synthesis of α-aminophosphonic acids, **4**, 462
use in cycloaddition reactions, **1**, 370
use in intermolecular radical additions, **1**, 326
use in the synthesis of aminonitriles, **3**, 616
1,3-Chirality transfer, in the synthesis of γ,δ-unsaturated esters, **5**, 159
Chiral ligands, reactions, with rhodium catalysts, **1**, 353
Chiral radicals
abstraction by, **1**, 326
reactions, with ethyl acrylate, **1**, 328

Chloral
reactions
with *S*-acetyl cyanophosphorothioite, **5**, 406
with ammonia, **4**, 297
with ethyl *N*-(dimethyloxyphosphoryl)benzimidate, **5**, 709
with phosgene, **4**, 58
Chloral, aryl hydrazones, as starting materials, for azo compounds, **4**, 799
Chloral hydrate
reactions, with hydroxylamine, **5**, 822
as starting material, for dichloroacetic acid, **4**, 14; **5**, 59
Chloramidites, synthesis of, **2**, 409
Chloramine
reactions
with arylsulfonamides, **1**, 35
with 10-chloro-5,10-dihydrophenarsazine, **2**, 892
with dialkyl sulfides, **2**, 164
with dimethylalkylboranes, **2**, 306
with 10-methyl-5,10-dihydrophenarsazine, **2**, 892
with organoboranes, **2**, 305
with oximes, **3**, 463
with phenoxide, **5**, 288
as reagent
for hydrazines, **2**, 374
for quaternary hydrazinium salts, **2**, 384
as starting material, for triazanium salts, **2**, 387
use in the synthesis of *N*-aminoamides, **5**, 354
Chloramine, diethyl, reactions, with phenylethynemagnesium bromide, **2**, 1041
Chloramines, for the *N*-amination of hydrazines, **2**, 379
Chloramine-T
reactions
with allyl sulfides, **2**, 244
with α-amino acids, **3**, 620
with (arylthio)dibenzoylmethane, **4**, 326
with cyclohexylisonitrile, **5**, 1064
with dimethyl sulfoxide, **2**, 163
with isonitriles, **5**, 1064
with phosphonium salts, **5**, 894
with selenium, **2**, 298
with sulfur mustards, **4**, 773
with triphenylarsine, **2**, 809
as reagent
for dithiane hydrolysis, **5**, 410
for the synthesis of thiocarbonyl *S*-imides, **3**, 370
use in the synthesis of dialkyl sulfoxides, **2**, 148
See also Sodium *N*-chlorotoluene-*p*-sulfonamide
Chloranil, reactions, with dialkyl acyl phosphoramidites, **5**, 373
o-Chloranil *See o*-Benzoquinone, 3,4,5,6-tetrachloro-
Chloranil
use in dehydrogenation reactions, **1**, 554
use in the synthesis of α,β-unsaturated ketones, **3**, 226
o-Chloranil, as reagent, for the synthesis of phenols, **2**, 659
Chloreal *See* Isocyanuric acid, trichloro-
Chloride displacement, for the synthesis of aryl fluorides, **2**, 614
Chloride fluoride, thiocarbonyl
from, dichlorofluoromethanesulfenyl chloride, **6**, 531
reactions, with alcohols, **6**, 533
synthesis of, **6**, 531
Chloride ions
for the α-chlorination of carbonyls, **2**, 22
reactions
with bis(pyridine)iodine(I) tetrafluoroborate, **2**, 24
with manganese(III) compounds, **2**, 24
Chlorides
from
allylic alcohols, **2**, 25
benzylic alcohols, **2**, 25
precursors for radical cyclization, **1**, 341
as radical precursors, **1**, 330
reactions, with water, **5**, 26
Chlorinated lime, reactions, with phosphonium salts, **5**, 894
Chlorinated solvents, manufacture of, **6**, 23
Chlorinating agents
for the synthesis of amidines, **5**, 751
for the synthesis of aryl chlorides, **2**, 619
Chlorinating reagents, reactions, with terminal alkynes, **2**, 1016
Chlorination
of alkanesulfonylthioformamides, **6**, 238
of alkyl halodithioformates, **6**, 233
of animal charcoal, **6**, 215
of aryl methyl sulfides, **6**, 232
of benzene, **6**, 215
of carbon disulfide, **6**, 214, 236
of carbonyl difluoride, **6**, 217
of chloromethane, **6**, 215
of *N*,*N*-diarylthioformamides, **6**, 238
of 1,1-difluoroethane, **6**, 225
of *N*,*N*-dimethylamine, **6**, 238
of dimethyl carbonate, **6**, 230
of *N*,*N*-dimethylchloroformamide, **6**, 239
of *N*,*N*-disubstituted chlorothioformamides, **6**, 238
of disulfides, **6**, 236
electrochemical, for dichlorodifluoromethane, **6**, 225
of graphite, **6**, 217
of hexafluoroethane, **6**, 218
of metal trihalomethanethiolates, **6**, 236
of methyl chloroformate, **6**, 230
of methyl isocyanate, **6**, 240
of methyl perchloro-3-butenoate, **6**, 230
of methyl sulfides, **6**, 233
of naphthalene, **6**, 215
of potassium bis(trifluoromethyl)amide, **6**, 240
regioselectivity of, studies, **4**, 67
of silver chlorodifluoroacetate, **6**, 225
of silver trifluoroacetate, **6**, 218
for the synthesis of hydroxamoyl halides, **5**, 664
of thiophosgene, **6**, 215
of trifluoromethane, **6**, 218
of (trifluoromethyl)trimethylstannane, **6**, 245
of trimethylsilyl, **6**, 243
Chlorination, allylic, reagents for, **2**, 21
Chlorination, benzylic, reagents for, **2**, 21
Chlorination, oxidative, of thiols, **2**, 173
o-Chlorination, selectivity, **2**, 620
para Chlorination, selectivity, **2**, 620
Chlorine
for the α-chlorination of ethers, **4**, 47
elimination of, **1**, 569
1,2-elimination of, **5**, 807
exchange for fluorine, **6**, 9
as impurity in commercial phosgene, **6**, 414
for the oxidation of alcohols to ketones, **3**, 118
reactions
with alcohols, **2**, 94
with alkenes, **2**, 23
with amides, **2**, 756
with *N*-(anilinothiocarbonyl)thiobenzamides, **5**, 657
with bis(acetyl) sulfide, **5**, 244
with bis(benzoyl) sulfide, **5**, 244
with carbon disulfide, **6**, 255
with carbon monosulfide, **6**, 529
with chloroalkenes, **4**, 14
with chlorodicyclohexylarsine, **2**, 492
with chloromethyldichlorophosphine, **2**, 476
with cyanogen, **5**, 662
with cyanogen chloride, **5**, 809
with cyclobutanone oximes, **4**, 108
with 1-diethylamino-2-phthalimidostyrenes, **4**, 793
with diiminosuccinonitrile, **5**, 818
with 3,5-dioxo-1,2,4-dithiazolidines, **6**, 449
with ditelluroacetals, **4**, 291

with dithiobissulfonamides, **2**, 365
with fluoroalkenes, **4**, 31
with isothiouronium salts, **2**, 177
with malononitrile, **5**, 662
with mercury fulminate, **5**, 822
with nitroethenes, **4**, 108
with pentaphenylantimony, **1**, 446
with perfluoronitriles, **5**, 808
with 1-phenylazo-1,2,2-trichloroethene, **5**, 682
with phenylhydrazones, **2**, 805
with phenyl isothiocyanate, **6**, 604
with phosphonium salts, **4**, 344
with silver acetate, **5**, 203
with silver fluoroacetate, **4**, 33
with sulfines, **5**, 489
with sulfur dioxide, **2**, 175
with thiobisamines, **2**, 360
with thiocyanates, **2**, 177
with thiol esters, **2**, 180
with thiophene-2-aldoxime, **3**, 683
with 1,2,3-triazolo[1,5-*a*]pyridine, **4**, 19
with trichloroacetonitrile, **5**, 662
with trimethylborane, **4**, 142
with ylides, **5**, 895
as reagent, for the synthesis of *N*-chloroamides, **5**, 315
sources of, for the chlorination of amines, **2**, 337
as starting material, for phosgene, **6**, 412
for the synthesis of aryl chlorides, **2**, 619
for the synthesis of carbonic difluoride, **6**, 409
for the synthesis of vinyl chlorides, **2**, 617
use in the synthesis of ^{13}C-labeled phosgene, **6**, 414
Chlorine compounds, of phosphorus functions, synthesis of, **5**, 876, 894
Chlorine fluoride
reactions
with perfluorohydrazones, **6**, 260
with trifluoromethyl isocyanide, **6**, 613
transhalogenation of, with chlorocarbonyl isocyanate, **6**, 442
Chlorine fluoro sulfate, reactions, with perfluorinated nitriles, **5**, 710
Chlorine gas, use in the synthesis of alkanesulfenyl chlorides, **2**, 166
Chlorine monofluoride
1,2-elimination of, **5**, 807
reactions
with *N*-acylimines, **5**, 315
with perfluoro ketones, **2**, 94
with trifluoroacyl iminosulfur difluorides, **5**, 317
as starting material, for carbonic chloride fluoride, **6**, 419
Chlorine oxide, as reagent, for the synthesis of alkyl chlorides, **2**, 21
Chlorine trifluoride
reactions
with carbonyl fluoride, **6**, 231
with perfluoro-*t*-butanol, **2**, 98
Chlorine trifluoromethanesulfonate, as starting material, for chloroformate esters, **6**, 428
Chlorinolysis, of disulfides, **2**, 716
α-Chloro-α-alkoxyacyl chlorides, α-chlorination of, **6**, 39
Chloroalkylation, of arenes, **2**, 27
Chloroamine, reactions, with thioamides, **5**, 736
Chloroamine-T
for the oxidation of tellurides, **1**, 611
reactions
with selenides, **1**, 621
with tellurides, **1**, 594, 621
Chlorocarbons
reactions
with alkylmercury compounds, **1**, 668
with hydrogen fluoride, **6**, 8
Chloro compounds, from, (*S*)-(-)-methyl phenylvinylphosphine oxide, **4**, 804
α-Chloro compounds, as starting materials, for α-iodo ethers, **4**, 51
Chlorodecarboxylation, of carboxylic acids, **2**, 27
Chloroethynylation, for the synthesis of alkynyl chlorides, **2**, 1015
Chlorofluorocarbons (CFC's)
replacements for, **6**, 4
synthesis of, **6**, 8
uses of, as refrigerants, **6**, 7
Chlorofluorohydrocarbons, synthesis of, **6**, 7
Chloroform
oxidation of, **6**, 412, 413
reactions
with acrylonitrile, **6**, 6
with allyltributyltin, **4**, 14
with dimethoxyboron chloride, **6**, 181, 206
with phenols, **3**, 91
with potassium *t*-butoxide, **4**, 16, 737
with silicon, **6**, 173, 267
with styrene, **4**, 17
with thiophene 1,1-dioxide, **4**, 742
with *p*-toluenesulfonic acid, **2**, 808
with trimethylstannyllithium, **6**, 208
with trimethyltinsodium, **4**, 699
with triorganylplumbyllithium, **6**, 209
as starting material
for fluorinated orthoformates, **6**, 69
for trialkyl orthoformates, **6**, 68
for triethyl orthoformates, **6**, 68
for tripentyl orthoformates, **6**, 68
for trithioorthoformate esters, **6**, 81
use in the synthesis of isoselenocyanates, **5**, 1051
Chloroiodination, of alkenes, **2**, 24
Chloromethylation
of arenes, **2**, 14
reagents for, **2**, 13
Chloromethyl compounds, aminosulfuration of, **5**, 601
Chloromethylenation, for the synthesis of vinyl chlorides, **2**, 618
Chloromethyl heterocycles, nitrosation of, **5**, 667
Chloromethylidyne complexes, synthesis of, **5**, 1097
1,2-Chlorotropic rearrangement, for the synthesis of benzoylbis(*t*-butyl)phosphide, **5**, 397
Chlorotropic shift, for the synthesis of carbodiimides, **5**, 1069
Cholest-5-ene, 3β-acetoxy-6-nitro-, reactions, with lithium organocuprates, **3**, 426
Cholestenethiol, synthesis of, **2**, 125
Cholesterine, synthesis of, **5**, 526
Cholesterol, 3-iodo-, synthesis of, **6**, 431
Cholines, synthesis of, **5**, 520
Chorismate, [3,3] sigmatropic rearrangement of, **1**, 826
2,4-Chromanedione, 3-diazo-, synthesis of, **3**, 469
2-Chromanols, synthesis of, **4**, 167
Chromanone, oxidation of, **1**, 554
Chromate, 2,2'-bipyridinium chloro-, as reagent, for aldehydes, **3**, 4
Chromate, bis(trimethylsilyl)-, as reagent, for aldehyde regeneration from oximes, **3**, 18
Chromate, *t*-butyl, synthesis of, **5**, 47
Chromate, cyanopyridinium chloro-, as reagent, for the synthesis of α-chloro ketones, **3**, 151
Chromate, 4-(dimethylamino)pyridinium chloro-, for the oxidation of benzylic alcohol, **3**, 88
Chromate, polyvinylpyridinium chloro-, as reagent, for aldehydes, **3**, 4
Chromate, pyridinium chloro- (pcc)
disadvantages of, **3**, 4
as oxidant, for the synthesis of ketones, **3**, 115
for the oxidation of benzylic alcohol, **3**, 88
reactions, with allylic alcohols, **3**, 78
as reagent
for aldehydes, **3**, 4

for oxime cleavage, **3**, 134
use in the synthesis of ketones, **3**, 134
use in the synthesis of α,β-unsaturated aldehydes, **3**, 72
Chromate, pyridinium fluoro-
for the oxidation of benzylic alcohol, **3**, 88
as reagent
for aldehydes, **3**, 4
for the synthesis of ketones, **3**, 115
Chromate, trimethylammonium chloro-, as reagent, for aldehydes, **3**, 4
Chromate, trimethylsilyl chloro-
as reagent
for aldehydes, **3**, 4
for the synthesis of ketones, **3**, 115
synthesis of, **3**, 4
Chromates, ammonium chloro-, as reagents, for aldehydes, **3**, 4
Chromenes
oxidation of, **2**, 654
synthesis of, **3**, 496
Chromenes, 3-nitro-, reduction of, **2**, 341
Chromia-based catalysts, for the synthesis of trihalides, **6**, 8
Chromia catalysts, use in vapor-phase processes, **6**, 17
Chromic acid
use in the synthesis of ketones, **3**, 115
See also Chromium trioxide
Chromic anhydride, reactions, with azidotrimethylsilane, **5**, 368
Chromic chloride, reactions, with trimethylsilylmethyl chloride, **1**, 677
Chromium
as catalyst, for the substitution of sulfur functions, **1**, 238
dimeric allyls of, synthesis of, **4**, 673
reactions, with arsabenzenes, **4**, 1041
Chromium, alkoxycarbene pentacarbonyl-, use in the synthesis of imino esters, **5**, 696
Chromium, benzene(pentafluorolithiobenzene)-, use in the synthesis of metal complexes, **4**, 720
Chromium, benzenetricarbonyl-, lithiation of, **4**, 707
Chromium, bisallyl-, decomposition of, **4**, 673
Chromium, bis(η^6-arsabenzene)-, synthesis of, **4**, 1041
Chromium, 2-chloroallylic, reactions, with aldehydes, **2**, 80
Chromium, dicumene-, as reducing agent, **5**, 808
Chromium, hexacarbonyl-
photolysis of, **5**, 1089
reactions, with dicarbonylcyclopentadienyl(substituted phenyl)iron complexes, **4**, 715
Chromium, lithiated tricarbonylthiophene-, reactions, with bromopentacarbonylmanganese, **4**, 720
Chromium, (lithiobenzene)tricarbonyl-, synthesis of, **4**, 705
Chromium, pentacarbonyl(cyano isocyanide)-, synthesis of, **5**, 1089
Chromium, pentacarbonyl[(dimethyloxosulfono)methanide]-, reactions, with bisdiphenylphosphino ethene, **4**, 1029
Chromium, pentacarbonyl(isocyanomethylene-triphenylphosphorane)-, reactions, with methyl isocyanate, **5**, 905
Chromium, pentacarbonyl(trichloromethyl isocyanide)-, use in the synthesis of carbodiimides, **5**, 1088
Chromium, tricarbonyl(benzene)-, as starting material, for chlorocarbonylchromium complexes, **6**, 456
Chromium, tricarbonyl(chlorobenzene)-, reactions, with palladium(0), **4**, 722
Chromium, tricarbonyl(lithiobenzene)-, reactions, with transition metal halides, **4**, 723
Chromium, tricarbonylthiophene-, metallation of, **4**, 707
Chromium, tricarbonyl(trisammonia)-, reactions, with tungsten hydride complexes, **4**, 715
Chromium, trimesityl-, reactions, with *t*-butylisonitrile, **5**, 798
Chromium(0), ([bromo(phenyl)methyleneamino]-dimethylaminocarbene)pentacarbonyl-, synthesis of, **5**, 660
Chromium(0), (methoxy)methylcarbenepentacarbonyl-, anion of, structure of, **4**, 955
Chromium(0), pentacarbonyl(dimethylamino[methoxy(phenyl)-methyleneamino]carbene)-, reactions, with boron tribromide, **5**, 660
Chromium(II), for the reduction of quinones, **2**, 657
Chromium(II) chloride
reactions
with alkynyl iodides, **1**, 1038
with haloalkynes, **1**, 1140
with α,β-unsaturated nitroalkenes, **3**, 426
Chromium(II) compounds, reactions, with electrophiles, **1**, 974
Chromium(II) salts
reactions, with alkyl halides, **2**, 571
use in the synthesis of alkenes, **1**, 515
Chromium(III) chloride, reactions, with methyllithium, **4**, 708
Chromium(IV) salts, transmetallation of, **2**, 572
Chromium(V) complexes, as reagents, for secondary alcohol oxidation, **3**, 116
Chromium(VI) compounds, use in the synthesis of α-amino ketones, **3**, 184
Chromium(VI) trioxide, for the oxidation of benzylic alcohol, **3**, 88
Chromium acetate, for the mediation of radical cyclizations, **1**, 346
Chromium acyl compounds
reactions
with alkyllithiums, **5**, 444
with aryllithiums, **5**, 444
Chromium alkynides, synthesis of, **1**, 1140
Chromium aminocarbene complexes, synthesis of, **5**, 953
Chromium-based oxidants, for the oxidation of benzyl halides, **3**, 89
Chromium carbene, reactions, with azobenzenes, **5**, 356
Chromium carbene complexes
reactions
with alkynes, **1**, 551
with cyclohexyl isocyanide, **3**, 597
with phosphines, **4**, 580
with vinylogous silylketenes, **3**, 545
synthesis of, **3**, 513; **6**, 327, 605, 721
Chromium carbene complexes, arylalkoxy-, reactions, with carbon monoxide, **1**, 551
Chromium carbene complexes, α,β-unsaturated alkoxy-, as starting materials, for dihydropyrrole complexes, **5**, 953
Chromium carbenes
α-alkylation of, **1**, 165
reactions, with aryllithiums, **3**, 513
synthesis of, **5**, 444
use in the synthesis of phenols, **2**, 670
Chromium carbenes, dienyl-, photolysis of, **2**, 670
Chromium carbonyl, reactions, with aryllithiums, **5**, 444
Chromium carbonyl complexes, synthesis of, **6**, 522
Chromium chloride
use in the synthesis of carbon-bridged dichromium complexes, **4**, 673
use in the synthesis of dimeric allyls of chromium, **4**, 673
Chromium complexes
as mediator of radical cyclizations, **1**, 346
reactions, with *n*-butyllithium, **4**, 1030
synthesis of, **6**, 456
use in the synthesis of alkenes, **1**, 507
Chromium complexes, alkoxycarbene-
reactions
with phosphoranes, **1**, 987
with selenium, **5**, 515, 520
Chromium complexes, alkyne-derived bridged, synthesis of, **4**, 676
Chromium complexes, alkynylcarbene-, as starting materials, for α-silyl enamines, **4**, 1014

Chromium complexes, bromomethyl-, synthesis of, **4**, 147
Chromium complexes, chlorocarbonyl-, from, tricarbonyl(benzene)chromium, **6**, 456
Chromium complexes, chloromethyl-, synthesis of, **4**, 147
Chromium complexes, cyclopentadienyl-, from, chromium trichloride, **2**, 571
Chromium complexes, haloarene-, nucleophilic substitution of, **4**, 715
Chromium complexes, (halomethyl)-, synthesis of, **4**, 147
Chromium complexes, pentacarbonyl-, reactions, with dichlorocyclopropenes, **3**, 513
Chromium complexes, thioaldehyde pentacarbonyl-, reactions, with ynamines, **5**, 605, 614
Chromium complexes, trialkylcyclopentadienyl-, synthesis of, **2**, 571
Chromium complexes, trisimidoyl-, synthesis of, **5**, 798
Chromium compounds
  for the oxidation of benzylic alcohol, **3**, 88
  reactions
    with 1-metalloalkylphosphine oxides, **4**, 579
    with phosphorus ylides, **4**, 580
  for the synthesis of aromatic ketones, **3**, 293
  use in the synthesis of ketones, **3**, 115
Chromium compounds, acyl-, synthesis of, **5**, 444
Chromium compounds, alkenyl-
  reactions, with aldehydes, **1**, 1099
  synthesis of, **1**, 1099
Chromium compounds, alkyl-
  chemoselectivity of, **2**, 571
  synthesis of, **2**, 570
    in situ, **2**, 571
Chromium compounds, alkynyl-
  reactions, with aldehydes, **1**, 1038
  synthesis of, **1**, 1140
Chromium compounds, allylic, from, vinyl epoxides, **2**, 86
Chromium compounds, bisaryl-, carbonylation of, **5**, 444
Chromium compounds, 1-haloalkyl-, reactions, with phosphines, **4**, 579
Chromium compounds, imidoyl-
  from
    isonitriles, **5**, 798
    metalloalkoxycarbenes, **5**, 798
  synthesis of, **5**, 798
Chromium compounds, 2-monosubstituted allylic, reactions, with aldehydes, **2**, 73
Chromium compounds, peroxy-, as reagents, for secondary alcohol oxidation, **3**, 116
Chromium compounds, trialkyl-, synthesis of, **2**, 571
Chromium(III) compounds, alkyl-
  as reagents, **2**, 570
  synthesis of, **2**, 571
Chromium dialkoxycarbene complexes, synthesis of, **6**, 723
Chromium dichloride, synthesis of, **2**, 571
Chromium dichloride, allyl-, Schlenk equilibrium of, **2**, 571
Chromium dichloride, di-*t*-imido-, reactions, with methylene(triphenyl)phosphorane, **5**, 918
Chromium dichloride, pentamethylcyclopentadienyl-, use in the synthesis of chromacyclopentadienes, **4**, 676
Chromium dihalides, alkyl-, as sources of carbon nucleophiles, **2**, 571
Chromium functions, doubly-bonded, synthesis of, **3**, 513
Chromium hexacarbonyl
  reactions
    with formamidines, **5**, 781
    with sodium cyclopentadienide, **6**, 522
  synthesis of, **6**, 522
Chromium oxidants, use in the synthesis of aldehydes, **3**, 4
Chromium pentacarbonyl, reactions, with diazirines, **5**, 943
Chromium pentacarbonyl complex, *N*-1,2-dichloroethylidenamino-, synthesis of, **5**, 658
Chromium pentacarbonyl complex, phenylarsine-, addition to alkynes, **2**, 875
Chromium pentacarbonyl complexes, vacuum thermolysis of, **3**, 716
Chromium pentacarbonyl complexes, chlorovinyl isocyanide-, synthesis of, **3**, 716
Chromium pentacarbonyl complexes, α-halogenated isocyanide-, synthesis of, **3**, 702
Chromium pentacarbonyl dianion, reactions, with acid chlorides, **5**, 444
Chromiumphosphinoalkene complexes, synthesis of, **4**, 1023, 1028
Chromiums, allyl-, reactions, with dialkyl acetals, **2**, 73
Chromium salts
  carbonylation of, **6**, 522
  reactions
    with allylic halides, **2**, 571
    with tosylates, **2**, 571
Chromium tricarbonylarene complexes, directed metallation of, **2**, 942
Chromium tricarbonyl complexes
  for the activation of aryl halides, **2**, 691
  Friedel–Crafts alkylation of, **1**, 202
Chromium tricarbonyl complexes, arene-, from, arenes, **1**, 202
Chromium trichloride
  reactions
    with alkyllithiums, **2**, 572
    with organometallics, **2**, 571
  as starting material, for cyclopentadienylchromium complexes, **2**, 571
Chromium trioxide
  reactions
    with *t*-butyl alcohol, **5**, 47
    with deuterium-labeled 3-phenyl-1-propanol, **5**, 47
  as reagent, for the synthesis of ketones, **3**, 115
  use in the synthesis of aromatic ketones, **3**, 295
  use in the synthesis of chlorotetrolic acid, **5**, 101
  use in the synthesis of α,β-unsaturated ketones, **3**, 215
  See also Chromic acid
Chromium vapor, reactions, with mesitylene, **4**, 694
Chromocenes, synthesis of, **6**, 522
Chromone-3-carboxaldehyde, oxidation of, **5**, 74
Chromone-3-carboxylic acid, synthesis of, **5**, 74
Chromyl chloride, for the oxidation of toluenes, **3**, 86
Chrysanthemic esters, synthesis of, **5**, 145
Chugaev elimination, in the synthesis of dienes, **1**, 788
Chugaev reaction
  for the pyrolysis of xanthate esters, **1**, 775
  of thiocarbonyls, **1**, 939
Chymotrypsin, inhibition by α-aminoboronic acids, **4**, 525
Cimetidine, synthesis of, **6**, 647
Cinammic acid anhydrides, synthesis of, **5**, 189
Cinchonidine, use in the synthesis of chiral phosphinites, **2**, 443
Cinchonine, use in the synthesis of chiral phosphinites, **2**, 443
Cinnamaldehyde
  reactions
    with benzenethiol, **2**, 138
    with diethyl trimethylsilyl phosphite, **4**, 918
    with ethyl cyanoacetate, **3**, 366
    with sodium hydrogen selenide, **3**, 39
    with titanium enolates, **5**, 607
Cinnamaldehyde, hydro-, reactions, with (methoxycarbonyl)sulfenyl chloride, **3**, 35
Cinnamaldehyde, α-(methoxycarbonylsulfenyl)hydro-, synthesis of, **3**, 35
trans-Cinnamaldehyde, 2-isothiocyanato-, synthesis of, **5**, 1033
Cinnamaldehydes, reduction of, **1**, 78, 86
Cinnamate, *t*-butyl
  reactions
    with lithium (*R*)-(α-methylbenzyl)benzylamide, **2**, 315
    with lithium (*S*)-(α-methylbenzyl)benzylamide, **2**, 315
Cinnamate, ethyl, synthesis of, **5**, 104
Cinnamate, (*E*)-methyl-β-benzylamino-α-isocyano-, synthesis of, **3**, 720

Cinnamate, (*Z*)-methyl-β-bromo-α-isocyano-, reactions, with benzylamine, **3**, 720
Cinnamates, (*Z*)-2-β-chloro-, reactions, with Grignard reagents, **1**, 469
Cinnamates, ethyl 2-bromo-, reduction of, **1**, 431
Cinnamhydroxamoyl chlorides, synthesis of, **5**, 664
Cinnamic acid
  from, 4-phenyl-3-buten-2-one, **5**, 103
  synthesis of, **5**, 101
Cinnamic acid, acetamido-, reduction of, **1**, 80
Cinnamic acid, α-acetamino-
  hydrolysis of, **5**, 66
  reduction of, **5**, 90
Cinnamic acid, α-benzoylamino-, azlactone, hydrolysis of, **5**, 87
Cinnamic acid, α-mercapto-, reactions, with nitrostyrenes, **3**, 250
Cinnamic acid, *m*-nitro-, synthesis of, **5**, 98
Cinnamic acid, *o*-nitro-, synthesis of, **2**, 782
Cinnamic acid, *o*-nitroso-, as intermediate, **2**, 782
Cinnamic acid, β-phenyl-, from, benzophenone, **5**, 103
Cinnamic acids
  as starting materials, for amides, **5**, 613
  synthesis of, **1**, 446
Cinnamic acids, *N*-acetyl-, reduction of, **1**, 76
trans-Cinnamic acids, synthesis of, **5**, 99
Cinnamic esters, synthesis of, **5**, 157
Cinnamides, *N*,*N*-disubstituted, reactions, with Grignard reagents, **1**, 310
trans-Cinnamoyl azides, as starting materials, for *trans*-β-styryl isocyanates, **5**, 989
Cinnamoyl bromide, synthesis of, **5**, 18
Cinnamoyl chloride, *p*-chloro-, reactions, with trimethyl phosphite, **4**, 922
Cinnamyl bromide
  reactions
    with methyl isocyanide, **3**, 702
    with palladium carbonyls, **5**, 463
Cinnoline, 4-amino-3-cyano-, synthesis of, **3**, 671
Circularly-polarized light, use in asymmetric [2+2] intermolecular additions, **1**, 357
Citraconate, diethyl, reactions, with *t*-butyl radicals, **1**, 327
Citraconic acid, synthesis of, **5**, 99
Citraconic anhydride, hydrolysis of, **5**, 99
(*E*)-Citral, reactions, with phosphine oxides, **1**, 754
Citric acid, as starting material, for aconitic acid, **5**, 98
(*R*)-Citronellol, synthesis of, **1**, 77
Civetone, synthesis of, **1**, 674
Claisen carbozincation, metalla-, of alkenyl organometallics, **2**, 1006
Claisen condensation
  of ketones, **3**, 349
  for the synthesis of diazo compounds, **3**, 472
  of thiocarboxylic esters, **5**, 237
Claisen–Cope rearrangement, of propargyl vinyl ethers, **1**, 979
Claisen–Johnson rearrangement, for the synthesis of γ,δ-unsaturated esters, **5**, 159
Claisen reaction, metallo-, for the synthesis of alkyl 1,1-bis(zinc) compounds, **2**, 594
Claisen rearrangement
  abnormal, **1**, 827
  aliphatic, **1**, 414
  of allyl aryl ethers, **1**, 414
  of allyl vinyl ethers, **1**, 190, 825
  aromatic, **1**, 414
  of 2,3-butene-1,4-diols, **5**, 166
  Carroll modification of, **1**, 416
  catalysis of, **1**, 826
  catalysts for, **1**, 415
  definition of, **1**, 410
  Eschenmoser modification, for the synthesis of bis(tributyltin) compounds, **4**, 702
  Eschenmoser variant, **1**, 825
  heteroatom variants, **1**, 836
  of imidate esters, **5**, 271
  Johnson variant, **1**, 825
  mechanism, **1**, 414, 825
  stereocontrolled, **1**, 830
  for the synthesis of acylsilanes, **5**, 421
  for the synthesis of amides, **5**, 271
  for the synthesis of α,α-difluoro ketones, **3**, 150
  for the synthesis of ketene silyl acetals, **4**, 831
  for the synthesis of keto esters, **5**, 145
  for the synthesis of natural products, **1**, 825
  for the synthesis of α-sulfonyl ketones, **3**, 171
  for the synthesis of α,β-unsaturated ketones, **3**, 217, 233
  use in the synthesis of phenols, **2**, 656
  use in the synthesis of α-stannylsilylalkanes, **4**, 661
  use in the synthesis of unsaturated aldehydes, **3**, 21
  use in the synthesis of unsaturated ketones, **3**, 141
  of vinyl ethers, **5**, 431
Claisen rearrangement, aromatic, mechanism, **1**, 826
Claisen rearrangement, aza-
  catalysis of, **1**, 835
  mechanism, **1**, 835
Claisen rearrangement, azo-
  for the synthesis of β-allenic nitriles, **3**, 624
  for the synthesis of nitriles, **3**, 623
Claisen rearrangement, charge-accelerated
  asymmetric, **1**, 832
  mechanism, **1**, 832
Claisen rearrangement, hetero-, for the synthesis of milbemycin-β_3, **6**, 473
Claisen rearrangement, retro-, mechanism, **1**, 826
Claisen rearrangement, thia-, mechanism, **1**, 833
Claisen rearrangement, thio-
  of *S*,*N*-ketene acetals, **5**, 616
  for the synthesis of β-hydroxy dithioesters, **5**, 557
  for the synthesis of thionoamides, **5**, 607
  for the synthesis of γ,δ-unsaturated aldehydes, **1**, 141
Claisen rearrangement, ynamine-, mechanism, **1**, 828
o-Claisen rearrangement, mechanism, **1**, 827
p-Claisen rearrangement
  of aryl allyl ethers, **1**, 808
  mechanism, **1**, 827, 840
Claisen–Tishchenko reaction, mechanism, **5**, 141
Clathrate compounds, for the handling of thiophosgene, **6**, 528
Claycop, as nitration agent, **4**, 431
Clayfen
  reactions, with aldehydes, **4**, 431
  as reagent, for the oxidative nitration of oximes, **4**, 429
Cleavage, base-induced, of phosphonium salts, **2**, 441
Clemmensen reduction
  for the reduction of aldehydes, **1**, 18
  for the reduction of ketones, **1**, 18
Clodronic acid, for the treatment of bone formation, **6**, 263
Clonidine, synthesis of, **6**, 645
Clovene, synthesis of, **3**, 536
Cluster framework, cobalt, **6**, 402
^{14}C NMR studies, of azoalkenes, **2**, 770
Cobalamin, methyl-, reactions, with potassium hexachloroplatinate, **2**, 584
Cobaloximes, synthesis of, **1**, 345
Cobaloximes, alkyl-, reactions, with trichloromethanesulfonyl chloride, **2**, 176
Cobaloximes, tertiary(alkylperoxy)(pyridine)-, photolysis of, **1**, 931
Cobalt
  for carbonylation reactions, **1**, 242
  as catalyst, for the carbonylation of benzylthiols, **1**, 239
  catalyzed substitution, with Grignard reagents, **1**, 970
  for the coupling of cyclobutenones with alkynes, **2**, 672
  reactions
    with carbon monoxide, **6**, 524

with diazoalkane, **4**, 686
with perfluorocyclooctatetraene, **4**, 687
Cobalt, acyl-, as acyl radical precursors, for the synthesis of ketones, **3**, 136
Cobalt, carbonylcyclopentadienyldiiodo-, reactions, with sodium amalgam, **4**, 690
Cobalt, chlorotris(trimethylphosphine)-, reactions, with methyllithium, **2**, 576
Cobalt, cyclopentadienyl(1-phenylborole)-, addition reactions of, **4**, 717
Cobalt, dicarbonylcyclopentadienyl-
reactions
with disilylalkynes, **4**, 689
with photopyrone, **4**, 690
with tolan, **4**, 689
Cobalt, dimesityl-
dimeric structure of, **4**, 690
synthesis of, **4**, 690
Cobalt, tetrakis(trialkylphosphine)-, alkylation of, **2**, 575
Cobalt(I), chlorotris(triphenylphosphine)-, use in the synthesis of ketones, **3**, 114
Cobalt(I) complexes
alkylation of, **2**, 578
reactions, with alkenes, **2**, 579
Cobalt(II) chloride
as catalyst
for the reduction of alkyl halides, **1**, 4
for the synthesis of anhydrides, **5**, 184, 195
for the synthesis of geminal diacetates, **4**, 203
for the synthesis of thiocyanates, **5**, 1108
for the mediation of radical cyclizations, **1**, 345
reactions
with arsaalkenes, **6**, 155
with arylmagnesium bromides, **1**, 117
with lithium compounds, **6**, 195
with mesityllithium, **4**, 690
with sodium borohydride, **1**, 75
with α,β-unsaturated aldehydes, **3**, 214
for the reduction of chloroalkanes, **1**, 7
use in the synthesis of phosphoramides, **5**, 370
Cobalt(II) complexes, reactions, with sodium borohydride, **1**, 457
Cobalt(II) complexes, *N,N'*-bis-(ethoxycarbonyl-3-oxobutylidene)ethylenediaminato-, use in the synthesis of α-hydroiminocarbonyls, **3**, 428
Cobalt(II) complexes, bis(trimethylsilyl)methyl-, synthesis of, **6**, 195
Cobalt(II) dichromate, tetrakis(pyridino)-
as oxidant, **3**, 4
reactions, with 4-nitrobenzyl alcohol, **3**, 99
Cobalt(II) ethyl hexanoate, reactions, with hydrogen, **6**, 525
Cobalt(III) complexes, hexacoordinate, synthesis of, **5**, 780
Cobalt(III) compounds, alkyl-, as starting materials, for carbon-centered radicals, **1**, 946
Cobalt(III) ions, pentacyano-, reactions, with hydrogen, **1**, 74
Cobalt(III) salen complexes, synthesis of, **1**, 346
Cobalt(III) trifluoroacetate, for the oxidation of arenes, **2**, 652
Cobalt(acac)$_3$, reduction of, **2**, 576
Cobaltacyclopentadiene complexes
from
nonacarbonyldiiron, **4**, 689
phenylethyne, **4**, 689
sodium tetracarbonylcobaltate, **4**, 689
trifluoroacetic acid, **4**, 689
reactions
with aluminum hydride triethylamine, **4**, 726
with nonacarbonyldiiron, **4**, 714
Cobalt acyl compounds, allylation of, **5**, 457
Cobalt alkynes, reactions, with boron enolates, **1**, 199
Cobalt anions, acyl-, thermal decomposition of, **5**, 421
Cobalt carbene complexes, reactions, with carbon disulfide, **6**, 324
Cobalt carbonyl
reactions, with Wittig reagents, **1**, 225
use in the synthesis of thiocarboxylic esters, **5**, 237
Cobalt carbonyl complexes, synthesis of, **6**, 524, 525
Cobalt–carbonylcyclopentadienyl complexes, synthesis of, **1**, 94
Cobalt carbonyls
as catalysts, for haloalkane addition, **6**, 7
reactions
with alkynic esters, **4**, 712
with vinyliron complexes, **4**, 711
Cobalt catalysis, for the synthesis of diols, **1**, 186
Cobalt catalysts, for hydroformylation reactions, **1**, 296
Cobalt(III) chelate complexes, alkyl-, oxidation of, **2**, 579
Cobalt chloride, reactions, with sodium borohydride, **2**, 312
Cobalt cluster compounds, use in the synthesis of monochalcogenomethane, **6**, 348
Cobalt complexes
as mediator of radical cyclizations, **1**, 346
reactions
with alkyl halides, **1**, 335, 345
with allylsilanes, **1**, 199
with aluminum alkyne, **1**, 199
with aryl halides, **1**, 345
with germanes, **6**, 404
with oxiranes, **1**, 345
with secondary alkylarsines, **2**, 512
with succinimide, **5**, 387
with tosylates, **1**, 335, 345
use in the synthesis of α,β-arylnitriles, **3**, 662
Cobalt complexes, alkyl-, synthesis of, **1**, 335
Cobalt complexes, allyl-, use in radical fragmentations, **1**, 334
Cobalt complexes, (chloroacetyl)-, synthesis of, **4**, 148
Cobalt complexes, (chloromethyl)-, synthesis of, **4**, 148
Cobalt complexes, cyclopentadienyl-
alkylation of, **2**, 579
reactions
with alkyl iodides, **2**, 579
with triphenylstannane, **4**, 726
as starting materials, for organocobalts, **6**, 341
Cobalt complexes, germanium-bridged, reactions, with alkynes, **4**, 687
Cobalt complexes, maleoyl-
reactions, with alkynes, **5**, 457
as starting materials, for quinones, **5**, 457
Cobalt complexes, metal–metal double-bonded
reactions
with alkynes, **4**, 689
with 3,3-dimethylcyclopropene, **4**, 689
Cobalt complexes, methyl-, synthesis of, **2**, 576
Cobalt complexes, mono-bridged, synthesis of, **4**, 686
Cobalt complexes, tetrakis(1-norbornyl)-, synthesis of, **2**, 579
Cobalt complexes, tri-bridged, synthesis of, **4**, 686
Cobalt complexes, tungstacyclopentadiene-, synthesis of, **4**, 714
Cobalt(III) complexes, alkyl-, as radical precursors, **2**, 578
Cobalt compounds
reactions
with carbon disulfide, **6**, 352
with 1-metalloalkylphosphines, **4**, 582
with phosphorus ylides, **4**, 582
with thiocarbyne complexes, **6**, 352
Cobalt compounds, acyl-
as starting materials, for acid chlorides, **5**, 456
synthesis of, **5**, 456
Cobalt compounds, alkyl-
reactions
with carbonyls, **2**, 577
with vinyl halides, **2**, 577
for the substitution of vinyl halides, **1**, 466
synthesis of, **2**, 575

Cobalt compounds, anionic, alkylation of, **2**, 575
Cobalt compounds, imidoyl-, synthesis of, **5**, 799
Cobalt(I) compounds, alkyl-, synthesis of, **2**, 575
Cobalt(II) compounds, alkyl-, synthesis of, **2**, 576
Cobalt hydrides
  elimination of, **1**, 667
  synthesis of, **1**, 74
Cobalt isocyanide complexes, reactions, with diaryl diazoalkanes, **3**, 599
Cobalt–molybdate–alumina catalyst, for the synthesis of benzene, **1**, 56
Cobalt multidecker sandwich compounds, from, pentamethylcyclopentadiene, **4**, 695
Cobaltocene, oxidation of, **6**, 463
Cobalt octacarbonyl
  in the cyclization of alkynylsilanes, **2**, 926
  reactions, with azobenzenes, **5**, 356
Cobalt phthalocyanin, use in the synthesis of alkylchromium(III) compounds, **2**, 572
Cobalt(III) porphyrin complexes, alkyl-, synthesis of, **2**, 578
Cobalt radicals, coupling procedures, **1**, 495
Cobalt salophens, acyl-
  as starting materials, for acyl radicals, **5**, 457
  synthesis of, **5**, 457
Cobalt salophens, alkyl-, as starting materials, for alkyl radicals, **1**, 667
Cobalt salophens, allylacyl-, homolysis of, **1**, 933
Cobalt salophens, arylmethyl-, homolysis of, **1**, 933
Cobalt sulfides, as catalysts, for the synthesis of alkanethiols, **2**, 130
Cobalt tetracarbonyl
  acylation of, **2**, 575
  reactions
    with alkenes, **5**, 456
    with ketenes, **5**, 456
Cobalt tetracarbonyls, alkyl-, stability of, **2**, 575
Cobalt trifluoride
  as fluorinating agent, **4**, 3
  use in the synthesis of trifluoromethyls, **6**, 19
Cobicyclization, enyneisonitrile, nickel-promoted, **2**, 985
Cocyclotrimerization, of alkynes, co-catalyzed, **2**, 984
Coenzyme A dithioic acid esters, synthesis of, **5**, 551
Coenzyme B_{12}, synthesis of, **2**, 578
Colchicine compounds, elimination of, **1**, 637
Collidine, for the removal of dihydroquinols, **1**, 558
2,4,6-Collidine, reactions, with lithium iodide, **5**, 39, 78
Collins' reagent
  synthesis of, **3**, 4
  use in the synthesis of aldehydes, **3**, 4
  use in the synthesis of ketones, **3**, 115
Collman's reagent
  for the conjugate reduction of enones, **3**, 128
  reactions, with alkyl halides, **2**, 575
  use in the synthesis of aldehydes, **3**, 20
  use in the synthesis of ketones, **3**, 137
  See also Disodium tetracarbonylferrate
Colnelic acid, synthesis of, **1**, 601
Compactin, synthesis of, **1**, 633, 688
Complexation
  for organometallics, **2**, 952
  for the synthesis of chiral phosphines, **2**, 442
π-Complexes
  metallation of, **4**, 705
  synthesis of, **2**, 577
Complex salts, from, dimethylcyanamide, **6**, 728
Concerted cycloadditions
  regioselectivity of, **1**, 350
  stereochemistry of, **1**, 351
Concerted reaction, mechanism, **1**, 350
Condensation
  chromium-mediated, **1**, 676
  for the synthesis of alkynylantimony compounds, **2**, 1081
  for the synthesis of alkynylstannanes, **2**, 1100
  for the synthesis of carbon–carbon double bonds, **1**, **719–770**
  for the synthesis of *N*-carbon-substituted ketimines, **3**, 407
  for the synthesis of one or more carbon–carbon double bonds, **1**, **673–717**
  for the synthesis of 3-thio α,β-unsaturated ketones, **3**, 251
  via Peterson reactions, **4**, 936
  zinc-mediated, **1**, 678
Condensation, dehydrogenative, for the synthesis of alkylsilathianes, **2**, 540
Condensation, intramolecular, for the synthesis of alkylidenebisamides, **4**, 425
Condensation, seven-component, of isocyanides, **3**, 694
Conducted tour mechanism, definition of, **1**, 858
Cone angle, of the silylmethyl group, **6**, 390
(+)-Conhydrine, synthesis of, **2**, 326
Conjugate addition
  of amines to α,β-unsaturated aldehydes, **3**, 44
  of aryl nucleophiles, to α,β-unsaturated ketones, **3**, 145
  frontier molecular orbital analysis of, **1**, 301
  for the synthesis of alkyltins, **2**, 600
  for the synthesis of dialkyl sulfones, **2**, 159
  to alkynyl epoxides, **1**, 277
  to carbon–carbon bonds, **1**, 300
  to enals, **3**, 12, 24
  to α,β-enolates, **5**, 151
  to enones
    carbon nucleophiles, **3**, 128
    of unsaturated nucleophiles, **3**, 140
  to α,β-unsaturated aldehydes, **1**, 273
  to α,β-unsaturated carboxylic acids, **1**, 276
  to α,β-unsaturated ketones, **1**, 273
  to vinyl epoxides, **1**, 277
  of unsaturated nucleophiles, to α,β-unsaturated ketones, **3**, 147
Conjugate addition, asymmetric
  chiral catalysts for, **1**, 306
  of lithium diorganocuprates, **1**, 257
Conjugate addition, charge-directed, mechanism, **1**, 303
Conjugate addition, nucleophilic, to metal carbide, **5**, 1097
Conjugate reaction, of carbene complexes, **5**, 958
Conjugate reduction
  of enals, **3**, 12
  of enones, **3**, 127, 145
Conjugation
  for the stabilization of carbanions, **1**, 847
  for the synthesis of carbanions, **1**, 849, 859
Contact ion pairs (CIP), equilibrium, with solvent separated ion pairs, **1**, 302
Cooligomerization, of phosphaalkynes, **5**, 885
Coordinating reagents, use in carbolithiation reactions, **2**, 555
Coordination, intramolecular, for the stabilization of alkyl metals, **2**, 564
Cope elimination
  method, **1**, 782
  regiochemistry of, **1**, 783
  stereochemistry of, **1**, 782
  for the synthesis of alkenes, **1**, 782
  for the synthesis of dienes, **1**, 788
Cope elimination, sila-, **1**, 596
  for the synthesis of dienes, **1**, 789
Cope–Mannich cyclization, aza-, of iminium ions, **3**, 419
Cope rearrangement
  of amine *N*-oxides, **2**, 346
  of bis(thiobenzoyl)hydrazine, **5**, 736
  catalysis of, **1**, 808
  choice of solvent for, **1**, 411
  definition of, **1**, 410, 410
  of divinyl cycloalkanes, **1**, 811
  of hydroxylamines, **2**, 345
  mechanism, **1**, 411, 623, 806
  regioselectivity of, **1**, 624

stereoselectivity of, **1**, 624
thermal, mechanism, **1**, 806
use with Claisen rearrangement, for the synthesis of aldehydes, **3**, 21
use in the synthesis of natural products, **1**, 811
of vinylbicycloalkenes, **1**, 814
Cope rearrangement, anionic oxy-, **1**, 411
acceleration of, **1**, 413
mechanism, **1**, 810
use in the synthesis of natural products, **1**, 810
Cope rearrangement, aza-
of iminium salts, **3**, 422
zwitterionic, **1**, 835
Cope rearrangement, 1-aza-, for the synthesis of amines, **2**, 314
Cope rearrangement, 3-aza-
of *N*-alkyl-*N*-allylanilines, **1**, 835
of *N*-allyl enamines, **3**, 413
Cope rearrangement, 1,3-diphospha-, for the synthesis of allylic isocyanates, **1**, 837
Cope rearrangement, dithia-, for the synthesis of thionoamides, **5**, 607
Cope rearrangement, metallo-, mechanism, **1**, 836
Cope rearrangement, oxy-, **1**, 411
catalysis of, **1**, 413
disadvantages of, **3**, 141
mechanism, **1**, 412, 809
for the synthesis of unsaturated aldehydes, **3**, 22, 24
for the synthesis of α,β-unsaturated ketones, **3**, 217
use in the synthesis of unsaturated ketones, **3**, 141
use in the synthesis of δ,ε-unsaturated ketones, **3**, 144
of vinyl alkynyl carbinols, **3**, 276
Cope rearrangement, 3-phospha-, **1**, 836
Cope rearrangement, retro-
for the synthesis of amine *N*-oxides, **2**, 349
for the synthesis of *N*,*N*-dialkylhydroxylamines, **2**, 346
Cope rearrangement, seleno-, of allyl butynyl selenides, **3**, 553
Cope rearrangement, sila-, mechanism, **1**, 626
Cope rearrangement, 3-sila-, **1**, 836
Cope rearrangement, thia-, for the synthesis of allylthioketenes, **3**, 548
Cope rearrangement–Mannich cyclization, aza-, mechanism, **3**, 420
Copper
as catalyst, in [2,3] Wittig rearrangements, **1**, 234
for the displacement of sulfur, **1**, 237
highly activated, synthesis of, **2**, 585
reactions
with alkyl halides, **2**, 585
with α-amino acids, **5**, 385
with perfluoroalkanes, **6**, 63
use in the displacement of allylic alcohols, **1**, 187
use in the synthesis of 1,1-bis(silyl)alkanes, **4**, 609
Copper, alkenyl-, as reagent, for the synthesis of copper salts of dithioalkenoic acids, **5**, 548
Copper, bis(hexafluoroacetylacetonato)-, as homogeneous catalyst, **6**, 708
Copper, *n*-butyl-, reactions, with 1-hexyne, **1**, 540
Copper, cyanomethyl-
reactions, with allyl halides, **3**, 623
synthesis of, **3**, 640
Copper, α-iodomethyl-, synthesis of, **2**, 998
Copper, methyl-
reactions
with alkenyldibromoboranes, **1**, 488
with dialkenylchloroboranes, **1**, 488
synthesis of, **2**, 586
Copper, silylallyl-, reactions, with electrophiles, **2**, 935
Copper, vinyl-, as intermediate, **2**, 960
Copper, zerovalent, synthesis of, **1**, 121
Copper(0), reactions, with 1,2-dichloropropene, **2**, 74
Copper(I), 1-methoxyvinyl-, from, methoxyvinyllithium, **4**, 953
Copper(I), (trifluoromethyl)-, synthesis of, **6**, 246
Copper(I) aldimines, synthesis of, **3**, 418
Copper(I) alkynides
alkylation of, **1**, 1002
oxidation of, **1**, 1034
reactions
with alkenyl halides, **1**, 1005
with allenyl halides, **1**, 1005
with aryl iodides, **1**, 1004
with (bromoethynyl)trimethylsilane, **1**, 1019
with 2-hydroxybenzothiazole, **1**, 1025
synthesis of, **1**, 1004; **3**, 273
Copper(I) ate complexes, α-bromoalkenyl-, as intermediates, in the synthesis of radialenes, **4**, 820
Copper(I) bromide
reactions
with 2-lithioanisole, **4**, 696
with 2-lithio-1-(dimethylaminomethyl)benzenes, **4**, 695
with vinylcadmiums, **4**, 821
with vinylzincs, **4**, 821
use in the synthesis of 1,3-dioxalanes, **4**, 207
Copper(I) carbene complexes, synthesis of, **6**, 723
Copper(I) catalysts, synthesis of, **1**, 119
Copper(I) chloride
as catalyst, for the synthesis of amidines, **5**, 750
reactions
with mesitylmagnesium bromide, **4**, 696
with tris(trimethylsilyl)benzamidine, **5**, 781
Copper(I) cocatalysis, for the synthesis of alkynes, **1**, 1008
Copper(I) complexes
reactions, with carbon disulfide, **6**, 544
reduction of, **1**, 121
Copper(I) compounds, uses of, **4**, 953
Copper(I) compounds, alkenyl-, from, alkenylboranes, **1**, 1018
Copper(I) compounds, alkyl-, reactions, with stannylalkynes, **1**, 968
Copper(I) compounds, perfluoroalkyl-, nucleophilic substitution with, **1**, 967
Copper(I) compounds, silyl-, reactions, with terminal alkynes, **2**, 917
Copper(I) compounds, trimethylsilylmethyl-, nucleophilic substitution with, **1**, 967
Copper(I) compounds, vinyl-, reactions, with *S*-methyl methanethiosulfonate, **4**, 940
Copper(I) cyanide
reactions
with aryl halides, **3**, 661
with diethylzinc, **2**, 69
with pent-4-enylzinc–magnesium compounds, **4**, 724
with phenylethyne, **5**, 1140
Copper(I) cyanoborohydride, bis(triphenylphosphine)-, use in the synthesis of aldehydes, **3**, 13
Copper(I) enolates, nucleophilic substitution with, **1**, 968
Copper(I) halides
reactions
with Grignard reagents, **1**, 967
with organolithiums, **2**, 1095
with terminal alkynes, **1**, 1004
Copper(I) hydrides, for the reduction of alkynes, **1**, 458
Copper(I) iodide
reactions
with alkenyllithiums, **1**, 1093
with methyl fluorosulfonyldifluoroacetate, **6**, 15
with potassium tri-*s*-butylborohydride, **1**, 999
with sulfoxides, **4**, 396
Copper(I) ketenides, synthesis of, **3**, 537
Copper(I) oxide, as catalyst, for the synthesis of imidoylphosphines, **5**, 785
Copper(I) phenoxides, for the tetrasubstitution of tetrachloromethane, **6**, 296
Copper(I) salts
reactions

with alkyl iodides, **2**, 72
with alkyllithiums, **2**, 586
with alkynes, **2**, 1094
with alkynyllithiums, **2**, 1095
with alkynylpotassiums, **2**, 1095
with allyl bromide, **1**, 1004
with Grignard reagents, **1**, 311
with lithium biphenylide, **2**, 585
with lithium methyl(trialkyl)borates, **1**, 125
with magnesium compounds, **2**, 586
with organometallics, **2**, 586
as starting materials, for mixed dialkylcuprates, **2**, 587
use in the synthesis of dialkyl carbonates, **6**, 465
use in the synthesis of imidoylcoppers, **5**, 802
Copper(I) systems, alkynyl-, synthesis of, **2**, 1094
Copper(I) tetrahydroborate, bis(triphenylphosphine)-, for the reduction of hydrazones, **1**, 41
Copper(I) thiocyanates, as catalyst, for the synthesis of thiocyanates, **5**, 1108
Copper(I) triflate, use in the synthesis of aromatic ketones, **3**, 282
Copper(I) trimethylsilylalkynide, stability of, **1**, 1006
Copper(II), as catalysts, in the synthesis of trihalides, **6**, 7
Copper(II) acetate
for the decarboxylation of acids, **1**, 561
reactions, with terminal alkynes, **1**, 1034
Copper(II) acetyl acetone, as catalyst, for the insertion of isonitriles into silanes, **5**, 792
Copper(II) bromide
for the dibromination of alkenes, **2**, 30
use in the synthesis of enones, **1**, 556
Copper(II) chloride
for the chlorination of alkenes, **2**, 23
for the chlorination of carbonyls, **2**, 21
reactions
with 1-phenylpropyne, **1**, 1142
with quadricyclane, **3**, 608
with triisopropylsilane, **2**, 538
with ylides, **2**, 447
for the ring-opening of thiirane sulfoxides, **2**, 222
use in coupling reactions, **5**, 279
use in the synthesis of ^{11}C-labeled phosgene, **6**, 414
Copper(II) chloride, benzil monohydrazone-, reactions, with pyridine, **3**, 470
Copper(II) cyanide, reactions, with bis(triphenylgermanium)mercury(II) complexes, **5**, 1136
Copper(II) halides, reactions, with alkenes, **5**, 1115
Copper(II) nitrate
for the oxidation of alcohols, **3**, 118
use in the synthesis of *N*-nitroamides, **5**, 361
Copper(II) nitrate, montmorillonite-supported, as nitration agent, **4**, 431
Copper(II) perchlorate, use in the synthesis of allylic amines, **2**, 308
Copper(II) salts
reactions, with *β*,*γ*-unsaturated diazo ketones, **3**, 532
as starting materials, for alkylcoppers, **2**, 586
Copper(III) nitrate hydrate, as oxidant, for the synthesis of phenols, **2**, 647
Copper alkynides
reactions, with alkyllithiums, **2**, 587
synthesis of, **1**, 1135
use in coupling reactions, **1**, 1136
use in the synthesis of organocuprates, **1**, 1136
Copper alkynylides
iodination of, **2**, 1019
as starting materials, for iodoalkynes, **2**, 1019
Copper boranes, vinyl-, synthesis of, **1**, 1094
Copper borohydrides, complex, for the reduction of acyl chlorides, **3**, 83
Copper bromide, as reagent, for the synthesis of alkyl bromides, **2**, 28
Copper bromide–dimethyl sulfide complexes, synthesis of, **2**, 586
Copper bromide–dimethyl sulfur, reactions, with alkyltitanium complexes, **2**, 587
Copper *t*-butoxide, use in the synthesis of copper alkynides, **1**, 1136
Copper carbenoids, synthesis of, **1**, 109
Copper carbonyl complexes, synthesis of, **6**, 525
Copper catalysis, for epoxide opening reactions, **1**, 478
Copper catalysts
in cyclopropanation reactions, **1**, 353
reactions, with seleno esters, **3**, 307
synthesis of, **1**, 468
for the synthesis of ylides, **1**, 54
Copper-catalyzed compounds, use in the displacement of alkyl alcohols, **1**, 177
Copper-catalyzed couplings, **1**, 116
Copper chloride, use in the synthesis of homocuprates, **2**, 586
Copper chloride, lithiated, use in organocuprate chemistry, **1**, 119
Copper chromite, use in the reduction of esters, **2**, 52
Copper complexes
as mediator of radical cyclizations, **1**, 346
for the oxidation of thiols, **2**, 215
Copper complexes, alkyl-
reactions
with alkynes, **5**, 100
with imides, **5**, 387
Copper complexes, octanuclear, synthesis of, **4**, 696
Copper cyanide
reactions
with alkyllithiums, **2**, 588
with alkylzinc iodides, **2**, 588
with alkylzirconiums, **2**, 588
with alkynes, **3**, 675
reduction of, **2**, 585
Copper dimethyl sulfide, trimethylstannyl-
reactions
with esters, **2**, 600
with ketones, **2**, 600
Copper enolates, synthesis of, **2**, 702
Copper groups, anti-elimination of, **1**, 668
Copper halides
reactions
with 1-metalloalkylphosphine oxides, **4**, 585
with 1-metalloalkylphosphines, **4**, 585
with 1-metalloalkyl phosphonates, **4**, 585
Copper hydrides
for the conjugate reduction of enones, **3**, 128
reactions, with 1-alkynyl halides, **1**, 999
Copper iodide
as catalyst, for carbozincation, **2**, 973
reactions
with 1-ethoxyvinyllithium, **4**, 953
with sodium trifluoroacetate, **6**, 14
as starting material, for homocuprates, **2**, 586
Copper iodide–lithium chloride, reactions, with alkyltitanium complexes, **2**, 587
Copper iodide triethyl phosphite complexes, reactions, with alkylsamarium(III) compounds, **2**, 587
Copperlithium, dimethyl-
reactions
with acetals, **4**, 388
with *α*,*β*-alkynic nitriles, **3**, 676
with (*α*-phenylseleno) *α*,*β*-unsaturated ketones, **4**, 401
with vinyl sulfones, **4**, 964
Copperlithium, diphenyl-, reactions, with vinyl iodonium salts, **1**, 431
Copper–lithium complexes, synthesis of, **2**, 588
Copper–lithium compounds, dialkyl-, use in the synthesis of tertiary amines, **2**, 306
Copper–lithium compounds, dimethyl-, reactions, with allylic stannanes, **2**, 589

Copper manganate(VII), for alcohol oxidation, **3**, 116
Copper nucleophiles, reactions, with allyloxybenzthiazoles, **1**, 187
Coppers
from, cadmium compounds, **6**, 15
for the opening of oxetanes, **1**, 229
for the opening of vinyl epoxides, **1**, 226
reactions
with γ-mesyloxy-α,β-enoate, **1**, 188
with phosphorus ylides, **4**, 585
with propargylic alcohols, **1**, 197
synthesis of, **4**, 963
for transmetallation reactions, **1**, 464
Coppers, alkenyl-
cross-coupling of, **2**, 978
reactions
with alkynyl(phenyl)iodonium tosylates, **1**, 1018
with electrophiles, **2**, 978
synthesis of, **1**, 1093, 1107; **2**, 969
Coppers, (*Z*)-alkenyl-, from, carbocupration of alkynes, **2**, 977
Coppers, α-alkoxy-, synthesis of, **4**, 384
Coppers, alkyl-
conjugate addition to ethynylsilanes, **1**, 160
deuterium-labeled, elimination of, **1**, 665
from, copper(II) salts, **2**, 586
Lewis acid-modified, **1**, 119
reactivity of, **2**, 586
stability of, **2**, 586
for the substitution of vinyl halides, **1**, 466
synthesis of, **1**, 121; **2**, 585
via carbometallation, **2**, 589
via halogen-metal exchange, **2**, 585
via transmetallation, **2**, 586
Coppers, alkynyl-
palladium-catalyzed cross-coupling reactions of, **1**, 1006
reactions
with acid chlorides, **1**, 1045
with imidoyl chlorides, **1**, 1047
reactivity of, **2**, 587
synthesis of, **1**, 1006; **2**, 1094
Coppers, allyl-
reactions, with allyl bromide, **1**, 121
synthesis of, **1**, 121
Coppers, allylic, from, vinylcoppers, **2**, 594
Coppers, aryl-
reactions, with alkyl halides, **1**, 121
use in the synthesis of biaryls, **1**, 473
Coppers, carbon-bridged, from, metal(I) halides, **4**, 695
Coppers, carbon-bridged pentameric, synthesis of, **4**, 696
Coppers, (*Z*,*Z*)-dienyl-, from, carbocupration of alkynes, **2**, 977
Coppers, α-haloalkenyl-, synthesis of, **4**, 821
Coppers, imidoyl-
from, imidoyllithiums, **5**, 802
as starting materials, for 1,4-diketones, **5**, 802
synthesis of, **5**, 802
Coppers, monoalkyl-, synthesis of, **2**, 586
Coppers, perfluoroisopropyl-, synthesis of, **4**, 151
Coppers, phenyl-
reactions
with α-bromoenamines, **1**, 464
with α-halotosylhydrazones, **1**, 464
Coppers, silylated, for the synthesis of arylhydrazines, **2**, 769
Coppers, silylvinyl-, as reagents, in the synthesis of alkenylsilanes, **2**, 931
Coppers, trifluoromethyl-, use in the synthesis of trifluoromethyls, **6**, 22
Coppers, trihalomethyl-, synthesis of, **6**, 246
Coppers, triorganyl-, characteristics of, **2**, 960
Coppers, vinyl-
activation of, **2**, 918
reactions
with electrophiles, **2**, 917
with iodomethylzinc iodide, **2**, 74
as starting materials, for allyliccoppers, **2**, 594
synthesis of, **2**, 958
N-Coppers, reactions, with aryl halides, **2**, 753
Copper salts
as catalysts, for 2,3 rearrangements, **1**, 409
use in the substitution of vinyl halides, **1**, 469
use in the synthesis of biaryls, **1**, 472
Copper salts, vinyl-, dimerization of, **1**, 491
Copper sulfate
as catalyst, for the synthesis of phenols, **2**, 647
use in the cleavage of carbon–silicon bonds, **1**, 63
Copper thiolates, synthesis of, **2**, 273
Copper triflate catalysts, use in metal-catalyzed [2+2] additions, **1**, 357
Copper–zinc compounds, synthesis of, via transmetallation, **1**, 292
Copper–zinc organometallics, reactions, with 1-nitroalkenes, **2**, 785
Corey–Fuchs sequence, for the synthesis of α,β-alkynoic esters, **5**, 160
Corey–Kim reagents
reactions, with 1,3-dicarbonyls, **3**, 168
for the synthesis of sulfonium ylides, **3**, 373
use in the synthesis of α-amino ketones, **3**, 184
Corey–Winter reaction, for the deoxygenation of diols, **1**, 585
Coriolin, synthesis of, via selenoxide pyrolysis, **1**, 782
Coronafacic acid, synthesis of, **1**, 760
Corynantheine alkaloids, synthesis of, **1**, 149
Cottonseed oil, uses, for the removal of chlorine in commercial phosgene, **6**, 414
Coumarin, 4-chloro-3-nitro-
reactions
with potassium *N*,*N*-dialkyl dithiocarbamates, **6**, 562
with potassium *N*-phenyl dithiocarbamate, **6**, 562
Coumarin, 3,4-dihydro-, reactions, with 1,2-bis-*O*-trimethylsilyloxyethane, **6**, 72
Coumarin, 4-hydroxy-, reactions, with benzylidineaniline, **4**, 304
Coumarins, synthesis of, **1**, 61
Coumarins, 3-acetyl-, reactions, with dialkyl phosphinates, **6**, 117
Coumarins, dihydro-
reduction of, **4**, 170
as starting materials, for ketenes, **3**, 543
Coumarin–silicon bonds, cleavage of, **1**, 61
Counter-nucleophile reaction, of iminium ions, **3**, 419
Coupling, oxidative, for organometallics, **2**, 9527
Coupling, palladium-catalyzed
for the synthesis of acylsilanes, **5**, 419
of terminal alkynes with chloromethyldithioformate, **5**, 560
sp^2-sp^2 Coupling, mechanism, **1**, 335
Coupling reaction
from bimolecular reactions, **1**, 395
of methylene carbenes, **1**, 1144
reagents for, **6**, 378
for the synthesis of *N*-arylarylamides, **5**, 303
for the synthesis of tetrasilylmethanes, **6**, 378
Coupling reaction, carbonylative, of alkynyltins with alkynes, **1**, 1029
Coupling reaction, reductive, for the synthesis of alcohols, **2**, 80
Coupling reagents
for the acylation of amines, **5**, 265
for the synthesis of *N*-alkylamides, **5**, 278
alternatives, **5**, 280
for the synthesis of *N*-alkyl arylamides, **5**, 299
for the synthesis of *N*-arylalkanoamides, **5**, 291
for the synthesis of *N*-arylamides, **5**, 291
use in the synthesis of amides, **5**, 261

Covalent hydration, mechanism, **6**, 109
Cram's rule, Felkin–Anh modification, **2**, 345
anti-Cram isomers, synthesis of, **2**, 54
p-Cresol, reactions, with 3,3,3-trichloro-2-methylpropene, **4**, 745
(*R*,*R*)-Crinan, synthesis of, **1**, 837
(+)-Crinipellin B, synthesis of, **1**, 452
Criss-cross addition, of hydrazines, **2**, 383
Cristol–Firth modification, of the Hunsdiecker reaction, **2**, 627
Cross-coupling, carbonylative, for the synthesis of aryl ketones, **3**, 289
Cross coupling, nickel-catalyzed, of alkenylalanes, **2**, 976
Cross-coupling reaction
  of alkenylalanes, **2**, 976
  of alkenylcoppers, **2**, 978
    palladium-catalyzed, **2**, 978
  of alkenylzirconiums, **2**, 969
  of alkynes, **1**, 1036
  of diorganylcuprates, with vinyl halides, **2**, 960
  palladium-catalyzed, **2**, 992
  for the synthesis of alkynes, **1**, 1006
Cross-coupling reaction, palladium-catalyzed, mechanism, **1**, 1013
Crotonaldehydes
  reactions, with trialkylboranes, **1**, 274
  reduction of, **1**, 85
Crotonate, ethyl 2-methyl-, from, ethyl 2-methyl-3-bromocrotonate, **1**, 431
Crotonate, ethyl 2-methyl-3-bromo-, as starting material, for ethyl 2-methylcrotonate, **1**, 431
Crotonate, ethyl *β*-uramino-, as starting material, for 6-methyluracil, **5**, 93
Crotonate, methyl, reactions, with *α*-methylbenzylamine, **2**, 315
Crotonates, Reformatsky reaction of, **4**, 736
Crotonates, ethyl 2-isothiocyanato-, synthesis of, **5**, 1040
Crotonic acid, synthesis of, **1**, 80
Crotonic aldehydes, 4-*aci*-nitro-, reactions, with pyrrolidinium acetate, **2**, 789
Crotonyl chloride, synthesis of, **5**, 10
Crotyl chloride, reactions, with lithium hexynide, **1**, 1001
2-Crotyl-1,3-dithiane-1-oxide, synthesis of, **4**, 263
18-Crown-6
  as catalyst
    for alkyl fluoroformates, **6**, 422
    for fluoroformates, **6**, 423
  use in enhancing solubility of potassium fluoride, **5**, 4
  use in the rearrangement of benzyl isothiocyanates, **5**, 1039
  use in the synthesis of alkynes, **1**, 1065
  use in the synthesis of *N*-carbonyl fluorides, **6**, 438
  use in the synthesis of cyanohydrins, **3**, 615
  use in the synthesis of nitriles, **3**, 613
  use in the synthesis of purple benzene, **5**, 47
18-Crown-6-complexes, use in the synthesis of phosphonium ylides, **1**, 727
Crown ethers, organometallic, synthesis of, **4**, 453
[2.2.2]Cryptands, use in the synthesis of *N*-alkylphthalimides, **2**, 299
Cryptaustoline, synthesis of, **1**, 149
Cryptocarbene, electrophilic, as starting material, for germaethenes, **6**, 711
Cryptocarbenes, reactions, with stannylenes, **6**, 724
(±)-Cryptofauronol, synthesis of, **3**, 325
Cubane, reactions, with rhodium tetracarbonyl dichloride, **5**, 460
Cucurbitine, synthesis of, **2**, 331
Cumene, benzylic oxidation of, **2**, 655
Cumulatrienes, 1,6-dibromo-, synthesis of, **1**, 995
Cumulenes
  from, 4,5-dien-2-yn-1-ols, **1**, 962
  metal-catalyzed homocoupling of, **1**, 992
  oxidation of, **1**, 983
  reactions
    with hexafluoro-2-butyne, **4**, 613
    with phenyl isocyanate, **4**, 421
  as starting materials
    for allenyl acylsilanes, **5**, 427
    for 1,3-diarsacyclobutanes, **4**, 594
  synthesis of, **1**, **953–995**; **3**, 584; **4**, 598, 851
Cumulenes, bicyclic, synthesis of, **1**, 988
Cumulenes, cyclopropyl, synthesis of, **1**, 988
Cumulenes, hetero-
  reactions, with silylphosphines, **6**, 687
  as starting materials, for thioimidic esters, **5**, 729
cis-Cumulenes, synthesis of, **1**, 989
Cuprate, bis(phenyldimethylsilyl)-, reactions, with allene, **2**, 929
Cuprate, bis(silylvinyl)-, as reagents, in the synthesis of alkenylsilanes, **2**, 931
Cuprate, butyl-, reactions, with 1,3-dibromo-1-trimethylsilyl-1-propene, **2**, 931
Cuprate, methyl-, reactions, with epoxy propargylic alcohols, **1**, 966
Cuprate, tributylstannyl-
  reactions
    with alkynic acylsilanes, **5**, 428
    with phenylthioethynes, **4**, 962
Cuprates
  additives for, **1**, 465
  conjugate addition to enones, **3**, 128
  from, bromides, **3**, 286
  higher order, coupling reactions, **1**, 118
  higher-order, synthesis of, **2**, 588
  higher order, use in the substitution of alkyl halides, **1**, 465
  low-order, synthesis of, **2**, 586
  reactions
    with alkenones, **1**, 120
    with 1-alkoxyallenes, **1**, 496
    with alkyl bromides, **1**, 465
    with alkyl iodides, **1**, 465
    with alkynones, **3**, 271
    with alkynyl phenyl iodosyl salts, **1**, 541
    with *β*-chloroenones, **3**, 271, 273
    with cyclopropenone acetals, **2**, 589
    with (1,2-dialkoxyethylene)iron complexes, **1**, 481
    with *γ*,*δ*-dihydroxy-*α*,*β*-enoates, **5**, 158
    with enones, **3**, 162; **4**, 385
    with haloacetates, **1**, 465
    with haloalcohols, **1**, 465
    with *β*-haloenones, **1**, 469
    with *β*-iodo-enones, **3**, 272
    with 8-phenylmenthyls, **5**, 152
    with *α*-phenylselenoheptyllithium, **4**, 401
    with 2-pyridyl esters, **3**, 287
    with *N*-tosylaziridines, **1**, 486
  as reagents, for the opening of epoxides, **1**, 206
  synthesis of, **4**, 1061
  use in the displacement of alkyl alcohols, **1**, 177
  use in the displacement of triflates, **1**, 179
Cuprates, acyl-, synthesis of, **3**, 136
Cuprates, alkenyl-, protonation of, **1**, 541
Cuprates, (*Z*)-alkenyl-, synthesis of, **1**, 468
Cuprates, *α*-alkoxy-
  reactions, with *α*,*β*-unsaturated ketones, **4**, 384
  synthesis of, **4**, 384
Cuprates, alkyl-
  from, alkenes, **2**, 587
  reactions, with terminal alkynes, **1**, 541
Cuprates, alkyl alkynyl-, for selective ligand transfer, **2**, 587
Cuprates, alkynic, as starting materials, for tertiary ynamines, **2**, 1041
Cuprates, alkynyl-
  reactions
    with acid chlorides, **3**, 273
    with cyanogen bromide, **3**, 675

Cuprates, alkynylcyano-, reactions, with alkynyl(phenyl)iodonium tosylates, **1**, 1018
Cuprates, allyl-, synthesis of, **2**, 74
Cuprates, allylic, synthesis of, **2**, 589
Cuprates, α-amino-, synthesis of, **2**, 1002
Cuprates, α-aminoalkylcyano-, from, organostannyl-*t*-butylcarbamates, **2**, 315
Cuprates, aryl-, Michael addition of, **3**, 626
Cuprates, α-boryl-, synthesis of, **2**, 1005
Cuprates, cyano-
  reactions
    with allenyl halides, **1**, 1015
    with lithioalkenes, **4**, 953
    with selenocyanogen, **4**, 947
    with stannylalkenes, **4**, 953
  reactivity of, **2**, 588
  stability of, **2**, 588
  structure of, **2**, 588; **5**, 1140
  synthesis of, **1**, 312; **2**, 588; **4**, 947
Cuprates, α-cyano-, as intermediates, **3**, 642
Cuprates, dialkyl-
  for the cleavage of α-keto sulfoxides, **1**, 236
  reactions
    with acid chlorides, **3**, 286
    with π-alkene complexes, **2**, 580
    with alkynyl(phenyl)iodonium tosylates, **1**, 1018
    with α,β-epoxysilanes, **1**, 759, 761
  reactivity of, **1**, 177
Cuprates, diaryl-
  reactions
    with iodoarenes, **1**, 473
    with vinyl halides, **1**, 469
Cuprates, divinyl-, reactions, with esters, **1**, 540
Cuprates, (*Z*)-divinyl-, alkylation of, **1**, 1107
Cuprates, divinylmagnesio-, use in the substitution of vinyl halides, **1**, 468
Cuprates, ethyl-, reactions, with propargylic ethers, **1**, 961
Cuprates, α-halo-, synthesis of, **2**, 998
Cuprates, hetero-, synthesis of, **2**, 588
Cuprates, lithio-, synthesis of, **4**, 953
Cuprates, lithiodialkenyl-, synthesis of, **1**, 479
Cuprates, lithiodivinyl-, use in the substitution of vinyl halides, **1**, 468
Cuprates, methyl-, use in the synthesis of transition metal amides, **5**, 384
Cuprates, methyl alkyl-, as starting materials, for alkylallenes, **1**, 966
Cuprates, mixed
  synthesis of, **2**, 587
  use in the synthesis of aromatic ketones, **3**, 287
Cuprates, mixed dialkyl-, from, copper(I) salts, **2**, 587
Cuprates, mononuclear, synthesis of, **6**, 194
Cuprates, α-selenyl-, synthesis of, **4**, 401
Cuprates, silyl-
  reactions
    with acyl chlorides, **5**, 417
    with 3-bromo 1-silyl 1-butynes, **4**, 1046
Cuprates, α-silylvinyl-, as intermediates, **1**, 1094
Cuprates, β-silylvinyl-, reactions, with alkynes, **3**, 271
Cuprates, α-stannyl-, synthesis of, **1**, 164
Cuprates, 2-thiophenylalkylcyano-, reactions, with aldehydes, **2**, 69
Cuprates, trialkylsilyl-, addition to enols, **3**, 52
Cuprates, vinyl-
  acylation of, **1**, 422
  alkylation of, **1**, 123
  chlorination of, for vinyl chlorides, **2**, 617
  from, organo cuprates, **2**, 629
  reactions
    with alkyl iodides, **1**, 465
    with cyanogen chloride, **3**, 642
  as starting materials
    for vinyl bromides, **2**, 624
    for vinyl chlorides, **2**, 616
    for vinyl halides, **2**, 607
Cuprates, vinylidene-, reactions, with acylaminophosphinic esters, **4**, 1001
Cuprate systems, alkynyl, synthesis of, **2**, 1095
Cupric bromide, for the synthesis of aryl bromides, **2**, 626
Cupric chloride
  for the *para* chlorination of arenes, **2**, 620
  reactions
    with arenediazoniums, **2**, 798
    with organostannanes, **4**, 774
  for the synthesis of aryl iodides, **2**, 631
Cupric triflate, use in the synthesis of vinyl sulfides, **2**, 710
Cuprous alkynides, reactions, with aryl halides, **1**, 472
Cuprous chloride, reactions, with oxygen, **5**, 54
Cuprous halides
  choice of, **1**, 276
  reactions, with alkenes, **3**, 698
Cuprous iodide
  reactions
    with 1-methoxyvinyllithium, **4**, 953
    with potassium naphthalenide, **1**, 472
Cuprous oxide, reactions, with butanethiol, **2**, 273
Curcumenes, synthesis of, **1**, 683
Curtius acyl azide rearrangement, for the synthesis of isocyanates, **4**, 446
Curtius reaction
  for the synthesis of amines, **2**, 319
  for the synthesis of α-aminophosphonic acids, **2**, 457
Curtius rearrangement
  of acyl azides, **2**, 757
  of amino acid azides, **4**, 426
  for the decomposition studies of phenylpropioyl azides, **2**, 1055
  mechanism, **5**, 989
  modified, mechanism, **5**, 993
  and Schmidt rearrangement compared, **5**, 1001
  for the synthesis of *N*-acylated α-ureido enamines, **4**, 982
  for the synthesis of α-aminoalkylphosphorus compounds, **4**, 476, 492
  for the synthesis of carbodiimides, **5**, 1076
  for the synthesis of diisocyanates, **5**, 990
  for the synthesis of *gem*-diisocyanates, **4**, 433
  for the synthesis of enamides, **5**, 288
  for the synthesis of isocyanates, **5**, 989
  for the synthesis of ketenimines, **3**, 564
  of trichloroacetyl azide, **6**, 606
2-Cyanacrylate, 3,3-diazido-, reactions, with triphenylphosphine, **4**, 993
Cyanamide
  chlorination of, **6**, 621
  reactions
    with alcohols, **6**, 731
    with alkenes, **5**, 1080, 1118
    with alkyl halides, **2**, 308; **5**, 1118
    with amidines, **5**, 1121
    with amine salts, **6**, 642
    with dithioesters, **6**, 630
    with electrophiles, **5**, 1118
    with hydrogen bromide, **6**, 623
    with hydrogen chloride, **6**, 621, 728
    with hydrogen iodide, **6**, 623
    with *N*-hydroxysuccinimide esters, **5**, 1120
    with isocyanates, **5**, 1120
    with 2-methanesulfonylpyrimidines, **5**, 1120
    with phthalocyanatogermanium dihydro compounds, **5**, 1083
    with trichloromethyl carbinols, **5**, 1121
    with trimethylsilyl cyanide, **5**, 1076
  as reagent, for the dehydrosulfurization of dithiocarbamates, **5**, 1023
  as starting material
    for acylcyanamides, **5**, 1120

for aminoguanidine, **5**, 847
for *N*-aryl cyanamides, **5**, 1120
for iminium salts, **6**, 623
for monomethyl cyanamides, **5**, 1118
use in the synthesis of bicyclic α-amino organophosphorus compounds, **4**, 458
Cyanamide, *N*-*t*-butyl-*N*-chloro-, reactions, with cyclohexene, **5**, 1120
Cyanamide, diethyl-, from, *N*,*N*-diethylurea, **5**, 1122
Cyanamide, difluoro-, reactions, with caesium fluoride, **6**, 261
Cyanamide, dimethyl-, as starting material, for complex salts, **6**, 728
Cyanamide, *N*,*N*-dimethyl-, reactions, with bistrimethylsilyl selenide, **6**, 597
Cyanamide, disodium, reactions, with phenylisocyanide dichloride, **5**, 1070
Cyanamide, *N*-methyl-*N*-phenyl-
reactions
with phenyllithium, **6**, 648
with sodium, **6**, 648
Cyanamide, methyl(trimethylsilyl)-, from, bis(trimethylsilyl)methylamine, **5**, 1127
Cyanamide, phenyl-, reactions, with metal chlorides, **5**, 1083
Cyanamides
acylation of, **6**, 620
bromination of, **6**, 615
from
alkyl amines, **5**, 1125
amides, **5**, 1125
amines, **5**, 1125
cyanogen halides, **5**, 1124
sulfonamides, **5**, 1125
tertiary amines, **5**, 1126
thioureas, **5**, 1122
[1,2,3]triazolo[1,5-a]pyrimidinium bromide, **5**, 1122
hydrolysis of, **6**, 504
as isomers of carbodiimides, **5**, 1062
reactions
with alcohols, **6**, 632
with aziridinones, **5**, 1118
with dichlorosilanes, **5**, 1077
with thiols, **6**, 635
with water, **6**, 504
as starting materials
for *N*-acyliminocarbonyls, **6**, 646
for *N*-alkyliminocarbonyls, **6**, 643
for arylguanidines, **6**, 645
for *N*-aryliminocarbonyls, **6**, 645
for carbamimidic chlorides, **6**, 620
for *N*-chlorocarbonyls, **6**, 445
for cyanimides, **3**, 409
for 2-cyanoguanidines, **6**, 648
for *N*-cyanoiminocarbonyls, **6**, 648
for guanidines, **6**, 642
for isothioureas, **6**, 635
for isoureas, **6**, 632
for ketene aminals, **4**, 975
for nitriles, **3**, 614
for selenoureas, **6**, 596
synthesis of, **5**, 1117
via dehydration reactions, **5**, 1121
via reaction with electrophilic cyanide, **5**, 1124
via reaction with nucleophilic cyanide, **5**, 1124
via reactions on the carbon α to NCN, **5**, 1124
via rearrangement and fragmentation reactions, **5**, 1122
via von–Braun reaction, **5**, 1126
Cyanamides, acyl-
from, cyanamide, **5**, 1120
synthesis of, **5**, 1120
Cyanamides, *N*-acyl-, hydrolysis of, **6**, 506
Cyanamides, *N*-acyl-*N*-(trimethylstannyl)-, synthesis of, **5**, 1121
Cyanamides, adamantyl-, synthesis of, **5**, 1118
Cyanamides, alkyl-
from, cyanogen azide, **5**, 1119
as starting materials
for *N*-alkyl-*N'*-silyl carbodiimides, **5**, 1082
for guanidines, **6**, 643
synthesis of, **5**, 1118
Cyanamides, alkylidene-, synthesis of, **6**, 643
Cyanamides, alkylnitro-
as intermediates, **5**, 963
isomerization of, **5**, 963
synthesis of, **5**, 1119
thermolysis of, **5**, 964
Cyanamides, *N*-alkyl-*N*-trimethylsilyl-, reactions, with 1-adamantyl chloride, **5**, 1118
Cyanamides, allyl-, synthesis of, **5**, 1118
Cyanamides, *N*-amino-*N*-phenyl-, synthesis of, **5**, 1125
Cyanamides, aryl-
from, anilines, **5**, 1125
synthesis of, **5**, 1120
Cyanamides, *N*-aryl-, from, cyanamide, **5**, 1120
Cyanamides, *N*-aryl-*N*-hydroxy-, synthesis of, **5**, 1124
Cyanamides, *N*-aryl-*N*-methyl-, synthesis of, **5**, 1122
Cyanamides, β-bromoselenenylalkyl-, synthesis of, **5**, 1118
Cyanamides, β-chloroalkyl-, synthesis of, **5**, 1120
Cyanamides, dialkyl-
reactions
with *t*-alkyl chlorides, **5**, 1146
with carbon disulfide, **6**, 575
with phosphoryl chloride, **6**, 656
with trimethyloxonium tetrafluoroborate, **5**, 1120
as starting materials, for carbamimidic chlorides, **6**, 620
synthesis of, **5**, 1118, 1119
Cyanamides, diaryl-, synthesis of, **5**, 1122
Cyanamides, heteroaryl-, from, hydroxyformamidines, **5**, 1121
Cyanamides, imino-
from
amidines, **5**, 1125
guanidines, **5**, 1125
Cyanamides, isocyanatomethyl-, synthesis of, **4**, 441
Cyanamides, α-methoxy-, synthesis of, **5**, 1124
Cyanamides, monoalkyl-
as starting materials
for 1,1-dialkylguanidines, **6**, 642
for 1,1,2-trialkylguanidines, **6**, 642
Cyanamides, monomethyl-
from, cyanamide, **5**, 1118
reactions, with arenediazonium salts, **5**, 1118
Cyanamides, perfluoro-, as starting materials, for methanediamines, **6**, 259
Cyanamides, phenyl-, synthesis of, **5**, 1123
Cyanamides, β-phenylselenenylalkyl-, synthesis of, **5**, 1118
Cyanamides, *N*-silyl-, use in the synthesis of carbodiimides, **5**, 1082
Cyanamides, *N*-substituted-*N*-chloro-, reactions, with alkenes, **5**, 1120
Cyanamides, *N*-substituted *N*-(trialkylstannyl)-, synthesis of, **5**, 390
Cyanamides, *N*-sulfonyl-, acylation of, **5**, 344
Cyanamidinium hexachloroantimonates, reactions, with amines, **6**, 643
Cyanamidinium salts, acyl, synthesis of, **5**, 1147
Cyanamidinium salts, trialkyl, synthesis of, **5**, 1146
Cyanamidium metallates, synthesis of, **5**, 1149
Cyanamidium salts, synthesis of, **5**, 1146
Cyanate, alkyl, reactions, with hydrogen selenide, **6**, 593
Cyanate, allyl, stability of, **5**, 1101
Cyanate, benzyl, stability of, **5**, 1101
Cyanate, butyl, synthesis of, **5**, 1104
Cyanate, ethyl, synthesis of, **5**, 1101
Cyanate, *n*-pentyl, synthesis of, **5**, 1104
Cyanate, phenethyl, stability of, **5**, 1101

Cyanate, phenyl, reactions, with acetylenides, **3**, 675
Cyanates
  alkylation of, **5**, 995
  chemistry of, **5**, 1100
  from
    alcohols, **5**, 1101
    O-alkyl *N*-hydroxythiocarbamates, **5**, 1104
    O-alkyl thiocarbamates, **5**, 1103
    cyanic acid, **5**, 1104
    1,4-dihydroxybicyclo[2.2.2]octane, **5**, 1102
    3-ethoxy-2,2,4,4-tetramethylcyclobutan-1-ol, **5**, 1102
    metal cyanates, **5**, 1104
    neopentyl alcohol, **5**, 1102
    phenols, **5**, 1101
    thiatriazoles, **5**, 1101
    thiocyanates, **5**, 1103
  reactions
    with amines, **6**, 632
    with thiols, **6**, 629
  as starting materials
    for iminocarbonates, **6**, 624
    for iminothiocarbonates, **6**, 629
    for isothiocyanates, **5**, 1037
    for isoureas, **6**, 632
  synthesis of, **5**, 1100
Cyanates, alkyl
  from
    thallium alkoxides, **5**, 1103
    thiatriazoles, **5**, 1101
  synthesis of, **5**, 1102
Cyanates, allylic, rearrangement of, **1**, 837
Cyanates, aryl
  from, 5-aryloxy-1,2,3,4-thiatriazoles, **5**, 1101
  reactions
    with acyl chlorides, **6**, 616
    with methylene groups, **4**, 889
    with phosphorus pentachloride, **6**, 728
  as starting materials
    for isothiocyanates, **5**, 1038
    for ketene hemiaminals, **4**, 889
    for orthocarbonates, **6**, 296
    for tetraalkyl orthocarbonates, **6**, 300
Cyanates, *o*-formylphenyltellurenyl, synthesis of, **5**, 1116
Cyanates, methyl, from, *O*-methyl thiocarbamate, **5**, 1103
Cyanates, polyfluoro-, synthesis of, **5**, 1102
Cyanation
  of allylic acetates, **1**, 192
  for the synthesis of heterocyclic aromatic nitriles, **3**, 669
Cyanic acid
  reactions
    with alkoxydichlorophosphines, **5**, 1018
    with diazoisobutane, **5**, 1104
    with hydrogen fluoride, **6**, 439
  as starting material, for cyanates, **5**, 1104
Cyanic anhydrides, carboxylic, as intermediates, in the synthesis of anhydrides, **5**, 191
Cyanide
  alkylation of, **1**, 178
  for the displacement of hydroxyl groups, **1**, 200
  for the displacement of hydroxyls, **1**, 185
  Michael addition to β-nitroalkenes, **3**, 638
  as nucleophile, **1**, 179
  for the opening of thiiranes, **1**, 242
  reactions
    with acylsilanes, **4**, 353
    with aldehydes, **5**, 85
    with arylalkenes, **3**, 626
    with bis(pentafluorophenyl)bis(dioxane)nickel(II) compounds, **5**, 1138
    with β-chloro-alcohols, **3**, 632
    with chlorotrifluoroethene, **4**, 5
    with dialkyltin halides, **5**, 1144
    with dihydropyridinium salts, **3**, 637
    with epoxides, **3**, 632
    with ketones, **5**, 85
    with selenium compounds, **5**, 1116
    with *C*-sulfonylthionoformamides, **5**, 592
    with vinyl sulfones, **3**, 636
  and reductive deaminations, **1**, 38
  use in substitution reactions, **3**, 612
Cyanide, benzoyl, reactions, with β-diketones, **3**, 268
Cyanide, benzyl
  reactions, with sulfuric acid, **5**, 34
  use in the synthesis of amidines, **5**, 742
Cyanide, *t*-butyldimethylsilyl
  reactions
    with adamantane-1-carboxylic acid, **5**, 226
    with 4-nitrobenzoic acid, **5**, 226
Cyanide, cuprous, reactions, with tetrakis(1-methoxyvinyl)tin, **4**, 954
Cyanide, mercuric, reactions, with diphenylthallium(III) chloride, **5**, 1143
Cyanide, *p*-nitrobenzyl
  reactions, with hydriodic acid, **5**, 33
  as starting material, for *p*-nitrophenylacetic acid, **5**, 94
Cyanide, pentafluorobenzyl, reactions, with sulfuric acid, **5**, 34
Cyanide, phenylselenyl
  reactions
    with alkenes, **3**, 637
    with enamines, **3**, 637
Cyanide, tetrabutylammonium, reactions, with diazo sulfides, **3**, 664, 667
Cyanide, thioformyl, synthesis of, via thermolysis of sulfenyl chlorides, **3**, 337
Cyanide, *p*-toluenesulfonyl (TsCN)
  reactions
    with acetylenides, **3**, 675
    with aryl organozinc halides, **3**, 670
    with organozinc halides, **3**, 646
Cyanide, tosyl
  for cyanation processes, **4**, 815
  reactions
    with alcohols, **2**, 721
    with benzylic organozinc halides, **3**, 664
Cyanide, trimethylacetyl, reactions, with trialkylmethoxysilanes, **5**, 1133
Cyanide, 2,4,6-trimethylbenzyl
  hydrolysis of, **5**, 34
  reactions, with sulfuric acid, **5**, 34
Cyanide, trimethylene, as starting material, for glutaric acid, **5**, 34
Cyanide, trimethylgermanyl, reactions, with trimethylgallium, **5**, 1143
Cyanide, trimethylsilyl (TMS-CN)
  asymmetric addition to aldehydes, **3**, 628
  as cyanide source, for Strecker reaction, **3**, 616
  reactions
    with acid chlorides, **3**, 627
    with acyloxystannanes, **5**, 1144
    with aldehydes, **3**, 646
    with alkyl halides, **3**, 613
    with alkynes, **3**, 671
    with allenes, **3**, 660
    with allylic methoxymethyl ethers, **1**, 191
    with arylalkynes, **3**, 660
    with aryl iodides, **3**, 661
    with aryl sulfinate salts, **5**, 1112
    with benzimidazole, **4**, 426
    with bistrifluoroacetates, **3**, 701
    with chiral acetals, **3**, 631
    with cyanamide, **5**, 1076
    with dialkylaluminums, **5**, 1142
    with di-*n*-butylboron chloride, **5**, 1136
    with *N*,*N*′-diisopropylcarbodiimide, **5**, 777
    with diketones, **3**, 667

with epoxides, **3**, 632
with halotrialkylsilanes, **5**, 1133
with α-methoxyamides, **3**, 638
with 6-methoxytetralone, **3**, 667
with oxetanes, **3**, 633
with phosphorus halides, **5**, 1128
with pyridine *N*-oxide, **3**, 671
with quinoline *N*-oxides, **3**, 671
with selenenyl halides, **5**, 1116
with selenoacetals, **3**, 637
with seleno ortho esters, **3**, 637
with thioacetals, **3**, 634
with *N*-tosylaziridines, **3**, 638
with trifluoromethyl isocyanate, **5**, 721
with α,β-unsaturated carbonyls, **3**, 633
synthesis of, **4**, 369; **5**, 1134
use in the synthesis of cyano ethers, **3**, 627
use in the synthesis of cyanohydrins, **3**, 615
use in the synthesis of heteroarylnitriles, **3**, 671
use in the synthesis of isocyanides, **3**, 698
use in the synthesis of α,β-unsaturated ketones, **3**, 213
use in the synthesis of β,γ-unsaturated nitriles, **3**, 622
Cyanide, vinyl, Michael addition of, **2**, 829
Cyanide anions
reactions
with alkylphosphonyl halides, **5**, 1131
with arylphosphonyl halides, **5**, 1131
Cyanide ions
oxidation of, **5**, 698
reactions
with carbonyls, **3**, 667
with fluorenes, **3**, 667
with tetrafluoroethene, **4**, 1076
with vinyl halides, **3**, 658
as starting materials, for thiocyanates, **5**, 1111
use in the synthesis of benzonitriles, **3**, 664
use in the synthesis of heterocyclic aromatic nitriles, **3**, 669
Cyanide ligands, π-acidic nature of, **2**, 588
Cyanides
acylation of, **3**, 694
alkylation of, **3**, 694
for the opening of epoxides, **1**, 208
reactions
with amines, **6**, 505
with nitrogen, **5**, 1125
Cyanides, acyl
as *O*-acylation reagents, for the synthesis of arylhydroxylamines, **5**, 212
decarbonylation of, **5**, 1139
as starting materials, for α-amino ketones, **3**, 183
synthesis of, **5**, 1144
Cyanides, alkoxydimethylsilyl
reactions
with phosphorus(V) dichlorides, **5**, 1131
with phosphorus halides, **5**, 1128
Cyanides, alkylsilyl, as silylating agents, for carboxylic acids, **5**, 226
Cyanides, allenyl acyl, synthesis of, **1**, 978
Cyanides, aryl
reactions, with diethyl methyl phosphonate, **4**, 927
as starting materials, for arylamidines, **5**, 755
synthesis of, **3**, 661
Cyanides, formimidoyl, photodissociation of, **3**, 699
Cyanides, imidoyl, as starting materials, for ketenimines, **3**, 563
Cyanides, phenylmercuric, synthesis of, **5**, 1142
Cyanides, silyl
from
alkylhalosilanes, **5**, 1134
bis(trimethylsilyl) sulfate, **5**, 1135
halosilanes, **5**, 1134
for the opening of epoxides, **1**, 208
synthesis of, **5**, 1134
Cyanides, sulfonyl
chlorination of, **5**, 809
from, sulfonyl chlorides, **5**, 1112
synthesis of, **5**, 1112
Cyanides, trialkyl
synthesis of, **5**, 1144
use in the synthesis of acyl cyanides, **5**, 1144
Cyanides, trialkylsilyl, synthesis of, **5**, 1133
Cyanides, trimethylsilyl, as starting materials, for thiocyanates, **5**, 1111
Cyanides, β-(trimethylsilyl)oxy, synthesis of, **3**, 698
Cyanides, α,β-unsaturated acyl
reactions
with alkenes, **3**, 652
with 3,3-dimethyl-1-trimethylsilylallene, **1**, 1061
with trimethylsilylallene, **1**, 1061
Cyanide salts, reactions, with alkyl halides, **3**, 612
Cyanimides, from, cyanamides, **3**, 409
Cyanine iodides, reactions, with triethylamine, **2**, 1046
Cyanines, azamonomethine-, synthesis of, **3**, 439
Cyanines, betaine-, synthesis of, **3**, 258
Cyanine salts, use in the synthesis of ynamine amidines, **2**, 1046
Cyanoalkyl radicals, use in the synthesis of ketenimines, **3**, 573
β-Cyano bromides, synthesis of, **3**, 657
α-Cyano carbanions, stability of, **1**, 850
Cyano carbocations, synthesis of, via bond scission, **1**, 904
Cyano complexes, alkylation of, **5**, 1087
Cyano compounds, reactions, with ketones, **3**, 268
α-Cyano compounds, cyclization of, **1**, 244
Cyanoform, synthesis of, **3**, 565
Cyanogen
chlorination of, **5**, 809
as electrophilic cyanide sources, **3**, 612
reactions
with alanates, **1**, 534
with alanes, **1**, 534
with chlorine, **5**, 662
with hydrazine, **5**, 847
as starting material, for diamidoximes, **5**, 833
use in the synthesis of nitriles, **3**, 614
Cyanogen, iodo-
reactions
with dimethylboron iodide, **5**, 1136
with hexaphenyldilead, **5**, 1145
Cyanogen azide
reactions
with *N*-acyl-1,2-dihydropyridines, **5**, 1119
with acylphosphoranes, **3**, 476
with alkanes, **5**, 1119
with alkynes, **3**, 476
with 1,2-dihydropyridines, **5**, 1119
as starting material, for alkyl cyanamides, **5**, 1119
Cyanogen bromide
as activating agent, for the synthesis of anhydrides, **5**, 191
as electrophilic cyanide sources, **3**, 612
for the oxidation of thiols, **2**, 214
reactions
with *N*-alkylanilines, **5**, 1126
with alkylidene aminoguanidines, **5**, 1125
with alkynyl cuprates, **3**, 675
with amines, **4**, 413; **5**, 1125; **6**, 632
with anilines, **5**, 1125
with antimony(V) chloride, **6**, 613
with *N*-aryl thioformamides, **3**, 721
with enamines, **2**, 748; **3**, 657
with methyldifluorophosphine, **5**, 1129
with phenyl-1,4-bis(hexafluoroisopropanol), **5**, 1102
with silver diazomethyl phosphoryls, **3**, 478
with tertiary amines, **2**, 33
as starting material, for nitriles, **3**, 614
Cyanogen chloride

reactions
with acetylenides, **3**, 675
with alkenes, **6**, 606
with alkyl chlorides, **5**, 962
with alkyl halides, **5**, 962
with alkynyllithiums, **1**, 1048
with amines, **6**, 632
with chlorine, **5**, 809
with chlorodimethylphosphine, **5**, 1131
with dimethylaminomethylbis(isopropyl)-phosphine, **5**, 1130
with dithiols, **6**, 627
with enol ethers, **3**, 627
with 2,2,3,3,4,4-hexafluoro-1,6-hexanediol, **5**, 1102
with 1,6-hexanediol, **5**, 1103
with imidazoles, **5**, 1127
with ketone enolates, **3**, 633
with 2-methylhexafluoroisopropanol, **5**, 1102
with peroxydisulfuryl difluoride, **6**, 277
with silylamines, **5**, 1082
with sulfinic acids, **5**, 1112
with thallium(I) ethoxide, **5**, 1103
with thiocarboxylate salts, **5**, 1112
with *N*-trimethylsilylacylamides, **5**, 1082
with vinyl cuprates, **3**, 642
as starting material
for carbamoyl fluorides, **6**, 441
for carbonimidic dichlorides, **6**, 606
for *N*-chlorocarbonyl isocyanide dichloride, **6**, 451
for dichloroiminium cations, **6**, 610
for iminodithiocarbonates, **6**, 627
for iminothiocarbonates, **6**, 629
for nitriles, **3**, 614
use in the synthesis of cyanates, **5**, 1102
Cyanogen *N*,*N*-dioxide
as starting material
for tetraoximino-1,4-dithianes, **5**, 828
for thiooxalic bis(hydroxamic acids, **5**, 828
Cyanogen fluoride
reactions, with bromine, **5**, 809
as starting material, for *N*-bromocarbonimidic difluoride, **6**, 603
Cyanogen halides
reactions
with alcohols, **5**, 1101
with dialkylaminomethylphosphines, **5**, 1129
with dialkylaminophosphines, **5**, 1129
with diols, **5**, 1102
with nitrogen, **5**, 1127
with phenols, **5**, 1101
with tertiary amines, **5**, 1126
with triarylarsines, **5**, 1132
with triarylstibines, **5**, 1132
as starting materials
for cyanamides, **5**, 1124
for iminocarbonates, **6**, 624
for isoureas, **6**, 632
for thiocyanates, **5**, 1111
Cyanogen iodide
reactions
with antimony(V) chloride, **6**, 613
with triarylbismuth compounds, **5**, 1132
Cyanogen oxide, synthesis of, **3**, 686
Cyanogens, diiso-
characterization of, **3**, 710
from, norbornadienoneazine, **3**, 710
Cyanogermylation, of alkynes, **2**, 946
Cyano groups
acidity of, **1**, 155
characteristics of, **1**, 154
Cyanohydrin, acetone
as cyanide source, for the synthesis of cyanohydrins, **3**, 616
reactions, with epoxides, **3**, 632
as starting material, for γ-ketonitriles, **3**, 633
use in the synthesis of asymmetric cyanohydrins, **3**, 628
use in the synthesis of nitriles, **3**, 613
Cyanohydrin, bis(trimethylsilyl)-, synthesis of, **3**, 667
Cyanohydrin, *O*-trimethylsilyl-α-trimethylsilyl-, synthesis of, **3**, 640
Cyanohydrin anions, silylation of, **3**, 568
Cyanohydrins
addition reactions of, **1**, 309
arylation of, **3**, 292
carbanions of, **3**, 627
cleavage of, **1**, 884
from
aldehydes, **3**, 615
ketones, **3**, 615
α,β-unsaturated carbonyls, **3**, 633
α,β-unsaturated ketones, **3**, 616
vinyl acetates, **3**, 627
vinyl ethers, **3**, 627
inversion of configuration, **3**, 631
optically active
from chiral acetals, **3**, 631
synthesis of, **3**, 628
reactions
with Grignard reagents, **2**, 330
with organometallics, **3**, 157
reductive silylation of, **5**, 420
as starting materials
for α-aminonitriles, **3**, 616
for α-chloronitriles, **3**, 627
for α-fluoronitriles, **3**, 627
for hydroxy aldehydes, **3**, 30
for imidoyl halides, **5**, 654
for imidoylsilanes, **5**, 794
for α-selenonitriles, **3**, 636
for α-silyloxy aldehydes, **3**, 33
synthesis of, **3**, 615, 627, 667
use in the synthesis of aromatic ketones, **3**, 292
use in the synthesis of furanyl ketones, **3**, 303
use in the synthesis of phenyl ketones, **3**, 295
See also Nitriles, α-hydroxy-
Cyanohydrins, acetylsilane *O*-silyl-
as starting materials
for β-siloxy-*N*,*N*-bis-silylenamines, **2**, 814
for α-silyl ketones, **2**, 814
Cyanohydrins, chiral, from, sulfoxides, **3**, 631
Cyanohydrins, β-fluoro-, synthesis of, **3**, 627
Cyanohydrins, *O*-silylated, use in the synthesis of nitriles, **3**, 651
(*R*)-Cyanohydrins, synthesis of, **2**, 321
(S)-Cyanohydrins, synthesis of, **3**, 628
Cyano ligands, in coupling reactions, **1**, 120
Cyanomethyl anion, structure of, **1**, 154
4-Cyanophenoxy-substituted complexes, isolation of, **5**, 462
4-Cyanophenyls, synthesis of, **6**, 606
Cyanophosphorothioite, *S*-acetyl, reactions, with chloral, **5**, 406
2-Cyanopropyl radicals, reactions, with anthrone, **1**, 265
α-Cyano sulfones, α-alkylated-, synthesis of, **3**, 635
Cyanuric chloride
as dehydrating reagent, **3**, 697
fluorination of, **6**, 258
reactions, with carbodiimides, **6**, 622
use in the synthesis of acid chlorides, **5**, 15
use in the synthesis of alkyl chlorides, **2**, 26
use in the synthesis of carbodiimides, **5**, 1064, 1078
Cyanuric fluoride
fluorination of, **6**, 261
reactions, with carboxylic acids, **5**, 5
as reagent, for acid fluorides, **5**, 5
Cyclialkylation, for the substitution of alkyl halides, **1**, 464
Cyclic radicals
reactions, with alkenes, **1**, 325

stereoselectivity in intermolecular additions, **1**, 325
Cyclic systems
elimination of, **1**, 625
elimination reactions in, **1**, 602
regioselectivity of, **1**, 598
Cyclization
of alkynylsilanes, **2**, 926
for the displacement of alkyl alcohols, **1**, 183
of 5-hexenyl radicals, **1**, 336
of iminium ions, **3**, 420
manganese-mediated, **1**, 346
ring-size selectivities of, **1**, 214
of squalene oxide, **1**, 294
for the synthesis of alkynes, **1**, 1038
for the synthesis of prolines, **2**, 323
tandem, **1**, 316
tandem radical, **1**, 548
to cyclohexanes, **1**, 229
to cyclopropanes, **1**, 229
transition state in, **1**, 336
Cyclization, 5-*exo-trig*-, for the synthesis of alkyllithiums, **2**, 556
Cyclization, 6-endo, mechanism, **1**, 548
Cyclization, 7-endo, mechanism, **1**, 547
Cyclization, 5-exo, mechanism, **1**, 547
Cyclization, 6-exo, mechanism, **1**, 547
Cyclization, intramolecular
for the synthesis of radicals, **1**, 948
for the synthesis of selenides, **2**, 286
Cyclization, palladium-catalyzed, of germylenes, **2**, 949
Cyclization, reductive, of nitro compounds, **5**, 324
Cyclization, tandem
mechanism, **1**, 366
use of vinyl radicals in, **1**, 1131
Cycloaddition
of alkynylsilanes, **2**, 927
with allenes, **1**, 497
frontier orbital approach to, **1**, 350
with germaethenes, **6**, 711
mechanism, **1**, 349
of nitrile ylides with diazaphospholes, **5**, 785
for the synthesis of α-aminoalkylarsenic compounds, **4**, 502
for the synthesis of α-aminoalkylphosphorus compounds, **4**, 488, 493
for the synthesis of aminocarbene complexes, **5**, 954
for the synthesis of arylsilanes, **2**, 943
for the synthesis of cyclic monothioacetals, **4**, 231
for the synthesis of dialkyl sulfones, **2**, 158
for the synthesis of thionoamides, **5**, 604, 622
of unsaturated carbene complexes, **5**, 959
Cycloaddition, [1+3] dipolar
factors influencing the stereoisomeric ratios in, **2**, 354
of nitronic esters onto alkenes, **2**, 355
Cycloaddition, intramolecular, of carbonyl ylides, **3**, 315, 317
Cycloaddition, intramolecular diyl, mechanism, **1**, 359
Cycloaddition, intramolecular thermal alkene, mechanism, **1**, 355
Cycloaddition, photochemical
with allenes, **1**, 497
for the synthesis of alkenes, **1**, 516
Cycloaddition, photochemical alkene
in diastereofacial selectivity, **1**, 357
regioselectivity of, **1**, 356
Cycloaddition, retro-
for the synthesis of carbon–carbon bonds, **1**, 771
for the synthesis of enols, **2**, 636
Cycloaddition, thermal, with allenes, **1**, 497
[2+2] Cycloaddition
for the synthesis of alkenes, **1**, 516
for the synthesis of *N*-aryl β-lactams, **5**, 294
for the synthesis of β-lactams, **5**, 286
for the synthesis of α,β-unsaturated ketones, **3**, 216, 247
[3+2] Cycloaddition
mechanism, **1**, 527
for the synthesis of α,β-unsaturated ketones, **3**, 225
[4+2] Cycloaddition, for the synthesis of phenols, **2**, 664
Cycloaliphatic acids, homologation of, **5**, 144
1-Cycloalkanecarboxamides, *N*-substituted 1-isocyano-, synthesis of, **3**, 706
Cycloalkanes, 1-azido-1-methylthiol-, rearrangement of, **5**, 731
Cycloalkanes, benzenesulfinyl substituted, rate of elimination in, **1**, 595
Cycloalkanes, bromomethylene-, synthesis of, **2**, 622, 625
Cycloalkanes, 2-diphenylphosphino-1-cyano-, synthesis of, **2**, 439
Cycloalkanes, divinyl-
anionic oxy-Cope rearrangement of, **1**, 813
Cope rearrangements of, **1**, 811
equilibrium with dienes, **1**, 811
from, 1,5-cyclononadienes, **1**, 812
Cycloalkanes, methylene-, reactions, with iodine isocyanate, **5**, 985
Cycloalkanes, α,β-unsaturated diazo-, synthesis of, **3**, 489
Cycloalkane sodium sulfide, 1-iodo-2-isothiocyanato-, as starting material, for 1,3-thiazolidine-2-thiones, **6**, 565
Cycloalkanethiols, synthesis of, **2**, 125
Cycloalkanoic acid esters, synthesis of, **5**, 145
Cycloalkanols, *cis*-2-alkyl-, from, epoxides, **2**, 83
Cycloalkanols, 1,2-dialkyl-, synthesis of, **2**, 69
Cycloalkanols, 2-methylene-, synthesis of, **2**, 81
Cycloalkanols, 2-vinyl-, [1,5] hydrogen shifts of, **1**, 795
Cycloalkanone ketals, rhenium-mediated oxidation of, **2**, 44
Cycloalkanones
from, cycloalkenols, **1**, 401
reactions, with silicon tetraisothiocyanate, **5**, 1043
as starting materials
for isothiocyanates, **5**, 1043
for lactones, **5**, 161
Cycloalkanones, 2-chloro-2-(benzenesulfonyl)-, synthesis of, **4**, 80
Cycloalkanones, 2,3-cyclopropyl-, reactions, with Grignard reagents, **2**, 67
Cycloalkanones, α-diazo-
reactions
with benzeneselenenyl bromide, **4**, 90
with benzeneselenenyl chloride, **4**, 90
Cycloalkanones, 2-(3'-ethoxy-1'-hydroxy-2'-propenyl)-, synthesis of, **2**, 759
Cycloalkanones, 2-exomethylene-, from, 2-(nitromethyl)cyclohexenones, **1**, 257
Cycloalkanones, α-halo-, reactions, with Brady's reagent, **3**, 451
Cycloalkanones, (*S*)-2-hydroxy-, synthesis of, **2**, 44
Cycloalkanones, 3-nitro-, reactions, with 1-methoxy-3-trimethylsilyloxy-1,3-butadiene, **2**, 665
Cycloalkanones, α-vinyl-, from, cycloalkenes, **3**, 142
Cycloalkenals, reactions, with sodium azide, **3**, 77
Cycloalkenals, azido-, synthesis of, **3**, 77
2-Cycloalkene-1-ones, nitro-, synthesis of, **3**, 258
Cycloalkenes
from, cyclic sulfones, **1**, 613
reactive, synthesis of, **1**, 707
as starting materials
for bisalkylidyne complexes, **3**, 691
for ω-cyanoaldehydes, **3**, 634
for methylenecycloalkenes, **1**, 654
for α-vinyl cycloalkanones, **3**, 142
strained, synthesis of, **1**, 707
synthesis of, via Cope reactions, **1**, 625
Cycloalkenes, diazo-, synthesis of, via diazo group transfer, **3**, 490
Cycloalkenes, 1-hetero-, reactions, with monoisopinocampheylborane, **2**, 42

Cycloalkenes, methylene-, from, cycloalkenes, **1**, 654
Cycloalkenes, 1-nitro-, as starting materials, for nitriles, **3**, 624
Cycloalkenes, polyfluoro-, reactions, with potassium hydroxide, **5**, 59
Cycloalkenes, 3-(trimethylsilyl)-, oxidation of, **3**, 272
Cycloalkenols, as starting materials, for cycloalkanones, **1**, 401
Cycloalkenones
  reactions, with diazoalkanes, **3**, 473
  synthesis of, **1**, 489
Cycloalkenones, 3-phenylsulfinyl-, reactions, with 1-methoxy-3-trimethylsilyloxy-1,3-butadiene, **2**, 665
Cycloalkenones, β-silyl-, synthesis of, **3**, 272
Cycloalkenones, β-substituted, reactions, with vinylcyclopropyl organometallic compounds, **1**, 811
Cycloalk-2-enones, 3-halo-, reactions, with organobis(cuprates), **1**, 120
Cycloalk-3-enylboranes, from, vinylboranes, **2**, 43
Cycloalkylamine oxides, elimination of, **1**, 625
Cycloalkylmethylenes, diazo-, synthesis of, **6**, 53
Cyclobutadienes, synthesis of, **1**, 517, 789
Cyclobutadienes, metalla-, from, reaction of metal carbynes with alkynes, **2**, 979
Cyclobutadienes, tetramethyl-, ring-opening of, **4**, 715
Cyclobuta-1,3-dithione, 2,2,4,4-tetramethyl-, flash vacuum pyrolysis of, **3**, 550
Cyclobutan-1-carbinols, tetrahydropyranyloxy-2,2,4,4-tetramethyl-, as starting materials, for hydroxyacids, **5**, 73
Cyclobutane, β,β-dialkyltitana-, reactions, with trialkylphosphine, **3**, 508
Cyclobutane, 1,3-dizirconia-, synthesis of, **6**, 707
Cyclobutane, octahydroxy-, reactions, with triethylborane, **1**, 60
Cyclobutane, perfluoro-, synthesis of, **4**, 7
Cyclobutane, 1-sila-3-titana-, synthesis of, **4**, 657
Cyclobutane, thora-, reactions, with tetramethylsilane, **4**, 658
Cyclobutane, titana-, reactions, with alkynes, **2**, 979
Cyclobutane alkoxides, divinyl-, oxy-Cope rearrangement of, **1**, 810
Cyclobutanecarbonitrile, 2-chloro-, dehydrochlorination of, **3**, 645
Cyclobutane-1-carboxylic acid, 1-bromo-, displacement of, **5**, 1113
Cyclobutane complexes, oxaplatina-, reactions, with acetyl chloride, **4**, 151
Cyclobutane complexes, titana-, as starting materials, for titanium methylene complexes, **3**, 508
1,2-Cyclobutanedicarbonitrile, dehydrocyanation of, **3**, 645
Cyclobutane-1,3-diisocyanide, 2,2,4,4-tetramethyl-, synthesis of, **3**, 700
Cyclobutanedinitriles, dehydrochlorination of, **1**, 571
Cyclobutanedione, sulfuration of, **3**, 345
1,3-Cyclobutanedione, tetramethyl-, reactions, with difluoroamine, **5**, 314
Cyclobutane-1,3-diones
  cycloreversion of, **3**, 538
  as starting materials, for ketenes, **3**, 538
Cyclobutanedithione, synthesis of, **3**, 346
Cyclobutane pyrolysis, for the synthesis of alkenes, **1**, 784
Cyclobutanes
  photochemical cleavage of, **1**, 789
  reactivity of, **1**, 848
  ring-fused, **1**, 357
  as starting materials, for perfluoro 2,3-dimethylbuta-1,3-diene, **4**, 730
  synthesis of, **1**, 354; **6**, 399
    via radical cyclization, **1**, 340
Cyclobutanes, *exo*-alkylidene-, synthesis of, **1**, 161
Cyclobutanes, 1,3-bis(dimethylamino)dititana-, synthesis of, **6**, 706
Cyclobutanes, 1,2-bis(phosphaallenyl)-, synthesis of, **3**, 606
Cyclobutanes, bis(phosphoranylidene)dititana-, synthesis of, **6**, 706
Cyclobutanes, bromo-, reactions, with vinylmagnesium bromide, **1**, 117
Cyclobutanes, chlorofluoro-, synthesis of, **4**, 32
Cyclobutanes, 1,2-diethylene-, synthesis of, **1**, 497
Cyclobutanes, 1,3-digermyl-, synthesis of, **4**, 626
Cyclobutanes, 1,3-dimetallo-, synthesis of, **4**, 656
Cyclobutanes, 1,3-disilyl-, ring-openings of, **4**, 622
Cyclobutanes, divinyl-
  oxy-Cope rearrangement of, **1**, 812
  rearrangement of, **1**, 412
Cyclobutanes, *trans*-1,2-divinyl-, Cope rearrangement of, **1**, 812
Cyclobutanes, metalla-
  fragmentation of, **1**, 666
  as starting materials
    for alkenes, **1**, 666
    for carbene complexes, **1**, 666
Cyclobutanes, metallo-, synthesis of, **2**, 584
Cyclobutanes, oxaplatina(IV), reactions, with actinium chloride, **4**, 151
Cyclobutanes, photochemical, stereoselectivity of, **1**, 356
Cyclobutanes, 1-sila-3-thora-, synthesis of, **4**, 658
Cyclobutanes, titana-
  from, Tebbe reagent, **3**, 508
  reactions, with alkylphosphines, **3**, 508
  stability of, **1**, 666
Cyclobutanes, tris(phosphoranylidene)dititana-, as synthetic equivalents for carbon atoms, **6**, 706
Cyclobutanes, vinyl-
  rearrangement of, **1**, 805
    anion-accelerated, **1**, 805
  ring-expansion of, **1**, 402
  as starting materials, for cyclohexenes, **1**, 805
Cyclobutanethiones
  as starting materials, for thioketenes, **3**, 550
  synthesis of, **3**, 353
Cyclobutanethiones, 2,2-dimethyl-, synthesis of, **3**, 343
Cyclobutanol, divinyl-, synthesis of, **1**, 812
Cyclobutan-1-ol, 3-ethoxy-2,2,4,4-tetramethyl-, as starting material, for cyanates, **5**, 1102
Cyclobutanols, as starting materials, for ethanol, **2**, 636
Cyclobutanols, α-halo-, from, perfluorocyclobutanone, **4**, 43
Cyclobutanols, 2-vinyl-, as starting materials, for 3-hydroxycyclohexene, **1**, 805
Cyclobutanone
  reactions, with vanadium compounds, **1**, 931
  synthesis of, **3**, 119
Cyclobutanone, perfluoro-, as starting material, for α-halocyclobutanols, **4**, 43
Cyclobutanones
  electrocyclic ring-opening of, **3**, 539
  from, 2-hydroxy sulfides, **1**, 390
  reactions, with carbonyl fluoride, **4**, 59
  ring-opening of, **3**, 269
  as starting materials, for ketenes, **3**, 538
Cyclobutanones, α-chloro-, synthesis of, **5**, 145
Cyclobutanones, α,α-dichloro-, synthesis of, **3**, 151
Cyclobutanones, 2-halo-, as starting materials, for cyclopropanecarboxylic acids, **1**, 393
Cyclobutanones, methylene-, synthesis of, **1**, 498
Cyclobutanones, α-methylene-, from, α-chloro-β-silyl ketones, **1**, 653
Cyclobutanones, 2-vinyl-, synthesis of, **1**, 812
Cyclobutene, ring-opening of, **3**, 233
Cyclobutene, 1-benzylthioperfluoro-, chlorination of, **2**, 168
Cyclobutene, 1-chloro-2,3,3-trifluoro-, reactions, with potassium permanganate, **5**, 59
Cyclobutene, 3,4-dimethyl-, from, 2,4-hexadiene, **1**, 418
Cyclobutene, 1,2-dimethyl-3,4-bismethylene-, metallation of, **1**, 879

Cyclobutene, hexafluoro-, as starting material, for hexafluoro-2-butyne, **6**, 18
Cyclobutene, 3-methylene-, synthesis of, **1**, 517
Cyclobutene, 1-silyl-2-methyl-, synthesis of, **2**, 926
Cyclobutene, 3,3,4,4-tetrafluoro-, synthesis of, **4**, 5
Cyclobutenedione, 3,4-dichloro-, reactions, with sodium azide, **3**, 539
Cyclobutene-1,2-dione, 3,4-di(methylseleno)-, thermal decarbonylation of, **2**, 1038
Cyclobutenediones, silyl-, synthesis of, **2**, 928
Cyclobutenes
from
1,3-dienes, **1**, 418
quinones, **1**, 517
synthesis of, **1**, 516, 690
via electrocyclic reactions, **1**, 418
Cyclobutenes, alkylidene-, synthesis of, **1**, 498
Cyclobutenes, dichloro-, ring-opening of, **4**, 742
Cyclobutenes, metalla-
from
carbene complexes, **2**, 979
reaction of metal carbene with alkynes, **2**, 979
as intermediates, for vinyl metals, **2**, 986
Cyclobutenes, perfluoro-
reactions
with trimethylgermane, **4**, 141
with trimethylstannane hydride, **4**, 153
Cyclobutenium carbocations, synthesis of, **1**, 901
Cyclobutenones, synthesis of, **1**, 551; **3**, 216
Cyclobutenones, metalla-, synthesis of, **5**, 465
Cyclobutenones, methylene-, as starting materials, for ketenes, **3**, 539
Cyclobutenones, silyldichloro-, from, alkynylsilanes, **2**, 928
2-Cyclobutenones, 4-alkenyl-, synthesis of, **2**, 669
2-Cyclobutenones, 3-alkoxy-, synthesis of, **3**, 247
Cyclobutenyl metals, 2-methyl-, from, zirconium-catalyzed methylalumination reaction, **2**, 976
Cyclobutylcarbinyls, use in the synthesis of radicals, **1**, 928
Cyclobutyl complexes, as starting materials, for ferracyclopentenes, **5**, 937
Cyclocarbosilanes, synthesis of, **6**, 178
α-Cyclocitral, reactions, with benzaldehyde, **3**, 66
Cyclocondensation, for the synthesis of 1,3-aminoalkylarsenic compounds, **4**, 495
1,3-Cyclodecadiene, as starting material, for bicyclo[6,2,0]decene, **1**, 418
(*Z,Z*)-Cyclodeca-1,6-diene, synthesis of, **1**, 459
Cyclodeca-1,2,6-triene, reduction of, **1**, 459
Cyclodecatrienes, use in the retardation of electrocyclization, **1**, 421
Cyclodecenones, synthesis of, via oxy-Cope rearrangements, **1**, 813
Cyclodextrin, amino-, synthesis of, **2**, 859
β-Cyclodextrin, use in the synthesis of thiotetronic acids, **6**, 473
Cyclodextrin azides, reactions, with triphenylphosphine, **2**, 859
β-Cyclodextrin–epichlorohydrin copolymers, use in the synthesis of alkyl iodides, **2**, 35
Cyclodextrins, reactions, with amine–borane complexes, **2**, 57
Cyclodimerization, nickel(0)-catalyzed, of tetraarylhexapentaenes, **1**, 992
Cyclodiphosphazenes, synthesis of, **2**, 468
Cyclododecanone, ring-contraction of, **5**, 40
Cyclododecyl radicals, synthesis of, **1**, 939
Cycloetherification, for the synthesis of carboxylic esters, **5**, 142
Cyclofunctionalization, use of term, **2**, 286
Cycloheptadienes, synthesis of, **1**, 812
1,3-Cycloheptadienes, 1,5 sigmatropic rearrangement of, **1**, 402
1,3-Cycloheptadienyls, use in the synthesis of vinyl cations, **1**, 1113
Cycloheptane, synthesis of, **1**, 340
Cycloheptane, trimethyl-, synthesis of, **1**, 72
Cycloheptane-1-carboxylic acid, 1-hydroxy-, as starting material, for acid fluorides, **5**, 5
Cycloheptanes, divinyl-, thermal rearrangement of, **1**, 813
Cycloheptanone, 2-chloro-, as starting material, for cyclohexanecarboxylic acid, **1**, 391
Cycloheptanones, synthesis of, **3**, 124, 319
Cycloheptatriene
acidity of, **1**, 865
reactions, with benzylpotassium, **1**, 865
synthesis of, **4**, 557
Cycloheptatriene complexes, chroma-, synthesis of, **4**, 676
Cycloheptatrienes
1,5 sigmatropic rearrangement of, **1**, 402
1,7 sigmatropic rearrangements of, **1**, 404
as starting materials, for norcaradienes, **1**, 420
Cycloheptatrienium bromide, synthesis of, **1**, 906
Cycloheptatrienyl anions, synthesis of, **1**, 865
Cycloheptatrienyl carbocations, synthesis of, **1**, 906
Cycloheptene, metathetic dimerization of, **1**, 674
Cycloheptene, 1-fluoro-, synthesis of, **2**, 610
Cycloheptene, 1-nitro-, synthesis of, **2**, 785
Cyclohept-2-enes, 7-vinyl-, rearrangement of, **1**, 799
Cycloheptenones
dehydrogenation of, **1**, 554
reactions, with 2,3-dichloro-5,6-dicyano-1,4-benzoquinone, **1**, 554
Cycloheptenyl chloride, synthesis of, **1**, 1120
1,4-Cyclohexadienamines, synthesis of, **2**, 740
Cyclohexadiene
hydroboration of, **2**, 519
reactions
with bis(η^3-allyl) complexes, **4**, 694
with phenylethyne, **1**, 529
Cyclohexadiene, 1,3,5,5-tetramethyl-, for the freezing point depression of sulfuric acid, **1**, 911
Cyclohexadiene, trimethylsilyloxy-, [4+2] cycloaddition of, **2**, 665
1,3-Cyclohexadiene
from, 1,3,5-hexatriene, **1**, 419
reactions, with diethyl fumarate, **6**, 343
as starting material, for 1,5-enynes, **1**, 1036
1,3-Cyclohexadiene, 1-methyl-2,5,6-tris(trimethylsilyl)-, synthesis of, **2**, 938
1,3-Cyclohexadiene, perfluoro-, reactions, with octacarbonyldicobalt, **4**, 687
1,4-Cyclohexadiene, synthesis of, **1**, 95
1,4-Cyclohexadiene, 1,3-bis(trimethylsilyl)-, synthesis of, **2**, 938
1,4-Cyclohexadiene, 1-methoxy-, synthesis of, **1**, 890
1,4-Cyclohexadiene, 1-methyl-, synthesis of, **1**, 96
1,4-Cyclohexadiene, 2-trimethylsilyl-, synthesis of, **2**, 938
Cyclohexa-3,4-diene-1-al, oxidation of, **5**, 51
2,4-Cyclohexadieneone, 5-allyl-, as starting material, for phenols, **1**, 808
Cyclohexadienes
synthesis of, via electrocyclic reactions, **1**, 419
uses of, **1**, 98
Cyclohexa-1,3-dienes, synthesis of, **1**, 736
Cyclohexa-1,4-dienes, synthesis of, **1**, 101
1,4-Cyclohexadienes, synthesis of, via elimination of β-nitro sulfones, **1**, 617
1,4-Cyclohexadienes, 1-alkoxy-, reactions, with arenesulfonyl azides, **5**, 702
1,4-Cyclohexadienes, 1-nitro-, synthesis of, **1**, 600
2,4-Cyclohexadienes, 1-bora-, thermal cycloreversion of, **6**, 713
Cyclohexadienimines, as starting materials, for ketenimines, **3**, 582
Cyclohexadienols, reactions, with sulfuric acid, **3**, 70
Cyclohexadienone, 2,4,4,6-tetrabromo-, for the dibromination of alkenes, **2**, 31

Cyclohexa-2,5-dienone, 2,4,4,6-tetrabromo-, for the *para* bromination of phenols, **2**, 626
Cyclohexa-2,5-dien-1-one *N*-oxide, 4-(2-methoxycarbonyl-4,6-dimethoxyphenyl)imino-3-methoxycarbonyl-5-methoxy-, synthesis of, **3**, 433
Cyclohexadienones
 aromatization of, **2**, 658, 658
 from, cyclohexenones, **2**, 658
 synthesis of, **3**, 233
Cyclohexadienones, chloromethyl-, synthesis of, **4**, 17
Cyclohexadienones, tetrabromo-, synthesis of, **4**, 23
Cyclohexa-2,4-dienones, as starting materials, for vinylketenes, **3**, 540
Cyclohexadienyl cations, heptamethyl-, synthesis of, **1**, 912
3-Cyclohexanal-1-one, 2,4,4-trimethyl-, oxidation of, **5**, 102
Cyclohexane
 reactions, with azoalkynes, **2**, 1054
 reactivity of, **1**, 848
 as starting material, for cyclohexanecarboxylic acid chloride, **5**, 16
 synthesis of, **1**, 93
Cyclohexane, 1-acetoxy-1-nitroso-, synthesis of, **4**, 314
Cyclohexane, 1,1'-azocyano-, thermal decomposition of, **3**, 573
Cyclohexane, 1,1-bis(methylthio)-, as starting material, for 1-azido-1-methylthiocyclohexane, **4**, 329
Cyclohexane, bistriflamido-*trans*-1,2-diamino-, reactions, with titanium tetraisopropoxide, **2**, 72
Cyclohexane, bis(trimethylenedioxyboryl)methylene-, reactions, with iodine, **4**, 750
Cyclohexane, 1-bromo-4-*t*-butyl-, reactions, with lithium dimethylcuprate, **1**, 118
Cyclohexane, 1-bromo-2-chloro-, elimination of, for vinyl bromides, **2**, 622
Cyclohexane, 1-cyano-1-(*m*-methoxyphenyl)-4-hydroxy-, synthesis of, **5**, 64
Cyclohexane, 1,2-dihydroxy-1-methyl-4-(2-propenyl)-, synthesis of, **2**, 84
Cyclohexane, 1,3,5-(dimethylamino)-1,3,5-triboro-, synthesis of, **4**, 632
Cyclohexane, 5,5-dimethyl-1,3-dioxo-, oxidation of, **4**, 206
Cyclohexane, 1-fluoro-, synthesis of, **2**, 610
Cyclohexane, hexadeuterio-, synthesis of, **1**, 93
Cyclohexane, perfluoro-
 pyrolysis of, **4**, 737
 synthesis of, **4**, 3
Cyclohexane, 1,1,3,3-tetramethyl-2-(thiocarbonyl)-, reactions, with methanol, **2**, 202
Cyclohexane, *cis*-1,3,5-trihydroxy-, reactions, with ortho esters, **6**, 79
Cyclohexane, 1-trimethylsilyloxy-, reactions, with tin tetrachloride, **1**, 251
Cyclohexanecarbaldehyde, *trans*-2-methyl-, from, 1-methylcyclohexene, **1**, 297
Cyclohexane carboxaldehyde, *trans*-2-trialkylsilyloxy-, synthesis of, **3**, 33
Cyclohexanecarboxylate, (*E*)-methyl 2-acetoxy, pyrolysis of, **1**, 774
Cyclohexanecarboxylate, methyl 2-oxo-, as starting material, for methyl 2-methyl-1-cyclohexene-1-carboxylate, **5**, 103
Cyclohexanecarboxylic acid, from, 2-chlorocycloheptanone, **1**, 391
Cyclohexanecarboxylic acid chloride, from, cyclohexane, **5**, 16
Cyclohexane-1,3-dione, reactions, with *N*-methylhydrazone, **3**, 264
Cyclohexane-1,3-dione, 2-bromo-, as starting material, for ketenes, **3**, 536
Cyclohexane-3,5-dione, 1,1-dimethyl-, reactions, with ethyl diazonitroacetate, **3**, 473
Cyclohexane-1,2-diones
 reactions, with chiral amines, **3**, 260
 use in the synthesis of thiocarbamic acid *O*-esters, **6**, 557
Cyclohexane-1,3-diones, reactions, with amines, **3**, 263
Cyclohexane 1,1-dithiol, as starting material, for dithionooxamides, **5**, 580
1,2-Cyclohexanedithiones, as intermediates, for the synthesis of tetrathioorthocarbonates, **6**, 306
Cyclohexane oxide, ring-opening of, **1**, 1139
Cyclohexanes
 cyclization to, **1**, 229
 synthesis of, **1**, 175, 822
Cyclohexanes, 1-acyloxy-1-nitroso-, synthesis of, **4**, 314
Cyclohexanes, alkyl-substituted, synthesis of, **1**, 89
Cyclohexanes, bis(halomercurio)methylene-, transmetallation of, **1**, 454
Cyclohexanes, (*R*,*R*)-diamino-, as starting materials, for (*R*)-α-chloroalkylphosphonic acids, **4**, 129
Cyclohexanes, diazo-, reactions, with 2,2-dimethylpropylidynylphosphine, **4**, 480
Cyclohexanes, dilithiomethylene-, synthesis of, **4**, 1064
Cyclohexanes, 1-hydroxy-1-azo-, as starting materials, for 1-benzoyl-2-phenylhydrazines, **5**, 357
Cyclohexanes, isothiocyanato-, synthesis of, **5**, 1044
Cyclohexanes, methylene-, synthesis of, **1**, 1127
Cyclohexanesulfonyl chlorides, dehydrochlorination of, **4**, 331
Cyclohexanetriamine, reactions, with triethyl orthoformate, **6**, 140
cis,cis-1,3,5-Cyclohexanetriisocyanide, synthesis of, **3**, 700
Cyclohexanoates, synthesis of, **5**, 146
Cyclohexanol, *trans*-2-trimethylsilyl(phenylthio)methyl-, synthesis of, **4**, 943
Cyclohexanols
 oxidation of, **5**, 46, 765
 reactions, with 4-morpholinosulfur trifluoride, **2**, 19
 synthesis of, **1**, 89
Cyclohexanols, 1,2-divinyl-, oxy-Cope rearrangements of, **1**, 813
Cyclohexanone
 reactions
  with acetonitrile, **3**, 644
  with bromochloromethyllithium, **4**, 758
  with ceric ammonium nitrate, **5**, 95
  with cyanoacetic acid, **3**, 644
  with di-*t*-butyl(dimethylamino)stibine, **2**, 495
  with di-*t*-butylmethoxystibine, **2**, 495
  with dimethoxyamine, **5**, 331
  with ethanedithiol, **4**, 246
  with α-silylcarbanions, **1**, 763
  with tribromomethyllithium, **6**, 31
 as starting material
  for enethiol ethers, **4**, 247
  for ethyl cyclohexylideneacetate, **5**, 103
  for 2-nitro-2-nitroso compounds, **4**, 439
Cyclohexanone, 4-*t*-butyl-, reactions, with methyllithium, **2**, 68
Cyclohexanone, *trans*-4-*t*-butyl-, reduction of, **2**, 61
Cyclohexanone, 2-chloro-, ring-contraction of, **1**, 393
Cyclohexanone, 1-methyl-, bromination of, **2**, 28
Cyclohexanone, 2-methyl-
 alkylation of, **2**, 675
 sulfenylation of, **2**, 658
Cyclohexanone, methylene-, synthesis of, **1**, 765
Cyclohexanone, 2-methylene-, reactions, with ylides, **1**, 736
Cyclohexanone, 2-methyl-2-nitro-, synthesis of, **3**, 190
Cyclohexanone, 2-morpholino-, synthesis of, **3**, 183
Cyclohexanone, 3-propargyl-, synthesis of, **1**, 1061
Cyclohexanone, 3,3,5-trimethyl-, reduction of, **2**, 61
Cyclohexanones
 aromatization of, **2**, 646, 658
  for the synthesis of phenols, **2**, 659
 cleavage of, **3**, 448
 reactions
  with alkynyl metals, **1**, 1039

with organoiron(II) compounds, **2**, 70
with phosphonic diamides, **1**, 645
with sulfuric acid, **2**, 658
with thiolethanol, **4**, 229
reduction of, **2**, 60
synthesis of, **1**, 86, 89; **3**, 124
Cyclohexanones, 2-acyl-, reactions, with cyanoacetamide, **3**, 672
Cyclohexanones, alkyl-, reduction of, **2**, 60
Cyclohexanones, 2-alkyl-, synthesis of, **3**, 415
Cyclohexanones, 2-fluoro-2-alkyl-, synthesis of, **3**, 415
Cyclohexanones, substituted
reactions, with pyrrolidones, **3**, 262
reduction of, **2**, 59
Cyclohexasilane, dodecamethyl-, as starting material, for dimethylsilylene, **3**, 502
Cyclohex-3,5-diene, 1,2-*cis*-dihydroxy-, from, benzene, **2**, 39
Cyclohex-5-en-1-carboxylic acid, *cis*-2-hydroxy-1-methyl-, synthesis of, **5**, 63
Cyclohexene
addition of tetrachloromethane to, **6**, 7
cyclization of, **1**, 1126
reactions
with *N*-*t*-butyl-*N*-chlorocyanamide, **5**, 1120
with *N*,*N*-dibromobenzenesulfonamide, **5**, 701
with phenyl selenyl chloride, **2**, 785
with phosphorus trichloride, **2**, 448
with potassium selenocyanate, **5**, 1115
with tetranitromethane, **6**, 149
reduction of, **1**, 79
as starting material
for adipic acid, **5**, 52
for methylcyclopentyl cations, **1**, 895
Cyclohexene, 1-bromo-, synthesis of, **2**, 622
Cyclohexene, 1-chloro-
reactions, with hydrogen fluoride, **4**, 30
reduction of, **1**, 430
Cyclohexene, 3-hydroxy-, from, 2-vinylcyclobutanols, **1**, 805
Cyclohexene, methyl-, reduction of, **1**, 76
Cyclohexene, 1-methyl-, as starting materials, for *trans*-2-methylcyclohexanecarbaldehyde, **1**, 297
Cyclohexene, 4-methyl-1-chloro-, reduction of, **1**, 430
Cyclohexene, morphilino-, reactions, with methyl α-chloroacrylate, **2**, 757
Cyclohexene, morpholino-, reactions, with chlorosulfene, **5**, 497
Cyclohexene, 1-morpholino-, reactions, with 1,2-benzoquinone, **4**, 307
Cyclohexene, 1-pyrrolidino-, reactions, with *o*-quinonemethides, **4**, 307
1-Cyclohexene, 4-*t*-butyl-, reactions, with trimethylsilylethyne, **1**, 1029
1-Cyclohexene, 1-nitro-, synthesis of, **2**, 786
1-Cyclohexene, 1,3,4,5,6-pentakis(trimethylsilyl)-, synthesis of, **2**, 938
Cyclohexenebiscarboxylic acids, synthesis of, **1**, 103
1-Cyclohexene-1-carboxylate, methyl 2-methyl-, from, methyl 2-oxocyclohexanecarboxylate, **5**, 103
Cyclohexene oxide
reactions, with trimethylsilyl(phenylthio)methyllithium, **4**, 943
as starting material, for alcohols, **1**, 388
Cyclohexene oxide, 1-methyl-, rearrangement of, **1**, 386
Cyclohexene oxides, ring-opening of, **2**, 82
Cyclohexenes
from, vinylcyclobutanes, **1**, 805
reactions
with ozone, **5**, 76
with thiol radicals, **2**, 137
synthesis of, **1**, 709
Cyclohexenes, 1-alkoxy-, reactions, with arenesulfonyl azides, **5**, 702
Cyclohexenes, alkyl-substituted-1-nitro-, synthesis of, **2**, 785
Cyclohexenes, 1-dialkylamino-
reactions
with amyl nitrite, **2**, 745
with ethyl nitrite, **2**, 745
Cyclohexenes, 4-imino-, Diels–Alder reaction of, **3**, 582
Cyclohexenes, trimethylsilyl-, reactions, with iodine, **4**, 139
syn-1,4-Cyclohex-2-enes, from, 1,2-diox-4-enes, **2**, 52
Cyclohexene sulfide, synthesis of, **2**, 136
Cyclohexenol, reactions, with diazophosphonates, **6**, 118
Cyclohex-2-en-1-ol, dihydroxylation of, **2**, 46
Cyclohexenone
reactions
with α-alkoxy ester enolates, **1**, 303
with bis(phenylthio)methane, **1**, 303
with *t*-butylmagnesium chloride, **1**, 301
with isochromans, **3**, 325
with methyl isocyanide, **3**, 707
synthesis of, **1**, 99
Cyclohexenone, 3-iodo-, reactions, with *N*-benzyl-1,4-dihydronicotinamide, **1**, 431
Cyclohexenone, 2-methyl-, synthesis of, **1**, 99
Cyclohexenone, 2-(nitromethyl)-, reactions, with ethyl sodio-2-cyanobutanoate, **1**, 251
2-Cyclohexenone
1,4 addition of, **1**, 1061
reactions
with 3-methyl-1-(triphenylstannyl)allene, **1**, 1061
with trimethylsilylallene, **1**, 1061
Cyclohex-2-enone, 3-ethoxy-, use in the synthesis of monothioacetals, **4**, 218
Cyclohex-2-enone, 3-methyl-, from, 1,3-dibromopropane, **1**, 141
2-Cyclohexenone, 6-phenyl selenyl-, synthesis of, **3**, 173
2-Cyclohexen-1-one, (*S*)-2-(2-(methoxymethyl)-1-pyrrolidinyl)methyl-, from, 2-(nitromethyl)-2-cyclohexen-1-one, **1**, 257
2-Cyclohexen-1-one, 2-(nitromethyl)-, as starting material, for (*S*)-2-(2-(methoxymethyl)-1-pyrrolidinyl)methyl-2-cyclohexen-1-one, **1**, 257
Cyclohexenones
aromatization of, **2**, 658
for the synthesis of phenols, **2**, 659
reactions, with alkynyl metals, **1**, 1039
reduction of, **1**, 86
as starting materials, for cyclohexadienones, **2**, 658
synthesis of, **3**, 219
Cyclohexenones, alkenyl-, reactions, with rhodium(III) chloride, **2**, 660
Cyclohexenones, homochiral, synthesis of, **1**, 558
Cyclohexenones, α-methylene-, synthesis of, **3**, 223
Cyclohexenones, 2-(nitromethyl)-, as starting materials, for 2-exomethylenecycloalkanones, **1**, 257
Cyclohexenyls, synthesis of, **1**, 1124
1,3-Cyclohexenyne, synthesis of, **1**, 1080
Cyclohexenyne complexes, hexafluoro-, synthesis of, **4**, 687
Cyclohexylamines, synthesis of, **1**, 89
Cyclohexyl analogues, 4-*t*-butyl-, alkylation of, problems, **1**, 139
Cyclohexyl compounds, elimination of, **1**, 625
Cyclohexyl halides, substitutions of, **1**, 112
Cyclohexylidenes, reactions, with lithium 1-dimethylaminonaphthalenide, **4**, 945
Cyclohexylides, synthesis of, **3**, 500
Cyclohexyl iodide, as alkylating agent, **1**, 136
2-Cyclohexylpentane, 1,1-dicyano-, reactions, with sulfuric acid, **5**, 35
Cyclohexylphosphine, dichloro-, synthesis of, **2**, 448
Cyclohexyl radicals
reactions, with methyl acylate, **1**, 331
synthesis of, **1**, 118
trapping of, **1**, 330
Cyclohexyne, synthesis of, **1**, 1084

Cyclohexyne *p*-dioxine, 1,4-dioxa-, synthesis of, **2**, 1027
Cyclols
  structure of, **6**, 104
  synthesis of, **6**, 104
Cyclonona-1,2-diene
  as starting material
    for (*Z*)-cyclononene, **1**, 459
    for cyclotetradeca-1,2-diene, **1**, 459
Cyclonona-1,2-diene, 1-methyl-, as starting material, for (*Z*)-1-methylcyclonon-1-ene, **1**, 460
(*Z*,*Z*)-Cyclonona-1,5-diene
  from, cyclonona-1,2,6-triene, **1**, 459
  synthesis of, **1**, 459
1,5-Cyclononadienes, as starting materials, for divinylcycloalkanes, **1**, 812
Cyclononatetraenide anion, from, bicyclo[6.1.0]nona-2,4,6-triene, **1**, 863
Cyclononatetraenyl 10 π-electron anion, 1,5-methano-, synthesis of, **1**, 864
Cyclononatriene
  as starting material
    for bicyclo[4,3,0]nonadiene, **1**, 421
    for dienes, **1**, 421
Cyclonona-1,2,3-triene, synthesis of, **1**, 990
Cyclonona-1,2,6-triene
  reduction of, **1**, 459
  as starting material, for (*Z*,*Z*)-cyclonona-1,5-diene, **1**, 459
1-Cyclononene, 1,9-bis(trimethylsilyl)-, synthesis of, **2**, 929
Cyclonon-1-ene, (*Z*)-1-methyl-, from, 1-methylcyclonona-1,2-diene, **1**, 460
(*Z*)-Cyclononene, from, cyclonona-1,2-diene, **1**, 459
1,3-Cyclooctadiene, as starting material, for bicyclo[4,2,0]octene, **1**, 418, 419
1,5-Cyclooctadiene
  reactions, with phosphine, **2**, 430
  as starting material, for 9-borabicyclononane, **2**, 516
cis,*trans*-1,5-Cyclooctadiene, synthesis of, **1**, 640
1,5-Cyclooctadiene ligands, for the synthesis of alkyl metal(I) compounds, **2**, 576
Cyclooctadienes, synthesis of, **1**, 412
2,4-Cyclooctadienol, rearrangement of, **1**, 799
1,3-Cyclooctadienyls, use in the synthesis of vinyl cations, **1**, 1113
Cyclooctane, synthesis of, **1**, 340
Cyclooctanes, 1,2-divinyl-, Cope rearrangement of, **1**, 813
Cyclooctatetraene
  protonation of, **1**, 917
  reduction of, **1**, 890
  as starting material
    for bridged iron complexes, **4**, 683
    for pentalene-bridged ruthenium complexes, **4**, 683
    for ruthenium complexes, **4**, 683
  use in the synthesis of carbon-bridged complexes, **4**, 703
  use in the synthesis of tris(cyclooctatetraene)dititanium, **4**, 671
Cyclooctatetraene, 1,4-dibromo-
  reactions
    with dodecacarbonyltriruthenium, **4**, 681
    with nonacarbonyldiiron, **4**, 681
    with tetracarbonyldimethylosmium, **4**, 681
Cyclooctatetraene, perfluoro-
  reactions
    with cobalt, **4**, 687
    with dicobalt, **4**, 687
Cyclooctatetraene-bridged complexes, synthesis of, **4**, 672
Cyclooctatetraenes, synthesis of, **1**, 255, 517
Cyclooctatetraenyl dicarbanions, synthesis of, **1**, 890
Cyclooctatetraenyls, use in the synthesis of vinyl cations, **1**, 1113
Cyclooctatriene, synthesis of, **1**, 421
Cyclooctene, metathetic dimerization of, **1**, 674
Cyclooctene, 1-bromo-, reduction of, **1**, 431
Cyclooctene, 1-nitro-, synthesis of, **2**, 786
trans-Cyclooctene, synthesis of, **1**, 640
Cyclooctyldimethylamine oxide, pyrolysis of, **1**, 624
Cyclooctyne
  reactions, with octacarbonyldicobalt, **4**, 689
  synthesis of, **1**, 1075
Cyclooligomerization
  of alkynes, **1**, 530
    via carbopalladation, **2**, 980
Cyclopalladation
  of amines, **4**, 720
  in the synthesis of butan-2-oxime complexes, **1**, 63
Cyclopentaarsine, pentakis(chloromethyl)-, synthesis of, **2**, 507
Cyclopentaarsine, pentamethyl-
  oxidation of, **2**, 496
  reactions, with alkylarsines, **2**, 507
Cyclopentadiene
  acidic properties of, **1**, 861
  mercuration of, **4**, 697
  metallation of, **4**, 625
  reactions
    with aminobismuthines, **2**, 485
    with benzoylethyne, **1**, 519
    with dimethylaminodimethylarsine, **2**, 484
    with dimethyl(dimethylamino)stibines, **2**, 485
    with methyl propiolate, **1**, 520
    with tris(dimethylamino)stibines, **2**, 485
  as starting material, for 2-phosphabicyclo[2.2.1]heptenes, **4**, 112
  use in the synthesis of carbon-bridged ditins, **4**, 701
Cyclopentadiene, 1,1-bis(trimethylgermyl)-, synthesis of, **4**, 625
Cyclopentadiene, 1,1-bis(trimethylsilyl)-, synthesis of, **4**, 603
Cyclopentadiene, 5-diazo-1,2,3,4-tetraphenyl-, synthesis of, **3**, 490
Cyclopentadiene, 1,3-di-*t*-butyl-5-vinylidene-, from, 6-chloro-6-methylpentafulvene, **1**, 982
Cyclopentadiene, 5,5-dichloro-1,2,3,4-tetracyano-, synthesis of, **1**, 862
Cyclopentadiene, 1-[dimethyl(methoxy)silyl]-1-trimethylsilyl-, synthesis of, **4**, 603
Cyclopentadiene, hexabromo-, for the *para* bromination of phenols, **2**, 626
Cyclopentadiene, 1,2,3,4,5-pentakis(trifluoromethyl)-, trifluoromethyl groups of, **1**, 862
Cyclopentadiene, pentamethyl-, as starting material, for cobalt multidecker sandwich compounds, **4**, 695
Cyclopentadiene, pentaphenylalumina-, reactions, with nickel(II) bromide, **4**, 693
Cyclopentadiene, 1,2,3,4-tetraphenyl-, reactions, with *p*-toluenesulfonyl azide, **3**, 490
Cyclopentadiene complexes
  reactions, with isonitriles, **5**, 799
  of zirconium and titanium, synthesis of, **5**, 797
Cyclopentadiene complexes, binuclear metalla-, synthesis of, **4**, 689
Cyclopentadiene complexes, rhena-, synthesis of, **4**, 677, 678
Cyclopentadiene complexes, rhodia-, synthesis of, **4**, 689
Cyclopentadiene complexes, ruthena-, synthesis of, **4**, 681, 683
Cyclopentadiene complexes, trinuclear osma-, as intermediates, **4**, 682
Cyclopentadienes
  from
    cyclopentadienes, **1**, 403
    vinylcyclopropenes, **1**, 802
  pyrolysis of, **1**, 785
  reactions
    with fluorocarbenes, **2**, 615
    with tetramethylchloroformamidinium chloride, **4**, 974
  1,5 sigmatropic rearrangement of, **1**, 402
  as starting materials
    for alkenes, **1**, 785

for cyclopentadienes, **1**, 403
Cyclopentadienes, benzoferra-
from
2-bromostyrene, **4**, 681
dodecacarbonyltriiron, **4**, 681
pentacarbonyliron, **4**, 681
photolysis of, **4**, 681
Cyclopentadienes, chroma-, synthesis of, **4**, 676
Cyclopentadienes, ferra-
from, monoalkyne complexes, **4**, 682
ruthenium analogue of, synthesis of, **4**, 683
synthesis of, **4**, 681, 714
via alkyne insertion, **4**, 683
use in the synthesis of diiron complexes, **4**, 681
Cyclopentadienes, metalla-
of aluminum, via transmetallation reaction, **2**, 989
of boron, via transmetallation reaction, **2**, 989
of germanium, via transmetallation reaction, **2**, 989
intermediacy of, **2**, 984
of silicon, via transmetallation reaction, **2**, 989
of tin, via transmetallation reaction, **2**, 989
Cyclopentadienes, metallo-, synthesis of, **3**, 272
Cyclopentadienes, tetracyano-, synthesis of, problems, **1**, 862
Cyclopentadienes, tetra(phenylsulfonyl)-, synthesis of, **4**, 281
Cyclopentadienes, zircona-
formation prevention of, **2**, 986
synthesis of, **2**, 927
Cyclopentadienide, tetrakis(benzenesulfonyl)-, protonation of, **4**, 281
Cyclopentadienide anion, chemistry of, **1**, 861
Cyclopentadienides, hetero-, synthesis of, **2**, 896
Cyclopentadienide salts, synthesis of, **1**, 862
Cyclopentadienones, synthesis of, **1**, 500
Cyclopentadienones, bis(silyl)-, from, α,ω-silyldiynes, **2**, 926
Cyclopentadienyl anion, tetracyano-
reactions of, **1**, 862
reactivity of, **1**, 862
Cyclopentadienyl anions
basicity of, **1**, 861
reactions, with trithiocyclopropenium salts, **4**, 852
Cyclopentadienyl–cobaltacyclopentadiene complexes, synthesis of, **4**, 690
Cyclopentadienyl complexes, synthesis of, **1**, 94
Cyclopentadienyl complexes, iridia-, synthesis of, **4**, 690
Cyclopentadienylide, dimethylsulfonium, structure of, **3**, 373
Cyclopentadienylide, triphenylstibonium tetraphenyl-
reactions, with aryl aldehydes, **1**, 647
structure of, **1**, 742
Cyclopentadienylides, synthesis of, **3**, 373, 500
Cyclopentadienyllanthanide complexes
reactions
with alkyllithiums, **2**, 565
with lithium tetraalkylaluminate, **2**, 565
with organolithiums, **2**, 564
Cyclopentane, methylene-, synthesis of, **1**, 547
Cyclopentanecarboxylic acids, hydroxy-, disilylation of, **5**, 223
Cyclopentane-1,3-dione, as starting material, for 3-aminocyclopent-2-en-1-one, **2**, 311
Cyclopentane-1,3-dione, 2-methyl-, reactions, with amines, **3**, 263
Cyclopentane-1,3-diones, 2-methyl-2-alkyl-, reduction of, **3**, 161
Cyclopentane-1,3-diyls, 2-alkylidene-, cycloaddition of, **1**, 359
Cyclopentanes
from
allyl anions, **1**, 363
trimethylenemethane, **1**, 358
synthesis of, **1**, 175, 183, 231, 360; **4**, 654
Cyclopentanes, acetyl-, alkylation of, **1**, 131
Cyclopentanes, 1,2-divinyl-, oxy-Cope rearrangement of, **1**, 812
Cyclopentanes, ethynylidene-, synthesis of, **1**, 1020
Cyclopentanes, methylene-
from, vinyllithiums, **2**, 556
synthesis of, **1**, 362, 1127
Cyclopentanes, silyl alkylidene-, synthesis of, **2**, 925
Cyclopentanes, zircona-, ring-contraction of, **2**, 987
Cyclopentanoates, synthesis of, **5**, 146
Cyclopentanoic esters, synthesis of, **5**, 146
Cyclopentanol (A), epoxidation of, **2**, 91
Cyclopentanone
from, cyclic titanium alkyl compounds, **5**, 441
reactions
with ceric ammonium nitrate, **5**, 95
with tungsten(II) complexes, **3**, 517
Cyclopentanone, 2-chloro-, reactions, with potassium *O*-isopropyl dithiocarbonate, **6**, 546
Cyclopentanone, 2-cyclopentylidene-, acidity of, **1**, 878
Cyclopentanone, 3-(3-hydroxypropyl)-, as starting material, for β-(3-oxocyclopentyl)propionic acid, **5**, 73
Cyclopentanone, 2-methoxy-, reactions, with lithium alkynide, **1**, 1039
Cyclopentanone, 2-methyl-, reactions, with carbon disulfide, **6**, 324
Cyclopentanone, sodio-2-carboethoxy-, reactions, with 1-(nitromethyl)-3,4-dihydronaphthalene, **1**, 250
Cyclopentanone-3,4-dicarboxylate, *trans*-dimethyl-, reactions, with 1,3-hydroxy thiols, **4**, 229
Cyclopentanone ring synthesis, mechanism, **3**, 129
Cyclopentanones
reactions, with phenylsulfenyl chloride, **3**, 249
synthesis of, **1**, 386; **3**, 113
Pauson–Khand reaction for, **3**, 114
via ring-contraction, **1**, 388
Cyclopentanones, 2-alkyl-, reactions, with ethynyl metals, **1**, 1039
Cyclopentanones, 2-alkylidene-, as starting materials, for 5-oxoalkanoic acids, **5**, 75
Cyclopentanones, 2-halo-, synthesis of 4-membered rings from, **1**, 393
Cyclopentanones, α-methylene-, from, α-chloro-β-silyl ketones, **1**, 653
Cyclopentanones, 2-silyl-, from, alkynylsilanes, **2**, 926
7*H*-Cyclopenta[*a*]pentalene-7,8-dicarboxylate, dimethyl 3,3*a*,3*b*,4,6*a*,7*a*-hexahydro-3,4,7-metheno-, saponification of, **5**, 30
Cyclopentenals, synthesis of, **3**, 59
Cyclopentene, reactions, with diazadienes, **3**, 453
Cyclopentene, 1-chloro-, synthesis of, **2**, 615
Cyclopentene, *cis*-3,5-dibromo-, reactions, with sodium trithiocarbonate, **6**, 554
Cyclopentene, metalla-, as intermediates, for vinyl metals, **2**, 986
3-Cyclopentene, 1-gerana-1,1-dichloro-, reactions, with 2-lithio-1,3-dithiane, **6**, 132
3-Cyclopentene, 1-oxa-2-alumina-, from, *anti*-hydroalumination of propargyl alcohols, **2**, 970
Cyclopentene annulation, trimethylsilyl-, use of term, **2**, 930
Cyclopent-1-enecarboxamides, synthesis of, **1**, 613
Cyclopentenes
from, vinylcyclopropanes, **1**, 801
cis-hydroxylation of, **2**, 47
as starting materials, for prostaglandins, **1**, 818
synthesis of, **1**, 431, 708
Cyclopentenes, diboro-, synthesis of, **4**, 632
Cyclopentenes, methylene-, synthesis of, via allylzincation of alkynylsilanes, **2**, 925
Cyclopentenes, 4-methylene-, synthesis of, **1**, 500
Cyclopentenes, zircona-, ring-contraction of, **2**, 987
Cyclopentenethiones, synthesis of, **3**, 353
cis-3,4-Cyclopenteno-1,3-dithiolane-2-thiones, synthesis of, **6**, 554

Cyclopentenone, reactions, with mixed cuprates, **1**, 1093
Cyclopentenone, 5,5-dimethyl-, as starting material, for
cyclopropylketenes, **3**, 538
2-Cyclopentenone, 2-(6-carbomethoxyhexyl)-4-hydroxy-,
reactions, with alkynyldimethylalanes, **1**, 1049
Cyclopent-2-en-1-one, 3-amino-
from, cyclopentane-1,3-dione, **2**, 311
synthesis of, **2**, 311
Cyclopent-2-en-1-one, 3-methyl-2-
[(dimethylthiocarbamoyl)oxy]-3-methyl-, synthesis
of, **6**, 556
Cyclopentenones
from
alkynic diols, **1**, 422
α-phenylsulfinyl ketones, **1**, 596
sulfur-substituted, synthesis of, **3**, 249
synthesis of, **1**, 421, 528, 708
via Pauson–Khand reaction, **3**, 218
via Ramberg–Bäcklund reactions, **1**, 619
via retro-Diels–Alder reaction, **1**, 786
Cyclopentenones, alkoxy-, reactions, with ammonia, **3**, 264
Cyclopentenones, amino-, synthesis of, **3**, 264
Cyclopentenones, 2-silyl-, synthesis of, **3**, 270
Cyclopentenones, α-silyl-, synthesis of, **2**, 931
Cyclopentenones, substituted, synthesis of, **1**, 556
Cyclopentenone synthesis, co-promoted, **2**, 985
Cyclopentenyl iodide, reactions, with
pentynyltributylstannane, **1**, 1015
Cyclopentenylsilanes, 4-methyl-, synthesis of, **2**, 925
Cyclopentyl cations, methyl-, from, cyclohexene, **1**, 895
1-Cyclopentyl cations, 1-methyl-, synthesis of, **1**, 895
Cyclophanes
synthesis of, **1**, 788
via rearrangements, **1**, 233
[2.2]-Cyclophanes, from, bisbenzyl sulfides, **1**, 232
Cyclophosphazenes, synthesis of, **2**, 864
Cycloplatinated complexes, reactions, with
iodomethane, **1**, 127
Cyclopolyarsines
from
arylarsenic dihalides, **2**, 893
arylarsonic acids, **2**, 893
primary arylarsines, **2**, 894
oxidation of, **2**, 495
synthesis of, **2**, 893
Cyclopolyphosphazenes, synthesis of, **2**, 468
Cyclopolyphosphines, metallated, reactions, with
1,1-dihaloalkanes, **4**, 545
Cyclopropabenzene, 7-lithio-7-trimethylsilyl-, synthesis
of, **4**, 647
Cyclopropa[*c*]chromene, ring-opening of, **4**, 742
1*H*-Cyclopropa[*b*]naphthalene, 1,1-bis(trimethylsilyl)-,
synthesis of, **4**, 604
Cyclopropa[*b*]naphthalene, reactions, with
n-butyllithium, **4**, 604
Cyclopropanation
of alkenes, **1**, 304
mechanism, **1**, 298
for the synthesis of 3,5-cyclovitamin D ring A
synthon, **1**, 109
Cyclopropanation, asymmetric, **1**, 299
of alkenes, **1**, 353
Cyclopropanation, intramolecular, of alkenes, **1**, 354
Cyclopropane
as intermediate, **1**, 1127
reactions, with silver nitrate, **3**, 235
synthesis of, **1**, 109
thermal rearrangement of, **1**, 804
Cyclopropane, 1-benzoyl-2-vinyl-, reactions, with cuprous
methyltrialkylboranes, **1**, 275
Cyclopropane, 1,1-bis(dibenzylamino)-, hydrolysis
of, **4**, 309
Cyclopropane, 1,1-bis(diphenylphosphino)-, synthesis
of, **4**, 548
Cyclopropane, bis(phenylseleno)-, synthesis of, **4**, 289
Cyclopropane, 1-bromo-1-ethoxy-, synthesis of, **4**, 50
Cyclopropane, chloro-, synthesis of, **2**, 615
Cyclopropane, divinyl-
Cope rearrangement of, **1**, 812
[3,3] rearrangement of, **1**, 550
Cyclopropane, 1-ethoxy-1-(trimethylsiloxy)-, reactions, with
phosphorous tribromide, **4**, 50
Cyclopropane, hexamethoxy-, thermolysis of, **4**, 1080
Cyclopropane, 1-lithio-2,2,3,3-tetramethyl-, pyrolysis
of, **2**, 1005
Cyclopropane, metalla-, synthesis of, **2**, 960
Cyclopropane, methyl-, reactions, with acyl chlorides, **1**, 285
Cyclopropane, methylene-, rearrangement of, **1**, 801
Cyclopropane, perfluoro-, synthesis of, **4**, 7
Cyclopropane, phenyl-, peroxymercuration of, **2**, 98
Cyclopropane, 1-siloxy-2-vinyl-, cleavage of, **3**, 224
Cyclopropane, tetrabromo-, ring-opening of, **4**, 747
Cyclopropane, trimethylsilyl-, reactions, with trimethylsilyl
chlorosulfonate, **2**, 190
Cyclopropane, 1-trimethylsilyl-2-methyl-, acylation
of, **3**, 140
Cyclopropanecarboxamide, reactions, with oxalyl
chloride, **5**, 995
Cyclopropanecarboxylate, ethyl 2-methyl-1-isocyano-,
synthesis of, **3**, 707
Cyclopropanecarboxylates, synthesis of, **4**, 146
Cyclopropanecarboxylates, 1-isocyano-
from, ethyl isocyanoacetate, **3**, 704
synthesis of, **3**, 704
1-Cyclopropanecarboxylates, trimethylsilyl 2-isocyanato-,
synthesis of, **5**, 992
Cyclopropanecarboxylic acids, from,
2-halocyclobutanones, **1**, 393
Cyclopropane-1,2-dicarboxaldehyde, reactions, with
hydrazine, **3**, 455
Cyclopropane 1,2-dicarboxylic acid, saponification of, **5**, 30
1,2-Cyclopropanedicarboxylic anhydride, 1,2-dimethyl-,
reactions, with trimethylsilyl azide, **5**, 992
Cyclopropane diesters, synthesis of, **5**, 148
Cyclopropanes
cyclization to, **1**, 229
desilylation of, **1**, 284
from
bismuthonium ylides, **1**, 743
dibromodifluoromethane, **4**, 1083
1,3-diphenylthiopropanes, **1**, 240
γ-epoxy sulfones, **1**, 214
α-halosilanes, **1**, 282
phosphorus ylides, **1**, 268
sulfur dioxide, **1**, 232
sulfur ylides, **1**, 811
reactions
with acyl chlorides, **1**, 285
with platinum(II) salts, **2**, 584
rearrangement of, **1**, 801
ring-expansion of, **1**, 390
ring-opening of, **1**, 841; **3**, 65, 271; **4**, 742, 747
as starting materials
for cyclopropyl acylsilanes, **5**, 428
for dibromocarbene, **4**, 1083
for dichlorocarbene, **4**, 1083
for α,β-unsaturated aldehydes, **3**, 65, 66
for α,β-unsaturated ketones, **3**, 232, 235
synthesis of, **1**, 183, 231, 282, 298, 352, 740, 812,
4, 1082; **6**, 351
via radical cyclization, **1**, 340
thermolysis of, **3**, 65, 232
use in the substitution of silicon functions, **1**, 284
use in the synthesis of carbenes, **4**, 1083
use in the synthesis of radicals, **1**, 928
use in the synthesis of α,β-unsaturated ketones, **3**, 224
Cyclopropanes, 1-acyl-2-vinyl-, reactions, with
trialkylboranes, **1**, 275

Cyclopropanes, alkenylidene-, synthesis of, **1**, 988
Cyclopropanes, alkoxychlorofluoro-, synthesis of, **3**, 66
Cyclopropanes, alkynic, from, 1-alkynylcyclopropyls, **1**, 975
Cyclopropanes, alkynyl-, synthesis of, **1**, 549
Cyclopropanes, allylic, hydrogenolysis of, **1**, 72
Cyclopropanes, aryl-, photochemical irradiation of, **1**, 841
Cyclopropanes, arylvinyl-, rearrangement of, **1**, 808
Cyclopropanes, bis(*N*-nitrosourethane)-, decomposition of, **3**, 489
Cyclopropanes, bis(silyl), synthesis of, **4**, 620
Cyclopropanes, chloro-, reactions, with potassium *t*-butoxide, **1**, 571
Cyclopropanes, 1-chloro-1-fluoro-, as starting materials, for fluorodienes, **2**, 612
Cyclopropanes, 1,1-dialkoxy-
  reactions, with alcohols, **6**, 76
  as starting materials, for ortho esters, **6**, 76
Cyclopropanes, 1,1-dialkoxy-2,2-dichloro-, as starting materials, for alkynyl ortho esters, **6**, 76
Cyclopropanes, dibromo-, metallation of, **3**, 271
Cyclopropanes, *gem*-dibromo-
  reactions
    with alkyllithiums, **4**, 146
    with silver acetate, **4**, 57
Cyclopropanes, *gem*-dibromo methylene-, reactions, with butyllithium, **1**, 990
Cyclopropanes, dichloro-, synthesis of, **4**, 16
Cyclopropanes, *gem*-dichloro-, reactions, with trimethylsilyl chloride, **4**, 605
Cyclopropanes, 1,1-dihalo-, as starting materials, for carbenoids, **1**, 164
Cyclopropanes, *gem*-dihalo-
  dehalogenation of, **1**, 990
  reactions, with alkyllithium, **1**, 990
Cyclopropanes, 1,1-dihalo-2-siloxy-, synthesis of, **3**, 235
Cyclopropanes, divinyl-
  rearrangement of, **1**, 411
  stability of, **1**, 411
Cyclopropanes, 1,2-divinyl-, Cope rearrangements of, **1**, 811
Cyclopropanes, imino-, synthesis of, **3**, 406
Cyclopropanes, 1-lithio-1-trimethylsilyl-, reactions, with dichloromethyl methyl ether, **5**, 429
Cyclopropanes, metalla-
  reactions
    with alkenes, **2**, 984
    with alkynes, **2**, 984
    with carbonyls, **2**, 984
    with nitriles, **2**, 984
  synthesis of, **2**, 984
Cyclopropanes, methylene-
  as byproducts, in the synthesis of alkynes, **1**, 111
  reactions, with alkenes, **1**, 362
  synthesis of, **1**, 145, 500, 1144
Cyclopropanes, siloxy-
  reactions
    with aryl triflates, **1**, 483
    with stannic chloride, **1**, 663
  ring-opening of, **1**, 663
  as starting materials
    for α-methylene aldehydes, **1**, 663
    for α-methylene ketones, **1**, 663
Cyclopropanes, silyl-, desilylation of, **1**, 284
Cyclopropanes, 1-silylketo-2-carbonyl-
  from
    α-haloacylsilanes, **5**, 430
    sulfur ylides, **5**, 430
Cyclopropanes, silyloxy-, ring-opening of, **3**, 177
Cyclopropanes, spirovinyl-, thermal rearrangement of, **1**, 801
Cyclopropanes, trihalo-, ring-opening of, **4**, 742
Cyclopropanes, (trimethylsilylethynyl)-, from, arsonium ylides, **1**, 270
Cyclopropanes, vinyl-
  charge-accelerated rearrangement of, **1**, 803
  [1,5] hydrogen shifts of, **1**, 795
  metal-catalyzed rearrangements of, **1**, 803
  photochemical rearrangement of, **1**, 802
  reactions
    with methacrylate esters, **4**, 830
    with sulfonyl isocyanates, **5**, 344
  rearrangement of, **1**, 402, 801
  ring-expansion of, **1**, 402
  [3,5] sigmatropic rearrangement of, **1**, 840
  as starting materials, for cyclopentenes, **1**, 801
  synthesis of, **1**, 459, 460
    via radical cyclization, **1**, 340
  thermal rearrangement of, **1**, 801
  use in the synthesis of natural products, **1**, 804
trans-Cyclopropanes, synthesis of, **4**, 109
Cyclopropanesulfonic acid, synthesis of, **2**, 190
Cyclopropane systems, oxy-, ring-opening of, **3**, 224
Cyclopropanoic esters
  as insecticides, **5**, 145
  synthesis of, **5**, 145
Cyclopropanols, from, dithioesters, **4**, 255
Cyclopropanols, α-bromo-, synthesis of, theoretical, **4**, 42
Cyclopropanone
  as starting material
    for aminals, **4**, 406
    for hemiaminals, **4**, 299
  synthesis of, **4**, 162
Cyclopropanone aminal, use in the synthesis of acylhydrazines, **5**, 359
Cyclopropanone enolates, synthesis of, **5**, 438
Cyclopropanone hydrates, synthesis of, **4**, 162
Cyclopropanone ketals, cleavage of, **1**, 122
Cyclopropanones, as intermediates, in Favorskii rearrangements, **1**, 392
Cyclopropanones, alkylidene-, synthesis of, **1**, 983
Cyclopropene, 3,3-dimethyl-
  reactions
    with metal–metal double-bonded cobalt complexes, **4**, 689
    with rhodium complexes, **4**, 689
    with Tebbe reagent, **3**, 508
  ring-opening of, **4**, 710
Cyclopropene, diphenyl-, reactions, with tungsten complexes, **3**, 517
Cyclopropene, 3,3-diphenyl-, reactions, with ruthenium(II) complexes, **3**, 520
Cyclopropene, tetrachloro-, thermal ring-opening of, **1**, 1069
Cyclopropenecarboxylic acid, 1-amino-, synthesis of, **5**, 989
Cyclopropene-3-carboxylic acids, synthesis of, **1**, 549
Cyclopropene intermediates, zircona-, in the synthesis of zirconacyclopentadienes, **2**, 927
Cyclopropenes
  oxidation of, **6**, 124
  oxidative coupling of, **2**, 984
  reactions
    with benzenesulfenyl chloride, **4**, 775
    with difluorocarbene, **4**, 737
    with titanocenebis(trimethylphosphane), **3**, 509
    with zirconocene-alkene complexes, **3**, 509
  as starting materials
    for alkynylphosphine oxides, **2**, 1068
    for ketenimines, **3**, 593
  synthesis of, **1**, 549, 690
Cyclopropenes, acyl-, ring-opening of, **2**, 671
Cyclopropenes, alkylidene-, synthesis of, **1**, 550
Cyclopropenes, bicyclic, as intermediates, **1**, 550
Cyclopropenes, dichloro-, reactions, with pentacarbonylchromium complexes, **3**, 513
Cyclopropenes, 1,1-dimethoxy-, as starting materials, for unsaturated ortho esters, **6**, 76
Cyclopropenes, metalla-
  characteristics of, **2**, 952

reactions
with alkenes, **2**, 984
with alkynes, **2**, 984
with carbonyls, **2**, 984
with nitriles, **2**, 984
synthesis of, **2**, 927, 984
Cyclopropenes, methylene-, ring-opening of, **3**, 490
Cyclopropenes, trichloro-, as starting materials, for ketene aminals, **4**, 979
Cyclopropenes, trimethylsilyl-, oxidation of, **3**, 271
Cyclopropenes, vinyl-, as starting materials, for cyclopentadienes, **1**, 802
Cyclopropenes, zircona-, as intermediates, **2**, 987
Cyclopropenethione, diphenyl, as starting material, for pyridin-4-thiones, **3**, 357
Cyclopropenethiones, synthesis of, **3**, 356
Cyclopropenethiones, amino, synthesis of, **3**, 356
Cyclopropenium bromides, triaryl-
as starting materials
for trinuclear palladium complexes, **4**, 693
for trinuclear platinum complexes, **4**, 693
Cyclopropenium perchlorates, *syn*-triphenyl-, synthesis of, **1**, 909
Cyclopropenium salts
stability of, **1**, 908
synthesis of, **2**, 1008
Cyclopropenium salts, halo-, synthesis of, **2**, 1006
Cyclopropenium salts, selenyl-, synthesis of, **2**, 1008
Cyclopropenium salts, sulfenyl-, synthesis of, **2**, 1008
Cyclopropenium salts, telluryl-, synthesis of, **2**, 1008
Cyclopropenone, diphenyl-
reactions
with dodecacarbonyltriiron, **4**, 681
with triphenylphosphine, **3**, 543
as starting material, for ketene phosphoranes, **3**, 540
Cyclopropenone, phenylhydroxy-, flash photolysis of, **2**, 1022
Cyclopropenone acetals, reactions, with cuprates, **2**, 589
Cyclopropenones
reactions, with tetrakis(triphenylphosphine)platinum, **5**, 465
as starting materials, for phosphatrifulvenes, **5**, 394
synthesis of, **1**, 550
Cyclopropenones, diamino-, as starting materials, for ynediamines, **2**, 1050
Cyclopropenones, phenylamino-, flash photolysis of, **2**, 1040
Cyclopropenyl anions, compounds of, **1**, 865
Cyclopropenyl cations, aromatic stability of, **1**, 907
Cyclopropenyl cations, tri-*t*-butyl-, stability of, **1**, 909
Cyclopropenylium cation, tris-1-(5-isopropyl-3,8-dimethylazulenyl)-, stabilization of, **1**, 916
Cyclopropenylium cations, reactions, with lithium salts, **5**, 874
Cyclopropenylium ions, reactions, with sodium azide, **4**, 483
Cyclopropenyl salts, synthesis of, **1**, 907
Cyclopropyl anions, reactions, with carbonyls, **1**, 284
Cyclopropyl anions, isocyano-, compared, with cyanocyclopropyl carbanions, **1**, 155
Cyclopropyl anions, triphenyl-, synthesis of, **1**, 865
Cyclopropyl carbanions, stability of, **1**, 846
Cyclopropyl carbanions, cyano-, compared, with isocyanocyclopropyl anions, **1**, 155
Cyclopropyl carbenes, synthesis of, **5**, 937
Cyclopropyl carbinols
rearrangement of, **2**, 11
as starting materials, for homoallylic halides, **2**, 11
Cyclopropylcarbinols, dichloro-, hydrolysis of, **3**, 224
Cyclopropylcarbinyl cations, synthesis of, **1**, 900
Cyclopropylcarbinyls, use in the synthesis of radicals, **1**, 928
Cyclopropyl-α-cyano sulfoxones, synthesis of, **3**, 635
Cyclopropyl groups, for the stabilization of carbocations, **1**, 894
Cyclopropylide, synthesis of, **3**, 494
Cyclopropyllithiums, 1-silyl-, synthesis of, **4**, 652
Cyclopropylmethyl radical, reviews, **1**, 919
Cyclopropylmethyl systems, solvolysis of, **1**, 894
Cyclopropyl radicals, reactions, with iron(III) chloride, **1**, 929
Cyclopropyls, 1-alkynyl-, as starting materials, for alkynic cyclopropanes, **1**, 975
Cyclopropyl tosylates, 1-alkenyl-, reduction of, **2**, 309
Cycloreversion
for the synthesis of carbonyl ylides, **3**, 323
for the synthesis of thioaldehydes, **3**, 334
[2+2] Cycloreversion, for the synthesis of silenes, **3**, 501
[2+4] Cycloreversion
for the synthesis of boraethenes, **3**, 504
for the synthesis of germenes, **3**, 503
Cyclotetradeca-1,2-diene, from, cyclonona-1,2-diene, **1**, 459
Cyclotetramerization, of phosphaalkynes, **5**, 885
Cyclotetramers, of phosphaalkynes, synthesis of, **5**, 885
Cyclotetraphosphine, reactions, with tetramethyldiarsine, **2**, 508
Cyclotetraphosphines, thermal decomposition of, **2**, 461
Cyclotetrastibane, synthesis of, **6**, 169
Cyclotetrastibine, tetra-*t*-butyl-, synthesis of, **2**, 507
Cyclotrigermanes, photolysis of, **2**, 949
Cyclotrimerization, of alkynes, **1**, 530; **5**, 461
Cyclotriphosphine, triphenyl-, synthesis of, **2**, 865
Cyclotriphosphines, synthesis of, **4**, 114
Cyclotriphosphines, tri-*t*-butyl-, cleavage of, **2**, 472
Cyclotrisilane, hexa-*t*-butyl-, use in the synthesis of carbodiimides, **5**, 1085
Cyclotrisilane, hexakis(*t*-butyl)-, photolysis of, **3**, 581
Cyclotrisilanes, reactions, with phosphaalkynes, **5**, 888
Cyclotrisilathiane, hexamethyl-, reactions, with diphenylphosphinous chloride, **2**, 851
Cycloundecanecarboxylic acid, synthesis of, **5**, 40
3,5-Cyclovitamin D ring A synthon, synthesis of, **1**, 109
Cylindrocarpon radicicola, use in the synthesis of amines, **2**, 331
Cymantrene, dimethylaminomethyl- *See* Manganese, (dimethylaminomethyl)cyclopentadienyltricarbonyl-
Cymantrenes *See* Manganese, cyclopentadienyl-
Cymantrenes, synthesis of, **4**, 1042
Cymantrenes, hetero-, halide displacement of, **2**, 896
p-Cymene, as starting material, for *p*-toluic acid, **5**, 115
p-2-Cymenes, as catalysts, **2**, 53
Cysteine
protecting groups for the sulfur of cysteine, **4**, 218, 223
protection of, **6**, 480
reactions, with *N*-thiophthalimide, **2**, 217
silylation of, **2**, 262
Cysteine, (*S*)-*N*-benzoyl-, use in the synthesis of thioboranes, **2**, 269
L-Cysteine, oxidation of, **2**, 213
Cysteine disulfide, oxidation of, **2**, 232
Cysteine methyl ester, *N*-(*t*-butoxycarbonyl)-, carbamoylation of, **6**, 494
Cysteines, as starting materials, for cystines, **2**, 230
Cysteines, *S*-(*N*,*N*-dimethylcarbamoyl)-, synthesis of, **6**, 493
Cystine-^{35}S-hydrochloride, reduction of, **2**, 126
Cystines
cleavage of, **4**, 218
from, cysteines, **2**, 230
Cytochalasin, synthesis of, **1**, 821
Cytochrome P-450
reactions, with alkyliron(III) compounds, **2**, 578
synthesis of, **2**, 579

D

DAIB, use in asymmetric ketone alkylation, **2**, 71
Dakin reaction, definition of, **2**, 655
Dakin–West reaction
mechanism, **3**, 180
for the synthesis of enamides, **2**, 747
Dammaradienol, synthesis of, **1**, 294
Danishefsky's diene, use in the synthesis of α,β-unsaturated ketones, **3**, 221, 243
Daphnanes, synthesis of, **3**, 325
Darzens–Claisen reaction, for the synthesis of vitamin A, **3**, 62
Darzens condensation
of chloro sulfoxides, **3**, 167
for the synthesis of alkanoic acids, **5**, 46
use in the synthesis of epoxynitriles, **3**, 627
Darzens reaction
of sulfinimines, **3**, 437
for the synthesis of α,β-epoxy sulfones, **4**, 236
for the synthesis of α,β-epoxy sulfoxides, **4**, 233
for the synthesis of α-sulfenyl aldehydes, **3**, 36
use in the synthesis of α-bromoacylsilanes, **5**, 423
Davis chiral oxaziridine oxidation, for the synthesis of vinyl sulfoxides, **2**, 713
Davis reagents
comparisons, with 2,2-dimethyldioxirane, **2**, 344
for the oxidation of amines, **2**, 344
for the synthesis of amine *N*-oxides, **2**, 349
use in the synthesis of hydroxylamines, **2**, 342
Davy's reagent, as reagent, for the synthesis of thioketones, **3**, 353
Deacetylation, of tris(acetylthio)methane, **6**, 89
Deacylation, of ketene aminals, **4**, 979
Dealkylation
of alkylbenzenes, **1**, 56
of sulfides, **2**, 127
for the synthesis of amines, **2**, 302
Deamination, of diazonium salts, **1**, 1118
Deamination, aromatic, mechanisms, **1**, 36
Deamination, reductive
types of, **1**, 28
via loss of amines, **1**, 32
via loss of cyanide, **1**, 38
via loss of nitrogen gas, **1**, 35
via loss of nitro groups, **1**, 36
via loss of pyridines, **1**, 30
via loss of succinimide, **1**, 29
via loss of sulfonamide anion, **1**, 28
Dean–Stark apparatus, for the synthesis of acetals, **4**, 177
Deca-2,4-dienoate, methyl 3-methyl-, synthesis of, **1**, 735
Decalin, *cis*-9-amino-, synthesis of, **2**, 298
cis-Decalin, synthesis of, **1**, 93
Decalinols, synthesis of, **1**, 475
Decalins, synthesis of, **1**, 46, 78, 101, 814
1-Decalone, failure to react with triphenylphosphine and tetrabromomethane, **4**, 745
2-Decalone, reactions, with triphenylphosphine, **4**, 745
cis-1-Decalone, methylenation of, **1**, 664
Decalones
reactions, **1**, 115
reduction of, **2**, 696
Decane, 1-bromo-, reactions, with dimethyl sulfide, **1**, 140
Decane, 1-trimethylsilyl-, synthesis of, **1**, 459
Decanoic acid, α-hydroxy-, synthesis of, **5**, 63
Decanol, synthesis of, **1**, 89
Decarbonylation
of aldehydes, **5**, 459
for the synthesis of radicals, **1**, 928, 933
Decarbonylation, photochemical, for the synthesis of alkyne-bridged diruthenium complexes, **4**, 681
Decarboxylation
of carboxylic acids, **1**, 560
of β-lactones, **1**, 681, 683
for the synthesis of carbanions, **1**, 883
for the synthesis of radicals, **1**, 933
for the synthesis of α-sulfenyl ketones, **3**, 166
Decarboxylation, catalytic, for the synthesis of enones, **1**, 561
Decarboxylation, nitrosative, mechanism, **2**, 781
Decarboxylation, oxidative, for the synthesis of vinyl cations, **1**, 1121
Decarboxylation–dehalogenation, for the synthesis of alkenes, **1**, 568
Decarboxylation–dehydration, for the synthesis of alkenes, **1**, 565
trans-Dec-2-en-4,6-diynal, oxidation of, **5**, 102
trans-Dec-2-en-4,6-diynoic acid
from, *trans*-undec-3-en-5,7-diyn-2-one, **5**, 102
synthesis of, **5**, 102
1-Decene, 4-methyl-, synthesis of, **1**, 113, 114
3-Decene, (*E*)-4-ethyl-, synthesis of, **1**, 278
(*Z*)-5-Decene, synthesis of, via hydrocupration, **2**, 969
(*E*)-Dec-2-ene oxide, synthesis of, **1**, 269
Dechlorination, reductive, for the synthesis of alkenes, **1**, 569
Decomplexation, in the synthesis of alkylaluminums, **2**, 595
Decomposition, for the synthesis of alkylmercury compounds, **2**, 594
3-Decyn-1-ol, synthesis of, **1**, 1024
Dediazonization, of diazonium ions, **1**, 36
Defluorination, for the synthesis of alkenes, **1**, 570
Dehalogenation
methods, for the synthesis of carbonimidic difluorides, **6**, 603
selective, **1**, 5
for the synthesis of alkynes, **1**, 1069
for the synthesis of alkynyl halides, **2**, 1012
for the synthesis of allenes and cumulenes, **1**, 990
for the synthesis of trialkylaluminums, **2**, 595
Dehydrating agents
for the synthesis of carbodiimides, **5**, 1062
for the synthesis of enamines, **2**, 739
Dehydration
of alcohols, **1**, 692
in elimination reactions, **1**, 575
Dehydration, acid-catalyzed, of tertiary benzylic alcohols, **1**, 577
Dehydroabietate, methyl, saponification of, **5**, 31
Dehydroabietic acid, synthesis of, reviews, **5**, 24
Dehydroalumination, mechanism, **2**, 596
Dehydrobromination
of bromohydrins, **3**, 114
of α-bromo ortho esters, **6**, 80
in elimination reactions, **1**, 572
for the synthesis of phenols, **2**, 658
Dehydrobromination, thermal, for the synthesis of *N*-borylimidoyls, **5**, 863
Dehydrochlorination
in elimination reactions, **1**, 571
for the synthesis of phosphaalkenes, **6**, 683
Dehydrocobaltation, mechanism, **1**, 335
Dehydrofluorination
of 1,1-difluorothiols, **5**, 474
in elimination reactions, **1**, 570

of thiols, **5**, 484
of trifluoromethylamines, **6**, 602
Dehydrogenation
of acylhydrazines, reagents for, **5**, 366
of hydrazones, **3**, 463
of hydrocarbons, **1**, 554
of α-ketohydrazones, **3**, 469
palladium-catalyzed, of alkenes, **1**, 1063
Dehydrohalogenation
of α- and β-haloalkyl ethers, **2**, 677
of halohydrins, **3**, 114
as side reaction, in metal–halogen exchange, **1**, 111
of sulfonyl halides, **5**, 488, 496
for the synthesis of alkynes, **1**, 1063
for the synthesis of alkynyl chlorides, **2**, 1014
for the synthesis of alkynyl halides, **2**, 1012
for the synthesis of tantalum carbene complexes, **3**, 511
1,2-Dehydrohalogenation, for the synthesis of alkynes, **1**, 1063
Dehydroiodination, in elimination reactions, **1**, 574
Dehydrolinloöl, cyclization of, **1**, 525
Dehydrometallation, for the synthesis of alkynes, **1**, 1081
3,5-Dehydronoriceane
reactions, with rhodium tetracarbonyl dichloride, **5**, 460
synthesis of, **5**, 460
Délépine reaction
mechanism, **2**, 308
for the synthesis of primary amines, **2**, 302
for the synthesis of 3-pyrrolines, **2**, 325
Demercuration, of metal hydride, **1**, 66
Demercuration, decarboxylative, of (sulfonyl)ethenylmercury, **2**, 1035
(+)-4-Demethoxydaunomycinone, synthesis of, **1**, 433
Demethoxylation, of 1,2,3-trimethoxybenzene, **1**, 432
Demjanov rearrangement, mechanism, **1**, 383
(±)-Dendrobine, synthesis of, **2**, 330
Deoxyanisoin, synthesis of, **3**, 317
Deoxygenation, of allylic compounds, **1**, 940
Deoxymercuration, for the synthesis of alkenes, **1**, 668
11-Deoxy-prostaglandin E_2, synthesis of, **1**, 1093
3'-Deoxythymidine, 3'-isocyanato-, synthesis of, **5**, 975
Dephosphorylation, of 2-fluoroalkenyl phosphonates, **1**, 1079
Deplancheine, synthesis of, **1**, 149; **4**, 532
(*S*)-(-)-Deplancheine, synthesis of, **1**, 594
Deprotection
of acetals, **4**, 175
of acylhydrazines, **5**, 351
of *N*-alkoxyimides, **5**, 332
of hydroxamates, **5**, 326
in modified Gabriel syntheses, **2**, 299
for the synthesis of amines, **2**, 300, 304
Deprotonation
vs. desilylation, **1**, 422
reagents for, *n*-butyllithium, **6**, 127
for the synthesis of alkyllithiums, **2**, 554
for the synthesis of α-lithioborylalkanes, **4**, 664
for the synthesis of α-lithiosilylalkanes, **4**, 646
for the synthesis of 1-metallo-1-silylalkanes, **4**, 646
Deprotonation, asymmetric, for the synthesis of organolithiums, **1**, 138
Deprotonation, enantioselective, lithium amide bases for, **1**, 135
Deprotonation–alkylation, of dithiocarboxylic acids and esters, **4**, 842
Dequaternization, of quaternary ammonium salts, **2**, 303
Dermadin, synthesis of, **3**, 712
Desilylation
of bis(silyl)benzene, **2**, 943
for the cleavage of carbon–silicon bonds, **1**, 448
vs. deprotonation, **1**, 422
metal-mediated, **1**, 63
reagents for, **1**, 448
for the synthesis of anions, **1**, 399
for the synthesis of azaallyl anions, **1**, 372
for the synthesis of azomethine ylides, **1**, 370
via thermal decomposition, **1**, 63
of vinylsilanes, **1**, 450
Desilylchlorination, fluoride ion induced, **1**, 653
Desoxypodocarpic acid, saponification of, **5**, 31
Dess–Martin periodinane, use in the synthesis of allenones, **1**, 957
Dess–Martin reagent
for the oxidation of alcohols to ketones, **3**, 119
for the synthesis of aldehydes, **3**, 6
for the synthesis of α-sulfenyl ketones, **3**, 165
use in the synthesis of dienals, **3**, 64
Destabilization, steric, of ketones, **2**, 642
Destannylselenenylation, for the synthesis of alkenes, **1**, 594
Desulfurization
mechanism, **1**, 438
for the synthesis of crown ethers, **5**, 511
Desulfurization, reductive, for the reduction of aryl sulfides, **1**, 4260
Desymmetrization, in elimination reactions, **1**, 573, 580
Deuteration, of triaryl trithioorthoacetates, **6**, 89
Deuteriodethallation, of aryl thallium(III) compounds, **1**, 452
2-Deuteriopentose compounds, 2-deoxy-, synthesis of, **4**, 253
Deuterium
reactions, with buta-1,2-diene, **1**, 459
use in hydroboration reactions, **2**, 42
Deuterium chloride, reactions, with trimethylsilylstyrene, **1**, 450
Deuteroaldehydes, synthesis of, **3**, 15
Deuterothionoformate, *O*-cholesteryl, reactions, with aryllithium, **3**, 341
1-Dewar-1,3-azaphosphinines, synthesis of, **5**, 883
Dewar benzene, perfluoro, use in the synthesis of fluoro compounds, **6**, 250
Dewar benzenes, synthesis of, **1**, 522
Dewar hexafluorobenzene
reactions, with bis(fluorooxy)difluoromethane, **4**, 46
as starting material, for α-fluoro epoxides, **4**, 46
Dextrorphan, synthesis of, **4**, 532
Diacetamide, *N*-methoxy-, synthesis of, **5**, 332
Diacetate, acrolein, use in the synthesis of acetals, **4**, 190
Diacetate, iodobenzene-, reactions, with α-cyano acetic esters, **3**, 573
Diacetate, phenyl iodoso-
as co-oxidant, with bis(triphenylphosphine)ruthenium(II) chloride, **3**, 5
reactions, with thioamides, **5**, 735
Diacetate, pyridine-4-carbaldehyde, synthesis of, **4**, 206
Diacetates
pyrolysis of, **1**, 787
synthesis of, **4**, 206
Diacetates, geminal
from
aldehydes, **4**, 203
ketones, **4**, 204
gem-Diacetates, allylic, synthesis of, **4**, 206
Diacetic acids, naphthalene, synthesis of, **5**, 43
Diacetyl telluride, synthesis of, **5**, 244
Diacid chlorides, from, dicarboxylic acids, **5**, 11
Diacid diesters, synthesis of, **4**, 553
1,3-Diacids, decarboxylation of, **1**, 772
1,8-Diacids, synthesis of, **5**, 109
Diacyl compounds, synthesis of, **4**, 203
gem Diacyl compounds, synthesis of, via oxidation of aromatic methyl and methylene groups, **4**, 206
O,*O*-Diacyl compounds, synthesis of, **4**, 206
gem-Diacyloxy compounds, synthesis of, **4**, 203
Dialdehydes, reductive cyclization of, **2**, 81
1,3-Dialdehydes, synthesis of, **4**, 250

1,5-Dialdehydes, reactions, with ammonia, **2**, 330
Dialkali metal arylphosphides, hydrolysis of, for primary phosphines, **2**, 821
Dialkenyl compounds, reactions, with phenylphosphine, **2**, 430
Dialkoxymethyl, as masked hydrogen substituent for phosphorus, **5**, 401
Dialkylamino compounds, reactions, with *t*-butoxybis(dimethylamino)methane, **2**, 745
β-Dialkylamino groups, elimination of, **1**, 656
Dialkylation, mechanism, **1**, 145
α,*α*'-Dialkylation, of saturated heterocycles, **1**, 149
Dialkyl complexes, reactions, with carbon monoxide, **5**, 462
Dialkyl compounds, synthesis of, **2**, 569
Dialkynes, macrocyclic, in dinuclear nickel complexes, **4**, 692
Diallenes, synthesis of, **1**, 959, 969
Diallyl, 2,3 Wittig rearrangement of, **1**, 404
Dialuminum, hexamethyl-
  reactions, with dichloro(pentamethylcyclopentadienyl)rhodium dimers, **4**, 687
  as reagent, for the reduction of aldehydes and ketones, **4**, 699
Dialuminum alkyls, reactions, with mercury(II) chloride, **4**, 697
Dialuminums, carbon-bridged, synthesis of, **4**, 698
Dialuminum tetrabromide, methylene-, reactions, with butyllithium, **4**, 708
4,9-Diamantyl dications, synthesis of, **1**, 900
Diamide, *N*,*N*,*N*',*N*'-tetramethylthiophosphoric, as thionating reagent, **5**, 575
1,3-Diamide, 2-keto-, reactions, with phosphorus pentasulfide, **3**, 345
Diamides, dehydration of, **3**, 674
Diamides, acyl phosphonate, synthesis of, **5**, 406
Diamides, *α*-aminophosphonic, synthesis of, **4**, 475
Diamides, phosphonic
  from, alkyl(diamino)phosphines, **4**, 133
  reactions, with cyclohexanones, **1**, 645
Diamides, phosphonothioic, synthesis of, **6**, 668
Diamides, phosphonous, oxidation of, **4**, 133
Diamidines, silylation of, **5**, 777
Diamidinium iodides, synthesis of, **3**, 419
Diamidoximes, from, cyanogen, **5**, 833
Diamine, N^1,N^2-bis(benzylidene)phenylmethylene-, synthesis of, **4**, 437
Diamine, *N*,*N*-dimethylethylene-, use in the synthesis of nitroalkenes, **2**, 784
Diamine, *N*,*N*'-dimethylethylene-, reactions, with glyoxal, **4**, 301
Diamine, *N*,*N*'-diphenylmethylene-, reactions, with phosgene, **5**, 971
Diamine, *N*,*N*-diphenyl trimethylene-, reactions, with hydrobromic acid, **1**, 258
Diamine, ethylene-, reactions, with lithium, **2**, 420
Diamine, *m*-phenylene-, reactions, with carbon disulfide, **6**, 565
Diamine, *N*,*N*,*N*,*N*-tetraacetylethylene-, as starting material, for peracetic acid, **2**, 390
Diamine, tetramethylene- (TMEDA), use in the synthesis of alkenes, **1**, 756
Diamine complexes, ethylene-
  alkylation of, **1**, 1001
  reactions, with carbonyls, **1**, 1037
Diamines
  reactions
    with alkali metal cyanates, **6**, 508
    with aminosulfenyl chlorides, **2**, 361
    with bistrifluoromethylketene, **4**, 886
    with carbon monoxide, **6**, 510
    with carbon oxide sulfide, **6**, 509
    with dialkyl carbonates, **6**, 507
    with dibenzoyltellurium diiodide, **5**, 734
    with hexamethyldisilazane, **6**, 507
    with ketene dithioacetals, **4**, 970
    with ketones, **4**, 410
    with phosgene, **6**, 506
    with vinylidene dihalides, **4**, 970
  silylmethylation of, **4**, 507
  as starting materials
    for aminals, **4**, 407
    for *α*-amino acetals, **3**, 41
    for azoxy compounds, **2**, 404
    for *α*-stannio enamines, **4**, 1020
    for vinyl sulfoximines, **2**, 793
  synthesis of, **1**, 261; **6**, 258
  use in polyazamacrocycle synthesis, **1**, 261
Diamines, alkylene-, phosgenation of, **5**, 981
Diamines, aromatic, reactions, with thionyl chloride, **2**, 796
Diamines, bis(3,5-dimethyl-1-pyrazolylthioformyl)-, synthesis of, **6**, 578
Diamines, chiral, as starting materials, for imidazolidines, **4**, 408
Diamines, *N*-chloro-, photolysis of, **6**, 259
Diamines, *N*,*N*'-disilylated, phosgenation of, **5**, 969
Diamines, methylene-
  reactions, with halides, **2**, 1008
  as starting materials, for *α*-haloamines, **4**, 97
Diamines, *o*-phenylene-
  for the condensation of *O*-ethyl selenocarboxylates, **5**, 522
  synthesis of, **5**, 711
Diamines, primary, use in the synthesis of thioureas, **6**, 578
Diamines, C_2-symmetric chiral, use in asymmetric *cis*-dihydroxylation reactions, **2**, 50
Diamines, *o*-tolylene-, phosgenation of, **5**, 966
Diamines, vicinal, synthesis of, **2**, 307, 311
1,2-Diamines
  from, aziridines, **2**, 319
  reactions, with dichloroglyoxime, **3**, 429
  synthesis of, **3**, 416
1,3-Diamines, from, 4-amino-1-aza-1,3-dienes, **2**, 312
gem-Diamines
  reactions, with thiols, **4**, 249
  synthesis of, **6**, 259
gem-Diamines, N^1,N^2-dihydroxy-, synthesis of, **4**, 435
gem-Diamines, halogenated, synthesis of, **6**, 260
S-Diamines, reactions, with 3-methoxybenzaldehyde, **4**, 408
vic-Diamines, synthesis of, via lead–aluminum bimetallic redox system, **2**, 312
α,*ω*-Diamines, reactions, with carbon disulfide, **6**, 575
Diamino acids, synthesis of, **5**, 88
α,*α*-Diamino acids, synthesis of, **4**, 417
Diaminocarbene complexes, acyclic, synthesis of, **6**, 663
α,*ω*-Diamino components, use in the synthesis of phosphacycloheptanes, **4**, 456
Dianils, benzil
  reactions
    with 2,4-bis(phenylthio)-1,3-dithia-2,4-diphosphetane-2,4-disulfide, **5**, 731
    with 2,4-bis(phenylthio)-1,3,2,4-dithiadiphosphetane-2,4-dithione, **4**, 320
Dianion equivalents, formyl, as acyl radical precursors, for the synthesis of ketones, **3**, 136
Dianionic complexes, alkylation of, **2**, 572
Dianions
  alkylation of, **1**, 633
  chemistry of, reviews, **1**, 890
  double alkylation of, **1**, 134
  from, *β*-keto ester enolates, **1**, 165
  as synthons, **1**, 634
Dianions, aromatic, synthesis of, **1**, 890
C,*S*-Dianions
  from, phenylmethanethiol(benzylthiol), **1**, 140
  reactions, with 1,3-dibromopropane, **1**, 140
1,3-Diarsacyclobutanes
  from, cumulenes, **4**, 594

synthesis of, **4**, 594
Diarsadigalletanes, synthesis of, **2**, 895
1,1'-Diarsaferrocene, as starting material, for 2,2',5,5'-tetradeuterio-1,1'-arsaferrocene, **4**, 1039
1,1'-Diarsaferrocene, 2-acetyl-, synthesis of, **4**, 1039
1,1'-Diarsaferrocenes, synthesis of, **4**, 1039
Diarsane, synthesis of, **6**, 169
Diarsanes, tetraorganyl-, reactions, with diorganyl dichalcogenides, **2**, 293
Diarsa selenides, nucleophilic substitution reactions of, **2**, 889
Diarsa tellurides, nucleophilic substitution reactions of, **2**, 889
Diarsenes, synthesis of, **6**, 168
Diarsenes, tris(trimethylsilyl)methyl-substituted, synthesis of, **6**, 375
Diarsenides, tetraaryl-, as starting materials, for sulfur compounds, **2**, 890
1,3-Diarsetane, synthesis of, **3**, 609
1,3-Diarsetanes, dialkyl-, from, arsacumulenes, **2**, 878
Diarsine, dimethyldiphenyl-, synthesis of, **2**, 893
Diarsine, tetracyclohexyl-, from, lithium dicyclohexylarsenide, **2**, 506
Diarsine, tetraethyl-
reactions, with bisdiethylarsine oxide, **2**, 494
synthesis of, **2**, 506
Diarsine, tetrakistrifluoromethyl-
reactions, with dimethylphosphine, **2**, 508
as starting material, for bistrifluoromethylarsines, **2**, 482
synthesis of, **2**, 506
Diarsine, tetramethyl-
cleavage of, **2**, 491, 510
as component of Cadet's liquid, **2**, 494
from, dimethylarsine, **2**, 506
reactions
with chloroethene, **4**, 134
with cyclotetraphosphine, **2**, 508
with fluoroethene, **4**, 134
with hexafluoropropene, **6**, 60
with trifluoroethene, **4**, 134
synthesis of, **2**, 505
Diarsine, tetraphenyl-, metathesis of, **2**, 893
Diarsine oxides, from, aminoarsines, **2**, 493
Diarsine oxides, tetraalkyl-
from, methylenebis(methyldimethylaminoarsine), **2**, 493
synthesis of, **2**, 493
Diarsines
from
alkylarsines, **2**, 506
diorganometal(III) alkali metal salts, **2**, 893
polyarsines, **2**, 506
reduction of, **1**, 445
synthesis of, **2**, 505
Diarsines, tetraalkyl-
as starting materials, for thioarsenates, **2**, 261
synthesis of, **2**, 505
Diarsine sulfides, synthesis of, **2**, 499
Diarylamine
reactions
with *N*-(chlorosulfinyl)iminocarbonyl chloride, **5**, 1122
with *N*-phosphoryliminocarbonyl chloride, **5**, 1122
Diarylamines, reactions, with 2,4-dinitrobenzenesulfenyl chloride, **2**, 768
Diastereofacial selection, acyclic, in intermolecular radical reactions, **1**, 320
Diastereofacial selectivity, with chiral substrates, **1**, 122
Diastereomers
of ketenimines, **3**, 556, 564
of 1-phosphaallenes, **3**, 608
Diastereoselection, for the synthesis of cyanohydrins, **3**, 615
Diastereoselective reductions, of *N*-phosphinoylimines, **3**, 439
Diastereoselectivity
in the addition of Grignard reagents to nitrones, **2**, 345
in allylations of carbonyls, **2**, 74
definition of, **1**, 351
in thermal [2 + 2] additions, **1**, 355
Diazaadamantane, synthesis of, **4**, 408
Diazaallyl anions, synthesis of, **4**, 536
Diazaarsoles, use in the synthesis of imidoylarsenic compounds, **5**, 791
Diazabicycloalkanes, synthesis of, **1**, 369
2,3-Diazabicyclo[2.2.1]heptanes, ozonolysis of, **4**, 162
Diazabicyclo[4.3.0]nonanes, synthesis of, **4**, 414
1,5-Diazabicyclo[4.3.0]non-5-ene (dbn), reactions, with oxaziridines, **3**, 404
1,4-Diazabicyclo[2.2.2]octane bis(tetrafluoroborate), 1-(chloromethyl)-4-fluoro-
as fluorinating reagent, **2**, 336
use in the synthesis of α-fluoro ketones, **3**, 149
1,4-Diazabicyclo[2.2.2]octane (dabco)
reactions
with acyldiazines, **5**, 354
with diazoacetyl chloride, **3**, 489
with oxaziridines, **3**, 404
use in the synthesis of methylenephosphines, **6**, 688
1,4-Diazabicyclo[2.2.2]octane (dabco) salts, 1-alkyl-4-fluoro-, synthesis of, **2**, 336
1,5-Diazabicyclo[5.4.0]undec-5-ene (dbu)
reactions, with nitro carboxylates, **4**, 108
use in dehydrobromination reactions, **1**, 573
use in the elimination of alcohols, **1**, 579
use in the synthesis of carboxylic esters, **5**, 125
use in the synthesis of methylenephosphines, **6**, 688
use in the synthesis of *N*-silylamides, **5**, 374
2,7-Diazabiphenylene, ring-opening of, **5**, 1033
Diazaboracyclohexane, reactions, with phenyl isothiocyanate, **6**, 572
Diazaborolane, *B*-allyl-, as starting material, for homoallylic alcohols, **2**, 76
1,3,2-Diazaborolidines, 2-chloro-, reactions, with bistrimethyl carbodiimide, **5**, 1085
2,3-Diazabutadiene, reactions, with lithium enolates, **5**, 364
1,3-Diazabutadienes, Diels–Alder reaction of, **4**, 421
1,4-Diaza-1,3-butadienes, 1,4-diaryl-2-cyano-3-chloro-, synthesis of, **5**, 657
1,3-Diazacyclododecane, 2-oxo-, synthesis of, **6**, 510
Diazadiborane ring, synthesis of, **2**, 811
1,3-Diaza-2,4-diboranes, synthesis of, **2**, 811
Diazadienes
reactions, with cyclopentene, **3**, 453
synthesis of, **5**, 1068
See also Diimines
Diazadienes, dichloro-, use in the synthesis of carbodiimides, **5**, 1075
1,4-Diaza-1,3-dienes, hydrosilylation of, **2**, 813
1,5-Diaza-3,7-diphosphabicyclo[3.3.0]octanes, synthesis of, **4**, 482
1,5-Diaza-3,7-diphosphacyclooctanes, synthesis of, via heteroatom exchange, **4**, 457
Diazadiphosphacyclopentanes, synthesis of, **2**, 466
Diazadiphosphetidines, synthesis of, **4**, 478
1,4-Diaza-2,5-diphosphorinanes, synthesis of, **4**, 483
Diazadithiapentalenes, synthesis of, **3**, 376
1,3-Diaza-2-germacyclopentane, 1,3-dimethyl-2,2-diethyl-, reactions, with thiophosgene, **6**, 577
Diazaheterocycles, thermolysis of, **3**, 323
2,5-Diazahexa-2,4-diene, perfluoro-, reactions, with caesium fluoride, **6**, 262
2,5-Diaza-1,3,5-hexatrienes, synthesis of, **3**, 678
Diazanes, synthesis of, **6**, 259
1,5-Diazapenta-1,4-diene, perfluoro-, synthesis of, **5**, 808
2,4-Diazapenta-1,4-diene, hexafluoro-, isomerization of, **5**, 1080
Diazapentalenes
from, bisoximes, **3**, 378

synthesis of, **3**, 378
Diazaphosphaboretidines, synthesis of, **2**, 467
1,5-Diaza-3-phosphacycloheptanes, synthesis of, **4**, 456
1,2-Diaza-4-phosphacyclopentanes, synthesis of, **4**, 482
1,3,5-Diazaphosphanes, from, tetrakis(aminomethyl)phosphonium salts, **4**, 458
3,5-Diazaphosphanes, synthesis of, via heteroatom exchange, **4**, 457
Diazaphospholanes, five-membered cyclic, synthesis of, **2**, 860
2,4-Diazaphospholanes, synthesis of, **4**, 493
Diazaphospholes
reactions
with diazo compounds, **4**, 549
with nitrilimines, **6**, 156
as starting materials, for 1,3-azaphospholines, **5**, 785
synthesis of, **4**, 483
use in the synthesis of imidoylarsenic compounds, **5**, 791
1,2,3-Diazaphospholes, reactions, with diazo compounds, **4**, 558
1,2,4-Diazaphospholes
synthesis of, **5**, 880
via rearrangement reactions, **4**, 479
3*H*-Diazaphospholes, 4,5-dihydro-, synthesis of, **4**, 484
3*H*-1,2,4-Diazaphospholes, synthesis of, **4**, 479
1,3,4-Diazaphospholidin-2-one 4-oxides, synthesis of, **4**, 488
1,4,2-Diazaphospholines, synthesis of, **4**, 493
Diazaphosphoranes, dimethyloxy-, synthesis of, **2**, 476
Diazaphosphorinanes, lithio, synthesis of, **2**, 1003
1,4-Diaza-2,5-phosphorinanes, synthesis of, **4**, 486
1*H*-1,2,4-Diazarsole, reactions, with sulfur ylides, **4**, 502
1,2,3-Diazarsoles, 2,5-diphenyl-, reactions, with diarylnitrones, **4**, 502
1,2,4-Diazarsoles, reactions, with diphenylnitrones, **4**, 502
Diazaselenuranes, synthesis of, **2**, 369
1,2-Diaza-3-sila-5-cyclopentenes, lithiation of, **4**, 649
Diazene, *N*,*N*-bis(trifluoromethyl)-, synthesis of, **6**, 242
Diazene-2-oxides, 2-(tribromophenyl)-, reactions, with ethyl vinyl ether, **5**, 680
Diazene-*N*-oxides, as starting materials, for hydrazonyl halides, **5**, 680
Diazenes
from, hydroxylamine-*O*-sulfonic acid, **2**, 402
as intermediates, **1**, 635
synthesis of, **2**, 402
use in the synthesis of diyls, **1**, 359
Diazenes, *N*-acyl-*N'*-alkyl-, synthesis of, **5**, 366
Diazenes, acyl arylsulfonyl-, synthesis of, **5**, 209
Diazenes, propargylic, rearrangement of, **1**, 962
Diazenes, *N*-trifluoromethyl-, synthesis of, **6**, 242
1,1-Diazenes, allylic, rearrangement of, **2**, 770
Diazenium salts, as starting materials, for azo compounds, **2**, 403
Diazenium salts, aryl-, as starting materials, for arylstibonic acids, **2**, 884
Diazenyl radicals, as intermediates, **1**, 944
Diazepines, synthesis of, **3**, 455
1,3-Diazetidine-2,4-dione, as starting material, for *N*-chlorocarbonyls, **6**, 450
1,2-Diazetidinone, tetraaryl-, decomposition of, **5**, 1011
Diazides, quinone, synthesis of, **3**, 490
gem-Diazides
from, bromide azide, **4**, 433
synthesis of, **4**, 432
2,5-Diazidoquinones, as starting materials, for cyanoketenes, **3**, 539
Diazines, acyl-
alkylation of, **5**, 354
from, carboxylic acids, **5**, 366
reactions
with 1,4-diazabicyclo[2.2.2]octane, **5**, 354
with enamines, **5**, 354
with Grignard reagents, **5**, 354
with *β*-keto esters, **5**, 354
reduction of, **5**, 354
synthesis of, **5**, 366
via condensation of acylhydrazines, **5**, 367
via dehydrogenation of acylhydrazines, **5**, 366
via oxidation of aldehydehydrazones, **5**, 366
Diazines, bis(alkoxycarbonyl)-, reactions, with amines, **6**, 506
Diazines, *α*-haloalkenyl-, synthesis of, **4**, 799
Diazines, 1,1'-(hydroxyimidoyl)bis(2-aryl)-, synthesis of, **6**, 650
1,2-Diazines
as starting materials, for dihydro indoles, **1**, 523
synthesis of, **4**, 105
1,3-Diazinones, synthesis of, **4**, 421
1*H*-Diazirenes, as intermediates, in the synthesis of nitrileimides, **3**, 680
Diaziridine, dibenzoyl-, thermal rearrangement of, **5**, 365
Diaziridines
from, pentafluoroguanidine, **6**, 285
as starting materials
for hydrazines, **2**, 374, 377
for ketene aminals, **4**, 979
Diaziridines, 1,2-diacyl-, rearrangement of, **5**, 365
Diaziridines, N^1,N^2-diacyl-, homolytic fission of, **4**, 427
Diaziridines, fluorinated, synthesis of, **6**, 285
Diaziridines, perfluoro-, stability of, **6**, 261
Diaziridinone, *N*,*N'*-di-*t*-butyl-, use in the synthesis of acyl hydrazines, **5**, 359
Diaziridinones
as intermediates, in the synthesis of hydrazines, **2**, 374
ring-opening of, **5**, 359
as starting materials, for hydrazines, **2**, 381
Diazirine, bromo-, reactions, with fluoride anions, **5**, 866
Diazirine, chlorofluoro-, from, dichlorobis(tetrafluorodiamino)methane, **6**, 261
Diazirine, 3-chloro-3-phenyl-, thermolysis of, **4**, 138
Diazirine, difluoro-
from, tetrafluoroformamidine, **6**, 261
as starting material, for difluorodinitromethane, **6**, 261
synthesis of, **6**, 261
vapor phase pyrolysis of, **6**, 604
Diazirine, (difluoroamino)fluoro-, production of, **6**, 285
Diazirine, fluoro-, synthesis of, **5**, 866
Diazirine, perfluoroamino-, thermolysis of, **6**, 616
Diazirine, phenylchloro-, thermal decomposition of, **4**, 154
Diazirines
from
amidines, **6**, 280
fluoroimines, **6**, 280
isoureas, **6**, 280
tris(difluoroamino)fluoromethane, **6**, 285
reactions, with chromium pentacarbonyl, **5**, 943
as starting materials, for halocarbenes, **6**, 280
thermolysis of, **6**, 614
Diazirines, bromoalkoxy-, as starting materials, for fluoroalkoxydiazirines, **6**, 280
Diazirines, chloro-, as starting materials, for vinyl chlorides, **2**, 619
Diazirines, chloroalkoxy-, as starting materials, for fluoroalkoxydiazirines, **6**, 280
Diazirines, fluoroalkoxy-, from, alkoxydiazirines, **6**, 280
Diazirines, halo-
stability of, **6**, 55
synthesis of, **5**, 866
Diazirines, phenyl halo-, use in the synthesis of acetals, **4**, 190
Diazo compounds
cycloaddition of, with thioketones, **1**, 685
from
N-acetylnitrosamides, **1**, 1077
aminoaziridinehydrazones, **1**, 1077
arylsulfonylhydrazones, **1**, 1077
carbonyls, **3**, 464

nitrosocarbamates, **5**, 867
N-nitrosoureas, **5**, 873
triphenylphosphoranylhydrazones, **1**, 685
germylation of, **5**, 874
irradiation of, **4**, 1082
lead compounds, **5**, 873, 874
mercury compounds, **5**, 871, 874
nitration of, **3**, 479
phosphorus-substituted, **6**, 636
photolysis of, **4**, 557; **5**, 925, 927
reactions
with allylic amines, **1**, 823
with benzeneselenenyl chloride, **4**, 90
with bromide, **4**, 90
with carbon monosulfide, **3**, 550
with diazaphospholes, **4**, 549
with 1,2,3-diazaphospholes, **4**, 558
with diazonium salts, **5**, 677
with 1,3-dienes, **5**, 145
with 3,5-diphenylphosphabenzene, **4**, 549
with Grignard reagents, **2**, 376
with iodonitroform, **4**, 995
with γ-ketonitriles, **3**, 453
with manganese complexes, **3**, 518
with mercury(I) oxide, **3**, 488
with phosphoranes, **3**, 445
with rhodium(II) complexes, **3**, 520
with selenium, **1**, 246; **3**, 391
with sulfinyl chlorides, **4**, 76
with sulfinyl halides, **4**, 75
with sulfur dioxide, **2**, 251
with triphenylantimony, **1**, 54
reduction of, **2**, 769
rhodium-catalyzed dimerization of, **1**, 717
silver compounds, **5**, 871, 874
stability of, **5**, 865
as starting materials
for alkanesulfonic esters, **2**, 207
for alkylidenes, **1**, 298
for alkynes, **1**, 1077
for bis(arenesulfonyl)methylenearsoranes, **6**, 708
for carbenes, **1**, 196; **4**, 1082
for *gem*-dinitroalkenes, **4**, 995
for germenes, **5**, 927
for hydrazines, **2**, 376
for isoselenocyanates, **5**, 1056
for ketocarbenoids, **1**, 352
for methylenes, **1**, 44
for silagermenes, **5**, 927
for silasilenes, **5**, 925
for silenes, **5**, 925
for thioketenes, **3**, 550
for 1,2,3-triazoles, **5**, 869
synthesis of, **3**, 461; **4**, 799; **5**, 866; **6**, 623
general methods, **3**, 461
reagents for, **5**, 867
thermolysis of, **5**, 925
tin compounds, **5**, 873, 874
uses of, **5**, 865
use in the synthesis of carbodiimides, **5**, 1087
Diazo compounds, aliphatic, as starting materials, for hydrazonoyl halides, **5**, 843
Diazo compounds, alkyl-, synthesis of, **3**, 462
Diazo compounds, antimony, synthesis of, **5**, 873
Diazo compounds, arsenic, synthesis of, **5**, 873
Diazo compounds, aryl-, synthesis of, **3**, 462
Diazo compounds, bismuth, synthesis of, **5**, 873
Diazo compounds, bis(thioxophosphoranyl)-, synthesis of, **6**, 667
Diazo compounds, germanium, synthesis of, **5**, 873
Diazo compounds, halo-, displacement of, **5**, 866
Diazo compounds, hydroxy-, synthesis of, **2**, 1066
Diazo compounds, α-lithio-, reactions, with palladium(II) complexes, **3**, 487
Diazo compounds, metallated, alkylation of, **3**, 474
Diazo compounds, nitro-
decomposition of, **5**, 326
synthesis of, **5**, 868
Diazo compounds, phosphinyl-, photolysis of, **4**, 558
Diazo compounds, silyl-
meta-chloroperbenzoic acid oxidation of, **5**, 413
photosensitized oxygenation of, **5**, 413
Diazo compounds, trialkylstannyl-, as starting materials, for nitrile imides, **3**, 682
N-Diazo compounds, synthesis of, **2**, 470
α-Diazo compounds
as starting materials, for α-diazophosphorus compounds, **5**, 870
synthesis of, **5**, 865
Diazo functions, with an attached heteroatom, synthesis of, **5**, **865–874**
Diazo groups, transfer of, **2**, 802
Diazo group transfer
deformylative, **3**, 479
mechanism, **3**, 472
mechanism, **3**, 467
for the synthesis of α-diazo phosphoryl compounds, **3**, 480
for the synthesis of α-diazo sulfinyls, **3**, 479
to carbonyls, **3**, 471
to ynamines, **3**, 477
2-Diazo-1,3-indanediones, protonation of, **2**, 799
Diazole, arsena-, reactions, with diphenylnitrileimine, **6**, 161
Diazoles, vinylidene-, as starting materials, for ketene aminals, **4**, 979
1,2-Diazoles, synthesis of, **4**, 105
1,3-Diazolidine, 2,2-dibromodimethyl-, synthesis of, **6**, 262
1,3-Diazolidine, 2,2-dichlorodimethyl-, synthesis of, **6**, 262
1,3-Diazolidine, perfluoro-, synthesis of, **6**, 261
1,3-Diazolines, fluorinated, synthesis of, **6**, 262
Diazomethyls
alkylation of, **3**, 474
lead-substituted, synthesis of, **3**, 484
reactions
with aldehydes, **3**, 474
with mercury, **3**, 488
stannylation of, **3**, 484
α-Diazomethyls, carbamoylation of, **3**, 482
Diazonabicyclo[2.2.2]octane bis(tetrafluoroborate), 1-chloromethyl-4-fluoro-, reactions, with vinylstannanes, **4**, 736
Diazonium chlorides, arene-
reactions
with carbon disulfide, **6**, 536
with sulfur dioxide, **2**, 717
Diazonium compounds, synthesis of, **4**, 799; **6**, 660
Diazonium compounds, arene-
reactions, with cupric chloride, **2**, 798
as starting materials, for diazonium salts, **2**, 803
Diazonium compounds, vinyl-, as intermediates, in the synthesis of phosphine oxides, **2**, 1067
Diazonium halides
reactions
with 2-chloroacetic acid esters, **5**, 676
with chloroacetone, **5**, 676
with 2-chloro-1-(2-furyl)-1,3-butanedione, **5**, 676
with chloromalonic acid, **5**, 676
with chloromalonic acid esters, **5**, 676
with chloromalonoaldehyde, **5**, 676
with 1-chloro-1-(4-nitrophenyl)acetone, **5**, 676
with 3-chloro-2,4-pentanedione, **5**, 676
with 2-chloro-1-phenyl-1,3-butanedione, **5**, 676
with 1-chloro-1-(4-toluenesulfonyl)acetone, **5**, 676
with dimethyl-[2-oxo-2-thienyl-(2)]ethyl sulfonium bromide, **5**, 676
as starting materials, for hydrazonyl halides, **5**, 676
Diazonium ions
dediazonization of, **1**, 36

as intermediates, in the synthesis of diazo compounds, **3**, 465
reactions, with formazans, **5**, 857
Diazonium ions, alkene-
from, diaryl triazenes, **2**, 801
as intermediates, **2**, 800
Diazonium ions, aminoarene-, synthesis of, **2**, 803
Diazonium ions, arene-, reactions, with ketene aminals, **3**, 260
Diazonium ions, α-chloroalkenyl-, synthesis of, **4**, 799
Diazonium ions, enol-, from, α-diazo monoketones, **2**, 799
Diazonium ions, β-oxygenated, synthesis of, **2**, 804
Diazonium nitrates, synthesis of, **2**, 806
Diazonium nitrate salts, synthesis of, **2**, 1056
Diazonium salts
deamination of, **1**, 1118
from
anthraquinone, **2**, 806
arenediazoniums, **2**, 803
arylhydrazines, **2**, 806
azofulvene, **2**, 800
α-diazocarbonyls, **2**, 799
hydroxylamines, **2**, 805
isocyanates, **2**, 800
nitro compounds, **2**, 805
nitroso acylamines, **2**, 805
nitroso compounds, **2**, 805
oximes, **2**, 805
sulfinylamines, **2**, 800
sulfonylhydrazones, **2**, 800
in Heck reactions, **1**, 494
in situ synthesis of, **1**, 486
as intermediates, in the synthesis of phenols, **2**, 647
as oxidizing agent, **5**, 1097
reactions
with acylaminomalonates, **4**, 448
with acylhydrazines, **5**, 365
with amide anions, **2**, 776
with amides, **5**, 365
with aryl tellurates, **2**, 731
with 2-bromo-1,3-diketones, **5**, 843
with 2-bromo-β-keto esters, **5**, 843
with 2-chloro-1,3-diketones, **5**, 843
with 2-chloro-β-keto esters, **5**, 843
with diazo compounds, **5**, 677
with formamide, **1**, 442
with heterocyclic pyruvic esters, **5**, 858
with hydrazines, **2**, 777
with hydrazones, **4**, 436
with hydroxylamines, **2**, 776
with malonic acid, **5**, 858
with methylenes, **5**, 857
with methylthio(thiocarbonyl)-methylenetriphenylphosphorane, **5**, 906
with phosphinyl aldehydes, **6**, 659
with primary amines, **2**, 777
with sodium hydrogen selenide, **2**, 734
with stannous chloride, **2**, 769
reduction of, **1**, 442
as starting materials
for acyl tetrazines, **5**, 365
for arylazo-functionalized ylides, **5**, 904
for aryl thiocyanates, **5**, 1108
for *gem*-diazoalkenes, **4**, 992
for hydrazones, **3**, 448
for organometallics, **2**, 876
synthesis of, **2**, 799, 1056
Diazonium salts, alkenyl(vinyl)-, synthesis of, **2**, 799
Diazonium salts, arene-
copper-catalyzed coupling of, **1**, 486
decomposition of, **2**, 305
from, aryl sulfinylamines, **2**, 800
reactions
with acylaminomalonates, **5**, 854
with alkynes, **1**, 1033
with alkynyl metals, **1**, 1033
with arylhydrazones, **5**, 856
with 2-chloroacetoacetanilide, **5**, 843
with chloroacetone, **5**, 843
with ω-chloroacetophenone, **5**, 843
with 3-chloroacetylacetone, **5**, 843
with chloromalondialdehyde, **5**, 843
with chloromalonic acid, **5**, 843
with chloromalonic dialdehyde, **5**, 843
with diazo esters, **5**, 843
with diazo ketones, **5**, 843
with ditellurides, **2**, 731
with enamines, **3**, 449
with ethyl 2-chloroacetate, **5**, 843
with methylamine, **4**, 444
with monomethyl cyanamides, **5**, 1118
with nitroalkanes, **3**, 449; **4**, 439
with nitroalkyls, **6**, 653
with phosphorus trichloride, **2**, 844
with silver alkynides, **1**, 1137; **2**, 1054
with silver phenylalkynide, **2**, 1054
with sulfide, **2**, 711
with sulfur dioxide, **2**, 719
with tetramethyltin, **3**, 290
with thiolates, **2**, 711
with thiols, **4**, 225
with vinylsilanes, **1**, 490
as starting materials
for arylarsonic acids, **2**, 884
for aryl bromides, **2**, 627
for aryl chlorides, **2**, 621
for aryl fluorides, **2**, 614
for arylhydrazones, **3**, 448
for aryl iodides, **2**, 632
for aryl metals, **2**, 956
for secondary amines, **2**, 305
for triarylarsine oxides, **2**, 885
for the synthesis of aryl halides, **2**, 609
use in the synthesis of arenethiols, **2**, 707
Diazonium salts, aromatic
reactions, with amines, **2**, 776
synthesis of, **2**, 802
Diazonium salts, naphthyl-, use in the synthesis of formazans, **5**, 857
Diazonium salts, perfluoroarene-, defluorination of, **1**, 426
Diazonium salts, triazole-, reactions, with trinitromethane salts, **6**, 145
Diazonium sulfates, synthesis of, **2**, 804
α-Diazosilyl compounds, synthesis of, **5**, 873
Diazo systems, reduction of, **1**, 44
Diazotates, synthesis of, **3**, 465
Diazotates, arene-, oxidation of, **2**, 798
Diazotization
of alkyl carbamates, **6**, 423
of amines, **1**, 1032; **2**, 802
of α-aminocarbonyls, **3**, 468
of α-amino compounds, **5**, 868
of 1,3-dicarbonyls, **3**, 193
of primary amines, **1**, 258
of sulfonylated 5-aminoisoxazoles, **2**, 1035
Diazo transfer reaction
reagents for, **5**, 869
for the synthesis of α-diazophosphorus compounds, **5**, 868
for the synthesis of α-diazosilicons, **5**, 873
DIBAL-H *See* Aluminum hydride, diisobutyl- (DIBAL-H)
5*H*-Dibenz-[*b,f*]azepine, 10-bromo-5-methyl-10,11-dihydro-, use in the synthesis of carboxylic acids, **5**, 95
Dibenzenesulfonamide, *p*-benzoquinone-, as activating agent, for the synthesis of anhydrides, **5**, 191
Dibenzenethiostibines, synthesis of, **2**, 501
Dibenzoates, reactions, with samarium diiodide, **1**, 587

7H-Dibenzo[*c,g*]fluorene, acidity of, **1**, 866
Dibenzophosphepin, reactions, with sodium hydride, **1**, 445
5H,7H-Dibenzo[*b,g*][1,5]selenazocine, *N*-methyl-, reactions, with *t*-butyl hypochlorite, **2**, 369
Dibenzoselenophene oxide, reduction of, **1**, 437
Dibenzothiophene, desulfurization of, **1**, 436
Dibenzothiophene, 4-chloro-, synthesis of, **2**, 621
Dibenzotriazole, 1,1'-(thiocarbonyldioxy)-, reactions, with amines, **5**, 1030
Dibenzotropylium ions, synthesis of, **1**, 906
Dibenzyl azodicarboxylate, reactions, with 3,4-dihydro-*2H*-pyran, **6**, 626
Dibenzyl carbonate, synthesis of, **6**, 465
Dibenzyl sulfides, reactions, with sulfuryl chloride, **6**, 28
Dibismaferrocenes, synthesis of, **4**, 1041
Dibismaferrocenes, 2,2',5,5'-tetramethyl-, synthesis of, **4**, 1041
Dibismuthine, tetramethyl-, synthesis of, **2**, 505
Dibismuthine, tetraphenyl-, from, diphenylbismuth iodide, **2**, 893
Dibismuthines
  from
    diorganometal(III) alkali metal salts, **2**, 893
    trialkylbismuthines, **2**, 505
  oxidation of, **2**, 493
  as starting materials, for dibismuth selenides, **2**, 890
  synthesis of, **2**, 505; **4**, 596
  thermochromic behavior of, **2**, 893
Dibismuthines, tetraalkyl-
  reactions, with chalcogens, **2**, 499
  synthesis of, **2**, 505, 510
1,1-Diboraalkanes, synthesis of, **1**, 1101
1,3-Diboracyclopent-4-ene, synthesis of, **4**, 632
Diboradioxanes, synthesis of, **4**, 363
1,3-Diborafulvenes, 1,3-dihydro-, synthesis of, **2**, 903
Diboraheterocycles, synthesis of, **2**, 903
1,4-Diboranaphtalenes, 1,4-dihydro-, synthesis of, **2**, 903
Diborane
  reactions
    with alcohols, **2**, 109
    with alkenes, **2**, 40, 514
    with amines, **2**, 418
    with *trans*-β-bromostyrene, **4**, 142
    with enamines, **1**, 443
    with nitriles, **3**, 458
    with tetrahydrofuran, **2**, 111
    with trialkoxyboranes, **2**, 109
    with vinylboronates, **4**, 631
  for the reduction of hydrazones, **2**, 374, 377
  for the reduction of nitro compounds, **2**, 341
  for transmetallation reactions, **2**, 652
  use in the synthesis of amines, **2**, 314
  use in the synthesis of thioboranes, **2**, 269
Diborane, 1,1-diaryl-2,2-difluoro-, reactions, with borataalkynes, **6**, 714
Diborane, 1,2-di *t*-butyl-1,2-dichloro-, reactions, with 1,2-bis(trimethylstannyl)ethyne, **4**, 1063
Diborane, tetrachloro-
  reactions
    with (dichloro)vinylborane, **4**, 632
    with trimethyl(vinyl)silane, **4**, 644
Diboranes
  dimerization of, **6**, 713
  reactions
    with alkenes, **2**, 521
    with secondary methylarsine, **2**, 509
    with silylalkynes, **4**, 1054
    with stannylalkynes, **4**, 1054
Diboranes, chloro-, reactions, with 1,2-bis(trimethylsilyl)ethyne, **4**, 1048
Diboranes, geminal, as starting materials, for α-boryl anions, **1**, 163
1,1-Diboranes, from, alkynes, **2**, 41
Diborapiperazines, synthesis of, **4**, 522
1,3-Diborataallenes, synthesis of, **6**, 201
1,3-Diboretanes
  from
    dilithiums, **4**, 633
    germaethenes, **6**, 188
  synthesis of, **4**, 632; **6**, 188, 202
  thermal cycloreversion of, **6**, 713
1,3-Diborete
  reactions
    with lithium, **6**, 188
    with potassium, **6**, 188
1,3-Diboretes, synthesis of, **5**, 928
Diborinanes, 1,4-diacylated, synthesis of, **5**, 383
Diboriranes
  oxidation of, **5**, 928
  as starting materials, for borenes, **5**, 928
Diboroles, synthesis of, **4**, 635
Diboronate, tetrabromomethylene-, synthesis of, **4**, 632
Diboronate, tetrabutyl ethylidene-, synthesis of, **4**, 631
Diboronate, tetramethylmethylene-, synthesis of, **4**, 632
Diboronation, of alkynes, **4**, 1054
Diboron dichloride, di-*t*-butyl, reactions, with alkynic compounds, **2**, 903
Diboronic acids, synthesis of, **4**, 631
1,1-Diboronic esters, alkene-, synthesis of, **6**, 182
gem-Diboronic esters, synthesis of, **6**, 199
Diboron tetrachloride
  reactions
    with alkynes, **6**, 181
    with alkynic compounds, **2**, 903
1,2-Diboronylethenes, trimethylsilyl, synthesis of, **4**, 1057
gem-Diboryls, synthesis of, **6**, 181
Dibromides
  reactions
    with mercuric oxide, **2**, 497
    with silver oxide, **2**, 497
  as starting materials
    for 1,2-dithianes, **2**, 225
    for dithiols, **2**, 119
Dibromides, vicinal, as starting materials, for vinyl bromides, **2**, 622
1,2-Dibromides
  from
    alkenes, **1**, 568
    epoxides, **2**, 245
  as starting materials, for bromoalkenes, **1**, 574
  telluration of, **4**, 876
1,4-Dibromides, debromination of, **1**, 569
gem-Dibromides, condensation of, **1**, 679
α,ω-Dibromides, reactions, with hexamethylphosphoramide, **1**, 574
Dibromo compounds, synthesis of, **6**, 251, 254
Dibromo compounds, acyclic, synthesis of, **4**, 622
1,1-Dibromovinyl groups
  reactions, with palladium [1,2-bis(diphenylphosphino)ethane], **5**, 38
  as starting materials, for carbocylic acids, **5**, 38
Di-*t*-butylmethyl radical, rate of dimerization, **1**, 922
trans-Dicalin, 10-carboethoxy-1,1-dimethyl-, deesterification of, **5**, 39
Dicarbaboranes, 1,2-bis(hydroxymethyl)-, reactions, with bromine, **4**, 143
1,2-Dicarbadodecaboranes, vinyl-, reactions, with bromine, **4**, 143
Dicarbanions, synthesis of, via proton loss, **1**, 879
gem-Dicarbodiimides, synthesis of, **4**, 436
Dicarbonate, di-*t*-butyl
  reactions
    with benzenethiol, **6**, 477
    with *O*-benzylhydroxylamine, **2**, 351
    with butanethiol, **6**, 477
Dicarbonate, dimethyl *N*-methylimino-, from, ethyl *N*-methylcarbamate, **6**, 625

Dicarbonates, synthesis of, **6**, 460
Dicarbonates, dialkyl
  reactions, with hydrazine hydrate, **6**, 489
  use in the synthesis of carbazates, **6**, 489
Dicarbonates, imido-, use in modified Gabriel syntheses, **2**, 299
Dicarbonothioates, *O,O*-dialkyl imido-, synthesis of, **6**, 558
1,3-Dicarbonyl anions, palladium catalysis of, **1**, 230
Dicarbonyls
  reactions
    with carbanions, **4**, 166
    with carbonyl oxides, **3**, 326
    with dialkylzincs, **4**, 166
    with organometallics, **4**, 166
    with phosphonium salts, **3**, 254
  reduction of, **1**, 25; **4**, 165
  for the stabilization of carbanions, **1**, 853
  as starting materials, for bisdithioacetals, **4**, 245
  synthesis of, **6**, 329
1,2-Dicarbonyls, as starting materials, for acetals, **4**, 183
1,3-Dicarbonyls
  alkoxy methylenation of, **3**, 247
  alkylation of, **1**, 187
  diazotization of, **3**, 193
  halogenation of, **3**, 239
  monoacetals of, **4**, 192
  NMR analysis of, **1**, 870
  reactions
    with arsine oxides, **3**, 48, 196
    with 1,3-bis(trimethylsilyloxy)-1-methoxy-1,3-butadiene, **2**, 662
    with Corey–Kim reagents, **3**, 168
    with ketones, **2**, 662
  as starting material, for α,β-unsaturated ketones, **3**, 246
  as starting materials
    for amino enals, **3**, 75
    for 2,2-bis-(4-methoxybenzoyloxy) compounds, **4**, 206
    for enols, **1**, 870
    for α,β-unsaturated ketones, **3**, 253
  synthesis of, **1**, 246, 387; **2**, 695
  use in the displacement of alcohols, **1**, 179
1,3-Dicarbonyls, 2-acyloxy-, as starting materials, for 2-acyloxy-2-chloro-1,3-dicarbonyls, **4**, 55
1,3-Dicarbonyls, 2-acyloxy-2-chloro-
  from
    2-acyloxy-1,3-dicarbonyls, **4**, 55
    2-diazo-1,3-dicarbonyls, **4**, 55
1,3-Dicarbonyls, 2-diazo-
  reactions, with arsines, **3**, 196
  as starting materials
    for 2-acyloxy-2-chloro-1,3-dicarbonyls, **4**, 55
    for α-bromoalkyl esters, **4**, 57
  synthesis of, **3**, 193
1,3-Dicarbonyls, α-ethynyl-, synthesis of, **1**, 1145
1,3-Dicarbonyls, 2-halo-, as starting materials, for hydrazonoyl halides, **5**, 843
1,4-Dicarbonyls
  from, alkynylborates, **1**, 126
  synthesis of, **1**, 245, 309
    via addition of acyl anion equivalents, **1**, 308
1,5-Dicarbonyls, monoacetals of, **4**, 193
α-Dicarbonyls, reactions, with lead(IV) acetate, **3**, 470
β-Dicarbonyls
  alk-1-ynylation of, **2**, 1102
  as ambident carbon nucleophiles, **1**, 879
  amination of, **3**, 263
  diazo transfer to, **3**, 472
  α-fluorination of, **2**, 16
  reactions, with *N*-trimethylsilyliminotriphenylphosphorane, **2**, 745
  selenenylation of, **1**, 607
  as starting materials, for selenoureas, **6**, 598
  synthesis of, **1**, 11; **2**, 740; **3**, 468
β-Dicarbonyls, cyclic, reactions, with trimethyl orthoformate, **2**, 746
β-Dicarbonyls, α-diazo-
  as starting materials, for α-diazocarbonyls, **3**, 475
  synthesis of, **3**, 468
    via diazo group transfer, **3**, 472
Dicarboxamide, *N,N,N',N'*-tetramethylazo-, synthesis of, **6**, 506
7,12-Dicarboxamide, as intermediate, **5**, 108
Dicarboxamides, azo-, synthesis of, **6**, 506
Dicarboxylate, azo-, reactions, with aniline, **2**, 777
Dicarboxylate, dibenzyl azo-, reactions, with α-tetralones, **2**, 310
Dicarboxylate, di *t*-butyl imino-, use in modified Gabriel syntheses, **2**, 299
Dicarboxylate, diethyl azo- (DEAD)
  for the oxidation of thiols, **2**, 213
  reactions
    with alcohols, **2**, 299
    with thioamides, **5**, 735
    with thiols, **2**, 217
  use in the synthesis of alkyl halides, **2**, 6
  use in the synthesis of hydrazines, **2**, 375, 378
  use in the synthesis of thioimidic esters, **5**, 726
Dicarboxylate, diethyl ethyne-, reactions, with dienes, **1**, 521
Dicarboxylate, diethyl hydrazo-
  as starting material, for 1,2-dimethylhydrazine, **2**, 377
  use in the synthesis of hydrazines, **2**, 376
Dicarboxylate, diisopropyl azo- (DIAD)
  use in the synthesis of acylgermanes, **5**, 431
  use in the synthesis of alkyl halides, **2**, 6
Dicarboxylate, dimethyl acetylene- (DMAD)
  reactions
    with azomethine ylides, **1**, 524
    with 3,4-dicyanofuran, **1**, 1083
    with dienes, **1**, 521
    with dihydropyrans, **1**, 518
    with dihydropyridines, **1**, 518
    with dimethyl malonate, **1**, 537
    with enamines, **1**, 518
    with ethyl cyanoacetate, **1**, 537
  use in ene reactions, **1**, 525
Dicarboxylate, dimethyl furan-, synthesis of, **1**, 1083
Dicarboxylate, ethyne-, aminosilylation of, **2**, 920
Dicarboxylate, methyl *t*-butyl imino-, use in modified Gabriel syntheses, **2**, 299
1,1-Dicarboxylate, diethyl 2-chloro-2-cyclopropyl-, hydrolysis of, **5**, 55
Dicarboxylate esters, azo-, reactions, with trinitromethane, **6**, 364
Dicarboxylates, azo-
  reactions
    with silyl enol ethers, **3**, 181
    with triphenylphosphine, **2**, 87
Dicarboxylates, dialkyl azo-, reactions, with 2-aza-1,3-butadienes, **3**, 414
trans-Dicarboxylates, synthesis of, **2**, 45
Dicarboxylation, mechanism, **1**, 298
Dicarboxylic acid, ethyne-, reactions, with dienes, **1**, 521
Dicarboxylic acid, 2,3-naphthalene-, from, 2,3-dimethylnaphthalene, **5**, 115
Dicarboxylic acid, 2,3-pyrazine-, synthesis of, **5**, 115
Dicarboxylic acid, pyrrolidine-, synthesis of, **1**, 831
4,4'-Dicarboxylic acid chloride, biphenyl-, synthesis of, **5**, 8
Dicarboxylic acid chlorides, reactions, with isocyanides, **3**, 417
Dicarboxylic acid dibromide fluoride, imino-, from, fluorocarbonyl isocyanate, **6**, 453
Dicarboxylic acid iodides, synthesis of, **5**, 19
Dicarboxylic acids
  for the acylation of amines, **5**, 305
  from, α,ω-diols, **5**, 46
  monooxyimidic amides of, synthesis of, **5**, 838

stability of, **5**, 25
as starting materials
for cyclic anhydrides, **5**, 199
for diacid chlorides, **5**, 11
Dicarboxylic acids, α,α'-imino-, synthesis of, **5**, 92
Dicarboxylic acids, monoester, from, unsaturated keto diesters, **5**, 52
Dicarboxylic acids, 2,6-naphthalene-, synthesis of, **5**, 109
trans-1,2-Dicarboxylic acids, reactivity of, **1**, 562
Dicarboxylic esters, azo-, as heterodienes, **6**, 626
Dications
of alkyne-bridged dimolybdenum complexes, synthesis of, **4**, 674
synthesis of, **2**, 1008
Dichalcogenides, acyl-substituted, synthesis of, **5**, 246
Dichalcogenides, bis(acyl), synthesis of, **5**, 251
Dichalcogenides, diorganyl, reactions, with tetraorganyl diphosphanes, **2**, 293
Dichalcogenides, monoacyl-substituted, synthesis of, **5**, 246
Dichlorides
reactions
with lithium, **6**, 398
with triethylamine, **3**, 526
reduction of, for the synthesis of borenes, **5**, 928
as starting materials
for boraborenes, **5**, 928
for borenes, **5**, 928
thermal rearrangement of, **5**, 928
Dichlorides, vicinal, from, alkenes, **2**, 24
1,2-Dichlorides, from, epoxides, **2**, 26
gem-Dichlorides
as starting materials
for diaryl thioketones, **3**, 363
for dithioacetals, **4**, 245, 249
for thioketones, **3**, 365
α,ω-Dichlorides, from, cyclic ethers, **2**, 26
Dichlorine heptoxide, reactions, with amines, **2**, 339
Dichlorine monoxide
as chlorinating agent, **6**, 28
for the synthesis of aryl chlorides, **2**, 619
Dichlorocerium anions, from, acetonitrile, **2**, 69
Dichloro complexes, bis(cyclopentadienyl)-, reactions, with 1,1'-dilithioferrocene, **4**, 720
Dichloro compounds
reactions, with potassium cyanate, **4**, 441
synthesis of, **6**, 251, 254
Di(chloromercurio) compounds, synthesis of, **4**, 1067
Dichloromethyl groups, from, trichloromethyl groups, **4**, 13
1,1-Dichlorovinyl groups, for the synthesis of carboxylic acids, **5**, 38
Dichromate, bis(benzyltriethylammonium)-, as oxidant, **3**, 4
Dichromate, bisphosphonium, as oxidant, **3**, 4
Dichromate, pyridinium
as oxidant, for the synthesis of ketones, **3**, 115
as reagent, for aldehydes, **3**, 4
Dichromate, tetrabutylammonium, as oxidizing agent, for aldehydes, **3**, 3
Dichromates, benzylpyridinium, as oxidants, for the synthesis of ketones, **3**, 116
Dichromium, tris(mesitylene)-, synthesis of, **4**, 694
Dichromium complexes, carbon-bridged, synthesis of, **4**, 672
Dichromium complexes, tetrakis(formamidinato)-, synthesis of, **5**, 781
Dicobalt, reactions, with perfluorocyclooctatetraene, **4**, 687
Dicobalt, alkynehexacarbonyl-, reactions, with metal carbonyls, **4**, 711
Dicobalt, dodecacarbonyl-, reactions, with alkynes, **4**, 687
Dicobalt, octacarbonyl-
reactions
with alkyl acetate, **3**, 20
with alkynes, **4**, 687
with cyclooctyne, **4**, 689
with enynes, **3**, 218
with perfluoro-1,3-cyclohexadiene, **4**, 687
as starting material, for hexafluoroisopropylidenedicobalt complexes, **4**, 685
synthesis of, **6**, 524
Dicobalt complexes, alkyne-bridged, synthesis of, **4**, 687
Dicobalt complexes, bridged, synthesis of, **4**, 690
Dicobalt complexes, carbon-bridged
from
diazoalkane, **4**, 685
diiodoalkanes, **4**, 686
synthesis of, **4**, 687
Dicobalt complexes, hexacarbonyl-
as alkyne protecting group, **4**, 688
of small and large-ring alkynes, **4**, 688
Dicobalt complexes, hexafluoroisopropylidene-
from, octacarbonyldicobalt, **4**, 685
synthesis of, **4**, 685
Dicobalt hexacarbonyl complexes
reactions, with nucleophiles, **1**, 198
as starting materials, for 1,4-diynes, **1**, 1025
Dicobalt octacarbonyl
reactions
with alkynes, **1**, 198
with carbon monoxide, **1**, 202
with tri(phenylethynyl)antimony, **2**, 1082
Dicobalt octacarbonyl complexes, reactions, with trihalomethanes, **6**, 401
Dictyopterenes, synthesis of, **1**, 811
Dicyanamide, reactions, with thionyl chloride, **5**, 810
Dicyanamides, synthesis of, **5**, 1118, 1125
Dicyanamides, alkyl-
reactions
with hydrogen sulfide, **5**, 1124
with sulfuric acid, **5**, 1124
Dicyanamides, *N*-alkyl-, as starting materials, for 1-cyano-3*H*-guanidines, **6**, 642
Dicyanates, synthesis of, **5**, 1102
Dicyandiamide, use in the synthesis of isothiocyanates, **5**, 1024
Dicyanides, dialkylsilyl, synthesis of, **5**, 1133
(Dicyano)iodonium triflate, reactions, with tin-substituted alkynes, **2**, 1021
Dicyclohexylarsine, chloro-, reactions, with chlorine, **2**, 492
Dicyclohexylcarbodiimide (dcc)
use in the synthesis of alkyl pyrophosphates, **2**, 108
use in the synthesis of sulfinamides, **2**, 797
1,3-Dicyclohexylisourea, 2-allyl-, reactions, with thioacetanilide, **5**, 726
endo-Dicyclopentadiene, reactions, with isothiocyanic acid, **5**, 1043
Dicyclopentadienyl, reactions, with iron pentacarbonyl, **6**, 524
Didecarboxylation, mechanism, **1**, 562
Didecarboxylation, electrochemical, examples, **1**, 563
Dideoxynucleosides, from, bis(xanthate) esters, **1**, 585
Dideuterio compounds, synthesis of, **4**, 618
Dieckmann condensation
for the synthesis of alkanoic acids, **5**, 46
for the synthesis of civetone, **1**, 674
Dieckmann cyclization
for annulation processes, **5**, 146
for the synthesis of α,β-unsaturated ketones, **3**, 222
Dieckmann–Michael reaction, for the synthesis of nitriles, **3**, 648
1,3-Dielectrophiles, reactions, with 1,3-diones, **3**, 246
Diels–Alder cyclization
of dienamides, **2**, 330
of 2-methyleneimidazolidines, **4**, 414
in the synthesis of arylsilanes, **2**, 944
for the synthesis of 2-silyl-α,β-unsaturated ketones, **3**, 271
Diels–Alder cycloaddition
of alkynylsilanes, **2**, 927
of allenes, **1**, 498

for the synthesis of alkenylsilanes, **2**, 938
for the synthesis of cyclohexanoates, **5**, 146
for the synthesis of phenols, **2**, 662, 664
Diels–Alder dimerization, for the synthesis of carboxylic acids, **5**, 93
Diels–Alder reaction
of acylnitroso compounds, **2**, 352
of 2-aza-1,3-butadienes, **3**, 414
diastereofacial selectivity in, **1**, 373
of 1,3-diazabutadienes, **4**, 421
of diselenoesters, **4**, 289
of ketene dithioacetal dioxides, **4**, 270
of ketene dithioacetals, **4**, 265
of ketene dithioacetal tetraoxides, **4**, 277
of ketene dithioacetal trioxides, **4**, 272
of maleic anhydride, **5**, 198
in nitrile synthesis, **3**, 621
of nitroalkenes, **2**, 396
regioselectivity of, **1**, 373
of selenothioester *S*,*S*-dioxides, **4**, 287
of silenes, **4**, 611
for the synthesis of alkanesulfonic esters, **2**, 206
for the synthesis of alkenes, **1**, 519
for the synthesis of α-aminoalkylphosphorus compounds, **4**, 489
for the synthesis of aza-1,3-dienes, **3**, 413
for the synthesis of bissilylated norbornenes, **4**, 619
for the synthesis of cyclic α-cyano sulfones, **3**, 635
for the synthesis of nitriles, **3**, 652
for the synthesis of six-membered rings, **1**, 372
for the synthesis of spiroaziridinium salts, **1**, 254
for the synthesis of sulfinamides, **2**, 366
for the synthesis of tetrahydrophosphorines, **4**, 459
for the synthesis of thioaldehydes, **3**, 331
for the synthesis of α-thiosulfoxides, **4**, 265
for the synthesis of α,β-unsaturated aldehydes, **3**, 61
for the synthesis of α,β-unsaturated ketones, **3**, 221, 243, 247
use in the synthesis of dithioacetals, **4**, 257
use in the synthesis of pyridines, **2**, 329
use in the synthesis of 1,4,5,6-tetrahydropyridazines, **3**, 453
Diels–Alder reaction, asymmetric intramolecular, mechanism, **1**, 375
Diels–Alder reaction, hetero-, use in the synthesis of arsenins, **2**, 879
Diels–Alder reaction, intermolecular, mechanism, **1**, 373
Diels–Alder reaction, intramolecular
of thioaldehydes, **3**, 331
uses of, **1**, 374
Diels–Alder reaction, inverse electron demand, with electron-rich dienophiles, **2**, 665
Diels–Alder reaction, retro-
for the synthesis of alkenes, **1**, 785
for the synthesis of alkynes, **1**, 1083
for the synthesis of α,β-unsaturated ketones, **3**, 233
Dienals
synthesis of, **3**, 64
via elimination reactions, **3**, 62
Dienals, δ-amino-, synthesis of, **3**, 66
2,4-Dienals, synthesis of, **1**, 1096
Dienamides
Diels–Alder cyclization of, **2**, 330
synthesis of, **5**, 991
use in the synthesis of amines, **2**, 330
Dienamines, synthesis of, **2**, 740
Dienamines, nitro
cycloadditions of, **2**, 789
as reagents, **2**, 789
synthesis of, **2**, 789
1,3-Diene, 1-chloro-, synthesis of, **2**, 617
1,4-Diene, 1-chloro-, synthesis of, **2**, 616
Diene magnesiation, for the synthesis of alkylmagnesiums, **2**, 561
Dienes
acylation of, **3**, 233
asymmetric dihydroxylation of, **2**, 49
[4+2] cycloaddition, **2**, 1020
[4+2] Diels–Alder additions of, **1**, 373
equilibrium with divinyl cycloalkanes, **1**, 811
formylation of, **3**, 64, 73
Friedel–Crafts acylation of, **3**, 230
from
allylic sulfoxides, **1**, 600
cyclononatriene, **1**, 421
1,5-dienes, **1**, 412
dienoyl triflates, **1**, 435
dithionosuccinates, **4**, 834
ethoxycarbonylallyl ylides, **1**, 739
α-pyrones, **1**, 789
vinylmercury compounds, **1**, 491
vinyl sulfide, **1**, 484
hydrometallation of, **2**, 568, 596
osmylation of, **2**, 47
ozonolysis of, **3**, 62, 215
protonation of, **1**, 911
reactions
with acetaldehyde, **2**, 140
with acid chlorides, **3**, 230
with *N*-acyl-*N*-sulfinylamines, **5**, 340
with alkylnitroso compounds, **2**, 354
with alkynes, **1**, 519
with arenediazonium tetrafluoroborates, **5**, 1044
with benzylic nitriles, **3**, 669
with but-3-yn-2-one, **1**, 519
with chlorophosphines, **2**, 453
with cyclopentadienylzirconium complexes, **5**, 442
with dialkenylchloroboranes, **1**, 1094
with dienophiles, **1**, 372
with diethyl acetylenedicarboxylate, **1**, 521
with dimethyl acetylenedicarboxylate, **1**, 521
with dithiophosphonic anhydrides, **2**, 256
with ethyne dicarboxylic acid, **1**, 521
with furans, **1**, 521
with iodine isocyanate, **5**, 985
with isocyanates, **5**, 386
with maleic anhydride, **5**, 199
with α-methylene-β-lactones, **1**, 566
with methyl phenylpropiolate, **1**, 522
with methyl propiolate, **1**, 521
with oxygen, **2**, 52
with perfluoro phosphaalkenes, **4**, 112
with peroxytrifluoroacetic acid, **4**, 46
with pyrroles, **1**, 521
with *N*-sulfinyl-*p*-toluenesulfonamide, **2**, 366
with sulfur dioxide, **2**, 201
with thexylborane, **2**, 518
with thionoethanal, **2**, 140
with tricarbonyliron, **1**, 291
with trifluoroperacetic acid, **4**, 46
with trimethylsilylphosphine, **2**, 428
with vinylcarbenes, **1**, 811
reactivity of, **1**, 74
synthesis of, **1**, 402, 488, 566, 569, 734, 741, 787
via alkoxide elimination, **1**, 579
via dimerization of vinylcopper salts, **1**, 491
via electrocyclic reactions, **1**, 418
via Hofmann elimination, **1**, 628
via Horner reactions, **1**, 754
via retro-cheletropic reactions, **1**, 790
use in the synthesis of disulfides, **2**, 221
Dienes, acyclic, rearrangement of, **1**, 413
Dienes, azido-, as starting materials, for 3-pyrrolines, **2**, 325
Dienes, bicyclic, synthesis of, **1**, 420
Dienes, bis(silyloxy)-, reactions, with alkynals, **1**, 520
Dienes, conjugated
reactions, with seleno radicals, **2**, 285

reduction of, **1**, 79
synthesis of, **1**, 644, 651, 700, 707
from sulfoxide pyrolysis, **1**, 788
Dienes, cyclic, reactions, with maleic anhydride, **5**, 199
Dienes, fluoro-, synthesis of, **2**, 612
Dienes, 3-hydroxy-, as starting materials, for ketones, **1**, 414
Dienes, 4-lithioalkoxy-, synthesis of, **3**, 63
Dienes, substituted, in Heck reactions, **1**, 494
Dienes, sulfinyl-, synthesis of, **2**, 713
Dienes, trichloroacetamido, synthesis of, **1**, 993
Dienes, tricyclic, synthesis of, **4**, 275
Dienes, zirconium-promoted, bicyclization of, **2**, 988
1,3-Dienes
hydroboration of, **2**, 43, 517
hydrometallation of, **2**, 579
oxidation of, **5**, 442
reactions
with calcium, **2**, 563
with catecholborane, **2**, 527
with diazo compounds, **5**, 145
with imino esters, **2**, 327
with magnesium metal, **2**, 561
as starting materials
for cyclobutenes, **1**, 418
for *β,γ*-unsaturated aldehydes, **5**, 442
synthesis of, **1**, 493, 531, 1095
via elimination of *β*-stannyl sulfides, **1**, 591
1,3-Dienes, *trans*-1-*N*-acylamino-, from, dienoic acids, **5**, 991
1,3-Dienes, 2-bromo-, solvolysis of, **1**, 1113
1,3-Dienes, 1-silyl-, synthesis of, **1**, 651
1,4-Dienes
as starting materials, for endo peroxides, **2**, 52
synthesis of, **1**, 123, 493, 524, 534, 1094, 1104
1,4-Dienes, 1-bromo-, from, allyl bromide, **2**, 624
1,4-Dienes, 1-iodo-1-silyl-, carbonylative cyclizations of, **2**, 931
1,5-Dienes
from, allylzinc halides, **1**, 127
interconversion of, **1**, 410, 806
as starting materials, for dienes, **1**, 412
synthesis of, **1**, 121, 157, 679
1,5-Dienes, 3-amino-, anion-accelerated Cope rearrangements of, **1**, 810
1,5-Dienes, hydroxy-, Cope rearrangement of, **1**, 809
1,6-Dienes, reactions, with zirconocenes, **2**, 568
cis,cis-Dienes, synthesis of, **1**, 1103
cis,trans-Dienes, synthesis of, **1**, 1103
(*E*)-Dienes, synthesis of, via elimination reactions, **1**, 616
(*E*)-1,3-Dienes, synthesis of, **4**, 737
(*E,E*)-Dienes, synthesis of, **1**, 490, 512; **3**, 228
trans,trans-Dienes, synthesis of, **1**, 1103
(*Z*)-1,4-Dienes, synthesis of, **4**, 809
1,5-Diene-3-ynes, synthesis of, **1**, 1007
Dienic esters, chloro, as starting materials, for phenols, **2**, 672
Dienoates, reactions, with lithiomethyl phosphonates, **3**, 229
α,β,γ,δ-Dienoates, reactions, with nucleophiles, **1**, 227
Dienoic acids
reduction of, **3**, 64
as starting materials, for *trans*-1-*N*-acylamino-1,3-dienes, **5**, 991
synthesis of, **1**, 476
Dienolates
s-cis vs. *s-trans*, structural effects, **1**, 877
reactions, with *N,N*-dimethyliminium chloride, **2**, 302
stability of, **1**, 878
Dienols
from, enynols, **1**, 513
synthesis of, **2**, 638
via elimination reactions, **1**, 643
Dienols, tertiary, oxidation of, **3**, 229
Dien-1-ols, oxidation of, **3**, 64
(*Z,Z*)-Dienols, synthesis of, **1**, 504
Dienone–phenol rearrangement, for the synthesis of phenols, **2**, 660
Dienones
reductive aromatization of, **2**, 660
seleno-substituted, synthesis of, **3**, 233
synthesis of, **1**, 1096; **3**, 227
via Wittig reaction, **3**, 231
Dienones, diazo, as starting materials, for resorcinol compounds, **2**, 668
Dienones, hydroxy, rearrangement of, **2**, 660
Dienones, steroidal, reactions, with trifluoroacetic anhydride, **2**, 660
Dienones, *α,β,α',β'*-unsaturated, synthesis of, **1**, 606
2,4-Dienones, synthesis of, **3**, 232
2,4-Dienones, 4-substituted, synthesis of, **3**, 232
α,β,γ,δ,-Dienones, synthesis of, **3**, 227
Dienophiles
alkynylsilanes as, **2**, 927
bis(iodonium)ethyne as, **2**, 1020
reactions
with dienes, **1**, 372
with heterodienes, **1**, 372
with hydroxypyran-2-ones, **2**, 665
with 1-methoxy-3-trimethylsilyloxy-1,3-butadiene, **2**, 665
with 1-thiocyanatobuta-1,3-dienes, **5**, 1041
Dienophiles, bisketene, from, benzocyclobutenediones, **3**, 539
Dienyl cations, synthesis of, **1**, 901
Dienylic cations
stability of, **1**, 912
synthesis of, **1**, 912
Dienynes
hydration-electrocyclization of, **1**, 421
as starting materials
for allyl vinyl ketones, **1**, 421
for divinyl ketones, **1**, 421
Dienynes, trimethylsilyl-, as starting materials, for 1,2,4,6-tetraenes, **1**, 963
1,2-Dien-4-ynes
from, 2-alkyl carbonates, **6**, 461
synthesis of, **1**, 976
1,2-Dien-4-yn-6-ols, synthesis of, **1**, 976
4,5-Dien-2-yn-1-ols, as starting materials, for cumulenes, **1**, 962
Diesters, acyl phosphonate, synthesis of, **5**, 404
Diesters, (chloromethyl)phosphonic, reactions, with aldimines, **4**, 473
Diesters, cumulene, from, bromoalkenylphosphonium bromides, **1**, 988
Diesters, *α*-diazo succinate, from, methyl ethynedicarboxylate, **3**, 471
Diesters, keto, as starting materials, for dithiolanes, **4**, 245
Diesters, 1,3-keto, synthesis of, **4**, 279
Diesters, *α*-ketophosphonic, as starting materials, for *α*-phosphoryl enamines, **4**, 1004
Diesters, unsaturated keto, as starting materials, for monoester dicarboxylic acids, **5**, 52
gem-Diesters, hydrolysis of, **5**, 30
β-Diesters, as radical precursors, **1**, 346
Diethanol, 2,2'-(ethylenedithio)-, use in the synthesis of alkenes, **1**, 503
Diether, resorcinol, nitrosation of, **2**, 781
1,3-Diethers, 2-hydroxy-, synthesis of, **2**, 53
gem-Diethols, synthesis of, **5**, 648
Diethylamine
reactions
with alkylthioalkynes, **4**, 905
with allenic sulfoxides, **3**, 171
with *N*-aryl *N*-thioacylthiocarbamates, **5**, 588
with *α*-bromo dithiocarboxylates, **4**, 843
with (1-chloro-2,2-dicyanovinyl)manganese compounds, **4**, 1019

with chlorotrifluoroethene, **6**, 54
with 1,1-cyano-*N*-fluoroformimidoyl fluoride, **5**, 817
with diethyl 2,4-dinitrobenzylidenemalonate, **4**, 306
with enamines, **2**, 308
with (*Z*)-fluoroimines, **5**, 761
with ketene acetals, **4**, 972
with 5-phenyl-1,2,4-dioxazolin-3-one, **6**, 109
with phenyliodinium ylides, **4**, 867
with 2,3,3,3-tetrafluorothiopropanoyl fluoride, **5**, 582
with tetrakis(*t*-butyl isocyanide)rhodium salt, **6**, 664
with thioacyl fluorides, **5**, 583
with trifluorothioacetyl chloride, **5**, 582
Diethylamine, *N*-hydroxymethyl-, reactions, with β-aminoalkylphosphines, **4**, 453
Diethylamine, *N*-trimethylsilyl-, reactions, with acetonitrile oxide, **5**, 765
Diethyne, bis(6-methyl-3-pyridyl)-, oxidation of, **5**, 116
Difluoride, reactions, with lithiophosphine, **6**, 167
Difluoro compounds, synthesis of, **6**, 250, 252
Difluoromethyl group, from, bromodifluoromethyl group, **4**, 5
1,1-Difluorothiols
dehydrofluorination of, **5**, 474
reactions
with potassium fluoride, **5**, 474
with secondary amines, **5**, 474
with sodium fluoride, **5**, 474
1,1-Difluorovinyl groups, as starting materials, for carboxylic acids, **5**, 38
Diformamides, as starting materials, for diisocyanides, **3**, 700
1,3-Digermacyclobutane
from
germenes, **4**, 627
trimethylsilyl(dichloromethyl)silane, **6**, 188
1,3-Digermacyclobutane, 1,1,3,3-tetramethyl-, synthesis of, **4**, 626
1,2-Digermacyclobut-3-ene, synthesis of, **2**, 947
1,4-Digermacyclohexadienes, synthesis of, **2**, 947
1,2-Digermacyclohex-4-enes, pyrolysis of, **1**, 656
Digermacyclopropanes
ring-expansion of, **4**, 629
synthesis of, **4**, 628
Digermanes, aryldialkyl-, cleavage of, **1**, 449
2,4-Digermapentane, 2,4-dimethyl-, reactions, with mercuracycles, **4**, 629
Digermathiacyclobutanes, synthesis of, **4**, 629
Digermene, tetrakis(2,6-diethylphenyl)-, reactions, with diazomethane, **4**, 628
Digermenes
addition to phenylethyne, **2**, 947
as starting materials, for 1,1-bisgermylalkanes, **4**, 627
synthesis of, **2**, 949
Diglyme
reactions
with 2-hydroxy-2-allyl-1-imines, **2**, 313
with 2-hydroxy-2-propargyl-1-imines, **2**, 313
as solvent, for the synthesis of thioketones, **3**, 344
use in the cleavage of carbon–boron bonds, **1**, 59
use in the reduction of alkyl halides, **1**, 4
use in the synthesis of dialkylmagnesiums, **2**, 559
Digold hexabromide, reactions, with Grignard reagents, **2**, 590
Dihafnium complexes, methylene-bridged, synthesis of, **4**, 670
Dihalide reduction, tellurium-mediated, **1**, 568
Dihalides
elimination of, **1**, 568
hydrolysis of, **5**, 26
reactions
with bis(sulfones), **4**, 275
with disulfones, **4**, 275
with sodium sulfide, **2**, 224
with zinc, **3**, 530
reduction of, for the synthesis of alkenes, **1**, 676
Dihalides, alkyl
reactions
with alkyldialkoxyphosphines, **2**, 453
with trialkylarsines, **4**, 135
Dihalides, dibromomethylene, synthesis of, **6**, 226
Dihalides, dichloromethylene, synthesis of, **6**, 224
Dihalides, difluoromethylene, synthesis of, **6**, 223
Dihalides, diiodomethylene, synthesis of, **6**, 227
Dihalides, vicinal, elimination of, for the synthesis of vinyl chlorides, **2**, 615
1,1-Dihalides, as starting materials, for 1-alkynes, **1**, 1065
1,2-Dihalides
from, alkenes, **1**, 568
as starting materials, for 1-alkynes, **1**, 1065
gem-Dihalides
condensation of, **1**, 676
reactions
with selenols, **4**, 288
with telluride anions, **4**, 290
as starting materials
for diselenoacetals, **4**, 288
for ditelluroacetals, **4**, 290
α,α-Dihalides, hydrolysis of, **5**, 410
α,ω-Dihalides, reactions, with methyl phenyl sulfone, **1**, 146
Dihalo compounds
reactions, with sodium azide, **4**, 432
as starting materials, for aminals, **4**, 411
Dihalo compounds, geminal
reactions, with amines, **4**, 410
reduction of, **2**, 14
as starting materials
for alkyl halides, **2**, 14
for aminals, **4**, 410
α,α-Dihalo compounds, reactions, with alkyllithiums, **1**, 164
Dihalo functions, synthesis of, **4**, 730
Dihalogens
elimination of, for vinyl halides, **2**, 606
reactions
with amines, **2**, 334
with enol ethers, **5**, 423
Dihalomethylenephosphines, kinetically-stabilized, **6**, 682
gem-Diheteroatomic compounds, cleavage of, **4**, 99
Dihydrazides, synthesis of, **5**, 849
Dihydrazidines
as byproducts, **5**, 848
from, primary amines, **5**, 854
1,2-Dihydrazones
oxidation of, **1**, 1075
reactions
with lead(IV) acetate, **1**, 1075
with mercury(II) oxide, **1**, 1075
with silver trifluoroacetate, **1**, 1075
as starting materials, for alkynes, **1**, 635
Dihydrogen phosphate, vinyl-, synthesis of, **2**, 693
Dihydrogen phosphates, monoalkyl
from, bis-(2,2,2-trichloroethyl)phosphochloridate, **2**, 105
synthesis of, **2**, 105
Dihydrogenphosphorane, tris(trifluoromethyl)-, synthesis of, **2**, 475
Dihydrogen trifluoride, tetrabutylammonium
as phase-transfer catalyst, **4**, 768
as reagent, for vinyl fluorides, **2**, 611
3,4-Dihydronapthalene, 1-bromo-, from, 1-tetralone, **2**, 624
Dihydropyran (DHP), cycloaddition of, **3**, 71
Dihydropyridazin-4-ones, *N*-cyano-, synthesis of, **5**, 1119
1,2-Dihydroquinaldamide, 1-benzoyl-, as starting material, for 1-benzoyl-1,2-dihydroquinaldinic acid, **5**, 95
1,2-Dihydroquinaldinic acid, 1-benzoyl-, from, 1-benzoyl-1,2-dihydroquinaldamide, **5**, 95
Dihydroxylation, mechanism, **2**, 44
cis-Dihydroxylation, asymmetric (ADH)

of alkenes, **2**, 48
mechanism, **2**, 50
mixes, availability of, **2**, 49
Diimidazole, carbonyl-
as coupling reagent, for the synthesis of sulfonamides, **5**, 342
as dehydrating agent, **5**, 206
in the dehydration of oximes, **3**, 617
reactions, with alkoxyamines, **5**, 327
as reagent, for the synthesis of amides, **5**, 277
use in the synthesis of *O*-acylated hydroxylamines, **5**, 211
Diimidazole, *N,N*-carbonyl-, as acylating agent, **5**, 196
Diimidazole, *N,N'*-carbonyl-
as coupling reagent, for the synthesis of hydroxylamines, **5**, 322
reactions
with amines, **5**, 972
with sulfonamides, **5**, 997
synthesis of, **5**, 972; **6**, 416
use as phosgene alternative, **6**, 416
uses of, as phosgene substitute, **5**, 972
use in the synthesis of symmetrical and unsymmetrical ureas, **6**, 416
Diimidazole, 1,1'-oxalyl-
as activating agent, for the synthesis of anhydrides, **5**, 189
reactions, with carboxylic acids, **5**, 196
Diimidazole, sulfonyl-, synthesis of, **1**, 578
Diimidazole, thiocarbonyl-
reactions, with alcohols, **6**, 543
as starting material, for thionocarbonates, **6**, 543
Diimidazole, 1,1'-thiocarbonyl-, reactions, with nitrones, **5**, 605
Diimidazole, *N,N'*-thiocarbonyl-
reactions
with alcohols, **6**, 557
with amines, **5**, 1029
as starting material, for thiocarbamic acid *O*-esters, **6**, 557
Diimidazolide, carbonyl- (CDI), use in the synthesis of carboxylic esters, **5**, 125
Diimidazolides, as starting materials, for imidazolides, **5**, 477
Diimide
for the reduction of alkenes, **1**, 74
for the reduction of alkynes, **1**, 458
for the reduction of allenes, **1**, 459
use in the synthesis of alkyl halides, **2**, 14
Diimide coupling reagents, for the synthesis of *N*-alkylamides, **5**, 279
Diimides, as coupling reagents, for the synthesis of amides, **5**, 265
Diimides, alkyl, as intermediates, in the synthesis of alkanes, **1**, 39
Diimides, *β*-chloroalkyl-, synthesis of, **5**, 1120
Diimides, water-soluble, as coupling reagents, for the synthesis of *N*-alkylamides, **5**, 279
S,S'-Diimides, from, sulfur monochloride, **5**, 336
Diimines
synthesis of, **3**, 417
See also Diazadienes
Diimines, silyl-, reactions, with hydrogen chloride, **4**, 107
1,2-Diimines, synthesis of, **3**, 417
1,3-Diimines
as starting materials, for *β*-amino ketones, **3**, 188
synthesis of, **3**, 417
1,4-Diimines, synthesis of, **3**, 418
Diindium, tetrachloro(*μ*-methylene)-, synthesis of, **4**, 698
Diiodides
as starting materials, for dichlorophosphines, **4**, 114
use in the synthesis of alkenes, **1**, 676
use in the synthesis of carbon-bridged diiron complexes, **4**, 679
1,3-Diiodides, reductive cyclization of, **5**, 146
gem-Diiodides, condensation of, **1**, 679
α,ω-Diiodides, synthesis of, **2**, 36
Diiodo compounds
from, *trans*-1,2-dinitrospiropentane, **4**, 108
synthesis of, **2**, 35; **6**, 254
Diiodonium salts, synthesis of, **2**, 1021
Diiridium complexes, dihydrido carbon-bridged
from
but-2-yne, **4**, 689
isopropyllithium, **4**, 689
Diiridium complexes, methylene-bridged, synthesis of, **4**, 685
Diiron, dicyclopentadienyltetracarbonyl-, reactions, with phospholes, **4**, 1034
Diiron, dodecacarbonyl-, use in the synthesis of carbon-bridged diiron complexes, **4**, 681
Diiron, nonacarbonyl-
reactions
with cobaltacyclopentadiene complexes, **4**, 714
with 1,4-dibromocyclooctatetraene, **4**, 681
with propargylruthenium complexes, **4**, 711
with tolan, **4**, 681
as starting material, for cobaltacyclopentadiene complexes, **4**, 689
use in the synthesis of carbon-bridged diiron complexes, **4**, 679, 681, 683
Diiron complexes, synthesis of, **4**, 679
Diiron complexes, bis-bridged, synthesis of, **4**, 681
Diiron complexes, carbon-bridged, synthesis of, **4**, 679, 683
Diiron complexes, cationic, synthesis of, **4**, 680
Diiron complexes, dichloro-substituted carbon-bridged, synthesis of, **4**, 685
Diiron complexes, methylene-, synthesis of, **4**, 679
Diiron complexes, mono-bridged, synthesis of, **4**, 681
Diiron dianion, octacarbonyl-, reactions, with diiodomethane, **4**, 679
Diiron nonacarbonyl, use in the synthesis of diiron complexes, **4**, 681
Diisocyanate, dibutyltin, synthesis of, **5**, 1016
Diisocyanate, hexamethylene, synthesis of, **5**, 994
Diisocyanate, pentamethylene, synthesis of, **5**, 971
Diisocyanate, sulfuryl, synthesis of, **5**, 1018
2,6-Diisocyanate, pyridine-, synthesis of, **5**, 990
Diisocyanates
as dehydrating reagents, **3**, 684
from
α,ω-diiodoperfluoroalkanes, **5**, 967
N-phenylcarbamates, **5**, 984
reactions, with diols, **6**, 488
synthesis of, **5**, 964
via Curtius rearrangement, **5**, 990
Diisocyanates, alkylene, industrial synthesis of, **5**, 1013
Diisocyanates, aryldisulfonyl, synthesis of, **5**, 981
Diisocyanates, diacyl, synthesis of, **5**, 995
Diisocyanates, ether, synthesis of, **5**, 990
Diisocyanates, phosphoryl halide, synthesis of, **5**, 1008
Diisocyanates, sulfido, synthesis of, **5**, 990
Diisocyanates, sulfonyl, synthesis of, **5**, 994
Diisocyanates, *o*-tolylene
stability of, **5**, 966
synthesis of, **5**, 966
gem-Diisocyanates
alcoholysis of, **4**, 447
reactions, with triphenylphosphine imides, **4**, 436
synthesis of, **4**, 433
o-Diisocyanates, synthesis of, **5**, 1013
Diisocyanato diesters, synthesis of, **5**, 1007
Diisocyanato esters, synthesis of, **5**, 1006, 1007
Diisocyanide, 1,4-butylene, synthesis of, **3**, 700
Diisocyanide, 1,2-ethylene, synthesis of, **3**, 700
Diisocyanide, 1,3-propylene, synthesis of, **3**, 700
Diisocyanides, from, diformamides, **3**, 700
gem-Diisocyanides, synthesis of, **4**, 437
Diisopropylamine
reactions
with phenylvinyl selenides, **4**, 964

with thioimidic esters, **5**, 757
with trifluoromethylphosphine, **5**, 1152
1-(Diisopropylamino)-3-borolenes, dimetallation of, **4**, 643
Di-*O*-isopropylidenes, as starting materials, for fructopyranose compounds, **4**, 222
Diisothiocyanate, carbonyl
chlorination of, **6**, 605
reactions, with nucleophiles, **6**, 565
Diisothiocyanates, alkyl, synthesis of, **5**, 1026
Diisothiocyanates, aromatic, synthesis of, **5**, 1026
Diisothiocyanates, polyethylene oxy-, synthesis of, **6**, 476
gem-Diisothiocyanates, synthesis of, **4**, 433, 434
Diketene
reactions
with alkyl Grignard reagents, **1**, 481
with benzaldehyde oxime, **5**, 217
ring-opening of, **2**, 701
Diketenes, reactions, with alkynylzinc chlorides, **1**, 1028
Diketones
p-acetoxybenzylation of, **3**, 145
acidity of, **1**, 853
amination of, **3**, 263
ozonolysis of, **3**, 326
photolysis of, **3**, 250
reactions
with hydrazones, **3**, 264
with hydroxylamines, **3**, 264
with nitromethane, **4**, 166
with phosphines, **4**, 339
with rhodium(I), **3**, 272
with trimethylsilyl cyanide, **3**, 667
as starting materials
for nonclassical thiophenes, **3**, 372
for thioketones, **3**, 343, 345
synthesis of, **1**, 141, 929; **5**, 462
1,2-Diketones
asymmetric hydrogenation of, **2**, 53
reactions
with allylic alcohols, **2**, 73
with cyanomethylenetriphenylphosphorane, **3**, 642
with diethoxyethenylidinetriphenylphosphorane, **6**, 75
with ylides, **3**, 220
as starting material, for cyclic ortho esters, **6**, 75
1,3-Diketones
fluorination of, **3**, 149
periodate oxidation of, **5**, 49
reactions
with amines, **2**, 744
with azides, **3**, 194
with dichloroaldehydes, **3**, 219
with nitrosyl cations, **3**, 260
with selenamidoacetonitriles, **3**, 656
with selenium dichlorides, **3**, 177
with selenoxides, **3**, 177
with telluroxides, **3**, 177
reduction of, **3**, 161
selenylation of, **3**, 174
1,3-Diketones, 2-azido-, synthesis of, **3**, 194
1,3-Diketones, 2-bromo-, reactions, with diazonium salts, **5**, 843
1,3-Diketones, 2-chloro-, reactions, with diazonium salts, **5**, 843
1,3-Diketones, 2-diazo-
reactions, with dialkyl hydrogen phosphites, **3**, 200
use in the synthesis of ketenimines, **3**, 586
1,3-Diketones, 2-hydroxy-, as starting materials, for α-chloro alcohols, **4**, 43
1,3-Diketones, monoprotected, synthesis of, **3**, 159
1,3-β-Diketones, as starting materials, for alkynones, **3**, 274
1,4-Diketones
from, imidoylcoppers, **5**, 802
reactions, with hydrazine, **3**, 453
synthesis of, **4**, 287
1,4-Diketones, nitromethyl-, cyclization of, **3**, 258
1,5-Diketones
cyclization of, **3**, 219
reactions, with hydrazine, **3**, 455
synthesis of, **1**, 254
1,5-Diketones, masked, synthesis of, **4**, 255
α-Diketones
reactions
with amines, **3**, 260
with hydrazines, **3**, 450
with tetrafluorohydrazine, **5**, 314
with tosylhydrazine, **3**, 446
as starting materials
for dibromo ketones, **4**, 27
for *N*,*N*-difluoroamides, **5**, 314
for osazones, **3**, 451
synthesis of, **3**, 470
α-Diketones, monotosylhydrazones, synthesis of, **3**, 446
β-Diketones
as radical precursors, **1**, 346
reactions
with benzoyl cyanide, **3**, 268
with bismuth dichloride, **3**, 196
with methanesulfonyl azide, **3**, 472
with sodium hydroxide, **1**, 852
with γ,δ-unsaturated β-keto esters, **3**, 453
as starting materials, for 1,2-dithioles, **3**, 371
β-Diketones, rhena-, reactions, with primary amines, **5**, 801
Dilead, hexaphenyl-
reactions
with hydrogen cyanide, **5**, 1145
with iodocyanogen, **5**, 1145
as starting material, for organolead cyanides, **5**, 1145
Dileads, carbon-bridged, synthesis of, **4**, 699
Dileads, hexaalkyl-, synthesis of, **2**, 602
C,*O*-Dilithiated compounds, synthesis of, **1**, 136
Dilithio compounds, configurational, stability of, **1**, 154
1,1-Dilithio compounds, synthesis of, **2**, 1005
C,*O*-Dilithio compounds, synthesis of, **1**, 154
Dilithioferrocene–tetramethylethylenediamine complexes, synthesis of, **4**, 1034
Dilithium, diphenylmethylene-, synthesis of, **4**, 668
Dilithium, methylene-
from
bis(mercury halides), **4**, 668
dibromomethane, **4**, 668
diiodomethane, **4**, 668
methyllithium, **4**, 668
structure of, **4**, 668
synthesis of, **4**, 668
Dilithium, phenylmethylene-, synthesis of, **4**, 668
Dilithium alkylarsenide, from, alkylarsines, **2**, 510
Dilithium alkylphosphides, reactions, with organogallium dichlorides, **2**, 475
Dilithium *t*-butylarsenide, reactions, with *t*-butyldichloroarsine, **2**, 507
Dilithium intermediates, stabilization of, **3**, 284
Dilithiums
reactions
with alkyl halides, **1**, 1001
with chlorophosphonates, **3**, 198
as starting materials, for 1,3-diboretanes, **4**, 633
synthesis of, **6**, 398
use in the synthesis of metalloids, **6**, 389
Dilithium salts
diboraallene structure of, **6**, 400
synthesis of, **6**, 400
Dilithium tetrabromocuprate, for the ring-opening of glycidyl tosylates, **2**, 84
Dilithium tetrachlorocuprate
as catalyst, **1**, 466
use in the substitution of alkyl halides, **1**, 465
Dilithium tetramethylferrate, synthesis of, **2**, 577

Dimagnesium dibromide, 1,4-butanediyl-, reactions, with alkyl phosphorodichloridates, **2**, 455
Dimagnesium salt, hydrazone, reactions, with selenium(I) chloride, **3**, 392
Dimanganese, decacarbonyl-
  reactions, with 2,5-dimethyl-1-phenylarsole, **4**, 1042
  for the synthesis of manganacyclopentadiene complexes, **4**, 677
  use in the synthesis of radicals, **1**, 937
Dimanganese complexes, carbon-bridged, synthesis of, **4**, 677
Dimanganese complexes, methylene-bridged, synthesis of, **4**, 676
Dimedone
  reactions, with phenyl isothiocyanate, **5**, 593
  as starting material, for thiodimedone, **3**, 347
  synthesis of, **1**, 11
Dimedones, 1,3-diketone, reactions, with benzenesulfenyl chloride, **4**, 70
Dimedonyliodone, phenyl-, reflux of, **1**, 11
Dimercuracycles, as starting materials, for carbosilanes, **4**, 624
Dimercuration, of alkynes, **4**, 1067
Dimercury complexes, carbon-bridged, synthesis of, **4**, 696
Dimeric, dimesityliron complexes, synthesis of, **4**, 685
Dimerization
  of alkenes, **1**, 294
  of carbonimidic difluorides, **6**, 602
  of substituted nitriles, **3**, 657
Dimerization, metathetic
  of cycloheptene, **1**, 674
  of cyclooctene, **1**, 674
Dimerization, oxidative, of ynamines, **2**, 1050
Dimerization, rhodium-catalyzed, of diazo compounds, **1**, 717
Dimerization, thermal, of acrylonitriles, **1**, 354
Dimesityl compounds, use in the synthesis of organoboranes, **2**, 525
1,1-Dimesityl compounds, synthesis of, **4**, 1048
Dimesylate synthons, reactions, with dilithiophenylphosphide, **2**, 438
Dimetal amides, groups 13 through 16, synthesis of, **5**, 391
Dimetal complexes, synthesis of, **4**, 719
1,1-Dimetal complexes, synthesis of, via transmetallation, **4**, 669
Dimetallated compounds, synthesis of, **1**, 146
α,α'-Dimetallated intermediates, reactions, with haloalkanes, **1**, 146
Dimetallic complexes, carbon-bridged, synthesis of, **4**, 686
Dimetallics
  reactions, with phenylsulfonyl chloride, **2**, 594
  transmetallation of, **4**, 656
Dimetals, reactions, with electrophiles, **4**, 809
Dimetals, carbon-bridged, from, diazoalkanes, **4**, 709
Dimethanes, *o*-quino-
  from, benzocyclobutenes, **1**, 418
  synthesis of, **1**, 569, 655
Dimethides, *o*-quinone, reactions, with propyne, **1**, 519
Dimethylamine
  alkylation of, **1**, 151
  electrocyclization of, **2**, 790
  reactions
    with fluoroimines, **5**, 762
    with hydrazonate esters, **5**, 773
    with ketene dithioacetals, **4**, 969
    with methyltetrafluorophosphorane, **2**, 477
Dimethylamine, *N*-iodo-, synthesis of, **2**, 339
Dimethylamine, perfluoro-*N*-bromo-, reactions, with perfluoro-1,1,3-tris(dimethylamino)propadiene, **4**, 794
Dimethylamine, trichloromethyl-, reactions, with dicyanoketene aminals, **4**, 987
Dimethylamine, trimethyltin-, reactions, with benzoyl *t*-butylcarbodiimide, **5**, 390
Dimethylamine hydrochloride, reactions, with primary alcohols, **2**, 309
Dimethylamine oxides, 2-phenylcyclohexyl-, use in the synthesis of alkenes, **1**, 624
Dimethyl chloride, thexyl-, for the silylation of amides, **5**, 375
Dimethylphenyl, reactions, with methyldiphenylphosphine, **2**, 1003
Dimethyl sulfoxide–iodine, for the oxidation of active methylenes, **4**, 161
Dimolybdenum, hexacarbonylbis(η^5-cyclopentadienyl)-
  reactions, with alkynes, **4**, 674
  use in the synthesis of alkyne-bridged dimolybdenum complexes, **4**, 674
Dimolybdenum, tris(cyclooctatetraene)-, synthesis of, **4**, 673
Dimolybdenum complexes, alkyne-bridged, α-carbocations of, **4**, 674
Dimolybdenum complexes, allene-bridged, synthesis of, **4**, 673
Dimolybdenum complexes, carbon-bridged, synthesis of, **4**, 672
Dimolybdenum complexes, phenylethyne–ethylene diamine, synthesis of, **4**, 675
Dimolybdenum complexes, sulfide-bridged dithiocarbamate, synthesis of, **4**, 675
Dimolybdenum complexes, tetrakis(formamidinato)-, synthesis of, **5**, 781
Dimolybdenum hexacarbonyl complexes, synthesis of, **2**, 496
Dimsyl anion
  as non-transferable ligand, **1**, 120
  reactions, with benzhydryl chloride, **1**, 142
  See also Sodium hydride
Dinickel, bis(pentadienyl)-, synthesis of, **4**, 694
Dinickel complexes, synthesis of, **4**, 693
Dinickel complexes, alkyne-bridged
  from
    nickel dinuclear carbonyls, **4**, 692
    nickelocene, **4**, 692
  palladium analogues of, synthesis of, **4**, 692
Dinickel salts, tris(cyclopentadienyl)-, synthesis of, **4**, 694
Dinitramine, methyl, as nitrating agent, **4**, 429
Dinitramine, methylene-, synthesis of, **4**, 435
Dinitramines, synthesis of, **4**, 435
gem-Dinitramines
  from, monocyclicamines, **4**, 435
  synthesis of, **4**, 434
Dinitrile, succinic, reactions, with hydroxylamine, **5**, 833
Dinitriles
  hydrolysis of, **5**, 305
  reactions
    with hydrogen sulfide, **5**, 580
    with sodium hydrazide, **5**, 848
  use in the synthesis of amidines, **5**, 756
Dinitriles, 3-hydroxy-2-methoxycarbonyl glutaconic, synthesis of, **3**, 653
Dinitro compounds
  cleavage of, **1**, 57
  reactions, with hydrochloric acid, **1**, 57
  synthesis of, **6**, 260
β-Dinitro compounds, elimination of, **1**, 638
Dinitrogen monoxide, reactions, with methyllithium, **3**, 485
Dinitrogen pentoxide
  for the oxidation of *gem*-dinitrosamines, **4**, 435
  reactions
    with ethyl diazoacetate, **3**, 479
    with mercury bis(*t*-butyl diazoacetate), **3**, 480
    with nitrodiazomethane, **3**, 480
Dinitrogen tetroxide
  for the oxidation of thiols, **2**, 214
  as oxidative nitrating agent, **4**, 429
  reactions
    with amines, **2**, 88

with nitrile oxides, **6**, 150
with nitronate salts, **5**, 838
with nitronic acids, **5**, 838
with oximes, **5**, 838
as starting material, for α-aminoalkylphosphorus compounds, **4**, 490
use in the synthesis of *gem*-dinitroalkenes, **4**, 993
use in the synthesis of nitrate esters, **2**, 104
use in the synthesis of *N*-nitrosoamides, **5**, 360
gem-Dinitrosamines
from, imidazolidines, **4**, 434
oxidation of, **4**, 435
synthesis of, **4**, 434
Dinitroso compounds, as starting materials, for α,ω-bis(diazo)alkanes, **3**, 467
Dinucleotides, synthesis of, **4**, 216
1,2-Diol, *cis*-*p*-menth-3-ene-, synthesis of, **2**, 52
Diolate salts, ethyne-, synthesis of, **2**, 1022
Diolides, synthesis of, **5**, 177
Diolithium salts, as starting materials, for ketene dithioacetals, **4**, 847
Diols
annulation reactions, **1**, 183
bisaminomethylation of, **4**, 300
cleavage of, **3**, 103
from
alkenes, **2**, 50
alkenylacetates, **1**, 67
2-keto-*N*-methoxy-*N*-methyl amides, **2**, 59
substituted benzenes, **2**, 39
monobenzoylation of, **5**, 138
oxidation of, **4**, 164
protection of, **2**, 540; **4**, 186
protection as cyclic acetals, **4**, 184
reactions
with boric acid, **2**, 111
with boron trichloride, **2**, 110
with cyanogen halides, **5**, 1102
with dichlorosilanes, **4**, 342
with 1,6-dihydroxy-2,4-hexadiynes, **1**, 986
with diisocyanates, **6**, 488
with dimethyl(dimethylamino)arsine, **2**, 494
with lactones, **6**, 72
with 2-methoxypropene, **4**, 186
with sulfur tetrafluoride, **2**, 100
rearrangement of, **1**, 384
as starting materials
for aldehydes, **3**, 7
for ketones, **3**, 121
for methylene acetals, **4**, 190
synthesis of, via the opening of oxetanes, **1**, 229
Diols, α,ω-alkane-, as starting materials, for spirocyclic orthocarbonates, **6**, 300
Diols, alkynic, as starting materials, for cyclopentanones, **1**, 422
Diols, amino-
as starting materials, for alkenes, **1**, 513
synthesis of, **4**, 507
Diols, vicinal
from, epoxides, **1**, 386
pinacol rearrangement of, **1**, 384
rearrangement of, for the synthesis of aldehydes, **1**, 386
as starting materials, for ketones, **1**, 384
1,2-Diols
deoxygenation of, **1**, 585
elimination of, **1**, 585
from
alkenes, **2**, 44
glycidols, **1**, 220
monobenzoylation of, **5**, 138
oxidation of, **3**, 30
oxidative cleavage of, **3**, 40
protection as acetals, **4**, 180
reactions, with ortho esters, **6**, 79
selective oxidation of, **3**, 158
as starting materials, for α-hydroxy aldehydes, **3**, 30
synthesis of, **4**, 231
1,2-Diols, 3-amino-, synthesis of, **2**, 81, 318
1,2-Diols, bis-*O*-trimethylsilyl-, as starting materials, for ortho esters, **6**, 72
1,3-Diols
acylation of, **5**, 135
from
alkenols, **2**, 50
2,4-diketo sulfoxides, **2**, 61
glycidols, **1**, 221
3-hydroxy ketones, **2**, 63
oxidation of, **3**, 31
protection as acetals, **4**, 180
reactions, with ortho esters, **6**, 79
selective oxidation of, **3**, 161
synthesis of, **2**, 41, 540
via the opening of epoxides, **1**, 212
via the reduction of aldols, **2**, 66
1,3-Diols, (*E*)-alkene-, from, 3-borolenes, **2**, 75
1,3-Diols, 2-dialkyl-, synthesis of, **2**, 86
1,4-Diols
oxidation of, **3**, 31
synthesis of, **2**, 42, 540
1,4-Diols, 2-alkyne-, hydrogenation of, **1**, 503
1,5-Diols
oxidation of, **3**, 31
synthesis of, **1**, 282
2,3-Diols, 1-amino-, synthesis of, **2**, 318
3,4-Diols, *anti*-alk-1-ene-, synthesis of, **2**, 75
anti-1,2-Diols, synthesis of, **2**, 61
cis-Diols, Swern oxidation of, **5**, 424
syn-1,3-Diols, synthesis of, via ring-opening of disubstituted epoxides, **2**, 85
trans-Diols, synthesis of, **2**, 40, 44
trans-1,2-Diols, synthesis of, **2**, 42
trans-1,3-Diols, synthesis of, **2**, 42
α-Diols
from, strained cyclic ketones, **4**, 162
synthesis of, **4**, 160
via hydration of carbonyls, **4**, 160
via oxidation, **4**, 161
α,ω-Diols
as starting materials
for alkyldialkoxyarsines, **2**, 496
for dicarboxylic acids, **5**, 46
1,3-Dione, indane-, oxidation of, **4**, 161
3,17-Dione, 5α-androstane-, reactions, with phosphonates, **3**, 651
Diones, diazo-, synthesis of, **3**, 470
1,2-Diones, reactions, with amines, **3**, 418
1,3-Diones
amination of, **3**, 264
reactions
with 1,3-dielectrophiles, **3**, 246
with α-halo acid chlorides, **3**, 246
as starting materials, for enones, **3**, 217
1,3-Diones, 2-allenyl-2-hydroxyindane-, synthesis of, **1**, 980
Diorganoantimony(III) halides, synthesis of, **1**, 446
Diorganoarsenic(III) halides, from, triorganoarsenic(IV) dihalides, **1**, 445
Diorganoarsenic halides, reduction of, **1**, 445
Diorganoarsenic trihalides, pyrolysis of, **2**, 489
Diorganoarsines
reactions, with alkynes, **2**, 875
as starting materials
for diaryl metal(V) trihalides, **2**, 881
for unsymmetrical tetraorganodiarsines, **2**, 893
Diorganobismuth(III) halides, synthesis of, **1**, 446
Diorganoboron halides, 1-alkynylation of, **2**, 1083
Diorganochalcogenoarsines, alkyl-, synthesis of, **2**, 500

Diorganocopperlithiums, reactions, with aziridines, **2**, 319
Diorganocuprates
reactions
with alkenyl triflates, **1**, 1027
with alkyl halides, **1**, 118
with aryl triflates, **1**, 1027
Diorganodichalcogenides, reactions, with tetraalkyldipnicogens, **2**, 499
Diorganogermoxanes, hexaalkyl-, attempted use in the synthesis of carbodiimides, **5**, 1081
Diorganolithiums
reactions, with dimethylgermanium dichloride, **4**, 636
synthesis of, **4**, 618
Diorganometal(III) alkali metal salts
reactions
with 1,2-dihaloethanes, **2**, 893
with diorganometal(III) halides, **2**, 893
as starting materials
for diarsines, **2**, 893
for dibismuthines, **2**, 893
for distibines, **2**, 893
Diorganometal(III) halides
hydrolysis of, **2**, 885
reactions, with diorganometal(III) alkali metal salts, **2**, 893
as starting materials, for diarylmetal(V) trihalides, **2**, 881
Diorganophosphides
alkylation of, **2**, 437
reactions, with alkyl halides, **2**, 437
synthesis of, **2**, 437
Diorganosilylenes, reactions, with carbon monoxide, **5**, 1019
Diorganostibines
reactions, with alkynes, **2**, 875
as starting materials, for diaryl metal(V) trihalides, **2**, 881
Diorganothalliums, synthesis of, **4**, 1068
Diorganozincs
reactions, with 1-alkynes, **2**, 1096
as reagents, for alkylation, **1**, 465
Diorganylcuprates, cross-coupling of, with vinyl halides, **2**, 960
Diorganyl dichalcogenides, reactions, with tetraorganyl diarsanes, **2**, 293
Diorganylgermylenes, reactions, with alkynes, **2**, 946
Diosmium complexes, carbon-bridged, synthesis of, **4**, 679
1,3,2-Dioxaborinane, 2-isocyanato-, synthesis of, **5**, 1015
Dioxaborinanes
homologation of, **4**, 364
reactions, with methoxy(phenylthio)methyllithium, **4**, 364
1,3,2-Dioxaborolane, *t*-butylperoxy-, for the oxidation of aryl organometallics, **2**, 652
Dioxaborolanes, chloro-, synthesis of, **2**, 110
Dioxaborolanes, hydroxy-, synthesis of, **2**, 111
1,3,2-Dioxaborolanes, 2-chloro-, synthesis of, **2**, 110
1,3,2-Dioxaborolanes, 2,2'-oxybis-, synthesis of, **2**, 111
Dioxacycles, synthesis of, **3**, 315
1,2,3,5-Dioxadisilolane, 3,3,5,5-tetramesityl-, synthesis of, **4**, 624
1,3-Dioxalanes, synthesis of, **4**, 207
Dioxan
as solvent, **1**, 1000
use in the synthesis of secondary phosphines, **2**, 432
Dioxane
fluorination of, **6**, 37
reactions, with phenoxides, **6**, 616
use in the synthesis of dialkylmagnesiums, **2**, 559
1,3-Dioxane, reactions, with trimethyltin chloride, **6**, 133
1,3-Dioxane, perfluoro-, synthesis of, **6**, 251
1,4-Dioxane, 2,3-dibromo-, stability of, **4**, 51
1,4-Dioxane, 2,3-dichloro-, synthesis of, **4**, 49
Dioxanes, 2,2,3-trifluoro-, synthesis of, **6**, 37
1,3-Dioxanes
hydrolysis of, **5**, 424
as starting materials, for alkynyl acylsilanes, **5**, 424
1,3-Dioxanes, bis(hydroxymethyl)-, as starting materials, for disubstituted carboxyl hydroxymethyl-1,3-dioxanes, **5**, 117
1,3-Dioxanes, disubstituted carboxyl hydroxymethyl-, from, bis(hydroxymethyl)-1,3-dioxanes, **5**, 117
1-3-Dioxanes, 2-trimethylammonium-, reactions, with isopropyl diphenylphosphinate, **6**, 116
3,6-Dioxanonanoyl fluoride, perfluoro-9-azido-2,5-dimethyl-8-n-propoxy-, pyrolysis of, **4**, 766
3,6-Dioxanon-1-ene, perfluoro-9-azido-5-methyl-8-n-propoxy-, synthesis of, **4**, 767
1,3-Dioxans, 5-substituted-2,2-dimethyl-4,6-dioxo-, synthesis of, **4**, 204
1,3,5,2-Dioxaphosphaborinanes, synthesis of, **4**, 342
1,3,5,2-Dioxaphosphaborinane salts, synthesis of, **4**, 342
1,3,2-Dioxaphosphacycloheptane, Arbuzov rearrangement of, **2**, 453
1,3,2-Dioxaphospholane, 2-phenyl-, reactions, with aryl 2-nitrophenyl sulfides, **2**, 856
1,3,5-Dioxaphosphorinanes, synthesis of, **4**, 337, 341
1,4,2-Dioxaphosphorinanes, synthesis of, **4**, 340
Dioxarsolane, chloro-, as starting material, for triorganoarsines, **2**, 872
1,3,5-Dioxathianes, synthesis of, **4**, 229
Dioxathiapentalenes
from, pyran-4-thiones, **3**, 377
as starting materials, for trithiapentalenes, **3**, 377
synthesis of, **3**, 377
1,3-Dioxa-2,4-thiazole 2-oxides, thermal rearrangement of, **5**, 1013
1,3,2,4-Dioxathiazol-*S*-oxides, synthesis of, **5**, 666
1,5,3-Dioxazepines, synthesis of, **4**, 300
Dioxazoles, from, bis(trifluoromethyl)ketene, **4**, 833
1,3,4-Dioxazoles, synthesis of, **5**, 665
1,2,4-Dioxazolin-3-one, 5-phenyl-, reactions, with diethylamine, **6**, 109
Dioxazolones, thermal rearrangement of, **5**, 1013
1,2-Dioxenes, synthesis of, **2**, 98
1,2-Diox-4-enes, as starting materials, for *syn*-1,4-cyclohex-2-enes, **2**, 52
Dioxenium salts, reactions, with alkoxides, **6**, 73
Dioxetane, 3-hydroxymethyl-, synthesis of, **6**, 426
1,2-Dioxetans, stability of, **2**, 98
Dioxide, ^{14}C-labeled, reactions, with Grignard reagents, **5**, 45
1,3-Dioxides, synthesis of, **4**, 268
S,S-Dioxides, as intermediates, **5**, 746
S,S'-Dioxides, optically pure, synthesis of, **4**, 269
trans-1,3-Dioxides, synthesis of, **4**, 268
Dioxime, succinamide, synthesis of, **5**, 833
Dioxime, succinimide, synthesis of, **5**, 833
Dioximes, oxidation of, **2**, 781
1,3-Dioximes, synthesis of, **3**, 429
Dioxindoles, 3-*t*-butyl-, synthesis of, **5**, 438
1,3-Dioxin-4-one, 2,2,6-trimethyl-4*H*-, reactions, with sultams, **5**, 343
Dioxins, stannyl, use in the synthesis of α,β-unsaturated ketones, **3**, 240
Dioxirane, difluoro-, synthesis of, **6**, 250
Dioxirane, dimethyl-
for the oxidation of diazo carbonyls, **4**, 162
for the oxidation of silyl enol ethers, **3**, 156
for the oxidation of sulfides, **1**, 780
as reagent
for oxime cleavage, **3**, 133
for the synthesis of sulfines, **3**, 368
as starting material, for nitro compounds, **2**, 397
use in the synthesis of epoxy esters, **4**, 209
use in the synthesis of glycoside bonds, **4**, 188
use in the synthesis of hydroxamic acids, **5**, 324
Dioxirane, 2,2-dimethyl-
comparisons
with *m*-chloroperbenzoic acid, **2**, 344

with Davis reagents, **2**, 344
for the oxidation of amines, **2**, 340
for the oxidation of secondary amines, **2**, 343
Dioxirane, methyl(trifluoromethyl)-, as reagent, for the synthesis of ketones from alcohols, **3**, 120
Dioxirane, trifluoromethyl-, for the oxidation of silyl enol ethers, **3**, 156
Dioxiranes
as epoxidizing agents, **2**, 90
for the oxidation of enolates, **2**, 44
reactions, with enolates, **3**, 156
1,3-Dioxo-2,4-disiletanes, synthesis of, **5**, 892
1,3-Dioxolan, 2,2-di(3-cyanopropyl)-, as starting material, for δ-oxoazelaic acid, **5**, 64
1,3-Dioxolane, 2-(2-bromoethyl)-, as starting material, for ω-functionalized aldehydes, **1**, 119
1,3-Dioxolane, 2-(methoxycarbonylcyanomethylene)-
reactions
with *O*-benzylhydroxylamine, **4**, 896
with 1,1-dimethylhydrazine, **4**, 897
1,3-Dioxolane, 2-(*N*-phenylimino)-, synthesis of, **6**, 624
1-3-Dioxolane, 2-trimethylammonium-, reactions, with triethyl phosphite, **6**, 116
Dioxolanes, amino-, hydrolysis of, **2**, 639
1,3-Dioxolanes
chlorination of, **6**, 43
synthesis of, **4**, 181
1,3-Dioxolanes, 2-acetoxy-, chlorination of, **6**, 43
1,3-Dioxolanes, 2-alkyl-, electrochemical oxidation of, **6**, 77
1,3-Dioxolanes, 2-chloro-, synthesis of, **6**, 43
1,3-Dioxolanes, 2-fluoro-2-trifluoromethyl-, synthesis of, **6**, 43
1,3-Dioxolanes, 4-oxo-
from
α-hydroxy acids, **4**, 213
β-hydroxy acids, **4**, 213
1,3-Dioxolane-2-thione, reactions, with *N*-methyl-*N*,*O*-bis(tributylstannyl)-2-aminoethanol, **6**, 313
Dioxolanones, as starting materials, for carbonyl ylides, **1**, 365
1,3-Dioxolan-4-ones, 5-alkyl-2-*t*-butyl-, synthesis of, **4**, 213
1,3-Dioxolan-4-ones, 5-ylidene-2,2-pentamethylene-, hydrolysis of, **5**, 79
1,3-Dioxole, perfluoro-, synthesis of, **6**, 250
Dioxolenium, reactions, with alkoxides, **6**, 73
1,3-Dioxol-2-one 5,5-dioxide, 4,6-diphenylthieno[3,4-*d*]-, reactions, with amines, **6**, 502
Dioxygens, synthesis of, **4**, 203
Di(oxyimidicamides), cyclization of, **5**, 833
1,3,2-Dioxyphosphole, 2,2,2-trimethyl-, synthesis of, **2**, 476
Dipalladium complexes, alkyne-bridged, from, hex-3-yne, **4**, 692
Dipeptides
chemoselective monothionation of, **5**, 569
isosteres, **1**, 817
thionation of, **5**, 515, 568
Dipeptides, cyclic, as catalysts, for the synthesis of asymmetric cyanohydrins, **3**, 628
Dipeptides, protected, thionation of, **5**, 574
Dipeptides, thionated, synthesis of, **5**, 586
Diphenol, α,α-dimethyl-4,4'-methylene-, reactions, with *N*-chlorocarbonyl isocyanate, **5**, 1007
Diphenylamine, 4-isothiocyanato-4'-nitro-, synthesis of, **5**, 1035
(η^2-Diphenylcyclopropene) complexes, synthesis of, **3**, 509
Diphenylphosphido–cyclopentadienyl compounds, lithiation of, **4**, 706
Diphosgene
decomposition of, **5**, 972
and phosgene compared, **5**, 972
reactions
with alcohols, **6**, 426
with amines, **5**, 965, 972
with amino acids, **5**, 972
with amino alcohols, **5**, 972
as reagent, for the synthesis of diisocyanates, **5**, 972
as starting material
for carbamoyl chlorides, **6**, 446
for chloroformate esters, **6**, 426
synthesis of, **5**, 972; **6**, 415
for the synthesis of isocyanates, **5**, 971
use in the synthesis of aryl isocyanides, **3**, 721, 722
use in the synthesis of isocyanides, **3**, 696
See also Formate, trichloromethyl chloro-
1,3-Diphosphaalkenes
cycloaddition to, **4**, 548
reactions of, **4**, 557
1,3-Diphosphaallenes
hydrolysis of, **4**, 559
isomerization of, **4**, 548
protonation of, **5**, 887
reactions, with sulfur, **4**, 559
as starting materials, for phosphaalkenes, **5**, 884
synthesis of, **5**, 1092
1,5-Diphosphabicyclo[3.3.1]nonane, synthesis of, **2**, 439
1,4-Diphosphabuta-1,3-diene, intramolecular cyclization, avoidance of, **5**, 880
Diphosphabutadienes, synthesis of, **6**, 689
Diphosphabutadiyne, synthesis of, **5**, 1159
1,4-Diphosphabutatriene, from, lithiated phosphaallene, **3**, 607
Diphosphachiropteradienes, synthesis of, **4**, 549
Diphosphacyclobutanes, synthesis of, **6**, 160, 166
Diphosphacyclopentadiene, reactions, with transition metals, **6**, 154
Diphosphacyclopentanes, synthesis of, **2**, 472
Diphosphaferrocene, diacetyltetraphenyl-, synthesis of, **4**, 1036
Diphosphaferrocenes
chemistry of, **4**, 1035
from, 1,2,3-triphenylphosphole, **4**, 1035
Diphosphaferrocenes, tetraphenyl-, synthesis of, **4**, 1035
1,3-Diphosphaindene, reactions, with 1,4-diphenylbutadiene, **4**, 548
Diphosphanes, tetraorganyl-, reactions, with diorganyl dichalcogenides, **2**, 293
1,3-Diphosphapropanes, 1,3-dichloro-
dehydrohalogenation of, **5**, 884
as starting materials, for *C*-phosphinomethylenephosphanes, **5**, 884
1,3-Diphosphapropenes
synthesis of, **5**, 1058
use in the synthesis of methylenephosphines, **6**, 688
Diphosphatricyclodecadienes, synthesis of, **2**, 472
Diphosphene, bis[2,4-di-*t*-butyl-6-(dimethylamino)phenyl]-, synthesis of, **2**, 865
trans-Diphosphene, reactions, with carbenes, **6**, 264
Diphosphenes
carbene addition to, **4**, 547
[2+2] cycloadditions of, **5**, 1092
reactions
with bis(trimethylsilyl)methane, **6**, 166
with halocarbenes, **6**, 688
as starting materials, for diphosphiranes, **3**, 607
synthesis of, **6**, 166
Diphosphenes, symmetrical, synthesis of, **5**, 1092
Diphosphenes, unsymmetrical, synthesis of, **5**, 1092
1,2-Diphosphetane, synthesis of, **5**, 1092
1,3-Diphosphetane, synthesis of, **6**, 679, 686
Diphosphetanediones, synthesis of, **5**, 1018
1,3-Diphosphetanes
from, bis(trifluoromethyl)phosphine, **6**, 264
reactions, with methanol, **6**, 265
synthesis
of, **3**, 606; **4**, 547, 549, 558, 1026; **5**, 882, 1086; **6**, 324
via dimerization of phosphaalkenes, **6**, 683

Diphosphetes, dihydro-, synthesis of, **5**, 885
1,2-Diphosphetes, reactions, with lithium, **4**, 545
1,3-Diphosphetes, hydrolysis of, **4**, 562
Diphosphinate esters, reduction of, **2**, 431
Diphosphinates
reactions, with carbonyls, **4**, 563
synthesis of, **2**, 453, 845
Diphosphine, 1,2-bis(iodocyclohexyl)-, synthesis of, **2**, 449
Diphosphine, 1,2-diphenyl-1,2-bis(trimethylsilyl)-, reactions, with phenylphosphonous dichloride, **2**, 865
Diphosphine, tetrafluoro-, reactions, with fluoroalkenes, **4**, 114
Diphosphine, tetrakistrifluoro-, reactions, with dimethylarsine, **2**, 508
Diphosphine, tetramethyl-, reactions, with hexafluoropropene, **4**, 802
Diphosphine, tetraphenyl-, synthesis of, **2**, 835, 864
vic-Diphosphine, reactions, with acylating agents, **5**, 400
Diphosphine dioxide, tetraphenyl-, reactions, with phenylselenenyl bromide, **2**, 857
Diphosphine dioxides, as starting materials, for diphosphine disulfides, **2**, 462
Diphosphine disulfides
from, diphosphine dioxides, **2**, 462
synthesis of, **2**, 461
Diphosphine ligands, tetrakis(hydroxymethyl)-, synthesis of, **2**, 440
Di(phosphine oxides), alkylidene-
alkylation on carbon, **4**, 563
synthesis of, **4**, 553
Diphosphines
carbene insertion into, **4**, 547
dehydrogenation of, **2**, 472
reactions
with bromine, **2**, 449
with disulfides, **2**, 460
synthesis of, **2**, 471, 864; **5**, 398
Diphosphines, alkyldialkoxy-, reactions, with alkyl halides, **2**, 453
Diphosphines, alkylidene-
oxidation of, **4**, 552, 562
synthesis of, **4**, 545
Diphosphines, bis(tris(trimethylsilyl)methyl)-, synthesis of, **6**, 373
Diphosphines, dialkyldisilyl-, synthesis of, **2**, 471
Diphosphines, *trans*-diaryl-, reactions, with isopropylidenecarbene, **4**, 1025
Diphosphines, 1,2-dimetallo-, reactions, with 1,1-dihaloalkanes, **4**, 545
Diphosphines, methylene-
reactions of, **4**, 550, 562
reactions, with mercuric acetate, **6**, 371
Diphosphines, silyl-, reactions, with 2,4,6-tri-*t*-butylphenylphosphaketene, **6**, 687
Diphosphines, silyl-substituted, synthesis of, **6**, 373
Diphosphines, tetraalkyl-
reactions, with dialkyl ditellurides, **2**, 465
synthesis of, **2**, 471
1,4-Diphosphines, synthesis of, **2**, 438
Di(phosphine sulfides), alkylidene-, synthesis of, **4**, 553
α,α-Diphosphino alcohols, synthesis of, **6**, 118
1,1-Diphosphinoalkanes
oxidation of, **4**, 552, 562
synthesis of, **4**, 545, 548
1,1-Diphosphinoalkenes, addition to, **4**, 548
Diphosphinocarbenes, intramolecular insertion of, **4**, 548
Diphosphinoethane, reactions, with formaldehyde, **2**, 440
Diphosphinomethane
reactions of, **4**, 550, 562
reactions, with diethylvinylphosphine, **2**, 439
Diphosphinomethanes
cyclization of, **2**, 439
deprotonation of, **1**, 158
Diphosphiranes
from, diphosphenes, **3**, 607
photoinduced rearrangement of, **6**, 688
synthesis of, **2**, 460; **4**, 545, 547; **6**, 264, 371
Diphosphiranes, di(chlorophenyl)methylene-, synthesis of, **4**, 1025
Diphosphiranes, halogenated, reductive ring-opening of, **5**, 886
Diphosphiranes, methylene-, synthesis of, **4**, 1025
1*H*-1,3-Diphosphole, 1-*t*-butyl-2-ethylidene-2,3-dihydro-4-methyl-3-phenyl-, synthesis of, **4**, 1026
Diphospholes, synthesis of, **4**, 1026, 1030
1,3-Diphospholes, synthesis of, **4**, 563
1,3-Diphospholides, as starting materials, for cyclic phosphaalkenes, **5**, 886
Diphosphonate, tetraethyl dialkylaminomethylene-, as starting material, for α-phosphoryl enamines, **4**, 1002
Diphosphonate, tetraethyl dichloromethane-, reactions, with *n*-butyllithium, **4**, 801
Diphosphonate, tetraethyl formamidomethylene-, as starting material, for *gem*-formamido phosphorylalkenes, **4**, 1004
Diphosphonates, alkene, synthesis of, **4**, 803
Diphosphonates, alkylidene
alkylation on carbon, **4**, 563
reduction of, **4**, 550
synthesis of, **4**, 553
Diphosphonates, alkyne, hydrogen halide hydrolysis of, **4**, 803
Diphosphonates, amino-, reactions, with sodium hydride, **4**, 1002
Diphosphonates, benzylidene, synthesis of, **4**, 559
Diphosphonates, methylene-
mercuration of, **6**, 371
reactions, with mercuric acetate, **6**, 371
Diphosphonates, tetraphenyl, synthesis of, **4**, 555
1,1-Diphosphonatoalkanes, reduction of, **4**, 550
Diphosphonic acids, dihalomethylene-, hydrolysis of, **6**, 518
Diphosphonic acids, (*Z*)-haloalkene-, synthesis of, **4**, 803
Diphosphonites, reactions, with aldehydes, **4**, 564
Diphosphonites, alkylidene-, Arbuzov reactions of, **4**, 552
Diphosphonium salts, synthesis of, **2**, 444
Diphosphonium salts, alkylidene-, synthesis of, **4**, 553
Diphosphoranes, alkylidene-, synthesis of, **4**, 564
1,1-Diphosphoranoalkanes, synthesis of, **4**, 564
Diphosphorines, synthesis of, **4**, 1027
Diphosphorous pentoxide, reactions, with hexamethylsiloxane, **2**, 106
Diphosphorus tetraiodide
as reagent, for the synthesis of alkyl iodides, **2**, 36
for the reduction of thioketones, **1**, 24
Diphosphorus tetrasulfide, diethyl-, use in the synthesis of thiophosphines, **2**, 254
Di(phosphoryl chloride), methylene-
reactions of, **4**, 563
reduction of, **4**, 550
Dipiperidine, 1,1'-(azodicarbonyl)-, as reagent, for the synthesis of aldehydes, **3**, 7
Dipivaloyl telluride, reactions, with bis(trimethylsilyl) telluride, **5**, 542
Diplatinum complexes, carbene-bridged, from, perfluoropropene, **4**, 691
Diplatinum complexes, methylene-bridged, synthesis of, **4**, 691
Dipnicogens, mixed tetraalkyl-, synthesis of, **2**, 508
Dipnicogens, tetraalkyl-
insertion reactions of, **2**, 499
reactions, with diorganodichalcogenides, **2**, 499
Dipnicogens, tetramethyl-, synthesis of, problems, **2**, 508
1,3-Dipolar addition, of diphenylsulfine and 2-diazopropane, **6**, 316
Dipolar cycloaddition, for the synthesis of α-aminoalkylphosphorus compounds, **4**, 479

[1+3] Dipolar cycloaddition
of diazo compounds, **1**, 298
mechanism, **1**, 523
of nitrile oxides, **3**, 600
of nitrile oxides to alkynes, **3**, 189
for the substitution of silicon functions, **1**, 285
for the synthesis of β-keto phosphonates, **3**, 200
for the synthesis of nitriles, **3**, 621
for the synthesis of nitrile ylides, **3**, 678
[2+3] Dipolar cycloaddition, in the synthesis of arylsilanes, **2**, 943
1,3-Dipolar cycloadducts, synthesis of, **3**, 680
Dipolarophiles
reactions
with 2-acetoxymethyl-3-allyltrimethylsilane, **1**, 287
with nitrile sulfides, **3**, 686
Dipole destabilization, of ketones, **2**, 642
1,3-Dipoles
[3+2] cycloaddition reactions, with 1,1-dialkoxyalkenes, **6**, 76
reactions
with aldehydes, **4**, 183
with ketones, **4**, 183
with sulfines, **2**, 209
1,1-Dipotassio compounds, synthesis of, **2**, 1005
Dipotassium cyclooctatetraenide
use in the synthesis of carbon-bridged dichromium complexes, **4**, 673
use in the synthesis of carbon-bridged diruthenium complexes, **4**, 683
use in the synthesis of cyclooctatetraene-bridged complexes, **4**, 672
use in the synthesis of tris(cyclooctatetraene)dimolybdenum complexes, **4**, 673
use in the synthesis of tris(cyclooctatetraene)ditungsten complexes, **4**, 673
1,2-Dipotassium-1,2-di-*t*-butyldiphosphide, reactions, with 1,1-dichloroalkenes, **4**, 1025
1,2-Dipotassium di-*t*-butyldiphosphide, reactions, with tetrachloromethane, **6**, 366
1,2-Dipotassium diorganylphosphides, reactions, with dichloromethylenephosphines, **6**, 685
Dipotassium disulfonate, hydroxylamine
O-alkylation of, **2**, 103
use in the synthesis of *O*-alkylhydroxylamines, **2**, 103
1,8-Dipotassium naphthalate, synthesis of, **5**, 109
Dipyrazole, 1,1-carbonyl-, thermolysis of, **6**, 365
Dipyridines, 2,2'-carbonyl-, use in the synthesis of carbodiimides, **5**, 1078
Di-2-pyridyl sulfites, use in the synthesis of carbodiimides, **5**, 1078
Dipyrrolidide, dithiomalonic acid, sulfhydrolysis of, **5**, 556
Diradical intermediates
in photochemical [2+2] additions, **1**, 356
in thermal [2+2] additions, **1**, 354
Directed metallation groups (DMG's)
definition of, **3**, 90
use in the synthesis of benzaldehydes, **3**, 90
Directing groups, in radical additions, **1**, 325
Direct synthesis, of alkylsilyl chlorides, **2**, 538
Dirhenium, decacarbonyl-, synthesis of, **4**, 677; **6**, 523
Dirhenium complexes, flyover, synthesis of, **4**, 678
Dirhenium heptaoxide, reactions, with alkylzincs, **2**, 585
Dirhodiacyclobutane complexes, cyclopentadienyl, use in the synthesis of alkyne-bridged dirhodium complexes, **4**, 688
Dirhodium, octakis(phosphorus trifluoride)-, reactions, with alkynes, **4**, 688
Dirhodium complexes, alkyne-bridged, reactions, with diazomethane, **4**, 689
Dirhodium complexes, bismethylene-bridged, synthesis of, **4**, 687
Dirhodium complexes, carbon-bridged, from, diazoalkanes, **4**, 686
Dirhodium complexes, carbon-bridged ketonic, from, hex-3-yne, **4**, 690
Dirhodium complexes, methylene-bridged, synthesis of, **4**, 689
Diruthenium, tetracarbonyldicyclopentadienyl-, use in the synthesis of diruthenium complexes, **4**, 679
Diruthenium complexes
synthesis of, **4**, 679
via photochemical decarbonylation, **4**, 681
Diruthenium complexes, alkyne-bridged, synthesis of, **4**, 680
Diruthenium complexes, carbon-bridged, synthesis of, **4**, 678, 683
Diruthenium complexes, chloride-bridged, synthesis of, **4**, 680
Diruthenium complexes, methylene-bridged, synthesis of, **4**, 680
Diruthenium complexes, trismethylene-, synthesis of, **4**, 680
Disaccharides, as starting materials, for dithiocarbamates, **6**, 561
1,3-Diselenacyclobutane, synthesis of, **3**, 394
Diselenacyclopentane, thionation of, **3**, 399
1,4-Diselenane, octafluoro-, synthesis of, **6**, 49
Diselenaphosphonous acid esters, phenyl-, synthesis of, **2**, 857
Diselenenamides, synthesis of, **2**, 369
1,3-Diselenetane, 1-trifluoromethyl-1,3,3-trifluoro-, synthesis of, **6**, 256
1,3-Diselenetanes
pyrolysis of, **3**, 395
synthesis of, **6**, 256
1,2-Diselenetene, as starting material, for metal selenocarbonyl complex, **3**, 397
Diselenide, reactions, with phosphine, **2**, 858
Diselenide, bis(4-biphenylcarbonyl), reactions, with sodium methoxide, **5**, 233
Diselenide, bis(4-methylbenzoyl), synthesis of, **5**, 542
Diselenide, bis(2-oxyethyl), synthesis of, **3**, 39
Diselenide, bis(perfluoroethyl), synthesis of, **6**, 49
Diselenide, bis(selenocarbamoyl), as intermediate, **6**, 594
Diselenide, bis(trifluoromethyl), synthesis of, **6**, 49
Diselenide, dilithium *o*-phenylene, reactions, with dibromo(methyl)bismuthine, **2**, 501
Diselenide, dimesityl
as catalyst
for the synthesis of aldehydes from alcohols, **3**, 7
for the synthesis of ketones from alcohols, **3**, 120
Diselenide, dimethyl
reactions
with sodium, **4**, 283
with tetrachloroethene, **4**, 786
Diselenide, diphenyl
reactions
with alkenes, **2**, 284
with trimethylsilylethyne, **4**, 947
with tris(phenylseleno)methanide ion, **6**, 307
with vinyl carbenes, **4**, 875
Diselenide, heptafluoropropyl, synthesis of, **6**, 49
Diselenide anions, alkylation of, **2**, 292
Diselenidecopper(I) triflate, diphenyl, reactions, with trimethylsilyl-substituted alkynes, **2**, 1037
Diselenide dianions, reactions, with aryl halides, **2**, 734
Diselenides
as by-products, **2**, 278
cleavage of, **6**, 356
from
areneselenols, **2**, 727
potassium organic seleno sulfates, **2**, 292
selenolate salts, **2**, 292
selenols, **2**, 292
halogenation of, **2**, 288
halolysis of, **2**, 732

oxidation of, **3**, 174
reactions
with alkynes, **4**, 871
with manganese pentacarbonyl hydride, **6**, 49
with selenium dioxide, **3**, 37
with vinyl halides, **2**, 727
reduction of, **2**, 279; **6**, 92
as starting materials
for alkaneselenenyl halides, **2**, 288
for alkaneseleninic acids, **2**, 291
for alkaneseleninic anhydrides, **2**, 291
for alkaneseleninyl chlorides, **2**, 289
for diselenoacetals, **4**, 290
for α-halo selenides, **4**, 88
for selenols, **2**, 278
for unsymmetrical dialkyl selenides, **2**, 279
synthesis of, **5**, 521
use in the synthesis of α-selenyl ketones, **3**, 173
Diselenides, alkali, alkylation of, **2**, 292
Diselenides, alkyl, synthesis of, **2**, 292
Diselenides, aroyl aryl, from, potassium selenocarboxylates, **5**, 250
Diselenides, aryl, synthesis of, **2**, 734
Diselenides, bis(acyl), synthesis of, **5**, 233, 244, 252
Diselenides, dialkyl
photolysis of, **2**, 465
as starting materials, for 1,2-bis(selenides), **2**, 286
synthesis of, **2**, 292
Diselenides, diaryl
disproportionation of, **2**, 735
from, arylamines, **2**, 734
reactions
with alkynes, **5**, 241
with aroylsulfenyl bromides, **5**, 248
with carbon monoxide, **5**, 241
with 1,3-dithiane, **6**, 102
with iodine, **2**, 732
with metallated arenes, **2**, 729
as starting materials, for aryl alkyl selenides, **2**, 279
synthesis of, **2**, 729, 734
Diselenides, diphenyl
reactions
with perfluoroalkyl iodides, **4**, 284
with sulfone carbanions, **4**, 286
Diselenides, monoacyl-substituted, synthesis of, **5**, 250
Diseleniums, synthesis of, **4**, 287, 874
Diselenoacetals
cleavage of, **4**, 284
deprotonation of, **4**, 289
elimination of selenols from, **2**, 728
from
aldehydes, **4**, 287
diselenides, **4**, 290
gem-dihalides, **4**, 288
ketones, **4**, 287
selenocarbonyls, **4**, 289
metal-selenium exchange in, **3**, 175
reactions, with butyllithium, **4**, 400
reduction of, **1**, 25
selenation of, **6**, 93
as starting materials
for thioselenoacetals, **4**, 284
for triselenoortho esters, **6**, 93
synthesis of, **4**, 287; **6**, 334
Diselenoacetals, ^{13}C-labeled, synthesis of, **4**, 289
Diselenoacetals, diphenyl, transmetallation of, **4**, 401
Diselenoacetals, ketene
from
gem-dibromoalkanes, **4**, 875
malononitrile, **4**, 875
reactions, with *n*-butyllithium, **4**, 964
synthesis of, **4**, 875
Diselenoacetic acid *Se*-methyl ester, α-naphthyl, synthesis of, **5**, 563
Diselenocarbamates
alkylation of, **6**, 594
from
carbon diselenide, **6**, 594
phosgeneiminium chloride, **6**, 594
reactions, with electrophilic reagents, **5**, 1055
stability of, **5**, 1055
as starting materials, for iminium salts, **6**, 591
Diselenocarbamates, *N*-alkyl, as intermediates, in the synthesis of isoselenocyanates, **5**, 1055
Diselenocarbazates, from, hydrazines, **6**, 594
Diselenocarbazic acids, stability of, **6**, 594
Diselenocarbonate, dibenzyl, synthesis of, **6**, 474
Diselenocarbonates
ammonolysis of, for selenocarbamates, **6**, 593
hydrazinolysis of, for selenocarbamates, **6**, 593
metal complexes of, **3**, 387
Diselenocarbonates, *O*-alkyl, from, diselenoxanthate salts, **6**, 592
Diselenocarboxylates, *Se*-methyl, synthesis of, **5**, 523
Diselenoesters, Diels–Alder reaction of, **4**, 289
Diselenofulvenes, from, 4-aryl-1,2,3-selenodiazoles, **3**, 553
Diselenoic acid esters, synthesis of, **5**, 562, 563
Diselenoic acids
structure of, **5**, 545
synthesis of, **5**, 549
1,1-Diselenolates, reactions, with alkyl halides, **4**, 875
1,3-Diselenolates, exchange reactions, with triethyl orthoformate, **6**, 100
1,3-Diselenole, 2-methoxy-, from, dibromomethyl methyl ether, **6**, 100
1,3-Diselenoles, 2-benzylidene, synthesis of, **4**, 874
1,3-Diselenoles, 2-methaneseleno-, synthesis of, **6**, 93
1,3-Diselenoles, 2-vinylidene, synthesis of, **4**, 874
Diseleno monothioortho esters, from, diselenoortho esters, **6**, 102
1,3-Diselenonium cations, reactions, with trialkyl phosphites, **6**, 123
Diselenoortho esters
from
α,α-dihalo ethers, **6**, 100
ortho esters, **6**, 100
reactions, with benzenethiol, **6**, 102
as starting materials, for diselenomonothioortho esters, **6**, 102
Diselenoorthoformate esters, from, dichloromethyl methyl ethers, **6**, 100
Diselenooxamides, synthesis of, **5**, 627
Diselenophosphinic acids, synthesis of, **2**, 465
Diselenotelluroorthocarbonates, theoretical, **6**, 311
Diselenotellurothioorthocarbonates, theoretical, **6**, 312
Diselenothiocarbonates, *S*-alkyl, from, diselenoxanthate salts, **6**, 592
Diselenothionocarbonate, bis(trifluoromethyl), photolysis of, **4**, 872
Diselenoxanthate salts, as starting materials, for selenocarbonyls, **6**, 592
1,3-Diseletane, 2,2,4,4-tetrachloro-, pyrolysis of, for selenophosgene, **6**, 588
β-Diselones, from, selenopyrone, **3**, 397
1,3-Disilabicyclo[1.1.0]butanes, synthesis of, **4**, 1049; **6**, 178
1,3-Disilabutane, 3-methyl-1,1,3-trichloro-, synthesis of, **4**, 611
1,3-Disilabutanes, synthesis of, **4**, 1054
1,2-Disilabut-3-enes, cleavage of, **4**, 1051
1,2-Disilacyclobutane, synthesis of, **6**, 178
1,3-Disilacyclobutane, synthesis of, **4**, 1053
1,3-Disilacyclobutane, 1,3-dichloro-1,3-dimethyl-, reactions, with organometallics, **4**, 621
1,3-Disilacyclobutane, 2-lithio-, synthesis of, **6**, 191
1,3-Disilacyclobutane, 1,1,3,3-tetrachloro-, reduction of, **4**, 621
1,3-Disilacyclobutane, 1,1,3,3-tetramethyl-

metallation of, **4**, 620
reactions
with alkyllithiums, **6**, 191
with bromine, **4**, 622
with methyllithium, **4**, 606
with organomercury carbenoid, **4**, 623
with sulfur dioxide, **4**, 624
with sulfur trioxide, **4**, 624
with trichlorosilane, **4**, 617
with trimethylsilanol, **4**, 623
with zirconium tetrachloride, **4**, 610
ring-opening of, **4**, 653
synthesis of, **4**, 608
1,3,Disilacyclobutane, 1,1,3,3-tetramethyl-2,4-dineopentyl-, synthesis of, **4**, 611
Disilacyclobutanes
from, bis(chloromethyl)dimethylsilane, **4**, 605
metallation of, **4**, 623
1,2-Disilacyclobutanes, thermolysis of, **6**, 383
1,3-Disilacyclobutanes
from
chloromethyldimethylsilyl halides, **4**, 654
disilenes, **4**, 614
α-lithiosilylalkanes, **4**, 605
silenes, **4**, 610
hydrosilylation of, **4**, 617
ring-opening of, **4**, 622, 658, 661
as starting materials
for α-borylsilylalkanes, **4**, 644
for tetrasilylethenes, **4**, 1049
synthesis of, **4**, 608, 1049; **6**, 179, 383
via pyrolysis of silacyclobutanes, **4**, 610
1,3-Disilacyclobutanes, 1,1,3,3-tetramethyl-, ring-opening of, **4**, 658
3,4-Disilacyclobut-3-ene, 3,4-bis(silyl), synthesis of, **4**, 1052
1,2-Disilacyclohexadienes, as starting materials, for silenes, **4**, 612
1,4-Disilacyclohexa-2,5-dienes, synthesis of, **4**, 1053
1,3-Disilacyclohexane, 1,1-dichloro-3,3-dimethyl-, synthesis of, **4**, 608
1,3-Disilacyclopentane, synthesis of, **6**, 268
1,4-Disilacyclopentanes, synthesis of, **4**, 621
1,3-Disilacyclopentene, 1,1,3,3-tetramethyl-, from, 1,2-diallyl-1,1,2,2-tetramethyldisilane, **4**, 617
Disilacyclopentenes, from, silirenes, **4**, 1052
1,3-Disilacyclopentenes
from
molybdenum complexes, **4**, 616
silacyclopropenes, **4**, 616
Disilacylobutane, synthesis of, **6**, 380
Disilane, 1,2-bis(bromomethyl)tetramethyl-, reactions, with magnesium, **4**, 654
Disilane, 1,2-bis(1-naphthyl)tetramethyl-, photolytic transformation of, **4**, 617
Disilane, chloromethylpentamethyl-
reactions
with alkoxides, **4**, 614
with aluminum chloride, **4**, 614
with potassium acetate, **4**, 614
Disilane, 1,2-diallyl-1,1,2,2-tetramethyl-, as starting material, for 1,1,3,3-tetramethyl-1,3-disilacyclopentene, **4**, 617
Disilane, 1,2-dichlorotetramethyl-
reactions
with methylene(triphenyl)phosphorane, **6**, 703
with ylides, **5**, 914
Disilane, hexachloro-
reactions, with diazomethane, **4**, 140
use in the synthesis of tertiary phosphines, **2**, 441
Disilane, hexamethyl-
cleavage of, **2**, 546
reactions, with methyllithium, **4**, 355
synthesis of, **2**, 544
thermolysis of, **4**, 616
use in the synthesis of acylsilanes, **5**, 419
Disilane, hexaphenyl-, synthesis of, **4**, 609
Disilane, pentamethylchloro-, use in the synthesis of carbodiimides, **5**, 1076
Disilane, 2-phenylethenylpentamethyl-, irradiation of, **4**, 614
Disilane, phenylphosphinylidene-, reactions, with phenylphosphine, **6**, 165
Disilane, 1-phenyl-2-vinyltetramethyl-, photolysis of, **4**, 613
Disilanes
cleavage of, **2**, 546
palladium-catalyzed addition of, **2**, 915
reactions
with alcohols, **2**, 539
with difluorocarbenes, **6**, 268
with isocyanides, **3**, 408
as starting materials
for alkenes, **4**, 1054
for 1,1-bis(silyl)alkanes, **4**, 614
synthesis of, **2**, 544
Disilanes, alkenyl-, 1,3-silyl migrations of, **4**, 613
Disilanes, alkyl-, photolysis of, **4**, 1052
Disilanes, alkyne-
from, 1,1-disilylallenes, **4**, 1046
rearrangement of, **4**, 1046
Disilanes, alkynic, rearrangement of, **1**, 983
Disilanes, divinyl-, as starting materials, for silenes, **3**, 502
Disilanes, hexaalkyl-, reactions, with 1-alkynes, **2**, 1090
Disilanes, methoxy-, cleavage of, **2**, 546
Disilanes, pentamethyl(3'-trimethylsilyloxy-1-alkynyl)-, as starting materials, for 1,1-bis(silyl)allenes, **2**, 940
Disilanes, polyhalogenated, reactions, with phosphorus ylides, **5**, 914
Disilanes, styryl-, rearrangement of, for the synthesis of 1,1-bis(silyl)alkanes, **4**, 614
Disilanyl chlorides, reactions, with ylides, **5**, 913
2,4-Disilapentane, 1-chloro-4-fluoro-2,2,4-trimethyl-, reactions, with magnesium, **4**, 608
2,4-Disilapentane, 1-lithio-2,2,4,4-tetramethyl-, reactions, with trimethylsilyl chloride, **4**, 606
2,4-Disilapentane, 2-methoxy-2,4,4-trimethyl-
from, methoxytrimethylsilane, **4**, 602
reactions, with *t*-butyllithium, **4**, 620
synthesis of, **4**, 623
2,4-Disilapentane, 2,2,4-trimethyl-, synthesis of, **4**, 616
1,4-Disilapentanes, synthesis of, **4**, 1054
2,4-Disilapentanes, 2-chloro-2,4,4-trimethyl-, synthesis of, **4**, 614
1,4-Disilapent-2-enes, cleavage of, **4**, 1051
1,3-Disilapropane, perchloro-, synthesis of, **6**, 267
1,3-Disilapropanes, 2,2-dichloro-, synthesis of, **6**, 267
Disilatanes, synthesis of, **5**, 1095
Disilathiane, hexamethyl-
reactions
with butanethiol, **2**, 263
with *N*,*N*-diethylcarbamoylmercury, **6**, 515
with nitroalkanes, **5**, 637
as reagent
for the synthesis of thioaldehydes, **3**, 336
for the synthesis of thioketones, **3**, 346, 353, 363
Disilazane, hexaalkyl-, reactions, with hydrogen cyanide, **5**, 1133
Disilazane, hexamethyl- (HMDS)
as ammonia substitute, in the synthesis of *gem*-imines, **4**, 438
reactions
with acid chlorides, **5**, 376, 379
with alkanethiols, **2**, 263
with carboxylic acids, **5**, 224, 376
with diamines, **6**, 507
with hydrogen sulfide, **2**, 541
with hydroxamic acids, **5**, 380

with ketene, **5**, 379
with primary thionoamides, **5**, 645
synthesis of, **2**, 603
uses of, **2**, 543
use in the synthesis of acyloxysilanes, **5**, 224
use in the synthesis of carbodiimides, **5**, 1081
use in the synthesis of imides, **5**, 306
Disilazane, *N*-methylhexamethyl-, reactions, with methyl perfluoromethacrylate, **3**, 588
Disilazanes, as *O*-silylation agents, **5**, 536
Disilazanes, hexaalkyl-, reactions, with urea, **5**, 1006
Disilazanes, *N*-substituted hexamethyl-
as starting materials
for sulfinylamines, **2**, 399
for sulfurdiimines, **2**, 399
Disilazides, hexamethyl-, for the enolization of carbonyls, **2**, 696
Disilene, tetramethyl-, reactions, with alkynes, **4**, 1052
Disilenes
as intermediates, in the synthesis of 1,1-bis(silyl)alkanes, **4**, 614
reactions, with methyl furan-2-carboxylate, **6**, 125
as starting materials
for 1,1-bis(silyl)alkanes, **4**, 609
for 1,3-disilacyclobutanes, **4**, 614
for 1-germyl-1-silylalkanes, **4**, 639
synthesis of, **4**, 614
1,3-Disiletanes
reactions, with methylenephosphoranes, **6**, 702
as starting materials, for silyl-functionalized ylides, **5**, 915
Disilethylene, 1,1,4,4-tetramethyl-1,4-bis(*N*,*N*-dimethylamino)-, use in the protection of amines, **2**, 543
Disiloxane, from, α-silyl silenes, **4**, 612
Disiloxane, chloromethylpentamethyl-, as starting material, for α-silyl Grignard reagents, **4**, 654
Disiloxane, di(chloromethyl)-, cyclization of, **4**, 609
Disiloxane, 1,3-dichloro-1,1,3,3-tetraisopropyl-, use in the protection of diols, **2**, 540
Disiloxane, hexamethyl-
as byproduct, **1**, 758
deprotonation of, **4**, 650
elimination of, for the synthesis of phosphaalkynes, **5**, 1154
reactions
with hydroxamic acids, **5**, 719
with phosphorus(III) bromide, **2**, 538
synthesis of, **5**, 542, 978
Disiloxanes, synthesis of, **4**, 623
Disilver complexes, dilithio-, synthesis of, **4**, 708
N,*N*-Disilyl compounds, synthesis of, **2**, 1057
Disodium dichloride, as starting material, for dithiooxophosphanylmethylenephosphoranes, **5**, 908
Disodium disulfide, reactions, with alkyl halides, **2**, 120
Disodium methylarsenide, reactions, with 1,4-dichlorobutane, **2**, 486
Disodium tetracarbonylferrate
for the reduction of anhydrides, **3**, 84
synthesis of, **2**, 575
See also Collman's reagent
(+)-Disparlure, synthesis of, **2**, 152
Disparlures, synthesis of, **2**, 49
Displacement, oxidative, for organometallics, **2**, 952
Displacement, solvolytic, use of term, **1**, 891
S_N2 Displacement–Michael addition, for the synthesis of pyrrolidines, **2**, 323
Displacement reaction, for the synthesis of ketene hemiaminals, **4**, 893
S_N2-Displacement reactions, with malonate anions, **5**, 158
Disproportionation
of arylphosphonous acids, for primary phosphines, **2**, 821
from bimolecular reactions, **1**, 395
for the synthesis of alkylaluminums, **2**, 596
for the synthesis of organocalciums, **1**, 115
for the synthesis of organometal(III) dihalides, **2**, 882
Dissociation mechanism, use of term, **1**, 891
Dissolving metal reduction
of allenes, **1**, 459
of arenes, **1**, 95
of carbonyls, **2**, 54
mechanism, **1**, 436
for the synthesis of alkenes, **1**, 456
Distannacyclobutanes, synthesis of, **6**, 399
Distannacycloheptanes, synthesis of, **4**, 700
Distannane, hexaalkyl-, reactions, with thiourea, **5**, 1082
Distannane, hexabutyl-, reactions, with acyl azides, **5**, 391
Distannanes, telluration of, **4**, 876
1,1'-Distibaferrocene, 2,2',5,5'-tetramethyl-, synthesis of, **4**, 1041
Distibaferrocenes, synthesis of, **4**, 1041
Distiba selenide, tetraphenyl-, from, tetraphenyldistibine, **2**, 890
Distibine, tetra-*t*-butyl-, synthesis of, **2**, 506
Distibine, tetraethyl-, reactions, with phenyllithium, **2**, 510
Distibine, tetramethyl-, cleavage of, **2**, 491
Distibine, tetraphenyl-
as starting material
for distiba selenides, **2**, 890
for tetraphenyldistiba selenide, **2**, 890
Distibine oxides, synthesis of, **2**, 493
Distibines
from, diorganometal(III) alkali metal salts, **2**, 893
synthesis of, **2**, 505; **4**, 595
thermochromic behavior of, **2**, 893
Distibines, symmetrical, from, diarylstibine halides, **2**, 893
Distibines, tetraalkyl-
reactions, with chalcogens, **2**, 499
synthesis of, **2**, 505
N,*N*'-Disuccinimidyl carbonate, as starting material, for carbamates, **6**, 417
Disulfane, [methoxy(thiocarbonyl)]chlorocarbonyl-, synthesis of, **6**, 548
Disulfane, methyl methylthiothiocarbonyl-, synthesis of, **6**, 553
Disulfanes, aryl-, reactions, with sulfenyl chlorides, **2**, 723
Disulfates, peroxy-, reactions, with sulfuric acid, **2**, 782
Disulfenamides
nomenclature, **2**, 767
synthesis of, **2**, 361
Disulfene
reactions
with trialkylsilyl perfluorobutanesulfonates, **6**, 128
with trimethylgermyl perfluorobutanesulfonate, **6**, 132
Disulfide, alkyl acetyl, reactions, with hydrochloric acid, **2**, 211
Disulfide, benzoyl *p*-methylbenzoyl, synthesis of, **5**, 252
Disulfide, bis(bromocarbonyl), from, bis(fluorocarbonyl) disulfide, **6**, 436
Disulfide, bis(chlorocarbonyl), from, bis(trichloromethyl) disulfide, **6**, 434
Disulfide, bisdiethylthiocarbamoyl, reactions, with 4-nitroaniline, **5**, 1029
Disulfide, bis(ferrocenecarbonyl), synthesis of, **5**, 252
Disulfide, bis(fluorocarbonyl), as starting material, for bis(bromocarbonyl) disulfide, **6**, 436
Disulfide, bis(trichloromethyl), as starting material, for bis(chlorocarbonyl) disulfide, **6**, 434
Disulfide, bis(trifluoromethyl)
synthesis of, **6**, 235
for the synthesis of trihalomethyl sulfides, **6**, 234
Disulfide, dibenzoyl, reactions, with cyclohexyl isocyanide, **5**, 1036
Disulfide, diethyl, use in the synthesis of sulfides, **6**, 316
Disulfide, dimethyl
reactions
with bis(benzenethio)methane, **6**, 86

with *n*-butyllithium, **6**, 98
with fluoroiodoalkanes, **6**, 45
with norbornene, **2**, 138
with thiols, **2**, 219
synthesis of, **2**, 185
Disulfide, dimorpholine, reactions, with acetophenones, **5**, 599
Disulfide, diphenyl
reactions
with diethyl ethyl phosphonate, **4**, 346
with sodium hydroxide, **4**, 284
with sulfolanes, **4**, 266
with trialkylboranes, **2**, 712
2,4-Disulfide, 2,4-bis(4-methoxyphenyl)-1,3,2,4-dithiaphosphetane
as thionating reagent, **5**, 570
See also Lawesson's reagent
2,4-Disulfide, 2,4-bis(phenylthio)-1,3-dithia-2,4-diphosphetane, reactions, with benzil dianils, **5**, 731
Disulfides
S-alkylation of, **6**, 316
chlorination of, **6**, 236
chlorinolysis of, **2**, 716
electrolytic reduction of, **2**, 126
from
alkanethiols, **2**, 215
alkylthiosilanes, **2**, 231
alkylthiostannanes, **2**, 231
dithioacetals, **4**, 260
sulfenyl halides, **2**, 229
thiocarbonyls, **2**, 230
thiocyanates, **2**, 231
thiols, **2**, 223
thiosulfinates, **2**, 218
thiosulfonates, **2**, 218
vinyl sulfides, **2**, 232
irridation of, **2**, 127
oxidation of, **2**, 177, 192, 198, 232, 724
reactions
with alkynes, **4**, 940
with amines, **2**, 358
with chloroamines, **2**, 360
with chlorotrimethylsilane, **2**, 265
with 1,1-dihaloalkenes, **6**, 46
with diphosphines, **2**, 460
with dithioacetals, **6**, 85
with enolate anions, **3**, 164
with enolates, **2**, 139
with haloimides, **5**, 336
with hexamethylphosphorus triamide, **2**, 143
with lithium dialkyl amides, **2**, 243
with mercury bis((ethyl diazoacetate), **3**, 483
with metallo enamines, **3**, 35
with methyl iodide, **5**, 830
with nitrile enolates, **3**, 634
with secondary phosphines, **2**, 463
with silyl sulfides, **2**, 228
with sulfenyl halides, **2**, 232
with sulfuryl chloride, **2**, 173
reduction of, **2**, 357; **5**, 338
reductive cleavage of, **2**, 126, 273
as starting materials
for alkanesulfinic esters, **2**, 198
for α-chloroalkanesulfenyl chlorides, **4**, 71
for dialkyl sulfides, **2**, 143
for sulfenamides, **2**, 243, 358, 768
for sulfenimines, **3**, 436
for sulfenyl chlorides, **2**, 166
for sulfides, **2**, 713
for thiocyanates, **5**, 1111
for thiols, **2**, 708; **3**, 165
for thiophosphates, **2**, 256
for thiosilanes, **2**, 265
sulfurization of, **2**, 233
synthesis of, **2**, 212; **6**, 308
via the oxidation of thiols, **2**, 212
via photochemical and thermal reactions, **2**, 228
via ring-opening of thiiranes, **2**, 222
thiolysis of, **2**, 219
use in the synthesis of β-substituted sulfides, **2**, 138
Disulfides, acetyl aryl, synthesis of, **5**, 248
Disulfides, acetyl α-chloro, reactions, with morpholine, **5**, 590
Disulfides, *S*-acetyl 1,1-dichloroalkyl
reactions, with morpholine, **5**, 483
as starting materials, for thioacyl chlorides, **5**, 483
Disulfides, acyl alkoxycarbonyl, reactions, with thiols, **5**, 247
Disulfides, alkyl, synthesis of, **2**, 212
Disulfides, alkyl hydro, from, thiosulfones, **2**, 211
Disulfides, alkylphthalimido, as starting materials, for thioalkylphthalimides, **5**, 338
Disulfides, *N*-alkyl pyridyl, reactions, with thioamides, **5**, 735
Disulfides, amino, synthesis of, **2**, 361
Disulfides, aryl
oxidation of, **2**, 724
synthesis of, **2**, 722
Disulfides, azetidinone imidoyl, synthesis of, **5**, 736
Disulfides, bis(acyl)
reactions, with phosphines, **5**, 243
synthesis of, **5**, 251
Disulfides, bis(alkoxythiocarbonyl), chlorolysis of, **6**, 534
Disulfides, bisamine, nomenclature, **2**, 766
Disulfides, bis(*N*-arylimidoyl), synthesis of, **5**, 735
Disulfides, bis(dialkoxyphosphoryl), reactions, with amines, **2**, 361
Disulfides, bishydrazonyl, reactions, with potassium cyanide, **5**, 641
Disulfides, bis(methylthio)divinyl, as starting materials, for tetrathiosuccimic acid esters, **5**, 556
Disulfides, diacyl, reactions, with isocyanides, **5**, 1036
Disulfides, dialkoxy
reactions
with sodium alkoxides, **2**, 99
with sulfur dichloride, **2**, 100
synthesis of, **2**, 100
Disulfides, dialkyl
reactions, with potassium enolates, **3**, 35
for the α-sulfenylation of enol ethers, **4**, 836
use in the synthesis of tetrathioorthocarbonates, **6**, 303
Disulfides, *S,S'*-dialkyl, reactions, with *N,N*-dichloroamides, **5**, 349
Disulfides, diaryl
N-bromosuccinimide oxidation of, **2**, 721
hydrogen peroxide oxidation of, **2**, 720
reactions
with alkynes, **5**, 238
with Grignard reagents, **2**, 711
reduction of, **2**, 708
synthesis of, via oxidation of thiols, **2**, 723
use in the synthesis of tetrathioorthocarbonates, **6**, 303
Disulfides, dibenzyl, reactions, with carbon monoxide, **5**, 237
Disulfides, di(disulfonamido), reactions, with alcohols, **2**, 100
Disulfides, diorganyl, reactions, with acyllithiums, **5**, 238
Disulfides, hydro, synthesis of, **2**, 211
Disulfides, imidoyl
from, thioamides, **5**, 734
synthesis of, **5**, 734
via oxidation of thioamides, **5**, 735
Disulfides, imidyl, synthesis of, **5**, 638
Disulfides, iminomethyl, oxidation of, **5**, 623
Disulfides, monoacyl-substituted, synthesis of, **5**, 246
Disulfides, phenyl, as starting materials, for thioketones, **3**, 365
Disulfides, α-silyl, as starting materials, for thioaldehydes, **3**, 333

Disulfides, α-sulfinyl, as starting materials, for
thioaldehydes, **3**, 333
Disulfides, α-sulfonyl, as starting materials, for
thioaldehydes, **3**, 340
Disulfides, tetraalkylthiuramide, as starting materials, for
thiocarbamoyl chlorides, **6**, 538
Disulfides, thiouram, as intermediates, in the synthesis of
isothiocyanates, **5**, 1025
Disulfides, trichloromethyl, chlorination of, **6**, 272
Disulfides, vinyl
from, vinylthiolates, **2**, 722
synthesis of, **2**, 722
Disulfonamides, synthesis of, **2**, 366
Disulfonate esters, deprotonation of, **2**, 207
Disulfone, trichloro-, synthesis of, **6**, 254
Disulfones
from
di(*p*-tolylsulfonyl)diazomethane, **6**, 307
1,1,3,3-tetraoxo-1,3-dithietane, **6**, 254
reactions
with dihalides, **4**, 275
with 2-ethoxyvinyl sulfones, **4**, 865
synthesis of, **6**, 307
Disulfones, alkylidene, as starting materials, for
bis(sulfones), **4**, 274
Disulfones, halogenated, synthesis of, **6**, 254
Disulfones, methylene
as starting materials, for bis(sulfones), **4**, 274
synthesis of, **4**, 272
Disulfones, spiro, synthesis of, **4**, 275
Disulfones, tetrachloro, hydrolysis of, **6**, 254
1,2-Disulfones, reductive desulfonylation of, **1**, 617
gem-Disulfones, dibromo, synthesis of, **6**, 254
Disulfonic acid salts, synthesis of, **2**, 190
Disulfonimides
from, sulfonamides, **1**, 28
pyrolysis of, **1**, 782
reductive displacement of, **1**, 28
N,*N*-Disulfonimides, pyrolysis of, **1**, 636
Disulfonium salts, synthesis of, **4**, 863
Disulfonyl fluoride, reactions, with
p-(dimethylamino)benzaldehyde, **4**, 864
Disulfoxides, synthesis of, **2**, 232; **6**, 307
Disulfur dichloride
as chlorinating agent, **6**, 215
reactions
with alcohols, **2**, 99, 100
with arenethiols, **2**, 723
with hydrazones, **3**, 363
with sodium ethoxide, **2**, 100
with sulfur, **2**, 712
as reagent
for the synthesis of diazapentalenes, **3**, 378
for the synthesis of thioketones, **3**, 347
Ditechnetium, decacarbonyl-, synthesis of, **6**, 523
1,3-Ditelluretane, tetrafluoro-, synthesis of, **6**, 256
1,3-Ditelluretanes, *trans*-dibenzylidene-, synthesis of, **4**, 877
Ditelluride, bis(benzoyl), synthesis of, **5**, 253
Ditelluride, bis(*N*,*N*-dimethylcarbamoyl), synthesis
of, **6**, 498
Ditelluride, bis(*o*-methoxybenzoyl), synthesis of, **5**, 253
Ditelluride, diferrocenyl, reduction of, **4**, 290
Ditelluride, diphenyl
reactions
with acylsulfenyl bromide, **5**, 249
with dichloromethane, **4**, 93
with lead(IV)acetate, **5**, 210
with vinyl carbenes, **4**, 877
as starting material, for phenyl tellurol, **1**, 437
Ditelluride, perfluoroalkyl, synthesis of, **6**, 49
Ditelluride anions
from, tellurium, **2**, 292
as starting materials, for alkyl ditellurides, **2**, 292
Ditellurides
as by-products, **2**, 278
from, diaryltellurium dichlorides, **2**, 730
reactions
with arenediazonium salts, **2**, 731
with diazomethane, **4**, 290
with Grignard reagents, **2**, 727
reduction of, **4**, 291
as starting materials
for alkyl aryl tellurides, **2**, 280
for ditelluroacetals, **4**, 290
for unsymmetrical tellurides, **2**, 280
Ditellurides, alkyl
from
ditelluride anions, **2**, 292
metal alkanetellurolates, **2**, 292
tellurium, **2**, 292
synthesis of, **2**, 292
Ditellurides, aryl, synthesis of, **2**, 735
Ditellurides, bis(acyl)
stability of, **5**, 253
synthesis of, **5**, 253
Ditellurides, dialkyl, reactions, with
tetraalkyldiphosphines, **2**, 465
Ditellurides, diaryl
reactions
with carbon monoxide, **5**, 242
with sodium borohydride, **5**, 241
with sodium hydroxide, **4**, 291
synthesis of, **2**, 735
Ditellurides, diphenyl, reactions, with potassium
cyanide, **5**, 1117
1,2-Ditellurides
condensation of, **4**, 877
synthesis of, **4**, 876
Ditelluriums, synthesis of, **4**, 290
Ditelluroacetals
deprotonation of, **4**, 290
from
gem-dihalides, **4**, 290
ditellurides, **4**, 290
reactions, with chlorine, **4**, 291
Ditelluroacetals, ketene, synthesis of, **4**, 876
Ditelluroic acid, structure of, **5**, 545
1,3-Ditelluroles, deprotonation of, **4**, 290
Ditelluromethane, diphenyl-, from, lithium tellurides, **4**, 290
Ditellurophosphines, synthesis of, **6**, 357
Ditellurothioorthocarbonates, theoretical, **6**, 312
Diterpenes, synthesis of, **1**, 810
Diterpenes, lathrane, synthesis of, **4**, 948
1,3,2-Dithiaborinane–dimethyl sulfide, reactions, with tin(II)
chloride, **4**, 257
Dithiaborinanes, reactions, with carboxylic acids, **3**, 83
Dithiacetals, bromo, from, aldehydes, **4**, 245
1,3,2,4-Dithiadiphosphetane, as reagent, for the synthesis of
thiocarboxylates, **5**, 512
1,3,2,4-Dithiadiphosphetane-2,4-dithione,
2,4-bis(phenylthio)-
reactions
with benzil dianils, **4**, 320
with benzil monoanils, **4**, 320
1,2-Dithia-3,7-diselenacycloheptane, 5,5-dimethyl-, use in
the synthesis of disulfides, **2**, 228
Dithiadiselenafulvalenes, synthesis of, **4**, 868
Dithiafulvalene, synthesis of, **6**, 698
1,3-Dithiagermetanes, synthesis of, **2**, 949
Dithiane, 2-benzoyloxy-, synthesis of, **6**, 99
Dithiane, silyl-, solvolysis of, **5**, 422
1,3-Dithiane
metallation of, **6**, 102
reactions
with diaryl diselenides, **6**, 102
with imines, **3**, 43

with sodium periodate, **4**, 268
with trityl fluoroborate, **4**, 1074
as starting material, for 2-seleno-1,3-dithianes, **6**, 102
uses of, **3**, 29
use in the synthesis of α-hydroxycarbonyls, **3**, 157
1,3-Dithiane, 2-carbethoxy-, oxidation of, **4**, 269
1,3-Dithiane, 2-chloro-4,6-dimethyl-, reactions, with sodium (*p*-methoxybenzene)selenolate, **6**, 101
1,3-Dithiane, 2,2-dimethyl-, synthesis of, **4**, 262
1,3-Dithiane, 5,5-dimethyl-, reactions, with sodium periodate, **4**, 268
1,3-Dithiane, 2-ethyl-, deprotonation of, **4**, 261
1,3-Dithiane, 2-lithio-
reactions
with 1-germana-1,1-dichloro-3-cyclopentene, **6**, 132
with germanium compounds, **6**, 132
with α-lithio sulfides, **3**, 169
with nitriles, **3**, 181
with tin compounds, **6**, 134
with titanium compounds, **6**, 134
with triethylgermyl chloride, **6**, 132
with triphenylgermyl chloride, **6**, 132
with α,β-unsaturated ketones, **3**, 169
use in the synthesis of aldehydes, **3**, 20
use in the synthesis of ketones, **3**, 137
1,3-Dithiane, 2-lithio-2-methyl-, reactions, with dichlorodimethylsilane, **6**, 127
1,3-Dithiane, 2-lithio-2-tributylstannyl-, alkylation of, **6**, 134
1,3-Dithiane, 2-lithio-2-trimethylsilyl-
1,2-addition of, **6**, 128
alkylation of, **1**, 282
1,3-Dithiane, 2-methoxy-, from, dichloromethyl methyl ether, **6**, 97
1,3-Dithiane, 2-methyl-, oxidation of, **4**, 260
1,3-Dithiane, 2-methylene-
reactions, with alkyllithiums, **4**, 254
as starting material, for dithioacetals, **4**, 254
1,3-Dithiane, 2-methyl-2-trimethylsilyl-, oxidation of, **4**, 260
1,3-Dithiane, 2-triethylgermyl-, reactions, with *n*-butyllithium, **6**, 520
1,3-Dithiane, 2-triethylgermyl-2-trimethylsilyl-, reactions, with mercuric chloride, **6**, 520
1,3-Dithiane, 2-trimethylsilyl-, acylation of, **6**, 128
1,3-Dithiane-*S,S'*-dioxide, enantiomerically pure, synthesis of, **4**, 269
1,3-Dithiane-1-oxides, 2-lithio-, reactions, with electrophiles, **4**, 263
Dithianes
as acyl anion equivalents, **1**, 165
from, alkynes, **4**, 258
glyoxalate-derived, **1**, 309
hydrolysis of, **5**, 410
oxidation of, with Sharpless reagent, **4**, 261
reactions, with methanesulfonic acid, **4**, 252
as starting materials
for dithioacetals, **4**, 250
for ketals, **1**, 309
uses of, **4**, 244
use in the synthesis of acylgermanes, **5**, 431
use in the synthesis of aromatic ketones, **3**, 291
use in the synthesis of α,β-unsaturated ketones, **3**, 214
Dithianes, acyl-, oxidation of, **2**, 147
Dithianes, α-halo-, as starting materials, for monoselenodithioortho esters, **6**, 101
Dithianes, keto-, reduction of, **2**, 66
Dithianes, lithiated, synthesis of, **6**, 134
Dithianes, lithio-, reactions, with alkoxycarbene complexes, **5**, 951
1,2-Dithianes, from, dibromides, **2**, 225
1,3-Dithianes
anodic oxidation of, **5**, 410
chlorination of, **6**, 48
hydrolysis of, **5**, 424
lithiation of, **6**, 134
reactions, with α,β-unsaturated carbonyls, **1**, 308
reductive cleavage of, **2**, 265
as starting materials, for phosphonium salts, **6**, 120
1,3-Dithianes, 2-alkyl-, lithiation of, **3**, 291
1,3-Dithianes, alkylidene-, reactions, with trifluoroacetic acid, **4**, 252
1,3-Dithianes, 2-aryl-, synthesis of, **3**, 291
1,3-Dithianes, 2-aryl-2-lithio-, arylation of, **3**, 291
1,3-Dithianes, 2,2-bis(triethylgermyl)-, synthesis of, **6**, 326
1,3-Dithianes, 2-chloro-
reactions, with areneselenolates, **6**, 101
synthesis of, **6**, 101
use in coupling reactions, **6**, 48
1,3-Dithianes, 2-germyl-, lithiation of, **6**, 326
1,3-Dithianes, 2-ithio-, reactions, with chlorodiphenylphosphine, **6**, 120
1,3-Dithianes, β-keto-, synthesis of, **4**, 258
1,3-Dithianes, 2-lithio-, reactions, with diethyl chlorophosphonate, **6**, 120
1,3-Dithianes, metallated, synthesis of, **6**, 134
1,3-Dithianes, silyl-, hydrolysis of, **5**, 410
1,3-Dithianes, 2-triethylgermyl-2-trimethylsilyl-, synthesis of, **6**, 326
1,4-Dithianes, tetraoximino-, from, cyanogen *N,N*-dioxide, **5**, 828
α,α-Dithianes, hydrolysis of, **5**, 410
1,3-Dithiane-2-thione, reactions, with amines, **6**, 305
2,9-Dithia-1-phosphabicyclo[4.3.0]-nona-3,7-diene 1-sulfides, synthesis of, **2**, 464
1,3,5-Dithiaphosphorinanes, synthesis of, **4**, 347
1,2,4-Dithiazol-3,5-dithione, *N*-methyl-, synthesis of, **6**, 564
1,2,3-Dithiazole, 4-chloro-5-tosyl-, reactions, with amines, **6**, 651
1,4,2-Dithiazoles, from, thiones, **3**, 686
5*H*-1,2,3-Dithiazoles, 5-(arylimino)-4-chloro-, reactions, with alkylamines, **5**, 736
1,2,4-Dithiazole-3-thione, 5-phenyl-
reactions
with alkenes, **5**, 634
with alkynes, **5**, 634
with benzyne, **5**, 634
synthesis of, **6**, 564
Dithiazolethiones, as starting materials, for nitrile sulfides, **3**, 687
1,2,4-Dithiazolidines, 3,5-dioxo-, reactions, with chlorine, **6**, 449
1,2,4-Dithiazoline, 3-aryl-5-arylimino-, synthesis of, **5**, 657
1,2,4-Dithiazolium cations, reactions, with amines, **6**, 114
1,2,3-Dithiazolium chloride, chlorination of, **5**, 674
1,2,3-Dithiazolium chloride, 4,5-dichloro-, synthesis of, **5**, 674
Dithiazolium salts
cleavage of, **5**, 622
synthesis of, **6**, 562
1,2,4-Dithiazol-5-ones, as starting materials, for nitrile sulfides, **3**, 686
1,2,4-Dithiazol-3-thiones, 5-amino-, desulfurization of, **5**, 1050
1,2,4-Dithiazol-3-thiones, 5-aryl-, desulfurization of, **5**, 1050
1,3,4-Dithiazol-2-thiones, synthesis of, **5**, 1049
3*H*-1,2,4-Dithiazol-3-thiones, 5-aryl-, from, aroyl isothiocyanates, **5**, 1049
Dithiepines, as starting materials, for dithioacetals, **4**, 250
Dithiet, di-*t*-butyl, synthesis of, **3**, 346
Dithietane
from, 1,3-dithietane-1-oxide, **4**, 258
synthesis of, **4**, 852
Dithietane, polycyclic, reactions, with triphenylphosphine, **3**, 351
Dithietane, 1,1,4,4-tetrafluoro-, as starting material, for difluorosulfine, **6**, 530

1,3-Dithietane, as starting material, for thioketenes, **5**, 511
1,3-Dithietane, 2-chloro-2,4,4-trifluoro-, pyrolysis of, **6**, 532
1,3-Dithietane, 2,4-dichloro-2,4-difluoro-, reactions, with trifluoroperacetic acid, **6**, 532
1,3-Dithietane, perfluoro-, oxidation of, **6**, 253
1,3-Dithietane, tetrabromo-
  oxidation of, **6**, 531
  synthesis of, **6**, 254
1,3-Dithietane, 2,2,4,4-tetrachloro-
  reactions, with trichloroperacetic acid, **6**, 530
  as starting material, for dichlorosulfine, **6**, 530
1,3-Dithietane, 2,2,4,4-tetrafluoro-
  oxidation of, **6**, 530
  as starting material, for thiocarbonyl difluoride, **6**, 528
1,3-Dithietane, 2,2,4,4-tetrakis[bis(trifluoromethyl)aminoxy]-, synthesis of, **6**, 312
1,3-Dithietane, 1,1,3,3-tetraoxo-, as starting material, for disulfones, **6**, 254
Dithietane dioxide, pyrolysis of, **5**, 476
Dithietane-1,1-dioxides, synthesis of, **5**, 496
1,2-Dithietane-1,1-dioxides, reactions, with lithium cyanide, **1**, 622
1,3-Dithietane-1-oxide, as starting material, for dithietane, **4**, 258
Dithietanes
  oxidation of, **4**, 854
  synthesis of, **5**, 474
Dithietanes, chlorofluoro-, detection of, **6**, 532
Dithietanes, trifluoro-, synthesis of, **6**, 253
1,3-Dithietanes
  from, bis(trifluoromethyl)thioketene, **6**, 312
  reactions, with 1,2-diaminobenzene, **4**, 415
  synthesis of, **6**, 49, 253, 311
1,3-Dithietanes, halogenated, systhesis of, **6**, 254
1,3-Dithietanes, 2-methylene-, as thioacylating reagent, **5**, 641
Dithietanones, as starting materials, for thioketenes, **3**, 550
1,3-Dithietanones, 2,4-bis(alkylidene)-, as starting materials, for thioketenes, **3**, 550
2,3-Dithietanones, ring-opening of, **2**, 211
1,2-Dithietes
  from, α-dithiones, **3**, 330
  synthesis of, **3**, 352
Dithiets, synthesis of, **3**, 366
1,2-Dithiin, 5,6-dihydro-, synthesis of, **2**, 229
Dithioacetal, benzaldehyde diethyl, as starting material, for triethyl trithioorthobenzoate, **6**, 86
Dithioacetal, formaldehyde, from, dithiols, **4**, 249
Dithioacetal, formaldehyde dimethyl, chlorination of, **2**, 168
Dithioacetal, hydroxyketene, selenocyclization of, **6**, 98
Dithioacetal, γ-tosyloxy, Grob fragmentation of, **4**, 846
Dithioacetal *S*,*S*-dioxides, ketene
  from
    aliphatic aldehydes, **4**, 858
    alkyl halides, **4**, 857
    methyl methylthiomethyl sulfones, **4**, 857
    methylthiomethyl *p*-tolyl sulfone, **4**, 857
    thiocarbonyls, **4**, 862
    vinyl sulfones, **4**, 861
  isomerization of, **4**, 857
  oxidation of, **4**, 854
  synthesis of, **4**, 853, 856
Dithioacetal *S*,*S'*-dioxides, ketene, synthesis of, **4**, 853, 862
Dithioacetalization
  of aldehydes, **4**, 247
  of ketones, **4**, 247
Dithioacetal monosulfoxides, ketene
  synthesis of, **4**, 856
    via elimination reactions, **4**, 858
Dithioacetal monoxides, 2-acetoxy, as intermediates, in the synthesis of ketene dithioacetal monoxides, **4**, 857
Dithioacetal monoxides, ketene, (*E*)–(*Z*) isomerization of, **4**, 856
Dithioacetal *S*-oxides, ketene
  from
    alkyl alkylthiomethyl sulfoxides, **4**, 857
    α-alkylthiosulfines, **4**, 859
    alkynes, **4**, 859
    methyl methylthiomethyl sulfoxides, **4**, 856
    phosphonate esters, **4**, 860
    α-substituted dithioacetals, **4**, 855
    vinyl sulfoxides, **4**, 858
  oxidation of, **4**, 854
Dithioacetals
  alkylation of, **2**, 141
  asymmetrically-substituted, synthesis of, **4**, 255
  chlorination of, **2**, 168
  cyclization of, **4**, 228
  enantiomerically enriched, synthesis of, **4**, 248
  enantioselective oxidation of, **4**, 263
  from
    acetals, **4**, 245
    aldehydes, **2**, 541; **4**, 245
    carboxylic acids, **4**, 257
    α-chloro sulfides, **4**, 258
    gem-dichlorides, **4**, 245, 249
    dideuteriodichloromethane, **4**, 249
    dithioacetals, **4**, 249
    ketene dithioacetals, **4**, 252
    ketones, **4**, 245
    2-methylene-1,3-dithiane, **4**, 254
    thiocarbonyls, **4**, 255
    thioglycolic acid, **4**, 272
    trithioorthoformates, **4**, 250
  as intermediates, in the syntheses of gingerols, **4**, 245
  C-lithium compounds, synthesis of, **6**, 327
  metallation of, **6**, 85
  methylation of, **4**, 216
  oxidation of, **4**, 853
    with peroxides, **6**, 99
    for the synthesis of α-thiosulfoxides, **4**, 258
  S-oxide, synthesis of, **4**, 261
  reactions
    with alkyllithiums, **2**, 141
    with *n*-butyllithium, **4**, 1075
    with disulfides, **6**, 85
    with Raney nickel, **1**, 11
    with succinimide, **6**, 140
    with tosyl azide, **4**, 860
    with trialkylboranes, **4**, 372
  as reagents, **4**, 269
  reduction of, **1**, 21, 24
  reviews, **4**, 244
  C-silicon compounds, synthesis of, **6**, 327
  as starting materials
    for α-alkylthiosulfoxides, **4**, 258
    for bis(sulfones), **4**, 272
    for bissulfoxides, **4**, 267
    for disulfides, **4**, 260
    for dithioacetals, **4**, 249
    for dithioketals, **4**, 250
    for dithioortho esters, **6**, 98
    for epoxy-bis(sulfone), **4**, 273
    for monoselenodithioortho esters, **6**, 102
    for α-sulfoxide sulfones, **4**, 271
    for trithioortho esters, **6**, 85
    for vinyl sulfides, **2**, 710
  synthesis of, **1**, 1058; **4**, 244, 245
    via sulfenation of methylenes, **6**, 325
  for the synthesis of radical centres flanked by two sulfur atoms, **4**, 1080
  C-tin compounds, synthesis of, **6**, 327
  use in the synthesis of phenyl ketones, **3**, 295
Dithioacetals, α-acetoxy, oxidation of, **4**, 855
Dithioacetals, acyclic, as starting materials, for cyclic dithioacetals, **4**, 228

Dithioacetals, acyl
as starting materials, for hydroxy aldehydes, **3**, 30
synthesis of, **4**, 247
Dithioacetals, α-acyloxy, as starting materials, for trithioortho esters, **6**, 82
Dithioacetals, α-alkoxy, as starting materials, for trithioortho esters, **6**, 82
Dithioacetals, allylketene, rearrangement of, **1**, 834
Dithioacetals, chloral, oxidation of, **4**, 855
Dithioacetals, α-chloro, reactions, with thiols, **6**, 82
Dithioacetals, crotonaldehyde, reactions, with benzaldehyde, **4**, 849
Dithioacetals, *S*-crotylketene, thia–Claisen rearrangement of, **1**, 834
Dithioacetals, cyclic, from, acyclic dithioacetals, **4**, 228
Dithioacetals, cyclopropyl, from, β-oxo dithio esters, **4**, 255
Dithioacetals, deuterium-labeled, synthesis of, **4**, 249, 250
Dithioacetals, diphenyl
reductive lithiation of, **4**, 380, 394
as starting materials, for α-thioalkylboranes, **4**, 372
Dithioacetals, fluoro, synthesis of, **4**, 252
Dithioacetals, α-halo, as starting materials, for trithioortho esters, **6**, 82
Dithioacetals, hydroxy, from, dihydropyran, **4**, 246
Dithioacetals, ketene
alkylation of, **4**, 860
chlorination of, **4**, 743
Diels–Alder reaction of, **4**, 265
from
acid chlorides, **4**, 846
carboxylic acids, **4**, 845
carboxylic esters, **4**, 845
dichloroalkyne, **4**, 851
gem-dihaloalkenes, **4**, 851
dithiocarboxylates, **4**, 843
dithiocarboxylic acids, **4**, 842
1,3-dithioles, **4**, 850
dithiolium salts, **4**, 847
methyl cyanoacetate, **4**, 843
β-oxo dithioic acids, **5**, 555
α-thiodithioacetals, **4**, 846
thioketenes, **4**, 852
p-tosylhydrazones, **4**, 852
trithioortho esters, **4**, 844
α,β-unsaturated dithioesters, **4**, 843
as intermediates, **4**, 969
for dithioortho esters, **6**, 98
monosulfoxidations of, **4**, 853
oxidation of, **4**, 853
polysulfoxidation of, **4**, 853
reactions
with alkenes, **1**, 358
with amines, **4**, 969
with diamines, **4**, 970
with dimethylamine, **4**, 969
with methyl fluorosulfonate, **4**, 863
with trifluoroacetic anhydride, **3**, 253
reduction of, **1**, 438
as starting materials
for *gem*-disulfoximidoalkenes, **4**, 997
for dithioacetals, **4**, 252
for α-hydrazino enamines, **4**, 983
for ketene aminals, **4**, 969
for ketene dithioacetal *S*-imides, **4**, 860
for ketene hemiaminals, **4**, 888
for ketene monothioacetals, **4**, 836
substitution of, **1**, 484
synthesis of, **1**, 140; **4**, 842
alkenation methods, **4**, 847
via dehydrogenation, **4**, 847
via double bond formation by elimination, **4**, 844
via Peterson alkenation reactions, **4**, 849
via rearrangement of α,β-unsaturated dithioacetals, **4**, 849
Dithioacetals, keto, from, dithioorthoformates, **4**, 250
Dithioacetals, lithiated
reactions
with benzophenenone, **6**, 98
with sulfur, **6**, 86
Dithioacetals, mono-*S*-trimethylsilylketene, reactions, with imines, **2**, 313
Dithioacetals, nitroketene, reactions, with hydrazines, **4**, 983
Dithioacetals, non-polar, synthesis of, **4**, 261
Dithioacetals, β-oxo, from, α-oxoketene dithioacetals, **4**, 253
Dithioacetals, oxoketene, reactions, with hydrogen, **4**, 253
Dithioacetals, α-oxoketene
reactions, with organometallics, **1**, 485
as starting materials, for β-oxo dithioacetals, **4**, 253
Dithioacetals, *S*-polyfluoroketene, synthesis of, **4**, 868
Dithioacetals, α-substituted, as starting materials, for ketene dithioacetal *S*-oxides, **4**, 855
Dithioacetals, theophylline, synthesis of, **4**, 259
Dithioacetals, trifluoromethyl
reactions
with lithium biphenylide, **4**, 775
with lithium naphthalenide, **4**, 775
Dithioacetals, α,β-unsaturated, rearrangement of, **4**, 849
Dithioacetals, unsymmetrical, synthesis of, **4**, 247
Dithioacetal tetroxides, ketene
cycloaddition reactions of, **4**, 277
Diels–Alder reaction of, **4**, 277
as electrophiles, **4**, 277
from
disulfonylmethanes, **4**, 864
vinyl ethers, **4**, 865
as starting materials, for bis(sulfones), **4**, 277
synthesis of, **4**, 855, 864
See also Alkenes, 1,1-disulfonyl-
Dithioacetal *S,S,S',S'*-tetroxides, ketene
synthesis of, **4**, 853
See also Alkenes, 1,1-disulfonyl-
Dithioacetal trioxides
deprotonation of, **4**, 271
synthesis of, **4**, 271
Dithioacetal trioxides, ketene, Diels–Alder reaction of, **4**, 272
Dithioacetal *S,S,S'*-trioxides, ketene, synthesis of, **4**, 853, 864
Dithioacetate, carboxymethyl, as thioacylating reagent, **5**, 586
Dithioacetate, methyl, reactions, with hydrazines, **5**, 639
Dithioacetate, methyl azido-, reactions, with piperidine, **5**, 586
Dithioacetate, methyl triphenylphosphoranylidene, reactions, with sodium hexamethyldisilylamide, **3**, 547
Dithioacetic acid
from, methylmagnesium halides, **5**, 546
as starting material, for thioacetyl chloride, **5**, 477
Dithioacetic acid, trifluoro-, synthesis of, **5**, 546
Dithioacetic acid *S*-allyl esters, oxidation of, **5**, 562
Dithioacetic acid esters, α-phenyl-, synthesis of, **5**, 550
Dithioacetic acid *S*-methyl ester, synthesis of, **5**, 550
Dithioacid, use in the synthesis of thionoamides, **5**, 583
Dithioacid amine salts, synthesis of, **5**, 583
Dithioacid anhydrides, for the thioacylation of amines, **5**, 620
Dithioacids
reactions of, **5**, 589
reactions
with amines, **5**, 614
with isonitriles, **5**, 609
as starting materials, for nitriles, **3**, 618
thioacylation of, **5**, 636
Dithioacids, hydrazinium salts, as starting materials, for hydrazides, **5**, 639
Dithioacids, metal salts, reactions, with hydrazines, **5**, 639

Dithioacrylic acid esters, β,β-dimethyl, synthesis of, **5**, 559
Dithioalkanoic acid esters, synthesis of, **5**, 550
Dithioalkanoic acids, synthesis of, **5**, 546
Dithioalkanoic acids, *S*-deuterio-, synthesis of, **5**, 547
Dithioalkanoic acids, β-imino-, synthesis of, **5**, 547
α,β-Dithioalkenals, synthesis of, **3**, 73
Dithioalkenoic acids, magnesium salts, synthesis of, **5**, 548
Dithioamide, keto-, synthesis of, **3**, 345
Dithioamides
  synthesis of, via acyclation of dithioamines, **5**, 338
  use in the synthesis of distibines, **4**, 595
Dithioanhydrides, as intermediates, in the synthesis of thioketenes, **3**, 547
Dithioanthraquinone, stability of, **3**, 358
Dithioate, hexafluoro, reactions, with trifluoromethanethiol, **4**, 256
Dithioates, alkane-, oxidation of, **5**, 248
Dithioates, *O*-alkylcarbonochloro-, reactions, with sodium isothiocyanate, **6**, 558
Dithioates, carboxymethyl
  reactions, with sodium azide, **5**, 634
  as thioacylating reagents, **5**, 640
Dithioates, furfuryl, reactions, with ethylmagnesium bromide, **4**, 255
Dithioates, methyl 3-alkyne, synthesis of, **1**, 1058
Dithiobenzoate, carboxymethyl, as thioacylating reagent, **5**, 586
Dithiobenzoate, methyl, reactions, with *N*-sulfinyldimethylammonium tetrafluoroborate, **5**, 731
Dithiobenzoate, phenyl, reactions, with phenyllithium, **4**, 255
Dithiobenzoate esters, as starting materials, for orthobenzoate esters, **6**, 73
Dithiobenzoates, methyl, reduction of, **1**, 24
Dithiobenzoic acid ester, 4-vinyl-, polymerization of, **5**, 561
Dithiobenzoic acid esters, 4-hydroxy-, synthesis of, **5**, 561
Dithiobenzoic acid esters, 4-siloxy, synthesis of, **5**, 560
Dithiobenzoic acid *S*-phenyl ester, synthesis of, **5**, 560
Dithiobenzoic acids, as starting materials, for thiobenzoyl bromides, **5**, 486
Dithiobenzoic acid *S*-trifluoromethyl esters, synthesis of, **5**, 561
Dithio-*p*-benzoquinone, synthesis of, via pyrolysis, **3**, 358
Dithio-*o*-benzoquinones
  structure of, **3**, 358
  trapping as Diels–Alder adducts, **3**, 358
Dithiobisamines, reactions, with sulfuryl chloride, **2**, 360
N,*N*-Dithiobisamines, synthesis of, **2**, 361
Dithiobissuccinimide, as reagent, for the synthesis of thioketones, **3**, 349, 358
Dithiobissulfonamides, reactions, with chlorine, **2**, 365
Dithiocarbamate
  from
    carbon disulfide, **6**, 563
    isothiocyanates, **6**, 565
Dithiocarbamate, ammonium, reactions, with sodium hypochlorite, **5**, 1025
Dithiocarbamate, *S*-benzyl *N*-phenyl, synthesis of, **6**, 565
Dithiocarbamate, carbethoxy, as intermediate, in the synthesis of isothiocyanates, **5**, 1023
Dithiocarbamate, carbomethoxy, as intermediate, in the synthesis of isothiocyanates, **5**, 1023
Dithiocarbamate, *N*,*N*-diethyl *O*-phenyl, from, phenol, **6**, 556
Dithiocarbamate, dimethylammonium *N*,*N*-dimethyl, reactions, with benzonitrile, **5**, 582
Dithiocarbamate, dimethylammonium *N*,*N*'-dimethyl, reactions, with carbon disulfide, **5**, 609
Dithiocarbamate, methyl *N*,*N*-dimethyl, reactions, with dimethyl sulfate, **6**, 314
Dithiocarbamate, *S*-methyl *N*,*N*-dimethyl, synthesis of, **6**, 567
Dithiocarbamate, tetra-*n*-butyl-*n*-thioformyl, reactions, with iodoalkanes, **6**, 561
Dithiocarbamate anions
  alkylation of, for dithio-substituted selenocarbonates, **6**, 589
  reactions, with α-halocarbonyls, **6**, 561
Dithiocarbamate esters
  chlorination of, **6**, 728
  decomposition of, **5**, 1026
Dithiocarbamates
  alkylation of, **6**, 559, 627, 730
  carboxymethyl esters of, synthesis of, **5**, 1026
  chlorination of, **6**, 605
  cleavage of, under oxidative conditions, **5**, 1025
  decomposition of, **5**, 1023
  from
    2-aminothiazole, **6**, 564
    disaccharides, **6**, 561
    dithiocarbamates, **6**, 561
    epoxides, **6**, 562
    thiocarbamoyl chlorides, **6**, 566
    thiuram disulfides, **6**, 566
    tosylates, **6**, 562
  reactions
    with haloalkanes, **6**, 561
    with isocyanates, **6**, 627
    with methylenes, **5**, 593
    with thiocarbamates, **2**, 218
  as starting materials
    for dithiocarbamates, **6**, 561
    for thiols, **2**, 124
  synthesis of, **2**, 889; **6**, 560
  Wittig rearrangement of, **5**, 614
Dithiocarbamates, alkyl, as starting materials, for thiols, **2**, 124
Dithiocarbamates, alkylammonium, synthesis of, **6**, 560
Dithiocarbamates, *S*-alkyl-*n*-formyl, synthesis of, **6**, 563
Dithiocarbamates, *S*-alkyl-*n*-formyl-*N*-methyl, synthesis of, **6**, 561
Dithiocarbamates, *S*-allyl *N*-aryl, synthesis of, **6**, 563
Dithiocarbamates, allylic, rearrangement of, **1**, 836
Dithiocarbamates, aryl, reactions, with phosgene, **5**, 1023
Dithiocarbamates, *S*-aryl, synthesis of, **6**, 562
Dithiocarbamates, aryl *N*,*N*-dimethyl, as starting materials, for cyclic bis(dithioorthocarbonates), **6**, 313
Dithiocarbamates, *N*-aryl *S*-methyl, from, anilines, **6**, 563
Dithiocarbamates, *N*-aryl *S*-nonyl, synthesis of, **6**, 563
Dithiocarbamates, *S*-cyclopentadienyl, synthesis of, **6**, 567
Dithiocarbamates, dialkyl, reactions, with *N*-iodoamides, **5**, 335
Dithiocarbamates, *N*,*N*-dialkyl, reactions, with α-bromocarboxylic acids, **6**, 561
Dithiocarbamates, ethyl, as starting materials, for alkyl isothiocyanates, **5**, 1026
Dithiocarbamates, *N*-hydroxy, synthesis of, **6**, 567
Dithiocarbamates, hydroxybenzyl dialkyl, synthesis of, **6**, 565
Dithiocarbamates, *N*-methyl-*N*-phenyl, synthesis of, **6**, 567
Dithiocarbamates, 1,3,4-oxadiazol-2-ylmethyl *N*-aryl, oxidation of, **6**, 564
Dithiocarbamates, *N*-substituted, as starting materials, for isothiocyanates, **5**, 1023
Dithiocarbamates, *N*,*N*,*S*-trialkyl, synthesis of, **6**, 563
Dithiocarbamate salts
  acylation of, **6**, 562
  alkylation of, **6**, 561
  as intermediates, in the synthesis of isothiocyanates, **5**, 1023
  oxidation of, **6**, 563
  reactions, with iodomethane, **6**, 561
  synthesis of, **6**, 560
Dithiocarbamic acid esters, synthesis of, **6**, 561
Dithiocarbamic acid esters, *N*-thioformyl, synthesis of, **6**, 561
Dithiocarbamic acids

reactions, with vinyl ethers, **6**, 565
synthesis of, **6**, 560
Dithiocarbamic esters
chlorination of, **6**, 617
reactions, with α-amino acids, **6**, 575
Dithiocarbamidate, reactions, with carboxylic acid chlorides, **5**, 537
Dithiocarbamidic acid, synthesis of, **6**, 560
2-Dithiocarbamoyldithioles, from, dithiolium salts, **6**, 85
Dithiocarbazates, acyl-, from, acylhydrazines, **6**, 563
Dithiocarbazate salts, synthesis of, **6**, 560
Dithiocarbazic acids, stability of, **6**, 594
Dithiocarbene, equilibrium with lithiotrithioorthoformates, **6**, 317
Dithiocarbene complexes, synthesis of, **6**, 720
Dithiocarbenes, trimethylene-, synthesis of, **4**, 250
Dithiocarbenium salts
reactions, with alkoxides, **6**, 98
as starting materials
for dithioortho esters, **6**, 98
for trithioortho esters, **6**, 85
Dithiocarbonate, *S,S'*-bis(1-phenyl-1*H*-tetrazol-5-yl), as reagent, for the synthesis of isothiocyanates, **5**, 1024
Dithiocarbonate, *O*-butyl *S*-ethoxycarbonyl, synthesis of, **6**, 546
Dithiocarbonate, *S*-cyclopentylmethyl *O*-ethyl, synthesis of, **6**, 547
Dithiocarbonate, *O,S*-dimethyl, reactions, with sulfuryl chloride, **6**, 276
Dithiocarbonate, *S*-ethoxycarbonyl *O*-ethyl, synthesis of, **6**, 476
Dithiocarbonate, *O*-ethyl *S*-(3-*O*-acetyl-2-azido-4-(benzyloxycarbonyl)amino-2,4,6-trideoxy-β-D-galactopyranosyl), synthesis of, **6**, 549
Dithiocarbonate, *O*-ethyl *S*-cyanomethyl, use in the synthesis of alkyl α,β-unsaturated nitriles, **3**, 644
Dithiocarbonate, *O*-ethyl *S*-(2-hydroxy-4-oxo-4-phenylbutyl), synthesis of, **6**, 552
Dithiocarbonate, *S*-ethyl *O*-methyl, synthesis of, **6**, 550
Dithiocarbonate, *O*-ethyl *S*-(2-oxoethyl), reactions, with lithium enolates, **6**, 552
Dithiocarbonate, *O*-ethyl *S*-triphenylmethyl, synthesis of, **6**, 546
Dithiocarbonate, *S*-5-hexenyloxy *O*-ethyl, irradiation of, **6**, 547
Dithiocarbonate, *O*-isopropyl (*Z*)-*S*-4-phenylbut-3-enyl, synthesis of, **6**, 548
Dithiocarbonate, *S*-methyl *O*-trimethylsilyl, synthesis of, **6**, 546
Dithiocarbonate, triethylammonium, alkylation of, **6**, 471
Dithiocarbonate esters, reactions, with *N,N*-dimethylhydrazine, **6**, 560
Dithiocarbonate rearrangement, for the synthesis of thiotetronic acids, **6**, 473
Dithiocarbonates
chlorine addition to, **6**, 276
from
alcohols, **6**, 550
dithiocarbonates, **6**, 551
dithiocarbonic acid *O*-ester salts, **6**, 545
reactions
with α,ω-alkanedithiols, **6**, 305
with carbamoyl chlorides, **6**, 493
with organic tellurium(II) chloride, **2**, 238
with trialkyl phosphites, **6**, 120
with zinc chloride, **2**, 273
as starting materials, for dithiocarbonates, **6**, 551
synthesis of, **2**, 889; **6**, 545
via alkenes, **6**, 551
Dithiocarbonates, *S*-(alkoxycarbonyl) *O*-alkyl, synthesis of, **6**, 547, 551
Dithiocarbonates, *O*-(1-alkoxy-2,2-dialkylvinyl) *S*-alkyl, synthesis of, **6**, 550
Dithiocarbonates, *O*-alkyl, reactions, with carbanions, **5**, 513
Dithiocarbonates, *O*-alkyl *S*-phthaloylglycyl, synthesis of, **6**, 546
Dithiocarbonates, *O*-allylic, rearrangement of, **1**, 836
Dithiocarbonates, *S*-allylic, synthesis of, **1**, 836
Dithiocarbonates, *S*-cyclopentylmethyl *S'*-methyl, synthesis of, **6**, 473
Dithiocarbonates, dialkyl
from
phosgene, **6**, 471
thiones, **6**, 471
synthesis of, via [3,3] sigmatropic rearrangement, **6**, 472
Dithiocarbonates, *O,S*-dialkyl
from
alcohols, **6**, 549
alkoxythiocarbonyl chlorides, **6**, 550
carbon disulfide, **6**, 549
thiocarbonyl chlorides, **6**, 550
reactions, with bis(trifluoromethyl)thioketene, **6**, 311
Dithiocarbonates, *S,S*-dialkyl
reactions, with alkyl halides, **2**, 134, 143
synthesis of, via [3,3] sigmatropic rearrangements, **6**, 472
Dithiocarbonates, *S,S'*-dialkyl, synthesis of, **6**, 471
Dithiocarbonates, *S,S*-dialkyl imino-, from, amino compounds, **6**, 627
Dithiocarbonates, *S*-glycosyl *O*-ethyl, synthesis of, **6**, 546
Dithiocarbonates, *N*-hydroxyimino, from, hydroxylamine, **6**, 628
Dithiocarbonates, imino-
N-alkylation of, **6**, 730
from
carbonimidic halides, **6**, 626
cyanogen chloride, **6**, 627
thiocyanates, **6**, 627
as starting materials, for isothioureas, **6**, 636
synthesis of, **6**, 113, 626
S-alkylation methods, **6**, 627
via functional group interconversion, **6**, 628
See also Carbonimidodithioic acids
Dithiocarbonates, α-phosphoryl, oxygenation of, **6**, 474
Dithiocarbonates, thiobenzoylimino-, reactions, with diphenylcarbene, **6**, 314
Dithiocarbonates, triphenyltelluronium, synthesis of, **6**, 547
Dithiocarbonate salts, *O*-alkyl
acylation of, **6**, 546
alkylation of, **6**, 546
reactions, with halides, **6**, 547
Dithiocarbonate salts, imino-, synthesis of, **6**, 730
Dithiocarbonic acid, stability of, **6**, 545
Dithiocarbonic acid *O*-alkyl esters, synthesis of, **6**, 545
Dithiocarbonic acid chloride, as starting material, for trithiocarbonic acid diesters, **6**, 553
Dithiocarbonic acid *O,S*-diesters, synthesis of, **6**, 545
Dithiocarbonic acid *O*-ester salts, as starting materials, for dithiocarbonates, **6**, 545
Dithiocarbonimidates, diethyl, reactions, with 2-acetylthiophene, **4**, 980
Dithiocarboxylate, phenyl diphenyl, pyrolysis of, **3**, 547
Dithiocarboxylate esters, as starting materials, for trifluoromethyls, **6**, 10
Dithiocarboxylates
reactions, with haloimides, **5**, 336
as starting materials
for 1,1-dichlorosulfenyl chlorides, **5**, 478
for ketene dithioacetals, **4**, 843
for thiocarboxylates, **5**, 509, 510
for trifluoromethyls, **6**, 10
synthesis of, *in situ*, **4**, 843
Dithiocarboxylates, α-bromo, reactions, with diethylamine, **4**, 843
Dithiocarboxylates, carboxymethyl, for the thioacylation of amines, **5**, 620

Dithiocarboxylates, ethyl, synthesis of, **5**, 575
Dithiocarboxylates, *O*-homoallyl, radical cyclization of, **5**, 513
Dithiocarboxylates, methyl
  use in the synthesis of Leu-enkephalins, **5**, 587
  use in the synthesis of thionoamides, **5**, 583
Dithiocarboxylates, piperidinium, synthesis of, **5**, 584
Dithiocarboxylation, of Grignard reagents, **4**, 843
Dithiocarboxylic acid dianions
  monoacylation of, **3**, 547
  reactions, with phosgene, **3**, 550
Dithiocarboxylic acid esters
  reactions, with hydrazines, **5**, 639
  for the thioacylation of hydroxylamines, **5**, 636
Dithiocarboxylic acids
  oxidation of, **3**, 254
  oxidative dimerization of, **4**, 852
  reactions
    with 9-borabicyclo[3.3.1]nonyl, **5**, 589
    with catecholborane, **5**, 589
    with dialkoxychlorophosphines, **2**, 464
  as starting materials
    for imidazolides, **5**, 477
    for ketene dithioacetals, **4**, 842
    for ortho esters, **6**, 73
    for thioacyl chlorides, **5**, 476
    for thioketenes, **3**, 547
  synthesis of, reviews, **5**, 81
  for the thioacylation of amines, **5**, 588
Dithiocarboxylic acid salts, for the thioacylation of amines, **5**, 620
Dithiocarboxylic esters, reactions, with dialkoxystannanes, **6**, 73
Dithio cations, synthesis of, **4**, 1073
Dithiocinnamic acid, synthesis of, **5**, 548
Dithiocyanamic acid, synthesis of, **5**, 559
Dithiocyanate, thiocarbonyl, synthesis of, **6**, 559
Dithiocyanates, bis(alkynyl), from, bis(alkynyl) bis iodonium salts, **2**, 1034
α,β-Dithiocyanates, synthesis of, **5**, 1110
1,1-Dithio-1,3-dienes
  reactions
    with ethyllithium, **4**, 255
    with organolithiums, **4**, 254
Dithiodiglycollic acid, reduction of, **2**, 126
2,2-Dithio-1,3-dithianes, synthesis of, **6**, 301
α-Dithio dithioesters, synthesis of, **5**, 558
2,2-Dithio-1,3-dithioles, synthesis of, **6**, 301
Dithioester, dienyl, synthesis of, **5**, 559
Dithioester *S*-oxides, reactions, with thallium ethoxide, **4**, 859
Dithioesters
  aminolysis of, **5**, 586
  bromination of, **2**, 171
  chlorination of, **5**, 490
  from
    alkyl chlorodithioformates, **1**, 1047
    1,1-dichloroethylene, **5**, 552
  reactions
    with allenylmagnesium bromide, **1**, 1058
    with cyanamide, **6**, 630
    with Grignard reagents, **4**, 255
    with maleic anhydride, **4**, 843
    with organolithiums, **4**, 255
    with organomagnesiums, **6**, 134
    with organometallics, **4**, 255
    with ynamines, **5**, 614
  as starting materials
    for cyclopropanols, **4**, 255
    for hemidithioacetals, **4**, 245, 256
  synthesis of, **5**, 550
  as thioacylating reagents, **5**, 586
  for the thioacylation of amines, **5**, 584
  use in the synthesis of distibines, **4**, 595
Dithioesters, β-amino, synthesis of, **2**, 313
Dithioesters, β-amino-α,β-unsaturated, synthesis of, **5**, 559, 586
Dithioesters, α-azido, synthesis of, **5**, 558
Dithioesters, α,α-bis(trifluoromethyl), synthesis of, **5**, 551
Dithioesters, α-carboxy, synthesis of, **5**, 555
Dithioesters, α-cyano, synthesis of, **5**, 558
Dithioesters, ferrocene, from, dithioic acid thioanhydrides, **5**, 561
Dithioesters, β-hydroxy
  dehydration of, **5**, 558
  synthesis of, **5**, 557
Dithioesters, imido, sulfhydrolysis of, **5**, 556
Dithioesters, α-oxo
  from
    halomethyl aryl ketones, **5**, 554
    α-sulfonyl ketones, **5**, 554
  synthesis of, **5**, 554
Dithioesters, β-oxo, as starting materials, for cyclopropyl dithioacetals, **4**, 255
Dithioesters, δ-oxo, synthesis of, **5**, 556
Dithioesters, α-oxy, synthesis of, **5**, 558
Dithioesters, perimidine, from, chloromethyl-1*H*-perimidinium chloride, **5**, 559
Dithioesters, α-silyl, reactions, with aldehydes, **5**, 559
Dithioesters, α-sulfinyl, synthesis of, **5**, 558
Dithioesters, α-sulfonyl, synthesis of, **5**, 558
Dithioesters, α-thioxo, synthesis of, **5**, 554
Dithioesters, γ-thioxo, synthesis of, **5**, 556
Dithioesters, δ-thioxo, synthesis of, **5**, 556
Dithioesters, *S*-trihalomethyl, synthesis of, **6**, 235
Dithioesters, α,β-unsaturated, as starting materials, for ketene dithioacetals, **4**, 843
Dithioesters, β,γ-unsaturated, isomerization of, **5**, 559
Dithioesters, γ,δ-unsaturated, synthesis of, **5**, 552
Dithioformate, ethyl chloro-, synthesis of, **6**, 535
Dithioformate, methyl chloro-
  reactions, with methylmagnesium, **6**, 306
  synthesis of, **6**, 535
Dithioformate, methyl cyano-, reactions, with amines, **5**, 587
Dithioformate, *S*-methyl (diisopropoxyphosphonyl), reactions, with carbon nucleophiles, **6**, 121
Dithioformate, perfluoromethyl fluoro-, from, trifluoromethanethiol, **6**, 534
Dithioformate, phenyl chloro-
  from, phenyl chlorodithioformate, **6**, 535
  as starting material, for phenyl chlorodithioformate, **6**, 535
Dithioformate, trichloromethyl chloro-, from, trichloromethanesulfenyl chloride, **6**, 535
Dithioformate, trifluoromethyl bromo-, synthesis of, **6**, 536
Dithioformate ester complexes, as starting materials, for organoirons, **6**, 332
Dithioformates, alkali metal chloro-, synthesis of, **6**, 535
Dithioformates, alkyl bromo-, synthesis of, **6**, 536
Dithioformates, alkyl chloro-
  as starting materials, for dithio esters, **1**, 1047
  synthesis of, **6**, 535
Dithioformates, alkyl fluoro-, synthesis of, **6**, 534
Dithioformates, alkyl halo-, chlorination of, **6**, 233
Dithioformates, alkyl trimethylsilyl, synthesis of, **6**, 584
Dithioformates, *S*-alkyl(triphenylstannyl), reactions, with platinum complexes, **6**, 332
Dithioformates, allyl chloro-, reactions, with 8-hydrazino-5-chloro-3,4-dihydro-2,6-dimethylbenzopyran, **6**, 567
Dithioformates, aryl chloro-, synthesis of, **6**, 535
Dithioformates, bromo-
  from, thiols, **6**, 536
  synthesis of, **6**, 536
Dithioformates, chloro-
  from
    carbon monosulfide, **6**, 535

thiols, **6**, 535
reactions, with Grignard reagents, **3**, 364
as starting materials, for fluorodithioformates, **6**, 534
synthesis of, **5**, 483; **6**, 535
Dithioformates, fluoro-
dimerization of, **6**, 275
from, chlorodithioformates, **6**, 534
synthesis of, **6**, 534
Dithioformates, pentachlorophenyl chloro-, synthesis of, **6**, 535
Dithioformates, *C*-sulfonyl, as starting materials, for 1,4,2-oxathiazoles, **6**, 309
Dithioformates, thiophosphinoyl
aminolysis of, **6**, 584
as starting materials, for phosphinothioformamides, **6**, 584
Dithioformates, trimethylsilyl, synthesis of, **6**, 585
Dithioformic acid, synthesis of, **5**, 546
Dithioformic acid esters, stability of, **5**, 550
Dithioformic acids, *S*-ethyl isothiocyano-, synthesis of, **6**, 565
Dithioformic acids, *S*-propyl isothiocyano-, synthesis of, **6**, 565
Dithioglyoxal, synthesis of, via cycloreversion, **3**, 335
Dithioglyoxal, *t*-butyl, synthesis of, via cycloreversion, **3**, 335
Dithioic acid, structure of, **5**, 545
Dithioic acid, ferrocene, synthesis of, **5**, 549
Dithioic acid *S*-allylic esters, as starting materials, for γ,δ-alkenyldithioic acids, **5**, 547
Dithioic acid ester, *t*-butyl-, synthesis of, **5**, 550
Dithioic acid esters
from
S-alkylphosphorus sulfides, **5**, 551
imido esters, **5**, 551
thioic acid esters, **5**, 551
synthesis of, **5**, 550
Dithioic acid esters, α,β-alkynic, synthesis of, **5**, 560
Dithioic acid esters, α-aryl-, synthesis of, **5**, 552
Dithioic acid esters, α,β-unsaturated, synthesis of, **5**, 558
Dithioic acid *S*-phenyl esters, from, chlorothioic acid *S*-phenyl ester, **5**, 550
Dithioic acids
from, aromatic aldehydes, **5**, 549
inner salts of, **5**, 548
quaternary ammonium salts of, **5**, 546
synthesis of, **5**, 545
Dithioic acids, γ,δ-alkenyl-, from, dithioic acid *S*-allylic esters, **5**, 547
Dithioic acids, *N*-alkyl-β-imino-, synthesis of, **5**, 547
Dithioic acids, α,β-alkynic, synthesis of, **5**, 549
Dithioic acids, β-imino-, synthesis of, **5**, 547
Dithioic acids, β-imino-β-cyano-, synthesis of, **5**, 547
Dithioic acids, β-oxo-
alkylation of, **1**, 140
as starting materials, for ketene dithioacetals, **5**, 555
synthesis of, **5**, 547
Dithioic acids, α,β-unsaturated, synthesis of, **5**, 548
Dithioic acid thioanhydrides
as starting materials, for ferrocenedithioesters, **5**, 561
as thioacylating agents, **5**, 550
Dithioic acid *S*-trimethylsilyl esters, deuterolysis of, **5**, 547
Dithioisobutyrate, methyl, reactions, with diethylphosphinous chloride, **4**, 927
Dithioketals
from
dithioacetals, **4**, 250
ketones, **2**, 541
reactions, with tungsten hexacarbonyl, **3**, 365
Dithioketals, trifluoromethyl, reactions, with methyllithium, **1**, 570
Dithio ketone, bis(alkynyl), isolation of, **3**, 366
Dithio ketones, synthesis of, **3**, 346, 355
Dithiolactones
reactions, with phosphonium salts, **3**, 254
synthesis of, **3**, 346
Dithiolane, 1,1-dioxo-, thermolysis of, **3**, 334
1,3-Dithiolane, 2-methoxy-, from, dichloromethyl methyl ether, **6**, 97
Dithiolane *S,S'*-dioxide, synthesis of, **4**, 268
1,3-Dithiolane 1,1-dioxides, 2-alkylidene-, as starting materials, for thioketenes, **3**, 550
1,3-Dithiolane 1,3-dioxides, reactions, with lithium diisopropylamide, **2**, 247
Dithiolanes
double oxidation of, **4**, 266
from
keto diesters, **4**, 245
ketones, **4**, 246
thietanes, **2**, 224
oxidation of, **4**, 260
as starting materials, for thioketones, **3**, 349
synthesis of, **2**, 224
1,3-Dithiolanes
reductive cleavage of, **2**, 265
as starting materials
for phosphonium salts, **6**, 120
for thioketones, **3**, 365
1,3-Dithiolanes, 2-alkylidene-, as starting materials, for thioketenes, **3**, 550
1,3-Dithiolanes, (haloalkyl)-, synthesis of, **4**, 249
1,3-Dithiolane-2-thione
alkylation of, **6**, 309
reactions
with bis(trifluoromethyl)thioketene, **6**, 305
with 2-phenyloxirane, **6**, 305
as starting material, for spirocyclic trithioorthocarbonates, **6**, 309
1,3-Dithiolane-2-thione, 4-vinylidene-, desulfurization of, **1**, 982
1,3-Dithiolans, 2-acylimino-, acylation of, **5**, 635
1,3-Dithiolans, 2-thionobenzoylimino-, synthesis of, **5**, 634
Dithiolate, disodiumethane-, reactions, with dichloromethyl methyl ether, **6**, 97
1,3-Dithiole, 2,2-dichloro-3,4-bis(trifluoromethyl)-, reactions, with ethane-1,2-dithiol, **6**, 306
1,2,-Dithiole-3-imine, as starting material, for ketenimines, **3**, 579
1,2-Dithiole-3-iminium salts, as starting materials, for thioketones, **3**, 357
1,2-Dithiolene-3-thiones, as starting materials, for thioketenes, **3**, 552
1,3-Dithiolen-2-ones, as starting materials, for thioketenes, **3**, 552
1,3-Dithiolen-2-thiolones, as starting materials, for thioketenes, **3**, 552
1,2-Dithiole-3-ones, as starting materials, for thioaldehydes, **3**, 339
Dithioles
reactions
with enamines, **3**, 255
with enolates, **3**, 254
Dithioles, acylmethylene-, as starting materials, for trithiapentalenes, **3**, 376
Dithioles, enamino-, as starting materials, for trithiapentalenes, **3**, 376
Dithioles, thiocarbonylmethylene-, synthesis of, **3**, 356
1,2-Dithioles
from
β-diketones, **3**, 371
1,2-dithiole-3-thiones, **3**, 371
reactions, with Grignard reagents, **4**, 905
as starting materials
for thioacylketene acetals, **3**, 361
for trithiapentalenes, **3**, 376
synthesis of, **3**, 73

1,2-Dithioles, 3-phenacylidene-, synthesis of, **3**, 254
1,3-Dithioles, as starting materials, for ketene dithioacetals, **4**, 850
3*H*-1,2-Dithioles, 3-acylmethylene-, synthesis of, **3**, 73
1,3-Dithiole-2-thione, as starting material, for spirocyclic trithioorthocarbonates, **6**, 309
1,3-Dithiole-2-thione, 4,5-bis(methylthio)-, synthesis of, **6**, 555
1,3-Dithiole-2-thione, 3,4-dicyano-, reactions, with trimethyl phosphite, **6**, 316
1,2-Dithiole-3-thiones
  reactions, with α-diazo ketones, **3**, 255
  as starting materials
    for 1,2-dithioles, **3**, 371
    for thioaldehydes, **3**, 339
    for thioketones, **3**, 356, 357
    for trithiapentalenes, **3**, 376
1,3-Dithiole-2-thiones, 4,5-disubstituted, synthesis of, **6**, 555
Dithiole-3-thiones, as starting materials, for thioketones, **3**, 361
1,2-Dithiolimines, as starting materials, for ketene aminals, **4**, 979
Dithiolium cation, 4-phenyl-, as starting material, for thioaldehydes, **3**, 339
1,3-Dithiolium cations, 2-(methylthio)-, as starting materials, for tetrathioorthocarbonates, **6**, 301
Dithiolium salts
  reactions
    with potassium ethyl xanthate, **6**, 85
    with thiolates, **6**, 85
  as starting materials
    for 2-dithiocarbamoyldithioles, **6**, 85
    for ketene dithioacetals, **4**, 850
    for trithiapentalenes, **3**, 376
  synthesis of, **3**, 371
  x-ray crystal structure of, **2**, 1008
1,2-Dithiolium salts
  as starting materials, for thioketones, **3**, 361
  synthesis of, **3**, 371
1,2-Dithiolium salts, 5-alkyl-, as starting materials, for trithiapentalenes, **3**, 376
Dithiolium tetrafluoroborate salts, as starting materials, for 2-phenylthio-1,3-dithioles, **6**, 85
1,3-Dithiol-2-one, as starting material, for dimethyl ester, **5**, 554
Dithiolones, reactions, with ynamines, **5**, 605
1,3-Dithiolones, as starting materials, for 2-methanethio-1,3-dithioles, **6**, 87
5*H*-1,3,4-Dithiol-5-ones, 2-chloro-, chlorination of, **5**, 675
Dithiolonium salts, reactions, with ynamines, **5**, 605
Dithiols
  as byproducts, **2**, 130
  cyclization of, **4**, 1073
  exchange reactions, with trialkyl orthoformates, **6**, 83
  from, dibromides, **2**, 119
  reactions
    with cyanogen chloride, **6**, 627
    with dibromo(methyl)stibine, **2**, 501
    with α,α-dichloro ethers, **6**, 97
    with sulfur, **2**, 233
    with triethyl trithioorthoformate, **6**, 89
    with trifluoroacetic acid, **6**, 84
  as starting materials
    for formaldehyde dithioacetal, **4**, 249
    for thioketones, **3**, 349
  use in the synthesis of dialkyl sulfides, **2**, 133
Dithiols, α,ω-alkane-
  reactions
    with dithiocarbonates, **6**, 305
    with trithiocarbonates, **6**, 305
Dithiols, geminal, synthesis of, **2**, 130
1,2-Dithiols, reactions, with seleneous acid, **2**, 237
1,3-Dithiols, reactions, with dichlorophosphines, **2**, 252
gem-Dithiols, synthesis of, **4**, 244
1,3-Dithiolylium cations, 2-alkylthio-, use in the synthesis of aminotrithioorthoformates, **6**, **295–318**
Dithiolylium salts, hydrolysis of, **3**, 254
1,2-Dithiolylium salts, reactions, with sodium benzoyl acetate, **3**, 254
Dithiomalonates, synthesis of, **4**, 842
Dithiomalonic acid, synthesis of, problems, **5**, 548
Dithiomethane, disulfonyl-, as starting material, for disulfonylthiomethane, **6**, 91
Dithiomethanide ions, sulfenylation of, **6**, 303
Dithione, synthesis of, **3**, 345, 351
Dithione, acenaphthene-, synthesis of, **3**, 365
α-Dithiones, tautomerism with 1,2-dithietes, **3**, 330
Dithioniadiboratacyclohexanes, synthesis of, **4**, 373
Dithionite, reactions, with sulfones, **1**, 439
Dithionite, diethyl phosphochlorido-, reactions, with alcohols, **2**, 106
α,α-Dithionitriles, synthesis of, **3**, 635
N,N'-Dithionoacylhydrazides, synthesis of, **5**, 639
Dithionoanhydride, 1,8-naphthalic, synthesis of, **5**, 528
Dithionoanhydrides, properties of, **5**, 528
Dithiono crowns, synthesis of, **5**, 511
Dithionooxalate, dimethyl, reactions, with 2,3-dimethylbutadiene, **4**, 229
Dithionooxamides
  from
    cyclohexane 1,1-dithiol, **5**, 580
    trichloroethylene, **5**, 602
  reactions, with primary amines, **5**, 591
  synthesis of, **5**, 587
Dithionosuccinates, as starting materials, for dienes, **4**, 834
Dithioorthocarbonates, synthesis of, **6**, 312
Dithioorthocarbonates, spirocyclic, synthesis of, **6**, 312
Dithioortho esters
  from
    α,α-dihalo ethers, **6**, 97
    dithioacetals, **6**, 98
    dithiocarbenium salts, **6**, 98
    monothioacetals, **6**, 98
    ortho esters, **6**, 97
    N-p-toluenesulfonylsulfilimines, **6**, 99
  as starting materials, for amide dithioacetal, **6**, 111
Dithioorthoformates
  reactions, with silyl enol ethers, **4**, 250
  as starting materials, for keto dithioacetals, **4**, 250
Dithiooxalates, dialkyl, from, dithiooxalic acid alkali metal salts, **5**, 554
Dithiooxalic acid alkali metal salts, as starting materials, for dialkyl dithiooxalates, **5**, 554
Dithiooxamides, synthesis of, **5**, 566
Dithioperoxy esters, synthesis of, **5**, 562
Dithiophosphate, *S*-carbamoyl *O,O'*-diethyl, thermal decomposition of, **5**, 1037
Dithiophosphate, diisopropyl, reactions, with benzonitrile, **5**, 737
Dithiophosphate, ethyl, reactions, with methanol, **2**, 135
Dithiophosphates, pyrolysis of, **5**, 475
Dithiophosphates, alkynyl, synthesis of, **2**, 1033
Dithiophosphates, dialkyl
  reactions
    with aryl isonitriles, **5**, 737
    with isonitriles, **5**, 737
    with nitriles, **5**, 737
Dithiophosphates, diethyl *S*-alkoxycarbonyl-, synthesis of, **6**, 476
Dithiophosphates, imidoyl, synthesis of, **5**, 644, 737
Dithiophosphates, *S*-imidoyl
  from, imidoyl chlorides, **5**, 738
  synthesis of, via dithiophosphorylation of nitriles and isonitriles, **5**, 737
Dithiophosphates, triethyl, synthesis of, **2**, 259
Dithiophosphinates

reactions, with butyllithium, **2**, 460, 462
synthesis of, **2**, 462; **5**, 737
Dithiophosphinates, *S*-imidoyl, from, imidoyl chlorides, **5**, 738
Dithiophosphinic acids
bifunctional, synthesis of, **2**, 854
reactions, with imidoyl chlorides, **5**, 738
synthesis of, **2**, 462
Dithiophosphinic esters, synthesis of, **2**, 462
Dithiophosphinous acids, synthesis of, **2**, 459
Dithiophosphonates, synthesis of, **2**, 119
Dithiophosphonates, *S*-alkyl alkylchloro-, reactions, with alcohols, **2**, 464
Dithiophosphonates, cyclic, synthesis of, **2**, 464
Dithiophosphonates, *O*,*S*-dialkyl alkyl-, synthesis of, **2**, 465
Dithiophosphonates, *S*,*S*-dialkyl alkyl-, synthesis of, **2**, 464
Dithiophosphonates, *S*,*S*-dialkyl(α-alkylthio)alkyl-, synthesis of, **2**, 464
Dithiophosphonic acid anhydrosulfides, reactions, with epoxides, **2**, 464
Dithiophosphonic anhydride, diphenyl-, reactions, with *N*,*N*-dimethyltrimethylsilylamine, **2**, 862
Dithiophosphonic anhydrides
alkylation of, **2**, 255
reactions, with dienes, **2**, 256
Dithiophosphonic dimethylamide, trimethylsilyl phenyl-, synthesis of, **2**, 862
Dithiophosphonous acid, phenyl-, diesters, synthesis of, **2**, 852
Dithiophosphoric acid, *O*,*O*-diethyl-
for the thionation of amides, **5**, 575
use in the synthesis of thionoamides, **5**, 619
Dithiophosphoric acid, diethyl-, *S*-alkynyl esters, synthesis of, **2**, 1033
Dithiophosphoric acid, *O*,*O*-diisopropyl-, use in the synthesis of thionoamides, **5**, 575
Dithiophosphoric acid, diphenyl-, use in the synthesis of thionoamides, **5**, 581
Dithiophosphoric acid, *O*,*O*-di-*n*-propyl-, use in the synthesis of thionoamides, **5**, 575
Dithiophosphoric acids
reactions
with imidoyl chlorides, **5**, 738
with imino esters, **5**, 644
use in the synthesis of dithioic acid esters, **5**, 551
Dithiophosphoric acids, *O*,*O*-dialkyl-
as insecticides, **6**, 476
reactions
with benzonitriles, **5**, 643
with nitriles, **5**, 581
Dithiophosphoric acid salts, *O*,*O*-diethyl-, reactions, with methyl chloroformate, **6**, 476
Dithiophosphoric esters, synthesis of, **2**, 253
Dithiophosphorus(V) acids
reactions
with imidoyl chlorides, **5**, 644
with nitriles, **5**, 643
Di(thiophosphoryl chloride), methylene-, reduction of, **4**, 550
Dithiopropanoic acid *S*-ethyl ester, reactions, with benzaldehyde, **5**, 557
Dithioselenocarbonate, production of, **6**, 589
Dithioselenocarbonate, dimethyl, as starting material, for ketene thioselenoacetals, **4**, 872
N,*N*'-Dithiosuccinimide, synthesis of, **5**, 338
Dithiosuccinoyl difluoride, tetrafluoro-, synthesis of, **5**, 473
Dithizone, dehydro-, reactions, with phosphorus ylides, **4**, 490
Ditin, hexabutyl-, reactions, with acid chlorides, **5**, 467
Ditin, hexamethyl-, reactions, with acid chlorides, **5**, 467
Ditin chlorides, synthesis of, **2**, 599
(Ditin)compounds, methylene-, halogen replacement of, **4**, 700
Ditins
reactions, with methyllithium, **6**, 399
synthesis of, **6**, 399
Ditins, carbon-bridged
synthesis of, **4**, 699
via replacement reactions, **4**, 701
Ditins, hexaalkyl-
cleavage of, **2**, 600
synthesis of, **2**, 599
Dititanium, tris(cyclooctatetraene)-, synthesis of, **4**, 670
Dititanium complexes, methylene-bridged
from, tetramethylenetitanocene, **4**, 670
synthesis of, **4**, 670
Di(1,2,4-triazole), 1,1'-thiocarbonyl-, reactions, with nitrones, **5**, 605
C,*N*-Ditrytyl analogues, synthesis of, **3**, 682
Ditungsten, hexakis(*t*-butoxy)-, reactions, with alkynes, **3**, 691
Ditungsten, tris(cyclooctatetraene)-, synthesis of, **4**, 673
Ditungsten complexes, alkyne-bridged, synthesis of, **4**, 674
Ditungsten complexes, carbon-bridged, synthesis of, **4**, 672
Diuretic drugs, Mercaptomerin, **2**, 274
Divanadium complexes, carbon-bridged
pentacoordinate carbon atoms in, **4**, 671
synthesis of, **4**, 671
Divinyl chlorides, as starting materials, for pentadienyl cations, **1**, 423
Divinyl sulfones
reactions
with imines, **2**, 158
with oximes, **2**, 158
Diylides, cumulenyl, as intermediates, in the synthesis of butatrienes, **1**, 988
Diyls, synthesis of, **1**, 359
Diyls, stannane-, reactions, with ketene ylides, **5**, 919
Diynes
deprotonation of, **1**, 111
isomerization of, **1**, 954
reactions
with α,ω-diiodoalkanes, **1**, 111
with organosilvers, **1**, 544
reduction of, **1**, 456
skipped, synthesis of, **1**, 121
as starting materials, for bisallenyl compounds, **1**, 954
synthesis of, **1**, 1030, 1069, 1079, 1142
Diynes, acyclic diterminal, cyclization of, **1**, 1001
Diynes, bis(iodonium)-, synthesis of, **2**, 1021
Diynes, *o*-diketo-, reactions, with transition metals, **3**, 272
Diynes, 1-silyl-, reduction of, **1**, 512
Diynes, α,ω-silyl-
cyclization of, **2**, 926
as starting materials, for bis(silyl)cyclopentadienones, **2**, 926
Diynes, sulfone, reduction of, **1**, 515
Diynes, tertiary amino-, reduction of, **1**, 512
Diynes, zirconium-promoted, bicyclization of, **2**, 988
1,3-Diynes
from, alkenyl thioethers, **1**, 1072
Mannich reactions of, **1**, 1041
synthesis of, **1**, 1017, 1018, 1033, 1035, 1065
1,4-Diynes
from, dicobalt hexacarbonyl complexes, **1**, 1025
as starting materials, for alkynylallenes, **1**, 954
synthesis of, **1**, 1002, 1025
ω-Diynes, intramolecular oxidative coupling of, **1**, 1034
1,1-Dizincalkene, from, alkynes, **4**, 1067
Dizincs, carbon-bridged, synthesis of, **4**, 696
Dizirconium complexes, methylene-bridged, synthesis of, **4**, 670
DMF *See* Formamide, dimethyl- (DMF)
Dodecahedryl carbocations, synthesis of, **1**, 900
1,16-Dodecahedryl dication, synthesis of, **1**, 900
1-Dodecane sulfinic acid, synthesis of, **2**, 273

Dodecanoic acid
from, 1-tridecene, **5**, 52
as starting material, for dodecanoyl iodide, **5**, 20
Dodecanoic acid, 2-methylene, synthesis of, **5**, 99
Dodecanol, oxidation of, **5**, 51
Dodecanoyl iodide, from, dodecanoic acid, **5**, 20
trans-6-Dodecene, as starting material, for *trans*-6-dodecenedioic acid, **5**, 117
trans-6-Dodecenedioic acid, from, *trans*-6-dodecene, **5**, 117
2-Dodecenoic acid, *trans*-2-methyl-, synthesis of, **5**, 98
Dodec-1-yne, reduction of, **1**, 457
6-Dodecyne, synthesis of, **1**, 1017
Doebner reaction, for the synthesis of α,β-unsaturated acids, **5**, 98
Double activation, mechanism, **5**, 134
Double bond migration, in heterogeneous hydrogenation, **1**, 71
Double bonds
oxidative cleavage, **3**, 294
ozonolytic cleavage of, **5**, 141
Double deprotonation, of nitroalkanes, **4**, 540
Douglas Fir Tussock Moth, pheromones, synthesis of, **1**, 438
Dowex-50®, use in the synthesis of ketones, **3**, 133
Duff formylation, for the synthesis of 1,4,8-trimethoxy-2-naphthaldehyde, **3**, 101
Duff reaction
for the formylation of ellipticine, **3**, 107
modified, for the synthesis of alkyl benzaldehydes, **3**, 94
for the synthesis of aryl aldehydes, **3**, 91
Durene, trichloromethylation of, **6**, 29
Dyes, synthesis of, **3**, 257
Dynemicin, synthesis of, **1**, 1009
1,1-Dyotropic rearrangement, use in the synthesis of 1,1-bis(silyl)alkanes, **4**, 614

E

Ebelactone A, synthesis of, **1**, 831
Ebselen PZ 51, use in the synthesis of sulfenyl selenides, **2**, 235
Ecdysone, 2,3 Wittig rearrangements of, **1**, 818
α-Effect, in phosphonous compounds, **5**, 402
β-Effect, mechanism, **1**, 447
Eglington reaction
  mechanism, **1**, 1034
  for the synthesis of 1,3-diynes, **1**, 1034
Eicosanoids, synthesis of, **1**, 1009
Einhorn reaction
  for the synthesis of *N*-acyl aminals, **4**, 416
  for the synthesis of monoacylated aminals, **4**, 417
Elaeokanines, synthesis of, **1**, 694
Elaeokanines A and B, synthesis of, **2**, 328
Elaiomycin, synthesis of, **2**, 773
Elaiomycin, L-dihydro-, synthesis of, **2**, 773
Elastase, inhibition via α-aminoboronic acids, **4**, 526
Elbs oxidation
  definition of, **2**, 651
  modification of, **2**, 651
(-)-Eldanolide, synthesis of, **1**, 1098
Electrocarbonylation, of halides, **5**, 44
Electrochemical fluorination (ECF), mechanism, **6**, 20
Electrochemical methods, for the synthesis of acetals, **4**, 197
Electrochemical oxidation
  of acetals, **6**, 77
  of alcohols, to aldehydes, **3**, 7
  of amines, to ketones, **3**, 132
  for the synthesis of ketones from alcohols, **3**, 120
  for the synthesis of α-sulfenyl aldehydes, **3**, 36
Electrochemical reduction
  for the synthesis of dialkyl sulfides, **2**, 133
  for the synthesis of hydroxylamines, **2**, 764
Electrochemical substitution, for the synthesis of isocyanates, **5**, 986
Electrochemical synthesis
  of alkylcadmiums, **2**, 594
  alkylmercury compounds, **2**, 594
Electrochemistry, for the synthesis of carbonates, **6**, 469
Electrocyclic reaction
  of dienylketenes, **2**, 662
  mechanism, **1**, 417
  Woodward–Hoffmann rules for, **1**, 419
Electrocyclization, of carbonyl ylides, **3**, 318
Electrofluorination, of dimethyl ether, **6**, 229
Electrolysis
  of chlorodifluoroacetic acid, **6**, 225
  for the synthesis of carbodiimides, **5**, 1078
Electron carriers
  in the oxidation of alcohols, **3**, 120
  for the oxidation of alcohols, to aldehydes, **3**, 7
Electron donors, as radical traps, **1**, 329
Electronic effects, and radical stability, **1**, 922
Electron spin resonance spectroscopy, of alkyl radicals, **1**, 923
Electron transfer
  in radical reactions, **1**, 336
  for the synthesis of cations, **4**, 1072
  in the synthesis of enamines, **1**, 149
Electron transfer photosensitization, for the synthesis of carbonyl ylides, **3**, 317
Electrooxidation, for the synthesis of vinyl cations, **1**, 1121
Electrophiles
  reactions
    with acylphosphines, **5**, 399
    with alkenes, **1**, 295
    with alkenyllithiums, **1**, 535
    with alkoxyvinyllithiums, **2**, 88
    with alkylethynyllithium cuprates, **1**, 1136
    with *N*-alkylthiomethylamides, **4**, 100
    with alkynes, **1**, 1122
    with alkynylborates, **2**, 904
    with alkynyl metals, **1**, 1044
    with alkynylsilanes, **1**, 1138
    with allenes, **1**, 1125
    with allenyl organometallics, **1**, 971
    with allylzirconiums, **1**, 197
    with anions, **4**, 917
    with arylgermanes, **1**, 448
    with benzylsamarium(III), **2**, 565
    with 1,2-bismetalloalkenes, **2**, 917
    with 2-borataallenes, **6**, 716
    with carbanions, **1**, 971; **3**, 620
    with carbenoids, **4**, 817
    with carbon–tin bonds, **1**, 70
    with chromium(II) compounds, **1**, 974
    with cyanamide, **5**, 1118
    with α-deprotonated alkanenitriles, **3**, 566
    with diazoalkanes, **5**, 871
    with diazomethanes, **5**, 871
    with α-diazophosphonates, **5**, 871
    with dichlorophosphines, **4**, 125
    with dimetals, **4**, 809
    with enamines, **1**, 296; **4**, 98
    with enol ethers, **1**, 296
    with germylenes, **6**, 187
    with α-haloamines, **4**, 100
    with imines, **4**, 98
    with 2-lithio allyl chloride, **2**, 74
    with α-lithiodifluorovinyl carbamates, **1**, 1089
    with 2-lithio-1,3-dithiane-1-oxides, **4**, 263
    with lithium (1-alkynyl)organoborates, **1**, 1033
    with metal alkynyl complexes, **1**, 1145
    with metallophosphides, **2**, 438
    with organometallics, **2**, 564; **3**, 289
    with organosamariums, **1**, 975
    with organozincs, **1**, 974
    with pentaorganoantimony compounds, **1**, 446
    with pentaorganoarsenic(V) compounds, **1**, 445
    with pentaorganobismuth compounds, **1**, 446
    with α-phenylthioalkylboronic esters, **6**, 124
    with phosphaferrocenes, **4**, 1035
    with phosphenium ylides, **5**, 893
    with phosphino ylides, **4**, 551
    with propargylmercury iodide, **1**, 974
    with propargyl organometallics, **1**, 971
    with propargyl trialkylsilanes, **1**, 977
    with silicon-substituted allyl metals, **2**, 935
    with silylallylcopper, **2**, 935
    with silyl allyllithiums, **2**, 935
    with silylallylmagnesium bromide, **2**, 935
    with silylallyltitanium, **2**, 935
    with silylallylzinc, **2**, 935
    with silyl enol ethers, **5**, 236
    with silyloxypropargyllithiums, **1**, 971
    with 2-silylpropenylzinc, **2**, 936
    with stable organotins, **1**, 973
    with α-sulfonyl carbanions, **4**, 236
    with titanium compounds, **1**, 973
    with trinuclear dianions, **6**, 338
    with tris(benzotriazolyl)methane, **6**, 143

with tris(trimethylsilyl)methyllithium, **4**, 1048; **6**, 373
with vinylboranes, **4**, 144
with vinylcoppers, **2**, 917
with vinylgermanes, **1**, 451
with vinylic organometallics, **1**, 463
with vinyllithiums, **1**, 633
with vinylsilanes, **4**, 139
with ynamines, **4**, 791
use in the synthesis of acetals, **4**, 187
use in the synthesis of trithiocarbonates, **6**, 553
See also Bifunctional electrophiles
Electrophiles, silyl
reactions
with α-metalloborylalkanes, **4**, 642
with α-metallogermylalkanes, **4**, 636
with α-metallomethyl transition metals, **4**, 657
with organolithiums, **4**, 602
Electrophilic complexes, reactions, with α-metallosilylalkanes, **4**, 656
1,2 Electrophilic migration, mechanism, **1**, 378
Electrophilic radicals
definition of, **1**, 920
from, iodomalonitriles, **1**, 340
intermolecular addition reactions, **1**, 325
radical cyclization of, **1**, 341
reactions, with alkenes, **1**, 331
Electrophilic reagents
reactions, with diselenocarbamates, **5**, 1055
for the synthesis of trialkylsilyl imidates, **5**, 722
Electropositive metals, for the chemical generation of active metals, **2**, 958
(-)-Elemol, from, (+)-hedycaryol, **1**, 812
Elfamycins, synthesis of, **1**, 581
Elimination
of alkoxy groups, **3**, 209
of alkylboranes, **1**, 657
of aluminum, **1**, 662
of amine oxides, **1**, 623
of amines, **1**, 635
of antimony groups, **1**, 647
of arsenic groups, **1**, 646
of bismuth groups, **1**, 647
of boron functions, **1**, 5965
of germanium groups, **1**, 656
of β-halosilanes, **1**, 652
of hydrogen bromide, **3**, 208
of hydrogen chloride, **3**, 208
of hydrogen halides, **4**, 730
of hydroxy groups, **3**, 209
of β-hydroxyphosphine oxides, **1**, 641
of β-hydroxyphosphinothioic amides, **1**, 643
of β-hydroxyphosphonic acid bisamides, **1**, 643
of β-hydroxysilanes, **1**, 648
of lead, **1**, 663
of metal functions, **1**, 660
of nitrogen functions, **1**, 623
of nitro groups, **1**, 637
of phosphine oxide, **1**, 639
of phosphonium groups, **1**, 640
of phosphonium ylides, **1**, 640
of phosphonothioate esters, **1**, 645
of phosphorus groups, **1**, 638
of quaternary ammonium salts, **1**, 627
in radical generation, **1**, 925
rate of, factors, **1**, 649
reaction rate, **1**, 1064
with secondary haloalkanes, **1**, 111
of selenides, **1**, 592
of selenones, **1**, 620
of selenoxides, **1**, 600; **3**, 207
silane induced, **1**, 587
of silicon groups, **1**, 648
stereochemistry of, **1**, 606
of β-substituted alkylboranes, **1**, 655
of β-substituted silanes, **1**, 653
of β-substituted sulfones, **1**, 614
of sulfones, **1**, 612; **3**, 209
of sulfoxides, **1**, 595; **3**, 209
of sulfur, **1**, 982
for the synthesis of alkynes, **1**, 1062
for the synthesis of allenes and cumulenes, **1**, 981
for the synthesis of aromatic nitriles, **3**, 662
for the synthesis of carbon–carbon bonds, **1**, **553–587**, **589–671**
for the synthesis of carbonyl ylides, **3**, 324
for the synthesis of ketenimines, **3**, 556
for the synthesis of ketones, **3**, 274
for the synthesis of α,β-unsaturated ketones, **3**, 206, 225, 234, 242
of tellurides, **1**, 594
of tellurones, **1**, 620
of telluroxides, **1**, 610
of tin, **1**, 663
of transition metals, **1**, 665
of triethyl phosphate, **1**, 646
Elimination, alkylative, of selenoxides, **1**, 610
Elimination, decarboxylative
for the synthesis of vinyl iodides, **2**, 628
to prepare vinyl fluorides, **2**, 610
Elimination, ElcB, for the synthesis of carbanions, **1**, 845
Elimination, exocyclic, in the synthesis of β-milbemycins, **1**, 602
Elimination, reductive, use in the cleavage of carbon–mercury bonds, **1**, 67
Elimination, regioselective, of acetates, **1**, 773
Elimination, stereoselective *anti*, of β-hydroxy selenides, **1**, 593
Elimination, stereospecific *cis*, mechanism, **1**, 595
Elimination, thermal, for the synthesis of carbon–carbon bonds, **1**, 771
syn-Elimination
of acetates, **1**, 773
pyrolytic, **1**, 773
α-Elimination
for the synthesis of carbanions, **1**, 845
for the synthesis of carbenes, **4**, 1081
α',β-Elimination
mechanism, **1**, 629
regioselectivity of, **1**, 629
stereoselectivity of, **1**, 629
β-Elimination, of radicals, **1**, 321
Elimination reaction
for the synthesis of alkynyl bromides, **2**, 1017
for the synthesis of alkynyl chlorides, **2**, 1015
for the synthesis of alkynyl halides, **2**, 1012
for the synthesis of azoalkenes, **2**, 769
for the synthesis of bromo iodoalkenes, **4**, 759
for the synthesis of chlorohaloalkenes, **4**, 756
for the synthesis of 1,1-dichloroalkenes, **4**, 742
for the synthesis of difluoroalkenes, **4**, 730
for the synthesis of fluorohaloalkenes, **4**, 750
for the synthesis of α-haloenols, **4**, 766
for the synthesis of 1-halo-1-thioalkylalkenes, **4**, 773
for the synthesis of ketene dithioacetals, **4**, 844
for the synthesis of perfluoropropyne, **2**, 1013
for the synthesis of tertiary ynamines, **2**, 1045
for the synthesis of tetracoordinate sulfur compounds, **4**, 779
for the synthesis of vinyl bromides, **2**, 622
for the synthesis of vinyl chlorides, **2**, 615
for the synthesis of vinyl fluorides, **2**, 610
for the synthesis of vinyl halides, **2**, 606
for the synthesis of vinyl iodides, **2**, 628
for the synthesis of vinyl sulfides, **2**, 710
Elimination reaction, decarboxylative, for the synthesis of vinyl bromides, **2**, 622

syn Elimination reaction, of selenoxides, **2**, 658
1,2-Elimination reactions, for the synthesis of metalloid functions, **3**, 501
α-Elimination reactions, of metal acyl complexes, **5**, 937
β-Elimination reactions, for the synthesis of phosphaalkynes, **3**, 688
Ellipticines
  formylation of, **3**, 107
  synthesis of, **4**, 164
Emde reduction, for the synthesis of tertiary amines, **2**, 303
Emetine, synthesis of, **1**, 149
Enals
  conjugate addition to, **3**, 12
  conjugate reduction of, **3**, 12
  reactions
    with arsonium ylides, **1**, 741
    with azides, **3**, 48
    with diethyl methyl phosphinate, **3**, 51
    with morpholinobenzoselenamide, **3**, 74
    with nitroalkanes, **3**, 47
  α-selenium-substituted, synthesis of, **3**, 74
  β-selenium-substituted, synthesis of, **3**, 74
  synthesis of, **3**, 57; **4**, 188
Enals, amino, from, 1,3-dicarbonyls, **3**, 75
Enals, chloro, reactions, with sodium sulfide, **3**, 72
Enals, β-chloro
  reactions, with selenols, **3**, 74
  reduction of, **3**, 72
Enals, halo, from, carbonyls, **3**, 68
Enals, β-sulfonyl, synthesis of, **3**, 37
4-Enals, synthesis of, **1**, 1096
(*E*)-Enals, synthesis of, **3**, 58
Enamides
  from
    allylamides, **5**, 290
    α,β-epoxysilanes, **2**, 747
    imines, **5**, 288
    oximes, **5**, 289
  hydrosilylation of, **4**, 510
  reduction of, **1**, 77
  synthesis of, **5**, 288
    via acylation, **5**, 288
    via alkylation of amides, **5**, 289
    via Beckmann rearrangement, **5**, 288
    via Curtius rearrangement, **5**, 288
    via rearrangement reactions, **5**, 288
  thionation of, **5**, 572
Enamides, chloro-, synthesis of, **1**, 571
Enamides, β-silyl, synthesis of, **2**, 927
Enamidines, metallation of, **1**, 1136
Enamimes, as starting materials, for α-aminoalkylphosphonites, **4**, 458
Enaminals, synthesis of, **3**, 76
Enamine, reactions, with sodium trichloroacetate, **3**, 235
Enamine, *N*-silyl, flash vacuum thermolysis of, **3**, 562
Enamines
  acylation of, **3**, 259, 261
  asymmetric hydroboration of, **2**, 519
  bromination of, **2**, 28
  cleavage of, **1**, 443
  cycloaddition of, **4**, 414
  fluorination of, **2**, 15
  from
    acetaldehyde, **3**, 49
    acetals, **2**, 739
    acetic acids, **2**, 742
    carbonyls, **2**, 739
    nitriles, **2**, 739
    2-nitro-5-carbomethoxy toluene, **2**, 740
    ynamines, **2**, 739
  halogenation of, **3**, 76
  Hofmann elimination of, **1**, 1075
  hydroboration of, **2**, 43, 514; **4**, 524
  Michael addition of, **1**, 307; **2**, 759
  nucleophilic addition to, **4**, 426
  protonation of, **2**, 757
  reactions
    with acid chlorides, **3**, 261
    with acrylates, **5**, 152
    with acyldiazines, **5**, 354
    with acyl isothiocyanates, **5**, 631
    with alanes, **1**, 443
    with alkyl halides, **2**, 757
    with amines, **4**, 971
    with arenediazonium salts, **3**, 449
    with arylsulfonyl peroxides, **3**, 163
    with azides, **3**, 467
    with 9-borabicyclo[3.3.1]nonyl, **1**, 443
    with boranedimethyl sulfide, **1**, 60
    with ω-bromoacetophenone semicarbazone, **2**, 770
    with bromosulfenes, **5**, 497, 499
    with chlorosulfene, **5**, 496
    with cyanogen bromide, **2**, 748; **3**, 657
    with diborane, **1**, 443
    with α,α'-dibromo ketones, **3**, 225
    with diethylamine, **2**, 308
    with *N,N*-dimethylacetamide diethyl acetal, **4**, 987
    with dimethyl acetylenedicarboxylate, **1**, 518
    with dimethyl ethynedicarboxylate, **3**, 233, 275
    with *S,S*-dioxide ketenethioselenoacetals, **4**, 286
    with diphenylphosphoryl azide, **5**, 775
    with dithioles, **3**, 255
    with electrophiles, **1**, 296; **4**, 98
    with 3-ethoxy-2-alkylacrolein, **2**, 759
    with formic acid, **1**, 82
    with hydrogen cyanide, **3**, 615
    with hydrogen sulfide, **2**, 130; **3**, 348
    with isothiocyanates, **5**, 597, 614
    with ketenes, **3**, 269
    with nitrilium salts, **3**, 408
    with phenyl selenocyanates, **2**, 284
    with phenyl selenyl chloride, **3**, 38
    with phenylselenyl cyanide, **3**, 637
    with piperidine, **2**, 308
    with propargylic ketones, **3**, 233
    with salicyl aldehyde, **4**, 307
    with Selectfluor reagents, **2**, 16
    with selenamides, **3**, 38
    with sodium borohydride, **1**, 82
    with sodium dimethylamide, **4**, 1017
    with sulfamoyl chlorides, **3**, 171
    with sulfonyl isocyanates, **5**, 344
    with sulfonyl oxiridines, **3**, 156
    with tetrachloromethane, **4**, 742
    with trichlorosilane, **4**, 510
    with trifluoroacetic acid, **2**, 759
    with vinyl isocyanates, **5**, 990
  reduction of, **1**, 82
  as starting materials
    for alkenes, **1**, 443
    for α-aminoalkylideneamino enamines, **4**, 987
    for α-bromo ketones, **3**, 154
    for α-chloro ketones, **3**, 152
    for dichloromethyl ketones, **4**, 15
    for α-fluoro ketones, **3**, 148
    for halo aldehydes, **3**, 26
    for α-haloamines, **4**, 100
    for imidoyl-tetracoordinated phosphorus compounds, **5**, 789
    for imines, **3**, 408
    for iminium salts, **2**, 757
    for α-iodo ketones, **3**, 155
    for ketene aminals, **4**, 979
    for β-ketonitriles, **3**, 633
    for α-selenocarbonyls, **2**, 282
    for α-sodio enamines, **4**, 1017

for thioaldehydes, **3**, 338
for thioketones, **3**, 348, 363
sulfonylation of, **3**, 170
super, **1**, 156
synthesis of, **1**, 149; **2**, 739; **3**, 43; **5**, 789
via aza-Cope rearrangements, **1**, 835
via elimination of hydrogen cyanide, **1**, 559
via retro-Diels–Alder reaction, **1**, 787
tautomers of, **4**, 880
use in the synthesis of *O*,*N*-acetals, **4**, 307
use in the synthesis of chlorosulfene, **5**, 496
use in the synthesis of ketones, **3**, 137
See also Amines, vinyl-
Enamines, *N*-acyl
from, 1-methoxylated cyclic *N*-formylamines, **2**, 748
synthesis of, **2**, 747, 748
Enamines, α-acyl, synthesis of, **2**, 745
Enamines, β-acyl, synthesis of, **2**, 744
Enamines, *N*-acylated α-hydrazino, from, α-hydrazino enamines, **4**, 986
Enamines, *N*-acylated α-ureido
from
acyl azides, **4**, 982
N-acylketenimines, **4**, 982
synthesis of, **4**, 982
Enamines, α-*O*-acyl-*N*,*N*-diacyl, from, ketoximes, **2**, 747
Enamines, 1-acyloxy, synthesis of, **4**, 889
Enamines, acylsilane, as starting materials, for imidoylsilicons, **5**, 794
Enamines, 1-alkoxy, tautomers of, **4**, 880
Enamines, α-alkoxycarbonylamino, from, *N*-acylketenimines, **4**, 981
Enamines, alkoxycyano, synthesis of, **3**, 658
Enamines, β-alkyl-α-phosphoryl, synthesis of, **4**, 1003
Enamines, *N*-allyl, 3-aza-Cope rearrangement of, **3**, 413
Enamines, allylic, as starting materials, for imines, **1**, 835
Enamines, α-amido
from
N-acylated ketene *N*,*S*-acetals, **4**, 981
N-acylketenimines, **4**, 980
isothioureas, **4**, 980
ketene aminals, **4**, 980
ynamines, **4**, 980
synthesis of, **4**, 980
Enamines, (*Z*)-γ-amino, synthesis of, **2**, 740
Enamines, β-amino, synthesis of, **2**, 741
Enamines, α-aminoalkylideneamino
from
acetals, **4**, 987
enamines, **4**, 987
iminium perchlorates, **4**, 987
ketene *N*,*S*-acetals, **4**, 988
ketene aminals, **4**, 987
as intermediates, in the synthesis of pyrimidines, **4**, 988
synthesis of, **4**, 987
Enamines, α-azido
from
N-alkyl-5-phenylisoxazolium fluoroborates, **4**, 990
α-chloro enamines, **4**, 990
gem-diazidoalkenes, **4**, 989
ketene *N*,*S*-acetals, **4**, 990
as starting materials, for α-phosphimino enamines, **4**, 990
synthesis of, **4**, 989
Enamines, α-azo, from, pyrazoles, **4**, 989
Enamines, α-bis(methylthio)methyleneamino, from, *N*,*N*-dimethylthiourea, **4**, 988
Enamines, *N*,*N*-bis(silyl), synthesis of, **2**, 544
Enamines, *trans*-*N*,*N*-bis(silyl), synthesis of, **2**, 813
Enamines, *N*,*N*-bis(trimethylsilyl), from, *N*,*N*-bis(trimethylsilyl)thioformamides, **2**, 814
Enamines, α-borio
from
trialkylboranes, **4**, 1015
ynamines, **4**, 1015
Enamines, bromo, synthesis of, **4**, 790
Enamines, α-bromo, reactions, with phenylcoppers, **1**, 464
Enamines, β-bromo, as starting materials, for tertiary ynamines, **2**, 1049
Enamines, β-carbonyl, synthesis of, **2**, 745
Enamines, chiral organotin, synthesis of, **2**, 816
Enamines, chloro
as intermediates, in the synthesis of ketene aminals, **4**, 970
reactions
with Grignard reagents, **4**, 15
with lithium amides, **4**, 971
as starting materials, for ketene aminals, **4**, 971
Enamines, 1-chloro
reactions
with metal salts of thiols, **4**, 904
with thiophenols, **4**, 904
as starting materials, for ketene *S*,*N*-acetals, **4**, 904
Enamines, *gem*-chloro
reactions, with sodium diphenylphosphide, **4**, 997
as starting materials, for α-isocyano enamines, **4**, 990
Enamines, α-chloro
alkylation of, **2**, 758
from, dimethylphosgeniminium chloride, **4**, 794
reactions
with carboxylic acids, **2**, 758
with diazoacetic acid, **3**, 475
as starting materials
for α-azido enamines, **4**, 990
for α-ferrio enamines, **4**, 1019
for keteniminium salts, **3**, 601
for α-lithio enamines, **4**, 1016
for α-magnesio enamines, **4**, 1017
for α-rhenio enamines, **4**, 1020
for α-sodio enamines, **4**, 1017
for tertiary ynamines, **2**, 1045
synthesis of, **2**, 1046; **4**, 791
Enamines, α,β-chloro, as starting materials, for tertiary ynamines, **2**, 1047
Enamines, β-chloro, as starting materials, for tertiary ynamines, **2**, 1049
Enamines, α-chloro-β-acyl, synthesis of, **2**, 746
Enamines, 2-chloro-2-nitro, from, nitroacetone, **4**, 796
Enamines, α-cuprio, from, ynamines, **4**, 1018
Enamines, α-cyano
from
α-chloroaldimines, **2**, 749
α-(*N*-methylanilino)-acetonitrile, **2**, 749
as starting materials
for ketenimines, **3**, 563
for trialkylketimines, **3**, 409
synthesis of, **2**, 748; **3**, 657
Enamines, β-cyano, synthesis of, **2**, 748, 749
Enamines, cyclic α-acyl, synthesis of, **2**, 745
Enamines, cyclic β,β-dicarbonyl, synthesis of, **2**, 746
Enamines, cyclic *N*-hydroxy, from, pyridine *N*-oxides, **2**, 765
Enamines, *N*,*N*-dialkyl α-halo, synthesis of, **4**, 790
Enamines, α-diaminomethyleneamino, from, *gem*-aminochloromethyleneamino chloroalkenes, **4**, 988
Enamines, dichloro, as starting materials, for tertiary ynamines, **2**, 1047
Enamines, α,β-dichloro
as starting materials, for tertiary ynamines, **2**, 1047
synthesis of, **2**, 1047
Enamines, β,β-difunctionalized
from, thioamides, **2**, 750
synthesis of, **2**, 749, 750
Enamines, β,β-dimethyl, reactions, with sulfonyl isothiocyanates, **5**, 638
Enamines, 1,2-disubstituted, reactions, with sulfonyl oxaziridines, **3**, 183
Enamines, α-ferrio
from

α-chloro enamines, **4**, 1019
iron vinylidene complexes, **4**, 1019
synthesis of, **4**, 1019
Enamines, fluoro, synthesis of, **4**, 790
Enamines, α-fluoro, synthesis of, **4**, 793
Enamines, formyl, reactions, with tosyl azide, **3**, 473
Enamines, halo
from, dialkylisobutyryl amides, **2**, 8
reactions, with hydrogen sulfide, **5**, 611
as starting materials, for α-trialkylphosphino enamines, **4**, 998
Enamines, 2-halo, as starting materials, for nitro enamines, **2**, 789
Enamines, *gem*-halo, as starting materials, for α-phosphino enamines, **4**, 997
Enamines, *N*-halo, synthesis of, **2**, 761
Enamines, α-halo
metallation of, **3**, 259
NMR spectra of, **4**, 792
as starting materials, for α-phosphoryl enamines, **4**, 1004
synthesis of, **4**, 790; **5**, 2
via addition of halide, **4**, 790
via amination, **4**, 792
via electrophilic attack, **4**, 793
Enamines, β-halo, reactions, with alkyllithiums, **2**, 1049
Enamines, heteroaryl, synthesis of, **2**, 741
Enamines, α-hydrazino
from
imino esters, **4**, 983
ketene acetals, **4**, 983
ketene *N,S*-acetals, **4**, 983
ketene dithioacetals, **4**, 983
nitriles, **4**, 984
thioamides, **4**, 984
thiosemicarbazides, **4**, 984
α-trichloromethyl enamines, **4**, 984
tricyanoquinodimethanes, **4**, 985
ynamines, **4**, 984
as intermediates, in the synthesis of heterocycles, **4**, 983
reactions
with carbon disulfide, **4**, 986
with methyl chloroformate, **4**, 986
with methyl iodide, **4**, 986
with methyl isocyanate, **4**, 986
as starting materials, for *N*-acylated α-hydrazino enamines, **4**, 986
synthesis of, **4**, 983, 986
Enamines, α-hydrazino β-nitro, use in the synthesis of α-hydrazino enamines, **4**, 986
Enamines, hydroxy, equilibrium, with nitrones, **2**, 764
Enamines, iodo, synthesis of, **4**, 790
Enamines, α-isocyano, from, *gem*-chloro enamines, **4**, 990
Enamines, β-lithiated, as enolate equivalents, **2**, 759
Enamines, (*Z*)-β-lithio, synthesis of, **3**, 76
Enamines, α-lithio
from
β-amino-α,β-unsaturated amides, **4**, 1016
β-amino-α,β-unsaturated esters, **4**, 1016
α-chloro enamines, **4**, 1016
synthesis of, **4**, 1016
Enamines, β-lithio, formylation of, **3**, 76
Enamines, α-magnesio, from, α-chloro enamines, **4**, 1017
Enamines, α-manganio, from, *gem*-chloro manganioalkenes, **4**, 1019
Enamines, γ-metallated silyl, synthesis of, **2**, 751
Enamines, γ-metallated stannyl, synthesis of, **2**, 751
Enamines, metallo
reactions, with disulfides, **3**, 35
use in the synthesis of carbonyls, **2**, 816
Enamines, α-metallo, synthesis of, **3**, 259
Enamines, 1-(methylseleno), synthesis of, **4**, 909
Enamines, α-molybdenio, as intermediates, in the synthesis of dicyanoketeneimmonium compounds, **4**, 1018
Enamines, nitro
from
2-halo enamines, **2**, 789
1-nitro-2-phenylthioalkenes, **2**, 789
reactions
with aldehyde enolates, **3**, 47
with aldehydes, **2**, 789
with benzenesulfonyl chloride, **4**, 907
substitution of, **1**, 486
Enamines, 2-nitro
reactions, with *N*-halosuccinimides, **4**, 796
synthesis of, **2**, 789
Enamines, *N*-nitro, acylation of, **1**, 983
Enamines, β-nitro
reactions, with sodium sulfide, **5**, 611
synthesis of, **2**, 787; **3**, 192
Enamines, α-palladio, from, isocyanides, **4**, 1017
Enamines, phenylhydrazino, synthesis of, **4**, 983
Enamines, α-phosphimino
from
α-azido enamines, **4**, 990
bisphosphiminoalkenes, **4**, 991
ketene aminals, **4**, 990
synthesis of, **4**, 990
Enamines, α-phosphinoyl
from
dialkyl phosphinous acids, **4**, 1000
malonitrile, **4**, 1000
Enamines, α-phosphorodiamidoyl
from
alkoxyphosphorus diamines, **4**, 1011
imino chlorides, **4**, 1011
Enamines, α-phosphoryl
alkylation of, **4**, 1004
from
acetals, **4**, 1004
aldehydes, **4**, 1002
1-aminoalkyl phosphonates, **4**, 1003
α-chloroalkanimines, **4**, 1004
1-dimethylamino-1-cyanomethanephosphonic acid, **4**, 1004
α-halo enamines, **4**, 1004
iminium esters, **4**, 1004
ketene *N,O*-acetals, **4**, 1003
ketene *N,S*-acetals, **4**, 1003
ketene aminals, **4**, 1003
α-ketophosphonic diesters, **4**, 1004
nitroalkenes, **4**, 1004
N-phosphorylated α-phosphoryl enamines, **4**, 1004
α-phosphoryl enamines, **4**, 1004
tertiary amides, **4**, 1003
tetraethyl dialkylaminomethylene diphosphonate, **4**, 1002
ynamines, **4**, 1003
reactions, with *t*-butyllithium, **4**, 1004
as starting materials, for α-phosphoryl enamines, **4**, 1004
synthesis of, **4**, 1002
Enamines, phosphorylated chloro, synthesis of, **5**, 656
Enamines, *N*-phosphorylated α-phosphoryl, as starting materials, for α-phosphoryl enamines, **4**, 1004
Enamines, α-platino, from, isocyanides, **4**, 1017
Enamines, polyhalogenated α-chloro, synthesis of, **4**, 793
Enamines, primary α-halo, synthesis of, **4**, 792
Enamines, α-rhenio, from, α-chloro enamines, **4**, 1020
Enamines, salts, synthesis of, **2**, 757
Enamines, secondary α-chloro
from, 2,2-dicyanovinyl azides, **4**, 792
reactions, with 2,2-dicyano-*N*-phenylacetamide, **4**, 792
Enamines, β-siloxy-*N,N*-bissilyl, from, acetylsilane *O*-silylcyanohydrins, **2**, 814
Enamines, α-silyl
from
alkynylcarbenechromium complexes, **4**, 1014

cyanohydrins, **4**, 1013
epoxynitriles, **4**, 1014
methacrylic amides, **4**, 1014
nitriles, **4**, 1014
propargylic alcohols, **4**, 1014
synthesis of, **4**, 1013
Enamines, α-sodio
from
α-chloro enamines, **4**, 1017
enamines, **4**, 1017
Enamines, α-stannio, from, diamines, **4**, 1020
Enamines, γ-stannylated, synthesis of, **2**, 751
Enamines, substituted, synthesis of, **4**, 927
Enamines, substituted trichloro, as starting materials, for ynediamines, **2**, 1047
Enamines, β-sulfonyl
from
alkynes, **2**, 743
α,β-epoxy sulfones, **2**, 743
synthesis of, via nucleophilic β-addition, **2**, 742
Enamines, tertiary α-cyano, from, α,β-unsaturated aldehydes, **2**, 748
Enamines, α-thiophosphinoyl
from
malonitrile, **4**, 1001
phosphinothious acid, **4**, 1001
Enamines, α-thiophosphoryl, from, ynamines, **4**, 1010
Enamines, β-tosyl, synthesis of, **2**, 743
Enamines, trichloro
as starting materials
for tertiary ynamines, **2**, 1047
for ynamines, **5**, 595
Enamines, α,β-trichloro, from, trichloroacetamides, **2**, 1047
Enamines, α-trichloromethyl, as starting materials, for α-hydrazino enamines, **4**, 984
Enamines, trifluoromethylated, synthesis of, **2**, 751
Enamines, α-(*N*,*N'*,*N''*-triphenylphosphorimidic triamido), from, phosphorimidic tribromide, **4**, 991
Enamines, α-tungstenio, as intermediates, in the synthesis of dicyanoketeneimmonium compounds, **4**, 1018
Enamines, α-ureido
from
imino esters, **4**, 982
ketene aminals, **4**, 981
1,3-oxazin-6-ones, **4**, 981
synthesis of, **4**, 981
Enamino esters, primary, synthesis of, **2**, 745
Enaminones
from, thioamides, **2**, 744; **3**, 268
hydrolysis of, **5**, 501
substitution of, **1**, 486
synthesis of, **2**, 744; **3**, 437
via acylation, **3**, 261
via acylation of enamines, **2**, 746
use in the Prelog synthesis of phenols, **2**, 662
Enaminones, *N*-β-carbomethoxyethyl, cyclization of, **3**, 262
Enaminones, *N*-hydroxy, synthesis of, **3**, 264
Enaminones, secondary
reactions
with acetyl chloride, **3**, 413
with benzoyl chloride, **3**, 413
with pivaloyl chloride, **3**, 413
Enaminones, trifluoromethyl, synthesis of, **3**, 261
(*E*)-Enaminones, synthesis of, **3**, 262
(*Z*)-Enaminones, synthesis of, **3**, 262
β-Enaminones, from, ketenes, **3**, 269
Enammonium salts, synthesis of, **2**, 757
Enantiodifferentiation, in monoacylation reactions, **5**, 138
Enantioface differentiation, for the 1,4-addition of tin enethiolates, **5**, 557
Enantiomeric forms, of ketenimines, **3**, 556
Enantiomers, of phosphaallenes, **3**, 605
(*R*)-Enantiomers, of 1-amino-2-methoxymethylpyrrolidine, synthesis of, **1**, 153
(*S*)-Enantiomers, of 1-amino-2-methoxymethylpyrrolidine, synthesis of, **1**, 153
Enantioselectivity
definition of, **1**, 351
in the synthesis of α-thiosulfoxides, **4**, 258
Enantiotopic differentiation, for the synthesis of δ-lactones, **5**, 172
NB-Enantride, use in the reduction of ketones, **2**, 60
Enazo compounds, synthesis of, **2**, 770
Enders hydrazone, alkylation of, **1**, 180
Endomyces reessii CBS 179.60, use in the synthesis of carboxylic acids, **5**, 118
Endothionopeptides, synthesis of, **5**, 587
γ,δ-Ene acids, halolactonizations of, **5**, 168
δ,ε-Ene acids
halolactonization of, **5**, 172
mercurylactonization of, **5**, 172
Eneamidines, as starting materials, for *gem*-iminomethylamino lithioalkenes, **4**, 1016
Eneaminones, chemistry of, **3**, 261
Enecarbamates, reactions, with acetyl chloride, **3**, 261
Ene–dicarbonyl compounds, synthesis of, **3**, 221
cis-2-Enediyn-1-als, oxidation of, **5**, 102
1,4-Enediyne, isomerization of, **1**, 955
Enediyne antibiotics, synthesis of, **1**, 1009
Enediynes, synthesis of, **1**, 111, 585, 617, 709
(*Z*)-Enediynes, synthesis of, **1**, 1029
Ene reaction
of acyl nitroso compounds, **5**, 326
in allylation reactions, **2**, 80
combined with Oppenauer oxidation, **3**, 119
of dimethylaminobis(trifluoromethyl)borane, **6**, 243
of maleic anhydride, **5**, 201
mechanism, **1**, 524
for the synthesis of α,β-unsaturated ketones, **3**, 222
of trifluoronitrosomethane, **6**, 241
Eneselenolate, metal complexes, **6**, 594
Enethiolate, *S*-trimethylsilyl
reactions
with acetals, **5**, 557
with aldehydes, **5**, 557
Enethiolates
from
O,*O'*-diethyl trithiopyro carbonate, **4**, 834
dimethyloxosulfonium methylide, **4**, 834
reactions
with β-alkenyl-β-propiolactones, **5**, 553
with α,β-disubstituted enones, **5**, 556
with α-methyl aldehydes, **5**, 557
with 2-phenylpropanal, **5**, 557
[3,3] sigmatropic rearrangement of, **5**, 547
synthesis of, **4**, 834
Enethiolates, *S*-trimethylsilyl, reactions, with α,β-unsaturated ketones, **5**, 557
Enethiols, protonation of, **5**, 552
4-Ene-2-ynes, 1-bromo-, as starting materials, for allenes, **1**, 958
Eniminium ions, use in the synthesis of ketene aminals, **4**, 976
α,β-Enoate, γ-mesyloxy-, reactions, with copper compounds, **1**, 188
Enoates, reactions, with alkynes, **1**, 1051
Enoates, γ-alkoxy-, reactions, with silylenol ethers, **1**, 196
Enoates, 4-amino-, synthesis of, **2**, 331
Enoates, β-halo-, use of in substitution reactions, **1**, 469
Enoates, 4-hydroxy-, osmylation of, **2**, 46
α,β-Enoates, synthesis of, **5**, 155
α,β-Enoates, γ,δ-dihydroxy-, reactions, with cuprates, **5**, 158
β,γ-Enoates, synthesis of, **1**, 192; **5**, 158
Enolate, acetaldehyde
alkylation of, theoretical studies, **1**, 130
synthesis of, **1**, 135
Enolate alkylation, for the synthesis of alkanoic esters, **5**, 148
Enolate anions

chlorination of, **2**, 22
for the opening of epoxides, **1**, 209
O-phosphorylation of, **3**, 198
reactions
with alkenyl sulfoxides, **3**, 172
with alkyl halides, **1**, 8475
with disulfides, **3**, 164
with epoxides, **5**, 166
with ethynyl-lead tetraacetate, **1**, 1145
with ethynyl(phenyl)iodonium tetrafluoroborate, **1**, 1145
with quaternary ammonium salts, **1**, 253
with α,β-unsaturated alkoxycarbene complexes, **5**, 958
stability of, **1**, 850
synthesis of, reviews, **1**, 873
Enolate anions, carbonyl, reactions, with germyl chlorides, **4**, 625
Enolate anions, chiral, reactions of, **1**, 878
Enolate carbanions, synthesis of, **1**, 853
Enolate Claisen rearrangement, mechanism, **1**, 825
Enolate equivalents, reactions, with nitroalkenes, **3**, 192
Enolate–imine condensation, for the synthesis of *N*-aryl β-lactams, **5**, 294
Enolate ions
α-iodination of, **2**, 34
protonation of, **1**, 888
reactions, with iodine, **2**, 34
stability of, **1**, 301
Enolates
O-acylation of, **2**, 674
addition reactions of, **1**, 306
alkylation of, **1**, 129, 187, 236, 653; **3**, 295
for the alkylation of aldehydes, **3**, 10
O-alkylation of, **2**, 674
bromination of, asymmetric, **2**, 29
equilibrium constants, determination of, **1**, 869
fluorination of, **2**, 16
from
dialkyl amides, **1**, 131
dibromomethyllithium, **2**, 1023
lithium amides, **1**, 135
unsymmetrical ketones, **1**, 872
geometry of, **1**, 129; **2**, 696
as intermediates, in the synthesis of silyl enol ethers, **5**, 415
isopropenylation of, **3**, 138
kinetic, **1**, 131
rate constants, determination of, **1**, 869
reactions of, **1**, 878
reactions
with alkylating reagents, **3**, 179
with *t*-alkyl halides, **1**, 133
with alkynyl chlorides, **1**, 1016
with alkynyl halides, **1**, 1016
with alkynyl(phenyl)iodonium salts, **1**, 1016
with aminocarbene complexes, **5**, 958
with bis(ethoxythiocarbonyl) sulfide, **5**, 513
with *t*-butyldimethylsilyl chloride, **1**, 417
with carbon disulfide, **3**, 251
with chlorodiethylphosphine, **2**, 457
with chlorosulfonates, **2**, 690
with dialkyl phosphorochloridate, **2**, 693
with *S,S*-dialkylsuccinimidosulfonium salts, **3**, 574
with diamidophosphorochloridate, **2**, 693
with dioxiranes, **3**, 156
with disulfides, **2**, 139
with dithioles, **3**, 254
with iminium ions, **1**, 255
with isocyanates, **5**, 307
with nitriles, **3**, 268
with ortho esters, **4**, 192
with phosphorochloridites, **2**, 693
with selenium, **3**, 174
with *N*-silylimines, **3**, 440
with sulfur dioxide, **2**, 157
with trimethylsilyl chloride, **1**, 417
with α,β-unsaturated sulfoxides, **3**, 169
scrambling of, **1**, 131
selectivity of, (*E*) vs. (*Z*), **1**, 874
[3,3] sigmatropic rearrangements of, **1**, 831
as starting materials
for acylhydrazones, **5**, 364
for γ-lactones, **5**, 166
structure of, **1**, 873, 876
synthesis of, **1**, 129; **2**, 701
stereoelectronic effects in, **1**, 878
thermodynamic, **1**, 131
transmetallation of, **1**, 132, 819
use in the synthesis of ketones, **3**, 124
Enolates, aldehyde
reactions
with diethyl chlorophosphite, **3**, 50
with nitro enamines, **3**, 47
reactivity of, **1**, 873
Enolates, β-alkyl, synthesis of, **1**, 313
Enolates, alkyl aryl ketone, fluorination of, **3**, 149
Enolates, δ-alkynic malonate, reactions, with alkynyl halides, **1**, 1020
Enolates, allyloxy, 2,3 Wittig rearrangement of, **1**, 818
Enolates, amide, synthesis of, factors affecting, **1**, 876
Enolates, bromo, as intermediates, **3**, 650
Enolates, α-bromo ester, reactions, with trialkylboranes, **4**, 832
Enolates, ester
silylation of, **4**, 831
stability of, **1**, 130
as starting materials, for ketene silyl acetals, **4**, 831
synthesis of, factors affecting, **1**, 876
2,3 Wittig rearrangements of, **1**, 819
Enolates, α-germyl, synthesis of, **4**, 662
Enolates, β-keto ester, as starting materials, for dianions, **1**, 165
Enolates, ketone
alkylation of, **1**, 178
amination of, **3**, 181
Michael addition of, **1**, 306
reactions
with alkynyl ketones, **1**, 538
with aminoalkyltitanium alkoxides, **3**, 186
with β-chloroacrylate, **1**, 470
with cyanogen chloride, **3**, 633
with ethyl propiolate, **1**, 537
with isothiocyanates, **5**, 593
with phenyltellurium iodide, **3**, 176
with phosphorus(III) halides, **3**, 195
with sulfinates, **3**, 170
with sulfonyl oxiridines, **3**, 156
stereochemical assignments, **1**, 874
trifluoroacetylation of, **3**, 472
Enolates, kinetic, **1**, 156
alkylation of, **1**, 873
benzylation of, **1**, 873
use in Michael reactions, **1**, 307
Enolates, metalloid, synthesis of, **2**, 696
Enolates, nitrile
reactions
with aryl halides, **3**, 626
with diethyl chlorophosphate, **3**, 639
with disulfides, **3**, 634
with sultines, **3**, 634
Enolates, 3-oxoalkanoic ester, as intermediates, **1**, 291
Enolates, phosphino silyl, alkaline hydrolysis of, **4**, 1026
Enolates, α-silyl, synthesis of, **4**, 648
Enolates, α-silyl ester, *C*-stannylation of, **4**, 660
Enolates, thioester, as starting materials, for β-lactones, **1**, 566
Enolates, thiol ester

O-alkylation of, **4**, 833
O-silylation of, **4**, 834
(*E*)-Enolates
geometry, **1**, 873
synthesis of, **1**, 873
(*Z*)-Enolates
from, tertiary carboxylic acid amides, **1**, 876
geometry, **1**, 873
reactions, with aldehydes, **1**, 878
synthesis of, **1**, 873
α,β-Enolates, conjugate addition to, **5**, 151
Enolate-*N*-silylimine condensation, studies, **3**, 440
Enolboration, mechanism, **2**, 532
Enol esters
bromination of, **2**, 28
fluorination of, **2**, 15
iodination of, **2**, 34
α-iodination of, **2**, 33
reactions
with acrylates, **5**, 152
with hydrogen halides, **5**, 3
Enol ether, trialkylsilyl, reactions, with fluoroformates, **6**, 462
Enol ethers
asymmetric *cis*-dihydroxylation of, **2**, 49
cleavage of, **2**, 88
fluorination of, **2**, 15
Enol ethers, benzyl
hydrolysis of, **2**, 641
as starting materials, for perfluoropropen-2-ol, **2**, 641
Enol ethers, cyclic, synthesis of, **2**, 1029
Enol ethers, ethyl, synthesis of, **5**, 416
Enol ethers, 2-hydroxy, as starting materials, for 2,3-*syn*-1,2,3-triols, **2**, 50
Enol ethers, silyl
alkylation of, **2**, 258
from, silyl triflate, **2**, 699
geometry of, **2**, 699
reactions
with aldehydes, **2**, 66
with methyllithium, **2**, 696
with *N*-methylsulfamoyl chloride, **2**, 249
with phenyl selenocyanate, **2**, 284
with titanium tetrachloride, **2**, 702
as starting materials
for enols, **2**, 639
for nitro compounds, **2**, 392
for α-selenocarbonyls, **2**, 282
synthesis of, **2**, 699, 1044
use in the synthesis of dialkyl sulfides, **2**, 142
Enol ethers, 3-silyl, alkylation of, **2**, 51
Enol ethers, *O*-substituted, hydrolysis of, **2**, 638
Enol ethers, trimethylsilyl, protonolysis of, **2**, 639
Enolization
of carbonyls, **2**, 645
factors affecting, **1**, 869
of ketones, **3**, 173
regiochemistry of, **1**, 606
for the synthesis of vinyl sulfides, **2**, 709
Enols
O-acylation of, **2**, 674
O-alkylation of, **2**, 674
definition of, **2**, 635
elimination of, **1**, 1071
fluorinated, synthesis of, **2**, 640
from
1,3-dicarbonyls, **1**, 870
glycols, **2**, 643
indanones, **2**, 637
ketenes, **2**, 645
silyl enol ethers, **2**, 639
α,β-unsaturated ketones, **2**, 643
ketonization of, **1**, 869
keto tautomerization of, **2**, 635
oxidation of, **3**, 31, 54
reactions
with alkynes, **1**, 525
with diethylaminosulfur trifluoride, **6**, 19
with difluorocarbene, **6**, 229
with triphenylbismuth bistrifluoroacetate, **2**, 675
simple aliphatic, synthesis of, **2**, 636
stability of
via electronic stabilization, **2**, 643
via steric hindrance, **2**, 643
as starting materials
for alkynes, **1**, 1069
for α-bromo ketones, **3**, 154
for α-chloro ketones, **3**, 152
for α-fluoro ketones, **3**, 148
for α-iodo ketones, **3**, 155
sterically congested, synthesis of, **2**, 642
substitution of, **1**, 1026
synthesis of, **1**, 133, 811; **2**, 635
via photoisomerization of α,β-unsaturated ketones, **2**, 638
via photolysis, **2**, 636
via retro-Diels–Alder reaction, **1**, 787
via thermolysis, **2**, 636
Enols, cyclic fluorinated, synthesis of, **2**, 641
Enols, α-halo
from
alkynes, **4**, 764
haloalkynes, **4**, 764
synthesis of, **4**, 760
Enols, *gem*-iodo phosphonamido, synthesis of, **4**, 804
Enols, trimethylsilyl, synthesis of, **2**, 930
Enol tautomers, incorporating intramolecular hydrogen bonds, **1**, 870
Enone, β-chloro-
reactions
with thiols, **3**, 252
with thiophenol, **3**, 252
Enones
acetalization of, **4**, 179
1,4-addition of cuprates to, **1**, 130
conjugate addition to, **3**, 128
conjugate reduction of, **3**, 127, 145
α-dialkoxyalkylation of, **4**, 193
from
1,3-diones, **3**, 217
enol ethers, **1**, 558
α-hydroxy ketones, **3**, 209
β-keto acids, **1**, 561
ketones, **1**, 556
α-phosphorus-substituted, synthesis of, **3**, 269
β-phosphorus-substituted, synthesis of, **3**, 270
protection of, **1**, 653
reactions
with alanes, **1**, 1049
with alkynes, **1**, 1051
with alkynylalanes, **1**, 1049
with alkynylborons, **1**, 1050
with alkynyldiisopropoxyboranes, **1**, 1050
with alkynyltrichlorotins, **1**, 1050
with alkynylzinc bromides, **1**, 998
with azomethine ylides, **2**, 331
with carbanions, **3**, 188
with cuprates, **3**, 162; **4**, 385
with dialkylboranes, **2**, 533
with diethylalkynylalanes, **1**, 1048
with dimethyl phosphite, **4**, 339
with hydroxy alkyl anions, **3**, 161
with lithium alkynide, **3**, 62
with methylenethiiranes, **1**, 242
with organocuprates, **1**, 313
with organosilyllithiums, **1**, 653

with phosphines, **3**, 195
with propargylsilanes, **1**, 978
with selenate ions, **3**, 175
with thiocarbonates, **1**, 176
with *p*-toluenesulfonylhydrazine, **1**, 1082
with tributylstannylmagnesium bromide, **2**, 600
with trimethylstannyllithium, **2**, 600
with trimethylstannylsodium, **2**, 600
with triphenylstannylpotassium, **2**, 600
with vinyl halides, **3**, 233
with vinyl metallics, **3**, 229
reduction of, **1**, 86
1,4-reduction of, **1**, 564
selenenylation of, **1**, 606
as starting materials
for β-alkyl enones, **1**, 610
for epoxides, **1**, 304
for epoxy ketones, **1**, 1082
from β-amino ketones, **3**, 187
for ketones, **1**, 86; **3**, 274
synthesis of, **1**, 86; **3**, 208
via dehydrochlorination reactions, **1**, 571
Enones, acyclic, synthesis of, via selenoxide pyrolysis, **1**, 781
Enones, alkoxy-
amination of, **3**, 264
reactions
with benzylamine, **3**, 265
with lithium alkynides, **3**, 228
Enones, (*E*)-β-alkoxy-, synthesis of, **3**, 245
Enones, α-alkoxy-, synthesis of, **3**, 242
Enones, β-alkoxy-, substitution of, **1**, 481
Enones, β-alkyl-, from, enones, **1**, 610
Enones, β-alkylated, synthesis of, **1**, 482
Enones, α-chloro-, synthesis of, **3**, 236
Enones, β-chloro-
reactions
with allylphosphorus ylides, **1**, 470
with cuprates, **3**, 271, 273
substitution of, **1**, 469
Enones, conjugated, reactions, with arsonium ylides, **1**, 740
Enones, β-cyano-, synthesis of, **3**, 653
Enones, β,β-dialkoxy-, synthesis of, **3**, 246
Enones, β-dialkylamino-, synthesis of, **3**, 266
Enones, diazo-, photochemical Wolff rearrangement of, **2**, 668
Enones, dichlorodifluoro-, reactions, with zinc, **1**, 569
Enones, β,β-difluoro-, synthesis of, **2**, 906
Enones, α,β-disubstituted, reactions, with enethiolates, **5**, 556
Enones, β,β-disubstituted, addition reactions of, **1**, 314
Enones, α-fluoro-, synthesis of, **1**, 561
Enones, halo-, elimination from, **3**, 274
Enones, α-halo-
reactions, with organometallics, **3**, 217
synthesis of, **4**, 90
Enones, β-halo-
1,2 addition of, **1**, 469
reactions, with cuprates, **1**, 469
Enones, iodo-, synthesis of, **3**, 274
Enones, β-iodo-, reactions, with cuprates, **3**, 272
Enones, α-methyl-, synthesis of, **3**, 224
Enones, β-methyl-, formylation of, **3**, 64
Enones, nitro-, synthesis of, **2**, 785
Enones, 3-oxy-, synthesis of, **3**, 244, 247
Enones, silyl-
from, allenyl ethers, **5**, 427
synthesis of, **3**, 270
Enones, β-silyl-, synthesis of, **3**, 271
2,3-Enones, 2-sulfonyl-, synthesis of, **3**, 249
4-Enones, synthesis of, **1**, 1096
cis Enones, as starting materials, for kainic acid, **1**, 369
α,β-Enones
α-alkoxyalkylation of, **1**, 629
from
β,γ-enones, **1**, 797
ketones, **3**, 173
as starting materials, for γ-alkynones, **1**, 1139
α,ω-Enones, synthesis of, **1**, 785
β,γ-Enones, as starting materials, for α,β-enones, **1**, 797
β,γ-Enones, γ,γ-difluoro-, synthesis of, **1**, 569
γ,δ Enones, synthesis of, **1**, 190
Enyl carbocations
stability of, **1**, 901
synthesis of, via bond scission, **1**, 901
Enyl chlorides, as byproducts, **4**, 738
Enyne, silyl-, cyclization of, **3**, 270
Enynes
1,6-addition to, **2**, 940
cyclization of, **1**, 546
cycloisomerizations of, palladium(0)-catalyzed, **2**, 927
hydroboration of, **5**, 425
hydroformylation of, **3**, 64
hydrogenation of, **1**, 504
hydrohalogenation of, **1**, 965
metathesis of, **1**, 675
reactions
with carbenes, **1**, 549
with ethanethiol, **4**, 872
with Grignard reagents, **1**, 960
with octacarbonyldicobalt, **3**, 218
with organolithiums, **1**, 960
with zirconocenes, **2**, 568
reduction of, **1**, 508
skipped, synthesis of, **1**, 121
as starting materials
for alkenylallenes, **1**, 955
for allenyl enol ethers, **1**, 970
for allenyl esters, **1**, 970
synthesis of, **1**, 472, 488, 496, 541, 739, 1016, 1027
zirconium-promoted, bicyclization of, **2**, 988
Enynes, alkoxy-, hydrostannylation of, **3**, 63
Enynes, conjugated
reduction of, **1**, 959
as starting materials, for alkenylalanes, **2**, 967
Enynes, halo-, synthesis of, **2**, 902
Enynes, α,ω-silyl-, cyclization of, **2**, 926
Enynes, terminal
reactions, with boranes, **2**, 900
as starting materials, for dienylboranes, **2**, 900
1,3-Enynes
from
alkynes, **1**, 1036
allenyl thioethers, **1**, 1072
Mannich reactions of, **1**, 1041
reactions, with amines, **3**, 413
synthesis of, **1**, 1018
use in the synthesis of carbon-bridged ditins, **4**, 701
1,4-Enynes
isomerization of, **1**, 955
synthesis of, **1**, 1002
1,5-Enynes
Cope rearrangement of, **1**, 808
from, 1,3-cyclohexadiene, **1**, 1036
synthesis of, **1**, 1052
3-En-1-ynes, 1-silyl-, monosilylation of, **4**, 1046
(*E*)-Enynes, synthesis of, **1**, 512, 1036
(*E*)-1,5-Enynes, from, methyl (*E*,*E*)-2,4-hexadienoate, **1**, 1036
trans-Enynes, synthesis of, **1**, 659
(*E*)-Enynoates, methyl, synthesis of, **1**, 1021
Enynols
as starting materials
for dienols, **1**, 513
for ruthenium carbene complexes, **5**, 944
2-En-4-yn-1-ols, conjugated, reduction of, **1**, 959
Enynones

cyclization of, **2**, 672
as starting materials, for phenols, **2**, 672
(*E*)-Enynones, methyl, synthesis of, **1**, 1021
Enynyl bromides, displacement of, **2**, 1101
Enzymatic oxidation, for the synthesis of dialkyl sulfoxides, **2**, 151
Enzymatic systems, for cyanohydrin resolution, **3**, 631
Enzyme inhibitors, synthesis of, **3**, 421
Enzymes
as electron-transfer agents, **6**, 20
glutathione peroxidase, **2**, 235
for the oxidation of dithioacetals, **4**, 262
for the oxidation of sulfides, **2**, 151
as starting materials
for amides, **5**, 270
for peptides, **5**, 262
use in the synthesis of acylhydrazines, **5**, 353
use in the synthesis of amines, **2**, 318
use in the synthesis of *N*-aminoimides, **5**, 359
use in the synthesis of asymmetric cyanohydrins, **3**, 628
use in the synthesis of carboxylic acids, **5**, 117
Enzymes, oxynitrilase, isolation of, **3**, 628
Enzymic couplings, for the synthesis of amides, **5**, 263
Enzymic hydrolysis, of nitriles, **5**, 270
Ephedrine
hydrogenation of, **1**, 88
reactions, with phenylphosphonous dichloride, **2**, 862
as starting material, for *N*-benzylephedrine, **2**, 300
Ephedrine, *N*,*N*-diallyl-, use in Reformatsky reactions, **2**, 69
(-)-Ephedrine, use in the synthesis of chiral selenophosphonates, **2**, 466
(±)-Epilupinine, synthesis of, **2**, 350
Epimerization, of halides, **2**, 5
Epiphosphine, synthesis of, **6**, 166
Episulfides
desulfurization of, **1**, 646, 983
reactions
with alkyl xanthates, **2**, 123
with dimethyldiazomalonate, **1**, 982
with metal phosphides, **2**, 430
with trialkyl phosphites, **1**, 646
with triethyl phosphite, **1**, 592
as starting materials, for alkenes, **1**, 592
Episulfone, 1-bromo-1-methylethene, from, α-bromoethanesulfonyl chloride, **4**, 83
Episulfone intermediates
in Ramberg–Bäcklund reactions, **1**, 704
stability of, **1**, 618
Episulfones, synthesis of, **1**, 618, 619
Episulfones, chloro, synthesis of, **4**, 82
Episulfonium ions
as intermediates, **1**, 252; **5**, 248
reactions, with silyl enol ethers, **3**, 169
synthesis of, **1**, 390
Episulfoxides
rearrangement of, **2**, 185
ring-opening of, **2**, 200
thermal decomposition of, **2**, 223
See also Thiirane sulfoxides
Epitellurides, as intermediates, in the synthesis of alkenes, **1**, 595
Epoxidation
for the synthesis of mixed anhydrides, **5**, 709
of vinylsilanes, **4**, 357
Epoxidation, asymmetric, mechanism, **2**, 91
Epoxidation, metal-catalyzed, mechanism, **2**, 90
Epoxidation, vanadium-mediated, stereochemistry, **2**, 91
Epoxide, nitro, synthesis of, **2**, 392
Epoxides
alkylation of, **1**, 478
cleavage of, **4**, 343
deoxygenation of, **1**, 587, 646
eliminative ring-opening of, **2**, 85
Friedel–Crafts alkylation of, **1**, 475
from
aldehydes, **1**, 268
arsonium ylides, **1**, 738; **3**, 48
α-diazophosphonamides, **6**, 125
enones, **1**, 304
β-hydroxyarsonium salts, **1**, 646
homolytic fission of, **1**, 929
isomerization of, **1**, 212; **3**, 652
opening of, **1**, 204
with alkynes, **1**, 204
reagents for, **1**, 204
photochemical ring-cleavage of, **2**, 83
propargylation of, **1**, 1053
reactions
with acetone cyanohydrin, **3**, 632
with aldehydes, **4**, 181
with alkanesulfonic acids, **2**, 203
with alkenylalanes, **1**, 1106
with alkyl xanthates, **2**, 123
with allenyllithium, **1**, 478, 971
with aniline, **2**, 318
with antimony(V) fluoride, **3**, 122
with arenes, **2**, 85
with bis(cyclopentadienyl)titanium(III) chloride, **1**, 942
with boron trichloride, **2**, 110
with *N*-chlorocarbonyl isocyanate, **5**, 1008
with cyanide, **3**, 632
with 1,1-dialkoxyalkenes, **6**, 75
with diethylaluminum cyanide, **3**, 632
with diethylaminosulfur trifluoride, **2**, 10
with dimethylhydrazine, **3**, 617
with dithiophosphonic acid anhydrosulfides, **2**, 464
with enolate anions, **5**, 166
with ethyl 2-isocyano-2-lithiopropionate, **3**, 704
with formate esters, **6**, 71
with hydride complexes, **2**, 576
with ketones, **4**, 181
with lactones, **6**, 72
with Lewis acids, **3**, 9
with lithiated triselenoorthoformates, **6**, 94
with lithiated trithioorthoformates, **6**, 88
with α-lithio isocyanides, **3**, 704
with lithium 4,4'-(di-*t*-butyl)biphenylide, **2**, 553
with metal phosphides, **2**, 430
with nucleophilic anions, **5**, 946
with organosilyl anions, **1**, 650
with phosphines, **2**, 445
with phosphorous trichloride, **2**, 105
with polyphosphonic acid anhydrides, **2**, 458
with stannyl anions, **2**, 600
with sulfenyl halides, **2**, 194
with thiols, **2**, 136
with trialkyl phosphites, **1**, 646
with 1-triisopropylsilylpropyne, **1**, 1053
with trimethylsilyl cyanide, **3**, 632
with vinylic carbanions, **1**, 1094
with vinylsilanes, **1**, 1094
rearrangement of, **1**, 386, 842
reductive cleavage of, **2**, 36, 83
reductive lithiation of, **2**, 83
regioselectivity reversal of, **1**, 218
ring-opening of, **1**, 386, 580, 608, 1092, **2**, 60, 135, 771; **3**, 33, 160, 698
as starting materials
for alcohols, **1**, 842
for aldehydes, **1**, 387; **3**, 9
for alkenes, **1**, 593
for *cis*-2-alkylcycloalkanols, **2**, 83
for alkyl iodides, **2**, 36
for amines, **2**, 318
for betaines, **1**, 640
for α-bromo ketones, **3**, 153

for carbon-centered radicals, **1**, 942
for carbonyl ylides, **1**, 365
for α-chloro ketones, **3**, 151
for cyclic sulfides, **2**, 136
for dialkyl sulfides, **2**, 135
for 1,2-dibromides, **2**, 245
for 1,2-dichlorides, **2**, 26
for dithiocarbamates, **6**, 562
for enamino esters, **2**, 746
for ester aminals, **6**, 110
for α-fluoro ketones, **3**, 148
for halohydrins, **1**, 386
for β-hydroxyphosphine oxides, **1**, 753
for hydroxy selenides, **2**, 281
for α-iodo ketones, **3**, 155
for ketones, **3**, 122
for oxetanes, **1**, 210
for thiocyanates, **5**, 1106
for thiophosphates, **2**, 259
for α,β-unsaturated aldehydes, **3**, 62
for vicinal diols, **1**, 386
substitution of, **1**, 478, 1021
synthesis of, **1**, 282; **4**, 146; **6**, 124
use in alkylation reactions, **1**, 282
use in Friedel–Crafts alkylations, **1**, 212
use in the synthesis of alkenes, **1**, 722
use in the synthesis of α-heterosubstituted ketones, **3**, 183
See also Oxiranes
Epoxides, acyl, rearrangement of, **1**, 387
Epoxides, alkyne, reduction of, **1**, 506
Epoxides, alkynic, reduction of, **1**, 962
Epoxides, alkynyl, conjugate addition to, **1**, 277
Epoxides, 1-alkynyl, reactions, with diisobutylaluminum hydride, **2**, 83
Epoxides, allylic, reactions, with Grignard reagents, **1**, 1053
Epoxides, aryl, reactions, with lithium aluminum amides, **2**, 318
Epoxides, δ-azido, reactions, with triphenylphosphine, **2**, 324
Epoxides, chloro, ring-cleavage of, **2**, 85
Epoxides, α-chloro
from, chloroalkenes, **4**, 50
reactions, with silver tetrafluoroborate, **3**, 148
Epoxides, *gem*-dicyano, use in the synthesis of 2-acetoxy-3-bromo-2-propenenitriles, **3**, 650
Epoxides, 1,1-disilylated, as starting materials, for 1,1-bis(silyl)alkanes, **4**, 619
Epoxides, *meso*-2,3-disubstituted, ring-opening of, **2**, 85
Epoxides, disulfonyl, synthesis of, **4**, 853
Epoxides, α-fluoro
from
Dewar hexafluorobenzene, **4**, 46
2,3-difluoroallylic alcohols, **4**, 45
hexafluorobenzene, **4**, 45
synthesis of, **4**, 45
Epoxides, α-halo, as starting materials, for allyl alcohols, **1**, 664
Epoxides, α-hydroxy, rearrangement of, **3**, 62
Epoxides, iodo, reactions, with anionic complexes, **5**, 955
Epoxides, β-oxosilyl, as starting materials, for 2-alkyl-4-trimethylsilyl furans, **2**, 945
Epoxides, silyl, selectivity of, **1**, 217
Epoxides, sulfinyl
use in the synthesis of α-alkoxy ketones, **3**, 162
use in the synthesis of α-amino ketones, **3**, 183
use in the synthesis of β-sulfenyl ketones, **3**, 168
Epoxides, trifluoromethyl, ring-opening of, **2**, 85
Epoxides, α,β-unsaturated, reactions, with organoboranes, **1**, 277
Epoxides, vinyl
as allylating agents, **1**, 228
conjugate addition to, **1**, 277
from, α,β-epoxy ketones, **1**, 479
opening of, **1**, 225
reagents for, **1**, 226
reactions
with allylsilanes, **1**, 226
with bis(phenylsulfonyl)methane, **4**, 280
with iron nonacarbonyl, **5**, 163
with organostannanes, **2**, 86
with sodium azide, **2**, 85
ring-opening of, **2**, 85
as starting materials
for acetals, **4**, 183
for allylic chromiums, **2**, 86
for 3-hydroxyhexa-1,5-dienes, **2**, 86
synthesis of, **3**, 318
Epoxides, vinylic
for the alkylation of carbanions, **1**, 479
from, allylic arsonium ylides, **1**, 269
reactions, with organocuprates, **2**, 86
as starting materials, for iron complexes, **1**, 196
synthesis of, **1**, 739
2,3-Epoxides, 1-tosyloxy, ring-opening of, **2**, 84
meso-Epoxides, ring-opening of, **2**, 527
trans-Epoxides, from, arsonium ylides, **1**, 738
γ,δ-Epoxides, reactions, with nucleophiles, **1**, 227
δ,ε-Epoxides, ring-opening of, **5**, 172
meso Epoxyacids, asymmetric hydrogenolysis of, **2**, 83
Epoxy alcohols
reactions, with lithium alkynides, **1**, 223, 1023
as starting materials, for thiocarbonylimidazolides, **6**, 557
See also Glycidols
2,3-Epoxy alcohols
reduction of, **2**, 83
ring-opening of, **2**, 82, 84; **4**, 516; **5**, 1106
synthesis of, **2**, 91
β,γ-Epoxy alcohols, substitution of, **1**, 478
α-Epoxy aldehydes, as starting materials, for α,β-unsaturated aldehydes, **3**, 57
β,γ-Epoxy aldehydes, synthesis of, **1**, 580
Epoxyalkenes, synthesis of, **1**, 506
2,3-Epoxyalkyl chlorides, cleavage of, **2**, 85
1,2-Epoxyalk-3-ynes, ring-opening of, **2**, 86
Epoxyamides, cyclization of, **1**, 215
Epoxybis(sulfone), from, dithioacetals, **4**, 273
1,2-Epoxybutane, reactions, with allenylmagnesium bromide, **1**, 1053
3,4-Epoxybut-1-ynes, reactions, with trialkylboranes, **1**, 277, 977
1,2-Epoxycarbonyls, electroreductive ring-opening of, **2**, 83
5,6-Epoxycyclodecene, Friedel–Crafts reaction of, **1**, 475
Epoxycyclohexane, 1-chloro-, rearrangement of, **4**, 50
1,2-Epoxycyclohexane, reactions, with butyrolactone, **6**, 72
1,2-Epoxycyclohexane, 2-ethynyl-, reactions, with Grignard reagents, **2**, 86
2,3-Epoxycyclohexanones, nucleophilic cleavage of, **3**, 249
3,4-Epoxycyclopentene, reactions, with diethylhexynylaluminum, **1**, 1024
Epoxy esters, synthesis of, **4**, 209
trans-2,3-Epoxy esters, cleavage of, **2**, 84
1,2-Epoxyethane, reactions, with butyrolactone, **6**, 72
Epoxy ethers
as byproducts, in Favorskii rearrangements, **1**, 393
deprotonation of, **3**, 34
Epoxy hydroxy ionones, photolysis of, **3**, 319
β-Epoxy isocyanides, as starting materials, for isocyanide spirolactones, **3**, 718
α-Epoxy isocyano group, characteristics of, **3**, 704
ω-Epoxy ketimines, rearrangement of, **2**, 326
Epoxy ketones
from, enones, **1**, 1082
reduction of, **2**, 39
as starting materials, for lactones, **1**, 841
use in Eschenmoser fragmentation reactions, **1**, 1082
α-Epoxy ketones, reactions, with alcohols, **3**, 242

α,β-Epoxy ketones
reactions, with tosyl hydrazide, **3**, 25
as starting materials
for alkynones, **1**, 1082
for β-hydroxy ketones, **3**, 160
for vinyl epoxides, **1**, 479
tosylhydrazones of, Eschenmoser fragmentations of, **1**, 1082
use in the synthesis of thiocyanates, **5**, 1107
Epoxynitriles, as starting materials, for α-silyl enamines, **4**, 1014
α-Epoxynitriles, synthesis of, **3**, 612
α,β-Epoxynitriles
reactions, with Grignard reagents, **3**, 459
synthesis of, **3**, 627
β,γ-Epoxynitriles, synthesis of, **3**, 633
β,γ-Epoxy nitro compounds, ring-opening of, **3**, 168
2,3-Epoxynorbornane, *exo*-2-chloro-, synthesis of, **4**, 50
1,2-Epoxy-3-ols, reactions, with amines, **2**, 318
1,2-Epoxypentane, 4-methyl-, reactions, with 1-lithiocyclopropylphenyl sulfide, **3**, 224
1,2-Epoxypenten-4-ol, reactions, with amines, **2**, 318
4,5-*trans*-Epoxypent-2-en-1-ols, 5-aryl-, synthesis of, **1**, 270
4,5-*trans*-Epoxy-(*E*)-pent-2-en-1-ols, 5-aryl-, synthesis of, **1**, 270
Epoxyphosphonates, synthesis of, **4**, 341
1,2-Epoxypropane, reactions, with butyrolactone, **6**, 72
Epoxysilanes
alcoholysis of, **4**, 357
from
chloromethyltrimethylsilane, **4**, 355
α-methyl-α-chlorotrimethylsilane, **4**, 355
vinylsilanes, **4**, 357
hydrolysis of, **4**, 357
regiospecific ring-opening of, **1**, 650
as starting materials, for vinyl fluorides, **2**, 610
α,β-Epoxysilanes
cleavage of, **4**, 139
reactions, with dialkylcuprates, **1**, 759, 761
as starting materials
for enamides, **2**, 747
for α-fluoro-β-hydroxysilanes, **4**, 137
Epoxystannanes, as intermediates, **1**, 664
Epoxysugars, regiochemistry of, **1**, 218
Epoxy sulfides, oxidation of, **4**, 237
Epoxy sulfones
spirocyclization of, **2**, 331
as starting materials, for hydroxy aldehydes, **3**, 29
2,3-Epoxy sulfones, reactions, with alkyllithiums, **2**, 85
α,β-Epoxy sulfones
as starting materials
for α-azido aldehydes, **3**, 48
for β-sulfonyl enamines, **2**, 743
synthesis of, **4**, 236
γ-Epoxy sulfones, as starting materials, for cyclopropanes, **1**, 214
Epoxy sulfoxides
as starting materials, for α-acetoxy ketones, **3**, 162
synthesis of, **2**, 147
α,β-Epoxy sulfoxides
as starting materials
for α-amino aldehydes, **3**, 44
for α-sulfenyl aldehydes, **3**, 36
synthesis of, **3**, 167; **4**, 233
Equilibration, for the synthesis of dialkylboranes, **2**, 522
Erbium compounds, reactions, with phosphorus ylides, **4**, 578
Erbstatin, synthesis of, **3**, 716
(+)-Eremantholide A, synthesis of, **1**, 618
Eremorphilone, reduction of, **1**, 85
Ergolines, electrochemical cyanation of, **3**, 670
Ergolines, 2-cyano-, synthesis of, **3**, 670
Ergothionene, synthesis of, **1**, 255
Erythro β-hydroxy acid, as starting material, for alkenyl ethers, **2**, 677
Erythronolide, (9*S*)-9-dihydro-, synthesis of, **5**, 199
Erythronolide A, synthesis of, **1**, 821
Eschenmoser's salt
characteristics of, **4**, 96
for the condensation of alkenes, **4**, 913
reactions, with 1-methoxy-3-trimethylsilyloxy-1,3-butadiene, **4**, 305
use in Mannich reactions, **3**, 185
Eschenmoser–Claisen rearrangement, mechanism, **1**, 827
Eschenmoser fragmentation
mechanism, **1**, 1082
for the synthesis of alkenynes, **1**, 1082
for the synthesis of alkynyl ketones, **3**, 147
for the synthesis of alkynyl-substituted alkyl aldehydes, **3**, 25
Eschenmoser modification, of Claisen rearrangement, for the synthesis of bis(tributyltin) compounds, **4**, 702
Eschenmoser sulfide ring–contraction, method, **2**, 750
Eschweiler–Clarke procedure, for the synthesis of amines, **2**, 300, 310
Eschweiler–Clarke reaction, for the synthesis of *S,N*-acetals, **4**, 318
Esermethole, synthesis of, **4**, 410
Esperamicin, synthesis of, **1**, 1009
Esperamicin–calicheamicin analogues, synthesis of, **1**, 818
Esperamycin, synthesis of, **2**, 233
Ester, allyl β-keto, decarboxylative allylation of, **3**, 139
Ester, N^2-benzoylarginine ethyl, bromodenitrogenation of, **5**, 1122
Ester, N^2-benzoyl-N^5-cyanoornithine ethyl, synthesis of, **5**, 1122
Ester, 2-butynedioic acid dimethyl, reduction of, **1**, 456
Ester, dimethyl
from, 1,3-dithiol-2-one, **5**, 554
synthesis of, **5**, 554
Ester, dioxosuccinic, as starting material, for hydrazonoyl bromides, **5**, 841
Ester, dithioacrylic acid methyl, synthesis of, **5**, 558
Ester, formate, reactions, with α-keto phosphonates, **3**, 245
Ester, isocyanomethyl phosphonate diethyl
reactions
with aldehydes, **3**, 717
with ketones, **3**, 717
Ester, malonic, reactions, with formaldehyde, **1**, 786
Ester, methyl
reactions
with bromotrimethylsilane, **5**, 404
with iodotrimethylsilane, **5**, 404
use in the synthesis of amides, **5**, 261
Ester, oxalosorbic, as starting material, for formazans, **5**, 858
Ester, perfluoroacrylic, reactions, with aniline, **4**, 894
Ester, phenylaniline methyl, reactions, with trimethylaminecarboxyborane, **6**, 514
Ester, polyphosphoric acid trimethylsilyl (PPSE), as condensation agent, for the synthesis of dithioacetals, **4**, 247
Ester aminals
from
epoxides, **6**, 110
isocyanates, **6**, 110
nitriles, **6**, 109
reactions, with amines, **5**, 747
synthesis of, **6**, 108
Ester auxiliaries, chiral, in photochemical [2+2] additions, **1**, 357
Ester enolate, *O*-chiral, use in the synthesis of alkanoic esters, **5**, 148
Ester enolate Carroll rearrangement, mechanism, **1**, 832
Ester enolates
cyclization of, **5**, 145
phosphorylation of, **2**, 682

reactions
with imines, **5**, 287
with oximes, **5**, 324
with *N*-(phenylsulfenyl)benzaldimine, **5**, 335
with *N*-silylimines, **5**, 377
as starting materials
for hydroxamic acids, **5**, 324
for β-lactams, **5**, 274
Ester enolates, α-alkoxy, reactions, with cyclohexenone, **1**, 303
Ester enolates, *C*-chiral, use in the synthesis of alkanoic esters, **5**, 148
Ester hydrochloride, glycine ethyl, as starting material, for ethyl diazoacetate, **3**, 468
Ester hydrochlorides, imidate, synthesis of, **5**, 515
Ester hydrochlorides, imino
reactions
with 2-aminoalkyl arsines, **5**, 790
with thiosemicarbazides, **5**, 705
Ester hydrochlorides, methyl, as starting materials, for α-substituted cyclic α-imino acids, **5**, 93
Esterification
of methanetrisulfonic acid, **6**, 91
via hydroxyl group activation, mechanism, **5**, 131
trans-Esterification
of bis(alkoxythiocarbonyls, **5**, 511
enzymatic, mechanism, **5**, 135
of imidates, **5**, 694
of *O*-methyl selenobenzoate, **5**, 520
for the synthesis of carboxylic esters, **5**, 134
Esterification reaction, for the synthesis of carboxylic esters, **5**, 122
Esters
as activating groups for alkenes, **1**, 300
for the acylation of amines, **5**, 263
alkenation of, **2**, 680
alkylidenation of, **1**, 770; **2**, 681
aminolysis of, **5**, 298
α-bromination of, **2**, 29
as chiral auxiliaries, **1**, 305
chlorination of, **4**, 54
α-chlorination of, **2**, 21
deprotonation of, **4**, 831
as electrophilic substrates, in Friedel–Crafts reactions, **3**, 281
Friedel–Crafts alkylation of, **1**, 181
from
α-chloro ketones, **1**, 392
α-diazo ketones, **1**, 394
pyridyl sulfides, **1**, 333
racemates, **4**, 390
zirconium acyl compounds, **5**, 442
hydrolysis of, **5**, 24, 29, 78
choice of reagents for, **5**, 25
hydrosilylation of, **2**, 82
as intermediates, for trithioortho esters, **6**, 83
α-iodination of, **2**, 34
methylenation of, **1**, 766
reactions
with alkoxyamines, **5**, 329
with aluminum sulfides, **5**, 235
with anhydrides, **6**, 39
with carbon monoxide, **5**, 192
with diazo ketones, **1**, 1078
with di-*t*-butyl peroxide, **1**, 925
with dimethylaluminum methylselenolate, **5**, 239
with divinylcuprates, **1**, 540
with hydrazine hydrate, **5**, 350
with hydroxylamines, **5**, 322
with α-lithioalkyl phosphonates, **3**, 220
with methanesulfonic acid, **5**, 29
with phenylthiotrimethylsilylmethyllithium, **3**, 166
with selenation reagents, **5**, 520
with sulfur, **5**, 600
with thiols, **5**, 236
with titanocene, **1**, 21
with trialkyl stannyllithiums, **5**, 467
with triethoxysilane, **2**, 82
with trimethylsilyl isothiocyanate, **5**, 1107
with trimethylstannylcopper dimethyl sulfide, **2**, 600
with triphenylphosphine dibromide, **5**, 17
with triphenylphosphineimide, **5**, 373
with tris(thiophenyl)aluminum, **5**, 415
with water, **5**, 24
reduction of, **1**, 15; **2**, 52, 54, 60; **4**, 169
for secondary phosphines, **2**, 822
reduction to aldehydes, **3**, 13
as starting materials
for acid bromides, **5**, 17
for acid iodides, **5**, 20
for aldehydes, **3**, 8
for *N*-alkylalkanoamides, **5**, 277
for *N*-alkylamides, **5**, 277
for alkylphosphine oxides, **2**, 452
for alkynoic amides, **5**, 304
for α-bromo ketones, **3**, 154
for α-chloro ketones, **3**, 152
for 1,2-difluoroethers, **2**, 89
for α-fluoro ketones, **3**, 150
for hemiacetals, **4**, 169
for ketene aminals, **4**, 972
for ketones, **3**, 121, 130
for nitriles, **3**, 618
for primary amides, **5**, 274
for selenocarboxylic esters, **5**, 239
for thiocarboxylic esters, **5**, 235
for α-thio α-sulfinyl acetophenones, **4**, 857
for toluenes, **1**, 21
for α,β-unsaturated acids, **5**, 98
sterically hindered, **1**, 137
sulfinylation–desulfinylation of, **1**, 598
synthesis of, **1**, 698; **3**, 530
via Wolff reactions, **1**, 394
use in the synthesis of aldehydes, via homologation, **3**, 15
Esters, acetic, as starting materials, for ketenes, **3**, 535
Esters, acetimidic, thermal rearrangement of, **5**, 288
Esters, acetoacetic
oxidation of, **4**, 162
use in the synthesis of 1,3-benzazarsolines, **4**, 496
Esters, 3-acetoxy, synthesis of, **2**, 74
Esters, *O*-acetyl thiohydroxamic
as starting materials, for methyl radicals, **1**, 943
use in the synthesis of radicals, **1**, 940
Esters, acrylic
reactions
with organocoppers, **5**, 151
with trimethylenemethanes, **5**, 146
synthesis of, **5**, 143
Esters, activated, synthesis of, **5**, 124
Esters, active
for the acylation of amines, **5**, 263
from
N-hydroxysuccinimide, **5**, 277
4-nitrophenol, **5**, 277
as starting materials
for *N*-alkylamides, **5**, 277
for imides, **5**, 306
use in the synthesis of amides, **5**, 261
Esters, *O*-acyl, rearrangement to ortho esters, **6**, 71
Esters, acylaminophosphinic
reactions, with vinylidene cuprates, **4**, 1001
as starting materials, for *gem*-amido phosphonoylalkenes, **4**, 1001
Esters, *O*-acyl selenon, synthesis of, **5**, 243
Esters, β-aldehydo, reactions, with hydrazines, **3**, 453
Esters, alkanoic, synthesis of, **5**, 148

Esters, alkene-1,1-diboronic, reactions, with mercuric chloride, **4**, 1067
Esters, alkenyl, synthesis of, **2**, 674
Esters, alkenyl boronate, use in the synthesis of α-hydroxy ketones, **3**, 157
Esters, *N*-alkenylimino, synthesis of, **5**, 690
Esters, alkenyl β-keto, as starting materials, for phenols, **2**, 672
Esters, *N*-alkoxycarbonyl-ω-amino stannyl, hydrolysis of, **5**, 993
Esters, *N*-alkoxyimino, synthesis of, **5**, 700
Esters, γ-alkoxy-α,β-unsaturated, reduction of, **1**, 15
Esters, 2-alkoxyvinyl phosphonic, reactions, with acetyl nitrate, **4**, 490
Esters, alkyl
  reactions
    with potassium cyanate, **5**, 964
    with trimethylsilyl iodide, **5**, 225
Esters, *n*-alkyl, synthesis of, **6**, 263
Esters, *O*-alkyl
  lithiation of, **4**, 382
  silylation of, **4**, 360
  as starting materials, for methylselenocarboxylic esters, **5**, 239
Esters, β-alkylaminobutyric, hydrolysis of, **5**, 24
Esters, alkyl β-fluoro-β-nitroacrylate, synthesis of, **4**, 795
Esters, alkylidene biscarbamic, synthesis of, **4**, 426
Esters, alkylidene malonic, synthesis of, **1**, 742
Esters, *O*-alkyl oxyimidic
  from, *O*-alkyl oxyimidic bromides, **5**, 827
  synthesis of, **5**, 826, 827
Esters, *O*-alkyl-substituted oxyimidic, synthesis of, **5**, 826
Esters, alkynic
  reactions
    with cobalt carbonyls, **4**, 712
    with zinc homoenolates, **3**, 225
  as starting materials, for alkynic ketones, **3**, 275
  use in the synthesis of *N*-alkylamides, **5**, 280
Esters, alkynoic, synthesis of, **5**, 154
Esters, 2-alkynoic, synthesis of, **1**, 1076
Esters, α,β-alkynoic, synthesis of, **5**, 160
Esters, alkynol, synthesis of, **2**, 1028
Esters, alkynyl
  from, alkynyl(phenyl)iodonium salts, **2**, 1028
  rearrangement of, **1**, 993
  research background, **2**, 1021
  synthesis of, **2**, 1028
Esters, *S*-alkynyl, of diethyl dithiophosphoric acid, **2**, 1033
Esters, α-alkynyl, synthesis of, **5**, 554
Esters, allenic
  from, propargylic carbonates, **1**, 975
  reactions, with alkenes, **1**, 498
  synthesis of, **5**, 159
Esters, α-allenic, synthesis of, **5**, 103
Esters, β-allenic, synthesis of, **1**, 979
Esters, allenyl, from, enynes, **1**, 970
Esters, α-allenyl, synthesis of, **5**, 554
Esters, β-allenyl, synthesis of, **1**, 968
Esters, allenylboronic
  reactions
    with aldehydes, **1**, 1055
    with β-hydroxyketones, **1**, 1055
Esters, allyl, reactions, with hydroxylamines, **2**, 308
Esters, allyl boronate, rearrangement of, **1**, 838
Esters, allyl-*N*,*N*-dimethylcarbamic, rearrangements of, **1**, 836
Esters, *S*-allyl dithioic, as starting materials, for γ,δ-unsaturated esters, **5**, 552
Esters, allylic, reactions, with lithium amides, **1**, 830
Esters, allylic glycolate, Wittig rearrangement of, **1**, 819
Esters, *S*-allylic γ,δ-unsaturated, synthesis of, **5**, 552
Esters, α-amido, synthesis of, **5**, 558
Esters, α-amido-β-keto, decarboxylation of, **3**, 182
Esters, amino, iron(III) oxidation of, **2**, 88
Esters, α-amino, as starting materials, for pyrrolidines, **2**, 324
Esters, β-amino
  cyclization of, **3**, 262
  synthesis of, **1**, 261; **2**, 319
Esters, (*E*)-3-amino-3-alkoxythioacrylic-*O*-alkyl, synthesis of, **4**, 895
Esters, β-amino-α,β-unsaturated, as starting materials, for α-lithio enamines, **4**, 1016
Esters, *N*-arenesulfonylimino, synthesis of, **5**, 703
Esters, arenethiol, use in the synthesis of arenesulfinyl chlorides, **2**, 717
Esters, aromatic
  hydrolysis of, **5**, 106
  reduction of, **3**, 84
Esters, arsinous, from, arsine oxides, **2**, 888
Esters, aryl
  cleavage of, **2**, 658
  as intermediates, in the synthesis of phenols, **2**, 654
  reactions, with Grignard reagents, **3**, 285
  reduction of, **1**, 21
  synthesis of, **2**, 683
Esters, *S*-aryl, synthesis of, **5**, 551, 828
Esters, aryl orthobis(trifluoroacetic), reactions, with sulfur tetrafluoride, **6**, 43
Esters, β-(arylseleno)-α,β-unsaturated selenocarboxylic, synthesis of, **5**, 241
Esters, arylsulfenyl alkyl boronate, deprotonation of, **3**, 167
Esters, *N*-aryl thioimidic, synthesis of, **5**, 730
Esters, β-arylthio-α,β-unsaturated thiocarboxylic, synthesis of, **5**, 238
Esters, azido, electrochemical reduction of, **5**, 214
Esters, azo, as starting materials, for hydrazines, **2**, 375
Esters, benzoate, synthesis of, **1**, 521
Esters, benzoic, reduction of, **1**, 21
Esters, benzyl, reduction of, **1**, 15
Esters, *N*-benzylidene-α-(diphenylphosphinoyl)glycine, as starting materials, for azomethine ylides, **4**, 489
Esters, bissulfonic, synthesis of, **4**, 282
Esters, bis(xanthate), as starting materials, for dideoxynucleosides, **1**, 585
Esters, borinic
  reactions, with phosphines, **4**, 342
  synthesis of, **4**, 815
Esters, boronate
  reactions
    with dichloromethyllithium, **4**, 526
    with ethyl formate, **3**, 35
Esters, boronic, synthesis of, **1**, 163
Esters, boronic ethyl, synthesis of, **4**, 815
Esters, bromo, as starting materials, for nitro compounds, **2**, 393
Esters, 2-bromo, reactions, with carbonyls, **2**, 69
Esters, α-bromo
  cyclization of, **1**, 346
  reactions, with phenoxides, **4**, 199
  as starting materials
    for α-acylimines, **3**, 417
    for *gem*-imino phosphiminoalkenes, **4**, 996
    for ketene silyl acetals, **4**, 831
  synthesis of, **2**, 29
Esters, β-bromo, as starting materials, for alkenes, **1**, 568
Esters, α-bromoalkyl
  from
    2-diazo-1,3-dicarbonyls, **4**, 57
    vinyl acetates, **4**, 57
  synthesis of, **4**, 56
Esters, bromoalkyl trimethylsilyl, synthesis of, **5**, 225
Esters, *N*-bromoimidic, as intermediates, **5**, 813
Esters, 2-bromo-β-keto, reactions, with diazonium salts, **5**, 843
Esters, bromomalonic, reactions, with aldehydes, **1**, 742
Esters, bromomethanediphosphonate, thallium salt of, **4**, 801

Esters, *S*-*t*-butyl, synthesis of, **5**, 551
Esters, *t*-butyl
  hydrolysis of, **5**, 29
  use in the synthesis of (*Z*)-alkenes, **1**, 751
Esters, *S*-carboxyalkyl, synthesis of, **5**, 550
Esters, carboxylic
  alkylation of, **4**, 824
  reactions
    with alkynyl metals, **1**, 1045
    with allenyl organometallics, **1**, 1060
    with propargyl organometallics, **1**, 1060
  as starting materials
    for ketene acetals, **4**, 824
    for ketene dithioacetals, **4**, 845
    for thioketenes, **3**, 547
Esters, carboxylic ortho
  reactions
    with benzenesulfonamide, **5**, 701
    with *N*,*N*-disubstituted hydrazines, **5**, 705
  as starting materials
    for hydrazonic acid esters, **5**, 705
    for imidic acid esters, **5**, 690
    for *N*-phosphorylated imino esters, **5**, 707
Esters, *S*-carboxylmethyl, from, chloroacetic acid sodium salt, **5**, 550
Esters, cephalosporin, synthesis of, **4**, 48
Esters, α-chloro, reactions, with *t*-butyl carbazate, **4**, 448
Esters, 1-chloroalkyl, synthesis of, **5**, 139
Esters, α-chloroalkyl
  from
    butenolide alcohols, **4**, 54
    α-phenylthioalkyl esters, **4**, 54
    phthalide alcohols, **4**, 54
  synthesis of, **4**, 53
Esters, α-chloroboronic, as starting materials, for α-alkoxystannanes, **4**, 389
Esters, 1-chloro-1-(chlorosulfenyl), synthesis of, **4**, 72
Esters, α-chloro-α-(chlorosulfenyl)carboxylic, as starting materials, for α-chloro sulfides, **4**, 70
Esters, chlorodihydrophenanthrene, dehalogenation of, **1**, 934
Esters, chloroformate
  reactions
    with allenyltins, **1**, 1060
    with propargyltins, **1**, 1060
Esters, *N*-chloroimidic, reactions, with tertiary phosphines, **5**, 860
Esters, ω-chloro-α-iodoalkenylboronate, synthesis of, **4**, 815
Esters, 2-chloro-β-keto, reactions, with diazonium salts, **5**, 843
Esters, chlorophosphine, reactions, with triethylamine, **4**, 483
Esters, α-chloro thiol, reactions, with triethyl phosphite, **4**, 836
Esters, 1-chlorovinylboronate, from, 1,1-dichloroethane, **4**, 813
Esters, chromate, as catalytic oxidants, for alcohols, **3**, 116
Esters, cyano, reactions, with alumina, **3**, 659
Esters, 2-cyano, oxidation of, **4**, 894
Esters, *S*-β-cyano, synthesis of, **5**, 560
Esters, α-cyano acetic, reactions, with iodobenzene diacetate, **3**, 573
Esters, α-cyano allyl, synthesis of, **3**, 623
Esters, cyanomethyl, saponification of, **5**, 30
Esters, cyclic allenic, synthesis of, **1**, 983
Esters, dialkenyl, from, β,γ-unsaturated esters, **5**, 553
Esters, dialkyl, synthesis of, **2**, 99
Esters, 2-dialkylaminoalkanoic, synthesis of, **1**, 261
Esters, dialkylamino(1-benzotriazolyl)acetic, nucleophilic displacement of, **1**, 261
Esters, dialkylborinic, synthesis of, **1**, 1102
Esters, diazo
  reactions
    with arenediazonium salts, **5**, 843
    with *N*-phenylmaleimide, **3**, 321
Esters, α-diazo
  acylation of, **3**, 475
  reactions
    with dialkylboranes, **2**, 533
    with triethyloxonium hexachloroantimonate, **2**, 799
  as starting materials
    for boryloxy compounds, **4**, 832
    for carbenes, **1**, 549
    for α-diazo amides, **3**, 475
  synthesis of, **3**, 468
Esters, *N*,*N*-dibenzylamino-α,β-unsaturated, oxidation of, **1**, 839
Esters, dichloromethyl, synthesis of, **6**, 38
Esters, dichloromethyl alkyl, as starting materials, for trithioorthoformate esters, **6**, 82
Esters, dienoic, from, allylic alcohols, **1**, 788
Esters, dienyl, isomerization of, **3**, 714
Esters, diethylene boronate, synthesis of, **4**, 634
Esters, difluoroaminoalkyl methyl, as intermediates, **5**, 814
Esters, α,α-digermylated, synthesis of, **4**, 625
Esters, α,α-dihalo, as starting materials, for ortho esters, **6**, 69
Esters, 1,2-dihydroxy, synthesis of, via asymmetric *cis*-dihydroxylation reactions, **2**, 49
Esters, 2,6-dimethylthiophenyl, use in the synthesis of β-lactones, **1**, 683
Esters, dioxobutyric, as starting materials, for hydrazonoyl bromides, **5**, 841
Esters, diphenyl, synthesis of, **4**, 467
Esters, α,α-disubstituted γ-keto, synthesis of, **5**, 79
Esters, disulfonate, reactions, with aldehydes, **4**, 864
Esters, α,α-dithioboronate, use in the synthesis of ketene dithioacetals, **4**, 849
Esters, dithiol, synthesis of, **1**, 822
Esters, dithionomalonic
  reactions
    with amines, **4**, 895
    with ammonia, **4**, 895
    with *t*-butoxy-bis(dimethylamino)methane, **4**, 896
Esters, α,α-dithiophosphonate, reactions, with ketones, **4**, 847
Esters, enamino
  from
    epoxides, **2**, 746
    iminium salts, **2**, 746
Esters, β-enamino, decarboxylation of, **3**, 411
Esters, enol
  epoxidation of, **4**, 209
  reactions, with peracids, **4**, 209
  as starting materials
    for acyl alkyl acetals, **4**, 209
    for aldehydes, **3**, 9
    for ketones, **3**, 123
Esters, 1-ethoxyethyl, synthesis of, **4**, 208
Esters, ethyl, as starting materials, for carboxylic acids, **5**, 92
Esters, ethynecarboxylate
  reactions
    with carbon diselenide, **4**, 874
    with selenolate salts, **4**, 874
Esters, α-fluoro, from, alkenes, **4**, 52
Esters, α-fluoro aliphatic, reactions, with Grignard reagents, **1**, 108
Esters, α-fluoroalkyl
  from
    fluoroalkenes, **4**, 51
    3-oxocarboxylic acids, **4**, 52
    perfluoroacyl fluorides, **4**, 52
    perfluoro alkoxides, **4**, 52
    vinyl acetates, **4**, 51
  as starting materials, for ethyl acetate, **4**, 52
  synthesis of, **4**, 51
Esters, α-fluorobenzyl phosphonate, synthesis of, **4**, 131

Esters, *N*-fluoroimidic, synthesis of, **5**, 813
Esters, α-fluoro-α-iodo, reactions, with *n*-butyllithium, **4**, 137
Esters, α-fluorophenylacetic, synthesis of, **1**, 108
Esters, fluoro sulfonate, use in the synthesis of alkoxycarbene complexes, **5**, 935
Esters, α-fluoro-α-trimethylsilyl, synthesis of, **4**, 137
Esters, 1-formamido-2,2,2-trichloroethanephosphonic, chlorination of, **5**, 790
Esters, glycolate, [2,3] Wittig rearrangement of, **1**, 819
Esters, α-halo
  reactions
    with arylzincs, **1**, 465
    with *N*-hydroxysuccinimide, **5**, 333
  Reformatsky–Claisen rearrangement of, **1**, 831
Esters, haloalkyl, synthesis of, **5**, 139
Esters, 2-haloalkyl
  from
    carboxylic acids, **5**, 139
    ketene acetals, **5**, 139
Esters, α-haloalkyl, synthesis of, **4**, 51
Esters, α-haloboronic
  chemistry of, **4**, 362
  reactions, with zinc, **1**, 163; **4**, 665
Esters, *N*-haloimidic
  from
    imidates, **5**, 813
    imidic esters, **5**, 813
    nitriles, **5**, 813
    N,*N*-perfluoroamidines, **5**, 813
  synthesis of, **5**, 813
  See also *N*-Haloimidates
Esters, halomethanediphosphonate, use in the synthesis of 2-haloalkenylphosphorus compounds, **4**, 801
Esters, heterocyclic pyruvic, reactions, with diazonium salts, **5**, 858
Esters, homochiral isocyano, synthesis of, **3**, 696
Esters, hydrazonic, as starting materials, for hydrazonamides, **5**, 852
Esters, hydroxamate, as starting materials, for radicals, **1**, 231
Esters, hydroxamic vinyl, reactions, with ketene acetals, **4**, 833
Esters, hydroxy, as starting materials, for *O*-trimethylsilyl ortho esters, **6**, 72
Esters, 2-hydroxy, synthesis of, via asymmetric hydrogenation, **2**, 53
Esters, 3-hydroxy, synthesis of, **2**, 69
Esters, 5-hydroxy, synthesis of, via ring-cleavage of epoxides, **2**, 85
Esters, *N*-hydroxy, as acylating agents, **5**, 266
Esters, β-hydroxy
  alkylation of, **1**, 682
  from, β-silyloxy esters, **1**, 283
  as starting materials, for β-lactones, **1**, 682
  synthesis of, **1**, 682
Esters, α-hydroxyalkyl phosphonic, synthesis of, **4**, 338
Esters, *o*-hydroxy aromatic, as starting materials, for ketenes, **3**, 536
Esters, α-hydroxybenzyl phosphonate, reactions, with diethylaminosulfur trifluoride, **4**, 129
Esters, *S*-2-hydroxyethyl, synthesis of, **5**, 828
Esters, *N*-hydroxyimino, synthesis of, **5**, 699
Esters, β-hydroxy-γ-imino, as azadiene precursors, **3**, 414
Esters, *N*-hydroxysuccinimide
  as acylating reagents, **5**, 322
  reactions, with cyanamide, **5**, 1120
Esters, ω-hydroxy thiocarboxylic, synthesis of, **5**, 235
Esters, β-hydroxythiol, cyclization of, **5**, 161
Esters, imidate
  Claisen rearrangement of, **5**, 271
  reactions, with hydrogen selenide, **5**, 516
  selenation of, **5**, 510
  as starting materials
    for *O*-alkyl selenocarboxylates, **5**, 515
    for selenoesters, **5**, 515
  sulfhydrolysis of, **5**, 512
  telluration of, **5**, 510
  thionation of, **5**, 510
Esters, imidic
  as intermediates, in the synthesis of amidines, **5**, 748
  reactions
    with acylhydrazines, **5**, 773
    with amines, **5**, 748
    with hydrazines, **5**, 769
  as starting materials
    for *N*-haloimidic esters, **5**, 813
    for hydrazonamides, **5**, 849
    for oxyimidic esters, **5**, 826
  synthesis of, via base-catalyzed reactions of nitriles, **5**, 749
  tautomers of, **4**, 880
  See also Imidates
Esters, imido
  as starting materials
    for dithioic acid esters, **5**, 551
    for tetrathiophthalic acid esters, **5**, 561
  sulfhydrolysis of, **5**, 551
Esters, imidophosphorane, synthesis of, **5**, 860
Esters, iminium
  reactions, with sodium diethyl phosphite, **4**, 1003
  as starting materials, for α-phosphoryl enamines, **4**, 1004
Esters, imino
  from
    alkylaminomethoxycarbenegold complexes, **5**, 696
    amines, **5**, 691
    benzylidencyanalkylamino oxides, **5**, 698
    α-chloroimines, **5**, 697
    ketene acetals, **5**, 691
    ortho esters, **5**, 690
  reactions
    with arenesulfonyl halides, **5**, 703
    with benzoylhydrazines, **4**, 986
    with 1,3-dienes, **2**, 327
    with dithiophosphoric acids, **5**, 644
    with phenylhydrazine hydrochloride, **5**, 705
  as starting materials
    for α-hydrazino enamines, **4**, 983
    for hydrazonic acid esters, **5**, 704
    for imidoyl-tetracoordinated phosphorus compounds, **5**, 787
    for ketene aminals, **4**, 972
    for α-ureido enamines, **4**, 982
  synthesis of, **5**, 688
Esters, *S*-β-imino, synthesis of, **5**, 560
Esters, α-iodo, use in atom transfer reactions, **1**, 344
Esters, α-iodoalkyl, synthesis of, **4**, 58
Esters, (iodomethyl)boronic, as starting materials, for boronic acids, **4**, 144
Esters, isocyanato, synthesis of, **5**, 969
Esters, isocyanide, synthesis of, **3**, 718
Esters, α-isocyano
  allylation of, **3**, 707
  use in the synthesis of isocyanides, **3**, 707
Esters, 2-isocyanoacrylic, reactions, with carbanions, **3**, 706
Esters, α-isocyanoalkanoic, metallation of, **3**, 704
Esters, 2-isocyano-2-methylalkanoic, synthesis of, **3**, 704
Esters, isocyanomethyl phosphonate diethyl, reactions, with steroidal ketones, **3**, 718
Esters, isopropenyl, as starting materials, for ketenes, **3**, 536
Esters, 2-isothiocyanatoacrylic, synthesis of, **5**, 1046
Esters, α-isothiocyanatoacrylic, synthesis of, **5**, 1040
Esters, keto
  allylation of, **1**, 231
  hydrogenation of, **2**, 52
  as intermediates, **5**, 145
Esters, 2-keto
  asymmetric hydrogenation of, **2**, 53

reduction of, **2**, 39
Esters, α-keto
pyrolysis of, **5**, 421
reactions
with phosphines, **5**, 1130
with sodium glyoxylate, **5**, 104
use in the synthesis of 1,3-azarsa compounds, **4**, 495
use in the synthesis of 1,3-benzazarsolines, **4**, 496
Esters, β-keto
base induced cleavage of, **1**, 884
dianions, alkylation of, **1**, 178
formylation of, **3**, 50
from
α-amino acid esters, **1**, 34
arsonium ylides, **1**, 54
ethyl diazoacetate, **1**, 389
methylenation of, **1**, 767
as radical precursors, **1**, 336, 346
reactions
with acyldiazines, **5**, 354
with hydrazines, **3**, 453
with sodium hydroxide, **1**, 852
rearrangement of, **1**, 831
as starting materials
for iodonium ylides, **1**, 109
for ketones, **3**, 131
synthesis of, **1**, 51, 132, 607; **3**, 461
thallium salts of, for the opening of bicyclic salts, **1**, 241
use in the synthesis of 1,3-azarsa compounds, **4**, 495
Esters, γ-keto, reactions, with hydrazine, **3**, 453
Esters, δ-keto, use in the synthesis of 1,3-benzazarsolines, **4**, 496
Esters, α-keto perfluoroalkane carboxylic, reactions, with hydrazine, **4**, 311
Esters, α-ketoxime, synthesis of, **3**, 428
Esters, malonic
reactions
with alkylarsenic dichlorides, **4**, 594
with arylarsenic dichlorides, **4**, 594
as starting materials, for ketenes, **3**, 536
Esters, malonoamide imido, synthesis of, **5**, 698
Esters, methacrylate, reactions, with vinyl cyclopropanes, **4**, 830
Esters, methane sulfonate (mesylate), synthesis of, **1**, 583
Esters, β-methoxy, synthesis of, **1**, 298
Esters, *N*-methoxybenzimino, synthesis of, **5**, 699
Esters, (methoxymethylene)bisphosphonic, reactions, with benzaldehyde, **4**, 916
Esters, methoxyvinyl, synthesis of, **4**, 827
Esters, methyl
cleavage of, **5**, 78
electrofluorination of, **6**, 230
as starting materials
for 2,2-diphenyl-3-hydroxypropionic acid, **5**, 62
for 2,4,6-trimethylbenzoic acid, **5**, 107
synthesis of, **5**, 132
Esters, *S*-methyl, synthesis of, **5**, 552
Esters, *p*-methylbenzyl, cleavage of, **4**, 478
Esters, *N*-methyldiethanolamine, as starting materials, for aryl fluorides, **2**, 613
Esters, methylidenemalonic, synthesis of, via retro-Diels–Alder reaction, **1**, 786
Esters, methyl malonate, synthesis of, **1**, 65
Esters, methylselenocarboxylic, from, *O*-alkyl esters, **5**, 239
Esters, methyl 4-tolyl, synthesis of, **6**, 624
Esters, β-methyl-α,β-unsaturated, synthesis of, **1**, 485
Esters, nitrite, as starting materials, for ketones, **3**, 122
Esters, nitro, synthesis of, **2**, 393
Esters, 2-nitro, reactions, with Lawesson's reagent, **5**, 603
Esters, α-nitro
from, acylimidazoles, **2**, 396
synthesis of, **2**, 391, 393
Esters, ω-nitro, from, nitro ketones, **2**, 396
Esters, α-nitroalkanoate, as starting materials, for nitrile oxides, **3**, 684
Esters, *p*-nitro-*t*-benzylic, displacement of, **1**, 200
Esters, nitronate, as intermediates, in the synthesis of nitrile oxides, **3**, 684
Esters, nitronic, reactions, with alkenes, **2**, 355
Esters, α-nitroso, production of, **2**, 391
Esters, *N*-organothioimino, synthesis of, **5**, 701
Esters, orthocarbamate, reactions, with methylene groups, **4**, 894
Esters, oxime, synthesis of, **3**, 431
Esters, oxiranylimino, synthesis of, **5**, 692
Esters, *S*-β-oxo, synthesis of, **5**, 560
Esters, oxyimidic
from
carboxylic acid acyloxyamides, **5**, 827
carboxylic acid alkoxyamides, **5**, 827
imidic esters, **5**, 826
nitrile oxides, **5**, 826
oxyimidic acids, **5**, 827
oxyimidic chlorides, **5**, 826
thiocarboxylic *O*-esters, **5**, 827
reactions
with ferric chloride, **5**, 864
with hydroxylamine, **5**, 835
as starting materials, for oxyimidic amides, **5**, 835
synthesis of, **5**, 826
Esters, penicillanic, hydrolysis of, **5**, 81
Esters, peptide, thionation of, **5**, 573
Esters, perfluoromethacrylic, reactions, with aniline, **3**, 588
Esters, phenolic, thermal rearrangement of, **3**, 297
Esters, phenolic phosphate, reduction of, **1**, 432
Esters, phenyl
Fries rearrangement of, **2**, 656
synthesis of, **2**, 656
Esters, *S*-phenyl, synthesis of, **5**, 551
Esters, phenylacetate, cleavage of, **5**, 107
Esters, phenylthioalkyl, bromination of, **4**, 57
Esters, α-phenylthioalkyl, as starting materials, for α-chloroalkyl esters, **4**, 54
Esters, *N*-phenyl thioimidic, synthesis of, **5**, 729, 731
Esters, phenylthiomethyl, synthesis of, **4**, 221
Esters, phosphate, pyrolysis of, **1**, 778
Esters, phosphinate, pyrolysis of, **1**, 778
Esters, phosphinic, use in the synthesis of amides, **5**, 264
Esters, phosphonate
as starting materials
for ketene dithioacetal *S*-oxides, **4**, 860
for thiaselenafulvenes, **4**, 870
Esters, *S*-α-phosphonio, synthesis of, **5**, 560
Esters, α-phosphono-α-oxo, hydration of, **6**, 117
Esters, phosphonothioate, elimination of, **1**, 645
Esters, phosphonous acid dialkyl, alkylation of, **2**, 456
Esters, *N*-phosphorylated imino
from
carboxylic ortho esters, **5**, 707
diethyl phosphoramidate, **5**, 707
diphenylphosphinic acid amide, **5**, 707
imidates, **5**, 708
imidoyl halides, **5**, 707
phosphorus compounds, **5**, 708
synthesis of, **5**, 707
Esters, pivalate, pyrolysis of, **1**, 582
Esters, propargylic, reactions, with alcohols, **3**, 72
Esters, propiolate, reactions, with tertiary amines, **1**, 835
Esters, β-pyridinesulfinyl, elimination of, **1**, 599
Esters, 2-pyridyl
reactions, with cuprates, **3**, 287
as starting materials, for ketones, **3**, 130
Esters, *S*-2-pyridyl, reactions, with tris(trimethylsilyl)aluminum, **5**, 418
Esters, pyruvate, photolysis of, for the synthesis of aldehydes, **3**, 8

Esters, quinoline, reduction of, **3**, 109
Esters, selenoimidic
  synthesis of, **5**, 732
    via alkylation of selenoamides, **5**, 732
Esters, silyl
  alkylidenation of, **2**, 700
  reactions, with trialkylsilanes, **4**, 832
  synthesis of, via dealkylative cleavage of esters and lactones, **5**, 225
Esters, *N*-silylimidoyl, from, imidates, **5**, 863
Esters, β-silyloxy, as starting materials, for β-hydroxy esters, **1**, 283
Esters, silylsalicylate, synthesis of, **5**, 227
Esters, *O*-silyl selenon, reactions, with acid chlorides, **5**, 244
Esters, *O*-silyl thion, reactions, with acid chlorides, **5**, 243
Esters, α-sily-α-stannyl, from, silylketenes, **4**, 661
Esters, (*E*)-α-stannyl-β-germylalkenyl, synthesis of, **2**, 947
Esters, stannylthioimidic, synthesis of, **5**, 739
Esters, *S*-stannylthioimidic, synthesis of, **5**, 739
Esters, α-substituted, from, 2-halo acetals, **1**, 391
Esters, *N*-substituted imidic, as starting materials, for *N*-substituted hydrazonamides, **5**, 849
Esters, α-sulfenyl, electrolysis of, **4**, 197
Esters, α-sulfinyl, elimination of, **1**, 598
Esters, sulfonate
  acylation of, **3**, 171
  Friedel–Crafts alkylation of, **1**, 475
  pyrolysis of, **1**, 778
  reactions, with alkynes, **1**, 1021
  substitution of, **1**, 1021
  synthesis of, **1**, 1021
Esters, sulfone ortho, reactions, with aldols, **5**, 166
Esters, sulfonyl, deprotonation of, **4**, 391
Esters, β-sulfonyl-α,β-unsaturated, reactions, with 1-octynylmagnesium bromide, **1**, 1031
Esters, sulfoxylic dialkyl, as starting materials, for alkanesulfinic esters, **2**, 201
Esters, α-sulfuranylidene, as starting materials, for boryloxy compounds, **4**, 832
Esters, tellurocarboxylic
  from
    acyl halides, **5**, 241
    tellurocarboxylates, **5**, 242
  synthesis of, **5**, 241, 526
Esters, telluroimidic
  from, imidoyl chlorides, **5**, 733
  synthesis of, **5**, 732
Esters, thiobenzimidate, synthesis of, **4**, 1001
Esters, thiocarboxylic
  Claisen condensation of, **5**, 237
  from
    acyl halides, **5**, 233
    carbon monoxide, **5**, 237
    carboxylic acids, **5**, 234
    esters, **5**, 235
    silver thiolates, **5**, 234
    thallium(I) thiolates, **5**, 234
    thiocarboxylic esters, **5**, 236
    trimethylsilyl sulfides, **5**, 234
    vinylic sulfides, **5**, 238
  reactions, with lithium 2,2,6,6-tetramethylpiperidide, **5**, 237
  as starting materials, for thiocarboxylic esters, **5**, 236
  synthesis of, **5**, 233
Esters, thiocarboxylic, enolate anions, synthesis of, **5**, 237
Esters, thiocyanato, synthesis of, **5**, 1107
Esters, thioformate, for the synthesis of radicals, **1**, 940
Esters, thiohydrazonic, from, thiohydrazides, **5**, 846
Esters, thiohydroxamate
  from, carboxylic acids, **1**, 332
  homolysis of, **1**, 332, 344
Esters, thiohydroxamic
  decarboxylation of, **5**, 727
  photolysis of, **1**, 935
  sonication of, **1**, 935
  use in the synthesis of radicals, **1**, 939
Esters, thioimidic
  alkylation of, **5**, 730
  from
    heterocumulenes, **5**, 729
    imidoyl halides, **5**, 728
    isonitriles, **5**, 727
    isothiocyanates, **5**, 729
    ketenimines, **5**, 729
    nitriles, **5**, 727
    N-phenyl phenylthioimidoyl chloride, **5**, 728
    thioamides, **5**, 726
    thiocyanates, **5**, 727
    thioimidic esters, **5**, 730
    thiolactams, **5**, 726
  reactions
    with amines, **5**, 742, 753, 757
    with ammonia, **5**, 753
    with ammonium acetate, **5**, 757
    with diisopropylamine, **5**, 757
    with hydrazine, **5**, 769
    with methylamine, **5**, 757
  S-allylation of, **5**, 726
  as starting materials
    for aliphatic amidines, **5**, 753
    for aromatic amidines, **5**, 757
    for hydrazonamides, **5**, 850
    for thioimidic esters, **5**, 730
  synthesis of, **5**, 726, 731, 827
    via alkylation, **5**, 726
    via radical arylation, **5**, 727
  use of term, **5**, 726
  See also Thioimidates
Esters, α-thioketo, tautomers of, **1**, 140
Esters, thiol
  hydrogenolysis of, **3**, 85
  oxidation of, **2**, 208; **4**, 259
  reactions
    with 1-alkynyltrimethylsilanes, **1**, 1045
    with chlorine, **2**, 180
    with sulfuryl chloride, **2**, 168, 173
    with Wittig reagents, **2**, 709
  reduction of, **3**, 85
  reduction to aldehydes, **3**, 13
  as starting materials
    for α-amino acids, **3**, 40
    for dialkyl sulfides, **2**, 143
    for thiols, **2**, 124
  synthesis of, **2**, 118, 124
  use in the synthesis of aromatic ketones, **3**, 287
Esters, thionoformate, reactions, with hydrazines, **5**, 640
Esters, thionoimino, hydrolysis of, **5**, 644
Esters, thionselenol, reactions, with *m*-chloroperbenzoic acid, **5**, 249
Esters, thiophenyl
  reactions, with lithium diisopropylamide, **4**, 932
  as starting materials, for trimethylsilyl enol ethers, **4**, 932
Esters, thiosulfonate
  from
    benzo-1,3-dithiole-2-thiones, **4**, 862
    benzo-1,3-dithiole-5-thiones, **4**, 862
  reactions, with ketones, **3**, 164
Esters, β-thioxo, synthesis of, **3**, 348
Esters, *O*-tosyl oxyimidic, from, tosylated malononitrile oxime, **5**, 826
Esters, *N*-tosyl phenylglycine, as starting materials, for *N*-benzoylsulfonamides, **5**, 346
Esters, trialkylsilyl
  reactions, with triphenylphosphine dibromide, **5**, 17
  as starting materials, for acid bromides, **5**, 17
  use in the synthesis of ketene disilyl acetals, **4**, 832

Esters, S-(tri-n-butylstannyl)thioimidic, synthesis of, **5**, 739
Esters, trichloromethyl, synthesis of, **6**, 230
Esters, triflate, use in the synthesis of alkoxycarbene complexes, **5**, 935
Esters, trifluoroalkane sulfonate, synthesis of, **1**, 433
Esters, trifluoromethyl, from, methyl perfluoroadamantaneacetate, **6**, 230
Esters, trihalomethyl, synthesis of, **6**, 230
Esters, trimethylaminecarboxyborane methyl, synthesis of, **6**, 491
Esters, trimethylsilyl
  hydrolysis of, **5**, 404
  methanolysis of, **4**, 129
  as starting materials, for anhydrides, **5**, 189, 196
  of strong acids, synthesis of, **1**, 64
  synthesis of, **5**, 978
Esters, α-trimethylsilyl, reactions, with carbonyls, **1**, 283
Esters, N-trimethylsilylacetimidic silyl, as starting materials, for N-halo-O-silyl imidates, **5**, 814
Esters, trimethylsilyl carboxylate, as starting materials, for lithium alkoxides, **5**, 414
Esters, 2-trimethylsilylethyl, elimination of, **1**, 5955
Esters, N-trimethylsilyl hydroxamic acid trimethylsilyl, synthesis of, **5**, 719
Esters, S-(trimethylsilyl)thioimidic, synthesis of, **5**, 739
Esters, α-trimethylstannyl boronic, destannylation of, **1**, 163
Esters, S-(tri-n-methylstannyl)thioimidic, synthesis of, **5**, 739
Esters, S-(triphenylstannyl)thioimidic, synthesis of, **5**, 739
Esters, unsaturated, reduction of, **1**, 77, 86
Esters, (E)-α,β-unsaturated, synthesis of, **1**, 534
Esters, α,β-unsaturated
  acidity of, **1**, 871
  allylation of, **1**, 314
  cycloaddition reactions of, **4**, 827
  from, β-hydroxy selenides, **1**, 594
  hydrostannation of, **2**, 601
  reactions
    with enaminostannanes, **3**, 406
    with hydrazine, **3**, 453
    with hydrazines, **5**, 351
    with magnesium salts, **5**, 550
    with malonate salts, **1**, 306
    with N-substituted hydroxylamines, **2**, 346
  as starting materials
    for α-aminophosphinic acids, **4**, 460
    for α-aminophosphonic acids, **4**, 460
    for ketene acetals, **4**, 827
  synthesis of, **1**, 542, 545, 596, 607, 763; **5**, 154
    via alkenation, **5**, 155
    via alkoxycarbonylation, **5**, 157
    via Heck coupling, **5**, 157
    via Ramberg–Bäcklund reactions, **1**, 619
    via redox reactions, **5**, 154
    via sulfoxide pyrolysis, **1**, 778
Esters, β,γ-unsaturated
  as starting materials, for dialkenyl esters, **5**, 553
  synthesis of, **1**, 467; **5**, 158, 553
    via S^N2'-reaction, **5**, 158
Esters, γ,δ-unsaturated
  from, S-allyl dithioic esters, **5**, 552
  synthesis of, **1**, 416, 827; **5**, 158
    via Claisen–Johnson rearrangement, **5**, 159
    via Tsuji–Trost reaction, **5**, 158
Esters, α,β-unsaturated β-amino, synthesis of, **4**, 496
Esters, γ,δ-unsaturated β-keto, reactions, with β-diketones, **3**, 453
Esters, α,β-unsaturated methyl, synthesis of, **1**, 531
Esters, γ,δ-unsaturated silyl, synthesis of, **1**, 830
Esters, α,β-unsaturated thioimidic, from, thioimidoyl chloride, **5**, 729
Esters, valine methyl, synthesis of, **4**, 536
Esters, vinyl
  trans-esterification of, **5**, 134
  lithiated compounds of, use in the synthesis of ketones, **3**, 137
  synthesis of, **3**, 526
Esters, S-vinyl, synthesis of, **5**, 552
Esters, vinylboronate, hydroboration of, **4**, 631
Esters, xanthate
  pyrolysis of, **1**, 775
  reduction of, **1**, 700
  as starting materials, for alkenes, **1**, 700
  use in the synthesis of radicals, **1**, 938
O-Esters, thiocarboxylic, as starting materials, for oxyimidic esters, **5**, 827
α-Esters, β-hydroxylated, stereoselectivity, in intermolecular radical additions, **1**, 327
Ester salts, imidic, reactions, with amines, **5**, 742
Estradiol
  bromination of, **2**, 685
  synthesis of, **1**, 604
Estrogens, 6,7-didehydro-, synthesis of, **1**, 435
Etard's reagent, use in the synthesis of α-chloro ketones, **3**, 151
Ethanal, benzoyl(phenyl)-, condensation of, **4**, 894
Ethanal, bis(4-methoxyphenyl)-, synthesis of, **3**, 317
Ethanamine, 2-chloro-N,N-diethyl-1,1,2-trifluoro-
  polymer supported, **2**, 19
  synthesis of, **6**, 54
  use in the synthesis of alkyl fluorides, **2**, 19
Ethane, electrochemical fluorination of, **4**, 3
Ethane, 1-(o-amidophenyl)-1-(8-methyl-1-naphthyl)-, reactions, with potassium hydroxide, **5**, 108
Ethane, 1-aryl-1,2,2-trichloro-2-fluoro-, as starting material, for 1-aryl-2-chloro-2-fluoroethenes, **4**, 750
Ethane, 1-benzoyloxy-1-ethoxy-, from, diethyl ether, **4**, 207
Ethane, 1,2-bis(chlorodimethylsilyl)-
  synthesis of, **2**, 538
  use in the protection of amines, **2**, 543
Ethane, 1,2-bis(dialkylarsino)-, synthesis of, **2**, 486
Ethane, 1,2-bis(dialkylstibo)-, synthesis of, problems, **2**, 486
Ethane, 1,2-bis(diazo)-, detection of, **3**, 467
Ethane, 1,2-bis(dichloroboryl)-1-(trimethylsilyl)-, synthesis of, **4**, 644
Ethane, 1,2-bis(dicyclohexylphosphino)-, nickel(0) complex of, use in the synthesis of carboxylic acids, **5**, 101
Ethane, 1,2-bis(difluoroamino)trifluoro-, as starting material, for perfluoro-1-amino-2-imino ethane, **5**, 807
Ethane, 1,1-bis(difluoroboryl)-, synthesis of, **4**, 632
Ethane, 1,2-bis(diphenylphosphino)- (DPPE)
  reactions, with acrolein and bis(cyclooctene)chlororhodium, **6**, 456
  synthesis of, **2**, 439
  use in the substitution of nitrogen functions, **1**, 250
Ethane, 1,2-bis(N-methylamino)-, reactions, with bis(dichloroarsino), **4**, 593
Ethane, 1,2-bis-O-trimethylsilyloxy-, reactions, with 3,4-dihydrocoumarin, **6**, 72
Ethane, bromochlorotrifluoro-, as anesthetic, **6**, 16
Ethane, 2-bromo-1,1-dichloro-, reactions, with arylalkynes, **2**, 307
Ethane, 1-(o-carboxyphenyl)-1-(8-methyl-1-naphthyl)-, synthesis of, **5**, 108
Ethane, chloro-, reactions, with selenium dioxide, **2**, 101
Ethane, 2-chloro-1-aminodioximino-, synthesis of, **5**, 822
Ethane, 1,2-diamino-, reactions, with bis(succinimido) carbonate, **6**, 507
Ethane, 1,2-dianilino-, use in the synthesis of aldehydes, **4**, 408
Ethane, diazo-, use in the synthesis of diruthenium complexes, **4**, 681
Ethane, 1,1-dibromo-, from, nitroethane, **6**, 54
Ethane, 1,2-dibromo-
  for the activation of magnesium, **2**, 558
  for chemical activation of metal surfaces, **2**, 958

reactions
with alkali metal dialkylstibides, **2**, 505
with cycloalkyllithiums, **1**, 110
with *o*-lithiomethylphenyl isocyanides, **3**, 724
with magnesium, **1**, 662
with zinc, **1**, 508
use in the synthesis of alkylzincs, **2**, 590
Ethane, 1,2-dibromo-1-ethoxy-, reactions, with triphenylphosphine, **4**, 912
Ethane, 1,2-dibromo-1-methoxy-, reactions, with trimethyl phosphite, **4**, 912
Ethane, 1,2-dibromotetrachloro-, use in the synthesis of alkyl bromides, **2**, 27
Ethane, 1,2-dibromotetrafluoro-, reactions, with *O*-(1-lithiovinyl) carbamates, **4**, 768
Ethane, 1,1-dichloro-
reactions
with lithium diisopropylamide, **4**, 813
with sodium diphenylarsenic, **4**, 592
with trimethylborate, **4**, 813
as starting material
for 1-chlorovinylboronate esters, **4**, 813
for orthoacetate esters, **6**, 75
Ethane, 1,2-dichloro-
reactions
with alkali metal dialkylstibides, **2**, 505
with magnesium, **2**, 558
Ethane, 1,2-dichloro-1,2-diisopropoxy-, reactions, with potassium *t*-butoxide, **4**, 767
Ethane, 1,2-dichloro-1,2-dimethoxy-, synthesis of, **4**, 767
Ethane, 1,1-dichlorophenyl-, synthesis of, **4**, 13
Ethane, 1,1-dideuterio-, synthesis of, **4**, 668
Ethane, 1,2-dideuteriophenyl-, synthesis of, **1**, 73
Ethane, 1,1-difluoro-, chlorination of, **6**, 225
Ethane, 1,2-dihydroxy-, reactions, with lactones, **6**, 72
Ethane, 2,2-diiodo-1,1-dinitro-, synthesis of, **4**, 993
Ethane, 1,2-dilithiophenyl[tris(trimethylsilyl)]-, synthesis of, **6**, 193
Ethane, 1,2-dilithio[tetrakis(trimethylsilyl)]-
reactions, with paraformaldehyde, **6**, 176
synthesis of, **6**, 193
Ethane, *N,N'*dimethyl-1,2-diamino-, reactions, with glyoxal, **4**, 409
Ethane, dinitro-, electrolysis of the sodium salt of, **6**, 151
Ethane, 1,1-dinitro-, alkylation of, **4**, 432
Ethane, 1,1-diphenyl-, chlorocarbonylation of, **5**, 16
Ethane, 1,2-diphenyl-, synthesis of, **1**, 110
Ethane, 1,1-disodio-1-phenyl-, synthesis of, **4**, 669
Ethane, 1,2-epoxy-1-phenyl-, reactions, with butyrolactone, **6**, 72
Ethane, hexachloro-
as chlorinating agent, for dichlorotriphenylphosphorane, **2**, 835
for the chlorination of alkenes, **2**, 24
pyrolysis of, **6**, 215
reactions, with alkyldiphenylphosphines, **4**, 112
Ethane, hexafluoro-
bromination of, **6**, 218
chlorination of, **6**, 218
iodination of, **6**, 219
radiolysis of, **2**, 1013
Ethane, hexalithio-, synthesis of, **6**, 204, 205
Ethane, hexanitro-, reactions, with benzene, **4**, 994
Ethane, iodopentafluoro-, reactions, with tetramethyllead, **6**, 65
Ethane, nitro-, as starting material, for 1,1-dibromoethane, **6**, 54
Ethane, 1-nitro-, deprotonation of, **4**, 335
Ethane, 1,1,1,2,2-pentafluoro-, as refrigerant, **6**, 8
Ethane, pentafluoroiodo-, synthesis of, **6**, 3
Ethane, perfluoro-1-amino-2-imino-, from, 1,2-bis(difluoroamino)trifluoroethane, **5**, 807
Ethane, perlithiated, from, diethylmercury, **6**, 205
Ethane, 1,1,2,2-tetrachloro-, reactions, with sodium thiophenoxide, **4**, 771
Ethane, 1,1,1,2-tetrafluoro-, as refrigerant, **6**, 8
Ethane, tetrakis(trimethylsilyl)-, synthesis of, **6**, 193
Ethane, *N,N,N',N'*-tetramethyldiamino- (TMDM), reactions, with boron enolates, **1**, 255
Ethane, 1,1,2,2-tetramorpholino-
reactions, with methanol, **4**, 309
synthesis of, **4**, 300
Ethane, 1,1,2,2-tetranitro-, fluorination of, **4**, 795
Ethane, 1,2,2-tribromo-1,2,2-trinitro-, pyrolysis of, **4**, 994
Ethane, 1,1,1-trichloro-
reactions, with allyltributyltin, **4**, 14
synthesis of, **6**, 23
Ethane, 1,1-trichlorodifluoro-, reactions, with alkenyllithiums, **4**, 758
Ethane, 1,1,2-trichloronitroso-, reactions, with dichloro-(2,2,3,3-tetrafluoropropyloxy)phosphine, **5**, 683
Ethane, 1-trichlorophosphazopentachloro-, reactions, with *p*-methoxybenzaldehyde oxime, **5**, 683
Ethane, 1,1,1-tricyano-, reactions, with tetrakis(triphenylphosphine)platinum(0), **5**, 1140
Ethane, 1,1,1-trifluoro-, as refrigerant, **6**, 8
Ethane, 1,2,2-trifluoro-1,2-dichloro-1-nitro-, as starting material, for difluorochloroacetoximidic fluoride, **5**, 822
Ethane, 1,1,1-trinitro-, as starting material, *gem*-dinitroalkanes, **4**, 431
Ethane, 1,1,2-tris(dichloroboryl)-, synthesis of, **4**, 632
Ethane, tris(trimethylsilyl)phenyl-, synthesis of, **6**, 193
Ethane, 1,1,1-tris(trimethylsilyl)-2-phenyl-, synthesis of, **6**, 175
1,2-Ethanebis(trithiocarbonic acid), synthesis of, **6**, 553
1,2-Ethanediamines, synthesis of, **2**, 314
Ethane-1,2-diamines, 1,2-diaryl-
as starting materials, for azaallyl carbanions, **3**, 408
synthesis of, **3**, 408
Ethanediol, reactions, with *P*-amino-oxazaphospholes, **2**, 476
Ethane-1,2-diol, reactions, with phosgene, **6**, 467
Ethanedithiol
reactions
with arsenic trichloride, **2**, 260
with cyclohexanone, **4**, 246
Ethane-1,2-dithiol, reactions, with 2,2-dichloro-3,4-bis(trifluoromethyl)-1,3-dithiole, **6**, 306
Ethane hexamercarbide
reactions, with carboxylic acids, **6**, 405
synthesis of, **6**, 405
Ethanephosphonate, diethyl 2-lithio-2-cyano-2-trimethylsilyl, condensation of, **3**, 646, 659
Ethanephosphonates, dialkyl 2-cyano-2-(trimethylsilyl), synthesis of, **3**, 640
Ethanes, phenyl-substituted, synthesis of, **1**, 108
Ethanes, 1-alkoxy-2,2,2-tribromo-1-chloro-, reactions, with sodium ethoxide, **4**, 767
Ethanes, 2-aryl-1-nitro-, deprotonation of, **1**, 156
Ethanes, bis(alkylthio)tetrachloro-, from, 1-chlorovinyl sulfides, **6**, 46
Ethanes, 1,2-bis(isocyanophenyl)-, synthesis of, **3**, 724
Ethanes, 1,1-bis(silyl)-, from, vinylsilanes, **4**, 617
Ethanes, diaryl-, synthesis of, **1**, 289
Ethanes, 1,1-diaryl-, as starting materials, for *gem*-dinitroalkenes, **4**, 993
Ethanes, 1,1-diaryl-1,2,2-trichloro-2-fluoro-, as starting materials, for 1,1-diaryl-2-chloro-2-fluoroethenes, **4**, 750
Ethanes, 1,1-diaryltrifluoro-, synthesis of, **1**, 475
Ethanes, 1,2-dihalo-
1,2 addition, **1**, 116
reactions
with diorganometal(III) alkali metal salts, **2**, 893
with metal dialkylstibides, **2**, 505

Ethanes, α,β-dihalo-, synthesis of, **4**, 804
Ethanes, dinitrotetraalkyl-, as starting materials, for 1,3-butadienes, **1**, 638
Ethanes, 1,2-disilyl-, as starting materials, for 1,1-bis(silyl)alkanes, **4**, 616
Ethanes, *N*-methyl-*N*-nitrosour-, decomposition of, **3**, 466
Ethanes, phosphaazo-, synthesis of, **4**, 110
Ethanes, polychlorinated, reactions, with sodium phenoxide, **4**, 761
Ethaneselenol, 2-amino-, reactions, with glyoxylic acid, **4**, 333
Ethanesulfenyl chloride, 2-chlorotetrafluoro-, as starting material, for chlorodifluorothioacetyl fluoride, **5**, 476
Ethanesulfenyl chloride, pentachloro-, as starting material, for trichlorothioacetyl chloride, **5**, 478
Ethanesulfenyl chlorides, pentachloro-, synthesis of, **5**, 492
Ethanesulfenyl fluoride, 1,2,2,2-tetrafluoro-, reactions, with pyridine, **5**, 502
Ethane sulfinates, alkyl-β-halo-, synthesis of, **2**, 200
Ethanesulfinic acid, 1-chloro-
  from, α-chlorosulfonyl chloride, **4**, 78
  reactions, with thionyl chloride, **4**, 78
Ethanesulfinyl chloride, 2,2,2-trifluoro-1,1-diphenyl-, synthesis of, **2**, 173
Ethanesulfonamides, alkylation of, **1**, 147
Ethanesulfonic esters, 2-aryl-, synthesis of, **2**, 203
Ethanesulfonyl chloride, bromo-, reactions, with sodium sulfite, **4**, 85
Ethanesulfonyl chloride, α-bromo-, as starting material, for 1-bromo-1-methylethene episulfone, **4**, 83
Ethanethiol
  reactions
    with 2-alkoxybenzodithioles, **6**, 82
    with bis(diethylamino)dimethylsilane, **2**, 263
    with bismuth nitrate, **2**, 261
    with chloroethoxyphenylphosphine, **2**, 851
    with diphenylselenophosphinic acid chloride, **2**, 857
    with enynes, **4**, 872
    with ethene, **2**, 137
    with ethyl acetate, **6**, 84
    with ethyl acrylate, **2**, 138
    with ethyl cyanoacetate, **4**, 905
    with *N*-(hexafluoroisopropylidene)-benzenesulfinamides, **4**, 325
    with penta-*O*-acetyl-β-D-glucopyranose, **4**, 224
    with 4-substituted 2-trifluoromethyl-5(4*H*)-oxazolones, **4**, 320
  synthesis of, **2**, 118
Ethanethiol, 2-amino-, reactions, with diethylgold bromide, **2**, 274
Ethanethiol, 1-chloro-1,2,2-trifluoro-, reactions, with potassium fluoride, **5**, 484
Ethanethiol, 2-chloro-1,1,2-trifluoro-, synthesis of, **5**, 484
Ethanethiol, 2-hydroxy-, reactions, with ortho esters, **6**, 95
Ethanethiol, 1,1,2-trifluoro-2-chloro-, reactions, with triethyl phosphite, **5**, 645
Ethanethiolate, reactions, with allyl bromides, **4**, 373
Ethanethiols, reactions, with boron trifluoride etherate, **2**, 304
Ethanethiols, 2-amino-, as antiradiation agents, **2**, 237
N-Ethanethiophthalimide, as thiolating agent, **6**, 90
Ethanol
  from
    cyclobutanols, **2**, 636
    ethylamine, **2**, 88
    3-thietanol, **2**, 636
  reactions
    with allylsulfonyl chloride, **2**, 203
    with 1-[bis-(2-chloroethyl)amino]-1,2,2-tricyanoethene, **4**, 893
    with bis(difluoroamino)fluoroaminomethyl isocyanate, **6**, 363
    with boron trifluoride, **2**, 110
    with dichlorotetrafluoroacetone, **4**, 164
    with diethoxyethenylidinetriphenylphosphorane, **6**, 75
    with ethyl propiolate, **4**, 195
    with mercuric oxide, **6**, 405
    with methane disulfonyl chloride, **4**, 282
    with nitriles, **4**, 883
    with oxime tosylates, **1**, 1119
    with Raney nickel, **1**, 47
    with selenium oxychloride, **2**, 101
    with silicon tetrachloride, **2**, 108
    with sodium ethoxide, **1**, 1065
    with trichlorosilane, **2**, 108, 108
    with *O*-trimethylsilyl thiocarboxylates, **5**, 539
    with ynamines, **4**, 885
  synthesis of, **2**, 636
Ethanol, 2-chloro-
  reactions
    with dipyridyl ketones, **4**, 183
    with methanesulfonyl chloride, **4**, 282
    with 1,1,1-trifluoroacetone, **4**, 164
  transesterification of orthoformate esters with, **6**, 78
Ethanol, 2-cyano-, reactions, with tristriethylamino phosphate, **2**, 105
Ethanol, 1,1-difluoro-1-nitro-, synthesis of, **6**, 54
Ethanol, 2,2-dinitro-2-fluoro-, dehydroxymethylation of, **6**, 57
Ethanol, 2-halo-2,2-dinitro-, as starting material, for potassium bromodinitromethides, **6**, 287
Ethanol, 2-(methylamino)-, reactions, with glyoxal, **4**, 302
Ethanol, *N*-methyl-*N*,*O*-bis(tributylstannyl)-2-amino-, reactions, with 1,3-dioxolane-2-thione, **6**, 313
Ethanol, 1-phenyl-, deprotonation of, **1**, 138
Ethanol, thiol-, reactions, with cyclohexanones, **4**, 229
Ethanol, 2-thiol-, reactions, with 1,1,3,3-tetramethoxypropane, **4**, 230
Ethanol, 2,2,2-trichloro-, reactions, with acetals, **4**, 184
Ethanol, trifluoro-, as solvent, in the synthesis of acylirons, **5**, 453
Ethanol, 2,2,2-tris(trimethylgermanyl)-, as starting material, for 1,1-bis(trimethylgermanyl)ethene, **4**, 1055
Ethanol, 2,2,2-tris(trimethylgermyl)-, synthesis of, **6**, 186
Ethanolamine
  reactions
    with potassium cyanate, **6**, 485
    with ruthenium complexes, **6**, 722
Ethanolamine, *O*,*N*-bis(tributylstannyl)-*N*-methyl-, reactions, with 3-methyl-2-oxazolidinethione, **6**, 322
Ethanolamines
  condensation of, **4**, 888
  reactions, with 7,7,8,8-tetracyanoquinodimethane, **4**, 893
Ethanols, α-alkyl-β-(*N*-methylamino)-, synthesis of, **2**, 311
Ethanols, 2-amino-1-aryl-, synthesis of, **2**, 330
Ethanols, β-amino-α-aryl-, synthesis of, **2**, 318
Ethanols, 2-arylthio-2,2-dichloro-, reactions, with zinc, **4**, 775
Ethanols, 2-aryl-1-trifluoromethyl-, synthesis of, **2**, 106
Ethanols, naphthyl-, as starting materials, for *gem*-dinitroalkenes, **4**, 993
Ethanols, triaryl-, synthesis of, **2**, 645
Ethanols, trifluoro-, dehydrofluorination of, **1**, 570
Ethanone, 2-methanesulfonyl-1-(2-chlorophenyl)-, reactions, with diethyl chlorothiophosphate, **4**, 784
1-Ethanone, 1-(*t*-butyldimethylsilyl)-2-chloro-, reactions, with trimethyl phosphite, **4**, 934
1-Ethanone, 2-diazo-1,2-diphenyl-, synthesis of, **3**, 470
Ethanthiol, for the hydrolysis of aromatic esters, **5**, 106
Ethene
  carbolithiation of, **2**, 555
  carbonylation of, **5**, 192
  reactions
    with alkynes, **1**, 528
    with aluminum hydride, **2**, 112
    with benzene, **1**, 364

with ethanethiol, **2**, 137
with platinum complexes, **2**, 582
with sulfur, **2**, 137
reduction of, **1**, 71
as starting material, for vinylpotassium, **2**, 907
See also Ethylene
Ethene, 2-anilino-1-bromo-1-nitro-, stability of, **4**, 796
Ethene, 1-benzenesulfinyl-1-nitro-2,2-diphenyl-, synthesis of, **4**, 907
Ethene, 1,1-bis(benzenesulfonyl)-, reactions, with 3*H*-indolium ylides, **4**, 277
Ethene, 1-[bis-(2-chloroethyl)amino]-1,2,2-tricyano-, reactions, with ethanol, **4**, 893
Ethene, 1,1-biscyclopropyl-, tandem rearrangements of, **1**, 801
Ethene, bis(dibutoxyphosphino)-, synthesis of, **4**, 913
Ethene, 1,1-bis(diphenylphosphino)-2,2-bistrifluoromethyl-, synthesis of, **4**, 1023
Ethene, bis(phenylthio)-, reactions, with butyllithium, **2**, 1031
Ethene, 1,1-bis(sulfonyl)-
Michael addition to, **4**, 277
reactions, with 2,3-dihydropyran-4-one, **4**, 277
Ethene, 1,1-bis-(toluene-*p*-sulfinyl)-, as chiral dienophile, **4**, 270
Ethene, 1,1-bis(trifluoromethylthio)-2,2-bis[(trifluoromethyl)thiocarbonyl]-, synthesis of, **6**, 235
Ethene, 1,1-bis(trimethylgermanyl)-, from, 2,2,2-tris(trimethylgermanyl)ethanol, **4**, 1055
Ethene, 1,1-bis(trimethylsilyl)-
cycloaddition of, **6**, 126
synthesis of, **4**, 1048
Ethene, 1,1-bis(trimethylsilyl)-2-phenyl-, from, phenylmethanal, **4**, 1049
Ethene, 1,1-bis(trimethylstannyl)-, synthesis of, **4**, 1069
Ethene, 1-bromo-1,2-diisopropoxy-, synthesis of, **4**, 767
Ethene, 1-(bromo)iodo-2-bromo(iodo)-1-phenoxy-, synthesis of, **4**, 765
Ethene, bromotrifluoro-, reactions, with tin halides, **4**, 812
Ethene, 1-bromo-1-trimethylsilyl-
reactions, with magnesium, **4**, 1060
synthesis of, **4**, 807
Ethene, chloro-, reactions, with tetramethyldiarsine, **4**, 134
Ethene, 1-chloro-1,2-diisopropoxy-, synthesis of, **4**, 767
Ethene, 2-chloro-1,1-dimethoxy-, [4+2] cycloaddition reactions, with α,β-unsaturated carbonyls, **6**, 75
Ethene, 1-chloro 2,2-diphenyl-, deprotonation of, **4**, 741
Ethene, chlorotrifluoro-
addition of iodine fluoride to, **6**, 3
reactions
with cyanide, **4**, 5
with diethylamine, **6**, 54
with hydrogen sulfide, **5**, 484
with silver trifluoroacetate, **6**, 63
with sulfur, **5**, 479
use in the synthesis of chlorofluoroalkanes, **4**, 32
Ethene, dibenzoyl-
reactions
with carboxylic acids, **5**, 190
with diphenylsulfilimine, **2**, 745
Ethene, *cis*-dibromo-, dehydrohalogenation of, **2**, 1017
Ethene, 2,2-dibromo-1,1-dinitro-, synthesis of, **4**, 994
Ethene, 1,2-dibromo-1-fluoro-, dehalogenation of, **2**, 1013
Ethene, 1,1-dibromo-2-phenoxy-, synthesis of, theoretical, **4**, 761
Ethene, 1,2-dibromo-1-phenoxy-, synthesis of, **4**, 761
Ethene, 1,1-dichloro-, condensation of, **5**, 38
Ethene, 1,2-dichloro-, reactions, with sodium methoxide, **2**, 1014
Ethene, *cis*-1,2-dichloro-, reactions, with benzeneselenol, **4**, 786
Ethene, (*Z*)-1,2-dichloro-, reactions, with 1-alkynes, **1**, 1065
Ethene, 1,1-dichlorodifluoro-, addition of iodine fluoride to, **6**, 3
Ethene, 1,1-dichloro-2,2-difluoro-
reactions, with organometallics, **1**, 1069
as starting material, for chlorodifluorothioacetyl chloride, **5**, 479
Ethene, *cis*-1,2-dichloro-1,2-di-(*p*-tolylthio)-, synthesis of, **4**, 770
Ethene, 1,2-dichloro-1-ethoxy-, synthesis of, **4**, 761
Ethene, dichlorofluoro-, dehydrochlorination of, **2**, 1013
Ethene, 1,1-dichloro-2-fluoro-, metallation of, **4**, 817
Ethene, 1,1-dichloro-2-nitro-, reactions, with sodium methoxide, **6**, 74
Ethene, 1,2-dichloronitro-, synthesis of, **4**, 797
Ethene, 1,1-dicyanobis(trifluoromethyl)-
reactions
with molybdenum chlorides, **4**, 800
with tungsten chlorides, **4**, 800
Ethene, 1,1-difluoro-
addition of iodine fluoride to, **6**, 3
dehydrofluorination of, **1**, 1067
reactions
with hydrogen bromide, **4**, 5
with hydrogen iodide, **4**, 5
with iodine, **6**, 2
Ethene, (*E*)-1,2-difluoro-1,2-dinitro-, synthesis of, **4**, 795
Ethene, 1,2-diisopropoxy-, synthesis of, **4**, 767
Ethene, 1,1-dilithio-, stability of, **4**, 1064
Ethene, 1-(dimesitylboryl)-1-trimethylsilyl-1-, nucleophilic addition to, for the synthesis of α-borylsilylalkanes, **4**, 644
Ethene, 1,1-dimethoxy-, [4+2] cycloaddition reactions, with α,β-unsaturated carbonyls, **6**, 75
Ethene, 1,1-dimethoxy-2,2-dimethyl-, [4+2] cycloaddition reactions, with α,β-unsaturated carbonyls, **6**, 75
Ethene, 1,2-dimethoxy-1,2-diphenoxy-, synthesis of, **4**, 830
Ethene, 1,2-dimethoxy-1-lithio-, reactions, with trialkylboranes, **4**, 935
Ethene, 1,1-dimethoxy-2-methyl-, [4+2] cycloaddition reactions, with α,β-unsaturated carbonyls, **6**, 75
Ethene, 1,1-dinitro-, synthesis of, *in situ*, **4**, 994
Ethene, 1,1-dinitro-2,2-diphenyl-
reactions
with potassium phenyl selenide, **4**, 911
with sulfur nucleophiles, **4**, 906
Ethene, 1,1-diphenyl-, synthesis of, **1**, 646, 752
Ethene, 1,2-diphenyl-, synthesis of, **1**, 753
Ethene, 1,1-diphenyl-2-fluoro-, halogenation of, **4**, 751
Ethene, 1-ethoxy-1,2-dichloro-, reactions, with triethyl phosphite, **4**, 912
Ethene, *cis*-1-ethyleneimino-2-(*p*-tolylsulfonyl)-, synthesis of, **2**, 742
Ethene, 1-ferrocenyl-2-nitro-
chloromercuriation of, **4**, 797
iodination of, **4**, 796
Ethene, fluoro-
reactions
with nitrogen tetrafluoride, **5**, 812
with tetramethyldiarsine, **4**, 134
Ethene, 1-fluoro-1-(benzenesulfonyl)-, as starting material, for α-fluoro sulfones, **4**, 79
Ethene, iodotrifluoro-
Arbuzov reaction of, **4**, 802
reactions
with nitric oxide, **4**, 798
with sulfur, **5**, 473
Ethene, nitro-, reactions, with secondary amines, **2**, 301
Ethene, 1-nitroso-1-phenyl-, reactions, with diphenyl trithiocarbonate, **6**, 309
Ethene, 1-nitroso-1,2,2-trichloro-
as starting material, for hydroxamoyl halides, **5**, 669
synthesis of, **5**, 670
Ethene, nitrosotrifluoro-, synthesis of, **4**, 798

Ethene, nitrotriiodo-, synthesis of, **4**, 797
Ethene, perfluoro-, reactions, with *N*-dichloroperfluoropropylamine, **5**, 659
Ethene, phenoxytriiodo-, synthesis of, **4**, 765
Ethene, 1-phenylazo-1,2,2-trichloro-, reactions, with chlorine, **5**, 682
Ethene, tetrachloro-
  reactions
    with benzeneselenol, **4**, 786
    with dimethyl diselenide, **4**, 786
    with sodium arenethiolates, **4**, 770
    with sodium thiolate, **4**, 770
Ethene, tetrafluoro-
  addition of iodine fluoride to, **6**, 3
  addition of iodine to, **6**, 2
  dimerization of, **4**, 5
  reactions
    with cyanide ions, **4**, 1076
    with ethyne, **4**, 5
    with methanol, **4**, 5
  reactivity of, **4**, 802
  for the synthesis of α-fluoro ethers, **4**, 44
Ethene, tetraiodo-, nitration of, **4**, 993
Ethene, tetrakis(dimethylamino)-
  in defluorination reactions, **1**, 570
  as starting material, for amide acetals, **6**, 105
Ethene, tetrakis(trimethylsilyl)-, reactions, with lithium, **4**, 618
Ethene, tetramethoxy-, as byproduct, in the synthesis of carbenes, **4**, 1080
Ethene, 1,1,2,2-tetramethoxy-, [4+2] cycloaddition reactions, with α,β-unsaturated carbonyls, **6**, 75
Ethene, tetramethyl-, reactions, with dibromodinitromethane, **4**, 995
Ethene, tetranitro-, synthesis of, **4**, 994
Ethene, tetraphenyl-, synthesis of, **1**, 142
Ethene, 1,1,2,2-tetrasilyl-, from, 1,2-bis(silyl)ethyne, **4**, 1045
Ethene, tetra(trimethylsilyl)-, synthesis of, **4**, 1050
Ethene, tribromo-, dehydrobromination of, **2**, 1017
Ethene, 1,2,2-tribromo-1-fluoro-, reactions, with thiophenol, **4**, 771
Ethene, 1-tribromogermanyl-1-trimethylsilyl-, synthesis of, **4**, 1058
Ethene, 1,1,2-tri(*t*-butyldimethylsilyl)-1-(trimethylsilyl)-, x-ray analysis of, **4**, 1051
Ethene, trichloro-
  reactions
    with lithium amides, **4**, 976
    with sodium arenethiolates, **4**, 772
  as starting material, for dichloroethyne, **2**, 1014
Ethene, 1,1,2-trichloro-, reactions, with sodium arenethiolates, **4**, 770
Ethene, 1,1,2-trichloro-2-nitro-, reactions, with sodium ethoxide, **6**, 74
Ethene, 1,2,2-trichloro-1-phenoxy-, synthesis of, **4**, 766
Ethene, trifluoro-
  reactions
    with fluoronitroso sulfates, **4**, 104
    with perfluorohydrazine, **5**, 663
    with tetramethyldiarsine, **4**, 134
  as starting material, for fluorotrifluoromethylsulfines, **5**, 494
  use in the synthesis of α-fluoroalkenyllithiums, **4**, 817
Ethene, 1,1,2-trifluoro-, fluorination of, **5**, 811
Ethene, trifluorochloro-, reactions, with *N*-dichloroperfluoropropylamine, **5**, 659
Ethene, 1,2,3-trifluoro-1-methoxy-, synthesis of, **4**, 761
Ethene, 1,1,2-trimethoxy-, [4+2] cycloaddition reactions, with α,β-unsaturated carbonyls, **6**, 75
Ethene, 1-trimethylsilyl-, synthesis of, **4**, 1061
Ethene, 1-trimethylsilyl-1-trimethylstannyl-, hydrostannylation of, **4**, 661
Ethene, triodonitro-, reactions, with trifluoromethylthiocopper, **4**, 771
Ethene, 1,1,2-tris(trimethylsilyl)-, synthesis of, **4**, 1049
Ethene, tris(trimethylsilyl)methyl-substituted, synthesis of, **6**, 176
1-Ethene-1,1-dicarboxylates, 2-substituted-2-chloro-, monoethyl esters of, hydrolysis of, **5**, 99
Ethene hexamer, tetrafluoro-, reactions, with amines, **3**, 589
Ethenes, 1-alkoxy-2,2-dibromo-1-chloro-, synthesis of, **4**, 767
Ethenes, 1-alkylthio-2-seleno-2-chloro-, synthesis of, **4**, 787
Ethenes, 1-aryl-2-chloro-2-fluoro-, from, 1-aryl-1,2,2-trichloro-2-fluoroethane, **4**, 750
Ethenes, 1,2-bisazo-, synthesis of, **3**, 680
Ethenes, bis(dialkylboryloxy)-, from, carboxylic acids, **4**, 832
Ethenes, 1,2-bis(silyl)-, palladium(0)-catalyzed coupling, **2**, 931
Ethenes, 1-chloro-2,2-diaryl-, metallation of, **4**, 817
Ethenes, chlorolithio-
  synthesis of, **2**, 998
  x-ray crystal structure of, **2**, 998
Ethenes, 1,1-dialkoxy-2-trimethylsilyl-, aminoalkylation of, **5**, 723
Ethenes, 2,2-diamino-1,1-disulfonyl-, synthesis of, **4**, 866
Ethenes, 1,1-diaryl-
  ozonolysis of, **2**, 643
  reactions
    with lithium aluminum hydride, **1**, 80
    with mercuric nitrate, **4**, 1067
  reduction of, **1**, 80
Ethenes, 1,1-diaryl-2-chloro-2-fluoro-, from, 1,1-diaryl-1,2,2-trichloro-2-fluoroethanes, **4**, 750
Ethenes, 1,1-diboro 2,2-bis(trimethylsilyl)-, synthesis of, **4**, 1048
Ethenes, dicyanodifluoro-, as starting materials, for α-fluoroamines, **4**, 104
Ethenes, α,α-dihalo-, as starting materials, for monothioortho esters, **6**, 95
Ethenes, 1,2-disilyl-, as starting materials, for 1-silyl-1,3-dienes, **2**, 939
Ethenes, 1,1-disilyl 2-phenyl-, synthesis of, **4**, 1048
Ethenes, fluoro-, as starting materials, for *N*-fluoroimino fluorides, **5**, 812
Ethenes, germanyl-substituted, synthesis of, **4**, 827
Ethenes, *gem*-halo(phosphono)-, synthesis of, **4**, 804
Ethenes, nitro-, reactions, with chlorine, **4**, 108
Ethenes, 1-phenylthio-1-triethylgermyl-, synthesis of, **4**, 940
Ethenes, polyamino-, as starting materials, for ortho amides, **6**, 141
Ethenes, polychlorinated, dehydrochlorination of, **2**, 1014
Ethenes, propargyl-, as starting materials, for allenylsilanes, **2**, 939
Ethenes, 1-silyl-2-stannyl-, palladium(0)-catalyzed coupling, **2**, 931
Ethenes, silyl-substituted, synthesis of, **4**, 827
Ethenes, tetraalkoxy-, synthesis of, **4**, 830
Ethenes, tetrasilyl-
  from
    bis(silyl) dichloromethanes, **4**, 1050
    1,3-disilacyclobutanes, **4**, 1049
Ethenes, 1-thioalkyl-1-nitro-2,2-diphenyl-, synthesis of, **4**, 906
Ethenes, 1,1,2-trisilyl-
  from, silirenes, **4**, 1051
  synthesis of, **4**, 1044
Ethene-1-sulfonyl fluoride, 2,2-difluoro-, reactions, with sulfur trioxide, **4**, 867
Ethenols, stable, criteria for, **2**, 643
Ethenols, 1-alkyl-2,2-diaryl-, synthesis of, **2**, 645
Ethenols, 2-alkyl-1,2-diaryl-, synthesis of, **2**, 643
Ethenols, 2,2-diaryl-
  from, diarylketenes, **2**, 645
  synthesis of, **2**, 643
Ethenols, 1,2-dimesityl-, from, α,β-unsaturated ketones, **2**, 643

Ethenols, isopropyl-, synthesis of, **2**, 643
Ethenols, polyaryl-substituted, synthesis of, **2**, 643
Ethenylation, of enolsilanes, **3**, 138
Ether, allylvinyl
  Claisen rearrangement of, **3**, 21, 141
  synthesis of, **3**, 141
Ether, β-azido enol, reactions, with carbon nucleophiles, **1**, 263
Ether, α-(*N*-benzotriazoyl)butyl methyl, reactions, with octyl-magnesium iodide, **1**, 262
Ether, benzyl chloromethyl, reactions, with tributylstannyllithium, **4**, 389
Ether, bis(chloromethyl)
  reactions, with amines, **4**, 411
  synthesis of, **4**, 47
Ether, bis(fluoromethyl)
  as intermediate, in the synthesis of α-fluoro ethers, **4**, 44
  synthesis of, **4**, 44
Ether, bis(methoxymethyl), synthesis of, **4**, 232
Ether, 2,2-bis[methylsulfonyl]vinyl ethyl, reactions, with amines, **2**, 743
Ether, bis(trichloromethyl), synthesis of, **6**, 229
Ether, bis(trifluoromethyl), synthesis of, **6**, 229
Ether, bis(trimethylsilyl)vinyl, synthesis of, **4**, 1049
Ether, α-bromocyanohydrin, synthesis of, **3**, 627
Ether, *t*-butyl methyl
  deprotonation of, **4**, 378
  reactions, with Schlosser's base, **1**, 136
Ether, *t*-butyl octyl, synthesis of, **1**, 574
Ether, butylvinyl, reactions, with oximes, **3**, 430
Ether, 1-chlorocyclopropyl methyl, synthesis of, **4**, 49
Ether, chloromethyl benzyl, as formaldehyde equivalent, **3**, 163
Ether, chloromethyl methyl
  reactions
    with lithium bis(trimethylsilyl)amide, **4**, 313
    with magnesium, **4**, 383
    with tributylstannylmagnesium chloride, **4**, 389
    with trichlorogermane, **4**, 361
  synthesis of, **4**, 48, 49
Ether, chloromethyl phenyl, from, sodium chloromethyl sulfonate, **4**, 49
Ether, cholesteryl benzyl, synthesis of, **5**, 526
Ether, 18-crown-6, as catalyst, for the oxidation of dithioacetals, **4**, 266
Ether, dibenzyl, cleavage of, **2**, 559
Ether, dibromomethyl methyl, reactions, with dilithium benzene-1,2-diselenolate, **6**, 100
Ether, 2,2-dibromovinyl ethyl, reactions, with *n*-butyllithium, **4**, 818
Ether, dichloromethyl methyl
  as formylating reagent, **3**, 92
  reactions
    with disodium ethanedithiolate, **6**, 97
    with 1-lithio-1-trimethylsilyl cyclopropanes, **5**, 429
  as starting material
    for 2-methoxy-1,3-dithiane, **6**, 97
    for triselenoorthoformate esters, **6**, 92
  use in the synthesis of alkylsulfonyl chlorides, **2**, 182
Ether, α,α-dichloromethyl methyl, as transfer agent, **6**, 40
Ether, di(2,2-dibromo-1-chlorovinyl), synthesis of, **4**, 767
Ether, diethyl
  reactions, with 2-chlorobutane, **2**, 558
  as starting material, for 1-benzoyloxy-1-ethoxyethane, **4**, 207
  use in the synthesis of silaethenes, **6**, 709
Ether, α,α-dihaloalkyl, reactions, with silver cyanate, **5**, 965
Ether, dimethyl
  electrofluorination of, **6**, 229
  lithiation of, **4**, 378
  photochlorination of, **6**, 229
Ether, enol, reactions, with benzyltrimethylammonium fluoride, **1**, 133
Ether, ethylvinyl
  reactions
    with azidocephalosporine, **5**, 697
    with phosphorus halides, **3**, 50
    with 2-(tribromophenyl)diazene-2-oxides, **5**, 680
Ether, α-halo-α-(trimethylsilyl)methyl, use in the synthesis of α,β-unsaturated ketones, **3**, 271
Ether, imino, as starting material, for ketenimines, **3**, 562
Ether, iodomethyl methyl, synthesis of, **4**, 51
Ether, lithiated trimethylsilylmethyl methyl, reactions, with bromotrialkylgermanes, **6**, 131
Ether, methylvinyl, reactions, with *t*-butyllithium, **1**, 1089; **4**, 948
Ether, phenylsulfonylmethyl *t*-butyl, as formyl anion equivalent, **3**, 31
Ether, propargylic methyl, reactions, with butyllithium, **1**, 981
Ether, silyl allyl, reactions, with *s*-butyllithium, **1**, 136
Ether, silyl enol, reactions, with α-chlorosilanes, **4**, 365
Ether, trialkylsilyl, silyl-Wittig rearrangement of, **5**, 413
Ether, trichloromethyl phenyl, reactions, with triethyl phosphite, **6**, 281
Ether, 1,2,2-trichlorovinylphenyl, synthesis of, **4**, 742
Ether, trimethylsilylmethyl methyl, lithiation of, **6**, 133
Etherate complexes, synthesis of, **2**, 595
Ether cleavage, for the synthesis of alcohols, **2**, 82
α-Ether complexes, as starting materials, for iron carbene complexes, **3**, 519
Ether groups, elimination of, **1**, 985
Ethers
  α-alkoxylation of, **4**, 201
  α-chlorination of, **4**, 47
  cleavage of, **2**, 559
  fluorination of, **6**, 36
  from
    alcohols, **6**, 36
    2-alkoxy alkylmercury compounds, **2**, 89
    benzyl alcohols, **2**, 89
    thiono esters, **2**, 89
  halogenation of, **6**, 36
  metallation of, **4**, 378
  oxyselenylation of, **3**, 175
  photochlorination of, **6**, 36
  reactions
    with acyl peroxides, **4**, 207
    with aluminum trichloride, **6**, 39
    with butylpotassium, **2**, 999
    with chlorotrialkylsilanes, **4**, 358
    with hydrogen iodide, **2**, 36
    with selenoxides, **6**, 256
  2,3 rearrangement of, **1**, 404
  reduction of, **1**, 15
  [1,5] sigmatropic rearrangements of, **1**, 838
  as starting materials
    for acetals, **4**, 201
    for acyl alkyl acetals, **4**, 207
    for aldehydes, **3**, 8
    for alkyl bromides, **2**, 32
    for alkyl halides, **2**, 10
    for alkyl iodides, **2**, 36
    for alkyllithiums, **2**, 553
    for arylamines, **2**, 756
    for α,α-dihalo ethers, **6**, 36
    for imidoyl-tricoordinated phosphorus compounds, **5**, 784
    for ketones, **3**, 121
  synthesis of, **2**, 88; **6**, 145
    via Markovnikov addition, **2**, 89
    via reduction of acetals, **2**, 89
  use in the synthesis of alkylhaloboranes, **2**, 525
Ethers, acyclic
  alkylation of, **1**, 475
  synthesis of, **2**, 88

Ethers, acyclic silyl enol, oxidation of, **1**, 556
Ethers, α-acyloxy, as starting materials, for acetals, **4**, 200
Ethers, acylselenenic, synthesis of, **5**, 246
Ethers, acylsulfenic, synthesis of, **5**, 246
Ethers, α-acyl α-sulfonyl, acylation of, **4**, 838
Ethers, acyltellurenic, synthesis of, **5**, 246
Ethers, alkenyl
  from
    acetals, **2**, 678
    alkynes, **2**, 676
    erythro β-hydroxy acid, **2**, 677
    ethanol ethers, **2**, 681
    ketones, **2**, 680
  synthesis of, **2**, 674
    via 1,2-elimination of substituted alkyl ethers, **2**, 677
Ethers, (*Z*)-alkenyl, synthesis of, **1**, 770
Ethers, α-alkoxy bromo, reactions, with *t*-butyllithium, **4**, 381
Ethers, alkoxy enol, elimination of, **3**, 54
Ethers, alkyl, cleavage of, **2**, 10
Ethers, α-*s*-alkyl, fluorination of, **4**, 44
Ethers, alkyl aryl
  synthesis of, **2**, 683
    via aromatic nucleophilic substitution, **2**, 684
Ethers, *t*-alkyl aryl, synthesis of, **2**, 684
Ethers, γ-alkylated, synthesis of, **1**, 136
Ethers, alkyl 1-azidoalkyl, synthesis of, **4**, 314
Ethers, alkyl chloromethyl
  reactions, with thiirane-*S*-oxide, **2**, 195
  as starting materials, for dialkylaminomethylalkyl ethers, **4**, 309
Ethers, alkyl enol
  of acylsilanes, synthesis of, **5**, 416
  reactions, with Grignard reagents, **1**, 480
Ethers, alkylmethyl, reduction of, **1**, 15
Ethers, alkylsilyl, from, chlorosilanes, **2**, 539
Ethers, alkylvinyl, reactions, with allylic alcohols, **1**, 414
Ethers, alkynic, reactions, with amines, **4**, 884
Ethers, alkynyl
  as electrophilic alkynylation reagents, **1**, 1030
  isomerization of, **1**, 1085
  reactions
    with azides, **3**, 477
    with lithium amides, **2**, 1043
  as starting materials, for tertiary ynamines, **2**, 1043
  synthesis of, **2**, 1025
Ethers, 1-alkynyl, reduction of, **1**, 999
Ethers, 2-alkynyl, as starting materials, for allenyl ethers, **1**, 955
Ethers, α-alkynyl, synthesis of, **1**, 1003
Ethers, β-alkynyl, reactions, with sulfonyl azides, **3**, 477
Ethers, β-*H*-alkynyl, reactions, with 4-dimethylaminobenzenesulfonyl azide, **3**, 477
Ethers, alkynyl silyl enol
  carbomercuration of, **1**, 544
  use in ene reactions, **1**, 526
Ethers, γ,δ-alkynyl silyl enol, synthesis of, **1**, 998
Ethers, β-alkyoxyalkyl allyl, rearrangement of, **1**, 397
Ethers, allenol silyl, synthesis of, **1**, 971
Ethers, allenyl
  from
    2-alkynyl ethers, **1**, 955
    oxonium compounds, **1**, 978
  as starting materials
    for α,β-alkynic acylsilanes, **5**, 427
    for silyl enones, **5**, 427
  synthesis of, **1**, 955
Ethers, allenyl enol, from, enynes, **1**, 970
Ethers, allyl
  alkylation of, **1**, 136
  isomerization of, **2**, 675
  oxidation of, **4**, 208
  reactions
    with nitroalkanes, **2**, 395
    with silyl enol ethers, **1**, 189
  rearrangement of, **1**, 397
  as starting materials, for β,γ-unsaturated cyanohydrins, **3**, 623
Ethers, *O*-allyl, synthesis of, **3**, 431
Ethers, allyl aryl
  aromatic Claisen rearrangement of, **2**, 656
  [3,3] sigmatropic rearrangement of, **1**, 826
Ethers, allyl 2-benzylaminoalkyl, synthesis of, **2**, 312
Ethers, allyl cyclohexenyl, rearrangement of, **1**, 832
Ethers, allyl enol, as starting materials, for acetoxy tetrahydrofurans, **4**, 208
Ethers, allylic
  lithiation of, **4**, 381
  ozonolysis of, **3**, 31
  reduction of, **1**, 15
  silylation of, **4**, 358
Ethers, allylic alkoxy thioenol, hydrolysis of, **3**, 54
Ethers, allylic methoxymethyl, reactions, with trimethylsilyl cyanide, **1**, 191
Ethers, allylic α-(trimethylsilyl)vinyl, Claisen rearrangement of, **5**, 421
Ethers, allyl methyl, as starting materials, for β,γ-unsaturated nitriles, **3**, 623
Ethers, allylpropargyl
  2,3 Wittig rearrangement of, **1**, 817
  2,3 Wittig ring-contraction of, **1**, 818
Ethers, allylsilyl
  alkylation of, **1**, 136
  substitution of, with vinyl nucleophiles, **1**, 477
Ethers, allylvinyl
  Claisen rearrangement of, **1**, 190, 414, 825
  1,3 sigmatropic rearrangement of, **1**, 402
  [3,3] sigmatropic rearrangement of, **1**, 825
  as starting materials
    for alkenes, **1**, 415
    for γ,δ-unsaturated aldehydes, **1**, 415
Ethers, amino, reactions, with organoaluminums, **1**, 973
Ethers, 3-amino, synthesis of, **2**, 320
Ethers, β-amino, synthesis of, **2**, 310
Ethers, aromatic, nitrosation of, **2**, 781
Ethers, aryl
  acylation of, **3**, 279, 298
  organolithiums of, synthesis of, **3**, 298
  para iodination of, for the synthesis of aryl iodides, **2**, 631
  reduction of, **1**, 98
  synthesis of, **2**, 683
Ethers, aryl alkyl
  Birch reduction of, **2**, 682
  cleavage of, **2**, 658
  as starting materials, for alkyl chlorides, **2**, 26
Ethers, arylallyl, *p*-Claisen rearrangement of, **1**, 808
Ethers, aryl imidate
  reactions
    with lithium dialkyl amides, **3**, 721, 722
    with magnesium diisopropyl amides, **3**, 721
Ethers, arylmethyl, reactions, with Grignard reagents, **1**, 482
Ethers, arylsilyl, from, chlorosilanes, **2**, 539
Ethers, 2-arylsulfonyl-substituted cyclic, reactions, with alkynylzincs, **1**, 1031
Ethers, aryltetrazolyl, displacement of, **1**, 482
Ethers, aryl trichloromethyl, synthesis of, **6**, 229
Ethers, aryl trifluoromethyl, synthesis of, **6**, 229
Ethers, azacrown, as catalysts, for the synthesis of α-aminoalkylphosphorus compounds, **4**, 469
Ethers, α-azido, synthesis of, **5**, 696
Ethers, azidoalkyl alkyl, synthesis of, **4**, 314
Ethers, α-azido oxime, reactions, with lithium aluminum hydride, **2**, 314
Ethers, azoxy, synthesis of, **2**, 405
Ethers, benzyl
  optically-active, rearrangements of, **1**, 396

rearrangements of, **1**, 396
reduction of, **1**, 15
synthesis of, **2**, 90
Ethers, benzylcrotyl, Wittig rearrangement of, **1**, 816
Ethers, benzylic
lithiation of, **4**, 381
silylation of, **4**, 358
Ethers, benzyl methyl, as starting materials, for benzylmagnesiums, **2**, 559
Ethers, *O*-benzyl oxime, use in the synthesis of *N*-alkylhydroxylamines, **2**, 343
Ethers, bisbenzyl, synthesis of, **1**, 202
Ethers, bis(formamidinium), as starting materials, for ketene aminals, **4**, 975
Ethers, bishomoallylic, carbolithiation of, **2**, 555
Ethers, bislactim, synthesis of, **4**, 535
Ethers, bispropadienyl, from, bispropargylic ethers, **1**, 955
Ethers, bispropargylic, as starting materials, for bispropadienyl ethers, **1**, 955
Ethers, bis(silyl), selectivity of, **1**, 149
Ethers, bis(trimethylsilyl) silyl enol, attempted isolation of, **4**, 1048
Ethers, bisvinyl, reduction of, **1**, 82
Ethers, 1-boro 1-germanyl alkenol, synthesis of, **4**, 1059
Ethers, boryl enol
of acylsilanes, synthesis of, **5**, 417
from, silylalkynes, **5**, 423
Ethers, α-bromo
from
acetals, **4**, 50
α-chloro ethers, **4**, 51
vinyl ethers, **4**, 51
synthesis of, **4**, 50
Ethers, bromochloro, reactions, with antimony trifluoride, **6**, 38
Ethers, *gem*-bromo lithio enol
as starting materials, for (*E*)-2-ethoxyvinyl bromide, **4**, 818
synthesis of, **4**, 818
Ethers, bromomethyldimethylsilyl, radical cyclization, **2**, 539
Ethers, α-bromo oxime, reactions, with sodium azide, **2**, 314
Ethers, α-bromo-β-selenenyl, reactions, with alcohols, **4**, 51
Ethers, (Z)-2-bromovinyl, dehydrohalogenation of, **2**, 1024
Ethers, 2-bromozincmethyl allylic
reactions
with imines, **1**, 194
with ketones, **1**, 194
Ethers, butatrienyl, from, 1-trimethylsilyloxy-4-alkoxy but-2-ynes, **1**, 981
Ethers, *t*-butoxymethyl, synthesis of, **4**, 201
Ethers, *t*-butyl
cleavage of, **4**, 176
synthesis of, **2**, 90
Ethers, *t*-butyl aryl, synthesis of, **2**, 684
Ethers, *t*-butyldimethylsilyl
synthesis of, **2**, 539
use in the synthesis of ketones, **3**, 121
Ethers, *t*-butylmethoxyphenylsilyl, use in the protection of alcohols, **2**, 539
Ethers, chloro, fluorination of, **6**, 36
Ethers, α-chloro
from
acetals, **4**, 48
alcohols, **4**, 47
aldehydes, **4**, 47
α-alkoxy acid chlorides, **4**, 49
α-aryloxy acid chlorides, **4**, 49
α-dichloro ethers, **4**, 49
ketones, **4**, 47
vinyl ethers, **4**, 49
reactions
with thiolates, **4**, 218
with trinitromethane salts, **6**, 145
replacement of, **4**, 340
as starting materials, for α-bromo ethers, **4**, 51
synthesis of, **4**, 46
via α-chlorination of ethers, **4**, 47
use in the synthesis of acetals, **4**, 198
use in the synthesis of nitriles, **3**, 613
Ethers, β-chloroalkyl trifluoromethyl, synthesis of, **6**, 229
Ethers, β-chlorocyanohydrin, synthesis of, **3**, 627
Ethers, α-chloro-β,β-dibromovinyl, reactions, with butyllithium, **2**, 1025
Ethers, 2-chloroethyl, synthesis of, **4**, 305
Ethers, chlorofluoro, synthesis of, **6**, 43
Ethers, chloromethyl
displacement of, **4**, 389
as starting materials
for α-alkoxystannanes, **4**, 389
for Grignard reagents, **4**, 383
Ethers, chloromethyl aryl, synthesis of, **4**, 47
Ethers, (1-chloro-1-methyl)ethyl methyl, synthesis of, **4**, 49
Ethers, chloromethyl methyl
metallation of, **4**, 381
synthesis of, **4**, 47
Ethers, chloromethyl-α-trimethylsilyl-α-arylmethyl, 1,3-elimination from, **3**, 326
Ethers, crown, synthesis of, via desulfurization, **5**, 511
Ethers, cyano, synthesis of, **3**, 627
Ethers, 2-cyano, from, allylic acetals, **2**, 89
Ethers, α-cyano
lithiation of, **4**, 381
synthesis of, **3**, 627
Ethers, α-cyano-α-amino, synthesis of, **3**, 627
Ethers, cyanohydrin silyl
synthesis of, **3**, 615
use in the synthesis of aromatic ketones, **3**, 292
Ethers, cyclic
cleavage of, **1**, 661; **2**, 553
reactions, with bis(1,5-cyclooctanediylbornyl) sulfide, **2**, 270
as starting materials
for α,ω-dichlorides, **2**, 26
for hydroxy selenides, **2**, 281
for isothiocyanates, **5**, 1042
synthesis of, **2**, 90, 93, 593
Ethers, cyclic alkenyl, synthesis of, **2**, 679
Ethers, cyclic benzylic, hydrogenation of, **1**, 15
Ethers, cyclic imino
ring-opening of, **5**, 1033
as starting materials, for ketenimines, **3**, 575
Ethers, cyclopropyl silyl
reactions, with cupric tetrafluoroborate, **1**, 929
thermolysis of, **4**, 616
Ethers, dialkyl
cleavage of, **2**, 26
as starting materials, for alkyl chlorides, **2**, 26
Ethers, dialkylaminomethylalkyl, from, alkyl chloromethyl ethers, **4**, 309
Ethers, diallylic, 2,3 Wittig rearrangements of, **1**, 818
Ethers, diaryl, synthesis of, **2**, 684
Ethers, di-*t*-butylsilyl, use in the protection of diols, **2**, 540
Ethers, dichloro
monofluorination of, **6**, 43
reactions, with antimony trifluoride, **6**, 38
Ethers, α-dichloro, as starting materials, for α-chloro ethers, **4**, 49
Ethers, α,α-dichloro, reactions, with dithiols, **6**, 97
Ethers, dichloromethyl
from, aryl formate esters, **6**, 40
reactions, with areneselenomagnesium halides, **6**, 100
use in the synthesis of benzaldehydes, **3**, 97
Ethers, α,α-dichloromethyl, as starting materials, for α-keto acid chlorides, **5**, 14
Ethers, dichloromethyl alkyl, use in the synthesis of alkyl benzaldehydes, **3**, 94

Ethers, dichloromethyl aryl, reactions, with thioesters, **6**, 48
Ethers, dichloromethyl methyl, reactions, with sodium benzeneselenolate, **6**, 100
Ethers, dichlorovinyl, structure of, **4**, 761
Ethers, 1,2-dichlorovinyl, as starting materials, for lithium alkoxyethynylides, **2**, 1026
Ethers, dienol, synthesis of, **2**, 682
Ethers, 1,2-difluoro, from, esters, **2**, 89
Ethers, α,α-difluoro, as starting materials, for ketene acetals, **4**, 826
Ethers, 1,1-difluoromethyl, synthesis of, **2**, 89
Ethers, [difluoro(methylthio)methyl], as starting materials, for trifluoromethyl ethers, **6**, 229
Ethers, dihalo, as starting materials, for α,α-dihalo ethers, **6**, 38
Ethers, α,α-dihalo
  from
    carbonyls, **6**, 39
    dihalo ethers, **6**, 38
    ethers, **6**, 36
    haloalkenes, **6**, 41
    haloalkoxyalkenes, **6**, 41
    trihalomethanes, **6**, 37
  as intermediates, for ortho esters, **6**, 74
  as starting materials
    for diselenoortho esters, **6**, 100
    for dithioortho esters, **6**, 97
    for ortho esters, **6**, 68
    for triselenoortho esters, **6**, 92
    for trithioortho esters, **6**, 92
Ethers, dihalomethyl methyl, reactions, with selenolates, **6**, 100
Ethers, diisocyanatophenyl, reactions, with phosgene, **5**, 967
Ethers, diisopropylsilyl, use in the protection of diols, **2**, 540
Ethers, dimethylaminomethyl alkyl, synthesis of, **4**, 305
Ethers, 1,1-diphenylmethyl(benzhydryl), synthesis of, **2**, 89
Ethers, *N,N*-disubstituted aminoalkyl
  reactions
    with 1-alkynes, **1**, 1041
    with alkynylmagnesium halides, **1**, 1041
Ethers, enethiol
  from, cyclohexanone, **4**, 247
  selenylation of, **4**, 871
Ethers, enol
  α-acylation of, **3**, 240
  alkoxymercuration of, **4**, 188
  cleavage of, **3**, 72
  as dienophiles, for the synthesis of acetals, **4**, 202
  formylation of, **3**, 71
  from
    carbonyl ylides, **3**, 323
    α-lactones, **1**, 785
  hydration of, **4**, 171, 174
  hydroboration of, **1**, 82
  hydrogenation of, **1**, 81
  hydrolysis of, for the synthesis of acylsilanes, **5**, 413
  hydroxy mercuration of, **4**, 171
  metallation of, **4**, 949
    for the synthesis of acylsilanes, **5**, 413
  osmylation of, **4**, 174
  oxidation of, **4**, 187
  as radical traps, for the synthesis of Corey lactones, **1**, 176
  reactions
    with alcohols, **4**, 186
    with allyl alcohols, **4**, 188
    with carboxylic acids, **4**, 208
    with α-cyanoacrylate, **4**, 827
    with cyanogen chloride, **3**, 627
    with dihalogens, **5**, 423
    with electrophiles, **1**, 296
    with *N*-halo amides, **3**, 43
    with hydrogen cyanide, **3**, 615
    with methyl nitrite, **3**, 192
    with phenyl selenyl bromide, **3**, 38
    with phenylsulfenyl chloride, **3**, 35; **5**, 413
    with prop-2-yn-1-ols, **1**, 979
    with sulfonyl isocyanates, **5**, 344
    with trimethylsilyl azide, **1**, 263
    with trinitromethanes, **6**, 145
    with (triseleno)orthoformates, **4**, 289
  reduction of, **1**, 434
  as starting materials
    for acetals, **4**, 186
    for acyl alkyl acetals, **4**, 208
    for aldehydes, **3**, 9
    for alkenes, **1**, 434
    for 1-alkoxyalkyl isocyanates, **4**, 314
    for enones, **1**, 558
    for α-functionalized acetals, **4**, 187
    for hemiacetals, **4**, 171
    for ketones, **3**, 123
    for monoselenoacetals, **4**, 238
    for monotelluroacetals, **4**, 238
    for α-selenocarbonyls, **2**, 282
    for thioketones, **3**, 347
    for α,β-unsaturated acylsilanes, **5**, 425
    for unsymmetrical acetals, **4**, 186, 188
  α-sulfenylation of, **4**, 836
  synthesis of, **1**, 1119; **4**, 749
    via elimination reactions, **1**, 579
    via hydrolysis of vinyl phosphonium salts, **1**, 446
Ethers, epoxy silyl
  Lewis acid-catalyzed rearrangement of, **3**, 33
  as starting materials
    for β-oxygenated aldehydes, **3**, 160
    for β-siloxy ketones, **3**, 160
Ethers, ethanol
  homologation of, **2**, 681
  as starting materials, for alkenyl ethers, **2**, 681
Ethers, ethoxyethyl (EE), use in the synthesis of acetals, **4**, 185
Ethers, ethyl, synthesis of, **5**, 521
Ethers, ethylene glycol, electrochemical fluorination of, **6**, 36
Ethers, ethyl enol, reactions, with carboxylic acids, **4**, 208
Ethers, ethylvinyl
  deprotonation of, **4**, 949
  nitration of, **4**, 208
Ethers, fluorinated 1-lithiovinyl, synthesis of, **4**, 949
Ethers, α-fluoro
  from
    acetyl hypofluorite, **4**, 44
    paraformaldehyde, **4**, 44
    perfluoroallyl fluoro sulfate, **4**, 44
    polypropylene oxide, **4**, 44
  stability of, **4**, 44
  synthesis of, **4**, 44
Ethers, β-fluoroalkyl trifluoromethyl, synthesis of, **6**, 229
Ethers, 1-fluoro enol, reactions, with amines, **4**, 894
Ethers, fluoromethyl aryl, synthesis of, **4**, 45
Ethers, 1-germanylalkyne, reactions, with germanium tetrachloride, **4**, 1055
Ethers, germyl dienol, as starting materials, for α,β-unsaturated acylgermanes, **5**, 431
Ethers, glycal alkenyl, from, glycosyl halides, **2**, 678
Ethers, glyceryl, asymmetric hydrogenation of, **2**, 53
Ethers, 1,2-halo, as starting materials, for alkynols, **1**, 1066
Ethers, α-halo
  from
    methoxyethoxymethyl ethers, **4**, 199
    methoxymethyl ethers, **4**, 199
  reactions
    with alcohols, **4**, 198
    with allenylmagnesium halides, **1**, 1052
    with *t*-butyllithium, **4**, 381
    with thiols, **4**, 218
  as starting materials

for acetals, **4**, 198
for monoselenoacetals, **4**, 239
for monotelluroacetals, **4**, 239
for monothioacetals, **4**, 218
synthesis of, **4**, 43, 760
Ethers, haloalkenyl, dehydrohalogenation of, **2**, 1025
Ethers, α-haloalkyl
dehydrohalogenation, **2**, 677
reactions, with carboxylic acids, **4**, 209
Ethers, β-haloalkyl, dehydrohalogenation, **2**, 677
Ethers, 1-halo enol, from, 1-alkoxyalkynes, **4**, 765
Ethers, halomethyl, reactions, with metal germanes, **4**, 360
Ethers, *gem*-halo silyl enol, synthesis of, **4**, 806
Ethers, α-halo α-sulfonyl, dehydrohalogenation of, **4**, 838
Ethers, homopropargyl thiomethyl, synthesis of, **1**, 1031
Ethers, hydrazonyl aryl, as starting materials, for 1-aroyl-2-*N*,*N*'-diarylhydrazines, **5**, 358
Ethers, 5-hydroxy, synthesis of, **2**, 57
Ethers, α-hydroxy enol, reactions, with phosphorus oxychloride, **3**, 208
Ethers, α-hydroxyoximo benzyl, diastereoselective reduction of, **2**, 350
Ethers, imidate, reactions, with organophosphorus compounds, **5**, 784
Ethers, imino
reactions, with arylhydrazines, **5**, 859
as starting materials, for nitriles, **3**, 620
Ethers, α-iodo
from, α-chloro compounds, **4**, 51
synthesis of, **4**, 51
Ethers, α,α'-isocyanato, synthesis of, **2**, 544
Ethers, γ-isocyano silyl enol, synthesis of, **3**, 708
Ethers, α-isothiocyanato, synthesis of, **5**, 1042
Ethers, 2-keto, synthesis of, **2**, 89
Ethers, lactim
from, lactams, **5**, 688
reactions, with sulfonyl chlorides, **5**, 346
sulfonylation of, **5**, 346
Ethers, lithiated enol, synthesis of, **4**, 817
Ethers, lithiated silylmethyl methyl
reactions
with aldehydes, **2**, 680
with ketones, **2**, 680
Ethers, lithiated sulfenyl methyl, as intermediates, for the synthesis of alkenyl ethers, **2**, 678
Ethers, (*Z*)-β-lithiated vinyl, synthesis of, **2**, 676
Ethers, lithiomethyl allylic, use in 2,3 Wittig rearrangements, **1**, 817
Ethers, macrocyclic, synthesis of, **3**, 315
Ethers, α-metallated, synthesis of, **1**, 135
Ethers, α-metallated enol, as starting materials, for α,β-unsaturated ketones, **3**, 240
Ethers, α-metallo, synthesis of, **2**, 999
Ethers, methoxybenzyl, reactions, with 2,3-dichloro-5,6-dicyano-1,4-benzoquinone, **4**, 201
Ethers, methoxyethoxymethyl (MEM), as starting materials, for α-halo ethers, **4**, 199
Ethers, methoxymethyl (MOM)
as starting materials, for α-halo ethers, **4**, 199
synthesis of, **4**, 185
use in the synthesis of acetals, **4**, 185
Ethers, methyl
reactions, with butylmagnesium iodide, **1**, 262
synthesis of, **1**, 262; **2**, 89
Ethers, methyl allyl, as starting materials, for 1-alkoxybut-3-enes, **1**, 816
Ethers, methyl(α,α-dichloromethyl), as starting materials, for pyruvoyl chloride, **5**, 14
Ethers, methyldiphenylsilyl, use in the protection of alcohols, **2**, 539
Ethers, methyl enol, synthesis of, **5**, 416
Ethers, methylthiomethyl (MTM)
protection of alcohols as, **4**, 219
reactions, with bromine, **4**, 216
synthesis of, **4**, 219
via oxidation, **4**, 220
via Pummerer reaction, **4**, 222
Ethers, α-monochloro, synthesis of, **4**, 49
Ethers, oxime
as radical traps, **1**, 175
reactions, with alkynyllithiums, **1**, 1042
Ethers, perfluoro, synthesis of, **4**, 44; **6**, 251
Ethers, perfluoro iodo, reactions, with aluminum chloride, **4**, 768
Ethers, phenol, formylation of, **3**, 92
Ethers, phenolic, reduction of, **1**, 432
Ethers, phenyl
fluorination of, **4**, 44
o-lithiation of, **1**, 213
Ethers, phenyl alkenyl, synthesis of, **2**, 675
Ethers, α-(phenylthio), reductive lithiation of, **2**, 999
Ethers, phenylthiomethyl, synthesis of, **4**, 219, 221
Ethers, 1-phosphaallene silyl, synthesis of, **3**, 608
Ethers, polyfluorinated, as starting materials, for acylketenes, **3**, 545
Ethers, propargyl
from, acetals, **1**, 1026
as starting materials, for chromanodihydrofurans, **3**, 315
synthesis of, **1**, 1026
Ethers, propargyl allyl, Claisen–Cope rearrangement of, **1**, 980
Ethers, propargylic
isomerization of, **3**, 61
reactions
with biscyclopentadienylzirconium, **1**, 1057
with ethylcuprates, **1**, 961
with organocoppers, **1**, 667
Ethers, propargylvinyl, Claisen–Cope rearrangements of, **1**, 979
Ethers, propynol, reactions, with allylzinc bromide, **4**, 809
Ethers, pyridylcarbonylsulfenic, synthesis of, **5**, 246
Ethers, resorcinol dimethyl, synthesis of, **1**, 433
Ethers, saturated cyclic, as starting materials, for cyclic Fischer complexes, **5**, 948
Ethers, silyl
oxidation of, for the synthesis of ketones, **3**, 121
reactions
with aldehydes, **4**, 181
with hydrochloric acid, **4**, 160
with ketones, **4**, 181
reduction of, **1**, 17
as starting materials
for aldehydes, **3**, 8
for alkyl bromides, **2**, 10
for alkyl fluorides, **2**, 10
synthesis of, **1**, 765; **4**, 620
Ethers, α-silyl, synthesis of, **4**, 358
Ethers, 1-silyl 1-alkyn-2-ol alkyl, reactions, with trihaloboranes, **4**, 1057
Ethers, silyl dienol, synthesis of, **3**, 38
Ethers, silyl enol
of acylsilanes, synthesis of, **5**, 414
alkylation of, **1**, 133, 200, 239
in the presence of a Lewis acid, **1**, 133
for the alkylation of ketones, **3**, 124
amino alkylation of, **3**, 186
arylation of, **1**, 471
benzylation of, **1**, 133
conjugate addition of, **1**, 304
dehydrogenation of, **1**, 556
from, aldehydes, **4**, 1049
nitration of, **3**, 190
oxidation of, **1**, 556; **3**, 33, 54, 156
reactions
with acetals, **3**, 220
with *N*-acyliminium salts, **3**, 186

with alcohols, **4**, 188
with aldehydes, **3**, 159, 220
with γ-alkoxyenoates, **1**, 196
with alkoxymethylthiomethylsilanes, **1**, 286
with alkyltellurium trichlorides, **3**, 177
with allyl ethers, **1**, 189
with arylsulfonyl azides, **5**, 346
with arylsulfonyl peroxides, **3**, 163
with aryltellurium trichlorides, **3**, 177
with azodicarboxylates, **3**, 181
with benzylic acetates, **1**, 202
with borane, **1**, 434
with bromomethanesulfonyl bromide, **4**, 84
with chloroalkyl sulfides, **3**, 168
with dihalocarbenes, **3**, 67, 235
with diiron salts, **4**, 680
with dithioorthoformates, **4**, 250
with electrophiles, **5**, 236
with ethyl azido formate, **3**, 181
with Grignard reagents, **1**, 480
with hydrazoic acid, **4**, 314
with iminium ions, **1**, 255
with ketones, **3**, 220
with Lewis acids, **1**, 189
with orthoformates, **3**, 39
with oxime sulfonates, **3**, 269
with phenylsulfenyl chloride, **3**, 35
with selenium compounds, **1**, 606
with sulfamoyl chlorides, **3**, 171
with sulfonimines, **3**, 437
with sulfonium ions, **3**, 169
with sulfonyl oxiridines, **3**, 156
with trimethylthio orthoformate, **3**, 36, 168
α-selenylation of, **3**, 173
as starting materials
for acetals, **4**, 188
for α,β-unsaturated acylsilanes, **5**, 425
substitution of, **1**, 482
sulfenylation of, **3**, 164
synthesis of, **3**, 38
use in Michael reactions, **1**, 308
use in the substitution of nitrogen functions, **1**, 251
use in the synthesis of cyclopentenones, **3**, 225
use in the synthesis of α,β-unsaturated ketones, **3**, 206
Ethers, (*E*)-silyl enol, synthesis of, **4**, 918
Ethers, *O*-silyl enol, as starting materials, for *N*-fluoroamides, **5**, 313
Ethers, 1-silyl ethyn-2-ol, reactions, with germanium tetrachloride, **4**, 1058
Ethers, silyloxyenol, dehydration of, **3**, 209
Ethers, silylpropargyl silyl, use in the synthesis of silyl acroleins, **2**, 938
Ethers, silyl ynol
desilylation of, **2**, 1023
from, α-diazocarbonyls, **2**, 1024
spectral properties, **2**, 1024
stability of, **2**, 1024
synthesis of, **2**, 1024
Ethers, stannylvinyl, synthesis of, via reverse Diels–Alder reaction, **4**, 952
Ethers, (*Z*)-β-stannylvinyl, synthesis of, **2**, 676
Ethers, α-substituted, as starting materials, for acetals, **4**, 198
Ethers, α-substituted alkenyl, synthesis of, **2**, 676
Ethers, α-sulfonyl
condensation of, with aldehydes, **4**, 838
reactions, with methyl formate, **4**, 838
C-silylation of, **4**, 838
as starting materials
for acetals, **4**, 200
for ketene monothioacetal *S,S*-dioxides, **4**, 838
use in Horner–Wittig reactions, **4**, 840
Ethers, α-sulfonyl β-halo, dehydrohalogenation of, **4**, 838
Ethers, sulfonyl methyl, as intermediates, for the synthesis of alkenyl ethers, **2**, 678
Ethers, 2-tetrahydrofuryl, synthesis of, **4**, 199
Ethers, tetrahydropyranyl, synthesis of, **1**, 976
Ethers, thio enol
elimination of, **3**, 54
oxidation of, **4**, 233
Ethers, thiolactim
metallation of, **5**, 731
synthesis of, **5**, 727
reviews, **5**, 726
Ethers, trialkyl silyl, as starting materials, for silyloxy aldehydes, **3**, 33
Ethers, 2-(trialkylsilyloxy)oxime *O*-trialkylsilyl, from, nitroalkanes, **3**, 436
Ethers, α,α,α'-trichloro, synthesis of, **6**, 39
Ethers, trichloromethyl
as starting materials, for trifluoromethyl ethers, **6**, 229
synthesis of, **6**, 229
Ethers, trifluoromethyl
from
fluoroformic acid esters, **6**, 229
trichloromethyl ethers, **6**, 229
synthesis of, **2**, 90
Ethers, trifluoromethyl alkyl, synthesis of, **4**, 1083
Ethers, trihalomethyl, synthesis of, **6**, 229
Ethers, triisopropyl silyl enol, α-azidonation of, **3**, 194
Ethers, trimethylammonium alkyl, elimination of, **1**, 629
Ethers, 1-trimethylgermanyl-1-alkynol, hydroboration of, **4**, 1059
Ethers, trimethylsilyl
elimination of, **1**, 983
as intermediates, in the synthesis of alkyl bromides, **2**, 32
reactions
with hydrogen chloride, **4**, 150
with iodonium triflate, **1**, 108
as starting materials, for aldehydes, **3**, 8
synthesis of, **2**, 539
Ethers, *O*-trimethylsilyl, as starting materials, for fluorenylmethyl chloroformate, **6**, 425
Ethers, trimethylsilyl enol
from, thiophenyl esters, **4**, 932
reactions
with imines, **3**, 187
with α-tributylstannyl thioacetals, **4**, 398
sulfenylation of, **3**, 227
use in the synthesis of α,β-unsaturated ketones, **3**, 224
Ethers, β-trimethylsilylethoxymethyl (SEM), elimination of, **1**, 654
Ethers, trimethylsilylpropargyl, as starting materials, for allenylsilanes, **2**, 940
Ethers, tris(trimethylsilyl)ethyl, synthesis of, **4**, 1048
Ethers, α,β-unsaturated
reactions
with hydroxymethyl isocyanate, **5**, 986
with isocyanic acid, **5**, 985
Ethers, vinyl
acylation of, **3**, 245
aminomethyleneation of, **3**, 76
Claisen rearrangement of, **5**, 431
deprotonation of, **2**, 908; **4**, 929
reagents for, **4**, 949
from, disulfonylmethanes, **4**, 865
geometry of, **4**, 949
Heck aroylation of, **3**, 245
metallation of, **4**, 929
pyrolysis of, **1**, 777
reactions
with 2-aminoalkanols, **4**, 306
with 1-azabuta-1,3-dienes, **4**, 305
with *t*-butyllithium, **4**, 929
with *N*-chlorocarbonyl isocyanate, **5**, 1008
with dialkylphosphinoyl azides, **5**, 709
with dithiocarbamic acids, **6**, 565
with hydrogen cyanide, **3**, 627

with hydrogen sulfide, **2**, 128
with nucleophiles, **4**, 306
with phosphorazidates, **5**, 861
with tetranitromethane, **6**, 149
reduction of, **1**, 74
solvolysis of, **1**, 1112
as starting materials
for alkenes, **1**, 480
for 2-alkoxy-2-alkenenitriles, **3**, 652
for α-bromo ethers, **4**, 51
for α-chloro ethers, **4**, 49
for cyanohydrins, **3**, 627
for ketene dithioacetal tetroxides, **4**, 865
synthesis of, **5**, 521
Ethers, vinylalkoxymethyl, reactions, with aluminum trichloride, **2**, 89
Ethers, vinyl allyl, reactions, with thexylborane, **2**, 41
Ethers, vinylic, synthesis of, **2**, 901
Ethoxide
reactions
with allenic phosphonates, **3**, 200
with vinyl phosphonium salts, **3**, 50
Ethoxyethyl compounds, from, α-hydroxy sulfones, **4**, 236
Ethylamine
as starting material
for ethanol, **2**, 88
for ethylmethylhydrazine, **2**, 379
Ethylamine, *N*,*N*-diiodo-, synthesis of, **2**, 339
Ethylamine, diisopropyl-
reactions, with trifluoromethanesulfonic anhydride, **3**, 696
as starting material, for quaternary ammonium salts, **1**, 254
Ethylamine, (*S*)-1-phenyl-
reactions, with formalin, **4**, 99
use in the synthesis of α-amino acetals, **3**, 41
Ethylamine, *N*-[1-(2-thienyl)cyclohexyl]-, isoselenocyanato compound of, **5**, 1052
Ethylamine hydrobromide, 1-ethylthio-2,2,2-trifluoro-, synthesis of, **4**, 320
Ethylamine hydrochloride, 1-methoxy-2,2,2-trifluoro-, synthesis of, **4**, 305
Ethylamine hydrochlorides, chloro-, synthesis of, **5**, 1033
Ethylamines, 2-aryl-, synthesis of, **2**, 320
Ethylamines, β-aryl-, synthesis of, **2**, 318
Ethylamines, β-chloro-, reactions, with metal phosphides, **2**, 430
Ethylamines, ferrocenyl-, synthesis of, **4**, 720
Ethylamines, β-nitro-, synthesis of, via Michael addition, **2**, 301
Ethylamines, phenyl-, synthesis of, via Heck arylation, **2**, 321
Ethylamines, *N*-substituted 1-(benzotriazolyl)-2,2-diethoxy-, reactions, with Grignard reagents, **1**, 261
Ethyl bromide, reactions, with hydroxamic acids, **5**, 701
Ethyl cations, 1,1-diphenyl-, synthesis of, **1**, 896
Ethyl cations, 1-phenyl-2,2,2-trifluoro-, synthesis of, **1**, 898
Ethyl 3-dimethylamino-2-isocyanoacrylate, synthesis of, **3**, 720
Ethylene
reactions, with tetranitromethane, **6**, 149
Wacker oxidation of, **2**, 983; **3**, 113
Ziegler telomerization of, **2**, 43
See also Ethene
Ethylene, 1,1-bis(diethylamino)-, reactions, with alcohols, **4**, 888
Ethylene, 1,2-bis(dimethylarsino)- (EDAS), synthesis of, **2**, 876
Ethylene, (*E*)-1,2-bis(tri-*n*-butylstannyl)-, as nucleophilic ethynyl group equivalent, **1**, 1048
Ethylene, *trans*-1,2-bis(tri-*n*-butylstannyl)-, reactions, with butyllithium, **1**, 1091
Ethylene, 1-bromo-2-alkoxy-, dehydrobromination of, **2**, 1026
Ethylene, 1-*t*-butylamino-2-methyl-2-nitro-, transamination of, **2**, 788
Ethylene, 1-chloro-2-iodo-, use in the synthesis of alkynes, **1**, 1005
Ethylene, diamino-, reactions, with sulfur, **5**, 548
Ethylene, 1,1-dichloro-
reactions, with carbonium ions, **5**, 26
as starting material, for dithioesters, **5**, 552
Ethylene, 1,2-dichloro-, use in the synthesis of natural products, **1**, 1009
Ethylene, 1,2-dichloro-1,2-difluoro-, as starting material, for chlorofluoronitromethane, **6**, 54
Ethylene, *trans*-1,2-dicyano-1,2-bis(trifluoromethyl)-, reactions, with stilbene bis(triphenylphosphine)platinum, **5**, 1139
Ethylene, 1,1-difluoro-, as starting material, for fluoroalkyne, **1**, 1017
Ethylene, 1,2-dimethoxy-, chlorination of, **4**, 767
Ethylene, 1,1-dimethylamino-, reactions, with 1,1-dichloroethylenes, **1**, 995
Ethylene, 1-dimethylamino-2-nitro-
from, dimethylformamide diethylacetal, **2**, 788
use in the synthesis of β-nitro enamines, **2**, 787
Ethylene, 1,1-dinitro-2,2-diphenyl-
reactions
with alkoxides, **4**, 898
with phenoxides, **4**, 898
Ethylene, 1,1-diphenyl-, synthesis of, **1**, 738
Ethylene, 1-ethoxy-1-diethylamino-, synthesis of, **4**, 888
Ethylene, 1-ethoxy-1-(*N*-ethylanilino)-, synthesis of, **4**, 887
Ethylene, fluorodichloro-, use in the synthesis of ynediamines, **2**, 1047
Ethylene, iodotrifluoro-, photolysis of, **2**, 778
Ethylene, nitrosotrifluoro-, synthesis of, **2**, 778
Ethylene, (*E*)-1-phenyl-2-(3-indolyl)-, synthesis of, **1**, 733
Ethylene, polytetrafluoro-, as starting material, for difluorocarbene, **6**, 12
Ethylene, tetrabromo-, fluorination of, **5**, 811
Ethylene, tetrachloro-
oxidation of, **6**, 413
as starting material, for phosgene, **6**, 413
Ethylene, tetracyano- (TCNE)
as activating agent, for the synthesis of anhydrides, **5**, 191
reactions
with alkenes, **1**, 355
with benzyltrimethylstannane, **5**, 1145
with bisbenzylmercury, **5**, 1142
reactivity of, **1**, 560
as starting material, for dicyanoketene acetals, **4**, 829
synthesis of, **3**, 656
use in the synthesis of aromatic anhydrides, **5**, 196
use in the synthesis of cyclobutanes, **1**, 355
use in the synthesis of keteniminato complexes, **3**, 570
Ethylene, tetrafluoro-
from, sodium perfluoropropionate, **4**, 731
oxidation of, **6**, 214
ozonation of, **6**, 251
photolysis of, **2**, 1013
reactions
with dialkyltin hydrides, **6**, 65
with selenium, **6**, 49
with sodium methoxide, **4**, 761
with sulfur, **6**, 528
with tetraselenium bis(hexafluoroarsenate(V)), **6**, 49
as starting material
for carbonic difluoride, **6**, 410
for carbonimidic difluorides, **6**, 604
thermolysis of, **6**, 46
Ethylene, tetrakis(trimethylsilyl)-, reactions, with lithium, **6**, 193
Ethylene, tetralithio-, synthesis of, **6**, 205
Ethylene, tetramethoxy-
ozonolysis of, **6**, 469
as starting material, for dimethyl carbonate, **6**, 469
Ethylene, tribromo-, reactions, with potassium phenolate, **4**, 761

Ethylene, trichloro-
  reactions
    with benzeneselenol, **4**, 786
    with phenols, **4**, 761
    with sodium ethoxide, **4**, 761
    with sodium phenolates, **4**, 761
  as starting material
    for 1,2-dichlorovinyl ethers, **2**, 1026
    for dithionooxamides, **5**, 602
Ethylene, 1-trimethylsilyl-2-benzenesulfonyl-, use in Diels–Alder cycloadditions, **1**, 654
Ethylene, tris(dimethylchlorosilyl)-, synthesis of, **6**, 178
Ethylene carbonate, as starting material, for *β*-chloroethyl chloroformate, **6**, 429
Ethylene cyanohydrin
  hydrolysis of, **5**, 56
  as starting material, for *β*-hydroxypropionic acid, **5**, 64
Ethylenediamine
  reactions
    with *N*-(2,4-dichlorophenyl)dichlorocarbiimides, **6**, 645
    with diethyl carbonate, **6**, 484
    with *S*-methylisothiouronium salts, **6**, 645
Ethylenediamine, *N*,*N*'-bis(tributylstannyl)-*N*,*N*'-dimethyl-, reactions, with ethylene thionocarbonate, **6**, 322
Ethylenediamine, tetramethyl- (TMEDA)
  for the deprotonation of vinyl ethers, **4**, 949
  for the enhanced reactivity of organolithium, **1**, 111
  lithiation of, **4**, 528
  reactions
    with alkyllithium complexes, **6**, 391
    with alkyllithiums, **6**, 190
    with allylic alcohols, **1**, 316
    with *t*-butyllithium, **4**, 623
    with diisopinocampheylborane, **2**, 515, 523
  use in the deprotonation of alkanes, **2**, 554
  use in the synthesis of carbon-bridged diindiums, **4**, 698
  use in the synthesis of dialkylmagnesiums, **2**, 559
  use in the synthesis of lithioferrocenes, **4**, 706
  use in the synthesis of stable imidoylboranes complexes, **5**, 795
Ethylenedithiol, reactions, with bis(dimethylamino)methylarsine, **2**, 501
Ethylene glycol
  dehydration of, **2**, 636
  reactions
    with monotins, **6**, 200
    with perthiophosphonic anhydrides, **2**, 856
Ethylene glycol bis(chloroformate), as starting material, for vinyl chloroformate, **6**, 429
Ethyleneimine, reactions, with ethyl chloroformate, **6**, 483
Ethylene oxide
  reactions
    with alkenylmagnesiocuprates, **1**, 1107
    with carboxylate anions, **5**, 138
    with methyl isocyanate, **6**, 324
    with methyl isothiocyanate, **6**, 324
    with silicon tetraisocyanate, **5**, 1016
    with trithioisocyanuric acid, **6**, 335
Ethylene oxide, 2-substituted 1,1-dicyano-, reactions, with hydrobromic acid, **5**, 56
Ethylene oxide, tetracyano-, thermal carbon–carbon bond scission of, **3**, 317
Ethylenes, reactions, with oxaziridines, **6**, 256
Ethylenes, aryl-, oxidation of, **3**, 87
Ethylenes, 1,1-bis(dimethylamino)-2-aryl-2-cyano-, synthesis of, **4**, 974
Ethylenes, 1,1-bis(silyl)-, reactions, with lithium, **4**, 618
Ethylenes, bromotrimethylsilyl-, monoalkynylation of, **1**, 1007
Ethylenes, 1-chloro-2-nitro-, as starting materials, for nitrovinylamines, **2**, 787
Ethylenes, diamino-, synthesis of, **6**, 141
Ethylenes, diaryl-, from, aryl diazoalkanes, **1**, 717
Ethylenes, 1,1-diaryl-, irradiation of, **1**, 1120
Ethylenes, 1,1-diaryl-2-chloro-, rearrangement of, **1**, 1144
Ethylenes, dibromo-, dialkynylation of, **1**, 1007
Ethylenes, dichloro-, selective substitution of, **1**, 468
Ethylenes, 1,1-dichloro-, reactions, with 1,1-dimethylaminoethylene, **1**, 995
Ethylenes, *trans*-1,2-dichloro-1-phenoxy-, synthesis of, **4**, 761
Ethylenes, 1,1-disubstituted, synthesis of, **1**, 650
Ethylenes, isocyanato-, synthesis of, **2**, 1055
Ethylenes, (nitroaryl)-, synthesis of, **1**, 613, 622
Ethylenes, polyhalogenated, reactions, with sodium phenoxide, **4**, 761
Ethylenes, tetraamino-
  reactions, with carbon disulfide, **5**, 548
  as starting materials
    for ortho amides, **6**, 141
    for telluroureas, **6**, 599
  synthesis of, **4**, 975
Ethylenes, 1-trimethylsilyl-1-dimesitylboryl-, as starting materials, for silyl(boryl)methanes, **6**, 201
Ethylene sulfoxide, thermal decomposition of, **2**, 153
Ethylenimine, reactions, with potassium seleno sulfate, **2**, 237
Ethyl group, migratory aptitudes of, **1**, 384
Ethyl halides, 2-(phenyl)-, substitutions of, **1**, 112
Ethylide, triphenylarsonium
  reactions
    with octanal, **1**, 269
    with *p*-tolualdehyde, **1**, 738
Ethylidenebis(mercury chloride), reactions, with lithium, **4**, 668
Ethylidene clusters, decomposition of, **6**, 347
Ethyl iodide
  reactions
    with carbanions, **1**, 858
    with silver cyanide, **3**, 694
    with triisopropylphosphine, **2**, 446
Ethyl iodide, 2,2,2-trifluoro-, bistrifluoroacetoxylation of, **5**, 205
Ethylthiols, electrooxidative cleavage of, **3**, 645
O-Ethyl triphenyltin dithiocarbonate, reactions, with bromoalkanes, **6**, 551
Ethyne
  reactions
    with antimony pentachloride, **2**, 881
    with arsenic trichloride, **2**, 881
    with bis(trimethylsilyl)methylstannylene, **2**, 1101
    with butadiene, **1**, 519
    with dinuclear bisimino complexes, **4**, 683
    with diphenylphosphine, **2**, 439
    with hexamethylguanidinium chloride, **6**, 139
    with hydrogen fluoride, **2**, 611
    with lithium diorganocuprates, **1**, 541
    with selenium oxychloride, **2**, 732
    with sodium hypoiodite, **2**, 1019
    with sulfur, **2**, 708
    with tetrafluoroethene, **4**, 5
    with trialkylalanes, **2**, 974
    with zinc amide, **2**, 1096
  as starting material, for trinitromethanes, **6**, 150
Ethyne, acetyl-, reactions, with oxygen, **1**, 274
Ethyne, benzoyl-
  reactions
    with cyclopentadiene, **1**, 519
    with salicyclic acids, **4**, 213
Ethyne, 1,2-bis(benzhydrylthio)-, synthesis of, **1**, 1040
Ethyne, bis(iodinium)-, reactions, with thiophenolate anion, **2**, 1033
Ethyne, bis(iodonium)-, synthesis of, **2**, 1020
Ethyne, 1,2-bis(methylthio)-, reactions, with thioketones, **5**, 559
Ethyne, bis(phenylthio)-, synthesis of, **2**, 1033

Ethyne, 1,2-bis(silyl)-, as starting material, for 1,1,2,2-tetrasilylethene, **4**, 1045
Ethyne, bis(trifluoromethyl)-, reactions, with *N*-benzoylpyrrole, **1**, 521
Ethyne, 1,2-bis(trimethylphenyl)-, synthesis of, **2**, 1101
Ethyne, bis(trimethylsilyl)-
  co-oligomerization of, **2**, 944
  reactions
    with *p*-bis(iodosyl)benzene, **2**, 1021
    with 1,1-dimethyl-2,3-bis(trimethylsilyl)silirane, **4**, 1052
  as starting material, for ethynyl(phenyl)iodonium tetrafluoroborate, **2**, 1020
  use in the synthesis of alkenes, **1**, 549
Ethyne, 1,2-bis(trimethylsilyl)-
  diboronation of, **4**, 1048
  reactions
    with chloro diboranes, **4**, 1048
    with hydrosilanes, **4**, 1045
    with triarylsilylboranes, **4**, 1058
Ethyne, 1,2-bis(trimethylstannyl)-, reactions, with 1,2-di *t*-butyl-1,2-dichloro diborane, **4**, 1063
Ethyne, bromocyano-, reactions, with trithiocarbonate, **5**, 487
Ethyne, *t*-butyl-
  dimerization of, **1**, 992
  reactions, with molybdenum carbyne complexes, **3**, 513
Ethyne, *t*-butyl fluoro-
  stability of, **2**, 1013
  synthesis of, **2**, 1014
Ethyne, *t*-butylnitro-
  oxidation of, **2**, 1052
  synthesis of, **2**, 1052
Ethyne, *t*-butyltrimethylstannyl-, reactions, with nitrogen pentoxide, **2**, 1051
Ethyne, chloro-, metal compounds of, **2**, 1015
Ethyne, cyano-, as starting material, for alkynyl chlorides, **2**, 1016
Ethyne, cyclopropyl-, reactions, with radicals, **1**, 1130
Ethyne, 2-deuterio-1-phenyl-, reactions, with trimethylaluminum, **2**, 975
Ethyne, dibromo-, synthesis of, **2**, 1017
Ethyne, di-*t*-butyl-, use in the synthesis of diiron complexes, **4**, 681
Ethyne, dichloro-
  reactions, with triphenylphosphine, **2**, 1062
  synthesis of, **2**, 1014, 1016
  use in the synthesis of ynediamines, **2**, 1047
Ethyne, dicyano-, synthesis of, **3**, 674
Ethyne, difluoro-
  detection of, **2**, 1013
  stability of, **2**, 1013
  synthesis of, **2**, 1013
Ethyne, dihydroxy-, trimerization of, **2**, 1023
Ethyne, diiodo-, synthesis of, **2**, 1019
Ethyne, di(methylseleno)-, synthesis of, **2**, 1038
Ethyne, diphenoxy-
  reactions, with sodium bis(trimethylsilyl)amine, **4**, 884
  synthesis of, **2**, 1028
Ethyne, diphenyl-
  as bridge, in binuclear nickel complexes, **4**, 692
  reactions
    with butadiene, **1**, 519
    with *t*-butyllithium, **1**, 539
    with Grignard reagents, **1**, 539
    with ruthenium cluster compounds, **6**, 344
    with sodium hydrazide, **3**, 447
    with triscarbonyl iridium complexes, **5**, 461
  reduction of, **1**, 515
  as starting material, for (*Z*)-stilbenes, **1**, 458
  synthesis of, **1**, 1077
Ethyne, ethoxy-
  availability of, **2**, 1027
  from, lithium diethylamide, **2**, 1025
  hydrostannation of, **2**, 676
  iodoboration of, **2**, 902
  reactions
    with allylboranes, **1**, 1030
    with sulfonyl azides, **5**, 827
    with triethylplumbyllithium, **4**, 955
    with triethylplumbylsodium, **4**, 955
  as starting material, for phenylethyne, **1**, 1030
Ethyne, 1-ethoxy-
  reactions
    with alkoxy(trimethylsiloxy)phosphines, **4**, 914
    with diethyl phosphite, **4**, 913
    with oximes, **6**, 74
    with tribromophosphine, **4**, 914
    with tributylphosphine, **4**, 914
    with triethyl phosphite, **4**, 913
  as starting material, for triethyl orthoacetate, **6**, 74
  as starting materials, for mixed orthoacetate esters, **6**, 74
Ethyne, ethoxytrimethylsilyl-, as activating agent, for the synthesis of anhydrides, **5**, 189
Ethyne, ethoxy(trimethylstannyl)-, thermolysis of, **3**, 540
Ethyne, fluoro-
  properties of, **2**, 1013
  stability of, **2**, 1013
  synthesis of, **2**, 1012, 1013
Ethyne, fluorochloro-
  stability of, **2**, 1013
  synthesis of, **2**, 1013
Ethyne, hydroxy-
  from, semisquaric acid, **2**, 1022
  isomerization of, **2**, 1022
  occurrence, **2**, 1022
Ethyne, methoxy-, as activating agent, for the synthesis of anhydrides, **5**, 189
Ethyne, 1-methoxy-, reactions, with carboxylic acids, **4**, 827
Ethyne, monobromo-, synthesis of, **2**, 1017
Ethyne, monochloro-
  properties of, **2**, 1014
  synthesis of, **2**, 1014
Ethyne, phenyl-
  carbalumination of, **4**, 1068
  carbonylation of, **3**, 276
  from, ethoxyethyne, **1**, 1030
  reactions
    with copper(I) cyanide, **5**, 1140
    with cyclohexadiene, **1**, 529
    with dimethylthiocarbamoyl chloride, **5**, 617
    with isothiocyanates, **5**, 617
    with tellurium, **5**, 627
    with tetraamines, **6**, 139
    with tetraorganodiarsines, **2**, 876
    with triphenylstannane, **4**, 701
  as starting material
    for cobaltacyclopentadiene complexes, **4**, 689
    for dialkynylzincs, **2**, 1096
  synthesis of, **1**, 485, 999, 1017
Ethyne, phenylazido-, synthesis of, problems, **2**, 1055
Ethyne, phenylchloro-, reactions, with sodium azide, **2**, 1055
Ethyne, phenylhydroxy-, from, phenylhydroxycyclopropenone, **2**, 1022
Ethyne, phenyl(*p*-methoxyphenyltelluro)-, oxidation of, **1**, 999
Ethyne, phenyl nitro-, synthesis of, **2**, 1052
Ethyne, phosphinyl-, reactions, with *N*-sulfinylarylsulfonamides, **5**, 645
Ethyne, siloxy-, synthesis of, **2**, 1024
Ethyne, *p*-toluenesulfonyl-, reactions, with butyllithium, **2**, 1035
Ethyne, (tolylsulfonyl)-, synthesis of, **2**, 1036
Ethyne, *p*-tolylsulfonyl-, reactions, with ethyleneimine, **2**, 742
Ethyne, 1-trimethylgermanyl-2-phenyl-, reactions, with diisobutylaluminum hydride, **4**, 1063

Ethyne, trimethylsilyl-
bromination of, **2**, 929
dimerization of, **1**, 992
reactions
with 4-*t*-butyl-1-cyclohexene, **1**, 1029
with diphenyl diselenide, **4**, 947
with organocoppers, **4**, 809
as starting material
for cumulenylsilanes, **2**, 940
for terminal alkynes, **1**, 1012
Ethyne, trimethylsilylethoxy-, as dehydrating agent, for the synthesis of acylated oximes, **5**, 215
Ethyne, trimethylsilylphenyl-, silyltitanation of, **2**, 924
Ethyne, 1-trimethylsilyl-2-phenyl-
carbatitanation of, **4**, 1061
reactions, with silyl silirenes, **4**, 1053
Ethyne, 1-(trimethylstannyl)-, reactions, with triethyl boron, **4**, 1068
Ethyne, trimethylstannyl(trimethylsilyl)-, reactions, with alkenyl iodides, **1**, 1007
Ethyne, triphenylsilyl-, reactions, with organocoppers, **4**, 809
1-Ethyne-1-carboxylates, ethyl 2-substituted-, synthesis of, **5**, 99
Ethyne complexes, diiridium diphenyl-, synthesis of, **4**, 688
Ethynedicarboxylate, bisdimethyl, as reagent, for selones, **3**, 394
Ethynedicarboxylate, dimethyl (DMAD)
reactions
with 6-amino-1,3-dimethyl uracil, **4**, 893
with α-aminosilanes, **4**, 510
with aryl isocyanides, **3**, 593
with carbon selenide sulfide, **4**, 869
with enamines, **3**, 233, 275
with rhenium pentacarbonyls, **5**, 451
with selone, **3**, 394
with silylated ynamines, **2**, 1045
Ethynedicarboxylate, methyl, as starting material, for α-diazo succinate diesters, **3**, 471
Ethynes, as starting materials, for borirenes, **4**, 1058
Ethynes, alkoxy-
hydrozirconation of, **2**, 969
oxidative dimerization of, **2**, 1027
Ethynes, alkoxyaryl-, synthesis of, **2**, 1026
Ethynes, aryl-, as starting materials, for alkynyl chlorides, **2**, 1016
Ethynes, arylated, as starting materials, for alkenylalanes, **2**, 967
Ethynes, arylsulfonyl-, deprotonation of, **2**, 1035
Ethynes, bis(alkylthio)-, reactions, with phosphorus pentachloride, **4**, 926
Ethynes, 1,2-bis(silyl)-
hydrosilylation of, **4**, 1044
reactions, with chloro methyl hydrosilanes, **4**, 1045
Ethynes, boron-substituted alkoxy-, synthesis of, **2**, 1027
Ethynes, phenoxy-, synthesis of, **2**, 1028
Ethynes, phosphorus-substituted, reactions, with benzenesulfenyl chlorides, **4**, 926
Ethynes, silicon-substituted alkoxy-, synthesis of, **2**, 1027
Ethynes, silyl-
carbocupration of, **2**, 924
reactions, with 2-aminobenzenethiol, **4**, 939
Ethynes, 1-silyl-2-alkoxy-, thermolysis of, **2**, 941
Ethynes, silylated chloro-, synthesis of, **2**, 1016
Ethynes, substituted alkoxy-, synthesis of, **2**, 1027
Ethynes, substituted chloro-, synthesis of, **2**, 1016
Ethynes, terminal alkoxy-, functionalization of, **2**, 1027
Ethynes, tin-substituted alkoxy-, synthesis of, **2**, 1027
Ethynolate, phenyl-, from, isoxazole, **2**, 1023
Ethynolate, trimethylsilyl-, from, ketene, **2**, 1023
Ethyn-2-ols, 1-germanyl-, reactions, with silicon tetrachloride, **4**, 1058
Ethynylation
of saturated ketones, **3**, 146
for the synthesis of phenylethyne, **1**, 1030
Ethynyl carbinols, aryloxymethyl, reduction of, **1**, 961
Ethynylides, alkoxy-
reactions
with carbonyls, **2**, 1027
with electrophiles, **2**, 1027
Ethynylides, chloro-, from, dichloroethenes, **2**, 1014
Ethynylides, halo-, alkylation of, **2**, 1012
Ethynyl metals, reactions, with 2-alkylcyclopentanones, **1**, 1039
Ethynylphosphonites, dialkyl, synthesis of, **2**, 838
Δ^5-Etienate, 3β-acetoxy-, use in the synthesis of alkanoic acids, **5**, 39
Etienic acid, 3β-hydroxy-, synthesis of, **5**, 74
Eucannabinolide, synthesis of, **1**, 813
Euparin, synthesis of, **1**, 1136
Europium, reactions, with alkyl iodides, **2**, 564
Europium compounds, as catalysts, for the synthesis of cyanohydrins, **3**, 615
Evans' oxazolidinones
as starting materials, for hydroxamates, **5**, 329
synthesis of, **5**, 148
Evans' oxazolinone auxiliary, use in alkylation reactions, **1**, 134
Exaltone, synthesis of, **1**, 1082
Exchange reaction, for the synthesis of mixed tetraalkyl dipnicogens, **2**, 508
Exomethylene, reactions, with hydrazines, **6**, 141
Extrusion reaction, of sulfur compounds, **1**, 231

F

Farnesol, as starting material, for agelasidine A, **6**, 472
Fatty acids, unsaturated
  synthesis of, **5**, 97
    reviews, **5**, 100
Favorskii rearrangement
  of *N*-chloroamides, **5**, 315
  of α-haloimines, **3**, 416
  mechanism, **1**, 391, 884
  for the synthesis of carboxylic esters, **5**, 145
Felkin–Anh model, extensions of, **1**, 305
Felkin–Anh rules, **1**, 761
Fenchane, 2-chloro-2-nitroso-, rearrangement of, **3**, 434
Feniculin, synthesis of, **1**, 641
Fenton's reagent, for the oxidation of arenes, **2**, 650
Ferracyclopentadiene–nickel complexes, synthesis of, **4**, 715
Ferracyclopentenes, from, cyclobutyl complexes, **5**, 937
Ferracyclopentenones, synthesis of, **5**, 937, 938
Ferrate anions, tetrahydridocarbonyl-, for the reduction of thioketones, **1**, 24
Ferrate complexes, trichloro-, reactions, with *t*-alkyl chlorides, **5**, 1146
Ferrate oxidants, use in asymmetric *cis*-dihydroxylation reactions, **2**, 48
Ferrates, as oxidants, for the synthesis of ketones, **3**, 117
Ferrates, carbonyl-, reactions, with manganese dioxide, **6**, 523
Ferrates, cyano-, use in hydroxylation reactions, **2**, 45
Ferrate salts, tetrachloro-, synthesis of, **3**, 687; **6**, 610
Ferric chloride
  as catalyst
    for chlorination, **2**, 608
    for coupling reactions, **1**, 117
    for the synthesis of aromatic ketones, **3**, 279
  for the oxidation of thiols, **2**, 214
  reactions
    with *N*-benzylsulfonyl acid phenylhydrazides, **5**, 87
    with *N*-carboalkyloxyamino acid phenylhydrazides, **5**, 87
    with oxyimidic esters, **5**, 864
    with oxyimidic halides, **5**, 864
    with potassium carbonate, **2**, 214
  reduction of, **2**, 577
Ferric compounds, for the oxidation of thiols, **2**, 214
Ferric iodide, for the synthesis of aryl iodides, **2**, 631
Ferrocene
  cyclopentadienyl anions of, substitution of, **1**, 862
  lithiation of, **4**, 706
  metallation of, **4**, 720
  reactions
    with acetyl chloride, **1**, 862
    with pentylsodium, **4**, 706
  stability of, **1**, 862
  use in the synthesis of difluorodiazirine, **6**, 261
  X-ray analyses of, **1**, 862
  See also Dicyclopentadienyliron
Ferrocene, bis(diphenylphosphine) (DPPF), use in the substitution of boron functions, **1**, 279
Ferrocene, bromo-
  reactions, with butyllithium, **4**, 706
  as starting material, for ferrocenylmagnesium bromide, **4**, 709
Ferrocene, chloro-, metallation of, **4**, 706
Ferrocene, 2-chlorolithio-, reactions, with chlorocyclopentadienyl(triphenylphosphine)-nickel, **4**, 719
Ferrocene, (chloromercuri)-, transmetallation of, **4**, 706
Ferrocene, 1,1'-dichloro-, lithiation of, **4**, 706
Ferrocene, 1',2-dichlorolithio-, use in the synthesis of metal complexes, **4**, 719
Ferrocene, 1,1'-dilithio-
  reactions, with bis(cyclopentadienyl)dichloro complexes, **4**, 720
  use in the synthesis of metal complexes, **4**, 719
Ferrocene, (dimethylaminomethyl)-, cyclometallation of, **4**, 720
Ferrocene, (*R*,*S*)-1-diphenylphosphino-2-(1-dimethylaminoethyl)-, reactions, with *n*-butyllithium, **4**, 1033
Ferrocene, disodio-, reactions, with magnesium bromide, **4**, 709
Ferrocene, iodo-
  mercuration of, **4**, 724
  reactions, with butyllithium, **4**, 706
Ferrocene, lithio-, reactions, with transition metal halides, **4**, 723
Ferrocene, 2-lithio(dimethylaminomethyl)-, reactions, with transition metal halides, **4**, 723
Ferrocenes
  reactions
    with *n*-butyllithium–tetramethylethylenediamine, **4**, 1034
    with trinitromethane salts, **6**, 145
Ferrocenes, acyl-, reactions, with pentacarbonyls, **4**, 720
Ferrocenes, ethyl-, synthesis of, **4**, 1033
Ferrocenes, hetero-, halide displacement of, **2**, 896
Ferrocenes, lithio-, synthesis of, **4**, 705
Ferrocenes, vinyl-, synthesis of, **4**, 1033
Ferrocenophanes
  reactions, with phenyllithium, **4**, 1034
  synthesis of, **4**, 1034
Ferrocenoyl chloride, reactions, with organometallic anions, **4**, 722
Ferrocenyl compounds, synthesis of, **6**, 338
Ferrocenylphosphine–transition metal complexes, use in the synthesis of amines, **2**, 331
Ferroles, tetrasubstituted, synthesis of, **4**, 681
Ferrous chloride, transmetallation of, **2**, 577
Fetizon's reagent, use in the synthesis of γ-lactones, **5**, 171
Finkelstein reaction
  for dichlorofluoroiodomethane, **6**, 226
  for the synthesis of alkyl iodides, **1**, 107; **2**, 35
  use in the synthesis of α-iodo alkyl esters, **4**, 58
First aid, for phosgene exposure, **6**, 411
Fischer carbene complexes
  alkylation of, **5**, 163
  reactions
    with alkynes, **1**, 551
    with methyl selenol, **5**, 952
    with pyridine, **5**, 953
  as starting materials, for α-alkoxystannanes, **4**, 388
Fischer carbenes
  in [3+2+1] benzannulation, **2**, 669
  synthesis of, limitations, **5**, 936
Fischer complexes
  definition of, **5**, 932
  reactions, with *n*-butyllithium, **5**, 955
  as starting materials, for thioether complexes, **5**, 951
Fischer complexes, cyclic, from, saturated cyclic ethers, **5**, 948
Fischer–Hepp reaction, for the synthesis of aromatic nitroso compounds, **2**, 780
Fischer reaction

for the synthesis of alkoxycarbene complexes, **5**, 933
for the synthesis of cyclic alkoxycarbene complexes, **5**, 943
for the synthesis of organoosmiums, **6**, 345
Fischer–Speier esterification
mechanism
$A_{AC}1$-type, **5**, 122
$A_{AC}2$-type, **5**, 122
Fischer-type carbenes, use of term, **3**, 507
FITS reagents, as perfluoroalkylating agents, **6**, 12
Five-component condensation, of isocyanides, **3**, 694
Five-membered rings, synthesis of, **1**, 358
Flame retardants, synthesis of, **4**, 456
Flash vacuum pyrolysis (FVP)
definition of, **1**, 772
of 2,3-dihydropyrrole-2,3-dione, **3**, 584
of Meldrum's acid compounds, **4**, 894
of metal clusters, **6**, 403
in retro-Diels–Alder reactions, **1**, 785
of sulfoxides, **2**, 718
for the synthesis of cyanamides, **5**, 1123
for the synthesis of nitrile imides, **3**, 681
for the synthesis of sulfenic acids, **2**, 184
of 1-vinylbenzotriazoles, **3**, 579
Flash vacuum thermolysis
of *N*-silyl enamine, **3**, 562
for the synthesis of carbenes, **4**, 1080
for the synthesis of ketenimines, **3**, 582
Flavanone, oxidation of, **1**, 554
Flavins, for the oxidation of thiols, **2**, 216
Flavobacterium oxyclans cells, mutant, use in the synthesis of carboxylic acids, **5**, 117
Flavones, synthesis of, **2**, 651
Flow pyrolyses, definition of, **1**, 771
Fluoradene, reactions, with sodium hydroxide, **1**, 862
Fluorene, acidity of, **1**, 862
Fluorene, 2-acetyl-, oxidation of, **5**, 114
Fluorene, 2-carboxamido-9-deuterio-9-methyl-, isoracemization of, **1**, 855
Fluorene, diazo-
photolysis of, **3**, 314
as starting material, for mixed 1,1-dihaloalkenes, **4**, 755
Fluorene, dichloromethylene-, synthesis of, **4**, 742
Fluorene, 2-nitro-7-carboxamido-9-deutero-9-methyl-, deuteration of, **1**, 858
Fluorenes, reactions, with cyanide ions, **3**, 667
Fluorenes, diphosphatetrahydro-, synthesis of, **4**, 548
Fluorenes, 9-silyl-, deprotonation of, **4**, 647
Fluorenes, zircona-, formation prevention of, **2**, 986
Fluorenone, reactions, with propylidenephosphoranes, **1**, 988
Fluorenylide, triphenylarsonium, reactions, with *p*-dimethylaminobenzaldehyde, **1**, 738
Fluorenylides, synthesis of, **3**, 500
Fluoride
displacement of, **1**, 107
nucleophilic displacement of, **6**, 615
reactions
with alkoxymethylthiomethylsilanes, **1**, 286
with α-halosilanes, **1**, 282
with perfluoromethanimine, **5**, 313
Fluoride anions
for the conjugate addition of allylsilanes, **1**, 314
reactions, with bromodiazirine, **5**, 866
Fluoride ions
reactions
with bismesitylboryl(trimethylsilyl)alkanes, **1**, 764
with β-silyl sulfones, **1**, 616
with β-stannyl sulfones, **1**, 616
with vinylsilanes, **1**, 63
use in elimination reactions, **1**, 653
Fluoride ions, naked, synthesis of, **2**, 19
Fluorides, primary, reduction of, **1**, 3
Fluorinating agents, reactions, with amines, **2**, 334
Fluorination
of activated graphite, **6**, 212
of *O*-alkyl chlorothioformic acid esters, **6**, 229
of amides, **5**, 313
of *N*,*N*-disubstituted fluorothioformamides, **6**, 238
electrochemical
of acetic acid, **6**, 214
of acetone, **6**, 214
of acetonitrile, **6**, 214
of acetyl chloride, **6**, 214
of dimethyl sulfide, **6**, 214
of methanol, **6**, 214
of trifluoroacetic acid, **6**, 214
of trimethylamine, **6**, 214
of fluoroformic acid esters, **6**, 229
of metal carbides, **6**, 212
of methyl esters, **6**, 230
for the synthesis of carbamoyl fluorides, **6**, 439
for the synthesis of α-halo nitro compounds, **4**, 108
of tetrabromomethane, **6**, 226
of tetrachloromethane, **6**, 212, 217, 225
of tetrafluoromethane, **6**, 212
of tetraiodomethane, **6**, 212, 226, 227
of 1,1,3,3-tetranitropropane, **4**, 795
of trichlorofluoromethane, **6**, 225
of trifluoromethane, **6**, 212
of trifluoromethyl isocyanide difluoride, **6**, 240
of xanthates, **6**, 229
Fluorination, benzylic, reagents for, **2**, 15
Fluorination, enantioselective, **2**, 16
Fluorine
addition to alkenes, **2**, 17
as electrophilic compound, **2**, 15
1,2-elimination of, **5**, 807
as fluorinating agent, for the synthesis of acylhypofluorites, **5**, 202
for the fluorination of amines, **2**, 334
reactions
with benzylidenealkylamines, **5**, 659
with bis(difluoroamino)(fluoroamino)-aminomethane, **6**, 362
with carbon disulfide, **6**, 46
with carbon monoxide, **6**, 409
with chloroalkenes, **4**, 31
with haloimines, **6**, 53
with heptafluorobutyramidine hydrochloride, **5**, 815
with isocyanides, **6**, 603
with perfluoro ketones, **2**, 94
with perfluorooxo compounds, **6**, 250
with polymethylene oxide, **6**, 251
with quinuclidine, **2**, 336
with sodium oxalate, **6**, 250
with sodium trifluoroacetate, **6**, 250
with tertiary amines, **2**, 336
with thiocarbonyls, **6**, 252
with *N*-(trimethylsilyl)pyridine, **5**, 313
Fluorine-18, use in the synthesis of radioactive acetylhypofluorite, **5**, 203
Fluorine compounds, of phosphorus functions, synthesis of, **5**, 876, 894
Fluorine compounds, organic, chemistry of, **6**, 212
Fluorine–fluorine bond strength, effect in elimination, **4**, 730
Fluorine–sodium fluoride, reactions, with potassium trinitromethanide, **6**, 261
Fluorocarbon ligands, for the stabilization of hypervalent iodine compounds, **1**, 108
Fluorocarbons, as starting materials, for aminals, **4**, 411
Fluorodecarboxylation, mechanism, **2**, 13
Fluorodestannylation, for the synthesis of alkanethiols, **2**, 131
Fluoromethylenation, of aldehydes and ketones, for vinyl fluorides, **2**, 611
Fluoroselenenylation, reagents for, **2**, 280

Fluoxetines, synthesis of, **2**, 65
Foam blowing agents, chlorofluorocarbons for, **6**, 7
Folic acid, synthesis of, reviews, **5**, 92
Formaldazine, tetrabromo-, reactions, with silver fluoride, **6**, 613
Formaldehyde
  reactions
    with *β*-allylaminoalkylphosphines, **4**, 458
    with amides, **4**, 416
    with amines, **1**, 255; **4**, 295
    with 2-aminopropan-1,3-diol, **4**, 302
    with ammonia, **4**, 407
    with arsines, **2**, 1079
    with arylhydrazines, **3**, 446
    with aryloxymagnesium halides, **3**, 97
    with benzenesulfonamide, **4**, 423
    with *N*-benzylglycine, **4**, 466
    with benzylthiol, **4**, 215
    with caesium(IV) ions, **5**, 49
    with chlorodiphenylphosphine, **4**, 119
    with dialkyl phosphinous acids, **4**, 119
    with dibenzylphosphine oxide, **4**, 471
    with dichlorophosphines, **4**, 125
    with difluoronitromethane, **6**, 54
    with 2,4-dimethyl-2-thiolpentan-3-one, **4**, 317
    with 1,3-dimethylurea, **4**, 301
    with diphosphinoethane, **2**, 440
    with formic acid, **2**, 310
    with hydrazines, **4**, 436
    with hydrofluoric acid, **4**, 44
    with *β*-hydroxy-thioamides, **4**, 230
    with isocyanic acid, **4**, 313
    with malonic ester, **1**, 786
    with methylamine hydrochloride, **4**, 317
    with α-(*N*-methylanilino)-α-(trimethylsilyl)acetonitrile, **2**, 748
    with *p*-methylphenylhydrazine, **3**, 446
    with methylsilylphosphines, **4**, 337
    with morpholine, **4**, 315
    with nitramines, **4**, 442
    with *N*-nitroamines, **4**, 311
    with phosphine, **2**, 440
    with styrene, **4**, 827
    with sulfonamides, **4**, 427
    with trialkylsilyl isocyanates, **4**, 313
    with trimethylhydrazine, **4**, 437
    with trinitromethanes, **6**, 145
  as starting material, for methylene aminals, **4**, 410
  use in aldol reactions, **3**, 163
Formaldehyde, nitro-, synthesis of, **3**, 449
Formaldehyde diaryl acetals, chlorination of, **6**, 252
Formaldehydehydrazones, dinitro-, from, nitroalkyls, **6**, 653
Formaldehyde oxime, dinitro-, synthesis of, **6**, 650
Formaldehyde oxime, *O*-(trifluoromethyl)difluoro-, synthesis of, **6**, 232
Formaldoxime, chlorofluoro-, polymer conversion of, **6**, 613
Formaldoxime, diiodo-, synthesis of, **6**, 612
Formalin, reactions, with (*S*)-1-phenylethylamine, **4**, 99
Formamide
  alkylation of, **6**, 138
  reactions
    with carbon disulfide, **6**, 560
    with carboxylic acids, **5**, 262
    with diazonium salts, **1**, 442
    with *N*,*N*-dibromobenzenesulfonamide, **5**, 701
    with macrocyclic triamines, **6**, 140
    with triethyl orthoformate, **6**, 140
    with trimethylsilyl chloride, **5**, 379
  silylmethylation of, **4**, 516
  as starting material
    for ortho esters, **6**, 70
    for triformamidomethane, **6**, 138
    for trithioorthoformate esters, **6**, 83
Formamide, bis(trimethylsilyl)-
  as *N*,*N*-disilylated amide, **5**, 720
  reactions, with organolithiums, **3**, 440
Formamide, *N*,*N*-bis(trimethylsilyl)-, reactions, with Schiff bases, **4**, 419
Formamide, *N*-chlorocarbonyl-*N*-phenyl-, synthesis of, **6**, 445
Formamide, *N*-(1-cyanocyclohexyl)-, as starting material, for cyclohexenyl isocyanide, **3**, 714
Formamide, *N*-diethoxyphosphoryl-, silylation of, **5**, 713
Formamide, *N*,*N*-diethyl-, thionation of, **5**, 577
Formamide, *N*,*N*-dimethyl-
  fluorination of, **5**, 808
  reactions, with carbene complexes, **6**, 723
  as starting material, for *O*-alkyl selenoformates, **5**, 516
  thionation of, **5**, 568
Formamide, *N*,*N*-dimethylchloro-, chlorination of, **6**, 239
Formamide, dimethyl- (DMF)
  1:1 adduct of, **6**, 513
  as catalyst, for acid chlorides, **5**, 7, 9, 11
  reactions
    with amides, **5**, 999
    with amines, **5**, 745
    with arylphosphanes, **5**, 882
    with bis(acetonitrile)palladium, **1**, 1007
    with carbonyl difluoride, **6**, 52
    with hydrogen sulfide, **1**, 46
    with isonitriles, **5**, 767
    with lithium iodide, **5**, 78
    with methoxynaphthalenes, **3**, 101
    with 1-naphthyl isocyanate, **5**, 745
    with phenyl isocyanate, **5**, 745; **6**, 143
    with phosphorus oxychloride, **3**, 91
    with phosphorus trichloride, **2**, 25
    with silylarsanes, **5**, 892
    with trifluoromethane sulfonic anhydride, **3**, 92
  as solvent
    for the synthesis of ketene aminals, **4**, 971
    for the synthesis of nitriles, **3**, 613
  use in the synthesis of aldehydes, **3**, 14
  use in the synthesis of aryl sulfides, **2**, 711
  use in the synthesis of *β*-formyl phosphonates, **3**, 49
Formamide, *N*-2-hydroxy-2-phenylethyl-, as starting material, for 2-chloro-2-phenylethyl isocyanide, **3**, 703
Formamide, *N*-methyl-, fluorination of, **5**, 808
Formamide, 1-substituted-*N*-(α-amino)-, synthesis of, **3**, 710
Formamide, *N*-((trimethylsilyl)methyl)-, synthesis of, **3**, 711
Formamide acetals, synthesis of, **6**, 105
Formamide azine, dimethyl-, reactions, with hydrogen selenide, **5**, 626
Formamide dimethyl acetal, *N*,*N*-dimethyl- (DMFDMA), as starting material, for amide acetals, **6**, 104
Formamide oximes, *N*-heteroaryl-, synthesis of, **5**, 835
Formamide oximes, substituted, synthesis of, **5**, 835
Formamides
  dehydration of, **3**, 693
  reactions
    with alkynyllithiums, **1**, 1046
    with *N*-allyl isocyanides, **5**, 760
    with benzotriazole, **3**, 709
    with bis(benzenesulfonyl)methane, **4**, 866
    with chloromethyltrimethylsilane, **3**, 711
    with Grignard reagents, **2**, 320
    with Lawesson's reagent, **5**, 608
    with phosgene, **6**, 52
    with *N*-sulfonyl isocyanates, **5**, 767
    with tetrakis(dimethylamino)titanium, **5**, 745
    with thionyl chloride, **6**, 52
  as starting materials
    for allylic unsaturated isocyanides, **3**, 700
    for 4-(*t*-butyldimethylsiloxy)-2,6-xylyl isocyanides, **3**, 723

for carbamoyl chlorides, **6**, 449
for formamidines, **5**, 744
for isocyanides, **3**, 695
for isonitriles, **4**, 908; **6**, 417
for polycyclic aromatic isocyanides, **3**, 724
for Vilsmeier reagents, **6**, 57
thionation of, **5**, 570, 608
Formamides, *N*-(α-alkylthioalkyl)-, as intermediates, **3**, 709
Formamides, allenic methyl-, synthesis of, **1**, 974
Formamides, *N*-(1-arenesulfonylalkenyl)-, synthesis of, **4**, 908
Formamides, aryl-
dehydration of, **3**, 721
reactions, with dimethyl sulfate, **5**, 744
as starting materials, for aryl isocyanides, **3**, 721
Formamides, *N*-(arylthiomethyl)-, synthesis of, **4**, 328
Formamides, cyano-, synthesis of, **5**, 747
Formamides, dialkyl-
reactions
with metallated pyridines, **3**, 108
with naphthyllithium halides, **3**, 101
with naphthylmagnesium halides, **3**, 101
with phenylcyanoacetaldehyde, **2**, 749
with thiophenes, **3**, 105
Formamides, dihydroxy-, as starting materials, for bis β-chloro isocyanides, **3**, 715
Formamides, *N*,*N*-disubstituted
as starting materials
for *N*,*N*-disubstituted carbamoyl bromides, **6**, 452
for trifluoromethylamines, **6**, 440
Formamides, *N*-hydroxy-, reactions, with acyl chlorides, **6**, 506
Formamides, monosubstituted
reactions, with carbamoyl chlorides, **5**, 744
as starting materials, for formamidines, **5**, 744
Formamides, *N*-monosubstituted, dehydration of, **3**, 695
Formamides, thiobenzoyl-, synthesis of, **5**, 760
Formamides, α,β-unsaturated
dehydration of, **3**, 712
epoxidation of, **3**, 704
from, sulfenimines, **3**, 713, 719
as starting materials, for α,β-unsaturated isocyanides, **3**, 712
N-Formamides, dehydration of, **3**, 700
Formamidic esters, chloro-, synthesis of, **6**, 616
Formamidine, *N*,*N*'-diisopropyl-*N*'-trimethylsilyl-1-cyano-, synthesis of, **5**, 777
Formamidine, *N*,*N*-dimethyl-*N*'-thiobenzoyl-, *S*-alkylation of, **4**, 328
Formamidine, *N*,*N*-diphenyl-, reactions, with phosgene, **6**, 445
Formamidine, *N*,*N*'-diphenyl-, monosilylation of, **5**, 777
Formamidine, methoxytrifluoro-, defluorination–cyclization of, **6**, 281
Formamidine, *N*-*n*-propyl-*N*'-chloro-1-cyano-, synthesis of, **5**, 818
Formamidine, tetrafluoro-
reactions, with alcohols, **5**, 814
as starting material, for difluorodiazirine, **6**, 261
synthesis of, **5**, 815
Formamidine, trimethylsilyl-, use in the synthesis of nitriles, **3**, 621
Formamidine esters, synthesis of, **5**, 746
Formamidines
configurational stability of, **1**, 150
for the deprotonation of organometallics, **1**, 149
from
acetals, **5**, 746
aminals, **5**, 746
α-amino acids, **5**, 746
O-*t*-butyl valinol, **1**, 149
formamides, **5**, 744
isonitriles, **5**, 747
monosubstituted formamides, **5**, 744
orthoformates, **5**, 746
thioformamides, **5**, 744
thioureas, **5**, 746
1,3,5-triazine, **5**, 747
Vilsmeier reagents, **5**, 744
lithiation of, **4**, 531
metallation of, **1**, 149
reactions
with chromium hexacarbonyl, **5**, 781
with hydroxylamine, **5**, 764
with molybdenum hexacarbonyl, **5**, 781
synthesis of, **5**, 743, 744
via reduction, **5**, 745
Formamidines, *N*-acyl-, synthesis of, **5**, 759
Formamidines, *N*'-acyl-*N*,*N*-dimethyl-, synthesis of, **5**, 759
Formamidines, chiral
from, isoquinoline, **5**, 746
reactions, with tetrahydroquinoline, **5**, 746
Formamidines, chloro-
reactions, with Grignard reagents, **5**, 755
synthesis of, **6**, 645
Formamidines, *N*-chloro-, from, *N*-chloroimines, **5**, 818
Formamidines, cyano-, synthesis of, **5**, 747, 777
Formamidines, 1-cyano-*N*,*N*-diethyl-*N*'-fluoro-, synthesis of, **5**, 817
Formamidines, diaryl-
reactions, with ethyl orthoformate, **5**, 746
synthesis of, **5**, 744
Formamidines, *N*,*N*'-diaryl-*N*-hydroxy-, as starting materials, for aryl isocyanides, **3**, 723
Formamidines, *N*-(dichloromethyl)-, synthesis of, **5**, 744
Formamidines, *N*,*N*-dimethyl-*N*'-alkyl-, synthesis of, **5**, 745
Formamidines, *N*,*N*-dimethyl-*N*'-aryl-, synthesis of, **5**, 745
Formamidines, *N*,*N*'-disubstituted, synthesis of, **5**, 747
Formamidines, dithiophosphonyl-, synthesis of, **5**, 775
Formamidines, (hetero)aryl-, reactions, with hydrogen selenide, **5**, 626
Formamidines, hydroxy-
dehydration of, **5**, 1121
as starting materials, for heteroaryl cyanamides, **5**, 1121
Formamidines, lead-substituted, synthesis of, **6**, 664
Formamidines, α-lithio-, alkylation of, **1**, 149
Formamidines, monosilyl-, synthesis of, **5**, 777
Formamidines, *N*,*N*-perfluoro-, reactions, with alkoxides, **5**, 813
Formamidines, phosphoryl-, synthesis of, **5**, 774
Formamidines, *N*-silyl-
reactions, with acid chlorides, **5**, 759
synthesis of, **5**, 759
Formamidines, *N*-silylated, reactions, with Grignard reagents, **3**, 406
Formamidines, thioacyl-, synthesis of, **5**, 760
Formamidines, *N*-thionoacyl-, reactions, with methylene(triphenyl)phosphorane, **5**, 633
Formamidines, thiophosphoryl-, synthesis of, **5**, 774
Formamidines, tin-substituted, synthesis of, **6**, 664
Formamidines, trialkylplumbyl-, synthesis of, **6**, 664
Formamidines, trialkylstannyl-, synthesis of, **6**, 664
Formamidines, *N*-trichloromethyl chloro-, synthesis of, **3**, 409
Formamidines, *N*-(triethyltin)-, synthesis of, **5**, 780
(*E*)-Formamidines, synthesis of, **5**, 747
(*Z*)-Formamidines, synthesis of, **5**, 747
Formamidinium chloride, as starting material, for triamines, **6**, 321
Formamidinium chloride, tetramethylchloro-
reactions
with cyclopentadienes, **4**, 974
with lithium dimethylamide, **6**, 361
as starting material, for ketene aminals, **4**, 974
Formamidinium chlorides, synthesis of, **6**, 313
Formamidinium chlorides, chloro-

reactions, with sodium isopropoxide, **4**, 974
as starting materials
for hydroxyguanidines, **6**, 650
for isopropoxymethanetriamines, **4**, 973
synthesis of, **4**, 974
Formamidinium chlorides, 1,3-dialkylchloro-, synthesis of, **5**, 980
Formamidinium salts
reactions, with secondary amines, **6**, 138
as starting materials, for aminals, **4**, 414
Formamidinium salts, chloro-
reactions
with amines, **6**, 641, 646
with aniline, **6**, 645
with benzenethiols, **6**, 732
synthesis of, **6**, 728
Formamidium salts, *N*,*N*,*N*',*N*'-tetrasubstituted, reactions, with sodium alkoxides, **6**, 105
Formamidoxime, as starting material, for oxyimidic amides, **5**, 837
Formamidrazone, alkylation of, **6**, 660
Formamidrazone, *N*[1]-phenyl-*N*[1]-methyl-*N*[3]-dimethyl-, methylation of, **5**, 773
Formamidrazones, synthesis of, **5**, 770
Formamidrazones, *N*-acyl-, synthesis of, **5**, 773
Formamidrazones, *N*'-aryl-*C*-(dialkoxyphosphoryl)-, synthesis of, **6**, 659
Formamimidinium chlorides
reactions
with alcoholates, **6**, 632
with phenolates, **6**, 632
Formanilide
as starting material
for amidines, **5**, 575
for thioformanilide, **5**, 566
Formanilide, *N*-methyl- (MFA)
reactions, with phosphorus oxychloride, **3**, 91
thionation of, **5**, 577
Formanilides
as starting materials, for carbonimidic dichlorides, **6**, 606
synthesis of, **5**, 723
Formanilides, *N*-methyl-, synthesis of, **2**, 305
Formanilides, silyl-, synthesis of, **5**, 376
Formate, alkyl chloro-, reactions, with trialkyl phosphite, **6**, 513
Formate, alkyl thiochloro-, reactions, with thioamides, **5**, 727
Formate, allyl chloro-, as starting material, for alkyl allyl carbonates, **6**, 461
Formate, ammonium, use in the reduction of nitriles, **1**, 47
Formate, benzyl chloro-, reactions, with thiocaprolactam, **5**, 727
Formate, *t*-butyl azido-, reactions, with thiols, **2**, 768
Formate, butyl chloro-, synthesis of, **6**, 425
Formate, *n*-butyl chloro-, reactions, with ammonia, **6**, 483
Formate, *t*-butyl chloro-, synthesis of, **6**, 425
Formate, t-butyl chlorothiol, stability of, **6**, 434
Formate, *t*-butyl fluoro-
from, *t*-butyl alcohol, **6**, 422
stability of, **6**, 422
uses of, for amine protection, **6**, 422
Formate, 1-chloroethyl chloro-, reactions, with benzimidazole, **4**, 60
Formate, α-chloroethyl chloro-
from, *N*-dealkylation of tertiary amines, **6**, 428
synthesis of, via radical chlorination, **6**, 428
Formate, *β*-chloroethyl chloro-, from, ethylene carbonate, **6**, 429
Formate, cyano-, from, phenyl iodoformate, **6**, 431
Formate, dichloromethyl chloro-, synthesis of, **6**, 428
Formate, 2,2-dichlorovinyl chloro-, from, trichloroacetaldehyde, **6**, 429
Formate, ethyl
hydrolysis of, **5**, 24
reactions
with arenethiomagnesium bromides, **6**, 84
with boronate esters, **3**, 35
with triphenylsilyllithium, **6**, 125
Formate, ethyl azido-
decomposition of, **2**, 161
reactions, with silyl enol ethers, **3**, 181
Formate, ethyl chloro-
as activating agent, for the synthesis of anhydrides, **5**, 189
reactions
with *O*-alkyl *N*-hydroxythiocarbamates, **5**, 1104
with *N*,*N*-dimethylacetamide, **4**, 882
with ethyleneimine, **6**, 483
with nitro alcohols, **6**, 477
with potassium *O*-butyl dithiocarbonate, **6**, 546
with potassium ethyl isocyanoacetate, **3**, 706
with sodium *O*-ethyl dithiocarbonate, **6**, 476
with tertiary amines, **2**, 27
as reagent
for the synthesis of amides, **5**, 277
for the synthesis of isothiocyanates, **5**, 1023
as starting material, for mixed anhydrides, **5**, 264
for the synthesis of acylated oximes, **5**, 217
Formate, ethyl cyano-, reactions, with nitromethane, **2**, 788
Formate, ethyl fluoro-, synthesis of, **6**, 422
Formate, ethyl thiochloro-, reactions, with thioamides, **5**, 727
Formate, fluorenylmethylchloro-, from, *O*-trimethylsilyl ethers, **6**, 425
Formate, (9-fluorenyl)methylchloro-, bromination of, **6**, 428
Formate, fluoromethyl fluoro-, from, carbonic difluoride, **6**, 423
Formate, methyl, reactions, with α-sulfonyl ethers, **4**, 838
Formate, methyl chloro-
chlorination of, **4**, 59
as dehydrating agent, **3**, 684
from, palladium bis(methoxycarbonyl) complexes, **6**, 428
photochlorination of, **5**, 972; **6**, 230
reactions
with acetone, **5**, 630
with didecyl selenide, **6**, 482
with *O*,*O*-diethyl dithiophosphoric acid salts, **6**, 476
with α-hydrazino enamines, **4**, 986
with lithium sulfide, **6**, 474
with thiols, **6**, 475
with trimethylaminecarboxyborane, **6**, 491
as reagent, for the synthesis of isothiocyanates, **5**, 1023
as starting material, for trichloromethyl chloroformate, **6**, 415
Formate, 4-methyl-3-nitrophenyl chloro-, synthesis of, **6**, 428
Formate, 4-methylphenyl chloro-, nitration of, **6**, 428
Formate, 4-nitrophenyl chloro-, as starting material, for carbamates, **6**, 417
Formate, peroxy chloro-, from, *t*-butyl hydroperoxide, **6**, 430
Formate, phenyl
reactions, with 3-hydroxy-2-amino pyridines, **5**, 290
saponification of, **5**, 506
as starting material, for thioformic acid, **5**, 506
Formate, phenyl bromo-, synthesis of, **6**, 431
Formate, phenyl chloro-
sulfonation of, **6**, 428
use in the synthesis of L-2-oxothiazolidine-4-carboxylic acid, **5**, 81
Formate, phenyl fluoro-
reactions, with phenol, **6**, 422
synthesis of, **6**, 422
Formate, phenyl iodo-, as starting material, for cyanoformate, **6**, 431
Formate, 2-propenyl chloro-, synthesis of, **6**, 429
Formate, 2-pyrrolylfluorothiol, from, fluorocarbonylsulfenyl chloride, **6**, 432

Formate, tertiary fluoro-, stability of, **6**, 425
Formate, trialkylammonium, use in the reduction of vinyl triflates, **1**, 435
Formate, 2,2,2-trichloroethyl bromo-, synthesis of, **6**, 431
Formate, trichloromethyl chloro-
  conversion to phosgene, **6**, 415
  as phosgene alternative, **6**, 428
  as phosgene substitute, **6**, 462
  safety precautions, **6**, 415
  as starting material
    for carbonic difluoride, **6**, 409
    for phosgene, **6**, 413
  synthesis of, **6**, 415
    via photochemical chlorination, **6**, 428
  use as phosgene alternative, **6**, 415
  See also Diphosgene
Formate, 2-(trifluoromethoxy)ethyl fluoro-, from, oxirane, **6**, 424
Formate, trifluoromethyl fluoro-, synthesis of, **6**, 424
Formate, 9-triptycyl iodo-, from, 9-triptycyl hydrogen oxalate, **6**, 431
Formate, vinyl chloro-, from, ethylene glycol bis(chloroformate), **6**, 429
Formate esters
  reactions
    with anhydrides, **6**, 39
    with epoxides, **6**, 71
    with phosphorus pentachloride, **6**, 39
  as starting materials, for trithioorthoformate esters, **6**, 84
Formate esters, aryl, as starting materials, for dichloromethyl ethers, **6**, 40
Formate esters, azido-, synthesis of, **6**, 489
Formate esters, bromo-, synthesis of, **6**, 431
Formate esters, chloro-
  from
    alcohols, **6**, 424
    diphosgene, **6**, 426
    electrophilic substitution of chloroformates, **6**, 427
    phosgene, **6**, 424
    triphosgene, **6**, 426
  reactions
    with alcohols, **6**, 461
    with imines, **6**, 507
    with lithiated trithioorthoformates, **6**, 88
    with primary phosphines, **2**, 429
  safety precautions for, **6**, 424
  synthesis of, **6**, 424
Formate esters, fluoro-
  from
    alcohols, **6**, 422
    alkyl carbamates, **6**, 423
    carbonic fluoride halides, **6**, 422
    1-chloroethyl carbonates, **6**, 423
    chloroformates, **6**, 422
    phenols, **6**, 422
  synthesis of, **6**, 422
Formates
  pyrolysis of, **1**, 774
  as starting materials, for dialkyl carbonates, **6**, 466
  synthesis of, **1**, 940
Formates, acyloxymethyl chloro-, synthesis of, **6**, 427
Formates, alkyl chloro-
  reactions
    with alkanethiols, **6**, 475
    with alkynyllithiums, **1**, 1047
    with alkynylzinc chlorides, **1**, 1047
    with monoalkylphosphines, **6**, 490
    with nitriles, **3**, 687
    with sodium sulfide nonahydrate, **6**, 475
    with sodium telluride, **6**, 481, 482
    with trialkyl phosphites, **6**, 490
    with tris(trimethylsilyl)phosphite, **6**, 491
  as starting materials, for alkyl chlorides, **2**, 25
  use in the synthesis of alkyl thiocarbonates, **6**, 475
  use in the synthesis of anhydrides of dithiocarbonic acids, **6**, 546
Formates, alkyl fluoro-
  from, alkyl carbamates, **6**, 423
  synthesis of, **6**, 422, 422
Formates, *N*-alkyl-*N*-(trifluoromethyl) fluoro-, from, isocyanates, **6**, 239
Formates, amidino-, from, cyanoformates, **5**, 753
Formates, aryl fluoro-, as starting materials, for aryl fluorides, **2**, 615
Formates, azido-
  reactions
    with molybdenum carbyne complexes, **5**, 955
    with tungsten carbyne complexes, **5**, 955
  synthesis of, **6**, 489
Formates, bromo-
  stability of, **6**, 421
  synthesis of, **6**, 431
Formates, chiral chloro-, from, chiral amines, **6**, 426
Formates, chloro-
  as coupling reagents, for the synthesis of sulfonamides, **5**, 342
  from
    alcohols via carbonates, **6**, 426
    O-alkyl *S*-ethyl thiocarbonate, **6**, 427
    2-bromo-2-methylpropanoyl phosphonates, **4**, 921
    17-hydroxysteroids, **6**, 489
  pyrolysis of, **1**, 774
  reactions
    with hydrazine hydrate, **6**, 488
    with sodium azide, **6**, 489
  as starting materials
    for carbamates, **6**, 421
    for fluoroformate esters, **6**, 422
    for urethanes, **6**, 483
  transhalogenation of
    with fluorides, **6**, 422
    with hydrogen bromide, **6**, 431
  use as phosgene alternatives, **6**, 417
  use in the synthesis of hydroxamates, **5**, 327
Formates, α-chloroalkyl chloro-, synthesis of, **4**, 58
Formates, β-chloroalkyl chloro-
  from, oxirane, **6**, 428
  synthesis of, **6**, 429
Formates, ω-chloroalkyl fluoro-, synthesis of, **6**, 423
Formates, α-chlorochloro-, from, bis(trichloromethyl) carbonate, **6**, 416
Formates, α-chloroethyl chloro-, reactions, with mesityl oxide, **4**, 59
Formates, chloromethyl chloro-
  reactions, with thiols, **4**, 60
  as starting materials, for α-chloroalkyl carbonates, **4**, 59
  synthesis of, **4**, 59; **6**, 428
Formates, cyano-, as starting materials, for amidinoformates, **5**, 753
Formates, enol chloro-, synthesis of, **6**, 429
Formates, fluoro-
  fluorination of, **6**, 421
  reactions, with trialkylsilyl enol ether, **6**, 462
  stability of, **6**, 421, 422
  synthesis of, **6**, 462
  uses of, **6**, 421
Formates, α-fluoroalkyl fluoro-, synthesis of, **4**, 59; **6**, 423
Formates, halo-, use of term, **6**, 421
Formates, α-haloalkyl chloro-, synthesis of, **6**, 428
Formates, haloalkyl fluoro-, synthesis of, **6**, 423
Formates, α-haloalkyl halo-, synthesis of, **4**, 58
Formates, imidate chloro-, from, secondary amides, **6**, 429
Formates, iodo-
  stability of, **6**, 431
  synthesis of, **6**, 431
Formates, mono-α-chloroalkyl chloro-, synthesis of, **6**, 428

Formates, oxime chloro-
  (*E*) and (*Z*) forms, synthesis of, **6**, 430
  as starting materials, for nitrilium ions, **3**, 687
  synthesis of, **6**, 429
Formates, perfluoroalkyl chloro-, from, perfluoroalkyl hypochlorites, **6**, 428
Formates, secondary alkyl chloro-, synthesis of, in the vapor phase, **6**, 425
Formates, tertiary chloro-
  decomposition of, to tertiary alkyl halides, **6**, 425
  stability of, **6**, 425
  synthesis of, **6**, 425
Formates, thiochloro-, reactions, with amidines, **5**, 760
Formates, trifluoromethyl thiocyano-, synthesis of, **6**, 477
o-Formates, reactions, with alkynylmagnesium halides, **1**, 1026
Formazan, 3-formyl-1,5-diphenyl-, as starting material, for *gem*-diazoalkenes, **4**, 992
Formazans
  from
    arylhydrazines, **5**, 859
    azo compounds, **5**, 858
    glyoxylic acid arylhydrazones, **5**, 856
    heteroarylacetonitriles, **5**, 858
    heterocyclic benzaldehydehydrazones, **5**, 857
    hydrazones, **5**, 856
    hydrazonoyl halides, **5**, 859
    methylenes, **5**, 857
    oxalocrotonic acid, **5**, 858
    oxalosorbic ester, **5**, 858
  reactions
    with diazonium ions, **5**, 857
    with hydrogen sulfide, **5**, 641
  reduction of, **5**, 847, 855
  synthesis of, **5**, 843, 856; **6**, 653
  unsubstituted at *C*-atom, synthesis of, **5**, 856
Formazans, 1,5-diaryl-3-amino-, synthesis of, **6**, 654
Formazans, 1,5-diaryl-3-arylazo-, synthesis of, **6**, 653
Formazans, 1,5-diaryl-3-nitro-, synthesis of, **6**, 653
Formazans, dihydro-, synthesis of, **5**, 850
Formazans, *C*-heterocyclic, synthesis of, **5**, 858
Formazans, 3-nitroso-, synthesis of, **6**, 655
Formhydrazidoyl bromides, synthesis of, **5**, 680
Formic acid
  oxidative elimination of, **3**, 210
  reactions
    with benzeneselenol, **6**, 92
    with *N,N*-bis(dichloromethylene)alkylphosphonamide, **5**, 1010
    with dialkoxyphosphoryl isocyanide dichlorides, **5**, 1005
    with *N*-dichloromethylenealkylphosphonic amide chloride, **5**, 1010
    with *N*-dichloromethylenephosphoric diester amides, **5**, 1009
    with enamines, **1**, 82
    with formaldehyde, **2**, 310
    with hydrogen sulfide and acetic anhydride, **6**, 85
  as reducing agent, **2**, 310
  as starting material, for trithioorthoformate esters, **6**, 84
  synthesis of, **5**, 49
    reviews, **5**, 24
Formic acid, benzoyl-, synthesis of, **5**, 63
Formic acid anhydrides, chloro-, synthesis of, **6**, 430
Formic acid *O*-benzyl oxime, as starting material, for formic acid oxyimidic chloride, **5**, 825
Formic acid esters, dithiochloro-, as starting materials, for α-dialkylphosphonodithioesters, **5**, 558
Formic acid esters, fluoro-, fluorination of, **6**, 229
Formic acid halothio-*S*-esters, nomenclature for, **6**, 432
Formic acid oxyimidic chloride, from, formic acid *O*-benzyl oxime, **5**, 825
Formic amides, halo-, nomenclature of, **6**, 437
Formic anhydride, acetic-, reactions, with triethyl orthoformate, **6**, 79
Formimidate, ethyl *N*-cyclohexyl-, reactions, with chlorocyanoketene, **4**, 304
Formimidate, ethyl *N*-phenyl-, deprotonation of, **3**, 722
Formimidates, reactions, with dinitromethane, **4**, 994
Formimidates, chloro-, thermal decomposition of, **5**, 977
Formimidates, ethyl *N*-aryl-, synthesis of, **5**, 746
Formimide chloride, *N*-chlorocyano-, as starting material, for *N*-chloroamidines, **5**, 816
Formimidic acids, *N*-(isocyanatocarbonyl)chloro-, synthesis of, **5**, 1008
Formimidoyl azides, 1-cyano-*n*-fluoro-, from, 1-cyano-*N*-fluoroformimidoyl fluorides, **5**, 819
Formimidoyl chloride, *N*-benzoylethoxy-, from, *N*-benzoylurethane, **5**, 656
Formimidoyl chloride, *N*-chlorochloro-, synthesis of, **5**, 809
Formimidoyl chloride, *N*-chloro-*N'*-chlorothio-*N'*-cyanoamino-, synthesis of, **5**, 810
Formimidoyl chloride, *N*-chlorocyano-, synthesis of, **5**, 809
Formimidoyl chloride, *N*-chlorothiochloro-, synthesis of, **5**, 809
Formimidoyl chloride, *N*-ethylamino cyano-, synthesis of, **5**, 658
Formimidoyl chloride, *N*-fluoro cyano-, synthesis of, **5**, 663
Formimidoyl chloride, *N*-fluorofluoro-, synthesis of, **5**, 811
Formimidoyl chloride, *N*-isopropylamino cyano-, synthesis of, **5**, 658
Formimidoyl chloride, *N*-methyl-*C*-(trichlorotitanio)-, synthesis of, **5**, 657
Formimidoyl chloride, *N*-phenyl-*C*-(trichlorotitanio)-, synthesis of, **5**, 657
Formimidoyl chlorides, reactions, with carbamates, **5**, 977
Formimidoyl chlorides, *N*-chloro-, synthesis of, **5**, 809
Formimidoyl fluoride, *N*-bromofluoro-, synthesis of, **5**, 809
Formimidoyl fluoride, 1,1-cyano-*N*-fluoro-, reactions, with diethylamine, **5**, 817
Formimidoyl fluoride, 1-cyano-*N*-fluoro-, synthesis of, **5**, 812
Formimidoyl fluoride, *N*-fluorodifluoroamino-, synthesis of, **5**, 811
Formimidoyl fluorides, 1-cyano-*N*-fluoro-, as starting materials, for 1-cyano-*n*-fluoroformimidoyl azides, **5**, 819
Formohydroximoyl iodide, as starting material, for fulminic acid, **3**, 685
Formonitrile oxide, benzenesulfonyl-, synthesis of, **3**, 686
Formonitrile oxide, bromo-, synthesis of, **3**, 686
Formonitrile oxide, ethoxycarbonyl-
  from
    diethyl nitromalonate, **3**, 685
    ethyl chloro(hydroxyimino) acetate, **3**, 684
Formonitrolic acid, use in the synthesis of fulminic acid, **3**, 685
Formoxyimidic chloride, ethoxycarbonyl, synthesis of, **5**, 819
α-(Formyl-amino)acrylates, ethyl, synthesis of, **3**, 717
Formylaminomethylenation, for the synthesis of α,β-unsaturated isocyanides, **3**, 713
Formyl anion equivalents, for the synthesis of aldehydes, **3**, 20
Formyl anions
  reactions, with ketones, **3**, 35
  reviews, **3**, 29
  as starting materials, for β-hydroxy aldehydes, **3**, 31
  synthesis of, **2**, 575
Formylation
  of arenes, **3**, 91
  of benzofurans, **3**, 103
  of dienes, **3**, 64
  of furans, **3**, 103
  of Grignard reagents, **3**, 90
  of organometallics, **3**, 14

palladium-catalyzed, **3**, 90
of phosphonate anions, **3**, 49
of unsaturated nucleophiles, **3**, 23
Formylation, decarboxylative, for the synthesis of enamines, **2**, 742
Formyl cations, reactions, with arenes, **3**, 101
Formyl chloride, (dimethoxyphosphinyl)-, synthesis of, **6**, 454
Formyl chromanones, oxidation of, **2**, 654
Formyl compounds, chloro-, tautomerization of, **6**, 606
Formyl fluoride
reactions
with arenes, **3**, 93
with perfluoroalkoxides, **4**, 59
Formyloxazetidine, 2-fluoro-, pyrolysis of, **5**, 1011
Formyl peroxide, bisfluoro-
reactions, with caesium fluoride, **6**, 424
synthesis of, **6**, 424
Formyl peroxide, fluoro-, decarboxylation of, **2**, 98
Formyl peroxide, trifluoromethyl fluoro-, synthesis of, **6**, 424
2-Formylphenethyl bromides, reactions, with D-phenylglycinol, **2**, 328
Formyl radicals
decarboxylation of, **1**, 939
as intermediates, **1**, 940
Forster reaction
of α-keto oximes, **3**, 469
of oximes, **3**, 463
for the synthesis of α-diazoalkane sulfones, **5**, 867
for the synthesis of α-diazo ketones, **3**, 193, 469
for the synthesis of α-diazophosphorus compounds, **5**, 870
Four-center transition state, mechanism, for carbon–mercury cleavage, **1**, 65
Four-component condensation, of isocyanides, **3**, 694
Four-membered rings, synthesis of, **1**, 354
Fractional distillation, for the synthesis of alkyl isocyanates, **5**, 980
Fragmentation
decarbonylative, **1**, 1081
decarboxylative, **1**, 1081
mechanism, **1**, 1081
for the synthesis of alkynes, **1**, 1081
for the synthesis of alkynyl-substituted alkyl aldehydes, **3**, 25
for the synthesis of vinyl cations, **1**, 1119
use in the synthesis of alkynyl ketones, **3**, 147
Fragmentation, silyl-directed, mechanism, **1**, 655
β-Fragmentation
of alkoxy radicals, **1**, 929
kinetic study of, **1**, 931
of oxygen-centered radicals, **1**, 929
for the synthesis of lactones, **1**, 931
Fragmentation method
for intermolecular radical addition, **1**, 334
for radical cyclization reactions, **1**, 344
in tandem cyclizations, **1**, 348
Fragmentation–rearrangement, for the synthesis of thiocyanates, **5**, 1112
Free carbenes, reactivity of, **1**, 352
Free radical addition
intermolecular, **1**, 546
intramolecular, **1**, 547
for the synthesis of alkenes, **1**, 546
for the synthesis of primary phosphines, **2**, 428
for the synthesis of secondary phosphines, **2**, 430
for the synthesis of tertiary phosphines, **2**, 439
Free radical-mediated reaction, for the substitution of silicon functions, **1**, 289
Free radicals, synthesis of, **1**, 395
Freon 11
synthesis of, **6**, 220
See also Methane, trichlorofluoro-
Freon 12
synthesis of, **6**, 220, 224
See also Methane, dichlorodifluoro-
Freon 13
synthesis of, **6**, 217
See also Methane, chlorotrifluoro-
Friedel–Crafts catalysts, reactions, with hydrochloric acid, **3**, 92
Friedel–Crafts reaction
acylation
of arenes, **2**, 654
with carboxylic acids, **3**, 280
of indoles, **3**, 309
of mesitylene, **5**, 116
with nitriles, **3**, 282
for the synthesis of aromatic ketones, **3**, 278
for the synthesis of α,β-unsaturated ketones, **3**, 230, 240
acylation–Baeyer–Villiger oxidation tandem protocol, for the synthesis of phenyl esters, **2**, 656
of acyl isothiocyanates, **5**, 631
alkylation
of alkyl–oxygen leaving groups, **1**, 475
of alkynes, **1**, 1025
of alkynylsilanes, **1**, 1003
of arenes, **1**, 486
for carbon–carbon bond synthesis, **1**, 106
of carbonium ions, **1**, 199
of chromium tricarbonyl complexes, **1**, 202
selectivity of, **1**, 181
for the substitution of hydrogen, **1**, 492
for the substitution of sulfur functions, **1**, 241
use of alkyl sulfones in, **1**, 484
use of epoxides in, **1**, 212
allylation, of 2-phenyloxetane, **1**, 229
aluminum trichloride-catalyzed, **2**, 718
for the arylation of aldehydes, **3**, 24
cyclizations, intramolecular, **1**, 216
in the displacement of alkyl alcohols, **1**, 180
methylation, of toluene, **1**, 247
of nitroalkanes, **1**, 251
of phosgene, for the synthesis of aromatic acid chlorides, **5**, 15
of phosphorus tribromide, with benzene, **2**, 832
of phosphorus trichloride, with benzene, **2**, 832
of pyrrole, with carbon disulfide, **5**, 561
for the substitution of alkyl halides, **1**, 462
sulfonylation, of arenes, **2**, 715
for the synthesis of aryl sulfones, **2**, 715
for the synthesis of thioketones, **3**, 364
for the synthesis of α,β-unsaturated ketones, **3**, 211
thioacylation, with *O*-phenyl chlorothioformate, **5**, 513
(trichloromethyl)thiothioacylations, **6**, 235
triflation, **6**, 235
of *S*-trifluoromethyl fluorodithioformic acid ester with benzenes, **5**, 561
(trifluoromethyl)thiothioacylations, with antimony(V) fluoride, **6**, 235
use in cleavage reactions, **1**, 57
Friedlander quinoline synthesis, for the synthesis of imines, **3**, 406
Fries rearrangement
for the synthesis of benzothiophenyl ketones, **3**, 306
for the synthesis of phenolic ketones, **3**, 297
for the synthesis of α,β-unsaturated ketones, **3**, 269
of thiophenes, **3**, 305
use in the synthesis of phenols, **2**, 656
Fries rearrangement, metallo-, definition of, **2**, 657
Fritsch–Buttenberg–Wiechell rearrangement
of alkenyl carbamates, **1**, 1071
of alkenyl triflates, **1**, 1071
bases for, **1**, 1068
mechanism, **1**, 1067, 1144
Frontier molecular orbital (FMO)

analysis
of conjugate addition, **1**, 301
of electrocyclic addition, **1**, 351
theory
in [3+2] additions, **1**, 365
and carbonyl ylides, **3**, 315
and radical addition, **1**, 323
Fructopyranose compounds, from, di-*O*-isopropylidenes, **4**, 222
β-L-Fructosylamine, use in the synthesis of *α*-aminophosphonic acids, **4**, 462
Fuel cells, electrolytes for, **6**, 278
Fujimoto–Belleau reaction, for the synthesis of *α*,*β*-unsaturated ketones, **3**, 219
Fullerenes, hydroboration of, **2**, 43
Fullerols, synthesis of, **2**, 43
Fulminates, as starting materials, for oxyimidic thioesters, **5**, 828
Fulminic acid
from, formohydroximoyl iodide, **3**, 685
synthesis of, **3**, 685
use in the synthesis of oximidic chlorides, **5**, 819
Fulvene, azo-, as starting material, for diazonium salts, **2**, 800
Fulvenes
from, alkynes, **1**, 1014
reactions, with lithium alkyls, **1**, 862
synthesis of, **4**, 835
Fulvenes, 6,6-bis(dimethylamino)-, synthesis of, **4**, 974
Fumarate, diethyl, reactions, with 1,3-cyclohexadiene, **6**, 343
Fumarate, dimethyl, synthesis of, **1**, 456, 507
Fumarate, dimethyl chloro-, reactions, with diphenyldiazomethane, **4**, 109
Fumarates, isothiocyanato-, synthesis of, **5**, 1044
Fumaric acid
from, furfural, **5**, 102
reactions, with bromine, **5**, 60
Fumaric acids, synthesis of, **1**, 446
Fumaric dichloride, difluoro-, reactions, with sodium tetracarbonylcobaltate, **4**, 821
Funabiki's reagent, for the oxidation of arenes, **2**, 650
Functional groups, oxygen-based, attached to *sp* carbon, synthesis of, **2**, 1021
Functional groups, selenium-based
attached to *sp* carbon, synthesis of, **2**, 1036
synthesis of, **2**, **277–295**, **705–736**
Functional groups, sulfur-based, synthesis of, **705–736**
Functional groups, tellurium-based
attached to *sp* carbon, synthesis of, **2**, 1036
synthesis of, **2**, **277–295**, **705–736**
Functionalization, of carbene complexes, **5**, 955
Fungal metabolites
from, sugars, **1**, 174
Marcfortine A, **5**, 572
Fungicides
carbamoyl fluorides as intermediates for, **6**, 437
trithioorthoformates as, **6**, 273
Furan, 2,5-bis(trimethylsilyl)-, photolysis of, **2**, 1091
Furan, 2,3-dibromo-, metallation of, **3**, 103
Furan, 2,5-dicarbazido-, reactions, with toluene, **5**, 990
Furan, 3,4-dicyano-, reactions, with dimethyl acetylenedicarboxylate, **1**, 1083
Furan, dihydro-, reactions, with butyllithium, **2**, 1023
Furan, 2,3-dihydro-, deprotonation of, **4**, 949
Furan, 2,5-diisocyanato-, synthesis of, **5**, 990
Furan, 2,5-dimethoxytetrahydro-, reactions, with benzotriazole, **4**, 309
Furan, 3-formyl-2-hydroseleno-2-phenyl-, condensation of, **5**, 521
Furan, 3-iodo-, palladium-catalyzed formylation of, **3**, 103
Furan, 2-methyl-, reactions, with iminium salts, **2**, 316
Furan, 3-nitro-2,5-dihydro-, reactions, with ammonia, **2**, 307
Furan, tetrahydro- (THF)
as electrophile, **3**, 163
fragmentation of, **2**, 56
reactions
with alkenes, **1**, 365
with *n*-butyllithium, **1**, 135
with carbonic chloride fluoride, **6**, 424
with ceric triethylammonium nitrate, **4**, 201
with dialkylmagnesiums, **1**, 305
with diborane, **2**, 111
with lithium aluminum hydride, **1**, 3
with lithium naphthalenide, **1**, 405
with pentyl nitrite, **1**, 442
ring-opening of, **1**, 186; **6**, 387
as solvent, for the synthesis of acylsilanes, **5**, 419
use in the synthesis of alkenylmagnesiums, **1**, 1092
use in the synthesis of alkynylantimony compounds, **2**, 1081
use in the synthesis of amino aldehydes, **3**, 44
use in the synthesis of lithium carbenoids, **6**, 62
use in the synthesis of organonickel cyanides, **5**, 1137
use in the synthesis of secondary phosphines, **2**, 432
use in the synthesis of tris(trimethylsilylmethyl)lithium, **6**, 390
Furan, 2-vinyl-, as starting material, for 2-furylpropionitrile, **3**, 615
3-Furanamine, *trans*-tetrahydro-4-nitro-, reduction of, **2**, 307
Furan-2-carboxylate, methyl, reactions, with disilenes, **6**, 125
Furan-2-carboxylic acid, reduction of, **1**, 100
3,4-Furandiamine, *trans*-tetrahydro-, from, 2,5-dihydrofurans, **2**, 307
Furan-2,3-dione, 4-benzoyl-5-phenyl-, reactions, with oximes, **5**, 217
Furan-2,3-dione, 4,4-dimethyl-5-(*i*-propyl)dihydro-, saponification of, **5**, 63
Furandiones, synthesis of, **3**, 246
Furan dithioic acid *S*-aryl esters, synthesis of, **5**, 561
Furanecarboxhydroxamoyl chlorides, synthesis of, **5**, 665
Furanocembranolides, synthesis of, **3**, 103
Furanoid epoxides, deoxygenation of, **1**, 587
Furanones, synthesis of, **1**, 568; **3**, 320; **4**, 824
Furanones, acetoxy, from, furans, **4**, 206
Furanones, alkoxy, synthesis of, **4**, 211
Furanones, alkylidene, synthesis of, **1**, 582
Furanones, hydroxy, reactions, with alcohols, **4**, 211
3-Furanones, reactions, with aldehydes, **3**, 241
Furan-3-ones, synthesis of, **3**, 246, 323
3(2*H*)-Furanones, synthesis of, **3**, 244, 245
3(2*H*)-Furanones, 5-disubstituted amino-2,2,4-trichloro-, synthesis of, **5**, 996
5(*H*)Furan-2-ones, 4-azido-3-halo-5-methoxy-, as starting materials, for halocyanoketenes, **3**, 543
Furanons, tetrahydro-, reactions, with *γ*-methoxy-*γ*-lithioallenes, **1**, 971
Furanoyl compounds, synthesis of, **5**, 468
Furans
acylation of, **3**, 302
deprotonation of, **1**, 278
in Diels–Alder reactions, **2**, 667
formylation of, **3**, 103
from
carbenes, **3**, 321
carbonyl ylides, **3**, 318
lithiation of, **3**, 103
oxidation of, **3**, 221; **4**, 206
photooxygenation of, **3**, 327
reactions
with butenolides, **5**, 171
with *n*-butyllithium, **4**, 846
with dienes, **1**, 521
with iminium chlorides, **2**, 316

with α-ketocarbenes, **3**, 65, 232
with lead(IV) acetate, **4**, 206
reduction of, **1**, 90
ring-opening of, **1**, 980; **3**, 62, 64
as starting materials
for acetoxy furanones, **4**, 206
for alkynylsilanes, **2**, 1091
synthesis of, **2**, 927; **3**, 321; **4**, 961
Furans, acetoxy tetrahydro-, from, allyl enol ethers, **4**, 208
Furans, 2-alkoxy-3,4-bisalkoxymethyl-, hydrolysis of, **3**, 227
Furans, alkyl-, reactions, with aluminum trichloride, **3**, 302
Furans, 3-alkyl-, acylation of, **3**, 303
Furans, α-alkylidene-, synthesis of, **2**, 982
Furans, 2-alkyl-4-trimethylsilyl-, from, β-oxosilyl epoxides, **2**, 945
Furans, alkynyl-, intramolecular Diels–Alder reaction of, **2**, 668
Furans, 2,3-biscarbomethoxy-, synthesis of, **3**, 318
Furans, 2,5-bis(trimethylsilyloxy)-, use in the synthesis of hydroquinones, **2**, 668
Furans, chromanodihydro-, from, propargyl ethers, **3**, 315
Furans, cyano-
reduction of, **3**, 102
synthesis of, **3**, 672
Furans, 5-cyano-, from, 2-substituted furans, **3**, 670
Furans, 2,5-dialkyl-, acylation of, **3**, 302
Furans, dihydro-
synthesis of, **4**, 961
via acetate pyrolysis, **1**, 774
use in the synthesis of acetals, **4**, 186
Furans, 2,5-dihydro-
nitromercuration of, **2**, 307
as starting materials, for *trans*-tetrahydro-3,4-furandiamine, **2**, 307
Furans, 4,5-dihydro-, synthesis of, **1**, 1058
Furans, 3,4-distannyl-, monoacylation of, **3**, 303
Furans, 2,5-disubstituted, ring-opening of, **3**, 221
Furans, epoxy tetrahydro-, ring-opening of, **2**, 84
Furans, 3-ketotetrahydro-, synthesis of, **1**, 410
Furans, lithiated, acylation of, **3**, 303
Furans, 5-lithio-2,3-dihydro-, reactions, with trialkylboranes, **4**, 935
Furans, methylene tetrahydro-, synthesis of, **1**, 548
Furans, silyl-, synthesis of, **2**, 945
Furans, 2-silyldihydro-, synthesis of, **2**, 936
Furans, 2-silyloxy-, oxidation of, **4**, 206
Furans, stannyl-, reactions, with allyl chlorides, **3**, 303
Furans, 2-substituted, as starting materials, for 5-cyanofurans, **3**, 670
Furans, tetrahydro-
cleavage of, **2**, 553
elimination of, **1**, 1072
synthesis of, **1**, 90, 289, 506; **2**, 93; **3**, 317; **4**, 168; **5**, 142
Furans, thiophenyl α-substituted tetrahydro-, reactions, with potassium 3-aminopropylamide, **1**, 1085
Furans, 2-tri-*n*-butylstannyl-, oxidation of, **4**, 206
Furan-2-selones, synthesis of, **5**, 521
Furan-2-selones, 3-aminomethylene-5-phenyl-, synthesis of, **5**, 521
Furan xanthate, tetrahydro-, cyclization of, **1**, 175
Furanyl acid chlorides, use in the synthesis of furanyl ketones, **3**, 303
Furanyl ethers, tetrahydro-, synthesis of, **4**, 188
Furanyl ethers, 2-tetrahydro-, synthesis of, **4**, 201
Furanyls, 2-diphenylacetoxytetrahydro-, reactions, with alcohols, **4**, 200
Furazan *N*-oxides
ring-opening of, **5**, 1013
as starting materials, for nitrile oxides, **3**, 685
synthesis of, **3**, 683, 685
Furazans, as starting materials, for nitrile oxides, **3**, 685
Furfural
reactions, with sodium hydroxide, **5**, 74
as starting material, for fumaric acid, **5**, 102
Furfural, 5-methyl-, oxidation of, **5**, 74
Furfuraldehyde, 5-methyl-, oxidation of, **5**, 102
Furfuryl chloride, tetrahydro-, reactions, with sodium amide, **1**, 1066
2-Furoate, methyl 3-methyl-, as starting material, for 3-methyl-2-furoic acid, **5**, 62
2-Furoates, ethyl, silylation of, **2**, 1091
Furofurans, synthesis of, **3**, 320
Furoic acid, 5-methyl-, synthesis of, **5**, 74, 102
2-Furoic acid, synthesis of, **5**, 74
2-Furoic acid, 3-methyl-, from, methyl 3-methyl-2-furoate, **5**, 62
3-Furoic acid, reactions, with bis(*N*-methylpiperazinyl)aluminum hydride, **3**, 102
Furoic acids, reduction of, **3**, 102
Furo[3,4-*d*]pyridines, 4-anilino-5-oxo-2-phenyl-7-thioxo-5,7-dihydro-, synthesis of, **5**, 529
Furoxans, synthesis of, **5**, 1003
Furylium salts, dihydro-, synthesis of, **2**, 1007
Fusaric acid, synthesis of, **5**, 116
Fused phenols, synthesis of, **2**, 668
Fused rings, synthesis of, via radical cyclization, **1**, 338
Fuson's enols
chemistry of, **2**, 643
synthesis of, **2**, 642

G

Gabriel synthesis
of amines, problems, **3**, 187
modification of, **2**, 299
for the synthesis of amines, **2**, 299
for the synthesis of α-aminoalkylphosphorus compounds, **4**, 473
for the synthesis of arylamines, **2**, 753
for the synthesis of primary amines, **2**, 308
Gadolinium compounds, reactions, with phosphorus ylides, **4**, 578
β-D-Galactoside, methyl 2,6-di-*O*-methyl-3,4-*O*-thiocarbonyl-, as starting material, for iodothiocarbonates, **6**, 477
Galactosylamine, use in the synthesis of α-aminophosphonic acids, **4**, 462
Gallane, chlorodi-*t*-butyl-, reactions, with lithium di-*t*-butylarsenide, **2**, 511
Gallane, tris(trifluoromethyl)-, synthesis of, **6**, 245
Gallium, (iodomethyl)diethyl-, synthesis of, **4**, 153
Gallium, trialkyl-, reactions, with *N*,*N*'-dimethylacetamidine, **5**, 780
Gallium, triisobutyl-, thermal decomposition of, **2**, 597
Gallium, trimethyl-
reactions
with hydrogen cyanide, **5**, 1143
with trimethylgermanyl cyanide, **5**, 1143
Gallium, tris(trimethylsilylmethyl)-, synthesis of, **4**, 659
Gallium(I) iodide, as starting material, for alkylgallium diiodides, **2**, 597
Gallium arsenide, use in the synthesis of semiconductors, **2**, 510
Gallium chloride, bis(trimethylsilylmethyl)-, synthesis of, **4**, 659
Gallium complexes, alkyl-, structure of, **6**, 395
Gallium cyanide, dimethyl-, synthesis of, **5**, 1143
Gallium dichloride, trimethylsilylmethyl-, synthesis of, **4**, 659
Gallium diiodides, alkyl-, from, gallium(I) iodide, **2**, 597
Gallium hydride, diisobutyl-, reactions, with alkenes, **2**, 597
Gallium hydride dihalide, reactions, with alkenes, **2**, 597
Galliums
for the opening of epoxides, **1**, 205
reactions, with phosphorus ylides, **4**, 587
Galliums, alkyl-, synthesis of, **2**, 597
Galliums, alkynyl-, reactions, with oxiranes, **1**, 1024
Galliums, imidoyl-, synthesis of, **5**, 803
Galliums, trihalomethyl-, synthesis of, **6**, 245
Gallium trichloride
reactions
with organolithiums, **2**, 597
with trimethylsilylmethylmagnesium chloride, **4**, 659
with ylides, **4**, 397
Gallium trihalides
reactions
with arsinosilanes, **2**, 511
with stibinosilanes, **2**, 511
Gascardic acid, synthesis of, **1**, 710
Gatterman amide synthesis, mechanism, **5**, 298
Gattermann–Koch reaction
of alkyl benzenes, **3**, 94
for the synthesis of aldehydes, **3**, 92
for the synthesis of benzaldehydes, **3**, 93
Gattermann reaction, for the synthesis of aldehydes, **3**, 92
Gatterman reaction
for the synthesis of *N*-arylarylamides, **5**, 304
for the synthesis of aryl chlorides, **2**, 621
for the synthesis of aryl halides, **2**, 609
Gauche effect, review, **1**, 849
Geminal difluoro compounds, synthesis of, **6**, 423
Geminal dihalides
reactions, with alkoxides, **4**, 189
as starting materials, for acetals, **4**, 189
synthesis of, via *N*-bromosuccinimide bromination, **5**, 410
Geometrical isomers, of 1-phosphabutatrienes, **3**, 607
(±)-Gephyrotoxin 223AB, synthesis of, **2**, 329
Geraniols
acylation of, **5**, 135
synthesis of, **1**, 77, 1105
7-Germabenzonorbornadienes
decomposition of, **2**, 947
reactions, with thiacycloheptynes, **2**, 947
Germacranolides, synthesis of, **1**, 813
Germacycles, synthesis of, **4**, 626
1-Germacyclobutane, 1,1-diethyl-
reactions, with
phenyl(bromodichloromethyl)mercury, **1**, 291
ring-expansion of, **1**, 291
Germacyclobutanes, pyrolysis of, **4**, 627
Germacyclopentadienes, synthesis of, **2**, 947
1-Germacyclopentane, 1,1-diethyl-2-dichloro-, synthesis of, **1**, 291
1-Germa-3-cyclopentenes, synthesis of, **2**, 949
Germaethenes
dimerization of, **6**, 711
from, electrophilic cryptocarbene, **6**, 711
reactions
with methanol, **6**, 188
with organolithiums, **6**, 192
as starting materials, for 1,3-diboretanes, **6**, 188
use in the synthesis of metalloid functions, **6**, 187
Germaethenes, *C*,*C*-dihetero-substituted
stability of, **6**, 711
synthesis of, **6**, 710
Germafulvene, diethyl-, as intermediate, in the synthesis of 1,1-bisgermylalkanes, **4**, 627
Germagermenes, synthesis of, **5**, 927
Germane, (acetoxymethyl)trimethyl-, synthesis of, **4**, 362
Germane, acetyl triethyl-, synthesis of, **5**, 432
Germane, acetyl triphenyl-, synthesis of, **5**, 431
Germane, amino-, use in the synthesis of *O*-germanium enolates, **2**, 700
Germane, benzoyl triethyl-, synthesis of, **5**, 432
Germane, benzoyl trimesityl-, synthesis of, **5**, 432
Germane, benzoyl trimethyl-, synthesis of, **5**, 431
Germane, benzylthiotrimethyl-, reactions, with *t*-butyllithium, **4**, 371
Germane, bis(chloromethyl)dimethyl-, reactions, with sodium acetate, **4**, 362
Germane, bis(hydroxymethyl)dimethyl-, chlorination of, **4**, 141
Germane, bis(iodomethyl)-, as starting material, for thiagermatanes, **4**, 372
Germane, bromotriethyl-
reactions
with imides, **5**, 391
with silver phthalimide, **5**, 391
Germane, bromotrimethyl-
reactions
with carbon nucleophiles, **4**, 361
with trimethylgermylacetonitrile, **6**, 185
Germane, bromo(trivinyl)-, synthesis of, **4**, 141

Germane, (but-2-en-1-ynyl)trimethyl-, reactions, with areneselenyl chlorides, **4**, 948
Germane, chlorodimethylphenyl-, reactions, with bromotris(trimethylsilyl)methyllithium, **6**, 292
Germane, chlorodiphenyl[tris(trimethylsilyl)methyl]-, reactions, with caesium fluoride, **6**, 710
Germane, chloromethyltrimethyl-, reactions, with dimethylbismuthinosodium, **4**, 599
Germane, chloromethyl(trimethyl)-, reactions, with lithium, **4**, 626
Germane, (chloromethyl)trimethyl-, reactions, with potassium acetate, **4**, 362
Germane, chlorotrimethyl-
  reactions
    with arsinosilanes, **2**, 509
    with di-Grignard reagents, **4**, 626
    with 1-methoxy-1-vinyllithium, **4**, 929
  as starting material, for α-alkoxygermanes, **4**, 361
Germane, diiodobis(trifluoromethyl)-, synthesis of, **6**, 244
Germane, dimethylaminotrimethyl-
  reactions
    with diazoacetates, **3**, 483
    with diazomethane, **3**, 483
Germane, dimethylarsinotrimethyl-, synthesis of, **2**, 509
Germane, dimethyl(divinyl)-
  hydroboration of, **4**, 646
  reactions, with borane dimethyl sulfide, **4**, 646
Germane, fluorovinyl-, reactions, with *t*-butyllithium, **3**, 503
Germane, (hydroxymethyl)trimethyl-, synthesis of, **4**, 362
Germane, (iodomethyl)trimethyl-
  reactions, with lithium phenyl selenide, **4**, 377
  synthesis of, **4**, 141
Germane, pivaloyl triethyl-, synthesis of, **5**, 432
Germane, tetrachloro-, as starting material, for hydroxymethylgermanes, **4**, 362
Germane, tetraisocyano-, synthesis of, **3**, 711
Germane, tetrakis(trimethylsilylmethyl)-, reactions, with bromine, **4**, 641
Germane, tetraphenyl-, reactions, with bromine, **1**, 449
Germane, tetravinyl-, chlorination of, **4**, 141
Germane, triacyl-, synthesis of, **5**, 432
Germane, tribromo(bromomethyl)-, synthesis of, **4**, 142
Germane, trichloro-
  reactions
    with alkenes, **6**, 62
    with chloromethyl methyl ether, **4**, 361
Germane, trichloro(chloromethyl)-
  reactions, with nucleophiles, **4**, 142
  reduction of, **4**, 141
  synthesis of, **4**, 142
Germane, trichloro(iodomethyl)-, synthesis of, **4**, 142
Germane, trichloro(methyl)-, chlorination of, **4**, 142
Germane, triethyl-, reactions, with phenylthioethynes, **4**, 940
Germane, trifluoroacetyl(triethyl)-, synthesis of, **5**, 432
Germane, triiodo(iodomethyl)-, synthesis of, **4**, 142
Germane, triiodo(trifluoromethyl)-, synthesis of, **6**, 244
Germane, trimethyl-, reactions, with perfluorocyclobutenes, **4**, 141
Germane, trimethylbromo-, reactions, with stannylacetone, **2**, 700
Germane, trimethylisocyano-, synthesis of, **3**, 711
Germane, triphenyl-, reactions, with silylallenes, **4**, 639
Germane, triphenylallenyl-, synthesis of, **2**, 946
Germane, triphenylgermylacetyl triphenyl-, synthesis of, **1**, 290; **5**, 432
Germane, tris(pentafluorophenyl)-, reactions, with triethylbismuthine, **2**, 509
Germane, tris(trimethylsilylmethyl)-, reactions, with diethylmercury, **4**, 641
Germane, tris(trimethylsilylmethyl)bromo-, synthesis of, **4**, 641
Germane, vinyltrimethyl-, reactions, with arenesulfenyl chlorides, **4**, 372
Germanediyls, use in the synthesis of *C*-germylphosphaalkenes, **6**, 689
Germane hydrides, reactions, with haloalkenes, **4**, 141
Germanes
  from
    bis(trimethylgermyl)diazomethane, **5**, 927
    germylenes, **6**, 187
  photolysis of, **5**, 927
  pyrolysis of, **5**, 927
  reactions, with cobalt complexes, **6**, 404
  silylation of, **6**, 292
  synthesis of, **5**, 431
Germanes, acyl-
  as acyl radical precursors, for the synthesis of ketones, **3**, 136
  from, germyl dienol ethers, **5**, 431
  synthesis of, **1**, 290; **2**, 950; **5**, 431
Germanes, alkanoyl-, synthesis of, **5**, 432
Germanes, alkenyl-
  from
    alkenyl metals, **2**, 946
    alkynes, **2**, 946
    alkynylgermanes, **2**, 948
  synthesis of, **2**, 945
Germanes, alkoxy-
  reverse Brook rearrangement of, **4**, 361
  as starting materials, for α-hydroxygermanes, **4**, 361
Germanes, α-alkoxy-
  from, chlorotrimethylgermane, **4**, 361
  synthesis of, **4**, 360
Germanes, alkyl arylsilyl-, as starting materials, for germyl anions, **4**, 361
Germanes, alkyl(α-haloalkyl)-, synthesis of, **4**, 141
Germanes, alkynyl-
  from, aminoalkynylstannanes, **2**, 1093
  hydroalumination of, **2**, 948, 967
  hydroboration of, **2**, 948
  hydrogenation of, **2**, 948
  reduction of, **1**, 1000
  as starting materials, for alkenylgermanes, **2**, 948
  synthesis of, **2**, 1092
    reviews, **2**, 1092
Germanes, allyl-, arylation of, **1**, 490
Germanes, α-amino-, synthesis of, **4**, 519
Germanes, aroyl-, synthesis of, **5**, 432
Germanes, arsino-, synthesis of, **2**, 509
Germanes, aryl-
  reactions, with electrophiles, **1**, 448
  reduction of, **1**, 448
  as starting materials, for arenes, **1**, 448
  synthesis of, **2**, 948
Germanes, arylalkyl-, photochemical behavior of, **1**, 449
Germanes, aryl(α-haloalkyl)-, synthesis of, **4**, 141
Germanes, aryltrimethyl-, halodegermylation of, **1**, 449
Germanes, bis(chloromethyl)-, synthesis of, **4**, 141
Germanes, bis(halomethyl)dialkyl-
  reactions
    with hydrogen sulfide, **4**, 372
    with sodium sulfide, **4**, 372
Germanes, bis(silyl)-, photolysis of, **2**, 949
Germanes, bromo-
  displacement of bromide from, **2**, 1093
  reactions, with chloromethyllithium, **4**, 141
Germanes, bromotrialkyl-, reactions, with lithiated trimethylsilylmethyl methyl ether, **6**, 131
Germanes, bromo(triphenyl)-, synthesis of, **4**, 663
Germanes, carbamoyl-, synthesis of, **6**, 515
Germanes, chloro-
  alkylation of, **4**, 141
  reactions, with silylphosphines, **2**, 473
  synthesis of, **4**, 519
Germanes, (chloromethyl)-
  halogen exchange in, **4**, 141

reactions, with amides, **4**, 521
as starting materials
for Grignard reagents, **4**, 371
for (iodomethyl)germanes, **4**, 141
Germanes, α,α-(dibromomethyl)-, reduction of, **4**, 141
Germanes, α,α-dichloroalkyl-, synthesis of, **6**, 62
Germanes, dienyl-, synthesis of, **2**, 946
Germanes, diiodobis(trifluoromethyl)-, synthesis of, **6**, 244
Germanes, furyl-, from, aryllithiums, **2**, 949
Germanes, halo-
nucleophilic displacement of, **4**, 361
reactions
with *t*-butyllithium, **3**, 503
with iminolithiums, **2**, 815
with lithium phosphides, **2**, 474
with phosphorus ylides, **4**, 574
with silylphosphines, **2**, 474
Germanes, β-halo-, decomposition of, **1**, 656
Germanes, α-haloalkyl-, synthesis of, **4**, 141
Germanes, halo(α-haloalkyl)-
alkylation of, **4**, 141
synthesis of, **4**, 142
Germanes, halomethyl-
reactions
with oxygen nucleophiles, **4**, 362
with sulfur nucleophiles, **4**, 371
Germanes, (halomethyl)trimethyl-, displacement of, **4**, 663
Germanes, hexaorganosilyl-, reactions, with aldehydes, **5**, 431
Germanes, hydro-, use in the synthesis of alkenylgermanes, **2**, 946
Germanes, α-hydroxy-
from, alkoxygermanes, **4**, 361
synthesis of, **4**, 360; **5**, 431
Germanes, β-hydroxyacyl-, synthesis of, **5**, 432
Germanes, hydroxymethyl-, from, tetrachlorogermane, **4**, 362
Germanes, (hydroxymethyl)-, reactions, with thionyl chloride, **4**, 141
Germanes, imidoyl-, synthesis of, **5**, 794
Germanes, (iodomethyl)-, from, (chloromethyl)germanes, **4**, 141
Germanes, oligo-, synthesis of, **2**, 949
Germanes, organobismuthino-, synthesis of, **2**, 509
Germanes, β-oxygenated, elimination of, **1**, 656
Germanes, pentafluorobenzoyl(triethyl)-, synthesis of, **5**, 432
Germanes, 1-phosphinoalkyl-, synthesis of, **4**, 568
Germanes, 1-phosphonatoalkyl-, synthesis of, **4**, 574
Germanes, 1-phosphonioalkyl-, synthesis of, **4**, 574
Germanes, 1-phosphonitoalkyl-, synthesis of, **4**, 568
Germanes, silyl-, reactions, with boron trichloride, **4**, 639
Germanes, β-substituted, as starting materials, for alkenes, **1**, 656
Germanes, tetraalkynyl-, synthesis of, **2**, 1092
Germanes, tetrakis(dialkylamino)-
from, amines, **2**, 418
synthesis of, **2**, 418
Germanes, thienyl-, from, aryllithiums, **2**, 949
Germanes, triiodo(trifluoromethyl)-, synthesis of, **6**, 244
Germanes, triphenyl-, synthesis of, **1**, 449
Germanes, vinyl-
bromination of, **4**, 141, 811
electrophilic sulfenylation of, **4**, 372
from, alkynes, **1**, 452
reactions
with arene diazonium tetrafluoroborates, **1**, 452
with bromine, **1**, 452
with electrophiles, **1**, 451
with halogens, **4**, 141
with iodine, **1**, 452
with organolithiums, **4**, 663
reduction of, **1**, 451
as starting materials
for alkenes, **1**, 451
for vinyl bromides, **2**, 624
for vinyl iodides, **2**, 630
Germanethiones, stability of, **2**, 949
Germanium
reactions, with trichloro(chloromethyl)silane, **4**, 639
substitution of, **1**, 489
synthesis of functions, **4**, 1063; **5**, **409–433**
bearing a germanium and a boron, **4**, 646
bearing germanium and metals, **6**, 202
bearing a silicon and a germanium, **4**, 636
bearing three germaniums, **6**, 185
bearing two germaniums, **4**, 625
two silicon functions, **6**, 292
Germanium, dichlorodimethyl-, reactions, with phenyl isocyanide, **2**, 949
Germanium, doubly-bonded, synthesis of functions, **6**, **677–724**
Germanium, *N*-phenylbenzimidoyl triphenyl-, from, *N*-phenylbenzimidoyl chloride, **5**, 794
Germanium, tetra(trifluoromethyl)-, reactions, with trimethyltin hydride, **6**, 65
Germanium(II) chloride, reactions, with bis(diphenylphosphinyl)methane, **6**, 169
Germanium(II) iodide, reactions, with trifluoroiodomethane, **6**, 244
Germanium acetamides, trialkyl-, as starting materials, for *N*-germylamides, **5**, 390
Germanium alkoxides, reactions, with carbon suboxide, **1**, 65
Germanium alkyls, perfluoro-, synthesis of, **6**, 62
Germanium amides, synthesis of, **5**, 389
Germanium amides, trialkyl-
rearrangement of, **5**, 390
use in the substitution of amides, **5**, 388
Germanium anions, as intermediates, for the synthesis of α-germylsilylalkanes, **4**, 638
Germanium bromide, tricyclohexyl-, reactions, with phenyllithium, **2**, 949
Germanium bromide, trimethyl-, use in the synthesis of germanium amides, **5**, 389
Germanium bromides, triaryl-, reactions, with lead cyanide, **5**, 1136
Germanium carbodiimide, synthesis of, **5**, 1081
Germanium–carbon bonds, synthesis of, **4**, 625
Germanium chloride, trimethyl-
reactions
with dilithiomethane, **6**, 388
with silenes, **6**, 387
Germanium chlorides, reactions, with α-silylated organometallics, **6**, 186
Germanium chlorides, alkyl-, reactions, with *N*-lithioimines, **3**, 457
Germanium compound, trimethyl-, for the synthesis of alkenes, **1**, 680
Germanium cumulenes, synthesis of, **1**, 988
Germanium cyanide, triethyl-, synthesis of, **5**, 1136
Germanium cyanides, trialkyl-, synthesis of, **5**, 1136
Germanium cyanides, triaryl-, synthesis of, **5**, 1136
Germanium cyanides, triphenyl-, synthesis of, **5**, 1136
Germanium dichloride, bis(trichlorosilylmethyl)-, synthesis of, **4**, 639
Germanium dichloride, dimethyl-, reactions, with diorganolithiums, **4**, 636
Germanium dihydro compounds, phthalocyanato-, reactions, with cyanamide, **5**, 1083
Germanium diiodide, reactions, with diiodomethane, **4**, 628
Germanium dimethylamine, trimethyl-, reactions, with benzoyl *t*-butylcarbodiimide, **5**, 390
O-Germanium enolates, synthesis of, **2**, 700
Germanium ethyl sulfonate, triethyl-, reactions, with potassium cyanide, **5**, 1136

Germanium functions
doubly-bonded, synthesis of, **3**, 503
elimination of, **1**, 1080
substitution of, **1**, 290, 1033
Germanium groups, elimination of, **1**, 656
Germanium halides
reactions
with lithium compounds, **6**, 388
with trifluorovinylmagnesium halides, **4**, 811
with trisilyllithiomethanes, **6**, 387
for the synthesis of germanium–nitrogen bonds, **2**, 814
Germanium halides, trialkyl-, alkynylation of, **2**, 1093
Germanium hydride, tributyl-, use in the synthesis of radicals, **1**, 936
Germanium hydrides
for the mediation of radical reactions, **1**, 331
reactions, with aldimines, **4**, 521
use in radical cyclization reactions, **1**, 341
Germanium hydrides, trialkyl-, use in radical cyclization, **1**, 342
Germanium isocyanate, tributyl-, as starting material, for tributylgermylphthalimide, **5**, 391
Germanium isocyanates, synthesis of, **5**, 1016
Germanium–nitrogen bonds
reactions
with isocyanates, **2**, 815
with isothiocyanates, **2**, 815
synthesis of compounds with, **2**, 814
Germanium oxides, trialkyl-, reactions, with hydrogen cyanide, **5**, 1136
Germanium peroxides, use in the synthesis of dialkyl peroxides, **2**, 98
Germanium–phosphorus heterocycles, synthesis of, **4**, 485
Germaniums
of phosphorus functions, synthesis of, **5**, 889
reactions
with 2-lithio-1,3-dithiane, **6**, 132
with α,β-unsaturated aldehydes, **1**, 290
synthesis
of, **1**, 290; **2**, 949; **4**, 599, 1031; **5**, 874, 927; **6**, 169
Germaniums, acyl-, synthesis of, **5**, 431
Germaniums, alkyl-, chlorination of, **6**, 62
Germaniums, alkyl(perfluoroalkyl)-, synthesis of, **6**, 62
Germaniums, alkynyl-, synthesis of, **2**, 1083, 1092
Germaniums, aryl-, synthesis of, **2**, **899–950**, 945
Germaniums, bis(carbodiimidoyl)phthalocyanato-, synthesis of, **5**, 1083
Germaniums, cyano-, synthesis of, **5**, 1136
Germaniums, cyclic imidoyl-, from, *N*-vinyl pyrrolidinone, **5**, 794
Germaniums, α-diazo-, synthesis of, **5**, 874
Germaniums, α-diazo phosphonothioic diamide triethyl-, synthesis of, **6**, 669
Germaniums, α-halo-, synthesis of, **4**, 141
Germaniums, α-haloalkenyl-, synthesis of, **4**, 811
Germaniums, singly-bonded, synthesis of, **5**, 650
Germaniums, tetracoordinate
from
imidoyl chlorides, **5**, 794
isonitriles, **5**, 794
lactones, **5**, 794
synthesis of, **5**, 794
Germaniums, tricoordinated, synthesis of, **6**, 709
Germaniums, trifluorovinyl-, synthesis of, **4**, 811
Germaniums, trihalo(perfluoroalkyl)-, reactions, with dialkylcadmiums, **6**, 62
Germaniums, vinyl-, synthesis of, **2**, **899–950**, 945
Germanium sulfides, synthesis of, **2**, 815
Germanium tetrachloride
alkynylation of, **2**, 1092
reactions
with 1-germanylalkyne ethers, **4**, 1055
with lithium alkynides, **2**, 1093
with 1-silyl ethyn-2-ol ethers, **4**, 1058
with sodium alkynides, **2**, 1093
Germanium tetrahalides, reactions, with bis(perfluoroalkyl)mercury compounds, **6**, 62
Germanium trichloride, (trichlorosilylmethyl)-, synthesis of, **4**, 639
Germanyl compounds, *N*-aryl, synthesis of, **2**, 814
Germaoxetane, synthesis of, **1**, 766
1-Germa-3-silacyclobutane, 1,1,3,3-tetramethyl-, synthesis of, **4**, 638
1-Germa-3-silacyclobutanes, synthesis of, **4**, 636
Germasilenes, synthesis of, **5**, 926
Germazenes, synthesis of, **4**, 1055
Germene, dimesitylneopentyl-, synthesis of, **3**, 503
Germenes
dimerization of, **4**, 627; **6**, 388, 397
from
diazo compounds, **5**, 927
α-germylcarbenes, **3**, 503
as intermediates, **3**, 501
reviews of, **5**, 927
stability of, **5**, 927
as starting materials
for 1,1-bisgermylalkanes, **4**, 627
for 1,3-digermacyclobutane, **4**, 627
synthesis of, **2**, 949; **3**, 501; **4**, 627; **5**, 926; **6**, 397
via dehydrohalogenation reactions, **3**, 503
Wittig reaction of, **1**, 765
Germirenes, synthesis of, **2**, 947
Germoles, synthesis of, **2**, 949
Germoxanes, synthesis of, **1**, 766
Germyl anions, from, alkyl arylsilylgermanes, **4**, 361
Germyl anions, trialkyl-, synthesis of, **4**, 626
Germylation, double, of alkynes, **2**, 946
C-Germylation, mechanism, **4**, 636
Germyl chlorides
reactions, with carbonyl enolate anions, **4**, 625
for the synthesis of germanium–carbon bonds, **4**, 625
Germylcupration, of alkynes, **2**, 946, 977
Germyl electrophiles, reactions, with α-metallosilylalkanes, **4**, 636
Germylene, bis(trimethylsilyl)methyl-(pentamethylcyclopentadienyl)-, synthesis of, **6**, 186
Germylenes
addition to alkynes, **2**, 947
palladium-catalyzed cyclization of, **2**, 949
reactions
with carbenes, **3**, 503
with cyanoallenes, **2**, 947
with electrophiles, **6**, 187
with phosphaalkynes, **5**, 889
as starting materials, for germanes, **6**, 187
transmetallation reactions with, **6**, 387
use in the synthesis of germaethenes, **6**, 711
Germylenes, bis[bis(trimethylsilyl)methyl]-, synthesis of, **6**, 186
Germyl halides, aryl-, reactions, with lithium phosphides, **2**, 867
Germyl halides, trialkyl-
reactions
with Grignard reagents, **2**, 948
with methyl alkanesulfonates, **6**, 132
O-Germyl isomer, as intermediate, **6**, 131
Germyl metallics, synthesis of, **5**, 431
Germyl radical, reactions, with carbon–carbon double bonds, **1**, 950
Gibberelin precursors, synthesis of, **1**, 678
Gibberellic acid, synthesis of, **1**, 523
Gibberellins, displacement of bridgehead mesylates in, **1**, 177
Giese reaction, for the synthesis of *N*-alkoxy secondary amines, **2**, 354
Gilman compounds, **1**, 124
for the opening of epoxides, **1**, 206

Gilman reagents, synthesis of, **6**, 396
Gingerols, synthesis of, **4**, 245
Ginkgolide B, synthesis of, **1**, 1029
Glaser reaction
 mechanism, **1**, 1034, 1142
 for the synthesis of 1,3-diynes, **1**, 1034
Glaucosterol, synthesis of, **6**, 461
Gloesporone, synthesis of, **5**, 177
D-Glucal, 3,4,6-tri-*O*-*t*-butyldimethylsilyl-, reactions, with *t*-butyllithium, **4**, 651
α-D-Glucofuranose, (-)-1,2:5.6-di-*O*-isopropylidene-, synthesis of, **2**, 534
Gluconolactone, 2,3,4,6-tetra-*O*-benzyl-, as starting material, for ortho esters, **6**, 72
β-D-Glucopyranose, penta-*O*-acetyl-, reactions, with ethanethiol, **4**, 224
Glucopyranoses, synthesis of, **2**, 261
β-D-Glucopyranosides, aryl 1-thio-, synthesis of, **4**, 225
Glucosamines, synthesis of, **4**, 224
Glucose
 for the reductive cleavage of disulfides, **2**, 127
 use in the synthesis of azoxy compounds, **2**, 776
Glucose, diacetone, reactions, with 9-borabicyclo[3.3.1]nonyl, **2**, 57
D-Glucose, 2-deoxy tri-*O*-benzyl, as starting materials, for isocyanides, **3**, 695
D-Glucose, diacetone-, use in the synthesis of aryl sulfoxides, **2**, 714
D-Glucose, 1-seleno-, synthesis of, **4**, 239
D-Glucose osazone, reactions, with benzenediazonium salts, **5**, 857
α-*C*-Glucosides, synthesis of, **5**, 152
Glutacondialdehyde, reactions, with isothiocyanates, **5**, 593
Glutamate *N*-carboxyanhydride, benzyl, for the acylation of phenylamidoxime, **5**, 220
Glutamic acids, as starting materials, for 5-hydroxypipecolic acids, **2**, 326
Glutaraldehyde
 condensation of, **2**, 331
 reactions, with Grignard reagents, **4**, 166
Glutarates, reactions, with acetoacetates, **2**, 662
Glutaric acid
 from
  δ-hydroxyvaleraldehyde, **5**, 46
  trimethylene cyanide, **5**, 34
Glutaric acid, α-chloro-, synthesis of, **5**, 57
Glutaric acid, β-(2,2'-dimethylbenzhydryl), from, 2-(2,2'-dimethylbenzhydryl)-1,3-propanedinitrile, **5**, 35
Glutaric acid, 3-ethyl-3-methyl-, from, α,α'-dicyano-β-ethyl-β-methylglutarimide, **5**, 37
Glutaric acid, α-keto-, from, ethyl oxalylsuccinate, **5**, 62
Glutaric acid monoamides, hydrolysis of, **5**, 34
Glutarimide, α,α'-dicyano-β-ethyl-β-methyl-, as starting material, for 3-ethyl-3-methylglutaric acid, **5**, 37
Glutarimide, perfluoro-, reactions, with phosphorus pentachloride, **5**, 656
Glutathione, oxidation of, **2**, 235
Glutathione peroxidase, action of, **2**, 235
Glutathiones, synthesis of, **6**, 494
Glycals
 azidonitration of, **4**, 174
 synthesis of, **1**, 584
Glycerol, reactions, with ortho esters, **6**, 79
Glycidol, 3-phenyl-, reactions, with 1-(benzenesulfonyl)alkynes, **4**, 840
Glycidols
 opening of, **1**, 220
 reactions, with aminonitrile anions, **1**, 224
 reduction of, **1**, 14
 as starting materials
  for 1,2-diols, **1**, 220
  for 1,3-diols, **1**, 221
 See also Epoxy alcohols
 See also Oxiranes, hydroxymethyl-
Glycinamide, reactions, with butoxycarbonylleucine, **5**, 587
Glycinate, *t*-butyl, alkylation of, **3**, 415
Glycinate, *t*-butyl *N*,*N*-bis(benzyloxycarbonyl)-α-fluoro-, synthesis of, **4**, 101
Glycinate, *t*-butyl *N*-*t*-butoxycarbonyl, halogenation of, **4**, 102
Glycinate, ethyl, as starting material, for 3-pyrrolidone, **2**, 324
Glycinate, ethyl *N*-benzoyl, reactions, with acid anhydrides, **3**, 182
Glycinate, methyl *N*-acetyl-α-hydroxy-, as starting material, for α-bromoamides, **4**, 99
Glycinate, 8-phenylmenthyl *N*-butoxycarbonyl, bromination of, **4**, 102
Glycinate anions, in methylating agents, **1**, 180
Glycinates, 2-alkyl, synthesis of, **4**, 102
Glycinates, benzyloxycarbonyl)-α-hydroxy-, chlorination of, **4**, 102
Glycinates, α-chloro, synthesis of, **4**, 101
Glycinates, ethyl fluoro, synthesis of, **4**, 101
Glycinates, (*S*)-2'-thiazinylidene, reactions, with *N*-chlorosuccinimide, **4**, 101
Glycine, synthesis of, **5**, 85
Glycine, *N*-benzyl-
 reactions
  with dichlorophosphines, **4**, 466
  with formaldehyde, **4**, 466
 as starting material, for α-aminophosphonites, **4**, 466
Glycine, benzyloxycarbonyl-, reactions, with 4-methylpentanoyl chloride, **5**, 193
Glycine, *N*-benzyloxycarbonyl-, reactions, with diphenylketene, **5**, 193
Glycine, *N*-*t*-butoxycarbonyl-protected, trianion of, **1**, 166
Glycine, *N*-phenyl-, reactions, with paraformaldehyde, **4**, 302
Glycine, α-phenyl-, synthesis of, **5**, 88
Glycine, *N*-phospho-, synthesis of, **2**, 414
Glycine *t*-butyl ester, from, *t*-butyl azidoacetate, **5**, 90
Glycine carbanion synthons, as lithiated imines of glycine, **4**, 535
Glycine equivalents, for the synthesis of α-amino acids, **1**, 166
Glycine ethyl ester hydrochloride, synthesis of, **5**, 85
Glycines
 reactions, with Belleau's reagent, **5**, 574
 thionation of, **5**, 573
Glycines, α-alkoxy-, reactions, with phosphorus trichloride, **4**, 465
Glycines, α-amino-, monoacyl compounds, synthesis of, **4**, 417
Glycines, aryl-, synthesis of, **4**, 466
Glycines, α-bromo-, reactions, with trimethyl phosphite, **4**, 465
Glycines, cysteinyl-, synthesis of, **6**, 494
Glycines, (*E*)-vinyl-, synthesis of, **1**, 1135
Glycine synthons, synthesis of, **4**, 535
Glycinol, (*R*)-*N*-methylphenyl-, reactions, with aldehydes, **2**, 313
Glycinol, D-phenyl-, reactions, with 2-formylphenethyl bromides, **2**, 328
Glycinol, (*R*)-phenyl-, reactions, with aldehydes, **2**, 313
Glycol, diethylene
 reactions, with benzenesulfonylhydrazones, **3**, 446
 as solvent, for the synthesis of nitriles, **3**, 613
Glycol, ethylene
 reactions
  with sodium, **1**, 630
  with tetrabutyl 2-phenylethylidene-1,1-bisboronate, **4**, 634
Glycol, tetramethylethylene, as starting material, for methyl *t*-butyl ketone, **1**, 384
Glycol, triethylene, use in the synthesis of isothiouronium salts, **2**, 121

Glycol-300, polyethylene, as solvent, for the synthesis of nitriles, **3**, 613
Glycolates, reactions, with 2-trimethylsilylbut-2-ene, **2**, 70
Glycolates, allylic ethers, rearrangement of, **2**, 88
Glycolates, propargyl, as starting materials, for allenic acids, **1**, 980
Glycolic acid, from, indol, **5**, 95
Glycolic acid esters, thionoacyl-, for the thioacylation of hydroxylamines, **5**, 636
Glycols
 cleavage of, **5**, 114
 dehydration of, for the synthesis of enols, **2**, 643
 hydrogenation of, **1**, 506
 oxidative cleavage of, **2**, 646
 reactions
  with *O,O*'-dimethyl 2,6-pyridinedicarbothioate, **5**, 511
  with potassium fluoride, **2**, 19
Glycols, ethylene, reactions, with *β*-chlorovinyl aldehydes, **4**, 196
Glycols, polyethylene, reactions, with dimethyl(dimethylamino)arsine, **2**, 494
Glycopyranosides, 1-*O*-acetyl, reactions, with thiols, **4**, 224
O-Glycoside, reactions, with ethynylmagnesium bromide, **1**, 1026
Glycoside bonds
 cleavage of, **4**, 176
 synthesis of, **4**, 188
Glycoside coupling
 for the synthesis of acetals, **4**, 198, 200
 for the synthesis of unsymmetrical acetals, **4**, 202
Glycosides, synthesis of, **1**, 1099
Glycosides, alkyl, deprotection of, **4**, 176
O-Glycosides, reactions, with dimethylboron bromide, **4**, 225
Glycosylation, use of sulfoxides for, **4**, 232
Glycosyl bromides, reactions, with sodium manganese pentacarbonyl, **5**, 446
Glycosyl fluorides
 from, phenyl thioglycosides, **4**, 45
 as starting materials, for thioglycosides, **4**, 225
Glycosyl halides
 reactions, with sulfur nucleophiles, **4**, 223
 as starting materials
  for glycal alkenyl ethers, **2**, 678
  for glycosyl sulfones, **4**, 236
Glycosyl radicals, from, glycosyl tellurides, **4**, 240
Glycosyl tellurides, synthesis of, **4**, 240
Glycyl chloride, phthaloyl-, reactions, with potassium *O*-alkyl dithiocarbonates, **6**, 546
Glyoxal
 reactions
  with amines, **4**, 298
  with benzylamines, **4**, 298
  with 1,3-diaminopropane, **4**, 409
  with *N,N*'dimethyl-1,2-diaminoethane, **4**, 409
  with *N,N*'-dimethylethylenediamine, **4**, 301
  with 2-(methylamino)ethanol, **4**, 302
  with morpholine, **4**, 300
 as starting material, for *α*-amino acetals, **3**, 40
 use in the synthesis of diamines, **1**, 261
Glyoxal, phenyl-
 reactions
  with amines, **4**, 300
  with primary amines, **3**, 417
  with toluene *p*-sulfinic acid, **4**, 234
Glyoxalate, phenyl, hydrogenation of, **2**, 53
Glyoxalates, reactions, with carbonyls, **2**, 81
Glyoxal bis(dialkoxyphosphorylhydrazones), bromination of, **5**, 679
Glyoxaldehyde, phenyl-, condensation of, **4**, 311
Glyoxal diphenylosazone, chlorination of, **5**, 840
Glyoxal hydrates, aryl-, from, phenacyl bromides, **4**, 161
Glyoxime, dichloro-, reactions, with 1,2-diamines, **3**, 429
Glyoximes, oxidation of, **3**, 685
Glyoxylate, methyl, reactions, with *p*-toluenesulfonyl isocyanate, **2**, 327
Glyoxylic acid
 hydrazonyl chloride of, synthesis of, **5**, 843
 reactions, with 2-aminoethaneselenol, **4**, 333
Glyoxylic acid aldoxime
 bromination of, **6**, 611
 chlorination of, **6**, 608
Glyoxylic acid arylhydrazones
 as starting materials, for formazans, **5**, 856
 synthesis of, **5**, 858
Glyoxylic acids, reactions, with nitrous acid, **5**, 325
Glyoxyloyl chlorides, phenyl-, reactions, with 3,3-(bisdimethylamino)acrylonitriles, **5**, 658
(*R*)-(+)-Glyphos, use in Pauson–Khand reactions, **1**, 529
Gold, alkyl-, chemistry of, **2**, 589
Gold, cyanodimethyl-
 from, iododimethylgolds, **5**, 1141
 reactions, with 1,8-naphthyridine, **5**, 1141
 synthesis of, **5**, 1141
Gold, trimethyl-, reactions, with thioglycolic acid, **2**, 274
Gold(I), cyano(triphenylphosphoniomethide)-, synthesis of, **5**, 1141
Gold(I), (trifluoromethyl)-, synthesis of, **6**, 246
Gold(I) amino(thio)carbene complexes, synthesis of, **6**, 723
Gold(I) chloride phosphine complexes, reactions, with alkylmagnesiums, **2**, 590
Gold(I) ketenides, synthesis of, **3**, 537
Gold(II), as catalyst, in the synthesis of methyl ketones, **3**, 114
Gold(III), cyanobis(pentafluorophenyl)triphenylphosphine)-, synthesis of, **5**, 1141
Gold(III), cyanodimethyl(triphenylphosphine)-, synthesis of, **5**, 1141
Gold(III) complex, tris(trifluoromethyl)-, synthesis of, **6**, 246
Gold(III) compounds, *trans*-cyanodihalo(triphenylphosphoniomethide)-, synthesis of, **5**, 1141
Gold(III) *μ*-halide dimers, bis(trifluoromethyl)-, synthesis of, **6**, 247
Gold(III) halides, reactions, with alkyllithiums, **2**, 590
Gold(III) halides, dialkyl-, synthesis of, **2**, 590
Gold(III) iodide, bis(trifluoromethyl)-, synthesis of, **6**, 246
Gold bromide, diethyl-, reactions, with 2-aminoethanethiol, **2**, 274
Gold carbonyl complexes, synthesis of, **6**, 525
Gold chloride complexes, characterization of, **6**, 539
Gold complexes, alkylaminomethoxycarbene-, as starting materials, for imino esters, **5**, 696
Gold complexes, (chloromethyl)-, synthesis of, **4**, 152
Gold complexes, imidoyl-, synthesis of, **5**, 802
Gold complexes, methylene-bridged, synthesis of, **4**, 695
Gold complexes, pentafluorophenyl-, displacement of chloride from, **5**, 1140
Gold complexes, thioacyl-, synthesis of, **5**, 651
Gold(I) complexes, alkyl-, synthesis of, **2**, 590
Gold(III) compounds, alkyl-, synthesis of, **2**, 590
Gold–copper compounds, synthesis of, **4**, 724
Gold cyanide, reactions, with triphenylphosphoniomethanide, **5**, 1141
Golds
 reactions
  with 1-metalloalkylphosphine sulfides, **4**, 586
  with phosphorus ylides, **4**, 586
Golds, alkyl-, synthesis of, **2**, 589
Golds, cyano-, synthesis of, via nucleophilic displacement, **5**, 1140
Golds, imidoyl-, synthesis of, **5**, 802
Golds, iododimethyl-, as starting materials, for cyanodimethylgold, **5**, 1141
Golds, monoalkyl-, stability of, **2**, 589

Golds, trialkyl-, synthesis of, **2**, 590
Golds, trihalomethyl-, synthesis of, **6**, 246
Golds, triorganyl-, characteristics of, **2**, 960
Gold salts, aryl-bridged, synthesis of, **4**, 696
Gold salts, ferrocenyl-bridged, synthesis of, **4**, 696
Gold salts, vinyl-bridged, synthesis of, **4**, 696
Gold thiolates, synthesis of, **2**, 274
Gold trichloride pyridine complexes, reactions, with Grignard reagents, **2**, 590
Gold–zinc compounds, synthesis of, **4**, 724
Goniomitine, synthesis of, **2**, 328
(+)-Goniothalamin, synthesis of, **1**, 596
β-Gorgonenes, methylenation of, **1**, 649
(±)-β-Gorgonenes, methylenation of, **1**, 759
Gossyplure, synthesis of, **1**, 755
Graham reaction, for the synthesis of α-diazo halo compounds, **5**, 866
Graham synthesis, of diazirines, **6**, 55
(-)-Grandisol, synthesis of, **1**, 357
Graphite
for the activation of metals, **2**, 958
chlorination of, **6**, 217
for the synthesis of active metals, **2**, 959
Graphite, activated, fluorination of, **6**, 212
Grignard addition, for the synthesis of enamides, **5**, 290
Grignard intermediates, organosilyl, use in the synthesis of imidoylsilicons, **5**, 793
Grignard reagents
acylation of, **3**, 285
in alkene cross-coupling, **1**, 468
aminoalkylation of, **2**, 304
autoxidation of, **2**, 96
carbometallation of, **2**, 973
for the carbometallation of allenes, **1**, 495
carbonation of, **5**, 43, 112
for the displacement of fluoride, **1**, 108
for the displacement of sulfur, **1**, 238
dithiocarboxylation of, **4**, 843
fluorination of, via *N*-fluoro-*O*-benzenedisulfonimide, **2**, 613
formylation of, **3**, 90
from
alkyl fluorides, **2**, 558
(1-bromovinyl)trimethylsilane, **4**, 947
chloromethyl ethers, **4**, 383
chloromethylgermanes, **4**, 371
chloromethyl methyl sulfide, **4**, 395
α-chlorophenylacetic acid, **4**, 147
chlorotrimethylsilane, **4**, 370
perfluorovinyl bromide, **4**, 820
sulfones, **4**, 395
trialkoxysilylmethyl chlorides, **1**, 160
trichloro(chloromethyl)silane, **4**, 606
halogenation of, **6**, 63
with β-hydrogens, **1**, 481
iodination of, for the synthesis of aryl iodides, **2**, 632
for the opening of oxetanes, **1**, 229
for the opening of thiiranes, **1**, 242
photolysis of, **1**, 662
reactions
with acid chlorides, **3**, 129
with *N*-acyl α-amino acid chlorides, **3**, 180
with acyldiazines, **5**, 354
with acyl isothiocyanates, **5**, 631
with aldehydes, **2**, 67; **5**, 468
with aldimines, **2**, 312
with aldonitrones, **2**, 345
with alkenes, **1**, 316
with alkenyl sulfides, **1**, 1031
with (*Z*)-alkenyl tellurides, **1**, 545
with alkenylzirconocene chlorides, **2**, 992
with α-alkoxyacrylonitrile, **3**, 243
with alkyl enol ethers, **1**, 480
with alkyl halides, **1**, 114
with alkylmercury bromide, **6**, 394
with alkylnitro compounds, **2**, 346
with alkylnitroso compounds, **2**, 346
with alkylthioarenes, **1**, 436
with alkyl tosylates, **1**, 1092
with alkynes, **1**, 539
with alkynyl chlorides, **1**, 1017
with alkynyl halides, **1**, 1017
with alkynylphosphonites, **2**, 1060
with alkynyl sulfones, **1**, 1032
with allenyl halides, **1**, 466
with allyl halides, **1**, 119
with allylic acetals, **2**, 679
with allylic epoxides, **1**, 1053
with allylic phosphates, **1**, 1053
with allyl sulfides, **1**, 238
with aminoarsenic halides, **2**, 504
with ammonium halide salts, **1**, 253
with antimony trihalides, **2**, 483
with arsenic trihalides, **2**, 483
with aryl esters, **3**, 285
with aryl methyl ethers, **1**, 482
with azides, **2**, 776
with azines, **2**, 377
with benzo-1,3,2-dioxaphospholes, **4**, 126
with benzothiophene, **1**, 484
with benzotriazoloiminophosphorane, **5**, 1067
with benzotriazolylmethyliminophosphorane, **5**, 1031
with α-1-benzotriazolyl sulfides, **1**, 263
with *N*-benzyl tetrahydro-1,3-oxazines, **2**, 320
with bis(benzotriazol-1-ylmethyl)amine, **1**, 260
with *N*,*N*-bis(benzotriazolylmethyl)arylamines, **2**, 304
with bis(benzotriazol-1-ylmethyl)hydroxylamine, **1**, 260
with bis(diethylamino)chloroarsine, **2**, 498
with bismuth trihalides, **2**, 483
with bis(trifluoromethyl) trisulfide, **6**, 528
with 1-bromoalkynes, **1**, 999
with *t*-butyl 2(*S*)-*N*-tosylaziridine carboxylate, **1**, 263
with carbodiimides, **5**, 743
with carbon disulfides, **5**, 546
with carbonium ions, **1**, 202
with carbon monoxide, **5**, 438
with carbon sulfide, **5**, 232
with carbonyl groups, **2**, 67
with carbonyls, **1**, 662
with carboxylic acids, **3**, 285
with carboxymethylrheniums, **5**, 451
with chiral silylallyl phosphates, **1**, 194
with chloroarsines, **4**, 597
with (*Z*)-2-β-chlorocinnamates, **1**, 469
with chlorodialkoxyphosphines, **2**, 451
with chlorodithioformates, **3**, 364
with chloro enamines, **4**, 15
with chloroformamidines, **5**, 755
with chloro(methyl)silanes, **4**, 137
with α-chloronitrosoadamantane, **3**, 432
with ^{14}C-labeled carbon dioxide, **5**, 71
with ^{14}C-labeled dioxide, **5**, 45
with copper(I) halides, **1**, 967
with copper(I) salts, **1**, 311
with cyanoamides, **5**, 743
with cyanohydrins, **2**, 330
with 2,3-cyclopropylcycloalkanones, **2**, 67
with dialkylaminochlorophosphines, **2**, 466
with 2-dialkylamino-2-(trimethylsilyl)acetonitriles, **2**, 305
with dialkyl phosphonates, **2**, 840
with dialkyl phosphorochloridites, **2**, 838
with dialkyl sulfites, **2**, 197
with dialkylthalliums, **2**, 598
with dialkyl thiophosphonates, **2**, 459
with diaminochlorophosphine, **2**, 859

with dianilido phosphorochloridate, **2**, 861
with diaryl disulfides, **2**, 711
with diaryl telluride, **1**, 437
with diazo compounds, **2**, 376
with α,ω-dibromoalkane, **1**, 119
with 1,3-dibromo-1-propenes, **1**, 115
with digold hexabromide, **2**, 590
with *N*-diisobutylaluminum imines, **2**, 316
with diphenylethyne, **1**, 539
with *N*,*N*-disubstituted cinnamides, **1**, 310
with ditellurides, **2**, 727
with dithioesters, **4**, 255
with 1,2-dithioles, **4**, 905
with enynes, **1**, 960
with α,β-epoxynitriles, **3**, 459
with ethyl diazoacetate, **3**, 486
with ethyl (trimethylsilyl) acetate, **1**, 650
with 2-ethynyl-1,2-epoxycyclohexane, **2**, 86
with α-fluoro aliphatic esters, **1**, 108
with formamides, **2**, 320
with glutaraldehyde, **4**, 166
with gold trichloride pyridine complexes, **2**, 590
with hafnium compounds, **5**, 443
with *N*-haloamines, **2**, 305
with α-halo ketones, **1**, 464
with hexahydropyrimidines, **4**, 410
with α-hydroxynitriles, **2**, 316
with imidazolidines, **4**, 410
with imidazolidinium salts, **3**, 95
with imides, **4**, 303
with iminium salts, **2**, 305
with indium trichloride, **2**, 597
with 1-iodoalkynes, **1**, 999
with iron pentacarbonyl, **3**, 90; **5**, 43
with isocyanates, **5**, 293, 301, 304
with isoselenoazoles, **3**, 256
with isothiocyanates, **5**, 595, 614
with ketenylidenetriphenylphosphoranes, **3**, 220
with ketones, **2**, 67
with lactones, **3**, 219
with lithium amides, **1**, 580
with manganese(II) chloride, **6**, 194
with nickel catalysts, **1**, 466
with nitric oxide, **2**, 798
with nitriles, **3**, 131, 405, 459
with nitroarenes, **2**, 318
with 1-octene, **2**, 561
with organometal(V) halides, **2**, 881
with orthocarbonate esters, **6**, 77
with ortho esters, **4**, 192
with orthoformates, **4**, 192
with oxime sulfonates, **1**, 1059; **3**, 407
with palladium catalysts, **1**, 466
with *S*-phenyl carbonochloridothioate, **3**, 285
with phenyl isocyanate, **5**, 293, 551
with phosphinic chlorides, **4**, 120
with phosphonic dichlorides, **2**, 453
with phosphonous acid monoalkyl esters, **2**, 450
with phosphorochloridites, **2**, 434
with phosphorous acid dialkyl esters, **2**, 450
with phosphorous acids, **2**, 452
with phosphorus chlorides, **2**, 434
with phosphorus trichloride, **2**, 433
with propargyl alcohol, **4**, 1065
with propargyl alcohols, **1**, 539
with propargyls, **1**, 970
with pyrroles, **3**, 308
with pyrylium salts, **3**, 232
with rhenium complexes, **3**, 521
with selenium, **2**, 727
with selenium dioxide, **2**, 238
with selenium disulfide, **2**, 238
with selenophene, **1**, 484
with silver bromide–lithium bromide, **2**, 589
with *N*-silylated formamidines, **3**, 406
with silyl enol ethers, **1**, 480
with sodium alkoxides, **2**, 112
with *N*-substituted 1-(benzotriazolyl)-2,2-diethoxyethylamines, **1**, 261
with sulfenylimines, **2**, 359
with *N*-sulfinylamines, **2**, 363
with *N*-sulfinylaniline, **2**, 797
with sulfinylimines, **2**, 363
with sulfonyl fluorides, **4**, 280
with sulfur, **2**, 707
with sulfur dioxide, **2**, 186, 273, 718
with sulfuryl chloride, **2**, 176
with tellurophene, **1**, 484
with thallium(I) bromide, **2**, 597
with thiophene, **1**, 484
with transition metal catalysts, **1**, 439
with trialkoxyacetonitriles, **6**, 78
with trialkoxyphosphines, **2**, 451
with trialkylgermyl halides, **2**, 948
with 1-trialkylsilyl-1-alkynes, **1**, 160
with trichloroacetaldehyde, **6**, 27
with triethoxyacetonitriles, **6**, 78
with trimethylsilyl isocyanate, **5**, 299
with trimethyl(vinyl)silane, **4**, 654
with triphenylphosphine thiocyanate, **5**, 595
with *N*-triphenylphosphorylidene-1-(benzotriazol-1-yl)methylamine, **2**, 306
with trisilacyclohexanes, **6**, 268
with trithiocarbonate *S*,*S*-dioxides, **4**, 862
with α,β-unsaturated carbonyls, **1**, 310
with vanadium(III) chloride, **3**, 127
with vanadium complexes, **3**, 510
with vinylsilanes, **4**, 654
for the rearrangement of epoxides, **1**, 388
and Schlenk equilibrium, **2**, 559
as starting materials, for carboxyls, **2**, 1079
storage of, as powders, **2**, 68
for the substitution of aryl halides, **1**, 470
synthesis of, **1**, 1092; **2**, 958; **4**, 1017; **6**, 393
for the synthesis of alkylmagnesiums, **2**, 558
for the synthesis of anions, **1**, 155
for the synthesis of aryl bromides, **2**, 627
for the synthesis of methyl ethers, **1**, 262
for the synthesis of phosphinous chlorides, **2**, 834
for the synthesis of secondary amines, **1**, 260
for the synthesis of tertiary alkylarsines, **2**, 483
for the synthesis of tertiary phosphines, **2**, 824
thiophilic addition of, **4**, 245
transition metal-catalyzed alkylation of, **1**, 476
transmetallation of, **1**, 677, 680; **2**, 990
use in the substitution of alkyl halides, **1**, 464
use in the synthesis of aldehydes, via formylation, **3**, 14
use in the synthesis of alkynylcuprates, **2**, 1095
use in the synthesis of alkynylsilicons, **2**, 1089
use in the synthesis of 1,1-bisgermylalkanes, **4**, 626
use in the synthesis of 1,1-bis(silyl)alkanes, **4**, 606
use in the synthesis of dialkyl sulfides, **2**, 134
use in the synthesis of dialkyl sulfoxides, **2**, 152
use in the synthesis of 1,1-dimetal complexes, **4**, 669
use in the synthesis of enols, **2**, 643
use in the synthesis of nitriles, **3**, 614
use in the synthesis of α-substituted tertiary amines, **2**, 310
use in the synthesis of thionoamides, **5**, 607
use in the synthesis of triazenes, **2**, 387
use in the synthesis of α,β-unsaturated ketones, **3**, 212
use in the synthesis of vinylsilanes, **2**, 924
See also Magnesium halides, alkyl-
Grignard reagents, α-(acylamino)alkyl, use in the synthesis of α-(acylamino)alkylmagnesiums, **2**, 1002
Grignard reagents, alkenyl
reactions, with trialkylboranes, **1**, 1100

reactivity of, **1**, 466
Grignard reagents, alkyl
reactions
with alkylphosphonous dihalides, **2**, 435
with diketene, **1**, 481
with iron catalysts, **1**, 467
with phenylphosphonous dihalides, **2**, 435
with phosphinous chlorides, **2**, 435
with silylalkynes, **1**, 1105
with vinyl halides, **1**, 1097
Grignard reagents, alkynic
reactions, with bis(amino)chlorophosphites, **2**, 1065
use in the synthesis of alkynylantimony compounds, **2**, 1081
Grignard reagents, alkynyl
for the alkynylation of boron, **2**, 1085
reactions
with alkenyl iodides, **1**, 1135
with chloro(amino)phosphine oxides, **2**, 1066
with chlorophosphine oxides, **2**, 1066
with dialkyl phosphorochloridates, **2**, 1070
with diaryl phosphorochloridates, **2**, 1070
with iodine, **1**, 1035
with organosiloxanes, **2**, 1090
with *N*-substituted isocyanide dichlorides, **1**, 1035
with thionyl chloride, **1**, 1035
with 2,4,6-trichloroborazine, **2**, 1085
for the synthesis of alkynylphosphines, **2**, 1058
for the synthesis of alkynyl phosphonates, **2**, 1072
use in the synthesis of chiral vinyl sulfoxides, **2**, 713
Grignard reagents, allylic
carbophilic addition to, **4**, 255
stability of, **2**, 73
Grignard reagents, α-amino, synthesis of, **4**, 531
Grignard reagents, aryl
alkylation of, **1**, 464
formylation of, **3**, 97
hydrolysis of, **1**, 452
reactions
with nitriles, **3**, 286
with oxazolinium salts, **4**, 306
with sulfur dioxide, **2**, 717
with sulfuryl chloride, **2**, 717
with trialkoxyboranes, **2**, 906
with vinyl halides, **1**, 469
as starting materials, for benzaldehydes, **3**, 90
for the synthesis of biaryls, **1**, 473
Grignard reagents, benzyl, as ambident nucleophiles, **1**, 115
Grignard reagents, bis-
reactions, with alkylphosphonous dichlorides, **2**, 435
use in the synthesis of carbon-bridged dimercury complexes, **4**, 696
Grignard reagents, butyl, for the synthesis of alkenes, **1**, 516
Grignard reagents, copper-catalyzed
for the displacement of phenylselenyl groups, **1**, 246
as nucleophiles, **1**, 185
reactivity of, **1**, 177
regiochemistry of, **1**, 187
Grignard reagents, di-
reactions
with *t*-butyldichloroarsenic, **2**, 484
with chlorotrimethylgermane, **4**, 626
with dimethyltin dichloride, **4**, 700
with tetrachlorosilane, **4**, 608
as starting materials, for 1,1-bis(silyl)alkanes, **4**, 608
synthesis of, **4**, 618, 669; **6**, 398
Grignard reagents, diisopropoxy(methyl)silylmethyl, development of, **1**, 160
Grignard reagents, α-haloalkyl, synthesis of, **4**, 147
Grignard reagents, primary, synthesis of, **2**, 562
Grignard reagents, α-silyl-
from, chloromethylpentamethyldisiloxane, **4**, 654
synthesis of, **4**, 653
Grignard reagents, α-silylalkenyl, synthesis of, **1**, 160
Grignard reagents, α-silylmethyl-
reactions, with aluminum trichloride, **4**, 659
synthesis of, **4**, 655
Grignard reagents, α-sulfenyl, synthesis of, **2**, 999
Grignard reagents, α-sulfonyl
reactions, with haloalkanes, **1**, 144
synthesis of, **1**, 144
Grignard reagents, (trialkylsilyl)methyl-, synthesis of, **4**, 654
Grignard reagents, vinyl
alkylation of, **1**, 464
hydrolysis of, **1**, 453
reactions
with allylic phosphates, **1**, 477
with chlorosilanes, **2**, 910
with selenenyl halides, **2**, 727
with sulfur, **2**, 722
use in the synthesis of chiral vinyl sulfoxides, **2**, 713
Griseoviridin, partial synthesis of, **3**, 350
Grob fragmentation
for the synthesis of alkenes, **1**, 683
of γ-tosyloxy dithioacetal, **4**, 846
Group 1 acyl metals
stability of, **5**, 436
synthesis of, **5**, 436
Group 1 metals
functions containing two, synthesis of, **4**, 667
organometallic compounds of, **1**, 110
synthesis of functions, **4**, 705
Group 2 acyl metals, synthesis of, **5**, 438
Group 2 metals
functions containing two, synthesis of, **4**, 669
organometallic compounds of, **1**, 114
synthesis of functions, **4**, 709
Group 2 organo metal cyanides, synthesis of, **5**, 1137
Group 13 acyl metals, synthesis of, **5**, 466
Group 13 elements, chemistry of, **2**, 564
Group 13 organo metal cyanides, synthesis of, **5**, 1142
Group 14 acyl metals, synthesis of, **5**, 467
Group 14 alkyl metals, chemistry of, **2**, 565
Group 14-centered radicals, reactions, with carbon–carbon double bonds, **1**, 950
Group 14 methylseleno compounds, synthesis of, **2**, 294
Group 14 organo metal cyanides, synthesis of, **5**, 1144
Group 15, synthesis of functions, **5**, 1091
Group 15 alkyl metals, chemistry of, **2**, 569
Group 15 elements, synthesis of functions, **6**, **137–170**, **359–375**
Group 15 heterobenzenes, synthesis of, reviews, **2**, 872
Group 16 metal(0) carbonyl complexes, chemistry of, **2**, 572
Group transfer, in radical additions to alkenes, **1**, 332
Group transfer, alkyl, from organometallics, **2**, 589
Grovenstein–Zimmerman rearrangement, mechanism, **1**, 886
Guanidine, amino-, from, cyanamide, **5**, 847
Guanidine, 1-[bis(difluoroamino)fluoromethyl]-1,2,3,3-tetrafluoro-, synthesis of, **5**, 815
Guanidine, 1-*t*-butyl-, synthesis of, **6**, 641
Guanidine, *N*-cyano-*N*′,*N*′,*N*″,*N*″-tetramethyl-, alkylation of, **6**, 733
Guanidine, pentafluoro-
reactions
with amines, **6**, 650
with isocyanic acid, **6**, 363
with semicarbazide hydrochloride, **5**, 811
as starting material
for diaziridines, **6**, 285
for tetraazamethanes, **6**, 362
synthesis of, **5**, 815
Guanidine, pentamethyl-, reactions, with isocyanates, **6**, 364
Guanidine, perchloro-, synthesis of, **6**, 649
Guanidine, 2-phenyl-1,1,2,2-tetramethyl-, reactions, with benzoyl isocyanates, **6**, 647

Guanidine, *N*-phenyltetramethyl-, reactions, with methyl isocyanate, **6**, 364
Guanidine, 1,1,3,3-tetraisopropyl-, synthesis of, **6**, 641
Guanidine, tetramethyl-, reactions, with alkyltitanium complexes, **6**, 140
Guanidine, 1,1,3,3-tetramethyl-
halogenation of, **6**, 649
reactions
with *N*,*N*-dimethylthiocarbonyl chloride, **6**, 647
with isobutyraldehyde, **6**, 644
use in the synthesis of phosphine oxides, **6**, 513
Guanidine nitrate, chlorination of, **6**, 649
Guanidines
acylation of, **6**, 646
alkylation of, **6**, 642
fluorination of, **6**, 649
from
alkylcyanamides, **6**, 643
N,*N'*-bisurethane-protected carboxamidines, **6**, 641
carbonimidic dichlorides, **6**, 619
cyanamides, **6**, 642
thioureas, **6**, 641
halogenation of, **6**, 648
nitration of, **6**, 655
reactions
with amines, **2**, 749
with aryl halides, **6**, 642
with phosgene, **6**, 444
with phosphoryl chlorides, **6**, 655
with tosylates, **6**, 642
as starting materials
for *N*-acyliminocarbonyls, **6**, 646
for *N*-alkenyliminocarbonyls, **6**, 644
for iminocyanamides, **5**, 1125
for ortho amides, **6**, 139
synthesis of, **5**, 1070; **6**, 640, 732
for the synthesis of vinyl iodides, **2**, 630
Guanidines, acetoxy-, synthesis of, **6**, 651
Guanidines, acyl-
from, carbonic acid ortho amides, **6**, 647
reduction of, **6**, 644
synthesis of, **6**, 646, 647
Guanidines, 1-acyl-2,3-dialkyl-, from, *S*-methylisothiouronium salts, **6**, 646
Guanidines, alkoxy-
from, *O*-alkylhydroxylamines, **6**, 651
synthesis of, **6**, 651
Guanidines, alkyl-, synthesis of, **6**, 487, 641, 642
Guanidines, 1-alkyl-
from, 1,3-di-*t*-butoxycarbonylthiourea, **6**, 641
synthesis of, **6**, 642
Guanidines, *N*-alkyl-, synthesis of, **6**, 642
Guanidines, (alkylaminosulfonyl)-, synthesis of, **6**, 651
Guanidines, alkylidene amino-, reactions, with cyanogen bromide, **5**, 1125
Guanidines, *N*-alkyl-*N'*-nitro-*N*-nitroso-, as starting materials, for diazoalkanes, **3**, 466
Guanidines, amino-, synthesis of, **6**, 653
Guanidines, *N*-amino-, synthesis of, **6**, 653
Guanidines, antimony-substituted, synthesis of, **6**, 656
Guanidines, aryl-
from
N-arylphosphimines, **6**, 645
carbodiimides, **6**, 645
cyanamides, **6**, 645
synthesis of, **6**, 642, 644
Guanidines, *N*-aryl-*N'*-cyano-, synthesis of, **5**, 1124
Guanidines, *N*-aryl-*N'*-(triphenylstannyl)-*N'*,*N''*-dicyano-, synthesis of, **5**, 1121
Guanidines, azido-, synthesis of, **6**, 653
Guanidines, benzoyloxy-, synthesis of, **6**, 651
Guanidines, 1,3-bis(alkoxycarbonyl)-, fluorination of, **6**, 649
Guanidines, bis(alkoxycarbonyl)fluoro-, synthesis of, **6**, 649
Guanidines, bismuth-substituted, synthesis of, **6**, 656
Guanidines, *N*-borated, synthesis of, **6**, 656
Guanidines, boron-substituted, synthesis of, **6**, 656
Guanidines, carboalkoxytetrafluoro-, synthesis of, **6**, 649
Guanidines, *N*-chloro-, reactions, with nitroalkynes, **4**, 796
Guanidines, 2-chloro-1,1-dialkyl-, synthesis of, **6**, 649
Guanidines, cyano-
fluorination of, **5**, 815; **6**, 649
from
carbodiimides, **6**, 648
isothiocyanates, **6**, 647
thioureas, **6**, 647
reactions, with amines, **5**, 1121
synthesis of, **5**, 1121; **6**, 642, 647
Guanidines, 2-cyano-
from, cyanamides, **6**, 648
synthesis of, **6**, 647
Guanidines, *N*-cyano-, synthesis of, **5**, 1121
Guanidines, 2-cyano-1-(4-pyridinyl)-, synthesis of, **6**, 648
Guanidines, dialkyl-, from, *N*,*N'*-dialkylcarbodiimides, **6**, 643
Guanidines, 1,1-dialkyl-
from, monoalkylcyanamides, **6**, 642
synthesis of, **6**, 642
Guanidines, 1,3-dialkyl-, from, 1-alkylthiourea, **6**, 641
Guanidines, 1,3-dialkyl-2-aryl-, synthesis of, **6**, 645
Guanidines, germanium-substituted, synthesis of, **6**, 656
Guanidines, *N*-germylated, synthesis of, **6**, 656
Guanidines, *N*-halo-, synthesis of, **6**, 648
Guanidines, (heteroarylmethyl)nitro-, synthesis of, **6**, 654
Guanidines, hydroxy-
alkylation of, **6**, 651
from
chloroformamidinium chlorides, **6**, 650
thioureas, **6**, 650
synthesis of, **6**, 650
Guanidines, imino-, synthesis of, **6**, 653
Guanidines, monoacyl-, synthesis of, **6**, 647
Guanidines, monoalkyl-, synthesis of, **6**, 641
Guanidines, monocarboalkoxy-, fluorination of, **6**, 649
Guanidines, nitro-, synthesis of, **6**, 653
Guanidines, nitroso-, synthesis of, **6**, 653
Guanidines, 2-nitroso-, stability of, **6**, 655
Guanidines, pentaalkyl-, from, dichlorocarbiimide, **6**, 643
Guanidines, pentafluoro-
reactions
with amines, **5**, 817
with ammonia, **5**, 817
synthesis of, **6**, 649
Guanidines, pentafluorosulfenyl-, synthesis of, **6**, 651
Guanidines, perfluoro-
synthesis of, **6**, 649
use in the synthesis of *N*-fluoroimidic esters, **5**, 814
Guanidines, phosphorus-substituted, synthesis of, **6**, 655
Guanidines, polyfluoro-, synthesis of, **6**, 362
Guanidines, silicon-substituted, synthesis of, **6**, 656
Guanidines, *N*-silylated, synthesis of, **6**, 656
Guanidines, substituted 1,1,2-trifluoro-, synthesis of, **5**, 817
Guanidines, sulfonyl-
from
amidinium chlorides, **6**, 651
carbodiimides, **6**, 652
S-methylisothiouronium salts, **6**, 651
synthesis of, **6**, 651
via [2+2] cycloaddition, **6**, 652
Guanidines, *N*-sulfonyl-, synthesis of, **6**, 651
Guanidines, sulfur-substituted, synthesis of, **6**, 651
Guanidines, tetraalkyl-
from
N,*N'*-dialkylcarbodiimides, **6**, 643
dichlorocarbiimide, **6**, 643
Guanidines, 2-(thiocarbamoyl)-, synthesis of, **6**, 647
Guanidines, tosyl-, synthesis of, **6**, 652

Guanidines, trialkyl-
  from
    N,*N*'-dialkylcarbodiimides, **6**, 643
    dichlorocarbiimide, **6**, 643
  synthesis of, **6**, 642
Guanidines, 1,1,2-trialkyl-, from,
    monoalkylcyanamides, **6**, 642
Guanidines, triaryl-, synthesis of, **6**, 646
Guanidines, *N*,*N*,*N*'-trisubstituted-*N*''-cyano-, synthesis
    of, **5**, 1121
3*H*-Guanidines, 1-cyano-, from,
    N-alkyldicyanamides, **6**, 642
Guanidine-*O*-sulfonate, hydroxy-, use in the synthesis of
    N-unsubstituted iminocarbonyls, **6**, 642
Guanidinium azide, tetramethyl-, use in the synthesis of acyl
    azides, **5**, 991
Guanidinium chloride, hexamethyl-, reactions, with
    ethyne, **6**, 139
Guanidinium salts
  from, *S*-methylisothiouronium salts, **6**, 641
  mixed hydride reduction of, **6**, 140
  synthesis of, **4**, 973; **6**, 732
Guanidino functions, protection of, **6**, 487
Guanylating agents, **6**, 641

H

H2 antagonist ranitidine, synthesis of, **4**, 907
Hafnium(IV) compounds, dialkylbis[bis(trimethylsilyl)amido]-
  reactions, with isonitriles, **5**, 797
  as starting materials, for imidoyl hafnium compounds, **5**, 797
Hafnium carbonyl complexes, synthesis of, **6**, 522
Hafnium complexes, biscyclopentadienyl bisalkyl-, reactions, with carbon monoxide, **5**, 443
Hafnium compounds, reactions, with Grignard reagents, **5**, 443
Hafnium compounds, acyl-, synthesis of, **5**, 443
Hafnium compounds, alkyl-, synthesis of, **2**, 565
Hafnium compounds, imidoyl-, synthesis of, **5**, 797
Hafnium dichloride, dicyclopentadienyl-, as starting material, for metallated ylides, **5**, 917
Half ester, reactions, with thionyl chloride, **5**, 13
Halide, selenoacyl, from, selenocarbonyl difluoride, **6**, 588
Halide displacement
  rate of, **1**, 107
  for the synthesis of alkynylalanes, **2**, 1098
Halide exchange
  phase-transfer catalysis of, **2**, 31
  for the synthesis of aryl halides, **2**, 609
  for the synthesis of organometal(III) halides, **2**, 882
  for the synthesis of sulfonyl halides, **2**, 183
Halide ions, for the α-halogenation of carbonyls, **2**, 3
Halides
  as borylating agents, **2**, 698
  condensation of, **1**, 676
  electrocarbonylation of, **5**, 44
  elimination of, **3**, 226
  epimerization of, **2**, 5
  from
    betylates, **2**, 9
    fluorosulfonates, **2**, 9
    imidazolyl sulfonates, **2**, 9
  hydrolysis of, **2**, 87
  nucleophilic displacement of, **2**, 647; **4**, 346
  nucleophilic interconversion of, **2**, 5
  reactions
    with alkali metal thiocyanates, **5**, 1038
    with *O*-alkyl dithiocarbonate salts, **6**, 547
    with amino ketones, **3**, 179
    with ammonium thiocyanates, **5**, 1038
    with hexamethylenetetramine, **2**, 308
    with β-ketostannanes, **3**, 195
    with Lewis acids, **2**, 1006
    with methylenediamines, **2**, 1008
    with organometallics, **6**, 186
    with phosphorus(III) compounds, **3**, 50
    with selenate anions, **3**, 174
    with selenocyanates, **5**, 1053
    with silver carboxylates, **2**, 12
    with silver thiocyanate, **5**, 1039
    with trialkylphosphines, **3**, 48
  reactivity of, **2**, 2
  reduction of, **2**, 505
  solvolysis of, **1**, 892
  as starting materials
    for aromatic carboxylic esters, **5**, 143
    for hydroperoxides, **2**, 96
    for iron complexes, **2**, 577
    for isocyanates, **5**, 1015
    for isoselenocyanates, **5**, 1053
    for isothiocyanates, **5**, 1038
    for ketones, **3**, 114
    for nickel complexes, **2**, 577
    for phosphonium salts, **3**, 492
    for radicals, **1**, 937
Halides, primary, reactions, with hexamethylenetetramine, **2**, 302
Halides, unsaturated, reduction of, **1**, 986
Haller–Bauer reaction, mechanism, **1**, 885
α-Halo acids
  as starting materials, for α-sulfenyl aldehydes, **3**, 35
  synthesis of, via diazotization of α-amino acids, **2**, 11
Haloalkylation
  of aldehydes, **4**, 47
  of arenes, **2**, 13
  of ketones, **4**, 47
β-Haloalkyl compounds, synthesis of, **2**, 583
(α-Haloalkylsilanes), alkyl-, synthesis of, **4**, 136
(α-Haloalkylsilanes), aryl-, synthesis of, **4**, 136
Halo atom transfer cyclization, mechanism, **1**, 344
Halobenzene–tricarbonylchromium complexes, reactions, with anionic organometallics, **4**, 722
Haloboration
  of alkynes, stereochemistry of, **2**, 902
  of terminal alkynes, for α-bromo aldehydes, **3**, 27
Haloboration reagents, for the synthesis of alkenylboranes, **2**, 900
α-Halo carbanions
  synthesis of, **1**, 164, 282
  See also Carbenoids
Halo compounds, α-diazo, synthesis of, **5**, 866
Halo compounds, mixed, synthesis of, **4**, 750
N-Halo compounds, reactions, with sulfides, **2**, 159
Halodecarboxylation, of carboxylic acids, **2**, 12
Halodestannylation, for the synthesis of *gem*-halo stannylalkenes, **4**, 812
Haloethynylation
  for the synthesis of alkynyl bromides, **2**, 1017
  for the synthesis of alkynyl chlorides, **2**, 1012
Haloform reaction
  for the synthesis of alkanoic acids, **5**, 48
  for the synthesis of carboxylic acids, **5**, 38
  for the synthesis of unsaturated acids, **5**, 102
Haloforms, mixed, synthesis of, **6**, 32
Halogen
  displacement of, for the synthesis of α-germylsilylalkanes, **4**, 640
  elimination of, **1**, 568
  and one nitrogen function, synthesis of functions, **6**, 437
  and one oxygen function, synthesis of functions, **6**, 421
  and one phosphorus function, synthesis of functions, **6**, 454
  and one sulfur function, synthesis of functions, **6**, 431
  substitution of, **1**, 462, 1052
  synthesis of functions, **4**, **41–93**
  and three other heteroatom substituents, synthesis of functions, **6**, **271–293**
Halogen, nucleophilic, reactions, with phosphinic acid esters, **4**, 125
Halogen abstraction, via tributyltin radicals, **1**, 330
Halogen acids, reactions, with thioureas, **2**, 119, 120
Halogenating reagents, electrophilic, use in the synthesis of alkynyl halides, **2**, 1012
Halogenation
  of aldehydes, **3**, 26
  of aldoximes, **5**, 664
  of alkynylsilanes, **2**, 929

of arsines, **2**, 881
of azines, **5**, 677
of bismuthines, **2**, 881
of carbonyls, **2**, 3
of hydrazones, **5**, 677
of ketones, **2**, 3
of nitroalkanes, **5**, 665
of organometallics, **4**, 774
of stibines, **2**, 881
of sulfoxides, **4**, 75
of sulfur ylides, **5**, 494
for the synthesis of, α-halo nitro compounds, **4**, 108
for the synthesis of α-haloamides, **4**, 98, 99
for the synthesis of α-halo imides, **4**, 98, 99
Halogenation, allylic
reagents for, **2**, 21
for the synthesis of alkyl halides, **2**, 3
Halogenation, benzylic, for the synthesis of alkyl halides, **2**, 3
Halogenation, reductive
of arsinic acids, **2**, 883
of arsonic acids, **2**, 883
of carbonyls, **2**, 14
of stibinic acids, **2**, 883
of stibonic acids, **2**, 883
Halogen atom transfer reaction, for the synthesis of alkyl halides, **2**, 14
Halogen-based oxidants
for the synthesis of aldehydes, **3**, 6
use in the synthesis of ketones, **3**, 118
Halogen carriers
as starting materials
for hydrazonyl halides, **5**, 675
for hydroxamoyl halides, **5**, 665
Halogen compounds
of arsenic functions, synthesis of, **5**, 891
of phosphorus functions, synthesis of, **5**, 876
synthesis of, **1**, 108; **4**, 769; **5**, 866; **6**, 36
use in the synthesis of dialkyl sulfoxides, **2**, 147
Halogen exchange
of chlorotrifluoromethane, **6**, 224
for the synthesis of alkyl bromides, **2**, 31
for the synthesis of aryl bromides, **2**, 627
for the synthesis of aryl chlorides, **2**, 621
for the synthesis of aryl fluorides, **2**, 614
for the synthesis of aryl iodides, **2**, 632
for the synthesis of aryl trifluoromethyl sulfides, **6**, 232
for the synthesis of bromotrifluoromethane, **6**, 219
for the synthesis of tribromoiodomethane, **6**, 223
for the synthesis of trichlorofluoromethane, **6**, 220
for the synthesis of trifluoromethyl ethers, **6**, 229
for the synthesis of trifluoromethyls, **6**, 16
of tetrachloromethane, for dichlorodifluoromethane, **6**, 224
of trihalomethanesulfenyl halides, **6**, 236
of trihalomethyl sulfides, **6**, 233
Halogen functions, substitution of, **1**, 1000
Halogen functions, singly-bonded, synthesis of, **5**, **1–21**
Halogenides, acyclic covalent, synthesis of, **6**, 273
Halogenides, cyclic covalent, synthesis of, **6**, 275
Halogen–lithium exchange
for the synthesis of α-alkoxycarbanions, **4**, 381
for the synthesis of alkyllithiums, **2**, 551
for the synthesis of α-lithiogermylalkanes, **4**, 662
for the synthesis of α-lithiosilylalkanes, **4**, 652
for the synthesis of manganese compounds, **4**, 707
for the synthesis of α-thiocarbanions, **4**, 395
Halogen–magnesium exchange, for the synthesis of α-magnesiosilylalkanes, **4**, 653
Halogen–metal exchange
scope of, **2**, 962
for the synthesis of alkylcoppers, **2**, 585
for the synthesis of alkyllithiums, **2**, 551
for the synthesis of alkynes, **1**, 999, 1017
for the synthesis of 1,1-bisgermylalkanes, **4**, 626
for the synthesis of metallostannanes, **2**, 600
for the synthesis of monoalkyl phenyl ketones, **3**, 295
for the synthesis of α-selenolithiums, **4**, 400
for the synthesis of triaryl compounds, **2**, 874
use in the synthesis of 1,1-bis(silyl)alkanes, **4**, 602
See also Metal–halogen exchange
Halogen–metal interconversion
for the synthesis of alkyllithiums, **2**, 554
for the synthesis of dialkylzincs, **2**, 591
Halogen migration, in carbonyl ylide reactions, **3**, 323
Halogen monofluorosulfates, reactions, with perfluoroalkenes, **6**, 42
Halogen nucleophiles, reactions, with enol sulfonates, **3**, 239
Halogen replacement, of methylene(ditin)compounds, **4**, 700
Halogens
additions to fluoroalkenes, **6**, 2
and any other elements, synthesis of functions, **6**, **35–66**
containing a carbonyl group and at least one halogen, synthesis of functions, **6**, **407–457**
containing two heteroatoms, synthesis of functions, **6**, **249–270**
elimination of, **1**, 1063
four similar, synthesis of functions, **6**, 212
and one other heteroatom substituent, synthesis of functions, **6**, **211–247**
for the oxidation of thiols, **2**, 212
reactions
with alkenylboranes, **1**, 1103
with alkylarsine halides, **2**, 492
with alkynes, **1**, 1123
with amides, **5**, 657
with dialkylhaloarsines, **2**, 492
with dinitroalkanes, **6**, 55
with 2-iodooctane, **1**, 108
with selenium bis(thiocarboxylates), **5**, 252
with selenyl-substituted alkynes, **4**, 786
with sulfenyl halogenides, **6**, 273
with tellurium bis(thiocarboxylates), **5**, 252
with tertiary arsines, **2**, 492
with thiocarbonyls, **6**, 273, 275
with trisulfonylmethanes, **6**, 274
with trithiocarbonates, **6**, 273
with vinylgermanes, **4**, 141
reactivity of, **2**, 961; **3**, 612
stabilization by, for the synthesis of carbenes, **4**, 1082
synthesis of functions, **4**, **789–822**
and another heteroatom group other than a chalcogen, **4**, **95–157**
incorporating two halogens or a halogen and a chalcogen, **4**, **729–788**
three and a chalcogen, synthesis of functions, **6**, 228
three and a Group 15 element, synthesis of functions, **6**, 237
three and a metalloid, synthesis of functions, **6**, 243
three and a phosphorus function, synthesis of functions, **6**, 242
three with a selenium or tellurium function, synthesis of functions, **6**, 237
three similar and one different, synthesis of functions, **6**, 217
three and a sulfur function, synthesis of functions, **6**, 232
two similar, synthesis of functions, **6**, 223
Halogen sources, active, **2**, 18
Halohydrins, as precursors of ketones, **3**, 114
Halohydrocarbons, reductive silylation of, **6**, 178
Halolactonization
of δ,ε-ene acids, **5**, 172
for the synthesis of γ-lactones, **5**, 171
Halolysis, of organoboranes, **2**, 4
Halomercuration, mechanism, **2**, 982
Halometallation
of alkynes, with heavy metals, **2**, 982

for organometallics, **2**, 952
with thallium, **2**, 982
for vinyl metals, **2**, 956
Halonium ions
in Markovnikov addition, **2**, 5
reactions, with *O*-trimethylsilyl thiocarboxylates, **5**, 539
α-Halo radicals, synthesis of, **2**, 1010
Halo salts, stability of, **5**, 403
Halosilane elimination, for the synthesis of carbonyl ylides, **1**, 366
Halothane, as anesthetic, **6**, 8
Halotropic rearrangement, of trichloromethylphosphines, **6**, 696
Handling precautions, sulfinyl chlorides, **2**, 172
Handling problems, hydrogen sulfide, **2**, 117
Hantzsch's ester, use in the synthesis of amines, **2**, 312
Hapalindole J, synthesis of, **3**, 701
Haptophilicity, definition of, **1**, 72
Hard and soft acids and bases (HSAB)
concept of, **1**, 886
theoretical basis for, **1**, 301
Harringtonine alkaloids, synthesis of, **3**, 420
Hay procedure, mechanism, **1**, 1034
HBTU, as coupling reagents, for the synthesis of *N*-alkylamides, **5**, 279
Heck aroylation, of vinyl ethers, **3**, 245
Heck arylation, of 3-ethenyl-4,5-diphenyl-2(3*H*)oxazolone, **2**, 321
Heck coupling
for the synthesis of alkenes, **1**, 546
for the synthesis of α,β-unsaturated esters, **5**, 157
of vinyl halides, **1**, 1096
Heck reaction
alkyl equivalent of, **1**, 495
of allenes, **1**, 496
of aryl triflates, **3**, 293
with silanes, **1**, 490
for the substitution of vinylhydrogen, **1**, 493
for the synthesis of alkynes, **1**, 1015
for the synthesis of amines, **2**, 320
for the synthesis of β-aryl-γ-butyrolactols, **4**, 168
for the synthesis of nitriles, **3**, 647
for the synthesis of phenylalkynes, **1**, 483
use in the substitution of boron, **1**, 489
use in the synthesis of alkenyl ethers, **2**, 682
(+)-Hedycaryol, as starting material, for (-)-elemol, **1**, 812
Helicoidal molecules, synthesis of, **1**, 971
Hell–Volhard–Zelinsky reaction
for the α-bromination of carboxylic acids, **2**, 28
for the synthesis of α-bromo acid bromides, **5**, 264
Helminthosporium, for the deprotection of arylamines, **2**, 304
Hemiacetal, cyclopropanone ethyl, synthesis of, **4**, 171
Hemiacetals
acylation of, **4**, 209
alkylation of, **4**, 183
from
acetals, **4**, 175
aldehydes, **4**, 164
enol ethers, **4**, 171
esters, **4**, 169
ketones, **4**, 164
silylation of, **4**, 202
as starting materials
for acetals, **4**, 200
for acyl alkyl acetals, **4**, 209
synthesis of, **4**, 163
via addition of carbanions, **4**, 171
via reduction of esters, **4**, 169
Hemiacetals, acyclic
from, α-halo ketones, **4**, 164
synthesis of, **4**, 164
Hemiacetals, 2-azido-, synthesis of, **4**, 174
Hemiacetals, cyclic
from, hydroxy carbonyls, **4**, 164
as starting materials, for acetals, **4**, 201
synthesis of
via addition of carbanions to dicarbonyls, **4**, 166
via oxidation of diols, **4**, 164
via reduction of dicarbonyls, **4**, 165
Hemiaminals
from
aldehydes, **4**, 294
cyclopropanone, **4**, 299
ketones, **4**, 294
reactions, with sodium cyanoborohydride, **2**, 310
synthesis of, **4**, 293
See also O,N-Acetals
Hemiaminals, *N*-aminoketene, synthesis of, **4**, 897
Hemiaminals, cyclic, reactions, with diethyl cyanophosphate, **2**, 328
Hemiaminals, *N*-hydroxylketene, synthesis of, **4**, 896
Hemiaminals, ketene
from
amides, **4**, 881
aryl cyanates, **4**, 889
dialkynic ynamines, **4**, 885
1-halo-1-dialkylaminoalkenes, **4**, 889
ketene acetals, **4**, 887
ketene aminals, **4**, 888
ketene dithioacetals, **4**, 888
ketene hemiaminals, **4**, 888
ketenes, **4**, 886, 887
ketenimines, **4**, 894
nitriles, **4**, 883
substituted alkynes, **4**, 884
thioesters, **4**, 895
as starting materials, for ketene hemiaminals, **4**, 888
synthesis of, **4**, 880
Hemiaminals, *N*-nitrosoketene
from, *N*-nitroamides, **4**, 896
synthesis of, **4**, 896
Hemiaminals, *N*-sulfonyl, synthesis of, **4**, 312
Hemidithioacetals
from
aldehydes, **4**, 245
dithioesters, **4**, 245, 256
synthesis of, **4**, 245
Hemiketals, silylated, synthesis of, **6**, 124
Hemimellitene, acylation of, **3**, 296
Hemiselenoacetals, *O*-silyl, synthesis of, **2**, 542
Hemithioacetals
from, dihydropyran, **4**, 223
with the oxygen in a ring, **4**, 223
with the sulfur in a ring, **4**, 226
synthesis of, **4**, 216
Hemithioacetals, *O*-silyl, synthesis of, **2**, 541
Hemithioaminals
from
aldehydes, **4**, 324
alkenes, **4**, 326
alkylthiomethyleneiminium salts, **4**, 321
imines, **4**, 322
ketones, **4**, 324
tricoordinate sulfur compounds, **4**, 329
N-functionalized, synthesis of, **4**, 323
synthesis of, **4**, 315
See also S,N-Acetals
Hemithioaminals, *N*-acyl, synthesis of, **4**, 325
Hemithioaminals, *N*-formamido, synthesis of, **4**, 327
Hemithioaminals, *N*-nitroso, synthesis of, **4**, 325
Hemithioaminals, *N*-tosyl, synthesis of, **4**, 326
Henry nitroaldol reaction
mechanism, **3**, 191
use in the synthesis of nitro compounds, **2**, 394
Henry reaction
of nitroalkanes, **4**, 540

for the synthesis of nitroalkenes, **2**, 784
See also Aldol reaction, nitro-
n-Heptadecafluorooctyl iodide, electrochemical addition to alkenes, **6**, 7
1,5-Heptadiene, from, 3-methyl-1,5-hexadiene, **1**, 411
Heptafluoropropyl sulfone, methyl, cleavage of, **2**, 193
Heptamethylated compounds, synthesis of, **2**, 570
Heptanals, reduction of, **1**, 18
1,7-Heptandioate, diethyl 2,4,6-trioxo-, as starting material, for chelidonic acid, **5**, 62
Heptane, 1-bromo-2-fluoro-, condensation of, **5**, 250
Heptane, (-)-3-methyl-, synthesis of, **1**, 112
Heptanoate, *s*-butyl 3-methyl-, hydrolysis of, **5**, 30
Heptanoic acid, from, 1-octene, **5**, 52
Heptanoic acid, 7-carboxymethyl-, reactions, with chloroacetyl chloride, **5**, 193
Heptanoic acid, 2-fluoro-, synthesis of, **5**, 250
Heptan-2-one, 7-bromo-, as starting material, for radicals, **1**, 546
Heptanoyl chloride, 7-isothiocyanato-, synthesis of, **5**, 1031
Heptanoyl fluoride, synthesis of, **5**, 6
Heptasila[4.4.4]propellane, synthesis of, **6**, 379
1,3,5-Heptatrienes, use in 1,7 sigmatropic rearrangements, **1**, 403
Heptene, (1*E*)-chloro-
  reactions
    with heptyne, **1**, 1012
    with heptynyl tributylstannane, **1**, 1012
(*E*)-Hept-2-enoate, methyl-7-hydroxy-2-methyl-, synthesis of, **1**, 725
Heptenone, methyl-, reactions, with *n*-butylidenetriphenylphosphorane, **1**, 730
6-Heptenyl radical, cyclization of, **1**, 339
Heptyne, reactions, with (1*E*)-chloroheptene, **1**, 1012
Heptyne, 1-bromo-, reactions, with pentylmagnesium bromide, **1**, 1017
1-Heptyne, reactions, with bromobenzenes, **1**, 1009
Herbicides
  aminophosphonic acids for, **6**, 286
  synthesis of, **5**, 993
Herz reaction, for the synthesis of arenethiols, **2**, 708
11(*R*)-HETE, synthesis of, **1**, 1016
Heterafulvalenes, synthesis of, **4**, 870
Heteroarenes, reduction of, **1**, 87
Heteroaromatic acids
  as starting materials, for amides, **5**, 298
  synthesis of, **5**, 105, 549
Heteroaryl dihalides, reactions, with alkynes, **1**, 1009
Heteroarylium salts, trimethylsilyl-, synthesis of, **5**, 226
Heteroaryl polyhalides, reactions, with alkynes, **1**, 1009
Heteroatom–carbon bonds, for the synthesis of allenes and cumulenes, **1**, 993
Heteroatom compounds, triple-bonded, synthesis of, **5**, **1151–1160**
Heteroatom donors, as radical traps, **1**, 329
Heteroatom exchange, for the synthesis of 3,5-diazaphosphanes, **4**, 457
Heteroatom radicals
  carbon–carbon bond scission of, **1**, 928
  reactions, with carbon–carbon double bonds, **1**, 948
  synthesis of, **1**, 927
Heteroatoms
  effect on 5-hexenyl radical cyclization, **1**, 337
  other than chalcogen or nitrogen, synthesis of functions, **5**, **1091–1098**
  reactions, with carbenes, **1**, 400
  synthesis of functions containing two halogens, **6**, **249–270**
α-Hetero carbanions, reactivity of, **1**, 301
Heterocuprates
  reactions, with allenyl halides, **1**, 466
  use in the substitution of alkyl halides, **1**, 465
Heterocycles
  from reactions involving Lawesson's reagent, **5**, 571
  halogenation of, **6**, 36
  hydrolysis of, **5**, 348
  pyrolysis of, **2**, 783
  reactions
    with dimethylketene, **4**, 886
    with hydroxylamine, **5**, 322
    with thiols, **2**, 235
  ring-cleavage of, **3**, 666
  ring-opening of
    for the synthesis of hydroxamic acids, **5**, 322
    for the synthesis of α,β-unsaturated isocyanides, **3**, 714
  solvolysis of, **5**, 354
  as starting materials
    for amines, **2**, 320
    for carbodiimides, **5**, 1073
    for hydrazonamides, **5**, 854
  synthesis of, via intramolecular Michael addition, **1**, 307
Heterocycles, *N*-benzoyl-, synthesis of, **5**, 336
Heterocycles, cyanophosphorus-substituted, synthesis of, **5**, 1128
Heterocycles, *N*-(silylmethyl), synthesis of, **4**, 519
Heterocyclic compounds
  directed metallation of, **2**, 942
  extrusion of nitrogen from, **1**, 367
  from, nitriles, **3**, 641
  reactions, with hydrazine, **3**, 453
  ring-cleavage of, **3**, 722
  as starting materials
    for imidic acid esters, **5**, 693
    for ketenes, **3**, 541
    for thioketenes, **3**, 550
Heterocyclic halides, reactions, with 1-alkynes, **1**, 1137
Heterocyclic rings, synthesis of, via carbodiimides, **5**, 1066
Heterocyclic systems, ring-opening of, **5**, 847
Hetero Diels–Alder reaction, for the synthesis of acetals, **4**, 202
Heterodienes
  [4+2] Diels–Alder additions of, **1**, 375
  reactions
    with alkenes, **1**, 375
    with dienophiles, **1**, 372
  as starting materials, for heterocyclic anhydrides, **5**, 201
  synthesis of, **1**, 375
Heterogeneous catalysis, in asymmetric hydrogenation reactions, **2**, 52
Heterogeneous catalysts
  for the addition of hydrogen to alkynes, **1**, 455
  for the synthesis of acetals, **4**, 178
Heterolysis mechanism, use of term, **1**, 891
Heterometallation
  of alkynes
    with heavy metals, **2**, 982
    with transition metals, **2**, 983
  for aryl metals, **2**, 956
  for organometallics, **2**, 952
  for vinyl metals, **2**, 956
Heteronitrilium compounds, synthesis of, **5**, 1145
Heteronitrilium salts, synthesis of, **5**, 1145
Hevea brasiliensis, as source of oxynitrilase enzymes, **3**, 628
Hexaalkylditin, as catalyst, for the synthesis of α-haloalkenylsilicons, **4**, 811
Hexaboraadamantane, hexabromo-, synthesis of, **6**, 183
Hexaboraadamantane, hexachloro-, synthesis of, **6**, 183
2,4,6,8,9,10-Hexaboraadamantane, synthesis of, **6**, 183
Hexabutylditin
  use in atom transfer reactions, **1**, 344
  use in mediating atom transfer reactions, **1**, 334
Hexachlorides, reactions, with *o*-(bistrimethylsilyl)benzene, **4**, 290
Hexachlorohomoenols, synthesis of, **1**, 868
Hexadecanoate, ethyl, at starting material, for hexadecanoyl iodide, **5**, 20
Hexadecanoic acid anhydride, synthesis of, **5**, 189

Hexadecanoyl iodide, from, ethyl hexadecanoate, **5**, 20
1,5-Hexadiene, 3-methyl-, as starting material, for
  1,5-heptadiene, **1**, 411
2,4-Hexadiene, as starting material, for
  3,4-dimethylcyclobutene, **1**, 418
1,5-Hexadienes, 3,3 sigmatropic rearrangement of, **1**, 411
Hexa-1,5-dienes, 3-hydroxy-, from, vinyl epoxides, **2**, 86
(*E*,*E*)-2,4-Hexadienoate, methyl, as starting material, for
  (*E*)-1,5-enynes, **1**, 1036
Hexa-1,5-dien-3-ols, oxy-Cope rearrangement of, **3**, 22
3,5-Hexadienones, synthesis of, **3**, 144
2,4-Hexadiyne, 1,6-dichloro-, 1,6-dehydrochlorination
  of, **1**, 1069
2,4-Hexadiyne-1,6-diol, as starting material, for silyl-
  substituted triynes, **2**, 1091
Hexa-1,5-diyn-3-enes, 3-alkyl-4-(1-alkynyl)-, synthesis
  of, **1**, 1035
2,4-Hexadiynes, 1,6-dihydroxy-, reactions, with diols, **1**, 986
Hexaene, protonation of, **1**, 912
Hexakis(phosphines), synthesis of, **4**, 547, 550
Hexamethyl compounds, reactions, with α,β-unsaturated
  carbonyls, **2**, 565
Hexamethylenetetramine (HMT)
  reactions
    with arenes, **3**, 91
    with benzyl halides, **3**, 88
Hexamethylphosphoric triamide (HMPTA), use in
  metallation reactions, **6**, 390
Hexanal
  reactions
    with allylic ylides, **1**, 735
    with *m*-chloroperbenzoic acid, **5**, 52
Hexanal, 2-carboxy-, from, tetrahydroterrein, **5**, 73
Hexanal, 6-phenylsulfenyl, synthesis of, **3**, 36
Hexanamidine, *N*,*N*,*N*′-trifluoro-, synthesis of, **5**, 815
Hexane, 1,6-bis(diazo)-, synthesis of, **3**, 467
Hexane, 1,1-bis(dichloroboryl)-, synthesis of, **4**, 631
Hexane, 1,6-bis(trimethylsilyl)decafluoro-, synthesis
  of, **6**, 61
Hexane, 1,6-dibromodecafluoro-, lithiation of, **6**, 61
Hexane, 1,6-dicyanato-, synthesis of, **5**, 1103
Hexane, 1,6-diisocyano-, reactions, with silaarsene, **6**, 691
Hexane, lithiobis(diethylalumino)-, from,
  1,1,1-tris(diethylalumino)hexane, **6**, 210
Hexane, 1,1,1-tris(diethylalumino)-, as starting material, for
  lithiobis(diethylalumino)hexane, **6**, 210
Hexane, 3,3,4-tris(diethylboryl)-, synthesis of, **4**, 632
1,6-Hexanediol
  reactions
    with *N*-chlorocarbonyl isocyanate, **5**, 1007
    with cyanogen chloride, **5**, 1103
1,6-Hexanediol, 2,2,3,3,4,4-hexafluoro-, reactions, with
  cyanogen chloride, **5**, 1102
Hexane-3,4-diol, (*S*,*S*)-2,5-dimethyl- (DIPED), reactions,
  with lithium benzyloxide, **4**, 362
Hexanenitrile, synthesis of, **2**, 1057
Hexane-1,3,4,6-tetraones, 1,6-diaryl-, as starting materials,
  for bis(diazo) compounds, **3**, 469
Hexanitratocerate, ammonium, for the electrophilic
  iodination of arenes, **2**, 631
Hexanoate, benzyl *N*-acetyl-*O*-benzoyl-6-hydroxylamino-,
  synthesis of, **5**, 214
Hexanoate, benzyl 6-amino-, reactions, with acetyl
  chloride, **5**, 214
Hexanoate, (*S*)-ethyl 2-fluoro-, enzyme catalysis of, **5**, 55
Hexanoic acid, synthesis of, **5**, 52
Hexanoic acid, 6-amino-, reactions, with hydrogen
  chloride, **5**, 970
Hexanoic acid, 6-bromo-, synthesis of, **5**, 59
Hexanoic acid, 3,3-dimethyl-5-keto-, from,
  isophorone, **5**, 75
Hexanol, 2-methyl-, as starting material, for δ-acetyl
  n-valeric acid, **5**, 73
Hexanol, 2-phenylsulfenyl chloro-, cleavage of, **3**, 36
Hexan-1-ol, synthesis of, **2**, 40
1-Hexanol, synthesis of, **4**, 630
1-Hexanol, 6-bromo-, oxidation of, **5**, 59
Hexan-2-ol, synthesis of, **2**, 40
Hexan-2-one, 3-bromo-5-methyl-, reactions, with potassium
  O-ethyl dithiocarbonate, **6**, 546
2-Hexanone, 1,1,1-tris(trimethylgermyl)-, synthesis
  of, **6**, 186
Hexanoyl chloride, 6-isocyanato-, synthesis of, **5**, 970
Hexaorganodistannanes, use in the synthesis of
  radicals, **1**, 924
1,1,3,3,5,5-Hexaoxo-1,3,5-trithiane, hexachloro-, from,
  trithiane, **6**, 254
Hexapentaenes, synthesis of, **1**, 986, 995
Hexapentaenes, tetraaryl-, nickel(0)-catalyzed
  cyclodimerization of, **1**, 992
Hexaphosphino complexes, synthesis of, **2**, 475
Hexatertiary phosphines, synthesis of, **2**, 439
Hexathiaadamantane, synthesis of, **5**, 546
Hexathiodiorthooxalates, from, trithioorthoformate
  esters, **6**, 88
Hexathioethane, dialkyl disulfide elimination from, **4**, 846
Hexatriazines, as starting materials, for chiral amides, **4**, 99
1,3,5-Hexatriazines, reactions, with acyl chlorides, **4**, 99
1,3,5-Hexatriene, as starting material, for
  1,3-cyclohexadiene, **1**, 419
Hexatriene oxides, thermolysis of, **3**, 318
Hexatrienes, synthesis of, **1**, 145
Hexatrienes, 1-dimethylamino-6-nitro-, synthesis of, **2**, 790
1,3,5-Hexatriynes, synthesis of, **1**, 1069
trans-2-Hexenal, oxidation of, **5**, 102
Hexene, 6-bromo-, cyclization of, **1**, 342
Hex-1-ene
  gas phase pyrolysis of, **1**, 772
  hydration of, **2**, 40
1-Hexene, 1,1-bis(trimethylsilyl)-, from, trimethylsilyl-1-
  hexyne, **2**, 923
Hex-1-ene, 1-bromo 1-trimethylsilyl 2-ethyl-, bromination–
  dehydrobromination of, **4**, 748
Hex-1-ene, 6-chloro-, synthesis of, via xanthate
  pyrolysis, **1**, 776
Hex-1-ene, (*E*)-1-iodo-, synthesis of, **1**, 654
1-Hexene, perfluoro-, reactions, with
  phenylhydrazine, **5**, 679
3-Hexene, *cis*-3,4-bis(diethylboryl)-, hydroboration
  of, **4**, 632
3-Hexene, *cis*-3,4-di(diethylboryl)-, pyrolysis of, **4**, 632
5-Hexene, 1-bromo-, cyclization of, **2**, 600
(*Z*)-Hex-2-ene, synthesis of, **1**, 514
Hexenes, 4,5,6-*syn*-5,6-dialkoxy-4-hydroxy-4-
  trimethylsilyl-, synthesis of, **2**, 79
Hexenoates, synthesis of, **1**, 819
2-Hexenoic acid, synthesis of, **5**, 99
trans-2-Hexenoic acid, synthesis of, **5**, 102
5-Hexenyl radicals
  effect of substituents on cyclization, **1**, 337
  regioselectivity of, **1**, 336
Hexoses, allylation of, **1**, 173
t-Hexylbromoborane–dimethyl sulfide complex, for the
  reduction of nitriles, **3**, 86
Hexyne, ethoxy-, reactions, with hexynyllithium, **1**, 1030
1-Hexyne
  photochemical [2+2] cycloaddition of, **1**, 516
  reactions
    with *n*-butylcopper, **1**, 540
    with dicyclohexylborane, **4**, 630
1-Hexyne, 1-ethoxy-, reactions, with organolithiums, **1**, 1030
1-Hexyne, trimethylsilyl-
  silylalumination of, **2**, 923
  as starting material, for 1,1-bis(trimethylsilyl)-1-
    hexene, **2**, 923
2-Hexyne, reduction of, **1**, 514

Hex-3-yne
as starting material
for alkyne-bridged dipalladium complexes, **4**, 692
for carbon-bridged ketonic dirhodium complexes, **4**, 690
for tetranuclear carbon-bridged rhodiacyclopentadiene complexes, **4**, 689
5-Hexyn-1-yl acetate, 6-phenyl-, solvolysis of, **1**, 1124
5-Hexynyl cyclizations, for the synthesis of vinyl radicals, **1**, 1131
Hexynyl metals, reactions, with alkenyl iodides, **1**, 1015
5-Hexynyl radical, cyclization of, **1**, 547
HF pyridine, reactions, with *N*-bromosuccinimide, **4**, 45
High dilution techniques, for the synthesis of macrolactones, **5**, 173
Higher fatty acids, natural unsaturated, synthesis of, **5**, 97
Highest occupied molecular orbital (HOMO)
of alkenes, **1**, 325, 350
of carbon-centered radicals, **1**, 920
of nucleophiles, **1**, 301
High pressure, use in Diels–Alder reactions, **1**, 373
β-Himachalene, synthesis of, **1**, 812
Hinesol, synthesis of, **1**, 804
Hinsberg synthesis, for the synthesis of amines, **2**, 300
Hippurate, methyl 2-methoxy-, as starting material, for *N*-benzoyl compounds, **4**, 327
Hippuric acid, synthesis of, **4**, 525
Hirsutene, synthesis of, **1**, 347
Histamines, H2 antagonist ranitidine, **4**, 907
Histidines, 2-methyl thiol-, synthesis of, **1**, 255
Histrionicotoxin, (±)-perhydro-, synthesis of, **1**, 263
Hiyama reagent, for the synthesis of allenes, **1**, 958
Hofmann's rule, definition of, **1**, 628
Hofmann carbylamine reaction
for the synthesis of aryl isocyanides, **3**, 721
for the synthesis of isocyanides, **3**, 693, 695
Hofmann degradation
for the conversion of amides to nitriles, **3**, 620
mechanism, **1**, 627
for the synthesis of bromochloroalkanes, **4**, 38
Hofmann elimination
of bromoamides, **5**, 318
of enamines, **1**, 1075
sila versions of, **1**, 596
silyl equivalent of, **1**, 629
stereoselectivity of, **1**, 628
for the synthesis of alkenes, **1**, 612
in the synthesis of carbon–carbon bonds, **1**, 399
for the synthesis of nitriles, **3**, 621
for the synthesis of tertiary amines, **2**, 303
Hofmann–Loeffler–Freytag reaction, **2**, 20
for the synthesis of piperidines, **2**, 326
for the synthesis of pyrrolidines, **2**, 323
Hofmann reaction
for the synthesis of α-aminoalkylorganophosphorus compounds, **4**, 472
for the synthesis of α,β-unsaturated ketones, **3**, 210
Hofmann rearrangement
of amides, **2**, 374
mechanism, **5**, 997
for the synthesis of amines, **2**, 320
for the synthesis of α-aminoalkylphosphorus compounds, **4**, 476
for the synthesis of arylamines, **2**, 756
for the synthesis of carbamates, **5**, 999
for the synthesis of isocyanates, **5**, 997
Holmberg, synthesis of trithioorthoformate esters, **6**, 84
Holmium compounds, reactions, with phosphorus ylides, **4**, 578
Homoallylic halides
from, cyclopropyl carbinols, **2**, 11
synthesis of, **1**, 825
Homoaromatic acids, synthesis of, **5**, 105
Homoaromaticity
concept of, **1**, 865
reviews, **1**, 917
Homoaromaticity, carbanionic, **1**, 866
Homobinuclear compounds, synthesis of, **6**, 331
Homocoupling reaction
of alkynes, **1**, 1036
in copper-catalyzed reactions, **1**, 119
in the synthesis of alkylcoppers, **2**, 586
Homocuprates
from
alkylmanganeses, **2**, 586
copper iodide, **2**, 586
dialkylzincs, **2**, 587
functionalized, synthesis of, **2**, 587
reactions, with allenyl halides, **1**, 466
synthesis of, **2**, 586
Homocysteine, *S*-adenosyl-, 6'-arylethynyl-6'-dethia analogues of, attempted synthesis of, **1**, 1074
Homoelimination, of mesylates, **1**, 583
Homoenolate anion equivalents, reactions, with sulfenylating reagents, **3**, 168
Homoenolate anions
reactions
with aldehydes, **3**, 161
with carbonyls, **3**, 31
with ketones, **3**, 161
sulfenylation of, **3**, 36
synthesis of, **1**, 867
Homoenolate carbanions, definition of, **1**, 868
Homoenolate equivalents, use in the synthesis of aldehydes, **3**, 21
Homoenolates, use in the synthesis of ketones, **3**, 137
γ-Homoenolates, synthesis of, **1**, 868
Homoenolization
mechanism, **1**, 868
reviews, **1**, 868
Homoenols, definition of, **1**, 868
Homogeneous catalysis
for the addition of hydrogen to alkynes, **1**, 456
in asymmetric hydrogenation reactions, **2**, 52
Homogeneous catalysts
for the hydrogenation of alkynes, **1**, 507
for the reduction of alkenes, **1**, 73
Homolaudanosine, synthesis of, **1**, 149
Homologation
of aldehydes, **3**, 10
of alkyl halides, **2**, 15
of aryl ketones, **3**, 146
of ethanol ethers, **2**, 681
of ketones, **3**, 125
for the synthesis of α-sulfenyl ketones, **3**, 166
of triselenoorthoformate esters, **6**, 94
of trithioformate esters, **6**, 88
with zinc compounds, **2**, 594
Homolysis, of organocobalt complexes, **1**, 335
Homopolymers, synthesis of, via ring-opening metathesis polymerization (ROMP), **1**, 675
Homopropargyl compounds
carbometallation of, **2**, 975
zirconium-catalyzed carboalumination of, **2**, 977
Homosecodaphniphyllate, (±)-methyl, synthesis of, **2**, 330
D-Homosteroids, synthesis of, **1**, 1138
Homotropylium carbocations, synthesis of, **1**, 917
Horner alkenation, for the synthesis of α,β-enoates, **5**, 155
Horner–Emmons reaction
of 4-nitrobenzaldehyde with β-phosphinyl dithioic acid esters, **5**, 559
of phosphononitriles, **3**, 646
for the synthesis of butatrienes, **1**, 988
for the synthesis of nitriles, **3**, 650
Horner–Emmons reagents, use in the synthesis of ketene dithioacetals, **4**, 847

Horner reaction
  mechanism, **1**, 744, 753
  modifications of, **1**, 756
  stereochemistry of, **1**, 754
    solvent effects, **1**, 755
  stereoselective synthesis of (*E*) and (*Z*)-alkenes, **1**, 754
  stereoselectivity of, **1**, 753
  for the synthesis of alkenes, **1**, 744, 752
Horner–Wadsworth–Emmons reaction
  mechanism, **1**, 639
  stereoselectivity of, **1**, 639
  use in the synthesis of α,β-unsaturated acylsilanes, **5**, 426
Horner–Wittig reaction
  of 2-phosphoryl benzo-1,3-diselenoles, **4**, 875
  for the synthesis of ketene monothioacetals, **4**, 835
  for the synthesis of thiaselenafulvenes, **4**, 870
  for the synthesis of α,β-unsaturated ketones, **3**, 236
  use of α-sulfonyl ethers in, **4**, 840
  use in the synthesis of ketene acetals, **4**, 828
  use in the synthesis of ketenimines, **3**, 593
Horse liver alcohol dehydrogenase (HLADH), use in the synthesis of L-α-hydroxy acids, **5**, 117
Houben–Hoesch reaction
  mechanism, **3**, 282
  for the synthesis of aromatic ketones, **3**, 282
  for the synthesis of pyrrolic ketones, **3**, 307
Huang–Minlon modification
  for the deamination of *N*-alkylimine systems, **1**, 45
  for the reduction of aldehydes, **1**, 18
  for the reduction of aldehydes and ketones, **1**, 11
  for the reduction of ketones, **1**, 19
Hückel molecular orbital (HMO) calculations, for the analysis of phenalenium anion, **1**, 866
Hückel ($4n + 2$) rule
  and carbanion synthesis, **1**, 861
    destabilized, **1**, 864
  and carbocation synthesis, **1**, 906
Hunig's base, for the elimination of hydrogen bromide, **5**, 790
Hunig's base, polymeric, use in the synthesis of 6-acylaminopenem-3-carboxylic acids, **5**, 905
Hunsdiecker reaction
  Cristol–Firth modification, **2**, 12, 627
  modification of, for aryl iodides, **2**, 633
  with silver bromochlorofluoroacetate, **6**, 228
  for the synthesis of aryl bromides, **2**, 627
  for the synthesis of bromodichlorofluoromethane, **6**, 225
  for the synthesis of bromodifluoroiodomethane, **6**, 224
  for the synthesis of bromofluoroalkanes, **4**, 35
  for the synthesis of chlorofluoroalkanes, **4**, 33
  for the synthesis of chlorotrifluoromethane, **6**, 218
  for the synthesis of dibromoalkanes, **4**, 27
  for the synthesis of fluoroiodoalkanes, **4**, 37
  for the synthesis of haloforms, **6**, 33
  for the synthesis of iodoperfluoroalkanes, **6**, 32
  for the synthesis of iodotrifluoromethane, **6**, 219
  for the synthesis of trichlorofluoromethane, **6**, 220
  for the synthesis of trifluoromethyls, **6**, 14
  for the synthesis of vinyl bromides, **2**, 625
  use in the synthesis of alkyl halides, **2**, 12
Hyalophora cecropia, juvenile hormones, synthesis of, **1**, 748
Hybridization, in the synthesis of carbanions, **1**, 848
Hydantoin, 5-chloro-, synthesis of, **4**, 103
Hydantoins, reduction of, **4**, 414
Hydantoins, *N*-cyano-, synthesis of, **5**, 1126
Hydantoins, *N*-sulfenyl-substituted, as starting materials, for thiobisamines, **2**, 361
Hydralumination
  of alkynes, **4**, 1068
  of 1-trimethylsilylalkynes, **4**, 1062
Hydratases, for the conversion of nitriles to amides, **5**, 118
Hydrates
  from, bicyclo[2.2.0]hex-5-en-2-ones, **4**, 160
  synthesis of, **2**, 731; **4**, 160
Hydrates, trifluoromethyl, synthesis of, **4**, 160
Hydration, of alkenes, **2**, 40
Hydration, mercury-catalyzed, for acetaldehyde, **2**, 982
Hydrazanoyl halides
  as starting materials, for amidrazones, **5**, 771
  synthesis of, **5**, 771
Hydrazide, mesitylenesulfonyl-, use in the synthesis of alkynyl-substituted alkyl aldehydes, **3**, 25
Hydrazide, tosyl-, reactions, with α,β-epoxy ketones, **3**, 25
Hydrazides
  bromination of, **5**, 771
  diazotization of, **6**, 511
  from, dithioacid hydrazinium salts, **5**, 639
  reactions
    with carbon disulfide, **6**, 558
    with phosphorus pentachloride, **5**, 842
  reduction of, **2**, 374, 377, 381, 382
  as starting materials
    for hydrazidoyl bromides, **5**, 675
    for hydrazines, **2**, 374, 377, 381, 382
    for hydrazonoyl halides, **5**, 842
  thionation of, **5**, 639
Hydrazides, acyl-, reactions, with sodium hydride, **5**, 1126
Hydrazides, alkyl-*p*-toluenesulfonyl-, thermal decomposition of, **1**, 39
Hydrazides, *N*,*N'*-diacyl-, thionation of, **5**, 639
Hydrazides, heteroaryl-, as starting materials, for hydrazonates, **5**, 845
Hydrazides, imino-, as intermediates, **3**, 445
Hydrazides, sulfonyl-, as starting materials, for tetrazenes, **2**, 388
Hydrazides, thioformic acid diethyl-, methylation of, **5**, 846
Hydrazidines
  reactions, with hydrogen sulfide, **5**, 641
  See also Amidrazones
Hydrazidoyl bromides, from, hydrazides, **5**, 675
Hydrazine
  acylation of, **5**, 350
  alkylation of, **2**, 373, 383
  oxidation of, **1**, 458
  reactions
    with 2-acetoxyvinyl isothiocyanate, **6**, 574
    with alkyl azides, **2**, 317
    with azines, **3**, 444
    with cyanogen, **5**, 847
    with cyclic nitrones, **3**, 445
    with cyclopentadienyliron tricarbonyl cation, **5**, 975
    with cyclopropane-1,2-dicarboxaldehyde, **3**, 455
    with *N*,*N*-dialkylhydrazones, **3**, 445
    with dialkyl ketones, **3**, 444
    with 1,4-diketones, **3**, 453
    with 1,5-diketones, **3**, 455
    with *O*-ethyl selenoacetate, **5**, 642
    with heterocyclic compounds, **3**, 453
    with ketenimines, **5**, 770
    with γ-keto esters, **3**, 453
    with ketones, **1**, 41
    with α-keto perfluoroalkane carboxylic esters, **4**, 311
    with naphthols, **2**, 769
    with nitriles, **5**, 768, 848
    with oxidizing agents, **1**, 74
    with 2-pyridinecarbonitrile, **5**, 769
    with selenoesters, **5**, 704
    with β-substituted ketones, **3**, 452
    with thiocarboxyamides, **5**, 769
    with thioimidic esters, **5**, 769
    with α,β-unsaturated esters, **3**, 453
  for the reduction of aldehydes, **1**, 18
  for the reduction of ketones, **1**, 19
  as starting material
    for azines, **3**, 444
    for hydrazones, **3**, 444

use in the synthesis of amines, **2**, 299
Hydrazine, 1-acetyl-2-benzyl-, oxidation of, **5**, 366
Hydrazine, *N*-acyl-*N'*-triflyl-
reactions, with alkyl halides, **5**, 363
as starting material, for acylhydrazones, **5**, 363
Hydrazine, benzyl-, alkylation of, **2**, 378
Hydrazine, *N*-benzyl-*N*-phenyl-, reactions, with mercury(I) oxide, **3**, 447
Hydrazine, *N*,*N*-bis(dibromomethylene)-, reactions, with silver(II) fluoride, **6**, 242
Hydrazine, bis(thiobenzoyl)-, Cope rearrangement of, **5**, 736
Hydrazine, *N*,*N'*-bis(trifluoroacetyl)-, as starting material, for bisimidoyl chlorides, **5**, 676
Hydrazine, bis(triphenylphosphoranediyl)-, synthesis of, **2**, 864
Hydrazine, *t*-butyl-, synthesis of, **3**, 452
Hydrazine, *N*-carboxymethyl-*N*-trimethylsilyl-, as starting material, for carbamoyl fluorides, **6**, 438
Hydrazine, *N*-chloroperfluoroalkyl-, reactions, with trifluoroacetonitrile, **5**, 679
Hydrazine, 1,1-diacyl-, synthesis of, via amination of imides, **5**, 360
Hydrazine, 1,2-diacyl-, reactions, with trimethylsilyl chloride, **5**, 380
Hydrazine, 1,2-diacyl bis(trimethylsilyl)-, reactions, with nitrogen pentoxide, **5**, 362
Hydrazine, dimethyl-, reactions, with epoxides, **3**, 617
Hydrazine, 1,1-dimethyl-
acylation of, **4**, 897
reactions
with 2-(methoxycarbonylcyanomethylene)-1,3-dioxolane, **4**, 897
with *N*-phenylbenzimidoyl chloride, **5**, 850
Hydrazine, 1,2-dimethyl-
from, diethyl hydrazodicarboxylate, **2**, 377
reactions, with *N*-phenylbenzimidoyl chloride, **5**, 770
Hydrazine, *N*,*N*-dimethyl-, reactions, with dithiocarbonate esters, **6**, 560
Hydrazine, (dinitromethyl)-, synthesis of, **6**, 369
Hydrazine, *o*,*p*-dinitrophenyl-, reactions, with dichloroacetaldehyde, **3**, 451
Hydrazine, diphenyl-, synthesis of, **5**, 848
Hydrazine, 1-ethoxycarbonyl-1-methyl-, synthesis of, **6**, 489
Hydrazine, ethyl-
dibenzoylation of, **2**, 376
synthesis of, **2**, 376
Hydrazine, ethylmethyl-, from, ethylamine, **2**, 379
Hydrazine, 1,2-ethylmethyl-, synthesis of, **2**, 376
Hydrazine, methyl-
acylation of, **5**, 351
reactions
with carbon disulfide, **6**, 560
with α-chloroaldehyde, **2**, 771
with diethyl carbonate, **6**, 489
Hydrazine, 1-methyl-1-phenyl-, reactions, with ethyl thioformate, **5**, 853
Hydrazine, *N*-methyl-*N*-phenyl-
reactions
with benzonitrile, **5**, 848
with sugars, **3**, 451
Hydrazine, *p*-methylphenyl-, reactions, with formaldehyde, **3**, 446
Hydrazine, perfluoro-, reactions, with trifluoroethene, **5**, 663
Hydrazine, phenyl-
decomposition of, **2**, 769
reactions
with alkynyl acids, **3**, 453
with carbon dioxide, **5**, 68
with dichloroacetaldehyde, **3**, 451
with *N*,*N*-disubstituted amidines, **5**, 851
with ketones, **3**, 452
with nitrous acid, **2**, 806
with oximes, **3**, 469
with perfluoro-1-alkenes, **4**, 799
with perfluoro-1-hexene, **5**, 679
synthesis of, **2**, 769
use in the synthesis of hydrazonoyl fluorides, **5**, 839
use in the synthesis of N^1-phenylhydrazonamide, **5**, 848
Hydrazine, tetrafluoro-
for addition to alkenes, **2**, 335
photolysis of, **2**, 170
reactions
with acyl fluorides, **6**, 52
with alkenes, **4**, 104
with alkynic compounds, **2**, 762
with allenes, **2**, 762
with cyclopentanone enol acetate, **5**, 314
with dihalothiocarbonyls, **2**, 166
with α-diketones, **5**, 314
with fluorotribromomethane, **6**, 613
with haloalkenes, **5**, 811
with hexafluoro-2-butyne, **2**, 762
Hydrazine, *p*-toluenesulfonyl-, reactions, with enones, **1**, 1082
Hydrazine, tosyl-
reactions
with aldehydes, **3**, 446
with α-diketones, **3**, 446
with ketones, **3**, 446
Hydrazine, 1-trifluoroacyl 2,2-dimethyl-, reactions, with boron trifluoride etherate, **5**, 383
Hydrazine, (2,4,6-triisopropylbenzenesulfonyl)-, decomposition of, **1**, 74
Hydrazine, trimethyl-
oxidation of, **3**, 447
reactions, with formaldehyde, **4**, 437
Hydrazine, 1,1,2-tri-*n*-propyl-, use in the synthesis of acylhydrazines, **5**, 356
Hydrazine aminals, oxidation of, **5**, 359
Hydrazinecarbodithioates, allyl-, synthesis of, **6**, 567
Hydrazine disulfonates, synthesis of, **2**, 769
Hydrazine hydrate
reactions
with *N*-aryl thioamides, **5**, 770
with chloroformates, **6**, 488
with dialkyl dicarbonates, **6**, 489
with esters, **5**, 350
with *O*-ethyl selenocarboxylates, **5**, 522
with fluoroalkenes, **5**, 839
with lactones, **5**, 350
Hydrazine hydrochloride, phenyl-, reactions, with imino esters, **5**, 705
Hydrazines
acetylation of, **4**, 897
N-acylated, **4**, 986
alkylation of, **2**, 378
criss-cross addition of, **2**, 383
from
alkenes, **2**, 376
amineimines, **2**, 384
amines, **2**, 374, 382
N-aminoimides, **2**, 379
azines, **2**, 377, 377
azo compounds, **2**, 377, 382, 383
azo esters, **2**, 375
azomethines, **2**, 374
diaziridines, **2**, 374, 377
diaziridinones, **2**, 381
diazo compounds, **2**, 376
hydrazides, **2**, 374, 377, 381
hydrazines, **2**, 376, 382
hydrazinium salts, **2**, 380
hydrazones, **2**, 374, 377, 380, 382
hydroxylamine-*O*-sulfonic acid, **2**, 384
O-mesitylenesulfonylhydroxylamine, **2**, 384
nitroamines, **2**, 379

    N-nitrosoamines, **2**, 379, 381
    oxaziridines, **2**, 374, 379
    sulfamides, **2**, 374, 378
    sydnones, **2**, 374
    tetrazenes, **2**, 383
    ureas, **2**, 374, 378
  Michael addition of, **2**, 380, 383, 384, 393
  oxidation of, **1**, 442, 962; **2**, 772; **3**, 447
  reactions
    with acid chlorides, **5**, 351
    with acyl azides, **5**, 351
    with aldehydes, **3**, 444
    with *β*-aldehydo esters, **3**, 453
    with *O*-alkyl thioesters, **5**, 640
    with amides, **5**, 772
    with carboxylic acids, **5**, 351
    with diazonium salts, **2**, 777
    with *α*-diketones, **3**, 450
    with dithiocarboxylic acid esters, **5**, 639
    with exomethylene, **6**, 141
    with formaldehyde, **4**, 436
    with *α*-halocarbonyls, **3**, 451
    with *α*-hydroxy ketones, **3**, 450
    with imidic esters, **5**, 769
    with imidoyl chlorides, **5**, 770
    with isothiocyanates, **6**, 573
    with ketene-*O*,*O*-acetals, **5**, 706
    with *β*-keto esters, **3**, 453
    with ketones, **3**, 444
    with mercury(I) oxide, **3**, 447
    with metal salts of dithioacids, **5**, 639
    with methyl dithioacetate, **5**, 639
    with nitriles, **5**, 770
    with nitroketene dithioacetals, **4**, 983
    with oxirane, **2**, 378
    with phosgene, **6**, 444
    with tetranitromethane, **4**, 995
    with thioamides, **4**, 984
    with thionoformate esters, **5**, 640
    with 1,3,5-triazine, **5**, 770
    with trimethylaluminum, **5**, 350
    with tris-(chloromethyl)amine, **4**, 437
    with *α*,*β*-unsaturated esters, **5**, 351
    with *α*,*β*-unsaturated ketones, **3**, 451
  as starting materials
    for alkyl halides, **2**, 12
    for amines, **2**, 315
    for azo compounds, **2**, 403, 772
    for azoxy compounds, **2**, 404
    for carbazates, **4**, 60
    for diselenocarbazates, **6**, 594
    for hydrazines, **2**, 376, 382
    for hydrazonic acid esters, **5**, 704, 705
    for nitrosohydrazines, **2**, 388
    for quaternary hydrazinium salts, **2**, 383, 384
    for selenosemicarbazides, **6**, 598
    for *N*-substituted amidrazones, **5**, 769
    for sulfinylhydrazines, **2**, 386
    for tetrazenes, **2**, 388
    for *N*-thionitrosoamines, **2**, 385
    for thioureas, **6**, 573
  synthesis of, **2**, 387, 769
    via *N*-amination, **2**, 379
  thioacylation of, **5**, 639
Hydrazines, acyl-
  chlorination of, **5**, 771
  condensation of, **5**, 367
  dehydrogenation of, **5**, 366
  deprotection of, **5**, 351
  from
    azo alcohols, **5**, 357
    N-heteroamides, **5**, 353
    hydroxamic acids, **5**, 353
    phenyldiazomethane, **5**, 358
  oxidation of, **5**, 365, 367
  reactions
    with diazonium salts, **5**, 365
    with dimethylformamide dimethyl acetal, **5**, 773
    with *α*-halo ketones, **5**, 367
    with imidic esters, **5**, 773
    with isothiocyanates, **6**, 573
    with nitriles, **5**, 773
  as starting materials
    for acyl dithiocarbazates, **6**, 563
    for *N*-acylhydrazonamides, **5**, 849
    for acylhydrazones, **5**, 363
  synthesis of
    via acylation of hydrazine, **5**, 350
    via acylation of *N*-silylhydrazines, **5**, 353
    via alkylation of acyldiazines, **5**, 354
    via amination of amides, **5**, 354
    via carbon monoxide insertion, **5**, 356
    via hydrolysis of thioacylhydrazines, **5**, 358
    via oxidation of hydrazine aminals, **5**, 359
    via oxidation of hydrazones, **5**, 356
    via oxidative cleavage of 3-pyrazolin-5-one, **5**, 357
    via 2+2 reactions of hydrazones, **5**, 355
    via rearrangement of phenoxyhydrazones, **5**, 358
    via reduction of acylhydrazones and acyldiazines, **5**, 354
    via reduction of acyl nitrosoamines, **5**, 353
    via ring-opening of diaziridinone, **5**, 359
    via solvolysis of heterocycles, **5**, 354
Hydrazines, 1-acyl-
  reactions
    with acetals, **5**, 363
    with imines, **5**, 363
    with sodium nitrate, **5**, 367
    with trichloromethylarenes, **5**, 363
Hydrazines, acylaminomethyl-, synthesis of, **4**, 448
Hydrazines, acyl arylsulfonyl-, oxidation of, **5**, 209
Hydrazines, 1-acyl-2-arylsulfonyl-, decomposition of, **3**, 85
Hydrazines, 2-acyl-1,1-disubstituted, synthesis of, **1**, 260
Hydrazines, acyl silyl-, synthesis of, **5**, 380
Hydrazines, alkoxy-, synthesis of, **2**, 387
Hydrazines, alkyl-, synthesis of, **2**, **371–423**
Hydrazines, *N*-alkyl-
  reactions
    with aldehydes, **3**, 446
    with ketones, **3**, 446
Hydrazines, 1-alkyl-1-(2,3-butadienyl)-, oxidation of, **2**, 770
Hydrazines, amino-, from, alkyl chlorides, **2**, 373
Hydrazines, *γ*-amino-, from, *β*-cyanoethylhydrazines, **2**, 381
Hydrazines, aminomethyl-, synthesis of, **4**, 441, 444
Hydrazines, arenesulfonyl-
  decomposition of, **1**, 74
  synthesis of, **2**, 726
Hydrazines, 1-aroyl-2-*N*,*N'*-diaryl-, from, hydrazonyl aryl ethers, **5**, 358
Hydrazines, *N*-aroyl-*N*-methyl-, as starting materials, for benzylmethylhydrazines, **2**, 379
Hydrazines, aryl-
  from, aromatic iodides, **2**, 769
  reactions
    with alcohols, **3**, 451
    with aldehydes, **3**, 446
    with aldoses, **3**, 450
    with carbon disulfide, **6**, 560
    with dimethylformamide dimethyl acetals, **5**, 770
    with formaldehyde, **3**, 446
    with *α*-hydroxy ketones, **3**, 451
    with imino ethers, **5**, 859
    with ketones, **3**, 446
    with ketoses, **3**, 450
    with *N*-phenylbenzimidoyl chloride, **5**, 850
    with *β*-substituted ketones, **3**, 452
    with *α*,*β*-unsaturated ketones, **3**, 451

as starting materials
for diazonium salts, **2**, 806
for formazans, **5**, 859
Hydrazines, *N*-aryl-*N*-benzyl-*N'*-tosyl-, as starting materials, for hydrazones, **3**, 446
Hydrazines, benzoyl-, reactions, with imino esters, **4**, 986
Hydrazines, 1-benzoyl-2-arenesulfonyl-, reactions, with thionyl chloride, **5**, 676
Hydrazines, 1-benzoyl-2-phenyl-, from, 1-hydroxy-1-azocyclohexanes, **5**, 357
Hydrazines, benzylmethyl-, from, *N*-aroyl-*N*-methylhydrazines, **2**, 379
Hydrazines, bis-, synthesis of, **4**, 437
Hydrazines, *N,N'*-bis(alkoxycarbonyl)-, synthesis of, **6**, 489
Hydrazines, *N,N'*-bis(aminocarbonyl)-, oxidation of, **6**, 506
Hydrazines, N^1,N^2-bisphosphinomethylene-N^1,N^2-dimethyl-, synthesis of, **4**, 482
Hydrazines, *N*-*t*-butyl-*N*-cyano-*N'*-toluenesulfonyl-, synthesis of, **5**, 1123
Hydrazines, *N*-chloro perfluoroalkyl-, as starting materials, for hydrazonyl halides, **5**, 679
Hydrazines, α-cyano-, synthesis of, **2**, 378
Hydrazines, β-cyanoethyl-
from, acrylonitriles, **2**, 376
as starting materials, for γ-aminohydrazines, **2**, 381
Hydrazines, 1,1-diacyl-, synthesis of, via acylation of amines, **5**, 359
Hydrazines, 1,2-diacyl-
from, 4-amino-3-pyrazolin-5-one, **5**, 357
synthesis of, **5**, 383
Hydrazines, *N,N*-dialkyl-
reactions
with aldehydes, **3**, 446
with ketones, **3**, 446
Hydrazines, N^1,N^2-dialkyl-, reactions, with bis(hydroxymethyl)phosphine, **4**, 482
Hydrazines, *N,N'*-di-*t*-butyl acyl-, synthesis of, **5**, 359
Hydrazines, *N',N'*-dimethyl-, cleavage of, **2**, 315
Hydrazines, dinitroso-, synthesis of, **2**, 389
Hydrazines, 1,1-disubstituted
reactions
with dicyanoketene acetals, **5**, 845
with *s*-triazine, **5**, 855
synthesis of, **1**, 260
Hydrazines, *N,N*-disubstituted
reactions
with carboxylic ortho esters, **5**, 705
with oxiranes, **2**, 384
Hydrazines, *N,N'*-disubstituted
as starting materials, for β-hydrazino esters, **2**, 381
synthesis of, **2**, 376, 378
Hydrazines, fluorocarbonyl-, synthesis of, **6**, 439
Hydrazines, homochiral, synthesis of, **2**, 376
Hydrazines, α-hydroxy acyl-, synthesis of, **5**, 356
Hydrazines, hydroxylamino-, synthesis of, **4**, 441
Hydrazines, hydroxylaminomethyl-, reduction of, **4**, 444
Hydrazines, methoxyaminocarbonyl-, synthesis of, **4**, 986
Hydrazines, methoxycarbonyl-, synthesis of, **4**, 986
Hydrazines, *N*-methoxymethyl-, chlorination of, **4**, 105
Hydrazines, monoalkyl-, synthesis of, **2**, 373
Hydrazines, naphthyl-, synthesis of, **2**, 769
Hydrazines, nitro-, synthesis of, **2**, 398
Hydrazines, nitromethylacyl-, diazotization of, **4**, 440
Hydrazines, nitroso-
from, hydrazines, **2**, 388
synthesis of, **2**, 388
Hydrazines, *N*-nitroso-1,2-diacyl-, synthesis of, **5**, 362
Hydrazines, phenyl-, cyanation of, **5**, 1125
Hydrazines, *N*-silyl-
acylation of, **5**, 353
reactions, with acyl fluorides, **5**, 353
Hydrazines, sulfinyl-
from, hydrazines, **2**, 386
as starting material, for *N*-thionitrosoamines, **2**, 385
synthesis of, **2**, 386
Hydrazines, sulfonyl-, reactions, with isothiocyanates, **6**, 573
Hydrazines, tetrasubstituted, synthesis of, **2**, 382
Hydrazines, thioacyl-
hydrolysis of, **5**, 358
reactions, with trimethyloxonium tetrafluoroborate, **5**, 358
Hydrazines, *N*-thionoacetyl-, synthesis of, **5**, 640
Hydrazines, *N*-thionoacyl-
as starting materials, for *N*-thionoacylhydrazones, **5**, 642
synthesis of, **5**, 639
Hydrazines, thionyl-, synthesis of, **2**, 386
Hydrazines, tosyl-, as starting materials, for alkyl halides, **2**, 12
Hydrazines, (trinitromethyl)-, synthesis of, **6**, 364
Hydrazines, triphenylphosphoranylidene-, as starting materials, for hydrazonyl halides, **5**, 680
Hydrazines, trisubstituted
alkylation of, **2**, 380
synthesis of, **2**, 380
Hydrazine sulfate, reactions, with bis(trimethylsilyl)amine, **6**, 626
Hydrazine tetrafluoride, reactions, with vinylsulfur pentafluoride, **5**, 814
Hydrazinium chloride, trimethyl-, reactions, with potassamide, **2**, 384
Hydrazinium iodide, trimethyl-, synthesis of, **2**, 384
Hydrazinium nitrates, dehydration of, **2**, 384
Hydrazinium salts, as starting materials, for hydrazines, **2**, 380, 382
Hydrazinium salts, quaternary
from
amines, **2**, 384
hydrazines, **2**, 383, 384
synthesis of, **2**, 383
α-Hydrazino acids
from, α-amino acids, **2**, 374
synthesis of, **5**, 87
Hydrazino compounds, aminomethyl, synthesis of, **4**, 444
α-Hydrazino esters, saponification of, **5**, 87
β-Hydrazino esters, from, *N,N'*-disubstituted hydrazines, **2**, 381
Hydrazoamides, cyano-, synthesis of, **5**, 847
Hydrazoic acid
reactions
with carboxylic acids, **5**, 1001
with imines, **4**, 444
with ketones, **5**, 283
with silyl enol ethers, **4**, 314
with trimethylbismuthine, **2**, 504
Hydrazonamide, N^1-phenyl-, synthesis of, **5**, 848
Hydrazonamide, N^2-substituted, synthesis of, **5**, 850
Hydrazonamides
from
acylaminomalonates, **5**, 854
amides, **5**, 851
amidines, **5**, 850
arylthiocarboxanilides, **5**, 852
carbodiimides, **5**, 849
formylhydrazonamides, **5**, 849
heterocycles, **5**, 854
hydrazonic esters, **5**, 852
hydrazonic thioesters, **5**, 852
hydrazonoyl halides, **5**, 854
imidic esters, **5**, 849
imidoyl halides, **5**, 850
ketimines, **5**, 849
nitriles, **5**, 847
1-nitroaldehydehydrazones, **5**, 854
oxalic acids, **5**, 852
thioamides, **5**, 852
thioimidic esters, **5**, 850

1,2,4-triazolium salts, **5**, 855
phosphorane compounds of, synthesis of, **5**, 862
as starting materials, for N^3-hydroxyhydrazonamides, **5**, 856
synthesis of, **5**, 847
via reduction, **5**, 855
Hydrazonamides, acryl-, synthesis of, **5**, 850
Hydrazonamides, *N*-acyl-, from, acylhydrazines, **5**, 849
Hydrazonamides, N^2-alkylated, synthesis of, **5**, 855
Hydrazonamides, N^3-alkylated, synthesis of, **5**, 855
Hydrazonamides, N^1,N^1-disubstituted, synthesis of, **5**, 850
Hydrazonamides, N^1,N^3-disubstituted, synthesis of, **5**, 850, 851
Hydrazonamides, formyl-, as starting materials, for hydrazonamides, **5**, 849
Hydrazonamides, N^3-hydroxy-
from, hydrazonamides, **5**, 856
synthesis of, **5**, 856
Hydrazonamides, silyl-, synthesis of, **5**, 850
Hydrazonamides, *N*-substituted, from, *N*-substituted imidic esters, **5**, 849
Hydrazonamides, N^1-substituted, synthesis of, **5**, 850
Hydrazonamides, N^2-substituted, synthesis of, **5**, 850
Hydrazonamides, N^3-substituted, from, N^3-unsubstituted hydrazonamides, **5**, 854
Hydrazonamides, N^1,N^1,N^3-trisubstituted, synthesis of, **5**, 850
Hydrazonamides, N^1,N^2,N^3-trisubstituted, synthesis of, **5**, 852
Hydrazonamides, N^3-unsubstituted, as starting materials, for N^3-substituted hydrazonamides, **5**, 854
Hydrazonate esters
reactions
with amines, **5**, 772
with ammonia, **5**, 772
with dimethylamine, **5**, 773
Hydrazonates
from
heteroarylhydrazides, **5**, 845
ketene acetals, **5**, 845
See also Acid esters, hydrazonic
Hydrazonates, aliphatic, from, nitrenes, **5**, 845
Hydrazonazides, as starting materials, for hydrazonimidophosphoranes, **5**, 862
Hydrazone, acetone 4-chlorophenyl-, bromination of, **4**, 329
Hydrazone, acetophenone dimethyl-, quaternization of, **2**, 384
Hydrazone, acetophenone phenyl-
bromination of, **5**, 842
oxidation of, **4**, 314
Hydrazone, *N*-aminoaziridine-, use in the synthesis of alkynyl-substituted alkyl aldehydes, **3**, 25
Hydrazone, benzaldehyde phenyl-, chlorination of, **5**, 840
Hydrazone, benzophenone-, reactions, with lead(IV) acetate, **3**, 464
Hydrazone, benzophenone 1-aminoethylidene-, reactions, with triphenylphosphine, **5**, 862
Hydrazone, cinnamaldehyde benzoyl-, reactions, with diethyl phosphite, **4**, 1008
Hydrazone, α,*β*-dichlorocrotonaldehyde 2,4,6-trichlorophenyl-, chlorination of, **5**, 841
Hydrazone, 2,4-dinitrophenyl-, oxidation of, **4**, 313
Hydrazone, *o*,*p*-dinitrophenyl-, reactions, with nitrones, **3**, 445
Hydrazone, 9-fluorenecarbaldehyde tosyl-, reactions, with antimony pentachloride, **2**, 800
Hydrazone, ketene-, synthesis of, **3**, 560
Hydrazone, *N*-methyl-, reactions, with cyclohexane-1,3-dione, **3**, 264
Hydrazone, *N*-methylene-*p*-methylphenyl-, synthesis of, **3**, 446
Hydrazone, nitroformaldehyde aryl-, as starting material, for *gem*-arylazo nitroalkenes, **4**, 995
Hydrazone, 2-phenyl-, chlorination of, **5**, 841
Hydrazone, phthaloyl-, synthesis of, **5**, 364
Hydrazone, tosyl-, use in the synthesis of α-diazophosphorus compounds, **5**, 870
5-Hydrazone, benzaldehyde tetrazolyl-, reactions, with bromine, **5**, 841
Hydrazone lithio anions, dimethyl-, synthesis of, **2**, 770
Hydrazone *N*-oxides, as starting materials, for hydroxylamines, **2**, 347
Hydrazones
as acyl anion equivalents, **1**, 165
acylation of, **5**, 362
bromination of, **3**, 242
dehydrogenation of, **3**, 463, 478
electrophilic halogenation of, **4**, 109
from
aniline, **3**, 446
arylalkyl ketones, **3**, 444
N-aryl-*N*-benzyl-*N'*-tosylhydrazines, **3**, 446
azoalkenes, **6**, 604
azo compounds, **3**, 448
diazonium salts, **3**, 448
hydrazine, **3**, 444
oximes, **3**, 446
halogenation of, **5**, 677
niobium-promoted reactions, with alkynes, **2**, 988
for the opening of epoxides, **1**, 212
oxidation of, **3**, 477, 666; **5**, 356, 870, 873
reactions
with acyl chlorides, **5**, 362
with anhydrides, **5**, 362
with *s*-butyllithium, **1**, 1091
with carboxyalkylphosphines, **4**, 482
with diketones, **3**, 264
with disulfur dichloride, **3**, 363
with hydrogen cyanide, **3**, 616
with iodine, **4**, 29
with mercury(II) oxide, **3**, 470
with *O*-mesitylenesulfonyl hydroxylamine, **4**, 444
with organoceriums, **1**, 1042
with organolithiums, **3**, 44
with selenium(I) bromide, **3**, 392
with silyl triflates, **1**, 186
with sulfones, **3**, 172
with sulfur dichloride, **5**, 641
with *p*-toluene sulfinate, **3**, 37
2+2 reactions of, **5**, 355
reactivity of, **1**, 1042
reduction of, **1**, 944; **2**, 374, 377, 381
as starting materials
for *O*,*N*-acetals, **4**, 313
for aldehydes, **3**, 18
for alkenes, **1**, 635
for alkynes, **1**, 1075
for azines, **3**, 444
for azo compounds, **2**, 403
for chiral auxiliaries, **1**, 153
for dibromoalkanes, **4**, 27
for formazans, **5**, 856
for hydrazines, **2**, 374, 377, 380, 381, 382
for hydrazonoyl halides, **5**, 839
for ketones, **3**, 133
for methylene, **1**, 43
for nitrile imides, **3**, 681
for thioaldehydes, **3**, 340
for thioketones, **3**, 347, 363
for vinyl iodides, **2**, 630
synthesis of, **3**, 444; **6**, 608, 623
tantalum-promoted reactions, with alkynes, **2**, 988
Hydrazones, α-acetoxy-, synthesis of, **3**, 451
Hydrazones, acyl-
from
acylhydrazines, **5**, 363

N-acyl-*N'*-triflylhydrazine, **5**, 363
azines, **5**, 364
enolates, **5**, 364
semicarbazones, **5**, 363
tetrazoles, **5**, 363
reduction of, **5**, 354
as starting materials, for methyl compounds, **1**, 42
synthesis of
via acylation of hydrazones, **5**, 362
via oxidation of conjugated hydrazones, **5**, 363
via sulfur dioxide extrusion by thiadiazolines, **5**, 364
Hydrazones, aldehyde-
chlorination of, **3**, 681
oxidation of, **5**, 366
Hydrazones, aldehyde *t*-butyl-, as acyl anion equivalents, **1**, 153
Hydrazones, alkyl-*p*-toluenesulfonyl-, as starting materials, for alkenes, **1**, 39
Hydrazones, arenesulfonyl-
as starting materials
for alkenes, **1**, 630
for alkenylsilanes, **2**, 911
Hydrazones, aryl-
from, arenediazonium salts, **3**, 448
halogenation of, **5**, 678
nitration of, **5**, 704
reactions
with arenediazonium salts, **5**, 856
with cetyltrimethylammonium permanganate, **3**, 134
as starting materials, for bisformazans, **5**, 856
synthesis of, **6**, 611, 634
Hydrazones, *N*-aryl-, synthesis of, **3**, 446
Hydrazones, arylsulfonyl-
cleavage of, **3**, 464
as starting materials
for diazo compounds, **1**, 1077
for vinyl anions, **1**, 1091
Hydrazones, azido-, synthesis of, **6**, 659
Hydrazones, benzaldehyde-, reactions, with bromine, **5**, 841
Hydrazones, benzenesulfonyl-, reactions, with diethylene glycol, **3**, 446
Hydrazones, benzoyl-
reactions, with lead tetraacetate, **6**, 107
as starting materials, for *gem*-phosphorylamino phosphorylalkenes, **4**, 1008
Hydrazones, bis(arylsulfonyl)formaldehyde-, dehydrogenation of, **3**, 478
Hydrazones, *t*-butyl-, metallation of, **4**, 536
Hydrazones, carbomethoxy-, rearrangement of, **1**, 832
Hydrazones, chloro-, reactions, with nitrogen nucleophiles, **6**, 659
Hydrazones, α-chloro tosyl-, reactions, with selenate anions, **3**, 174
Hydrazones, conjugated, oxidation of, **5**, 363
Hydrazones, cyclic, synthesis of, **3**, 451
Hydrazones, *N*,*N*-diacyl-
from
N-nitrenyl phthalimide, **5**, 364
semicarbazones, **5**, 364
synthesis of
via rearrangement of 1,2-diacyl diaziridines, **5**, 365
via ring-opening of *N*-aziridylphthalimides, **5**, 364
Hydrazones, *N*,*N*-dialkyl-
as carbonyl equivalents, **1**, 153
deprotonation of, **1**, 153
reactions, with hydrazine, **3**, 445
Hydrazones, *N*,*N*-diethylaminosulfonyl-, use in Shapiro reaction, **1**, 630
Hydrazones, dimethyl-
as starting materials
for ketones, **3**, 133
for nitriles, **3**, 617
synthesis of, **3**, 447
Hydrazones, *N*,*N*-dimethyl-, synthesis of, **6**, 611
Hydrazones, 2,4-dinitrobenzenesulfonyl-, use in the synthesis of alkynes, **1**, 1082
Hydrazones, dinitrophenyl-
as precursors of aldehydes, **3**, 18
as starting materials, for ketones, **3**, 133
Hydrazones, *N*,*N*-disubstituted, synthesis of, **3**, 446
Hydrazones, formyl-
as starting materials
for dialkylamino isocyanides, **3**, 709
for imino isocyanides, **3**, 710
Hydrazones, glyoxal β-aryl-, synthesis of, **3**, 449
Hydrazones, 2-halo-, synthesis of, **5**, 367
Hydrazones, α-halo-
reactions, with methanol, **3**, 451
as starting materials
for α-selenylhydrazones, **3**, 39
for α,β-unsaturated azo compounds, **2**, 771
synthesis of, **3**, 451
Hydrazones, α-halotosyl-, reactions, with phenylcoppers, **1**, 464
Hydrazones, ketene-, synthesis of, **3**, 586
Hydrazones, keto-, reactions, with α-diazo ketones, **5**, 362
Hydrazones, α-keto-, dehydrogenation of, **3**, 469
Hydrazones, ketone-
as α-amino carbanion equivalents, **1**, 154
synthesis of, **3**, 449
Hydrazones, ketone tosyl-, reductive cyclization of, **1**, 944
Hydrazones, α-keto tosyl-, Bamford–Stevens reaction of, **3**, 470
Hydrazones, metallated, reactions, with oxygen, **3**, 464
Hydrazones, α-methoxy-, synthesis of, **3**, 451
Hydrazones, methylene-, synthesis of, **3**, 446
Hydrazones, methylsulfinyl-, cleavage of, **3**, 464
Hydrazones, monoalkyl-, synthesis of, **3**, 446
Hydrazones, *N*-monoalkyl-, deprotonation of, **1**, 153
Hydrazones, nitro-, synthesis of, **4**, 440; **6**, 659
Hydrazones, α-nitro-, alkylation of, **3**, 190
Hydrazones, 1-nitroaldehyde-
reactions, with ammonia, **5**, 854
as starting materials, for hydrazonamides, **5**, 854
Hydrazones, α-nitroaldehyde-, as starting materials, for nitrile imides, **3**, 681
Hydrazones, *p*-nitrophenyl-, synthesis of, **3**, 449
Hydrazones, 2-nitrophenylsulfenyl-, cleavage of, **3**, 464
Hydrazones, perfluoro-
reactions
with chlorine fluoride, **6**, 260
with methyl iodide, **6**, 260
Hydrazones, phenoxy-, rearrangement of, **5**, 358
Hydrazones, phenyl-
flash vacuum pyrolysis of, **5**, 1123
oxidation of, **4**, 314; **5**, 357
reactions
with benzenediazonium salts, **5**, 857
with chlorine, **2**, 805
with hydroxylamines, **4**, 441
tautomerization of, **3**, 448
Hydrazones, α-phenyl-, synthesis of, **5**, 68
Hydrazones, phosphoranylidine-
as starting materials, for selones, **3**, 391
synthesis of, **3**, 391
Hydrazones, α-selenyl-, from, α-halohydrazones, **3**, 39
Hydrazones, *N*-substituted, synthesis of, **3**, 446
Hydrazones, α-sulfenyl-, synthesis of, **3**, 37
Hydrazones, sulfonyl-, as starting materials, for diazonium salts, **2**, 800
Hydrazones, tetrazolyl-, bromination of, **6**, 611
Hydrazones, *N*-thionoacyl-
from, *N*-thionoacylhydrazines, **5**, 642
synthesis of, **5**, 639
Hydrazones, toluene-*p*-sulfonyl-, decomposition of, **1**, 630
Hydrazones, *p*-toluenesulfonyl-

reactions
with lithium aluminumhydride, **1**, 39
with sodium borohydride, **1**, 39
Hydrazones, toluene-*p*-sulfonyl-, as starting materials, for vinyllithiums, **1**, 630
Hydrazones, tosyl-
Bamford–Stevens reaction of, **3**, 464, 489
cleavage of, **1**, 1082
as precursors of aldehydes, **3**, 18
reactions
with alkoxides, **1**, 1077
with alkyllithiums, **1**, 630, 632
with catecholborane, **1**, 635
with hydrides, **1**, 822
with lithium diisopropylamide, **1**, 630
with metal hydrides, **1**, 18
with potassium carbonate, **3**, 471
with sodium borohydride, **1**, 19
with sodium cyanoborohydride, **1**, 19, 635
reduction of, **1**, 962
as starting materials
for alkenes, **1**, 635
for ketones, **3**, 133
for pregnenynes, **1**, 1077
for vinyllithiums, **2**, 956
synthesis of, **3**, 446; **6**, 627
for the synthesis of alkynes, **1**, 1077
vacuum pyrolysis of, **3**, 465
Hydrazones, *N*-tosyl-, as starting materials, for alkenyllithiums, **2**, 946
Hydrazones, *p*-tosyl-, as starting materials, for ketene dithioacetals, **4**, 852
Hydrazones, triisopropylbenzenesulfonyl-, cleavage of, **3**, 465
Hydrazones, (2,4,6-triisopropylbenzenesulfonyl)-, use in Shapiro reaction, **1**, 630, 633
Hydrazones, triphenylphosphoranyl-
reactions, with selenium, **1**, 686
as starting materials, for diazo compounds, **1**, 685
Hydrazones, triphenylphosphoranylidene-, reactions, with selenium, **2**, 858
Hydrazones, triphenylphosphorylidene-, thermal decomposition of, **3**, 347
Hydrazones, α,β-unsaturated tosyl-
fragmentation of, **1**, 1082
synthesis of, **1**, 597
Hydrazones, *N*-unsubstituted, synthesis of, **3**, 444
Hydrazones, unsymmetrical, metallation of, **1**, 153
Hydrazone salts, tosyl-, vacuum pyrolysis of, **3**, 489
Hydrazonic acid esters, from, benzotrichlorides, **5**, 703
Hydrazonoates, aryl-, synthesis of, **5**, 707
Hydrazonoates, phenyl-, synthesis of, **5**, 704
Hydrazono compounds, synthesis of, **6**, 659
Hydrazono-2,4-pentanediones, 3-aryl-, reactions, with bromine, **5**, 679
Hydrazonoyl bromides
from
dioxobutyric esters, **5**, 841
dioxosuccinic ester, **5**, 841
Hydrazonoyl chloride, *N*-chloro-, synthesis of, **5**, 841
Hydrazonoyl chloride, polychlorinated, synthesis of, **5**, 841
Hydrazonoyl chlorides
reactions, with alkoxides, **5**, 703
as starting materials, for arylazomethylenetriphenylphosphoranes, **5**, 905
synthesis of, **3**, 681
Hydrazonoyl chlorides, 2,4,6-trichloro-, synthesis of, **5**, 840
Hydrazonoyl compounds, synthesis of, **5**, 839
Hydrazonoyl fluorides, synthesis of, **5**, 839
Hydrazonoyl halides
from
aliphatic diazo compounds, **5**, 843
fluoroalkenes, **5**, 839
2-halo-1,3-dicarbonyls, **5**, 843
hydrazides, **5**, 842
hydrazones, **5**, 839
as starting materials
for formazans, **5**, 859
for hydrazonamides, **5**, 854
for hydrazonic acid esters, **5**, 703
for nitrile imides, **3**, 678, 680
synthesis of, **5**, 839
Hydrazonoyl halides, aromatic, as starting materials, for hydrazonic acid esters, **5**, 844
Hydrazonyl bromides, synthesis of, **5**, 839
Hydrazonyl chloride, reactions, with thioketones, **6**, 128
Hydrazonyl chlorides, aryl-, as starting materials, for ylides, **5**, 905
Hydrazonyl halides
from
2-arylhydrazino-3-oxobutyric acid esters, **5**, 678
carboxylic acid hydrazides, **5**, 675
N-chlorinated perfluoroalkylamines, **5**, 679
N-chloro perfluoroalkylhydrazines, **5**, 679
diazene-*N*-oxides, **5**, 680
diazonium halides, **5**, 676
halogen carriers, **5**, 675
hydrazonyl halides, **5**, 681
perfluorinated alkenes, **5**, 679
trifluoroacetonitrile, **5**, 679
triphenylphosphoranylidenehydrazines, **5**, 680
reactions, with hydrogen sulfide, **5**, 641
as starting materials, for hydrazonyl halides, **5**, 681
synthesis of, **5**, 675
side reactions, **5**, 677
Hydride, trimethyltin, photochemical addition of, to propargyl acetate, **4**, 701
Hydride, tris(trimethylsilylmethyl)
metallation of, **6**, 391
reactions, with methyllithium, **6**, 391
Hydride abstraction
from tributyltin hydride, **1**, 177
mechanism, **1**, 663
in the substitution of sulfur functions, **1**, 232
Hydride acceptors, use in elimination reactions, **1**, 661
Hydride complexes
reactions
with alkenes, **2**, 576
with aziridines, **5**, 446
with epoxides, **2**, 576
with oxiranes, **2**, 576
Hydride complexes, alkyl, synthesis of, **2**, 582
Hydride complexes, cyclopentadienyl, use in Meerwein–Ponndorf–Verley reduction, **2**, 55
Hydride donors, reactions, with alkylboranes, **2**, 534
Hydride reducing agents, for the synthesis of alkenes, **1**, 509, 514
Hydride reduction
of amines, **1**, 32
of arenes, **1**, 102
of phosphonium salts, **2**, 442
Hydrides
for the cleavage of carbon–phosphorus bonds, **1**, 51
for the deamination of *N*-alkylimine systems, **1**, 45
reactions
with aryl iodides, **1**, 429
with tosylhydrazones, **1**, 822
for the reduction of nitriles, **1**, 48
as starting materials
for arylamines, **2**, 752
for dialkylmagnesiums, **2**, 561
use in the cleavage of nitriles, **1**, 55
Hydrides, nucleophilic, from, sodium borohydride, **1**, 28
Hydrides, silyl
for the reduction of bromoalkanes, **1**, 9
for the reduction of chloroalkanes, **1**, 7

Hydrides, triphenyl, for the reduction of bromoarenes, **1**, 428
Hydrido complexes, synthesis of, **6**, 340
Hydrido compounds, synthesis of, **6**, 331
Hydrin, epichloro-
  reactions
    with aluminate, **1**, 224
    with dimethyl malonate, **1**, 224
Hydrin, trimethylene chloro-, oxidation of, **5**, 59
Hydrindan compounds, synthesis of, **1**, 606
Hydrindane, from, vinylallenes, **1**, 499
Hydrines, silylalkenyl cyano-, synthesis of, **2**, 927
Hydrins, bromo-
  as precursors of ketones, **3**, 114
  synthesis of, **4**, 122, 139
Hydrins, chloro-, synthesis of, **2**, 74; **4**, 146
Hydrins, cyano-
  silylation of, **4**, 359
  as starting materials, for α-silyl enamines, **4**, 1013
Hydrins, epihalo-, use in the synthesis of acetals, **4**, 181
Hydrins, fluoro-, from, alcohols, **2**, 15
Hydrins, halo-
  as byproducts, **1**, 478
  from, epoxides, **1**, 386
  semipinacol rearrangement of, **1**, 390
  as starting materials, for ketones, **1**, 390
  synthesis of, **4**, 146
Hydrins, *O*-trimethylsilyl cyano-, synthesis of, **2**, 544
Hydrins, β,γ-unsaturated cyano-
  from, allyl ethers, **3**, 623
  synthesis of, **3**, 627
Hydriodic acid
  reactions
    with alcohols, **2**, 86
    with *p*-nitrobenzyl cyanide, **5**, 33
Hydroacylation, of alkenes, for the synthesis of ketones, **3**, 113
Hydroalumination
  of alkenes, **2**, 4, 970
  of alkynes, **2**, 967
    with diisobutylalane, **2**, 968
    regiochemistry, **2**, 968
  of alkynylgermanes, **2**, 948
    stereochemistry, **2**, 967
  of alkynylsilanes, **2**, 922; **4**, 808
    stereochemistry, **2**, 967
  vs. carboalumination, **2**, 974
  with diisobutylaluminum hydride, **1**, 161
  and *syn*-hydrozirconation compared, **2**, 969
  of internal alkynes, regioselectivity, **2**, 967
  mechanism, **1**, 453; **2**, 596
  side-reactions, **1**, 511
  for the synthesis of alkenes, **1**, 511, 533
  for the synthesis of alkyl bromides, **2**, 30
  for the synthesis of alkyl chlorides, **2**, 23
  for the synthesis of vinyl iodides, **2**, 629
  use in the synthesis of trialkylaluminums, **2**, 595
anti-Hydroalumination, of propargyl alcohols, **2**, 973
syn-Hydroalumination, mechanism, **2**, 967
Hydroaromatic compounds, polycyclic, aromatization of, **1**, 554
Hydroborating agents, from, monoisopinocampheylorganoboranes, **1**, 487
Hydroboration
  of alkenes, **2**, 4, 514, 525, 527; **2**, 970
  of alkynes, **1**, 449, 531, 1100; **2**, 900, 967
    for the synthesis of ketones, **3**, 114
  of alkynylgermanes, **2**, 948
  of alkynylsilanes, **1**, 161; **2**, 922
  of 1-bromoalkynes, **4**, 814
  enantiocontrol in, **2**, 41
  and *syn*-hydrozirconation compared, **2**, 969
  of 1-iodoalkynes, **4**, 814
  mechanism, **2**, 40
  mediation of, via transition metal catalysis, **2**, 527
  regiochemistry of, **2**, 514
  regioselectivity of, **1**, 1102
  of 1-silylalkynes, **4**, 1056
  stereocontrol in, **2**, 41
  for the synthesis of alkenes, **1**, 509
  for the synthesis of alkyl bromides, **2**, 30
  for the synthesis of dialkyl sulfides, **2**, 138
  for the synthesis of vinyl bromides, **2**, 623
  for the synthesis of vinyl iodides, **2**, 628
  of terminal alkynes, **3**, 3
  uses of, **2**, 966
  use in the synthesis of aldehydes, **3**, 19
  use in the synthesis of ketones, **3**, 134
  of vinylsilanes, **4**, 356, 643
Hydroboration, asymmetric
  for the synthesis of alcohols, **2**, 42
  for the synthesis of alkylboranes, **2**, 519
Hydroboration, chiral, for the synthesis of alkyl iodides, **2**, 34
Hydroboration, rhodium(I)-catalyzed, for the synthesis of alkoxyboranes, **2**, 527
Hydroboration–oxidation, for the synthesis of α,β-unsaturated acylsilanes, **5**, 425
Hydroboration reagents, for the synthesis of alkenylboranes, **2**, 900
Hydrobromic acid
  for the hydrolysis of nitriles, **5**, 25
  reactions
    with *N*,*N*-diphenyl trimethylenediamine, **1**, 258
    with 2-substituted 1,1-dicyanoethylene oxide, **5**, 56
  for silicon–phenyl bond cleavage, **6**, 379
Hydrobromination
  of alkenes, **2**, 29
  of 1-trimethylsilylalkynes, **2**, 929
Hydrocarbons
  acidity of, **1**, 848
  acylation of, **1**, 462
  amination of, **2**, 298
  of bond dissociation energies, **1**, 925
  chemistry of, reviews, **6**, 172
  dehydrogenation of, **1**, 554
  electrophilic mercuration of, **6**, 207
  highly strained, acidity of, **1**, 848
  hydrogen isotope exchange, **1**, 848
  microbial oxidation of, **2**, 38
  oxidation of, **2**, 86
  oxidative coupling of, **1**, 674
  reactions, with lithium vapor, **6**, 204
  as starting materials, for nitroso compounds, **2**, 391
Hydrocarbons, allenic, metallation of, **1**, 971
Hydrocarbons, polyunsaturated, reduction of, **1**, 959
Hydrocarbons, unactivated, halogenation of, **2**, 2
Hydrocarbons, unsaturated acidic, as starting materials, for carbanions, **1**, 866
Hydrocarbon salts, synthesis of, **1**, 867
Hydrocarbon solvents, use in the synthesis of alkyllithiums, **2**, 551
Hydrocarboxylation
  mechanism, **1**, 297
  for the synthesis of carboxylic acids, **1**, 296; **5**, 27
Hydrocarboxylation, asymmetric, for the synthesis of ibuprofen, **1**, 298
Hydrochloric acid
  for the hydrolysis of nitriles, **5**, 25
  reactions
    with alkyl acetyl disulfide, **2**, 211
    with alkylmercury halide, **1**, 65
    with dialkylboranes, **2**, 526
    with 2,2-dimorpholinoacetophenone, **4**, 309
    with dinitro compounds, **1**, 57
    with Friedel–Crafts catalysts, **3**, 92
    with potassium 4-cyanobutanoate, **5**, 34

with silyl ethers, **4**, 160
with thione *S*-imides, **5**, 336
use in the synthesis of alkyl chlorides, **2**, 25
See also Hydrogen chloride
Hydrochlorides, amine, as starting materials, for alkyliminotrihalophosphines, **2**, 407
Hydrochloride salts
reactions, with sodium nitrite, **1**, 259
synthesis of, **4**, 478
Hydrochloride salts, imino ester, reactions, with urea, **4**, 982
Hydrochloride salts, 2-pyrazoline, from, acetone azine, **3**, 452
Hydrochlorofluorocarbons (HCFC's), as replacements for chlorofluorocarbons, **6**, 8
Hydrocobaltation, mechanism, **1**, 335
Hydrocupration, for (*Z*)-5-decene, **2**, 969
Hydrocyanation, of alkynylsilanes, **2**, 928
Hydrofluoric acid
in elimination reactions, **1**, 648
reactions
with formaldehyde, **4**, 44
with potassium thiocyanate, **5**, 593
Hydrofluorination, of alkenes, **2**, 17
Hydrofluorocarbons (HFC's), as replacements for chlorofluorocarbons, **6**, 8
Hydroformylation
of alkenes, **2**, 576; **5**, 442
for aldehydes, **3**, 2
of alkynylsilanes, **2**, 928
mechanism, **1**, 297
of silylalkynes, **3**, 77
for the synthesis of aldehydes, **1**, 296
for the synthesis of γ,δ-unsaturated aldehydes, **3**, 23
Hydroformylation, asymmetric, **1**, 297
Hydroformylation, cobalt-catalyzed, of alkenes, **5**, 456
Hydrogen
addition to ketene dithioacetals, **4**, 252
elimination of, **1**, 1063
[1,*j*] migrations of, **1**, 794
oxidative elimination of, **3**, 225
reactions
with alkynes, **1**, 455
with allenes, **1**, 458
with cobalt(II) ethyl hexanoate, **6**, 525
with oxoketene dithioacetals, **4**, 253
with pentacyanocobalt(III) ions, **1**, 74
replacement of, **2**, 298
substitution of, **1**, 492
Hydrogen, vinyl-, substitution of, **1**, 493
γ-Hydrogen
abstraction of, for the synthesis of 1,4-biradicals, **4**, 1079
for the synthesis of biradicals, **4**, 1079
δ-Hydrogen, for the synthesis of biradicals, **4**, 1079
Hydrogen abstraction, in the substitution of oxygen functions, **1**, 172
Hydrogen abstraction, intermolecular, mechanism, **4**, 1078
α-Hydrogen abstraction, for the synthesis of tantalum carbene complexes, **3**, 511
Hydrogenation
of alkynes, **1**, 455
of alkynylgermanes, **2**, 948
catalytic, for the reduction of enones, **3**, 127
and catalytic transfer hydrogenation reactions compared, **1**, 457
for the reduction of acyl chlorides, **3**, 83
for the reduction of carbon–carbon double bonds, **1**, 71
for the synthesis of alkenes, **1**, 502
for the synthesis of *N*-imidoylhydroxylamines, **5**, 765
use of Lindlar catalyst in, **1**, 503
Hydrogenation, asymmetric, mechanism, **2**, 52
Hydrogenation, catalytic
of amines, **1**, 32
for the dechlorination of aryl halides, **1**, 427
for the reduction of aldehydes, **1**, 18
for the reduction of alkyl halides, **1**, 2
for the reduction of aryl bromides, **1**, 428
for the reduction of fluoroalkanes, **1**, 6
for the reduction of vinyl bromides, **1**, 431
for the synthesis of 2,6-disubstituted piperidines, **2**, 326
for the synthesis of hydroxylamines, **2**, 764
of vinyl fluorides, **1**, 430
Hydrogenation, catalytic transfer
and hydrogenation reactions compared, **1**, 457
for the reduction of alkenes, **1**, 74
for the synthesis of alkenes, **1**, 457
Hydrogenation, heterogeneous
of arenes, **1**, 87
mechanism, **1**, 71
Hydrogenation, heterogeneous catalytic
of alkynes, **1**, 502
for the synthesis of alkenes, **1**, 459
Hydrogenation, homogeneous
of arenes, **1**, 93
for the synthesis of alkenes, **1**, 459
Hydrogenation, homogeneous catalytic, **1**, 507
Hydrogenation, ionic
mechanism, **1**, 13
for the synthesis of α-aminoalkyl phosphonates, **4**, 491
1,4-Hydrogenation, of α,β-unsaturated ketones, **2**, 643
Hydrogen atom transfer reaction, mechanism, **1**, 331, 343
Hydrogen bromide
for the cleavage of trialkylboranes, **1**, 58
elimination of, **3**, 208
Hunig's base for the elimination of, **5**, 790
reactions
with bis(trifluoromethyl)thioketene, **5**, 487
with bromoalkenes, **4**, 23
with cyanamide, **6**, 623
with 1,1-difluoroethene, **4**, 5
with propyne, **4**, 24
with triacetates, **4**, 121
for the synthesis of vinyl bromides, **2**, 623
use in the synthesis of α-bromo ethers, **4**, 50
Hydrogen chloride
elimination of, **3**, 208
intermolecular eliminations of, **5**, 1153
reactions
with alkenes, **2**, 23
with alkyl complexes, **5**, 949
with alkyldiphenylstibine, **2**, 490
with 1-alkylselenobut-3-en-1-ynes, **4**, 786
with 4-aminobutanoic acid, **5**, 970
with 6-aminohexanoic acid, **5**, 970
with 5-aminopentanoic acid, **5**, 970
with bis[bis(trifluoromethyl)arsine]oxide, **2**, 492
with bisphosphines, **5**, 398
with boron trifluoride etherate, **4**, 885
with buta-1,2-diene, **1**, 1125
with *t*-butyl isocyanide, **4**, 894
with α-(*t*-butyl)D-α-phenoxymethylpenicilloate, **5**, 81
with α-(*t*-butyl)DL-α-phenoxymethylpenicilloate, **5**, 81
with carbodiimides, **6**, 728
with cyanamide, **6**, 621, 728
with dialkoxyvinyl orthoformates, **6**, 43
with dialkylethynylarsine, **2**, 876
with dimethylaminodiethylarsine, **2**, 492
with dimethylfluoroamine, **2**, 762
with ethyl dicyanoacetate, **4**, 792
with ethyl thiocyanate, **6**, 618
with ethynylplatinum complexes, **4**, 822
with germyl(silyl)methane, **4**, 641
with mercury(II) trifluoromethanethiolate, **6**, 232
with (methoxymethyl)iron complexes, **4**, 148
with methyl cellosolve, **5**, 82
with 1-methylselenohexa-1,3-diyne, **4**, 786
with methyl trimethylsilyl ketone, **4**, 944

with *aci*-nitro compounds, **5**, 819
with organylbis(diisopropylamino)boranes, **2**, 321
with potassium 1,2-ethanebis(trithiocarbonate), **6**, 553
with rhenium carbene complexes, **3**, 522
with silyl diimines, **4**, 107
with α-silyl ketones, **5**, 647
with tetramethylbismuthine, **2**, 491
with 1-thiobenzoyl imidazole, **5**, 477
with thioketenes, **5**, 480
with trimethoxyborohydride, **2**, 109
with trimethylbis(trimethylsiloxy)antimony, **2**, 492
with trimethylsilyl ethers, **4**, 150
with trimethylstibine oxide, **2**, 492
with vinylrhodium complexes, **4**, 149
with zinc cyanide, **3**, 92
with zirconium compounds, **4**, 547
as reagent, for the hydrolysis of nitriles, **5**, 269
as starting material, for dichloroiminium cations, **6**, 610
for the synthesis of vinyl chlorides, **2**, 616
See also Hydrochloric acid
Hydrogen cyanide
addition to aldehydes and ketones, **3**, 615
asymmetric addition of, to aldehydes, **3**, 629
elimination of, **1**, 559
reactions
with alkenes, **3**, 614
with alkoxydialkyl phosphinites, **5**, 1131
with alkynes, **1**, 545; **3**, 660
with chlorotrimethylsilane, **5**, 1134, 1135
with conjugated alkenes, **3**, 615
with diethylaluminum hydride, **5**, 1143
with dimethylberyllium, **5**, 1137
with dimethyldimethylaminophosphine, **5**, 1130
with enamines, **3**, 615
with enol ethers, **3**, 615
with hexaalkyldisilazane, **5**, 1133
with hexaphenyldilead, **5**, 1145
with hydrazones, **3**, 616
with imines, **3**, 616
with ortho esters, **3**, 627
with oximes, **3**, 616
with semicarbazones, **3**, 616
with thioboranes, **5**, 1136
with trialkylaluminums, **5**, 1142
with trialkylgermanium oxides, **5**, 1136
with trimethylgallium, **5**, 1143
with trimethylindium, **5**, 1143
with α,β-unsaturated carbonyls, **3**, 633
with vinyl ethers, **3**, 627
Hydrogen cyanides
reactions
with dialkylaminomethylphosphines, **5**, 1129
with dialkylaminophosphines, **5**, 1129
Hydrogen disulfide, acylation of, **5**, 252
Hydrogen dithiophosphate, *O,O'*-diethyl, reactions, with isocyanates, **5**, 1037
Hydrogen donors, as radical traps, **1**, 329
Hydrogen fluoride
addition to alkenes, **2**, 17
electrolysis of, **6**, 20
hazards relating to, **6**, 408
reactions
with antimony fluorides, **6**, 16
with antimony pentafluoride, **6**, 38
with *N*-arylcarbonimidic dichlorides, **6**, 602
with 1-chlorocyclohexene, **4**, 30
with cyanic acid, **6**, 439
with ethyne, **2**, 611
with melamine, **2**, 17
with metal fluorides, **6**, 16
with xenon difluoride, **2**, 613
reactivity of, **2**, 6
as starting material, for carbamoyl fluorides, **6**, 439
use in the synthesis of fluoro steroids, **4**, 9
Hydrogen fluoride, anhydrous (AHF)
for the cleavage of trialkylboranes, **1**, 58
as industrial source of fluorine, **6**, 9
reactions
with isopropenyl octanoate, **5**, 3
with organic halides, **6**, 9
as starting material, for carbonic chloride fluoride, **6**, 419
Hydrogen halides
Anti-Markovnikov addition of, **2**, 4
elimination of, **1**, 570, 981, 1063, 1069; **4**, 730
for the synthesis of phosphaalkynes, **5**, 1152
for vinyl halides, **2**, 606
Markovnikov addition of, **2**, 4
reactions
with alcohols, **2**, 6
with (alkoxymethyl) metal complexes, **4**, 147
with alkylvinyl selenides, **4**, 91
with arylvinyl selenides, **4**, 91
with enol esters, **5**, 3
with haloalkenes, **4**, 2
with haloiminium salts, **6**, 53
with sulfonium dinitro ylides, **6**, 56
Hydrogen iodide
reactions
with alkenes, **2**, 34
with 1,12-bis(hydroxymethyl)decahydrododecaborate caesium salt, **4**, 143
with cyanamide, **6**, 623
with 1,1-difluoroethene, **4**, 5
with ethers, **2**, 36
with phthalimides, **4**, 101
with (trifluoromethylthio)silane, **6**, 232
for the synthesis of vinyl iodides, **2**, 628
Hydrogen isocyanate
reactions
with ammonia, **6**, 500
with cyclopentadienyltungsten complexes, **6**, 517
Hydrogen isocyanate complexes, reactions, with cyclohexyl isocyanate, **5**, 1088
Hydrogenolysis, benzylic, **1**, 74
Hydrogen oxalate, 9-triptycyl-, as starting material, for 9-triptycyl iodoformate, **6**, 431
Hydrogen peroxide
for the oxidation of alkynylphosphines, **2**, 1066
for the oxidation of amines, **2**, 340
for the oxidation of dithioacetals, **4**, 259, 271, 272
for the oxidation of oximes, **4**, 429
for the oxidation of selenides, **2**, 287
for the oxidation of tertiary amines, **2**, 347
as oxidizing agent, for the synthesis of bisacetoxyiodobenzene, **5**, 204
reactions
with alcohols, **2**, 95
with aldehydes, **2**, 96
with alkenes, **2**, 96
with alkyl bromides, **2**, 96
with *N*-(benzyloxy)amines, **3**, 435
with bis(1-chloroethyl)chlorostibine, **2**, 498
with carbenium ions, **2**, 98
with carbonyls, **2**, 96
with carboxylic acids, **2**, 90
with 3,4-dimethoxyphenylpyruvic acid, **5**, 51
with 1-hydroxy-2-nitro-2-phenylselenononane, **4**, 335
with imines, **2**, 96
with ketones, **2**, 96
with 2-phenylselenocyclooctane, **1**, 607
with α,β-unsaturated sulfones, **4**, 237
as starting material, for alkyl hydroperoxides, **2**, 95
use in the synthesis of alkynylarsenic compounds, **2**, 1080
use in the synthesis of dialkyl sulfones, **2**, 155
use in the synthesis of dialkyl sulfoxides, **2**, 144
use in the synthesis of hydroperoxides, **2**, 95

use in the synthesis of ketones, **3**, 134
use in the synthesis of tertiary arylphosphine oxides, **2**, 841
Hydrogen phosphite, bis(*p*-methylbenzyl), reactions, with aldimines, **4**, 461
Hydrogen phosphites, dialkyl-
reactions
with 2-diazo-1,3-diketones, **3**, 200
with 2-diazo ketones, **3**, 200
with imidoyl chlorides, **5**, 787
synthesis of, **2**, 104
Hydrogen polysulfides, as reagents, for the synthesis of 1,2-dithioles, **3**, 371
Hydrogen selenide
reactions
with acetylacetone, **3**, 397
with aldehydes, **3**, 382
with alkyl cyanate, **6**, 593
with benzoyl chloride, **5**, 233
with carbodiimides, **6**, 596
with 3-cyanopyridine, **5**, 625
with dimethylformamide azine, **5**, 626
with (hetero)aryl formamidines, **5**, 626
with imidate esters, **5**, 516
with ketone, **3**, 391
with nitriles, **5**, 626
with 1,3,4-oxadiazolium salts, **5**, 642
with phosgeneiminium chloride, **6**, 594
with triphenylarsenic dichloride, **2**, 890
use in the synthesis of bis(acyl) selenides, **5**, 243
use in the synthesis of selenoamides, **5**, 625
Hydrogen selenite, methyl, synthesis of, **2**, 101
Hydrogen selenites, alkyl
reactions, with diazoalkanes, **2**, 101
synthesis of, **2**, 100
[1,3] Hydrogen shifts, mechanism, **1**, 794
[1,5] Hydrogen shifts
mechanism, **1**, 794
of vinyl cyclopropanes, **1**, 795
[1,7] Hydrogen shifts, mechanism, **1**, 794
Hydrogen sulfates, alkyl
alkylation of, **2**, 102
synthesis of, **2**, 101
Hydrogen sulfide
equivalents for, **5**, 580
handling problems, **2**, 117
reactions
with N^2-acyl-N^1,N^1-dimethylamidines, **5**, 631
with acylsilanes, **4**, 371
with alcohols, **2**, 118
with aldehydes, **2**, 223
with alkenes, **2**, 116, 137
with alkoxyalkylchlorophosphines, **2**, 461
with alkoxyaminoalkylphosphines, **2**, 461
with alkyldicyanamides, **5**, 1124
with alkyl halides, **2**, 119
with alkynes, **3**, 337
with amidines, **5**, 611
with aminodialkylphosphines, **2**, 460
with *o*-azido aldehydes, **3**, 342
with benzoyl chloride, **5**, 528
with bis(halomethyl)dialkylgermanes, **4**, 372
with chloroiminium salts, **5**, 611
with chlorotrifluoroethene, **5**, 484
with *N*-dialkyloxyphosphorylimino chlorides, **5**, 644
with dialkyl phenylphosphonites, **2**, 851
with *N*-dialkylphosphorylimino chlorides, **5**, 644
with dihalotrialkylphosphoranes, **2**, 461
with dimethylaminomalononitrile, **5**, 580
with dimethylformamide, **1**, 46
with dinitriles, **5**, 580
with dithio-*N*,*N*-dimethyliminium salts, **6**, 555
with enamines, **2**, 130; **3**, 348
with formazans, **5**, 641
with formic acid, **6**, 85
with haloenamines, **5**, 611
with hexafluoroacetone, **4**, 215
with hexamethyldisilazane, **2**, 541
with hydrazidines, **5**, 641
with hydrazonyl halides, **5**, 641
with imines, **2**, 224
with isonitriles, **5**, 608
with ketals, **3**, 347
with ketimines, **2**, 130
with ketones, **2**, 223; **3**, 343; **4**, 244
with methylarsine, **2**, 500
with 1,8-naphthaloyl chloride, **5**, 529
with nitriles, **5**, 579, 619
with organoarsenic(III) anhydrides, **2**, 890
with organoarsenic(III) dihalides, **2**, 890
with *N*-phenylimines, **3**, 348
with α-silyl ketones, **5**, 647
with sulfonyl chlorides, **2**, 223
with thioimidates, **5**, 611
with trichloroacetyl chloride, **5**, 506
with triethylorthoformate, **5**, 586
with trifluoroacetaldehyde, **4**, 215
with trifluoroacetonitrile, **5**, 546, 580
with 1-(trimethylsilyl)imidazole, **2**, 541
with trimethylstibine oxide, **2**, 502
with vinyl ethers, **2**, 128
as reagent
for the synthesis of 1,2-dithioles, **3**, 371
for the synthesis of thioaldehydes, **3**, 342
for the synthesis of thioketones, **3**, 343, 352, 356, 362
as starting material, for thiocarboxylic acids, **5**, 232
use in the reduction of carbon–mercury bonds, **1**, 68
use in the synthesis of disulfides, **2**, 223
use in the synthesis of thionoamides, **5**, 579
Hydrogen sulfite ions, use in the synthesis of sulfonic acids, **2**, 190
Hydrogen telluride
reactions
with ketone, **3**, 400
with nitrones, **3**, 408
use in the synthesis of amines, **2**, 311
Hydrogen thioperoxide (HOSH), synthesis of, **2**, 99
1,5-Hydrogen transfer, intramolecular, in radical cyclizations, **1**, 339
1,6-Hydrogen transfers, mechanism, **4**, 1079
Hydrogermylation
of alkynes, **2**, 946
of allenes, **2**, 946
for the synthesis of α-germylsilylalkanes, **4**, 639
Hydro halides, reactions, with phosphines, **2**, 445
Hydro halides, isothiosemicarbazide, as starting materials, for 3-(ω-aminoalkyl)-1,2,4-triazoles, **2**, 310
Hydrohalide salts, amine, use in the synthesis of alkyl halides, **2**, 6
Hydrohalogenation
of alkynylsilanes, **2**, 929
Markovnikov, mechanism, **2**, 4
Hydroiodic acid
reactions
with acetic acid, **1**, 432
with 1,3-bis(diphenylarsino)propane, **2**, 490
Hydroiodination, of alkenes, **2**, 34
Hydrolysis
of aromatic nitriles, **5**, 298
of coordinated nitriles, **5**, 385
of nitriles, **5**, 268, 281, 296
of sulfenyl compounds, **2**, 185
of sulfinic acids, **2**, 187
for the synthesis of amino acids, **5**, 85
for the synthesis of aromatic acids, **5**, 105
for the synthesis of carboxylic acids, **5**, 24

for the synthesis of haloalkanoic acids, **5**, 55
for the synthesis of heteroatom-substituted alkanoic acids, **5**, 96
for the synthesis of nitrogen-substituted alkanoic acids, **5**, 92
for the synthesis of oxygen-substituted alkanoic acids, **5**, 62
for the synthesis of sulfur-substituted alkanoic acids, **5**, 81
for the synthesis of α,β-unsaturated acids, **5**, 98
for the synthesis of unsubstituted alkanoic acids, **5**, 29
of thioimidate *N*-oxides, **5**, 327
Hydrolysis, enzymic, for the synthesis of α-aminophosphonic acids, **4**, 462
Hydrolysis, stereospecific enzymatic, for the synthesis of carboxylic acids, **5**, 118
Hydrolytic reaction, for the synthesis of alcohols, **2**, 87
Hydromagnesiation
of alkynylsilanes, **2**, 923
for the synthesis of alkylmagnesiums, **2**, 561
of 1-trimethylsilyl-1-alkynes, **1**, 161
Hydrometallation
of alkynes, **1**, 531
use in the synthesis of unsaturated ketones, **3**, 140
of alkynylsilanes, **2**, 921
anti-stereoselective, **2**, 965
for aryl metals, **2**, 965
mechanism, **2**, 596
non-stereoselective, **2**, 965
for organometallics, **2**, 952
stereo-random, **2**, 970
syn-stereoselective, **2**, 965
for the synthesis of alkenes, **1**, 458
for the synthesis of alkylcobalt(I) compounds, **2**, 576
for the synthesis of alkyl metal(II) compounds, **2**, 577
for the synthesis of alkylzirconiums, **2**, 568
for vinyl metals, **2**, 956, 965
Hydrometallation, stereo-random, of alkynes, **2**, 970
anti-Hydrometallation
of alkynes, **2**, 970
with aluminum-containing reagents, **2**, 970
as a barrier to *syn*-hydrometallation, **2**, 965
syn-Hydrometallation
of alkynes, **2**, 965, 969
with aluminum, **2**, 966
and carbometallation compared, **2**, 972
Hydrometallation–transmetallation
for the synthesis of alkyl cuprates, **2**, 587
for the synthesis of alkyl metals, **2**, 592
Hydroperoxide, allyl, synthesis of, **2**, 95
Hydroperoxide, aryl, synthesis of, **2**, 687
Hydroperoxide, *t*-butyl-, magnesium salt of, for the oxidation of aryl organometallics, **2**, 636
Hydroperoxide, *t*-butyl
for the oxidation of alkynylarsines, **2**, 1080
for the oxidation of arsines and stibines, **2**, 497
for the oxidation of selenides, **1**, 601
reactions
with aldoximes, **3**, 684
with alkenes, **2**, 90
with allylic alcohols, **2**, 91
with *t*-butylacetimidates, **2**, 98
with chloro-*o*-nitrophenoxyboranes, **2**, 99
with oxybis(diacetoxy)borane, **2**, 99
with trichloro-1,3,5-triazine, **2**, 687
with triflates, **2**, 98
as starting material
for dialkyl peroxides, **2**, 98
for peroxy chloroformate, **6**, 430
Hydroperoxide, cumene, synthesis of, **2**, 96
Hydroperoxide, cumyl, reactions, with alkenes, **2**, 90
Hydroperoxide, 2-methoxy, synthesis of, **2**, 96
Hydroperoxide, trichloromethyl, synthesis of, **6**, 231
Hydroperoxide, trifluoromethyl, synthesis of, **6**, 231
Hydroperoxide rearrangement, benzylic, for the synthesis of phenols, **2**, 655
Hydroperoxides
from
halides, **2**, 96
tertiary alcohols, **2**, 95
as intermediates, **2**, 87
for the oxidation of sulfoxides, **2**, 156
reactions, with acid chlorides, **5**, 206
synthesis of, via displacement of alcohols and ethers, **2**, 95
use in the synthesis of dialkyl sulfoxides, **2**, 146
Hydroperoxides, alkyl
alkylation of, **2**, 97
decomposition of, **2**, 97
from
alkanes, **2**, 96
alkyl dichloroboranes, **2**, 97
hydrogen peroxide, **2**, 95
organoboranes, **2**, 97
reactions
with acetoxyacyl chlorides, **6**, 71
with alkyl sulfates, **2**, 97
with boron trihalides, **2**, 99
with carbenium ions, **2**, 98
with haloalkanes, **2**, 98
as starting materials
for dialkyl peroxides, **2**, 97
for peroxyborates, **2**, 99
synthesis of, **2**, 95
Hydroperoxides, allylic, synthesis of, via autoxidation reactions, **2**, 96
Hydroperoxides, α-azo, synthesis of, **5**, 357
Hydroperoxides, benzylic, synthesis of, via autoxidation reactions, **2**, 96
Hydroperoxides, 1,1-hydroxy, as starting materials, for ω-thiocyanatocarboxylic acids, **5**, 1110
Hydroperoxides, 2-hydroxy, synthesis of, **2**, 96
Hydroperoxides, tertiary, synthesis of, via autoxidation reactions, **2**, 96
Hydroquinone, trifluoromethylation of, **6**, 13
Hydroquinones, alkyl-, synthesis of, **1**, 273
Hydrosilylation
of alkenes, **2**, 4, 50, 536
of alkynes, **4**, 1044
platinum-catalyzed, **2**, 969
of alkynylsilanes, **2**, 923; **4**, 1044
of carbodiimides, **5**, 777
catalysts for, **4**, 1044
of enamides, **4**, 510
of enamines, **4**, 510
of silyl(alkylsilyl)alkynes, **4**, 1054
for the synthesis of alkenes, **1**, 535
for the synthesis of 1,1-bis(silyl)alkanes, **4**, 617
for the synthesis of silicon–nitrogen bonds, **2**, 812
for the synthesis of tetraorganosilanes, **2**, 536
use of chlorosilanes for, **2**, 538
Hydrosilylation, transition metal-catalyzed, of alkynes, **2**, 913
Hydrostannation
of alkynes, **2**, 966, 970
of ethoxyethyne, with trialkyltin hydride, **2**, 676
of 1-stannylalkynes, **4**, 1068
of 1-stannyl-1-alkynes, **6**, 209
for the synthesis of alkyltins, **2**, 601
of terminal alkynes, **2**, 971
of 1-trimethylsilylpentyne, **4**, 1062
1,4-Hydrostannation, of α,β-unsaturated ketones, **2**, 701
Hydrostannylation, for the synthesis of alkenes, **1**, 535
Hydrosulfide, tetraethylammonium
reactions
with dichloroalkenes, **4**, 851
with nitriles, **5**, 581

Hydroxamates
acylation of, **5**, 332
deprotection of, **5**, 326
from
acyl nitroso compounds, **5**, 330
O-alkyloximes, **5**, 329
Evans oxazolidinones, **5**, 329
vinyl triflates, **5**, 331
as starting materials, for *N*-hetero-*N*-chloroamides, **5**, 317
synthesis of
via acylation of alkoxyamines, **5**, 327
via alkylation of *N,N'*-dimethoxyurea, **5**, 331
via oxidation of hydroxylamines, **5**, 325
via oxidation of nitrones, **5**, 331
via ring-expansion of ketones, **5**, 331
Hydroxamates, *N,O*-bis(trimethylsilyl)-, synthesis of, **5**, 380
Hydroxamates, cyclic, synthesis of, **5**, 325
Hydroxamates, *O*-(*N*-phenylcarbamyl), synthesis of, **5**, 1000
Hydroxamic acids
acylation of, **5**, 214, 332
O-acyl compounds of, **1**, 935
alkylation of, **2**, 103; **3**, 434; **5**, 701
from
acyl nitroso compounds, **5**, 326
2-alkoxypyridine *N*-oxides, **5**, 326
amines, **5**, 324
ester enolates, **5**, 324
as intermediates, in the synthesis of isocyanatoethylenes, **2**, 1055
reactions
with ethyl bromide, **5**, 701
with hexamethyldisilazane, **5**, 380
with hexamethyldisiloxane, **5**, 719
with ketene dimers, **5**, 1001
with polychlorosilanes, **5**, 1001
with sulfur trioxide, **5**, 1000
with triethyloxonium tetrafluoroborate, **5**, 701
with triphenylphosphine, **5**, 1000
as starting materials
for acylhydrazines, **5**, 353
for acyl nitroso compounds, **5**, 326
for hydroxamoyl halides, **5**, 665
for isocyanates, **5**, 999
for lactams, **5**, 324
synthesis of, **4**, 449; **5**, 322, 327
via acylation of hydroxylamine, **5**, 322
via deprotection of hydroxamates, **5**, 326
via hydrolysis of thioimidate *N*-oxides, **5**, 327
via oxidation reactions, **5**, 324
via photolytic rearrangement of nitro compounds, **5**, 325
via reduction of *O*-acyl oximes, **5**, 324
via reductive cyclization of nitro compounds, **5**, 324
via ring-opening of heterocycles, **5**, 322
thermal decomposition of, **5**, 1001
use in the synthesis of *O*-alkylhydroxylamines, **2**, 103
use in the synthesis of *N*-arylamides, **5**, 294
See also Amides, *N*-alkoxy-
Hydroxamic acids, acylated, synthesis of, **5**, 214
Hydroxamic acids, *N*-methyl-, synthesis of, **5**, 11, 327, 331
Hydroxamic acids, silylated, thermal treatment of, **5**, 1000
Hydroxamic acids, *O*-sulfinyl-, synthesis of, **5**, 347
Hydroxamoyl chlorides
reactions, with alkoxides, **5**, 699
synthesis of, **5**, 664, 667
Hydroxamoyl chlorides, heterocyclic, synthesis of, **5**, 670
Hydroxamoyl chlorides, *O*-phosphorylated, synthesis of, **5**, 669
Hydroxamoyl chlorides, *O*-silylated, reactions, with isocyanates, **5**, 672
Hydroxamoyl chlorides, thienoyl-, synthesis of, **5**, 664
Hydroxamoyl fluorides, synthesis of, **5**, 666
Hydroxamoyl fluorides, perfluoro-, as starting materials, for imidoyl halides, **5**, 671
Hydroxamoyl halides
from
carboxylic acid chlorides, **5**, 669
chlorosilicons, **5**, 669
dichloronitrosoalkanes, **5**, 669
1,1-difluoroalkenes, **5**, 666
halogen carriers, **5**, 665
hydroxamic acids, **5**, 665
hydroxamoyl halides, **5**, 671
nitrile oxides, **5**, 669
1-nitroso-1,2,2-trichloroethene, **5**, 669
phosphorus compounds, **5**, 669
trichloromethyls, **5**, 666
as starting materials, for hydroxamoyl halides, **5**, 671
synthesis of, **5**, 664
via halogenation of aldoximes, **5**, 664
via nitrosation, **5**, 667
Hydroxamoyl halides, 3-bromo-2-oxopropion-, nitrosation of, **5**, 667
Hydroxides
for the cleavage of allylphosphonium bromide, **1**, 49
reactions, with tertiary phosphine oxides, **2**, 454
for the reduction of carbon–phosphorus bonds, **1**, 49
Hydroxides, diazo
as intermediates, in the synthesis of diazo compounds, **3**, 465
reactions, with acetic acid, **3**, 462
Hydroxides, metallic, reactions, with monocarboxylic acids, **1**, 883
Hydroxides, quaternary ammonium
Hofmann elimination of, **1**, 627
thermal decomposition of, **1**, 627
Hydroxides, sulfonium, thermal decomposition of, **1**, 622
Hydroximates, *O*-tetrahydropyranyl-, [3,3] sigmatropic rearrangement of, **1**, 837
Hydroximic acid chlorides, reactions, with hydrosulfide anions, **5**, 637
Hydroximoyl chlorides, synthesis of, **3**, 683
Hydroximoyl halides
dehydrochlorination of, **3**, 686
dehydrohalogenation of, **3**, 683
as starting materials, for nitrile oxides, **3**, 678, 683
Hydroxy acids
from, tetrahydropyranyloxy-2,2,4,4-tetramethylcyclobutan-1-carbinols, **5**, 73
reactions, with dimethylformamide dimethylacetal, **1**, 565
synthesis of, reviews, **5**, 62
3-Hydroxy acids, 2-amino-, synthesis of, **2**, 84
cis-Hydroxy acids, synthesis of, **5**, 63
L-α-Hydroxy acids, synthesis of, **5**, 117
α-Hydroxy acids
decarboxylation, with thionyl chloride, **5**, 10
as starting materials, for 4-oxo-1,3-dioxolanes, **4**, 213
synthesis of, **4**, 231; **5**, 63
β-Hydroxy acids
decarboxylation of, **1**, 683
reactions, with phenylboronic acid, **5**, 229
as starting materials
for alkenes, **1**, 565, 681
for 4-oxo-1,3-dioxolanes, **4**, 213
synthesis of, **1**, 681; **5**, 63
δ-Hydroxy acids, cyclization of, **5**, 171
Hydroxy alkyl anions, reactions, with enones, **3**, 161
N-Hydroxyalkylation, mechanism, **4**, 296
Hydroxyalkyl compounds, synthesis of, **2**, 1082
trans-1,2-Hydroxy chloride, synthesis of, **2**, 44
1,2-Hydroxy chlorides, synthesis of, **2**, 44
Hydroxy complexes
dehydration of, **5**, 949
synthesis of, problems, **5**, 934
Hydroxy compounds, as starting materials, for oxyimidic halides, **5**, 823
N-Hydroxy compounds

as co-reagent, in coupling reactions, **5**, 266
use in the synthesis of *N*-halo enamines, **2**, 762
α-Hydroxy compounds
as starting materials
for α-aminoalkylphosphorus compounds, **4**, 475
for α-haloamides, **4**, 98
for α-haloimides, **4**, 98
Hydroxy dialkyl compounds, from, α-bromoboranes, **2**, 524
Hydroxy functions, hindered, acylation of, **5**, 129
Hydroxy group activation (HGA)
for the synthesis of β-lactones, **5**, 161
for the synthesis of macrolactones, **5**, 177
Hydroxy groups
activation of, **1**, 648
elimination of, **3**, 209, 226
oxidation of, **4**, 165
β- to ketones, elimination of, **3**, 210
δ-Hydroxy groups, elimination of, **3**, 63
Hydroxyl, phenolic, alkylation of, **2**, 683
Hydroxylactonization, for the synthesis of γ-lactones, **5**, 168
Hydroxylamine
O-acylation of, **5**, 322
reactions
with aldehydes, **3**, 425
with amine transfer agents, **2**, 691
with benzenes, **2**, 764
with *p*-bromobenzonitrile, **5**, 763
with carbamidic acid chloride, **6**, 505
with chloral hydrate, **5**, 822
with formamide acetals, **5**, 764
with formamidines, **5**, 764
with glutaronitrile, **5**, 833
with heterocycles, **5**, 322
with imidoyl chlorides, **5**, 764
with ketones, **3**, 425
with nitriles, **5**, 763, 833
with nitrosobenzene, **2**, 806
with nitrosotrifluoromethane, **5**, 360
with oxyimidic esters, **5**, 835
with (*E*)-3-phenylpropenitriles, **5**, 763
with phthalonitrile, **5**, 833
with pyridine-3-carbonitrile, **5**, 763
with succinic dinitrile, **5**, 833
with thioamides, **5**, 764
with trichloroacetonitrile, **5**, 822
with 2,4,6-trimethyl benzotrichloride, **5**, 666
with urethanes, **6**, 505
as starting material
for *N*-alkoxyimidates, **5**, 700
for *N*-hydroxyimidates, **5**, 700
for *N*-hydroxyiminodithiocarbonates, **6**, 628
for *N*-imidoylhydroxylamines, **5**, 763
use in the synthesis of alkynes, **1**, 1018
use in the synthesis of oxyimidic esters, **5**, 826
Hydroxylamine, *O*-alkenyl-, synthesis of, **2**, 691
Hydroxylamine, *O*-alkyl-, use in the synthesis of *O*-alkyl-substituted oxyimidic esters, **5**, 826
Hydroxylamine, 2-aminobenzoyl-, as reagent, for the synthesis of nitriles, **5**, 219
Hydroxylamine, β-aroylamino *N*,*N*-disubstituted, synthesis of, **2**, 347
Hydroxylamine, *N*-(2-benzoylstyryl)-, use in the synthesis of ketenimines, **3**, 578
Hydroxylamine, *O*-benzyl-
alkylation of, **2**, 351
N-chlorocarbonyls of, **6**, 446
reactions
with di-*t*-butyl dicarbonate, **2**, 351
with 2-(methoxycarbonylcyanomethylene)-1,3-dioxolane, **4**, 896
Hydroxylamine, bis(benzotriazol-1-ylmethyl)-, reactions, with Grignard reagents, **1**, 260
Hydroxylamine, *N*,*N*-bis(trifluoromethyl)-
reactions, with thiophosgene, **6**, 312
synthesis of, **6**, 241
Hydroxylamine, bis(trimethylsilyl)-, use in the synthesis of oximes, **5**, 219
Hydroxylamine, *N*-*t*-butyl-, reactions, with sulfonylalkyne, **2**, 691
Hydroxylamine, *N*-[1-chloro-2-diethoxyphosphoryl)ethylidene]-, synthesis of, **5**, 821
Hydroxylamine, *N*-[(*p*-chlorophenyl)diphenylmethyl]-, synthesis of, **2**, 342
Hydroxylamine, *N*-[2-(diethoxyphosphoryl)ethylidene]-, chlorination of, **5**, 821
Hydroxylamine, diethyl-, for the reduction of quinones, **2**, 657
Hydroxylamine, dimethyl-, use in the synthesis of ynamines, **2**, 1041
Hydroxylamine, *O*-diphenylphosphinyl-, reactions, with potassium phthalimide, **5**, 360
Hydroxylamine, *O*-mesitoyl-, for the *N*-amination of hydrazines, **2**, 379
Hydroxylamine, *O*-mesitylenesulfonyl- (MSH)
reactions, with hydrazones, **4**, 444
as starting material, for hydrazines, **2**, 384
use in the synthesis of *O*-aryl hydroxylamino, **2**, 691
Hydroxylamine, *N*-methyl-*N*,*O*-bis(trimethylsilyl)-, reactions, with carbonyls, **3**, 433
Hydroxylamine, phenyl-
condensation of, **4**, 894
reactions, with sodium dichromate, **2**, 781
as starting material, for β,γ-unsaturated amines, **2**, 307
Hydroxylamine, *N*-phenyl-, reactions, with imidoyl chlorides, **5**, 764
Hydroxylamine, *O*-phenyl-, reactions, with methyl isocyanate, **6**, 506
Hydroxylamine, *O*-silylated (*Z*)-imidoyl-, synthesis of, **5**, 765
Hydroxylamine, *N*-(trifluoromethyl)-, synthesis of, **6**, 241
Hydroxylamine-*O*-carbamates, *N*-acyl-, rearrangement of, **2**, 316
Hydroxylamine hydrochloride
for the hydrolysis of nitrones, **2**, 343
reactions
with ε-caprolactam, **5**, 971
with diphenyloxamidine, **5**, 833
with phosgene, **6**, 505
Hydroxylamine hydrochloride, *O*-methyl-
reactions, with phosgene, **6**, 505
use in the synthesis of *N*-heteroarylacetamide oximes, **5**, 836
Hydroxylamines
acylation of, **5**, 211, 322, 332
alkylation of, **2**, 349
bisacylation of, **5**, 214
classification of, **2**, 763
from
amine oxides, **1**, 839
amines, **2**, 340
hydrazone *N*-oxides, **2**, 347
nitro compounds, **2**, 341
oximes, **2**, 763
hydrolysis of, **2**, 347
monoalkylation of, **2**, 341, 344
neutralization of salts, **2**, 347
oxidation of, **2**, 781; **3**, 432; **5**, 324
reactions
with aldehydes, **4**, 435
with alkanesulfinyl chlorides, **2**, 250
with alkenes, **2**, 343
with allyl esters, **2**, 308
with allylic acetates, **2**, 344
with aryl imines, **4**, 442
with carbon disulfide, **6**, 575
with carboxylic acids, **5**, 322

with diazonium salts, **2**, 776
with diketones, **3**, 264
with esters, **5**, 322
with ethyl diphenylmethylene carbamate, **4**, 448
with hydroxymethylenes, **3**, 431
with isothiocyanates, **6**, 570
with ketones, **4**, 435
with phenylhydrazones, **4**, 441
with sulfides, **2**, 161
with sulfinyl chlorides, **2**, 365
with sulfur dioxide, **2**, 366
with thiocarboxylic acid esters, **5**, 636
with trichloromethylarenes, **3**, 429
with trimethylsilyl chloride, **5**, 323
as reagents, for the synthesis of nitriles, **3**, 617
as starting materials
for *N*-alkoxyimidates, **5**, 700
for azoxy compounds, **2**, 405
for diazonium salts, **2**, 805
for *N*-hydroxyimidates, **5**, 700
for *N*-hydroxyimides, **5**, 332
for nitrones, **3**, 431
for nitroso compounds, **2**, 390
for nitroso dimers, **2**, 405
for *N*-sulfinyloxyamines, **2**, 357
synthesis of, **2**, 340, 763
via Cope reactions, **1**, 623
thioacylation of, **5**, 635
Hydroxylamines, *N*-acyl-
Mannich-type reactions of, **4**, 442
thionation of, **5**, 635
Hydroxylamines, *O*-acyl-
reductive approaches to, **5**, 213
synthesis of, **5**, 211
via Polonovski rearrangements, **5**, 214
Hydroxylamines, acylaminomethyl-, synthesis of, **4**, 448
Hydroxylamines, acylated, synthesis of, via direct acylation, **5**, 211
Hydroxylamines, *O*-acylated
as intermediates, in the synthesis of hydroxamic acids, **5**, 324
reactions, with alcohols, **5**, 125
synthesis of, **5**, 211
Hydroxylamines, ω-alkenyl *N*,*N*-disubstituted, retro-Cope rearrangements of, **2**, 349
Hydroxylamines, *N*-alkyl-
from
nitrones, **2**, 342
nitroxides, **2**, 343
monoalkylation of, **2**, 344
reactions
with alkenes, **2**, 346
with oxiranes, **2**, 342
synthesis of, **2**, 340, 340, 764
via nitrones, **2**, 342
Hydroxylamines, *O*-alkyl-
from
N-hydroxyphthalimides, **2**, 102
N-hydroxyurethanes, **2**, 102
as starting materials, for alkoxyguanidines, **6**, 651
synthesis of, **1**, 1042; **2**, 102; **6**, 232
Hydroxylamines, *N*-alkyl-*O*-(arylsulfonyl)-, rearrangement of, **3**, 413
Hydroxylamines, *O*-alkylated, synthesis of, **1**, 624
Hydroxylamines, *O*-alkyl-*N*,*N*-bis(trifluoromethyl)-, synthesis of, **6**, 240
Hydroxylamines, *N*-allyl-, synthesis of, **2**, 308
Hydroxylamines, *O*-(2-aminobenzoyl)-*N*-arylidene-, synthesis of, **5**, 219
Hydroxylamines, *N*-aryl-, Mannich-type reactions of, **4**, 442
Hydroxylamines, *O*-aryl-, synthesis of, **2**, 691
Hydroxylamines, *N*-benzyl-, reactions, with thiomaronyl dichlorides, **5**, 533
Hydroxylamines, *O*-benzyl-, synthesis of, **2**, 103
Hydroxylamines, bis-(*N*-thiomethyl)-, synthesis of, **4**, 325
Hydroxylamines, *N*,*N*-bis(trifluoromethyl)-, synthesis of, **6**, 241
Hydroxylamines, *N*-α-chloroalkyl-*N*-alkenyl-, synthesis of, **3**, 432
Hydroxylamines, dialkyl-, synthesis of, via Cope reactions, **1**, 626
Hydroxylamines, *N*,*N*-dialkyl-
O-alkylation of, **2**, 353
from, nitroxides, **2**, 346
reactions, with alkyl halides, **2**, 349
synthesis of, **2**, 343
via alkylation, **2**, 344
via Cope rearrangements, **2**, 346
via reduction of nitrones, **2**, 344
via retro-Cope rearrangements, **2**, 346
Hydroxylamines, *N*,*O*-dimethyl-, for the reactivity enhancement of carbonyls, **1**, 1047
Hydroxylamines, *N*,*N*-disubstituted
reactions, with titanium trichloride, **3**, 409
synthesis of, **1**, 260
Hydroxylamines, *N*-formidoyl-, synthesis of, **5**, 764
Hydroxylamines, *N*-formyl-, as starting materials, for *N*-thionoformylhydroxylamines, **5**, 635
Hydroxylamines, *N*-imidoyl-
from
amines, **5**, 765
ammonia, **5**, 765
hydroxylamine, **5**, 763
reactions, with tris(dimethylamino)phosphine, **5**, 775
synthesis of, **5**, 763
via hydrogenation, **5**, 765
See also Amidoximes
Hydroxylamines, (*Z*)-imidoyl-, synthesis of, **5**, 765
Hydroxylamines, *N*-imidoyl-*N*-phenyl-, synthesis of, **5**, 764
Hydroxylamines, keto-, synthesis of, **3**, 189
Hydroxylamines, *O*-mesitylsulfonyl-, as starting materials, for α-aminoalkylphosphorus compounds, **4**, 490
Hydroxylamines, *N*-nitroso-, synthesis of, **2**, 798
Hydroxylamines, oximino-, monoacylation of, **5**, 212
Hydroxylamines, phenyl-, use in the synthesis of *N*-hydroxyenaminones, **3**, 264
Hydroxylamines, *N*-substituted
reactions
with aldehydes, **2**, 344
with 5,6,7,8-tetrahydroquinolines, **2**, 346
with α,β-unsaturated esters, **2**, 346
synthesis of, **2**, 341
Hydroxylamines, *O*-substituted *N*-alkyl-
protonation of, **2**, 356
synthesis of, **2**, 350
Hydroxylamines, *N*-thionoacyl-
from, primary nitroalkanes, **5**, 637
synthesis of, **5**, 635
via thioacylation of hydroxylamines, **5**, 635
See also Thiohydroxamic acids
Hydroxylamines, *O*-thionoacyl-, synthesis of, **5**, 531
Hydroxylamines, *N*-thionoformyl-, from, *N*-formyl hydroxylamines, **5**, 635
Hydroxylamines, *N*-(trifluoromethyl)-, synthesis of, **6**, 241
Hydroxylamines, *N*-(trihalomethyl)-, synthesis of, **6**, 240
N-Hydroxylamines, reactions, with phosgene, **6**, 505
Hydroxylamine salts, *N*-alkyl-
from, isoxazoles, **2**, 347
handling precautions, **2**, 347
synthesis of, **2**, 347
Hydroxylamine salts, *O*-substituted-*N*-alkyl-, synthesis of, **2**, 356
Hydroxylamine sulfonates, reactions, with lithium alkynides, **2**, 1058
Hydroxylamine-*O*-sulfonic acid (HOSA)

for the *N*-amination of hydrazines, **2**, 379
reactions
with arylsulfonamides, **1**, 35
with organoboranes, **2**, 305
with piperidine, **2**, 382
with primary amines, **1**, 35
with thiane, **2**, 161
with thioamides, **5**, 736
as reagent, for hydrazines, **2**, 374
as starting material
for azo compounds, **2**, 403
for diazenes, **2**, 402
for hydrazines, **2**, 384
for triazanium salts, **2**, 387
Hydroxylation
of arenes, **2**, 39, 649
for the substitution of silicon functions, **1**, 283
Hydroxyl groups, functionalization of, **1**, 615
Hydroxyl groups, anomeric
deprotection of, **4**, 176
protection of, **4**, 176
Hydroxyls
displacement of, **1**, 185
as starting materials, for amines, **2**, 308
Hydroxymercuration, of enol ethers, **4**, 171
Hydroxymethyl anion equivalents, acylation of, **3**, 157
Hydroxymethyl anions, ethers as equivalents for, **1**, 136
Hydroxymethylation agents, silyl compounds as, **1**, 160
2-Hydroxymethyl compounds, 1-bromomethyl-, synthesis of, **4**, 143
Hydroxymoyl bromides, synthesis of, **5**, 821
Hydroxymoyl chlorides, synthesis of, **5**, 821
Hydroxynitrilium ion equivalents, from, nitronate salts, **3**, 429
N-Hydroxy-2-pyridone, 3-allyl-, from, 2-allyloxypyridine *N*-oxide, **5**, 326
Hydroxyselenenylation, of alkenes, **2**, 281
Hydroxythiols, as starting materials, for oxathiolanes, **4**, 229
1,3-Hydroxythiols
reactions, with *trans*-dimethylcyclopentanone-3,4-dicarboxylate, **4**, 229
as starting materials
for 1,3-oxathianes, **4**, 229
for 1,3-oxathiolanes, **4**, 229
Hydrozirconation
of alkenes, **2**, 4
of alkynes, **2**, 967
of alkynylsilanes, **2**, 923
of internal alkynes, regiochemistry, **2**, 969
mechanism, **5**, 441
for the synthesis of alkenes, **1**, 536
for the synthesis of alkenylzirconiums, **1**, 1105
for the synthesis of alkyl cuprates, **2**, 587
for the synthesis of alkylzirconiums, **2**, 568
in the synthesis of nitriles, **3**, 614
for the synthesis of vinyl bromides, **2**, 623
for the synthesis of vinyl iodides, **2**, 629
syn-Hydrozirconation, of alkynes, **2**, 968
Hyperconjugation
of radicals, **1**, 921
for the stabilization of carbocations, **1**, 892
Hyperconjugation, isovalent, use of term, **1**, 892
Hyperconjugation, negative (anionic), use of term, **1**, 852
Hyperconjugation, sacrificial, use of term, **1**, 892
Hypervalent halogen compounds, and carbon–carbon bond synthesis, **1**, 108
Hypobromite, reactions, with amides, **5**, 997
Hypobromite, acetyl
reactions
with amides, **5**, 320
with difluoroalkenes, **4**, 52
synthesis of, **5**, 203
use in the synthesis of bromoamides, **5**, 318
Hypobromite, *t*-butyl
reactions, with nitromethane, **6**, 54
synthesis of, **2**, 95
Hypobromite, 3-chlorobenzoyl, reactions, with 3,4-dihydro-2*H*-pyran, **4**, 208
Hypobromites
as intermediates, **2**, 685
synthesis of, **2**, 95
Hypobromites, acyl, synthesis of, **5**, 203
Hypobromites, alkyl, synthesis of, **2**, 95
Hypobromous acid, reactions, with tertiary alcohols, **2**, 95
Hypochloric acid, use in the synthesis of oxyimidic chlorides, **5**, 821
Hypochlorite
reactions
with thioureas, **5**, 1078
with *N*-unsubstituted imidoates, **5**, 698
with ureas, **2**, 374, 380
Hypochlorite, acetyl, synthesis of, **5**, 203
Hypochlorite, *t*-butyl
as chlorinating reagents, **2**, 338
for the chlorination of amides, **5**, 317
for the chlorination of amines, **2**, 337
for the chlorination of imides, **5**, 316
for the oxidation of diazo ketones, **4**, 162
for the oxidation of sulfides, **2**, 149
for the *para* chlorination of arenes, **2**, 620
reactions
with bis(*t*-butyl)urea, **6**, 509
with crotononitrile, **3**, 650
with *N*-methyl-5*H*,7*H*-dibenzo[b,g][1,5]selenazocine, **2**, 369
with *O*-trimethylsilyl thiopivalate, **5**, 530
with ylides, **5**, 895
as reagent, for the synthesis of *N*-chloroamides, **5**, 315
stability of, **2**, 94
as starting material, for sulfuranes, **2**, 793
use in the synthesis of *N*-haloimidates, **5**, 698
use in the synthesis of hydroxamoyl halides, **5**, 664
use in the synthesis of oxyimidic chlorides, **5**, 821
Hypochlorite, ethyl
as chlorinating reagents, **2**, 338
reactions, with iodine, **2**, 95
Hypochlorite, 2,2,3,3,4,4,5,5-octafluoropentyl, reactions, with perfluorobutyronitrile, **5**, 699
Hypochlorite, trifluoromethyl
reactions
with alkenes, **6**, 229
with 1,1-dihaloalkenes, **6**, 43
with vinylsilanes, **4**, 140
synthesis of, **6**, 231
Hypochlorite, 1,1,5-trihydroperfluoroamyl, reactions, with *n*-perfluoropropionitrile, **5**, 813
Hypochlorite ions, reactions, with amines, **2**, 761
Hypochlorites
reactions, with acetamides, **3**, 668
stability of, **2**, 94
synthesis of, **2**, 94
Hypochlorites, acyl, synthesis of, **5**, 203
Hypochlorites, alkyl, synthesis of, **2**, 94
Hypochlorites, perfluoroalkyl
as starting materials, for perfluoroalkyl chloroformates, **6**, 428
synthesis of, **2**, 94
Hypochlorous acid
reactions
with alcohols, **2**, 94
with amines, **2**, 761
Hypofluorite, acetyl
as fluorinating agent, for the synthesis of acylhypofluorites, **5**, 203
as reagent, for aryl fluorides, **2**, 613
as starting material, for α-fluoro ethers, **4**, 44

synthesis of, **5**, 202
for the synthesis of aryl fluorides, **2**, 613
Hypofluorite, *t*-butyl, synthesis of, **2**, 94
Hypofluorite, fluorocarbonyl
reactions, with sulfur tetrafluoride, **6**, 424
synthesis of, **6**, 424
Hypofluorite, methyl, synthesis of, **2**, 94; **4**, 51
Hypofluorite, trifluoroacetyl, synthesis of, **5**, 202
Hypofluorite, trifluoromethyl
as fluorinating reagent, **6**, 217
reactions
with alkenes, **6**, 229
with carbon monoxide, **6**, 424
with carbonyl difluoride, **2**, 98
as starting material, for aziridine-*N*-carbonyl fluorides, **6**, 441
synthesis of, **2**, 94; **6**, 231
Hypofluorites, synthesis of, **2**, 94
Hypofluorites, acyl, synthesis of, **5**, 202
Hypofluorites, alkyl, perfluorinated, stability of, **2**, 94
Hypofluorites, *gem*-fluoro, synthesis of, **6**, 250
Hypohalides, alkoxy, reactions, with amidines, **5**, 816
Hypohalites
for the oxidation of sulfides, **2**, 149
reactions, with haloalkenes, **6**, 42
as starting materials, for *N*-chloroimides, **5**, 317
synthesis of, **2**, 685
Hypohalites, acyl, synthesis of, **2**, 12; **5**, 201
Hypohalites, alkenyl, synthesis of, **2**, 685
Hypohalites, alkyl, synthesis of, **2**, 94
Hypohalites, aryl
as intermediates, in the halogenation of phenolic compounds, **2**, 685
synthesis of, **2**, 685
Hypohalites, trihalomethyl, synthesis of, **6**, 231
Hypoiodides, use in the synthesis of alkynyl iodides, **2**, 1019
Hypoiodite, acetyl
reactions
with alcohols, **2**, 95
with alkanols, **2**, 95
Hypoiodite, *t*-butyl, synthesis of, **2**, 95
Hypoiodite, ethyl, synthesis of, **2**, 95
Hypoiodites, synthesis of, **2**, 95
Hypoiodites, alkyl, synthesis of, **2**, 95
Hypoiodites, *t*-butyl, synthesis of, **2**, 95
Hypoiodites, cyclohexyl, synthesis of, **2**, 95
Hypoiodites, cyclopentyl, synthesis of, **2**, 95
Hypoiodites, ethyl, synthesis of, **2**, 95
Hypoiodites, methyl, synthesis of, **2**, 95
Hypoiodites, 1-pentyl, synthesis of, **2**, 95
Hypoiodites, 2-propyl, synthesis of, **2**, 95
Hypomyces semitranslucens, as source of sibirinone, **3**, 535
Hypophosphite esters, reactions, with alkenes, **2**, 451
Hypophosphite hydrate, triethylammonium, use in the synthesis of aryl aldehydes, **3**, 86
Hypophosphorous acid
reactions
with alkenes, **2**, 451
with diphenylmethylimines, **2**, 451
with paraformaldehyde, **4**, 125
as reducing agent, **1**, 442
for the reductive cleavage of disulfides, **2**, 127
use in aromatic deaminations, **1**, 36
use in the synthesis of radicals, **1**, 936, 938, 945

I

Ibuprofen, synthesis of, **1**, 298
Idenes, cumylenyl-, as reactive intermediates, **1**, 1145
Ikarugamycin
  reactions, with alkenes, **1**, 434
  synthesis of, **1**, 815
Ilosvay's reagent, use in the synthesis of alkynylcoppers, **2**, 1094
Imidate, reactions, with sodium diethyl phosphite, **6**, 159
Imidate ester hydrochloric salts, from, nitriles, **6**, 69
Imidate ester hydrochlorides
  reactions, with alcohols, **6**, 70
  thermal decomposition, to alkyl chlorides and carboxamides, **6**, 70
Imidate ester salts, as starting materials, for ortho esters, **6**, 69
Imidates
  N-acylation of, **5**, 694
  trans-esterification of, **5**, 694
  reactions, with ethoxycarbonyl nitrene, **5**, 845
  silylation of, **5**, 863
  as starting materials
    for *N*-haloimidic esters, **5**, 813
    for imidic acid esters, **5**, 694
    for *N*-phosphorylated imino esters, **5**, 708
    for *N*-silylimidoyl esters, **5**, 863
  See also Esters, imidic
Imidates, alkenyl, synthesis of, **5**, 685
Imidates, *N*-alkoxy-
  from
    N-hydroxyimidic acid esters, **5**, 700
    hydroxylamine, **5**, 700
    hydroxylamines, **5**, 700
    thiocarboxylic acid esters, **5**, 700
  synthesis of, **5**, 699
  See also Acid esters, hydroxamic
Imidates, alkyl *N*-acyl
  synthesis of, **3**, 417
  See also Imines, *N*-acyl-
Imidates, alkyl-*N*-(diorganophosphin)(thi)oyl, synthesis of, **5**, 707
Imidates, amino-, as intermediates, in the synthesis of acylhydrazines, **5**, 356
Imidates, *N*-arenesulfonyl, synthesis of, **5**, 703
Imidates, aryl
  deprotonation of, **3**, 721
  synthesis of, **5**, 685
Imidates, *N*-(aryldichlorothio)-, synthesis of, **5**, 703
Imidates, aryl-*N*-(diorganophosphin)(thi)oyl, synthesis of, **5**, 707
Imidates, *N*-aryl-substituted, synthesis of, **5**, 694
Imidates, *N*-(arylthio)-
  chlorination of, **5**, 703
  synthesis of, **5**, 703
Imidates, *O*,*N*-bis(trialkylplumbyl), synthesis of, **5**, 717
Imidates, bis(trimethylsilyl), synthesis of, **5**, 379
Imidates, bromo-, as intermediates, in the synthesis of acylhydrazines, **5**, 356
Imidates, chloro-, as intermediates, in the synthesis of acylhydrazines, **5**, 356
Imidates, *N*-chloro-
  from, silylamides, **5**, 317
  reactions
    with diaryl sulfides, **5**, 703
    with sulfenyl chlorides, **5**, 703
  synthesis of, **6**, 56
Imidates, α-diazo-, synthesis of, **3**, 477
Imidates, *N*-fluorinated, synthesis of, **5**, 698
Imidates, *N*-fluoro-, synthesis of, **5**, 813
Imidates, *N*-halo-
  synthesis of, **5**, 698
  See also Esters, *N*-haloimidic
Imidates, *N*-halo-*O*-silyl, from, *N*-trimethylsilylacetimidic silyl esters, **5**, 814
Imidates, hydroperoxy-, as intermediates, **5**, 709
Imidates, *N*-hydroxy-
  from
    N-hydroxyimidic acid esters, **5**, 700
    hydroxylamine, **5**, 700
    hydroxylamines, **5**, 700
    thiocarboxylic acid esters, **5**, 700
  synthesis of, **5**, 699
Imidates, methyl, synthesis of, **5**, 699
Imidates, methylthiomethoxy-, synthesis of, **5**, 689
Imidates, 4-nitrophenyl, synthesis of, **5**, 707
Imidates, *N*-phenyl, as starting materials, for *N*-silylketene *N*,*O*-acetals, **1**, 828
Imidates, *N*-phenylaminochloro-, reactions, with quinoline, **5**, 1123
Imidates, *O*-silyl, oxidation of, **1**, 558
Imidates, stannylated, synthesis of, **5**, 716
Imidates, *C*-substituted, synthesis of, **3**, 687
Imidates, *N*-sulfonyl, synthesis of, **5**, 827
Imidates, *N*-(trialkylplumbyl), synthesis of, **5**, 717
Imidates, trialkylsilyl
  from
    carbonic acids, **5**, 721
    carboxylic acid amides, **5**, 718
    carboxylic acids, **5**, 721
    imidic acids, **5**, 722
    N-silylated amines, **5**, 721
  synthesis of, **5**, 718
Imidates, trichloro-, [3,3] sigmatropic rearrangement of, **1**, 836
Imidates, trimethylsilyl, synthesis of, **5**, 723
Imidates, *N*-unsubstituted, carboxylation of, **5**, 694
Imidate salts, *O*-methyl, alcoholysis of, **6**, 70
Imidazol-5-carboxylic acid, 4-nitro-1-methyl-, from, 4-nitro-1-methyl-5-cyanoimidazole, **5**, 94
Imidazole
  reactions, with phosgene, **6**, 444
  ring-opening of, **5**, 1033
  use in the synthesis of alkanoic acids, **5**, 31
  use in the synthesis of *N*-silylamides, **5**, 374
Imidazole, 1-(*N*-benzyl-*N*-methylthiocarbamoyl)-, reactions, with sodium hydride, **6**, 557
Imidazole, 1-[(benzyloxy)thiocarbonyl]-, synthesis of, **6**, 557
Imidazole, 2-chloro-4,5-dihydro-, reactions, with potassium *O*-alkyl dithiocarbonates, **6**, 547
Imidazole, 2-(2'-mercaptophenyl)dihydro-, reactions, with phthalolyl chloride, **6**, 111
Imidazole, *N*-methyl-, metallation of, **6**, 598
Imidazole, *N*-methyl-2-phenyl-4-benzamino-, as starting material, for ketones, **3**, 311
Imidazole, 2-methylthio-4-phenyl-1-triphenylphosphoranylidenamino-, reactions, with carboxylic acid chlorides, **5**, 680
Imidazole, 4-nitro-1-methyl-5-cyano-, as starting material, for 4-nitro-1-methylimidazol-5-carboxylic acid, **5**, 94
Imidazole, 1-thiobenzoyl-, reactions, with hydrogen chloride, **5**, 477
Imidazole, *N*-thiobenzoyl-
  reactions, with *N*-methylaniline, **5**, 587

as thioacylating reagent, **5**, 641
Imidazole, thiocarbonyl-, tributyltin cleavage of, **1**, 177
Imidazole, *N*-trifluoroacetyl-, as acylating agent, **5**, 196
Imidazole, triiodo-, use in the synthesis of alkyl iodides, **2**, 7
Imidazole, 1-(trimethylsilyl)-
reactions
with hydrogen sulfide, **2**, 541
with thiols, **2**, 263
Imidazole, *N*-trimethylsilyl-, reactions, with thiophosgene, **6**, 577
Imidazole-1-carbodithioate, *S*-methyl 2-methylthio-4,5-dihydro-, synthesis of, **6**, 564
Imidazoles
from, glyoxal α-oximes, **3**, 673
as phase-transfer catalysts, **4**, 793
reactions
with 2-alkylthio-1,3-thiaselenolylium cations, **6**, 317
with cyanogen chloride, **5**, 1127
with lithium, **3**, 109
with oxygen, **5**, 759
as starting materials
for ketene aminals, **4**, 973
for ketones, **3**, 311
use in the synthesis of aminotrithioorthoformates, **6**, **295–318**
Imidazoles, acyl-
as acylating reagents, **5**, 235
reactions
with chlorotrimethylsilane, **4**, 933
with nitromethane, **5**, 829
with nitronate salts, **3**, 46
reductive silylation of, **5**, 416, 419
as starting materials
for α-amino acids, **3**, 40
for α-nitro esters, **2**, 396
for α-nitro ketones, **2**, 396
use in the synthesis of α-nitro ketones, **3**, 190
use in the synthesis of selenocarboxylic esters, **5**, 239
Imidazoles, 1-acyl-, reduction of, **3**, 85
Imidazoles, 4-acyl-, synthesis of, **3**, 311
Imidazoles, *N*-acyl-, use in the synthesis of aromatic ketones, **3**, 281
Imidazoles, 1-cyano-, from, [1,2,3,4]tetrazolo[1,5-c]pyrimidines, **5**, 1122
Imidazoles, 2-cyano-, synthesis of, **3**, 672
Imidazoles, 4-cyano-, synthesis of, **3**, 673
Imidazoles, 4-cyano-1-alkyl-5-nitro-, synthesis of, **3**, 673
Imidazoles, 5-cyano-1-alkyl-4-nitro-, synthesis of, **3**, 673
Imidazoles, 2-cyano-1-hydroxy-, synthesis of, **3**, 673
Imidazoles, formyl-, synthesis of, **3**, 109
Imidazoles, 2-formyl-, synthesis of, **3**, 109
Imidazoles, 5-keto-, thermal rearrangement of, **3**, 311
Imidazoles, nitro-, reactions, with dimethylformamide dimethyl acetal, **3**, 673
Imidazoles, *N*-thiobenzoyl-, reactions, with amines, **5**, 587
Imidazoles, triaryl-, dimerization of, **6**, 143
Imidazoles, 4-(trifluoromethyl)-, reactions, with ammonium hydroxide, **3**, 673
Imidazoles, vinyl-, synthesis of, via retro-Diels–Alder reaction, **1**, 786
Imidazole-2-thione, from, Schiff bases, **6**, 575
Imidazolides
as acylating agents, **5**, 125
from
diimidazolides, **5**, 477
dithiocarboxylic acids, **5**, 477
Imidazolides, thiocarbonyl
from, epoxy alcohols, **6**, 557
reactions, with tributyltin hydride, **1**, 938
Imidazolidine, 1-benzoyl-2-oxo-3-(2-phenylethyl)-, synthesis of, **6**, 510
Imidazolidine, 1,3-dimethyl-2-oxo-, synthesis of, **6**, 507
Imidazolidine, 1,3-diphenyl-2-oxo-, synthesis of, **6**, 508
Imidazolidine, 2-oxo-, synthesis of, **6**, 507
Imidazolidine, 2-oxo-1,3,4,5-tetraaryl-, synthesis of, **6**, 507
4,5-Imidazolidinediones, 2,2-dichloro-, reactions, with amines, **6**, 360
Imidazolidines
from, chiral diamines, **4**, 408
reactions, with Grignard reagents, **4**, 410
as starting materials, for *gem*-dinitrosamines, **4**, 434
synthesis of, **4**, 408
Imidazolidines, N^1,N^2-diacyl-, synthesis of, **4**, 425
Imidazolidines, 1,3-dihydroxy-, synthesis of, **4**, 435
Imidazolidines, 2-methylene-, Diels–Alder cyclizations of, **4**, 414
Imidazolidine-2-thione, 1,3-dimethyl-, synthesis of, **6**, 577
Imidazolidinethiones, synthesis of, **4**, 421
Imidazolidinetrione, reduction of, **4**, 103
2-Imidazolidininone, 1,3-dimethyl-, use in the synthesis of alkenes, **1**, 568
Imidazolidinium salts, reactions, with Grignard reagents, **3**, 95
Imidazolidin-2-one, as starting material, for thiones, **6**, 579
Imidazolidinones
from
dimethylketene-*N*-phenylimine, **4**, 422
thiohydantoins, **4**, 420
synthesis of, **4**, 416, 419, 421
2-Imidazolidinones
reactions, with phosgene, **5**, 1013
synthesis of, **3**, 416
4-Imidazolidinones, *N*-hydroxy-, synthesis of, **4**, 442
Imidazolinediones, synthesis of, **6**, 115
Imidazolines
from, isocyanides, **4**, 538
as starting materials
for aminals, **4**, 414
for ketene aminals, **4**, 979
Imidazolines, 2-(arylimino)-, reactions, with alkyl halides, **6**, 642
Imidazolines, N^1,N^2-diacyl-, synthesis of, **4**, 426
Imidazolium salts, cycloaddition to, **4**, 414
Imidazol-4-ones, 5-substituted-1-cyano-2-imino-, synthesis of, **5**, 1121
Imidazol-2-ylidene, 1,3-dimesityl-, use in the synthesis of carbene complexes, **6**, 722
Imidazopyridines, perhydro-, from, 2-(methylamino)methylhexahydropyridine, **4**, 408
Imide, benzonitrile *N*-phenyl-, synthesis of, **3**, 680
Imide, bis(dichloroaluminum)phenyl-
as iminating reagent, **3**, 404
use in the synthesis of *N*-substituted imines, **3**, 408
Imide, *N*-fluorobis(benzenesulfonyl)-, reactions, with phosphonates, **4**, 131
Imide, *N*-fluoro bis(trifluoromethylsulfonyl)-, use in the synthesis of α-fluoro ketones, **3**, 149
Imide, triphenylphosphine benzotriazole-1-ylmethyl-, synthesis of, **4**, 329
Imide, triphenylphosphine *N*-trimethylsilyl-, reactions, with thiophosgene, **6**, 537
Imide radicals, reactions, with allylstannane, **1**, 328
Imides
alkylation of, **4**, 881
amination of, **5**, 360
as chiral auxiliary in radical reactions, **1**, 326
desilylation of, **1**, 375
from, active esters, **5**, 306
reactions
with alkylcopper complexes, **5**, 387
with antimony halides, **5**, 372
with arylpalladium complexes, **5**, 387
with bromotriethylgermane, **5**, 391
with dialkylnickel complexes, **5**, 387
with dimethyl sulfoxide, **2**, 163
with Grignard reagents, **4**, 303

with mercury(I) carbonate, **5**, 387
with metal alkoxides, **5**, 372
with metal halides, **5**, 372
with phosphorus halides, **5**, 369
with sulfene, **5**, 348
with tributyltin isocyanate, **5**, 391
silylation of, **5**, 378
as starting materials
for *O,N*-acetals, **4**, 303
for *N*-bromoimides, **5**, 318
for *N*-chloroimides, **5**, 316
for *N*-fluoroimides, **5**, 314
for imidoaluminates, **5**, 384
for *N*-iodo(III)imides, **5**, 322
for *N*-iodo succinimide, **5**, 321
for metal imides, **5**, 390
for *N*-phosphinoimides, **5**, 369
for phosphoroimides, **5**, 372
for *N*-selenoimides, **5**, 336
for *N*-silylimides, **5**, 378
for sulfides, **2**, 713
for *N*-sulfonylimides, **5**, 348
for thioimides, **5**, 336
for *N*-thioimides, **5**, 336
for transition metal imides, **5**, 387
synthesis of, **5**, 305
via oxidation of amides, **5**, 307
thionation of, **5**, 631, 632
Imides, acyclic, synthesis of, via oxidation of amides, **5**, 307
Imides, *N*-alkoxy-, deprotection of, **5**, 332
Imides, *β*-alkoxy-, as starting materials, for *β*-alkoxy aldehydes, **3**, 32
Imides, alkoxy-substituted, synthesis of, **5**, 332
Imides, *N*-amino-, as starting materials, for hydrazines, **2**, 379
Imides, α-azido-, synthesis of, **6**, 615
Imides, azo-, synthesis of, **6**, 616
Imides, bis(arylsulfen)-, nomenclature, **2**, 767
Imides, boron-substituted, synthesis of, **6**, 661, 669
Imides, *N*-bromo-
from
imides, **5**, 318
metallated imides, **5**, 318
synthesis of, **6**, 615
via electrolysis of sodium bromide, **5**, 318
Imides, *N*-chloro-
from
hypohalites, **5**, 317
imides, **5**, 316
synthesis of, **6**, 621
Imides, copper-substituted, synthesis of, **6**, 674
Imides, *N*-fluoro-, from, imides, **5**, 314
Imides, *N*-germylated, synthesis of, **5**, 391
Imides, halo-
reactions
with dialkyl sulfides, **5**, 336
with disulfides, **5**, 336
with dithiocarboxylates, **5**, 336
with tin sulfides, **5**, 337
as starting materials
for imido selenides, **5**, 341
for imido tellurides, **5**, 341
for metal imides, **5**, 391
for transition metal imides, **5**, 388
Imides, *N*-halo-
as starting materials
for *N*-iodo succinimide, **5**, 321
for phosphoroimides, **5**, 372
for *N*-silylimides, **5**, 378
for thioimides, **5**, 336
Imides, α-halo-
from, α-hydroxy compounds, **4**, 98
synthesis of, **4**, 98
via cleavage of *gem*-diheteroatomic compounds, **4**, 99
via halogenation, **4**, 99
Imides, *N*-haloalkyl-, synthesis of, **4**, 99
Imides, α-haloalkyl-, reactions, with zinc, **2**, 1002
Imides, hydroxy-, alkylation of, **5**, 333
Imides, *N*-hydroxy-
alkylation of, **5**, 333
from
hydroxylamines, **5**, 332
α-nitro ketones, **5**, 332
reactions, with diazomethane, **5**, 333
synthesis of, **5**, 332
via deprotection of *N*-alkoxyimides, **5**, 332
Imides, hydroxy-substituted, synthesis of, **5**, 332
Imides, *N*-iodo(I), from, *N*-metalimides, **5**, 320
Imides, *N*-iodo(III), from, imides, **5**, 322
Imides, metallated
reactions, with bromine, **5**, 318
as starting materials, for *N*-bromoimides, **5**, 318
Imides, metal-substituted, synthesis of, **6**, 663
Imides, *N*-methoxy-, synthesis of, **5**, 333
Imides, nitrile
dimerization of, **3**, 680
from
diazoalkanes, **3**, 682
hydrazones, **3**, 681
hydrazonoyl halides, **3**, 678, 680
α-nitro aldehydehydrazones, **3**, 681
oxadiazolinones, **3**, 681
oxathiadiazoles, **3**, 681
tetrazoles, **3**, 681
trialkylstannyldiazo compounds, **3**, 682
rearrangement of, **3**, 680
stable, synthesis of, **3**, 681
synthesis of, **3**, 680
Imides, *N*-nitro-, synthesis of, **5**, 362
Imides, *N*-phenylsulfenyl-, reactions, with alcohols, **2**, 135
Imides, phosphene-, synthesis of, **2**, 859
Imides, phosphine-
synthesis of, **2**, 863; **4**, 800
See also Phosphoranes, imino-
Imides, phosphorus-substituted, synthesis of, **6**, 656, 665
Imides, silicon-substituted, synthesis of, **6**, 661, 668, 671, 673
Imides, *N*-silyl-
from
N-haloimides, **5**, 378
imides, **5**, 378
as starting materials, for thioimides, **5**, 338
Imides, silylated phosphine-, synthesis of, **2**, 470
Imides, stilbine, synthesis of, **5**, 372
Imides, α-substituted-*β*-hydroxy-, synthesis of, **1**, 878
Imides, *N*-sulfenyl-, as sulfenylating reagents, **2**, 688
Imides, sulfinyl-, as starting materials, for transition metal imides, **5**, 388
Imides, *N*-sulfonyl-, from, imides, **5**, 348
Imides, sulfuryl fluoride alkyl-, synthesis of, **2**, 401
Imides, thioalkyl-, as transfer reagents, **5**, 336
Imides, thionophosphorylcarbonyl mixed, as starting materials, for *N*-monosubstituted thionoamides, **5**, 643
Imides, tin-substituted, synthesis of, **6**, 674
Imides, *N*(trialkylstannyl)-, reactions, with amines, **5**, 390
Imides, *N*-tributylstannyl-, synthesis of, **5**, 391
Imides, triphenylphosphine-
reactions
with *gem*-diisocyanates, **4**, 436
with *gem*-diisocyanatoalkanes, **4**, 441
S-Imides, ketene dithioacetal
from, ketene dithioacetals, **4**, 860
synthesis of, **4**, 860
S-Imides, thiocarbonyl-
from
α-chlorosulfenamides, **3**, 370

sulfines, **3**, 370
sulfinylamines, **3**, 370
thioketones, **3**, 370
synthesis of, **3**, 370
Imidic acid benzenesulfonic acid anhydrides, as intermediates, **5**, 710
Imidic acid esters
from
alkoxymethyleniminium salts, **5**, 686
amidines, **5**, 694
carboxylic ortho amides, **5**, 690
carboxylic ortho esters, **5**, 690
heterocyclic compounds, **5**, 693
imidates, **5**, 694
imidoyl halides, **5**, 685
isonitriles, **5**, 691
metal complexes, **5**, 696
nitriles, **5**, 691
nitrilium salts, **5**, 685
synthesis of, **5**, 685
Imidic acid esters, *N*-hydroxy-
alkylation of, **5**, 700
as starting materials
for *N*-alkoxyimidates, **5**, 700
for *N*-hydroxyimidates, **5**, 700
Imidic acid esters, *N*-unsubstituted, silylation of, **5**, 722
Imidic acids
as starting materials, for trialkylsilyl imidates, **5**, 722
synthesis of, **5**, 685
Imidic acids, *N*-fluorinated, reactions, with diazomethane, **5**, 699
Imidic acids, *N*-halo-, synthesis of, **5**, 812
Imidic acids, thiol, synthesis of, **5**, 725
Imidic compounds, halo-, synthesis of, **5**, 806
Imidic compounds, *N*-halo-, hydrolysis of, **5**, 812
Imidic fluorides, *N*-bromo-, synthesis of, **5**, 808
Imidic fluorides, *N*-fluoro-, synthesis of, **5**, 807
Imidic fluorides, halo-, synthesis of, **5**, 809
Imidic fluorides, perfluorinated *N*-fluoro-, synthesis of, **5**, 811
Imidic fluorides, perfluoro-*N*-chloro-, synthesis of, **5**, 808
Imidic fluorides, perfluoro *N*-fluoro-, hydrolysis of, **5**, 813
Imidic halides, halo-
from
halogenated alkenes, **5**, 811
nitriles, **5**, 808
Imidic halides, *N*-halo-, as starting materials, for *N*-haloamidines, **5**, 816
Imidium chloride salts, chloroalkyl-, synthesis of, **2**, 758
Imidoalkyl compounds, synthesis of, **3**, 514
Imidoates, alkyl-*N*-hydroxy-, synthesis of, **5**, 699
Imidoates, *N*-bromo-, synthesis of, **5**, 698
Imidoates, ethyl-*N*-ethoxy-, synthesis of, **5**, 701
Imidoates, *N*-iodo-, synthesis of, **5**, 698
Imidoates, methyl *N*-hydroxy-, synthesis of, **5**, 699
Imidoates, *N*-unsubstituted, reactions, with hypochlorite, **5**, 698
Imido groups, silylation of, **5**, 863
Imidophosphinates, (bromomethyl)-, synthesis of, **4**, 129
Imidoyl azides, *N*-halo-, synthesis of, **5**, 819
Imidoyl bromides
as intermediates, in the synthesis of ketenimines, **3**, 561
reactions, with methyl thiocyanate, **5**, 864
synthesis of, **3**, 557
Imidoyl bromides, *N*-alkoxy-, from, *N*-alkoxyamides, **5**, 666
Imidoyl bromides, hydroxy-, synthesis of, **5**, 667
Imidoyl bromides, *N*-hydroxy-, reactions, with alkoxides, **5**, 699
Imidoyl bromides, *N*-phenyl-, reactions, with silver cyanide, **3**, 710
Imidoyl chloride, reactions, with tris(trimethylsilyl)phosphane, **6**, 660
Imidoyl chlorides
dehydrochlorination of, for the synthesis of nitrile ylides, **3**, 678
from, phosgene, **5**, 516
hydrogenation of, **3**, 406
oxidative addition of, **5**, 797
reactions
with alcohols, **4**, 312; **5**, 685
with alkylarsines, **5**, 791
with 1-alkynes, **1**, 1047
with alkynylcoppers, **1**, 1047
with alkynyltins, **1**, 1047
with amines, **5**, 743
with *o*-arsinophenols, **5**, 791
with benzoic acids, **5**, 714
with carboxylic acids, **5**, 716
with 4-chlorothiophenol, **5**, 728
with chlorotrimethylsilane, **5**, 793
with dialkyl hydrogen phosphites, **5**, 787
with diethylphenylphosphoramidite, **5**, 787
with dithiophosphinic acids, **5**, 738
with dithiophosphoric acids, **5**, 738
with dithiophosphorus(V) acids, **5**, 644
with hydrazines, **5**, 770
with hydroxylamine, **5**, 764
with iron carbonyl anions, **5**, 799
with lithium arsenides, **5**, 791
with metal carbonyl anions, **5**, 797
with monothiophosphorus(V) acids, **5**, 644
with organophosphorus compounds, **5**, 784
with organotins, **3**, 409
with phenols, **5**, 685
with *N*-phenylhydroxylamine, **5**, 764
with potassium diphenyl dithiophosphinate, **5**, 738
with selenium nucleophiles, **5**, 732
with silver dibenzylphosphate, **5**, 712
with silver nitrate, **5**, 362
with silver nitrite, **5**, 361
with sodium benzeneselonate, **5**, 732
with sodium benzenetellurolate, **5**, 733
with sodium diethyl dithiophosphate, **5**, 738
with sodium diethylthiophosphate, **5**, 738
with sodium selenocyanate, **5**, 1057
with tetrakis(triphenylphosphine)palladium, **5**, 801
with thallous phenyl selenide, **5**, 732
with thallous thiophenoxide, **5**, 728
with thiols, **5**, 728
with trialkyl phosphinites, **5**, 787
with trialkylphosphites, **5**, 787
with trialkyl stannyllithium, **5**, 803
with triallyl plumbyllithium, **5**, 803
with triphenyl germyllithium, **5**, 794
with tris(trimethylsilyl)phosphine, **6**, 686
with tris(trimethylsilyl)phosphite, **5**, 787
as starting materials
for alkoxymethylenephosphanes, **5**, 880
for *gem*-alkylideneamino phosphorylalkenes, **4**, 1009
for *N*-aryl pivaloylimide chlorides, **5**, 891
for *N*-(2,2-dichloroethenyl)benzamide, **5**, 657
for imidoylarsenic compounds, **5**, 791
for *S*-imidoyldithiophosphates, **5**, 738
for *S*-imidoyldithiophosphinates, **5**, 738
for imidoyl fluorides, **5**, 660
for imidoylleads, **5**, 803
for imidoylpalladium(II) complexes, **5**, 801
for imidoylsilicons, **5**, 793
for imidoyl-tetracoordinated phosphorus compounds, **5**, 787
for *S*-imidoylthiophosphates, **5**, 738
for imidoyltins, **5**, 803
for imidoyl-tricoordinated phosphorus compounds, **5**, 784
for ketenimines, **3**, 559
for ketimines, **3**, 409

for *N*-nitroamides, **5**, 362
for *N*-nitrosoamides, **5**, 361
for phosphaalkenes, **5**, 881
for telluroimidic esters, **5**, 733
for *N*-(1,2,2,2-tetrachloroethyl)acetamides, **5**, 657
for tetracoordinate germanium compounds, **5**, 794
for *N*-(2,2,2-trichloro-1-hydroxyethyl)trichloroacetamide, **5**, 657
synthesis of, **3**, 558; **5**, 743; **6**, 617, 621
use in the synthesis of amidines, **5**, 751
Imidoyl chlorides, *N*-alkoxy-, use in the synthesis of mixed anhydrides, **5**, 713
Imidoyl chlorides, *N*-alkyl-, synthesis of, **5**, 655, 657
Imidoyl chlorides, *N*-arenesulfonyl-
reactions
with alkoxides, **5**, 703
with phenoxides, **5**, 703
Imidoyl chlorides, *N*-aryl-
reactions, with *O*-alkyldithiocarbonatepotassium salts, **5**, 632
synthesis of, **5**, 655, 657
Imidoyl chlorides, aryl *N*-hydroxy-, reactions, with hydridoundecacarbonyltriferrate anions, **3**, 666
Imidoyl chlorides, *N*-chloro-, synthesis of, **5**, 662
Imidoyl chlorides, α-chloro-
as intermediates, for ketenimines, **3**, 581
as starting materials, for ketenimines, **3**, 561
Imidoyl chlorides, *N*-(chlorocarbonyl)-, reactions, with methanesulfonic acid, **5**, 1004
Imidoyl chlorides, *N*-(chloromethylphosphinoyl)-, reactions, with sodium nitrophenoxide, **5**, 707
Imidoyl chlorides, *N*-(dichlorophosphinoyl)-
reactions
with alkoxides, **5**, 707
with phenoxides, **5**, 707
Imidoyl chlorides, *N*-(dichlorophosphinthioyl)-
reactions
with alkoxides, **5**, 707
with phenoxides, **5**, 707
Imidoyl chlorides, *N*-dichlorothiophosphoryl-, synthesis of, **5**, 683
Imidoyl chlorides, α-fluorinated, synthesis of, **5**, 675
Imidoyl chlorides, α-keto-, oxidative addition of, for the synthesis of imidoylplatinums, **5**, 802
Imidoyl chlorides, *N*-methanesulfonyloxy, synthesis of, **5**, 670
Imidoyl chlorides, perhalogenated, dehalogenation of, **5**, 659
Imidoyl chloride salts, as starting materials, for ortho esters, **6**, 70
Imidoyl compounds
of Group 13 metals, synthesis of, **5**, 803
of Group 14 metals, synthesis of, **5**, 803
of Group 1 metals, synthesis of, **5**, 795
of Group 2 metals, synthesis of, **5**, 796
as starting materials, for aminobis(phosphonato)methanes, **6**, 159
structure of, **5**, 709
synthesis of, **5**, 860
Imidoyl compounds, *N*-boryl-, synthesis of, **5**, 863
Imidoyl compounds, *N*-silyl-, synthesis of, **5**, 863
Imidoyl dichloride, *N*-ethoxycarbonyl-, use in the synthesis of carbodiimides, **5**, 1071
Imidoyl dichlorides
as starting materials, for carbodiimides, **5**, 1070
synthesis of, **5**, 1070
Imidoyl dichlorides, *N*-aryl-
reactions, with amines, **5**, 1070
synthesis of, **5**, 1070
Imidoyl ethers, as starting materials, for imidoylarsenic compounds, **5**, 790
Imidoyl fluoride, *N*-(trifluoromethyl)perfluorobutane-, synthesis of, **5**, 662
Imidoyl fluorides
from
1,1-difluoroalkenes, **3**, 588
imidoyl chlorides, **5**, 660
as starting materials, for ketenimines, **3**, 560
Imidoyl fluorides, *N*-acyl-, synthesis of, **5**, 662
Imidoyl fluorides, *N*-alkoxy-, synthesis of, **5**, 674
Imidoyl fluorides, *N*-bromo-, synthesis of, **5**, 662
Imidoyl fluorides, *N*-(2-bromo-1,1-difluoroalkyl)haloalkane-, synthesis of, **5**, 662
Imidoyl fluorides, *N*-fluorinated, reactions, with alkoxides, **5**, 698
Imidoyl fluorides, *N*-fluoro-, reactions, with sodium azide, **5**, 819
Imidoyl fluorides, *N*-halogenated, synthesis of, **5**, 662
Imidoyl fluorides, perfluorinated, hydrolysis of, **5**, 685
Imidoyl fluorides, *N*-perfluoroalkyl-, synthesis of, **5**, 674
Imidoyl fluorides, *N*-perhaloethylpropyl-, synthesis of, **5**, 659
Imidoyl halides
displacement of, for the synthesis of imidoyltungstens, **5**, 798
from
amides, **5**, 654
amines, **5**, 659
cyanohydrins, **5**, 654
imidoyl halides, **5**, 659
imines, **5**, 659
isocyanates, **5**, 657
isocyanides, **5**, 657
nitriles, **5**, 657
perfluorohydroxamoyl fluorides, **5**, 671
ureas, **5**, 654
reactions
with metal carbonyl anions, **5**, 798
with phosphorus acid, **6**, 159
with pyridine-*N*-oxides, **5**, 712
as starting materials
for hydrazonamides, **5**, 850
for imidic acid esters, **5**, 685
for imidoyl halides, **5**, 659
for imidoyl isothiocyanates, **5**, 1050
for ketenimines, **3**, 558
for nitrile ylides, **3**, 678
for nitrilium ions, **3**, 687
for *N*-phosphorylated imino esters, **5**, 707
for thioimidic esters, **5**, 728
synthesis of, **5**, 654
Imidoyl halides, *N*-alkoxy-
from, *N*-alkoxyamides, **5**, 824
as intermediates, **3**, 666
Imidoyl halides, *O*-benzyloxy-, synthesis of, **5**, 824
Imidoyl halides, *N*-borylated, synthesis of, **5**, 684
Imidoyl halides, *O*-*t*-butoxy-, synthesis of, **5**, 824
Imidoyl halides, *N*-halo-, synthesis of, **5**, 662
Imidoyl halides, *N*-methoxy-, hydrogenation of, **3**, 430
Imidoyl halides, *O*-methoxy-, synthesis of, **5**, 824
Imidoyl halides, *N*-organoelemental-substituted, synthesis of, **5**, 684
Imidoyl halides, *N*-phosphorylated
from, *N*-phosphorylated carboxylic acid amides, **5**, 683
synthesis of, **5**, 683
Imidoyl iodides
reactions, with lithium ethanethiolate, **5**, 728
synthesis of, **3**, 437
Imidoyl radicals, as intermediates, in the synthesis of isothiocyanates, **5**, 1037
Imination
of acid chlorides, **3**, 404
of aldehydes, **3**, 404
of ketones, **3**, 404
Imine, aryl aldehyde aryl-, reactions, with sodium, **6**, 507
Imine, benzophenone
as ammonia equivalent, **3**, 187

reactions, with acrolein, **3**, 44
as transiminating reagent, **3**, 406
Imine, *N*-chlorodifluoromethylene-, synthesis of, **5**, 807
Imine, dimethylketene-*N*-phenyl-, as starting material, for imidazolidinones, **4**, 422
Imine, dimethylsilyl-, synthesis of, **5**, 794
Imine, ethylene-, reactions, with *p*-tolylsulfonylethyne, **2**, 742
Imine, hexafluoroacetone, reactions, with acetic anhydride, **4**, 304
Imine, hexafluoroisopropylidene-, reactions, with methylphosphines, **2**, 440
Imine, *N*-phenyldichloro-, as starting material, for tris(phosphoryl)methylanilines, **6**, 367
Imine, *N*-trichlorovinylbenzyl-, synthesis of, **4**, 794
Imine, (trifluoroguanylazo)difluoromethylene-, reactions, with ammonia, **6**, 363
Imine anions, alkylation of, **1**, 152
Imine chlorides, synthesis of, **3**, 417
Imine complexes, *N*-metallo-, synthesis of, **3**, 460
Imine–enamine isomerization, for the synthesis of imidoylsilanes, **5**, 794
Imine functions, α-metallated, synthesis of, **4**, 535
Imines
arsenic-substituted, synthesis of, **3**, 456
beryllium-substituted, synthesis of, **3**, 459
catalytic asymmetric hydrogenation of, **2**, 330
α-chlorination of, **2**, 22
cycloaddition of, **4**, 414
from
alkyl azides, **3**, 407
allylic enamines, **1**, 835
aryl azides, **3**, 407
cycloalkyl azides, **3**, 407
enamines, **3**, 408
ketoximes, **3**, 408
N-(methoxymethyl)anilines, **4**, 412
nitro compounds, **3**, 408
nitrones, **3**, 408
the oxidation of amines, **3**, 17
gallium-substituted, synthesis of, **3**, 458
germanium-substituted, synthesis of, **6**, 669
halogenation of, **3**, 416
as heterodienophiles, **4**, 304
hydrogenation of, **2**, 331
hydrolysis of, **3**, 132
iron-substituted, synthesis of, **3**, 460
lithiation of, **1**, 371
Mannich reactions of, **3**, 186
C-metal compounds of
as acyl anion equivalents, **3**, 418
synthesis of, **3**, 418
nitrosation of, **2**, 805
nucleophilic addition to, **4**, 426
for the opening of epoxides, **1**, 212
optically pure, for the synthesis of α-aminophosphonic acids, **4**, 462
properties of, **3**, 404
reactions
with acetyl chloride, **4**, 100
with 1-acylhydrazines, **5**, 363
with 1-alkynes, **1**, 1040
with alkynyl chlorides, **2**, 1053
with alkynyl metals, **1**, 1040
with amines, **4**, 411
with 2-aryloxymethylallylzinc bromides, **2**, 324
with bistrifluoromethylketene, **4**, 886
with boron trichloride, **4**, 107
with 2-bromozincmethyl allylic ethers, **1**, 194
with *t*-butyl thiol, **4**, 325
with carbon nucleophiles, **2**, 313
with chloroacetyl chloride, **4**, 100
with chloroformate esters, **6**, 507
with *m*-chloroperbenzoic acid, **4**, 104
with chlorophosphonites, **4**, 486
with diaryl phosphites, **4**, 463
with dibutylzirconocene, **2**, 70
with dichlorophosphines, **4**, 463
with 1,3-dithiane, **3**, 43
with divinyl sulfones, **2**, 158
with electrophiles, **4**, 98
with ester enolates, **5**, 287
with ethyl phenylphosphonite, **4**, 463
with halocarbenes, **6**, 51
with hydrazoic acid, **4**, 444
with hydrogen cyanide, **3**, 616
with hydrogen peroxide, **2**, 96
with hydrogen sulfide, **2**, 224
with isocyanates, **6**, 504
with isothiocyanates, **6**, 571
with ketenes, **5**, 286
with mono-*S*-trimethylsilylketene dithioacetals, **2**, 313
with nitroacetone, **3**, 258
with nucleophilic phosphorus compounds, **4**, 455
with organolithiums, **1**, 310
with ortho formates, **5**, 690
with palladium diacetate, **5**, 464
with (perfluoroalkyl)lithiums, **2**, 311
with phenylphosphine, **4**, 455
with phosgene, **6**, 500
with phosphines, **3**, 455; **4**, 455
with phosphinic acids, **4**, 463
with phosphonites, **4**, 463
with phosphorus(III) compounds, **3**, 50
with phosphorus trichloride, **2**, 449
with semicarbazide, **3**, 450
with *S*-silyl alkynylthioethers, **5**, 605
with silylphosphines, **4**, 481
with sodium borohydride, **2**, 308
with sodium hydrogen selenide, **5**, 626
with sulfonyl chlorides, **2**, 249
with thioketenes, **4**, 905
with tin(II) carboxylic thioesters, **5**, 236
with tributyltin chloride, **2**, 816
with *S*-trimethylsilylalkynethiols, **5**, 614
with trimethylsilyl enol ethers, **3**, 187
with trimethylsilyl trifluoromethanesulfonate, **2**, 313
with trinitromethanes, **6**, 147
with vinylidene complexes, **5**, 954
reactivity of, **1**, 1041
reduction of, **1**, 46; **2**, 330
for the synthesis of *N*-acylaminals, **4**, 420
for the synthesis of amines, **2**, 312
silylation of, **4**, 518
silylmethylation of, **4**, 517
as starting materials
for *O*,*N*-acetals, **4**, 304, 312
for *S*,*N*-acetals, **4**, 319, 328
for aminals, **4**, 411
for amines, **2**, 311; **3**, 404
for α-aminoalkylphosphinic acids, **4**, 463
for α-aminoalkylphosphonic acids, **4**, 463
for α-aminophosphinic acids, **4**, 463
for α-aminophosphonic acids, **4**, 462
for azaallyl anions, **3**, 404
for azomethine ylides, **1**, 368
for *N*-chloroamides, **5**, 315
for *N*-chlorocarbonyls, **6**, 445
for chlorofluoroalkanes, **4**, 32
for dibromoalkanes, **4**, 27
for dichloroalkanes, **4**, 19
for difluoroalkanes, **4**, 10
for 4,5-dihydro-1,2,4-thiadiazoles, **3**, 686
for diiodoalkanes, **4**, 29
for enamides, **5**, 288
for *N*-fluoroamines, **2**, 335

for α-haloamines, **4**, 100
for hemithioaminals, **4**, 462
for imidoyl halides, **5**, 659
for imidoylphosphorus dichlorides, **5**, 785
for imidoyl-tetracoordinated phosphorus compounds, **5**, 789
for imidoyl-tricoordinated phosphorus compounds, **5**, 785
for β-lactams, **5**, 274
for nitrones, **3**, 432
for oxaziridines, **4**, 312
for pyrrolidines, **1**, 289
for sulfenimines, **3**, 436
for thioketones, **3**, 348, 362, 363, 365
for unsymmetrical alkanoic anhydrides, **5**, 193
α-sulfenylation of, **3**, 418
synthesis of, **3**, **403–423**, **443–490**
general methods, **3**, 404
via palladium-catalyzed reactions, **3**, 408
zinc-substituted, synthesis of, **3**, 461
Imines, acyl-
Mannich reactions of, **3**, 186
reactions, with alcohols, **4**, 304
synthesis of, **3**, 417
Imines, *N*-acyl-
pyrolysis of, **5**, 1002
reactions
with alkynyl metals, **1**, 1042
with chlorine monofluoride, **5**, 315
as starting materials, for α-aminophosphonic acids, **4**, 486
use in [4+2] cycloaddition reactions, **3**, 417
See also Imidates, alkyl *N*-acyl-
Imines, α-acyl-
from
α-bromo esters, **3**, 417
α-bromo ketones, **3**, 417
synthesis of, **3**, 417
Imines, *N*-acyl-3-alkyn-2-one, synthesis of, **1**, 1078
Imines, alkanesulfonyl-, hydrolysis of, **2**, 251
Imines, *N*-alkyl-
deprotonation of, **1**, 152
See also Schiff bases
Imines, *N*-alkylbis(silyl)-, synthesis of, **6**, 671
Imines, alkynyl-, synthesis of, **1**, 1047
Imines, allyl-, rearrangement of, **3**, 180
Imines, *N*-alumino-, synthesis of, **3**, 458
Imines, amine-, as starting materials, for hydrazines, **2**, 384
Imines, aminomethyl-
from, α-bromo glycine diphenylmethanimine, **4**, 445
synthesis of, **4**, 445
Imines, *N*-arenesulfonyl-, reactions, with secondary phosphines, **4**, 481
Imines, *N*-aroyl-, synthesis of, **4**, 304
Imines, arsino-, from, organoarsines, **2**, 892
Imines, aryl-
from, *N*-silylated amides, **3**, 412
reactions, with hydroxylamines, **4**, 442
use in the synthesis of *N*-aryl β-lactams, **5**, 294
use in the synthesis of resorcinols, **1**, 46
Imines, *N*-aryl-, synthesis of, **3**, 407
Imines, benzocyclobutanone, as intermediates, **3**, 665
Imines, benzophenone, lithiation of, **4**, 535
Imines, bisaryl-, from, aryl thiobenzamides, **3**, 412
Imines, bis(phosphoryl)-, synthesis of, **6**, 367
Imines, *N*-boryl-
as starting materials, for carbinamines, **2**, 316
synthesis of, **3**, 457
Imines, α-bromo-, dehydrodimerization of, **3**, 418
Imines, *N*-carbon-substituted
from, carbonyls, **3**, 405
synthesis of, **3**, 405
Imines, carbonyl-, synthesis of, **3**, 327
Imines, chiral
as intermediates
for the synthesis of chiral amines, **3**, 415
for the synthesis of chiral ketones, **3**, 415
synthesis of, **3**, 414
use in Michael reactions, **3**, 414
Imines, chloro-
reactions
with arenesulfenyl chlorides, **4**, 110
with sulfur dichloride, **4**, 110
Imines, *N*-chloro-
reactions, with Lewis acids, **4**, 110
as starting materials, for *N*-chloroformamidines, **5**, 818
Imines, α-chloro-
reactions, with trimethyl phosphite, **5**, 789
as starting materials
for imino esters, **5**, 697
for nitrilimine, **3**, 452
Imines, β-chloro-
as starting materials, for azetidines, **3**, 416
synthesis of, **3**, 416
Imines, *N*-(cyanomethyl)-, synthesis of, **3**, 409
Imines, *N*-cyanovinyl-, synthesis of, **5**, 1118
Imines, cyclic
from
amines, **3**, 411
nitrogen-centered radicals, **3**, 410
synthesis of, **1**, 152; **3**, 409
via Wacker oxidation, **3**, 411
Imines, cycloheptanone, alkylation of, **1**, 152
Imines, cyclohexanone, alkylation of, **1**, 152
Imines, 2-cyclopropyl-substituted piperidine, synthesis of, **3**, 410
Imines, diaryl-, reduction of, **1**, 46
Imines, α-diazo-, synthesis of, **3**, 476
Imines, α-diazo *N*-cyano-, synthesis of, **3**, 476
Imines, α,α-dichloro-, α-sulfenylation of, **3**, 418
Imines, diphenylmethyl-, reactions, with hypophosphorous acid, **2**, 451
Imines, *N*-diphenylphosphinoyl-, as starting materials, for phosphoramides, **2**, 331
Imines, fluoro-
reactions, with dimethylamine, **5**, 762
as starting materials, for diazirines, **6**, 280
Imines, *N*-fluoro-
reactions, with alcohols, **6**, 279
as starting materials, for *N*-fluoroamidines, **5**, 817
synthesis of, **5**, 807
use in the synthesis of *N*-fluoroimidic esters, **5**, 814
Imines, (*Z*)-fluoro-
reactions
with ammonia, **5**, 761
with 4-chloroaniline, **5**, 761
with diethylamine, **5**, 761
Imines, *N*-fluorosulfonyl-
reactions, with diphenylphosphine, **4**, 481
as starting materials, for α-aminophosphonic acids, **4**, 486
Imines, *N*-germyl-, synthesis of, **2**, 815; **3**, 457
Imines, halo-
reactions, with fluorine, **6**, 53
synthesis of, **3**, **403–423**
Imines, *N*-halo-, synthesis of, **3**, 419
Imines, α-halo-
from, amines, **3**, 415
synthesis of, **3**, 415
Imines, β-hydroxyamine-, production of, **2**, 384
Imines, isocyano-, synthesis of, **3**, 695
Imines, ketone-, reactions, with isothiocyanates, **5**, 597
Imines, *N*-lithio-
reactions
with alkylgermanium chlorides, **3**, 457
with alkyltin halides, **3**, 457
with haloboranes, **3**, 458
with silyl chlorides, **3**, 456

Imines, *N*-magnesio-, synthesis of, **3**, 459
Imines, manganese *N*-metallo-, synthesis of, **3**, 460
Imines, nitro-, synthesis of, **3**, 439
Imines, *N*-nitro-, reduction of, **3**, 439
Imines, *N*-nitroso-
  synthesis of, **3**, 439
  as synthetic intermediates, **3**, 439
  use in the synthesis of heterocycles, **3**, 439
Imines, perfluoro-, **6**, 279
  oxidation of, **6**, 279
  as starting materials, for oxaziridines, **6**, 279
Imines, *N*-phenyl-
  reactions, with hydrogen sulfide, **3**, 348
  as starting materials, for thioketones, **3**, 348
Imines, *N*-phosphinoyl-
  diastereoselective reductions of, **3**, 439
  from, phosphinamides, **3**, 439
  reduction of, **3**, 439
  as starting materials, for *N*-phosphinoyloxaziridines, **3**, 439
  synthesis of, **3**, 439
Imines, phosphosilyl-, reactions, with acid chlorides, **5**, 373
Imines, *N*-plumbyl-, synthesis of, **3**, 457
Imines, polyfluorinated phenyl-, reactions, with bis(trifluoromethyl)thioketene, **4**, 905
Imines, propargyl-, metallation of, **4**, 518
Imines, 2-propargyloxy-, Claisen–Cope rearrangement of, **1**, 980
Imines, selenyl-
  synthesis of, **2**, 251
  See also Selenimides, *N*-substituted
Imines, *N*-silicon-substituted, synthesis of, **3**, 439
Imines, silyl-, synthesis of, **2**, 812
Imines, *N*-silyl-
  reactions
    with enolates, **3**, 440
    with ester enolates, **5**, 377
    with Lewis acids, **4**, 110
    with silylketene acetals, **3**, 440
  as starting materials
    for *β*-lactams, **5**, 377
    for *N*-silylamides, **5**, 377
  structure of, **3**, 440
  synthesis of, **3**, 440, 456
  use as masked *N*-H imines, **3**, 439
  use in the synthesis of *β*-lactams, **3**, 440
  use in the synthesis of *N*-unsubstituted *β*-lactams, **3**, 440
Imines, α-silyl-
  acylation of, **4**, 514
  deprotonation of, **4**, 648
Imines, *N*-silylated bis(trimethylsilyl)-, synthesis of, **6**, 673
Imines, silylmethyl-
  from
    α-aminosilanes, **4**, 517
    carbonyls, **4**, 517
Imines, *N*-stannyl-, synthesis of, **3**, 457
Imines, α-stannyl-, synthesis of, **4**, 536
Imines, substituted, synthesis of, **3**, 455
Imines, *C*-substituted, synthesis of, **3**, 687
Imines, *N*-substituted
  from
    aldehydes, **3**, 405
    ketones, **3**, 405
    primary amines, **3**, 405
  synthesis of, problems, **3**, 406
Imines, *N*-substituted bis(silyl)-, synthesis of, **3**, 408
Imines, sulfenyl-
  reactions
    with Grignard reagents, **2**, 359
    with organolithiums, **2**, 359
Imines, α-sulfenyl-, synthesis of, **3**, 418
Imines, sulfinyl-, reactions, with Grignard reagents, **2**, 363
Imines, *N*-sulfinyl-, reduction of, **2**, 362
Imines, *N*-sulfonoyl-, from, sulfonamides, **3**, 439
Imines, *N*-sulfonyl-
  reactions, with water, **4**, 312
  synthesis of, **1**, 1042; **4**, 311, 427
Imines, *N*-thioacyl-, use in [4+2] cycloaddition reactions, **3**, 417
Imines, *N*-tosyl-
  acylation of, **5**, 344
  reactions, with cyclopropane acetals, **5**, 344
  as starting materials, for *N*-acylsulfonamides, **5**, 344
  synthesis of, **3**, 438
Imines, *N*-trialkylsilyl-
  reactions, with haloboranes, **3**, 458
  use in the synthesis of *N*-metalloketimines, **3**, 460
Imines, trichloroacetaldehyde, as starting materials, for *O*,*N*-acetals, **4**, 304
Imines, trimethylsilyl-, as starting materials, for primary amines, **3**, 439
Imines, *N*-(1-trimethylsilylallyl)-, protiodesilylation of, **3**, 414
Imines, *N*-(trimethylsilylmethyl)-, synthesis of, **3**, 440
Imines, α,*β*-unsaturated
  from, thiazolidine-4-carboxylic acid esters, **3**, 412
  synthesis of, **3**, 411
Imines, ynamine, synthesis of, **2**, 1044
Imines, *N*-(α-cyanobenzyl)-, synthesis of, **3**, 409
1-Imines, 2-hydroxy-2-allyl-, reactions, with diglyme, **2**, 313
1-Imines, 2-hydroxy-2-propargyl-, reactions, with diglyme, **2**, 313
N-H Imines
  from, oximes, **3**, 405
  synthesis of, **3**, 404
Imine systems, *N*-alkyl-
  deamination of, **1**, 45
  reduction of, **1**, 45
Imine tautomerism, vinylamine-, studies, **2**, 757
Iminium cations, acyl-, reactions, with propargylsilanes, **1**, 978
Iminium cations, alkoxy-, as starting materials, for amide thioacetals, **6**, 111
Iminium cations, dichloro-
  from
    cyanogen chloride, **6**, 610
    hydrogen chloride, **6**, 610
  synthesis of, **6**, 609
Iminium chloride, *N*,*N*-dimethyl-, reactions, with dienolates, **2**, 302
Iminium chloride, *N*,*N*-dimethylchloromethan-, as coupling reagent, for the synthesis of hydroxylamines, **5**, 322
Iminium chloride, *N*,*N*-dimethylchloromethylene-, reactions, with nitro compounds, **3**, 427
Iminium chloride, *N*,*N*-dimethyldichloro-
  reactions, with phenol, **6**, 727
  as reagent, **6**, 727
Iminium chloride, dimethylphosgen-
  reactions, with fluoroacetyl chloride, **4**, 794
  as starting material, for α-chloro enamines, **4**, 794
Iminium chloride (DFCC), *N*,*N*-dimethylchloroform-, use in the synthesis of formamidines, **5**, 836
Iminium chlorides
  reactions
    with furans, **2**, 316
    with trialkyl phosphite, **4**, 1003
  as starting materials, for amide acetals, **6**, 105
Iminium chlorides, carboxymethylen-
  as intermediates, in the synthesis of aryl aldehydes, **3**, 82
  synthesis of, **2**, 758
Iminium chlorides, *N*,*N*-dialkylchloromethylen-, use in the synthesis of imidoyl chlorides, **5**, 660
Iminium chlorides, dichlorophosgen-, synthesis of, **6**, 238
Iminium compounds, *N*,*N*-dialkyl-, synthesis of, **6**, 660
Iminium dyes, quinone-, uses of, **3**, 419

Iminium ions
cyclization of, **2**, 329; **3**, 420
from
α-methoxyamines, **3**, 421
perhydrotriazine, **4**, 455
photospirocyclization of, **3**, 420
reactions
with alkynes, **3**, 421
with enolates, **1**, 255
with nucleophilic phosphorus compounds, **4**, 455
with silyl enol ethers, **1**, 255
with tertiary phosphines, **4**, 462, 487
with trisubstituted phosphines, **4**, 478
structure of, **6**, 726
synthesis of, **3**, 404, 419; **4**, 533
via decarbonylation of α-amino acids, **3**, 419
via oxidative functionalization, **3**, 419
use in the synthesis of ketene hemiaminals, **4**, 894
See also Iminium salts
Iminium ions, *N*-acyl-
synthesis of, **3**, 422
use in the synthesis of alkaloids, **3**, 422
Iminium ions, dichloro-, synthesis of, **6**, 726
Iminium ions, 1-pyrroline-, synthesis of, **3**, 419
Iminium ion salts, synthesis of, **3**, 419
Iminium ion–vinylsilane cyclization, mechanism, **3**, 420
Iminium salts
aza-Cope (2-azonia-[3,3] sigmatropic) rearrangement of, **3**, 422
for converting alcohols to alkyl chlorides, **2**, 25
as electrophiles, **4**, 835
from
amine perchlorates, **3**, 422
cyanamide, **6**, 623
enamines, **2**, 757
hydrolysis of, **2**, 304
phosphorus analogues of, **6**, 733
photochemistry of, **3**, 420
reactions
with alkynyl metals, **1**, 1041
with allenyl organometallics, **1**, 1059
with amines, **4**, 412; **6**, 156
with arylsulfonamides, **6**, 156
with *N,N*-dichlorosulfonamides, **6**, 619
with *N,N*-dichlorourethanes, **6**, 619
with Grignard reagents, **2**, 305
with metal dithiocarbamates, **4**, 320
with 2-methylfuran, **2**, 316
with nucleophilic phosphines, **4**, 481
with organoaluminums, **1**, 973
with propargyl organometallics, **1**, 1059
with propargylsilanes, **1**, 978
as starting materials
for *O,N*-acetals, **4**, 304
for *S,N*-acetals, **4**, 319
for *Te,N*-acetals, **4**, 334
for aminals, **4**, 411
for amines, **2**, 316
for enamino esters, **2**, 746
for lactam acetals, **6**, 105
for thioketones, **3**, 356, 361
synthesis of, **2**, 1008; **3**, 185, 419; **4**, 517; **6**, 610, 612
via diselenocarbamates, **6**, 591
use in Mannich reactions, **3**, 185
See also Iminium ions
Iminium salts, *N*-acyl-
reactions
with alkynylmetals, **1**, 1042
with silyl enol ethers, **3**, 186
synthesis of, **2**, 1009; **3**, 422
Iminium salts, alkoxymethylen-
deprotonation of, **5**, 686
synthesis of, **3**, 419; **4**, 881; **5**, 688
Iminium salts, alkylthiolmethylen-, synthesis of, **5**, 726
Iminium salts, alkylthiomethylen-, as starting materials, for hemithioaminals, **4**, 321
Iminium salts, chloro-
reactions
with hydrogen sulfide, **5**, 611
with triethylamine, **2**, 758
as starting materials, for amide thioacetals, **6**, 111
synthesis of, **6**, 52
Iminium salts, chloromethylen-, hydrolysis of, **3**, 91
Iminium salts, dialkyl-, synthesis of, **2**, 1008
Iminium salts, *N,N*-dialkylalkoxymethylen-, reactions, with lithium alkoxides, **1**, 828
Iminium salts, *N,N*-dialkyldichloro-, synthesis of, **6**, 727
Iminium salts, dibromo-, synthesis of, **6**, 612
Iminium salts, dichloro-
reactions
with alcohols, **6**, 729
with amines, **6**, 619
with nitrogen nucleophiles, **6**, 728
as reagents, **6**, 727
as starting materials, for carbamimidic chlorides, **6**, 619
Iminium salts, diethoxy-, synthesis of, **6**, 624
Iminium salts, dihalo-, chemistry of, **6**, 727
Iminium salts, dimethyl-, synthesis of, **6**, 612
Iminium salts, dithio-*N,N*-dimethyl-, reactions, with hydrogen sulfide, **6**, 555
Iminium salts, halo-
reactions, with hydrogen halides, **6**, 53
as starting materials, for α-dihaloamines, **6**, 53
synthesis of, **6**, 50, 55
use in the synthesis of alkyl halides, **2**, 8
Iminium salts, heteroaromatic, synthesis of, **2**, 1009
Iminium salts, methylene-, reactions, with triphenylphosphine, **4**, 487
Iminium salts, oxo-, reactions, with diaryl phosphites, **4**, 485
Iminium salts, *N*-phenyl-*N*-fluor-, synthesis of, **2**, 763
Iminium salts, *N*-trimethylsilyl-, reactions, with organolithiums, **2**, 313
Iminium salts, trimethylsilylallyl-, synthesis of, **1**, 286
Iminium salts, *N*-(trimethylsilylmethyl)-, as starting materials, for azomethine ylides, **4**, 517
α-Imino acids
hydrolysis of, **5**, 68
reactions, with sulfur, **5**, 603
reduction of, **5**, 88
α-Imino acids, α-substituted cyclic, from, methyl ester hydrochlorides, **5**, 93
Imino chlorides
aminolysis of, **6**, 657
reactions
with phosphoramides, **5**, 776
with *N*-(trimethylsilyl)aniline, **5**, 777
as starting materials, for α-phosphorodiamidoyl enamines, **4**, 1011
Imino chlorides, *N*-dialkyloxyphosphoryl-, reactions, with hydrogen sulfide, **5**, 644
Imino chlorides, *N*-dialkylphosphoryl-, reactions, with hydrogen sulfide, **5**, 644
Imino compounds, dehydrohalogenation of, **3**, 678
Imino compounds, *N*-acyl-, synthesis of, **6**, 658
Imino compounds, *N*-alkyl-, synthesis of, **6**, 657
Imino compounds, *N*-aryl-, synthesis of, **6**, 657
Imino compounds, bis(diphenylphosphinothioyl)-, synthesis of, **6**, 666
Imino compounds, phenylsulfonyl-, synthesis of, **6**, 666
Imino fluorides, *N*-fluoro-
from
acetonitrile, **5**, 807
fluoroethenes, **5**, 812
synthesis of, **5**, 807
Imino germylene–chromium complexes, synthesis of, **3**, 457
Imino–rhenium complexes, cyclopentadienylidene-bridged, synthesis of, **4**, 678

Iminoxy, di-*t*-butyl-, reactions, with pentane, **3**, 407
Imminium salts, reactions, with sodium hydrogen telluride, **5**, 627
Imminium salts, silylmethyl-, desilylation of, **1**, 524
Immonium chloride, chloromethylene dimethyl-, reactions, with lithium (triphenylstannyl)phenyl phosphide, **5**, 881
Immonium compounds, dicyanoketene-, synthesis of, **4**, 1018
Immonium salts, from, *N*,*N*-dimethylthiourea, **4**, 988
Immonium salts, chloromethylene-
  reactions, with phosphaallylic cations, **5**, 882
  as starting materials, for phosphaalkenes, **5**, 881
Indan-1,3-diones, 2-amino-2-hydroxy-, synthesis of, **4**, 298
1,3-Indanedione, 2-thiomethyl-2-phenyl-, as starting material, for thiocarbonates, **6**, 475
Indane-1-one, 2-hydroxy-5,6-dimethoxy-, as starting material, for 2-carboxy-4,5-dimethoxyphenylacetaldehyde, **5**, 73
Indane-1,2,3-trione, reactions, with alkynes, **1**, 980
Indanone, chloro-, reduction of, **1**, 97
Indan-2-one
  from
    titanium metallaindanes, **5**, 443
    zirconium metallaindanes, **5**, 443
Indanones
  reduction of, **1**, 97
  as starting materials, for enols, **2**, 637
1-Indanones, synthesis of, **1**, 156
Indazole, 3-amino-2-phenyl, as starting material, for 2-cyanoazobenzene, **2**, 773
Indazoles, 1-bromocarbonyl-, from, 4-bromo-3-arylsydnones, **6**, 453
Indazoles, *N*-methyl, lithiation of, **4**, 539
Indazolines, 3-oxo-, synthesis of, **5**, 356
Indene, acidity of, **1**, 862
Indene, dihydro-, synthesis of, **1**, 840
Indenes
  1,5 sigmatropic rearrangements of, **1**, 403
  synthesis of, **1**, 255, 549, 814
Indenes, metallo-, stability of, **5**, 938
Indenone, 3-amino-2-phenyl-, reactions, with sodium nitrite, **2**, 801
Indenone oxides, pyrolysis of, **3**, 317
Indenones, hydroxy-, reduction of, **1**, 84
Indenones, silyl-, synthesis of, via cyclization of alkynylsilanes, **2**, 928
Indium
  highly reactive, synthesis of, **2**, 597
  reactions
    with alkyl iodides, **2**, 597
    with allylic bromides, **2**, 597
  use in Barbier reactions, **2**, 73
Indium, trialkyl-, reactions, with *N*,*N*'-dimethylacetamidine, **5**, 780
Indium, trimethyl-, reactions, with hydrogen cyanide, **5**, 1143
Indium(I) bromide, synthesis of, **4**, 153
Indium(I) chloride
  reactions
    with allylic iodides, **2**, 597
    with dichloromethane, **4**, 153
  use in the synthesis of carbon-bridged diindium compounds, **4**, 698
Indium(I) iodide
  synthesis of, **4**, 153
  use in Barbier reactions, **2**, 73
Indium(III) chloride, use in the synthesis of carbon-bridged diindium compounds, **4**, 698
Indium chloride, reactions, with tris(trimethylsilylmethyl)lithium, **6**, 395
Indium chlorides, reactions, with trimethylstannylmethyllithium, **4**, 726
Indium complexes, (bromomethyl)dibromo-, synthesis of, **4**, 153
Indium complexes, (chloromethyl)dichloro-, synthesis of, **4**, 153
Indium compounds, reactions, with phosphorus ylides, **4**, 587
Indium compounds, alkyl-, synthesis of, **2**, 597
Indium compounds, allyl-, use in the synthesis of carbohydrates, **2**, 597
Indium compounds, (bromomethyl)dibromo-, synthesis of, **4**, 153
Indium compounds, 1-haloalkyl-, reactions, with phosphines, **4**, 588
Indium compounds, imidoyl-, synthesis of, **5**, 803
Indium compounds, (iodomethyl)diiodo-, synthesis of, **4**, 153
Indium compounds, trialkyl-, synthesis of, **2**, 597
Indium compounds, trihalomethyl-, synthesis of, **6**, 245
Indium cyanide, dimethyl-, synthesis of, **5**, 1143
Indium halides, alkyl-, synthesis of, **6**, 395
Indium–magnesium alloy, reactions, with alkyl halides, **2**, 597
Indium phosphide, use in the synthesis of semiconductors, **2**, 510
Indium trichloride
  reactions
    with Grignard reagents, **2**, 597
    with organolithiums, **2**, 597
    with potassium, **2**, 597
    with ylides, **4**, 397
Indium trihalides
  reactions
    with arsinosilanes, **2**, 511
    with stibinosilanes, **2**, 511
Indol, as starting material, for glycolic acid, **5**, 95
Indole
  hydrogenation of, **1**, 90
  reactions
    with nitroso compounds, **3**, 722
    with oxalyl chloride, **5**, 15
Indole, 3-acetyl-, synthesis of, **3**, 291
Indole, *N*-chloro-
  detection of, **2**, 761
  as intermediates, in the synthesis of 3-chloroindolenines, **2**, 761
Indole, dihydro-, synthesis of, **6**, 111
Indole, 3-(dimethylaminomethyl)-, reactions, with triethyl tricarbethoxymethane, **1**, 256
Indole, 4-hydroxymethyl-, oxidation of, **3**, 106
Indole, 3-(2-iodoethyl)-, transmetallation of, **1**, 123
Indole, 3-(2-nitroethyl)-, reactions, with sodium methoxide, **5**, 665
Indole, octahydro-, synthesis of, **1**, 90
Indole, 2-phenyl-, reduction of, **1**, 92
1*H*-Indole, 2-(1-(dimethylamino)ethyl)-3-(2-hydroxyethyl)-, as starting material, for nitriloalcohols, **1**, 253
3*H*-Indole, 3-chloro-, detection of, **2**, 761
3*H*-Indole, 5,7-di-*t*-butyl-3,3-dimethyl-, oxidation of, **5**, 1011
Indole-3-acetic acid, synthesis of, **1**, 257; **5**, 95
Indole-3-aldehyde, reactions, with benzylidenetriphenylphosphorane, **1**, 733
Indole alkaloids, synthesis of, **1**, 55; **5**, 574
Indolenine *N*-oxides, equilibrium, with *N*-hydroxyindoles, **2**, 764
Indolenines, 3-chloro-, from, indoles, **2**, 761
Indolenines, 3-chloro-2,3-dimethyl-, reactions, with silver ions, **2**, 761
Indolenines, 3-methoxy-, synthesis of, **2**, 761
Indolenium ions, as starting materials, for hexahydropyrrolo[2,3-*b*]indole nucleus, **4**, 418
Indole-3-one, *N*-acetyl, reactions, with benzaldehydes, **3**, 257
Indole-3-ones, reactions, with thiophene aldehydes, **3**, 257

3*H*-Indole-1-oxide, 5,7-di-*t*-butyl-3,3-dimethyl-, photolysis of, **5**, 1011
Indoles
alkylation of, **1**, 256
cyanation of, **3**, 669
Friedel–Crafts acylation of, **3**, 309
reactions
with dichloroketene, **6**, 111
with nitrobenzenesulfenyl chloride, **2**, 767
with ozone, **6**, 107
as starting materials, for 3-chloroindolenines, **2**, 761
synthesis of, **4**, 653
via thermal Cope rearrangements, **1**, 807
Indoles, 3-acetyl-*N*-chloromethyl-, from, *N*-dimethylaminomethylamines, **4**, 97
Indoles, 3-acyl-, reactions, with diethyl phosphorocyanidate, **3**, 670
Indoles, *N*-alkyl, reduction of, **1**, 92
Indoles, 3-amino-, as starting materials, for 3-isothiocyanatoindoles, **5**, 1025
Indoles, 1-(*N*-aroylcarbamoyl)-, synthesis of, **3**, 309
Indoles, 3-(arylthio)-, synthesis of, **2**, 767
Indoles, azido-, thermal decomposition of, **3**, 673
Indoles, 3-cyano-, synthesis of, **3**, 673
Indoles, dihydro-, from, 1,2-diazines, **1**, 523
Indoles, 2,3-dihydro-, synthesis of, **1**, 100
Indoles, 4,7-dihydro-, synthesis of, **1**, 100
Indoles, *N*-hydroxy-, equilibrium, with indolenine *N*-oxides, **2**, 764
Indoles, 2-hydroxymethyl-, reactions, with manganese dioxide, **3**, 106
Indoles, 2-isothiocyanato-, synthesis of, **5**, 1045
Indoles, 3-isothiocyanato-, from, 3-aminoindoles, **5**, 1025
Indoles, keto-, synthesis of, **3**, 309
Indoles, metallated, reactions, with aldehydes, **3**, 309
Indoles, 1-methyl-, lithiation of, **3**, 309
Indoles, 3-methyl-, reactions, with selenium dioxide, **3**, 106
Indoles, 1-phenylsulfonyl-, lithiation of, **3**, 309
1*H*-Indoles, 2-aroyl-, synthesis of, **3**, 309
1*H*-Indoles, 5,6-dimethoxy-2,3-dimethyl1-(phenylmethyl)-, synthesis of, **1**, 47
Indoline
reactions, with carbon disulfide, **6**, 561
synthesis of, **1**, 90
Indoline, 2-phenyl-, synthesis of, **1**, 92
Indoline, 1,3,3-trimethyl-2-methylene-, reactions, with 3-azathiopyrylium salts, **5**, 634
(*S*)-2-Indoline carboxylic acid, as starting material, for oxazaborolidines, **2**, 66
Indolines, deprotonation of, **1**, 151
Indolinones, 3-aminoiso-, synthesis of, **4**, 419
Indolinones, hydroxy-, oxidation of, **2**, 146
Indolium salts, hexahydro-, synthesis of, **2**, 757
Indolizidine, reactions, with phosphoryl chloride, **3**, 341
Indolizidine, thioformyl, synthesis of, **3**, 341
Indolizidine alkaloids, *trans*-2,5-disubstituted, synthesis of, **2**, 325
Indolizidines, synthesis of, **1**, 370; **2**, 328, 354; **3**, 411
Indolizidinones, synthesis of, via acyliminiumvinylsilane cyclizations, **2**, 328
Indolizines, hexahydro, from, *N*-*β*-carboethoxyethylpyrrolidones, **3**, 262
5-Indolones, as starting materials, for ketene aminals, **4**, 979
Indoxyls, thienylidene-, synthesis of, **3**, 257
Indsocin, synthesis of, **3**, 717
Induction, for the synthesis of carbanions, **1**, 851
Induction, asymmetric
in cyclopalladation, **4**, 720
in Friedel–Crafts acylation, **1**, 463
in manganese acetate mediated-radical cyclizations, **1**, 346
in the synthesis of methylenecyclobutanones, **1**, 498
Induction, 1,4-asymmetric, in intermolecular radical additions, **1**, 326
1,2-Induction, in intermolecular radical additions, **1**, 326
Inert fluids, synthesis of, **6**, 20
Infrared spectroscopy, of alkyl radicals, **1**, 923
Ingenanes, synthesis of, **3**, 325
Initiation, for the synthesis of radicals, **1**, 924
Inner salts, synthesis of, **5**, 548
Inorganic halides, reactions, with triphenylmethyl chloride, **1**, 898
D-*myo*-Inositols, synthesis of, **1**, 828
Insecticides
bioresmethrin, synthesis of, **1**, 724
carbamoyl fluorides as intermediates for, **6**, 437
O,*O*-dialkyl dithiophosphoric acids, **6**, 476
pluridone, **4**, 268
thiophosphates, **2**, 252
trichlorofon, **4**, 912
Insect pheromones
gossyplure, **1**, 755
sex, synthesis of, **1**, 647
synthesis of, **1**, 842, 1107
Insertion, carbonyl migratory, in the synthesis of acyliron complexes, synthesis of, **2**, 575
Insertion reaction, for the synthesis of isoselenocyanates, **5**, 1056
Interconversion, and transmetallation, **2**, 992
Interconversion, nucleophilic, of halides, **2**, 5
Interconversion, organometallic, characteristics of, **2**, 957
Intermolecular radical addition
atom transfer method, **1**, 334
Barton method, **1**, 332
borane method, **1**, 333
effect of substituents, **1**, 323
fragmentation method, **1**, 334
use of mercury hydrides in, **1**, 332
Intermolecular radical reaction, reviews, **1**, 919
Intermolecular reaction, for the substitution of oxygen functions, **1**, 172
Internal alkenes, addition of iodine fluoride to, **6**, 3
Intramolecular addition, for the synthesis of enones, **1**, 538
Intramolecular migration, of silyl groups, **5**, 542
Intramolecular processes, for the addition of carbon nucleophiles to alkenes, **1**, 304
Intramolecular radical addition, reviews, **1**, 320
Intramolecular reaction
of alkenes, **1**, 357
for the substitution of oxygen functions, **1**, 174
for the substitution of sulfur functions, **1**, 240
Iodide
reactions
with aryl α-iodo ketones, **1**, 10
with Brönsted acids, **1**, 5
with cyclohexyldichlorophosphine, **2**, 449
with Lewis acids, **1**, 5
Iodides
from, tosylates, **2**, 9
precursors for radical cyclization, **1**, 341
as radical precursors, **1**, 334
reactions, with tosylates, **1**, 465
Iodinane, [*N*-(*p*-toluenesulfonyl)imino]phenyl-, reactions, with trialkylboranes, **2**, 311
Iodinanes, use in the synthesis of dialkyl sulfoxides, **2**, 148
Iodinanes, alkyl-, stability of, **1**, 108
Iodinating reagents, reactions, with metal alkynylides, **2**, 1019
Iodination
of aldehydes, **3**, 27
of bromotrichloromethane, **6**, 221
of mercury(II) bis(trifluoromethyl)amide, **6**, 240
of silver chlorodifluoroacetate, **6**, 224
of silver trifluoroacetate, **6**, 219
o-Iodination, of phenols, for the synthesis of aryl iodides, **2**, 631
para Iodination, of aryl ethers, for the synthesis of aryl iodides, **2**, 631

Iodine
for the activation of magnesium, **2**, 558
additions to tetrafluoroethene, **6**, 2
as catalyst, for the synthesis of aromatic ketones, **3**, 279
for chemical activation of metal surfaces, **2**, 958
for the electrophilic iodination of arenes, **2**, 631
oxidative acylation of, **5**, 204
as oxidizing agent, for the synthesis of imidoyl disulfides, **5**, 735
reactions
with alanates, **1**, 961
with alkenes, **2**, 44; **5**, 1107
with alkenylboranes, **1**, 531
with alkenyldialkylboranes, **1**, 658
with alkoxyiron tetracarbonylmagnesium bromide complexes, **5**, 43
with alkynyl Grignard reagents, **1**, 1035
with amides, **5**, 320
with arsonium bromide, **3**, 649
with bis(trialkylsilyl)amides, **5**, 377
with bis(trimethylenedioxyboryl)-methylenecyclohexane, **4**, 750
with *t*-butoxide, **2**, 95
with *gem*-chloronitrosoalkanes, **4**, 40
with diaryl diselenides, **2**, 732
with diazoalkanes, **4**, 29
with di(bistrifluoromethylamino)mercury, **2**, 339
with 1,1-difluoroethene, **6**, 2
with diiodoalkyne, **4**, 797
with 2-(dimethylsulfuranylidene)acetophenone, **5**, 495
with enolate ions, **2**, 34
with ethyl hypochlorite, **2**, 95
with fluorostannylalkenes, **1**, 455
with hydrazones, **4**, 29
with lithiated trithioorthoformates, **6**, 88
with lithium (1-alkynyl)organoborates, **1**, 1033
with phenoxyalkyne, **4**, 765
with phenoxyethynylmagnesium bromide, **4**, 765
with phenyliodine(III) bis(phthalimide), **5**, 321
with phosphine, **2**, 708
with silver cyanate, **5**, 984
with *N*-silver succinimide, **5**, 320
with sodium alkyl sulfinate, **2**, 182
with tetramethylbismuthine, **2**, 491
with trifluoromethylphenylmercury, **4**, 1083
with trimethylsilylcyclohexenes, **4**, 139
with vinylgermanes, **1**, 452
with zirconocycles, **4**, 1039
for the synthesis of aryl iodides, **2**, 630
for the synthesis of vinyl iodides, **2**, 629
use in the synthesis of disulfides, **2**, 212
use in the synthesis of phosphonates, **2**, 456
Iodine(I) tetrafluoroborate, bis(pyridine)-
for the bromoiodination of alkenes, **2**, 31
reactions, with chloride ions, **2**, 24
use in iodination reactions, **2**, 1019
Iodine(I) thiocyanate, reactions, with alkenes, **5**, 1044
Iodine(III), for the oxidation of ketones, **3**, 178
Iodine(III), phenyl-, reactions, with phthalimide, **5**, 322
Iodine(III) acetate, use in the synthesis of aldehydes, via diol cleavage, **3**, 8
Iodine(III) bis(phthalimide), phenyl-
reactions
with bromine, **5**, 318
with iodine, **5**, 321
Iodine(III) compounds, trifluoromethyl-, synthesis of, **6**, 220
Iodine(III) difluoride, trifluoromethyl-
reactions, with *N*-silylamides, **5**, 321
synthesis of, **6**, 220
Iodine(III) trifluoroacetate, use in the dihydroxylation of alkenes, **2**, 45
Iodine(V) compounds, for the synthesis of aldehydes, **3**, 6
Iodine bromide, reactions, with fluoroalkenes, **6**, 2
Iodine chloride, reactions, with fluoroalkenes, **6**, 2
Iodine compounds
of phosphorus functions, synthesis of, **5**, 878, 896
reactions
with aldazines, **4**, 198
with ketones, **2**, 44
Iodine compounds, bis(trifluoromethyl)-, synthesis of, **6**, 220
Iodine compounds, hypervalent, stabilization of, with fluorocarbon ligands, **1**, 108
Iodine compounds, trivalent, chemistry of, **1**, 108
Iodine fluoride
additions to chlorotrifluoroethene, **6**, 3
additions to 1,1-dichlorodifluoroethene, **6**, 3
additions to 1,1-difluoroethene, **6**, 3
additions to hexafluoropropene, **6**, 3
additions to internal alkenes, **6**, 3
additions to tetrafluoroethene, **6**, 3
Iodine isocyanate (INCO)
reactions
with alkenes, **3**, 703; **5**, 984
with alkynes, **5**, 985
with allenes, **5**, 985
with dienes, **5**, 985
with methylallenes, **5**, 985
with methylene cycloalkanes, **5**, 985
with styrenes, **5**, 985
synthesis of, **5**, 984
Iodine monobromide, as catalyst, for the cleavage of ethers, **2**, 33
Iodine monochloride
for chlorine–iodine addition to alkenes, **2**, 24
reactions
with bis(perfluoroethyl) monoselenide, **6**, 49
with organoaluminums, **4**, 758
with organomercury compounds, **4**, 750
with thiocyanates, **6**, 618
with tyrosine, **2**, 608
Iodine pentafluoride
as oxidizing agent, for aldehydes, **3**, 3
reactions
with fluoroalkenes, **6**, 2
with tetrabromomethane, **6**, 226
with tetrachloromethane, **6**, 220
with tetrahalomethane, **6**, 217
with tetraiodomethane, **6**, 219
as starting material, for carbonic fluoride iodide, **6**, 420
Iodine pentoxide, reactions, with phenylpropiolic acid, **4**, 749
Iodines, as radical precursors, **1**, 330
Iodoalkylation, of arenes, **2**, 36
β-Iodoalkylbis(trifluoromethyl)amines, synthesis of, **6**, 239
Iodoboration, of ethoxyethyne, **2**, 902
Iodocyclization, mechanism, **6**, 468
Iododecarboxylation
of carboxylic acids, **2**, 36
mechanism, **2**, 13
Iodofluorination, of alkenes, **2**, 18
Iodoform
reactions
with alkali metals, **6**, 32
with sodium sulfite, **4**, 281
reduction of, **4**, 29
as starting material
for iodomethylenephosphanes, **5**, 878
for methane bissulfonic acid, **4**, 281
for tris(trifluoromethyl) trithioorthoformate, **6**, 81
See also Methane, triiodo-
Iodolactonization
mechanism, **5**, 377
for the synthesis of dialkyl carbonates, **6**, 468
of *β*,*γ*-unsaturated carboxylic acids, **5**, 161
Iodomethylation, of arenes, **2**, 13
Iodonium bromide, diphenyl-, for the *O*-acylation of hydroxylamine, **2**, 691

Iodonium chloride, diphenyl-, reactions, with *N*-hydroxyphthalimide, **2**, 692
Iodonium compounds, substitution of, **1**, 108
Iodonium ions
 stability of, **1**, 1109
 synthesis of, **2**, 45
Iodonium oxides, rearrangement of, **2**, 88
Iodonium salts
 thermolysis of, for the synthesis of carbenes, **4**, 1081
 use in the synthesis of quaternary phosphonium salts, **2**, 445
Iodonium salts, alkenyl(phenyl)-, elimination of, **1**, 1080
Iodonium salts, alkynyl-
 reactions
  with nucleophilic compounds, **2**, 1012
  with organocopper(I) compounds, **1**, 1018
  with sodium sulfinate, **2**, 1036
 as starting materials, for alkynyl thiocyanates, **2**, 1033
 synthesis of, **2**, 1012, 1019
Iodonium salts, alkynyl(phenyl)-
 reactions, with enolates, **1**, 1016
 as starting materials
  for α-acetoxy ketones, **3**, 162
  for alkynyl esters, **2**, 1028
  for alkynyl phosphonium salts, **2**, 1062
 synthesis of, **2**, 1019
Iodonium salts, diaryl-, as starting materials, for aryl fluorides, **2**, 614
Iodonium salts, vinyl-, reactions, with diphenylcopperlithium, **1**, 431
Iodonium tetrafluoroborate, bis(pyridine)-, for the synthesis of vinyl iodides, **2**, 629
Iodonium tetrafluoroborates, alkynyl-, reactions, with triphenylphosphine, **2**, 1062
Iodonium tosylates, alkynyl-, synthesis of, **2**, 1020
Iodonium tosylates, alkynyl(phenyl)-
 reactions
  with alkenylcoppers, **1**, 1018
  with alkynylcyanocuprates, **1**, 1018
  with dialkylcuprates, **1**, 1018
 as starting materials, for alkynyl sulfones, **2**, 1036
 synthesis of, **2**, 1020
Iodonium transfer reaction, between phenyl(cyano)iodonium triflate and tin alkynes, **2**, 1021
Iodonium triflate, reactions, with sodium thiocyanate, **2**, 1033
Iodonium triflate, cyano-, as starting material, for ethynyl(phenyl)iodonium triflates, **2**, 1020
Iodonium triflate, phenyl(cyano)-, reactions, with tin alkynes, **2**, 1021
Iodonium triflates, β-alkylethynyl(phenyl)-, synthesis of, **2**, 1020
Iodonium triflates, alkynyl(phenyl)-, as starting materials, for alkynyl sulfones, **2**, 1036
Iodonium triflates, β-arylethynyl(phenyl)-, synthesis of, **2**, 1020
Iodonium triflates, dialkynyl-, from, iodosyl triflate, **2**, 1021
Iodonium triflates, ethynyl(phenyl)-, synthesis of, **2**, 1020
Iodosyl salts, alkynyl phenyl-, reactions, with cuprates, **1**, 541
Ion exchange resins
 for the dehydration of sterols, **1**, 578
 use in the synthesis of *N*-alkylamides, **5**, 276
 use in the synthesis of furanyl ketones, **3**, 302
Ionic additions, for the synthesis of alkenes, **1**, 531
Ionic reactions, and radical reactions compared, **1**, 320
Ionization, for the synthesis of iron carbene complexes, **3**, 519
(*R*)-Ionone, synthesis of, **1**, 582
α-Ionone, 1-hydroxy-4-oxo-, reactions, with ethoxycarbonylmethylidenetriphenylphosphorane, **1**, 726
β-Ionone
 reactions
  with cyano phosphonates, **1**, 752
  with diethyl ethoxycarbonylmethyl phosphonate, **1**, 748
Ionophore tetronasins, synthesis of, **1**, 614
Ions, synthesis of, **1**, **1087–1145**
Ions, dicoordinated, synthesis of, **4**, **1071–1083**
Ions, stabilized substituted, bearing one heteroatom, synthesis of, **2**, **997–1010**
Ions, tricoordinated, synthesis of, **4**, **1071–1083**
Ireland–Claisen rearrangement
 double, **1**, 831
 mechanism, **1**, 416, 825, 829
 for the ring-contraction of lactones, **1**, 831
 for the synthesis of acyloxysilanes, **5**, 227
 for the synthesis of macrolactones, **5**, 179
 for the synthesis of natural products, **1**, 831
Ireland–Claisen rearrangement, 3-aza, **1**, 835
Ireland enolization model, for lithium amides, **1**, 876
Iridium, use in Barbier reactions, **2**, 73
Iridium, allylic complexes, reactions, with carbon monoxide, **5**, 461
Iridium, bisphosphine complexes, reactions, with acid halides, **5**, 460
Iridium, chlorotris(triphenylphosphine)-, reactions, with methyllithium, **5**, 460
Iridium, cyclopentamethyl-, reduction of, **1**, 94
Iridium, dicarbonyl(μ-pyrazolyl)-, reactions, with diiodomethane, **4**, 685
Iridium(I), as catalyst, for the isomerization of allyl ethers, **2**, 675
Iridium(I) carbonyl azides, reactions, with carbon monoxide, **5**, 983
Iridium(I) cations, pentacoordinated
 reactions, with xenon difluoride, **6**, 457
 as starting materials, for fluorocarbonyl iridium(III) complexes, **6**, 457
Iridium(I) halide complexes, reactions, with alkyl halides, **2**, 583
Iridium(I) imide, synthesis of, **5**, 387
Iridium(III) complexes, fluorocarbonyl-, from, pentacoordinated iridium(I) cations, **6**, 457
Iridium(III) compounds, alkyl-, synthesis of, **2**, 583
Iridium(III) hydride complexes, use in the hydrogenation of imines, **2**, 312
Iridium(III) porphyrin complexes, synthesis of, **2**, 583
Iridium(V) complexes, synthesis of, **2**, 584
Iridium acyl compounds, synthesis of, **5**, 460
Iridium carbonyl complexes, synthesis of, **6**, 525
Iridium catalysts
 for the dehydrogenation of alkanes, **1**, 554
 in hydroboration reactions, **2**, 43
 use in the synthesis of vinylsilanes, **2**, 915
Iridium complexes
 as hydrogenation catalysts, **1**, 456
 reactions
  with carbon monoxide, **4**, 931
  with secondary alkylarsines, **2**, 512
 synthesis of, **2**, 53; **6**, 457
 use in the synthesis of enols, **2**, 639
Iridium complexes, alkyl-, reactions, with alkynes, **5**, 461
Iridium complexes, anionic
 reactions, with methyl iodide, **5**, 460
 synthesis of, **5**, 460
Iridium complexes, (halomethyl)-, synthesis of, **4**, 150
Iridium complexes, triscarbonyl-, reactions, with diphenyl ethyne, **5**, 461
Iridium compounds, reactions, with phosphorus ylides, **4**, 583
Iridium compounds, acyl-, synthesis of, **5**, 460
Iridium compounds, alkyl-, carbonylation of, **5**, 460
Iridium compounds, imidoyl-, synthesis of, **5**, 800
Iridium cyclic complexes, synthesis of, **5**, 948

Iridium halides, reactions, with carbon monoxide, **6**, 525
Iridium maleoyl compounds, synthesis of, **5**, 461
Iridium–nitrogen bonds, reactions, with carbon monoxide, **6**, 516
Iridium pentahydride, as catalyst, for the oxidation of alcohols, **3**, 118
Iron
catalyzed substitution, with Grignard reagents, **1**, 970
reactions
with allyl cations, **1**, 191
with nucleophiles, **1**, 196
Iron, (benzylideneacetone)tricarbonyl-, reactions, with alkynes, **4**, 681
Iron, dicarbonyl(η^5-cyclopentadienyl)-, photolysis of, **1**, 937
Iron, dicarbonylcyclopentadienylmethyl-, reactions, with bis(cyclopentadienyl)hydridoniobium, **4**, 717
Iron, dicyclopentadienyl-
as reducing agent, **5**, 808
synthesis of, **1**, 862
See also Ferrocene
Iron, pentacarbonyl-, as starting material, for benzoferracyclopentadiene complexes, **4**, 681
Iron, pentakis (*t*-butylisonitrile)-, oxidative addition, with alkyl halides, **5**, 799
Iron, tricarbonyl-, reactions, with dienes, **1**, 291
Iron, tricarbonylbenzocyclobutadiene-, as starting material, for benzoferracyclopentadiene complexes, **4**, 681
Iron, tricarbonylcyclobutadiene-
lithiation of, **4**, 706
mercuration of, **4**, 724
Iron, tricarbonyl(lithiocyclobutadiene)-, reactions, with transition metal halides, **4**, 723
Iron(0) compounds, alkyl-, synthesis of, **2**, 575
Iron(II), allyl(cyclopentadienyl)-, use in allylation of aldehydes, **2**, 73
Iron(II), tetraphenylporphyrino-, reactions, with tetrachloromethane, **6**, 717
Iron(II) chloride
reactions
with arsoles, **4**, 1039
with bismoles, **4**, 1039
with bismolyl anions, **4**, 1041
with mesitylmercuric chloride, **4**, 685
with phosphonyl anions, **4**, 1026, 1035
with potassium arsolyls, **4**, 1039
with stiboles, **4**, 1039
with stibolyl anions, **4**, 1041
Iron(II) compounds, alkyl-, synthesis of, **2**, 576
Iron(II) sulfate, as catalyst, for the synthesis of thiocyanates, **5**, 1108
Iron(II) tetraphenylporphyrin (TPP), reactions, with dichlorocarbene, **3**, 700
Iron(III), as catalysts, in the synthesis of trihalides, **6**, 7
Iron(III) chloride
as catalyst, for the synthesis of geminal diacetates, **4**, 203
σ-complexes, stability of, **4**, 820
reactions
with acrylonitrile, **5**, 770
with cyclopropyl radicals, **1**, 929
use in the synthesis of ketones, **3**, 123
Iron(III) complexes, acyl-, reactions, with sodium azide, **5**, 991
Iron(III) compounds, alkyl-, reactions, with cytochrome P-450; **2**, 578
Iron(III) nitrate
for the oxidation of alcohols, **3**, 118
as reagent, for the oxidative nitration of oximes, **4**, 429
Iron(III) porphyrin complexes, reactions, with alkenes, **2**, 579
Iron(acac)$_3$, reduction of, **2**, 576
Iron acyl complexes
reactions, with triflic anhydride, **5**, 950
as starting materials
for thiocarbene complexes, **5**, 951
for vinylidene complexes, **5**, 950
synthesis of, **5**, 452
Iron acyl complexes, homochiral
definition of absolute stereochemistry of, **5**, 453
x-ray crystal structure of, **5**, 453
Iron acyls, use in the synthesis of hydroxamates, **5**, 329
Iron alkoxycarbene complexes, reactions, with isocyanides, **5**, 954
Iron alkyl complexes, carbonylation of, **5**, 452
Iron alkyls, reactions, with triethyloxonium tetrafluoroborate, **5**, 942
Iron aminocarbene complexes, synthesis of, **5**, 953
Iron carbene complexes
from
α-ether complexes, **3**, 519
α-thio ester complexes, **3**, 519
η^1-vinyliron complexes, **3**, 520
reactions, with phosphines, **4**, 581
as starting materials, for organoiron complexes, **6**, 332
synthesis of, **3**, 519
Iron–carbon bonds, cleavage of, **2**, 577
Iron carbonyl anions
reactions
with imidoyl chlorides, **5**, 799
with ketenimines, **5**, 799
as starting materials, for imidoylirons, **5**, 799
Iron carbonyl complexes, synthesis of, **1**, 196; **6**, 523, 525
Iron carbonyls, as catalysts, for haloalkane addition, **6**, 7
Iron carbyne complexes, synthesis of, **3**, 690
Iron catalysts, reactions, with alkyl Grignard reagents, **1**, 467
Iron cation, tricarbonyl(cyclohexadienyl)-, as starting material, for iodocarbonyl complexes, **6**, 455
Iron cluster compounds, synthesis of, **6**, 340
Iron complexes
alkylation of, **2**, 578
from
cyclic sulfonates, **1**, 196
halides, **2**, 577
vinylic epoxides, **1**, 196
as mediator of radical cyclizations, **1**, 346
photolytic rearrangement of, **4**, 639
reactions
with anions, **5**, 951
with carbon monoxide, **6**, 518
with *N*-lithioketimines, **3**, 460
with succinimide, **5**, 387
synthesis of, **4**, 641; **6**, 455
thermal rearrangement of, **4**, 1032
use in the synthesis of acyl aldehydes, **3**, 83
use in the synthesis of alkoxy aldehydes, **3**, 29
Iron complexes, alkyl-
from, alkenes, **2**, 577
reactions, with trityl ions, **1**, 666
Iron complexes, alkynyl-, as intermediates, **1**, 545
Iron complexes, allyl-, reactions, with dialkylcadmiums, **1**, 291
Iron complexes, bridged, from, cyclooctatetraene, **4**, 683
Iron complexes, cyclopentadienyl-
reactions
with alkenes, **2**, 577
with carbon monoxide, **5**, 453
synthesis of, **2**, 576; **5**, 453
use in diastereoselective and enantioselective synthesis, **5**, 453
Iron complexes, (1,2-dialkoxyethylene)-, reactions, with cuprates, **1**, 481
Iron complexes, dicarbonylcyclopentadienyl(substituted phenyl)-, reactions, with hexacarbonylchromium, **4**, 715
Iron complexes, di-μ-(4,4'-dimethylbenzophenoniminato)bis(tricarbonyl)-, synthesis of, **3**, 460

Iron complexes, dienylallene tricarbonyl-, synthesis of, **1**, 966
Iron complexes, halo-, reactions, with ferrocenyllithium, **4**, 669, 684
Iron complexes, (halomethyl)-, synthesis of, **4**, 148
Iron complexes, menthyl ester, reactions, with allyllithiums, **5**, 453
Iron complexes, (methoxymethyl)-, reactions, with hydrogen chloride, **4**, 148
Iron complexes, propargyl-, rearrangement of, **4**, 683
Iron complexes, tetrakis(1-norbornyl)-, synthesis of, **2**, 579
Iron complexes, vinyl-, reactions, with cobalt carbonyls, **4**, 711
Iron complexes, η^1-vinyl-
  as starting materials, for iron carbene complexes, **3**, 520
  synthesis of, **3**, 520
Iron ethoxycarbene complexes, synthesis of, **5**, 942
Iron functions, doubly-bonded, synthesis of, **3**, 519
Iron hydride complexes, reactions, with tetracarbonylvanadium, **4**, 717
Iron imides, synthesis of, **5**, 387
Iron intermediates, acyl-, use in the synthesis of β-lactones, **5**, 161
Iron intermediates, allylic acyl-, use in the synthesis of β-lactones, **5**, 163
Iron iodide, cyclopentadienyldicarbonyl-, reactions, with arsolyl anions, **4**, 1039
Iron iodide, dicarbonylcyclopentadienyl-
  reactions
    with diferrocenylmercury, **4**, 685
    with tetraphenylphospholyl anion, **4**, 1034
Iron multidecker sandwich compounds, synthesis of, **4**, 694
Iron–nickel complexes, synthesis of, **4**, 713
Iron nitrenes, *N*-acyl-, synthesis of, **5**, 387
Iron nitrosyl complexes, methylene-bridged, synthesis of, **4**, 679
Iron nonacarbonyl
  reactions
    with α,α'-dibromoacetone, **1**, 364
    with α,α'-dibromoketones, **1**, 364
    with vinyl epoxides, **5**, 163
  use in the synthesis of α,β-unsaturated ketones, **3**, 225
Iron pentacarbonyl
  as catalyst, for the synthesis of α-haloalkenylsilicons, **4**, 810
  reactions
    with allyl halides, **5**, 452
    with dicyclopentadienyl, **6**, 524
    with dimethylgermyl(dimethylsilyl)methane, **4**, 641
    with Grignard reagents, **3**, 90; **5**, 43
    with organolithiums, **3**, 90
    with trithiocarbonate complexes, **6**, 333
  for the reduction of α-halocarbonyls, **1**, 2
  for the reduction of vinyl chlorides, **1**, 430
  as starting material, for iodocarbonyl complexes, **6**, 455
  use in the synthesis of carbon-bridged diiron complexes, **4**, 681
Iron–phosphinoalkene complexes, synthesis of, **4**, 1023
Iron phthalocyanine μ carbides, synthesis of, **5**, 1097
Iron phthalocyanine dimer, synthesis of, **5**, 1097
Iron–porphyrin complexes, use in the synthesis of aryl sulfoxides, **2**, 714
Irons
  reactions
    with alkynes, **1**, 545
    with phosphorus ylides, **4**, 581
    with tetramethylthiourea, **6**, 351
Irons, acyl-
  synthesis of, **5**, 451
  use in α-alkylation, **1**, 134
Irons, alkyl-
  reactions
    with carbonyls, **2**, 577
    with vinyl halides, **2**, 577
  for the substitution of vinyl halides, **1**, 466
  synthesis of, **2**, 575
Irons, 1-haloalkyl-, reactions, with phosphines, **4**, 582
Irons, homochiral alkyl-, reactions, with carbon monoxide, **5**, 453
Irons, imidoyl-
  from
    iron carbonyl anions, **5**, 799
    isonitriles, **5**, 799
  synthesis of, **5**, 799
Irons, monoimidoyl-, synthesis of, **5**, 799
Iron tetracarbonyl anions, hydrido-
  reactions
    with acyl chlorides, **3**, 102
    with *p*-nitrobenzoyl chloride, **3**, 98
Iron tetracarbonyls, synthesis of, **5**, 451
Iron tetrafluoroborate, 1-methoxycarbonyl-3-methyl-(η3-allyl)-tetracarbonyl-, reactions, with amines, **2**, 331
Iron tetraphenylporphyrin μ-carbides, synthesis of, **5**, 1097
Iron tetraphenylporphyrin complexes, μ-nitrido, synthesis of, **5**, 1097
Iron tetraphenylporphyrin complexes, μ-oxo, synthesis of, **5**, 1097
Iron tricarbonyl cation, cyclopentadienyl-, reactions, with hydrazine, **5**, 975
Iron tricarbonyl complexes, acyl-, reactions, with alkylating agents, **5**, 451
Iron tricarbonyl complexes, acyl diene-, reactions, with organolithiums, **3**, 30
Iron tricarbonyl phenylmethyl carbocations, biscyclobutadienyl-, structure of, **1**, 898
Iron tricarbonyls, cyclohexa-1,3-diene-, methoxycarbonyls of, **5**, 96
Iron vinylidene complexes, as starting materials, for α-ferrio enamines, **4**, 1019
Irradiation
  ultraviolet
    of bis(trifluoromethyl) sulfide, **6**, 219
    of methyl trifluoromethyl sulfide, **6**, 219
Isatin, 5-bromo-, as starting material, for nitrogen-functionalized ylides, **5**, 906
Isatin, *N*-vinyl-, use in the synthesis of α-haloamines, **4**, 101
Isatoic anhydride, synthesis of, **5**, 969, 970
Ishikawa's reagent
  for the fluorination of alkenes, **2**, 18
  use in the synthesis of alkyl fluorides, **2**, 19
Isobacteriochlorin macrocycles, synthesis of, **3**, 106
Isobenzofurans, synthesis of, **1**, 789; **3**, 322
Isobutene
  reactions, with silaethenes, **6**, 177
  as starting material, for trimethylenemethane dianions, **6**, 206
Isobutene, perfluoro-
  reactions, with benzenesulfonamide, **3**, 588
  as starting material, for ketenimines, **3**, 588
Isobutylamine, as starting material, for (dialkylamino)dihaloarsines, **2**, 414
Isobutylene oxide, reactions, with ethyl cyanoacetate, **4**, 883
Isobutylene sulfide, synthesis of, **2**, 136
Isobutyl hypochlorite, reactions, with benzal chlorides, **6**, 28
Isobutyraldehyde
  reactions, with 1,1,3,3-tetramethylguanidine, **6**, 644
  synthesis of, **1**, 387
Isocamphane, 2-isothiocyanato-, synthesis of, **5**, 1043
Isocaproic acid, as starting material, for β-hydroxyisocaproic acid, **5**, 118
Isocaproic acid, β-hydroxy-, from, isocaproic acid, **5**, 118
Isocarbostyrils, synthesis of, **5**, 989
Isochroman, 1-ethoxy-, reactions, with acetic anhydride, **4**, 212
Isochromans, reactions, with cyclohexenone, **3**, 325

Isocomene
  from, ketenes, **3**, 528
  synthesis of, **1**, 382, 832
Isocyanate, acetyl
  reactions, with methoxytripropylstannane, **5**, 716
  synthesis of, **5**, 1006
Isocyanate, adamantyl, synthesis of, **5**, 969
Isocyanate, (alkoxychlorophosphonyl)(alkoxyfluorophosphonyl)chloromethyl, production of, **6**, 285
Isocyanate, aryl, reactions, with phosphorus(III) acid anhydrides, **6**, 668
Isocyanate, benzoyl, reactions, with amines, **5**, 721
Isocyanate, benzyl
  photochlorination of, **5**, 658
  reactions, with 3-desacetylcephalosporins, **6**, 485
Isocyanate, benzyloxycarbonyl, synthesis of, **5**, 977
Isocyanate, *N*-benzyloxymethyl, synthesis of, **5**, 993
Isocyanate, bis(difluoroamino)fluoroaminomethyl
  reactions, with ethanol, **6**, 363
  as starting material, for tetraazamethanes, **6**, 363
  synthesis of, **6**, 363
Isocyanate, bis(phosphoryl)halomethyl, production of, **6**, 285
Isocyanate, bromocarbonyl, synthesis of, **5**, 1015
Isocyanate, *N*-bromocarbonyl, from, chlorocarbonyl isocyanate, **6**, 453
Isocyanate, butyl
  as starting material, for carbonimidic dichlorides, **6**, 606
  synthesis of, **5**, 995
Isocyanate, *t*-butyl
  reactions, with molybdenum complexes, **3**, 514
  synthesis of, **5**, 963
Isocyanate, *t*-butyldiphenylsiloxymethyl, synthesis of, **5**, 993
Isocyanate, chloroacetyl, reactions, with alkynylmagnesium bromides, **1**, 1043
Isocyanate, chlorobis(diethoxyphosphonyl)methyl, reactions, with amidines, **6**, 367
Isocyanate, chlorocarbonyl
  cycloaddition of, **6**, 450
  properties of, **6**, 451
  reactions, with nitrile oxides, **5**, 669
  as starting material, for *N*-bromocarbonyl isocyanate, **6**, 453
  synthesis of, **6**, 451
  transhalogenation of, with fluorides, **6**, 442
Isocyanate, *N*-chlorocarbonyl
  reactions
    with carboxylic acid chlorides, **5**, 1008
    with α,α-dimethyl-4,4'-methylenediphenol, **5**, 1007
    with epoxides, **5**, 1008
    with 1,6-hexanediol, **5**, 1007
    with nucleophiles, **5**, 1007
    with triethylene glycol, **5**, 1007
    with vinyl ethers, **5**, 1008
  as starting material, for isocyanates, **5**, 1007
  synthesis of, **5**, 1007; **6**, 451
Isocyanate, β-chloroethyl, synthesis of, **5**, 976, 1012
Isocyanate, 3-chloro-4-methoxyphenyl, synthesis of, **5**, 1001
Isocyanate, chloromethyl, synthesis of, **5**, 986
Isocyanate, 4-chlorophenyl, reactions, with stannylated ynamines, **2**, 1044
Isocyanate, chlorosulfonyl (CSI)
  as activating agent, for the synthesis of anhydrides, **5**, 189
  for the cyanation of indoles, **3**, 669
  cycloaddition of, **6**, 128
  as dehydrating agent, for the synthesis of aromatic anhydrides, **5**, 196
  for the dehydration of amides, **3**, 619
  reactions
    with alkynylsilanes, **1**, 1043
    with carboxylic acids, **3**, 618, 666
    with ketones, **3**, 633
    with perfluoroalkanesulfinic amides, **5**, 995
    with silver cyanate, **5**, 994, 1018
    with stannylated arenes, **2**, 726
    with vinylsilanes, **4**, 514
  use in the synthesis of *N*-arylalkenoamides, **5**, 297
  use in the synthesis of carboxylic esters, **5**, 125
Isocyanate, 1-chlorovinyl
  reactions, with triethyl phosphite, **4**, 1010
  as starting material, for *gem*-isocyanato phosphorylalkenes, **4**, 1010
Isocyanate, cinnamyl, synthesis of, **5**, 986
Isocyanate, cyclohexyl
  deoxygenation of, **6**, 515
  reactions
    with 2-aminobenzoic acid, **6**, 504
    with hydrogen isocyanate complexes, **5**, 1088
    with lithium arsides, **6**, 513
Isocyanate, 1,2-dibromoethyl
  reactions, with triethyl phosphite, **4**, 1009
  as starting material, for *gem*-isocyanato phosphorylalkenes, **4**, 1009
Isocyanate, dichlorofluoromethyl, reactions, with dialkyl chlorophosphites, **6**, 285
Isocyanate, dichloromethyl, reactions, with triethyl phosphite, **5**, 787
Isocyanate, dichlorophenyl, synthesis of, **5**, 1004
Isocyanate, difluorophosphoryl, synthesis of, **5**, 1007
Isocyanate, ethoxycyclopropyl, synthesis of, **5**, 1011
Isocyanate, ethoxy(thiocarbonyl)
  reactions
    with alcohols, **6**, 559
    with amines, **6**, 559
Isocyanate, ethyl, chlorination of, **4**, 102
Isocyanate, fluorocarbonyl
  as starting material, for iminodicarboxylic acid dibromide fluoride, **6**, 453
  synthesis of, **5**, 1007, 1015; **6**, 441
Isocyanate, fluorosulfonyl, synthesis of, **5**, 1018
Isocyanate, *n*-heptadecyl, synthesis of, **5**, 993
Isocyanate, *n*-hexyl, synthesis of, **5**, 1009
Isocyanate, hydroxymethyl
  reactions
    with thionyl chloride, **5**, 986
    with α,β-unsaturated ethers, **5**, 986
  stability of, **4**, 313
Isocyanate, *p*-isocyanatobenzenesulfonyl, synthesis of, **5**, 997
Isocyanate, isopropenyl, synthesis of, **5**, 997
Isocyanate, *m*-isopropenyl-α,α-dimethylbenzyl (*m*-TMI), hydrosilylation of, **5**, 1007
Isocyanate, isopropyl
  reactions
    with tris(trimethylsilyl)phosphane, **6**, 511
    with tris(trimethylsilyl)phosphine, **5**, 1157
Isocyanate, 2-isopropyl-4,6-di-*t*-butylphenyl, synthesis of, **5**, 1011
Isocyanate, methoxymethyl, synthesis of, **5**, 1013
Isocyanate, methyl
  bromination of, **6**, 240, 611
  photochlorination of, **6**, 240
  reactions
    with ethylene oxide, **6**, 324
    with α-hydrazino enamines, **4**, 986
    with nitrilimines, **5**, 681
    with pentacarbonyl(isocyanomethylenetriphenylphosphorane)chromium, **5**, 905
    with *O*-phenylhydroxylamine, **6**, 506
    with *N*-phenyltetramethylguanidine, **6**, 364
  as starting material
    for carbonimidic dichlorides, **6**, 606
    for *N*-chlorocarbonyl isocyanide dichloride, **6**, 451
Isocyanate, 2-methylallyl, synthesis of, **5**, 1009
Isocyanate, 1-naphthyl, reactions, with dimethylformamide, **5**, 745

Isocyanate, neopentyl, radical bromination of, **6**, 453
Isocyanate, β-(5-nitro-2-furyl)vinyl, synthesis of, **5**, 990
Isocyanate, pentafluoro-, as starting material, for pentafluorotellurium isocyanate, **5**, 348
Isocyanate, pentafluoroethyl, synthesis of, **5**, 1003
Isocyanate, perfluoroisopropyl, synthesis of, **5**, 987
Isocyanate, 4-phenanthrene, synthesis of, **5**, 1001
Isocyanate, phenyl
  reactions
    with alkylarsines, **6**, 513
    with alkynyltins, **1**, 1043
    with arylarsines, **6**, 513
    with benzophenone phenylimine, **6**, 504
    with cumulenes, **4**, 421
    with dimethylformamide, **5**, 745; **6**, 143
    with Grignard reagents, **5**, 293, 551
    with phosphomethyltriorganostannanes, **5**, 390
    with phosphorus pentaselenide, **5**, 1056
    with triphenylallyllead, **5**, 390
    with triphenylallyltin, **5**, 390
    with ylides, **3**, 591
  as starting material, for *N*,*N*-bisfluorocarbonylaniline, **6**, 438
  synthesis of, **5**, 971, 988, 1001
Isocyanate, 2-phenylthioethyl, synthesis of, **5**, 1010
Isocyanate, phenylthiomethyl, synthesis of, **4**, 328
Isocyanate, phosphoryl dichloride
  from, diethyl *N*-chloroiminocarbonate, **5**, 988
  synthesis of, **5**, 988
Isocyanate, phosphoryl difluoride, synthesis of, **5**, 1018
Isocyanate, *n*-propyl, fluorination of, **6**, 439
Isocyanate, 1,2,2,2-tetrachloroethyl, synthesis of, **5**, 977
Isocyanate, 2-thienyl, synthesis of, **5**, 992
Isocyanate, 3-thienyl, synthesis of, **5**, 992
Isocyanate, β-thienylvinyl, cyclization of, **5**, 990
Isocyanate, 4-toluenesulfonyl, reactions, with sulfoxides, **2**, 162
Isocyanate, *p*-toluenesulfonyl, reactions, with methyl glyoxylate, **2**, 327
Isocyanate, tosyl, reactions, with sulfoxide, **3**, 365
Isocyanate, tribromomethyl
  isomerization of, **6**, 453
  synthesis of, **6**, 240, 611
Isocyanate, trichloromethyl
  reactions
    with alkyl dihalophosphites, **6**, 666
    with dialkyl halophosphite, **6**, 285
    with triethyl phosphite, **6**, 285, 366
  synthesis of, **6**, 240
  tautomerization of, **6**, 606
Isocyanate, triethyltin, reactions, with acetyl bromide, **5**, 1006
Isocyanate, trifluoromethyl
  fluorination of, **6**, 440
  from, perfluoro-2-azapropene, **5**, 988
  hydrolysis of, **6**, 604
  reactions, with trimethylsilyl cyanide, **5**, 721
  synthesis of, **6**, 240
Isocyanate, trimethylsilyl
  reactions
    with alkynylmagnesium bromides, **1**, 1043
    with α-chloro isocyanates, **4**, 434
    with Grignard reagents, **5**, 299
    with lithioaminofluorosilane, **5**, 1083
    with silylcarboxylic acids, **5**, 1006
  use in the synthesis of isocyanates, **5**, 1007
Isocyanate, trimethylsilylmethyl, synthesis of, **4**, 516
Isocyanate, tris(difluoroamino)methyl, reactions, with alcohols, **6**, 363
3-Isocyanate, pyridine-, synthesis of, **5**, 990
Isocyanates
  chlorination of, **6**, 606
  from
    acyl azides, **5**, 989
    aldehydes, **5**, 986
    alkenes, **5**, 984
    alkoxycarbonyliminophosphoranes, **5**, 988
    alkyl halides, **5**, 962
    amides, **5**, 995
    amidines, **5**, 987
    amines, **5**, 965; **6**, 443
    anhydrides, **5**, 995
    arenes, **5**, 984
    (arylsulfonyliminoiodo)benzenes, **5**, 987
    azides, **5**, 983
    aziridines, **5**, 1010
    carbamates, **5**, 976
    carbamic acids, **5**, 976
    N,*N'*-carbonyldiimidazole, **6**, 417
    carboxylic acids, **5**, 989
    N-chlorocarbonyl isocyanate, **5**, 1007
    chlorosulfonyl isocyanates, **5**, 1005
    cyanomethylene amines, **5**, 1003
    halides, **5**, 1015
    hydroxamic acids, **5**, 999
    iminophosphoranes, **5**, 974
    isocyanates, **5**, 1005
    isocyanic acid, **5**, 1015
    isocyanide dichlorides, **5**, 1004
    isonitrile dichlorides, **5**, 1004
    isonitriles, **5**, 1003
    isothiocyanates, **5**, 1009
    ketones, **5**, 986
    metal cyanates, **5**, 1015
    N-monosubstituted carbamoyl chlorides, **6**, 443
    nitrile oxides, **5**, 1003
    nitriles, **5**, 1002
    nitro compounds, **5**, 984
    nitroso compounds, **5**, 984
    phosphoramidate anions, **5**, 974
    silicon halides, **5**, 968
    silylamines, **6**, 443
    sulfur hexafluoride, **5**, 975
    thiocarbamates, **5**, 979
    ureas, **5**, 979
  hazards in the use of, **6**, 415
  industrial synthesis of, problems, **5**, 991
  niobium-promoted reactions, with alkynes, **2**, 988
  nucleophilic addition to, **4**, 426
  phosphorus analogues of, **3**, 607
  reactions
    with acetals, **4**, 426
    with alcohols, **6**, 484, 622
    with aldoximes, **3**, 434
    with alkanes, **5**, 386
    with alkenes, **5**, 386
    with alkyllithiums, **3**, 276
    with alkynes, **5**, 297, 386
    with alkynide anions, **5**, 304
    with 9-(1-alkynyl)-9-borabicyclo[3.3.1]nonanes, **1**, 1043
    with alkynyl metals, **1**, 1043
    with allenes, **5**, 386
    with allophanoyl chlorides, **5**, 980
    with amides, **5**, 754
    with amines, **6**, 504
    with anhydrides, **5**, 305
    with aroylsulfenamides, **5**, 253
    with arylamines, **3**, 684
    with arylstannanes, **5**, 344
    with arylsulfonamides, **5**, 767
    with aziridines, **6**, 510
    with bis(ethylthio)phenylborane, **2**, 267
    with *t*-butanol, **4**, 1007
    with carboxylic acids, **5**, 293
    with cyanamide, **5**, 1120
    with dienes, **5**, 386

with *O,O'*-diethyl hydrogen dithiophosphate, **5**, 1037
with dimethoxymethane, **4**, 310
with dithiocarbamates, **6**, 627
with enolates, **5**, 307
with germanium–nitrogen bonds, **2**, 815
with Grignard reagents, **5**, 293, 301, 304
with *S*-imidoyl sulfenamides, **5**, 736
with imines, **6**, 504
with iminophosphoranes, **5**, 1066
with isothiocyanates, **5**, 1009
with ketene acetals, **6**, 106
with nitrosyl hexachloroantimonate, **2**, 800
with pentamethylguanidine, **6**, 364
with phosphatrifulvene, **6**, 511
with phosphine, **6**, 511
with phosphine oxides, **6**, 512
with phosphoramidates, **5**, 1069
with phosphorus pentachloride, **6**, 606
with *O*-silylated hydroxamoyl chlorides, **5**, 672
with silylphosphines, **6**, 684
with sulfonamides, **5**, 981
with thioacetals, **4**, 426
with thiols, **6**, 494
with thionoamides, **5**, 631
with triarylarsine oxides, **2**, 809
with trimethyl chlorosilane, **5**, 1016
with water, **6**, 504
reduction of, **3**, 698, 721
as starting materials
for *O*-alkyl-*N*-(trifluoromethyl)carbamates, **6**, 239
for *N*-alkyl-*N*-(trifluoromethyl)fluoroformates, **6**, 239
for amide acetals, **6**, 107
for *N*-arylamides, **5**, 293
for carbamoyl bromides, **6**, 452
for carbodiimides, **5**, 1061, 1065
for carbonimidic dichlorides, **6**, 606
for *N*-carbonyl chlorides, **6**, 448
for diazonium salts, **2**, 800
for ester aminals, **6**, 110
for imidoyl halides, **5**, 657
for isocyanates, **5**, 1005
for isocyanides, **3**, 698
for isoselenocyanates, **5**, 1056
for isothiocyanates, **5**, 1037
for ketenimines, **3**, 591
for metal amides, **5**, 390
for nitro compounds, **2**, 397
for ortho amides, **6**, 143
for sulfamic acids, **2**, 366
for thiocarbamates, **6**, 494
for transition metal amides, **5**, 386
for 1,3,5-triazines, **4**, 426
for urethanes, **6**, 484
synthesis of, **2**, 725; **5**, 962, 1016
tantalum-promoted reactions, with alkynes, **2**, 988
uses of, **5**, 962
use in the synthesis of formamidines, **5**, 745
Isocyanates, acyl
reactions
with diazomethylcarbonyls, **3**, 475
with thionoamides, **5**, 631
as starting materials, for *N*-(α-chloroalkyliden) carbamidic acid chlorides, **5**, 658
synthesis of, **5**, 994
Isocyanates, acylaminomethyl, synthesis of, **4**, 446
Isocyanates, alkenyl, use in the synthesis of carbodiimides, **5**, 1078
Isocyanates, 2-alkenyl, synthesis of, **5**, 1013
Isocyanates, α-alkoxy, synthesis of, **5**, 985
Isocyanates, 3-alkoxyacryloyl, synthesis of, **5**, 1008
Isocyanates, 1-alkoxyalkyl, from, enol ethers, **4**, 314
Isocyanates, alkyl
hydrosilylation of, **6**, 515
reactions
with 1-diethylaminopropyne, **3**, 600
with β-ribofuranose phosphorus ylides, **5**, 1068
with water, **6**, 504
as starting materials, for carbamoyl fluorides, **6**, 439
synthesis of, **5**, 963
via fractional distillation, **5**, 980
Isocyanates, alkylethyl, synthesis of, **5**, 991
Isocyanates, alkylphosphonic chloride, synthesis of, **5**, 988
Isocyanates, alkynyl, synthesis of, **2**, 1055
Isocyanates, allenyl, from, 5-propargyloxythiatriazole, **5**, 1101
Isocyanates, allylic, synthesis of, via rearrangement of allylic cyanates, **1**, 837
Isocyanates, amino aryl, synthesis of, via phosgenation of amines, **5**, 965
Isocyanates, arenesulfinyl, synthesis of, **2**, 726
Isocyanates, arenesulfonyl, synthesis of, **2**, 726; **5**, 997, 1005, 1014
Isocyanates, aroyl, synthesis of, **5**, 994
Isocyanates, aryl
chloroformic acid esters of, **5**, 966
reactions
with amide thioacetals, **6**, 115
with iminophosphoranes, **3**, 194; **5**, 1067
with triethylamine, **4**, 446
as starting materials
for aryl isocyanides, **3**, 721
for bis(phosphinyl)iminocarbonyls, **6**, 668
for carbamoyl fluorides, **6**, 439
synthesis of, via phosgenation of amines, **5**, 965
use in the synthesis of carbodiimides, **5**, 1066
Isocyanates, aryloxysulfonyl, synthesis of, **5**, 1005
Isocyanates, arylsulfenyl, synthesis of, **5**, 995
Isocyanates, arylsulfonyl, synthesis of, **5**, 981, 987
Isocyanates, azomethyl, addition of Grignard reagents to, **4**, 448
Isocyanates, azomethylene, synthesis of, **4**, 440, 441
Isocyanates, benzoyl, reactions, with 2-phenyl-1,1,2,2-tetramethylguanidine, **6**, 647
Isocyanates, bis, reactions, with benzenemethanethiol, **6**, 121
Isocyanates, bis(silyl)methyl, synthesis of, **4**, 517
Isocyanates, bromo-, synthesis of, **5**, 1014
Isocyanates, α-bromoalkyl, reactions, with potassium thiocyanate, **4**, 434
Isocyanates, butyl, synthesis of, **5**, 964
Isocyanates, carbamoyl, synthesis of, **5**, 1008
Isocyanates, carbamoylmethyl, synthesis of, **4**, 447
Isocyanates, carbodiimide, synthesis of, **5**, 1079
Isocyanates, α-chloro, reactions, with trimethylsilyl isocyanate, **4**, 434
Isocyanates, 1-chloroalkyl, use in the synthesis of functionalized alkyl isocyanates, **5**, 1069
Isocyanates, 2-chloroalkyl, synthesis of, **5**, 1010
Isocyanates, α-chloroalkyl
chlorotropic shift of, **5**, 1069
reactions
with alkoxides, **4**, 314
with phenoxides, **4**, 314
with *N*-silylated carbodiimides, **4**, 441
as starting materials
for α-chloroalkylcarbodiimides, **5**, 1068
for imidoyl-tetracoordinated phosphorus compounds, **5**, 787
tautomerization of, **6**, 606
tautomers of, **4**, 329
Isocyanates, β-chloroalkyl, synthesis of, **5**, 998
Isocyanates, γ-chlorobutyroyl, synthesis of, **5**, 995
Isocyanates, 1-(chloromethyl)alkoxycarbonyl, synthesis of, **5**, 1008
Isocyanates, chlorosulfonyl (CSI)
reactions
with arenes, **5**, 1005

with phenols, **5**, 1005
as starting materials, for isocyanates, **5**, 1005
Isocyanates, *β*-cyano-*α*,*β*-unsaturated, synthesis of, **5**, 967
Isocyanates, dialkylacetyl, synthesis of, **5**, 994
Isocyanates, dialkylcarbamoyl, synthesis of, **5**, 1006
Isocyanates, *α*,*α*-dichloro-, reactions, with phosphinimines, **5**, 1068
Isocyanates, dichloromethyl, as starting materials, for *N*-acylimino compounds, **6**, 658
Isocyanates, dihaloethyl, as starting materials, for *gem*-dihalophosphoryl isocyanatoalkenes, **4**, 1012
Isocyanates, ethyl, synthesis of, **5**, 964
Isocyanates, fluoroalkyl, from, fluoroazaalkanes, **5**, 989
Isocyanates, *β*-fluoroalkyl, synthesis of, **6**, 441
Isocyanates, fluorocarbonylsulfenyl, synthesis of, **5**, 1018
Isocyanates, functionalized alkyl, synthesis of, **5**, 1069
Isocyanates, glycosyl, synthesis of, **5**, 1068
Isocyanates, haloalkyl, as starting materials, for *gem*-dihalophosphoryl isocyanatoalkenes, **4**, 1012
Isocyanates, *α*-haloalkyl aryl, use in the synthesis of isocyanates, **5**, 1008
Isocyanates, halocarbonyl, synthesis of, **6**, 437
Isocyanates, *α*-hydroxy, synthesis of, **5**, 986
Isocyanates, iminoacyl, reactions, with phosgene, **5**, 967
Isocyanates, iminoalkyl, synthesis of, **5**, 967, 987
Isocyanates, isopropenyl, synthesis of, **5**, 989
Isocyanates, isopropyl, synthesis of, **5**, 963
Isocyanates, *α*-(isothiocyanato), synthesis of, **4**, 434
Isocyanates, 2-keto-, synthesis of, **3**, 194
Isocyanates, monophosphorylated dichloromethyl, use in the synthesis of isocyanates, **6**, 286
Isocyanates, monosulfonyl, synthesis of, **5**, 994, 995
Isocyanates, 2-naphthalenesulfonyl, synthesis of, **5**, 1005
Isocyanates, *gem*-nitro-, synthesis of, **4**, 440
Isocyanates, organic, uses of, **5**, 962
Isocyanates, organodisulfonyl, synthesis of, **5**, 997
Isocyanates, perfluoroacyl, synthesis of, **5**, 995
Isocyanates, perfluoroalkanesulfonyl, synthesis of, **5**, 1005
Isocyanates, phenylazomethyl, synthesis of, **5**, 986
Isocyanates, phosphoryl
chlorination of, **6**, 606
synthesis of, **5**, 1017
Isocyanates, pyrazole, synthesis of, **5**, 973
Isocyanates, silyloxymethyl, synthesis of, **4**, 313
Isocyanates, *trans*-*β*-styryl
from, *trans*-cinnamoyl azides, **5**, 989
isomerization of, **5**, 989
Isocyanates, sulfonyl
chlorination of, **6**, 606
reactions
with alkenes, **5**, 344
with alkynes, **5**, 344
with allenes, **5**, 344
with aryl compounds, **5**, 344
with benzocyclopropene, **5**, 344
with enamines, **5**, 344
with enol ethers, **5**, 344
with ketones, **5**, 344
with thioketenes, **5**, 344
with thionoamides, **5**, 631
with vinylcyclopropanes, **5**, 344
with vinylsilanes, **5**, 344
as starting materials
for *N*-acyl sulfonamides, **5**, 344
for sulfonamides, **2**, 249
synthesis of, **5**, 979, 1005
Isocyanates, *N*-sulfonyl
reactions
with acetamides, **5**, 767
with formamides, **5**, 767
Isocyanates, tertiary cyclo(polycyclo)alkyl, synthesis of, **5**, 992
Isocyanates, tetrachloroethyl
reactions, with triethyl phosphite, **4**, 1010
as starting materials
for *gem*-dihalophosphoryl isocyanatoalkenes, **4**, 1013
for *gem*-isocyanato phosphorylalkenes, **4**, 1010
Isocyanates, thienyl, synthesis of, **5**, 990
Isocyanates, thioacyl, synthesis of, **5**, 996
Isocyanates, *α*-thioalkyl, synthesis of, **4**, 329
Isocyanates, thiocarbonyl, synthesis of, **5**, 996
Isocyanates, thiocyano aryl, synthesis of, **5**, 966
Isocyanates, trialkylsilyl, reactions, with formaldehyde, **4**, 313
Isocyanates, trichloromethylthiosulfenyl, synthesis of, **5**, 1018
Isocyanates, trifluoroacetyl, synthesis of, **5**, 995
Isocyanates, trihalomethyl, synthesis of, **6**, 239
Isocyanates, trimethylsiloxy, reactions, with isocyanato alkanoyl chlorides, **5**, 1007
Isocyanates, trimethylsiloxyalkyl
reactions
with alkanedioic acid dichlorides, **5**, 1006
with isocyanatoalkanoyl chlorides, **5**, 1006
Isocyanates, trimethylsiloxy-substituted, synthesis of, **5**, 1007
Isocyanates, tris(silyl)methyl, synthesis of, **4**, 517
Isocyanates, *α*,*β*-unsaturated
as starting materials
for *α*-bromoalkylcarbamoyl bromides, **6**, 452
for *α*-iodoalkylcarbamoyl iodides, **6**, 454
Isocyanates, vinyl
reactions, with enamines, **5**, 990
as starting materials, for bifunctional electrophiles, **6**, 448
synthesis of, **5**, 989, 993
Isocyanates, vinylidene, as starting materials, for *gem*-isocyano lithioalkenes, **4**, 1015
5-Isocyanates, Δ^2-isoxazoline, use in the synthesis of isoxazoles, **5**, 990
Isocyanatoalkenes, *gem*-alkylideneamino, from, *gem*-diisocyanatoalkanes, **4**, 992
Isocyanato groups, hydrolysis of, **4**, 1006
Isocyanic acid
reactions
with alkenes, **5**, 985
with alkyl chlorides, **5**, 964
with alkyl halides, **5**, 964
with carbonyl chlorides, **5**, 994
with carbonyls, **5**, 986
with carboxylic acid chlorides, **5**, 994
with diisopropenylbenzenes, **5**, 985
with formaldehyde, **4**, 313
with isopropenylbenzenes, **5**, 985
with pentafluoroguanidine, **6**, 363
with perfluorocarboxylic anhydrides, **5**, 995
with trialkylantimony oxides, **5**, 1017
with triethyltin oxide, **5**, 1017
with trifluoroacetic anhydride, **5**, 995
with *α*,*β*-unsaturated ethers, **5**, 985
as starting material, for isocyanates, **5**, 1015
Isocyanide, all *trans* retinyl, synthesis of, **3**, 702
Isocyanide, aminomethyl, synthesis of, **4**, 446
Isocyanide, aryl, reactions, with octamethyltrisilane, **6**, 163
Isocyanide, azidophenyl, synthesis of, **3**, 722
Isocyanide, 1*H*-benzotriazol-1-ylmethyl (BetMIC), synthesis of, **3**, 710
Isocyanide, benzyl
deprotonation of, **1**, 155
reactions, with *n*-butyllithium, **5**, 88
synthesis of, **3**, 702
Isocyanide, bis(trimethylsilyl)amino, reactions, with sulfenyl chlorides, **6**, 617
Isocyanide, bis(trimethylsilyl)methyl
synthesis of, **3**, 711
via silylation of methyl isocyanide, **6**, 162
Isocyanide, *t*-butyl

reactions
with acetals, **3**, 658
with fluorenylideneborane, **3**, 595
with hydrogen chloride, **4**, 894
as trapping agent, **4**, 1082
use in the synthesis of carbon-bridged diiron cation complexes, **4**, 680
use in the synthesis of carboxylic esters, **5**, 125
use in the synthesis of nitriles, **3**, 614
Isocyanide, 4-chlorophenyl, from, *N*-(4-chlorophenyl)dichloromethylsulfonamide, **3**, 723
Isocyanide, 2-chloro-2-phenylethyl, from, *N*-2-hydroxy-2-phenylethyl formamide, **3**, 703
Isocyanide, cyclohexenyl, from, *N*-(1-cyanocyclohexyl) formamide, **3**, 714
Isocyanide, cyclohexyl
reactions
with chlorocarbonylsulfenyl chloride, **5**, 1036
with chromium carbene complexes, **3**, 597
with dibenzoyl disulfide, **5**, 1036
with dichlorocarbene, **3**, 596
Isocyanide, diphenyl methyl, reactions, with 1,3-dibromopropane, **3**, 702
Isocyanide, disilanyl, synthesis of, **3**, 711
Isocyanide, 2-fluoroethyl, synthesis of, **3**, 703
Isocyanide, β-ionyl, synthesis of, **3**, 702
Isocyanide, α-lithiomethyl
reactions, with arenesulfonyl fluorides, **3**, 709
silylation of, **3**, 711
Isocyanide, lithio-α-(trimethylsilyl)tosylmethyl, use in the synthesis of α,β-unsaturated isocyanides, **3**, 714
Isocyanide, mesityl, reactions, with diarylstannadiyl, **5**, 1086
Isocyanide, methoxyisobutyl, synthesis of, **3**, 707
Isocyanide, methyl
deprotonation of, **1**, 155
lithiation of, **4**, 516
reactions
with cinnamyl bromide, **3**, 702
with cyclohexenone, **3**, 707
Isocyanide, *o*-methyl phenyl, deprotonation of, **3**, 723
Isocyanide, pentafluorophenyl, synthesis of, **3**, 722
Isocyanide, *t*-pentyl, reactions, with nickel chloride, **5**, 1072
Isocyanide, phenoxy, synthesis of, **3**, 703
Isocyanide, phenyl
reactions
with dichlorodimethylgermanium, **2**, 949
with tributylborane, **4**, 522
synthesis of, **3**, 680, 722
Isocyanide, (*E*)-4-phenyl-3-butenyl, synthesis of, **3**, 702
Isocyanide, phenylthiomethyl, reactions, with nitroarenes, **3**, 702
Isocyanide, *n*-propyl, from, *N*-*n*-propyl-1,8-di-*t*-butyl-5,6-dioxo-2,3-benzobicycloocta-2,7 diene, **3**, 699
Isocyanide, 2,2,5,5-tetramethyl cyclopentyl, synthesis of, **3**, 699
Isocyanide, theonellin, synthesis of, **3**, 700
Isocyanide, tosylmethyl (TosMIC)
alkylation of, **3**, 709
synthesis of, **4**, 331, 538
use in the synthesis of isocyanides, **3**, 694
Isocyanide, trichloromethyl, reactions, with secondary amines, **6**, 605
Isocyanide, 2,2,2-trifluoroethyl, synthesis of, **3**, 703
Isocyanide, trifluoromethyl
reactions, with chlorine fluoride, **6**, 613
synthesis of, **3**, 699, 703
Isocyanide, trimethylsilyl, use in the synthesis of nitriles, **3**, 614
Isocyanide, trimethylsilylmethyl
deprotonation of, **3**, 714
as starting material
for amidines, **4**, 518
for 1-aminoalkylphosphonium salts, **4**, 492
for thioimidates, **4**, 518
for trimethylsilylmethyl isothiocyanate, **4**, 517
synthesis of, **3**, 711; **4**, 516
Isocyanide, α-(trimethylsilyl)tosylmethyl, reactions, with aldehydes, **3**, 720
Isocyanide, tris(trimethylsilyl)methyl, synthesis of, **3**, 711
Isocyanide, vinyl, synthesis of, **1**, 574
Isocyanide, 2-vinyloxyethyl, synthesis of, **3**, 707
Isocyanide anions, for the opening of epoxides, **1**, 209
Isocyanide complexes, as starting materials, for carbene complexes, **6**, 721
Isocyanide complexes, alkynyl, synthesis of, **3**, 716
Isocyanide complexes, cyclobutenyl, synthesis of, **3**, 716
Isocyanide complexes, hydrido alkyl, reactions, with azetidine, **6**, 663
Isocyanide dibromide, bromocarbonyl, synthesis of, **6**, 453
Isocyanide dibromides *See* Carbonimidic dibromides
Isocyanide dichloride, chlorocarbonyl, synthesis of, **6**, 449
Isocyanide dichloride, *N*-chlorocarbonyl
from
cyanogen chloride, **6**, 451
methyl isocyanate, **6**, 451
Isocyanide dichloride, dichlorophenyl, reactions, with phosphorus pentoxide, **5**, 1004
Isocyanide dichloride, phenyl
reactions
with disodium cyanamide, **5**, 1070
with silver oxide, **5**, 1004
use in the synthesis of isocyanates, **5**, 1005
Isocyanide dichloride, trichloromethyl
reactions
with ammonium *O*,*O*-diethylthiophosphate, **6**, 316
with sodium *N*,*N*-diethyl dithiocarbamate, **6**, 316
with sodium *N*,*N*-dimethyl dithiocarbamate, **6**, 316
as starting material
for bis(trifluoromethyl)amines, **6**, 238, 239
for orthocarbonates, **6**, 296
for tetraaryl orthocarbonates, **6**, 296
See also Carbonimidic dichloride, trichloromethyl-
Isocyanide dichloride complexes, *t*-butyl
reactions
with dimethyl sulfoxide, **5**, 963
with trifluoroacetic acid, **5**, 963
Isocyanide dichloride complexes, isopropyl, reactions, with zinc oxide, **5**, 963
Isocyanide dichlorides
reactions
with acids, **5**, 1004
with alkynes, **1**, 1047
with silylphosphines, **6**, 686
with transition metal complexes, **3**, 699
as starting materials
for isocyanates, **5**, 1004
for isocyanides, **3**, 699
for secondary trihalomethylamines, **6**, 238
See also Carbonimidic dichlorides
See also Carbonimidoyl dichlorides
Isocyanide dichlorides, aryl, reactions, with phenyl bis-(trimethylsilyl)phosphines, **5**, 1086
Isocyanide dichlorides, *N*-aryl, as starting materials, for aryl isocyanides, **3**, 721
Isocyanide dichlorides, ω-chloroalkyl, synthesis of, **3**, 409
Isocyanide dichlorides, dialkoxyphosphoryl, reactions, with formic acid, **5**, 1005
Isocyanide dichlorides, *N*-substituted, reactions, with alkynyl Grignard reagents, **1**, 1035
Isocyanide difluoride, as starting material, for tris(trifluoromethyl)amine, **6**, 239
Isocyanide difluoride, trifluoromethyl
for bis(trifluoromethyl)amines, **6**, 239
fluorination of, **6**, 240

as starting material
for bis(trifluoromethyl)amines, **6**, 238, 239
for metal (trihalomethyl)amides, **6**, 242
for tris(trifluoromethyl)amine, **6**, 239
Isocyanide difluorides *See* Carbonimidic difluorides
Isocyanide dihalides, reduction of, **3**, 698, 721
Isocyanide dihalides *See* Carbonimidic dihalides
Isocyanide dihalides, as starting materials, for isocyanides, **3**, 699
Isocyanide dihalides, aryl, as starting materials, for aryl isocyanides, **3**, 721
Isocyanide dihalides, halocarbonyl, synthesis of, **6**, 437
Isocyanide dihalides, trihalomethyl, synthesis of, **6**, 239
Isocyanide diiodides *See* Carbonimidic diiodides
Isocyanides
addition of sulfenyl chlorides to, **6**, 617
bromination of, **6**, 610
chemistry of, reviews, **3**, 694
chlorine addition to, **6**, 605
condensation of, **3**, 707
deprotonation of, **3**, 697, 714
from
5-alkyl(aryl)aminotetrazoles, **3**, 699
amines, **3**, 693
carbodiimides, **3**, 695
2-deoxy tri-*O*-benzyl D-glucose, **3**, 695
difluoromethanimines, **3**, 702
formamides, **3**, 695
isocyanates, **3**, 698
isocyanide dichlorides, **3**, 699
isocyanide dihalides, **3**, 699
isothiocyanates, **3**, 698
N-sulfinylamines, **3**, 695
halo-substituted, synthesis of, **3**, 702
metallation of, **3**, 697
natural occurrence of, **3**, 694
nickel-catalyzed polymerization of, **5**, 463
reactions
with alkyllithiums, **3**, 418; **5**, 88
with azetidine, **6**, 663
with boranes, **6**, 637
with *N*-bromosuccinimide, **6**, 506
with *N*-chlorodialkylamines, **5**, 1146
with diacyl disulfides, **5**, 1036
with dicarboxylic acid chlorides, **3**, 417
with diphenylphosphine, **5**, 1130
with disilanes, **3**, 408
with fluorine, **6**, 603
with iron alkoxycarbene complexes, **5**, 954
with metal chlorides, **6**, 622
with nitrogen oxides, **5**, 1004
with nucleophiles, **6**, 721
with octamethyltrisilane, **6**, 163
with oligosilanes, **6**, 384
with organic azides, **5**, 1071
with organometallics, **5**, 68
with organozincs, **3**, 418
with selenium, **5**, 1051
with sulfur, **5**, 1036
with thiocarboxylic acids, **5**, 632
with thiols, **5**, 1036
with trialkylboranes, **4**, 522
with tributyltin hydride, **1**, 945
with tris(trimethylsilyl)silane, **5**, 1135
safety precautions for, **3**, **693–726**
as starting materials
for aminobis(phosphonato)methanes, **6**, 160
for aryl aldimines, **3**, 411
for carbon-centered radicals, **1**, 945
for carbonimidic dibromides, **6**, 610
for carbonimidic dichlorides, **6**, 605
for carbonimidic difluorides, **6**, 603
for imidazolines, **4**, 538
for imidoyl halides, **5**, 657
for iminocarbonyl chlorides, **6**, 617
for isoselenocyanates, **5**, 1051
for isothiocyanates, **5**, 1036
for *gem*-isothiocyanato silylalkenes, **4**, 1014
for ketenimines, **3**, 593
for oxazolines, **4**, 538
for α-palladio enamines, **4**, 1017
for α-platino enamines, **4**, 1017
for pyrroles, **4**, 538
for thiazolines, **4**, 538
for ureas, **6**, 505
for urethanes, **6**, 485
structure of, **3**, 694
synthesis of, **3**, **693–726**
historical background, **3**, 693
methods, **3**, 694
photochemical method, **3**, 699
reagent-free method, **3**, 699
via Hofmann carbylamine reaction, **3**, 695
uses of, **3**, 694
use in the synthesis of aryl ketones, **3**, 290
See also Isonitriles
Isocyanides, acyl, synthesis of, **3**, 704
Isocyanides, α-acyloxy, synthesis of, **3**, 703
Isocyanides, alkyl
from
silver hexacyanocobaltates, **3**, 695
silver hexacyanoferrates, **3**, 695
reactions
with aminoaziridines, **6**, 337
with *n*-butyllithium, **3**, 704
with triethylamine, **5**, 1052
as starting materials, for *N*-alkyliminocarbonyls, **6**, 671, 674
Isocyanides, *N*-alkyl, synthesis of, **3**, 699
Isocyanides, allyl, isomerization of, **3**, 714
Isocyanides, *N*-allyl, reactions, with formamides, **5**, 760
Isocyanides, allylic unsaturated, from, formamides, **3**, 700
Isocyanides, α-amino, synthesis of, **3**, 710
Isocyanides, aralkyl, synthesis of, **3**, 702
Isocyanides, arenesulfonylmethyl
reactions
with aldehydes, **4**, 908
with ketones, **4**, 908
with 17-oxosteroids, **4**, 908
with pyrrolidone-4-aldehydes, **4**, 908
synthesis of, **3**, 709
use in the synthesis of substituted alkenes, **4**, 908
Isocyanides, aryl
from
amines, **3**, 721
5-aryl aminotetrazoles, **3**, 723
aryl formamides, **3**, 721
aryl isocyanates, **3**, 721
N-aryl isocyanide dichlorides, **3**, 721
aryl isocyanide dihalides, **3**, 721
aryl isothiocyanates, **3**, 721
benzoxazoles, **3**, 723
N,*N'*-diaryl-*N*-hydroxyformamidines, **3**, 723
N-sulfinylarylamines, **3**, 721
reactions
with alkyl halides, **3**, 158
with dimethyl ethynedicarboxylate, **3**, 593
with sulfur, **5**, 1036
as starting materials, for *N*-aryliminocarbonyls, **6**, 671, 674
synthesis of, **3**, 721, 722
via Beckmann rearrangement, **3**, 722
via Hofmann carbylamine reaction, **3**, 721
Isocyanides, *N*-(arylalkyl)amino, synthesis of, **3**, 710
Isocyanides, aryl amino, synthesis of, **3**, 709
Isocyanides, aryl diazo, from, formyl triazenes, **3**, 710

Isocyanides, 1-(arylthio)alkenyl, synthesis of, **3**, 720
Isocyanides, (arylthio)methyl, reactions, with trimethylsilyl chloride, **3**, 720
Isocyanides, α-(benzotriazolyl)alkyl, synthesis of, **3**, 710
Isocyanides, bis β-chloro, from, dihydroxy formamides, **3**, 715
Isocyanides, 4-(*t*-butyldimethylsiloxy)-2,6-xylyl, from, formamides, **3**, 723
Isocyanides, 1-chloro α,β-unsaturated, synthesis of, **3**, 717
Isocyanides, coordinated
 nucleophilic addition to, **6**, 721
 synthesis of, **6**, 732
Isocyanides, cyclopropyl, deprotonation of, **1**, 155
Isocyanides, dialkylamino
 from, formylhydrazones, **3**, 709
 synthesis of, **3**, 709
Isocyanides, dienyl
 from, β-tosyl isocyanides, **3**, 718
 synthesis of, **3**, 714
Isocyanides, disubstituted aryl, synthesis of, **3**, 721
Isocyanides, ethyl
 structural formula, **3**, 694
 synthesis of, **3**, 694
Isocyanides, ethynyl, synthesis of, **3**, 694, 716
Isocyanides, fluorinated
 reviews, **3**, 694
 synthesis of, **3**, 702
Isocyanides, α-haloalkyl, stability of, **5**, 1087
Isocyanides, heteroaryl, from, 4-iminoisoxazalones, **3**, 722
Isocyanides, heterocyclic aromatic, from, iminoisoxazolones, **3**, 724
Isocyanides, β-hydroxy, synthesis of, **3**, 698, 704
Isocyanides, γ-hydroxy
 from, dihydro-1,3-oxazines, **3**, 707
 synthesis of, **3**, 698
Isocyanides, *N*-imidoyl, synthesis of, **3**, 710
Isocyanides, imino, from, formylhydrazones, **3**, 710
Isocyanides, β-iodo, synthesis of, **3**, 703
Isocyanides, keto, reactions, with lithium (*Z*)-β-lithioacrylate, **3**, 718
Isocyanides, β-keto, synthesis of, **3**, 194
Isocyanides, lithiated, cyclization of, **4**, 538
Isocyanides, α-lithiated, synthesis of, **4**, 537
Isocyanides, α-lithio
 alkylation of, **3**, 704
 reactions
  with epoxides, **3**, 704
  with oxetanes, **3**, 704
 synthesis of, **3**, 704
Isocyanides, α-lithioalkyl, synthesis of, **2**, 1002
Isocyanides, *o*-lithiomethylphenyl, reactions, with 1,2-dibromoethane, **3**, 724
Isocyanides, α-lithiovinyl, silylation of, **2**, 9000
Isocyanides, α-metallated
 as α-amino anion equivalents, **3**, 698
 condensation of, **3**, 697
 reactions
  with aldehydes, **3**, 713
  with ketones, **3**, 713
 synthesis of, **4**, 537
 uses of, **3**, 694
Isocyanides, α-metalloalkyl, synthesis of, **2**, 1002
Isocyanides, α-metallomethyl, reactions, with α,β-unsaturated ketones, **3**, 708
Isocyanides, β-methylthioether, synthesis of, **3**, 709
Isocyanides, multidentate, reviews, **3**, 694
Isocyanides, multidentate aryl, chemistry of, **3**, 723
Isocyanides, non-racemic oxygenated, synthesis of, **3**, 707
Isocyanides, organometallic, uses of, **3**, 698
Isocyanides, α-oxygenated, synthesis of, **3**, 703
Isocyanides, phosphine functionalized, synthesis of, **3**, 712
Isocyanides, polycyclic aromatic, from, formamides, **3**, 724
Isocyanides, saturated, synthesis of, **3**, 699
Isocyanides, 2-siloxyaryl, synthesis of, **3**, 722
Isocyanides, silyl enol ether, synthesis of, **3**, 707
Isocyanides, steroidal, synthesis of, **3**, 706
Isocyanides, sulfoximidoylmethyl, synthesis of, **4**, 330
Isocyanides, tertiary alkyl, reactions, with *t*-butyllithium, **3**, 408
Isocyanides, thiomethyl, synthesis of, **3**, 708
Isocyanides, thionaphthylmethyl, synthesis of, **3**, 708
Isocyanides, thiophene, from, thiopheneamines, **3**, 725
Isocyanides, β-tosyl, as starting materials, for dienyl isocyanides, **3**, 718
Isocyanides, trichloromethyl
 chromium complexes of, **6**, 610
 synthesis of, **3**, 703
Isocyanides, tridentate aryl, from, triformamides, **3**, 723
Isocyanides, unsaturated, synthesis of, **3**, 700
Isocyanides, α,β-unsaturated
 from, α,β-unsaturated formamides, **3**, 712
 synthesis of, **3**, 712
Isocyanides, β,γ-unsaturated, synthesis of, **3**, 702
Isocyanides, vinyl, synthesis of, **3**, 712, 714, 715
Isocyanides, vinylidene
 reactions, with chlorotrimethylsilane, **4**, 1014
 as starting materials, for α-isocyano silylalkenes, **4**, 1014
Isocyanoates, aryl-, as starting materials, for nitriles, **3**, 614
Isocyanogen, use in the synthesis of carbodiimides, **5**, 1089
Isocyano groups, characteristics of, **1**, 154
Isocyanomethyl anion, structure of, **1**, 154
Isocyanurate, triphenyl-, pyrolysis of, **5**, 1015
Isocyanurates, synthesis of, **5**, 980
Isocyanuric acid, dibromo-, reactions, with amides, **5**, 318
Isocyanuric acid, dichloro-, for the chlorination of amines, **2**, 338
Isocyanuric acid, trichloro-
 alcohol oxidations with, **3**, 119
 for the chlorination of amines, **2**, 338
 for the chlorination of imides, **5**, 316
 for the chlorination of ketones, **2**, 22
 for the chlorination of sulfides, **4**, 68
 as oxidant, for the synthesis of aldehydes from alcohols, **3**, 6
 for the synthesis of aryl chlorides, **2**, 619
 thermal decomposition of, **5**, 1015
Isocyanuric acids, trialkoxy-, synthesis of, **5**, 973
Isocyaphides, as intermediates, in the synthesis of phosphaalkynes, **3**, 689
Isodisulfonate, hydroxylamine, alkylation of, **2**, 342
Isodurene, trichloromethylation of, **6**, 29
Isoflavones, jamaicin, synthesis of, **3**, 278
Isoflurane, synthesis of, **4**, 49
Isofulminic acid, synthesis of, **3**, 703
Isoindole-1-ones, 3-substituted 3-hydroxy-1,3-dihydro-, synthesis of, **4**, 303
Isoindoles, 1-cyano-2-substituted, synthesis of, **3**, 671
Isoindolines, 1,2-dihydro-, from, allylfurfurylamine hydrochlorides, **2**, 329
Isoindolones, synthesis of, **4**, 299
Isoinversion, definition of, **1**, 858
Isomeric bromides, solvolysis of, **1**, 1114
Isomerization
 of alkynes, **1**, 954
 of allyl ethers, **2**, 675
 of allylic alcohols, **2**, 639
 of diynes, **1**, 954
 of 1,4-enediyne, **1**, 955
 of 1,4-enynes, **1**, 955
 (*E*)- to (*Z*)-, **4**, 808
 in Friedel–Crafts acylation, **1**, 462
 in heterogeneous hydrogenation, **1**, 71
 of ketene dithioacetal *S*,*S*-dioxides, **4**, 857
 of nitroalkenes, (*Z*) to (*E*), **4**, 796
 of organoboranes, **2**, 522
 of oxyimidic halides, **5**, 825

for the synthesis of alkynes, **1**, 1084
for the synthesis of tertiary ynamines, **2**, 1049
for the synthesis of α,β-unsaturated aldehydes, **3**, 61
for the synthesis of α,β-unsaturated ketones, **3**, 218, 229
π to σ Isomerization, for the synthesis of alkyl metal(II) compounds, **2**, 577
Isomunchones, synthesis of, **3**, 324
Isonitramine, synthesis of, **2**, 331
Isonitrile, *t*-butyl-
reactions
with alkenylrutheniums, **5**, 455
with *n*-butyllithium, **5**, 792, 795
with dichlorotrimethyltantalum, **5**, 797
with osmium methylimide complexes, **5**, 801
with trimesitylchromium, **5**, 798
use in the synthesis of imidoylboranes, **5**, 795
Isonitrile, caverno-, synthesis of, **3**, 704
Isonitrile, cyclohexyl-
reactions, with chloramine-T, **5**, 1064
as starting material, for *N*-cyclohexylformimidoyltrimethylsilane, **5**, 792
Isonitrile, methyl-, reactions, with trimethylaluminum, **5**, 1147
Isonitrile, phenyl-
reactions
with *t*-butyllithium, **5**, 796
with 2-(dialkylboryl)aminopyridine, **6**, 662
with dialkylboryls, **6**, 662
with piperidine, **5**, 747
Isonitrile, 1,1,3,3-tetramethylbutyl- (TMBI), use in the synthesis of imidoyllithium intermediates, **5**, 796
Isonitrile, vinyl-
reactions, with phosphines, **5**, 784
as starting material, for 1,3-azaphospholine, **5**, 784
Isonitrile, 2,6-xylyl-, reactions, with samarium iodide, **5**, 803
Isonitrile 270, synthesis of, **3**, 718
Isonitrile compounds, use in the synthesis of carbodiimides, **5**, 1071
Isonitrile dichlorides, as starting materials, for isocyanates, **5**, 1004
Isonitrile–nitrile rearrangement, via flash-pyrolysis, **1**, 266
Isonitriles
as by-products, in nitrile synthesis, **3**, 612
conversion to formimidoylphosphines, **5**, 785
deamination of, **1**, 38
from
formamides, **4**, 908; **6**, 417
silver cyanide, **3**, 612
oxidation of, **5**, 1003
reactions
with alcohols, **5**, 692
with alkyllithiums, **5**, 796
with alkylmanganese pentacarbonyl complexes, **5**, 799
with alkyltantalum complexes, **5**, 797
with amines, **5**, 747
with arylmanganese pentacarbonyl complexes, **5**, 799
with boranes, **5**, 1147
with *s*-butyllithium, **5**, 796
with chloramine-T, **5**, 1064
with cyclopentadiene complexes, **5**, 799
with dialkylbis[bis(trimethylsilyl)amido]hafnium(IV) compounds, **5**, 797
with dialkyl dithiophosphates, **5**, 737
with diisobutylaluminum, **5**, 803
with dimethylformamide, **5**, 767
with dimethyl sulfoxide, **5**, 1003
with dithioacids, **5**, 609
with hydrogen sulfide, **5**, 608
with nitrile oxides, **5**, 1004
with nitrines, **5**, 1071
with ozone, **5**, 1003
with phosphonic acids, **5**, 712
with silanes, **5**, 792
with thiols, **5**, 728
with thiophenols, **5**, 728
with toluenesulfonic acid, **5**, 710
with trimethylsilane, **5**, 792
as starting materials
for diethylformimidoylphosphines, **5**, 785
for formamidines, **5**, 747
for formimidoylsilanes, **5**, 792
for imidic acid esters, **5**, 691
for imidoylchromiums, **5**, 798
for imidoylirons, **5**, 799
for imidoyllithium intermediates, **5**, 796
for imidoylmagnesiums, **5**, 796
for imidoylnickels, **5**, 800
for imidoylosmiums, **5**, 801
for imidoylpalladiums, **5**, 801
for imidoylplatinums, **5**, 801
for imidoylsilicons, **5**, 792
for imidoyltantalums, **5**, 797
for imidoyl transition metals, **5**, 797
for imidoyl-tricoordinated phosphorus compounds, **5**, 784
for imidoyltungstens, **5**, 798
for imidoylzincs, **5**, 802
for isocyanates, **5**, 1003
for mixed anhydrides, **5**, 714
for nitriles, **3**, 621
for nitrilium ions, **3**, 687
for tetracoordinate germanium compounds, **5**, 794
for thioimidic esters, **5**, 727
structure of, **5**, 1089
See also Isocyanides
Isonitriles, alkyl-, reactions, with alkyllithiums, **5**, 795
Isonitriles, aryl-
reactions
with amines, **5**, 747
with dialkyl dithiophosphates, **5**, 737
Isonitriles, phenyl-, reactions, with alkyllithiums, **5**, 795
Isonitriles, silyloxylalkyl-, reactions, with triphenylborane, **5**, 1147
Isonitriles, tolyl-, reactions, with lithium diisopropylamide, **5**, 620
Isonitrin A, synthesis of, **3**, 718
Isonitrin B, synthesis of, **3**, 718
Isonitrin C, synthesis of, **3**, 704
Isonitrins, synthesis of, **3**, 712
Isopagodane skeleton, synthesis of, **1**, 574
Isophorone
addition reactions of, **1**, 311
as starting material, for 3,3-dimethyl-5-ketohexanoic acid, **5**, 75
Isophthalic acid, 2-hydroxy-, from, 2-hydroxy-3-methylbenzoic acid, **5**, 115
Isopinocamphone, methylenation of, **1**, 654
Isoprenoids, synthesis of, **1**, 754
Isoprenyl compounds, synthesis of, **1**, 292
Isopropanol, use in the synthesis of amines, **2**, 312
Isopropenylation, of enolates, **3**, 138
2-Isopropenyl-2-enoate, methyl, synthesis of, **1**, 735
Isopropenyl octanoate, reactions, with anhydrous hydrogen fluoride, **5**, 3
Isopropylamine
condensation with perfluorophosphapropene, **5**, 786
reactions
with chlorosilane, **2**, 417
with perfluoro-2-phosphapropene, **5**, 1087
Isopropyl cations, synthesis of, **1**, 899
N^1,N^2-Isopropylidene protected dipeptides, synthesis of, **4**, 417
Isopropyl iodide, reactions, with silver cyanate, **5**, 1104
Isopropyl iodide, heptafluoro-, reactions, with phosphorus, **4**, 114
Isopyrazoles, 4-hydroxy-, synthesis of, **4**, 948

Isopyrocalciferol, from, precalciferol, **1**, 419
Isoquinoline
  reactions
    with allenyltins, **1**, 1060
    with propargyltins, **1**, 1060
  reduction of, **1**, 90, 99
  ring-opening of, **5**, 1033
  as starting material, for chiral formamidines, **5**, 746
Isoquinoline, 2-chloro-, synthesis of, **5**, 657
Isoquinoline, 1-chloromethyl-3,4-dihydro-, α-sulfenylation of, **3**, 418
Isoquinoline, 3-cyano-, use in the synthesis of hydrazonamides, **5**, 847
Isoquinoline, 1-cyano-1-skatyl-2-benzoyl-1,2-dihydro-, synthesis of, **1**, 257
Isoquinoline, tetrahydro-, synthesis of, **3**, 422
Isoquinoline alkaloids, synthesis of, **1**, 55, 149
Isoquinoline-3-carboxylic acid, 1,2,3,4-tetrahydro-, reactions, with carbon disulfide, **6**, 564
Isoquinolines, synthesis of, **1**, 257
Isoquinolines, 1-alkoxycarbonyl-2-(2-alkynyl)-1,2-dihydro-, synthesis of, **1**, 1060
Isoquinolines, 1-benzyl-, synthesis of, **1**, 77
Isoquinolines, cyano-, synthesis of, **3**, 672
Isoquinolines, 3,4-dihydro-, as starting materials, for 1-phosphonic acids, **4**, 464
Isoquinolines, 8-methoxytetrahydro-, synthesis of, **3**, 419
Isoquinolines, oxazalo-[2,3-*a*]tetrahydro-, synthesis of, **2**, 328
Isoquinolines, phthalidyl-, synthesis of, **1**, 282
Isoquinolines, tetrahydro-
  alkylation of, **1**, 149, 150
  synthesis of, **4**, 531
Isoquinolines, 1,2,3,4-tetrahydro-, synthesis of, **1**, 99; **2**, 327
[2,3,4-*i,j*]Isoquinolines, fluoreno-, synthesis of, **3**, 585
Isoquinolones, synthesis of, **5**, 301
Isoracemization, mechanism, **1**, 855
Isoretronecanol, synthesis of, **2**, 326
Isosafroles, synthesis of, **1**, 756
Isoselenoazoles, reactions, with Grignard reagents, **3**, 256
Isoselenocyanate, phenyl, reactions, with amines, **5**, 1056
Isoselenocyanates
  from
    alcohols, **5**, 1053
    amines, **5**, 1055
    carbonimidoyl dichlorides, **5**, 1056
    diazo compounds, **5**, 1056
    halides, **5**, 1053
    isocyanates, **5**, 1056
    isocyanides, **5**, 1051
    selenocyanates, **5**, 1053
  as intermediates, in the synthesis of isothiocyanates, **5**, 1036
  as starting materials
    for carbon-centered radicals, **1**, 945
    for selenocarbamates, **6**, 593
    for selenosemicarbazides, **6**, 598
    for selenoureas, **6**, 596
  synthesis of, **5**, 761, 1051
Isoselenocyanates, *O*-acetylglycosyl, synthesis of, **5**, 1053
Isoselenocyanates, acyl
  from, acyl chlorides, **5**, 1056
  synthesis of, **5**, 1056
Isoselenocyanates, alkyl, synthesis of, **5**, 1052
Isoselenocyanates, arylimidoyl, synthesis of, **5**, 1057
Isoselenocyanates, imidoyl, synthesis of, **5**, 1056
Isoselenocyanates, silyl, synthesis of, **2**, 544
Isoselenoureas, synthesis of, **6**, 637
Isosteres, synthesis of, **1**, 817
Isosydnones, ring-opening of, **5**, 641
Isotellurocyanates
  synthesis of, **5**, 1051
    problems, **5**, 1057
Isothiazole 1,1-dioxide, 3-methylthio-1,2-benz-, reactions, with 1-diethylaminoprop-1-yne, **4**, 905
Isothiazoles
  as starting materials, for thioketenes, **3**, 552
  synthesis of, **3**, 686
2-Isothiazolines, from, alkenes, **3**, 686
4-Isothiazolin-3-ones, 5-halo-, reactions, with nucleophiles, **4**, 775
Isothiazolium salts
  as starting materials
    for thioaldehydes, **3**, 339
    for thioketones, **3**, 356, 361
Isothiazol-5-thiones, as starting materials, for thioketenes, **3**, 552
Isothiocyanate, 2-acetoxyvinyl, reactions, with hydrazine, **6**, 574
Isothiocyanate, acetyl, synthesis of, **5**, 1047
Isothiocyanate, benzoyl
  from, benzoyl chloride, **5**, 1047
  reactions, with fluorenone oximes, **6**, 558
  synthesis of, **5**, 1047
Isothiocyanate, benzyl, synthesis of, **5**, 1023, 1031
Isothiocyanate, bromocarbonyl, from, oxalyl bromide thiocyanate, **6**, 453
Isothiocyanate, *t*-butyl, synthesis of, **5**, 1038
Isothiocyanate, carbonyl isocyanate
  reactions, with nucleophiles, **6**, 565
  synthesis of, **6**, 559
Isothiocyanate, chlorocarbonyl, synthesis of, **6**, 451
Isothiocyanate, *N*-chlorocarbonyl
  from, oxalyl chloride isothiocyanate, **6**, 452
  synthesis of, **6**, 451
Isothiocyanate, 2-chloro-5-nitrobenzoyl, reactions, with sodium hydrogen sulfide, **6**, 566
Isothiocyanate, ethoxycarbonyl
  reactions
    with aromatics, **5**, 620
    with aryltributylstannanes, **5**, 593
    with benzenes, **5**, 592
    with pyrroles, **5**, 592
    with thiophenes, **5**, 592
  as reagent, for the synthesis of thioketones, **3**, 355
Isothiocyanate, 4-hydroxyphenyl, synthesis of, **5**, 1023
Isothiocyanate, iodo-, reactions, with *n*-butyllithium, **5**, 729
Isothiocyanate, methanesulfonyl, use in the synthesis of thiocyanates, **5**, 1110
Isothiocyanate, methyl
  reactions
    with 2-alkylthiiranes, **6**, 314
    with amines, **6**, 647
    with bis(trifluoromethyl)thioketene, **6**, 314
    with ethylene oxide, **6**, 324
    with *N*-tributylstannyldiphenylmethyleneamine, **6**, 571
Isothiocyanate, α-methylallyl, from, crotyl thiocyanate, **5**, 1040
Isothiocyanate, α-naphthyl, synthesis of, **5**, 1035
Isothiocyanate, 4-nitrophenyl
  from, 3-(4-nitrophenyl)-1,4,2-oxathiazole, **5**, 1046
  synthesis of, **5**, 1029
Isothiocyanate, oxalyl chloride, as starting material, for *N*-chlorocarbonyl isothiocyanate, **6**, 452
Isothiocyanate, phenethyl, synthesis of, **5**, 1025
Isothiocyanate, phenyl
  reactions
    with aminothiols, **6**, 570
    with 2,3-aziridines, **6**, 571
    with benzenethiol, **6**, 565
    with benzyne, **5**, 596
    with bis(trimethylsilyl)phosphines, **6**, 582
    with 2-*t*-butylamino-1,3,2-benzodioxaborole, **5**, 1038
    with carbon nucleophiles, **5**, 729
    with chlorine, **6**, 604
    with diazaboracyclohexane, **6**, 572

with dimedone, **5**, 593
with dimethylformamide diethyl acetal, **6**, 630
with diphenylmethyleneamine, **6**, 571
with germylphosphines, **6**, 582
with methylmagnesium chloride, **5**, 739
with 2-oxetanone, **5**, 729
with phosphorus(V) acid *t*-butyl esters, **6**, 583
with silylphosphines, **6**, 582
with stannylamines, **6**, 571
with trimethylsilyl tetraethylphosphorodiamidite, **6**, 582
Isothiocyanate, [^{35}S]-phenyl, synthesis of, **5**, 1038
Isothiocyanate, picryl, synthesis of, **5**, 1040
Isothiocyanate, *n*-propyl, synthesis of, **5**, 1024
Isothiocyanate, *p*-toluenesulfonyl, reactions, with substituted pyrroles, **5**, 638
Isothiocyanate, tri-*n*-butylstannyl, use in the synthesis of thiocyanates, **5**, 1107
Isothiocyanate, trimethylsilyl
chlorination of, **6**, 605
reactions
with alkyl aldehydes, **5**, 1042
with esters, **5**, 1107
as starting material, for adamantanes, **5**, 1045
as thiocyanating agent, **5**, 1106
Isothiocyanate, trimethylsilylmethyl
from, trimethylsilylmethyl isocyanide, **4**, 517
synthesis of, **4**, 516
Isothiocyanate, triphenylphosphine
use in the synthesis of acyl isothiocyanates, **5**, 1048
use in the synthesis of isothiocyanates, **5**, 1041
Isothiocyanate, triphenylsilyl, reactions, with phenyllithium, **5**, 595
Isothiocyanate, 2-(vinyloxy)ethyl, synthesis of, **5**, 1026
Isothiocyanates
chlorination of, **6**, 617
dimerization of, **6**, 572
in elimination reactions, **1**, 603
from
alcohols, **5**, 1041
aldehydes, **5**, 1042
alkenes, **5**, 1043
alkylamines, **5**, 1024
2-alkyloxaziridines, **5**, 1032
alkynes, **5**, 1043
amides, **5**, 1030
amines, **5**, 1022, 1030
amino acid trimethylsilyl esters, **5**, 1027
aromatic nitro compounds, **5**, 1033
aryl cyanates, **5**, 1038
N-arylsulfinylamines, **5**, 1032
carbonimidoyl dichlorides, **5**, 1032
cyanates, **5**, 1037
cyclic ethers, **5**, 1042
cycloalkanones, **5**, 1043
2,6-dichlorothiobenzamide, **5**, 1037
halides, **5**, 1038
iminophosphoranes, **5**, 1031
isocyanates, **5**, 1037
isocyanides, **5**, 1036
isothiocyanates, **5**, 1038
ketimines, **5**, 1032
ketones, **5**, 1042
lactams, **5**, 1031
nitrile oxides, **5**, 1038
oxetanes, **5**, 1042
phosphoramidates, **5**, 1031
N-substituted dithiocarbamates, **5**, 1023
thioamides, **5**, 1031, 1037
thiobenzamides, **5**, 1037
thiocarbonyl tetrachloride, **5**, 1028
thiocyanates, **5**, 1038
thiocyanogen, **5**, 1045
thiophosphoramidates, **5**, 1031
thiosemicarbazides, **5**, 1036
thioureas, **5**, 1035
trifluoroacetamides, **5**, 1030
N-trimethylsilylamines, **5**, 1030
trimethylsilylmethyl azide, **5**, 1031
reactions
with acylhydrazines, **6**, 573
with alcohols, **6**, 558
with alkynyl metals, **1**, 1044
with amidines, **5**, 760
with amines, **6**, 569
with ω-amino alcohols, **6**, 570
with ω-aminophenols, **6**, 570
with ammonia, **6**, 569
with benzaldehyde, **4**, 1014
with 1-(carbazol-9-ylmethyl)benzotriazole anions, **5**, 596
with diethyl malonate, **5**, 593
with enamines, **5**, 597, 614
with ethyl acetoacetate, **5**, 593
with ethyl cyanoacetate, **5**, 593
with germanium–nitrogen bonds, **2**, 815
with glutacondialdehyde, **5**, 593
with Grignard reagents, **5**, 595, 614
with halo ketones, **3**, 269
with hydrazines, **6**, 573
with hydroxylamines, **6**, 570
with imines, **6**, 571
with isocyanates, **5**, 1009
with ketene *S,N*-acetals, **5**, 597
with ketone enolates, **5**, 593
with ketoneimines, **5**, 597
with 4-methyl-1,2-benzenedithiol, **2**, 307
with methylenes, **5**, 593
with organolithiums, **5**, 595
with phenylethyne, **5**, 617
with phosphatrifulvenes, **6**, 582
with phosphorus(III) acid amides, **6**, 583
with phosphorus(V) compounds, **6**, 581
with phosphorus ylides, **5**, 597
with silylphosphines, **6**, 684
with stannylated ynamines, **2**, 1044
with sulfonylhydrazines, **6**, 573
with thiols, **6**, 565
with triethylaluminum, **5**, 595
with trimethylsilylamines, **6**, 571
with trimethylsilyldiphenylphosphine, **6**, 582
with (trimethylsilylmethylene)dimethyl phenylphosphorane, **5**, 645
with triphenylphosphine, **6**, 573
with tris(trimethylsilyl)phosphite, **6**, 582
reduction of, **3**, 698, 721; **5**, 609
as starting materials
for *O*-alkyl thiocarbamates, **6**, 558
for carbodiimides, **5**, 1062, 1065
for carbon-centered radicals, **1**, 945
for carbonimidic dichlorides, **6**, 604
for carbonimidic difluorides, **6**, 603
for cyanoguanidines, **6**, 647
for dithiocarbamate, **6**, 565
for iminocarbonates, **6**, 624
for iminocarbonyl chlorides, **6**, 617
for isocyanates, **5**, 1009
for isocyanides, **3**, 698
for isothiocyanates, **5**, 1038
for keteniminylidenephosphoranes, **3**, 605
for phosphinothioformamides, **6**, 581
for thioimidic esters, **5**, 729
for thionoamides, **5**, 592
for thiosemicarbazides, **6**, 573
for thioureas, **6**, 569
synthesis of, **2**, 273, 307; **5**, 1022, 1105

reviews, **5**, 1022
via aromatic electrophilic substitution, **5**, 1045
via cleavage of nitrogen heterocycles, **5**, 1033
via cleavage of nitrogen–sulfur-containing heterocycles, **5**, 1045
use in the synthesis of imidoyl dichlorides, **5**, 1070
Isothiocyanates, acyl
chlorination of, **6**, 605
from, acyl halides, **5**, 1047
reactions
with acetylacetone, **5**, 631
with enamines, **5**, 631
with ethyl cyanoacetate, **5**, 631
with Grignard reagents, **5**, 631
with malonitrile, **5**, 631
with phosphonium ylides, **5**, 631
with propargyl alcohols, **6**, 629
with substituted benzenes, **5**, 631
synthesis of, **5**, 1022, 1047
Isothiocyanates, acylaminomethyl, synthesis of, **4**, 446
Isothiocyanates, alkanoyl
reactions, with secondary amines, **6**, 572
synthesis of, **5**, 1047
Isothiocyanates, alkenyl
intramolecular tin-mediated radical cyclizations of, **5**, 593
synthesis of, **5**, 1032
Isothiocyanates, alkoxyalkyl, synthesis of, **5**, 1042
Isothiocyanates, alkoxycarbonyl, reactions, with amines, **6**, 572
Isothiocyanates, 3-alkoxypropyl, synthesis of, **5**, 1025
Isothiocyanates, alkyl
chlorination of, **6**, 604
from, ethyl dithiocarbamates, **5**, 1026
reactions
with acyl anions, **5**, 596
with thionoamides, **5**, 633
as starting materials, for α-oxothioimidates, **5**, 729
synthesis of, **5**, 1023
Isothiocyanates, alkynyl, synthesis of, **2**, 1056
Isothiocyanates, allyl, synthesis of, via [3,3] sigmatropic rearrangement, **5**, 1040
Isothiocyanates, allylic, synthesis of, **5**, 1110
Isothiocyanates, amino
reactions, with 3-aminopyridine, **6**, 574
as starting materials, for thiosemicarbazides, **6**, 574
Isothiocyanates, arenesulfonyl, synthesis of, **2**, 726
Isothiocyanates, aroyl
as starting materials, for 5-aryl-3*H*-1,2,4-dithiazol-3-thiones, **5**, 1049
synthesis of, **5**, 1047
Isothiocyanates, aryl
chlorination of, **6**, 617
reactions
with nitrones, **5**, 605
with (stannylethynyl)amines, **5**, 739
with thionoamides, **5**, 633
with ynamines, **5**, 618
as starting materials
for aryl isocyanides, **3**, 721
for carbonimidic difluorides, **6**, 603
Isothiocyanates, *N*-arylimidoyl, as starting materials, for 4(3*H*)-quinazoline thiones, **5**, 1050
Isothiocyanates, α-aryloxyalkyl, stability of, **4**, 314
Isothiocyanates, arylsulfonyl, reactions, with diphenyldiazomethane, **5**, 638
Isothiocyanates, azomethylene, synthesis of, **4**, 440
Isothiocyanates, 1-bromoalkyl, reactions, with secondary amines, **6**, 571
Isothiocyanates, α-bromoalkyl, reactions, with potassium thiocyanate, **4**, 434
Isothiocyanates, α-chloroalkyl, synthesis of, **5**, 1032
Isothiocyanates, 2-chloroethyl, synthesis of, **5**, 1033
Isothiocyanates, cinnamyl, from, cinnamyl thiocyanates, **5**, 1040
Isothiocyanates, 1-cycloalkenyl, synthesis of, **5**, 1043
Isothiocyanates, dialkoxythiophosphoryl, reactions, with sodium dialkylphosphites, **6**, 583
Isothiocyanates, ω-dialkylaminoalkyl, synthesis of, **5**, 1025
Isothiocyanates, glycosyl, synthesis of, via phase-transfer catalysis, **5**, 1039
Isothiocyanates, heterocyclic, synthesis of, **5**, 1022
Isothiocyanates, imidoyl
from, imidoyl halides, **5**, 1050
stability of, **5**, 1050
synthesis of, **5**, 1050
Isothiocyanates, *N*-imidoyl, as starting materials, for ketenimines, **3**, 593
Isothiocyanates, *vic*-iodo-, synthesis of, **5**, 1044
Isothiocyanates, 2-iodoalkyl, reactions, with sulfur nucleophiles, **6**, 314
Isothiocyanates, 2-keto, synthesis of, **3**, 194; **5**, 1046
Isothiocyanates, lithiated, synthesis of, **4**, 538
Isothiocyanates, α-metallated, synthesis of, **4**, 537
Isothiocyanates, *N*-methylbenzimidoyl, synthesis of, **5**, 1051
Isothiocyanates, phosphoryl
chlorination of, **6**, 605
use in the synthesis of acyl isothiocyanates, **5**, 1048
use in the synthesis of isothiocyanates, **5**, 1041
Isothiocyanates, 2-polyfluoroalkylethyl, synthesis of, **5**, 1022
Isothiocyanates, propadienyl, synthesis of, **1**, 993
Isothiocyanates, sesquiterpene, synthesis of, **5**, 1044
Isothiocyanates, silyl, reactions, with acyl halides, **5**, 1048
Isothiocyanates, *N*-substituted benzimidoyl, from, benzimidoyl chlorides, **5**, 1051
Isothiocyanates, substituted phenoxyphenyl, synthesis of, **5**, 1035
Isothiocyanates, sulfonyl
chlorination of, **6**, 605
reactions, with β,β-dimethyl enamines, **5**, 638
Isothiocyanates, thienyl, synthesis of, **5**, 1037
Isothiocyanates, thioacyl, synthesis of, **5**, 1022, 1049
Isothiocyanates, thioaroyl, synthesis of, **5**, 1050
Isothiocyanates, thiobenzoyl, synthesis of, **5**, 1050
Isothiocyanates, thiocarbamoyl, synthesis of, **5**, 1050
Isothiocyanates, thiocarbonyl, chlorination of, **2**, 169
Isothiocyanates, vinyl, from, Δ^3-thiazoline-2-thiones, **5**, 1046
9-Isothiocyanato compounds, synthesis of, **5**, 1040
Isothiocyanatophosphazenes, reactions, with alcohols, **6**, 558
Isothiocyanic acid
reactions
with alkenes, **2**, 307; **5**, 1043
with alkynes, **5**, 1108
with bicyclo[2.2.1]heptene systems, **5**, 1043
with camphene, **5**, 1043
with *endo*-dicyclopentadiene, **5**, 1043
with α,β-unsaturated carbonyls, **5**, 1107
use in the synthesis of thiocyanates, **5**, 1105
Isothiosemicarbazides, synthesis of, **4**, 984
Isothiothiocyanate, carbonyl isocyanate, chlorination of, **6**, 605
Isothiourea, 1,3-bis(methoxycarbonyl)-S-methyl-
reactions, with 2-aminobenzamides, **6**, 364
as reagent, for the synthesis of benzimidazole-2-carbamates, **6**, 363
Isothiourea, *N*-cyano-*N'*,*S*-dimethyl-, reactions, with amines, **6**, 647
Isothiourea, *N*-cyano-*S*-methyl-, reactions, with sodium hydroxide, **5**, 1118
Isothiourea, *S*-methyl-, as starting material, for *N*-acyliminocarbonyls, **6**, 646
Isothioureas
acylation of, **6**, 636
as dienes, **6**, 636
from
carbodiimides, **6**, 635

carbonimidic halides, **6**, 635
cyanamides, **6**, 635
iminodithiocarbonates, **6**, 636
thiocyanates, **6**, 635
thioureas, **6**, 635
reactions, with acetophenones, **4**, 980
as starting materials
for α-amido enamines, **4**, 980
for ketene aminals, **4**, 973
for tetrathioorthocarbonates, **6**, 301
synthesis of, **6**, 634
cycloaddition methods, **6**, 636
Isothioureas, *S*-alkyl-*N*-hydroxy-, synthesis of, **5**, 831
Isothioureas, *N*,*N*'-dinitro-*S*-alkyl-, from, isothioureas, **6**, 301
Isothioureas, *N*,*N*'-dinitro-*S*-aryl-, from, isothioureas, **6**, 301
Isothioureas, *S*-methyl-, chlorination of, **6**, 619
Isothioureas, *S*-methyl-*N*-nitro-, reactions, with amines, **6**, 654
Isothiouronium salts
displacement of, **1**, 238
from, alcohols, **2**, 120
hydrolysis of, **4**, 346
reactions, with chlorine, **2**, 177
synthesis of, **2**, 119; **6**, 731
use in the synthesis of alkanethiols, **2**, 120
Isothiouronium salts, alkyl-, oxidation of, **2**, 177
Isothiouronium salts, *S*-alkyl-
reactions, with anilines, **6**, 644
as starting materials, for *N*-alkylguanidines, **6**, 642
Isothiouronium salts, *S*-aryl-, synthesis of, **6**, 732
Isothiouronium salts, *S*-methyl-
acylation of, **6**, 646
reactions
with amines, **6**, 641
with ethylenediamine, **6**, 645
as starting materials
for 1-acyl-2,3-dialkylguanidines, **6**, 646
for sulfonylguanidines, **6**, 651
Isotopic exchange, rate constant for, **1**, 854
Isoureas
N-alkylation of, **6**, 633
from
carbodiimides, **6**, 633
carbonimidic esters, **6**, 631
carbonimidic halides, **6**, 631
carbonimidic thioesters, **6**, 631
cyanamides, **6**, 632
cyanates, **6**, 632
cyanogen halides, **6**, 632
ureas, **6**, 633
as starting materials
for diazirines, **6**, 280
for ureas, **6**, 503
synthesis of, **6**, 631
via [1+3] dipolar cycloaddition, **6**, 634
via functional group interconversion, **6**, 633
Isouronium salts, synthesis of, **6**, 632
Isouronium salts, *O*-phosphoryl-
reactions
with amines, **6**, 641
with aniline, **6**, 645
Isovaleric acid, as starting material, for β-hydroxyisovaleric acid, **5**, 118
Isovaleric acid, β-hydroxy-, from, isovaleric acid, **5**, 118
n-Isovaleric acid, 2-oxazolidone-, synthesis of, **6**, 485
Isoxazalones, 4-imino-, as starting materials, for heteroaryl isocyanides, **3**, 722
Isoxazilines
photolysis of, **3**, 77
reduction of, **3**, 269
Isoxazole
as starting material
for lithium phenylethynolate, **1**, 683
for phenylethynolate, **2**, 1023
Isoxazole, 3,4-diphenyl-, as starting material, for phenyltrimethylsilylketene, **3**, 543
Isoxazole, 5-mercapto-, characteristics of, **5**, 531
Isoxazole, triphenyl-, as starting material, for ketenimines, **3**, 576
Isoxazoles
N-alkylation of, **2**, 342
deprotonation of, **5**, 162
hydrogenation of, **2**, 744
photolysis of, **4**, 892
reactions, with alkynes, **3**, 683
reduction of, **3**, 269
as starting materials
for *N*-alkylhydroxylamine salts, **2**, 347
for ketene aminals, **4**, 979
for ketenimines, **3**, 576
synthesis of, **5**, 990
thiolic forms of, **5**, 532
Isoxazoles, 5-amino-, as starting materials, for alkynes, **1**, 1076
Isoxazoles, chloro-, reactions, with potassium hydrosulfide, **5**, 531
Isoxazoles, dihydro-, synthesis of, **6**, 118, 126
Isoxazoles, 5-hydroxymethyl-, hydrogenation of, **3**, 247
Isoxazoles, 3-methyl-4-nitro-5-(2-arylstyryl)-, hydrolysis of, **5**, 99
Isoxazoles, silyl-, synthesis of, **2**, 943
Isoxazoles, 5-substituted, reactions, with lithium diisopropylamide, **3**, 653
Isoxazoles, sulfonylated 5-amino-, diazotization of, **2**, 1035
Isoxazole salts, ring-opening of, **4**, 892
5-Isoxazolethiols, synthesis of, **5**, 531
Isoxazolidine, hydroxy-, use in the synthesis of ketenimines, **3**, 578
Isoxazolidines
reactions, with molybdenum hexacarbonyl, **2**, 320
synthesis of, **4**, 307
via cycloaddition reactions, **2**, 354
Isoxazolidines, *N*-alkoxyl-, synthesis of, **2**, 355
Isoxazolidines, *N*-silyloxy-, synthesis of, **2**, 356
Isoxazolidine-3-thiones, 2-benzyl-5-phenyl-, reactions, with phosphorus pentasulfide, **5**, 637
Isoxazolidin-3-one, 2-benzyl-5-phenyl-, thiation of, **5**, 637
Isoxazoline *N*-oxides
reactions, with alkenes, **2**, 356
use in the synthesis of *N*,*N*-dialkoxylamines, **2**, 356
Isoxazolines
as intermediates, for the synthesis of β-hydroxy ketones, **3**, 160
as starting materials, for 2-sulfonyl-α,β-unsaturated ketones, **3**, 250
2-Isoxazolines
reactions
with alkenes, **3**, 683
with alkynyllithiums, **1**, 1042
4-Isoxazolines, cleavage of, **3**, 250
2-Isoxazoline-5-thiones, characteristics of, **5**, 531
2-Isoxazoline-5-thiones, 4-arylhydrazono-
reactions, with diazomethane, **5**, 532
synthesis of, **5**, 532
3-Isoxazoline-5-thiones, characteristics of, **5**, 531
3-Isoxazoline-5-thiones, *N*-methyl-, synthesis of, **5**, 532
3-Isoxazoline-5-thiones, 2-substituted, from, 5-chloroisoxazolium tetrafluoroborates, **5**, 532
Isoxazoline-5-thiones, synthesis of, **5**, 531
Isoxazolium salts
as starting materials
for ketene aminals, **4**, 979
for ketenimines, **3**, 576
Woodward's Reagent K, **3**, 577

Isoxazol-5(4H)-one, 3-phenyl-, reactions, with nitroso compounds, **3**, 724
Isoxazolones, imino-
  as starting materials, for heterocyclic aromatic isocyanides, **3**, 724
  synthesis of, **3**, 724
Isoxazol-5-ones, flash vacuum pyrolysis of, **3**, 686
5-Isoxazolones, as starting materials, for alkynes, **1**, 1076
Isoxazol-5-ones, as starting materials, for ketenimines, **3**, 578
5-Isoxazolones, 2-nitroso-, as starting materials, for alkynes, **1**, 1076
Isoxazol-5(4*H*)-ones, 4-(anilinomethylene)-3-methyl-, flash vacuum pyrolysis of, **2**, 1040
Isoxazol-5(4*H*)-ones, 4-hydrazono-, pyrolysis of, **3**, 710
Itaconic acid, from, itaconic anhydride, **5**, 99
Itaconic anhydride, as starting material, for itaconic acid, **5**, 99
Itsuno's reagent, for the reduction of ketimines, **2**, 312
Ivanov reagent, use in the synthesis of α-aryl-β-hydroxypropionic acids, **5**, 76

J

Jacobson rearrangement, for the synthesis of
2,4-dimethyltrichloromethylbenzene, **6**, 29
Japp–Klingemann reaction
mechanism, **3**, 448
for the synthesis of formazans, **5**, 857
for the synthesis of hydrazones, **3**, 448
for the synthesis of 1,4,5,6-tetrahydropyridazin-4-ones, **3**, 453
Johnson–Claisen rearrangement, mechanism, **1**, 416, 827
Johnson (*N*-methylphenylsulfonimidoyl)methyllithium
method, for the synthesis of alkenes, **1**, 710
Jones' reagent
for the oxidation of silyl ethers, **3**, 121
reactions, with alcohols, **1**, 957
synthesis of, **5**, 47
use in the synthesis of carboxylic acids, **5**, 51, 74
use in the synthesis of ketones, **3**, 113, 115
Jones oxidation, for the synthesis of α,β-unsaturated
ketones, **3**, 249
Juglone, reduction of, **1**, 85
Julia reaction
mechanism, **1**, 614
for the synthesis of alkenes, **1**, 144, 696
for the synthesis of alkynes, **1**, 1073
use in natural product synthesis, **1**, 702
Julia synthesis, of retinoic acids, **1**, 612
Juvenile hormones
of *cecropia*, **1**, 747
of *Hyalophora cecropia*, **1**, 748
synthesis of, **5**, 154

K

K10-Montmorillonite, as catalyst, for the alkylation of alcohols, **1**, 180
Kabachnik–Fields reaction
  mechanism, **4**, 469
  for the synthesis of α-aminoalkylorganophosphorus compounds, **4**, 469
  for the synthesis of α-aminoalkylphosphorus compounds, **4**, 487
Kainic acid
  from, *cis* enones, **1**, 369
  synthesis of, **1**, 345
Kainic acid, allo-, synthesis of, **1**, 369
(-)-Kainic acid, synthesis of, **1**, 831
Kaluza reaction, for the synthesis of isothiocyanates, **5**, 1023
Kaplan–Shechter reaction, for the synthesis of *gem*-dinitroalkanes, **4**, 428
Kaurenoide, synthesis of, **6**, 472
Keck's procedure, for the synthesis of macrolactones, **5**, 173
Ketal, 3-buten-2-one ethylene, reactions, with vinyl halides, **1**, 1096
Ketals
  bromination of, **2**, 29
  from, dithianes, **1**, 309
  halogenation of, **2**, 3
  reactions
    with allenyltrialkylstannanes, **1**, 1053
    with hydrogen sulfide, **3**, 347
    with methyllithium–lithium bromide, **4**, 961
    with pyrazole, **4**, 437
    with trimethylgermyl chloride, **6**, 130
  reductive halogenation of, **2**, 14
  reductive lithiation of, **4**, 394
  as starting materials, for amines, **2**, 310
Ketals, diethyl, reactions, with peroxycarboxylic acid, **6**, 466
Ketals, silyl, synthesis of, **6**, 124
Ketazines
  as starting materials
    for α-aminoalkylorganophosphorus compounds, **4**, 460
    for azo compounds, **2**, 404
Ketene
  as activating and dehydrating agent, in the synthesis of anhydrides, **5**, 185
  deprotonation of, **2**, 1023
  reactions
    with *N*-*t*-alkylazomethines, **4**, 421
    with *t*-butyl diethyl phosphite, **4**, 915
    with chloroacetic acid, **5**, 193
    with hexamethyldisilazane, **5**, 379
    with mercuric halide complexes of trialkyl phosphites, **4**, 915
    with trimethylsilylamines, **5**, 376
  synthesis of, via Wolff reactions, **1**, 394
Ketene, 2,2-bis(fluorosulfonyl)-, synthesis of, **4**, 867
Ketene, 2,2-bis(trifluoromethanesulfonyl)-, synthesis of, **4**, 867
Ketene, bis(trifluoromethyl)-
  reactions
    with dialkylphosphines, **4**, 1023
    with diamines, **4**, 886
    with imines, **4**, 886
  as starting material
    for cyclic sulfates, **4**, 833
    for dioxazoles, **4**, 833
Ketene, bis(trifluoromethylseleno)-, synthesis of, **6**, 237
Ketene, bis(trimethylsilyl)-
  from, *t*-butyl bis(trimethylsilyl) acetate, **3**, 537
  synthesis of, **2**, 941
Ketene, bis(trimethylstannyl)-, synthesis of, **3**, 540
Ketene, bromo-, synthesis of, **4**, 807
Ketene, bromotrimethylsilyl-, from, trimethylsilylketene, **2**, 941
Ketene, *t*-butyl cyano-
  reactions
    with *S*,*S*-diphenylsulfilimine, **5**, 349
    with triphenylphosphine sulfide, **3**, 546
Ketene, *n*-butyl trimethylsilyl-, synthesis of, **2**, 941
Ketene, chlorocyano-
  reactions, with ethyl *N*-cyclohexylformimidate, **4**, 304
  synthesis of, **3**, 539
Ketene, cyclopentadienyl-, from, *o*-phenylene carbonates, **3**, 543
Ketene, di-*t*-butyl-
  reactions, with phosphorus pentasulfide, **3**, 546
  synthesis of, via dehydrohalogenation with sodamide, **3**, 526
Ketene, dichloro-
  reactions
    with alkynes, **3**, 237
    with indoles, **6**, 111
  use in the synthesis of carbenes, **4**, 1083
Ketene, dimethyl-
  reactions
    with acridine, **4**, 886
    with carbon disulfide, **4**, 836
    with heterocycles, **4**, 886
    with quinoline, **4**, 886
    with thiophosgene, **5**, 481
    with trivalent phosphorus compounds, **4**, 914
  synthesis of, **3**, 534
Ketene, dimethylmethylene-, from, α-methylene-β-lactams, **3**, 541
Ketene, diphenyl-
  reactions
    with azetines, **4**, 829
    with azirines, **4**, 829
    with *N*-benzyloxycarbonyl glycine, **5**, 193
    with diphenylacetic anhydride, **4**, 829
    with oxiranes, **4**, 829
    with silanes, **2**, 1044
    with silylphosphines, **6**, 684
    with sulfur diimides, **2**, 244
    with tetramethylallene-1,2-sulfide, **4**, 836
    with triethyl phosphite, **4**, 914
  synthesis of, **3**, 530
  use in the synthesis of carbon-bridged dirhodium complexes, **4**, 686
Ketene, phenyl-, synthesis of, **3**, 536
Ketene, phenyltrimethylsilyl-, from, 3,4-diphenylisoxazole, **3**, 543
Ketene, silylvinyl-, Diels–Alder cyclization of, **2**, 944
Ketene, trimethylsilyl-
  bromination of, **4**, 807
  reactions, with butyllithium, **3**, 537
  as starting material, for bromotrimethylsilylketene, **2**, 941
Ketene, trimethylsilylvinyl-, from, (*Z*)-2-trimethylsilyl-2-butenoic acid, **2**, 941
Ketene–anthracene cycloaddition adducts, as starting materials, for alkynes, **1**, 1084
Ketene dimers, reactions, with hydroxamic acids, **5**, 1001
Ketenes
  addition of dimethylzirconocene to, **2**, 703
  analogues of, synthesis of, **3**, 602

bearing heteroatom constituents, synthesis of, **3**, 527
characteristics of, **3**, 525
cycloaddition of, for iminodithiocarbonates, **6**, 113
[2+2] cycloadditions of, **5**, 162
dimerization of, **3**, 525
from
acetaldehyde, **3**, 535
acetic acid, **3**, 535
acetic amides, **3**, 535
acetic anhydrides, **3**, 534
acetic esters, **3**, 535
acetone, **3**, 535
acid chlorides, **3**, 527
acid halides, **3**, 526
acyclic ketones, **3**, 535
alkoxyalkynes, **3**, 540
alkylphthalimides, **3**, 536
aroyl halides, **3**, 527
biacetyl, **3**, 535
2-bromocyclohexane-1,3-dione, **3**, 536
butanone, **3**, 535
carbocyclic ketones, **3**, 537
carboxylic acid anhydrides, **3**, 534
carboxylic acids, **3**, 535, 536
cyclobutane-1,3-diones, **3**, 538
cyclobutanones, **3**, 538
α-diazocarbonyls, **3**, 530
diethyl ketone, **3**, 535
dihydrocoumarins, **3**, 543
ethyl acetate, **3**, 535
α-germyl-α-diazocarbonyls, **3**, 533
2-haloacyl halides, **3**, 530
heterocyclic materials, **3**, 541
o-hydroxy aromatic esters, **3**, 536
isopropenyl esters, **3**, 536
ketones, **3**, 535
malonic acid anhydrides, **3**, 534
malonic esters, **3**, 536
o-mercaptobenzoic acid, **3**, 536
methyl alkyl ketones, **3**, 535
methylenecyclobutenones, **3**, 539
mixed anhydrides, **3**, 534
mono esters, **3**, 536
1,3,4-oxadiazines, **3**, 543
α-silyl-α-diazocarbonyls, **3**, 533
sulfonium methylides, **3**, 536
transition metal complexes, **3**, 544
α,β-unsaturated acid halides, **3**, 527
photolytic generation of, **5**, 174
reactions
with *N*-acyl-*N*-sulfinylamines, **5**, 340
with alkoxyalkynes, **3**, 247
with alkylidenephosphoranes, **1**, 986
with allenes, **1**, 498; **3**, 216
with cobalt tetracarbonyl, **5**, 456
with enamines, **3**, 269
with imines, **5**, 286
with iminotriphenylphosphoranes, **3**, 584
with ketene acetals, **3**, 246
with lithiated silylphosphine, **3**, 605
with silylamines, **4**, 886
with silylated amides, **4**, 886
with stannous thiolates, **5**, 236
with thionoamides, **5**, 630
with triethylgermylphenylamine, **5**, 389
stability of, **3**, 525
as starting materials
for *N*-acyl sulfonamides, **5**, 343
for alkenylsilanes, **2**, 930
for benzofurans, **3**, 529
for 1,1-bisgermylalkanes, **4**, 628
for dichlorocarbene, **4**, 1083
for β-enaminones, **3**, 269
for enols, **2**, 645
for isocomene, **3**, 528
for ketene acetals, **4**, 829
for ketene hemiaminals, **4**, 886, 887
for ketenimines, **3**, 584
for β-lactams, **3**, 528; **5**, 274; **6**, 113, 115
for β-lactones, **1**, 683
for prostaglandins, **3**, 529
for 6-protoilludene, **3**, 528
for retigeranic acid, **3**, 528
for terpenes, **3**, 528
for thioacyl chlorides, **5**, 480
for thioketenes, **3**, 546
for tricyclo[5.3.0.0]decane, **3**, 528
for unsymmetrical alkanoic anhydrides, **5**, 193
for vinylsilanes, **2**, 929
synthesis of, **3**, **525–554**, **4**, 831
in situ, **3**, 530
in situ from acid halides, **3**, 526
overviews, **3**, 526
problems, **3**, 540
reagents for, **3**, 530
via dehalogenation, **3**, 530
via Norrish type I cleavage, **3**, 537
via vinylogous Wolf rearrangement, **3**, 532
via Wolff rearrangement, **3**, 530
for the synthesis of manganese acyl compounds, **5**, 449
tautomerization of, **2**, 1022
Ketenes, acyl-
from, polyfluorinated ethers, **3**, 545
reactions, with diazo ketones, **3**, 475
synthesis of, **3**, 527, 536, 542
Ketenes, alkenyl-
[4+2] annulation of, **2**, 668
from, α,β-unsaturated acid chlorides, **2**, 668
reactions, with alkynes, **2**, 668
Ketenes, alkyl (trimethylsilyl)-, from, alkyl alkoxyalkynes, **3**, 540
Ketenes, allyl-, synthesis of, **3**, 527
Ketenes, arsino-
from, arsino ynethers, **2**, 876
synthesis of, **2**, 878, 896
Ketenes, arsinyl-, synthesis of, **2**, 876
Ketenes, aryl-, synthesis of, **3**, 527
Ketenes, bis(silyl)-, as starting materials, for 1,1-bis(silyl)alkanes, **4**, 619
Ketenes, bis(thiophosphinyl)-, dimerization of, **4**, 559
Ketenes, chiral, synthesis of, **3**, 528
Ketenes, cyano-
from
azidotropones, **3**, 539
2,5-diazidoquinones, **3**, 539
synthesis of, **3**, 539
vinylogous, synthesis of, **3**, 539
Ketenes, cyclopentadienyl-, synthesis of, **3**, 536
Ketenes, cyclopropyl-, from, 5,5-dimethylcyclopentenone, **3**, 538
Ketenes, diacyl-, synthesis of, **4**, 829
Ketenes, dialkyl-
from
alkylphthalimides, **3**, 541
malonic acid anhydrides, **3**, 541
Meldrum's acid compounds, **3**, 541
organophosphorus compounds, **3**, 541
ortho esters, **3**, 541
oxiranes, **3**, 541
oxirenes, **3**, 541
stability of, **3**, 525
synthesis of, **3**, 527, 534, 536
Ketenes, diaryl-, as starting materials, for 2,2-diarylethenols, **2**, 645
Ketenes, diazo-, synthesis of, **3**, 489
Ketenes, dienyl-, electrocyclic reactions of, **2**, 662

Ketenes, 1-dihaloboronyl 1-silyl-, synthesis of, **4**, 1057
Ketenes, dimetallo-, synthesis of, **3**, 540
Ketenes, 2,2-disulfonyl-, synthesis of, **4**, 867
Ketenes, fluoro-, thermolysis of, **4**, 736
Ketenes, germyl-, synthesis of, **3**, 534
Ketenes, halocyano-, from, 4-azido-3-halo-5-methoxy-5(*H*)furan-2-ones, **3**, 543
Ketenes, imino-, synthesis of, **3**, 542
Ketenes, metallated, reactions, with dialkylphosphine oxides, **4**, 915
Ketenes, metallo-, synthesis of, **3**, 537, 545
Ketenes, methylene-
  from, α-diazo-γ-lactones, **3**, 541
  as intermediates, in the synthesis of ketenes, **3**, 527
  synthesis of, **3**, 541
Ketenes, monoalkyl-, synthesis of, **3**, 527, 536
Ketenes, phosphoryl-
  as starting materials, for ketenimines, **3**, 587
  synthesis of, **3**, 543
Ketenes, *o*-quinonoid methylene-, synthesis of, **3**, 542
Ketenes, silyl-
  from
    ethoxyalkynes, **2**, 941
    lithium ynolates, **2**, 941
  as intermediates, for the synthesis of silenes, **5**, 925
  as starting materials, for α-sily-α-stannyl esters, **4**, 661
  synthesis of, **3**, 534, 545
Ketenes, thioacyl-, synthesis of, **3**, 542
Ketenes, vinyl-
  from
    crotonic anhydrides, **3**, 535
    cyclohexa-2,4-dienones, **3**, 540
  as intermediates, in the synthesis of ketenes, **3**, 527
  synthesis of, **3**, 527, 543
Ketenes, vinylogous silyl-, reactions, with chromium carbene complexes, **3**, 545
Ketene–sulfur dioxide adducts, reactions, with dihydrothiazoles, **6**, 113
Ketenimides, reactions, with lithium trimethylsilyldiazomethanide, **2**, 944
Keteniminato complexes
  carbon–nitrogen sigma bonding in, **3**, 568
  synthesis of, **3**, 570
Ketenimine, *N*-alkoxy-, synthesis of, **3**, 559
Ketenimine, 1,2-azaborolidine-, synthesis of, **3**, 595
Ketenimine, *C*-benzoyl-*C*,*N*-diphenyl-, synthesis of, **3**, 578
Ketenimine, *N*,*C*-bis(*t*-butyl)-, from, imino ether, **3**, 562
Ketenimine, *N*-(*t*-butyl)benzoyl-, synthesis of, **3**, 577
Ketenimine, cyclic, synthesis of, **3**, 564, 572
Ketenimine, *N*-cyclohexyl-*C*,*C*-diphenyl-, synthesis of, **3**, 596
Ketenimine, dichloro-, synthesis of, **3**, 596
Ketenimine, *C*,*N*-diphenyl-, deprotonation of, **3**, 582
Ketenimine, *N*-isopropyldiphenyl-, synthesis of, **3**, 561
Ketenimine, ketene-, synthesis of, **3**, 582
Ketenimine, *N*-metallo-, synthesis of, **3**, 570
Ketenimine, *N*-methyl-, synthesis of, **3**, 575
Ketenimine, *N*-methyldiphenyl-, synthesis of, **3**, 561
Ketenimine, *N*-(methylthiomethyl)-, synthesis of, **3**, 574
Ketenimine, *N*-(*p*-nitrophenyl)diphenyl-, synthesis of, **3**, 561
Ketenimine, *N*-phenyl-, synthesis of, **3**, 562
Ketenimine, *N*-phenyl-*C*,*C*-bis(ethoxycarbonyl)-, synthesis of, **3**, 562
Ketenimine, *N*-phenyl-*C*-methyl-, synthesis of, **3**, 562
Ketenimine, *C*-phenylthio-*C*-vinyl-, synthesis of, **3**, 597
Ketenimine, *C*-phthalimide, synthesis of, **3**, 559
Ketenimine, *N*-(*p*-tolyl)diphenyl-, synthesis of, **3**, 561
Ketenimine, triphenyl-, reactions, with pentacarbonylbenzylidene tungsten, **4**, 1018
Ketenimine, tris(trimethylsilyl)-
  as starting material, for ketenimines, **3**, 581
  synthesis of, **3**, 565
Ketenimine, *C*,*C*,*N*-tris(trimethylsilyl)-, synthesis of, **3**, 567
η^1-Ketenimine complex, of transition metals, **3**, 597
Ketenimines
  axial dissymmetry of, **3**, 556, 564
  canonical forms of, **3**, 556, 565
  characteristics of, **3**, 555
  with chiral substituents, **3**, 584
    synthesis of, **3**, 559
  diastereomers of, **3**, 564
  from
    alkanenitriles, **3**, 565
    amines, **3**, 588, 590
    azlactones, **3**, 578
    carbenes, **3**, 596
    chloroalkenes, **3**, 590
    chloroalkynes, **3**, 590
    chloroethynyl phosphonates, **3**, 591
    α-cyanoalkylsulfonium salts, **3**, 574
    α-cyano enamines, **3**, 563
    diphenylcarbodiimide, **3**, 593
    1,2,-dithiole-3-imine, **3**, 579
    haloalkenes, **3**, 588
    haloalkynes, **3**, 588
    imidoyl cyanides, **3**, 563
    N-imidoyl isothiocyanates, **3**, 593
    iminophosphoranes, **3**, 584
    iminosulfuranes, **3**, 587
    5-iminothiazolines, **3**, 579
    imino thioethers, **3**, 562
    isocyanates, **3**, 591
    isocyanides, **3**, 593
    isoxazol-5-ones, **3**, 578
    ketenes, **3**, 584
    ketenimines, **3**, 580
    oximes, **3**, 563
    oxime sulfonates, **3**, 564
    perchlorobutenyne, **3**, 590
    perfluoroisobutene, **3**, 588
    phosphonate anions, **3**, 593
    phosphoramidate anions, **3**, 587
    phosphorus ylides, **3**, 591, 592
    phosphorylketenes, **3**, 587
    potassium tricyanomethanide, **4**, 977
    thioketene, **3**, 587
    α,β-unsaturated nitriles, **3**, 570
    vinyl azides, **3**, 564
  reactions
    with alkylamines, **4**, 977
    with alkyl thiols, **5**, 730
    with amines, **5**, 753
    with aryl thiols, **5**, 730
    with hydrazine, **5**, 770
    with iron carbonyl anions, **5**, 799
    with lithium salts, **5**, 874
    with *N*-sulfinyl sulfonamides, **5**, 767
    with sulfur dioxide, **2**, 247
  as starting materials
    for amides, **2**, 1040
    for ketene aminals, **4**, 979
    for ketene hemiaminals, **4**, 894
    for ketenimines, **3**, 580
    for thioimidic esters, **5**, 729
  synthesis of, **3**, **555–610**
    via alkylation of alkanenitriles, **3**, 566
    via cleavage of aziridines, **3**, 574
    via cleavage of cyclic imino ethers, **3**, 575
    via cleavage of isoxazoles and isoxazolium salts, **3**, 576
    via cleavage of triazoles, **3**, 579
    via conjugate addition to alkylidenemalonodinitriles, **3**, 570
    via dechlorination of α-chloro imidoyl chlorides, **3**, 561
    via dehydration of secondary amides, **3**, 556, 557
    via dehydrohalogenation of imidoyl halides, **3**, 558, 559
    via [1+3] dipolar cycloaddition, **3**, 600

via elimination reactions, **3**, 556
via eliminations from thioamides, **3**, 561
via pericyclic processes, **3**, 582
via Perkow reaction of α-haloalkanenitriles, **3**, 568
via reaction of α-deprotonated alkanenitriles with electrophiles, **3**, 566
tautomerism of, **3**, 565
x-ray diffraction of, **3**, 556
Ketenimines, *C*-acyl-, synthesis of, **3**, 557, 562, 576, 584, 586, 592
Ketenimines, *N*-acyl-
reactions
with carbodiimides, **4**, 982
with *p*-chloroaniline, **4**, 980, 981
as starting materials
for *N*-acylated α-ureido enamines, **4**, 982
for α-alkoxy carbonylamino enamines, **4**, 981
for α-amido enamines, **4**, 980
synthesis of, **3**, 567, 578, 580, 582, 585, 592
Ketenimines, *N*-adamantyl-, synthesis of, **3**, 584
Ketenimines, *N*-alkoxy-, synthesis of, **3**, 559
Ketenimines, *N*-alkyl-, synthesis of, **3**, 563
Ketenimines, *C*-alkylthio-, synthesis of, **3**, 563
Ketenimines, *C*-allenyl-, synthesis of, **3**, 597
Ketenimines, *N*-allyl-, rearrangement of, **1**, 835
Ketenimines, *C*-amino-, synthesis of, **3**, 597
Ketenimines, *N*-amino-, synthesis of, **3**, 560, 586
Ketenimines, aryl-, synthesis of, **3**, 599
Ketenimines, *C*-azido-, synthesis of, **3**, 577
Ketenimines, bis(ethanesulfonyl)-, synthesis of, **3**, 561
Ketenimines, bis(phosphoryl)-
cycloaddition of, **4**, 559
synthesis of, **3**, 557
Ketenimines, bis(trifluoromethyl)-, synthesis of, **3**, 560
Ketenimines, *C*-bromo-
reactions, with methyllithium, **3**, 582
synthesis of, **3**, 557, 574
Ketenimines, *N*-bromo-, synthesis of, **3**, 567
Ketenimines, *N*-(*t*-butyl)-, synthesis of, **3**, 566
Ketenimines, *C*-carbamoyl-, synthesis of, **3**, 601
Ketenimines, cephalosporin, synthesis of, **3**, 559, 581
Ketenimines, *C*-chloro-, synthesis of, **3**, 566
Ketenimines, *C*-cyano-, synthesis of, **3**, 557, 564, 565, 592
Ketenimines, *N*-cyano-, synthesis of, **3**, 579
Ketenimines, *N*-cyanoalkyl-, synthesis of, **3**, 585
Ketenimines, cyclic, synthesis of, **3**, 564, 565, 572
Ketenimines, cyclopropropenyl-
from, hexafluoro-2-butyne, **3**, 593
synthesis of, **3**, 593
Ketenimines, *C*-dienyl-, synthesis of, **3**, 582
Ketenimines, *C*,*N*-dilithiated, synthesis of, **1**, 155
Ketenimines, dimeric, synthesis of, **3**, 573
Ketenimines, *C*,*C*-diphenyl-, synthesis of, **3**, 584
Ketenimines, ferrocenyl-, synthesis of, **3**, 567
Ketenimines, fluorenylidene-, synthesis of, **3**, 585, 592
Ketenimines, fluorinated, synthesis of, **3**, 570
Ketenimines, fluoro-, synthesis of, **3**, 590
Ketenimines, *N*-imidoyl-, synthesis of, **3**, 567, 593
Ketenimines, imino-
from, 3-aminoacryl amides, **3**, 558
synthesis of, **3**, 579
use in the synthesis of amidines, **3**, 562
Ketenimines, *C*-imino-, synthesis of, **3**, 588
Ketenimines, *N*-iodo-, synthesis of, **3**, 573
Ketenimines, *C*-lithio-, synthesis of, **3**, 580
Ketenimines, *N*-lithio-, synthesis of, **3**, 566
Ketenimines, metallated, synthesis of, **3**, 599
Ketenimines, *C*-metallated, synthesis of, **3**, 582
Ketenimines, *N*-metallated, synthesis of, **3**, 568, 570, 580, 597
Ketenimines, α-methoxy-
from, penicillin ketenimines, **3**, 581
synthesis of, **3**, 581
Ketenimines, *N*-(methylthiomethyl)-, synthesis of, **3**, 574
Ketenimines, *N*-nitroso-, synthesis of, **3**, 580
Ketenimines, *C*-oxygenated, synthesis of, **3**, 568
Ketenimines, penicillin
as starting materials, for α-methoxy ketenimines, **3**, 581
synthesis of, **3**, 559
Ketenimines, *N*-phenyl-, synthesis of, **3**, 579
Ketenimines, *C*-phosphinoyl-, from, halophosphoranes, **3**, 592
Ketenimines, *C*-phosphoryl-, synthesis of, **3**, 557, 585, 591
Ketenimines, *N*-phosphoryl-, synthesis of, **3**, 568
Ketenimines, *N*-phosphorylated, reactions, with triethyl phosphite, **4**, 1008
Ketenimines, *C*-phthalimide, synthesis of, **3**, 559
Ketenimines, *N*-(2-pyridyl)-, from, *N*-pyridyl iminophosphorane, **3**, 587
Ketenimines, silyl-, synthesis of, **3**, 565, 576
Ketenimines, *C*-silyl-, synthesis of, **3**, 562
Ketenimines, *N*-silyl-, synthesis of, **3**, 567, 580
Ketenimines, silylated, synthesis of, **3**, 568
Ketenimines, *C*-sulfenyl-, synthesis of, **3**, 557, 563, 597
Ketenimines, *C*-sulfonyl-, synthesis of, **3**, 557, 561, 566, 587
Ketenimines, *N*-sulfonyl-, synthesis of, **3**, 557, 568
Ketenimines, thioacyl-, synthesis of, **3**, 579
Ketenimines, *N*-thionoacyl-, synthesis of, **5**, 634
Ketenimines, trialkyl-, synthesis of, **3**, 557, 563
Ketenimines, triaryl-, synthesis of, **3**, 556
Ketenimines, trifluoromethyl-, synthesis of, **3**, 584
Ketenimines, *N*-trifluoromethyl-, synthesis of, **3**, 600
Ketenimines, trisilyl, synthesis of, **2**, 1058
Ketenimines, tris(trimethylsilyl)-, reactions, with aldehydes, **3**, 660
Ketenimines, vinyl-, synthesis of, **3**, 593, 599
Ketenimines, *C*-vinyl-, synthesis of, **3**, 557, 592
Ketenimines, *N*-vinyl-, synthesis of, **3**, 585, 592
Keteniminium halides, use in the synthesis of α-halo enamines, **4**, 790
Keteniminium ions, as intermediates, in the synthesis of ketene aminals, **4**, 971
Keteniminium salts
from
amides, **3**, 601
α-chloro enamines, **3**, 601
ynamines, **3**, 602
reactions, with alkynes, **3**, 216
synthesis of, **3**, 601; **4**, 790
X-ray diffraction of, **3**, 602
*η*2-Ketenyl complexes, as starting materials, for tungsten complexes, **2**, 878
Ketimines
alkylation of, **2**, 331
enantioselective hydrogenation of, **2**, 330
from, imidoyl chlorides, **3**, 409
as intermediates, in the synthesis of amines, **1**, 1041
reactions
with germylphosphines, **4**, 998
with hydrogen sulfide, **2**, 130
with organolithiums, **2**, 312
with silylphosphines, **4**, 998
with ytterbium, **2**, 312
as starting materials
for alkyl arylketiminoboranes, **2**, 811
for 1-aminoalkyl phosphinates, **4**, 463
for 1-aminoalkyl phosphonates, **4**, 463
for hydrazonamides, **5**, 849
for isothiocyanates, **5**, 1032
for α-phosphino enamines, **4**, 998
synthesis of, **3**, 404
Ketimines, *N*-alumino-, synthesis of, **3**, 458
Ketimines, aryl-, synthesis of, **3**, 412
Ketimines, *N*-aryl *β*-chloro, as starting materials, for 2-methyleneazetidines, **2**, 322
Ketimines, α-bromo-, cyclization of, **3**, 406

Ketimines, *N*-carbon-substituted
synthesis of, **3**, 407
via condensation reactions, **3**, 407
via oxidation reactions, **3**, 407
via rearrangement reactions, **3**, 407
via reduction reactions, **3**, 407
Ketimines, chiral pinanone, alkylation of, **3**, 415
Ketimines, α-chloro-, synthesis of, **3**, 416
Ketimines, α,α-dichloro-, from, α,α-dichloro-β-iminocarbonyls, **3**, 409
Ketimines, α-halo-, alkylation of, **3**, 409
Ketimines, *N*-lithio-
reactions
with beryllium chloride, **3**, 459
with iron complexes, **3**, 460
with titanocene dichloride, **3**, 460
with zirconocene dichloride, **3**, 460
synthesis of, **3**, 459
Ketimines, *N*-magnesio-, synthesis of, **3**, 459
Ketimines, molybdenum *N*-metallo-, synthesis of, **3**, 460
Ketimines, *N*-phosphinyl-, synthesis of, **3**, 455
Ketimines, phosphorus-substituted, synthesis of, **3**, 455
Ketimines, platinum-substituted
from, *N*-stannylketimines, **3**, 461
synthesis of, **3**, 461
Ketimines, *N*-silyl-, synthesis of, **3**, 456
Ketimines, *N*-stannyl-, as starting materials, for platinum-substituted ketimines, **3**, 461
Ketimines, tertiary alkyl-, synthesis of, **3**, 408
Ketimines, *N*-thionoacyl-, synthesis of, **5**, 634
Ketimines, *N*-thiophosphinyl-, synthesis of, **3**, 455
Ketimines, titanium *N*-metallo-, synthesis of, **3**, 460
Ketimines, trialkyl-, from, α-cyano enamines, **3**, 409
Ketimines, trimethylsilyl, reactions, with titanocene dichloride, **3**, 460
Ketimines, tungsten *N*-metallo-, synthesis of, **3**, 460
Ketimines, α,β-unsaturated, synthesis of, **3**, 413
Ketimines, *N*-unsubstituted, deprotonation of, **3**, 459
N-H Ketimines
from
N-alkyloxaziridines, **3**, 405
nitriles, **3**, 405
synthesis of, **3**, 405
N-H-Ketimines, diaryl-, synthesis of, **3**, 405
α-Ketimines, synthesis of, **3**, 417
α-Ketimines, *N*-phenyl-, synthesis of, **3**, 417
β-Ketimines, synthesis of, **3**, 417
α-Keto acid dithioacetals, synthesis of, **4**, 320
Keto acids
as starting materials, for α-silyloxy propenals, **3**, 33
synthesis of, **6**, 25
α-Keto acids
arylhydrazones of, **5**, 857
cleavage of, **5**, 114
from, 2,2-bis(ethylthio)carboxylic acids, **1**, 265
reactions, with bromine, **4**, 28
synthesis of, **5**, 62, 68, 79
use in the synthesis of 1,3-azarsa compounds, **4**, 495
β-Keto acids
decarboxylation of, **1**, 772
decarboxylative allylation of, **3**, 139
from, methylenes, **5**, 70
as radical precursors, **1**, 336
as starting materials, for enones, **1**, 561
synthesis of, **1**, 832
β-Keto acids, aliphatic, synthesis of, **5**, 68
Ketocarbenoids, from, diazo compounds, **1**, 352
Ketocarbenoids, metal-stabilized, synthesis of, **1**, 352
Keto diacids, reactions, with acetyl chloride, **4**, 204
Keto–enol equilibrium, constants, **1**, 869
Keto esters, 2-(but-4-enyl), diastereoselective reductive cyclization reactions of, **2**, 81
3-Keto esters, 4-(3-hydroxyketo)-, stereoselective reduction of, **2**, 62
Ketone
reactions
with 2,2-diethoxyvinylidenetriphenylphosphorane, **3**, 245
with hydrogen selenide, **3**, 391
with hydrogen telluride, **3**, 400
Ketone, 2-chloroethyl)phenyl, reduction of, **2**, 65
Ketone, cyclopropyl ethyl, synthesis of, **1**, 1116
Ketone, diethyl, as starting material, for ketenes, **3**, 535
Ketone, diethylaminomethyl phenyl, oxidation of, **5**, 114
Ketone, isoartemisia, synthesis of, **4**, 256
Ketone, mesityl 2-pyridyl, reactions, with amines, **2**, 302
Ketone, methyl *t*-butyl, from, tetramethylethylene glycol, **1**, 384
Ketone, methyl β-naphthyl, haloform oxidation of, **5**, 114
Ketone, methyl trimethylsilyl, reactions, with hydrogen chloride, **4**, 944
Ketone, methylvinyl (MVK)
reactions
with difluoroamine, **4**, 439
with nitromethane, **3**, 192
with organoboranes, **1**, 273
with trialkynylborons, **1**, 1050
with triethyl phosphite, **3**, 200
Ketone, phenyl triethylgermyl, reactions, with triethylsilyllithium, **6**, 131
Ketone, phenyl trimethylsilyl, reactions, with dimethyl(trimethylgermyl)phosphine, **6**, 131
Ketone, α-stannyl, reactions, with dichlorobis(tri-o-tolylphosphine)palladium(II), **3**, 144
Ketone, triethylgermyl trimethylsilyl, synthesis of, **6**, 520
Ketone, trimethylsilylvinyl, reactions, with thiocarbonyls, **4**, 232
p-Ketone, synthesis of, **3**, 298
Ketone complexes, 3-carbon-bridged, synthesis of, **4**, 676
Ketone enolates, β-lithio, synthesis of, **2**, 555
Ketone hydrates, synthesis of, **4**, 162
Ketoneimines, reactions, with Vilsmeier salts, **6**, 644
Ketones
acetalization of, **4**, 178
acidity of, **1**, 877
as activating groups for alkenes, **1**, 300
alkenation of, **2**, 680; **3**, 642
alkylation of, **3**, 123
asymmetric hydrogenations of, **2**, 52
bearing an α,β-alkenic bond, **3**, **205–276**
bearing an antimony function, synthesis of, **3**, 195
bearing an arsenic function, synthesis of, **3**, 195
bearing an α,β-aryl or -hetaryl substituent, synthesis of, **3**, **277–312**
bearing an oxygen function, synthesis of, **3**, 156
bearing an α,β-triple bond, synthesis of, **3**, 273
bearing a bismuth function, synthesis of, **3**, 195
bearing a nitrogen function, synthesis of, **3**, 178
bearing a phosphorus function, synthesis of, **3**, 195
bearing a selenium function, synthesis of, **3**, 173
bearing a sulfur function, synthesis of, **3**, 164
bearing a tellurium function, synthesis of, **3**, 173
boron-functionalized, synthesis of, **3**, 203
bromination of, **1**, 853; **4**, 20
α-bromination of, **2**, 28; **3**, 154
bromoacylation of, **4**, 56
α-chlorination of, **2**, 21; **3**, 152
chloroacylation of, **4**, 53
Claisen condensation of, **3**, 349
cleavage of, **1**, 884
dehydrogenation of, **1**, 554
deoxygenation of, **1**, 18, 24
deuteration of, **1**, 853
diazotization of, **3**, 193
α,α-dibromination of, **4**, 21
dithioacetalization of, **4**, 247
elimination of, **1**, 606

enantioselective alkylation of, **2**, 71
enolization of, **1**, 1038; **2**, 646; **3**, 173
  direct, **2**, 699
γ-enolization of, **1**, 868
α-fluorination of, **2**, 16; **3**, 149
from
  acetals, **3**, 123
  acid anhydrides, **3**, 130
  acyl anions, **3**, 136
  alcohols, **3**, 114
  aldehydes, **3**, 123, 127
  alkanes, **3**, 112
  alkenes, **1**, 296; **3**, 113
  alkenyllithiums, **2**, 911
  alkynes, **3**, 114
  allylic alcohols, **3**, 122
  amides, **3**, 131
  amines, **3**, 132
  benzylic methylene groups, **3**, 293
  carbonates, **3**, 122
  carboxylic acids, **3**, 129
  diols, **3**, 121
  enol esters, **3**, 123
  enol ethers, **3**, 123
  enones, **1**, 86; **3**, 274
  epoxides, **3**, 122
  esters, **3**, 121, 130
  ethers, **3**, 121
  α-functionalized ketones, **3**, 128
  halides, **3**, 114
  halohydrins, **1**, 390
  hydrazones, **3**, 133
  3-hydroxy-dienes, **1**, 414
  2-hydroxy thiazoles, **3**, 311
  imidazoles, **3**, 311
  β-keto esters, **3**, 131
  ketones, **3**, 123
  N-methyl-2-phenyl-4-benzaminoimidazole, **3**, 311
  nitriles, **3**, 134
  nitroalkanes, **3**, 134
  organoboranes, **3**, 134
  organocuprates, **3**, 130
  organomanganese halides, **3**, 130
  organosilanes, **3**, 134
  oxazoles, **3**, 311
  oximes, **3**, 133
  pyridyl sulfides, **1**, 333
  selenoacetals, **3**, 132
  thiazoles, **3**, 311
  thioacetals, **3**, 131
  triethylgermyllithium, **6**, 131
  vicinal diols, **1**, 384
  vinyl sulfides, **3**, 131
β-functionalization of, **3**, 137
C-germylation of, **4**, 360
haloalkylation of, **4**, 47
homologation of, **3**, 125; **5**, 873
hydrogenation of, **2**, 52
imination of, **3**, 404
α-iodination of, **3**, 155
long chain synthesis of, **1**, 677
with a metal function, synthesis of, **3**, 204
with a metalloid function, synthesis of, **3**, 202
methylation of, **1**, 765
α-methylation of, **1**, 133
methylenation of, **1**, 591, 643, 664, 767
monoalkylation of, **1**, 194
monobromination of, **3**, 208
niobium-promoted reactions, with alkynes, **2**, 988
nitration of, **3**, 190
non-enolizable, cleavage of, **1**, 884
for the opening of epoxides, **1**, 212
oxidation of, **1**, 554, 1077; **3**, 156, 162, 178; **5**, 48
oximes of, synthesis of, **3**, 425
perphenylation of, **3**, 145
pyrolysis of, **1**, 772
racemization of, **1**, 853
reactions
  with acyl anions, **3**, 157
  with acyllithiums, **3**, 158
  with alcohols, **2**, 678; **4**, 177
  with alkenylaluminums, **1**, 534
  with alkenylceriums, **1**, 1098
  with *N*-alkylhydrazines, **3**, 446
  with alkylphosphine oxides, **1**, 641
  with 1-alkynes, **1**, 1038
  with alkynezirconocenes, **2**, 987
  with *B*-1-alkynyl-9-borabicyclo[3.3.1]nonanes, **1**, 1037
  with alkynylcerium dichlorides, **1**, 1038
  with alkynyldiethylalanes, **1**, 1037
  with alkynyl metals, **1**, 1037
  with alkynyltitanium triisopropoxides, **1**, 1038
  with 9-allenyl-9-borabicyclo[3.3.1]nonyl, **1**, 1055
  with allenyllithium, **1**, 971
  with allenylmagnesiums, **1**, 1055
  with allenyl metals, **1**, 1054
  with allenylsilanes, **1**, 1056
  with allylboranes, **1**, 281
  with allylboronates, **1**, 281
  with Alpine-Borane, **2**, 57
  with aluminums, **1**, 1055
  with amines, **4**, 294, 405
  with (ω-aminoalkyl)-phenylphosphines, **4**, 453
  with aminoiminoboranes, **5**, 381
  with 1-aminopyrazoles, **2**, 316
  with ammonia, **3**, 405
  with ammonium acetate, **5**, 600
  with arenesulfonylmethyl isocyanides, **4**, 908
  with arylhydrazines, **3**, 446
  with aryltellurium trichlorides, **3**, 177
  with benzeneselenenyl chloride, **4**, 90
  with 1,3-benzodithiolylium, **3**, 251
  with benzynezirconocenes, **2**, 987
  with bishydroxylamines, **4**, 436
  with bis thiols, **4**, 347
  with bis(trimethylsilyl)methyllithiums, **4**, 936
  with 1,3-bis(trimethylsilyl)propenyl anions, **1**, 651
  with 2-bromozincmethyl allylic ethers, **1**, 194
  with carbon-centered radicals, **1**, 931
  with chloro(diethyl)phosphine, **4**, 120
  with α-chloro-α-isocyanomethyl phosphonates, **3**, 717
  with chlorophosphines, **4**, 119, 121, 345
  with chloro sulfones, **3**, 29
  with chlorosulfonyl isocyanate, **3**, 633
  with 1,1-chloro sulfoxides, **3**, 36
  with cyanide, **5**, 85
  with cyano compounds, **3**, 268
  with cyanodibutyl telluronium ylides, **3**, 647
  with cyanothioacetamide, **5**, 615
  with cyclopentadienyltantalum complexes, **4**, 355
  with dialkylformamide acetals, **3**, 266
  with *N*,*N*-dialkylhydrazines, **3**, 446
  with dialkyl phosphites, **4**, 338
  with dialkylzincs, **2**, 68
  with diamines, **4**, 410
  with diazoalkanes, **1**, 389
  with dibromomethyllithium, **4**, 746
  with dibromotriphenylphosphines, **1**, 983
  with di-*t*-butyl peroxide, **1**, 925
  with 1,3-dicarbonyls, **2**, 662
  with diethoxycarbenium tetrafluoroborate, **4**, 193
  with diethyl cyanomethyl phosphonate, **1**, 749
  with diethyl ethoxycarbonylmethyl phosphonate, **1**, 744
  with diethyl phosphorocyanide, **3**, 647
  with diethyl succinate, **5**, 98
  with dimethylphenylsilyllithium, **4**, 355

with 1,3-dipoles, **4**, 183
with α,α-dithiophosphonate esters, **4**, 847
with epoxides, **4**, 181
with ethynyl ylides, **1**, 739
with fluoromethylsulfoximine, **4**, 86
with formyl anions, **3**, 35
with Grignard reagents, **2**, 67
with homoenolate anions, **3**, 161
with hydrazine, **1**, 41
with hydrazines, **3**, 444
with hydrazoic acid, **5**, 283
with hydrogen peroxide, **2**, 96
with hydrogen sulfide, **2**, 223; **3**, 343; **4**, 244
with hydroxylamine, **3**, 425
with hydroxylamines, **4**, 435
with (3-hydroxypropyl)phenylphosphines, **4**, 337
with iodine compounds, **2**, 44
with isocyanomethyl phosphonate diethyl ester, **3**, 717
with lithiated silylmethyl methyl ethers, **2**, 680
with lithiated trithioorthoformates, **6**, 88
with lithiodichloromethane, **4**, 741
with 1-lithio-1-phenylpropa-1,2-diene, **3**, 250
with α-lithio phosphonamides, **1**, 644
with 2-lithio thiazole, **3**, 29
with lithium aluminum amides, **3**, 405
with lithium hydroxide, **6**, 556
with magnesium halide carbon dioxide complexes, **5**, 70
with mercaptoalcohols, **4**, 229
with α-metallated ethyl isocyanoacetate, **3**, 717
with α-metallated isocyanides, **3**, 713
with 2-methyl-1-aza-1-cycloalkenes, **3**, 413
with methylene(triphenyl)phosphorane, **1**, 759
with nitro compounds, **1**, 252
with organoboranes, **2**, 531
with organolithiums, **1**, 693; **2**, 68
with ortho esters, **4**, 190
with orthoformates, **4**, 190
with pentaorganostiboranes, **1**, 272
with phenylhydrazine, **3**, 452
with phenylselenoalkyllithium, **3**, 125
with phenylselenyl chloride, **3**, 207
with *N*-(phenylselenyl)phthalimide, **3**, 173
with phosphacumulenes, **1**, 988
with phosphines, **2**, 446
with phosphonates, **1**, 744; **3**, 60
with phosphoramidites, **4**, 475
with phosphorous acid dialkyl esters, **2**, 457
with phosphorous oxybromide, **3**, 69
with phosphorus(III) chlorides, **2**, 454
with phosphorus(III) esters, **2**, 454
with potassium *t*-butoxide, **1**, 43
with potassium hydroxide, **5**, 48
with propargyl metals, **1**, 1054
with Selectrides, **2**, 60
with semicarbazide, **3**, 449
with silicon compounds, **1**, 1038
with silirenes, **4**, 1052
with silyl enol ethers, **3**, 220
with silyl ethers, **4**, 181
with β-silylvinylceriums, **1**, 1098
with sodium alkynide, **3**, 230
with sodium bisulfite, **4**, 237
with 1-substituted allenylboranes, **1**, 1056
with α-sulfinyl carbanions, **4**, 77
with sulfonyl carbanions, **1**, 614
with sulfonyl isocyanates, **5**, 344
with telluronium salts, **1**, 623
with telluronium ylides, **1**, 623
with tetrabromomethane, **4**, 745
with thio esters, **1**, 682
with α-thiol carboxylic acids, **4**, 230
with thiols, **2**, 709; **4**, 247
with thionyl chloride, **5**, 485
with thiosemicarbazide, **3**, 449
with thiosulfonate esters, **3**, 164
with titanium complexes, **1**, 991
with tosylhydrazine, **3**, 446
with trialkylalanes, **1**, 1037
with trialkylaluminum, **2**, 699
with trialkyl orthoacetates, **6**, 80
with *N*-trialkylsilyl iminophosphoranes, **3**, 457
with tri(alkylthio)phosphines, **2**, 464
with triborylmethide anions, **6**, 182
with tributylstannylmagnesium chloride, **4**, 386
with tributylstibonium ylides, **1**, 647, 743
with trichloromethyl anions, **6**, 25
with trimethylsilyl chlorosulfonate, **2**, 209
with trimethylsilyl sulfides, **4**, 216
with α-trimethylsilylvinylceriums, **1**, 1098
with trimethylstannylcopper dimethyl sulfide, **2**, 600
with triphenylphosphine, **4**, 745
with tungsten complexes, **3**, 517
with Vilsmeier reagent, **4**, 196
with vinylidene triphenylphosphoranes, **1**, 988
rearrangement of, **1**, 832
reduction of, **1**, 11, 18; **2**, 60; **3**, 31, 159; **4**, 699
reductive amination of, **2**, 331; **3**, 188
reductive halogenation of, **2**, 14
regioselective propargylation of, **1**, 1055
ring-expansion of, **3**, 125
ring expansion of, **5**, 331
samarium-mediated methylenation of, **1**, 680
selenation of, **3**, 392, 394
selenenylation of, **1**, 606
α-selenylation of, **3**, 173
silyl anion addition to, **4**, 354
α-silylation of, **3**, 202
C-stannylation of, **4**, 386
as starting materials
for acetals, **4**, 177
for *O,As*-acetals, **4**, 344
for *O,N*-acetals, **4**, 311
for *Se,N*-acetals, **4**, 333
for *S,N*-acetals, **4**, 315, 328
for *S,P*-acetals, **4**, 345
for acyl alkyl acetals, **4**, 212
for *N*-acylsulfonamides, **5**, 345
for aldehydes, **3**, 10
for alkanethiols, **2**, 130
for alkenes, **1**, 593, 630
for α,β-alkenic nitriles, **3**, 644
for alkenyl ethers, **2**, 680
for alkynyl-substituted dialkyl ketones, **3**, 146
for amides, **5**, 276
for aminals, **4**, 405
for amines, **2**, 310
for aromatic acids, **5**, 114
for bromochloroalkanes, **4**, 38
for carboxylic esters, **5**, 141
for α-chloro ethers, **4**, 47
for chloroiodoalkanes, **4**, 40
for cyanohydrins, **3**, 615
for dialkyl carbonates, **6**, 466
for dialkyl ketones, **3**, 138, 144
for *gem*-diazidoalkanes, **4**, 432
for dibromoalkanes, **4**, 25
for dichloroalkanes, **4**, 18
for difluoroalkanes, **4**, 8
for diselenoacetals, **4**, 287
for dithioacetals, **4**, 245
for dithioketals, **2**, 541
for dithiolanes, **4**, 246
for enones, **1**, 556
for α,β-enones, **3**, 173
for geminal diacetates, **4**, 204
for hemiacetals, **4**, 164

for hemiaminals, **4**, 294
for hemithioaminals, **4**, 324
for isocyanates, **5**, 986
for isothiocyanates, **5**, 1042
for ketene monothioacetals, **4**, 834
for ketenes, **3**, 535
for ketones, **3**, 123
for nitriles, **3**, 621
for nitroalkenes, **2**, 785
for phosphorus compounds, **4**, 336
for selones, **3**, 391
for α-silyl aldehydes, **3**, 51
for *N*-substituted imines, **3**, 405
for sulfides, **2**, 139
for symmetrical acetals, **4**, 190
for tellones, **3**, 400
for thioacyl chlorides, **5**, 484
for thioketones, **3**, 342
for α,β-unsaturated aldehydes, **3**, 54
for α,β-unsaturated ketones, **3**, 266
for vinyl bromides, **2**, 624
for vinyl chlorides, **2**, 617
for vinyl fluorides, **2**, 610, 611
for vinyl halides, **2**, 607
for vinyl iodides, **2**, 630
in Strecker reaction, **3**, 616
sulfenylation of, **3**, 164
sulfinylation of, **3**, 170
sulfinylation-desulfinylation of, **1**, 598
synthesis of, **1**, 126, 1119; **4**, 936
factors affecting, **1**, 873
via alcohol oxidation with chromium compounds, **3**, 115
via alcohol oxidation with manganese compounds, **3**, 116
via alcohol oxidation with miscellaneous metal oxidants, **3**, 117
via alcohol oxidation with ruthenium compounds, **3**, 116
via electrochemical oxidation of alcohols, **3**, 120
via Oppenauer oxidation of alcohols, **3**, 119
via oxidation of alcohols with dimethyl sulfoxides, **3**, 118
via oxidation of alcohols with halogen-based oxidants, **3**, 118
via oxidation with ruthenium tetroxide, **3**, 117
via oxidation with tetrapropylammonium perruthenate, **3**, 117
via photochemical oxidation of alcohols, **3**, 120
synthesis of functions, **3**, **313–327**
tantalum-promoted reactions, with alkynes, **2**, 988
thiocyanation of, **5**, 1110
transposition of, **3**, 126
use in Beckmann rearrangements, **5**, 270
use in the synthesis of 1,3-azarsa compounds, **4**, 495
use in the synthesis of 1,3-benzazarsolines, **4**, 496
Ketones, acetamido, synthesis of, **3**, 183
Ketones, α-acetoxy
from
alkynylphenyl iodonium salts, **3**, 162
epoxy sulfoxides, **3**, 162
propargylic acetates, **3**, 162
reduction of, **1**, 16
Ketones, α-acetoxy-α-methoxy, synthesis of, **4**, 161
Ketones, acyclic
ring-contractions of, **5**, 51
as starting materials, for ketenes, **3**, 535
Ketones, α-acyloxy, synthesis of, **3**, 162
Ketones, δ-acyloxy, synthesis of, **3**, 163
Ketones, aldehydro, reduction of, **2**, 55
Ketones, aliphatic
reduction of, **1**, 19
synthesis of, **3**, 350
Ketones, alkenic, synthesis of, **1**, 831
Ketones, α,β-alkenic, synthesis of, **3**, 247
Ketones, alkenyl, reduction of, **1**, 506
Ketones, α-alkenyl, synthesis of, **1**, 481, 488
Ketones, α-alkenyl α,β-unsaturated, synthesis of, **3**, 233
Ketones, alkoxy, reactions, with zinc borohydride, **2**, 61
Ketones, 2-alkoxy, synthesis of, **2**, 89
Ketones, 3-alkoxy, stereoselectivity of, **2**, 68
Ketones, α-alkoxy
Baeyer–Villiger oxidation of, **4**, 212
as byproducts, in Favorskii rearrangements, **1**, 393
from, nitroalkenes, **3**, 162
reduction of, **1**, 15
as starting materials, for α,β-unsaturated ketones, **3**, 241
synthesis of, **3**, 162
Ketones, γ-alkoxy, synthesis of, **3**, 162
Ketones, alkoxyaryl, synthesis of, **3**, 298
Ketones, α-alkoxy-α,β-unsaturated, synthesis of, **3**, 241
Ketones, β-alkoxy-α,β-unsaturated, reactions, with amines, **3**, 264
Ketones, α-alkoxyvinyl, synthesis of, **3**, 243
Ketones, α-alkyl-α-amino, synthesis of, **3**, 179
Ketones, alkyl aryl, synthesis of, **3**, 278, 285, 290
Ketones, β-alkylated vinyl, synthesis of, **1**, 469
Ketones, alkyl benzyl, synthesis of, **3**, 146
Ketones, β-(*N*-alkyl-*N*-nitrosoamino), cleavage of, **3**, 465
Ketones, alkyl perfluoroalkyl, synthesis of, **3**, 150
Ketones, α-alkyl selenyl, synthesis of, **3**, 37
Ketones, alkynic
from, alkynic esters, **3**, 275
isomerization of, **3**, 229
reactions, with propanedithiol, **4**, 258
synthesis of, **3**, 274
Ketones, α-alkynic
pyrolysis of, **3**, 223
reactions
with organocoppers, **3**, 222
with trialkylboranes, **3**, 223
reduction of, **3**, 223
Ketones, α,β-alkynic
reactions
with Alpine-Borane, **2**, 57
with organoboranes, **1**, 274
synthesis of, **1**, 1137
Ketones, β-alkynic, rearrangement of, **1**, 957
Ketones, alkynyl
hydration of, **3**, 244
reactions
with amines, **3**, 267
with ketone enolates, **1**, 538
synthesis of, **1**, 1135
via fragmentation reactions, **3**, 147
Ketones, α,β-alkynyl, reduction of, **2**, 57
Ketones, allenic
hydrochlorination of, **3**, 238
synthesis of, **1**, 971
Ketones, α-allenic
from, homopropargylic alcohols, **1**, 957
synthesis of, **3**, 230
via rearrangements, **1**, 957
Ketones, β-allenic, synthesis of, **1**, 979
Ketones, allenyl, synthesis of, **1**, 980
Ketones, γ-allenyl, synthesis of, **1**, 971
Ketones, allenyl diazomethyl, synthesis of, **1**, 981
Ketones, allylic, hydroboration of, **2**, 42
Ketones, allyl pentenyl
reductive tandem cyclization of, **2**, 81
as starting materials, for bicyclo[3.3.0]octanols, **2**, 81
Ketones, allylvinyl
from, dienynes, **1**, 421
Nazarov cyclization of, **1**, 422
Ketones, α-amido, synthesis of, **3**, 182
Ketones, amino

diazotization of, **3**, 193
α,α-disubstituted, synthesis of, **3**, 186
reactions
with halides, **3**, 179
with thiophosgene, **3**, 194
synthesis of, **3**, 43, 178
via Gabriel synthesis, **3**, 187
Ketones, 3-amino, enantioselective reduction of, **2**, 64
Ketones, α-amino
α-alkylation of, **3**, 179
asymmetric hydrogenation of, **2**, 305
diazotization of, **3**, 468
from
acyl anions, **3**, 182
acyl cyanides, **3**, 183
azido ketones, **3**, 184
α-azido ketones, **3**, 179
as starting materials
for ω-aminonitriles, **3**, 638
for nitriles, **3**, 622
synthesis of, **2**, 313
via Neber rearrangement, **3**, 435
Ketones, β-amino
deamination of, **1**, 635
from
aminals, **1**, 255
1,3-diimines, **3**, 188
enones, **3**, 187
synthesis of, **1**, 255
Ketones, γ-amino, synthesis of, **2**, 315
Ketones, α-amino-α-allyl, synthesis of, **3**, 180
Ketones, aminomethylene, reactions, with halosulfenes, **5**, 501
Ketones, α-amino-α,β-unsaturated, synthesis of, **3**, 260
Ketones, aminovinyl, synthesis of, **3**, 259
Ketones, anthryl, synthesis of, **3**, 301
Ketones, aralkyl, reactions, with *gem*-diiminoalkanes, **4**, 445
Ketones, aromatic
from
acyl halides, **3**, 278
thioindoxyls, **3**, 247
as intermediates, **3**, 278
reduction of, **1**, 19
as starting materials, for arylalkanes, **1**, 19
synthesis of, **3**, 278
via Friedel–Crafts acylation, **3**, 278
via Houben–Hoesch reaction, **3**, 282
via oxidation, **3**, 293
Vilsmeier–Haack reaction of, **3**, 69
Ketones, aryl
from, enolsilanes, **3**, 144
homologation of, **3**, 146
reactions
with 3,3-dialkyl-substituted allenyllithiums, **1**, 1056
with dimethyl(diazomethyl) phosphonate, **1**, 1078
with methyl methylthiomethyl sulfoxides, **4**, 857
with trimethylsilyldiazomethane, **1**, 1078
sulfuration of, **3**, 362
synthesis of, **1**, 489; **3**, 288
via carbonylative cross-coupling, **3**, 289
via Friedel–Crafts acylation, **1**, 462
Ketones, α-aryl, synthesis of, **1**, 464, 474, 479
Ketones, aryl alkyl
Meerwein–Ponndorf–Verley reduction of, **2**, 56
reactions
with amines, **5**, 599
with ammonium sulfide, **5**, 599
reduction of, **1**, 19; **2**, 64
as starting materials, for hydrazones, **3**, 444
Ketones, α-arylated, synthesis of, **1**, 492
Ketones, arylcyclopropyl, synthesis of, **1**, 1115
Ketones, aryl diazo, reactions, with thiophosgene, **5**, 482
Ketones, α-(aryldichlorotelluro), synthesis of, **3**, 177
Ketones, aryl α-iodo
reactions
with benzeneselenolate ions, **1**, 10
with benzenethiolate ions, **1**, 10
with iodide, **1**, 10
Ketones, aryl methyl
nitrosation of, **5**, 667
reactions, with paraformaldehyde, **1**, 256
Reformatsky reaction of, **2**, 69
synthesis of, **3**, 350
Ketones, aryloxy, reactions, with zinc, **4**, 167
Ketones, aryl pyridyl, synthesis of, **3**, 310
Ketones, α-aryl selenyl, synthesis of, **3**, 37
Ketones, aryl-substituted, from, carboxylic acids, **3**, 146
Ketones, azido, as starting materials, for α-amino ketones, **3**, 184
Ketones, α-azido
decomposition of, **3**, 194
from, α-diazo ketones, **3**, 194
as starting materials, for α-amino ketones, **3**, 179
Ketones, β-azido, synthesis of, **3**, 194
Ketones, α-azidovinyl, synthesis of, **3**, 258, 259
Ketones, benzofuranyl, synthesis of, **3**, 304
Ketones, benzothiophenyl, synthesis of, **3**, 306
Ketones, benzyl
reactions, with mercury(II) trifluoroacetate, **1**, 1075
synthesis of, **1**, 272
Ketones, α-benzyl, from, enolsilanes, **3**, 145
Ketones, benzylic
reduction of, **1**, 11
synthesis of, **3**, 146
Ketones, biaryl
selectivity of, **1**, 879
synthesis of, **3**, 290
Ketones, bicyclic, synthesis of, **2**, 568
Ketones, α,α'-bis(diazo), synthesis of, **3**, 469
Ketones, β-bisselenyl, synthesis of, **3**, 176
Ketones, bis(silyl), synthesis of, **6**, 519
Ketones, α,α-bissulfenyl, synthesis of, **3**, 164
Ketones, bis(trimethylsilyl), synthesis of, **6**, 325, 520
Ketones, bis(triphenylsilyl), synthesis of, **6**, 519
Ketones, boronate, synthesis of, **3**, 204
Ketones, bromo
reactions, with sodium azide, **3**, 259
as starting materials, for silphiperfil-6-ene, **1**, 1130
Ketones, 1-bromo, *C*-phosphorylation of, **2**, 456
Ketones, α-bromo
as electrophiles, for dithiocarboxylic enethiolates, **4**, 842
Favorskii rearrangements of, **1**, 392
from
acids, **3**, 154
alkenes, **3**, 153
enamines, **3**, 154
enols, **3**, 154
epoxides, **3**, 153
esters, **3**, 154
ketones, **3**, 154
halide exchange of, for the synthesis of α-fluoro ketones, **3**, 149
reactions
with arylboranes, **1**, 464
with dialkylphosphines, **3**, 195
with hexamethylene tetramine, **3**, 179
with magnesium metal, **2**, 697
with oximes, **3**, 189
with sodium hydrogen sulfide, **3**, 164
reduction of, **1**, 9, 10
as starting materials
for α-acylimines, **3**, 417
for α-telluro ketones, **3**, 176
synthesis of, **2**, 28
Ketones, bromoaliphatic, synthesis of, **3**, 153
Ketones, α-bromoaliphatic, synthesis of, **3**, 153

Ketones, bromochloromethyl, synthesis of, **3**, 152
Ketones, bromomethanesulfonyl, synthesis of, **4**, 84
Ketones, bromomethyl, synthesis of, **3**, 153, 154
Ketones, α-bromovinyl, substitution of, **3**, 259
Ketones, α-t-butyldimethylsilyl, synthesis of, **3**, 202
Ketones, carbocyclic, as starting materials, for ketenes, **3**, 537
Ketones, carbohydrate, synthesis of, **3**, 122
Ketones, chloro, synthesis of, **1**, 285; **3**, 235
Ketones, α-chloro
- Favorskii rearrangements of, **1**, 392
- from
  - acids, **3**, 152
  - alcohols, **3**, 151
  - alkenes, **3**, 151
  - alkynes, **3**, 151
  - enamines, **3**, 152
  - enols, **3**, 152
  - epoxides, **3**, 151
  - esters, **3**, 152
  - ketones, **3**, 152
- halide exchange of, for the synthesis of α-fluoro ketones, **3**, 149
- reactions
  - with potassium salts, **6**, 627
  - with trimethylstannylmethyllithium, **1**, 664
- as starting materials
  - for esters, **1**, 392
  - for α-oxothionoamides, **5**, 601
  - for oxyimidic chlorides, **5**, 823

Ketones, β-chloro, use in the synthesis of β-amino ketones, **3**, 187
Ketones, chloroaliphatic, synthesis of, **3**, 151
Ketones, α-chloroaliphatic, synthesis of, **3**, 151
Ketones, β-chloroaliphatic, synthesis of, **3**, 153
Ketones, chlorodifluoromethyl, as starting materials, for enol phosphates, **4**, 735
Ketones, chlorofluoromethyl, synthesis of, **4**, 32
Ketones, α-chloro-α-methoxy, synthesis of, **4**, 161
Ketones, chloromethyl aryl, as starting materials, for oxyimidic chlorides, **5**, 823
Ketones, α-chloro-α-phenylselenyl, synthesis of, **4**, 88
Ketones, α-chloro-β-silyl
- as starting materials
  - for α-methylenecyclobutanones, **1**, 653
  - for α-methylenecyclopentanones, **1**, 653

Ketones, (*E*)-α-chloro-α,β-unsaturated, synthesis of, **3**, 236
Ketones, α-chloro-α,β-unsaturated, synthesis of, **3**, 237
Ketones, cyano, electrolytic reduction of, **2**, 81
Ketones, α-cyano, as starting materials, for alkynic nitriles, **3**, 676
Ketones, cyclic
- photochemical ring-opening of, **1**, 841
- reactions
  - with 2,6-dimethylaniline, **5**, 292
  - with lithium alkoxide, **5**, 557
  - with methylvinyl ketones, **3**, 220
  - with piperidine, **2**, 310
- reduction of, **2**, 57
- ring-contraction of, **1**, 391
- ring-expansion of, **3**, 144
- synthesis of, **3**, 130; **5**, 460, 464

Ketones, cyclic α-diazo, ring-contraction of, **1**, 395
Ketones, cyclic unsaturated, synthesis of, **3**, 220
Ketones, cyclopentenylvinyl, Nazarov cyclization of, **3**, 211
Ketones, cyclopropyl
- from, cyclopropyltrimethylsilane, **1**, 284
- reactions
  - with diethylaluminum cyanide, **3**, 634
  - with phenyl selenide, **3**, 176
  - with phosphines, **3**, 195
- synthesis of, **1**, 268
  - via oxidation of cyclopropylmethyl groups, **3**, 113

Ketones, 2-cyclopropyl, reactions, with lithium triethylborohydride, **2**, 61
Ketones, dialkyl
- alkynyl-substituted, synthesis of, **3**, 146
- from, ketones, **3**, 138, 144
- halo-substituted, synthesis of, **3**, 148
- reactions
  - with 1,4-dibromobut-2-ene, **1**, 982
  - with hydrazine, **3**, 444
- synthesis of, **3**, **111–204**, 138

Ketones, dialkyl phenyl, synthesis of, **3**, 296
Ketones, diaryl
- acetalization of, **4**, 179
- as starting materials, for α,α-diarylnitriles, **3**, 626
- synthesis of, **3**, 305

Ketones, diaryl alkyl, reduction of, **1**, 19
Ketones, diazo
- reactions
  - with acylketenes, **3**, 475
  - with aldehydes, **1**, 1078
  - with ammonia, **5**, 276
  - with arenediazonium salts, **5**, 843
  - with esters, **1**, 1078
  - with propanal, **3**, 321
  - with thiophosgene, **5**, 482
- reactivity of, **3**, 532
- rearrangement of, **3**, 250; **5**, 271
- reduction of, **1**, 36
  - for the synthesis of antimalarial drugs, **1**, 45
- as starting materials
  - for α-halo ketones, **2**, 12
  - for methyl ketones, **1**, 45
  - for thioacyl chlorides, **5**, 482
  - for α,β-unsaturated ketones, **3**, 206
- synthesis of, **2**, 1056
- Wolff rearrangement of, **5**, 144

Ketones, 2-diazo, reactions, with dialkyl hydrogen phosphites, **3**, 200
Ketones, α-diazo
- decomposition of, **1**, 820
- from
  - 1-aminophenazine, **3**, 468
  - diazoalkanes, **3**, 474
- insertion reactions, **3**, 129
- reactions
  - with benzenesulfenyl chloride, **4**, 69
  - with 1,2-dithiole-3-thione, **3**, 255
  - with ketohydrazones, **5**, 362
  - with thiophenol, **3**, 166
- as starting materials
  - for amides, **1**, 395
  - for α-azido ketones, **3**, 194
  - for boron enolates, **2**, 697
  - for esters, **1**, 394
  - for Wolff rearrangements, **1**, 393
- synthesis of, **3**, 192, 446, 470
  - via diazo group transfer, **3**, 473
- use in the synthesis of *N*-acyl sulfonamides, **5**, 343
- use in the synthesis of nitrile ylides, **3**, 679

Ketones, diazomethyl, reactions, with trialkylsilyl triflates, **3**, 483
Ketones, 2-diazo-3-sulfonyl, use in the synthesis of ketenimines, **3**, 587
Ketones, α-diazovinyl, synthesis of, **3**, 260
Ketones, dibromo, from, α-diketones, **4**, 27
Ketones, α-dibromo, as starting materials, for alkynolate anions, **2**, 1023
Ketones, α,α'-dibromo
- reactions
  - with enamines, **3**, 225
  - with iron nonacarbonyl, **1**, 364
  - with primary amines, **3**, 417

Ketones, dibromomethyl, synthesis of, **3**, 154

Ketones, α,α-dichloro, as starting materials, for
α-oxothionoamides, **5**, 602
Ketones, β,β-dichloroalkenyl aryl, reactions, with
pyrrolidine, **4**, 793
Ketones, dichloromethyl
from
enamines, **4**, 15
terminal alkynes, **4**, 16
synthesis of, **3**, 151, 152
Ketones, α-dichloromethylene, synthesis of, **4**, 742
Ketones, dichlorovinyl, reactions, with amines, **4**, 970
Ketones, α-(diethoxymethyl), synthesis of, **4**, 193
Ketones, α,α-difluoro, synthesis of, **3**, 148, 150
Ketones, *N*,*N*-diformylamino, from, sodium
diformylamide, **2**, 308
Ketones, α,α-digermyl, synthesis of, **4**, 628
Ketones, α,α-digermylated, synthesis of, **4**, 625
Ketones, dihalo, as starting materials, for
α-ketoacetals, **4**, 189
Ketones, α,α'-dihalo
reduction to ketones, **3**, 129
synthesis of, **3**, 152
Ketones, dihydroxy, synthesis of, **2**, 77
Ketones, α,α'-dihydroxy, synthesis of, **3**, 156, 158
Ketones, α-dimethylphenylsilyl, synthesis of, **3**, 202
Ketones, 3,3-dimethyl unsaturated, synthesis of, **3**, 223
Ketones, dipyridyl, reactions, with 2-chloroethanol, **4**, 183
Ketones, α,α-disilyl, reactions, with 1,1,4,4-
tetrakis(trimethylsilyl)butatriene, **4**, 619
Ketones, β,β-disulfenyl, synthesis of, **3**, 168
Ketones, dithiolidene, synthesis of, **3**, 254
Ketones, 3,3-dithio α,β-unsaturated, synthesis of, **3**, 251
Ketones, divinyl
electrocyclization of, **1**, 421
from, dienynes, **1**, 421
Nazarov cyclization of, **1**, 422
stability of, **1**, 421
synthesis of, **1**, 1096; **3**, 214
Ketones, enamino
cycloaddition of, **5**, 501
synthesis of, **2**, 744
Ketones, α,β-epoxy silyl, synthesis of, **5**, 430
Ketones, ethyl, asymmetric α-amination of, **2**, 331
Ketones, ethynyl, reactions, with amines, **6**, 565
Ketones, α-fluoro
from
enamines, **3**, 148
enols, **3**, 148
epoxides, **3**, 148
ketones, **3**, 149
synthesis of, **1**, 108; **2**, 3, 12; **3**, 237
Ketones, fluoroaliphatic, synthesis of, **3**, 148
Ketones, α-fluoroaliphatic
from, alcohols, **3**, 148
synthesis of, **3**, 148
Ketones, β-fluoroaliphatic, synthesis of, **3**, 150
Ketones, α-fluoro-α,β-unsaturated, synthesis of, **3**, 237
Ketones, γ-formyloxy, synthesis of, **3**, 163
Ketones, α-functionalized, as starting materials, for
ketones, **3**, 128
Ketones, furanyl, synthesis of, **3**, 302
Ketones, furyl, bromination of, **2**, 28
Ketones, α-furyl, tosylhydrazones of, thermolysis of, **1**, 1077
Ketones, germanium-functionalized, synthesis of, **3**, 203
Ketones, germanium-substituted, synthesis of, **6**, 520
Ketones, germyl silyl, synthesis of, **6**, 520
Ketones, halo
reactions
with isothiocyanates, **3**, 269
with phosphorus(III) esters, **3**, 196
Ketones, 2-halo, as starting materials, for carboxylic
acids, **1**, 391
Ketones, α-halo
from, diazo ketones, **2**, 12
reactions
with acyl hydrazines, **5**, 367
with alkoxides, **4**, 183
with dilithioalkyl sulfones, **3**, 224
with Grignard reagents, **1**, 464
with nitrogen nucleophiles, **3**, 178
with phosphines, **3**, 195
with sodium diethyl phosphite, **3**, 197
with sodium thiosulfate, **5**, 601
with sulfur nucleophiles, **3**, 165
with thiolate anions, **3**, 164
with trialkyl phosphites, **3**, 196
rearrangement of, **1**, 115
reduction of, **1**, 2
reduction to ketones, **3**, 129
as starting materials
for acyclic hemiacetals, **4**, 164
for alkynolate anions, **2**, 1023
for α,β-unsaturated ketones, **3**, 259
synthesis of, **2**, 3
use in the synthesis of nitriles, **3**, 612
Ketones, β-halo, reactions, with nitrite anions, **3**, 191
Ketones, α-halo-α-benzeneselenenyl, synthesis of, **4**, 90
Ketones, halogenated, as starting materials, for
acetals, **4**, 183
Ketones, α-halogenated, nitrosation of, **5**, 667
Ketones, 3-halogenated α,β-unsaturated, synthesis of, **3**, 238
Ketones, halogenated α,β-unsaturated, synthesis of, **3**, 234
Ketones, halomethyl aryl, as starting materials, for α-oxo
dithioesters, **5**, 554
Ketones, halophenyl, synthesis of, **3**, 296
Ketones, α-halo-α,β-unsaturated, synthesis of, **3**, 235, 237
Ketones, hetaryl, synthesis of, **3**, 302
Ketones, hexynylcyano, rearrangement of, **3**, 653
Ketones, α-hydrazino, synthesis of, **3**, 181
Ketones, hydroxy
diastereoselective reductions of, **2**, 61
synthesis of, **2**, 74; **5**, 438
Ketones, 2-hydroxy
reactions
with allyltrialkoxysilanes, **2**, 78
with zinc borohydride, **2**, 61
Ketones, 3-hydroxy
as starting materials, for 1,3-diols, **2**, 63
stereoselective reduction of, **2**, 62
synthesis of, **2**, 73
Ketones, α-hydroxy
from
enolsilanes, **3**, 156
α-sulfonyloxy ketones, **3**, 156
periodate oxidation of, **5**, 49
reactions
with aryl hydrazines, **3**, 451
with chlorodiethylphosphine, **2**, 457
with diethyl chlorophosphite, **3**, 198
with hydrazines, **3**, 450
with malononitrile, **3**, 671
reduction to ketones, **3**, 129
as starting materials
for enones, **3**, 209
for osazones, **3**, 451
synthesis of, **3**, 30, 156; **4**, 536; **5**, 437
Ketones, β-hydroxy
from
α,β-epoxy ketones, **3**, 160
β-silyl ketones, **3**, 160
hydrosilylation reductions of, **2**, 62
reactions, with allenylboronic esters, **1**, 1055
synthesis of
via aldol reactions, **3**, 159
via nitrile oxide-alkene cycloaddition, **3**, 160
Ketones, γ-hydroxy, synthesis of, **1**, 241; **3**, 161

Ketones, δ-hydroxy, synthesis of, **3**, 163
Ketones, hydroxylamino, reactions, with hydroxylamino alcohols, **3**, 189
Ketones, α-hydroxymethyl, from, terminal alkenes, **3**, 158
Ketones, 4-imino, synthesis of, **3**, 418
Ketones, α-imino, reactions, with lithium ester enolates, **3**, 414
Ketones, iodo, reactions, with aldehydes, **2**, 74
Ketones, α-iodo
  from
    alkenes, **3**, 155
    enamines, **3**, 155
    enols, **3**, 155
    epoxides, **3**, 155
    ketones, **3**, 155
  reduction of, **1**, 10
Ketones, iodoaliphatic, synthesis of, **3**, 155
Ketones, α-iodoaliphatic, synthesis of, **3**, 155
Ketones, β-iodoaliphatic, synthesis of, **3**, 156
Ketones, iodomethyl, synthesis of, **3**, 155
Ketones, iodovinyl, synthesis of, **3**, 238
Ketones, *trans* β-iodovinyl, synthesis of, **3**, 238
Ketones, β-keto, from, arsonium ylides, **1**, 54
Ketones, macrocyclic, ring-opening of, **1**, 841
Ketones, α-mercapto, synthesis of, **3**, 164
Ketones, α-methoxy, synthesis of, **3**, 162
Ketones, α-methoxycarbonylsulfenyl, synthesis of, **3**, 165
Ketones, methyl
  from, diazo ketones, **1**, 45
  oxidation of, **5**, 102
  reactions
    with diiminosuccinonitrile, **4**, 438
    with nitrosyl chloride, **5**, 823
    with oxaziridines, **6**, 256
    with thionyl chloride, **5**, 485
    with trimethyl phosphonacetate, **1**, 748
    with trimethyl thiophosphite, **4**, 345
  as starting materials, for terminal alkynes, **1**, 1070
  synthesis of, **1**, 273; **3**, 113, 114, 180
Ketones, methyl alkyl, as starting materials, for ketenes, **3**, 535
Ketones, α-methylene
  from, siloxycyclopropanes, **1**, 663
  reactions, with dialkylamines, **3**, 187
Ketones, α-methyl selenyl, synthesis of, **3**, 174
Ketones, α-methylsulfenyl
  as intermediates, **3**, 164
  synthesis of, **3**, 165
Ketones, α-methylsulfenyl methyl, synthesis of, **3**, 165
Ketones, methylvinyl
  reactions
    with alkynes, **1**, 1051
    with cyclic ketones, **3**, 220
  synthesis of, **2**, 938
  use in the synthesis of phenols, **2**, 663
Ketones, monoalkyl phenyl, synthesis of, **3**, 295
Ketones, monobromo, as starting materials, for alkynolate anions, **2**, 1023
Ketones, monochloro, as starting materials, for alkynolate anions, **2**, 1023
Ketones, β-monoselenyl, synthesis of, **3**, 176
Ketones, naphthyl, synthesis of, **3**, 300
Ketones, nitro
  as starting materials
    for ω-nitroacids, **2**, 396
    for ω-nitro esters, **2**, 396
Ketones, 2-nitro, reactions, with Lawesson's reagent, **5**, 603
Ketones, α-nitro
  alkylation at the α and α' positions, **3**, 190
  dehydration of, **5**, 829, 830
  from
    acylimidazoles, **2**, 396
    β-nitrostyrenes, **2**, 393
  reactions, with 2-propanol, **3**, 427
  reduction of, **3**, 184
  as starting materials, for *N*-hydroxyimides, **5**, 332
  synthesis of, **2**, 391, 392; **3**, 46, 190
Ketones, β-nitro
  as starting materials, for cyclic nitroalkenes, **2**, 785
  synthesis of, **2**, 394; **3**, 191
Ketones, γ-nitro
  reduction of, **3**, 434
  synthesis of, **2**, 396; **3**, 191
Ketones, nitrogen-substituted, reactions, with carbonyls, **3**, 257
Ketones, nitroso, isomerization of, **3**, 192
Ketones, α-nitroso, synthesis of, **3**, 192
Ketones, oximino
  fragmentation of, **3**, 672
  synthesis of, **3**, 672
Ketones, α-oximino, synthesis of, **3**, 192
Ketones, pentenyl, intramolecular reductive cyclization reactions of, **2**, 81
Ketones, perfluoro
  reactions
    with chlorine monofluoride, **2**, 94
    with fluorine, **2**, 94
Ketones, phenanthryl, synthesis of, **3**, 301
Ketones, phenolic
  synthesis of, **3**, 296
    via Fries rearrangement, **3**, 297
Ketones, phenyl
  Beckmann rearrangement of, **5**, 292
  α,α-dichlorination of, **4**, 12
  from, benzene, **3**, 295
  reactions, with sulfuryl chloride, **4**, 12
  synthesis of, **3**, 295
Ketones, α-phenyl, alkylation of, **1**, 131
Ketones, phenylethynyl, synthesis of, **1**, 1015
Ketones, α-(phenylselenenyl), synthesis of, **1**, 1105, 1139
Ketones, α-phenylseleno, elimination of hydrogen chloride from, **3**, 256
Ketones, (α-phenylseleno) α,β-unsaturated, reactions, with dimethylcopperlithium, **4**, 401
Ketones, α-phenyl selenyl, synthesis of, **3**, 173
Ketones, α-phenyl selenyl-α,β-unsaturated, synthesis of, **3**, 174
Ketones, α-phenylsulfenyl, as intermediates, **3**, 164
Ketones, α-phenylsulfinyl, as starting materials, for cyclopentenones, **1**, 596
Ketones, α-(phenylthio)
  α-acetoxylation of, **4**, 220
  synthesis of, **4**, 220
Ketones, β-phenylthio-β-tributylstannyl, synthesis of, **4**, 398
Ketones, 2-phenylthio-α,β-unsaturated, from, sulfoxides, **3**, 249
Ketones, β-phenyl-α,β-unsaturated, conjugate reduction of, **3**, 145
Ketones, α-phosphorus-substituted, reactions, with aldehydes, **3**, 269
Ketones, α-phosphoryloxy, synthesis of, **3**, 163
Ketones, polyalkylphenyl, synthesis of, **3**, 296
Ketones, polycyclic aryl, synthesis of, **3**, 300, 302
Ketones, propargylic
  reactions
    with arsabenzenes, **3**, 269
    with enamines, **3**, 233
Ketones, α-propargylic, synthesis of, **1**, 1062
Ketones, purinyl, synthesis of, **3**, 283
Ketones, pyridyl
  from, alcohols, **3**, 311
  synthesis of, **3**, 310
Ketones, pyrrolic, synthesis of, **3**, 306
Ketones, saturated
  allylation of, **3**, 139
  ethynylation of, **3**, 146

propargylation of, **3**, 147
vinylation of, **3**, 138
Ketones, saturated unsubstituted, synthesis of, **3**, 112
Ketones, α-selenocyano, cyclization of, **6**, 630
Ketones, selenoester, reactions, with alkoxides, **1**, 246
Ketones, α-selenyl
synthesis of, **3**, 173
See also Selenoxides, β-keto
Ketones, γ-selenyl, synthesis of, **3**, 176
Ketones, β-siloxy, from, epoxy silyl ethers, **3**, 160
Ketones, silyl
from, halovinylsilanes, **3**, 271
reactions, with magnesium, **6**, 125
as starting materials, for acetals, **6**, 124
Ketones, α-silyl
from
acetylsilane *O*-silylcyanohydrins, **2**, 814
acids, **3**, 202
alcohols, **3**, 202
aldehydes, **3**, 202
ketones, **3**, 202
reactions
with bis(trimethylsilyl) sulfide, **5**, 648
with hydrogen chloride, **5**, 647
with hydrogen sulfide, **5**, 647
as starting materials, for α-silyl thioketones, **5**, 647
synthesis of, **3**, 202
Ketones, β-silyl
Baeyer–Villiger oxidation of, **1**, 650
as starting materials, for β-hydroxy ketones, **3**, 160
synthesis of, **3**, 203
Ketones, 2-silyl-3-chloro-α,β-unsaturated, synthesis of, **3**, 270
Ketones, 3-silyldivinyl, synthesis of, **3**, 271
Ketones, silyloxy
from, aldehydes, **3**, 163
as intermediates, **3**, 163
Ketones, 2-silyl α,β-unsaturated
synthesis of, **3**, 270
via Diels–Alder cyclization, **3**, 271
Ketones, 3-silyl α,β-unsaturated, synthesis of, **3**, 271
Ketones, silylvinyl, reactions, with trimethylphenylthiosilane, **4**, 944
Ketones, stannyl, synthesis of, **3**, 273
Ketones, β-stannyl, synthesis of, **3**, 204
Ketones, steroidal, reactions, with isocyanomethyl phosphonate diethyl esters, **3**, 718
Ketones, strained cyclic, as starting materials, for α-diols, **4**, 162
Ketones, (*E*)-γ-substituted, from, titanium ate complexes, **4**, 918
Ketones, β-substituted
reactions
with aryl hydrazines, **3**, 452
with hydrazine, **3**, 452
Ketones, 1-substituted-2-hydroxy, synthesis of, **2**, 51
Ketones, *N*-substituted phenyl, synthesis of, **3**, 299
Ketones, α-sulfenyl
from, alkenes, **3**, 165
oxidation of, **3**, 170, 171
synthesis of, **3**, 164
Ketones, γ-sulfenyl, synthesis of, **3**, 169
Ketones, α-sulfenylated, synthesis of, **6**, 478
Ketones, α-sulfenyl-α'-oxygenated, synthesis of, **3**, 167
Ketones, α-sulfinyl
synthesis of, **3**, 170
See also Sulfoxides, β-keto
Ketones, β-sulfinyl, synthesis of, **3**, 172
Ketones, γ-sulfinyl, synthesis of, **3**, 172
Ketones, α-sulfonamido
reactions, with benzaldehydes, **3**, 257
synthesis of, **3**, 171
Ketones, sulfonyl, alkylation of, **3**, 170
Ketones, α-sulfonyl
reduction to ketones, **3**, 129
as starting materials, for α-oxo dithioesters, **5**, 554
synthesis of, **3**, 170
Ketones, β-sulfonyl, synthesis of, **3**, 172
Ketones, γ-sulfonyl, synthesis of, **3**, 172
Ketones, α-sulfonyloxy
as starting materials, for α-hydroxy ketones, **3**, 156
synthesis of, **3**, 163
Ketones, 2-sulfonyl-α,β-unsaturated, from, isoxazolines, **3**, 250
Ketones, symmetrical, synthesis of, **4**, 523
Ketones, α-thiol, reactions, with amines, **4**, 318
Ketones, α-thiomethyl, tosylhydrazones of, **1**, 1077
Ketones, thiophenyl, synthesis of, **3**, 299
Ketones, 2-thio α,β-unsaturated, synthesis of, **3**, 247
Ketones, 3-thio α,β-unsaturated, synthesis of, **3**, 251
Ketones, tin-functionalized, synthesis of, **3**, 204
Ketones, trialkylsilyl, synthesis of, **3**, 202
Ketones, β-trialkylstannyl, synthesis of, **3**, 204
Ketones, β-tributylstannyl α,β-unsaturated, synthesis of, **4**, 398
Ketones, trichloromethyl
as acylating reagents, **5**, 345
synthesis of, **3**, 152; **4**, 195; **6**, 26
Ketones, β-trichlorostannyl, dehydrostannation of, **1**, 663
Ketones, tricyclic, synthesis of, **1**, 422
Ketones, α-triethylsilyl, synthesis of, **3**, 203
Ketones, α-trifluoromethanesulfonyloxy, synthesis of, **3**, 163
Ketones, trifluoromethyl
from, ethyl trifluoroacctate, **3**, 150
reactions, with diphenyl(trimethylsilyl)phosphine, **4**, 338
synthesis of, **3**, 253
Ketones, trifluorovinyl, synthesis of, **3**, 237
Ketones, α,α,α'-trihalo, synthesis of, **3**, 152
Ketones, α-triisopropylsilylmethyl, synthesis of, **3**, 203
Ketones, trimethylplumbyl α-diazo, synthesis of, **3**, 484
Ketones, β-trimethylsilyl, desilylation of, **1**, 62
Ketones, trimethylsilylmethyl, synthesis of, **3**, 202
Ketones, β-trimethylstannyl-α,β-unsaturated, synthesis of, **3**, 272
Ketones, β-triphenylgermyl, synthesis of, **3**, 203
Ketones, α-trisubstituted, synthesis of, **1**, 385
Ketones, unsaturated
as byproducts, in Favorskii rearrangements, **1**, 393
oxidation of, **5**, 102
reduction of, **1**, 86
synthesis of, **3**, 138
Ketones, α,β-unsaturated
acidity of, **1**, 871
addition reactions of, **1**, 302
aminoalkylation of, **2**, 302
aresenic-substituted, synthesis of, **3**, 269
conjugate addition of aryl nucleophiles to, **3**, 145
conjugate addition to, **1**, 273
conjugate addition of unsaturated nucleophiles to, **3**, 147
1,4-disilylation of, **2**, 51
from
acid chlorides, **3**, 261
alkenes, **3**, 211
α-alkoxy ketones, **3**, 241
alkynes, **3**, 222, 238, 252
carbonyls, **3**, 268
cyclopropanes, **3**, 232, 235
diazo ketones, **3**, 206
1,3-dicarbonyls, **3**, 246, 253
α-halo ketones, **3**, 259
ketones, **3**, 266
α-metallated enol ethers, **3**, 240
oxime sulfonates, **3**, 269
pyrylium salts, **3**, 232
α,β-unsaturated acid chlorides, **3**, 212
vinyl compounds, **3**, 210

1,4-hydrostannation of, **2**, 701
metal reduction of, **1**, 889
nitration of, **3**, 269
2-nitrogen-substituted, synthesis of, **3**, 257
3-nitrogen-substituted, synthesis of, **3**, 261
α-nitro-substituted, synthesis of, **3**, 258
oxidation of, **1**, 560
oxygen-substituted, synthesis of, **3**, 240
2-oxygen-substituted, synthesis of, **3**, 240
3-oxygen-substituted, synthesis of, **3**, 243
phosphorus-substituted, synthesis of, **3**, 269
reactions
with alkenylboranes, **1**, 1103
with α-alkoxycuprates, **4**, 384
with aryl hydrazines, **3**, 451
with diethyl methyl phosphonite, **3**, 200
with hydrazines, **3**, 451
with (2-hydroxyaryl)mercury chlorides, **4**, 168
with 2-lithio-1,3-dithiane, **3**, 169
with α-metallomethyl isocyanides, **3**, 708
with nitrite anions, **3**, 191
with organoboranes, **1**, 274
with phenylseleno cyanide, **3**, 656
with phosphorus trichloride, **2**, 456
with selenol anions, **3**, 175
with tertiary phosphines, **2**, 446
with trialkyl allylsilanes, **1**, 314
with 3-(trimethyllsilyl)-2-propynylidenetriphenylarsorane, **1**, 270
with *S*-trimethylsilyl enethiolates, **5**, 557
with vinyl nucleophiles, **3**, 228
with vinylzirconium(IV) complexes, **1**, 1105
reduction of, **1**, 20
1,4-reduction of, **2**, 59
selenium-substituted, synthesis of, **3**, 256
as starting materials
for acyl alkyl acetals, **4**, 213
for boron enolates, **2**, 698
for cyanohydrins, **3**, 616
1,2-dimesityl ethenols, **2**, 643
for enols, **2**, 638, 643
for 4-ketonitriles, **3**, 616
synthesis of, **1**, 395, 546, 596, 607; **3**, **205–276**
via aldol condensation, **3**, 219, 230
via carbonylation reactions, **3**, 218, 233
via Claisen rearrangement, **3**, 217, 233
via [3+2] cycloaddition, **3**, 225
via Diels–Alder reaction, **3**, 221, 243
via elimination reactions, **3**, 206, 225, 234
via Friedel–Crafts acylation, **3**, 230
via isomerization, **3**, 218
via oxidation reactions, **3**, 215
via oxy-Cope rearrangement, **3**, 217
via Pummerer reaction, **3**, 249
via rearrangement reactions, **3**, 216
via Wittig reaction, **3**, 220, 254
tellurium-substituted, synthesis of, **3**, 256
via sulfoxide pyrolysis, **1**, 778
Wolff–Kishner reduction of, **1**, 887
Ketones, β,γ-unsaturated
enolization of, **1**, 555
isomerization of, **3**, 218
synthesis of, **1**, 292
Ketones, γ,δ-unsaturated
isomerization of, **3**, 218
synthesis of, **1**, 189, 275, 1103
Ketones, α,β-unsaturated cyclic, sulfenylation of, **2**, 138
Ketones, α,β-unsaturated cyclopropyl, synthesis of, **1**, 285
Ketones, α,β-unsaturated diazo, synthesis of, **3**, 473
Ketones, β,γ-unsaturated diazo, reactions, with copper(II) salts, **3**, 532
Ketones, unsymmetrical
as starting materials, for enolates, **1**, 872
synthesis of, **5**, 800
Ketones, uretidinethione, as intermediates, in the synthesis of isothiocyanates, **5**, 1037
Ketones, vinyl
cyclization of, **3**, 242
synthesis of, **3**, 212
Ketones, α-vinyl, synthesis of, **1**, 479, 489; **3**, 142
Ket-2-ones, 1-chloroalkyl-, as starting materials, for 2-oxoalkyl phosphonates, **2**, 457
β-γ-Ketones, synthesis of, **4**, 742
Ketonic acids, synthesis of, reviews, **5**, 62
Ketoses, reactions, with arylhydrazines, **3**, 450
6-Ketosteric acid, 4-methyl-, synthesis of, **5**, 76
Ketosteroids, bromination of, **1**, 878
20-Ketosteroids, 17-bromo-, rearrangement of, **1**, 391
β-Ketothiol, reactions, with phenylselenyl chloride, **2**, 235
Ketoximes
acetylation of, **2**, 747
Beckmann fragmentation of, **3**, 663
chlorination of, **5**, 821
from, α,β-unsaturated nitroalkenes, **3**, 426
oxidation of, **4**, 314
reactions, with hexamethylphosphorus triamide, **5**, 755
as starting materials
for α-*O*-acyl-*N*,*N*-diacyl enamines, **2**, 747
for imines, **3**, 408
for nitroalkanes, **2**, 397
synthesis of, **5**, 215
Ketoximes, acylated, use in *trans*-esterification reactions, **5**, 134
Ketoximes, α-nitro, dehydration of, **3**, 685
Ketoximes, *O*-silylated, synthesis of, **3**, 436
Ketoximes, β-trimethylsilyl, Beckmann fragmentation of, **3**, 663
α-Ketoximes, silylation of, **3**, 436
Ketyl radicals, synthesis of, **1**, 548; **2**, 1010
Khellin analogues, synthesis of, **3**, 87
Kinetic resolution, of allylic alcohols, **2**, 44
Kirsanov reaction, for the synthesis of alkylphosphineimides, **2**, 468
Knoevenagel condensation
of bis(sulfones), **4**, 275
of *o*-hydroxybenzaldehyde, **3**, 671
for the synthesis of ketene dithioacetal tetroxides, **4**, 864
for the synthesis of nitriles, **3**, 653
for the synthesis of β-sulfonyl enamines, **2**, 742
for the synthesis of α,β-unsaturated acids, **5**, 104
use of methyl methylthiomethyl sulfoxides in, **4**, 856
Koch carboxylic acid synthesis, mechanism, **5**, 27
Koenigs–Knorr reaction, mechanism, **4**, 198
Kolbe decarboxylation, for the synthesis of alkyl radicals, **1**, 934
Kolbe electrolysis
reviews of, **5**, 28
for the synthesis of α,ω-dinitroalkanes, **2**, 396
Kolbe reaction
for the synthesis of radicals, **1**, 925
for the synthesis of trifluoromethyls, **6**, 20
Kornblum oxidation
for the synthesis of aldehydes, **3**, 3, 8
for the synthesis of benzaldehydes, **3**, 88
for the synthesis of ketones, **3**, 114
for the synthesis of pyridine carboxaldehydes, **3**, 108
Kornblum reaction, mechanism, **3**, 57
Koser's reagent, use in the synthesis of bromo iodoalkenes, **4**, 760
Kröhnke procedure, for the synthesis of benzaldehydes, **3**, 88
Kugelrohr distillation
of esters, **1**, 582
for the synthesis of alkenes, **1**, 566
Kuhn's anion, synthesis of, **1**, 867
Kumada's ligands, synthesis of, **4**, 1032

L

Lactam acetals
from, iminium salts, **6**, 105
reactions, with nitromethane, **2**, 788
γ-Lactam acetals, synthesis of, **6**, 107
Lactams
alkylation of, **4**, 881
bromination of, **5**, 318
chlorination of, **5**, 315
fluorination of, **5**, 314
from, hydroxamic acids, **5**, 324
hydrolysis of, **5**, 86
metallation of, **5**, 385
phosphorolation of, **5**, 369
reactions
with alkoxyamines, **5**, 329
with alkynyl metals, **1**, 1047
with sulfur transfer reagents, **5**, 333
selenenylation of, **1**, 607
silylation of, **5**, 720
as starting materials
for isothiocyanates, **5**, 1031
for lactim ethers, **5**, 688
for the synthesis of radicals, **1**, 925
thionation of, comparison of methods, **5**, 576
use in the synthesis of thionoamides, **5**, 569
Lactams, *N*-alkyl-, reactions, with trialkyl oxonium tetrafluoroborates, **5**, 576
Lactams, allenyl-, synthesis of, **1**, 978
Lactams, *N*-aryl-, synthesis of, **5**, 292
Lactams, bicyclic, **1**, 134
synthesis of, **4**, 496
Lactams, α-bromo-, synthesis of, **4**, 99
Lactams, *N*-chloro-, reactions, with benzenethiol, **5**, 335
Lactams, *N*-chlorosulfonyl-, reductive cleavage of, **4**, 306
Lactams, enamino-, synthesis of, **2**, 746
Lactams, *N*-fluoro-, for the synthesis of aryl fluorides, **2**, 613
Lactams, hydroxy-, synthesis of, **4**, 303
Lactams, α-methylene-, synthesis of, **5**, 296
Lactams, *N*-sulfonyl-, synthesis of, **5**, 344, 345, 346
Lactams, *N*-tosyl-, synthesis of, **5**, 344
Lactams, *N*-trimethylsilyloxymethyl-, from, α-chloroamides, **4**, 100
Lactams, α,β-unsaturated, synthesis of, **1**, 596, 607
β-Lactams
acetoxylations of, **6**, 111
analogues, synthesis of, **4**, 274
from
ester enolates, **5**, 274
imines, **5**, 274
ketene acetals, **6**, 106
ketenes, **3**, 528; **5**, 274; **6**, 113, 115
N-silylimines, **5**, 377
trimethylsilylallenes, **6**, 128
optically active, synthesis of, **1**, 264
oxidation of, **6**, 111
pyrolysis of, **1**, 785
reactions
for amide bond formation, **5**, 260
with phosphochloridates, **5**, 370
with xenon difluoride, **4**, 778
rearrangement of, **5**, 998
as starting materials
for amides, **4**, 825
for monoacyl aminals, **4**, 419
stereochemistry, **1**, 215
sulfonation of, **5**, 345
synthesis of, **1**, 410; **4**, 906; **5**, 273
reviews, **5**, 286
via asymmetric *cis*-dihydroxylation reactions, **2**, 49
via Mitsunobu conditions, **5**, 273
via photochemical extrusion, **1**, 232
thionation of, **5**, 570, 575
See also Azetidinones
β-Lactams, *N*-alkoxy-, from, *O*-alkyloximes, **5**, 329
β-Lactams, *N*-alkyl alkanoamide, synthesis of, **5**, 287
β-Lactams, allenic, from, 4-amino-2-alkynyl carbonates, **1**, 976
β-Lactams, *N*-amino-, synthesis of, **5**, 355
β-Lactams, *N*-aryl-, synthesis of, **5**, 294
β-Lactams, *N*-butoxycarbonyl-, ring-opening of, **5**, 329
β-Lactams, chloro-, as starting materials, for oxazepinones, **4**, 824
β-Lactams, *N*-hydroxy-, synthesis of, **5**, 324
β-Lactams, *N*-imino-, synthesis of, **5**, 364
β-Lactams, α-methylene-, as starting materials, for dimethylmethyleneketene, **3**, 541
β-Lactams, selenium-substituted, synthesis of, **6**, 116
β-Lactams, *N*-silyl-, synthesis of, **2**, 544
β-Lactams, *N*-tosyl-, from, (*N*-tosyl)propargylamines, **5**, 346
β-Lactams, *N*-unsubstituted, synthesis of, **3**, 440
δ-Lactams, reactions, with Lawesson's reagent, **5**, 573
Lactic acid, esterification of, **5**, 122
Lactim ethers, reactions, with nitromethane, **2**, 788
Lactols
from, lactones, **4**, 171
reactions, with mercury(II) oxide, **1**, 931
synthesis of, **4**, 164
Lactols, tribromo-, synthesis of, **6**, 31
Lactols, trichloromethyl-, synthesis of, **6**, 25
δ-Lactols, synthesis of, **4**, 166
Lactone, aristo-, synthesis of, **1**, 818
Lactone, hybrida-, synthesis of, **1**, 1016
Lactone complexes, synthesis of, via carbon monoxide insertion, **4**, 688
β-Lactone decarboxylation, tungsten catalyst, **1**, 568
β-Lactone enolates, alkylation of, **5**, 162
α-Lactone pyrolysis, stereochemistry of, **1**, 785
Lactones
alkenation of, **2**, 680
alkylidenation of, **1**, 770; **2**, 681; **3**, 161
C,*O*-disilylation of, **4**, 831
Friedel–Crafts alkylation of, **1**, 181, 476
from
cycloalkanones, **5**, 161
epoxy ketones, **1**, 841
nitrosation of, **3**, 469
reactions
with alkoxyamines, **5**, 329
with alkynyllithiums, **1**, 1045
with bis-*O*-trimethylsilyl-1,2-diols, **6**, 72
with carbanions, **4**, 171
with di-*t*-butyl peroxide, **1**, 925
with 1,2-dihydroxyethane, **6**, 72
with diols, **6**, 72
with epoxides, **6**, 72
with Grignard reagents, **3**, 219
with hydrazine hydrate, **5**, 350
with lithium alkynide, **4**, 171
with lithium alkynides, **3**, 244
with selenolates, **2**, 282
with trimethylsilyl bromide, **5**, 225
reduction of, **4**, 170

ring-contraction of, **1**, 831
ring-opening of, **5**, 83
as starting materials
for alkynones, **1**, 1045
for lactols, **4**, 171
for ortho esters, **6**, 72
for tetracoordinate germanium compounds, **5**, 794
for thionolactones, **5**, 512
synthesis of, **1**, 205, 822; **5**, **121–179**, 161
via Baeyer–Villiger oxidation, **5**, 161
Lactones, aldohydroximino-, synthesis of, **3**, 428
Lactones, cephem spiroacetal, synthesis of, **4**, 208
Lactones, β-chloro-, synthesis of, **1**, 1108
Lactones, corey, synthesis of, **1**, 176
Lactones, enamino-, synthesis of, **2**, 746
Lactones, enol, acylation of, **3**, 246
Lactones, ω-ethenyl-, substitution of, **1**, 476
Lactones, fluoro-, synthesis of, **4**, 52
Lactones, furo-, synthesis of, **3**, 321, 322
Lactones, iodo-, synthesis of, **1**, 931
Lactones, isocyano-, synthesis of, **3**, 704
Lactones, lignan, synthesis of, **3**, 291
Lactones, α-methylene-
reactions, with acetonitrile, **3**, 652
synthesis of, **1**, 546, 547, 607
Lactones, α-methylene-γ-valero-, synthesis of, **1**, 455
Lactones, β-propio-, reactions, with dimethylalkynylaluminums, **1**, 1024
Lactones, α,β-unsaturated, synthesis of, **1**, 561, 596, 607
Lactones, ynenol-, synthesis of, **1**, 1020
α-Lactones
pyrolysis of, **1**, 785
as starting materials
for allenes, **1**, 785
for enol ethers, **1**, 785
β-Lactones
decarboxylation of, **1**, 681
from
(2-alkenyl)triphenoxytitaniums, **1**, 683
β-bromocarboxylic acids, **5**, 161
β-hydroxy esters, **1**, 682
ketenes, **1**, 683
thio ester enolates, **1**, 566
opening of, **1**, 230
pyrolysis of, **1**, 789
ring-expansion of, **5**, 162, 170
as starting materials
for alkenes, **1**, 681
for 1,3-thiazine-2-thiones, **6**, 563
for thiocyanates, **5**, 1106
substitution of, **1**, 476
synthesis of, **1**, 681; **5**, 161
via carbonylation reactions, **5**, 163
thermolysis of, **1**, 681
use in the synthesis of alkenes, **1**, 565
β-Lactones, α-alkylidene-, synthesis of, via selenoxide pyrolysis, **1**, 781
β-Lactones, α-methylene-, reactions, with dienes, **1**, 566
γ-Lactones
from
allylic alcohols, **5**, 170
butane-1,4-diols, **5**, 171
carboxylates, **5**, 168
enolates, **5**, 166
homopropargylic alcohols, **5**, 170
(*Z*)-vinyl iodides, **5**, 170
synthesis of, **5**, 164
via enolate additions, **5**, 166
via radical cyclization, **1**, 342
via redox reactions, **5**, 171
γ-Lactones, bicyclic, saponification of, **5**, 63
γ-Lactones, γ-chloro-, from, levulinic acid, **4**, 54
γ-Lactones, α-diazo-, as starting materials, for methyleneketenes, **3**, 541
γ-Lactones, γ-fluoro-, synthesis of, **4**, 52
γ-Lactones, hydroxy-, synthesis of, **1**, 1139
γ-Lactones, α-methylene-, synthesis of, **1**, 1136
δ-Lactones, synthesis of, **1**, 1093; **5**, 171
δ-Lactones, bicyclic, saponification of, **5**, 63
δ-Lactones, 3-oxo-, synthesis of, **2**, 39
Lactonium tetrafluorborate salts, *O*-alkyl-, reactions, with alkoxides, **6**, 73
Lactonization, for the synthesis of radicals, **1**, 938
Laevulinic acid, 2-mercapto-, synthesis of, **2**, 124
Lanceol, Julia's synthesis of, **4**, 254
Lanthanide(III) complexes, synthesis of, **5**, 782
Lanthanide chlorides, as starting materials, for lanthanides, **2**, 564
Lanthanide complexes
reactions, with lithium trimethylsilyldiazomethane, **3**, 487
as starting materials, for bis(trimethylsilyl)methyls, **6**, 194
Lanthanide hydrides, dicyclopentadienyl-, reactions, with alkenes, **2**, 43
Lanthanide metals, reactions, with alkyl iodides, **2**, 564
Lanthanides
dimeric bis(cyclopentadienyl) metal–methyl complexes of, synthesis of, **4**, 702
from
lanthanide chlorides, **2**, 564
triflates, **2**, 564
for the opening of epoxides, **1**, 205
for the oxidation of alcohols, **3**, 118
oxidation states of, **2**, 564
reactions, with methyl 3-iodopropionate, **2**, 564
synthesis of, **5**, 650
use in organic synthesis, **2**, 564
Lanthanides, acyl, synthesis of, **5**, 439
Lanthanides, dialkylated, stability of, **2**, 565
Lanthanides, imidoyl, synthesis of, **5**, 803
Lanthanides, trialkylated
from
t-butoxides, **2**, 565
lanthanide trichlorides, **2**, 565
stability of, **2**, 565
Lanthanide salts, reactions, with organometallics, **2**, 564
Lanthanide trichlorides, as starting materials, for trialkylated lanthanides, **2**, 565
Lanthanide trihalide, as catalyst, for Friedel–Crafts acylation, **1**, 462
Lanthanum aryl oxides, reactions, with lithium bis(trimethylsilyl)methanide, **6**, 194
Lanthanum chloride, as catalyst, for the synthesis of dithioacetals, **4**, 246
Lanthanum compounds
reactions, with phosphorus ylides, **4**, 578
with two carbon-bridges, **4**, 703
(+)-Laurencin, synthesis of, **1**, 829
Laurenene, synthesis of, **1**, 49
Lauric acid, synthesis of, **5**, 39
Lauroate, 2-bromoethyl, cleavage of, **5**, 39
Lawesson's reagent
modifications to, **5**, 573
reactions
with acid chlorides, **5**, 590, 619
with acrylamides, **5**, 571
with amino amides, **5**, 571
with formamides, **5**, 608
with hexamethylphosphoramide, **5**, 575
with 2-hydroxybenzamides, **5**, 571
with δ-lactams, **5**, 573
with 2-nitroamides, **5**, 603
with 2-nitro esters, **5**, 603
with 2-nitro ketones, **5**, 603
with pseudo acid chlorides, **5**, 528
with 1,4,6-trisubstituted 2(1H)-pyrimidones, **6**, 579
as reagent
for the synthesis of alkenyl aryl thioketones, **3**, 360

for the synthesis of enaminothiones, **3**, 355
for the synthesis of thioaldehydes, **3**, 342
for the synthesis of
thioketones, **3**, 346, 353, 360, 361; **3**, 363
for the synthesis of thiocarboxylates, **5**, 512
for the synthesis of thionoalkenoamides, **5**, 613
for the thionation of carbonyls, **5**, 513
for the thionation of imides, **5**, 631
use of hexamethylphosphoramide as solvent with, problems, **5**, 571
use in the synthesis of dithioic acid esters, **5**, 551
use in the synthesis of thionoalkanoamides, **5**, 610
use in the synthesis of thionoamides, **5**, 566, 570
use in the synthesis of thiophosphines, **2**, 256
See also 2,4-Disulfide, 2,4-bis(4-methoxyphenyl)-1,3,2,4-dithiaphosphetane-
Lead
alkynylation of, **2**, 1101
elimination of, **1**, 663
Lead, bis(ethylthio)-, reactions, with alkylarsenic halides, **2**, 499
Lead, bis(trifluoromethyl)dimethyl-, bromination of, **6**, 245
Lead, pentafluoroethyltrimethyl-, synthesis of, **6**, 65
Lead, tetraethyl-, as petrol additive, **2**, 601
Lead, tetrakis(trifluoromethyl)-, synthesis of, **6**, 245
Lead, tetrakis(trifluorovinyl)-, synthesis of, **4**, 813
Lead, tetrakis(trimethylsilylmethyl)-, synthesis of, **4**, 660
Lead, tetramethyl-, reactions, with iodopentafluoroethane, **6**, 65
Lead, triphenylallyl-, reactions, with phenyl isocyanate, **5**, 390
Lead, tris(trifluoromethyl)methyl-, synthesis of, **6**, 245
Lead(II) chloride
reactions
with lithium compounds, **2**, 602
with organometallics, **2**, 602
Lead(II) cyanamide, reactions, with alkylchlorohydridosilanes, **5**, 1081
Lead(II) thiocyanate, reactions, with phosgene, **6**, 451
Lead(IV) acetate
for the dehydrogenation of hydrazones, **3**, 464
for the oxidative cleavage of alkenes, with osmium(VIII) oxide, **3**, 113
reactions
with alkenylstannanes, **1**, 1081
with benzophenone hydrazone, **3**, 464
with α-dicarbonyls, **3**, 470
with 1,2-dihydrazones, **1**, 1075
with diphenyl ditelluride, **5**, 210
with furans, **4**, 206
with (1-halo-1-alkenyl)dialkylboranes, **1**, 660
with malonic acids, **4**, 206
with 2-thioxo-1,3-benzodithiole, **6**, 309
as starting material, for aldehydes, **3**, 2
use in the synthesis of aldehydes, via diol cleavage, **3**, 8
Lead(IV) halides, use in the synthesis of tetraalkylleads, **2**, 602
Lead(IV) salts, reactions, with trifluorovinylmagnesium iodide, **4**, 813
Lead bromide, bis(trifluoromethyl)methyl-, synthesis of, **6**, 245
Lead bromides, (trifluoromethyl)dialkyl-, as starting materials, for bis(trifluoromethyl)dialkyllead complexes, **6**, 245
Lead chloride, trimethyl-, reactions, with silver cyanate, **5**, 1017
Lead chlorides, alkyl-, synthesis of, **2**, 602
Lead complexes, bis(trifluoromethyl)dialkyl-, from, (trifluoromethyl)dialkyllead bromides, **6**, 245
Lead cyanide, reactions, with triarylgermanium bromides, **5**, 1136
Lead cyanides, triphenyl-, synthesis of, **5**, 1145
Lead dichloride, reactions, with trimethylsilylmethylmagnesium chloride, **4**, 660
Lead halides, reactions, with tris(trimethylsilylmethyl)lithium, **6**, 396
Lead halides, (trifluoromethyl)dialkyl-, synthesis of, **6**, 245
Lead hydride, diphenyl-, reactions, with diphenylethynylstibine, **2**, 876
Lead imides, synthesis of, **5**, 391
Lead isocyanate, trimethyl-, synthesis of, **5**, 1017
Lead isocyanates, synthesis of, **5**, 1017
Lead nitrate, use in the synthesis of phenyl isothiocyanate, **5**, 1023
Leads
reactions, with phosphorus ylides, **4**, 589
synthesis of, **6**, 208
Leads, α-alkoxy-
reactions, with aldehydes, **4**, 390
synthesis of, **4**, 390
Leads, alkyl-
reactions, with silver nitrate, **2**, 589
synthesis of, **2**, 601
uses of, **2**, 601
Leads, alkynyl-, synthesis of, **2**, 1101
Leads, amino-, synthesis of, **2**, 816
Leads, α-haloalkenyl-, synthesis of, **4**, 812
Leads, imidoyl-
from, imidoyl chlorides, **5**, 803
synthesis of, **5**, 803
Leads, perfluoro organo-, synthesis of, **6**, 65
Leads, tetraalkyl-
from, lead tetraacetate, **2**, 602
reactions
with aldehydes, **4**, 390
with *N*-chlorosuccinimide, **5**, 391
with trityl cations, **1**, 663
synthesis of, **2**, 601
via transmetallation, **2**, 602
Leads, trihalomethyl-, synthesis of, **6**, 245
Lead selenocarboxylates, use in the synthesis of monoacyl diselenides, **5**, 250
Lead tetraacetate
for the oxidation of aryl organometallics, **2**, 652
for the oxidation of semicarbazones, **4**, 441
for the oxidation of thiols, **2**, 215
reactions
with *S*,*N*-acetals, **5**, 731
with amides, **5**, 997
with benzoylhydrazones, **6**, 107
with bis(alk-1-ynyl)mercury compounds, **2**, 1102
with α-chloro carboxylic acids, **4**, 40
with diaryl sulfides, **2**, 198
with pyridine, **5**, 999
with trialkyl(alk-1-ynyl)stannanes, **2**, 1102
as reagent, for the decarboxylation of carboxylic acids, **1**, 560
as starting material, for tetraalkylleads, **2**, 602
use in dehydrogenation reactions, **5**, 366
use in didecarboxylation reactions, **1**, 562
use in the synthesis of allylic hemiacetals, **4**, 174
use in the synthesis of carboxylic acids, **5**, 28
use in the synthesis of hydroxamic acids, **5**, 325
Lead tetraacetate, ethynyl-, reactions, with enolate anions, **1**, 1145
Lead tetraacetate–copper(II) acetate, for the dehydrogenation of β-ketoamides, **1**, 556
Lead tetraacetyl, use in the synthesis of α,β-unsaturated ketones, **3**, 210
Lead tetra(trifluoroacetate), reactions, with alkanes, **2**, 86
Lead thiolates, silylation of, **2**, 264
Lead triacetates, alk-1-ynyl-, synthesis of, **2**, 1102
Lead triacetates, aryl-, displacement of lead from, **1**, 490
Lead triacetates, vinyl-, displacement of lead from, **1**, 490
Leaving groups
comparison of, **1**, 107
in nitrile synthesis, **3**, 612

β-Leaving groups, displacement of, **3**, 59, 217
Leinamycin, synthesis of, **4**, 948
Lemieux–von Rudloff reagent, use in the synthesis of alkanoic acids, **5**, 49
Leucine, butoxycarbonyl-, reactions, with glycinamide, **5**, 587
Leu-enkephalins, synthesis of, **5**, 587; **6**, 487
Leukotrienes, synthesis of, **1**, 111, 659; **2**, 136
Levoglucosenone, as starting material, for α-bromo alcohols, **4**, 42
Levulinic acid, as starting material, for γ-chloro γ-lactones, **4**, 54
Lewis acid
  as catalyst
    for the synthesis of cyanohydrins, **3**, 615, 616
    use in the synthesis of β-oxygenated nitriles, **3**, 632
  catalyzed cycloaddition of allenes, **1**, 498
  use in the synthesis of aminonitriles, **3**, 616
  use in the synthesis of thioketones, **3**, 353
Lewis acid–aldehyde complexes, synthesis of, **3**, 26
Lewis acid catalysis
  in Diels Alder reactions, **1**, 499
  in ene reactions, **1**, 524
Lewis acid-catalyzed addition, of acetals to enol ethers, **4**, 188
Lewis acid-catalyzed rearrangement, for the synthesis of β-silyloxy aldehydes, **3**, 33
Lewis acid halides, reactions, with alkyl chlorides, **1**, 900
Lewis acids
  in allylation reactions, **2**, 78
  as catalysts
    for aryl halides, **2**, 608
    in Cope rearrangements, **1**, 809
    in Diels–Alder cycloadditions, **1**, 350
    for Friedel–Crafts acylation, **1**, 462
    in Friedel–Crafts alkylations, **1**, 475
    for the substitution of sulfur functions, **1**, 239
    for the synthesis of acetals, **4**, 178
  choice of, **1**, 386, 678
  in Friedel–Crafts alkylations, choice of, **1**, 212
  for halotrinitromethanes, **6**, 283
  for the insertion of carbon monoxide into the manganese–carbon bonds, **5**, 447
  for the opening of thiiranes, **1**, 242
  reactions
    with acetals, **2**, 1007
    with alkyl halides, **1**, 106
    with allylic alcohols, **1**, 364
    with *S*-benzyl thioethers, **5**, 481
    with Brönsted acids, **2**, 1007
    with *N*-chloroimines, **4**, 110
    with epoxides, **3**, 9
    with halides, **2**, 1006
    with iodide, **1**, 5
    with methylene chloride, **5**, 81
    with peroxy acids, **2**, 650
    with silyl enol ethers, **1**, 189
    with *N*-silylimines, **4**, 110
    with ureas, **5**, 979
  for the rearrangement of epoxides, **1**, 386
  for the substitution of sulfur functions, **1**, 235
  for the synthesis of organometal(III) halides, **2**, 883
  for the synthesis of ortho esters, **6**, 72
  use in the cleavage of carbon–silicon bonds, **1**, 61
  use in cleavage reactions, **1**, 57
  use in the conjugate addition of cuprates, **1**, 312
  use in Diels–Alder reactions, **1**, 373
  use in the opening of epoxides, **1**, 209
  use in the rearrangement of vinylcyclopropanes, **1**, 803
  use in the substitution of epoxides, **1**, 478
  use in the synthesis of alkanoic acids, **5**, 40
  use in the synthesis of alkenes, **1**, 697
  use in the synthesis of aromatic ketones, **3**, 278
  use in the synthesis of carbonyls, **6**, 588
  use in the synthesis of 1,1-diacetates, **4**, 203
  use in the synthesis of dialkyl sulfides, **2**, 133
  use in the synthesis of isocyanates, **5**, 984
  use in the synthesis of β-lactones, **5**, 170
  use in the synthesis of β-sulfenyl ketones, **3**, 168
  use in the synthesis of vinyl sulfides, **2**, 709
Lewis acids, chiral, use in Diels–Alder reactions, **1**, 374
Lewis base adducts
  of alkynylboranes, synthesis of, **2**, 1084
  as starting materials, for tetraorganoborates, **2**, 1087
Lewis bases
  definition of, **1**, 845
  reactions, with boron trifluoride, **2**, 1084
Lewisite
  synthesis of, **2**, 881
  uses of, **4**, 805
LIDAKOR, use in the synthesis of dienes and trienes, **1**, 579
Ligands
  dummy, in substitution reactions, **1**, 119
  non-transferable, alkynes as, **2**, 587
  reaction rates, **1**, 873
  sterically hindered, **1**, 163
Ligands, hetero, as residual ligands, **2**, 588
Ligands, imino, stability of, **2**, 570
π-Ligands, for the stabilization of transition metal complexes, **2**, 564
Ligated metals, for the activation of metals, **2**, 958
Limonene
  reduction of, **1**, 75, 78
  as starting material, for (*R*)-(+)-perrillaldehyde, **1**, 580
Lincosamine, synthesis of, **3**, 244
Lindestrene, synthesis of, via xanthate pyrolysis, **1**, 776
Lindlar catalyst
  as heterogeneous catalyst, **1**, 455
  stereoselectivity of, **1**, 506
  synthesis of, **1**, 502
  use in hydrogenation reactions, **1**, 503
  use in the synthesis of alkenylgermanes, **2**, 948
Lindlar hydrogenation, use in the synthesis of α,β-unsaturated esters, **5**, 154
Linear multi-sulfur systems, structure of, **3**, 354
Lipase P
  reactions, with oximes, **5**, 217
  use in the synthesis of α-alkoxystannanes, **4**, 389
  use in the synthesis of carboxylic acids, **5**, 55
Lipases
  for cyanohydrin resolution, **3**, 631
  for the resolution of β-hydroxynitriles, **3**, 632
  use in *trans*-esterification reactions, **5**, 135
Lipids, synthesis of, **2**, 84
Lipoic acid, synthesis of, **2**, 224
Lipoxin B_4, synthesis of, **1**, 1009
Lipstatin, tetrahydro-, synthesis of, **3**, 541
(-)-Lipstatin, synthesis of, **3**, 541
Liquid ammonia, reactions, with alkali metals, **1**, 55
Liquid crystals, ferroelectric, **5**, 246
Lithiation
  of alkenes and arenes, **2**, 965
  heteroatom-assistance in, **2**, 554
  of pyridines, **3**, 108
  via deprotonation, **4**, 535
o-Lithiation
  of arenes, with heteroatom-containing substituents, **2**, 952
  of phenyl ethers, **1**, 213
  for the synthesis of 2-fluorobenzaldehydes, **3**, 96
  for the synthesis of phenyl ketones, **3**, 300
Lithiation–germylation, of aminoacetonitriles, **4**, 521
Lithiation–silylation reactions, for the synthesis of tetrasilylmethanes, **6**, 378
2-Lithioanisole, reactions, with copper(I) bromide, **4**, 696
Lithio compounds, reactions, with tributylchlorostannane, **4**, 961

α-Lithio compounds, from, 1,1-bis(silyl)alkanes, **4**, 653
Lithiotrithioorthoformates, isotopically labeled, synthesis of, **6**, 317
Lithium
and metal-halogen exchange, **2**, 961
methyl-, lead-substituted, synthesis of, **6**, 405
oxidative addition of, **2**, 958
reactions
with acylsilanes, **1**, 985
with alkylamines, **1**, 508
with alkyl halides, **2**, 551
with arenes, **2**, 551
with 1,1-bissilylethylenes, **4**, 618
with (*Z*)-2-bromobut-2-ene, **1**, 1088
with chloromethyl(trimethyl)germane, **4**, 626
with chloromethyltrimethylsilane, **4**, 602
with chlorotriphenylsilane, **2**, 545
with dialkylmercury, **2**, 555
with 1,3-diborete, **6**, 188
with dichlorides, **6**, 398
with dichloromethane, **4**, 632
with dichloro(trimethylsilyl)methane, **6**, 192
with *gem*-dihaloalkanes, **4**, 667
with 1,2-diphosphetes, **4**, 545
with ethylenediamine, **2**, 420
with ethylidenebis(mercury chloride), **4**, 668
with hexachloro-2-butyne, **6**, 173
with imidazoles, **3**, 109
with oxazoles, **3**, 109
with pentaphenylphosphole, **4**, 112
with tetrachloromethane, **6**, 404
with tetrakis(trimethylsilyl)ethene, **4**, 618
with tetrakis(trimethylsilyl)ethylene, **6**, 193
with thiazoles, **3**, 109
with trichlorosilane, **2**, 545
with vinyl sulfides, **2**, 722
for the reduction of alkyl halides, **1**, 3
for the reduction of selenides, **1**, 25
for the reduction of sulfones, **1**, 24
for the substitution of sulfur functions, **1**, 236
Lithium, alkenyl-, reactions, with bistrimethylsilyl peroxide, **2**, 700
Lithium, α-alkoxy-, synthesis of, **1**, 397
Lithium, alkyl-
reactions
with *gem*-dihalocyclopropanes, **1**, 990
with ethyl thionoformate, **3**, 332
with fluoroalkenes, **4**, 785
with trialkylantimony dibromide, **2**, 488
Lithium, alkyl bis(trimethylsilyl)silyl-, reactions, with acyl chlorides, **5**, 418
Lithium, alkynyl-
oxidative coupling of, **1**, 1035
reactions
with 1-(dialkylaminoalkyl)benzotriazoles, **1**, 261
with diethylaluminum chloride, **1**, 314
Lithium, allenyl-
reactions
with alkyl halides, **1**, 971
with amides, **1**, 971
with carbon dioxide, **1**, 971
with epoxides, **1**, 478, 971
with ketones, **1**, 971
with ureas, **1**, 971
Lithium, 9-anthryl-, reactions, with bis(chlorodimethylsilyl)methane, **4**, 621
Lithium, aryl-
addition to diarylketenes, **2**, 645
reactions
with aryl halides, **2**, 717
with carbon disulfide, **5**, 560
with *O*-cholesteryl deuterothionoformate, **3**, 341
Lithium, benzoyl-, as intermediate, **5**, 436
Lithium, benzyl-
from, toluene, **2**, 554
reactions, with cyclopentadienyltitanium complexes, **2**, 566
Lithium, *exo,exo*-[1,3-bis(trimethylsilyl)allyl]-, synthesis of, **4**, 603
Lithium, bis(trimethylsilyl)bromomethyl-, reactions, with aldehydes, **4**, 805
Lithium, bis(trimethylsilyl)methyl-
as starting material, for chloroarsines, **4**, 597
synthesis of, **6**, 190
Lithium, bromochloromethyl-
reactions, with cyclohexanone, **4**, 758
use in the synthesis of 1-bromo-1-chloroalkenes, **4**, 758
Lithium, 1-bromo-2-methoxyvinyl-, use in the synthesis of 2-bromo-alkenals, **3**, 67
Lithium, bromomethyl-, synthesis of, **4**, 146
Lithium, bromotris(trimethylsilyl)methyl-, reactions, with chlorodimethylphenylgermane, **6**, 292
Lithium, butyl-
for the cleavage of tetrathioorthoformates, **6**, 317
as metallating agent, **4**, 705
reactions
with alkenylalanes, **1**, 479
with α-alkoxystannanes, **4**, 378
with alkyltelluriums, **1**, 437
with 1-alkynes, **1**, 1134
with (*Z*)-allyltriphenylphosphonium bromides, **1**, 735
with anisole complexes, **4**, 707
with arsonium salts, **1**, 738
with aryltelluriums, **1**, 437
with bis(phenyltelluro)methanes, **4**, 402
with bis(phenylthio)ethene, **2**, 1031
with *trans*-1,2-bis(tri-*n*-butylstannyl)ethylene, **1**, 1091
with 2-bis(trimethylsilyl)methylpyridine, **6**, 195
with 1,3-bis(trimethylsilyl)propyne, **4**, 1059
with borane–amine complexes, **2**, 60
with bromoferrocene, **4**, 706
with 3-bromothiophene, **3**, 105
with bromotrichloromethane, **4**, 16
with 1-*t*-butyl-1-sila-2,4-cyclohexadiene, **4**, 603
with carbosilanes, **6**, 178
with chlorobenzene complexes, **4**, 707
with α-chloro-*β*,*β*-dibromovinyl ethers, **2**, 1025
with 1,1-dibromoalkenes, **4**, 757
with *gem*-dibromo methylene cyclopropanes, **1**, 990
with dichlorobis(trimethylsilyl)methane, **6**, 288
with diethyl trichloromethyl phosphonate, **6**, 266
with 1,2-difluorovinyl chloride, **4**, 817
with dihydrofuran, **2**, 1023
with 1-dimethylamino-4-methoxy-4-methyl-2-pentyne, **2**, 940
with 1,1-dimethyl-2,5-diphenyl-1-silacyclohex-3,5-diene, **4**, 616
with diselenoacetals, **4**, 400
with dithiophosphinates, **2**, 460, 462
with iodoferrocene, **4**, 706
with ketophosphine oxides, **1**, 757
with methylenedialuminum tetrabromide, **4**, 708
with onium salts, **6**, 708
with (pentachlorocyclopentadienyl) (1,5-cyclooctadiene)rhodium, **4**, 707
with propargylic methyl ether, **1**, 981
with selenoketals, **4**, 400
with 1-silylalkynes, **4**, 1058
with sulfones, **6**, 541
with tertiary thionoamides, **5**, 606
with tetrakis(dimethoxyboryl)methane, **6**, 200
with tetravinyltin, **1**, 1091
with 1,4-thiatellurins, **1**, 441
with thietanonium salts, **1**, 234
with thioureas, **5**, 1112
with *p*-toluenesulfonylethyne, **2**, 1035

with trihalostannanes, **4**, 380
with triisopropylsilyldiazomethane, **5**, 1148
with trimethylaluminum, **1**, 220
with trimethylsilyldiazomethane, **6**, 674
with trimethylsilylketene, **3**, 537
with triphenyl(triseleno)orthoformate, **4**, 289
with vinylic phosphirenes, **1**, 985
with vinyl iodides, **1**, 278
as starting material, for (trichloromethyl)lithium, **6**, 244
use in the reduction of carbon–arsenic bonds, **1**, 53
Lithium, *n*-butyl-
reactions
with alkenyl alanes, **1**, 543
with α-alkoxysilylstannanes, **4**, 358
with alkyl isocyanides, **3**, 704
with alkynes, **5**, 101
with allyl thiolates, **1**, 140
with α-aminostannanes, **2**, 1001
with benzotriazoles, **3**, 29
with *N*-(benzotriazol-1-ylmethyl)pyrrole, **6**, 370
with benzyl alcohol, **4**, 383
with benzyl chloride, **1**, 110
with benzyl isocyanide, **5**, 88
with benzyl thiolates, **1**, 140
with benzyltrimethylsilane, **2**, 1005
with bis(benzenethio)methane, **6**, 86
with bis(benzotriazoyl)methane, **6**, 364
with 1,1-bismethylselenoalkene, **1**, 440
with bis(phenylseleno)methane, **4**, 400
with bis(trimethylsilyl)methane, **6**, 190
with bis(trimethylsilyl) selenide, **6**, 481
with bromoanilines, **2**, 813
with (+)-2-bromobutane, **1**, 112
with α-bromo selenides, **4**, 400
with *t*-butylisonitrile, **5**, 792, 795
with *n*-butyl tellurobenzoate, **5**, 437
with 1-butyne, **4**, 605
with 2-butyne, **4**, 605
with carbene complexes, **5**, 946
with carbon monoxide, **5**, 437
with chlorofluoromethyl phenyl sulfide, **4**, 773
with α-chlorophosphonates, **4**, 132
with 3-chloropropyne, **5**, 59
with chloro(trihalomethyl)phosphines, **6**, 682
with 1-chloro-3-(trimethylsilyl)-1-propyne, **4**, 648
with chromium complexes, **4**, 1030
with cyclopropa[*b*]naphthalene, **4**, 604
with 2,2-dibromovinyl ethyl ether, **4**, 818
with 1,1-dichloro-2-fluoroalkenes, **1**, 1069
with dichloromethylenephosphines, **6**, 690
with diethyl α-methoxy phosphonate, **6**, 119
with 1,2-diiodoadamantane, **1**, 661
with diisopropyl trichloromethanephosphonate, **5**, 1152
with (dimethylaminomethyl)cyclopentadienyl-tricarbonylmanganese, **4**, 1036
with dimethyl disulfide, **6**, 98
with (*R*,*S*)-1-diphenylphosphino-2-(1-dimethylaminoethyl)ferrocene, **4**, 1033
with dithioacetals, **4**, 1075
with (*E*)-ethoxyvinyl chlorides, **4**, 817
with (*Z*)-2-ethoxyvinyl chlorides, **4**, 817
with Fischer complexes, **5**, 955
with α-fluoro-α-iodo esters, **4**, 137
with fluoro sulfonate salts, **4**, 369
with furans, **4**, 846
with *o*-haloaryl triflates, **1**, 1071
with (*o*-halobenzyl)phosphines, **4**, 575
with β-hydroxysilanes, **2**, 547
with iodoisothiocyanate, **5**, 729
with ketene diselenoacetals, **4**, 964
with manganese carbene complexes, **5**, 946
with methoxysilanes, **6**, 393
with α-methylbenzyl(phenyl) sulfoxide, **2**, 1000
with methyl-substituted nitrogen heteroaryls, **2**, 741
with nitroalkanes, **2**, 784
with organomercury compounds, **6**, 205
with organostannanes, **4**, 964
with 1,3-pentadiyne, **6**, 206
with phenylbenzyl sulfide, **2**, 999
with α-(phenylseleno)methyl phosphonate, **6**, 136
with phenylthio(trimethylsilyl)methane, **6**, 129
with phenylvinyl selenides, **4**, 401, 964
with phosphorus pentasulfide, **5**, 569
with potassium *t*-butoxide, **2**, 557
with propyne, **6**, 206
with sulfones, **4**, 391
with tellurium, **6**, 498
with tetraethyl dichloromethanediphosphonate, **4**, 801
with tetrahydrofuran, **1**, 135
with tetraphenyl tetraselenoorthocarbonate, **6**, 93
with tetraphenyl tetrathioorthocarbonate, **6**, 87
with tributylstannylmethanol, **4**, 378
with tributyl(trimethylsilylmethyl)tin, **4**, 652
with 2-triethylgermyl-1,3-dithiane, **6**, 520
with trifluoromethyltriphenylsilane, **4**, 735
with trimethylphosphine, **2**, 1003
with trimethylsilylmethanol, **4**, 383
with tris(ethylenedioxyboryl)methane, **2**, 905
with tris(methylseleno)methane, **6**, 135
with vinyl sulfides, **2**, 1000
as reagent, for the synthesis of isothiocyanates, **5**, 1024
as reagents, for deprotonation, **6**, 127
as starting material, for acyllithiums, **5**, 437
use in the cleavage of carbon–phosphorus bonds, **1**, 51
use in the synthesis of imidoylarsenic compounds, **5**, 791
use in the synthesis of ketenimines, **3**, 574
Lithium, *s*-butyl-
reactions
with alkyl benzoates, **4**, 382
with amidines, **4**, 107
with carbamates, **1**, 138
with cycloheptanone 2,4,6-triisopropylbenzenesulfonylhydrazone, **2**, 785
with (*E*)-1,3-bis(trimethylsilyl)propene, **4**, 603
with hydrazones, **1**, 1091
with isonitriles, **5**, 796
with methyltriphenylphosphonium bromide, **6**, 705
with 1,3-oxathianes, **6**, 126
with silyl allyl ether, **1**, 136
with (*E*)-*N*-(3-tosyl-2-propenyl)morpholine, **1**, 1090
with triphenylsilyloxirane, **4**, 944
as starting material, for acyllithiums, **5**, 437
Lithium, *t*-butyl-
as deprotonating agent, **4**, 358
for the deprotonation of vinyl ethers, **4**, 949
reactions
with acetonitrile, **2**, 1058
with α-alkoxy bromo ethers, **4**, 381
with allylic amines, **1**, 1090
with benzylthiotrimethylgermane, **4**, 371
with benzylthiotrimethylsilane, **4**, 368
with boranes, **4**, 524
with *C*-borylborataalkyne, **6**, 201
with *t*-butylnitrile, **3**, 459
with carbon monoxide, **5**, 729
with chloro(dimethyl)vinylsilane, **4**, 611
with chloro(methyl)phenyl(vinyl)silane, **4**, 611
with chlorovinylsilanes, **4**, 611; **5**, 926
with dichlorodimethylsilane, **2**, 538
with dichlorodiphenylsilane, **2**, 538
with diphenylethyne, **1**, 539
with fluorovinylgermane, **3**, 503
with α-halo ethers, **4**, 381
with halogermanes, **3**, 503
with halosilanes, **2**, 536
with 2-methoxy-2,4,4-trimethyl-2,4-disilapentane, **4**, 620

with methylvinyl ether, **1**, 1089; **4**, 948
with phenyl isonitrile, **5**, 796
with α-phosphoryl enamines, **4**, 1004
with pinacol(1-iodoethyl)boronate, **4**, 664
with tertiary alkyl isocyanides, **3**, 408
with tetramethylethylenediamine, **4**, 623
with 3,4,6-tri-*O*-*t*-butyldimethylsilyl-D-glucal, **4**, 651
with vinyl chlorosilanes, **3**, 502
with vinyl ethers, **4**, 929
with vinyltrimethylsilane, **4**, 606
as starting material, for acyllithiums, **5**, 437
use in the synthesis of alkyllithiums, **2**, 554
use in the synthesis of lithioferrocenes, **4**, 706
use in the synthesis of methylenedilithium, **4**, 668
Lithium, (*t*-butylcarbamoyl)-, reactions, with trimethylgermyl chloride, **6**, 515
Lithium, *t*-butyldiphenylsilyl-, synthesis of, **2**, 545
Lithium, *N*-*t*-butylimino pentylidin-, reactions, with *n*-hexyl-9-Borabicyclo[3.3.1.]nonane, **5**, 795
Lithium, 1-chloro-2,2-bis(4-chlorophenyl)vinyl-, structure of, **4**, 816
Lithium, chlorobis(trimethylsilyl)methyl-, synthesis of, **4**, 146
Lithium, 1-chloro-2,2-difluorovinyl-, use in the synthesis of 2-chloro-2-alkenals, **3**, 67
Lithium, chlorodiphenylmethyl-, synthesis of, **4**, 146
Lithium, chloromethyl-
reactions
with bromogermanes, **4**, 141
with chloro(methyl)phenylsilane, **4**, 136
with triphenylsilyl chlorides, **4**, 138
synthesis of, **1**, 681; **4**, 146
Lithium, 3-chloro-3-methylbut-2-yn-1-yl-, for the synthesis of allenic thioethers, **1**, 980
Lithium, cyanomethyl-, synthesis of, **3**, 640
Lithium, cyclopentadienyl-, reactions, with chlorodimethylarsine, **2**, 484
Lithium, cyclopropyl-, reactions, with triphenyl phosphite, **2**, 434
Lithium, diazomethyl-
reactions
with cadmium(II) chloride, **3**, 488
with chlorophosphanes, **6**, 668
as starting material
for bissilyldiazomethanes, **6**, 672
for α-diazophosphonothioic diamides, **6**, 667
synthesis of, **3**, 485; **6**, 670, 674
Lithium, dibromomethyl-
for the homologation of ketones, **3**, 125
reactions, with ketones, **4**, 746
as starting material, for enolates, **2**, 1023
use in the synthesis of dibromoalkanes, **4**, 22
Lithium, 1',2-dichloroferrocenyl-, use in the synthesis of dichloro-substituted carbon-bridged diiron complexes, **4**, 685
Lithium, dichloromethyl-
reactions
with alkyldialkoxyboranes, **2**, 529
with boronate esters, **4**, 526
use in the synthesis of dichloroalkanes, **4**, 19
Lithium, diethylphosphinyldifluoromethyl-, reactions, with ortho-α-bromoxylene, **6**, 58
Lithium, diisopropylphosphinyldifluoromethyl-, reactions, with *o*-tolualdehyde, **6**, 58
Lithium, dimesitylborylmethyl-, reactions, with trimethylsilyl chloride, **4**, 642
Lithium, α-dimethylaminobenzyl-, synthesis of, **2**, 1001
Lithium, *o*-*N*,*N*-dimethyl benzyl-, use in the synthesis of carbon-bridged dimanganese complexes, **4**, 677
Lithium, dimethylphenylsilyl-
reactions
with aldehydes, **4**, 355
with chloromethyl(dimethyl)phenylsilane, **4**, 609
with chlorosilanes, **4**, 609
with ketones, **4**, 355
synthesis of, **2**, 545; **4**, 355
Lithium, diphenylarsinomethyl-
from
bis(diphenylarsino)methane, **4**, 599
iodomethyldiphenylarsine, **4**, 599
reactions, with tributyl chloride, **4**, 600
Lithium, diphenyl-*t*-butylsilyl-
for the synthesis of isocyanides, **3**, 698
use in the synthesis of aryl isocyanides, **3**, 721
Lithium, diphenylmethyl-, reactions, with α-phenyl neopentyl chloride, **1**, 110
Lithium, diphenylstibinomethyl-, from, bis(diphenylstibino)methane, **4**, 600
Lithium, diphenylstibinylmethyl-, from, bis(diphenylstibine)methane, **1**, 271
Lithium, 1-ethoxyvinyl-
reactions
with copper iodide, **4**, 953
with zinc chloride, **4**, 954
Lithium, (*Z*)-2-ethoxyvinyl-, synthesis of, **4**, 818
Lithium, ethyl-
reactions
with 2,5-dichloro-3-iodoselenophene, **4**, 785
with 1,1-dithio-1,3-dienes, **4**, 255
with tetrabenzyl metals, **2**, 566
with 2,3,5-tribromoselenophene, **4**, 785
with trimethylvinylsilane, **1**, 761
Lithium, ethylthioethynyl-, synthesis of, **2**, 1031
Lithium, ferrocenyl-, reactions, with haloiron complexes, **4**, 669, 684
Lithium, α-fluorovinyl-, synthesis of, **4**, 812
Lithium, formyl-, stability of, **5**, 436
Lithium, heptafluoroisopropyl-, synthesis of, **4**, 146
Lithium, hexynyl-, reactions, with ethoxyhexyne, **1**, 1030
Lithium, iodomethyl-, synthesis of, **4**, 146
Lithium, isopropyl-
reactions, with pentachlorophosphorane, **2**, 434
as starting material, for dihydrido diiridium carbon-bridged complexes, **4**, 689
Lithium, mesityl-
reactions, with cobalt(II) chloride, **4**, 690
use in the synthesis of anions, **4**, 1077
Lithium, methoxydimethylsilyl(trimethylsilyl)methyl-, synthesis of, **6**, 190
Lithium, methoxy(phenylthio)methyl-, reactions, with dioxaborinanes, **4**, 364
Lithium, methoxy(trimethylsilyl)methyl-, synthesis of, **4**, 650
Lithium, methoxyvinyl-
as starting material
for lithium di(1-methoxyvinyl)cuprate, **4**, 953
for 1-methoxyvinylcopper(I), **4**, 953
Lithium, 1-methoxyvinyl-
for the displacement of fluoride, **4**, 621
pure, synthesis of, **4**, 935
reactions
with chlorometal complexes, **4**, 955
with cuprous iodide, **4**, 953
with trialkylboranes, **4**, 934
with triisobutylborane, **4**, 935
with triisopropylborate, **4**, 934
synthesis of, **4**, 949
Lithium, 1-methoxy-1-vinyl-
reactions
with chlorotrimethylgermane, **4**, 929
with chlorotrimethylsilane, **4**, 929
Lithium, methyl-
for the lithiation of metals, **4**, 706
for the metallation of tris(silyl)methanes, **6**, 174
pyrolysis of, **6**, 206
reactions
with alkenyl acetates, **2**, 696

with alkenylaluminums, **1**, 534
with 1-*O*-benzyl-2,3-isopropylidene-L-lyxose, **6**, 541
with bis(diisopropylamino)boryls, **3**, 682
with bis(silyl) sulfide, **5**, 538
with *C*-bromo ketenimines, **3**, 582
with 4-*t*-butylcyclohexanone, **2**, 68
with carbyne complexes, **3**, 518
with chloroiodomethane, **1**, 681
with chlorotris(trimethylphosphine)cobalt, **2**, 576
with chlorotris(triphenylphosphine)iridium, **5**, 460
with chromium(III) chloride, **4**, 708
with cyclopentadienylmanganese complexes, **5**, 448
with cyclopentadienylrhenium complexes, **5**, 450
with dinitrogen monoxide, **3**, 485
with ditins, **6**, 399
with fluorosilane, **6**, 392
with hexamethyldisilane, **4**, 355
with hexamethyltungsten, **2**, 572
with β-ketosilane, **1**, 761
with manganese pentacarbonyls, **5**, 448
with metal(II) chlorides, **2**, 572
with phenylvinyl sulfide, **4**, 366
with phosphonium salts, **2**, 475
with rhenium pentacarbonyls, **5**, 450
with silver cyanide, **2**, 589
with silyl enol ethers, **2**, 696
with sulfonylsulfines, **4**, 272
with 1,1,3,3-tetramethyl-1,3-disilacyclobutane, **4**, 606
with thiols, **2**, 272
with trifluoroiodomethane, **6**, 244
with trifluoromethyl dithioketals, **1**, 570
with trimethylarsenic dichloride, **2**, 488
with trimethyl metal dichlorides, **2**, 570
with tris(dialkoxyboryl)methanes, **6**, 199
with tris(dimethoxyboryl)methane, **4**, 633
with tris(trimethylsilyl)arsine, **5**, 1160
with tris(trimethylsilylmethyl) hydride, **6**, 391
stability of, **1**, 851
as starting material, for methylenedilithium, **4**, 668
use in the synthesis of bismethylene-bridged dirhodium complexes, **4**, 687
use in the synthesis of carbon-bridged diruthenium complexes, **4**, 680
use in the synthesis of lithiotrithioorthoformates, **6**, 317
use in the synthesis of trismethylenediruthenium complexes, **4**, 680
Lithium, methyl bis(trimethylsilyl)silyl-, reactions, with acyl chlorides, **5**, 418
Lithium, methylthiomethyl-
reactions
with trimethylamine–borane complexes, **4**, 373
with triphenyl phosphite, **2**, 434
Lithium, α-nitrobenzyl-, synthesis of, **2**, 1003
Lithium, 2-octyl-, racemization of, **1**, 846
Lithium, organotin-, reactions, with halophosphines, **2**, 868
Lithium, pentadienyl-, reactions, with phenylphosphonous dichloride, **2**, 436
Lithium, phenyl-
reactions
with alkoxysilanes, **2**, 942
with aminotriphenylphosphonium salt, **2**, 830
with benzyldiphenylphosphine oxide, **1**, 752
with carbodiimides, **5**, 777
with carbon monoxide, **5**, 436
with α-diazoacetates, **1**, 486
with dicyclohexylarsine, **2**, 510
with *N*,*N*'-dimethoxyurea, **5**, 331
with diphenyl trithiocarbonate, **6**, 87
with ferrocenophanes, **4**, 1034
with *N*-methyl-*N*-phenylcyanamide, **6**, 648
with osmium complexes, **3**, 690
with phenyl dithiobenzoate, **4**, 255
with phospholes, **4**, 1031
with quaternary ammonium halides, **1**, 628
with selenoamides, **4**, 334, 909
with stannaethenes, **6**, 193
with tetraethyldistibine, **2**, 510
with tetravinyltin, **1**, 1091
with thiobenzophenone, **4**, 394
with tricyclohexylgermanium bromide, **2**, 949
with triphenylsilyl isothiocyanate, **5**, 595
with tris(triphenylplumbyl)methane, **6**, 204
with tungsten hexacarbonyl, **5**, 933
as reagent, for the synthesis of borenes, **5**, 928
as starting material, for carbon-bridged triosmium complexes, **4**, 683
Lithium, phenyldimethylsilyl-, cuprate of, **2**, 546
Lithium, (phenyldimethylsilyl)chloromethyl-, reactions, with organoboranes, **4**, 357
Lithium, (phenyldimethylsilyl)dichloromethyl-, use in the synthesis of ketones, **3**, 135
Lithium, phenylethynyl-, reactions, with *N*,*N*'-dimethoxy-*N*,*N*'-dimethylurea, **1**, 1047
Lithium, phenylselenoalkyl-, reactions, with ketones, **3**, 125
Lithium, α-phenylselenoheptyl-, reactions, with cuprates, **4**, 401
Lithium, [2-phenyl-1-(silyl)vinyl]-, reactions, with carbon monoxide, **5**, 438
Lithium, phenylthio(diphenylphosphinoxido)methyl-, reactions, with β-silylacroleine, **2**, 939
Lithium, phenylthiomethyl-
reactions, with trimethylborate, **4**, 372
titanation of, **4**, 396
Lithium, phenylthiotrimethylsilylmethyl-, reactions, with esters, **3**, 166
Lithium, 2-phenyl-2-(*p*-tolyl)propyl-, rearrangements of, **1**, 378
Lithium, *iso*-propyl-, reactions, with phenylvinyl selenide, **4**, 401
Lithium, ruthenocenyl-, as starting material, for diruthenocenylmercury, **4**, 723
Lithium, α-silyl acyl-, from, α-lithiosilanes, **5**, 415
Lithium, α-silylchloromethyl-, synthesis of, **4**, 357
Lithium, [1-(silyl)vinyl]-, reactions, with carbon monoxide, **5**, 438
Lithium, stibinomethyl-, reactions, with diphenylchloroarsine, **4**, 596
Lithium, tetrathiafulvalenyl-, formylation of, **3**, 73
Lithium, *p*-tolylthiomethyl-, reactions, with menthyl toluene-*p*-sulfinate, **4**, 265
Lithium, trialkyl stannyl-, reactions, with imidoyl chlorides, **5**, 803
Lithium, triallyl plumbyl-, reactions, with imidoyl chlorides, **5**, 803
Lithium, tribromomethyl-
reactions, with cyclohexanone, **6**, 31
as starting material, for tribromomethylcarbinols, **6**, 31
synthesis of, **6**, 244
Lithium, 2,4,6-tri-*t*-butylphenyl-
reactions
with *O*-*t*-butyl selenoformates, **5**, 524
with *O*-cholesteryl selenoformates, **5**, 524
Lithium, tri-*t*-butylphenyl-, reactions, with selenoformates, **3**, 383
Lithium, tributylstannyl-
addition to aldehydes and ketones, **4**, 386
reactions
with aminomethyl phenyl sulfides, **4**, 533
with benzyl chloromethyl ether, **4**, 389
with (chloromethyl)trimethylsilane, **4**, 660
with thionolactones, **6**, 134
Lithium, tri-*n*-butylstannyl-, reactions, with bicyclo[2,2,1]hept-5-en-2-one, **4**, 952
Lithium, tributylstannylmethyl-
reactions
with diphenylchloroarsine, **4**, 600

with trimethylsilyl chloride, **4**, 660
Lithium, tributylzinc-, reactions, with bromochloroalkene, **1**, 431
Lithium, (trichloromethyl)-
from, butyllithium, **6**, 244
reactions
with carbonyls, **6**, 25
with iodoalkanes, **6**, 23
synthesis of, **6**, 244
Lithium, triethylgermyl-
reactions
with amides, **6**, 515
with *N,N*-dialkyl carboxamides, **5**, 432
with dimethylamides, **6**, 131
with pyrrolidinones, **5**, 794
as starting material
for bis(triethylgermyl)carbinol, **6**, 131
for ketones, **6**, 131
Lithium, triethylplumbyl-, reactions, with ethoxyethyne, **4**, 955
Lithium, triethylsilyl-, reactions, with phenyl triethylgermyl ketone, **6**, 131
Lithium, (trifluoromethyl)-
stability of, **6**, 13
synthesis of, **6**, 244
Lithium, trifluorovinyl-
reactions
with carbonyls, **3**, 67
with phosphorus(III) halides, **4**, 802
Lithium, trimethylgermylmethyl-, synthesis of, **4**, 663
Lithium, trimethylsilyl-
cuprate of, **2**, 546
reactions, with thiobenzaldehydes, **4**, 376
synthesis of, **2**, 546
Lithium, (trimethylsilyl)bromomethyl-, reactions, with organoboranes, **4**, 357
Lithium, trimethylsilylcyclopentadienyl-
reactions
with dichlorodimethylsilane, **4**, 603
with trimethylsilyl chloride, **4**, 603
Lithium, trimethylsilyldiazomethyl-, as starting material, for trimethylgermyltrimethylsilyldiazomethane, **3**, 483
Lithium, (trimethylsilyl)dichloromethyl-, reactions, with aldehydes, **4**, 742
Lithium, trimethylsilylmethyl-
from, butyl trimethylsilyl telluride, **4**, 652
reactions
with chloroarsines, **4**, 597
with stiboranes, **4**, 599
with *β*-tetralone, **1**, 759
with thallium trichloride, **4**, 659
with trialkylboranes, **4**, 642
with triphenyltin chloride, **4**, 660
synthesis of, **4**, 652
Lithium, trimethylsilyl(phenylthio)methyl-, reactions, with cyclohexene oxide, **4**, 943
Lithium, trimethylsilylpinacolborylmethyl-, as starting material, for alkenylpinacolboronates, **2**, 905
Lithium, 1-trimethylsilyl propynyl-, transmetallation of, **1**, 973
Lithium, *α*-(trimethylsilyl)vinyl-, use in the synthesis of coronafacic acid, **1**, 760
Lithium, trimethylstannyl-
reactions
with benzimidoyl chlorides, **5**, 803
with chloroform, **6**, 208
with enones, **2**, 600
with pinacol(1-chloroethyl)boronate, **4**, 665
Lithium, trimethylstannylmethyl-
reactions
with *α*-chloro ketones, **1**, 664
with indium chlorides, **4**, 726
Lithium, (trimethylstannylmethyl)-, synthesis of, **4**, 708
Lithium, trimethyltin-, reactions, with dihaloalkanes, **4**, 699
Lithium, triorganylplumbyl-, reactions, with chloroform, **6**, 209
Lithium, triphenylgermyl-
reactions
with acyl chlorides, **5**, 432
with carboxylic esters, **5**, 432
with imidoyl chlorides, **5**, 794
with trifluorovinylsilanes, **4**, 811
as starting material, for bis(triphenylgermyl)methanol, **6**, 131
Lithium, triphenylgermylmethyl-, synthesis of, **4**, 663
Lithium, triphenyllead-, reactions, with dibromomethane, **4**, 699
Lithium, triphenylsilyl-
cuprate of, **2**, 546
reactions
with acetyl chloride, **5**, 417
with ethyl formate, **6**, 125
with sulfur, **5**, 254
synthesis of, **2**, 545
Lithium, triphenylstannyl-, reactions, with benzimidoyl chlorides, **5**, 803
Lithium, (triphenylstannyl)methyl-, reactions, with benzeneselenenyl bromide, **4**, 401
Lithium, tris(phenylseleno)methyl-, reactions, with triphenylphosphine, **6**, 697
Lithium, tris(phenylthio)methyl-, reactions, with phosphines, **6**, 697
Lithium, tris(trimethylsilylmethyl)-, reactions, with aluminum chloride, **6**, 395
Lithium, tris(trimethylsilyl)methyl-, reactions, with electrophiles, **4**, 1048
Lithium, tris(trimethylsilylmethyl)-
reactions
with electrophiles, **6**, 373
with *O*-ethyl thioformate, **6**, 176
with indium chloride, **6**, 395
with lead halides, **6**, 396
with manganese chloride, **6**, 396
with mercury chloride, **6**, 394
with methyl iodide, **6**, 174
with phosphorus trichloride, **6**, 165
Lithium, tris(trimethylsilyl)methyl-, reactions, with styrene oxide, **4**, 620
Lithium, tris(trimethylsilylmethyl)-
reactions
with thallium chloride, **6**, 395
with (trimethylsilyl)aminodifluorophosphines, **6**, 372
with zinc chloride, **6**, 393
as starting material, for aliphatic thioaldehydes, **6**, 176
as starting materials, for tetrasilylmethanes, **6**, 379
synthesis of, **6**, 174, 390
use in the synthesis of monochalcogenomethanes, **6**, 349
Lithium, tri*o*-tolylsilyl-, synthesis of, **2**, 545
Lithium, vinyl-
reactions, with selenenyl halides, **2**, 727
as starting material, for vinyl fluorides, **2**, 611
synthesis of, **1**, 1091
Lithium, (2,6-xylylimino)(trimethylsilyl)methyl-, as starting material, for (2,6-xylylimino)bissilanes, **6**, 672
Lithium acyl metallates, as starting materials, for carbene complexes, **6**, 719
Lithium alanates, synthesis of, **4**, 816
Lithium alanates, propargylic, from, methylalkynes, **1**, 973
Lithium aldimines
carbonation of, **5**, 88
as intermediates, **3**, 408
synthesis of, **3**, 418; **5**, 795, 796
See also Lithium intermediates, imidoyl-
Lithium alkoxide
reactions
with chlorodialkylaminoalkylphosphines, **2**, 451

with cyclic ketones, **5**, 557
Lithium alkoxides
from, trimethylsilyl carboxylate esters, **5**, 414
reactions
with *N*,*N*-dialkylalkoxymethyleneiminium salts, **1**, 828
with methanesulfenyl chloride, **2**, 194
with sulfones, **4**, 235
use in the synthesis of alkenes, **1**, 693
Lithium alkoxyethynylides
from, 1,2-dichlorovinyl ethers, **2**, 1026
reactions, with phosphine halides, **2**, 1027
synthesis of, **2**, 1025
Lithium alkyls
covalent character of, **1**, 885
reactions
with fulvenes, **1**, 862
with methylenetrialkylphosphoranes, **6**, 706
with methyltriphenylphosphorane, **6**, 705
with tin tetrachloride, **4**, 949
Lithium alkynide
alkylation of, **1**, 1000
reactions
with dialkylalkenylboranes, **1**, 488
with enones, **3**, 62
with lactones, **4**, 171
with 2-methoxycyclopentanone, **1**, 1039
with oxiranes, **1**, 1022
as starting material, for 1-alkynes, **1**, 1034
Lithium alkynide complexes, alkylation of, **1**, 1001
Lithium alkynide–ethylenediamine complex, alkylation of, **1**, 1000
Lithium alkynides
alkoxycarbonylation of, **5**, 103
from
organolithiums, **1**, 1069
terminal alkynes, **1**, 111
reactions
with alkoxyenones, **3**, 228
with aluminum trichloride, **1**, 1003
with ammonia, **1**, 1000
with carbon disulfide, **5**, 549
with dialkylaluminum halides, **1**, 1139
with dialkylchlorophosphines, **2**, 1059
with dialkyl phosphorochloridites, **2**, 1064
with dichloro(phenyl)phosphine, **2**, 1060
with epoxy alcohols, **1**, 223, 1023
with germanium tetrachloride, **2**, 1093
with hydroxylamine sulfonates, **2**, 1058
with lactones, **3**, 244
with 1,1,1-triphenyl-3-penten-2-one, **1**, 1048
with zinc chloride, **1**, 1139
synthesis of, **1**, 1063, 1133
transmetallation of, **1**, 1133
Lithium alkynolates
reactions
with acid chlorides, **2**, 1030
with silyl chlorides, **2**, 1024
synthesis of, **2**, 1023
Lithium alkynolate salts, stability of, **2**, 1023
Lithium alkynylides
bromination of, **2**, 1018
reactions
with arylselenium bromide, **2**, 1037
with aryltellurium bromide, **2**, 1037
with bis(trimethylsilyl) peroxide, **2**, 1019
with sulfur, **2**, 1032
Lithium (1-alkynyl)organoborates
reactions
with electrophiles, **1**, 1033
with iodine, **1**, 1033
with methanesulfinyl chloride, **1**, 1033
with tributyltin chloride, **1**, 1033
Lithium alkynylthiolates, synthesis of, **1**, 1077
Lithium alkynyltrifluoroborates, reactions, with carboxylic anhydrides, **1**, 1045
Lithium allyl compounds, reactions, with alkenes, **1**, 363
Lithium aluminum amides
reactions
with aldehydes, **3**, 405
with aryl epoxides, **2**, 318
with ketones, **3**, 405
Lithium aluminum deuteride (LAD), for the reduction of ketene dithioacetals, **4**, 253
Lithium aluminum hydride (BINAL-H), 2,2'-dihydroxy-1,1'-binaphthyl-modified, for the reduction of acylstannanes, **4**, 386
Lithium aluminum hydride (LAH)
for the cleavage of carbon–phosphorus bonds, **1**, 51
reactions
with alkenes, **2**, 596
with alkoxyboranes, **2**, 535
with alkynenitriles, **3**, 642
with allylic alcohols, **1**, 429
with amides, **1**, 34
with α-azido oxime ethers, **2**, 314
with (1,1'-bi-2,2'-naphthol, **2**, 63
with chloroalkynes, **4**, 816
with 1-chloro 1-alkynes, **4**, 758
with 3-chloro-3,3-difluoro-2-trifluoromethylprop-1-ene, **4**, 735
with chlorosilanes, **2**, 536
with 1,1-diarylethenes, **1**, 80
with 1-phenyl-1-propynes, **1**, 512
with propargyl alcohols, **1**, 513
with propargyl amines, **1**, 514
with tetrahydrofuran, **1**, 3
with titanium tetrachloride, **4**, 1082
with *p*-toluenesulfonyl hydrazones, **1**, 39
with transition metal salts, **1**, 72
with trichloride, **2**, 571
with ureas, **4**, 414
as reducing agent, for the synthesis of allyl-bridged palladium complexes, **4**, 693
in reductions, **2**, 643
use in the reduction of aldehydes, **1**, 18
use in the reduction of alkylarsenic halides, **2**, 481
use in the reduction of alkyl iodides and bromides, **1**, 3
use in the reduction of alkynes, **1**, 512
use in the reduction of amides, **2**, 319; **3**, 85
use in the reduction of α-amino acids, **3**, 40
use in the reduction of aryl bromides, **1**, 428
use in the reduction of aryl chlorides, **1**, 427
use in the reduction of aryl fluorides, **1**, 426
use in the reduction of benzylic sulfones, **1**, 24
use in the reduction of benzylic sulfoxides, **1**, 23
use in the reduction of bromoalkanes, **1**, 8
use in the reduction of carbonyls, **2**, 60
use in the reduction of chlolesteryl thiocyanate, **2**, 127
use in the reduction of chloroalkanes, **1**, 7
use in the reduction of esters, **1**, 17
use in the reduction of fluoroalkanes, **1**, 6
use in the reduction of glycidols, **1**, 14
use in the reduction of 3-hydroxypropynylsilanes, **2**, 921
use in the reduction of iodoalkanes, **1**, 10
use in the reduction of ketene dithioacetals, **4**, 253
use in the reduction of ketones, **1**, 19
use in the reduction of nitroalkanes, **1**, 37
use in the reduction of oxiranes, **1**, 14
use in the reduction of phosphine oxides, **2**, 441
use in the reduction of phosphonium salts, **2**, 442
use in the reduction of vinyl bromides, **1**, 431
use in the reduction of vinyl chlorides, **1**, 430
use in the reduction of vinyl fluorides, **1**, 429
use in the synthesis of alkylphosphines, **2**, 428
use in the synthesis of amines, **2**, 314
use in the synthesis of arenethiols, **2**, 708

use in the synthesis of aziridines, **2**, 322
use in the synthesis of benzylamines, **1**, 57
use in the synthesis of dialkylchloroboranes, **2**, 522
use in the synthesis of 2,2-diaryl ethenols, **2**, 645
use in the synthesis of formamidines, **5**, 746
use in the synthesis of secondary phosphines, **2**, 430, 822
use in the synthesis of sulfinic acids, **2**, 187
use in the synthesis of thiols, **2**, 126
use in the synthesis of α,β-unsaturated aldehydes, **3**, 67
Lithium amide bases
enantiomerically pure, **1**, 144
for enantioselective deprotonations, **1**, 135
Lithium amides
Ireland enolization model for, **1**, 876
reactions
with aldehydes, **4**, 299
with alkynyl chlorides, **2**, 1041
with alkynyl ethers, **2**, 1043
with allylic esters, **1**, 830
with amide chlorides, **2**, 1045
with arenetellurenyl iodides, **2**, 370
with chloro enamines, **4**, 971
with Grignard reagents, **1**, 580
with phosphorus trichloride, **2**, 466
with trichloroethene, **4**, 976
for regioselective enolate formation, **1**, 131
as starting materials, for enolates, **1**, 135
synthesis of, **2**, 315, 420
use in the synthesis of amides, **5**, 263
Lithium amides, *N*-alkyl-, synthesis of, **2**, 420
Lithium amides, *N*-silyl-, reactions, with organoarsenic halides, **2**, 891
Lithium aminoborohydrides, use in the reduction of carbonyls, **2**, 59
Lithium anilide, reactions, with (-)-(*S*)-methylmethylphenylphosphinate, **2**, 807
Lithium–arene complexes, as starting materials, for alkyllithiums, **2**, 551
Lithium arsenides
reactions, with imidoyl chlorides, **5**, 791
synthesis of, **5**, 791
Lithium arsides, reactions, with cyclohexyl isocyanate, **6**, 513
Lithium arsolides, bonding of, **2**, 895
Lithium aryl tellurides, as starting materials, for perfluoroalkyl tellurides, **6**, 50
Lithium bases, reactions, with organic halides, **2**, 554
Lithium benzyloxide
reactions
with boronates, **4**, 362
with (*S*,*S*)-2,5-dimethylhexane-3,4-diol, **4**, 362
Lithium biphenylide
reactions
with barium(II) iodide, **1**, 116
with barium iodide, **2**, 563
with calcium bromide, **2**, 563
with copper(I) salts, **2**, 585
with magnesium halides, **2**, 561
with trifluoromethyl dithioacetals, **4**, 775
use in the reduction of calcium halides, **1**, 115
Lithium bis(2-adamantyl)amide (LBAA), synthesis of, **1**, 131
Lithium bis(boryl)methides, synthesis of, **6**, 201
Lithium bis(dialkoxyboryl)methides, reactions, with organoelement halides, **6**, 200
Lithium bis(plumbyl)methides, synthesis of, **6**, 204
Lithium bis(stannyl)methide, synthesis of, **6**, 210
Lithium bis(trialkylsilyl)amides, reactions, with carbonyls, **3**, 456
Lithium bis(trifluoroacetate), aryl-, as starting material, for biaryls, **1**, 491
Lithium bis(trimethylgermyl)methide, synthesis of, **6**, 202
Lithium 1,1-bis(trimethylsilyl) alkoxides, reactions, with benzophenone, **4**, 932
Lithium bis(trimethylsilyl)amide
reactions
with benzonitriles, **5**, 777
with chloromethyl methyl ether, **4**, 313
with diphenyltrifluorophosphorane, **2**, 861
Lithium bis(trimethylsilylamide)amide, reactions, with benzoic acid esters, **5**, 721
Lithium bis(trimethylsilyl)arsine, reactions, with 2,4,6-tris(*t*-butyl)benzoyl chloride, **5**, 1160
Lithium bis(trimethylsilyl)arsine complex, from, tris(trimethylsilyl)arsine, **2**, 509
Lithium bis(trimethylsilyl)chloromethane, reactions, with dichlorophosphines, **6**, 692
Lithium bis(trimethylsilyl)methanide
reactions
with aluminum chloride, **6**, 196
with lanthanum aryl oxides, **6**, 194
with metallic halides, **6**, 194
transmetallation reactions of, **6**, 195
Lithium bis(trimethylsilyl)phosphide, use in the synthesis of phosphaalkynes, **5**, 1155
Lithium bis(triphenylplumbyl)methides, synthesis of, **6**, 210
Lithium borohydride
as catalyst, for hydroboration, **2**, 43
as hydroborating agent, for vinyl iodides, **2**, 628
use in the reduction of carbonyls, **2**, 59
use in the reduction of glycidols, **1**, 14
use in the ring-opening of 2,3-epoxy alcohols, **2**, 83
Lithium-boron *ate* complexes, ^{11}B NMR spectrum of, **4**, 935
Lithium borylmethides, synthesis of, **6**, 201
Lithium borylmethide salts, synthesis of, **6**, 201
Lithium bromide
reactions
with *N*-aziridylphthalimides, **5**, 364
with 3-pentanone, **2**, 31
with silver compounds, **2**, 589
Lithium–bromine exchange, for the synthesis of *gem*-dibromoalkenes, **4**, 818
Lithium bromoethynylides
reactions, with triphenylchlorosilane, **2**, 1017
synthesis of, **2**, 1017
Lithium α-bromomethyl acrylate, reactions, with alkyl *p*-tolyl sulfoxides, **1**, 143
Lithium butanetellurolate, reactions, with *N*,*N*-dimethylmethyleneammonium chloride, **4**, 334
Lithium *t*-butylamide, reactions, with carbon monoxide, **6**, 515
Lithium *t*-butylbis(trimethylsilyl)acetate, synthesis of, **6**, 190
Lithium butyl(diisobutyl)aluminum hydride, reactions, with phenyl (1-trimethylsilyl)propadienyl sulfide, **4**, 946
Lithium *n*-butyl diisobutylaluminum hydride, for the reduction of amides, **3**, 85
Lithium *t*-butyl peroxide, for the oxidation of alkenes, **6**, 99
Lithium *n*-butyl selenide, synthesis of, **6**, 481
Lithium butyltriallylborates, synthesis of, **1**, 280
Lithium carbanions, synthesis of, **1**, 129
Lithium carbenoids
reactions, with metal salts, **4**, 820
and transmetallation reactions, **4**, 821
use of, **6**, 62
Lithium carbenoids, α-haloalkyl-
in asymmetric synthesis, **4**, 146
synthesis of, **4**, 146
Lithium carbonate, use in dehydrochlorination reactions, **1**, 571
Lithium cations, chelation of, **5**, 150
Lithium chloride
reactions, with acetone, **2**, 25
use in the synthesis of alkynes, **1**, 1027
Lithium (*E*)-1-chloro-1-alkenyl alanates, synthesis of, **4**, 758
Lithium chloroalkynide, use in the synthesis of disubstituted alkynes, **1**, 1034
Lithium chlorodifluoroacetate, use in the synthesis of difluorocarbenes, **4**, 732

Lithium chloroethynylide, condensation of, **2**, 1015
Lithium–chromium hexamethyl compounds, synthesis of, **2**, 571
Lithium complexes, synthesis of, **6**, 398
Lithium complexes, alkyl-, reactions, with tetramethylethylenediamine, **6**, 391
Lithium complexes, alkynyl-, synthesis of, **1**, 1134
Lithium cuprates
  reactions, with arene tricarbonylmanganese complexes, **5**, 448
  use in the synthesis of allenes, **1**, 966
Lithium cuprates, alkenyl-, synthesis of, **1**, 1093
Lithium cuprates, alkylethynyl-, reactions, with electrophiles, **1**, 1136
Lithium cuprates, dialkyl-
  reactions, with alkynes, **1**, 1107
  use in the synthesis of ylides, **3**, 496
Lithium cyanide
  reactions
    with 1,2-dithietane-1,1-dioxides, **1**, 622
    with vinylsulfoximines, **1**, 622
  use in the synthesis of nitriles, **3**, 613
  use in the synthesis of trimethylsilyl cyanide, **5**, 1134
Lithium cyclopentadienyliron alkyldicarbonyl, carbonation of, **5**, 97
Lithium dialkenylcuprates, reactions, with vinyl halides, **1**, 1097
Lithium dialkylamides
  reactions
    with 4-aryl-1,1-dichloro-1-buten-3-ynes, **2**, 307
    with aryl imidate ethers, **3**, 721, 722
    with 1-chloro-1*H*-phosphirine, **2**, 859
    with disulfides, **2**, 243
    with polyfluoroalkenes, **4**, 793
  use in the synthesis of enolates, **1**, 874
Lithium dialkylarsenides, reactions, with dichloroalkanes, **2**, 486
Lithium dialkylcuprates
  reactions
    with allenyl halides, **1**, 1015
    with primary iodoalkanes, **1**, 120
  as reagents, for displacement reactions, **1**, 117
Lithium dialkylphosphides, reactions, with trialkoxycarbenium ions, **6**, 314
Lithium dialkynyldiethylaluminates, reactions, with episulfonium tetrafluoroborates, **1**, 1031
Lithium diarylphosphides, hydrolysis of, for diarylphosphines, **2**, 823
Lithium diazoacetate, use in the synthesis of diazoacetates, **3**, 484
Lithium di-*t*-butylarsenide, reactions, with chlorodi-*t*-butylgallane, **2**, 511
Lithium 4,4'-(di-*t*-butyl)biphenylide (LDBB)
  reactions
    with epoxides, **2**, 553
    with oxetanes, **2**, 553
    with vinyl sulfides, **2**, 68
  use in metallation reactions, **2**, 551
Lithium dibutylcopper, reactions, with acrylaziridines, **1**, 263
Lithium dibutylcuprate, reactions, with phospholenes, **1**, 159
Lithium di-*n*-butylcuprate, reactions, with *β*-styrylmethyl thioether, **1**, 484
Lithium dibutylphosphide, synthesis of, **2**, 475
Lithium di-*t*-butylstibide dioxane complex, reactions, with 1,4-dichlorobutane, **2**, 506
Lithium dichloromethyl methyl ether, use in the synthesis of ketones, **3**, 135
Lithium dicyclohexamide, for the deprotonation of phenylvinyl tellurides, **4**, 965
Lithium dicyclohexylamide (LDCA)
  reactions, with bis(trimethylstannyl)methane, **6**, 210
  use in the synthesis of anions, **4**, 1077
Lithium dicyclohexylarsenide
  from, dicyclohexylarsine, **2**, 485
  as starting material, for tetracyclohexyldiarsine, **2**, 506
  synthesis of, **2**, 510
Lithium diethylamide
  reactions, with allyl(phenylseleno)silane, **6**, 136
  as starting material, for ethoxyethyne, **2**, 1025
  use in the synthesis of imidic acid esters, **5**, 696
Lithium diethylarsenide, reactions, with bis(dimethylamino)chloroborane, **2**, 509
Lithium diethylphosphide, reactions, with alkyltrichlorostannane, **2**, 474
Lithium diethylstibide, synthesis of, **2**, 510
Lithium difluoroalkylsulfinates, from, iodofluoroalkanes, **6**, 45
Lithium dihydrogen arsenide, reactions, with mesitoyl chloride, **5**, 891
Lithium dihydrogen phosphide, reactions, with aromatic acids, **5**, 879
Lithium diisobutylamide, reactions, with bis(phenylseleno)methane, **4**, 401
Lithium diisopropylamide (LDA)
  as deprotonation agent, **4**, 398
  for the enolization of carbonyls, **2**, 696
  for the generation of anions, **1**, 155
  for the lithiation of triselenoorthoformates, **6**, 317
  reactions
    with *α*-aminonitriles, **2**, 741
    with aryl triflates, **1**, 1071
    with bromoform, **4**, 746
    with carboxylic esters, **6**, 550
    with chloromethyl phenyl sulfoxide, **3**, 170
    with 1,1-dichloroethane, **4**, 813
    with diethylbenzyl phosphonate, **2**, 1003
    with diethyl difluoromethyl phosphonate, **6**, 266
    with (*E*)-dimethyl 2-phenylvinyl phosphonate, **4**, 927
    with di(phenyltelluro)methane, **4**, 290
    with 1,3-dithiolane 1,3-dioxides, **2**, 247
    with *α*-methoxyallylphosphine oxide, **4**, 917
    with *α*-silyloxyphosphonates, **6**, 119
    with 5-substituted isoxazoles, **3**, 653
    with sulfimides, **6**, 326
    with sulfonates, **1**, 183
    with 2,3,4,5-tetrahydro-6-methylpyridine, **2**, 743
    with thiocarboxylic acids, **5**, 232
    with thiophenyl esters, **4**, 932
    with tolyl isonitriles, **5**, 620
    with tosylhydrazones, **1**, 630
    with *α*-trimethylsilyl-*α*-phenylselenyltoluene, **6**, 130
  use in dehydrofluorination reactions, **1**, 570
  use in the synthesis of alkanoic esters, **5**, 148
  use in the synthesis of enolates, **1**, 131
  use in the synthesis of *α*,*β*-unsaturated ketones, **3**, 231
  use in 2,3 Wittig rearrangements, **1**, 404
Lithium diisopyramide, for the deprotonation of phenylvinyl selenides, **4**, 965
Lithium dimesitylborylmethide, synthesis of, **6**, 199
Lithium di(1-methoxyvinyl)cuprate, from, methoxyvinyllithium, **4**, 953
Lithium dimethylamide, reactions, with tetramethylchloroformamidinium chloride, **6**, 361
Lithium 1-(dimethylamino)naphthalenide (LDMAN)
  reactions, with cyclohexylidenes, **4**, 945
  for the reductive lithiation of *α*-phenyl thioethers, **4**, 380
  use in metallation reactions, **2**, 551
  use in the synthesis of *α*-lithio sulfides, **4**, 394
  use in the synthesis of *α*-thiosilanes, **4**, 366
Lithium dimethylarsenide, synthesis of, **2**, 510
Lithium dimethylcuprate
  reactions
    with *N*,*N*-bistrifluoromethanesulfonimides, **1**, 257
    with 1-bromo-4-*t*-butylcyclohexane, **1**, 118
    with chloroselenurane salts, **2**, 369

with diethyl chlorophosphate, **5**, 103
with methyl phenylpropiolate, **1**, 540
Lithium dimethyl phosphonate, reactions, with vinylsulfoximines, **1**, 622
Lithium diorganocuprates
asymmetric conjugate addition of, **1**, 257
as nucleophiles, **1**, 253
reactions
with alkynyl halides, **1**, 1018
with ethyne, **1**, 541
Lithium diorganophosphides, synthesis of, **2**, 432
Lithium diphenylcuprate
reactions
with alkyl bromides, **1**, 119
with chloroselenurane salts, **2**, 369
with α-fluoro sulfoxides, **4**, 777
Lithium diphenylphosphide
reactions
with alkyl halides, **1**, 52
with *N*-alkylthioformamidoyl chloride, **6**, 584
with aryl halides, **1**, 52
with *N*-arylthioformamidoyl chloride, **6**, 584
with α-chloro tertiary amines, **4**, 455
use in the reduction to ketones, **3**, 129
Lithium diphosphinomethanides, as starting materials, for phosphorus ylides, **5**, 909
Lithium disulfide, acylation of, **5**, 252
Lithium dithiocarbamates
from, amines, **5**, 1024
synthesis of, **6**, 560
Lithium di(trimethylsilyl)amide, use in the synthesis of anions, **4**, 1076
Lithium divinylcuprates
reactions, with alkynes, **3**, 222
synthesis of, **1**, 1093
Lithium enolates
alkylation of, with vinyl halides, **1**, 467
for the alkylation of aldehydes, **3**, 10
conjugate addition of, **1**, 304
from, 1-silyloxyvinyltin, **5**, 415
metal exchange of, **2**, 701
reactions
with alkenyl bromides, **1**, 467
with amino methyl benzotriazoles, **3**, 186
with carbon disulfide, **6**, 550
with 2,3-diazabutadiene, **5**, 364
with dimethylzinc, **1**, 132
with *O*-ethyl *S*-(2-oxoethyl) dithiocarbonate, **6**, 552
with haloborane, **2**, 698
with nitroalkenes, **3**, 435
with trialkylsilyl chlorides, **4**, 832
with vinyl ether–iron complexes, **1**, 481
synthesis of, **2**, 696
transmetallation of, **2**, 701, 702
vinylation of, **3**, 138
Lithium enolates, tertiary, reactions, with haloalkynes, **1**, 1108
Lithium ester enolates
reactions
with *N*-(alkylamino)benzotriazoles, **2**, 319
with α-iminoketones, **3**, 414
synthesis of, **1**, 130, 876
Lithium ethanethiolate, reactions, with imidoyl iodides, **5**, 728
Lithium ethoxide
reactions
with 2,3-di(*t*-butylimino)propiononitrile dihydrochloride, **4**, 894
with phenylnitromethane, **2**, 1003
Lithium (ethyl diazoacetate)
reactions
with allyl iodide, **3**, 474
with 2-chloro-1,2,3-benzodioxaborole, **3**, 485
with methyl iodide, **3**, 474
Lithium ethyneselenolates, dimerization of, **4**, 874
Lithium ethynyltrihexylborate, deprotonation of, **2**, 1088
Lithium formate, as formylating reagent, **3**, 90
Lithium halides, elimination of, **1**, 985
Lithium halides, bis(trimethylsilyl)-, synthesis of, **6**, 293
Lithium halides, naphthyl-, reactions, with dialkylformamides, **3**, 101
Lithium halide salts, synthesis of, **1**, 478
Lithium–halogen exchange
for the synthesis of α-fluoroalkenyllithiums, **4**, 817
for the synthesis of *gem*-halo lithioalkenes, **4**, 816
tertiary alkyllithiums as reagents for, **2**, 961
Lithium heptynide, reactions, with aziridines, **1**, 1032
Lithium hexamethyldisilazide (LHMDS), use in the synthesis of azidomethylboronates, **4**, 526
Lithium hexynide, reactions, with crotyl chloride, **1**, 1001
Lithium hydrazide, use in the synthesis of hydrazonamides, **5**, 848
Lithium hydridoalkoxyaluminates, use in the reduction of carbonyls, **2**, 60
Lithium (μ-hydrido)trialkylalanate, synthesis of, **6**, 197
Lithium hydridotri-*t*-butoxyaluminate, use in the reduction of carbonyls, **2**, 60
Lithium hydroxide, reactions, with ketones, **6**, 556
Lithium intermediates, imidoyl-
from, isonitriles, **5**, 796
synthesis of, **5**, 796
See also Lithium aldimines
Lithium iodide
reactions
with chloromethyl methyl sulfide, **4**, 71
with 2,4,6-collidine, **5**, 39, 78
with dimethylformamide, **5**, 78
with 2,6-lutidine, **5**, 78
use in the reduction of hydroxy ketones, **2**, 61
Lithium–iodine exchange, for the synthesis of iodo compounds, **1**, 110
Lithium *N*-isopropylcyclohexylamide, reactions, with aryl benzyl sulfoxides, **1**, 142
Lithium isopropylcyclohexylamide (LICA)
for the enolization of carbonyls, **2**, 696
use in the synthesis of alkanoic esters, **5**, 148
Lithium ketimines, reactions, with phosphorus chlorides, **3**, 456
Lithium (*Z*)-β-lithioacrylate, reactions, with keto isocyanides, **3**, 718
Lithium–manganese complexes, use in the synthesis of organomanganese(II) compounds, **2**, 574
Lithium metal
reactions
with acetals, **1**, 397
with triphenylacetic benzoic anhydride, **3**, 84
for the reduction of thiol esters, **3**, 85
Lithium metal vapor, reactions, with dichlorobis(trimethylsilyl)methane, **4**, 618
Lithium methaneselenate, reactions, with aryl halides, **2**, 729
Lithium methanetellurolate, as starting material, for methyltelluro(trimethylsilane), **2**, 294
Lithium methide, synthesis of, **6**, 203
Lithium methoxide, reactions, with tin compounds, **6**, 200
Lithium methoxy alkynyldialkylborinates, reactions, with boron trifluoride diethyletherate, **2**, 1084
Lithium methylamide, reactions, with benzonitrile, **5**, 777
Lithium (*R*)-(α-methylbenzyl)benzylamide, reactions, with *t*-butyl cinnamate, **2**, 315
Lithium (*S*)-(α-methylbenzyl)benzylamide, reactions, with *t*-butyl cinnamate, **2**, 315
Lithium methylcyanamide, synthesis of, **5**, 1123
Lithium 4-methyl-2,6-di(*t*-butyl) phenoxide, use in the synthesis of carboxylic acids, **5**, 97
Lithium methylphenylphosphonium dimethanide, reactions, with tricyclopentadienyluranium chloride, **5**, 920

Lithium methyl(trialkyl)borates, reactions, with copper(I) salts, **1**, 125
Lithium *N*-methyl-*N*-trimethylsilyl amides, reactions, with diazomethane, **3**, 485
Lithium methyl tris(trimethylsilyl)aluminum, reactions, with acyl chlorides, **5**, 418
Lithium naphthalene, reactions, with benzophenones, **1**, 136
Lithium naphthalenide (LN)
  reactions
    with bis(phenylthio)bis(trimethylsilyl)methane, **4**, 945
    with ketene thioacetals, **4**, 956
    with tetrahydrofuran, **1**, 405
    with trifluoromethyl dithioacetals, **4**, 775
    with zinc chloride, **2**, 590
    with zinc halides, **1**, 123
  for the reductive lithiation of α-(phenylthio) ethers, **4**, 380
  use in metallation reactions, **2**, 551
  use in the synthesis of alkylmagnesiums, **2**, 558
  use in the synthesis of α-lithio sulfides, **4**, 394
  use in the synthesis of α-thiosilanes, **4**, 366
Lithium naphthalide
  reactions, with sulfides, **4**, 1026
  for the synthesis of active metals, **2**, 958
Lithium nitronate, reactions, with phenyl selenenyl bromide, **2**, 784
Lithium nucleophiles, α-alkoxy-, reactions, with tributyltin chloride, **4**, 388
Lithium *t*-octyl butyl amide, use of, **1**, 874
Lithium organocuprates
  reactions
    with 3β-acetoxy-6-nitrocholest-5-ene, **3**, 426
    with 1-(4-chlorophenyl)-2-nitropropene, **3**, 426
Lithium pentamethylcyclopentadienide, use in the synthesis of carbon-bridged dicobalt complexes, **4**, 687
Lithium perchlorate, use for carbocation formation, **1**, 386
Lithium phenylalkynide, reactions, with *N*-chlorodiethylamine, **4**, 976
Lithium phenylcyanamide, synthesis of, **5**, 1123
Lithium phenylethynolate, from, isoxazole, **1**, 683
Lithium phenylethynylide, reactions, with (dichloro)iodobenzene, **2**, 1019
Lithium phenyl selenide
  reactions
    with (iodomethyl)trimethylgermane, **4**, 377
    with (iodomethyl)trimethylsilane, **4**, 377
Lithium phenyl telluride, reactions, with diiodomethane, **4**, 290
Lithium(phenylthio)alkylcuprates, in substitution reactions, **1**, 469
Lithium phosphide dimethoxyethane complex, reactions, with acyl chlorides, **5**, 1155
Lithium phosphides
  reactions
    with alkylgermyl halides, **2**, 867
    with arylgermyl halides, **2**, 867
    with dialkyldichlorosilanes, **2**, 473
    with haloboranes, **2**, 474
    with halogermanes, **2**, 474
  as starting materials, for methylenephosphanes, **5**, 879
  synthesis of, **5**, 397
Lithium phospholides, synthesis of, **4**, 112
Lithium–platinum hexamethyl compounds, synthesis of, **2**, 584
Lithium poly(boryl)methides, from, methanetriboronic esters, **6**, 198
Lithium propynoate, reactions, with aldehydes, **5**, 154
Lithium pyrrolidinoborohydride, for the 1,2-reduction of enones, **2**, 60
Lithiums
  from
    arylsilanes, **6**, 392
    propanediolboronic esters, **6**, 400
  for the opening of oxetanes, **1**, 229
  reactions
    with cerium chloride, **1**, 1140
    with cerium iodide, **1**, 1140
    with chlorosilanes, **6**, 392
    with cobalt(II) chloride, **6**, 195
    with *N*-diisobutylaluminum imines, **2**, 316
    with germanium halides, **6**, 388
    with lead(II) chloride, **2**, 602
    with metal halides, **6**, 396
    with metalloid halides, **6**, 396
    with perfluoroacylcyanomethylene-triphenylphosphoranes, **3**, 649
    with silicon halides, **6**, 388
    with silyl chlorides, **6**, 399
  stability of, **4**, 956
  as starting materials, for carbamates, **4**, 934
  structure of, **4**, 949
  synthesis of, **4**, 599, 956; **6**, 136, 204, 391
  use in the synthesis of 1,1-dimetal complexes, **4**, 669
Lithiums, acyl-
  cyclization reactions of, **5**, 438
  from
    n-butyllithium, **5**, 437
    s-butyllithium, **5**, 437
    t-butyllithium, **5**, 437
  reactions
    with diorganyl disulfides, **5**, 238
    with ketones, **3**, 158
  silylation of, **5**, 418
  synthesis of, **3**, 136
    via addition of organolithiums to carbon monoxide, **5**, 436
    via deprotonation of the formyl group of aldehydes, **5**, 437
    via desilylation of acylsilanes, **5**, 437
    via intramolecular trapping reactions, **5**, 438
    via lithium–tellurium exchange of acyl tellurides, **5**, 437
Lithiums, alkenyl-
  from, *N*-tosyl hydrazones, **2**, 946
  protonolysis of, **1**, 453
  reactions
    with benzeneselenenyl bromide, **4**, 785
    with copper(I) iodide, **1**, 1093
    with electrophiles, **1**, 535
    with 1,1-trichlorodifluoroethane, **4**, 758
  as starting materials
    for alkenylboronic esters, **2**, 904
    for ketones, **2**, 911
  synthesis of, **1**, 1090, 1108
Lithiums, α-alkoxy-
  synthesis of, **4**, 378
    via transmetallation reactions, **4**, 378
Lithiums, α-alkoxyalkenyl-, synthesis of, **2**, 999
Lithiums, α-alkoxyalkyl-
  as configurationally stable carbanions, **4**, 379
  synthesis of, **2**, 999
Lithiums, alkoxyvinyl-, reactions, with electrophiles, **2**, 88
Lithiums, α-alkoxyvinyl-
  silylation of, **2**, 908
  synthesis of, **1**, 455
Lithiums, alkyl-
  for the deprotonation of amides, **5**, 384
  elimination of, **1**, 660, 661
  from
    1-alkoxystannanes, **2**, 68
    alkyl halides, **2**, 551
    alkyl iodides, **2**, 554
    ethers, **2**, 553
    lithium–arene complexes, **2**, 551
  reactions
    with alkenes, **2**, 555
    with α-alkoxystannanes, **1**, 138
    with alkylaluminum halides, **4**, 698

with alkylisonitriles, **5**, 795
with aluminum chloride, **4**, 698
with aluminum trichloride, **2**, 595
with boryl(stannyl)alkanes, **6**, 201
with carbon monoxide, **5**, 238
with carboxylic acid, **3**, 131
with cerium chloride, **2**, 313
with chloroboranes, **2**, 526
with chromium acyl compounds, **5**, 444
with chromium trichloride, **2**, 572
with copper(I) salts, **2**, 586
with copper alkynides, **2**, 587
with copper cyanide, **2**, 588
with cyclopentadienyllanthanide complexes, **2**, 565
with cyclopentadienyl metal complexes, **2**, 566
with cyclopentadienyltungsten complexes, **2**, 573
with dialkylaluminum chlorides, **2**, 595
with dialkylthalliums, **2**, 598
with *gem*-dibromocyclopropanes, **4**, 146
with dihaloalkanes, **2**, 998; **6**, 62
with α,α-dihalo compounds, **1**, 164
with 6-dimethylamino-1-azafulvene, **1**, 256
with dithioacetals, **2**, 141
with 2,3-epoxy sulfones, **2**, 85
with gold(III) halides, **2**, 590
with β-haloenamines, **2**, 1049
with isocyanates, **3**, 276
with isocyanides, **3**, 418; **5**, 88
with isonitriles, **5**, 796
with metal(II) chlorides, **2**, 577
with metal halides, **2**, 578
with methoxyamine, **2**, 305
with methyleneboranes, **6**, 201
with 2-methylene-1,3-dithiane, **4**, 254
with 2-methyl-2-thione-1,2,3-phosphilidine, **2**, 463
with molybdenum hexacarbonyl, **5**, 445
with nickelocene, **6**, 403
with nitriles, **3**, 459
with nitrones, **2**, 343
with organosilanes, **6**, 379
with osmium complexes, **2**, 581
with oximes, **2**, 343
with phenylisonitriles, **5**, 795
with selenides, **2**, 554
with sulfones, **4**, 936
with sulfonium *n*-alkylides, **1**, 623
with sulfoxides, **4**, 391
with tellurides, **2**, 554
with 1,1,3,3-tetramethyl-1,3-disilacyclobutane, **6**, 191
with tetramethylethylenediamine, **6**, 190
with tetraorganophosphonium salts, **2**, 830
with thioketenes, **4**, 956
with tosylhydrazones, **1**, 630, 632
with trialkylstannylmethanimines, **1**, 372
with trihaloalkenes, **6**, 62
with *N*-(trimethylsilyl)benzaldimine, **2**, 313
with tungsten hexacarbonyl, **5**, 446
with vinylboranes, **6**, 201
with vinylphosphines, **4**, 575
with vinyl selenides, **4**, 400
with vinylsilanes, **6**, 135
with vinylstannanes, **2**, 911
as reagents, for the synthesis of dialuminum alkyl compounds, **4**, 698
as starting materials, for lithiation of alkenes and arenes, **2**, 965
synthesis of, **2**, 551; **6**, 392
use in the synthesis of aldehydes, via formylation, **3**, 15
use in the synthesis of alkylcoppers, **2**, 586
use in the synthesis of vinyllithiums, **1**, 630
x-ray crystallography studies, **4**, 647
Lithiums, 1-alkylthiovinyl-
from
ketene thioacetals, **4**, 956
2-methoxyalkyl sulfides, **4**, 958
Lithiums, alkynyl-
hydrolysis of, **1**, 998
hydroxymethylation of, **1**, 1037
reactions
with aldimines, **1**, 1041
with alkyl chloroformates, **1**, 1047
with *S*-alkylthioamidinium salts, **1**, 1041
with boron trifluoride etherate, **1**, 1023, 1045
with chlorophosphaalkenes, **6**, 370
with copper(I) salts, **2**, 1095
with cyanogen chloride, **1**, 1048
with *N,N*-dialkyl amides, **1**, 1046
with dialkylhalophosphine sulfides, **2**, 1069
with dialkylphosphinous chlorides, **2**, 1060
with dialkyl phosphorochloridothionates, **2**, 1072
with 2,2-dinitropropane, **1**, 1033
with formamides, **1**, 1046
with halodialkylphosphine sulfides, **2**, 1070
with 2-isoxazolines, **1**, 1042
with lactones, **1**, 1045
with methoxy dialkylborane, **2**, 1084
with *N*-methoxy-*N*-methyl amides, **1**, 1047
with organosilyl halides, **2**, 1089
with oxetanes, **1**, 1024
with oxime ethers, **1**, 1042
with oxiranes, **1**, 1023
with *N*-sulfonyl aldimines, **1**, 1042
with trimethylaluminum, **2**, 1099
as starting materials
for alkynylalanes, **2**, 1098
for alkynylantimony compounds, **2**, 1080
for 1-phosphaallenes, **3**, 608
synthesis of, **2**, 1076
Lithiums, 1-alkynyl-, reactions, with boronic esters, **2**, 1085
Lithiums, 5-alkyn-1-yl-, reactions, with aldehydes, **2**, 68
Lithiums, allenyl-
alkylation of, **1**, 464, 1052
reactions
with acylsilanes, **1**, 1057
with alkyl bromides, **1**, 464
with alkyl iodides, **1**, 464
with carbon disulfide, **1**, 1058
with α,β-unsaturated carbonyls, **1**, 1061
Lithiums, allyl-
from, allylic selenides, **1**, 245
reactions
with alkyl halides, **1**, 112
with menthyl esteriron complexes, **5**, 453
silyl-substituted, synthesis of, **4**, 648
synthesis of, **2**, 555
Lithiums, allylic, use in the synthesis of α,β-unsaturated ketones, **3**, 210
Lithiums, α-aminoalkenyl-, synthesis of, via deprotonation, **2**, 1002
Lithiums, aminoalkyl-
electrophilic silylation of, **4**, 511
stability of, **4**, 528
Lithiums, aminomethyl-, silylation of, **4**, 511
Lithiums, (aminosilyl)-, synthesis of, **2**, 546
Lithiums, aryl-
carbonation of, **5**, 110
oxidation of, **2**, 652
reactions
with chromium acyl compounds, **5**, 444
with chromium carbenes, **3**, 513
with chromium carbonyl, **5**, 444
with molybdenum hexacarbonyl, **5**, 445
with oxime ethers, **2**, 351
with sulfur dioxide, **2**, 718
with tellurium, **5**, 241
with tetraorganophosphonium salts, **2**, 830

with thioketenes, **4**, 956
as starting materials
for aryl bromides, **2**, 627
for aryl chlorides, **2**, 621
for benzaldehydes, **3**, 90
for furylgermanes, **2**, 949
for thienylgermanes, **2**, 949
synthesis of, via lithium-halogen exchange, **2**, 961
for the synthesis of biaryls, **1**, 473
Lithiums, arylgermyl-, reactions, with aldehydes, **4**, 360
Lithiums, arylthiomethyl-, reactions, with trialkyltin chlorides, **4**, 397
Lithiums, benzyl-
from, benzylic selenides, **1**, 245
reactions, with alkyl halides, **1**, 112
Lithiums, bis(trimethylsilyl)methyl-
reactions
with aldehydes, **4**, 936
with ketones, **4**, 936
Lithiums, α-borylalkyl-, synthesis of, **2**, 1005
Lithiums, α-bromo-, reactions, with mercuric chloride, **4**, 152
Lithiums, α-bromoalkenyl-
synthesis of, **4**, 818
See also Alkenes, *gem*-bromo lithio-
Lithiums, *t*-butoxymethyl-, transmetallation of, **4**, 384
Lithiums, butyl-, reactions, with α-nitro styrenes, **3**, 190
Lithiums, chiral, synthesis of, **2**, 999
Lithiums, chiral α-amino-, synthesis of, **4**, 531
Lithiums, α-chloroalkenyl-
synthesis of, **4**, 817
via metallation of 1-chloro-2,2-diarylethenes, **4**, 817
Lithiums, chloromethyl-, synthesis of, **6**, 379
Lithiums, cycloalkyl-
reactions
with 1,2-dibromoethane, **1**, 110
with iodomethane, **1**, 110
Lithiums, cyclopentenyl-, synthesis of, **1**, 1091
Lithiums, cyclopentylmethyl-, synthesis of, **2**, 556
Lithiums, cyclopropyl-, synthesis of, **2**, 555
Lithiums, 3,3-dialkyl-substituted allenyl-
reactions
with aldehydes, **1**, 1056
with aryl ketones, **1**, 1056
Lithiums, α,α-dibromoalkyl-, synthesis of, **4**, 22
Lithiums, dimethoxyvinyl-, reactions, with aldehydes, **3**, 71
Lithiums, 3,3-disubstituted allenyl-, reactions, with carbonyls, **1**, 1056
Lithiums, α-fluoroalkenyl-
synthesis of, **4**, 817
via metallation of 1,1-dichloro-2-fluoroethene, **4**, 817
Lithiums, germyl-
reactions, with aldehydes, **5**, 431
synthesis of, **4**, 641; **5**, 431
Lithiums, α-haloalkenyl-, synthesis of, **2**, 998; **4**, 816
Lithiums, α-haloalkyl-
configurational stability of, **4**, 146
synthesis of, **2**, 997; **4**, 145, 146
Lithiums, α-(halomethyl)-, synthesis of, **4**, 146
Lithiums, α-halovinyl-, synthesis of, **2**, 908
Lithiums, 5-hexenyl-
anionic cyclization of, **2**, 556
intramolecular carbolithiation of, **2**, 556
Lithiums, imidoyl-
reactions, with chlorotrimethylsilane, **5**, 792
as starting materials, for imidoylcoppers, **5**, 802
transmetallation of, **5**, 802
Lithiums, imino-, reactions, with halogermanes, **2**, 815
Lithiums, α-iodoalkenyl-, synthesis of, **4**, 819
Lithiums, methyl-, reactions, with metal(I) chloride, **2**, 579
Lithiums, α-methylthio alkyl-, synthesis of, **4**, 394
Lithiums, organosilyl-, reactions, with enones, **1**, 653
Lithiums, 1-oxyalkenyl-, synthesis of, **4**, 948
Lithiums, perfluoroalkyl-, elimination of, **1**, 661
Lithiums, (perfluoroalkyl)-, reactions, with imines, **2**, 311
Lithiums, α-phenylthioalkyl-
copper promoted allylation of, **4**, 396
synthesis of, **4**, 394
Lithiums, phenylthiomethyl-, alkylation of, **1**, 140
Lithiums, α-phosphinoalkyl-, synthesis of, **2**, 1003
Lithiums, 1-plumbylvinyl-
from
1,1-diplumbylalkenes, **4**, 1065
1-plumbyl 1-stannylalkenes, **4**, 1065
Lithiums, propargyl-, reactions, with α,β-unsaturated carbonyls, **1**, 1061
Lithiums, α-selenenyl-, synthesis of, via halogen–metal exchange, **4**, 400
Lithiums, α-selenoalkyl-
dimerization of, **4**, 401
synthesis of, **4**, 375
Lithiums, α-silyl-, synthesis of, **6**, 192
Lithiums, α-silylalkyl-, synthesis of, **2**, 1004
Lithiums, silylallyl-, reactions, with electrophiles, **2**, 935
Lithiums, α-silylbenzyl-, synthesis of, **2**, 1005
Lithiums, silylmethyl-, reactions, with carbon monoxide, **4**, 931
Lithiums, α-silyloxyalkyl-, crystal structure of, **2**, 999
Lithiums, silyloxypropargyl-, reactions, with electrophiles, **1**, 971
Lithiums, silylvinyl-, as reagents, in the synthesis of alkenylsilanes, **2**, 931
Lithiums, 1-stannylvinyl-, synthesis of, **4**, 1065
Lithiums, α-sulfenylalkenyl-, synthesis of, **2**, 1000
Lithiums, α-sulfenylalkyl-, synthesis of, **2**, 999
Lithiums, α-sulfinylalkyl-, synthesis of, **2**, 1000
Lithiums, 1-sulfonyl-1-alkenyl-, synthesis of, **2**, 85
Lithiums, α-sulfonylalkyl-, synthesis of, **2**, 1000
Lithiums, 1-sulfonylvinyl-, from, sulfonylalkenes, **4**, 961
Lithiums, α-sulfonylvinyl-, from, sulfonylalkenes, **1**, 145
Lithiums, tertiary alkyl-, as reagents, for lithium–halogen exchange, **2**, 961
Lithiums, thienyl-
reactions, with Weinreb amides, **3**, 305
synthesis of, **3**, 305
Lithiums, trialkylsilylethynyl-, reactions, with aldehydes, **1**, 1039
Lithiums, trialkyl stannyl-
reactions
with acid chlorides, **5**, 467
with esters, **5**, 467
with thioesters, **5**, 467
Lithiums, α-(trimethylsilyl)benzyl-, synthesis of, **2**, 1005
Lithiums, 1-trimethylsilylcyclopropyl-, synthesis of, **4**, 652
Lithiums, trisilyl-, synthesis of, **6**, 393
Lithiums, trisilylmethyl-
reactions, with halosilanes, **6**, 379
as starting materials, for carbosilanes, **6**, 385
synthesis of, **6**, 380
Lithiums, vinyl-
alkylation of, **1**, 141
from
1-iodo-1-alkenes, **1**, 1090
toluene-*p*-sulfonylhydrazones, **1**, 630
tosylhydrazones, **2**, 956
vinyl tellurides, **1**, 441
as intermediates, for the synthesis of methyl enol ethers, **5**, 416
as nucleophiles, **1**, 316
reactions
with carbon dioxide, **5**, 101
with carboxylic acids, **3**, 210
with α,ω-dibromoalkanes, **1**, 464
with electrophiles, **1**, 633
with trialkylboranes, **1**, 278
silylation of, **2**, 910
as starting materials

for methylene cyclopentanes, **2**, 556
for vinyl bromides, **2**, 624
for vinyl halides, **2**, 607
stereoisomerization of, **2**, 961
for the substitution of alkyl halides, **1**, 464
synthesis of, **1**, 1090; **4**, 667
via lithium-halogen exchange, **2**, 961
via Shapiro reaction, **3**, 58
Lithiums, (*E*)-vinyl-, synthesis of, **1**, 633
Lithiums, vinylic
configurationally pure, **1**, 111
reactions, with primary haloalkanes, **1**, 111
Lithium salts
alkylation of, **2**, 308
as byproducts, in the synthesis of ylides, **3**, 493
as intermediates, **6**, 331
reactions
with cyclopropenylium cations, **5**, 874
with ketenimines, **5**, 874
with trimethylamine hydrochloride, **4**, 373
with trimethylsilyl chloride, **6**, 163
rearrangement of, **1**, 378
synthesis of, **5**, 400; **6**, 349
transmetallation of, **5**, 780
use in Wittig reactions, **1**, 726
Lithium salts, diazo-, synthesis of, **5**, 1148; **6**, 670
Lithium salts, 2-trialkylboronthiophene-, synthesis of, **5**, 82
Lithium selenoates, reactions, with alkynyl bromides, **2**, 1037
Lithium silaenolate anions, synthesis of, **5**, 924
Lithium silaneselenolate
from, disilyl selenide, **5**, 541
reactions, with acetyl chloride, **5**, 541
Lithium silanethiolate, reactions, with acetyl chloride, **5**, 538
Lithium silyl arsenides, reactions, with benzophenone, **5**, 892
Lithium silyldiazomethanes, synthesis of, **6**, 674
Lithium silylphosphide, as starting material, for lithium sulfidomethylenephosphines, **6**, 684
Lithium sulfide, reactions, with methyl chloroformate, **6**, 474
Lithium sulfides, acylation of, **5**, 242
Lithium sulfinates, synthesis of, **2**, 272
Lithium tellurides, as starting materials, for diphenylditelluromethane, **4**, 290
Lithium–tellurium exchange, of acyl tellurides, **5**, 437
Lithium telluroates, reactions, with alkynyl bromides, **2**, 1037
Lithium tellurolates, as intermediates, for alkynyl tellurides, **2**, 1037
Lithium tetraalkylaluminate, reactions, with cyclopentadienyllanthanide complexes, **2**, 565
Lithium tetraalkynylaluminate, use in the synthesis of alkynylarsenic compounds, **2**, 1076
Lithium tetraethylaluminate
reactions
with antimony trichlorides, **2**, 483
with bismuth trichlorides, **2**, 483
Lithium tetrahydridoaluminate, for the reduction of propargylics, **1**, 961
Lithium tetrakis(trimethylsilyl)aluminate, use in the synthesis of α-hydroxysilanes, **4**, 355
Lithium tetrakis(trimethylsilyl)aluminum, reactions, with acyl chlorides, **5**, 418
Lithium tetramethylaluminate, reactions, with chlorobis(cyclopentadienyl) metal complexes, **4**, 726
Lithium 2,2,6,6-tetramethylpiperidide
deprotonation of, **1**, 162
for the deprotonation of phenylvinyl tellurides, **4**, 965
as metallating reagent, **6**, 203
reactions
with 2-benzoboroles, **4**, 642
with thiocarboxylic esters, **5**, 237
use in the synthesis of alkynes, **1**, 1070
Lithium tetramethylpiperidide (LITMP)
as deprotonation agent, **4**, 398
for the deprotonation of phenylvinyl selenides, **4**, 965
for the enolization of carbonyls, **2**, 696
reactions
with aldehydes, **5**, 437
with carbamates, **1**, 1089
with (*Z*)-2-phenyltetrafluoropropenes, **4**, 755
Lithium tetra(methylseleno)aluminate, reactions, with trimethylsilyl halides, **2**, 294
Lithium 2-thienylcyanocuprate, as reagent, **2**, 589
Lithium thiocarbamates, synthesis of, **6**, 494
Lithium thiolate, equilibrium with lithiotrithioorthoformates, **6**, 317
Lithium thiolates, synthesis of, **2**, 272
Lithium thiolates, alkynyl-, synthesis of, **2**, 1032
Lithium trialkylalkynylborates
reactions, with carbon dioxide, **5**, 101
synthesis of, **5**, 101
Lithium trialkylaluminum hydrides, reactions, with nitriles, **3**, 458
Lithium trialkylgermanium, reactions, with alkyn-2-on-1-yl chlorides, **4**, 1055
Lithium trialkylmanganates, reactions, with haloalkanes, **1**, 125
Lithium trialkyl zincates, reactions, with 1-bromo-1-chloroalkenes, **1**, 431
Lithium tribiphenylenephosphate, reactions, with *N*-methylbrucinium iodide, **2**, 869
Lithium tri(*t*-butoxy)aluminum hydride, for the reduction of acyl chlorides, **3**, 83
Lithium tri-*s*-butylborohydride
as reagent, for the reduction of amides to aldehydes, **3**, 13
See also L-Selectride
Lithium triethoxyaluminum hydride, reactions, with nitriles, **3**, 458
Lithium triethylborohydride
reactions
with cyclopentadienyltungsten complexes, **5**, 446
with 2-cyclopropyl ketones, **2**, 61
with sulfur, **5**, 242, 252
for the reduction of alkyl halides, **1**, 4
for the reduction of allylic ethers, **1**, 15
for the reduction of allylic sulfides, **1**, 23
for the reduction of allylic sulfones, **1**, 24
for the reduction of bromoalkanes, **1**, 9
for the reduction of chloroalkanes, **1**, 7
for the reduction of esters, **1**, 17
for the reduction of iodoalkanes, **1**, 10
for the reduction of oxiranes, **1**, 14
for the reduction of primary fluorides, **1**, 6
for the reduction of selenides, **1**, 25
use in the synthesis of carbon-bridged diruthenium complexes, **4**, 679
use in the synthesis of carbon-bridged trinuclear osmium complexes, **4**, 683
Lithium triisopropoxides, alkynyl-, reactions, with aldehydes, **1**, 1039
Lithium trimethylsilanolate
elimination of, **5**, 1155
for the synthesis of phosphaalkynes, **5**, 1155
Lithium trimethylsilylacetonitrile, reactions, with α-chlorocarbonyls, **3**, 650
Lithium trimethylsilylalkylamide, reactions, with alkynes, **2**, 306
Lithium trimethylsilyldiazomethane, reactions, with lanthanide complexes, **3**, 487
Lithium trimethylsilyldiazomethanide, reactions, with ketenimides, **2**, 944
2-Lithium-2-trimethylsilyl-1,3-dithiane, reactions, with phosphenimidous amides, **6**, 684
Lithium trimethylsilylphosphides, reactions, with thioformamidoyl chlorides, **6**, 584
Lithium triphenylsilanethiolate, synthesis of, **5**, 254

Lithium (triphenylstannyl)phenyl phosphide, reactions, with chloromethylene dimethylimmonium chloride, **5**, 881
Lithium(tris *t*-butoxy)aluminum hydride, use in the reduction of hydroxy ketones, **2**, 61
Lithium tris(dialkylamino)aluminum hydrides, for the reduction of benzoic acids, **3**, 82
Lithium tris(diethylamino)aluminum hydride
  for the reduction of amides, **3**, 85
  for the reduction of aromatic esters, **3**, 84
  use in the synthesis of aldehydes, **3**, 13
Lithium tris(dihexylamino)aluminum hydride, for the reduction of nitriles, **3**, 86
Lithium–tungsten octamethyls, synthesis of, **2**, 572
Lithium vapor
  reactions
    with dichloro(trimethylsilyl)methane, **6**, 205
    with halocarbons, **6**, 204
    with hydrocarbons, **6**, 204
    with perchlorocarbons, **6**, 204
    with polyhaloalkanes, **6**, 204
Lithium ynolates
  as intermediates, in the synthesis of aldehydes, **3**, 15
  as starting materials, for silylketenes, **2**, 941
Lithium–zirconium exchange, in 1,2 Wittig rearrangement, **1**, 397
Lobesia botrana, sex pheromone, **1**, 253, 1102
Long chain fatty acids, for the cleavage of carbon–boron bonds, **1**, 58
Longifolene
  reactions, with mercuric acetate, **4**, 748
  synthesis of, **1**, 389, 802
    via acetate pyrolysis, **1**, 774
(+)-Longifolene, use in the synthesis of alkylboranes, **2**, 520
Lossen rearrangement
  mechanism, **5**, 999
  for the synthesis of α-aminoalkylphosphorus compounds, **4**, 492
  for the synthesis of isocyanates, **5**, 999
  use in the synthesis of amino acids, **5**, 90
Lowest unoccupied molecular orbital (LUMO)
  of alkenes, **1**, 323, 350
  of carbon-centered radicals, **1**, 920
  of conjugate acceptors, **1**, 301
Low temperature method, for the synthesis of radicals, **1**, 946
Low-valent metal reduction
  for the reduction of alkyl halides, **1**, 2
  for the synthesis of alkenes, **1**, 457
Low-valent metals, for the reduction of chloroalkanes, **1**, 6
Lutetium
  chemistry of, **2**, 564
  pentamethylcyclopentadienyl complexes of, synthesis of, **4**, 702
Lutetium compounds, reactions, with phosphorus ylides, **4**, 578
Lutetium compounds, acyl, synthesis of, **5**, 440
2,6-Lutidine, reactions, with lithium iodide, **5**, 78
Lutidinium *p*-toluene sulfonate, as catalyst, for the synthesis of cyclic alkenyl ethers, **2**, 679
(±)-α-Lycorane, synthesis of, **2**, 346
Lycorine, synthesis of, **6**, 487
Lysergic acids, synthesis of, **3**, 299; **5**, 166
L-Lyxose, as starting material, for pseudomonic acid C, **1**, 173
L-Lyxose, 1-*O*-benzyl-2,3-isopropylidene-, reactions, with methyllithium, **6**, 541

M

Macrocyclization, chiral induction in, **1**, 340
Macrocyclization–transannular cyclization, mechanism, **1**, 347
Macrolactones
  synthesis of, **5**, 173
  use in the synthesis of natural products, **5**, 173
Macrolactonization, enzymatic, for the synthesis of macrolactones, **5**, 179
Macrolides, synthesis of, **5**, 277
Magic methyl *See* Sulfonate, methyl fluoro-
Magnesiocuprates, from, alkynes, **1**, 479
Magnesiocuprates, alkenyl-
  reactions, with ethylene oxide, **1**, 1107
  synthesis of, **1**, 1107
Magnesio–ene reaction, for the synthesis of alkylmagnesiums, **2**, 562
Magnesium
  activation of, **2**, 558
  oxidative addition of, **2**, 958
  reactions
    with anthracene, **2**, 561, 959
    with 1,2-bis(bromomethyl)tetramethyldisilane, **4**, 654
    with 1-bromo-1-trimethylsilylethene, **4**, 1060
    with 1-chloro-4-fluoro-2,2,4-trimethyl-2,4-disilapentane, **4**, 608
    with chloromethyl methyl ether, **4**, 383
    with α-chloro sulfides, **2**, 999
    with chlorotrimethylsilane, **2**, 473
    with dibromodi(trimethylsilyl)methane, **4**, 618
    with 1,2-dibromoethane, **1**, 662
    with 1,2-dichloroethane, **2**, 558
    with 1,2-dichloro-1,2,2-trifluoroethyl phenyl selenide, **4**, 785
    with ethyl α-bromobutyrate, **1**, 117
    with iodonitriles, **3**, 459
    with methanol, **1**, 617
    with organic halides, **2**, 558
    with phenylthiomethyl azide, **2**, 753
    with phosphoryl azide, **2**, 753
    with polyhalogenated ethyl phenyl selenide, **4**, 785
    with silyl ketones, **6**, 125
    with tosyl azide, **2**, 753
    with trichloro(chloromethyl)silane, **4**, 654
    with trimethylsilyl chloride, **5**, 793
    with trimethylsilylmethyl azide, **2**, 753
    with trimethylsilylmethyl chloride, **4**, 653
    with trimethylstannylmethyl halides, **4**, 709
  reduction of, for the synthesis of borenes, **5**, 928
  for the reduction of γ-alkoxy-α,β-unsaturated esters, **1**, 15
  for the reduction of dithioacetals, **1**, 25
  for the reduction of sulfones, **1**, 24
  use in the synthesis of alkenes, **1**, 679
  use in the synthesis of carbon-bridged tritins, **4**, 700
Magnesium, bis(trimethylsilylmethyl)-
  from, trimethylsilylmethylmagnesium chloride, **4**, 655
  use in the synthesis of chloride-bridged diruthenium complexes, **4**, 680
  use in the synthesis of methylene-bridged diruthenium complexes, **4**, 680
Magnesium, bis(trimethylsilylmethylene)-, use in the synthesis of carbon-bridged dichromium complexes, **4**, 672
Magnesium, butadienyl-, use in the synthesis of dihydrido diiridium carbon-bridged complexes, **4**, 689
Magnesium, *n*-butyl-*s*-butyl-, as starting material, for magnesium diisopropylamide, **2**, 56
Magnesium, dibenzyl-
  reactions
    with tungsten tetrachlorides, **2**, 572
    with vanadium tetrachloride, **2**, 570
Magnesium, diethyl-, reactions, with triethylantimony dichloride, **2**, 488
Magnesium, dimethoxy-, reactions, with carbon dioxide, **5**, 95
Magnesium, dimethyl-, reactions, with trimethylaluminum, **4**, 709
Magnesium, methyl-, reactions, with methyl chlorodithioformate, **6**, 306
Magnesium, (trifluoromethyl)-, synthesis of, **6**, 244
Magnesium, trifluorovinyl-, reactions, with phosphorus(III) halides, **4**, 802
Magnesium, trimethylsilylmethyl-, reactions, with bis(diethylamino)chloroarsine, **4**, 597
Magnesium aldimines, synthesis of, **3**, 418
Magnesium alkoxides, synthesis of, **2**, 112
Magnesium alkoxides, alkyl-, synthesis of, via displacement reactions, **2**, 558
Magnesium alkoxides, halo-, oxidation of, **3**, 119
Magnesium alkynides
  acylation of, **3**, 273
  reactions, with α-silyl-α,β-unsaturated sulfones, **1**, 1048
  synthesis of, **1**, 1134
  transmetallation of, **1**, 1133
Magnesium amalgam, use in the synthesis of di-Grignard reagents, **4**, 669
Magnesium amides
  as intermediates, **1**, 636
  for the ring-opening of epoxides, **1**, 580
Magnesium–anthracene, use in the synthesis of Grignard reagents, **4**, 654
Magnesium–anthracene complexes
  as source of active magnesium, **2**, 959
  as source of soluble magnesium, **2**, 559
  use in the synthesis of alkylmagnesiums, **2**, 559
Magnesium bisdialkyl amides, synthesis of, **2**, 697
Magnesium bromide
  reactions, with disodioferrocene, **4**, 709
  use in the synthesis of *N*-magnesioketimines, **3**, 459
Magnesium bromide, alkyl-, nucleophilic substitution with, **1**, 967
Magnesium bromide, alkylamide-, reactions, with *O*-ethyl selenocarboxylates, **5**, 522
Magnesium bromide, allenyl-
  reactions
    with benzothiazoles, **1**, 1061
    with benzoxazoles, **1**, 1061
    with di-*t*-butyl thioketone, **1**, 1058
    with dithio esters, **1**, 1058
    with 1,2-epoxybutane, **1**, 1053
Magnesium bromide, allyl-
  reactions
    with 2-bromooctane, **1**, 114
    with vanadium oxytrichloride, **2**, 570
  use in the synthesis of allyl-bridged palladium complexes, **4**, 693
  use in the synthesis of dimeric allyls of chromium, **4**, 673
Magnesium bromide, β-aminoethyl-, synthesis of, **2**, 558
Magnesium bromide, aryl-
  reactions
    with aminodichlorophosphines, **2**, 859
    with diethyl phosphonate, **2**, 842

Magnesium bromide, benzyl-, reactions, with 2-bromobutane, **1**, 114
Magnesium bromide, *t*-butoxy-, reactions, with thietane dioxides, **2**, 200
Magnesium bromide, *t*-butyl-, reactions, with oxazaphospholidine-2-thione, **2**, 257
Magnesium bromide, (diphenylphosphino)ethynyl-, reactions, with carbon dioxide, **2**, 1061
Magnesium bromide, ethyl-
  carbonylation of, **5**, 439
  reactions
    with alkanethiols, **2**, 273
    with alkynes, **1**, 1134
    with 1-alkynes, **2**, 1076
    with furfuryl dithioates, **4**, 255
    with methyl diethylphosphonodithioformate, **6**, 316
    with methyl phenyl sulfone, **4**, 395
    with 1-octyne, **1**, 278
    with zirconocene dichloride, **2**, 562
  as reagent, for the synthesis of ketenimines, **3**, 561
  use in the synthesis of isothiocyanates, **5**, 1024
Magnesium bromide, ethynyl-
  reactions
    with *t*-butylphosphonous dichloride, **2**, 435
    with *O*-glycoside, **1**, 1026
Magnesium bromide, ferrocenyl-, from, bromoferrocene, **4**, 709
Magnesium bromide, heptafluoropropyl-, synthesis of, **6**, 63
Magnesium bromide, mesityl-
  reactions
    with acid chlorides, **2**, 646
    with boron trifluoride, **2**, 520
    with copper(I) chloride, **4**, 696
    with silver chloride, **4**, 696
Magnesium bromide, 3-methoxyphenyl-, reactions, with keto-aldehydes, **4**, 166
Magnesium bromide, methyl-
  reactions
    with acylarenes, **2**, 655
    with ethyl diazoacetate, **3**, 486
    with titanium trichloride, **2**, 566
    with triphenyl phosphite, **2**, 434
Magnesium bromide, *n*-octyl-, synthesis of, **2**, 561
Magnesium bromide, 1-octynyl-, reactions, with *β*-sulfonyl-*α*,*β*-unsaturated esters, **1**, 1031
Magnesium bromide, pentyl-
  reactions
    with 1-bromoheptyne, **1**, 1017
    with bromophenylalkyne, **1**, 1017
Magnesium bromide, phenoxyethynyl-, reactions, with iodine, **4**, 765
Magnesium bromide, phenyl-
  reactions
    with ethoxythietanes, **4**, 731
    with 1-iodoheptafluoropropane, **6**, 63
    with pyridyl sulfoxide, **3**, 310
    with tetraalkyl orthocarbonate esters, **6**, 77
    with trifluoroacetonitrile, **3**, 627
Magnesium bromide, phenylethyne-, reactions, with diethyl chloramine, **2**, 1041
Magnesium bromide, phenylethynyl-, reactions, with thionyl chloride, **4**, 863
Magnesium bromide, propargyl-, reactions, with trimethoxyborane, **2**, 904
Magnesium bromide, pyrryl-, reactions, with 2-bromooctane, **1**, 114
Magnesium bromide, silylallyl-, reactions, with electrophiles, **2**, 935
Magnesium bromide, tributylstannyl-, reactions, with enones, **2**, 600
Magnesium bromide, 1-trimethylsilylvinyl-, reactions, with (-)-menthyl *p*-toluene sulfinate, **4**, 946
Magnesium bromide, triphenylsilylallyl-, synthesis of, **4**, 654
Magnesium bromide, tris(trimethylsilyl)methyl-, synthesis of, **6**, 174
Magnesium bromide, vinyl-, reactions, with bromocyclobutanes, **1**, 117
Magnesium bromide complexes, alkoxyiron tetracarbonyl-, reactions, with iodine, **5**, 43
Magnesium bromide etherate, reactions, with oxiranes, **4**, 783
Magnesium bromides, alkenyl-, reactions, with alkynes, **1**, 1105
Magnesium bromides, alkynyl-
  reactions
    with alkyl chlorothionoformates, **1**, 1047
    with arylselenium halides, **2**, 1037
    with chloroacetyl isocyanate, **1**, 1043
    with trimethylsilyl isocyanate, **1**, 1043
Magnesium bromides, aryl-
  addition to diarylketenes, **2**, 645
  oxidation of, **2**, 652
  reactions
    with *t*-butyl peroxybenzoate, **2**, 684
    with cobalt(II) chloride, **1**, 117
    with haloalkanes, **1**, 117
    with silver(I) bromide, **1**, 117
Magnesium bromides, ethynyl-, as starting materials, for alkynylborons, **2**, 1083
Magnesium bromides, imino-, acylation of, **5**, 290
Magnesium bromides, *α*-silylvinyl-, synthesis of, **1**, 1105
Magnesium carbonate, methyl-, synthesis of, **5**, 95
Magnesium chloride
  catalyzed exchange reactions, with alcohols, **6**, 79
  reactions, with potassium, **2**, 558
  reduction of, **2**, 558
Magnesium chloride, allyl-, use in the synthesis of dimeric allyls of molybdenum, **4**, 673
Magnesium chloride, benzylthiomethyl-, from, chloromethyl sulfide, **4**, 395
Magnesium chloride, *t*-butyl-
  reactions
    with antimony trichlorides, **2**, 484
    with *t*-butyldimethylsilylamines, **5**, 376
    with cyclohexenone, **1**, 301
    with di-*t*-butylphosphinous chloride, **2**, 434
    with phosphorus trichloride, **2**, 434
  as reagent, for the synthesis of thioketones, **3**, 351
Magnesium chloride, 2,2-dimethylpropyl-, reactions, with tantalum chloride complexes, **5**, 444
Magnesium chloride, isobutenyl-, reactions, with ketoketene thioacetals, **2**, 663
Magnesium chloride, isopropyl-
  reactions
    with chlorobenzene, **1**, 470
    with dibromomethane, **4**, 147
Magnesium chloride, methyl-
  reactions
    with ferrocenylacetonitrile, **3**, 567
    with phenyl isothiocyanate, **5**, 739
    with 2,3,3-trimethyl-2-chlorobutane, **1**, 114
Magnesium chloride, phenylethynyl-, use in the synthesis of alkynes, **1**, 1025
Magnesium chloride, propargyl-, synthesis of, **2**, 959
Magnesium chloride, tribromomethyl-, synthesis of, **6**, 244
Magnesium chloride, tributylstannyl-
  reactions
    with aldehydes, **4**, 386
    with chloromethyl methyl ether, **4**, 389
    with ketones, **4**, 386
Magnesium chloride, (trichloromethyl)-, synthesis of, **6**, 244
Magnesium chloride, triethoxysilylmethyl-, stability of, **4**, 607
Magnesium chloride, trimethylsilylmethyl-
  availability of, **4**, 653
  reactions

with antimony trichloride, **4**, 598
with arsenic trichloride, **4**, 597
with bismuth trichloride, **4**, 599
with gallium trichloride, **4**, 659
with lead dichloride, **4**, 660
with thioacetals, **2**, 936
with tin tetrachloride, **4**, 660
with trimethylborate, **4**, 642
as starting material, for bis(trimethylsilylmethyl)magnesium, **4**, 655
synthesis of, **4**, 653
Magnesium chlorides, trialkylvinylborate, carbonation of, **5**, 71
Magnesium diisopropylamide (MDA)
from, *n*-butyl-*s*-butylmagnesium, **2**, 56
use in Meerwein–Ponndorf–Verley reduction, **2**, 56
Magnesium diisopropylamides, reactions, with aryl imidate ethers, **3**, 721
Magnesium 2,3-dimethylbutane sulfinate, synthesis of, **2**, 273
Magnesium enolates, synthesis of, **2**, 697
Magnesium enolates, bromo-, synthesis of, **2**, 697
Magnesium fluoride, *p*-tolyl-, synthesis of, **2**, 958
Magnesium halide, *i*-propyl-, as reducing agent, **2**, 466
Magnesium halide carbon dioxide complexes, reactions, with ketones, **5**, 70
Magnesium halides
reactions
with alkenylzirconocene chlorides, **2**, 991
with dialkylmagnesiums, **2**, 559
with lithium biphenylide, **2**, 561
Magnesium halides, alkyl-
reactions
with alkynyl bromides, **1**, 1017
with phosphorus oxychloride, **2**, 452
synthesis of, **2**, 557
See also Grignard reagents
Magnesium halides, *t*-alkyl-, reactions, with anthrone, **1**, 265
Magnesium halides, alkynyl-
reactions
with *o*-acetates, **1**, 1026
with *o*-carbonates, **1**, 1026
with *N*,*N*-disubstituted aminoalkyl ethers, **1**, 1041
with *o*-formates, **1**, 1026
with oxiranes, **1**, 1022
with propargyl toluene sulfonate, **1**, 1025
with pyridine, **1**, 1042
with succinimide, **1**, 1045
Magnesium halides, allenyl-
reactions
with allenylaluminums, **1**, 1055
with allylic halides, **1**, 1052
with carbonyls, **1**, 1055
with α-halo ethers, **1**, 1052
with propargylic halides, **1**, 1052
Magnesium halides, aryl-
reactions
with dialkyl phosphorochloridites, **2**, 838
with phosphoramidic dichlorides, **2**, 844
Magnesium halides, aryloxy-, reactions, with formaldehyde, **3**, 97
Magnesium halides, bis(trimethylsilyl)methyl-, synthesis of, **6**, 192
Magnesium halides, 1,2-butadienyl-, reactions, with aldehydes, **1**, 1055
Magnesium halides, ethyl-
pyrolysis of, **1**, 662
reactions, with alcohols, **2**, 112
Magnesium halides, 1-indolyl-, acylation of, **3**, 309
Magnesium halides, methyl-
reactions, with phosphonous dichlorides, **2**, 435
as starting materials, for dithioacetic acid, **5**, 546
Magnesium halides, naphthyl-, reactions, with dialkylformamides, **3**, 101
Magnesium halides, α-silylbenzyl-, synthesis of, **4**, 654
Magnesium halides, trifluorovinyl-
reactions, with germanium halides, **4**, 811
as starting materials, for α-fluorovinyltin, **4**, 812
Magnesium halohydride, use in the synthesis of alkylmagnesiums, **2**, 561
Magnesium hydride
dehydrogenation of, **2**, 558
use in the synthesis of alkylmagnesiums, **2**, 561
Magnesium hydrides, alkyl-
as magnesium sources, **2**, 561
synthesis of, via reduction reactions, **2**, 558
Magnesium iodide, butyl-, reactions, with methyl ethers, **1**, 262
Magnesium iodide, methyl-
reactions
with acylphosphine oxides, **4**, 343
with (*R*)-(+)-α-[*t*-butyldimethylsilyl)oxyl]-phenylacetonitrile, **2**, 316
with platinum salts, **2**, 584
Magnesium iodide, octyl-, reactions, with α-(*N*-benzotriazoyl)butyl methyl ether, **1**, 262
Magnesium iodide, trifluorovinyl-
reactions, with lead(IV) salts, **4**, 813
synthesis of, **4**, 820
Magnesium metal
reactions
with α-bromo ketones, **2**, 697
with 1,3-dienes, **2**, 561
Magnesium metallocycle, for the synthesis of carbocycles, **1**, 116
Magnesium methoxide, reactions, with methyl cyanoacetate, **3**, 653
Magnesium monoperoxyphthalate (MMPP)
for the cleavage of hydrazones, **3**, 18, 133
for the oxidation of sulfoxides, **2**, 156
Magnesiums
elimination of, **1**, 661
hydroxymethylation of, **1**, 1037
reactions
with aluminum trichloride, **2**, 595
with copper(I) salts, **2**, 586
with metal halides, **2**, 578
with phosphorus ylides, **4**, 577
with platinum(IV) tetrachloride, **2**, 584
with potassium hexachloroplatinate, **2**, 584
Magnesiums, acyl-, synthesis of, **5**, 438
Magnesiums, α-(acylamino)alkyl-, synthesis of, **2**, 1002
Magnesiums, alkenyl-
as starting materials, for alkenylboronic esters, **2**, 904
synthesis of, **1**, 1092, 1105
Magnesiums, α-alkoxy-, synthesis of, **4**, 383
Magnesiums, alkyl-
from, alkyl halides, **2**, 558
reactions
with alkylmercury compounds, **1**, 846
with gold(I) chloride phosphine complexes, **2**, 590
stability of, **2**, 558
synthesis of, **2**, 557
Magnesiums, alkynyl-
reactions
with aldehydes, **1**, 1039
with organosilyl halides, **2**, 1089
synthesis of, **2**, 1076
Magnesiums, allenyl-
reactions
with aldehydes, **1**, 1055
with ketones, **1**, 1055
Magnesiums, α-amino-, transmetallation of, **4**, 531
Magnesiums, aryl-
reactions
with benzoyl chlorides, **3**, 285
with vinylborinic esters, **2**, 907

Magnesiums, arylimino-, reactions, with
benzophenones, **3**, 412
Magnesiums, benzyl-, from, benzyl methyl ethers, **2**, 559
Magnesiums, 2-butene 1,4-diyl-, synthesis of, **2**, 561
Magnesiums, dialkyl-
from
dialkylmercury compounds, **2**, 560
hydrides, **2**, 561
reactions
with magnesium halides, **2**, 559
with tetrahydrofuran, **1**, 305
synthesis of, **2**, 557, 559
Magnesiums, α-haloalkenyl-, synthesis of, **4**, 820
Magnesiums, α-haloalkyl-, synthesis of, **4**, 147
Magnesiums, (halomethyl)-, synthesis of, **4**, 147
Magnesiums, imidoyl-, from, isonitriles, **5**, 796
Magnesiums, methyl-, reactions, with metal(I)
chloride, **2**, 579
Magnesiums, monosilyl-, for the synthesis of
vinylsilanes, **2**, 918
Magnesiums, α-silyl-, synthesis of, **6**, 192
Magnesiums, α-silyl alkyl-, synthesis of, **4**, 653
Magnesiums, silylvinyl-, as reagents, in the synthesis of
alkenylsilanes, **2**, 931
Magnesiums, α-sulfenylalkyl-, synthesis of, **2**, 999
Magnesiums, vinyl-
alkylation of, **2**, 923
silylation of, **2**, 923
stannylation of, **2**, 923
Magnesium salts
acid hydrolysis of, **5**, 549
reactions, with α,β-unsaturated esters, **5**, 550
Magnesium sulfinates, synthesis of, **2**, 273
Magnesium thiolates, synthesis of, **2**, 273
Magnesium turnings, dry stirring of, **2**, 559
Magnesium vapor, condensation of, **2**, 558
Makaiyama rearrangement, for the synthesis of ketene silyl
acetals, **4**, 831
Malachite green, compounds of, **1**, 898
Maleate, diethyl, as starting material, for
hexakis(trimethylsilyl)-2-butene, **6**, 174
Maleate, dimethyl, synthesis of, **1**, 507
Maleates, ynamine silyl, synthesis of, **2**, 1045
Maleic anhydride
Diels–Alder reaction of, **5**, 198
ene reaction of, **5**, 201
reactions
with aniline, **5**, 95
with arylalkanes, **5**, 201
with benzonitrile *N*-sulfide, **2**, 240
with cyclic dienes, **5**, 199
with dienes, **5**, 199
with dithio esters, **4**, 843
with osmium complexes, **5**, 201
Maleic anhydride, fluoro-, pyrolysis of, **2**, 1013
Maleic anhydrides, reactions, with tributyltin
isocyanate, **5**, 391
Maleimide, *N*-phenyl-, reactions, with diazo esters, **3**, 321
Malenanilic acid, synthesis of, **5**, 95
Malodinitrile, dichloro-, reactions, with phenylsubstituted
benzotrichlorides, **5**, 658
Malonaldehydes, reactions, with acetamides, **3**, 75
Malonaldehydes, methyl-, synthesis of, **1**, 979
Malonanitriles, reactions, with dialkylselenium
dichlorides, **3**, 637
Malonate, diazo-, reactions, with thietanes, **1**, 233
Malonate, di(*t*-butyl), reactions, with
paraformaldehyde, **5**, 104
Malonate, di(*t*-butyl)methylene-, synthesis of, **5**, 104
Malonate, diethyl
reactions
with aldehydes, **4**, 559
with isothiocyanates, **5**, 593
with *N*-sulfonylaziridines, **5**, 345
use in the synthesis of alkanoic acids, **5**, 32
Malonate, diethyl amino-, from, diethyl
acetamidomalonate, **5**, 90
Malonate, diethyl chloro-, as starting material, for α-
chlorophosphines, **4**, 115
Malonate, diethyl 2,4-dinitrobenzylidene, reactions, with
diethylamine, **4**, 306
Malonate, diethyl nitro-, as starting material, for
ethoxycarbonylformonitrile oxide, **3**, 685
Malonate, dimethyl
dichlorination of, **4**, 12
reactions
with dimethyl acetylenedicarboxylate, **1**, 537
with epichlorohydrin, **1**, 224
Malonate, dimethyl benzenesulfamido, reactions, with
bromine, **4**, 105
Malonate, dimethyl diazo-, reactions, with
di-*t*-butylthioketene, **4**, 837
Malonate, dimethyldiazo-, reactions, with episulfides, **1**, 982
Malonate, ethyl cyano-, synthesis of, **3**, 659
Malonate, ethyl 2,5-dimethylphenylhydroxy, as starting
material, for 2,5-dimethylmandelic acid, **5**, 62
Malonate, monoethyl acetamido-, oxidation of, **4**, 311
Malonate anions, S_N2-displacement reactions with, **5**, 158
Malonate anions, dialkyl, for the opening of epoxides, **1**, 209
Malonate enolates, chiral, cycloalkylation of, **5**, 148
Malonates
reactions
with dialkoxymethylammonium salts, **4**, 193
with toluic acid chlorides, **2**, 664
synthesis of, **4**, 824
Malonates, acylamino-
reactions
with arenediazonium salts, **5**, 854
with diazonium salts, **4**, 448
as starting materials, for hydrazonamides, **5**, 854
Malonates, alkylidene
reactions
with alkynylalanes, **1**, 1050
with propargylsilanes, **1**, 978
Malonates, α-aryl, as intermediates, for nitriles, **3**, 625
Malonates, diethyl α-alkyl-α-isocyano-, synthesis of, **3**, 706
Malonate salts, reactions, with α,β-unsaturated esters, **1**, 306
Malondialdehyde, chloro-, reactions, with arenediazonium
salts, **5**, 843
Malonic acid
from, sodium α-cyanoethanoate, **5**, 35
reactions
with aldehydes, **5**, 98
with diazonium salts, **5**, 858
Malonic acid, chloro-
reactions
with arenediazonium salts, **5**, 843
with diazonium halides, **5**, 676
Malonic acid esters, chloro-, reactions, with diazonium
halides, **5**, 676
Malonic acids
diiodination of, **4**, 28
reactions
with isopropenyl acetate, **4**, 204
with lead(IV) acetate, **4**, 206
Malonic dialdehyde, chloro-, reactions, with arenediazonium
salts, **5**, 843
Malonic esters, carbonylative cyclization of, **5**, 464
Malonic ester synthesis, for the synthesis of amino
acids, **5**, 89
Malonimidoyl dichloride, tetrachloro-, synthesis of, **5**, 809
Malonitrile
reactions, with acyl isothiocyanates, **5**, 631
as starting material
for α-phosphinoyl enamines, **4**, 1000
for α-thiophosphinoyl enamines, **4**, 1001

Malonitriles, reactions, with sodium alkylthiolate, **3**, 655
Malonitriles, 3,3-bis(alkylthio)methylene-, reduction of, **3**, 655
Malonitriles, iodo-, as starting materials, for electrophilic radicals, **1**, 340
Malonoaldehyde, chloro-, reactions, with diazonium halides, **5**, 676
Malonodinitriles, alkylidene-, as starting materials, for ketenimines, **3**, 570
Malononitrile
 chlorination of, **5**, 809
 reactions
  with bianthrone, **3**, 668
  with chlorine, **5**, 662
  with α-chloroacetylacetone, **3**, 672
  with α-hydroxy ketones, **3**, 671
 as starting material, for ketene diselenoacetals, **4**, 875
Malononitrile, amino-, from, malononitriles, **3**, 638
Malononitrile, benzylidene-, reactions, with β-ketothionoamides, **5**, 607
Malononitrile, α-bromo-, as starting material, for *gem*-bromo iminophosphoranes, **4**, 800
Malononitrile, α-dichloroarsine methyl, synthesis of, **3**, 639
Malononitrile, dimethylamino-, reactions, with hydrogen sulfide, **5**, 580
Malononitriles
 as starting materials, for aminomalononitrile, **3**, 638
 synthesis of, **4**, 836
Malononitriles, bis(dialkylamino)-, as starting materials, for ketene aminals, **4**, 975
Malonyl chloride, reactions, with thiodiimides, **5**, 340
Malonyl chlorides, dimethyl-, reactions, with monoalkylamides, **4**, 881
Mandelamides, synthesis of, **2**, 53
Mandelates, synthesis of, **2**, 53
Mandelic acid
 reduction of, **1**, 88
 synthesis of, **5**, 63, 66
Mandelic acid, *p*-bromo-, synthesis of, **5**, 66
Mandelic acid, 2,5-dimethyl-, from, ethyl 2,5-dimethylphenylhydroxymalonate, **5**, 62
Mandelic acid, hexahydro-, synthesis of, **1**, 88
Mandelic acid, pentafluoro-, synthesis of, **5**, 63
Mandelic acids, α-alkyl-, synthesis of, **5**, 63
Mandelonitrile, synthesis of, **3**, 615
Mandelonitrile, *o*-nitro-, as starting material, for *o*-nitrobenzoic acid, **2**, 783
(*R*)-Mandelonitrile, synthesis of, **3**, 628
Manganacyclopentadiene complexes, synthesis of, **4**, 677
Manganate, dilithio(1-norbornyl)-, oxidation of, **2**, 575
Manganate(VI) salts, as oxidants, **3**, 5
Manganate(VII), benzyltriethylammonium, for alcohol oxidation, **3**, 116
Manganate(VII) salts, as oxidants, **3**, 5
Manganates, trialkyl-
 reactions
  with enol phosphates, **1**, 480
  with enol triflates, **1**, 480
Manganese
 reactivity of, **2**, 574
 use in Barbier reactions, **2**, 74
 use in Reformatsky reactions, **2**, 74
 use in the synthesis of radicals, **1**, 926
Manganese, bis[bis(trimethylsilyl)methyl]-, synthesis of, **6**, 194
Manganese, bromopentacarbonyl-
 reactions
  with dimethylstibolyl anion, **4**, 1042
  with lithiated tricarbonylthiophenechromium, **4**, 720
  with succinimide, **5**, 388
Manganese, cyclopentadienyl-, synthesis of, **4**, 1036
Manganese, cyclopentadienyldicarbonyl-, use in the synthesis of methylene-bridged dimanganese complexes, **4**, 676
Manganese, cyclopentadienyltricarbonyl-, deprotonation of, **4**, 1036
Manganese, decacarbonyl-, synthesis of, **6**, 523
Manganese, (dimethylaminomethyl)cyclopentadienyltricarbonyl-, reactions, with *n*-butyllithium, **4**, 1036
Manganese, tetra(1-norbornyl)-, synthesis of, **2**, 575
Manganese, tricarbonyl(cyclopentadienyl)-, lithiation of, **4**, 706
Manganese, tricarbonyl(lithiocyclopentadienyl)-
 reactions
  with chlorocyclopentadienyl-(triphenylphosphine)nickel, **4**, 719
  with chlorotriphenyltin, **4**, 725
  with titanocene dichloride, **4**, 719
  with transition metal halides, **4**, 723
 use in the synthesis of dimetal complexes, **4**, 719
Manganese(I) compounds, alkyl-, synthesis of, **2**, 573
Manganese(II) chloride
 reactions, with Grignard reagents, **6**, 194
 use in the synthesis of carbon-bridged dimanganese complexes, **4**, 677
 use in the synthesis of manganese η^5-pentadienyl complexes, **4**, 678
Manganese(II) complexes, (salen)-, as catalysts, in asymmetric epoxidation, **2**, 438
Manganese(II) compounds, alkyl-, synthesis of, **2**, 574
Manganese(II) enolates, for the α-methylation of ketones, **1**, 133
Manganese(II) iodide, use in the synthesis of carbon-bridged dimanganese complexes, **4**, 677
Manganese(II) salts, reactions, with trialkylaluminums, **6**, 523
Manganese(III) compounds, reactions, with chloride ions, **2**, 24
Manganese(III) compounds, alkyl-, synthesis of, **2**, 574
Manganese(III) enolate, synthesis of, **1**, 346
Manganese(III) Schiff-base complexes, synthesis of, **2**, 574
Manganese(IV) compounds, alkyl-, synthesis of, **2**, 574
Manganese(IV) compounds, tetraalkyl-, synthesis of, **2**, 575
Manganese(IV) oxide
 use in the synthesis of diazoketones, **3**, 464
 use in the synthesis of α-keto hydrazones, **3**, 470
Manganese acetate
 for the mediation of radical cyclizations, **1**, 346
 reactions, with alkenes, **1**, 336
Manganese alkyl compounds, reactions, with carbon monoxide, **5**, 446
Manganese bis(acyl) complexes, synthesis of, **5**, 448
Manganese bromide, use in the synthesis of alkylzinc halides, **2**, 591
Manganese carbene complexes
 reactions, with *n*-butyllithium, **5**, 946
 synthesis of, **3**, 518
Manganese carbonyl complexes, synthesis of, **6**, 523, 525
Manganese carbyne complexes, reactions, with trimethylphosphane, **5**, 919
Manganese chloride
 reactions, with tris(trimethylsilylmethyl)lithium, **6**, 396
 as starting material, for organomanganese(II) compounds, **2**, 574
Manganese chloride, alkyl-, synthesis of, **6**, 396
Manganese chloride, phenyl-, as starting material, for 1-tetradecene, **1**, 125
Manganese complexes
 from, manganese pentacarbonyls, **5**, 447
 as mediator of radical cyclizations, **1**, 346
 reactions
  with diazo compounds, **3**, 518
  with organomolybdenum complexes, **6**, 328
 synthesis of, **4**, 955
Manganese complexes, arene tricarbonyl-, reactions, with lithium cuprates, **5**, 448

Manganese complexes, cyclopentadienyl-, reactions, with methyllithium, **5**, 448
Manganese complexes, (halomethyl)-, synthesis of, **4**, 148
Manganese complexes, iodomethyl-, synthesis of, **4**, 148
Manganese complexes, α-keto acyl-, synthesis of, **5**, 449
Manganese complexes, methylene-bridged, synthesis of, **4**, 676
Manganese compounds
lithiation of, **4**, 706
reactions
with 1-metalloalkylphosphine oxides, **4**, 581
with phosphorus ylides, **4**, 581
with vinyl halides, **1**, 469
synthesis of, **4**, 1037
use in the synthesis of ketones, **3**, 116
Manganese compounds, acyl-, synthesis of, **5**, 446
Manganese compounds, alkyl-
as starting materials, for homocuprates, **2**, 586
synthesis of, **2**, 573
Manganese compounds, (1-chloro-2,2-dicyanovinyl)-, reactions, with diethylamine, **4**, 1019
Manganese compounds, dialkyl-, oxidation of, **2**, 575
Manganese compounds, imidoyl-, synthesis of, **5**, 799
Manganese compounds, silyl-
additions to alkynes, **2**, 919
use in the synthesis of 1,1,2-trisilylalkenes, **4**, 1045
Manganesedicarbonyl, methylcyclopentadienyl-, use in the synthesis of methylene-bridged dimanganese complexes, **4**, 676
Manganese diiodide, as starting material, for organomanganese(II) compounds, **2**, 574
Manganese dioxide
for the oxidation of hydroxymethyl furans, **3**, 102
reactions
with benzylic alcohols, **3**, 98
with carbonylferrates, **6**, 523
with 2-hydroxymethyl indoles, **3**, 106
with pyridoxol, **4**, 164
use in the synthesis of aromatic azo compounds, **2**, 772
use in the synthesis of aromatic ketones, **3**, 294
use in the synthesis of cyclic hemiacetals, **4**, 164
use in the synthesis of formylpyridines, **3**, 108
use in the synthesis of α,β-unsaturated aldehydes, **3**, 57
use in the synthesis of α,β-unsaturated ketones, **3**, 215, 247
Manganese dioxide, activated, for the oxidation of benzylic alcohol, **3**, 87
Manganese functions, doubly-bonded, synthesis of, **3**, 518
Manganese hexacarbonyl, reactions, with sodium amalgam, **5**, 446
Manganese oxidants
for the oxidation of benzylic alcohol, **3**, 88
use in the synthesis of aldehydes, **3**, 5
Manganese pentacarbonyl
alkylation of, **2**, 573
reactions, with 1-phenylphosphole, **4**, 1037
Manganese pentacarbonyl, acetyl-, synthesis of, **5**, 446
Manganese pentacarbonyl, bromo-, transmetallation of, **2**, 573
Manganese pentacarbonyl, ethyl-, stability of, **2**, 573
Manganese pentacarbonyl, methyl-
in metallation reactions, **4**, 721
migratory insertion reactions, **5**, 447
reactions
with amines, **5**, 447
with carbon monoxide, **5**, 446
with triphenylphosphine, **5**, 447
stability of, **2**, 573
Manganese pentacarbonyl complexes, acyl-, from, acid chlorides, **5**, 446
Manganese pentacarbonyl complexes, alkyl-
reactions
with alkynes, **3**, 218
with isonitriles, **5**, 799
Manganese pentacarbonyl complexes, aryl-, reactions, with isonitriles, **5**, 799
Manganese pentacarbonyl hydride, reactions, with diselenides, **6**, 49
Manganese pentacarbonyls
acylation of, **2**, 573
alkylation of, **2**, 573
for the opening of epoxides, **1**, 208
reactions
with aziridines, **5**, 447
with 2-chloroethyl methyl sulfide, **5**, 448
with methyllithium, **5**, 448
as starting materials, for manganese complexes, **5**, 447
Manganese pentacarbonyls, glycosyl-, synthesis of, **5**, 446
Manganese η^5-pentadienyl complexes, synthesis of, **4**, 678
Manganese radials, synthesis of, **1**, 231
Manganese (salen) complexes, use in the synthesis of aryl sulfoxides, **2**, 714
Manganesesilanes, 1,2-addition to aldehydes, **5**, 449
Manganese tetracarbonyls, synthesis of, **5**, 447, 448
Manicone, synthesis of, **4**, 255
Mannich bases, reactions, with phosphines, **4**, 452
Mannich reaction
of alkynes, **1**, 1041
of benzo-1,3-dithiolane *S,S'*-dioxides, **4**, 863
of boron enolates, **2**, 302
of carbonyls, **1**, 255
mechanism, **3**, 184
for nitroamines, **2**, 394
for the synthesis of *S,N*-acetals, **4**, 315
for the synthesis of aminals, **4**, 406
for the synthesis of amines, **2**, 302, 316
for the synthesis of β-amino aldehydes, **3**, 45
for the synthesis of 1,3-aminoalkylarsenic compounds, **4**, 499
for the synthesis of 1-aminoalkylarsines, **4**, 499
for the synthesis of β-amino ketones, **3**, 184
for the synthesis of α-amino organophosphorus compounds, **4**, 452
for the synthesis of α-aminosilanes, **4**, 509
for the synthesis of dialkyl α-aminoalkyl phosphonates, **2**, 457
for the synthesis of enamines, **2**, 739
for the synthesis of formyl phosphonates, **3**, 50
for the synthesis of hemiaminals, **4**, 299
for the synthesis of α-methylene aldehydes, **3**, 60
for the synthesis of nitraminoalkanes, **4**, 440
for the synthesis of *N*-nitroso *O,N*-acetals, **4**, 312
for the synthesis of piperidines, **2**, 326
for the synthesis of tertiary phosphines, **2**, 440
for the synthesis of tricoordinate phosphorus functions, **4**, 481
for the synthesis of α,β-unsaturated ketones, **3**, 210
Mannich reaction, intramolecular, for the synthesis of saturated nitrogen heterocycles, **3**, 187
Mannich reactions, use in the synthesis of 1,1,1-trinitroalkanes, **6**, 147
Mannitol, reactions, with boric acid, **2**, 111
Marcfortine A, uses of, **5**, 572
Marine alkaloids, Sarain A, **1**, 369
Marine metabolites, (+)-laurencin, **1**, 829
(+)-Maritidine, synthesis of, **1**, 55
Markovnikov's rule, **2**, 676
Markovnikov addition
of hydrogen halides, **2**, 4
methods, **2**, 30
for the synthesis of alcohols, **2**, 40
for the synthesis of alkanethiols, **2**, 117
in the synthesis of alkylcobalt(I) compounds, **2**, 576
for the synthesis of alkyl iodides, **2**, 34
for the synthesis of ethers, **2**, 89
for the synthesis of tertiary phosphines, **2**, 439
to alkenes, **3**, 614

use in the synthesis of alkyl chlorides, **2**, 23
Markovnikov adducts, synthesis of, **2**, 280
anti-Markovnikov adducts, as intermediates, **2**, 280
anti-Markovnikov amination, for the synthesis of primary amines, **2**, 306
anti-Markovnikov hydration, mechanism, **2**, 40
Markovnikov reaction, for the synthesis of bromo selenides, **3**, 174
Martin sulfurane reagent
synthesis of, **1**, 576
use in dehydration reactions, **1**, 576
Mass spectra, of vinyl halides, **1**, 1122
Maylansinoids, synthesis of, **4**, 936
Maysine, synthesis of, **4**, 255
Maysine, 4,5-deoxy-, synthesis of, **4**, 255
McFadyen–Stevens procedure
for the synthesis of benzaldehydes, **3**, 85
for the synthesis of polyaromatic aldehydes, **3**, 100
McMurry coupling, of carbonyls, **1**, 687
McMurry reaction, mechanics of, **1**, 687
Mediterranean fruit fly, sex pheromone, synthesis of, **1**, 756
Medium sized rings, synthesis of, via radical cyclization, **1**, 340
Meerwein's salt, as alkylating agents, **2**, 674
Meerwein arylation
mechanism, **1**, 486
use in the synthesis of thiocyanates, **5**, 1108
Meerwein–Ponndorf–Verley reduction
mechanism, **2**, 55
as side reaction, **2**, 67
Meerwein–Ponndorf–Verley reduction, enantioselective, **2**, 56
Meisenheimer complexes, oxidation of, **6**, 30
Meisenheimer rearrangement, mechanism, **2**, 355
Melamine, reactions, with hydrogen fluoride, **2**, 17
Meldrum's acid
reactions, with silylated amines, **5**, 227
synthesis of, **4**, 204
use in the synthesis of ketene acetals, **4**, 827
Meldrum's acid, 5,5-dibromo, use in the synthesis of α-bromo ketones, **3**, 154
Meldrum's acids
flash vacuum pyrolysis of, **4**, 894
pyrolysis of, **2**, 1032
reduction of, **2**, 39
as starting materials, for dialkylketenes, **3**, 541
Melodinus alkaloids, synthesis of, **3**, 422
6-Membered rings, synthesis of, via radical cyclization, **1**, 339
Menschutkin reaction, use in the synthesis of primary amines, **2**, 298
2-Menthene, synthesis of, **1**, 625
Menthols, (-)-8-phenyl-, as chiral auxiliaries, in radical reactions, **1**, 328
Menthyls, 8-phenyl-, reactions, with cuprates, **5**, 152
Mercaptals, ketene, *C*-acylation of, **3**, 253
Mercaptans *See* Thiols
Mercaptides
reactions
with alkyl alkylchlorophosphonates, **2**, 464
with *S*-alkyl alkylhalothiophosphonates, **2**, 464
with alkyldichlorophosphines, **2**, 461
with phosphonic dihalides, **2**, 464
synthesis of, **5**, 739
3-Mercapto-2,3-enone anions, alkylation of, **3**, 254
Mercaptomerin, synthesis of, **2**, 274
Mercuracycles, reactions, with 2,4-dimethyl-2,4-digermapentane, **4**, 629
Mercuration
of acetone, **6**, 208
of aldehydes, **6**, 207
of alkenes, **2**, 40; **4**, 1067
of carboxylic acids, **4**, 697
of cyclopentadiene, **4**, 697
for the synthesis of alkylmercury compounds, **2**, 593
for the synthesis of transition metal complexes, **4**, 724
Mercuration, electrophilic, of hydrocarbons, **6**, 207
Mercuretin
structure of, **6**, 207
synthesis of, **6**, 207
Mercuribenzoic acid, *p*-chloro-, from, *p*-tolylmercuric chloride, **5**, 115
Mercuric amides, (trinitromethyl)-, from, bis(trinitromethyl)mercury, **6**, 368
Mercuric bromide, phenyl-, as by-product, in the synthesis of α,α-dihalo ethers, **6**, 38
Mercuric bromides, phenyldihalomethyl-, as starting materials, for halomethylides, **3**, 497
Mercuric chloride
reactions
with alkene 1,1-diboronic esters, **4**, 1067
with alkynes, **1**, 1108
with α-bromolithiums, **4**, 152
with nitrones, **2**, 274
with sodium acetate, **6**, 207
with 2-triethylgermyl-2-trimethylsilyl-1,3-dithiane, **6**, 520
as reagent, in complete mercuration of cyclopentadiene, **4**, 697
Mercuric chloride, mesityl-, reactions, with iron(II) chloride, **4**, 685
Mercuric chloride, *p*-tolyl-, as starting material, for *p*-chloromercuribenzoic acid, **5**, 115
Mercuric compounds, aryl-, as starting materials, for carbenes, **3**, 321
Mercuric dichloride, reactions, with benzylbis(boryl)methane, **6**, 201
Mercuric fluoride
reactions
with dibromochloromethane, **6**, 33
with haloalkenes, **6**, 64
use in the synthesis of difluoroalkanes, **4**, 3
Mercuric halides, reactions, with alkynes, **2**, 1097
Mercuric halides, β-nitro, elimination of, **1**, 668
Mercuric nitrate, reactions, with 1,1-diarylethenes, **4**, 1067
Mercuri compounds, acetoxy-, synthesis of, **4**, 724
Mercuric oxide
for the oxidation of arsines and stibines, **2**, 497
reactions
with *O*-alkyl thiocarbamates, **5**, 1103
with arsenous acids, **1**, 446
with boron trifluoride, **4**, 195
with dibromides, **2**, 497
with ethanol, **6**, 405
with *O*-ethyl selenocarbamate, **5**, 1103
with imidothiocarbamates, **5**, 1104
with succinimide, **5**, 321
for the synthesis of aryl bromides, **2**, 627
for the synthesis of aryl iodides, **2**, 631
for the synthesis of carbodiimides, **5**, 1078
use in the synthesis of carbodiimides, **5**, 1062
Mercuric salts
as catalysts, for the synthesis of alkenyl phosphates, **2**, 693
reactions
with methyl compounds, **6**, 207
with tris(dimethoxyboryl)methane, **6**, 206
Mercuronium ions, as intermediates, **2**, 593
Mercury
reactions, with diazomethyl compounds, **3**, 488
use in the detection of chlorine, in commercial phosgene, **6**, 414
use in the synthesis of alkyl radicals, **1**, 948
use in the synthesis of carbon-bridged dimercury complexes, **4**, 696
use in the synthesis of *N*-fluoroimines, **5**, 807
use in Wittig reactions, **4**, 734

Mercury, aryl(trinitromethyl)-, from, diphenylmercury, **6**, 368
Mercury, bis(alkylseleno)-, reactions, with dialkylhaloarsines, **2**, 500
Mercury, bisbenzyl-, reactions, with tetracyanoethylene, **5**, 1142
Mercury, bis[bis(trimethylsilyl)methyl]-, synthesis of, **6**, 193
Mercury, bis(chlorodinitromethyl)-, production of, **6**, 287
Mercury, bis(diazo(dimethylarsino)methyl)-, synthesis of, **6**, 671
Mercury, bis(diazomethyl)-, synthesis of, **3**, 488; **6**, 670
Mercury, bis(diazonitromethyl)-, synthesis of, **6**, 664
Mercury, bis(difluorosulfoximidoyl)-, reactions, with tetrachloromethane, **6**, 362
Mercury, bis(fluorodinitromethyl)-, characteristics of, **6**, 287
Mercury, bis(heptafluoropropylseleno)-, synthesis of, **6**, 49
Mercury, bis(iodomethyl)-, synthesis of, **4**, 153
Mercury, bis(perfluoroisopropyl)-
  reactions
    with diethylaluminum iodide, **3**, 395
    with sulfur, **3**, 348
Mercury, bis(phenylethyne)-, reactions, with mercury(II) cyanide, **5**, 1142
Mercury, bis(*p*-tolyl)-, reactions, with haloalkanes, **1**, 128
Mercury, bis(trialkylgermyl)-, reactions, with bis(haloalkyl)mercury compounds, **6**, 62
Mercury, bis(trichloromethyl)-, synthesis of, **6**, 246
Mercury, bis(triethylgermyl)-
  reactions, with triethylgermyldiazoacetone, **6**, 185
  as starting material, for haloalkyl(triethylgermyl)mercury compounds, **6**, 62
Mercury, bis(2,2,2-trifluoro-1,1-dichloroethyl)-, reactions, with sulfur, **5**, 480
Mercury, bis(trifluoromethyl)-
  reactions of, **6**, 718
  reactions, with arsenic triiodide, **2**, 483
  synthesis of, **6**, 246
Mercury, bis(trifluoromethylthio)-, reactions, with alkylarsenic halides, **2**, 499
Mercury, bis(trifluorovinyl)-, reactions, with alane complexes, **4**, 815
Mercury, bis(trimethylsilyl)-
  reactions
    with alkali metals, **4**, 354
    with *N*,*N*'-diphenylcarbodiimide, **6**, 661
    with 1,2,3,3,3-pentafluoro-1-trimethylsilylpropene, **4**, 1048
    with pyridine, **2**, 814
    with trifluorovinyl bromide, **4**, 821
Mercury, bis(trimethylsilylmethyl)-, reactions, with potassium sand, **4**, 653
Mercury, bis(trimethylsilyl)methyl-, transmetallation of, **2**, 557
Mercury, bis(trinitromethyl)-
  metathesis reactions, of, **6**, 368
  as starting material, for (trinitromethyl)mercuric amides, **6**, 368
  synthesis of, **6**, 368
Mercury, (bromochloromethyl)phenyl-, reactions, with aldehydes, **2**, 618
Mercury, (chloromethyl)phenyl-, synthesis of, **4**, 152
Mercury, dialkyl-
  reactions
    with lithium, **2**, 555
    with mercury nitrate, **1**, 65
  synthesis of, **6**, 395
  transmetallation of, **2**, 557
Mercury, dibenzyl-, reactions, with primary alkylarsines, **2**, 507
Mercury, di(bistrifluoromethylamino)-, reactions, with iodine, **2**, 339
Mercury, dibutyl-
  reactions, **1**, 128
    with trinitromethane, **6**, 368
Mercury, diethyl-
  reactions, with tris(trimethylsilylmethyl)germane, **4**, 641
  as starting material, for perlithiated ethane, **6**, 205
Mercury, *N*,*N*-diethylcarbamoyl-, reactions, with hexamethyldisilathiane, **6**, 515
Mercury, diferrocenyl-
  reactions
    with dicarbonylcyclopentadienyliron iodide, **4**, 685
    with tin chloride, **4**, 726
  transmetallation of, **4**, 706
Mercury, di-1-hexynyl-, reactions, with nitrosyl chloride, **2**, 1056
Mercury, dimethyl-, use in monomethylation reaction, **2**, 570
Mercury, diphenyl-
  reactions, with acetic acid, **1**, 65
  as starting material, for aryl(trinitromethyl)mercury, **6**, 368
Mercury, diruthenocenyl-, from, ruthenocenyllithium, **4**, 723
Mercury, disilyl-, reactions, with chlorodialkylphosphine, **2**, 471
Mercury, phenyl(bromodichloromethyl)-
  reactions
    with carboxylic acids, **6**, 38
    with 1,1-diethyl-1-germacyclobutane, **1**, 291
  use in the synthesis of dichlorocarbene, **4**, 16
Mercury, phenyl(fluorodibromomethyl)-, use in the synthesis of bromofluorocarbene, **4**, 35
Mercury, phenyl tribromomethyl-, as starting material, for dibromocarbene, **4**, 748
Mercury, phenyl(tribromomethyl)-, use in the synthesis of dibromocarbene, **4**, 25
Mercury, phenyltrifluoromethyl-, use in the synthesis of difluorocarbene, **4**, 8
Mercury, (sulfonyl)ethenyl-, decarboxylative demercuration of, **2**, 1035
Mercury, trifluoromethylphenyl-
  reactions, with iodine, **4**, 1083
  as starting material, for difluorocarbene, **4**, 1083
Mercury(I), use in the synthesis of transition metal amides, **5**, 384
Mercury(I) carbonate, reactions, with imides, **5**, 387
Mercury(I) iodide, as catalyst, for the synthesis of carbon-bridged dimercury complexes, **4**, 696
Mercury(I) nitrate, reactions, with benzyl bromides, **3**, 89
Mercury(I) oxide
  reactions
    with *N*-benzyl-*N*-phenylhydrazine, **3**, 447
    with diazo compounds, **3**, 488
    with hydrazines, **3**, 447
Mercury(II), as catalyst, in the synthesis of methyl ketones, **3**, 114
Mercury(II), bis(bistrifluoromethylaminooxy)-, reactions, with thiophosgene, **6**, 436
Mercury(II) acetamide, as dehydrogenating reagent, **3**, 463
Mercury(II) acetate
  reactions, with diazomethane, **3**, 488
  use in the synthesis of trinuclear carbon-bridged mercury compounds, **4**, 697
Mercury(II) bis(trifluoromethanethiolate)
  as agent for trifluoromethylthiolation, **6**, 236
  reactions, with acyl chlorides, **6**, 235
  synthesis of, **6**, 236
  trifluoromethylthiolation with, **6**, 233
Mercury(II) bis(trifluoromethyl)amide
  bromination of, **6**, 240
  iodination of, **6**, 240
  reactions, with alkanesulfenyl chlorides, **6**, 241
α-Mercury(II) carbonyls, demercuration of, **1**, 67
Mercury(II) chloride
  reactions
    with amines, **5**, 1055
    with butanal, **4**, 697

with dialuminum alkyls, **4**, 697
with pentan-2,4-dione, **4**, 697
with propanal, **4**, 697
as reagent, for dithiane hydrolysis, **5**, 410
Mercury(II) chloride, (trichloromethyl)-, synthesis of, **6**, 246
Mercury(II) complexes, bis(triphenylgermanium)-, reactions, with copper(II) cyanide, **5**, 1136
Mercury(II) compounds, bisphenyl-, synthesis of, **5**, 1142
Mercury(II) cyanide
reactions
with bis(phenylethyne)mercury, **5**, 1142
with iodotrifluoromethylphosphine, **5**, 1129
with phenyltrifluorosilanes, **5**, 1142
Mercury(II) cyanide, benzyl-, synthesis of, **5**, 1142
Mercury(II) cyanide, phenylethyne-, synthesis of, **5**, 1142
Mercury(II) cyanide, trifluoromethyl-, synthesis of, **5**, 1141
Mercury(II) cyanides
reactions
with chlorotrialkylsilane, **5**, 1135
with organogermanium halides, **5**, 1136
synthesis of, **5**, 1141, 1142, 1143
Mercury(II) cyanides, chloromethyl-, synthesis of, **5**, 1142
Mercury(II) iodide, as starting material, for alkynylmercury compounds, **2**, 1098
Mercury(II) iodide, trifluoromethyl-
reactions, with silver cyanide, **5**, 1141
synthesis of, **5**, 1141
Mercury(II) iodides, allylic, reactions, with acid chlorides, **3**, 140
Mercury(II) nitrate, reactions, with bromoalkanes, **2**, 103
Mercury(II) oxide
for the dehydrogenation of hydrazones, **3**, 463
reactions
with ammonium hydroxide, **5**, 1026
with 1,2-dihydrazones, **1**, 1075
with hydrazones, **3**, 470
with lactols, **1**, 931
with nitrodiazomethane, **6**, 664
as reagent, for dithiane hydrolysis, **5**, 410
use in the synthesis of α-keto hydrazones, **3**, 470
Mercury(II) salt, bis(diazomethylene)-, synthesis of, **6**, 670
Mercury(II) selenocyanates, reactions, with alkyl halides, **5**, 1053
Mercury(II) trifluoroacetamide, as dehydrogenating reagent, **3**, 463
Mercury(II) trifluoroacetate
reactions, with benzyl ketones, **1**, 1075
use in the synthesis of polymeric mercury compounds, **4**, 697
Mercury(II) trifluoromethaneselenolate
reactions
with tetrabromomethane, **6**, 307
with *S*-trifluoromethyl bromothioformate, **6**, 237
Mercury(II) trifluoromethanethiolate
reactions
with dialkyl aminosulfenyl chlorides, **2**, 361
with hydrogen chloride, **6**, 232
Mercury acetate
reactions
with alkynes, **2**, 982
with sodium borohydride, **2**, 43
with trialkylboranes, **2**, 592
Mercury acetate, 1-propenyl-, reactions, with mercuric acetate, **4**, 697
Mercury acetates, alkenyl-, protonolysis of, **1**, 66
Mercury acetates, alkyl-, reduction of, **1**, 332
Mercury acetate ylide adducts, as starting materials, for mercury-bridged bisylides, **5**, 919
Mercury alkoxides, reactions, with trialkylboranes, **2**, 592
Mercury alkyls, reactions, with *N*-bromosuccinimide, **5**, 388
Mercury alkynylides, iodination of, **2**, 1019
Mercury bis(alkynides), reactions, with nitrosyl chloride, **2**, 1051
Mercury (bis)alkynides, as starting materials, for trialkynylphosphines, **2**, 1060
Mercury bis[bis(trimethylsilyl)amide], reactions, with ethyl diazoacetate, **3**, 488
Mercury bis(*t*-butyl diazoacetate), reactions, with dinitrogen pentoxide, **3**, 480
Mercury bis(diazocarbonyl) compounds, reactions, with disilylmercury compounds, **3**, 483
Mercury bis(ethyl diazoacetate)
reactions
with disulfides, **3**, 483
with iodo trialkylsilanes, **3**, 483
with metal sulfides, **3**, 484
with sulfuryl chloride, **3**, 478
synthesis of, **3**, 488
Mercury bis(phenylalkyne), reactions, with thiobenzanilide, **5**, 739
Mercury bromide, alkyl-, reactions, with Grignard reagents, **6**, 394
Mercury carboxylates, decarboxylation of, **2**, 594
Mercury carboxylates, alkyl-, synthesis of, **2**, 594
Mercury chloride
reactions
with tris(trimethylsilylmethyl)lithium, **6**, 394
with ylides, **5**, 919
as starting material, for α-mercurated ylides, **5**, 919
Mercury chloride, alkenyl-, synthesis of, **1**, 1108
Mercury chloride, hexenyl-, reactions, with benzene thiol, **1**, 68
Mercury chlorides, (2-hydroxyaryl)-, reactions, with α,β-unsaturated ketones, **4**, 168
Mercury complexes, synthesis of, **6**, 732
Mercury complexes, trinuclear carbon-bridged, synthesis of, **4**, 697
Mercury compounds
of phosphorus functions, synthesis of, **5**, 890
reactions
with alkynes, **4**, 687
with 1-metalloalkylphosphine oxides, **4**, 587
with 1-metalloalkylphosphine sulfides, **4**, 587
with phosphorus ylides, **4**, 587
synthesis of, **6**, 206
in situ, **2**, 715
toxicity of, **2**, 590
transmetallation of, **1**, 493
Mercury compounds, α-acetoxy-, synthesis of, **4**, 385
Mercury compounds, β-acetoxyalkenyl-, from, reaction of alkynes with mercury acetate, **2**, 982
Mercury compounds, alkenyl-
dimerization of, **1**, 1096
synthesis of, **1**, 1095, 1108
Mercury compounds, α-alkoxy-, synthesis of, **4**, 385
Mercury compounds, 2-alkoxy alkyl-, as starting materials, for ethers, **2**, 89
Mercury compounds, alkyl-
electrochemical synthesis of, **2**, 594
protonolysis of, **1**, 65
reactions
with alkylmagnesiums, **1**, 846
with chlorocarbons, **1**, 668
with platinum(IV) tetrachloride, **2**, 584
with potassium hexachloroplatinate, **2**, 584
with sodium thiolates, **2**, 274
stability of, **1**, 668
synthesis of, **2**, 590
transmetallation of, **2**, 595
Mercury compounds, alkynyl-
cleavage of, **1**, 1000
from
alkynyl metals, **2**, 1097
mercuric thiocyanate, **2**, 1098
mercury(II) iodide, **2**, 1098
organomercury halides, **2**, 1097

peroxyalkynes, **2**, 1098
synthesis of, **2**, 1097
Mercury compounds, aryl-
carbonylation of, **3**, 290
cleavage of, **1**, 453
dimerization of, **1**, 492
fluorination of, for aryl fluorides, **2**, 613
from, metal carboxylates, **2**, 956
in Heck reactions, **1**, 493
reactions, with boron trihalides, **2**, 906
synthesis of, **1**, 493
for the synthesis of diaryl selenides, **2**, 729
Mercury compounds, bis(alk-1-ynyl)-, reactions, with lead tetraacetate, **2**, 1102
Mercury compounds, bis(carbonyl)-, methylation of, **6**, 719
Mercury compounds, bis(haloalkyl)-, reactions, with bis(trialkylgermyl)mercury, **6**, 62
Mercury compounds, bis(perfluoroalkyl)-, reactions, with germanium tetrahalides, **6**, 62
Mercury compounds, (bromomethyl)-, synthesis of, **4**, 152
Mercury compounds, carbon-bridged bischloro-, synthesis of, **4**, 697
Mercury compounds, (α-chloroalkyl)-, synthesis of, **4**, 152
Mercury compounds, dialkyl
cleavage of, **1**, 66
reactions, with dialkynyl mercury compounds, **2**, 1097
Mercury compounds, dialkyl-
reactions, with trityl cations, **1**, 668
as starting materials
for dialkylberylliums, **2**, 563
for dialkylmagnesiums, **2**, 560
for potassium salts, **1**, 378
synthesis of, **2**, 592, 597
Mercury compounds, dialkynyl-, reactions, with dialkyl mercury compounds, **2**, 1097
Mercury compounds, diazo-, reactions, with bis(trialkylphosphine)palladium(II) halides, **3**, 487
Mercury compounds, α-diazo-, synthesis of, **4**, 385
Mercury compounds, 1,1-dihalo-2,2,2-trifluoroethyl, reactions, with sulfur, **5**, 474
Mercury compounds, disilyl-, reactions, with mercury bis(diazocarbonyl) compounds, **3**, 483
Mercury compounds, divinyl-, dimerization of, **1**, 491
Mercury compounds, haloalkyl(triethylgermyl)-, from, bis(triethylgermyl)mercury, **6**, 62
Mercury compounds, 1-halo-2,2,2-trifluoroethyl, reactions, with sulfur, **5**, 474
Mercury compounds, imidoyl-, synthesis of, **5**, 802
Mercury compounds, *S*-imidoyl-, synthesis of, **5**, 739
Mercury compounds, (iodomethyl)-, synthesis of, **4**, 153
Mercury compounds, perfluoroisopropyl-, synthesis of, **4**, 152
Mercury compounds, perhalodialkyl-, synthesis of, **6**, 64
Mercury compounds, polymeric, synthesis of, **4**, 697
Mercury compounds, silyl-
stability of, **2**, 1097
synthesis of, **4**, 821
Mercury compounds, trihalomethyl-
synthesis of, **6**, 245
use in the synthesis of α,α-dihalo ethers, **6**, 38
Mercury compounds, vinyl
cleavage of, **1**, 454
demercuration of, **1**, 454
Mercury compounds, vinyl-
reactions, with selenenyl halides, **2**, 727
as starting materials, for dienes, **1**, 491
Mercury compounds, vinylic, reactions, with palladium chloride, **1**, 1095
Mercury difluoride, use in the synthesis of chlorofluoroalkanes, **4**, 30
N-Mercury enamines, synthesis of, **2**, 817
Mercury ethoxides, alkyl-, reactions, with diazoalkanes, **3**, 488
Mercury ethoxides, aryl-, reactions, with diazoalkanes, **3**, 488
Mercury fulminate
reactions
with chlorine, **5**, 822
with trimethylsilyl bromide, **3**, 686
use in the synthesis of fulminic acid, **3**, 685
Mercury halide, alkyl-, reactions, with hydrochloric acid, **1**, 65
Mercury halides, alkyl-, reduction of, **1**, 332
Mercury hydride reagents, synthesis of, **1**, 332
Mercury hydrides
for the mediation of radical reactions, **1**, 332
synthesis of, **1**, 343
use in radical cyclization reactions, **1**, 343
Mercury hydrides, alkyl-, synthesis of, **1**, 332
Mercury imides, synthesis of, **5**, 387
Mercury iodide, propargyl-
reactions
with acyl chlorides, **1**, 974
with electrophiles, **1**, 974
Mercury iodide, (trifluoromethyl)-, synthesis of, **6**, 246
Mercurylactonization
of δ,ε-ene-acids, **5**, 172
for the synthesis of γ-lactones, **5**, 168
Mercury–lithium exchange
for the synthesis of alkyllithiums, **2**, 555
for the synthesis of lithiomethanes, **6**, 205
Mercurymethylene complexes, synthesis of, via transmetallation, **4**, 669
Mercury nitrate, reactions, with dialkylmercury, **1**, 65
Mercury organometallics, synthesis of, **6**, 258
Mercury oxides, reactions, with trinitromethane, **6**, 368
Mercury salts
as catalysts, in Cope rearrangements, **1**, 809
reactions
with alkenes, **2**, 593
with chloromethyltrifluorosilanes, **5**, 1142
with dialkylthallium acetates, **2**, 598
with sulfinic acid, **2**, 594
use in the synthesis of furanyl ketones, **3**, 302
Mercury thiocarboxylates, phenyl-
iodination of, **5**, 245
reactions, with *N*-bromosuccinimide, **5**, 245
Mercury thiolates, synthesis of, **2**, 273
Mercury trifluoroacetate, reactions, with perfluoropropen-2-ol, **6**, 64
Merker–Scott reaction, for the synthesis of tris(organylsilyl)methanes, **6**, 173
Mesaconic acid, synthesis of, **5**, 99
Mesembranol, synthesis of, **1**, 828
Mesitoates, alkyl, saponification of, **5**, 106
Mesitoic acid, synthesis of, **5**, 112, 116
Mesitonitrile oxides, isolation of, **3**, 683
Mesitoyl chloride, reactions, with lithium dihydrogen arsenide, **5**, 891
Mesitylene
acylation of, **3**, 296
Friedel–Crafts acylation of, **5**, 116
reactions, with chromium vapor, **4**, 694
trichloromethylation of, **6**, 29
Mesitylene, trichloromethyl-, synthesis of, **6**, 29
Mesityl oxide
reactions
with α-chloroethyl chloroformates, **4**, 59
with dibutylphosphine oxides, **4**, 339
with primary amines, **3**, 465
Meso-epoxides, enantioselective rearrangement of, **1**, 842
Mesylate elimination, palladium-catalyzed, use in the synthesis of glycals, **1**, 584
Mesylates
cleavage of, **2**, 553
displacement of, **1**, 188, 198; **2**, 95

Friedel–Crafts alkylation of, **1**, 475
reactions
with potassium halides, **2**, 9
with sodium sulfide, **2**, 224
synthesis of, **1**, 700
use in the alkylation of alcohols, **1**, 182
use in the synthesis of alkyl halides, **2**, 9
Mesylates, alkyl, as starting materials, for dialkyl peroxides, **2**, 97
Mesylates, allylic, as starting materials, for allenes, **1**, 983
Mesylates, bridgehead, displacement of, **1**, 177
Mesylates, primary, alkylation of, **5**, 140
Mesyl azide, as diazo transfer reagent, **5**, 869
Metal, pentacarbonyl-1-alkoxy-2-aza-3,3-diarylallenyl, synthesis of, **5**, 1088
Metal(0) complexes, reactions, with alkyl halides, **2**, 582
Metal(0) compounds, alkyl, synthesis of, **2**, 575
Metal(I) chloride
reactions
with methyllithiums, **2**, 579
with methylmagnesiums, **2**, 579
Metal(I) complexes, anionic, alkylation of, **2**, 583
Metal(I) compounds, alkyl, synthesis of, **2**, 575
Metal(I) halides
as starting materials
for carbon-bridged coppers, **4**, 695
for carbon-bridged silver compounds, **4**, 695
Metal(II) chlorides
reactions
with alkyllithiums, **2**, 577
with methyllithium, **2**, 572
Metal(II) compounds, alkyl, synthesis of, **2**, 576
Metal(II) cyclopentadienylcarbonyl complexes, alkylation of, **2**, 572
Metal(II) cyclopentadienylcarbonyl complexes, alkyl, synthesis of, **2**, 572
Metal(III) alkoxides, synthesis of, via nucleophilic substitution reactions, **2**, 885
Metal(III) compounds, alkyl, synthesis of, **2**, 578
Metal(III) dihalides, aryl, synthesis of, via disproportionation reactions, **2**, 882
Metal(III) dihalides, vinyl, synthesis of, via disproportionation reactions, **2**, 882
Metal(III) halides, synthesis of, **2**, 880
Metal(III) halides, diaryl, synthesis of, via disproportionation reactions, **2**, 882
Metal(III) halides, divinyl, synthesis of, via disproportionation reactions, **2**, 882
Metal(III) oxides
synthesis of
via oxidative processes, **2**, 886
via reductive methods, **2**, 888
Metal(IV) complexes, alkyl, structure of, **2**, 572
Metal(IV) complexes, dialkyl, oxygen oxidation of, **2**, 573
Metal(IV) compounds
reduction of, **2**, 565
synthesis of, **2**, 584
Metal(IV) compounds, alkyl
synthesis of, **2**, 578
problems, **2**, 579
Metal(V) halides, tetraphenyl, synthesis of, **2**, 881
Metal(V) oxides, synthesis of, via oxidative processes, **2**, 886
Metal(V) trihalides, diaryl
from
diorganoarsines, **2**, 881
diorganometal(III) halides, **2**, 881
diorganostibines, **2**, 881
Metal acyl complexes, α-elimination reactions of, **5**, 937
Metal acyl complexes, pentacarbonyl, reactions, with oxaloyl halides, **3**, 691
Metal alanates, reactions, with alkynes, **2**, 1099
Metal alkanetellurolates, as starting materials, for alkyl ditellurides, **2**, 292
Metal alkoxides
reactions
with alcohols, **2**, 87
with imides, **5**, 372
with phosphonium salts, **2**, 441
synthesis of, **2**, 112
Metal *N*-alkylcarbamato complexes, decomposition of, **5**, 975
Metal alkyls
as catalysts, for Friedel–Crafts acylation, **1**, 462
reactions, with carbonyl ligands, **5**, 441
as starting materials, for tertiary phosphine sulfides, **2**, 461
Metal alkynides
reactions
with antimony halides, **2**, 1080
with arsenic halides, **2**, 1076
with boron halides, **2**, 1083
with chlorophosphine oxides, **2**, 1065
with phosphorus halides, **2**, 1058
Metal alkynyl complexes, reactions, with electrophiles, **1**, 1145
Metal alkynylides, iodination of, **2**, 1019
Metal amides
from
amides, **5**, 388
bis(trialkylstannyl)carbodiimide, **5**, 389
carbodiimides, **5**, 390
isocyanates, **5**, 390
tin isocyanates, **5**, 391
groups 1 and 2, synthesis of, **5**, 384
groups 13 through 16, synthesis of, **5**, 388
reactions
with alkylantimony halides, **2**, 502
with alkylarsenic halides, **2**, 502
with alkylbismuth halides, **2**, 502
with diazomethanes, **6**, 669
synthesis of, **2**, 423
via acylation of metallated amines, **5**, 389
via nitrile hydrolysis, **5**, 389
via rearrangement of trialkylgermanium amides, **5**, 390
via ring-opening of metal imides, **5**, 390
via transmetallation, **5**, 389
use in the synthesis of tertiary ynamines, **2**, 1041
Metal amides, *N*-alkyl, synthesis of, **2**, 423
Metalates, tetraalkylammonium pentacarbonylacyl-, reactions, with acyl halides, **5**, 949
Metal azides
reactions
with alkylhaloarsines, **2**, 504
with alkylhalostibines, **2**, 504
Metal-based oxidants, for the synthesis of ketones, **3**, 115
Metal bis(trifluoromethyl)amides, as starting materials, for bis(trifluoromethyl)alkylamines, **6**, 238
Metal carbene
catalytic reactions, with alkynes, **2**, 981
reactions, with alkynes, **2**, 979
Metal carbene complexes
reactions
with silylalkynes, **2**, 927
with tellurocyanate, **3**, 400
synthesis of, **3**, 507
See also Alkylidene complexes
Metal carbenoids
additions of, **1**, 352
reactions, with alkenes, **1**, 299
Metal carbides
characteristics of, **5**, 1096
fluorination of, **6**, 212
Metal carbonyl anions
reactions
with imidoyl chlorides, **5**, 797
with imidoyl halides, **5**, 798
Metal carbonyl complexes

nucleophilic attack on, **6**, 517
synthesis of, **2**, 1082; **6**, 521
Metal carbonyl complexes, alkyl, synthesis of, **2**, 573
Metal carbonyls
as desulfurization reagents, **1**, 437
reactions
with alkynehexacarbonyldicobalt, **4**, 711
with dihydroboroles, **4**, 695
with organolithiums, **6**, 719
with 1-phenyl-1-boracyclohepta-2,6-diene, **4**, 695
as starting materials, for carbene complexes, **6**, 719
synthesis of, **6**, 522
Metal carboxylates
reactions
with acyl chlorides, **5**, 198
with dichlorophenylphosphine, **5**, 221
as starting materials
for anhydrides, **5**, 182
for arylmercury compounds, **2**, 956
for symmetrical alkanoic anhydrides, **5**, 185
for symmetrical aromatic anhydrides, **5**, 194
for unsymmetrical alkanoic anhydrides, **5**, 192
Metal carbyne, reactions, with alkynes, **2**, 979
Metal carbyne complexes, synthesis of, **3**, 690
Metal catalysis, for the oxidation of tertiary amines, **2**, 348
Metal catalysts, for the synthesis of cyclopropanes, **1**, 299
Metal catalysts, ligands, carbon–hydrogen activation of, **2**, 963
Metal-catalyzed rearrangements, of vinylcyclopropanes, **1**, 803
Metal cations, effects on Wittig reactions, **1**, 727
Metal chelating agents, use in enhancing solubility of potassium fluoride, **5**, 4
Metal chlorides
reactions
with alcohols, **2**, 112
with isocyanides, **6**, 622
with phenylcyanamide, **5**, 1083
Metal chlorides, trihexamethyldisilazide, transmetallation of, **2**, 603
Metal chloroethynylides, synthesis of, **2**, 1015
Metal clusters
flash vacuum pyrolysis of, **6**, 403
from, carbyne clusters, **6**, 402
Metal clusters, mixed, synthesis of, **6**, 402
Metal complexes, **6**, 593
as catalysts, in Cope rearrangements, **1**, 809
containing a metal–halocarbonyl bond, **6**, 455
reactions
with acyl azides, **5**, 386
with alkynes, **4**, 712
of selenobenzaldehydes, **3**, 386
as starting materials, for imidic acid esters, **5**, 696
Metal complexes, acyl
decarbonylation of, **2**, 572; **4**, 722
synthesis of
via addition of a metal alkyl reagent to a coordinated carbonyl ligand, **5**, 441
via migratory insertion of alkyl ligands, **5**, 441
via oxidative addition of acid chlorides to transition metal complexes, **5**, 441
Metal complexes, (alkoxymethyl), reactions, with hydrogen halides, **4**, 147
Metal complexes, alkyl, synthesis of, **2**, 572
Metal complexes, alkyloxo, synthesis of, **2**, 572
Metal complexes, alkyne
characteristics of, **2**, 952
as intermediates, **2**, 989
Metal complexes, benzylidene, reactions, with selenocyanate, **3**, 386
Metal complexes, chloro, reactions, with 1-methoxyvinyllithium, **4**, 955
Metal complexes, chlorobis(cyclopentadienyl), reactions, with lithium tetramethylaluminate, **4**, 726
Metal complexes, cyclopentadienyl
alkylation of, **2**, 602
reactions, with alkyllithiums, **2**, 566
synthesis of, **2**, 566
Metal complexes, functionalized isocyanide, reviews, **3**, 694
Metal complexes, halo, reactions, with diazoalkanes, **4**, 147
Metal complexes, ketenimine, synthesis of, **3**, 596
Metal complexes, nitrogen ylide pentacarbonyl, synthesis of, **5**, 953
Metal complexes, selone pentacarbonyl, synthesis of, **3**, 396
Metal π-complexes, alkyne, characteristics of, **2**, 952
Metal compounds, of phosphorus functions, synthesis of, **5**, 889
Metal counterions, use in the synthesis of carbanions, **1**, 1088
Metal cyanates
reactions, with carbonyl chlorides, **5**, 994
as starting materials
for cyanates, **5**, 1104
for isocyanates, **5**, 1015
Metal cyanides
reactions
with aryl halides, **3**, 661
with selenenyl halides, **5**, 1116
with thallium salts, **5**, 1143
as starting materials, for thiocyanates, **5**, 1111
use in the synthesis of phosphorus(III) cyanides, **5**, 1129
Metal dialkylstibides, reactions, with 1,2-dihaloethanes, **2**, 505
Metal diarylphosphides
reactions
with alkyl halides, **2**, 826
with aryl halides, **2**, 826
Metal dichlorides, trimethyl, reactions, with methyllithium, **2**, 570
Metal dithiocarbamates, reactions, with iminium salts, **4**, 320
Metal enolates
alkylation of, **3**, 124
reactions, with silyl halides, **2**, 699
rearrangement of, **1**, 829
as starting materials, for α-selenocarbonyls, **2**, 282
synthesis of, **2**, 696
use in the synthesis of α-selenyl ketones, **3**, 173
Metal exchange reaction
for the substitution of selenium and tellurium, **1**, 245
for the substitution of sulfur functions, **1**, 236
Metal fluorides
as Lewis acids, **6**, 16
reactions
with hydrogen fluoride, **6**, 16
with perfluorobicyclobutylidene, **6**, 18
Metal fulminates, reactions, with alkyl halides, **3**, 683
Metal functions
elimination of, **1**, 660, 1080
substitution of, **1**, 291
synthesis of, **5**, 1096
Metal functions, acyl, synthesis of, **5**, **435–469**
Metal functions, alkynyl, reduction of, **1**, 1000
Metal functions, doubly-bonded, synthesis of, **3**, **507–523**
Metal functions, imidoyl, synthesis of, **5**, 795
Metal germanes, reactions, with halomethyl ethers, **4**, 360
Metal halides
reactions
with alkyllithiums, **2**, 578
with imides, **5**, 372
with lithiums, **6**, 396
with magnesium compounds, **2**, 578
with methanes, **6**, 370
with organometals, **2**, 990
with transition metal carbene complexes, **4**, 722
Metal halides, triaryl, thermal stability of, **2**, 881
Metal–halogen exchange
of aryl halides, **2**, 961

for aryl metals, **2**, 956
vs. metal–hydrogen exchange, **1**, 1144
of organic halides, **2**, 956
for organometallics, **2**, 952
reactions, **1**, 110
for the reduction of vinyl bromides, **1**, 431
side reactions of, **2**, 961
for the synthesis of alkenes, **4**, 1048
for the synthesis of carbanions, **1**, 885
for the synthesis of diazohalogenmethylcarbonyls, **3**, 478
for the synthesis of 1-lithio 1-silylalkenes, **4**, 1059
for the synthesis of α-metallated 1,1-bis(silyl)alkanes, **4**, 618
for the synthesis of transition metals, **4**, 821
use in the substitution of vinyl halides, **1**, 466
of vinyl halides, **2**, 952, 961
for vinyl metals, **2**, 956
See also Halogen–metal exchange
Metal hydride bonds, reactions, with alkenes, **1**, 296
Metal hydride nucleophilic displacement, for the reduction of alkyl halides, **1**, 2
Metal hydrides
demercuration of, **1**, 66
for the deprotonation of amides, **5**, 384
reactions
with aromatic acids, **3**, 82
with tosylhydrazones, **1**, 18
with transition metal salts, **1**, 515
for the reduction of bromoalkanes, **1**, 9
for the reduction of carbon–halogen bonds, **1**, 3
for the reduction of carbonyls, **2**, 56
for the reduction of iodoalkanes, **1**, 10
for the reduction of nitriles, **3**, 86
for the reduction of oxiranes, **1**, 13
use in the reductive cleavage of epoxides, **2**, 83
use in the synthesis of alcohols, **2**, 52
Metal–hydrogen exchange
of arenes, **2**, 956
vs. metal–halogen exchange, **1**, 1144
for organometallics, **2**, 952
of silylmethanes, **6**, 190
synthesis via, **6**, 198
Metal-hydrogen exchange, non-redox, of alkenes and arenes, **2**, 963
Metal–hydrogen exchange, non-redox
for aryl metals, **2**, 956
for vinyl metals, **2**, 956
Metal hydroxides, reactions, with phosphonium salts, **2**, 441
Metal hypobromides, reactions, with terminal alkynes, **2**, 1017
Metal hypochlorites, as starting materials, for alkynyl chlorides, **2**, 1016
Metal imides
from
haloimides, **5**, 391
imides, **5**, 390
groups 1 and 2, synthesis of, **5**, 384
groups 13 through 16, synthesis of, **5**, 390
ring-opening of, **5**, 390
synthesis of, via transmetallation, **5**, 391
N-Metalimides, as starting materials, for *N*-iodo(I)imides, **5**, 320
Metal imine complexes, reactions, with alkynes, **4**, 1017
Metal iodides, alkyl, synthesis of, **2**, 564
Metallacycles
five-membered, **2**, 989
synthesis of, **2**, 561, 581; **5**, 465
three-membered and five-membered, **2**, 984, 986
vinyl metals, **2**, 953
Metallacyclic complexes, stability of, **1**, 666
Metallaferrocenophanes, synthesis of, **4**, 720
Metalla-heterocycles, synthesis of, **4**, 1029
Metallaindane compounds, synthesis of, **5**, 443
Metallaphosphiranes, synthesis of, **4**, 581
Metallates, acyl, alkylation of, **6**, 719
Metallates, carbonyl
for the nucleophilic metallation of halides, **2**, 512
reactions, with halocarbonyl metals, **6**, 525
Metallation
of alkenes and arenes, **2**, 965
via deprotonation with bases, **2**, 964
of arenes, via electrophilic substitution, **2**, 963
of π-complexes, **4**, 705
of nitrosamines, **4**, 529
of organic compounds, **1**, 881
of substituted benzenes, regiochemistry, **2**, 964
for the synthesis of alkyllithiums, **2**, 551
for the synthesis of *gem*-halo lithioalkenes, **4**, 816
Metallation, acyl-, example of, **2**, 978
Metallation, oxidative
of alkynes, **2**, 983, 989
for aryl metals, **2**, 956
for vinyl metals, **2**, 956
with alloys and amalgams, **2**, 959
of aryl halides, **2**, 957
using activated metals, **2**, 959
for aryl metals, **2**, 956
with metals activated by arenes and graphite, **2**, 959
with metal vapors, **2**, 961
of organic halides, **2**, 956
for organometallics, **2**, 952
for the synthesis of metal(III) halides, **2**, 880
with transition metal complexes, **2**, 960
of unsaturated compounds, for organometallics, **2**, 952
of vinyl halides, **2**, 957
using activated metals, **2**, 959
for vinyl metals, **2**, 956
o-Metallation
for the synthesis of aryl ethers, **3**, 298
for the synthesis of benzaldehydes, **3**, 97
Metallations, directed
of aromatic compounds, **2**, 942
of chromium tricarbonylarene complexes, **2**, 942
of heterocyclic compounds, **2**, 942
Metallation–silylation reactions, for the synthesis of tetrasilylmethanes, **6**, 378
Metallic complexes, for the mediation of radical reactions, **1**, 336
Metallic elements, synthesis of functions containing two atoms of, **4**, **667–703**
Metallic halides, reactions, with lithium bis(trimethylsilyl)methanide, **6**, 194
Metallics, acyl
silylation of, **5**, 418
as starting materials, for acylsilanes, **5**, 417
Metallics, alkenyl, from, vinyl halides, **1**, 1090
Metallics, aryl
as starting materials
for aryl bromides, **2**, 627
for aryl chlorides, **2**, 620
for aryl fluorides, **2**, 613
for aryl halides, **2**, 608
for aryl iodides, **2**, 632
Metallics, silyl
reactions, with acyl chlorides, **5**, 417
as starting materials, for acylsilanes, **5**, 417
Metallics, vinyl
reactions, with enones, **3**, 229
as starting materials
for vinyl bromides, **2**, 624
for vinyl chlorides, **2**, 617
for vinyl fluorides, **2**, 611
for vinyl iodides, **2**, 630
use in the synthesis of α,β-unsaturated ketones, **3**, 210
Metallic salts, diazo, as starting materials, for bis(phosphinyl)iminocarbonyls, **6**, 667

Metalloboration, of alkynes, **2**, 903
Metallocene dichlorides, reactions, with methylenebis(magnesium bromide), **4**, 670
Metallocenes, synthesis of, **1**, 862
Metallocenes, decamethyl-, use in the synthesis of triple decker sandwich complexes, **4**, 717
Metallocycles
  reactions, with α,ω-dihaloalkanes, **1**, 116
  synthesis of, **4**, 1067
Metallo-ene reaction, mechanism, **1**, 526
Metallogermylation, of alkynes, **2**, 946
Metalloidal halides
  reactions
    with organolithiums, **6**, 186
    with organomagnesiums, **6**, 186
Metalloid functions, synthesis of, **5**, 1019, 1095
Metalloid functions, alkynyl, reduction of, **1**, 999
Metalloid functions, doubly-bonded, synthesis of, **3**, **501–505**, **923–930**
Metalloid functions, imidoyl, synthesis of, **5**, 792
Metalloid halides, reactions, with lithium compounds, **6**, 396
Metalloids
  of arsenic functions, synthesis of, **5**, 892
  with mixed metalloids and metals, synthesis of functions, **6**, 203
  synthesis of functions, **4**, **1043–1070**
  synthesis of functions containing at least one, **6**, **171–210**, **377–406**
    with another metalloid or metal, **4**, **601–665**
Metalloids, halo-, reactions, with lithiomethanes, **6**, 389
Metalloids, mixed, synthesis of functions, **6**, 186
Metalloids, trifluoromethyl-, synthesis of, **6**, 22
N-Metalloids, synthesis of, **2**, 810
Metalloles, synthesis of, **2**, 949; **4**, 822
Metallonitrene complexes, synthesis of, **4**, 800
Metal–metal exchange, for the synthesis of carboxylic esters, **5**, 142
α-Metal nitriles, synthesis of, **3**, 640
Metal nitrocyanamides, reactions, with alkyl halides, **5**, 963
Metal nucleophile, addition elimination of, **5**, 1097
Metal oxidants, use in the synthesis of ketones, **3**, 117
Metal oxides
  for the oxidation of thiols, **2**, 214
  as starting materials, for carbodiimides, **5**, 1062
Metal oxides, alkenyl, synthesis of, via oxidation with manganese dioxide, **2**, 887
Metal oximes, as starting materials, for transition metal amides, **5**, 386
Metal oxo complexes, synthesis of, **2**, 572
Metal oxygen bonds
  synthesis of
    via esterification processes, **2**, 886
    via substitution reactions, **2**, 885
Metal pentacarbonyl, bromo, transmetallation of, **2**, 572
Metal pentacarbonyl complexes, synthesis of, **2**, 572
Metal pentachlorides, reactions, with methyl metals, **2**, 570
Metal perchlorates, reactions, with triethyl thiophosphate, **2**, 94
Metal phenolates, synthesis of, **2**, 703
Metal phosphides
  alkylation of, **2**, 437
    for secondary phosphines, **2**, 823
  hydrolysis of, for secondary phosphines, **2**, 823
  reactions
    with alkyl halides, **2**, 427
    with carbonyls, **6**, 686
    with β-chloroethylamines, **2**, 430
    with 1,1-dihaloalkanes, **4**, 545
    with episulfides, **2**, 430
    with epoxides, **2**, 430
    with 1-haloalkylphosphines, **4**, 546
    with 1-haloalkylsilanes, **4**, 566
    with halocarboxylic acids, **2**, 430
    with organohalides, **2**, 429, 444
Metal–phosphido bonds, reactions, with carbon monoxide, **6**, 519
Metal polysulfides, use in the synthesis of arenethiols, **2**, 706
Metal porphyrin complexes, alkyl, synthesis of, **2**, 578
Metal reduction
  of amines, **1**, 32
  of annulenes, **1**, 890
  dissolving, for the synthesis of alkenes, **1**, 508
  of α,β-unsaturated ketones, **1**, 889
Metals
  bearing borons and metals, synthesis of functions, **6**, 198
  bearing germanium and metals, synthesis of functions, **6**, 202
  as catalysts, in Cope rearrangements, **1**, 808
  doubly-bonded, synthesis of functions, **6**, **677–724**
  with mixed metalloids and metals, synthesis of functions, **6**, 203
  oxidative metallation with, **2**, 959
  reactions
    with alcohols, **2**, 112
    with carbonyls, **1**, 1057
    with organic halides, **2**, 957
  for the reduction of nitriles, **1**, 48
  substitution of, **1**, 490
  synthesis of functions, **4**, **1043–1070**
  synthesis of functions containing three, **6**, **171–210**
  synthesis of functions doubly-bonded to, **5**, **931–959**
  two dissimilar, synthesis of functions, **4**, **705–727**
  use in the synthesis of vinylic carbanions, **1**, 1088
Metals, acyl, characteristics of, **2**, 952
Metals, η^1-acyl, from, transition metals, **5**, 440
Metals, α-(acylamino)alkyl, synthesis of, **2**, 1002
Metals, acyloxy, synthesis of, **5**, 229
Metals, alkenyl
  dimerization of, **1**, 491
  from, alkynes, **2**, 876
  silylation of, **2**, 907
  as starting materials
    for alkenylboranes, **2**, 904
    for alkenylgermanes, **2**, 946
    for alkenylsilanes, **2**, 907
    for vinylsilanes, **2**, 910
  synthesis of
    via β-elimination, **2**, 877
    via metal–metal exchange, **1**, 1090
Metals, alkenylidene, characteristics of, **2**, 952
Metals, α-alkoxy, synthesis of, **4**, 381
Metals, α-alkoxyalkyl, synthesis of, **2**, 999
Metals, alkyl
  chemistry of, **2**, 550
  elimination of, **1**, 660
  from
    alkenes, **2**, 592
    terpenoids, **2**, 56
  reactions
    with alcohols, **2**, 112
    with boron halides, **2**, 520
    with metal salts, **2**, 566
    with tetra(dialkylamide), **2**, 566
    with titanium tetraisopropoxide, **2**, 566
  as starting materials, for chlorophosphines, **2**, 447
  synthesis of, **2**, **549–603**
    methods, **2**, 551
    reviews, **2**, 550
Metals, α-alkylthioalkyl, synthesis of, **2**, 999
Metals, alkynyl
  carboxylation of, **1**, 1040
  dimerization of, **1**, 1035
  reactions
    with *N*-acylimines, **1**, 1042
    with *N*-acyliminium salts, **1**, 1042
    with aldehydes, **1**, 1037

with arenediazonium salts, **1**, 1033
with borates, **2**, 1085
with carbodiimides, **1**, 1044
with carbonyls, **1**, 1037
with carboxylic acid chlorides, **1**, 1045
with carboxylic amides, **1**, 1046
with carboxylic anhydrides, **1**, 1045
with carboxylic esters, **1**, 1045
with cyclohexanones, **1**, 1039
with cyclohexenones, **1**, 1039
with *N*,*N*-dialkylcarbamoyl chlorides, **1**, 1048
with electrophiles, **1**, 1044
with halotins, **2**, 1100
with imines, **1**, 1040
with iminium salts, **1**, 1041
with isocyanates, **1**, 1043
with isothiocyanates, **1**, 1044
with ketones, **1**, 1037
with lactams, **1**, 1047
with organogermanium halides, **2**, 1092
with pyridine, **1**, 1042
with triorganoboranes, **2**, 1086
as starting materials, for alkynylmercury compounds, **2**, 1097
synthesis of, **2**, 1094
use in substitution reactions, **1**, 1021
Metals, allenyl
1,4 addition reactions of, **1**, 1061
reactions
with aldehydes, **1**, 1054
with aldimines, **1**, 1059
with ketones, **1**, 1054
Metals, allyl
reactions
with allenes, **1**, 496
with trialkoxyboranes, **2**, 529
Metals, η^3-allyl, characteristics of, **2**, 952
Metals, α-aminoalkyl, synthesis of, **2**, 1001
Metals, aryl
characterization, identification, and quantitative analysis of, **2**, 957
classification of, **2**, 952
definitions of, **2**, 952
from
arenediazonium salts, **2**, 956
aroyl chlorides, **2**, 957
silylation of, **2**, 941
synthesis of, **2**, **951–995**, 956
general discussion, **2**, 954
methods, **2**, 956
via transmetallation, **2**, 990
Metals, η^2-coordinated acyl, synthesis of, **5**, 441
Metals, cyclic vinyl
structure of, **2**, 954
types of, **2**, 953
Metals, η^5-cyclopentadienyl
characteristics of, **2**, 952
σ-metallation of, **2**, 952
Metals, α-diazo, synthesis of, **5**, 874
Metals, *cis*-α,β-disubstituted vinyl, synthesis of, via hydrometallation, **2**, 965
Metals, *trans*-α,β-disubstituted vinyl, synthesis of, via hydrometallation, **2**, 965
Metals, α-fluoroalkenyl, synthesis of, **4**, 821
Metals, α-halo, synthesis of, **4**, 145
Metals, α-haloalkenyl
from
haloalkynes, **4**, 822
metalloalkynes, **4**, 822
Metals, α-haloalkyl, synthesis of, **2**, 997
Metals, hexaalkyl, synthesis of, **2**, 565
Metals, β-hydroxyalkyltriphenyl, acid-catalyzed elimination of, **1**, 664
Metals, imidoyl, as starting materials, for imidoylsilicons, **5**, 792
Metals, imidoyloxy, synthesis of, **5**, 723
Metals, methyl, reactions, with metal pentachlorides, **2**, 570
Metals, phenyl, use in the synthesis of phenyl ketones, **3**, 295
Metals, propargyl
reactions
with aldehydes, **1**, 1054
with ketones, **1**, 1054
Metals, silicon-substituted allyl, reactions, with electrophiles, **2**, 935
Metals, α-silylalkyl, synthesis of, **2**, 1004
Metals, singly-bonded, synthesis of, **5**, 650
Metals, (*E*)-β-substituted vinyl, synthesis of, via hydrometallation, **2**, 965
Metals, (*Z*)-β-substituted vinyl, synthesis of, via hydrometallation, **2**, 965
Metals, α-substituted vinyl, synthesis of, via hydrometallation, **2**, 965
Metals, tetraalkyl, synthesis of, **2**, 565
Metals, tetrabenzyl, reactions, with ethyllithium, **2**, 566
Metals, tribenzyl, synthesis of, **2**, 566
Metals, (trifluoroalkyl), synthesis of, **6**, 244
Metals, (trifluoromethyl), synthesis of, **6**, 22, 244
Metals, trimethyliodo, reactions, with silver cyanide, **3**, 711
Metals, vinyl
characterization, identification, and quantitative analysis of, **2**, 957
classification of, **2**, 952
containing one vinyl group, **2**, 953
definitions of, **2**, 952
reactions, with allylzinc bromide, **4**, 709
regiodefined, **2**, 953
stereodefined, **2**, 953
synthesis of, **2**, **951–995**, 956
general discussion, **2**, 954
methods, **2**, 956
via transmetallation, **2**, 990
Metal salts
reactions
with alkali metals, **2**, 958
with alkyl metals, **2**, 566
with lithium carbenoids, **4**, 820
Metal salts, alkenyl, from, ylides, **2**, 877
Metal salts, nitronate, reactions, with alkyl halides, **4**, 540
Metal salts, thiophilic, reactions, with anomeric sulfides, **4**, 200
Metal salts of thiols, reactions, with 1-chloro enamines, **4**, 904
Metal selenocarbonyl complex, from, 1,2-diselenetene, **3**, 397
Metal selenyl dihalomethyls, transmetallation of, **6**, 49
Metal silanes, use in the synthesis of arsinosilanes, **2**, 509
Metal silicides, reactions, with silicon–metal bonds, **2**, 535
Metal stabilized isocyanocarboxamides, synthesis of, **5**, 1088
Metal sulfides
reactions
with alkyl halides, **2**, 132
with aryl halides, **2**, 711
with carbon disulfide, **6**, 552
with dialkylhaloarsines, **2**, 499
with dialkylhalostibines, **2**, 499
with mercury bis(ethyl diazoacetate), **3**, 484
as starting materials, for dialkyl sulfides, **2**, 132
use in the synthesis of arenethiols, **2**, 706
Metal sulfides, isothiocyanatobis(tributyl), synthesis of, **5**, 1082
Metal sulfur pentafluoride complexes, as starting materials, methylenesulfur tetrafluoride, **3**, 380
Metal tellurolates
from, tellurium, **2**, 278
as starting materials, for unsymmetrical tellurides, **2**, 280
Metal thiocyanates, reactions, with alkyl halides, **5**, 1105

Metal thiolates
alkylation of, **2**, 134
reactions, with chlorosilanes, **2**, 540
synthesis of, **2**, 125, 270
Metal trihalo methoxides, synthesis of, **6**, 232
Metal (trihalomethyl)amides, synthesis of, **6**, 242
Metal trihalomethyls, as starting materials, for dihalocarbene complexes, **6**, 717
Metal vapor deposition, use in the synthesis of triorganoarsines, **2**, 872
Metal vapors, oxidative metallation with, **2**, 961
Metal–vinylidene intermediates, dimerization of, **1**, 991
Meta photocycloaddition
of arenes, **1**, 364
intermolecular, **1**, 364
intramolecular, **1**, 364
Metathesis
of alkenes, **1**, 674
of carbonic difluoride and phosgene, **6**, 409
of diphenylarsine, **2**, 893
for the synthesis of tungsten carbene complexes, **3**, 515, 518
of tetraphenyldiarsine, **2**, 893
of vinylsilanes, **2**, 938
Metathetical reaction
for the synthesis of antimony–gallium bonds, **2**, 511
for the synthesis of antimony–indium bonds, **2**, 511
for the synthesis of arsenic–gallium bonds, **2**, 511
for the synthesis of arsenic–indium bonds, **2**, 511
Methacrolein, reactions, with phosphorus pentasulfide, **3**, 338
Methacrylate, methyl perfluoro-, reactions, with *N*-methylhexamethyldisilazane, **3**, 588
Methacrylic acid, asymmetric chlorination of, **2**, 24
Methacrylimide, trimethylamine, pyrolysis of, **5**, 997
Methacryloyl fluoride, perfluoro-, as starting material, for 1,3-oxazin-2-ones, **3**, 590
Methallylthiols, double deprotonation of, **4**, 367
Methanal, phenyl-, as starting material, for 1,1-bis(trimethylsilyl)-2-phenylethene, **4**, 1049
Methanamines, perfluoro-, reactions, with acyl fluorides, **5**, 313
Methane, reactions, with phosphorus trichloride, **2**, 448
Methane, (arylthio)dibenzoyl-, reactions, with chloramine-T, **4**, 326
Methane, benzeneselenotrinitro-, synthesis of, **6**, 355
Methane, benzenesulfonylnitro-, monoalkylation of, **4**, 333
Methane, benzenethiobis(trimethylsilyl)lithio-, use in the synthesis of dichalcogenomethanes, **6**, 325
Methane, benzenethiochlorodinitro-, synthesis of, **6**, 323
Methane, benzoylnitro-, reactions, with alkynyltin, **4**, 541
Methane, benzylbis(boryl)-, reactions, with mercuric dichloride, **6**, 201
Methane, benzyloxynitro-, reactions, with aldehydes, **4**, 897
Methane, bis(acetoxymercury)-, as starting material, for bis(cyanomercury)methane, **5**, 1141
Methane, bis(2-acetylpyrrol-1-yl)-, synthesis of, **4**, 410
Methane, bis(benzenesulfinyl)-, dithiocarboxylation of, **4**, 863
Methane, bis(benzenesulfonyl)-, reactions, with formamides, **4**, 866
Methane, bis(benzenethio) dinitro-
from
iodonium ylides, **6**, 323
nitro ylides, **6**, 323
selenolium ylides, **6**, 323
sulfonium ylides, **6**, 323
ylides, **6**, 323
Methane, bis(benzenethio)lithiotrialkylsilyl-, nuclear magnetic resonance study of, **6**, 327
Methane, bis(benzenethio)lithiotrimethylsilyl-, synthesis of, **6**, 327
Methane, bis(benzotriazoyl)-
reactions
with *n*-butyllithium, **6**, 364
with tosyl azide, **6**, 364
Methane, bis(boryl)bis(stannyl)-, synthesis of, **6**, 200
Methane, bis(bromomagnesio)trimethylsilyl-
reactions, with chlorotrimethylstannane, **6**, 198
synthesis of, **6**, 192
Methane, bis(chlorodimethylsilyl)-
reactions
with 9-anthryllithium, **4**, 621
with 2,3-dilithio-1,3-butadiene, **4**, 621
as starting material, for α-silyl ylides, **5**, 914
Methane, bis(cyanomercury)-, from, bis(acetoxymercury)methane, **5**, 1141
Methane, bis(dialkylamino)-1-methoxy-, as starting material, for *gem*-dihydrazinoalkenes, **4**, 985
Methane, bis(dibromoarsino)-, synthesis of, **4**, 593
Methane, bis(dichloroarsino)-, as starting material, for bisarsinomethanes, **4**, 593
Methane, bis(dichlorophosphino)-, reactions, with sodium bis(trimethylsilyl)amide, **5**, 884
Methane, 1,1-bis(dichlorophosphino)-, reactions of, **4**, 550, 562
Methane, bis(dichlorostibino)-, synthesis of, **4**, 595
Methane, bis(difluoroamino)(fluoroamino)amino-, reactions, with fluorine, **6**, 362
Methane, bis(difluoroamino)-[(trifluoromethyl)fluoroamino]fluoro-, defluorination–cyclization of, **6**, 285
Methane, [bis(diisopropylamino)boranyl]diazo-, synthesis of, **3**, 485
Methane, bis(dimesitylboryl)-, as starting material, for potassium bis(boryl)methanide, **6**, 199
Methane, bis(dimethylamino)-
reactions
with acid anhydrides, **4**, 309
with secondary arsines, **4**, 501
synthesis of, **4**, 405
Methane, bis(dimethylamino)diethoxy-, synthesis of, **6**, 321
Methane, bis(dimethylamino)dimethoxy-, synthesis of, **6**, 321
Methane, bis[dimethyl(9-anthryl)silyl]-, synthesis of, **4**, 621
Methane, bis(dimethylantimony)diazo-, synthesis of, **3**, 482
Methane, bis(dimethylarsano)diazo-, synthesis of, **3**, 482
Methane, bis(dimethylbismuth)diazo-, synthesis of, **3**, 482
Methane, 1,1-bis(dimethylsilyl)-, synthesis of, **4**, 614
Methane, bis(diphenylarsanyl)-, lithium salts, synthesis of, **6**, 170
Methane, bis(diphenylarsino)-
as starting material
for arsines, **4**, 593
for diphenylarsinomethyllithium, **4**, 599
for methylene bisphenylarsenic acid, **4**, 592
synthesis of, **4**, 592
Methane, bis(diphenylbismutino)-, synthesis of, **1**, 54
Methane, bis(diphenylphosphino)-, reactions of, **4**, 563
Methane, 1,1-bis(diphenylphosphino)-, reactions of, **4**, 550
Methane, bis(diphenylphosphinyl)-, reactions, with germanium(II) chloride, **6**, 169
Methane, bis(diphenylstibanyl)-, lithium salts, synthesis of, **6**, 170
Methane, bis(diphenylstibine)-, as starting material, for diphenylstibinylmethyllithium, **1**, 271
Methane, bis(diphenylstibino)-, as starting material, for diphenylstibinomethyllithium, **4**, 600
Methane, bisdiphenylthiophosphinoyl-, reactions, with aldehydes, **4**, 1022
Methane, bis(ethanesulfonyl)-, condensation of, **4**, 864
Methane, bis(fluorooxy)difluoro-, reactions, with Dewar hexafluorobenzene, **4**, 46
Methane, bis(fluorosulfonyl)-, reactions, with carbodiimides, **4**, 866

Methane, bis(fluoroxy)difluoro-
for the fluorination of amines, **2**, 335
reactions, with alkenes, **6**, 250
as reagent, for aryl fluorides, **2**, 612
synthesis of, **6**, 250
Methane, bis(formylamino)-, reactions, with phosgene, **4**, 439
Methane, bis(germyl)diazo-, synthesis of, **6**, 672
Methane, bis(methanesulfonyl)-
from, methyl methylsulfinylmethyl sulfide, **4**, 274
reactions, with benzenediazonium chloride, **6**, 114
Methane, bismethanesulfonyl diazo-, reactions, with α-morpholinostyrene, **2**, 802
Methane, bis(methanethio)bis(trimethylsilyl)-, synthesis of, **6**, 325
Methane, bis(methanethio)bis(trimethylstannyl)-, from, bis(methanethio)lithiotrimethylstannylmethane, **6**, 326
Methane, bis(methanethio)lithiotrimethylsilyl-, synthesis of, **6**, 327
Methane, bis(methanethio)lithiotrimethylstannyl-, as starting material, for bis(methanethio)bis(trimethylstannyl)methane, **6**, 326
Methane, bismethoxazonyl-, as starting material, for *gem*-dimethoxazonylalkenes, **4**, 996
Methane, bis(methylchloroarsino)-, as starting material, for tetraarsacyclohexanes, **4**, 594
Methane, bis(methylsulfonyl)-, synthesis of, **4**, 272
Methane, bis(phenylseleno)-
lithiation of, **6**, 135
reactions
with *n*-butyllithium, **4**, 400
with lithium diisobutylamide, **4**, 401
Methane, bis(phenylsulfonyl)-
metallation of, **4**, 274
reactions
with quinoline-1-oxide, **4**, 281
with vinyl epoxides, **4**, 280
as starting material, for bis(sulfones), **4**, 274
Methane, bis(phenylthio)-, reactions, with cyclohexenone, **1**, 303
Methane, bis(phenylthio)bis(trimethylsilyl)-, reactions, with lithium naphthalenide, **4**, 945
Methane, bis(phosphonato)-, synthesis of, **6**, 157
Methane, bisstannyldiazo-
reactions
with chloroboranes, **3**, 682
with chlorophosphanes, **3**, 682
Methane, (+)-(*S*,*S*)-bis(toluene-*p*-sulfinyl)-, condensation reactions of, **4**, 863
Methane, (*S*,*S'*)-bis-*p*-tolylsulfinyl-, synthesis of, **4**, 270
Methane, bis(trichlorosilyl)trimethylsilyl-, production of, **6**, 290
Methane, bis(triiodogermyl)-
as starting material, for bis(trimethylgermyl)methane, **4**, 629
synthesis of, **4**, 628
Methane, bis(trimethylenedioxyboryl)-, deprotonation of, **4**, 634
Methane, bis(trimethylgermyl)-
from, bis(triiodogermyl)methane, **4**, 629
metallation of, **4**, 628
synthesis of, **4**, 626
Methane, bis(trimethylgermyl)diazo-
photolysis of, **4**, 627
as starting material, for germanes, **5**, 927
synthesis of, **3**, 483; **6**, 202
Methane, bis(trimethylplumbyl)diazo-, synthesis of, **3**, 484
Methane, bis(trimethylsilyl)-
metallation of, **4**, 620
reactions
with antimony trichlorides, **6**, 168
with arsenic trichlorides, **6**, 168
with *n*-butyllithium, **6**, 190
with diphosphines, **6**, 166
with phosphorus trichloride, **6**, 164
with sodium methoxide, **4**, 653
synthesis of, **4**, 602; **6**, 165
Methane, bis(trimethylsilyl)chloro-
from, bis(trimethylsilyl)dichloromethane, **6**, 60
reactions, with sodium azide, **6**, 163
Methane, bis(trimethylsilyl)diazo-
photolysis of, **6**, 178
synthesis of, **3**, 482; **5**, 1148; **6**, 672
thermolysis of, **4**, 612; **6**, 178
Methane, bis(trimethylsilyl)dichloro-, as starting material, for bis(trimethylsilyl)chloromethane, **6**, 60
Methane, bis(trimethylsilyl)dilithio-, use in the synthesis of dichalcogenomethanes, **6**, 325
Methane, bis(trimethylsilyl)phenyl-, synthesis of, **4**, 608
Methane, bis(trimethylstannyl)-, reactions, with lithium dicyclohexylamide, **6**, 210
Methane, bis(trimethylstannyl)diazo-, reactions, with chlorotriisopropylsilane, **5**, 1148
Methane, bis(trimethylstannyl)trimethylsilyl-, synthesis of, **6**, 198
Methane, bis(triphenylstannyl)ethylenedioxyboryl-, synthesis of, **6**, 200
Methane, (boryl)bis(stannyl)-substituted, synthesis of, **6**, 200
Methane, bromo[bromobis(isopropyl)aminoboryl]-bis(trimethylsilyl)-, synthesis of, **6**, 292
Methane, bromochloro-, reactions, with phosphorothioates, **2**, 255
Methane, bromochlorodifluoro-
as starting material, for dibromodifluoromethane, **6**, 226
synthesis of, **6**, 223
Methane, bromochlorodiiodo-, synthesis of, theoretical, **6**, 228
Methane, bromochlorodinitro-, synthesis of, **6**, 261
Methane, bromochlorofluoro-, synthesis of, **6**, 33
Methane, bromochlorofluorodiiodo-, synthesis of, theoretical, **6**, 228
Methane, bromochloroiodo-
oxidation of, **6**, 420
synthesis of, **6**, 33
Methane, bromodiazo-, synthesis of, **3**, 478; **5**, 844
Methane, bromodichlorofluoro-
as starting material, for dichlorofluoroiodomethane, **6**, 226
synthesis of, **6**, 225
Methane, bromodichloroiodo-, synthesis of, **6**, 226
Methane, bromodifluoro-, vapor phase chlorination of, **6**, 223
Methane, bromodifluoroiodo-, synthesis of, **6**, 224
Methane, bromodinitro-, synthesis of, **6**, 56
Methane, bromofluorodiiodo-, synthesis of, **6**, 228
Methane, bromonitro-, reactions, with aldehydes, **4**, 798
Methane, bromosilyl-, from, phenylsilylmethane, **6**, 379
Methane, bromotriboryl-, production of, **6**, 292
Methane, bromotrichloro-
electrochemical addition to alkenes, **6**, 7
iodination of, **6**, 221
photolytic addition of, to ethyl enol ether, **5**, 426
reactions
with butyllithium, **4**, 16
with α-cholorovinyltrimethylsilanes, **4**, 758
with enol sulfonates, **4**, 742
with ethynyltins, **4**, 813
with ketene silyl acetates, **4**, 743
as starting material, for trihalocarbinols, **6**, 26
synthesis of, **6**, 221
use in the synthesis of dibromoalkanes, **4**, 21
use in the synthesis of 1,1-dichloroalkenes, **4**, 738
use in Wittig reactions, **4**, 738
Methane, bromotricyano-, reactions, with triethyl phosphite, **4**, 1008

Methane, bromotrifluoro-
  reactions
    with hexafluorobenzene, **6**, 12
    with triphenylphosphine, **4**, 754
  as starting material, for trifluoromethyl radicals, **6**, 20
  synthesis of, **6**, 218
    via halogen exchange, **6**, 219
  trifluoromethylation with, **6**, 233
Methane, bromotriiodo-, synthesis of, **6**, 223
Methane, bromotrinitro-, as starting material, for trinitro compounds, **6**, 145
Methane, bromotriphenyl-, reactions, with potassium *O*-ethyl dithiocarbonate, **6**, 546
Methane, bromotris(trimethylsilyl)-, synthesis of, **6**, 373
Methane, *t*-butoxybis(dimethylamino)-
  reactions
    with dialkylamino compounds, **2**, 745
    with dithionomalonic esters, **4**, 896
Methane, *t*-butylsulfonyldiazo-, synthesis of, **3**, 479
Methane, chiral bromochlorofluoro-, synthesis of, **6**, 33
Methane, chloro-, chlorination of, **6**, 215
Methane, chlorobis(methanethio)-, reactions, with methanethiol, **6**, 82
Methane, chlorobis(trimethylsilyl)-, synthesis of, **6**, 191
Methane, chlorodiazo-, synthesis of, **3**, 478; **5**, 844
Methane, chlorodifluoro-
  bromination of, **6**, 223
  copyrolysis of, with pentafluorobenzotrifluoride, **4**, 737
  reactions
    with (arylalkylidene)triphenylphosphoranes, **4**, 735
    with sodium benzyl sulfide, **2**, 179
    with sodium diethylphosphonate, **6**, 58
    with sulfur, **6**, 528
  as starting material
    for α,α-dichloro sulfides, **6**, 45
    for difluorocarbene, **4**, 8; **6**, 38
    for difluoromethylenes, **6**, 37
    for triaryl trithioortho esters, **6**, 81
  synthesis of, **6**, 32
Methane, chlorodifluoroiodo-, synthesis of, **6**, 224
Methane, chlorodifluoronitroso-, reactions, with diethyl phosphite, **6**, 603
Methane, chlorodiphenoxy-, as starting material, for triphenyl monothioorthoformates, **6**, 95
Methane, chlorofluoro-
  reactions, with sodium benzyl sulfide, **2**, 179
  synthesis of, **4**, 30
Methane, chlorofluorodiiodo-, synthesis of, theoretical, **6**, 228
Methane, chlorofluoronitro-, from, 1,2-dichloro-1,2-difluoroethylene, **6**, 54
Methane, chloroiodo-
  reactions
    with methyllithium, **1**, 681
    with triphenylphosphine, **2**, 618
Methane, chloronitro-
  synthesis of, **4**, 798
    via chlorination of nitromethane, **6**, 55
Methane, chlorotricyano-
  for the chlorination of imides, **5**, 316
  reactions, with triethyl phosphite, **4**, 1008
Methane, chlorotrifluoro-
  halogen exchange of, **6**, 224
  synthesis of, **6**, 217
Methane, chlorotriiodo-, synthesis of, theoretical, **6**, 223
Methane, chlorotris(dimethylamino)-, synthesis of, **6**, 321
Methane, chlorotris(methoxydimethylsilyl)chloro-, synthesis of, **6**, 291
Methane, cyanodinitro-, rubidium salt of, **6**, 170
Methane, diazo-
  acylation of, **3**, 474
  as alkylating agent, **4**, 833
  as carbene sources, **4**, 669
  chlorination of, **3**, 478
  cycloaddition of, **6**, 122
  from, *N*-methyl-*N*-nitrosourea, **3**, 466
  for the homologation of ketones, **3**, 125
  reactions
    with acid chlorides, **1**, 393; **3**, 474
    with A-frame platinum complexes, **4**, 691
    with alkaloids, **1**, 627
    with alkyne-bridged dirhodium complexes, **4**, 689
    with allylic selenides, **4**, 277
    with 4-arylhydrazono-2-isoxazoline-5-thiones, **5**, 532
    with benzeneselenenyl thiocyanate, **4**, 335
    with bis(diisopropylamino)chloroborane, **3**, 485
    with chlorosulfenes, **4**, 82
    with 1-cyclobutenyldiphenylphosphine oxide, **4**, 488
    with diazophosphaalkenes, **2**, 470
    with dichloro(phenyl)thallium, **4**, 153
    with diethylaluminum iodide, **4**, 153
    with dimethylaminotrimethylgermane, **3**, 483
    with dimethylaminotrimethylplumbane, **3**, 484
    with ditellurides, **4**, 290
    with *N*-fluorinated imidic acids, **5**, 699
    with hexachlorodisilane, **4**, 140
    with *N*-hydroxyimides, **5**, 333
    with lithium *N*-methyl-*N*-trimethylsilyl amides, **3**, 485
    with mercury(II) acetate, **3**, 488
    with 1,6-methano[10]annulene, **1**, 910
    with methyl 2-chloro-2-cyclopropylideneacetate, **4**, 109
    with 4-nitrobenzenediazonium chloride, **5**, 843
    with osmium complexes, **3**, 522
    with *N*-phenylselenophthalimide, **4**, 335
    with phenylsulfinyl chloride, **3**, 479; **5**, 867
    with potassium selenocyanates, **5**, 1056
    with ruthenium complexes, **3**, 520
    with selenenyl bromides, **4**, 291
    with selenium complexes, **3**, 387
    with seleno esters, **3**, 174
    with sulfinyl halides, **4**, 76
    with tetrabromosilane, **4**, 140
    with tetrachlorosilane, **4**, 140
    with tetrakis(2,6-diethylphenyl)digermene, **4**, 628
    with thioamides, **5**, 726
    with thioketones, **2**, 243; **6**, 128
    with thiolactams, **5**, 726
    with thionoamide *S*-oxides, **5**, 623
    with thionyl chloride, **4**, 76
    with tin tetrabromide, **4**, 155
    with tin tetrachloride, **4**, 155
    with trichlorofon, **4**, 912
    with trimethylstannyl chloride, **3**, 482
    with trimethylthallium, **3**, 485
    with vinyl sulfones, **6**, 122
    with zinc benzoate, **4**, 385
  synthesis of, **3**, 462, 465
    via Forster reaction of oximes, **3**, 463
  for the synthesis of methylene-bridged dimanganese complexes, **4**, 676
  use in the synthesis of allene-bridged dimolybdenum complexes, **4**, 673
  use in the synthesis of carboxylic esters, **5**, 144
  use in the synthesis of methylene-bridged diruthenium complexes, **4**, 681
  use in the synthesis of methylene-bridged iron nitrosyl complexes, **4**, 679
  use in the synthesis of methyl esters, **5**, 132
  use in the synthesis of methyl ethers, **2**, 89
Methane, diazo bis(dimethylthallium)-, synthesis of, **3**, 485
Methane, diazo di(trimethylgermanyl)-, flash vacuum pyrolysis of, **4**, 1055
Methane, diazo-, nickel complex, synthesis of, **6**, 675
Methane, diazophenyl-
  from, azibenzil, **3**, 475
  reactions, with tosyl chloride, **3**, 479

Methane, diazophenyltosyl-, synthesis of, **3**, 479
Methane, diazo-, rhodium complex, synthesis of, **6**, 675
Methane, dibismuthino-, synthesis of, **4**, 596
Methane, dibromo-
reactions
with aluminum, **4**, 698
with (diethoxy)phenylphosphine, **4**, 127
with isopropylmagnesium chloride, **4**, 147
with phenylarsenic oxide, **4**, 592
with sulfinate anions, **4**, 83
with triphenylleadlithium, **4**, 699
with triphenylphosphine, **5**, 1093
as starting material, for methylenedilithium, **4**, 668
Methane, dibromobis(trimethylsilyl)-, metallation of, **6**, 290
Methane, dibromochloro-
reactions
with mercuric fluoride, **6**, 33
with sodium iodide, **6**, 33
with triphenylphosphine, **4**, 757
Methane, dibromochlorofluoro-, synthesis of, **6**, 227
Methane, dibromochloroiodo-
from, potassium bromodifluoroacetate, **6**, 227
synthesis of, **6**, 227
Methane, dibromochloronitroso-, bromination of, **6**, 222, 227
Methane, dibromodichloro-, synthesis of, **6**, 226
Methane, dibromodifluoro-
electrochemical addition to alkenes, **6**, 7
reactions, with ketene silyl acetals, **4**, 736
as starting material
for carbonic difluoride, **6**, 410
for cyclopropanes, **4**, 1083
synthesis of, **6**, 226
use in the synthesis of difluorocarbenes, **4**, 7, 733
use in the synthesis of trifluoromethyls, **6**, 15
Methane, dibromodiiodo-, synthesis of, theoretical, **6**, 228
Methane, dibromodinitro-
from, 2,4,6-trichloroaniline, **6**, 260
reactions, with tetramethylethene, **4**, 995
synthesis of, **6**, 260
Methane, dibromodi(trimethylsilyl)-, reactions, with magnesium, **4**, 618
Methane, dibromofluoro-, synthesis of, **6**, 32
Methane, dibromofluoroiodo-, synthesis of, **6**, 227
Methane, dibromonitro-, synthesis of, **6**, 54
Methane, dibromo(trimethylsilyl)-, reactions, with zinc–copper couples, **6**, 194
Methane, di-*t*-butylthioxophosphoranyldiazo-, reactions, with chlorobis(diisopropylamino)phosphane, **5**, 1148
Methane, dichloro-
as alternative to tetrachloromethane, for phosgene, **6**, 413
bromination of, **6**, 226
reactions
with acid chlorides, **5**, 129
with aluminum, **1**, 680; **4**, 698
with aminohaloboranes, **6**, 243
with diphenylbismuthsodium, **4**, 596
with diphenyl ditelluride, **4**, 93
with indium(I) chloride, **4**, 153
with lithium, **4**, 632
with phenylmethylantimonysodium, **4**, 595
with pyrrole, **4**, 410
with sodium iodide, **4**, 28
with triethylgermylpotassium, **6**, 185
with trimethyltinsodium, **4**, 699
with trivinyltinsodium, **4**, 699
with vinylic dialkoxyboranes, **2**, 530
as starting material, for tetrachlorobisphosphines, **2**, 448
use in the synthesis of carbon-bridged diindium compounds, **4**, 698
use in the synthesis of carbon-bridged diruthenium complexes, **4**, 678
use in the synthesis of chlorofluoroalkanes, **4**, 30
Methane, dichlorobis(tetrafluorodiamino)-, as starting material, for chlorofluorodiazirine, **6**, 261
Methane, dichlorobis(trimethylsilyl)-
hydride reduction of, **4**, 619
reactions
with butyllithium, **6**, 288
with lithium metal vapor, **4**, 618
Methane, dichlorodiazo-, metal complexes of, **6**, 610
Methane, dichlorodifluoro-
reactions, with sodium alkoxides, **6**, 69
synthesis of, **6**, 224
Methane, dichlorodiiodo-, synthesis of, **6**, 227
Methane, dichlorodinitro-
from, 2,4,6-trichloroaniline, **6**, 260
synthesis of, **6**, 260
Methane, dichlorodiphenoxy-, synthesis of, **6**, 251
Methane, dichlorodiphenyl-, use in the synthesis of diphenylmethylenedilithium, **4**, 668
Methane, dichlorofluoro-
oxidation of, **6**, 420
reactions
with 2,4,6-tri-*t*-butylphenylphosphane, **5**, 876
with potassium *t*-butoxide, **4**, 751, 753
as starting material, for chlorofluorocarbene, **4**, 1083
synthesis of, **6**, 32
vapor phase bromination of, **6**, 227
Methane, dichlorofluoroiodo-, synthesis of, **6**, 225
Methane, dichlorofluoronitroso-, reactions, with phosphites, **6**, 613
Methane, dichlorophenyl-, use in the synthesis of phenylmethylenedilithium, **4**, 668
Methane, dichloro(trimethylsilyl)-
reactions
with lithium, **6**, 192
with lithium vapor, **6**, 205
Methane, dicyanodiazo-, synthesis of, **3**, 477
Methane, dideuteriodichloro-, as starting material, for dithioacetals, **4**, 249
Methane, difluoro-, vapor phase bromination of, **6**, 226
Methane, difluoro(bisisocyanato)-, synthesis of, **6**, 260
Methane, difluorochloro-, as starting material, for trialkyl orthoformates, **6**, 69
Methane, difluorodiiodo-, synthesis of, **6**, 227
Methane, difluoro(dimethylamino)-, synthesis of, **6**, 52
Methane, difluorodinitro-
from, difluorodiazirine, **6**, 261
reactions
with *p*-nitroaniline, **6**, 260
with sodium ethoxide, **6**, 469
Methane, difluorodiphenyl-, synthesis of, **4**, 4
Methane, difluoronitro-, reactions, with formaldehyde, **6**, 54
Methane, diiodo-
photolysis of, **4**, 668
reactions
with calcium, **4**, 669
with dicarbonyl(μ-pyrazolyl)iridium, **4**, 685
with germanium diiodide, **4**, 628
with lithium phenyl telluride, **4**, 290
with octacarbonyldiiron dianion, **4**, 679
with sodium benzene sulfinate, **4**, 84
with sodium phenyl sulfinate, **4**, 281
with thiols, **4**, 249
with tin(II) bromide, **4**, 699
with trinitromethane salts, **6**, 145
with zinc, **4**, 696
as starting material, for methylenedilithium, **4**, 668
synthesis of, **4**, 28
use in the synthesis of carbon-bridged dimercury complexes, **4**, 696
use in the synthesis of methylenebis(magnesium bromide), **4**, 669
Methane, diiododi(trimethylsilyl)-, reactions, with zinc, **4**, 618

Methane, 1,1-diisonitrilo-, synthesis of, **4**, 439
Methane, dilithio-
reactions
with bis(dimethylamino)boron chloride, **4**, 632
with trimethylgermanium chloride, **6**, 388
with trimethylsilyl chloride, **4**, 603
with trimethyltin chloride, **4**, 699
synthesis of, **6**, 206
Methane, dilithiodifluoro-, structure of, **6**, 270
Methane, dilithiodi(trimethylsilyl)-, synthesis of, **4**, 618
Methane, dilithio(trimethylsilyl)-, synthesis of, **6**, 192
Methane, (dimesitylboryl)-, metallation of, **4**, 634
Methane, dimesitylboryl(trimethylsilyl)-, synthesis of, **4**, 642
Methane, dimethoxy-
reactions
with benzeneselenol, **4**, 238
with isocyanates, **4**, 310
with trimethylsilyl iodide, **4**, 51
as starting material, for methoxymethyl acetate, **4**, 211
Methane, dimethylaminotrimethylsilyloxy-, synthesis of, **4**, 295
Methane, dimethyldiazo-, use in the synthesis of tellones, **3**, 401
Methane, (+)-(*S*)-(1,1-dimethylethanesulfonyl)toluene-*p*-sulfinyl-, Mannich reaction of, **4**, 864
Methane, dimethylgermyl(dimethylsilyl)-, reactions, with iron pentacarbonyl, **4**, 641
Methane, di(methylthio)-, as starting material, for dithioacetals, **4**, 250
Methane, dinitro-
reactions
with amido nitriles, **4**, 995
with dimethylformamide dimethyl acetal, **4**, 995
with formimidates, **4**, 994
as starting material, for *gem*-dinitroalkenes, **4**, 994
Methane, dinitrodiazo-, synthesis of, **3**, 480
Methane, diphenyl-, synthesis of, **1**, 847
Methane, diphenyldiazo-
[1+4] cycloaddition of, **6**, 314
photolysis of, **3**, 596
reactions
with alcohols, **2**, 89; **4**, 198
with aryl sulfonyl isothiocyanates, **5**, 638
with dimethyl chlorofumarate, **4**, 109
with iminoboranes, **4**, 145
with methyl 2-chloro-2-cyclopropylideneacetate, **4**, 109
with rhodium complexes, **3**, 520
with thioketone, **3**, 363
as starting material, for mixed dihaloalkenes, **4**, 755
synthesis of, **3**, 464
Methane, di(phenyltelluro)-, reactions, with lithium diisopropylamide, **4**, 290
Methane, di(phenylthio)-, as starting material, for dithioacetals, **4**, 250
Methane, (diphenylthiophosphinoyl)phenyldiazo-, Pummerer rearrangement of, **4**, 346
Methane, dipyrryl-, synthesis of, **3**, 341
Methane, disilver diazo-, synthesis of, **3**, 487
Methane, dithiobenzoyl-, synthesis of, **3**, 366
Methane, di(*p*-tolylsulfonyl)diazo-, as starting material, for disulfones, **6**, 307
Methane, ethoxytris(dimethylamino)-, synthesis of, **6**, 337
Methane, fluorochlorodinitro-, synthesis of, **6**, 261
Methane, fluorodichloro-, as starting material, for trialkyl orthoformates, **6**, 69
Methane, fluorodinitro-, reactions, with silver oxide, **6**, 287
Methane, fluoronitro-, characteristics of, **6**, 284
Methane, fluoropentakis(difluoroamino)azo-, synthesis of, **6**, 363
Methane, fluorotribromo-, reactions, with tetrafluorohydrazine, **6**, 613
Methane, fluorotrichloro-
reactions
with alkenes, **6**, 4
with polychlorinated benzenes, **6**, 30
as starting material, for carbonic chloride fluoride, **6**, 419
Methane, fluorotriiodo-, synthesis of, theoretical, **6**, 223
Methane, fluorotrinitro-
reactions, with nucleophiles, **6**, 279
as starting material, for fluorodinitromethide, **6**, 287
Methane, fluoroxytrifluoro-
as fluorinating reagent, **2**, 335
as reagent
for aryl fluorides, **2**, 612
for aryl halides, **2**, 608
for the synthesis of aryl fluorides, **2**, 613
Methane, germylbis(stannyl)-, synthesis of, **6**, 203
Methane, germyl(silyl)-, reactions, with hydrogen chloride, **4**, 641
Methane, halodinitro-, nitration of, **6**, 284
Methane, hexafluoroazo-, reactions, with carbon monoxide, **5**, 989
Methane, iodo-
reactions
with allenes, **4**, 1024
with bisdiphenylphosphino ethene, **4**, 1024
with cycloalkyllithiums, **1**, 110
with cycloplatinated complexes, **1**, 127
with dithiocarbamate salts, **6**, 561
with ozone, **4**, 42
with sodium *N*-formyl dithiocarbamate, **6**, 563
Methane, iodotrifluoro-
from
methyl bromodifluoroacetate, **6**, 219
methyl chlorodifluoroacetate, **6**, 219
reactions, with pyridines, **6**, 12
as starting material, for trifluoromethyl radicals, **6**, 20
synthesis of, **6**, 219
Methane, iodotrinitro-
methylation of, **6**, 145
as starting material, for trinitro compounds, **6**, 145
Methane, lithiodichloro-, reactions, with ketones, **4**, 741
Methane, lithiodichloromethoxy-, reactions, with trisubstituted alkylboranes, **2**, 530
Methane, lithiotrichloro-, reactions, with *n*-heptylphenyl sulfone, **4**, 741
Methane, methanethiotrinitro-, synthesis of, **6**, 348
Methane, methanethiotris(trimethylsilyl)-, reactions, with *m*-chloroperbenzoic acid, **6**, 325
Methane, methoxy(arylthio)-, synthesis of, **4**, 218
Methane, methoxy bis(trimethylsilyl), as starting material, for silicon stabilized anions, **5**, 417
Methane, methoxy(phenylthio)-
as starting material, for silanes, **4**, 370
synthesis of, **4**, 218
Methane, methoxytrifluoromethylphenylchloro-, synthesis of, **5**, 12
Methane, methoxytris(dimethylamino)-, synthesis of, **6**, 337
Methane, methylsulfonyldibenzoyl-, reactions, with 4-nitrobenzenediazonium tetrafluoroborate, **4**, 332
Methane, nitro-
alkylation of, **4**, 540
chlorination of, **6**, 55
reactions
with acylimidazoles, **5**, 829
with amide acetals, **2**, 788
with benzenediazonium salts, **3**, 449
with *t*-butyl hypobromite, **6**, 54
with diketones, **4**, 166
with ethyl cyanoformate, **2**, 788
with 3*β*-hydroxyandrost-5-en-17-one, **2**, 785
with lactam acetals, **2**, 788
with lactim ethers, **2**, 788
with methylvinyl ketone, **3**, 192
with pentamethinium salts, **2**, 790

Methane, nitrodiazo-
reactions
with dinitrogen pentoxide, **3**, 480
with mercury(II) oxide, **6**, 664
synthesis of, **3**, 480
Methane, nitro(phenylseleno)-, reactions, with aldehydes, **4**, 910
Methane, nitro(phenylthio)-, reactions, with aldehydes, **4**, 906
Methane, nitrosotrifluoro-
reactions, with hydroxylamine, **5**, 360
use in the synthesis of trifluoromethyls, **6**, 20
Methane, nitrotrichloro-, reduction of, **5**, 822
Methane, nitrotrifluoro-, synthesis of, **6**, 241
Methane, oxyiminodibromo-, synthesis of, **5**, 821
Methane, perfluorodimethoxy-, synthesis of, **6**, 251
Methane, phenylbis(diethyldiamino)-, reactions, with acetyl chloride, **4**, 455
Methane, phenyldiazo-
reactions
with bis(trifluoromethyl)sulfine, **5**, 358
with phosphonium ylides, **3**, 445
as starting material, for acyl hydrazines, **5**, 358
synthesis of, **3**, 465
Methane, phenylnitro-
reactions
with lithium ethoxide, **2**, 1003
with triphenylarsenic dichloride, **5**, 921
Methane, *aci*-phenylnitro-, as starting material, for oxyimidic acid, **5**, 819
Methane, phenylsilyl-
as starting material, for bromosilylmethane, **6**, 379
synthesis of, **6**, 378
Methane, phenylsulfinyldiazo-, synthesis of, **3**, 479
Methane, phenylthiobis(trimethylsilyl)-, synthesis of, **6**, 129
Methane, (phenylthio)nitro-
deprotonation of, **1**, 157
reactions, with 1,5-dibromopentane, **1**, 157
Methane, phenylthio(phenylseleno)-, lithiation of, **6**, 135
Methane, phenylthio(trimethylsilyl)-, reactions, with *n*-butyllithium, **6**, 129
Methane, α-phenyl(*p*-toluenesulfonyl)diazo-, synthesis of, **3**, 479
Methane, phenyltrimethylgermyldiazo-, pyrolysis of, **3**, 503
Methane, phosphinodiazo-, photolysis of, **6**, 704
Methane, phosphinyldiazo-, reactions, with methyl propiolate, **2**, 861
Methane, polychlorotetrasilyl-, synthesis of, **6**, 383
Methane, polyethylene glycoldichloro-, for the synthesis of aryl bromides, **2**, 627
Methane, silylated trinitro-, reactions, with silylaminal, **6**, 147
Methane, 1-silyl-1-chloro-1-lithio-, reactions, with transition metal complexes, **4**, 658
Methane, stannyldiazo-, synthesis of, **6**, 670
Methane, tetrabromo-
addition to aldehydes, **6**, 31
bromination of, **6**, 222
electrochemical addition to alkenes, **6**, 7
electroreduction of, **4**, 38
fluorination of, **6**, 218, 226
from, tribromomethane, **6**, 215
oxidation of, **6**, 418
reactions
with aldehydes, **4**, 745
with bromine trifluoride, **6**, 217, 218, 222
with ketones, **4**, 745
with mercury(II) trifluoromethaneselenolate, **6**, 307
with triethyl phosphite, **6**, 286
with triphenylphosphine, **4**, 745
as starting material, for carbonic dibromide, **6**, 417
synthesis of, **6**, 215
Methane, tetrachloro-
addition to cyclohexene, **6**, 7
bromination of, **6**, 221, 222, 226
for the chlorination of benzyl sulfide, **4**, 64
fluorination of, **6**, 212, 220, 225
for chlorotrifluoromethane, **6**, 217
halogen exchange of, for dichlorodifluoromethane, **6**, 224
oxidation of, **6**, 413
pyrolysis of, **2**, 1014
reactions
with acrylonitrile, **6**, 6
with alkanethiolates, **6**, 81
with alkyldiphenylphosphines, **4**, 112
with alkyl halides, **6**, 24
with azides, **2**, 773
with azosilanes, **6**, 608
with bis(arenesulfonyl)methylenephosphines, **6**, 699
with bis(difluorosulfoximidoyl)mercury, **6**, 362
with bis(trimethylsilyl)aminophosphines, **4**, 572
with bromine trifluoride, **6**, 212, 217, 220, 226
with diamino(alkyl)phosphines, **4**, 116
with di-*t*-butyltrimethylsilylmethylphosphane, **5**, 915
with 1,2-dipotassium di-*t*-butyldiphosphide, **6**, 366
with enamines, **4**, 742
with iodine pentafluoride, **6**, 220
with ketene silyl acetates, **4**, 743
with lithium, **6**, 404
with *N*-methoxy amides, **3**, 666
with pentafluorophenol, **6**, 272
with phosphanes, **5**, 898
with sodium alkoxides, **6**, 69
with sodium hexafluorosilicate, **6**, 225
with sulfonylmethylphosphanes, **5**, 903
with tetraalkyl diamidofluorophosphites, **6**, 696
with tetrachloroantimony(V) azide, **6**, 733
with tetraphenylporphyrinoiron(II), **6**, 717
with thienoyl azides, **5**, 990
with trialkylaluminums, **2**, 596
with triphenylphosphine, **4**, 737; **5**, 1024
with tris(trimethylsilyl)silane, **2**, 541
with ylides, **5**, 895
as starting material
for carbonimidic dichlorides, **6**, 608
for dichlorocarbene, **4**, 1083
for orthocarbonates, **6**, 296
for phosgene, **6**, 412, 413
for tetraiodomethane, **6**, 216
for tetrakis(difluorosulfoximidoyl)methanes, **6**, 362
for tetrakis(1-pyrazolyl)methanes, **6**, 362
for tetrathioorthocarbonates, **6**, 301
for trihalocarbinols, **6**, 26
synthesis of, **6**, 214, 404
tetrasilylation of, **6**, 379
use in the synthesis of acid chlorides, **5**, 13
use in the synthesis of ^{13}C-labeled phosgene, **6**, 414
Methane, tetra(cyanomercury)-, synthesis of, **5**, 1141
Methane, tetra(dimethoxyboryl)-, use in the synthesis of 1,1-bisborylalkanes, **4**, 633
Methane, tetraethyl 1-amino-1,1-bis(phosphonato)-1-phenyl-, synthesis of, **6**, 159
Methane, tetraethynyl-, synthesis of, **1**, 1077
Methane, tetrafluoro-
reactions, with phosphorus, **6**, 58
synthesis of, **6**, 212
Methane, tetraiodo-
fluorination of, **6**, 219, 226, 227
from, tetrachloromethane, **6**, 216
purification of, **6**, 217
reactions
with aldehydes, **4**, 749
with bromine trifluoride, **6**, 212, 217, 219
with iodine pentafluoride, **6**, 219
as source of carbon, **5**, 1097

synthesis of, **6**, 216
Methane, tetrakis(bromomethyl)-, reactions, with sodium diphenylphosphide, **2**, 437
Methane, tetrakis(dichlorodimethylphosphoranyl)-, from, chlorodimethylphosphines, **6**, 366
Methane, tetrakis(difluoroamino)-, synthesis of, **6**, 362, 363
Methane, tetrakis(dimethoxyboryl)-, reactions, with butyllithium, **6**, 200
Methane, tetrakis(dimethylamino)-
as starting material, for ketene aminals, **4**, 975
synthesis of, **6**, 337, 361
See also Tetraamines
Methane, tetrakis(diphenylphosphinomethyl)-, synthesis of, **2**, 437
Methane, tetrakis[(phenylsodiophosphino)methyl]-, reactions, with carbon disulfide, **5**, 1058
Methane, tetralithio-
coupling of, **6**, 379
reactions of, **6**, 405
synthesis of, **6**, 204
Methane, *N*,*N*,*N*',*N*'-tetramethyldiamino-, reactions, with boron enolates, **2**, 302
Methane, tetranitro-
addition to alkenes, **6**, 147
methylation of, **6**, 145
as nitrating agent, **4**, 429, 431; **6**, 150
for the oxidation of sulfides, **2**, 150
reactions
with anhydrous metal halides, **6**, 283
with butadiene, **6**, 149
with cyclohexene, **6**, 149
with ethylene, **6**, 149
with hydrazines, **4**, 995
with nitroalkanes, **4**, 429
with vinyl ethers, **6**, 149
as starting material, for anthracene, **6**, 147
synthesis of, **6**, 364, 365
Methane, tetrastannyl-, synthesis of, **6**, 405
Methane, thionitroso-, as starting material, for dimethylsulfurdiimide, **2**, 795
Methane, tosyl diazo-, reactions, with sulfonic acids, **4**, 235
Methane, tosylnitro-
alkylation of, **4**, 332
synthesis of, **4**, 331
Methane, tribromo-
fluorination of, **6**, 32
reactions
with sodium hypobromite, **6**, 215
with sodium hypoiodite, **6**, 223
as starting material, for tetrabromomethane, **6**, 215
See also Bromoform
Methane, tribromochloro-, synthesis of, **6**, 222
Methane, tribromofluoro-
as starting material
for carbonic bromide fluoride, **6**, 420
for dibromodifluoromethane, **6**, 226
synthesis of, **6**, 222
Methane, tribromoiodo-
synthesis of, **6**, 223
via halogen exchange, **6**, 223
Methane, trichloro-
bromination of, **6**, 221, 222, 227
fluorination of, **6**, 32
reactions, with sodium hypoiodite, **6**, 221
as starting material, for trichloromethyl anions, **6**, 25
Methane, trichlorofluoro-
fluorination of, **6**, 225
reactions
with triphenylphosphine, **4**, 753
with tris(dimethylamino)phosphine, **4**, 753
as starting material, for chlorofluorocarbene, **4**, 1082
synthesis of, **6**, 220
via halogen exchange, **6**, 220
Methane, trichlorogermyl(trimethylsilyl)-, synthesis of, **4**, 636
Methane, trichloroiodo-, synthesis of, **6**, 221
Methane, trichloronitro-
reactions, with potassium *O*,*O*-diethyldithiophosphate, **6**, 316
reduction of, **6**, 608
as starting material, for orthocarbonates, **6**, 296
synthesis of, **6**, 241
Methane, trichloronitroso-, synthesis of, **6**, 241
Methane, trichlorosilyldichloro-, as starting material, for trimethylsilyldichloromethane, **6**, 61
Methane, tricyano-
chlorination of, **5**, 810
potassium salt of, for the synthesis of primary α-halo enamines, **4**, 792
Methane, triethylsilyldiazo-, for the *O*-methylation of alcohols, **2**, 89
Methane, triethyl tricarbethoxy-, reactions, with 3-(dimethylaminomethyl)indole, **1**, 256
Methane, trifluoro-
bromination of, **6**, 218
chlorination of, **6**, 218
fluorination of, **6**, 212
Methane, trifluoroiodo-
reactions
with germanium(II) iodide, **6**, 244
with methyllithium, **6**, 244
with selenium, **6**, 49
trifluoromethylation with, **6**, 233
Methane, trifluoromethanesulfonyl(phenylthio)-, oxidation of, **4**, 273
Methane, trifluoronitroso-
ene reaction of, **6**, 241
iodination of, **6**, 219
pyrolysis of, **6**, 613
reactions
with amines, **6**, 242
with bis(trifluoromethyl)aminoallenes, **3**, 600
with perfluorobutadiene, **6**, 54
synthesis of, **6**, 241
vapor phase pyrolysis of, **6**, 604
Methane, triformamido-
from, formamide, **6**, 138
synthesis of, **6**, 140
Methane, triiodo-
reactions
with potassium hypoiodite, **6**, 216
with sodium hypobromite, **6**, 223
See also Iodoform
Methane, triisopropylsilyldiazo-
reactions
with butyllithium, **5**, 1148
with chlorobis(diisopropylamino)borane, **5**, 1148
Methane, trilithio-
from, tris(chloromercuri)methane, **6**, 205
synthesis of, **6**, 205
Methane, trimethylarsinediazo-, synthesis of, **5**, 871
Methane, trimethylene- (TMM)
complexes, synthesis of, **1**, 500
cycloadditions of, regiochemical outcomes, **1**, 361
palladium(0) complexes of, **1**, 360
reactions
with alkynes, **1**, 527
with tropone, **1**, 288
with vinyl sulfones, **1**, 288
as starting material, for cyclopentanes, **1**, 358
structure of, **1**, 358
synthesis of, **1**, 287
synthon, synthesis of, **1**, 193
Methane, trimethylgermyldiazo-, pyrolysis of, **5**, 926
Methane, trimethylgermyltrimethylsilyldiazo-, from, trimethylsilyldiazomethyllithium, **3**, 483

Methane, trimethylplumbyltrimethylsilyldiazo-, synthesis of, **3**, 484
Methane, trimethylsilyl-, reactions, with tosyl azide, **3**, 483
Methane, (trimethylsilyl)[bis(diisopropylamino)-phosphino]diazo-, as starting material, for alkynylphosphonium oxides, **2**, 1067
Methane, trimethylsilyldiazo-
acylation of, **3**, 475, 483
photolytic decomposition of, **3**, 502
reactions
with aldehydes, **1**, 1078
with aryl ketones, **1**, 1078
with butyllithium, **6**, 674
with chlorotrimethylsilane, **5**, 1148
with trimethyl bis[(trimethylsilyl)amino]plumbane, **3**, 484
synthesis of, **3**, 482, 483
use in the synthesis of methyl aryl ethers, **2**, 683
Methane, trimethylsilyldichloro-, from, trichlorosilyldichloromethane, **6**, 61
Methane, trimethylsilyl lithiobis(methylthio)-, as starting material, for acylsilanes, **5**, 411
Methane, trimethylsilyl(methylthio)-, bromination of, **6**, 129
Methane, (trimethylsilyl)trifluoro- (TMSTFM), reactions, with tetra-*n*-butylammonium fluoride, **2**, 70
Methane, trimethylsilyl(trimethylgermyl)-, lithiation of, **6**, 203
Methane, trinitro-
fluorination of, **6**, 282
reactions
with azodicarboxylate esters, **6**, 364
with dibutylmercury, **6**, 368
with mercury oxides, **6**, 368
with nitroxyl fluoride, **6**, 364
with sulfenyl chlorides, **6**, 652
Methane, trinitro-, salts
stability of, **6**, 369
synthesis of, **6**, 369
Methane, triphenyl-, synthesis of, **1**, 128
Methane, triphenylsilylphenyldiazo-, synthesis of, **3**, 483
Methane, (triphenylstannyl)bis(ethylenedioxyboryl)-, synthesis of, **6**, 200
Methane, tris(acetylthio)-
deacetylation of, **6**, 89
from, formic acid, **6**, 85
Methane, tris(benzotriazolyl)-, reactions, with electrophiles, **6**, 143
Methane, tris(biphenylenevinyl)-, acidity of, **1**, 866
Methane, tris(boryl)stannyl-, synthesis of, **6**, 200
Methane, tris(chloromercuri)-, as starting material, for trilithiomethane, **6**, 205
Methane, tris-(7*H*-dibenzo[*c*,*g*]fluorenylidenemethyl)-, synthesis of, **1**, 867
Methane, tris(diethylboryl)-, pyrolysis of, **6**, 183
Methane, tris(difluoroamino)fluoro-, as starting material, for (difluoroamino)fluorodiazirine, **6**, 285
Methane, tris(dimethoxyboryl)-
reactions
with boron trichloride, **6**, 181
with mercuric salts, **6**, 206
with methyllithium, **4**, 633
synthesis of, **6**, 181
transborylation of, **6**, 182
transesterification of, **6**, 181
Methane, tris(dimethylphosphino)-, methylation of, **6**, 151
Methane, tris(diphenylarsino)-, synthesis of, **6**, 155
Methane, tris(diphenylphosphino)-, reactions, with sulfur, **6**, 151
Methane, tris(ethylenedioxyboryl)-
reactions, with *n*-butyllithium, **2**, 905
use in the synthesis of 1,1-bisborylalkanes, **4**, 633
Methane, trisformamido-, reactions, with heterocyclic amines, **5**, 836
Methane, tris(methylseleno)-, reactions, with *n*-butyllithium, **6**, 135
Methane, tris(phenylsilyl)-, synthesis of, **6**, 173
Methane, tris(phenylthio)-, reactions, with arenes, **3**, 98
Methane, tris(phosphino)-, synthesis of, problems, **6**, 372
Methane, tris(trifluoromethylsulfonyl)-, lithium salt of, as electrolyte, **6**, 317
Methane, tris(trimethylsilyl)-
desilylation of, **6**, 194
halogenation of, **6**, 289
reactions, with sodium methoxide, **4**, 619
as starting material
for bis(trimethylsilyl)thioketone, **6**, 585
for tris(trimethylsilyl)methanethiol, **6**, 349
synthesis of, **4**, 1049; **6**, 173
Methane, tris(trimethylstannyl)-, synthesis of, **6**, 208
Methane, tris(triphenylplumbyl)-, reactions, with phenyllithium, **6**, 204
Methane, tris(triphenylstannyl)-, synthesis of, **6**, 210
Methane, zinc diiodo-, use in the synthesis of vinyl sulfides, **2**, 710
Methane anions, bis(arylseleno)-, condensation of, **2**, 728
Methane anions, triboryl-, condensation of, **4**, 814
Methane anions, tris(phenylthio)-, reactions, with trialkylboranes, **4**, 373
Methane bissulfonic acid
from, iodoform, **4**, 281
as starting material, for methane disulfonyl chloride, **4**, 281
Methane bi(sulfinic acids), difluoro-, synthesis of, **6**, 252
Methaneboronates, iodo-, reactions, with primary amines, **4**, 525
Methane cations, diarylhalo-, synthesis of, **2**, 1006
Methane complexes, dichlorodiazo-, synthesis of, **6**, 610
Methane complexes, trimethylsilyldiazo-, synthesis of, **6**, 675
Methanediamines, from, perfluorocyanamides, **6**, 259
Methanedisulfonic acid, disodio-, diphenyl ester, synthesis of, **6**, 326
Methanedisulfonyl chloride
from, methane bissulfonic acid, **4**, 281
reactions, with ethanol, **4**, 282
Methanedisulfonyl fluoride, synthesis of, **4**, 282
Methanehydrazonyl azides, *N*-aryl-1-nitro-, synthesis of, **6**, 654
Methane isocyanates, chlorodifluoro-, synthesis of, **5**, 1018
Methanepersulfonic acid, trifluoro-, as oxidizing agent, **6**, 530
Methanephosphonate, diisopropyl trichloro-, reactions, with *n*-butyllithium, **5**, 1152
Methanephosphonate, dimethyl trichloro-, synthesis of, **6**, 242
Methanephosphonates, dioxy-, electrochemical methoxylation of, **6**, 314
Methanephosphonic acid, 1-dimethylamino-1-cyano-, as starting material, for α-phosphoryl enamines, **4**, 1004
Methanephosphonic acid, trifluoro-, synthesis of, **6**, 242
Methanephosphonic acid diesters, trithio-substituted, synthesis of, **6**, 316
Methanephosphonic acid monoalkyl esters, trifluoro-, synthesis of, **6**, 242
Methanes
bearing four metal functions, synthesis of, **6**, 404
bearing four metalloid functions, synthesis of, **6**, 378
bearing at least one metalloid function, synthesis of, **6**, 378
bearing one metalloid and three metal functions, synthesis of, **6**, 401
bearing three metalloid functions and a metal function, synthesis of, **6**, 390
bearing two metalloid and two metal functions, synthesis of, **6**, 398
reactions
with alkoxides, **1**, 759
with metal halides, **6**, 370

synthesis of, **6**, 360
with three phosphorus functions, synthesis of, **6**, 369
Methanes, alkoxy tris(dimethylamino)-, as starting materials, for ketene aminals, **4**, 973
Methanes, aminobis(phosphonato)-
from
imidoyl compounds, **6**, 159
isocyanides, **6**, 160
ortho amides, **6**, 157
Methanes, aminotris(phosphonato)-, synthesis of, via phosphite addition to imines, **6**, 159
Methanes, arenesulfonyl diazo-, photolysis of, **4**, 235
Methanes, aryldinitro-, reactions, with benzenediazonium ions, **3**, 449
Methanes, aryl hydrazono aryl nitro-, reactions, with phenoxide, **5**, 704
Methanes, arylselenotrinitro-, as starting materials, for halotrinitromethanes, **6**, 283
Methanes, arylthiotrinitro-, as starting materials, for halotrinitromethanes, **6**, 283
Methanes, azinodialkoxyphenyl-, synthesis of, **5**, 685
Methanes, (benzoyloxy)iodo-, stability of, **4**, 58
Methanes, bis(3-valent organometallic)diazo-, synthesis of, **6**, 668
Methanes, bis(alkylsulfonyl)diazo-, synthesis of, **3**, 479
Methanes, bis(alkylthio)-
as starting materials, for bis(alkylthio)trialkylsilylmethanes, **6**, 127
synthesis of, **5**, 550
Methanes, bis(alkylthio)trialkylsilyl-, from, bis(alkylthio)methanes, **6**, 127
Methanes, bis(alkylthio)trimethylsilyl-, lithiation of, **6**, 128
Methanes, bis(arylsulfonyl)-, synthesis of, **4**, 272
Methanes, bis(arylsulfonyl)chloro-, synthesis of, **6**, 277
Methanes, bis(arylsulfonyl)diazo-, synthesis of, **3**, 478, 479
Methanes, bis(arylsulfonyl)(methylsulfonyl)-, as starting materials, for bis(3-nitroarylsulfonyl)bromonitromethanes, **6**, 277
Methanes, bis(benzenesulfonyl)halo-, reactions, with chlorophosphines, **6**, 699
Methanes, bis(benzoylamino)-, reactions, with phosphorus pentachloride, **5**, 657
Methanes, bis(benzoylamino)aryl-, reactions, with phosphorus pentachloride, **5**, 657
Methanes, bis(boryl)(germyl)-, synthesis of, **6**, 200
Methanes, bis(dialkoxyboryl)-, metallation of, **6**, 199
Methanes, bis(dialkylamino)difluoro-, as starting materials, for phosphaalkenes, **6**, 685
Methanes, bis(dichloroarsino)-, synthesis of, **4**, 592
Methanes, bis(dithianyl)-, deprotonation of, **4**, 845
Methanes, bis(halophosphino)-, cyclodehalogenation of, **2**, 460
Methanes, bis(3-nitroarylsulfonyl)bromonitro-, from, bis(arylsulfonyl)(methylsulfonyl)methanes, **6**, 277
Methanes, bis(perfluoroalkanesulfonyl)-, as nucleophiles, **4**, 864
Methanes, bis(phenyltelluro)-, reactions, with butyllithium, **4**, 402
Methanes, bis(phosphino)-
reactions, with sulfur, **6**, 167
silylation of, **6**, 164
Methanes, bis(phosphino)trimethylsilyl-, synthesis of, **6**, 164
Methanes, bis(silyl)-
metallation of, **6**, 190
synthesis of, **6**, 173
Methanes, 1,1-bis(silyl)-, synthesis of, **4**, 609
Methanes, bissilyldiazo-
from
diazomethyllithium, **6**, 672
lithiodiazomethanes, **6**, 672
Methanes, bis(silyl) dichloro-, as starting materials, for tetrasilyl ethenes, **4**, 1050
Methanes, bis(sulfonyl)-
reactions
with alkylthioalkyl sulfones, **6**, 90
with sulfonyl chlorides, **6**, 90
Methanes, bis(sulfonyl)halo-, metallation of, **6**, 278
Methanes, bis(trichloromethylsilyl)dichloro-, cleavage of, **6**, 61
Methanes, bis(trimethylsilyl)amino-
from
amides, **6**, 162
trimethylsilyl cyanide, **6**, 162
Methanes, boranylsilyldiazo-, synthesis of, **6**, 673
Methanes, bromodinitro-, reactions, with α,α'-dicarbonylazo compounds, **6**, 284
Methanes, bromotris(trialkylsilyl)-, production of, **6**, 291
Methanes, chiral halo-, synthesis of, **6**, 228
Methanes, chlorobis(trimethylsilyl)-, reactions, with dichlorophosphines, **6**, 693
Methanes, chlorodiaryloxy-
as starting materials, for diaryloxymethyltriethylammonium chlorides, **6**, 105
synthesis of, **6**, 44
Methanes, chlorodinitro-, reactions, with α,α'-dicarbonylazo compounds, **6**, 284
Methanes, diacyldiazo-, synthesis of, **3**, 473
Methanes, dialkylaminophosphanylsilyldiazo-, synthesis of, **6**, 668
Methanes, dialkylphosphanylsilyldiazo-, synthesis of, **6**, 668
Methanes, diaminoalkoxy-
as starting materials
for amide acetals, **6**, 140
for triaminomethanes, **6**, 140
Methanes, diaminodihalo-, alcoholysis of, **6**, 321
Methanes, diaryl-
as starting materials, for benzophenones, **3**, 293
synthesis of, **1**, 46, 475
Methanes, diaryldiazo-
reactions
with dichlorocarbene, **4**, 740
with diphenyl trithiocarbonate, **6**, 87
Methanes, diaryldichloro-, synthesis of, **6**, 29
Methanes, diarylnitro-, reactions, with benzenediazonium ions, **3**, 449
Methanes, diazo-
reactions
with electrophiles, **5**, 871
with metal amides, **6**, 669
with phosphaalkynes, **4**, 479
synthesis of, **6**, 672, 674
Methanes, diazo-, antimony-substituted, synthesis of, **6**, 668
Methanes, diazo-, arsenic-substituted, synthesis of, **6**, 668
Methanes, diazo-, bismuth-substituted, synthesis of, **6**, 668
Methanes, diazo-, bis(silyl-substituted), synthesis of, **6**, 673
Methanes, diazo-, lithium salts, use in the preparation of silyldiazomethanes, **6**, 668
Methanes, α-diazo β-oxosulfonyldiazo-, synthesis of, **3**, 479
Methanes, diazo-, silicon-substituted, synthesis of, **6**, 668
Methanes, diboryldilithio-, synthesis of, **4**, 634
Methanes, dichalcogeno-, synthesis of, **6**, 320
Methanes, dichlorobis(aryloxy)-
as starting materials, for iminocarbonates, **6**, 624
synthesis of, **6**, 252
Methanes, difluorodialkoxy-, synthesis of, **6**, 251
Methanes, dihalo-
displacement of, **4**, 190
double quaternization of, **5**, 1094
reactions
with alkali metal salts, **4**, 147
with selenide anions, **4**, 88
with zinc, **4**, 152
with zirconocene dichloride, **1**, 680
use in the synthesis of methylene-bridged gold complexes, **4**, 695
Methanes, dihalobis(trimethylsilyl)-, metallation of, **6**, 293

Methanes, dihalodifluoro-
reactions
with cadmium, **6**, 15
with zinc, **6**, 15
Methanes, diiodo-, use in Simmons–Smith reactions, **1**, 300
Methanes, diiododinitro-, synthesis of, **6**, 261
Methanes, dinitro-
alkali metal salts of, **6**, 170
reactions, with triethylthallium, **6**, 170
synthesis of, **6**, 150
Methanes, dioxodihalo-, aminolysis of, **6**, 321
Methanes, disilyl-, reactions, with organolithiums, **6**, 379
Methanes, disilyl(boryl)-, synthesis of, **6**, 188
Methanes, disulfonyl-
condensation reactions of, **4**, 864
reactions, with ortho esters, **4**, 865
as starting materials
for ketene dithioacetal tetroxides, **4**, 864
for vinyl ethers, **4**, 865
Methanes, fluorochloro-, synthesis of, **6**, 217
Methanes, fluorohalo-, reactions, with sulfur trioxide, **6**, 410
Methanes, germyldiazo-, synthesis of, **4**, 521
Methanes, germyldisilyl-, thermal decomposition of, **6**, 711
Methanes, germylsilyldiazo-, synthesis of, **6**, 673
Methanes, halo-
reactions
with organogermanium halides, **6**, 386
with phosphines, **2**, 444
with silylpotassium, **6**, 173
sequential metallations of, **6**, 386
as source of carbon, **5**, 1097
Methanes, halodisulfonyl-, characteristics of, **6**, 279
Methanes, halonitro-, reactions, with aldehydes, **4**, 798
Methanes, halo(silyl)-
metallation of, **6**, 189
oxidative metallation of, **6**, 191
Methanes, halotriamino-, production of, **6**, 284
Methanes, halo tricyano-, as starting materials, for *gem*-phosphorylamino phosphorylalkenes, **4**, 1008
Methanes, halotrinitro-
from, trinitromethyl sulfones, **6**, 283
production of, **6**, 283
synthesis of, **6**, 282
Methanes, halotris(trialkylsilyl)-
production of, **6**, 290
synthesis of, **6**, 292
Methanes, hydrazono alkoxyphenyl-, synthesis of, **5**, 685
Methanes, lithiated bis(silyl)-, synthesis of, **6**, 192
Methanes, lithio-
reactions, with halometalloids, **6**, 389
synthesis of, via mercury-lithium exchange, **6**, 205
Methanes, lithiodiazo-, as starting materials, for bissilyldiazomethanes, **6**, 672
Methanes, lithio(dimethylphosphino)-, synthesis of, **2**, 1003
Methanes, α-metallated boryl-, synthesis of, **6**, 201
Methanes, monoboryl-, synthesis of, **6**, 199
Methanes, monochalcogeno-, synthesis of, **6**, 336
Methanes, perfluorodiamino-, synthesis of, **6**, 258
Methanes, phosphoryldiazo-, reactivity of, **3**, 480
Methanes, plumbyl(trimethylsilyl)diazo-, synthesis of, **6**, 675
Methanes, polyborylated, reactions, with organolithiums, **6**, 199
Methanes, poly(fluoroamino)halo-, uses, **6**, 284
Methanes, polyhalo-, reactions, with aldehydes, **2**, 81
Methanes, polymercurated, synthesis of, **6**, 206
Methanes, polysilylated, as starting materials, for silaethenes, **6**, 177
Methanes, selenium-substituted, synthesis of, **6**, 49
Methanes, silyl-
alkali metal compounds of, synthesis of, **6**, 189
magnesium compounds of, synthesis of, **6**, 189
metal–hydrogen exchange reactions of, **6**, 189
Methanes, silylbis(plumbyl)-, transmetallation of, **6**, 203
Methanes, silyl(boryl)-, from, 1-trimethylsilyl-1-dimesitylborylethylenes, **6**, 201
Methanes, silyldiazo-, synthesis of, **6**, 668
Methanes, silyltrihalo-, synthesis of, problems, **6**, 401
Methanes, (stannyl)bis(plumbyl)-, synthesis of, **6**, 210
Methanes, stannyl(trimethylsilyl)diazo-, synthesis of, **6**, 675
Methanes, sulfonyldiazo-
from, *N*-nitrosamines, **5**, 867
synthesis of, **3**, 479
Methanes, tellurium-substituted, synthesis of, **6**, 49
Methanes, tetraamino-
as aminating agents, **6**, 361
from, 2-chloro-1,3-dimethylbenzoimidazolium tetrafluoroborate, **6**, 361
synthesis of, **6**, 360
Methanes, tetrachalcogeno-, synthesis of, **6**, 295
Methanes, tetragermyl-
rarity of, **6**, 386
synthesis of, **6**, 386
Methanes, tetrahalo-
reactions
with di(*t*-butyl)fluorophosphine, **6**, 695
with iodine pentafluoride, **6**, 217
with phosphines, **6**, 695
with silylalkylphosphines, **6**, 703
as starting materials, for phosphonium ylides, **3**, 495
synthesis of, **6**, 212
Methanes, tetrakis(dialkylamino)-, synthesis of, **6**, 360
Methanes, tetrakis(difluorosulfoximidoyl)-, from, tetrachloromethane, **6**, 362
Methanes, tetrakis(1-pyrazolyl)-
from, tetrachloromethane, **6**, 362
synthesis of, **6**, 365
Methanes, tetrakis(silyl)-, synthesis of, **6**, 173
Methanes, tetramercurio-, synthesis of, **6**, 405
Methanes, tetrametallo-, synthesis of, **6**, 395, 405
Methanes, tetrasilyl-
from
carbosilanes, **6**, 385
in situ coupling reactions, **6**, 378
silenes, **6**, 380
trisilyllithiomethane, **6**, 379
tris(trimethylsilylmethyl)lithium, **6**, 379
vinylsilanes, **6**, 392
range of, **6**, 378
synthesis of, **6**, 378
photolytic methods, **6**, 383
thermolytic methods, **6**, 383
via transition metal-catalyzed reactions, **6**, 384
Methanes, thiophosphinoyldiazo-, reactions, with chlorophosphanes, **3**, 681
Methanes, triacyl-, as starting materials, for ketene acetals, **4**, 824
Methanes, trialkylsulfonyl-, reactions, with halogens, **6**, 274
Methanes, triamino-
from, diaminoalkoxymethanes, **6**, 140
synthesis of, **6**, 140, 141
Methanes, triarylsulfonyl-, reactions, with halogens, **6**, 274
Methanes, triazulenyl-, reactions, with 2,3-dichloro-5,6-dicyano-1,4-benzoquinone, **1**, 899
Methanes, triboryl-, as intermediates, in the synthesis of 1,1,1-tris(dialkylboryl)methanes, **6**, 182
Methanes, trichalcogeno-, synthesis of, **6**, 313
Methanes, trifluoro-, as starting materials, for trifluoromethyl radicals, **6**, 11
Methanes, trifluorosulfonyl-, reactions, with halogens, **6**, 274
Methanes, trihalo-
reactions
with alkoxides, **4**, 826
with dicobaltoctacarbonyl complexes, **6**, 401
with phenoxides, **4**, 826
reductive silylation of, **6**, 172

as starting materials
for α,α-dihalo ethers, **6**, 37
for ketene acetals, **4**, 826
for ketene aminals, **4**, 979
Methanes, trihalonitro-, synthesis of, **6**, 240
Methanes, trihalonitroso-, synthesis of, **6**, 240
Methanes, trimercurated, synthesis of, **6**, 206
Methanes, trimethylene-, reactions, with acrylic esters, **5**, 146
Methanes, trimethylsilyl-, Michael addition of, **6**, 128
Methanes, trimethylsilyl(ethylenedioxyboryl)-, metallation of, **6**, 203
Methanes, trinitro-
O-alkylation of, **6**, 145
in cycloaddition reactions, **6**, 149
from, ethyne, **6**, 150
methylation of, **6**, 145
reactions
with enol ethers, **6**, 145
with formaldehyde, **6**, 145
with imines, **6**, 147
with pyridinium salts, **6**, 145
O-silylation of, **6**, 149
synthesis of, **6**, 368
via nitration reactions, **6**, 150
Methanes, trinitro-, potassium salt, phase-transfer catalyzed methylation of, **6**, 145
Methanes, trinitro-, silver salt, alkylation of, **6**, 145
Methanes, triplumbyl-, synthesis of, **6**, 209
Methanes, tris(boryl)-, synthesis of, **6**, 180
Methanes, tris(dialkoxyboryl)-
reactions, with methyllithium, **6**, 199
synthesis of, **6**, 181
See also Methanetriboronic esters
Methanes, 1,1,1-tris(dialkylboryl)-, synthesis of, **6**, 182
Methanes, tris(dichoroboryl)-, synthesis of, **6**, 181
Methanes, tris(dihaloboryl)-
from, boron halides, **6**, 181
synthesis of, **6**, 181
Methanes, tris(dimethylsilyl)-, from, dimethylchlorosilane, **6**, 174
Methanes, trisilyl-
from
halosilanes, **6**, 174
silaethenes, **6**, 178
tris(trimethylsilyl)methylsilicon halides, **6**, 177
metallation of, **6**, 391
synthesis of, **4**, 1050
methods, **6**, 173
Methanes, trisilyllithio-
reactions
with germanium halides, **6**, 387
with halo- or pseudohalosilanes, **6**, 379
as starting materials, for tetrasilylmethanes, **6**, 379
Methanes, tris(organosilyl)-, synthesis of, **6**, 173
Methanes, tris(organylsilyl)-, synthesis of, **6**, 173
Methanes, tris(phosphino)-
from
bisphosphines, **6**, 151
phosphaalkynes, **6**, 152
modification of phosphorus, **6**, 151
synthesis of, **6**, 151
Methanes, tris(phosphonato)-, synthesis of, **6**, 151
Methanes, tris(silyl)-
metallation of, **6**, 174
synthesis of, **6**, 173
Methanes, tris(trialkylsilyl)-, metallation of, **6**, 289
Methanes, tris(triorganylplumbyl)-, synthesis of, **6**, 209
Methanes, trisulfonyl-
reactions
with halogens, **6**, 274
with nitric acid, **6**, 316
with xenon difluoride, **6**, 274
Methane salts, dinitro-, reactions, with diethylboranyl, **6**, 169
Methane salts, trinitro-
halogenation of, **6**, 282
reactions
with bromoadamantane, **6**, 145
with α-chloroamines, **6**, 145
with α-chloro ethers, **6**, 145
with diiodomethane, **6**, 145
with ferrocenes, **6**, 145
with triazolediazonium salts, **6**, 145
Methaneselanol, as starting material, for methylselenosilanes, **2**, 294
Methaneselenenic acid, from, methaneseleninic acid, **2**, 290
Methaneseleninic acid, as starting material, for methaneselenenic acid, **2**, 290
Methaneselenonic acid, trifluoro-, synthesis of, **6**, 237
Methanesulfenamide, *N*,*N*-diethyl tris[(trifluoromethyl)thio]-, synthesis of, **6**, 307
Methanesulfenamides, triphenyl-, synthesis of, via reduction of tritylsulfenimines, **2**, 359
Methane sulfenates, *O*-alkyl-, synthesis of, **2**, 194
Methane sulfenates, trifluoro-, synthesis of, **2**, 194
Methanesulfenates, trihalo-, synthesis of, **2**, 194
Methanesulfenic acid, elimination reactions of, **4**, 924
Methanesulfenic acids, trihalo-, stability of, **6**, 236
Methanesulfenic compounds, trihalo-, synthesis of, **6**, 236
Methanesulfenyl bromide, as starting material, for methyl bromothiolformate, **6**, 437
Methanesulfenyl bromide, tris(trimethylsilyl)-, as intermediate, **6**, 585
Methanesulfenyl chloride, reactions, with lithium alkoxides, **2**, 194
Methanesulfenyl chloride, chlorodifluoro-, synthesis of, **2**, 166
Methanesulfenyl chloride, dichloro(difluoroamino)-, synthesis of, **2**, 170
Methanesulfenyl chloride, dichlorofluoro-, as starting material, for thiocarbonyl chloride fluoride, **6**, 531
Methanesulfenyl chloride, methoxydichloro-, reactions, with potassium *O*-methyl dithiocarbonate, **6**, 548
Methanesulfenyl chloride, trichloro-
hydrolysis of, **6**, 531
for chlorocarbonylsulfenyl chloride, **6**, 434
reactions
with alcohols, **2**, 194
with sodium trithiocarbonate, **6**, 554
reduction of, **6**, 529
as starting material
for dichlorosulfine, **6**, 531
for orthocarbonates, **6**, 296
for tetraalkyl orthocarbonates, **6**, 308
for tetrathioorthocarbonates, **6**, 301
for thiophosgene, **6**, 529
for trichloromethyl chlorodithioformate, **6**, 535
trichloromethylthiolation with, **6**, 233
Methanesulfenyl chloride, trifluoro-
reactions, with silver cyanate, **5**, 1018
synthesis of, **2**, 169
trifluoromethylthiolation with, **6**, 233
Methanesulfenyl chloride, tris[(trifluoromethyl)thio]-, synthesis of, **6**, 307
Methanesulfenyl chlorides, trichloro-, synthesis of, **6**, 236
Methanesulfenyl compounds, alkoxydichloro-, as starting materials, for chlorothiolformate esters, **6**, 435
Methanesulfenyl fluoride, synthesis of, problems, **2**, 166
Methanesulfenyl fluoride, dichloro(difluoroamino)-, synthesis of, **2**, 166
Methanesulfenyl fluoride, trichloro-, isomerization of, **2**, 166
Methanesulfenyl fluoride, trifluoro-, dimerization of, **2**, 166
Methanesulfenyl halides, trihalo-
halogen exchange in, **6**, 236
synthesis of, **6**, 236
Methanesulfenyl iodide, triphenyl-, synthesis of, **2**, 172
Methanesulfenyl isocyanate, trifluoro-, synthesis of, **5**, 1018

Methane sulfinate, methyl-, oxidation of, **2**, 178
Methane sulfinate, sodium hydroxy-, for the reduction of organic halides, **4**, 234
Methane sulfinate, trifluoro-, displacement of, **4**, 309
Methanesulfinic acid, aminoimino-, fluorination of, **6**, 258
Methanesulfinic acid, chloro-, from, chloromethanesulfinyl chloride, **4**, 78
Methanesulfinic compounds, trihalo-, synthesis of, **6**, 236
Methanesulfinyl bromide, trichloro-, as starting material, for α-bromo sulfoxides, **4**, 77
Methanesulfinyl chloride
  reactions
    with lithium (1-alkynyl)organoborates, **1**, 1033
    with silver acetate, **5**, 209
Methanesulfinyl chloride, chloro-
  as starting material, for chloromethanesulfinic acid, **4**, 78
  synthesis of, **4**, 78
Methanesulfinyl chloride, perfluoro-, synthesis of, **2**, 172
Methanesulfinyl chloride, phenyl-, reactions, with ylides, **5**, 900
Methanesulfinyl chloride, trichloro-, reactions, with silver cyanate, **5**, 995
Methanesulfinyl halides, trihalo-, synthesis of, **6**, 236
Methanesulfonamide, use in asymmetric *cis*-dihydroxylation reactions, **2**, 48
Methanesulfonamide, *N*,*N*-dimethyl-, synthesis of, **1**, 147
Methanesulfonamide, *N*-methyltrifluoro-, reactions, with triphenylphosphine, **2**, 309
Methanesulfonamide, *N*-(trifluoromethyl)-*N*-nitrosotrifluoro-, as trifluoromethylation agent, **6**, 233
Methanesulfonamides, *N*,*N*-disubstituted, methylation of, **2**, 365
Methanesulfonanilides, fluoro-, synthesis of, **4**, 86
Methanesulfonate, chlorine trifluoro-, as starting material, for chlorocarbonyltrifluoromethanesulfonate, **6**, 430
Methanesulfonate, chloro-, reactions, with phosphorus pentachloride, **4**, 85
Methanesulfonate, chlorocarbonyltrifluoro-, from, chlorine trifluoromethanesulfonate, **6**, 430
Methane sulfonate, 2-methylpropyl bromo-, synthesis of, **4**, 86
Methanesulfonate, (*Z*)-4-phenylbut-3-enyl-, reactions, with potassium *O*-isopropyl dithiocarbonate, **6**, 548
Methanesulfonate, silver trifluoro-, as alkylating agent, **4**, 881
Methanesulfonate, trichloromethyl trifluoro-, synthesis of, **6**, 232
Methane sulfonate, trifluoromethyl-, alkylation of, **2**, 157
Methanesulfonate, *Se*-(trifluoromethyl)dibenzoselenophenium trifluoro-, synthesis of, **6**, 237
Methanesulfonate, *Te*-(trifluoromethyl)dibenzotellurophenium trifluoro-, synthesis of, **6**, 237
Methanesulfonate, trifluoromethyl trifluoro-, synthesis of, **6**, 232
Methane sulfonate, trimethylsilyl trifluoro-, reactions, with imines, **2**, 313
Methanesulfonates, acyloxytrifluoro-, synthesis of, **5**, 209
Methanesulfonates, alkyl-, reactions, with sodium hexamethyldisilazide, **4**, 605
Methane sulfonates, trifluoro-
  synthesis of, **2**, 689
  See also Triflates
Methanesulfonate salts, reactions, with potassium tricyanomethanide, **4**, 977
Methane sulfonate(triflate) salts, trifluoro-, synthesis of, **2**, 1020
Methanesulfonic acid
  reactions
    with *N*-(chlorocarbonyl)imidoyl chlorides, **5**, 1004
    with dithianes, **4**, 252
    with esters, **5**, 29
    with oxomethylenebis(imidazole), **3**, 696
    with phosphorus pentoxide, **5**, 282
Methanesulfonic acid, aminoimino-, from, thioureas, **6**, 641
Methanesulfonic acid, methylsulfonyl-, sodium salt of, synthesis of, **4**, 282
Methanesulfonic acid, trifluoro-
  reactions
    with acid chlorides, **5**, 209
    with bis(trimethylsilyl) peroxide, **2**, 650
  as starting material, for trifluoromethyl radicals, **6**, 11
  synthesis of, **6**, 20
  use in the synthesis of alkenyl triflates, **2**, 689
  use in the synthesis of aromatic ketones, **3**, 279
  use in the synthesis of mixed anhydrides, **5**, 710
Methanesulfonic anhydride, trifluoro-
  reactions
    with diisopropylethylamine, **3**, 696
    with dimethylformamide, **3**, 92
  use in the synthesis of alkenyl triflates, **2**, 689
  See also Triflic anhydride
Methanesulfonic compounds, trihalo-, synthesis of, **6**, 236
Methane sulfonimide, *N*-phenyltrifluoro-, use in the synthesis of alkenyl triflates, **2**, 689
Methane sulfonimides, *N*,*N*-bistrifluoro-, reactions, with lithium dimethylcuprate, **1**, 257
Methanesulfonyl azide, trifluoro-, as starting material, for α-aminoalkylphosphorus compounds, **4**, 491
Methanesulfonyl bromide, synthesis of, **2**, 178
Methanesulfonyl bromide, bromo-
  reactions
    with alkenes, **4**, 83
    with alkynes, **4**, 84
    with 2-methylpropanol, **4**, 86
    with silyl enol ethers, **4**, 84
  as starting material, for alkyl bromomethyl sulfones, **4**, 83
Methanesulfonyl bromide, chloro-
  reactions
    with alkenes, **4**, 82
    with alkynes, **4**, 82
  synthesis of, **4**, 85
Methanesulfonyl bromide, iodo-
  as starting material, for α-iodo sulfones, **4**, 84
  synthesis of, **4**, 85
Methanesulfonyl chloride
  as activating agent, for the synthesis of anhydrides, **5**, 189
  reactions
    with alcohols, **4**, 780
    with 2-chloroethanol, **4**, 282
  as reagent, for the synthesis of symmetrical aromatic anhydrides, **5**, 196
  synthesis of, **2**, 181
  use in the synthesis of alkenes, **1**, 592
Methanesulfonyl chloride, bromo-, reactions, with ammonia, **4**, 86
Methanesulfonyl chloride, chloro-
  reactions
    with alkenes, **4**, 82
    with ketene diethyl acetal, **5**, 496
    with sodium sulfite, **4**, 85
    with triethylamine, **5**, 496
  synthesis of, **4**, 85
Methanesulfonyl chloride, dichloro-, synthesis of, **2**, 177
Methanesulfonyl chloride, iodo-
  reactions
    with ammonia, **4**, 86
    with benzamidine, **4**, 86
  synthesis of, **4**, 85
Methanesulfonyl chloride, trichloro-
  reactions
    with alkylcobaloximes, **2**, 176
    with potassium cyanide, **2**, 272
Methanesulfonyl chloride, trifluoro-

reactions
with haloalkenes, **6**, 46
with vinyl benzoate, **4**, 56
use in the synthesis of dichloroalkanes, **4**, 12
use in the synthesis of trifluoromethyls, **6**, 14
Methanesulfonyl chloride, trimethylsilyl-, synthesis of, **2**, 177
Methanesulfonyl chlorides, trihalo-, synthesis of, **6**, 236
Methanesulfonyl halides, chloro-, as starting materials, for α-chloro sulfones, **4**, 82
Methanesulfonyl halides, trihalo-, synthesis of, **6**, 236
Methanetellurenyl iodide, tris(trimethylsilyl)-, reactions, with silver cyanide, **5**, 1117
Methanetellurocyanate, tris(trimethylsilyl)-, synthesis of, **5**, 1117
Methanetetraboronic esters, synthesis of, **6**, 386
Methanetetraphosphonic acid, synthesis of, **6**, 365
2-Methanethio-1,3-dithioles
from
1,3-dithiolones, **6**, 87
2-methanethio-1,3-dithiolium salts, **6**, 85
2-Methanethio-1,3-dithiolium salts, reduction of, to 2-methanethio-1,3-dithioles, **6**, 85
Methanethiol
reactions
with chlorobis(methanethio)methane, **6**, 82
with chlorodimethylarsine, **2**, 499
with ethyl acrylate, **2**, 138
Methanethiol, sodium salt, for the reduction of nitroalkanes, **1**, 37
Methanethiol, trichloro-
reduction of, **6**, 529
synthesis of, **6**, 232
Methanethiol, trifluoro-
reactions, with hexafluoro dithioate, **4**, 256
as starting material, for perfluoromethyl fluorodithioformate, **6**, 534
synthesis of, **6**, 232
trifluoromethylthiolation with, **6**, 233
Methanethiol, tris(trimethylsilyl)-
from, tris(trimethylsilyl)methane, **6**, 349
as intermediate, **6**, 585
Methanethiolate, copper(I) trifluoro-
synthesis of, **6**, 236
trifluoromethylthiolation with, **6**, 233
Methanethiolates, metal trihalo-
chlorination of, **6**, 236
synthesis of, **6**, 236
Methanethiol(benzylthiol), phenyl-, as starting material, for *C,S*-dianions, **1**, 140
Methanethiol carbanions, synthesis of, **1**, 141
Methanethiols, trichloro-, hydrolysis of, for chlorothiolformate esters, **6**, 434
Methanethiols, trihalo-, synthesis of, **6**, 232
2-Methanethio-1,3-oxathiane, from, 1,3-oxathiane, **6**, 98
Methanethiosulfonate, *S*-methyl
reactions
with vinylcopper(I) compounds, **4**, 940
with vinyl sulfones, **4**, 858
synthesis of, **2**, 185
Methanethiosulfonic acid esters, reactions, with potassium *S*-methyl trithiocarbonate, **6**, 553
Methanetriboronic esters
as starting materials, for lithium poly(boryl)methides, **6**, 198
synthesis of, **6**, 181
See also Methanes, tris(dialkoxyboryl)-
Methanetrisulfonic acid, esterification of, **6**, 91
Methanetrisulfonyl trifluoride, reactions, with trimethylchlorosilane, **6**, 317
Methanetrisulfonyl trifluoride, (trimethylsilyl)-, synthesis of, **6**, 317
Methanetrithiol, from, tris(acetylthio)methane, **6**, 89
Methanide, acetonitrile, synthesis of, **3**, 679
Methanide, benzonitrile bis(trifluoromethyl)-, synthesis of, **3**, 680
Methanide, bis(thiophosphinyl)-, synthesis of, **6**, 371
Methanide, bromodinitro-, synthesis of, **6**, 261
Methanide, tricyano-, as starting material, for ketene aminals, **4**, 977
Methanide ion, tris(phenylseleno)-, reactions, with diphenyl diselenide, **6**, 307
Methanides, bis(phosphino)-, synthesis of, **6**, 371
Methanides, fluorodinitro-
reactions
with aldehydes, **6**, 57
with alkenes, **6**, 57
Methanide salts, synthesis of, **4**, 978
Methanide salts, ammonium tricyano-, as intermediates, in the synthesis of ketene aminals, **4**, 977
Methanide salts, tricyano-, as intermediates, in the synthesis of ketene aminals, **4**, 971
Methanimine, α-bromo glycine diphenyl-, as starting material, for aminomethyl imines, **4**, 445
Methanimine, dibromo-*N*-(pentafluorophenyl)-, dehalogenation of, **3**, 722
Methanimine, diphenyl-, as starting material, for bisimines, **4**, 438
Methanimine, *N*-methyl-, reactions, with nitrosyl chloride, **4**, 325
Methanimine, perfluoro-
reactions, with fluoride, **5**, 313
as starting material, for *N*-fluoroamides, **5**, 313
Methanimines, bis(methylthio)-, reactions, with carbanions, **4**, 901
Methanimines, diaryl-, Mannich-type reaction of, **4**, 445
Methanimines, difluoro-, as starting materials, for isocyanides, **3**, 702
Methanimines, trialkylstannyl-, reactions, with alkyllithiums, **1**, 372
Methaniminium chloride, *N,N*-dimethylchloro-, reactions, with diethyl tetrathiomalonate, **4**, 775
Methaniminium chloride, *N,N*-dimethyl dichloro-, use in the synthesis of isothiocyanates, **5**, 1035
1,6-Methano[10]annulene, reactions, with diazomethane, **1**, 910
Methanodibenzo[*a,e*]cyclooctenes, imino-, synthesis of, **2**, 328
Methanol
electrochemical fluorination of, **6**, 214
electrolytic carbonylation of, **6**, 469
elimination of, **1**, 981
photochemical gas phase chlorination of, **4**, 42
reactions
with allenes, **4**, 736
with benzonitrile, **5**, 515
with bis(arenesulfonyl)alkynes, **4**, 840
with boron trifluoride, **2**, 110
with chlorocarbonylsulfenyl chloride, **6**, 478
with chlorodialkylphosphines, **2**, 450
with β,β-diacylketene aminals, **4**, 979
with 3-*N,N*-dimethylaminopropynal, **3**, 76
with 1,3-diphosphetanes, **6**, 265
with ethyl dithiophosphate, **2**, 135
with germaethenes, **6**, 188
with α-halohydrazones, **3**, 451
with magnesium, **1**, 617
with nitrile oxides, **5**, 699
with perfluoroamidines, **5**, 814
with potassium hydroxide, **4**, 748, 906
with potassium methoxide, **2**, 344
with samarium(II) iodide, **1**, 7
with selenium dioxide, **2**, 101
with sodium amalgam, **1**, 615, 617
with sodium carbonate, **5**, 951
with tetrafluoroethene, **4**, 5

with 1,1,3,3-tetramethyl-2-(thiocarbonyl)cyclohexane, **2**, 202
with 1,1,2,2-tetramorpholinoethane, **4**, 309
with trimethylaminecarboxyborane, **6**, 491
with trimethylsilyl ethynyl triethylsilyl sulfide, **3**, 548
with tri α-naphthyl boron, **1**, 447
use in the cleavage of carbon–silicon bonds, **1**, 63
Methanol, bis(triphenylgermyl)-, from, triphenylgermyllithium, **6**, 131
Methanol, 1,1-bis(triphenylsilyl)-, oxidation of, **6**, 519
Methanol, chloro-, synthesis of, **4**, 42
Methanol, fluoro-, synthesis of, **4**, 42
Methanol, iodo-, synthesis of, **4**, 42
Methanol, 2-methoxy-, reactions, with sodium hydroxide, **1**, 42
Methanol, phenyldimethylsilyl-, for the protection of anomeric hydroxyl groups, **4**, 176
Methanol, tributylstannyl-, reactions, with *n*-butyllithium, **4**, 378
Methanol, trifluoro-, synthesis of, **6**, 228
Methanol, trimethylsilyl-
reactions, with *n*-butyllithium, **4**, 383
Swern oxidation of, **5**, 422
Methanol–hydrochloric acid, reactions, with 4-methyl-2-trifluoromethyl-5(2*H*)-oxazolone, **4**, 305
Methanols, reactions, with dicyclohexylcarbodiimide, **6**, 519
Methanols, dialkylamino-, reactions, with tristrimethylsilylphosphine, **4**, 459
Methanols, furan-
oxidation of, **3**, 303
as starting materials, for 6-hydroxypyran-2-ones, **3**, 221
Methanols, 2-furan-, Sharpless epoxidation of, **3**, 222
Methanols, trihalo-, synthesis of, **6**, 228
Methide, bis(perfluoroalkanesulfonyl), use in the synthesis of methylenephosphoranes, **6**, 699
Methide, fluorodinitro, from, fluorotrinitromethane, **6**, 287
Methide anions, bis(dimethoxyboryl), synthesis of, **4**, 633
Methide anions, triboryl
alkylation of, **6**, 182
metallation of, **6**, 183
reactions
with aldehydes, **6**, 182
with ketones, **6**, 182
Methides, bis(phosphonyl)halo, production of, **6**, 287
Methides, germyl(plumbyl), reactions, with trimethylsilyl chloride, **6**, 203
Methides, polyboryl, from, methyleneboranes, **6**, 201
Methides, thionaphthoquinone, synthesis of, **3**, 359
Methides, thioquinone, synthesis of, **3**, 358
Methide salts, from, tins, **6**, 200
Methide salts, diboryl, alkylation of, **6**, 199
Methide salts, halodinitro, production of, **6**, 286
Methide salts, triboryl, synthesis of, **6**, 292
Methine, azo-, as starting material, for hydrazines, **2**, 374
Methine carbocations, diphenyl-, synthesis of, **1**, 902
Methines, *N*-aryldifluoroazo-, as starting materials, for carbamoyl fluorides, **6**, 438
Methines, azo-
reactions, with *N*-chloro primary amines, **2**, 377
reduction of, **2**, 311
as starting materials, for nitroso compounds, **2**, 390
Methiodides, from, methyl iodide, **2**, 760
DL-Methionine, synthesis of, **5**, 90
Methionines, phospho-, synthesis of, **4**, 466
Methoxide
reactions
with halomethylsilanes, **4**, 352
with tetrafluoroborate salts, **4**, 892
Methoxide anion, reactions, with phenylethynyl halides, **2**, 1028
Methoxylamine hydrochloride, use in the synthesis of *N*-alkylamides, **5**, 282
Methoxylation, of dioxymethanephosphonates, **6**, 314
Methyl, silylation of, **6**, 263
Methyl, alkyl trichloro-, use in the synthesis of aromatic ketones, **3**, 283
Methyl, tris(trifluoromethylthio)-, synthesis of, **6**, 734
Methyl, tris(trimethylsilyl)-, synthesis of, **6**, 734
Methylal
reactions, with polyols, **4**, 185
use in the synthesis of acetals, **4**, 185
Methylal, dibromo-, synthesis of, **6**, 252
Methylalumination
scope of, **2**, 975
zirconium-catalyzed
groups tolerance, **2**, 975
mechanism, **2**, 976
Methylamine
reactions
with arenediazonium salts, **4**, 444
with *N*,*N*-dimethyl-*N*,*N*-dinitrosooxalamide, **3**, 467
with nitrosyl chloride, **3**, 462
with thioimidic esters, **5**, 757
Methylamine, *N*-acetoxymethyl-*N*-nitro-, cleavage of, **4**, 106
Methylamine, bis(trimethylsilyl)-, as starting material, for methyl(trimethylsilyl)cyanamide, **5**, 1127
Methylamine, *N*,*N*-bis(trimethylsilyl)methoxy-, use in the synthesis of amines, **1**, 1041
Methylamine, cyano-, bisulfate salt, use in the synthesis of glycine, **5**, 85
Methylamine, *N*-iodobistrifluoro-, synthesis of, **2**, 339
Methylamine, *N*-methyl-*N*-nitrosochloro-, reactions, with thiols, **4**, 325
Methylamine, *N*-nitro-, reactions, with acid chlorides, **5**, 361
Methylamine, *N*-nitroso(1-methoxy-2-chloroethyl)-, from, *N*-nitroso(methyl)vinylamine, **4**, 896
Methylamine, trimethylsilyl-, synthesis of, **4**, 506, 512
Methylamine, *N*-triphenylphosphorylidene-1-(benzotriazol-1-yl)-
from, 1-azidomethylbenzotriazole, **1**, 262
use in the synthesis of amines, **1**, 1041
Methylamine hydrochloride
reactions
with formaldehyde, **4**, 317
with monothiooxamides, **5**, 591
Methylamines, *N*-amino-, reactions, with acid halides, **4**, 99
Methylamines, bis(difluoroamino)fluoroamino-, as intermediates, **5**, 817
Methylamines, bis(silyl)-, from, *N*,*N*-dimethylbenzamide, **4**, 511
Methylamines, chloro-, reactions, with tris(trimethylsilyl)phosphine, **2**, 433
Methylamines, *N*-dimethylamino , as starting materials, for 3-acetyl-*N*-chloromethylindoles, **4**, 97
Methylamines, germyl-, synthesis of, **4**, 519
Methylamines, halo-, reactions, with metallosilanes, **4**, 510
Methylamines, *N*-hydroxy-, use in the synthesis of aminomethylarsines, **4**, 501
Methylamines, tertiary, demethylation of, **2**, 304
Methyl anions, reactivity of, **1**, 847
Methyl anions, diphenyl-, stability of, **1**, 847
Methyl anions, α-silyl-, reactions, with transition metals, **4**, 655
Methylation, of arsines, **2**, 1079
α-Methylbenzyl groups, migration of, **1**, 398
Methyl bromide, cyclopropyl-, metallation of, **2**, 558
Methyl bromide, cyclopropylidene-, solvolysis of, **1**, 1114
Methyl carbanions, amino-, from, thioaminals, **4**, 528
Methyl carbocations, diphenyl-
stability of, **1**, 895
synthesis of, **1**, 894
Methyl carbocations, substituted tri(1-azulenyl)-, synthesis of, **1**, 899
Methyl carbocations, tri-(*p*-aminophenyl)-, structure of, **1**, 897
Methyl carbocations, tri(1-azulenyl)-, synthesis of, **1**, 898

Methyl carbocations, triphenyl-
stability of, **1**, 895, 898
synthesis of, **1**, 894, 897
Methyl cations, cyanodiaryl-, nuclear magnetic resonance analysis of, **1**, 904
Methyl cations, cyclopropyl-, ring-expansion of, **1**, 383
Methyl chloride, methoxy-, reactions, with potassium, **4**, 360
Methyl chloride, 2-pyridyl-, synthesis of, **1**, 861
Methyl chloride, trimethylsilyl-
reactions
with chromic chloride, **1**, 677
with magnesium, **4**, 653
Methyl chloride, β-trimethylsilylethoxy-, as formaldehyde equivalent, **3**, 163
Methyl chloride, triphenyl-
ionization of, **1**, 898
reactions
with inorganic halides, **1**, 898
with potassium cyanotriphenylborate, **5**, 1147
Methyl chlorides, trialkoxysilyl-, as starting materials, for Grignard reagents, **1**, 160
Methyl complexes
reactions
with carbon monoxide, **5**, 463
with trityl cations, **2**, 580
stability of, **2**, 581
as starting materials
for tantalum carbene complexes, **3**, 512
for triethylphosphine analogues, **5**, 463
Methyl complexes, halo-, from, formylmolybdenum complexes, **4**, 148
Methyl compounds
from
acylhydrazones, **1**, 42
semicarbazones, **1**, 42
reactions
with alkynes, **5**, 466
with mercuric salts, **6**, 207
Methyl compounds, active
reactions, with thionyl chloride, **5**, 484
as starting materials, for thioacyl chlorides, **5**, 484
Methyl compounds, cyano-, synthesis of, **4**, 509
Methyl compounds, hydroxy-, synthesis of, **4**, 1078
Methyl compounds, *N*-hydroxy-, as intermediates, in the synthesis of α-haloamides, **4**, 99
Methyl compounds, *N*-iodo-, synthesis of, **4**, 103
Methyl compounds, *N*-isocyanido-, synthesis of, **4**, 441
Methyl compounds, *N*-methoxy-, as starting materials, for *N*-chloromethyltriazacyclohexanes, **4**, 97
Methyl compounds, nitro-, as starting materials, for nitrile oxides, **3**, 684
Methyl compounds, tris(difluoroamino)-, as starting materials, for perfluoroamidines, **5**, 818
Methyl dianion, hydroxy-, synthon for, **4**, 383
Methylenating agents
organozirconium-based, synthesis of, **1**, 680
synthesis of, **1**, 680
Methylenating reagents, synthesis of, **4**, 669
Methylenation
of aldehydes, **1**, 769
of amides, **1**, 766
of carbonyls, **1**, 678, 767
of esters, **1**, 766
of β-keto esters, **1**, 767
of ketones, **1**, 767
metal-induced, **1**, 766
via Peterson reaction, **1**, 759
Methylene
from, hydrazones, **1**, 43
as starting material, for selenoformaldehyde, **3**, 382
Methylenebis(methyldimethylaminoarsine), as starting material, for tetraalkyldiarsine oxides, **2**, 493
Methylenebis(triphenylphosphonium) bromide, as starting material, for hexaphenyl carbodiphosphoranes, **5**, 1093
Methylene blue, for the oxidation of disulfides, **2**, 232
Methylene chloride
lithiation of, **6**, 267
reactions
with Lewis acids, **5**, 81
with trimethylsilyl chloride, **6**, 267
Methylenecyclopropanes, aryl-substituted, photolysis of, **1**, 1143
Methylene groups
nitrosation of, **5**, 838
reactions
with aryl cyanates, **4**, 889
with orthocarbamate esters, **4**, 894
Methylene groups, acidic, use in Ramberg–Bäcklund reactions, **1**, 706
Methylene groups, allylic, oxidation of, **3**, 215
Methylene groups, benzylic
oxidation of, **3**, 293
as starting materials, for ketones, **3**, 293
Methylene halides
as starting materials
for ethylene-bridged palladium complexes, **4**, 691
for methylene-bridged palladium complexes, **4**, 691
Methylenephosphonate, tetraethyl dimethylamino, use in the synthesis of alkanoic acids, **5**, 42
Methylenes
from
carbonyls, **1**, 767
diazocompounds, **1**, 44
reactions
with diazonium salts, **5**, 857
with dichlorodiphenylphosphorane, **3**, 499
with dithiocarbamates, **5**, 593
with haloarene transition metal complexes, **1**, 474
with isothiocyanates, **5**, 593
with polychlorophosphoranes, **6**, 698
with sulfinyl chloride, **4**, 266
with *C*-sulfonylthionoformamides, **5**, 593
with tertiary arsenic oxide, **5**, 920
with *N*,*N*,*N*',*N*'-tetramethylmethylthioformamidinium iodide, **4**, 973
with thiapyrylium iodide, **3**, 255
as starting materials
for formazans, **5**, 857
for β-keto acids, **5**, 70
synthesis of, **1**, 758
Methylenes, active, reactions, with thionyl chloride, **4**, 71
Methylenes, alkoxy-, synthesis of, **2**, 787
Methylenes, 1,1-bis(trimethylsilyl)-, synthesis of, **4**, 1049
Methylenes, difluoro-
from, chlorodifluoromethane, **6**, 37
synthesis of, **6**, 39
Methylenes, diiodo-, synthesis of, **4**, 750
Methylenes, α-halogenated, reactions, with thioamides, **2**, 749
Methylenes, hydroxy-, reactions, with hydroxylamines, **3**, 431
Methylenes, 2-nitro-, synthesis of, **2**, 788
exo-Methylenes, synthesis of, **4**, 827
Methylene transfer, for the synthesis of carbon-bridged dirhodium complexes, **4**, 686
Methyleniminium salts, alkoxy-, as starting materials, for imidic acid esters, **5**, 686
Methylenomycin B, synthesis of, **1**, 527
Methyl ester hydrochloride, acetimino, reactions, with phenyl-2-aminoethylarsine, **5**, 790
Methyl groups
displacement of, **4**, 641
metallation of, **6**, 392
migratory aptitudes of, **1**, 384
Methyl groups, allylic, oxidation of, **3**, 58

Methyl groups, aromatic, oxidation of, **4**, 206
Methyl groups, cyclopropyl-, oxidation of, for the synthesis of cyclopropyl ketones, **3**, 113
Methyl halide, amino-, reactions, with azides, **4**, 444
Methyl halides, cyclopropyl-, fragmentation of, **2**, 600
Methyl halides, dialkoxy-, from, vinyl orthoformates, **6**, 43
Methyl halides, germyl-
reactions
with amines, **4**, 519
with potassium cyanate, **4**, 521
Methyl halides, silyl-, reactions, with amides, **4**, 513
Methyl halides, tribromo-, synthesis of, **6**, 222
Methyl halides, trichloro-, synthesis of, **6**, 220
Methyl halides, trifluoro-, synthesis of, **6**, 217
Methyl halides, triiodo-, synthesis of, **6**, 223
Methyl halides, trimethylstannyl-, reactions, with magnesium, **4**, 709
Methyl halides, trisilyl-, metallation of, **6**, 393
N-Methyl heterocycles, metallation of, **4**, 539
Methylide, dimethyloxosulfonium, as starting material, for enethiolates, **4**, 834
Methylide, dimethylsulfonium
as reagent, in cyclopropanation reactions, **1**, 304
uses of, **3**, 372
Methylide, dimethylsulfonium dinitro-, nitration of, **6**, 283
Methylide, dimethylsulfoxonium
reactions
with acid compounds, **3**, 171
with β-substituted ethyl α-isocyanoacrylates, **3**, 704
with thiobenzoyl chloride, **3**, 364
as reagent
in cyclopropanation reactions, **1**, 304
for the synthesis of thiabenzene *S*-oxides, **3**, 380
stability of, **3**, 378
as starting material, for sulfoxonium ylides, **3**, 378
synthesis of, via deprotonation of a trimethylsulfoxonium salt, **3**, 378
Methylide, lithium tribromo-, synthesis of, **4**, 746
Methylide, phenoxyfluoroxonium (fluorosulfonyl), reactions, with aniline, **4**, 282
Methylide, triethoxyphosphonium bis(phenylsulfonyl), reactions, with bromine, **6**, 278
Methylides, synthesis of, **3**, 500
Methylides, ethoxycarbonyl
reactions
with aldehydes, **1**, 724
with benzaldehyde, **1**, 722
Methylides, halo-, from, phenyldihalomethylmercuric bromides, **3**, 497
Methylides, iodonium dinitro-, halogenation of, **6**, 261
Methylides, monochloro-, synthesis of, **3**, 495
Methylides, monofluoro-, synthesis of, **3**, 495
Methylides, phenyliodonium bis(arenesulfonyl), as starting materials, for arsonium ylides, **6**, 708
Methylides, sulfonium, as starting materials, for ketenes, **3**, 536
Methylides, sulfoxonium, synthesis of, **2**, 363
S-Methylides, thiocarbonyl, synthesis of, **3**, 367
Methylidynes, germyl-substituted, synthesis of, **6**, 404
Methylimines, reduction of, **1**, 46
Methyl iodide
for chemical activation of metal surfaces, **2**, 958
hydrolysis of, **1**, 891
reactions
with anionic iridium complexes, **5**, 460
with π-arene complexes, **2**, 582
with 3-(α-dialkylaminoarylidene)-1-alkylthiotriazenes, **3**, 419
with 4,4-dimethyl-1-piperidino-pentan-3-one, **1**, 254
with disulfides, **5**, 830
with α-hydrazino enamines, **4**, 986
with β-hydroxyphosphinothioic amides, **1**, 643
with lithium (ethyl diazoacetate), **3**, 474
with organoborate salts, **2**, 524
with perfluorohydrazones, **6**, 260
with phosphines, **2**, 430, 1063
with silver alkynoates, **1**, 1137
with tertiary alkylstibines, **2**, 487
with thiosemicarbazides, **4**, 984
with trimethyl phosphite, **4**, 485
with triphenyl phosphite, **2**, 8
with tris(diethylamino)sulfonium-trimethyldifluorosilicate, **1**, 282
with tris(trifluoromethyl) triselenoorthoformate, **6**, 94
with tris(trimethylsilyl)methyllithium, **6**, 174
as reagents, in quaternization reactions, **2**, 1079
as starting material, for methiodides, **2**, 760
Methyl iodide, bis(trifluoroacetoxy)trifluoro-, synthesis of, **5**, 205
Methyl iodide, tributylstannyl-
displacement of iodide from, **4**, 387
as starting material, for α-alkoxystannanes, **4**, 387
Methyl iodide, trifluoro-, reactions, with tris(dimethylamino)arsine, **2**, 504
Methylium cations, tris(alkoxy)-, synthesis of, **6**, 729
Methylium ions, synthesis of, **6**, 726
Methylium salts, tris(alkoxy)-, synthesis of, **6**, 729
Methyl ligands, bis(trimethylsilyl)-, lithiation of, **4**, 656
Methyl ligands, silyl-, as starting materials, for transition metal complexes, **4**, 656
Methyllithium–lithium bromide, reactions, with ketals, **4**, 961
Methyl radicals, from, *O*-acetyl thiohydroxamic esters, **1**, 943
1-Methylthioethenamine, *N*,*N*-dimethyl-, use of term, **4**, 898
Methylthiomethyl (MTM) group, for the protection of carboxylic acids, **4**, 219
[[2-(Methylthio)phenyl]thio]methyl (MPTM) protecting group, development of, **4**, 219
Methylylide, triphenylstibonium, reactions, with benzophenone, **1**, 743
Methynolide, synthesis of, **1**, 821
Mevalonolactone, synthesis of, **4**, 167
Mevinolin
reduction of, **1**, 79
synthesis of, **1**, 688
Meyer reaction
for the synthesis of alkylarsinic acids, **2**, 497
for the synthesis of alkylarsonic acids, **2**, 497
Meyers' oxazoline chemistry, for the synthesis of γ-lactones, **5**, 166
Micellar catalysis, cationic, use in Williamson reactions, **2**, 89
Michael acceptors
carbene as, **5**, 958
from, 1,1,1-trifluoroacetone, **4**, 766
nitroalkenes as, **2**, 393
reactions, with carbene complexes, **5**, 958
for the synthesis of nitriles, **3**, 621
use in the synthesis of thionoamides, **5**, 606
Michael addition
of allylsilanes, **3**, 624
of arylphosphines, to vinyl cyanide, **2**, 829
of enamines, **2**, 759
of hydrazines, **2**, 380, 381, 383, 384, 393
intramolecular, **1**, 307
of nitroalkanes to enals, **3**, 47
of nitroamines, **2**, 386
of organometallics to nitroalkenes, **2**, 784
of phosphites, to perfluoromethacrylate esters, **4**, 803
for the synthesis of aldimines, **3**, 406
for the synthesis of alkenes, **1**, 537
for the synthesis of amines, **2**, 315
for the synthesis of carbanions, **1**, 888
for the synthesis of carboxylic esters, **5**, 145
for the synthesis of nitro ketones, **3**, 192

for the synthesis of pyrrolidines, **2**, 324
for the synthesis of (-)-tylophorine, **2**, 328
to bis(sulfonyl)ethene, **4**, 277
to nitroalkenes, **4**, 431
of trimethylsilylmethanes, **6**, 128
Michael donors
carbene as, **5**, 958
nitroalkanes as, **2**, 394
reactions, with bisdiphenylphosphinoethene, **4**, 1028
Michaelis–Arbuzov reaction
for *N*-alkylimino compounds, **6**, 657
between dialkyl halophosphite and trichloromethyl isocyanate, **6**, 285
between triethyl phosphite and trichloromethyl isocyanate, **6**, 285
between triethyl phosphite and trichloromethyl phenyl ether, **6**, 281
of phosphonic acid esters, with alkyl halides, **2**, 842
for the synthesis of alkylphosphine oxides, **2**, 452
for the synthesis of alkylphosphine sulfides, **2**, 462
for the synthesis of amidinium salts, **6**, 661
for the synthesis of amidinophosphonates, **6**, 657
for the synthesis of bis(phosphinyl)iminocarbonyls, **6**, 666
for the synthesis of bis(phosphoryl)imines, **6**, 367
for the synthesis of dialkyl alkyl phosphonates, **4**, 920
for the synthesis of isocyanates, **5**, 1008
for the synthesis of phenylsulfonyliminos, **6**, 666
for the synthesis of phosphonates, **2**, 456
use in the synthesis of acylphosphorus compounds, **5**, 402
See also Arbuzov reaction
Michaelis–Becker reaction
of dialkyl phosphonate ions, with aryl halides, **2**, 848
for the synthesis of α-aminoalkylphosphorus compounds, **4**, 464
for the synthesis of α-aminophosphorus compounds, **4**, 487
for the synthesis of tetraesters, **6**, 264
Michael reaction
for the addition of carbon nucleophiles to carbon–carbon bonds, **1**, 300
of α-aminosilanes, **4**, 510
for the displacement of sulfur, **1**, 240
historical background, **1**, 307
mechanism, **1**, 888
retrograde, **1**, 884
reviews, **1**, 306
for the synthesis of alkanoic acids, **5**, 46
for the synthesis of *N*-alkylhydroxylamines, **2**, 343
use of chiral imines in, **3**, 414
Microbial agents, for the synthesis of carboxylic acids, **5**, 117
Microbial hydrolysis, for the synthesis of carboxylic acids, **5**, 117
Microbial oxidation
of hydrocarbons, **2**, 38
for the synthesis of carboxylic acids, **5**, 117
Microemulsions, use in the synthesis of alkyl bromides, **2**, 31
Microwave irradiation, for the synthesis of dithioacetals, **4**, 257
Migration
of double bonds, **2**, 4
oxygen to nitrogen, of chlorocarbonyl group, **6**, 446
Migration, anion-accelerated, of hydrogen, **1**, 799
Migration, base-catalyzed, of hydrogen, **1**, 795
Migration, Lewis acid-catalyzed, of hydrogen, **1**, 795
Migration, metal-catalyzed, of hydrogen, **1**, 797
Migration, sigmatropic, of hydrogen, **1**, 794
[1,2]-Migration, for the synthesis of α-amino aldehydes, **3**, 43
H-Migration, homo [1,5] sigmatropic, in the synthesis of α,β-unsaturated ketones, **3**, 232
[1,7] Migrations, of carbons, **1**, 800
[1,*j*] Migrations
of double bonds, **1**, 794
of hydrogen, **1**, 794
Migratory aptitude
in dienone-phenol rearrangement, **2**, 660
in pinacol rearrangement, **1**, 384
in the synthesis of carboxylic esters, **5**, 142
Milbemycin-β_3
synthesis of, **6**, 472
via hetero-Claisen rearrangement, **6**, 473
Milbemycin B, synthesis of, **1**, 602, 754
Mineral acids
reactions
with amines, **2**, 759
with arylsilanes, **1**, 448
use in the cleavage of carbon–silicon bonds, **1**, 61
Mitomycins, synthesis of, **1**, 648
Mitsunobu conditions
for the displacement of alcohols, **1**, 179, 200
in elimination reactions, **3**, 561
for the monobenzoylation of diols, **5**, 138
for the synthesis of β-lactams, **5**, 273
Mitsunobu coupling, of hydroxy bis(sulfones), **1**, 183
Mitsunobu reaction
mechanism, **2**, 87
for the synthesis of alkoxyimides, **5**, 333
for the synthesis of amines, **2**, 299, 309
for the synthesis of α-aminoalkylphosphorus compounds, **4**, 475, 491
for the synthesis of aryl ethers, **2**, 683
for the synthesis of carbodiimides, **5**, 1078
for the synthesis of carboxylic esters, **5**, 123, 131
for the synthesis of guanidines, **6**, 487
for the synthesis of imidic acid esters, **5**, 689
for the synthesis of thioglycosides, **4**, 225
for the synthesis of thioimidic esters, **5**, 726
for the synthesis of thiolesters, **2**, 118
for the synthesis of thiols, **2**, 124
use in the synthesis of alkyl halides, **2**, 6
use in the synthesis of nitriles, **3**, 613
Mitsunobu reagents
for the synthesis of β-lactones, **5**, 161
for the synthesis of macrolactones, **5**, 177
Mixed cuprates
development of, **1**, 312
reactions, with cyclopentenone, **1**, 1093
MK 801, synthesis of, **2**, 328
MMI (antithyroid drug), selenium analogue, **6**, 598
Modhephene, synthesis of, **3**, 223, 245
Modified neglect of diatomic overlap (MNDO)
Hamiltonian calculations, in oxazaborolidine ketone reduction, **2**, 66
studies
of oxiranes, **1**, 225
of silicon functions, **5**, 1019
Moenocinol, C-11 methylene unit of, **1**, 654
Molecular orbital (MO) calculations, in the opening of epoxides, **1**, 217
Molecular sieves, use in the synthesis of carboxylic esters, **5**, 122
Moleozonides, synthesis of, **6**, 251
Molozonides
synthesis of, **2**, 98
use in the synthesis of carbonyl oxides, **3**, 326
Molybdate, benzyltrimethylammonium tetrabromooxo-, as catalyst, for the oxidation of alcohols, **3**, 117
(Molybdenacyclononatriene) complexes, 8-carbon-bridged, synthesis of, **4**, 676
Molybdenacyclopentadienes, synthesis of, **4**, 676
Molybdenum
for the allylation of seleno phenoxide, **1**, 246
as catalyst
in allyl acetate reactions, **1**, 196
for the substitution of sulfur functions, **1**, 238
dimeric allyls of, synthesis of, **4**, 673
reactions, with allyl cations, **1**, 191

Molybdenum, dimesityl dioxo-, as starting material, for ylides, **5**, 918
Molybdenum, hexacarbonyl-, reactions, with tetramethylurea, **2**, 125
Molybdenum(III) compounds, alkyl-, synthesis of, **2**, 572
Molybdenum(V) chloride, use in the synthesis of dimeric allyls of molybdenum, **4**, 673
Molybdenum(VI) complexes, dioxy-, use in the synthesis of amines, **2**, 307
Molybdenum acyl compounds, cyclopentadienyl-, from, acetyl chloride, **5**, 445
Molybdenum-alkylidyne complexes, synthesis of, **3**, 691
Molybdenum alkyne complexes, reactions, with pentacarbonylrhenate, **4**, 710
Molybdenum carbene complex, difluoro-, detection of, **6**, 717
Molybdenum carbene complexes
  reactions
    with alkenes, **3**, 514
    with phosphines, **4**, 580
  synthesis of, **3**, 513
Molybdenum carbenes, use in the synthesis of phenols, **2**, 670
Molybdenum carbonyl complexes
  as catalysts, for haloalkane addition, **6**, 7
  synthesis of, **6**, 522
Molybdenum carbyne complexes
  reactions
    with azidoformates, **5**, 955
    with *t*-butylethyne, **3**, 513
Molybdenum catalysis, in epoxidation reactions, **2**, 91
Molybdenum chlorides, reactions, with 1,1-dicyanobis(trifluoromethyl)ethene, **4**, 800
Molybdenum complexes
  reactions
    with alkynic alcohols, **5**, 944
    with *t*-butyl isocyanate, **3**, 514
  as starting materials, for 1,3-disilacyclopentenes, **4**, 616
  use in the synthesis of α-hydroxy ketones, **3**, 156
Molybdenum complexes, alkyne-bridged, synthesis of, **4**, 674
Molybdenum complexes, η3-allyl-, reactions, with allylate aldehydes, **2**, 76
Molybdenum complexes, cyclopenytadienyl-, alkylation of, **2**, 572
Molybdenum complexes, flyover 6-carbon-bridged, synthesis of, **4**, 676
Molybdenum complexes, formyl-, as starting materials, for halomethyl complexes, **4**, 148
Molybdenum complexes, (halomethyl)-, synthesis of, **4**, 147
Molybdenum complexes, peroxy-, use in the synthesis of dialkyl sulfones, **2**, 155
Molybdenum complexes, phosphine-bridged, synthesis of, **4**, 675
Molybdenum cyclopentadienyl complexes, synthesis of, **5**, 446
Molybdenum dichloride, di-*t*-butylimido-, reactions, with methylene(triphenyl)phosphorane, **5**, 918
Molybdenum difluorocarbene complexes, detection of, **6**, 717
Molybdenum functions, doubly-bonded, synthesis of, **3**, 513
Molybdenum halides, use in the synthesis of carbon-bridged dimolybdenum complexes, **4**, 672
Molybdenum hexacarbonyl
  for the displacement of phenylthio, **1**, 238
  reactions
    with alkyllithiums, **5**, 445
    with aryllithiums, **5**, 445
    with formamidines, **5**, 781
    with isoxazolidines, **2**, 320
    with phosphaferrocenes, **4**, 1026
    with 1,1,4,4-tetrakisdiphenylphosphino-1,3-butadiene, **4**, 1023
  for the reduction of alkyl thiols, **1**, 22
  for the reductive coupling of sulfonyl chlorides, **2**, 221
  synthesis of, **6**, 522
  use in the synthesis of disulfides, **2**, 228
Molybdenum hexafluoride, use in the synthesis of difluoroalkanes, **4**, 10
Molybdenum pentoxide–pyridine complexes, use in the hydroxylation of enolate anions, **2**, 44
Molybdenum pentoxide–pyridine–hexamethylphosphoramide complex, use in the synthesis of α-hydroxy sulfones, **4**, 234
Molybdenum peroxide–pyridine–hexamethylphosphoramide (MoOPH), for the oxidation of aryl organometallics, **2**, 652
Molybdenum–phosphinoalkene complexes, synthesis of, **4**, 1028
Molybdenums, reactions, with phosphorus ylides, **4**, 580
Molybdenums, acyl-, synthesis of, **5**, 445
Molybdenums, alkyl-, synthesis of, **2**, 570
Molybdenums, bromomethyl-, synthesis of, **4**, 148
Molybdenums, chloromethyl-, synthesis of, **4**, 148
Molybdenums, imidoyl-, synthesis of, **5**, 798
Molybdenum salts, as catalysts, for the oxidation of alcohols, **3**, 117
Molybdenum sulfides, as catalysts, for the synthesis of alkanethiols, **2**, 130
Molybdenum tetrachlorides, use in the synthesis of tris(cyclooctatetraene)dimolybdenum complexes, **4**, 673
Molybdenum trichloride, methyl-, synthesis of, **2**, 572
Monel reactors, for the synthesis of bisfluoroethyl monotelluride, **6**, 49
Monoacylation, of polyols, **5**, 138
Monoacyl complexes, synthesis of, **5**, 443
Monoalkali phosphides, reactions, with alkyl halides, **2**, 429
Monoalkenes, classification of, **2**, 952
Monoalkylation, for the synthesis of amines, **2**, 300
Monoalkyne complexes, as starting materials, for ferracyclopentadienes, **4**, 682
Monoalkynes, reactions, with bromoalkynes, **1**, 1142
Monoanils, benzil, reactions, with 2,4-bis(phenylthio)-1,3,2,4-dithiadiphosphetane-2,4-dithione, **4**, 320
Monoanions, monoalkylation of, **1**, 155
Monoarsaferrocenes, synthesis of, **4**, 1039
1-Monoazadienes, hydrosilylation of, **2**, 813
Monobactam antibiotics, *N*-acyl sulfamate, synthesis of, **2**, 366
Monobenzoates, synthesis of, **5**, 138
Monobenzoylation, of diols, **5**, 138
Monobismaferrocenes, synthesis of, **4**, 1041
Monocarbometallation, of alkynes, **2**, 978
syn-Monocarbometallation, control reactions, **2**, 971
Monocarboxylic acids, reactions, with metallic hydroxides, **1**, 883
Mono(chloromercurio) compounds, synthesis of, **4**, 1067
Monoenes
  synthesis of
    via retro-cycloaddition reactions, **1**, 784
    via retro-ene reactions, **1**, 772
Monoepoxide, 1,3-butadiene, reactions, with trialkylboranes, **1**, 277
Monoesters, as starting materials, for ketenes, **3**, 536
Monoesters, acylphosphonate, synthesis of, **5**, 404
Monoimines, synthesis of, **3**, 417
Monoimines, *N*-thionoacyldiphosphane-, synthesis of, **5**, 635
Monoisocyanates, phosphorylated, synthesis of, **5**, 1009
Monoketones, α-diazo, as starting materials, for enoldiazonium ions, **2**, 799
Monolithium alkynide, reactions, with carbonyls, **1**, 1037
Monometal chalcogens, synthesis of, **2**, 891
Monometalloalkenyls, configurational stability of, **2**, 909
Monomethylation, for the synthesis of aromatic amines, **1**, 29

cis,cis-Monomethylmuconate, synthesis of, **5**, 54
Monoorganocoppers
  reactions, with boron trifluoride etherate, **1**, 312
  synthesis of, **1**, 311
Monoorganometal(III) oxides, synthesis of, **2**, 885
(*E*)-Monooxime, benzil, reactions, with *N*-chlorosuccinimide, **3**, 433
Monooximes, quinone, synthesis of, **3**, 427
Monoperoxides, reactivity of, **2**, 687
Monoperphthalic acid, reactions, with aryl methoxymethyl sulfides, **4**, 235
Monophosphacarbodiimides, synthesis of, **5**, 1086
Monophosphazenes, reactions, with tris(dimethylamino)phosphonium chloride, **2**, 860
Monophosphines
  reactions, with organodihalides, **2**, 444
  synthesis of, **2**, 827
Monophosphinic acids, synthesis of, **4**, 470
Monophosphonium salts, synthesis of, **4**, 555
Monophosphoric acid, peroxy-
  in Baeyer–Villiger oxidation, **2**, 6367
  for the oxidation of arenes, **2**, 650
Monopotassium phosphide, reactions, with allyl (2-chloroethyl)amine, **2**, 427
Monosaccharides, as sources of chiral auxiliaries, **1**, 134
Monosaccharides, amino-, difluorination of, **2**, 335
Monoselenide, bis(pentafluoroethyl), synthesis of, **6**, 49
Monoselenide, bis(perfluoroethyl), reactions, with iodine monochloride, **6**, 49
Monoselenoacetals
  from
    acetals, **4**, 238
    carbonyls, **4**, 238
    enol ethers, **4**, 238
    α-halo ethers, **4**, 239
Monoselenoacetals, ketene, synthesis of, **4**, 841
Monoselenocarbamates, synthesis of, **6**, 593
Monoselenodithioortho esters
  from
    dithioacetals, **6**, 102
    α-halodithianes, **6**, 101
    selenodithiocarbenium salts, **6**, 101
Monoseleno monotelluro compounds, syntheses of, **4**, 876
Monoseleno monothioortho esters, from, monothioacetals, **6**, 100
Monoselenoortho esters, from, 2-alkoxyselenophenes, **6**, 100
Monostibaferrocenes, synthesis of, **4**, 1041
Monosulfides, thiuram, as thiocarbamoylating reagents, **5**, 595
Monosulfones, synthesis of, **4**, 854
Monosulfoxides, synthesis of, **2**, 232
Monosulfoxides, 2-aminoketene dithioacetal
  from, nitriles, **4**, 857
  as starting materials, for α-amino acids, **4**, 857
Monotelluride, bisfluoroethyl, synthesis of, **6**, 49
Monotellurides, reactions, with tellurium, **6**, 356
Monotelluroacetals
  from
    acetals, **4**, 238
    carbonyls, **4**, 238
    enol ethers, **4**, 238
    α-halo ethers, **4**, 239
Monotelluroacetals, ketene, synthesis of, **4**, 841
Monoterpenes, synthesis of, **1**, 683
Monothalliums, stability of, **2**, 597
Monothiatriselenafulvalenes, from, 1,3-diselenole 2-thiones, **4**, 869
Monothioacetal *S,S*-dioxides, ketene
  from
    1-sulfonylalkynes, **4**, 840
    α-sulfonyl ethers, **4**, 838
  synthesis of, via sulfoxidation, **4**, 838
  See also Alkoxyvinyl sulfones
Monothioacetals
  from
    acetals, **4**, 215
    O,O-acetals, **4**, 216
    aldehydes, **4**, 217
    amides, **6**, 111
    carbonyls, **4**, 215
    α-halo ethers, **4**, 218
    α-halo sulfides, **4**, 218
    sulfoxides, **4**, 221
  oxidation of, **4**, 232
  as starting materials
    for α-alkoxy sulfones, **4**, 235
    for dithioortho esters, **6**, 98
    for monoselenomonothioortho esters, **6**, 100
    for monothioortho esters, **6**, 97
    for thioimidates, **6**, 111
  synthesis of, **4**, 215; **6**, 111
    via oxidation of sulfides, **4**, 220
Monothioacetals, acyclic, synthesis of, **4**, 215
Monothioacetals, cyclic
  from, carbonyls, **4**, 229
  synthesis of, **4**, 229
    via cycloaddition reactions, **4**, 231
    via intramolecular alkylation, **4**, 231
Monothioacetals, ketene
  from
    aldehydes, **4**, 834
    ketene dithioacetals, **4**, 836
    ketones, **4**, 834
    monothiocarboxylic acids, **4**, 833
    α,β-unsaturated *O,S*-acetals, **4**, 836
  synthesis of
    via Horner–Wittig reactions, **4**, 835
    via Peterson alkenation, **4**, 834
Monothioacetals, ketene *O*-silyl, synthesis of, **4**, 833
Monothioacetals, silyloxy, as starting materials, for dialkyl sulfides, **2**, 142
Monothioanthraquinone, synthesis of, **3**, 358
Monothiobenzils, synthesis of, **3**, 365
Monothiobenzoate esters, as starting materials, for monothioorthobenzoate esters, **6**, 96
Monothio-*p*-benzoquinone, synthesis of, via pyrolysis, **3**, 358
Monothio-*o*-benzoquinones, trapping as Diels–Alder adducts, **3**, 358
Monothiocarbonates, oxidation of, **6**, 479
Monothiocarboxylic acids
  as starting materials, for ketene monothioacetals, **4**, 833
  synthesis of, reviews, **5**, 81
Monothiocarboxylic esters, as starting materials, for monothioortho esters, **6**, 96
Monothiodiketone, synthesis of, **3**, 343
Monothiodione, synthesis of, **3**, 345
Monothioesters, hydroxy, as starting materials, for monothioortho esters, **6**, 96
Monothioglyoxal, synthesis of, via cycloreversion, **3**, 335
Monothionoamides, synthesis of, **5**, 580
Monothionoanhydride, 1,8-naphthalic, synthesis of, **5**, 529
Monothionocarbonates, synthesis of, **6**, 541
Monothionocarbonates, diol, synthesis of, **6**, 542
Monothionocarbonates, propargylic diol, use in the synthesis of allenes, **1**, 993
Monothionoimides, photolysis of, **5**, 609
Monothioorthoacetates, synthesis of, **6**, 95
Monothioorthobenzoate esters, from, monothiobenzoate esters, **6**, 96
Monothioorthobenzoates, synthesis of, **6**, 95
Monothioortho esters
  from
    O-acyl orthoformate ester, **6**, 95
    2-alkoxythiophenes, **6**, 97

α,α-dihaloethenes, **6**, 95
dioxocarbenium salts, **6**, 96
hydroxy monothioesters, **6**, 96
monothioacetals, **6**, 97
monothiocarboxylic esters, **6**, 96
ortho esters, **6**, 95
oxothiocarbenium salts, **6**, 96
Monothioorthoformates
from, *N*-*p*-toluenesulfonylsulfilimines, **6**, 97
synthesis of, **6**, 95
Monothioorthoformates, triphenyl, from, chlorodiphenoxymethane, **6**, 95
Monothiooxamides
reactions, with methylamine hydrochloride, **5**, 591
synthesis of, **5**, 566
Monothiophosphoric acids, *O*,*O*-dialkyl-, reactions, with nitriles, **5**, 581
Monothiophosphorus(V) acids
reactions
with imidoyl chlorides, **5**, 644
with nitriles, **5**, 643
Monotin chlorides, synthesis of, **2**, 599
Monotins, reactions, with ethylene glycol, **6**, 200
Monoxime, γ-benzil, reactions, with benzenesulfonyl chloride, **3**, 722
Montmorillonite
as catalyst, for the synthesis of dithioacetals, **4**, 246
as dehydrating agent, in the synthesis of cyclic anhydrides, **5**, 199
Montreal Protocol, and the replacement of chlorofluorocarbons, **6**, 8
Mordenite, use in the synthesis of trichloromethyls, **6**, 29
Morpholides, dichlorothioacetyl-, synthesis of, **5**, 582
Morpholides, trichlorothioacetyl-, synthesis of, **5**, 582
Morpholine
reactions
with acetophenones, **5**, 599
with acetyl α-chloro disulfides, **5**, 590
with *S*-acetyl 1,1-dichloroalkyl disulfides, **5**, 483
with benzaldehydes, **4**, 406
with 2,2-dichloro-1,3-benzodioxole, **6**, 321
with 4-fluorobenzaldehyde, **4**, 298
with formaldehyde, **4**, 315
with glyoxal, **4**, 300
with trichlorothioacetyl chloride, **5**, 582
with trithione, **5**, 559
use in dehydrobromination reactions, **1**, 574
Morpholine, 4-methyl-, fluorination of, **6**, 37
Morpholine, perfluoro-, pyrolysis of, **6**, 257
Morpholine, *N*-styryl-, reactions, with salicyl aldehyde, **4**, 307
Morpholine, (*E*)-*N*-(3-tosyl-2-propenyl)-, reactions, with *s*-butyllithium, **1**, 1090
Morpholine-*N*-oxide, *N*-methyl- (NMO)
as co-oxidant, with bis(triphenylphosphine)ruthenium(II) chloride, **3**, 5
for the oxidation of benzyl halides, **3**, 89
for the oxidation of sulfides, **2**, 155
as reoxidant, **4**, 357
Morpholinium nitrate, use in the synthesis of *N*-nitroamides, **5**, 361
Morpholinones, reactions, with *t*-butoxy chloride, **4**, 102
Morpholinones, benzyloxycarbonyl-, bromination of, **4**, 102
Mortierella isabellina, for the deprotection of arylamines, **2**, 304
Muconates, 3-fluoro-, cyclization of, **4**, 53
Muconic acid, from, diethyl α,δ-dibromoadipate, **5**, 103
Mukaiyama–Michael reaction, mechanism, **1**, 308
Mukaiyama reaction
mechanism, **3**, 162
for the synthesis of β-sulfenyl ketones, **3**, 168
for the synthesis of α,β-unsaturated ketones, **3**, 220
Multiphoton irradiation, for the synthesis of carbenes, **4**, 1083
Multiple substituents, use in the synthesis of carbanions, **1**, 852
Muscarine analogues, synthesis of, **4**, 232
Muscone, synthesis of, **1**, 1082
Muscopyridine, synthesis of, **3**, 437
Myrtanyl compounds, dithianyl, reactions, with borane, **2**, 42

N

NAD, use in the synthesis of L-α-hydroxy acids, **5**, 117
Nafion-H
as catalyst, for the benzylation of benzenes, **1**, 201
use in the synthesis of benzophenones, **3**, 279
use in the synthesis of diacetates, **4**, 203
use in the synthesis of α-diols, **4**, 161
Nalbuphine, synthesis of, **2**, 303
Nametkin rearrangement, **1**, 380
mechanism, **1**, 383
2-Naphthaldehyde, 1,4,8-trimethoxy-, synthesis of, **3**, 101
Naphthaldehydes
as starting materials, for polycyclic aryl aldehydes, **3**, 99
synthesis of, **3**, 100
Naphthalene, 1-acetyl-, reduction of, **1**, 96
Naphthalene, 1,8-bis(dimethylamino)-, use in the synthesis of isocyanates, **5**, 993
Naphthalene, 1,8-bis(trimethylsilymethyl)-, reactions, with butyllithium, **4**, 603
Naphthalene, 2-bromomethyl-, reactions, with triphenylarsine, **1**, 268
Naphthalene, 2-bromomethylheptafluoro-, as starting material, for heptafluoro-2-naphthoic acid, **5**, 116
Naphthalene, chloro-, reduction of, **1**, 427
Naphthalene, α-cyano-, hydrolysis of, **5**, 108
Naphthalene, α-diazo-, oxidation of, **2**, 804
Naphthalene, dicyano-, use in the synthesis of carbonyl ylides, **3**, 317
Naphthalene, 1,8-dicyano-, synthesis of, **3**, 668
Naphthalene, dihydro-, synthesis of, **3**, 322
Naphthalene, 1,2-dihydro-, photochemical ring-opening of, **1**, 420
Naphthalene, 1,4-dihydro-, synthesis of, **1**, 98
Naphthalene, 2,3-dimethyl-, as starting material, for 2,3-naphthalenedicarboxylic acid, **5**, 115
Naphthalene, fluoro-, reduction of, **1**, 426
Naphthalene, α-fluoro-, reduction of, **1**, 426
Naphthalene, 1-methoxy-, reduction of, **1**, 98
Naphthalene, 2-methoxy-, reduction of, **1**, 98, 103
Naphthalene, 1-methyl-, reduction of, **1**, 96
Naphthalene, 2-methyl-, reduction of, **1**, 96
Naphthalene, 1-(nitromethyl)-3,4-dihydro-, reactions, with sodio-2-carboethoxycyclopentanone, **1**, 250
Naphthalenes
acylation of, **3**, 300, 300
chlorination of, **6**, 215
electrochemical reduction of, **1**, 103
formylation of, **3**, 101
nitration of, **2**, 791
reactions, with potassium permanganate, **5**, 74
reduction of, **1**, 94
as starting materials, for carboxylic acids, **5**, 116
synthesis of, **1**, 523; **2**, 708
Naphthalenes, alkyl-, acylation of, **3**, 300
Naphthalenes, dicyano-, synthesis of, **3**, 667
Naphthalenes, dihydro-, synthesis of, **1**, 96, 98, 103
Naphthalenes, dimethyl-, dimetallation of, **5**, 43
Naphthalenes, 1,4-epoxytetrahydro-, flash vapor pyrolysis of, **1**, 789
Naphthalenes, methoxy-
reactions
with dimethylformamide, **3**, 101
with pyrophosphoryl chloride, **3**, 101
Naphthalenes, tetrahydro-, synthesis of, **1**, 98
1-Naphthalenesulfonic acid, 5-hydroxy-, reactions, with tris(2-aminoethyl)amine, **2**, 309
1-Naphthalene sulfonic acid, 5-tris(2-aminoethyl)amino-, synthesis of, **2**, 309
2-Naphthalenesulfonyl azide, as starting material, for (diazomethylene)bisphosphonates, **6**, 666
2-Naphthalenetellurenyl iodide, synthesis of, **2**, 732
2-Naphthalenethiol, reduction of, **1**, 436
Naphthalenethiols, synthesis of, **2**, 711
1,8-Naphthaloyl chloride
reactions, with hydrogen sulfide, **5**, 529
thionation of, **5**, 528
Naphthamide, 2-*N*,*N*-dimethyl-, reduction of, **3**, 100
Naphthamidines, synthesis of, **5**, 758
1-Naphthamidines, synthesis of, **5**, 757, 758
1,4-Naphthaquinones, 2-alkyl-, from, 1,4-naphthaquinones, **1**, 275
1,4-Naphthaquinones, 2,3-dialkyl-, synthesis of, **1**, 275
Naphthenic acid, synthesis of, reviews, **5**, 24
2-Naphthoic acid, heptafluoro-, from, 2-bromomethylheptafluoronaphthalene, **5**, 116
α-Naphthoic acid, synthesis of, **5**, 112
1-Naphthol, reduction of, **1**, 89
Naphthols
from
alkynyl acetophenones, **2**, 672
benzamides, **2**, 673
reactions
with hydrazine, **2**, 769
with thiophosgene, **6**, 533
as starting materials, for chlorothioformates, **6**, 533
synthesis of, **1**, 551
Naphthols, tetrahydro-, synthesis of, **2**, 662
1-Naphthonitrile
reactions, with arylamines, **5**, 756
as starting material, for arylamidines, **5**, 755
1-Naphthonitrile, 2-methyl-, reactions, with amines, **5**, 756
1-Naphthonitriles, as starting materials, for amidines, **5**, 758
Naphthoquinone diols, rearrangement of, **2**, 660
Naphthoquinones, synthesis of, **4**, 901
2-Naphthoyl chloride, reduction of, **3**, 100
1-Naphthylamine, reactions, with dimethylthioformamide, **5**, 745
1-Naphthylamine, *N*-ethyl-*N*-methyl-, oxidative azidation of, **4**, 443
Naphthyl ligands, isoquinolinyl-, in asymmetric hydroboration, **2**, 43
1,8-Naphthyridine, reactions, with cyanodimethylgold, **5**, 1141
Naproxen, synthesis of, **1**, 77; **2**, 29; **3**, 626
(*S*)-Naproxen, synthesis of, **1**, 389
1-Napthol, 2-fluoro-, synthesis of, **2**, 613
Natural products
synthesis of, **1**, 558, 702, 1132
limitations, **1**, 1008
via alkylation reactions, **1**, 1134
via imino esters, **5**, 692
via Ireland–Claisen rearrangement, **1**, 831
via macrolactones, **5**, 173
via rearrangement of vinylcyclopropanes, **1**, 804
via Wadsworth–Emmons reaction, **1**, 747
Nazarov cyclization
of cyclopentenylvinyl ketones, **3**, 211
mechanism, **1**, 421
for the synthesis of α,β-unsaturated ketones, **3**, 224, 253
Nazarov reaction, for the synthesis of α,β-unsaturated ketones, **3**, 216

Neber rearrangement, for the synthesis of α-amino ketones, **3**, 184, 435
Nef reaction
 of nitroalkanes, **4**, 540
 in the synthesis of aldehydes, **3**, 18
 for the synthesis of ketones, **3**, 134
 for the synthesis of α,β-unsaturated aldehydes, **3**, 57
 use in the synthesis of γ,δ-unsaturated ketones, **3**, 143
Neocarzinostatin chromophores, synthesis of, **1**, 818, 1015, 1029
Neodymium compounds, reactions, with phosphorus ylides, **4**, 578
Neoflavonoids, synthesis of, **1**, 435
Neopentane, cleavage of, **1**, 901
Neopentyl chloride, α-phenyl, reactions, with diphenylmethyllithium, **1**, 110
Neopentyl chlorides, reduction of, **1**, 7
Neopentyl halides, from, tosylates, **2**, 9
Nephromopsinic acid, synthesis of, **5**, 166
Nerol, acylation of, **5**, 135
Neuraminic acid, synthesis of, **1**, 751
Newman–Kwart rearrangement
 for the synthesis of arenethiols, **2**, 707
 for the synthesis of *S*-aryl thiocarbamates, **6**, 495
Nicholas reaction
 in the displacement of propargylic alcohols, **1**, 198
 for the synthesis of tertiary amines, **2**, 301
Nickel
 as catalyst, for the substitution of sulfur functions, **1**, 238
 catalyzed substitution, with Grignard reagents, **1**, 970
 for the coupling of cyclobutenones with alkynes, **2**, 672
 reactions
  with allyl cations, **1**, 191
  with allylic alcohols, **1**, 195
  with α-amino acids, **5**, 385
Nickel, bis(cyclooctadienyl)-, use in the synthesis of enediynes, **1**, 585
Nickel, *trans*-chlorobis(triphenylphosphine)-, use in the synthesis of α,β-arylnitriles, **3**, 661
Nickel, chlorocyclopentadienyl(triphenylphosphine)-
 reactions
  with 2-chlorolithioferrocene, **4**, 719
  with tricarbonyl(lithiocyclopentadienyl)-manganese, **4**, 719
Nickel, dicyclooctadiene-, as starting material, for alkyne-bridged binuclear nickel complexes, **4**, 692
Nickel, pentacarbonylcyclopentadienylcobalt-, reactions, with alkynes, **4**, 713
Nickel, tetrakis(triethylphosphine)-, reactions, with aryl nitriles, **5**, 1138
Nickel(0), as catalyst, for the synthesis of aromatic ketones, **3**, 289
Nickel(0) complexes, alkyl-, synthesis of, via reductive alkylation, **2**, 575
Nickel(II), complex formation equilibria, **5**, 864
Nickel(II) bromide
 as catalyst, for the synthesis of aldehydes, **3**, 5
 reactions, with pentaphenylaluminacyclopentadiene, **4**, 693
Nickel(II) catalysts, for the synthesis of alkenes, **1**, 700
Nickel(II) chloride
 as catalyst, for the reduction of alkyl halides, **1**, 4
 reactions
  with 1,4-pentadiene, **4**, 694
  with sodium borohydride, **1**, 75
 for the reduction of chloroalkanes, **1**, 7
Nickel(II) complex, (diazomethyl)trimethylsilane-, synthesis of, **6**, 675
Nickel(II) compounds, alkyl-, synthesis of, **2**, 576
Nickel(II) compounds, bis(pentafluorophenyl)bis(dioxane)-, reactions, with cyanide, **5**, 1138
Nickel(II) cyanide, bis(pentafluorophenyl)triphenylphosphine-, synthesis of, **5**, 1138
Nickel acetate complexes, pentahalophenyl-, reactions, with sodium cyanide, **5**, 1137
Nickel acetoacetate, as catalyst, for the addition of trimethylane to an α,β-unsaturated ketone, **2**, 699
Nickel acetylacetonate
 reactions, with dialkylzincs, **1**, 314
 use in the synthesis of ketene aminals, **4**, 975
Nickel acyl compounds, synthesis of, **5**, 461
Nickel acyl compounds, pentacoordinated, synthesis of, **5**, 463
Nickel alkyl complexes, reactions, with nickelocene, **6**, 403
Nickel alkynides, synthesis of, **1**, 1141
Nickel–aluminum oxide catalyst, for the synthesis of benzene, **1**, 56
Nickelasilacyclobutenes, as intermediates, **4**, 1053
Nickel bis(ethene), alkyl-, synthesis of, **2**, 575
Nickel boride
 for the reduction of aryl sulfides, **1**, 436
 for the reduction of selenides, **1**, 25
 for the reduction of vinyl selenides, **1**, 440
Nickel carbonyl
 reactions, with allyl halides, **1**, 526
 use in the synthesis of α,β-unsaturated ketones, **3**, 218
Nickel carbonyl anions, as starting materials, for imidoylnickels, **5**, 800
Nickel carbonyl complexes, synthesis of, **6**, 525
Nickel catalysis, for the synthesis of alkynes, **1**, 1006
Nickel catalysts
 for the dealkylation of alkylbenzenes, **1**, 56
 for hydrocarboxylation reactions, **1**, 297
 for the hydrogenation of alkynes, **1**, 455
 reactions
  with enol triflates, **1**, 1099
  with Grignard reagents, **1**, 466
 for the substitution of aryl halides, **1**, 470
 for the substitution of vinyl ethers, **1**, 480
 for the synthesis of alkenes, **1**, 502
 use in the substitution of vinyl halides, **1**, 468
 use in the synthesis of alkylmagnesiums, **2**, 561
Nickel chloride
 for the catalyzed coupling of vinyl halides, **1**, 677
 reactions, with *t*-pentyl isocyanide, **5**, 1072
Nickel chloride, bis(triethylphosphine)pentachlorophenyl-, reactions, with silver cyanide, **5**, 1137
Nickel–cobalt alkyne complexes, synthesis of, **4**, 713
Nickel complexes
 as catalysts, for alkylation reactions, **1**, 465
 from, halides, **2**, 577
 reactions
  with alkynes, **1**, 545
  with carbon disulfide, **6**, 332
 synthesis of, **4**, 955
Nickel complexes, acyl-, synthesis of, **5**, 462
Nickel complexes, alkyne-bridged binuclear, from, dicyclooctadienenickel, **4**, 692
Nickel complexes, chlorovinyl-, reactions, with potassium cyanide, **5**, 1137
Nickel complexes, cyclic, synthesis of, **5**, 948
Nickel complexes, dialkyl-, reactions, with imides, **5**, 387
Nickel complexes, diazomethyl-, synthesis of, **3**, 487
Nickel complexes, ethyl-, reactions, with carbon monoxide, **5**, 461
Nickel complexes, methyl-, reactions, with carbon monoxide, **5**, 461
Nickel complexes, vinyl-, cleavage of, **1**, 453
Nickel complex reducing agents, for the reduction of vinyl sulfides, **1**, 438
Nickel dicarbonyl, bis(triphenylphosphine)-
 reactions, with cyclic anhydrides, **1**, 564
 use in the synthesis of cyclic systems, **1**, 564
Nickel dinuclear carbonyls, as starting materials, for alkyne-bridged dinickel complexes, **4**, 692
Nickel dioxide, use in the synthesis of α,β-unsaturated ketones, **3**, 215

Nickelocene
reactions
with alkyllithiums, **6**, 403
with fluoroboric acid, **4**, 694
with nickel alkyl complexes, **6**, 403
as starting material, for alkyne-bridged dinickel complexes, **4**, 692
Nickelocene–lithium aluminum hydride complex, for the reduction of thiols, **1**, 436
Nickel peroxide
for the oxidation of thiocarboxylic acids, **5**, 251
use in the synthesis of alkanoic acids, **5**, 48
use in the synthesis of aromatic acids, **5**, 114
use in the synthesis of diazo compounds, **3**, 464
use in the synthesis of monocarboxylic acids, **5**, 28
Nickel-promoted reaction, of alkynes, with carbon dioxide, **2**, 988
Nickels
reactions
with benzoyl chloride, **5**, 463
with 1-haloalkylphosphines, **4**, 583
with phosphorus ylides, **4**, 583
Nickels, acyl-, synthesis of, **5**, 461
Nickels, alkyl-, synthesis of, **2**, 575
Nickels, π-allyl-, for the allylation of haloalkanes, **1**, 127
Nickels, cyano-, synthesis of, **5**, 1137
Nickels, imidoyl-
from
isonitriles, **5**, 800
nickel carbonyl anions, **5**, 800
synthesis of, **5**, 799
Nickel salts, use in Ullmann reactions, **1**, 472
Nickel tetracarbonyl, synthesis of, **6**, 525
Nicotinamide, reduction of, **1**, 103
Nicotinamide, *N*-benzyl-1,4-dihydro- (BNAH)
reactions, with 3-iodocyclohexenone, **1**, 431
for the reduction of nitroalkanes, **1**, 37
use in the reduction of carbon–hydrogen bonds, **1**, 69
use in the reduction of vinyl bromides, **1**, 431
Nicotinamides, for the reduction of thioketones, **1**, 24
Nicotinic acid, reduction of, **1**, 91
Nicotinic acid, 6-hydroxy-, synthesis of, **5**, 106
Ninhydrin
acylation of, **4**, 55
bromoacetylation of, **4**, 57
hemithioacetal of, **4**, 221
synthesis of, **4**, 160
Niobium
in the synthesis of metallacyclopropenes, **2**, 927
use for the deoxygenation of epoxides, **1**, 587
Niobium, bis(cyclopentadienyl)hydrido-, reactions, with dicarbonylcyclopentadienylmethyliron, **4**, 717
Niobium(III) chloride, reactions, with alkynes, **2**, 672
Niobium bis(cyclopentadienyl)trihydrides, use in the synthesis of dimeric niobocene, **4**, 672
Niobium carbene complexes, synthesis of, **3**, 510
Niobium carbenes, reactions, with phosphines, **3**, 510
Niobium carbonyl complexes, synthesis of, **6**, 522
Niobium complexes, dinuclear, synthesis of, **4**, 672
Niobium compounds, alkyl-, synthesis of, **2**, 569
Niobium compounds, imidoyl-, synthesis of, **5**, 797
Niobium functions, doubly-bonded, synthesis of, **3**, 510
Niobium-promoted reaction
of alkynes
with aldehydes, **2**, 988
with hydrazones, **2**, 988
with isocyanates, **2**, 988
with ketones, **2**, 988
Niobocene, dimeric, synthesis of, **4**, 672
Nitramine, *N*-chloromethyl-*N*-ethyl-, reactions, with sodium alkoxides, **4**, 313
Nitramine, cyclotrimethylene-, synthesis of, **4**, 434
(±)-Nitramine, synthesis of, **2**, 331
Nitramines
from, primary aromatic amines, **2**, 798
Mannich-type reaction of, **4**, 442
reactions, with formaldehyde, **4**, 442
synthesis of, **6**, 260
Nitramines, acylaminomethyl-, synthesis of, **4**, 447
Nitramines, alkyl-, reactions, with acylamines, **4**, 447
Nitramines, *N*-alkyl-*N*-chloromethyl-, displacement of, **4**, 327
Nitramines, aminomethyl-, synthesis of, **4**, 442
Nitramines, aminomethylene-, nitrosation of, **4**, 440
Nitramines, aromatic, synthesis of, **2**, 798
Nitramines, *N*-chloro-, reduction of, **2**, 798
Nitramines, primary, synthesis of, **2**, 798
Nitrate, acetyl
as nitrating agent, **4**, 435
reactions, with 2-alkoxyvinyl phosphonic esters, **4**, 490
safety precautions, **4**, 208
as starting material, for α-aminoalkylphosphorus compounds, **4**, 490
Nitrate, 3,5-dimethyl-1*H*-pyrazole-1-carboxamidine, as guanylating agent, **6**, 641
Nitrate, ethyl, as starting material, for α-aminoalkylphosphorus compounds, **4**, 490
Nitrate, thionyl, synthesis of, **2**, 104
Nitrate esters, synthesis of, **2**, 103
Nitrates
for the oxidation of sulfides, **2**, 150
as starting materials
for amide acetals, **6**, 105
for nitro compounds, **2**, 397
synthesis of, safety precautions, **2**, 104
Nitrates, acetyl, as reagents, for the synthesis of nitrate esters, **2**, 104
Nitrates, alkyl
reactions
with aldimines, **2**, 788
with amines, **2**, 798
as starting materials, for nitroalkanes, **4**, 428
Nitrates, metallic, use in nitration reactions, **2**, 792
Nitration
reviews, **2**, 790
for the synthesis of arylamines, **2**, 752
Nitration reactions, for the synthesis of trinitromethanes, **6**, 150
4-Nitrazaalkanoic acids, from, 4-nitrazaalkanonitriles, **5**, 94
Nitrazones, reduction of, **5**, 855
Nitrene, ethoxycarbonyl, reactions, with imidates, **5**, 845
Nitrenes
reactions, with alkenes, **4**, 515
as starting materials, for aliphatic hydrazonates, **5**, 845
Nitrenes, acyl, reactions, with dimethyl sulfoxide, **2**, 163
Nitrenes, alkynyl, as intermediates, in the synthesis of dicyanostilbenes, **2**, 1055
Nitrenes, cyano-, as intermediates, in the synthesis of phosphaalkynes, **5**, 1159
Nitrenium ion, as intermediate, in Bamberger rearrangement, **2**, 649
Nitrenoids, from, (arenesulfonylimino)iodoarenes, **2**, 726
Nitric acid
for the oxidation of dithioacetals, **4**, 260
reactions
with acetic anhydride, **6**, 365
with alkenes, **4**, 797
with dinitroalkenes, **6**, 150
with pentenes, **1**, 895
with trisulfonylmethanes, **6**, 316
use in the synthesis of alkanoic acids, **5**, 46
use in the synthesis of *N*-nitroamides, **5**, 361
Nitric oxide
reactions
with Grignard reagents, **2**, 798
with iodotrifluoroethene, **4**, 798

with vinyl radicals, **2**, 778
Nitrides, hydrolysis of, **5**, 33
Nitrilamines, reactions, with amines, **6**, 659
Nitrilases, for the hydrolysis of nitriles, **5**, 118
Nitrile, α-amino-β-trifluoro-, synthesis of, **3**, 627
Nitrile, α-amino-α-trimethylsilyl-, synthesis of, **3**, 640
Nitrile, arylmethine-, synthesis of, **3**, 625
Nitrile, 3-benzyl-3-methylpentane-, as starting material, for 3-benzyl-3-methylvaleric acid, **5**, 34
Nitrile, 5-bromovalero-, use in the synthesis of amines, **2**, 316
Nitrile, *t*-butyl-, reactions, with *t*-butyllithium, **3**, 459
Nitrile, cinnamo-, reduction of, **1**, 48
Nitrile, croto-, synthesis of, **3**, 643
Nitrile, crotono-, reactions, with *t*-butyl hypochlorite, **3**, 650
Nitrile, β-diethylamino-, synthesis of, **3**, 638
Nitrile, dithyrea-
from, amides, **4**, 245
synthesis of, **3**, 636
Nitrile, α-ethyl-β-hydroxy-, synthesis of, **3**, 632
Nitrile, glutaro-, reactions, with hydroxylamine, **5**, 833
Nitrile, isoamyl-, reactions, with α-amino acid esters, **3**, 468
Nitrile, α-lithioamino-, crystal structure of, **2**, 1001
Nitrile, maleic, chlorination of, **5**, 809
Nitrile, phenylpropyne-, synthesis of, **3**, 675
Nitrile, β-tributylstannylproponyl-, synthesis of, **3**, 640
Nitrile anions, amino-, reactions, with glycidols, **1**, 224
Nitrile carbanions, reactions, with alkyl halides, **3**, 620
Nitrileimine, diphenyl-, reactions, with arsenadiazole, **6**, 161
Nitrileimines
from, chlorophosphanes, **6**, 667
reactions, with diazaphospholes, **6**, 156
rearrangement, to diazo compounds, **6**, 667
Nitrile oxide, bromo-, from, oximes, **6**, 611
Nitrile oxide-alkene cycloaddition, for the synthesis of β-hydroxy ketones, **3**, 160
Nitrile oxide oximes, as starting materials, for oxyimidic amides, **5**, 837
Nitrile oxides
from
aldoximes, **3**, 683
furazan *N*-oxides, **3**, 685
furazans, **3**, 685
hydroximoyl halides, **3**, 678, 683, 684
α-nitroalkanoate esters, **3**, 684
nitrolic acids, **3**, 684
nitromethyl compounds, **3**, 684
reactions
with alcohols, **5**, 699, 826
with alkenes, **3**, 45
with alkynes, **3**, 189
with allylsilanes, **3**, 142
with carboxylic acid chlorides, **5**, 669
with chlorocarbonyl isocyanate, **5**, 669
with (2,2-dimethylpropylidyne)phosphane, **5**, 839
with dinitrogen tetroxide, **6**, 150
with *O,O*-disubstituted thiocarbonates, **6**, 309
with isonitriles, **5**, 1004
with methanol, **5**, 699
with sulfur ylides, **2**, 779
with thioketones, **6**, 127
with thiols, **5**, 826, 828
with thiophenols, **5**, 828
with triphenylphosphine, **5**, 1131
rearrangement of, **5**, 1003
as starting materials
for hydroxamoyl halides, **5**, 669
for isocyanates, **5**, 1003
for isothiocyanates, **5**, 1038
for ketenimines, **3**, 600
for oxyimidic acids, **5**, 828
for oxyimidic esters, **5**, 826
for oxyimidic halides, **5**, 819
synthesis of, **3**, 683
via nitronate esters, **3**, 684
use in the synthesis of oximidic chlorides, **5**, 819
use in the synthesis of β,γ-unsaturated ketones, **3**, 142
Nitrile oxides, arene-, synthesis of, **3**, 685
Nitrile oxides, ω-cyano-, synthesis of, **3**, 685
Nitrile oxides, hetero-, synthesis of, **5**, 1148
Nitrile *N*-oxides, use in the synthesis of 1,4,2-oxathiazoles, **6**, 309
Nitriles
acidic hydrolysis of, **5**, 269
acidity of adjacent protons, **3**, 620
as activating groups for alkenes, **1**, 300
addition of amides to, **4**, 418
additions to, **1**, 174
alkaline hydrolysis of, **5**, 269
alkylation of, **3**, 86
bearing α,β-vinylic bonds, synthesis of, **3**, 643
α-bromination of, **2**, 29
cyclization of, **3**, 264
deuterium exchange of, **1**, 851
enzymic hydrolysis of, **5**, 270
formylation of, **3**, 43
in Friedel–Crafts acylation, **3**, 282
from
acid chlorides, **3**, 618
acids, **3**, 617
alcohols, **3**, 613
aldehydes, **3**, 617, 621
alkyl halides, **3**, 612
amides, **3**, 619
amines, **3**, 620
amino acids, **3**, 620
α-amino acids, **3**, 620
α-amino ketones, **3**, 622
arylisocyanoates, **3**, 614
N-*t*-butyl amides, **3**, 619
carboxylic acids, **3**, 617
cyanamides, **3**, 614
cyanogen bromide, **3**, 614
cyanogen chloride, **3**, 614
dimethylhydrazones, **3**, 617
dithioacids, **3**, 618
esters, **3**, 618
imino ethers, **3**, 620
isonitriles, **3**, 621
ketones, **3**, 621
nitriles, **3**, 620
nitro compounds, **3**, 620
1-nitrocycloalkenes, **3**, 624
α-nitrosophosphonium salts, **5**, 904
organoboranes, **3**, 622
oximes, **3**, 617; **5**, 215
primary amides, **5**, 569
thioamides, **3**, 619
halogenation of, **2**, 3; **6**, 53
hydration of, for the synthesis of primary alkenoamides, **5**, 295
hydrolysis of, **5**, 25, 34, 268, 281, 296, 385
reagents for, **5**, 275, 281
for the synthesis of metal amides, **5**, 389
isoracemization of, **1**, 858
peroxide hydrolysis of, **5**, 269
Pinner reaction of, **5**, 727
reactions
with acylhydrazines, **5**, 773
with alcohols, **4**, 883; **5**, 691, 742
with alkyl chloroformates, **3**, 687
with alkyllithiums, **3**, 459
with alkynezirconocenes, **2**, 987
with amidoalcohols, **4**, 425
with amines, **2**, 749
with ammonia, **5**, 780
with ammonium alkyl sulfonates, **5**, 749

with ammonium aryl sulfonates, **5**, 749
with antimony pentachloride, **5**, 389
with aryl Grignard reagents, **3**, 286
with benzynezirconocenes, **2**, 987
with carbonyl fluoride, **5**, 1003
with dialkyl dithiophosphates, **5**, 737
with *O,O*-dialkyldithiophosphoric acids, **5**, 581
with *O,O*-dialkylmonothiophosphoric acids, **5**, 581
with dialkylphosphinous acids, **5**, 786
with diborane, **3**, 458
with diisobutylaluminum hydride, **3**, 458
with dithiophosphorus(V) acids, **5**, 643
with enolates, **3**, 268
with ethanol, **4**, 883
with Grignard reagents, **3**, 131, 405, 459
with hydrazine, **5**, 768, 848
with hydrazines, **5**, 770
with hydrogen selenide, **5**, 626
with hydrogen sulfide, **5**, 579, 619
with hydroxylamine, **5**, 763, 833
with 2-lithio-1,3-dithiane, **3**, 181
with lithium trialkylaluminum hydrides, **3**, 458
with lithium triethoxyaluminum hydride, **3**, 458
with metallacyclopropanes, **2**, 984
with metallacyclopropenes, **2**, 984
with methyl triflate, **3**, 282
with monoalkylboranes, **3**, 457
with monothiophosphorus(V) acids, **5**, 643
with organoceriums, **2**, 312, 316
with organometallics, **3**, 440
with organosilylstannanes, **3**, 418
with phenylphosphine, **4**, 481
with phosgene, **5**, 1002
with phosphorus trihalides, **6**, 159
with potassium amide, **5**, 750
with Raney nickel, **3**, 86
with silver fluoride, **6**, 53
with sodamide, **5**, 750
with sodium trialkylaluminum hydrides, **3**, 458
with tetraethylammonium hydrosulfide, **5**, 581
with thiolates, **2**, 240
with thiols, **5**, 727
with thionoamides, **5**, 630
with trialkylaluminums, **3**, 458
with trioxan, **4**, 425
with tri-*n*-propylamine, **1**, 858
with α-unsubstituted bromoacetates, **3**, 461
with zinc ester enolates, **3**, 461
reduction of, **1**, 46; **2**, 312, 316
reduction to aldehydes, **3**, 13
rhodium-catalyzed hydrogenation of, **2**, 316
saponification of, **5**, 709
saturated unsubstituted, synthesis of, **3**, 622
selenenylation of, **1**, 607
as starting materials
for aliphatic amidines, **5**, 748
for *N*-alkylalkanoamides, **5**, 281
for *N*-alkylalkenoamides, **5**, 296
for *N*-alkylamides, **5**, 281
for amides, **5**, 268
for amidines, **5**, 742
for amidrazones, **5**, 768
for amines, **2**, 316
for α-aminoalkylorganophosphorus compounds, **4**, 464
for aminobis(phosphonates), **6**, 159
for 2-aminoketene dithioacetal monosulfoxides, **4**, 857
for aromatic amidines, **5**, 755
for azines, **3**, 445
for carbonimidic difluorides, **6**, 603
for carboxylic acids, **5**, 25
for α-chloroalkanesulfenyl chlorides, **4**, 72
for *N*-chlorocarbonyls, **6**, 445
for enamines, **2**, 739
for ester aminals, **6**, 109
for *N*-haloimidic esters, **5**, 813
for haloimidic halides, **5**, 808
for heterocyclic compounds, **3**, 641
for α-hydrazino enamines, **4**, 984
for hydrazonamides, **5**, 847
for imidate ester hydrochloric salts, **6**, 69
for imidic acid esters, **5**, 691
for imidoyl halides, **5**, 657
for imidoyl-tetracoordinated phosphorus compounds, **5**, 786
for isocyanates, **5**, 1002
for ketene *S,N*-acetals, **4**, 905
for ketene aminals, **4**, 979
for ketene hemiaminals, **4**, 883
for *N*-H ketimines, **3**, 405
for ketones, **3**, 134
for nitriles, **3**, 620
for ortho esters, **6**, 70
for oxyimidic amides, **5**, 833
for primary amides, **5**, 275
for selenoamides, **5**, 625
for α-silyl enamines, **4**, 1014
for thioimidic esters, **5**, 727
for thionoamides, **5**, 579, 610
for trifluoromethyl, **6**, 10
synthesis of, **1**, 252; **3**, **611–640**; **4**, 512
choice of solvent for, **3**, 613
general methods, **3**, 641
with a heteroatom attached to the cyanocarbon, **5**, **1099–1149**
reducing agents for, **3**, 621
via addition reactions, **3**, 614
via electrophilic substitutions, **3**, 614
via elimination reactions, **3**, 617
via nucleophilic substitutions, **3**, 612
via substitution reactions, **3**, 612
use in atom transfer reactions, **1**, 344
Nitriles, acyl-, reactions, with iminophosphoranes, **3**, 478
Nitriles, adamantane-, reduction of, **1**, 48
Nitriles, aliphatic
reactions, with alkali amide bases, **5**, 750
synthesis of, **3**, 622
Nitriles, alkane-
alkylation of, **3**, 566
as starting materials, for ketenimines, **3**, 565
tautomerism of, **3**, 565
Nitriles, (*E*)-2-alkene-, synthesis of, **3**, 642
Nitriles, 2-alkenic, synthesis of, **3**, 642
Nitriles, (*E*)-2-alkenic, synthesis of, **3**, 645
Nitriles, α,β-alkenic
from
α,β-alkynenitriles, **3**, 642
ketones, **3**, 644
Nitriles, alkenyl-, synthesis of, **1**, 534
Nitriles, 2-alkoxy-2-alkene-, from, vinyl ethers, **3**, 652
Nitriles, β-alkoxy-α-imino-, synthesis of, **3**, 658
Nitriles, alkyl-
from, trialkylboranes, **3**, 622
reactions, with dialkylboranes, **3**, 457
as starting materials, for α-(tosyloxyimino)alkylnitriles, **3**, 427
synthesis of, **1**, 276
electrochemical methods, **1**, 276
Nitriles, α-alkyl-α-aryl-, synthesis of, **3**, 625
Nitriles, β-alkylthio-α,β-alkenic, synthesis of, **3**, 655
Nitriles, alkyne-
reactions
with lithium aluminum hydride, **3**, 642
with organocopper(I) compounds, **3**, 642
Nitriles, 2-alkyne-, reduction of, **1**, 514
Nitriles, α,β-alkyne-, as starting materials, for α,β-alkenic nitriles, **3**, 642

Nitriles, alkynic, from, α-cyano ketones, **3**, 676
Nitriles, α,β-alkynic
  reactions, with dimethylcopperlithium, **3**, 676
  synthesis of, **3**, 674
Nitriles, alkynyl-
  reactions, with organosilvers, **1**, 544
  synthesis of, **1**, 1048
Nitriles, β-allenic, synthesis of, **3**, 624
Nitriles, allenyl-, reactions, with thiols, **3**, 655
Nitriles, amido-, reactions, with dinitromethane, **4**, 995
Nitriles, amino-
  cleavage of, **1**, 184
  synthesis of, **3**, 616
Nitriles, α-amino-
  deprotonation of, **3**, 638, 657
  from
    amines, **3**, 638
    cyanohydrines, **3**, 616
  reactions, with lithium diisopropylamide, **2**, 741
  reductive decyanation of, **2**, 305
  as starting materials
    for *N*-acylaminals, **4**, 418
    for α-aminooxyimidic amides, **5**, 833
    for α-diazonitriles, **3**, 477
  synthesis of, **2**, 310; **3**, 408, 616, 637
Nitriles, β-amino-
  from, amines, **3**, 638
  synthesis of, **1**, 264
Nitriles, ω-amino-, from, α-amino ketones, **3**, 638
Nitriles, α-amino-α,β-alkenic, synthesis of, **3**, 657
Nitriles, γ-amino-α,β-alkenic, synthesis of, **3**, 658
Nitriles, α-amino-β-alkoxy-α,β-unsaturated, synthesis of, **3**, 658
Nitriles, α-amino-β-(phenylseleno)-, synthesis of, **2**, 284
Nitriles, β-amino-a,β-unsaturated, synthesis of, **3**, 658
Nitriles, aromatic
  hydrolysis of, **5**, 108, 300
    reagents for, **5**, 298
  reactions, with sodium hydrazide, **5**, 848
  reduction of, **3**, 86
  synthesis of
    via elimination reactions, **3**, 662
    via substitution, **3**, 661
Nitriles, β-aroylpropio-, synthesis of, **1**, 256
Nitriles, aryl-
  reactions
    with dialkylboranes, **3**, 457
    with platinum, **5**, 1140
    with tetrakis(triethylphosphine)nickel, **5**, 1138
    with tris(triarylphosphine)palladium(II) compounds, **5**, 1139
  synthesis of, **3**, **641–676**
Nitriles, α-aryl-
  from, aryl halides, **3**, 626
  synthesis of, **3**, 625
Nitriles, α,β-aryl-, synthesis of, **3**, 661
Nitriles, β-aryl-, synthesis of, **3**, 626
Nitriles, α-aryl-α-alkyl-, synthesis of, **3**, 625
Nitriles, arylmethyl-, reactions, with alcohols, **3**, 625
Nitriles, arylmethylene-
  deprotonation of, **3**, 625
  reactions, with alcohol, **3**, 625
Nitriles, α-(arylsulfonyl)-, synthesis of, **3**, 635
Nitriles, benzyl-, synthesis of, **3**, 625
Nitriles, benzylic, reactions, with dienes, **3**, 669
Nitriles, bis(alkylthio)-α,β-unsaturated, synthesis of, **3**, 655
Nitriles, bisaryl-, synthesis of, **3**, 668
Nitriles, bisgermyl-, synthesis of, **4**, 625
Nitriles, bromo-, synthesis of, **3**, 652
Nitriles, α-bromo-, synthesis of, **3**, 627
Nitriles, α-*t*-butoxy-α,β-unsaturated, synthesis of, **3**, 650
Nitriles, α-carboxy-, synthesis of, **3**, 621
Nitriles, α-carboxy-α,β-alkenic, synthesis of, **3**, 653
Nitriles, α-carboxyethyl-α,β-unsaturated, synthesis of, **3**, 653
Nitriles, α-carboxy-β-hydroxy-α,β-unsaturated, synthesis of, **3**, 653
Nitriles, chiral amino-, synthesis of, **3**, 616
Nitriles, α-chloro-
  from, cyanohydrins, **3**, 627
  synthesis of, **3**, 627
Nitriles, coordinated, hydrolysis of, **5**, 385
Nitriles, cyclic-α-aryl-, synthesis of, **3**, 626
Nitriles, α-deprotonated alkane-, reactions, with electrophiles, **3**, 566
Nitriles, dialkylamino-, use in the synthesis of ketones, **3**, 137
Nitriles, dialkynic, synthesis of, **3**, 674
Nitriles, α,α-diamino-, synthesis of, **3**, 637
Nitriles, α,α-diaryl-, from, diaryl ketones, **3**, 626
Nitriles, diazo-, Diels–Alder reactions of, **5**, 1119
Nitriles, α-diazo-
  from, α-aminonitriles, **3**, 477
  synthesis of, **3**, 477
Nitriles, diene-, synthesis of, **3**, 646
Nitriles, α,α-digermylated, synthesis of, **4**, 625
Nitriles, α,α-dihalogenated, reactions, with bromine, **5**, 662
Nitriles, enamino-, reactions, with phosgene, **5**, 967
Nitriles, β-enamino-, synthesis of, **3**, 656
Nitriles, α-ethylenic, chlorination of, **3**, 650
Nitriles, α-ethylthio-α,β-unsaturated, synthesis of, **3**, 655
Nitriles, α-fluorinated, reactions, with boron trichloride, **5**, 684
Nitriles, α-fluoro-, from, cyanohydrins, **3**, 627
Nitriles, fluoroimino-
  reactions
    with amines, **5**, 761
    with ammonia, **5**, 761
Nitriles, α-formylamino-, as starting materials, for α-isocyanonitriles, **3**, 710
Nitriles, α-germyl-, deprotonation of, **4**, 662
Nitriles, halo-, synthesis of, **3**, 627
Nitriles, α-halo-
  reactions
    with arylzincs, **1**, 465
    with dibutyltellurium, **3**, 637
Nitriles, α-haloalkane-, as starting materials, for ketenimines, **3**, 568
Nitriles, α-halo-α,β-alkenic, synthesis of, **3**, 649
Nitriles, halo-substituted aliphatic, synthesis of, **3**, 627
Nitriles, α-halo-α,β-unsaturated, as intermediates, **3**, 649
Nitriles, heteroaryl-, synthesis of, **3**, 669
Nitriles, heterocyclic, as starting materials, for *C*-heteroaryl-substituted oxyimidic amides, **5**, 834
Nitriles, heterocyclic aromatic, synthesis of, **3**, 669
Nitriles, α-hydrazono-, reactions, with aluminum chloride, **3**, 671
Nitriles, α-hydroxy-
  reactions, with Grignard reagents, **2**, 316
  synthesis of, **3**, 627
  See also Cyanohydrins
Nitriles, β-hydroxy-
  from
    alkenes, **3**, 632
    bromoacetonitrile, **3**, 632
  resolution of, **3**, 632
  synthesis of, **3**, 632
Nitriles, γ-hydroxy-
  from, oxetanes, **3**, 633
  synthesis of, **3**, 620
Nitriles, 4-hydroxyalk-2-ene-, synthesis of, **3**, 651
Nitriles, γ-hydroxy-α,β-alkenic, synthesis of, **3**, 651
Nitriles, (*E*)-2-(1-hydroxyalkyl)alk-2-ene-, synthesis of, **3**, 660
Nitriles, α-hydroxylamino-, from, secondary amines, **3**, 638
Nitriles, α-imino-, synthesis of, **3**, 417
Nitriles, iodo-, reactions, with magnesium, **3**, 459
Nitriles, γ-iodo-, synthesis of, **3**, 627

Nitriles, α-isocyano-, from, α-formylaminonitriles, **3**, 710
Nitriles, 4-keto-, from, α,β-unsaturated ketones, **3**, 616
Nitriles, 5-keto-, synthesis of, **3**, 634
Nitriles, β-keto-
  from
    cyanoacetic acid, **3**, 633
    enamines, **3**, 633
  synthesis of, **3**, 620, 633
Nitriles, γ-keto-
  from
    acetone cyanohydrin, **3**, 633
    calcium cyanide, **3**, 634
    diethylaluminum cyanide, **3**, 634
    α,β-unsaturated carbonyls, **3**, 633
  reactions, with diazo compounds, **3**, 453
  synthesis of, **3**, 633
Nitriles, metallated alkyl-, use in the synthesis of benzonitriles, **3**, 665
Nitriles, γ-methoxy-α,β-alkenic, synthesis of, **3**, 654
Nitriles, α-methoxy-α,β-unsaturated, synthesis of, **3**, 652
Nitriles, α-methylthio-α,β-unsaturated, synthesis of, **3**, 654
Nitriles, 4-nitrazaalkano-, as starting materials, for 4-nitrazaalkanoic acids, **5**, 94
Nitriles, α-nitro-
  reactions, with alcohols, **5**, 692
  synthesis of, **2**, 391
Nitriles, β-nitro-, from, nitroalkenes, **3**, 638
Nitriles, β-oxy-, synthesis of, **3**, 632
Nitriles, α-oxygenated, synthesis of, **3**, 627
Nitriles, β-oxygenated, synthesis of, **3**, 632
Nitriles, perfluorinated
  reactions
    with acetoxime, **5**, 711
    with chlorine fluoro sulfate, **5**, 710
Nitriles, perfluoro-
  reactions
    with bromine, **5**, 808
    with chlorine, **5**, 808
    with phosphines, **5**, 786
  as starting materials, for imidoyl-tricoordinated phosphorus compounds, **5**, 786
Nitriles, perfluoroalkyl-, reactions, with caesium fluoride, **3**, 419
Nitriles, β-perfluoroalkyl-α,β-alkenic, synthesis of, **3**, 649
Nitriles, perfluorooctano-, reactions, with phenylphosphine, **5**, 786
Nitriles, perhalogenated, as starting materials, for *N*-bromoperhalo-1-alkanimines, **5**, 808
Nitriles, γ-phenoxy-α,β-unsaturated, synthesis of, **3**, 653
Nitriles, γ-phenylseleno-α,β-unsaturated, synthesis of, **3**, 656
Nitriles, β-phenylselenyl-, synthesis of, **3**, 637
Nitriles, β-phenylthio-γ-keto-, synthesis of, **3**, 636
Nitriles, phosphono-, Horner–Emmons reaction of, **3**, 646
Nitriles, γ-phosphonoxy-α,β-alkenic, synthesis of, **3**, 659
Nitriles, polycyclic aromatic, synthesis of, **3**, 667
Nitriles, β-seleno-α-amino-, synthesis of, **3**, 637
Nitriles, α-silyl-, synthesis of, **3**, 639
Nitriles, (*E*)-silylalk-2-ene-, synthesis of, **3**, 660
Nitriles, α-(*N*-silylamino)-, dehydrocyanation of, **3**, 457
Nitriles, 2-silyloxy-, as starting materials, for 1,2-amino alcohols, **2**, 317
Nitriles, β-silyloxy-
  from, trimethylsilylacetonitrile, **3**, 632
  synthesis of, **3**, 632
Nitriles, (*Z*)-β-silyloxy-α,β-alkenic, synthesis of, **3**, 653
Nitriles, γ-silyloxy enol ethers, synthesis of, **3**, 634
Nitriles, substituted, dimerization of, **3**, 657
Nitriles, *N*-substituted, synthesis of, **3**, **677–692**
Nitriles, 3-substituted-3-methylthio-2-alkenic, synthesis of, **3**, 654
Nitriles, ω-tellurio-α,β-unsaturated, as starting materials, for β-cyclic-α-pyridylthio-nitriles, **3**, 635
Nitriles, α-thio-α,β-alkenic, synthesis of, **3**, 654
Nitriles, β-thio-α,β-alkenic, synthesis of, **3**, 655
Nitriles, β-thio-β,γ-alkenic, synthesis of, **3**, 655
Nitriles, α-thio-α-amino-, from, thioamides, **3**, 638
Nitriles, α-thioether-, bromination of, **3**, 655
Nitriles, α-thiophenyl-α,β-unsaturated, synthesis of, **3**, 655
Nitriles, α-tosyl-, cyclization of, **2**, 322
Nitriles, *N*-tosyl β-amino-, from, *N*-tosylaziridines, **2**, 319
Nitriles, β-*N*-tosylamino-, from, *N*-tosylaziridines, **3**, 638
Nitriles, α-(tosyloxyimino)alkyl-, from, alkylnitriles, **3**, 427
Nitriles, (*E*)-2-trimethylsilylalk-2-ene-, synthesis of, **3**, 660
Nitriles, β-trimethylsilyl-β-γ-unsaturated, synthesis of, **3**, 660
Nitriles, unsaturated, synthesis of, **3**, 622
Nitriles, α,β-unsaturated
  acidity of, **1**, 871
  allylation of, **1**, 314
  deprotonation of, **3**, 623
  in Diels–Alder reactions, **3**, 621
  in [1 + 3] dipolar cycloaddition reactions, **3**, 621
  epoxidation of, **3**, 627
  from
    allyl α-cyanocarboxylates, **3**, 645
    vinylboranes, **3**, 623
  as Michael acceptors, **3**, 621
  reactions
    with amines, **3**, 638
    with diphenylphosphine, **2**, 439
  reduction of, **3**, 61
  as starting materials, for ketenimines, **3**, 570
  synthesis of, **1**, 536, 545, 597, 622; **3**, 621, 650, **641–676**
Nitriles, β,γ-unsaturated
  deprotonation of, **3**, 623
  epoxidation of, **3**, 633
  etherification of, **3**, 653
  from
    allyl methyl ethers, **3**, 623
    allyl phosphonates, **3**, 622
  as starting materials, for β-thionitriles, **3**, 636
  synthesis of, **1**, 191, 835; **3**, 622
Nitriles, γ,δ-unsaturated, from, allyl halides, **3**, 623
Nitriles, δ,ε-unsaturated, synthesis of, **3**, 623
Nitriles, β,γ-unsaturated-α-amino-, synthesis of, **3**, 617, 637
Nitriles, α-unsubstituted, synthesis of, **3**, 621
Nitriles, vinyl-, synthesis of, **3**, 642
Nitrile sulfides
  [1 + 3] dipolar cycloaddition of, **6**, 129
  reactions, with phosphaalkenes, **6**, 129
Nitrile ylides, metallo-, synthesis of, **5**, 1089
Nitrilimine
  from
    α-chloroimines, **3**, 452
    tetrazoles, **3**, 452
Nitrilimine, bis(triisopropylsilyl)-, synthesis of, **5**, 1148
Nitrilimines
  from, nitrines, **5**, 1071
  as intermediates, in the synthesis of arylazomethylenetriphenylphosphoranes, **5**, 905
  reactions, with methyl isocyanate, **5**, 681
  for the synthesis of phosphorus-containing heterocycles, **4**, 488
  use in the synthesis of carbodiimides, **5**, 1071
Nitrilimines, *N*-boryl-, synthesis of, **5**, 1148
Nitrilimines, phosphoranyl-, synthesis of, **5**, 1148
Nitrilimines, silyl-, synthesis of, **5**, 1148
Nitrilium ions
  from
    imidoyl halides, **3**, 687
    isonitriles, **3**, 687
    oxime chloroformates, **3**, 687
  as intermediates
    in reaction process, **3**, 687
    in the synthesis of silyloxazolines, **4**, 519
  reactions, with thiols, **5**, 728

synthesis of, **3**, 687; **6**, 515
use in the synthesis of ketenimines, **3**, 566
Nitrilium ions, *N*-alkyl-, reactions, with triethylsilane, **3**, 86
Nitrilium organometallates, synthesis of, **5**, 1147
Nitrilium salts
reactions
with dimethyl sulfoxide, **5**, 339, 689
with enamines, **3**, 408
as starting materials
for *N*-acyl sulfonium salts, **5**, 339
for imidic acid esters, **5**, 685
synthesis of, **2**, 1009; **3**, 687; **5**, 1147
use in the synthesis of amidines, **5**, 751
Nitrilium salts, *N*-methyl-, synthesis of, **3**, 282
Nitrimines, as starting materials, for *N*-nitroamines, **2**, 386
Nitrines
reactions, with isonitriles, **5**, 1071
as starting materials, for nitrilimines, **5**, 1071
Nitrite, amyl
reactions
with 1-dialkylaminocyclohexenes, **2**, 745
with 1,1-disubstituted alkenes, **2**, 786
for the synthesis of aryl iodides, **2**, 632
Nitrite, butyl, reactions, with 1,1-disubstituted alkenes, **2**, 786
Nitrite, *n*-butyl, reactions, with amides, **5**, 35
Nitrite, *t*-butyl
for the synthesis of aryl bromides, **2**, 627
for the synthesis of aryl chlorides, **2**, 621
use in the synthesis of nitrite esters, **2**, 104
Nitrite, ethyl
reactions
with 1-dialkylaminocyclohexenes, **2**, 745
with styrenes, **3**, 429
Nitrite, isoamyl, reactions, with α-chloroacetoacetates, **5**, 824
Nitrite, methyl, reactions, with enol ethers, **3**, 192
Nitrite, pentyl, reactions, with tetrahydrofuran, **1**, 442
Nitrite anions
reactions
with cationic carbyne complexes, **5**, 449
with β-halo ketones, **3**, 191
with α,β-unsaturated ketones, **3**, 191
Nitrite esters, synthesis of, **2**, 103
Nitrites
photolysis of, **2**, 391
reactions, with acrolein, **3**, 46
Nitrites, alkyl
for the diazotization of anthranilic acid, **6**, 99
oxidation of, for the synthesis of aldehydes, **3**, 8
reactions, with arylamines, **1**, 36
synthesis of, **2**, 104
use in the synthesis of acetals, **4**, 193
Nitro acids, as starting materials, for nitro compounds, **2**, 393
ω-Nitroacids, from, nitro ketones, **2**, 396
Nitroalkyl compounds
reactions, with arenediazonium salts, **6**, 653
as starting materials, for dinitroformaldehydehydrazones, **6**, 653
α-Nitroalkyl enones, substitution of, **1**, 251
Nitro anions, Michael addition of, **1**, 309
α-Nitro anions, as intermediates, for the synthesis of nitroalkanes, **4**, 540
α-Nitro carbanions, stability of, **1**, 850
Nitro compounds
acylation of, **3**, 161
C-acylation of, **3**, 190
in aldol reactions, **2**, 394
deoxygenation of, **2**, 782
displacement of, **1**, 486
from
alkenes, **2**, 392
alkyl halides, **2**, 393
amines, **2**, 397
bromoalkanes, **2**, 393
bromo esters, **2**, 393
dimethyldioxirane, **2**, 397
enol acetates, **2**, 392
isocyanates, **2**, 397
nitrates, **2**, 397
nitro acids, **2**, 393
nitroalkenes, **2**, 393
nitrones, **2**, 397
oximes, **2**, 397
phosphazenes, **2**, 397
silyl enol ethers, **2**, 392
urea peroxide, **2**, 397
halogenation of, **4**, 108
as intermediates, **1**, 250
photolysis of, **5**, 325
photolytic rearrangement of, **5**, 325
reactions
with acetone, **3**, 427
with aldehydes, **1**, 252
with carbon disulfide, **3**, 643
with carbon monoxide, **5**, 984
with *N*,*N*-dimethylchloromethyleneiminium chloride, **3**, 427
with ketones, **1**, 252
with organometallics, **2**, 346
with rhodium carbonyl complexes, **5**, 984
with thiols, **5**, 829
with tin(II) complexes, **3**, 426
with tributyltin hydride, **1**, 945
reduction of, **2**, 764, 782; **5**, 324
reductive cyclization of, **5**, 324
as starting materials
for alkyl iodides, **2**, 36
for amines, **2**, 318
for arylamines, **2**, 752
for diazonium salts, **2**, 805
for hydroxylamines, **2**, 341
for imines, **3**, 408
for isocyanates, **5**, 984
for nitriles, **3**, 620
for oximes, **3**, 426
for oxyimidic halides, **5**, 822
for oxyimidic thioesters, **5**, 829
synthesis of, **2**, 391, 783; **5**, 868
via direct nitration, **2**, 391
Nitro compounds, aliphatic
halogenation of, **4**, 108
as starting materials, for oxyimidic amides, **5**, 838
Nitro compounds, alkyl-, reactions, with Grignard reagents, **2**, 346
Nitro compounds, allyl-, synthesis of, **2**, 394
Nitro compounds, allylic
reactions
with allyltrimethylsilane, **1**, 251
with cyanotrimethylsilane, **1**, 252
as reactive substrates, **1**, 250
as starting materials, for allylic oximes, **3**, 430
Nitro compounds, aromatic
reactions, with carbonyl sulfide, **5**, 1033
reduction of, **2**, 772, 775
as starting materials
for azoxy compounds, **2**, 775
for isothiocyanates, **5**, 1033
synthesis of, **2**, 790
Nitro compounds, bromo-
from
2-adamantanone oxime, **4**, 108
triazines, **4**, 109
Nitro compounds, *gem*-chloro-, synthesis of, **4**, 108
Nitro compounds, α-chloro-, as starting materials, for *N*,*O*-dialkylamines, **2**, 353

Nitro compounds, *gem*-fluoro-, from, 2-nitropropane, **4**, 108
Nitro compounds, α-halo-
synthesis of, **4**, 108
via addition of halogens to nitroalkenes, **4**, 108
via halogenation, **4**, 108
via halogenation of nitro compounds, **4**, 108
Nitro compounds, primary
reactions
with ethylthiotrimethylsilane, **5**, 829
with phenylthiotrimethylsilane, **5**, 829
Nitro compounds, β-substituted, reactions, with tributyltin hydride, **1**, 638
Nitro compounds, α,β-unsaturated, synthesis of, **1**, 608
aci-Nitro compounds
reactions, with hydrogen chloride, **5**, 819
as starting materials, for oxyimidic chlorides, **5**, 819
Nitroform, reactions, with azocarboxamides, **6**, 364
Nitroform, iodo-, reactions, with diazo compounds, **4**, 995
Nitrogen
reactions
with cyanides, **5**, 1125
with cyanogen halides, **5**, 1127
and reductive deaminations, **1**, 35
substitution of, **1**, 486
synthesis of functions, **4**, **967–1020**
with another group 15 element, **4**, **451–504**
bearing at least one nitrogen and no chalcogens, **5**, **1061–1089**
one halogen, **6**, 282
one halogen and one oxygen function, **6**, 279
one halogen and one sulfur function, **6**, 281
one halogen and two chalcogens, **6**, 277
one halogen and two phosphorus functions, **6**, 285
one metal, **4**, **505–541**
one metalloid, **4**, **505–541**
with one phosphorus, **4**, 451
two, synthesis of functions, **4**, **403–449**
α-Nitrogen anions, reactions, with acylating reagents, **3**, 182
Nitrogen–antimony functions, synthesis of, **2**, 809
Nitrogen–arsenic functions, synthesis of, **2**, 809
Nitrogen–bismuth functions, synthesis of, **2**, 809
Nitrogen bonds, synthesis of systems containing, **2**, 776
Nitrogen-centered radicals, as starting materials, for cyclic imines, **3**, 410
Nitrogen chloride, bis(triphenylphosphine)-, reactions, with ruthenium carbonyl, **6**, 456
Nitrogen compounds
for the alkylation of aldehydes, **3**, 10
for the alkylation of ketones, **3**, 124
of arsenic functions, synthesis of, **5**, 891
of phosphorus functions, synthesis of, **5**, 880
reactions, with *O*-ethyl selenocarboxylates, **5**, 522
reduction of, **2**, 764
Nitrogen compounds, alkyl-, synthesis of, **2**, **297–332**, **333–370**, **371–423**
Nitrogen compounds, alkynyl-, synthesis of, **2**, **1039–1074**, 1057
Nitrogen compounds, aryl-, synthesis of, **2**, **737–817**
Nitrogen compounds, α-diazo-, synthesis of, **5**, 868
Nitrogen compounds, electrophilic, as starting materials, for α-aminoalkylphosphorus compounds, **4**, 490
Nitrogen compounds, nucleophilic, reactions, with organometal halides, **2**, 891
Nitrogen compounds, vinyl-, synthesis of, **2**, **737–817**
Nitrogen dioxide, reactions, with dichloroalkyne, **4**, 797
Nitrogen fluorides, uses of, **2**, 335
Nitrogen functions
condensation of, **1**, 717
elimination of, **1**, 623, 1075
substitution of, **1**, 250, 1032
synthesis of, **5**, 962, 1022
Nitrogen heteroaryls, methyl-substituted, reactions, with *n*-butyllithium, **2**, 741
Nitrogen heterocycles
ring-opening of, **5**, 1010
silylmethylation of, **4**, 519
Nitrogen heterocycles, saturated, synthesis of, **3**, 187
Nitrogen–hydrogen insertion, of aminoiminoboranes, **5**, 381
Nitrogen–lead bonds, synthesis of, **2**, 815
Nitrogen ligands, use in orthometallation reactions, **2**, 582
Nitrogen–mercury bonds, synthesis of, **2**, 815
Nitrogen–metal bonds, synthesis of compounds with, **2**, 815
Nitrogen–metalloid bonds, synthesis of, **2**, 810
Nitrogen nucleophile anions, reactions, with carbon monoxide, **6**, 517
Nitrogen nucleophiles
reactions
with alkenes, **3**, 187
with chlorohydrazones, **6**, 659
with dichloroalkyne, **4**, 792
with dichloroiminium salts, **6**, 728
with α-halo ketones, **3**, 178
Nitrogenous compounds, use in the synthesis of dialkyl sulfoxides, **2**, 150
Nitrogen oxides, reactions, with isocyanides, **5**, 1004
Nitrogen–oxygen bonds, for the synthesis of *N*-alkyl primary amines, **2**, 351
Nitrogen pentoxide
reactions
with *t*-butyltrimethylsilylethyne, **2**, 1051
with 1,2-diacyl bis(trimethylsilyl)hydrazine, **5**, 362
Nitrogen–phosphorus functions, synthesis of, **2**, 807
Nitrogen-substituted carbene complexes, synthesis of, **6**, 719
Nitrogen substitution, and elimination reactions, **1**, 603
Nitrogen tetrafluoride, reactions, with fluoroethene, **5**, 812
Nitrogen tetroxide, reactions, with *N*-propylthiobenzoylamide, **5**, 361
Nitrogen–tin bonds, synthesis of, **2**, 815
Nitrogen trifluoride
reactions, with phosgene, **6**, 409
as starting material, for carbonic chloride fluoride, **6**, 419
Nitrogen ylides
reactions
with trialkylboranes, **4**, 522
with triarylboranes, **4**, 522
2,3 rearrangements of, **1**, 823
2,3 Stevens rearrangement of, **1**, 823
synthesis of, **2**, 1003
Nitro groups
acidity of, **1**, 155
as activating groups for alkenes, **1**, 300
elimination of, **1**, 637
reduction of, **2**, 752; **3**, 189
and reductive deaminations, **1**, 36
as starting materials, for radicals, **1**, 548
Nitro groups, aryl
displacement of, **1**, 486
reduction of, **1**, 441
Nitrolic acids
as starting materials
for nitrile oxides, **3**, 684
for oxyimidic amides, **5**, 838
synthesis of, **5**, 765, 838
Nitromethyl compounds, dehydration of, **3**, 683
Nitronate anions
reactions, with α,β-unsaturated carbonyls, **3**, 191
reduction of, **2**, 341
Nitronate esters, as starting materials, for thiocyanates, **5**, 1106
Nitronates
as ambident nucleophiles, **1**, 156
hydrolysis to carbonyls, **4**, 541
reactions, with thioketones, **4**, 232
synthesis of, **1**, 155; **4**, 540
Nitronates, acyl, synthesis of, **3**, 410

Nitronates, silyl
  cycloaddition of, **6**, 118
  as starting materials, for β-nitro alcohols, **2**, 394
Nitronate salts
  reactions
    with acylimidazoles, **3**, 46
    with dinitrogen tetroxide, **5**, 838
  as starting materials
    for hydroxynitrilium ion equivalents, **3**, 429
    for oximes, **3**, 429
Nitrone, *N*-benzyl-*C*-phenyl-, oxidation of, **4**, 312
Nitrone, *C*-*t*-butyl-*N*-phenyl-, synthesis of, **3**, 433
Nitrone, *C*,*N*-diphenyl-, reactions, with 1-phenyl-1-pyrrolidinoethene, **4**, 307
Nitrone, fluorenone methyl-, synthesis of, **3**, 432
Nitrones
  [1 + 3] cycloadditions of, **2**, 204
  [1 + 3] dipolar cycloadditions of, **3**, 431
  equilibrium, with hydroxy enamines, **2**, 764
  formylation of, **3**, 43
  from
    hydroxylamines, **3**, 431
    imines, **3**, 432
    oximes, **3**, 433
    secondary amines, **3**, 432
  hydrolysis of, **2**, 342
  nucleophilic addition of, **2**, 345
  oxidation of, **5**, 331
  reactions
    with alkenes, **2**, 352; **3**, 45
    with alkyllithiums, **2**, 343
    with allylamines, **2**, 345
    with aryl isothiocyanates, **5**, 605
    with *o*,*p*-dinitrophenylhydrazone, **3**, 445
    with hydrogen telluride, **3**, 408
    with mercuric chloride, **2**, 274
    with 1,1'-thiocarbonyldiimidazole, **5**, 605
    with 1,1'-thiocarbonyldi(1,2,4-triazole), **5**, 605
    with thioketones, **4**, 232
    with trialkyl phosphites, **5**, 789
  reduction of, **2**, 344
  as starting materials
    for *O*,*N*-acetals, **4**, 312
    for *N*-alkylhydroxylamines, **2**, 342
    for imidoyl-tetracoordinated phosphorus compounds, **5**, 789
    for imines, **3**, 408
    for nitro compounds, **2**, 397
    for oxaziridines, **6**, 110
    for thioketones, **3**, 365
  synthesis of, **2**, 764; **3**, 431
Nitrones, acyclic, synthesis of, **3**, 431
Nitrones, acyl-, synthesis of, **2**, 765
Nitrones, adamantylidene-, synthesis of, **3**, 432
Nitrones, α-aroyl-*N*-phenyl-, synthesis of, **3**, 433
Nitrones, *C*-aryl-*N*-(1-carboxyalkyl)-, synthesis of, **3**, 434
Nitrones, α-aryl-*N*-phenyl-, reduction of, **3**, 412
Nitrones, *N*-carbamoyl-, synthesis of, **3**, 434
Nitrones, chiral
  reactions, with dialkyl phosphites, **4**, 485
  use in the synthesis of *N*-alkoxy secondary amines, **2**, 354
Nitrones, chloro-, synthesis of, **3**, 434
Nitrones, cyclic
  from, allenic oximes, **2**, 765
  reactions, with hydrazine, **3**, 445
  synthesis of, **3**, 434
Nitrones, diaryl-, reactions, with 2,5-diphenyl-1,2,3-diazarsoles, **4**, 502
Nitrones, diphenyl-, reactions, with 1,2,4-diazarsoles, **4**, 502
Nitrones, *N*-glycosyl *C*-dialkoxyphosphoryl-, as starting materials, for α-aminophosphonic acids, **4**, 489
Nitrones, methyl-, from, α-chloronitroso compounds, **3**, 433
Nitrones, *N*-methyl-, synthesis of, **3**, 433
Nitrones, methylthiomethyl-, cyclization of, **3**, 433
Nitrones, phenyl-*t*-butyl-, synthesis of, **3**, 433
Nitrones, polycyclic, synthesis of, **2**, 765
(*E*)-Nitrones, synthesis of, **3**, 434
(*Z*)-Nitrones, synthesis of, **3**, 434
Nitronic acids
  reactions, with dinitrogen tetroxide, **5**, 838
  silyl esters of, **2**, 356
Nitronium salts, for the oxidation of sulfoxides, **2**, 156
Nitronium tetrafluoroborate, use in the synthesis of halogenotrinitromethanes, **6**, 283
4-Nitrophenyl compounds, 2,6-dichloro-, synthesis of, **6**, 606
1-Nitrophex-1-yne, synthesis of, **2**, 1056
Nitrosamides, *N*-acetyl-, as starting materials, for diazo compounds, **1**, 1077
Nitrosamines
  from, nitrosyl chloride, **2**, 798
  isomerization of, **1**, 258
  metallation of, **4**, 529
Nitrosamines, aminomethyl-, synthesis of, **4**, 442
Nitrosamines, aromatic, synthesis of, **2**, 798
Nitrosamines, α-chloroalkyl-, as intermediates, **4**, 312
Nitrosamines, β-keto-, synthesis of, **3**, 194
Nitrosamines, α-metalloalkyl-, synthesis of, **2**, 1002
Nitrosamines, methoxyalkyl-, lithiation of, **4**, 529
N-Nitrosamines, as starting materials, for sulfonyldiazomethanes, **5**, 867
p-Nitrosated compounds, synthesis of, **2**, 780
Nitrosation
  of amines, **2**, 779
  of aromatic ethers, **2**, 781
  of carbonyls, **3**, 469
  of phenols, **2**, 780
  for the synthesis of carboxylic acids, **5**, 108
  for the synthesis of hydroxamoyl halides, **5**, 667
Nitrosation, oxidative, for the synthesis of aromatic nitroso compounds, **2**, 779
Nitrosative deamination, of oxyimidic amides, **5**, 824
Nitroselenation, mechanism, **2**, 785
β-Nitroselenenylation, of alkenes, **2**, 284
Nitroselenylation, process, **2**, 392
o-Nitrosoacetophenone, synthesis of, **2**, 782
Nitrosoamines
  reduction of, **5**, 354
  as starting materials, for *N*-nitrosoamides, **5**, 360
Nitrosoamines, acyl-, reduction of, **5**, 353
N-Nitrosoamines, synthesis of, **1**, 148
Nitroso chloride, methyloleate, reactions, with thujene nitroso chloride, **2**, 778
Nitroso compounds
  carbonylation of, **5**, 984
  dimerization of, **2**, 389
  from
    alkenes, **2**, 391
    amines, **2**, 389
    azo compounds, **2**, 390
    azomethines, **2**, 390
    3,5-dibromo-4-hydroxybenzoic acid, **2**, 781
    hydrocarbons, **2**, 391
    hydroxylamines, **2**, 390
  reactions
    with *N*,*N*-dichloroalkylamines, **2**, 775
    with indole, **3**, 722
    with organometallics, **2**, 346
    with 3-phenylisoxazol-5(4H)-one, **3**, 724
    with sodium cyanide, **5**, 1124
  as starting materials
    for azoxy compounds, **2**, 405
    for diazonium salts, **2**, 805
    for isocyanates, **5**, 984
    for oxyimidic halides, **5**, 822
  synthesis of, **2**, 389, 777; **4**, 314

tautomerization of, **2**, 389
Nitroso compounds, acyl
as chiral auxiliaries, in cycloaddition reactions with alkenes, **2**, 352
ene reaction of, **5**, 326
from, hydroxamic acids, **5**, 326
as starting materials
for hydroxamates, **5**, 330
for hydroxamic acids, **5**, 326
use in Diels–Alder reactions, **5**, 330
Nitroso compounds, alkyl
reactions
with alkenes, **2**, 352, 354
with diazoalkanes, **2**, 765
with dienes, **2**, 354
with Grignard reagents, **2**, 346
with sulfonium ylides, **2**, 765
Nitroso compounds, arenesulfonyl, synthesis of, **2**, 726
Nitroso compounds, aromatic
as starting materials, for azoxy compounds, **2**, 775
synthesis of, **2**, 779
Nitroso compounds, aryl
reactions
with diazoalkanes, **2**, 765
with sulfonium ylides, **2**, 765
Nitroso compounds, chloro, from, 2-bornanones, **4**, 107
Nitroso compounds, α-chloro, as starting materials, for methyl nitrones, **3**, 433
Nitroso compounds, α-halo, synthesis of, via chlorination of oximes, **4**, 107
Nitroso compounds, α,β-unsaturated
from
α-halooximes, **2**, 778
nitroso halides, **2**, 778
vinyl radicals, **2**, 778
synthesis of, **2**, 777
Nitroso compounds, vinyl, synthesis of, **2**, 778
1-Nitroso compounds, 1-chloro-, as intermediates, **5**, 821
2-Nitroso compounds, 2-nitro-, from, cyclohexanone, **4**, 439
4-Nitroso compounds, synthesis of, **2**, 780
C-Nitroso compounds
as intermediates, for oxime carbonates, **6**, 429
isomerization to imines, **3**, 427
α-Nitroso compounds, α-chloro-, use in the synthesis of amines, **2**, 331
Nitrosodecarboxylation, mechanism, **2**, 781
Nitroso dimers
from
azoxy compounds, **2**, 405
hydroxylamines, **2**, 405
as starting materials, for azoxy compounds, **2**, 404
synthesis of, **2**, 405
Nitroso halides, as starting materials, for α,β-unsaturated nitroso compounds, **2**, 778
Nitrosolic acids
reduction of, **5**, 838
as starting materials, for oxyimidic amides, **5**, 838
Nitrosonate anion
synthesis of, **2**, 1002
See also Carbanions, α-nitro
Nitrosonium ions, reactions, with benzylic amines, **3**, 665
aci-Nitro structures, synthesis of, **1**, 156
Nitrosyl cations, reactions, with 1,3-diketones, **3**, 260
Nitrosyl chloride
for the hydrolysis of oximes, **3**, 133
reactions
with amines, **3**, 462
with di-1-hexynylmercury, **2**, 1056
with 2,2-diphenylvinyl amine, **1**, 1118
with ethyl oxalic acid oxyimidic amide, **5**, 825
with mercury bis(alkynides), **2**, 1051
with methylamine, **3**, 462
with methyl ketones, **5**, 823
with *N*-methylmethanimine, **4**, 325
with 2-propanol, **5**, 823
with *N*-silylated (diazomethyl)phosphine sulfides, **5**, 1131
with silyl enol ethers, **3**, 192
as reagent, for the synthesis of nitrile oxides, **3**, 683
as starting material, for nitrosamines, **2**, 798
synthesis of, **3**, 18
use in the synthesis of hydroxamoyl halides, **5**, 665
use in the synthesis of nitrite esters, **2**, 104
use in the synthesis of oxyimidic chlorides, **5**, 821
use in the synthesis of phosphides, **6**, 518
Nitrosyl fluoride, as nitrating agent, **6**, 150
Nitrous acid
for the deamination of amines, **1**, 637
elimination of, **1**, 637
reactions
with alcohols, **2**, 104
with amines, **2**, 798; **3**, 462
with 2-aminomethyl pyrroles, **1**, 259
with 9-amino-10-nitrophenanthrene, **3**, 468
with 18-amino-5α-pregnane-3β,20-diols, **1**, 259
with camphor-3-glyoxylic acid, **5**, 325
with carboxylic acid hydrazides, **5**, 989
with diiodoalkyne, **4**, 797
with glyoxylic acids, **5**, 325
with phenylhydrazine, **2**, 806
with tertiary amines, **2**, 779
with thiosemicarbazides, **5**, 1046
Nitroxide radicals, synthesis of, **1**, 946
Nitroxides
disproportionation of, **2**, 343
as starting materials
for *N*-alkoxyl secondary amines, **2**, 355
for *N*-alkylhydroxylamines, **2**, 343
for *N*,*N*-dialkylhydroxylamines, **2**, 346
Nitroxides, β-phosphorylated, synthesis of, **4**, 485
Nitroxyl fluoride, reactions, with trinitromethane, **6**, 364
Nitryl fluoride
as nitrating agent, **6**, 284
reactions
with ammonium trinitromethanide, **6**, 364
with bis(trimethylsilyl)alkyne, **4**, 807
Noble metals, use in the synthesis of isocyanates, **5**, 984
Nomenclature
carbocations, **1**, 890
of thiohydroxylamines, **2**, 766
Nonactin, synthesis of, **6**, 469
Nona-1,2-diene
reduction of, **1**, 459
as starting material, for 2-(*E*)-alkenes, **1**, 459
Nonaflates, aryl, synthesis of, **1**, 433
Nonane, synthesis of, **1**, 122
n-Nonanes, synthesis of, **1**, 118
Nona-1,2,6-triene, reduction of, **1**, 459
Non-chain reactions
in radical additions to alkenes, **1**, 335
in radical cyclizations, **1**, 345
Non-concerted mechanism
for allyl cation additions, **1**, 363
in cycloaddition reactions, **1**, 351
Non-ethereal solvents, reactions, with zinc compounds, **2**, 566
Non-Kolbé electrolysis, of carboxylic acids, **1**, 561
Norbonyl cations, synthesis of, **2**, 1007
Norbonyl iodides, reactions, with peracetic acid, **1**, 108
Norbornadiene
reactions
with alkynes, **1**, 530
with dichloromethylphosphine, **2**, 448
Norbornadiene, trimethylsilyl-, synthesis of, **2**, 908
2,5-Norbornadiene, synthesis of, **1**, 625
Norbornadiene–molybdenum tetracarbonyl, reactions, with phosphaferrocenes, **4**, 1026

Norbornadienes
synthesis of, **1**, 675
thermolysis of, **4**, 1081
Norbornadienoneazine, as starting material, for diisocyanogens, **3**, 710
Norbornane, *exo*-2-cyano-, synthesis of, **3**, 614
Norbornane, tetranitro-, synthesis of, **4**, 429
Norbornane epoxides, fragmentation of, **4**, 169
Norbornanes, synthesis of, **1**, 67
Norbornanone, reactions, with ceric ammonium nitrate, **5**, 95
Norbornene
reactions
with alkynes, **1**, 529
with dimethyl disulfide, **2**, 138
2-Norbornene, synthesis of, **1**, 625
Norborn-5-ene-2-carbaldehyde, reactions, with chlorosulfene, **5**, 496
Norbornenes, hydrogenation of, **1**, 72
Norbornenes, alkylidene, synthesis of, **1**, 566
Norbornenes, bissilylated, synthesis of, via Diels–Alder reactions, **4**, 619
2-Norbornenes, 7-*syn*-amino-, synthesis of, **5**, 991
Norbornenols, retro-Diels–Alder reaction of, use in the synthesis of enols, **2**, 636
2-Norbornen-7-yl cations, 2,3-dimethyl-7-phenyl-, synthesis of, **1**, 906
2-Norborneol, *endo*-2-methyl-, as starting material, for alkenes, **1**, 576
Norbornyl cations, crystal structure of, **1**, 906
2-Norbornyl cations, studies, **1**, 905
2-Norbornyl system, solvolysis of, **1**, 905
Norcaradienes
from, cycloheptatrienes, **1**, 420
synthesis of, **4**, 557
Norcarane, 7,7-dichloro-, synthesis of, **4**, 16
Norephedrine, use in the synthesis of amines, **2**, 331
(-)-Norephedrine, reactions, with borane, **2**, 314
Norescurinine, synthesis of, **1**, 749
Nornadiene, 7-isopropylidene-2,3-dicarbomethoxy-, use in the synthesis of molybdenum carbene complexes, **3**, 514
Norrish type I cleavage, for the synthesis of ketenes, **3**, 537
Norrish type I fission, of acetones, **4**, 1080
Norrish type II process
intermolecular, use in the synthesis of enols, **2**, 637
involving 1,5-hydrogen transfer, **4**, 1079
use in the synthesis of enols, **2**, 637
Norsecularines, synthesis of, **1**, 626
Nortricyclane, as nonclassical structure, **1**, 906
Noyori reaction, for the synthesis of enolates, **1**, 132
NPP (optical material), use in the synthesis of carboxylic esters, **5**, 124
N-TBDP imines, use in the synthesis of β-lactams, **5**, 377
N-TPS imines, use in the synthesis of β-lactams, **5**, 377
Nucleofuges, uses in photochemical processes, **4**, 1075
Nucleophiles
copper-catalyzed Grignard reagents as, **1**, 185
for the displacement of benzylic sulfones, **1**, 238
for the displacement of sulfur, **1**, 240
hard, **1**, 225
reactions
with acylsilanes, **4**, 930
with acylstannanes, **4**, 951
with alkyl halides, **4**, 761
with alkylidene halides, **4**, 761
with alkynyl thiocarboxylates, **3**, 548
with aminosulfenyl chlorides, **2**, 361
with borenes, **5**, 929
with (bromomethyl)chlorosilanes, **4**, 140
with carbene complexes, **2**, 580
with carbonyl diisothiocyanate, **6**, 565
with carbonyl isocyanate isothiocyanate, **6**, 565
with *N*-chlorocarbonyl isocyanate, **5**, 1007
with chloro(chloromethyl)silanes, **4**, 140
with *N*-cyanocarboimidates, **5**, 1124
with dicobalt hexacarbonyl complexes, **1**, 198
with $\alpha,\beta,\gamma,\delta$-dienoates, **1**, 227
with α,α-dihaloalkylboranes, **6**, 61
with γ,δ-epoxides, **1**, 227
with fluorotrinitromethane, **6**, 279
with *N*-α-haloalkylamides, **4**, 96
with halo(halomethyl)silanes, **4**, 140
with 5-halo-4-isothiazolin-3-ones, **4**, 775
with halosulfenamides, **2**, 239
with iron, **1**, 196
with isocyanides, **6**, 721
with organoboranes, **2**, 523
with ortho esters, **4**, 192, 193
with osmium complexes, **2**, 583
with palladium complexes, **2**, 583
with perfluorinated oxaziridine, **6**, 441
with perfluoroalkenes, **4**, 1076
with platinum complexes, **2**, 583
with propenoyl trimethylsilane, **5**, 416
with ruthenium complexes, **2**, 583
with selenide chloride, **2**, 768
with sulfines, **4**, 859
with trichloro(chloromethyl)germane, **4**, 142
with α,β-unsaturated aldehydes, **1**, 303
with vinyl ethers, **4**, 306
with vinylsilanes, **1**, 451
with ynamines, **4**, 885
soft, **1**, 225
sources of, **2**, 550
use in the substitution of nitrogen functions, **1**, 250
use in the synthesis of ylides, **3**, 496
Nucleophiles, alkenyl, for the substitution of vinyl halides, **1**, 467
Nucleophiles, alkyl
for the substitution of aryl halides, **1**, 470
use in the substitution of vinyl halides, **1**, 466
Nucleophiles, aryl, for the substitution of vinyl halides, **1**, 469
Nucleophiles, heteroatomic, for the ring-opening of epoxides, **2**, 84
Nucleophiles, unsaturated, conjugate addition of, to α,β-unsaturated ketones, **3**, 147
Nucleophiles, vinyl, as starting materials, for butadienes, **1**, 496
Nucleophilic attack, on alkynyl halides, **2**, 1055
Nucleophilic compounds, for the synthesis of organometal(III) compounds, **2**, 889
Nucleophilic displacement, of alkyl alcohol compounds, **1**, 177
Nucleophilic displacement reaction, for the synthesis of aminals, **4**, 417
Nucleophilic substitution, alkynic, for the synthesis of alkynes, **2**, 1011
Nucleosides
protection of, **2**, 540
reactions, with phosphonium salts, **2**, 135
Nucleosides, fluoro, synthesis of, **4**, 79
Nucleotide analogues, physiological properties of, **6**, 263
Nucleotides, hydroxylamino, synthesis of, **2**, 341
Nylon 66, nitrosation of, **3**, 467

O

Octacarbonyl complexes, bis(alkyne)-, synthesis of, **4**, 682
trans-9-Octadecanoic acid, as starting material, for *trans*-9-octandecenedioic acid, **5**, 117
Octa-1,2-diene, as starting material, for 2-(*E*)-alkenes, **1**, 459
1,5-Octadiene, *trans*-1,5-dimethyl-7-*p*-toluenesulfonyl-, carbonation of, **5**, 82
2,4-Octadiyne, metallation of, **6**, 206
Octahydronaphthalene products, unwanted production of, **1**, 498
Octalins
  metallation of, **4**, 532
  reduction of, **1**, 78
  synthesis of, **1**, 101
Octalins, hydroxy-, synthesis of, **1**, 99
cis-Octalones, synthesis of, **1**, 1126
Octanal
  methylenation of, **1**, 759
  reactions, with triphenylarsonium ethylide, **1**, 269
trans-9-Octandecenedioic acid, from, *trans*-9-octadecanoic acid, **5**, 117
Octane, 1-bromo-, as starting material, for 1-octene, **1**, 574
Octane, 2-bromo-
  reactions
    with allylmagnesium bromide, **1**, 114
    with allylsodium, **1**, 113
    with pyrrylmagnesium bromide, **1**, 114
Octane, (+)-2-chloro-, reactions, with ethylsodium, **1**, 112
Octane, 1,8-diamino-, reactions, with carbon disulfide, **6**, 575
Octane, 1,1-difluoro-, synthesis of, **4**, 9
Octane, 2-iodo-, reactions, with halogens, **1**, 108
2,7-Octanedione, from, 1,4-dibromobutane, **1**, 141
Octanoic acid, heptadecafluoro-, synthesis of, **5**, 59
4-Octanone, 3-hydroxy-3-ethyl-, synthesis of, **5**, 437
Octanoyl bromide, synthesis of, **5**, 17
Octanoyl fluoride, synthesis of, **5**, 3
2,4,6-Octatriene, electrocyclization of, **1**, 419
1-Octene
  from, 1-bromooctane, **1**, 574
  reactions, with Grignard reagents, **2**, 561
Oct-1-ene, reactions, with phosphine, **2**, 428
1-Octene, as starting material, for heptanoic acid, **5**, 52
1-Octenes, (*E*)-2-ethyl-1-iodo-, synthesis of, **1**, 278
Oct-4-enes, synthesis of, **1**, 761
2-Octenoic acid, 4-ethyl-2-methyl-, synthesis of, **5**, 98
n-Octyl bromide, reactions, with vinyl selenides, **1**, 1089
Octyne, 1-chloro-, reactions, with tributyltin hydride, **4**, 812
1-Octyne
  hydroalumination of, **1**, 128
  reactions, with ethylmagnesium bromide, **1**, 278
1-Octyne, 1-ethoxy-, reactions, with iodotrimethylsilane, **4**, 765
1-Octyne, 1-trimethylsilyl-, reactions, with allylzinc bromide, **4**, 809
16-Octynyl compounds, synthesis of, **1**, 1050
Oestrone, synthesis of, **1**, 64
(±)-Oestrone, synthesis of, **1**, 448
Olah's reagent
  use in the synthesis of alkyl halides, **2**, 6
  See also Pyridine poly(hydrogen fluoride) (PPHF)
Oleanolate, methyl, synthesis of, **1**, 938
Oleum
  use in the synthesis of carbonic chloride fluoride, **6**, 419
  use in the synthesis of phosgene, **6**, 412, 413
Oligocarbometallation, of alkynes, **2**, 978
Oligomerization
  of thioaldehydes, **3**, 330
  of thioketones, **3**, 330
Oligomerization, cyclic, of alkynes, catalyzed by transition metals, **2**, 984
Olivetol, prepartion of, **1**, 432
One-carbon homologation, for the synthesis of α,β-alkynoic esters, **5**, 160
Onium rearrangement
  in the synthesis of tertiary ynamines, **2**, 1043
  in the synthesis of ynediamines, **2**, 1047
Onium salts
  cleavage of, **1**, 943
  reactions, with butyllithium, **6**, 708
  use in the synthesis of ylides, **3**, 500
Ophibolins, synthesis of, **1**, 1099
Opoponax oil, constituents of, **1**, 757
Oppenauer oxidation
  for the synthesis of aldehydes, **3**, 7
  for the synthesis of ketones, **3**, 119
Organic acids, reactions, with amines, **2**, 759
Organic bases, for the synthesis of anhydrides, **5**, 184
Organic compounds, metallation of, **1**, 881
Organic halides
  metal–halogen exchange of, **2**, 956
  oxidative metallation of, **2**, 952, 956
  reactions
    with alkenyl alanes, **1**, 542
    with alkynes, **1**, 1137
    with anhydrous hydrogen fluoride, **6**, 9
    with bis(triorganotin) sulfides, **2**, 131
    with lithium bases, **2**, 554
    with magnesium, **2**, 558
    with metals, **2**, 957
    with trimethylsilylalkyne, **1**, 1137
Organic molecules, as synthetic metals, **3**, 648
Organic reaction prediction software (OGOR2), for the study of enamine hydroboration, **4**, 524
Organoalanes
  elimination of, **1**, 662
  reactions, with alkynes, **2**, 974
  synthesis of, **4**, 659
Organoalkali compounds, x-ray crystallographic studies of, **1**, 885
Organoaluminum cyanides, synthesis of, **5**, 1142
Organoaluminum-promoted Beckmann rearrangement, of sulfonimines, **3**, 437
Organoaluminums
  carboxylation of, **5**, 44
  β-elimination of, **1**, 663
  for the hydroalumination of alkynes, **1**, 128
  for the opening of epoxides, **1**, 204
  reactions
    with alkynes, **1**, 542
    with allyl halides, **1**, 973
    with amino ethers, **1**, 973
    with bromine, **4**, 758
    with carbon dioxide, **1**, 973
    with carbonyls, **1**, 971
    with enol phosphates, **1**, 480
    with iminium salts, **1**, 973
    with iodine monochloride, **4**, 758
    with triethyl orthoformate, **1**, 971
  synthesis of, **6**, 373
  use in Claisen rearrangements, **1**, 826
  use in conjugate addition reactions, **1**, 311
  use in the synthesis of alkyl aryl ketones, **3**, 289

Organoantimony, chemistry of, **1**, 271
Organoantimony alkoxides, reactions, with aldehydes, **4**, 198
Organoantimony chlorides, reactions, with ureas, **5**, 1017
Organoantimony compounds
  reactions
    with aldehydes, **2**, 73
    with carbonyls, **1**, 272
  safety precautions, **2**, 481
  synthesis of, **6**, 340
    reviews, **2**, 481
Organoarsenic(III) anhydrides, reactions, with hydrogen sulfide, **2**, 890
Organoarsenic(III) dihalides, reactions, with hydrogen sulfide, **2**, 890
Organoarsenic(III) diiodides, synthesis of, **1**, 445
Organoarsenic compounds
  from, aminoarsines, **2**, 504
  safety precautions, **2**, 481
  synthesis of, reviews, **2**, 481
Organoarsenic dihalides
  reactions
    with amides, **2**, 892
    with primary amines, **2**, 892
Organoarsenic halides
  reactions
    with alkynes, **2**, 875
    with *N*-silyllithium amides, **2**, 891
  synthesis of, **2**, 504
Organoarsines
  as starting materials
    for arsenoimines, **2**, 892
    for sulfur compounds, **2**, 890
Organoarsinic acids, synthesis of, **2**, 504
Organoarsonic acids, synthesis of, **2**, 504
Organobariums
  chemistry of, **2**, 562
  selectivity of, **1**, 116
  synthesis of, **1**, 116
Organoberylliums
  as carbon nucleophiles, **2**, 563
  chemistry of, **2**, 562
Organobis(cuprates), reactions, with 3-halocycloalk-2-enones, **1**, 120
Organobismuth compounds
  alkylation of, **6**, 340
  safety precautions, **2**, 481
  synthesis of, reviews, **2**, 481
Organo *gem*-biszincs, reactions, with phenylsulfonyl chloride, **2**, 998
Organoboranes
  carbonylation of, **4**, 363
  cleavage of, **1**, 60
  halolysis of, **2**, 4
  in Heck reactions, **1**, 493
  hydroboration of, **2**, 40
  as intermediates, in the synthesis of primary alcohols, **1**, 316
  isomerization of, **2**, 522
  oxidation of, **2**, 97, 523
  palladium-catalyzed coupling of, **1**, 488
  reactions
    with acetic acid, **1**, 58
    with acrylates, **1**, 276
    with aldehydes, **2**, 531
    with α,β-alkynic ketones, **1**, 274
    with alkynyltins, **4**, 665
    with aryl triflates, **1**, 482
    with chloramine, **2**, 305
    with ^{13}C-labeled carbon monoxide, **5**, 45
    with hydroxylamine-*O*-sulfonic acid, **2**, 305
    with ketones, **2**, 531
    with methyl vinyl ketone, **1**, 273
    with nucleophiles, **2**, 523
    with (phenyldimethylsilyl)chloromethyllithium, **4**, 357
    with potassium hydride, **2**, 534
    with propargylides, **1**, 976
    with (trimethylsilyl)bromomethyllithium, **4**, 357
    with α,β-unsaturated epoxides, **1**, 277
    with α,β-unsaturated ketones, **1**, 274
  redistribution of, **2**, 522
  as starting materials
    for aldehydes, **1**, 273; **3**, 19
    for alkyl hydroperoxides, **2**, 97
    for carbon-centered radicals, **1**, 946
    for ketones, **3**, 134
    for nitriles, **3**, 622
  synthesis of, **2**, 1085; **5**, 380
    reviews, **2**, 514
    via alkylation α-to boron, **2**, 525
  transmetallation of, **1**, 1020
  use in the substitution of boron functions, **1**, 273
  use in the synthesis of ^{13}C-labeled oxygen-substituted alkanoic acid, **5**, 71
  use in the synthesis of radicals, **1**, 925
Organoboranes, β-alkoxy-, *syn* elimination of, **1**, 655
Organoboranes, α-lithio-, synthesis of, **4**, 664
Organoboranes, monoisopinocampheyl-, as starting materials, for hydroborating agents, **1**, 487
Organoboranes, propargylic
  reactions
    with allylic halides, **1**, 977
    with carbon dioxide, **1**, 977
    with carbonyls, **1**, 977
Organoboranes, trialkyl-, reactions, with bis(trimethylsilylethynyl)dimethylplumbane, **2**, 1102
Organoborate salts, reactions, with methyl iodide, **2**, 524
Organoboron molecules, formation of, **5**, 1096
Organoborons
  reactions
    with alkynes, **1**, 542
    with enol triflates, **1**, 480
  synthesis of, **1**, 125
Organocadmium halides, as reagents, for the synthesis of ketones, **3**, 130
Organocadmiums
  for the alkylation of antimony and bismuth trihalides, **2**, 483
  reactions, with haloalkanes, **1**, 127
  synthesis of, **6**, 64
  use in the synthesis of aromatic ketones, **3**, 289
Organocalciums
  chemistry of, **1**, 115; **2**, 562
  from, fluorobenzene, **1**, 116
  synthesis of, **1**, 115
Organoceriums
  from, organolithiums, **2**, 69
  reactions
    with hydrazones, **1**, 1042
    with nitriles, **2**, 312, 316
  synthesis of, **2**, 564; **4**, 954
Organochromiums, synthesis of, **6**, 340
Organocobaloximes, as starting materials, for sugars, **1**, 335
Organocobalt complexes
  photolysis of, **1**, 345
  as starting materials, for alkenes, **1**, 335
Organocobalt group transfer
  in radical cyclizations, **1**, 345
  for the synthesis of radicals, **1**, 335
Organocobalts
  from, cyclopentadienylcobalt complexes, **6**, 341
  as intermediates, **2**, 985
  synthesis of, **6**, 333, 340, 346, 352
Organocopper(I) compounds
  reactions
    with alkynenitriles, **3**, 642

with alkynyl halides, **1**, 1017
with alkynyl iodonium salts, **1**, 1018
synthesis of, **1**, 967
Organocopper chemistry, reviews, **1**, 116
Organocopper cyanides, synthesis of, **5**, 1140
Organocopper reactions, catalytic, **1**, 119
Organocoppers
acylation of, **3**, 286
chemistry of, **2**, 585
conjugate addition, to α,β-unsaturated carbonyls, **2**, 702
from, alkyl iodides, **3**, 287
nucleophilic substitution with, **1**, 966
for the opening of epoxides, **1**, 204
reactions
with acrylic esters, **5**, 151
with aldehydes, **2**, 69
with alkoxyalkynes, **2**, 676
with alkynes, **1**, 539; **2**, 976
with α-alkynic ketones, **3**, 222
with allenyl halides, **1**, 1015
with 1-ethynylsilanes, **4**, 940
with propargylic ethers, **1**, 667
with selenocyanogen, **5**, 1115
with trimethyl silylethyne, **4**, 809
with triphenyl silylethyne, **4**, 809
with vinylphosphorus compounds, **1**, 159
reactivity of, **1**, 311
selectivity of, **1**, 301
as starting materials, for allenylsilanes, **2**, 940
substitution reactions, **1**, 115
synthesis of, **1**, 116; **3**, 128
thienyl-based, **1**, 121
use in the synthesis of allenes, **1**, 966
Organocoppers, thienyl-based, synthesis of, **1**, 121
Organocuprates
commercial availability of, **1**, 312
for conjugate addition to enals, **3**, 12
conjugate addition to enones, **3**, 128
reactions
with acid chlorides, **3**, 130
with 1,3-alkadien-2-yl diethyl phosphates, **1**, 970
with alkynes, **3**, 642; **4**, 772
with alkynic sulfides, **4**, 963
with 9-borabicyclo[3.3.1]nonyl, **2**, 522
with enones, **1**, 313
with ortho esters, **4**, 192
with (*E*)-3-penten-2-one, **1**, 311
with propargylic acetates, **1**, 960
with unactivated alkynes, **1**, 540
with α,β-unsaturated amides, **1**, 313
with α,β-unsaturated carbonyls, **1**, 540
with vinylic epoxides, **2**, 86
as starting materials
for ketones, **3**, 130
for vinyl cuprates, **2**, 629
synthesis of, **1**, 311, 1136
tolerance of functional groups, **1**, 120
use in the synthesis of furanyl ketones, **3**, 303
Organocuprates, chiral α-alkoxy-, synthesis of, **4**, 385
Organodiarsines, synthesis of, **2**, 893
Organodihalides
reactions
with bisphosphines, **2**, 444
with monophosphines, **2**, 444
Organodisilanes, reactions, with *N*-bromosuccinimide, **5**, 378
Organodistibines, synthesis of, **2**, 893
Organoelement halides, reactions, with lithium bis(dialkoxyboryl)methides, **6**, 200
Organofluorine, synthesis of, **2**, 803
Organogallium cyanides, synthesis of, **5**, 1143
Organogallium dichlorides, reactions, with dilithium alkylphosphides, **2**, 475
Organogermanium cyanides, synthesis of, **5**, 1136
Organogermanium halides
reactions
with alkynyl metals, **2**, 1092
with halomethanes, **6**, 386
with mercury(II) cyanides, **5**, 1136
with silver cyanide, **5**, 1136
Organogermaniums, reactions, with 1-alkynes, **2**, 1093
Organogermaniums, alkyl-, reactions of, **1**, 290
Organogold cyanides, synthesis of, **5**, 1140
Organogolds, synthesis of, **6**, 343, 349, 355
Organohafnium compounds, synthesis of, **6**, 340
Organohalides, reactions, with metal phosphides, **2**, 429, 444
Organohalosilanes, Wurtz coupling of, **2**, 544
Organoindium cyanides, synthesis of, **5**, 1143
Organoiron(II) compounds, reactions, with cyclohexanones, **2**, 70
Organoiron complexes, from, iron carbene complexes, **6**, 332
Organoirons
from, dithioformate ester complexes, **6**, 332
synthesis of, **6**, 332, 346, 351
Organolanthanides
as catalysts for hydroboration, **2**, 43
reactions, with carbonyls, **2**, 69
synthesis of, **2**, 564, 565
Organolead alkoxides, as starting materials, for alkynylplumbanes, **2**, 1101
Organolead cyanides
from, hexaphenyldilead, **5**, 1145
synthesis of, **5**, 1145
Organolead halides, reactions, with amines, **2**, 816
Organolead hydroxides, as starting materials, for alkynylplumbanes, **2**, 1101
Organoleads, synthesis of, **2**, 601, 868
Organoleads, β-substituted, elimination of, **1**, 664
Organolithium, reactions, with chlorodialkylalane, **2**, 990
Organolithium nucleophiles, reactions, with vinyl sulfides, **4**, 366
Organolithiums
acylation of, **3**, 284
alkylation of, with vinyl halides, **1**, 466
amination of, **2**, 305
as bases, for the synthesis of ylides, **3**, 493
for the carbometallation of allenes, **1**, 495
chemistry of, **2**, 551
deuterolysis of, **1**, 1090
elimination of, **1**, 661
β-elimination of, **1**, 660
from, tributylstannanes, **4**, 652
for the generation of anions, **1**, 155
for halogen–lithium exchange reactions, **4**, 652
in Heck reactions, **1**, 493
for the opening of epoxides, **1**, 204
reactions
with acid chlorides, **3**, 212
with acyl dieneiron tricarbonyl complexes, **3**, 30
with aldimines, **2**, 312
with alkynes, **1**, 539; **2**, 972
with alkynyl halides, **1**, 1017
with alkynyl sulfones, **1**, 1032
with aminals, **3**, 42
with arsines, **2**, 510
with betaine–lithium complexes, **1**, 729
with 1,1-bis(silyl)alkenes, **4**, 618
with bis(trimethylsilyl)formamide, **3**, 440
with α-bromophosphonium salts, **5**, 895
with caprothionolactone, **4**, 226
with carbodiimides, **5**, 743
with carbon dioxide, **3**, 284
with carbon monoxide, **5**, 436
with carboxylic acids, **3**, 284
with ceric chloride, **4**, 954
with cerium chloride, **1**, 1098

with cerium iodide, **1**, 1098
with chloroboranes, **2**, 520
with chloro(vinyl)silanes, **4**, 651
with copper(I) halides, **2**, 1095
with cyanoamides, **5**, 743
with cyclopentadienyllanthanide complexes, **2**, 564
with dialkyl phosphorochloridates, **2**, 848
with *gem*-diborylalkanes, **4**, 665
with disilylmethanes, **6**, 379
with 1,1-dithio-1,3-dienes, **4**, 254
with dithioesters, **4**, 255
with enynes, **1**, 960
with 1-ethoxy-1-hexyne, **1**, 1030
with gallium trichloride, **2**, 597
with germaethenes, **6**, 192
with 2-halo-2-alkenoyl silanes, **1**, 985
with hydrazones, **3**, 44
with imines, **1**, 310
with indium trichloride, **2**, 597
with iron pentacarbonyl, **3**, 90
with isonitriles, **5**, 796
with isothiocyanates, **5**, 595
with ketimines, **2**, 312
with ketones, **1**, 693; **2**, 68
with metal carbonyls, **6**, 719
with metalloidal halides, **6**, 186
with oximes, **2**, 764
with α-(phenylthio)vinylsilanes, **4**, 366
with phosphinous halides, **2**, 435
with phosphonous halides, **2**, 435
with phosphorus trichloride, **2**, 434
with polyborylated methanes, **6**, 199
with rhenium pentacarbonyls, **5**, 449
with silaethenes, **6**, 192
with silanes, **2**, 536
with silyl electrophiles, **4**, 602
with stibines, **2**, 510
with sulfenylimines, **2**, 359
with β-sulfinyl nitroalkenes, **2**, 784
with thioketones, **4**, 392
with thionolactones, **4**, 226
with *N*-trimethylsilyliminium salts, **2**, 313
with trimethyl(vinyl)silane, **4**, 651
with triphenyl(vinyl)silane, **4**, 651
with trithiocarbonates, **6**, 317
with α,β-unsaturated aldehydes, **1**, 310
with α,β-unsaturated amidate anions, **4**, 651
with vinyldialkoxyboranes, **1**, 487
with vinylgermanes, **4**, 663
with vinylic sulfones, **1**, 310
with vinyl phosphonates, **3**, 197
with vinyl selenides, **4**, 375, 400
with vinyl selenones, **4**, 400
with vinyl selenoxides, **4**, 400
with vinylsilanes, **1**, 310; **4**, 366, 651
with vinyl sulfides, **4**, 392
reactivity of, **1**, 310
as reagents, for the dehydrohalogenation of dichloroalkanes, **1**, 1066
selectivity of, **1**, 301
as starting materials
for alkenylzincs, **1**, 468
for α-aluminosilylalkanes, **4**, 659
for lithium alkynides, **1**, 1069
for organoceriums, **2**, 69
for thiocyanates, **5**, 1110
stereochemistry of coupling, **1**, 112
synthesis of, **1**, 398; **2**, 961; **6**, 373
for the synthesis of 1,1-bis(silyl)alkanes, **4**, 602
for the synthesis of sulfinic acids, **2**, 186
for the synthesis of tertiary phosphines, **2**, 434
transmetallation of, **2**, 990
use in the synthesis of alkanethiols, **2**, 125
use in the synthesis of carbon-bridged copper and silver compounds, **4**, 695
use in the synthesis of carboxylic acids, **5**, 27
use in the synthesis of imidoyllithium intermediates, **5**, 796
1,2 Wittig rearrangement of, **1**, 398
x-ray crystal structures of, **1**, 110
Organolithiums, α-alkoxy-
configurational stability of, **1**, 138
synthesis of, **4**, 378
Organolithiums, benzylic, reactions, with benzyl fluorides, **1**, 108
Organolithiums, geminal dihalo-, synthesis of, **1**, 164
Organolithiums, α-oxygenated, synthesis of, **1**, 138
Organolithiums, β-substituted, elimination of, **1**, 661
Organomagnesiocuprates, reactions, with silylalkynes, **1**, 1094
Organomagnesium halides
reactions
with alkynyl halides, **1**, 1018
with 3-bromo-1-(phenylsulfonyl)propene, **1**, 144
stereochemistry of reactions, **1**, 114
Organomagnesiums
α,α-addition, to isonitriles, **5**, 796
for the opening of epoxides, **1**, 204
pyrolysis of, **1**, 662
reactions
with alkynes, **2**, 972
with dithioesters, **6**, 134
with metalloidal halides, **6**, 186
with phosphine oxides, **1**, 988
with silanes, **2**, 536
reactivity of, **1**, 310
selectivity of, **1**, 301
as starting materials, for α-aluminosilylalkanes, **4**, 659
synthesis of, **2**, 558
transmetallation of, **2**, 991
use in nucleophilic substitution reactions, **1**, 970
use in the synthesis of carbon-bridged copper and silver compounds, **4**, 695
use in the synthesis of carboxylic acids, **5**, 27
use in the synthesis of trifluorovinylsilanes, **4**, 806
Organomagnesiums, α-germyl, as starting materials, for α-thiogermanes, **4**, 371
Organomagnesiums, β-substituted, elimination of, **1**, 662
Organomanganates, use in the synthesis of furanyl ketones, **3**, 303
Organomanganese(II) chloride, copper-catalyzed substitution of, **1**, 125
Organomanganese(II) compounds
from
manganese chloride, **2**, 574
manganese diiodide, **2**, 574
synthesis of, via transmetallation reactions, **2**, 574
Organomanganese chlorides, for the alkylation of primary bromides, **1**, 465
Organomanganese complexes, synthesis of, **2**, 574
Organomanganese compounds
synthesis of, **6**, 330, 347
use in the synthesis of α-chloro ketones, **3**, 152
Organomanganese halides
reactions, with acid chlorides, **3**, 130
as starting materials, for ketones, **3**, 130
Organomercury(II) halides, reduction of, **1**, 68
Organomercury carbenoid, reactions, with 1,1,3,3-tetramethyl-1,3-disilacyclobutane, **4**, 623
Organomercury compounds
exchange reactions between, **6**, 270
handling problems, **2**, 1060
in Heck reactions, **1**, 493
palladium catalysis of, **1**, 230
photolysis of, **1**, 1143
protonolysis of, **1**, 65

reactions
with allyl halides, **1**, 128
with aluminum, **4**, 698
with *n*-butyllithium, **6**, 205
with haloalkanes, **1**, 128
with iodine monochloride, **4**, 750
with thiols, **1**, 68
with trityl chloride, **1**, 128
reduction of, **1**, 454
safety precautions, **2**, 1097
as starting materials
for alkylpotassiums, **2**, 556
for alkylsodiums, **2**, 556
for α-bromo ylides, **5**, 896
for the synthesis of organometallics, **2**, 590
toxicity of, **2**, 1097
Organomercury compounds, α-alkoxy-, synthesis of, **4**, 385
Organomercury compounds, allenic, reactions, with acid chlorides, **1**, 1060
Organomercury compounds, hydroperoxy-, use in the synthesis of *t*-butylalkylperoxides, **2**, 98
Organomercury cyanides, synthesis of, **5**, 1141
Organomercury halides
reactions, with diethylvinyl phosphonate, **4**, 587
as starting materials, for alkynylmercury compounds, **2**, 1097
Organometal(III) compounds
synthesis of, **2**, 872
via nucleophilic displacement reactions, **2**, 889
via reductive processes, **2**, 879
Organometal(III) dihalides
synthesis of
via disproportionation reactions, **2**, 882
via transmetallation reactions, **2**, 882
Organometal(III) halides, synthesis of, via halogenation, **2**, 881
Organometal(V) chlorides, synthesis of, **2**, 883
Organometal(V) compounds, synthesis of, via displacement reactions, **2**, 890
Organometal(V) halides
reactions, with Grignard reagents, **2**, 881
as starting materials
for pentaorganoantimony, **2**, 873
for pentaorganoarsenic, **2**, 873
for pentaorganobismuth, **2**, 873
Organometal(V) imines, *N*-sulfonyl
as starting materials
for pentaorganoantimony, **2**, 873
for pentaorganoarsenic, **2**, 873
for pentaorganobismuth, **2**, 873
Organometal halides, reactions, with nucleophilic nitrogen compounds, **2**, 891
Organometallic addition, for the synthesis of α-methylene lactams, **5**, 296
Organometallic carbanions, substitution by, **1**, 479, 482
Organometallic carbodiimides
reactions of, **5**, 1081
synthesis of, **5**, 1081
Organometallic catalysts
for hydrogen cyanide addition, to alkenes, **3**, 614
for the synthesis of (*Z*)-alkenes, **1**, 456
Organometallic chemistry, reviews, **1**, 885
Organometallic complexes
use in the synthesis of alkenyl ethers, **2**, 681
use in the synthesis of allenes and cumulenes, **1**, 991
Organometallic complexes, η^2-acyl, as starting materials, for ketenimines, **3**, 598
Organometallic complexes, triene, dihydroxylation of, **2**, 47
Organometallic groups, displacement of, **3**, 664
Organometallics
acylation of, **3**, 289
alkynylation of, **1**, 1019
amination of, **2**, 305
and carbanions, **1**, 845
carbonation of
for the synthesis of amino acids, **5**, 88
for the synthesis of aromatic acids, **5**, 110
for the synthesis of carboxylic acids, **5**, 27
for the synthesis of halo alkanoic acids, **5**, 58
for the synthesis of heteroatom-substituted alkanoic acids, **5**, 97
for the synthesis of nitrogen-substituted alkanoic acids, **5**, 95
for the synthesis of oxygen-substituted alkanoic acids, **5**, 68
for the synthesis of sulfur-substituted alkanoic acids, **5**, 82
for the synthesis of α,β-unsaturated acids, **5**, 100
for the synthesis of unsubstituted alkanoic acids, **5**, 42
as catalysts, for the synthesis of asymmetric cyanohydrins, **3**, 628
chemistry of, **2**, 550
chlorination of, **6**, 270
electrophilic acylation of, **3**, 284
equilibration of, **1**, 1054
formylation of, **3**, 14
from
carbon–halogen bonds, **1**, 106
diazonium salts, **2**, 876
halogenation of, **4**, 774
number of theoretical combinations of, **6**, 378
preparative methods, reviews, **6**, 172
quenching of, **4**, 757, 760
reactions
with acid chlorides, **3**, 198
with aldonitrones, **2**, 345
with alkenyldimesitylboranes, **6**, 201
with alkoxyboranes, **2**, 520, 529
with alkylantimony halides, **2**, 483
with alkylarsenic halides, **2**, 483
with alkylbismuth halides, **2**, 483
with alkynes, **1**, 538
with alkynyl fluorides, **1**, 1017
with allyl halides, **1**, 119
with amino acids, **3**, 180
with α-amino aldehydes, **3**, 39
with aminoarsenic halides, **2**, 503
with aminosulfenyl chlorides, **2**, 359
with antimony(III) halides, **2**, 872
with antimony halides, **2**, 483
with antimony trihalides, **2**, 488
with arsenic(III) halides, **2**, 872
with arsenic halides, **2**, 483
with arsenic trihalides, **2**, 488
with bisdialkylaminochlorophosphines, **2**, 467
with bismuth(III) halides, **2**, 872
with bismuth halides, **2**, 483
with bismuth trihalides, **2**, 488
with bisphosphinoalkenes, **4**, 1029
with boranes, **2**, 529
with carbodiimides, **5**, 754
with carbonyl groups, **2**, 67
with carbonyls, **1**, 692
with chloroarsenic amides, **2**, 891
with chromium trichloride, **2**, 571
with copper(I) salts, **2**, 586
with cyanohydrins, **3**, 157
with dialkylaminoboranes, **2**, 520
with dialkylaminochlorophosphines, **2**, 448
with dicarbonyls, **4**, 166
with 1,1-dichloro-2,2-difluoroethene, **1**, 1069
with 1,3-dichloro-1,3-dimethyl-1,3-disilacyclobutane, **4**, 621
with dithioesters, **4**, 255
with electrophiles, **2**, 564; **3**, 289
with furanyl aldehydes, **3**, 303

with halides, **6**, 186
with haloalkanes, **1**, 109
with haloboranes, **2**, 520, 526
with α-haloenones, **3**, 217
with isocyanides, **5**, 68
with lanthanide salts, **2**, 564
with lead(II) chloride, **2**, 602
with nitriles, **3**, 440
with nitro compounds, **2**, 346
with nitroso compounds, **2**, 346
with ortho esters, **4**, 192
with oxime ethers, **2**, 351
with α-oxoketene dithioacetals, **1**, 485
with oxygen, **2**, 96
with palladium salts, **2**, 581
with phenylphosphonous acid dibornyl esters, **2**, 443
with phenylphosphonous acid dimenthyl esters, **2**, 443
with phenyl (1-trimethylsilyl)propadienyl sulfide, **4**, 946
with phosphinous halides, **2**, 436
with phosphonous dihalides, **2**, 436
with phosphorous halides, **2**, 841
with phosphorus, **2**, 433
with platinum salts, **2**, 581
with selenium, **2**, 278
with β-silylvinyl sulfones, **4**, 393
with sulfines, **4**, 272
with sulfinyl compounds, **2**, 363
with sulfur, **2**, 125
with sulfur dioxide, **2**, 176, 186
with trialkylantimony dihalides, **2**, 488
with vinylsilanes, **1**, 759; **6**, 192
reactivity of, **1**, 885
selectivity of, **1**, 301
as starting materials
for α-amino acetals, **3**, 40
for carbanions, **1**, 885
for ketenimines, **3**, 596
substitution of, **1**, 291
sulfur insertion reactions of, **2**, 125
synthesis of, **5**, 874; **6**, 186
classification of methods, **2**, 954
reviews, **2**, 872
use in the synthesis of alkylaminoboranes, **2**, 533
use in the synthesis of amines, **2**, 312
use in the synthesis of dialkyl sulfides, **2**, 134
use in the synthesis of phosphinic acids, **2**, 453
use in the synthesis of tertiary phosphines, **2**, 436, 824
Organometallics, alkenyl, metalla–Claisen carbozincation of, **2**, 1006
Organometallics, alkyl, alkynylation of, **1**, 1000
Organometallics, allenyl
reactions
with carboxylic esters, **1**, 1060
with electrophiles, **1**, 971
with iminium salts, **1**, 1059
Organometallics, allyl, reactions, with aldehydes, **3**, 31
Organometallics, anionic, reactions, with halobenzene–tricarbonylchromium complexes, **4**, 722
Organometallics, aryl
oxidation of, **2**, 652
reactions, with alkenyl halides, **1**, 471
as starting materials, for benzaldehydes, **3**, 89
Organometallics, α-boryl, alkylation of, **1**, 163
Organometallics, fluorinated, synthesis of, **6**, 266
Organometallics, α-hetero-substituted, synthesis of, **1**, 310
Organometallics, propargyl
reactions
with carboxylic esters, **1**, 1060
with electrophiles, **1**, 971
with iminium salts, **1**, 1059
Organometallics, α-silyl, advantages of, **1**, 759
Organometallics, α-silylated
reactions, with germanium chlorides, **6**, 186
synthesis of, **6**, 189
Organometallics, vinyl
alkylation of, **1**, 465
reactions, with vinyl halides, **1**, 1096
Organometallics, vinylcyclopropyl, reactions, with β-substituted cycloalkenones, **1**, 811
Organometallics, vinylic
reactions, with electrophiles, **1**, 463
as starting materials, for vinyl halides, **2**, 607
Organometal–mercury compounds, synthesis of, **4**, 724
Organometals
oxidative addition of, **2**, 957
oxidative displacement of, **2**, 957
reactions
with alkynes, **2**, 875
with arenes, **2**, 875
with metal halides, **2**, 990
starting materials for, **2**, 954
synthesis of, **2**, 953
transmetallation of, **2**, 956
Organometals, allylic, synthesis of, **2**, 964
Organometals, benzylic, synthesis of, **2**, 964
Organomolybdenum complexes
reactions
with manganese complexes, **6**, 328
with organorutheniums, **6**, 329
Organomolybdenums, synthesis of, **6**, 327
Organonickel cyanides
synthesis of, **5**, 1137
via oxidative addition reactions, **5**, 1138
Organonickels, synthesis of, **6**, 332, 346
Organoniobiums, reactions, with carbodiimides, **5**, 781
Organoosmiums, synthesis of, **6**, 344
Organopalladium complexes, synthesis of, **6**, 336
Organopalladium cyanides
synthesis of, **5**, 1138
via oxidative insertion reactions, **5**, 1139
Organopalladiums
for the substitution of alkenes, **1**, 493
synthesis of, **6**, 334
Organophosphides, dimetallo-, use in the synthesis of tertiary phosphines, **2**, 438
Organophosphines, 1,2-elimination reactions, of, **6**, 678
Organophosphorus bis(isothiocyanates), use in the synthesis of acyl isothiocyanates, **5**, 1048
Organophosphorus compounds
reactions
with imidate ethers, **5**, 784
with imidoyl chlorides, **5**, 784
reviews, **4**, 451
as starting materials, for dialkylketenes, **3**, 541
for the synthesis of alkylphosphines, **1**, 157
as thionating agents, **5**, 570
for the thionation of amides, **5**, 575
Organophosphorus compounds, α-amino-, synthesis of, via Mannich type reactions, **4**, 452
Organophosphorus compounds, α-aminoalkyl-
from
azines, **4**, 460
hydrobenzamide, **4**, 460
ketazines, **4**, 460
nitriles, **4**, 464
oximes, **4**, 464
α-oxophosphorus compounds, **4**, 471
Schiff's bases, **4**, 461
synthesis of
via conjugate addition, **4**, 460
via Hofmann reactions, **4**, 472
via Kabachnik–Fields reactions, **4**, 469
Organophosphorus compounds, α-nitroalkyl-, synthesis of, **4**, 493
Organophosphorus isothiocyanates, use in the synthesis of isothiocyanates, **5**, 1041

Organophosphorus selenocyanates, reactions, with alcohols, **5**, 1055
Organoplatinum cyanides
synthesis of, **5**, 1139
via displacement of neutral ligands, **5**, 1139
via oxidative insertion reactions, **5**, 1140
Organoplatinums, synthesis of, **6**, 332, 351
Organoplumbanes, 2-alkoxy-, reactions, with aldehydes, **2**, 70
Organopotassiums
elimination of, **1**, 663
stability of, **2**, 556
substitution reactions, **1**, 113
Organorhodiums, synthesis of, **6**, 347
Organoruthenium
reactions, with organomolybdenum complexes, **6**, 329
synthesis of, **6**, 334, 343, 348
Organosamarium(III) compounds, as intermediates, **2**, 565
Organosamariums, reactions, with electrophiles, **1**, 975
Organoselenium electrophiles, reactions, with alkenes, **2**, 286
Organoseleniummanganese complexes, synthesis of, **6**, 49
Organoseleniums
synthesis of, **6**, 334, 354
transmetallation of, **1**, 245
uses of, **1**, 437
Organoselenoarsines, synthesis of, **2**, 500
Organosilanes
elimination of, **1**, 648
in Heck reactions, **1**, 493
metallation of, **1**, 758
α-metallation of, **1**, 759
reactions
with alkyllithiums, **6**, 379
with carboxylic acids, **5**, 225
with thionocarbonates, **1**, 938
for the reduction of alkyl halides, **1**, 4
for the reduction of thionoethers, **1**, 17
reviews, **1**, 919
as starting materials
for aldehydes, **3**, 18
for ketones, **3**, 134
for tertiary ynamines, **2**, 1043
synthesis of, **6**, 289
transmetallation of, **6**, 189
use in the synthesis of acyloxysilanes, **5**, 225
use in the synthesis of alcohols, **2**, 70
use in the synthesis of radicals, **1**, 945
Organosilanes, amido, rearrangement of, **5**, 376
Organosilanes, polyhalogenated, reactions, with phosphorus ylides, **5**, 914
Organosilicon chemistry, reviews, **2**, 535
Organosilicon hydrides, use in radical cyclizations, **1**, 342
Organosilicons
from, α-halosilanes, **4**, 136
reactions, with arsenic trichloride, **2**, 489
synthesis of, **6**, 335
use in the synthesis of isocyanates, **5**, 1006
Organosiloxanes, reactions, with alkynyl Grignard reagents, **2**, 1090
Organosilvers
reactions
with alkynylnitriles, **1**, 544
with diynes, **1**, 544
synthesis of, **6**, 63, 343, 349
Organosilyl anions, reactions, with epoxides, **1**, 650
Organosilyl halides
reactions
with alkynyllithiums, **2**, 1089
with alkynylmagnesiums, **2**, 1089
Organosodiums
elimination of, **1**, 663
stability of, **2**, 556
substitution reactions, **1**, 113
use in the synthesis of 1,1-bis(silyl)alkanes, **4**, 609
use in the synthesis of carboxylic acids, **5**, 27
Organostannanes
reactions
with aryl triflates, **1**, 482; **3**, 290
with *n*-butyllithium, **4**, 964
with cupric chloride, **4**, 774
with enol triflates, **1**, 480
with vinyl epoxides, **2**, 86
for the reduction of alkyl halides, **1**, 4
transmetallation of, **4**, 390, 652
Organostibine halides, as starting materials, for trimethylsilylstibides, **2**, 894
Organostrontiums, chemistry of, **2**, 562
Organosulfur trichlorides, as starting materials, for sulfinyl chlorides, **2**, 174
Organotantalums
reactions, with carbodiimides, **5**, 781
synthesis of, **6**, 338
Organotelluriums
from, aryltellurium trichlorides, **2**, 732
synthesis of, **6**, 354, 356
transmetallation of, **1**, 245
Organotellurium ylides, reactions, with carbonyls, **3**, 647
Organothallium cyanides, synthesis of, **5**, 1143
Organothalliums, for the synthesis of ketones, **3**, 130
N-Organothioimidoyl halides, synthesis of, **5**, 674
Organotin carbodiimide, synthesis of, **5**, 1081
Organotin cations, synthesis of, **2**, 1088
Organotin chlorides
reactions
with silver cyanate, **5**, 1016
with ureas, **5**, 1017
Organotin cyanides, synthesis of, **5**, 1144
Organotin enolates, reactions, with stannazanes, **2**, 816
Organotin halides, reactions, with amines, **2**, 815
Organotin hydrides
disadvantages of, **1**, 342
toxicity of, **1**, 331
use in the synthesis of radicals, **1**, 939
Organotin hydroxides, reactions, with alkynes, **2**, 1100
Organotin imines, reactions, with trimethylchlorosilane, **2**, 813
Organotin oxides
reactions
with alkynes, **2**, 1100
with ureas, **5**, 1017
Organotins
acylation of, **3**, 287
oxidation of, **3**, 703
reactions
with acid halides, **3**, 288
with alkenylpalladium(II) chlorides, **1**, 1027
with arylpalladium(II) chlorides, **1**, 1027
with ceric ammonium nitrate, **4**, 252
with imidoyl chlorides, **3**, 409
reviews, **1**, 919
as starting materials, for trialkynylphosphines, **2**, 826
synthesis of, **1**, 279; **2**, 816; **4**, 755; **6**, 327, 351
use in the synthesis of isocyanates, **5**, 1006
use in the synthesis of radicals, **1**, 935, 948
Organotins, cyclic, synthesis of, **4**, 665
Organotins, stable, reactions, with electrophiles, **1**, 973
Organotins, β-substituted, elimination of, **1**, 664
Organotin selenocarboxylates, use in the synthesis of monoacyl diselenides, **5**, 250
Organotitanium, use in the synthesis of alcohols, **2**, 70
Organotitaniums
copper-catalyzed allylation of, **1**, 124
synthesis of, **6**, 340
use for the deoxygenation of epoxides, **1**, 587
use in organic synthesis, **2**, 565
Organotitaniums, perfluoroalkyl-, use in the fluoroalkylative amination of aldehydes, **2**, 311

Organotransition metals
  carbometallation with, **2**, 978
  anti-carbometallation with, **2**, 981
  in catalytic reactions, **2**, 979
Organotungstens
  from, tungsten carbene complexes, **6**, 354
  synthesis of, **6**, 330, 342, 348, 353
  use for the deoxygenation of epoxides, **1**, 587
Organovanadium dichlorides, use in the synthesis of ketones, **3**, 127
Organovanadiums, for the synthesis of ketones, **3**, 130
Organozinc halides
  reactions
    with acid chlorides, **3**, 130
    with *p*-toluenesulfonyl cyanide, **3**, 646
  as reagents, for the synthesis of ketones, **3**, 130
Organozinc halides, aryl, reactions, with *p*-toluenesulfonyl cyanide, **3**, 670
Organozinc halides, benzylic, reactions, with tosyl cyanide, **3**, 664
Organozincs
  acylation of, **3**, 288
  for the alkylation of carbonyls, **2**, 68
  for the methylenation of carbonyls, **1**, 678
  for the opening of vinyl epoxides, **1**, 227
  reactions
    with alkynes, **1**, 543; **2**, 972
    with alkynylzincs, **2**, 1095
    with allyl halides, **1**, 123
    with electrophiles, **1**, 974
    with isocyanides, **3**, 418
    with trichloroacetyl chloride, **3**, 152
    with vinyl halides, **1**, 467
    with vinyl iodides, **4**, 954
  reactivity of, **1**, 122
  stability of, **2**, 69
  synthesis of, **1**, 122, 313; **4**, 665
    in situ, **2**, 69
  for the synthesis of amino esters, **1**, 261
  for the synthesis of organometallics, **2**, 590
  use in conjugate addition reactions, **1**, 311
Organozirconiums
  protonation of, **1**, 536
  synthesis of, **6**, 340
    via transmetallation, **2**, 991
  use in organic synthesis, **2**, 565
Organyl bromides, reactions, with triethylarsine, **2**, 1079
Ortho:para ratio, in the chlorination of arenes, **2**, 620
Orthoacetate, ethyl, reactions, with amines, **5**, 753
Orthoacetate, triethyl
  exchange reactions, with 2-methyl-2-thiohydroxymethylpropane-1,3-dithiol, **6**, 83
  from, 1-ethoxyethyne, **6**, 74
  reactions
    with 2-aminobenzenecarboxamide, **6**, 109
    with phenol, **6**, 79
    with trifluoromethylselenenyl chloride, **6**, 80
Orthoacetate, triethyl α,α-bis(trifluoromethylselenenyl), from, triethyl α-(trifluoromethylselenenyl)orthoacetate, **6**, 80
Orthoacetate, triethyl α-(trifluoromethylselenenyl), from, trifluoromethylselenenyl chloride, **6**, 80
Orthoacetate, trimethyl
  reactions
    with aluminum compounds, **4**, 192
    with boron tris(benzene selenide), **6**, 92
    with boron tris(methane selenide), **6**, 92
  as starting material, for trimethyl triselenoorthoacetate, **6**, 92
Orthoacetate, triphenyl, synthesis of, **6**, 74
Orthoacetate esters, synthesis of, **6**, 74
Orthoacetate esters, mixed, from, 1-ethoxyethyne, **6**, 74
Orthoacetates, trialkyl
  condensation of, **4**, 912
  reactions
    with perhalogenated aldehydes, **6**, 80
    with perhalogenated ketones, **6**, 80
Orthoacetates, tris(β-alkoxyalkyl), from, 1,1-dichloroethane, **6**, 75
Ortho acids, as starting materials, for ortho amides, **6**, 140
Orthoacrylates, trialkyl, synthesis of, **6**, 69
Orthoacrylates, triaryl, synthesis of, **6**, 69
Ortho amides
  from
    alkyl halides, **6**, 140
    amide salts, **6**, 138
    amidinium salts, **6**, 138
    guanidines, **6**, 139
    haloalkanes, **6**, 140
    isocyanates, **6**, 143
    ortho acids, **6**, 140
    polyaminoethenes, **6**, 141
    tetraaminoethylenes, **6**, 141
    tetrakis(dimethylamino)titanium, **6**, 141
    titanium salts, **6**, 141
    triamines, **6**, 143
  as starting materials, for aminobis(phosphonato)methanes, **6**, 157
  synthesis of, **6**, 137, 337
    via heterocyclization, **6**, 142
Ortho amides, carboxylic, as starting materials, for imidic acid esters, **5**, 690
Orthobenzoate, triethyl, from, imidoyl chloride salts, **6**, 70
Orthobenzoate, triethyl 2-6-difluoro-, from, trialkoxyacetonitriles, **6**, 78
Orthobenzoate, trimethyl, from, trialkoxyacetonitriles, **6**, 78
Orthobenzoate, trimethyl 2-methyl, from, trialkoxyacetonitriles, **6**, 78
Orthobenzoate esters
  from
    benzimidate ester hydrochlorides, **6**, 70
    dithiobenzoate esters, **6**, 73
Orthobenzoate esters, fluorinated, synthesis of, **6**, 69
Orthobenzoate esters, trialkyl, from, trialkoxyacetonitriles, **6**, 78
Orthobenzoates, trialkyl
  from, tetraalkyl carbonates, **6**, 77
  synthesis of, **6**, 69
  transesterification reactions, with benzene-1,2-diol, **6**, 79
Orthoborates, symmetrical, synthesis of, **2**, 111
Orthoborates, unsymmetrical, synthesis of, **2**, 112
Ortho-α-bromoxylene, reactions, with diethylphosphinyldifluoromethyllithium, **6**, 58
Orthocarbonate, alkyl, as alkylating agent, **2**, 683
Orthocarbonate, diethylene, use in the synthesis of ethylene acetals, **4**, 190
Orthocarbonate, ethyl, reactions, with 1,4,7,10-tetraazacyclododecane, **6**, 365
Orthocarbonate, tetraethyl
  fluorination of, **6**, 301
  reactions
    with amines, **6**, 625
    with azulene, **1**, 898
Orthocarbonate, tetrakis(2,2,2-trichloroethyl), synthesis of, **6**, 299
Orthocarbonate, tetramethyl
  fluorination of, **6**, 301
  use in the synthesis of spirocyclic orthocarbonates, **6**, 300
Orthocarbonate, tetrapropyl, fluorination of, **6**, 301
Orthocarbonate, triethyl, reactions, with arenesulfonamides, **6**, 625
Orthocarbonate esters
  reactions, with Grignard reagents, **6**, 77
  as starting materials, for ortho esters, **6**, 77
Orthocarbonate esters, tetraalkyl, reactions, with phenylmagnesium bromide, **6**, 77

Orthocarbonates
from
aryl cyanates, **6**, 296
chloroorthoformates, **6**, 272
tetrachloromethane, **6**, 296
trialkoxyacetonitrile, **6**, 296
trialkoxycarbenium salts, **6**, 296
trichloroacetonitrile, **6**, 296
trichloromethanesulfenyl chloride, **6**, 296
trichloromethyl isocyanide dichloride, **6**, 296
trichloronitromethane, **6**, 296
reactions
with alkynes, **1**, 1025
with boron trifluoride etherate, **6**, 729
as starting materials, for iminocarbonates, **6**, 624
transesterification of, **6**, 300
use for the synthesis of polymers, **6**, 300
Orthocarbonates, aliphatic
from, alkanols, **6**, 296
synthesis of, **6**, 299
Orthocarbonates, oligospirocyclic, synthesis of, **6**, 300
Orthocarbonates, spirocyclic
from
α,ω-alkanediols, **6**, 300
thiocarbonates, **6**, 300
synthesis of, **6**, 296
transesterification of, **6**, 300
Orthocarbonates, symmetrical, reactions, with dialkylphosphorus acid anhydrides, **6**, 314
Orthocarbonates, tetraalkyl
from
aryl cyanates, **6**, 300
trialkoxycarbenium tetrafluoroborates, **6**, 300
trichloromethanesulfenyl chloride, **6**, 308
synthesis of, **6**, 296
Orthocarbonates, tetraaryl
from
carbonic acid diesters, **6**, 300
trichloromethyl isocyanide dichloride, **6**, 296
Orthocarbonic acid esters, synthesis of, **6**, 296
Orthocarbonic acids
as starting materials, for ureas, **6**, 503
synthesis of, **6**, 320
Orthocarbonic acids, substituted, synthesis of, **6**, 296
Orthocinnamate esters, triethyl, synthesis of, **6**, 75
Ortho esters
acid-catalyzed hydrolysis of, **6**, 70
α-bromination of, **6**, 80
as dehydrating agents, for the synthesis of acetals, **4**, 177
elimination of alcohols from, **4**, 825
exchange reactions
with boron triselenides, **6**, 92
with thiols, **6**, 98
with triols, **6**, 79
from
acetals, **6**, 77
1-alkoxyalkynes, **6**, 74
amide acetals, **6**, 80
carboxylic acids, **6**, 71
carboxylic esters, **6**, 71
1,1-dialkoxyalkenes, **6**, 74
2,2-dialkoxyaziridimes, **6**, 77
2,2-dialkoxy-2*H*-azirimes, **6**, 77
1,1-dialkoxycyclopropanes, **6**, 76
1,1-dihaloalkenes, **6**, 74
α,α-dihalo ethers, **6**, 68
dioxocarbenium salts, **6**, 73
dithiocarboxylic acids, **6**, 73
α-halo acetals, **6**, 68
imidate ester salts, **6**, 69
lactones, **6**, 72
nitriles, **6**, 70
orthocarbonate esters, **6**, 77
trialoxyacetonitriles, **6**, 77
1,1,1-trihaloalkanes, **6**, 68
trithioortho esters, **6**, 80
hydrolysis of, **2**, 638
irradiation of, for oxygen–carbon bond fission, **4**, 1075
reactions
with aldehydes, **4**, 190
with alkynes, **1**, 1025; **4**, 192
with allylic alcohols, **1**, 416; **5**, 159
with allylsilanes, **4**, 192
with azides, **6**, 108
with bis(dimethylaluminum) selenide, **5**, 520
with 1,3-dihydroxy-2-hydroxymethyl-2-methylpropane, **6**, 79
with 1,2-diols, **6**, 79
with 1,3-diols, **6**, 79
with disulfonylmethanes, **4**, 865
with enolates, **4**, 192
with glycerol, **6**, 79
with Grignard reagents, **4**, 192
with hydrogen cyanide, **3**, 627
with 2-hydroxybenzenethiol, **6**, 95
with 2-hydroxyethanethiol, **6**, 95
with ketones, **4**, 190
with nucleophiles, **4**, 192, 193
with organocuprates, **4**, 192
with organometallics, **4**, 192
with peroxides, **6**, 79
with phenol, **6**, 79
with phosphines, **6**, 116
with selenation reagents, **5**, 520
with sulfonamides, **5**, 827
with 2-thiohydroxyacetic acid, **6**, 95
with thiols, **6**, 95
with trichloro-acetic acid, **4**, 1081
with trifluoro-acetic acid, **4**, 1081
with 1,2,4-trihydroxybutane, **6**, 79
with *cis*-1,3,5-trihydroxycyclohexane, **6**, 79
reduction of, **1**, 21
reviews, **6**, 36
stability of, **6**, 68
as starting materials
for acetals, **4**, 192
for aliphatic amidines, **5**, 753
for amide acetals, **6**, 104
for amidines, **5**, 743
for dialkylketenes, **3**, 541
for diselenoortho esters, **6**, 100
for dithioortho esters, **6**, 97
for imino esters, **5**, 690
for ketene acetals, **4**, 825
for ketene aminals, **4**, 975, 979
for monothioortho esters, **6**, 95
for thiophosphates, **2**, 259
for triselenoortho esters, **6**, 92
synthesis of, **6**, 68, 296
via solvolysis, **6**, 71
transesterification reactions of, **6**, 78
use in the synthesis of carboxylic esters, **5**, 126
Ortho esters, *O*-acyl
from, triethyl orthoformate, **6**, 79
as starting materials, for mixed ortho esters, **6**, 79
Ortho esters, acyloxy
synthesis of, **6**, 79
via solvolysis of acetoxyacyl chlorides, **6**, 71
Ortho esters, alkynyl, from, 1,1-dialkoxy-2,2-dichlorocyclopropanes, **6**, 76
Ortho esters, α-amino, synthesis of, **6**, 77
Ortho esters, aminoxy, from, 1,1-dialkoxyalkenes, **6**, 74
Ortho esters, α-bromo
dehydrobromination of, **6**, 80
reactions, with sodium sand, **4**, 827
synthesis of, **6**, 80

Ortho esters, carboxylic
  from, other ortho esters, **6**, 78
  as starting materials, for trithioortho esters, **6**, 82
  synthesis of, **6**, 68
Ortho esters, cyclic
  from, 1,2-diketones, **6**, 75
  ring-opening of, **4**, 827
  synthesis of, **6**, 308
    via thermal [4+2] cycloaddition reactions, **6**, 75
Ortho esters, α-halo, as starting materials, for ketene acetals, **4**, 827
Ortho esters, α-imino, synthesis of, **6**, 77
Ortho esters, mixed
  from
    O-acyl ortho esters, **6**, 79
    1,1-dialkoxyalkenes, **6**, 74
    1,1-dialkoxycyclopropanes, **6**, 76
    imidate ester salts, **6**, 70
  synthesis of
    via exchange reactions of trialkyl ortho esters with simple alcohols, **6**, 78
    via the magnesium chloride-catalyzed exchange reaction with alcohols, **6**, 79
Ortho esters, mixed ester, synthesis of, **6**, 296
Ortho esters, monoperoxy, from, peroxides, **6**, 79
Ortho esters, perfluorinated, synthesis of, **6**, 301
Ortho esters, trialkenyl, from, tris(β-chloroalkyl) ortho esters, **6**, 80
Ortho esters, trialkyl
  exchange reactions, with alcohols, **6**, 78
  transesterification reactions, with β-hydroxycarboxylic acids, **6**, 79
Ortho esters, *O*-trimethylsilyl, from, hydroxy esters, **6**, 72
Ortho esters, tris(β-chloroalkyl), dehydrochlorination of, **6**, 80
Ortho esters, unsaturated, from, 1,1-dimethoxycyclopropenes, **6**, 76
Orthoformate, *t*-butyldiethyl, synthesis of, via transesterification of triethyl orthoformate with *t*-butanol, **6**, 78
Orthoformate, di-*t*-butylethyl, synthesis of, via transesterification of triethyl orthoformate with *t*-butanol, **6**, 78
Orthoformate, ethyl
  reactions
    with cyanoacetic acid, **2**, 749
    with diarylformamidines, **5**, 746
    with phosphorus pentasulfide, **2**, 259
  use in the synthesis of formamidines, **5**, 746
Orthoformate, tri-*O*-acetyl, from, ozonolysis of tris(2-propenyl) orthoformate, **6**, 81
Orthoformate, triethyl
  exchange reactions
    with 1,3-diselenolates, **6**, 100
    with 2-methyl-2-thiohydroxymethylpropane-1,3-dithiol, **6**, 83
  for the formylation of enol ethers, **3**, 71
  reactions
    with acetic formic anhydride, **6**, 79
    with boron tris(benzene selenide), **6**, 92
    with cyclohexanetriamine, **6**, 140
    with dichlorophosphines, **4**, 126
    with formamide, **6**, 140
    with hydrogen sulfide, **5**, 586
    with hydroxyalkylphosphorus compounds, **4**, 342
    with 2-mercaptobenzamide, **6**, 111
    with 3-methyl-2-aminopyridine, **6**, 157
    with *N*-methylaniline, **6**, 140
    with 1-nitroalkanes, **2**, 789
    with organoaluminums, **1**, 971
    with phenol, **6**, 79
  as reagent, for the synthesis of aldehydes, **3**, 14
  as starting material, for *O*-acyl ortho esters, **6**, 79
Orthoformate, trimethyl, reactions, with cyclic β-dicarbonyls, **2**, 746
Orthoformate, tris(2-propenyl), ozonolysis of, **6**, 81
Orthoformate ester, *O*-acyl, as starting material, for monothioortho esters, **6**, 95
Orthoformate esters
  exchange reactions
    with alcohols, **6**, 78
    with selenols, **6**, 92
    with thiols, **6**, 82
Orthoformate esters, fluorinated, synthesis of, **6**, 69
Orthoformate esters, mixed, from, formamide, **6**, 70
Orthoformates
  dimerization of, **4**, 830
  reactions
    with aldehydes, **4**, 190
    with aluminum compounds, **4**, 192
    with dibutoxyphosphine, **2**, 451
    with Grignard reagents, **4**, 192
    with imines, **5**, 690
    with ketones, **4**, 190
    with phosphorus(III) halides, **4**, 340
    with Reformatsky reagents, **4**, 193
    with silyl enol ethers, **3**, 39
  as starting materials
    for dialkyl thiocarbonates, **6**, 478
    for formamidines, **5**, 746
Orthoformates, alkyl, reactions, with nitroacetic esters, **2**, 787
Orthoformates, chloro-
  as intermediates, for orthocarbonates, **6**, 272
  isolation of, **6**, 276
  as starting materials, for fluoroorthoformates, **6**, 272
Orthoformates, dialkoxyvinyl, reactions, with hydrogen chloride, **6**, 43
Orthoformates, enamine, synthesis of, **2**, 740
Orthoformates, fluoro-, from, chloroorthoformates, **6**, 272
Orthoformates, polyhalogenated, synthesis of, **6**, 273
Orthoformates, trialkyl
  exchange reactions, with dithiols, **6**, 83
  from
    chloroform, **6**, 68
    difluorochloromethane, **6**, 69
    fluorodichloromethane, **6**, 69
  use in the synthesis of thioketones, **3**, 343
Orthoformates, triaryl
  from, ethyl formate, **6**, 84
  reactions, with acetyl chloride, **6**, 44
  synthesis of, **6**, 69
Orthoformates, triethyl
  from, chloroform, **6**, 68
  synthesis of, **6**, 68
Orthoformates, tripentyl, from, chloroform, **6**, 68
Orthoformates, (triseleno)-, reactions, with enol ethers, **4**, 289
Orthoformates, vinyl, as starting materials, for dialkoxymethyl halides, **6**, 43
Orthohydrogen esters, stability of, **6**, 68
Orthometallation
  for the synthesis of alkyl actinides, **2**, 602
  for the synthesis of palladium complexes, **2**, 581
  for the synthesis of platinum complexes, **2**, 581
  for the synthesis of rhodium complexes, **2**, 581
Orthopropionate, triethyl-, from, imidoyl chloride salts, **6**, 70
Ortho thioesters
  chlorination of, **6**, 48
  from, thio compounds, **6**, 10
  stability of, **1**, 1031
Orthotoluic acids, annulation of, **2**, 664
Osazones
  from
    α-diketones, **3**, 451

α-hydroxy ketones, **3**, 451
synthesis of, **3**, 450
Osazones, glyoxal, synthesis of, **3**, 451
Osazones, sugar, structure of, **3**, 450
Osazones, sugar phenyl-, synthesis of, **3**, 450
Osazonization, of sugars, **3**, 451
Oshima's reagent
synthesis of, **1**, 767
use in methylenation reactions, **1**, 766
Osmacyclopentadiene complexes, synthesis of, **4**, 681
Osmium, tetracarbonyldimethyl-, reactions, with 1,4-dibromocyclooctatetraene, **4**, 681
Osmium, tetracarbonylethyne-, reactions, with carbon(η^5-cyclopentadienyl)phosphinerhodium, **4**, 710
Osmium(0) complexes, reactions, with alkyl halides, **2**, 582
Osmium(I) carbene complex, difluoro-, synthesis of, **6**, 717
Osmium(I) difluorocarbene complexes, synthesis of, **6**, 717
Osmium(II) complexes, synthesis of, **2**, 582
Osmium(VIII) oxide
for the oxidative cleavage of alkenes, **3**, 113
reactions, with sodium periodate, **3**, 2
Osmium carbene complexes
reactions, with sulfur, **3**, 553
synthesis of, **3**, 522
Osmium carbonyl complexes, synthesis of, **6**, 524
Osmium cluster compounds, synthesis of, **6**, 344
Osmium complexes
alkylation of, **6**, 720
reactions
with alkylaluminums, **2**, 581
with alkyllithiums, **2**, 581
with diazomethane, **3**, 522
with maleic anhydride, **5**, 201
with nucleophiles, **2**, 583
with phenyllithium, **3**, 690
with tellurium, **3**, 391
synthesis of, **3**, 391; **5**, 650
use in the synthesis of organometallics, **6**, 593
Osmium complexes, π-arene-, alkylation of, **2**, 581
Osmium complexes, carbon-bridged trinuclear, synthesis of, **4**, 679
Osmium complexes, (α-chloromethyl)-, synthesis of, **4**, 150
Osmium complexes, α-diazomethyl-, synthesis of, **3**, 486
Osmium complexes, difluorocarbene-, synthesis of, **6**, 718
Osmium complexes, methylimido-, reactions, with carbon monoxide, **5**, 456
Osmium complexes, trinuclear allyl-bridged, synthesis of, **4**, 683
Osmium complexes, trinuclear carbon-bridged, synthesis of, **4**, 682
Osmium complexes, trinuclear carbon-bridged benzyne, synthesis of, **4**, 682
Osmium functions, doubly-bonded, synthesis of, **3**, 522
Osmium methylene complex, reactions, with selenium, **3**, 387
Osmium methylimide complexes, reactions, with *t*-butylisonitrile, **5**, 801
Osmiums, reactions, with pyridine, **6**, 345
Osmiums, acyl-, synthesis of, **5**, 455
Osmiums, dialkyl-
reactions, with carbon monoxide, **5**, 455
synthesis of, **5**, 455
Osmiums, imidoyl-
from, isonitriles, **5**, 801
synthesis of, **5**, 800
Osmiums, methylene-bridged, synthesis of, **4**, 683
Osmium salts
reactions, with alkyl iodides, **5**, 455
synthesis of, **5**, 455
Osmium tetracarbonyl dianion, alkylation of, **2**, 579, 582
Osmium tetroxide
for the cleavage of alkanes, **4**, 168
for the *cis*-hydroxylation of alkenes, **2**, 45
for the oxidation of dialkyl sulfides, **2**, 155
reactions
with alkenes, **2**, 50
with allylic alcohols, **2**, 46
with (*E*,*E*)-1,4-diphenylbutadiene, **2**, 47
with vinylsilanes, **4**, 357
as reagent, for the synthesis of aldehydes from alcohols, **3**, 5
Osmylation
of enol ethers, **4**, 174
reagent control in, **2**, 47
substrate control in, **2**, 46
for the synthesis of hemiacetals, **4**, 168
Oudemansins, synthesis of, **1**, 643
1,3,2-Oxaazaphospholane, reactions, with alkyl halides, **2**, 861
Oxaazaphospholanes, five-membered cyclic, synthesis of, **2**, 860
Oxabicycles, synthesis of, **3**, 315
7-Oxabicyclo[2.2.1]hept-5-enes, ring-opening metathesis polymerization, choice of catalysts for, **1**, 675
5-Oxabicyclo[2:1:0]pentane, photolysis of, **3**, 314
1,2-Oxaboretanes, synthesis of, **6**, 713
1,2-Oxaboretane-4-thione, synthesis of, **5**, 534
Oxaboretane thiones, properties of, **5**, 534
Oxaborolidines, synthesis of, **2**, 528
Oxacycloalkenylidene complexes, synthesis of, **5**, 944
Oxacyclohexylidene complexes, synthesis of, **5**, 944
Oxacyclopentylidene complexes
from, carbene complexes, **5**, 946
synthesis of, **5**, 944
1,2,5-Oxadiazinanes, synthesis of, **2**, 307, 345
Oxadiazines, synthesis of, **4**, 886
Oxadiazines, dihydro, synthesis of, **4**, 982
1,3,4-Oxadiazines, as starting materials, for ketenes, **3**, 543
1,3,5-Oxadiazines, tetrahydro-, synthesis of, **4**, 428
Oxadiazoles, thermolysis of, **3**, 323
1,2,4-Oxadiazoles, pyrolysis of, **5**, 1013
1,3,4-Oxadiazoles
pyrolysis of, **5**, 1013
ring-opening of, **5**, 854
1,3,4-Oxadiazoles, 2,5-dihydro-, synthesis of, **6**, 107
1,3,4-Oxadiazole-2-thiones, synthesis of, **6**, 558
1,2,4-Oxadiazolidine, 2,2'-carbonylbis(3,5-dioxo-4-methyl-, reactions, with carboxylic acids, **5**, 333
Oxadiazolines
as starting materials, for carbonyl ylides, **1**, 365
thermolysis of, **5**, 707
Oxadiazolinones, as starting materials, for nitrile imides, **3**, 681
Oxadiazolium perchlorate, triphenyl-, hydrogen sulfide opening of, **5**, 642
Oxadiazolium salts, use in the synthesis of carbodiimides, **5**, 1073
1,3,4-Oxadiazolium salts
reactions
with amines, **6**, 109
with hydrogen selenide, **5**, 642
1,2,4-Oxadiazol-5-ones, synthesis of, **6**, 450
1,3,4-Oxadiazol-2(3H)-ones, 5-aryl-, reactions, with phosgene, **6**, 444
1,3,4-Oxadiazolo[3,2-*d*][1,3,4]thiadiazine-6(5*H*)-thiones, 5-aryl-, synthesis of, **6**, 564
Oxadiazonium salts, synthesis of, **5**, 1146
Oxadi-π-methanes, rearrangement of, **1**, 800
1,3,5-Oxadithianes, synthesis of, **4**, 229
4-Oxa-β-lactam, as starting material, for amide acetals, **6**, 107
Oxalamide, *N*,*N*-dimethyl-*N*,*N*-dinitroso-, reactions, with methylamine, **3**, 467
Oxalate, bis(trifluoromethyl), synthesis of, **6**, 230
Oxalate, dialkyl, as alkylating agent, **2**, 683
Oxalates, thermolysis of, **1**, 172
Oxalates, ethyl, hydrolysis of, **5**, 24

Oxalates, methyl, hydrolysis of, **5**, 24
Oxalate salts, synthesis of, **2**, 347
Oxaldihydrazonamides, synthesis of, **5**, 847
Oxaldiimidate, dimethyl, synthesis of, **5**, 698
Oxalic acid, use in the synthesis of radicals, **1**, 940
Oxalic acid hydrazonoyl chlorides, synthesis of, **5**, 843
Oxalic acids, as starting materials, for hydrazonamides, **5**, 852
Oxalimide chloride, *N,N'*-dichloro-, as starting material, for *N,N'*-dichloroamidines, **5**, 816
Oxalimidoyl dichloride, *N,N'*-dichloro-, synthesis of, **5**, 809
Oxaline, occurrence, in natural products, **6**, 138
Oxalocrotonic acid, as starting material, for formazans, **5**, 858
Oxaloyl chloride, use in the synthesis of dichloroalkanes, **4**, 18
Oxaloyl halides, reactions, with pentacarbonyl metal acyl complexes, **3**, 691
Oxalyl bromide
  reactions, with benzophenone imine, **6**, 453
  as starting material, for carbonic dibromide, **6**, 417
  synthesis of, **5**, 17
Oxalyl chloride
  for the dehydration of amides, **3**, 619
  properties of, **5**, 11
  reactions
    with amides, **5**, 576, 743, 995
    with amines, **5**, 973
    with 2-aminobenzothiazole, **5**, 974
    with aromatic thioamides, **5**, 996
    with carbamates, **5**, 977
    with carboxylic acid amides, **5**, 654
    with cyclopropanecarboxamide, **5**, 995
    with *N,N*-disubstituted acetamides, **5**, 996
    with indole, **5**, 15
    with isoindolecarboxylic acid, **5**, 11
    with methylphosphonic acid dimethyl ester, **2**, 469
    with phosphoric ester amides, **5**, 1017
    with potassium salt, **5**, 12
    with silver salts, **5**, 183
    with sulfonamides, **2**, 726; **5**, 996
    with *s*-triazinamines, **5**, 973
  as reagent
    for acid chlorides, **5**, 1, 10
    for the synthesis of aldehydes, **3**, 6
  as starting material, for phosgene, **6**, 412
Oxalyl compounds, reactions, with pivaloyl chloride, **5**, 396
Oxamide, *N*-phenyl-, thionation of, **5**, 566
Oxamides, sulfonyl-, synthesis of, **5**, 996
Oxamidine, diphenyl-, reactions, with hydroxylamine hydrochloride, **5**, 833
Oxamidines
  reactions, with boranes, **5**, 779
  See also Amides, oxyimidic
Oxamidines, bis(dimethylmetal)-, synthesis of, **5**, 780
Oxaminium salts, as oxidants, for the synthesis of aldehydes, **3**, 7
Oxamoyl chlorides, sulfonyl-, synthesis of, **5**, 996
Oxanes, electrochemical fluorination of, **6**, 37
Oxaphosphetanes
  as intermediates
    vs. betaines, **1**, 639
    in the synthesis of alkenes, **1**, 721
1,2-Oxaphosphetanes, synthesis of, **2**, 831
1,2-Oxaphosphetane-4-thiones, synthesis of, **5**, 534
1-Oxa-2-phospholane, 2-chloro-2-thiono-5-methyl-, as thionating reagent, **5**, 576
Oxaphospholanes
  synthesis of, **4**, 339
  thermal decomposition of, **1**, 268
1,3-Oxaphospholanes, synthesis of, **4**, 337, 337
Oxaphospholans, synthesis of, **2**, 473
1,3-Oxaphospholene oxides, synthesis of, **4**, 344
Oxaphospholenes, synthesis of, **1**, 993
1,2-Oxaphospholenes, 2-oxo-, synthesis of, **4**, 129
1,2-Oxaphosphol-3-ene-5-thione, synthesis of, via carbonyl–thiocarbonyl rearrangements, **5**, 534
1,3,2-Oxaphospholidine, 2-phenyl-3-methyl-, use in the synthesis of aryl isocyanides, **3**, 721
1,3-Oxaphosphorinanes, synthesis of, **4**, 337
1,2-Oxarsolane, 2-chloro-
  alkylation of, **2**, 494
  from, dichloro(3-hydroxypropyl)arsine, **2**, 494
  synthesis of, **2**, 497
1,2-Oxarsolane, 1-methyl-, synthesis of, **2**, 494
Oxasilenes
  dimerization of, **5**, 924
  from, acylsilanes, **5**, 923
  as reactive intermediates, **5**, 924
  synthesis of, **5**, 923
    via retro dimerization, **5**, 924
Oxasiletanides, synthesis of, **2**, 547
1,2-Oxasiletanides, as intermediates, **2**, 547
11-Oxasteroids, synthesis of, **3**, 322
1,4,3,5-Oxathiadiazines, hydrolysis of, **5**, 348
Oxathiadiazoles, as starting materials, for nitrile imides, **3**, 681
1,3-Oxathiafulvenes, synthesis of, **4**, 835
1,3-Oxathiane
  metallation of, with *n*-butyllithium, **6**, 98
  as starting material, for 2-methaneseleno-1,3-oxathiane, **6**, 100
1,3-Oxathiane, 2-methaneseleno-, from, 1,3-oxathiane, **6**, 100
1,3-Oxathiane, 2-trimethylsilyl-, methylation of, **6**, 126
Oxathiane dioxides, acylation of, **1**, 241
1,3-Oxathiane dioxides, synthesis of, **4**, 235
Oxathianes
  acylation of, **3**, 161
  use in the synthesis of alkanethiols, **2**, 128
1,3-Oxathianes
  as chiral auxiliaries, **4**, 230
  from
    1,3-hydroxy thiols, **4**, 229
    10-mercapto-*exo*-borneol, **4**, 231
  metallation of, **1**, 141
  reactions
    with *s*-butyllithium, **6**, 126
    with trimethylgermyl chloride, **6**, 132
  synthesis of, **4**, 229
1,3-Oxathianes, 2-acyl-, hydride reduction of, **4**, 231
1,3-Oxathianes, 2-lithio-
  reactions
    with dimethyl thiophosphonyl chloride, **6**, 119
    with trimethyllead acetate, **6**, 133
1,3-Oxathian-4-ones, synthesis of, **4**, 230
1,2-Oxathian-2-oxide, synthesis of, **2**, 201
1,4,3-Oxathiazine 4,4-dioxides, 6-methyl-5,6-dihydro-, hydrolysis of, **5**, 348
Oxathiazines, from, thiodiimides, **5**, 340
4*H*-1,5,2-Oxathiazines, synthesis of, **4**, 232; **6**, 546
6*H*-1,3,5-Oxathiazines, synthesis of, **5**, 633
4*H*-1,5,2-Oxathiazine-6-thione, 3-phenyl-, synthesis of, **6**, 546
1,4,2-Oxathiazole, 3-(4-nitrophenyl)-, as starting material, for 4-nitrophenyl isothiocyanate, **5**, 1046
1,3,4-Oxathiazoles, from, aldehydes, **3**, 686
1,4,2-Oxathiazoles
  as intermediates, for the synthesis of isothiocyanates, **5**, 1046
  synthesis of, **6**, 309
1,4,2-Oxathiazoles, 3,5,5-triaryl-, as intermediates, in the synthesis of isothiocyanates, **5**, 1038
1,4,2-Oxathiazolidines, synthesis of, **4**, 232
1,2,4-Oxathiazoline, use in the synthesis of carbodiimides, **5**, 1073

1,3,4-Oxathiazolines, synthesis of, **4**, 232
1,4,2-Oxathiazolines, synthesis of, **4**, 232
Oxathiazolinone, *N*-tosyl-, hydrolysis of, **5**, 348
1,3,4-Oxathiazol-2-ones, as starting materials, for nitrile sulfides, **3**, 686
4*H*-1,3-Oxathiins, synthesis of, **4**, 232
Oxathioacetals, synthesis of, **2**, 120
1,3-Oxathiolane, 2-ethoxy-, reactions, with *t*-butyl peroxide, **6**, 478
Oxathiolane dioxides
  metallation of, **4**, 236
  synthesis of, **4**, 235
Oxathiolane nucleosides, synthesis of, **4**, 231
Oxathiolanes
  from, hydroxythiols, **4**, 229
  oxidative hydrolysis of, **5**, 411
  reactions, with sodium periodate, **5**, 411
  reductive cleavage of, **2**, 265
  ring-contraction of, **2**, 140
  as starting materials, for thiocarbamates, **6**, 496
  synthesis of, **4**, 229; **6**, 628
1,3-Oxathiolanes
  from, 1,3-hydroxy thiols, **4**, 229
  synthesis of, **1**, 286; **4**, 229
Oxathiolanones, as starting materials, for thioketones, **3**, 350
1,3-Oxathiolan-5-ones
  reactions, with *m*-chloroperbenzoic acid, **4**, 837
  as starting materials, for sulfoxides, **4**, 837
  synthesis of, **4**, 230
Oxathiolan-5-ones, as starting materials, for alkenes, **1**, 686
Oxathiolanyl nucleosides, synthesis of, **4**, 229
1,3-Oxathiole, synthesis of, **6**, 111
1,3-Oxathioles, 2-alkylidene-, as starting materials, for thioketenes, **3**, 551
Oxathiolium salts
  reactions, with alcohols, **6**, 97
  synthesis of, **6**, 50
1-Oxa-4,6,9-trithiaspiro[4.4]non-2-enes, synthesis of, **6**, 309
Oxazaborolidines
  as catalysts, for the reduction of arylalkyl ketones, **2**, 66
  from, (*S*)-2-indoline carboxylic acid, **2**, 66
  use in the reduction of ketones, **2**, 64
1,3,2-Oxazaborolidines, synthesis of, **2**, 66
Oxazaborolines, from, cyanoborate salts, **4**, 523
1,2-Oxazaheterocycles, synthesis of, via ring-expansion reactions, **2**, 355
1,3,5-Oxazaphosphanes, synthesis of, **4**, 455
1,3,2-Oxazaphospholane, Arbuzov rearrangement of, **2**, 453
1,3,2-Oxazaphospholans, reactions, with alkyl halides, **2**, 455
1,3,2-Oxazaphospholenes, synthesis of, **2**, 467
Oxazaphospholes, as starting materials, for nitrile ylides, **3**, 680
Oxazaphospholes, *P*-amino-, reactions, with ethanediol, **2**, 476
1,2,4-Oxazaphospholes, synthesis of, **5**, 839
1,3,2-Oxazaphospholidine, 2-phenyl-3-methyl-, as reducing agent, **3**, 698
Oxazaphospholidines, synthesis of, **2**, 443
Oxazaphospholidine-2-thione, reactions, with *t*-butylmagnesium bromide, **2**, 257
Oxazaphospholimines
  as starting materials, for nitrile ylides, **5**, 789
  thermolysis of, **5**, 789
Oxazaphospholines, synthesis of, via cycloaddition reactions, **4**, 493
Oxazepinones, from, chloro *β*-lactams, **4**, 824
Oxazetidines
  as starting materials
    for ketenimines, **3**, 600
    for perfluoroalkylcarbonimidic difluorides, **6**, 604
Oxazetidinone *N*-oxides, as intermediates, for the synthesis of nitrile oxides, **3**, 685
1,2-Oxazetines, perfluoro-2-alkyl-, pyrolysis of, **5**, 812
Oxazine, tetrahydro-, synthesis of, **4**, 306
1,3-Oxazine, dihydro-, thermal rearrangement of, **5**, 1014
1,3-Oxazine-2,4-(6H)-dione, 1-trimethylsilyl-, synthesis of, **5**, 992
2*H*-1,4-Oxazine-2-ones, 5-chloro-, reactions, with alkenes, **5**, 661
Oxazines
  alkylation of, **1**, 151
  reduction of, **3**, 61
  synthesis of, via the opening of epoxides, **1**, 209
  use in the synthesis of *α,β*-unsaturated ketones, **3**, 211
1,2,-Oxazines, *N*-trichloromethyl-3,6-dihydro-, hydrolysis of, **6**, 450
1,3-Oxazines
  as starting materials, for *gem*-imino phosphiminoalkenes, **4**, 996
  synthesis of, **4**, 302, 304
1,3-Oxazines, *N*-benzyl tetrahydro-, reactions, with Grignard reagents, **2**, 320
1,3-Oxazines, dihydro-
  deprotonation of, for the synthesis of ketenimines, **3**, 575
  as starting materials, for *γ*-hydroxy isocyanides, **3**, 707
1,3-Oxazines, 2-lithio(silyl)methyl-, synthesis of, **4**, 649
1,3-Oxazines, *N*-substituted tetrahydro-, reactions, with chlorodiphenylphosphine, **4**, 468
1,3-Oxazines, tetrahydro-
  ring-opening of, **2**, 320
  synthesis of, **4**, 302, 306
2*H*-1,3-Oxazines, 5,6-dihydro-, synthesis of, **4**, 313
4*H*-1,3-Oxazines, 5,6-dihydro-, reduction of, **4**, 306
6*H*-1,2-Oxazines, as starting materials, for aziridines, **2**, 322
1,3-Oxazine-2-thiones, synthesis of, **6**, 557
2*H*-1,3-Oxazin-2-one, tetrahydro-
  reactions
    with aniline salts, **2**, 320
    with thiophenols, **2**, 320
Oxazinones, bromination of, **1**, 1135
1,3-Oxazin-2-ones, from, perfluoromethacryloyl fluoride, **3**, 590
1,3-Oxazin-6-ones
  reactions, with butylamine, **4**, 981
  as starting materials, for *α*-ureido enamines, **4**, 981
Oxaziranes, *N*-acyl-, reactions, with amines, **5**, 353
Oxaziranes, *N*-tosyl-, rearrangement of, **5**, 347
Oxaziridine, perfluorinated, reactions, with nucleophiles, **6**, 441
Oxaziridine, *trans*-2-phenylsulfonyl-3-phenyl-, synthesis of, **4**, 312
Oxaziridine–amide rearrangement, for the synthesis of *N*-alkylamides, **5**, 283
Oxaziridine rearrangement, for the synthesis of *N*-arylamides, **5**, 292
Oxaziridines
  from
    imines, **4**, 312
    nitrones, **6**, 110
    perfluoroimines, **6**, 279
  as intermediates, **2**, 765
  for the oxidation of dithianes, **4**, 262
  for the oxidation of selenides, **2**, 287
  for the oxidation of sulfides, **2**, 150
  reactions
    with *t*-butanethiol, **2**, 185
    with 1,5-diazabicyclo[4.3.0]non-5-ene, **3**, 404
    with 1,4-diazabicyclo[2.2.2]octane, **3**, 404
    with ethylenes, **6**, 256
    with methyl ketones, **6**, 256
  rearrangement of, **5**, 272
  ring-opening of, **5**, 283, 1011
  as starting materials
    for *N*-H aldimines, **3**, 404
    for amides, **5**, 272

for aryl aldimines, **3**, 411
for hydrazines, **2**, 374, 379
for oxyimidic halides, **5**, 824
for thiocyanates, **5**, 1107
synthesis of, **6**, 256
via peroxidation of imines, **2**, 343
use in the synthesis of hydroxylamines, **2**, 342
Oxaziridines, 2-alkyl-, as starting materials, for isothiocyanates, **5**, 1032
Oxaziridines, *N*-alkyl-, as starting materials, for *N*-H ketimines, **3**, 405
Oxaziridines, 3-fluoro-, ring-opening of, **5**, 674
Oxaziridines, perfluorinated, thermal rearrangement of, **6**, 614
Oxaziridines, perfluoro-, synthesis of, **4**, 104
Oxaziridines, perfluoro-2,3-dialkyl-
reactions, with chloropentafluoroacetone, **5**, 825
as starting materials, for oxyimidic fluorides, **5**, 824
Oxaziridines, 2-(phenylsulfonyl)-3-aryl-
for the oxidation of amines, **2**, 344
for the oxidation of tertiary amines, **2**, 349
Oxaziridines, *N*-phosphinoyl-, from, *N*-phosphinoylimines, **3**, 439
Oxaziridines, sulfonyl-
for the oxidation of aryl organometallics, **2**, 652
reactions, with 1,2-disubstituted enamines, **3**, 183
Oxazole, 4,5-dihydro-2-methyl-, as starting material, for *N*-acyloxyethyl-*N*-cyanoacetamides, **5**, 1123
Oxazole, 5-(1,3-dithianyl)-, acylation of, **3**, 311
Oxazole, 2-formyl-, synthesis of, **3**, 109
Oxazole, 2-lithio-, use in the synthesis of formyl oxazoles, **3**, 109
5-Oxazole, 3-methyl-5-oxo-4-phenyloximino-4,5-dihydro-, flash vacuum pyrolysis of, **3**, 703
Oxazoles
for the opening of epoxides, **1**, 212
ozonolysis of, **5**, 197
photooxygenation of, **5**, 174
reactions
with alkynes, **1**, 522
with lithium, **3**, 109
reduction of, **4**, 306
reductive cleavage of, **3**, 189
as starting materials
for azomethine ylides, **1**, 524
for ketene aminals, **4**, 979
for ketones, **3**, 311
for the synthesis of anhydrides, **5**, 197
Oxazoles, diaryl-, ozonolysis of, **5**, 197
Oxazoles, dihydro-
reactions, with benzonitrile oxide, **6**, 107
synthesis of, **6**, 107, 634
Oxazoles, 2,5-dihydro-, synthesis of, **4**, 313
Oxazoles, 4,5-dihydro-, reduction of, **4**, 306
Oxazoles, 4-keto-, synthesis of, **3**, 311
Oxazoles, trimethylsilyl-, synthesis of, **6**, 634
Oxazoles, trimethylstannyl-, synthesis of, **6**, 634
1,2-Oxazoles, arylhydrazono-3-methyl-5-oxo-4,5-dihydro-, thermolysis of, **3**, 709
Oxazolidine-2,5-diones, silylation of, **5**, 1012
2,5-Oxazolidinediones, synthesis of, **5**, 970
Oxazolidine-2,5-diones, *N*-(trimethylsilyl)-, synthesis of, **5**, 1012
Oxazolidines
as auxiliaries, in radical reactions, **1**, 326
lithiation of, **4**, 531
reduction of, **3**, 35
as starting materials, for azomethine ylides, **1**, 368
synthesis of, **4**, 301
use in the synthesis of peptides, **5**, 259
Oxazolidines, chiral, as intermediates, **3**, 422
1,3-Oxazolidines, ring-opening of, **2**, 313
1,3-Oxazolidines, perfluoro-, synthesis of, **6**, 257
Oxazolidine-2-selones
from
amino alcohols, **6**, 595
metallation and selenation, **6**, 595
2-Oxazolidinethione, 3-methyl-, reactions, with *O*,*N*-bis(tributylstannyl)-*N*-methylethanolamine, **6**, 322
Oxazolidine-2-thiones, *N*-phosphorylated, ring-opening of, **5**, 1046
Oxazolidine-5-thiones, synthesis of, **5**, 533
Oxazolidine-5-thiones, 3-oxo-, synthesis of, **5**, 533
1,3-Oxazolidinine-2-thiones, reactions, with alkoxyamines, **5**, 329
Oxazolidinone, reactions, with tosyl chloride, **5**, 1012
1,3-Oxazolidin-2-one, reactions, with phosphorus pentachloride, **5**, 1225
Oxazolidin-2-one, 4-carboxy-, synthesis of, **6**, 486
2-Oxazolidinone, (*S*)-3-(1-oxo-2-methyl-3-hydroxy-3-phenylpropyl)-4-(phenylmethyl)-, hydrolysis of, **5**, 80
Oxazolidinones
alkylation of, **2**, 321
lithiation of, **4**, 531
mesylation of, **2**, 309
Oxazolidinones, 5,5-dialkyl-*N*-nitroso-, as starting materials, for vinyl iodides, **2**, 630
Oxazolidinones, *N*-sulfinyl-, reactions, with amides, **2**, 363
2-Oxazolidinones
pyrolysis of, **5**, 1012
synthesis of, **5**, 999
Oxazolidin-2-ones
synthesis of, **6**, 486
thermal rearrangement of, **5**, 1012
2-Oxazolidone *N*-acetic acid, synthesis of, **6**, 485
Oxazolidones, synthesis of, **6**, 485
Oxazolidones, 5,5-disubstituted *N*-nitroso-, decomposition of, **1**, 1143
Oxazolidones, *N*-nitroso-
cleavage of, **3**, 488
decomposition of, **2**, 801
2-Oxazolidones, 5,5-alkyl aryl-3-nitroso-, as starting materials, for alkynes, **1**, 1076
2-Oxazolidones, 5-aryl-3-nitroso-, as starting materials, for alkynes, **1**, 1076
2-Oxazolidones, 5,5-diaryl-3-nitroso-, as starting materials, for alkynes, **1**, 1076
2-Oxazolidones, 3-nitroso-, decomposition of, **1**, 1118
Oxazolines
cleavage of, **5**, 622
from, isocyanides, **4**, 538
lithiation of, **4**, 531
for the stabilization of carbanions, **1**, 405
as starting materials, for *gem*-formamido phosphorylalkenes, **4**, 1005
synthesis of, **6**, 142
use in the reduction of carbonyls, **2**, 64
Oxazolines, 2-alkyl-, ring-opening of, **5**, 1033
Oxazolines, 2-(1-piperidinyl)-, metallation of, **1**, 152
Oxazolines, silyl-, synthesis of, **4**, 519
Oxazolines, α-substituted α-hydroxy-, hydrolysis of, **5**, 80
Oxazolines, α,β-unsaturated, synthesis of, **1**, 597
2-Oxazolines, 5-alkyl-4-tosyl-, synthesis of, **2**, 311
Oxazolinethiones, synthesis of, **4**, 538
Oxazolinium salts, reactions, with aryl Grignard reagents, **4**, 306
Oxazolinones
for the alkylation of amides, **1**, 134
from, amino acids, **1**, 134
Oxazolinones, allenyl, synthesis of, **1**, 985
2-3-Oxazolin-5-ones, as starting materials, for nitrile ylides, **3**, 680
Oxazolium salts, ring-opening of, **5**, 627
2(3*H*)Oxazolone, 3-ethenyl-4,5-diphenyl-, Heck arylation of, **2**, 321
5(2*H*)-Oxazolone, 4-methyl-2-trifluoromethyl-, reactions, with methanol–hydrochloric acid, **4**, 305

Oxazolones
reactions, with 1-phenylpropynone, **1**, 537
reduction of, **1**, 76
2*H*-Oxazol-5-ones, 4-alkyl-2,2-bis(trifluoromethyl)-, hydrolysis of, **5**, 79
5(4*H*)-Oxazolones, 4-substituted 2-trifluoromethyl-, reactions, with ethanethiol, **4**, 320
Oxazolo[3,4-*c*]oxazoles, synthesis of, **4**, 302
Oxepines, from, carbonyl ylides, **3**, 318
Oxetane
reactions
with carbonic chloride fluoride, **6**, 424
with tertiary phosphines, **2**, 445
Oxetane, 2-phenyl-, Friedel–Crafts allylation of, **1**, 229
Oxetanes
alkylation of, **1**, 478
allylation of, **1**, 229
Friedel–Crafts alkylation of, **1**, 475
from, epoxides, **1**, 210
opening of, **1**, 229
pyrolysis of, **1**, 785
reactions
with alkynes, **1**, 1024
with alkynyllithiums, **1**, 1024
with α-lithio isocyanides, **3**, 704
with lithium 4,4'-(di-*t*-butyl)biphenylide, **2**, 553
with trimethylsilyl cyanide, **3**, 633
with trimethylsilyl selenocyanate, **5**, 1114
ring-opening of, **2**, 121; **3**, 698, 707
as starting materials
for γ-hydroxynitriles, **3**, 633
for isothiocyanates, **5**, 1042
for γ-lithio alkoxides, **5**, 943
synthesis of, **2**, 92
Oxetanes, vinyl-, opening of, **1**, 230
Oxetanone, tetraphenyl-, irradiation of, **3**, 314
2-Oxetanone, reactions, with phenyl isothiocyanate, **5**, 729
Oxetan-2-ones, 3-trimethylsilyl-, cleavage of, **1**, 282
Oxidants
for the oxidation of sulfides to sulfoxides, **2**, 150
for the synthesis of alkylphosphine oxides, **2**, 452
for the synthesis of aryl aldehydes, **3**, 87
Oxidants, metal-based, for the synthesis of ketones, **3**, 115
Oxidants, solid-supported
for aldehydes, **3**, 4
for the synthesis of ketones, **3**, 115
Oxidases, for the oxidation of arenes, **2**, 39
Oxidation
of alkylzirconiums, **2**, 568
of allylic alcohols, **3**, 215, 229
of aryl carbon–hydrogen bonds, **2**, 649
of aryl ethylenes, **3**, 87
of aryl trifluoromethyl selenides, **6**, 237
of benzylamines, **3**, 89
of benzyl halides, **3**, 88
of benzylic alcohols, **3**, 87
of benzylic methylene groups, **3**, 293
of furans, **3**, 221
of ketones, **3**, 156
of nitrones, **5**, 331
of secondary benzylic alcohols, **3**, 294
of sulfides, **2**, 144
of sulfur compounds, **2**, 177
for the synthesis of alcohols, **2**, 86
for the synthesis of *N*-alkylhydroxylamines, **2**, 340
for the synthesis of alkynylantimony compounds, **2**, 1082
for the synthesis of alkynylarsenic compounds, **2**, 1080
for the synthesis of aromatic acids, **5**, 113
for the synthesis of aromatic ketones, **3**, 293
for the synthesis of *N*-arylamides, **5**, 293
for the synthesis of *N*-carbon-substituted ketimines, **3**, 407
for the synthesis of carboxylic acids, **5**, 28
for the synthesis of carboxylic esters, **5**, 141
for the synthesis of *N,N*-dialkylhydroxylamines, **2**, 343
for the synthesis of α-diols, **4**, 161
for the synthesis of haloalkanoic acids, **5**, 59
for the synthesis of nitrogen-substituted alkanoic acids, **5**, 95
for the synthesis of primary alkenoamides, **5**, 295
for the synthesis of primary arylamides, **5**, 299
for the synthesis of sulfonic acids, **2**, 192
for the synthesis of sulfur-substituted alkanoic acids, **5**, 83
for the synthesis of α,β-unsaturated acids, **5**, 101
for the synthesis of α,β-unsaturated ketones, **3**, 215
for the synthesis of unsubstituted alkanoic acids, **5**, 46
of tetrafluoroethylene, **6**, 214
of thioacyl halides, **5**, 490
of thiols, **2**, 212
of thioureas, **6**, 502
of toluenes, **3**, 86
of trifluoromethyl sulfides, **6**, 234
use of ultrasound in, **3**, 4
Oxidation, benzylic, of cumene, **2**, 655
Oxidation, electrochemical
of acylsilanes, **5**, 343
of bis(sulfones), **4**, 277
Oxidation, electrophilic, of arenes, **2**, 649
Oxidation, enantioselective, **2**, 147
Oxidation, osmium-catalyzed, of (*E*)-alkenylsilanes, **5**, 424
Oxidations, enzyme-catalyzed, use in the synthesis of aryl sulfoxides, **2**, 714
Oxidative addition
of aroyl chlorides, **2**, 957
of aryl halides, with transition metal complexes, **2**, 960
of imidoyl chloride, **5**, 797
intermolecular, **2**, 963
of lithium, **2**, 958
of magnesium, **2**, 958
of organometals, **2**, 957
for the synthesis of alkylmanganese(II) compounds, **2**, 574
for the synthesis of samarium compounds, **2**, 565
of vinyl halides, **2**, 952
with transition metal complexes, **2**, 960
Oxidative cleavage, of silicon–carbon bonds, **2**, 536
Oxidative coupling
of alkynes, **2**, 983
for aryl metals, **2**, 956
of cyclopropenes, **2**, 984
for vinyl metals, **2**, 956
Oxidative coupling–ring-expansion
with difunctional organic compounds, **2**, 984
mechanism, **2**, 984
of zirconium, **2**, 986
Oxidative displacement
of aryl halides, with transition metal complexes, **2**, 960
for aryl metals, **2**, 956
of organometals, **2**, 957
of vinyl halides, with transition metal complexes, **2**, 960
for vinyl metals, **2**, 956
Oxidative homocoupling reaction, of terminal alkynes, **1**, 1034
Oxide, bis[bis(trifluoromethyl)arsine]-, reactions, with hydrogen chloride, **2**, 492
Oxide, bis(di-*t*-butylstibine), synthesis of, **2**, 493
Oxides, for the reduction of nitriles, **1**, 48
Oxides, bis(dialkylbismuthine), synthesis of, **2**, 493
Oxides, bis(diorganometal), synthesis of, **2**, 885
Oxides, tertiary, as starting materials, for ylides, **3**, 500
N-Oxides, synthesis of, **2**, 765
N-Oxide sulfates, synthesis of, **2**, 350
Oxidizing agents
reactions, with hydrazine, **1**, 74
for the synthesis of disulfides, **2**, 212
Oxidizing reagents, for the conversion of aldehydes to alkanoic acids, **5**, 74
2,3(*S*)-Oxidosqualene, synthesis of, **1**, 116

trans-Oximation reaction, for the synthesis of oximes, **3**, 426
Oxime, 2-adamantanone, as starting material, for bromo nitro compounds, **4**, 108
Oxime, benzaldehyde, reactions, with diketene, **5**, 217
Oxime, cycloheptanone, rearrangement of, **3**, 564
Oxime, formamide, reactions, with aromatic aldehydes, **5**, 837
Oxime, silyl protected, as starting material, for oxime chloroformates, **6**, 430
Oxime, tosylated malononitrile, as starting material, for *O*-tosyl oxyimidic esters, **5**, 826
Oxime, *O*-(trifluoromethyl)difluoroformaldehyde, synthesis of, **6**, 232
Oxime dianions, for the alkylation of aldehydes, **1**, 154
Oxime dianions, acetaldehyde, alkylation of, **1**, 154
Oxime esters, synthesis of, **5**, 217
Oxime ethers
  cyclization of, **2**, 351
  reactions
    with aryllithiums, **2**, 351
    with organometallics, **2**, 351
Oximes
  acylation of, **5**, 215
  alkylation of, **2**, 342
  N-alkylation of, **3**, 430, 433
  O-alkylation of, **3**, 430
  arylation of, **2**, 692
  borane-mediated reduction of, **5**, 324
  chlorination of, **4**, 107
  dehydration of, **3**, 617, 643, 662
  deprotonation of, **1**, 154
  Forster reaction of, **3**, 463
  from
    carbonyls, **3**, 425
    nitro compounds, **3**, 426
    nitronate salts, **3**, 429
    α,β-unsaturated nitroalkanes, **3**, 426
  α-functionalization of, **3**, 429
  N-heteroatom analogues of, synthesis of, **3**, 436
  hydrolysis of, **3**, 18
  as intermediates, for oxime carbonates, **6**, 429
  metallation of, **3**, 426, 427
  oxidation of, **4**, 108; **5**, 324
  oxidative nitration of, **4**, 429
  photochlorination of, **5**, 664
  reactions
    with alkyllithiums, **2**, 343
    with 4-benzoyl-5-phenylfuran-2,3-dione, **5**, 217
    with α-bromo ketones, **3**, 189
    with butylvinyl ether, **3**, 430
    with chloramine, **3**, 463
    with 1,1-diethoxy-1-propene, **3**, 722
    with dinitrogen tetroxide, **5**, 838
    with divinyl sulfones, **2**, 158
    with ester enolates, **5**, 324
    with 1-ethoxyethyne, **6**, 74
    with formic acetic anhydride, **3**, 712
    with haloalkenes, **6**, 42
    with hydrogen cyanide, **3**, 616
    with lipase P, **5**, 217
    with organolithiums, **2**, 764
    with phenylhydrazine, **3**, 469
    with phosgene, **6**, 444
    with phosphinic acid, **4**, 464
    with sulfonyl halides, **3**, 437
  reactivity of, **1**, 1042
  [2,3] rearrangement of, **1**, 839
  reduction of, **2**, 341, 350
    for the synthesis of α-aminoalkyl phosphonates, **4**, 472
  as starting materials
    for *O*,*N*-acetals, **4**, 313
    for aldehydes, **3**, 18
    for *N*-alkoxy-α-amino phosphonates, **4**, 485
    for amines, **2**, 314
    for α-aminoalkylorganophosphorus compounds, **4**, 464
    for bromonitrile oxide, **6**, 611
    for diazonium salts, **2**, 805
    for enamides, **5**, 289
    for hydrazones, **3**, 446
    for hydroxylamines, **2**, 763
    for *N*-H imines, **3**, 405
    for ketenimines, **3**, 563
    for ketones, **3**, 133
    for nitriles, **3**, 617; **5**, 215
    for nitro compounds, **2**, 397
    for nitrones, **3**, 433
    for pseudonitroles, **4**, 439
    for thioketones, **3**, 365
  synthesis of, **2**, 578; **3**, **425–441**
    via Peterson reactions, **3**, 431
  use in the synthesis of *O*-benzylhydroxylamines, **2**, 103
Oximes, acyclic *O*-thionoacyl, stability of, **5**, 531
Oximes, acyl
  reduction of, **5**, 213
  uses of, **5**, 215
Oximes, *O*-acyl
  reduction of, **5**, 324
  synthesis of, **3**, 431; **5**, 215
Oximes, aldehyde, acylation of, **5**, 214
Oximes, α-alkoxy, synthesis of, **3**, 426
Oximes, *O*-alkyl
  deprotonation of, **3**, 431
  as starting materials
    for *N*-alkoxy β-lactams, **5**, 329
    for hydroxamates, **5**, 329
Oximes, allenic, as starting materials, for cyclic nitrones, **2**, 765
Oximes, β-allenic *O*-methyl, hydrostannylation of, **2**, 351
Oximes, allylic, from, allylic nitro compounds, **3**, 430
Oximes, *O*-arsenic-substituted, synthesis of, **3**, 435
Oximes, aryl, synthesis of, **3**, 429
Oximes, *O*-aryl, synthesis of, **2**, 692; **3**, 430
Oximes, azido, acetylation of, **5**, 220
Oximes, azo, synthesis of, **5**, 858
Oximes, *C*,*N*-bisamide, synthesis of, **5**, 836
Oximes, α-bromo, oxidation of, **2**, 786
Oximes, *O*-(*t*-butoxycarbonyl), synthesis of, **3**, 431
Oximes, *O*-(1-butoxyethyl), synthesis of, **3**, 430
Oximes, *O*-carbon-substituted, synthesis of, **3**, 430
Oximes, *O*-chalcogen-substituted, synthesis of, **3**, 435
Oximes, chiral, dehydration of, **3**, 617
Oximes, chloro, synthesis of, **6**, 622
Oximes, α chloro
  oxidation of, **2**, 786
  as starting materials, for *gem*-chloro nitrosoalkenes, **4**, 798
Oximes, cyclic, synthesis of, **3**, 430
Oximes, cyclobutanone, reactions, with chlorine, **4**, 108
Oximes, α,α-dialkylated aldehyde, synthesis of, **1**, 154
Oximes, fluorenone, reactions, with benzoyl isothiocyanate, **6**, 558
Oximes, furan-2-carboxaldehyde, reactions, with triethyl phosphite, **4**, 485
Oximes, α-halo
  reactions, with bases, **2**, 778
  as starting materials, for α,β-unsaturated nitroso compounds, **2**, 778
  synthesis of, **3**, 429
Oximes, hetero-substituted acyl, synthesis of, **5**, 219
Oximes, α-hydroxy, resolution of, **3**, 427
Oximes, hydroxyiminoethyl phosphonate, reactions, with chlorodiphenylphosphine, **4**, 1008
Oximes, α-hydroxylamino, synthesis of, **2**, 340; **3**, 427
Oximes, α-keto
  Forster reaction of, **3**, 469
  synthesis of, **3**, 192
Oximes, *O*-mesyl, reduction of, **2**, 314

Oximes, *O*-methyl, electroreduction of, **2**, 314
Oximes, *O*-silicon-substituted, synthesis of, **3**, 436
Oximes, *O*-stannyl, as intermediates, in the synthesis of nitrile oxides, **3**, 684
Oximes, substituted acetamide, synthesis of, **5**, 835
Oximes, α-sulfenyl, synthesis of, **3**, 165
Oximes, *O*-thionoacyl, synthesis of, **5**, 531
Oximes, *O*-tosyl, use in the synthesis of α-amino ketones, **3**, 184
Oximes, *O*-trimethylsilyl, alkylation of, **3**, 433
Oximes, α,β-unsaturated, synthesis of, **3**, 429
α-Oximes, glyoxal
  reactions, with trifluoroacetic anhydride, **3**, 673
  as starting materials, for imidazoles, **3**, 673
6-Oximinocholest-4-ene-4-thiocarboxylic anhydrides, synthesis of, **5**, 533
Oximino pyruvates, thermal decomposition of, **3**, 672
Oxindoles, thionation of, **5**, 567
Oxindoles, arylimino-, reactions, with trialkyl phosphites, **4**, 464
Oxindoles, α,β-unsaturated, thionation of, **5**, 613
Oxirane
  reactions
    with azulenes, **1**, 212
    with carbonic chloride fluoride, **6**, 424
    with carbonic difluoride, **6**, 424
    with hydrazines, **2**, 378
    with phosgene, **6**, 429
  as starting material, for β-chloroalkyl chloroformates, **6**, 429
Oxirane, 2-lithio-2-(trimethylstannyl)-, from, 2-(tributylstannyl)oxirane, **6**, 133
Oxirane, phenyl-, reactions, with diphenylcarbodiimide, **6**, 510
Oxirane, 2-phenyl-, reactions, with 1,3-dithiolane-2-thione, **6**, 305
Oxirane, tetraphenyl-, irradiation of, **3**, 314
Oxirane, 2-(tributylstannyl)-, as starting material, for 2-lithio-2-(trimethylstannyl)oxirane, **6**, 133
Oxirane, triphenylsilyl-, reactions, with *s*-butyllithium, **4**, 944
Oxirane, triphenylsilylsilyl-, deprotonation of, **6**, 126
Oxiranes
  addition of nucleophiles to, theoretical study, **1**, 217
  cleavage of, **3**, 315; **4**, 208
  reactions
    with *N*-alkylhydroxylamines, **2**, 342
    with alkynes, **1**, 1022
    with alkynylalanates, **1**, 1024
    with alkynylaluminums, **1**, 1023
    with alkynylgalliums, **1**, 1024
    with alkynyllithiums, **1**, 1023
    with alkynylmagnesium halides, **1**, 1022
    with ammonia, **1**, 1022
    with carbodiimides, **6**, 510
    with cobalt complexes, **1**, 345
    with diphenylketene, **4**, 829
    with *N,N*-disubstituted hydrazines, **2**, 384
    with hydride complexes, **2**, 576
    with lithium alkynide, **1**, 1022
    with magnesium bromide etherate, **4**, 783
    with methoxymethylaluminum alkynides, **1**, 1024
    with sodium alkynides, **1**, 1022
    with triisopropoxytitanium alkynides, **1**, 1024
    with trimethylsilyl selenocyanate, **5**, 1114
  reduction of, **1**, 13
  ring-opening of, **1**, 264; **2**, 129; **3**, 698, 707
  as starting materials
    for carbonyl ylides, **3**, 315
    for dialkylketenes, **3**, 541
    for monoalkylhydrazines, **2**, 373
    for oxygen heterocycles, **3**, 315
    for tetrasubstituted hydrazines, **2**, 382
  synthesis of, **2**, 90; **4**, 146
  See also Epoxides
Oxiranes, 2-aminomethyl-, synthesis of, **2**, 318
Oxiranes, bis(silyl)-, synthesis of, **6**, 126
Oxiranes, 2-(1-bromoalkyl)-, reactions, with amines, **2**, 318
Oxiranes, α,α-dichloro-, as starting materials, for α-azido acids, **6**, 42
Oxiranes, α,α-difluoro-, synthesis of, **6**, 42
Oxiranes, diol, reactions, with dimethylaluminum alkynides, **1**, 1024
Oxiranes, 2,3-disubstituted, reduction of, **1**, 14
Oxiranes, hydroxymethyl-
  reduction of, **1**, 14
  See also Glycidols
Oxiranes, 2-phenylsulfonyl-, synthesis of, **5**, 423
Oxiranes, silyl-
  reactions
    with acetonitriles, **4**, 519
    with amines, **4**, 506
    with *N*-silylamides, **4**, 513
  as starting materials, for β-hydroxy-α-silyl azides, **4**, 516
  use in the synthesis of α-aminosilanes, **4**, 508
Oxiranes, 2-substituted, use in the synthesis of spirocyclic trithioorthocarbonates, **6**, 309
Oxiranes, *N*-tosyl-, ring-opening of, **5**, 347
Oxiranes, 2-trimethylsilyl-, reactions, with trimethyltin bromide, **6**, 133
Oxiranes, vinyl-, synthesis of, via elimination reactions, **1**, 664
cis-Oxiranes, as starting materials, for α-aminosilanes, **4**, 508
Oxirenes
  as intermediates, in the synthesis of ketenes, **3**, 530
  as starting materials, for dialkylketenes, **3**, 541
Oxiridines, sulfonyl-
  reactions
    with enamines, **3**, 156
    with ketone enolates, **3**, 156
    with silyl enol ethers, **3**, 156
2-Oxoacethydroxamoyl chlorides, 2-aryl-, synthesis of, **5**, 668
Oxo acid chlorides, as starting materials, for (chloromethyl)thiophosphinic chlorides, **4**, 127
α-Oxoacids, reactions, with secondary amines, **5**, 603
5-Oxoalkanoic acids, from, 2-alkylidenecyclopentanones, **5**, 75
δ-Oxoazelaic acid, from, 2,2-di(3-cyanopropyl)-1,3-dioxolan, **5**, 64
Oxocarbene complexes, synthesis of, **3**, 521
Oxocarbonium ion, chloro-, synthesis of, **6**, 428
4-Oxocinnolin-3-nitriles, 1-ethyl-1,4-dihydro-, use in the synthesis of carboxylic acids, **5**, 94
Oxo compounds, as starting materials, for alkylrhenium(VI) compounds, **2**, 585
Oxone
  for the oxidation of dithioacetals, **4**, 272
  for the oxidation of sulfides, **2**, 150
  reactions, with alkynes, **3**, 151
  as reagent, for the synthesis of ketones from alcohols, **3**, 120
  use in the synthesis of dialkyl sulfones, **2**, 155
Oxonium compounds, as starting materials, for allenyl ethers, **1**, 978
Oxonium intermediates, use in the synthesis of carboxylic esters, **5**, 126
Oxonium ions, as intermediates, in the synthesis of acetals, **4**, 177
Oxonium salts
  synthesis of, **2**, 1007
  use in the synthesis of thioimidic esters, **5**, 726
Oxonium salts, trialkyl-, reactions, with *N*-methyleneamides, **3**, 422
2-Oxopenams, alkylation of, **4**, 833
α-Oxophosphinates, as starting materials, for α-aminoalkylphosphinates, **4**, 471

Oxo reaction, of alkenes, for aldehydes, **3**, 2
Oxo synthesis
  mechanism, **2**, 576
  with rhodium catalysts, **2**, 580
Oxovanadium(V), for the ring-opening of dichlorocyclobutenes, **4**, 743
Oxovanadiums
  reactions
    with benzylicsilanes, **1**, 290
    with cinnamyltrimethylsilane, **1**, 290
Oxovanadiums, triallyl-, synthesis of, **2**, 570
Oxyallyl cation, as intermediate, in Favorskii rearrangements, **1**, 392
Oxyamide oximes
  reduction of, **5**, 838
  as starting materials, for oxyimidic amides, **5**, 838
Oxyanions, stability of, **1**, 301
Oxy functions, synthesis of, **5**, **653–723**
Oxygen
  reactions
    with acetylethyne, **1**, 274
    with boron, **2**, 56
    with cuprous chloride, **5**, 54
    with dienes, **2**, 52
    with imidazoles, **5**, 759
    with metallated hydrazones, **3**, 464
    with methylenephosphines, **6**, 692
    with organometallics, **2**, 96
    with silenes, **6**, 382
    with thiols, **2**, 214
    with trialkylboranes, **2**, 97
    with vinyl radicals, **1**, 1127
  substitution of, **1**, 475
    nucleophilic aromatic substitution, **1**, 483
  and sulfur, synthesis of functions, **4**, 833
  synthesis of functions, **6**, 68
    with another chalcogen, **4**, **215–241**
    containing at least one oxygen, **5**, **961–1020**
    containing selenium or tellurium, **4**, 841
    and one halogen, **6**, 272
    one halogen and two nitrogen functions, **6**, 279
    one halogen and two sulfur functions, **6**, 276
  synthesis of functions bearing two oxygens, **4**, **159–214**, 824
  use in the synthesis of monocarboxylic acids, **5**, 28
Oxygen, sulfinyl-, reactions, with tin compounds, **2**, 210
Oxygen, vinyl-, displacement of, via addition–elimination, **1**, 481
Oxygen anions, reactions, with nitroalkenes, **3**, 162
Oxygen bases, reactions, with polyhaloalkenes, **2**, 1012
Oxygen compounds
  of arsenic functions, synthesis of, **5**, 891
  of phosphorus functions, synthesis of, **5**, 878, 897
  synthesis of, **6**, 36, 256
Oxygen compounds, α-diazo, synthesis of, **5**, 867
Oxygen difluoride, reactions, with carbonyl difluoride, **2**, 99
β-Oxygen effect, in radical generation, **1**, 945
Oxygen functions
  elimination of, **1**, 575, 1069
  substitution of, **1**, 172, 1021, 1053
Oxygen functions, alkyl-, substitution of, **1**, 1021
Oxygen heterocycles
  from
    carbenes, **3**, 319, 320
    oxiranes, **3**, 315
Oxygen leaving groups, alkenyl-, substitution of, **1**, 479
Oxygen leaving groups, alkyl-, substitution of, **1**, 475
Oxygen leaving groups, aryl-, displacement of, **1**, 482
Oxygen nucleophiles
  reactions
    with halomethylgermanes, **4**, 362
    with halomethylsilanes, **4**, 352
Oxygen-substituted carbene complexes, synthesis of, **6**, 719
2-Oxygen substitution, for the synthesis of α-alkoxyacroleins, **3**, 71
3-Oxygen substitution, for the synthesis of β-alkoxyacroleins, **3**, 71
Oxyimidic acids
  from
    nitrile oxides, **5**, 828
    aci-phenylnitromethane, **5**, 819
  as starting materials, for oxyimidic esters, **5**, 827
Oxyimidic azides, *O*-aroyl-substituted, reactions, with triphenylphosphine, **5**, 862
Oxyimidic bromides, *O*-acyl, synthesis of, **5**, 820
Oxyimidic bromides, *O*-alkyl, as starting materials, for *O*-alkyl oxyimidic esters, **5**, 827
Oxyimidic chloride, *O*-benzoyl
  from, 2-(nitrobromo)ethyl quinoline, **5**, 823
  synthesis of, **5**, 823
Oxyimidic chloride, (*Z*)-*O*-methyl-4-nitrophenyl, substitution of chlorine in, **5**, 834
Oxyimidic chlorides
  from
    amides, **5**, 834
    benzoic acid *O*-ethyl oxyimidic amide, **5**, 825
    chloroacetone, **5**, 823
    α-chloro ketones, **5**, 823
    chloromethyl aryl ketones, **5**, 823
    nitroalkenes, **5**, 819
    aci-nitro compounds, **5**, 819
    oxyimidic amides, **5**, 825
    thioamides, **5**, 834
  reactions
    with ammonia, **5**, 765
    with aziridine, **5**, 765
    with benzamide, **5**, 765
    with selenophenol, **5**, 832
    with thiol ions, **5**, 828
  as starting materials, for oxyimidic esters, **5**, 826
Oxyimidic chlorides, glyoxylic acidoxime, from, 2-nitroacetaldehyde oxime, **5**, 822
Oxyimidic compounds
  salts of, synthesis of, **5**, 864
  synthesis of, **5**, 819
Oxyimidic fluoride, 2,3,3,3-tetrafluoro-, synthesis of, **5**, 821
Oxyimidic fluorides, from, perfluoro-2,3-dialkyloxaziridines, **5**, 824
Oxyimidic fluorides, *O*-perfluoroalkyl, synthesis of, **5**, 825
Oxyimidic halides
  from
    aldoximes, **5**, 821
    alkenes, **5**, 821
    carbonyls, **5**, 823
    hydroxy compounds, **5**, 823
    nitrile oxides, **5**, 819
    nitro compounds, **5**, 822
    nitroso compounds, **5**, 822
    oxaziridines, **5**, 824
  isomerization of, **5**, 825
  reactions, with ferric chloride, **5**, 864
  as starting materials
    for oxyimidic amides, **5**, 834
    for oxyimidic thioesters, **5**, 828
  synthesis of, **5**, 819
Oxyimidic iodides
  reactions, with thiol ions, **5**, 828
  synthesis of, **5**, 822
Oxymercuration
  of alkynes, intramolecular, **2**, 982
  mechanism, **2**, 982
Oxymercuration, enantioselective, **2**, 40
Oxymetallation
  of alkenes, **6**, 209
  of alkynes
    with heavy metals, **2**, 982

with transition metals, **2**, 983
for organometallics, **2**, 952
with palladium, **2**, 983
with thallium, **2**, 982
for vinyl metals, **2**, 956
D-Oxynitrilase, as catalyst, for the synthesis of asymmetric cyanohydrins, **3**, 628
Oxypalladation
of alkenes, **2**, 983
intramolecular, of alkynoic acids, **2**, 983
Oxyphosphonium intermediates
synthesis of, **2**, 5
use in the synthesis of alkyl halides, **2**, 6
Oxyphosphonium salts, as intermediates, in the synthesis of carboxylic esters, **5**, 123
Oxy radicals, use in the synthesis of cyclic ethers, **2**, 93
Oxythallation, for the synthesis of alkylthalliums, **2**, 598
Ozazine, dihydro-, for the stabilization of carbanions, **1**, 405
Ozazines, halo-, metallation of, **3**, 27
Ozazines, 2-vinyl-, synthesis of, **3**, 576
Ozone
destruction of, **6**, 8
for the oxidation of selenides, **2**, 287
for the oxidation of styrenes, **3**, 87
for the oxidation of tertiary amines, **2**, 348
for the oxidative cleavage of alkenes, **3**, 113
reactions
with alkenes, **2**, 98
with cyclohexenes, **5**, 76
with dibromothiophosgene, **6**, 254
with indoles, **6**, 107
with iodomethane, **4**, 42
with isonitriles, **5**, 1003
with 4-methylbenzaldehyde, **5**, 208
with phenanthrene, **5**, 115
with pyrene, **5**, 115
with selenenyl chlorides, **2**, 733
with selenides, **4**, 241
with triphenyl phosphite, **6**, 520
as reagent
for aldehydes, **3**, 2
for the synthesis of sulfines, **3**, 368
use in the synthesis of 2-methoxy hydroperoxide, **2**, 96
use in the synthesis of monocarboxylic acids, **5**, 28
Ozonides
from, carbonyl ylides, **3**, 319
hydrogenolysis of, **4**, 162
synthesis of, **3**, 326; **4**, 214; **6**, 107
Ozonides, allylic, reduction of, **1**, 84
Ozonolysis
of 1,1-diarylethenes, **2**, 643
of oxazoles, **5**, 197
of phosphorus–phosphorus double bonds, **6**, 373
for the synthesis of carbonates, **6**, 469
for the synthesis of carbonyl oxides, **3**, 326
for the synthesis of α-hydroxy aldehydes, **3**, 30
of vinyl sulfones, **6**, 480

P

Paget's disease, agents for, **6**, 157
Pagodyl dications, synthesis of, **1**, 918
Palladacyclopentadiene, tetrakis(carbomethoxy)-, reactions, with tris(2,6-dimethoxyphenyl)phosphine, **1**, 1051
Palladacyclopentadienes, 2,3,4,5-tetrakis(methoxycarbonyl)-, use in enyne metathesis reactions, **1**, 676
Palladium
  on carbon with hydrogen, as reducing agent, **1**, 43
  as catalyst
    for the oxidation of alcohols, **3**, 118
    for the reduction of nitriles, **1**, 46
    for the substitution of sulfur functions, **1**, 238
    for the synthesis of aldehydes from alcohols, **3**, 5
    use in the synthesis of aryl-substituted aldehydes, **3**, 24
  oxymetallation of, **2**, 983
  reactions
    with alkyl halides, **2**, 582
    with allyl cations, **1**, 191
    with α-amino acids, **5**, 385
    with barium sulfate, **1**, 503
    with *t*-butyldimethylsilane, **5**, 225
    with silk fibroin, **1**, 76
    with siloxane polymers, **1**, 456
    with tellurides, **1**, 247
  for the reduction of iodoalkanes, **1**, 10
  use in the reduction of alkenes, **1**, 74
  use in the synthesis of α-methylene lactams, **5**, 296
Palladium, alkenyl-, cascade carbopalladation of, **2**, 980
Palladium, bis(acetonitrile)-, reactions, with dimethylformamide, **1**, 1007
Palladium, bis(triphenylphosphine)-, use in the synthesis of alkynes, **1**, 1007
Palladium, 2,3,4,5-tetrakis(methoxycarbonyl)cyclopentadiene biphenylphosphinoethane-, reactions, with *N,N*-dichloro-*para*-toluamide, **5**, 387
Palladium, tetrakis(triphenylphosphine)-
  reactions, with imidoyl chlorides, **5**, 801
  for the substitution of enol phosphates, **1**, 479
  use in the synthesis of alkynes, **1**, 1007
  use in the synthesis of thiocarboxylic esters, **5**, 238
Palladium(0)
  catalyzed substitution, with Grignard reagents, **1**, 970
  reactions
    with tributyltin hydride, **1**, 86
    with tricarbonyl(chlorobenzene)chromium, **4**, 722
Palladium(0), tetrakis(triphenylphosphine)-
  reactions, with phosphaalkenes, **5**, 1158
  use in the synthesis of α,β-arylnitriles, **3**, 661
Palladium(0) catalysis
  for the isomerization of allyl ethers, **2**, 675
  for the oxidation of allyl chlorides, **1**, 156
  for the synthesis of amines, **2**, 304
  for the synthesis of aryl metals, **2**, 983
  for the synthesis of bis(sulfones), **4**, 279
  for the synthesis of phenols, **2**, 659
  for the synthesis of vinyl metals, **2**, 983
Palladium(0)-catalysts, use in the synthesis of α,β-unsaturated ketimines, **3**, 413
Palladium(0)-catalyzed coupling
  with acid bromide, **2**, 931
  with acid chlorides, **2**, 931
  with arylstannanes, **2**, 931
  with Grignard reagents, **2**, 931
  with vinyl and aryl halides, **2**, 931
  with vinyl and aryl triflates, **2**, 931
  with zinc compounds, **2**, 931
Palladium(0) complexes
  reactions, with allylic carbonates, **6**, 470
  of trimethylenemethane, **1**, 360
Palladium(II)
  as catalyst, for the isomerization of allyl ethers, **2**, 675
  as oxidant, for the synthesis of enones, **1**, 556
  as reagent, for the synthesis of phenols, **2**, 659
Palladium(II), benzoylchlorobis(triphenylphosphine)-, use in the synthesis of aromatic ketones, **3**, 288
Palladium(II), chloro(cyanomethyl)bis(triphenylphosphine)-, reactions, with potassium cyanide, **5**, 1138
Palladium(II), dichlorobis(tri-o-tolylphosphine)-, reactions, with α-stannyl ketone, **3**, 144
Palladium(II) acetate
  synthesis of, **5**, 229
  use in the synthesis of α,β-arylnitriles, **3**, 662
Palladium(II) bis(acetonitrile) dichloride
  as reagent
    for the synthesis of aldehydes, **3**, 9
    for the synthesis of ketones from acetals, **3**, 123
Palladium(II) catalysts
  for dicarboxylation reactions, **1**, 298
  for the synthesis of alkenes, **1**, 700
  for the synthesis of α-haloalkenylsilicons, **4**, 810
Palladium(II) chlorides, alkenyl-, reactions, with organotins, **1**, 1027
Palladium(II) chlorides, aryl-, reactions, with organotins, **1**, 1027
Palladium(II) complex, interlamellar montmorillonite diphenylphosphine-, use in the synthesis of amines, **2**, 317
Palladium(II) complexes
  as catalysts, in Cope rearrangements, **1**, 413, 809
  reactions, with α-lithiodiazo compounds, **3**, 487
  synthesis of, **2**, 581
Palladium(II) complexes, diaminocarbene, synthesis of, **5**, 1072
Palladium(II) complexes, imidoyl-, from, imidoyl chlorides, **5**, 801
Palladium(II) compounds, tris(triarylphosphine)-, reactions, with aryl nitriles, **5**, 1139
Palladium(II) diacetate, reactions, with triphenylphosphine, **5**, 220
Palladium(II) halides, bis(trialkylphosphine)-, reactions, with diazomercury compounds, **3**, 487
Palladium(II) iodides, alkenyl-, synthesis of, via oxidative addition, **1**, 1013
Palladium(II) salts
  reactions, with aryl halides, **3**, 662
  transmetallation of, **1**, 493
Palladium(II) salts, tetrachloro-, reactions, with sulfur ylides, **4**, 396
Palladium(II) salts. as catalysts, in Claisen rearrangement, **1**, 415
Palladium(II) trifluoroacetate, synthesis of, **5**, 229
Palladium acetate
  as oxidant, **1**, 558
  reactions
    with alkenyldialkylboranes, **1**, 660
    with triphenylphosphine, **1**, 1051
  use in the synthesis of vinyl sulfides, **2**, 710
Palladium acetate, phenyl-, synthesis of, **5**, 220

Palladium acyl compounds
from, chiral alkyl bromides, **5**, 463
synthesis of, stereoselectivity, **5**, 463
Palladium alkene complexes, reactions, with amines, **5**, 464
Palladium alkynides, synthesis of, **1**, 1137
N-Palladium amides, *N*-chloro-, synthesis of, **5**, 387
Palladium azoaryls, cyclization of, **5**, 356
Palladium–barium sulfate catalyst
for the synthesis of alkenes, **1**, 503
for the synthesis of (*Z*)-alkenes, **1**, 506
Palladium bis(acetylacetonate), reactions, with alkylaluminums, **2**, 581
Palladium [1,2-bis(diphenylphosphino)ethane], reactions, with 1,1-dibromovinyl groups, **5**, 38
Palladium bis(methoxycarbonyl) complexes, as starting materials, for methyl chloroformate, **6**, 428
Palladium on carbon
for the deprotection of amines, **2**, 304
for the synthesis of amines, **2**, 318
use in the synthesis of aryl aldehydes, **3**, 85
Palladium–carbon catalysis
for the reduction of esters, **1**, 16
for the reduction of ketones, **1**, 19
Palladium carbonyl complexes
reactions, with alkyl halides, **5**, 463
synthesis of, **6**, 525
Palladium carbonyls, reactions, with cinnamyl bromide, **5**, 463
Palladium catalysis
in ene reactions, **1**, 525
for the reduction of vinyl triflates, **1**, 435
for the synthesis of alkynes, **1**, 1006
Palladium catalysts
in elimination reactions, **1**, 613
for the hydrogenation of alkynes, **1**, 455
reactions, with Grignard reagents, **1**, 466
selectivity of, **1**, 502
for the substitution of alkenyl halides, **1**, 466
for the substitution of aryl halides, **1**, 470
for the synthesis of alkenes, **1**, 502
for the synthesis of aryl ketones, **3**, 289
for the synthesis of enynes, **1**, 468
for trimethylenemethane reactions, **1**, 287
use in carbon monoxide insertion, **5**, 346
use in the substitution of vinyl halides, **1**, 468
use in the synthesis of acyl aldehydes, **3**, 83
use in the synthesis of α,β-unsaturated ketones, **3**, 206
water soluble, **1**, 1013
Palladium-catalyzed cross-coupling, of alkenylalanes, **2**, 976
Palladium-catalyzed reaction
of alkenyl bromides
with monoalkyl phosphonites, **2**, 845
for tertiary phosphine oxides, **2**, 843
of aryl halides, with dialkyl phosphonates, **2**, 848
Palladium-catalyzed reactions, for the synthesis of imines, **3**, 408
Palladium chloride
reactions, with vinylic mercury compounds, **1**, 1095
use in the alkylation of alkenes, **1**, 315
use in the synthesis of alkenes, **1**, 507
use in the synthesis of dienes, **1**, 491
Palladium chloride, π-allyl-, synthesis of, **5**, 974
Palladium complexes
as catalysts, for alkylation reactions, **1**, 465
reactions, with nucleophiles, **2**, 583
stability of, **2**, 581
as starting materials, for ketenimines, **3**, 599
synthesis of, **4**, 955; **5**, 1089
use in the synthesis of alkylzinc halides, **2**, 591
use in the synthesis of chiral phosphines, **2**, 442
Palladium complexes, acyl-, synthesis of, **5**, 463
Palladium complexes, acylcyanol-, synthesis of, theoretical, **5**, 1139
Palladium complexes, alkyl-, bearing dithiocarbamate ligands, **2**, 581
Palladium complexes, alkynyl-, synthesis of, **1**, 1138
Palladium complexes, π-allyl-, synthesis of, **1**, 192
Palladium complexes, (π-allyl)-, synthesis of, **6**, 470
Palladium complexes, allyl-bridged
from, cyclopentadienylpalladiums, **4**, 693
platinum analogues, synthesis of, **4**, 693
synthesis of, **4**, 693
Palladium complexes, aryl-, reactions, with imides, **5**, 387
Palladium complexes, bisallyl-bridged, synthesis of, **4**, 693
Palladium complexes, bis(cyclopentadienyl)-bridged, synthesis of, **4**, 693
Palladium complexes, cationic trinuclear, synthesis of, **4**, 692
Palladium complexes, α-diazo-, synthesis of, **3**, 487
Palladium complexes, ethylene-bridged, from, methylene halides, **4**, 691
Palladium complexes, (halomethyl)-, synthesis of, **4**, 150
Palladium complexes, methylene-bridged, from, methylene halides, **4**, 691
Palladium complexes, monoalkyl-, synthesis of, **2**, 582
Palladium complexes, trinuclear, from, triarylcyclopropenium bromides, **4**, 693
Palladium complexes, trinuclear carbon-bridged, from, thallium acetylacetonate, **4**, 693
Palladium complexes, tris(*t*-butyl)phosphine-, orthometallation of, **2**, 581
Palladium diacetate, reactions, with imines, **5**, 464
Palladium intermediates, allenyl-, hydrogenolysis of, **1**, 962
Palladium-on-carbon catalysis
for the reduction of alkyl halides, **1**, 2
for the reduction of benzyl chlorides, **1**, 6
for the reduction of benzylic alcohols, **1**, 12
for the reduction of bromoalkanes, **1**, 8
for the reduction of fluoroalkanes, **1**, 6
Palladium oxide, as catalyst, for decarbonylation and replacement by isonitriles, **5**, 799
Palladiumphosphine catalysts, for the synthesis of carboxylic esters, **5**, 142
Palladiumphosphine complexes
as catalysts
for haloalkane addition, **6**, 7
for the substitution of vinyl halides, **1**, 468
reactions, with acetyl chloride, **5**, 463
use in the substitution of nitrogen functions, **1**, 250
Palladium–phosphinoalkene complexes, synthesis of, **4**, 1023, 1028
Palladiums
reactions
with 1-haloalkyl phosphonates, **4**, 583
with phosphorus ylides, **4**, 583
Palladiums, acyl-, synthesis of, **5**, 463
Palladiums, alkenyl-, synthesis of, **1**, 1096, 1107
Palladiums, alkyl-
β-hydride elimination from, **1**, 1015
as intermediates, **2**, 579
Palladiums, π-alkyl-, carbonylation of, **1**, 315
Palladiums, aryl-, as starting materials, for benzaldehydes, **3**, 89
Palladiums, cyano-, synthesis of, **5**, 1138
Palladiums, cyclopentadienyl-, as starting materials, for allyl-bridged palladium complexes, **4**, 693
Palladiums, imidoyl-
from, isonitriles, **5**, 801
synthesis of, **5**, 800
via oxidative addition, **5**, 800
Palladiums, α-sulfinyl-, synthesis of, **4**, 396
Palladium salts
reactions
with alkynes, **1**, 1107
with organometallics, **2**, 581
Palladium–trimethylenemethane complexes
applications of, **1**, 362

intramolecular cycloadditions of, **1**, 362
Palladium–zeolite catalysts, for the reduction of alkynes, **1**, 456
Palmitate, methyl, synthesis of, **5**, 18
Palytoxin, synthesis of, **1**, 1099
β-Panasinsene, synthesis of, **1**, 711
Papain, use in the synthesis of peptides, **5**, 263
Papaverine, tetrahydro-, synthesis of, **1**, 77
Parabanates, pyrolysis of, **5**, 996
Parabanates, sulfonyl-, synthesis of, **5**, 996
Paraconic acids, synthesis of, **5**, 166
Paracyclophanes, synthesis of, **1**, 243
Paraformaldehyde
reactions
with aryl methyl ketones, **1**, 256
with di(*t*-butyl)malonate, **5**, 104
with diethyl phosphite, **4**, 338
with 1,2-dilithio[tetrakis(trimethylsilyl)]ethane, **6**, 176
with hypophosphorous acid, **4**, 125
with perfluoroalkyldichlorophosphines, **4**, 125
with *N*-phenylglycine, **4**, 302
with phosphinic acid chlorides, **4**, 128
with γ-thiol amides, **4**, 325
as starting material
for ethyl α-(hydroxymethyl)acrylate, **5**, 103
for α-fluoro ethers, **4**, 44
for phosgene, **6**, 413
for primary chloro sulfides, **4**, 69
use in hydroxymethylation reactions, **1**, 1037
use in the synthesis of amines, **2**, 300
Paraldehyde, as starting material, for secondary chloro sulfides, **4**, 69
Parikh–Doering conditions, in the synthesis of α-amino aldehydes, **3**, 40
Passerini reaction, for the synthesis of isocyanides, **3**, 694
β-Patchouline oxide, reactions, with boron trifluoride etherate, **1**, 842
Paterno–Büchi photocyclization, for the synthesis of oxetanes, **2**, 92
Patuolide A, synthesis of, **5**, 199
Pauson–Khand [2+2+1] cycloaddition, of 1-silyl-1-alkynes, **2**, 926
Pauson–Khand reaction
mechanism, **1**, 528
for the synthesis of cyclopentanones, **3**, 114
for the synthesis of α,β-unsaturated ketones, **3**, 218, 243, 270
Payne rearrangement
in the opening of epoxides, **1**, 223
for the synthesis of δ,ε-epoxides, **5**, 172
Pedersen thionation reaction, for the synthesis of 1,8-naphthalic monothionoanhydride, **5**, 529
Peduncularine, synthesis of, via acetate pyrolysis, **1**, 774
Penem-3-carboxylic acids, 6-acylamino-, synthesis of, **5**, 905
Penicillacylases, use in the synthesis of α-aminophosphonic acids, **4**, 462
Penicillanic acid, 6-amino-, synthesis of, **4**, 898
Penicillanic acid *S,S*-dioxide, synthesis of, **4**, 898
Penicillin, synthesis of, **4**, 897
Penicillin, α-phenoxyethyl-, as starting material, for acid chlorides, **5**, 9
Penicillin acid chloride, synthesis of, **5**, 9
Penicillin carboxylic acids, activation of, with ethyl chloroformate, **5**, 217
Penicillin G acylase, use in the synthesis of *N*-acyl-α-aminosilanes, **4**, 513
Penicillinic acids, 6-diazo-, synthesis of, **3**, 468
Penicillins
as starting materials, for acid chlorides, **5**, 9
synthesis of, **6**, 488
via carbon–sulfur bond cleavage, **2**, 128
Penicillins, 6,6-diacylamino-, synthesis of, **4**, 426
Penicillins, α-diazo, from, acetamides, **3**, 471
Penicillin sulfoxide esters, use in the synthesis of thiosilanes, **2**, 267
Penicillin sulfoxides
reactions, with thioamides, **5**, 736
as starting materials, for cephalosporins, **2**, 184
thionation of, **5**, 577
Penicilloate, α-(*t*-butyl)D-α-phenoxymethyl-, reactions, with hydrogen chloride, **5**, 81
Penicilloate, α-(*t*-butyl)DL-α-phenoxymethyl-, reactions, with hydrogen chloride, **5**, 81
Penicilloic acid hydrochloride, D-phenoxymethyl-, synthesis of, **5**, 81
Penicilloic acid hydrochloride, DL-phenoxymethyl-, synthesis of, **5**, 81
Pentaarsacyclopentane, pentakis(chloromethyl)-, synthesis of, **4**, 135
Pentacarbonyl complexes, reactions, with propargylic alcohols, **5**, 954
Pentacarbonyl complexes, (pyridinato)-, synthesis of, **5**, 953
Pentacarbonyls, reactions, with acylferrocenes, **4**, 720
Pentachlorophosphate, ethyl, reactions, with silver cyanide, **2**, 476
Pentacoordinate compounds, dihalo, as starting materials, for ylides, **3**, 500
Pentacyclopropylethyl carbocations, synthesis of, **1**, 900
2,4-Pentadienal, 2-chloro-, synthesis of, **3**, 68
Penta-1,2-diene, 3-ethyl-, reduction of, **1**, 459
1,3-Pentadiene, reduction of, **1**, 79
Penta-1,3-diene, 1-trimethylsilyl-, reactions, with methyl propiolate, **1**, 520
1,4-Pentadiene, reactions, with nickel(II) chloride, **4**, 694
Penta-1,4-diene, synthesis of, **1**, 788
1,4-Pentadiene, 3-methoxymethylene-2,4-bis(trimethylsilyloxy)-, as starting material, for phenols, **2**, 667
1,3-Pentadienes, rearrangement of, **1**, 402
1,4-Pentadienes, synthesis of, **1**, 496
2,4-Pentadienhydroxamoyl chlorides, 5-dialkylamino-2-nitro-, synthesis of, **5**, 674
2,3-Pentadienoate, methyl, reactions, with alkynes, **1**, 1051
Pentadienoic acid nitriles, perchloro-, as starting materials, for triphenoxy compounds, **4**, 829
2,4-Pentadienoic acids, 5-aryl-2,3,4,5-tetrafluoro-, synthesis of, **5**, 58
Penta-2,3-dien-1-ol, 5-trimethylsilyl-2-methyl-, synthesis of, **1**, 967
Pentadienyl cations
from
divinyl chlorides, **1**, 423
vinylallene oxides, **1**, 423
synthesis of, **1**, 421
1,3-Pentadiyne, reactions, with *n*-butyllithium, **6**, 206
1,4-Pentadiyne, as starting material, for 3-stanna-1,4-cyclohexadienes, **2**, 971
1,3-Pentadiynes, 1,5,5,5-tetrakis(trimethylgermyl)-, synthesis of, **6**, 185
Pentaenes, synthesis of, **1**, 754
Pentaenes, bishexa-, synthesis of, **1**, 986
(*E*)-Pentaenes, synthesis of, **1**, 677
Pentafluoroethyl radical, synthesis of, **4**, 1080
Pentafulvene, 6-chloro-6-methyl-, as starting material, for 1,3-di-*t*-butyl-5-vinylidene cyclopentadiene, **1**, 982
Pentalenes, synthesis of, **1**, 789, 804; **3**, 356, 376
Pentalenes, *tetra*-azo-, synthesis of, **2**, 1054
(+/-)-Pentalenolactone (*E*)-methyl ester, synthesis of, **6**, 461
Pentamethinium salts, reactions, with nitromethane, **2**, 790
Pentamethylated compounds, synthesis of, **2**, 570
Pentanal, reactions, with diphenylphosphine oxides, **1**, 755
Pentan-2,4-dione, reactions, with mercury(II) chloride, **4**, 697
Pentane, reactions, with di-*t*-butyliminoxy, **3**, 407
Pentane, 1,5-dibromo-, reactions, with (phenylthio)nitromethane, **1**, 157

Pentane, 3-methyl- (3-MP)
reactions, with 2-methylbutane, **5**, 1020
use in the synthesis of silylenes, **5**, 1019
Pentane 1,5-dicarboxylate, ethyl *t*-butyl 2-carbobenzyloxyamino-, use in the synthesis of alkanoic acids, **5**, 32
Pentane-2,4-dione, reactions, with 5-methyl-2-morpholino-1,3-dithiolylium-4-thiolate, **3**, 255
2,4-Pentanedione, 3-chloro-, reactions, with diazonium halides, **5**, 676
Pentane-2,4-dione, 1,1,1-trifluoro-, as starting material, for α-bromo alcohols, **4**, 42
Pentanenitrile, synthesis of, **2**, 1057
Pentanoate, methyl 5-fluoro-, hydrolysis of, **5**, 55
Pentanoic acid, 5-amino-, reactions, with hydrogen chloride, **5**, 970
Pentanoic acid, 4-methyl-2-oxo-, synthesis of, **5**, 63
5-Pentanol, 1,1,1-trichloro-, oxidation of, **5**, 59
n-Pentanol, synthesis of, **5**, 439
2-Pentanone, 5-phenyl selenyl-, from, 2-acetyl 4-butyrolactone, **3**, 176
3-Pentanone
deprotonation of, **1**, 874
enolization of, **1**, 131
reactions, with lithium bromide, **2**, 31
as starting material, for diethylacetic acid, **5**, 85
Pentan-3-one, decafluoro-, irradiation of, **4**, 1080
Pentan-3-one, 2,4-dimethyl-2-thiol-, reactions, with formaldehyde, **4**, 317
Pentan-4-ones, β-(*N*-alkyl-*N*-nitrosoamino)methyl-, cleavage of, **3**, 465
Pentanoyl chloride, isocyanato-, synthesis of, **5**, 970
Pentanoyl chloride, 4-methyl-, reactions, with benzyloxycarbonylglycine, **5**, 193
Pentaorganoantimony
from
organometal(V) halides, **2**, 873
N-sulfonyl organometal(V) imines, **2**, 873
Pentaorganoantimony compounds, reactions, with electrophiles, **1**, 446
Pentaorganoarsenic
from
organometal(V) halides, **2**, 873
N-sulfonyl organometal(V) imines, **2**, 873
Pentaorganoarsenic(V) compounds, reactions, with electrophiles, **1**, 445
Pentaorganobismuth
from
organometal(V) halides, **2**, 873
N-sulfonyl organometal(V) imines, **2**, 873
Pentaorganobismuth compounds
reactions, with electrophiles, **1**, 446
synthesis of, problems, **2**, 1082
Pentaorganophosphoranes, synthesis of, **2**, 830
Pentaorganostiboranes
from, benzyltributylstibonium bromide, **1**, 272
reactions
with acyl chlorides, **1**, 272
with ketones, **1**, 272
synthesis of, **1**, 272
Pentatetraenes, synthesis of, **1**, 986
1,2,3,4-Pentatetraenes, synthesis of, **1**, 983
Pentatraenes, synthesis of, **1**, 987
Penta-1,2,3-triene, 4-((benzyloxy)*t*-(butyl)(methyl)silyloxy)-, synthesis of, **1**, 985
1,2,3-Pentatriene, 1-dimethylamino-1-trimethylsilyl-4-methyl-, synthesis of, **2**, 940
Pentazadienes, synthesis of, **2**, 777
Pent-3-enal, 4-methyl-3-phenyl-, oxime acetate of, **1**, 800
4-Pentenal, synthesis of, **1**, 414
(*E*)-Pent-2-en-1-al, reactions, with 9-acetoxynonyltriphenylphosphonium bromide, **1**, 734
Pent-2-enals, (*Z*)-3-(trimethylstannyl)-, synthesis of, **3**, 78
Pentene, aryl-, as starting material, for (-)-retigeranic acid, **1**, 365
Pentene, phenyl-, use in meta photocycloaddition, **1**, 364
Pent-1-ene, hydroboration of, **2**, 517
Pent-1-ene, 4,4-dichloro-, synthesis of, **4**, 14
Pent-2-ene, ozonolysis of, **3**, 326
Pent-2-ene, perfluoro-2-methyl-, use in the synthesis of ketenimines, **3**, 588
5-Pentene carboximidoyl chlorides, synthesis of, **5**, 655
Pent-4-ene-2,3-diols, 1-amino-, synthesis of, **2**, 318
Pentenes, reactions, with nitric acid, **1**, 895
Pentenes, 1,1-bis(silyl)-, synthesis of, **4**, 1050
2-Pentenoates, *t*-butyl 5,5-disubstituted-5-hydroxy-, use in the synthesis of α,β-unsaturated acids, **5**, 98
4-Pentenoic acid, 5,5-dichloro-, hydrolysis of, **5**, 57
Pent-4-en-1-oic acids, synthesis of, **5**, 227
Pent-3-enols, 5-alkoxy-, synthesis of, **2**, 79
3-Penten-2-one, 4-methyl-, as starting material, for β,β-dimethylacrylic acid, **5**, 103
3-Penten-2-one, 1,1,1-triphenyl-, reactions, with lithium alkynides, **1**, 1048
(*E*)-3-Penten-2-one, reactions, with organocuprates, **1**, 311
4-Pentenyl radicals, cyclization of, **1**, 340
Pent-4-enylzinc–magnesium compounds
reactions
with chlorotrimethylstannane, **4**, 724
with copper(I) cyanide, **4**, 724
1,4-Pentenynes, synthesis of, **1**, 1030
n-Pentyl chloride, reactions, with sodium cyanate, **5**, 1104
Pent-3-yn-1,5-diols, as starting materials, for β-allenic alcohols, **1**, 961
Pentyne, 1-trimethylsilyl-, hydrostannation of, **4**, 1062
Pent-1-yne, 1-(phenylseleno)-, reactions, with bromine, **4**, 786
2-Pentyne, 1-dimethylamino-4-methoxy-4-methyl-, reactions, with butyllithium, **2**, 940
Pent-2-ynes, 1-dialkylamino-4-methoxy-4-methyl-, as starting materials, for allenyl amines, **1**, 981
4-Pentynes, 2-(*N*-succinimidyl)-, aza-Cope rearrangement of, **1**, 980
Pentynoates, reactions, with propargylic acetates, **1**, 975
4-Pentyn-1-ol, synthesis of, **1**, 1066
Peptide coupling
with acid chlorides, **5**, 9
for the synthesis of amides, **5**, 278
for the synthesis of anhydrides, **5**, 189
use of amino acid chlorides in, **5**, 15
Peptides
from
amino acids, **5**, 259
enzymes, **5**, 262
reactions, with Belleau's reagent, **5**, 574
solid-phase synthesis of, **5**, 259, 262
synthesis of, **3**, 687; **5**, 185, 221
methods, **5**, 259
reviews, **5**, 84
via carbon–sulfur bond cleavage, **2**, 128
thionation of, **5**, 515
Peptides, phosphoro-, synthesis of, **6**, 487
Peracetate, *t*-butyl
reactions
with sulfides, **4**, 220
with tetrahydrothiophene, **4**, 226
Peracetic acid
from, *N*,*N*,*N*,*N*-tetraacetylethylenediamine, **2**, 390
as oxidizing agent, **6**, 410
reactions
with 2-iodobenzene-1,3-dicarboxylic acid, **5**, 204
with norbonyl iodides, **1**, 108
as reagent, for the synthesis of 1,2-dithioles, **3**, 371
Peracetic acid, trichloro-, reactions, with 2,2,4,4-tetrachloro-1,3-dithietane, **6**, 530

Peracetic acid, trifluoro-
reactions
with *trans*-2,3-dibromobut-2-ene, **4**, 24
with 2,4-dichloro-2,4-difluoro-1,3-dithietane, **6**, 532
with dienes, **4**, 46
Peracids
as oxidation reagents, **4**, 854
reactions
with alkenes, **2**, 90
with alkyl iodides, **1**, 108
with allyl iodides, **2**, 88
with dicyclohexylcarbodiimide, **5**, 208
with enol esters, **4**, 209
synthesis of, *in situ*, **2**, 90
for the synthesis of diacyl peroxides, **5**, 208
use in the synthesis of dialkyl sulfoxides, **2**, 147
See also Peroxides, monoacyl
Peracids, organic, for the oxidation of alkynyl thioethers, **2**, 1034
Peralkylated compounds, synthesis of, **2**, 571
Perbenzoic acid
for the oxidation of disulfides, **2**, 232
use in the synthesis of arsenic oxide, **2**, 1080
Perbenzoic acid, *m*-chloro- (mcpba)
comparisons, with 2,2-dimethyldioxirane, **2**, 344
as epoxidizing agent, **2**, 90
for the oxidation of alkanedithioates, **5**, 248
for the oxidation of alkynyl sulfones, **2**, 1034
for the oxidation of arenes, **2**, 650
for the oxidation of α-chloro sulfides, **4**, 74
for the oxidation of disulfides, **2**, 232
for the oxidation of α-fluoro sulfides, **4**, 73
for the oxidation of silyl diazo compounds, **5**, 413
for the oxidation of sulfines, **6**, 306
for the oxidation of tertiary amines, **2**, 348
reactions
with *N*-alkylaziridines, **1**, 636
with alkylphenyl tellurides, **1**, 611
with 2,2-bis(methylthio)-1,3-diphenylpropane, **4**, 268
with hexanal, **5**, 52
with imines, **4**, 104
with methanethiotris(trimethylsilyl)methane, **6**, 325
with 1,3-oxathiolan-5-ones, **4**, 837
with thionselenol esters, **5**, 249
with thiophosgene, **6**, 531
as reagent
for the synthesis of ketones from alcohols, **3**, 120
for the synthesis of sulfines, **3**, 368
for the stepwise oxidation of aromatic dithioesters, **5**, 562
use in monoxidation reactions, **4**, 853
use in the synthesis of acyldiazines, **5**, 366
use in the synthesis of α-alkoxy sulfoxides, **4**, 232
use in the synthesis of 1-benzenesulfinyl-1-nitro-2,2-diphenylethene, **4**, 907
use in the synthesis of α-chloro sulfones, **4**, 80
use in the synthesis of α-diazosilicons, **5**, 873
use in the synthesis of α-hydroxyketones, **3**, 156
use in the synthesis of imidosulfonamides, **5**, 342
use in the synthesis of isocyanates, **5**, 964
use in the synthesis of α-ketoacylsilanes, **5**, 424
use in the synthesis of nitroso compounds, **2**, 389
Percarboxylic acid, reactions, with acyl chloride, **5**, 207
Perchlorate, benzo[*a*]phenalenium, synthesis of, **1**, 910
Perchlorate, *t*-butyldimethylsilyl, use in the protection of alcohols, **2**, 539
Perchlorate, cyclobutenephosphonium, hydrolysis of, **1**, 444
Perchlorate, ethyl, synthesis of, **2**, 94
Perchlorate, triethylsilyl, as silylating agent, **4**, 831
Perchlorates
from, benzyl chlorides, **1**, 904
safety precautions, **4**, 250
Perchlorates, acylium, reduction of, **4**, 425
Perchlorates, alkyl
stability of, **2**, 94
synthesis of, **1**, 910; **2**, 94
Perchlorates, amine, as starting materials, for iminium salts, **3**, 422
Perchlorates, benzothiophenium, deprotonation of, **1**, 985
Perchlorates, iminium
hydrolysis of, **4**, 987
as starting materials, for α-aminoalkylideneamino enamines, **4**, 987
Perchlorates, thiopyrylium, as starting materials, for 3-oxidopyrylium ylides, **3**, 326
Perchlorates, trialkylsilyl, use in the synthesis of (*Z*)-silylketene acetals, **2**, 699
Perchlorate salts, synthesis of, **4**, 987
Perchloro 1,3-dioxolane, synthesis of, **6**, 252
Perchlorodiphenylmethyl radical, synthesis of, **2**, 1010
Perchloryl fluoride
as fluorinating agent, **4**, 3, 754
for the fluorination of amines, **2**, 336
for the fluorination of carbon–hydrogen bonds, **4**, 780
precautions in the use of, **2**, 336
reactions
with 2,2,6,6-tetramethylpiperidin-4-ones, **2**, 336
with triethylamine, **2**, 339
for the synthesis of vinyl fluorides, **2**, 612
use in the synthesis of dinitroalkanes, **6**, 56
Perfluoroacyl anhydrides, decomposition of, **6**, 32
Perfluoroacyl chlorides, decomposition of, **6**, 32
Perfluoro alkoxides, reactions, with formyl fluoride, **4**, 59
Perfluoroalkyl analogues, synthesis of, **1**, 1130
Perfluoroalkylation
of alkenes, **1**, 1130
of alkynes, **1**, 1130
Perfluoroalkyl iodides, synthesis of, **6**, 5
Perfluoroalkyl radicals, reactions, with tellurium, **6**, 49
N-Perfluoroalkyls, synthesis of, via photochemical dechlorination, **6**, 603
Perfluoroalkyl transfer, selectivity reversal, **4**, 810
Perfluoro compounds, organic-, applications of, **6**, 212
Perfluorooctyl bromide (PFOB)
synthesis of, **6**, 32
use in diagnostic imaging, **6**, 32
Perfluorooctyl iodide, reactions, with tetraalkylphosphonium bromide, **6**, 32
Perfluorooxo compounds, reactions, with fluorine, **6**, 250
Perfluorophenyl groups, in anionic platinum and palladium complexes, **4**, 691
Perhydrates, hydroxy, synthesis of, **2**, 288
Pericyclic cleavage reaction, retro-, for the synthesis of carbon–carbon bonds, **1**, 771
Pericyclic processes, for the synthesis of ketenimines, **3**, 582
Pericyclic reaction, for the synthesis of alkynes, **1**, 1083
Perillane, synthesis of, **4**, 255
1*H*-Perimidinium chloride, chloromethyl-, as starting material, for perimidine dithioesters, **5**, 559
Perinaphthenylium cations, stability of, **1**, 910
Periodate
as starting material, for aldehydes, **3**, 2
use in the synthesis of aldehydes, via diol cleavage, **3**, 8
Periodate, tetrabutylammonium, as oxidizing agent, for aldehydes, **3**, 3
Periodate oxidation, of acetoin, **5**, 49
Periodic acid, as oxidizing agent, **6**, 410
Periodinane, alkoxyaryltrifluoro-
as oxidant, for the synthesis of aldehydes from alcohols, **3**, 6
for the oxidation of alcohols to ketones, **3**, 119
Periodinanes
for the oxidation of alcohols to ketones, **3**, 119
as reagent, for the synthesis of aldehydes, **3**, 6
(-)-Periplanone-B, synthesis of, **1**, 591
Periplanone B, synthesis of, **1**, 813

Perkin reaction, for the synthesis of α,β-unsaturated acids, **5**, 98
Perkow–Arbusov process, for the synthesis of phosphonium salts, **3**, 492
Perkow reaction
 of α-haloalkanenitriles, **3**, 568
 for the synthesis of alkenyl phosphates, **2**, 693
 for the synthesis of enol phosphates, **4**, 920
Permanganate, cetyltrimethylammonium, for the oxidation of benzylic alcohol, **3**, 88
Pernitrate, trichloromethyl-, synthesis of, **6**, 232
Peroxide, acetyl, synthesis of, **5**, 205
Peroxide, acetyl trichloromethyl, synthesis of, **6**, 232
Peroxide, allylic, rearrangement of, **1**, 838
Peroxide, benzoyl
 as catalyst, for the synthesis of α-haloalkenylsilicons, **4**, 811
 reactions
  with benzoquinuclidine, **2**, 765
  with triethyl phosphite, **6**, 154
 synthesis of, **5**, 205
 use in the synthesis of *N*-oxides, **2**, 765
Peroxide, bisperfluorodi-*t*-butyl, synthesis of, **2**, 98
Peroxide, bis(trifluoroacetyl), thermal decomposition of, **6**, 12
Peroxide, bis(trifluoromethyl), synthesis of, **2**, 98; **6**, 231
Peroxide, bis(trimethylsilyl)
 as co-oxidant, with bis(triphenylphosphine)ruthenium(II) chloride, **3**, 5
 as oxasilylating agent, **4**, 358
 for the oxidation of aryl organometallics, **2**, 639
 reactions
  with alkenyllithium, **2**, 700
  with lithium alkynylides, **2**, 1019
  with terminal alkynes, **2**, 1018
  with trialkyl phosphites, **2**, 108
  with trifluoromethanesulfonic acid, **2**, 650
 use in oxysilylation reactions, **6**, 335
 use in the synthesis of α-hydroxy sulfones, **4**, 234
 use in the synthesis of phosphine oxides, **2**, 452
Peroxide, *t*-butyl, reactions, with 2-ethoxy-1,3-oxathiolane, **6**, 478
Peroxide, *t*-butyl-(2-phenyl) ethyl, synthesis of, **2**, 98
Peroxide, chloro trifluoromethyl, synthesis of, **6**, 231
Peroxide, dibenzoyl
 reactions
  with alcohols, **4**, 220
  with ethyl phenyl selenide, **4**, 240
  with methylselenobenzene, **5**, 210
  with 1-methylselenonaphthalene, **5**, 210
  with secondary amines, **2**, 344
  with thioamides, **5**, 735
 synthesis of, **5**, 206
 use in the synthesis of radicals, **1**, 925
Peroxide, dibenzyl, reactions, with amines, **2**, 340, 351
Peroxide, di-*t*-butyl
 as catalyst, for the synthesis of α-haloalkenylsilicons, **4**, 810
 irradiation of, **4**, 1078
 reactions
  with carboxylic acids, **1**, 925
  with esters, **1**, 925
  with ketones, **1**, 925
  with lactones, **1**, 925
 use in the synthesis of dialkyl carbonates, **6**, 465
Peroxide, diethyl, reactions, with tributylphosphine, **2**, 476
Peroxide, dipropionyl, synthesis of, **5**, 208
Peroxide, disuccinoyl, use in the synthesis of 1,2-diols, **2**, 45
Peroxide, urea, as starting materials, for nitro compounds, **2**, 397
Peroxide functions, synthesis of, **2**, 686
Peroxide hydrolysis, of nitriles, **5**, 269
Peroxide oxidation, of tris(3,3-dimethyl-1-butynyl)arsines, **2**, 1076
Peroxides
 for the oxidation of thiols, **2**, 213
 potential hazards with, **2**, 830
 as radical initiators, **1**, 330
 reactions
  with ortho esters, **6**, 79
  with secondary amines, **2**, 343
 synthesis of, **1**, 68
 thermal decomposition of, safety precautions, **2**, 643
 use in the synthesis of dialkyl sulfoxides, **2**, 146
 use in the synthesis of radicals, **1**, 924
Peroxides, acyl
 from
  acid chlorides, **5**, 205
  2-methoxyprop-2-yl acyl peroxides, **5**, 206
 reactions, with ethers, **4**, 207
 synthesis of, **5**, 205
  via phase-transfer conditions, **5**, 205
Peroxides, alkenyl, synthesis of, **2**, 686
Peroxides, alkyl
 for the oxidation of arenes, **2**, 650
 for the oxidation of tertiary amines, **2**, 348
 reactions, with triphenylphosphine, **2**, 830
Peroxides, alkyl acyl
 hydrolysis of, **5**, 205
 synthesis of, **5**, 206
Peroxides, alkyl *t*-butyl, synthesis of, **2**, 98
Peroxides, alkyl monoacyl, hydrolysis of, **5**, 206
Peroxides, alkynoyl, synthesis of, **5**, 206
Peroxides, aryl, synthesis of, **2**, 687
Peroxides, *t*-butylalkyl, synthesis of, **2**, 98
Peroxides, diacyl
 hydrolysis of, **5**, 205
 as starting materials, for alkyl radicals, **1**, 934
 synthesis of, **5**, 206
  via oxidative dimerization, **5**, 208
 use in the synthesis of radicals, **1**, 924
Peroxides, dialkyl
 from
  alkyl hydroperoxides, **2**, 97
  alkyl mesylates, **2**, 97
  alkyl sulfates, **2**, 97
  t-butyl hydroperoxide, **2**, 98
  poly(alkylperoxy)silanes, **2**, 98
 synthesis of, **2**, 97
Peroxides, dialkyl, unsymmetrical, synthesis of, **2**, 98
Peroxides, 1,1-dihydroxydialkyl, synthesis of, **2**, 96
Peroxides, 2,2'-dihydroxydialkyl, synthesis of, **2**, 96
Peroxides, endo, from, 1,4-dienes, **2**, 52
Peroxides, heteroaryl, synthesis of, **2**, 687
Peroxides, 2-methoxyprop-2-yl acyl, as starting materials, for acyl peroxides, **5**, 206
Peroxides, monoacyl
 reactions, with acid chlorides, **5**, 206
 synthesis of, **5**, 205
  via oxidation of acids, **5**, 205
  via oxidation of anhydrides, **5**, 205
 for the synthesis of diacyl peroxides, **5**, 206
 See also Peracids
Peroxides, perfluoro, synthesis of, **6**, 251
Peroxides, *O*-trihalomethyl, synthesis of, **6**, 231
Peroxides, tris-, reactivity of, **2**, 687
Peroxides, vinyl, as intermediates, **2**, 686
Peroxide transfer reaction, for the synthesis of dialkyl peroxides, **2**, 98
Peroxidic compounds, synthesis of, **5**, 530
Peroxo compounds, use in decarboxylation reactions, **1**, 934
Peroxy acids
 for the oxidation of arenes, **2**, 650
 for the oxidation of tertiary amines, **2**, 348
 reactions, with Lewis acids, **2**, 650
 use in the synthesis of dialkyl sulfones, **2**, 155
Peroxybenzoate, *t*-butyl, reactions, with arylmagnesium bromide, **2**, 684

Peroxy compounds, safety precautions, **2**, 687
Peroxymercury compounds, demercuration of, **1**, 68
Perphenylation, of ketones, **3**, 145
Perrhenate, tetrabutylammonium, as catalyst, for the synthesis of α,β-unsaturated acylsilanes, **5**, 427
(*R*)-(+)-Perrillaldehyde, from, limonene, **1**, 580
Perruthenate, tetrapropylammonium (TPAP)
as catalytic oxidant, for the synthesis of aldehydes, **3**, 5
use in the synthesis of ketones, **3**, 117
Perruthenates, tetraalkylammonium
as oxidants, for the synthesis of aldehydes, **3**, 5
use in the synthesis of ketones, **3**, 117
Persistent radicals
classification of, **6**, 725
definition of, **6**, 734
Persulfate oxidation, of phenols, **2**, 651
Perthiophosphonic anhydrides
reactions
with ethylene glycol, **2**, 856
with phenyltetrafluorophosphorane, **2**, 855
Peterson alkenation
chemoselectivity of, **1**, 649
for elimination reactions, **1**, 648
of α-silyl α-sulfonyl ethers, **4**, 838
for the synthesis of alkenes, **1**, 678
for the synthesis of alkenyl ethers, **2**, 680
for the synthesis of alkenylsilanes, **2**, 936
for the synthesis of ketene monothioacetals, **4**, 834
for the synthesis of α,β-unsaturated aldehydes, **3**, 60
for the synthesis of α,β-unsaturated isocyanides, **3**, 714
use of lithium *t*-butylbis(trimethylsilyl)acetate for, **6**, 190
use of α-silyl carbanions in, **1**, 160
use in the synthesis of ketene dithioacetals, **4**, 849
use in the synthesis of sulfenylsulfines, **5**, 562
use in the synthesis of α-thiosulfoxides, **4**, 263
Peterson elimination
for the synthesis of acylsilanes, **5**, 417, 426
for the synthesis of biaryls, **1**, 489
for the synthesis of *gem*-halo phosphonoalkenes, **4**, 801
use in the synthesis of β,γ-unsaturated ketones, **3**, 142
vinylogous, **1**, 651
Peterson protocol, use in the synthesis of thioketenes, **3**, 549
Peterson reaction
for aldehyde homologation, **3**, 11
for condensation, **4**, 936
effect of Lewis-acid catalysis, **1**, 759
intermediates in, **2**, 547
mechanism, **1**, 758
for the synthesis of alkenes, **1**, 758
for the synthesis of α-cyano enamines, **2**, 748
for the synthesis of methylenes, **1**, 759
for the synthesis of nitriles, **3**, 650
for the synthesis of oximes, **3**, 431
for the synthesis of phosphaallenes, **3**, 605
for the synthesis of phosphacumulenes, **3**, 605
for the synthesis of phosphorylsulfines, **4**, 926
for the synthesis of silylalkenes, **4**, 805
for the synthesis of sulfur functions, **4**, 925
for the synthesis of tin compounds, **4**, 962
for the synthesis of vinyl selenides, **2**, 728
for the synthesis of vinyl sulfones, **2**, 715
use in the synthesis of sulfines, **3**, 368
Peterson-type elimination, for the synthesis of silenes, **3**, 502
Peters reaction, mechanism, **2**, 594
PFP esters, use in the synthesis of amides, **5**, 264
PFP thioesters, use in the synthesis of amides, **5**, 264
Phase-transfer catalysis
for acid chlorides, **5**, 8
in alkyl chloride synthesis, **2**, 25
for the cleavage of vinyl mercury compounds, **1**, 454
of halide exchange, **2**, 25, 31
for the synthesis of acyl isothiocyanates, **5**, 1047
for the synthesis of alkyl bromides, **2**, 31
for the synthesis of alkyl fluorides, **2**, 19
for the synthesis of alkyl iodides, **2**, 35
for the synthesis of alkynes, **1**, 1002, 1013
for the synthesis of anhydrides, **5**, 183
for the synthesis of arylalkynes, **1**, 472
for the synthesis of benzaldehydes, **3**, 91
for the synthesis of bis(acyl) disulfides, **5**, 252
for the synthesis of bis(acyl) sulfides, **5**, 242
for the synthesis of carbodiimides, **5**, 1065
for the synthesis of dialkyl sulfides, **2**, 133
for the synthesis of isocyanates, **5**, 998
for the synthesis of isocyanides, **3**, 695
for the synthesis of nitriles, **3**, 613
for the synthesis of thiocyanates, **5**, 1106
use in enhancing solubility of potassium fluoride, **5**, 4
use in the synthesis of arsenic compounds, **2**, 891
Phase-transfer catalysts
for the elimination of hydrogen halide from α-haloacetals, **4**, 826
in Ramberg–Bäcklund reactions, **1**, 707
for the synthesis of trichloromethyls, **6**, 25
use in dehydrobromination reactions, **1**, 572
Phase-transfer conditions, for the synthesis of enolates, **1**, 135
Phase-transfer environments, carbenes in, **4**, 1080
Phase-transfer reagents, for the avoidance of deoxymercuration reactions, **1**, 67
Phenacyl bromide
reactions
with bis(triphenylstannyl)telluride, **3**, 176
with phenylpropiolic thioanilide, **6**, 111
Phenacyl bromides, as starting materials, for arylglyoxal hydrates, **4**, 161
Phenacyl chlorides, reactions, with selenium oxychloride, **3**, 177
Phenacylides, reactions, with alkyl halides, **2**, 877
Phenalene
as starting material
for phenalenium anion, **1**, 866
for phenalenium cations, **1**, 866
for phenalenium radicals, **1**, 866
Phenalene anion, synthesis of, via proton loss, **1**, 866
Phenalene cations, synthesis of, via bond scission, **1**, 910
Phenalenes, chemistry of, **1**, 866
Phenalenium anion, from, phenalene, **1**, 866
Phenalenium cations
from, phenalene, **1**, 866
stability of, **1**, 910
Phenalenium radicals, from, phenalene, **1**, 866
Phenalenium salts, synthesis of, **1**, 910
Phenanthraquinone, use in the synthesis of α-diazocarbonyls, **3**, 470
Phenanthrene
reactions, with ozone, **5**, 115
synthesis of, **1**, 934
Phenanthrene, 9-amino-10-nitro-, reactions, with nitrous acid, **3**, 468
Phenanthrenes
acylation of, **3**, 301
reduction of, **1**, 99
synthesis of, **1**, 709
Phenanthrenone, 9-diazo-10-nitro-, synthesis of, **3**, 468
Phenanthroline, oxidative addition to, **2**, 584
Phenarsazine, 10-chloro-5,10-dihydro-, reactions, with chloramine, **2**, 892
Phenarsazine, 10-methyl-5,10-dihydro-, reactions, with chloramine, **2**, 892
Phenazine, 1-amino-, as starting material, for α-diazo ketones, **3**, 468
Phenethyl chloride, racemization of, **1**, 112
Phenol
acylation of, **3**, 296
O-amination of, **2**, 691

deoxygenation of, **1**, 432
o-formylation of, **3**, 97
industrial synthesis of, **2**, 655
reactions
with 2-chlorobenzoxazole, **1**, 432
with *N*,*N*-dimethyldichloroiminium chloride, **6**, 727
with ortho esters, **6**, 79
with 1-phenyl-5-chlorotetrazole, **1**, 432
with phenyl fluoroformate, **6**, 422
with potassium phenoxide, **1**, 856
with Viehe's salt, **6**, 593
reduction of, **1**, 87
as starting material
for *N*,*N*-diethyl *O*-phenyl dithiocarbamate, **6**, 556
for phenyl fluoroformate, **6**, 422
Phenol, *o*-amino-
for the condensation of *O*-ethyl selenocarboxylates, **5**, 522
use in the synthesis of α-halophosphinic acids, **4**, 128
Phenol, 3-amino-4-methyl-, reactions, with styrylphosphonic dichloride, **5**, 969
Phenol, 1-bromo-4-nitro-, as starting material, for 4-aminothiophenol, **2**, 706
Phenol, 2-*t*-butyl-, alkylation of, **1**, 201
Phenol, 2,6-dibromo-
reactions, with sodium nitrite, **2**, 780
synthesis of, **2**, 626
Phenol, 2,6-di-*t*-butyl-, reactions, with 3-azathiopyrylium salts, **5**, 634
Phenol, 2,5-dichloro-, synthesis of, **1**, 57
Phenol, 2-hydroxythio-, reactions, with Viehe's salt, **6**, 593
Phenol, 4-nitro-
as starting material, for active esters, **5**, 277
synthesis of, **2**, 648
Phenol, *p*-allyl-, synthesis of, **2**, 660
Phenol, pentachloro-, dechlorination of, **1**, 427
Phenol, pentafluoro-, reactions, with tetrachloromethane, **6**, 272
Phenol, 2,4,6-tribromo-, bromination of, **4**, 23
Phenol, 2,4,6-trichloro-, synthesis of, **2**, 619
Phenolates
O-acylation of, **2**, 683
O-alkylation of, **2**, 683
phosphorylation of, **2**, 694
reactions
with chlorosulfonates, **2**, 690
with diamidophosphorochloridate, **2**, 694
with formamimidinium chlorides, **6**, 632
Phenolates, metalloid, synthesis of, **2**, 703
Phenolic ethers, *O*-substituted, as starting materials, for phenols, **2**, 657
Phenols
O-acylation of, **2**, 683
alkylation of, **2**, 683
O-alkylation of, **2**, 683
Birch reduction of, **3**, 218
o-bromination of, **2**, 626
cleavage of, **1**, 70
cyclization of, **4**, 167
definition of, **2**, 646
formylation of, **3**, 92
from
acylarenes, **2**, 654
alkenyl β-keto esters, **2**, 672
5-allyl-2,4-cyclohexadieneone, **1**, 808
chlorodienic esters, **2**, 672
enynones, **2**, 672
iodovanilin, **2**, 647
3-methoxymethylene-2,4-bis(trimethylsilyloxy)-1,4-pentadiene, **2**, 667
nitroarenes, **2**, 648
O-substituted phenolic ethers, **2**, 657
iodination of, for aryl iodides, **2**, 631
o-iodination of, for the synthesis of aryl iodides, **2**, 631
monochlorination of, for aryl chlorides, **2**, 619
mononitration of, **2**, 791
nitrosation of, **2**, 780
para bromination of, **2**, 626
reactions
with alkali cyanides, **5**, 1103
with aminoarsine, **2**, 695
with aminostilbine, **2**, 695
with borenes, **5**, 930
with 1-chloro-1-dialkylaminoalkenes, **4**, 889
with chloroform, **3**, 91
with chlorosulfonyl isocyanates, **5**, 1005
with cyanogen halides, **5**, 1101
with dialkyl phosphite, **2**, 694
with diethylchlorophosphidate, **1**, 432
with imidoyl chlorides, **5**, 685
with phosphorus oxychloride, **2**, 694
with phosphorus trichloride, **2**, 693
with sulfamoyl chlorides, **2**, 367
with thiocyanates, **5**, 728
with thiophosgene, **6**, 533
with trichloroethylene, **4**, 761
with triflic anhydride, **2**, 689
with ynamines, **4**, 885
reduction of, **1**, 89, 432
as starting materials
for arenes, **1**, 432
for arenethiols, **2**, 707
for arylamines, **2**, 753
for aryl bromides, **2**, 627
for aryl chlorides, **2**, 621
for chlorothioformates, **6**, 533
for cyanates, **5**, 1101
for fluoroformate esters, **6**, 422
O-sulfonylation of, **2**, 689
synthesis of, **2**, 646
from aromatic substrates, **2**, 646
via, alkene hydroformylation process, **1**, 52
via Claisen rearrangement, **2**, 656
via dienonephenol rearrangement, **2**, 660
via Fries rearrangement, **2**, 656
use in reduction reactions, **1**, 99
Phenols, 2-acyl-4-alkoxy-, from, oxacycloalkylidene carbene complexes, **2**, 670
Phenols, 2-alkoxy-, synthesis of, **2**, 670
Phenols, alkynyl-, from, polyhaloethyl, **2**, 657
Phenols, amino-
oxidation of, **2**, 772
reactions
with phosgene, **5**, 966, 969
with phosphorus trihalides, **5**, 969
Phenols, ω-amino-, reactions, with isothiocyanates, **6**, 570
Phenols, annulated, synthesis of, **2**, 670
Phenols, *o*-arsino-, reactions, with imidoyl chlorides, **5**, 791
Phenols, 2,6-bis(sulfenyl), synthesis of, **2**, 658
Phenols, bromo-, synthesis of, **2**, 647
Phenols, *o*-cyano-, synthesis of, **3**, 664
Phenols, hindered, reactions, with 1-alkoxyalkynes, **4**, 827
Phenols, isocyanato-, synthesis of, **5**, 966
Phenols, nitro-, reactions, with phosphorus pentahalides, **5**, 969
Phenols, pentachloro-, use in the synthesis of amides, **5**, 277
Phenols, pentafluoro-, use in the synthesis of amides, **5**, 277
Phenols, *O*-silylated, sulfinylation of, **2**, 688
Phenone imines, lithiation of, **2**, 311
Phenonium ion, bridged, use in the synthesis of ketones, **1**, 384
Phenonium ions, synthesis of, **1**, 915
Phenothiazine, reactions, with propargyl bromide, **2**, 1049
Phenoxathiin, desulfurization of, **1**, 436
Phenoxide
reactions
with arylhydrazono aryl nitromethanes, **5**, 704

with chloramine, **5**, 288
Phenoxide anions, as starting materials, for biaryls, **1**, 474
Phenoxides
reactions
with *N*-arenesulfonylimidoyl chlorides, **5**, 703
with α-bromo esters, **4**, 199
with α-chloroalkyl isocyanates, **4**, 314
with *N*-(dichlorophosphinoyl)imidoyl chlorides, **5**, 707
with *N*-(dichlorophosphinthioyl)imidoyl chlorides, **5**, 707
with 1,1-dinitro-2,2-diphenylethylene, **4**, 898
with dioxane, **6**, 616
with trihalomethanes, **4**, 826
Phenoxy compounds, *O*-allyl-, [3,3] sigmatropic rearrangement of, **5**, 326
Phenyl, 2,4,6-tris(2'-methylbutyl)-, use in the synthesis of carbodiphosphines, **5**, 1091
Phenylarsine, dilithio anion, reactions, with *N*-(2-chloroethyl)benzimidoyl chloride, **5**, 791
Phenylation, of saturated ketones, **3**, 144
Phenyl azide, 4-nitro-, as diazo transfer reagent, **5**, 869
Phenyldimethylsilyl groups, in elimination reactions, **1**, 648
o-Phenylenebis(lithiophenylphosphine), reactions, with dichloro(methyl)arsine, **2**, 508
o-Phenylenebis(phenylphosphine)
reactions, with alkylphosphonous dichlorides, **2**, 865
as starting material, for 1,3,2-benzodiphosphaarsoles, **2**, 865
o-Phenylenedithiabis(trimethylsilane), reactions, with phenyltetrafluorophosphorane, **2**, 856
o-Phenylenedithiol, reactions, with diethoxy(methyl)stibine, **2**, 501
Phenylethynyl halides, reactions, with methoxide anion, **2**, 1028
β-Phenyl group migration, in the synthesis of selenones, **2**, 287
Phenyl groups
as chromophores, in photochemical reactions, **1**, 232
displacement of, **4**, 640
migration of, **1**, 886
reactions, with allylic carbocations, **1**, 901
Phenyl radicals, from, *N*-nitrosoacetanilide, **6**, 323
Phenyl telluroxides, *n*-alkyl, stability of, **2**, 287
Phenylthio, displacement of, **1**, 238
Phenyselenation, of alkynes, **2**, 1037
Pheromones
bombykol, synthesis of, **1**, 659
from, allenes, **1**, 966
insect sex, synthesis of, **1**, 647
muscalure, synthesis of, **1**, 659
synthesis of, **1**, 253, 455, 626, 652, 790, 1097; **4**, 1079
See also Insect pheromones
Pheromones, spiroketal, synthesis of, **1**, 165
Phorbol, synthesis of, **3**, 325
Phosgenation, of primary amines, **5**, 965
Phosgene
alternatives to, **6**, 415
bis(trichloromethyl) carbonate, **6**, 416
carbonates, **6**, 417
N,N'-carbonyldiimidazole, **6**, 416
chloroformates, **6**, 417
trichloromethyl chloroformate, **6**, 415
Arbuzov reaction, with trimethyl phosphite, **6**, 455
chlorination of, **6**, 217
detection of, **6**, 411
and diphosgene compared, **5**, 972
disassociation of, **6**, 409
as electrophile, for dithiocarboxylic enethiolates, **4**, 842
exposure to, first aid, **6**, 411
fluorination of, **6**, 225
from
aluminum trichloride, **6**, 413
carbon monoxide, **6**, 412
chlorine, **6**, 412
oxalyl chloride, **6**, 412
paraformaldehyde, **6**, 413
sulfuric acid, **6**, 412
sulfur trioxide, **6**, 412
tetrachloroethylene, **6**, 413
tetrachloromethane, **6**, 412
trichloromethyl chloroformate, **6**, 413
gas phase of, **6**, 443
handling of, **6**, 411
imino analogues, use in the synthesis of 1-phosphacumulenes, **3**, 604
industrial production of, **6**, 412
purification of, **6**, 414
reactions
with acylamines, **6**, 506
with *N*-acyl amino acids, **1**, 985
with alcohols, **6**, 425, 460
with aldehydes, **4**, 58; **6**, 428
with alkoxycarbonyl thioureas, **5**, 1063
with *N*-alkylamidines, **5**, 987
with 1,1'-alkylenebis-(3-arylsulfonylureas, **5**, 981
with amides, **2**, 746; **4**, 971; **5**, 995, 996
with amine hydrochloride, **5**, 965
with amines, **5**, 965; **6**, 500
with amino acid ester hydrochlorides, **5**, 971
with α-amino acids, **5**, 970
with 4-aminobenzylaniline, **5**, 971
with aminophenols, **5**, 966, 969
with ammonium hydrogen fluoride, **6**, 409
with ammonium thiocyanate, **6**, 559
with anthranilic acid, **5**, 969
with antimony trifluoride, **6**, 409
with arylacetonitriles, **5**, 1002
with *N*-arylamidines, **5**, 987
with aryl dithiocarbamates, **5**, 1023
with 5-aryl-1,3,4-oxadiazol-2(3H)-ones, **6**, 444
with 1-arylsulfonyl-3-alkylureas, **5**, 980
with aziridines, **5**, 1010
with benzimidazole, **6**, 444
with bis(formylamino)methane, **4**, 439
with bisphenol A, **6**, 470
with bis(trimethylsilyl)amides, **5**, 997
with bis(trimethylsilylamino)alkane, **6**, 507
with t-butylbis(trimethylsilyl)phosphine, **6**, 454
with calcium fluoride, **6**, 409
with carbamates, **5**, 977
with carbazates, **6**, 444
with carbodiimides, **6**, 621
with carboxylic acid amides, **5**, 654
with chloral, **4**, 58
with 1,3-dialkylureas, **5**, 980
with diamines, **6**, 506
with diisocyanatophenyl ethers, **5**, 967
with *N,N*-diphenylformamidine, **6**, 445
with *N,N'*-diphenylmethylenediamine, **5**, 971
with diphenyl(trimethylsilyl)phosgene, **6**, 517
with dithiocarboxylic acid dianions, **3**, 550
with enaminonitriles, **5**, 967
with ethane-1,2-diol, **6**, 467
with formamides, **6**, 52
with guanidines, **6**, 444
with hydrazines, **6**, 444
with hydroxylamine hydrochloride, **6**, 505
with *N*-hydroxylamines, **6**, 505
with 3-hydroxymethylpiperidine, **6**, 445
with 3-hydroxypiperidine, **6**, 446
with imidazole, **6**, 444
with 2-imidazolidinones, **5**, 1013
with imines, **6**, 500
with iminoacyl isocyanates, **5**, 967
with lead(II) thiocyanate, **6**, 451
with mesitylbis(trimethylsilyl)phosphine, **6**, 454

with *O*-methyl-4-butyrolactim, **5**, 998
with *N*-methylene-2,6-diethyl benzenamine, **4**, 102
with *O*-methylhydroxylamine hydrochloride, **6**, 505
with nitriles, **5**, 1002
with 4-nitrocatechol, **6**, 467
with nitrogen trifluoride, **6**, 409
with oximes, **6**, 444
with oxirane, **6**, 429
with *N*-phenylpropionamide, **5**, 655
with phosphanes, **5**, 1019
with primary amines, **6**, 443
with secondary amines, **5**, 971; **6**, 443, 500
with silylphosphanes, **5**, 1018
with silyl primary amines, **5**, 968
with sodium fluoride, **6**, 409
with sulfamoyl chlorides, **6**, 444
with sulfonamides, **2**, 726
with tertiary amines, **3**, 696; **6**, 443, 500
with thiols, **6**, 434, 471
with thionyl fluoride, **6**, 409
with thioureas, **6**, 619
with triethylamine, **4**, 790
with triethylamine trishydrogen fluoride complex, **6**, 409
with *N*-trimethylsilylamine, **6**, 443
with trimethylsilylphosphines, **6**, 454
with ureas, **5**, 980; **6**, 444
with*N*,*N*,*N'*,*N'*-tetrasubstituted ureas, **6**, 313
as reagent, for the synthesis of isocyanides, **3**, 694
safety precautions, **6**, 411, 412
solvents for
toluene, **6**, 425
water, **6**, 425
as starting material
for carbamoyl chlorides, **6**, 443
for carbonic bromide chloride, **6**, 420
for carbonic chloride fluoride, **6**, 419
for *N*-chlorocarbonyls, **6**, 443
for chloroformate esters, **6**, 424
for chlorothiolformate esters, **6**, 434
for dialkyl carbonates, **6**, 460
for dialkyl dithiocarbonates, **6**, 471
for imidoyl chlorides, **5**, 516
synthesis of, **6**, 412
laboratory methods, **6**, 412
with oxalyl chloride, **5**, 15
thio analogues of, **3**, 604
toxicity of, **6**, 411, 462
transhalogenation of, for carbonic difluoride, **6**, 409
use in dehydrating reagents, **3**, 696
uses of, **6**, 415
use in the synthesis of aromatic acid chlorides, **5**, 15
use in the synthesis of aryl isocyanides, **3**, 721
use in the synthesis of carbodiimides, **5**, 1062
use in the synthesis of dicarbonates, **6**, 460
use in the synthesis of formamidines, **5**, 744
use in the synthesis of ^{14}C-labeled phosgene, **6**, 415
use in the synthesis of thioacyl chlorides, **5**, 476
use of term, **6**, 408, 411
See also Carbonic dichloride
Phosgene, bromo-
synthesis of, **6**, 417
See also Carbonic dibromide
Phosgene, difluoro-, use in the synthesis of oxyimidic halides, **5**, 825
Phosgene, diphenyl(trimethylsilyl)-, reactions, with phosgene, **6**, 517
Phosgene, isotopically labeled, synthesis of, **6**, 414
Phosgene, ^{11}C-labeled, synthesis of, **6**, 414
Phosgene, ^{13}C-labeled, synthesis of, **6**, 414
Phosgene dehydration protocol, for the synthesis of heterocyclic aromatic isocyanides, **3**, 725
Phosgeneiminium chloride
reactions, with diols, **6**, 589
as starting material
for diselenocarbamates, **6**, 594
for selenocarbonates, **6**, 589
Phosgene oxime, synthesis of, **5**, 822
Phosgenes, halo-, use of term, **6**, 408
Phosgeniminium salts
reactivity of, **4**, 794
Van Braun dechloroalkylation of, **3**, 409
1-Phosphaadamantane, synthesis of, **2**, 441
Phosphaalkene, *C*-chloro-, synthesis of, **5**, 1152
Phosphaalkene, perchloro-, synthesis of, **6**, 679
Phosphaalkene complexes, from, phosphaalkynes, **5**, 893
η^1-Phosphaalkene complexes
from, η^3-phosphaallyl complexes, **5**, 893
synthesis of, **5**, 893
Phosphaalkenes
chemistry of, **6**, 678
reviews, **5**, 876
cycloaddition reactions of, **4**, 112
as dienophiles, **4**, 112
dimerization of, **4**, 549, 558; **5**, 879, 882
from
amides, **5**, 882
aminoiminophosphines, **6**, 689
azetes, **5**, 883
bis(dialkylamino)difluoromethanes, **6**, 685
carbenium salts, **5**, 883
carboxylic amide acetals, **5**, 883
chloromethylene immonium salts, **5**, 881
P-chloro(methylene)phosphorane, **5**, 888
diazoalkanes, **5**, 880
dichloro(2,4,6-tri-*t*-butylphenyl)phosphine, **6**, 682
1,3-diphosphaallenes, **5**, 884
disilylphosphanes, **5**, 881
α-halophosphines, **4**, 111
imidoyl chlorides, **5**, 881
iminophosphanes, **5**, 888
lithiated acyl phosphides, **5**, 880
phosphaalkynes, **5**, 883, 885, 888, 889
potassium phosphides, **5**, 883
silylphosphines, **6**, 684
trifluoromethylphosphines, **6**, 684
ylides, **5**, 888
as intermediates
in the synthesis of amino (α-haloalkyl)phosphines, **4**, 116
in the synthesis of chlorotetrahydrophosphorins, **4**, 111
lithiation of, **5**, 889
nucleophilic behavior of, **5**, 395
oxidation of, **6**, 661
palladium-catalyzed rearrangements of, for the synthesis of phosphaalkynes, **5**, 1158
reactions
with amines, **4**, 116; **6**, 683
with bromine, **6**, 689
with chlorocarbenes, **4**, 115
with diazoalkanes, **4**, 484
with nitrile sulfides, **6**, 129
with *o*-quinones, **4**, 340
with secondary amines, **2**, 467
with tetrakis(triphenylphosphine)palladium(0), **5**, 1158
with transition metal complexes, **5**, 893
stability of, **6**, 678, 681
as starting materials
for *N*,*N*-dialkyliminiums, **6**, 660
for phosphaalkynes, **3**, 688
synthesis of, **5**, 876
via dehydrochlorination reactions, **6**, 683
trimerization of, **4**, 558
See also Phosphanes, methylene-
See also Phosphines, methylene-
Phosphaalkenes, *C*-amino-, from, carboxylic amides, **5**, 880

Phosphaalkenes, chloro-
  reactions, with alkynyllithiums, **6**, 370
  as starting materials, for phosphaalkynes, **3**, 688
Phosphaalkenes, chlorolithio-, as starting materials, for phosphaalkynes, **3**, 689
Phosphaalkenes, cyclic, from, 1,3-diphospholides, **5**, 886
Phosphaalkenes, cyclopropenyl-, photochemical rearrangement of, **4**, 341
Phosphaalkenes, diazo-, reactions, with diazomethane, **2**, 470
Phosphaalkenes, 2,2-dichloro-, dechlorination of, **3**, 689
Phosphaalkenes, *P*-(ethynyl)-, synthesis of, **6**, 689
Phosphaalkenes, fluoro-, reactions, with tertiary phosphines, **6**, 700
Phosphaalkenes, germyl methylene-, from, phosphaalkynes, **5**, 889
Phosphaalkenes, *P*-halo-, nucleophilic substitution at, **5**, 889
Phosphaalkenes, *C*-lithio-
  germylation of, **5**, 889
  silylation of, **5**, 889
Phosphaalkenes, mercurated, synthesis of, **5**, 890
Phosphaalkenes, *C*-metallated, from, phosphaalkynes, **5**, 890
Phosphaalkenes, perfluoro-, reactions, with dienes, **4**, 112
Phosphaalkenes, phosphinomethyl-substituted, synthesis of, **6**, 370
Phosphaalkenes, 2-silyl-, addition to, **4**, 566, 572
Phosphaalkenes, *P*-silyl-, chlorination of, **5**, 880
Phosphaalkenes, stannylated, from, phosphaalkynes, **5**, 890
Phosphaalkenes, trifluoromethyl-, synthesis of, **6**, 242
(*E*)-Phosphaalkenes, rearrangement of, **5**, 879
Phosphaalkene transition metal complexes, synthesis of, **6**, 692
Phosphaalkynes
  1,2-additions to, **5**, 887
  characteristics of, via spectroscopy, **5**, 1152
  cooligomerization of, **5**, 885
  cycloaddition reactions, with heterocycles, **6**, 154
  cyclotetramerization of, **5**, 885
  from
    alkynylphosphines, **3**, 689
    chlorolithiophosphaalkenes, **3**, 689
    chlorophosphaalkenes, **3**, 688
    P-chloro(trimethylsilyl)phosphaethenes, **5**, 1154
    dichloromethylenephosphines, **6**, 690
    phosphaalkenes, **3**, 688
    phosphaalkynes, **5**, 1159
    stannylated phosphaalkenes, **5**, 890
    tris(trimethylsilyl)methyldichlorophosphines, **5**, 1154
  oligomerization of, in presence of metal complexes, **6**, 154
  reactions of, **4**, 549
  reactions
    with chlorocarbenes, **4**, 110
    with cyclotrisilanes, **5**, 888
    with α-diazo carbonyls, **4**, 479
    with diazomethanes, **4**, 479
    with germylenes, **5**, 889
    with tolan, **5**, 885
    with 1,2,5-triphospha–Dewar benzenes, **5**, 885
  reduction of, **5**, 893
  short-lived, **5**, 1152
  spirocyclotrimerization of, **5**, 885
  as starting materials
    for germyl methylene phosphaalkenes, **5**, 889
    for *C*-metallated phosphaalkenes, **5**, 890
    for phosphaalkene complexes, **5**, 893
    for phosphaalkenes, **5**, 883, 885, 888, 889
    for phosphaalkynes, **5**, 1159
    for phosphinodiazoalkanes, **5**, 1159
    for *C*-phosphinomethylenephosphanes, **5**, 884
    for 1,2,4-thiazaphospholes, **3**, 686
    for tris(phosphino)methanes, **6**, 152
  synthesis of, **3**, 687, 726; **5**, 1151, 1155
    via dehydrohalogenation of halophosphanes, **3**, 688
    via β-elimination reactions, **3**, 688
    via flash vacuum pyrolysis, **5**, 1152
Phosphaalkynes, amino-substituted, synthesis of, **5**, 1157
Phosphaalkynes, kinetically stabilized
  cyclooligomerization of, **5**, 1151
  stability of, **5**, 1154
$\lambda^3\sigma^1$-Phosphaalkynes, synthesis of, **3**, 687
λ^5-Phosphaalkynes, synthesis of, **5**, 1159
$\lambda^5\sigma^3$-Phosphaalkynes, synthesis of, **3**, 689
Phosphaallene, lithiated, as starting material, for 1,4-diphosphabutatriene, **3**, 607
Phosphaallenes
  axial dissymmetry of, **3**, 605
  canonical forms of, **3**, 603
  cyclodimerization of, **4**, 1026
  enantiomers of, **3**, 605
  reactions, with 2,4,6-tri(*t*-butyl)phenylchlorophosphine, **3**, 607
  synthesis of, **5**, 1091
Phosphaallenes, cyclic, synthesis of, **3**, 603
1-Phosphaallenes
  diastereomers of, **3**, 608
  from
    alkynyllithiums, **3**, 608
    lithiated λ^3-phosphaethenes, **3**, 605
  homologation of, **3**, 607
  synthesis of, **3**, 603, 605
    via Wittig reactions, **3**, 604
1λ^3-Phosphaallenes
  synthesis of, **3**, 605
  transition metal complexes of, **3**, 609
1λ^5-Phosphaallenes
  synthesis of, **3**, 603
  See also Phosphoranes, alkenylidene-
3-Phosphaallenes, *t*-butyl-3-(2,4,6-tri-*t*-butylphenyl)-1-aza-, synthesis of, **5**, 1086
2-Phosphaallyl chlorides, synthesis of, **5**, 882
η^3-Phosphaallyl complexes, as starting materials, for η^1-phosphaalkene complexes, **5**, 893
2-Phosphaallylic salts, synthesis of, **6**, 660
Phosphaamidines, synthesis of, **5**, 784
1-Phospha-4-arsacyclohexa-2,5-dienes, synthesis of, **2**, 896
Phosphaarsenes, synthesis of, **2**, 866; **6**, 375
Phosphaarsenyl complexes, reactions, with sulfur, **2**, 866
Phospha-λ^5-azenes, *N*-acrylic, Wittig reaction of, **3**, 414
Phosphabenzene, 3,5-diphenyl-, reactions, with diazo compounds, **4**, 549
Phosphabenzene, 2,4,6-triphenyl-, use in the synthesis of phosphacyclohexadienyl ligands, **4**, 1036
2-Phosphabicyclo[2.2.1]heptenes, from, cyclopentadiene, **4**, 112
Phosphabicyclooctanes, synthesis of, **4**, 339
Phosphaborazirines, synthesis of, **2**, 474
Phosphabutadiynes, synthesis of, **5**, 1153
1-Phosphabutatriene, 4,4-bis(trimethylsilyl)-, from, 1,3-dilithiated bis(trimethylsilyl) allene, **3**, 607
1-Phosphabutatrienes
  from
    alkynylphosphine, **3**, 606
    β-ethoxyallylidenephosphoranes, **3**, 604
    lithiated trimethylsilylallenes, **3**, 607
  geometrical isomers of, **3**, 607
  synthesis of, via the elimination of trimethylsilanol, **3**, 608
1λ^5-Phosphabutatrienes, synthesis of, **3**, 603
Phosphacumulenes
  canonical forms of, **3**, 603
  reactions
    with ketones, **1**, 988
    with 2,4,6-tri(*t*-butyl)phenylchlorophosphine, **3**, 607
1-Phosphacumulenes
  from
    1,1-dichloroalkenes, **3**, 604

hexaphenylcarbodiphosphorane, **3**, 604
methylene(triphenyl)phosphorane, **3**, 604
synthesis of, **3**, 603, 607
1λ^3-Phosphacumulenes
synthesis of, **3**, 605
transition metal complexes of, **3**, 609
1λ^5-Phosphacumulenes, synthesis of, **3**, 603
Phosphacyclohexadienyl ligands, synthesis of, **4**, 1036
Phosphacymantrenes, synthesis of, **4**, 1037
1-Phosphadamantane, synthesis of, **1**, 445
1,2,4-Phosphadiazetidin-3-ones, decomposition of, **5**, 982
Phosphadiazoles, synthesis of, **6**, 667, 668
Phosphaethenes, homologation of, **3**, 607
Phosphaethenes, *P*-chloro(trimethylsilyl)-, as starting materials, for phosphaalkynes, **5**, 1154
Phosphaethenes, *P*-cyano-, from, cyanochlorotrimethylsilylmethylphosphine, **5**, 1130
λ^3-Phosphaethenes, lithiated, as starting materials, for 1-phosphaallenes, **3**, 605
Phosphaethyne, 1-adamantyl-, synthesis of, **5**, 1155
Phosphaethyne, amino-
from, cyanophosphine, **5**, 1157
synthesis of, **5**, 1157
Phosphaethyne, *t*-butyl-
cleavage of, **5**, 1159
flash vacuum pyrolysis of, **3**, 689
synthesis of, **5**, 1154
Phosphaethyne, *t*-butylamino-, synthesis of, **3**, 689
Phosphaethyne, diisopropylamino-, synthesis of, **5**, 1152
Phosphaethyne, fluoro-
characterization of
via microwave spectroscopy, **5**, 1152
via photoelectron, **5**, 1152
from, trifluoromethylphosphine, **5**, 1152
Phosphaethyne, *N*-trimethylsilyl-*N*-isopropylamino-, synthesis of, **5**, 1157
Phosphaethyne, 1-tripticyl-, synthesis of, **5**, 1155
Phosphaethyne, 2,4,6-tris(*t*-butyl)phenyl-, synthesis of, **5**, 1155
Phosphaethynes
flash vacuum pyrolysis of, **5**, 1159
from, triallylphosphine, **5**, 1158
synthesis of, **5**, 1152
Phosphaferrocenes
reactions
with electrophiles, **4**, 1035
with molybdenum hexacarbonyl, **4**, 1026
with norbornadiene–molybdenum tetracarbonyl, **4**, 1026
synthesis of, **4**, 1026, 1032, 1034
Phosphaferrocenes, acetyl-, synthesis of, **4**, 1035
Phosphaferrocenes, tetraphenyl-, synthesis of, **4**, 1034
Phosphagermirenes, synthesis of, **5**, 889
Phosphaisocyanides
as intermediates, **3**, 726
in the synthesis of phosphaalkynes, **5**, 1158
Phosphaketene, *t*-butyl-
as intermediate, detection of, **6**, 686
stability of, **5**, 1018
Phosphaketene, 2,4,6-tri-*t*-butylphenyl-, reactions, with silyldiphosphines, **6**, 687
Phosphaketenes
addition reactions, **5**, 1092
synthesis of, **5**, 1018; **6**, 454
Phosphametallocenes, synthesis of, **4**, 1034
Phosphamethine cyanines, synthesis of, **5**, 882
Phosphanates, α-hydroxyimino-, Beckmann rearrangement of, **5**, 371
Phosphane, bromomethyldibromo-, as starting material, for bromomethylenephosphanes, **5**, 877
Phosphane, 2,4,6-tri-*t*-butylphenyl-, reactions, with dichlorofluoromethane, **5**, 876
Phosphane, di-*t*-butyltrimethylsilylmethyl-, reactions, with tetrachloromethane, **5**, 915
Phosphane, chloro-, as starting material, for methylenephosphanes, **5**, 887
Phosphane, chlorobis(diisopropylamino)-, reactions, with di-*t*-butylthioxophosphoranyldiazomethane, **5**, 1148
Phosphane, chloromethyldichloro-, reactions, with thiobenzamide, **5**, 635
Phosphane, (2,2-dimethylpropylidyne)-, reactions, with nitrile oxides, **5**, 839
Phosphane, ethynyldiphenyl-, as starting material, for triorganoboranes, **2**, 1088
Phosphane, *N*-isocyanoiminotriphenyl-, synthesis of, **3**, 696, 710
Phosphane, methylidyne-
characterization of, **3**, 688
synthesis of, **3**, 689
Phosphane, trifluoromethyl-, reactions, with diisopropylamine, **3**, 689
Phosphane, trimethyl-
reactions
with carbyne complexes, **5**, 919
with manganese carbyne complexes, **5**, 919
Phosphane, triphenyl-, reactions, with sulfonylcarbamimidic azides, **6**, 651
Phosphane, trishydroxymethyl-, reactions, with 2-chloroquinolinium salts, **5**, 882
Phosphane, tris(trimethylsilyl)-
acylation of, **3**, 688
reactions
with imidoyl chloride, **6**, 660
with isopropyl isocyanate, **6**, 511
Phosphanes
reactions
with phosgene, **5**, 1019
with tetrachloromethane, **5**, 898
as starting materials
for oxyimidic phosphorus compounds, **5**, 839
for trichloromethylphosphonium salts, **5**, 898
for ylides, **5**, 902
Phosphanes, acyl-
1,3-silatropic rearrangement of, **5**, 878
silatropy of, **5**, 879
Phosphanes, alkoxymethylene-
from
hydroxyalkylphosphanes, **5**, 880
imidoyl chlorides, **5**, 880
Phosphanes, aryl-, reactions, with dimethylformamide, **5**, 882
Phosphanes, bis(trimethylsilyl)-
reactions, with pivaloyl chloride, **5**, 878
as starting materials, for silyloxymethylenephosphanes, **5**, 878
Phosphanes, bromomethylene-
from
bromomethyldibromophosphane, **5**, 877
bromomethylenephosphanes, **5**, 878
primary phosphanes, **5**, 877
as starting materials, for bromomethylenephosphanes, **5**, 878
Phosphanes, chloro-
reactions
with bisstannyldiazomethane, **3**, 682
with diazomethyllithium, **6**, 668
with thiophosphinoyldiazomethanes, **3**, 681
as starting materials, for nitrilimines, **6**, 667
Phosphanes, *P*-chloro disilylmethyl-, thermolysis of, **5**, 887
Phosphanes, chloromethyldichloro-, as starting materials, for chloromethylenephosphanes, **5**, 876
Phosphanes, chloromethylene-
deprotonation of, **5**, 877
from
chloromethyldichlorophosphanes, **5**, 876
dichloromethylenephosphanes, **5**, 877
primary phosphanes, **5**, 876

Phosphanes, dichloro-, flash pyrolysis of, **3**, 688
Phosphanes, α,α-dichloro-, dehydrochlorination of, **3**, 688
Phosphanes, dichloromethylene-
lithiation of, **5**, 877
as starting materials, for
chloromethylenephosphanes, **5**, 877
Phosphanes, diiodomethylene-
lithiation of, **5**, 878
as starting materials, for
iodomethylenephosphanes, **5**, 878
Phosphanes, disilyl-, as starting materials, for
phosphaalkenes, **5**, 881
Phosphanes, halo-, reactions, with α-silylated ylides, **5**, 908
Phosphanes, α-halomethyl-, synthesis of, **5**, 898
Phosphanes, *P*-halo(silylmethyl)-
dehydrohalogenation of, **5**, 887
as starting materials, for silyl
methylenephosphoranes, **5**, 887
Phosphanes, hydroxyalkyl-, as starting materials, for
alkoxymethylenephosphanes, **5**, 880
Phosphanes, imino-
as starting materials
for 1,2,4-azadiphosphetines, **5**, 885
for phosphaalkenes, **5**, 888
Phosphanes, iminomethylene-, synthesis of, **5**, 1087
Phosphanes, iodomethylene-
from
diiodomethylenephosphanes, **5**, 878
iodoform, **5**, 878
Phosphanes, methylene-
from
carboxylic esters, **5**, 880
chlorophosphane, **5**, 887
lithium phosphides, **5**, 879
synthesis of, **5**, 876
See also Phosphaalkenes
Phosphanes, monochloro-, reactions, with ylides, **5**, 908
Phosphanes, organodisilyl-, reactions, with sodium
hydroxide, **5**, 882
Phosphanes, phenyl bis(trimethylsilyl)-, reactions, with
thiophosgene, **5**, 1058
Phosphanes, *C*-phosphinomethylene-
from
1,3-dichloro-1,3-diphosphapropanes, **5**, 884
phosphaalkynes, **5**, 884
Phosphanes, phosphoniomethylene-, from, ylides, **5**, 886
Phosphanes, primary
as starting materials
for bromomethylenephosphanes, **5**, 877
for chloromethylenephosphancs, **5**, 876
Phosphanes, selenylmethylene-, from, phosphonium
salts, **5**, 903
Phosphanes, silyl-
reactions, with phosgene, **5**, 1018
as starting materials, for
silyloxymethylenephosphanes, **5**, 878
Phosphanes, silyloxymethylene-
from
acid chlorides, **5**, 878
bis(trimethylsilyl)phosphanes, **5**, 878
silylphosphanes, **5**, 878
Phosphanes, sulfonylmethyl-
disproportionation of, **5**, 902
reactions, with tetrachloromethane, **5**, 903
Phosphanes, tertiary, as starting materials, for *P*-halo
ylides, **5**, 894
Phosphanes, trialkyl-, as starting materials, for ylides, **5**, 908
Phosphanes, triaryl-, as starting materials, for ylides, **5**, 908
1,3-Phosphanes, synthesis of, **4**, 453
Phosphanylcarbene, synthesis of, **6**, 668
2-Phosphapropene, perfluoro-
from, hexafluorodimethylphosphine, **5**, 786
reactions
with alcohols, **6**, 683
with *t*-butylamine, **5**, 1087
with isopropylamine, **5**, 1087
as starting material
for formimidoylphosphines, **5**, 786
for sulfinic salts, **6**, 258
synthesis of, **6**, 679
Phosphapropyne
from
divinylphosphine, **5**, 1158
1-vinylphosphirane, **5**, 1158
1*H*-1,3-Phospharsoles, 1-substituted 2-ethylidene-2,3-
dihydro-4-methyl-3-phenyl-, synthesis of, **4**, 1030
Phosphasilirenes, synthesis of, **5**, 888
1-Phospha-4-stannabenzene, 1,4-dihydro-, synthesis
of, **4**, 1038
1*H*-1,3-Phosphastannole, 1-*t*-butyl-2-ethylidene-2,3-
dihydro-4-methyl-3,3-di-*n*-butyl-, synthesis
of, **4**, 1038
Phosphastibenes, aromatic, synthesis of, **2**, 866
Phosphate, cyanohydrin diethyl, synthesis of, **3**, 627
Phosphate, diethyl chloro-
reactions
with lithium dimethyl cuprate, **5**, 103
with nitrile enolates, **3**, 639
Phosphate, diethyl cyano-, reactions, with cyclic
hemiaminals, **2**, 328
Phosphate, diethylvinyl, bromination of, **4**, 768
Phosphate, dihydroxy acetone, reactions, with
aldehydes, **3**, 159
Phosphate, mono(2-cyanoethyl), reactions, with
alcohols, **2**, 105
Phosphate, phenylbis(dimethylamino)fluorophosphonium
phenylpentafluoro-, from,
phenyldimethylaminotriflurophosphorane, **2**, 869
Phosphate, phenyl dichloro-
reactions, with carboxylic acids, **5**, 239
as reagent, for the hydrolysis of acetals, **3**, 123
use in the synthesis of acyl azides, **5**, 367
Phosphate, phenylpentafluoro-, synthesis of, **2**, 869
Phosphate, tetraethyl hypo-, reactions, with aluminum
ethoxide, **2**, 107
Phosphate, tributyl, reactions, with potassium
cyanate, **5**, 995
Phosphate, tri(2-chloroethyl), reactions, with chloroacetyl
chloride, **4**, 768
Phosphate, triethyl, elimination of, **1**, 646
Phosphate, tristriethylamino-, reactions, with
2-cyanoethanol, **2**, 105
Phosphate anions, elimination of, **1**, 639
Phosphate esters, synthesis of, **2**, 1072
Phosphate esters, alkynyl
from, alkynyl(phenyl)iodonium phosphates, **2**, 1028
research background, **2**, 1022
Phosphates
chirality transfer of, **1**, 187
displacement of, **1**, 187
from, phosphoramidites, **2**, 106
rearrangement of, **1**, 923
reduction of, **1**, 17
regioselectivity of, **1**, 187
as starting materials, for *Se*,*P*-acetals, **4**, 348
Phosphates, acyl trimethylsilyl, reactions, with sodium
hydrogen carbonate, **5**, 189
Phosphates, 1,3-alkadien-2-yl diethyl, reactions, with
organocuprates, **1**, 970
Phosphates, alkenyl
elimination of, **1**, 1070
reactions
with diethylalkynylalanes, **1**, 1027
with metallated alkynes, **1**, 1027
with ylides, **4**, 737
synthesis of, **2**, 693

See also Phosphates, enol
Phosphates, *N*-alkoxybenzimidoyl, synthesis of, **5**, 713
Phosphates, alkyl
as starting materials, for alkylselenophosphonium salts, **2**, 293
synthesis of, **2**, 105
Phosphates, *S*-alkyl chloro-, synthesis of, **2**, 257
Phosphates, alkyl chloroisocyanato-, synthesis of, **5**, 1008
Phosphates, alkyl hydrido-, synthesis of, **2**, 477
Phosphates, alkynyl
reactions, with phosphorus pentachloride, **2**, 1071
synthesis of, **2**, 1030
Phosphates, alkynyl(phenyl)iodonium, as starting materials, for alkynyl phosphate esters, **2**, 1028
Phosphates, allenyl dialkyl, synthesis of, **1**, 993
Phosphates, allyl, carbonylation of, **1**, 192
Phosphates, allylic
reactions
with Grignard reagents, **1**, 1053
with vinyl Grignard reagents, **1**, 477
use in the synthesis of allylchromiums, **2**, 73
Phosphates, arenediazonium tris(perfluoroalkyl) trifluoro-, as starting materials, for aryl fluorides, **2**, 614
Phosphates, aryl
hydrolysis of, **1**, 445
reactions, with metallated alkynes, **1**, 1027
Phosphates, arylpentahalo-, synthesis of, **2**, 869
Phosphates, bromo enol, synthesis of, **4**, 768
Phosphates, carbamate methyltris(trifluoromethyl), synthesis of, **2**, 476
Phosphates, chiral silylallyl, reactions, with Grignard reagents, **1**, 194
Phosphates, 1-chlorothiol, synthesis of, **4**, 72
Phosphates, cyano-, reactions, with boron trifluoride etherate, **3**, 647
Phosphates, cyanovinyl, as starting materials, for monoalkyl phosphates, **2**, 106
Phosphates, cyclohexadienyl, synthesis of, **3**, 198
Phosphates, dialkyl
alkylation of, **2**, 107
reactions, with dialkyl phosphochloridates, **2**, 108
synthesis of, **2**, 106
via dicyclohexylcarbodiimide-coupling reactions, **2**, 106
use in the synthesis of (*N*-alkylthiocarbamoyl)phosphonic acid esters, **6**, 581
Phosphates, diaryl ethyl, metallation-induced rearrangement of, **2**, 845
Phosphates, di-*t*-butyl alkyl, from, di-*t*-butyl-*N*,*N*-diethylphosphoramidite, **2**, 107
Phosphates, diphenyl, synthesis of, **2**, 870
Phosphates, dithiocarbamate methyltris(trifluoromethyl), synthesis of, **2**, 476
Phosphates, enol
elimination of, **1**, 1073
from, chlorodifluoromethyl ketones, **4**, 735
reactions
with organoaluminums, **1**, 480
with trialkylmanganates, **1**, 480
reduction of, **1**, 434
substitution of, **1**, 479, 482
synthesis of, **4**, 836, 920
thermal decomposition of, **2**, 640
See also Phosphates, alkenyl
Phosphates, hexavalent, synthesis of, **2**, 476
Phosphates, monoalkyl
from, cyanovinylphosphates, **2**, 106
synthesis of, **2**, 105
Phosphates, monoaryl, synthesis of, **2**, 694
Phosphates, monophenyl, synthesis of, **2**, 870
Phosphates, pentachloro-, synthesis of, **2**, 477
Phosphates, α-phosphorylated enol, from, α-halo acid halides, **4**, 920
Phosphates, propargylic, rearrangement of, **1**, 993
Phosphates, 1-sulfinylvinyl, synthesis of, **4**, 837
Phosphates, 1-sulfonylvinyl, synthesis of, **4**, 838
Phosphates, tetraarylphosphonium hexaaryl, synthesis of, **2**, 869
Phosphates, thiocarbamate methyltris(trifluoromethyl)-, synthesis of, **2**, 476
Phosphates, 1-thiovinyl, oxidation of, **4**, 837, 838
Phosphates, trialkyl
from
alkanes, **2**, 108
dialkyl phosphites, **2**, 107
as starting materials, for dialkyl alkylphosphonates, **2**, 456
synthesis of, **2**, 107
Phosphates, triaryl, synthesis of, **2**, 694
Phosphates, vinyl, synthesis of, **3**, 568
Phosphate salts, hexafluoro-, as starting materials, for aryl fluorides, **2**, 614
Phosphathiaarsiranes, synthesis of, **2**, 866
Phosphathioketenes, as intermediates, **5**, 1058
Phosphatrifulvene, reactions, with isocyanates, **6**, 511
Phosphatrifulvenes
from, cyclopropenones, **5**, 394
reactions, with isothiocyanates, **6**, 582
Phospha–Wittig synthesis, for the synthesis of methylenephosphines, **6**, 692
Phosphazene bases, reactive, **1**, 574
Phosphazenes
acylation of, **5**, 371
as starting materials, for nitro compounds, **2**, 397
synthesis of, **2**, 468, 469
Phosphazenes, arsino-, synthesis of, **2**, 504
Phosphazenes, cyclic, synthesis of, **2**, 468
Phosphazenes, *N*-phosphorylated, synthesis of, **2**, 470
Phosphazenes, polyamino-, synthesis of, **2**, 860
Phosphazenes, sulfonyltrichloro-, synthesis of, **5**, 982
Phosphazenes, trichloro-, synthesis of, **5**, 982
Phosphazenium fluoride, use in dehydrobromination reactions, **1**, 574
Phosphazenium naked fluorides, reactions, with benzene sulfonates, **1**, 584
Phosphazines, synthesis of, **3**, 445
Phosphazines, chloro-, reactions, with acid halides, **3**, 619
Phosphazines, thiocarbonyl-, from, thioalkylphosphines, **6**, 581
Phosphazo compounds, chloro-, decomposition of, **5**, 988
Phosphenes, bis-, synthesis of, **6**, 167
1,3-Phosphepanes, synthesis of, **4**, 453
Phosphetanes, synthesis of, **4**, 114
Phosphetan-1-oxide, reduction of, **4**, 114
Phosphidate, diethylchloro-, reactions, with phenols, **1**, 432
Phosphide, benzoylbis(*t*-butyl), synthesis of, via 1,2-chlorotropic rearrangements, **5**, 397
Phosphide, dilithiophenyl, reactions, with dimesylate synthons, **2**, 438
Phosphide, diphenyl, use in the synthesis of heterocuprates, **2**, 588
Phosphide, sulfur-substituted, reactions, with acyl chlorides, **5**, 398
Phosphides, acyl
spectroscopic studies, **5**, 395
synthesis of, **6**, 687
Phosphides, dialkyl, synthesis of, **2**, 475
Phosphides, lithiated acyl, as starting materials, for phosphaalkenes, **5**, 880
Phosphides, metallo, reactions, with electrophiles, **2**, 438
Phosphido complexes, deprotonation of, **5**, 400
Phosphimines, *N*-aryl-, reactions, with thioureas, **6**, 645
Phosphimines, *N*-(trimethylsilyl)-, reactions, with carbonyls, **3**, 440
Phosphinamidates, enamides of, synthesis of, **2**, 808
Phosphinamide, *N*-(*p*-nitrophenyl)-*p*-methoxyphenylphenyl-, synthesis of, **2**, 807

Phosphinamides
  reactions, with phosphorus pentachloride, **2**, 470
  as starting materials, for *N*-phosphinoylimines, **3**, 439
  synthesis of, **2**, 455
Phosphinamides, (-)-(*S*)-*N*-(phenyl)methylphenyl-, synthesis of, **2**, 807
Phosphinanilides, hydrolysis of, **2**, 807
Phosphinate, diethyl methyl, reactions, with enals, **3**, 51
Phosphinate, ethyl(chloromethyl)methyl, synthesis of, **4**, 126
Phosphinate, isopropyl diphenyl, reactions, with 2-trimethylammonium-1-3-dioxanes, **6**, 116
Phosphinate, methyl chloromethyl, synthesis of, **4**, 125
Phosphinate, methyl(chloromethyl)methyl, synthesis of, **4**, 126
Phosphinate, methyl(diazobenzyl), synthesis of, **3**, 480
Phosphinate, (-)-(*S*)-methylmethylphenyl, reactions, with lithium anilide, **2**, 807
Phosphinate, phenyl bis(chloromethyl), reactions, with potassium iodide, **4**, 127
Phosphinate acetals, Arbuzov reaction of, **6**, 117
Phosphinate anions, elimination of, **1**, 639, 641
Phosphinate anions, α-fluoroalkyl, alkylation of, **4**, 131
Phosphinate esters, use in the synthesis of chiral phosphines, **2**, 443
Phosphinate esters, menthyl, use in the synthesis of chiral phosphines, **2**, 443
Phosphinates
  from
    phosphinic chlorides, **4**, 126
    vinylphosphorus compounds, **4**, 561
  synthesis of, **2**, 454
Phosphinates, 2-alkoxyalkenyl, as starting materials, for α-nitrophosphinic acid, **4**, 490
Phosphinates, alkyl, reduction of, **2**, 431
Phosphinates, alkyl alkenylphenyl, synthesis of, **2**, 846
Phosphinates, α-amino-, reactions, with *N*-bromosuccinimide, **5**, 790
Phosphinates, 1-aminoalkyl, from, ketimines, **4**, 463
Phosphinates, α-aminoalkyl
  from
    α-oxophosphinates, **4**, 471
    phosphorus esters, **4**, 464
Phosphinates, α-aminobenzyl, synthesis of, **4**, 461
Phosphinates, bis(chloromethyl), synthesis of, **4**, 127
Phosphinates, α-bromo-α^1-amino-
  as starting materials, for formimidoyl phosphinates, **5**, 790
  synthesis of, **5**, 790
Phosphinates, bromomethyl, synthesis of, **4**, 127
Phosphinates, chiral alkylmethylphenyl, synthesis of, **2**, 455
Phosphinates, chloromethyl, reactions, with aldehydes, **4**, 126
Phosphinates, cyclic, synthesis of, **2**, 454
Phosphinates, dialkyl, reactions, with 3-acetylcoumarins, **6**, 117
Phosphinates, α-diazo-, from, α-ketophosphoryls, **3**, 480
Phosphinates, diphenylcyanamido-, reactions, with triorganometal chlorides, **5**, 1084
Phosphinates, ethyl, as starting materials, for phosphinic chlorides, **4**, 125
Phosphinates, ethyl bis(2-hydroxyaryl), synthesis of, **2**, 846
Phosphinates, ethyl(diethoxy)methyl, synthesis of, **4**, 126
Phosphinates, formimidoyl, from, α-bromo-α^1-aminophosphinates, **5**, 790
Phosphinates, α-halo-α^1-amino-, as starting materials, for imidoyl-tetracoordinated phosphorus compounds, **5**, 790
Phosphinates, hydroxyimino-, rearrangement of, **5**, 371
Phosphinates, α-hydroxyimino-, Beckmann rearrangement of, **5**, 371
Phosphinates, 1,4-keto, synthesis of, **3**, 201
Phosphinates, 1,5-keto, synthesis of, **3**, 201
Phosphinates, γ-keto, synthesis of, **3**, 200
Phosphinates, methyl, synthesis of, **2**, 455
Phosphinates, methyl chloromethyl(1-hydroxyalkyl), synthesis of, **4**, 126
Phosphinates, methyl α-diazo, reactions, with trimethylsilyl bromide, **3**, 482
Phosphinates, silyloxy-, reactions, with thiocarbonyls, **2**, 266
Phosphinate salts, diphenyl-, synthesis of, **1**, 641
Phosphine
  oxidation of, **6**, 151
  reactions
    with acrylonitrile, **2**, 428
    with 1,5-cyclooctadiene, **2**, 430
    with diselenide, **2**, 858
    with formaldehyde, **2**, 440
    with iodine, **2**, 708
    with isocyanates, **6**, 511
    with oct-1-ene, **2**, 428
    with *p*-tolyl azide, **6**, 151
  tellurylation of, **2**, 858
Phosphine, acetoxydiethylaminophenyl-, synthesis of, **5**, 221
Phosphine, alkynyl-, as starting material, for 1-phosphabutatrienes, **3**, 606
Phosphine, allyldibromo-, synthesis of, **2**, 449
Phosphine, 1-azidocycloprop-2-inylmethyl)-, synthesis of, **4**, 483
Phosphine, benzyldichloro-, synthesis of, **2**, 448
Phosphine, bis(arylazo)methylene-, synthesis of, **6**, 659
Phosphine, bis(benzoyloxy)methyl-, synthesis of, **5**, 221
Phosphine, bis(benzoyloxy)phenyl-, synthesis of, **5**, 221
Phosphine, bis(2-cyanoethyl)-, synthesis of, **2**, 428
Phosphine, bis(diethylamino)phenyl-, acylation of, **5**, 221
Phosphine, bis(dimethylamino)trichloromethyl-, synthesis of, **6**, 696
Phosphine, bis(hydroxymethyl)-, reactions, with N^1,N^2-dialkylhydrazines, **4**, 482
Phosphine, bis(trifluoromethyl)-, as starting material, for 1,3-diphosphetanes, **6**, 264
Phosphine, bromochloromethylene-, synthesis of, **6**, 683
Phosphine, bromocyanomethyl-, synthesis of, **5**, 1129
Phosphine, bromo difluoro-, reactions, with bis(trimethylsilyl)acetamide, **5**, 713
Phosphine, t-butylbis(trimethylsilyl)-, reactions, with phosgene, **6**, 454
Phosphine, *t*-butyldiethynyl-
  deprotonation of, **2**, 435
  reactions, with phenylphosphine, **4**, 1027
Phosphine, *t*-butyldipropynyl-
  reactions
    with di-*n*-butyltin hydride, **4**, 1038
    with phenylarsine, **4**, 1030
Phosphine, chloro-, reactions, with diazomethylsilver, **6**, 667
Phosphine, chlorobis(*t*-butyl)-, use in the synthesis of phosphaalkynes, **5**, 1159
Phosphine, *P*-chlorobis(trimethylsilyl)methylene-, reactivity of, **6**, 689
Phosphine, chlorodialkyl-
  reactions
    with dialkylphosphine, **2**, 471
    with dialkylphosphinites, **2**, 471
    with disilylmercury, **2**, 471
    with trialkylphosphine, **2**, 471
Phosphine, chlorodiethoxy-, reactions, with ω-aminoalkyl bromides, **2**, 457
Phosphine, chlorodiethyl-
  reactions
    with acetic anhydride, **4**, 911
    with enolates, **2**, 457
    with α-hydroxy ketones, **2**, 457
Phosphine, chloro(diethyl)-, reactions, with ketones, **4**, 120
Phosphine, chlorodimethyl-
  reactions
    with cyanogen chloride, **5**, 1131
    with *p*-toluenesulfonamide, **2**, 469
Phosphine, chlorodiphenyl-

reactions
with acetic acid, **6**, 118
with carbon suboxide, **4**, 550
with formaldehyde, **4**, 119
with hydroxyiminoethyl phosphonate oximes, **4**, 1008
with 3-lithio-1-diphenylphosphinopropyne, **4**, 1023
with 2-lithio-1,3-dithiane, **6**, 120
with sodium diphenyldithioarsinate, **2**, 891
with *N*-substituted tetrahydro-1,3-oxazines, **4**, 468
Phosphine, chlorodi-*i*-propyl-, reactions, with alkyldialkoxyphosphines, **2**, 471
Phosphine, chloroethoxyphenyl-, reactions, with ethanethiol, **2**, 851
Phosphine, chloromethyldichloro-
reactions, with chlorine, **2**, 476
synthesis of, **2**, 448
Phosphine, cyano-, as starting material, for aminophosphaethyne, **5**, 1157
Phosphine, cyanobis(isopropyl)-, synthesis of, **5**, 1130
Phosphine, cyanobis(pentafluorophenyl)-, synthesis of, **5**, 1129
Phosphine, cyanochlorotrimethylsilylmethyl-
from, dichlorotrimethylsilylmethylphosphine, **5**, 1130
as starting material, for *P*-cyanophosphaethenes, **5**, 1130
Phosphine, cyanodimethylamino-, synthesis of, **5**, 1130
Phosphine, 2-cyanoethyl-, synthesis of, **2**, 428
Phosphine, cyclohexyldichloro-, reactions, with iodide, **2**, 449
Phosphine, dialkoxychloro-, reactions, with carboxylic acid amides, **5**, 713
Phosphine, dialkyl-, reactions, with chlorodialkylphosphine, **2**, 471
Phosphine, diaminochloro-, reactions, with Grignard reagents, **2**, 859
Phosphine, dibenzoyl-, methylation of, **5**, 395
Phosphine, dibromo(bromomethyl)-, synthesis of, **4**, 115
Phosphine, dibromomethyl-, reactions, with dicyanomethylphosphine, **5**, 1129
Phosphine, dibromo(trifluoromethyl)-, as starting material, for (trifluoromethyl)phosphinylbistriazolide, **6**, 242
Phosphine, dibromo(tris(trimethylsilyl)methyl)-, synthesis of, **6**, 373
Phosphine, dibutoxy-, reactions, with orthoformates, **2**, 451
Phosphine, di-*t*-butyl-, reactions, with aminostannanes, **2**, 474
Phosphine, di(*t*-butyl)fluoro-, reactions, with tetrahalomethanes, **6**, 695
Phosphine, dichloromethyl-
reactions
with 2,3-dimethylbutadiene, **4**, 111
with norbornadiene, **2**, 448
as starting material, for fluorophosphines, **2**, 448
Phosphine, dichloromethyldichloro-, reactions, with triethylamine, **6**, 679
Phosphine, dichloromethylenetriphenyl-, from, dichlorocarbene, **4**, 737
Phosphine, dichloro(phenyl)-
reactions
with lithium alkynide, **2**, 1060
with metal carboxylates, **5**, 221
with *N*-methyl(*o*-hydroxybenzylidene)amine, **4**, 478
with phosphine sulfide, **4**, 114
Phosphine, dichloro-(2,2,3,3-tetrafluoropropyloxy)-, reactions, with 1,1,2-trichloronitrosoethane, **5**, 683
Phosphine, dichloro(2,4,6-tri-*t*-butylphenyl)-, as starting material, for phosphaalkenes, **6**, 682
Phosphine, dichlorotrimethylsilylmethyl-, as starting material, for cyanochlorotrimethylsilylmethylphosphine, **5**, 1130
Phosphine, dicyanomethyl-
reactions, with dibromomethylphosphine, **5**, 1129
synthesis of, **5**, 1129
Phosphine, dicyanotrifluoromethyl-, synthesis of, **5**, 1129
Phosphine, dicyclopropylchloro-, synthesis of, **2**, 448
Phosphine, (diethoxy)phenyl-, reactions, with dibromomethane, **4**, 127
Phosphine, diethylaminofluoromethyl-, reactions, with biacetyl, **2**, 476
Phosphine, diethylchloro-, synthesis of, **2**, 448
Phosphine, di(ethylthio)-, reactions, with aldehydes, **2**, 461
Phosphine, diethyl trimethylsilyl-, reactions, with *N*-acetyl benzophenoneimine, **5**, 720
Phosphine, diethylvinyl-, reactions, with diphosphinomethane, **2**, 439
Phosphine, difluoro(iodomethyl)-, synthesis of, **4**, 115
Phosphine, diisopropylchloro-, reactions, with thallium compounds, **5**, 400
Phosphine, dilithio phenyl-, as starting material, for 1,3-azaphospholine, **5**, 784
Phosphine, dimethoxy-, synthesis of, **6**, 165
Phosphine, dimethyl-
reactions
with hexafluoropropene, **4**, 802
with tetrakistrifluoromethyldiarsine, **2**, 508
with tris(but-3-enyl)phosphine, **2**, 439
as starting material, for dimethylphosphinic acid, **2**, 454
synthesis of, **2**, 430
Phosphine, dimethylaminomethylbis(isopropyl)-, reactions, with cyanogen chloride, **5**, 1130
Phosphine, dimethyldimethylamino-
reactions, with hydrogen cyanide, **5**, 1130
synthesis of, **5**, 1130
Phosphine, dimethyl(7-norbornyl)-, synthesis of, **2**, 435
Phosphine, 2,2-dimethylpropylidynyl-, reactions, with diazocyclohexanes, **4**, 480
Phosphine, dimethyl(trimethylgermyl)-, reactions, with phenyl trimethylsilyl ketone, **6**, 131
Phosphine, diphenyl-
reactions
with 1,2-bisdiphenylphosphino ethyne, **4**, 1023
with diphenylphosphinous chloride, **2**, 864
with ethyne, **2**, 439
with *N*-fluorosulfonyl imines, **4**, 481
with isocyanides, **5**, 1130
with perfluoronitriles, **5**, 786
with quaternary ammonium salts, **4**, 483
with α,β-unsaturated nitriles, **2**, 439
with vinylsulfonyl fluoride, **2**, 440
Phosphine, diphenyl(trimethylsilyl)-
reactions
with *gem*-dihaloalkenes, **4**, 997
with trifluoromethyl ketones, **4**, 338
Phosphine, dipivaloyl-, x-ray crystal structure of, **5**, 396
Phosphine, divinyl-, as starting material, for phosphapropyne, **5**, 1158
Phosphine, ethoxytrimethylsilyloxy(diethoxymethyl)-, use in the synthesis of aminoalkylphosphonous acids, **2**, 451
Phosphine, ethyldichloro-, synthesis of, **2**, 448
Phosphine, *P*-fluoro-*C*,*C*-bis(trimethylsilyl)methylene-, synthesis of, **6**, 689
Phosphine, fluorodiphenyl-, reactions, with benzoyl fluoride, **5**, 407
Phosphine, hexafluorodimethyl-, as starting material, for perfluoro-2-phosphapropene, **5**, 786
Phosphine, iminomethylene-, synthesis of, **6**, 242
Phosphine, iodobis(pentafluoroethyl)-, reactions, with silver cyanide, **5**, 1129
Phosphine, iodotrifluoromethyl-, reactions, with mercury(II) cyanide, **5**, 1129
Phosphine, lithiated silyl-, reactions, with ketenes, **3**, 605
Phosphine, lithio-, reactions, with difluoride, **6**, 167
Phosphine, lithio(chloro)methylene-, synthesis of, **6**, 690
Phosphine, lithio diethyl-, reactions, with *N*-phenylbenzimidoyl chloride, **5**, 784
Phosphine, lithiomethylene-, stability of, **6**, 690
Phosphine, mesitylbis(trimethylsilyl)-, reactions, with phosgene, **6**, 454

Phosphine, *N*-metallated amino-
reactions
with chloroarsine, **2**, 472
with chlorodialkylstibine, **2**, 472
Phosphine, methyl-
reactions, with alkyl halides, **2**, 429
synthesis of, **2**, 427
Phosphine, methylbis(trimethylsilyl)-, reactions, with carbon disulfide, **6**, 324
Phosphine, methyldifluoro-, reactions, with cyanogen bromide, **5**, 1129
Phosphine, methyldiphenyl-
deprotonation of, **1**, 157
reactions, with dimethylphenyl, **2**, 1003
Phosphine, monophenyl-, reactions, with perfluoronitriles, **5**, 786
Phosphine, (7-norcaranyl)diphenyl-, synthesis of, **2**, 438
Phosphine, phenyl-
reactions
with *t*-butyldiethynylphosphine, **4**, 1027
with *t*-butyldipropynylphosphorine, **4**, 1026
with dialkenyls, **2**, 430
with imines, **4**, 455
with nitriles, **4**, 481
with perfluorooctanonitriles, **5**, 786
with phenylphosphinylidene disilane, **6**, 165
Phosphine, phenyldichloro-
methylation of, **2**, 448
reactions, with *N*-phenylurea, **4**, 488
Phosphine, polymer-supported triphenyl-, use in the synthesis of acid chlorides, **5**, 13
Phosphine, stannyl-, as starting material, for 2-phospha-1-perfluorobutene, **6**, 679
Phosphine, tetrachloro(phenyl)-, use in the synthesis of alkyl chlorides, **2**, 7
Phosphine, trialkyl-
reactions
with chlorodialkylphosphine, **2**, 471
with *β*,*β*-dialkyltitanacyclobutane, **3**, 508
as starting material, for phosphonium salts, **3**, 492
Phosphine, triallyl-, as starting material, for phosphaethynes, **5**, 1158
Phosphine, triaryl-, amination of, **2**, 859
Phosphine, tribromo-, reactions, with 1-ethoxyethyne, **4**, 914
Phosphine, tributyl-
reactions
with alcohols, **2**, 135
with chlorodifluoroacetate, **6**, 695
with 2,3-dichlorobuta-1,3-diene, **2**, 446
with diethyl peroxide, **2**, 476
with 1-ethoxyethyne, **4**, 914
with *o*-nitrophenyl selenocyanate, **1**, 604
with quaternary ammonium salts, **2**, 447
with thiophosphinic acid thioester, **2**, 851
with trichloroacetamides, **4**, 791
for the reductive cleavage of disulfides, **2**, 127
use in Wittig reactions, **4**, 732
Phosphine, tri-*n*-butyl-, use in the synthesis of carbodiimides, **5**, 1073
Phosphine, tri-*t*-butyl-, synthesis of, **2**, 434
Phosphine, 2,4,6-tri(*t*-butyl)phenylchloro-
reactions
with phosphaallenes, **3**, 607
with phosphacumulenes, **3**, 607
Phosphine, trichloro-
reactions
with 3-chloropropyne, **5**, 1153
with 1-cyanopropyne, **5**, 1153
with dialkylcadmium, **2**, 447
as starting material, for chlorophosphines, **2**, 447
use in the synthesis of *α*-(benzyloxycarbonylamino)benzylphosphonic acids, **4**, 467
Phosphine, trichloromethyl-substituted acyl-, synthesis of, **5**, 398
Phosphine, tri(cyanomethyl)-, synthesis of, **3**, 639
Phosphine, tricyclopropyl-, synthesis of, **2**, 434
Phosphine, triethyl-, synthesis of, **2**, 434
Phosphine, trifluoromethyl-
reactions, with diisopropylamine, **5**, 1152
as starting material, for fluorophosphaethyne, **5**, 1152
Phosphine, (trifluoromethyl)silver(I) trimethyl-, synthesis of, **6**, 246
Phosphine, triisopropyl-
reactions, with ethyl iodide, **2**, 446
synthesis of, **2**, 434
Phosphine, trimethyl-
reactions
with *n*-butyllithium, **2**, 1003
with tribromobismuth, **2**, 473
with tungsten complexes, **3**, 514
synthesis of, **2**, 434
use in the synthesis of carbon-bridged dichromium complexes, **4**, 672
use in the synthesis of trisimidoylchromium complexes, **5**, 798
use in the synthesis of trismethylenediruthenium complexes, **4**, 680
Phosphine, (trimethylsiloxy)methylidene-, use in the synthesis of phosphaalkyne, **5**, 1155
Phosphine, trimethylsilyl-
reactions
with alkenes, **2**, 428
with dienes, **2**, 428
Phosphine, trimethylsilyldiphenyl-, reactions, with isothiocyanates, **6**, 582
Phosphine, trioctyl-, reactions, with phosphorus tribromide, **2**, 449
Phosphine, triphenyl-
elimination of, **2**, 1038
fluorination of, for diflourotriphenylphosphorane, **2**, 835
reactions
with acid chloride, **5**, 402
with alcohols, **2**, 299
with aldehydes, **4**, 745, 749
with alkyl aldehydes, **2**, 446
with alkyl peroxides, **2**, 830
with alkynyliodonium tetrafluoroborates, **2**, 1062
with *O*-aroyl-substituted oxyimidic azides, **5**, 862
with 2-arylacrylonitriles, **3**, 649
with azides, **2**, 324
with 3'-azido-3'-deoxythymidine (AZT), **5**, 975
with *δ*-azido epoxides, **2**, 324
with azidotrithioorthoformates 6: 295–31865
with azodicarboxylates, **2**, 87
with benzophenone 1-aminoethylidene hydrazone, **5**, 862
with bromotrifluoromethane, **4**, 754
with chloroacetonitrile, **3**, 639
with *N*-chloroarylamidines, **5**, 861
with *N*-chlorobenzamidines, **5**, 775
with chlorodiphenylacetonitrile, **5**, 860
with chloroiodomethane, **2**, 618
with cyclodextrin azides, **2**, 859
with 2-decalone, **4**, 745
with 3,3-diazido-2-cyanacrylate, **4**, 993
with dibromochloromethane, **4**, 757
with 1,2-dibromo-1-ethoxyethane, **4**, 912
with dibromomethane, **5**, 1093
with dichloroethyne, **2**, 1062
with dimethoxyphosphoryl chloroalkynes, **2**, 1062
with diphenylcyclopropenone, **3**, 543
with 1,3-dithiole-2-thiones, **4**, 850
with *α*-haloalkyl aryl selenides, **4**, 347
with hydroxamic acids, **5**, 1000
with 2-hydroxy azides, **2**, 860

with isothiocyanates, **6**, 573
with ketones, **4**, 745
with methyleneiminium salts, **4**, 487
with methylmanganese pentacarbonyl, **5**, 447
with *N*-methyltrifluoromethanesulfonamide, **2**, 309
with nitrile oxides, **5**, 1131
with palladium(II) diacetate, **5**, 220
with palladium acetate, **1**, 1051
with polycyclic dithietane, **3**, 351
with sodium chlorodifluoroacetate, **6**, 694
with sulfenyl chlorides, **5**, 477
with sulfuryl chloride, **2**, 182
with tetrabromomethane, **4**, 745
with tetrachloromethane, **4**, 737; **5**, 1024
with thiocyanogen, **5**, 1041
with trichlorofluoromethane, **4**, 753
with tris(phenylseleno)methyllithium, **6**, 697
with ylides, **6**, 264
as reagent, for the synthesis of borenes, **5**, 928
as reducing agents, **2**, 769
for the ring-cleavage of epoxides, **2**, 84
as starting material, for vinyl chlorides, **2**, 617
Staudinger phosphorylation reaction with, **6**, 365
synthesis of, **3**, 391; **5**, 860
for the synthesis of ylides, **3**, 492
use in the synthesis of acid chlorides, **5**, 13
use in the synthesis of carbodiimides, **5**, 1073
use in the synthesis of 1,1-dichloroalkenes, **4**, 737
use in the synthesis of 1,1-difluoroalkenes, **4**, 732
use in the synthesis of fluoro haloalkenes, **4**, 751
use in the synthesis of *N*-methoxyamides, **5**, 327
use in the synthesis of thioimides, **5**, 338
Phosphine, tris(but-3-enyl)-, reactions, with dimethylphosphine, **2**, 439
Phosphine, tris(chloromethyl)-, from, tetrakis(hydroxymethyl)phosphonium chloride, **4**, 113
Phosphine, tris(2-cyanoethyl)-, synthesis of, **2**, 428
Phosphine, trisdialkylamino-, reactions, with chlorotrimethylsilane, **2**, 467
Phosphine, tris(diethylamino)-, use in the synthesis of thioimides, **5**, 338
Phosphine, tris(2,6-dimethoxyphenyl)-, reactions, with tetrakis(carbomethoxy)-palladacyclopentadiene, **1**, 1051
Phosphine, tris(dimethylamino)-
reactions
with aldehydes, **2**, 469
with *N*-imidoylhydroxylamines, **5**, 775
with trichlorofluoromethane, **4**, 753
use in the synthesis of alkyl halides, **2**, 7
use in Wittig reactions, **4**, 733, 738, 745
Phosphine, tris(methylthiomethyl)-, synthesis of, **2**, 434
Phosphine, tris(trifluoromethyl)-, synthesis of, **2**, 434
Phosphine, tris(2,4,6-trimethoxyphenyl)-, as starting material, for (chloromethyl)phosphonium salts, **4**, 118
Phosphine, tris(trimethylsilyl)-
desilylation of, **6**, 155
reactions
with acyl chlorides, **5**, 1154
with *N*-(chloroacetamide), **4**, 483
with chloromethylamines, **2**, 433
with dialkylaminomethanols, **4**, 459
with imidoyl chlorides, **6**, 686
with isopropyl isocyanate, **5**, 1157
synthesis of, **5**, 1154
Phosphine, tris(trimethylsilyl)methyldifluoro-, synthesis of, **6**, 372
Phosphine, tri-*o*-tolyl-, use in the synthesis of alkenylpalladiums, **1**, 1096
Phosphine acetate, diphenyl-, oxidation of, **5**, 221
Phosphine–aluminum chloride complexes
monoalkylation of, **2**, 427
reactions, with 1,3-disubstituted haloadamantanes, **2**, 427
Phosphine analogues, triethyl-, from, methyl complexes, **5**, 463
Phosphine–borane adducts
deprotonation of, **4**, 557
reactions of, **4**, 564, 570, 575
Phosphine–borane complexes, synthesis of, **1**, 158
Phosphineboranes, reactions, with amines, **2**, 444
Phosphinecarboxylate oxide, methyl di-*n*-butoxy-, synthesis of, **6**, 490
Phosphinecarboxylate oxides, alkoxy(methyl)-, synthesis of, **6**, 490
Phosphinecarboxylate oxides, alkyl dialkoxy-, synthesis of, **6**, 490
Phosphinecarboxylates
synthesis of, **6**, 489
via Arbuzov reaction, **6**, 490
Phosphinecarboxylates, alkyl(trimethylsilyl)-, synthesis of, **6**, 491
Phosphinecarboxylates, bis(trimethylsilyloxy)-, synthesis of, **6**, 491
Phosphinecarboxylates, monoalkyl-, synthesis of, **6**, 490
Phosphine chalcogenides, alkynyl-, synthesis of, **2**, 1065
Phosphine complexes
order of reactivity for, **2**, 582
synthesis of, **5**, 952
Phosphine complexes, alkyliridium-, synthesis of, **2**, 579
Phosphine complexes, alkylrhodium-, synthesis of, **2**, 579
Phosphine complexes, bistriethyl-, reactions, with acetyl bromide, **5**, 460
Phosphine complexes, copper(I) iodide triphenyl-, reduction of, **3**, 286
Phosphine complexes, iridium(I), synthesis of, **2**, 579
Phosphine dibromide, triphenyl-
reactions
with amines, **5**, 975
with carboxylic acids, **5**, 17
with esters, **5**, 17
with primary amines, **5**, 1065
with sulfonamides, **2**, 863
with trialkylsilyl esters, **5**, 17
for the synthesis of isocyanides, **3**, 696
Phosphinedicarboxylates, synthesis of, **6**, 490
Phosphine dihalides, triaryl-, reactions, with amines, **2**, 859
Phosphine dihalides, triphenyl-, reactions, with amines, **2**, 860
N'-Phosphineformamidrazone, synthesis of, **6**, 659
Phosphinegold chloride, triphenyl-, reactions, with ylides, **4**, 397
Phosphine halides, reactions, with lithium alkoxyethynylides, **2**, 1027
Phosphineimide, *N*-trimethylsilyl-
reactions
with acid anhydrides, **2**, 860
with acyl chlorides, **2**, 860
Phosphineimides, *N*-acyl-, synthesis of, **2**, 860
Phosphineimides, alkyl-, synthesis of, **2**, 468
Phosphineimides, alkynyl-, synthesis of, **2**, 1068
Phosphineimides, aryl-, synthesis of, **2**, 859
Phosphineimides, triaryl-, as aza-Wittig reagents, **2**, 859
Phosphineimides, trichloro-, synthesis of, **5**, 373
Phosphineimides, triphenyl-
reactions
with acyl chlorides, **5**, 373
with anhydrides, **5**, 373
with esters, **5**, 373
Phosphineimines, synthesis of, **6**, 151
Phosphineimines, 1-lithioalkyl-, synthesis of, **4**, 576
Phosphineimines, *N*-substituted, synthesis of, **2**, 470
Phosphine ligands, for the stabilization of alkylvanadium(II) compounds, **2**, 569
Phosphine ligands, chiral, synthesis of, **2**, 829

Phosphine oxide
as catalyst, in the synthesis of phosgene, **6**, 413
synthesis of, **5**, 1094; **6**, 116, 167
Phosphine oxide, (*o*-aminobenzyl)phenyl-, reactions, with benzaldehyde, **4**, 471
Phosphine oxide, benzyldiphenyl-
deprotonation of, **4**, 349
reactions, with phenyllithium, **1**, 752
Phosphine oxide, bis-, synthesis of, **6**, 118
Phosphine oxide, bis(acetoxy)phenyl-, synthesis of, **5**, 221
Phosphine oxide, bis(benzoyloxy)phenyl-, synthesis of, **5**, 221
Phosphine oxide, bisbenzyl(phenyl)-, bromination of, **4**, 122
Phosphine oxide, bis(chloromethyl)methyl-, synthesis of, **4**, 120
Phosphine oxide, bis(chloromethyl)phenyl-, as starting material, for (trimethylsilyloxy phosphonium iodides, **4**, 122
Phosphine oxide, bis(chloromethyl)*n*-propyl-, synthesis of, **4**, 121
Phosphine oxide, bis(iodomethyl)methyl-, synthesis of, **4**, 122
Phosphine oxide, bis(iodomethyl)phenyl-, synthesis of, **4**, 122
Phosphine oxide, (bromomethyl)dimethyl-, synthesis of, **4**, 121
Phosphine oxide, (chloromethyl)dimethyl-, from, tetrakis(hydroxymethyl)phosphonium chloride, **4**, 119
Phosphine oxide, (chloromethyl)diphenyl-, synthesis of, **4**, 119
Phosphine oxide, cyanodimethyl-, synthesis of, **5**, 1131
Phosphine oxide, cyanomethyldiphenyl-, use in the synthesis of nitriles, **3**, 642
Phosphine oxide, 1-cyclobutenyldiphenyl-, reactions, with diazomethane, **4**, 488
Phosphine oxide, *N*,*N*-dialkylanilinomethyldiphenyl-, as reagent, **2**, 739
Phosphine oxide, diazomethyldiphenyl-, synthesis of, **3**, 480
Phosphine oxide, dibenzyl-, reactions, with formaldehyde, **4**, 471
Phosphine oxide, dichlorophenyl-, reactions, with silver benzoate, **5**, 221
Phosphine oxide, (diethoxymethyl)-, reactions, with benzaldehyde, **4**, 828
Phosphine oxide, difluoromethyl-, use in Wittig–Horner reactions, **4**, 734
Phosphine oxide, ethyl methyl phenyl-, deprotonation of, **1**, 158
Phosphine oxide, (α-fluorobenzyl)diphenyl-, synthesis of, **4**, 118
Phosphine oxide, α-methoxyallyl-
reactions
with acetaldehyde, **4**, 917
with benzaldehyde, **4**, 917
with lithium diisopropylamide, **4**, 917
Phosphine oxide, *N*-methyl-*N*-anilinomethyldiphenyl-, as reagent, **2**, 739
Phosphine oxide, methyldiphenyl-, reactions, with sodamide, **1**, 752
Phosphine oxide, (*S*)-(-)-methyl phenylvinyl-
as starting material
for (*R*)-(-)-α-bromo compounds, **4**, 804
for chloro compounds, **4**, 804
Phosphine oxide, α-nitro-, from, 2-alkoxyalkenylphosphine oxides, **4**, 490
Phosphine oxide, trialkyl-, use in Pauson–Khand reactions, **1**, 529
Phosphine oxide, tri-*n*-butyl-, as catalyst, in the synthesis of phosgene, **6**, 413
Phosphine oxide, triphenyl-
elimination of, **1**, 983
reactions, with triflic anhydride, **3**, 662; **5**, 196
Phosphine oxide, tris(bromomethyl)-, synthesis of, **4**, 122
Phosphine oxide, tris(chloromethyl)-, synthesis of, **4**, 121
Phosphine oxide, tris(iodomethyl)-, synthesis of, **4**, 122
Phosphine oxide, tris(pentafluoroethyl)-, reactions, with caesium fluoride, **2**, 476
Phosphine oxides
deprotonation of, **1**, 158, 639
elimination of, **1**, 639, 640
from, phosphonium salts, **1**, 444
metallation of, **1**, 754; **6**, 374
reactions
with aldehydes, **1**, 757; **3**, 182
with alkenes, **2**, 450
with boron trisulfide, **2**, 462
with (*E*)-citral, **1**, 754
with haloalkyl phosphonates, **4**, 554
with isocyanates, **6**, 512
with organomagnesiums, **1**, 988
with phosphorus pentachloride, **6**, 699
with potassium phthalimide, **4**, 473
with triflic anhydrides, **5**, 123
reactivity of, **4**, 120
reduction of, **2**, 440, 443, 444
for tertiary phosphines, **2**, 827
as starting materials
for *Se*,*P*-acetals, **4**, 348
for arylphosphinic acids, **2**, 843
for β-ketophosphine oxides, **3**, 197
for phosphine-boranes, **1**, 158
synthesis of, **1**, 159; **3**, 50
use in the synthesis of alkenyl ethers, **2**, 680
use in the synthesis of chiral phosphines, **2**, 444
Phosphine oxides, acetoxydiphenyl-, synthesis of, **5**, 221
Phosphine oxides, (acetoxymethyl)-
chlorination of, **4**, 119
as starting materials, for (chloromethyl)phosphine oxides, **4**, 119
Phosphine oxides, acyl-
hydrolysis of, **6**, 118
reactions, with methylmagnesium iodide, **4**, 343
Phosphine oxides, acyl diazo-, synthesis of, **3**, 482
Phosphine oxides, alkenylalkylphenyl-, synthesis of, **2**, 843
Phosphine oxides, alkenyldiphenyl-, synthesis of, **2**, 843
Phosphine oxides, 2-alkoxyalkenyl-, as starting materials, for α-nitrophosphine oxide, **4**, 490
Phosphine oxides, alkyl-
from
alkoxydialkylphosphines, **2**, 452
esters, **2**, 452
phosphorus halides, **2**, 452
reactions, with ketones, **1**, 641
synthesis of, **2**, 451
Phosphine oxides, alkyl arylphenyl-, synthesis of, **2**, 843
Phosphine oxides, alkyldiphenyl-, metallation of, **4**, 346
Phosphine oxides, alkyl diphenyl-, reactions, with aldehydes, **1**, 641
Phosphine oxides, alkynyl-
from
aldehydes, **2**, 1067
cyclopropenes, **2**, 1068
synthesis of, **2**, 1065
Phosphine oxides, allenyl dihalo-, synthesis of, **1**, 993
Phosphine oxides, allyl-
epoxidation of, **1**, 754
reactions, with acetaldehyde, **1**, 755
Phosphine oxides, α-amino-, from, tertiary phosphines, **4**, 477
Phosphine oxides, α-aminoalkyl-, from, phosphorus esters, **4**, 464
Phosphine oxides, aminomethyl-, synthesis of, **4**, 473
Phosphine oxides, anilinomethyl-, synthesis of, **4**, 471
Phosphine oxides, aryl-, synthesis of, **2**, 839
Phosphine oxides, bis(arenesulfonyl)methyl-, as starting materials, for methylenephosphoranes, **6**, 699

Phosphine oxides, bis(chloromethyl) (1-hydroxyalkyl)-, synthesis of, **4**, 121
Phosphine oxides, α-bromo-, synthesis of, **4**, 121
Phosphine oxides, 1-cerioalkyl-, synthesis of, **4**, 577
Phosphine oxides, chiral
  from, *O*-isopropylmethylphosphorothioate, **2**, 453
  synthesis of, **2**, 453
Phosphine oxides, chloro-
  reactions
    with alkynyl Grignard reagents, **2**, 1066
    with 1,1-dimetalloalkanes, **4**, 556
    with metal alkynides, **2**, 1065
    with 1-metalloalkylphosphorus compounds, **4**, 557
    with 1-metalloalkylsilanes, **4**, 569
    with phosphorus ylides, **4**, 557
Phosphine oxides, α-chloro-, synthesis of, **4**, 119
Phosphine oxides, (2-chloroalkyl)-, synthesis of, **2**, 452
Phosphine oxides, (α-chloroalkyl)-
  phosphinylmethylating agents, **4**, 119
  synthesis of, **4**, 120
Phosphine oxides, chloro(amino)-, reactions, with alkynyl Grignard reagents, **2**, 1066
Phosphine oxides, (chloromethyl)-, from, (acetoxymethyl)phosphine oxides, **4**, 119
Phosphine oxides, chloromethyl(divinyl)-, synthesis of, **4**, 120
Phosphine oxides, chlorovinyl-, reactions, with thiophenol, **4**, 120
Phosphine oxides, 1-cycloalkenyldiphenyl-, from, 1-cycloalkenyltriphenylphosphonium salts, **2**, 843
Phosphine oxides, 1-cyclopropenyl-, reactions, with diazoalkanes, **3**, 481
Phosphine oxides, dialkyl-, reactions, with metallated ketenes, **4**, 915
Phosphine oxides, dialkyl(chloromethyl)-, synthesis of, **4**, 119
Phosphine oxides, diaryl-
  chlorination of, for arylphosphinic chlorides, **2**, 845
  sulfenylation of, **2**, 854
  synthesis of, **2**, 839
Phosphine oxides, α-diazo-, synthesis of, **3**, 480
Phosphine oxides, α-diazoalkyl-, for the synthesis of phosphorus-containing heterocycles, **4**, 489
Phosphine oxides, (diazomethyl)diphenyl-, reactions, with aldehydes, **2**, 1066
Phosphine oxides, dibutyl-, reactions, with mesityl oxide, **4**, 339
Phosphine oxides, dichloro-, synthesis of, **4**, 478
Phosphine oxides, 2,2-difluorovinyl-, reduction of, **1**, 429
Phosphine oxides, diphenyl-, reactions, with pentanal, **1**, 755
Phosphine oxides, α-fluoro-, synthesis of, **4**, 118
Phosphine oxides, α-halo-, synthesis of, **4**, 118
Phosphine oxides, heteroaromatic-substituted alkynyl-, synthesis of, **2**, 1067
Phosphine oxides, *threo*-β-hydroxy-, synthesis of, **1**, 756
Phosphine oxides, α-hydroxy-, chlorination of, **4**, 119
Phosphine oxides, β-hydroxy-
  chromatographic separation of, **1**, 755
  elimination of, **1**, 641
  from
    epoxides, **1**, 753
    β-ketophosphine oxides, **1**, 641
  isomerization of, **1**, 753
  reactions, with sodium hydride, **1**, 641
  as starting materials, for alkenes, **1**, 641
  synthesis of, **1**, 752, 755
Phosphine oxides, (hydroxymethyl)-
  bromination of, **4**, 121
  chlorination of, **4**, 119, 120, 121
Phosphine oxides, imino-, synthesis of, **5**, 786
Phosphine oxides, α-iodo-, synthesis of, **4**, 122
Phosphine oxides, (iodomethyl)-, synthesis of, **4**, 122
Phosphine oxides, keto-, reactions, with butyllithium, **1**, 757
Phosphine oxides, α-keto-, reduction of, **1**, 756
Phosphine oxides, β-keto-
  from, phosphine oxides, **3**, 197
  as starting materials, for β-hydroxyphosphine oxides, **1**, 641
  synthesis of, **1**, 641
Phosphine oxides, 1-lithioalkyl-, synthesis of, **4**, 575
Phosphine oxides, 1-metalloalkyl-
  reactions
    with chlorophosphines, **4**, 551
    with chlorosilanes, **4**, 570
    with chlorostannanes, **4**, 588
    with chromium compounds, **4**, 579
    with copper halides, **4**, 585
    with manganese compounds, **4**, 581
    with mercury compounds, **4**, 587
    with tantalum compounds, **4**, 579
Phosphine oxides, monofluorovinyl-, synthesis of, **1**, 429
Phosphine oxides, perfluoroacyldiphenyl-, synthesis of, **5**, 220
Phosphine oxides, 1-phosphinoalkyl-, synthesis of, **4**, 551
Phosphine oxides, 1-potassioalkyl-, synthesis of, **4**, 577
Phosphine oxides, secondary, unsymmetrical, synthesis of, **2**, 840
Phosphine oxides, 1-silylalkyl-, synthesis of, **4**, 569
Phosphine oxides, 1-sodioalkyl-, synthesis of, **4**, 576
Phosphine oxides, stibinomethyl-, synthesis of, **4**, 565
Phosphine oxides, tertiary
  alkaline hydrolysis of, for phosphinic acids, **2**, 844
  reactions
    with hydroxides, **2**, 454
    with sodium hydride, **2**, 454
  as starting materials, for tertiary phosphine sulfides, **2**, 462
  synthesis of, **2**, 452, 841
Phosphine oxides, tertiary aryl-
  reduction of, **2**, 827
    for tertiary phosphines, **2**, 827
  synthesis of, **2**, 840
Phosphine oxides, trialkyl-, synthesis of, **2**, 452
Phosphine oxides, α,β-unsaturated, addition to, **4**, 561
Phosphine oxides, unsymmetrical, synthesis of, **2**, 452
Phosphine oxides, vinyl-
  alkylation of, **1**, 159
  reactions
    with alcohols, **2**, 453
    with amines, **2**, 453
Phosphinephosphite–rhodium complexes, use in hydroformylation reactions, **1**, 297
Phosphines
  addition, to a metal carbide, **5**, 1096
  as additives, to cuprates, **1**, 465
  alkylation of, **2**, 427
  chloromethylation of, **4**, 112
  choice of, for the synthesis of ylides, **3**, 492
  deprotonation of, **1**, 157
  from
    phosphinites, **2**, 436
    phosphonium iodide, **2**, 427
  methylene-, mesomerically-stabilized, **6**, 686
  optically active
    synthesis of, **2**, 442
    for the synthesis of ylides, **3**, 492
  reactions
    with acrylonitrile, **3**, 498
    with acyl azides, **5**, 373
    with alcohols, **2**, 445
    with aldehydes, **2**, 446; **5**, 1130
    with alkalis, **2**, 475
    with alkenes, **2**, 438, 440; **6**, 58
    with alkyl azides, **5**, 373
    with alkyl compounds, **5**, 452
    with alkynes, **2**, 438
    with alkynyl halides, **2**, 1062

with 1-aminoalkylphosphines, **4**, 546
with azides, **2**, 1068
with bis(acyl) disulfides, **5**, 243
with borinic esters, **4**, 342
with carbenes, **3**, 497
with carbonimidic dichlorides, **6**, 686
with carbon–nitrogen bonds, **2**, 440
with carbonyls, **4**, 548; **6**, 686
with chlorotrimethylstannane, **2**, 474
with chromium carbene complexes, **4**, 580
with cyclopropyl ketones, **3**, 195
with 1,1-dihaloalkanes, **4**, 546
with diketones, **4**, 339
with enones, **3**, 195
with epoxides, **2**, 445
with fluoroalkenes, **4**, 111
with fluoro(trifluoromethyl)carbene, **4**, 111
with 1-haloalkylchromiums, **4**, 579
with 1-haloalkylindiums, **4**, 588
with 1-haloalkylirons, **4**, 582
with 1-haloalkylphosphines, **4**, 546
with 1-haloalkylphosphorus compounds, **4**, 554
with 1-haloalkylplatinums, **4**, 585
with 1-haloalkylrhodiums, **4**, 582
with 1-haloalkylsilanes, **4**, 569
with 1-haloalkylstannanes, **4**, 588
with 1-haloalkylzincs, **4**, 586
with α-halo ketones, **3**, 195
with halomethanes, **2**, 444
with hexafluoroacetone, **4**, 337
with hydrohalides, **2**, 445
with imines, **3**, 455; **4**, 455
with iron carbene complexes, **4**, 581
with α-keto esters, **5**, 1130
with ketones, **2**, 446
with Mannich bases, **4**, 452
with methyl iodide, **2**, 430, 1063
with molybdenum carbene complexes, **4**, 580
with niobium carbenes, **3**, 510
with ortho esters, **6**, 116
with perfluoronitriles, **5**, 786
with phosphonium hydrohalides, **2**, 445
with rhenium carbene complexes, **4**, 581
with selenium, **2**, 1070; **6**, 151
with sulfur, **2**, 1069
with tertiary amines, **2**, 474
with tetrachloro-*o*-benzoquinone, **2**, 1074
with tetrahalomethanes, **6**, 695
with tris(phenylthio)methyllithium, **6**, 697
with tungsten carbene complexes, **4**, 580
with α,β-unsaturated carbonyls, **4**, 339
with vinylisonitrile, **5**, 784
reduction of, **1**, 444
as starting materials
for *Se*,*P*-acetals, **4**, 347
for phosphonium ylides, **3**, 497
for quaternary phosphonium salts, **2**, 444
for secondary alkylphosphines, **2**, 429
for tertiary phosphines, **2**, 432
synthesis of, **2**, 820; **6**, 242
water soluble, **2**, 7
Phosphines, acetyl-, synthesis of, **5**, 395
Phosphines, acyl-
Chapman rearrangement of, **5**, 396
chlorination of, **5**, 398
oxidation of, **5**, 403
quaternization of, **5**, 402
reactions
with electrophiles, **5**, 399
with sulfur, **5**, 406
reactivity of, **5**, 395
self-condensation of, **5**, 395
tautomerization of, **5**, 396
Phosphines, acyl-, metal derivatives, synthesis of, **5**, 400
Phosphines, adamantyl-, synthesis of, **2**, 427
Phosphines, adamantylmethylhalo-, synthesis of, **2**, 448
Phosphines, alkoxyalkyl(alkylthio)-, synthesis of, **2**, 461
Phosphines, alkoxyalkylamino-
alcoholysis of, **2**, 451
hydrolysis of, **2**, 451
Phosphines, alkoxyalkylchloro-
reactions
with hydrogen sulfide, **2**, 461
with thiols, **2**, 461
synthesis of, **2**, 451
Phosphines, alkoxyaminoalkyl-, reactions, with hydrogen sulfide, **2**, 461
Phosphines, alkoxybis(dialkylamino)-
Arbuzov rearrangement of, **2**, 469
from
alkoxydichlorophosphines, **2**, 410
bis(dialkylamino)chlorophosphines, **2**, 411
synthesis of, **2**, 410
Phosphines, alkoxydialkyl-
Arbuzov reactions of, **2**, 453
from, alkyldialkoxyphosphines, **2**, 450
reactions, with sulfur, **2**, 462
as starting materials, for alkylphosphine oxides, **2**, 452
synthesis of, **2**, 450
Phosphines, alkoxydialkylaminoalkyl-, synthesis of, **2**, 451
Phosphines, alkoxydichloro-
reactions, with cyanic acid, **5**, 1018
as starting materials
for alkoxybis(dialkylamino)phosphines, **2**, 410
for chloro-*N*,*N*-dialkylaminoalkoxyphosphines, **2**, 409
Phosphines, alkoxydicyanato-, synthesis of, **5**, 1018
Phosphines, alkoxy(α-haloalkyl)-, synthesis of, **4**, 117
Phosphines, alkoxy(trimethylsiloxy)-, reactions, with 1-ethoxyethyne, **4**, 914
Phosphines, alkyl-
chlorination of, **2**, 448
reactions, with titanacyclobutanes, **3**, 508
synthesis of, **2**, 426
Phosphines, alkyl(alkylthio)chloro-
reactions
with alcohols, **2**, 461
with α,β-unsaturated carboxylic acids, **2**, 463
Phosphines, alkyl(alkylthio)halo-, synthesis of, **2**, 461
Phosphines, alkylamino-, as starting materials, for selenides, **2**, 1070
Phosphines, alkylbis(chloromethyl)-, from, chlorophosphines, **4**, 112
Phosphines, alkylbis(trimethylsiloxy)-, reactions, with sulfur, **2**, 464
Phosphines, alkyldialkoxy-
Arbuzov reactions of, **2**, 453
hydrolysis of, **2**, 451
reactions
with alkyldichlorophosphines, **2**, 451
with alkyl dihalides, **2**, 453
with alkyldihalophosphines, **2**, 461
with chlorodi-*i*-propylphosphine, **2**, 471
as starting materials, for alkoxydialkylphosphines, **2**, 450
Phosphines, alkyldi(alkylthio)-, synthesis of, **2**, 460
Phosphines, alkyldiamino-
alcoholysis of, **2**, 451
hydrolysis of, **2**, 451
Phosphines, alkyl(diamino)-, as starting materials, for phosphonic diamides, **4**, 133
Phosphines, alkyldichloro-
alcoholysis of, **2**, 451
displacement of, **2**, 467
hydrolysis of, **2**, 451
reactions
with alkyldialkoxyphosphines, **2**, 451
with benzaldehydes, **4**, 123

with mercaptides, **2**, 461
with phosphorus, **2**, 471
with primary phosphines, **2**, 471
with sulfenyl chlorides, **2**, 464
with thiols, **2**, 461
as starting materials
for alkylaminomethylphosphinic acids, **4**, 465
for diiodophosphines, **2**, 449
synthesis of, **2**, 447
Phosphines, alkyldihalo-, reactions, with alkyldialkoxyphosphines, **2**, 461
Phosphines, alkyldiphenyl-
reactions
with hexachloroethane, **4**, 112
with tetrachloromethane, **4**, 112
synthesis of, **2**, 436; **6**, 490
Phosphines, *N*-alkylformidoyldialkyl-, synthesis of, **5**, 1130
Phosphines, alkylgermyl-, synthesis of, **2**, 867
Phosphines, alkyliminohalo-, synthesis of, **2**, 406
Phosphines, alkyliminotrihalo-
from, amine hydrochlorides, **2**, 407
synthesis of, **2**, 407
Phosphines, alkylthiodialkyl-
reactions, with sulfur, **2**, 462
synthesis of, **2**, 460
Phosphines, alkynyl-
from
alkynyl chlorides, **2**, 1060
alkynylphosphines, **2**, 1061
alkynylphosphonites, **2**, 1060
oxidation of, **2**, 1065
with hydrogen peroxide, **2**, 1066
reactions, with allyl bromide, **2**, 1063
as starting materials
for alkynylphosphines, **2**, 1061
for alkynylphosphine sulfides, **2**, 1069
for alkynyl phosphonium salts, **2**, 1063
for phosphaalkynes, **3**, 689
synthesis of, **2**, 1058
via alkynic Grignard reagents, **2**, 1059
via the nucleophilic substitution of alkynylphosphonites, **2**, 1060
via the nucleophilic substitution of phosphorus halides by metal alkynides, **2**, 1058
Phosphines, 1-alkynyl-
Lewis base-induced rearrangement of, for the synthesis of phosphaalkynes, **5**, 1157
reduction of, **1**, 999
Phosphines, alkynylamino-
reactions, with sulfur tetrafluoride, **2**, 1073
synthesis of, **2**, 1065
Phosphines, allyl-, synthesis of, **2**, 473
Phosphines, β-allylaminoalkyl-, reactions, with formaldehyde, **4**, 458
Phosphines, amidino-, synthesis of, **6**, 658
Phosphines, amino-
reduction of, **2**, 431
as starting materials, for chlorophosphines, **2**, 448
synthesis of, **2**, 471
Phosphines, aminoalkyl-, synthesis of, **2**, 433, 440
Phosphines, 1-aminoalkyl-
reactions
with diethyl phosphite, **4**, 551
with phosphines, **4**, 546
Phosphines, α-aminoalkyl-
alkylation of, **4**, 478
from, azacrown compounds, **4**, 453
optically pure, synthesis of, **4**, 455
as starting materials, for α-aminoalkyl phosphonium salts, **4**, 478
synthesis of, **4**, 452, 456
Phosphines, β-aminoalkyl-, reactions, with *N*-hydroxymethyldiethylamine, **4**, 453
Phosphines, (ω-aminoalkyl)phenyl-
reactions
with aldehydes, **4**, 453
with ketones, **4**, 453
Phosphines, amino(chloromethyl)-, synthesis of, **4**, 116
Phosphines, aminodialkyl-, reactions, with hydrogen sulfide, **2**, 460
Phosphines, aminodichloro-
reactions, with arylmagnesium bromide, **2**, 859
synthesis of, **2**, 466
Phosphines, amino(α-haloalkyl)-, synthesis of, **4**, 116
Phosphines, aminoimino-, as starting materials, for phosphaalkenes, **6**, 689
Phosphines, aminomethyl-, synthesis of, via Mannich reactions, **2**, 440
Phosphines, α-aminomethyl-, synthesis of, **4**, 452
Phosphines, aminomethylene-, reactions, with alcohols, **6**, 685
Phosphines, α-aminomethyl tertiary, synthesis of, **4**, 455
Phosphines, aromatic, synthesis of, **2**, 829
Phosphines, 1-arsinoalkyl-, synthesis of, **4**, 564
Phosphines, aryl-
reactions, with amide acetals, **6**, 686
use in the synthesis of primary phosphines, **2**, 821
Phosphines, arylbis(trimethylgermyl)-, synthesis of, **2**, 867
Phosphines, arylbis(trimethylsilyl)-, synthesis of, **2**, 866, 867
Phosphines, aryldialkyl-
cleavage of, **2**, 475
reactions, with ethyl *N*-chlorobenzimidate, **5**, 708
Phosphines, aryldichloro-
amination of, **2**, 859
reactions, with benzaldehydes, **4**, 123
Phosphines, aryldihalo-, synthesis of, **2**, 832
Phosphines, arylgermyl-, synthesis of, **2**, 867
Phosphines, aryloxybis(dialkylamino)-
from, bis(dialkylamino)chlorophosphines, **2**, 411
synthesis of, **2**, 410
Phosphines, aryl(trimethylsilyl)-, synthesis of, **2**, 866
Phosphines, azido-, reactions, with chlorotrimethylsilane, **2**, 470
Phosphines, 1-barioalkyl-, synthesis of, **4**, 577
Phosphines, benzoyl-
methylation of, **5**, 395
synthesis of, **5**, 395
Phosphines, bid(dialkylamino)chloro-, synthesis of, **2**, 410
Phosphines, bidentate-, use in the synthesis of carbodiimides, **5**, 1072
Phosphines, binaphthyl-, in asymmetric hydrogenation reactions, **2**, 53
Phosphines, bis(acyl)-, synthesis of, **5**, 395, 396
Phosphines, bisadamantanoyl-, synthesis of, **5**, 396
Phosphines, bis(alkylthio)methylene-, from, chlorophosphines, **6**, 683
Phosphines, bis(amino)methylene-
as intermediates, theoretical, **6**, 685
synthesis of, **6**, 685
Phosphines, bis(arenesulfonyl)methylene-, reactions, with tetrachloromethane, **6**, 699
Phosphines, bis(chloromethyl)-, synthesis of, **4**, 115
Phosphines, biscyclohexanoyl-, synthesis of, **5**, 396
Phosphines, bis(dialkylamino)-, synthesis of, **2**, 414
Phosphines, bis(dialkylamino)alkoxy-, as starting materials, for *N*,*N*-dialkylphosphoramidites, **2**, 410
Phosphines, bisdialkylaminochloro-, reactions, with organometallics, **2**, 467
Phosphines, bis(dialkylamino)chloro-
as starting materials
for alkoxybis(dialkylamino)phosphines, **2**, 411
for aryloxybis(dialkylamino)phosphines, **2**, 411
Phosphines, bis(ethylthio)alkyl-, reactions, with dibutyltin dihydride, **6**, 581
Phosphines, bis-α-hydroxyalkyl-, reactions, with aldehydes, **4**, 341

Phosphines, bis(α-hydroxybenzyl)-, synthesis of, **4**, 337
Phosphines, bistriazolylalkoxy-, as starting materials, for *N,N*-dialkylphosphoramidites, **2**, 410
Phosphines, bis(trichloromethyl)-, synthesis of, **6**, 242
Phosphines, bis(trimethylsilyl)-, reactions, with phenyl isothiocyanate, **6**, 582
Phosphines, bis(trimethylsilyl)acyl-, synthesis of, **5**, 1154
Phosphines, bis(trimethylsilyl)amino-, reactions, with tetrachloromethane, **4**, 572
Phosphines, bis(trimethylsilyl)methylene-, synthesis of, **6**, 688
Phosphines, bromo-, synthesis of, **2**, 448
Phosphines, bromodialkyl-, synthesis of, **2**, 449
Phosphines, (bromomethyl)-, oxidation of, **4**, 122
Phosphines, *t*-butylchloro-, synthesis of, **2**, 448
Phosphines, carbamoyl-
  synthesis of, **6**, 511
  use in the synthesis of carbodiimides, **5**, 1086
Phosphines, carboxyalkyl-
  reactions, with hydrazones, **4**, 482
  as starting materials, for 1-amino-1,3-azaphospholan-5-one, **4**, 482
Phosphines, chiral
  in Heck reactions, **1**, 495
  synthesis of, **2**, 442
Phosphines, chloro-
  dehydrochlorination of, **6**, 688
  from
    alkyl metals, **2**, 447
    aminophosphines, **2**, 448
    trichlorophosphine, **2**, 447
  reactions
    with alcohols, **2**, 450
    with aldehydes, **4**, 119, 121
    with amines, **4**, 116
    with bis(benzenesulfonyl)halomethanes, **6**, 699
    with diazoalkanes, **4**, 114
    with dienes, **2**, 453
    with 1,1-dimetalloalkanes, **4**, 546
    with α-hydroxyamides, **4**, 465
    with iminophosphines, **2**, 471
    with ketones, **4**, 119, 121, 345
    with 1-metalloalkylphosphine oxides, **4**, 551
    with 1-metalloalkylphosphines, **4**, 546
    with 1-metalloalkylphosphine sulfides, **4**, 551
    with 1-metalloalkylphosphonamidates, **4**, 551
    with 1-metalloalkyl phosphonates, **4**, 551
    with 1-metalloalkylphosphorus compounds, **4**, 551
    with 1-metalloalkylsilanes, **4**, 565
    with methylenebisphosphines, **6**, 151
    with sulfur, **2**, 463
    with 1-trialkylsilylalkylphosphines, **4**, 547
    with triethylamine, **6**, 265
  reduction of, **6**, 167
  as starting materials
    for alkylbis(chloromethyl)phosphines, **4**, 112
    for bis(alkylthio)methylenephosphines, **6**, 683
    for (di-*t*-butylamino)phosphines, **4**, 116
    for fluorophosphines, **2**, 449
Phosphines, 2-chloro-, synthesis of, **6**, 679
Phosphines, α-chloro-, from, diethyl chloromalonate, **4**, 115
Phosphines, α-chloroalkyl-
  from, dichlorophosphoranes, **4**, 112
  synthesis of, **4**, 111
Phosphines, (α-chloroalkyl)diphenyl-, synthesis of, **4**, 112
Phosphines, *P*-chloro-α-chloroalkyl-, dechlorination of, **6**, 689
Phosphines, chlorodialkoxy-, reactions, with Grignard reagents, **2**, 451
Phosphines, chlorodialkyl-
  hydrolysis of, **2**, 450
  reactions
    with methanol, **2**, 450
    with propanol, **2**, 450
Phosphines, chloro-*N,N*-dialkylaminoalkoxy-, from, alkoxydichlorophosphines, **2**, 409
Phosphines, chlorodialkylaminoalkyl-, reactions, with lithium alkoxide, **2**, 451
Phosphines, chlorodimethyl-, as starting materials, for tetrakis(dichlorodimethylphosphoranyl)-methane, **6**, 366
Phosphines, (chloromethyl)-, oxidation of, **4**, 120
Phosphines, chloromethylbis(octyloxy)-, synthesis of, **4**, 117
Phosphines, chloromethyldihalo-, synthesis of, **4**, 115
Phosphines, (chloromethyl)dimethyl-, synthesis of, **4**, 112
Phosphines, chloro(trihalomethyl)-, reactions, with *n*-butyllithium, **6**, 682
Phosphines, cyano-
  from
    dialkylaminomethylphosphines, **5**, 1129
    dialkylaminophosphines, **5**, 1129
Phosphines, β-cyano-, synthesis of, **3**, 639
Phosphines, cyanoalkyl-, synthesis of, **2**, 433
Phosphines, 2-cyanoethyl-, synthesis of, **2**, 829
Phosphines, cyanomethyl-, synthesis of, **3**, 639
Phosphines, cyclic amino-, synthesis of, **2**, 466
Phosphines, cyclic α-aminoalkyl-, synthesis of, **4**, 453
Phosphines, cyclic aminomethyl-
  from
    orthodisubstituted aromatic phosphines, **4**, 453
    secondary amines, **4**, 453
Phosphines, cyclic halo-, synthesis of, **2**, 449
Phosphines, dialkoxyalkylamino-, reactions, with dialkylchlorophosphines, **2**, 471
Phosphines, dialkoxychloro-, reactions, with dithiocarboxylic acids, **2**, 464
Phosphines, dialkoxyhalo-, reactions, with alkyl halides, **2**, 458
Phosphines, dialkyl-
  reactions
    with bistrifluoromethylketene, **4**, 1023
    with α-bromo ketones, **3**, 195
  synthesis of, **2**, 429
Phosphines, dialkylalkoxy-, hydrolysis of, **2**, 450
Phosphines, dialkylalkynyl-, reactivity of, **2**, 1059
Phosphines, dialkylamino-
  reactions
    with cyanogen halides, **5**, 1129
    with hydrogen cyanides, **5**, 1129
  as starting materials, for cyanophosphines, **5**, 1129
Phosphines, dialkylaminoalkylimino-
  from, amines, **2**, 406
  synthesis of, **2**, 406
Phosphines, dialkylaminochloro-
  as intermediates, in the synthesis of *p*-chlorophospholanes, **2**, 449
  reactions
    with Grignard reagents, **2**, 466
    with organometallics, **2**, 448
Phosphines, dialkylaminodialkoxy-, synthesis of, **2**, 410
Phosphines, dialkylaminodichloro-
  from, amines, **2**, 409
  reactions, with alkoxyamino alcohols, **2**, 410
Phosphines, dialkylaminodihalo-, synthesis of, **2**, 409
Phosphines, dialkylaminomethyl-
  reactions
    with cyanogen halides, **5**, 1129
    with hydrogen cyanides, **5**, 1129
  as starting materials, for cyanophosphines, **5**, 1129
Phosphines, dialkylchloro-
  reactions
    with dialkoxyalkylaminophosphines, **2**, 471
    with lithium alkynides, **2**, 1059
    with thiols, **2**, 460
Phosphines, dialkylcyano-, synthesis of, **5**, 1130
Phosphines, dialkylhalo-

reactions
with alkoxyalkynes, **4**, 914
with amines, **2**, 466
Phosphines, dialkyl(α-haloalkyl)-, synthesis of, **4**, 111
Phosphines, dialkylphenyl-, synthesis of, **2**, 435
Phosphines, dialkynyl-, synthesis of, **2**, 1059
Phosphines, dialkynylamino-, synthesis of, **2**, 1065
Phosphines, diallylaminomethyl-, as starting materials, for 1,3-azaphosphanes, **4**, 458
Phosphines, diamino(alkyl)-, reactions, with tetrachloromethane, **4**, 116
Phosphines, diaminochloro-, synthesis of, **2**, 467
Phosphines, diarylalkynyl-, synthesis of, **2**, 1060
Phosphines, diarylhalo-, synthesis of, **2**, 834
Phosphines, diaryl(α-haloalkyl)-, synthesis of, **4**, 111
Phosphines, diaryl-, symmetrical, synthesis of, **2**, 823
Phosphines, diaryl(trimethylsilyl)-
synthesis of, **2**, 866
transmetallation of, **2**, 868
Phosphines, diaryl(triorganostannyl)-, synthesis of, **2**, 868
Phosphines, dibromomethylene-, synthesis of, **6**, 682, 689
Phosphines, dibromotriphenyl-, reactions, with ketones, **1**, 983
Phosphines, dibutoxy(dialkoxymethyl)-, synthesis of, **2**, 451
Phosphines, dibutoxy(dialkylaminomethyl)-, synthesis of, **2**, 451
Phosphines, (di-*t*-butylamino)-, from, chlorophosphines, **4**, 116
Phosphines, dichloro-
acid hydrolysis of, **4**, 123
alkylation of, **4**, 124
from, diiodides, **4**, 114
reactions
with *N*-benzylglycine, **4**, 466
with chlorobis(trimethylsilyl)methanes, **6**, 693
with 1,3-dithiols, **2**, 252
with electrophiles, **4**, 125
with formaldehyde, **4**, 125
with imines, **4**, 463
with lithium bis(trimethylsilyl)chloromethane, **6**, 692
with *N*-methyl thioacetamide, **2**, 252
with triethyl orthoformate, **4**, 126
synthesis of, **2**, 448
Phosphines, (dichloroalkyl)-
from, diisopropyl dichloroalkanephosphonates, **5**, 1152
reactions, with potassium carbonate, **5**, 1152
Phosphines, dichloro(chloromethyl)-
chlorination of, **4**, 133
synthesis of, **4**, 114
Phosphines, dichloromethylene-
reactions
with *n*-butyllithium, **6**, 690
with 1,2-dipotassiumdiorganylphosphides, **6**, 685
synthesis of, **6**, 682
via dechlorination, **6**, 681
via dehydrochlorination, **6**, 679
Phosphines, diethoxy(heptafluoroisopropyl)-, synthesis of, **4**, 117
Phosphines, diethylaminomethyl-, substitution of, **4**, 546
Phosphines, diethylformimidoyl-, from, isonitriles, **5**, 785
Phosphines, difluoromethylene-
synthesis of, **6**, 682
via pyrolysis, **6**, 679
Phosphines, digermyl-, synthesis of, **2**, 474
Phosphines, dihalo-2-methylbut-2-enyl-, synthesis of, **2**, 448
Phosphines, dihalomethylene-, synthesis of, **6**, 678
Phosphines, *C*,*C*-dihalomethylene-, reactions of, **6**, 684
Phosphines, diiodo-
from, alkyldichlorophosphines, **2**, 449
synthesis of, **4**, 114
Phosphines, diiodomethylene-, synthesis of, **6**, 682
Phosphines, (dimethylamino)-, synthesis of, **4**, 116
Phosphines, diphenyl-, synthesis of, **6**, 490
Phosphines, ethynyl-, synthesis of, **2**, 820
Phosphines, fluoro-
from
chlorophosphines, **2**, 449
dichloromethylphosphine, **2**, 448
synthesis of, **2**, 448; **4**, 116
Phosphines, α-fluoro-, synthesis of, **4**, 111
Phosphines, fluoro(alkoxy)methylene-
reactions, with trialkylphosphines, **6**, 700
synthesis of, **6**, 683
Phosphines, α-fluoroalkyl-, synthesis of, **4**, 111
Phosphines, α-fluoroalkyl(trifluoromethyl)iodo-, synthesis of, **4**, 114
Phosphines, fluoro(amino)alkynyl-, as starting materials, for alkynyl phosphoranes, **2**, 1073
Phosphines, fluoro(amino)methylene-, synthesis of, **6**, 684
Phosphines, formimidoyl-, from, perfluoro-2-phosphapropene, **5**, 786
Phosphines, gallyl-, synthesis of, **2**, 475
Phosphines, germyl-
reactions
with ketimines, **4**, 998
with phenyl isothiocyanate, **6**, 582
synthesis of, **2**, 474
Phosphines, halo-
reactions
with aldehydes, **4**, 123
with organotinlithium, **2**, 868
with phosphorus ylides, **4**, 551
Phosphines, (halo)-, reactions, with thioformaldehyde, **4**, 123
Phosphines, α-halo-
reactions, with sulfur, **4**, 128
stability of, **4**, 111
as starting materials, for phosphaalkenes, **4**, 111
Phosphines, 1-haloalkyl-
reactions
with metal phosphides, **4**, 546
with nickel compounds, **4**, 583
with phosphines, **4**, 546
Phosphines, α-haloalkyl-
stability of, **4**, 110
as starting materials, for methylenephosphoranes, **6**, 694
synthesis of, **6**, 699
Phosphines, α-haloalkylbis(diisopropylamino)-, synthesis of, **4**, 116
Phosphines, α-haloalkyl-, trimethylsilyl-substituted, halotropic rearrangements of, **6**, 703
Phosphines, (*o*-halobenzyl)-, reactions, with *n*-butyllithium, **4**, 575
Phosphines, halo-*N*,*N*-dialkylaminoalkoxy-, synthesis of, **2**, 409
Phosphines, *P*-halogenated acyl-, synthesis of, **5**, 398
Phosphines, halo(α-haloalkyl)-, synthesis of, **4**, 114
Phosphines, (1-hydroxyalkyl)di(alkylthio)-, synthesis of, **2**, 461
Phosphines, (α-hydroxyalkyl)phenyl-, disproportionation of, **4**, 341
Phosphines, α-hydroxymethyl-, synthesis of, **4**, 336
Phosphines, (3-hydroxypropyl)phenyl-
reactions
with aldehydes, **4**, 337
with ketones, **4**, 337
Phosphines, imidoyl-, synthesis of, **5**, 785
Phosphines, imino-
reactions
with alkenes, **2**, 468
with carbenoids, **6**, 693
with chlorophosphines, **2**, 471
as starting materials, for methylenephosphoranes, **6**, 693
Phosphines, (iminomethylidene)-, stability of, **5**, 1086
Phosphines, iodo-, synthesis of, **2**, 448
Phosphines, keto-, as starting materials, for phosphonium salts, **3**, 195

Phosphines, α-keto-, synthesis of, **4**, 1023
Phosphines, β-keto-, synthesis of, **3**, 195
Phosphines, α-lithio-, synthesis of, **4**, 575
Phosphines, 1-lithioalkyl-, synthesis of, **4**, 575
Phosphines, lithium dialkyl(aryl)-, reactions, with α-chloro tertiary amines, **4**, 455
Phosphines, lithiumsulfidomethylene-, from, lithium silylphosphide, **6**, 684
Phosphines, 1-lutetioalkyl-, synthesis of, **4**, 577
Phosphines, mesityl(diphenylmethylene)-, reactions, with phenyl azide, **4**, 483
Phosphines, metallated
  as starting materials
    for primary phosphines, **2**, 427
    for secondary alkylphosphines, **2**, 429
    for tertiary phosphines, **2**, 437
Phosphines, *C*-metallated methylene-, synthesis of, **6**, 690
Phosphines, 1-metalloalkyl-
  reactions
    with aluminum compounds, **4**, 587
    with chlorophosphines, **4**, 546
    with chlorosilanes, **4**, 566
    with chlorostannanes, **4**, 588
    with cobalt compounds, **4**, 582
    with copper halides, **4**, 585
    with tantalum compounds, **4**, 579
    with thorium compounds, **4**, 589
    with zirconium compounds, **4**, 579
Phosphines, α-metalloalkyl-, synthesis of, **2**, 1003
Phosphines, methoxy-, synthesis of, **4**, 117
Phosphines, methyl-
  reactions, with hexafluoroisopropylideneimine, **2**, 440
  synthesis of, **6**, 151
Phosphines, methylbromocyano-, synthesis of, **5**, 1129
Phosphines, methyldichloro-, use in the synthesis of phosphaalkynes, **5**, 1152
Phosphines, methylene-
  characteristics of, **6**, 678
  photolysis of, **4**, 459
  reactions
    with carbenoids, **6**, 693
    with oxygen, **6**, 692
    with selenium, **6**, 692
    with sulfur, **6**, 692
  stability of, **6**, 665
  as starting materials
    for α-aminoalkylphosphonites, **4**, 459
    for methylenephosphoranes, **6**, 693
    for tetrahydrophosphorines, **4**, 459
  synthesis of
    via condensation reactions, **6**, 685
    via elimination reactions, **6**, 688
    via intramolecular rearrangements, **6**, 688
    via [1,3]-silyl migration, **6**, 684
    via trimethylsilyl migrations, **6**, 686
  See also Phosphaalkenes
Phosphines, methylene-, *P*-arsino-substituted, synthesis of, **6**, 689
Phosphines, methylene-, η^1(P)-coordinated, synthesis of, **6**, 692
Phosphines, methylene-, *C*,*C*-dihalo-substituted, **6**, 682
Phosphines, methylene-, *C*-germanium-substituted, synthesis of, **6**, 688
Phosphines, methylene-, *C*-germyl-*C*-silyl-substituted, synthesis of, **6**, 689
Phosphines, methylene-, *C*-nitrogen-substituted, synthesis of, **6**, 684
Phosphines, methylene-, *C*-oxygen-substituted, synthesis of, **6**, 683
Phosphines, methylene-, palladium-substituted, stability of, **6**, 690
Phosphines, methylene-, *C*-phosphorus-substituted, synthesis of, **6**, 684
Phosphines, methylene-, *C*-silicon-substituted, synthesis of, **6**, 688
Phosphines, methylene-, *C*-sulfur-substituted, synthesis of, **6**, 683
Phosphines, methylsilyl-, reactions, with formaldehyde, **4**, 337
Phosphines, monoalkyl-, reactions, with alkyl chloroformates, **6**, 490
Phosphines, monoalkynylamino-, synthesis of, **2**, 1065
Phosphines, nucleophilic, reactions, with iminium salts, **4**, 481
Phosphines, oragnodichloro-, as starting materials, for methylenephosphoranes, **6**, 693
Phosphines, organoarsino-, synthesis of, **2**, 893
Phosphines, ortho-disubstituted aromatic, as starting materials, for cyclic aminomethylphosphines, **4**, 453
Phosphines, (3-oxoalkyl)phenyl-
  reactions
    with aldehydes, **4**, 454
    with aldimines, **4**, 454
Phosphines, *t*-pentylchloro-, synthesis of, **2**, 448
Phosphines, perfluoro-, as starting materials, for imidoyl-tricoordinated phosphorus compounds, **5**, 786
Phosphines, perfluoroalkyldichloro-
  reactions
    with paraformaldehyde, **4**, 125
    with trioxane, **4**, 125
Phosphines, perfluorotrialkyl-, synthesis of, **4**, 112
Phosphines, phenyl bis(trimethylsilyl)-, reactions, with aryl isocyanide dichlorides, **5**, 1086
Phosphines, phenyldihalo-, reactions, with methylselenylsilane, **2**, 857
Phosphines, (phosphoranomethyl)-, synthesis of, **4**, 552
Phosphines, pivaloyl-, synthesis of, **5**, 395
Phosphines, polycyclic-, synthesis of, **2**, 471
Phosphines, polycycloalkyl-, synthesis of, **2**, 471
Phosphines, 1-potassioalkyl-, synthesis of, **4**, 577
Phosphines, primary
  alkylation of, **2**, 429, 430
  from, metallated phosphines, **2**, 427
  reactions
    with acyl chlorides, **5**, 396
    with alkyldichlorophosphines, **2**, 471
    with carbonyls, **2**, 450
    with chloroformate esters, **2**, 429
    with α,ω-dichloroalkanes, **2**, 433
    with sulfenyl chlorides, **2**, 460
  synthesis of, **2**, 427, 820
Phosphines, primary α-halo-, synthesis of, **4**, 111
Phosphines, secondary
  oxidation of, for arylphosphinic acids, **2**, 843
  reactions
    with *N*-(alkoxymethyl)-*N*-arylamines, **4**, 453
    with *N*-arenesulfonyl imines, **4**, 481
    with disulfides, **2**, 463
  synthesis of, **2**, 429, 822
Phosphines, secondary acyl-, synthesis of, **5**, 396
Phosphines, secondary alkyl-
  from
    metallated phosphines, **2**, 429
    phosphines, **2**, 429
Phosphines, secondary α-halo-, synthesis of, **4**, 111
Phosphines, silyl-
  condensation of, **6**, 685
  reactions
    with carbonic acid halides, **6**, 686
    with chlorogermanes, **2**, 473
    with chlorosilanes, **2**, 473
    with chlorostannanes, **2**, 473, 474
    with diphenylketene, **6**, 684
    with halogermanes, **2**, 474
    with heterocumulenes, **6**, 687
    with imines, **4**, 481

with isocyanates, **6**, 684
with isocyanide dichlorides, **6**, 686
with isothiocyanates, **6**, 684
with ketimines, **4**, 998
with phenyl isothiocyanate, **6**, 582
with thiophosgene, **6**, 687
as starting materials, for phosphaalkenes, **6**, 684
synthesis of, **2**, 473
Phosphines, silylalkyl-, reactions, with tetrahalomethanes, **6**, 703
Phosphines, silylated, use in the synthesis of acylphosphides, **5**, 396
Phosphines, 1-sodioalkyl-, synthesis of, **4**, 576
Phosphines, stanno-, use in the synthesis of phosphines, **2**, 474
Phosphines, stannyl-, synthesis of, **2**, 868
Phosphines, stibinomethyl-, synthesis of, **4**, 565
Phosphines, substituted, use in alkylation reactions, **1**, 489
Phosphines, *N*-substituted aminomethyl-, synthesis of, **4**, 481
Phosphines, sulfonated, use in the synthesis of alkynes, **1**, 1013
Phosphines, tertiary
aromatic, synthesis of, **2**, 829
from
carbonyls, **2**, 440
metallated phosphines, **2**, 437
phosphines, **2**, 432
phosphorus, **2**, 433
oxidation of, **2**, 452
for tertiary arylphosphine oxides, **2**, 840
reactions
with alkali metals, **2**, 432, 437
with alkyl halides, **2**, 444
with alkynes, **2**, 446
with borohydride, **2**, 474
with carbon disulfide, **3**, 549
with *N*-chloroimidic esters, **5**, 860
with fluorophosphaalkenes, **6**, 700
with iminium ions, **4**, 462, 487
with oxetane, **2**, 445
with quaternary ammonium salts, **2**, 446
with α,β-unsaturated ketones, **2**, 446
ring-opening of, **2**, 445
as starting materials
for α-aminophosphine oxides, **4**, 477
for α-aminophosphine sulfides, **4**, 477
for quaternary phosphonium salts, **2**, 444
synthesis of, **1**, 52; **2**, 432, 824
via addition of phosphorus–hydrogen bonds to unsaturated compounds, **2**, 438
via phosphonium salts, **2**, 441
via photostimulated radical reactions, **2**, 437
via reduction reactions, **2**, 440
Phosphines, tertiary acyl-, synthesis of, **5**, 397
Phosphines, tertiary alkyl-, synthesis of, **2**, 437
Phosphines, tertiary, asymmetrical, synthesis of, **2**, 825
Phosphines, tertiary benzoyl-, chemistry of, **5**, 397
Phosphines, tertiary α-halo-, synthesis of, **4**, 111
Phosphines, thioalkyl-, as starting materials, for thiocarbonylphosphazines, **6**, 581
Phosphines, transition metal-bound, synthesis of, **5**, 400
Phosphines, trialkoxy-
reactions
with Grignard reagents, **2**, 451
with α-methoxyurethanes, **2**, 457
Phosphines, (trialkoxymethyl)-, synthesis of, **6**, 314
Phosphines, trialkyl-
as coupling reagents, for the synthesis of amides, **5**, 268
deprotonation of, **4**, 575
oxidation of, **4**, 123
reactions
with carbon diselenide, **6**, 596
with carbon disulfide, **6**, 121
with dichloroacetamides, **2**, 1047
with fluoro(alkoxy)methylenephosphines, **6**, 700
with halides, **3**, 48
as starting materials, for ylides, **6**, 121
synthesis of, **2**, 433
Phosphines, trialkyl(aryl)-, as starting materials, for 1-aminoalkylphosphonium salts, **4**, 492
Phosphines, trialkylphenyl-, synthesis of, **2**, 435
Phosphines, 1-trialkylsilylalkyl-, reactions, with chlorophosphines, **4**, 547
Phosphines, tri(alkylthio)-, reactions, with ketones, **2**, 464
Phosphines, trialkynyl-
from
mercury (bis)alkynides, **2**, 1060
organotins, **2**, 826
synthesis of, **2**, 1059
Phosphines, triaryl-, as starting materials, for phosphonium salts, **3**, 492
Phosphines, triaryl-, symmetrical, synthesis of, **2**, 825
Phosphines, trichloromethyl-
halotropic rearrangement of, **6**, 696
as intermediates, in the synthesis of dichloromethylenephosphines, **6**, 681
Phosphines, trifluoromethyl-, as starting materials, for phosphaalkenes, **6**, 684
Phosphines, trimethylsilyl-, reactions, with phosgene, **6**, 454
Phosphines, (trimethylsilyl)aminodifluoro-, reactions, with tris(trimethylsilyl)methyllithium, **6**, 372
Phosphines, triphenyl-
polymer-supported, use in the synthesis of alkyl halides, **2**, 7
reactions
with carbon tetrahalide, **2**, 7
with *N*-halosuccinimide, **2**, 8
use in the synthesis of alkyl halides, **2**, 6
Phosphines, tris(aminomethyl)-, synthesis of, **2**, 433; **4**, 456
Phosphines, trisaminomethyl-
synthesis of, **4**, 459
use in the synthesis of α-aminophosphine sulfides, **4**, 477
Phosphines, tris(dialkylamino)-
as starting materials, for alkoxybis(dialkylamino)phosphines, **2**, 411
synthesis of, **2**, 411
Phosphines, tris(dialkylaminomethyl)-, synthesis of, **4**, 456
Phosphines, tris(dimethylaminomethyl)-
reactions, with secondary amines, **4**, 456
thermal displacement of, **4**, 453
Phosphines, tris(hydroxymethyl)-
reactions, with secondary amines, **4**, 456
synthesis of, **4**, 336
Phosphines, tris(trimethylsilyl)methyldichloro-, as starting materials, for phosphaalkynes, **5**, 1154
Phosphines, trisubstituted, reactions, with iminium ions, **4**, 478
Phosphines, trithioalkyl-, use in the synthesis of disulfides, **2**, 227
Phosphines, vinyl-
reactions, with alkyllithiums, **4**, 575
as starting materials
for alkenes, **1**, 446
for allenes, **1**, 446
synthesis of, **2**, 820
via retro-Diels–Alder reaction, **1**, 787
Phosphines, (vinyldimethylsilylmethyl)-, synthesis of, **4**, 566
Phosphine selenide, triphenyl-, synthesis of, **2**, 858
Phosphine selenides
reduction of, **2**, 441
synthesis of, **2**, 465, 858
Phosphine selenides, alkyl-, synthesis of, **2**, 858
Phosphine selenides, alkynyl-, synthesis of, **2**, 1070
Phosphine selenides, aryl-, synthesis of, **2**, 858
Phosphine selenides, tertiary, synthesis of, **2**, 465

Phosphine selenides, triaryl-, synthesis of, **2**, 858
Phosphine sulfide, synthesis of, **6**, 163
Phosphine sulfide, bis-, reactions, with red selenium, **6**, 151
Phosphine sulfide, dimethylbromo-, synthesis of, **2**, 1070
Phosphine sulfide, dimethyl-1-butynyl-, synthesis of, **2**, 1070
Phosphine sulfide, dimethylethyl-, synthesis of, **2**, 462
Phosphine sulfide, dimethylphenylethynyl-, synthesis of, **2**, 1070
Phosphine sulfide, diphenyl-, synthesis of, **2**, 851
Phosphine sulfide, tributyl-, synthesis of, **2**, 461
Phosphine sulfide, triethyl-, synthesis of, **2**, 462
Phosphine sulfides
 alkylation of, **2**, 441
 methylation of, **6**, 121
 reduction of, **2**, 440
 synthesis of, **6**, 151, 242
Phosphine sulfides, alkyl-, synthesis of, **2**, 461
Phosphine sulfides, alkyl aryl-, synthesis of, **2**, 853
Phosphine sulfides, alkynyl-
 from, alkynylphosphines, **2**, 1069
 synthesis of, **2**, 1069
Phosphine sulfides, α-aminoalkyl tertiary, synthesis of, **2**, 462
Phosphine sulfides, aryl-, synthesis of, **2**, 853
Phosphine sulfides, bis(dialkylamino)alkynyl-, synthesis of, **2**, 1069
Phosphine sulfides, dialkylhalo-, reactions, with alkynyllithiums, **2**, 1069
Phosphine sulfides, α-diazo-, synthesis of, **6**, 669
Phosphine sulfides, halodialkyl-, reactions, with alkynyllithiums, **2**, 1070
Phosphine sulfides, α-hydroxyalkyl tertiary, synthesis of, **2**, 462
Phosphine sulfides, *N*-silylated (diazomethyl)-, reactions, with nitrosyl chloride, **5**, 1131
Phosphine sulfides, tertiary
 from
  metalalkyls, **2**, 461
  tertiary phosphine oxides, **2**, 462
 synthesis of, **2**, 461
Phosphine sulfides, triaryl-, synthesis of, **2**, 853
Phosphine sulfides, vinyl-, as reagents, for polyphosphines, **2**, 853
Phosphine telluride, triphenyl-
 stability of, **3**, 389
 synthesis of, **2**, 858
Phosphine tellurides, synthesis of, **2**, 465
Phosphine tellurides, monophenyl-, synthesis of, **2**, 858
Phosphinic acetate, diphenyl-, synthesis of, **5**, 220
Phosphinic acid
 reactions, with oximes, **4**, 464
 use in the alkylation of amines, **2**, 300
Phosphinic acid, amidino-
 [2+2] cycloaddition reaction of, **6**, 366
 as intermediate, **6**, 366
Phosphinic acid, (1-aminomethyl) (chloromethyl)-, synthesis of, **4**, 124
Phosphinic acid, bis(chloromethyl)-, reactions, with thiourea, **4**, 346
Phosphinic acid, α-chloromethyl-, synthesis of, **4**, 123
Phosphinic acid, (chloromethyl)phenyl-, synthesis of, **4**, 123
Phosphinic acid, (2-cyano-1-fluoroethenyl)*n*-butyl-, hydrogenation of, **4**, 123
Phosphinic acid, di(chloromethyl)-, reactions, with ammonia, **4**, 473
Phosphinic acid, dimethyl-, from, dimethylphosphine, **2**, 454
Phosphinic acid, diphenyl-
 elimination of, **1**, 755
 synthesis of, **1**, 752
Phosphinic acid, (fluoromethyl)phenyl-, from, phenyltetrafluorophosphorane, **4**, 123
Phosphinic acid, imidetetraphenyl-, synthesis of, **2**, 864
Phosphinic acid, α-nitro-, from, 2-alkoxyalkenyl phosphinates, **4**, 490
Phosphinic acid amide, diphenyl-, as starting material, for *N*-phosphorylated imino esters, **5**, 707
Phosphinic acid amides, dialkyl-, synthesis of, **2**, 468
Phosphinic acid anhydrides, reactions, with phosphorus pentachloride, **4**, 125
Phosphinic acid chlorides
 reactions, with paraformaldehyde, **4**, 128
 thiation of, **2**, 854
Phosphinic acid esters
 optically active, synthesis of, **2**, 846
 synthesis of, **2**, 845
Phosphinic acid esters, aryl-, synthesis of, **2**, 845
Phosphinic acid esters, diaryl-, synthesis of, **2**, 846
Phosphinic acid esters, phenyl-, reactions, with alkanes, **2**, 845
Phosphinic acid *O*-esters, phenyl-, sulfenylation of, **2**, 855
Phosphinic acid halides, alkyl-, synthesis of, **2**, 453
Phosphinic acids
 from
  dialkylphosphinous acids, **2**, 454
  ω-haloalkylphosphonites, **2**, 453
  phosphorinanes, **4**, 124
 reactions
  with acid chlorides, **6**, 118
  with imines, **4**, 463
  with perfluorocarboxylic acid anhydrides, **5**, 221
 reduction of, **2**, 450
 as starting materials, for thiophosphinic chlorides, **2**, 463
 synthesis of, **2**, 843; **5**, 404
Phosphinic acids, acyl-, synthesis of, **5**, 404
Phosphinic acids, alkyl-, synthesis of, **2**, 453
Phosphinic acids, alkylaminomethyl-
 from
  alkyldichlorophosphines, **4**, 465
  α-hydroxyamides, **4**, 465
Phosphinic acids, α-amino-
 from
  alkyl phosphonites, **4**, 460
  imines, **4**, 463
  α,β-unsaturated esters, **4**, 460
 synthesis of, **2**, 454
Phosphinic acids, 1-aminoalkyl-, synthesis of, **4**, 461
Phosphinic acids, α-aminoalkyl-, from, imines, **4**, 463
Phosphinic acids, α-aminobenzyl-, synthesis of, **4**, 460, 467
Phosphinic acids, aryl-
 alkylation of, **2**, 845
 amides of, synthesis of, **2**, 860
 reduction of, for secondary phosphines, **2**, 822
 synthesis of, **2**, 843
Phosphinic acids, bis(chloromethyl)-, synthesis of, **4**, 125
Phosphinic acids, α-chloroalkyl-, synthesis of, **4**, 124
Phosphinic acids, α-chlorobenzyl-, synthesis of, **4**, 123
Phosphinic acids, (chloromethyl)-, synthesis of, **4**, 124
Phosphinic acids, chloromethyl(1-hydroxyalkyl)-, synthesis of, **4**, 124
Phosphinic acids, diaryl-
 chlorination of, for arylphosphinic chlorides, **2**, 845
 synthesis of, **2**, 843, 844
Phosphinic acids, diheteroaryl-, synthesis of, **2**, 844
Phosphinic acids, α-halo-, synthesis of, **4**, 123
Phosphinic acids, halomethyl-, as starting materials, for α-aminoalkylphosphorus compounds, **4**, 473
Phosphinic acids, *p*-methoxyphenylphenyl-, reactions, with thionyl chloride, **2**, 807
2-Phosphinic acids, 2,3-dihydropyranyl-, synthesis of, **4**, 339
Phosphinic acid salts
 chlorination of, **4**, 125
 as starting materials, for (chloromethyl)methylphosphinic chloride, **4**, 125
Phosphinic amides, optically active, synthesis of, **2**, 861
Phosphinic amides, aryl-
 hydrolysis of, for phosphinic acids, **2**, 843
 synthesis of, **2**, 860

Phosphinic amides, diaryl-, hydrolysis of, for phosphinic acids, **2**, 844
Phosphinic amides, α-halo-, synthesis of, **4**, 127
Phosphinic (2-aminophosphonic) acids, 2-amino-, synthesis of, **2**, 454
Phosphinic anhydrides, synthesis of, **2**, 455
Phosphinic anhydrides, (chloromethyl)perfluoroalkyl-, synthesis of, **4**, 128
Phosphinic anhydrides, α-halo-, synthesis of, **4**, 128
Phosphinic azide, methylphenyl-, photolysis of, **2**, 862
Phosphinic azides, diaryl-, synthesis of, **2**, 863
Phosphinic chloride, reactions, with L-proline ethyl ester, **2**, 455
Phosphinic chloride, (chloromethyl)methyl-, from, phosphinic acid salts, **4**, 125
Phosphinic chloride, diphenyl-
  reactions
    with carboxylic acids, **5**, 220
    with diphenylthiophosphinic chloride, **5**, 910
    with trifluoroacetic anhydride, **5**, 220
Phosphinic chlorides
  alcoholysis of, **4**, 126
  alkylation of, **4**, 119
  from
    dialkyltrihalophosphoranes, **2**, 455
    ethyl phosphinates, **4**, 125
  hydrolysis of, **4**, 124
  reactions
    with amines, **4**, 127
    with carboxylic acids, **5**, 234
    with Grignard reagents, **4**, 120
    with thiols, **4**, 128
  as starting materials
    for (chloromethyl)methylthiophosphinic chloride, **4**, 128
    for phosphinates, **4**, 126
    for phosphinico acrylic acids, **4**, 124
Phosphinic chlorides, alkyl chloromethyl-, synthesis of, **4**, 125
Phosphinic chlorides, aryl-
  alcoholysis of, **2**, 845
  synthesis of, **2**, 845
Phosphinic chlorides, aryl chloromethyl-, synthesis of, **4**, 125
Phosphinic chlorides, bis(chloromethyl)-
  as starting materials, for amides, **4**, 127
  synthesis of, **4**, 125
Phosphinic chlorides, 2-cyanoethyl-, synthesis of, **4**, 125
Phosphinic chlorides, dialkyl-, synthesis of, **2**, 455
Phosphinic compounds, reduction of, **2**, 430
Phosphinic compounds, alkyl-, synthesis of, **2**, 453
Phosphinic dihalides, *N,N*-dialkylamino-, synthesis of, **2**, 409
Phosphinic fluorides, aryl-, synthesis of, **2**, 845
Phosphinic halides
  reactions
    with alcohols, **2**, 454
    with sodium hydrosulfide, **2**, 462
  reduction of, **2**, 450
Phosphinic halides, alkyl-, reduction of, **2**, 430
Phosphinic halides, aryl-
  reactions, with amines, **2**, 860
  reduction of, for secondary phosphines, **2**, 822
Phosphinic halides, α-halo-, synthesis of, **4**, 125
Phosphinic pyrazolides, synthesis of, **2**, 861
Phosphinic trifluoroacetate, diphenyl-, synthesis of, **5**, 220
Phosphinidenebis(diphenylborane), aryl-, synthesis of, **2**, 867
Phosphinidene complexes, use in the synthesis of methylenephosphines, **6**, 692
Phosphinidenes, monocyano-, synthesis of, **5**, 1128
Phosphinides, chalcogeno-, reactions, with tellurium, **2**, 465
Phosphinimidate, trimethylsilylfluorodiphenyl-, synthesis of, **2**, 861
Phosphinimines
  for the deprotection of amines, **2**, 304
  reactions
    with carbon dioxide, **5**, 1079
    with α,α-dichloro isocyanates, **5**, 1068
  synthesis of, **4**, 552
Phosphinimines, bis(trimethylsilyl)-, synthesis of, **4**, 563
Phosphiniminium chloride, triphenyl-, reactions, with thiophosgene, **6**, 537
Phosphinite, diethyl phenyl-, reactions, with *N*-tetrachloroethylamides, **4**, 1002
Phosphinite, ethyl diphenyl-
  reactions
    with *N*-chloromethylamide, **4**, 487
    with chloromethyl methyl sulfide, **4**, 347
    with *N*-tetrachloroethylamides, **4**, 1000
  as starting material, for *gem*-methoxycarbonylamino diphenylphosphinoylalkenes, **4**, 1000
Phosphinites
  as starting materials, for phosphines, **2**, 436
  synthesis of, **2**, 443
    via methanolysis of phosphines, **4**, 117
  use in the synthesis of chiral phosphines, **2**, 443
Phosphinites, alkoxydialkyl-, reactions, with hydrogen cyanide, **5**, 1131
Phosphinites, alkyl-
  reactions
    with 1-haloalkylphosphorus compounds, **4**, 556
    with 1-haloalkylsilanes, **4**, 570
Phosphinites, alkynyl, rearrangement of, **1**, 993
Phosphinites, cyclic aminoethyl-, synthesis of, **2**, 859
Phosphinites, dialkyl-, reactions, with chlorodialkylphosphine, **2**, 471
Phosphinites, diethyl-, alkylation of, **2**, 435
Phosphinites, diphenyl-, alkylation of, **2**, 435
Phosphinites, trialkyl-, reactions, with imidoyl chlorides, **5**, 787
O-Phosphinoacetates, equilibrium of, **5**, 535
Phosphinoacetimidate, *O*-trimethylsilyl-*N*-difluoro-, synthesis of, **5**, 713
Phosphinoalkenes, deprotonation of, **4**, 1031
Phosphinoalkenes, *gem*-amido-, from, α-phosphino enamines, **4**, 998
Phosphinoalkenes, *gem*-chloro-, synthesis of, **4**, 997
1-Phosphinoalkenes, 1-silyl-, hydrostannylation of, **4**, 567
α-Phosphinoalkyl radicals, synthesis of, **2**, 1010
Phosphinoallene, tetrakisdiphenyl-, synthesis of, **4**, 1023
N-Phosphinoamides
  from
    amides, **5**, 368
    silylamides, **5**, 369
Phosphinoamidines, synthesis of, **5**, 775
Phosphinoarsines, synthesis of, **2**, 472, 508
Phosphino-1,3-butadiene, 1,1,4,4-tetrakisdiphenyl-
  detection of, **4**, 1023
  reactions, with molybdenum hexacarbonyl, **4**, 1023
Phosphinobutatrienes, synthesis of, **1**, 985
α-Phosphino carbanions, synthesis of, **2**, 1003
Phosphinocarbene complexes
  from, carbyne complexes, **5**, 952
  synthesis of, **5**, 952
Phosphinocarbenes
  as starting materials, for methylenephosphoranes, **6**, 704
  use in the synthesis of phosphaalkenes, **6**, 689
Phosphino–carbonyl complexes, synthesis of, **6**, 690
Phosphinocymantrenes, diphenyl-, synthesis of, **4**, 1036
Phosphinodithioic acids, alkylation of, **2**, 255
Phosphinoenamines, α-trialkyl, from, halo enamines, **4**, 998
α-Phosphinoenamines
  from
    gem-dihaloalkenes, **4**, 997
    gem-halo enamines, **4**, 997
    ketimines, **4**, 998

as starting materials, for *gem*-amido phosphinoalkenes, **4**, 998
synthesis of, **4**, 997
Phosphinoethane adducts, 1,2-(bis)dimethyl-, synthesis of, **2**, 570
Phosphinoethene, bisdiphenyl-
reactions
with cyclopentadienylbisdiphenylphosphinoruthenium chloride, **4**, 1029
with iodomethane, **4**, 1024
with Michael donors, **4**, 1028
with pentacarbonyl[(dimethyloxosulfono)-methanide]chromium, **4**, 1029
with platinum(II) chloride, **4**, 1028
Phosphinoethene, 1,1-bisdiphenyl-
alkylation of, **4**, 1023
reactions, with cycloocta-1,5-dienyldimethylplatinum, **4**, 1028
Phosphinoethene, trisdiphenyl-, synthesis of, **4**, 1023
2-Phosphinoethylamines, as starting materials, for 1,3-azaphospholine, **5**, 784
Phosphinoethyne, 1,2-bisdiphenyl-, reactions, with diphenylphosphine, **4**, 1023
Phosphinoferrocenes
as asymmetric catalysts, in Grignard cross-couplings, **4**, 1033
synthesis of, **4**, 1032
Phosphinoguanidines, imino-, synthesis of, **6**, 655
N-Phosphinoimides, from, imides, **5**, 369
Phosphinolithium, dialkyl-, reactions, with chlorosilanes, **2**, 473
Phosphinometallocenes, synthesis of, **4**, 1032
Phosphinomethanes, bischlorodiethylamino-, synthesis of, **2**, 467
Phosphinomethanes, bis(trimethylsilyl)-, metallation of, **6**, 373
Phosphinomethyllithium, bis(trimethylsilyl)dimethyl-, exchange reactions of, **6**, 374
Phosphinomethyl ylides, synthesis of, **4**, 552
2-Phosphinophospholanes, synthesis of, **4**, 547
Phosphinophosphoranes, synthesis of, **6**, 371
Phosphinophosphorus ylides, protonation of, **4**, 551
Phosphinopropene, bisdiphenyl-, synthesis of, **4**, 1022
Phosphinopropyne, 3-lithio-1-diphenyl-, reactions, with chlorodiphenylphosphine, **4**, 1023
Phosphino propyne, tetrakisdiphenyl-, non-detection of, **4**, 1023
Phosphinorutheniumcarbene complexes, rearrangement of, **4**, 582
Phosphinoruthenium chloride, cyclopentadienylbisdiphenyl-
reactions
with bischlorodicarbonylrhodium, **4**, 1029
with bisdiphenylphosphino ethene, **4**, 1029
Phosphinosilanes, synthesis of, **2**, 473
Phosphinothioacetate, *O*-difluoro-, synthesis of, **5**, 535
S-Phosphinothioacetates, equilibrium of, **5**, 535
Phosphinothioates, dimethyl-, synthesis of, **5**, 221
Phosphinothiochloridates, *O*-alkyl
nucleophilic displacement of, **5**, 1131
reactions, with potassium cyanide, **5**, 1131
Phosphinothioformamide, trimethylsilyl-, synthesis of, **6**, 584
Phosphinothioformamides
from
chlorothioamides, **6**, 584
halothioamides, **6**, 584
isothiocyanates, **6**, 581
phosphorus(III) esters, **6**, 584
thiophosphinoyldithioformates, **6**, 584
synthesis of, **6**, 583
uses of, **6**, 580
Phosphinothioformamides, chiral, synthesis of, **6**, 581
Phosphinothioformamides, diphenyl-, synthesis of, **6**, 582, 584
Phosphinothioformamides, *N*-thiophosphonate, synthesis of, **6**, 583
Phosphinothioic mixed anhydrides, dimethyl-, synthesis of, **5**, 221
Phosphinothious acid, as starting material, for α-thiophosphinoyl enamines, **4**, 1001
Phosphinothioyl chloride, dimethyl-, reactions, with peptide carboxylic acids, **5**, 221
Phosphinotins, synthesis of, **2**, 868
Phosphinous acid, aryl-, amides of, synthesis of, **2**, 859
Phosphinous acid, diphenyl-, reactions, with 4-nitrobenzonitrile, **5**, 786
Phosphinous acid amides, alkyl-, synthesis of, **2**, 466
Phosphinous acid amides, dialkyl-, synthesis of, **2**, 466
Phosphinous acid amides, monoalkyl-, synthesis of, **2**, 466
Phosphinous acid chlorides, reactions, with selenium, **2**, 857
Phosphinous acid esters, synthesis of, **2**, 436
Phosphinous acid *S*-esters, oxidation of, **2**, 854
Phosphinous acids, synthesis of, **2**, 450
Phosphinous acids, alkyl-
reduction of, **2**, 431
synthesis of, **2**, 449
Phosphinous acids, aryl-
reduction of, for secondary phosphines, **2**, 822
synthesis of, **2**, 839
Phosphinous acids, dialkyl-
reactions
with formaldehyde, **4**, 119
with nitriles, **5**, 786
with sulfur, **2**, 462
as starting materials
for phosphinic acids, **2**, 454
for α-phosphinoyl enamines, **4**, 1000
Phosphinous acids, diaryl-, synthesis of, **2**, 839
Phosphinous acids chloride, diphenyl-, reactions, with sulfinic acid, **2**, 854
Phosphinous amide, aryl-, synthesis of, **2**, 859
Phosphinous amide, diphenyl-*N*-propyl-, reactions, with benzoyl, **2**, 835
Phosphinous amides, synthesis of, **2**, 859
Phosphinous amides, *P*,*P*-diaryl-*N*,*N*-diethyl-, as starting materials, for diarylphosphinous chlorides, **2**, 834
Phosphinous bromides, diaryl-, synthesis of, **2**, 835
Phosphinous chloride, di-*t*-butyl-, reactions, with *t*-butylmagnesium chloride, **2**, 434
Phosphinous chloride, diethyl-, reactions, with methyl dithioisobutyrate, **4**, 927
Phosphinous chloride, diphenyl-
reactions
with ammonium chloride, **2**, 864
with diphenylphosphine, **2**, 864
with hexamethylcyclotrisilathiane, **2**, 851
with phenylselenomagnesium chloride, **2**, 856
with trialkylboranes, **2**, 436
Phosphinous chlorides
reactions, with alkyl Grignard reagents, **2**, 435
synthesis of, using Grignard reagents, **2**, 834
Phosphinous chlorides, aryl-, synthesis of, one-step, **2**, 834
Phosphinous chlorides, dialkyl-, reactions, with alkynyllithiums, **2**, 1060
Phosphinous chlorides, diaryl-
synthesis of, **2**, 834
one-step, **2**, 834
Phosphinous chlorides, diaryl-, unsymmetrical, synthesis of, **2**, 834
Phosphinous compounds, reduction of, **2**, 430
Phosphinous compounds, alkyl-, synthesis of, **2**, 449
Phosphinous fluoride, diphenyl-, synthesis of, **2**, 835
Phosphinous halides
metallation of, **2**, 437
reactions
with organolithiums, **2**, 435
with organometallics, **2**, 436

Phosphinous halides, alkyl-, reduction of, **2**, 430
Phosphinous halides, aryl-
  amination of, **2**, 859
  reduction of, for secondary phosphines, **2**, 822
  use in the synthesis of tertiary phosphines, **2**, 824
Phosphinous halides, diaryl-
  hydrolysis of, for diarylphosphine oxides, **2**, 840
  reactions
    with alcohol, **2**, 840
    with water, **2**, 840
  synthesis of, **2**, 834
Phosphinous iodides, diaryl-, synthesis of, **2**, 835
Phosphinoyl azides, dialkyl-, reactions, with vinyl ethers, **5**, 709
Phosphinoyl compounds, chloro-, reactions, with phosphorus pentachloride, **4**, 1013
Phosphino ylides
  reactions, with electrophiles, **4**, 551
  synthesis of, **4**, 551
Phosphinthioyl chlorides, reactions, with thiols, **2**, 252
Phosphinyl aldehydes, reactions, with diazonium salts, **6**, 659
Phosphinyl chloride, diphenyl-, reactions, with bisphosphines, **6**, 151
Phosphinyl chlorides, diaryl-, reactions, with sodium azide, **2**, 863
Phosphinyl compounds, synthesis of, **5**, 397
Phosphinyl groups, and acylphosphorus functions, chemical behavior of, **5**, 393
Phosphinyl halides, aryl-, hydrolysis of, for phosphinic acids, **2**, 843
Phosphinylidenes, reduction of, **6**, 165
Phosphinylidenes, amino-, dimerization of, **6**, 166
Phosphinylmethyllithium, bis(trimethylsilyl)dimethyl-, structure of, **6**, 373
Phosphinyl peroxide, diphenyl-
  for the oxidation of primary amines, **2**, 340
  reactions, with amines, **2**, 344, 357
Phosphinyl radicals, synthesis of, **6**, 167
Phosphirane, 1-vinyl-, as starting material, for phosphapropyne, **5**, 1158
Phosphirane oxides, synthesis of, **4**, 924
Phosphiranes, synthesis of, **2**, 430
Phosphiranes, dichloro-
  cleavage of, with organolithiums, **3**, 607
  synthesis of, **4**, 115
Phosphiranes, disilyl-, synthesis of, **4**, 567
Phosphiranes, 2-silyl-, synthesis of, **4**, 567
Phosphiranes, 2-vinyl-, reactions, with methylenephosphoranes, **5**, 908
Phosphirenes, ring-expansion of, **2**, 894
Phosphirenes, chloro-, synthesis of, **2**, 449
Phosphirenes, vinylic, reactions, with butyllithium, **1**, 985
1*H*-Phosphirenes, 1-chloro-, synthesis of, **4**, 110
2*H*-Phosphirenes, 2-chloro-, as intermediates, in the synthesis of 1-chloro-1*H*-phosphirenes, **4**, 110
Phosphiridines, synthesis of, **2**, 470
1*H*-Phosphirine, 1-chloro-, reactions, with lithium dialkyl amides, **2**, 859
Phosphite, bis(diethylamino)chloro-, reactions, with alcohols, **6**, 234
Phosphite, *t*-butyl diethyl, reactions, with ketene, **4**, 915
Phosphite, chloro-, reactions, with *N*-acetyltrifluoroacetamide, **4**, 1009
Phosphite, dialkyl
  reactions
    with phenols, **2**, 694
    with tetraaminoalkene, **6**, 156
Phosphite, dialkyl halo-, reactions, with trichloromethyl isocyanate, **6**, 285
Phosphite, diethyl
  reactions
    with aldimines, **4**, 461
    with aldimine salts, **4**, 464
    with amidinium salts, **6**, 156
    with 1-aminoalkylphosphines, **4**, 551
    with chlorodifluoronitrosomethane, **6**, 603
    with cinnamaldehyde benzoylhydrazone, **4**, 1008
    with diphenylcarbodiimide, **6**, 366
    with 1-ethoxyethyne, **4**, 913
    with paraformaldehyde, **4**, 338
    with uronium salts, **6**, 156
  as starting material
    for α-aminoalkylphosphonic acids, **4**, 464
    for bis(amino)methyl phosphonate, **6**, 156
  use in the synthesis of dichloromethyls, **4**, 13
Phosphite, diethyl chloro-
  reactions
    with aldehyde enolates, **3**, 50
    with α-hydroxy ketones, **3**, 198
    with *N*-methylthiobenzamide, **5**, 643
    with potassium phthalimide, **5**, 369
Phosphite, *S*,*S*-diethyl dithiochloro-, reactions, with *N*,*O*-bis(trimethylsilyl)acetamide, **5**, 369
Phosphite, diethyl trimethylsilyl, reactions, with cinnamaldehyde, **4**, 918
Phosphite, dimethyl
  for the aromatization of bromo dienones, **2**, 660
  reactions, with enones, **4**, 339
Phosphite, halo-
  reactions
    with acid anhydrides, **5**, 403
    with sodium salts, **5**, 403
Phosphite, hexaethyl, reactions, with amidines, **5**, 863
Phosphite, hexamethyl triamido-, as dechlorinating reagent, **6**, 696
Phosphite, methyl dicyano-, synthesis of, **5**, 1128
Phosphite, sodium diethyl, reactions, with alkynyl bromides, **2**, 1072
Phosphite, tetraethyl diamidophthalimido-, synthesis of, **5**, 369
Phosphite, thioacetyl, synthesis of, **5**, 535
Phosphite, trialkyl
  reactions
    with alkylchloroformate, **6**, 513
    with benzoyl phosphonates, **4**, 557
    with iminium chlorides, **4**, 1003
    with thiols, **1**, 942
Phosphite, tribenzyl, reactions, with *N*-chlorobenzamidines, **5**, 774
Phosphite, triethyl
  as additive, to cuprates, **1**, 465
  reactions
    with *t*-alkyl azides, **2**, 317
    with benzo-1,3-dithiole-2-thione, **6**, 314
    with benzoyl peroxide, **6**, 154
    with bis(trifluoromethyl)thioketenes, **4**, 926
    with bromotricyanomethane, **4**, 1008
    with chlorodiphenylacetonitrile, **3**, 568
    with α-chloro thiol esters, **4**, 836
    with chlorotricyanomethane, **4**, 1008
    with 1-chlorovinyl isocyanate, **4**, 1010
    with cyanoborane, **6**, 514
    with 1,2-dibromoethyl isocyanate, **4**, 1009
    with dichloromethyl isocyanate, **5**, 787
    with diphenylacetyl chloride, **4**, 920
    with diphenylketene, **4**, 914
    with episulfides, **1**, 592
    with 1-ethoxy-1,2-dichloroethene, **4**, 912
    with 1-ethoxyethyne, **4**, 913
    with furan-2-carboxaldehyde oximes, **4**, 485
    with methylthioacetone, **4**, 775
    with methylvinyl ketone, **3**, 200
    with *N*-phosphorylated ketenimines, **4**, 1008
    with tetrabromomethane, **6**, 286
    with tetrachloroethyl isocyanates, **4**, 1010

with trichloromethylamines, **6**, 366
with trichloromethyl isocyanate, **6**, 285, 366
with trichloromethyl phenyl ether, **6**, 281
with 1,1,2-trifluoro-2-chloroethanethiol, **5**, 645
with 2-trimethylammonium-1-3-dioxolane, **6**, 116
use in the synthesis of aryl isocyanides, **3**, 721
Phosphite, trimethyl
Arbuzov reaction, with phosgene, **6**, 455
reactions
with β-alkoxy alkylthalliums, **2**, 598
with azides, **2**, 317
with α-bromoglycines, **4**, 465
with 1-(*t*-butyldimethylsilyl)-2-chloro-1-ethanone, **4**, 934
with *p*-chlorocinnamoyl chloride, **4**, 922
with α-chloroimines, **5**, 789
with 1,2-dibromo-1-methoxyethane, **4**, 912
with 3,4-dicyano-1,3-dithiole-2-thione, **6**, 316
with methyl iodide, **4**, 485
with phosphorus tricyanide, **5**, 1128
for the reduction of quinones, **2**, 657
as starting material, for serines, **4**, 487
as trapping agent, **1**, 598
use in the synthesis of thiophosphates, **2**, 257
Phosphite, trimethylsilyl dialkyl, reactions, with acrylonitrile, **3**, 640
Phosphite, triphenyl
reactions
with bromine, **4**, 25
with cyclopropyllithium, **2**, 434
with methyl iodide, **2**, 8
with methylmagnesium bromide, **2**, 434
with methylthiomethyllithium, **2**, 434
with ozone, **6**, 520
Phosphite, tris(*N*-ethyl-*N*-phenylamido), reactions, with *N*,*N*-dichlorobenzamide, **5**, 379
Phosphite, tris(trimethylsilyl)
reactions
with alkyl chloroformates, **6**, 491
with imidoyl chlorides, **5**, 787
with isothiocyanates, **6**, 582
Phosphite reduction, trimethyl, of arenesulfonyl chlorides, **2**, 721
Phosphites
reactions
with acid chlorides, **6**, 118
with amide acetals, **6**, 116
with dichlorofluoronitrosomethane, **6**, 613
with polyfluoroalkenes, **4**, 803
with thiophosgene, **6**, 123
as starting materials
for *N*-alkylphosphorimidates, **2**, 407
for imidoyl-tetracoordinated phosphorus compounds, **5**, 788
Phosphites, acyloxy-, synthesis of, **5**, 221
Phosphites, alkenyl, synthesis of, **2**, 693
Phosphites, alkyl
reactions, with sulfenyl chlorides, **2**, 257
synthesis of, **2**, 104
Phosphites, alkyl dihalo-, reactions, with trichloromethyl isocyanate, **6**, 666
Phosphites, amido-, synthesis of, **5**, 788
Phosphites, aryl, synthesis of, **2**, 693
Phosphites, bis(amino)chloro-, reactions, with alkynic Grignard reagents, **2**, 1065
Phosphites, chloro-, reactions, with silylamides, **5**, 369
Phosphites, ω-chlorotrialkyl, synthesis of, **2**, 105
Phosphites, dialkyl
reactions
with acetic anhydride, **4**, 921
with aldimines, **4**, 462
with amines, **5**, 975
with aryl aldazines, **4**, 460
with carbonyls, **4**, 469, 559
with chiral nitrones, **4**, 485
with N^1,N^2-dialkylidene 1,1-diaminoalkanes, **4**, 461
with fluoroalkyl halides, **4**, 131
with hydrobenzamide, **4**, 460
with ketene *N*,*X*-acetals, **4**, 1003
with ketones, **4**, 338
as starting materials
for α-aminoalkyl phosphonates, **4**, 462
for α-aminophosphonic acids, **4**, 460
for trialkyl phosphates, **2**, 107
synthesis of, **2**, 104
use in the synthesis of radicals, **1**, 936, 938
Phosphites, dialkyl chloro-
reactions
with dichlorofluoromethyl isocyanate, **6**, 285
with secondary amines, **5**, 788
Phosphites, dialkyl halo-, reactions, with trichloromethyl isocyanate, **6**, 666
Phosphites, dialkyltrimethylsilyl, reactions, with alkyl thiocyanates, **5**, 1135
Phosphites, diaryl
reactions
with imines, **4**, 463
with oxoiminium salts, **4**, 485
Phosphites, dihalopropargyl, rearrangement of, **1**, 993
Phosphites, imidoyl, synthesis of, **5**, 788
Phosphites, monoalkyl, synthesis of, **2**, 104
Phosphites, monoaryl, synthesis of, **2**, 693
Phosphites, *N*-phthalimido bis(diethylamino)-, synthesis of, **5**, 369
Phosphites, propargyl, as starting materials, for alkynyl phosphonates, **2**, 1073
Phosphites, tetraalkyl diamidofluoro-, reactions, with tetrachloromethane, **6**, 696
Phosphites, trialkyl
alkylation of, **4**, 920
Arbuzov reaction of, **6**, 120
reactions
with alkyl chloroformates, **6**, 490
with *S*-alkylthiophthalimides, **2**, 259
with *S*-alkylthiosuccinimides, **2**, 259
with alkynyl halides, **2**, 1071
with aryliminooxindoles, **4**, 464
with bis(trimethylsilyl) peroxide, **2**, 108
with carbonimidic dichloride, *N*-phenyl-, **6**, 367
with 2,4-diisocyanatobenzyl chloride, **5**, 1009
with 1,3-diselenonium cations, **6**, 123
with 1,3-dithiole-2-thiones, **4**, 850
with episulfides, **1**, 646
with epoxides, **1**, 646
with α-halo aldehydes, **3**, 50
with 1-haloalkylphosphorus compounds, **4**, 555
with 1-haloalkylsilanes, **4**, 569
with 1-haloalkylstannanes, **4**, 588
with α-halo ketones, **3**, 196
with imidoyl chlorides, **5**, 787
with *N*-methoxymethyl arylamines, **4**, 467
with nitrones, **5**, 789
with phosphorus tricyanide, **5**, 1128
with phosphorus triisocyanate, **5**, 1008
with phosphoryl triisocyanate, **5**, 1008
with thiocarbonyls, **4**, 345
with thiolsulfonates, **2**, 199
with 2,2,2-trichloroacetic amidines, **6**, 58
with trichloromethyl isocyanate, **6**, 666
synthesis of, **2**, 105
use in the Arbuzov reaction of aryl halides, **2**, 848
Phosphites, trialkyl(aryl)
reactions, with α-haloamines, **4**, 464
as starting materials, for α-aminoalkyl phosphonates, **4**, 466
Phosphites, trialkyl, mercuric halide complexes, reactions, with ketene, **4**, 915

Phosphites, triaryl
oxidation of, **2**, 694
synthesis of, **2**, 693
Phosphites, trifluoromethylethyl, reactions, with aldehydes, **2**, 476
Phosphites, vinyl
oxidation of, **2**, 693
synthesis of, **2**, 693
Phosphite triesters, reactions, with *N,N'*-diphenylcarbodiimide, **6**, 657
Phosphoacyl functions, synthesis of, **5**, **875–922**
N-Phosphoamides, *N*-silyl-
from
N,O-bis(trimethylsilyl)acetamide, **5**, 369
monosilylamides, **5**, 369
Phosphoamines, acylation of, **5**, 370
Phosphochloridate, bis-(4-nitrobenzyl)-, stability of, **2**, 105
Phosphochloridate, bis-(2,2,2-trichloroethyl)-, as starting material, for monoalkyldihydrogenphosphates, **2**, 105
Phosphochloridate, dibenzyl, use in the synthesis of monoalkylphosphates, **2**, 105
Phosphochloridate, diphenyl, reactions, with potassium phthalimide, **5**, 372
Phosphochloridates, reactions, with β-lactams, **5**, 370
Phosphochloridates, amido-, synthesis of, **2**, 412
Phosphochloridates, dialkyl, reactions, with dialkyl phosphates, **2**, 108
Phosphochloridite, phenylene, use in the synthesis of alkenes, **1**, 693
N-Phospho compounds, synthesis of, **2**, 470
Phosphofluoridates, for the alkylation of alcohols, **2**, 108
Phosphoimines, acylation of, **5**, 371, 373
Phosphoimines, *N*-silyl-, acylation of, **5**, 373
Phospholane 1-oxides, 1-hydroxy-, synthesis of, **2**, 455
Phospholanes, synthesis of, **2**, 438, 473
Phospholanes, *p*-chloro-, synthesis of, **2**, 449
Phospholanes, dimethylfluoro-, synthesis of, **2**, 475
Phospholanes, dimethylfluoro-1,3,2-dioxy-, synthesis of, **2**, 476
Phospholanic acids, synthesis of, **2**, 455
Phosphole, 1-butyl-3,4-dimethyl-, synthesis of, **4**, 1037
Phosphole, pentaphenyl-, reactions, with lithium, **4**, 112
Phosphole, 1-phenyl-, reactions, with manganese pentacarbonyl, **4**, 1037
Phosphole, 1,2,3-triphenyl-, as starting material, for diphosphaferrocenes, **4**, 1035
Phosphole complexes, 2,3,4,5-tetraphenyl-, synthesis of, **4**, 1037
2-Phospholene 1-oxide, 1-phenyl-, reactions, with bromine, **4**, 122
Phospholenes, reactions, with lithium dibutylcuprate, **1**, 159
3-Phospholenes, *p*-chloro-, synthesis of, **2**, 449
3-Phospholene sulfides, synthesis of, **2**, 462
Phospholes
dimerization of, **2**, 472
reactions
with benzoyl chloride, **4**, 341
with dicyclopentadienyltetracarbonyldiiron, **4**, 1034
with phenyllithium, **4**, 1031
synthesis of, **4**, 1026
Phospholes, 1,3,2-dioxo-, reactions, with secondary amines, **6**, 572
Phospholes, thiooxo-, use in Diels–Alder reactions, **4**, 489
Phospholidines, synthesis of, **2**, 252
2-Phospholine, 1-phenyl-, hydrozirconation of, **4**, 547, 579
3-Phospholine oxide, 3-methyl-1-phenyl-, use in the synthesis of carbodiimides, **5**, 1078
Phospholines, trichloro-, reactions, with alkanethiols, **2**, 253
Phosphol-1-oxides, 3-keto-, synthesis of, **2**, 452
Phospholyl anion, tetraphenyl-, reactions, with dicarbonylcyclopentadienyliron iodide, **4**, 1034
1*H*-Phosphol-1-yl anions, synthesis of, **1**, 53
Phosphonacetate, trimethyl, reactions, with methyl ketones, **1**, 748
Phosphonamidate, methyl *N*-methylphenyl-, synthesis of, **2**, 862
Phosphonamidate, methyl *N*-phenylmethyl-, synthesis of, **2**, 862
Phosphonamidates, enamides of, synthesis of, **2**, 808
Phosphonamidates, *N*-*t*-butyl α-chloro-, synthesis of, **4**, 133
Phosphonamidates, α-halo-, synthesis of, **4**, 133
Phosphonamidates, 1-metalloalkyl-, reactions, with chlorophosphines, **4**, 551
Phosphonamide, *N,N*-bis(dichloromethylene)alkyl-, reactions, with formic acid, **5**, 1010
Phosphonamides
phosgenation of, **5**, 997
as starting materials, for α-aminophosphonic acid compounds, **4**, 473
use in the synthesis of alkenyl ethers, **2**, 680
Phosphonamides, *N*-acyl-, synthesis of, **5**, 371
Phosphonamides, chloromethyl-, reactions, with sodium azide, **4**, 492
Phosphonamides, 1,2-diaryl-2-hydroxyethane-, elimination of, **1**, 644
Phosphonamides, α-diazo-, as starting materials, for epoxides, **6**, 125
Phosphonamides, 2-ethoxyvinyl-, halogenation of, **4**, 804
Phosphonamides, β-hydroxy-
synthesis of, **1**, 644
via epoxide ring-opening, **1**, 644
Phosphonamides, β-keto-, reduction of, **1**, 644
Phosphonamides, α-lithio-
reactions
with aldehydes, **1**, 644
with ketones, **1**, 644
Phosphonamides, α-silyloxyalkyl-, deprotonation of, **4**, 917
Phosphonamidic chlorides, *N,N*-dialkyl α-haloalkyl-, synthesis of, **4**, 131
Phosphonamidic fluorides, *N,N*-dialkyl α-haloalkyl-, synthesis of, **4**, 131
Phosphonamidic halides, α-halo-, synthesis of, **4**, 131
Phosphonamidites, as starting materials, for imidoyl phosphonates, **5**, 788
Phosphonate, 1-alkynyl, as starting material, for enamine phosphonates, **2**, 739
Phosphonate, bis(amino)methyl, from, diethyl phosphite, **6**, 156
Phosphonate, bis(3-isocyanato-4-methylphenyl) styryl, synthesis of, **5**, 969
Phosphonate, carboxychloromethyl, as carboxychloro alkenation reagent, **5**, 103
Phosphonate, α-cyanomethyl, as starting material, for arylacetonitrile, **3**, 626
Phosphonate, diazomethyl, use in the synthesis of alkenyl ethers, **2**, 680
Phosphonate, diethyl
condensation of, **4**, 912
reactions, with arylmagnesium bromide, **2**, 842
Phosphonate, diethyl benzenesulfonylmethyl, reactions, with aldehydes, **4**, 925
Phosphonate, diethyl 1-(benzenesulfonyl)methyl, fluorination of, **4**, 79
Phosphonate, diethylbenzyl, reactions, with lithium diisopropylamide, **2**, 1003
Phosphonate, diethyl α,α-bis(methylthio)-, as starting material, for alcohols, **6**, 121
Phosphonate, diethyl α-bromomethyl, hydrolysis of, **4**, 130
Phosphonate, diethyl α-bromovinyl, reactions, with ammonia, **4**, 473
Phosphonate, diethyl *t*-butoxycyanomethyl, reactions, with sodium hydride, **3**, 650
Phosphonate, diethyl chloro-, reactions, with 2-lithio-1,3-dithiane, **6**, 120
Phosphonate, diethyl chloro(phenylseleno)methyl, synthesis of, **4**, 90

Phosphonate, diethyl cyanomethyl, reactions, with ketones, **1**, 749
Phosphonate, diethyl cyanomethylene, use in the synthesis of nitriles, **3**, 642
Phosphonate, diethyl(diazomethyl), synthesis of, **3**, 481
Phosphonate, diethyl dibromomethyl, use in Wittig–Horner reaction, **4**, 745
Phosphonate, diethyl dichloromethyl, use in Wittig–Horner reaction, **4**, 740
Phosphonate, diethyl difluoromethyl
  metallation of, **6**, 541
  reactions, with lithium diisopropylamide, **6**, 266
  synthesis of, **6**, 58
  use in Wittig–Horner reaction, **4**, 734
Phosphonate, diethyl diiodomethyl
  from
    diethyl iodomethyl phosphonate, **4**, 749
    diethyl methyl phosphonate, **4**, 749
Phosphonate, diethyl ethoxycarbonylmethyl
  reactions
    with aldehydes, **1**, 745
    with β-ionone, **1**, 748
    with ketones, **1**, 744
Phosphonate, diethyl ethyl, reactions, with diphenyl disulfide, **4**, 346
Phosphonate, diethyl 1-fluoro-1-(phenylsulfonyl)methyl, reactions, with carbonyls, **4**, 780
Phosphonate, diethyl 1-hydroxy-2,2-diphenylethenyl, synthesis of, **4**, 920
Phosphonate, diethyl iodomethyl
  reactions
    with sodium benzeneselenolate, **4**, 348
    with thioacetate, **4**, 346
  as starting material
    for diethyl diiodomethylphosphonate, **4**, 749
    for diethyl phenylselenomethylphosphonate, **4**, 784
Phosphonate, diethyl(isocyano(arylthio)methyl), use in the synthesis of 1-(arylthio)alkenyl isocyanides, , **3**, 720
Phosphonate, diethyl(isocyanomethyl), reactions, with aldehydes, **3**, 714
Phosphonate, diethyl lithiodichloromethyl, use in Wittig–Horner reaction, **4**, 740
Phosphonate, diethyl α-methoxy, reactions, with *n*-butyllithium, **6**, 119
Phosphonate, diethyl methyl
  deprotonation of, **4**, 929
  reactions, with aryl cyanides, **4**, 927
  as starting material, for diethyl diiodomethyl phosphonate, **4**, 749
Phosphonate, diethyl(methylthio)methyl
  deprotonation of, **6**, 129
  use in Peterson reaction, **4**, 925
Phosphonate, diethyl (2-oxoethyl), deprotonation of, **3**, 481
Phosphonate, diethyl phenylselenomethyl
  from, diethyl iodomethyl phosphonate, **4**, 784
  use in Wittig reaction, **4**, 784
Phosphonate, diethyl trichloromethyl
  chlorine–lithium exchange, **6**, 289
  reactions
    with alkenes, **6**, 58
    with butyllithium, **6**, 266
  use in Wittig–Horner reaction, **4**, 740
Phosphonate, diethyl (2-trimethylsilylethoxy)methyl, Peterson reaction of, **4**, 916
Phosphonate, diethyl(trimethylsilyl)methyl, deprotonation of, **6**, 129
Phosphonate, diethylvinyl
  addition of cuprates to, **4**, 585
  halogenation of, **4**, 804
  reactions
    with organomercury halides, **4**, 587
    with tetraethyl pyrophosphite, **4**, 552
Phosphonate, diisopropyl fluoromethyl, from, sodium diisopropylphosphites, **4**, 131
Phosphonate, dimethyl(diazomethyl)
  reactions
    with aldehydes, **1**, 1078; **3**, 11
    with aryl ketones, **1**, 1078
Phosphonate, dimethyl 1-methoxyvinyl, synthesis of, **4**, 912
Phosphonate, dimethyl 2-oxoheptyl, reactions, with aldehydes, **1**, 747
Phosphonate, (*E*)-dimethyl 2-phenylvinyl, reactions, with lithium diisopropylamide, **4**, 927
Phosphonate, diphenyl cyanomethyl, reactions, with benzaldehyde, **1**, 751
Phosphonate, diphenyl[(iodomethyl)phenoxyphosphinyl]methyl, synthesis of, **4**, 127
Phosphonate, formimidoyl, from, α-phosphonyl-α¹-chloroalkyl ureas, **5**, 787
Phosphonate, lithio diethyl(isocyanomethyl), reactions, with 2,5-dimethoxybenzaldehyde, **3**, 716
Phosphonate, α-(phenylseleno)methyl, reactions, with *n*-butyllithium, **6**, 136
Phosphonate, Still's fluorinated, use in alkenation reactions, **5**, 155
Phosphonate, α-sulfinyl, reactions, with phenylselenyl bromide, **6**, 123
Phosphonate, trifluoroethyl, reactions, with benzaldehyde, **1**, 752
Phosphonate anions
  elimination of, **1**, 639
  formylation of, **3**, 49
  reactions
    with amines, **5**, 603
    with benzeneselenyl bromide, **4**, 929
    with selenium, **5**, 627
  as starting materials, for ketenimines, **3**, 593
  use in the synthesis of thionoamides, **5**, 622
Phosphonate anions, alkyl, electrophilic fluorination of, **4**, 131
Phosphonate anions, α-silyl, reactions, with sulfur dioxide, **5**, 645
Phosphonate carbanions
  acylation of, **3**, 197
  reactions
    with acid compounds, **3**, 196
    with aldehydes, **1**, 750
    with carbonyls, **1**, 744
  synthesis of, **1**, 749
α-Phosphonate carbanions, synthesis of, **3**, 201
Phosphonate dichlorides, acyl, synthesis of, **5**, 403
Phosphonate diester acetals, synthesis of, **6**, 116
Phosphonate diesters, synthesis of, **2**, 456; **6**, 120
Phosphonate ester, thioethynyl, oxidation of, **2**, 1034
Phosphonate ester, α-(trimethylsilyloxy), deprotonation of, **6**, 125
Phosphonate esters
  reactions, with amines, **6**, 512
  synthesis of, **6**, 116
Phosphonate esters, alkynyl, synthesis of, **2**, 1071
Phosphonate monoamides, acyl, synthesis of, **5**, 404
Phosphonate monochlorides, acyl, synthesis of, **5**, 403
Phosphonates
  as activating groups for alkenes, **1**, 300
  deprotonation of, **1**, 158
  lithiation of, **4**, 129
  for the opening of epoxides, **1**, 212
  reactions
    with 5α-androstane-3,17-dione, **3**, 651
    with *N*-fluorobis(benzenesulfonyl)imide, **4**, 131
    with ketones, **1**, 744; **3**, 60
    with metallic cadmium, **6**, 266
    with metallic zinc, **6**, 266
    with sodium selenide, **4**, 348
  reduction of, **4**, 342
  synthesis of, **4**, 802; **6**, 313, 622, 659

use in the synthesis of alkenyl ethers, **2**, 680
use in the synthesis of *N*-alkylalkenoamides, **5**, 295
Phosphonates, 1-acetoxyvinyl, synthesis of, **4**, 921
Phosphonates, acetyl, synthesis of, **4**, 921
Phosphonates, acyl
reactions, with acylating agents, **4**, 920
reduction of, **2**, 428
Phosphonates, α-acylamino-, electrochemical oxidation of, **6**, 117
Phosphonates, (*N*-acyliminomethyl), use in Diels–Alder reactions, **4**, 489
Phosphonates, γ-(acyloxy)-β-keto, Wadsworth–Emmons condensation of, **3**, 244
Phosphonates, alkenyl, synthesis of, **2**, 849
Phosphonates, α-alkoxyalkyl, synthesis of, **4**, 340
Phosphonates, 1-alkoxyallyl, isomerization of, **4**, 918
Phosphonates, *N*-alkoxy-α-amino, from, oximes, **4**, 485
Phosphonates, alkyl
bromination of, **4**, 132
reduction of, **2**, 428
as starting materials, for alkylselenophosphonium salts, **2**, 293
Phosphonates, alkyl alkylchloro-, reactions, with mercaptides, **2**, 464
Phosphonates, alkylchloro-, synthesis of, **2**, 458
Phosphonates, alkyl hydrogen (1-benzamido-2,2-dichlorovinyl), synthesis of, **4**, 1006
Phosphonates, α-alkylthioalkyl
reactions, with alkyl halides, **4**, 347
synthesis of, **4**, 346
Phosphonates, 2-(alkylthio)vinyl, bromination of, **4**, 804
Phosphonates, alkynyl
from
alkynyl halides, **2**, 1071
propargyl phosphites, **2**, 1073
synthesis of, **2**, 849, 1070
via non-Arbuzov mechanisms, **2**, 1071
Phosphonates, allenic, reactions, with ethoxide, **3**, 200
Phosphonates, allyl
deprotonation of, **1**, 159
as starting materials, for β,γ-unsaturated nitriles, **3**, 622
Phosphonates, α-amino
diazotization of, **5**, 868
synthesis of, **4**, 476
Phosphonates, aminoalkyl
acylation of, **4**, 493
quaternization of, **4**, 493
Phosphonates, 1-aminoalkyl
from, ketimines, **4**, 463
as starting materials, for α-phosphoryl enamines, **4**, 1003
synthesis of, **4**, 466
Phosphonates, 2-aminoalkyl, synthesis of, **2**, 457
Phosphonates, α-aminoalkyl
from
aldehydes, **4**, 466
dialkyl phosphites, **4**, 462
diethyl phosphoramidate, **4**, 470
α-oxophosphonates, **4**, 471
phosphorus esters, **4**, 464
thioureas, **4**, 466
trialkyl(aryl) phosphites, **4**, 466
reactions, with perfluoro-2-azapropene, **5**, 1076
synthesis of, **2**, 457
via ionic hydrogenation, **4**, 491
Phosphonates, α-aminobenzyl, synthesis of, **4**, 461
Phosphonates, aryl
hydrolysis of, **1**, 445
reduction of, for primary phosphines, **2**, 821
synthesis of, **2**, 850
Phosphonates, *N*-arylaminomethyl, synthesis of, **4**, 467
Phosphonates, α-aryl-α-cyano, synthesis of, **3**, 626, 639
Phosphonates, (arylhydrazono)(arylazomethyl), synthesis of, **6**, 659
Phosphonates, 1-azidoalkyl, synthesis of, via Mitsunobu reaction, **4**, 491
Phosphonates, benzoyl
reactions, with trialkyl phosphite, **4**, 557
as starting materials, for α-diazobenzyl phosphonates, **5**, 404
synthesis of, **5**, 404
Phosphonates, bisamido, synthesis of, **5**, 406
Phosphonates, bis(trifluoroethyl), reactions, with aldehydes, **1**, 751
Phosphonates, bis(trimethylsilyl)methyl, metallation of, **6**, 374
Phosphonates, α-bromo-, synthesis of, **4**, 132
Phosphonates, α-bromoalkyl, synthesis of, **4**, 132
Phosphonates, 2-bromo-2-methylpropanoyl, as starting materials, for chloroformates, **4**, 921
Phosphonates, *t*-butylammonium α-diazo, synthesis of, **3**, 482
Phosphonates, *t*-butyl ester, condensation of, **1**, 749
Phosphonates, 1-cesioalkyl, synthesis of, **4**, 577
Phosphonates, chiral, synthesis of, **2**, 459
Phosphonates, chloro-, reactions, with dilithiums, **3**, 198
Phosphonates, α-chloro-, reactions, with *n*-butyllithium, **4**, 132
Phosphonates, α-chloroalkyl, from, trichloromethyl phosphonates, **4**, 131
Phosphonates, chloroethynyl, as starting materials, for ketenimines, **3**, 591
Phosphonates, α-chloro-α-isocyanomethyl
reactions
with aldehydes, **3**, 717
with ketones, **3**, 717
Phosphonates, 1-chloromethyl, azidation of, **4**, 492
Phosphonates, α-chloromethyl, as starting materials, for bis(phosphonic acids), **6**, 288
Phosphonates, 1-chlorovinyl, displacement of, **4**, 927
Phosphonates, cyano-, reactions, with β-ionone, **1**, 752
Phosphonates, α-cyano-
enolate of, synthesis of, **3**, 639
synthesis of, **3**, 639
use in the synthesis of alkenes, **1**, 748
Phosphonates, cyclopropene, as starting materials, for diazo phosphonates, **5**, 869
Phosphonates, 1-cyclopropenyl, reactions, with diazoalkanes, **3**, 481
Phosphonates, (1,1-dialkoxyethyl), synthesis of, **4**, 912
Phosphonates, dialkyl
reactions
with Grignard reagents, **2**, 840
with *N*-methylbenzimidoyl chloride, **5**, 713
as starting materials, for phosphonic acid esters, **2**, 848
Phosphonates, dialkylalkoxy, synthesis of, **6**, 513
Phosphonates, dialkyl alkyl-
from, trialkyl phosphates, **2**, 456
halogenation of, **2**, 458
metallation of, **4**, 346
reactions, with dichlorotriphenylphosphorane, **2**, 458
synthesis of, **4**, 920
Phosphonates, dialkyl α-aminoalkyl, synthesis of, **2**, 457
Phosphonates, dialkyl aryl, synthesis of, **2**, 848, 850
Phosphonates, dialkyl(bromodifluoromethyl), as starting materials, for dialkyl(iodofluoromethyl)phosphonates, **6**, 58
Phosphonates, dialkyl 1-hydroxyalkyl, synthesis of, **2**, 457
Phosphonates, dialkyl 2-hydroxyalkyl, synthesis of, **2**, 457
Phosphonates, dialkyl(iodofluoromethyl), from, dialkyl(bromodifluoromethyl)phosphonates, **6**, 58
Phosphonates, dialkyl(iodotetrafluoroethyl), dimerization of, **6**, 58
Phosphonates, dialkyl-1-isocyanomethyl, synthesis of, **3**, 706
Phosphonates, dialkylmethyl, metallation of, **6**, 374
Phosphonates, dialkylvinyl, synthesis of, **2**, 850

Phosphonates, diazo-
  from
    cyclopropene phosphonates, **5**, 869
    ynamine phosphonates, **5**, 869
  reactions, with cyclohexenol, **6**, 118
  synthesis of, **6**, 636
  use in the synthesis of α-aminophosphonates, **4**, 472
Phosphonates, α-diazo-
  cleavage of, **3**, 482
  photochemical oxidation of, **6**, 118
  reactions
    with boron trifluoride etherate, **6**, 124
    with electrophiles, **5**, 871
  solvolysis of, **6**, 119
  synthesis of, **3**, 480; **6**, 668
Phosphonates, (1-diazo-2-alken-1-yl), synthesis of, **3**, 489
Phosphonates, α-diazoalkyl
  reactions, with alcohols, **4**, 913
  for the synthesis of phosphorus-containing heterocycles, **4**, 489
Phosphonates, α-diazobenzyl, from, benzoyl phosphonates, **5**, 404
Phosphonates, diazomethyl, synthesis of, **3**, 480; **5**, 868
Phosphonates, α-diazomethyl, deprotonation of, **5**, 871
Phosphonates, dibutyl, synthesis of, **2**, 1072
Phosphonates, diethyl 1-alkoxyvinyl, synthesis of, **4**, 912
Phosphonates, diethyl alkynyl, synthesis of, **2**, 850
Phosphonates, diethyl α-bromoalkyl, synthesis of, **4**, 133
Phosphonates, diethyl α-chloroalkyl, synthesis of, **4**, 132
Phosphonates, diethyl 1-dimethylaminoalkenyl, synthesis of, **5**, 42
Phosphonates, diethyl ethoxycarbonylalkyl, reactions, with aliphatic aldehydes, **1**, 747
Phosphonates, diethyl 3-hydroxy-1-isocyanoalkyl, synthesis of, **3**, 706
Phosphonates, diethyl iodomethyl
  hydrolysis of, **4**, 130
  synthesis of, **4**, 133
Phosphonates, diethyl 1-isocyanocyclopropyl, synthesis of, **3**, 706
Phosphonates, diethyllithiobenzyl, synthesis of, **2**, 1003
Phosphonates, diethyl α-lithio-α-(trimethylsilyl)(chloromethyl), alkylation of, **4**, 131
Phosphonates, 1,1-difluoroalkyl, synthesis of, **6**, 542
Phosphonates, difluoromethylene-, synthesis of, **6**, 264
Phosphonates, diisopropyl α-bromoalkyl, synthesis of, **4**, 132
Phosphonates, diisopropylfluoromethyl, hydrolysis of, **4**, 129
Phosphonates, dimeric, synthesis of, **4**, 478
Phosphonates, dimethyl alkyl, synthesis of, **2**, 457
Phosphonates, dimethyl α-diazo, reactions, with trimethylsilyl bromide, **3**, 482
Phosphonates, dimethyl α-tosylhydrazono, cleavage of, **3**, 480
Phosphonates, diphenyl (1-benzamido-2,2-dichlorovinyl), synthesis of, **4**, 1006
Phosphonates, enamine, from, 1-alkynylphosphonate, **2**, 739
Phosphonates, ethyl, reactions, with 2-phenylpropanal, **1**, 748
Phosphonates, α-fluoro-, synthesis of, **4**, 131
Phosphonates, 2-fluoroalkenyl, dephosphorylation of, **1**, 1079
Phosphonates, α-fluoro-α-cyano-, synthesis of, **3**, 639
Phosphonates, α-fluoromethyl, treatment of, **6**, 289
Phosphonates, formyl
  reactions, with amines, **3**, 49
  as starting materials, for β-keto phosphonates, **3**, 198
Phosphonates, β-formyl, synthesis of, **3**, 49
Phosphonates, haloalkyl, reactions, with phosphine oxides, **4**, 554
Phosphonates, 1-haloalkyl
  reactions
    with palladiums, **4**, 583
    with platinums, **4**, 585
Phosphonates, α-halomethyl, reactions, with dimethyl sulfide, **4**, 346
Phosphonates, heteroaryl, synthesis of, **2**, 849
Phosphonates, α-hydroxy
  reactions, with phthalimide, **4**, 475
  synthesis of, **5**, 404
Phosphonates, β-hydroxy, synthesis of, **1**, 639
Phosphonates, α-hydroxyalkyl
  reactions, with diethylaminosulfur trifluoride, **4**, 131
  synthesis of, **4**, 343
Phosphonates, 1-*N*-hydroxyiminoalkyl, oxidation of, **4**, 491
Phosphonates, imidoyl
  from
    phosphonamidites, **5**, 788
    N-trifluoroethylidenylbenzylamine, **5**, 789
  synthesis of, **5**, 789
Phosphonates, isocyanato-, synthesis of, **5**, 977
Phosphonates, α-isocyanomethyl, use in the synthesis of α,β-unsaturated isocyanides, **3**, 717
Phosphonates, keto, synthesis of, **3**, 49
Phosphonates, α-keto
  reactions, with formate ester, **3**, 245
  use in the synthesis of α,β-unsaturated ketones, **3**, 220
Phosphonates, β-keto
  alkylation of, **3**, 197
  from
    allene oxides, **3**, 200
    formyl phosphonates, **3**, 198
  hydrolysis of, **3**, 198
  synthesis of, **3**, 196
Phosphonates, γ-keto, synthesis of, **3**, 200
Phosphonates, lithiated α-diazo-, synthesis of, **6**, 668
Phosphonates, 1-lithioalkyl, synthesis of, **4**, 576
Phosphonates, α-lithioalkyl, reactions, with esters, **3**, 220
Phosphonates, lithiomethyl, reactions, with dienoates, **3**, 229
Phosphonates, 1-magnesioalkyl, synthesis of, **4**, 577
Phosphonates, α-mercapto-, as starting materials, for thionoamides, **5**, 603
Phosphonates, metallated α-diazomethyl, alkylation of, **3**, 482
Phosphonates, 1-metalloalkyl
  reactions
    with chlorophosphines, **4**, 551
    with chlorosilanes, **4**, 570
    with chlorostannanes, **4**, 588
    with copper halides, **4**, 585
    with titanium compounds, **4**, 578
Phosphonates, α-metalloalkyl, synthesis of, **2**, 1003
Phosphonates, *N*-methoxy, synthesis of, **4**, 485
Phosphonates, methyl, synthesis of, **2**, 467
Phosphonates, methylsulfinyl, synthesis of, **4**, 347
Phosphonates, monoalkyl aryl, synthesis of, **2**, 848
Phosphonates, 1-nitroalkyl
  from, α-oxophosphonates, **4**, 491
  synthesis of, **4**, 490
Phosphonates, α-oxo
  as starting materials
    for *gem*-amido phosphorylalkenes, **4**, 1005
    for α-aminoalkyl phosphonates, **4**, 471
    for 1-nitroalkyl phosphonates, **4**, 491
Phosphonates, 2-oxoalkyl
  from, 1-chloroalkylket-2-ones, **2**, 457
  synthesis of, **2**, 457
Phosphonates, 2-oxoazetidin-4-yl, from, 4-acetoxyazetidin-2-ones, **4**, 487
Phosphonates, α-phenylthio-α-cyano-, synthesis of, **3**, 639
Phosphonates, 1-phosphinoalkyl, synthesis of, **4**, 551
Phosphonates, 1-rubidioalkyl, synthesis of, **4**, 577
Phosphonates, 1-silylalkyl, synthesis of, **4**, 569
Phosphonates, silylmethyl, synthesis of, **4**, 569

Phosphonates, α-silyloxy, reactions, with lithium diisopropylamide, **6**, 119
Phosphonates, substituted phenyl, synthesis of, **2**, 849
Phosphonates, α-sulfinyl, Pummerer rearrangement of, **6**, 119
Phosphonates, α-sulfonyl, use in the synthesis of vinyl sulfones, **2**, 715
Phosphonates, thioacyl, synthesis of, **5**, 406
Phosphonates, α-thioalkyl, synthesis of, **4**, 346
Phosphonates, α-thiocyanatoalkyl, synthesis of, **4**, 346
Phosphonates, trichloromethyl, as starting materials, for α-chloroalkyl phosphonates, **4**, 131
Phosphonates, α,β-unsaturated, addition to, **4**, 561
Phosphonates, α-unsubstituted, synthesis of, **5**, 869
Phosphonates, vinyl
  reactions
    with alkyl thiols, **3**, 166
    with organolithiums, **3**, 197
  as starting materials, for 1-bromo-2-aminoethylphosphonic acids, **4**, 130
  synthesis of, **2**, 850
  for the synthesis of phosphorus-containing heterocycles, **4**, 489
Phosphonates, ynamine, as starting materials, for diazo phosphonates, **5**, 869
Phosphonic acid, 1-amino-1-phenylethyl-, synthesis of, **4**, 473
Phosphonic acid, bis(methanethio)trimethylsilylmethyl-, diethyl ester, synthesis of, **6**, 326
Phosphonic acid, chloromethyl-, reactions, with amines, **4**, 473
Phosphonic acid, formyl methyl-, synthesis of, **3**, 49
Phosphonic acid, methylenearsonic-, synthesis of, **4**, 565
Phosphonic acid, thiocarbamoyl-, synthesis of, **6**, 582
Phosphonic acid amides, alkyl-, synthesis of, **2**, 469
Phosphonic acid amides, 2-dimethylamino-(2-hydroxymethyl)-, synthesis of, **2**, 469
Phosphonic acid amido esters, alkyl-, synthesis of, **2**, 469
Phosphonic acid amido halides, alkyl-, synthesis of, **2**, 469
Phosphonic acid anhydrides, cleavage of, **2**, 456
Phosphonic acid bisamides, β-hydroxy-
  elimination of, **1**, 643
  as starting materials, for alkenes, **1**, 644
Phosphonic acid diamides
  from, phosphonic acid dihalides, **2**, 469
  synthesis of, **2**, 469; **6**, 659
Phosphonic acid dichlorides
  hydrolysis of, **2**, 456
  synthesis of, **2**, 458
Phosphonic acid diethyl ester, α-oxo-, reactions, with N-ethoxymethyl dialkylamines, **4**, 458
Phosphonic acid dihalides, as starting materials, for phosphonic acid diamides, **2**, 469
Phosphonic acid dimethyl ester, methyl-, reactions, with oxalyl chloride, **2**, 469
Phosphonic acid esters
  dealkylation of, with iodotrimethylsilane, **2**, 847
  Michaelis–Arbuzov reaction of, for tertiary phosphine oxides, **2**, 842
  synthesis of, **2**, 848
Phosphonic acid esters, (N-alkylthiocarbamoyl)-, synthesis of, **6**, 581
Phosphonic acid esters, (arylcarbonimidoyl)bis-, synthesis of, **6**, 665
Phosphonic acid esters, N-methyliminobenzyl-, synthesis of, **5**, 713
Phosphonic acid halides, alkyl-, synthesis of, **2**, 455
Phosphonic acid methyl ester amides, α-diazo-, reactions, with trimethylsilyl bromide, **3**, 482
Phosphonic acid monochlorides, synthesis of, **2**, 458
Phosphonic acid monoesters, synthesis of, **2**, 456
Phosphonic acids
  fluorination of, for arylphosphonic difluorides, **2**, 848
  reactions, with isonitriles, **5**, 712
  as starting materials, for alkyltetrafluorophosphoranes, **2**, 476
  synthesis of, **2**, 847; **5**, 404; **6**, 242
Phosphonic acids, acyl-, synthesis of, **5**, 404
Phosphonic acids, alkenyl-, synthesis of, **2**, 837
Phosphonic acids, N-alkoxy-α-amino-, synthesis of, **4**, 485
Phosphonic acids, alkyl-, synthesis of, **2**, 455
Phosphonic acids, amino-, for herbicides, **6**, 286
Phosphonic acids, 1-amino-
  from, nitrilotriacetic acid, **4**, 492
  synthesis of, via sonochemical activation, **4**, 463
Phosphonic acids, α-amino-
  from
    N-acylimines, **4**, 486
    aldazines, **4**, 460
    aldimines, **4**, 461
    chiral phosphono alcohols, **4**, 475
    dialkyl phosphites, **4**, 460
    N-fluorosulfonylimines, **4**, 486
    N-glycosyl C-dialkoxyphosphoryl nitrones, **4**, 489
    imines, **4**, 462
    phosphonamides, **4**, 473
    α-phosphonocarboxylic acids, **4**, 476
    α,β-unsaturated esters, **4**, 460
  optically pure, synthesis of, **4**, 462
  synthesis of, **2**, 457
    via hydrolysis, **4**, 469, 476
Phosphonic acids, 1-aminoalkyl-, from, phosphorus acid, **4**, 470
Phosphonic acids, 2-aminoalkyl-, synthesis of, **2**, 459
Phosphonic acids, α-aminoalkyl-
  from
    diethyl phosphite, **4**, 464
    imines, **4**, 463
  optically active, synthesis of, **4**, 492
Phosphonic acids, ω-aminoalkyl-, synthesis of, **2**, 457
Phosphonic acids, α-aminobenzyl-, synthesis of, **4**, 467
Phosphonic acids, 1-aminomethyl-, synthesis of, via hydrolysis, **4**, 469
Phosphonic acids, α-aminomethyl-, synthesis of, via Kabachnik–Fields reactions, **4**, 469
Phosphonic acids, aryl-
  amides of, synthesis of, **2**, 861
  fluorination of, for aryltetrafluorophosphoranes, **2**, 836
  reduction of, for primary phosphines, **2**, 820
  synthesis of, **2**, 837, 847
Phosphonic acids, aziridinyl-, synthesis of, **4**, 473
Phosphonic acids, 1-benzamido-(2,2-dichlorovinyl)-, synthesis of, **4**, 1006
Phosphonic acids, α-(benzyloxycarbonylamino)benzyl-, synthesis of, **4**, 467
Phosphonic acids, 1-bromo-2-aminoethyl-, from, vinyl phosphonates, **4**, 130
Phosphonic acids, bromomethyl-, synthesis of, **4**, 130
Phosphonic acids, chiral α-aminoalkyl-, synthesis of, **2**, 459
Phosphonic acids, chiral 1-chloroalkyl-, synthesis of, **2**, 459
Phosphonic acids, chiral 1-hydroxyalkyl-, synthesis of, **2**, 459
Phosphonic acids, (R)-α-chloroalkyl-, from, (R,R)-diaminocyclohexanes, **4**, 129
Phosphonic acids, α-chloroalkyl-, synthesis of, **4**, 129
Phosphonic acids, cyclic, synthesis of, **2**, 458
Phosphonic acids, α-diazomethyl-, synthesis of, **3**, 482
Phosphonic acids, fluoromethyl-, synthesis of, **4**, 129
Phosphonic acids, N-formylaminoalkyl-, synthesis of, **4**, 488
Phosphonic acids, guanidinoalkyl-, synthesis of, **4**, 487
Phosphonic acids, α-halo-, synthesis of, **4**, 129
Phosphonic acids, halomethyl-, as starting materials, for α-aminoalkylphosphorus compounds, **4**, 473
Phosphonic acids, α-hydrazino-, from, aldazines, **4**, 460
Phosphonic acids, 1-hydroxyalkyl-, synthesis of, **2**, 456
Phosphonic acids, iodomethyl-, synthesis of, **4**, 130
Phosphonic acids, β-keto-, synthesis of, **3**, 198

Phosphonic acids, 3-oxoalkyl-, synthesis of, **2**, 456
Phosphonic acids, unsaturated, synthesis of, **2**, 847
1-Phosphonic acids, from, 3,4-dihydroisoquinolines, **4**, 464
Phosphonic amide chloride, *N*-dichloromethylenealkyl-, reactions, with formic acid, **5**, 1010
Phosphonic amides, aryl-, synthesis of, **2**, 861
Phosphonic amides, cyclic aryl- synthesis of, **2**, 862
Phosphonic bis(dimethylamide), [aryl(dimethylamino)methyl]-, deprotonation of, **2**, 740
Phosphonic chlorides, diaryl-, synthesis of, **2**, 845
Phosphonic compounds, reduction of, **2**, 428
Phosphonic compounds, alkyl-, synthesis of, **2**, 455
Phosphonic compounds, allenyl-, synthesis of, **1**, 993
Phosphonic diamide, diazo-, synthesis of, **6**, 668
Phosphonic diamides, aryl-, acidic hydrolysis of, for arylphosphonic acids, **2**, 847
Phosphonic diamides, α-halo-, synthesis of, **4**, 133
Phosphonic dichloride, alkynyl-, synthesis of, **2**, 1071
Phosphonic dichloride, α-chloroethyl-, condensation of, **4**, 132
Phosphonic dichloride, chloromethyl-, reactions, with sodium fluoride, **4**, 130
Phosphonic dichloride, fluoromethyl-, synthesis of, **4**, 130
Phosphonic dichloride, iodomethyl-, synthesis of, **4**, 131
Phosphonic dichloride, styryl-, reactions, with 3-amino-4-methylphenol, **5**, 969
Phosphonic dichlorides
  alcoholysis of, for phosphonic acid esters, **2**, 848
  from, chlorophosphonium complexes, **2**, 458
  monoesterification of, **4**, 131
  reactions
    with alcohols, **2**, 456
    with amines, **4**, 133
    with Grignard reagents, **2**, 453
  synthesis of, **2**, 458
Phosphonic dichlorides, alkenyl-, synthesis of, **2**, 848
Phosphonic dichlorides, alkyl-, reduction of, **2**, 428
Phosphonic dichlorides, aryl-, synthesis of, **2**, 847
Phosphonic dichlorides, α-chloro-, from, dichloroalkanes, **4**, 130
Phosphonic dichlorides, chloroalkyl-, synthesis of, **2**, 458
Phosphonic diester chloride, reactions, with methylene(triphenyl)phosphorane, **5**, 910
Phosphonic difluorides, aryl-, synthesis of, **2**, 848
Phosphonic dihalides
  alcoholysis of, **2**, 456
  reactions, with mercaptides, **2**, 464
Phosphonic dihalides, α-chloro-, from, *gem*-dihaloalkanes, **4**, 130
Phosphonic dihalides, α-halo-, synthesis of, **4**, 130
Phosphonic dihalides, α-haloalkyl-, synthesis of, **4**, 130
Phosphonic fluoroamides, aryl-, synthesis of, **2**, 862
Phosphonic halides, hydrolysis of, for phosphonic acids, **2**, 847
Phosphonioalkenes, *gem*-alkoxycarbonylamido-, from, tetrachloroethyl urethanes, **4**, 999
Phosphonioalkenes, *gem*-amido-
  from
    gem-amido carboxyalkenes, **4**, 998
    gem-amido phosphonioalkenes, **4**, 999
    N-tetrachloroethylamides, **4**, 999
    N-trichloroethylamides, **4**, 999
  N-methylation of, **4**, 999
  as starting materials, for *gem*-amido phosphonioalkenes, **4**, 999
  synthesis of, **4**, 998
Phosphonioalkenes, *gem*-chloroarylideneamino-, from, *gem*-amido phosphonioalkenes, **4**, 1000
Phosphoniomethanide, triphenyl-
  from, methyltriphenylphosphonium iodide, **5**, 1141
  reactions, with gold cyanide, **5**, 1141
Phosphonite, bis(trimethylsilyl) (BTSP), synthesis of, **3**, 200
Phosphonite, diethyl methyl, reactions, with α,β-unsaturated ketones, **3**, 200
Phosphonite, dimethyl phenyl, as starting material, for phenylphosphorimidates, **2**, 862
Phosphonite, ethyl phenyl
  reactions, with imines, **4**, 463
  as starting material, for α-aminoalkylphosphonites, **4**, 460
Phosphonites, reactions, with imines, **4**, 463
Phosphonites, alkyl, as starting materials, for α-aminophosphinic acids, **4**, 460
Phosphonites, alkynyl
  reactions
    with Grignard reagents, **2**, 1060
    with pyruvic acid esters, **2**, 1074
  as starting materials, for alkynylphosphines, **2**, 1060
  synthesis of, **2**, 1064
Phosphonites, α-amino-, from, *N*-benzylglycine, **4**, 466
Phosphonites, α-aminoalkyl
  from
    enamimes, **4**, 458
    methylenephosphines, **4**, 459
Phosphonites, bis(trimethylsilyl), reactions, with *N*-trityl alkanamine, **4**, 461
Phosphonites, chloro-, reactions, with imines, **4**, 486
Phosphonites, dialkyl
  reactions
    with 1-haloalkylphosphorus compounds, **4**, 556
    with 1-haloalkylsilanes, **4**, 570
    with 1-haloalkylstannanes, **4**, 588
Phosphonites, dialkyl aryl
  reactions, with alkyl halides, **2**, 845
  synthesis of, **2**, 838
Phosphonites, dialkyl phenyl, reactions, with hydrogen sulfide, **2**, 851
Phosphonites, dialkylvinyl, synthesis of, **2**, 838
Phosphonites, ethyl, reactions, with aldazines, **4**, 460
Phosphonites, ω-haloalkyl, as starting materials, for phosphinic acids, **2**, 453
Phosphonites, monoalkyl
  with palladium-catalyzed reaction of alkenyl bromides, **2**, 845
  synthesis of, **2**, 838
Phosphonium anhydrides, synthesis of, **5**, 196
Phosphonium betaine, as starting material, for alkenes, **1**, 639
Phosphonium bromide, hydrolysis of, **1**, 444
Phosphonium bromide, 9-acetoxynonyltriphenyl-, reactions, with (*E*)-pent-2-en-1-al, **1**, 734
Phosphonium bromide, allyl-, cleavage of, **1**, 49
Phosphonium bromide, benzyl[2-(*N*,*N*-dimethylamino)phenyl]diphenyl-, cleavage of, **1**, 50
Phosphonium bromide, benzyl[4-(*N*,*N*-dimethylamino)phenyl]diphenyl-, cleavage of, **1**, 50
Phosphonium bromide, (bromodifluoromethyl)triphenyl-, as starting material, for difluorocarbene, **4**, 737
Phosphonium bromide, (bromomethyl)triphenyl-, synthesis of, **4**, 118
Phosphonium bromide, (dibromofluoromethyl)triphenyl-, reactions, with iodine, **6**, 227
Phosphonium bromide, hexadecyltributyl-
  as phase-transfer catalyst, **6**, 475
  use in the synthesis of alkanoic acids, **5**, 29
Phosphonium bromide, 2-methoxybenzyltriphenyl-, reactions, with aldehydes, **1**, 732
Phosphonium bromide, (2-methyl-2-pyrrolin-3-yl) triphenyl-, *N*-alkylation of, **1**, 267
Phosphonium bromide, methyltriphenyl-, reactions, with *s*-butyllithium, **6**, 705
Phosphonium bromide, (*E*)-pent-2-entyltriphenyl-, reactions, with 9-oxonon-1-yl acetate, **1**, 734
Phosphonium bromide, prop-2-ynyl triphenyl-, reactions, with aziridines, **1**, 267
Phosphonium bromide, tetraalkyl-, reactions, with perfluorooctyl iodide, **6**, 32

Phosphonium bromide, tetrakis(bromomethyl)-, synthesis of, **4**, 118
Phosphonium bromide, triphenyl-, reactions, with phosphonium ylides, **2**, 1051
Phosphonium bromides, (*Z*)-allyltriphenyl-, reactions, with butyllithium, **1**, 735
Phosphonium bromides, bromoalkenyl-, as starting materials, for cumulene diesters, **1**, 988
Phosphonium bromides, bromotrialkyl-, reactions, with sodium fluoride, **2**, 475
Phosphonium bromides, α-halo-, synthesis of, **4**, 118
Phosphonium bromides, (α-iodopropyl)-, synthesis of, **4**, 118
Phosphonium chloride, isocyanomethyl-, use in the synthesis of metallonitrile ylides, **5**, 1089
Phosphonium chloride, tetrakis(chloromethyl)-, synthesis of, **4**, 118
Phosphonium chloride, tetrakis(hydroxymethyl)-
  as starting material
    for (chloromethyl)dimethylphosphine oxide, **4**, 119
    for tris(chloromethyl)phosphine, **4**, 113
Phosphonium chloride, triphenyl(cyanomethyl)-, synthesis of, **3**, 639
Phosphonium chloride, tris(dimethylamino)-, reactions, with monophosphazenes, **2**, 860
Phosphonium chlorides, hexacoordinated, synthesis of, **2**, 870
Phosphonium chlorides, (hydroxymethyl)-, chlorination of, **4**, 119
Phosphonium chloromethylide, triphenyl-, synthesis of, **5**, 894
Phosphonium complexes, chloro-, as starting materials, for phosphonic dichlorides, **2**, 458
Phosphonium compounds, acyl-, synthesis of, **5**, 402
Phosphonium compounds, iodo-, synthesis of, **6**, 373
Phosphonium compounds, (*Z*)-perfluorovinyl-, synthesis of, **4**, 801
Phosphonium dibromide, triphenyl-, reactions, with phosphonium ylides, **2**, 1063
Phosphonium diphenylarsinomethylide, triphenyl-, synthesis of, **5**, 911
Phosphonium fluorenylide ylide, triphenyl-, reactions, with chlorofluorocarbene, **4**, 753
Phosphonium groups
  elimination of, **1**, 640
  structural alteration of, **1**, 732
Phosphonium halides, synthesis of, **3**, 492
Phosphonium halides, benzoyl tris(dialkylamino)-, synthesis of, **5**, 402
Phosphonium hydrogen difluoride, tetraphenyl-, as starting material, for aryl fluorides, **2**, 614
Phosphonium hydrohalides, reactions, with phosphines, **2**, 445
Phosphonium hydroxides, decomposition of, **2**, 452
Phosphonium iodide
  reactions, with alkyl halides, **2**, 427
  as starting material, for phosphines, **2**, 427
Phosphonium iodide, σ-chlorophenyltriphenyl-, hydrolysis of, **1**, 445
Phosphonium iodide, methyl(triphenoxy)-, reactions, with (α-hydroxyalkyl)trimethylsilanes, **4**, 139
Phosphonium iodide, methyltriphenyl-, as starting material, for triphenylphosphoniomethanide, **5**, 1141
Phosphonium iodide, tetrakis-*t*-butyl-, synthesis of, **2**, 447
Phosphonium iodide, tetrakis(isopropyl)-, synthesis of, **2**, 446
Phosphonium iodides, synthesis of, **2**, 1063
Phosphonium iodides, (trimethylsilyloxy, from, bis(chloromethyl)phenylphosphine oxide, **4**, 122
Phosphonium ions, methylene-
  from, carbenes, **6**, 691
  synthesis of, **5**, 893; **6**, 691
Phosphonium oxides, alkynyl-
  from
    acid chlorides, **2**, 1067
    (trimethylsilyl)[bis(diisopropylamino)-phosphino]diazomethane, **2**, 1067
Phosphonium pentaiodide, hexaalkyliodophosphorane-, as intermediate, in the synthesis of trialkylphosphine oxide, **2**, 452
Phosphonium perchlorate, chlorotrisdimethylamino-, as activating agent, for the synthesis of anhydrides, **5**, 191
Phosphonium salt, aminotriphenyl-, reactions, with phenyllithium, **2**, 830
Phosphonium salts
  acidity of, **3**, 493
  as alkylating reagents, **2**, 299
  base-induced cleavage of, **2**, 441
  bromination of, **5**, 896
  chlorination of, **5**, 894
  condensation of, **4**, 925
    with sulfoxides, **5**, 900
  decomposition of, **1**, 640; **5**, 860
  dehalogenation of, **3**, 495; **6**, 695
  deprotonation of, **3**, 493
  desilylation of, **3**, 496
  electrolysis of, for the synthesis of ylides, **3**, 497
  elimination, **3**, 494
  from
    alkyl halides, **3**, 492
    P-chloro-*P*,*P*-bis(dialkylamino)-substituted ylides, **6**, 691
    1,3-dithianes, **6**, 120
    1,3-dithiolanes, **6**, 120
    halides, **3**, 492
    ketophosphines, **3**, 195
    trialkylphosphine, **3**, 492
    triarylphosphines, **3**, 492
  hydride reduction of, **2**, 442
  hydrolysis of, for tertiary phosphine oxides, **2**, 842
  as intermediates, **5**, 861
    in the synthesis of tertiary phosphines, **2**, 433
  ligand exchange, **3**, 494
  mechanisms for decomposition of, **1**, 50
  proton abstraction of, **5**, 894
  reactions
    with carbanionic compounds, **3**, 494
    with chloramine-T, **5**, 894
    with chlorinated lime, **5**, 894
    with chlorine, **4**, 344
    with dicarbonyls, **3**, 254
    with dithio lactones, **3**, 254
    with isocyanomethylenephosphoranes, **5**, 904
    with metal alkoxides, **2**, 441
    with metal hydroxides, **2**, 441
    with methyllithium, **2**, 475
    with nucleosides, **2**, 135
    with sodium bis(trimethylsilyl)amide, **6**, 702
    with tetravalent selenium compounds, **5**, 903
    with zinc, **4**, 753
  reduction of, **2**, 441
  as starting materials
    for arylazo-functionalized ylides, **5**, 904
    for carbodiphosphoranes, **5**, 1094
    for α-chloro ylides, **5**, 894
    for ketene acetals, **4**, 825
    for methylenephosphoranes, **6**, 694
    for phosphine oxides, **1**, 444
    for phosphinomethylphosphonium salts, **5**, 907
    for phosphonium ylides, **3**, 491
    for selenylmethylenephosphanes, **5**, 903
    for α-sulfonylphosphonium ylides, **5**, 901
    for sulfurimino difluorides, **2**, 400
    for thioalkoxymethylenephosphoranes, **5**, 897
  synthesis of, **1**, 157; **2**, 860, **3**, 455, 492, 639, **4**, 112, 336, 347, 571, 753, 1023, **5**, 952

thermal decomposition of, **2**, 442
transylidation of, **5**, 894
use in the synthesis of alkenes, **3**, 48
use in the synthesis of alkenyl ethers, **2**, 680
use in the synthesis of chiral phosphines, **2**, 443
use in Wittig reactions, **3**, 195
Phosphonium salts, acyl-, synthesis of, **4**, 341
Phosphonium salts, (acylmethyl)triphenyl-, reactions, with azidinium salts, **3**, 481
Phosphonium salts, α-alkoxyalkyl-, synthesis of, **4**, 340
Phosphonium salts, alkynyl-
from
alkynylbis(phosphines), **2**, 1063
alkynyl halides, **2**, 1062
alkynyl(phenyl)iodonium salts, **2**, 1062
alkynylphosphines, **2**, 1063
reduction of, **1**, 999
synthesis of, **2**, 1062
via quaternization of alkynylphosphines, **2**, 1063
Phosphonium salts, allyl-, deprotonation of, **1**, 157
Phosphonium salts, (allyl)arylalkynyl-, synthesis of, **2**, 1063
Phosphonium salts, amido-, reduction of, for tertiary phosphines, **2**, 829
Phosphonium salts, α-amino-, synthesis of, **4**, 487
Phosphonium salts, 1-aminoalkyl-
from
N-alkoxymethylamides, **4**, 492
N-hydroxymethylamides, **4**, 492
trialkyl(aryl)phosphines, **4**, 492
trimethylsilylmethyl isocyanide, **4**, 492
Phosphonium salts, α-aminoalkyl-
from, α-aminoalkylphosphines, **4**, 478
synthesis of, **4**, 462
Phosphonium salts, 1-arsinoalkyl-, synthesis of, **4**, 564
Phosphonium salts, aryl-, cleavage of, **1**, 444
Phosphonium salts, arylazo-functionalized, as starting materials, for azomethylenephosphoranes, **5**, 904
Phosphonium salts, benzoylmethyl-, reactions, with selenoxides, **4**, 928
Phosphonium salts, benzyl-, reactions, with benzaldehyde, **1**, 733
Phosphonium salts, benzyltriphenyl-, reactions, with propionaldehyde, **1**, 732
Phosphonium salts, α-bromo-
deprotonation of, **5**, 896
reactions, with organolithiums, **5**, 895
Phosphonium salts, bromodifluoromethyl-, as starting materials, for difluorocarbene, **4**, 7
Phosphonium salts, chloro-, as intermediates, in the synthesis of ylides, **5**, 895
Phosphonium salts, α-chloro-
from, iodoarene dichlorides, **5**, 897
synthesis of, **5**, 895
Phosphonium salts, (chloromethyl)-, from, tris(2,4,6-trimethoxyphenyl)phosphine, **4**, 118
Phosphonium salts, 1-cycloalkenyl-, for the synthesis of phosphorus-containing heterocycles, **4**, 488
Phosphonium salts, 1-cycloalkenyltriphenyl-, as starting materials, for 1-cycloalkenyldiphenylphosphine oxides, **2**, 843
Phosphonium salts, cyclopropyl-
nucleophilic ring-opening of, **3**, 496
as starting materials, for ylides, **3**, 496
Phosphonium salts, α-diazo-, synthesis of, **3**, 481
Phosphonium salts, diazomethyl-, synthesis of, **3**, 481
Phosphonium salts, (dichloromethyl)-, synthesis of, **4**, 112
Phosphonium salts, enol, elimination of, **1**, 1071
Phosphonium salts, (ethoxycarbonylfluoromethyl)-, synthesis of, **4**, 118
Phosphonium salts, furfuryl-, reduction of, **2**, 442
Phosphonium salts, germylmethyl-
deprotonation of, **5**, 916
as starting materials, for α-germyl ylides, **5**, 916
Phosphonium salts, α-halo-
from, phosphoranes, **4**, 117
as starting materials, for Wittig reagents, **4**, 117
synthesis of, **4**, 117
transylidation of, **3**, 499
Phosphonium salts, hydroxyalkyl-, use in the synthesis of alcohols, **2**, 84
Phosphonium salts, α-hydroxyalkyl-, synthesis of, **4**, 336
Phosphonium salts, β-hydroxyalkyl-, synthesis of, **2**, 445
Phosphonium salts, 3-hydroxypropyl-, reactions, with potassium *t*-butoxide, **1**, 268
Phosphonium salts, β-keto-, synthesis of, **3**, 195
Phosphonium salts, γ-keto-, synthesis of, **3**, 195
Phosphonium salts, mercurated, deprotonation of, **5**, 919
Phosphonium salts, α-nitroso-, as starting materials, for nitriles, **5**, 904
Phosphonium salts, organothiomethyl-, deprotonation of, **5**, 897
Phosphonium salts, 1-phosphinoalkyl-, synthesis of, **4**, 551
Phosphonium salts, phosphinomethyl-
deprotonation of, **5**, 907
from
bisphosphinomethanes, **5**, 907
phosphonium salts, **5**, 907
Phosphonium salts, quaternary
from
phosphines, **2**, 444
tertiary phosphines, **2**, 444
ylides, **2**, 446
synthesis of, via transquaternization, **2**, 446
uses of, **2**, 444
Phosphonium salts, quaternary alkyl-, synthesis of, **2**, 444
Phosphonium salts, α-selenyl, deprotonation of, **5**, 903
Phosphonium salts, 1-silylalkyl-, synthesis of, **4**, 569
Phosphonium salts, silylmethyl-
deprotonation of, **5**, 912
pyrolysis of, **5**, 913
Phosphonium salts, silyloxyalkyl-, synthesis of, **2**, 446
Phosphonium salts, stiboniomethyl-, synthesis of, **4**, 565
Phosphonium salts, tetraalkyl-, reactions, with sodium hydroxide, **2**, 452
Phosphonium salts, tetrakis(aminomethyl)-, as starting materials, for 1,3,5-diazaphosphanes, **4**, 458
Phosphonium salts, tetrakis(hydroxymethyl)-
reactions
with primary amines, **4**, 456
with sodium ethoxide, **4**, 458
Phosphonium salts, tetrakishydroxymethyl-, as starting materials, for α-aminoalkylphosphorus compounds, **4**, 475
Phosphonium salts, tetramethyl-, deprotonation of, **1**, 157
Phosphonium salts, tri-*n*-butyl-, deprotonation of, **4**, 118
Phosphonium salts, tri-*n*-butyl(fluoromethyl)-, synthesis of, **4**, 117
Phosphonium salts, trichloromethyl-, from, phosphanes, **5**, 898
Phosphonium salts, triphenyl-, as starting materials, for thiaselenafulvenes, **4**, 870
Phosphonium salts, vinyl-
hydrolysis of, **1**, 446
reactions, with ethoxide, **3**, 50
as starting materials
for alkenylidenephosphoranes, **3**, 603
for diethoxyvinylidinetriphenylphosphorane, **2**, 1004
for ylides, **3**, 496
synthesis of, **1**, 640
use in the synthesis of cyclopentanoid natural products, **4**, 925
Phosphonium ylide, tris(ethoxy)-, synthesis of, **6**, 699
Phosphonium ylides, difluoro-, reactions, with alcohols, **2**, 476
Phosphonium ylides, triphenyl-, synthesis of, **4**, 773
Phosphonoacetamides, as starting materials, for α-aminoalkylphosphorus compounds, **4**, 476

Phosphonoacetate, triethyl, use in the synthesis of α,β-unsaturated acids, **5**, 103
Phosphono alcohols, chiral, as starting materials, for α-aminophosphonic acids, **4**, 475
Phosphonoalkenes, *gem*-chloro-, synthesis of, **4**, 801
Phosphonoalkenes, *gem*-fluoro-, synthesis of, **4**, 801, 803
Phosphonoalkenes, *gem*-halo-, synthesis of, via Peterson elimination, **4**, 801
α-Phosphono carbanions, synthesis of, **2**, 1003
α-Phosphonocarboxylic acids, as starting materials, for α-aminophosphonic acids, **4**, 476
Phosphonochloridates, α-halo-, synthesis of, **4**, 131
Phosphono compounds, α-oxo, synthesis of, **5**, 870
Phosphonodichloridate, 2-chloro-2-(1-chloroethenyloxy)ethyl-, synthesis of, **4**, 766
Phosphonodithioate, *S,S*-diethyl 1-chlorovinyl, reactions, with sodium methanethiolate, **4**, 927
Phosphonodithioate, *S,S*-diethyl 1-ethylthiovinyl, synthesis of, **4**, 927
Phosphonodithioesters, α-dialkyl-, from, dithio chloroformic acid esters, **5**, 558
Phosphonodithioformate, methyl diethyl-, reactions, with ethylmagnesium bromide, **6**, 316
Phosphonodithioformates
  reactions, with alkanethiols, **6**, 316
  use in the synthesis of ketene dithioacetals, **4**, 847
Phosphonofluoridous acid esters, phenyl-, synthesis of, **2**, 839
Phosphonoketenes, *gem*-chloro-, synthesis of, **4**, 802
Phosphonothioates, dimethyl *n*-alkyl, reactions, with aldehydes, **1**, 646
Phosphonothioates, β-hydroxy, elimination of, **1**, 645
Phosphonothioic acid *O*-esters, phenyl-, isomerization of, **2**, 855
Phosphonothioic diamides, α-diazo-
  from, diazomethyllithium, **6**, 667
  synthesis of, **6**, 669
Phosphonothioic dichloride, phenyl-, reactions, with sodium fluoride, **2**, 855
Phosphonothioic difluoride, phenyl-, synthesis of, **2**, 855
Phosphonothionates, rearrangement of, **2**, 257
N-Phosphonothionoamides, synthesis of, **5**, 642
N-Phosphonothionoselenoamides, synthesis of, **5**, 642
N-Phosphonothionotelluroamides, synthesis of, **5**, 642
Phosphonous acid, aryl-, amides of, synthesis of, **2**, 859
Phosphonous acid, dibutyl-, reactions, with alkyl chlorothioformates, **6**, 497
Phosphonous acid amides, alkyl-, synthesis of, **2**, 467
Phosphonous acid amido esters, alkyl-, synthesis of, **2**, 467
Phosphonous acid diamides, alkyl-, synthesis of, **2**, 467
Phosphonous acid diamides, aryl-, synthesis of, **2**, 859
Phosphonous acid dibornyl esters, phenyl-, reactions, with organometallics, **2**, 443
Phosphonous acid diesters, oxidation of, for phosphonic acid esters, **2**, 848
Phosphonous acid dimenthyl esters, phenyl-, reactions, with organometallics, **2**, 443
Phosphonous acid ethyl ester, ethyl-, reactions, with dialkylamines, **2**, 469
Phosphonous acid halides, alkyl-, synthesis of, **2**, 450
Phosphonous acid methyl esters, reactions, with *N*-chlorodialkylamine, **2**, 469
Phosphonous acid monoalkyl esters, reactions, with Grignard reagents, **2**, 450
Phosphonous acid monoester monochlorides, aryl-, synthesis of, **2**, 838
Phosphonous acid monoesters, synthesis of, **2**, 451
Phosphonous acids
  esterification of, **2**, 451
  reactions
    with alcohols, **2**, 456
    with alkenes, **2**, 454
    with phosphorus pentasulfide, **2**, 461
  synthesis of, **2**, 451
Phosphonous acids, alkyl-, synthesis of, **2**, 450
Phosphonous acids, aminoalkyl-, synthesis of, **2**, 451
Phosphonous acids, aryl-
  disproportionation of, for primary phosphines, **2**, 821
  reduction of, for primary phosphines, **2**, 820
  synthesis of, **2**, 837
Phosphonous acids, unsaturated, synthesis of, **2**, 837
Phosphonous amide chloride, synthesis of, **2**, 859
Phosphonous amide chlorides, aryl-, synthesis of, **2**, 859
Phosphonous amides, aryl-, oxidation of, for arylphosphonic amides, **2**, 862
Phosphonous amido halides, alkyl-, synthesis of, **2**, 467
Phosphonous chlorides, reactions, with alcohols, **2**, 838
Phosphonous compounds
  reactions, with amino alcohols, **2**, 443
  reduction of, **2**, 428
Phosphonous compounds, alkyl-, synthesis of, **2**, 450
Phosphonous dibromide, phenyl-, synthesis of, **2**, 832
Phosphonous dichloride, *t*-butyl-, reactions, with ethynylmagnesium bromide, **2**, 435
Phosphonous dichloride, phenyl-
  reactions
    with 1,2-diphenyl-1,2-bis(trimethylsilyl)diphosphine, **2**, 865
    with ephedrine, **2**, 862
    with pentadienyllithium, **2**, 436
  as starting material, for tetrachlorophenylphosphorane, **2**, 836
  synthesis of, **2**, 832
Phosphonous dichlorides, reactions, with methylmagnesium halides, **2**, 435
Phosphonous dichlorides, alkenyl-
  as starting materials, for alkenylphosphonic acids, **2**, 837
  synthesis of, **2**, 833
Phosphonous dichlorides, alkyl-
  reactions
    with bis-Grignard reagents, **2**, 435
    with *o*-phenylenebis(phenylphosphine), **2**, 865
Phosphonous dichlorides, aryl-
  disproportionation of, for diarylphosphinous chlorides, **2**, 834
  fluorination of, for aryltetrafluorophosphoranes, **2**, 836
  reduction of, for primary phosphines, **2**, 821
  as starting materials, for arylphosphonic acids, **2**, 837
  synthesis of, **2**, 833
Phosphonous difluoride, phenyl-, synthesis of, **2**, 832
Phosphonous dihalides
  reactions, with organometallics, **2**, 436
  reduction of, **2**, 428
Phosphonous dihalides, alkenyl-, synthesis of, **2**, 832
Phosphonous dihalides, alkyl-, reactions, with alkyl Grignard reagents, **2**, 435
Phosphonous dihalides, aryl-, synthesis of, **2**, 832
Phosphonous dihalides, phenyl-, reactions, with alkyl Grignard reagents, **2**, 435
Phosphonous diiodide, phenyl-, as starting material, for tetrabromophenylphosphorane, **2**, 836
Phosphonous halides, reactions, with organolithiums, **2**, 435
Phosphonous halides, aryl-, use in the synthesis of tertiary phosphines, **2**, 824
Phosphonyl anions
  reactions, with iron(II) chloride, **4**, 1026, 1035
  synthesis of, **4**, 1031
Phosphonyl compounds, α-diazo, synthesis of, **5**, 868
Phosphonyl groups, and acylphosphorus functions, chemical behavior of, **5**, 393
Phosphonyl halides, alkyl-, reactions, with cyanide anions, **5**, 1131
Phosphonyl halides, aryl-, reactions, with cyanide anions, **5**, 1131
Phosphoramidate, *N*-(*t*-butoxycarbonyl)-, reactions, with alcohols, **2**, 299

Phosphoramidate, *N,N*-dibromo-, reactions, with terminal alkenes, **2**, 307
Phosphoramidate, diethyl-
as starting material
for α-aminoalkyl phosphonates, **4**, 470
for *N*-phosphorylated imino esters, **5**, 707
Phosphoramidate, diethyl *N*-(*t*-butoxycarbonyl)-, use in modified Gabriel syntheses, **2**, 299
Phosphoramidate, methyl *N,N*-dimethylphenyl-, synthesis of, **2**, 862
Phosphoramidate anions
reactions, with carbon dioxide, **5**, 974, 975
as starting materials
for isocyanates, **5**, 974
for ketenimines, **3**, 587
Phosphoramidates
from, amines, **2**, 412
reactions
with alkoxides, **2**, 108
with carbonyls, **5**, 1069
with isocyanates, **5**, 1069
with sodium hydride, **5**, 975
reduction of, **1**, 17
as starting materials, for isothiocyanates, **5**, 1031
synthesis of, **2**, 412
via hydrolysis of *N*-acyl iminophosphoranes, **5**, 371
via rearrangement of hydroxyiminophosphinates, **5**, 371
via ring-opening of phosphoroimides, **5**, 371
use in the synthesis of phosphates, **2**, 807
Phosphoramidates, *N*-acyl-, synthesis of, **5**, 371
Phosphoramidates, alkyl *N,N*-dialkyl-, synthesis of, **2**, 414
Phosphoramidates, *N*-α-chloro-*N*-β-phenylsulfenyl-, synthesis of, **4**, 106
Phosphoramidates, dialkyl *N*-alkyl-, synthesis of, **5**, 975
Phosphoramidates, *N,N*-dialkylhalo-, synthesis of, **2**, 412
Phosphoramidates, diethyl alkyl-, from, alkyl bromides, **5**, 1031
Phosphoramidates, diethyl *N*-(β-bromoalkyl)-, synthesis of, **2**, 307
Phosphoramidates, enol, reduction of, **1**, 434
Phosphoramidates, α-hydroxy, chlorination of, **4**, 106
Phosphoramide, hexamethyl- (HMPA)
as additive, to cuprates, **1**, 465
carcinogenic nature of, **1**, 1000
for the deprotonation of phenylvinyl selenides, **4**, 965
influence on addition reactions, **1**, 308
reactions, **5**, 406
with *N*-chlorosuccinimide, **5**, 379
with α,ω-dibromides, **1**, 574
with Lawesson's reagent, **5**, 575
with *n*-propylthiol, **5**, 39
with sodium, **1**, 508
with sodium cyanoborohydride, **1**, 9
as solvent
for the generation of organosilyl Grignard intermediates, **5**, 793
for the synthesis of nitriles, **3**, 613
for thionyl chloride, **5**, 10
use in conjugate addition reactions, **1**, 302
use in the opening of epoxides, **1**, 204, 225
use in the reduction of chloroalkanes, **1**, 7
use in the synthesis of alkanoic esters, **5**, 148
use in the synthesis of carboxylic esters, **5**, 141
use in the synthesis of α-silyl enamines, **4**, 1013
use in thionation reactions, **5**, 570
Phosphoramides
from, *N*-diphenylphosphinoylimines, **2**, 331
reactions
with imino chlorides, **5**, 776
with sodium hydride, **2**, 1057
synthesis of
via acylation of phosphoamines, **5**, 370
via acylation of phosphoaminosilanes, **5**, 370
via acylation of phosphoimines, **5**, 371
Phosphoramides, dialkylaminochloro-, as starting materials, for *N,N*-dialkylphosphoramidites, **2**, 410
Phosphoramides, fluoro-, synthesis of, **2**, 412
Phosphoramides, α-metalloalkyl, synthesis of, **2**, 1003
Phosphoramides, *N*-propargyl-, synthesis of, **2**, 1057
Phosphoramidic acids, *N*-acyl-, alkylation of, **2**, 808
Phosphoramidic acids, *N*-alkyl-, alkylation of, **2**, 808
Phosphoramidic compounds, dichloromethylene-, reactions, with dialkylamines, **6**, 655
Phosphoramidic dichlorides, reactions, with arylmagnesium halides, **2**, 844
Phosphoramidic difluorides, dialkyl-, synthesis of, **2**, 411
Phosphoramidic dihalides, *N,N*-dialkyl-, synthesis of, **2**, 411
Phosphoramidic esters, synthesis of, **6**, 656
Phosphoramidic esters, bis(dialkylamino)methylene-, synthesis of, **6**, 656
Phosphoramidite, di-*t*-butyl-*N,N*-diethyl-, as starting material, for di-*t*-butyl alkyl phosphates, **2**, 107
Phosphoramidite, diethyl *N*-acyl(trimethylsilyl)-, deprotection of, **5**, 369
Phosphoramidite, diethylphenyl-, reactions, with imidoyl chlorides, **5**, 787
Phosphoramidites
reactions
with aldehydes, **4**, 475
with ketones, **4**, 475
as starting materials
for α-aminoalkylphosphorus compounds, **4**, 475
for phosphates, **2**, 106
synthesis of, **2**, 106, 410
use in the synthesis of thiophosphines, **2**, 253
Phosphoramidites, acyl, oxidation of, **5**, 373
Phosphoramidites, *N,N*-dialkyl-
from
bis(dialkylamino)alkoxyphosphines, **2**, 410
bistriazolylalkoxyphosphines, **2**, 410
dialkylaminochlorophosphoramides, **2**, 410
synthesis of, **2**, 410
Phosphoramidites, dialkyl acyl
reactions
with 1,4-benzoquinone, **5**, 373
with chloranil, **5**, 373
with1,4-napthaquinone, **5**, 373
Phosphoramidites, *N,N*-dialkylchloro-, synthesis of, **2**, 409
Phosphoramidites, *N,N*-dialkylhalo-, synthesis of, **2**, 409
Phosphoramidites, *N*-phenyl-, reactions, with benzaldehyde, **4**, 476
Phosphoramidochloridate, *N*-phenyl-, as activating agent, for the synthesis of anhydrides, **5**, 189
Phosphoramidothioic acids, synthesis of, **6**, 656
Phosphorane, 1-adamantylmethylidene-, attempted reactions, with tosyl fluoride, **5**, 901
Phosphorane, allylidenetriphenyl-, reactions, with α,β-unsaturated β-chlorocarbonyls, **1**, 736
Phosphorane, arylsulfonylmethylenetriphenyl-, reactions, with *p*-carboxybenzenesulfonyl azide, **3**, 479
Phosphorane, benzotriazoloimino-, reactions, with Grignard reagents, **5**, 1067
Phosphorane, benzotriazolylmethylimino-, reactions, with Grignard reagents, **5**, 1031
Phosphorane, benzylidenetriphenyl-
reactions
with indole-3-aldehyde, **1**, 733
with α,β-unsaturated aldehydes, **1**, 733
Wittig reactions of, **1**, 733
Phosphorane, bis(η^5-cyclopentadienyl)chlorozirconylmethylenetriphenyl-, synthesis of, **6**, 706
Phosphorane, bis(diisopropylamino)chlorodiazomethylene-, reactions, with boron trifluoride etherate, **3**, 485

Phosphorane, bis(methylene)chloro-, synthesis of, **6**, 693
Phosphorane, bis(phenylseleno)methylene-, synthesis of, **6**, 697
Phosphorane, bromomethylenetriphenyl-, use in the synthesis of alkynes, **1**, 1064
Phosphorane, *n*-butylidenetriphenyl-, reactions, with methylheptenone, **1**, 730
Phosphorane, *t*-butyltetraiodo-, synthesis of, **2**, 476
Phosphorane, *P*-chloro(methylene)-
  as starting material, for phosphaalkenes, **5**, 888
  thermolysis of, **5**, 888
Phosphorane, chloromethyltetrachloro-, synthesis of, **2**, 476
Phosphorane, 1-chlorovinylimino-, reactions, with amines, **5**, 861
Phosphorane, cyanomethylenetriphenyl-
  reactions
    with acid chlorides, **3**, 674
    with 1,2-diketones, **3**, 642
    with succinimides, **3**, 658
  use in the synthesis of nitriles, **3**, 642
Phosphorane, cyclopentadienylidenetriphenyl-, reactions, with dodecacarbonyltriruthenium, **4**, 685
Phosphorane, diazomethylene-
  reactions
    with benzaldehyde, **2**, 1068
    with borane, **5**, 872
    with boron trifluoride, **5**, 872
Phosphorane, dibromomethylenetriphenyl-, use in the synthesis of alkynes, **1**, 1067
Phosphorane, dibromotriphenyl-
  reactions
    with ketene aminals, **4**, 990
    with secondary amides, **3**, 557
  use in the synthesis of ketenimines, **3**, 563
Phosphorane, 2,4-di-*t*-butyl-6-(dimethylamino)phenyldiselenoxy-, deselenation of, **2**, 857
Phosphorane, dichlorocyanodimethyl-, synthesis of, **5**, 1131
Phosphorane, dichlorodiphenyl-, reactions, with methylenes, **3**, 499
Phosphorane, dichloromethylene-, acylation of, **5**, 895
Phosphorane, dichloromethylenetriphenyl-, use in the synthesis of 2-aryl-3,3-dichloroacrylonitriles, **3**, 649
Phosphorane, dichlorotriphenyl-
  reactions, with dialkyl alkylphosphonates, **2**, 458
  synthesis of, **2**, 835
  use in the synthesis of ketenimines, **3**, 558
Phosphorane, diethoxyethenylidinetriphenyl-
  reactions
    with 1,2-diketones, **6**, 75
    with ethanol, **6**, 75
Phosphorane, 2,2-diethoxyvinylidenetriphenyl-, reactions, with ketone, **3**, 245
Phosphorane, diethoxyvinylidinetriphenyl-, from, vinyl phosphonium salts, **2**, 1004
Phosphorane, diflourotriphenyl-, synthesis of, **2**, 835
Phosphorane, difluoromethylenetributyl-, synthesis of, **6**, 694
Phosphorane, difluorotrimethyl-
  reactions, with trimethylsilylmethylene trimethylphosphorane, **5**, 911
  as starting material, for fluorotrimethylsilane, **5**, 1093
  synthesis of, **6**, 700
Phosphorane, difluorotris(trifluoromethyl)-, use in the synthesis of difluorocarbene, **4**, 7
Phosphorane, dimethyltrichloro-, reactions, with ammonium chloride, **2**, 468
Phosphorane, diphenyltrifluoro-, reactions, with lithium bis(trimethylsilyl)amide, **2**, 861
Phosphorane, diphenyltris(2,2,2-trifluoroethoxy)-, reactions, with sodium 2,2,2-trifluoro ethoxide, **2**, 870
Phosphorane, ethoxycarbonylmethylenetriphenyl-, acylation of, **1**, 1079
Phosphorane, ethoxycarbonylmethylidenetriphenyl-
  reactions
    with aldehydosugars, **1**, 725
    with 1-hydroxy-4-oxo-α-ionone, **1**, 726
Phosphorane, ethylidene-, sulfonylation of, **5**, 901
Phosphorane, ethylidenetriphenyl-
  reactions
    with 2,2-dimethylbut-3-enal, **1**, 729
    with pivalaldehyde, **1**, 729
Phosphorane, fluorenylidenetriphenyl-, reactions, with aldehydes, **1**, 723
Phosphorane, hexylidenetriphenyl-, reactions, with *endo*-aldehydes, **1**, 729
Phosphorane, imidoyl-, as reactive intermediate, for the synthesis of imidoyl phosphonates, **5**, 790
Phosphorane, iodomethylenetriphenyl-
  from, ylides, **5**, 896
  synthesis of, *in situ*, **5**, 896
Phosphorane, α-lithiomethylenetriphenyl-, synthesis of, **6**, 705
Phosphorane, 4-methoxybenzylidenetris(*p*-chlorophenyl)-, use in the synthesis of stilbenes, **1**, 732
Phosphorane, methoxycarbonylethylidenetriphenyl-, reactions, with acetaldehyde, **1**, 725
Phosphorane, methoxymethylenetriphenyl-
  acylation of, **5**, 897
  reactions
    with acyl chloride, **5**, 897
    with ethyl carboxylate, **5**, 897
  synthesis of, *in situ*, **5**, 897
  transylidation of, **5**, 897
Phosphorane, 1-methyl-, reactions, with 1,4-dibromobutane, **2**, 444
Phosphorane, methylenetrimethyl-
  reactions, with dichlorodimethylsilane, **6**, 702
  synthesis of, **3**, 496
Phosphorane, methylene(triphenyl)-
  condensation of, **3**, 364
  reactions
    with alkynyl bromides, **1**, 1016
    with benzaldehyde, **1**, 639
    with 3-borahomoadamantane, **2**, 447
    with di-*t*-butylimidomolybdenum dichloride, **5**, 918
    with tri-*t*-butylimidorhenium chloride, **5**, 919
    with chlorodiphenylarsane, **5**, 911
    with chlorotrimethylstannanes, **5**, 920
    with 1,2-dichlorotetramethyldisilane, **6**, 703
    with dipyridinedi-*t*-butylimidotungsten dichloride, **5**, 918
    with di-*t*-imidochromium dichloride, **5**, 918
    with ketones, **1**, 759
    with pentacarbonyl bromorhenium, **5**, 919
    with phosphonic diester chloride, **5**, 910
    with selenocarboxylates, **5**, 521
    with *N*-thionoacylformamidines, **5**, 633
  as starting material
    for disubstituted ylides, **6**, 701
    for 1-phosphacumulenes, **3**, 604
Phosphorane, methylenetris(dimethylamino)-, reactions of, **6**, 706
Phosphorane, methyltetrachloro-
  reactions
    with ammonium chloride, **2**, 468
    with quaternary ammonium chlorides, **2**, 477
Phosphorane, methyltetrafluoro-, reactions, with dimethylamine, **2**, 477
Phosphorane, methylthio(thiocarbonyl)methylenetriphenyl-, reactions, with diazonium salts, **5**, 906
Phosphorane, methyltriphenyl-, reactions, with lithium alkyls, **6**, 705
Phosphorane, monosilylmethylene-, reactions, with stannyl chlorides, **6**, 702

Phosphorane, neopentylidene-, attempted reactions, with tosyl fluoride, **5**, 901
Phosphorane, 4-nitrobenzylidenetriphenyl-, reactions, with anisaldehyde, **1**, 731
Phosphorane, pentaalkoxy-, as alkylating agent, **2**, 683
Phosphorane, pentachloro-, reactions, with isopropyllithium, **2**, 434
Phosphorane, pentaethoxy-, as alkylating agents, **2**, 674
Phosphorane, pentaphenyl-, synthesis of, **2**, 830
Phosphorane, phenacylidenetriphenyl-, reactions, with *p*-nitrobenzaldehyde, **1**, 724
Phosphorane, phenoxytetraphenyl-, synthesis of, **2**, 832
Phosphorane, phenyldimethylaminotrifluro-, as starting material, for phenylbis(dimethylamino)fluorophosphonium phenylpentafluorophosphate, **2**, 869
Phosphorane, 1,2-phenylenedioxytribromo-, use in the synthesis of thiobenzoyl bromide, **5**, 486
Phosphorane, phenyltetrafluoro-
  reactions
    with perthiophosphonic anhydrides, **2**, 855
    with *o*-phenylenedithiabis(trimethylsilane), **2**, 856
  as starting material, for (fluoromethyl)phenylphosphinic acid, **4**, 123
Phosphorane, phenyltetrakis(2,2,2-trifluoroethoxy)-, reactions, with sodium 2,2,2-trifluoroethoxide, **2**, 870
Phosphorane, *N*-pyridylimino-, as starting material, for *N*-(2-pyridyl)ketenimines, **3**, 587
Phosphorane, spiropentaoxy-, reactions, with pyridine, **2**, 869
Phosphorane, tetrabromophenyl-, synthesis of, **2**, 836
Phosphorane, tetrachlorophenyl-, synthesis of, **2**, 836
Phosphorane, tetraoxy-, synthesis of, **2**, 476
Phosphorane, tribromodiphenyl-, synthesis of, **2**, 836
Phosphorane, tributyldiethoxy-, synthesis of, **2**, 476
Phosphorane, 2,4,6-tri-*t*-butylphenyl-, use in the synthesis of 1,3-diphosphapropene, **5**, 1058
Phosphorane, trichlorodiphenyl-, synthesis of, **2**, 836
Phosphorane, trifluorodiphenyl-, synthesis of, **2**, 835
Phosphorane, triiododiphenyl-, synthesis of, **2**, 836
Phosphorane, trimethyl-, reactions, with rhenium carbyne complexes, **5**, 919
Phosphorane, trimethylbis(trimethylsilyl)methylene-, use in the synthesis of 1,1-bis(silyl)alkanes, **4**, 624
Phosphorane, *N*-trimethylsilyliminotriphenyl-, reactions, with *β*-dicarbonyls, **2**, 745
Phosphorane, (trimethylsilylmethylene)dimethyl phenyl-, reactions, with isothiocyanates, **5**, 645
Phosphorane, trimethylsilylmethylene trimethyl-, reactions, with difluorotrimethylphosphorane, **5**, 911
Phosphorane amides, imido-, synthesis of, **5**, 861
Phosphorane chlorides, imido-, synthesis of, **5**, 860
Phosphorane complexes, imino-, synthesis of, **4**, 800
Phosphoranes
  carbonation of, **5**, 43
  reactions
    with acid halides, **3**, 274
    with acyl halides, **1**, 986
    with alkoxycarbenechromium complexes, **1**, 987
    with benzophenone, **1**, 763
    with carbon dioxide, **1**, 987
    with carbon disulfide, **4**, 843
    with diazo compounds, **3**, 445
    with phthalic anhydride, **2**, 1064
  reduction of, **1**, 52
  resonance-stabilized, **1**, 157
  as starting materials
    for 1,3-aminophosphoranes, **4**, 478
    for α-halophosphonium salts, **4**, 117
    for phosphonium ylides, **3**, 499
  sulfur-containing spirocyclic, synthesis of, **2**, 856
  synthesis of, **2**, 435, 475, 830; **3**, 674; **4**, 825
  use in the synthesis of 1,5-dienes, **1**, 157
  use in the synthesis of α,β-unsaturated aldehydes, **3**, 60
Phosphoranes, acyl-
  O-alkylation of, **4**, 824
  reactions
    with azides, **3**, 476
    with cyanogen azide, **3**, 476
Phosphoranes, *N*-acylimino-
  from
    acyl azides, **5**, 373
    amides, **5**, 373
  hydrolysis of, **5**, 371
  synthesis of
    via acylation of phosphoimines, **5**, 373
    via acylation of *N*-silylphosphoimines, **5**, 373
    via oxidation of acyl phosphoramidites, **5**, 373
Phosphoranes, acylmethylenetriphenyl-, deprotonation of, **3**, 603
Phosphoranes, alkenylidene-
  from, vinyl phosphonium salts, **3**, 603
  See also 1λ^5-Phosphaallenes
Phosphoranes, 1-alkoxyallyltriphenyl-, isomerization of, **4**, 918
Phosphoranes, alkoxycarbonylimino-
  reactions, with phosphorus pentachloride, **5**, 988
  as starting materials, for isocyanates, **5**, 988
Phosphoranes, alkoxycarbonyliminotrichloro-, as intermediates, **5**, 988
Phosphoranes, [(alkoxycarbonyl)methylidene]-, reactions, with acyl halides, **1**, 986
Phosphoranes, alkyl(dichloro)diphenyl-, synthesis of, **4**, 112
Phosphoranes, alkylidene-
  reactions
    with ketenes, **1**, 986
    with trifluoroacetamides, **2**, 751
  as starting materials, for dialkylborylalkylidenetriphenylphosphoranes, **6**, 704
Phosphoranes, (alkylidene)chloro-, 1,2-chlorotropic rearrangement of, **4**, 112
Phosphoranes, alkylidenetriphenyl-, synthesis of, **4**, 348
Phosphoranes, alkyltetrachloro-
  hydrolysis of, **2**, 456
  reduction of, for alkenylphosphonous dichlorides, **2**, 833
Phosphoranes, alkyltetrafluoro-, from, phosphonic acids, **2**, 476
Phosphoranes, alkylthiomethylene-, from, organothiomethylenephosphoranes, **5**, 898
Phosphoranes, *N*-(alkylthiomethyl)imino-, as intermediates, **4**, 329
Phosphoranes, alkynyl-
  from, fluoro(amino)alkynylphosphines, **2**, 1073
  synthesis of, **2**, 1073
Phosphoranes, 1,3-amino-
  from
    phosphoranes, **4**, 478
    Schiff bases, **4**, 478
    vinylphosphoranes, **4**, 478
Phosphoranes, aminomethylene-, *ab initio* calculations, **6**, 700
Phosphoranes, *C*-aminomethylene-, synthesis of, **6**, 700
Phosphoranes, α-arsino methylene-, from, dimethylarsinotrimethylsiloxane, **5**, 911
Phosphoranes, aryl-, with sulfur function, synthesis of, **2**, 856
Phosphoranes, (arylalkylidene)triphenyl-, reactions, with chlorodifluoromethane, **4**, 735
Phosphoranes, arylazo-, as starting materials, for nitrogen-functionalized ylides, **5**, 906
Phosphoranes, arylazomethylenetriphenyl-, from, hydrazonoyl chlorides, **5**, 905
Phosphoranes, arylchloro-, reactions, with iodotrimethylsilane, **2**, 837
Phosphoranes, arylhalo-, synthesis of, **2**, 835
Phosphoranes, aryliodo-, synthesis of, **2**, 837
Phosphoranes, aryltetrachloro-, as starting materials, for arylpentahalophosphates, **2**, 869

Phosphoranes, aryltetrafluoro-
as starting materials, for arylpentahalophosphates, **2**, 869
synthesis of, **2**, 836
Phosphoranes, *N*-aziridinylimino-, as starting materials, for *N*-aminoketenimines, **3**, 586
Phosphoranes, azomethylene-, from, arylazo-functionalized phosphonium salts, **5**, 904
Phosphoranes, benzoylimino-, synthesis of, **5**, 708
Phosphoranes, bicyclictetraoxy-, synthesis of, **2**, 476
Phosphoranes, bis(dialkylamino)bromo-*N*-alkylimino-, from, tris(dialkylamino)dibromophosphoranes, **2**, 408
Phosphoranes, bis(methylene)-, tetrakis(trimethylsilyl)-substituted, synthesis of, **6**, 692
Phosphoranes, bis(phenylthio)methylene-, synthesis of, **6**, 697
Phosphoranes, *gem*-bromoimino-, from, α-bromomalononitrile, **4**, 800
Phosphoranes, bromomethylene-
reactions, with *t*-butyllithium, **6**, 705
synthesis of, **5**, 895
Phosphoranes, (α-chlorovinyl)tetrachloro-, synthesis of, **4**, 804
Phosphoranes, cyanomethylene-, deprotonation of, **5**, 559
Phosphoranes, dialkoxy-, synthesis of, **2**, 830
Phosphoranes, dialkyl-, synthesis of, **2**, 476
Phosphoranes, dialkylborylalkylidenetriphenyl-
from, alkylidenephosphoranes, **6**, 704
synthesis of, **6**, 704
Phosphoranes, dialkyltrihalo-, as starting materials, for phosphinic chlorides, **2**, 455
Phosphoranes, dibromo-, reactions, with benzaldehyde, **4**, 745
Phosphoranes, dibromomethylene-, synthesis of, **6**, 695
Phosphoranes, dichloro-
reactions, with amines, **2**, 468
as starting materials, for α-chloroalkylphosphines, **4**, 112
synthesis of, **4**, 112
Phosphoranes, dichloromethylene(amino)difluoro-, synthesis of, **6**, 696
Phosphoranes, dichloromethylenebis(amino)fluoro-, synthesis of, **6**, 696
Phosphoranes, dichloromethylenetriphenyl-, synthesis of, **6**, 695
Phosphoranes, dichlorophosphanylmethylene-, thiolysis of, **5**, 907
Phosphoranes, difluoromethylene-, synthesis of, **6**, 695
Phosphoranes, difluorooxy-, synthesis of, **2**, 476
Phosphoranes, dihalo-, hydrolysis of, **2**, 452
Phosphoranes, *C*,*C*-dihalomethylene-, synthesis of, **6**, 694
Phosphoranes, dihalomethylenetriphenyl-, synthesis of, **6**, 695
Phosphoranes, dihalotrialkyl-, reactions, with hydrogen sulfide, **2**, 461
Phosphoranes, dihalotriphenyl-, synthesis of, **2**, 835
Phosphoranes, diiodomethylene-, synthesis of, **6**, 695
Phosphoranes, dimethylamino bicyclic, reactions, with catechol, **2**, 477
Phosphoranes, diphenoxy-, synthesis of, **2**, 831
Phosphoranes, diselenoxo-, as starting materials, for selenadiphosphiranes, **2**, 857
Phosphoranes, dithiooxophosphanylmethylene-, from, disodium dichloride, **5**, 908
Phosphoranes, β-ethoxyallylidene-, as starting materials, for 1-phosphabutatrienes, **3**, 604
Phosphoranes, [(ethoxycarbonyl)methylidene]-, as starting materials, for ethyl allenyl carboxylates, **1**, 986
Phosphoranes, ethylidenetriphenyl-, reactions, with acetophenones, **1**, 730
Phosphoranes, fluoro-, reactions, with *O*-silylated carbamates, **2**, 476
Phosphoranes, *P*-fluoro-, synthesis of, **4**, 803
Phosphoranes, formyl-
reactions
with acid chlorides, **3**, 79
with aldehydes, **3**, 60, 65
Phosphoranes, *o*-formylarylazomethylene-, intramolecular cyclization of, **5**, 906
Phosphoranes, formylmethylene-, halogenation of, **3**, 68
Phosphoranes, halo-, as starting materials, for *C*-phosphinoylketenimines, **3**, 592
Phosphoranes, α-halo-
reactions, with carbonyls, **3**, 649
synthesis of, **4**, 133
Phosphoranes, haloformylmethylene-, synthesis of, **3**, 68
Phosphoranes, hydrazonimido-, from, hydrazonazides, **5**, 862
Phosphoranes, imino-
from, *o*-azidobenzamides, **5**, 1066
homologation of, **5**, 1067
reactions
with acylnitriles, **3**, 478
with aryl isocyanates, **3**, 194; **5**, 1067
with carbon dioxide, **5**, 974
with carbonyls, **4**, 518
with isocyanates, **5**, 1066
as starting materials
for isocyanates, **5**, 974
for isothiocyanates, **5**, 1031
for ketenimines, **3**, 584
synthesis of, **5**, 975; **6**, 651
uses of, **5**, 974
use in the synthesis of thionoamides, **5**, 622
See also Imides, pohosphine
Phosphoranes, imino(thioxo)-, synthesis of, **6**, 702
Phosphoranes, iminotriphenyl-, reactions, with ketenes, **3**, 584
Phosphoranes, iodo-, reactions, with aldehydes, **3**, 274
Phosphoranes, isocyanomethylene-, reactions, with phosphonium salts, **5**, 904
Phosphoranes, ketene-, from, diphenylcyclopropenone, **3**, 540
Phosphoranes, keteniminylidene-
from, isothiocyanates, **3**, 605
as starting materials, for ketenimines, **3**, 581
Phosphoranes, ketenylidene-, synthesis of, **3**, 603
Phosphoranes, ketenylidenetriphenyl-, reactions, with Grignard reagents, **3**, 220
Phosphoranes, α-keto-
reactions
with aldehydes, **3**, 220
with α,β-unsaturated aldehydes, **3**, 220
Phosphoranes, β-ketoalkylidene-, pyrolysis of, **3**, 674
Phosphoranes, β-ketoalkylidenetriphenyl-, pyrolysis of, **3**, 674
Phosphoranes, α-methoxybenzylidene-, as reactive intermediates, **5**, 897
Phosphoranes, α-methoxyethylidene-, as reactive intermediates, **5**, 897
Phosphoranes, methylene-
chemistry of, **6**, 694
effect of heterosubstituents on stability, **6**, 694
from
bis(arenesulfonyl)methylphosphine oxides, **6**, 699
α-haloalkylphosphines, **6**, 694
iminophosphines, **6**, 693
methylenephosphines, **6**, 693
oragnodichlorophosphines, **6**, 693
phenyliodonium ylides, **6**, 699
phosphinocarbenes, **6**, 704
phosphonium salts, **6**, 694
lithium-carbon, synthesis of, **6**, 705
reactions
with 1,3-disiletanes, **6**, 702
with hexamethylsiliranes, **6**, 702
with siletanes, **6**, 702

with 2-vinylphosphiranes, **5**, 908
synthesis of, **5**, 894
with a transition metal–carbon bond, synthesis of, **6**, 706
Phosphoranes, $\sigma^3\lambda^5$methylene-, synthesis of, **6**, 692
Phosphoranes, methylene-, antimony-substituted, synthesis of, **6**, 700
Phosphoranes, methylene-, arsenic-substituted, synthesis of, **6**, 700
Phosphoranes, methylene-, boron-substituted ylides, synthesis of, **6**, 702
Phosphoranes, methylene-, *C*-gallyl-substituted, synthesis of, **6**, 704
Phosphoranes, methylene-, germanium-substituted ylides, synthesis of, **6**, 702
Phosphoranes, methylene(imino)-, synthesis of, **6**, 692, 693
Phosphoranes, methylene-, metal-substituted, synthesis of, **6**, 705
Phosphoranes, methylene-, nitrogen-substituted, synthesis of, **6**, 700
Phosphoranes, methylene(oxo)-, synthesis of, **6**, 694
Phosphoranes, methylene-, oxygen-substituted, synthesis of, **6**, 697
Phosphoranes, methylene-, phosphorus-substituted, synthesis of, **6**, 700
Phosphoranes, methylene-, selenium-substituted, synthesis of, **6**, 697
Phosphoranes, methylene(selenoxo)-, synthesis of, **6**, 694
Phosphoranes, methylene-, silicon-substituted ylides, synthesis of, **6**, 702
Phosphoranes, methylene-, sulfonyl-stabilized, synthesis of, **6**, 699
Phosphoranes, methylene-, sulfur-substituted, synthesis of, **6**, 697
Phosphoranes, methylene(thioxo)-, synthesis of, **6**, 694
Phosphoranes, methylenetrialkyl-, reactions, with lithium alkyls, **6**, 706
Phosphoranes, α-methylthio-β-keto-, synthesis of, **5**, 898
Phosphoranes, monoalkyl-, synthesis of, **2**, 476
Phosphoranes, organothiomethylene-
reactions, with acyl halides, **5**, 898
as starting materials, for alkylthiomethylene phosphoranes, **5**, 898
Phosphoranes, oxoalkylidenetriphenyl-, as intermediates, **1**, 1079
Phosphoranes, 3-oxoindazolinylmethylene-, synthesis of, **5**, 906
Phosphoranes, oxyimidic imido-, synthesis of, **5**, 862
Phosphoranes, pentaalkyl-, synthesis of, **2**, 475
Phosphoranes, perfluoroacylcyanomethylenetriphenyl-, reactions, with lithium compounds, **3**, 649
Phosphoranes, perfluoroacylmethylenetriphenyl-, stability of, **3**, 649
Phosphoranes, phenylfluoro-, for the fluorodehydroxylation of alcohols, **2**, 20
Phosphoranes, phenylseleno(phenylthio)methylene-, synthesis of, **6**, 697
Phosphoranes, polychloro-, reactions, with methylenes, **6**, 698
Phosphoranes, propargylidene(triphenyl)-, synthesis of, **1**, 1016
Phosphoranes, propylidene-, reactions, with fluorenone, **1**, 988
Phosphoranes, α-silylated methylene-, synthesis of, **6**, 702
Phosphoranes, silyl methylene-, from, *P*-halo(silylmethyl)phosphanes, **5**, 887
Phosphoranes, α-stibinomethylene-
from
chlorodimethylstibane, **5**, 912
metallostibanes, **5**, 912
Phosphoranes, α-sulfoniomethylene-, dealkylation of, **5**, 898
Phosphoranes, tetraalkyl-, synthesis of, **2**, 475
Phosphoranes, tetraoxyhydrido-, reactions, with aldehydes, **2**, 476
Phosphoranes, α-thioacyl-, *S*-alkylation of, **4**, 834
Phosphoranes, thioalkoxymethylene-
from
phosphonium salts, **5**, 897
ylides, **5**, 897
synthesis of, **5**, 897
Phosphoranes, thiocarbonyl-, synthesis of, **5**, 1058
Phosphoranes, *N*-thionoacyliminoethylidenetriphenyl-, synthesis of, **5**, 633
Phosphoranes, thioxophosphanylmethylene-, sulfurization of, **5**, 908
Phosphoranes, thioxophosphanylmethylenetriphenyl-, synthesis of, **5**, 907
Phosphoranes, trialkyl-, synthesis of, **2**, 475
Phosphoranes, trialkyldifluoro-, synthesis of, **2**, 475
Phosphoranes, *N*-trialkylsilyl imino-, reactions, with ketones, **3**, 457
Phosphoranes, triaryldioxy-, polycyclic, synthesis of, **2**, 831
Phosphoranes, trichlorodiaryl-, reactions, with phosphorus pentoxide, **2**, 845
Phosphoranes, tricyclic, synthesis of, **4**, 478
Phosphoranes, triflourodiphenyl-, synthesis of, **2**, 836
Phosphoranes, tris(dialkylamino)dibromo-, as starting materials, for bis(dialkylamino)bromo-*N*-alkyliminophosphoranes, **2**, 408
Phosphoranes, vinyl-
as starting materials, for 1,3-aminophosphoranes, **4**, 478
synthesis of, **2**, 1074
Phosphoranes, vinylidene-, reactions, with carbonyls, **1**, 986
Phosphoranes, vinylidene triphenyl-, reactions, with ketones, **1**, 988
Phosphoranesulfenyl bromides, reactions, with carbon–carbon double bonds, **2**, 256
Phosphoranium salts, bis-, synthesis of, **6**, 265
Phosphorazidate, diphenyl (DPPA)
as diazo transfer reagent, **5**, 869
reactions, with carboxylic acids, **5**, 993
Phosphorazidates, reactions, with vinyl ethers, **5**, 861
Phosphordiamidic chloride, *N*,*N*-bis(2-oxo-3-oxazolidinyl)-
as activating agent, for the synthesis of anhydrides, **5**, 187
reactions, with carboxylic acids, **5**, 188
Phosphoric acid
reactions
with potassium iodide, **2**, 34
with vinyl acetate, **2**, 693
Phosphoric acid bisamides, use in the synthesis of alkenes, **1**, 752
Phosphoric acid dianilides, aryl-, synthesis of, **2**, 861
Phosphoric acids, synthesis of, **2**, 411
Phosphoric acids, alkylbenzyl-, synthesis of, **2**, 105
Phosphoric amide, *N*,*N*-dialkyl-, synthesis of, **2**, 414
Phosphoric anhydride, *t*-butylcarbonic diethyl-, reactions, with potassium azide, **6**, 489
Phosphoric anhydrides, alkylcarbonic diethyl-, as chloroformate substitute, **6**, 489
Phosphoric diamides, *N*,*N*,*N'*,*N'*-tetraalkyl-, synthesis of, **2**, 414
Phosphoric diester amides, *N*-dichloromethylene-, reactions, with formic acid, **5**, 1009
Phosphoric ester amides, reactions, with oxalyl chloride, **5**, 1017
Phosphoric triamide, heaxamethyl- (HMPT), use in the synthesis of acylmagnesiums, **5**, 439
Phosphoric trichloride, reactions, with alkanols, **2**, 105
Phosphorimidates, *N*-alkyl
from
azides, **2**, 407
phosphites, **2**, 407
synthesis of, **2**, 407
Phosphorimidates, alkylaminochloro-, synthesis of, **2**, 408
Phosphorimidates, phenyl-, from, dimethyl phenylphosphonite, **2**, 862
Phosphorimidic acid amides, *N*-alkyldichloro-, synthesis of, **2**, 408

Phosphorimidic acid triamides, *N*-alkyl-
from
amines, **2**, 407
azides, **2**, 407
synthesis of, **2**, 407
Phosphorimidic tribromide, as starting material, for α-(*N,N',N''*-triphenylphosphorimidic triamido) enamines, **4**, 991
Phosphorinane, reactions, with poly(HF-pyridine), **2**, 412
Phosphorinanes, as starting materials, for phosphinic acids, **4**, 124
Phosphorinanes, *p*-chloro-, synthesis of, **2**, 449
Phosphorinanes, 2,6-dihydroxy-, synthesis of, **4**, 339
Phosphorine, *t*-butyldipropynyl-, reactions, with phenylphosphine, **4**, 1026
Phosphorines, tetrahydro-, from, methylenephosphines, **4**, 459
Phosphorins, chlorotetrahydro-, synthesis of, **4**, 111
Phosphoriodites, dialkyl, reactions, with aldimines, **4**, 4525
Phosphoroamidates, alkenyl, synthesis of, **2**, 693
Phosphoroamidates, aryl, synthesis of, **2**, 694
Phosphorochloramidates, hydrolysis of, **2**, 414
Phosphorochloridate, dialkyl
in situ synthesis of, **2**, 694
reactions, with enolates, **2**, 693
Phosphorochloridate, diamido-
reactions
with enolates, **2**, 693
with phenolates, **2**, 694
Phosphorochloridate, dianilido, reactions, with Grignard reagents, **2**, 861
Phosphorochloridate, diphenyl, as activating agent, for the synthesis of anhydrides, **5**, 189
Phosphorochloridates, dialkyl
reactions
with alkynyl Grignard reagents, **2**, 1070
with amines, **5**, 975
with organolithiums, **2**, 848
Phosphorochloridates, diaryl, reactions, with alkynyl Grignard reagents, **2**, 1070
Phosphorochloridite, *o*-phenylene, reactions, with β-hydroxy sulfides, **1**, 590
Phosphorochloridites
reactions
with enolates, **2**, 693
with Grignard reagents, **2**, 434
with sodium phenolates, **2**, 693
Phosphorochloridites, dialkyl
reactions
with arylmagnesium halides, **2**, 838
with Grignard reagents, **2**, 838
with lithium alkynides, **2**, 1064
Phosphorocyanatidites, reactions, with *N*-alkylideneacetamide, **4**, 493
Phosphorocyanidate, diethyl
as cyanide source, for Strecker reaction, **3**, 616
reactions
with 3-acylindoles, **3**, 670
with α,β-unsaturated carbonyls, **3**, 659
Phosphorocyanide, diethyl, reactions, with ketones, **3**, 647
Phosphorodiamidates, *N,N,N',N'*-tetraalkyl-
from, amines, **2**, 413
synthesis of, **2**, 413
Phosphorodiamidic chloride, *N,N*-bis[2-oxo-3-oxazolidinyl]-
as dehydrating agent, in the synthesis of cyclic anhydrides, **5**, 199
as reagent, for the synthesis of symmetrical aromatic anhydrides, **5**, 195
Phosphorodiamidic chlorides, as starting materials, for anhydrides, **5**, 188
Phosphorodiamidic halides, *N,N,N',N'*-tetraalkyl-, synthesis of, **2**, 413
Phosphorodiamidite, trimethylsilyl tetraethyl-, reactions, with phenyl isothiocyanate, **6**, 582
Phosphorodichloridates, reactions, with silver nitrate, **2**, 106
Phosphorodichloridates, alkyl
hydrolysis of, **2**, 105
reactions, with 1,4-butanediyldimagnesium dibromide, **2**, 455
Phosphorodiimidates, Wittig rearrangement of, **4**, 923
Phosphorodithioate, sodium *O,O*-diethyl, reactions, with benzohydroximoyl chloride, **5**, 644
Phosphorodithioates, from, thiophosphoramidites, **2**, 410
Phosphorodithioates, trialkyl-, saponification of, **2**, 144
Phosphoro halides, dialkyl, reactions, with alkynide ions, **2**, 1072
Phosphoroimides
from
N-haloimides, **5**, 372
imides, **5**, 372
ring-opening of, **5**, 371
Phosphorolation, of amides, **5**, 368
Phosphorophthalimide, diphenyl, synthesis of, **5**, 372
Phosphorothioate, *S*-cyanomethyl diethyl-, use in the synthesis of alkyl α,β-unsaturated nitriles, **3**, 644
Phosphorothioate, *O*-isopropylmethyl-, as starting material, for chiral phosphine oxides, **2**, 453
Phosphorothioates
alkylation of, **2**, 254
reactions, with bromochloromethane, **2**, 255
Phosphorothioates, *S*-alkyl, synthesis of, **2**, 254
Phosphorothioite, *S*-acetyl, synthesis of, **5**, 535
Phosphorothioites, *S*-acetyl, from, thioalkanoic acids, **5**, 406
Phosphorothiolates
oxidation of, **2**, 210
reactions, with alkane halides, **2**, 120
as starting materials, for alkanethiols, **2**, 120
Phosphorothiolates, dichloro-, reactions, with butanethiol, **2**, 252
Phosphorothionates, rearrangement of, **2**, 257
Phosphorous
reactions
with alkanols, **2**, 108
with alkenes, **2**, 104
Phosphorous(III) compounds, reactions, with *N*-chloroamines, **2**, 860
Phosphorous acid
reactions
with alcohols, **2**, 106
with alkanols, **2**, 104
Phosphorous acid dialkyl esters
reactions
with aldehydes, **2**, 457
with Grignard reagents, **2**, 450
with ketones, **2**, 457
Phosphorous acids
reactions
with acylating agents, **5**, 402
with aziridines, **2**, 457
with Grignard reagents, **2**, 452
synthesis of, **2**, 409
Phosphorous halides, reactions, with organometallics, **2**, 841
Phosphorous oxybromide, reactions, with ketones, **3**, 69
Phosphorous oxychloride, reactions, with alkanols, **2**, 107
Phosphorous pentachloride, reactions, with sulfamic acids, **2**, 368
Phosphorouss
reactions, with organic azides, **5**, 1065
as starting materials, for carbodiimides, **5**, 1063
Phosphorous–selenium bonds, chemistry of, **2**, 856
Phosphorous–tellurium bonds, chemistry of, **2**, 856
Phosphorous tribromide, reactions, with 1-ethoxy-1-(trimethylsiloxy)cyclopropane, **4**, 50
Phosphorous trichloride, reactions, with epoxides, **2**, 105
Phosphorous trihalides
reactions
with alcohols, **2**, 104

with alkanols, **2**, 104
Phosphorus
aminoalkylation of, **6**, 117
cleavage reactions, chirality, **1**, 53
α-halo oxo acids of, **4**, 123
reactions
with alkyldichlorophosphines, **2**, 471
with alkyl iodides, **2**, 452
with bromine, **2**, 28
with heptafluoroisopropyl iodide, **4**, 114
with α-hydroxyamines, **4**, 466
with *N*-hydroxymethyl dialkylamines, **4**, 452
with organometallics, **2**, 433
with tetrafluoromethane, **6**, 58
with thiols, **2**, 252
stabilization by, **4**, 1082
stabilizing effects on ylide functions, **6**, 700
as starting material, for tertiary phosphines, **2**, 433
synthesis of functions, **4**, **543–589**
with nitrogen, **4**, 451
with no halogen, chalcogen or nitrogen, **4**, **1021–1042**
one halogen, **6**, 286
one halogen and one chalcogen, **6**, 281
one halogen and one nitrogen function, **6**, 285
one halogen and two sulfur functions, **6**, 278
Phosphorus, bromodifluoro-, reactions, with silver cyanamides, **5**, 1085
Phosphorus, doubly-bonded, synthesis of functions, **6**, **677–724**
Phosphorus(II) coupling systems, for the synthesis of *N*-alkylamides, **5**, 280
Phosphorus(III) acid amides, reactions, with isothiocyanates, **6**, 583
Phosphorus(III) acid anhydrides, reactions, with aryl isocyanate, **6**, 668
Phosphorus(III) bromide, reactions, with hexamethyldisiloxane, **2**, 538
Phosphorus(III) chloride, reactions, with (methylseleno)trimethylsilane, **2**, 293
Phosphorus(III) chlorides
reactions
with aldehydes, **2**, 454
with ketones, **2**, 454
with thionoamides, **5**, 643
Phosphorus(III) chlorocarbonyls, synthesis of, **6**, 454
Phosphorus(III) compounds
reactions
with aldehydes, **5**, 404
with carboxylic chlorides, **5**, 403
with halides, **3**, 50
with imines, **3**, 50
use in the synthesis of nitriles, **3**, 620
Phosphorus(III) compounds, acyloxy-
rearrangement of, **5**, 403
synthesis of, **5**, 220, 403
Phosphorus(III) compounds, chlorocarbonyl-, synthesis of, **6**, 454
Phosphorus(III) compounds, α-diazo-, oxidation of, **5**, 872
Phosphorus(III) compounds, trimethylsilyloxy-, reactions, with *N*-benzylidineallylamines, **4**, 461
Phosphorus(III) cyanides, synthesis of, **5**, 1128
Phosphorus(III) esters
reactions
with acid chlorides, **5**, 870
with aldehydes, **2**, 454
with halo ketones, **3**, 196
with ketones, **2**, 454
as starting materials, for phosphinothioformamides, **6**, 584
Phosphorus(III) halides
reactions
with chloroimino carbonates, **5**, 988
with ketone enolates, **3**, 195
with orthoformates, **4**, 340
with trifluorovinyllithium, **4**, 802
with trifluorovinylmagnesium, **4**, 802
Phosphorus(III) oxyacids, use in the synthesis of acylphosphorus compounds, **5**, 402
Phosphorus(III) trimethylsilyl esters, Arbuzov reactions of, **2**, 453
Phosphorus(V) acid *t*-butyl esters, reactions, with phenyl isothiocyanate, **6**, 583
Phosphorus(V) acids, synthesis of, **4**, 106
Phosphorus(V) chlorocarbonyls, synthesis of, **6**, 454
Phosphorus(V) compounds
as coupling reagents
for the synthesis of *N*-alkylamides, **5**, 279
for the synthesis of amides, **5**, 267
reactions, with isothiocyanates, **6**, 581
for the synthesis of *N*,*N*-dimethylamides, **5**, 280
Phosphorus(V) compounds, acyloxy-, synthesis of, **5**, 221
Phosphorus(V) compounds, chlorocarbonyl-, synthesis of, **6**, 454
Phosphorus(V) compounds, vinyl-, halogenation of, **4**, 804
Phosphorus(V) cyanides, synthesis of, **5**, 1130
Phosphorus(V) dichlorides
reactions
with alkoxydimethylsilyl cyanides, **5**, 1131
with silver cyanide, **5**, 1131
Phosphorus(V) isocyanates, from, phosphorus(V) isothiocyanates, **5**, 1009
Phosphorus(V) isothiocyanates, as starting materials, for phosphorus(V) isocyanates, **5**, 1009
Phosphorus acetals, oxidation of, **4**, 343
Phosphorus acid
reactions
with cyclohexylamine, **4**, 470
with imidoyl halides, **6**, 159
with phosphorus trichloride, **6**, 158
as starting material, for 1-aminoalkylphosphonic acids, **4**, 470
use in the synthesis of α-aminophosphonic acids, **4**, 462
Phosphorus acid, tristrimethylsilyl-, reactions, with acid chlorides, **5**, 404
Phosphorus acid anhydrides, dialkyl-, reactions, with symmetrical orthocarbonates, **6**, 314
Phosphorus acid esters, α-aminoalkyl-, synthesis of, **4**, 478
Phosphorus acids, in Friedel–Crafts alkylations, **1**, 181
Phosphorus acids, α-aminoalkyl-
deesterification of, **4**, 478
esterification of, **4**, 478
transesterification of, **4**, 478
Phosphorus acids, tervalent, use in the synthesis of α-aminoalkylphosphorus compounds, **4**, 470
Phosphorus allenes, *gem*-halo-, from, propargyl alcohols, **4**, 802
Phosphorus amides
as starting materials
for α-aminoalkylphosphorus compounds, **4**, 475
for 3-aminobenzophosphoranes, **4**, 478
N-Phosphorus amidines, synthesis of, **5**, 775
Phosphorus anions, reactions, with 1-haloalkylphosphorus compounds, **4**, 554
Phosphorus–arsenic analogues, synthesis of, **5**, 1094
Phosphorus bromide, tetraphenyl-, for the synthesis of aryl fluorides, **2**, 614
Phosphorus bromide difluoride, reactions, with *O*-silyl thioacetate, **5**, 535
Phosphorus–bromine complexes, triphenyl-, reactions, with ureas, **5**, 1063
Phosphorus bromodifluoride, reactions, with *N*,*O*-bis(trimethylsilyl)acetamide, **5**, 369
Phosphorus–carbon bonds, reductive cleavage of, **2**, 432
Phosphorus chemistry, reviews, **4**, 336
Phosphorus chloride, use in the synthesis of aromatic amidines, **5**, 757

Phosphorus chloride, diisopropyl-, reactions, with trimethylsilylamidines, **5**, 775
Phosphorus chlorides
reactions
with Grignard reagents, **2**, 434
with lithium ketimines, **3**, 456
Phosphorus compounds
from
aldehydes, **4**, 336
ketones, **4**, 336
reactions
with alkoxyalkynes, **4**, 913
with carbonimidic dichlorides, **6**, 665
as starting materials
for hydroxamoyl halides, **5**, 669
for *N*-phosphorylated imino esters, **5**, 708
synthesis of, **4**, 345; **5**, 868, 1092
as thionating reagents, **5**, 575
use in the synthesis of acid chlorides, **5**, 13
Phosphorus compounds, acyl-
organometallics of, **5**, 407
reactions, with amines, **5**, 403
structural and chemical diversity of, **5**, 394
Phosphorus compounds, acyloxy-, synthesis of, **5**, 220
Phosphorus compounds, α-alkoxy alkyl-, as starting materials, for anions, **4**, 917
Phosphorus compounds, alkyl-
reactions
with selenium, **2**, 465
with tellurium, **2**, 465
synthesis of, **2**, **425–477**
Phosphorus compounds, alkyl-, hexavalent, synthesis of, **2**, 476
Phosphorus compounds, alkyl-, pentacoordinated, synthesis of, **2**, 475
Phosphorus compounds, alkynyl-, synthesis of, **2**, 1058, **1039–1074**
Phosphorus compounds, amino-, oxidation of, **4**, 493
Phosphorus compounds, α-amino-
synthesis of
via Arbuzov reactions, **4**, 487
via Michaelis–Becker reactions, **4**, 487
for the synthesis of tricoordinate phosphorus functions, **4**, 480
Phosphorus compounds, α-aminoalkyl-
from
acetyl nitrate, **4**, 490
amines, **4**, 466
α-aminoalkylphosphorus compounds, **4**, 493
benzyl carbamates, **4**, 467
carbamates, **4**, 466
carbinolamines, **4**, 465
dinitrogen tetroxide, **4**, 490
electrophilic nitrogen compounds, **4**, 490
ethyl carbamates, **4**, 467
ethyl nitrate, **4**, 490
halomethylphosphinic acids, **4**, 473
halomethylphosphonic acids, **4**, 473
α-hydroxyalkylphosphorus compounds, **4**, 456
α-hydroxyamides, **4**, 465
α-hydroxy compounds, **4**, 475
O-mesitylsulfonylhydroxylamines, **4**, 490
methyl carbamates, **4**, 467
α-oxophosphorus compounds, **4**, 491
phenyl azide, **4**, 490
phosphonoacetamides, **4**, 476
phosphoramidites, **4**, 475
phosphorus amides, **4**, 475
tetrakishydroxymethylphosphonium salts, **4**, 475
thioureas, **4**, 466
trifluoromethanesulfonyl azide, **4**, 491
ureas, **4**, 466
as starting materials, for α-aminoalkylphosphorus compounds, **4**, 493
synthesis of
via Arbuzov reactions, **4**, 464
via Curtius rearrangement, **4**, 476, 492
via cycloaddition reactions, **4**, 488, 493
via Diels–Alder reactions, **4**, 489
via dipolar cycloaddition reactions, **4**, 479
via Hofmann rearrangement, **4**, 476
via Kabachnik–Fields reactions, **4**, 487
via Lossen rearrangement, **4**, 492
via Michaelis–Becker reactions, **4**, 464
via Mitsunobu reactions, **4**, 491
Phosphorus compounds, aminoalkyl hexacoordinate, synthesis of, **4**, 494
Phosphorus compounds, aryl-, synthesis of, **2**, **819–870**, 863
Phosphorus compounds, α-carbonyl-, enolization of, **4**, 920
Phosphorus compounds, α-chloroalkyl-, replacement of, **4**, 339
Phosphorus compounds, α-diazo-
from, α-diazo compounds, **5**, 870
synthesis of, **5**, 868
Phosphorus compounds, α-diazomethyl-, synthesis of, **5**, 872
Phosphorus compounds, dicoordinated, synthesis of, **5**, 1091
Phosphorus compounds, diphenyl-, fluorination of, for triflourodiphenylphosphoranes, **2**, 836
Phosphorus compounds, germyl-, synthesis of, **2**, 474
Phosphorus compounds, α-halo-, synthesis of, **4**, 110
Phosphorus compounds, 2-haloalkenyl-
from
alkenes, **4**, 802
carbonyls, **4**, 801
Phosphorus compounds, α-haloalkenyl-
synthesis of, **4**, 801
halogenation methods, **4**, 804
Phosphorus compounds, 1-haloalkyl-
Arbuzov reactions of, **4**, 555
reactions
with alkyl phosphinites, **4**, 556
with dialkyl phosphonites, **4**, 556
with phosphines, **4**, 554
with phosphorus anions, **4**, 554
with trialkyl phosphites, **4**, 555
Phosphorus compounds, hexa-coordinated, synthesis of, **2**, 868
Phosphorus compounds, hydroxyalkyl-
reactions
with acrylonitrile, **4**, 342
with triethyl orthoformate, **4**, 342
Phosphorus compounds, α-hydroxyalkyl-
as starting materials, for α-aminoalkylphosphorus compounds, **4**, 456
synthesis of, **4**, 339
Phosphorus compounds, imidoyl-dicoordinated, synthesis of, **5**, 784
Phosphorus compounds, imidoyl-tetracoordinated
from
α-chloroalkyl isocyanates, **5**, 787
enamines, **5**, 789
α-halo-α[1]-aminophosphinates, **5**, 790
imidoyl chlorides, **5**, 787
imines, **5**, 789
nitriles, **5**, 786
nitrile ylides, **5**, 789
nitrones, **5**, 789
phosphites, **5**, 788
synthesis of, **5**, 786
Phosphorus compounds, imidoyl-tricoordinated
from
amines, **5**, 786
imidoyl chlorides, **5**, 784
imines, **5**, 785
isonitriles, **5**, 784
nitrile ylides, **5**, 785

perfluoronitriles, **5**, 786
perfluorophosphines, **5**, 786
synthesis of, **5**, 784
Phosphorus compounds, low-coordinated, phosphaalkynes as, **5**, 1151
Phosphorus compounds, 1-metalloalkyl-
reactions
with chlorophosphine oxides, **4**, 557
with chlorophosphines, **4**, 551
Phosphorus compounds, α-methylene-, oxidation of, **5**, 394
Phosphorus compounds, nucleophilic
reactions
with acylating agents, **5**, 394
with imines, **4**, 455
with iminium ions, **4**, 455
Phosphorus compounds, α-oxo-
as starting materials
for α-aminoalkylorganophosphorus compounds, **4**, 471
for α-aminoalkylphosphorus compounds, **4**, 491
Phosphorus compounds, oxyimidic
from, phosphanes, **5**, 839
synthesis of, **5**, 839
Phosphorus compounds, tetracoordinated, synthesis of, **4**, 569
Phosphorus compounds, α-trifluorovinyl-, synthesis of, **4**, 802
Phosphorus compounds, trivalent, reactions, with dimethylketene, **4**, 914
Phosphorus compounds, β,γ-unsaturated α-oxy-, isomerization of, **4**, 917
Phosphorus compounds, vinyl-
reactions, with organocoppers, **1**, 159
as starting materials, for phosphinates, **4**, 561
synthesis of, **2**, **819–870**
Phosphorus-containing heterocycles, synthesis of, **4**, 488
Phosphorus diamines, alkoxy-, as starting materials, for α-phosphorodiamidoyl enamines, **4**, 1011
Phosphorus dichlorides, imidoyl-, from, imines, **5**, 785
Phosphorus esters
as starting materials
for α-aminoalkyl phosphinates, **4**, 464
for α-aminoalkylphosphine oxides, **4**, 464
for α-aminoalkyl phosphonates, **4**, 464
Phosphorus esters, trivalent, reactions, with *N*-bromomethylphthalimide, **4**, 464
Phosphorus functional groups, properties of, **5**, 394
Phosphorus functions
elimination of, **1**, 1079
substitution of, **1**, 268, 1032
Phosphorus functions, acyl-, synthesis of, **5**, **393–407**
Phosphorus functions, dicoordinated
synthesis of, **4**, 110, 479, 565; **5**, 394, 876; **6**, 678
theoretical, **4**, 452
Phosphorus functions, imidoyl-, synthesis of, **5**, 784
Phosphorus functions, tetracoordinated, synthesis of, **4**, 117, 460, 485; **5**, 402, 786, 894; **6**, 694
Phosphorus functions, tricoordinated
synthesis of, **4**, 110, 452, 480, 565; **5**, 395, 893; **6**, 691
via Mannich reactions, **4**, 481
Phosphorus groups, elimination of, **1**, 638
Phosphorus halides
reactions
with alkoxydimethylsilyl cyanides, **5**, 1128
with carbamates, **5**, 977
with ethylvinyl ether, **3**, 50
with imides, **5**, 369
with metal alkynides, **2**, 1058
with silver cyanate, **5**, 1017
with trimethylsilyl cyanide, **5**, 1128
as reagents, **2**, 426
as starting materials, for alkylphosphine oxides, **2**, 452
synthesis of, **2**, 832
Phosphorus halides, alkyl-, synthesis of, **2**, 447
Phosphorus halides, cyclic alkyl-, synthesis of, **2**, 449
Phosphorus hexafluoride, nitroso-, as reagent, for the hydrolysis of nitriles, **5**, 270
Phosphorus–hydrogen bonds
for the synthesis of primary phosphines, **2**, 428
for the synthesis of secondary phosphines, **2**, 430
use in the synthesis of tertiary phosphines, **2**, 438
Phosphorus isocyanates
synthesis of, **5**, 1017
use in the synthesis of isocyanates, **5**, 1008
Phosphorus–nitrogen bonds, reactions, with carbon disulfide, **6**, 564
Phosphorus nucleophiles
reactions
with alkynes, **4**, 561
with 1,1-dihaloalkanes, **4**, 553
with vinyl chlorides, **4**, 553
synthesis of, **1**, 52
for the synthesis of tetracoordinate phosphorus functions, **4**, 460
Phosphorus oxides, α-haloalkenyl-, synthesis of, **4**, 804
Phosphorus oxychloride
for the decomposition of dithiocarbamates, **5**, 1023
as dehydrating agent, **3**, 684
for the dehydration of amides, **3**, 619
reactions
with activated aromatic compounds, **2**, 845
with alkylmagnesium halides, **2**, 452
with amides, **5**, 576
with *O*-aryl-*O*,*O*-dialkylthiophosphates, **2**, 257
with dimethylformamide, **3**, 91
with α-hydroxy enol ethers, **3**, 208
with *N*-methylformamilide, **3**, 91
with phenols, **2**, 694
with tetrazoles, **5**, 774
use in the synthesis of acyl chlorides, **5**, 264
use in the synthesis of isocyanides, **3**, 696
Phosphorus–oxygen double bonds, for the synthesis of alkyl halides, **2**, 6
Phosphorus oxytrichloride, use in the synthesis of sulfonyl chlorides, **2**, 181
Phosphorus pentabromide
reactions, with sodium iodomethanesulfonate, **4**, 85
use in the synthesis of dibromoalkanes, **4**, 25
use in the synthesis of sulfonyl bromides, **2**, 181
Phosphorus pentachloride
for the α-chlorination of ethers, **4**, 47
reactions
with acetoxy groups, **4**, 766
with acetylated sugars, **4**, 766
with *N*-acetylcarbazole, **4**, 791
with *N*-acetylthiourea, **6**, 619
with alkoxycarbonylimino phosphoranes, **5**, 988
with *N*-alkoxycarboxylic acid amides, **5**, 665
with (alkylthio)alkyne chlorides, **4**, 804
with alkynes, **4**, 772
with alkynylphosphates, **2**, 1071
with amides, **5**, 743
with *gem*-amido phosphinoylalkenes, **4**, 1001
with *N*-(1-arenesulfonyl-2-chloroethenyl)amides, **5**, 656
with aryl cyanates, **6**, 728
with *N*-(1-aryldiphenylphosphonyl-2-dichloroethenyl)amides, **5**, 656
with 1-arylsulfonyl-3-alkylureas, **5**, 982
with *N*,*N*'-bis(alkylidene)ureas, **5**, 1063
with bis(alkylthio)ethynes, **4**, 926
with bis(benzoylamino)arylmethanes, **5**, 657
with bis(benzoylamino)methanes, **5**, 657
with carbamate esters, **6**, 617
with carbamates, **5**, 976
with carbodiimides, **6**, 622
with α-chloro amides, **3**, 561
with chloromethanesulfonate, **4**, 85

with chlorophosphinoyl compounds, **4**, 1013
with diarylacetamides, **5**, 656
with α,ω-dicyanoalkanes, **5**, 57
with 1,3-disubstituted ureas, **5**, 982
with formate esters, **6**, 39
with hydrazides, **5**, 842
with isocyanates, **6**, 606
with 1,3-oxazolidin-2-one, **5**, 126
with perfluoroglutarimide, **5**, 656
with perfluorosuccinimide, **5**, 656
with phenyl acetate, **4**, 766
with phosphinamides, **2**, 470
with phosphine oxides, **6**, 699
with phosphinic acid anhydrides, **4**, 125
with primary amides, **5**, 373
with secondary amides, **3**, 560
with sodium iodomethanesulfonate, **4**, 85
with tertiary amide, **4**, 790
with thiocyanates, **5**, 861
with thiolformates, **6**, 46
with ureas, **5**, 982
with vinyl acetate, **4**, 766
as reagent, for the synthesis of thiocarboxylates, **5**, 512
for the synthesis of vinyl chlorides, **2**, 617
use in the synthesis of alkyl chlorides, **2**, 25
use in the synthesis of dichloroalkanes, **4**, 18
use in the synthesis of formamidines, **5**, 744
use in the synthesis of hydrazonyl halides, **5**, 675
use in the synthesis of sulfonyl chlorides, **2**, 181
use in the synthesis of trichloromethyls, **6**, 28
Phosphorus pentahalides
reactions
with amides, **5**, 655
with nitrophenols, **5**, 969
Phosphorus pentaselenide
reactions
with amides, **5**, 626
with carboxamides, **5**, 624
with phenyl isocyanate, **5**, 1056
as starting material, for selenoureas, **6**, 596
use in the synthesis of selenals, **3**, 389
Phosphorus pentasulfide
reactions
with acyl chlorides, **5**, 484
with alcohols, **2**, 119, 258
with amides, **4**, 424; **5**, 613
with 2-benzyl-5-phenylisoxazolidine-3-thiones, **5**, 637
with di-*t*-butylketene, **3**, 546
with *n*-butyllithium, **5**, 569
with ethyl orthoformate, **2**, 259
with 2-keto-1,3-diamide, **3**, 345
with methacrolein, **3**, 338
with perfluoroalkyl iodides, **3**, 348; **5**, 474
with phosphonous acids, **2**, 461
with sodium carbonate, **5**, 568
with sodium sulfide, **5**, 582
as reagent
for the synthesis of alkenyl aryl thioketones, **3**, 360
for the synthesis of 1,2-dithioles, **3**, 371
for the synthesis of nonclassical thiophenes, **3**, 372
for the synthesis of thioaldehydes, **3**, 338, 341
for the synthesis of thioketones, **3**, 344, 351, 352, 357; **3**, 360, 361, 365
for the synthesis of trithiapentalenes, **3**, 376
use in the synthesis of nitriles, **5**, 570
use in the synthesis of thionoalkanoamides, **5**, 610
use in the synthesis of thionoamides, **5**, 566
disadvantages, **5**, 569
use in the synthesis of thiophosphines, **2**, 254
Phosphorus pentasulfide–alkyllithium compound, for the synthesis of thionoamides, **5**, 569
Phosphorus pentoxide
as catalyst, for the synthesis of anhydrides, **5**, 197
reactions
with carbamates, **5**, 977
with dichlorophenyl isocyanide dichloride, **5**, 1004
with methanesulfonic acid, **5**, 282
with pyridine, **5**, 1078
with secondary amides, **3**, 556
with trichlorodiarylphosphoranes, **2**, 845
Phosphorus sulfides, *S*-alkyl-, as starting materials, for dithioic acid esters, **5**, 551
Phosphorus sulfoxybromide, as reagent, for the synthesis of thioketones, **3**, 365
Phosphorus triamide, hexaethyl-, reactions, with phthalimide, **5**, 369
Phosphorus triamide, hexamethyl-
reactions
with aryl aldehyde, **2**, 740
with aryl aldehydes, **4**, 475
with benzoyl chloride, **5**, 407
with disulfides, **2**, 143
with ketoximes, **5**, 755
synthesis of, **2**, 411
use as dechlorinating agent, **5**, 1093
Phosphorus triamides, hexaalkyl-, reactions, with benzoyl halides, **5**, 402
Phosphorus tribromide
Friedel–Crafts reaction, with benzene, **2**, 832
reactions, with trioctylphosphine, **2**, 449
for the synthesis of vinyl bromides, **2**, 624
use in the synthesis of alkyl bromides, **2**, 31
use in the synthesis of dibromoalkanes, **4**, 25
Phosphorus trichloride
Friedel–Crafts reaction, with benzene, **2**, 832
reactions
with aldehydes, **2**, 456; **4**, 130
with aldimines, **4**, 488
with alkoxyalkynes, **4**, 765
with α-alkoxyglycines, **4**, 465
with arenediazonium salts, **2**, 844
with arenediazonium tetrafluoroborates, **2**, 844
with benzylideneaniline, **5**, 785
with *N*,*O*-bis(trimethylsilyl)acetamide, **5**, 369
with bis(trimethylsilyl)methane, **6**, 164
with butadiene, **2**, 449
with *t*-butylmagnesium chloride, **2**, 434
with *N*-chlorobenzamidines, **5**, 775
with cyclohexene, **2**, 448
with 1,3-dimethoxybenzene, **2**, 829
with *N*,*N*-dimethylaniline, **2**, 829
with dimethylarsinic acid, **2**, 492
with dimethylformamide, **2**, 25
with Grignard reagents, **2**, 433
with hydroxymethylamides, **4**, 469
with imines, **2**, 449
with lithium amides, **2**, 466
with methane, **2**, 448
with organolithiums, **2**, 434
with phenols, **2**, 693
with phenyl acetate, **4**, 742
with phosphorus acid, **6**, 158
with potassium phthalimide, **5**, 369
with Schiff bases, **5**, 785
with silver cyanate, **5**, 1017
with sulfonate esters, **2**, 182
with trialkylaluminums, **2**, 434
with tributyl(cyanomethyl)tin, **3**, 639
with 1,3,5-trimethoxybenzene, **2**, 829
with tris(trimethylsilyl)methyllithium, **6**, 165
with α,β-unsaturated ketones, **2**, 456
with zirconium complexes, **6**, 153
use in the synthesis of acyl chlorides, **5**, 264
use in the synthesis of nitriles, **3**, 620
use in the synthesis of polyphosphines, **2**, 471
Phosphorus trichloride, catechyl-, for the synthesis of vinyl chlorides, **2**, 617

Phosphorus trichloride bisphosphonic acids, reactions, with carboxylic acids, **6**, 118
Phosphorus tricyanide
reactions
with trialkyl phosphites, **5**, 1128
with trimethylphosphite, **5**, 1128
Phosphorus trihalides
reactions
with 1-alkenes, **2**, 448
with amides, **5**, 655
with aminophenols, **5**, 969
with nitriles, **6**, 159
as starting materials, for aminobis(phosphonates), **6**, 158
use in the synthesis of tertiary phosphines, **2**, 824
Phosphorus triiodide, reactions, with bis(trifluoromethyl)cadmium, **2**, 434
Phosphorus triisocyanate, reactions, with trialkyl phosphites, **5**, 1008
Phosphorus trisulfide, use in the synthesis of thionoamides, **5**, 566
Phosphorus ylides
formation of borane adducts, **4**, 574
from, lithium diphosphinomethanides, **5**, 909
lithiation of, **4**, 576
reactions
with alkanesulfonyl chlorides, **5**, 902
with allenyl acyl chloride, **1**, 987
with aluminum compounds, **4**, 587
with amines, **5**, 603
with benzene sulfenyl chloride, **5**, 898
with beryllium chloride, **4**, 577
with boron compounds, **4**, 574
with *n*-butane sulfenyl chloride, **5**, 898
with cadmium compounds, **4**, 586
with carbon disulfide, **3**, 550
with carboxylic acid chlorides, **1**, 1079
with chlorophosphine oxides, **4**, 557
with chromium compounds, **4**, 580
with cobalt compounds, **4**, 582
with copper compounds, **4**, 585
with dehydrodithizone, **4**, 490
with dichloroalkylboranes, **5**, 916
with *gem*-diisocyanatoalkanes, **4**, 992
with erbium compounds, **4**, 578
with gadolinium compounds, **4**, 578
with gallium compounds, **4**, 587
with gold compounds, **4**, 586
with halogermanes, **4**, 574
with halophosphines, **4**, 551
with halosilanes, **4**, 570
with halostannanes, **4**, 588
with holmium compounds, **4**, 578
with indium compounds, **4**, 587
with iridium compounds, **4**, 583
with iron compounds, **4**, 581
with isothiocyanates, **5**, 597
with lanthanum compounds, **4**, 578
with lead compounds, **4**, 589
with lutetium compounds, **4**, 578
with magnesium compounds, **4**, 577
with manganese compounds, **4**, 581
with mercury compounds, **4**, 587
with molybdenum compounds, **4**, 580
with neodymium compounds, **4**, 578
with nickel compounds, **4**, 583
with palladium compounds, **4**, 583
with platinum compounds, **4**, 585
with polyhalogenated disilanes, **5**, 914
with polyhalogenated organosilanes, **5**, 914
with praseodymium compounds, **4**, 578
with rhodium compounds, **4**, 582
with ruthenium compounds, **4**, 582
with samarium compounds, **4**, 578
with scandium compounds, **4**, 578
with Schiff bases, **2**, 860
with silver compounds, **4**, 585
with tantalum compounds, **4**, 579
with thallium compounds, **4**, 587
with thorium compounds, **4**, 589
with titanium compounds, **4**, 578
with transition metal compounds, **4**, 578
with tungsten compounds, **4**, 580
with uranium compounds, **4**, 589
with vanadium compounds, **4**, 579
with zinc compounds, **4**, 586
with zirconium compounds, **4**, 579
silyl shifts in, **5**, 912
as starting materials
for *S*,*P*-acetals, **4**, 346
for *Te*,*P*-acetals, **4**, 348
for allenes, **1**, 986
for cyclopropanes, **1**, 268
for ketenimines, **3**, 591
synthesis of, **3**, **491–500**, **5**, 894, 897
use in the synthesis of thionoamides, **5**, 622
Phosphorus ylides, allyl-, reactions, with β-chloroenones, **1**, 470
Phosphorus ylides, β-oxido-, synthesis of, **1**, 729
Phosphorus ylides, β-ribofuranose, reactions, with alkyl isocyanates, **5**, 1068
Phosphorus ylides, silicon-stabilized, **2**, 936
Phosphorus ylides, silyl-, protonation of, **4**, 571
N-Phosphorylated imidates, synthesis of, **5**, 708
Phosphorylation
of amides, **5**, 370, 643
of ester enolates, **2**, 682
of thioacetic acid, **5**, 535
of thioamides, **5**, 737
of triethylammonium salts, **5**, 535
of *O*-trimethylsilyl thioacetate, **5**, 535
of *O*-trimethylsilyl thiobenzoate, **5**, 535
Phosphoryl azide, diphenyl- (DPPA)
reactions, with enamines, **5**, 775
for the synthesis of azides, **5**, 278
use in the synthesis of amides, **5**, 265
Phosphoryl chloride
reactions
with dialkylcyanamides, **6**, 656
with indolizidine, **3**, 341
Phosphoryl chloride, (dichlorophosphinomethyl)-, reactions of, **4**, 553
Phosphoryl chloride, diethyl, reactions, with 2-imino-5,6-dihydro-4*H*-1,3-thiazine, **5**, 1112
Phosphoryl chloride, pyrocatechyl, use in the synthesis of acyl thiocyanates, **5**, 1112
Phosphoryl chloride, (trimethylsilylmethyl)-, reactions of, **4**, 574
Phosphoryl chlorides, reactions, with guanidines, **6**, 655
Phosphoryl chlorides, aryl-, reactions, with amines, **2**, 861
Phosphoryl compounds, α-diazo-
synthesis of
via Bamford–Stevens reaction, **3**, 480
via diazo group transfer, **3**, 480
Phosphoryl compounds, α-diazomethyl-, acylation of, **3**, 482
Phosphoryl compounds, α-halo-, synthesis of, **4**, 123
Phosphoryl compounds, halodiazomethyl-, synthesis of, **3**, 478
Phosphoryl compounds, α-keto-, as starting materials, for α-diazophosphinates, **3**, 480
Phosphoryl compounds, γ,δ-unsaturated α-diazo-, synthesis of, **3**, 481
Phosphoryl fluoride, reactions, with tetraisocyanatosilane, **5**, 1018
Phosphoryl groups, chloro-, hydrolysis of, **4**, 1006
Phosphoryl halides, bis(dialkylamino)-
from, amines, **2**, 413

synthesis of, **2**, 413
Phosphoryl trichloride, reactions, with silver cyanate, **5**, 1017
Photochemical oxidation
of alcohols, to aldehydes, **3**, 7
for the synthesis of ketones from alcohols, **3**, 120
Photochemical reaction
radical, for the synthesis of primary amides, **5**, 276
for the substitution of sulfur functions, **1**, 231
for the synthesis of disulfides, **2**, 228
for the synthesis of metalloid functions, **3**, 501
Photochemical rearrangement, for the synthesis of silenes, **5**, 925
Photochlorination
of dimethyl ether, **6**, 229
of ethers, **6**, 36
of methyl ethers, **6**, 229
for the synthesis of imidoyl halides, **5**, 659
Photofluorination, for the synthesis of aryl fluorides, **2**, 613
Photo-Fries rearrangement, mechanism, **3**, 297
Photogermylation, use in the synthesis of germanes, **5**, 432
Photohydration, of alkynes, for the synthesis of enols, **2**, 637
Photoinitiation, for the synthesis of radicals, **1**, 925
Photolysis
for the synthesis of carbenes, **4**, 1081
for the synthesis of radicals, **1**, 925, 935
for the synthesis of vinyl cations, **1**, 1120
Photolytic rearrangement, of nitro compounds, **5**, 325
Photooxidation, 9,10-dicyanoanthracene-sensitized, for the synthesis of amines, **2**, 304
Photooxygenation, of siloxyalkenes, **5**, 227
Photopyrone, reactions, with dicarbonylcyclopentadienylcobalt, **4**, 690
Photoreactions, for the synthesis of arsines, **4**, 598
Photorearrangement, for the synthesis of allenes, **1**, 957
Photoreduction, of thiocarbonyls, **6**, 115
Photoresists, iminosilanes as, **4**, 507
Photosensitized oxygenation, of silyl diazo compounds, **5**, 413
Phthalaldehyde
reactions
with ammonia, **4**, 298
with thiourea, **4**, 298
o-Phthalaldehyde, reactions, with amines, **3**, 671
Phthalaldoxime, tere-, as starting material, for terephthalohydroxymoyl chloride, **5**, 821
Phthalamides, phospho-, synthesis of, **5**, 371
Phthalate, diester dimethyl 2-nitro-, partial hydrolysis of, **5**, 105
Phthalic acid, reduction of, **1**, 103
Phthalic acid, 4-chloro-, synthesis of, **5**, 116
Phthalic acid, homo-, from, 2α-thiohomophthalimide, **5**, 37
Phthalic acid, monoperoxy-, as reagent, for the synthesis of sulfines, **3**, 368
Phthalic acid, 4-nitro-, from, 4-nitrophthalimide, **5**, 109
Phthalic acids, nitro-, from, phthalic anhydride, **5**, 109
Phthalic anhydride
reactions, with phosphoranes, **2**, 1064
as starting material, for nitrophthalic acids, **5**, 109
Phthalic anhydride, homo-, as a chemical equivalent of 6-hydroxy-pyran-2-ones, **2**, 665
Phthalic anhydrides
reactions, with tributyltin isocyanate, **5**, 391
synthesis of, **5**, 199
Phthalide, bromination of, **4**, 56
Phthalide, α,α-dichloro-, reactions, with 3-hydroxybutanoic acid, **6**, 69
Phthalides, 3-chloro-, synthesis of, **4**, 54
Phthalides, 3-fluoro-, synthesis of, **4**, 52
Phthalimide
irradiation of, **4**, 1077
reactions
with aluminum chloride, **5**, 384
with hexaethylphosphorus triamide, **5**, 369
with α-hydroxyphosphonates, **4**, 475
with phenyliodine(III), **5**, 322
with tributyltin oxide, **5**, 390
silylmethylation of, **4**, 513
use in the synthesis of radicals, **4**, 1077
Phthalimide, alkylselenyl-, as starting material, for alkane selenenamides, **2**, 292
Phthalimide, *N*-amino-
oxidation of, **5**, 364
synthesis of, **5**, 360
Phthalimide, *N*-bromo-, synthesis of, **5**, 318
Phthalimide, *N*-bromomethyl-
reactions
with silver nitrocyanamide, **4**, 447
with trivalent phosphorus esters, **4**, 464
synthesis of, **4**, 99
Phthalimide, *N*-chloro-, synthesis of, **5**, 316
Phthalimide, *N*-chloromethyl-, displacement of chlorine from, **4**, 499
Phthalimide, *N*-germyl-, synthesis of, **5**, 391
Phthalimide, *N*-hydroxy-
reactions
with aryl halides, **2**, 692
with diphenyliodonium chloride, **2**, 692
Phthalimide, *N*-iodo-, from, trialkylstannylphthalimide, **5**, 320
Phthalimide, *N*-nitrenyl-, as starting material, for *N*,*N*-diacyl hydrazones, **5**, 364
Phthalimide, 4-nitro-, as starting material, for 4-nitrophthalic acid, **5**, 109
Phthalimide, *N*-phenylseleno- (NPSP)
reactions
with alkynes, **4**, 787
with diazomethane, **4**, 335
with ethyl diazoacetate, **4**, 335
as starting material, for unsymmetrical selenides, **2**, 279
use in the synthesis of selenides, **2**, 280
Phthalimide, *N*-(phenyl selenyl)-
reactions, with ketones, **3**, 173
as selenylating reagent, **3**, 38
Phthalimide, 2α-thiohomo-, as starting material, for homophthalic acid, **5**, 37
Phthalimide, trialkylstannyl-, as starting material, for *N*-iodophthalimide, **5**, 320
Phthalimide, tributylgermyl-, from, tributylgermanium isocyanate, **5**, 391
Phthalimide, *N*-vinyl-, use in the synthesis of α-haloamines, **4**, 101
Phthalimide anions, alkylation of, **2**, 299
Phthalimide esters, hydroxy-, use in the synthesis of alkyl halides, **2**, 13
Phthalimides
deprotection of, **2**, 304
reactions
with amines, **2**, 358
with hydrogen iodide, **4**, 101
as starting materials, for thioaldehydes, **3**, 333
Phthalimides, *N*-(alkanesulfenyl)-, oxidation of, **2**, 363
Phthalimides, *N*-(alkanethio)-, as starting materials, for alkanesulfenic esters, **2**, 195
Phthalimides, alkyl-
as starting materials
for dialkylketenes, **3**, 541
for ketenes, **3**, 536
Phthalimides, *N*-alkyl-, synthesis of, **2**, 299
Phthalimides, *N*-(alkylsulfinyl)-, reactions, with platinum π-allyl compounds, **5**, 388
Phthalimides, *N*-alkylsulfinyl-, synthesis of, **2**, 688
Phthalimides, *N*-alkylsulfonyl-, synthesis of, **5**, 348
Phthalimides, *N*-(arenesulfenyl)-, oxidation of, **2**, 363
Phthalimides, *N*-(arylaminothio)-, as starting materials, for thionitrosoarenes, **2**, 792

Phthalimides, *N*-aryloxy-, synthesis of, **2**, 692
Phthalimides, *N*-(arylsulfinyl)-, reactions, with platinum π-allyl compounds, **5**, 388
Phthalimides, *N*-arylsulfinyl-, synthesis of, **2**, 688
Phthalimides, *N*-aziridyl-
 reactions, with lithium bromide, **5**, 364
 ring-opening of, **5**, 364
Phthalimides, α-bromo-, synthesis of, **4**, 99
Phthalimides, *N*-chloromethyl-, synthesis of, **4**, 483
Phthalimides, *N*-hydroxy-, as starting materials, for *O*-alkylhydroxylamines, **2**, 102
Phthalimides, *N*-phenoxy-, synthesis of, **2**, 692
Phthalimides, *N*-phosphinomethyl-, synthesis of, **4**, 483
Phthalimides, *N*-sulfenyl-, use in the synthesis of disulfides, **2**, 217
Phthalimides, sulfinyl-
 as sulfinylating reagents, **2**, 688
 synthesis of, **2**, 363
Phthalimides, thioalkyl-, from, alkylphthalimido disulfides, **5**, 338
Phthalimidines, synthesis of, **4**, 298, 303
Phthalimidoalkenes, *gem*-halo, synthesis of, **4**, 794
Phthalimidomalonic acid synthesis, use in the synthesis of amino acids, **5**, 90
Phthalimidomethyl fluorides, synthesis of, **4**, 101
Phthalohydroxymoyl chloride, tere-, from, terephthalaldoxime, **5**, 821
Phthalolyl chloride, reactions, with 2-(2'-mercaptophenyl)dihydroimidazole, **6**, 111
Phthalonitrile, reactions, with hydroxylamine, **5**, 833
Phthaloyl chloride
 as starting material, for thionophthalic anhydride, **5**, 528
 use in the synthesis of acid chlorides, **5**, 15
o-Phthaloyl chloride, as starting material, for benzotellurophenedione, **5**, 244
Phthaloyl dichloride, as starting materials, for bisphosphonium salts, **2**, 1063
Phtholic acid, tetrahydro-, synthesis of, reviews, **5**, 24
Phyllanthocin ring system, synthesis of, **1**, 289
Physovenine, synthesis of, **1**, 345
2-Picoline, reactions of, **1**, 860
4-Picoline, oxidation of, **4**, 206
α-Picoline, as starting material, for picolinic acid, **5**, 115
α-Picoline, trichloro-, reduction of, **4**, 13
Picolinic acid, from, α-picoline, **5**, 115
Picolinic acids, 5-alkyl, synthesis of, **5**, 116
Picryl chloride, reactions, with pyridine, **5**, 1040
Picryl fluoride, reactions, with potassium thiocyanate, **5**, 1040
Pictet–Spengler cyclization, for the synthesis of 8-methoxytetrahydroisoquinolines, synthesis of, **2**, 328
Pictet–Spengler reaction, use in the synthesis of iminium salts, **3**, 419
Pimelic acid, synthesis of, **5**, 31
Pinacol esters, synthesis of, **6**, 386
Pinacol rearrangement
 mechanism, **1**, 384
 modification of, **1**, 389
 silicon variant of, **3**, 202
 for the synthesis of enols, **2**, 643
 for the synthesis of ketones, from 1,2-diols, **3**, 121
 use in the synthesis of α-aryl ketones, **3**, 146
 use in the synthesis of unsaturated ketones, **3**, 142
Pinacols, synthesis of, via electroreduction, **2**, 80
Pinanediol, reactions, with diisopropyl (1-methoxyvinyl)boronate, **4**, 934
Pinanone, reactions, with benzylamine, **2**, 331
Pinene
 regeneration of, **2**, 531
 as starting material, for (*Z*)-3-alkoxyallylboranes, **2**, 75
 use in the reduction of carbonyls, **2**, 58
2-Pinene, reactions, with borane, **2**, 41
α-Pinene
 oxidation of, **5**, 74
 reactions
  with borane, **2**, 514
  with thexylborane, **2**, 515
 as starting material, for bornyl chloride, **1**, 380
Pinner, synthesis of ortho esters, **6**, 70
Pinner reaction
 side reactions, **5**, 748
 for the synthesis of amidines, **5**, 748
 for the synthesis of thioimidic esters, **5**, 727
 use in the synthesis of amidines, **5**, 742
Pinner synthesis, for the synthesis of imidic acid esters, **5**, 687
Pipecolic acid, reactions, with carbon disulfide, **6**, 564
Pipecolic acid hydrochloride, 4-keto-, synthesis of, **5**, 92
Pipecolic acids, synthesis of, **2**, 326
Pipecolic acids, 5-hydroxy-, from, glutamic acids, **2**, 326
Pipecolic esters, synthesis of, **2**, 327
α-Pipecoline oxide, *N*-methyl-, elimination of, **1**, 625
Piperazine, diketo-, synthesis of, **4**, 218
Piperazine, *N*-methyl-, reactions, with sodium bis(methoxyethoxy)aluminum dihydride, **3**, 84
Piperazines, synthesis of, **4**, 301
Piperazines, diketo-
 as catalysts, for the synthesis of asymmetric cyanohydrins, **3**, 628
 use in the synthesis of asymmetric cyanohydrins, **3**, 628
Piperazines, 2,3-dioximino-, synthesis of, **3**, 429
1-Piperideine, perfluoro-, synthesis of, **5**, 656, 662
1-Piperideine, 2,2,6-trichloro-3,3,4,4,5,5-hexafluoro-, synthesis of, **5**, 656
Piperidenes, 4-hydroxy-, synthesis of, **3**, 422
Piperidine
 reactions
  with *N*-benzoylthionoglycinate, **5**, 586
  with bis(acyl) tellurides, **5**, 250
  with carbonic difluoride, **6**, 438
  with cyclic ketones, **2**, 310
  with 2,2-dichloro-1,3-benzodioxole, **6**, 321
  with enamines, **2**, 308
  with 4-fluorobenzaldehyde, **4**, 298
  with hydroxylamine-*C*-sulfonic acid, **2**, 382
  with methyl azidodithioacetate, **5**, 586
  with phenyl isonitrile, **5**, 747
  with pyrrole-2-carbaldehyde, **4**, 406
  with selenium, **2**, 369
  with sodium diethyldihydroaluminate, **3**, 83
 reduction of, **1**, 89
 synthesis of, **1**, 91
 use in the synthesis of aminotrithioorthoformates, **6**, 314
Piperidine, *N*-acetyl-, synthesis of, **4**, 99
Piperidine, *N*-amino-, oxidation of, **2**, 385
Piperidine, 1,1'-azodicarbonyl-, as reagent, for the synthesis of ketones from alcohols, **3**, 120
Piperidine, 1-benzhydryl-, synthesis of, **4**, 334
Piperidine, *N*-cyano-, from, *N*-cyanomethylpiperidine, **5**, 1127
Piperidine, *N*-cyanomethyl-, as starting material, for *N*-cyanopiperidine, **5**, 1127
Piperidine, 2,6-diphenyl-, chlorination of, **2**, 338
Piperidine, *N*-formyl-, as formylating reagent, **3**, 90
Piperidine, *N*-(*N*-formyl-*N*-methylamino)-, as reagent, for the synthesis of aldehydes, **3**, 14
Piperidine, 3-hydroxy-
 reactions, with phosgene, **6**, 446
 synthesis of, **1**, 91
Piperidine, 4-hydroxy-, synthesis of, **1**, 91
Piperidine, 3-hydroxymethyl-, reactions, with phosgene, **6**, 445
Piperidine, *N*-methyl-
 deprotonation of, **1**, 148
 metallation of, **4**, 527

Piperidine, *N*-nitroso-, lithiation of, **4**, 335
Piperidine, perfluoro-, defluorination of, **5**, 662
Piperidine, perfluoro-*N*-fluoro-, for the fluorination of amines, **2**, 334
Piperidine, 1-(1-propenyl)-, reactions, with betaine, **6**, 314
Piperidine alkaloids, synthesis of, **2**, 938
3-Piperidinecarboxylic acid, *N*-(4,4-diphenylbut-3-enyl)-, synthesis of, **5**, 92
Piperidine ketals, synthesis of, **1**, 91
Piperidine oxide, *N*-methyl-, cleavage of, **1**, 625
Piperidine-*N*-oxides
  Polonovski rearrangement of, **5**, 214
  synthesis of, **1**, 626
Piperidines
  alkylation of, **1**, 151
  α-alkylation of, **1**, 149
  reactions, with aldehydes, **2**, 310
  synthesis of, **2**, 320, 326; **3**, 421
    via reduction, **2**, 327
Piperidines, *N*-aryl-, synthesis of, **2**, 326
Piperidines, benzo-, synthesis of, **2**, 327
Piperidines, 2-cyano-1,2-dimethyl-, as starting materials, for dimethyl piperidines, **1**, 55
Piperidines, dimethyl-, from, 2-cyano-1,2-dimethylpiperidines, **1**, 55
Piperidines, 2,6-disubstituted, synthesis of, via catalytic hydrogenation, **2**, 326
Piperidines, 2-silyl-, synthesis of, **4**, 511
Piperidinium salts
  of dithioalkanoic acids, **5**, 546
  storage of, **5**, 549
Piperidinoborane, *t*-butyliminotetramethyl-, as starting material, for diborated carbodiimide, **5**, 1085
Piperidin-4-ones, 2,2,6,6-tetramethyl-, reactions, with perchloryl fluoride, **2**, 336
1-Piperidino-pentan-3-one, 4,4-dimethyl-, reactions, with methyl iodide, **1**, 254
Piperidinyl-1-oxy, 2,2,6,6-tetramethyl (TEMPO)
  electrooxidation of, **3**, 665
  for the oxidation of benzylic alcohol, **3**, 88
  use in the synthesis of 3-formyl benzofuran, **3**, 104
4-Piperidone, *N*-*t*-butoxycarbonyl-, ring-expansion of, **2**, 324
4-Piperidone, 1-methyl-, reactions, with chlorosulfene, **5**, 496
Piperonal, as starting material, for piperonylic acid, **5**, 114
Piperonylic acid, from, piperonal, **5**, 114
Pishchimuka reaction, ylide version of, **6**, 699
Pisonic acid, synthesis of, **5**, 74
Pivalaldehyde, reactions, with ethylidenetriphenylphosphorane, **1**, 729
Pivalamide, *N*,*N*-dimethyl, as starting material, for *t*-butylchloromethylenedimethylammonium chloride, **5**, 526
Pivalate, iodomethyl, reactions, with zinc, **4**, 385
Pivalates, properties of, **5**, 526
Pivaldehyde
  condensation of, **4**, 302
  reactions, with α-hydroxycarboxylic acids, **4**, 213
Pivalic acid, demethoxycarbonylation of, **4**, 846
Pivalimidoyl chloride, *N*-phenyl-, as starting material, for imidoylsilanes, **5**, 793
Pivalimidoyl chloride, *N*-*p*-tolyl-, condensation of, **5**, 880
Pivaloyl chloride
  acylation problems, **3**, 278
  reactions
    with alkylbis(trimethylsilyl)arsanes, **5**, 891
    with arylbis(trimethylsilyl)arsanes, **5**, 891
    with bis(trimethylsilyl)phosphanes, **5**, 878
    with bis(trimethylsilyl) telluride, **5**, 255, 542
    with oxalyl compounds, **5**, 396
    with secondary enaminones, **3**, 413
Pivaloyl fluoride, synthesis of, **5**, 4
Pivaloylimide chlorides, *N*-aryl
  from
    disilylarsanes, **5**, 891
    imidoyl chlorides, **5**, 891
Pivaloylimidoyl chloride, reactions, with phenylbis(trimethylsilyl)arsine, **5**, 791
Plant hormone, antheridiogen-An, **1**, 804
Platinic(IV) acid, hexachloro-
  as catalyst, for the synthesis of 1,1-bis(silyl)alkanes, **4**, 617
  as hydrosilylation catalyst, **4**, 1044
  See also Speier's catalyst
Platinum
  as catalyst, in the synthesis of phosgene, **6**, 413
  reactions
    with air, **5**, 48
    with alkyl halides, **2**, 582
    with aryl nitriles, **5**, 1140
  as reagent, for the synthesis of phenols, **2**, 659
Platinum, cycloocta-1,5-dienyldimethyl-, reactions, with 1,1-bisdiphenylphosphino ethene, **4**, 1028
Platinum, ethylenebis(triphenylphosphine)-, reactions, with rhodium bis(triflate) complexes, **4**, 716
Platinum, iodopentachlorophenylbis(triethylphosphine)-, reactions, with potassium cyanide, **5**, 1139
Platinum, stilbene bis(triphenylphosphine)-, reactions, with *trans*-1,2-dicyano-1,2-bis(trifluoromethyl)ethylene, **5**, 1139
Platinum, tetrakis(triphenylphosphine)-
  reactions
    with acid chlorides, **5**, 464
    with aldehydes, **5**, 466
    with cyclopropenones, **5**, 465
Platinum(0), tetrakis(triphenylphosphine)-
  as hydrosilylation catalyst, **4**, 1044
  reactions, with 1,1,1-tricyanoethane, **5**, 1140
Platinum(II) chloride, reactions, with bisdiphenylphosphino ethene, **4**, 1028
Platinum(II) complexes
  reactions, with iodomethane, **4**, 1029
  synthesis of, **2**, 581
Platinum(II) complexes, alkyl-, reactions, with alkyl halides, **2**, 584
Platinum(II) complexes, dimethyl-, synthesis of, **4**, 1028
Platinum(II) salts
  reactions
    with alkynes, **5**, 939
    with cyclopropanes, **2**, 584
Platinum(IV) chloride, use in the synthesis of ^{11}C-labeled phosgene, **6**, 414
Platinum(IV) complexes, alkylation of, **2**, 584
Platinum(IV) compounds, synthesis of, **2**, 584; **4**, 151
Platinum(IV) oxide catalysis
  for the reduction of esters, **1**, 16
  for the reduction of tertiary alcohols, **1**, 12
Platinum π-allyl compounds
  reactions
    with *N*-(alkylsulfinyl)phthalimides, **5**, 388
    with *N*-(arylsulfinyl)phthalimides, **5**, 388
Platinum carbamoyl complexes, synthesis of, **6**, 516, 517
Platinum carbene complexes, synthesis of, **5**, 939
Platinum carbonates, (π-allyl)-, synthesis of, **6**, 470
Platinum carbonyl complexes, synthesis of, **6**, 525
Platinum carbonyls, oxidative addition of succinic anhydride to, **5**, 465
Platinum catalysts, for the synthesis of dialkyl sulfides, **2**, 134
Platinum complexes
  in hydroformylation reactions, **2**, 583
  reactions
    with *S*-alkyl(triphenylstannyl)dithioformates, **6**, 332
    with ethene, **2**, 582
    with nucleophiles, **2**, 583
    with (*E*)-vinyl-1,2-dimethyl-1-trimethylsilane, **1**, 63

stability of, **2**, 581
synthesis of, **3**, 461; **4**, 955; **5**, 1089; **6**, 732
Platinum complexes, acylcyano-, synthesis of, **5**, 1140
Platinum complexes, alkyl-, bearing dithiocarbamate ligands, **2**, 581
Platinum complexes, anionic, with perfluorophenyl-bridged groups, **4**, 691
Platinum complexes, cationic trinuclear, synthesis of, **4**, 692
Platinum complexes, (chloromethyl)-, synthesis of, **4**, 151
Platinum complexes, dialkyl-, elimination of, **1**, 665
Platinum complexes, *cis*-dialkyltriphenylphosphine-, as starting materials, for imidoylplatinums, **5**, 801
Platinum complexes, dichloro-, synthesis of, **4**, 1028
Platinum complexes, dihalo-, reactions, with ethyl diazoacetate, **4**, 151
Platinum complexes, dinuclear, synthesis of, **4**, 692
Platinum complexes, ethyl-, synthesis of, **2**, 582
Platinum complexes, ethynyl-, reactions, with hydrogen chloride, **4**, 822
Platinum complexes, ferrocenyl-, synthesis of, **4**, 722
Platinum complexes, α-haloalkyl-, synthesis of, **4**, 151
Platinum complexes, (halomethyl)-, synthesis of, **4**, 151
Platinum complexes, hydroxo-, reactions, with methyl cyanoacetate, **3**, 568
Platinum complexes, methylene-bridged, palladium analogues of, synthesis of, **4**, 691
Platinum complexes, monoalkyl-, synthesis of, **2**, 582
Platinum complexes, phosphaalkenyl-, synthesis of, **6**, 690
Platinum complexes, trinuclear
from, triarylcyclopropenium bromides, **4**, 693
synthesis of, **4**, 692
Platinum halides, on silica, for the reduction of amines, **1**, 32
Platinum hexafluorides, synthesis of, **5**, 464
Platinum hydride complexes, cationic, synthesis of, **4**, 691
Platinum iodide, trimethyl-, synthesis of, **2**, 584
Platinum oxide
as catalyst, for the reduction of benzylic fluoride, **1**, 2
reactions, with acetic acid, **1**, 88
Platinumphosphine complexes, reactions, with 1,1-dihaloalkanes, **4**, 584
Platinum–phosphinoalkene complexes, synthesis of, **4**, 1023, 1028
Platinum–potassium complexes, synthesis of, **2**, 584
Platinums
reactions
with 1-haloalkyl phosphonates, **4**, 585
with phosphorus ylides, **4**, 585
Platinums, acyl-, synthesis of, **5**, 464
Platinums, 1-haloalkyl-, reactions, with phosphines, **4**, 585
Platinums, imidoyl-
from
cis-dialkyltriphenylphosphineplatinum complexes, **5**, 801
isonitriles, **5**, 801
synthesis of, **5**, 800
via oxidative addition with α-ketoimidoyl chlorides, **5**, 802
Platinum salts
reactions
with but-1-yn-4-ol, **5**, 944
with methylmagnesium iodide, **2**, 584
with organometallics, **2**, 581
Platinum tetrachloride
reactions
with alkylmercury compounds, **2**, 584
with magnesium compounds, **2**, 584
Platinum tetrakis(phenyldimethylphosphine), as starting material, for imidoylplatinums, **5**, 802
Platinum tetrakis(triphenylphosphine), as starting material, for imidoylplatinums, **5**, 802
Plaunotol, synthesis of, **3**, 646
Plumbane, bis(trimethylsilylethynyl)dimethyl-, reactions, with trialkylorganoboranes, **2**, 1102
Plumbane, chlorotriphenyl-, as starting material, for (iodomethyl)triphenylplumbane, **4**, 156
Plumbane, dimethylaminotrimethyl-, reactions, with diazomethane, **3**, 484
Plumbane, (iodomethyl)triphenyl-, from, chlorotriphenylplumbane, **4**, 156
Plumbane, trimethyl bis[(trimethylsilyl)amino]-, reactions, with trimethylsilyldiazomethane, **3**, 484
Plumbanes
reactivity of, **2**, 1101
synthesis of, **2**, 70
Plumbanes, alkynyl-
from
organolead alkoxides, **2**, 1101
organolead hydroxides, **2**, 1101
synthesis of, reviews, **2**, 1101
Plumbanes, aminotrialkyl-, uses of, **2**, 1101
Plumbanes, α-bromo-, synthesis of, **4**, 156
Plumbanes, enynyl-, synthesis of, **2**, 1101
Plumbanes, tetraalkyl-, synthesis of, **2**, 1102
Plumbanes, tetraalkynyl-, synthesis of, **2**, 1101
Plumbanes, trialkylmethoxy-, reactions, with trichloroacetonitrile, **5**, 717
Plumbyl bromide, tributyl-, reactions, with 2-methoxy stannanes, **2**, 70
Pluridone, synthesis of, **4**, 268
Pnicogen iodides, dimethyl, synthesis of, **2**, 491
Pnicogens, use of term, **2**, 481
PO-activated reactions
advantages of, **1**, 744
mechanism, **1**, 744
Podocarpate, *O*-methyl-, use in the synthesis of alkanoic acids, **5**, 39
Podocarpate, methyl *O*-methyl-, saponification of, **5**, 31
Polonovski reaction
of sulfitoamines, **2**, 357
for the synthesis of iminium ions, **3**, 419
of trimethylamine *N*-oxide, **4**, 311
Polonovski rearrangement, for the synthesis of *O*-acylhydroxylamines, **5**, 214
Polyacids, synthesis of, reviews, **5**, 24
Poly(acrylamidoxime) resins, synthesis of, **5**, 864
Polyalkylation, in the synthesis of enolates, **1**, 131
Polyalkynes, synthesis of, via Cadiot–Chodkiewicz reactions, **1**, 1018
Polyamidines, synthesis of, **5**, 753
Polyamidoxime complexes, synthesis of, **5**, 864
Polyamines, synthesis of, **2**, 344
Polyamino acids, synthesis of, reviews, **5**, 84
Polyanions, chemistry of, reviews, **1**, 890
Polyarsines
oxidation of, **2**, 495
as starting materials, for diarsines, **2**, 506
synthesis of, **2**, 506
Polyarsines, 2,2,2-trideuterioethyl, synthesis of, **2**, 507
Polyazapolycycles, synthesis of, **4**, 298
Polycarbonates, synthesis of, **6**, 470
Polycarbosilanes, synthesis of, **6**, 190
Polycondensation, for the synthesis of polycarbonates, **6**, 470
Polycyclics, tetrahydrofuran-based, synthesis of, **1**, 367
Polydibenzo-18-crown-6, use in the reduction of carbonyls, **2**, 61
Poly(dimethylsilene), metallation of, **6**, 190
Polyenamines, nitro
cycloadditions of, **2**, 789
synthesis of, **2**, 789
Polyene isoprenoids, synthesis of, **1**, 754
Polyenes
protonation of, **1**, 912
reactions, with perfluorobutanesulfinic acid, **2**, 156
synthesis of, **1**, 470, 489, 707, 787
Polyenes, conjugated, synthesis of, **1**, 736

8-Polyenes, synthesis of, **1**, 754
Polyene synthesis, sulfone elimination in, **1**, 612
Polyenyl cations, synthesis of, **1**, 901
Polyenylic cations, synthesis of, **1**, 912
Polyethers, use in the cleavage of carbon–boron bonds, **1**, 58
Polyhaloalkyls, sulfurization of, **5**, 479
Polyhaloaromatic compounds, peralkynylation of, **1**, 1009
Polyhaloethyl, as starting material, for alkynyl phenols, **2**, 657
Polyhaloheteroaromatic compounds, peralkynylation of, **1**, 1009
Polyhalo substrates, selectivity of, **1**, 107
Poly(HF-pyridine), reactions, with phosphorinane, **2**, 412
Polyisocyanates, synthesis of, **5**, 974
Polyisocyanates, phosphorylated, synthesis of, **5**, 1009
Polyisocyanides, synthesis of, **3**, 702
Polyketides, biomimetic synthesis of, **2**, 662
Polylithium organic compounds, synthesis of, **6**, 204
Polymer-bound catalysts, use in the alkylation of aldehydes, **2**, 71
Polymer-bound reagents, for diazo transfer reactions, **5**, 869
Polymerization
  avoidance in radical additions, **1**, 329
  in borane mediated radical reactions, **1**, 333
  of thioaldehydes, **3**, 330
  of thioketones, **3**, 330
Polymers
  synthesis of, **2**, 561, 565
  use in the synthesis of carbodiimides, **5**, 1065
Polymers, perfluorinated, synthesis of, **6**, 36
Polymer-supported phosphines, use in the synthesis of alkyl halides, **2**, 7
Polymethylene oxide, reactions, with fluorine, **6**, 251
Polyols
  monoacylation of, **5**, 138
  reactions, with methylal, **4**, 185
  synthesis of, **2**, 85
Polyphenols, synthesis of, **1**, 473
Polyphosphate, trimethylsilyl- (PPSE)
  synthesis of, **2**, 106
  use in the synthesis of benzylic iodides, **2**, 36
Polyphosphazenes
  silylation of, **4**, 570
  synthesis of, **2**, 468
Polyphosphines
  synthesis of, **2**, 827
    vinylphosphine sulfides as reagent for, **2**, 853
Polyphosphines, acyclic, synthesis of, **2**, 471
Polyphosphines, cyclic, synthesis of, **2**, 471
Polyphosphonic acid anhydrides, reactions, with epoxides, **2**, 458
Polyphosphoric acid
  as catalyst, for the synthesis of aromatic ketones, **3**, 280
  for the hydrolysis of nitriles, **5**, 25
  as reagent, for the hydrolysis of nitriles, **5**, 269
Polypropylene oxide, as starting material, for α-fluoro ethers, **4**, 44
Polysilacycloalkynes
  photolysis of, **2**, 1089
  synthesis of, **2**, 1089
Polysilanes, rearrangement of, **4**, 624
Poly(sila(*N*-substituted)imines), synthesis of, **6**, 671
Poly(styryldiphenylphosphine dibromide), as reagent, for the synthesis of ketenimines, **3**, 558
Polysulfide, ammonium
  reactions, with aldehydes, **5**, 639
  use in the synthesis of α-aminophosphine sulfides, **4**, 477
Polysulfides
  as byproducts, **2**, 130
  synthesis of, **2**, 233
Polysulfides, aryl, synthesis of, **2**, 723
Polysulfides, cyclic
  reactions, with phosphonium ylides, **3**, 363
  as reagents
    for the synthesis of thioaldehydes, **3**, 336
    for the synthesis of thioketones, **3**, 363
Polysulfides, dialkoxy, synthesis of, **2**, 100
Polysulfides, hydro, synthesis of, **2**, 211
N,N-Polythiobisamines, chemistry of, **2**, 361
Polyurethanes, synthesis of, **6**, 488
Poly(4-vinylpyridine), as acid acceptor, **6**, 425
Poly(4-vinylpyridinium) poly(hydrogen fluoride) (PVPHV), use in the synthesis of alkyl fluorides, **2**, 17
Polyynes, synthesis of, **1**, 1034
Ponzio reaction, for the synthesis of aryldinitromethanes, **4**, 429
(±)-Porantheridine, synthesis of, **2**, 345
Porcine pancreatic lipase, use in the synthesis of amines, **2**, 318
Porphyrin complexes
  chemistry of, **2**, 578
  synthesis of, **2**, 578
Porphyrin complexes, alkyl, synthesis of, **2**, 583
Potassamide, reactions, with trimethylhydrazinium chloride, **2**, 384
Potassiophosphides, alkylation of, **2**, 437
Potassium
  reactions
    with alkyl halides, **2**, 556
    with 1,3-diborete, **6**, 188
    with 2,5-dimethyl-1-phenylarsole, **4**, 1039
    with indium trichloride, **2**, 597
    with magnesium chloride, **2**, 558
    with methoxymethyl chloride, **4**, 360
  for the reduction of alkyl fluorides, **1**, 3
  for the reduction of alkyl sulfones, **1**, 24
  for the reduction of fluoroalkanes, **1**, 6
  for the synthesis of active metals, **2**, 958
Potassium, benzyl-, reactions, with cycloheptatriene, **1**, 865
Potassium, bis(trimethylsilyl)methyl-, synthesis of, **6**, 193
Potassium, butyl-, reactions, with ethers, **2**, 999
Potassium, *n*-butyl-, synthesis of, **2**, 557
Potassium, crotyl-, reactions, with chlorobis-(dimethylamino)borane, **2**, 529
Potassium, (2,4-dimethylpentadienyl)-, reactions, with cyclopentadienylsodium, **4**, 678
Potassium, silyl-, reactions, with halomethanes, **6**, 173
Potassium, triethylgermyl-, reactions, with dichloromethane, **6**, 185
Potassium, trimethylsilylmethyl-
  synthesis of, **4**, 653
  for the synthesis of allylpotassiums, **2**, 557
Potassium, triphenylsilyl-
  reactions, with benzoyl chloride, **5**, 417
  synthesis of, **2**, 545
Potassium, triphenylstannyl-, reactions, with enones, **2**, 600
Potassium, vinyl-
  from, ethene, **2**, 907
  reactions, with chlorosilanes, **2**, 907
Potassium acetate
  reactions
    with acyl chlorides, **5**, 182
    with arsenic(III) oxide, **2**, 494
    with chloromethylpentamethyldisilane, **4**, 614
    with (chloromethyl)trimethylgermane, **4**, 362
Potassium alkoxides
  reactions, with thiophosgene, **6**, 534
  rearrangements of, **1**, 800
  use in sigmatropic rearrangements, **1**, 402
Potassium *O*-alkyl dithiocarbonates
  acylation of, **6**, 546
  alkylation of, **6**, 546
  reactions
    with *O*-alkyl chlorothiocarbonates, **6**, 547
    with carboxylic acid chlorides, **6**, 546
    with 2-chloro-4,5-dihydroimidazole, **6**, 547

with phthaloylglycyl chloride, **6**, 546
Potassium *O*-alkyl monothiocarbonates, reactions, with *O*-alkyl chlorothiocarbonates, **6**, 551
Potassium *O*-alkyl thiocarbonate, synthesis of, **6**, 479
Potassium *O*-alkyl xanthates, reactions, with dimethyl sulfate, **6**, 472
Potassium alkynyl selenolates, reactions, with alcohols, **5**, 519
Potassium amide
reactions
with aminotrialkylarsonium halides, **2**, 504
with nitriles, **5**, 750
with quaternary ammonium halides, **1**, 628
Potassium amides, synthesis of, **2**, 423
Potassium amides, *N*-alkyl-, synthesis of, **2**, 423
Potassium 3-aminopropylamide (KAPA)
reactions, with vinyl sulfides, **1**, 590
synthesis of, **2**, 423
use in the synthesis of alkynes, **1**, 1072, 1085
Potassium 9-*sec*-amyl-9-boratabicyclo[3.3.1]nonane, reactions, with 4-methylbenzonitrile, **3**, 93
Potassium arsolyls, reactions, with iron(II) chloride, **4**, 1039
Potassium azide, reactions, with *t*-butylcarbonic diethylphosphoric anhydride, **6**, 489
Potassium azodicarboxylate, decomposition of, **1**, 74
Potassium bicarbonate, reactions, with resorcinol, **5**, 110
Potassium bis(boryl)methanide, from, bis(dimesitylboryl)methane, **6**, 199
Potassium bis(trifluoromethyl)amide, chlorination of, **6**, 240
Potassium bis(trimethylsilyl)amide, reactions, with selenocarboxylates, **5**, 522
Potassium bromodifluoroacetate, as starting material, for difluorodiiodomethane, **6**, 227
Potassium bromodinitromethides, from, 2-halo-2,2-dinitroethanol, **6**, 287
Potassium *t*-butoxide
for the deprotonation of vinyl ethers, **4**, 949
for the elimination of sulfones, **1**, 612
for the migration of double bonds, **1**, 795
as a nucleophile, **6**, 425
reactions
with aldehydes, **1**, 43
with aliphatic sulfoxides, **1**, 600
with alkenes, **1**, 882
with *n*-butyllithium, **2**, 557
with (+)-camphenilone, **1**, 868
with chlorocyclopropanes, **1**, 571
with chloroform, **4**, 16, 737, 740
with 1,2-dichloro-1,2-diisopropoxyethane, **4**, 767
with dichlorofluoromethane, **4**, 751, 753
with 3-hydroxypropylphosphonium salts, **1**, 268
with ketones, **1**, 43
with tricyclo-[4.1.0.0^{2,7}]-heptane, **1**, 848
with vinyl triflates, **1**, 1143
use in dehydrobromination reactions, **1**, 572
use in the elimination of alcohols, **1**, 579
use in the elimination of hydrogen cyanide, **1**, 560
use in the synthesis of alkynes, **1**, 1065
use in the synthesis of phosphonium ylides, **1**, 727
Potassium *O*-butyl dithiocarbonate, reactions, with ethyl chloroformate, **6**, 546
Potassium butyrate, oxidation of, **6**, 469
Potassium carbonate
reactions
with acetophenone, **5**, 70
with alkyl bromides, **6**, 462
with dibromo(methyl)arsine, **2**, 495
with (dichloroalkyl)phosphines, **5**, 1152
with ferric chloride, **2**, 214
with tosylhydrazones, **3**, 471
use in the synthesis of hemiaminals, **4**, 300
Potassium chloride
reactions
with acetamidine hydrochlorides, **5**, 763
with butanoamidine hydrochlorides, **5**, 763
with heptafluorobutanamidine hydrochloride, **5**, 763
Potassium chlorodinitromethides, from, 2-halo-2,2-dinitroethanol, **6**, 287
Potassium 18-crown-6 salts, synthesis of, **5**, 1140
Potassium cyanate
reactions
with alkyl esters, **5**, 964
with anisole, **5**, 986
with carbonyl fluoride, **5**, 1015
with dichloro compounds, **4**, 441
with ethanolamine, **6**, 485
with germylmethyl halides, **4**, 521
with sulfur trioxide, **5**, 994
with tributyl phosphate, **5**, 995
Potassium cyanide
reactions
with alkenyldialkylboranes, **2**, 748
with *O*-alkylphosphinothiochloridates, **5**, 1131
with aryl azides, **5**, 1124
with bishydrazonyl disulfides, **5**, 641
with bis(pentachlorophenyl)thallium(III) perchlorates, **5**, 1143
with bis(trimethylsilyl) sulfate, **5**, 1135
with chloroarsines, **5**, 1132
with chloro(cyanomethyl)bis(triphenylphosphine)palladium(II), **5**, 1138
with chlorovinylnickel complexes, **5**, 1137
with 4-(dimethylamino) phenylmercuric acetate, **5**, 1141
with diphenyl ditellurides, **5**, 1117
with iodopentachlorophenylbis(triethylphosphine)platinum, **5**, 1139
with *o*-nitrobenzaldehyde, **2**, 783
with phenyl cyanate, **6**, 624
with phenyltellurium halides, **5**, 1117
with 2-propynols, **3**, 649
with pyridazine 1-oxide, **3**, 648
with sulfonamides, **5**, 1125
with triarylbismuth(V) halides, **5**, 1133
with trichloromethanesulfonyl chloride, **2**, 272
with triethylgermanium ethyl sulfonate, **5**, 1136
with vinyl bromide, **3**, 658
Potassium 4-cyanobutanoate, reactions, with hydrochloric acid, **5**, 34
Potassium cyanotriphenylborate, reactions, with triphenylmethyl chloride, **5**, 1147
Potassium cyclooctaldraenide, use in the synthesis of cyclooctatetraene-bridged complexes, **4**, 672
Potassium *N*,*N*-dialkyl dithiocarbamates, reactions, with 4-chloro-3-nitrocoumarin, **6**, 562
Potassium dicarbonylcyclopentadienylium, use in the synthesis of carbon-bridged diiron complexes, **4**, 679
Potassium dichromate, use in the synthesis of *α*,*β*-unsaturated ketones, **3**, 215
Potassium *O*,*O*-diethyldithiophosphate, reactions, with trichloronitromethane, **6**, 316
Potassium dimsylate, reactions, with benzothiazines, **1**, 622
Potassium diorganophosphides, synthesis of, **2**, 432
Potassium diphenyl dithiophosphinate, reactions, with imidoyl chlorides, **5**, 738
Potassium diphenylphosphine, reactions, with *N*-phenylbenzimidoyl chloride, **5**, 784
Potassium disulfide, use in the synthesis of disulfides, **2**, 225
Potassium dithiobenzoate, reactions, with sodium chloroacetate, **5**, 83
Potassium enolates
nitration of, **3**, 190
reactions, with dialkyl disulfides, **3**, 35
synthesis of, **4**, 838
triethylboron *ate* complexes of, **1**, 133
Potassium 1,2-ethanebis(trithiocarbonate), reactions, with hydrogen chloride, **6**, 553

Potassium ethoxide, reactions, with thiadiazoles, **5**, 480
Potassium *O*-ethyl dithiocarbonate
 reactions
  with 3-bromo-5-methylhexan-2-one, **6**, 546
  with bromotriphenylmethane, **6**, 546
  with chloroacetaldehyde, **6**, 546
  with 2-chloromethylpyridine, **6**, 545
  with 2-chloromethylquinoline, **6**, 545
  with *V.harveyi* oxidoreductase, **6**, 549
Potassium ethyl isocyanoacetate, reactions, with ethyl chloroformate, **3**, 706
Potassium ethyl xanthate
 reactions, with dithiolium salts, **6**, 85
 use in the synthesis of alkyl xanthates, **2**, 123
Potassium ferrate, for the oxidation of benzylic alcohol, **3**, 88
Potassium ferricyanide
 reactions, with sodium sulfide, **3**, 377
 use in asymmetric *cis*-dihydroxylation reactions, **2**, 48
Potassium fluoride
 as coreagent, **6**, 602
 enhancing solubility of, **5**, 4
 reactions
  with acid chlorides, **5**, 4
  with benzoyl trimethylsilane, **5**, 437
  with 1-chloro-1,2,2-trifluoroethanethiol, **5**, 484
  with difluoroaminocarbonyl fluoride, **6**, 257
  with 1,1-difluorothiols, **5**, 474
  with 2,2-dihalovinyl sulfones, **4**, 783
  with glycols, **2**, 19
  with α-keto alkyl chlorides, **4**, 3
  with perfluoro acetone, **5**, 987
  with potassium trinitromethanide, **6**, 261
  with thioacyl bromides, **5**, 476
 as starting material, for fluoroformate esters, **6**, 422
 use in Hofmann rearrangements, **5**, 998
 use in the synthesis of chlorofluoroalkanes, **4**, 30
Potassium fluoro sulfinate, reactions, with sulfonyl chlorides, **2**, 183
Potassium graphite
 reactions, with zinc chloride, **2**, 590
 as reagent, for the synthesis of alkenes, **1**, 439
 as a source of active potassium, **2**, 959
Potassium halides, reactions, with mesylates, **2**, 9
Potassium hexachloroplatinate
 reactions
  with alkylmercury compounds, **2**, 584
  with magnesium compounds, **2**, 584
  with methylcobalamin, **2**, 584
Potassium hexacyanoferrate(III), reactions, with thioamides, **5**, 735
Potassium hexamethyldisilazane (KHMDS), use in the synthesis of alkanoic esters, **5**, 148
Potassium hexamethyldisilazide
 reactions, with selenoesters, **4**, 841
 use in the elimination of alcohols, **1**, 579
 use in the synthesis of alkenes, **1**, 751
Potassium hydride
 reactions
  with alkoxyboranes, **2**, 534
  with *threo*-β-hydroxysilanes, **1**, 761
  with organoboranes, **2**, 534
 for the reduction of bromoalkanes, **1**, 9
 use in the synthesis of alkenes, **1**, 63
 use in the synthesis of oximes, **5**, 219
Potassium hydroborates, synthesis of, **2**, 534
Potassium hydrogen carbonate, reactions, with benzyl bromide, **6**, 462
Potassium hydrogen difluoride, reactions, with *N,N*-diethyl-1-propynamine, **4**, 790
Potassium hydrogen sulfide, as reagent, for the synthesis of thioketones, **3**, 357
Potassium hydrosulfide, reactions, with chloroisoxazoles, **5**, 531
Potassium hydroxide
 reactions
  with aldehydes, **1**, 42
  with 1-(*o*-amidophenyl)-1-(8-methyl-1-naphthyl) ethane, **5**, 108
  with aryl selenocyanates, **2**, 734
  with α-bromoethyl ethyl sulfone, **1**, 704
  with 7,12-dicyanobenz[*k*]fluoroanthene, **5**, 108
  with 3-(2-hydroxyethyl)benzothiazolin-2-ones, **6**, 545
  with 3-(3-hydroxypropyl)benzothiazolin-2-ones, **6**, 545
  with ketones, **5**, 48
  with methanol, **4**, 748, 906
  with *N*-methyl-*N*-nitrosoacetamide, **3**, 467
  with 1,8-naphthalic anhydride, **5**, 109
  with polyfluorocycloalkenes, **5**, 59
  with *N*-(trimethylsilyl)methyl-*N*-nitrosourea, **3**, 483
 as reagent, for the hydrolysis of nitriles, **5**, 269
 for the reduction of nitroalkancs, **1**, 37
 use in the cleavage of organoboranes, **1**, 60
 use in the synthesis of alkynes, **1**, 1065
Potassium hydroxide, methanolic, reactions, with bis(acyl) selenides, **5**, 240
Potassium hypobromite, reactions, with succinimide, **5**, 86
Potassium hypoiodite
 reactions
  with acetone, **6**, 216
  with triiodomethane, **6**, 216
Potassium iodate, use in the dihydroxylation of alkenes, **2**, 45
Potassium iodide
 reactions
  with *N*-alkylcarbonimidic dichlorides, **6**, 612
  with arenesulfonyl chlorides, **2**, 724
  with benzoyl chloride, **5**, 19
  with α-bromo ylides, **5**, 897
  with phenyl bis(chloromethyl)phosphinate, **4**, 127
  with phosphoric acid, **2**, 34
 use in the synthesis of Rieke's magnesium, **2**, 558
Potassiumiron(III) cyanide, for the nitration of nitroalkanes, **4**, 428
Potassium *O*-isopropyl dithiocarbonate
 reactions
  with 2-chlorocyclopentanone, **6**, 546
  with (*Z*)-4-phenylbut-3-enyl methanesulfonate, **6**, 548
Potassium isothiocyanate, acylation of, **6**, 565
Potassium manganate(VII), for alcohol oxidation, **3**, 116
Potassiummercury(II) cyanide, reactions, with halotrialkylsilanes, **5**, 1133
Potassium metabisulfite, reactions, with sugars, **4**, 237
Potassium methoxide
 reactions
  with 2,4-dinitro-5-methoxyselenophene, **6**, 100
  with 2,4-dinitro-5-methoxythiophene, **6**, 97
  with methanol, **2**, 344
Potassium 4-methyl-2,6-di-*t*-butyl phenoxide, use in the synthesis of aliphatic β-keto acids, **5**, 68
Potassium *O*-methyl dithiocarbonate, reactions, with methoxydichloromethanesulfenyl chloride, **6**, 548
Potassium *N*-methyl *N*-thioformyl dithiocarbamate, oxidation of, **6**, 564
Potassium *S*-methyl trithiocarbonate, reactions, with methanethiosulfonic acid esters, **6**, 553
Potassium monothiocarbonates, oxidation of, **6**, 479
Potassium naphthalenide
 reactions, with cuprous iodide, **1**, 472
 use in the synthesis of carbon-bridged dititanium complexes, **4**, 671
 use in the synthesis of carbon-bridged zirconium complexes, **4**, 671
Potassium nitrate, reactions, with alcohols, **2**, 104
Potassium nitrite, reactions, with arylamine, **1**, 442
Potassium nitronates
 reactions
  with allyl halides, **1**, 156

with *p*-toluenesulfonyl iodide, **4**, 331
with tosyl azide, **4**, 331
Potassium *O*-alkyl dithiocarbonates, reactions, with *N*-arylbenzimidoyl chlorides, **6**, 559
Potassium *O*-ethyl dithiocarbonate, as starting material, for thiocarbamates, **6**, 559
Potassium organic seleno sulfates, as starting materials, for diselenides, **2**, 292
Potassium perfluoro 2-or 3-aminopropanoates, as starting materials, for *N*-trifluorovinyl secondary amines, **4**, 794
Potassium permanganate
as dihydroxylating agent, **2**, 44
for the oxidation of dithioacetals, **4**, 265, 272
for the oxidation of sulfinic acids, **2**, 193
reactions
with benzene, **5**, 47
with 1-chloro-2,3,3-trifluorocyclobutene, **5**, 59
with homoallylic alcohols, **4**, 168
with naphthalenes, **5**, 74
use in the synthesis of monosulfones, **4**, 854
Potassium persulfate, as oxidizing agent, for the synthesis of acyl peroxides, **5**, 205
Potassium phenolate, reactions, with tribromoethylene, **4**, 761
Potassium phenoxide, reactions, with phenol, **1**, 856
Potassium *N*-phenyl dithiocarbamate, reactions, with 4-chloro-3-nitrocoumarin, **6**, 562
Potassium phenyl selenide, reactions, with 1,1-dinitro-2,2-diphenylethene, **4**, 911
Potassium phosphides, as starting materials, for phosphaalkenes, **5**, 883
Potassium phthalimide
reactions
with aryl halides, **2**, 753
with diethyl chlorophosphite, **5**, 369
with *O*-diphenylphosphinylhydroxylamine, **5**, 360
with diphenyl phosphochloridate, **5**, 372
with phosphine oxides, **4**, 473
with phosphorus trichloride, **5**, 369
with pyridinium salts, **4**, 448
silylmethylation of, **4**, 513
use in the synthesis of α-amino ketones, **3**, 179
Potassiums, α-alkoxyalkyl-, synthesis of, **2**, 999
Potassiums, alkyl-
from, organomercury compounds, **2**, 556
synthesis of, **2**, 556; **6**, 393
Potassiums, alkynyl-, reactions, with copper(I) salts, **2**, 1095
Potassiums, allyl-
alkylation of, **1**, 113
synthesis of, **2**, 557
use in the synthesis of allylalkylboranes, **2**, 521
Potassiums, benzyl-, from, toluene, **2**, 557
Potassiums, α-fluoroalkenyl-, synthesis of, **4**, 819
Potassiums, prenyl-, reactions, with bromoalkanes, **1**, 113
Potassiums, unsymmetrical allyl-, use in the synthesis of α-santalol, **1**, 113
Potassium salt, reactions, with oxalyl chloride, **5**, 12
Potassium salt, dibenzyl iminodicarboxylate-, *N*-alkylation of, **4**, 101
Potassium salts
from, dialkylmercury compounds, **1**, 378
reactions
with amines, **4**, 977
with α-chloroketones, **6**, 627
rearrangement of, **1**, 378
structural determination of, **2**, 547
Potassium salts, *O*-alkyldithiocarbonate-, reactions, with *N*-arylimidoyl chlorides, **5**, 632
Potassium salts, cyanothioformamide-, alkylation of, **5**, 592
Potassium salts, ethyl xanthate-, reactions, with alkyl iodides, **6**, 478
Potassium sand, reactions, with bis(trimethylsilylmethyl)mercury, **4**, 653
Potassium selenobenzoate, reactions, with trimethylsilyl chloride, **5**, 255
Potassium selenocarboxylates
oxidation of, **5**, 252
reactions of, **4**, 288
reactions, with chlorosilanes, **5**, 541
as starting materials, for aroyl aryl diselenides, **5**, 250
synthesis of, **5**, 240
Potassium selenocyanate
reactions
with alkenes, **5**, 1115
with cyclohexene, **5**, 1115
with selenides, **5**, 1114
use in the synthesis of aryl selenocyanates, **5**, 1113
Potassium selenocyanates
reactions
with alkyl halides, **5**, 1053
with diazomethane, **5**, 1056
with ethyl diazoacetate, **5**, 1056
with trityl chloride, **5**, 1053
Potassium seleno sulfate
alkylation of, **2**, 292
from, selenium, **2**, 292
reactions, with ethylenimine, **2**, 237
synthesis of, **2**, 237
Potassium sulfinate, synthesis of, **2**, 272
Potassium superoxide
for the oxidation of sulfides, **2**, 150
for the oxidation of sulfinic acids, **2**, 193
reactions
with alkyl bromides, **2**, 98
with *N*-benzylamide, **2**, 649
with *N*-chloroamines, **3**, 407
with 1,4-dinitrobenzene, **2**, 648
use in the synthesis of azo compounds, **2**, 772
Potassium *t*-butoxide
for the isomerization of allyl ethers, **2**, 675
reactions, with butynes, **4**, 1028
Potassium tellurocyanate
reactions, with benzylhalides, **5**, 1117
as starting material, for benzyl tellurocyanates, **2**, 290
Potassium tetrafluorocobaltate, reactions, with methyl benzoate, **4**, 3
Potassium 2,2,6,6-tetramethyl-2,6-disilapiperidide, as starting material, for primary amines, **2**, 300
Potassium thiobenzoate, reactions, with *p*-methylbenzoylsulfenyl bromide, **5**, 252
Potassium thiocarboxylates
oxidation of, **5**, 251
reactions
with *N*-chlorosuccinimide, **5**, 253
with diphenyltin dichloride, **5**, 245
with phenyltellurenyl bromides, **5**, 249
Potassium thiocyanate
reactions
with acetophenones, **3**, 165
with acid chlorides, **5**, 1047
with alkynes, **5**, 1044
with benzyl halides, **5**, 1039
with bromine, **5**, 1107
with α-bromo alkyl isocyanates, **4**, 434
with α-bromo alkyl isothiocyanates, **4**, 434
with 9-chloroacridines, **5**, 1040
with hydrofluoric acid, **5**, 593
with β-mannosylchitobiose trisaccharides, **5**, 1042
with *o*-nitrobenzene sulfonates, **5**, 1108
with picryl fluoride, **5**, 1040
with pyridinium chloride, **5**, 1040
use in the synthesis of cyclohexene sulfide, **2**, 136
use in the synthesis of isothiocyanates, **5**, 1032
Potassium thiocyanates, for the synthesis of alkyl thiocyanates, **5**, 1105
Potassium tribiphenylenephosphates, synthesis of, **2**, 869

Potassium tri-*s*-butylborohydride
for the conjugate reduction of enones, **3**, 127
reactions, with copper(I) iodide, **1**, 999
Potassium trichloroacetate, bromination of, **6**, 221
Potassium trichlorodicyanoantimonate(III), reactions, with sodium cyclopentadienide, **5**, 1132
Potassium tricyanomethanide
reactions, with methane sulfonate salts, **4**, 977
as starting material, for ketenimines, **4**, 977
Potassium trifluoro methoxide, synthesis of, **6**, 232
Potassium trifluoromethylfluoroborate, pyrolysis of, **4**, 7
Potassium triisopropoxyborohydride (KIPBH), use in the synthesis of thiols, **2**, 126
Potassium trinitromethanide
reactions
with fluorine-sodium fluoride, **6**, 261
with potassium fluoride, **6**, 261
with pyridinium salts, **4**, 995
Potassium triphenylborohydride
for the conjugate reduction of enones, **3**, 127
for the reduction of enones, **1**, 86
Potassium triphenylcyanoborate, reactions, with benzenediazonium tetrafluoroborate, **5**, 1148
Potential-energy distribution (PED), of thioformic *O*-acids, **5**, 507
Praseodymium compounds
reactions
with phosphorus ylides, **4**, 578
with sodiumiron tetracarbonyl, **6**, 518
synthesis of, **6**, 518
Precalciferol
1,7 sigmatropic rearrangements of, **1**, 403
as starting material
for isopyrocalciferol, **1**, 419
for pyrocalciferol, **1**, 419
Precapnelladiene, synthesis of, **1**, 632
Precorrins, synthesis of, **2**, 578
Pregnane, 20-keto-, synthesis of, **3**, 643
5α-Pregnane-3β,20-diols, 18-amino-, reactions, with nitrous acid, **1**, 259
5α-Pregnene-3β,20-diol, 14β,18-cyclo-, synthesis of, **1**, 259
Pregnenolone, construction of side chains on, **4**, 939
5-Pregnen-20-one, 3β-acetoxy-, oxidation of, **5**, 74
Pregnenynes, from, tosylhydrazones, **1**, 1077
Prelog condensation, use in the synthesis of phenols, **2**, 662
Prelog–Djerassi lactone framework, synthesis of, **1**, 832
Prephenate, synthesis of, **1**, 826
Prévost reaction, mechanism, **2**, 44
Primary cation instability, **1**, 894
Prins addition, of formaldehyde to tetrafluoroethene, **4**, 44
Prins reaction, for the synthesis of trihalides, **6**, 4
Prodrugs, synthesis of, **6**, 427
Prodrugs, ester, of β-lactam antibiotics, synthesis of, **4**, 205
Progesterones, synthesis of, **3**, 643
Proline, *N*-chlorocarbonyl compounds of, **6**, 446
L-Proline ethyl ester
reactions, with phosphinic chloride, **2**, 455
use in the synthesis of chiral phosphonates, **2**, 459
Prolinol compounds, synthesis of, **2**, 324
Propadiene, 1-bromo-1-(trimethylgermyl)-, synthesis of, **4**, 811
Propadiene, perfluoro-1,1,3-tris(dimethylamino)-, reactions, with perfluoro-*N*-bromodimethylamine, **4**, 794
Propa-1,2-diene, 1-lithio-1-phenyl-, reactions, with ketones, **3**, 250
Propadienes, 1,3-bis(trimethylsilyl) 1-silyl 3-phenyl-, synthesis of, **4**, 1047
Propadienes, *gem*-bromosilyl-, synthesis of, **4**, 807
Propadienes, *gem*-halo(trimethylstannyl)-, synthesis of, **4**, 813
Propa-1,2-dienes, polyalkyl-, synthesis of, **1**, 967
Propanal
reactions
with diazo ketones, **3**, 321
with mercury(II) chloride, **4**, 697
synthesis of, **1**, 85
Propanal, aldehyde 3-nitro-, synthesis of, **3**, 46
Propanal, *S*-2-benzyloxy-, use in the synthesis of chiral dihydropyrimidines, **4**, 408
Propanal, 2,2-dimethyl 3,3-bis(methylthio)-, synthesis of, **3**, 36
Propanal, 2-phenyl-
reactions
with enethiolate, **5**, 557
with ethyl phosphonates, **1**, 748
Propanamidrazone, methylation of, **5**, 773
Propanamidrazone, N^1-phenyl-N^1,N^2,N^3-trimethyl-, structure of, **5**, 768
2-Propanamine, *N*,*N*-dimethyl-1-ferrocenyl-, deprotonation of, **4**, 1032
Propanamines, (arylthio)-, synthesis of, **2**, 320
Propan-1,3-diol, 2-amino-, reactions, with formaldehyde, **4**, 302
Propane, 1,3-bis(diazo)-, synthesis of, **3**, 467
Propane, 1,1-bis(diethylboryl)-3-chloro-, synthesis of, **4**, 631
Propane, 2,2-bis(difluoroamino)-, pyrolysis of, **5**, 812
Propane, 1,3-bis(diiodoarsino)-, synthesis of, **2**, 490
Propane, 1,3-bis(diphenylarsino)-, reactions, with hydroiodic acid, **2**, 490
Propane, 1,2-bis(diphenylphosphino)-, reactions, with 1,3-diiodopropane, **2**, 444
Propane, 2,2-bis(methylthio)-1,3-diphenyl-, reactions, with *m*-chloroperbenzoic acid, **4**, 268
Propane, 1-bromo-3-chloro-, reactions, with benzhydrylamine, **2**, 322
Propane, 3-bromo-1-chloro-, as reagent, **1**, 168
Propane, 1-bromo-1-fluoro-1-nitro-, synthesis of, **6**, 55
Propane, 2-bromo-2-nitro-, reactions, with thiolates, **4**, 327
Propane, 1,3-diamino-
reactions
with carbon oxide sulfide, **6**, 509
with glyoxal, **4**, 409
Propane, 2-diazo-, reactions, with diphenylsulfine, **6**, 316
Propane, 1,3-dibromo-
reactions
with *C*,*S*-dianions, **1**, 140
with diphenyl methyl isocyanide, **3**, 702
as starting material, for 3-methylcyclohex-2-enone, **1**, 141
use in the synthesis of cyclic alkoxycarbene complexes, **5**, 943
Propane, 2,2-dibromo-, synthesis of, **4**, 24
Propane, 1,3-dihydroxy-2-hydroxymethyl-2-methyl-, reactions, with ortho esters, **6**, 79
Propane, 1,3-diiodo-, reactions, with 1,2-bis(diphenylphosphino)propane, **2**, 444
Propane, 1,3-diisocyano-, deprotonation of, **3**, 720
Propane, 2,2-dimethoxy-
as dehydrating agent, for the synthesis of acetals, **4**, 178
use in the synthesis of acetals, **4**, 184
Propane, 2,2-dinitro-, reactions, with alkynyllithiums, **1**, 1033
Propane, (-)-1,2-diphenyl-, synthesis of, **1**, 463
Propane, 3,3-diphenyl-3-iisocyano-1-bromo-, synthesis of, **3**, 703
Propane, 1,2-epoxy-3-(4-methoxyphenoxy)-, reactions, with butyrolactone, **6**, 72
Propane, 1,2-epoxy-3-phenoxy-, reactions, with butyrolactone, **6**, 72
Propane, 2-(*N*-fluorimino)-, synthesis of, **5**, 812
Propane, 1-iodoheptafluoro-, reactions, with phenylmagnesium bromide, **6**, 63
Propane, 2-iodo-2-nitro-, reactions, with sodium benzene sulfinate, **4**, 332
Propane, 2-methyl-2-nitro-, reactions, with *p*-substituted benzaldehydes, **3**, 433

Propane, 2-methyl-2-nitroso-, synthesis of, **2**, 389
Propane, nitro-, as starting material, for propionitrile, **3**, 620
Propane, 2-nitro-
  reactions
    with benzyl halides, **3**, 88
    with thiolates, **4**, 327
  as starting material, for *gem*-fluoronitro compounds, **4**, 108
Propane, 1,1,3,3-tetramethoxy-, reactions, with 2-thiolethanol, **4**, 230
Propane, 1,1,3,3-tetranitro-, fluorination of, **4**, 795
Propane, 1,1,3,3-tetraphenyl-1,3-diisocyano-, synthesis of, **3**, 702
Propane anions, 2-nitro-, reactions, with nitrobenzyl halides, **1**, 156
1,3-Propanediamine, *N,N'*-dibenzoyl-*N,N'*-dinitroso-, reactions, with sodium hydroxide, **3**, 467
1,3-Propanediamine, *N*-methyl-, silylation of, **2**, 543
1,3-Propanedinitrile, 2-(2,2'-dimethylbenzhydryl)-, as starting material, for *β*-(2,2'-dimethylbenzhydryl) glutaric acid, **5**, 35
Propanediol, 2,2-dinitro-, synthesis of, **4**, 428
Propane-1,3-diol, 2,2-dinitro-, cleavage of, **6**, 56
Propanediolboronic esters, as starting materials, for lithiums, **6**, 400
Propanedithiol, reactions, with alkynic ketones, **4**, 258
1,3-Propanedithiol, reactions, with aldehydes, **5**, 410
Propane-1,3-dithiol, reactions, with 3,3-dichloropropenal, **3**, 72
1,3-Propanedithiol, reactions, with hexachlorobutadiene, **4**, 770
Propane-1,3-dithiol, reactions, with *α,β*-ynones, **3**, 168
Propane-1,3-dithiol, 2-methyl-2-thiohydroxymethyl-
  exchange reactions
    with triethyl orthoacetate, **6**, 83
    with triethyl orthoformate, **6**, 83
Propanenitrile, 2-ethoxy-2-phenyl-, synthesis of, **3**, 559
Propaneone, *N*-isopropyl-1-phenoxy-3-amino-, hydrogenation of, **2**, 54
Propanes, synthesis of, **6**, 4
Propanes, 2-alkyl-2-hydroxymethyl-1,3-dihydroxy-, reactions, with carboxylic acids, **6**, 71
Propanes, 3-chloro-1-cyano-2-hydroxy-, 1,2-elimination from, **3**, 650
Propanes, 3-substituted-1,2-diamino-, from, *N,N'*-bis-protected 2-(aminomethyl)aziridines, **2**, 309
Propanesulfenyl fluoride, perfluoro-, synthesis of, **2**, 166
Propanesulfinic acid, synthesis of, **2**, 188
Propane-2-sulfonyl fluorides, 1-alkylthio-2-*H*-pentafluoro-, synthesis of, **6**, 46
Propane 2 thiol, 2 methyl , irradiation of, **4**, 1077
Propanoate, methyl 2,3-dibromo-, hydrobromic acid hydrolysis of, **5**, 55
Propanoates, 2-chloro-, synthesis of, **1**, 182
Propanoates, 2,3-epoxy-3-trialkylstannyl, ring-opening of, **2**, 85
Propanoates, 2-sulfonyloxy-, optically active, reactions, **1**, 182
Propanoic acid, 2-aryl-, hydrolysis of, **5**, 30
Propanoic acid, 3-chloro-2,2,3-trifluoro-, synthesis of, **4**, 5
Propanoic acid, 2,3-dibromo-, synthesis of, **5**, 55
Propanoic acid, 3,3-dichloro-2,2-dimethyl-, synthesis of, **4**, 11
Propanoic acid, (2*S*,3*S*)-3-hydroxy-3-phenyl-2-methyl-, synthesis of, **5**, 80
Propanoic acid, (*R*)-(-)-2-(4'-isobutylphenyl)-, synthesis of, **5**, 118
Propanol, reactions, with chlorodialkylphosphines, **2**, 450
Propanol, 1-chloro-2-, transesterification of orthoformate esters with, **6**, 78
Propanol, 2-methyl-, reactions, with bromomethanesulfonyl bromide, **4**, 86
Propanol, 2-methylhexafluoroiso-, reactions, with cyanogen chloride, **5**, 1102
Propanol, *β*-nitro-, synthesis of, **2**, 392
Propanol, 3,3,3-tris(trimethylsilyl)-, deprotonation of, **4**, 620
1-Propanol, carbonylation of, **6**, 464
Propan-1-ol, (1*S*,2*R*)-2-(di-*n*-butyl)amino-1-phenyl-, use in the synthesis of chiral amines, **2**, 331
1-Propanol, 3-phenyl-, deuterium-labeled, reactions, with chromium trioxide, **5**, 47
Propan-2-ol, hydrogen abstraction from, **1**, 546
2-Propanol
  reactions
    with *α*-nitro ketones, **3**, 427
    with nitrosyl chloride, **5**, 823
Propan-2-ol, as reducing agent, **2**, 320
Propan-2-ol, 2-(1-adamantyl)-, as starting material, for adpoc fluoride, **6**, 422
2-Propanol, 2-(methylarsino)hexafluoro-, synthesis of, **2**, 482
Propanols, *α*-alkoxy-, synthesis of, **3**, 32
Propanols, 3-amino-, cyclization of, **2**, 323
Propanols, 3-*t*-amino-, synthesis of, **2**, 320
Propanols, *α*-arylsulfenyl-, synthesis of, **3**, 35
Propanols, 2,3-epoxy-3-aryl-, synthesis of, **1**, 269
Propanone, 1-phenyl-2-methylsulfenyl, synthesis of, **3**, 165
Propanothiohydroxamic acids, *N*-benzyl-3-aryl-3-phenyl-, synthesis of, **5**, 637
Propanoyl bromide, reactions, with nitroalkenes, **5**, 820
Propanoyl chloride, 1-chloro-1-(chlorosulfenyl)-, reactions, with sodium ethoxide, **4**, 72
Propanoyl chloride, 1,1-dimethyl-, reactions, with thallium 2-methylpropanoate, **5**, 183
Propanoyl chloride, 2,2-dimethyl-, reactions, with tetrazolylacetic acid, **5**, 193
Propanoyl chloride, 3-isocyanato-, synthesis of, **5**, 970
Propanoyl chlorides, synthesis of, **4**, 125
Propargyl, [3,3]-sigmatropic rearrangements of, **5**, 1041
Propargylation, of saturated ketones, **3**, 147
Propargylation, enantioselective, of aldehydes, **1**, 1055
Propargyl bromide, reactions, with phenothiazine, **2**, 1049
Propargyl bromide, trimethylsilyl-, reactions, with aluminum, **1**, 971
Propargyl bromides, *α*-substituted, reactions, with aldehydes, **1**, 1056
Propargyl carbanions
  1,2 addition–elimination reactions of, **1**, 1060
  addition reactions of, **1**, 1054
  substitution of, **1**, 1051
Propargyl cations, synthesis of, **1**, 904
Propargyl chloride
  reactions, with diethylborane, **4**, 631
  as starting material, for methyl 4-chloro-2-butynoate, **5**, 103
Propargyl halides
  carbonylation of, **1**, 976
  reactions, with terminal alkynes, **1**, 121
  as starting materials, for allenes, **1**, 958
Propargylic alcohols, homo-, as starting materials, for *γ*-lactones, **5**, 170
Propargylic bromides, as starting materials, for *α*-allenic alcohols, **1**, 967
Propargylic halides
  metallation of, **2**, 559
  reactions
    with alkynes, **1**, 1002
    with 1-alkynes, **1**, 1004
    with alkynides, **1**, 1001
    with allenylmagnesium halides, **1**, 1052
    with stannous chloride, **1**, 973
    with transition metals, **1**, 1004
  as starting materials, for allenynes, **1**, 975
Propargylics
  palladium-catalyzed reduction of, **1**, 962
  reduction of, **1**, 961
  as starting materials, for allenes, **1**, 966

synthesis of, **2**, 1056
Propargylides, reactions, with organoboranes, **1**, 976
Propargyl metals
1,4 addition reactions of, **1**, 1061
reactions, with aldimines, **1**, 1059
Propargyloxy groups, reactions, with aryl radicals, **3**, 104
Propargyls
carbometallation of, **2**, 975
reactions, with Grignard reagents, **1**, 970
zirconium-catalyzed carboalumination of, **2**, 977
Propargyl synthons, chlorofluoro-, reduction of, **1**, 958
[4.4.4]Propellahexaene, synthesis of, **1**, 634
Propellane, synthesis of, via xanthate pyrolysis, **1**, 776
Propenal, 3,3-dichloro-, reactions, with propane-1,3-dithiol, **3**, 72
Propenals, 2-acyloxy 3-heterosubstituted, enzymatic resolution of, **3**, 32
Propenals, bis-2,3-(*N*-alkylamino)-, synthesis of, **3**, 75
Propenals, α-silyloxy-, from, keto acids, **3**, 33
Propenamidrazones
reactions, with thiols, **5**, 770
synthesis of, **5**, 770
1-Propen-1-amine, 1-chloro-*N*,*N*-2-trimethyl-, reactions, with *N*-benzyloxycarbonyl amino acids, **5**, 15
Propene
chlorinolysis of, **6**, 215
metallation of, **4**, 668
reactions, with tripropylaluminum, **1**, 662
Propene, 1-azidopentafluoro-, synthesis of, **4**, 799
Propene, 2-bromo-, synthesis of, **2**, 622
Propene, (*Z*)-1-bromo-, synthesis of, **2**, 622
Propene, 3-bromo-1-(phenylsulfonyl)-, reactions, with organomagnesium halides, **1**, 144
Propene, 3-chloro-1-methoxy-, reactions, with alcohols, **4**, 199
Propene, 1-chloro-2-methyl-, hydroboration of, **4**, 144
Propene, 1-(4-chlorophenyl)-2-nitro-, reactions, with lithium organocuprates, **3**, 426
Propene, 1-(cyclohex-1-enyl)-3-cyclohexylidene-, synthesis of, **1**, 736
Propene, 1-diazo-2-methyl-, synthesis of, **3**, 488
Propene, (*Z*)-1,3-dibromo-1-trimethylsilyl-, synthesis of, **4**, 807
Propene, 1,2-dichloro-, reactions, with copper(0), **2**, 74
Propene, 2,3-dichlorotetrafluoro-, reactions, with dimethylarsine, **4**, 805
Propene, 1,2-dimetallated, synthesis of, **2**, 74
Propene, (*E*)-1,3-bis(trimethylsilyl)-, reactions, with *s*-butyllithium, **4**, 603
Propene, hexafluoro-
addition of iodine fluoride to, **6**, 3
reactions
with azide anions, **4**, 799
with benzyl azide, **6**, 54
with carboxylic acids, **5**, 6
with dimethylarsine, **6**, 60
with dimethylphosphine, **4**, 802
with *N*-iodobis(ditrifluoromethyl)amine, **6**, 54
with selenium, **3**, 395
with tetramethyldiarsine, **6**, 60
with tetramethyldiphosphine, **4**, 802
reduction of, **1**, 429
use in the synthesis of 2,3,3,3-tetrafluorooxyimidic fluoride, **5**, 821
Propene, 2-methoxy-, reactions, with diols, **4**, 186
Propene, 1-nitro-2-nitro-, as starting material, for 2-chloro-2-methylpropaneoxyimidic chloride, **5**, 820
Propene, 1,2,3,3,3-pentafluoro-1-trimethylsilyl-, reactions, with bis(trimethylsilyl)mercury, **4**, 1048
Propene, perfluoro-
oxidation of, **6**, 410
reactions
with sodium methoxide, **6**, 74
with sulfur dioxide, **2**, 175
with sulfur trioxide, **4**, 86
as starting material
for carbene-bridged diplatinum complexes, **4**, 691
for carbonic difluoride, **6**, 410
Propene, 3,3,3-trichloro-2-methyl-, reactions, with *p*-cresol, **4**, 745
Propene, 2-trifluoromethyl perfluoro-, reactions, with sulfur trioxide, **2**, 208
1-Propene, 2,3-bis(benzenesulfonyl)-, displacement of, **4**, 275
Prop-1-ene, 3-chloro-3,3-difluoro-2-trifluoromethyl-, reactions, with lithium aluminum hydride, **4**, 735
1-Propene, 1,3-dibromo-1-trimethylsilyl-, reactions, with butylcuprate, **2**, 931
1-Propene, 1,1-diethoxy-, reactions, with oximes, **3**, 722
1-Propene, 1,1-dilithio-2-methyl-, synthesis of, **4**, 1064
1-Propene, 3-dimethylphenylsilyl-3-triphenylgermyl-, synthesis of, **4**, 639
Prop-1-ene, (*E*)-(1)-morpholino-1-phenyl-, hydroboration of, **1**, 443
Prop-1-ene, (*E*)-1-phenyl-, detection of, **1**, 443
1-Propene, 1-phenylthio-3-(trimethylsilyl)-, deprotonation of, **4**, 938
1-Propene, 1,3,3,3-tetrafluoro-2-trifluoromethyl-1-methoxy-, synthesis of, **4**, 762
1-Propene, (*E*)-1,3,3-tris(trimethylsilyl)-, synthesis of, **4**, 603
2-Propenenitriles, 2-acetoxy-3-bromo-, synthesis of, **3**, 650
Propenes, pentafluoro-, synthesis of, **1**, 429
Propenes, phenylfluoro-, reduction of, **1**, 429
Propenes, (*Z*)-2-phenyltetrafluoro-, reactions, with lithium tetramethylpiperidide, **4**, 755
1-Propenes, 2-aryl-3-(*N*,*N*-dimethylamino)-, as starting materials, for quaternary ammonium salts, **1**, 253
Prop-1-enes, 3-chloro-1-cyano-, synthesis of, **3**, 650
1-Propenes, 1,3-dibromo-, reactions, with Grignard reagents, **1**, 115
Propene-2-sulfonyl fluoride, pentafluoro-, reactions, with alkyl hydrosulfides, **6**, 46
1-Propene systems, 2-aryl-, synthesis of, **1**, 253
Prop-1-ene systems, 2-aryl-, synthesis of, **1**, 253
Propenethial, 3-dimethylamino-, synthesis of, **3**, 338
2-Propenethiol, deprotonation of, **1**, 140
Propenitriles, (*E*)-3-phenyl-, reactions, with hydroxylamine, **5**, 763
2-Propenoate, ethyl 3-amino-3-(ethylthio)-, synthesis of, **4**, 905
2-Propenoate, methyl-, reactions, with alkynyl iodides, **1**, 1021
Propenoic acid anions, synthesis of, **1**, 634
Propen-2-ol, synthesis of, **2**, 636
Propen-2-ol, perfluoro-
from, benzyl enol ether, **2**, 641
reactions, with mercury trifluoroacetate, **6**, 64
Prop-2-en-1-ol, 2-ethyl-, isomerization of, **2**, 639
Prop-2-en-1-ol, 2-methyl-, isomerization of, **2**, 639
Prop-2-en-1-ol, 1,1,3,3-tetra-*t*-butyl-, as starting material, for tetra-*t*-butylallene, **1**, 981
Propenolic acid, 3,3-dichloro-, from, chloropropiolic acid, **4**, 743
Propen-2-ols, perfluoro-, synthesis of, **2**, 640
Propenylamine, 1-chloro-*N*,*N*-2-trimethyl-
as starting material, for acid chlorides, **5**, 1
synthesis of, **5**, 1
Propenylamine, *N*,*N*-dimethyl-1-chloro-2-methyl-, reactions, with zinc chloride, **2**, 758
Propenylamines, 1-halo-*N*,*N*-2-trimethyl-, as starting materials, for acid halides, **5**, 1
Propenyl anions, 1,3-bis(trimethylsilyl)-
reactions
with aldehydes, **1**, 651
with ketones, **1**, 651
1-Propenyllithium, 1-ethoxy-, isomerization of, **4**, 949

β-Propiolactones, β-alkenyl-, reactions, with enethiolates, **5**, 553
Propiolate, ethyl
hydroboration of, **1**, 532
reactions
with anthranilamides, **4**, 422
with diphenylacetonitrile, **1**, 537
with ethanol, **4**, 195
with iodopyrimidines, **1**, 1009
with ketone enolates, **1**, 537
use in the synthesis of (*E*)-3-[(*E*)-3-phenyl-2-propenoxy]acrylic acid, **5**, 105
Propiolate, methyl
reactions
with alkenes, **1**, 524
with alkenylketene acetals, **1**, 521
with 2-azabutadienes, **1**, 522
with *t*-butyl radicals, **1**, 546
with cyclopentadiene, **1**, 520
with dienes, **1**, 521
with phosphinyldiazomethane, **2**, 861
with 2-silyloxybutadienes, **1**, 520
with 1-trimethylsilylpenta-1,3-diene, **1**, 520
Propiolate, methyl bromo-, reactions, with aziridine, **4**, 793
Propiolate, methyl phenyl
reactions
with dienes, **1**, 522
with lithium dimethylcuprate, **1**, 540
Propiolate, triethyl orthophenyl
from
tetraalkyl orthocarbonate, **6**, 77
triethoxyacetonitriles, **6**, 78
Propiolic acid, chloro-
as starting material, for 3,3-dichloropropenolic acid, **4**, 743
synthesis of, **2**, 1016
Propiolic acid, phenyl
from, ethyl α,β-dibromo-β-phenylpropionate, **5**, 103
reactions, with iodine pentoxide, **4**, 749
Propiolic acids, synthesis of, **2**, 1061
Propiolyl fluoride, fluoro-
detection of, **2**, 1013
stability of, **2**, 1013
Propionaldehyde, reactions, with benzyltriphenylphosphonium salts, **1**, 732
Propionaldehyde, β-chloro-, as starting material, for β-chloropropionic acid, **5**, 59
Propionamide, *N*-phenyl-, reactions, with phosgene, **5**, 655
Propionamides, (1'*R*,3*R*)-3-hydroxy-*N*-(2-hydroxy-1-phenylethyl)-3-substituted, hydrolysis of, **5**, 66
Propionamides, (1'*R*,3*S*)-3-hydroxy-*N*-(2-hydroxy-1-phenylethyl)-3-substituted, hydrolysis of, **5**, 66
Propionamides, (1'*S*,3*R*)-3-hydroxy-3-phenyl-*N*-(1-substituted), hydrolysis of, **5**, 66
Propionamidine, 2-chloro-*N*,*N*,*N'*-trifluoro-, synthesis of, **5**, 815
Propionamidine, *N*,*N*,*N'*-trifluoro-, synthesis of, **5**, 815
Propionate, 1,2-dichloropropenyl-α-chloro-, synthesis of, **4**, 768
Propionate, ethyl 3-chloro-, reactions, with sodium, **4**, 171
Propionate, ethyl α,β-dibromo-β-phenyl-, as starting material, for phenylpropiolic acid, **5**, 103
Propionate, ethyl 3,3-diethoxy-, synthesis of, **4**, 195
Propionate, ethyl 2-isocyano-2-lithio-, reactions, with epoxides, **3**, 704
Propionate, ethyl β-phenyl-β-cyano-, as starting material, for phenylsuccinic acid, **5**, 34
Propionate, 2-isocyanato-, alkylation of, **5**, 634
Propionate, methyl (*R*)-(+)-2-chloro-, hydrolysis of, **5**, 118
Propionate, methyl (*S*)-(-)-2-chloro-, hydrolysis of, **5**, 118
Propionate, methyl 3-iodo-, reactions, with lanthanides, **2**, 564
Propionate, methyl 3-methoxy-, fluorination of, **6**, 251
Propionate, octyl (*R*)-(+)-2-chloro-, hydrolysis of, **5**, 118
Propionic acid, β-bromo-, synthesis of, **5**, 56
Propionic acid, (*R*)-(+)-2-chloro-, synthesis of, **5**, 118
Propionic acid, (*S*)-(-)-2-chloro-, synthesis of, **5**, 118
Propionic acid, β-chloro-
from, β-chloropropionaldehyde, **5**, 59
synthesis of, **5**, 59
Propionic acid, 3-chloro-2,2,3-trifluoro-, synthesis of, **5**, 56
Propionic acid, β-dialkylamino-, reactions, with trimethylchlorosilane, **5**, 863
Propionic acid, 2,2-diphenyl-3-hydroxy-, from, methyl esters, **5**, 62
Propionic acid, β-hydroxy-, from, ethylene cyanohydrin, **5**, 64
Propionic acid, 2-hydroxy-2-phenyl-, reactions, with thionyl chloride, **5**, 10
Propionic acid, (±)-α-(isopropylideneaminooxy)-, synthesis of, **5**, 93
Propionic acid, β-(3-oxocyclopentyl)-, from, 3-(3-hydroxypropyl)cyclopentanone, **5**, 73
Propionic acid, α-phenyl-, from, acetophenone, **5**, 85
Propionic acids, synthesis of, **1**, 81
Propionic acids, β-alkynyl-, synthesis of, **1**, 1024
Propionic acids, α-aryl-β-hydroxy-, synthesis of, **5**, 76
Propionic acids, 2-halo-, synthesis of, **4**, 129
Propionic acids, (*R*)-3-hydroxy-3-phenyl-, synthesis of, **5**, 66
Propionic acids, 3-hydroxy-3-substituted, synthesis of, **5**, 66
Propionic acids, phosphinico-, synthesis of, **4**, 124
Propionimidate, trimethylsilyl *N*-ethyl-2-oxo-, synthesis of, **5**, 719
Propionimidoyl chloride, *N*-(2-phenoxyethyl)-, synthesis of, **5**, 655
Propionimidoyl chloride, *N*-phenyl-, synthesis of, **5**, 655
Propionitrile, from, nitropropane, **3**, 620
Propionitrile, 2-furyl-, from, 2-vinyl furan, **3**, 615
Propionitrile, 3-hydroxy-, vacuum pyrolysis of, **3**, 582
Propionitrile, 3-iodo-, reactions, with zinc, **1**, 122
Propionitrile, β-methylamino-, as starting material, for *N*-nitroamines, **2**, 386
Propionitrile, *n*-perfluoro-, reactions, with 1,1,5-trihydroperfluoroamyl hypochlorite, **5**, 813
Propionitrile, 2-phenylseleno-3-tosyl-, synthesis of, **3**, 637
Propionitrile, 2-phenylthio-3-trimethylsilyl-, synthesis of, **3**, 639
Propiononitrile dihydrochloride, 2,3-di(*t*-butylimino)-, reactions, with lithium ethoxide, **4**, 894
Propionyl chloride, reactions, with ruthenium salts, **5**, 453
Propionyl chloride, α-chloro-, reactions, with triethylamine, **4**, 768
Propionyl compounds, synthesis of, **5**, 461
Propionyl fluoride, synthesis of, **5**, 5
Propionyl fluoride, 3-chloro-, synthesis of, **5**, 4
Propiophenone, reactions, with chlorosulfene, **5**, 498
Propiophenone, 2,4,6-trimethyl-, haloform oxidation of, **5**, 114
Propiophenones, synthesis of, **4**, 365
Proponate, methyl 2-methyl-, deesterification of, **5**, 39
(*S*)-Propranolol, synthesis of, **2**, 54
Propylamide, potassium 3-amino (KAPA), reactions, with thiophenyl α-substituted tetrahydrofurans, **1**, 1085
Propylamine, *N*-dichloroperfluoro-
reactions
with perfluoroethene, **5**, 659
with trifluorochloroethene, **5**, 659
Propylamine, *N*-(isocyanomethyl)-*N*-nitro-, synthesis of, **3**, 710
1-Propylamine, 1-lithio-3-hydroxy-, synthetic equivalent to, **1**, 151
n-Propylamine
reactions
with carbon disulfide, **5**, 1024
with *N*,*N'*-dichlorodiiminosuccinonitrile, **5**, 818
Propylamines, 1,2-diphenyl-, Hofmann degradations of, **1**, 627

Propyl bromide, reactions, with sodium butylarsenite, **2**, 498
Propyl bromide, 3-chloro-, as starting material, for 4-chlorobutyronitrile, **3**, 612
Propylenimine, *N*-chloro-, pyrolysis of, **3**, 575
n-Propylthiol, reactions, with hexamethylphosphoramide, **5**, 39
Propynal, 3-*N*,*N*-dimethylamino-, reactions, with methanol, **3**, 76
Propynals
  hydroselenation of, **3**, 74
  synthesis of, **3**, 78
Propynals, 3-amino-, reactions, with alcohols, **3**, 76
1-Propynamine, *N*,*N*-diethyl-, reactions, with potassium hydrogen difluoride, **4**, 790
Propyne
  reactions
    with *n*-butyllithium, **6**, 206
    with hydrogen bromide, **4**, 24
    with *o*-quinone dimethides, **1**, 519
    with sodium amide, **5**, 100
  as starting material, for ethyl tetrolate, **5**, 103
Propyne, 1,3-bis(trimethylsilyl)-, reactions, with butyllithium, **4**, 1059
Propyne, chloro-, from, *cis*-1,2 dichloropropyne, **2**, 1015
Propyne, 3-chloro-
  reactions
    with *n*-butyllithium, **5**, 59
    with trichlorophosphine, **5**, 1153
Propyne, 1-chlorotrifluoro-, reactions, with silver trifluoro methoxide, **4**, 764
Propyne, 1-cyano-, reactions, with trichlorophosphine, **5**, 1153
Propyne, diazo-
  from
    N-methyl-*N*-nitrosoureas, **3**, 490
    N-nitrosoacetamides, **3**, 490
Propyne, *cis*-1,2 dichloro-, as starting material, for chloropropyne, **2**, 1015
Propyne, diethylamino-, as activating agent, for the synthesis of anhydrides, **5**, 189
Propyne, 1-diethylamino-, reactions, with alkyl isocyanates, **3**, 600
Propyne, 1,3-dilithio-, alkylation of, **1**, 1053
Propyne, lithiated 1-(triisopropylsilyl)-, 1,4 addition of, **1**, 1061
Propyne, perfluoro-
  stability of, **2**, 1013
  synthesis of, **2**, 1013
Propyne, perlithio-, synthesis of, **6**, 205, 206
Propyne, 1-phenyl-
  dilithiation of, **4**, 668
  reactions, with copper(II) chloride, **1**, 1142
  reduction of, **1**, 507
Propyne, 3,3,3-triethoxy-, use in the synthesis of alkynes, **1**, 1009
Propyne, 1-(triethylsilyl)-, dihydroboration of, **4**, 632
Propyne, 1-triisopropylsilyl-, reactions, with epoxides, **1**, 1053
Propyne, 1-trimethylsilyl-, reactions, with zirconocene dichloride, **4**, 1038
Propyne, 1,1,3-triphenyl-, as starting material, for triphenylallene, **1**, 957
Propyne, 1,3,3-tris(trimethylsilyl)-, as starting material, for lithioallenes, **4**, 1047
1-Propyne, 1-chloro-3-(trimethylsilyl)-, reactions, with *n*-butyllithium, **4**, 648
1-Propyne, 1-cyclohexyl-, hydroboration of, **1**, 1102
Prop-1-yne, 1-diethylamino-
  reactions
    with *N*-benzenesulfonyl-*S*-phenylthiobenzimidate, **4**, 905
    with *N*-benzoyl-*S*-phenylthiobenzimidate, **4**, 905
    with 3-methylthio-1,2-benzisothiazole 1,1-dioxide, **4**, 905
    with 3-phenylseleno-1,2-benzisothiazole 1,1-dioxide, **4**, 911
Prop-1-yne, 1-phenoxy-, reactions, with diethylsilane, **4**, 933
1-Propyne, 1-phenyl-, hydroboration of, **1**, 1102
Prop-1-yne, 3-(phenylthio)-, alkylation of, **1**, 141
1-Propyne, 1-trimethylsilyl-, monohydroboration of, **4**, 644
Propyne complexes, use in the synthesis of α-carbocations of alkyne-bridged dimolybdenum complexes, **4**, 674
Propynes, 1,3-bis(trimethylstannyl)-, γ-halodestannylation of, **4**, 813
Propynes, 1-silyl 3-phenyl-, reactions, with trimethylsilyl chloride, **4**, 1047
Propynes, 3-titano 1-trimethylsilyl-, synthesis of, **4**, 1061
Propynes, 1-(triisopropylsilyl)-, alkylation of, **1**, 1052
Propynes, 1-(trimethylsilyl)-, alkylation of, **1**, 1052
Propynes, 3-trimethylsilyl 1-silyl 3-phenyl-, synthesis of, **4**, 1047
1-Propynes, 1-phenyl-, reactions, with lithium aluminum hydride, **1**, 512
Propynethial, synthesis of, via pyrolysis of dipropynal sulfide, **3**, 342
Prop-1-ynide, bromomagnesium 3-tetrahydropyranyloxy-, reactions, with acylsilanes, **1**, 985
Propynoate, methyl, organocupration of, **1**, 974
Propynol, aminomercuration of, **3**, 75
1-Propyn-2-ol, 1-trimethylsilyl-, hydromagnesiation of, **4**, 1062
Prop-2-yn-1-ol, reactions, with vinylzinc bromide, **4**, 1067
2-Propynols, reactions, with potassium cyanide, **3**, 649
Prop-2-yn-1-ols
  reactions
    with amide acetals, **1**, 979
    with enol ethers, **1**, 979
Propynone, 1-phenyl-, reactions, with oxazolones, **1**, 537
Propynylides, synthesis of, **3**, 500
Prop-2-ynyls, isomerization of, **1**, 956
(±) Prosopinine, synthesis of, **5**, 572
Prostacyclins, synthesis of, **1**, 173
Prostaglandin
  synthesis of, **1**, 111, 1016; **5**, 171
    three-component coupling, **1**, 132
Prostaglandin D_1, synthesis of, **3**, 254
Prostaglandin I_1, synthesis of, **4**, 279
Prostaglandins
  from
    cyclopentenes, **1**, 818
    ketenes, **3**, 529
  synthesis
    of, **1**, 218, 525, 659, 747; **1**, 1024, 1093, 1103; **6**, 467
Protected metal, use of term, **5**, 1087
N-Protected nucleosides, synthesis of, **6**, 488
Protecting groups, for sulfur containing amino acids, **6**, 480
Protein iodination, use of sulfenyl iodides in, **2**, 172
Proteins
  structure of, reviews, **5**, 84, 85
  synthesis of, reviews, **5**, 83, 85
Protic acids
  as catalysts
    for Friedel–Crafts acylation, **1**, 462
    for the synthesis of acetals, **4**, 178
  reactions, with arsenide anions, **1**, 445
  for the rearrangement of epoxides, **1**, 386
  use in the synthesis of alkenes, **1**, 765
  use in the synthesis of aromatic ketones, **3**, 278
Protiodegermylation, mechanism, **1**, 449
Protocatechuic acid, synthesis of, **5**, 114
Protodesilylation
  mechanism, **1**, 447, 448
  rate of, **1**, 1000
6-Protoilludene, from, ketenes, **3**, 528
Protonation
  of alcohols, **2**, 5
  of enamines, **2**, 757

for the synthesis of tricoordinate phosphorus functions, **4**, 480
Proton loss, for the synthesis of vinylic carbanions, **1**, 1089
Protonolysis
of carbon–metal bonds, **1**, 69
mechanisms, **1**, 65
for the reduction of carbon–mercury bonds, **1**, 65
for the reduction of vinylboranes, **1**, 449
studies, **1**, 59
of trimethylsilyl enol ethers, **2**, 639
1,2-Prototopy, in aliphatic oximes, **2**, 352
Proximity effect, in the synthesis of arylamidines, **5**, 755
Prozac, synthesis of, **2**, 65
Pseudo acid chlorides, reactions, with Lawesson's reagent, **5**, 528
Pseudocumene
acylation of, **3**, 296
trichloromethylation of, **6**, 29
Pseudomonas
use in the synthesis of alcohols, **2**, 39
use in the synthesis of cyanohydrins, **3**, 631
Pseudomonic acid C, from, L-lyxose, **1**, 173
Pseudomonic acids, synthesis of, **1**, 763
Pseudonitroles
from, oximes, **4**, 439
synthesis of, **4**, 439
(-)-Ptilocaulin, synthesis of, **1**, 434
Pulegone, use in the synthesis of 1,3-hydroxythiols, **4**, 229
Pulsed laser assisted reaction, of vaporized boron atoms with methane, **5**, 1096
Pumiliotoxins, synthesis of, **3**, 420
Pummerer reaction
mechanism, **4**, 221
for the synthesis of α-halo selenides, **4**, 88
for the synthesis of monothioacetals, **4**, 226
for the synthesis of 2-(phenylsulfinyl)acylals, **4**, 207
for the synthesis of selenoacetals, **2**, 286
for the synthesis of α,β-unsaturated ketones, **3**, 249
use in the synthesis of aldehydes, **3**, 16, 20
Pummerer rearrangement
of (diphenylthiophosphinoyl)phenyldiazomethane, **4**, 346
of α-sulfinyl phosphonates, **6**, 119
of sulfoxides, **5**, 334
for the synthesis of α-alkoxy aldehydes, **3**, 32
for the synthesis of α-cyano sulfoxide, **3**, 634
for the synthesis of dialkyl sulfides, **2**, 142
for the synthesis of α,α-dichloro sulfides, **6**, 46
for the synthesis of enals, **3**, 57
for the synthesis of phenols, **2**, 660
for the synthesis of sulfur functions, **4**, 923
for the synthesis of *N*-thioamides, **5**, 335
for the synthesis of α,β-unsaturated ketones, **3**, 254
use in the synthesis of α-fluoro aldehydes, **3**, 27
Pummerer rearrangement, sila-, for the synthesis of dichalcogenomethanes, **6**, 325
Punctatin A, synthesis of, **1**, 817
Pupukeanane, 9-isocyano-, synthesis of, **3**, 700
Purdie modification, for the synthesis of ethers, **2**, 88
Purine nucleoside phosphorylase inhibitors, research, **6**, 58
Purines
from, adenines, **1**, 36
reduction of, **4**, 413
Purines, 6-substituted, reactions, with 3,4-dihydro-2*H*-pyran, **4**, 306
Purple benzene
reactions, with styrene, **5**, 115
synthesis of, **5**, 47
Pyradones, *N*-hydroxy-, synthesis of, **5**, 327
Pyran, thermolysis of, **3**, 68
Pyran, 2,3-dihydro-, deprotonation of, **4**, 949
Pyran, dihydro- (DHP)
reactions
with benzenesulfinic acid, **4**, 236
with carboxylic acids, **4**, 208
ring-opening substitution of, **1**, 481
as starting material
for hemithioacetals, **4**, 223
for hydroxy dithioacetals, **4**, 246
synthesis of, **4**, 827
use in the synthesis of acetals, **4**, 186
Pyran, 6-lithio-2,3-dihydro-, synthesis of, **1**, 1089
Pyran, 2-methoxytetrahydro-, reactions, with benzenesulfinic acid, **4**, 236
2*H*-Pyran, 3,4-dihydro-
reactions
with 3-chlorobenzoylhypobromite, **4**, 208
with dibenzyl azodicarboxylate, **6**, 626
with 6-substituted purines, **4**, 306
Pyranone, synthesis of, **1**, 581
2-Pyranone, 6-hydroxy-, synthesis of, **3**, 222
2*H*-Pyran-2-one, 3,4-dihydro-4,4-dimethyl-, reactions, with primary amines, **2**, 320
Pyran-4-one, 2,3-dihydro-, reactions, with 1,1-bis(sulfonyl)ethene, **4**, 277
Pyranones, dihydro-, synthesis of, **3**, 246
Pyran-2-ones, 3-carbomethoxy-, inverse electron demand Diels–Alder reactions, **2**, 665
Pyran-2-ones, hydroxy-, reactions, with dienophiles, **2**, 665
Pyran-2-ones, 6-hydroxy-, from, furanmethanols, **3**, 221
2*H*-Pyran-2-ones, hydroxy-, in Diels–Alder cycloaddition, **2**, 665
Pyranosides, reactions, with phenyl chlorothionoformate, **6**, 541
Pyranosyl nitrate, 3-*O*-acetyl-2-azido-4-(benzyl oxycarbonyl)amino-2,4,6-trideoxy-D-galacto-, synthesis of, **6**, 549
Pyrans, 2-alkoxy-, synthesis of, **4**, 202
Pyrans, 2-(benzenesulfonyl)tetrahydro-, synthesis of, **4**, 236
Pyrans, α-bromotetrahydro-, stability of, **4**, 51
Pyrans, 2-chlorotetrahydro-, synthesis of, **4**, 49
Pyrans, dihydro-
reactions, with dimethyl acetylenedicarboxylate, **1**, 518
synthesis of, via xanthate pyrolysis, **1**, 776
Pyrans, *cis*-2,3-dihydro-, synthesis of, **3**, 243
Pyrans, 2-hydroxy-3-phenyl selenyl tetrahydro-, synthesis of, **3**, 38
Pyrans, 3-substituted-3,4-dihydro-, cleavage of, **3**, 163
Pyrans, tetrahydro-, synthesis of, **2**, 93
Pyrans, 2-[[(trimethylsilyl)methyl]thio]tetrahydro-, reactions, with trimethylsilyl chloride, **6**, 349
Pyran-4-thiones
from, pyrones, **3**, 357
reactions, with sodium hydrogen sulfide, **3**, 255
as starting materials, for dioxathiapentalenes, **3**, 377
Pyranyl ethers, tetrahydro-
reduction of, **2**, 57
as starting materials
for alkyl bromides, **2**, 10
for alkyl halides, **2**, 10
for alkyl iodides, **2**, 10
synthesis of, **4**, 186
Pyrazine-1-oxides, 2-azido-, thermal decomposition of, **3**, 673
Pyrazines, synthesis of, **5**, 973
Pyrazines, bis(trimethylsilyl)-, synthesis of, **2**, 943
Pyrazines, 2-cyanamido-, methylation of, **5**, 1118
Pyrazines, diboradihydro-, synthesis of, **4**, 522
Pyrazines, dihydro-, photolysis of, **3**, 699
Pyrazines, 6-phenyl-2,5-dihydro-, synthesis of, **2**, 770
Pyrazole
reactions
with acetals, **4**, 437
with dichlorocarbenes, **6**, 141
with ketals, **4**, 437
use in the synthesis of amino acids, **5**, 85
Pyrazole, dihydro-, synthesis of, **6**, 122

Pyrazole, *N,N'*-thiocarbonylbis(3,5-dimethyl)-
reactions
with aniline, **6**, 578
with 1,2-diaminobenzene, **6**, 578
1*H*-Pyrazole-1-carboxamidine hydrochloride, as guanylating agent, **6**, 641
Pyrazoles
as starting matcrials, for α-azo cnamines, **4**, 989
synthesis of, **3**, 680
via thermal Cope rearrangements, **1**, 807
Pyrazoles, acyl-, as starting materials, for α-amino acids, **3**, 40
Pyrazoles, *N*-acyl-3,5-dimethyl-, use in the synthesis of aromatic ketones, **3**, 281
Pyrazoles, alkenylamino-, hydrogenolysis of, **2**, 316
Pyrazoles, 1-amino-, reactions, with ketones, **2**, 316
Pyrazoles, 4-amino-3-trimethylsilyl-, synthesis of, **5**, 874
Pyrazoles, silyl-, synthesis of, **2**, 943
Pyrazoles, *N*-trimethylsilyl-, synthesis of, **6**, 444
3*H*-Pyrazoles, 5-acyl-, Wolff rearrangement of, **3**, 543
Pyrazoline
as intermediate, in the synthesis of cyclopropanes, **1**, 298
photolysis of, **1**, 1120
synthesis of, **1**, 19
Pyrazolines
reactions
with bromine, **4**, 105
with *N*-bromosuccinimide, **4**, 109
rearrangement of, **5**, 873
synthesis of, **3**, 479; **4**, 488
Pyrazolines, 3-bis(phenylsulfonyl)-, synthesis of, **4**, 277
Pyrazolines, 3-chloro-, synthesis of, **4**, 109
Pyrazolines, 3-nitro-, synthesis of, **4**, 440
2-Pyrazolines
from, alkenes, **3**, 452
synthesis of, **3**, 451, 680
2-Pyrazolines, *N*-formyl-, synthesis of, **3**, 452
3-Pyrazolin-5-one, oxidative cleavage of, **5**, 357
3-Pyrazolin-5-one, 4-amino-, as starting material, for 1,2-diacylhydrazines, **5**, 357
2-Pyrazolin-5-ones, synthesis of, **3**, 451
2-Pyrazolin-5-ones, 3,4-polymethylene-, oxidation of, **1**, 983
Pyrazolones, reactions, with thallium(III) nitrate, **1**, 1076
5-Pyrazolones, reactions, with ynamines, **4**, 986
5-Pyrazolones, 4,4-dihalo-, as starting materials, for alkynes, **1**, 1076
5-Pyrazolones, 3,4-disubstituted-4-monohalo-, use in the synthesis of alkynes, **1**, 1076
Pyrene
acylation of, **3**, 302
reactions, with ozone, **5**, 115
Pyrethroid acids, synthesis of, reviews, **5**, 28
Pyrethroids, as insecticides, **5**, 145
Pyridazine, 2-amino-3-cyano-5-methyl-, reactions, with formamide acetals, **5**, 746
1,2-Pyridazine-3,6-diones, dehydrogenation of, **5**, 366
Pyridazine 1-oxide, reactions, with potassium cyanide, **3**, 648
Pyridazines, 2-aminocarbonyl-2,5-dihydro-, as starting materials, for 2,5-dihydropyridazines, **2**, 770
Pyridazines, cyano-, synthesis of, **3**, 648
Pyridazines, 2,5-dihydro-, from, 2-aminocarbonyl-2,5-dihydropyridazines, **2**, 770
Pyridazines, 4,5-dihydro-, synthesis of, **3**, 453
Pyridazines, *cis*-2,5-diphenyl tetrahydro-, irradiation of, **4**, 105
Pyridazines, silyl-, synthesis of, **2**, 944
Pyridazines, tetrahydro-, synthesis of, **3**, 453
Pyridazines, 1,4,5,6-tetrahydro-, synthesis of, via [4+2] cycloaddition reactions, **3**, 453
Pyridazin-6-one, 1,4,5,6-tetrahydro-, synthesis of, **3**, 453
Pyridazinones, dihydro-, synthesis of, **3**, 453
Pyridazin-4-ones, 1,4,5,6-tetrahydro-, synthesis of, via Japp–Klingemann reaction, **3**, 453
Pyridine
acylation of, **3**, 310
as an added base in the synthesis of acid chlorides, **5**, 9
as catalyst, for chloroformates, **6**, 426
for the chlorination of sulfides, **4**, 65
electrochemical reduction of, **1**, 103
reactions
with acetic anhydride, **2**, 747
with acyl radicals, **3**, 311
with alkynylmagnesium halides, **1**, 1042
with alkynyl metals, **1**, 1042
with benzil monohydrazonecopper(II) chloride, **3**, 470
with bis(trimethylsilyl)mercury, **2**, 814
with 1,1-dimethyl-2,2-bis(trimethylsilyl)silaethenes, **6**, 177
with Fischer carbene complexes, **5**, 953
with lead tetraacetate, **5**, 999
with osmium compounds, **6**, 345
with phosphorus pentoxide, **5**, 1078
with picryl chloride, **5**, 1040
with pyridinium chloride, **5**, 81
with selenium(IV) dichlorides, **4**, 88
with spiropentaoxyphosphorane, **2**, 869
with 1,2,2,2-tetrafluoroethanesulfenyl fluoride, **5**, 502
with tosyl chloride, **5**, 282
with trimethylsilyl iodide, **5**, 226
with trimethylsilyltriflate, **5**, 226
reduction of, **1**, 99, 102
for the reduction of piperidine, **1**, 89
ring-opening of, **5**, 1033
silylmethylation of, **4**, 519
as solvent, for the synthesis of thionoamides, **5**, 566
use in the substitution of boron functions, **1**, 274
use in the synthesis of alkenes, **1**, 503
use in the synthesis of anhydrides, **5**, 183
use in the synthesis of bis(cyclopentadienyl) metal–methyl complexes of lanthanides, **4**, 702
use in the synthesis of dialkylmagnesiums, **2**, 559
Pyridine, 2-acetyl-, reduction of, **1**, 91
Pyridine, 3-acetyl-, reduction of, **1**, 91
Pyridine, 3-amino-, reactions, with amino isothiocyanates, **6**, 574
Pyridine, 2-bis(trimethylsilyl)methyl-
reactions, with butyllithium, **6**, 195
synthesis of, **4**, 604
Pyridine, bromo-, reactions, with 9-alkyl-9-borabicyclo[3.3.1]nonyl, **3**, 311
Pyridine, 2-bromo-, use in the synthesis of carboxylic esters, **5**, 125
Pyridine, 4-*t*-butyl-, reactions, with borenes, **5**, 929
Pyridine, 2-chloromethyl-, reactions, with potassium *O*-ethyl dithiocarbonate, **6**, 545
Pyridine, 2-cyano-, use in the synthesis of hydrazonamides, **5**, 847
Pyridine, 3-cyano-
reactions, with hydrogen selenide, **5**, 625
reduction of, **1**, 102
Pyridine, 2-(dialkylboryl)amino-, reactions, with phenylisonitrile, **6**, 662
Pyridine, 2-dibromomethyl-, synthesis of, **4**, 27
Pyridine, 2-dichloromethyl-, synthesis of, **4**, 19
Pyridine, 2,6-dimethyl-, use in the synthesis of dimethyl carbonate, **6**, 465
Pyridine, 4-dimethylamino-
reactions, with 2,4,6-trichlorobenzoyl chloride, **5**, 173
use in esterification reactions, **1**, 583
use in the synthesis of carboxylic esters, **5**, 124
use in the synthesis of *N*-silylamides, **5**, 374
Pyridine, *N,N*-dimethylamino- (DMAP), use in the synthesis of carboxylic esters, **5**, 129
Pyridine, 3-hydroxy-, reduction of, **1**, 91
Pyridine, iodo-, reactions, with boronic acids, **3**, 311
Pyridine, 2-methyl-

reactions, with dichlorocarbene, **3**, 722
reduction of, **1**, 99
Pyridine, 3-methyl-
deprotonation of, **1**, 861
as starting material, for 3-trichloromethylpyridine, **6**, 28
Pyridine, 3-methyl-2-amino-, reactions, with triethyl orthoformate, **6**, 157
Pyridine, 2-(methylamino)methylhexahydro-, as starting material, for perhydroimidazopyridines, **4**, 408
Pyridine, 2-(*N*-methyl-*N*-formyl) amino-, as formylating reagent, **3**, 90
Pyridine, poly-4-vinyl- (Poly-4-VP)
as catalyst
for the synthesis of anhydrides, **5**, 184, 198
for the synthesis of hydroxylamines, **5**, 213
Pyridine, tetrahydro-, from, dihydropyrrole, **1**, 400
Pyridine, 2,3,4,5-tetrahydro-6-methyl-, reactions, with lithium diisopropylamide, **2**, 743
Pyridine, 1,2,3-triazolo[1,5-*a*]-
reactions
with bromine, **4**, 27
with chlorine, **4**, 19
Pyridine, 3-trichloromethyl-, from, 3-methylpyridine, **6**, 28
Pyridine, *N*-(trimethylsilyl)-, reactions, with fluorine, **5**, 313
Pyridine, 2,4,6-triphenyl-, synthesis of, **1**, 30
Pyridine, 2-vinyl-, reduction of, **1**, 99
Pyridine-3-acetic acids, hydrogenation of, **1**, 91
Pyridine bromide perbromide, 4-(dimethylamino)-, use in the synthesis of α-bromo ketones, **3**, 154
2,6-Pyridinedicarbothioate, *O,O'*-dimethyl-, reactions, with glycols, **5**, 511
Pyridine dichromate, 3-carboxy-, as oxidant, **3**, 4
Pyridine hydrotribromide, use in the synthesis of alkyl bromides, **2**, 28
Pyridine methiodide, 2-bromo-, reactions, with sodium hydrogen selenide, **5**, 625
Pyridine *N*-oxide
for the oxidation of benzyl halides, **3**, 89
reactions
with benzofuran-3-acetic acids, **3**, 103
with trimethylsilyl cyanide, **3**, 671
use in the synthesis of xanthates, **2**, 123
Pyridine *N*-oxide, 2-allyloxy-, as starting material, for 3-allyl-*N*-hydroxy-2-pyridone, **5**, 326
Pyridine *N*-oxide, dimethylamino-, for the oxidation of benzyl halides, **3**, 89
Pyridine *N*-oxide, 4-dimethylamino-
as reagent
for aldehydes, **3**, 3
for the synthesis of ketones, **3**, 114
Pyridine *N*-oxide, 2-methyl-, reactions, with *p*-toluenesulfonyl chloride, **1**, 861
Pyridine-1-oxides, 2-azido-, thermal decomposition of, **3**, 673
Pyridine *N*-oxides
acylation of, **3**, 310
reactions
with alkynylsilvers, **1**, 1043
with imidoyl halides, **5**, 712
as starting materials
for cyclic *N*-hydroxy enamines, **2**, 765
for 2-lithiopyridine *N*-oxides, **2**, 765
Pyridine *N*-oxides, 2-alkoxy-, as starting materials, for hydroxamic acids, **5**, 326
Pyridine *N*-oxides, 2-lithio-, from, pyridine *N*-oxides, **2**, 765
Pyridine *N*-oxides, 2-substituted, hydrolysis of, **5**, 327
Pyridine *N*-oxides, 2-sulfonyl, reactions, with sodium ethoxide, **2**, 271
Pyridine poly(hydrogen fluoride) (PPHF)
for the fluorination of adamantane, **2**, 15
reactions, with Selectfluor reagent, **2**, 17
synthesis of, **2**, 18
use in the synthesis of alkyl fluorides, **2**, 17
use in the synthesis of alkyl halides, **2**, 6, 11
use in the synthesis of α-fluorocarboxylic acids, **2**, 11
See also Olah's reagent
Pyridines
acidity of, **1**, 861
elimination of, **1**, 629
hydrogenolysis of, **1**, 91
lithiation of, **3**, 108, 310
reactions
with acid chlorides, **5**, 129
with alkynyltins, **1**, 1043
with amides, **3**, 310
with iodotrifluoromethane, **6**, 12
with sodium borohydride, **1**, 102
reduction of, **3**, 107
and reductive deaminations, **1**, 30
ring-opening of, **3**, 66
as starting materials, for carboxylic acids, **5**, 116
synthesis of, **5**, 973
reviews, **5**, 90
trichloromethylation of, **6**, 28
use in the synthesis of monochalcogenomethanes, **6**, 338
Pyridines, *N*-acyl-1,2-dihydro-, reactions, with cyanogen azide, **5**, 1119
Pyridines, 2-alkoxy-, pyrolysis of, **1**, 777
Pyridines, 2-alkoxytetrahydro-, synthesis of, **4**, 305
Pyridines, alkyl-, trifluoromethylation of, **6**, 22
Pyridines, 2-alkyl-, substitution of, **1**, 860
Pyridines, 4-alkyl-, reduction of, **1**, 99
Pyridines, 1-alkyl-4,4-dimethyl-1,4,5,6-tetrahydro-, synthesis of, **2**, 320
Pyridines, amino-
as starting materials, for ketene aminals, **4**, 979
synthesis of, **6**, 157
Pyridines, 6-anilino-5-carboxy-4-methyl-2-phenyl-, reactions, with thionyl chloride, **5**, 529
Pyridines, bis(silylmethyl)-, synthesis of, **4**, 604
Pyridines, 6-*t*-butyl-2-aryl-, synthesis of, **1**, 254
Pyridines, 2-chloro-3-cyano-, reactions, with selenamidoacetonitriles, **3**, 656
Pyridines, 2-cyano-, synthesis of, **3**, 671
Pyridines, 2,6-dialkylated tetrahydro-, synthesis of, **2**, 326
Pyridines, dihydro-
as intermediates, **1**, 102
reactions, with dimethyl acetylenedicarboxylate, **1**, 518
Pyridines, 1,2-dihydro-
reactions, with cyanogen azide, **5**, 1119
synthesis of, **1**, 102
Pyridines, 1,4-dihydro-, synthesis of, **1**, 31, 99
Pyridines, 2,3-disubstituted 1,2-dihydro-, synthesis of, **1**, 1043
Pyridines, 2-formyl-, synthesis of, **3**, 107
Pyridines, 3-formyl-
synthesis of, **3**, 107
via palladium-catalyzed formylation, **3**, 108
Pyridines, 4-formyl-, synthesis of, **3**, 107
Pyridines, 4-halo-, as starting materials, for pyridin-4-thiones, **3**, 357
Pyridines, haloalkyl-, phosphinylation of, **2**, 430
Pyridines, hydroxy-, hydrogenation of, **1**, 91
Pyridines, 3-hydroxy-2-amino-, reactions, with phenyl formate, **5**, 290
Pyridines, hydroxymethyl-, oxidation of, **3**, 108
Pyridines, metallated, reactions, with dialkylformamides, **3**, 108
Pyridines, methyl-, acidity of, **1**, 861
Pyridines, 2-methyl-, trichlorination of, **6**, 28
Pyridines, 4-methyl-, trichlorination of, **6**, 28
Pyridines, 2-methyl-5-alkyl-, oxidation of, **5**, 116
Pyridines, quaternized, amination of, **4**, 413
Pyridines, 2-sulfonyl-, cleavage of, **2**, 188
Pyridines, tetrahydro-
reduction of, **1**, 85

synthesis of, **3**, 410
Pyridines, 1,2,5,6-tetrahydro-, synthesis of, **2**, 329
Pyridines, 2,3,4,5-tetrahydro-, synthesis of, **3**, 410
Pyridines, trimethylsilylmethyl-, metallation of, **4**, 647
Pyridines, 2-trimethylsilylmethyl-
deprotonation of, **4**, 648
synthesis of, **4**, 604
Pyridines, trimethylstannyl-, acylation of, **3**, 311
Pyridineselenones, *N*-methyl-, synthesis of, **5**, 625
2-Pyridine sulfinate, methyl-, as reagent, in elimination reactions, **1**, 598
Pyridine-2-thione, *N*-(cyclopentyloxy)-, as source of radicals, **1**, 931
Pyridine-2-thione, 1-(2-hydroxyethyl)-4,6-diphenyl-, reactions, with alkyl chlorides, **2**, 121
Pyridine-2-thione, 1-methyl-, reactions, with alkyl halides, **2**, 121
Pyridine-2-thione carbamates, *N*-hydroxy-, irradiation of, **2**, 323
Pyridine thiones, synthesis of, **5**, 593
Pyridinium bromide, phenacyl-, reactions, with nitrosoarenes, **2**, 765
Pyridinium cations, as starting materials, for alkenes, **1**, 629
Pyridinium chloride
reactions
with potassium thiocyanate, **5**, 1040
with pyridine, **5**, 81
use in the synthesis of carboxylic acids, **5**, 107
Pyridinium chloride, 2-benzoylthio-1-methyl-, as phase-transfer activating agent, for the synthesis of alkyl aryl anhydrides, **5**, 199
Pyridinium chloride, *N*-benzyloxy-, synthesis of, **1**, 1043
Pyridinium chloride, cetyl-, as catalyst, for the synthesis of cyanates, **5**, 1104
Pyridinium dichromates, benzyl-, as oxidants, **3**, 4
Pyridinium hydrobromide perbromide, development of, **2**, 30
Pyridinium hydrobromide perbromide, 4-dimethylamino-, use in the synthesis of alkyl bromides, **2**, 30
Pyridinium iodide, 2-chloro-, use in the synthesis of sulfinamides, **2**, 363
Pyridinium iodide, 2-chloro-1-methyl-
reactions, with thioureas, **5**, 1064
as reagent, for the synthesis of isothiocyanates, **5**, 1024
Pyridinium iodides, 2,4,6-triphenyl-, as starting materials, for aryl iodides, **2**, 632
Pyridinium ions, synthesis of, **3**, 420
Pyridinium perchlorate, 2-chloro-1-methoxy-3-nitro-, reactions, with amines, **5**, 674
Pyridinium poly(hydrogen fluoride)
as fluorinating agent, **6**, 423
reactions
with acid anhydrides, **5**, 4
with acid chlorides, **5**, 4
as solvent, **6**, 423
as solvent and reagent, for carbamoyl fluorides, **6**, 439
Pyridiniums, *N*-methyl-, reduction of, **1**, 102
Pyridinium salt, 1-amino-2,4,6-triphenyl-, use in the synthesis of isocyanates, **5**, 1002
Pyridinium salt, 2-chloro-1-methyl-, reactions, with *N,N'*-disubstituted thioureas, **5**, 1078
Pyridinium salts
ammonolysis of, **4**, 448
elimination of, **1**, 629
reactions
with potassium phthalimide, **4**, 448
with potassium trinitromethanide, **4**, 995
with trinitromethanes, **6**, 145
reduction of, **1**, 90
ring-opening of, **3**, 66
as starting materials, for alkyl halides, **2**, 11
synthesis of, **1**, 31; **4**, 519
as thioacylating agents, **5**, 561
Pyridinium salts, acyl-, as intermediates, **5**, 130
Pyridinium salts, *N*-acylamino-, thermal decomposition of, **5**, 1002
Pyridinium salts, chloro-, use in the synthesis of cyclic aminocarbene complexes, **5**, 954
Pyridinium salts, dihydro-, reactions, with cyanide, **3**, 637
Pyridinium salts, 2-halo-, as coupling reagents, for the synthesis of amides, **5**, 268
Pyridinium salts, *N*-substituted 2,4,6-triphenyl-, reduction of, via sodium borohydride, **1**, 30
Pyridinium salts, tetrahydro-, synthesis of, **4**, 305
Pyridinium salts, thioaroyl-, as thioacylating agents, **5**, 588
α-Pyridinium salts, reactions, with aldehydes, **3**, 258
Pyridin-3(2H)-one, 1,2,4-triazolo[4,3-a], synthesis of, **6**, 444
Pyridin-4-thiones
from
diphenyl cyclopropenethione, **3**, 357
4-halopyridines, **3**, 357
pyridones, **3**, 357
4-thiopyrones, **3**, 358
molecular structure of, **3**, 352
synthesis of, via rearrangement, **3**, 357
Pyridone, *N*-fluoro-, synthesis of, **5**, 313
2-Pyridone, 1-methyl-, lithiation of, **4**, 540
2-Pyridone, 1-trimethylsilyl-, synthesis of, **4**, 540
2,2'-Pyridone, 1,1'-thiocarbonyl-, reactions, with amines, **5**, 1030
2-Pyridone-5-carboxylate, methyl, reactions, with sodium hydroxide, **5**, 106
Pyridone *N*-oxide, as reagent, for aldehydes, **3**, 3
Pyridones
as starting materials, for pyridin-4-thiones, **3**, 357
thermal decomposition of, **5**, 1002
Pyridones, *N*-alkyl, as starting materials, for thioketones, **3**, 357
Pyridones, dihydro-, synthesis of, **3**, 264
Pyridones, hydroxy-, reactions, with alkyl thiocyanates, **6**, 558
Pyridones, silylmethyl, desilylation of, **4**, 540
2(1*H*)-Pyridones, 3-substituted-, synthesis of, **5**, 992
4-Pyridones, dihydro-, synthesis of, **3**, 262
Pyrido[2,3-c]pyridines, 1,2-addition to, **5**, 1127
2*H*-Pyrido[1,2-a]pyrimidines, 2-amino-, as starting materials, for ketene aminals, **4**, 979
Pyridoxal, synthesis of, **4**, 164
Pyridoxal prototropic rearrangement, for the synthesis of ketones, **3**, 132
Pyridoxol, reactions, with manganese dioxide, **4**, 164
Pyridyl bromides, transmetallation of, **3**, 310
Pyridyl compounds, synthesis of, **6**, 168
Pyrimidine, reduction of, **1**, 90
Pyrimidine, 2,4-dichloro-, use in the synthesis of carbodiimides, **5**, 1078
Pyrimidine, 2,4-dichloro-5-nitro-, use in the synthesis of carbodiimides, **5**, 1078
Pyrimidine, 2,4-dimethyl-6-phenyl-, oxidation of, **5**, 116
Pyrimidine, 2-oxoperhydro-, synthesis of, **6**, 509
Pyrimidine, 2,4,6-trichloro-, use in the synthesis of carbodiimides, **5**, 1078
[3,2a]Pyrimidine-6-carboxylic acid, 5-oxothiazolo-, synthesis of, **5**, 92
Pyrimidine-4-carboxylic acid, 2-methyl-6-phenyl-, synthesis of, **5**, 116
Pyrimidinediones, reactions, with amines, **6**, 360
Pyrimidinediones, 2,2-(dichloro)dihydro-4,6-, reactions, with amines, **6**, 360
Pyrimidinediones, hexahydro-, synthesis of, **4**, 425
Pyrimidines
ring-opening of, **4**, 978
as starting materials, for ketene aminals, **4**, 978
synthesis of, **4**, 421, 988; **5**, 973
Pyrimidines, chloro-, use in the synthesis of carbodiimides, **5**, 1065

Pyrimidines, 2-cyanamido-, synthesis of, **5**, 1120
Pyrimidines, 4,6-dichloro-5-substituted, reactions, with amines, **4**, 978
Pyrimidines, dihydro-, synthesis of, **4**, 408
Pyrimidines, ethynyl-, synthesis of, **1**, 1009
Pyrimidines, hexahydro-
  reactions, with Grignard reagents, **4**, 410
  synthesis of, **4**, 408
Pyrimidines, *N*-hydroxy-, reactions, with benzimidoyl chlorides, **5**, 711
Pyrimidines, iodo-
  reactions
    with alkylalkynes, **1**, 1009
    with arylalkynes, **1**, 1009
    with ethylpropiolate, **1**, 1009
Pyrimidines, 2-methanesulfonyl-, reactions, with cyanamide, **5**, 1120
Pyrimidines, tetrahydro-, synthesis of, **1**, 90; **4**, 408
[1,5-c]Pyrimidines, [1,2,3,4]tetrazolo-, as starting materials, for 1-cyanoimidazoles, **5**, 1122
Pyrimidine-2-thiones, synthesis of, **6**, 579
Pyrimidine-2(1*H*)-thiones, from, 4-amino-1-azabutadienes, **6**, 575
Pyrimidine-4-thiones, synthesis of, **4**, 421
[1,5-a]Pyrimidinium bromide, [1,2,3]triazolo-, as starting material, for cyanamides, **5**, 1122
Pyrimidones, synthesis of, **4**, 978
2(1*H*)-Pyrimidones, 1,4,6-trisubstituted-, reactions, with Lawesson's reagent, **6**, 579
Pyrocalciferol, from, precalciferol, **1**, 419
Pyrocarbonate, di-*t*-butyl, as activating agent, for the synthesis of anhydrides, **5**, 189
Pyrolysis
  of hexachloroethane, **6**, 215
  for the synthesis of metalloid functions, **3**, 501
  of thioesters, **5**, 475
Pyrolysis reactions, for the synthesis of 1,1,1-trilithioalkanes, **6**, 206
Pyrones
  as starting materials, for pyran-4-thiones, **3**, 357
  synthesis of, **6**, 616
Pyrones, dihydro-, synthesis of, **3**, 243, 244
4-Pyrones, 2,3-dihydro-, synthesis of, **3**, 244
α-Pyrones, as starting materials, for dienes, **1**, 789
γ-Pyrones, synthesis of, **1**, 957
Pyrophosphates, synthesis of, **5**, 712
Pyrophosphates, alkyl, synthesis of, **2**, 108
Pyrophosphite, tetraethyl
  reactions, with diethylvinyl phosphonate, **4**, 552
  use in the synthesis of diethoxy(heptafluoroisopropyl)phosphines, **4**, 117
Pyrophosphoramide, octamethyl-, synthesis of, **2**, 413
Pyrophosphoryl chloride
  reactions
    with alcohols, **2**, 105
    with methoxynaphthalenes, **3**, 101
  use in the synthesis of aryl aldehydes, **3**, 92
Pyrrole
  cyanation of, **3**, 669
  reactions, with dichloromethane, **4**, 410
  reduction of, **1**, 90
  as starting material, for *N*-chlorocarbonylpyrrole, **6**, 450
  Vilsmeier reaction of, **3**, 106
Pyrrole, *N*-(benzotriazol-1-ylmethyl)-, reactions, with *n*-butyllithium, **6**, 370
Pyrrole, *N*-benzoyl-, reactions, with bis(trifluoromethyl)ethyne, **1**, 521
Pyrrole, bis-, deprotonation, **6**, 170
Pyrrole, *N*-chlorocarbonyl-, from, pyrrole, **6**, 450
Pyrrole, 1-chloroformyl-5-chloro-2,3-dihydro-, synthesis of, **5**, 970
Pyrrole, 2-cyano-, synthesis of, **3**, 669
Pyrrole, dihydro-, as starting material, for tetrahydropyridine, **1**, 400
Pyrrole, 3,5-dimethyl-2-(trimethylsilyloxy)-, isomerization of, **5**, 992
Pyrrole, *N*-methyl-3,4-dinitro-, reactions, with secondary amines, **4**, 415
Pyrrole, *N*-phenyl-substituted, reactions, with sodium, **1**, 100
Pyrrole, *N*-triisopropylsilyl-, as starting material, for β-silylpyrroles, **2**, 942
Pyrrole-2-carboxylic acid, 4-hexadecylamino-1-methyl-, synthesis of, **5**, 92
Pyrrole complexes, dihydro-, from, α,β-unsaturated alkoxychromium carbene complexes, **5**, 953
Pyrrole-2,3-dione, 2,3-dihydro-, flash vacuum pyrolysis of, **3**, 584
Pyrroles
  acylation of, **3**, 306
  deprotonation of, **1**, 278
  from, isocyanides, **4**, 538
  oxidation of, **6**, 107
  reactions
    with dienes, **1**, 521
    with ethoxycarbonyl isothiocyanate, **5**, 592
    with Grignard reagents, **3**, 308
  synthesis of, **2**, 320; **4**, 534; **5**, 924
Pyrroles, acyl-, as starting materials, for aldehydes, **3**, 106
Pyrroles, 2-acyl-, as starting materials, for 2-formylpyrroles, **3**, 106
Pyrroles, *N*-alkyl-, metallation of, **2**, 942
Pyrroles, 2-aminomethyl-, reactions, with nitrous acid, **1**, 259
Pyrroles, *N*-(*t*-butoxycarbonyl)-, reduction of, **1**, 92
Pyrroles, cyano-, from, diarylalkynes, **3**, 671
Pyrroles, 4-cyano-, synthesis of, **3**, 670
Pyrroles, 2-cyano-5-amino-, synthesis of, **3**, 671
Pyrroles, 2-cyano-1-hydroxy-, synthesis of, **3**, 673
Pyrroles, 2,5-disubstituted, reduction of, **1**, 92
Pyrroles, 2-formyl-, from, 2-acyl pyrroles, **3**, 106
Pyrroles, 2-keto-
  isomerization of, **3**, 307
  synthesis of, **3**, 307
Pyrroles, lithio-, x-ray crystal structure of, **2**, 1002
Pyrroles, magnesio-, reactions, with acid chlorides, **3**, 308
Pyrroles, metallated, reactions, with aldehydes, **3**, 308
Pyrroles, 2-methyl-, reactions, with sulfuryl chloride, **3**, 106
Pyrroles, *N*-phthaloylamino-, acylation of, **3**, 307
Pyrroles, β-silyl-, from, *N*-triisopropylsilylpyrrole, **2**, 942
Pyrroles, substituted, reactions, with *p*-toluenesulfonyl isothiocyanate, **5**, 638
Pyrroles, 2-substituted, reduction of, **1**, 92
Pyrroles, *N*-substituted, lithiation of, **3**, 308
Pyrroles, 3-(thiol ester)-, reduction of, **3**, 106
Pyrroles, vinyl-, thionation of, **5**, 572
Pyrrolidine
  potassium compounds of, **2**, 423
  reactions
    with Angeli's salt, **3**, 453
    with β,β-dichloroalkenyl aryl ketones, **4**, 793
    with triethoxyacetonitrile, **6**, 140
  ring-expansion of, **3**, 453
  synthesis of, **1**, 626
Pyrrolidine, *N*-allyl-, lithiation of, **4**, 511
Pyrrolidine, 1-amino-2-methoxymethyl-, enantiomers of, **1**, 153
Pyrrolidine, 1(R)-amino-2-methoxymethyl- (RAMP), use in the synthesis of aldehydes, **3**, 10
Pyrrolidine, 1(S)-amino-2-methoxymethyl- (SAMP), use in the synthesis of aldehydes, **3**, 10
Pyrrolidine, bis(phosphonato)-, synthesis of, **6**, 160
Pyrrolidine, *N*-cyano-, as starting material, for squaric acids, **6**, 623
Pyrrolidine, (2*S*,5*S*)-2,5-dimethyl-, from, L-alanine, **2**, 325
Pyrrolidine, *trans*-2,5-dimethyl-
  from

D-alanine, **2**, 331
L-alanine, **2**, 331
Pyrrolidine, *N*-isobutyl-, reactions, with thione *S*-imides, **5**, 336
Pyrrolidine, 1-methoxy carbonyl-, methoxylation of, **4**, 303
Pyrrolidine, *N*-methyloctafluoro-, chlorination of, photochemical, **6**, 450
Pyrrolidine, (*S*)-(-)-2-(di-β-naphthylhydroxymethyl)-, reactions, with boronic acids, **2**, 65
Pyrrolidine, *N*-phenyl-, synthesis of, **2**, 320
Pyrrolidine carboxylic acids, 3-substituted, synthesis of, **2**, 331
Pyrrolidine complexes, *N*-methyl-, use in the reduction of carbonyls, **2**, 60
Pyrrolidinedione, *N*-fluoro-, synthesis of, **5**, 314
Pyrrolidine ligands, bis(phosphino)-, synthesis of, **2**, 437
Pyrrolidines
  α-alkylation of, **1**, 149
  from
    δ-halo amines, **2**, 323
    imines, **1**, 289
  synthesis of, **1**, 256, 285, 368; **2**, 320, 323; **4**, 536
Pyrrolidines, allenic, synthesis of, **1**, 979
Pyrrolidines, bis(methoxymethyl)-, as chiral auxiliary in radical reactions, **1**, 326
Pyrrolidines, 2,6-dialkyl-, synthesis of, **2**, 323
Pyrrolidines, 2,5-dimethoxymethyl-, as chiral auxiliaries, in radical reactions, **1**, 328
Pyrrolidines, 2,5-dimethyl-, as chiral auxiliaries, in radical reactions, **1**, 328
Pyrrolidines, 2-(diphenylphosphinoyl)-, synthesis of, **4**, 468
Pyrrolidines, 3-methylene-, synthesis of, **2**, 324
Pyrrolidines, *N*-sulfonyl-, synthesis of, **2**, 365
Δ^1-Pyrrolidines, synthesis of, **3**, 410
Δ^2-Pyrrolidines, 5-trimethylsilyl-, synthesis of, **3**, 409
Pyrrolidinoalkenes, *gem*-chloro-, synthesis of, **4**, 793
1-Pyrrolidinoethene, 1-phenyl-, reactions, with *C,N*-diphenylnitrone, **4**, 307
Pyrrolidino iodides, synthesis of, **2**, 353
Pyrrolidinone, *N*-vinyl-, as starting material, for cyclic imidoylgermaniums, **5**, 794
Pyrrolidin-2-one, *N*-chloromethyl-, phosphonylation of, **4**, 465
2-Pyrrolidinone, 5-ethoxy-, reactions, with 1-methyl-1-(trimethylsilyl)allene, **1**, 1060
2-Pyrrolidinone, *N*-methyl-, as solvent, use in oxy-Cope rearrangement, **1**, 413
Pyrrolidinones
  cyclization of, **1**, 547
  mesylation of, **2**, 309
  reactions, with triethylgermyllithium, **5**, 794
  synthesis of, **3**, 659
4-Pyrrolidinopyridine (PPY), use in the synthesis of carboxylic esters, **5**, 129
Pyrrolidone, *N*-methyl-, reactions, with sodium formate, **2**, 753
2-Pyrrolidone, electroreduction of, **6**, 24
2-Pyrrolidone, *N*-methyl-, as solvent, in the synthesis of ketones, **3**, 113
3-Pyrrolidone, from, ethyl glycinate, **2**, 324
Pyrrolidone complexes, metalla-, reactions, with ruthenium cluster compounds, **6**, 344
2-Pyrrolidone hydrotribromide, use in the synthesis of alkyl bromides, **2**, 28
Pyrrolidones, reactions, with substituted cyclohexanones, **3**, 262
Pyrrolidones, *N*-β-carboethoxyethyl-, as starting materials, for hexahydro indolizines, **3**, 262
5-Pyrrolidones, 2-cyanomethylene-, synthesis of, **3**, 658
Pyrroline, perfluoro-, synthesis of, **5**, 656
1-Pyrroline, 5-ethyl-, synthesis of, **3**, 411
Δ'-Pyrroline-*N*-oxides, use in the synthesis of pyrrolizidines, **2**, 354
Pyrrolines, synthesis of, **2**, 325; **4**, 415
1-Pyrrolines, 3-halo-, α-sulfenylation of, **3**, 418
3-Pyrrolines
  from, azidodienes, **2**, 325
  synthesis of, via Délépine reaction, **2**, 325
Δ^1-Pyrrolines, synthesis of, **3**, 410
3-Pyrrolin-2-one, 3,5-dimethyl-1-trimethylsilyl-, synthesis of, **5**, 992
Pyrrolin-5-one, 2-ethoxy-, irradiation of, **5**, 1011
Pyrrolizidine alkaloids, synthesis of, **1**, 829; **2**, 325, 765
Pyrrolizidines, synthesis of, **1**, 90, 370; **2**, 325, 354
Pyrrolo[1,2-*a*]imidazolinones, synthesis of, **4**, 422
Pyrrolo[2,3-*b*]indole nucleus, hexahydro-, from, indolenium ions, **4**, 418
Pyrrolo[2,3-b]pyridine-5-carboxylate, 1-benzyl-2,3-dimethyl-4-alkoxy-, hydrolysis of, **5**, 106
Pyruvate, methyl trichloro-, reactions, with amides, **3**, 417
Pyruvates, alkyl-, reactions, with amines, **2**, 745
Pyruvic acid
  from, tartaric acid, **5**, 73
  reactions, with sulfuryl chloride, **5**, 60
Pyruvic acid, chloro-, synthesis of, **5**, 60
Pyruvic acid, 3,4-dimethoxyphenyl-, reactions, with hydrogen peroxide, **5**, 51
Pyruvic acid, phenyl-, synthesis of, **5**, 66
Pyruvic acid esters, reactions, with alkynylphosphonites, **2**, 1074
Pyruvic acids, 3-(*N*-benzoylamino)-3-substituted, from, *N*-benzoylaminoacrylic acid nitriles, **5**, 94
Pyruvoyl chloride, from, methyl(α,α-dichloromethyl) ethers, **5**, 14
Pyrylium cations, 2,3,5,6-tetraphenyl-
  reactions
    with primary alkylamines, **1**, 31
    with primary arylamines, **1**, 31
Pyrylium ions
  reactions
    with ammonia, **3**, 420
    with primary amines, **3**, 420
Pyrylium nitrate, triphenyl-, reactions, with amines, **2**, 104
Pyrylium salts
  reactions
    with amines, **1**, 30
    with dimethyl sulfoxide, **4**, 863
    with Grignard reagents, **3**, 232
  ring-opening of, **3**, 64
  as starting materials, for α,β-unsaturated ketones, **3**, 232
  synthesis of, **2**, 1008
Pyrylium salts, 2,4,6-triphenyl-, use in the synthesis of isocyanates, **5**, 1002
Pyrylium salts, 2,4,6-trisubstituted, reactions, with sodium borohydride, **3**, 232

Q

Quadricyclane
reactions
with copper(II) chloride, **3**, 608
with silenes, **5**, 925
Quadrone, synthesis of, **1**, 382
Quantum mechanics, molecular orbitals, **1**, 848
Quassinoid, synthesis of, via selenoxide pyrolysis, **1**, 781
Quaternary centers, synthesis of, via radical additions, **1**, 320
Quaternary halides, thermolysis of, **1**, 446
Quaternary salts, synthesis of, **4**, 597
Quaternary salts, *N*-chloro, synthesis of, **2**, 338
Quaternary salts, *N*-fluoro, synthesis of, **2**, 336
Quaternary salts, trimethyl, synthesis of, **2**, 760
Quaternization
for the synthesis of alkynylantimony compounds, **2**, 1082
for the synthesis of alkynylarsenic compounds, **2**, 1079
Quinaldine, tribromo-, reduction of, **4**, 23
Quinane, synthesis of, via retro-Diels–Alder reaction, **1**, 786
Quinazoline(benzo[*d*]pyrimidine) compounds, use in the synthesis of arylamines, **2**, 753
Quinazoline-2,4-dione, 3-cyano-, synthesis of, **5**, 1120
Quinazolinediones, from, bis(trichloromethyl) carbonate, **6**, 416
Quinazoline-3-oxide, reactions, with acetic anhydride, **3**, 722
Quinazolinones, synthesis of, **4**, 422; **6**, 109
Quinazolones, acetoxy, as starting materials, for aziridines, **4**, 515
4-Quinazolones, synthesis of, **5**, 987
Quinidine, dihydro-, use in asymmetric *cis*-dihydroxylation reactions, **2**, 48
Quinodimethane, 7,7,8,8-tetracyano-, reactions, with ethanolamines, **4**, 893
2,3-Quinodimethane, *N*-benzoylindole-, use in the synthesis of bis(sulfones), **4**, 277
Quinodimethanes, tetracyano-, synthesis of, **3**, 648
o-Quinodimethanes, from, imidoylsilanes, **5**, 793
Quinodimethans, tricyano-
as starting materials
for *gem*-dihydrazinoalkenes, **4**, 985
for α-hydrazino enamines, **4**, 985
Quinoline
oxidation of, **2**, 782
reactions
with allenyltins, **1**, 1060
with dimethylketene, **4**, 886
with *N*-phenylaminochloroimidates, **5**, 1123
with propargyltins, **1**, 1060
with sodium methoxide, **4**, 195
with thiophosgene, **5**, 1033
reduction of, **1**, 99
use in the synthesis of alkenes, **1**, 502
Quinoline, 2-chloro-, reactions, with triethylamine, **2**, 301
Quinoline, 2-chloromethyl-
reactions, with potassium *O*-ethyl dithiocarbonate, **6**, 545
use in the synthesis of dithiocarbamates, **6**, 561
Quinoline, 2-cyano-, synthesis of, **3**, 671
Quinoline, 6-cyano-4-methoxy-, hydrolysis of, **5**, 108
Quinoline, 4,7-dichloro-, ring-opening of, **5**, 1033
Quinoline, 2-(diethylamino)-, synthesis of, **2**, 301
Quinoline, 1,4-dihydro-, synthesis of, **1**, 99
Quinoline, 6-methoxy-8-cyano-, hydrolysis of, **5**, 108
Quinoline, 4-methyl-, reactions, with benzoyl chloride, **1**, 861
Quinoline, 2-(nitrobromo)ethyl-, as starting material, for *O*-benzoyl oxyimidic chloride, **5**, 823
Quinoline, nitromethyl-, reactions, with benzoyl chloride, **5**, 823
Quinoline, 4-phenacyl-, synthesis of, **1**, 861
Quinoline, tetrahydro-, reactions, with chiral formamidines, **5**, 746
Quinoline, 1,2,3,4-tetrahydro-, acetylation of, **3**, 299
Quinoline alkaloids, *trans*-decahydro-, synthesis of, **2**, 326
Quinoline ligands, 8-alkyl-, orthometallation of, **2**, 582
Quinoline-1-oxide, reactions, with bis-(phenylsulfonyl)methane, **4**, 281
Quinoline *N*-oxide, 6-amino-, from, quinoline *N*-oxides, **2**, 767
Quinoline *N*-oxides
reactions, with trimethylsilyl cyanide, **3**, 671
as starting materials, for 6-aminoquinoline *N*-oxide, **2**, 767
Quinolines
metallation of, **3**, 109
reduction of, **2**, 327
trichloromethylation of, **6**, 28
Quinolines, 1-alkoxycarbonyl-2-(2-alkynyl)-1,2-dihydro-, synthesis of, **1**, 1060
Quinolines, 1-aroyl-2-cyano-1,2-dihydro-, synthesis of, **3**, 85
Quinolines, 4-arylamino-, synthesis of, **3**, 588
Quinolines, cyano-, synthesis of, **3**, 672
Quinolines, 2-(*N*-cyanophenylamino)-, synthesis of, **5**, 1123
Quinolines, decahydro-
deprotonation of, **1**, 151
synthesis of, **1**, 90
Quinolines, dihydro-, synthesis of, **3**, 407
Quinolines, methyl-, reactions, with selenium dioxide, **3**, 109
Quinolines, 2-methyl-, trichlorination of, **6**, 28
Quinolines, 4-methyl-, trichlorination of, **6**, 28
Quinolines, *cis*-perhydro-, alkylation of, **1**, 151
Quinolines, quaternized, amination of, **4**, 413
Quinolines, 1,2,3,4-tetrahydro-
deprotonation of, **1**, 151
from, *N*-butoxycarbonyl arylamines, **2**, 328
synthesis of, **1**, 258; **2**, 327
Quinolines, 5,6,7,8-tetrahydro-, reactions, with *N*-substituted hydroxylamines, **2**, 346
2-Quinolines, 3,4-dihydro-, synthesis of, **3**, 414
Quinolinium cation, 4,4-diiodo-1,1-dimethyl-1,4-dihydro-, as starting material, for α-iodo alcohols, **4**, 43
Quinolinium salts, 2-chloro-, reactions, with trishydroxymethylphosphane, **5**, 882
Quinolizidinones, synthesis of, via acyliminiumvinylsilane cyclizations, **2**, 328
Quinolizines, silyl, synthesis of, **2**, 944
Quinols, dihydro-, removal of, **1**, 554
Quinone, anthra-, as starting material, for diazonium salts, **2**, 806
Quinone, 1,4-naptha-, reactions, with dialkyl acyl phosphoramidites, **5**, 373
Quinone diazides
acylation of, **2**, 803
protonation of, **2**, 803
p-Quinone–iron complexes, isolation of, **2**, 670
o-Quinonemethides, reactions, with 1-pyrrolidinocyclohexene, **4**, 307
Quinones
alkylation of, **1**, 273
condensation of, **3**, 648
from, maleoylcobalt complexes, **5**, 457
reactions
with alkyl radicals, **1**, 275
with trialkylborane, **1**, 275

reduction of, **2**, 657
as starting materials
for cyclobutenes, **1**, 517
for hydroquinones, **2**, 657
synthesis of, **1**, 551; **4**, 901; **5**, 973
Quinones, allyloxyanthra-, rearrangement of, **2**, 656
Quinones, halohydro-, synthesis of, **2**, 657
Quinones, hydro-
from, quinones, **2**, 657
synthesis of, **2**, 651, 655
Quinones, 1,4-naphtha-, as starting materials, for 2-alkyl-1,4-naphthaquinones, **1**, 275
o-Quinones, reactions, with phosphaalkenes, **4**, 340
Quinonoids, synthesis of, **4**, 855
Quinoxaline, use in the synthesis of aromatic acids, **5**, 115
Quinoxalines, 2-(bromodifluoromethoxy)-, 6-substituted, synthesis of, **6**, 229
Quinuclidine
irradiation of, **4**, 98
reactions, with fluorine, **2**, 336
Quinuclidine ligands, use in the synthesis of alkylrheniums, **2**, 585
Quinuclidinium fluoride, *N*-fluoro-, synthesis of, **2**, 336
Quinuclidones, reactions, with quinoline aldehydes, **3**, 258

R

Racemates
  as starting materials
    for alcohols, **4**, 390
    for esters, **4**, 390
Racemization
  in the dehydration of oximes, **3**, 617
  of ketenimines, **3**, 556
  with phenethyl chloride, **1**, 112
  rate constant for, **1**, 854
Radialenes, synthesis of, **4**, 820
Radical addition
  asymmetric induction, **1**, 326
  asymmetric 1,4-induction, **1**, 328
  factors affecting, **1**, 323
  FMO treatment, **1**, 323
  mediated by mercury hydrides, **1**, 332
  reactivity and selectivity requirements, **1**, 329
  for the synthesis of carbon–carbon bonds, **1**, 320
  to alkenes, transition state, **1**, 322
Radical addition, intermolecular, to carbon–carbon multiple bonds, **1**, 322
Radical addition, intramolecular, to carbon–carbon multiple bonds, **1**, 336
Radical additions, acyclic, stereoselectivity in intermolecular additions, **1**, 326
Radical allylation, in asymmetric reactions, **1**, 328
Radical anions, rearrangements for, **1**, 395
Radical cation, via electron transfer, **4**, 1072
Radical centres
  flanked by nitrogen and oxygen, **4**, 1077
  flanked by nitrogen and sulfur, **4**, 1077
  flanked by two halogens, **4**, 1080
  flanked by two oxygens, **4**, 1078
  flanked by two sulfurs, **4**, 1080
Radical chemistry, for the synthesis of α,β-unsaturated ketones, **3**, 223
Radical chlorination, for the synthesis of α-chloroethyl chloroformate, **6**, 428
Radical cyclization
  and intermolecular radical additions, compared, **1**, 336
  methods, **1**, 341
  for the synthesis of 6-membered rings, **1**, 339
Radical cyclization, aryl
  6-*endo*, **1**, 339
  7-*endo*, **1**, 340
  8-*endo*, **1**, 340
Radical cyclization, metal-mediated, mechanisms, **1**, 346
Radical cyclization, sequential, mechanism, **1**, 347
Radical delocalization, mechanism, **1**, 922
Radical diastereofacial selection, reviews, **1**, 919
Radical frontier molecular orbital theory, **1**, 920
Radical initiators, in the synthesis of alkylmercury carboxylates, **2**, 594
Radical intermediates, in coupling reactions, **1**, 119
Radical macrocyclization, mechanism, **1**, 340
Radical-mediated processes, for the cleavage of carbon–nitrogen bonds, **1**, 265
Radical nitrogen, for the synthesis of heteroatom radicals, **1**, 927
Radical oxygen, for the synthesis of heteroatom radicals, **1**, 927
Radical–radical coupling, mechanism, **1**, 921
Radical reactions
  β-elimination of, **1**, 321
  intermolecular–intermolecular addition, **1**, 348
  intermolecular–intramolecular addition, **1**, 348
  intramolecular–intermolecular processes, **1**, 348
  intramolecular–intramolecular sequences, **1**, 347
  and ionic reactions compared, **1**, 320
  methods, **1**, 329
  for the substitution of oxygen functions, **1**, 172
  tandem processes, **1**, 347
Radical reactions, transition metal-mediated, reviews, **1**, 320
Radical retrosynthetic planning, reviews, **1**, 919
Radical ring-expansions, reviews, **1**, 919
Radicals
  adjacent to chalcogen, synthesis of, **2**, 1010
  adjacent to halogen, synthesis of, **2**, 1010
  adjacent to nitrogen, synthesis of, **2**, 1010
  adjacent to phosphorus, synthesis of, **2**, 1010
  adjacent to silicon, synthesis of, **2**, 1010
  definition of, **1**, 920
  dicoordinate, synthesis of, **1**, 1126
  electronic nature of, **1**, 323
  from
    alcohols, **1**, 172
    7-bromoheptan-2-one, **1**, 546
    halides, **1**, 937
    hydroxamate esters, **1**, 231
    nitro groups, **1**, 548
    xanthates, **1**, 231
    ynones, **1**, 546
  general literature survey, **1**, 919
  plasma generation of, **6**, 21
  reactions
    with α-chloroacrylonitriles, **3**, 627
    with cyclopropylethyne, **1**, 1130
    for the substitution of sulfur, **1**, 230
    with trialkylboranes, **1**, 546
    with trifluoroamidines, **5**, 818
  regioselectivity of, **1**, 925
  reversible addition to alkenes, **1**, 949
  solvent and concentration effects, **1**, 923
  stabilization of, **2**, 141
  as starting materials, for carbocations, **1**, 336
  substituent migrates as, **1**, 395
  synthesis of, **1**, **843–951**, 919, **1087–1145**; **4**, 1077, **1071–1083**
    reviews, **1**, 1126
    via bond scission, **1**, 925
    via homolysis of organocobalt complexes, **1**, 345
  thermolytic generation of, **1**, 173
  vicinal substituents, **1**, 925
Radicals, acylamino, synthesis of, **1**, 341
Radicals, α-acylamino, cyclization of, **2**, 325
Radicals, carbon-centered
  characteristics of, **1**, 919
  synthesis of, **1**, 106
Radicals, conjugated, conformationally-restricted, in asymmetric induction, **1**, 327
Radicals, ester-substituted, lowest energy conformation of, **1**, 327
Radicals, α-heteroatom-stabilized, stability of, **2**, 1010
Radicals, hetero-substituted, synthesis of, via intramolecular 1,5-hydrogen transfer, **1**, 339
Radicals, nitrogen-centered, reactions, with carbon–carbon double bonds, **1**, 950
Radicals, nucleophilic
  definition of, **1**, 920
  intermolecular addition reactions, **1**, 323
  reactions, with alkynes, **1**, 324

Radicals, oxygen-centered
  β-fragmentation of, **1**, 929
  reactions, with carbon–carbon double bonds, **1**, 948
Radicals, persistent
  definition of, **1**, 922
  reviews, **1**, 919
Radicals, persistent heteroatom-stabilized, synthesis of, **2**, 1010
Radicals, primary, synthesis of, **1**, 330
Radicals, secondary, synthesis of, **1**, 330
Radicals, α-silylalkyl, synthesis of, **2**, 1010
Radicals, sulfur-centered, reactions, with carbon–carbon double bonds, **1**, 949
Radicals, thiol, reactions, with cyclohexenes, **2**, 137
Radicals, tricoordinate, synthesis of, **1**, 919
Radicals, tricoordinated stabilized, synthesis of, **6**, **725–734**
Radicals, vinylic, synthesis of, **1**, 1126
Radical substitution, for the reduction of alkyl halides, **1**, 2
Radiolabeled compounds, synthesis of, **1**, 449
Ramberg–Bäcklund reaction
  bases for, **1**, 618
  Michael induced, **1**, 707
  for ring synthesis, **1**, 707
  stereoselectivity problems, **1**, 707
  for the synthesis of alkenes, **1**, 144, 617
  for the synthesis of 1,1-dichloroalkenes, **4**, 742
  for the synthesis of α-halo sulfones, **4**, 79
  for the synthesis of vinyl chlorides, **2**, 619
  use in the synthesis of alkenes, **1**, 704
  vinylogous, **1**, 707
RAMP hydrazones, use in the synthesis of α-hydroxy ketones, **3**, 156
Raney cobalt, reactions, with carbon monoxide, **6**, 524
Raney copper, use in the synthesis of secondary phosphines, **2**, 431
Raney nickel
  as catalyst
    for the reduction of alkyl halides, **1**, 2
    for the reduction of nitriles, **1**, 46
  catalysts, for the semihydrogenation of alkynylsilanes, **2**, 921
  for the cleavage of dinitro compounds, **1**, 57
  dehydrogenation of alcohols with, **3**, 118
  reactions
    with aryl sulfides, **1**, 435
    with aryl thiols, **1**, 435
    with dithioacetals, **1**, 11
    with ethanol, **1**, 47
    with nitriles, **3**, 86
  as reagent, for oxime cleavage, **3**, 134
  for the reduction of aldehydes, **1**, 18
  for the reduction of amines, **1**, 34
  for the reduction of aryl chlorides, **1**, 428
  for the reduction of aryl fluorides, **1**, 426
  for the reduction of aryl iodides, **1**, 429
  for the reduction of bromoalkanes, **1**, 8
  for the reduction of chloroalkanes, **1**, 6
  for the reduction of dithioacetals, **1**, 24
  for the reduction of esters, **1**, 16
  for the reduction of fluoroalkanes, **1**, 6
  for the reduction of iodoalkanes, **1**, 10
  for the reduction of ketones, **1**, 19
  for the reduction of sulfides, **1**, 23
  for the reduction of sulfoxides, **1**, 23
  for the reduction of tertiary alcohols, **1**, 12
  for the reduction of vinyl sulfides, **1**, 438
  use in reduction reactions, **1**, 81
  use in the synthesis of acyl aldehydes, **3**, 85
  use in the synthesis of alkanes, **1**, 21
  use in the synthesis of alkenes, **1**, 503
  use in the synthesis of semicarbazones, **3**, 450
Raney nickel reagent, for the reduction of thiol esters to aldehydes, **3**, 13
Rapamycin, synthesis of, **1**, 818, 828
Rathke guanidine synthesis
  mechanism, **6**, 641, 642
  using *S*-methylisothiouronium salts, **6**, 651
  variants, **6**, 641
S^N2'-Reaction, for the synthesis of β,γ-unsaturated esters, **5**, 158
Reaction rate, effect of changes of halogen on, **1**, 107
Reactions, types of, **1**, 377
Reactivity, in the addition of nucleophiles to carbon–carbon multiple bonds, **1**, 300
Reagent control, in the reduction of carbonyls, **2**, 63
Reagent-controlled asymmetric synthesis, for the synthesis of alcohols, **2**, 42
Reagents
  for coupling reactions, in the synthesis of tetrasilylmethanes, **6**, 378
  for the hydroboration of alkynes, **1**, 509
  for the opening of epoxides, **1**, 204
  for the oxidation of hydroxylamines, **2**, 782
  for the oxidation of selenides, **1**, 601
  polymer supported, **1**, 496
  for the synthesis of aromatic ketones, **3**, 294
  for the synthesis of arylamines, **2**, 752
  for the synthesis of tertiary phosphines, choice of, **2**, 440
Rearrangement
  of amine *N*-oxides, **2**, 355
  categorization of, **1**, 793
  involving [1,2] shifts and [1, *j*] migration, **1**, 794
  of ketoamines, **5**, 272
  migration of an aromatic ring, **1**, 378
  of oxaziridines, **5**, 272
  rate of, **1**, 396
  stereoelectronic control of, **1**, 115
  for the substitution of sulfur functions, **1**, 233
  for the synthesis of alcohols, **2**, 88
  for the synthesis of alkanesulfenic esters, **2**, 195
  for the synthesis of *N*-alkylamides, **5**, 282
  for the synthesis of amides, **5**, 270
  for the synthesis of *N*-arylamides, **5**, 292
  for the synthesis of *N*-arylarylamides, **5**, 303
  for the synthesis of carbon–nitrogen bonds, **1**, 266
  for the synthesis of *N*-carbon-substituted ketimines, **3**, 407
  for the synthesis of dialkyl sulfones, **2**, 159
  for the synthesis of dialkyl sulfoxides, **2**, 154
  for the synthesis of 1,1-dichloroalkenes, **4**, 742
  for the synthesis of enamides, **5**, 288
  for the synthesis of ketene hemiaminals, **4**, 894
  for the synthesis of one or more carbon–carbon and carbon–hydrogen bonds, **1**, **377–423**, **793–842**
  for the synthesis of primary amides, **5**, 275
  for the synthesis of primary arylamides, **5**, 298
  for the synthesis of α,β-unsaturated ketones, **3**, 216
  of triazolines, **3**, 419
  of vinylcyclobutanes, **1**, 805
  of vinylcyclopropanes, **1**, 801
  in vinyl radical cyclizations, **1**, 338
Rearrangement, anionic, **1**, 378
  of carbanions, **1**, 886
Rearrangement, cationic, **1**, 380
Rearrangement, charge-accelerated, **1**, 413
  mechanism, **1**, 401
  of vinylcyclopropanes, **1**, 803
Rearrangement, chiral, of ylides, **1**, 235
Rearrangement, electrophilic, mechanism, **1**, 378
Rearrangement, germanotropic, **1**, 766
Rearrangement, nucleophilic, mechanism, **1**, 379
Rearrangement, photochemical
  of hydrogen, **1**, 794
  of vinylcyclopropanes, **1**, 802
Rearrangement, pinacol-type, mechanism, **1**, 115
Rearrangement, propargylic, for the synthesis of allenes, **1**, 955

Rearrangement, prototropic, mechanism, **1**, 797
Rearrangement, semibenzilic, **1**, 393
Rearrangement, [3,3] sigmatropic
for the synthesis of chlorothiolformates, **6**, 436
for the synthesis of dialkyl dithiocarbonates, **6**, 472
for the synthesis of thiocarbamates, **6**, 495
Rearrangement, thermal [1,3], of alkyl and aryl groups, **1**, 799
Rearrangement, thermal [1,5], of alkyl and aryl groups, **1**, 799
[1,3] Rearrangement, anion-accelerated, **1**, 800
[1,3] Rearrangement, photochemical, of alkyl and aryl groups, **1**, 799
[1,*j*] Rearrangement, involving carbon migration, **1**, 799
[2,3] Rearrangement
of nitrogen ylides, **1**, 823
of oxonium ylides, **1**, 820
of sulfur ylides, **1**, 821
Redistribution
of organoboranes, **2**, 522
for the synthesis of alkylarsine halides, **2**, 490
for the synthesis of alkynylmercury compounds, **2**, 1097
for the synthesis of haloallylboranes, **2**, 526
for the synthesis of mixed alanes, **2**, 1099
Redox
for the synthesis of γ-lactones, **5**, 171
for the synthesis of oximes, **3**, 426
for the synthesis of radicals, **1**, 925, 926
for the synthesis of α,β-unsaturated esters, **5**, 154
Reducing agents, for the synthesis of hydroxylamines, **2**, 764
Reduction
of *O*-acyl oximes, **5**, 324
of alkenes, **1**, 71
of alkenyl-substituted alkenes, **1**, 79
of alkyl halides, **1**, 1
of alkyl-substituted alkenes, **1**, 78
of alkynyl-substituted alkenes, **1**, 79
of amides, **3**, 85
of anhydrides, **3**, 84
of aromatic esters, **3**, 84
of aromatic nitriles, **3**, 86
of aryl halides, **1**, 425
of aryl-substituted alkenes, **1**, 79
of benzoic acids, **3**, 82
of benzoyl halides, **3**, 83
of carbon–antimony bonds, **1**, 444
of carbon–arsenic bonds, **1**, 444
of carbon–bismuth bonds, **1**, 444
of carbon–boron bonds, **1**, 447
of carbon–germanium bonds, **1**, 447
of carbon–halogen bonds, **1**, 425
of carbon–metal bonds, **1**, 452
of carbon–nitrogen bonds, **1**, 441
of carbon–oxygen double bonds, **1**, 432
of carbon–phosphorus bonds, **1**, 444
of carbon–selenium bonds, **1**, 435
of carbon–silicon bonds, **1**, 447
of carbon–sulfur bonds, **1**, 435
of carbon–tellurium bonds, **1**, 435
of carboxylic acids, **3**, 82
of disulfides, **5**, 338
of heteroalkyl-substituted alkenes, **1**, 81
of heteroaryl-substituted alkenes, **1**, 79
of nitriles, **2**, 316
of remotely-substituted alkenes, **1**, 84
of sulfones, **2**, 154
of sulfonyl halides, **2**, 187
for the synthesis of amines, **2**, 312
for the synthesis of aryl sulfides, **2**, 713
for the synthesis of *N*-carbon-substituted ketimines, **3**, 407
for the synthesis of tertiary phosphines, **2**, 440
for the synthesis of α,β-unsaturated aldehydes, **3**, 61
of thiol esters, **3**, 85
of vinyl halides, **1**, 429
Reduction, cathodic
of carbonyls, **2**, 80
for the synthesis of amines, **2**, 318
Reduction, electrochemical
of arenes, **1**, 103
of aryl bromides, **1**, 428
for the reduction of carbon–mercury bonds, **1**, 68
for the synthesis of dichloromethyls, **4**, 13
Reduction, electrolytic
of amines, **1**, 32, 34
of disulfides, **2**, 126
of phosphonium salts, **2**, 441
for the synthesis of alcohols, **2**, 80
Reduction, enzymatic, for the synthesis of cyclic hemiacetals, **4**, 165
Reduction, palladium-catalyzed, of propargylic compounds, **1**, 962
Reductive amination
of aldehydes and ketones, **3**, 188
for the synthesis of amino acetals, **3**, 44
Reductive cleavage, of disulfides, **2**, 126
Reformatsky–Claisen rearrangement, of α-halo esters, **1**, 831
Reformatsky reaction
mechanism, **2**, 69
for the synthesis of β-lactones, **5**, 161
for the synthesis of zinc enolates, **2**, 702
Reformatsky reagents
reactions
with orthoformates, **4**, 193
with phenyl fluoroalkyl sulfonates, **1**, 483
synthesis of, **2**, 697; **3**, 461
for the synthesis of sulfinamides, **2**, 247
for the synthesis of β,γ-unsaturated esters, **1**, 467
use in the synthesis of carbonyls, **1**, 471
Refrigerants
chlorofluorocarbons for, **6**, 7
1,1,1,2,2-pentafluoroethane, **6**, 8
1,1,1,2-tetrafluoroethane, **6**, 8
1,1,1-trifluoroethane, **6**, 8
Regioisomers, in cycloaddition reactions, **1**, 350
Regioselectivity
in the addition of nucleophiles to carbon–carbon multiple bonds, **1**, 300
of electrocyclic additions, **1**, 350
of 5-hexenyl radicals, **1**, 336
of radical additions, **1**, 323
of radicals, **1**, 925
of sulfoxide elimination, **1**, 595
γ-Regioselectivity, of organocuprates, **1**, 123
Reimer–Tiemann reaction
mechanism, **6**, 38
for the synthesis of aromatic ketones, **3**, 283
for the synthesis of aryl aldehydes, **3**, 91
for the synthesis of dichloroalkanes, **4**, 17
for triaryl orthoformates, **6**, 69
Reissert compounds
alkylation of, **1**, 257
synthesis of, **3**, 85
Reissert–Henze reaction, for the synthesis of heteroarylnitriles, **3**, 671
Renin inhibitors, synthesis of, **3**, 171
Reppe alcohol synthesis, mechanism, **2**, 44
Resins, synthesis of, **5**, 864
Resonance, for the stabilization of anions, **1**, 850
Resonance, no-bond, stabilizing mechanism, **3**, 399
Resorcinol
reactions
with -2-chloro-3-methyl-1,3-benzothiazolium tetrafluroborate, **6**, 318
with potassium bicarbonate, **5**, 110
Resorcinols
from, diazo dienones, **2**, 668

synthesis of, **1**, 46; **2**, 664, 668
via thermal Wolff rearrangements, **2**, 668
β-Resorcyclic acid, synthesis of, **5**, 110
Reticuline, synthesis of, **1**, 149
(+)-Reticuline, synthesis of, **2**, 328
Retigeranic acid, from, ketenes, **3**, 528
(-)-Retigeranic acid, from, arylpentene, **1**, 365
cis-Retinal, synthesis of, **1**, 752
Retinoic acids, Julia synthesis of, **1**, 612
Retro-aldol reaction, for the synthesis of carbanions, **1**, 883
Retro-cheletropic reaction
for the synthesis of dienes, **1**, 790
for the synthesis of dienes and polyenes, **1**, 771
Retro-cycloaddition, for the synthesis of monoenes, **1**, 784
Retro Diels–Alder reaction
for the synthesis of silasilenes, **5**, 926
for the synthesis of silenes, **5**, 926
Retrodiene cleavage reaction, mechanism, **1**, 1083
Retro dimerization
for the synthesis of oxasilanes, **5**, 924
for the synthesis of silenes, **5**, 924
Retro-ene reaction
as side-reaction in the oxy-Cope rearrangement, **1**, 412
for the synthesis of monoenes, **1**, 772
Retrohydroboration, mechanism, **1**, 657
Retrosynthetic planning, reviews, **1**, 320
Reverse Brook rearrangement
of alkoxygermanes, **4**, 361
of alkoxysilanes, **4**, 358
of germyl thioethers, **4**, 371
of silyl thioethers, **4**, 368
See also Brook rearrangement
Reverse Cope elimination, mechanism, **1**, 626
Reverse Diels–Alder reaction, for the synthesis of stannylvinyl ethers, **4**, 952
Reversible addition, mechanism, **1**, 950
L-Rhamnose compounds, synthesis of, **4**, 235
Rhenacyclobutenone ring system, construction of, **5**, 451
Rhenate, pentacarbonyl-, reactions, with molybdenum alkyne complexes, **4**, 710
Rhenium, hexamethyl-, synthesis of, **2**, 585
Rhenium, pentacarbonyl bromo-, reactions, with methylene(triphenyl)phosphorane, **5**, 919
Rhenium, tetracarbonyl-, synthesis of, **6**, 523
Rhenium, tricarbonyl(cyclopentadienyl)-, lithiation of, **4**, 706
Rhenium(I), tricarbonyl(cyclopentadienyl)-, bromination of, **6**, 457
Rhenium(II) oxide, for the reduction of carboxylic acids, **1**, 21
Rhenium(III) compounds, alkyl-, synthesis of, **2**, 583
Rhenium(IV) clusters, alkyl-, synthesis of, **2**, 584
Rhenium(V) compounds, alkyl-, synthesis of, **2**, 585
Rhenium(VI) compounds, alkyl-, from, oxo compounds, **2**, 585
Rhenium(VII) compounds, alkyl-, synthesis of, **2**, 585
Rhenium acyl compounds, from, carbynes, **5**, 451
Rhenium acyl compounds, anionic, synthesis of, **5**, 450
Rhenium carbene complexes
from, rhenium carbene complexes, **3**, 522
reactions
with hydrogen chloride, **3**, 522
with phosphines, **4**, 581
as starting materials, for rhenium carbene complexes, **3**, 522
synthesis of, **3**, 521
Rhenium carbonyl complexes, synthesis of, **6**, 523
Rhenium carbyne complexes, reactions, with trimethyl phosphorane, **5**, 919
Rhenium chloride, tri-*t*-butylimido-, reactions, with methylene(triphenyl)phosphorane, **5**, 919
Rhenium clusters, synthesis of, **2**, 583
Rhenium complex, bromocarbonyl-, as intermediate, **6**, 457
Rhenium complexes
dehydrohalogenation of, **3**, 521
reactions
with η^2-acylzirconiums, **3**, 521
with *t*-butylamine, **3**, 522
with Grignard reagents, **3**, 521
synthesis of, **6**, 457
Rhenium complexes, alkyl-, synthesis of, via alkylation, **2**, 579
Rhenium complexes, (α-chlorobenzyl)-, synthesis of, **4**, 150
Rhenium complexes, cyclopentadienyl-
reactions, with methyllithium, **5**, 450
synthesis of, **5**, 451
Rhenium complexes, α-haloalkyl-, synthesis of, **4**, 150
Rhenium *cis*-dioxo complexes, photolysis of, **3**, 521
Rhenium functions, doubly-bonded, synthesis of, **3**, 521
Rhenium heptoxide, reactions, with carbon monoxide, **6**, 523
Rhenium nitrosyl complexes, synthesis of, **2**, 580
Rhenium oxide, tetramethyl-, alkylation of, **2**, 585
Rhenium pentacarbonyls
reactions
with acid chlorides, **5**, 449
with dimethyl ethynedicarboxylate, **5**, 451
with methyllithium, **5**, 450
with organolithiums, **5**, 449
Rhenium–phosphine complexes, as catalysts, for haloalkane addition, **6**, 7
Rheniums, transylidation of, **5**, 919
Rheniums, acyl-, synthesis of, **5**, 449
Rheniums, carboxymethyl-, reactions, with Grignard reagents, **5**, 451
Rheniums, imidoyl-
from, functionalized amines, **5**, 801
synthesis of, **5**, 800
Rheniums, monoalkyl trioxo-, synthesis of, **2**, 585
Rheniums, pentacarbonyl acyl-, synthesis of, **5**, 449
Rhenium vinyl compounds, reactions, with allyl bromide, **5**, 450
Rhizoxin, synthesis of, **1**, 1134
Rhodanine, as starting material, for thioketones, **3**, 350
Rhodium
reactions
with carbon monoxide, **6**, 525
with diazoalkane, **4**, 686
with titanium dioxide, **1**, 88
use in acylation reactions, **3**, 289
use in allylation reactions, **1**, 196
use in hydrogenation reactions, **1**, 87
Rhodium, bischlorodicarbonyl-, reactions, with cyclopentadienylbisdiphenylphosphinoruthenium chloride, **4**, 1029
Rhodium, bis(cyclooctene)chloro-, reactions, with acrolein and 1,2-bis(diphenylphosphino)ethane, **6**, 456
Rhodium, chlorotris(triphenylphosphine)-, reactions, with quinoline aldehydes, **5**, 459
Rhodium, cyclopentamethyl-, reduction of, **1**, 94
Rhodium, dicarbonylcyclopentadienyl-
decarbonylation of, **4**, 686
use in the synthesis of carbon-bridged dirhodium complexes, **4**, 686
Rhodium, (pentachlorocyclopentadienyl) (1,5-cyclooctadiene)-, reactions, with butyllithium, **4**, 707
Rhodium(I)
as catalyst, for the isomerization of allyl ethers, **2**, 675
reactions, with diketones, **3**, 272
Rhodium(I), chlorotris(triphenylphosphine)-, as hydrosilylation catalyst, **4**, 1044
Rhodium(I) carbonyl azides, reactions, with carbon monoxide, **5**, 983
Rhodium(I) catalysts, use in carbonylation reactions, **3**, 290
Rhodium(I) chloride, [1,2-bis(diphenylphosphino)butane]-, as catalyst, for the synthesis of anhydrides, **5**, 192

Rhodium(I) complex, (diazomethyl)trimethylsilane-, synthesis of, **6**, 675
Rhodium(I) complexes, Π-alkene-, isomerization of, **2**, 583
Rhodium(I) halide complexes, reactions, with alkyl halides, **2**, 583
Rhodium(I) imide, synthesis of, **5**, 387
Rhodium(I)phosphine complexes, synthesis of, **2**, 579
Rhodium(II) acetate, as catalyst, for the elimination of amines, **1**, 637
Rhodium(II) carboxylates, as catalysts, for the synthesis of cyclopropanes, **1**, 299
Rhodium(II) complexes, reactions, with diazo compounds, **3**, 520
Rhodium(II) salts, use in the synthesis of carbenes, **3**, 320
Rhodium(III) chloride, reactions, with alkenyl cyclohexenones, **2**, 660
Rhodium(III) complexes, (chloromethyl)-, synthesis of, **4**, 149
Rhodium(III) complexes, (iodomethyl)-, synthesis of, **4**, 149
Rhodium(III) compounds, alkyl-, synthesis of, **2**, 583
Rhodium(III) hydride complexes, reactions, with alkenes, **2**, 583
Rhodium(III)phosphine complexes, orthometallation of, **2**, 581
Rhodium(III) porphyrin complexes, synthesis of, **2**, 583
Rhodium acyl compounds, synthesis of, via incorporation of homochiral phosphine ligands, **5**, 459
Rhodium–aluminum oxide, for the dealkylation of xylene, **1**, 56
Rhodium–amine catalysts, for the synthesis of alkoxy aldehydes, **3**, 33
Rhodium analogues, synthesis of, **4**, 690
Rhodium bis(triflate) complexes, reactions, with ethylenebis(triphenylphosphine)platinum, **4**, 716
Rhodium carbene complexes, synthesis of, **3**, 520
Rhodium carbonyl, reactions, with titanium dioxide, **1**, 72
Rhodium carbonyl complexes
  reactions, with nitro compounds, **5**, 984
  synthesis of, **6**, 525
  use in the synthesis of amides, **2**, 318
  use in the synthesis of dienes, **1**, 491
Rhodium carbonylcyclopentadienyl complexes, synthesis of, **1**, 94
Rhodium carbonyls, use in the synthesis of amines, **2**, 314
Rhodium carbonyls, alkyl-, selectivity of, **2**, 580
Rhodium catalysis, for the reduction of 2-amino ketones, **2**, 54
Rhodium catalysts
  in cyclopropanation reactions, **1**, 353
  for the dealkylation of alkylbenzenes, **1**, 56
  for the dehydrogenation of alkanes, **1**, 554
  for hydroformylation reactions, **1**, 296
  reactions
    with chiral ligands, **1**, 353
    with silanes, **4**, 831
  for the substitution of alkenyl halides, **1**, 466
  synthesis of, **1**, 72, 88
  for the synthesis of alkenes, **1**, 507
  use in carbon monoxide insertion, **5**, 346
  use in the synthesis of vinylsilanes, **2**, 915
Rhodium chlorides, aroyl-, synthesis of, **2**, 957
Rhodium complex, diiodo-, reactions, with sodium hydrogen telluride, **3**, 391
Rhodium complexes
  for the decarbonylation of aldehydes, **5**, 459
  as hydrogenation catalysts, **1**, 456
  reactions
    with alkynes, **4**, 689
    with 3,3-dimethylcyclopropene, **4**, 689
    with diphenyldiazomethane, **3**, 520
  for the reduction of chloroarenes, **1**, 427
  synthesis of, **3**, 461; **6**, 456
Rhodium complexes, alkyl-, use in acylation reactions, **3**, 289
Rhodium complexes, allyl-, for the reduction of benzene, **1**, 93
Rhodium complexes, 2,2'-bis(diphenylphosphino)-1,1'-binaphthyl-, use in the synthesis of enamines, **2**, 816
Rhodium complexes, chlorocarbonyl-, synthesis of, **6**, 456
Rhodium complexes, (α-chloroethyl)-, synthesis of, **4**, 149
Rhodium complexes, diazomethyl-, synthesis of, **3**, 486
Rhodium complexes, (halomethyl)-, synthesis of, **4**, 149
Rhodium complexes, mono-bridged, synthesis of, **4**, 686
Rhodium complexes, tri-bridged, synthesis of, **4**, 686
Rhodium complexes, vinyl-, reactions, with hydrogen chloride, **4**, 149
Rhodium complexes, vinylidene-, reactions, with sulfur, **3**, 553
Rhodium compound, cyclopentadienyl-, reactions, with acid chlorides, **5**, 458
Rhodium dimers, dichloro(pentamethylcyclopentadienyl)-, reactions, with hexamethyldialuminum, **4**, 687
Rhodium functions, doubly-bonded, synthesis of, **3**, 520
Rhodium hydride, imidoyl-, use in the synthesis of unsymmetrical ketones, **5**, 800
Rhodium hydride, tetrakis(triphenylphosphine)-, use in the synthesis of α-silyl ketones, **3**, 202
Rhodium hydride complexes, allyl-, synthesis of, **1**, 94
Rhodium hydrides, acyl-, synthesis of, **5**, 459
Rhodiums
  reactions
    with acid chlorides, **5**, 458
    with phosphorus ylides, **4**, 582
  synthesis of, via cyclotrimerization of tolan, **4**, 689
Rhodiums, acyl-, synthesis of, **5**, 458
Rhodiums, dimeric, reactions, with polycyclic alkanes, **5**, 460
Rhodiums, 1-haloalkyl-, reactions, with phosphines, **4**, 582
Rhodiums, imidoyl-
  from, 3-methyl-2-aminopyridyl aldimines, **5**, 800
  synthesis of, **5**, 800
    via oxidative addition, **5**, 800
Rhodium salt, tetrakis(*t*-butyl isocyanide)-
  reactions
    with diethylamine, **6**, 664
    with dimethylamine, **6**, 664
Rhodium salts, use in the synthesis of enols, **2**, 639
Rhodium tetraacetate, for the reduction of diazocompounds, **1**, 44
Rhodium tetracarbonyl dichloride
  reactions
    with cubane, **5**, 460
    with 3,5-dehydronoriceane, **5**, 460
Rhodium trichloride trihydrate, reactions, with sodium borohydride, **1**, 73
Rhodium tristriphenylphosphine monochloride, use for the reduction of alkenes, **1**, 73
Rhodococcus s.p. BPM 1613, use in the synthesis of ibuprofen, **5**, 118
Ribonucleosides, oligo-, synthesis of, **4**, 219
L-(+)-Ribose, synthesis of, **4**, 362
Ricinolenic acid, as starting material, for azelaric acid, **5**, 52
Rieke barium, development of, **1**, 116
Rieke calcium, development of, **1**, 115
Rieke copper, development of, **1**, 121
Rieke magnesium, synthesis of, **2**, 558
Rieke zinc, development of, **1**, 123
Rifamycin, synthesis of, **1**, 438
(R)-Imidazolidinones, reactions, with *N*-bromosuccinimide, **4**, 102
Ring-cleavage, heterocyclic, for the synthesis of alkanethiols, **2**, 129
Ring-contraction
  in 2-amino alcohols, **1**, 389
  of bicyclo[3,2,1]octane, **1**, 382
  of cyclic α-diazo ketones, **1**, 395
  of cyclic ketones, **1**, 391
  in Favorskii rearrangements, **1**, 393

for the substitution of sulfur functions, **1**, 233
for the synthesis of cyclopentanones, **1**, 388
via 2,3 sigmatropic rearrangement, **1**, 410
Ring-contraction–expansion, for five-membered zirconacycles, **2**, 987
Ring-enlargement
dyotropic, **5**, 170
of β-lactones, **5**, 170
Ring-expansion
in aryl radical cyclization, **1**, 339
of bicyclo[4,2,0]octane, **1**, 382
of cyclic ketones, **3**, 144
of cyclopropanes, **1**, 390
of 2-hydroxy sulfides, **1**, 390
of ketones, via free-radical chemistry, **3**, 125
reactions for, **1**, 389
via 1,3 sigmatropic rearrangement, **1**, 401
via 2,3 sigmatropic rearrangement, **1**, 410
Ring-opening
of aziridines, **1**, 263
of diketene, with tributyltin alkoxides, **2**, 701
of epoxides, **2**, 82
for the synthesis of ketene *S,N*-acetals, **4**, 905
for the synthesis of ketene hemiaminals, **4**, 891
Ring-opening metathesis polymerization (ROMP), reviews, **1**, 675
Ring-size, and epoxides, **1**, 214
Ritter reaction
of nitriles, **4**, 883
for the synthesis of amides, **5**, 281
for the synthesis of amines, **2**, 308
for the synthesis of mixed anhydrides, **5**, 710
Robinson annelation, mechanism, **1**, 306
Robinson annulation reaction, for the synthesis of α,β-unsaturated ketones, **3**, 220, 241
Robinson-type annulation, for the synthesis of phenols, **2**, 663
Rongalite
use in the synthesis of selenides, **2**, 279
use in the synthesis of tellurides, **2**, 279
Rosenmund reduction
of acyl chlorides, **3**, 83
for the synthesis of aldehydes, **3**, 13
for the synthesis of amino aldehydes, **3**, 44
for the synthesis of arylarsonic acids, **2**, 884
Rosenmund–von Braun reaction, for the synthesis of α,β-arylnitriles, **3**, 661
Royal jelly, synthesis of, **1**, 745
Royleanone, synthesis of, **5**, 457
RSOX functions, synthesis of, **2**, 208
Rubidium, triphenylsilyl-, synthesis of, **2**, 545
Rubidiums, synthesis of, **2**, 557
Rubidiums, alkyl-, synthesis of, **2**, 557
Rubidium selenocarboxylates, from, *O*-trimethylsilyl selenocarboxylates, **5**, 542
Rupe rearrangement
mechanism, **3**, 216
for the synthesis of α,β-unsaturated ketones, **3**, 216
Ruthenacyclopentadiene complexes, from, dodecacarbonyl triruthenium, **4**, 683
Ruthenium
use in allylation reactions, **1**, 196
use in hydrogenation reactions, **1**, 87
Ruthenium(0), for the addition of carboxylic acids to alkynes, **2**, 676
Ruthenium(0) complexes, reactions, with alkyl halides, **2**, 582
Ruthenium(0)–trialkylphosphine complexes, use in the synthesis of alkynes, **1**, 1036
Ruthenium(II), as catalyst, for the isomerization of allyl ethers, **2**, 675
Ruthenium(II)–BINAP complexes, for the hydrogenation of alkenes, **1**, 77
Ruthenium(II) chloride, bis(triphenylphosphine)-
as catalyst, for the oxidation of alcohols, **3**, 117
as oxidant, for aldehydes, **3**, 5
Ruthenium(II) complexes
for the addition of carboxylic acids to alkynes, **2**, 676
reactions, with 3,3-diphenylcyclopropene, **3**, 520
synthesis of, **1**, 1137; **2**, 582
Ruthenium(II) halides, as catalysts, for the oxidation of alcohols, **3**, 119
Ruthenium(III)acetylacetone, reactions, with carbon monoxide, **6**, 524
Ruthenium(III) chloride, use in the synthesis of alkanoic acids, **5**, 49
Ruthenium(III) porphyrin complexes, alkyl-, synthesis of, **2**, 583
Ruthenium(IV) complexes, dialkyl-, reduction of, **2**, 583
Ruthenium acyl complexes, synthesis of, **5**, 453
Ruthenium–amide bonds, reactions, with carbon monoxide, **6**, 516
Ruthenium–bisphosphinoalkene complexes, synthesis of, **4**, 1029
Ruthenium carbene complexes
from
alkyl complexes, **5**, 938
enynols, **5**, 944
synthesis of, **3**, 520
Ruthenium carbonyl, reactions, with bis(triphenylphosphine)nitrogen chloride, **6**, 456
Ruthenium carbonyl complexes, synthesis of, **6**, 524
Ruthenium catalysts
for the reduction of 1,4-dimethylbenzene, **1**, 93
for the substitution of alkenyl halides, **1**, 466
use in the synthesis of pyridyl ketones, **3**, 311
Ruthenium cluster compounds
reactions
with diphenylethyne, **6**, 344
with metallapyrrolidone complexes, **6**, 344
reduction of, **6**, 344
Ruthenium complexes
alkylation of, **2**, 581
as catalysts, for the substitution of alkenes, **1**, 493
from
cyclooctatetraene, **4**, 683
terminal alkynes, **4**, 683
as mediator of radical cyclizations, **1**, 346
reactions
with diazomethane, **3**, 520
with ethanolamine, **6**, 722
with nucleophiles, **2**, 583
synthesis of, **6**, 456
use in the synthesis of arylamines, **2**, 753
use in the synthesis of butatrienes, **1**, 992
use in the synthesis of trichloromethyls, **6**, 24
Ruthenium complexes, chlorocarbonyl-, synthesis of, **6**, 456
Ruthenium complexes, cyclopentadienyl-
reactions, with alkyl halides, **2**, 582
synthesis of, **6**, 516
Ruthenium complexes, difluorocarbene-, synthesis of, **6**, 718
Ruthenium complexes, (halomethyl)-, synthesis of, **4**, 149
Ruthenium complexes, neopentyl-, for the metallation of ferrocene, **4**, 720
Ruthenium complexes, nitrosyl-, reactions, with benzyl halides, **2**, 582
Ruthenium complexes, pentalene-bridged, from, cyclooctatetraene, **4**, 683
Ruthenium complexes, propargyl-, reactions, with nonacarbonyldiiron, **4**, 711
Ruthenium complexes, tetranuclear carbon-bridged, synthesis of, **4**, 683
Ruthenium complexes, trinuclear, synthesis of, **4**, 683, 685
Ruthenium functions, doubly-bonded, synthesis of, **3**, 520
Ruthenium hydride complexes, alkyne-bridged, synthesis of, **4**, 681

Ruthenium hydridocarbonyls, trinuclear hydrazino-bridged, use in the synthesis of trinuclear bridged ruthenium complexes, **4**, 683
Ruthenium multidecker sandwich compounds, synthesis of, **4**, 694
Ruthenium oxidants, for the synthesis of aldehydes, **3**, 5
Ruthenium oximate, thermal rearrangement of, **5**, 386
Ruthenium–phosphine complexes, as catalysts, for haloalkane addition, **6**, 7
Ruthenium porphyrin complexes, dimethyl-, synthesis of, **2**, 584
Rutheniums
  as chiral catalysts, for haloalkane addition, **6**, 7
  reactions, with phosphorus ylides, **4**, 582
  use in the synthesis of ketones, **3**, 116
Rutheniums, acyl-, synthesis of, **5**, 453
Rutheniums, alkenyl-, reactions, with *t*-butylisonitrile, **5**, 455
Rutheniums, imidoyl-, synthesis of, **5**, 800
Ruthenium salts, reactions, with propionyl chloride, **5**, 453
Ruthenium tetroxide
  as catalyst, **6**, 410
  for the oxidation of amides, **5**, 307
  for the oxidation of sulfides, **2**, 154
  use in the synthesis of carboxylic acids, **5**, 52
  use in the synthesis of ketones, **3**, 117
Ruthenium trichloride, use in the synthesis of ketones, **3**, 117
Ruthenocene
  lithiation of, **4**, 706
  mercuration of, **4**, 724
Ruthenocene, 1,1'-dilithio-, use in the synthesis of metal complexes, **4**, 719
Ruthenocene, pentamethyl-, mercuration of, **4**, 724

S

Saccharin, as catalyst, for the silylation of amides, **5**, 375
Saccharin, *N*-bromo-, use in the synthesis of alkyl bromides, **2**, 27
Saegusa reaction, mechanism, **5**, 154
Safety precautions
for acetyl nitrate, **4**, 208
for alkynylantimony compounds, **2**, 1076
for alkynylarsenic compounds, **2**, 1076
for alkynylbismuth compounds, **2**, 1076
for bis(trichloromethyl) carbonate, **6**, 416
for carbonic dibromide, **6**, 417
for carbonic dichloride, **6**, 411
for chloroformate esters, **6**, 424
N-fluoroamidines, **5**, 761
for isocyanides, **3**, **693–726**
for organomercury compounds, **2**, 1097
for perchlorates, **4**, 250
for phosgene, **6**, 411
for the synthesis of nitrates, **2**, 104
for trichloromethyl chloroformate, **6**, 415
Sakurai cyclization, for the synthesis of γ-lactones, **5**, 166
Salicyclic acids, reactions, with benzoylethyne, **4**, 213
Salicyloyl chloride, pyrolysis of, **3**, 527
Salsolidine, synthesis of, **1**, 149
Salts, cyclic, synthesis of, **4**, 554
Salts, quaternary, synthesis of, **5**, 397
Samarium
chemistry of, **2**, 564
reactions, with alkyl iodides, **2**, 564
use for the deoxygenation of epoxides, **1**, 587
Samarium(II), for the reduction of phosphaalkynes, **5**, 893
Samarium(II) iodide
reactions
with alkyl halides, **1**, 937
with allylic acetates, **1**, 195
with methanol, **1**, 7
as reagent, for the synthesis of ketones, **3**, 130
for the reduction of α-acetoxyketones, **1**, 16
for the reduction of bromoalkanes, **1**, 8
for the reduction of α-functionalized ketones, **3**, 128
for the reduction of α-halocarbonyls, **1**, 2
for the reduction of α-hydroxycarbonyls, **1**, 13
for the reduction of iodoalkanes, **1**, 10
for the reduction of oxiranes, **1**, 14
for the reduction of primary iodides and bromides, **1**, 3
for the reduction of sulfides, **1**, 23
for the reduction of sulfonates, **1**, 17
for the reduction of sulfones, **1**, 24
for the reduction of sulfoxides, **1**, 23
use in the synthesis of alkenes, **1**, 698
Samarium(II) salts, reactions, with alkyl halides, **2**, 565
Samarium(III), benzyl-, reactions, with electrophiles, **2**, 565
Samarium(III) chloride, use in the synthesis of ketones, **3**, 123
Samarium(III) compounds, from, samariumcyclopentadienyl complexes, **2**, 565
Samarium(III) compounds, alkyl-, reactions, with copper iodide triethyl phosphite complexes, **2**, 587
Samarium(III) diiodides, alkyl-, as intermediates, **2**, 565
Samarium complexes, cyclopentadienyl-
reactions, with acid chlorides, **5**, 439
as starting materials, for samarium(III) compounds, **2**, 565
Samarium dialkylcuprates, diiodo-, synthesis of, **2**, 587
Samarium diiodide
as catalyst, for haloalkane addition, **6**, 7
for the cyclization of oxime ethers, **2**, 352
reactions
with alkyl radicals, **1**, 347
with aryl radicals, **1**, 347
with dibenzoates, **1**, 587
with vinyl radicals, **1**, 347
as reducing agent, **5**, 253
for the reduction of nitro compounds, **2**, 341
use in the displacement of benzoate groups, **1**, 179
use in radical cyclizations, **1**, 347
Samarium iodide
for the elimination of β-hydroxy imidazolyl sulfones, **1**, 615
reactions, with 2,6-xylylisonitrile, **5**, 803
Samariums
for the opening of epoxides, **1**, 205
reactions, with phosphorus ylides, **4**, 578
synthesis of, via oxidative addition, **2**, 565
with two carbon-bridges, **4**, 703
Samariums, acyl-, synthesis of, **5**, 439
Samariums, alkyl-, synthesis of, **1**, 347
Samariums, imidoyl-, synthesis of, **5**, 803
SAMP hydrazones, use in the synthesis of α-hydroxy ketones, **3**, 156
Sandmeyer reaction
for the synthesis of aryl bromides, **2**, 627
for the synthesis of aryl chlorides, **2**, 621
for the synthesis of aryl halides, **2**, 609
for the synthesis of thiocyanates, **5**, 1108
Sandwich compound, use of term, **1**, 862
(-)-β-Santalene, synthesis of, **1**, 381
α-Santalenes
synthesis of, **1**, 113
via π-allylnickels, **1**, 127
α-Santalol, synthesis of, **1**, 113
Saponification
of nitriles, **5**, 709
for the synthesis of alkanoic acids, **5**, 30
of trialkylphosphorodithioates, **2**, 144
Sarain A, synthesis of, **1**, 369; **5**, 572
Sarkomycin, synthesis of, **1**, 795, 821
Scandium, chemistry of, **2**, 564
Scandium compounds, reactions, with phosphorus ylides, **4**, 578
Scheller reaction, use in the synthesis of arylarsonic acids, **2**, 884
Schiemann reaction
for the synthesis of aryl fluorides, **2**, 614
for the synthesis of aryl halides, **2**, 609
See also Balz–Schiemann reaction
Schiff base condensation, use in the synthesis of imidoylrheniums, **5**, 801
Schiff bases
N-alkylation of, **3**, 412
from
N,*N*-bis(silyl)amines, **3**, 406
hexahydro-1,3,5-triazines, **4**, 463
metallation of, **1**, 881
reactions
with *N*,*N*-bis(trimethylsilyl)formamide, **4**, 419
with *N*-diphenylmethylenebenzylamine, **3**, 408
with phosphorus trichloride, **5**, 785
with phosphorus ylides, **2**, 860
as starting materials
for α-aminoalkylorganophosphorus compounds, **4**, 461
for 1,3-aminophosphoranes, **4**, 478
for imidazole-2-thione, **6**, 575

synthesis of, **2**, 578; **5**, 736
See also *N*-Alkylimines
Schlenk equilibrium, definition of, **2**, 559
Schlosser's base
reactions, with *t*-butyl methyl ether, **1**, 136
for the synthesis of delocalized carbanions, **1**, 161
use in deprotonation methods, **1**, 148
Schlosser's superbases, for the synthesis of allylpotassiums, **1**, 113
Schmidt reaction, for the synthesis of amines, **2**, 319
Schmidt rearrangement
and Curtius rearrangement compared, **5**, 1001
mechanism, **5**, 1001
for the synthesis of *N*-alkylamides, **5**, 283
for the synthesis of amides, **5**, 270
for the synthesis of isocyanates, **5**, 1001
Schoenberg rearrangements, for the synthesis of dialkyl dithiocarbonates, **6**, 472
Schotten–Baumann procedure, for the synthesis of amides, **5**, 264
Schrock complexes, definition of, **5**, 932
Schrock-type carbenes, use of term, **3**, 507
Schwartz's reagent
characteristics of, **2**, 568
See also Zirconocene chloride hydride
Scopulariopsis brevicaulis, as a source of *R*-ethylmethyl-*n*-propylarsine, **2**, 487
SE1 carbanionic mechanism, **1**, 883
Sebasic acid, synthesis of, reviews, **5**, 24
Secondary orbital effects, in cycloaddition reactions, **1**, 351
Selane, trimethylstannylpentafluoroethyl-, as starting materials, for trifluoromethylselenocarbonyl fluoride, **6**, 49
Selanes, structure of, **2**, 369
Selectfluor reagents
development of, **2**, 15
reactions
with enamines, **2**, 16
with enol acetates, **2**, 16
with enolsilanes, **2**, 16
with pyridine poly(hydrogen fluoride), **2**, 17
See also 1,4-Diazabicyclo[2.2.2]octane bis(tetrafluoroborate), 1-chloromethyl-4-fluoro-
Selectivity
diastereofacial, **1**, 353
effect of temperature on, **1**, 505
ring-size, **1**, 214
endo–exo Selectivity, in cycloaddition chemistry, **1**, 351
α-Selectivity, of allylbariums, **1**, 116
γ-Selectivity, in the displacement of allylic alcohols, **1**, 187
K-Selectride, for the conjugate reduction of enones, **3**, 127
L-Selectride
for the reduction of tertiary amides, **3**, 85
use in the reduction of carbonyls, **2**, 61
See also Lithium tri-*s*-butylborohydride
Selectrides, reactions, with ketones, **2**, 60
Selenadiazole, pyrolysis of, **3**, 382
1,2,3-Selenadiazoles
fragmentation of, **4**, 874
from, selenolates, **5**, 519
1,3,4-Selenadiazoles, 2,5-dihydro-, synthesis of, **4**, 333
1,2,4-Selenadiazolidine-3,5-dione, 4-methyl-2-phenyl-, synthesis of, **6**, 498
1,3,4-Selenadiazolines
as starting materials, for alkenes, **1**, 684, 686
synthesis of, **1**, 686
1,3,4-Selenadiazolium salts, synthesis of, **5**, 642
Selenadiphosphiranes, three-membered cyclic, synthesis of, **2**, 857
Selenal, propyne-, synthesis of, **3**, 382
Selenals
Diels–Alder adducts
intramolecular, **3**, 385
synthesis of, **3**, 384
from
amines, **3**, 388
bis-(dimethylaluminum)selenide, **3**, 385
bis-(trimethylsilyl) selenide, **3**, 385
trimethyl selenophosphate, **3**, 385
metal complexes
bimolecular, **3**, 386
isomeric forms, **3**, 386
synthesis of, **3**, 382
via displacement reactions, **3**, 387
via 1,2-elimination, **3**, 384
via phenylselenophosphonic dichloride, **3**, 388
via phosphorus pentaselenide, **3**, 388
via α-silyl selenocyanates, **3**, 383
via Vilsmeier–Haack transformation, **3**, 388
via Wittig reagents, **3**, 384
use of term, **3**, 381
See also Selenoaldehydes
Selenals, conjugatively-stabilized, synthesis of, **3**, 388
Selenals, cyclic trimers, synthesis of, **3**, 382
Selenals, metal-stabilized, synthesis of, **3**, 386
Selenals, stable, synthesis of, **3**, 383
Selenamides
reactions, with enamines, **3**, 38
use in the synthesis of α-phenyl selenyl aldehydes, **3**, 37
Selenamides, arene-, synthesis of, **2**, 736
Selena-γ-pyrone, synthesis of, **3**, 398
Selenate, chlorocarbonyl pentafluoro-, synthesis of, **6**, 430
Selenate anions
reactions
with α-chloro tosylhydrazones, **3**, 174
with halides, **3**, 174
Selenate ions, reactions, with enones, **3**, 175
1,3-Selenatelluretane, tetrafluoro-, synthesis of, **6**, 256
Selenates
reactions
with α,β-unsaturated aldehydes, **3**, 39
with vinyl halides, **2**, 727
Selenates, dialkyl, synthesis of, **2**, 102
Selenation
of aldehydes, **3**, 385
of imidate esters, **5**, 510
of ketones, **3**, 392, 394
α-Selenation, electrophilic, of acetone, **6**, 93
Selenation reagents
reactions
with esters, **5**, 520
with ortho esters, **5**, 520
Selenazaphospholes, synthesis of, via cycloaddition reactions, **4**, 493
Selenazinones, synthesis of, **6**, 618
Selenazole
from, selenazolium cations, **6**, 115
synthesis of, **6**, 115
Selenazolidines, synthesis of, **4**, 333
Selenazolium cations, as starting materials, for selenazole, **6**, 115
Selenenamides
reactions, with sulfuryl chloride, **2**, 369
synthesis of, **1**, 781; **2**, 368
Selenenamides, alkane-
from
alkylselenyl amines, **2**, 292
alkylselenyl halides, **2**, 292
alkylselenylphthalimide, **2**, 292
Selenenamides, diamino-, from, diaminoseleninamides, **2**, 369
Selenenamides, dichloro-, synthesis of, **2**, 369
Selenenic acid, 2-aminoethane-, synthesis of, **2**, 290
Selenenic acid esters, alkane-, synthesis of, *in situ*, **2**, 291
Selenenic acids
disproportionation of, **2**, 290

elimination of, **1**, 600, 604
Selenenic acids, alkane-
stability of, **2**, 290
synthesis of, **2**, 290
Selenenic acids, arene-, reactions, with thiols, **2**, 735
Selenenic esters, arene-, synthesis of, **2**, 734
Selenenothiocyanate, phenyl, as starting material, for thiocyanates, **5**, 1108
Selenenylation
of enones, **1**, 606
for the synthesis of phenols, **2**, 658
Selenenylation, electrophilic, of alkenes, **2**, 284
Selenenyl bromide, phenyl-
reactions
with lithium nitronate, **2**, 784
with tetraphenyldiphosphine dioxide, **2**, 857
Selenenyl bromides
reactions
with alkenes, **2**, 284
with diazomethane, **4**, 291
Selenenyl bromides, arene-
reactions, with selenocarboxylates, **5**, 250
synthesis of, **2**, 732
Selenenyl bromides, aryl-, reactions, with alkyl amines, **2**, 368
Selenenyl chloride, phenyl-, reactions, with ethyl acetate, **1**, 606
Selenenyl chloride, trifluoromethyl-, reactions, with triethyl orthoacetate, **6**, 80
Selenenyl chlorides
reactions
with alkenes, **2**, 44
with ozone, **2**, 733
as starting materials, for alkaneseleninyl chlorides, **2**, 289
Selenenyl chlorides, arene-
reactions, with arylvinyl selenides, **4**, 91
synthesis of, **2**, 732
Selenenyl halides
reactions
with alkynes, **1**, 1123
with chloroamines, **2**, 369
with metal cyanides, **5**, 1116
with trimethylsilyl cyanide, **5**, 1116
with vinyl Grignard reagents, **2**, 727
with vinyllithium, **2**, 727
with vinylmercury compounds, **2**, 727
Selenenyl halides, alkane-
from
diselenides, **2**, 288
selenocyanates, **2**, 288
synthesis of, **2**, 288
Selenenyl halides, arene-
from, aryl alkyl selenides, **2**, 732
reactions
with alcohols, **2**, 734
with ylides, **4**, 348
synthesis of, **2**, 732
Selenenyl halides, aryl-, reactions, with amines, **2**, 369
Selenenyl iodides, arene-, synthesis of, **2**, 732
Selenenyl phenyl tellurides, aroyl-, synthesis of, **5**, 250
Selenenyl sulfinates, alkyl-, as starting materials, for selenocyanates, **5**, 1116
Selenenyl tellurides, acyl-, synthesis of, **5**, 250
Seleneous acid
reactions, with 1,2-dithiols, **2**, 237
use in the synthesis of sulfenyl selenides, **2**, 237
1,2-Selenetane, 2,2,4,4-tetrafluoro-, as starting material, for bistrifluoromethyl triselenocarbonate, **6**, 592
Selenic acid elimination, for the synthesis of macrolactones, **5**, 179
Selenide, acyl, as acyl radical precursors, for the synthesis of ketones, **3**, 136
Selenide, alkylphenyl, reduction of, **1**, 437
Selenide, alkynyl phenyl, as starting material, for α-thioxo selenothioic acid *Se*-phenyl ester, **5**, 563
Selenide, allenyl, as starting material, for allenylsilanes, **2**, 939
Selenide, bis(1,5-cyclooctanediylboryl), as reagent, for the selenation of esters, **5**, 520
Selenide, bis(dimethylaluminum)
reactions
with acetals, **3**, 385; **5**, 520
with ketene silyl acetals, **3**, 553
with ortho esters, **5**, 520
as starting material, for selenals, **3**, 385
Selenide, bis(tricyclohexyltin), as starting material, for selones, **3**, 392
Selenide, bis(trimethylsilyl)
reactions
with aldehydes, **2**, 542
with boron trifluoride etherate, **5**, 625
with *n*-butyllithium, **6**, 481
with *N,N*-dimethylcyanamide, **6**, 597
as starting material, for selenals, **3**, 385
synthesis of, **2**, 542
Selenide, 2-bromoethylphenyl, as starting material, for aryl selenocyanates, **5**, 1114
Selenide, bromomethyl, synthesis of, **4**, 291
Selenide, *t*-butylthiophenyl, synthesis of, **2**, 237
Selenide, *cis*-β-chlorovinyl phenyl, synthesis of, **4**, 786
Selenide, cyanomethyl phenyl, oxidation of, **4**, 240
Selenide, 1,2-dichloro-1,2,2-trifluoroethyl phenyl, reactions, with magnesium, **4**, 785
Selenide, didecyl, reactions, with methyl chloroformate, **6**, 482
Selenide, (*E*)-1,2-difluoro-2-chlorovinyl phenyl, synthesis of, **4**, 785
Selenide, dimethyl, flash vacuum pyrolysis of, **3**, 382
Selenide, disilyl, as starting material, for lithiumsilaneselenolate, **5**, 541
Selenide, ethyl phenyl
oxidation of, **4**, 240
reactions, with dibenzoyl peroxide, **4**, 240
Selenide, hydroxy, synthesis of, **3**, 125
Selenide, methyl(difluoromethyl), synthesis of, **6**, 49
Selenide, phenyl, reactions, with cyclopropyl ketones, **3**, 176
Selenide, phenyl trimethylsilyl, reactions, with aldehydes, **4**, 239
Selenide, phenylvinyl
reactions, with *iso*-propyllithium, **4**, 401
as starting material, for α-selenosilanes, **4**, 375
synthesis of, **4**, 401
Selenide, polyhalogenated ethyl phenyl, reactions, with magnesium, **4**, 785
Selenide, thallous phenyl, reactions, with imidoyl chlorides, **5**, 732
Selenide, 1,2,2-trifluorovinyl phenyl, synthesis of, **4**, 785
Selenide anions, reactions, with dihalomethanes, **4**, 88
Selenide anions, phenyl, reactions, with chlorotrimethylsilane, **2**, 542
Selenide α-anions, alkylation of, **1**, 712
Selenide chloride, reactions, with nucleophiles, **2**, 768
Selenide dianions, use in the synthesis of areneselenols, **2**, 727
Selenides
alkylation of, **2**, 288
deprotonation of, **3**, 175; **4**, 398
elimination of, **1**, 592
fluorination of, **4**, 87
from
alkylaminophosphines, **2**, 1070
terminal alcohols, **1**, 604
β-functionalization of, **2**, 280
irridation of, **1**, 943
migrations of, **1**, 246
oxidation of, **1**, 621; **2**, 287, 731; **4**, 241
to selenoxides, **1**, 601

oxidative deselenenylation of, **4**, 92
as radical precursors, **1**, 330
reactions
with alkyllithiums, **2**, 554
with chloroamine-T, **1**, 621
with ozone, **4**, 241
with potassium selenocyanate, **5**, 1114
with tributyltin hydride, **1**, 594
reduction of, **1**, 25
as starting materials
for alkenes, **1**, 712
for alkyl bromides, **2**, 32
for selenones, **2**, 287
synthesis of, **2**, 278
tandem selenoxide elimination–Claisen rearrangement of, **1**, 829
vacuum pyrolysis of, **1**, 243
Selenides, α-acyloxy, synthesis of, via seleno-Pummerer reactions, **4**, 240
Selenides, acylsulfenyl
from
acylsulfenyl halides, **5**, 248
thiocarboxylic acids, **5**, 248
synthesis of, **5**, 248
Selenides, alkali, reactions, with alkyl halides, **2**, 278
Selenides, alkenyl, alkylation of, **1**, 1073
Selenides, β-alkoxy, as starting materials, for alkenes, **1**, 716
Selenides, alkoxy alkyl, synthesis of, **3**, 175
Selenides, alkyl
as starting materials, for *N*-substituted selenimides, **2**, 292
synthesis of, **2**, 278
Selenides, alkyl aryl
alkylation of, **2**, 288
from, aryl halides, **2**, 729
oxidation of, **2**, 287
as starting materials, for alkyl aryl selenones, **2**, 287
synthesis of, **2**, 729
Selenides, alkyl phenyl, synthesis of, **2**, 730
Selenides, alkyl silyl, synthesis of, **6**, 481
Selenides, alkylvinyl, reactions, with hydrogen halides, **4**, 91
Selenides, alkynyl
oxidation of, **2**, 1038
research background, **2**, 1036
synthesis of, **2**, 1038
Selenides, 1-alkynyl, as starting materials, for 1-alkynes, **1**, 999
Selenides, (alkynyl)alkyl, synthesis of, **2**, 1037, 1038
Selenides, allyl butynyl, seleno-Cope rearrangement of, **3**, 553
Selenides, allylic
reactions, with diazomethane, **4**, 277
as starting materials, for allyllithiums, **1**, 245
Selenides, allyl phenyl, as starting materials, for α-selenosilanes, **4**, 374
Selenides, aryl
reduction of, **1**, 437
as starting materials, for selenoarenes, **1**, 437
substitution of, **1**, 484
synthesis of, **2**, 729
Selenides, aryl alkyl
from
areneselenolate ions, **2**, 279
diaryl diselenides, **2**, 279
as starting materials, for areneselenenyl halides, **2**, 732
Selenides, aryl methyl, synthesis of, **2**, 729
Selenides, aryl trifluoromethyl
oxidation of, **6**, 237
synthesis of, **6**, 237
Selenides, arylvinyl
deprotonation of, **4**, 947
reactions
with areneselenenyl chlorides, **4**, 91
with arenesulfenyl chlorides, **4**, 91
with hydrogen halides, **4**, 91
silylation of, **2**, 908
sulfenylation of, **4**, 871
Selenides, β-azido, synthesis of, **2**, 284
Selenides, benzylic, as starting materials, for benzyllithiums, **1**, 245
Selenides, bis(acyl)
reactions, with methanolic potassium hydroxide, **5**, 240
synthesis of, **5**, 243
Selenides, bis(alkylthiol), synthesis of, **2**, 236
Selenides, bisallyl, reactions, with *N*-chlorobenzotriazole, **2**, 236
Selenides, bissugar, synthesis of, **4**, 239
Selenides, bromo, reactions, with sodium benzylselenide, **4**, 288
Selenides, α-bromo
reactions, with *n*-butyllithium, **4**, 400
synthesis of, **4**, 88, 90
Selenides, β-bromo, synthesis of, **2**, 280
Selenides, 1-butenyl-3-methyl, deprotonation of, **4**, 965
Selenides, chloro, synthesis of, **4**, 287
Selenides, α-chloro
from, benzeneselenenyl bromide, **4**, 90
synthesis of, **4**, 88, 90
Selenides, β-chloro
chlorination of, **1**, 610
synthesis of, **2**, 280
Selenides, 1-chlorobutadienyl, synthesis of, **4**, 786
Selenides, 1-chlorovinyl phenyl, synthesis of, **4**, 784
Selenides, cyano, synthesis of, **3**, 644
Selenides, α-cyano, synthesis of, **3**, 637
Selenides, dialkyl
reactivity of, **2**, 288
as starting materials, for trialkylselenonium halides, **2**, 288
synthesis of, **2**, 278
Selenides, diaryl
from
arenes, **2**, 729
arylamines, **2**, 729
aryl halides, **2**, 729
reactions, with alkanethiols, **2**, 237
as starting materials, for biaryls, **1**, 484
use in the synthesis of sulfenyl selenides, **2**, 237
Selenides, dibismuth, from, dibismuthines, **2**, 890
Selenides, distiba, from, tetraphenyldistibine, **2**, 890
Selenides, α-fluoro
alkylation of, **4**, 87
synthesis of, **4**, 87
Selenides, β-D-glucopyranosyl phenyl, synthesis of, **4**, 239
Selenides, α-halo
from
diselenides, **4**, 88
selenols, **4**, 88
selenophenols, **4**, 88
synthesis of, **4**, 400
via substitution reactions, **4**, 88
Selenides, β-halo
reactions, with phenylselenolate, **1**, 594
as starting materials, for alkenes, **1**, 594
Selenides, α-haloalkyl aryl, reactions, with triphenylphosphine, **4**, 347
Selenides, halomethyl, synthesis of, **4**, 88
Selenides, α-halovinyl, reactions, with alkenes, **4**, 92
Selenides, hydroxy
from
cyclic ethers, **2**, 281
epoxides, **2**, 281
selenocyanates, **2**, 281
as intermediates, in the synthesis of allylic alcohols, **2**, 85
synthesis of, **2**, 44
Selenides, 2-hydroxy, semipinacol rearrangement of, **1**, 390
Selenides, β-hydroxy
from

alkenes, **2**, 281
α-lithio selenoxides, **1**, 716
selenium-stabilized carbanions, **2**, 281
reductive elimination of, **1**, 592, 712
as starting materials
for alkenes, **1**, 592, 716
for allenes, **1**, 440
for α,β-unsaturated esters, **1**, 594
synthesis of, **1**, 608, 712, 713
Selenides, 2-hydroxy alkyl, synthesis of, **3**, 174
Selenides, 3-hydroxy alkyl, synthesis of, **3**, 174
Selenides, imido, from, haloimides, **5**, 341
Selenides, β-isothiocyanatoalkyl phenyl, synthesis of, **5**, 1107
Selenides, α-lithio
from, α-selenocarboxylic esters, **2**, 282
as starting materials
for α-selenocarbonyls, **2**, 282
for α-selenocarboxylic acids, **2**, 282
synthesis of, **4**, 375, 399
Selenides, α-lithiovinyl, from, 1,1-bis(seleno)alkenes, **2**, 911
Selenides, nitrile, characterization of, **3**, 678
Selenides, nitro, synthesis of, **2**, 784
Selenides, β-nitro, from, alkenes, **2**, 785
Selenides, β-nitrogen-substituted, from, alkenes, **2**, 284
Selenides, β-nitro-α-silyl, from, vinylsilanes, **2**, 284
Selenides, β-oxygenated, elimination of, **1**, 594
Selenides, perfluoroalkyl, synthesis of, **6**, 49
Selenides, phenyl
oxidation of, **1**, 780
precursors for radical cyclization, **1**, 341
Selenides, phenyl alkyl, cleavage of, **1**, 437
Selenides, phenyl alkynyl, synthesis of, **2**, 1037
Selenides, phenylvinyl
deprotonation of, **4**, 964
reactions
with *n*-butyllithium, **4**, 401, 964
with diisopropylamine, **4**, 964
Selenides, selenocarbamoyl, as starting materials, for selenoureas, **6**, 597
Selenides, silyl, synthesis of, **6**, 49
Selenides, β-stannyl, destannylselenylation of, **1**, 665
Selenides, sulfenyl
from, sulfides, **2**, 237
synthesis of, **2**, 234
via oxidation of thiols, **2**, 237
Selenides, symmetrical dialkyl, synthesis of, **2**, 278
Selenides, tertiary alkylarsine, synthesis of, **2**, 502
Selenides, β-thiocyanatoalkyl phenyl, synthesis of, **5**, 1107
Selenides, (thioethynyl), from, (chloroethynyl)thioethers, **2**, 1037
Selenides, trimethylsilyl, reactions, with acyl halides, **5**, 238
Selenides, α-(trimethylsilyloxy), synthesis of, **4**, 239
Selenides, tris(trimethylsilyl)methyl, synthesis of, **6**, 355
Selenides, unsymmetrical
from
N-phenylselenophthalimide, **2**, 279
selenols, **2**, 279
Selenides, unsymmetrical dialkyl
from
alkaneselenolates, **2**, 279
alkaneselenols, **2**, 279
diselenides, **2**, 279
selenocyanates, **2**, 279
Selenides, vinyl
deselenization of, **1**, 440
formylation of, **3**, 74
from
alkynes, **2**, 728
vinyl halides, **2**, 727
oxidation of, **1**, 441, 983; **3**, 175
reactions
with alkyllithiums, **4**, 400
with *n*-octyl bromide, **1**, 1089
with organolithiums, **4**, 375, 400
reduction of, **1**, 438, 440
as starting materials
for alkenes, **1**, 438
for selenoalkenes, **1**, 440
for vinyl bromides, **2**, 624
synthesis of, **2**, 727
via elimination reactions, **2**, 728
via Wittig and Peterson reactions, **2**, 728
use in the synthesis of vinyl radicals, **1**, 1129
α-Selenides, β-hydroxy-α-silyl-, as starting materials, for alkenylsilanes, **2**, 936
Seleniiranium salts, as intermediates, in the synthesis of alkenes, **1**, 716
Selenimides, elimination of, **1**, 621
Selenimides, *N*-substituted
from
alkyl selenides, **2**, 292
alkyl selenoxides, **2**, 292
See also Selenylimines
Seleninamides, synthesis of, **2**, 369
Seleninamides, alkane-, synthesis of, **2**, 292
Seleninamides, diamino-
as starting materials, for diaminoselenenamides, **2**, 369
synthesis of, **2**, 369
Seleninamides, trimethylsilyl esters, synthesis of, **2**, 370
Seleninate esters, aryl, as intermediates, **2**, 690
Seleninic acid esters, alkane-
from
seleninic acid salts, **2**, 291
seleninic anhydrides, **2**, 291
seleninyl chlorides, **2**, 291
Seleninic acids
esterification of, **2**, 734
reduction of, **2**, 281
as starting materials
for alkaneseleninic anhydrides, **2**, 291
for alkaneselenonic acids, **2**, 291
Seleninic acids, alkane-
from
diselenides, **2**, 291
selenium(III) halides, **2**, 291
selenocyanates, **2**, 291
selenols, **2**, 291
synthesis of, **2**, 291
Seleninic acids, arene-, synthesis of, **2**, 733
Seleninic acid salts, as starting materials, for alkaneseleninic acid esters, **2**, 291
Seleninic anhydrides, as starting materials, for alkaneseleninic acid esters, **2**, 291
Seleninic anhydrides, alkane-
from
diselenides, **2**, 291
seleninic acids, **2**, 291
selenols, **2**, 291
Seleninyl bromides, hydrolysis of, **2**, 733
Seleninyl chlorides, as starting materials, for alkaneseleninic acid esters, **2**, 291
Seleninyl chlorides, alkane-
from
diselenides, **2**, 289
selenenyl chlorides, **2**, 289
Seleninyl chlorides, arene-
reactions, with ammonia, **2**, 736
synthesis of, **2**, 732
Seleninyl halides, arene-, synthesis of, **2**, 732
Seleninyl halides, vinyl-, synthesis of, **2**, 732
Seleniolactones, synthesis of, **2**, 292
Selenious acid
alkylation of, **2**, 101
diesters of, **2**, 101
Selenirenium ions, stability of, **1**, 1109
Selenite, ethyl chloro-, synthesis of, **2**, 101

Selenites
reactions, with thiols, **2**, 236
use in the synthesis of sulfenyl selenides, **2**, 236
Selenites, alkyl chloro-, synthesis of, **2**, 101
Selenites, dialkyl, synthesis of, **2**, 101
Selenium
amidines of, **5**, 767
extrusion of, **1**, 246
and heteroatomic chemistry, **2**, 856
as oxidizing agent, **6**, 466
reactions
with alkoxycarbenechromium complexes, **5**, 515, 520
with alkylphosphorus compounds, **2**, 465
with chloramine-T, **2**, 298
with *N*-chloroamides, **5**, 341
with diazo compounds, **1**, 246; **3**, 391
with enolates, **3**, 174
with Grignard reagents, **2**, 727
with hexafluoropropene, **3**, 395
with isocyanides, **5**, 1051
with methylenephosphines, **6**, 692
with organometallics, **2**, 278
with osmium methylene complex, **3**, 387
with pentafluoroethylsilver salts, **6**, 49
with phosphines, **2**, 1070; **6**, 151
with phosphinous acid chlorides, **2**, 857
with phosphonate anions, **5**, 627
with piperidine, **2**, 369
with silver heptafluorobutyrate, **6**, 49
with sodium alkynylide, **2**, 1036
with sodium hydroxide, **5**, 253
with tetrafluoroethylene, **6**, 49
with 2-thia-spiro-[3.5]-nonane, **2**, 237
with trialkylarsines, **2**, 502
with trifluoroiodomethane, **6**, 49
with triphenylphosphoranylhydrazones, **1**, 686
with triphenylphosphoranylidenehydrazones, **2**, 858
with vinylidene transition metal complexes, **3**, 554
as reagent, for the synthesis of phenols, **2**, 659
as starting material, for potassium seleno sulfate, **2**, 292
substitution of, **1**, 243
synthesis of functions, **6**, 91
containing oxygen, **4**, 841
one halogen and sulfur functions, **6**, 276
with sulfur functions, **6**, 101
use in the synthesis of bis(acyl) diselenides, **5**, 253
use in the synthesis of triselenides, **6**, 594
Selenium, difluoro(perfluoroethyl)-, synthesis of, **6**, 49
Selenium, diphenyldimethoxy-, reactions, with trichloroacetamide, **5**, 340
Selenium, hypervalent, **3**, 399
Selenium, red-
reactions
with acetone, **6**, 93
with bisphosphine sulfide, **6**, 151
Selenium, trimethyl-, for the methylation of toluene, **1**, 247
Selenium(I) bromide, reactions, with hydrazones, **3**, 392
Selenium(I) chloride
reactions, with hydrazone dimagnesium salt, **3**, 392
as starting material, for aryl selones, **3**, 392
Selenium(I) halides, as starting materials, for selones, **3**, 392
Selenium(II) compounds, reactions, with aldehydes, **3**, 37
Selenium(III) halides, as starting materials, for alkaneseleninic acids, **2**, 291
Selenium(IV) compounds, amino-, synthesis of, **2**, 369
Selenium(IV) dichlorides, reactions, with pyridine, **4**, 88
Selenium(IV) dichlorides, α-halo-, synthesis of, **4**, 92
Selenium bis(thiocarboxylates), reactions, with halogens, **5**, 252
Selenium bromide, aryl-, reactions, with lithium alkynylides, **2**, 1037
Selenium bromide, 4-methoxyanthraquinon-1-yl-, acetoxylation of, **5**, 210
Selenium chloride, aryl-, reactions, with *N*-chloroamides, **5**, 341
Selenium chloride, phenyl-
reactions
with amides, **5**, 335
with *O*-cholestannyl selenobenzoate, **5**, 526
Selenium complexes, reactions, with diazomethane, **3**, 387
Selenium dichloride, benzylphenyl-
from, benzylphenylselenium dichloride, **5**, 903
as starting material, for α-selenyl ylides, **5**, 903
Selenium dichloride, methylphenyl-, as starting material, for α-selenyl ylides, **5**, 903
Selenium dichlorides, reactions, with 1,3-diketones, **3**, 177
Selenium dichlorides, aryl(acylamino)-, synthesis of, **5**, 341
Selenium dichlorides, bis(acylamino)-, synthesis of, **5**, 341
Selenium dichlorides, dialkyl-, reactions, with malononitriles, **3**, 637
Selenium difluorides, diamino-
stability of, **2**, 369
synthesis of, **2**, 369
Selenium diimide, *N*-phenylsulfonyl-*N'*-benzoyl-, use in the synthesis of sulfinamides, **5**, 341
Selenium dioxide
as dehydrating agent, for the synthesis of acetals, **4**, 178
for the oxidation of allylic carbon–hydrogen bonds, **2**, 86
for the oxidation of arenes, **2**, 654
for the oxidation of sulfides, **2**, 145
reactions
with alcohols, **3**, 109
with chloroethane, **2**, 101
with diselenides, **3**, 37
with Grignard reagents, **2**, 238
with methanol, **2**, 101
with 3-methylindoles, **3**, 106
with methylquinolines, **3**, 109
use in the synthesis of carboxylic acids, **5**, 28
use in the synthesis of formylpyridines, **3**, 108
use in the synthesis of α,β-unsaturated aldehydes, **3**, 58
Selenium disulfide, reactions, with Grignard reagents, **2**, 238
Selenium functions
elimination of, **1**, 590, 1072
substitution of, **1**, 1031
Selenium functions, acyl-, synthesis of, **5**, **231–255**
Selenium groups, elimination of, **3**, 227
Selenium halides, aryl-
reactions
with alkynylmagnesium bromides, **2**, 1037
with alkynylstannane, **2**, 1037
Seleniumimides, phenyl-, as transfer reagents, **5**, 336
Selenium–lithium exchange
for the synthesis of α-thiocarbanions, **4**, 394
in 1,2 Wittig rearrangement, **1**, 398
Selenium–metal exchange, for the synthesis of α-selenolithiums, **4**, 399
Selenium nucleophiles
reactions
with halomethylsilanes, **4**, 377
with imidoyl chlorides, **5**, 732
Selenium–organoarsanyl selenocarboxylates, use in the synthesis of acylselenenyl tellurides, **5**, 250
Selenium oxide, reactions, with *N*-trimethylsilyl amines, **2**, 370
Selenium oxide, diphenyl-, condensation of, with arsonium salts, **5**, 920
Selenium oxides, condensation of, **5**, 904
Selenium oxychloride
reactions
with arenes, **2**, 729
with ethanol, **2**, 101
with ethyne, **2**, 732
with phenacyl chlorides, **3**, 177
Selenium reactions, for the substitution of selenium and tellurium, **1**, 243

Seleniums
  cyclization of, **1**, 244
  reactions
    with alkynes, **1**, 545
    with carbonyls, **1**, 605
    with cyanide, **5**, 1116
    with enol acetates, **1**, 606
    with silyl enol ethers, **1**, 606
  reduction of, **1**, 21
  synthesis of, **4**, 785, 868, 964
    via oxidative processes, **2**, 890
Seleniums, acyloxy-, synthesis of, **5**, 210
Seleniums, alkynyl-, coupling reactions of, **1**, 1035
Seleniums, allyl-, use in the synthesis of carbon-bridged ditins, **4**, 701
Seleniums, α-halo-, synthesis of, **4**, 87
Seleniums, imido-, reactions, with alkenes, **2**, 298
Seleniums, imidoyl-, synthesis of, **5**, **725–740**
Seleniums, oxyimidic, synthesis of, **5**, 832
Seleniums, phenyl-, as radical precursors, **1**, 243
Seleniums, tetravalent, reactions, with phosphonium salts, **5**, 903
Seleniums, trifluoromethyl-, synthesis of, **6**, 237
N-Seleniums, synthesis of, **2**, 768
O-Seleniums, alkenyl-, synthesis of, **2**, 690
O-Seleniums, aryl-, synthesis of, **2**, 690
Selenium tetrachloride, reactions, with arenes, **2**, 729
Selenium tetrafluoride
  reactions
    with amines, **2**, 369
    with carboxylic acids, **5**, 5
  as reagent, for acid fluorides, **5**, 6
  use in the synthesis of difluoroalkanes, **4**, 10
Selenium transfer reagents, use in the synthesis of sulfenyl selenides, **2**, 236
Selenium trichloride, aryl-, reactions, with trimethylsilylamides, **5**, 341
Selenium trichloride, phenyl-, as starting material, for vinyl chlorides, **2**, 616
Selenium trifluorides, amino-, synthesis of, **2**, 369
Selenium ylides, synthesis of, **2**, 1001
Selenium ylides, α-cyano-, synthesis of, **3**, 637
Selenoacetaldehyde, trimeric, synthesis of, **3**, 382
Selenoacetals
  from, selenoxides, **2**, 286
  precursors of aldehydes, **3**, 16
  reactions, with trimethylsilyl cyanide, **3**, 637
  silylation of, **4**, 375
  as starting materials, for ketones, **3**, 132
  transmetallation of, **4**, 375
Selenoacetals, α-cyano-, synthesis of, **3**, 637
Selenoacetals, α-cyano-β-phenyl-, synthesis of, **2**, 284
1,1-Selenoacetals, cleavage of, **1**, 713
Selenoacetate, *O*-benzyl, synthesis of, problems, **5**, 516
Selenoacetate, *O*-ethyl, reactions, with hydrazine, **5**, 642
Selenoacetate, ethyl phenyl, fluorination of, **4**, 92
Selenoacetate, silyl, synthesis of, **5**, 541
Selenoacetates
  equilibrium of, **5**, 535
  reduction of, **5**, 521
Selenoacetic acid *Se*-alkynyl esters, as starting materials, for selenothioic acid *S*-alkyl esters, **5**, 563
Selenoacetone, hexafluoro-, synthesis of, **3**, 394
Selenoacridone, from, sodium hydrogen selenide, **3**, 398
Selenoacridones, from, acridones, **3**, 398
Selenoacylation, of amines, **5**, 626
Selenoacyl compounds, silyl, synthesis of, **5**, 541
Selenoacyl halides, synthesis of, **5**, **471–503**
P-Selenoacyloxy compounds, synthesis of, **5**, 534
Selenoacyloxy functions, synthesis of, **5**, **505–543**
Selenoaldehydes
  from, α-silylalkyl selenocyanates, **4**, 376
  nomenclature of, **3**, 381
  reactions
    with alkanethiols, **2**, 234
    with seleno anions, **4**, 289
  as starting materials, for aminals, **4**, 415
  synthesis of, **2**, 542; **3**, **381–401**
  See also Selenals
Selenoaldehydes, stable, synthesis of, **3**, 383
β-Selenoaldehydes, use in the synthesis of α,β-unsaturated ketones, **3**, 228
Selenoalkene, 1,1-bismethyl-, reactions, with *n*-butyllithium, **1**, 440
Selenoalkenes, from, vinyl selenides, **1**, 440
1-Selenoalkenes, 1-lithio-, synthesis of, **1**, 440; **4**, 964
Selenoalkynes
  lithium aluminum hydride reduction of, **2**, 728
  oxidative elimination in, **2**, 1035
  reactions, with amines, **4**, 910
Selenoalkynes, 1-alkyl-, addition of ethanethiol to, **4**, 872
1-Selenoalkynes, reactions, with acetic acid, **4**, 841
Selenoamides
  alkylation of, **5**, 732
  from, nitriles, **5**, 625
  IR spectra of, **5**, 726
  reactions
    with alkyl halides, **5**, 732
    with phenyllithium, **4**, 334, 909
    with sulfur, **5**, 623
  as starting materials, for 1-alkyl selenoenamines, **4**, 909
  synthesis of, **5**, 341, 522, 624
Selenoamides, *N*-acyl-, synthesis of, **5**, 629
Selenoamides, *N*-phenyl-, synthesis of, **5**, 335
Selenoanhydrides, synthesis of, **5**, 530
Selenoanions, reactions, with selenoaldehydes, **4**, 289
Selenoanisole, cleavage of, **4**, 398
Selenoarenes, from, aryl selenides, **1**, 437
Selenoarsines, alkyl-, synthesis of, **2**, 500
Selenoarsinous acid esters, alkyl-, synthesis of, **2**, 293
Selenobarbiturates, from, selenourea, **6**, 598
Selenobenzaldehyde, 2,4,6-tri-*t*-butyl-, synthesis of, **3**, 383; **5**, 1115
Selenobenzaldehyde, trimeric, synthesis of, **3**, 382
Selenobenzaldehydes
  metal complexes of, **3**, 386
  synthesis of, **3**, 386; **5**, 524
Selenobenzene, methyl-, reactions, with dibenzoyl peroxide, **5**, 210
Seleno-1,2-benzisothiazole 1,1-dioxide, 3-phenyl-, reactions, with 1-diethylaminoprop-1-yne, **4**, 911
Selenobenzoate, *O*-*t*-butyl, synthesis of, problems, **5**, 516
Selenobenzoate, *O*-cholestannyl
  reactions
    with benzeneseleninic anhydride, **5**, 523
    with bis(*p*-methoxyphenyl) telluroxide, **5**, 523
    with phenylselenium chloride, **5**, 526
Selenobenzoate, *O*-cholesteryl, reduction of, to cholesterine, **5**, 526
Selenobenzoate, *O*-ethyl, from, ethyl benzoate, **5**, 520
Selenobenzoate, *O*-methyl
  trans-esterification of, **5**, 520
  synthesis of, **5**, 515
Selenobenzoate, *O*-methyl 4-methyl, synthesis of, **5**, 515
Selenobenzoate, *O*-trimethylsilyl
  IR spectrum of, **5**, 541
  synthesis of, **5**, 541
Selenobenzoate, *O*-trimethylsilyl 4-methyl, hydrolysis of, **5**, 542
Selenobenzoates
  reactions, with *N*-acetoxy-2-thiopyridone, **5**, 526
  reduction of, **5**, 521
Selenobenzoic acid, from, benzoyl chloride, **5**, 508
Selenobiurets, from, alkylation-selenation procedures, **6**, 597
α-Selenoboranes, synthesis of, **4**, 377

Selenobuta-1,3-dienes, 3-phenyl-, as starting materials, for α-hydroxyallenes, **1**, 985
Selenobut-3-en-1-yne, methyl-, reactions, with bromine, **4**, 786
Selenobut-3-en-1-ynes, 1-alkyl-, reactions, with hydrogen chloride, **4**, 786
Selenocarbamate, aryl, production of, **6**, 593
Selenocarbamate, *N*-aryl-*O*-alkyl, pyrolysis of, **1**, 777
Selenocarbamate, *O*-ethyl, reactions, with mercuric oxide, **5**, 1103
Selenocarbamates
from
diselenocarbonates, **6**, 593
isoselenocyanates, **6**, 593
reduction of, **3**, 713
synthesis of, **6**, 497
Selenocarbamates, dialkyl, as starting materials, for bisselenocarbamoyl triselenide, **6**, 594
α-Selenocarbanions
reactions, with sulfur nucleophiles, **4**, 284
synthesis of, reviews, **4**, 398
Selenocarbene complexes, synthesis of, **5**, 952
α-Selenocarbenium ions, synthesis of, **2**, 1008
Selenocarbonates
metallation of, **6**, 590
organometallics, **6**, 593
synthesis of, **6**, 481
use in the synthesis of radicals, **1**, 939
via *cis*-diols, **6**, 589
via dithiocarbamates, **6**, 589
Selenocarbonates, alkoxy-substituted, synthesis of, **6**, 588
Selenocarbonates, dithio-substituted
from, phosgeneiminium chloride, **6**, 589
synthesis of, **6**, 589
Selenocarbonyl difluoride
dimerization of, **6**, 588
from, aluminum triiodide, **6**, 588
pyrolysis of, **6**, 588
reactions, with amines, **6**, 49
as starting material, for selenoacyl halide, **6**, 588
Selenocarbonyl fluoride, trifluoromethyl-, from, trimethylstannylpentafluoroethylselane, **6**, 49
Selenocarbonyl group
synthesis of functions, **6**, **587–599**
flanked by two different chalcogen atoms, **6**, 592
flanked by two selenium atoms, **6**, 590
at least one attached chalcogen, **6**, 588
at least one attached halogen, **6**, 588
at least one nitrogen function, **6**, 596
Selenocarbonyl halides, synthesis of, **6**, 588
Selenocarbonyls
from, thiocarbonyls, **6**, 592
reagents for, **6**, 587
scarcity of, **6**, 587
stability of, **3**, 381; **6**, 587
as starting materials
for *Se,N*-acetals, **4**, 333
for diselenoacetals, **4**, 289
use in the synthesis of sulfenyl selenides, **2**, 234
Selenocarbonyls, phosphorus-substituted, synthesis of, **6**, 596
Selenocarbonyls, zwitterionic, synthesis of, **6**, 596
α-Selenocarbonyls
from
dialkylbenzeneselenenamides, **2**, 282
enamines, **2**, 282
enol acetates, **2**, 282
enol ethers, **2**, 282
enolizable carbonyls, **2**, 282
α-lithio selenides, **2**, 282
metal enolates, **2**, 282
silyl enol ethers, **2**, 282
Selenocarboxylate, *O*-silyl, synthesis of, **5**, 536
Selenocarboxylates
from
selenoketenes, **5**, 510
sodium hydrogen selenide, **5**, 516
reactions
with areneselenenyl bromides, **5**, 250
with methylene(triphenyl)phosphorane, **5**, 521
with potassium bis(trimethylsilyl)amide, **5**, 522
reduction of, **5**, 521
as starting materials, for selenocarboxylic esters, **5**, 240
synthesis of, **5**, 510
Selenocarboxylates, *O*-alkyl
from, imidate esters, **5**, 515
NMR spectra of, **5**, 521
reactions, with sulfur, **5**, 523
synthesis of, **5**, 514
UV-Vis spectra of, **5**, 521
Selenocarboxylates, *Se*-alkyl, reactions, with alcohols, **5**, 519
Selenocarboxylates, *O*-aryl
synthesis of, **5**, 514
UV-Vis spectra of, **5**, 521
Selenocarboxylates, *t*-butyldimethylsilyl, synthesis of, **5**, 541
Selenocarboxylates, *O*-ethyl
reactions
with alkylamide magnesium bromide, **5**, 522
with hydrazine hydrate, **5**, 522
with nitrogen compounds, **5**, 522
Selenocarboxylates, *O*-halo, synthesis of, **5**, 530
Selenocarboxylates, *O*-methyl arene, synthesis of, **5**, 520
Selenocarboxylates, *O*-silyl, reactions, with benzenetellurenyl iodide, **5**, 250
Selenocarboxylates, *O*-trimethylsilyl
as starting materials
for caesium selenocarboxylates, **5**, 542
for rubidium selenocarboxylates, **5**, 542
synthesis of, **5**, 541
Selenocarboxylates, *O*-triorganosilyl, synthesis of, **5**, 542
Selenocarboxylic acid, 1-adamantyl-, synthesis of, **5**, 233
Selenocarboxylic acid, 4-biphenyl-, synthesis of, **5**, 233
Selenocarboxylic acid *O*-esters, synthesis of, **5**, 514
Selenocarboxylic acids
amine salts of, **5**, 240
from
sodium 1-adamantylselenocarboxylate, **5**, 508
sodium 4-biphenylselenocarboxylate, **5**, 508
oxidation of, **5**, 233, 252
synthesis of, **5**, 233, 506
Selenocarboxylic *O*-acids, synthesis of, **5**, 508
α-Selenocarboxylic acids, from, α-lithio selenides, **2**, 282
ω-Selenocarboxylic acids, synthesis of, **2**, 282
Selenocarboxylic esters
from
acyl halides, **5**, 238
carboxylic acids, **5**, 239
esters, **5**, 239
selenocarboxylates, **5**, 240
synthesis of, **5**, 238
α-Selenocarboxylic esters, as starting materials, for α-lithio selenides, **2**, 282
Selenocholines, synthesis of, **5**, 520
Seleno complexes, synthesis of, **5**, 952
α-Selenocoppers, synthesis of, **4**, 401
Selenocuprates, alkyl-, for the displacement of β-seleno sulfones, **1**, 486
Selenocyanate
reactions
with benzylidene metal complexes, **3**, 386
with chloroacetonitrile, **3**, 636
as starting material, for selenoureas, **6**, 596
Selenocyanate, ferric, reactions, with trialkylboranes, **5**, 1114
Selenocyanate, *o*-nitrophenyl, reactions, with tributylphosphine, **1**, 604

Selenocyanate, phenyl, reactions, with silyl enol ethers, **2**, 284
Selenocyanate, trimethylsilyl
reactions
with oxetanes, **5**, 1114
with oxiranes, **5**, 1114
Selenocyanates
from
alkylselenenyl sulfinates, **5**, 1116
arylselenenyl selenocyanates, **5**, 1116
dicyanogen triselenide, **5**, 1116
selenenyl thiocyanates, **5**, 1116
halolysis of, **2**, 732
isomerization of, **5**, 1053
reactions
with halides, **5**, 1053
with thiols, **2**, 735
reduction of, **2**, 279
as selenating reagents, **5**, 239
as starting materials
for alkaneselenenyl halides, **2**, 288
for alkaneseleninic acids, **2**, 291
for hydroxy selenides, **2**, 281
for isoselenocyanates, **5**, 1053
for selenols, **2**, 278
for selones, **3**, 392
for unsymmetrical dialkyl selenides, **2**, 279
synthesis of, **5**, 1113; **6**, 130
via additions to double bonds, **5**, 1115
via decarboxylative selenocyanation, **5**, 1115
via homolytic reactions, **5**, 1115
via nucleophilic substitution reactions, **5**, 1113
via reaction with electrophilic selenium, **5**, 1115
Selenocyanates, alkoxy, synthesis of, **5**, 1115
Selenocyanates, alkyl
from
alcohols, **5**, 1114
selenocyanogen, **2**, 289
synthesis of, **2**, 289; **5**, 1113
via nucleophilic substitution reactions, **2**, 289
Selenocyanates, allyl, isomerization of, **5**, 1054
Selenocyanates, allylic, synthesis of, **5**, 1115
Selenocyanates, aryl
from, 2-bromoethylphenyl selenide, **5**, 1114
reactions
with alcohols, **2**, 734
with metallated arenes, **2**, 729
with potassium hydroxide, **2**, 734
with thiols, **2**, 735
synthesis of, **5**, 1113
via selenocyanatodethallation, **5**, 1113
Selenocyanates, aryl selenenyl, as starting materials, for selenocyanates, **5**, 1116
Selenocyanates, benzyl, isomerization of, **5**, 1054
Selenocyanates, bis(4-toluyl)methyl, reactions, with benzene, **5**, 1054
Selenocyanates, halo-, synthesis of, **5**, 1115
Selenocyanates, iodo-, synthesis of, **5**, 1115
Selenocyanates, phenyl, reactions, with enamines, **2**, 284
Selenocyanates, phosphonium, oxidation of, **5**, 903
Selenocyanates, α-silyl
from, α-silyl tosylates, **5**, 1113
synthesis of, **3**, 383
Selenocyanates, α-silylalkyl
as starting materials, for selenoaldehydes, **4**, 376
synthesis of, **4**, 376
Selenocyanates, trimethylsiloxyalkyl, synthesis of, **5**, 1114
Selenocyanates, α-trimethylsilyl, synthesis of, **5**, 1115
Selenocyanation, of alkenes, **2**, 284
β-Selenocyanato propionic acids, synthesis of, **5**, 1114
Selenocyanide, synthesis of, **6**, 637
Selenocyanide, phenyl, reactions, with α,β-unsaturated ketones, **3**, 656
Selenocyanogen
reactions
with cyanocuprates, **4**, 947
with organocoppers, **5**, 1115
as starting material, for alkyl selenocyanates, **2**, 289
use in the synthesis of selenocyanates, **6**, 130
Selenocyclization, of hydroxyketene dithioacetal, **6**, 98
Selenocyclooctane, 2-phenyl-, reactions, with hydrogen peroxide, **1**, 607
L-Selenocystine, synthesis of, **5**, 1115
1,2,3-Selenodiazoles
as starting materials
for alkynes, **1**, 1077
for selenoketenes, **3**, 553
1,2,3-Selenodiazoles, 4-aryl-, as starting materials, for diselenofulvenes, **3**, 553
Selenodinitriles, α-phenyl-, synthesis of, **3**, 637
Selenodinitriles, γ-phenyl-, synthesis of, **3**, 637
Seleno-1,3-diselenoles, 2-alkane-, synthesis of, **6**, 93
Selenoditelluroorthocarbonates, theoretical, **6**, 312
Selenoditellurothioorthocarbonates, theoretical, **6**, 312
Seleno-1,3-dithianes, 2-arene-, from, 2-chloro-1,3-dithianes, **6**, 101
2-Seleno-1,3-dithianes, from, 1,3-dithiane, **6**, 102
Selenodithiocarbenium salts, as starting materials, for monoselenodithioortho esters, **6**, 101
Selenodithioorthocarbonates, theoretical, **6**, 311
Selenoenamines, 1-alkyl, from, selenoamides, **4**, 909
α-Selenoenones, synthesis of, **3**, 256
β-Selenoenones, synthesis of, **3**, 256
Selenoester, as starting material, for acyl radicals, **1**, 933
Selenoester, oxyimidic, synthesis of, **5**, 832
Selenoesters
from
alkynylselenolate anions, **5**, 515
alkynyl selenols, **5**, 515
imidate esters, **5**, 515
reactions
with copper catalysts, **3**, 307
with diazomethane, **3**, 174
with hydrazine, **5**, 704
with potassium hexamethyldisilazide, **4**, 841
with semicarbazides, **5**, 704
with thiosemicarbazides, **5**, 704
for the selenoacylation of amines, **5**, 626
as starting materials
for hydrazonic acid esters, **5**, 704
for ketones, **3**, 130
synthesis of, **5**, 514; **6**, 536
use in the synthesis of aromatic ketones, **3**, 282
use in the synthesis of furanyl ketones, **3**, 302
Selenoesters, *O*-alkyl
Se-alkylation of, **4**, 841
synthesis of, **5**, 514
Selenoesters, *O*-aryl, synthesis of, **5**, 514
Selenoethene, 1-phenylthio-1-phenyl-, oxidation of, **4**, 873
Selenoethers, synthesis of, **2**, 1036
Selenoethers, alkynyl, synthesis of, **2**, 1032, 1036
Selenoethers, allylic, displacement of, **1**, 483
Selenoethers, difluoromethyl-, synthesis of, **6**, 49
Selenofenchone, synthesis of, **3**, 392
Selenofluorenone, synthesis of, **3**, 392
Selenoformaldehyde
as bridging ligand, **3**, 387
from, methylene, **3**, 382
metal complexes of, **3**, 387
metal complexes, synthesis of, **3**, 387
Selenoformaldehyde, polymeric, synthesis of, **3**, 382
Selenoformaldehyde, trimeric, synthesis of, **3**, 382
Selenoformaldehyde metal complex, synthesis of, **3**, 387
Selenoformamides, synthesis of, **5**, 626
Selenoformates
reactions, with tri-t-butylphenyllithium, **3**, 383

reduction of, **5**, 521
Selenoformates, *O*-alkyl, from, *N*,*N*-dimethylformamide, **5**, 516
Selenoformates, *O*-*t*-butyl, reactions, with 2,4,6-tri-*t*-butylphenyllithium, **5**, 524
Selenoformates, *O*-cholesteryl
reactions, with 2,4,6-tri-*t*-butylphenyllithium, **5**, 524
reduction of, **5**, 521
α-Selenogermanes, synthesis of, **4**, 377, 377
1-Seleno-β-D-glucopyranose, tetra-*O*-acetyl-, as starting material, for selenoglucosinates, **5**, 832
Selenoglucosinates, from, tetra-*O*-acetyl-1-seleno-β-D-glucopyranose, **5**, 832
Selenoglycosides, synthesis of, **4**, 238, 239
Selenoglycosides, L-rhamnopyranosyl, synthesis of, **4**, 238
Seleno groups, *o*-nitrophenyl-, for the elimination of selenoxides, **1**, 601
Seleno groups, 2-pyridyl-
elimination of, **1**, 604
for the elimination of selenoxides, **1**, 602
Selenoguanosine, synthesis of, **6**, 598
Selenohexa-1,3-diyne, 1-methyl-, reactions, with hydrogen chloride, **4**, 786
Selenohydrazides, synthesis of, **5**, 642
Selenoic acid, arene-, elimination of, **1**, 983
N-Selenoimides, from, imides, **5**, 336
Selenoketals
reactions, with butyllithium, **4**, 400
silylation of, **4**, 375
transmetallation of, **4**, 375
Selenoketenes
from
benzo-1,2,3-selenodiazole, **3**, 553
1,2,3-selenodiazoles, **3**, 553
as intermediates, in the synthesis of selenocarboxylates, **5**, 519
irradiation of, **3**, 382
reactions, with amines, **5**, 626
for the selenoacylation of amines, **5**, 626
as starting materials
for selenocarboxylates, **5**, 510
for thiocarboxylates, **5**, 510
synthesis of, **3**, 553
transition metal complexes of, **3**, 554
Selenoketenes, silyl, synthesis of, **2**, 941
Selenoketones
nomenclature of, **3**, 381
as starting materials, for thioketones, **3**, 348
synthesis of, **1**, 686; **3**, **381–401**
α-Selenoketones, as starting materials, for alkenes, **1**, 593
β-Selenoketones, oxidation of, **3**, 207
Selenol, benzyl, reactions, with dibenzyltriseleno carbonate, **5**, 626
Selenol, methyl, reactions, with Fischer carbene complexes, **5**, 952
Selenolane, octafluoro-, synthesis of, **6**, 49
Selenol anions, reactions, with α,β-unsaturated ketones, **3**, 175
Selenolate, reactions, with carbon diselenide, **6**, 591
Selenolate, alkynyl, from, sodium alkynylide, **2**, 1036
Selenolate, phenyl, reactions, with β-halo selenides, **1**, 594
Selenolate anions, alkynyl
reactions, with alcohols, **5**, 519
as starting materials, for selenoesters, **5**, 515
Selenolate ions
as intermediates, **2**, 278
nucleophilicity of, **2**, 282
Selenolate ions, arene-, as starting materials, for aryl alkyl selenides, **2**, 279
Selenolates
reactions
with dihalomethyl methyl ethers, **6**, 100
with lactones, **2**, 282
as starting materials, for 1,2,3-selenadiazoles, **5**, 519
synthesis of, **2**, 279
Selenolates, alkane-
alkylation of, **2**, 279
as starting materials
for tris(alkylseleno)boranes, **2**, 294
for unsymmetrical dialkyl selenides, **2**, 279
Selenolates, alkynyl, synthesis of, **2**, 1038
Selenolates, arene-, reactions, with 2-chloro-1,3-dithianes, **6**, 101
Selenolate salts
reactions, with ethynecarboxylate esters, **4**, 874
as starting materials, for diselenides, **2**, 292
γ-Selenolbutyrolactone, synthesis of, **5**, 239
Selenolithium, phenyl-, reactions, with (iodomethyl)trimethylstannane, **4**, 401
α-Selenolithiums
synthesis of, **4**, 401
via selenium–metal exchange, **4**, 399
Selenols
acylation of, **5**, 238, 239
aminomethylation of, **4**, 333
boron derivatives of, **4**, 288
condensation with ketones, **4**, 287
exchange reactions, with orthoformate esters, **6**, 92
from
diselenides, **2**, 278
selenocyanates, **2**, 278
reactions
with acyl chlorides, **5**, 238
with alkyldihaloarsines, **2**, 500
with alkyldihalobismuthines, **2**, 500
with alkyldihalostibines, **2**, 500
with alkylthiochloroalkynes, **4**, 787
with alkynes, **2**, 728
with β-chloroenals, **3**, 74
with *gem*-dihalides, **4**, 288
silicon derivatives of, **4**, 288
as starting materials
for alkaneseleninic acids, **2**, 291
for alkaneseleninic anhydrides, **2**, 291
for diselenides, **2**, 292
for α-halo selenides, **4**, 88
for unsymmetrical selenides, **2**, 279
synthesis of, **5**, 238
Selenols, alkane-
reactions
with aminoarsines, **2**, 500
with dialkylhaloarsines, **2**, 499, 500
with sodium ethoxide, **2**, 279
with thiophosgene, **6**, 536
safety precautions, **2**, 278
as starting materials
for chloroselenothioformates, **6**, 536
for tris(alkylseleno)boranes, **2**, 294
for unsymmetrical dialkyl selenides, **2**, 279
synthesis of, **2**, 277
Selenols, alkynyl-
reactions, with alcohols, **5**, 519
as starting materials, for selenoesters, **5**, 515
Selenols, arene-
from
arylamines, **2**, 727
aryl halides, **2**, 727
as starting materials, for diselenides, **2**, 727
synthesis of, **2**, 727
Selenols, aryl-
oxidation of, **2**, 734
reactions, with aroyl halides, **5**, 238
Selenomagnesium chloride, aminoethyl-, as starting material, for selenothiobenzoic acid *Se*-aminoethyl ester, **5**, 562
Selenomagnesium chloride, phenyl-, reactions, with diphenylphosphinous chloride, **2**, 856

Selenomagnesium halides, arene-, reactions, with dichloromethyl ethers, **6**, 100
Selenomalonitrile, 2-methyl-2-phenyl-, reactions, with alkenes, **4**, 284
Selenomethanes, aryl-, fluorination of, **4**, 92
L-Selenomethionine, synthesis of, **5**, 1115
Selenonamides, synthesis of, **2**, 370
Selenonamides, alkane-, synthesis of, **2**, 292
Selenonaphthalene, 1-methyl-, reactions, with dibenzoyl peroxide, **5**, 210
Selenones
  deprotonation of, **4**, 399
  elimination of, **1**, 620
  from, selenides, **2**, 287
  nucleophilic displacement of, **1**, 620
  reactions, with diazoalkanes, **4**, 333
  synthesis of, **1**, 620; **2**, 732; **4**, 343
  use of term, **3**, 382; **6**, 587
Selenones, alkyl, synthesis of, **2**, 287
Selenones, alkyl aryl, from, alkyl aryl selenides, **2**, 287
Selenones, α-halo, synthesis of, **4**, 92
Selenones, α-metallo, synthesis of, **4**, 399
Selenones, vinyl
  reactions, with organolithiums, **4**, 400
  substitution of, **1**, 485
2-Selenones, 1,3-diselenole-, as starting materials, for tetraselenafulvalenes, **4**, 874
2-Selenones, 1,3-thiaselenole-
  dimerization of, **4**, 868
  as starting materials, for thiaselenafulvenes, **4**, 870
2(1*H*)-Selenones, 4,6-disubstituted-3-cyanopyridine-, synthesis of, **3**, 656
Selenon esters, stability of, **5**, 243
Selenonic acid esters, alkane-, synthesis of, **2**, 292
Selenonic acids, as oxidizing agents, **2**, 734
Selenonic acids, alkane-, from, seleninic acids, **2**, 291
Selenonic acids, arene-, synthesis of, **2**, 734
Selenonic esters, methyl-, reactions, with amines, **2**, 370
Selenonitriles, α-fluoro-α-aryl-, synthesis of, **3**, 637
Selenonitriles, α-phenyl-, oxidative elimination of, **3**, 643, 644
α-Selenonitriles
  deprotonation of, **3**, 637
  from, cyanohydrins, **3**, 636
  synthesis of, **3**, 636
Selenonitrosamines, α-phenyl-, synthesis of, **4**, 335
N-Selenonitrosimine, as intermediate, for selones, **3**, 392
Selenonium halides, trialkyl-, from, dialkyl selenides, **2**, 288
Selenonium salts
  elimination of, **1**, 623
  synthesis of, **1**, 716
Selenonium salts, alkyl, synthesis of, **2**, 288
Selenononane, 1-hydroxy-2-nitro-2-phenyl-, reactions, with hydrogen peroxide, **4**, 335
α-Selenonyl carbanions, synthesis of, **2**, 1001
Selenonyl halides, alkane-, synthesis of, **2**, 289
Selenoorthocarbonates, synthesis of, **6**, 308
Selenoortho esters, reactions, with trimethylsilyl cyanide, **3**, 637
Selenophenates, as starting materials, for thioselenoacetals, **4**, 283
Selenophene
  reactions, with Grignard reagents, **1**, 484
  reactivity of, **3**, 106
Selenophene, 2,5-dibromo-, chlorination of, **6**, 49
Selenophene, 2,5-dichloro-3-iodo-, reactions, with ethyllithium, **4**, 785
Selenophene, 2,4-dinitro-5-methoxy-, reactions, with potassium methoxide, **6**, 100
Selenophene, 2,3,5-tribromo-, reactions, with ethyllithium, **4**, 785
Selenophene-1,1-dichloride, tetrahydro-, reactions, with ylides, **5**, 903
Selenophenes, ring-opening of, **4**, 785
Selenophenes, 2-alkoxy-, as starting materials, for monoselenoortho esters, **6**, 100
Selenophenes, formyl, synthesis of, **3**, 106
Selenophenes, 3-halo-, as starting materials, for alkynes, **1**, 1072
Selenophene-2,3,4,5-tetrachloride, 2,5-dibromo-, synthesis of, **6**, 49
Selenophenols
  condensation with aldehydes, **4**, 287
  reactions, with oxyimidic chlorides, **5**, 832
  as starting materials, for α-halo selenides, **4**, 88
Selenophenoxide, allylation of, **1**, 246
Selenophosgene, from, 2,2,4,4-tetrachloro-1,3-diseletane, **6**, 588
Selenophosphate, trimethyl, reactions, with ω-*H*-perfluoroalkanals, **3**, 385
Selenophosphinates, synthesis of, **2**, 465
Selenophosphinates, alkyl-, as starting materials, for alkylselenophosphonium salts, **2**, 293
Selenophosphinic acid chloride, diphenyl-, reactions, with ethanethiol, **2**, 857
Selenophosphinic acid esters, synthesis of, **2**, 857
Selenophosphinic acid *Se*-phenyl ester, diphenyl-, synthesis of, **2**, 857
Selenophosphinic acids, synthesis of, **2**, 465
Selenophosphinic acids, α-halo-, synthesis of, **4**, 127
Selenophosphinic chlorides
  reactions
    with sodium hydroselenide, **2**, 465
    with sodium hydrosulfide, **2**, 465
  synthesis of, **4**, 128
Selenophosphinous acid esters, alkyl-, synthesis of, **2**, 293
Selenophosphinous acid *Se*-phenyl ester, diphenyl-, synthesis of, **2**, 856
Selenophospholes, use in Diels–Alder reactions, **4**, 489
Selenophosphonates, synthesis of, **2**, 465, 466
Selenophosphonic acids, alkyl ethyl-, synthesis of, **2**, 466
Selenophosphonic amides, aryl-, synthesis of, **2**, 862
Selenophosphonic dichloride, phenyl-, use in the synthesis of selenals, **3**, 388, 389
Selenophosphonic dihalides, reactions, with alcohols, **2**, 466
Selenophosphonium salts, as intermediates, **5**, 239
Selenophosphonium salts, alkyl-
  from
    alkylphosphates, **2**, 293
    alkylphosphonates, **2**, 293
    alkylselenophosphinates, **2**, 293
    selenophosphoranes, **2**, 293
Selenophosphoranes
  as starting materials, for alkylselenophosphonium salts, **2**, 293
  vacuum pyrolysis of, **2**, 1038
Selenophthalide, synthesis of, **5**, 515
Selenophthalimide, *N*-phenyl- (NPSP)
  reactions, with thiocarboxylic acids, **5**, 248
  as selenating reagents, **5**, 239
  synthesis of, **1**, 604
Selenopropane, 2-phenylthio-2-phenyl-, synthesis of, **4**, 284
Seleno–Pummerer reaction, for the synthesis of α-acyloxy selenides, **4**, 240
Selenopyridones, synthesis of, **3**, 398
Selenopyrimidines, from, selenourea, **6**, 598
Selenopyrone
  as starting material, for β-diselones, **3**, 397
  use in the synthesis of selones, **3**, 399
Seleno radicals
  reactions
    with alkenes, **2**, 285
    with conjugated dienes, **2**, 285
Selenosemicarbazides
  from
    hydrazines, **6**, 598

isoselenocyanates, **6**, 598
studies of, **5**, 1051
Selenosemicarbazides, phenyl-, synthesis of, **5**, 1056
Selenosemicarbazones, from, selenosemicarbazides, **6**, 598
Selenosilanes, methyl-, from, methaneselanol, **2**, 294
Selenosilanes, phenyl-, synthesis of, **4**, 375
α-Selenosilanes
deprotonation of, **4**, 375
from
allyl phenyl selenides, **4**, 374
(halomethyl)trimethylsilanes, **4**, 377
phenylvinyl selenide, **4**, 375
vinylsilanes, **4**, 376
synthesis of, **4**, 374
Selenosugars, synthesis of, **4**, 238
Selenosulfates, use in the synthesis of sulfenyl selenides, **2**, 237
Selenosulfides, synthesis of, **2**, 235
Selenosulfonates, reactions, with alkenes, **2**, 285
Selenosulfonation
of alkenes, **2**, 285
of alkynes, **2**, 285
of allenes, **2**, 285
Selenosulfones, reduction of, **1**, 440
α-Selenosulfones, deprotonation of, **4**, 286
β-Selenosulfones, displacement of, **1**, 486
Selenosulfoxides, α-phenyl, synthesis of, **4**, 285
α-Selenosulfoxides, from, α-sulfoxide carbanions, **4**, 285
Selenosulfuric acid, 2-aminoethane
synthesis of, **2**, 237
use in the synthesis of sulfenyl selenides, **2**, 237
Selenotelluroacetals, from, telluride anions, **4**, 291
Selenotellurodithioorthocarbonates, theoretical, **6**, 311
Selenotelluroic *Se*-acid, structure of, **5**, 545
Selenotelluroic *Te*-acid, structure of, **5**, 545
Selenotelluroorthocarbonates, theoretical, **6**, 311
Selenothioacetals, synthesis of, **6**, 336
Selenothioacetic acid *S*-alkyl esters, synthesis of, **5**, 563
Selenothiobenzoic acid *Se*-aminoethyl ester, from, aminoethylselenomagnesium chloride, **5**, 562
Selenothiocarbamates, via thioureas, alkylation of, **6**, 594
Selenothiocyanates, reactions, with amines, **2**, 368
Selenothioesters
S,S-dioxides, Diels–Alder reaction of, **4**, 287
reactions, with amines, **5**, 626
for the selenoacylation of amines, **5**, 626
Selenothioformate, *Se*-trifluoromethyl fluoro-, synthesis of, **6**, 536
Selenothioformates, bromo-, synthesis of, **6**, 537
Selenothioformates, chloro-
from, alkaneselenols, **6**, 536
synthesis of, **6**, 536
Selenothioformates, fluoro-, synthesis of, **6**, 536
Selenothioic *S*-acid, structure of, **5**, 545
Selenothioic *Se*-acid, structure of, **5**, 545
Selenothioic acid *S*-alkyl esters, from, selenoacetic acid *Se*-alkynyl esters, **5**, 563
Selenothioic acid esters, synthesis of, **5**, 562
Selenothioic acid *S*-phenyl ester, synthesis of, **5**, 563
Selenothioic acid *Se*-phenyl ester, α-thioxo, from, alkynyl phenyl selenide, **5**, 563
Selenothioic acid *Se*-phenyl esters
as starting materials
for selenothioperoxy esters, **5**, 563
for selenylsulfines, **5**, 563
synthesis of, **5**, 563
Selenothioic acids, synthesis of, **5**, 549
Selenothioorthocarbonates, theoretical, **6**, 311
Selenothioorthoformates, halo-, synthesis of, **6**, 276
Selenothioperoxy esters, from, selenothioic acid *Se*-phenyl esters, **5**, 563
Selenothiophosphinic acid *S*-ethyl ester, diphenyl-, synthesis of, **2**, 857
Selenothiophosphinic acids, synthesis of, **2**, 465
Selenothiopyrane *S*-oxides, 2-phenyl-, synthesis of, **4**, 285
α-Selenotins, synthesis of, **4**, 401
Seleno(trimethylsilane), methyl-, synthesis of, **2**, 294
Selenotritelluroorthocarbonates, synthesis of, **6**, 308
Selenotrithioorthocarbonates, synthesis of, **6**, 308
Selenourea, quasi-, synthesis of, **3**, 399
Selenoureas
alkylation of, **6**, 637, 732
from
bistrimethylsilyl selenide, **6**, 597
carbodiimides, **6**, 596
cyanamides, **6**, 596
isoselenocyanates, **6**, 596
phosphorus pentaselenide, **6**, 596
selenocarbamoyl selenides, **6**, 597
selenocyanate, **6**, 596
thioureas, **6**, 596
triselenides, **6**, 597
triselenocarbonates, **6**, 597
production of, **6**, 594, 596
as starting materials, for carbenes, **6**, 143
use in the synthesis of selones, **3**, 397
Selenoureas, *N*-benzoyl-, reactions, with thiophosgene, **5**, 655
Selenoureas, heterocyclic, synthesis of, **6**, 598
Selenourethane, from, phosgeneiminium chloride, **6**, 589
Selenoxide, reactions, with phenyl selinyl chloride, **3**, 177
Selenoxide, aryl, oxidative elimination of, **3**, 644
Selenoxide, bis(4-methoxyphenyl), use in Kornblum oxidation, **3**, 89
Selenoxide, 1-chloroethyl phenyl, stability of, **4**, 92
Selenoxide, 1,1-difluoromethyl phenyl, synthesis of, **4**, 92
Selenoxide, dimethyl
reactions
with benzyl bromides, **3**, 98
with 4-nitrobenzyl bromide, **3**, 99
use in Kornblum oxidation, **3**, 89
use in the synthesis of α-acyloxy selenides, **4**, 240
Selenoxide, β-hydroxy, use in the synthesis of alkenes, **1**, 716
Selenoxide, 1-lithiovinyl phenyl, half life of, **4**, 965
Selenoxide, 1,1,1-trifluoromethyl phenyl, synthesis of, **4**, 92
Selenoxide elimination
regiochemistry of, **1**, 602
regiospecificity of, **1**, 607
side reactions, **1**, 601
stereoselectivity of, **1**, 604
and sulfoxide elimination, compared, **1**, 601
for the synthesis of α,β-unsaturated nitro compounds, **1**, 608
and telluroxide elimination, compared, **1**, 611
Selenoxide elimination, asymmetric, for the synthesis of chiral allenes, **1**, 609
Selenoxide pyrolysis
regiochemistry of, **1**, 781
stereochemistry of, **1**, 781
Selenoxides
deprotonation of, **1**, 713; **4**, 399
elimination of, **1**, 600; **3**, 207
syn-elimination of, **2**, 658
intermediates in, **2**, 290
elimination reactions of, **4**, 924
exo elimination of, **1**, 607
from, alkenes, **1**, 712
hydrogen bonding of, **2**, 286
as intermediates, in the synthesis of alkynic acylsilanes, **5**, 427
metallation of, **1**, 715
oxidation of, **1**, 620
oxidative elimination of, **3**, 656
pyrolysis of, **1**, 780
reactions
with benzoylmethylphosphonium salts, **4**, 928

with 1,3-diketones, **3**, 177
with ethers, **6**, 256
with thionyl chloride, **2**, 369
with trimethylsilyl triflate, **2**, 369
stability of, **4**, 240
as starting materials
for alkenes, **1**, 716
for selenoacetals, **2**, 286
synthesis of, **2**, 731; **4**, 241
Selenoxides, alkenyl, elimination of, **1**, 1073
Selenoxides, alkyl
elimination rate of, **1**, 601
as starting materials
for alkenes, **2**, 286
for *N*-substituted selenimides, **2**, 292
synthesis of, **2**, 286
Selenoxides, alkynyl, stability of, **2**, 1038
Selenoxides, 1-alkynyl phenyl, reduction of, **1**, 999
Selenoxides, allyl, [2,3] sigmatropic rearrangement of, **2**, 286, 291
Selenoxides, allylic, as starting materials, for allylic alcohols, **1**, 608
Selenoxides, aryl
elimination rate of, **1**, 601
stability of, **1**, 601
Selenoxides, dienyl, [2,3] sigmatropic rearrangement of, **1**, 609
Selenoxides, α-halo, synthesis of, **4**, 92
Selenoxides, β-keto
as intermediates, **3**, 177
synthesis of, **3**, 177
See also α-Selenyl ketones
Selenoxides, α-lithio, as starting materials, for β-hydroxy selenides, **1**, 716
Selenoxides, α-metallo, synthesis of, **4**, 399
Selenoxides, phenylvinyl, deprotonation of, **4**, 965
Selenoxides, propargyl, [2,3] sigmatropic rearrangement of, **2**, 286
Selenoxides, β-substituted, elimination of, regioselectivity, **1**, 603
Selenoxides, vinyl, reactions, with organolithiums, **4**, 400
Selenoxides, vinylic
elimination of, **1**, 605
as starting materials
for alkynes, **1**, 605
for allenes, **1**, 605
(*Z*)-Selenoxides, as starting materials, for allenes, **1**, 1073
Selenoxide *syn*-elimination, mechanism, **1**, 600
2-Selen trisulfide, ditrityl-, synthesis of, **2**, 236
Selenuranes, synthesis of, **4**, 240
Selenuranes, hydroxylamino-, synthesis of, **2**, 369
Selenuranes, *Se*-methylammonio-, synthesis of, **2**, 369
Selenuranes, *Se*-phenylammonio-, synthesis of, **2**, 369
Selenurane salts, chloro-
reactions
with lithium dimethyl cuprate, **2**, 369
with lithium diphenyl cuprate, **2**, 369
synthesis of, **2**, 369
Selenylating reagents, for the synthesis of α-selenyl ketones, **3**, 173
Selenylation
of aldehydes, **3**, 37
of carbanions, **6**, 123
α-Selenylation, of carbonyl groups, **3**, 173
Selenyl bromide, phenyl
reactions
with dimethyl sulfide, **2**, 238
with enol ethers, **3**, 38
with α-sulfinyl phosphonate, **6**, 123
with ylides, **6**, 123
as starting material, for vinyl bromides, **2**, 623
Selenyl bromide, 2-pyridyl
use in selenylation reactions, **3**, 174
use in the synthesis of α-phenyl selenyl aldehydes, **3**, 37
α-Selenyl carbanions
formylation of, **3**, 39
synthesis of, **3**, 175; **4**, 398
Selenyl chloride, phenyl
for quenching, **6**, 130
reactions
with cyclohexene, **2**, 785
with enamines, **3**, 38
with β-keto aldehydes, **3**, 38
with ketones, **3**, 207
with β-ketothiol, **2**, 235
with vinyl sulfones, **4**, 287
with ylides, **6**, 123
use in the synthesis of α-selenyl aldehydes, **3**, 37
use in the synthesis of thio selenides, **2**, 235
Selenyl chlorides, reactions, with thiocarbonyls, **6**, 276
Selenyl chlorides, arene
reactions
with (but-2-en-1-ynyl)trimethylgermane, **4**, 948
with (but-2-en-1-ynyl)trimethylsilane, **4**, 948
Selenyls, reactions, with haloalkynes, **4**, 786
Selenyls, phenyl, as starting materials, for carbon-centered radicals, **1**, 942
Self-condensation, problems, **4**, 557
Self-oligomerization, synthesis of, **5**, 546
(-)β-Selinene, synthesis of, **1**, 805
Selinyl chloride, phenyl, reactions, with selenoxide, **3**, 177
Selone
reactions
with dimethyl ethynedicarboxylate, **3**, 394
with Wittig reagents, **3**, 394
use of term, **6**, 587
via β-diketone analogues, synthesis of, **3**, 397
Selones
Diels–Alder adducts, **3**, 392
from
bis(tricyclohexyltin) selenide, **3**, 392
ketones, **3**, 391
phosphoranylidinehydrazones, **3**, 391
selenium(I) halides, **3**, 392
selenocyanates, **3**, 392
synthesis of, **2**, 858; **3**, 391
via dimethylaluminum selenide, **3**, 394
via metal carbene complexes, **3**, 396
via Wittig reagents, **3**, 394
use of term, **3**, 381
Selones, alkyl, synthesis of, **3**, 394
Selones, aryl
from, selenium(I) chloride, **3**, 392
synthesis of, **3**, 392, 394
Selones, hindered, comparison of syntheses, **3**, 392
Selones, metal-stabilized, synthesis of, **3**, 396
Selones, resonance-stabilized, synthesis of, **3**, 398
Selones, sterically hindered, synthesis of, **3**, 391
Semicarbazide
reactions
with aldehydes, **3**, 449
with imines, **3**, 450
with ketones, **3**, 449
use in the synthesis of hydrazonoyl fluorides, **5**, 839
Semicarbazide, benzyl, oxidation of, **5**, 1036
Semicarbazide hydrochloride, reactions, with pentafluoroguanidine, **5**, 811
Semicarbazides
for the condensation of *O*-ethyl selenocarboxylates, **5**, 522
reactions, with selenoesters, **5**, 704
Semicarbazone, ω-bromoacetophenone, reactions, with enamines, **2**, 770
Semicarbazones
as precursors of aldehydes, **3**, 18
reactions, with hydrogen cyanide, **3**, 616
as starting materials

for acylhydrazones, **5**, 363
for alkanes, **1**, 42
for *N,N*-diacylhydrazones, **5**, 364
for ketones, **3**, 133
for methyl compounds, **1**, 42
synthesis of, **3**, 449
Semicarbazones, 2-phenyl-, oxidation of, **5**, 986
Semiconductors, precursors for, **2**, 510
Semicorrin ligands, for copper catalysts, **1**, 353
Semipinacol rearrangement, mechanism, **1**, 388
Semisquaric acid, as starting material, for hydroxyethyne, **2**, 1022
Semmelhack reaction, for the synthesis of alkoxycarbene complexes, **5**, 937
Semmler–Wolff aromatization, for the synthesis of *N*-arylamides, **5**, 293
Senepoxide, synthesis of, **1**, 573
Serinals, dipeptide amino, synthesis of, **3**, 40
Serine, α-diazo esters, synthesis of, **3**, 468
Serine, *N*-tosyl-, disilylation of, **5**, 223
L-Serine, reactions, with triphosgene, **6**, 486
L-Serine β-lactone, *N*-(benzoyloxycarbonyl), ring-opening of, **5**, 85
Serines
O-acylation of, **5**, 140
from, trimethyl phosphite, **4**, 487
Sesquibromides, synthesis of, **2**, 597
Sesquicarene, synthesis of, **1**, 1082
Sesquihalides of aluminum, reduction of, **4**, 698
Sesquiiodides, synthesis of, **2**, 597
Sesquiterpenes
aristolactone, **1**, 818
cleavage of, **1**, 450
isocomene, **1**, 832
punctatin A, **1**, 817
synthesis of, **1**, 176, 548
Sevoflurane, synthesis of, **4**, 45
Sex pheromones
from *Lobesia botrana*, **1**, 1102
from *Spodoptera littoralis*, **1**, 734
synthesis of, **1**, 747
Seychellene, synthesis of, **1**, 1132
(±)-Seychellene, synthesis of, **1**, 434
Shapiro reaction
limitations of, **1**, 630
mechanism, **1**, 630, 630
regioselectivity of, **1**, 632
stereoselectivity of, **1**, 633
for the synthesis of nitroalkenes, **2**, 785
for the synthesis of vinyllithiums, **3**, 58
use in the synthesis of alkenylsilanes, **2**, 911
use of tosylhydrazones in, **3**, 446
for vinyllithiums, **2**, 956
Sharpless asymmetric dihydroxylation, for the oxidation of enolsilanes, **3**, 156
Sharpless asymmetric epoxidation, for the reduction of oxiranes, **1**, 14
Sharpless catalyst
for the synthesis of asymmetric cyanohydrins, **3**, 628
use in the synthesis of optically active α-cyano sulfoxides, **3**, 634
Sharpless epoxidation
of allylic alcohols, **2**, 318
of 2-furanmethanols, **3**, 222
for the synthesis of amine *N*-oxides, **2**, 348
for the synthesis of dialkyl sulfoxides, **2**, 147
use of alkenylsilanols in, **1**, 1091
Sharpless oxidation, Kagan-modified, for the synthesis of vinyl sulfoxides, **2**, 713
Sharpless reagent
Di Furia-modified, **4**, 261
Kagan-modified, **4**, 261
for the oxidation of dithianes, **4**, 269
for the oxidation of dithioacetals, **4**, 261
[1,2] Shifts
of double bonds, **1**, 794
of sulfur compounds, **1**, 233
Sibirinone, synthesis of, **3**, 535
Sigmatropic rearrangement
of carbon, **1**, 799
mechanism, **1**, 400
for the synthesis of 1,2-dihydrotriazines, **5**, 1075
[1,3] Sigmatropic rearrangement, mechanism, **1**, 400
[1,5] Sigmatropic rearrangement
of ethers, **1**, 838
mechanism, **1**, 402
of tertiary amines, **1**, 838
[1,7] Sigmatropic rearrangement, mechanism, **1**, 403
[2,3] Sigmatropic rearrangement
mechanism, **1**, 404
of selenoxides, **5**, 427
for the synthesis of β,γ-unsaturated esters, **5**, 553
[3,3] Sigmatropic rearrangement
of allylic acetates, **1**, 836
of *O*-allylphenoxy compounds, **5**, 326
of allyl vinyl thioethers, **1**, 833
of chorismate, **1**, 826
1,3-diheteroanalogues of, **1**, 836
mechanism, **1**, 410, 807
of silylketene acetals, **1**, 829
for the synthesis of β,γ-unsaturated esters, **5**, 553
[3,5] Sigmatropic rearrangement, of vinylcyclopropanes, **1**, 840
[5,5] Sigmatropic rearrangement, mechanism, **1**, 840
[7,7] Sigmatropic rearrangement, mechanism, **1**, 841
[2,3] Sigmatropic selenoxide–selenate rearrangement, mechanism, **1**, 839
Sigmatropic-shift processes, retro-, for the synthesis of carbon–carbon bonds, **1**, 771
Silaacyl complexes, synthesis of, **6**, 337
1-Sila-3-alanatetanes, synthesis of, **6**, 197
Silaarsene, reactions, with 1,6-diisocyanohexane, **6**, 691
Silabicyclo[2.2.2]octadienes, as starting materials, for silenes, **3**, 502
Silaboralanes, synthesis of, **4**, 643
Silacyclobutane, 1,1-dimethyl-, pyrolysis of, **4**, 610
1-Silacyclobutane, 1,1-difluoro-, as starting material, for α-silyl ylides, **5**, 915
1-Silacyclobutane, 1,1-dimethyl-, thermolysis of, **4**, 610
Silacyclobutanes
pyrolysis of, **4**, 610
reactions
with silacyclobutanes, **4**, 610
with sulfur, **2**, 267
as starting materials
for carbosilanes, **4**, 610
for silenes, **3**, 501
1-Sila-2,4-cyclohexadiene, 1-*t*-butyl-, reactions, with butyllithium, **4**, 603
Silacyclohexanes, reactions, with trityl fluoroborate, **1**, 656
1-Silacyclohex-3,5-diene, 1,1-dimethyl-2,5-diphenyl-, reactions, with butyllithium, **4**, 616
Silacyclopentane, 1,1-dimethyl-, attempted metallation of, **4**, 650
1-Silacyclopentene, 1,1-dimethyl-2,3-benzo-5-trimethylsilyl-, synthesis of, **4**, 614
Silacyclopropanes, reactions, with sulfur, **2**, 267
Silacyclopropenes
from, alkynes, **4**, 1048
as starting materials, for 1,3-disilacyclopentenes, **4**, 616
Siladiazoles, reactions, with carbon disulfide, **6**, 576
Siladihydrotriazoles, synthesis of, **6**, 673
Silaethene, reactions, with silyl azides, **6**, 673
Silaethene, 1,1-dimethyl-, trimerization of, **4**, 610
Silaethene, 1-methyl-1-phenyl-2-neopentyl-, synthesis of, **4**, 610

1-Silaethene, synthesis of, **4**, 610
Silaethenes
additions of, **6**, 373
cycloaddition of, **6**, 672
diazo group transfer to, **3**, 483
ene reactions of, **6**, 177
from, polysilylated methanes, **6**, 177
reactions
with carbonyls, **6**, 178
with isobutene, **6**, 177
with organolithiums, **6**, 192
stability of, **6**, 709
as starting materials, for trisilylmethanes, **6**, 178
synthesis of, choice of solvent for, **6**, 709
Silaethenes, 1,1-dimethyl-2,2-bis(trimethylsilyl)-, reactions, with pyridine, **6**, 177
Silaethenes, disilyl-substituted, synthesis of, **6**, 709
Silaethenes, kinetically-stabilized, **6**, 709
Silaethylene, 1,1,2-trimethyl-, synthesis of, **3**, 502
Silagermenes
from, diazo compounds, **5**, 927
synthesis of, **5**, 927
Silaketenes
stability of, **5**, 1020
structure of, **5**, 1020
Silaketenimine, synthesis of, **5**, 1085
Silane, 2-acetoxymethyl-3-allyltrimethyl-, reactions, with dipolarophiles, **1**, 287
Silane, acetoxymethyldimethylacetoxy-, synthesis of, **5**, 224
Silane, acetyl-, synthesis of, **5**, 412
Silane, alkyne-, as starting material, for 1,1-disilylalkenes, **4**, 1048
Silane, alkynic acyltriphenyl-, as starting material, for α,β-unsaturated acylsilanes, **5**, 428
Silane, allyl(phenylseleno)-, reactions, with lithium diethylamide, **6**, 136
Silane, allyltrimethyl-
mercuration of, **6**, 369
reactions
with allylic nitro compounds, **1**, 251
with *N*-bromosuccinimide, **4**, 807
with β-nitro sulfides, **1**, 252
synthesis of, **4**, 369
Silane, aminoalkyl-, stability of, **4**, 506
Silane, azidotrimethyl-, reactions, with chromic anhydride, **5**, 368
Silane, (benzoylthio)triphenyl-, synthesis of, **5**, 254
Silane, benzoyltrimethyl-, reactions, with potassium fluoride, **5**, 437
Silane, benzylthiotrimethyl-, reactions, with *t*-butyllithium, **4**, 368
Silane, benzyl(trimethyl)-
bromination of, **4**, 138
reactions, with *n*-butyllithium, **2**, 1005
Silane, bis(chloromethyl)dimethyl-
as starting material
for bis(lithiomethyl)dimethylsilane, **4**, 602, 652
for disilacyclobutanes, **4**, 605
Silane, bis(diethylamino)dimethyl-, reactions, with ethanethiol, **2**, 263
Silane, bis(lithiomethyl)dimethyl-, from, bis(chloromethyl)dimethylsilane, **4**, 602, 652
Silane, bis(trifluoromethyl)-, synthesis of, **6**, 243
Silane, bis(trimethylsilyl)methyl-, for the reduction of chloroalkanes, **1**, 7
Silane, bromo-, reactions, with *S*-tributylstannyl thioacetate, **5**, 537
Silane, (α-bromobenzyl)trimethyl-, synthesis of, **4**, 138
Silane, (bromoethynyl)trimethyl-, reactions, with copper(I) alkynides, **1**, 1019
Silane, bromomethyl-, as starting material, for thiocarbonyl ylides, **4**, 369
Silane, bromotrimethyl-
loss of, in the synthesis of arsoranes, **4**, 597
reactions, with methyl ester, **5**, 404
synthesis of, **2**, 538
Silane, (1-bromovinyl)trimethyl-
displacement of bromine from, **4**, 946
reactions
with benzeneselenenyl bromide, **4**, 947
with trialkylstannyl phenyl sulfide, **4**, 946
as starting material, for Grignard reagents, **4**, 947
Silane, (but-2-en-1-ynyl)trimethyl-, reactions, with areneselenyl chlorides, **4**, 948
Silane, *t*-butylchlorodiphenyl-, use in the protection of amines, **2**, 543
Silane, *t*-butyldimethoxy(trimethylsilylmethyl)-, synthesis of, **4**, 602
Silane, *t*-butyldimethyl-, reactions, with palladium, **5**, 225
Silane, *N*-*t*-butyl *o*-(dimethylaminomethyl)benzimidoyl-, synthesis of, **5**, 793
Silane, *t*-butyldimethylchloro-, synthesis of, **2**, 538
Silane, *t*-butyldimethyl(vinyl)-, synthesis of, predicted, **4**, 611
Silane, *t*-butyldiphenylchloro-, synthesis of, **2**, 538
Silane, *n*-butyl(trimethyl)-, metallation of, **4**, 650
Silane, chloro-
elimination of, **1**, 982
reactions, with isopropylamine, **2**, 417
Silane, chloro(chloromethyl)dimethyl-
reactions, with sodium, **4**, 609
use in the synthesis of carbodiimides, **5**, 1074
Silane, chlorodimethyl-, reactions, with 2,3-dimethylbut-2-ene, **2**, 538
Silane, chlorodimethylvinyl-, hydrosilylation of, **2**, 538
Silane, chloro(dimethyl)vinyl-, reactions, with *t*-butyllithium, **4**, 611
Silane, chloroethyl-, reactions, with ylides, **5**, 913
Silane, (α-chloroethyl)triethyl-, synthesis of, **4**, 137
Silane, [chloro(methoxy)methyl]trimethyl-, from, [methoxy(methylthio)methyl]trimethylsilane, **6**, 61
Silane, (chloromethyl)cyclohexyl(phenyl)-, synthesis of, **4**, 136
Silane, chloromethyldimethylacetoxy-, synthesis of, **5**, 224
Silane, chloromethyldimethylchloro-, reactions, with sodium acetate, **5**, 224
Silane, chloromethyl(dimethyl)phenyl-, reactions, with dimethylphenylsilyllithium, **4**, 609
Silane, chloro(methyl)phenyl-, reactions, with chloromethyllithium, **4**, 136
Silane, chloro(methyl)phenyl(vinyl)-, reactions, with *t*-butyllithium, **4**, 611
Silane, [chloro(methylthio)methyl]trimethyl-, synthesis of, **6**, 61
Silane, chloromethyltrimethyl-
reactions
with dimethylbismuthinosodium, **4**, 599
with formamides, **3**, 711
with lithium, **4**, 602
with trialkylarsines, **4**, 597
with tributylstannyllithium, **4**, 660
with trimethylarsine, **4**, 597
as starting material, for epoxysilanes, **4**, 355
synthesis of, **2**, 537
Silane, chlorotrialkyl-
reactions
with mercury(II) cyanides, **5**, 1135
with metalloaldimines, **5**, 792
with thallium(III) cyanides, **5**, 1135
use in the synthesis of carbodiimides, **5**, 1074
Silane, chlorotriethyl-, synthesis of, **2**, 538
Silane, chlorotriisopropyl-
reactions, with bis(trimethylstannyl)diazomethane, **5**, 1148
synthesis of, **2**, 538
Silane, chlorotrimethyl-
elimination of, for the synthesis of phosphaalkynes, **5**, 1154

reactions
with acylimidazoles, **4**, 933
with alkali metal cyanides, **5**, 1134
with alkyl phenyl sulfoxides, **4**, 945
with azidophosphines, **2**, 470
with dihydrothiazine, **4**, 370
with dimethyltetraphenylcarbodiphosphorane, **6**, 370
with disulfides, **2**, 265
with hydrogen cyanide, **5**, 1134, 1135
with imidoyl chlorides, **5**, 793
with imidoyllithiums, **5**, 792
with magnesium, **2**, 473
with 1-methoxy-1-vinyllithium, **4**, 929
with methyl benzene sulfinate, **4**, 370
with phenylselenide anions, **2**, 542
with sodium sulfide, **5**, 582
with thiocarboxylic acids, **5**, 537
with thionylanilide, **4**, 370
with trimethylsilyldiazomethane, **5**, 1148
with trisdialkylaminophosphine, **2**, 467
with vinylidene isocyanides, **4**, 1014
as starting material
for aryl(trimethylsilyl)phosphines, **2**, 866
for Grignard reagents, **4**, 370
synthesis of, **2**, 538
thermal elimination of, **6**, 688
use in the synthesis of alkylzincs, **2**, 590
Silane, chlorotriphenyl-, reactions, with lithium, **2**, 545
Silane, α-chlorovinyl-, as starting material, for silenes, **4**, 611
Silane, cinnamyltrimethyl-, reactions, with oxovanadiums, **1**, 290
Silane, cyano(methylthio)methyltrimethyl-, reactions, with carbonyls, **3**, 654
Silane, cyanotrimethyl-
reactions
with acetonitrile, **2**, 305
with aldonitrones, **3**, 417
with allylic nitro compounds, **1**, 252
Silane, *N*-cyclohexylformimidoyltrimethyl-, from, cyclohexylisonitrile, **5**, 792
Silane, cyclopentadienyltrimethyl-, as starting material, for cyclopentadienyl(dimethyl)arsine, **2**, 484
Silane, cyclopropyltrimethyl-, as starting material, for cyclopropyl ketones, **1**, 284
Silane, diazomethyltrimethyl-
synthesis of, **5**, 873
use in the synthesis of α-diazosilicons, **5**, 873
Silane, dichloro(chloromethyl)[trimethylsilyl]methyl-, reactions, with acetic acid, **5**, 224
Silane, dichlorodimethyl-
reactions
with *t*-butyllithium, **2**, 538
with 2-lithio-2-methyl-1,3-dithiane, **6**, 127
with methylenetrimethylphosphorane, **6**, 702
with trimethylsilylcyclopentadienyllithium, **4**, 603
synthesis of, **2**, 538, 544
Silane, dichlorodiphenyl-, reactions, with *t*-butyllithium, **2**, 538
Silane, di(chloromethyl)dimethyl-, use in the synthesis of α-germylsilylalkanes, **4**, 639
Silane, diethyl-
reactions, with 1-phenoxyprop-1-yne, **4**, 933
for the synthesis of bismethylene diruthenium complexes, **4**, 679
Silane, *N,N*-diethylcarbamoyl-, synthesis of, **6**, 515
Silane, diiodo-
reactions, with carboxylic acids, **5**, 20
use in the synthesis of acid iodides, **5**, 20, 20
Silane, diiododimethyl-, as reagent, for the hydrolysis of acetals, **3**, 123
Silane, dimethylbis(pentafluorophenyl)-, reactions, with *N*-acetylacetamide, **5**, 719
Silane, dimethylchloro-, as starting material, for tris(dimethylsilyl)methanes, **6**, 174
Silane, dimethyldiisocyano-, synthesis of, **3**, 711
Silane, dimethylphenyl-, reactions, with alkynes, **3**, 77
Silane, diphenyl-
as radical-based reducing agent, **2**, 536
for the reduction of thionoethers, **1**, 17
use in the synthesis of alkylphosphines, **2**, 428
use in the synthesis of amines, **2**, 314
Silane, diphenyldiazido-, reactions, with carbonyl acid chlorides, **5**, 992
Silane, ethylthiotrimethyl-, reactions, with primary nitro compounds, **5**, 829
Silane, ethynyltrimethyl-, reactions, with sodium triethylborohydride, **2**, 1088
Silane, fluoro-, reactions, with methyllithium, **6**, 392
Silane, fluorotrimethyl-, from, difluorotrimethylphosphorane, **5**, 1093
Silane, formyl triisopropyl-, synthesis of, **5**, 422
Silane, formyl trimethyl-
in situ generated, **5**, 413
as intermediate, in the synthesis of acylsilanes, **5**, 422
synthesis of, **5**, 421
Silane, formyl triphenyl-, synthesis of, **5**, 422
Silane, formyl tris(trimethylsilyl)-, synthesis of, **5**, 422
Silane, fulmidotrimethyl-, hydrolysis of, **3**, 685
Silane, (1-germylpropargyl)-, bromodesilylation of, **4**, 811
Silane, *syn*-β-hydroxyl acyl-, synthesis of, **5**, 415
Silane, iodo-, as starting material, for dialkylaminosilanes, **2**, 416
Silane, α-iodoacyl-, as starting material, for α-phosphonoacylsilane, **5**, 426
Silane, iodoethynyltrimethyl-, use in the synthesis of alkynes, **1**, 1018
Silane, (iodomethyl)trimethyl-
reactions, with lithium phenyl selenide, **4**, 377
synthesis of, **4**, 139
Silane, iodotrimethyl-
reactions
with arylchlorophosphoranes, **2**, 837
with ethoxyalkynes, **2**, 941
with 1-ethoxy-1-octyne, **4**, 765
with methyl ester, **5**, 404
as reagent
for acetal hydrol, **3**, 9
for the hydrolysis of acetals, **3**, 123
synthesis of, **2**, 538
use in the synthesis of phosphonic acids, **2**, 847
Silane, isoselenocyanatotrimethyl-, synthesis of, **2**, 544
Silane, isothiocyanatotrimethyl-, reactions, with aldehydes, **2**, 544
Silane, β-keto-, reactions, with methyllithium, **1**, 761
Silane, lithioaminofluoro-, reactions, with trimethylsilyl isocyanate, **5**, 1083
Silane, [methoxy(methylthio)methyl]trimethyl-, as starting material, for [chloro(methoxy)methyl]trimethylsilane, **6**, 61
Silane, [methoxy(phenylthio)methyl]trimethyl-
reactions
with alkenes, **4**, 370
with arenes, **4**, 370
Silane, methoxytrimethyl-
reactions, with silenes, **5**, 925
as starting material, for 2-methoxy-2,4,4-trimethyl-2,4-disilapentane, **4**, 602
Silane, α-methyl-α-chlorotrimethyl-, as starting material, for epoxysilanes, **4**, 355
Silane, (methylseleno)trimethyl-
reactions
with arsenous(III) chloride, **2**, 293
with phosphorus(III) chloride, **2**, 293
Silane, methylselenyl-, reactions, with phenyldihalophosphines, **2**, 857
Silane, (methylthiomethyl)trimethyl-, chlorination of, **6**, 61
Silane, methyl trichloro-, reactions, with sodium iodide, **5**, 39

Silane, phenyl-
as radical-based reducing agent, **2**, 536
use in the synthesis of tertiary phosphines, **2**, 441
Silane, phenyldimethyl-, as reducing agent, use in synthesis of acylhydroxylamines, **5**, 213
Silane, (phenylseleno)*t*-butyldimethyl-, synthesis of, **2**, 542
Silane, (phenylseleno)*t*-butyldiphenyl-, synthesis of, **2**, 542
Silane, (phenylseleno)triethyl-, synthesis of, **2**, 542
Silane, (phenylseleno)trimethyl-
reactions, with aldehydes, **2**, 542
synthesis of, **2**, 542
Silane, phenylthiotrimethyl-
reactions
with acylstannanes, **4**, 951
with primary nitro compounds, **5**, 829
use in the synthesis of thioglycosides, **4**, 224
Silane, 1-phenylthiovinyl-, reactions, with titanium tetrachloride, **4**, 369
Silane, α-phosphonoacyl-
from, α-iodoacylsilanc, **5**, 426
as starting material, for α,β-unsaturated acylsilanes, **5**, 426
Silane, propenoyltrimethyl-, reactions, with nucleophiles, **5**, 416
Silane, propionyltriphenyl-, synthesis of, **5**, 412
Silane, (selenobenzoyloxy)trimethyl-, synthesis of, **5**, 255
Silane, tetrabromo-, reactions, with diazomethane, **4**, 140
Silane, tetrachloro-
as catalyst, for the synthesis of dithioacetals, **4**, 246
reactions
with diazomethane, **4**, 140
with di-Grignard reagents, **4**, 608
Silane, tetra-2-(2-chloroethoxy)-, synthesis of, **2**, 108
Silane, 1,1,2,2-tetrafluoroethyl-, synthesis of, **6**, 61
Silane, tetraisocyanato-, reactions, with phosphoryl fluoride, **5**, 1018
Silane, tetrakis(*t*-butylamino)-, synthesis of, **2**, 417
Silane, tetrakis(trichloromethyl)-, synthesis of, **6**, 243
Silane, tetrakis(trimethylsilyl)-
as radical-based reducing agent, **2**, 545
synthesis of, **2**, 545
Silane, tetramethyl-
cleavage of, **1**, 62
partial deprotonation of, **4**, 649
pyrolysis of, **6**, 383
reactions
with *n*-butyllithiumpentamethyldiethylenetriamine, **4**, 602
with thoracyclobutane, **4**, 658
Silane, thexyldimethylchloro-, synthesis of, **2**, 537, 538
Silane, thiovinyl-, reactions, with α,β-unsaturated acid chloride, **3**, 253
Silane, trichloro-
reactions
with enamines, **4**, 510
with ethanol, **2**, 108, 108
with lithium, **2**, 545
with 1,1,3,3-tetramethyldisilacyclobutane, **4**, 617
for the reduction of benzoic acid, **1**, 21
for the reduction of nitrones, **2**, 344
as starting material
for bis(dialkylamino)silanes, **2**, 416
for dialkylaminosilanes, **2**, 416
for hydridosilanes, **2**, 535
synthesis of, **2**, 535
use in the synthesis of tertiary phosphines, **2**, 440
Silane, trichloro(chloromethyl)-
reactions
with germanium, **4**, 639
with magnesium, **4**, 654
as starting material, for Grignard reagents, **4**, 606
Silane, trichloro(dichlorofluoromethyl)-, as starting material, for α-fluoromethylsilanes, **4**, 136
Silane, trichloromethyl-, synthesis of, **2**, 538
Silane, trichloromethyltrimethyl-
reactions, with aldehydes, **6**, 26
as starting material, for trichloromethyl anions, **6**, 24
Silane, triethoxy-
reactions
with 1-alkyl-4-aryl-1-aza-1,3-dienes, **2**, 813
with esters, **2**, 82
synthesis of, **2**, 108
Silane, triethyl-
as radical-based reducing agent, **2**, 536
reactions
with *N*-alkylnitrilium ions, **3**, 86
with aluminum tribromide, **1**, 61
with boron trifluoride, **1**, 13
with chlorobenzene, **1**, 428
with vinyldipropargylamine, **2**, 920
for the reduction of aldehydes, **1**, 18
for the reduction of alkyl halides, **1**, 5
for the reduction of benzylic alcohols, **1**, 13
for the reduction of chloroalkanes, **1**, 7
for the reduction of ketones, **1**, 20
for the reduction of oxiranes, **1**, 13
for the reduction of thiol esters, **3**, 85
for the reduction of thionoethers, **1**, 17
use in the synthesis of acyl aldehydes, **3**, 83
Silane, (trifluoroacetylthio)trimethyl-, synthesis of, **5**, 254
Silane, trifluoromethyl-, synthesis of, **6**, 243
Silane, (trifluoromethylthio)-, reactions, with hydrogen iodide, **6**, 232
Silane, trifluoromethyltriphenyl-, reactions, with *n*-butyllithium, **4**, 735
Silane, triisopropyl-, reactions, with copper(II) chloride, **2**, 538
Silane, trimethyl-
reactions, with isonitriles, **5**, 792
as reducing agent, **1**, 202
Silane, trimethylazido-, as starting material, for imidoylsilicons, **5**, 794
Silane, trimethylchloro-
reactions
with acetamides, **5**, 719
with amides, **5**, 718
with β-dialkylaminopropionic acid, **5**, 863
with dichloronitrosoalkanes, **5**, 669
with isocyanates, **5**, 1016
with methanetrisulfonyl trifluoride, **6**, 317
with organotin imines, **2**, 813
for the silylation of trithioorthoformates, **6**, 317
Silane, trimethyldiethyl-, use in the synthesis of *N*-silylimidoyls, **5**, 863
Silane, trimethylgermyldimethylphenyl-, reactions, with aldehydes, **4**, 361
Silane, trimethyliodo-, reactions, with silver cyanide, **3**, 711
Silane, trimethylisocyano-, synthesis of, **3**, 711
Silane, trimethylmethoxy-, synthesis of, **4**, 610
Silane, trimethylpentafluorophenyl-, as silylating reagent, **5**, 719
Silane, trimethylsilyl(dichloromethyl)-, as starting material, for 1,3-digermacyclobutane, **6**, 188
Silane, trimethyltrifluoromethyl-
displacement of trifluoromethyl from, **6**, 22
synthesis of, **6**, 13
Silane, trimethylvinyl-
reactions
with arenesulfonyl chlorides, **4**, 942
with ethyllithium, **1**, 761
Silane, trimethyl(vinyl)-
reactions
with Grignard reagents, **4**, 654
with organolithiums, **4**, 651
with tetrachlorodiborane, **4**, 644
Silane, triphenyl-
for the conjugate reduction of enones, **3**, 128

as radical-based reducing agent, **2**, 536
Silane, triphenylchloro-, reactions, with lithium bromoethynylides, **2**, 1017
Silane, triphenyl-(*p*-tolyl)-, as starting material, for *p*-(triphenylsilyl)benzoic acid, **5**, 115
Silane, triphenyl(vinyl)-, reactions, with organolithiums, **4**, 651
Silane, tris(isopropylthio)-, synthesis of, **2**, 541
Silane, tris(methylthio)-, synthesis of, **2**, 541
Silane, tris(trimethylsilyl)- (TTMSS)
as radical-based reducing agent, **2**, 536
reactions
with isocyanides, **5**, 1135
with tetrachloromethane, **2**, 541
as reagent, for mediating radical additions, **1**, 331
for the reduction of bromoalkanes, **1**, 9
for the reduction of chloroalkanes, **1**, 7
for the reduction of thionoethers, **1**, 17
synthesis of, **2**, 545
use in radical cyclizations, **1**, 342
use in the synthesis of acyl radicals, **1**, 933
use in the synthesis of radicals, **1**, 936, 943
use in the synthesis of thiosilanes, **2**, 265
Silane, (*E*)-vinyl-1,2-dimethyl-1-trimethyl-, reactions, with platinum complexes, **1**, 63
Silane, vinyltriethyl-, reactions, with trimethylsilyl azide, **2**, 813
Silane, vinyltrimethyl-
hydroboration of, **4**, 356, 643
reactions
with arylsulfenyl chlorides, **4**, 369
with *t*-butyllithium, **4**, 606
Silaneamines, synthesis of, **6**, 163
Silanecarboximidamide, synthesis of, **6**, 661
Silaneimine, reactions, with dimethylethylamine, **6**, 163
Silane nitrones, allyl-, cyclization of, **2**, 326
Silanes
chlorination of, **4**, 137
cleavage of, **1**, 447
elimination from, **1**, 367
from, methoxy(phenylthio)methane, **4**, 370
photolysis of, **4**, 1048
reactions
with carbenes, **4**, 137
with dichlorocarbene, **6**, 60
with diphenylketene, **2**, 1044
with isonitriles, **5**, 792
with methylidyne tricobaltnonacarbonyl complexes, **6**, 401
with organolithiums, **2**, 536
with organomagnesiums, **2**, 536
with rhodium catalysts, **4**, 831
with silver cyanamides, **5**, 1076
with transition metal catalysts, **1**, 86
with trichloroacetaldehyde, **6**, 27
with trifluoroacetic acid, **1**, 281
with vinyl triflates, **1**, 490
for the reduction of iodoalkanes, **1**, 10
for the reduction of tertiary phosphine oxides, **2**, 440
for the reduction of thionoethers, **1**, 17
as starting materials, for carbonyl ylides, **1**, 366
synthesis of, **5**, 409
use in allylation reactions, **2**, 74
use in the synthesis of alkenes, **1**, 515, 763
use in the synthesis of secondary phosphines, **2**, 431
use in the synthesis of tertiary phosphines, **2**, 827
Silanes, acyl-
asymmetric reduction of, **2**, 58; **4**, 353
catalytic hydrogenation of, for the synthesis of aldehydes, **3**, 19
desilylation of, **5**, 437
electrochemical oxidation of, **5**, 343
enolization of, **4**, 930
from
acetylenylcarbinols, **5**, 413
acyl metallics, **5**, 417
aldehydes, **5**, 410
α-hydroxysilanes, **5**, 412
silyl metallics, **5**, 417
α-silyloxyallylsilanes, **5**, 419
silyloxycarbenes, **5**, 421
trimethylsilyl lithiobis(methylthio)methane, **5**, 411
organometallic addition to, **4**, 353
reactions
with allenyllithiums, **1**, 1057
with bromomagnesium 3-tetrahydropyranyloxy-prop-1-ynide, **1**, 985
with cyanide, **4**, 353
with hydrogen sulfide, **4**, 371
with lithium, **1**, 985
with nucleophiles, **4**, 930
with propargylzincs, **2**, 74
rearrangement of, **1**, 283
reduction of, **4**, 353
reviews, **5**, 409
as starting materials
for *N*-acyl sulfonamides, **5**, 343
for alkenylsilanes, **2**, 936
for benzocyclobutane, **1**, 238
for α-hydroxysilanes, **4**, 353
for oxasilenes, **5**, 923
for silenes, **5**, 923, 924
stereoselective addition to, **4**, 354
synthesis of, **2**, 902; **4**, 929; **5**, 409; **6**, 520
via Claisen rearrangement, **5**, 421
via reductive silylation, **5**, 419
thermolysis of, **5**, 924
Silanes, acyl-, alkyl enol ethers, synthesis of, **5**, 416
Silanes, *N*-acyl-α-amino-, synthesis of, **4**, 513
Silanes, acyl-, boryl enol ethers, synthesis of, **5**, 417
Silanes, acyl-, enol ethers, **5**, 413
synthesis of, **5**, 414
Silanes, acyloxy-
from, carboxylic acids, **5**, 223
as intermediates, for the synthesis of carboxylic acids, **5**, 222
synthesis of, **5**, 222
Silanes, α-acyloxy-, synthesis of, **4**, 352
Silanes, acyl-, silyl enol ethers, synthesis of, **5**, 414
Silanes, acyltrimethyl-, synthesis of, **5**, 418
Silanes, alkanethiotrimethyl-, reactions, with benzoate esters, **6**, 84
Silanes, alkanoyl-, synthesis of, **5**, 410
Silanes, alkenyl-
dehydrosilylation of, **1**, 1080
elimination of, **1**, 1080
from
acylsilanes, **2**, 936
alkenyl metals, **2**, 907
alkynes, **2**, 913
alkynylsilanes, **2**, 921
allenes, **2**, 929
allenylsilanes, **2**, 930
allylsilanes, **2**, 931
arenesulfonylhydrazones, **2**, 911
α,α-bis(silyl)carbanions, **2**, 936
α-bromovinylsilanes, **2**, 931
β-bromovinylsilanes, **2**, 931
carbonyls, **2**, 936
cyclopentadienyltitanium complexes, **2**, 936
electrophilic silicons, **2**, 907
β-hydroxy-α-silyl-α-selenides, **2**, 936
ketenes, **2**, 930
silyl epoxides, **2**, 939
thioacetals, **2**, 936
from vinyl and allylsilanes

via palladium-catalyzed rearrangements, **2**, 932
via S_N2 substitution, **2**, 932
as intermediates, **1**, 535
isomerization of, **4**, 808
oxidation of, **3**, 165
as starting materials
for alkynes, **1**, 1080
for α-sulfenyl aldehydes, **3**, 36
substitution of, **1**, 490
synthesis of, **2**, 907
via halogen–metal exchange of alkenyl metals, **2**, 909
via hydrogen–metal exchange of alkenyl metals, **2**, 907
via silylalkenation, **2**, 936
via silylmetallation, **2**, 929
via transmetallation of alkenyl metals, **2**, 910
Silanes, (*E*)-alkenyl-, osmium-catalyzed oxidation of, **5**, 424
Silanes, (*Z*)-alkenyl-, synthesis of, **1**, 512
Silanes, alkoxy-
deprotonation of, **4**, 358
oxidation of, **3**, 32
reactions
with hydrosilanes, **2**, 942
with phenyllithium, **2**, 942
Silanes, α-alkoxy-
from, halomethylsilanes, **4**, 352
synthesis of, **4**, 352
Silanes, 2,3-*syn*-alkoxy acyl-, reactions, with allyltributyltin, **2**, 79
Silanes, alkoxy(alkyl)-, deprotonation of, **4**, 650
Silanes, alkoxymethylthiomethyl-
reactions
with fluoride, **1**, 286
with silyl enol ethers, **1**, 286
Silanes, *syn*-α-alkoxy-β-(siloxy)acyl-, diastereoselective addition to, **4**, 354
Silanes, alkoxytrimethyl-
reactions, with thiophosgene, **6**, 534
as starting materials, for chlorothioformates, **6**, 533
Silanes, alkyl-
deprotonation of, **4**, 646, 649
direct deprotonation of, **1**, 160
synthesis of, **2**, 535
Silanes, alkylchlorohydrido-, reactions, with lead(II) cyanamide, **5**, 1081
Silanes, alkylhalo-, as starting materials, for silyl cyanides, **5**, 1134
Silanes, alkylthiotrimethyl-, reactions, with aldehydes, **2**, 541
Silanes, alkyltrimethyl-, cleavage of, **2**, 537
Silanes, alkynic, hydrolysis of, **1**, 963
Silanes, alkynic acyl-, reactions, with tributylstannyl cuprate, **5**, 428
Silanes, α,β-alkynic acyl-, from, allenyl ethers, **5**, 427
Silanes, alkynyl-
allylzincation of, **2**, 925
bromination of, **4**, 807; **6**, 60
carboalumination of, **2**, 925
carbocupration of, **2**, 924
carbomagnesiation of, **2**, 924
carbometallation of, **2**, 924; **4**, 809
cyclizations of, **2**, 926
[2+2] cycloaddition of, **2**, 927
cycloadditions of, **2**, 927
desilylation of, **1**, 999
as dienophiles, **2**, 927
double silylation of, **2**, 923
Friedel–Crafts alkylation of, **1**, 1003
from
alkynylsilanes, **2**, 1092
furans, **2**, 1091
propargylstannanes, **2**, 1090
halogenation of, **2**, 929
hydride reduction of, **2**, 921
hydroalumination of, **2**, 922, 967; **4**, 808
hydroboration of, **1**, 161; **2**, 922
hydroboration–oxidation of, **5**, 417
hydrocyanation of, **2**, 928
hydroformylation of, **2**, 928
palladium-catalyzed, **2**, 929
hydrohalogenation, **2**, 929
hydromagnesiation of, **2**, 923
hydrometallation of, **2**, 921
hydrosilylation of, **2**, 923; **4**, 1044
hydrozirconation of, **2**, 923
reactions
with acid chlorides, **1**, 1045
with alkylaluminum chlorides, **2**, 925
with allylic zincs, **2**, 925
with boron trifluoride etherate, **4**, 808
with chlorosulfonyl isocyanate, **1**, 1043
with electrophiles, **1**, 1138
with iodosylbenzene, **4**, 808
with tertiary alkyl halides, **1**, 1003
semihydrogenation of, **2**, 921
silylmetallation of, **2**, 923
silylzincation of, **2**, 923
as starting materials
for alkenylsilanes, **2**, 921
for alkynylsilanes, **2**, 1092
for 1,2-bis(silyl)alkene, **2**, 923
for 1,2-bis(silyl)alkenes, **4**, 1044
for 2-silylcyclopentanones, **2**, 926
for silyldichlorocyclobutenones, **2**, 928
Silanes, alkynyl acyl-, from, 1,3-dioxanes, **5**, 424
Silanes, alkynyltrimethyl-
reactions
with acetals, **1**, 1026
with alkenyl triflates, **1**, 1028
with D-glucal acetates, **1**, 1026
Silanes, 1-alkynyltrimethyl-
reactions
with *S*-ethyl carbothioates, **1**, 1045
with thiol esters, **1**, 1045
Silanes, allenyl-
from
alkoxyallenes, **2**, 939
allenes, **2**, 939
allenyl selenide, **2**, 939
allenylsilanes, **2**, 939
organocoppers, **2**, 940
propargyl ethenes, **2**, 939
propargyl sulfides, **2**, 939
1-silyl-1-alkynes, **2**, 940
1-silyl-2-stannyl-3-hydroxy-1-alkenes, **2**, 940
trimethylsilylpropargyl ethers, **2**, 940
reactions
with aldehydes, **1**, 1056, 1057
with ketones, **1**, 1056
with α,β-unsaturated acylsilanes, **3**, 272
reduction of, **1**, 459
as starting materials
for alkenylsilanes, **2**, 930
for allenylsilanes, **2**, 939
synthesis of, **1**, 963; **2**, 939
via 1,6-addition to enynes, **2**, 940
Silanes, α-allenyl-, synthesis of, **1**, 664
Silanes, allenyl acyl-, from, cumulenes, **5**, 427
Silanes, allyl-
acylation of, **3**, 140, 213
for the alkylation of oxiranes, **1**, 213
for the allylation of oxetanes, **1**, 229
arylation of, **1**, 490
cyclization of, **1**, 216
deprotonation of, **4**, 648
electrophilic addition of, **1**, 295
from, vinylsilanes, **1**, 652
Michael addition of, **3**, 624

reactions
with cobalt complexes, **1**, 199
with nitrile oxides, **3**, 142
with nitrosium tetrafluoroborate, **3**, 160
with ortho esters, **4**, 192
with vinyl epoxides, **1**, 226
as starting materials, for alkenylsilanes, **2**, 931
synthesis of, **1**, 565
via hydroxysilane elimination, **1**, 652
thermal 1,3-rearrangement of, **1**, 838
use in the synthesis of unsaturated ketones, **3**, 140
Silanes, allylic-, use in the synthesis of allylic selenocyanates, **5**, 1115
Silanes, allyltrialkoxy-
for the allylation of carbonyls, **2**, 77
reactions, with 2-hydroxy ketones, **2**, 78
Silanes, allyltrialkyl-, use in the synthesis of alcohols, **2**, 70
Silanes, allyltrichloro-, reactions, with aldehydes, **2**, 78
Silanes, allyltrimethyl-, synthesis of, **1**, 593
Silanes, amino-
acylation of, **4**, 513
reactions, with aldehydes, **4**, 509
Silanes, α-amino-
chemistry of, **4**, 506
from
α-aminosilanes, **4**, 508
cis-oxiranes, **4**, 508
Michael reactions of, **4**, 510
reactions
with benzynes, **4**, 508
with dimethyl ethynedicarboxylate, **4**, 510
as starting materials
for α-aminosilanes, **4**, 508
for silylmethyl imines, **4**, 517
synthesis of, **4**, 506
Silanes, amino bistrimethyl-, reactions, with aminosulfenyl chlorides, **2**, 361
Silanes, aminomethyl-, stability of, **4**, 506
Silanes, amino propargyl trimethyl-, as starting materials, for vinylidene cyclic amines, **1**, 978
Silanes, aromatic, as sources of alkenylsilanes, **2**, 937
Silanes, aroyl-, synthesis of, **5**, 410, 419
Silanes, arsino-
reactions
with chlorotrimethylgermane, **2**, 509
with gallium trihalides, **2**, 511
with indium trihalides, **2**, 511
Silanes, aryl-
reactions
with carboxylic acids, **1**, 448
with mineral acids, **1**, 448
reactivity of, **1**, 447
reduction of, **1**, 99, 447
reduction to cycloalkenylsilanes, **2**, 937
as starting materials
for arenes, **1**, 447
for aryl bromides, **2**, 627
for aryl iodides, **2**, 632
for biaryls, **1**, 489
for lithium compounds, **6**, 392
synthesis of, **2**, 941
via cycloaddition reactions, **2**, 943
via electrophilic substitution, **2**, 942
via silylation of aryl metals, **2**, 941
via transition metal catalysis, **2**, 912
via transition metal-catalyzed reactions, **2**, 944
via Würtz–Fittig type coupling, **2**, 941
use in Pictet–Spengler reactions, **3**, 419
Silanes, arylthiotrimethyl-, reactions, with aldehydes, **2**, 541
Silanes, α-arylthiovinyl-, electrophilic sulfenylation of, **4**, 368
Silanes, azo-, reactions, with tetrachloromethane, **6**, 608
Silanes, benzenoid aryl-, synthesis of, **2**, 945
Silanes, benzoyl-
from, benzylsilanes, **5**, 410
as starting materials, for α-silylstyrenes, **2**, 936
synthesis of, **5**, 421
Silanes, benzoyl triphenyl-, synthesis of, **5**, 410
Silanes, benzyl-
bromination of, **4**, 138
deprotonation of, **4**, 647
metallation of, **4**, 653
as starting materials, for benzoylsilanes, **5**, 410
Silanes, benzylic, reactions, with oxovanadiums, **1**, 290
Silanes, benzyltrimethyl-, reactions, with silver sulfate, **1**, 289
Silanes, *trans*-bis-1-alkynyl(pthalocyaninato)-, synthesis of, **2**, 1089
Silanes, bis(dialkylamino)-
from
amines, **2**, 416
trichlorosilane, **2**, 416
synthesis of, **2**, 416
Silanes, α-boryl-
oxidation of, **4**, 357
as starting materials, for α-hydroxysilanes, **4**, 357
Silanes, bromo-
reactions, with 1-alkynylstannanes, **2**, 1089
synthesis of, **4**, 140
Silanes, α-bromo-, synthesis of, **4**, 136, 138
Silanes, α-bromoacyl-, synthesis of, **4**, 783; **5**, 422, 423
Silanes, (bromomethyl)chloro-, reactions, with nucleophiles, **4**, 140
Silanes, (bromomethyl)oxy-, synthesis of, **4**, 140
Silanes, bromovinyl-, alkylation with cuprates, **2**, 931
Silanes, α-bromovinyl-
palladium(0)-catalyzed coupling, **2**, 931
as starting materials, for alkenylsilanes, **2**, 931
Silanes, β-bromovinyl-
palladium(0)-catalyzed coupling, **2**, 931
as starting materials, for alkenylsilanes, **2**, 931
Silanes, carbamoyl-, synthesis of, **6**, 515
Silanes, carbimino-, as starting materials, for *gem*-aminoborylamino silylalkenes, **4**, 1015
Silanes, α-carboxyacyl-, synthesis of, **5**, 430
Silanes, chloro-
bromination of, **4**, 140
direct synthesis of, **6**, 383
reactions
with alkali metals, **2**, 545
with alkanethiols, **2**, 262
with amides, **2**, 811
with amines, **2**, 811
with arsenides, **2**, 508
with chloroalkenes, **4**, 806
with dialkylphosphinolithium, **2**, 473
with dimethylphenylsilyllithium, **4**, 609
with fluoroalkenes, **4**, 140
with lithium aluminum hydride, **2**, 536
with lithium compounds, **6**, 392
with 1-metalloalkylphosphine oxides, **4**, 570
with 1-metalloalkylphosphines, **4**, 566
with 1-metalloalkylphosphine sulfides, **4**, 570
with 1-metalloalkyl phosphonates, **4**, 570
with metal thiolates, **2**, 540
with organometallic carbon nucleophiles, **6**, 172
with potassium selenocarboxylates, **5**, 541
with silver cyanate, **5**, 1016
with silylphosphines, **2**, 473
with stibides, **2**, 508
with ureas, **5**, 983
with vinyl Grignard reagents, **2**, 910
with vinylpotassium, **2**, 907
reactivity of, **2**, 537
reduction of, **2**, 536
as silylating agents, **2**, 942
as starting materials
for alkyl silyl ethers, **2**, 539

for aryl silyl ethers, **2**, 539
for silyloxycarbene complexes, **6**, 719
Silanes, α-chloro-
from, vinylsilanes, **4**, 137
reactions, with silyl enol ether, **4**, 365
synthesis of, **4**, 136, 137
Silanes, α-chloroacyl-, synthesis of, **5**, 422
Silanes, chloro(chloromethyl)-, reactions, with nucleophiles, **4**, 140
Silanes, 2-(chlorodifluoromethyl)allyl-, reactions, with aldehydes, **2**, 77
Silanes, (chloro)dimethyl-, synthesis of, **4**, 513
Silanes, chloro(α-fluoro)-, synthesis of, **4**, 140
Silanes, chloromethoxy-, use in the synthesis of alkyl silyl acetals, **4**, 202
Silanes, chloromethyl-
reactions
with aromatic amines, **4**, 507
with Grignard reagents, **4**, 137
use in the synthesis of 1,1-bis(silyl)alkanes, **4**, 609
Silanes, chloromethylchloro-, metal-induced coupling of, **6**, 178
Silanes, chloromethyl hydro-, reactions, with 1,2-bis(silyl) ethynes, **4**, 1045
Silanes, (chloromethyl)oxy-, synthesis of, **4**, 140
Silanes, chloromethyltrifluoro-, reactions, with mercury salts, **5**, 1142
Silanes, *m*-chlorophenyl-
lithiation of, **4**, 508
as starting materials, for azasilolines, **4**, 508
Silanes, chlorotrialkyl-
reactions
with alkanethiols, **2**, 262
with ethers, **4**, 358
Silanes, chlorotrimethoxy-, as starting materials, for ethynylmethoxysilanes, **2**, 1090
Silanes, chlorotriorganyl-, reactions, with thiocarboxylic acids, **5**, 254
Silanes, chlorovinyl-, reactions, with *t*-butyllithium, **4**, 611; **5**, 926
Silanes, chloro(vinyl)-, reactions, with organolithiums, **4**, 651
Silanes, α-cholorovinyltrimethyl-, reactions, with bromotrichloromethane, **4**, 758
Silanes, crotyl trifluoro-
for the allylation of carbonyls, **2**, 78
synthesis of, **2**, 78
Silanes, cumulenyl-
from, trimethylsilylethyne, **2**, 940
synthesis of, **2**, 939, 940
Silanes, cyano-, synthesis of, **5**, 1133
Silanes, β-cyano-β-arylalkenyl-, synthesis of, **3**, 660
Silanes, cyclic acyl-, from, 4-bromo-1-butene, **5**, 411
Silanes, cyclic vinyl-, synthesis of, **2**, 920
Silanes, cyclopropyl-, acylation of, **1**, 284
Silanes, cyclopropyl acyl-
from
cyclopropanes, **5**, 428
α,β-unsaturated acylsilanes, **5**, 429
ring-expansion of, **4**, 931
synthesis of, **5**, 412, 418, 428
Silanes, (*Z*)-1,2-dialkylalkenyl-, synthesis of, **1**, 161
Silanes, dialkylamino-
from
amines, **2**, 416
iodosilane, **2**, 416
trichlorosilane, **2**, 416
synthesis of, **2**, 416
Silanes, *N*,*N*-dialkylamino-, reactions, with amines, **2**, 358
Silanes, dialkylaminotrimethyl-, reactions, with sulfur tetrafluoride, **2**, 362
Silanes, dialkyldichloro-, reactions, with lithium phosphides, **2**, 473
Silanes, dialkylhydro-, synthesis of, **5**, 379
Silanes, 3,3-diborylpropyl-, synthesis of, **4**, 632
Silanes, α,β-dibromo-, synthesis of, **4**, 138
Silanes, dibromofluoromethyl-, synthesis of, **6**, 243
Silanes, dichloro-
reactions
with cyanamides, **5**, 1077
with diols, **4**, 342
Silanes, dichloromethyl-, synthesis of, **6**, 60
Silanes, α-dichloromethyl-, synthesis of, **6**, 61
Silanes, dienoyl-, synthesis of, **5**, 428
Silanes, dienyl-
from, thioacetals, **2**, 936
synthesis of, **2**, 939
Silanes, α,α-dihalo-, reduction of, **4**, 138
Silanes, α,α-dihaloalkyl-, synthesis of, **6**, 60
Silanes, β,β-disubstituted alkenoyl-, synthesis of, **5**, 426
Silanes, enamino-, α-alkylation of, **3**, 417
Silanes, enol-
allylation of, **3**, 139
bromination of, **2**, 28
chlorination of, **2**, 22
ethenylation of, **3**, 138
fluorination of, **2**, 15
iodination of, **2**, 34
α-iodination of, **2**, 33
oxidation of, **3**, 163
via Sharpless asymmetric dihydroxylation, **3**, 156
reactions, with Selectfluor reagents, **2**, 16
as starting materials
for aryl ketones, **3**, 144
for α-benzyl ketones, **3**, 145
for α-bromo ketones, **3**, 154
for α-chloro ketones, **3**, 152
for α-fluorocarbonyls, **2**, 16
for α-fluoro ketones, **3**, 148
for halo aldehydes, **3**, 26
for α-hydroxy ketones, **3**, 156
for α-iodo ketones, **3**, 155
Silanes, enynyl-, synthesis of, **2**, 939
Silanes, (*E*)-β-ethoxycarbonylvinyl-, synthesis of, **2**, 929
Silanes, ethynyl-
carbometallation of, **4**, 809
cross-conjugated, synthesis of, **2**, 1091
Silanes, 1-ethynyl-, reactions, with organocoppers, **4**, 940
Silanes, ethynylmethoxy-, from, chlorotrimethoxysilanes, **2**, 1090
Silanes, α-fluoro-
characterization of, **4**, 136
synthesis of, **4**, 137
Silanes, α-fluoro-β-hydroxy-, from, α,β-epoxysilanes, **4**, 137
Silanes, α-fluoromethyl-, from, trichloro(dichlorofluoromethyl)silane, **4**, 136
Silanes, (α-fluoromethyl)-, synthesis of, **4**, 137
Silanes, fluorovinyl-
reactions, with bromine, **1**, 653
reduction of, **1**, 429
Silanes, formamido-, synthesis of, **4**, 516
Silanes, formimidoyl-, from, isonitriles, **5**, 792
Silanes, formyl-
organometallic addition to, **4**, 353
synthesis of, **5**, 421
Silanes, functionalized vinyl-, synthesis of, **2**, 918
Silanes, halo-
reactions
with *t*-butyllithium, **2**, 536
with phosphorus ylides, **4**, 570
with selenium-stabilized carbanions, **4**, 374
with trisilyllithiomethanes, **6**, 379
with trisilylmethyllithiums, **6**, 379
as starting materials
for silyl cyanides, **5**, 1134
for trisilylmethanes, **6**, 174

Silanes, α-halo-
from, α-hydroxysilanes, **4**, 138
reactions, with fluoride, **1**, 282
as starting materials
for cyclopropanes, **1**, 282
for organosilicons, **4**, 136
synthesis of, **4**, 136
Silanes, β-halo-
elimination of, **1**, 652
as starting materials, for alkenes, **1**, 652
Silanes, haloacyl-, synthesis of, **5**, 411
Silanes, α-haloacyl-
as starting materials, for 1-silylketo-2-carbonyl cyclopropanes, **5**, 430
synthesis of, **5**, 422
Silanes, ω-haloacyl-, cyclization of, **2**, 936
Silanes, 2-halo-2-alkenoyl-, reactions, with organolithium, **1**, 985
Silanes, 1-haloalkyl-
Arbuzov reactions of, **4**, 569
reactions
with alkyl phosphinites, **4**, 570
with dialkyl phosphonites, **4**, 570
with metal phosphides, **4**, 566
with phosphines, **4**, 569
with sodium dialkylphosphites, **4**, 569
with trialkyl phosphites, **4**, 569
Silanes, (α-haloalkyl)oxy-, synthesis of, **4**, 140
Silanes, halo(α-haloalkyl)-, synthesis of, **4**, 139
Silanes, halo(halomethyl)-
condensation of, **4**, 608
reactions, with nucleophiles, **4**, 140
Silanes, halomethyl-
reactions
with amines, **4**, 506
with methoxide, **4**, 352
with oxygen nucleophiles, **4**, 352
with selenium nucleophiles, **4**, 377
with stannyl anions, **4**, 660
with sulfur nucleophiles, **4**, 367
as starting materials
for α-alkoxysilanes, **4**, 352
for α-thiosilanes, **4**, 367
use in the synthesis of α-aminosilanes, **4**, 506
Silanes, (halomethyl)trimethyl-, as starting materials, for α-selenosilanes, **4**, 377
Silanes, haloorgano-, reactions, with alkanes, **6**, 173
Silanes, halotrialkyl-
reactions
with potassiummercury(II) cyanide, **5**, 1133
with silver cyanide, **5**, 1133
with trimethylsilyl cyanide, **5**, 1133
Silanes, halovinyl-, as starting materials, for silyl ketones, **3**, 271
Silanes, hydrido-
as byproducts, **2**, 538
from, trichlorosilane, **2**, 535
reactions
with alcohols, **2**, 539
with alkenes, **2**, 536
synthesis of, **2**, 535
uses of, **2**, 536
Silanes, hydro-
radical addition to alkynes, **2**, 913
reactions
with alkanethiols, **2**, 264
with alkoxysilanes, **2**, 942
with alkynes, **2**, 913
with 1,2-bis(trimethylsilyl) ethyne, **4**, 1045
use in the synthesis of primary phosphines, **2**, 821
use in the synthesis of secondary phosphines, **2**, 822
Silanes, hydroxy-, as intermediates, in the synthesis of (*E*)- alkenyl ethers, **2**, 682
Silanes, *threo*-β-hydroxy-
reactions
with boron trifluoride etherate, **1**, 761
with potassium hydride, **1**, 761
Silanes, α-hydroxy-
from
acylsilanes, **4**, 353
α-borylsilanes, **4**, 357
vinylsilanes, **4**, 356
as starting materials
for acylsilanes, **5**, 412
for α-halosilanes, **4**, 138
Silanes, β-hydroxy-
as alkene protecting groups, **1**, 652
elimination of, **1**, 648
stereospecificity of, **1**, 649
from, β-ketosilanes, **1**, 650
as intermediates, **1**, 758
oxidation of
to α-silyl aldehydes, **3**, 51
to α-silyl ketones, **3**, 202
reactions, with *n*-butyllithium, **2**, 547
stability of, **1**, 649
synthesis of, **1**, 649; **4**, 357, 619
via regiospecific ring-opening of epoxysilanes, **1**, 650
Silanes, (α-hydroxyalkyl)-, synthesis of, **5**, 413
Silanes, (α-hydroxyalkyl)trimethyl-, reactions, with methyl(triphenoxy)phosphonium iodide, **4**, 139
Silanes, 2-hydroxycrotyl-
hydroxylation of, **2**, 46
as starting materials, for triols, **2**, 46
Silanes, hydroxymethyl-, from, acetate, **4**, 352
Silanes, 3-hydroxypropynyl-, reduction of, **2**, 921
Silanes, imidoyl-
from
C-chlorobenzylideneanilines, **5**, 793
cyanohydrins, **5**, 794
N-phenylpivalimidoyl chloride, **5**, 793
as starting materials, for *o*-quinodimethanes, **5**, 793
Silanes, α-imino-, synthesis of, **4**, 517
Silanes, α-iodo-, synthesis of, **4**, 139
Silanes, (*E*)-α-iodoalkenyl-
isomerization of, **4**, 808
reactions, with triethylborane, **4**, 811
Silanes, (α-iodoalkyl)-, synthesis of, **4**, 139
Silanes, (iodomethyl)-, synthesis of, **4**, 136
Silanes, iodo trialkyl-, reactions, with mercury bis(ethyl diazoacetate), **3**, 483
Silanes, isocyanato-
exchange reactions of, for the synthesis of isocyanates, **5**, 1006
reactions, with carboxylic acid anhydrides, **5**, 1006
synthesis of, **5**, 1006, 1016
Silanes, (iso)cyano-, use of term, **5**, 1133
Silanes, isoprenyl-, as nucleophilic reagents, **1**, 253
Silanes, β-keto-
organometallics, **1**, 759
reduction of, **1**, 761
as starting materials, for β-hydroxysilanes, **1**, 650
Silanes, α-ketoacyl-, synthesis of, **5**, 423
Silanes, δ-ketoacyl-, synthesis of, **5**, 430
Silanes, α-lithio-
as intermediates, in the synthesis of 1,1-bis(silyl)alkanes, **4**, 611
as starting materials, for α-silyl acyllithium, **5**, 415
Silanes, β-lithio-, synthesis of, **1**, 1091
Silanes, α-metallated, synthesis of, **1**, 761
Silanes, metallo-
reactions
with halomethylamines, **4**, 510
with ylides, **5**, 913
Silanes, 1-metalloalkyl-
reactions

with chlorophosphine oxides, **4**, 569
with chlorophosphines, **4**, 565
Silanes, α-metalloalkyl-, synthesis of, **2**, 1004
Silanes, methoxy-
reactions, with *n*-butyllithium, **6**, 393
reduction of, **4**, 136
Silanes, α-methoxy-, oxidation of, **4**, 197
Silanes, β-methoxy-, thermolysis of, **1**, 654
Silanes, methoxymethyl amino-, synthesis of, **4**, 509
Silanes, methylchloro-
chlorination of, **4**, 140
pyrolysis of, **6**, 173
synthesis of, **6**, 383
Silanes, 4-methyl-substituted aryl-, reduction of, **1**, 99
Silanes, monochloro(α-fluoro)-, synthesis of, **4**, 140
Silanes, oligo-, reactions, with isocyanides, **6**, 384
Silanes, organofluoro-, reactions, with aryl halides, **3**, 290
Silanes, β-oxido-
syn β-elimination of, **1**, 650
synthesis of, **1**, 761
Silanes, β-oxoacyl-, tautomerism of, **4**, 931
Silanes, α-phenylsulfinylacyl-, thermal decomposition of, **5**, 426
Silanes, α-(phenylthio)alkyl-, synthesis of, **4**, 366
Silanes, α-(phenylthio)cyclopropyl-, synthesis of, **4**, 366
Silanes, phenylthiovinyl-, oxidation of, **2**, 909
Silanes, α-(phenylthio)vinyl-, reactions, with organolithiums, **4**, 366
Silanes, phenyltrifluoro-, reactions, with mercury(II) cyanide, **5**, 1142
Silanes, 1-phosphinoalkyl-
oxidation of, **4**, 573
synthesis of, **4**, 565
Silanes, phosphoamino-, acylation of, **5**, 370
Silanes, 1-phosphoranoalkyl-, synthesis of, **4**, 575
Silanes, poly(alkylperoxy)-, as starting materials, for dialkyl peroxides, **2**, 98
Silanes, polyazaheterocyclic, synthesis of, **2**, 944
Silanes, polychloro-, reactions, with hydroxamic acids, **5**, 1001
Silanes, polymeric alkynyl-, synthesis of, **2**, 1089
Silanes, propargyl-
reactions
with acyl iminium cations, **1**, 978
with alkylidene malonates, **1**, 978
with enones, **1**, 978
with iminium salts, **1**, 978
reactivity of, **1**, 977
reduction of, **1**, 512
Silanes, propargylic
desilylation of, **1**, 963
reactions, with acid chlorides, **3**, 230
Silanes, propargyltrialkyl-, reactions, with electrophiles, **1**, 977
Silanes, propargyltrimethyl-, reactions, with aldehydes, **1**, 978
Silanes, propenoyl-, synthesis of, **5**, 428
Silanes, pseudohalo-, reactions, with trisilyllithiomethanes, **6**, 379
Silanes, ring-contracted, synthesis of, **2**, 1089
Silanes, α-seleno-β-nitro-, synthesis of, **4**, 376
Silanes, α-selenyl-, synthesis of, **4**, 374
Silanes, silyl acety-
as starting materials, for α,β-unsaturated acylsilanes, **5**, 426
synthesis of, **5**, 426
Silanes, α-silyloxyallyl-
isomerization of, **5**, 419
as starting materials, for acylsilanes, **5**, 419
Silanes, styryl-, use in Heck reactions, **1**, 490
Silanes, β-substituted, elimination of, **1**, 653
Silanes, α-telluryl-, synthesis of, **4**, 374
Silanes, tetraalkoxy-
as dehydrating agents, for the synthesis of acetals, **4**, 178
synthesis of, **2**, 108, 108
Silanes, tetraalkyl-, stability of, **2**, 537
Silanes, tetrabromo-, synthesis of, **6**, 60
Silanes, tetrakis(dialkylamino)-
from, amines, **2**, 417
synthesis of, **2**, 417
Silanes, tetramethyl-, as starting materials, for carbosilanes, **4**, 617
Silanes, α-thioacyl-, synthesis of, **5**, 426
Silanes, thionoacyloxy-, synthesis of, **5**, 254
Silanes, trialkoxy-, synthesis of, **2**, 108
Silanes, trialkyl-
reactions
with methylene carbenes, **1**, 1144
with silyl esters, **4**, 832
with sulfides, **2**, 265
as reagents, for mediating radical additions, **1**, 331
use in the synthesis of vinyl radicals, **1**, 1129
Silanes, trialkylallyl-
reactions
with aldehydes, **2**, 78
with α,β-unsaturated ketones, **1**, 314
Silanes, trialkylchloro-, use in the synthesis of *N*-silylimidoyls, **5**, 863
Silanes, trialkyl isocyano-, synthesis of, **3**, 711
Silanes, trialkylmethoxy-, reactions, with trimethylacetyl cyanide, **5**, 1133
Silanes, 2-(trialkylsilyloxy)allyl-, as acetone 2,2'-dianion equivalents, **2**, 77
Silanes, trialkylstannyltrialkyl-, reactions, with terminal alkynes, **2**, 916
Silanes, triaryl isocyano-, synthesis of, **3**, 711
Silanes, trichloro(halomethyl)-, synthesis of, **4**, 140
Silanes, trienyl-, from, thioacetals, **2**, 936
Silanes, trifluoro-, cleavage of, **4**, 152
Silanes, β-trifluoroacetyl-, elimination of, **1**, 654
Silanes, trifluorosilyl(vinyl)-, reactions, with bromine, **4**, 138
Silanes, trifluorovinyl-, reactions, with triphenylgermyllithium, **4**, 811
Silanes, trimethylsilyl(vinyl)-, reactions, with bromine, **4**, 138
Silanes, trimethyl(vinyl)-
reactions
with iododifluoroacetates, **4**, 139
with tosyl iodide, **4**, 139
Silanes, tris(alkylthio)-
as radical-based reducing agent, **2**, 536
for the reduction of chloroalkanes, **1**, 7
use in the synthesis of radicals, **1**, 936
Silanes, tris(dialkylamino)-
from, amines, **2**, 417
synthesis of, **2**, 417
Silanes, α,β-unsaturated acyl-
from
alkynic acyltriphenylsilane, **5**, 428
enol ethers, **5**, 425
α-phosphonoacylsilane, **5**, 426
silyl acetysilane, **5**, 426
silyl enol ethers, **5**, 425
reactions, with allenylsilanes, **3**, 272
as starting materials, for cyclopropyl acylsilanes, **5**, 429
synthesis of, **5**, 424
via hydroboration–oxidation, **5**, 425
via Michael addition reactions, **5**, 427
via oxidation reactions, **5**, 425
Silanes, vinyl-
acylation of, **1**, 422; **3**, 271
as anion sources, **2**, 931
carbometallation of, **1**, 161
in cyclization reactions, **1**, 316
desilylation of, **1**, 450
electrophilic selenylation of, **4**, 376
electrophilic sulfenylation of, **4**, 368

epoxidation of, **4**, 357
formylation of, **3**, 58
from
alkenyl metals, **2**, 910
alkynes, **2**, 913
allenes, **2**, 929
ketenes, **2**, 929
silylalkynes, **2**, 921
vinylstannanes, **2**, 915
hydroboration of, **4**, 356, 643
hydrometallation of, **1**, 161
hydrosilylation of, **4**, 617
metathesis of, **2**, 938
nucleophilic addition to, for the synthesis of α-metallo silylalkanes, **4**, 606
organometallic addition to, **6**, 189
oxidation of, **6**, 124
reactions
with acid chlorides, **3**, 211
with alkyllithiums, **6**, 135
with aniline, **4**, 507
with arenediazonium salts, **1**, 490
with arenesulfenyl chlorides, **4**, 138
with aryl halides, **1**, 490
with bromide azide, **4**, 514
with bromine, **4**, 138, 751
with chlorosulfonyl isocyanate, **4**, 514
with electrophiles, **4**, 139
with epoxides, **1**, 1094
with fluoride ions, **1**, 63
with Grignard reagents, **4**, 654
with nucleophiles, **1**, 451
with organolithiums, **1**, 310; **4**, 366, 651
with organometallics, **1**, 759; **6**, 192
with osmium tetroxide, **4**, 357
with sulfonyl isocyanates, **5**, 344
with trifluoromethyl hypochlorite, **4**, 140
with xenon difluoride, **2**, 617
reduction of, **1**, 450
as starting materials
for aldehydes, **3**, 18
for alkenes, **1**, 450
for allylsilanes, **1**, 652
for 1,2-bis(silyl)alkanes, **4**, 617
for 1,1-bissilylethanes, **4**, 617
for α-borylsilylalkanes, **4**, 643
for α-chlorosilanes, **4**, 137
for epoxysilanes, **4**, 357
for α-hydroxysilanes, **4**, 356
for β-nitro-α-silyl selenides, **2**, 284
for α-selenosilanes, **4**, 376
for α-silyl carbanions, **1**, 160
for tetrasilylmethanes, **6**, 392
for vinyl bromides, **2**, 624
for vinyl chlorides, **2**, 617
for vinyl halides, **1**, 653; **2**, 607
for vinyl iodides, **2**, 630
synthesis of, **1**, 161, 440, 676
via hydrosilylation of alkynes, **2**, 913
via transition metal catalysis, **2**, 912
use in Heck reactions, **1**, 490
use in the synthesis of ketones, **3**, 137
vicinal hydroxylation of, **3**, 19
Silanes, vinyl chloro-, reactions, with *t*-butylithium, **3**, 502
Silanes, vinyl-, exocyclic, synthesis of, **2**, 927
Silanes, vinyl-, heterosubstituted, synthesis of, **2**, 908
Silanes, vinylic, synthesis of, **2**, 902
γ-Silanes, synthesis of, **4**, 511
Silanethiol, hexyloxydimethyl-, reactions, with benzoyl chloride, **5**, 538
Silanethiol, triphenyl-
synthesis of, **2**, 541
use in the synthesis of alkanethiols, **2**, 117
Silanol, trimethyl-
elimination of, **1**, 981
for 1-phosphabutatrienes, **3**, 608
reactions
with bromodimethylstibine, **2**, 494
with chloro(dimethyl)arsine, **2**, 494
with 1,1,3,3-tetramethyl-1,3-disilacyclobutane, **4**, 623
Silanols, for the cleavage of 1,3-disilacyclobutanes, **4**, 623
Silanols, alkenyl-, use in Sharpless epoxidation, **1**, 1091
Silanols, alkynyl-, synthesis of, **2**, 1090
1-Silaphenalene, 2,3-dihydro-1,1-dimethyl-, reactions, with butyllithium–tetramethylethylenediamine [1,2-bis(dimethylamino)ethane], **4**, 603
Silaphosphiranes, synthesis of, **4**, 567
Silapinacol rearrangement
for the synthesis of aldehydes, **3**, 19
use in the synthesis of α-silyl aldehydes, **3**, 51
Sila-Pummerer reaction, mechanism, **4**, 222
Silasilenes
from
alkenes, **5**, 926
diazo compounds, **5**, 925
silenes, **5**, 925
synthesis of, **5**, 925
via retro Diels–Alder reaction, **5**, 926
1-Sila-1-stanna-1-alkenes, as starting materials, for α-stannylsilylalkanes, **4**, 661
Silathiacyclopropanes, synthesis of, **2**, 266
Silathianes, alkyl-, synthesis of, **2**, 540
3-Silathietane, 3,3-diethyl-2,4-dimethyl-, pyrolysis of, **2**, 266
Silatranes
ethynyl compounds of, synthesis of, **2**, 1091
structure of, **2**, 547
Silatriene, use in the synthesis of tetrasilylmethanes, **6**, 383
1,3-Silatropic rearrangement, of acylphosphanes, **5**, 878
Silatropy, of acylphosphanes, **5**, 879
Silazane, heptamethyl-, reactions, with trifluoroacetic anhydride, **5**, 377
Silazanes, alkyl-, synthesis of, **2**, 543
Silene, 1-methyl-, from, dimethylsilylene, **3**, 502
Silenes
addition reactions of, **6**, 380
Diels–Alder reactions of, **4**, 611
dimerization of, **5**, 924; **6**, 380
from
acylsilanes, **5**, 923
alkenes, **5**, 926
anthracene adducts, **5**, 926
α-chlorovinylsilane, **4**, 611
diazo compounds, **5**, 925
1,2-disilacyclohexadienes, **4**, 612
divinyldisilanes, **3**, 502
silabicyclo[2.2.2]octadienes, **3**, 502
silacyclobutanes, **3**, 501
silenes, **5**, 925
silylcarbenes, **3**, 502
silylenes, **3**, 502
as intermediates, **3**, 501
reactions
with boron trifluoride, **6**, 387
with methoxytrimethylsilane, **5**, 925
with oxygen, **6**, 382
with quadricyclane, **5**, 925
with silicon tetrafluoride, **6**, 382
with trimethylgermanium chloride, **6**, 387
with trimethyltin chloride, **6**, 395
reviews of, **5**, 923
as starting materials
for 1,1-bis(silyl)alkanes, **4**, 609
for 1,3-disilacyclobutanes, **4**, 610
for 1-germyl-1-silylalkanes, **4**, 639
for silasilenes, **5**, 925
for silenes, **5**, 925

for tetrasilylmethanes, **6**, 380
synthesis of, **3**, 501
via irradiation of acylsilanes, **5**, 924
via pyrolysis of diazo compounds, **5**, 926
via pyrolysis of silacyclobutanes, **4**, 610
via retro Diels–Alder reaction, **5**, 926
via retro dimerization, **5**, 924
via thermolysis of the dimer, **5**, 924
trapping of, **5**, 924
X-ray structure of, **5**, 923, 924
Silenes, dichloro-
as intermediates, in the synthesis of silenes, **5**, 925
rearrangement of, **5**, 926
Silenes, dimethyl-, synthesis of, **5**, 925
Silenes, neopentyl-, synthesis of, **5**, 926
Silenes, α-silyl-, as starting materials, for disiloxane, **4**, 612
Siletanes
cleavage of, via ylides, **5**, 914
reactions, with methylenephosphoranes, **6**, 702
Siletanes, 1-chloro-, reactions, with ylides, **5**, 913
Silica, reactions, with dialkyl carbonates, **2**, 108
Silica gel, as reagent, for the synthesis of *N*-alkylamides, **5**, 276
Silica gel catalysis, for the synthesis of alkenes, **1**, 566
Silicate, ammonium methylpentafluoro-, reactions, with thallium(III) acetate, **5**, 1143
Silicate, tris(diethylamino)sulfoniumtrimethyldifluoro- (TASF)
as methylating reagent, **1**, 490
reactions, with methyl iodide, **1**, 282
use in coupling reactions, **1**, 490
use in the synthesis of alkynes, **1**, 1007
Silicates, allyl, for the allylation of carbonyls, **2**, 77
Silicates, allylic, as intermediates, **1**, 314
Silicates, organofluoro-, synthesis of, **1**, 490
Silicates, tris(dialkylamino)sulfonium difluorotrimethyl, synthesis of, **2**, 362
Silicates, tris(diethylamino)sulfonium difluorotrimethyl, reactions, with 2-silylalkyl halides, **2**, 70
Silicon
Brook-type migration of, **5**, 793
characteristics of, **1**, 159
effect on 5-hexenyl radical cyclization, **1**, 337
ipso substitution of, **1**, 447
reactions
with alkyl chlorides, **2**, 538
with aryl chlorides, **2**, 538
with chloroform, **6**, 173, 267
stabilization by, **4**, 1082
substitution of, **1**, 489
synthesis of functions
bearing a silicon and a boron, **4**, 642
bearing a silicon and a germanium, **4**, 636
bearing three silicons, **6**, 172
bearing two silicons, **4**, 602
one boron function, **6**, 292
one germanium function, **6**, 292
one halogen, **6**, 289
one halogen and one sulfur function, **6**, 282
use in the transposition of ketones, **3**, 126
Silicon, acyl-, synthesis of functions, **5**, **409–433**
Silicon, doubly-bonded, synthesis of functions, **6**, **677–724**
Silicon(IV) isothiocyanate, reactions, with secondary amines, **6**, 572
Silicon alkynides, synthesis of, **1**, 1138
Silicon carbide
fluorination of, **6**, 212
precursors to, **6**, 379
Silicon–carbon bond fission, of tetraalkylsilanes, **2**, 537
Silicon–carbon bonds
cleavage of, **4**, 653
oxidative cleavage of, **2**, 536, 540
Silicon–carbon–silicon bonds, synthesis of, **4**, 602
Silicon chloride, reactions, with silver cyanate, **5**, 1016
Silicon cumulenes, synthesis of, **1**, 988
Silicon disulfide
reactions, with alcohols, **2**, 108
as reagent, for the synthesis of thioketones, **3**, 357, 361
Silicon functions
elimination of, **1**, 1080
substitution of, **1**, 281, 1033
Silicon functions, doubly-bonded, synthesis of, **3**, 501
Silicon groups, elimination of, **1**, 648
Silicon halides
reactions
with lithiums, **6**, 388
with silver cyanate, **5**, 1016
as starting materials, for isocyanates, **5**, 968
trifluoromethylation with, **6**, 243
Silicon halides, tris(trimethylsilyl)methyl-
reactions, with methanolic sodium methoxide, **6**, 177
as starting materials, for trisilylmethanes, **6**, 177
Silicon–halogen bonds, use in the synthesis of 1,1-bis(silyl)alkanes, **4**, 621
Silicon hydrides
halogenation of, **4**, 139
for the mediation of radical reactions, **1**, 331
reactions
with tin(IV) chloride, **4**, 140
with transition metal complexes, **6**, 402
use in radical cyclization reactions, **1**, 342
Silicon hydrides, hypervalent
reactions, with acyl chlorides, **3**, 107
for the reduction of 2-furoic acid, **3**, 102
use in the synthesis of acyl aldehydes, **3**, 83
use in the synthesis of 2-formyl thiophene, **3**, 104
Silicon isocyanates, synthesis of, **5**, 1016
Silicon isocyanates, fluoro-, synthesis of, **5**, 1016
Silicon–metal bonds, reactions, with metal silicides, **2**, 535
Silicon–methyl bonds, cleavage of, **1**, 61
Silicon–nitrogen bonds, synthesis of compounds with, **2**, 811
Silicon–oxygen bonds, cleavage of, **2**, 539
Silicon–phenyl bond cleavage, via hydrobromic acid, **6**, 379
Silicons
of phosphorus functions, synthesis of, **5**, 887
reactions
with aldehydes, **1**, 1038
with ketones, **1**, 1038
synthesis of, **2**, 907; **4**, 1031; **5**, 873, 923
Silicons, acyl-, synthesis of, **5**, 409
Silicons, alkenyl-, synthesis of, **1**, 1106
Silicons, alkyl-, synthesis of, **2**, 535, **513–547**
Silicons, alkynyl-, synthesis of, **2**, 1083, 1089
Silicons, allenyl-, for the synthesis of alcohols, **1**, 1055
Silicons, aryl-, synthesis of, **2**, **899–950**
Silicons, chloro-, as starting materials, for hydroxamoyl halides, **5**, 669
Silicons, α-diazo-, synthesis of, **5**, 873
Silicons, electrophilic, as starting materials, for alkenylsilanes, **2**, 907
Silicons, α-halo-, synthesis of, **4**, 136
Silicons, α-haloalkenyl-
from
alkyl halides, **4**, 810
carbonyls, **4**, 805
synthesis of, **4**, 805
halogenation methods, **4**, 807
silylation methods, **4**, 806
via addition of alkyl halides to alkynylsilanes, **4**, 810
via halogenation of α-silylalkenyl metals, **4**, 808
Silicons, hypervalent
reactions, with benzoic acids, **3**, 82
use in the synthesis of fluorine-substituted benzaldehydes, **3**, 95
Silicons, imidoyl-
from

acylsilane enamines, **5**, 794
imidoyl chlorides, **5**, 793
imidoyl metals, **5**, 792
isonitriles, **5**, 792
metalloaldimines, **5**, 792
trimethylazidosilane, **5**, 794
synthesis of, **5**, 792
Silicons, pentacovalent, synthesis of, **2**, 547
Silicons, singly-bonded, synthesis of, **5**, 647
Silicons, tricoordinated, synthesis of, **6**, 709
Silicons, vinyl-, synthesis of, **2**, **899–950**
1,2-Silicon shifts, in the synthesis of cyclopropyl acylsilanes, **5**, 429
Silicon tetrachloride
reactions
with ethanol, **2**, 108
with 1-germanyl ethyn-2-ols, **4**, 1058
as starting material, for alkylsilyl halides, **2**, 538
synthesis of, **2**, 538
Silicon tetrafluoride
reactions, with silenes, **6**, 382
as starting material, for carbonic chloride fluoride, **6**, 419
use in the synthesis of trifluoromethylamines, **6**, 440
Silicon tetraisocyanate
reactions
with antimony trifluoride, **5**, 1016
with ethylene oxide, **5**, 1016
Silicon tetraisothiocyanate, reactions, with cycloalkanones, **5**, 1043
Silirane, 1,1-dimethyl-2,3-bis(trimethylsilyl)-, reactions, with bis(trimethylsilyl)ethyne, **4**, 1052
Siliranes
reactions, with alkynes, **4**, 1052
as starting materials, for silyl silirenes, **4**, 1052
Siliranes, hexamethyl-
reactions, with methylenephosphoranes, **6**, 702
ring-opening of, **5**, 914
Silirenes
dimerization of, **4**, 1053
hydrolysis of, **4**, 1051
reactions
with acetic acid, **4**, 1051
with alcohols, **4**, 1051
with aldehydes, **4**, 1052
with alkynes, **4**, 1051, 1052
with ketones, **4**, 1052
as starting materials
for disilacyclopentenes, **4**, 1052
for 1,1,2-trisilylethenes, **4**, 1051
synthesis of, **4**, 1051
Silirenes, silyl
from, siliranes, **4**, 1052
hydrolysis of, **4**, 1051
methanolysis of, **4**, 1051
reactions, with 1-trimethylsilyl-2-phenylethyne, **4**, 1053
Silk fibroin, reactions, with palladium, **1**, 76
Silkworms, *Antheraea polyphemus*, sex pheromones, **1**, 455
Siloles, synthesis of, via pyrolysis, **1**, 777
Siloxane, dimethylarsinotrimethyl-, as starting material, for α-arsino methylenephosphoranes, **5**, 911
Siloxane, hexamethyl-, reactions, with diphosphorous pentoxide, **2**, 106
Siloxane, hexamethylgermano-, as starting material, for α-germyl ylides, **5**, 916
Siloxane, polymethylhydro-, for the reduction of quinones, **2**, 657
Siloxane polymers, reactions, with palladium, **1**, 456
Siloxanes
metallation of, **4**, 650
synthesis of, **4**, 623
Siloxanes, alkyl-, synthesis of, **2**, 539
Silphenene, synthesis of, **1**, 364
Silphinenes, synthesis of, **1**, 176
Silphiperfil-6-ene, from, bromo ketones, **1**, 1130
Silver, as additive, in Heck reactions, **1**, 494
Silver, diazomethyl-
reactions, with chlorophosphine, **6**, 667
synthesis of, **6**, 670
Silver, mesityl-, synthesis of, **4**, 696
Silver, perfluoroisopropyl-, synthesis of, **4**, 152
Silver, phenylethynyl-, as alkynyl transfer reagent, **2**, 1077
Silver(I) bis(trimethylsilyl)methyls, synthesis of, **6**, 195
Silver(I) bromide, reactions, with arylmagnesium bromides, **1**, 117
Silver(I) catalysis
for the benzylation of silyl enol ethers, **1**, 133
in elimination reactions, **1**, 646
Silver(I) cyanamide, use in the synthesis of carbodiimides, **5**, 1081
Silver(I) cyanide, reactions, with chlorocarbonylsulfenyl chloride, **6**, 435
Silver(I) ketenides, synthesis of, **3**, 537
Silver(I) nitrate, use in decarboxylation reactions, **1**, 934
Silver(I) oxide
for the addition of carboxylic acids to alkynes, **2**, 676
use in the synthesis of *trans*-dec-2-en-4,6-diynoic acid, **5**, 102
Silver(I) trifluoromethanethiolate
synthesis of, **6**, 237
trifluoromethylthiolation with, **6**, 233
Silver(II) fluoride
reactions, with *N*,*N*-bis(dibromomethylene) hydrazine, **6**, 242
for the synthesis of carbonic difluoride, **6**, 409
Silver(II) oxide
for the dehydrogenation of hydrazones, **3**, 463
use in the synthesis of α-keto hydrazones, **3**, 470
Silver acetate
reactions
with acetic anhydride, **3**, 537
with bromine, **5**, 203
with chlorine, **5**, 203
with *gem*-dibromocyclopropanes, **4**, 57
with dichloro(propyl)arsine, **2**, 496
with methanesulfinyl chloride, **5**, 209
Silver acrylate, reactions, with bromine, **4**, 27
Silver alkynides
reactions
with arenediazonium salts, **1**, 1137; **2**, 1054
with thiocyanogen, **5**, 1110
synthesis of, **1**, 1137
Silver alkynoates, reactions, with methyl iodide, **1**, 1137
Silver alkynylides, iodination of, **2**, 1019
Silver azide
for the nucleophilic displacement of halides, **4**, 432
reactions, with dimethyliodoarsine, **2**, 504
Silver benzoate
reactions, with dichlorophenylphosphine oxide, **5**, 221
synthesis of, **5**, 229
Silver bromide, reactions, with 2-lithio(dimethylaminomethyl)benzene, **4**, 708
Silver bromide–lithium bromide, reactions, with Grignard reagents, **2**, 589
Silver bromochlorofluoroacetate
bromination of, **6**, 227
chlorination of, for bromodichlorofluoromethane, **6**, 225
Hunsdiecker reaction with, **6**, 228
Silver bromodifluoroacetate, bromination of, **6**, 226
Silver carbonate, for the oxidation of hydroxymethyl furans, **3**, 102
Silver carbonyl complexes, synthesis of, **6**, 525
Silver carboxylates
bromination of, **5**, 203
reactions
with 1-chloro-1-dialkylaminoalkenes, **4**, 889
with halides, **2**, 12

Silver chloride, reactions, with mesitylmagnesium bromide, **4**, 696
Silver chlorodifluoroacetate
bromination of, **6**, 223
chlorination of, **6**, 225
iodination of, **6**, 224
Silver complexes, tetranuclear alkenyl, synthesis of, **4**, 696
Silver cyanamides
reactions
with bromodifluorophosphorus, **5**, 1085
with silanes, **5**, 1076
Silver cyanate
reactions
with acyl chlorides, **5**, 994
with alkyl halides, **5**, 965
with arylsulfenyl chlorides, **5**, 995
with bromine, **5**, 1015
with *s*-butyl iodide, **5**, 1104
with chlorosilanes, **5**, 1016
with chlorosulfonyl isocyanate, **5**, 994, 1018
with α,α-dihaloalkyl ether, **5**, 965
with iodine, **5**, 984
with isopropyl iodide, **5**, 1104
with organotin chlorides, **5**, 1016
with phosphorus halides, **5**, 1017
with phosphorus trichloride, **5**, 1017
with phosphoryl trichloride, **5**, 1017
with silicon chloride, **5**, 1016
with silicon halides, **5**, 1016
with sulfonic anhydrides, **5**, 995
with sulfonyl chlorides, **5**, 994
with trichloromethanesulfinyl chloride, **5**, 995
with trifluoromethanesulfenyl chloride, **5**, 1018
with trimethyllead chloride, **5**, 1017
Silver cyanide
alkylation of, **3**, 693
reactions
with acyl iodides, **3**, 704
with alkyl iodides, **3**, 694
with allyl iodide, **3**, 693
with benzoyl bromide, **3**, 703
with bis(triethylphosphine)pentachlorophenylnickel chloride, **5**, 1137
with di-*n*-butylboron chloride, **5**, 1136
with (dimethylamino)organo haloboranes, **2**, 811
with ethyl iodide, **3**, 694
with ethylpentachlorophosphate, **2**, 476
with halotrialkylsilanes, **5**, 1133
with iodobis(pentafluoroethyl)phosphine, **5**, 1129
with methyllithium, **2**, 589
with organogermanium halides, **5**, 1136
with *N*-phenylimidoyl bromides, **3**, 710
with phosphorus(V) dichlorides, **5**, 1131
with trifluoromethylmercury(II) iodide, **5**, 1141
with trimethyliodo metals, **3**, 711
with trimethyliodosilane, **3**, 711
with tris(trimethylsilyl)methanetellurenyl iodide, **5**, 1117
as starting material, for isonitriles, **3**, 612
use in the synthesis of phosphorus(III) cyanides, **5**, 1129
Silver (α-diazocarbonyl)compounds
stability of, **3**, 487
synthesis of, **3**, 487
Silver diazomethyl phosphoryls, reactions, with cyanogen bromide, **3**, 478
Silver dibenzylphosphate, reactions, with imidoyl chlorides, **5**, 712
Silver dichlorofluoroacetate
bromination of, for bromodichlorofluoromethane, **6**, 225
chlorination of, **6**, 220
iodination of, for dichlorofluoroiodomethane, **6**, 225
Silver dimesylamine, reactions, with acyl chlorides, **5**, 348
Silver (ethyl diazoacetate)
reactions
with bromides, **3**, 474
with bromine, **3**, 478
synthesis of, **3**, 487
Silver ferrate
for the oxidation of alcohols, **3**, 118
for the oxidation of benzylic alcohol, **3**, 88
Silver fluoride
as fluorinating reagent, **6**, 222
reactions
with benzeneselenenyl chloride, **4**, 52
with hexafluoro-2-butyne, **4**, 696
with nitriles, **6**, 53
with tetrabromoformaldazine, **6**, 613
with trialkylamines, **1**, 370
use in the synthesis of chlorofluoroalkanes, **4**, 30
use in the synthesis of difluoroalkanes, **4**, 4
Silver fluoroacetate, reactions, with chlorine, **4**, 33
Silver α-fluoro carboxylate, reactions, with bromine, **4**, 35
Silver fulminate, use in the synthesis of oximidic chlorides, **5**, 819
Silver heptafluorobutyrate, reactions, with selenium, **6**, 49
Silver hexacyanocobaltates, as starting materials, for alkyl isocyanides, **3**, 695
Silver hexacyanoferrates, as starting materials, for alkyl isocyanides, **3**, 695
Silver ions, reactions, with 3-chloro-2,3-dimethylindolenines, **2**, 761
Silver isocyanate, reactions, with sulfinyl chlorides, **2**, 726
Silver nitrate
reactions
with alkyl halides, **2**, 103
with alkylleads, **2**, 589
with *N*-arylbenzimidoyl chlorides, **5**, 711
with bromides, **1**, 893
with 5-bromo methyl-11*H*-benzo[*a*]fluoren-11-one, **3**, 100
with cyclopropane, **3**, 235
with (2-hydroxyaryl)alkynones, **3**, 242
with imidoyl chlorides, **5**, 362
with nitroalkanes, **4**, 428
with phosphorodichloridates, **2**, 106
Silver nitrite, reactions, with imidoyl chlorides, **5**, 361
Silver nitrocyanamide
reactions
with alkyl halides, **5**, 963
with *N*-bromomethylphthalimide, **4**, 447
with trityl chloride, **5**, 963
Silver oxide
reactions
with *O*-alkyl thiocarbamates, **5**, 1103
with *O*-aryl thiocarbamates, **5**, 1103
with diazoacetophenone, **3**, 530
with dibromides, **2**, 497
with fluorodinitromethane, **6**, 287
with phenyl isocyanide dichloride, **5**, 1004
with 3-thenaldehyde, **5**, 83
with vanillin, **5**, 114
use in the synthesis of carbodiimides, **5**, 1072
use in the synthesis of carboxylic acids, **5**, 51
use in the synthesis of ethers, **2**, 88
Silver perchlorate
reactions, with alkyl iodides, **2**, 94
use in the synthesis of alcohols, **2**, 70
Silver phenylalkynide, reactions, with arenediazonium salts, **2**, 1054
Silver phenylcyanomethylene nitronate, reactions, with trityl chloride, **5**, 963
Silver phthalimide, reactions, with bromotriethylgermane, **5**, 391
Silvers
reactions
with lithium bromide, **2**, 589

with phosphorus ylides, **4**, 585
Silvers, alkyl-, synthesis of, **2**, 589
Silvers, alkynyl-, reactions, with pyridine *N*-oxides, **1**, 1043
Silvers, carbon-bridged, from, metal(I) halides, **4**, 695
Silvers, imidoyl-, synthesis of, **5**, 802
Silvers, trihalomethyl-, synthesis of, **6**, 246
Silver salts
as catalysts, in Cope rearrangements, **1**, 809
reactions
with acyl halides, **1**, 1137
with chlorodiphenylstibine, **6**, 369
with dimethyl sulfoxide, **3**, 88
with oxalyl chloride, **5**, 183
use in the synthesis of alcohols, **2**, 87
use in the synthesis of styrenes, **1**, 490
Silver salts, pentafluoroethyl-, reactions, with selenium, **6**, 49
Silver salts, vinyl-, dimerization of, **1**, 491
Silver selenate, reactions, with alkyl iodides, **2**, 102
Silver succinimide, reactions, with bis(η-5-cyclopentadienyl)vanadium, **5**, 388
N-Silver succinimide, reactions, with iodine, **5**, 320
Silver sulfate
reactions
with alkyl iodides, **2**, 102
with benzyltrimethylsilanes, **1**, 289
Silver sulfide, reactions, with bis(trifluoromethyl)iodoarsine, **2**, 499
Silver tetrafluoroborate
reactions, with α-chloro epoxides, **3**, 148
use in the synthesis of ketones, **3**, 273
Silver thiocarboxylates, use in the synthesis of acylsulfenyl bromides, **5**, 245
Silver thiocyanate
reactions
with 9-chloroacridines, **5**, 1040
with halides, **5**, 1039
Silver thiolates
as starting materials, for thiocarboxylic esters, **5**, 234
synthesis of, **2**, 275
Silver tricyanomethanide, reactions, with alkylamine hydrobromides, **4**, 977
Silver triflate
as catalyst, for the synthesis of α-benzyl ketones, **3**, 145
use in the synthesis of multidecker borole complexes of ruthenium, **4**, 695
Silver trifluoroacetate
bromination of, **6**, 218
chlorination of, **6**, 218
iodination of, **6**, 219
reactions
with chlorotrifluoroethene, **6**, 63
with 1,2-dihydrazones, **1**, 1075
with dimethylchloroarsine, **5**, 222
with fluorocarbonylsulfenic acid chloride, **5**, 246
use in the synthesis of enolates, **1**, 133
Silver trifluoromethoxide
reactions, with 1-chlorotrifluoropropyne, **4**, 764
synthesis of, **4**, 764
Silver trimetaphosphate, alkylation of, **2**, 107
Silyl, *O*-trimethyl-, synthesis of, **5**, 542
Silyl, trimethyl- (TMS), use in the synthesis of acyloxysilanes, **5**, 222
Silylalkenation, for the synthesis of alkenylsilanes, **2**, 936
Silyl allenes, hydrolysis of, **5**, 427
Silylallyl anions, reactions, with carbonyls, **1**, 651
α-Silylallyl carbanions, use in the synthesis of alkenes, **1**, 762
Silylalumination
of alkynes, **2**, 918
of trimethylsilyl-1-hexyne, **2**, 923
Silylamines, metallated, oxidation of, **3**, 17
Silyl anions
reactions, with carbon electrophiles, **2**, 536
stable functional, synthesis of, **2**, 546
α-Silyl anions, reactions, with α,β-unsaturated aldehydes, **1**, 651
Silylating reagents
for the synthesis of trialkylsilyl imidates, **5**, 718
for the synthesis of ylides, **5**, 913
Silylation
of alkenyl metals, **2**, 907
of amides, **5**, 374
of cyanohydrin anions, **3**, 568
of imides, **5**, 378
of thioamides, **5**, 739
Silylation, double, of alkynylsilanes, **2**, 923
Silylation, electrophilic, of aminoalkyllithiums, **4**, 511
Silylation, reductive
of aromatic compounds, **2**, 945
for the synthesis of imidoylsilanes, **5**, 794
C-Silylation
of aldehydes, **4**, 354
examples of, **4**, 365
of ketones, **4**, 354
O-Silylation, mechanism, **4**, 636
Silylboration, of 1-alkynes, **2**, 903
Silyl bromide, trimethyl-
elimination of, for vinyl bromides, **2**, 622
reactions
with acid chlorides, **5**, 17
with α-diazophosphonic acid methyl ester amides, **3**, 482
with lactones, **5**, 225
for the synthesis of aryl bromides, **2**, 627
use in the synthesis of alkyl bromides, **2**, 32
α-Silyl carbanions
alkylation of, **1**, 160
from, vinylsilanes, **1**, 160
reactions
with aldehydes, **1**, 763
with carbonyls, **1**, 649, 758, 761
with cyclohexanone, **1**, 763
synthesis of, **1**, 160, 759
use of, **1**, 160
use in the synthesis of alkenes, **1**, 758
Silyl chloride, *t*-butyldimethyl-
reactions
with enolates, **1**, 417
with sodium 2-methyltellurobenzoate, **5**, 543
with sodium tellurocarboxylates, **5**, 255
as reagent, for the protection of alcohols, **5**, 223
for the *O*-silylation of ester enolates, **4**, 831
Silyl chloride, *t*-butyldiphenyl-, as protecting group for carboxylic acids, **5**, 223
Silyl chloride, dimethyl 1,1,2-trimethylpropyl-
as protecting group for carboxylic acids, **5**, 223
as silylation reagent, **5**, 224
Silyl chloride, thexyldimethyl-, reactions, with butanethiol, **2**, 262
Silyl chloride, triethyl-, for the silylation of amides, **5**, 375
Silyl chloride, trimethyl- (TMS-Cl)
as electrophile, for dithiocarboxylic enethiolates, **4**, 842
reactions
with acetamide, **5**, 379
with (arylthio)methyl isocyanides, **3**, 720
with carboxylic acids, **5**, 223
with 1,2-diacylhydrazine, **5**, 380
with *gem*-dichlorocyclopropanes, **4**, 605
with diisopropyl sulfides, **2**, 99
with dilithiomethane, **4**, 603
with dimesitylborylmethyllithium, **4**, 642
with enolates, **1**, 417
with formamide, **5**, 379
with hydroxylamines, **5**, 323
with 1-lithio-2,2,4,4-tetramethyl-2,4-disilapentane, **4**, 606
with magnesium, **5**, 793
with monolithioacetonitrile, **4**, 604

with potassium selenobenzoate, **5**, 255
with primary amines, **2**, 300
with 1-silyl 3-phenylpropynes, **4**, 1047
with tributylstannylmethyllithium, **4**, 660
with trifluoroacetylsulfenyl iodide, **5**, 254
with trimethylsilylcyclopentadienyllithium, **4**, 603
for the reduction of bromoalkanes, **1**, 9
use in conjugate addition reactions, **1**, 313
use in the displacement of alcohols, **1**, 179
use in Reformatsky reactions, **2**, 69
use in the synthesis of alkenes, **1**, 593
use in the synthesis of carboxylic esters, **5**, 123, 141
use in the synthesis of nitriles, **3**, 613
use in the synthesis of *N*-silylamides, **5**, 374
Silyl chlorides
effectiveness in the synthesis of tetrasilylmethanes, **6**, 379
reactions
with *N*-lithioimines, **3**, 456
with lithium alkynolates, **2**, 1024
with lithium compounds, **6**, 399
as *O*-silylation agents, **5**, 536
use in the synthesis of acyloxysilanes, **5**, 223
Silyl chlorides, alkyl-, large scale production of, **2**, 538
Silyl chlorides, alkynyl-, reduction of, **2**, 1092
Silyl chlorides, trialkyl-
reactions, with lithium enolates, **4**, 832
as silylating agents, **4**, 831
Silyl chlorides, triphenyl-, reactions, with chloromethyllithium, **4**, 138
Silyl compounds, as hydroxymethylation agents, **1**, 160
N-Silyl compounds, synthesis of, **2**, 470
Silylcupration
of alkynes, **2**, 917, 977
of allenes, **2**, 929
Silyl dichlorides, reactions, with carboxylic acids, **5**, 224
Silyl dienes, Diels–Alder reaction of, **2**, 938
1-Silyl-1,3-dienes
from
aldehydes, **2**, 936
1,2-disilylethenes, **2**, 939
Silylene, dimethyl-
from, dodecamethylcyclohexasilane, **3**, 502
reactions, with carbon monoxide, **5**, 1019
as starting material, for 1-methylsilene, **3**, 502
Silylenes
[2+1] cycloaddition of, **5**, 888
detection of, **5**, 1019
reactions, with carbon monoxide, **5**, 1020
as starting materials, for silenes, **3**, 502
Silyl epoxides, as starting materials, for alkenylsilanes, **2**, 939
Silylethoxymethyl (SEM) group, as protecting group for anomeric hydroxyl groups, **4**, 176
Silyl fluorides, effectiveness in the synthesis of tetrasilylmethanes, **6**, 379
Silylformylation
of alkynes, **3**, 77
for the synthesis of α-silyloxy aldehydes, **3**, 33
Silyl groups, intramolecular migration of, **5**, 542
Silyl groups, trimethyl-
metallation of, **4**, 651
1,2-shift of, **4**, 615
α-Silyl groups, stabilizing effect of, **6**, 392
Silyl halides
elimination of
for the synthesis of vinyl chlorides, **2**, 615
for vinyl halides, **2**, 606
reactions
with amides, **6**, 162
with metal enolates, **2**, 699
Silyl halides, alkyl-
from, silicon tetrachloride, **2**, 538
synthesis of, **2**, 538
Silyl halides, chloromethyldimethyl-, as starting materials, for 1,3-disilacyclobutanes, **4**, 654
Silyl halides, trimethyl-
reactions
with lithium tetra(methylseleno)aluminate, **2**, 294
with sultones, **2**, 209
α-Silyl heterocycles, as starting materials, for ylides, **4**, 519
Silyl iodide, trimethyl-
as catalyst, for the synthesis of acetals, **4**, 181
reactions
with acid chlorides, **5**, 19
with alkyl esters, **5**, 225
with dimethoxymethane, **4**, 51
with pyridine, **5**, 226
with vinyl sulfones, **4**, 84
use in the synthesis of acid iodides, **5**, 20
use in the synthesis of alkyl iodides, **2**, 36
Silyllithiation, of allenes, **2**, 929
Silylmagnesiation
of alkynes, **2**, 918
of allenes, **2**, 929
Silylmetallation
of alkynes, **2**, 917
of alkynylsilanes, **2**, 923
regiochemistry of, **2**, 923
Silylmethylation
of amines, **4**, 507
of formamide, **4**, 516
of nitrogen heterocycles, **4**, 519
of pyridine, **4**, 519
Silylmethyl group, cone angle of, **6**, 390
[1,3]-Silyl migration, for the synthesis of methylenephosphines, **6**, 684
Silyloxy compounds, synthesis of, **2**, 887
Silyloxy groups, displacement of, **1**, 197
Silyl radical, reactions, with carbon–carbon double bonds, **1**, 950
Silylstannylation, of allenes, **2**, 930
β-Silyl-substituted compounds, fragmentation of, **1**, 654
Silyltitanation, of trimethylsilylphenylethyne, **2**, 924
Silyl transfer reagents, for the synthesis of *N*-silylamides, **5**, 375
Silylzincation
of alkynes, **2**, 918
of alkynylsilanes, **2**, 923
Simmons–Smith cyclopropanation, of α-silyl allylic alcohols, **5**, 429
Simmons–Smith cyclopropanation reagent, synthesis of, experimental procedures, **1**, 300
Simmons–Smith reaction
asymmetric, enantioselective, **1**, 352
mechanism, **1**, 299, 352
for the synthesis of metallated carbenes, **4**, 1083
for the synthesis of sulfur ylides, **1**, 235
Simmons–Smith reagents
reactions
with chlorotrimethylstannane, **4**, 154
with dichlorodimethylstannane, **4**, 154
with trimethylammonia, **4**, 534
synthesis of, **4**, 152
Simple enols, use of term, **2**, 636
β-Sinesal, synthesis of, **1**, 840
Single electron transfer (SET)
in coupling mechanisms, **1**, 152
mechanism, **1**, 313, 884
in the synthesis of thiyl radicals, **2**, 215
Singlet carbenes, use in the synthesis of cyclopropanes, **1**, 298
Singlet oxygen, for the oxidation of dithioacetals, **4**, 260
Singly occupied molecular orbital (SOMO)
of carbon-centered radicals, **1**, 920
of nucleophilic radicals, **1**, 323
Sinularene, synthesis of, via acetate pyrolysis, **1**, 774
Six-membered rings, synthesis of, **1**, 372
Skeletal rearrangement, of ylides, **5**, 908
SMEAH, *See* Sodium bis(2-methoxyethoxy)aluminum hydride (Red-Al)

Smiles rearrangement
  reviews, **1**, 379
  for the synthesis of arenesulfinic acids, **2**, 719
  for the synthesis of *N*-arylamides, **5**, 292
SN1 mechanism, and nucleophilic substitution
    reactions, **1**, 891
SN2' reaction, for the synthesis of 1,1-difluoroalkenes, **4**, 735
SN2 mechanism, and nucleophilic substitution
    reactions, **1**, 891
Sodamide
  reactions
    with methyldiphenylphosphine oxide, **1**, 752
    with nitriles, **5**, 750
  use in the synthesis of ketenes, **3**, 526
  use in the synthesis of phosphonium ylides, **1**, 727
  use in the synthesis of primary amines, **2**, 298
Sodiates, dialkyl-, synthesis of, **6**, 393
Sodium
  for the deprotonation of amides, **5**, 384
  in liquid ammonia
    for the cleavage of nitriles, **1**, 55
    for the reduction of amine salts, **1**, 34
  reactions
    with alkylbismuthines, **2**, 510
    with alkyl chlorides, **2**, 599
    with alkyl halides, **2**, 556
    with amides, **5**, 723
    with ammonia, **1**, 508
    with aryl aldehyde arylimine, **6**, 507
    with bisdimethylarsine sulfide, **2**, 510
    with chloro(chloromethyl)dimethylsilane, **4**, 609
    with diarylsulfur diimide, **2**, 773
    with dimethyl diselenide, **4**, 283
    with ethyl 3-chloropropionate, **4**, 171
    with ethylene glycol, **1**, 630
    with hexamethylphosphoramide, **1**, 508
    with *N*-methyl-*N*-phenylcyanamide, **6**, 648
    with α-methylstyrene, **5**, 42
    with *N*-phenyl-substituted pyrrole, **1**, 100
    with styrene, **5**, 42
    with tetrathianes, **2**, 130
    with thiols, **2**, 271
    with toluene, **4**, 609
    with vinyl sulfides, **2**, 722
  for the reduction of alkyl halides, **1**, 3
  use in the synthesis of alkyllithiums, **2**, 551
Sodium, alk-3-en-1-ynyl-, reactions, with
    triphenylborane, **2**, 1086
Sodium, alkynyl-, reactions, with chloroborates, **2**, 1085
Sodium, allyl-, reactions, with 2-bromooctane, **1**, 113
Sodium, benzoyl-, synthesis of, theoretical, **5**, 436
Sodium, cyclopentadienyl-
  alkylation of, **2**, 483
  reactions, with (2,4-
    dimethylpentadienyl)potassium, **4**, 678
  use in the synthesis of carbon-bridged dichromium
    complexes, **4**, 673
  use in the synthesis of carbon-bridged dirhodium
    complexes, **4**, 686
  use in the synthesis of cyclooctatetraene-bridged
    complexes, **4**, 672
  use in the synthesis of cyclopentadienyl-bridged imino–
    rhenium complexes, **4**, 678
Sodium, dimethylbismuthino-
  reactions
    with chloromethyltrimethylgermane, **4**, 599
    with chloromethyltrimethylsilane, **4**, 599
    with chloromethyltrimethyltin, **4**, 600
Sodium, diphenylbismuth-, reactions, with
    dichloromethane, **4**, 596
Sodium, ethyl-, reactions, with (+)-2-chlorooctane, **1**, 112
Sodium, ethynyl-, use in the synthesis of alkynylantimony
    compounds, **2**, 1081
Sodium, methyl-, use in the synthesis of
    dialkylsodiates, **6**, 393
Sodium, pentyl-, reactions, with ferrocene, **4**, 706
Sodium, *n*-pentyl-
  for the deprotonation of alkenes, **2**, 557
  reactions, with camphene, **1**, 113
Sodium, phenylmethylantimony-, reactions, with
    dichloromethane, **4**, 595
Sodium, triethylplumbyl-, reactions, with
    ethoxyethyne, **4**, 955
Sodium, trimethylsilylmethyl-, synthesis of, **4**, 653
Sodium, trimethylstannyl-, reactions, with enones, **2**, 600
Sodium, trimethyltin-
  reactions
    with chloroform, **4**, 699
    with dichloromethane, **4**, 699
Sodium, trivinyltin-, reactions, with dichloromethane, **4**, 699
Sodium acetate
  reactions
    with acyl chlorides, **5**, 182
    with bis(chloromethyl)dimethylgermane, **4**, 362
    with chloromethyldimethylchlorosilane, **5**, 224
    with diaryltellurium dichlorides, **5**, 210
    with 3,3-dibromo-1,1,1-trifluoro acetone, **4**, 161
    with 2,4-dinitrobenzenesulfenyl chlorides, **5**, 208
    with mercuric chloride, **6**, 207
  use in dehydrochlorination reactions, **1**, 571
Sodium *N*-acylamides, synthesis of, **5**, 723
Sodium 1-adamantylselenocarboxylate, as starting material,
    for selenocarboxylic acids, **5**, 508
Sodium alcoholates, reactions, with carbon dioxide, **6**, 463
Sodium alkoxides
  for halogen displacement reactions, **2**, 494
  reactions
    with alkylantimony dihalides, **2**, 496
    with alkylarsenic dihalides, **2**, 496
    with alkylbismuth dihalides, **2**, 496
    with 3-alkyl-2-(*N*-cyanoimino)thiazolidines, **5**, 1120
    with *N*-*t*-butylthiophthalimide, **2**, 195
    with 1-chloro-1-dialkylaminoalkenes, **4**, 889
    with chloroform, **6**, 68
    with *N*-chloromethyl-*N*-ethylnitramine, **4**, 313
    with dialkoxy disulfides, **2**, 99
    with dichlorodifluoromethane, **6**, 69
    with difluorochloromethane, **6**, 69
    with 1,1-dihaloalkenes, **6**, 74
    with fluorodichloromethane, **6**, 69
    with Grignard reagents, **2**, 112
    with tetrachloromethane, **6**, 69
    with *N*,*N*,*N*',*N*'-tetrasubstituted formamidium
      salts, **6**, 105
  as starting materials, for 1,2-dichlorovinyl ethers, **2**, 1026
  use in Bamford–Stevens reactions, **1**, 630
Sodium alkoxyethynylide, synthesis of, **2**, 1025
Sodium *O*-alkyl dithiocarbonates
  reactions
    with 3-chloromethylbenzothiazolin-2-thione, **6**, 545
    with dimethyltellurium diiodide, **6**, 547
    with triphenyltelluronium chloride, **6**, 547
Sodium alkyl selenides, synthesis of, **6**, 481
Sodium alkyl sulfinate, reactions, with iodine, **2**, 182
Sodium alkylthiolate, reactions, with malonitriles, **3**, 655
Sodium alkylthio sulfates, oxidation of, **2**, 177
Sodium *S*-alkyl trithiocarbonates, reactions, with
    triphenyltellurium chloride, **6**, 553
Sodium alkynes, reactions, with diarylbismuth
    halides, **2**, 1082
Sodium alkynide
  carbonation of, **5**, 100
  reactions
    with carbon diselenide, **6**, 590
    with ketones, **3**, 230
    with thiobenzophenone, **1**, 1040

Sodium alkynides
reactions
with alkyl halides, **1**, 1000
with allylic halides, **1**, 1001
with germanium tetrachloride, **2**, 1093
with oxiranes, **1**, 1022
as starting materials, for alkynylstannanes, **2**, 1100
use in the synthesis of substituted alkynes, **1**, 113
Sodium alkynylide
reactions
with selenium, **2**, 1036
with tellurium, **2**, 1036
Sodium alkynylides
bromination of, **2**, 1018
iodination of, **2**, 1019
Sodium aluminum hydride, for the reduction of aromatic esters, **3**, 84
Sodium amalgam
for the cleavage of alkylmercury compounds, **1**, 69
reactions
with carbonylcyclopentadienyldiiodocobalt, **4**, 690
with manganese hexacarbonyl, **5**, 446
with methanol, **1**, 615, 617
for the reduction of benzoic acids, **3**, 96
for the reduction of sulfones, **1**, 24
for the reductive elimination of β-hydroxy sulfones, **1**, 698
use in the reduction of carbonyls, **2**, 54
use in the reduction of cyclopentadienylzirconium complexes, **2**, 566
use in the synthesis of aryl aldehydes, **3**, 83
Sodium amide
dehydrobromination of, **2**, 1031
reactions
with ammonia, **1**, 1065
with aryl chlorides, **1**, 1066
with benzazocine, **2**, 742
with benzazonine, **2**, 742
with *t*-butylacetonitrile, **3**, 567
with methyltriphenylarsonium bromide, **2**, 1004
with propyne, **5**, 100
with tetrahydrofurfuryl chloride, **1**, 1066
use in the synthesis of sodiotrithioformates, **6**, 317
use in the synthesis of trifluorovinyl anions, **4**, 819
Sodium amides, synthesis of, **2**, 422
Sodium amides, *N*-alkyl-, synthesis of, **2**, 422
Sodium 3-aminopropylamide (NAPA), synthesis of, **2**, 422
Sodium–arene complexes, stability of, **2**, 556
Sodium arene sulfinates, use in the synthesis of ketenimines, **3**, 568
Sodium arenethiolates
reactions
with dichloroalkyne, **4**, 772
with tetrachloroethene, **4**, 770
with trichloroethene, **4**, 772
with 1,1,2-trichloroethene, **4**, 770
Sodium arsenes, synthesis of, **4**, 499
Sodium-*N*-aryl hydroxamates, iminoacylation of, **5**, 711
Sodium aryl tellurides, synthesis of, **4**, 291
Sodium azide
acylation of, **5**, 367
reactions
with acrylic acids, **5**, 989
with acyl chlorides, **2**, 757
with acyliron(III) complexes, **5**, 991
with bis(trimethylsilyl)chloromethane, **6**, 163
with bromo compounds, **4**, 109
with bromo ketones, **3**, 259
with α-bromo oxime ethers, **2**, 314
with carbamoyl chloride, **6**, 511
with carboxylic acid chlorides, **5**, 989
with carboxymethyl dithioates, **5**, 634
with chloroformates, **6**, 489
with chloromethylphosphonamides, **4**, 492
with cycloalkenals, **3**, 77
with cyclopropenylium ions, **4**, 483
with diarylphosphinyl chlorides, **2**, 863
with 3,4-dichlorocyclobutenedione, **3**, 539
with dihalo compounds, **4**, 432
with ethyl dibromofluoroacetate, **6**, 54
with ethyl dibromonitroacetate, **6**, 57
with *N*-fluoroimidoyl fluorides, **5**, 819
with fluoromethyl phenyl sulfoxide, **4**, 86
with mixed anhydrides, **5**, 991
with 4-phenanthrenecarboxylic acid, **5**, 1001
with phenylchloroethyne, **2**, 1055
with vinyl epoxides, **2**, 85
with vinylidene dichlorides, **4**, 993
Sodium azodicarboxylate, decomposition of, **1**, 74
Sodium benzeneselenolate
reactions
with diethyl iodomethyl phosphonate, **4**, 348
with thioacyl chlorides, **5**, 563
Sodium benzeneselonate, reactions, with imidoyl chlorides, **5**, 732
Sodium benzene sulfinate
reactions
with *N*-diethylacrylamide, **3**, 657
with diiodomethane, **4**, 84
with 2-iodo-2-nitropropane, **4**, 332
Sodium benzenesulfonamide, reactions, with aldehydes, **5**, 345
Sodium benzenetellurolate, reactions, with imidoyl chlorides, **5**, 733
Sodium benzoaldoximate, reactions, with thiobenzoyl chloride, **5**, 531
Sodium–benzophenone ketal, use in the synthesis of alkenes, **1**, 570
Sodium benzoyl acetate, reactions, with 1,2-dithiolylium salts, **3**, 254
Sodium benzylselenide, reactions, with bromo selenides, **4**, 288
Sodium benzyl sulfide
reactions
with chlorodifluoromethane, **2**, 179
with chlorofluoromethane, **2**, 179
Sodium bicarbonate, use in the synthesis of carboxylic esters, **5**, 141
Sodium 4-biphenylselenocarboxylate, as starting material, for selenocarboxylic acids, **5**, 508
Sodium bis(2-methoxy)aluminum hydride, use in the synthesis of vinylsilanes, **2**, 921
Sodium bis(methoxyethoxy)aluminum dihydride
reactions, with *N*-methylpiperazine, **3**, 84
for the reduction of aromatic esters, **3**, 84
Sodium bis(methoxyethoxy)aluminum hydride (Red-Al)
reactions, with propargylic alcohols, **4**, 948
reduction of, **2**, 83
for the reduction of aldehydes, **1**, 18
for the reduction of esters, **1**, 21
for the reduction of oxiranes, **1**, 14
use in the reduction of carbonyls, **2**, 60
use in the reduction of nitriles, **1**, **27–70**
use in the synthesis of allylic alcohols, **1**, 514
Sodium bis(trialkylsilyl)amides, reactions, with carbonyls, **3**, 456
Sodium bis(trimethylsilyl)amide
reactions
with benzoic acid esters, **5**, 721
with bis(dichlorophosphino)methane, **5**, 884
with phosphonium salts, **6**, 702
use in the synthesis of phosphonium ylides, **1**, 727
Sodium bis(trimethylsilyl)amine, reactions, with diphenoxyethyne, **4**, 884
Sodium bisulfite
reactions
with aldehydes, **4**, 237

with ketones, **4**, 237
with 3-nitro-4-dimethylaminotoluene, **2**, 782
for the reduction of iodoalkanes, **1**, 10
Sodium borate, for the oxidation of thiols, **2**, 216
Sodium borodeutride, use in the reduction of organomercury compounds, **1**, 67
Sodium borohydride
reactions
with alcohols, **2**, 532
with alkynyl tellurides, **1**, 999
with aluminum trichloride, **2**, 59
with cobalt(II) chloride, **1**, 75
with cobalt(II) complexes, **1**, 457
with cobalt chloride, **2**, 312
with diaryl ditellurides, **5**, 241
with 1,4-dinitrobenzene, **2**, 773
with enamines, **1**, 82
with imines, **2**, 308
with mercury acetate, **2**, 43
with mixed anhydrides, **2**, 305
with nickel(II) chloride, **1**, 75
with pyridines, **1**, 102
with rhodium trichloride trihydrate, **1**, 73
with sulfur, **2**, 227; **3**, 620
with *p*-toluenesulfonyl hydrazones, **1**, 39
with tosylhydrazones, **1**, 19
with transition metal salts, **1**, 72
with 2,4,6-trisubstituted pyrylium salts, **3**, 232
with trithiocarbenium salts, **6**, 317
as reagent
for the synthesis of *N*-acylaminals, **4**, 420
for the synthesis of nitriles, **3**, 621
for the synthesis of pyrimidines, **4**, 420
as reducing agent, **2**, 310
for the reduction of acyl chlorides, **3**, 83
for the reduction of aldehydes, **1**, 18
for the reduction of alkyl halides, **1**, 4
for the reduction of benzylic alcohols, **1**, 13
for the reduction of bromoalkanes, **1**, 8
for the reduction of chloroalkanes, **1**, 7
for the reduction of diarylimines, **1**, 46
for the reduction of diazonium salts, **1**, 442
for the reduction of esters, **1**, 17
for the reduction of iodoalkanes, **1**, 10
for the reduction of ketones, **1**, 19
for the reduction of nitroalkenes, **2**, 341
for the reduction of quinones, **2**, 657
as starting material
for dialkoxyboranes, **2**, 109
for nucleophilic hydrides, **1**, 28
synthesis of, **2**, 422
use in the reduction of acid chlorides, **5**, 14
use in the reduction of acyclic esters, **4**, 170
use in the reduction of carbonyls, **2**, 59
use in the reduction of nitriles, **1**, 55
use in the reduction of nitro compounds, **2**, 318
use in the reduction of organomercury compounds, **1**, 66
use in the synthesis of alkenes, **1**, 514
use in the synthesis of amines, **2**, 300, 314
use in the synthesis of carbon-bridged diiron complexes, **4**, 679
use in the synthesis of cyclic hemiacetals, **4**, 165
use in the synthesis of diruthenium complexes, **4**, 679
use in the synthesis of formamidines, **5**, 746
use in the synthesis of thiols, **2**, 126
Sodium borohydride, alkaline, use in the demercuration of organomercury compounds, **1**, 67
Sodium borohydride–iodine, use in the synthesis of amines, **2**, 316
Sodium borohydride trisulfide, use in the synthesis of disulfides, **2**, 226
Sodium bromate
as reagent, for the synthesis of *N*-bromoimides, **5**, 318
use in the synthesis of ketones, **3**, 117
Sodium bromite
for the oxidation of alcohols to ketones, **3**, 119
for the oxidation of sulfides, **2**, 149
as reagent, for the synthesis of *N*-bromoimides, **5**, 318
use in the synthesis of alkanoic acids, **5**, 48
Sodium bromoethanesulfonate, chlorination of, **4**, 85
Sodium bromoethynylides, synthesis of, **2**, 1017
Sodium butylarsenite, reactions, with propyl bromide, **2**, 498
Sodium carbonate
reactions
with 1-ethynylarsine, **5**, 1160
with methanol, **5**, 951
with phosphorus pentasulfide, **5**, 568
Sodium carboxylates
reactions
with acyl chlorides, **5**, 183
with acyl halides, **5**, 183
with arenediazonium tetrafluoroborates, **5**, 198
with water, **1**, 373
synthesis of, **5**, 232
Sodium chloride, use in the synthesis of *N*-chloroimides, **5**, 317
Sodium chlorite, use in the synthesis of *trans*-2-hexenoic acid, **5**, 102
Sodium chloroacetate
reactions
with potassium dithiobenzoate, **5**, 83
with sodium cyanide, **5**, 95
Sodium chlorodifluoroacetate
reactions, with triphenylphosphine, **6**, 694
use in the synthesis of 1,1-difluoroalkenes, **4**, 732
Sodium chlorodifluoroacetates, pyrolysis of, **4**, 8
Sodium chloroethynylide, as starting material, for alkynyl chlorides, **2**, 1015
Sodium chloromethyl sulfonate, as starting material, for chloromethyl phenyl ether, **4**, 49
Sodium *N*-chlorotoluene-*p*-sulfonamide
reactions, with triphenylarsine, **2**, 892
See also Chloramine-T
Sodium cyanamide, reactions, with 2-amino-1,3-benzoxazin-4-one, **5**, 1120
Sodium cyanate, reactions, with *n*-pentyl chloride, **5**, 1104
Sodium cyanide
as catalyst, **5**, 818
reactions
with aryl halides, **3**, 661
with 9-benzoylanthracene, **3**, 667
with carbon disulfide, **5**, 589
with nitroso compounds, **5**, 1124
with pentahalophenylnickel acetate complexes, **5**, 1137
with sodium chloroacetate, **5**, 95
use in the displacement of alcohols, **1**, 179
Sodium cyanoacetate, hydrolysis of, **5**, 95
Sodium cyanoborohydride
reactions
with hemiaminals, **2**, 310
with hexamethylphosphoramide, **1**, 9
with tosylhydrazones, **1**, 19, 635
as reducing agent, **2**, 310
for the reduction of aldehydes, **1**, 18
for the reduction of alkyl bromides, **1**, 4
for the reduction of benzylic alcohols, **1**, 13
for the reduction of chloroalkanes, **1**, 7
for the reduction of esters, **1**, 17
for the reduction of iodoalkanes, **1**, 10
for the reduction of ketones, **1**, 20
for the reduction of oxiranes, **1**, 14
for the reduction of quaternary ammonium salts, **1**, 33
use in the synthesis of amines, **2**, 314
use in the synthesis of hydroxylamines, **2**, 763
Sodium α-cyanoethanoate, as starting material, for malonic acid, **5**, 35

Sodium cyanothioformate, as starting material, for thioureas, **6**, 580
Sodium 1,2-cyclohexenedithiolate, as starting material, for tetrathioorthocarbonates, **6**, 306
Sodium cyclopentadienide
reactions
with chlorodicyanoantimony(III), **5**, 1132
with chromium hexacarbonyl, **6**, 522
with potassium trichlorodicyanoantimonate(III), **5**, 1132
Sodium cyclopentadienyldicarbonylruthenium, use in the synthesis of carbon-bridged diruthenium complexes, **4**, 678
Sodium dialkylbismuthides, synthesis of, **2**, 510
Sodium dialkyl phosphites
reactions
with dialkoxythiophosphoryl isothiocyanates, **6**, 583
with 1-haloalkylsilanes, **4**, 569
with trialkoxycarbenium ions, **6**, 313
Sodium dialkylstibides
from, trialkylstibines, **2**, 510
as starting materials, for tertiary stibines, **2**, 485
Sodium dialkyl(thio)phosphates, use in the synthesis of mixed anhydrides, **5**, 713
Sodium diarylphosphides, hydrolysis of, for diarylphosphines, **2**, 823
Sodium dichlorofluoroacetate, decarboxylation of, **4**, 752
Sodium dichromate
reactions
with phenylhydroxylamine, **2**, 781
with sulfuric acid, **5**, 115
Sodium diethyldihydroaluminate
reactions
with acyl chlorides, **3**, 107
with piperidine, **3**, 83
Sodium *N*,*N*-diethyl dithiocarbamate, reactions, with trichloromethyl isocyanide dichloride, **6**, 316
Sodium diethyl dithiophosphate, reactions, with imidoyl chlorides, **5**, 738
Sodium diethyl phosphite
reactions
with α-halo ketones, **3**, 197
with imidate, **6**, 159
with iminium esters, **4**, 1003
with thioimidates, **6**, 122
Sodium diethylphosphonate
reactions
with alkynyl halides, **2**, 850
with chlorodifluoromethane, **6**, 58
with difluorocarbene, **6**, 58
Sodium diethylpiperidinoaluminum hydrides, for the reduction of aromatic esters, **3**, 84
Sodium diethylpiperidinohydroaluminate (SDPA)
for the reduction of nitriles, **3**, 86
use in the reduction of carbonyls, **2**, 60
Sodium diethyl thiophosphate, reactions, with imidoyl chlorides, **5**, 738
Sodium difluoroacetate, decarboxylation of, **3**, 497
Sodium diformylamide
as starting material, for *N*,*N*-diformylamino ketones, **2**, 308
use in Gabriel synthesis, **2**, 308
use in the synthesis of α-amino ketones, **3**, 179
Sodium dihydrobis(2-methoxyethanolato)aluminate, use in the synthesis of alkylphosphines, **2**, 429
Sodium dihydrobis(2-methoxyethoxy)aluminate, for the reduction of phosphinates, **2**, 431
Sodium diisopropyl phosphites, as starting materials, for diisopropyl fluoromethyl phosphonate, **4**, 131
Sodium dimethylamide, reactions, with enamines, **4**, 1017
Sodium dimethylarsenide, synthesis of, **2**, 510
Sodium *N*,*N*-dimethyl dithiocarbamate
reactions
with chloroacetonitrile, **6**, 561
with trichloromethyl isocyanide dichloride, **6**, 316
Sodium dimethylhydro-1-propynylborate, synthesis of, **2**, 1088
Sodium dimethyl malonate, alkylation of, **1**, 254
Sodium dinitromethanide, reactions, with bromine, **6**, 261
Sodium diphenylarsenic, reactions, with 1,1-dichloroethane, **4**, 592
Sodium diphenyldithioarsinate, reactions, with chlorodiphenylphosphine, **2**, 891
Sodium diphenylphosphide
reactions
with *gem*-chloro enamines, **4**, 997
with tetrakis(bromomethyl)methane, **2**, 437
as starting material, for tertiary phosphines, **2**, 827
Sodium disulfide, acylation of, **5**, 252
Sodium disulfite, reactions, with alkyl halides, **2**, 176
Sodium dithionate, alkylation of, **2**, 157
Sodium dithionite
for the conjugate reduction of enones, **3**, 128
reactions, with aldehydes, **4**, 234
as reducing agent, **5**, 1097
for the reduction of quinones, **2**, 657
use in desulfonation reactions, **1**, 439
use in the reduction of carbon–mercury bonds, **1**, 68
Sodium enolates, reactions, with diphenylantimony(III) salt, **2**, 695
Sodium ethanedithiolate, use in the synthesis of alkanoic acids, **5**, 39
Sodium ethanethiolate
reactions, with methoxythiocarbonyl chloride, **6**, 550
reduction of, **6**, 91
Sodium ethoxide
reactions
with alkaneselenols, **2**, 279
with 1-alkoxy-2,2,2-tribromo-1-chloroethanes, **4**, 767
with *t*-butanesulfenyl iodide, **2**, 194
with 1-chloro-1-(chlorosulfenyl)propanoyl chloride, **4**, 72
with difluorodinitromethane, **6**, 469
with disulfur dichloride, **2**, 100
with ethanol, **1**, 1065
with 2-ethoxythiolenium tetrafluoroborate, **6**, 96
with sulfones, **1**, 612
with 2-sulfonyl pyridine-*N*-oxides, **2**, 271
with tetrakis(hydroxymethyl)phosphonium salts, **4**, 458
with trichloroethylene, **4**, 761
with 1,1,2-trichloro-2-nitroethene, **6**, 74
with trimethylamineimidazolecarbonylborane, **6**, 492
as reagent, for the synthesis of α-iminophosphine oxides, **5**, 786
as starting material, for triethyl orthoacetate, **6**, 74
Sodium *O*-ethyl dithiocarbonate, reactions, with ethyl chloroformate, **6**, 476
Sodium ethylmethylphosphinate, reactions, with *N*-sulfonylaziridines, **2**, 454
Sodium ethyl selenides, reactions, with *S*-ethyl thiocarbonate, **6**, 482
Sodium fluoride
reactions
with bromotrialkylphosphonium bromides, **2**, 475
with chloromethylphosphonic dichloride, **4**, 130
with 1,1-difluorothiols, **5**, 474
with phenylphosphonothioic dichloride, **2**, 855
with phosgene, **6**, 409
with thionyl chloride, **6**, 409
Sodium formate, reactions, with *N*-methylpyrrolidone, **2**, 753
Sodium *N*-formyl dithiocarbamate
reactions, with iodomethane, **6**, 563
synthesis of, **6**, 565
Sodium *N*-formyl thiocarbamate, synthesis of, **6**, 560
Sodium glyoxylate, reactions, with α-ketoesters, **5**, 104

Sodium hexafluorosilicate, reactions, with tetrachloromethane, **6**, 225
Sodium hexamethyldisilazide
nitrile deprotonation of, **3**, 640
reactions
with alkylmethanesulfonates, **4**, 605
with methyl alkanesulfonates, **6**, 132
Sodium hexamethyldisilylamide, reactions, with methyl triphenylphosphoranylidene dithioacetate, **3**, 547
Sodium hydrazide
reactions
with alkynes, **3**, 447
with aromatic nitriles, **5**, 848
with dinitriles, **5**, 848
with diphenylethyne, **3**, 447
Sodium hydride
reactions
with acylhydrazides, **5**, 1126
with *O*-alkyl-4-nitrobenzaldoximes, **3**, 666
with aminodiphosphonates, **4**, 1002
with 1-(*N*-benzyl-*N*-methylthiocarbamoyl)imidazole, **6**, 557
with benzyltriphenylarsonium bromide, **1**, 740
with *p*-chlorobenzyltriphenylarsonium bromide, **1**, 740
with dibenzophosphepin, **1**, 445
with diethyl *t*-butoxycyanomethyl phosphonate, **3**, 650
with dimethyl sulfoxide, **4**, 391
with *β*-hydroxyarsonium salts, **1**, 646
with *β*-hydroxyphosphine oxides, **1**, 641
with methyl dichlorofluoroacetate, **4**, 752
with phosphoramidates, **5**, 975
with phosphoramides, **2**, 1057
with tertiary phosphine oxides, **2**, 454
with thionoamides, **5**, 632
with thionocarbamates, **5**, 632
with trialkoxyboranes, **2**, 109
with trimethoxyborane, **2**, 109
See also Dimsyl anion
Sodium hydrogen carbonate
reactions
with acyl trimethylsilyl phosphates, **5**, 189
with tetra-*n*-butylammonium bromide, **5**, 183, 195
use in the synthesis of thionoamides, **5**, 568
Sodium hydrogen selenide
reactions
with aryl aldehydes, **2**, 735
with cinnamaldehyde, **3**, 39
with diazonium salts, **2**, 734
with imines, **5**, 626
as starting material
for selenoacridone, **3**, 398
for selenocarboxylates, **5**, 516
use in the synthesis of selenals, **3**, 388
use in the synthesis of selenoamides, **5**, 625
use in the synthesis selenocarbonates, **6**, 589
use in the synthesis of selenocarbonyls, **6**, 593
use in the synthesis of selenothiocarbamates, **6**, 594
Sodium hydrogen sulfide
reactions
with alkyl chlorides, **2**, 119
with α-bromo ketones, **3**, 164
with calcium chloride, **2**, 764
with 2-chloro-5-nitrobenzoyl isothiocyanate, **6**, 566
with *N*-(2,6-dimethylphenyl)-*N*-isothiocyanatomethyl chloroacetamide, **6**, 565
with pyran-4-thiones, **3**, 255
as reagent
for the synthesis of thioketones, **3**, 357, 360, 362, 363
for the synthesis of trithiapentalenes, **3**, 376
Sodium hydrogen telluride
reactions
with benzoyl chloride, **5**, 253
with Bunte salts, **2**, 122
with diiodorhodium complex, **3**, 391
with imminium salts, **5**, 627
with methyl carbamates, **2**, 304
with α-methyl thiosulfones, **1**, 439
for the reductive cleavage of disulfides, **2**, 127
use in the reduction of nitroalkanes, **1**, 37
Sodium hydroselenide, reactions, with selenophosphinic chlorides, **2**, 465
Sodium hydrosulfide
reactions
with dialkoxycarbenium tetrafluoroborates, **5**, 512
with dioxocarbenium salts, **6**, 96
with phosphinic halides, **2**, 462
with selenophosphinic chlorides, **2**, 465
with thiophosphinic halides, **2**, 462
Sodium hydroxide
reactions
with benzamidines, **5**, 763
with benzene sulfonic acid, **2**, 647
with 3-chloro-4-cyanobiphenyl, **5**, 108
with 1-(2-chloroethyl)-1,3-diphenylurea, **6**, 508
with *N*-cyano-*S*-methylisothiourea, **5**, 1118
with diaryl ditellurides, **4**, 291
with *N,N'*-dibenzoyl-*N,N'*-dinitroso-1,3-propanediamine, **3**, 467
with di-*t*-butylchlorostibine, **2**, 493
with *β*-diketones, **1**, 852
with 6,7-dimethyloxytetralin-2,3-dicarboxylic anhydride, **5**, 37
with diphenyl disulfide, **4**, 284
with fluoradene, **1**, 862
with furfural, **5**, 74
with *β*-keto esters, **1**, 852
with 2-methoxymethanol, **1**, 42
with methyl 2-pyridone-5-carboxylate, **5**, 106
with organodisilylphosphanes, **5**, 882
with propargylbenzotriazole, **1**, 956
with selenium, **5**, 253
with tetraalkylphosphonium salts, **2**, 452
with tetra-*n*-butylammonium chloride, **5**, 183
with 1,2,4-thiadiazolium salts, **5**, 759
with thiols, **2**, 270
with zinc, **2**, 772
use in the synthesis of phosphaalkynes, **5**, 1154
Sodium hydroxylamine-*O*-sulfonate, reactions, with sodium thiocarboxylates, **5**, 253
Sodium hypobromite
Hofmann rearrangement, with trifluoroacetamide, **6**, 218
reactions
with acetone, **6**, 215
with fluorinated alkenes, **4**, 45
with 2-fluoro-2-nitro-1-butanol, **6**, 55
with tribromomethane, **6**, 215
with triiodomethane, **6**, 223
as reagent, for the synthesis of *N*-bromoimides, **5**, 318
use in the synthesis of *N*-bromoamines, **2**, 761
Sodium hypochlorite
alcohol oxidations with, **3**, 118
for the chlorination of amides, **5**, 317
for the chlorination of amines, **2**, 337
as oxidizing agent, **6**, 410
reactions
with amine salts, **2**, 337
with ammonium dithiocarbamate, **5**, 1025
with 1,3,5-trithiane, **6**, 254
Sodium hypoiodite
reactions
with ethyne, **2**, 1019
with tribromomethane, **6**, 223
with trichloromethane, **6**, 221
Sodium intermediates, alkynyl-, synthesis of, **2**, 1076
Sodium iodide
reactions

with alkyl chlorides, **3**, 612
with dibromochloromethane, **6**, 33
with dichloromethane, **4**, 28
with methyl trichlorosilane, **5**, 39
with sulfenyl chlorides, **2**, 172
use in the synthesis of acid iodides, **5**, 19
use in the synthesis of alkenes, **1**, 593
Sodium iodomethanesulfonate
reactions
with phosphorus pentabromide, **4**, 85
with phosphorus pentachloride, **4**, 85
Sodium–iron complexes
reactions, with aryl iodides, **5**, 452
synthesis of, **5**, 451
Sodiumiron tetracarbonyl, reactions, with praseodymiums, **6**, 518
Sodium–iron tetracarbonyl complexes
reactions, with alkyl halides, **5**, 451
synthesis of, **5**, 451
Sodium isopropoxide
reactions
with chloroformamidinium chlorides, **4**, 974
with thiocarbamoyl chlorides, **5**, 609
Sodium isothiocyanate, reactions, with *O*-alkyl carbonochlorodithioates, **6**, 558
Sodium–lead alloys, reactions, with alkyl halides, **2**, 601
Sodium manganese pentacarbonyl, reactions, with glycosyl bromides, **5**, 446
Sodium–mercury amalgams, in defluorination reactions, **1**, 570
Sodium metaperiodate
for the oxidation of dithioacetal oxides, **4**, 268
for the oxidation of sulfides, **2**, 149
Sodium methaneselenolate, synthesis of, **6**, 92
Sodium methanethiolate, reactions, with *S*,*S*-diethyl 1-chlorovinylphosphonodithioate, **4**, 927
Sodium methoxide
reactions
with amines, **2**, 760
with bis(4-biphenylcarbonyl) diselenide, **5**, 233
with bis(trimethylsilyl)methane, **4**, 653
with bis(trimethylsilyl) sulfide, **5**, 582
with α-bromoboronate, **4**, 363
with *t*-butyl diazoacetoacetate, **3**, 475
with α-chlorosulfenyl chloride, **3**, 375
with 1,2-dichloroethene, **2**, 1014
with 1,1-dichloro-2-nitroethene, **6**, 74
with ethynyl azaarenes, **4**, 195
with halogenated styrenes, **4**, 762
with hexachlorophenylalkyne, **4**, 764
with 3-(2-nitroethyl)indole, **5**, 665
with perfluoropropene, **6**, 74
with quinoline, **4**, 195
with tetrafluoroethylene, **4**, 761
with 1,2,2-trichloroacrylonitrile, **4**, 762
with tris(trimethylsilyl)methane, **4**, 619
with 4-twistanone, **1**, 879
for the reduction of amines, **1**, 34
Sodium methoxide, methanolic-, reactions, with tris(trimethylsilyl)methylsilicon halides, **6**, 177
Sodium (*p*-methoxybenzene)selenolate, reactions, with 2-chloro-4,6-dimethyl-1,3-dithiane, **6**, 101
Sodium *O*-methyl dithiocarbonate, reactions, with α-chloroacetophenone oxime, **6**, 546
Sodium methyl selenide, synthesis of, **4**, 283
Sodium naphthalenide
reactions, with benzoates, **1**, 590
use in the synthesis of alkylmagnesiums, **2**, 558
Sodium naphthelenide, for the cleavage of tertiary diphosphines, **1**, 445
Sodium nitrate
as diazotizing agent, for aryl iodides, **2**, 632
reactions, with 1-acyl hydrazines, **5**, 367
for the synthesis of aryl bromides, **2**, 627
Sodium nitrite
electrolysis of, **2**, 786
reactions
with amides, **5**, 36
with amines, **2**, 803
with 3-amino-2-phenylindenone, **2**, 801
with arylamine, **1**, 442
with 2,6-dibromophenol, **2**, 780
with fluoroalkenes, **6**, 54
with hydrochloride salts, **1**, 259
with *N*-methyl-*N'*-(2-phenylethyl)urea, **6**, 506
with nitroalkanes, **4**, 428
with substituted ureas, **5**, 980
with terpenoids, **1**, 1063
Sodium nitrocyanamide, reactions, with alkyl iodides, **5**, 964, 1119
Sodium nitronates, reactions, with allyl halides, **1**, 156
Sodium nitro phenoxide, reactions, with *N*-(chloromethylphosphinoyl)imidoyl chlorides, **5**, 707
Sodium oxalate, reactions, with fluorine, **6**, 250
Sodium oximate, reactions, with nitroarenes, **2**, 648
Sodium perborate
as hydroborating reagent, **2**, 43
for the oxidation of arenes, **2**, 650
as oxidizing agent, for the synthesis of acyloxyiodoarenes, **5**, 204
use in the synthesis of dialkyl sulfones, **2**, 155
Sodium perfluoropropionate
as starting material, for tetrafluoroethylene, **4**, 731
thermal decarboxylation of, **4**, 731
Sodium periodate
for the cleavage of alkanes, **4**, 168
for the oxidation of alkynyl thioethers, **2**, 1034
for the oxidation of dithioacetals, **4**, 259, 267
for the oxidation of selenides, **2**, 287
for the oxidative cleavage of alkenes, with osmium(VIII) oxide, **3**, 113
reactions
with 5,5-dimethyl-1,3-dithiane, **4**, 268
with 1,3-dithiane, **4**, 268
with osmium(VIII) oxide, **3**, 2
with oxathiolanes, **5**, 411
as reagent, for hydrazone cleavage, **3**, 133
use in the synthesis of aldehydes, via 1,2-diol cleavage, **3**, 8
use in the synthesis of aromatic acids, **5**, 114
Sodium phenolates
reactions
with bismuth(III) halides, **2**, 695
with bismuth(V) halides, **2**, 695
with phosphorochloridites, **2**, 693
with trichloroethylene, **4**, 761
transmetallation of, **2**, 695
Sodium phenoxide
reactions
with polychlorinated ethanes, **4**, 761
with polyhalogenated ethylenes, **4**, 761
Sodium phenylalkynide, reactions, with tosyl azide, **2**, 1055
Sodium phenylmethanethiolate, reactions, with polychlorobutene, **4**, 770
Sodium phenyl selenide, reactions, with α-brominated sulfone, **4**, 286
Sodium phenyl sulfinate, reactions, with diiodomethane, **4**, 281
Sodium phosphides, reactions, with carbeniumtetrafluoroborates, **6**, 686
Sodium–potassium alloy, reactions, with dichloro(diisopropylamino)borane, **6**, 184
Sodium ruthenate, use in the synthesis of cinnamic acid, **5**, 101
Sodiums, alkyl-
from, organomercury compounds, **2**, 556
synthesis of, **2**, 556, 599

Sodiums, alkynyl-
  reactions
    with aluminum trichloride, **2**, 1098
    with chloroaminoborane, **2**, 1085
  for the synthesis of alkynylborons, **2**, 1083
  use in the synthesis of alkynylsilicons, **2**, 1089
Sodiums, vinyl-
  silylation of, **2**, 910
  synthesis of, **1**, 113
Sodium salts
  for the displacement of aryl bromides, **2**, 874
  reactions
    with *t*-butyldimethylsilyl chloride, **5**, 542
    with halophosphite, **5**, 403
  rearrangement of, **1**, 378
  synthesis of, **4**, 545
    in situ, **5**, 183
Sodium sand, reactions, with α-bromo ortho esters, **4**, 827
Sodium selenide
  alkylation of, **2**, 278
  reactions, with phosphonates, **4**, 348
Sodium selenite–cysteine reduction, use in the synthesis of alkenes, **1**, 568
Sodium selenocarboxylates, oxidation of, **5**, 244, 252
Sodium selenocyanate, reactions, with imidoyl chlorides, **5**, 1057
Sodium selonates, synthesis of, **6**, 92
Sodium succinimide, reactions, with tetraphenylarsenic chloride, **5**, 372
Sodium sulfamates, synthesis of, **2**, 366
Sodium sulfide
  reactions
    with biphosphine disulfides, **2**, 462
    with bis(halomethyl)dialkylgermanes, **4**, 372
    with *t*-butyldichlorostibine, **2**, 500
    with chlorodi-*t*-butylstibine, **2**, 499
    with chloroenals, **3**, 72
    with 3-chloro-2-methyl-but-2-enal, **3**, 72
    with chlorotrimethylsilane, **5**, 582
    with α-cyano-β-nitro sulfones, **3**, 645
    with dialkoxycarbenium tetrafluoroborates, **5**, 512
    with dihalides, **2**, 224
    with dimethyliodoarsine, **2**, 499
    with dioxocarbenium salts, **6**, 96
    with mesylates, **2**, 224
    with β-nitroenamines, **5**, 611
    with phosphorus pentasulfide, **5**, 582
    with potassium ferricyanide, **3**, 377
    with tosylates, **2**, 224
    with tricyclohexylantimony dibromides, **2**, 502
    with triethylcyclohexylantimony dibromides, **2**, 502
  as reagent, for the synthesis of thioketones, **3**, 356, 357, 360
  for the reductive cleavage of disulfides, **2**, 127
  use in the synthesis of arylamines, **2**, 753
  use in the synthesis of disulfides, **2**, 224
  use in the synthesis of thionoamides, **5**, 580
Sodium sulfide nonahydrate, reactions, with alkyl chloroformates, **6**, 475
Sodium sulfides, acylation of, **5**, 242
Sodium sulfinate
  reactions
    with alkynyliodonium salts, **2**, 1036
    with thionyl chloride, **2**, 688
Sodium sulfinates
  reactions, with zinc chloride, **2**, 273
  synthesis of, **2**, 271
Sodium sulfite
  reactions
    with bromoethanesulfonyl chloride, **4**, 85
    with chloromethanesulfonyl chloride, **4**, 85
    with iodoform, **4**, 281
    with sulfonyl chlorides, **5**, 1112
Sodium telluride
  alkylation of, **4**, 377
  reactions
    with alkyl chloroformates, **6**, 481, 482
    with alkyl halides, **2**, 279
    with aroyl chlorides, **5**, 244
    with aryl iodides, **2**, 735
  as starting material, for symmetrical tellurides, **2**, 279
  use in the synthesis of *p*-methoxytoluenes, **1**, 46
Sodium tellurite, for the oxidation of thiols, **2**, 216
Sodium tellurocarboxylates, reactions, with *t*-butyldimethylsilyl chloride, **5**, 255
Sodium tetraalkynylborates, hydrolysis of, **2**, 1084
Sodium tetracarbonylcobaltate
  reactions, with difluorofumaric dichloride, **4**, 821
  as starting material, for cobaltacyclopentadiene complexes, **4**, 689
Sodium tetrachloropalladate, use in metallation reactions, **4**, 720
Sodium tetrahydroborate, use in the synthesis of α-aminoalkyl phosphonates, **4**, 472
Sodium tetraphenylborate
  reactions
    with alkyl iodides, **3**, 290
    with amino acids, **5**, 228
Sodium thiocarboxylates, reactions, with sodium hydroxylamine-*O*-sulfonate, **5**, 253
Sodium thiocyanate
  reactions
    with 2-acetamidothiophene, **5**, 1108
    with iodonium triflate, **2**, 1033
    with phenyliodoniumalkynes, **5**, 1108
    with thioaroyl chlorides, **5**, 1050
  use in the synthesis of α,β-unsaturated ketones, **3**, 252
Sodium thioethoxide, for the hydrolysis of aromatic esters, **5**, 106
Sodium thiolate, reactions, with tetrachloroethene, **4**, 770
Sodium thiolate methyl esters, alkylation of, **2**, 270
Sodium thiolates
  reactions, with alkylmercury compounds, **2**, 274
  synthesis of, **2**, 270
Sodium thiophenoxide
  for the hydrolysis of aromatic esters, **5**, 106
  reactions, with 1,1,2,2-tetrachloroethane, **4**, 771
  synthesis of, **4**, 284
Sodium thiophosphate, reactions, with alkyl halides, **2**, 255
Sodium thiosulfate
  reactions, with α-haloketones, **5**, 601
  use in the synthesis of Bunte salts, **2**, 122
  use in the synthesis of disulfides, **2**, 227
Sodium–titanium compounds, use in the synthesis of amines, **2**, 314
Sodium toluene-*p*-sulfinates
  reactions
    with acrylonitriles, **4**, 783
    with alkenyl halides, **4**, 781
Sodium triacetoxyborohydride, use in the synthesis of amines, **2**, 310
Sodium trialkylaluminum hydrides, reactions, with nitriles, **3**, 458
Sodium tri(*t*-butoxy)aluminum hydride
  reactions
    with acyl chlorides, **3**, 98
    with 4-methylbenzoyl chloride, **3**, 93
  as reducing agent, **3**, 83
Sodium trichloroacetate
  decomposition of, **4**, 740
  reactions
    with enamine, **3**, 235
    with thiofluorenone *S*-oxide, **4**, 742
  as starting material, for dichlorocarbenes, **6**, 25
  thermal decomposition of, **6**, 25
  thermolysis of, **3**, 721

Sodium triethylborohydride, reactions, with ethynyltrimethylsilane, **2**, 1088
Sodium trifluoroacetate
reactions
with copper iodide, **6**, 14
with fluorine, **6**, 250
Sodium 2,2,2-trifluoro ethoxide, reactions, with phenyltetrakis(2,2,2-trifluoroethoxy)phosphorane, **2**, 870
Sodium trimethoxyborohydride
for the reduction of acid chlorides, **2**, 59
as starting material, for dialkoxyboranes, **2**, 109
thermal decomposition of, **2**, 109
Sodium trimethylsilanethiolate, synthesis of, **5**, 582
Sodium tris(pentafluorophenoxy)borohydride, use in the reduction of substituted cyclohexanones, **2**, 59
Sodium trithiocarbonate
alkylation of, **6**, 554
reactions
with alkyl halides, **2**, 123
with chloroacetonitrile, **3**, 634
with *cis*-3,5-dibromocyclopentene, **6**, 554
with pentachlorobenzenesulfenyl chloride, **6**, 535
with sulfenyl chlorides, **6**, 554
with trichloromethanesulfenyl chloride, **6**, 554
use in the synthesis of alkanoic acids, **5**, 39
(*S*)-(-)-Solanone, synthesis of, **1**, 1097
Solenopsin B, synthesis of, **3**, 437
Solid-phase synthesis, of peptides, **5**, 259, 262
Solid-supported oxidants
for the synthesis of aldehydes, **3**, 4
for the synthesis of ketones, **3**, 115
Solvent effects, in radical reactions, **1**, 320, 325
Solvent polarity, effect on regioselectivity, **1**, 356
Solvents
aprotic, for lithiation, **6**, 134
chlorofluorocarbons for, **6**, 7
choice of
for radicals, **1**, 923
for 1,2 Stevens rearrangements, **1**, 399
for the synthesis of alkynide ions, **1**, 1133
for the synthesis of alkynylantimony compounds, **2**, 1081
for the synthesis of nitriles, **3**, 613
for the synthesis of ylides, **3**, 493
for Claisen rearrangement, **1**, 826
effects of, in the opening of vinyl epoxides, **1**, 227
role in transmetallation, **2**, 990
for the synthesis of alkenylmagnesiums, **1**, 1092
for the synthesis of alkynes, choice of, **1**, 1000, 1037
for the synthesis of 2-alkynoic acids, choice of, **1**, 1040
for the synthesis of isocyanates, **5**, 964, 984, 985
use in the synthesis of carbanions, **1**, 845
use in the synthesis of thionoamides, choice of, **5**, 566
in Wittig reactions, **1**, 727
Solvent separated ion pairs (SSIP), equilibrium, with contact ion pairs, **1**, 302
Solvolysis
of allenes, **1**, 1126
of β-bromostyrenes, **1**, 1110
factors affecting, **1**, 1113
rates of, **1**, 1111
for the synthesis of vinyl cations, **1**, 1111
Solvomercuration
of alkynes, **1**, 66
for the synthesis of radicals, **1**, 948
Sommelet–Hauser products
from, dibenzyltrimethylammonium iodide, **1**, 399
synthesis of, **1**, 399
Sommelet–Hauser rearrangement
[1,3] hydrogen shift in, **1**, 797
mechanism, **1**, 824
reviews, **1**, 407
in the synthesis of ketenimines, **3**, 574
Sommelet oxidation
for the synthesis of formyl thiophenes, **3**, 104
for the synthesis of indole carboxaldehydes, **3**, 106
Sommelet reaction
for the oxidation of benzylamines, **3**, 89
for the synthesis of benzaldehydes, **3**, 88
for the synthesis of furan carboxaldehydes, **3**, 102
for the synthesis of pyridine carboxaldehydes, **3**, 108
Sonication
for the activation of magnesium, **2**, 558
for activation of metal surfaces, **2**, 958
in elimination reaction, **1**, 569
for the synthesis of alkylaluminums, **2**, 595
for the synthesis of alkylzirconiums, **2**, 568
in the synthesis of nitriles, **3**, 613
for the synthesis of radicals, **1**, 925
use in the Strecker reaction, **3**, 616
use in the synthesis of alkyllithiums, **2**, 551
use in the synthesis of lithium diisopropylamide, **2**, 422
Sonn and Müller procedure, for the synthesis of benzaldehydes, **3**, 85
Sonochemical activation, for the synthesis of 1-aminophosphonic acids, **4**, 463
Sorghum bicolour, as source of oxynitrilase enzymes, **3**, 628
Sparsomicin, synthesis of, **4**, 264
(-)-Sparteine
use in Reformatsky reactions, **2**, 69
use in the synthesis of α-alkoxylithiums, **4**, 382
use in the synthesis of carbanions, **1**, 138
Speier's catalyst
for the hydrosilylation of 1,2-bis(silyl) ethynes, **4**, 1044
for the synthesis of 1,3-disilabutanes, **4**, 1054
use in hydrosilylation reactions, **2**, 536
uses of, **4**, 1045
See also Hexachloroplatinic(IV) acid
Spiroacetals, synthesis of, **4**, 171; **6**, 321
Spiroaminals, synthesis of, **4**, 413
Spiroarsonium salts, synthesis of, **2**, 488
Spiroaziridinium salts, synthesis of, **1**, 254
Spirobenzopyrans, synthesis of, **4**, 307
Spirobicyclophosphonium salts, synthesis of, **2**, 444
Spirocarbene complexes, dimetalla-, synthesis of, **6**, 324
Spirocycles, synthesis of, **1**, 184
Spirocyclic compounds, synthesis of, **1**, 120
Spirocyclotrimerization, of phosphaalkynes, **5**, 885
Spiro[4,5]decane, synthesis of, **1**, 384
Spiro[4.5]decane, 1,4,6,10-tetraoxa-, synthesis of, **6**, 300
Spiro[5.6]dodecane, 1,5,7,12-tetraoxa-, synthesis of, **6**, 300
2,2'-Spiroimidazolidine-4',5'-diones, as stabilizers, **6**, 360
Spiroketals
ring-opening of, **1**, 572
studies, **1**, 572
Spirolactones, synthesis of, **4**, 204
Spirolactones, isocyanide, from, β-epoxy isocyanides, **3**, 718
Spiro[4.4]nonane, 1,4,6,9-tetraoxa-, synthesis of, **6**, 300
Spirooxazines, synthesis of, **4**, 307
Spiropentane, *trans*-1,2-dinitro-, as starting material, for diiodo compounds, **4**, 108
Spirophosphoranes
reactions, with *N*-methylbenzylimine, **4**, 478
synthesis of, **4**, 134
Spiro–Reformatsky addition, for the synthesis of γ-lactones, **5**, 166
Spirosystems, synthesis of, **1**, 193
Spiro[2.2]tetraphosphapentane, synthesis of, **6**, 366
Spiro[6.6]tridecane, 1,6,8,13-tetraoxa-, synthesis of, **6**, 300
Spodoptera littoralis, sex pheromones from, **1**, 734
Squalene, synthesis of, **1**, 822
trans-Squalene, synthesis of, **1**, 721
Squalene oxide, cyclization of, **1**, 216, 294
Squarate dianions, synthesis of, **1**, 879
Squaric acid diamides, monothio compounds, synthesis of, **3**, 355

Squaric acids
from, *N*-cyanopyrrolidine, **6**, 623
synthesis of, **3**, 354
SRN1 mechanism, **1**, 250
Stabase compounds, use in the protection of amines, **2**, 543
Stability
of alkyl and vinyl radicals, **1**, 321
of alkynyl compounds, **2**, 1076
of carbanions, **1**, 846
Stabilization, for the synthesis of carbenes, **4**, 1082
Stabilized radicals, cyclization of, **1**, 343
Stannaalkenes, synthesis of, **6**, 724
1-Stanna-3-boracyclopentenes, synthesis of, **4**, 665
3-Stanna-1,4-cyclohexadienes, synthesis of, via hydrostannation, of 1,4-pentadiyne, **2**, 971
Stannacyclopentadienes, 2-stannyl, synthesis of, **4**, 1068
Stannadiyl, diaryl-, reactions, with mesityl isocyanide, **5**, 1086
Stannaethenes
reactions, with phenyllithium, **6**, 193
synthesis of, **6**, 202
Stannane, alkyltrichloro-, reactions, with lithium diethylphosphide, **2**, 474
Stannane, alkynyl-, reactions, with arylselenium halides, **2**, 1037
Stannane, allyl-
reactions, with imide radicals, **1**, 328
use in intermolecular–intermolecular tandem radical additions, **1**, 348
Stannane, benzyloxymethyl-, synthesis of, **4**, 389
Stannane, benzyltrimethyl-, reactions, with tetracyanoethylene, **5**, 1145
Stannane, bis(bromomethyl)dimethyl-, synthesis of, **4**, 154
Stannane, bis(α-chloroethyl)diethyl-, synthesis of, **4**, 154
Stannane, bis(dimethylamino)dimethyl-, reactions, with ethyl diazoacetate, **3**, 484
Stannane, bis(silylmethyl)-, synthesis of, **4**, 660
Stannane, bis(trifluoromethyl)-, synthesis of, **6**, 245
Stannane, (bromomethyl)trimethyl-, synthesis of, **4**, 154
Stannane, α-chloroalkoxy-, as starting material, for α-alkoxystannanes, **4**, 388
Stannane, α-chlorobenzyltri-*n*-butyl-, synthesis of, **4**, 154
Stannane, chlorotrimethyl-
reactions
with pent-4-enylzinc–magnesiums, **4**, 724
with phosphines, **2**, 474
with Simmons–Smith reagents, **4**, 154
Stannane, dibutyl bis(phenylthio)-, as reagent, for the synthesis of monothioacetals, **4**, 216
Stannane, di-*t*-butyldichloro-, use in the synthesis of diphosphadistannylbutanes, **2**, 474
Stannane, α,α-dichloro-, reduction of, **4**, 154
Stannane, dichlorodimethyl-, reactions, with Simmons–Smith reagents, **4**, 154
Stannane, ((2,6-dimethylphenylamino)(trimethylsilyl)methyl)-, use in the synthesis of imidoylsilanes, **5**, 793
Stannane, heptynyl tributyl-, reactions, with (1*E*)-chloroheptene, **1**, 1012
Stannane, (iodomethyl)trimethyl-, reactions, with phenylselenolithium, **4**, 401
Stannane, methoxy tributyl-, use in the synthesis of carbodiimides, **5**, 1081
Stannane, methoxytripropyl-, reactions, with acetyl isocyanate, **5**, 716
Stannane, mono(trifluoromethyl)-, synthesis of, **6**, 245
Stannane, pentynyltributyl-, reactions, with cyclopentenyl iodide, **1**, 1015
Stannane, 1,2-propadienyl tributyl-, reactions, with aldehydes, **3**, 30
Stannane, tetrakis(trifluoromethyl)-, as starting material, for tris(trifluoromethyl)tin iodide, **6**, 245
Stannane, tributyl-
reactions, with 4-bromobenzoyl chloride, **3**, 95
use in the synthesis of acyl aldehydes, **3**, 83
Stannane, tributylchloro-, reactions, with lithio compounds, **4**, 961
Stannane, tributyl(tribromomethyl)-, reactions, with aldehyde, **6**, 31
Stannane, tributyl(trichloromethyl)-, reactions, with aldehydes, **6**, 26
Stannane, tributyl(trimethylsilylmethyl)-
deprotonation of, **4**, 662
synthesis of, **4**, 660
Stannane, triethylethoxy-, reactions, with *O*-trimethylsilyl thiocarboxylates, **5**, 540
Stannane, triethylmethoxy-, reactions, with diazoacetates, **3**, 484
Stannane, trifluoromethyl(trimethyl)-, synthesis of, **6**, 245
Stannane, (trifluoromethyl)trimethyl-
chlorination of, **6**, 245
synthesis of, **6**, 244
Stannane, triphenyl-
reactions
with cyclopentadienylcobalt complexes, **4**, 726
with phenylethyne, **4**, 701
for the reduction of tellurides, **1**, 25
Stannane, tris(trifluoromethyl)-, synthesis of, **6**, 245
Stannane, vinyl-
as starting material, for vinyl bromides, **2**, 623
synthesis of, **5**, 428
Stannane–borane–air method, for the cyclization of selenium compounds, **1**, 244
Stannane hydride, trimethyl-, reactions, with perfluorocyclobutenes, **4**, 153
Stannane hydrides, trialkyl-, reactions, with fluoroalkenes, **4**, 153
Stannanes
for the alkylation of oxiranes, **1**, 213
for the allylation of oxetanes, **1**, 229
characteristics of, **4**, 951
cleavage of, **2**, 555
cyclization of, **1**, 216
dimerization of, **1**, 491
from, aminals, **4**, 533
in Heck reactions, **1**, 494
reactions
with sulfur dioxide, **2**, 210
with vinyl halides, **4**, 8855
for the reduction of selenides, **1**, 25
as starting materials
for dialkylaminoboranes, **2**, 418
for tertiary ynamines, **2**, 1043
for tris(dialkylamino)arsines, **2**, 415
for tris(dialkylamino)boranes, **2**, 419
Still–Wittig rearrangement of, **1**, **793–842**
transmetallation of, **1**, 405; **4**, 531
use in allylation reactions, **2**, 74
Stannanes, acyl-
asymmetric reduction of, **4**, 380, 386
reactions
with allyl halides, **3**, 142
with nucleophiles, **4**, 951
with phenylthiotrimethylsilane, **4**, 951
reduction of, **1**, 138; **4**, 386
synthesis of, **5**, 467
via oxidation of stannyl carbinols, **5**, 467
Stannanes, acylamino-, synthesis of, **4**, 533
Stannanes, acyloxy-, reactions, with trimethylsilyl cyanide, **5**, 1144
Stannanes, α-acyloxy-, hydrolysis of, **4**, 389
Stannanes, acyltributyl-, enantioselective reduction of, **4**, 386
Stannanes, alkenyl-, reactions, with lead(IV) acetate, **1**, 1081
Stannanes, (*E*)-alkenyl-
as intermediates, **1**, 535

synthesis of, **4**, 813
Stannanes, (*Z*)-alkenyl-, transmetallation of, **1**, 536
Stannanes, alkoxy-
reactions, with enol acetates, **2**, 701
use in the synthesis of disulfides, **2**, 215
Stannanes, 1-alkoxy-
as starting materials, for alkyllithiums, **2**, 68
transmetallation of, **2**, 68
Stannanes, α-alkoxy-
from
carbonyls, **4**, 386
α-chloroalkoxystannane, **4**, 388
α-chloroboronic esters, **4**, 389
chloromethyl ethers, **4**, 389
Fischer carbene complexes, **4**, 388
tributylstannyl acetals, **4**, 388
tributylstannylmethyl iodide, **4**, 387
tributyltin chloride, **4**, 387
reactions
with alkyllithiums, **1**, 138
with butyllithium, **4**, 378
as reagents, for the synthesis of α-alkoxy carbanions, **1**, 138
synthesis of, **1**, 138
transmetallation of, **2**, 999; **4**, 384
Stannanes, 4-alkoxyallyl-, use in the allylation of aldehydes, **2**, 79
Stannanes, 5-alkoxyallyl-, reactions, with aldehydes, **2**, 79
Stannanes, (alkoxymethyl)trihalo-, as intermediates, in the synthesis of α-alkoxylithiums, **4**, 389
Stannanes, α-alkoxysilyl-, reactions, with *n*-butyllithium, **4**, 358
Stannanes, 1-alkoxyvinyl-, synthesis of, **4**, 951
Stannanes, α-alkoxyvinyl-
reactions, with vinyl triflates, **3**, 243
transmetallation of, **1**, 455
Stannanes, β-alkoxyvinyl-, reactions, with acid chlorides, **3**, 245
Stannanes, alkyl-, reactions, with aryl iodide, **1**, 471
Stannanes, alkynyl-
as alkynyl transfer reagent, **2**, 1077
from, sodium alkynides, **2**, 1100
reactions
with aryl triflates, **1**, 1030
with diisobutylaluminum hydride, **2**, 967
synthesis of, reviews, **2**, 1099
Stannanes, 1-alkynyl-, reactions, with bromosilanes, **2**, 1089
Stannanes, alkynyltributyl-, reactions, with acetals, **1**, 1026
Stannanes, alkynyltrichloro-, synthesis of, **2**, 1100
Stannanes, alkynyltrimethyl-, use in the synthesis of alk-1-ynyllead triacetates, **2**, 1102
Stannanes, allenyl-, reactions, with aldehydes, **1**, 1057
Stannanes, allenyltrialkyl-
reactions
with acetals, **1**, 1053
with allylic acetates, **1**, 1053
with ketals, **1**, 1053
Stannanes, allyl-
alkylation of, **1**, 239
oxidation of, **1**, 664
use in the allylation of carbonyls, **2**, 78
use in radical cyclizations, **1**, 344
use in radical fragmentations, **1**, 334
use in the synthesis of allylboranes, **2**, 529
use in the synthesis of dialkyl sulfides, **2**, 141
use in the synthesis of unsaturated ketones, **3**, 140
Stannanes, *N*-allyl aminomethyl-, as starting materials, for homoallylic amines, **1**, 406
Stannanes, allylic, reactions, with dimethylcopper–lithium compounds, **2**, 589
Stannanes, allyl-, tartrate-modified, synthesis of, **2**, 79
Stannanes, allyltrialkyl-, reactions, with aldehydes, **2**, 78
Stannanes, amino-
reactions
with carbonyls, **2**, 701
with di-*t*-butylphosphine, **2**, 474
with silylalkynes, **2**, 1092
with terminal alkynes, **2**, 1100
synthesis of, **4**, 531
Stannanes, α-amino-
condensation of, with aldehydes and ketones, **4**, 537
reactions, with *n*-butyllithium, **2**, 1001
synthesis of, **4**, 528, 533
transmetallation of, **2**, 1001
Stannanes, aminoalkynyl-, as starting materials, for alkynylgermanes, **2**, 1093
Stannanes, aminomethyl-
as amino alkyl anion equivalents, **3**, 182
synthesis of, **2**, 322
transmetallation of, **4**, 530
Stannanes, aryl-
reactions
with acyl halides, **3**, 287
with aryl triflates, **1**, 482
with isocyanates, **5**, 344
as starting materials
for arenes, **1**, 452
for aryl bromides, **2**, 627
for aryl fluorides, **2**, 613
for aryl iodides, **2**, 632
use in the synthesis of biaryls, **1**, 473
Stannanes, aryltrialkyl-, use in the synthesis of benzaldehydes, **3**, 92
Stannanes, aryltributyl-, reactions, with ethoxycarbonyl isothiocyanate, **5**, 593
Stannanes, aryltrimethyl-, as starting materials, for arenes, **1**, 452
Stannanes, α-azido-, rearrangement of, **3**, 457
Stannanes, azidomethyl-
Staudinger reaction of, **4**, 537
synthesis of, **4**, 537
Stannanes, *gem*-bromo-, synthesis of, **4**, 813
Stannanes, α-bromo-
from, aldehydes, **4**, 155
synthesis of, **4**, 154
Stannanes, bromo(bromomethyl)-, synthesis of, **4**, 155
Stannanes, chiral α-alkoxy-, synthesis of, **4**, 386
Stannanes, chloro-
reactions
with 1-metalloalkylphosphine oxides, **4**, 588
with 1-metalloalkylphosphines, **4**, 588
with 1-metalloalkylphosphine sulfides, **4**, 588
with 1-metalloalkyl phosphonates, **4**, 588
with silylphosphines, **2**, 473, 474
Stannanes, α-chloro-, synthesis of, **4**, 154
Stannanes, chlorotrimethyl-, reactions, with methylene(triphenyl)phosphorane, **5**, 920
Stannanes, dialkoxy-, reactions, with dithiocarboxylic esters, **6**, 73
Stannanes, dialkoxydibutyl-, reactions, with carbon disulfide, **6**, 300
Stannanes, disubstituted allyl-, synthesis of, **1**, 566
Stannanes, enamino-, reactions, with α,β-unsaturated esters, **3**, 406
Stannanes, enol-, allylation of, **3**, 139
Stannanes, α-fluoro-, synthesis of, **4**, 153
Stannanes, halo-, reactions, with phosphorus ylides, **4**, 588
Stannanes, 1-haloalkyl-
reactions
with dialkyl phosphonites, **4**, 588
with phosphines, **4**, 588
with trialkyl phosphites, **4**, 588
Stannanes, halo(α-haloalkyl)-, synthesis of, **4**, 155
Stannanes, halo(halomethyl)-, synthesis of, **4**, 156
Stannanes, β-hydoxylalkyl trialkyl-, synthesis of, **2**, 600
Stannanes, hydroxy-, bromination of, **4**, 154

Stannanes, α-hydroxy-, kinetic resolution of, **4**, 389
Stannanes, β-hydroxyvinyl-, elimination of, **1**, 664
Stannanes, iodo-, reactions, with iodo(iodomethyl)zinc, **4**, 155
Stannanes, α-iodo-
from, aldehydes, **4**, 155
synthesis of, **4**, 155
Stannanes, α-iodopropyl-, synthesis of, **4**, 155
Stannanes, β-keto-, reactions, with halides, **3**, 195
Stannanes, metallo-
alkylation of, **2**, 600
reactions, with ylides, **5**, 920
synthesis of, **2**, 600
Stannanes, 2-methoxy-, reactions, with tributylplumbyl bromide, **2**, 70
Stannanes, nitro-, characterization of, **4**, 541
Stannanes, organosilyl-, reactions, with nitriles, **3**, 418
Stannanes, organosilyl(*N*-alkylimino)-, synthesis of, **6**, 674
Stannanes, 2-oxygenated allyl-, as allylating reagents, **2**, 79
Stannanes, phenylselenyl-, for the synthesis of selenoglycosides, **4**, 238
Stannanes, 1-phosphonatoalkyl-, reduction of, **4**, 588
Stannanes, phthalimidomethyl-, synthesis of, **4**, 533
Stannanes, polyfluorovinyl-, cleavage of, **1**, 455
Stannanes, propargyl-, as starting materials, for alkynylsilanes, **2**, 1090
Stannanes, α-selenenyl-, synthesis of, **4**, 401
Stannanes, tetraalkyl-, reactions, with carbon monoxide, **3**, 15
Stannanes, tetraalkynyl-, reactions, with triethylborane, **2**, 1088
Stannanes, 3-(tetrahydropyranyloxy)propenyl-, use in the allylation of aldehydes, **2**, 79
Stannanes, trialkyl(alk-1-ynyl)-, reactions, with lead tetraacetate, **2**, 1102
Stannanes, trialkyl(α-haloalkyl)-, synthesis of, **4**, 153
Stannanes, trialkyl(iodomethyl)-, synthesis of, **4**, 155
Stannanes, trialkyl(trifluoromethyl)-, reactions, with trimethyltin hydride, **6**, 65
Stannanes, triaryl(α-haloalkyl)-, synthesis of, **4**, 153
Stannanes, tributyl-, as starting materials, for organolithiums, **4**, 652
Stannanes, tributylallyl-, reactions, with aldehydes, **2**, 79
Stannanes, trihalo-
reactions, with butyllithium, **4**, 380
transmetallation of, **4**, 380
Stannanes, vinyl-
arylation of, **1**, 471
destannylation of, **1**, 440
hydrostannation of, **4**, 701
protonolysis of, **1**, 455
reactions
with alkyllithiums, **2**, 911
with arenesulfenyl chlorides, **4**, 154
with arylazo compounds, **1**, 486
with 1-chloromethyl-4-fluorodiazonabicyclo[2.2.2]octane bis(tetrafluoroborate), **4**, 736
reduction of, **1**, 83
as starting materials
for vinyl bromides, **2**, 624
for vinyl fluorides, **2**, 611
for vinyl halides, **2**, 607
for vinyl iodides, **2**, 630
for vinylsilanes, **2**, 915
synthesis of, **4**, 386, 534
use in radical cyclizations, **1**, 344
use in radical fragmentations, **1**, 334
Stannanes, vinyl trialkyl-, protonolysis of, **1**, 455
Stannates
reactivity of, **2**, 582
synthesis of, **1**, 132
Stannazanes, reactions, with organotin enolates, **2**, 816
Stannic chloride
as catalyst, for the synthesis of acetals, **4**, 181
reactions, with siloxycyclopropanes, **1**, 663
Stannichloride salts, aldimine, synthesis of, **3**, 86
Stannins, dihydro-, as starting materials, for heterobenzenes, **2**, 879
Stannole, 1,1-di-*n*-butyl-2,5-dimethyl-, reactions, with phenylantimony dichloride, **4**, 1041
Stannous chloride
reactions
with diazonium salts, **2**, 769
with propargylic halides, **1**, 973
Stannyl anions
reactions
with alkyl halides, **2**, 600
with epoxides, **2**, 600
with halomethylsilanes, **4**, 660
Stannylation, of alkynes, **1**, 342
Stannyl azide, butyl-, as reagent, in the synthesis of isocyanates, **5**, 992
Stannyl azides, reactions, with carbonyl acid chlorides, **5**, 992
Stannylboration, of 1-alkynes, **2**, 903
Stannyl carbinols
oxidation of, **5**, 467
synthesis of, **5**, 467
Stannyl chloride, trialkyl-, for the transmetallation of acyllithiums, **5**, 468
Stannyl chloride, trimethyl-, reactions, with diazomethane, **3**, 482
Stannyl chlorides, reactions, with monosilylmethylenephosphorane, **6**, 702
Stannyl compounds, synthesis of, **4**, 735
Stannyl compounds, anti-7-trimethyl-, synthesis of, **4**, 155
Stannylcupration, of alkynes, **2**, 977
Stannylene, bis[2-pyridylbis(trimethylsilyl)methyl]-, reactions, with aryl azides, **6**, 196
Stannylene, bis(trimethylsilyl)methyl-, reactions, with ethyne, **2**, 1101
Stannylenes
reactions
with boranediylboriranes, **6**, 202
with cryptocarbenes, **6**, 724
Stannylgermylation, of alkynes, **2**, 947
β-Stannyl groups, elimination of, **1**, 665
Stannyl halide, elimination of, **1**, 664
Stannyl radicals
reactions
with alkynes, **1**, 548
with carbon–carbon double bonds, **1**, 950
Stannylsilylation, of alkynes, **2**, 916
Stannylthiols, reactions, with aryl iodides, **2**, 711
Staudinger phosphorylation, for the synthesis of *N*-triphenylphosphorylidene-1-(benzotriazol-1-yl) methylamine, **1**, 262
Staudinger phosphorylation reaction, with triphenylphosphine, **6**, 365
Staudinger reaction
of azidomethylstannanes, **4**, 537
for the synthesis of *N*-acyl iminophosphoranes, **5**, 373
for the synthesis of alkylphosphineimides, **2**, 468
for the synthesis of alkynylphosphineimides, **2**, 1068
for the synthesis of amines, **2**, 308
for the synthesis of arsinophosphazenes, **2**, 504
for the synthesis of 1-azabicyclo[3.1.0]hexanes, **2**, 324
for the synthesis of iminophosphoranes, **5**, 862
for the synthesis of ketenimines, **3**, 591
for the synthesis of phenylphosphorimidates, **2**, 862
Staudinger reaction, modified, for the synthesis of primary amines, **2**, 309
Stearoyl fluoride, synthesis of, **5**, 3
Stegobiol, synthesis of, **2**, 529
Stephen's method, for the synthesis of naphthaldehydes, **3**, 100

Stephen reaction
for the synthesis of aryl aldehydes, **3**, 86
for the synthesis of 3-formyl pyridines, **3**, 107
Stephens–Castro coupling reaction, in the synthesis of arylalkynes, **1**, 472
Stepwise mechanisms
regioselectivity of, **1**, 350
in the synthesis of azaallyl anions, **1**, 371
Stereoelectronic characteristics, of addition reactions, **1**, 304
Stereoelectronic factors, in asymmetric radical additions, **1**, 327
Stereoisomerization, of vinyllithiums, in the α position, **2**, 961
Stereoisomers, in cycloaddition reactions, **1**, 351
Stereoselectivity
in conjugate addition reactions, **1**, 305
of electrocyclic addition, **1**, 351
in hydrogenation reactions, **1**, 505
in intermolecular radical additions, **1**, 325
in substituted 5-hexenyl radical cyclizations, **1**, 337
Steric effects, in radical additions, **1**, 323
Steroidal compounds, chloroethynyl, synthesis of, **2**, 1016
Steroids
chlorination of, regioselective, **2**, 20
conformational analysis of, **1**, 773
dehydrobromination of, **4**, 748
modification of, **1**, 993
(±)-oestrone, **1**, 448
synthesis of, **1**, 113, 578, 1122; **3**, 718
via ring-expansion of *N*-thioacylaziridine, **6**, 571
Steroids, bromoethynyl-substituted, synthesis of, **2**, 1018
Steroids, *gem*-dinitro-, synthesis of, **4**, 429
Steroids, 17-ethynyl-17-hydroxy-, cobalt carbonyl complexes of, **4**, 718
Steroids, fluoro-, synthesis of, **4**, 9
Steroids, 17-hydroxy-, as starting materials, for chloroformates, **6**, 489
Steroids, 17-oxo-, reactions, with arenesulfonylmethyl isocyanides, **4**, 908
Steroid side chains, synthesis of, **4**, 948
Sterols, dehydration of, **1**, 578
Stevens rearrangement
for the synthesis of α-aminosilanes, **4**, 509
for the synthesis of tertiary amines, **2**, 303
of ylides, **1**, 233
1,2 Stevens rearrangement, mechanism, **1**, 398
2,3 Stevens rearrangement, mechanism, **1**, 407
Stibacymantrenes, synthesis of, **4**, 1042
Stibaferrocenes, synthesis of, **4**, 1038, 1039
Stibane, bis(nitro)methyl-, synthesis of, **6**, 161
Stibane, chlorodimethyl-, as starting material, for α-stibino methylenephosphoranes, **5**, 912
Stibanes, synthesis of, **6**, 168
Stibanes, metallo-, as starting materials, for α-stibino methylenephosphoranes, **5**, 912
Stibaoxane, cyclo-*t*-butyl-, synthesis of, **2**, 495
Stibaoxanes, alkyl-, synthesis of, **2**, 495
Stibathiolanes, synthesis of, **2**, 888
Stibenes, aryl-, as starting materials, for arenes, **1**, 445
Stibides
from, stibines, **2**, 510
reactions, with chlorosilanes, **2**, 508
use in the synthesis of antimony–gallium bonds, **2**, 511
use in the synthesis of antimony–indium bonds, **2**, 511
Stibides, alkyl-
alkylation of, **2**, 485
synthesis of, reviews, **2**, 509
Stibides, dialkyl-, synthesis of, **2**, 510
Stibides, trimethylsilyl-, from, organostibine halides, **2**, 894
Stibine, alkyldiphenyl-, reactions, with hydrogen chloride, **2**, 490
Stibine, benzylbis(trimethylstannyl)-, synthesis of, **2**, 511
Stibine, bis(1-chloroethyl)chloro-, reactions, with hydrogen peroxide, **2**, 498
Stibine, bromodimethyl-
hydrolysis of, **2**, 493
reactions, with trimethylsilanol, **2**, 494
Stibine, butyldichloro-, hydrolysis of, **2**, 495
Stibine, *t*-butyldichloro-
reactions, with sodium sulfide, **2**, 500
reduction of, **2**, 507
Stibine, *t*-butyldiethoxy-, transesterification of, **2**, 496
Stibine, *t*-butyldimethyl-, synthesis of, **2**, 484
Stibine, butylethylpropyl-, synthesis of, **2**, 485
Stibine, chlorodialkyl-, reactions, with *N*-metallated aminophosphine, **2**, 472
Stibine, chlorodi-*t*-butyl-
reactions, with sodium sulfide, **2**, 499
synthesis of, **2**, 489
Stibine, chlorodiphenyl-, reactions, with silver salts, **6**, 369
Stibine, dibromo(methyl)-, reactions, with dithiols, **2**, 501
Stibine, di-*t*-butylchloro-, reactions, with sodium hydroxide, **2**, 493
Stibine, di-*t*-butyl(dimethylamino)-, reactions, with cyclohexanone, **2**, 495
Stibine, di-*t*-butylmethoxy-, reactions, with cyclohexanone, **2**, 495
Stibine, dichloro-, synthesis of, **2**, 490
Stibine, dichloro(chloromethyl)-, synthesis of, **4**, 136
Stibine, diethoxy(methyl)-, reactions, with *o*-phenylenedithiol, **2**, 501
Stibine, diiodoethyl-, synthesis of, **2**, 489
Stibine, diphenylethynyl-
reactions
with diphenyllead hydride, **2**, 876
with triphenyltin hydride, **2**, 876
Stibine, diphenyl(trinitromethyl)-, synthesis of, **6**, 369
Stibine, methyl(benzene-1,2-dithio)-, synthesis of, **2**, 501
Stibine, triethyl-, synthesis of, **2**, 483
Stibine, trimethyl-
purification of, **2**, 487
synthesis of, wet method, **2**, 487
Stibine, triphenyl-, reactions, with trimethyloxonium tetrafluoroborate, **1**, 271
Stibine, tris(chloromethyl)-, synthesis of, **4**, 136
Stibine, tris(trifluoromethyl)-, synthesis of, **2**, 486
Stibine, tris(trimethylsilylmethyl)-, synthesis of, **4**, 598
Stibine, tris(*N*-trimethylsilyl-*N*-methylamino)-, synthesis of, **2**, 416
Stibine amides, diaryl-, synthesis of, **2**, 891
Stibine halides, diaryl-, as starting materials, for symmetrical distibines, **2**, 893
Stibine oxide, bisdimethyl-, synthesis of, **2**, 493
Stibine oxide, trimethyl-
reactions
with hydrogen chloride, **2**, 492
with hydrogen sulfide, **2**, 502
Stibine oxides, alkyl-, synthesis of, **2**, 497
Stibine oxides, bisdialkyl-, synthesis of, **2**, 493
Stibine oxides, trialkyl-, synthesis of, **2**, 497
Stibines
alkylation of, **2**, 509
halogenation of, **2**, 881
reactions
with organolithiums, **2**, 510
with stiboranes, **4**, 599
with tertiary arsines, **2**, 492
as starting materials
for antimony ylides, **3**, 500
for stibides, **2**, 510
synthesis of, **4**, 598
Stibines, alkenyl-, reactions, with alkynes, **2**, 875
Stibines, alkoxy-, synthesis of, **2**, 494
Stibines, alkoxydialkyl-
reactions
with enol acetates, **2**, 494
with enol trifluoroacetates, **2**, 494

synthesis of, **2**, 494
Stibines, alkyl-
stability of, **2**, 481
synthesis of, **2**, 481
Stibines, alkylaminohalo-, synthesis of, **2**, 409
Stibines, alkyldialkoxy-, synthesis of, **2**, 496
Stibines, alkyldihalo-
hydrolysis of, **2**, 495
reactions
with selenols, **2**, 500
with tertiary alkylstibines, **2**, 490
with thiols, **2**, 500
synthesis of, **2**, 489, 490
use in the synthesis of antimony–transition metal bonds, **2**, 512
Stibines, alkylhalo-
reactions, with metal azides, **2**, 504
synthesis of, **2**, 489
Stibines, amino-
as starting materials, for tertiary alkylarsines, **2**, 484
synthesis of, **2**, 502, 503
Stibines, aminodialkyl-
reactions
with enol acetates, **2**, 494
with enol trifluoroacetates, **2**, 494
Stibines, aryl-
from
alkali metal stibides, **2**, 895
halostibines, **2**, 895
Stibines, azido-, synthesis of, **2**, 504
Stibines, dialkylethynyl-, reactions, with tributyltin hydrides, **2**, 876
Stibines, dialkylhalo-
hydrolysis of, **2**, 493
oxidative hydrolysis of, **2**, 498
reactions
with alcohols, **2**, 494
with alkali metals, **2**, 510
with antimony trihalides, **2**, 490
with metal sulfides, **2**, 499
Stibines, dichalcogeno-, from, organochalcogeno(dialkyl)stibines, **2**, 501
Stibines, dimethyl(dimethylamino)-, reactions, with cyclopentadiene, **2**, 485
Stibines, halo-, as starting materials, for arylstibines, **2**, 895
Stibines, organochalcogeno(dialkyl)-, as starting materials, for dichalcogenostibines, **2**, 501
Stibines, phenyl-, from, antimony, **2**, 872
Stibines, polyalkyl-, synthesis of, **2**, 506
Stibines, tertiary
alkylation of, **2**, 487
from, sodium dialkylstibides, **2**, 485
Stibines, tertiary alkyl-
oxidation of, **2**, 497
oxidative fluorination of, **2**, 492
reactions
with alkyldihalostibines, **2**, 490
with alkyl halides, **2**, 487
with antimony trihalides, **2**, 490
with chalcogens, **2**, 502
with methyl iodide, **2**, 487
synthesis of, **2**, 483
Stibines, trialkyl-
reactions
with allyl bromide, **2**, 73
with benzyl bromide, **1**, 272
as starting materials, for sodium dialkylstibides, **2**, 510
synthesis of, **2**, 483
use in the alkylation of aldehydes, **2**, 73
Stibines, triaryl-
cleavage of, **2**, 510
photochemical synthesis of, **2**, 874
reactions
with carboxylic acids, **2**, 888
with cyanogen halides, **5**, 1132
Stibines, tris(dimethylamino)-, reactions, with cyclopentadiene, **2**, 485
Stibines, unsymmetrical tertiary alkyl-, synthesis of, **2**, 483, 484
Stibines, vinyl-, from, antimony, **2**, 872
Stibine selenides, trialkyl-, synthesis of, **2**, 502
Stibine sulfide, *t*-butyl-, synthesis of, **2**, 500
Stibine sulfide, trimethyl-, synthesis of, **2**, 502
Stibine sulfides, trialkyl-, synthesis of, **2**, 502
Stibinic acid, bis(1-chloroethyl)-, synthesis of, **2**, 498
Stibinic acid, dimethyl-, synthesis of, **2**, 498
Stibinic acids, reductive halogenation of, **2**, 883
Stibinosilanes
reactions
with gallium trihalides, **2**, 511
with indium trihalides, **2**, 511
synthesis of, **2**, 508
Stibolane, 1-chloro-, synthesis of, **2**, 491
Stibole, 1-phenyl-2,5-dimethyl-, synthesis of, **4**, 1041
Stiboles
reactions, with iron(II) chloride, **4**, 1039
synthesis of, **4**, 1038
Stiboles, 2,5-bis(trimethylsilyl)-3,4-dimethyl-, synthesis of, **4**, 1038
Stibolyl anion, dimethyl-, reactions, with bromopentacarbonylmanganese, **4**, 1042
Stibolyl anions
reactions, with iron(II) chloride, **4**, 1041
synthesis of, **4**, 1039
Stibonic acids, reductive halogenation of, **2**, 883
Stibonic acids, aryl-, from, aryldiazenium salts, **2**, 884
Stibonium bromide, benzyltributyl-, as starting material, for pentaorganostiboranes, **1**, 272
Stibonium bromides, benzyltrialkyl-, synthesis of, **1**, 272
Stibonium chloride, tetraethyl-, alkylation of, **2**, 488
Stibonium iodide, synthesis of, **2**, 487
Stibonium iodide, tetraalkyl-, synthesis of, **2**, 487
Stibonium iodide, triphenylmethyl-, synthesis of, **1**, 271
Stibonium salts
synthesis of, **1**, 743
via halogen anion exchange, **2**, 488
Stibonium salts, alkyl-
alkylation of, **2**, 488
synthesis of, **2**, 487
Stibonium salts, triphenylmethyl-, reactions, with benzophenone, **1**, 271
Stibonium triflate, tetraphenyl-, use in the synthesis of amines, **2**, 318
Stibonium ylides, synthesis of, **2**, 1004
Stiborane, azidotetramethyl-, synthesis of, **2**, 504
Stiborane, benzyltetrabutyl-, reactions, with acyl chlorides, **1**, 272
Stiborane, pentacyclopropyl-, synthesis of, **2**, 488
Stiborane, pentaethyl-, synthesis of, **2**, 488
Stiborane, pentakis(trimethylsilylmethyl)-, synthesis of, **2**, 488
Stiborane, pentamethyl-
cleavage of, **2**, 504
synthesis of, **2**, 488
Stiborane, tetrakis (trimethylsilylmethyl)chloro-, dehydrochlorination of, **5**, 922
Stiborane, tetramethyliodo-, as starting material, for tetramethylstiboranes, **4**, 599
Stiboranes
reactions
with stibines, **4**, 599
with trimethylsilylmethyllithuim, **4**, 599
synthesis of, **4**, 599
Stiboranes, alkyl-, synthesis of, **2**, 488
Stiboranes, chloroalkoxytri-*n*-butyl-, as intermediates, **3**, 644

Stiboranes, diethoxytrialkyl-, synthesis of, **2**, 497
Stiboranes, methylene-
  rearrangement of, **5**, 922
  synthesis of, **5**, 922
Stiboranes, pentaalkyl-
  cleavage of, **2**, 488
  synthesis of, **2**, 488
Stiboranes, tetramethyl-, from, tetramethyliodostiborane, **4**, 599
Stibosilanes, alkyl-, synthesis of, **2**, 508
Stieglitz rearrangement, of trityl *N*-haloamines, **3**, 407
cis-Stilbene, synthesis of, **1**, 507
trans-Stilbene
  osmylation of, **2**, 50
  synthesis of, **1**, 142, 507; **4**, 81
Stilbene oxide, isomerization of, **3**, 317
(*Z*)-Stilbene oxides, irradiation of, **3**, 315
Stilbenes
  reduction of, **1**, 80
  synthesis of, **1**, 141, 282, 732, 745
Stilbenes, dicyano-, synthesis of, **2**, 1055
Stilbenes, 4-nitro-4'-methoxy-, synthesis of, **1**, 731
(*Z*)-Stilbenes
  from, diphenylethyne, **1**, 458
  synthesis of, **1**, 756
Stilbine, tris(trifluoromethyl)-, synthesis of, **6**, 243
Stilbines, alkoxy-, reactions, with enol trifluoroacetates, **2**, 695
Stilbines, amino-
  reactions
    with enol acetates, **2**, 695
    with phenols, **2**, 695
Stiles' reagent, for the synthesis of carboxylic acids, **5**, 95, 112
Still's phosphonate, use in the synthesis of streptovaricin, **1**, 752
Stille coupling, for the synthesis of α,β-unsaturated ketones, **3**, 271
Stille cross-coupling reactions, **2**, 689
Stille-type formylation, of methoxy-substituted iodobenzenes, **3**, 97
Stobbe condensation
  for the synthesis of γ-lactones, **5**, 166
  for the synthesis of α,β-unsaturated acids, **5**, 98
Stork enamine reaction
  alternatives to, **3**, 415
  mechanism, **2**, 758
$A^{1,3}$ Strain, in asymmetric intermolecular radical additions, **1**, 327
Strecker reaction
  of aldehydes, **3**, 137
  chiral version, for the synthesis of nitriles, **3**, 616
  with hindered ketones, **3**, 616
  for the synthesis of α-aminonitriles, **3**, 637
  for the synthesis of cyanomethyls, **4**, 509
  for the synthesis of nitriles, **3**, 616
Strecker synthesis, mechanism, **2**, 191
Streptogramin, synthesis of, via selenoxide pyrolysis, **1**, 781
Streptovaricin, synthesis of, **1**, 752
Strontiums, alkyl-, synthesis of, **2**, 563
Styrene
  diastereoselectivity of, **1**, 325
  reactions
    with azaallyl anions, **1**, 350
    with 1,2-bis(dimethylsilyl)benzene, **4**, 618
    with chloroform, **4**, 17
    with formaldehyde, **4**, 827
    with purple benzene, **5**, 115
    with sodium, **5**, 42
    with sulfur, **5**, 599
    with Wilkinson's catalyst, **1**, 73
  synthesis of
    via reduction of vinylsilanes, **1**, 450
    via xanthate pyrolysis, **1**, 776
Styrene, 2-bromo-, as starting material, for benzoferracyclopentadiene complexes, **4**, 681
Styrene, *trans*-β-bromo-, reactions, with diborane, **4**, 142
Styrene, 2-bromo-2-nitro-, NMR spectra of, **4**, 795
Styrene, *trans* 1-chloro-, deprotonation of, **4**, 754
Styrene, (*E*)-1-chloro-1-fluoro-, synthesis of, **4**, 754
Styrene, a-chloro-b-iodo-, synthesis of, **2**, 629
Styrene, 2-chloro-2-nitro-, NMR spectra of, **4**, 795
Styrene, 1-chloroperfluoro-, synthesis of, **4**, 737
Styrene, β-deuterio-, synthesis of, **1**, 450
Styrene, 1,1-dibromo-, synthesis of, **4**, 745
Styrene, a,b-dibromo-, synthesis of, **2**, 624
Styrene, 1,1-dibromo-2-fluoro-, synthesis of, **4**, 748
Styrene, 1,1-dichloro-, from, benzaldehyde, **4**, 738
Styrene, 2,2-difluoro-1-chloro-, synthesis of, **4**, 737
Styrene, α-methyl-, reactions, with sodium, **5**, 42
Styrene, β-methyl-, synthesis of, **1**, 48
Styrene, α-morpholino-, reactions, with bismethanesulfonyl diazomethane, **2**, 802
Styrene, *p*-nitro-, reduction of, **2**, 772
Styrene, β-nitro-, reduction of, **1**, 82
Styrene, nitroso-, synthesis of, **2**, 778
Styrene, α-nitroso-, reactions, with thiobenzophenone, **4**, 232
Styrene, octafluoro-, from, decafluoroethylbenzene, **4**, 730
Styrene, trimethylsilyl-, reactions, with deuterium chloride, **1**, 450
Styrene, β-trimethylsilyl-, cleavage of, **1**, 450
Styrene nitroso chloride, reactions, with triethylamine, **2**, 778
Styrene oxide
  reactions
    with carbon nucleophiles, **1**, 204
    with tris(trimethylsilyl)methyllithium, **4**, 620
Styrenes
  chlorination of, **2**, 24
  hydroboration of, **2**, 43
  hydroformylation of, **2**, 580
  hydrometallation of, **2**, 579
  as intermediates, **1**, 529
  nitrosation of, **3**, 429
  oxidative cleavage of, **3**, 87
  reactions
    with ethyl nitrite, **3**, 429
    with iodine isocyanate, **5**, 985
  reduction of, **1**, 80
  as starting materials, for arylacetaldehyde acetals, **4**, 197
  synthesis of, **1**, 490
Styrenes, a-bromo-, synthesis of, **2**, 623
Styrenes, α-bromo-, solvolysis of, **1**, 1111
Styrenes, β-bromo-
  solvolysis of, **1**, 1110
  use in the synthesis of alkynes, **1**, 1005
Styrenes, bromonitro-, synthesis of, **4**, 798
Styrenes, 2-bromo-2-nitro-, synthesis of, **4**, 798
Styrenes, 2-chloro-2-nitro-, synthesis of, **4**, 798
Styrenes, 2-chloro-2-tosylazo-, from, arylacetic acids, **4**, 799
Styrenes, 1-diethylamino-2-phthalimido-
  reactions
    with bromine, **4**, 793
    with chlorine, **4**, 793
Styrenes, halogenated, reactions, with sodium methoxide, **4**, 762
Styrenes, nitro-
  catalytic hydrogenation of, **2**, 318
  reactions, with α-mercaptocinnamic acid, **3**, 250
Styrenes, α-nitro-, reactions, with butyllithiums, **3**, 190
Styrenes, β-nitro-, as starting materials, for α-nitro ketones, **2**, 393
Styrenes, α-silyl-, from, benzoylsilanes, **2**, 936
Styrenes, β-trimethylstannyl-, destannylation of, **1**, 454

Substitution
of alkynyl carbanions, **1**, 1000
of allenyl carbanions, **1**, 1051
of antimony functions, **1**, 271, 1032
of arsenic functions, **1**, 268, 1032
of bismuth functions, **1**, 1032
of boron functions, **1**, 272, 1033
effects of substrate structure on, **1**, 107
of germanium functions, **1**, 290, 1033
of halogen, **1**, 1052
of halogen functions, **1**, 1000
of metal functions, **1**, 291
of nitrogen functions, **1**, 250, 1032
of oxygen functions, **1**, 1053
of phosphorus functions, **1**, 268, 1032
of propargyl carbanions, **1**, 1051
of selenium functions, **1**, 1031
of silicon functions, **1**, 281, 1033
of sulfur functions, **1**, 1031
for the synthesis of aromatic nitriles, **3**, 661
for the synthesis of carbon–carbon bonds, **1**, **171–247**, **249–292**, **461–500**
for the synthesis of carbon–hydrogen bonds, **1**, **425–460**
for the synthesis of diazoalkanes, **3**, 478
for the synthesis of hydrazonamides, **5**, 847
for the synthesis of tertiary ynamines, **2**, 1041
for the synthesis of tetracoordinate sulfur compounds, **4**, 781
of tellurium functions, **1**, 1031
Substitution, aliphatic, in organoboron reactions, **1**, 280
Substitution, alkenic, in organoboron reactions, **1**, 278
Substitution, allylic, enantioselective, **1**, 194
Substitution, aromatic, in organoboron reactions, **1**, 278
Substitution, electrophilic
of arsenic halides, **2**, 875
of (9-fluorenyl)methyl chloroformate, **6**, 428
of 4-methylphenyl chloroformate, **6**, 428
of phenyl chloroformate, **6**, 428
for the substitution of boron functions, **1**, 278
for the synthesis of aryl halides, **2**, 608
for the synthesis of arylsilanes, **2**, 942
for the synthesis of carbon–nitrogen bonds, **1**, 267
Substitution, electrophilic aromatic
for the synthesis of aryl bromides, **2**, 625
for the synthesis of aryl chlorides, **2**, 619
for the synthesis of aryl fluorides, **2**, 612
for the synthesis of aryl iodides, **2**, 630
for the synthesis of aryl sulfides, **2**, 712
for the synthesis of metal(III) halides, **2**, 880
for the synthesis of phenols, **2**, 646
Substitution, nucleophilic
at the carbene carbon, **6**, 722
of *P*-halophosphaalkenes, **5**, 889
internal, **5**, 1095
of nitrogen electrophiles, with alkynic anions, **2**, 1041
for the synthesis of allenes and cumulenes, **1**, 966
for the synthesis of α-carbocations of alkyne-bridged dimolybdenum complexes, **4**, 674
use of a halogen as a leaving group in, **1**, 106
Substitution, nucleophilic aromatic
of aryl halides, **1**, 473
for the synthesis of alkyl aryl ethers, **2**, 684
for the synthesis of arenethiols, **2**, 706
for the synthesis of aryl sulfides, **2**, 711
for the synthesis of isothiocyanates, **5**, 1039
for the synthesis of phenols, **2**, 646
cine Substitution, for the synthesis of tertiary amines, **2**, 301
ipso-Substitution, of 2-trimethylsilyl benzofurans, **3**, 304
Substrate control, in intermolecular radical additions, **1**, 326
Substrate preorganization, effect on regioselectivity, **1**, 356
Substrate–reagent ion-pair (SRIP) reaction, in the synthesis of alkyl halides, **2**, 9
Succinate, diethyl
reactions
with aldehydes, **5**, 98
with ketones, **5**, 98
Succinate, ethyl oxalyl, as starting material, for α-ketoglutaric acid, **5**, 62
Succinate, methyl, reactions, with *N*-(dimethylaminopropyl)-*N'*-ethylcarbodiimide hydrochloride, **5**, 185
Succinic acid, α-alkylidene-, monoethyl esters of, synthesis of, **5**, 98
Succinic acid, α-aralkylidene-, monoethyl esters of, synthesis of, **5**, 98
Succinic acid, α,β-dibromo-
as starting material, for ethynedicarboxylic acid, **5**, 103
synthesis of, **5**, 60
Succinic acid, dioxo-, bisphenylhydrazones of, **5**, 857
Succinic acid, phenyl-, from, ethyl β-phenyl-β-cyanopropionate, **5**, 34
Succinic acids, 2,2-difluoro-, synthesis of, **5**, 59
Succinic anhydride
cleavage of, **5**, 37
reactions, with carboxylic acid chlorides, **5**, 538
Succinic anhydrides, reactions, with tributyltin isocyanate, **5**, 391
Succinic anhydrides, substituted, synthesis of, **5**, 201
Succinimide
reactions
with alkynylmagnesium halides, **1**, 1045
with aluminum chloride, **5**, 384
with bromopentacarbonylmanganese, **5**, 388
with cobalt complexes, **5**, 387
with diphenylantimony chloride, **5**, 369
with dithioacetals, **6**, 140
with iron complexes, **5**, 387
with mercuric oxide, **5**, 321
with potassium hypobromite, **5**, 86
Succinimide, *N*-bromo- (NBS)
bromination, for the synthesis of geminal dihalides, **5**, 410
for the oxidation of thiols, **2**, 213
reactions
with acetonitriles, **6**, 143
with aldehydes, **5**, 18
with aldoximes, **5**, 821
with alkenyl amines, **2**, 339
with allyltrimethylsilane, **4**, 807
with amides, **4**, 138
with α-aminophosphinates, **5**, 790
with benzopyranones, **4**, 174
with *N,O*-bis(trimethylsilyl)acetamide, **5**, 378
with dibenzoyltellurium, **5**, 341
with diethylaminosulfur trifluoride, **4**, 45
with HF pyridine, **4**, 45
with (*R*)-imidazolidinones, **4**, 102
with 1-iodoalkynes, **4**, 760
with isocyanides, **6**, 506
with mercury alkyls, **5**, 388
with organodisilanes, **5**, 378
with phenylmercury thiocarboxylates, **5**, 245
with pyrazolines, **4**, 109
with sulfides, **4**, 71
with thiocarboxylic acid *O*-silyl esters, **5**, 245
with *O*-trimethylsilyl thiopivalate, **5**, 530
as reagent
for hydrazone cleavage, **3**, 134
for the synthesis of nitrile oxides, **3**, 683
use in the synthesis of alkanesulfinic esters, **2**, 198
use in the synthesis of alkyl bromides, **2**, 27
use in the synthesis of α-bromoacylsilanes, **5**, 423
use in the synthesis of carboxylic esters, **5**, 141
use in the synthesis of dialkyl sulfoxides, **2**, 148
use in the synthesis of dibromoalkanes, **4**, 21
use in the synthesis of *N*-iodo succinimide, **5**, 321

Succinimide, *N*-chloro- (NCS)
as chlorinating reagents, for the synthesis of arenesulfenyl chlorides, **2**, 716
for the chlorination of amines, **2**, 337
for the α-chlorination of enamines, **2**, 22
for the chlorination of enolsilanes, **2**, 22
for the chlorination of sulfides, **4**, 66
for the chlorination of sulfoxides, **4**, 74
as oxidant, for the synthesis of aldehydes, **3**, 6
reactions
with aldoximes, **5**, 821
with alkyl t-butyl sulfoxides, **2**, 199
with benzaldehyde, **5**, 15
with benzil (*E*)-monooxime, **3**, 433
with *N*,*N*-dialkylsulfenamides, **5**, 339
with diisopropyl sulfide, **3**, 118
with dimethyl sulfide, **5**, 339
with hexamethylphosphoramide, **5**, 379
with 1-phenylthio-1-silylalkanes, **4**, 943
with potassium thiocarboxylates, **5**, 253
with sulfides, **4**, 283
with sulfur, **5**, 336
with tetraalkylleads, **5**, 391
with tetraalkyltins, **5**, 391
with (*S*)-2'-thiazinylideneglycinates, **4**, 101
as reagent
for the synthesis of *N*-chloroamides, **5**, 315
for the synthesis of nitrile oxides, **3**, 683
as starting material, for *N*-iodo succinimide, **5**, 321
synthesis of, **5**, 316
use in the chlorination of ketones, **3**, 152
use in the synthesis of alkanesulfinic esters, **2**, 199
use in the synthesis of *N*-chloroamines, **2**, 337
use in the synthesis of α-chloro sulfides, **4**, 66
use in the synthesis of hydroxamoyl halides, **5**, 664
use in the synthesis of sulfinyl chlorides, **2**, 179
Succinimide, *N*-(chlorothio)-, synthesis of, **5**, 336
Succinimide, *N*-halo-, reactions, with triphenylphosphines, **2**, 8
Succinimide, 1-hydroxy-, as co-reagent, in coupling reactions, **5**, 266
Succinimide, *N*-hydroxy-
active esters of, use in the synthesis of amides, **5**, 264
reactions
with α-halo esters, **5**, 333
with triphosgene, **6**, 483
as starting material, for active esters, **5**, 277
use in coupling reactions, **5**, 279
Succinimide, *N*-iodo- (NIS)
alcohol oxidations with, **3**, 119
from
N-chloro succinimide, **5**, 321
N-haloimides, **5**, 321
imides, **5**, 321
trialkylstannylsuccinimide, **5**, 320
for the iodination of enol esters, **2**, 34
as oxidant, for the synthesis of aldehydes from alcohols, **3**, 6
reactions
with alcohols, **2**, 95
with 1-iodoalkynes, **4**, 760
synthesis of, **3**, 155
use in the dihydroxylation of alkenes, **2**, 45
Succinimide, perfluoro-, reactions, with phosphorus pentachloride, **5**, 656
Succinimide, (*N*-phenylthio)-, reactions, with ylides, **4**, 346
Succinimide, trialkylstannyl-, as starting material, for *N*-iodosuccinimide, **5**, 320
Succinimide, *N*-trimethylsilyl-, reactions, with *N*-halocarboxamideamides, **5**, 814
Succinimide, *N*-(trimethylsilyl)-
reactions
with sulfur dichloride, **5**, 338
with sulfur monochloride, **5**, 338
synthesis of, **5**, 378
Succinimides
reactions, with cyanomethylenetriphenylphosphorane, **3**, 658
and reductive deaminations, **1**, 29
use in the synthesis of sulfinamides, **2**, 363
Succinimides, *O*-acyl-*N*-hydroxy-, cleavage of, **5**, 332
Succinimides, *N*-alkyl-, reduction of, **4**, 303
Succinimides, *N*-(aroylthio)-
as aroylsulfenylating reagents, **5**, 253
use in the synthesis of bis(acyl) disulfides, **5**, 252
Succinimides, *N*-aryl-, reduction of, **4**, 303
Succinimides, *O*-benzyl-, deprotection of, **5**, 332
Succinimides, *N*-chloromethyl-, synthesis of, **4**, 483
Succinimides, *N*-(dialkylphosphonyl)-, synthesis of, **5**, 372
Succinimides, *N*-halo-
reactions
with aldimines, **3**, 416
with 2-nitro enamines, **4**, 796
Succinodinitrile, tetramethyl-, synthesis of, **3**, 573
Succinonitrile, *N*,*N*'-dichlorodiimino-, reactions, with *n*-propylamine, **5**, 818
Succinonitrile, diimino-
reactions
with chlorine, **5**, 818
with methyl ketones, **4**, 438
Sugar lactones, reactions, with trimethylsilyl azide, **4**, 432
Sugars
bearing isocyanides, synthesis of, **3**, 708
from, organocobaloximes, **1**, 335
osazonization of, **3**, 451
reactions
with *N*-methyl-*N*-phenylhydrazine, **3**, 451
with potassium metabisulfite, **4**, 237
as starting materials, for fungal metabolites, **1**, 174
synthesis of
via radical cyclization, **1**, 341
via radical reactions, **1**, 331
Sugars, acetylated, reactions, with phosphorus pentachloride, **4**, 766
Sugars, bis(benzylseleno), synthesis of, **4**, 287
Sulfamate, methyl *N*-phenyl, synthesis of, **2**, 794
Sulfamate esters
alkylation of, **2**, 367
reactions, with sulfonyl halides, **2**, 368
synthesis of, **2**, 366
uses of, **2**, 367
Sulfamate esters, 2-hydroxyphenyl, reactions, with amines, **2**, 367
Sulfamate esters, 2-(hydroxyphenyl), synthesis of, **2**, 366
Sulfamate esters, *N*-sulfonyl, synthesis of, **2**, 368
Sulfamates, acylation of, **5**, 342
Sulfamates, amine, synthesis of, **2**, 357
Sulfamate thiol esters, from, alkylsulfamoyl halides, **2**, 367
Sulfamic acids
from
amines, **2**, 366
isocyanates, **2**, 366
urethanes, **2**, 366
reactions, with phosphorous pentachloride, **2**, 368
synthesis of, **2**, 366
Sulfamides
reactions
with alkylamines, **2**, 367
with sulfides, **2**, 162
as starting materials
for azo compounds, **2**, 403
for *N*,*N*-difluoroamines, **2**, 335
for hydrazines, **2**, 374, 378
synthesis of, **2**, 797
Sulfamides, *N*,*N*'-bisalkyl-
synthesis of, **2**, 367

transamination reactions of, **2**, 367
Sulfamides, bis-*N*,*N*'-dichloro-*N*,*N*'-bisalkyl-, synthesis of, **2**, 367
Sulfamides, symmetrical, synthesis of, **2**, 797
Sulfamines, *S*-imidoyl hydro-, synthesis of, **5**, 736
Sulfamoyl chloride, carbomethoxy-, as reagent, for the synthesis of symmetrical aromatic anhydrides, **5**, 196
Sulfamoyl chloride, *N*-methyl-, reactions, with silyl enol ethers, **2**, 249
Sulfamoyl chlorides
dehydrohalogenation of, **2**, 793
from, sulfuryl chlorofluoride, **2**, 368
reactions
with alcohols, **2**, 367
with alkoxides, **2**, 366
with amines, **2**, 367
with arenes, **2**, 365
with enamines, **3**, 171
with phenols, **2**, 367
with phosgene, **6**, 444
with silyl enol ethers, **3**, 171
Sulfamoyl halides
reactions, with alkali azides, **2**, 368
synthesis of, **2**, 368
Sulfamoyl halides, alkyl-, as starting materials, for sulfamate thiol esters, **2**, 367
Sulfamyl chlorides, as starting materials, for sulfonylamines, **2**, 401
Sulfane, methoxythiocarbonyl methyl-, synthesis of, **6**, 548
Sulfanes, alkoxy-, from, sulfoxylic acid, **2**, 99
Sulfanes, aryl-
from
arenethiols, **2**, 722
arylchlorosulfanes, **2**, 725
synthesis of, **2**, 722
Sulfanes, arylchloro-
as starting materials, for arylsulfanes, **2**, 725
synthesis of, **2**, 725
Sulfanes, bis(methoxythiocarbonyl)-, synthesis of, **6**, 548
Sulfanes, chloro-, reactions, with titanocene pentasulfide, **2**, 100
Sulfate, bis(trimethylsilyl)
as catalyst, for the synthesis of dithioacetals, **4**, 246
reactions, with potassium cyanide, **5**, 1135
as starting material, for silyl cyanides, **5**, 1135
Sulfate, caesium fluoroxy, reactions, with alcohols, **5**, 6
Sulfate, dimethyl
reactions
with amines, **2**, 760
with arylformamides, **5**, 744
with methyl *N*,*N*-dimethyldithiocarbamate, **6**, 314
with 4-(3-methylthioureido)phenyl *N*-methoxycarbonyl-*N*-methylcarbamate, **6**, 335
with potassium *O*-alkyl xanthates, **6**, 472
with thioamides, **3**, 657
synthesis of, **2**, 102
Sulfate, *N*,*N*-dimethylmethoxymethaneiminium methyl, reactions, with alkanoic acids, **3**, 76
Sulfate, perfluoroallyl fluoro-, as starting material, for α-fluoro ethers, **4**, 44
Sulfate adducts, dialkyl, use in the synthesis of acetals, **4**, 184
Sulfate radical anion, as oxidising agent, **4**, 1072
Sulfates
cleavage of, **2**, 553
reactions
with *t*-butanethiol, **2**, 134
with primary amines, **2**, 322
reduction of, **1**, 17
Sulfates, alkenyl, synthesis of, **2**, 690
Sulfates, alkyl
for the alkylation of amides, **5**, 284
displacement of, via nucleophiles, **1**, 177
reactions, with alkyl hydroperoxides, **2**, 97
as starting materials
for dialkyl peroxides, **2**, 97
for dialkyl sulfides, **2**, 134
Sulfates, alkylfluoro-, reactions, with amines, **2**, 366
Sulfates, β-amino-, synthesis of, **2**, 322
Sulfates, aryl, synthesis of, **2**, 690
Sulfates, arylfluoro-, reactions, with amines, **2**, 366
Sulfates, cyclic
from, bis(trifluoromethyl)ketene, **4**, 833
ring-strain of, **1**, 185
as starting materials, for *N*-unsubstituted aziridines, **2**, 322
use in the displacement of alkyl alcohols, **1**, 185
Sulfates, dialkyl
as electrophiles, for dithiocarboxylic enethiolates, **4**, 842
reactions, with alkynes, **1**, 1021
synthesis of, **2**, 102
use in the synthesis of thioimidic esters, **5**, 726
Sulfates, fluoronitroso-, reactions, with trifluoroethene, **4**, 104
Sulfates, 2-haloalkylfluoroalkyl fluoro-, synthesis of, **6**, 42
Sulfates, monoalkyl
as detergents, **2**, 102
synthesis of, **2**, 101
Sulfenamide, as sulfenylating reagents, **2**, 688
Sulfenamide, *S*-(*N*,*N*-diethylthiocarbamoyl)-, reactions, with acid chlorides, **5**, 335
Sulfenamides
S-amination of, **2**, 360
from
alkylthionitroso compounds, **2**, 359
arenesulfenyl chlorides, **2**, 358
arenesulfinic esters, **2**, 358
azides, **2**, 359
N-chloroamines, **2**, 359
disulfides, **2**, 243, 358, 768
α-halomethyl sulfoxides, **2**, 360
sulfenimines, **2**, 244
sulfenyl halides, **2**, 239, 767
sulfenyl thiocyanates, **2**, 243
sulfides, **2**, 244
thiiranes, **2**, 244
thiocarbonyls, **2**, 243
thiols, **2**, 240, 768
thionitrites, **2**, 243
N-thiophthalimides, **2**, 241
oxidation of, **2**, 250, 362, 365
reactions, with acetic anhydride, **2**, 361
as starting materials
for sulfenimines, **3**, 436
for sulfenyl halides, **2**, 169
for sulfinamides, **2**, 362
for the sulfenylation of thiocarboxylic acids, **5**, 247
synthesis of, **2**, 239, 358, 766; **4**, 843
reviews, **2**, 767
via oxidation of thiols, **2**, 358
via transamination reactions, **2**, 359
thiolalkylation of, **2**, 217
See also Thiohydroxylamines
Sulfenamides, acyl-
reactions, with thiocarboxylates, **5**, 251
stability of, **5**, 254
synthesis of, **5**, 253
Sulfenamides, alkane-, synthesis of, **2**, 239
Sulfenamides, *S*-alkoxy-, synthesis of, **2**, 363
Sulfenamides, *N*-(alkylthio)-, nomenclature, **2**, 767
Sulfenamides, arene-, synthesis of, **2**, 725
Sulfenamides, aroyl-, reactions, with isocyanates, **5**, 253
Sulfenamides, *N*-(arylthio)-, nomenclature, **2**, 767
Sulfenamides, *N*,*N*-bis(trifluoromethyl)alkane-, synthesis of, **6**, 241
Sulfenamides, *N*,*N*-bis(trifluoromethyl)-2-chloroalkane-, synthesis of, **6**, 241

Sulfenamides, 1-chloro-, synthesis of, **4**, 72
Sulfenamides, α-chloro-, as starting materials, for thiocarbonyl *S*-imides, **3**, 370
Sulfenamides, β-chloro-, synthesis of, **2**, 359
Sulfenamides, cyclic, synthesis of, **2**, 766
Sulfenamides, *N,N*-dialkyl-, reactions, with *N*-chlorosuccinimide, **5**, 339
Sulfenamides, halo-, reactions, with nucleophiles, **2**, 239
Sulfenamides, *N*-imidoyl-, synthesis of, **5**, 765
Sulfenamides, *S*-imidoyl-
  reactions
    with isocyanates, **5**, 736
    with salicyl aldehyde, **5**, 736
  synthesis of, **5**, 736
Sulfenamides, primary, synthesis of, **2**, 359
Sulfenamides, secondary, synthesis of, **2**, 359
Sulfenamides, *N*-sulfonyl-
  oxidation of, **2**, 366
  synthesis of, **2**, 365
Sulfenamides, *N*-(trihalomethyl)-, synthesis of, **6**, 241
Sulfenamides, vinyl-, synthesis of, **2**, 725
Sulfenamide salts, synthesis of, **2**, 360
Sulfenate esters, allyl alkyl, rearrangement of, **2**, 154
Sulfenate esters, aryl, synthesis of, **2**, 688
Sulfenates, as starting materials, for thiophosphates, **2**, 257
Sulfenates, alkenyl, synthesis of, **2**, 688
Sulfenates, arene
  reactions, with amines, **2**, 358
  synthesis of, **2**, 720
Sulfenates, aryl, synthesis of, **2**, 688
Sulfenates, β-chloroalkyl, synthesis of, **2**, 194
Sulfenates, *S*-imidoyl
  reactions, with amines, **5**, 736
  synthesis of, **5**, 734
Sulfenates, silyl
  syn elimination of, **1**, 655
  as starting materials, for β-silyl sulfoxides, **1**, 655
Sulfenates, unstable propargyl, rearrangement of, **1**, 993
Sulfene
  as dienophile, **6**, 636
  reactions, with imides, **5**, 348
  synthesis of, **2**, 177
Sulfene, chloro-
  reactions
    with acetophenone, **5**, 499
    with enamines, **5**, 496
    with ketene-*N,N*-acetals, **5**, 497
    with 1-methyl-4-piperidone, **5**, 496
    with morpholinocyclohexene, **5**, 497
    with norborn-5-ene-2-carbaldehyde, **5**, 496
    with propiophenone, **5**, 498
  synthesis of, **5**, 496
Sulfene, methanesulfonyl-, as starting material, for thietane dioxides, **4**, 281
Sulfene, methylchloro-, synthesis of, **5**, 497, 499
Sulfene, phenyl-, reactions, with carbethoxymethylene triphenylarsorane, **5**, 920
Sulfene adducts, synthesis of, **3**, 375
Sulfeneamides, aroyl-, synthesis of, **5**, 253
Sulfenes
  as intermediates, in the synthesis of ylides, **5**, 902
  reactions
    with alkenes, **5**, 497
    with amines, **4**, 282
    with sulfonyl halides, **5**, 488
  as starting materials, for alkanesulfonic esters, **2**, 204
  synthesis of, **3**, 375; **5**, 495
  in situ trapping of, **3**, 375
Sulfenes, amine-stabilized, synthesis of, **5**, 502
Sulfenes, bromo-, reactions, with enamines, **5**, 497, 499
Sulfenes, chloro-, reactions, with diazomethane, **4**, 82
Sulfenes, fluoro-, synthesis of, **5**, 502
Sulfenes, halo-
  reactions, with aminomethylene ketones, **5**, 501
  as sulfur(VI) thioacyl halides, **5**, 472
  synthesis of, **5**, 495
    via dehydrohalogenation of sulfonyl halides, **5**, 496
Sulfenes, imino-, as reactive intermediates, **3**, 375
Sulfenes, iodo-, synthesis of, **5**, 497
Sulfenic acid chloride, fluorocarbonyl-, reactions, with silver trifluoroacetate, **5**, 246
Sulfenic acids
  from, sulfur monoxide, **2**, 184
  reactions, with alkynes, **2**, 184
  as starting materials, for thiohydroxylamines, **2**, 766
  synthesis of, **1**, 595, 779; **2**, 184; **6**, 307
Sulfenic acids, *O*-acyl-, synthesis of, **5**, 208
Sulfenic acids, alkane-, synthesis of, **2**, 184
Sulfenic acids, alkyl-, reactions, with alkenes, **2**, 153
Sulfenic acids, allylic, synthesis of, **2**, 185
Sulfenic acids, arene-, synthesis of, **2**, 718
Sulfenic esters, as starting materials, for sulfenyl halides, **2**, 169
Sulfenic esters, alkane-
  from
    alkanesulfenic esters, **2**, 195
    N-(alkanethio)phthalimides, **2**, 195
    thiirane-*S*-oxide, **2**, 195
  as starting materials, for alkanesulfenic esters, **2**, 195
  synthesis of, **2**, 194
Sulfenic esters, β-chloroethyl-, synthesis of, **2**, 195
Sulfenimines
  from
    carbonyls, **3**, 436
    disulfides, **3**, 436
    imines, **3**, 436
    sulfenamides, **3**, 436
  oxidation of, **3**, 436
  reduction of, **3**, 436
  as starting materials
    for sulfenamides, **2**, 244
    for thioamides, **5**, 335
    for α,β-unsaturated formamides, **3**, 713, 719
  stereoisomers of, **3**, 436
  synthesis of, **2**, 240; **3**, 436
Sulfenimines, chiral, synthesis of, **3**, 436
Sulfenimines, trityl-, reduction of, **2**, 359
Sulfenylating reagents
  reactions
    with alkenes, **2**, 221
    with homoenolate anion equivalents, **3**, 168
    with thioamides, **5**, 734
  for the synthesis of α-sulfenyl ketones, **3**, 164
Sulfenylation
  of aldehydes, **3**, 35
  of carbanions, **6**, 122, 123
  of ketones, **3**, 164
  sequential, **6**, 121
  of α,β-unsaturated cyclic ketones, **2**, 138
α-Sulfenylation, of imines, **3**, 418
Sulfenyl bromide, acyl-, reactions, with diphenyl ditelluride, **5**, 249
Sulfenyl bromide, fluorocarbonyl-
  as starting material, for fluorothiolformate esters, **6**, 432
  synthesis of, **6**, 432
Sulfenyl bromide, *p*-methylbenzoyl-, reactions, with potassium thiobenzoate, **5**, 252
Sulfenyl bromides, synthesis of, **2**, 171
Sulfenyl bromides, acyl-
  stability of, **5**, 245
  synthesis of, **5**, 245
Sulfenyl bromides, alkane-, synthesis of, **2**, 170
Sulfenyl bromides, arene-, synthesis of, **2**, 716
Sulfenyl bromides, aroyl-, reactions, with diaryl diselenides, **5**, 248
Sulfenyl bromides, α-bromo-, synthesis of, **2**, 171

Sulfenyl bromides, α,α-dibromo-, synthesis of, **2**, 171
Sulfenyl chloride
pyrolysis of, **5**, 476
reactions, with propargylic alcohols, **1**, 993
Sulfenyl chloride, acetyl-
reactions, with aniline, **5**, 254
synthesis of, **5**, 244
Sulfenyl chloride, arene-, reactions, with chloroalkyne, **4**, 772
Sulfenyl chloride, benzoyl-, synthesis of, **5**, 244
Sulfenyl chloride, benzyl-, synthesis of, **2**, 168
Sulfenyl chloride, *n*-butane-, reactions, with phosphorus ylides, **5**, 898
Sulfenyl chloride, α-chloro-, reactions, with sodium methoxide, **3**, 375
Sulfenyl chloride, chlorocarbonyl-
from, trichloromethanesulfenyl chloride, **6**, 434
nomenclature for, **6**, 432
reactions
with *O*-alkyl chlorothioformates, **6**, 480
with carboxamides, **3**, 686
with cyclohexyl isocyanide, **5**, 1036
with methanol, **6**, 478
with silver(I) cyanide, **6**, 435
with thiophene, **6**, 432
reactivity of, **6**, 432
as starting material
for chlorocarbonyl thiocyanate, **6**, 435
for fluorocarbonylsulfenyl chloride, **6**, 432
Sulfenyl chloride, dimethylamino-, reactions, with alkenes, **2**, 359
Sulfenyl chloride, fluorocarbonyl-
reactions, with trimethylsilyl bromide, **6**, 432
reactivity of, **6**, 432
as starting material
for fluorothiolformate esters, **6**, 432
for 2-pyrrolylfluorothiolformate, **6**, 432
synthesis of, **6**, 432
Sulfenyl chloride, methoxycarbonyl-
reactions
with benzenemethanethiol, **6**, 478
with hydrocinnamaldehyde, **3**, 35
synthesis of, **5**, 244; **6**, 478
as thiocarboxylating reagents, **6**, 478
Sulfenyl chloride, perchloromethyl-, reduction of, **6**, 529
Sulfenyl chloride, phenyl-
reactions
with cyclopentanones, **3**, 249
with enol ethers, **3**, 35; **5**, 413
with silyl enol ethers, **3**, 35
with thioamide, **5**, 727
Sulfenyl chloride, toluene-, reactions, with benzamidine, **5**, 766
Sulfenyl chloride, *N*-2,2-trichlorocyanoacetimidoyl-, synthesis of, **5**, 832
Sulfenyl chloride, trityl-, reactions, with alkenes, **2**, 221
N-Sulfenyl chloride, *N*-(amido)-, synthesis of, **5**, 335
Sulfenyl chlorides
addition to alkenes, **6**, 609
from
acetanilides, **5**, 491
S-benzyl thioethers, **5**, 492
disulfides, **2**, 166
isothiocyanates, **6**, 605
oxidation of, **4**, 78
reactions
with alkyl aldehydes, **3**, 35
with alkyldichlorophosphines, **2**, 464
with alkyl phosphites, **2**, 257
with aryl disulfanes, **2**, 723
with bis(trimethylsilyl)amino isocyanide, **6**, 617
with carbon monosulfide, **6**, 535
with *N*-chloroimidates, **5**, 703
with iodonium nitro ylides, **6**, 281
with isocyanides, **6**, 617
with primary amines, **2**, 361
with primary phosphines, **2**, 460
with sodium iodide, **2**, 172
with sodium trithiocarbonate, **6**, 554
with thioamides, **5**, 734
with α-thiolcarboxylic acids, **2**, 126
with thiosulfonic acids, **2**, 220
with tin, **6**, 528
with trinitromethane, **6**, 652
with triphenylphosphine, **5**, 477
with vinyl sulfones, **4**, 861
as starting materials
for α-chloro sulfides, **4**, 69
for chlorothiolformate esters, **6**, 434
for thioaldehydes, **3**, 337
for thiocarbonates, **6**, 273
for thiophosphates, **2**, 257
for trithioorthoformates, **6**, 273
for the sulfenylation of thiocarboxylic acids, **5**, 247
synthesis of, **2**, 166; **6**, 273, 617
via chlorination of isothiocyanates, **6**, 604
Sulfenyl chlorides, acyl-
alcoholysis of, **5**, 246
from, diphenyltin bis(thiocarboxylates, **5**, 245
reactions, with thiols, **5**, 246
stability of, **5**, 245
synthesis of, **5**, 244
Sulfenyl chlorides, alkane-
reactions
with alcohols, **2**, 194
with *N,N*-diethyl ynamines, **4**, 791
with mercury(II) bis(trifluoromethyl)amide, **6**, 241
synthesis of, **2**, 166
Sulfenyl chlorides, alkoxycarbonyl-
as starting materials, for dialkyl thiocarbonates, **6**, 478
as thiocarboxylating reagents, **6**, 478
Sulfenyl chlorides, alkyl-, as starting materials, for sulfenyl halides, **2**, 358
Sulfenyl chlorides, amino-
reactions
with allylic alcohols, **2**, 363
with amino bistrimethylsilanes, **2**, 361
with diamines, **2**, 361
with nucleophiles, **2**, 361
with organometallics, **2**, 359
synthesis of, **2**, 360
Sulfenyl chlorides, arene-
reactions
with arylvinyl selenides, **4**, 91
with chloroimines, **4**, 110
with *N,N*-diethyl ynamines, **4**, 791
with tertiary amines, **2**, 360
with triorganogermanium selenocarboxylates, **5**, 249
with vinylsilanes, **4**, 138
with vinylstannanes, **4**, 154
with vinyltrimethylgermane, **4**, 372
as starting materials, for sulfenamides, **2**, 358
synthesis of, **2**, 716
use in the synthesis of 6-aminoquinoline *N*-oxide, **2**, 767
use in the synthesis of diaryl sulfides, **2**, 712
Sulfenyl chlorides, aryl-
reactions
with alkenes, **2**, 712
with silver cyanate, **5**, 995
with vinyltrimethylsilane, **4**, 369
Sulfenyl chlorides, aryldichloro-, synthesis of, **5**, 490
Sulfenyl chlorides, carbamoyl-, from, triazines, **4**, 103
Sulfenyl chlorides, carbamoylcarbonyl-, synthesis of, **5**, 245
Sulfenyl chlorides, α-carbamoyl dichloro-, as starting materials, for chlorosulfines, **5**, 491
Sulfenyl chlorides, α-carbamoyl-1,1-dichloro-, from, *S*-benzyl thioethers, **5**, 477

Sulfenyl chlorides, α-carbamoyl-α,α-dichloro-, hydrolysis of, **5**, 245
Sulfenyl chlorides, α-chloro-
as intermediates, **5**, 485
reactions, with thioacetic acid, **5**, 590
as starting materials, for thioketones, **3**, 365
Sulfenyl chlorides, α-chloroalkane-
from
disulfides, **4**, 71
α-mercaptoacetanilides, **4**, 71
nitriles, **4**, 72
reactions, with amines, **4**, 72
synthesis of, **4**, 71
Sulfenyl chlorides, ω-chloroalkane-, synthesis of, **2**, 167
Sulfenyl chlorides, dialkyl amino-, reactions, with mercury(II) trifluoromethanethiolate, **2**, 361
Sulfenyl chlorides, dichloro-
as starting materials
for *N,N*-disubstituted α-carbamoyl chlorosulfines, **5**, 490
for halosulfines, **5**, 491
Sulfenyl chlorides, 1,1-dichloro-
from, dithiocarboxylates, **5**, 478
as starting materials
for chlorosulfines, **5**, 492
for thioacyl chlorides, **5**, 477
Sulfenyl chlorides, 1,1-dihalo-, hydrolysis of, **5**, 492
N-Sulfenyl chlorides, sulfonamide, synthesis of, **2**, 365
Sulfenyl fluoride, trifluoromethyl-, reactions, with bis(trifluoromethyl) thiocarbonate, **6**, 274
Sulfenyl fluorides, synthesis of, **2**, 165
Sulfenyl fluorides, alkane-, synthesis of, **2**, 165
Sulfenyl halides
from
alkylsulfenyl chlorides, **2**, 358
sulfenamides, **2**, 169
sulfenic esters, **2**, 169
nomenclature, **2**, 767
oxidation of, **2**, 178
reactions of, **4**, 940
reactions
with alkynes, **1**, 1123
with amines, **2**, 358, 767
with epoxides, **2**, 194
with thiocarbonyls, **6**, 275
stability of, **2**, 165
as starting materials
for chlorosulfines, **5**, 492
for disulfides, **2**, 229
for sulfenamides, **2**, 239, 767
for thiocyanates, **5**, 1111
as sulfenylating reagents, **2**, 688
thiolysis of, **2**, 217
transylidation reactions of, **6**, 697
use in the synthesis of β-substituted sulfides, **2**, 138
Sulfenyl halides, acyl-
reactions, with thiocarboxylates, **5**, 251
spectral analysis of, **5**, 246
as starting materials, for acylsulfenyl selenides, **5**, 248
synthesis of, **5**, 244
Sulfenyl halides, alkane-, synthesis of, **2**, 165, 194
Sulfenyl halides, alkyl-, reactions, with alkenes, **2**, 715
Sulfenyl halides, arene-
reactions
with alkenes, **2**, 715
with alkoxides, **2**, 720
with vinyl acetates, **4**, 56
synthesis of, **2**, 716
Sulfenyl halides, aryl-, reactions, with allenes, **1**, 1125
Sulfenyl halides, fluorocarbonyl-, as starting materials, for fluorothiolformate esters, **6**, 432
Sulfenyl halides, imidoyl-, synthesis of, **5**, 734
Sulfenyl halides, trihalomethane-, synthesis of, **6**, 236
Sulfenyl halogenides, reactions, with halogens, **6**, 273
Sulfenyl iodide, trifluoroacetyl-, reactions, with trimethylsilyl chloride, **5**, 254
Sulfenyl iodides
synthesis of, **2**, 172
use in protein iodination, **2**, 172
use in the synthesis of alkanesulfenic esters, **2**, 194
Sulfenyl iodides, acyl-
from, diphenyltin bis[thiocarboxylates], **5**, 246
stability of, **5**, 246
synthesis of, **5**, 245
Sulfenyl iodides, alkane-, synthesis of, **2**, 172
Sulfenyl nitroalkane, 1-chloro-1-chloro-, synthesis of, **6**, 57
Sulfenyls
hydrolysis of, **2**, 185
oxidation of, **2**, 250
Sulfenyls, chlorocarbonyl-, from, alkoxydichloromethanesulfenyls, **6**, 435
Sulfenyl selenides, alkane-, synthesis of, **2**, 234
Sulfenyl tellurides, synthesis of, **2**, 238
Sulfenyl tellurides, acyl-, synthesis of, **5**, 249
Sulfenyl tellurides, alkane-, synthesis of, **2**, 234, 238
Sulfhaloform reaction, mechanism, **2**, 179
Sulfhydrolysis
of imidate esters, **5**, 512
of imido esters, **5**, 551
Sulfidamides, β-hydroxy
synthesis of, **1**, 695
thermal elimination of, **1**, 695
Sulfide, reactions, with arenediazonium salts, **2**, 711
Sulfide, acetonitrile, synthesis of, **3**, 686
Sulfide, allyl benzyl, reactions, with dichlorocarbene, **4**, 775
Sulfide, allyl methyl
reactions, with carbenes, **1**, 409
reduction of, **1**, 85
Sulfide, allyl phenyl, reactions, with carbenes, **1**, 409
Sulfide, allyl tris(trimethylsilyl), synthesis of, **2**, 541
Sulfide, allylvinyl, alkylation of, **1**, 141
Sulfide, ammonium
reactions
with aryl alkyl ketones, **5**, 599
with carbonimidoyl dichlorides, **5**, 1032
use in the synthesis of thionoamides, **5**, 580
Sulfide, benzonitrile, synthesis of, **3**, 686
Sulfide, benzyl methyl, metallation of, **4**, 347
Sulfide, benzyl trityl, chlorination of, **2**, 168
Sulfide, bis(acetyl), reactions, with chlorine, **5**, 244
Sulfide, bis(benzoyl), reactions, with chlorine, **5**, 244
Sulfide, bis(1,5-cyclooctanediylbornyl), reactions, with cyclic ethers, **2**, 270
Sulfide, bis(di-*t*-butylstibine), synthesis of, **2**, 499
Sulfide, bis(diethylaluminum), as reagent, for the synthesis of thioketones, **3**, 363
Sulfide, bisdiethylthiocarbamoyl, reactions, with 4-nitroaniline, **5**, 1029
Sulfide, bis(dimethylaluminum), as reagent, for the synthesis of thioaldehydes, **3**, 336
Sulfide, bis(ethoxythiocarbonyl), reactions, with enolates, **5**, 513
Sulfide, bis(methoxycarbonyl), synthesis of, **6**, 474
Sulfide, bis(methoxythiocarbonyl), synthesis of, **6**, 551
Sulfide, bis(silyl), reactions, with methyllithium, **5**, 538
Sulfide, bis(tributyltin), use in the synthesis of bis(acyl) sulfides, **5**, 242
Sulfide, bis(tricyclohexyltin), reactions, with boron trichloride, **5**, 575
Sulfide, bis(trifluoromethyl), ultraviolet irradiation of, **6**, 219
Sulfide, bis(trimethylsilyl)
reactions
with carbonimidoyl dichlorides, **5**, 1032
with carboxylic acid chlorides, **5**, 538
with α-silyl ketones, **5**, 648
with sodium methoxide, **5**, 582

synthesis of, **2**, 265, 541
use in the synthesis of bis(acyl) sulfides, **5**, 242
use in the synthesis of thionoamides, **5**, 576
Sulfide, bis(trimethylsilylmethyl), reactions, with carbonyls, **2**, 139
Sulfide, bis(trimethylsilyl)methyl (trimethylsilyl)methyl, bromination of, **6**, 282
Sulfide, borane dimethyl
as hydroborating agent, **5**, 417
reactions
with ω-alkoxycarbonyl alkyl amides, **2**, 326
with dimethyl(divinyl)germane, **4**, 646
with enamines, **1**, 60
with 1-(trimethylsilyl)-1,3-butadiene, **4**, 643
as reducing agent, **2**, 320
use in the synthesis of aryl aldehydes, **3**, 82
Sulfide, borane methyl, use in the reduction of imines, **2**, 312
Sulfide, 2-butenyl *t*-butyl, crystal structure of, **2**, 999
Sulfide, *t*-butyl chloromethyl, synthesis of, **4**, 66
Sulfide, carbonyl
reactions
with aromatic nitro compounds, **5**, 1033
with ethyl cyanoacetate, **4**, 833
Sulfide, chlorocyclopropyl, synthesis of, **4**, 68
Sulfide, 2-chloroethyl methyl, reactions, with manganese pentacarbonyls, **5**, 448
Sulfide, chlorofluoromethyl phenyl, reactions, with *n*-butyllithium, **4**, 773
Sulfide, chloromethyl, as starting material, for benzylthiomethylmagnesium chloride, **4**, 395
Sulfide, β-chloro methyl, use in the synthesis of amino acids, **5**, 90
Sulfide, chloromethyl methyl
reactions
with ethyl diphenylphosphinite, **4**, 347
with lithium iodide, **4**, 71
as starting material, for Grignard reagents, **4**, 395
Sulfide, chloromethyl *p*-nitrobenzyl, chlorination of, **4**, 64
Sulfide, chloromethyl phenyl
oxidation of, **4**, 74
as starting material, for *N*-acyl-*S*,*N*-acetals, **4**, 329
synthesis of, **4**, 68
Sulfide, diallyl, as starting material, for thioacrolein, **3**, 337
Sulfide, dibenzyl
α-acetoxylation of, **4**, 220
reactions, with sulfuryl chloride, **4**, 64
Sulfide, dibromoborane dimethyl, reactions, with alkynes, **1**, 1102
Sulfide, dibutyl, synthesis of, **2**, 137
Sulfide, di(chloromethyl)
from
dimethyl sulfide, **4**, 63
1,3,5-trithiane, **4**, 63
synthesis of, **4**, 65
Sulfide, diethyl, synthesis of, **2**, 135
Sulfide, diethynyl, synthesis of, **2**, 1031
Sulfide, diisopropyl
as oxidant, for the synthesis of aldehydes, **3**, 6
reactions, with *N*-chlorosuccinimide, **3**, 118
Sulfide, dimethyl
electrochemical fluorination of, **6**, 214
lithiation of, **4**, 391
reactions
for *N*-bromoamides, **5**, 338
with 1-bromodecane, **1**, 140
with *N*-chloroamides, **5**, 338, 689
with *N*-chlorosuccinimide, **5**, 339
with α-halomethyl phosphonates, **4**, 346
with phenylselenyl bromide, **2**, 238
with α,β-unsaturated acetals, **4**, 201
as starting material, for di(chloromethyl) sulfide, **4**, 63
use in the synthesis of alkylcoppers, **2**, 586
Sulfide, dipropynal, pyrolysis of, **3**, 342
Sulfide, (*Z*,*Z*)-divinyl, synthesis of, **1**, 441
Sulfide, ethylmethyl, synthesis of, **2**, 135
Sulfide, iodomethyl phenyl, synthesis of, **4**, 71
Sulfide, 1-lithiocyclopropylphenyl, reactions, with 4-methyl-1,2-epoxypentane, **3**, 224
Sulfide, methyl
reactions
with fluoroiodoalkanes, **6**, 45
with thionyl chloride, **4**, 65
Sulfide, methyl butyl, oxidation of, **2**, 145
Sulfide, methyl methylsulfinylmethyl, as starting material, for bis(methanesulfonyl)methane, **4**, 274
Sulfide, methyl norbornyl, synthesis of, **2**, 138
Sulfide, methyl phenyl
acidity of, **1**, 142
as starting material, for
α-acetoxyphenylthiomethane, **4**, 220
Vilsmeier reaction of, **3**, 98
Sulfide, methylpropyl, synthesis of, **1**, 85
Sulfide, methyl trifluoromethyl, ultraviolet irradiation of, **6**, 219
Sulfide, octafluoroethyl methyl, synthesis of, **6**, 46
Sulfide, phenacyl neopentyl, photolytic cleavage of, **3**, 331
Sulfide, phenylbenzyl, reactions, with *n*-butyllithium, **2**, 999
Sulfide, phenylchloro, as intermediate, **5**, 335
Sulfide, phenyl crotyl, chlorination of, **4**, 68
Sulfide, phenyl(trimethylsilyl), use in the synthesis of monothioacetals, **4**, 217
Sulfide, phenyl trimethylsilylmethyl, anodic oxidation of, **4**, 221
Sulfide, phenyl (1-trimethylsilyl)propadienyl
reactions
with lithium butyl(diisobutyl)aluminum hydride, **4**, 946
with organometallics, **4**, 946
Sulfide, phenylvinyl
cycloaddition of, **4**, 226
reactions, with methyllithium, **4**, 366
silylation of, **2**, 908
as starting material, for α-lithio sulfides, **4**, 392
Sulfide, phosphine, reactions, with dichloro(phenyl)phosphine, **4**, 114
Sulfide, propargyl phenyl, silylcupration of, **2**, 940
Sulfide, trialkylstannyl phenyl, reactions, with (1-bromovinyl)trimethylsilane, **4**, 946
Sulfide, trichloromethyl chloromethyl, chlorination of, **4**, 85
Sulfide, 1,1,1-trichloromethyl methyl, synthesis of, **4**, 65
Sulfide, trifluoroacetonitrile, synthesis of, **3**, 686
Sulfide, trifluoromethyl methyl, synthesis of, **6**, 233
Sulfide, trimethylsilyl ethynyl triethylsilyl, reactions, with methanol, **3**, 548
Sulfide, 1,1,1-trinitrodimethyl, synthesis of, **6**, 348
Sulfide, triphenylphosphine, reactions, with *t*-butyl cyanoketene, **3**, 546
Sulfide, vinyl
reactions, with alcohols, **3**, 634
as starting material
for dienes, **1**, 484
for trienes, **1**, 484
Sulfide, vinylethynyl, synthesis of, **2**, 1031
1,2-Sulfide, tetramethylallene-, reactions, with diphenylketene, **4**, 836
Sulfide anions, hydro-, reactions, with hydroximic acid chlorides, **5**, 637
Sulfide complexes, bromoalkylborane dimethyl, use in hydroboration reactions, **1**, 531
Sulfide complexes, bromoborane dimethyl, reactions, with haloalkynes, **4**, 815
β-Sulfide groups, elimination of, **1**, 665
Sulfide ions, hydro-
reactions
with 2,5-dichlorobenzonitrile oxide, **5**, 832
with 2,3,5,6-tetramethylbenzonitrile oxide, **5**, 832

Sulfides
as additives, to cuprates, **1**, 465
asymmetric oxidation of, **2**, 714
bromination of, **2**, 171; **4**, 71
chlorination of, **2**, 167
α-chlorination of, **6**, 44
cleavage of, **2**, 553
dealkylation of, **2**, 127
deprotonation of, **4**, 391
electrochemical fluorination of, **4**, 62
elimination of, **1**, 590
fluorination of, **2**, 165; **4**, 61
α-fluorination of, **4**, 61
from
aldehydes, **2**, 139
disulfides, **2**, 713
imides, **2**, 713
ketones, **2**, 139
sulfones, **2**, 713
sulfoxides, **2**, 713
thiocyanates, **2**, 713
oxidation of, **2**, 144, 192; **4**, 220
electrochemical, **4**, 220
oxidative coupling of, **2**, 230
reactions
with amines, **2**, 160
with azides, **2**, 161
with bromine, **4**, 71
with *N*-bromosuccinimide, **4**, 71
with *t*-butyl peracetate, **4**, 220
with *N*-chlorosuccinimide, **4**, 283
with dichlorocarbene, **4**, 775
with *N*-haloamines, **2**, 160
with *N*-halo compounds, **2**, 159
with hydroxylamines, **2**, 161
with lithium naphthalide, **4**, 1026
with sulfamides, **2**, 162
with sulfuryl chloride, **4**, 80
with trialkylsilanes, **2**, 265
reduction of, **1**, 21, 23
as starting materials
for α-chloro sulfoxides, **4**, 74
for dialkyl sulfones, **2**, 154
for α,α-dichloro sulfides, **6**, 44
for sulfenamides, **2**, 244
for sulfenyl selenides, **2**, 237
for sulfonium ylides, **3**, 373
synthesis of, **4**, 956; **6**, 316
via hydroboration, **1**, 85
Sulfides, α-acetoxy
synthesis of, **4**, 220
via Pummerer reaction, **4**, 221
Sulfides, acylselenenyl, synthesis of, **5**, 249
Sulfides, acyltellurenyl, synthesis of, **5**, 250
Sulfides, acyl thioacyl, use in the synthesis of thionoamides, **5**, 584
Sulfides, alkenyl
deprotonation of, **4**, 956
elimination of, **1**, 1085
nickel-catalyzed cross-coupling reactions of, **1**, 1031
oxidation of, **3**, 165
reactions, with Grignard reagents, **1**, 1031
as starting materials, for α-sulfenyl aldehydes, **3**, 36
Sulfides, alkyl
chlorination of, **4**, 63
as starting materials
for α-chloro sulfides, **4**, 63
for thiosilanes, **2**, 265
Sulfides, *t*-alkyl, synthesis of, **1**, 263
Sulfides, alkyl aryl
chlorination of, **4**, 63
from, haloarenes, **2**, 711
as starting materials, for aldehydes, **3**, 16
synthesis of, **2**, 135
Sulfides, alkyl dialkylaminomethyl, synthesis of, **4**, 315
Sulfides, alkyl hydro, reactions, with pentafluoropropene-2-sulfonyl fluoride, **6**, 46
Sulfides, alkyl phenyl
anions derived from, **1**, 140
chlorination of, **4**, 68
Sulfides, alkylvinyl, deprotonation of, **1**, 141
Sulfides, alkynic, reactions, with organocuprates, **4**, 963
Sulfides, alkynyl
reactions
with bis(pyridine)iodine(1) tetrafluoroborate, **4**, 772
with bromine, **4**, 772
as starting materials, for thioketenes, **3**, 548
synthesis of, **2**, 1031
Sulfides, α-alkynyl, synthesis of, **1**, 1003
Sulfides, alkynyl alkyl
pyrolysis of, **3**, 548
reactions, with amines, **5**, 590
Sulfides, alkynyl allenyl
[3,3] sigmatropic rearrangement of, **5**, 618
thermolysis of, **3**, 548
Sulfides, alkynyl allyl, thio-Claisen rearrangement of, for the synthesis of γ,δ-unsaturated thionoamides, **5**, 614
Sulfides, alkynyl propargyl
thermolysis of, **3**, 548
use in the synthesis of thionoamides, **5**, 614
Sulfides, allenyl, from, alkynyl thioethers, **1**, 954
Sulfides, allyl
metallation of, **1**, 141
reactions
with chloramine-T, **2**, 244
with Grignard reagents, **1**, 238
reduction of, **1**, 85
as starting materials
for alkenes, **1**, 236
for thioaldehydes, **3**, 335
for thioketones, **3**, 351
synthesis of, **1**, 643
Sulfides, allyl (1-arylvinyl), Thio–Claisen rearrangement of, **3**, 365
Sulfides, allylic
chlorination of, **4**, 66, 68
diketene thia-Claisen rearrangement of, **1**, 834
as radical traps, **1**, 231
reactions, with alkynes, **1**, 235
reduction of, **1**, 23
as starting materials
for allenic thioethers, **1**, 980
for homoallylic sulfides, **1**, 408
[2,3] Wittig rearrangement of, **1**, 820
Sulfides, allyl 2-pyridyl, alkylation of, **1**, 591
Sulfides, allylvinyl
synthesis of, **5**, 552
thio-Claisen rearrangement of, **5**, 552
Sulfides, aminomethyl phenyl, reactions, with tributylstannyllithium, **4**, 533
Sulfides, α-aminophosphine, from, tertiary phosphines, **4**, 477
Sulfides, anomeric, reactions, with thiophilic metal salts, **4**, 200
Sulfides, aryl
reactions, with Raney nickel, **1**, 435
reduction of, **1**, 435
synthesis of, **2**, 710
via addition reactions, **2**, 712
via electrophilic aromatic substitution, **2**, 712
via nucleophilic aromatic substitution, **2**, 711
via reduction reactions, **2**, 713
Sulfides, aryl alkyl, desulfurization of, **1**, 22
Sulfides, aryl methoxymethyl, reactions, with monoperphthalic acid, **4**, 235
Sulfides, aryl methyl, photochlorination of, **6**, 232

Sulfides, aryl 2-nitrophenyl, reactions, with 2-phenyl-1,3,2-dioxaphospholane, **2**, 856
Sulfides, arylselenyl, synthesis of, **2**, 735
Sulfides, aryl trichloromethyl, synthesis of, **6**, 232
Sulfides, aryl trifluoromethyl, synthesis of, **6**, 232
Sulfides, α-azido, oxidation of, **4**, 330
Sulfides, α-azo, synthesis of, **4**, 329
Sulfides, α-1-benzotriazolyl, reactions, with Grignard reagents, **1**, 263
Sulfides, benzyl, chlorination of, **2**, 168
Sulfides, benzyl alkyl, chlorination of, **5**, 561
Sulfides, benzyl haloethyl, reactions, with titanium tetrafluoride, **5**, 582
Sulfides, bis(acyl), synthesis of, **5**, 242
Sulfides, bis(alkoxycarbonyl), synthesis of, **6**, 475
Sulfides, bisamine, nomenclature, **2**, 766
Sulfides, bisbenzyl
  photochemical desulfurization of, **1**, 232
  as starting materials, for [2.2]-cyclophanes, **1**, 232
Sulfides, bis(dialkoxycarbonyl), synthesis of, **6**, 474
Sulfides, bis(fluorophenylthiophosphinyl), synthesis of, **2**, 855
Sulfides, bis(thioacyl), as starting materials, for thiocarboxylates, **5**, 509, 510
Sulfides, bis(triorganotin), reactions, with organic halides, **2**, 131
Sulfides, α-bromo
  from
    aldehydes, **4**, 71
    thiols, **4**, 71
  as starting materials, for α-bromo sulfones, **4**, 83
  synthesis of, **4**, 71
Sulfides, β-bromo, as starting materials, for alkenes, **1**, 591
Sulfides, butadiynyl, synthesis of, **2**, 1031
Sulfides, *t*-butyl, synthesis of, **2**, 135
Sulfides, carbonyl, synthesis of, **3**, 327
Sulfides, α-chloro
  from
    aldehydes, **4**, 69
    alkyl aryl sulfoxides, **4**, 69
    alkyl sulfides, **4**, 63
    alkyl sulfoxides, **4**, 69
    benzoates, **4**, 69
    α-chloro-α-(chlorosulfenyl)carboxylic esters, **4**, 70
    sulfenyl chlorides, **4**, 69
    thiols, **4**, 69
  oxidation of, **4**, 74
  reactions, with magnesium, **2**, 999
  safety precautions, **4**, 63
  stability of, **4**, 62
  as starting materials
    for α-chloro sulfones, **4**, 80
    for dithioacetals, **4**, 258
  synthesis of, **3**, 16; **4**, 62
  use in the synthesis of monothioacetals, **4**, 227
  use in the synthesis of α-selenyl ketones, **3**, 175
Sulfides, β-chloro
  as starting materials, for alkenes, **1**, 591
  synthesis of, **2**, 138
Sulfides, chloroalkyl, reactions, with silyl enol ethers, **3**, 168
Sulfides, α-chloroalkyl phenyl, reactions, with zinc, **4**, 396
Sulfides, (α-chloroalkyl)phosphine, synthesis of, **4**, 123
Sulfides, 1-chloro-1,1-dinitromethyl, synthesis of, **6**, 281
Sulfides, chlorofluoro, synthesis of, **6**, 47
Sulfides, (chloromethyl)phosphine, synthesis of, **4**, 123
Sulfides, 1-chlorovinyl, as starting materials, for bis(alkylthio)tetrachloroethanes, **6**, 46
Sulfides, α-chlorovinyl, synthesis of, **4**, 773
Sulfides, cyanomethyl, electrochemical oxidation of, **4**, 221
Sulfides, cyclic
  deprotonation of, **1**, 234
  from, epoxides, **2**, 136
Sulfides, cyclic phosphine, phosphinylation of, **4**, 1026
Sulfides, cyclopropyl, as starting materials, for thioacrolein, **3**, 337
Sulfides, cyclopropylphenyl, ring-opening of, **3**, 224
Sulfides, dialkyl
  as byproducts, **2**, 116
  from
    alcohols, **2**, 135
    alkenes, **2**, 137
    alkyl halides, **2**, 132
    alkyl sulfates, **2**, 134
    alkyl thiolates, **2**, 132
    carbonyls, **2**, 139
    carboxyls, **2**, 140
    disulfides, **2**, 143
    epoxides, **2**, 135
    metal sulfides, **2**, 132
    silyloxy monothioacetals, **2**, 142
    α-sulfenyl anions, **2**, 143
    sulfones, **2**, 143
    sulfonic esters, **2**, 135
    sulfonium salts, **2**, 143
    sulfoxides, **2**, 142
    thioacetals, **2**, 141
    thiocarbonyls, **2**, 140
    thiol esters, **2**, 143
  oxidation of, **2**, 146
  reactions
    with *N*-bromoacetamide, **2**, 159
    with chloramine, **2**, 164
    with *N*-chloroamidines, **5**, 766
    with ethyl *N*-chlorocarbamate, **2**, 160
    with haloimides, **5**, 336
  synthesis of, **2**, 132, 210
  use in the synthesis of α-(phenylthio) ketones, **4**, 220
Sulfides, diaryl
  reactions
    with *N*-chloroimidates, **5**, 703
    with lead tetraacetate, **2**, 198
  synthesis of, **2**, 711
Sulfides, diazo, reactions, with tetrabutylammonium cyanide, **3**, 664, 667
Sulfides, α-diazo phoshpine, photolysis of, **4**, 346
Sulfides, dibenzyl, synthesis of, **4**, 345
Sulfides, dichloro, reactions, with antimony trifluoride, **6**, 47
Sulfides, α,α-dichloro
  from
    alkenes, **6**, 46
    carbon disulfide, **6**, 46
    chlorodifluoromethane, **6**, 45
    α,α-dichloroalkyl sulfides, **6**, 45
    α-fluoroalkyl sulfoxides, **6**, 46
    iodofluoroalkanes, **6**, 45
    sulfides, **6**, 44
    thiolformates, **6**, 46
Sulfides, dichloroalkyl, reactions, with antimony trifluoride, **6**, 45
Sulfides, α,α-dichloroalkyl, as starting materials, for α,α-dichloro sulfides, **6**, 45
Sulfides, α,α-dichlorodibenzyl, as starting materials, for diaryl-substituted alkynes, **1**, 1074
Sulfides, α,α-dichloromethyl, synthesis of, **6**, 46
Sulfides, di(cyanomethyl), from, thioamides, **3**, 634
Sulfides, dienyl, synthesis of, **2**, 901
Sulfides, difluoro, synthesis of, **6**, 45
Sulfides, α,α-difluoroalkyl, synthesis of, **6**, 46
Sulfides, diisopropyl, reactions, with trimethylsilyl chloride, **2**, 99
Sulfides, divinyl, synthesis of, **2**, 708, 722
Sulfides, enediyne, oxidation of, **1**, 956
Sulfides, ethyl, reactions, with benzyne, **2**, 713
Sulfides, α-fluoro
  from, sulfoxides, **4**, 61
  oxidation of, **4**, 73

as starting materials, for α-fluoro sulfones, **4**, 79
synthesis of, **4**, 61, 218
Sulfides, α-halo
pyrolysis of, **1**, 592
reactions, with alcohols, **4**, 218
as starting materials, for monothioacetals, **4**, 218
synthesis of, **4**, 61
Sulfides, α-halophosphine, synthesis of, **4**, 123
Sulfides, 1-halovinyl, synthesis of, **4**, 773
Sulfides, α-heteroarylthionitro, synthesis of, **4**, 327
Sulfides, homoallylic
from, allylic sulfides, **1**, 408
synthesis of, **1**, 407
Sulfides, hydroxy, as byproducts, **2**, 130
Sulfides, 2-hydroxy
ring-expansion of, **1**, 390
semipinacol rearrangement of, **1**, 390
Sulfides, β-hydroxy
elimination of, stereoselectivity of, **1**, 591
oxidation of, **3**, 165
reactions, with *o*-phenylene phosphorochloridite, **1**, 590
as starting materials, for alkenes, **1**, 590
synthesis of, **1**, 693; **4**, 220, 391
Sulfides, β-hydroxy-substituted, synthesis of, **2**, 139
Sulfides, imidyl, synthesis of, **5**, 638
Sulfides, α-iodo, synthesis of, **4**, 71
Sulfides, α-keto, synthesis of, **2**, 139
Sulfides, α-lithio
from
phenylvinyl sulfide, **4**, 392
vinyl sulfides, **4**, 392
reactions, with 2-lithio-1,3-dithiane, **3**, 169
stability of, **2**, 999
synthesis of, **2**, 999; **4**, 366
via thiophilic addition, **4**, 394
Sulfides, 1-lithioalkylphosphine, synthesis of, **4**, 576
Sulfides, α-lithiophenyl, synthesis of, **4**, 394
Sulfides, 1-lithiovinyl phenyl, reactions, with tin chlorides, **4**, 961
Sulfides, α-metallo, synthesis of, **2**, 999
Sulfides, 1-metalloalkylphosphine
reactions
with chlorophosphines, **4**, 551
with chlorosilanes, **4**, 570
with chlorostannanes, **4**, 588
with gold compounds, **4**, 586
with mercury compounds, **4**, 587
with tantalum compounds, **4**, 579
Sulfides, α-methoxy, synthesis of, **4**, 222
Sulfides, 2-methoxyalkyl, as starting materials, for 1-alkylthiovinyllithiums, **4**, 958
Sulfides, methyl, chlorination of, **6**, 233
Sulfides, methyl benzyl, synthesis of, **1**, 24
Sulfides, nitrile
cycloadditions of, **4**, 232
from
(alkylimino)sulfur difluorides, **3**, 686
4,5-dihydro-1,2,4-thiadiazoles, **3**, 686
dithiazolethiones, **3**, 687
1,2,4-dithiazol-5-ones, **3**, 686
1,3,4-oxathiazol-2-ones, **3**, 686
thiatriazoles, **3**, 687
N-thioaroyl diphenylsulfimides, **3**, 687
reactions, with dipolarophiles, **3**, 686
synthesis of, **3**, 686
Sulfides, nitro, as starting materials, for sulfones, **4**, 908
Sulfides, α-nitro, synthesis of, **4**, 326
Sulfides, β-nitro
elimination of, **1**, 638
reactions
with allyl trimethylsilane, **1**, 252
with tributyltin hydride, **1**, 592
synthesis of, **1**, 252
Sulfides, β-oxygenated, elimination of, **1**, 591
Sulfides, phenacyl
photolysis of, **3**, 331
as starting materials
for thioaldehydes, **3**, 331
for thioketones, **3**, 351
Sulfides, phenyl
lithiation of, **3**, 161
precursors for radical cyclization, **1**, 341
use in the synthesis of radicals, **1**, 942
Sulfides, phenyl fluoromethyl, synthesis of, **4**, 61
Sulfides, 1-phosphinoalkylphosphine
reduction of, **4**, 550
synthesis of, **4**, 551
Sulfides, phosphole, reduction of, **4**, 1026
Sulfides, primary chloro, from, paraformaldehyde, **4**, 69
Sulfides, propargyl
metallation of, **1**, 141
as starting materials, for allenylsilanes, **2**, 939
Sulfides, pyridyl
as starting materials
for acids, **1**, 333
for aldehydes, **1**, 333
for amides, **1**, 333
for esters, **1**, 333
for ketones, **1**, 333
synthesis of, via Barton reactions, **1**, 332
Sulfides, secondary chloro, from, paraldehyde, **4**, 69
Sulfides, selenyl, synthesis of, **2**, 234
Sulfides, silyl
as intermediates, **2**, 264
reactions, with disulfides, **2**, 228
use in the synthesis of thiocarboxylic esters, **5**, 236
Sulfides, 1-silylalkylphosphine, synthesis of, **4**, 569
Sulfides, α-silyloxy, synthesis of, **3**, 16; **4**, 223
Sulfides, β-stannyl, elimination of, **1**, 591
Sulfides, β-substituted
elimination of, **1**, 590
synthesis of, **2**, 138
Sulfides, tertiary alkylarsine, synthesis of, **2**, 502
Sulfides, thioacyl diphenylselenophosphinoyl
as thioacylating reagents, **5**, 584
use in the synthesis of thionoamides, **5**, 584
Sulfides, thioacyl diphenylthiophosphinoyl
as thioacylating reagents, **5**, 584
use in the synthesis of thionoamides, **5**, 584
Sulfides, trifluoromethyl
oxidation of, **6**, 234
synthesis of, **6**, 46
Sulfides, trihalomethyl
from, alcohols, **6**, 234
synthesis of, **6**, 232
via halogen exchange, **6**, 233
Sulfides, trimethylsilyl
reactions
with aldehydes, **4**, 216
with ketones, **4**, 216
as starting materials, for thiocarboxylic esters, **5**, 234
Sulfides, trimethylsilyl alkynyl, as starting materials, for trimethylsilyl thioketenes, **3**, 548
Sulfides, α-(trimethylsilyl)methyl, metallation of, **6**, 135
Sulfides, α-(trimethylsilyloxy), synthesis of, **4**, 216
Sulfides, 1-trimethylsilylvinyl, synthesis of, **4**, 945
Sulfides, unsymmetrical, chlorination of, **4**, 66
Sulfides, vinyl
cleavage of, **3**, 337
from
alkynes, **2**, 710
dithioacetals, **2**, 710
thioacetals, **1**, 590
vinyl halides, **2**, 709
hydrolysis of, for the synthesis of aldehydes, **3**, 15
as intermediates, in the synthesis of methylene, **1**, 41

lithiated compounds of, use in the synthesis of ketones, **3**, 137
oxidation of, **2**, 714
reactions
with *n*-butyllithium, **2**, 1000
with lithium, **2**, 722
with lithium 4,4'-(di-*t*-butyl)biphenylide, **2**, 68
with organolithium nucleophiles, **4**, 366
with organolithiums, **4**, 392
with potassium 3-aminopropylamide, **1**, 590
with sodium, **2**, 722
reduction of, **1**, 438
as starting materials
for alkenes, **1**, 438
for disulfides, **2**, 232
for ketones, **3**, 131
for α-lithio sulfides, **4**, 392
synthesis of, **1**, 572, 676; **2**, 708, 901
via elimination of halosilanes, **1**, 653
via elimination reactions, **2**, 710
via enolization, **2**, 709
via Wittig-type reactions, **2**, 709
Sulfides, vinylic
reduction of, **1**, 83
as starting materials, for thiocarboxylic esters, **5**, 238
Sulfides, vinylphosphine, synthesis of, **4**, 1023
Sulfilimine, diphenyl-, reactions, with dibenzoylethene, **2**, 745
Sulfilimine, *S*,*S*-diphenyl-, reactions, with *t*-butylcyanoketene, **5**, 349
Sulfilimines
elimination of, **1**, 621
from, amines, **2**, 401
synthesis of, **2**, 401
See also Sulfimines
Sulfilimines, allylic, rearrangement of, **2**, 365
Sulfimide complexes, synthesis of, **4**, 800
Sulfimides
reactions
with benzimidoyl chlorides, **5**, 766
with lithium diisopropylamide, **6**, 326
Sulfimides, *N*-acyl dichloro-, synthesis of, **5**, 349
Sulfimides, *S*,*S*-diphenyl-, synthesis of, **5**, 766
Sulfimides, α-halo alkenyl dichloro-, synthesis of, **4**, 800
Sulfimides, *N*-imidoyl-, synthesis of, **5**, 766
Sulfimides, *N*-thioaroyl diphenyl-, as starting materials, for nitrile sulfides, **3**, 687
Sulfimine, 4-toluenesulfonyl dimethyl-, synthesis of, **2**, 162
Sulfimines
from, sulfoxides, **2**, 162
reactions, with acid chlorides, **5**, 339
as starting materials
for *N*-acyl sulfonium salts, **5**, 339
for sulfoximines, **2**, 164
synthesis of, **2**, 159
See also Sulfilimines
Sulfimines, *N*-acyl-, synthesis of, **2**, 159
Sulfimines, *N*-ethoxycarbonyl-, synthesis of, **2**, 160
Sulfimines, *N*-sulfonyl-, synthesis of, **2**, 161
Sulfinamides
alkylation of, **1**, 695
deprotonation of, **4**, 391
from
N-acylsulfinamides, **2**, 363
N-acyl-*N*-sulfinylamines, **5**, 340
amides, **5**, 340
silylamides, **5**, 341
sulfenamides, **2**, 362
sulfinates, **2**, 363
sulfoximines, **2**, 245
oxidation of, **2**, 250; **5**, 347
as starting materials
for alkanesulfinic esters, **2**, 200
for alkenes, **1**, 621, 695
for *N*-haloamides, **5**, 341
for sulfonimidoyl chlorides, **2**, 401
synthesis of, **2**, 245, 362, 363, 796
use in the synthesis of arene thiosulfinate esters, **2**, 724
Sulfinamides, *N*-acyl-, as starting materials, for sulfinamides, **2**, 363
Sulfinamides, alkane-, synthesis of, **2**, 245
Sulfinamides, alkenyl-, synthesis of, **1**, 622
Sulfinamides, *N*-alkoxy-, synthesis of, **2**, 363
Sulfinamides, allylic
from, allylic alcohols, **2**, 247
synthesis of, **2**, 331, 363
Sulfinamides, arene-
optically active, for the synthesis of sulfinates, **2**, 721
reactions, with 1,3-dichloro-1,1,3,3-tetrafluoroacetone, **4**, 311
synthesis of, **2**, 726
Sulfinamides, arene *N*-alkoxy-, synthesis of, **2**, 364
Sulfinamides, aryl-, reactions, with alcohols, **2**, 200
Sulfinamides, β-hydroxy-, *cis*- elimination of, **1**, 621
Sulfinamides, *N*-imidoyl-, synthesis of, **5**, 767
Sulfinamides, α-lithio-, reactions, with carbonyls, **1**, 695
Sulfinamides, primary, *N*-alkylation of, **2**, 363
Sulfinamides, propargylic, hydrolysis of, **1**, 963
Sulfinamides, *N*-tosyl cyclic, synthesis of, via Diels–Alder reactions, **2**, 366
Sulfinamides, vinyl-, synthesis of, **2**, 726
Sulfinamidines, *N*-acyl-*N'*-aryl(alkyl)sulfonyl-, synthesis of, **5**, 340
Sulfinanilides, *trans*-2-hydroxycyclohexyl-, *cis*-elimination of, **1**, 695
Sulfinate, (-)-menthyl *p*-toluene-, reactions, with 1-trimethylsilylvinylmagnesium bromide, **4**, 946
Sulfinate, *p*-toluene-, reactions, with hydrazones, **3**, 37
Sulfinate, trifluoromethane, as leaving group in Ramberg–Bäcklund reactions, **1**, 705
p-Sulfinate, menthyl toluene-, reactions, with *p*-tolylthiomethyllithium, **4**, 265
p-Sulfinate, (-)-menthyl (*S*)-toluene-, use in the synthesis of dithioacetal trioxides, **4**, 271
Sulfinate anions
reactions
with alkyl halides, **4**, 235
with dibromomethane, **4**, 83
Sulfinate esters
from, sulfinyl chlorides, **2**, 721
oxidation of, **2**, 207
as starting materials, for alkanesulfonic esters, **2**, 207
use in the synthesis of dialkyl sulfoxides, **2**, 152
Sulfinate esters, allylic, rearrangement of, **2**, 159
Sulfinate esters, arene, synthesis of, **2**, 721
Sulfinates
alkylation of, **2**, 721
in Heck reactions, **1**, 494
reactions
with ketone enolates, **3**, 170
with sulfones, **1**, 619
as starting materials, for sulfinamides, **2**, 363
use in elimination reactions, **1**, 981
Sulfinates, alkenyl, synthesis of, **2**, 688
Sulfinates, arene
as nucleophiles, **2**, 716
synthesis of, **2**, 720
Sulfinates, aryl, synthesis of, **2**, 688
Sulfinates, chiral
nucleophilic substitution of, **2**, 714
synthesis of, **2**, 721
Sulfinates, chloro-, stability of, **2**, 100
Sulfinates, α-hydroxy, synthesis of, **4**, 234
Sulfinates, menthyl, use in the synthesis of sulfoxides, **2**, 720
Sulfinates, β-silylalkenyl, elimination of, **1**, 1080
Sulfinates, thiol, as sulfenylating reagents, **5**, 247

Sulfinates, trimethylsilyl propargylic, as starting materials, for silylated allenes, **1**, 967
Sulfinate salts, arene
reactions
with alkyl halides, **2**, 715
with aryl halides, **2**, 715
Sulfinate salts, aryl, reactions, with trimethylsilyl cyanide, **5**, 1112
Sulfine, bis(trifluoromethyl)-
reactions
with phenyldiazomethane, **5**, 358
with thiophosgene, **6**, 436
Sulfine, chlorofluoro-, synthesis of, **6**, 532
Sulfine, dibromo-, synthesis of, **6**, 531
Sulfine, dichloro-
from
allyl trichloromethyl sulfoxide, **6**, 531
2,2,4,4-tetrachloro-1,3-dithietane, **6**, 530
trichloromethanesulfenyl chloride, **6**, 531
synthesis of, **6**, 530
See also Thiophosgene-*S*-oxide
Sulfine, difluoro-
from
allyl difluorochloromethyl sulfoxide, **6**, 530
1,1,4,4-tetrafluorodithietane, **6**, 530
synthesis of, **6**, 530
Sulfine, diphenyl-, reactions, with 2-diazopropane, **6**, 316
Sulfine, methylchloro-, synthesis of, **5**, 489
Sulfine, phenyl-, reactions, with *p*-bromobenzoylmethylene triphenylarsorane, **5**, 920
Sulfine, phenylchloro-
from, thiobenzoyl chloride, **5**, 490
synthesis of, **5**, 489
Sulfines
m-chloroperbenzoic acid oxidation of, **6**, 306
from
allylvinyl sulfoxides, **3**, 369
sulfinyl chlorides, **3**, 368
thiophenes, **3**, 369
as intermediates, in the synthesis of ylides, **5**, 900
reactions
with chlorine, **5**, 489
with 1,3-dipoles, **2**, 209
with nucleophiles, **4**, 859
with organometallics, **4**, 272
stability of, **3**, 369
as starting materials
for bissulfoxides, **4**, 270
for thiocarbonyl *S*-imides, **3**, 370
for thioketones, **3**, 351
synthesis of, **3**, 368; **6**, 530, 537
via 1,2-elimination reactions, **3**, 368
via oxidation of thioketones, **3**, 368
via retro [2+2] cyclization, **3**, 369
via retro Diels–Alder reaction, **3**, 369
via sulfur dioxide, alkylidenation of, **3**, 368
Sulfines, α-(allylthio)-, tautomerization of, **4**, 859
Sulfines, chloro-
from
α-carbamoyldichlorosulfenyl chlorides, **5**, 491
α-chlorosulfinyl chlorides, **5**, 489
1,1-dichlorosulfenyl chlorides, **5**, 492
sulfenyl halides, **5**, 492
stability of, **5**, 488
synthesis of, **5**, 490
Sulfines, chlorotrifluoromethyl, synthesis of, **5**, 492
Sulfines, *N*,*N*-disubstituted α-carbamoyl chloro-, from, dichlorosulfenyl chlorides, **5**, 490
Sulfines, dithioester-, [4π+2π] cycloaddition of, **4**, 265
Sulfines, fluoro-, stability of, **5**, 488
Sulfines, fluorotrifluoromethyl-
from, trifluoroethene, **5**, 494
synthesis of, **5**, 492
Sulfines, halo-
from
camphor-10-sulfonyl chloride, **5**, 488
dichlorosulfenyl chlorides, **5**, 491
synthesis of, **5**, 488
via dehydrohalogenation of sulfonyl halides, **5**, 488
via oxidation of thioacyl halides, **5**, 490
as thioacyl halides, **5**, 472
Sulfines, α-oxo-, synthesis of, **3**, 171; **5**, 562
Sulfines, perhalogenated aliphatic halo-, synthesis of, **5**, 492
Sulfines, phosphoryl-, synthesis of, **4**, 926; **5**, 645
Sulfines, selenyl-, from, selenothioic acid *Se*-phenyl esters, **5**, 563
Sulfines, sulfenyl-, synthesis of, **5**, 562
Sulfines, sulfinyl-, synthesis of, **5**, 562
Sulfines, sulfonyl-
reactions, with methyllithium, **4**, 272
synthesis of, **5**, 562
Sulfines, γ,δ-unsaturated, synthesis of, **1**, 834
Sulfinic acid
reactions
with diphenylphosphinous acids chloride, **2**, 854
with mercury salts, **2**, 594
with thionyl chloride, **5**, 340
Sulfinic acid, *o*-benzylarene-, from, aryl sulfones, **1**, 379
p-Sulfinic acid, toluene-, reactions, with phenylglyoxal, **4**, 234
Sulfinic acids
alkylation of, **2**, 156, 197
elimination of, **1**, 612
esterification of, **2**, 197, 721
from, sulfones, **2**, 188
hydrolysis of, **2**, 187
oxidation of, **2**, 193
reactions
with acetals, **4**, 236
with amines, **2**, 363
with cyanogen chloride, **5**, 1112
with thionyl chloride, **2**, 172, 717
as starting materials
for alkanesulfinic esters, **2**, 197
for α-alkoxy sulfones, **4**, 235
for dialkyl sulfones, **2**, 156
synthesis of, **2**, 186; **4**, 270
use in the synthesis of arene thiosulfinate esters, **2**, 724
Sulfinic acids, *O*-acyl-, synthesis of, **5**, 209
Sulfinic acids, alkane-, reactions, with thionyl chloride, **2**, 172
Sulfinic acids, alkyl-
desulfination of, **2**, 594
reactions, with diazoalkanes, **2**, 100
Sulfinic acids, arene-
from
arenes, **2**, 718
arenethiols, **2**, 719
oxidation of, **2**, 724
synthesis of, **2**, 718
Sulfinic acids, aryl-, as starting materials, for aryl thiocyanates, **5**, 1112
Sulfinic acids, 2-chloroalkane-, synthesis of, **2**, 188
Sulfinic acids, β-substituted, synthesis of, **2**, 188
Sulfinic acids, *p*-toluene-, for the cleavage of sesquiterpenes, **1**, 450
Sulfinic anhydrides, synthesis of, **2**, 208
Sulfinic esters
hydrolysis of, **2**, 187
synthesis of, **2**, 196
Sulfinic esters, alkane-
from
alcohols, **2**, 202
alkyl sulfites, **2**, 197
disulfides, **2**, 198

hydroxythiols, **2**, 201
sulfinamides, **2**, 200
sulfinic acids, **2**, 197
sulfinyl halides, **2**, 196
sulfones, **2**, 200
sulfoxides, **2**, 199
sulfoxylic dialkyl esters, **2**, 201
sulfur dioxide, **2**, 201
thioketenes, **2**, 202
thiosulfonates, **2**, 199
synthesis of, **2**, 196
Sulfinic esters, arene-
reactions, with primary amines, **2**, 361
as starting materials, for sulfenamides, **2**, 358
Sulfinic esters, trimethylsilyl-, synthesis of, **2**, 209
Sulfinic salts, from, perfluoro-2-phosphapropene, **6**, 258
Sulfinimides, synthesis of, **6**, 252
Sulfinimidoyl chlorides, *N*-acyl, from, *N*-haloamides, **5**, 349
Sulfinimidoyl chlorides, *N*-α-chloroalkyl, synthesis of, **4**, 110
Sulfinimines
Darzens reaction of, **3**, 437
reduction of, **3**, 437
synthesis of, **3**, 437
Sulfinimines, alkyl aryl-, synthesis of, **3**, 437
Sulfinimines, chiral
synthesis of, **3**, 437
use in the synthesis of amino acids, **3**, 437
Sulfinyl, phenyl-, as radical propagator, **1**, 231
N-Sulfinylamines, α-chloro-α-methoxy-, synthesis of, **5**, 340
Sulfinylation, of ketones, **3**, 170
α-Sulfinyl carbanions
alkylation of, **4**, 77
chirality of, **4**, 391
diastereoselective alkylation of, **1**, 143
from
α-bromo sulfoxides, **4**, 77
phenyl fluoromethyl sulfoxides, **4**, 74
reactions
with aldehydes, **4**, 77
with ketones, **4**, 77
use in the synthesis of α-chloro sulfoxides, **4**, 77
Sulfinyl chloride, reactions, with methylenes, **4**, 266
Sulfinyl chloride, alkane-, reactions, with lithiated alkyne, **2**, 1035
Sulfinyl chloride, *t*-butyl-, synthesis of, **2**, 173
Sulfinyl chloride, phenyl-, reactions, with diazomethane, **3**, 479; **5**, 867
Sulfinyl chloride, toluene-, as starting material, for α-sulfinyl ylides, **5**, 900
Sulfinyl chlorides
from
organosulfur trichlorides, **2**, 174
thioacetals, **2**, 179
handling precautions, **2**, 172
reactions
with alcohols, **2**, 196
with anilines, **2**, 796
with diazo compounds, **4**, 76
with hydroxylamines, **2**, 365
with silver isocyanate, **2**, 726
reductive coupling of, **2**, 221
stability of, **4**, 78
as starting materials
for α-iodo sulfoxides, **4**, 78
for sulfinate esters, **2**, 721
for sulfines, **3**, 368
synthesis of, **2**, 172, 688; **4**, 78
Sulfinyl chlorides, alkane-
hydrolysis of, **2**, 187
reactions
with *N*-alkoxyamines, **2**, 363
with hydroxylamines, **2**, 250
Sulfinyl chlorides, alkyl-, reactions, with amines, **2**, 362
Sulfinyl chlorides, arene-
reactions, with thiols, **2**, 723
synthesis of, **2**, 717
Sulfinyl chlorides, α-chloro-
as starting materials, for chlorosulfines, **5**, 489
synthesis of, **5**, 489
Sulfinyl chlorides, α-chloroalkane-, synthesis of, **4**, 78
Sulfinyl chlorides, α-chloroalkyl-, from, trithianes, **4**, 78
Sulfinyl compounds
oxidation of, **2**, 250
reactions, with organometallics, **2**, 363
synthesis of, **4**, 958
Sulfinyl compounds, α-diazo-, synthesis of, via diazo group transfer, **3**, 479
N-Sulfinyl compounds
phosgenation of, **5**, 968
synthesis of, **2**, 725
Sulfinyl fluoride, trichloromethyl-, synthesis of, **2**, 175
Sulfinyl fluorides, synthesis of, **2**, 172
Sulfinyl fluorides, arene-, use of, **2**, 716
Sulfinyl group transfer, for the synthesis of dialkyl sulfoxides, **2**, 151
Sulfinyl halides
dechlorination of, **5**, 492
reactions
with diazo compounds, **4**, 75
with diazomethane, **4**, 76
as starting materials, for alkanesulfinic esters, **2**, 196
as sulfinylating reagents, **2**, 688
Sulfinyl halides, alkane-, synthesis of, **2**, 172
Sulfinyl halides, arene-, synthesis of, **2**, 716
Sulfinyl halides, trihalomethane-, synthesis of, **6**, 236
β-Sulfinyl-substituted compounds, elimination of, **1**, 599
Sulfinyl sulfones, arene-, synthesis of, **2**, 724
Sulfite, 2,2'-dipyridyl, use in the synthesis of nitriles, **3**, 619
Sulfites
extrusion of sulfur dioxide from, **1**, 232
reactions
with alkyl halides, **2**, 191
with diazosulfonates, **2**, 769
Sulfites, alkyl, as starting materials, for alkanesulfinic esters, **2**, 197
Sulfites, alkyl chloro-
reactions, with alkoxides, **2**, 102
stability of, **2**, 101
synthesis of, **2**, 101
factors influencing, **2**, 101
Sulfites, allylic, as starting materials, for alkanesulfonic esters, **2**, 206
Sulfites, amino, synthesis of, **2**, 364
Sulfites, cyclic, synthesis of, **2**, 100
Sulfites, dialkyl
as dehydrating agents, for the synthesis of acetals, **4**, 177
reactions
with Grignard reagents, **2**, 197
with sulfuryl chloride, **2**, 102
synthesis of, **2**, 100, 102
Sulfites, fluoro-, from, difluoroamines, **2**, 763
o-Sulfobenzoic acid, ammonium salt, synthesis of, **5**, 109
o-Sulfobenzoic imide, hydrolysis of, **5**, 109
Sulfobetaines, synthesis of, **5**, 902
Sulfodiimides, synthesis of, **2**, 402
Sulfodiimines, reactions, with bis(1,1-alkylthio) alkenes, **4**, 899
Sulfolane, 3-methyl-, as solvent, in the synthesis of ketones, **3**, 113
Sulfolanes
reactions, with diphenyl disulfide, **4**, 266
use in the reduction of chloroalkanes, **1**, 7
Sulfolene pyrolysis, for the synthesis of alkaloids, **1**, 791
Sulfolenes, alkylation of, **1**, 790
3-Sulfolenes, pyrolysis of, **1**, 790
Sulfonamide, *N*-acyl-

synthesis of
via hydrolysis of *N*,*N*-bis(sulfonyl)amides, **5**, 348
via hydrolysis of heterocycles, **5**, 348
via hydrolysis of 2-iminoazetidine, **5**, 348
Sulfonamide, *N*-chloro-*N*-methylsulfonylmethyl-, reactions, with sulfur, **2**, 99
Sulfonamide, *N*-(4-chlorophenyl)dichloromethyl-, as starting material, for 4-chlorophenyl isocyanide, **3**, 723
Sulfonamide, *N*,*N*-dibromophenyl-, for the electrophilic bromination of arenes, **2**, 626
Sulfonamide, ditellurenyl-, synthesis of, **2**, 769
Sulfonamide, *N*-methyl-*N*-nitroso-*p*-toluene-, use in the synthesis of diazomethane, **3**, 467
Sulfonamide, *N*-sulfinyl-*p*-toluene-, reactions, with *β*-keto sulfoxides, **4**, 326
Sulfonamide, *p*-toluene-, condensation of, **4**, 311
Sulfonamide anions, and reductive deaminations, **1**, 28
Sulfonamidebisimides
as starting materials, for sulfonimidamides, **2**, 402
synthesis of, **2**, 402
α,*N*-Sulfonamide dianions, synthesis of, **1**, 147
Sulfonamides
acylation of, **5**, 342
alkylation of, **2**, 365, 797
deprotonation of, **1**, 147; **4**, 391
elimination of, **1**, 622
fluorination of, **2**, 365
from
sulfonamides, **2**, 251
sulfonyl azides, **2**, 365
sulfonyl isocyanates, **2**, 249
α-oxidation of, **5**, 347
photolysis of, **5**, 343
reactions
with acyl halides, **5**, 342
with aldehydes, **4**, 423
with aminols, **4**, 427
with anhydrides, **5**, 342
with carbon disulfide, **6**, 627
with *N*,*N'*-carbonyldiimidazole, **5**, 997
with carboxylic acids, **5**, 342
with formaldehyde, **4**, 427
with isocyanates, **5**, 981
with ortho esters, **5**, 827
with oxalyl chloride, **2**, 726; **5**, 996
with phosgene, **2**, 726
with potassium cyanide, **5**, 1125
with sulfur dichloride, **2**, 365
with 2,2,2-trichloroacetophenone, **5**, 345
with trimethylaluminum, **5**, 342
with trioxan, **4**, 427
with triphenylphosphine dibromide, **2**, 863
N-stannylation of, **2**, 815
as starting materials
for alkanesulfonic esters, **2**, 205
for cyanamides, **5**, 1125
for disulfonimides, **1**, 28
for sulfonamides, **2**, 251
for *N*-sulfonoylimines, **3**, 439
synthesis of, **2**, 248, 364, 797
via transamination reactions, **2**, 364
Sulfonamides, acyl-, reduction of, **2**, 319
Sulfonamides, *N*-acyl-
from
acylsilanes, **5**, 343
aldehydes, **5**, 345
arylsulfonyl azides, **5**, 346
N-chlorosulfonamides, **5**, 345
ketenes, **5**, 343
ketones, **5**, 345
sulfonyl isocyanates, **5**, 344
N-tosylimines, **5**, 344
reactions, with *α*,*β*-unsaturated acids, **5**, 342
synthesis of
via acylation of *N*-substituted sulfonamides, **5**, 343
via acylation of sulfonamides, **5**, 342
via acylation of *N*-tosylimines, **5**, 344
via acylation of *N*-trimethylsilyl sulfonamides, **5**, 343
via carbon monoxide insertion, **5**, 346
via oxidation of sulfinamides, **5**, 347
via oxidative rearrangement of *N*-tosylamino acids, **5**, 346
via ring-opening of *N*-sulfonylaziridines, **5**, 345
via ring-opening of *N*-tosyloxiranes, **5**, 347
via sulfonylation of amides, **5**, 345
via sulfonylation of lactim ethers, **5**, 346
Sulfonamides, alkane-, synthesis of, **2**, 248
Sulfonamides, *N*-alkyl-, cyclization of, **2**, 365
Sulfonamides, *N*-alkyl aryl-, as starting materials, for bissulfonamides, **4**, 428
Sulfonamides, *N*-alkyl-*N*-nitroso-*p*-toluene-, as starting materials, for diazoalkanes, **3**, 467
Sulfonamides, allylic, synthesis of, **2**, 298
Sulfonamides, *β*-amino-*N*-acyl-, from, 2-*N*-(*p*-toluenesulfonyl)iminoazetidines, **5**, 348
Sulfonamides, arene-
reactions
with iminium salts, **6**, 156
with triethyl orthocarbonate, **6**, 625
as starting materials, for *N*-sulfonyl aminals, **4**, 423
synthesis of, **2**, 365, 726
uses of, **2**, 797
Sulfonamides, aryl-
reactions
with aluminum trichloride, **5**, 674
with chloramine, **1**, 35
with hydroxylamine-*O*-sulfonic acid, **1**, 35
with isocyanates, **5**, 767
Sulfonamides, *N*-aryl-, from, sulfonimines, **3**, 438
Sulfonamides, *N*-benzoyl-, from, *N*-tosyl phenylglycine esters, **5**, 346
Sulfonamides, benzyl-
alkylation of, problems, **1**, 147
synthesis of, **1**, 147
Sulfonamides, *N*-bromo-, synthesis of, **2**, 365
Sulfonamides, *α*-bromo-, synthesis of, **4**, 105
Sulfonamides, bromoalkane-, synthesis of, **4**, 86
Sulfonamides, *N*-chloro-
decomposition of, **3**, 666
reactions, with acetophenone, **5**, 345
as starting materials, for *N*-acylsulfonamides, **5**, 345
synthesis of, **2**, 365
Sulfonamides, chloroalkane-, synthesis of, **4**, 86
Sulfonamides, *N*,*N*-dichloro-, reactions, with iminium salts, **6**, 619
Sulfonamides, *N*-dichloromethylene-, reactions, with amine hydrochlorides, **5**, 1071
Sulfonamides, *N*,*N*-disubstituted arene-, *N*-methylation of, **2**, 365
Sulfonamides, *N*-fluoro-
as fluorinating agents, **2**, 16
synthesis of, **2**, 365
Sulfonamides, *N*-fluoro-*N*-alkyl-, for the synthesis of aryl fluorides, from aryl metallics, **2**, 613
Sulfonamides, *N*-fluoroperfluoroalkyl-, as reagents, for aryl fluorides, **2**, 613
Sulfonamides, *N*-halo-, synthesis of, **2**, 365
Sulfonamides, *α*-halo-, synthesis of, **4**, 104
Sulfonamides, *α*-haloalkane-, from, *α*-halosulfonyl halides, **4**, 86
Sulfonamides, imidazolyl-, use in the synthesis of alkanesulfonic esters, **2**, 205
Sulfonamides, imido-, synthesis of, via oxidation of thioimides, **5**, 342
Sulfonamides, imidoyl-, from, vinylsulfonyl chlorides, **5**, 767

Sulfonamides, *N*-imidoyl-
from, *N*-sulfonylcarboxyamides, **5**, 767
synthesis of, **5**, 767
Sulfonamides, iodoalkane-, synthesis of, **4**, 86
Sulfonamides, *β*-ketoalkane-, synthesis of, **2**, 249
Sulfonamides, *N*-methyl alkyl-, use in the synthesis of *N*-thio α-haloamines, **4**, 104
Sulfonamides, monosubstituted, reactions, with sulfonyl chlorides, **2**, 366
Sulfonamides, *N*-nitroso-, acylation of, **5**, 344
Sulfonamides, non-racemic α-substituted, synthesis of, **1**, 147
Sulfonamides, silyl-, synthesis of, **2**, 251
Sulfonamides, *N*-substituted
acylation of, **5**, 343
synthesis of, **2**, 365
Sulfonamides, *N*-sulfinyl-
reactions
with alcohols, **2**, 100
with aldehydes, **1**, 1042; **4**, 311
with amides, **5**, 767
with ketenimines, **5**, 767
with ynamines, **5**, 767
as starting materials, for sulfinylamines, **2**, 399
Sulfonamides, *N*-sulfinylaryl-, reactions, with phosphinylethyne, **5**, 645
Sulfonamides, *N*-sulfinyl perfluoroalkane-, reactions, with sulfoxides, **2**, 162
Sulfonamides, tertiary alkyl-, synthesis of, **2**, 365
Sulfonamides, *N*-trimethylsilyl-, acylation of, **5**, 343
Sulfonamides, vinyl-, synthesis of, **2**, 726
N-Sulfonamides, *N*-sulfinyl-, oxidation of, **2**, 366
Sulfonamide salts, synthesis of, **2**, 365
Sulfonamidium salts, synthesis of, **2**, 365
Sulfonamyl chlorides, reactions, with dichloroketene acetals, **6**, 106
Sulfonate, alkylalkane, deprotonation of, **5**, 867
Sulfonate, methyl fluoro-
as alkylating agents, **2**, 674
reactions, with ketene dithioacetals, **4**, 863
Sulfonate, 3-pentyn-1-yl *m*-nitrobenzene, solvolysis of, **1**, 1123
Sulfonate, propargyltoluene, reactions, with alkynylmagnesium halides, **1**, 1025
Sulfonate, trimethylsilyl chloro-
reactions
with aldehydes, **2**, 209
with ketones, **2**, 209
with trimethylsilylcyclopropane, **2**, 190
Sulfonate esters
as intermediates, in the synthesis of alkyl bromides, **2**, 32
reactions
with alkali metal halides, **2**, 9
with phosphorus trichloride, **2**, 182
with tetrabutylammonium halides, **2**, 9
with thioacetic acid, **2**, 128
Sulfonate esters, alkynyl, research background, **2**, 1022
Sulfonate esters, oxime, as starting materials, for *n*-methyleneynamines, **2**, 1053
Sulfonates
acid strengths of, **1**, 177
in Heck reactions, **1**, 494
rate of solvolysis, **1**, 177
reactions, with lithium diisopropylamide, **1**, 183
reduction of, **1**, 11, 17
regioselectivity of, **1**, 224
as starting materials, for bridgehead halides, **2**, 9
use in the synthesis of alkyl halides, **2**, 9
use in the synthesis of quaternary phosphonium salts, **2**, 445
Sulfonates, acyloxy, synthesis of, **5**, 209
Sulfonates, alkenyl
reactions
with metallated alkynes, **1**, 1027
with terminal alkynes, **1**, 1028
synthesis of, **2**, 688
Sulfonates, alkyl
as alkylating agents, **2**, 272
displacement of, via nucleophiles, **1**, 177
Sulfonates, alkyl chloro-, synthesis of, **2**, 102
Sulfonates, alkyl fluoro-, use in the synthesis of thioimidic esters, **5**, 726
Sulfonates, alkyl 1-formylalkane
reactions
with 4-carboxybenzenesulfonyl azide, **5**, 867
with tosyl azide, **5**, 867
Sulfonates, alkynyl, synthesis of, **2**, 1028
Sulfonates, ammonium alkyl, reactions, with nitriles, **5**, 749
Sulfonates, ammonium aryl, reactions, with nitriles, **5**, 749
Sulfonates, arene-, synthesis of, **2**, 721, 722
Sulfonates, arenesulfonyl, synthesis of, **2**, 724
Sulfonates, aryl
reactions
with metallated alkynes, **1**, 1027
with terminal alkynes, **1**, 1028
synthesis of, **2**, 688
Sulfonates, aryl fluoroalkane
reactions
with alkynylzinc chlorides, **1**, 1028
with terminal alkynes, **1**, 1029
Sulfonates, aryl perfluoroalkane, synthesis of, **2**, 689
Sulfonates, aryl perfluoroalkene, reduction of, **1**, 433
Sulfonates, aryl trifluoroalkane, displacement of, **1**, 433
Sulfonates, benzene, reactions, with phosphazenium naked fluorides, **1**, 584
Sulfonates, α,α-bis(silyl), synthesis of, **4**, 605
Sulfonates, chloro-
reactions
with enolates, **2**, 690
with phenolates, **2**, 690
Sulfonates, cyclic, as starting materials, for iron complexes, **1**, 196
Sulfonates, diazo-
reactions, with sulfites, **2**, 769
reduction of, **2**, 769
Sulfonates, α-diazoalkane, synthesis of, **5**, 867
Sulfonates, enol
reactions
with bromotrichloromethane, **4**, 742
with halogen nucleophiles, **3**, 239
as starting materials, for alkynes, **1**, 1081
Sulfonates, ethylalkane, synthesis of, **2**, 207
Sulfonates, fluoro-, as starting materials, for halides, **2**, 9
Sulfonates, α-fluoroalkane, synthesis of, **4**, 86
Sulfonates, imidazolyl, as starting materials, for halides, **2**, 9
Sulfonates, iminium, synthesis of, **5**, 689
Sulfonates, iodonium, reactions, with *N*-silylamides, **5**, 321
Sulfonates, lanthanide trifluoromethane, use in the synthesis of aromatic ketones, **3**, 281
Sulfonates, methylalkane
reactions
with sodium hexamethyldisilazide, **6**, 132
with trialkylgermyl halides, **6**, 132
Sulfonates, oxime
Beckmann rearrangement of, **1**, 1044, 1059; **3**, 407; **5**, 731
reactions
with Grignard reagents, **1**, 1059; **3**, 407
with silyl enol ethers, **3**, 269
as starting materials
for ketenimines, **3**, 564
for α,β-unsaturated ketones, **3**, 269
See also Sulfonimines
Sulfonates, perfluoroalkane, synthesis of, **2**, 689
Sulfonates, phenyl fluoroalkyl, reactions, with Reformatsky reagents, **1**, 483
Sulfonates, phenylselenenyl, homolytic decomposition of, **2**, 285

Sulfonates, pyridyl, pyrolysis of, **1**, 778
Sulfonates, 2-pyridylthiol, reactions, with amines, **2**, 364
Sulfonates, quinolyl, pyrolysis of, **1**, 778
Sulfonates, secondary alkyl
  reactions
    with alkynylaluminums, **1**, 1021
    with trihexynylaluminum, **1**, 1021
Sulfonates, silyl, synthesis of, **2**, 209
Sulfonates, α-silyl, use in the synthesis of vinyl sulfones, **2**, 715
Sulfonates, silylalkane, synthesis of, **4**, 397
Sulfonates, thiol
  reactions, with trialkyl phosphites, **2**, 199
  as sulfenylating reagents, **5**, 247
Sulfonates, toluene, reactions, with terminal alkynes, **1**, 1025
Sulfonates, trifluoro-, use in elimination reactions, **1**, 981
Sulfonates, trihalomethyl, synthesis of, **6**, 231
Sulfonates, trimethylsilyl 2-oxo-, synthesis of, **2**, 209
Sulfonates, vinyl, synthesis of, **2**, 721
O-Sulfonates, *t*-ammonium *N*-acylhydroxylamine-, Lossen rearrangement of, **5**, 1000
Sulfonate salts, fluoro-, reactions, with *n*-butyllithium, **4**, 369
Sulfonate–sulfate anhydrides, synthesis of, **2**, 208
Sulfonation
  of alcohols, **2**, 5
  for the synthesis of sulfonic acids, **2**, 190
Sulfone, benzyl phenyl, use in the synthesis of *S,N*-acetals, **4**, 332
Sulfone, α-brominated, reactions, with sodium phenyl selenide, **4**, 286
Sulfone, α-bromoethyl ethyl, reactions, with potassium hydroxide, **1**, 704
Sulfone, 1-bromovinyl methyl, synthesis of, **4**, 779
Sulfone, *t*-butoxymethyl phenyl, alkylation of, **4**, 200
Sulfone, cyclopropyl 1-chloro-1-phenylmethyl, synthesis of, **4**, 81
Sulfone, dibenzyl, synthesis of, **2**, 157
Sulfone, dimethyl
  acidity of, **1**, 142
  deprotonation of, **1**, 144; **4**, 391
  metallation of, **4**, 332
Sulfone, di(*n*-pentyl), synthesis of, **1**, 144
Sulfone, 1-fluoro-2-chloroethyl phenyl, dehydrochlorination of, **4**, 79
Sulfone, fluoromethyl phenyl, synthesis of, **4**, 780
Sulfone, α-fluoromethyl phenyl, as starting material, for α-sulfonyl carbanions, **4**, 79
Sulfone, *n*-heptylphenyl, reactions, with lithiotrichloromethane, **4**, 741
Sulfone, iodomethyl phenyl
  as starting material, for α-sulfonyl carbanions, **4**, 84
  synthesis of, **4**, 84
Sulfone, lithiomethyl phenyl, formylation of, **3**, 37
Sulfone, methoxymethyl *p*-tolyl, synthesis of, **4**, 235
Sulfone, methyl phenyl
  acidity of, **1**, 142
  alkylation of, **1**, 144
  deprotonation of, **4**, 271; **6**, 130
  reactions
    with α,ω-dihalides, **1**, 146
    with ethylmagnesium bromide, **4**, 395
Sulfone, methylthiomethyl *p*-toluene, as formyl anion equivalent, **3**, 31
Sulfone, methylthiomethyl *p*-tolyl, as starting material, for ketene dithioacetal *S,S*-dioxides, **4**, 857
Sulfone, methylvinyl, bromination–dehydrobromination of, **4**, 779
Sulfone, γ-morpholinovinyl, deprotonation of, **1**, 145
Sulfone, phenylselenomethyl phenyl, synthesis of, **4**, 286
Sulfone, phenyl (*p*-tolylsulfonylmethyl), condensation of, **4**, 858
Sulfone, phenylvinyl, bromination–dehydrobromination of, **4**, 779
Sulfone, propyl benzyl, electrochemical reduction of, **2**, 188
Sulfone, α-substituted vinyl, synthesis of, **1**, 145
Sulfone, vinyl chloro, reduction of, **1**, 430
β-Sulfone, ring-opening of, **6**, 15
Sulfone anions
  alkylation of, **1**, 654
  for the opening of epoxides, **1**, 225
  synthesis of, **1**, 697
Sulfone carbanions, reactions, with diphenyl diselenides, **4**, 286
Sulfone α,α'-dianions, oxidation of, **1**, 706
Sulfones
  as activating groups for alkenes, **1**, 300
  alkylation of, **1**, 612
  base induced cleavage of, **2**, 188
  α-chlorination of, **4**, 80
  deprotonation of, **4**, 391
  displacement of, **1**, 483
  elimination of, **1**, 612; **3**, 63, 209
  extrusion of sulfur dioxide from, **1**, 232
  free-radical bromination of, **4**, 83
  from, nitro sulfides, **4**, 908
  oxidation of, **2**, 192
  oxidation to ketones, **3**, 132
  oxidative desulfonylation of, **4**, 234
  as phase-transfer catalysts, **2**, 35
  precursors of aldehydes, **3**, 16
  reactions
    with alkyllithiums, **4**, 936
    with butyllithium, **6**, 541
    with *n*-butyllithium, **4**, 391
    with dithionite, **1**, 439
    with hydrazones, **3**, 172
    with lithium alkoxides, **4**, 235
    with sodium ethoxide, **1**, 612
    with sulfinates, **1**, 619
  reduction of, **1**, 24; **2**, 154
    using sodium dithionite, **1**, 700
  reductive cleavage of, **2**, 188
  as starting materials
    for *S,N*-acetals, **4**, 332
    for alkanesulfinic esters, **2**, 200
    for alkanesulfonic esters, **2**, 205
    for alkenes, **1**, 617; **4**, 840
    for dialkyl sulfides, **2**, 143
    for Grignard reagents, **4**, 395
    for α-lithiated alkenes, **4**, 960
    for sulfides, **2**, 713
    for sulfinic acids, **2**, 188
  synthesis of, **4**, 960; **6**, 253
Sulfones, β-acetoxy
  elimination of, **1**, 1073
    for the synthesis of alkenes, **1**, 698
Sulfones, β-acetoxy-substituted, elimination of, **1**, 613
Sulfones, acyl, use in the synthesis of aromatic ketones, **3**, 283
Sulfones, α-acyloxy, synthesis of, **4**, 235
Sulfones, alkenyl
  elimination of, **1**, 1073
  reactions, with alkynes, **1**, 545
Sulfones, α-alkoxy
  from
    α-diazo sulfones, **4**, 235
    monothioacetals, **4**, 235
    sulfinic acids, **4**, 235
  synthesis of, **4**, 235
    via carbon–carbon bond formation, **4**, 236
Sulfones, β-alkoxy, elimination of, **1**, 1073
Sulfones, β-alkoxy-substituted, elimination of, **1**, 613
Sulfones, alkoxyvinyl
  synthesis of, **4**, 840
  See also Ketene monothioacetal *S,S*-dioxides

Sulfones, alkyl
dimetallation of, **1**, 146
for Friedel–Crafts alkylation reactions, **1**, 484
reactions, with vinyl alanes, **1**, 483
reduction of, **1**, 24
Sulfones, alkyl bromomethyl, from, bromomethanesulfonyl bromide, **4**, 83
Sulfones, alkyl chloromethyl, synthesis of, **4**, 81
Sulfones, alkyl haloalkyl, synthesis of, **4**, 81
Sulfones, alkyl halomethyl, synthesis of, **4**, 81
Sulfones, alkylthioalkyl, reactions, with bis(sulfonyl)methanes, **6**, 90
Sulfones, alkynic
reactions
with amines, **2**, 742
with thiols, **4**, 861
Sulfones, alkynyl
displacement of, **1**, 485
reactions
with Grignard reagents, **1**, 1032
with organolithiums, **1**, 1032
synthesis of
via elimination reactions, **2**, 1034
via oxidation of alkynyl thioethers, **2**, 1034
via substitution at *sp* carbon, **2**, 1035
Sulfones, alkynyl trifluoromethyl, synthesis of, **6**, 235
Sulfones, allenic, synthesis of, **1**, 983
Sulfones, allenyl, from, 2,3-bis(phenylsulfonyl)-1,3-butadiene, **1**, 980
Sulfones, allenyl chloromethyl, synthesis of, **1**, 993
Sulfones, allyl, deprotonation of, **1**, 145
Sulfones, allylic
elimination of, **1**, 612
reduction of, **1**, 24
substitution of, **1**, 1031
Sulfones, α-amidoalkyl, from, aminals, **4**, 332
Sulfones, aryl
displacement of, **1**, 486
reduction of, **1**, 437
reductive lithiation of, **4**, 381
as starting materials, for *o*-benzylarenesulfinic acid, **1**, 379
substitution of, **1**, 485
synthesis of, **2**, 715
Sulfones, aryl alkyl
base-catalyzed rearrangement of, **2**, 719
synthesis of, **2**, 716
Sulfones, aryl arylazo, as starting materials, for aryl iodides, **2**, 632
Sulfones, aryl azoxy, rearrangement of, **2**, 804
Sulfones, aryl chloromethyl, synthesis of, **4**, 81
Sulfones, aryl haloalkyl, synthesis of, **4**, 81
Sulfones, aryl halomethyl, synthesis of, **4**, 81
Sulfones, aryl methoxymethyl, synthesis of, **4**, 235
Sulfones, aryl trifluormethyl, from, difluorocarbene, **6**, 235
Sulfones, arylvinyl, synthesis of, **4**, 797
Sulfones, α-azido, synthesis of, **4**, 331
Sulfones, β-benzoyloxy, elimination of, for the synthesis of alkenes, **1**, 698
Sulfones, benzylic
displacement of, via nucleophiles, **1**, 238
reduction of, **1**, 24
Sulfones, α-bromo
alkylation of, **4**, 83
from, α-bromo sulfides, **4**, 83
as starting materials, for α-sulfonyl carbanions, **4**, 83
synthesis of, **4**, 83
via halogenative decarboxylation, **4**, 83
Sulfones, α-bromoalkyl, synthesis of, **4**, 83
Sulfones, bromomethyl, synthesis of, **4**, 83
Sulfones, bromomethyl alkenyl, synthesis of, **1**, 619
Sulfones, bromomethyl dienyl, synthesis of, **4**, 84
Sulfones, α-bromomethylvinyl, synthesis of, **4**, 84
Sulfones, (*Z*)-α-bromovinyl, synthesis of, **4**, 783
Sulfones, *t*-butyl, for the synthesis of α-lithio compounds, **1**, 146
Sulfones, chloro, reactions, with ketones, **3**, 29
Sulfones, α-chloro
from
chloromethanesulfonyl halides, **4**, 82
α-chloro sulfides, **4**, 80
reactions, with carbonyls, **4**, 236
as starting materials, for α-sulfonyl carbanions, **4**, 82
synthesis of, **4**, 80
Sulfones, 1-chloroalkenyl phenyl, synthesis of, **4**, 781
Sulfones, 1-chlorocyclopropyl, synthesis of, **4**, 80
Sulfones, chloromethyl, synthesis of, **4**, 82
Sulfones, α-cyano, synthesis of, **3**, 635
Sulfones, β-cyano, synthesis of, **3**, 636
Sulfones, cyanomethyl(aryl), from, bromoacetonitrile, **3**, 635
Sulfones, α-cyano-β-nitro, reactions, with sodium sulfide, **3**, 645
Sulfones, cyclic
in Ramberg–Bäcklund reactions, **1**, 618
as starting materials, for cycloalkenes, **1**, 613
Sulfones, dialkyl
from
sulfides, **2**, 154
sulfinic acids, **2**, 156
sulfonic acids, **2**, 157
α-sulfonyl carbanions, **2**, 158
sulfoxides, **2**, 154
as starting materials, for vinylsulfonic acids, **2**, 719
synthesis of, **2**, 154
Sulfones, diallylic, as starting materials, for trienes, **1**, 618
Sulfones, diaryl
displacement of, **1**, 485
synthesis of, **2**, 716
Sulfones, diazo, synthesis of, **6**, 628
Sulfones, α-diazo
hydrolysis of, **4**, 234
as starting materials, for α-alkoxy sulfones, **4**, 235
synthesis of, **3**, 478
Sulfones, α-diazo-β-keto, cleavage of, **5**, 867
Sulfones, diazovinyl, synthesis of, **3**, 479
Sulfones, α,α-dichlorodibenzyl, Ramberg–Bäcklund rearrangement of, **1**, 1074
Sulfones, 2,2-dihalovinyl, reactions, with potassium fluoride, **4**, 783
Sulfones, dilithioalkyl, reactions, with α-halo ketones, **3**, 224
Sulfones, enyne allene, synthesis of, **1**, 956
Sulfones, 2-ethoxyvinyl, reactions, with disulfones, **4**, 865
Sulfones, fluorinated, synthesis of, **6**, 253
Sulfones, α-fluoro
from
1-fluoro-1-(phenylsulfonyl)ethene, **4**, 79
α-fluoro sulfides, **4**, 79
synthesis of, **4**, 79
Sulfones, α-fluoroalkenyl phenyl, reactions, with tributyltin hydride, **4**, 812
Sulfones, (*E*)-1-fluoro 2-aryl, synthesis of, **4**, 780
Sulfones, α-fluoro β-substituted ethyl phenyl, synthesis of, **4**, 79
Sulfones, glycosyl
from, glycosyl halides, **4**, 236
synthesis of, **4**, 235
Sulfones, α-halo
from
α-sulfonylcarboxylic acids, **4**, 81
sulfonyl halides, **2**, 157
Ramberg–Bäcklund reactions on, **1**, 706
as starting materials, for alkenes, **1**, 704
synthesis of, **1**, 704; **4**, 79
Sulfones, α-haloalkyl, synthesis of, **4**, 81
Sulfones, halogenated, use in the synthesis of organometallics, **6**, 258

Sulfones, β-halovinyl, substitution of, **1**, 470
Sulfones, homoallylic, elimination of, **1**, 612
Sulfones, 2-hydroxy, semipinacol rearrangements of, **1**, 390
Sulfones, α-hydroxy
as starting materials, for ethoxyethyls, **4**, 236
synthesis of, **4**, 234
Sulfones, β-hydroxy
electrochemical reductive elimination of, **1**, 615
elimination of, **1**, 614
stereoselectivity of, **1**, 615
reductive alkenation of, **5**, 526
reductive elimination of, **1**, 696
Sulfones, β-hydroxyimidazolyl, elimination of, **1**, 615
Sulfones, α-iodo
from, iodomethanesulfonyl bromide, **4**, 84
synthesis of, **4**, 84
Sulfones, β-iodoalkenyl, dehydrohalogenation of, **2**, 1034
Sulfones, α-iodoalkyl, synthesis of, **4**, 84
Sulfones, isochromanyl, synthesis of, **4**, 235
Sulfones, keto
reduction of, **2**, 39
thermal rearrangement of, **2**, 205
Sulfones, β-keto
reactions, with amines, **2**, 743
as starting materials, for 1-(phenylsulfonyl)alkynes, **2**, 1035
synthesis of, **3**, 171
Sulfones, 2-lithio, synthesis of, **2**, 1000
Sulfones, α-lithio
from, α-tributylstannyl sulfones, **4**, 394
synthesis of, **4**, 391
Sulfones, α-lithiovinyl, synthesis of, **1**, 1089
Sulfones, α-metallo, synthesis of, **2**, 1000
Sulfones, methoxymethyl *p*-tolyl, synthesis of, **4**, 235
Sulfones, 1-methoxyvinyl, synthesis of, **4**, 838
Sulfones, methyl methylthiomethyl
condensation reactions of, **4**, 857
as starting materials, for ketene dithioacetal *S,S*-dioxides, **4**, 857
Sulfones, methyl perfluoroalkyl, cleavage of, **2**, 193
Sulfones, α-nitro
α-fluorination of, **4**, 108
from, α-halonitroalkanes, **4**, 332
reactions, with thiolates, **4**, 328
synthesis of, **4**, 331
Sulfones, β-nitro, elimination of, **1**, 616, 638
Sulfones, β-nitro-α-cyano, synthesis of, **3**, 635
Sulfones, β-pivaloyloxy, substitution of, **1**, 482
Sulfones, 2-pyridyl, synthesis of, **1**, 616
Sulfones, β-silyl
elimination of, **1**, 616, 654
reactions, with fluoride ions, **1**, 616
synthesis of, **1**, 654
Sulfones, α-silyl-α,β-unsaturated, reactions, with magnesium alkynides, **1**, 1048
Sulfones, β-silylvinyl, reactions, with organometallics, **4**, 393
Sulfones, β-stannyl
elimination of, **1**, 616, 665
reactions, with fluoride ions, **1**, 616
Sulfones, (*E*)-styryl, synthesis of, **4**, 275
Sulfones, β-substituted, elimination of, **1**, 612, 614
Sulfones, β-sulfonyl, elimination of, **1**, 616
Sulfones, α-sulfonyloxy, synthesis of, **4**, 235
Sulfones, *gem*-sulfoxide, synthesis of, **4**, 272
Sulfones, α-sulfoxide, from, dithioacetals, **4**, 271
Sulfones, tertiary allylic, displacement of, **1**, 483
Sulfones, trialkyl, synthesis of, **6**, 91
Sulfones, triaryl, synthesis of, **6**, 91
Sulfones, tributylstannyl, β-elimination of, **1**, 440
Sulfones, α-tributylstannyl, as starting materials, for α-lithio sulfones, **4**, 394
Sulfones, trihalomethyl, synthesis of, **6**, 234
Sulfones, α-(trimethylsilyl)vinyl phenyl, addition reactions of, **1**, 310
Sulfones, trinitromethyl, as starting materials, for halotrinitromethanes, **6**, 283
Sulfones, unsaturated, use in the synthesis of dialkyl sulfones, **2**, 158
Sulfones, α,β-unsaturated
reactions, with hydrogen peroxide, **4**, 237
synthesis of, **1**, 707
Sulfones, β-α-unsaturated, synthesis of, **4**, 780
Sulfones, β,γ-unsaturated δ-cyano, synthesis of, **3**, 636
Sulfones, vinyl
alkylation of, **3**, 254
cycloaddition of, **2**, 158
displacement of, via Grignard reagents, **1**, 485
lithiation of, **1**, 1090
Michael additions to, **4**, 366
ozonolysis of, **6**, 480
reactions
with cyanide, **3**, 636
with diazomethane, **6**, 122
with dimethylcopperlithium, **4**, 964
with *S*-methyl methanethiosulfonate, **4**, 858
with sulfenyl chlorides, **4**, 861
with trialkylboranes, **1**, 485
with trimethylenemethane, **1**, 288
with trimethylsilyl iodide, **4**, 84
reduction of, **1**, 438
as starting materials
for alkenes, **1**, 439
for ketene dithioacetal *S,S*-dioxides, **4**, 861
structure of, **4**, 960
substitution of, **1**, 485
substitution, using Grignard reagents, **1**, 700
synthesis of, **1**, 572; **2**, 714; **4**, 858
via elimination of halosilanes, **1**, 653
via elimination reaction, **2**, 715
via Peterson and Wittig reactions, **2**, 715
Sulfones, vinylic, reactions, with organolithiums, **1**, 310
Sulfone sulfonates, alkane, synthesis of, **2**, 232
Sulfonic acid, reactions, with anilines, **6**, 645
Sulfonic acid, alkyl-, reactions, with alkynes, **2**, 689
Sulfonic acid, chloro-
reactions
with alkanols, **2**, 101
with amines, **2**, 366
with triethylamine, **2**, 368
use in the synthesis of arenesulfonyl chlorides, **2**, 717
Sulfonic acid, toluene-
reactions
with alcohols, **1**, 578
with isonitriles, **5**, 710
Sulfonic acid, *p*-toluene-
as dehydrating agent, **3**, 684
reactions, with tertiary amines, **1**, 256
p-Sulfonic acid, toluene-, as catalyst, for the condensation of thiols, **4**, 245
Sulfonic acid anhydrides, *N*-chlorocarboximidic fluoro-, synthesis of, **5**, 710
Sulfonic acids
as desilylating reagents, **1**, 448
elimination of, **1**, 583
from, sulfonic acids, **2**, 194
reactions
with amines, **2**, 364
with calcium cyanamide, **5**, 1119
with tosyl diazomethane, **4**, 235
as starting materials
for *S,N*-acetals, **4**, 331
for alkanesulfonic esters, **2**, 203
for dialkyl sulfones, **2**, 157
for sulfonic acids, **2**, 194
for sulfonyl halides, **2**, 181
for thiols, **2**, 708
for vinylsulfonyl chlorides, **2**, 717

synthesis of, **2**, 189; **4**, 267
via oxidation, **2**, 192
Sulfonic acids, *O*-acyl-, synthesis of, **5**, 209
Sulfonic acids, alkane-
alkylation of, **2**, 203
esterification of, **2**, 203
reactions, with epoxides, **2**, 203
Sulfonic acids, arene-
chlorination of, **2**, 718
synthesis of
via oxidation, **2**, 720
via sulfonation of arenes, **2**, 719
use in the synthesis of aryl sulfones, **2**, 715
Sulfonic acids, halo-, use in the synthesis of alkanesulfonyl chlorides, **2**, 176
Sulfonic acids, α-haloalkane-, salts of, synthesis of, **4**, 86
Sulfonic acids, α-hydroxy-, synthesis of, **4**, 237
Sulfonic acids, α-keto-, synthesis of, **2**, 190
Sulfonic acids, vinyl-, from, dialkyl sulfones, **2**, 719
Sulfonic acid salts, alkane-, synthesis of, **2**, 190
Sulfonic acid salts, hydroxy-, synthesis of, **2**, 191
Sulfonic acid salts, vinyl-, synthesis of, **1**, 618
Sulfonic anhydrides
as byproducts, **2**, 181
reactions, with silver cyanate, **5**, 995
Sulfonic anhydrides, acyl, synthesis of, **2**, 208
Sulfonic esters
as starting materials
for alkanesulfonic esters, **2**, 206
for dialkyl sulfides, **2**, 135
Sulfonic esters, alkane-
from
allylic sulfites, **2**, 206
diazo compounds, **2**, 207
sulfenes, **2**, 204
sulfinate esters, **2**, 207
sulfonamides, **2**, 205
sulfones, **2**, 205
sulfonic acids, **2**, 203
sulfonic esters, **2**, 206
sulfonyl chlorides, **2**, 202
sulfonyl esters, **2**, 207
sulfur trioxide, **2**, 206
synthesis of, **2**, 202
Sulfonic esters, hydroxyalkyl, synthesis of, **2**, 203
Sulfonimidamides
from
amines, **2**, 402
sulfonamidebisimides, **2**, 402
synthesis of, **2**, 401
Sulfonimidoyl chlorides
from, sulfinamides, **2**, 401
reactions, with alcohols, **2**, 365
as starting materials, for sulfoximines, **2**, 164
Sulfonimidoyl fluorides, synthesis of, **2**, 401
Sulfonimidoyl halides, synthesis of, **2**, 401
Sulfonimines
Beckmann rearrangement of, **3**, 437
nucleophilic reactions of, **3**, 438
reactions
with Grignard reagents, **3**, 438
with silyl enol ethers, **3**, 437
rearrangement, to imidoyl iodides, **3**, 437
reduction of, **3**, 438
as starting materials, for *N*-arylsulfonamides, **3**, 438
synthesis of, **3**, 437
use in Diels–Alder reactions, **3**, 438
See also Sulfonates, oxime
Sulfonium bromide, bromo-, synthesis of, **2**, 171
Sulfonium bromide, dimethylbromo-, use in the synthesis of alkyl bromides, **2**, 32
Sulfonium bromide, dimethyl-[2-oxo-2-thienyl-(2)]ethyl-, reactions, with diazonium halides, **5**, 676
Sulfonium bromides, synthesis of, **2**, 159
Sulfonium chloride, chloro-, use in the synthesis of α-chloro ketones, **3**, 151
Sulfonium chloride, chlorodimethyl-, for the *para* chlorination of arenes, **2**, 620
Sulfonium hydroxide, trimethyl-, as alkylating agent, **2**, 683
Sulfonium salts
as alkylating agents, **1**, 241
deprotonation of, **1**, 407, 821
elimination of, **1**, 622
reactions, with amidines, **5**, 766
rearrangement of, **2**, 524
as starting materials
for dialkyl sulfides, **2**, 143
for sulfonium ylides, **3**, 373
for thioaldehydes, **3**, 333
for ylides, **1**, 241
for the substitution of sulfur functions, **1**, 241
synthesis of, **3**, 519; **4**, 347
use in the synthesis of α-aminoalkylphosphines, **4**, 456
Sulfonium salts, *N*-acyl-
from
N-haloamides, **5**, 338
nitrilium salts, **5**, 339
sulfimines, **5**, 339
Sulfonium salts, acyloxy-, synthesis of, **4**, 323
Sulfonium salts, *N*-acyl-*N*-toluenesulfonylamino-, deprotonation of, **5**, 702
Sulfonium salts, alkenyl-, elimination of, **1**, 1072
Sulfonium salts, alkyldiphenyl-
for the *O*-alkylation of amides, **3**, 419
use in the synthesis of alkoxycarbene complexes, **5**, 936
Sulfonium salts, amino-, synthesis of, **5**, 339
Sulfonium salts, *S*-amino-, synthesis of, **2**, 160
Sulfonium salts, α-cyanoalkyl-, as starting materials, for ketenimines, **3**, 574
Sulfonium salts, *S*,*S*-dialkylsuccinimido-, reactions, with enolates, **3**, 574
Sulfonium salts, *S*-halo-, as intermediates, **2**, 160
Sulfonium salts, *N*-succinimido-, synthesis of, **5**, 339
Sulfonium salts, 1-thiovinyl-, synthesis of, **4**, 860
Sulfonium salts, trimethylsilyl-, as starting materials, for sulfur ylides, **1**, 821
Sulfonyl anhydrides, alkane-, reactions, with ylides, **5**, 901
Sulfonylation, of enamines, **3**, 170
Sulfonylation, electrophilic, of arenes, **2**, 708
Sulfonyl azide, chloro-, reactions, with amines, **2**, 368
Sulfonyl azides, arene-
reactions
with 1-alkoxy-1,4-cyclohexadienes, **5**, 702
with 1-alkoxy cyclohexenes, **5**, 702
synthesis of, **2**, 726
Sulfonyl azides, aryl-
reactions, with silyl enol ethers, **5**, 346
as starting materials, for *N*-acylsulfonamides, **5**, 346
Sulfonyl bromides, alkane-, synthesis of, **2**, 177, 182
Sulfonyl bromides, tertiary alkane-, synthesis of, **2**, 179
Sulfonyl carbanions
reactions
with aldehydes, **1**, 614
with ketones, **1**, 614
α-Sulfonyl carbanions
alkylation of, **4**, 82
bromination of, **4**, 83
configurational stability of, **1**, 146
from
α-bromo sulfones, **4**, 83
α-chloro sulfones, **4**, 82
α-fluoromethyl phenyl sulfone, **4**, 79
iodomethyl phenyl sulfone, **4**, 84
iodination of, **4**, 84
reactions, with electrophiles, **4**, 236
structure of, **4**, 391

synthesis of, **1**, 144
use of, **1**, 144
Sulfonyl chloride, allyl-, reactions, with ethanol, **2**, 203
Sulfonyl chloride, α-chloro-, as starting material, for 1-chloroethanesulfinic acid, **4**, 78
Sulfonyl chloride, phenyl-
reactions
with dimetallics, **2**, 594
with organo *gem*-biszincs, **2**, 998
Sulfonyl chloride, *p*-toluene-
for the dehydration of formamides, **3**, 696
reactions, with 2-methylpyridine-*N*-oxide, **1**, 861
use in the synthesis of carbodiimides, **5**, 1065
Sulfonyl chloride, trichloromethyl-, reactions, with 1-alkenes, **6**, 24
Sulfonyl chlorides
as coupling reagents, for the synthesis of amides, **5**, 268
from, sulfuryl chloride, **2**, 182
in Heck reactions, **1**, 494
reactions
with alcohols, **2**, 202
with amines, **2**, 364, 797
with ammonia, **2**, 726
with antimony trifluoride, **2**, 183
with bis(sulfonyl)methanes, **6**, 90
with diazoalkanes, **3**, 479
with hydrogen sulfide, **2**, 223
with imines, **2**, 249
with lactim ethers, **5**, 346
with monosubstituted sulfonamides, **2**, 366
with potassium fluoro sulfinate, **2**, 183
with silver cyanate, **5**, 994
with sodium sulfite, **5**, 1112
with triethylamine, **1**, 619
with xenon difluoride, **2**, 183
reduction of, **2**, 271
reductive coupling of, **2**, 220
as starting materials
for alkanesulfonic esters, **2**, 202
for sulfonyl cyanides, **5**, 1112
for thiocyanates, **5**, 1111
Sulfonyl chlorides, alkane-
reactions
with amines, **4**, 331
with phosphorus ylides, **5**, 902
use in the synthesis of aryl sulfones, **2**, 715
Sulfonyl chlorides, alkyl-
reactions, with amidines, **5**, 767
synthesis of, **2**, 182
Sulfonyl chlorides, arene-
reactions
with potassium iodide, **2**, 724
with trimethylvinylsilane, **4**, 942
with ylides, **5**, 898
reduction of, **2**, 724
synthesis of, **2**, 717
trimethyl phosphite reduction of, **2**, 721
use in the synthesis of arenesulfinic acids, **2**, 718
use in the synthesis of aryl sulfones, **2**, 715, 716
use in the synthesis of diaryl disulfides, **2**, 723
Sulfonyl chlorides, aryl-
reactions
with alkynylides, **2**, 1016
with amidines, **5**, 767
as starting materials, for aryl chlorides, **2**, 622
Sulfonyl chlorides, styryl-, as starting materials, for (*E*),(*E*)-1,4-diaryl-1,3-butadienes, **1**, 485
Sulfonyl chlorides, tertiary alkane-, synthesis of, **2**, 179
Sulfonyl chlorides, vinyl-
from, sulfonic acids, **2**, 717
reactions, with alcohols, **2**, 721
as starting materials, for imidoylsulfonamides, **5**, 767
Sulfonyl compounds, reactions, with amines, **2**, 364
Sulfonyl compounds, alkyl aryl-, as starting materials, for alkyl compounds, **1**, 39
Sulfonyl compounds, allylic-*p*-nitrophenyl-, rearrangement of, **1**, 840
Sulfonyl esters
as starting materials, for alkanesulfonic esters, **2**, 207
synthesis of, **2**, 202
Sulfonyl fluoride, vinyl-, reactions, with diphenylphosphine, **2**, 440
Sulfonyl fluorides
reactions, with Grignard reagents, **4**, 280
as starting materials, for bis(sulfones), **4**, 280
synthesis of, **2**, 175, 182
Sulfonyl fluorides, alkane-
reactions, with ylides, **5**, 901
synthesis of, **2**, 177
use in the synthesis of aryl sulfones, **2**, 715
Sulfonyl fluorides, arene-
reactions
with α-lithiomethyl isocyanide, **3**, 709
with ylides, **5**, 901
synthesis of, **5**, 4
Sulfonyl fluorides, (2-nitrosotrifluoroethyl)oxy-, synthesis of, **4**, 104
Sulfonyl fluorides, vinyl-, synthesis of, **2**, 717
Sulfonyl groups, addition of, for the synthesis of sulfonic acids, **2**, 190
Sulfonyl halides
dehydrohalogenation of, **5**, 472, 488, 496
from
sulfonic acids, **2**, 181
sulfur compounds, **2**, 177
reactions
with amines, **2**, 248
with ammonia, **2**, 248
with oximes, **3**, 437
with sulfamate esters, **2**, 368
with sulfenes, **5**, 488
reduction of, **2**, 187; **3**, 170
as starting materials
for alkanethiols, **2**, 125
for α-halo sulfones, **2**, 157
synthesis of, **2**, 175
use in the synthesis of carboxylic esters, **5**, 125
Sulfonyl halides, alkane-
hydrolysis of, **2**, 194
synthesis of, **2**, 175
Sulfonyl halides, arene-
reactions, with imino esters, **5**, 703
synthesis of, **2**, 717
Sulfonyl halides, bromoalkane-, synthesis of, **4**, 85
Sulfonyl halides, α-halo-
as starting materials, for α-haloalkanesulfonamides, **4**, 86
synthesis of, **4**, 85
Sulfonyl halides, α-haloalkane-, reactions, with alcohols, **4**, 86
Sulfonyl halides, trihalomethane-, synthesis of, **6**, 236
Sulfonyl halides, vinyl-, synthesis of, **2**, 717
N-Sulfonylimines, camphor, deprotonation of, **1**, 147
N-Sulfonylimines, fenchone, deprotonation of, **1**, 147
Sulfonyl iodide, *p*-toluene-, reactions, with potassium nitronates, **4**, 331
Sulfonyl iodides, synthesis of, **2**, 182
Sulfonyloximes, *O*-aryl-, rearrangement of, **5**, 689
Sulfonyloxyamine, mesitylene-
reactions
with α-bromo sulfoxides, **4**, 86
with α-chloro sulfoxides, **4**, 86
Sulfonyl peroxides, aryl-
reactions
with enamines, **3**, 163
with enol acetates, **3**, 163
with silyl enol ethers, **3**, 163

Sulfonyl peroxides, diaryl-, reactions, with amines, **2**, 357
Sulfonyl radicals, reactions, with carbon–carbon double bonds, **1**, 949
o-Sulfoperbenzoic acid, use in the synthesis of 1,2-diols, **2**, 45
Sulforaphane, analogues of, synthesis of, **5**, 1044
Sulfoxidation, for the synthesis of ketene monothioacetal *S,S*-dioxides, **4**, 838
Sulfoxidation, asymmetric, for the synthesis of chiral vinyl sulfoxides, **2**, 713
Sulfoxide, allyl difluorochloromethyl, as starting material, for difluorosulfine, **6**, 530
Sulfoxide, allyl trichloromethyl, as starting material, for dichlorosulfine, **6**, 531
Sulfoxide, benzhydryl benzyl, reactions, with sulfuryl chloride, **4**, 74
Sulfoxide, benzyl phenyl, methylation of, **1**, 143
Sulfoxide, (+)-(*S*)-1-bromovinyl *p*-tolyl, synthesis of, **4**, 776
Sulfoxide, chloromethyl methyl, as starting material, for ethoxymethyl methyl sulfoxide, **4**, 233
Sulfoxide, chloromethyl phenyl
  alkylation of, for 1-chloroalkene, **2**, 616
  deprotonation of, **4**, 77
  metallation of, **4**, 233
  reactions
    with aldehydes, **4**, 776
    with lithium diisopropylamide, **3**, 170
  synthesis of, **5**, 867
Sulfoxide, diallyl, as starting material, for thioacrolein, **3**, 337
Sulfoxide, α-diazomethyl, synthesis of, **5**, 867
Sulfoxide, dibenzyl, synthesis of, **2**, 150
Sulfoxide, dimethyl (DMSO)
  acidity of, **1**, 142
  deprotonation of, **1**, 142
  as oxidant, for the synthesis of alkyl chlorides, **2**, 24
  for the oxidation of benzylamines, **3**, 89
  for the oxidation of benzylic alcohol, **3**, 88
  as phase-transfer catalyst, **4**, 793
  reactions
    with acetamidines, **6**, 55
    with acyl nitrenes, **2**, 163
    with arylamines, **2**, 793
    with 1,7-bis(tosyl)trisulfur tetranitride, **2**, 162
    with *t*-butyl isocyanide dichloride complexes, **5**, 963
    with calcium(II) halides, **2**, 9
    with carbonic difluoride, **6**, 423
    with chloramine-T, **2**, 163
    with imides, **2**, 163
    with isonitriles, **5**, 1003
    with nitrilium salts, **5**, 339, 689
    with pyrylium salts, **4**, 863
    with silver salts, **3**, 88
    with sodium hydride, **4**, 391
    with trifluoroacetic acid, **5**, 209
  as reagent
    for the synthesis of aldehydes, **3**, 6
    for the synthesis of ketones, **3**, 114, 118
  as solvent, for the synthesis of nitriles, **3**, 613
  as starting material, for sulfoxonium ylides, **3**, 378
  for the synthesis of aryl bromides, **2**, 627
  thiophilic addition of, **4**, 270
  use in the opening of epoxides, **1**, 204
  use in the reduction of chloroalkanes, **1**, 7
  use in the synthesis of alkynes, **1**, 1000, 1066
  use in the synthesis of α-bromo ketones, **3**, 154
  use in the synthesis of α-chloro ketones, **3**, 151
  use in the synthesis of disulfides, **2**, 228
  use in the synthesis of ketones, **3**, 118, 123
Sulfoxide, divinyl, chlorination of, **4**, 77
Sulfoxide, ethoxymethyl methyl, from, chloromethyl methyl sulfoxide, **4**, 233
Sulfoxide, ethyl phenyl, reactions, with (dichloroiodo)benzene, **4**, 75
Sulfoxide, 1-fluoro-2-(4-biphenylyl)vinyl phenyl, synthesis of, **4**, 777
Sulfoxide, fluoromethyl phenyl, reactions, with sodium azide, **4**, 86
Sulfoxide, lithio bromomethyl phenyl, reactions, with aldehydes, **3**, 154
Sulfoxide, lithio chloromethyl phenyl, reactions, with aldehydes, **3**, 153
Sulfoxide, methoxymethyl phenyl, rearrangement of, **4**, 232
Sulfoxide, methoxymethyl *p*-tolyl, synthesis of, **4**, 233
Sulfoxide, α-methylbenzyl(phenyl), reactions, with *n*-butyllithium, **2**, 1000
Sulfoxide, methyl methylthiomethyl, use in the synthesis of *S,N*-acetals, **4**, 326
Sulfoxide, (+)-methyl *p*-tolyl, halogenation of, **4**, 75
Sulfoxide, phenyl, use in elimination reactions, **1**, 597
Sulfoxide, phenyl methyl, reactions, with (dimethylamino)tributyltin, **4**, 397
Sulfoxide, pyridyl, reactions, with phenylmagnesium bromide, **3**, 310
Sulfoxide, trimethylsilylmethyl phenyl, thermal rearrangement of, **4**, 222
α-Sulfoxide carbanions, as starting materials, for α-selenosulfoxides, **4**, 285
Sulfoxide elimination
  regioselectivity of, **1**, 595, 598
  and selenoxide elimination, compared, **1**, 601
  for the synthesis of cyclophanes, **1**, 788
Sulfoxide groups, displacement of, **1**, 237
Sulfoxide pyrolysis
  regiochemistry of, **1**, 779
  stereochemistry of, **1**, 779
  for the synthesis of dienoic esters, **1**, 788
Sulfoxides
  as activating groups for alkenes, **1**, 300
  alkylation of, **1**, 693
  bromination of, **4**, 77
  as chiral auxiliaries, **1**, 305
  chlorination of, **4**, 74, 75
  deprotonation of, **4**, 391
  elimination of, **1**, 595; **3**, 63, 209, 227
  enantiomerically enriched, synthesis of, **2**, 713, 714
  flash vacuum pyrolysis of, **2**, 718
  from, 1,3-oxathiolan-5-ones, **4**, 837
  halogenation of, **4**, 75
    stereochemistry, **4**, 75
  as precursors of aldehydes, **3**, 16
  Pummerer elimination of, **4**, 924
  Pummerer rearrangement of, **5**, 334
  pyrolysis of, **1**, 778
  reactions
    with alkyllithiums, **4**, 391
    with *N*-alkyl-*N*-sulfinylalkaminium salts, **2**, 360
    with copper(I) iodide, **4**, 396
    with diethylaminosulfur trifluoride, **4**, 73
    with *N*-sulfinyl perfluoroalkanesulfonamides, **2**, 162
    with 4-toluenesulfonyl isocyanate, **2**, 162
    with tosyl isocyanate, **3**, 365
    with zinc bromide, **1**, 305
  reduction of, **1**, 21
  reductive coupling of, **2**, 221
  as starting materials
    for alkanesulfinic esters, **2**, 199
    for alkenes, **1**, 693
    for chiral cyanohydrins, **3**, 631
    for dialkyl sulfides, **2**, 142
    for dialkyl sulfones, **2**, 154
    for α-fluoro sulfides, **4**, 61
    for monothioacetals, **4**, 221
    for nonclassical thiophenes, **3**, 372
    for 2-phenylthio-α,β-unsaturated ketones, **3**, 249
    for sulfides, **2**, 713
    for sulfimines, **2**, 162

for sulfonium ylides, **3**, 373
for sulfoximines, **2**, 163
for *N*-thioamides, **5**, 335
synthesis of, **1**, 143; **4**, 959
thermal elimination of, applications, **1**, 596
thermolysis of, **2**, 184
use in glycosylation, **4**, 232
use in the synthesis of disulfides, **2**, 227
Sulfoxides, aliphatic, reactions, with potassium *t*-butoxide, **1**, 600
Sulfoxides, alkenyl
deprotonation of, **4**, 961
elimination of, **1**, 1072
reactions, with enolate anions, **3**, 172
Sulfoxides, (*E*)-2-alkenyl, synthesis of, **4**, 958
Sulfoxides, α-alkoxy
from, α-chloro sulfoxides, **4**, 232
synthesis of, **4**, 232
Sulfoxides, β-alkoxyethyl, synthesis of, **4**, 233
Sulfoxides, alkyl
desulfurization of, **1**, 23
as starting materials, for α-chloro sulfides, **4**, 69
Sulfoxides, alkyl alkylthiomethyl, as starting materials, for ketene dithioacetal *S*-oxides, **4**, 857
Sulfoxides, alkyl aryl
as starting materials, for α-chloro sulfides, **4**, 69
thermal elimination of, **1**, 595
Sulfoxides, alkyl *t*-butyl, reactions, with *N*-chlorosuccinimide, **2**, 199
Sulfoxides, alkyl phenyl, reactions, with chlorotrimethylsilane, **4**, 945
Sulfoxides, alkylvinyl, use in the synthesis of dialkyl sulfoxides, **2**, 153
Sulfoxides, alkynyl
synthesis of
via oxidation of alkynyl thioethers, **2**, 1034
via substitution at *sp* carbon, **2**, 1035
Sulfoxides, 1-alkynyl, as starting materials, for 1-alkynes, **1**, 999
Sulfoxides, allenic, reactions, with diethylamine, **3**, 171
Sulfoxides, allenyl, synthesis of, **4**, 837
Sulfoxides, allyl, deprotonation of, **1**, 143
Sulfoxides, allylic
[2,3] sigmatropic rearrangement of, **2**, 195
as starting materials, for dienes, **1**, 600
Sulfoxides, allylvinyl
rearrangement of, **1**, 834
as starting materials, for sulfines, **3**, 369
Sulfoxides, aryl
reactions, with trialkyltinamines, **4**, 397
reduction of, **1**, 437
synthesis of, **2**, 713
Sulfoxides, aryl alkyl, synthesis of, **2**, 151
Sulfoxides, α-aryl-α-cyano, synthesis of, **3**, 635
Sulfoxides, aryl fluoromethyl, synthesis of, **4**, 73
Sulfoxides, benzylic, reduction of, **1**, 23
Sulfoxides, bis-
from, methylene bissulfoxides, **4**, 269
metallation of, **4**, 269
synthesis of, **4**, 267
Sulfoxides, bis(alkynyl), synthesis of, **2**, 1035
Sulfoxides, α-bromo
from, trichloromethanesulfinyl bromide, **4**, 77
reactions, with mesitylenesulfonyloxyamine, **4**, 86
as starting materials, for α-sulfinyl carbanions, **4**, 77
synthesis of, **4**, 77
Sulfoxides, β-bromo, stereoselective elimination of, **1**, 600
Sulfoxides, chiral vinyl, synthesis of, **2**, 713
Sulfoxides, chloro, Darzens condensation of, **3**, 167
Sulfoxides, 1,1-chloro, reactions, with ketones, **3**, 36
Sulfoxides, α-chloro
from, sulfides, **4**, 74
reactions
with amines, **2**, 322
with mesitylenesulfonyloxyamine, **4**, 86
as starting materials, for α-alkoxy sulfoxides, **4**, 232
synthesis of, **2**, 153; **4**, 74
Sulfoxides, α-chloromethyl, synthesis of, **4**, 76
Sulfoxides, α-chloro-α'-polychloro, chlorination of, **4**, 78
Sulfoxides, α-cyano
deoxygenation of, **3**, 655
from, α-cyanovinylic sulfoxides, **3**, 634
optically active, synthesis of, **3**, 634
oxidation of, **3**, 635
reactions, with arenes, **3**, 635
synthesis of, **3**, 634
Sulfoxides, α-cyanovinylic, as starting materials, for α-cyano sulfoxides, **3**, 634
Sulfoxides, cyclic, alkylation of, **1**, 144
Sulfoxides, dialkyl
from, sulfur-stabilized carbanions, **2**, 153
pyrolysis of, **1**, 595
as starting materials, for thioaldehydes, **3**, 336
synthesis of, **2**, 144
via oxidation of sulfides, **2**, 144
via reduction of sulfones, **2**, 154
use in the alkylation of amines, **2**, 300
Sulfoxides, α-diazo, synthesis of, **3**, 478
Sulfoxides, α-dichloro, reactions, with titanium(IV) chloride, **4**, 773
Sulfoxides, dicyclohexyl, deprotonation of, **4**, 264
Sulfoxides, 2,4-diketo, as starting materials, for 1,3-diols, **2**, 61
Sulfoxides, 1-fluoro, synthesis of, **4**, 73
Sulfoxides, α-fluoro, reactions, with lithium diphenylcuprate, **4**, 777
Sulfoxides, α-fluoroalkyl, as starting materials, for α,α-dichloro sulfides, **6**, 46
Sulfoxides, α-fluoroalkyl aryl, synthesis of, **4**, 73
Sulfoxides, 1-fluoro-2-hydroxyalkyl phenyl, synthesis of, **4**, 778
Sulfoxides, a-fluoro-a-lithio, alkylation of, for vinyl fluorides, **2**, 612
Sulfoxides, α-germyl, synthesis of, **4**, 371
Sulfoxides, α-halo
reactions, with trimethylsilyl triflate, **4**, 773
synthesis of, **4**, 73
Sulfoxides, 1-haloalkyl aryl, synthesis of, **4**, 73
Sulfoxides, α-halomethyl, as starting materials, for sulfenamides, **2**, 360
Sulfoxides, 3-hydroxy, as starting materials, for homoallylic alcohols, **2**, 85
Sulfoxides, β-hydroxy
pyrolysis of, **1**, 779
as starting materials, for alkenes, **1**, 600
Sulfoxides, *S*-imidoyl
synthesis of, **5**, 734
via *S*-alkylation of thioamide *S*-oxides, **5**, 734
Sulfoxides, α-iodo
from, sulfinyl chlorides, **4**, 78
synthesis of, **4**, 77
Sulfoxides, iodomethyl aryl, synthesis of, **4**, 78
Sulfoxides, keto, elimination of, **1**, 598
Sulfoxides, α-keto, cleavage of, **1**, 236
Sulfoxides, β-keto
deprotonation of, **4**, 961
reactions, with *N*-sulfinyl-*p*-toluenesulfonamide, **4**, 326
synthesis of, **3**, 631
See also Ketones, α-sulfinyl
Sulfoxides, α-lithio
synthesis of, **1**, 143; **4**, 391
transmetallation of, **4**, 397
Sulfoxides, metallated, synthesis of, **4**, 391
Sulfoxides, α-metallo, synthesis of, **2**, 1000
Sulfoxides, methyl, pyrolysis of, **1**, 597
Sulfoxides, methyl alkyl, chlorination of, **4**, 76

Sulfoxides, methyl methylthiomethyl
condensation of, **4**, 856
reactions, with aryl ketones, **4**, 857
as starting materials, for ketene dithioacetal *S*-oxides, **4**, 856
uses of, **4**, 856
Sulfoxides, optically active, from, chiral cyanohydrins, **3**, 631
Sulfoxides, phenyl fluoromethyl
alkylation of, **4**, 74
as starting materials, for α-sulfinyl carbanions, **4**, 74
Sulfoxides, phenylvinyl, as starting materials, for 2-(phenylsulfinyl)acylals, **4**, 207
Sulfoxides, pyrazinyl, elimination of, **1**, 595
Sulfoxides, pyridyl, elimination of, **1**, 595
Sulfoxides, α-silyl
synthesis of, **3**, 16
use in sila-Pummerer reactions, **4**, 222
Sulfoxides, β-silyl, from, silylsulfenates, **1**, 655
Sulfoxides, trihalomethyl, synthesis of, **6**, 234
Sulfoxides, β-trimethylsilyl, elimination of sulfenic acid from, **1**, 600
Sulfoxides, 2-trimethylsilylethyl, reactions, with sulfuryl chloride, **2**, 174
Sulfoxides, α,β-unsaturated, reactions, with enolates, **3**, 169
Sulfoxides, unsymmetrical, chlorination of, **4**, 75
Sulfoxides, vinyl
Michael additions to, **4**, 366
oxidation of, **2**, 714
reactions, with trimethylenemethane, **1**, 288
reduction of, **1**, 438
as starting materials
for alkenes, **1**, 439
for ketene dithioacetal *S*-oxides, **4**, 858
synthesis of, **2**, 713
δ-Sulfoxides, elimination of, **3**, 63
[2,3] Sulfoxide–sulfenate rearrangement, mechanism, **1**, 839
Sulfoximes, *N*-aryl-, reduction of, **2**, 796
Sulfoximide, vinyl-, α-sulfenylation of, **4**, 862
Sulfoximides, *N*-alkyl-, chlorination of, **4**, 86
Sulfoximides, α-aminoalkyl-, synthesis of, theoretical, **4**, 330
Sulfoximides, *N*-chloro-, chlorination of, **4**, 86
Sulfoximides, α-chloroalkyl *N*-alkyl-, synthesis of, **4**, 86
Sulfoximides, chloromethyl phenyl *N*-chloro-, reduction of, **4**, 86
Sulfoximides, fluoromethyl-, synthesis of, **4**, 86
Sulfoximides, α-haloalkyl-, synthesis of, **4**, 86
Sulfoximides, *N*-methyl-, fluorination of, **4**, 86
Sulfoximine, fluoromethyl-
reactions
with aldehydes, **4**, 86
with ketones, **4**, 86
Sulfoximine, *N*-(phenylethynyl)-*S*,*S*-dimethyl-, synthesis of, as evidence for alkynyl azides, **2**, 1055
Sulfoximine anions
as alternatives to Wittig reagents, **1**, 710
as reagents, **3**, 379
Sulfoximine group, substitution via nickel-catalyzed cross-coupling reactions, **1**, 711
Sulfoximines
alkylation of, **1**, 710; **2**, 363
axially chiral alkenyl, synthesis of, **1**, 711
characteristics of, **1**, 146
deprotonation of, **4**, 391
displacement of, **1**, 483
from
amines, **2**, 402
sulfimines, **2**, 164
sulfonimidoyl chlorides, **2**, 164
sulfoxides, **2**, 163
homochiral, for the synthesis of alkenes, **1**, 711
reactions, with bis(1,1-alkylthio)alkenes, **4**, 899
reduction of, **1**, 24; **2**, 362
as starting materials
for alkenes, **1**, 710
for sulfinamides, **2**, 245
for sulfoxonium ylides, **3**, 378
synthesis of, **2**, 163, 402, 793; **2**, 1055
use in the synthesis of cyclopropanes, **1**, 304
Sulfoximines, *S*-allyl-, deprotonation of, **1**, 146
Sulfoximines, aromatic, synthesis of, **2**, 793
Sulfoximines, *N*-aryl-, synthesis of, **2**, 793
Sulfoximines, β-hydroxy-
reductive elimination of, **1**, 622, 710
synthesis of, **1**, 710
Sulfoximines, *N*-nitro-, alkylation of, **1**, 146
Sulfoximines, (*S*)-(sulfinylmethyl)-, synthesis of, **4**, 272
Sulfoximines, vinyl-
displacement of, **1**, 485
from, diamines, **2**, 793
reactions
with lithium cyanide, **1**, 622
with lithium dimethyl phosphonate, **1**, 622
synthesis of, **2**, 793
N-Sulfoximines, use in cycloaddition reactions, **2**, 793
Sulfoxonium salt, trimethyl-, deprotonation of, **3**, 378
Sulfoxonium salts, as starting materials, for ylides, **1**, 210
Sulfoxylate, *d*-*n*-butyl-, synthesis of, **2**, 99
Sulfoxylates, synthesis of, **2**, 99
Sulfoxylates, dialkyl, synthesis of, **2**, 100
Sulfoxylic acid, as starting material, for alkoxysulfanes, **2**, 99
Sulfur
alkylations on, **1**, 140
cyanation of, nucleophilic, **5**, 1111
displacement of, **1**, 237
electrophilic cyanation of, **5**, 1111
elimination of, **1**, 982
higher oxidation states of, **5**, 488
oxidation of, in alkynyl thioethers, **2**, 1034
and oxygen, synthesis of functions, **4**, 833
reactions
with acylphosphines, **5**, 406
with alkanes, **2**, 144
with alkenes, **2**, 131, 137
with alkoxydialkylphosphines, **2**, 462
with alkylarsines, **2**, 499
with alkylbis(trimethylsiloxy)phosphines, **2**, 464
with alkyl bromides, **2**, 226
with *O*-alkyl selenocarboxylates, **5**, 523
with alkylthiodialkylphosphines, **2**, 462
with alkynes, **5**, 603
with aminals, **5**, 599
with aryl isocyanides, **5**, 1036
with 1,4-bis(chloromethylbenzene), **5**, 561
with bis(perfluoroisopropyl)mercury, **3**, 348
with bis(phosphino)methanes, **6**, 167
with bis(2,2,2-trifluoro-1,1-dichloroethyl)mercury, **5**, 480
with *N*-chloroamides, **5**, 349
with *N*-chlorocarbonimidic dichloride, **6**, 609
with chlorodifluoromethane, **6**, 528
with *N*-chloro-*N*-methylsulfonylmethylsulfonamide, **2**, 99
with chlorophosphines, **2**, 463
with *N*-chlorosuccinimide, **5**, 336
with chlorotrifluoroethene, **5**, 479
with dialkylphosphinous acids, **2**, 462
with diaminoethylene, **5**, 548
with 1,1-dihalo-2,2,2-trifluoroethyl mercury compounds, **5**, 474
with 1,3-diphosphaallenes, **4**, 559
with disulfur dichloride, **2**, 712
with dithiols, **2**, 233
with esters, **5**, 600
with ethene, **2**, 137
with ethyne, **2**, 708

with Grignard reagents, **2**, 707
with α-halophosphines, **4**, 128
with 1-halo-2,2,2-trifluoroethyl mercury compounds, **5**, 474
with α-imino acids, **5**, 603
with iodotrifluoroethene, **5**, 473
with isocyanides, **5**, 1036
with lithiated dithioacetals, **6**, 86
with lithium alkynylides, **2**, 1032
with lithium triethylborohydride, **5**, 242, 252
with methylarsaoxane, **2**, 500
with methylenephosphines, **6**, 692
with organometallics, **2**, 125
with osmium carbene complexes, **3**, 553
with phosphaarsenyl complexes, **2**, 866
with phosphines, **2**, 1069
with polyhaloalkenes, **5**, 472
with selenoamides, **5**, 623
with silacyclobutanes, **2**, 267
with silacyclopropanes, **2**, 267
with sodium borohydride, **2**, 227; **3**, 620
with styrene, **5**, 599
with tetrafluoroethylene, **6**, 528
with thiocarbonyls, **2**, 226
with trialkylarsines, **2**, 502
with α,α,α-trichloroacetamides, **5**, 602
with trifluoroalkenes, **5**, 472
with 1,1,1-trihaloalkanes, **5**, 480
with triphenylsilyllithium, **5**, 254
with tris(diphenylphosphino)methane, **6**, 151
with vinyl Grignard reagents, **2**, 722
with vinylidenerhodium complexes, **3**, 553
with ynamines, **5**, 554, 605
as reagent
for the synthesis of phenols, **2**, 659
for the synthesis of thioaldehydes, **3**, 336
for the synthesis of thioketones, **3**, 347, 351, 358, 362
reductive cyanation of, **5**, 1112
sources of, **2**, 225
as starting material, for aldehydes, **3**, 15
substitution of, via intramolecular reactions, **1**, 240
synthesis of functions, **6**, 81
and one halogen, **6**, 273
one halogen and a metal function, **6**, 278
one halogen and one oxygen function, **6**, 276
one halogen and one phosphorus function, **6**, 278
one halogen and selenium functions, **6**, 276
one halogen and two nitrogen functions, **6**, 281
one halogen and two silicon functions, **6**, 282
with selenium functions, **6**, 101
synthesis of functions containing two sulfurs, **4**, 244, 842
use in the synthesis of disulfides, **2**, 225
Sulfur, allylic, displacement of, **1**, 237
Sulfur, benzylic, displacement of, **1**, 237
Sulfur, electrophilic, reactions, with arenes, **2**, 712
Sulfur(IV) compounds, synthesis of, **5**, 488
Sulfurane, bis(methylene)-, reactions, with aldehydes, **4**, 231
Sulfuranes, from, *t*-butylhypochlorite, **2**, 793
Sulfuranes, aminofluoro-, use in the synthesis of difluoroalkanes, **4**, 10
Sulfuranes, dialkoxy diphenyl-, reactions, with carboxylic acid amides, **5**, 689
Sulfuranes, difluoro-, pyrolysis of, **2**, 166
Sulfuranes, imino-, as starting materials, for ketenimines, **3**, 587
Sulfuration, for the synthesis of thionoamides, **5**, 599
Sulfur carbonyl fluoride, pentafluoro-, synthesis of, **6**, 433
Sulfur chloride, reactions, with alcohols, **2**, 101
Sulfur compounds
extrusion reactions of, **1**, 231
from
organoarsines, **2**, 890
tetraaryldiarsenides, **2**, 890
hydrolysis of, problems, **3**, 20
oxidation of, **2**, 177
reactions, with alkynes, **1**, 545
[1,2] shifts of, **1**, 233
as starting materials, for sulfonyl halides, **2**, 177
synthesis of, **4**, 345, 769; **6**, 257
via oxidative processes, **2**, 890
Sulfur compounds, allylic, as starting materials, for ylides, **1**, 234
Sulfur compounds, α-diazo-, synthesis of, **5**, 867
Sulfur compounds, dicoordinated, synthesis of, **4**, 283
Sulfur compounds, (fluorocarbonyl)imido-, synthesis of, **6**, 440
Sulfur compounds, α-halo-, synthesis of, **4**, 61
Sulfur compounds, imidoyl-, synthesis of, **5**, **725–740**
Sulfur compounds, tetracoordinated, synthesis of, **4**, 286, 778, 838, 873
Sulfur compounds, tetracoordinate α-halo-, synthesis of, **4**, 79
Sulfur compounds, tricoordinated
as starting materials, for hemithioaminals, **4**, 329
synthesis of, **4**, 285, 776, 837, 873; **4**, 907
Sulfur compounds, tricoordinate α-halo-, synthesis of, **4**, 78
Sulfur dichloride
reactions
with acetophenones, **5**, 602
with alkyl amines, **2**, 360
with amines, **2**, 361
with *t*-butylamine, **2**, 361
with chloroimines, **4**, 110
with dialkoxy disulfides, **2**, 100
with hydrazones, **5**, 641
with *N*-phenyl-*N'*-trimethylsilylsulfurdiimide, **2**, 796
with sulfonamides, **2**, 365
with *N*-(trimethylsilyl)benzamide, **5**, 335
with *N*-(trimethylsilyl)succinimide, **5**, 338
with trithiocarbonates, **6**, 274
as reagent, for the synthesis of diazapentalenes, **3**, 378
use in the synthesis of alkanesulfenyl chlorides, **2**, 170
use in the synthesis of disulfides, **2**, 226
Sulfur difluoride, (benzylimino)-, synthesis of, **3**, 686
Sulfur difluoride oxide, alkylidene-, synthesis of, **3**, 379
Sulfur difluorides, (alkylimino)-, as starting materials, for nitrile sulfides, **3**, 686
Sulfur difluorides, bis(dialkylamino)-, synthesis of, **2**, 362
Sulfur difluorides, perfluoroacyl imino-, as starting materials, for *N,N*-dichloroamides, **5**, 317
Sulfur difluorides, trifluoroacyl imino-, reactions, with chlorine monofluoride, **5**, 317
Sulfur diimide, diaryl-, reactions, with sodium, **2**, 773
Sulfur diimide, dimethyl-, from, thionitrosomethane, **2**, 795
Sulfur diimide, *N*-phenyl-*N'*-trimethylsilyl-, reactions, with sulfur dichloride, **2**, 796
Sulfur diimides
reactions, with diphenylketene, **2**, 244
synthesis of, **2**, 794; **6**, 358
Sulfurdiimides, aromatic, synthesis of, **2**, 795
Sulfurdiimides, aryl-, synthesis of, **2**, 795
Sulfur diimides, diaryl-, synthesis of, **2**, 796
Sulfur diimides, vinyl-
isolation of, **2**, 795
synthesis of, **2**, 794
Sulfur diimines
from
amines, **2**, 399
N-substituted hexamethyldisilazanes, **2**, 399
sulfinylamines, **2**, 399
sulfurimino dichlorides, **2**, 399
sulfurimino difluorides, **2**, 399
sulfurimino dihalides, **2**, 399
as starting materials, for sulfinylamines, **2**, 399
synthesis of, **2**, 164, 399
Sulfur diimines, dialkyl-, synthesis of, **2**, 164

Sulfur dioxide
elimination of, **1**, 983
extrusion of, from cyclic sulfite, **1**, 232
reactions
with alkanes, **2**, 191
with alkenes, **2**, 192
with amine *N*-oxides, **2**, 357
with arenediazonium chlorides, **2**, 717
with arenediazonium salts, **2**, 719
with aryl Grignard reagents, **2**, 717
with aryllithiums, **2**, 718
with chlorine, **2**, 175
with *N*-chloroamines, **2**, 368
with diazoalkanes, **2**, 207
with diazo compounds, **2**, 251
with dienes, **2**, 201
with enolates, **2**, 157
with Grignard reagents, **2**, 186, 273, 718
with hydroxylamines, **2**, 366
with ketenimines, **2**, 247
with organometallics, **2**, 176, 186
with perfluoropropene, **2**, 175
with pyridine 2-carboxaldehyde, **4**, 237
with α-silyl phosphonate anions, **5**, 645
with stannanes, **2**, 210
with 1,1,3,3-tetramethyl-1,3-disilacyclobutane, **4**, 624
with trimethylsilylketene acetals, **2**, 209
as reducing agents, **2**, 769
as starting material
for alkanesulfinic esters, **2**, 201
for cyclopropanes, **1**, 232
use in the synthesis of carbocations, **1**, 898
use in the synthesis of carbodiimides, **5**, 1078
use in the synthesis of disulfides, **2**, 226
use in the synthesis of sulfenylsulfines, **5**, 562
use in the synthesis of sulfonic acids, **2**, 191
use in the synthesis of sulfonyl halides, **2**, 175
Sulfur fluoroformate, pentafluoro-, synthesis of, **6**, 424
Sulfur functions
elimination of, **1**, 590, 1072
substitution of, **1**, 230, 1031
Sulfur functions, acyl-
from, thiocarboxylic acids, **5**, 232
synthesis of, **5**, **231–255**
Sulfur functions, dicoordinated, synthesis of, **4**, 833, 868
Sulfur groups, elimination of, **3**, 227
Sulfur halides, reactions, with primary amines, **2**, 796
Sulfur heterocycles, functionalization of, **4**, 226
Sulfur hexafluoride
as starting material, for isocyanates, **5**, 975
for the synthesis of carbonic difluoride, **6**, 410
Sulfuric acid
as dehydrating agent, for the synthesis of acetals, **4**, 178
for the hydrolysis of nitriles, **5**, 25
reactions
with alcohols, **1**, 894
with alkanols, **2**, 101
with alkyldicyanamides, **5**, 1124
with aryl-substituted alkenes, **1**, 895
with benzyl cyanide, **5**, 34
with cyclohexadienols, **3**, 70
with cyclohexanones, **2**, 658
with 1,1-dicyano-2-cyclohexylpentane, **5**, 35
with pentafluorobenzyl cyanide, **5**, 34
with peroxydisulfates, **2**, 782
with sodium dichromate, **5**, 115
with 2,4,6-trimethylbenzyl cyanide, **5**, 34
with urea nitrate, **2**, 791
as reagent, for the hydrolysis of nitriles, **5**, 269
as starting material, for phosgene, **6**, 412
use in pinacol rearrangements, **1**, 384
use in the synthesis of ^{13}C-labeled phosgene, **6**, 414
Sulfuric acid, chloro-, reactions, with acetic acid, **4**, 282
Sulfuric acid, concentrated, use in the cleavage of alkyl–silicon bonds, **1**, 61
Sulfuric acid, nitrosyl, reactions, with α-naphthylamine, **2**, 780
Sulfuric acid catalysts, use in dehydration reactions, **1**, 578
Sulfuric acids, use in the synthesis of sulfonic acids, **2**, 190
Sulfur imide, hepta-, reactions, with acetyl chloride, **5**, 338
Sulfurimino dichlorides
from, sulfinylamines, **2**, 400
as starting materials, for sulfurdiimines, **2**, 399
synthesis of, **2**, 400
Sulfurimino difluorides
from, phosphonium salts, **2**, 400
as starting materials, for sulfur diimines, **2**, 399
synthesis of, **2**, 400
Sulfurimino dihalides
from, amines, **2**, 400
as starting materials, for sulfur diimines, **2**, 399
Sulfur isocyanates, synthesis of, **5**, 1018
Sulfurization
of polyhaloalkanes, **5**, 479
of polyhaloalkenes, **5**, 472, 479
of polyhaloalkyls, **5**, 479
Sulfur–lithium exchange
in [1,2] Wittig rearrangement, **1**, 398
in [2,3] Wittig rearrangement, **1**, 405
Sulfur monochloride
reactions
with alkenes, **2**, 221
with *N*-(trimethylsilyl)succinimide, **5**, 338
as starting material, for *S,S'*-diimide compounds, **5**, 336
use in the synthesis of disulfides, **2**, 226
Sulfur monoxide
reactions, with diazoalkanes, **3**, 368
as reagent, for the synthesis of sulfines, **3**, 368
as starting material, for sulfenic acids, **2**, 184
use in the synthesis of dialkyl sulfoxides, **2**, 153
Sulfur mustards, reactions, with chloramine-T, **4**, 773
Sulfur nucleophiles
reactions
with carbonimidic dichlorides, **6**, 617
with 3-chloro-2-alkenals, **3**, 72
with 1,1-dinitro-2,2-diphenylethene, **4**, 906
with glycosyl halides, **4**, 223
with α-halo ketones, **3**, 165
with halomethylgermanes, **4**, 371
with halomethylsilanes, **4**, 367
with 2-iodoalkyl isothiocyanates, **6**, 314
with α-selenocarbanions, **4**, 284
use in the synthesis of alkanethiols, **2**, 121
Sulfur pentafluoride, cyanofluoroimino-, synthesis of, **5**, 814
Sulfur pentafluoride, fluoroxy-, for the fluorination of amines, **2**, 335
Sulfur pentafluoride, vinyl-
reactions, with hydrazine tetrafluoride, **5**, 814
as starting material, for *N*-fluoroimidic thio compounds, **5**, 814
Sulfur pentafluoride chloride
use in the synthesis of alkanesulfenyl fluorides, **2**, 165
UV irradiation of, **5**, 675
Sulfur pentafluoride isocyanide, synthesis of, **3**, 703
Sulfur-substituted carbene complexes, synthesis of, **6**, 719
Sulfur tetrafluoride
reactions
with alkynylaminophosphines, **2**, 1073
with amines, **6**, 51
with anhydrides, **6**, 39
with aryl orthobis(trifluoroacetic) esters, **6**, 43
with benzoic acids, **6**, 10
with carbonyls, **6**, 39
with dialkylaminotrimethylsilanes, **2**, 362
with diols, **2**, 100
with fluorocarbonyl hypofluorite, **6**, 424

use in the fluorination of fluoroformates, **6**, 421
use in the synthesis of alkyl fluorides, **2**, 19
use in the synthesis of difluoroalkanes, **4**, 8
use in the synthesis of trifluoromethyls, **6**, 19
Sulfur tetrafluoride, methylene-
from, metal sulfur pentafluoride complexes, **3**, 380
structure of, **3**, 380
synthesis of, **3**, 380
Sulfur tetrafluorides, alkylidene-
hydrolysis of, **3**, 379
synthesis of, **3**, 380
Sulfur transfer reagents
reactions
with amides, **5**, 333
with lactams, **5**, 333
Sulfur trifluoride, diethylamino- (DAST)
for the fluorination of sulfoxides, **4**, 61
reactions
with alcohols, **4**, 97
with benzoic acids, **6**, 10
with *N*-bromosuccinimide, **4**, 45
with enols, **6**, 19
with epoxides, **2**, 10
with α-hydroxyalkyl phosphonates, **4**, 131
with α-hydroxybenzyl phosphonate esters, **4**, 129
with (hydroxymethyl)triphenylphosphonium tetrafluoroborate, **4**, 117
with sulfoxides, **4**, 73
synthesis of, **2**, 362
use in the synthesis of alkyl fluorides, **2**, 19
use in the synthesis of α,α-dichloro sulfides, **6**, 46
use in the synthesis of difluoroalkanes, **4**, 10
Sulfur trifluoride, morpholino-, reactions, with acid chlorides, **5**, 5
Sulfur trifluoride, 4-morpholino-, reactions, with cyclohexanols, **2**, 19
Sulfur trifluoride, phenyl-, use in the synthesis of difluoroalkanes, **4**, 9
Sulfur trifluoride, trifluoromethyl-, hydrolysis of, **2**, 175
Sulfur trifluorides, dialkylamino-
as reagents, for acid fluorides, **5**, 5
synthesis of, **2**, 362
use in the synthesis of sulfonyl fluorides, **2**, 183
Sulfur trifluorides, *N*,*N*-dialkylamino-, use in the synthesis of difluoro compounds, **4**, 10
Sulfur trifluorides, methylidyne-, synthesis of, **3**, 689
Sulfur triimines, synthesis of, **2**, 401
Sulfur trioxide
handling problems, **2**, 720
reactions
with alkanes, **2**, 176
with alkanols, **2**, 101
with alkenes, **2**, 206, 208; **6**, 42
with carboxylic acids, **2**, 176
with 2,2-difluoroethene-1-sulfonyl fluoride, **4**, 867
with fluorohalomethanes, **6**, 410
with hydroxamic acids, **5**, 1000
with perfluoropropene, **4**, 86
with potassium cyanate, **5**, 994
with 1,1,3,3-tetramethyl-1,3-disilacyclobutane, **4**, 624
with 2-trifluoromethyl perfluoropropene, **2**, 208
with tris(silylalkyl)aluminums, **4**, 397
with ylides, **5**, 902
as starting material
for carbonic bromide fluoride, **6**, 420
for carbonic chloride fluoride, **6**, 419
for phosgene, **6**, 412
as starting materials, for alkanesulfonic esters, **2**, 206
use in the synthesis of sulfonic acids, **2**, 190
Sulfuryl chloride
as chlorinating agent, **6**, 606
as chlorinating agents, for aryl chlorides, **2**, 620
as chlorinating reagents, for the synthesis of arenesulfenyl chlorides, **2**, 716
for the chlorination of alkanes, **2**, 21
for the chlorination of sulfides, **4**, 63
for the chlorination of sulfoxides, **4**, 74
for the chlorination of thiocarbonate, **6**, 427
reactions
with alkanols, **2**, 102
with alkynyl tellurides, **2**, 1038
with amides, **5**, 657
with amines, **2**, 367, 368, 797
with *N*-(anilinothiocarbonyl)thiobenzamide, **5**, 657
with anisole, **4**, 47
with arenethiols, **2**, 717
with aryl Grignard reagents, **2**, 717
with benzhydryl benzyl sulfoxide, **4**, 74
with dialkyl sulfites, **2**, 102
with dibenzyl sulfide, **4**, 64
with dibenzyl sulfides, **6**, 28
with *O*,*S*-dimethyl dithiocarbonate, **6**, 276
with disulfides, **2**, 173
with dithiobisamines, **2**, 360
with Grignard reagents, **2**, 176
with mercury bis(ethyl diazoacetate), **3**, 478
with 2-methylpyrroles, **3**, 106
with methylthioacetone, **4**, 775
with phenylketones, **4**, 12
with pyruvic acid, **5**, 60
with selenenamides, **2**, 369
with sulfides, **4**, 80
with β-thioalkoxyamides, **2**, 168
with thiol esters, **2**, 168, 173
with toluene, **4**, 11
with 2-trimethylsilylethyl sulfoxides, **2**, 174
with triphenylphosphine, **2**, 182
as starting material, for sulfonyl chlorides, **2**, 182
for the synthesis of aryl chlorides, **2**, 619
for the synthesis of vinyl chlorides, **2**, 617
use in the synthesis of acylsulfenyl chlorides, **5**, 244
use in the synthesis of alkanesulfenyl chlorides, **2**, 166
use in the synthesis of arenesulfinyl chlorides, **2**, 717
use in the synthesis of aromatic acid chlorides, **5**, 15
use in the synthesis of α-carbamoyl-1,1-dichlorosulfenyl chlorides, **5**, 477
use in the synthesis of chloroformates, **6**, 427
Sulfuryl chlorofluoride, as starting material, for sulfamoyl chlorides, **2**, 368
Sulfuryl difluoride, *N*-(fluorothiocarbonyl)imido-, synthesis of, **6**, 537
Sulfuryl difluoride, peroxydi-, reactions, with cyanogen chloride, **6**, 277
Sulfuryl halides, use in the synthesis of sulfonyl halides, **2**, 176
Sulfur ylides
from
benzynes, **1**, 821
carbenes, **1**, 821
trimethylsilyl sulfonium salts, **1**, 821
halogenation of, **5**, 494
metal complexes of, **3**, 379
reactions
with 1*H*-1,2,4-diazarsole, **4**, 502
with nitrile oxides, **2**, 779
with tetrachloropalladium(II) salts, **4**, 396
as reagents, in cyclopropanation reactions, **1**, 304
rearrangement of, **1**, 821
representation of, **3**, 366
as starting materials
for cyclopropanes, **1**, 811
for 1-silylketo-2-carbonyl cyclopropanes, **5**, 430
synthesis of, **2**, 1000; **3**, 168
Sulfur ylides, allylic, displacement of, **1**, 483
Sulphenyl halides, as starting materials, for thioaldehydes, **3**, 333
Sulphur, use in the transposition of ketones, **3**, 126

Sultams
as chiral auxiliaries, **1**, 305
reactions, with 2,2,6-trimethyl-4*H*-1,3-dioxin-4-one, **5**, 343
synthesis of, **2**, 797
via oxidation of amino sulfides, **5**, 347
Sultams, *N*-fluoro-
for the fluorination of carbonyls, **2**, 16
synthesis of, **2**, 365
use in the synthesis of α-fluoro ketones, **3**, 149
Sultenes, detection of, **2**, 195
γ-Sultenes, rearrangement of, **5**, 604
Sultines
reactions, with nitrile enolates, **3**, 634
synthesis of, **2**, 200
β-Sultines, as intermediates, **1**, 600
Sultones
reactions, with trimethylsilyl halides, **2**, 209
synthesis of, **2**, 204
Sultones, alkane, use in the synthesis of thioimidic esters, **5**, 726
2-β-Sultones, perfluoro-, synthesis of, **6**, 42
Superacid media, for the synthesis of cations, **1**, 900
Superacid systems, for oxidation, **2**, 650
Superacid technology, for the synthesis of carbocations, **1**, 898
Superbase
as metallating agent, **4**, 650
use in the synthesis of 1,1-bis(silyl)alkanes, **4**, 620
See also n-Butyllithium–potassium *t*-butoxide
Superhalogens, characteristics of, **6**, 228
Super Lewis acid, use of term, **1**, 893
Superoxide radical anions, for the nucleophilic displacement of leaving groups, **2**, 87
Supinidine, synthesis of, **2**, 326
Suprafacial addition, mechanism, **1**, 349
Surfactants, manufacture of, **1**, 674
Sustmann Type II dipoles, nitrile imides as, **3**, 680
Suzuki reaction, use in the substitution of boron, **1**, 489
Swart's catalyst, for the synthesis of trihalides, **6**, 8
Swern conditions, for the synthesis of α-amino ketones, **3**, 184
Swern–Moffatt oxidation
of silyl ethers, to aldehydes, **3**, 8
for the synthesis of aldehydes, **3**, 6, 13
for the synthesis of ketones, **3**, 118
Swern oxidation
of *cis*-diols, **5**, 424
for the oxidation of α-hydroxysilanes to acylsilanes, **5**, 412
of silyl ethers, **3**, 121
for the synthesis of furan carboxaldehydes, **3**, 102
of trimethylsilyl methanol, **5**, 422
use in the synthesis of acylgermanes, **5**, 431
Sydnones
as starting materials, for hydrazines, **2**, 374
thionation of, **5**, 642
Sydnones, 4-bromo-3-aryl-, as starting materials, for 1-bromocarbonylindazoles, **6**, 453
Symbioramide, synthesis of, via selenoxide pyrolysis, **1**, 782
Synergistic effect, use of term, **1**, 922
Synergistic radicals, reviews, **1**, 919
Synthons, synthetic equivalents for, **1**, 165
umpoled Synthons, use in the synthesis of aryl alkyl ketones, **3**, 291

T

Talopyranosides, synthesis of, **6**, 549
Tamoxifen, σ-stannylated, destannylation of, **1**, 452
3*H*-Tamoxifen, synthesis of, **1**, 452
Tamoxifen analogue, synthesis of, **1**, 576
Tandem rearrangement
of 1,1-bis-cyclopropylethene, **1**, 801
of 7-vinylcyclohept-2-enes, **1**, 799
Tandem selenoxide elimination–Claisen rearrangement
mechanism, **1**, 826
of selenides, **1**, 829
Tandem sigmatropic rearrangement, mechanism, **1**, 840
Tandem sulfoxide elimination–Claisen rearrangement, mechanism, **1**, 826
Tantalocene, dimeric, synthesis of, **4**, 672
Tantalum
in the synthesis of metallacyclopropenes, **2**, 927
use in the synthesis of α-hydroxysilanes, **4**, 355
Tantalum, dichlorotrimethyl-, reactions, with *t*-butylisonitrile, **5**, 797
Tantalum alkyl complexes, as starting materials, for carbene hydride complexes, **3**, 512
Tantalum bis(cyclopentadienyl)trihydrides, use in the synthesis of dimeric tantalocene, **4**, 672
Tantalum carbene complexes
from
methyl complexes, **3**, 512
tantalum carbene complexes, **3**, 512
ylides, **3**, 512
as starting materials
for tantalum carbene complexes, **3**, 512
for tungsten oxocarbene complexes, **3**, 517
synthesis of, **3**, 511
Tantalum carbonyl complexes, synthesis of, **6**, 522
Tantalum chloride complexes, reactions, with 2,2-dimethylpropylmagnesium chloride, **5**, 444
Tantalum complexes, alkyl-, reactions, with isonitriles, **5**, 797
Tantalum complexes, cyclopentadienyl-
reactions
with aldehydes, **4**, 355
with ketones, **4**, 355
Tantalum functions, doubly-bonded, synthesis of, **3**, 511
Tantalum-promoted reaction
of alkynes
with aldehydes, **2**, 988
with hydrazones, **2**, 988
with isocyanates, **2**, 988
with ketones, **2**, 988
Tantalums
reactions
with 1-metalloalkylphosphine oxides, **4**, 579
with 1-metalloalkylphosphines, **4**, 579
with 1-metalloalkylphosphine sulfides, **4**, 579
with phosphorus ylides, **4**, 579
Tantalums, acyl-, synthesis of, **5**, 444
Tantalums, alkyl-, synthesis of, **2**, 569; **5**, 444
Tantalums, imidoyl-, synthesis of, **5**, 797
Tantalums, pentabenzyl-, synthesis of, **2**, 570
Tartaric acid
as starting material
for boronates, **2**, 75
for pyruvic acid, **5**, 73
L-Tartaric acid anhydride, dibenzoyl-, use in the synthesis of α-aminophosphonic acids, **4**, 462
Tartrate, (-)-dialkyl, use in asymmetric epoxidation, **2**, 91
Tartrate, diisopropyl, use in the synthesis of amine *N*-oxides, **2**, 348
Tartrate, dimethyl, bromination of, **2**, 29
Tartrate esters, use in asymmetric epoxidation, **2**, 91
Tartrates, dialkyl, as starting materials, for homochiral allydialkoxyboranes, **2**, 529
D-Tartrates, use in asymmetric epoxidation, **2**, 92
L-Tartrates, use in asymmetric epoxidation, **2**, 92
Tartrate salts, dibenzoyl-, use in the synthesis of chiral phosphines, **2**, 443
Taurine, synthesis of, **2**, 191
Tautomeric equilibrium
of iminoesters and 1-alkoxyenamines, **4**, 880
of ketene *S,N*-acetals, **4**, 898
Tautomerism
of alkanenitriles, **3**, 565
of thiolimidic acids, **5**, 725
Tautomerization, imine-secondary enamine, **3**, 414
Taxane, synthesis of, **4**, 220
Taxol, synthesis of, **1**, 842
Taxol analogues, oxetane rings for, **2**, 642
Tazane skeleton, synthesis of, via macrocyclization–transannular cyclization, **1**, 347
TBTU, as coupling reagents, for the synthesis of *N*-alkylamides, **5**, 279
TDBTU, as coupling reagents, for the synthesis of *N*-alkylamides, **5**, 280
TEA, reactions, with tetrazolium salts, **5**, 1073
Tebbe reagent
reactions
with alkynes, **2**, 979
with 3,3-dimethylcyclopropene, **3**, 508
as starting material, for titanacyclobutanes, **3**, 508
synthesis of, **1**, 767
for the synthesis of diborocyclopentenes, **4**, 632
use in methylenation reactions, **1**, 766
use in the synthesis of alkenyl ethers, **2**, 680
use in the synthesis of transition metal complexes, **4**, 716
Technetium carbonyl complexes, synthesis of, **6**, 523
Tellone
bridged metal complex, synthesis of, **3**, 401
cycloaddition, with diazo compound, **3**, 400
Tellone, 1,1,3,3-tetramethylindan-, synthesis of, **3**, 400
Tellones
Diels–Alder adducts, **3**, 400
from
bis-(dimethylaluminum) telluride, **3**, 400
ketones, **3**, 400
synthesis of, **3**, 400
via metal carbenes, **3**, 400
use of term, **3**, 381
Tellones, stable, synthesis of, **3**, 400
Telluradiazoles, synthesis of, **4**, 334
1,3,4-Telluradiazoline, thermolysis of, **3**, 400
Telluradiazolines, synthesis of, **3**, 400
Tellurals
Diels–Alder adducts, **3**, 389
metal complexes of, **3**, 390
synthesis of, **3**, 389
via bis-(dimethylaluminum) telluride, **3**, 389
via metal carbenes, **3**, 390
use of term, **3**, 381
See also Telluroaldehydes
Telluramides, alkane-
synthesis of, **2**, 293

See also Tellurenyl amides
Tellurane, tetrakis(trifluoromethyl)-, synthesis of, **6**, 237
Tellurates, synthesis of, **2**, 735
Tellurates, arene, *in situ* synthesis of, **2**, 735
Tellurates, aryl
reactions
with aryl halides, **2**, 731
with diazonium salts, **2**, 731
Telluration
of imidate esters, **5**, 510
for the synthesis of tellones, **3**, 400
Tellurazoles, synthesis of, **2**, 769
Tellurenamides, synthesis of, **2**, 370, 769
Tellurenamides, arene-, synthesis of, **2**, 370
Tellurenamide tetrafluorides, diamino-, reactions, with silyl amines, **2**, 370
Tellurenic acid esters, alkane-, synthesis of, **2**, 292
Tellurenic acids, in elimination reactions, **1**, 611
Tellurenic acids, alkane-
disproportionation of, **2**, 290
synthesis of, **2**, 290
Tellurenyl bromides, phenyl-
reactions
with piperidinium thiocarboxylates, **5**, 249
with potassium thiocarboxylates, **5**, 249
Tellurenyl halides, as starting materials, for benzyl tellurocyanates, **2**, 290
Tellurenyl halides, alkane-, complexes of, **2**, 289
Tellurenyl halides, arene-
synthesis of, **2**, 732
thiourea adducts of, synthesis of, **2**, 732
Tellurenyl halides, benzyl-, as starting materials, for 2-carboethoxybenzyl tellurocyanate, **5**, 1117
Tellurenyl iodides, arene-, reactions, with lithium amides, **2**, 370
Tellurenyl trichlorides, alkane-, from, tellurium tetrachloride, **2**, 289
Telluride, bis(*t*-butyldimethylsilyl), synthesis of, **2**, 542
Telluride, bis(dimethylaluminum)
as starting material, for tellones, **3**, 400
synthesis of, **3**, 390
use in the synthesis of tellurals, **3**, 389
Telluride, bis(pivaloyl), synthesis of, **5**, 244
Telluride, bis(trifluoromethyl), reactions, with dialkyl tellurides, **6**, 237
Telluride, bis(trimethylsilyl)
reactions
with acetyl chloride, **5**, 244
with dipivaloyl telluride, **5**, 542
with pivaloyl chloride, **5**, 255, 542
synthesis of, **2**, 542
as tellurating reagent, **3**, 390
Telluride, bis(trimethyltin), transmetallation of, **3**, 390
Telluride, bis(triphenylstannyl), reactions, with phenacyl bromide, **3**, 176
Telluride, α-bromo, as intermediate, in the synthesis of α-bromotellurones, **4**, 93
Telluride, *n*-butyl *t*-butyl, reactions, with alkynes, **4**, 947
Telluride, *t*-butylthiophenyl, synthesis of, **2**, 238
Telluride, butyl trimethylsilyl, as starting material, for trimethylsilylmethyllithium, **4**, 652
Telluride, chloromethyl phenyl, synthesis of, **4**, 93
Telluride, dialkyl, reactions, with thiols, **2**, 238
Telluride, diaryl, reactions, with Grignard reagents, **1**, 437
Telluride, dimethyl
oxidation of, **2**, 288
reactions
with bis(trifluoromethyl)cadmium, **6**, 49
with trifluoromethylzinc bromide, **6**, 49
Telluride, phenyltrimethylsilyl, reactions, with acid chlorides, **5**, 242
Telluride, phenylvinyl, silylation of, **2**, 908
Telluride anions
alkylation of, **4**, 93
reactions, with *gem*-dihalides, **4**, 290
as starting materials, for selenotelluroacetals, **4**, 291
synthesis of, **4**, 291
Telluride dichlorides, reduction of, **1**, 25
Tellurides
elimination of, **1**, 594
in situ synthesis of, **2**, 731
migrations of, **1**, 246
oxidation of, **1**, 611; **2**, 288
reactions
with alkyllithiums, **2**, 554
with chloroamine-T, **1**, 594, 621
with palladium, **1**, 247
reduction of, **1**, 25
as starting materials, for alkane tellurinic acids, **2**, 291
vacuum pyrolysis of, **1**, 243
Tellurides, acyl, lithium–tellurium exchange of, **5**, 437
Tellurides, (*Z*)-alkenyl, reactions, with Grignard reagents, **1**, 545
Tellurides, alkyl
as starting materials, for alkyl aryl telluroxides, **2**, 287
synthesis of, **2**, 278
Tellurides, alkyl aryl
from, ditellurides, **2**, 280
oxidation of, **2**, 288
synthesis of, via Williamson reactions, **2**, 731
Tellurides, alkyl phenyl
oxidation of, **2**, 287
reactions, with *m*-chloroperbenzoic acid, **1**, 611
Tellurides, alkynyl
reactions
with sodium borohydride, **1**, 999
with sulfuryl chloride, **2**, 1038
research background, **2**, 1036
synthesis of, **2**, 1037
Tellurides, alkynyl(alkyl), synthesis of, **2**, 1037
Tellurides, aryl
reduction of, **1**, 437
as starting materials, for telluroarenes, **1**, 437
synthesis of, **2**, 730
Tellurides, arylvinyl, deprotonation of, **4**, 947
Tellurides, bis(acyl)
reactions
with piperidine, **5**, 250
with secondary amines, **5**, 242
reactivity of, **5**, 244
synthesis of, **5**, 244
Tellurides, bis(aroyl), synthesis of, **5**, 244
Tellurides, diacyl, from, tellurocarboxylic *O*-acids, **5**, 509
Tellurides, dialkyl
reactions, with bis(trifluoromethyl) telluride, **6**, 237
as starting materials, for alkane tellurimides, **2**, 293
synthesis of, **2**, 278; **4**, 377
Tellurides, diaryl
acetoxylation of, **5**, 210
from
arylamines, **2**, 730
aryl halides, **2**, 730
Tellurides, diphenacyl, synthesis of, **3**, 176
Tellurides, glycosyl, as starting materials, for glycosyl radicals, **4**, 240
Tellurides, imido, from, haloimides, **5**, 341
Tellurides, perfluoroalkyl
from, lithium aryl tellurides, **6**, 50
synthesis of, **6**, 49
Tellurides, phenylvinyl, deprotonation of, **4**, 965
Tellurides, primary, as starting materials, for alkenes, **1**, 594
Tellurides, symmetrical, from, sodium telluride, **2**, 279
Tellurides, unsymmetrical
from
ditellurides, **2**, 280

metal tellurolates, **2**, 280
Tellurides, vinyl
from, alkynes, **2**, 730
reduction of, **1**, 438, 441
silylation of, **2**, 911
as starting materials
for alkenes, **1**, 438
for telluroalkenes, **1**, 441
for vinyllithiums, **1**, 441
synthesis of, **2**, 730
Tellurimides, use in the synthesis of carbodiimides, **5**, 1078
Tellurimides, alkane-, from, dialkyl tellurides, **2**, 293
Tellurinic acid esters, alkane-, synthesis of, **2**, 292
Tellurinic acids, oxidation of, **2**, 290
Tellurinic acids, alkane-
from, tellurides, **2**, 291
synthesis of, **2**, 291
Tellurinic acids, arene-, from, aryltellurium trihalides, **2**, 733
Tellurinyl halides, alkane-, from, tellurium trihalides, **2**, 289
Tellurium
amidines of, **5**, 767
difluoromethyl compounds, synthesis of, **6**, 49
and heteroatomic chemistry, **2**, 856
reactions
with alkali metal cyanides, **2**, 290
with alkylphosphorus compounds, **2**, 465
with aryllithiums, **5**, 241
with *n*-butyllithium, **6**, 498
with chalcogenophosphinides, **2**, 465
with monotellurides, **6**, 356
with osmium complexes, **3**, 391
with perfluoroalkyl radicals, **6**, 49
with phenylethyne, **5**, 627
with sodium alkynylide, **2**, 1036
reduction of, **2**, 279
as starting material
for alkyl ditellurides, **2**, 292
for ditelluride anions, **2**, 292
for metal tellurolates, **2**, 278
substitution of, **1**, 243
synthesis of functions, containing oxygen, **4**, 841
Tellurium, alkyldibutylcyanomethyl-, synthesis of, **3**, 637
Tellurium, bis(trifluoromethyl)-
as alkylation agent, **2**, 483
use in the synthesis of tris(trifluoromethyl)arsine, **2**, 486
Tellurium, diaryl-, reactions, with bromoamides, **5**, 341
Tellurium, dibenzoyl-, reactions, with *N*-bromosuccinimide, **5**, 341
Tellurium, dibutyl-
reactions
with chloroacetonitrile, **3**, 637
with α-halonitriles, **3**, 637
Tellurium, dimethyl-, reactions, with thiuram disulfides, **6**, 566
Tellurium, dimethyldimethoxy-, synthesis of, **5**, 211
Tellurium, (*N*,*N*-dimethylmethanimamidato-*N*′)pentafluoro-, synthesis of, **5**, 767
Tellurium, tetrahalo-, reactions, with *N*,*O*-bis(trimethylsilyl)acetamide, **5**, 349
Tellurium, trimethyl-, for the methylation of toluene, **1**, 247
Tellurium(II) chloride, organic, reactions, with dithiocarbonates, **2**, 238
Tellurium(IV) chloride
reactions
with acetylacetone, **2**, 690
with alkenes, **1**, 595
Tellurium acetates, diaryl-, synthesis of, **5**, 210
Tellurium amides, synthesis of, **5**, 341
Tellurium bis(*O*-alkyl dithiocarbonates), dimethyl-, synthesis of, **6**, 547
Tellurium bis(thiocarboxylates), reactions, with halogens, **5**, 252
Tellurium bromide, aryl-, reactions, with lithium alkynylides, **2**, 1037
Tellurium chloride, triphenyl-, reactions, with sodium *S*-alkyl trithiocarbonates, **6**, 553
Tellurium complexes, use in the synthesis of arylamines, **2**, 753
Tellurium dichlorides, diaryl-
reactions, with sodium acetate, **5**, 210
as starting materials, for ditellurides, **2**, 730
Tellurium dihalides, aryl-, as starting materials, for aryl telluroxides, **2**, 731
Tellurium dihalides, diaryl-, as intermediates, in the synthesis of aryl telluroxides, **2**, 731
Tellurium diiodide, dibenzoyl-, reactions, with diamines, **5**, 734
Tellurium diiodide, dimethyl-
reactions, with sodium *O*-alkyl dithiocarbonates, **6**, 547
as reagent, for the synthesis of dimethyldimethoxytellurium, **5**, 211
Tellurium diiodides, synthesis of, **5**, 734
Tellurium enolates, synthesis of, **2**, 690
Tellurium functions
elimination of, **1**, 590, 1072
substitution of, **1**, 1031
Tellurium functions, acyl-, synthesis of, **5**, **231–255**
Tellurium halides, phenyl-, reactions, with potassium cyanide, **5**, 1117
Tellurium hexafluoride, reactions, with silyl amines, **2**, 370
Tellurium imides, elimination of, **1**, 621
Tellurium imines, *N*-acyl-
from
amides, **5**, 349
N-silylamides, **5**, 349
Tellurium iodide, phenyl-, reactions, with ketone enolates, **3**, 176
Tellurium isocyanate, pentafluoro-, from, pentafluoroisocyanate, **5**, 348
Tellurium oxide, diaryl-, reactions, with trifluoroacetamide, **5**, 349
Tellurium pentafluoride, dialkylamino-, synthesis of, **2**, 370
Telluriums
reactions, with alkynes, **1**, 545
reduction of, **1**, 21
synthesis of, **4**, 872
via oxidative processes, **2**, 890
Telluriums, acyloxy-, synthesis of, **5**, 210
Telluriums, alkyl-, reactions, with butyllithium, **1**, 437
Telluriums, alkynyl-, coupling reactions of, **1**, 1035
Telluriums, aryl-, reactions, with butyllithium, **1**, 437
Telluriums, α-halo-, synthesis of, **4**, 93
Telluriums, imidoyl-, synthesis of, **5**, **725–740**
Telluriums, tetraalkyl-, reactions, with alkynes, **1**, 247
Telluriums, trifluoromethyl-, synthesis of, **6**, 237
N-Telluriums, synthesis of, **2**, 768
O-Telluriums, alkenyl-, synthesis of, **2**, 690
O-Telluriums, aryl-, synthesis of, **2**, 690
Tellurium salts, synthesis of, **3**, 176, 637
Tellurium tetrachloride
reactions, with acetic anhydride, **4**, 291
as starting material, for alkane tellurenyl trichlorides, **2**, 289
Tellurium triacetate, phenyl-, synthesis of, **5**, 210
Tellurium trichlorides, alkyl-, reactions, with silyl enol ethers, **3**, 177
Tellurium trichlorides, aryl-
reactions
with ketones, **3**, 177
with silyl enol ethers, **3**, 177
reduction of, **2**, 735
as starting materials, for organotelluriums, **2**, 732
synthesis of, **2**, 732
Tellurium trihalides, as starting materials, for alkanetellurinyl halides, **2**, 289
Tellurium trihalides, aryl-, as starting materials, for arenetellurinic acids, **2**, 733

Tellurium trithiocarbonates, *S*-alkyl *S*-triphenyl-, synthesis of, **6**, 553
Tellurium ylides, synthesis of, **3**, 177
Telluroacetamide, *N*,*N*-diethylphenyl-, synthesis of, theoretical, **5**, 628
Telluroacetate, *O*-trimethylsilyl, synthesis of, **5**, 543
Telluroacetone, bridged metal complex, synthesis of, **3**, 401
Telluroacyl compounds, silyl, synthesis of, **5**, 542
Telluroacyl halides, synthesis of, **5**, **471–503**
Telluro acyloxy functions, synthesis of, **5**, **505–543**
Telluroaldehyde, trimeric, synthesis of, **3**, 389
Telluroaldehydes
  nomenclature of, **3**, 381
  synthesis of, **3**, **381–401**
  See also Tellurals
Telluroalkenes, from, vinyl tellurides, **1**, 441
Telluroalkynes, dichloro-, synthesis of, **2**, 1038
1-Telluroalkynes, reactions, with acetic acid, **4**, 841
Telluroamides
  IR spectra of, **5**, 726
  stability of, **5**, 627
  synthesis of, **5**, 627
Telluroamides, *N*-acyl-, synthesis of, **5**, 629
Telluroanhydrides, synthesis of, **5**, 530
Telluroarenes, from, aryl tellurides, **1**, 437
Telluroarsinous acid esters, alkyl-, synthesis of, **2**, 293
Tellurobenzaldehyde, synthesis of, **3**, 389, 390
Tellurobenzoate, *n*-butyl, reactions, with *n*-butyllithium, **5**, 437
Tellurobenzoate, *O*-cholesteryl, synthesis of, problems, **5**, 526
Tellurobenzoate, *Te*-ferrocenyl, synthesis of, **5**, 241
Tellurobenzoate, sodium 2-methyl-, reactions, with *t*-butyldimethylsilyl chloride, **5**, 543
Tellurobenzophenone, metal complex of, **3**, 400
α-Telluroboranes, synthesis of, **4**, 377
Tellurocarbamate, *Te*-butyl *N*,*N*-diethyl, synthesis of, **6**, 498
Tellurocarbamate, *Te*-*n*-hexyl *N*,*N*-dimethyl, synthesis of, **6**, 498
Tellurocarbamates, synthesis of, **6**, 498
α-Tellurocarbanions, synthesis of, reviews, **4**, 398
Tellurocarbonates, synthesis of, **6**, 481
Tellurocarbonyl difluoride, from, diethylaluminum iodide, **6**, 588
Tellurocarbonyl group
  synthesis of functions, **6**, **587–599**
    at least one attached halogen, **6**, 588
Tellurocarbonyls
  reagents for, **6**, 587
  scarcity of, **6**, 587
  stability of, **3**, 381; **6**, 587
Tellurocarboxylate, *O*-silyl, synthesis of, **5**, 536
Tellurocarboxylate, *Te*-triorganosilyl, as intermediate, in the synthesis of tellurocarboxylate, **5**, 543
Tellurocarboxylates
  alkylation of, **5**, 242
  from, tellurolate anions, **5**, 242
  as starting materials, for tellurocarboxylic esters, **5**, 242
  synthesis of, **5**, 510, 526, 543
Tellurocarboxylates, *O*-alkyl, synthesis of, **5**, 526
Tellurocarboxylates, *O*-halo, synthesis of, **5**, 530
Tellurocarboxylates, piperidinium, reactions, with 2,4-dinitrobenzenesulfenyl chloride, **5**, 250
Tellurocarboxylates, trimethylsilyl, synthesis of, **5**, 542
Tellurocarboxylic acids, synthesis of, **5**, 506
Tellurocarboxylic *O*-acids
  as starting materials, for diacyl tellurides, **5**, 509
  synthesis of, **5**, 509
Tellurochalcogenophosphinic acids, dialkyl-, synthesis of, **2**, 465
Tellurocyanate, reactions, with metal carbene complexes, **3**, 400
Tellurocyanate, 2-carboethoxybenzyl, from, benzyltellurenyl halides, **5**, 1117
Tellurocyanates, synthesis of, **5**, 1116
Tellurocyanates, alkyl
  stability of, **2**, 290
  synthesis of, **2**, 289
Tellurocyanates, benzyl
  from
    potassium tellurocyanate, **2**, 290
    tellurenyl halides, **2**, 290
  synthesis of, **5**, 1117
Tellurocyanates, phenyl
  synthesis of, **5**, 1117
    via reductive cleavage of tellurium–tellurium bonds, **5**, 1117
Tellurodicarbonic acid diesters, synthesis of, **6**, 481, 482
Tellurodithioorthocarbonates, theoretical, **6**, 311
Telluroethers, synthesis of, **2**, 1036
Telluroethers, alkynyl, synthesis of, **2**, 1032, 1036
Telluroformaldehyde
  bridged metal complex, synthesis of, **3**, 390
  metal complexes of, **3**, 391
  osmium complex of, **3**, 391
Telluroformamide, *N*,*N*-dimethyl-, synthesis of, **5**, 627
Tellurogermane, methyl-, from, methyltellurosilane, **2**, 294
α-Tellurogermanes, synthesis of, **4**, 377
Telluro groups, phenyl, conjugate addition to unsaturated ketones, **3**, 176
Telluro groups, β-phenyl-, oxidative elimination of, **1**, 611
Tellurohalogenides, synthesis of, **6**, 358
Tellurohydrazides, from, thionoamides, **5**, 642
Telluroketenes, as intermediates, **5**, 628
Telluroketones
  nomenclature of, **3**, 381
  synthesis of, **3**, **381–401**
Telluroketones, α-phenyl
  elimination of hydrogen chloride from, **3**, 256
  synthesis of, **3**, 176
Telluroketones, β-trichloro, synthesis of, **3**, 177
α-Telluroketones, from, α-bromo ketones, **3**, 176
β-Telluroketones, synthesis of, **3**, 39, 176
Tellurol, phenyl, from, diphenyl ditelluride, **1**, 437
γ-Tellurolactones, synthesis of, via phase-transfer catalysis, **5**, 242
Tellurolate, alkynyl, from, sodium alkynylide, **2**, 1036
Tellurolate anions
  acylation of, **5**, 241
  as starting materials, for tellurocarboxylates, **5**, 242
Tellurolate anions, aryl-, aroylation of, **5**, 241
α-Tellurolithiums, synthesis of, **4**, 402
Tellurols, oxidation of, **2**, 278
Tellurols, alkane-
  from, aluminum telluride, **2**, 278
  synthesis of, **2**, 277
Tellurols, arene-, synthesis of, **2**, 727
Tellurols, thienyl, synthesis of, **4**, 872
Tellurones
  elimination of, **1**, 620
  nucleophilic displacement of, **1**, 620
  synthesis of, **2**, 288
    theoretical, **2**, 732
  use of term, **3**, 382
Tellurones, alkyl, synthesis of, **2**, 287
Tellurones, alkyl aryl, pyrolysis of, **1**, 620
Tellurones, α-bromo, synthesis of, **4**, 93
Telluronic acid, alkane-, synthesis of, **2**, 291
Telluronic acid esters, alkane-, synthesis of, **2**, 292
N-Telluronitrosimine, use in the synthesis of tellones, **3**, 400
Telluronium alkoxides, triphenyl-, reactions, with carbon disulfide, **6**, 547
Telluronium chloride, dibutyl(cyanomethyl)-, synthesis of, **3**, 637
Telluronium chloride, triphenyl-, reactions, with sodium *O*-alkyl dithiocarbonates, **6**, 547
Telluronium iodide, dimethyl(trifluoromethyl)-, synthesis of, **6**, 237

Telluronium salts
reactions
with aldehydes, **1**, 623
with ketones, **1**, 623
Telluronium salts, alkyl, synthesis of, **2**, 288
Telluronyl halides, alkane-, synthesis of, **2**, 289
α-Telluroorganometallics, reactivity of, **1**, 717
Telluroorthocarbonates, synthesis of, **6**, 308
Tellurophene
reactions, with Grignard reagents, **1**, 484
reactivity of, **3**, 106
Tellurophenes, formyl, synthesis of, **3**, 106
Tellurophenes, 3-halo-, as starting materials, for alkynes, **1**, 1072
Tellurophosphines, synthesis of, **6**, 357
Tellurophosphines, dialkylalkyl-, synthesis of, **2**, 465
Tellurophosphinous acid esters, alkyl-, synthesis of, **2**, 293
Tellurophosphonates, synthesis of, **2**, 465
Tellurophosphonium salts, alkyl-, from, tellurophosphoranes, **2**, 293
Tellurophosphoranes, as starting materials, for alkyltelluro phosphonium salts, **2**, 293
Tellurophthalide, synthesis of, **5**, 242
Telluropivalates
properties of, **5**, 542
synthesis of, **5**, 526
Telluropivalates, *O*-alkyl, synthesis of, **5**, 526
Telluropivalates, *Te*-trimethylsilyl, synthesis of, **5**, 542
Tellurosilane, methyl-
as starting material
for methyltellurogermane, **2**, 294
for methyltellurostannane, **2**, 294
α-Tellurosilanes, synthesis of, **4**, 377
Tellurostannane, methyl-, from, methyltellurosilane, **2**, 294
Tellurotetrathiafulvalenes, monoalkyl, synthesis of, **4**, 873
Tellurothioic *S*-acid, structure of, **5**, 545
Tellurothioic *Te*-acid, structure of, **5**, 545
Tellurothioorthocarbonates, theoretical, **6**, 311
Telluro(trimethylsilane), methyl-, from, lithium methanetellurolate, **2**, 294
Tellurotrithiooorthocarbonates, synthesis of, **6**, 308
Telluroureas
from, 1,3-dimethylbenzimidazoline, **6**, 599
metal complexes, **6**, 599
synthesis of, **6**, 599
Telluroxide, bis(*p*-methoxyphenyl), reactions, with *O*-cholestannyl selenobenzoate, **5**, 523
Telluroxide elimination
regioselectivity of, **1**, 611
stereoselectivity of, **1**, 611
Telluroxide hydrates, synthesis of, **2**, 287
Telluroxides
elimination of, **1**, 610
syn elimination of, **1**, 610
oxidation of, **1**, 620
reactions, with 1,3-diketones, **3**, 177
as starting materials, for alkenes, **1**, 716
synthesis of, **1**, 611; **2**, 731
Telluroxides, alkyl, synthesis of, **2**, 286
Telluroxides, alkyl aryl, from, alkyl tellurides, **2**, 287
Telluroxides, alkynyl, stability of, **2**, 1038
Telluroxides, allylic, [2,3] sigmatropic rearrangement of, **1**, 611
Telluroxides, aryl, from, aryltellurium dihalides, **2**, 731
Telluroxides, cyclohexyl phenyl, stability of, **2**, 287
Telluroxides, 4-methoxyphenyl, as starting materials, for terminal alkenes, **1**, 611
α-Telluryl carbanions, synthesis of, **4**, 398
Telomerization, of alkenes, **6**, 5
Telomers, synthesis of, **6**, 4
Terephthalaldehyde
as starting material
for aminals, **4**, 406
for bishemithioaminals, **4**, 317
Terephthalic acid, from, *p*-methylacetophenone, **5**, 114
Ter Meer reaction, with chloronitromethane, **6**, 55
Terpene hydrogen donators, for the reduction of nitriles, **1**, 47
Terpenes
from, ketenes, **3**, 528
as sources of chiral auxiliaries, **1**, 134
Terpenoids
reactions, with sodium nitrite, **1**, 1063
as starting materials, for alkyl metals, **2**, 56
Terpenoid systems, Wagner–Meerwein rearrangements, **1**, 380
Terphenylene, synthesis of, **1**, 448
Terpinolines, synthesis of, **1**, 683
2,2':5',2"-Terthiophene, carbonation of, **5**, 82
Tertiary iodides, reactions, with zinc, **4**, 730
Tetraalkyl compounds, synthesis of, **4**, 660
Tetraamines
reactions, with phenylethyne, **6**, 139
as starting materials, for ketene aminals, **4**, 979
synthesis of, **6**, 360
See also Methane, tetrakis(dimethylamino)-
Tetraarsaadamantanes, synthesis of, **4**, 593
Tetraarsacubanes, synthesis of, **5**, 1160; **6**, 155
Tetraarsacyclohexanes, from, bis(methylchloroarsino)methane, **4**, 594
2,4,6,8-Tetraazabicyclo[3.3.0]octane, 3,7-dioxo-, synthesis of, **6**, 508
1,4,7,10-Tetraazacyclododecane, reactions, with ethyl orthocarbonate, **6**, 365
1,2,4,5-Tetraaza-3,6-diborane, 3,6-diphenyl-, reactions, with acetic anhydride, **5**, 383
Tetraazafluorenes, perhydro-, synthesis of, **4**, 409
Tetraazamethanes
from
bis(difluoroamino)fluoroaminomethyl isocyanate, **6**, 363
pentafluoroguanidine, **6**, 362
synthesis of, **6**, 362
2,4,6,8-Tetraaza-1,3,5,7-tetraarsaadamantanes, tetra-*N*-alkyl-, synthesis of, **2**, 503
Tetraazathiapentalenes, from, bishydrazones, **3**, 378
2,3,5,6-Tetraborabicyclo[2.1.1]hexane, synthesis of, **6**, 183
Tetracarbonyl complexes, alkylidyne(halo)-, synthesis of, **3**, 691
Tetracoordinated compounds, synthesis of, **5**, 1092
29,29,30,30-Tetracyanobionthraquinodimethane (TBAQ), synthesis of, **3**, 668
Tetracyano compounds, synthesis of, **4**, 850
1-Tetradecene
from, phenylmanganese chloride, **1**, 125
synthesis of, **1**, 125
Tetradec-7-yne, reduction of, **1**, 507
Tetra(dialkylamide), reactions, with alkyl metals, **2**, 566
1,2,4,6-Tetraenes
from, trimethylsilyldienynes, **1**, 963
synthesis of, **1**, 967
Tetraester, isopropyl-, as starting material, for bisphosphonates, **6**, 263
Tetraesters, synthesis of, via hypohalite method, **6**, 263
Tetraesters, difluoro-, synthesis of, **6**, 263
Tetrafluoroborate, triethyloxonium, reactions, with 3-methyl-1,3-benzothiazol-2-one, **6**, 317
Tetrafluoroborates, carbenium, reactions, with sodium phosphides, **6**, 686
Tetragallium tetrahedrane, synthesis of, **6**, 395
Tetrahalides, reactions, with antimony pentafluoride, **6**, 726
Tetrahedral intermediates, stability of, **3**, 284
Tetraheterafulvalenes, synthesis of, **4**, 868
Tetrahydro-2,2'-bipyridyl, *N*,*N*-bis(trimethylsilyl)-, synthesis of, **2**, 814
Tetrahydro-4,4'-bipyridyl, *N*,*N*-bis(trimethylsilyl)-, synthesis of, **2**, 814

9-(Tetrahydropyran-2-yl) compounds, synthesis of, **4**, 306
Tetrahydroquinazoline, 2-hydroxy-, synthesis of, **6**, 110
Tetrahydroterrein, as starting material, for
2-carboxyhexanal, **5**, 73
Tetrakis(acetodimolybdenum), synthesis of, **1**, 437
Tetrakis(phosphines), synthesis of, **4**, 546
Tetralin, synthesis of, **1**, 94
Tetralin, hydroxy-, reduction of, **1**, 99
Tetralin, 1-methoxy-, reduction of, **1**, 98
Tetralin, 2-methoxy-, reduction of, **1**, 98
Tetralin-2,3-dicarboxylic anhydride, 6,7-dimethyloxy-,
reactions, with sodium hydroxide, **5**, 37
Tetralone, 6-methoxy-, reactions, with trimethylsilyl
cyanide, **3**, 667
1-Tetralone, as starting material, for 1-bromo-
3,4-dihydronapthalene, **2**, 624
2-Tetralone, acidity of, **1**, 877
β-Tetralone, reactions, with
trimethylsilylmethyllithium, **1**, 759
Tetralones, synthesis of, **1**, 476
α-Tetralones, reactions, with dibenzyl
azodicarboxylate, **2**, 310
Tetra(methylthio) compounds, synthesis of, **4**, 842
Tetramine, hexamethylene- (HMT)
reactions
with (*Z*)-1,4-dichloro-2-butene, **2**, 325
with halides, **2**, 308
with primary halides, **2**, 302
with thiophenols, **4**, 317
as starting material, for tris(chloromethyl)amine, **4**, 307
synthesis of, **4**, 294, 407
for the synthesis of *N,N'*-methylenebisamides, **4**, 424
Tetranuclear carbon-bridged, rhodiacyclopentadiene
complexes, from, hex-3-yne, **4**, 689
Tetraorganoborates
from, Lewis base adducts, **2**, 1087
synthesis of, **2**, 1086
Tetraorganodiarsines
reactions
with hexafluorobut-2-yne, **2**, 876
with phenylethyne, **2**, 876
Tetraorganodiarsines, unsymmetrical, from,
diorganoarsines, **2**, 893
Tetraorgano metal(V) halides, reduction of, with
hydrides, **2**, 879
Tetraorganophosphonium salts
reactions
with alkyllithiums, **2**, 830
with aryllithiums, **2**, 830
Tetraorganophosphoranes, alkoxy-, synthesis of, **2**, 830
Tetraorganosilanes, synthesis of, **2**, 536
Tetraorganostannanes, reactions, with aryl halides, **1**, 471
Tetraorganostibonium salts, synthesis of, **2**, 876
Tetraoxaspirononanes, reactions, with
benzoxaphospholes, **4**, 478
1,1,3,3-Tetraoxide, 1,3-dithiane-, synthesis of, **4**, 277
Tetraphosphabicyclopentanes, synthesis of, **2**, 472
Tetraphosphabishomoprismane, synthesis of, **5**, 885
Tetraphosphacubanes, synthesis of, **5**, 1152; **6**, 153
Tetraphosphaspirosilanes, synthesis of, **2**, 473
Tetraphosphetanes, synthesis of, **2**, 472
Tetraphosphine ligands, synthesis of, **2**, 439
Tetraphosphinostannanes, synthesis of, **2**, 474
Tetraphospholanes, synthesis of, **5**, 881
Tetraquinane diterpenoids, (+)-Crinipellin B, synthesis
of, **1**, 452
Tetraselanes, amino-, synthesis of, **2**, 369
Tetraselenafulvalenes
from
1,3-diselenole-2-selenones, **4**, 874
1,3-diselenole-2-thiones, **4**, 874
synthesis of, **4**, 874
via triselenocarbonates, **6**, 590
Tetraselenafulvalenes, (methoxycarbonyl)-, synthesis
of, **4**, 874
Tetraselenium bis(hexafluoroarsentate(V)), reactions, with
tetrafluoroethylene, **6**, 49
Tetraselenoorthocarbonate, tetrakis(trifluoromethyl),
synthesis of, **6**, 307
Tetraselenoorthocarbonate, tetraphenyl
reactions, with *n*-butyllithium, **6**, 93
synthesis of, **6**, 307
Tetraselenoorthocarbonates, as starting materials, for
triselenoorthoformates, **6**, 93, 94
Tetraselenoorthocarbonic acid, substituted compounds,
synthesis of, **6**, 296
Tetraselenoorthocarbonic esters, synthesis of, **6**, 307
Tetrasilanes, as starting materials, for
tetrasilylmethanes, **6**, 384
Tetrasilylation
reactions, **6**, 378
of tetrachloromethane, **6**, 379
Tetrastannacyclohexane, synthesis of, **4**, 700
Tetrasulfanes, desulfurization of, **6**, 349
Tetrasulfide, diethyl, reactions, with alkenes, **2**, 137
Tetrasulfides, synthesis of, **2**, 237; **4**, 866
Tetrasulfides, aryl, synthesis of, **2**, 723
Tetrasulfides, diaryl, synthesis of, **2**, 723
Tetrasulfide tetroxides, synthesis of, **2**, 725
Tetrasulfones
reactions, with amines, **4**, 866
synthesis of, **4**, 277
Tetrasulfurtetranide, reactions, with bishydrazoles, **2**, 794
Tetratellurafulvalenes, synthesis of, **4**, 876
Tetratelluranaphthalenes, synthesis of, **4**, 877
Tetratelluroorthocarbonates
structure of, **6**, 296
synthesis of, **6**, 308
Tetratelluroorthocarbonates, perfluoroalkyl, synthesis
of, **6**, 308
Tetrathiafulvalene, bis(ethylenedithio)-, analogues of, **6**, 698
Tetrathiafulvalenes
from, dithio-substituted selenocarbonates, **6**, 589
synthesis of, **4**, 850, 851
telluration of, **4**, 873
Tetrathiafulvalenes, tetra(alkyltelluro)-, synthesis of, **4**, 873
Tetrathiafulvalenes, tetra(methylthio)-, synthesis of, **4**, 846
Tetrathiafulvenes, synthesis of, **1**, 570
Tetrathianes, reactions, with sodium, **2**, 130
1,4,6,9-Tetrathiaspiro[4.4]nona-2,7-diene, 2,3,7,8-
tetrakis(trifluoromethyl)-, from, 1,1,1,3,3,3-
hexafluoro-2-butyne, **6**, 306
1,4,6,9-Tetrathiaspiro[4.4]nonane, synthesis of, **6**, 305
1,4,6,9-Tetrathiaspiro[4.4]nonanes, synthesis of, **6**, 306
Tetrathiepane, synthesis of, **6**, 306
Tetrathiobisamines, cyclic, synthesis of, **2**, 361
Tetrathiocarbonates, as starting materials, for trithioortho
esters, **6**, 88
Tetrathiofulvalenes, synthesis of, **4**, 847
Tetrathiomalonate, diethyl, reactions, with
N,N-dimethylchloromethaniminium chloride, **4**, 775
Tetrathiomalonic acid esters, synthesis of, **5**, 555
Tetrathiomalonic acid trianion, methylation of, **5**, 559
Tetrathiometallates, use in the synthesis of disulfides, **2**, 228
Tetrathiomolybdate, benzyltriethylammonium
reactions, with acyl halides, **5**, 252
as sulfur transfer reagent, **2**, 228
Tetrathioorthocarbonate, tetra-*t*-butyl, synthesis of,
problems, **6**, 301
Tetrathioorthocarbonate, tetraethyl, from,
bis(ethanethio)methane, **6**, 86
Tetrathioorthocarbonate, tetramethyl, *S*-iodination
of, **6**, 306
Tetrathioorthocarbonate, tetraphenyl, reactions, with
n-butyllithium, **6**, 87
Tetrathioorthocarbonate *S,S'*-dioxide, tetraphenyl,
synthesis of, **6**, 307

Tetrathioorthocarbonates
from
2-azido-1,3-dithiole, **6**, 305
cyclic trithiocarbonates, **6**, 305
2-(methylthio)-1,3-dithiolium cations, **6**, 301
sodium 1,2-cyclohexenedithiolate, **6**, 306
tetrachloromethane, **6**, 301
trichloromethanesulfenyl chloride, **6**, 301
tris(methylthio)carbenium cations, **6**, 301
trithiocarbenium salts, **6**, 306
octaoxides of, problems in the synthesis of, **6**, 307
as starting materials, for trithioorthoformates, **6**, 87
synthesis of, **6**, 306
transesterification of, with thiols, **6**, 306
Tetrathioorthocarbonic acid, substituted compounds, synthesis of, **6**, 296
Tetrathioorthocarbonic acid esters, synthesis of, **6**, 301
Tetrathioorthoformates, cleavage of, with butyllithium, **6**, 317
Tetrathiooxalate, synthesis of, **5**, 554
Tetrathiophthalic acid esters
from, imido esters, **5**, 561
synthesis of, **5**, 561
Tetrathiosuccimic acid esters, from, bis(methylthio) divinyl disulfides, **5**, 556
Tetrathiotungstate, piperidinium, use in the synthesis of disulfides, **2**, 228
Tetraynes, synthesis of, **1**, 1069
Tetrazene, *N*-nitroso-, synthesis of, **2**, 389
Tetrazenes
from
hydrazines, **2**, 388
sulfonylhydrazides, **2**, 388
as starting materials, for hydrazines, **2**, 383
synthesis of, **2**, 388
1-Tetrazenes, synthesis of, **2**, 777
Tetrazenium salts, synthesis of, **2**, 388
s-Tetrazine, 1,4-dimethyl-1,4-dihydro-, quaternary salts of, ring-opening of, **5**, 855
1,2,4,5-Tetrazine-3,6-dicarboxylic acid, dihydro-, from, ethyldiazoacetate, **5**, 93
1,2,4,5-Tetrazine dimers, synthesis of, **3**, 680
Tetrazines
as byproducts, **5**, 848
reactions, with ethyl thionoformate, **4**, 229
synthesis of, **5**, 642
Tetrazines, acyl
from, diazonium salts, **5**, 365
synthesis of, via oxidation of acylhydrazines, **5**, 365
Tetrazines, dehydro-, from, arylthiocarboxanilides, **5**, 852
Tetrazines, 1,4-diacyl-, synthesis of, **5**, 365
Tetrazines, dihydro-, as byproducts, **5**, 848
Tetrazines, 1,4-dihydro-, synthesis of, **5**, 846
Tetrazines, tetrahydro-, synthesis of, **4**, 437
1,2,4,5-Tetrazines, 1,2-dihydro-, synthesis of, **5**, 839
1,2,4,5-Tetrazines, tetrahydro-, synthesis of, **4**, 436
s-Tetrazines, synthesis of, **5**, 849
Tetrazole, 1-phenyl-5-chloro-, reactions, with phenol, **1**, 432
Tetrazoles
reactions
with phosphorus oxychloride, **5**, 774
with Vilsmeier reagents, **5**, 363
as starting materials
for acylhydrazones, **5**, 363
for alkynes, **1**, 1078
for nitrile imides, **3**, 681
for nitrilimine, **3**, 452
thermolysis of, **5**, 358
Tetrazoles, *N*-alkyl-, lithiation of, **4**, 539
Tetrazoles, 5-alkyl(aryl)amino-, as starting materials, for isocyanides, **3**, 699
Tetrazoles, 5-aryl amino-, as starting materials, for aryl isocyanides, **3**, 723
Tetrazoles, arylidene hydrazino-, bromination of, **3**, 710
Tetrazoles, 2,5-diphenyl-, photolysis of, **5**, 707
Tetrazolin-5-ones, thermal cleavage of, **5**, 1014
Tetrazolium salts
reactions, with TEA, **5**, 1073
reduction of, **5**, 855
Tetrazolothiones, use in the synthesis of carbodiimides, **5**, 1073
Tetrazolylcephalosporins, synthesis of, **5**, 193
Tetrolate, ethyl, from, propyne, **5**, 103
Tetrolic acid, synthesis of, **5**, 100
Tetrolic acid, chloro-, synthesis of, **5**, 59
Tetroxides, synthesis of, **2**, 98
Thallium
halometallation of, **2**, 982
oxymetallation of, **2**, 982
Thallium, (chloromethyl)chloro(phenyl)-, synthesis of, **4**, 153
Thallium, dichloro(phenyl)-, reactions, with diazomethane, **4**, 153
Thallium, triethyl-, reactions, with dinitromethanes, **6**, 170
Thallium, trimethyl-, reactions, with diazomethane, **3**, 485
Thallium(I) alkoxides, reactions, with carbon disulfide, **6**, 300
Thallium(I) bromide, reactions, with Grignard reagents, **2**, 597
Thallium(I) ethoxide, reactions, with cyanogen chloride, **5**, 1103
Thallium(I) salts, reactions, with bromine, **4**, 27
Thallium(I) thiolates, as starting materials, for thiocarboxylic esters, **5**, 234
Thallium(III) acetate
reactions
with alkenes, **5**, 1115
with ammonium methylpentafluorosilicate, **5**, 1143
as reagent
for aldehyde regeneration from hydrazones, **3**, 18
for hydrazone cleavage, **3**, 134
synthesis of, **5**, 229
Thallium(III) acetate, cyanomethyl-, synthesis of, **5**, 1143
Thallium(III) chloride, reactions, with *N,N,N'*-trimethylsilylbenzamidine, **5**, 780
Thallium(III) chloride, diphenyl-, reactions, with mercuric cyanide, **5**, 1143
Thallium(III) compounds, aryl-
deuteriodethallation of, **1**, 452
ipso substitution of, **1**, 452
Thallium(III) compounds, vinyl-, protiodethallation of, **1**, 453
Thallium(III) cyanides
reactions, with chlorotrialkylsilane, **5**, 1135
synthesis of, **5**, 1143
Thallium(III) halides, as starting materials, for dialkylthallium halides, **2**, 597
Thallium(III) nitrate
reactions
with alkenes, **4**, 194
with methyl 3,5-dimethoxybenzoate, **3**, 433
with pyrazolones, **1**, 1076
Thallium(III) oxide, reactions, with acetic anhydride, **5**, 229
Thallium(III) perchlorates, bis(pentachlorophenyl)-, reactions, with potassium cyanide, **5**, 1143
Thallium(III) salts, reactions, with alkenes, **2**, 598
Thallium(III) thiocarboxylates, synthesis of, **5**, 251
Thallium acetate
for the oxidation of thiocarboxylic acids, **5**, 251
reactions, with alkynes, **4**, 1068
Thallium acetates, dialkyl-, reactions, with mercury salts, **2**, 598
Thallium acetylacetonate, as starting material, for trinuclear carbon-bridged palladium complexes, **4**, 693
Thallium alkoxides, as starting materials, for alkyl cyanates, **5**, 1103

Thallium bromide, dialkyl-, from, thallium tribromide, **4**, 659
Thallium carboxylates, use in the dihydroxylation of alkenes, **2**, 45
Thallium carboxylates, monoalkyl-, synthesis of, **2**, 598
Thallium chloride, reactions, with tris(trimethylsilylmethyl)lithium, **6**, 395
Thallium chloride, bis(trimethylsilylmethyl)-, synthesis of, **4**, 659
Thallium compounds
  reactions
    with diisopropylchlorophosphine, **5**, 400
    with phosphorus ylides, **4**, 587
Thallium compounds, β-alkoxy alkyl-, reactions, with trimethyl phosphite, **2**, 598
Thallium compounds, alkyl-
  synthesis of, **2**, 597
    via transmetallation, **2**, 597
Thallium compounds, (chloromethyl)-, synthesis of, **4**, 153
Thallium compounds, dialkyl-
  reactions
    with alkyllithiums, **2**, 598
    with Grignard reagents, **2**, 598
Thallium compounds, (halomethyl)-, synthesis of, **4**, 153
Thallium compounds, imidoyl-, synthesis of, **5**, 803
Thallium compounds, trialkyl-
  from, thallium salts, **2**, 598
  stability of, **2**, 597
  synthesis of, **2**, 597
Thallium compounds, trihalomethyl-, synthesis of, **6**, 245
Thallium cyanides, alkyl-, synthesis of, **5**, 1143
Thallium dichlorides, alkyl-, stability of, **2**, 597
Thallium ethoxide
  reactions, with dithioester *S*-oxides, **4**, 859
  use in the synthesis of ethers, **2**, 89
Thallium halides, dialkyl-
  from, thallium(III) halides, **2**, 597
  stability of, **2**, 597
Thallium hydroxide, as promoter for cross-coupling, **1**, 489
Thallium 2-methylpropanoate, reactions, with 1,1-dimethylpropanoyl chloride, **5**, 183
Thallium salts
  as additives, in Heck reactions, **1**, 494
  for the oxidation of flavanones, **1**, 556
  reactions
    with metal cyanides, **5**, 1143
    with trialkyltin cyanides, **5**, 1143
  as starting materials, for trialkylthalliums, **2**, 598
Thallium sulfate, reactions, with alkenes, **2**, 40
Thallium thioacetate, use in the synthesis of isothiocyanates, **5**, 1032
Thallium triacetate, for the synthesis of aryl bromides, **2**, 626
Thallium tribromide, as starting material, for dialkylthallium bromide, **4**, 659
Thallium trichloride, reactions, with trimethylsilylmethyllithium, **4**, 659
Thallium trinitrate trihydrate, as reagent, for dithiane hydrolysis, **5**, 410
Thallium tris(trifluoroacetate), use in the synthesis of benzonitriles, **3**, 664
Thapsane sesquiterpenes, synthesis of, **1**, 828
Thebaine, reactions, with thioaldehydes, **2**, 140
3-Thenaldehyde, reactions, with silver oxide, **5**, 83
2-Thenoic acid, synthesis of, **5**, 83
3-Thenoic acid, synthesis of, **5**, 83
Thenoyl chloride, reactions, with trimethylsilyl azide, **5**, 992
Thermal decarboxylation, for the synthesis of alkynylstannanes, **2**, 1100
Thermal decomposition
  of acetyl chloride, **6**, 215
  of methyl bromodifluoroacetate, **6**, 227
  of methyl chlorodifluoroacetate, **6**, 227
  of phosphonium salts, **2**, 442
  of trifluoroacetyl bromide, **6**, 218
  of trifluoroacetyl chloride, **6**, 219
  of trifluoroacetyl fluoride, **6**, 219
Thermal reaction
  for the substitution of selenium and tellurium, **1**, 243
  for the substitution of sulfur functions, **1**, 231
  for the synthesis of disulfides, **2**, 228
Thermal rearrangement
  of hydrogen, **1**, 794
  of phenolic esters, **3**, 297
  of vinylcyclopropanes, **1**, 801
Thermal Wolff rearrangement, for the synthesis of resorcinols, **2**, 668
Thermolysis
  of oxazaphospholimines, **5**, 789
  of sulfoxides, **2**, 185
  for the synthesis of radicals, **1**, 924, 935
  of ureas, **5**, 982
Thexylbromoborane–DMS complexes, for the synthesis of halobenzaldehydes, **3**, 94
Thexylchloroborane–dimethyl sulfide complex, as reagent, for the reduction of carboxylic acids to aldehydes, **3**, 13
Thexylhaloborane–dimethyl sulfide complexes, use in the synthesis of aryl aldehydes, **3**, 82
Thexyliodoborane–dimethyl sulfide complexes, hydroboration of, **3**, 3
$\lambda^6\sigma^4$-Thiaalkynes, synthesis of, **3**, 689
1-Thia-3-azabutadienes, synthesis of, **3**, 414
Thiabenzene *S*-oxides
  structure of, **3**, 379
  synthesis of, **3**, 379
Thiabenzenes
  from, thiopyrylium salts, **3**, 375
  synthesis of, **3**, 375
    via addition to thiopyrylium salts, **3**, 375
    via deprotonation of thiinium salts, **3**, 376
Thiacepham compounds, oxidation of, **4**, 274
Thiacrown compounds, synthesis of, **2**, 133
Thiacumulenes, synthesis of, **5**, 481
Thiacyclobutadiene, synthesis of, **3**, 376
1-Thiacyclohept-4-yne, 3,3,6,6-tetramethyl-, use in the synthesis of diiron complexes, **4**, 681
Thiacycloheptynes, reactions, with 7-germabenzonorbornadienes, **2**, 947
Thiacyclopentane, bromination of, **4**, 65
Thiacyclopentane, α-bromo-, synthesis of, **4**, 65
Thiacyclopentane, 2,3-dichloro-, synthesis of, **4**, 65
1,3,5-Thiadiazepine, 2-(2,6-dimethylphenyl)-6-oxo-2-thioxohexahydro-, synthesis of, **6**, 565
Thiadiazine, synthesis of, **4**, 229
Thiadiazine dioxides, synthesis of, **6**, 636
4*H*-1,3,5-Thiadiazine-4-one, 2,3-dihydro-6-methoxy-2-thioxo-, reactions, with dibenzylamine, **6**, 559
1,3,5-Thiadiazines, synthesis of, **6**, 575
1,3,4-Thiadiazole, dihydro-, synthesis of, **6**, 128
1,2,5-Thiadiazole-1,1-dioxides, pyrolysis of, **3**, 668
Thiadiazoles
  flash photolysis of, **2**, 792
  from, diazocyanoacetamide, **5**, 571
  pyrolysis of, **5**, 480
  reactions, with potassium ethoxide, **5**, 480
  synthesis of, **5**, 483, 641; **6**, 316, 627
Thiadiazoles, 2-cyanamido-, synthesis of, **5**, 1120
1,2,3-Thiadiazoles
  as starting materials
    for alkynes, **1**, 1077
    for thioketenes, **3**, 551; **5**, 511
  synthesis of, **4**, 852
1,2,4-Thiadiazoles, synthesis of, **3**, 686; **5**, 735
1,2,4-Thiadiazoles, 4,5-dihydro-
  from, imines, **3**, 686
  as starting materials, for nitrile sulfides, **3**, 686

1,2,4-Thiadiazoles, 3,5-disubstituted, synthesis of, **5**, 635
1,3,4-Thiadiazoles, synthesis of, **5**, 846; **6**, 114, 128, 535
1,3,4-Thiadiazoles, 2-alkylidene-2,5-dihydro-, synthesis of, **4**, 328
1,3,4-Thiadiazole-2(3*H*)-3-thiones, synthesis of, **6**, 564
1,3,4-Thiadiazolidines
stability of, **1**, 686
as starting materials, for 1,3,4-thiadiazones, **1**, 685
synthesis of, **4**, 318
Thiadiazolinediones, thermolysis of, **5**, 996
Thiadiazolines
synthesis of, **2**, 243
for the synthesis of acylhydrazones, **5**, 364
thermolysis of, **3**, 570
Thiadiazolines, imino-, use in the synthesis of carbodiimides, **5**, 1073
1,3,4-Thiadiazolines
stability of, **1**, 686
as starting materials
for alkenes, **1**, 684
for amide monothioacetals, **6**, 111
1,3,4-Thiadiazolines, 2-alkylidene-, as starting materials, for thioketenes, **3**, 551
1,2,4-Thiadiazolium salts, reactions, with sodium hydroxide, **5**, 759
1,3,4-Thiadiazones, from, 1,3,4-thiadiazolidines, **1**, 685
Thiadiphosphetanes, synthesis of, **4**, 559
Thiaenolization, of aldehydes, **2**, 709
Thiagermatanes, from, bis(iodomethyl)germane, **4**, 372
1,3-Thiagermatanes, 3,3-dialkyl-, synthesis of, **4**, 372
Thiagermetanes, pyrolysis of, **4**, 627
2-Thianaphthalenes, synthesis of, **3**, 376
Thiane, reactions, with hydroxylamine-*O*-sulfonic acid, **2**, 161
Thiaphosphetanes, as starting materials, for thioketenes, **3**, 550
Thiaphosphiranes, synthesis of, **6**, 350
1,2-Thiaphosphol-3-en-5-one 2-sulfide, synthesis of, **5**, 534
Thiaphosphonium salts, synthesis of, **2**, 254
1,2-Thiaphosphorin 2-sulfide, 4,5-dimethyl-2-phenyl-3,6-dihydro-, synthesis of, **2**, 851
2-Thiapropane, 1-chloro-, as starting material, for 2-thia-4-selenapentane, **4**, 283
Thiapyrans, synthesis of, **3**, 72
Thiapyrones, synthesis of, **3**, 251
Thiapyrones, dihydro-, synthesis of, **3**, 251
Thiapyrones, *trans*-dihydro-, synthesis of, **3**, 252
Thiapyrylium iodide, reactions, with methylenes, **3**, 255
1-Thia-2-selenacyclopentane-4-carboxylic acids, synthesis of, **2**, 237, 238
Thiaselenafulvenes
from
phosphonate esters, **4**, 870
1,3-thiaselenole-2-selenones, **4**, 870
1,3-thiaselenole-2-thiones, **4**, 870
1,3-thiaselenolium salts, **4**, 870
triphenylphosphonium salts, **4**, 870
synthesis of, **4**, 870
2-Thia-4-selenapentane, from, 1-chloro-2-thiapropane, **4**, 283
2-Thia-3-selenaspiro-[4.5]-decane, synthesis of, **2**, 238
Thiaselenazines, as starting materials, for thiaselenoliminium salts, **6**, 630
Thiaselenoliminium salts, from, thiaselenazines, **6**, 630
1,3-Thiaselenolium salts, as starting materials, for thiaselenafulvenes, **4**, 870
2-Thiaspiro-[3.5]-nonane, reactions, with selenium, **2**, 237
3-Thiatanones, synthesis of, **3**, 249
1,4-Thiatellurins, reactions, with butyllithium, **1**, 441
Thiation, for the synthesis of α-silyl thioketones, **5**, 647
Thiatriazine *S*-oxides, synthesis of, **6**, 633
Thiatriazole, 5-propargyloxy-, as starting material, for allenyl isocyanates, **5**, 1101
1,2,3,4-Thiatriazole, 5-mercapto-, alkylation of, **5**, 1046
Thiatriazoles
as starting materials
for alkyl cyanates, **5**, 1101
for cyanates, **5**, 1101
for nitrile sulfides, **3**, 687
synthesis of, **5**, 634; **6**, 114
thermolysis of, **5**, 1101
Thiatriazoles, α-diazo-, fragmentation of, **3**, 477
1,2,3,4-Thiatriazoles, thermal decomposition of, **5**, 1013
1,2,3,4-Thiatriazoles, 5-alkoxy-, synthesis of, **5**, 1101
1,2,3,4-Thiatriazoles, 5-amino-, as intermediates, in the synthesis of isothiocyanates, **5**, 1046
1,2,3,4-Thiatriazoles, 5-aryloxy-, as starting materials, for aryl cyanates, **5**, 1101
1,2,3,4-Thiatriazoline-5-thiones, cleavage of, for the synthesis of isothiocyanates, **5**, 1046
Thiazaphospholes
as starting materials, for nitrile ylides, **3**, 680
synthesis of, via cycloaddition reactions, **4**, 493
1,2,4-Thiazaphospholes, from, phosphaalkynes, **3**, 686
1,3,4-Thiazaphospholes, synthesis of, **5**, 635
Thiazepines, synthesis of, **4**, 325
2*H*-1,3-Thiazetes, thermal equilibrium of, **5**, 633
1,3-Thiazetidines, synthesis of, **4**, 905; **5**, 604
1,2-Thiazetidines 1-oxides, 3-alkoxy-, synthesis of, **4**, 312
Thiazine, dihydro-, reactions, with chlorotrimethylsilane, **4**, 370
4*H*-1,3-Thiazine, 2-imino-5,6-dihydro-, reactions, with diethyl phosphoryl chloride, **5**, 1112
Thiazine-2-ones, thiocarbazoylation of, **6**, 540
6*H*-1,3-Thiazine-6-ones, synthesis of, **3**, 414
2*H*-1,2,-Thiazine-1-oxides, 3,6-dihydro-2-(*p*-toluenesulfonyl)-, synthesis of, via Diels–Alder reactions, **2**, 366
Thiazines, from, β-thiol amides, **4**, 325
1,3-Thiazines, 2-alkyl, as starting materials, for thioketones, **3**, 350
4*H*-1,3-Thiazines, 2-substituted 5,6-dihydro-, reduction of, **4**, 321
1,3-Thiazine-2-thione, *N*-(ethoxyacetyl)tetrahydro-, photo-rearrangement of, **6**, 114
1,3-Thiazine-2-thiones
from, β-lactones, **6**, 563
synthesis of, **6**, 565
1,3-Thiazin-4-ones, 2-thioxo-, synthesis of, **6**, 565
1,4-Thiazin-3-ones, as starting materials, for α-benzoyloxy compounds, **4**, 226
Thiazirenes, synthesis of, **3**, 687
Thiaziridine-3-thiones, decomposition of, **5**, 1032
Thiazole
metallation of, **5**, 595
reactions, with carbonyls, **3**, 42
Thiazole, 2-amino-, as starting material, for dithiocarbamates, **6**, 564
Thiazole, 2-chloro-4-(trifluoromethyl)-5-isocyanato-, synthesis of, **5**, 990
Thiazole, 2-cyano-, use in the synthesis of hydrazonamides, **5**, 847
Thiazole, 2,5-dichloro-4-isocyanato-, synthesis of, **5**, 990
Thiazole, 4,5-dihydro-2-methyl-, as starting material, for *N*-acylthioethyl-*N*-cyanoacetamides, **5**, 1123
Thiazole, 2-lithio-
reactions
with aldehydes, **3**, 29
with ketones, **3**, 29
Thiazole, 2-trimethylsilyl-
reactions, with aldehydes, **3**, 29
synthesis of, **6**, 637
4-Thiazolecarboxamidine, *N*-chloro-*N'*-phenyl-, synthesis of, **5**, 816
4-Thiazolecarboxamidine, *N*-chloro-*N'*-(3-pyridyl)-, synthesis of, **5**, 816

Thiazoles
as by-products, **2**, 121
reactions, with lithium, **3**, 109
as starting materials, for ketones, **3**, 311
synthesis of, **6**, 637
use in the synthesis of α-hydroxy ketones, **3**, 157
Thiazoles, dihydro-
reactions, with ketene–sulfur dioxide adducts, **6**, 113
synthesis of, **6**, 629
Thiazoles, 2,3-dihydro-, synthesis of, **4**, 318
Thiazoles, 2,5-dihydro-, synthesis of, **4**, 317, 328
Thiazoles, 4,5-dihydro-, reduction of, **4**, 321
Thiazoles, formyl-, synthesis of, **3**, 109
Thiazoles, 2-formyl-, synthesis of, **3**, 109
Thiazoles, 2-hydroxy-, as starting materials, for ketones, **3**, 311
Thiazoles, imidazo-, from, 2-thiohydantoins, **6**, 575
Thiazoles, trimethylsilyldihydro-, synthesis of, **6**, 637
Thiazoles, trimethylstannyldihydro-, synthesis of, **6**, 637
Thiazolidindiones, as starting materials, for thioketones, **3**, 350
Thiazolidine-4-carboxylic acid, L-2-oxo-, synthesis of, **5**, 81
Thiazolidine-2-ones, thiocarbazoylation of, **6**, 540
Thiazolidines
from, aldehydes, **4**, 318
oxidation of, **4**, 226
reductive cleavage of, **2**, 265
Thiazolidines, 3-alkyl-2-(*N*-cyanoimino)-, reactions, with sodium alkoxides, **5**, 1120
Thiazolidines, 4-oxo-, synthesis of, **4**, 319
1,3-Thiazolidine-2-thione, reactions, with 5-chloro-2-nitroaniline 6: 295–31830
Thiazolidine-2-thione, reactions, with thiocarbazic acid chlorides, **6**, 566
Thiazolidine-2-thione, 4,4'-bis(3-alkyl-4-hydroxy-, synthesis of, **6**, 564
1,3-Thiazolidine-2-thiones, from, 1-iodo-2-isothiocyanatocycloalkanes sodium sulfide, **6**, 565
Thiazolidine-2-thiones
reactions, with alkoxyamines, **5**, 329
synthesis of, **6**, 567
Thiazolidine-2-thiones, 3-acyl-, reduction of, **3**, 85
Thiazolidine-2-thiones, 5-hydroxy-, cleavage of, for the synthesis of 2-keto isothiocyanates, **5**, 1046
Thiazolidinones, synthesis of, **6**, 114
Thiazoline, 2-thiol, use in the synthesis of *N*-alkylalkenoamides, **5**, 295
Thiazolinediones, thermolysis of, **5**, 996
Thiazolines, from, isocyanides, **4**, 538
Thiazolines, amino-, synthesis of, **5**, 593
Thiazolines, 2-aryloxy-, ring-opening of, **5**, 1033
Thiazolines, 5-imino-, as starting materials, for ketenimines, **3**, 579
Thiazoline thione esters, use in the synthesis of amides, **5**, 277
Thiazoline thiones
reactions, with ynamines, **5**, 605
synthesis of, **4**, 538
1,3-Thiazoline-2-thiones, synthesis of, **6**, 564
Δ^3-Thiazoline-2-thiones, as starting materials, for vinyl isothiocyanates, **5**, 1046
Thiazolium cations
reactions
with amines, **6**, 114
with azide anions, **6**, 113
Thiazolium perchlorate, 4-aryl-, reactions, with benzylamine, **3**, 339
Thiazolium salts
use in the synthesis of carboxylic esters, **5**, 125
use in the synthesis of α-hydroxy ketones, **3**, 157
Thiazolium salts, 2-acyl-, as acylating agents, for the synthesis of hydroxylamines, **5**, 212
Thiazolo-[2,3-*b*]thiazolium bromide, 3,5-dimethyl-, as starting material, for aminotrithioorthoformates, **6**, 314
Thiazolo-[2,3-*b*]thiazolium bromide, 2,3,5,6-tetrahydro-, as starting material, for bicyclic aminotrithioorthoformates, **6**, 314
Thieno compounds, bis-, synthesis of, **4**, 850
2-Thienohydroximoyl chloride, 5-chloro-, synthesis of, **3**, 683
2-Thienyl, for the synthesis of higher-order cuprates, **2**, 589
2-Thienyl groups, as dummy ligands, **1**, 312
2-Thienyl ligands, in coupling reactions, **1**, 120
Thiepin, alkylation of, **1**, 145
Thiepin-1,1-dioxides, 2,7-dihydro-
elimination of, **1**, 620
pyrolysis of, **1**, 791
Thiepin-*S*,*S*-dioxides, 2,7-dihydro-, alkylation of, **1**, 145
Thietane, difluorooxa-, ring-opening of, **2**, 183
Thietane dioxides
from, methanesulfonylsulfene, **4**, 281
reactions, with *t*-butoxymagnesium bromide, **2**, 200
use in the synthesis of sulfinic acids, **2**, 188
Thietane-1,1-dioxides, 2,4-dialkyl-3-oxo-, synthesis of, **5**, 500
Thietanes
reactions, with diazomalonate, **1**, 233
as starting materials, for dithiolanes, **2**, 224
Thietanes, ethoxy-, reactions, with phenylmagnesium bromide, **4**, 731
3-Thietanol, as starting material, for ethanol, **2**, 636
Thietanone, synthesis of, **3**, 247
Thietanonium salts, reactions, with butyllithium, **1**, 234
Thietes, synthesis of, **1**, 784
2*H*-Thiin, 3,4-dihydro-, synthesis of, **4**, 65
Thiinium salts, as starting materials, for thiabenzenes, **3**, 376
Thiirane, bis(trimethylsilyl)-, synthesis of, **6**, 129
Thiirane-1,1-dioxide, ring-opening of, **2**, 272
Thiirane dioxides, cleavage of, **2**, 188
Thiirane-*S*-oxide
reactions, with alkyl chloromethyl ethers, **2**, 195
as starting material, for alkanesulfenic esters, **2**, 195
Thiiranes
cleavage of, **2**, 265
desulfurization of, **4**, 862
from, aldehydes, **2**, 140
as intermediates, in the synthesis of alkenes, **1**, 685
opening of, **1**, 242
reactions
with alcohols, **2**, 128
with triphenylphosphonium ylides, **2**, 853
ring-opening of, **2**, 129, 222
as starting materials
for chlorothiolformate esters, **6**, 434
for sulfenamides, **2**, 244
for thiophosphates, **2**, 259
synthesis of, **2**, 136; **3**, 32
use in the synthesis of alkanesulfenyl chlorides, **2**, 167
use in the synthesis of alkanethiols, **2**, 125
Thiiranes, 2-alkyl-, reactions, to methyl isothiocyanate, **6**, 314
Thiiranes, methylene-, reactions, with enones, **1**, 242
Thiirane sulfoxides
ring-opening of, **2**, 222
See also Episulfoxides
Thiiren-1,1-dioxides, as intermediates, **1**, 1074
Thiirene dioxides, synthesis of, **1**, 620
Thiirenium ions, as intermediates, **1**, 1112
Thiiriene-1,1-dioxides, reactions, with diazoalkanes, **3**, 479
Thioacetaldehyde
synthesis of
via elimination of dithiolanes, **3**, 333
via photolysis of sulfenyl chlorides, **3**, 337
Thioacetaldehyde, diethyl-, as starting material, for α,α-diethylthionitriles, **3**, 636
Thioacetaldehyde, trifluoro-, synthesis of, via thermolysis of 1,1-dioxodithiolane, **3**, 334

Thioacetals
alcoholysis of, **4**, 196
cleavage of, **2**, 128
reactions
with 2-aminothiophenols, **4**, 899
with cyanogen iodide, **3**, 634
with dichlorocarbene, **4**, 775
with isocyanates, **4**, 426
with trimethylsilyl cyanide, **3**, 634
with trimethylsilylmethylmagnesium chloride, **2**, 936
stability of, **1**, 1031
as starting materials
for acetals, **4**, 196
for alkenylsilanes, **2**, 936
for dialkyl sulfides, **2**, 141
for dienylsilanes, **2**, 936
for ketones, **3**, 131
for sulfinyl chlorides, **2**, 179
for α-thionitriles, **3**, 634
for trienylsilanes, **2**, 936
for vinyl sulfides, **1**, 590
Thioacetals, α-chloro, as starting materials, for amide thioacetals, **6**, 111
Thioacetals, dibenzyl, synthesis of, **2**, 138
Thioacetals, diphenyl, reductive lithiation of, **4**, 366
Thioacetals, ketene
reactions, with lithium naphthalenide, **4**, 956
as starting materials, for 1-alkylthiovinyllithiums, **4**, 956
synthesis of, via hydrolysis of vinyl phosphonium salts, **1**, 446
Thioacetals, ketoketene, reactions, with isobutenylmagnesium chloride, **2**, 663
Thioacetals, silylated, reactions, with thiols, **4**, 247
Thioacetals, α-tributylstannyl, reactions, with trimethylsilyl enol ethers, **4**, 398
Thioacetamide, cyano-
condensation of, **5**, 615
reactions, with ketones, **5**, 615
use in the synthesis of heterocyclic aromatic nitriles, **3**, 669
Thioacetamide, *N*-methyl-, reactions, with dichlorophosphines, **2**, 252
Thioacetamide, *N*-methyl-*N*-phenyl-, reactions, with ylides, **5**, 898
Thioacetamides, phenyl-
phosphorylation of, **5**, 643
synthesis of, **5**, 600
Thioacetanilide
reactions
with 2-allyl-1,3-dicyclohexylisourea, **5**, 726
with triphenyltin iodide, **5**, 739
Thioacetate
reactions, with diethyl iodomethyl phosphonate, **4**, 346
use in the synthesis of α-thiol ketones, **3**, 164
Thioacetate, ethyl α-methyl, reactions, with acid chlorides, **3**, 166
Thioacetate, ethyl trimethylsilyl, as reagent, for the synthesis of thioketones, **3**, 350
Thioacetate, silyl
electron diffraction of, **5**, 539
NMR of, **5**, 539
X-ray analyses of, **5**, 539
Thioacetate, *O*-silyl
reactions, with phosphorus bromide difluoride, **5**, 535
synthesis of, **5**, 538
Thioacetate, *S*-tributylstannyl
reactions, with bromosilane, **5**, 537
use in the synthesis of *O*-difluorophosphino thioacetate, **5**, 535
Thioacetate, *O*-trimethylsilyl, phosphorylation of, **5**, 535
Thioacetates
hydrolysis of, **3**, 34
oxidation of, **2**, 192
pyrolysis of, **1**, 777, 778
Thioacetates, dicyano-, as starting materials, for ketene aminals, **4**, 977
Thioacetates, sugar, reactions, with dimethylarsenic iodide, **2**, 260
Thioacethydrazide, synthesis of, **5**, 641
Thioacetic acid
ab initio calculations of, **5**, 507
geometries of, **5**, 507
IR spectrum of, **5**, 506
phosphorylation of, **5**, 535
reactions
with α-chlorosulfenyl chlorides, **5**, 590
with dichlorodimethyltin, **5**, 541
with diethoxytrimethylantimony, **2**, 502
with sulfonate esters, **2**, 128
as reagent, for the synthesis of thioketones, **3**, 353
as starting material, for thioketenes, **3**, 547
use in the synthesis of thionoamides, **5**, 570, 581
Thioacetic acid, phenyl-, decarboxylation of, **4**, 68
Thioacetic acid, trichloro-
equilibrium of, **5**, 506
from, trichloroacetyl chloride, **5**, 506
IR spectrum of, **5**, 506
synthesis of, **5**, 506
Thioacetimidate hydroiodide, *S*-methyl, use in the synthesis of hydrazonamides, **5**, 862
Thioacetone, hexafluoro-
synthesis of, **3**, 348, 351
use in the synthesis of alkanesulfenyl chlorides, **2**, 170
Thioacetone, methyl-
reactions
with sulfuryl chloride, **4**, 775
with triethyl phosphite, **4**, 775
Thioacetonitrile, methyl-
reactions, with carbonyls, **3**, 654
as starting material, for arylacetonitrile, **3**, 626
Thioacetonitrile, phenyl-, enolate of, **3**, 639
Thioacetonitriles, α-oxygenated-α-phenyl-, synthesis of, **3**, 636
α-Thio-α-acetonitriles, synthesis of, **3**, 634
Thioacetophosphinic acid, fluorochloro-, diethyl ester, synthesis of, **5**, 645
Thioacetyl bromide, trifluoro-, from, *S*-methyl thioethers, **5**, 487
Thioacetyl chloride, from, dithioacetic acid, **5**, 477
Thioacetyl chloride, chlorodifluoro-, from, 1,1-dichloro-2,2-difluoroethene, **5**, 479
Thioacetyl chloride, difluoro-, synthesis of, **5**, 484
Thioacetyl chloride, trichloro-
from, pentachloroethanesulfenyl chloride, **5**, 478
reactions, with morpholine, **5**, 582
Thioacetyl chloride, trifluoro-
reactions, with diethylamine, **5**, 582
synthesis of, **5**, 473, 479
Thioacetyl fluoride, 2,2-bis(trifluoromethyl)-, synthesis of, **5**, 475
Thioacetyl fluoride, chlorodifluoro-
from, 2-chlorotetrafluoroethanesulfenyl chloride, **5**, 476
synthesis of, **5**, 479
Thioacetyl fluoride, dichlorofluoro-, synthesis of, **5**, 476
Thioacetyl fluoride, iododifluoro-, as intermediate, in the synthesis of tetrafluorodithiosuccinoyl difluoride, **5**, 473
Thioacetyl fluoride, trifluoro-, synthesis of, **5**, 474, 476
Thio acids
as reagents, for the synthesis of thioketones, **3**, 365
as starting materials, for alkanethiols, **2**, 130
Thio acids, ω-alkyl-, synthesis of, **5**, 83
Thio acids, ω-aryl-, synthesis of, **5**, 83
Thio acids, oxyimidic
acylation of, **5**, 831
oxidation of, **5**, 830
reduction of, **5**, 830

as starting materials, for oxyimidic thioesters, **5**, 831
Thioacrolein
Diels–Alder dimers of, **3**, 337
synthesis via sulfuration of acrolein, **3**, 337
synthesis of, **3**, 337
via 1,2-elimination, **3**, 337
via photolysis, **3**, 337
via thermolysis, **3**, 337
Thioacrolein dianions, transmetallation of, **4**, 395
Thioacrylamides, 3-amino-, use in the synthesis of ketenimines, **3**, 562
Thioacrylonitriles, α-alkyl-, synthesis of, **3**, 655
Thioacylating reagents, for the synthesis of thiocarboxylates, **5**, 510
Thioacylation, of amines, **5**, 582, 584, 611
Thioacyl bromides
from
S-benzyl thioethers, **5**, 487
bromofluoro *S*-benzyl thioethers, **5**, 481
cyclic trithiocarbonates, **5**, 487
thioethers, **5**, 487
thioketenes, **5**, 487
reactions
with amines, **5**, 582
with potassium fluoride, **5**, 476
stability of, **5**, 486
synthesis of, **5**, 486
Thioacyl bromides, α,β-unsaturated, thioacylation of, **5**, 614
Thioacyl chlorides
from
S-acetyl 1,1-dichloroalkyl disulfides, **5**, 483
active methyl compounds, **5**, 484
acyl chlorides, **5**, 484
alkenes, **5**, 480
S-benzyl thioethers, **5**, 481
diazo ketones, **5**, 482
1,1-dichlorosulfenyl chlorides, **5**, 477
dithiocarboxylic acids, **5**, 476
ketenes, **5**, 480
ketones, **5**, 484
thioketenes, **5**, 480
as intermediates, in the synthesis of thiacumulenes, **5**, 481
reactions
with amines, **5**, 583
with sodium benzeneselenolate, **5**, 563
as starting materials
for thiocarboxylates, **5**, 509
for thioketenes 316030.665
synthesis of, **5**, 476
via dehydrofluorination of thiols, **5**, 484
via sulfurization, **5**, 479
for the thioacylation of hydroxylamines, **5**, 636
use in the thioacylation of thiols, **5**, 550
Thioacyl chlorides, α-chloro-, synthesis of, **5**, 480, 482
Thioacyl chlorides, α-keto-
detection of, **5**, 485
as intermediates, in the synthesis of thionoamides, **5**, 583
Thioacyl fluorides
dimerization of, **5**, 474
from, trifluoroalkenes, **5**, 472
reactions, with diethylamine, **5**, 583
stability of, **5**, 472
synthesis of, **5**, 472
via acidic cleavage of *S*-benzyl thioethers, **5**, 475
via dehydrofluorination of 1,1-difluorothiols, **5**, 474
via pyrolysis of thioesters, **5**, 475
via sulfurization of polyhaloalkenes, **5**, 472
Thioacyl fluorides, perfluoro-, synthesis of, **5**, 474
Thioacyl functions, linked to metalloids or metals, **5**, **647–651**
Thioacyl halides
oxidation of, **5**, 490
reactions, with anthracene, **5**, 492
as starting materials, for thiocarboxylates, **5**, 510
synthesis of, **5**, **471–503**
general methods, **5**, 472
reviews, **5**, 471
use in the synthesis of thionoamides, **5**, 582
Thioacyl iodides, synthesis of, **5**, 488
Thioacyloxy functions, synthesis of, **5**, **505–543**
Thioaldehyde, diselenovinyl, synthesis of, **3**, 339
Thioaldehyde, α-keto, synthesis of, **3**, 333
Thioaldehyde *S*-oxides, aryl, synthesis of, **3**, 369
Thioaldehyde *S*-oxides, vinyl, synthesis of, **3**, 369
Thioaldehydes
desulfurization of, **1**, 24
from
acetals, **3**, 340
aldehyde dimethyl acetals, **3**, 336
aldehydes, **3**, 340, 341
Bunte salts, **3**, 333
dialkyl sulfoxides, **3**, 336
1,2-dithiole-3-ones, **3**, 339
enamines, **3**, 338
hydrazones, **3**, 340
isothiazolium salts, **3**, 339
4-phenyldithiolium cation, **3**, 339
phosphonium ylides, **3**, 336
phthalimides, **3**, 333
α-silyl disulfides, **3**, 333
sulfenyl chlorides, **3**, 337
sulfenyl halides, **3**, 333
α-sulfinyl disulfides, **3**, 333
sulfonium salts, **3**, 333
thiocyanohydrins, **3**, 337
Vilsmeier salts, **3**, 341
in situ trapping of, **3**, 331, 334
as intermediates, **1**, 833
in the synthesis of thionoamides, **5**, 603
intramolecular Diels–Alder reaction of, **3**, 331, 334, 336
metal complexes of, **3**, 370, 375
oligomerization of, **3**, 330
polymerization of, **3**, 330
reactions
with carbanions, **1**, 1040
with thebaine, **2**, 140
stability of, **1**, 1058; **3**, 331
as starting materials, for aldehydes, **1**, 834
synthesis of, **2**, 541; **3**, **329–380**, 331; **4**, 1049
via [3+2] cycloaddition, **3**, 339
via [2+2] cycloreversion, **3**, 334
via [3+2] cycloreversion, **3**, 335
via [4+1] cycloreversion, **3**, 335
via cycloreversion of heterocycles, **3**, 340
via cycloreversion reactions, **3**, 334
via elimination of dithiolanes, **3**, 333
via 1,2-elimination reactions, **3**, 333, 340
via fluoride induced cleavage of α-silyl disulfides, **3**, 333
via formation of α carbon–carbon bonds, **3**, 332
via modified Vilsmeier–Haack reaction, **3**, 341
via photolysis, **3**, 337
via photolysis of phenacyl sulfides, **3**, 331
via reaction of phosphonium ylides with elemental sulfur, **3**, 336
via retro Diels–Alder reaction, **3**, 335
via sulfuration of aldehydes, **3**, 336, 340, 341
via thermal rearrangement of *S*-alkyl thiosulfinates, **3**, 334
via thermolysis, **3**, 337
via Vilsmeier–Haack reaction, **3**, 338
via zirconation of thioketenes, **3**, 337
trapping as Diels–Alder adducts, **3**, 331, 334, 336, 340
trapping with phosphonium ylides, **3**, 336
trapping with secondary amines, **3**, 336
Thioaldehydes, aliphatic, from, tris(trimethylsilyl)methyllithium, **6**, 176

Thioaldehydes, alkenyl, synthesis of, **3**, 337
Thioaldehydes, alkyl, synthesis of, **3**, 331
Thioaldehydes, aryl, synthesis of, **3**, 340
Thioaldehydes, α-cyano-β-imino, synthesis of, **3**, 338
Thioaldehydes, enamino
 synthesis of
  from 4-phenyldithiolium cation, **3**, 339
  via sulfuration of aldehydes, **3**, 338
 tautomerism of, **3**, 339
Thioaldehydes, phosphonium ylide-stabilized alkyl, synthesis of, **3**, 332
Thioaldehydes, α,β-unsaturated, synthesis of, **3**, 337
1-Thioaldose, per-*O*-acetyl-, synthesis of, **4**, 223
Thioalkanes, 1,1-bis(trimethylsilyl)-1-phenyl-, reductive lithiation of, **4**, 618
Thioalkanoic acids, as starting materials, for *S*-acetyl phosphorothioites, **5**, 406
Thioalkanols, *O,S*-bis(9-borabicyclo[3.3.1]nonyl), synthesis of, **2**, 270
Thioalkanoyl chlorides, perhalo-, synthesis of, **5**, 481
α-Thioalkanoyl chlorides, as starting materials, for acyl azides, **4**, 328
2-Thio-2-alkenals, synthesis of, **3**, 72
α-Thioalkenals, synthesis of, **3**, 72
β-Thioalkenals, synthesis of, **3**, 72, 73
Thioalkenes, 1-alkoxy 1-halo 2-alkyl-, synthesis of, **4**, 765
Thioalkenes, aryl-, silylation of α-carbanions, **4**, 938
Thioalkenes, 1,2-dialkoxy 1-alkyl-, synthesis of, **4**, 765
Thioalkenes, 1-halo 1-alkyl-, oxidation of, **4**, 776
Thioalkenes, 1-halo 1-phenyl-, synthesis of, via Wittig–Horner reaction, **4**, 773
Thioalkenes, nitrophenyl-, synthesis of, **4**, 907
Thioalkenes, 1-nitro-2-phenyl-, as starting materials, for nitro enamines, **2**, 789
1-Thioalkenes, 1-amino-, synthesis of, **4**, 898
1-Thioalkenes, 1-halo-, synthesis of, **4**, 774
1-Thioalkenes, 1-nitro-, synthesis of, **4**, 906
β-Thioalkenones, synthesis of, **3**, 252
Thioalkoxides, use in the synthesis of heterocuprates, **2**, 588
Thioalkoxythiols, synthesis of, **2**, 128
Thioalkylation
 mechanism, **2**, 216
 of thiols, **2**, 216
 See also Thiolysis
α-Thioalkylation, mechanism, **4**, 315
Thioalkynes, hydrosilation of, **4**, 940
Thioalkynes, alkyl-, reactions, with diethylamine, **4**, 905
Thioalkynes, 1-alkyl-, addition of diselenides to, **4**, 871
Thioalkynes, phenyl-, reactions, with tributyltin hydride, **4**, 962
Thio-1-alkynes, 1-alkyl-, hydroboration of, **2**, 901
5-Thio-D-allose, synthesis of, **4**, 227
Thio allyl anions, bismethyl-, as starting materials, for α,β-unsaturated aldehydes, **3**, 54
α-Thioaluminums, synthesis of, **4**, 397
Thioamide
 reactions
  with *t*-butylthiyl radicals, **4**, 1077
  with phenylsulfenyl chloride, **5**, 727
Thioamide acetals, from, amide acetals, **6**, 111
Thioamide *S*-oxides
 O-alkylation of, **5**, 734
 deprotonation of, **5**, 734
 reactions, with triethyloxonium tetrafluoroborate, **5**, 734
Thioamides
 acylation of, **5**, 733
 alkylation of, **4**, 899; **5**, 753, 861
 S-allylation of, **1**, 834
 S-amination of, **5**, 736
 arylation of, **5**, 727
 from
  amides, **2**, 1049; **5**, 333
  carboxylic acids, **5**, 618
  N-silylamides, **5**, 335
  sulfenimines, **5**, 335
 oxidation of, **5**, 340, 341, 735
 phosphorylation of, **5**, 737
 reactions
  with alkylating agents, **5**, 726
  with alkyl halides, **3**, 638
  with *N*-alkylpyridyl disulfides, **5**, 735
  with alkyl thiochloroformate, **5**, 727
  with amines, **5**, 753
  with ammonia, **5**, 753
  with chloroamine, **5**, 736
  with diazomethane, **5**, 726
  with dibenzoyl peroxide, **5**, 735
  with diethyl azodicarboxylate, **5**, 735
  with dimethyl sulfate, **3**, 657
  with ethyl thiochloroformate, **5**, 727
  with formamide acetals, **5**, 760
  with α-halogenated methylenes, **2**, 749
  with hydrazines, **4**, 984
  with hydroxylamine, **5**, 764
  with hydroxylamine-O-sulfonic acid, **5**, 736
  with penicillin sulfoxides, **5**, 736
  with phenyl iodosodiacetate, **5**, 735
  with potassium hexacyanoferrate(III), **5**, 735
  with sulfenylating reagents, **5**, 734
  with sulfenyl chlorides, **5**, 734
  with *N*-thiosuccinimides, **5**, 735
  with tribenzenesulfenamide, **5**, 735
 silylation of, **5**, 739
 as starting materials
  for aliphatic amidines, **5**, 753
  for amidines, **5**, 743
  for aromatic amidines, **5**, 757
  for di(cyanomethyl) sulfides, **3**, 634
  for β,β-difunctionalized enamines, **2**, 750
  for enaminones, **2**, 744; **3**, 268
  for α-hydrazino enamines, **4**, 984
  for hydrazonamides, **5**, 852
  for imidoyl disulfides, **5**, 734
  for isothiocyanates, **5**, 1031, 1037
  for ketene *S,N*-acetals, **4**, 899
  for ketenimines, **3**, 561
  for nitriles, **3**, 619
  for *N*-nitrosoamides, **5**, 361
  for oxyimidic chlorides, **5**, 834
  for tertiary ynamines, **2**, 1049
  for α-thio-α-aminonitriles, **3**, 638
  for thioimidic esters, **5**, 726
 synthesis of, **1**, 1044, 1048
  via acylation of thioamines, **5**, 335
  via 1,4-addition to thione *S*-imides, **5**, 336
  via cycloaddition of thione *S*-imides, **5**, 336
 thiolimidic acid tautomerism, **5**, 725
 vinylogous, synthesis of, **3**, 355
 See also Amides, thiocarboxy-
Thioamides, *N*-acyl-, synthesis of, **5**, 733
Thioamides, *S*-acyl-, synthesis of, **5**, 733
Thioamides, allylic, oxidation of, **3**, 57
Thioamides, aromatic, reactions, with oxalyl chloride, **5**, 996
Thioamides, *N*-aryl-
 reactions
  with hydrazine hydrate, **5**, 770
  with triphenyltin iodide, **5**, 739
Thioamides, chloro-, as starting materials, for phosphinothioformamides, **6**, 584
Thioamides, cyclic, synthesis of, via cyclization of amides, **5**, 334
Thioamides, dialkyl-, enolization of, **1**, 876
Thioamides, α,α-dimethyl-, dehydration of, **3**, 636
Thioamides, halo-, as starting materials, for phosphinothioformamides, **6**, 584
Thioamides, β-hydroxy-, reactions, with formaldehyde, **4**, 230

Thioamides, *N*-(hydroxymethyl)-, synthesis of, **4**, 296
Thioamides, α-keto-
  from, 4-methoxy-3-nitroacetophenone, **5**, 485
  synthesis of, **5**, 485
Thioamides, tertiary
  alkylation of, **5**, 726
  reactions, with alkyl halides, **4**, 899
Thioamides, *N*-(trimethylsilyl)-, synthesis of, **5**, 739
Thioamides, α,β-unsaturated, as starting materials, for ketene *S*,*N*-acetals, **4**, 901
N-Thioamides, from, sulfoxides, **5**, 335
Thioamidinium salts, *S*-alkyl-, reactions, with alkynyllithiums, **1**, 1041
Thioaminals, as starting materials, for aminomethyl carbanions, **4**, 528
Thioamines, acylation of, **5**, 335
Thioanilide, phenylpropiolic, reactions, with phenacyl bromide, **6**, 111
α-Thioanions, synthesis of, **4**, 365
Thioanisole
  deprotonation of, **4**, 365
  lithiation of, **4**, 391
  metallation of, **1**, 140
Thioantimonates, synthesis of, **2**, 261
Thioantimonates, alkane-, synthesis of, **2**, 261
Thioarenes, alkyl-, reactions, with Grignard reagents, **1**, 436
Thioaroyl chlorides, reactions, with sodium thiocyanate, **5**, 1050
Thioarsenates
  from
    aminoarsines, **2**, 260
    arsenic halides, **2**, 260
    arsenoxides, **2**, 260
    tetraalkyldiarsines, **2**, 261
  synthesis of, **2**, 259
Thioarsenates, alkane-, synthesis of, **2**, 259
Thioarsenic esters
  from
    arsenic amides, **2**, 889
    arsenic halides, **2**, 889
    arsenic oxides, **2**, 889
Thioarsines, alkyl-, synthesis of, **2**, 499
Thioarsinic esters
  from
    arsenic amides, **2**, 889
    arsenic halides, **2**, 889
    arsenic oxides, **2**, 889
Thioarsonous acid anhydrides, alkyl-, synthesis of, **2**, 500
Thioate, *S*-phenyl carbonochlorido-, reactions, with Grignard reagents, **3**, 285
Thioates, phosphinamido-, enamides of, synthesis of, **2**, 808
Thioates, phosphoroamido-, reactions, with benzaldehydes, **4**, 467
Thioates, *S*-(2-pyridyl)-, as acylating agents, **3**, 286
2-Thiobarbituric acids, reduction of, **4**, 426
Thiobenzaldehyde, 4-methyl-, synthesis of, **3**, 98
Thiobenzaldehyde, 2,4,6-tri-(*t*-butyl)-, synthesis of, **3**, 340
Thiobenzaldehydes
  reactions, with trimethylsilyllithium, **4**, 376
  synthesis of, **3**, 340
Thiobenzamide
  reactions
    with acetonitrile, **5**, 760
    with aldehydes, **5**, 606
    with chloromethyldichlorophosphane, **5**, 635
    with dimethylformamide dimethyl acetal, **5**, 760
  synthesis of, **5**, 595
Thiobenzamide, *o*-amino-, for the condensation of *O*-ethyl selenocarboxylates, **5**, 522
Thiobenzamide, *N*-(anilinothiocarbonyl)-, reactions, with sulfuryl chloride, **5**, 657
Thiobenzamide, 2,6-dichloro-, as starting material, for isothiocyanates, **5**, 1037
Thiobenzamide, *N*,*N*-dimethyl-, synthesis of, **5**, 582
Thiobenzamide, *N*-methyl-, reactions, with diethyl chlorophosphite, **5**, 643
Thiobenzamide-*S*-oxide, *N*-phenyl-, oxidation of, **5**, 697
Thiobenzamides
  reactions, with thiophosgene, **5**, 1049
  as starting materials, for isothiocyanates, **5**, 1037
  synthesis of, **5**, 600
Thiobenzamides, 2-amino-, synthesis of, **5**, 584
Thiobenzamides, *N*-(anilinothiocarbonyl)-, reactions, with chlorine, **5**, 657
Thiobenzamides, aryl-, as starting materials, for bisarylimines, **3**, 412
Thiobenzamides, *N*,*N*-disubstituted, reactions, with bromine, **5**, 734
Thiobenzanilide
  amination of, **5**, 736
  reactions, with mercury bis(phenylalkyne), **5**, 739
Thiobenzanilide, *N*-trimethylsilyl-, reactions, with dibromophenylborane, **5**, 739
Thiobenzhydroxamic acids, use in the synthesis of oxyimidic thioesters, **5**, 831
Thiobenzimidate, *N*-benzenesulfonyl-*S*-phenyl-, reactions, with 1-diethylaminoprop-1-yne, **4**, 905
Thiobenzimidate, *N*-benzoyl-*S*-phenyl-, reactions, with 1-diethylaminoprop-1-yne, **4**, 905
Thiobenzimidates, *S*-methyl *N*-methyl-, hydrolysis of, **5**, 327
Thiobenzoate, *O*-benzylideneamino, synthesis of, **5**, 531
Thiobenzoate, *O*-trimethylsilyl, phosphorylation of, **5**, 535
Thiobenzoates, reactions, with tributyltin hydride, **1**, 938
Thiobenzoates, vinyl-, synthesis of, **1**, 603
Thio-1,3-benzodithiole, 2-ethyl-, from, 2-alkoxy-1,3-benzodithiole, **6**, 82
Thiobenzoic acid
  reactions
    with benzenetellurenic acid, **5**, 249
    with diethoxytrimethylantimony, **2**, 502
Thiobenzophenone
  reactions
    with azomethine ylides, **4**, 319
    with α-nitrosostyrene, **4**, 232
    with phenyllithium, **4**, 394
    with sodium alkynide, **1**, 1040
  stability of, **3**, 342
  synthesis of, **3**, 363
Thiobenzophenone *S*-oxide, thermolysis of, **3**, 365
Thiobenzophenones, reactions, with phenyliodinium ylides, **4**, 866
o-Thiobenzoquinone methides
  dimerization of, **3**, 359
  synthesis of, **3**, 359
  trapping as Diels–Alder adducts, **3**, 359
Thiobenzothiazoles, displacement of, **1**, 238
Thiobenzoyl bromides, from, dithiobenzoic acids, **5**, 486
Thiobenzoyl chloride
  reactions
    with dimethylsulfoxonium methylide, **3**, 364
    with sodium benzoaldoximate, **5**, 531
    with thionocarbamates, **5**, 632
  as starting material, for phenylchlorosulfine, **5**, 490
α-Thioberylliums, synthesis of, **4**, 395
Thiobisamines
  from, *N*-sulfenyl-substituted hydantoins, **2**, 361
  reactions, with chlorine, **2**, 360
N,*N*-Thiobisamines, synthesis of, **2**, 361
N,*N*'-Thiobisamines, synthesis of, **2**, 361
N,*N*-Thiobis(arylamines), as starting materials, for thionitrosoarenes, **2**, 792
Thiobismuthates, synthesis of, **2**, 261
Thiobismuthates, alkane-, synthesis of, **2**, 261
Thiobismuthates, α-carboxyl, synthesis of, **2**, 261
Thioborane complexes, trimethylamine–methyl-, synthesis of, **4**, 373

Thioboranes
from
alkylboranes, **2**, 268
trimethylamineboranes, **2**, 269
reactions, with hydrogen cyanide, **5**, 1136
synthesis of, **2**, 267
Thioboranes, alkane-, synthesis of, **2**, 267
Thioborates, synthesis of, **2**, 268
Thioborates, alkyl, synthesis of, **2**, 268
Thioboration
of alkynes, stereochemistry of, **2**, 902
for the synthesis of alkenylboranes, **2**, 900
Thioboric acid esters, dialkyl-, as starting materials, for alkyl arylketiminoboranes, **2**, 811
Thioborinates, dialkyl-, reactions, with amidines, **5**, 778
Thioboronates, α-alkyl-, synthesis of, **4**, 373
Thioboronates, dialkyl alkyl-, synthesis of, **2**, 269
Thioboronates, α-phenyl-, alkylation of, **4**, 373
Thioboronates, trialkyl-, synthesis of, **2**, 269
Thio-1,4-butadiene, 1-silyl-4-phenyl-, synthesis of, **2**, 939
Thiobut-3-enoyl fluoride, 2,2,3,4,4-pentafluoro-, from, 1,2,2,3,4,4-hexafluoro-1,1-diiodobut-3-ene, **5**, 473
Thiobutynylide, synthesis of, **2**, 1031
Thio-γ-butyrolactones, γ-phenyl-, synthesis of, **4**, 226
Thiocamphor, synthesis of, **3**, 349
Thiocaprolactam, reactions, with benzyl chloroformate, **5**, 727
Thiocarbamate, *O*,*N*-dibenzyl *N*-methyl, synthesis of, **6**, 557
Thiocarbamate, *N*,*S*-diethyl, synthesis of, **6**, 492
Thiocarbamate, *O*-ethyl *N*-phenyl, synthesis of, **6**, 559
Thiocarbamate, *S*-ethyl *N*-phenyl, synthesis of, **6**, 492
Thiocarbamate, *O*-methyl, as starting material, for methyl cyanates, **5**, 1103
Thiocarbamate, *N*,*N*,*O*-trimethyl, synthesis of, **6**, 556
Thiocarbamates
as amine protective groups, **6**, 496
from
alkylamide salts, **6**, 494
carbamoyl chlorides, **6**, 493
chlorothioformates, **6**, 492
1-chlorothioformimidates, **6**, 494
isocyanates, **6**, 494
oxathiolanes, **6**, 496
potassium *O*-ethyl dithiocarbonate, **6**, 559
reactions, with dithiocarbamates, **2**, 218
as starting materials, for isocyanates, **5**, 979
synthesis of, via [3,3] sigmatropic rearrangement, **6**, 495
uses of, **6**, 495
use in the synthesis of distibines, **4**, 595
use in the synthesis of disulfides, **2**, 218
use in the synthesis of α-thiol ketones, **3**, 164
Thiocarbamates, *N*-acyl imido-, use in the synthesis of α,β-unsaturated ketones, **3**, 267
Thiocarbamates, *O*-alkyl
from, isothiocyanates, **6**, 558
reactions
with mercuric oxide, **5**, 1103
with silver oxide, **5**, 1103
as starting materials, for cyanates, **5**, 1103
Thiocarbamates, *S*-alkyl, synthesis of, **6**, 494
Thiocarbamates, *S*-alkyldimethyl, pyrolysis of, **1**, 778
Thiocarbamates, *O*-alkyl *N*-hydroxy
reactions, with ethyl chloroformate, **5**, 1104
as starting materials, for cyanates, **5**, 1104
Thiocarbamates, ammonium, synthesis of, **6**, 495
Thiocarbamates, *O*-aryl, reactions, with silver oxide, **5**, 1103
Thiocarbamates, *S*-aryl, synthesis of, via Newman-Kwart rearrangement, **6**, 495
Thiocarbamates, *N*-aryl *N*-thioacyl, reactions, with diethylamine, **5**, 588
Thiocarbamates, *N*-aryl-*N*-thionoacyl, synthesis of, **5**, 632
Thiocarbamates, *S*-2-chloroalkyl *N*,*N*-dialkyl, from, dialkylcarbamoyl chlorides, **6**, 493
Thiocarbamates, *N*,*N*-disubstituted, synthesis of, **6**, 495
Thiocarbamates, imido-
reactions, with mercuric oxide, **5**, 1104
synthesis of, **5**, 1104
Thiocarbamates, methyl, reactions, with 2-chloro-1-ethylbenzoxazolium tetrafluoroborate, **5**, 979
Thiocarbamates, *O*-phenyl, synthesis of, **6**, 556
Thiocarbamic acid, structure of, **6**, 555
Thiocarbamic acid, *O*-ethyl, synthesis of, **6**, 559
Thiocarbamic acid, *O*-methyl, synthesis of, **6**, 559
Thiocarbamic acid esters, *N*-aryl-*N*-thiobenzoyl, synthesis of, **6**, 559
Thiocarbamic acid *O*-esters
from
O-alkyl chlorothioformates, **6**, 556
O-alkyl thiocarbamic acids, **6**, 559
amino alcohols, **6**, 557
O-aryl chlorothioformates, **6**, 556
N,*N*-dialkylthiocarbamoyl chlorides, **6**, 556
N,*N'*-thiocarbonyldiimidazole, **6**, 557
synthesis of, **6**, 556
Thiocarbamic acids, synthesis of, **6**, 555
Thiocarbamic acids, *O*-alkyl, as starting materials, for thiocarbamic acid *O*-esters, **6**, 559
Thiocarbamic acid thioanhydrides, as starting materials, for *O*-trimethylsilyl thiocarboxylates, **5**, 537
Thiocarbamic esters, chlorination of, **6**, 617
Thiocarbamoylation
of carbanions, **4**, 901
for the synthesis of thionoamides, **5**, 592, 620
Thiocarbamoylation, nucleophilic, for the synthesis of thionoamides, **5**, 597
Thiocarbamoyl bromides, synthesis of, **6**, 540
Thiocarbamoyl chloride, diethyl-
reactions
with arylamines, **5**, 1029
with tertiary amides, **5**, 578
Thiocarbamoyl chloride, *N*,*N*-diethyl-, reactions, with benzimidazole-2-thiol, **6**, 566
Thiocarbamoyl chloride, dimethyl-, reactions, with phenylethyne, **5**, 617
Thiocarbamoyl chloride, *N*,*N*-dimethyl-, synthesis of, **6**, 538
Thiocarbamoyl chlorides
from
amines, **6**, 537
tetraalkylthiuramide disulfides, **6**, 538
thioformamides, **6**, 539
as intermediates, **6**, 576
in the synthesis of isothiocyanates, **5**, 1027
reactions
with alcohols, **6**, 557
with aromatics, **5**, 593
with sodium isopropoxide, **5**, 609
with thiols, **6**, 566
as starting materials, for dithiocarbamates, **6**, 566
synthesis of, **5**, 1033; **6**, 537
Thiocarbamoyl chlorides, *N*,*N*-dialkyl-
reactions, with alkynes, **1**, 1048
as starting materials, for thiocarbamic acid *O*-esters, **6**, 556
Thiocarbamoyl fluoride, *N*,*N*-diethyl-, synthesis of, **6**, 537
Thiocarbamoyl fluoride, *N*,*N*-dimethyl-, synthesis of, **6**, 537
Thiocarbamoyl fluorides, synthesis of, **6**, 537
Thiocarbamoyl halides, reactions, with alkynes, **5**, 597
Thiocarbamoylphosphonates, synthesis of, **6**, 582
α-Thiocarbanions
reactions
with trialkylboranes, **4**, 372
with trialkylborate, **4**, 372
synthesis of, **4**, 390
via lithium exchange reactions, **4**, 394
Thiocarbazic acid chlorides
reactions, with thiazolidine-2-thione, **6**, 566
synthesis of, **6**, 540

Thiocarbene complexes
  from, iron acyl complexes, **5**, 951
  synthesis of, **5**, 949, 950
Thiocarbene complexes, cyclic, synthesis of, **5**, 951
Thiocarbenes, as intermediates, **5**, 951
α-Thiocarbenium ions, synthesis of, **2**, 1008
Thiocarbenium salts, oxo-, as starting materials, for monothioortho esters, **6**, 96
Thiocarbonate, *O*-alkyl *S*-ethyl
  as starting material, for chloroformates, **6**, 427
  synthesis of, **6**, 427
Thiocarbonate, bis(trifluoromethyl), reactions, with trifluoromethylsulfenyl fluoride, **6**, 274
Thiocarbonate, *S*-butyl *O*-methyl, synthesis of, **6**, 475
Thiocarbonate, *O*,*S*-diethyl, synthesis of, **6**, 478
Thiocarbonate, dimethyl, as dithiocarboxylating agent, **5**, 555
Thiocarbonate, di-2-pyridyl
  reactions, with amines, **5**, 1030
  use in the synthesis of carbodiimides, **5**, 1078
Thiocarbonate, *S*-ethyl, reactions, with sodium ethyl selenides, **6**, 482
Thiocarbonate, *S*-ethyl *O*-methyl, synthesis of, **6**, 475
Thiocarbonate, *S*-isopropyl *O*-methyl, synthesis of, **6**, 475
Thiocarbonate, *S*-vinyl *O*-*t*-butyl, synthesis of, **6**, 476
Thiocarbonates
  chlorination of, **6**, 427
  from
    carbonothioic acid salts, **6**, 479
    sulfenyl chlorides, **6**, 273
    2-thiomethyl-2-phenyl-1,3-indanedione, **6**, 475
  as hormone analogues, **6**, 477
  pyrolysis of, **1**, 777
  reactions, with enones, **1**, 176
  reduction of, **1**, 16
  selectivity of, **1**, 176
  as starting materials, for spirocyclic orthocarbonates, **6**, 300
Thiocarbonates, alkanesulfenyl methyl, synthesis of, **6**, 478
Thiocarbonates, alkyl, synthesis of, **6**, 475
Thiocarbonates, *O*-alkyl chloro-
  reactions
    with potassium *O*-alkyl dithiocarbonates, **6**, 547
    with potassium *O*-alkyl monothiocarbonates, **6**, 551
Thiocarbonates, *O*-alkyl *S*-ethyl, synthesis of, **6**, 477
Thiocarbonates, *S*-aryl, synthesis of, **6**, 478
Thiocarbonates, chloromethyl, synthesis of, **4**, 60
Thiocarbonates, dialkyl
  from
    alkoxycarbonylsulfenyl chlorides, **6**, 478
    carbonates, **6**, 474
    ethyl xanthatepotassium salts, **6**, 478
    orthoformates, **6**, 478
Thiocarbonates, *O*,*O*'-dialkyl, synthesis of, **6**, 300
Thiocarbonates, *O*,*O*-disubstituted
  as intermediates, **6**, 312
  reactions, with nitrile oxides, **6**, 309
Thiocarbonates, imino-
  from
    carbonimidic halides, **6**, 628
    cyanates, **6**, 629
    cyanogen chloride, **6**, 629
    thiocyanates, **6**, 629
  synthesis of, **6**, 628
    via *S*-alkylation methods, **6**, 629
    via functional group interconversion, **6**, 630
Thiocarbonates, iodo-, from, methyl 2,6-di-*O*-methyl-3,4-*O*-thiocarbonyl β-D-galactoside, **6**, 477
Thiocarbonates, iodomethyl, synthesis of, **4**, 60
Thiocarbonates, *S*-methyl, reactions, with tributyltin hydride, **1**, 590
Thiocarbonates, mixed alkyl, synthesis of, **6**, 475
Thiocarbonates, sulfenyl
  reactions, with thiols, **2**, 220
  use in the synthesis of disulfides, **2**, 220
Thiocarbonates, thioacyl *O*-ethyl, thermolysis of, **6**, 476
Thiocarbonyl bonds, polarization of, **3**, 330
Thiocarbonyl chloride, *S*-chloro-
  reactions
    with amines, **6**, 501
    with 1*H*-benzotriazole, **6**, 501
Thiocarbonyl chloride, *N*,*N*-dimethyl-, reactions, with 1,1,3,3-tetramethylguanidine, **6**, 647
Thiocarbonyl chloride, methoxy-, reactions, with sodium ethanethiolate, **6**, 550
Thiocarbonyl chlorides
  chlorination of, **6**, 727
  reactions, with thiols, **6**, 550
  as starting materials, for *O*,*S*-dialkyl dithiocarbonates, **6**, 550
Thiocarbonyl chlorides, alkoxy-
  as starting materials
    for *O*,*S*-dialkyl dithiocarbonates, **6**, 550
    for thiocarboxylates, **5**, 510
Thiocarbonyl complexes, desulfurization of, **5**, 900
Thiocarbonyl dibromide
  from
    carbon monosulfide, **6**, 529
    thiocarbonyl difluoride, **6**, 529
    thiophosgene, **6**, 529
  synthesis of, **6**, 529
Thiocarbonyl dichloride
  synthesis of, **6**, 529
  See also Thiophosgene
Thiocarbonyl difluoride
  from, 2,2,4,4-tetrafluoro-1,3-dithietane, **6**, 528
  as starting material, for thiocarbonyl dibromide, **6**, 529
  synthesis of, **6**, 528
Thiocarbonyl diiodide
  from, carbon monosulfide, **6**, 530
  synthesis of, **6**, 530
Thiocarbonyl fluoride, trifluoromethylthiolation with, **6**, 233
Thiocarbonyl groups
  reactions, with carbenes, **1**, 367
  synthesis of functions, **6**, **527–567**
    bearing two heteroatoms other than a halogen or chalcogen, **6**, **569–585**
Thiocarbonyl halides, synthesis of, **6**, 528
Thiocarbonyl *S*-oxide, synthesis of, **2**, 247
Thiocarbonyls
  bromination of, **2**, 171
  classes of, for the synthesis of isothiocyanates, **5**, 1022
  [2+2] cycloadditions of, **4**, 852
  dimerization of, **6**, 49
  photolysis of, **3**, 352
  photoreduction of, **6**, 115
  reactions
    with amines, **5**, 1022, 1028
    with carbenes, **4**, 862
    with fluorine, **6**, 252
    with halogens, **6**, 273, 275
    with selenyl chlorides, **6**, 276
    with silyloxyphosphinates, **2**, 266
    with sulfenyl halides, **6**, 275
    with sulfur, **2**, 226
    with trialkyl phosphites, **4**, 345
    with tributyltin hydride, **1**, 938
    with trimethylsilylvinyl ketone, **4**, 232
  as starting materials
    for *S*,*N*-acetals, **4**, 319, 328
    for *S*,*P*-acetals, **4**, 345
    for dialkyl sulfides, **2**, 140
    for disulfides, **2**, 230
    for dithioacetals, **4**, 255
    for ketene dithioacetal *S*,*S*-dioxides, **4**, 862

for selenocarbonyls, **6**, 592
for sulfenamides, **2**, 243
for thiosilanes, **2**, 266
synthesis of, **1**, 1078
tautomerism of, **3**, 330
thione-enethiol tautomerism, **3**, 330
use in the synthesis of alkanesulfenyl chlorides, **2**, 169
use in the synthesis of radicals, **1**, 939
Thiocarbonyls, alkoxy-, reactions, with carbanions, **5**, 510
Thiocarbonyls, dihalo-, reactions, with tetrafluorohydrazine, **2**, 166
Thiocarbonyls, ethoxy-, synthesis of, **6**, 4625
Thiocarbonyls, imidazoyl-, reduction of, **1**, 16
Thiocarbonyl tetrachloride, as starting material, for isothiocyanates, **5**, 1028
Thiocarbonyl transfer reagents, as starting materials, for thioureas, **6**, 577
Thiocarboxanilides, aryl-
as starting materials
for dehydrotetrazines, **5**, 852
for hydrazonamides, **5**, 852
Thiocarboxylate, *β*-isocyanatoethyl, synthesis of, **5**, 1012
Thiocarboxylate, ^{33}S-labeled, synthesis of, **5**, 523
Thiocarboxylate, *S*-trimethylsilyl, rearrangement of, to *O*-trimethylsilyl thiocarboxylate, **5**, 538
Thiocarboxylate *S*-oxide, *O*-methyl, synthesis of, problems, **5**, 527
Thiocarboxylates
from
alkoxythiocarbonyl chlorides, **5**, 510
O-alkyl thiocarboxylates, **5**, 509
bisthioacyl sulfides, **5**, 509
bis(thioacyl) sulfides, **5**, 510
dithiocarboxylates, **5**, 509, 510
selenoketenes, **5**, 510
thioacyl chlorides, **5**, 509
thioacyl halides, **5**, 510
thioketenes, **5**, 509, 511
reactions
with acylsulfenamides, **5**, 251
with acylsulfenyl halides, **5**, 251
synthesis of, **5**, 510
Thiocarboxylates, *O*-alkyl
as starting materials, for thiocarboxylates, **5**, 509
synthesis of, **5**, 510, 513
Thiocarboxylates, alkynyl, reactions, with nucleophiles, **3**, 548
Thiocarboxylates, *O*-aryl, synthesis of, **5**, 510
Thiocarboxylates, *O*-ethyl *β*-oxo-, synthesis of, **5**, 513
Thiocarboxylates, *O*-germanyl, synthesis of, **5**, 540
Thiocarboxylates, *S*-germanyl, synthesis of, **5**, 540
Thiocarboxylates, *O*-halo-, synthesis of, **5**, 530
Thiocarboxylates, *O*-methyl, oxidation of, with *m*-chloroperbenzoic acid, **5**, 527
Thiocarboxylates, piperidinium, reactions, with phenyltellurenyl bromides, **5**, 249
Thiocarboxylates, *O*-silyl
properties of, **5**, 539
synthesis of, **5**, 536
Thiocarboxylates, *S*-silyl, as intermediates, in the synthesis of *O*-silyl thiocarboxylates, **5**, 539
Thiocarboxylates, *O*-stannyl, synthesis of, **5**, 540
Thiocarboxylates, *S*-stannyl, synthesis of, **5**, 540
Thiocarboxylates, *O*-trimethylsilyl
from
acid chlorides, **5**, 539
thiocarbamic acid thioanhydrides, **5**, 537
reactions
with acetic acid, **5**, 539
with acyl chlorides, **5**, 539
with aldehydes, **5**, 540
with alkyl halides, **5**, 540
with ethanol, **5**, 539
with halonium ions, **5**, 539
with triethylethoxystannane, **5**, 540
synthesis of, **5**, 537, 538
Thiocarboxylate salts, reactions, with cyanogen chloride, **5**, 1112
Thiocarboxylic acid esters
as biological acylating agents, **5**, 199
reactions, with hydroxylamines, **5**, 636
as starting materials
for *N*-alkoxyimidates, **5**, 700
for *N*-hydroxyimidates, **5**, 700
Thiocarboxylic acids
acylation of, **5**, 243
equilibrium of, **5**, 506
from, hydrogen sulfide, **5**, 232
properties of, **5**, 506
reactions
with benzenesulfonyl chloride, **5**, 252
with chlorotrimethylsilane, **5**, 537
with chlorotriorganylsilanes, **5**, 254
with dicyclohexylcarbodiimide, **5**, 243
with isocyanides, **5**, 632
with lithium diisopropylamide, **5**, 232
with *N*-phenylselenophthalimide, **5**, 248
as starting materials
for acylsulfenyl selenides, **5**, 248
for acylsulfur functions, **5**, 232
sulfenylation of, **5**, 247
synthesis of, **5**, 232
Thiocarboxylic acids, *N'*-alkylhydrazonide esters, synthesis of, **5**, 846
Thiocarboxylic acids, *N'*-arylhydrazonide esters, synthesis of, **5**, 846
Thiocarboxylic acids, *β*-hydroxy-, synthesis of, **5**, 232
Thiocarboxylic acids, *α*-phenyl-, decarboxylation of, **4**, 71
Thiocarboxylic *O*-acids, synthesis of, **5**, 506
Thiocarboxylic acid *O*-silyl esters, reactions, with *N*-bromosuccinimide, **5**, 245
Thiocarboxylic esters, as starting materials, for trithioortho esters, **6**, 84
Thiocarbyne complexes, reactions, with cobalt compounds, **6**, 352
Thiocetals, hydrolysis of, for the synthesis of aldehydes, **3**, 15
Thiochroman-4-one 1-oxide, 2-azido-, from, thiochromone 1-oxide, **4**, 330
Thiochromone 1-oxide, as starting material, for 2-azidothiochroman-4-one 1-oxide, **4**, 330
Thio compounds, as starting materials, for ortho thioesters, **6**, 10
Thio compounds, *N*-fluoroimidic, from, vinylsulfur pentafluoride, **5**, 814
Thiocopper, trifluoromethyl-
reactions
with dichloroacrylonitriles, **4**, 771
with triodonitroethene, **4**, 771
α-Thiocoppers, synthesis of, **4**, 396
Thiocoumarin, synthesis of, **5**, 512
Thiocyanate, alkylammonium, use in the synthesis of amidines, **5**, 750
Thiocyanate, ammonium
reactions
with carbonyl chloride isocyanate, **6**, 559
with phosgene, **6**, 559
Thiocyanate, benzeneselenenyl, reactions, with diazomethane, **4**, 335
Thiocyanate, butadien-1-yl, Diels–Alder reaction of, **6**, 558
Thiocyanate, *t*-butyl, from, *t*-butyl bromide, **5**, 1105
Thiocyanate, chlolesteryl, reduction of, **2**, 127
Thiocyanate, (-)-4-chlorobenzhydryl, isomerization of, **5**, 1038
Thiocyanate, chlorocarbonyl, from, chlorocarbonylsulfenyl chloride, **6**, 435
Thiocyanate, crotyl, as starting material, for *α*-methylallyl isothiocyanate, **5**, 1040

Thiocyanate, *N,N*-dimethylthiocarbamoyl, synthesis of, **5**, 1112
Thiocyanate, ethyl, reactions, with hydrogen chloride, **6**, 618
Thiocyanate, mercuric, as starting material, for alkynylmercury compounds, **2**, 1098
Thiocyanate, methyl
  fluorination of, **5**, 808; **6**, 257
  reactions, with imidoyl bromides, **5**, 864
  as starting material, for iminocarbonyl chlorides, **6**, 618
Thiocyanate, oxalyl bromide, as starting material, for bromocarbonyl isothiocyanate, **6**, 453
Thiocyanate, phenyl selenenyl, reactions, with alkenes, **5**, 1107
Thiocyanate, phosphonium, oxidation of, **5**, 898
Thiocyanate, 2-propoxycarbonyl, synthesis of, **6**, 476
Thiocyanate, thiocarbonyl chloride, synthesis of, **6**, 559
Thiocyanate, *p*-toluene, reactions, with 4-amino-3,5-di(methylthio)-1,2,4-triazolium iodide, **5**, 1122
Thiocyanate, trialkylstannyl, synthesis of, **5**, 1082
Thiocyanate, triphenylphosphine, reactions, with Grignard reagents, **5**, 595
Thiocyanate anions
  reactions, with alkynyliodonium salts, **2**, 1033
  as starting materials, for thiocyanates, **5**, 1105
Thiocyanate ions, reactions, with alkynyl halides, **2**, 1056
Thiocyanates
  from
    alcohols, **5**, 1107
    alkenes, **5**, 1110
    alkynes, **5**, 1110
    amines, **5**, 1107
    N-arylthiohydantoins, **5**, 1111
    N-arylthioisatins, **5**, 1111
    aziridines, **5**, 1107
    Barton esters, **5**, 1110
    cyanide ions, **5**, 1111
    cyanogen halides, **5**, 1111
    disulfides, **5**, 1111
    epoxides, **5**, 1106
    β-lactones, **5**, 1106
    metal cyanides, **5**, 1111
    nitronate esters, **5**, 1106
    2-nitro-5-thiocyanatobenzoic acid, **5**, 1111
    organolithiums, **5**, 1110
    oxaziridines, **5**, 1107
    phenyl selenenothiocyanate, **5**, 1108
    sulfenyl halides, **5**, 1111
    sulfenyl thiocyanates, **5**, 1111
    sulfonyl chlorides, **5**, 1111
    thiocyanate anions, **5**, 1105
    thiocyanogen, **5**, 1109
    thiocyanogen halides, **5**, 1109
    thiosulfates, **5**, 1111
    trimethylsilyl cyanides, **5**, 1111
  reactions
    with acyl chlorides, **6**, 618
    with amines, **6**, 635
    with chlorine, **2**, 177
    with iodine monochloride, **6**, 618
    with phenols, **5**, 728
    with phosphorus pentachloride, **5**, 861
  rearrangement of, **5**, 1038
  as starting materials
    for alkanethiols, **2**, 127
    for cyanates, **5**, 1103
    for disulfides, **2**, 231
    for iminocarbonyl chlorides, **6**, 618
    for iminodithiocarbonates, **6**, 627
    for iminothiocarbonates, **6**, 629
    for isothiocyanates, **5**, 1038
    for isothioureas, **6**, 635
    for sulfides, **2**, 713
    for thioimidic esters, **5**, 727
    for thioketones, **3**, 365
    for thiophosphates, **2**, 259
  synthesis of, **5**, 1104
    via conjugate addition, **5**, 1107
    via homolytic reactions, **5**, 1110
  use in the synthesis of disulfides, **2**, 220
  use in the synthesis of sulfonic acids, **2**, 192
Thiocyanates, acyl, synthesis of, **5**, 1112
Thiocyanates, alkyl
  reactions
    with bromine, **6**, 623
    with dialkyltrimethylsilyl phosphites, **5**, 1135
    with hydroxypyridones, **6**, 558
    with thiols, **6**, 627
  synthesis of, **5**, 1105
Thiocyanates, alkynyl
  from, alkynyliodonium salts, **2**, 1033
  protonolysis of, **3**, 548
  synthesis of, **2**, 1033
Thiocyanates, ammonium
  reactions
    with *t*-butyl chloride, **5**, 1038
    with halides, **5**, 1038
  for the synthesis of alkyl thiocyanates, **5**, 1105
Thiocyanates, aryl
  from
    arylsulfinic acids, **5**, 1112
    N-arylthioformamides, **5**, 1112
    diazonium salts, **5**, 1108
    thiocyanogen, **5**, 1109
    thiocyanogen halides, **5**, 1109
  synthesis of, **5**, 1112
Thiocyanates, benzhydryl, isomerization of, **5**, 1038
Thiocyanates, benzyl, rearrangement of, **5**, 1039
Thiocyanates, benzylic, synthesis of, **5**, 1107
Thiocyanates, 2,3-butadienyl, [3,3] sigmatropic rearrangements of, **5**, 1041
Thiocyanates, cinnamyl, as starting materials, for cinnamyl isothiocyanates, **5**, 1040
Thiocyanates, *β*-halo-, synthesis of, **5**, 1107
Thiocyanates, *β*-hydroxy, synthesis of, **5**, 1106
Thiocyanates, propynyl, thermolysis of, **1**, 993
Thiocyanates, selenenyl, as starting materials, for selenocyanates, **5**, 1116
Thiocyanates, sulfenyl
  reactions, with amines, **2**, 360
  as starting materials
    for sulfenamides, **2**, 243
    for thiocyanates, **5**, 1111
  use in the synthesis of disulfides, **2**, 219
Thiocyanates, tertiary alkyl, synthesis of, **5**, 1105
Thiocyanates, vinyl, cleavage of, **3**, 337
Thiocyanation, electrochemical, for the synthesis of isothiocyanates, **5**, 1045
3-Thiocyanato compounds, synthesis of, **5**, 1106
Thiocyanic acid, reactions, with alkynes, **5**, 1044
Thiocyanogen
  chlorination of, **5**, 809
  reactions
    with alkenes, **5**, 1109
    with benzonorbornadiene, **5**, 1044
    with silver alkynides, **5**, 1110
    with triphenylphosphine, **5**, 1041
    with ylides, **5**, 898
  as starting material
    for aryl thiocyanates, **5**, 1109
    for isothiocyanates, **5**, 1045
    for thiocyanates, **5**, 1109
    for *α*-thiocyanatoarylalkanes, **5**, 1110
  synthesis of, **5**, 1109
    in situ, **5**, 1045
Thiocyanogen bromide, synthesis of, **5**, 1109
Thiocyanogen halides

as starting materials
for aryl thiocyanates, **5**, 1109
for thiocyanates, **5**, 1109
for α-thiocyanatoarylalkanes, **5**, 1110
Thiocyanohydrins, as starting materials, for thioaldehydes, **3**, 337
Thiocyclohexane, 1-azido-1-methyl-, from, 1,1-bis(methylthio)cyclohexane, **4**, 329
Thio-2-cyclohexenones, 2-alkyl-, synthesis of, **3**, 249
Thiocyclopropanes, alkoxybromo-, ring-opening of, **3**, 72
Thiodiimide, bis(diphenylarsino)-, synthesis of, **2**, 891
Thiodiimides
reactions, with malonyl chloride, **5**, 340
as starting materials, for oxathiazines, **5**, 340
Thiodimedone, as starting material, for dimedone, **3**, 347
Thiodisulfides, α-alkyl, synthesis of, **4**, 256
α-Thiodithioacetals, as starting materials, for ketene dithioacetals, **4**, 846
Thiodithioacrylate, (*E,Z*)-3-chloro-3-ethyl-, synthesis of, **4**, 775
Thiodithioacrylic acid ester, β,β-dimethyl, synthesis of, **5**, 559
α-Thio dithioesters, synthesis of, **5**, 558
Thio-1,3-dithioles, 2-phenyl-, from, dithiolium tetrafluoroborate salts, **6**, 85
Thio-1,3-dithiolium salts, 2-alkyl-, electrophilic reactions of, **4**, 852
Thioenamines, 2-nitro-2-phenyl, synthesis of, **4**, 907
β-Thioenamines, synthesis of, **2**, 744
Thioenones, β-alkyl-, substitution of, **1**, 485
Thioenones, α-phenyl-, synthesis of, **3**, 250
Thioepoxides, as nucleophilic compounds, **2**, 890
Thioester, buteneyne, synthesis of, **2**, 1032
α-Thioester complexes, as starting materials, for iron carbene complexes, **3**, 519
Thioesters
aminolysis of, **5**, 586
condensation of, **1**, 682
pyrolysis of, **5**, 475
reactions
with dichloromethyl aryl ethers, **6**, 48
with ketones, **1**, 682
with trialkyl stannyllithiums, **5**, 467
as starting materials
for amidines, **5**, 743
for ketene aminals, **4**, 979
for ketene hemiaminals, **4**, 895
synthesis of, **4**, 1074
via Wolff reactions, **1**, 394
use in the synthesis of amides, **5**, 277
Thioesters, *O*-acyl oxyimidic, synthesis of, **5**, 826
Thioesters, α-acyl oxyimidic, synthesis of, **5**, 829
Thioesters, alkyl, use in the synthesis of acylsilanes, **5**, 418
Thioesters, *O*-alkyl, reactions, with hydrazines, **5**, 640
Thioesters, alkylidene biscarbamic, synthesis of, **4**, 426
Thioesters, alkynyl, synthesis of, **2**, 1030, 1033
Thioesters, β-aminocarboxylic, synthesis of, **5**, 236
Thioesters, *S*-aryl oxyimidic, synthesis of, **5**, 828
Thioesters, carboxylic, as starting materials, for trithioortho esters, **6**, 84
Thioesters, chloroformamidic, synthesis of, **6**, 617
Thioesters, cumulenic, synthesis of, **1**, 990
Thioesters, diethylhydrazonic, synthesis of, **5**, 846
Thioesters, hydrazonic, as starting materials, for hydrazonamides, **5**, 852
Thioesters, imidic
synthesis of, **5**, 863
use in the synthesis of oxyimidic thioesters, **5**, 826
Thioesters, imidophosphorane, synthesis of, **5**, 861
Thioesters, iodomethyl, reactions, with zinc, **4**, 396
Thioesters, α-keto, from, 4-methoxy-3-nitroacetophenone, **5**, 485
Thioesters, α-keto oxyimidic, synthesis of, **5**, 830
Thioesters, lithiomethyl, synthesis of, **5**, 237
Thioesters, oxyimidic
from
alkyl thioimidates, **5**, 830
carbonyls, **5**, 830
fulminates, **5**, 828
nitro compounds, **5**, 829
oxyimidic halides, **5**, 828
oxyimidic thioacids, **5**, 831
oxidation of, **5**, 832
synthesis of, **5**, 826, 826
Thioesters, (*E*)-oxyimidic, synthesis of, **5**, 829
Thioesters, (*Z*)-oxyimidic, synthesis of, **5**, 829
Thioesters, *S*-phenyl, synthesis of, **4**, 906
Thioesters, pyridyl, use in the synthesis of pyrrolic ketones, **3**, 308
Thioesters, ring-opened, synthesis of, **4**, 1075
Thioesters, *S*-trihalomethyl, synthesis of, **6**, 235
Thioesters, vinylogous, synthesis of, **1**, 485
Thioethanethiols, β-alkyl-, synthesis of, **2**, 128
Thioether, bis(trimethylsilyl) silyl, synthesis of, via Brook type rearrangement, **4**, 1049
Thioether, bromo, dehydrobromination of, **2**, 1030
Thioether, 1-chloro-1,2,2,2-tetrafluoroethyl *S*-benzyl, as intermediate, **5**, 492
Thioether, ethynyl, from, bromothioether, **2**, 1030
Thioether, β-styrylmethyl, reactions, with lithium di-*n*-butylcuprate, **1**, 484
Thioether complexes
from
carbene complexes, **5**, 951
Fischer complexes, **5**, 951
synthesis of, **5**, 950
Thioethers
cleavage of, **1**, 241
deprotonation of, **1**, 140
desulfurization of, **1**, 436
lithiation of, **1**, 693
as starting materials
for thioacyl bromides, **5**, 487
for α-thio organometallics, **1**, 140
use in the synthesis of thiophenyl ketones, **3**, 299
Thioethers, alkenyl
displacement of, **1**, 484
elimination of, **1**, 1072
for the elimination of thiols, **1**, 1072
solvolysis of, **5**, 475
as starting materials
for 1,3-diynes, **1**, 1072
for 1,3,5-triynes, **1**, 1072
Thioethers, alkynyl
oxidation of, **2**, 1034
reactions, with amines, **5**, 611
as starting materials, for allenyl sulfides, **1**, 954
synthesis of, **2**, 1030
Thioethers, 1-alkynyl, cleavage of, **1**, 999
Thioethers, allenic, from, allylic sulfide, **1**, 980
Thioethers, allenyl, as starting materials, for 1,3-enynes, **1**, 1072
Thioethers, allylic, displacement of, **1**, 483
Thioethers, allylvinyl, [3,3] sigmatropic rearrangement of, **1**, 833
Thioethers, aryl, displacement of, **1**, 484
Thioethers, aryl alkyl, formylation of, **3**, 98
Thioethers, *S*-benzyl
acidic cleavage of, **5**, 475
reactions
with Brönsted acids, **5**, 481
with Lewis acids, **5**, 481
as starting materials
for α-carbamoyl-1,1-dichlorosulfenyl chlorides, **5**, 477
for sulfenyl chlorides, **5**, 492
for thioacyl bromides, **5**, 487

for thioacyl chlorides, **5**, 481
Thioethers, benzyl germyl, as starting materials, for germylbenzylthiols, **4**, 371
Thioethers, benzyl silyl, as starting materials, for α-silylbenzylthiols, **4**, 368
Thioethers, bromofluoro-*S*-benzyl, as starting materials, for thioacyl bromides, **5**, 481
Thioethers, β-bromo-β-phosphonovinyl, synthesis of, **4**, 804
Thioethers, (chloroethynyl), as starting materials, for (thioethynyl) selenides, **2**, 1037
Thioethers, 1-chloro-2,2,2-trifluoroethyl *S*-benzyl, synthesis of, **5**, 492
Thioethers, 1,1-dichloro-2,2,2-trifluoroethyl *S*-benzyl, synthesis of, **5**, 492
Thioethers, ethynyl, synthesis of, **2**, 1031
Thioethers, famesyl cysteinyl, synthesis of, **2**, 133
Thioethers, germyl, reverse Brook rearrangement of, **4**, 371
Thioethers, α-germyl, synthesis of, **4**, 371
Thioethers, halo, dehydrohalogenation of, **2**, 1030
Thioethers, imino, as starting materials, for ketenimines, **3**, 562
Thioethers, *S*-methyl, as starting materials, for trifluorothioacetyl bromide, **5**, 487
Thioethers, methyl phenyl, as starting materials, for α-thioalkylboranes, **4**, 372
Thioethers, nitroso, from, thionitrites, **2**, 391
Thioethers, α-phenyl, reductive lithiation of, **4**, 380
Thioethers, silyl
  reverse Brook rearrangement of, **4**, 368
  Wittig rearrangement of, **2**, 131
Thioethers, *S*-silyl alkynyl, reactions, with imines, **5**, 605
Thioethers, α-stannyl, synthesis of, **4**, 397
Thioethers, trimethylsilyl, synthesis of, **2**, 263, 541
Thioethylamine hydrobromide, 2,2-difluoro-1-ethyl-, synthesis of, **4**, 320
Thioethynes, phenyl-
  reactions
    with tributylstannylcuprate, **4**, 962
    with triethylgermane, **4**, 940
  synthesis of, **2**, 1031
Thioethynylide, reactions, with carbonyls, **2**, 1031
4-Thioflavone, synthesis of, **3**, 362
Thiofluorenone *S*-oxide, reactions, with sodium trichloroacetate, **4**, 742
Thiofluorenone-*S*-oxide, use in the synthesis of sulfinates, **2**, 209
Thioformaldehyde
  reactions, with (halo)phosphines, **4**, 123
  synthesis of
    via cycloreversion, **3**, 334
    via thermolysis of sulfenyl chlorides, **3**, 337
    via thermolysis of 1,2,4-trithiolane, **3**, 337
Thioformaldehyde, cyano, synthesis of, from dibromoacetonitrile, **3**, 334
Thioformaldehyde, dideutero-, synthesis of, via cycloreversion, **3**, 334
Thioformamide, dimethyl-
  reactions, with 1-naphthylamine, **5**, 745
  as reagent, for the synthesis of thioketones, **3**, 357, 363
Thioformamide, phenyl-, bromination of, **6**, 610
Thioformamides
  reactions, with bromine, **6**, 540
  as starting materials
    for formamidines, **5**, 744
    for thiocarbamoyl chlorides, **6**, 539
Thioformamides, acyl-, synthesis of, **5**, 632
Thioformamides, alkanesulfonyl-, chlorination of, **6**, 238
Thioformamides, alkyl-, synthesis of, **5**, 632
Thioformamides, *N*-aryl-
  reactions, with cyanogen bromide, **3**, 721
  as starting materials, for aryl thiocyanates, **5**, 1112
Thioformamides, *N*,*N*-bis(trimethylsilyl)-, as starting materials, for *N*,*N*-bis(trimethylsilyl) enamines, **2**, 814
Thioformamides, *N*,*N*-diaryl-, chlorination of, **6**, 238
Thioformamides, *N*,*N*-disubstituted chloro-, chlorination of, **6**, 238
Thioformamides, *N*,*N*-disubstituted fluoro-, fluorination of, **6**, 238
Thioformamidinium iodide, *N*,*N*,*N'*,*N'*-tetramethylmethyl-, reactions, with methylenes, **4**, 973
Thioformamidinium salts, as starting materials, for ketene aminals, **4**, 973
Thioformamidinium salts, chloro-, reactions, with aromatic compounds, **5**, 731
Thioformamidoyl chloride, *N*-alkyl-, reactions, with lithium diphenylphosphide, **6**, 584
Thioformamidoyl chloride, *N*-aryl-, reactions, with lithium diphenylphosphide, **6**, 584
Thioformamidoyl chlorides, reactions, with lithium trimethylsilylphosphides, **6**, 584
Thioformanilide, from, formanilide, **5**, 566
Thioformate, ethyl, reactions, with 1-methyl-1-phenylhydrazine, **5**, 853
Thioformate, *O*-ethyl, reactions, with tris(trimethylsilyl)methyllithium, **6**, 176
Thioformate, ethyl chloro-, reactions, with amines, **6**, 492
Thioformate, *O*-phenyl chloro-
  Friedel–Crafts thioacylation of, **5**, 513
  reactions, with (nitrophenylamino)aniline, **6**, 556
Thioformate, propargyl chloro-, synthesis of, **6**, 533
Thioformate, *S*-trifluoromethyl bromo-, reactions, with mercury(II) trifluoromethaneselenolate, **6**, 237
Thioformates, alkyl chloro-
  reactions
    with dibutylphosphonous acid, **6**, 497
    with thionoamides, **5**, 632
  synthesis of, **6**, 534
Thioformates, *O*-alkyl chloro-
  reactions
    with amines, **6**, 556
    with carbanions, **5**, 513
    with chlorocarbonylsulfenyl chloride, **6**, 480
  as starting materials, for thiocarbamic acid *O*-esters, **6**, 556
Thioformates, aryl chloro-
  synthesis of, **6**, 533
  use in the synthesis of thiocarbonic acid, **6**, 533
Thioformates, *O*-aryl chloro-
  reactions
    with amines, **6**, 556
    with bis(trimethylsilylamines), **6**, 556
  as starting materials, for thiocarbamic acid *O*-esters, **6**, 556
Thioformates, chloro-
  from
    alcoholates, **6**, 533
    alcohols, **6**, 533
    alkoxytrimethylsilanes, **6**, 533
    naphthols, **6**, 533
    phenols, **6**, 533
  as starting materials, for thiocarbamates, **6**, 492
  synthesis of, **6**, 533
Thioformates, fluoro-, synthesis of, **6**, 533
Thioformates, phenyl chloro-, synthesis of, **6**, 533
Thioformates, phosphinyl alkyl, synthesis of, **6**, 497
Thioformic acid
  ab initio calculation of, **5**, 507
  conforms of, **5**, 507
  from, phenyl formate, **5**, 506
  IR spectrum of, **5**, 506
  tautomers of, **5**, 507
Thioformic acid, *S*-benzyl imidate, use in the synthesis of oxyimidic thioesters, **5**, 831
Thioformic acid esters, *O*-alkyl chloro-, fluorination of, **6**, 229
Thioformic acid esters, guanidino-, synthesis of, **6**, 561
Thioformic acids, equilibrium of, **5**, 506
Thioformic *O*-acids

geometries of, **5**, 507
potential-energy distribution of, **5**, 507
vibrational frequencies of, **5**, 507
Thioformimidates, chloro-, hydrolysis of, **6**, 494
Thioformimidates, 1-chloro-, as starting materials, for thiocarbamates, **6**, 494
Thioformimidates, 1-halo-, synthesis of, **6**, 617
Thioformyl chlorides, synthesis of, **4**, 71
Thioformyl compounds, from, heterocyclic amines, **5**, 836
Thioformyl ligand, in osmium complexes, **5**, 650
1-Thio-D-galactose, 2,3,4,6-tetra-*O*-acetyl-, synthesis of, **4**, 223
α-Thiogalliums, synthesis of, **4**, 397
α-Thiogermanes
from, α-germyl organomagnesiums, **4**, 371
synthesis of, **4**, 371
1-Thioglucopyranoses, *S*-alkylation of, **4**, 225
Thioglucosides, *S*-ethyl, synthesis of, **4**, 224
1-Thioglycerol, reactions, with acetone, **4**, 229
Thioglycolic acid
reactions, with trimethylgold, **2**, 274
as starting material, for dithioacetals, **4**, 272
use in the synthesis of thioboranes, **2**, 268
Thioglycolic acid, thiobenzoyl-, synthesis of, **5**, 83
Thioglycosides
from, glycosyl fluorides, **4**, 225
oxidation of, **4**, 232, 235
synthesis of, **4**, 223
Thioglycosides, phenyl, as starting materials, for glycosyl fluorides, **4**, 45
Thioglycosides, pyridyl, synthesis of, **4**, 224
1-Thioglycosides, synthesis of, **4**, 223
Thioglyoxal, *t*-butyl, synthesis of, **3**, 351
α-Thiogolds, synthesis of, **4**, 396
Thio group, trihalomethyl-, as superhalogens, **6**, 228
Thio groups, alkyl-, nucleophilic displacement of, **4**, 899
Thio halides, α-phenyl-, for the alkylation of ketones, **3**, 124
Thiohippurate, methyl 2-benzyl-, oxidative cleavage of, **4**, 309
Thiohydantoins
as starting materials, for imidazolidinones, **4**, 420
synthesis of, **6**, 575
Thiohydantoins, *N*-aryl-, as starting materials, for thiocyanates, **5**, 1111
2-Thiohydantoins, as starting materials, for imidazothiazoles, **6**, 575
Thiohydrazides
alkylation of, **5**, 846
oxidation of, **5**, 846
as starting materials, for thiohydrazonic esters, **5**, 846
Thiohydroxamate esters, as starting materials, for alkyl halides, **2**, 13
Thiohydroxamates, *O*-acyl, synthesis of, **4**, 326
Thiohydroxamic acid anhydrides, aryl-, Beckmann rearrangement of, **5**, 832
Thiohydroxamic acids
reactions, with benzophenone dichloride, **5**, 1046
reduction of, **1**, 935
use in the synthesis of radicals, **1**, 940
See also Hydrozylamines, *N*-thionoacyl-
Thiohydroxamic acids, *N*-alkyl-, reactions, with alkyl iodides, **3**, 435
Thiohydroximates, synthesis of, **3**, 426
Thiohydroxylamines
from, sulfenic acids, **2**, 766
nomenclature, **2**, 766
selenium analogues of, synthesis of, **2**, 768
synthesis of, **2**, 357, 360, 766
via reduction of disulfides, **2**, 357
tellurium analogues of, synthesis of, **2**, 768
See also Sulfenamides
Thioic acid *O*-alkyl esters, reactions, with aluminum selenide, **5**, 563
Thioic acid esters, as starting materials, for dithioic acid esters, **5**, 551
Thioic acid *S*-ethyl ester, chloro-, use in the synthesis of dithioic acid esters, **5**, 550
Thioic acid *S*-phenyl ester, chloro-, as starting material, for dithioic acid *S*-phenyl esters, **5**, 550
5-Thio-L-idose, synthesis of, **4**, 228
Thioimidate ester hydrochloride salts, as starting materials, for trithioortho esters, **6**, 83
Thioimidate ester salts, as starting materials, for trithioortho esters, **6**, 83
Thioimidate *N*-oxides
hydrolysis of, **5**, 327
synthesis of, **3**, 435
Thioimidates
from
amides, **4**, 517
monothioacetals, **6**, 111
trimethylsilylmethyl isocyanide, **4**, 518
reactions
with hydrogen sulfide, **5**, 611
with sodium diethyl phosphite, **6**, 122
as starting materials
for imidic acid esters, **5**, 694
for thiols, **3**, 165
synthesis of, **4**, 518
as tautomers of ketene *S*,*N*-acetals, **4**, 898
use in the synthesis of oxyimidic thioesters, **5**, 830
See also Esters, thioimidic
Thioimidates, alkyl, as starting materials, for oxyimidic thioesters, **5**, 830
Thioimidates, *S*-allyl, thermolysis of, **5**, 614
Thioimidates, α-oxo-, from, alkyl isothiocyanates, **5**, 729
Thioimidates, *S*-phosphoryl, synthesis of, **5**, 643
Thioimidate salts, *N*-alkyl, synthesis of, **5**, 728
Thioimidazoles, 2-allyl-, [3,3] sigmatropic rearrangement of, **1**, 837
Thioimides
from
N-haloimides, **5**, 336
imides, **5**, 336
N-silylimides, **5**, 338
oxidation of, **5**, 342
synthesis of, **5**, 336
via reduction of disulfides, **5**, 338
N-Thioimides, from, imides, **5**, 336
Thioimidic acid esters, α-aryl-, synthesis of, **5**, 691
Thioimidic anhydrides, synthesis of, **5**, 727
Thioimidoyl chloride, as starting material, for α,β-unsaturated thioimidic esters, **5**, 729
Thioimidoyl chloride, *N*-phenyl phenyl-, as starting material, for thioimidic esters, **5**, 728
Thioimidoyl chlorides, reactions, with carbon nucleophiles, **5**, 728
α-Thioiminium salts, cyclization of, **2**, 749
Thioindigo, synthesis of, **3**, 251
α-Thioindiums, synthesis of, **4**, 397
Thioindoxyls, as starting materials, for aromatic ketones, **3**, 247
Thioisatins, *N*-aryl-, as starting materials, for thiocyanates, **5**, 1111
Thioisocoumarins, reactions, with amines, **5**, 586
Thioisocyanides, α-alkyl, synthesis of, **3**, 709
α-Thioisocyanides, synthesis of, **3**, 708
Thioketals, polyfluorinated, reactions, with thiolates, **4**, 775
Thioketene, as starting material, for ketenimines, **3**, 587
Thioketene, bis(trifluoromethyl)-
cycloaddition of, for the synthesis of 1,3-dithietanes, **6**, 312
reactions
with benzeneselenol, **5**, 563
with *O*,*S*-dialkyl dithiocarbonates, **6**, 311
with 1,3-dithiolane-2-thione, **6**, 305
with hydrogen bromide, **5**, 487

with methyl isothiocyanate, **6**, **295–318**
with polyfluorinated phenyl imines, **4**, 905
Thioketene, di-*t*-butyl-
reactions, with dimethyl diazomalonate, **4**, 837
synthesis of, **3**, 547
Thioketene, dimethyl-, synthesis of, **3**, 550
Thioketene, diphenyl-, synthesis of, **3**, 547
Thioketene, methylene-, synthesis of, **3**, 550
Thioketene, triphenylphosphoranylidene, synthesis of, **3**, 546, 547
Thioketenes
[2+2] cycloadditions of, **5**, 604
with isocyanates, **5**, 639
dimerization of, **3**, 546
electronic factors, **3**, 546
from
acid chlorides, **3**, 550
2-alkylidene-1,3-dithiolane 1,1-dioxides, **3**, 550
2-alkylidene-1,3-dithiolanes, **3**, 550
2-alkylidene-1,3-oxathioles, **3**, 551
2-alkylidene-1,3,4-thiadiazolines, **3**, 551
alkynyl sulfides, **3**, 548
alkynyl thiolates, **3**, 548
2,4-bis(alkylidene)-1,3-dithietanones, **3**, 550
carbon disulfide, **3**, 549
carbon monosulfide, **3**, 550
carboxylic acids, **3**, 547
carboxylic esters, **3**, 547
cyclobutanethiones, **3**, 550
diazetinethiones, **3**, 550
diazo compounds, **3**, 550
1,3-dithietane, **5**, 511
dithietanones, **3**, 550
dithiocarboxylic acids, **3**, 547
1,2-dithiolene-3-thiones, **3**, 552
1,3-dithiolen-2-ones, **3**, 552
1,3-dithiolen-2-thiolones, **3**, 552
heterocyclic materials, **3**, 550
2,2,4,4,6,6-hexamethyl-1,3,5-trithiane, **3**, 551
isothiazoles, **3**, 552
isothiazol-5-thiones, **3**, 552
ketene-*N*,*S*-acetals, **3**, 550
ketene-*O*,*S*-acetals, **3**, 550
ketene-*S*,*X*-acetals, **3**, 550
ketenes, **3**, 546
1,2,3-thiadiazoles, **3**, 551; **5**, 511
thiaphosphetanes, **3**, 550
thioacetic acid, **3**, 547
thioacyl chlorides, **3**, 550
transition metal complexes, **3**, 552
as intermediates, in the synthesis of thionoamides, **5**, 589
oligomerization of, **3**, 546
oxidation of, **2**, 202
reactions
with alcohols, **5**, 511
with alkyllithiums, **4**, 956
with amines, **5**, 589
with aryllithiums, **4**, 956
with hydrogen chloride, **5**, 480
with imines, **4**, 905
with sulfonyl isocyanates, **5**, 344
with trithiocarbonates, **6**, 305
with ynamines, **5**, 605
stability of, **3**, 546
as starting materials
for alkanesulfinic esters, **2**, 202
for ketene dithioacetals, **4**, 852
for thioacyl bromides, **5**, 487
for thioacyl chlorides, **5**, 480
for thiocarboxylates, **5**, 509, 511
substituent effects, **3**, 546
synthesis of, **3**, 546; **5**, 589
overviews, **3**, 546
techniques, **3**, 546
via thermolysis of *O*-trimethylsilyl thiocarboxylates, **5**, 539
via Wolff rearrangement, **3**, 551
as thioacylating agents, **5**, 550
zirconation of, **3**, 337
Thioketenes, allenic, synthesis of, **3**, 548
Thioketenes, allyl-, synthesis of, via thia-Cope rearrangement, **3**, 548
Thioketenes, bis(trifluoromethyl)-, reactions, with triethyl phosphite, **4**, 926
Thioketenes, cumulated, synthesis of, **3**, 547
Thioketenes, propargylic, synthesis of, **3**, 548
Thioketenes, silyl-, synthesis of, **2**, 941
Thioketenes, trimethylsilyl-, from, trimethylsilyl alkynyl sulfides, **3**, 548
Thioketenimines, *C*-alkyl-, synthesis of, **3**, 563
Thioketenoids
reactions, with amines, **5**, 589
synthesis of, **5**, 589
Thioketone, reactions, with diphenyldiazomethane, **3**, 363
Thioketone, bis(trimethylsilyl)
from, tris(trimethylsilyl)methane, **6**, 585
stability of, **6**, 584
synthesis of, **6**, 584
use in the synthesis of alkyltrimethylsilyl dithioformates, **6**, 584
Thioketone, t-butyl α-trimethylsilyl, synthesis of, **5**, 648
Thioketone, di-*t*-butyl
reactions, with allenylmagnesium bromide, **1**, 1058
synthesis of, **3**, 351
Thioketone, methyl α-trimethylsilyl, synthesis of, **5**, 648
Thioketone, 2,4,6-trimethylphenyl α-trimethylsilyl, attempted synthesis of, **5**, 648
Thioketone *S*-oxides, reactions, with ynamines, **5**, 623
Thioketones
bromination of, **2**, 171
cycloaddition of, with diazo compounds, **1**, 685
desulfurization of, **1**, 24
from
2-alkyl 1,3-thiazines, **3**, 350
allyl sulfides, **3**, 351
Bunte salts, **3**, 349, 365
diketones, **3**, 343, 345
1,2-dithiole-3-iminium salts, **3**, 357
1,2-dithiole-3-thiones, **3**, 356, 357
dithiole-3-thiones, **3**, 361
1,2-dithiolium salts, **3**, 361
dithiols, **3**, 349
enamines, **3**, 363
imines, **3**, 348, 362
iminium salts, **3**, 356, 361
iminovinyl chlorides, **3**, 360
isothiazolium salts, **3**, 356, 361
ketones, **3**, 342
oxathiolanones, **3**, 350
rhodanine, **3**, 350
seleno ketones, **3**, 348
thiazolidindiones, **3**, 350
vinylic chlorides, **3**, 356
intramolecular Diels–Alder reaction of, **3**, 360
metal complexes of, **3**, 370, 375
oligomerization of, **3**, 330
polymerization of, **3**, 330
reactions
with arsonium ylides, **1**, 741
with 1,2-bis(methylthio)ethyne, **5**, 559
with carbanions, **1**, 1040
with diazoalkanes, **4**, 328
with diazomethane, **2**, 243; **6**, 128
with hydrazonyl chloride, **6**, 128
with nitrile oxides, **6**, 127
with nitronates, **4**, 232

with nitrones, **4**, 232
with organolithiums, **4**, 392
with ynamines, **5**, 614
reductive coupling of, for the synthesis of alkenes, **1**, 690
stability of, **3**, 342
as starting materials
for thiocarbonyl *S*-imides, **3**, 370
for thiocarbonyl ylides, **3**, 367
synthesis of, **3**, **329–380**
via [3+2] cycloaddition, **3**, 356, 361
via [2+2] cycloreversion, **3**, 365
via cycloreversion of dithiolanes, **3**, 349
via cycloreversion reactions, **3**, 351, 365
via 1,2-elimination reactions, **3**, 349, 365
via formation of α carbon–carbon bonds, **3**, 349, 364
via Friedel–Crafts reaction, **3**, 364
via oxidative sulfuration, **3**, 363
via photolysis, **3**, 351
via photolysis of phenacyl sulfides, **3**, 351
via rearrangement, **3**, 351, 353
via reduction of sulfines, **3**, 351, 365
via retro Diels–Alder reaction, **3**, 353
via ring-opening of heterocycles, **3**, 350
via sulfuration of acetals, **3**, 347
via sulfuration of enamines, **3**, 348
via sulfuration of enol ethers, **3**, 347
via sulfuration of hydrazones, **3**, 347
via sulfuration of imines, **3**, 348, 363
via sulfuration of ketones, **3**, 343, 352, 354, 362, 378
via sulfur–selenium exchange, **3**, 348
via thermolysis, **3**, 351, 353
via Vilsmeier–Haack reaction, **3**, 363
via Willgerodt–Kindler reaction, **3**, 361
trapping as Diels–Alder adducts, **3**, 349, 349, 360, 365
trimer formation of, **3**, 343
use in the synthesis of disulfides, **2**, 228
See also Thiones
Thioketones, aliphatic, synthesis of, **3**, 345
Thioketones, alkenyl
Diels–Alder dimerization of, **3**, 360
Diels–Alder dimerization, stability of, **3**, 352
stability of, **3**, 352
synthesis of, **3**, 352, 366
via retro Diels–Alder reaction, **3**, 360
Thioketones, α,β-alkenyl, synthesis of, **3**, 353
Thioketones, alkenyl aryl, synthesis of, **3**, 360
Thioketones, alkyl aryl, synthesis of, **3**, 360
Thioketones, alkyl α-silyl, synthesis of, **5**, 648
Thioketones, α-allyl, synthesis of, **3**, 351
Thioketones, aryl
reactions, with amines, **3**, 361
synthesis of, via photolysis, **3**, 366
Thioketones, aryl hetaryl, synthesis of, **3**, 366
Thioketones, aryl α-silyl, synthesis of, **5**, 648
Thioketones, biaryl, reduction of, **1**, 24
Thioketones, α-boro, synthesis of, **5**, 650
Thioketones, chiral α-silyl, synthesis of, **5**, 648
Thioketones, dialkyl, synthesis of, **3**, 342, 347, 349
Thioketones, diaryl
from, *gem*-dichlorides, **3**, 363
stability of, **3**, 342
Thioketones, di-*t*-butyl, synthesis of, **3**, 348
Thioketones, *S*,*S*-dimethoxy, synthesis of, **3**, 375
Thioketones, dithiovinyl, synthesis of, **3**, 356
Thioketones, enamino
from, β-thioxo aldehydes, **3**, 356
synthesis of, **3**, 354, 360
Thioketones, α-germyl, synthesis of, **5**, 650
Thioketones, hetaryl, synthesis of, **3**, 363
Thioketones, heteroaryl α-silyl, synthesis of, **5**, 648
Thioketones, β-keto, synthesis of, **3**, 348
Thioketones, α-oxo, synthesis of, **3**, 349
Thioketones, β-oxo
synthesis of, **3**, 349, 363
from β-oxo ketones, **3**, 344
tautomerism of, **3**, 344
Thioketones, β-oxo aryl, synthesis of, **3**, 364
Thioketones, perfluoro, synthesis of, **3**, 348, 349
Thioketones, phenyl trimethylsilyl, synthesis of, **6**, 127
Thioketones, α-silyl
catalysts for the synthesis of, **5**, 648
from, α-silyl ketones, **5**, 647
synthesis of, **5**, 647
use in the synthesis of Lawesson's reagent, **5**, 648
Thioketones, β-thioketo, synthesis of, **3**, 363
Thioketones, α-trialkylsilyl, synthesis of, **5**, 648
Thioketones, α-triarylsilyl, synthesis of, **5**, 648
Thioketones, α,β-unsaturated, synthesis of, **3**, 352
2-Thioketones, aldol condensation of, **3**, 247
Thiol, trichloromethyl-
reduction of, **6**, 529
as starting material, for thiophosgene, **6**, 529
β-Thiol acids, synthesis of, **1**, 242
Thiolactams
from
amides, **5**, 333
N-haloamides, **5**, 335
N-silylamides, **5**, 335
reactions
with alkylating agents, **5**, 726
with diazomethane, **5**, 726
as starting materials, for thioimidic esters, **5**, 726
N-Thio-β-lactams, synthesis of, **5**, 335
Thiolane, 2,2-diethoxy-, from, 2-ethoxythiolenium tetrafluoroborate, **6**, 96
Thiolane 1,1-dioxide, dibromo-, reactions, with bromoform, **4**, 748
Thiolanes, 3-hydroxy-, synthesis of, **1**, 286
Thiolate, 1,1-dimethylethane, reactions, with allyl bromides, **4**, 373
4-Thiolate, 5-methyl-2-morpholino-1,3-dithiolylium-, reactions, with pentane-2,4-dione, **3**, 255
Thiolate anions
reactions
with α-halo ketones, **3**, 164
with nitroalkenes, **3**, 165
Thiolates
reactions
with alkynyl halides, **2**, 1033
with arenediazonium salts, **2**, 711
with 2-bromo-2-nitropropane, **4**, 327
with α-chloro ethers, **4**, 218
with dithiolium salts, **6**, 85
with nitriles, **2**, 240
with 2-nitropropane, **4**, 327
with α-nitro sulfones, **4**, 328
with polyfluorinated thioketals, **4**, 775
with prop-2-yniminium triflates, **4**, 321
as starting materials, for selenocarbonyls, **6**, 592
synthesis of, **5**, 480
use in the synthesis of dialkyl sulfides, **2**, 132
Thiolates, alkane, reactions, with tetrachloromethane, **6**, 81
Thiolates, alkyl, as starting materials, for dialkyl sulfides, **2**, 132
Thiolates, alkyne, reactions, with tungsten carbene complexes, **5**, 651
Thiolates, alkynyl, as starting materials, for thioketenes, **3**, 548
Thiolates, allyl, reactions, with *n*-butyllithium, **1**, 140
Thiolates, arene, reactions, with perchloro-1,3-butadiene, **4**, 851
Thiolates, benzyl, reactions, with *n*-butyllithium, **1**, 140
Thiolates, dithiocarboxylic ene-, electrophiles for, **4**, 842
Thiolates, silyl, as starting materials, for thiosilanes, **2**, 264
Thiolates, stannous, reactions, with ketenes, **5**, 236
Thiolates, thionoester ene-, *S*-acylation of, **4**, 834

Thiolates, vinyl
as starting materials, for vinyl disulfides, **2**, 722
synthesis of, **2**, 722
Thiolate salts, thiazolyl, oxidation of, **2**, 358
Thiolating reagents, reactions, with alkynylides, **2**, 1032
Thiol carbamates, *S*-vinyl, synthesis of, **6**, 496
Thiol carbonate *S,S*-dioxides, *O*-methyl-, synthesis of, **6**, 480
Thiolenium tetrafluoroborate, 2-ethoxy-, reactions, with sodium ethoxide, **6**, 96
Thioles, 2-methanethio-1,3-seleno-, from, 2-methanethio-1,3-selenothiolium salts, **6**, 101
Thiolformate, β-chloroalkyl chloro-, synthesis of, **6**, 434
Thiolformate, methyl bromo-, from, methanesulfenyl bromide, **6**, 437
Thiolformate, trifluoromethyl bromo-, from, trifluoromethylfluorothiolformate, **6**, 436
Thiolformate, trifluoromethyl fluoro-, transhalogenation of, with boron tribromide, **6**, 436
Thiolformate esters, bromo-, synthesis of, **6**, 436
Thiolformate esters, chloro-
from
alkoxydichloromethanesulfenyls, **6**, 435
benzenethiols, **6**, 434
carbon monoxide, **6**, 434
phosgene, **6**, 434
sulfenyl chlorides, **6**, 434
thiiranes, **6**, 434
thiols, **6**, 434
trichloromethanethiols, **6**, 434
synthesis of, **6**, 434
transhalogenation of, with hydrogen fluoride, **6**, 432
Thiolformate esters, fluoro-
from
carbonic dihalides, **6**, 432
chlorothiolformates, **6**, 432
fluorocarbonylsulfenyl bromide, **6**, 432
fluorocarbonylsulfenyl chloride, **6**, 432
rearrangement of aryl fluorothionoformates, **6**, 432
synthesis of, **6**, 432
via halogen exchange, **6**, 432
Thiolformate esters, halo-, nomenclature for, **6**, 432
Thiolformate esters, iodo-, synthesis of, **6**, 437
Thiolformates
reactions, with phosphorus pentachloride, **6**, 46
as starting materials, for α,α-dichloro sulfides, **6**, 46
Thiolformates, chloro-
stability of, **6**, 434
as starting materials, for fluorothiolformate esters, **6**, 432
Thiolformates, β-fluoroalkyl fluoro-, synthesis of, **6**, 433
Thiolic acid esters, α-keto, as starting materials, for α-alkoximinocarboxylic acids, **3**, 427
Thiol ions
reactions
with oxyimidic chlorides, **5**, 828
with oxyimidic iodides, **5**, 828
α-Thiolithiums, synthesis of, **4**, 391
Thiolium salts, halo-, synthesis of, **6**, 48
Thiolium salts, 2-methanethio-1,3-seleno-, reduction of, to 2-methanethio-1,3-selenothioles, **6**, 101
Thiols
boron derivatives, for the synthesis of dithioacetals, **4**, 248
cleavage of, **1**, 70
condensation of, **4**, 244
dehydrofluorination of, **5**, 484
deprotonation of, **1**, 140
desulfurization of, **1**, 436, 942
disproportionation reactions, with trithioortho esters, **6**, 89
elimination of, **1**, 1072
exchange reactions
with ortho esters, **6**, 98
with orthoformate esters, **6**, 82
from
Bunte salts, **2**, 122
disulfides, **2**, 708; **3**, 165
dithiocarbamates, **2**, 124
sulfonic acids, **2**, 708
thioimidates, **3**, 165
thiol esters, **2**, 124
oxidation of, **2**, 177, 185, 186, 192; **2**, 212, 237
oxidative chlorination of, **2**, 173
reactions of, **4**, 940
reactions
with acetals, **4**, 249
with 1-*O*-acetyl glycopyranosides, **4**, 224
with acyl alkoxycarbonyl disulfides, **5**, 247
with acyl chlorides, **5**, 233
with acylsulfenyl chlorides, **5**, 246
with aldehydes, **4**, 247
with alkenes, **2**, 137, 712
with alkoxyalkylchlorophosphines, **2**, 461
with alkoxycarbene complexes, **5**, 951
with alkylboranes, **2**, 533
with alkyldichlorophosphines, **2**, 461
with alkyldihaloarsines, **2**, 500
with alkyldihalobismuthines, **2**, 500
with alkyldihalostibines, **2**, 500
with alkyl thiocyanates, **6**, 627
with alkynes, **2**, 710; **4**, 939
with alkynic sulfones, **4**, 861
with allenylnitriles, **3**, 655
with amide acetals, **6**, 111
with amines, **2**, 240, 358
with antimony trichloride, **2**, 261
with arenediazonium salts, **4**, 225
with areneselenenic acids, **2**, 735
with arenesulfinyl chlorides, **2**, 723
with arsenoxides, **2**, 260
with aryl selenocyanates, **2**, 735
with aziridine, **4**, 899
with benzoyloxytrithiane, **6**, 82
with benzylideneanilines, **4**, 319
with boron monochlorides, **2**, 267
with bromoacetaldehyde, **3**, 35
with *t*-butyl azidoacetate, **2**, 768
with *t*-butyl azidoformate, **2**, 768
with carbamoyl chlorides, **6**, 493
with carbodiimides, **6**, 635
with carbon disulfide, **6**, 552
with carbonic bromide fluoride, **6**, 432
with carbonic chloride fluoride, **6**, 432
with carbonimidic dichlorides, **6**, 626
with carbon monosulfide, **5**, 550
with α-chloro dithioacetals, **6**, 82
with β-chloroenone, **3**, 252
with chlorofluoroalkenes, **6**, 47
with 1-(chloromethyl)benzotriazole, **4**, 322
with chloromethyl chloroformates, **4**, 60
with cyanamides, **6**, 635
with cyanates, **6**, 629
with dialkylchlorophosphines, **2**, 460
with dialkyl(diethylamino)arsines, **2**, 500
with dialkyl telluride, **2**, 238
with *gem*-diamines, **4**, 249
with diaminoarsines, **2**, 501
with diethyl azodicarboxylate, **2**, 217
with difluoroalkenes, **6**, 46
with difluorocarbene, **6**, 45
with 1,1-dihaloalkenes, **6**, 46
with diiodomethane, **4**, 249
with dimethyl disulfide, **2**, 219
with dioxocarbenium salts, **6**, 96
with epoxides, **2**, 136
with esters, **5**, 236
with fluoroalkyne, **4**, 772
with α-halo ethers, **4**, 218

with heterocycles, **2**, 235
with imidoyl chlorides, **5**, 728
with isocyanates, **6**, 494
with isocyanides, **5**, 1036
with isonitriles, **5**, 728
with isothiocyanates, **6**, 565
with α-ketoaldehydes, **4**, 215
with ketones, **2**, 709; **4**, 247
with methyl chloroformate, **6**, 475
with methyllithium, **2**, 272
with *N*-methyl-*N*-nitrosochloromethylamine, **4**, 325
with nitrile oxides, **5**, 826, 828
with nitriles, **5**, 727
with nitrilium ions, **5**, 728
with nitro compounds, **5**, 829
with organomercury compounds, **1**, 68
with ortho esters, **6**, 95
with oxygen, **2**, 214
with phosgene, **6**, 434, 471
with phosphacumulene ylides, **5**, 730
with phosphinic chlorides, **4**, 128
with phosphinthioyl chlorides, **2**, 252
with phosphorus, **2**, 252
with propenamidrazones, **5**, 770
with prop-2-yniminium triflates, **4**, 321
with selenites, **2**, 236
with selenocarbonyl *N*-benzenesulfonylimide, **2**, 235
with selenocyanates, **2**, 735
with silylated thioacetals, **4**, 247
with sodium, **2**, 271
with sodium hydroxide, **2**, 270
with sulfenyl thiocarbonates, **2**, 220
with thiocarbamoyl chlorides, **6**, 566
with thiocarbonyl chlorides, **6**, 550
with thiophosgene, **6**, 552
with thiosulfinates, **2**, 219
with trialkyl phosphite, **1**, 942
with trichloroacetaldehyde, **4**, 215
with 2-trichloromethylbenzimidazole, **6**, 81
with trimethylantimony dichloride, **2**, 502
with 1-(trimethylsilyl)imidazole, **2**, 263
with trityl chloride, **2**, 133
with α,β-unsaturated aldehydes, **2**, 138
reduction of, **1**, 21
silicon derivatives, for the synthesis of dithioacetals, **4**, 248
silylation of, reagents for, **2**, 263
as starting materials
for alkylthiophosphonium salts, **2**, 299
for bromodithioformates, **6**, 536
for α-bromo sulfides, **4**, 71
for chlorodithioformates, **6**, 535
for α-chloro sulfides, **4**, 69
for chlorothiolformate esters, **6**, 434
for disulfides, **2**, 223
for fluorothiolformate esters, **6**, 432
for sulfenamides, **2**, 240, 768
for thiophosphines, **2**, 252
for thiyl radicals, **1**, 949
storage as sulfides, **2**, 127
synthesis of, **1**, 1058
thioacylation of, **5**, 550
thioalkylation of, **2**, 216
tin derivatives, for the synthesis of dithioacetals, **4**, 248
for the transesterification of tetrathioorthocarbonates, **6**, 306
use in the synthesis of sulfenyl selenides, **2**, 234
Thiols, alkane-
reactions
with alkyl chloroformates, **6**, 475
with phosphonodithioformates, **6**, **295–318**
Thiols, alkoxy-, synthesis of, **2**, 128
Thiols, alkyl-, reduction of, **1**, 21
Thiols, amino-, reactions, with phenyl isothiocyanate, **6**, 570
Thiols, aryl-
reactions
with ketenimines, **5**, 730
with Raney nickel, **1**, 435
Thiols, benzylic, synthesis of, **2**, 119
Thiols, chlolesteryl, synthesis of, **2**, 127
Thiols, hydroxy-, as starting materials, for alkanesulfinic esters, **2**, 201
Thiols, β-hydroxy-, as starting materials, for alkenes, **1**, 591
Thiols, ω-hydroxy-, synthesis of, **2**, 132
Thiols, α-silyloxy-, alkylation of, **2**, 120
Thiols, stannylated, use in the synthesis of arene thiosulfinate esters, **2**, 723
Thiols, tertiary, synthesis of, **2**, 118
Thiols, trihalomethane-, synthesis of, **6**, 232
Thiolysis
for the synthesis of disulfides, **2**, 216
See also Thioalkylation
Thiomagnesium bromides, arene-, reactions, with ethyl formate, **6**, 84
α-Thiomagnesiums, synthesis of, **4**, 395
Thiomalononitriles, alkyl-, as intermediates, **3**, 655
Thiomaronyl dichlorides, reactions, with *N*-benzylhydrozylamines, **5**, 533
Thio metalloids, imidoyl-, synthesis of, **5**, 739
Thio metals, imidoyl-, synthesis of, **5**, 739
α-Thio metals, synthesis of, **4**, 397
Thiomethane, α-acetoxyphenyl-, from, methyl phenyl sulfide, **4**, 220
Thiomethane, disulfonyl-, from, disulfonyldithiomethane, **6**, 91
Thiomethane, tris(trimethylsilyl)methyl-, oxidation of, **6**, 520
Thiomethanes, disulfonyl-, from, bis(sulfonyl)methanes, **6**, 90
Thiomides, as starting materials, for thiophosphates, **2**, 259
Thiomuscarones, synthesis of, **2**, 235
Thionacetamide, 2-cyano-, chlorination of, **5**, 832
Thionacetates, pyrolysis of, **1**, 777
Thionamides, *N*-phosphorylated, synthesis of, **5**, 643
Thionamides, *N*-thiophosphorylated, *N*-methylation of, **5**, 644
Thionates, dialkyl phosphorochlorido-, reactions, with alkynyllithiums, **2**, 1072
Thionation
of amides, **5**, 610, 613
of carbonyls, **5**, 513
of esters, reagents for, **5**, 512
of formamides, **5**, 608
of imidate esters, **5**, 510
of imides, **5**, 631
mechanism, **5**, 568
methods, reviews, **5**, 565
of 1,8-naphthaloyl chloride, **5**, 528
of tertiary amides, **5**, 617
Thione, enamino-, synthesis of, **3**, 356
Thione, fluorene-, synthesis of, **3**, 363
Thione, methyl phenylethynyl-, synthesis of, **3**, 366
Thione, trimethylsilyl-*t*-butyl-, synthesis of, **4**, 371
Thione, xanthane-, synthesis of, **3**, 362
2-Thione, *N*-hydroxypyridine-, use in the synthesis of radicals, **1**, 935
3-Thione, 1,2-dithiole-, synthesis of, **4**, 834
Thione *S*-imides
cycloaddition of, **5**, 336
reactions
with hydrochloric acid, **5**, 336
with *N*-isobutylpyrrolidine, **5**, 336
Thione-*S*-imines, synthesis of, **2**, 400
2-Thione-1,2,3-phosphilidine, 2-methyl-, reactions, with alkyllithiums, **2**, 463
Thiones
from, imidazolidin-2-one, **6**, 579

photocycloaddition of, **5**, 634
as starting materials
for dialkyl dithiocarbonates, **6**, 471
for 1,4,2-dithiazoles, **3**, 686
synthesis of, **3**, 345; **4**, 343
See also Thioketones
Thiones, diazetine-, as starting materials, for thioketenes, **3**, 550
Thiones, α,α-dioxo-, synthesis of, **3**, 349
Thiones, enamino-, synthesis of, **2**, 744; **3**, 355
Thiones, *N*-hydroxymethylthiadiazole-, reactions, with thionyl chloride, **4**, 105
Thiones, 4(3*H*)-quinazoline, from, *N*-arylimidoyl isothiocyanates, **5**, 1050
Thiones, quinoline-, synthesis of, **3**, 362
Thiones, trimethylsilyl-
S-alkylation of, **4**, 371
cycloaddition of, **4**, 371
2-Thiones, *N*-allyl-, synthesis of, **1**, 837
2-Thiones, 1,3-diselenole-
as starting materials
for monothiatriselenafulvalenes, **4**, 869
for tetraselenafulvalenes, **4**, 874
2-Thiones, 1,3-dithiole-
1,3-dipolar additions of, **4**, 850
reactions
with trialkyl phosphites, **4**, 850
with triphenylphosphine, **4**, 850
2-Thiones, 1,3-thiaselenole-
dimerization of, **4**, 868
as starting materials, for thiaselenafulvenes, **4**, 870
Thione–thiol rearrangement
for the synthesis of bis(methylthiocarbonyl) polysulfanes, synthesis of, **6**, 474
for the synthesis of dialkyl dithiocarbonates, **6**, 472
Thionitrates, reactions, with alkanethiols, **2**, 220
Thionitriles, from, haloacetonitriles, **3**, 634
Thionitriles, bromo-, reactions, with triethylamine, **3**, 655
Thionitriles, α-chloro-β-alkyl-, rearrangement of, **3**, 655
Thionitriles, β-cyclic-α-pyridyl-, from, ω-tellurio-α,β-unsaturated nitriles, **3**, 635
Thionitriles, α,α-diethyl-, from, diethylthioacetaldehyde, **3**, 636
Thionitriles, α-pyridyl-, synthesis of, **3**, 635
α-Thionitriles
from
chloroacetonitrile, **3**, 634
thioacetals, **3**, 634
oxidation of, **3**, 634
synthesis of, **3**, 634
β-Thionitriles
from, β,γ-unsaturated nitriles, **3**, 636
synthesis of, **3**, 636
Thionitrilium salts, alkyl-, synthesis of, **5**, 1145
Thionitrite, *t*-butyl
for the synthesis of aryl bromides, **2**, 627
for the synthesis of aryl chlorides, **2**, 621
Thionitrites
as starting materials
for nitrosothioethers, **2**, 391
for sulfenamides, **2**, 243
Thionitronates, use in thiolysis reactions, **2**, 220
Thionitroso compounds
reactions, with alkenes, **2**, 244
synthesis of, **2**, 398
Thionitroso compounds, alkyl-, as starting materials, with sulfenamides, **2**, 359
Thionitroso compounds, aryl-, synthesis of, **2**, 792
Thionium ions, bisphenyl-, synthesis of, **3**, 283
Thionoacetamide, *N*,*N*-dimethyl-, deprotonation of, **5**, 616
Thionoacetamide, *N*-thiophosphorylated 2-benzoyl-2-cyano-, synthesis of, **5**, 644
Thionoacetate, phenyl, reactions, with cyclic amines, **5**, 586
Thionoacyl compounds, silyl, synthesis of, **5**, 536
Thionoacyloxy compounds, silyl, synthesis of, **5**, 536
B-Thionoacyloxy compounds, synthesis of, **5**, 534
P-Thionoacyloxy compounds, synthesis of, **5**, 534
Thionoamide *S*,*S*-dioxides
as intermediates, theoretical, **5**, 624
synthesis of, **5**, 623
Thionoamide *S*-oxides
reactions, with diazomethane, **5**, 623
reduction of, **5**, 623
synthesis of, **5**, 623
Thionoamides
acylation of, **5**, 630, 630
[4+2]-dimerization of, **5**, 615
from
alkanoamides, **5**, 610
amides, **5**, 566
α-amino-α-cyanocarboxylic acid lactate esters, **5**, 580
carboxylic acids, **5**, 570
isothiocyanates, **5**, 592
α-mercaptophosphonates, **5**, 603
nitriles, **5**, 579, 610
thionoamides, **5**, 606
oxidation of, **5**, 623
reactions
with acyl isocyanates, **5**, 631
with alkyl chlorothioformates, **5**, 632
with alkyl isothiocyanates, **5**, 633
with amide dialkyl acetals, **5**, 633
with aryl isothiocyanates, **5**, 633
with benzenesulfinyl chloride, **5**, 638
with benzenesulfonyl chloride, **5**, 638
with bis acyl halides, **5**, 630
with hexafluoroacetone, **5**, 633
with isocyanates, **5**, 631
with ketenes, **5**, 630
with nitriles, **5**, 630
with phosphorus(III) chlorides, **5**, 643
with sodium hydride, **5**, 632
with sulfonyl isocyanates, **5**, 631
as starting materials
for *N*-acylthionoamides, **5**, 630
for tellurohydrazides, **5**, 642
for thionoamides, **5**, 606
synthesis of, **5**, **565–628**
reviews, **5**, 565
via aminosulfuration, **5**, 599
via cycloaddition reactions, **5**, 604
via sulfuration, **5**, 599
via thiocarbamoylation, **5**, 592
as thioacylating reagents, **5**, 591
Thionoamides, *N*-acyl-
from, thionoamides, **5**, 630
synthesis of, **5**, 629
Thionoamides, *N*-alkenyl-, synthesis of, **5**, 606
Thionoamides, alkyl-, synthesis of, **5**, 595
Thionoamides, alkynyl-, synthesis of, **5**, 597
Thionoamides, allenyl-, synthesis of, **5**, 605
Thionoamides, *N*-aminomethyl-, synthesis of, **5**, 606
Thionoamides, β-amino-α,β-unsaturated, synthesis of, **5**, 586, 614
Thionoamides, α-chloro-, synthesis of, **5**, 577
Thionoamides, α,α-dichloro-, use in the synthesis of thionoalkenoamides, **5**, 617
Thionoamides, *N*,*N*-diethyl-, synthesis of, **5**, 583
Thionoamides, *N*,*N*-dimethyl-, synthesis of, **5**, 600, 603
Thionoamides, *N*,*N*-dimethyl α-oxo-, synthesis of, **5**, 602
Thionoamides, enamino-, synthesis of, **5**, 614
Thionoamides, *N*-ethoxycarbonyl-, synthesis of, **5**, 631
Thionoamides, *N*-hydroxymethyl-, synthesis of, **5**, 606
Thionoamides, β-hydroxy γ,δ-unsaturated, synthesis of, **5**, 616
Thionoamides, indole, synthesis of, **5**, 620

Thionoamides, β-keto-
  reactions
    with acrylonitrile, **5**, 607
    with benzylidenemalononitrile, **5**, 607
  synthesis of, **5**, 586, 597
Thionoamides, β-mercapto-, synthesis of, **5**, 568, 613
Thionoamides, *N*-monosubstituted, from, thionophosphorylcarbonyl mixed imides, **5**, 643
Thionoamides, α-nitro-, synthesis of, **5**, 611
Thionoamides, α-oxo-
  from
    α-chloro ketones, **5**, 601
    α,α-dichloro ketones, **5**, 602
  synthesis of, **5**, 596
Thionoamides, β-oxo-, synthesis of, **5**, 593
Thionoamides, primary, reactions, with hexamethyldisilazane, **5**, 645
Thionoamides, pyridinethione, synthesis of, **5**, 588
Thionoamides, *N*-substituted, synthesis of, **5**, **629–645**
Thionoamides, *N*-sulfonyl-, synthesis of, **5**, 638
Thionoamides, *N*-sulfonyl-α,α-dimethyl-β-dialkylamino-, synthesis of, **5**, 638
Thionoamides, *N*-sulfonyl pyrrole, synthesis of, **5**, 638
Thionoamides, tertiary
  deprotonation of, **5**, 607
  reactions, with butyllithium, **5**, 606
Thionoamides, *N*-thionoacyl-, synthesis of, **5**, 632
Thionoamides, triazole, from, thiadiazole aldehydes, **5**, 622
Thionoamides, *N*-(trimethylsilyl)-, synthesis of, **5**, 645
Thionoamides, α,β-unsaturated, from, unsaturated thionoesters, **5**, 586
Thionoamides, γ,δ-unsaturated, synthesis of, **5**, 614
Thionoamides, ynamino-, synthesis of, **5**, 618
Thionoanhydrides, synthesis of, **5**, 528
Thionobenzamides
  phosphorylation of, **5**, 643
  synthesis of, **5**, 586
Thionobenzamides, *N*-(aryl sulfonyl)-, synthesis of, **5**, 638
Thionobenzoic anhydride, synthesis of, problems, **5**, 528
Thionocarbamates
  reactions
    with sodium hydride, **5**, 632
    with thiobenzoyl chloride, **5**, 632
  synthesis of, **6**, 560
Thionocarbazates, synthesis of, **6**, 560
Thionocarbonate, bis(2-fluoro-2,2-dinitroethyl), synthesis of, **6**, 541
Thionocarbonate, ethylene
  reactions, with *N,N'*-bis(tributylstannyl)-*N,N'*-dimethylethylenediamine, **6**, 322
  use in the synthesis of monochalcogenomethanes, **6**, 337
Thionocarbonate, 5-oxohexyl phenyl, synthesis of, **6**, 542
Thionocarbonate, 5-oxopentyl phenyl, synthesis of, **6**, 542
Thionocarbonate group, as a protective group, **6**, 542
Thionocarbonates
  from
    chlorothionoformates, **6**, 541
    thiocarbonylbis(azoles), **6**, 543
    thiocarbonyldiimidazole, **6**, 543
    thiophosgene, **6**, 540
  reactions, with organosilanes, **1**, 938
  synthesis of, **6**, 540
    via cyclization reactions, **6**, 544
  use in the synthesis of radicals, **1**, 940
Thionocarbonates, diaryl, synthesis of, **6**, 544
Thionocarbonates, diethyl, chlorination of, **6**, 251
Thionocarbonic acid, *O,O*-diesters of, synthesis of, **6**, 540
Thionocarboxylates, thermal rearrangement of, **2**, 124
Thionocarboxylic esters, synthesis of, **5**, 510
Thionoester, peroxy-, homolytic decomposition of, **5**, 530
Thionoester, β-thioketo, synthesis of, **3**, 350
Thionoesters
  S-alkylation of, **4**, 833, 834
  aminolysis of, **5**, 586
  chlorination of, for chloroformates, **6**, 426
  cycloadditions of, **4**, 229
  from, protected amino acids, **5**, 586
  as starting materials
    for ethers, **2**, 89
    for thioketones, **3**, 349
  for the thioacylation of amines, **5**, 584
Thionoesters, cyanoacetic, reactions, with benzylamine, **4**, 896
Thionoesters, unsaturated, as starting materials, for α,β-unsaturated thionoamides, **5**, 586
Thionoethanal, reactions, with dienes, **2**, 140
Thionoethers, reduction of, **1**, 11, 16, 16
Thionoformamide, *N,N*-diethyl-, synthesis of, **5**, 590
Thionoformamide, *N,N*-dimethyl-, reactions, with secondary amines, **5**, 591
Thionoformamides
  alkylation of, **5**, 597
  synthesis of, **5**, 580, 586, 608
Thionoformamides, cyano-, synthesis of, **5**, 592
Thionoformamides, *C*-sulfonyl-
  reactions
    with cyanide, **5**, 592
    with methylenes, **5**, 593
Thionoformate, ethyl
  reactions
    with alkyllithium, **3**, 332
    with amines, **5**, 586
    with tetrazines, **4**, 229
  as reagent, for the synthesis of thioaldehydes, **3**, 332, 340
  synthesis of, *in situ*, **5**, 586
Thionoformate, phenyl chloro-, reactions, with pyranosides, **6**, 541
Thionoformate, *p*-tolylchloro-, reactions, with 1-(3',5'-*O*-(1,1,3,3-tetraisopropyldisiloxane-1,3-diyl)-β-D-ribofuranosyl)-1*H*-benzimidazole, **6**, 541
Thionoformates, transamination of, **5**, 609
Thionoformates, alkyl chloro-
  reactions
    with 1-alkynes, **1**, 1047
    with alkynylmagnesium bromides, **1**, 1047
Thionoformates, aryl fluoro-, rearrangement of, for fluorothiolformate esters, **6**, 432
Thionoformates, chloro-
  reactions, with alcohols, **6**, 541
  as starting materials, for thionocarbonates, **6**, 541
Thionoformohydrazides, synthesis of, **5**, 640
Thionoglycinate, *N*-benzoyl-, reactions, with piperidine, **5**, 586
Thionohydrazides, stability of, **5**, 640
Thionohydrazides, aroyl-, synthesis of, **5**, 640
Thionohydroximic acid anhydrides, synthesis of, **5**, 637
Thionohydroxylamines, *O*-phosphorylated, synthesis of, **5**, 644
Thionolactams
  oxidative dimerization of, **5**, 607
  synthesis of, **5**, 593
Thiono β-lactams, synthesis of, **5**, 604
Thionolactones
  cycloadditions of, **4**, 229
  from, lactones, **5**, 512
  reactions
    with organolithiums, **4**, 226
    with tributylstannyllithium, **6**, 134
γ-Thionolactones, synthesis of, **5**, 513
Thionophosphates, allylic, rearrangement of, **1**, 840
Thionophosphine oxide, trimeric, synthesis of, **2**, 855
Thionophthalic anhydride
  from, phthaloyl chloride, **5**, 528
  synthesis of, **5**, 528
Thionopropanoamides, synthesis of, **5**, 595
Thionopropionates, synthesis of, **4**, 834

Thionoselenoesters, alkylation of, **4**, 870
Thionyl chloride
for the chlorination of sulfides, **4**, 65
as dehydrating agent, in the synthesis of cyclic anhydrides, **5**, 199
for the dehydration of amides, **3**, 619
hexamethylphosphoramide as solvent for, **5**, 10
reactions
with active methyl compounds, **5**, 484
with active methylenes, **4**, 71
with adamantane, **2**, 174
with alcohols, **2**, 101; **4**, 97
with alkanesulfinic acids, **2**, 172
with alkenes, **2**, 188
with alkoxy acroleins, **3**, 69
with *N*-alkoxycarboxylic acid amides, **5**, 665
with alkoxyquinolinecarboxylic acids, **5**, 9
with 2-(alkylsulfinyl)benzamides, **5**, 336
with alkynyl Grignard reagents, **1**, 1035
with amides, **5**, 657
with amines, **5**, 968
with aminoamides, **5**, 340
with 6-anilino-5-carboxy-4-methyl-2-phenylpyridines, **5**, 529
with aromatic amines, **2**, 793
with aromatic diamines, **2**, 796
with 1-benzoyl-2-arenesulfonylhydrazines, **5**, 676
with benzylic amides, **2**, 27
with *t*-butanesulfinic acid, **2**, 173
with butanones, **3**, 249
with carboxylic acids, **5**, 184
with 1-chloroethanesulfinic acid, **4**, 78
with diazomethane, **4**, 76
with dicyanamide, **5**, 810
with dicyclohexylammonium salt, **5**, 9
with 1,3-disubstituted ureas, **5**, 981
with fluorenonecarboxylic acids, **5**, 246
with formamides, **6**, 52
with half ester, **5**, 13
with *N*-hydroxyamides, **5**, 666
with α-hydroxy aziridines, **2**, 364
with (hydroxymethyl)germanes, **4**, 141
with hydroxymethyl isocyanate, **5**, 986
with *N*-hydroxymethylthiadiazolethiones, **4**, 105
with 2-hydroxy-2-phenylpropionic acid, **5**, 10
with ketones, **5**, 485
with *p*-methoxyphenylphenylphosphinic acids, **2**, 807
with methyl ketones, **5**, 485
with methyl sulfide, **4**, 65
with *aci*-nitroacetamide, **5**, 819
with phenylethynylmagnesium bromide, **4**, 863
with selenoxides, **2**, 369
with sodium fluoride, **6**, 409
with sodium sulfinate, **2**, 688
with sulfinic acid, **5**, 340
with sulfinic acids, **2**, 172, 717
with thioureas, **5**, 657
with 1,3,5-trithiane, **4**, 65
with ureas, **5**, 981
as reagent, for acid chlorides, **5**, 1, 7
use in the synthesis of acyl chlorides, **5**, 264
use in the synthesis of alkanesulfinic esters, **2**, 198
use in the synthesis of alkyl chlorides, **2**, 25
use in the synthesis of α-chloroalkyl esters, **4**, 54
use in the synthesis of α-chloro ethers, **4**, 48, 49
use in the synthesis of dichloroalkanes, **4**, 18
use in the synthesis of α,α-dichloro sulfides, **6**, 44
use in the synthesis of imidoyl chlorides, **5**, 657
use in the synthesis of sulfonyl halides, **2**, 181
use in the synthesis of thioacyl chlorides, **5**, 476
Thionyl chloride nitrate, synthesis of, **2**, 104
Thionyl fluoride
reactions, with phosgene, **6**, 409
as starting material, for carbonic chloride fluoride, **6**, 419
α-Thioorganometallics, from, thioethers, **1**, 140
Thioorthocarbonates, synthesis of, **6**, 308
Thioorthocarbonic acid tetraalkyl esters, as intermediates, for the synthesis of tetraalkyl orthocarbonates, **6**, 308
Thioorthoformate, trimethyl, reactions, with silyl enol ethers, **3**, 36, 168
Thioorthoformates, chloro-, production of, **6**, 273
Thiooximes, α-alkyl, synthesis of, **3**, 426
Thiophen, 2-vinyltetrahydro-, synthesis of, **1**, 141
Thiophen dioxide, alkylation of, **1**, 145
Thiophen 1,1-dioxide, 3-dichloromethylene 2,3-dihydro-, synthesis of, **4**, 742
Thiophen-*S*,*S*-dioxides, 2,5-dihydro-, alkylation of, **1**, 145
Thiophen-2-dithiocarboxylic acid, as starting material, for thiophen-2-thiocarbonyl chloride, **5**, 476
Thiophene
characteristics of, **3**, 104
reactions
with chlorocarbonylsulfenyl chloride, **6**, 432
with dodecacarbonyltriiron, **4**, 681
with dodecacarbonyltriruthenium, **4**, 683
with Grignard reagents, **1**, 484
reduction of, **1**, 100
ring-opening of, **1**, 1072
use in the synthesis of carbon-bridged dicobalt complexes, **4**, 690
Thiophene, 2-acetamido-, reactions, with sodium thiocyanate, **5**, 1108
Thiophene, 2-acetyl-
haloform oxidation of, **5**, 83
reactions, with diethyl dithiocarbonimidates, **4**, 980
Thiophene, 2-azidotetrahydro-, oxidation of, **4**, 330
Thiophene, 3-bromo-
reactions, with butyllithium, **3**, 105
synthesis of, **1**, 428
Thiophene, 2-carbamyl-4-hexadecylamino-, hydrolysis of, **5**, 82
Thiophene, 4-cyano-3-oxotetrahydro-, reactions, with alkyl halides, **3**, 648
Thiophene, 2,5-dihydro-, synthesis of, **2**, 201
Thiophene, 2,3-dilithio-, synthesis of, **3**, 105
Thiophene, 2,4-dinitro-5-methoxy-, reactions, with potassium methoxide, **6**, 97
Thiophene, 2-ethyldifluorosilyl-, reactions, with aryl halides, **3**, 305
Thiophene, 2-formyl-, synthesis of, **3**, 104
Thiophene, 3-ketotetrahydro-, aldol condensation of, **3**, 247
Thiophene, 2-lithio-, reactions, with trialkylboron, **5**, 82
Thiophene, 2-methyl-
reactions
with dodecacarbonyltriruthenium, **4**, 683
with methyl *N*-(α-benzotriazol-1-ylalkyl)carbamates, **1**, 262
Thiophene, 2-nitro-, reactions, with secondary amines, **4**, 907
Thiophene, tetrabromo-, reduction of, **1**, 428
Thiophene, tetrahydro-, reactions, with *t*-butyl peracetate, **4**, 226
Thiophene, thieno-, aldol condensation of, **3**, 247
Thiophene, 2,3,5-tribromo-, reduction of, **1**, 428
Thiophene, 2-vinyl-, synthesis of, **1**, 629
Thiopheneamines, as starting materials, for thiophene isocyanides, **3**, 725
Thiophene-2-carboxylic acids, 5-alkyl-, synthesis of, **5**, 82
Thiophene 1,1-dioxide, reactions, with chloroform, **4**, 742
Thiophene-1,1-dioxide, 2,5-dihydro-, elimination of, **1**, 620
Thiophenes
acylation of, **3**, 279, 305
hydrogenation of, **1**, 91
metallation of, **3**, 305
nonclassical, synthesis of, **3**, 371
reactions

with dialkyl formamides, **3**, 105
with ethoxycarbonyl isothiocyanate, **5**, 592
as starting materials, for sulfines, **3**, 369
synthesis of, **1**, 428; **3**, 72
Thiophenes, 2-alkoxy-, as starting materials, for monothioortho esters, **6**, 97
Thiophenes, azido-, synthesis of, **4**, 856
Thiophenes, dihydro-
synthesis of, **1**, 103
via acetate pyrolysis, **1**, 774
Thiophenes, 2,3-dihydro-, synthesis of, **1**, 100
Thiophenes, formyl-, synthesis of, **3**, 104
Thiophenes, 3-formyl-, synthesis of, **3**, 104, 105
Thiophenes, 3-halo-, ring-opening of, **1**, 1072
Thiophenes, halomethyl-, oxidation of, **3**, 104
Thiophenes, 3-hydroxy-, dimerization of, **3**, 251
Thiophenes, 2-iodo-, carbonylation of, **3**, 305
Thiophenes, keto-, synthesis of, **3**, 305
Thiophenes, α-keto tetrahydro-, use in the synthesis of disulfides, **2**, 229
Thiophenes, 2-lithio-, synthesis of, **3**, 105, 305
Thiophenes, 2-lithiobeno-, synthesis of, **2**, 1000
Thiophenes, 3-methyl-, dibromination of, **3**, 104
Thiophenes, nonclassical
from
diketones, **3**, 372
sulfoxides, **3**, 372
synthesis of, **3**, 371
Thiophenes, tetrahydro-, synthesis of, **1**, 91, 242; **6**, 129
Thiophenium ions, synthesis of, **3**, 371
Thiophenol
reactions
with acrolein, **3**, 36
with benzaldehyde, **4**, 216
with bis(thiophenyl)tin, **2**, 317
with β-chloroenone, **3**, 252
with chlorovinylphosphine oxides, **4**, 120
with α-diazo ketones, **3**, 166
with α,α-di(morpholino)xylene, **4**, 322
with *N*-(hexafluoroisopropylidene)-benzenesulfinamides, **4**, 325
with 1,2,2-tribromo-1-fluoroethene, **4**, 771
Thiophenol, 2-amino-, reactions, with aldehydes, **4**, 318
Thiophenol, 4-amino-, from, 1-bromo-4-nitrophenol, **2**, 706
Thiophenol, 4-chloro-, reactions, with imidoyl chlorides, **5**, 728
Thiophenol, 2-hydroxy-, reactions, with acid chlorides, **4**, 1073
Thiophenol, 3-methyl-
reactions
with *N*-benzoyl aminals, **4**, 323
with carbon disulfide, **5**, 1033
Thiophenolate, reactions, with bishydrazonoyl halides, **5**, 681
Thiophenolate anion, reactions, with bis(iodinium)ethyne, **2**, 1033
Thiophenolates, reactions, with alkyl halides, **2**, 710
Thiophenols
reactions
with 1-chloro enamines, **4**, 904
with hexamethylenetetramine, **4**, 317
with isonitriles, **5**, 728
with nitrile oxides, **5**, 828
with tetrahydro-2*H*-1,3-oxazin-2-one, **2**, 320
Thiophenols, 2-amino-, reactions, with thioacetals, **4**, 899
Thiophenols, *o*-amino-, for the condensation of *O*-ethyl selenocarboxylates, **5**, 522
Thiophen-3-ones, 2-alkylidene-4,5-dihydro-, synthesis of, **3**, 250
Thiophenoxide, thallous, reactions, with imidoyl chlorides, **5**, 728
Thiophenoxide ligands, in coupling reactions, **1**, 120
Thiophen-2-thiocarbonyl chloride, from, thiophen-2-dithiocarboxylic acid, **5**, 476
Thiophilic addition
of dimethyl sulfoxide, **4**, 270
of organo metallics, **4**, 255
Thiophosgene
chlorination of, **2**, 169; **6**, 215
dimerization of, **6**, 253
from
carbon monosulfide, **6**, 529
trichloromethanesulfenyl chloride, **6**, 529
trichloromethyl thiol, **6**, 529
photolysis of, **2**, 170
reactions
with alcohols, **6**, 533
with alkaneselenols, **6**, 528
with alkoxides, **6**, 540
with alkoxytrimethylsilanes, **6**, 534
with allenic alcohols, **6**, 436
with amines, **5**, 1022, 1026; **6**, 576
with amino ketones, **3**, 194
with arylamines, **5**, 1027
with aryl diazo ketones, **5**, 482
with aziridine, **5**, 1033
with *N*-benzoylselenoureas, **5**, 655
with *N*-benzoylthioureas, **5**, 655
with bis(bistrifluoromethylaminooxy)mercury(II), **6**, 436
with *N*,*N*-bis(trifluoromethyl)hydroxylamine, **6**, 312
with bis(trifluoromethyl)sulfine, **6**, 436
with bis(trimethylsilyl)-*N*-methylbenzamidine, **5**, 1051
with carboxylic acid amides, **5**, 654
with *m*-chloroperbenzoic acid, **6**, 531
with diazocarbonyls, **6**, 535
with diazo ketones, **5**, 482
with 1,3-dimethyl-2,2-diethyl-1,3-diaza-2-germacyclopentane, **6**, 577
with dimethylketene, **5**, 481
with naphthols, **6**, 533
with phenols, **6**, 533
with phenyl bis(trimethylsilyl)phosphanes, **5**, 1058
with phosphites, **6**, 123
with potassium alkoxides, **6**, 534
with quinoline, **5**, 1033
with secondary amines, **6**, 537
with silylphosphines, **6**, 687
with thiobenzamides, **5**, 1049
with thiols, **6**, 552
with trimethylsilylamines, **6**, 537
with *N*-trimethylsilylimidazole, **6**, 577
with 1-trimethylsilyloxybenzotriazole, **5**, 1030
with triphenylphosphine *N*-trimethylsilylimide, **6**, 537
with triphenylphosphiniminium chloride, **6**, 537
as reagent, for the synthesis of thioketones, **3**, 364
as starting material
for thiocarbonyl dibromide, **6**, 529
for thionocarbonates, **6**, 540
for thiophosgene-*S*-oxide, **6**, 531
for thioureas, **6**, 576
for trithiocarbonic acid diesters, **6**, 552
See also Thiocarbonyl dichloride
Thiophosgene, chlorofluoro-, dimerization of, **6**, 254
Thiophosgene, dibromo-, reactions, with ozone, **6**, 254
Thiophosgene-*S*-oxide
from, thiophosgene, **6**, 531
See also Sulfine, dichloro-
Thiophosphate, ammonium *O*,*O*-diethyl-, reactions, with trichloromethyl isocyanide dichloride, **6**, **295–318**
Thiophosphate, diethyl chloro-, reactions, with 2-methanesulfonyl-1-(2-chlorophenyl)ethanone, **4**, 784
Thiophosphate, dimethyl chloro-, reactions, with amines, **5**, 614
Thiophosphate, triethyl, reactions, with metal perchlorates, **2**, 94

Thiophosphates
  alkylation of, **2**, 252
  from
    alcohols, **2**, 258
    carbonyls, **2**, 258
    disulfides, **2**, 256
    epoxides, **2**, 259
    ortho esters, **2**, 259
    sulfenates, **2**, 257
    sulfenyl chlorides, **2**, 257
    thiiranes, **2**, 259
    thiocyanates, **2**, 259
    thiomides, **2**, 259
  as insecticides, **2**, 252
  synthesis of, **5**, 737
Thiophosphates, *N*-alkoxybenzimidoyl, synthesis of, **5**, 713
Thiophosphates, *O*-aryl-*O,O*-dialkyl, reactions, with phosphorus oxychloride, **2**, 257
Thiophosphates, *S*-imidoyl, from, imidoyl chlorides, **5**, 738
Thiophosphates, vinyl, synthesis of, **2**, 693
Thiophosphinate *O*-esters, synthesis of, **2**, 462
Thiophosphinates, reactions, with thiophosphinyl chlorides, **2**, 463
Thiophosphinates, *O*-alkyl, synthesis of, **4**, 128
Thiophosphinates, *S*-alkyl, synthesis of, **4**, 128
Thiophosphinates, *O*-aryl, synthesis of, **4**, 128
Thiophosphinates, *S*-aryl, synthesis of, **4**, 128
Thiophosphinates, *S*-*n*-butyl(chloromethyl)-, synthesis of, **4**, 128
Thiophosphinates, (α-hydroxyalkyl)-, synthesis of, **2**, 463
Thiophosphines
  from
    alkyl halides, **2**, 254
    thiols, **2**, 252
  synthesis of, **2**, 252
Thiophosphines, alkane-, synthesis of, **2**, 252
Thiophosphinic acid, alkylation of, **2**, 854
Thiophosphinic acid amides, synthesis of, **2**, 468
Thiophosphinic acid chloride, diphenyl-, synthesis of, **2**, 854
Thiophosphinic acid *S*-esters, synthesis of, **2**, 854
Thiophosphinic acids, synthesis of, **2**, 462
Thiophosphinic acids, alkyl-, synthesis of, **2**, 462
Thiophosphinic acids, aryl-, synthesis of, **2**, 853
Thiophosphinic acids, α-halo-, synthesis of, **4**, 127
Thiophosphinic acids, methylphenyl-, synthesis of, **2**, 463
Thiophosphinic acid thioester, reactions, with tributylphosphine, **2**, 851
Thiophosphinic *N*-amidate chelates, synthesis of, **2**, 861
Thiophosphinic amides, aryl-, synthesis of, **2**, 860
Thiophosphinic chloride, bis(chloromethyl)-, reduction of, **4**, 115
Thiophosphinic chloride, (chloromethyl)methyl-, from, phosphinic chlorides, **4**, 128
Thiophosphinic chloride, diphenyl-, reactions, with diphenylphosphinic chloride, **5**, 910
Thiophosphinic chlorides
  from, phosphinic acids, **2**, 463
  reactions
    with alcohols, **2**, 462; **4**, 128
    with amines, **4**, 128
  synthesis of, **4**, 128
Thiophosphinic chlorides, (chloromethyl)-, from, oxo acid chlorides, **4**, 127
Thiophosphinic chlorides, dialkyl-, synthesis of, **2**, 463
Thiophosphinic compounds, alkyl-, synthesis of, **2**, 462
Thiophosphinic esters, synthesis of, **2**, 462
Thiophosphinic esters, dialkyl-, synthesis of, **2**, 255
Thiophosphinic halides, reactions, with sodium hydrosulfide, **2**, 462
Thiophosphinic iodides, dialkyl-, synthesis of, **2**, 463
Thiophosphinite, methyl *t*-butylphenyl-, optically active, synthesis of, **2**, 851
Thiophosphino sulfide, bisdiphenyl-, synthesis of, **2**, 851
Thiophosphinothioformamides, alkyl-, synthesis of, **6**, 581
Thiophosphinous acid, *S*-acetyldiphenyl-, reactions, with benzaldehyde, **4**, 345
Thiophosphinous acid arylthioester, phenyl-, synthesis of, **2**, 851
Thiophosphinous acid *O*-esters, phenyl-, synthesis of, **2**, 851
Thiophosphinous acids, synthesis of, **2**, 459
Thiophosphinous acids, alkyl-, synthesis of, **2**, 459
Thiophosphinous acids, aryl-, synthesis of, **2**, 851
Thiophosphinous acids, dialkyl-
  reactions
    with alkenes, **2**, 462
    with carbonyls, **2**, 462
Thiophosphinous compounds, alkyl-, synthesis of, **2**, 459
Thiophosphinyl chlorides, reactions, with thiophosphinates, **2**, 463
Thiophosphite, dialkyl, reactions, with ynamines, **4**, 1010
Thiophosphite, trimethyl, reactions, with methyl ketones, **4**, 345
Thiophosphites, *S*-imidoyl, synthesis of, via phosphorylation of thioamides, **5**, 737
Thiophospholipids, synthesis of, **2**, 257
Thiophosphonate, acyl, synthesis of, **5**, 406
Thiophosphonate, menthyl *S*-methylphenyl, synthesis of, **2**, 855
Thiophosphonates, alkyl alkylchloro-, hydrolysis of, **2**, 463
Thiophosphonates, *O*-alkyl alkylchloro-, synthesis of, **2**, 464
Thiophosphonates, *S*-alkyl alkylchloro-
  reactions, with alcohols, **2**, 464
  synthesis of, **2**, 464
Thiophosphonates, *S*-alkyl alkylhalo-, reactions, with mercaptides, **2**, 464
Thiophosphonates, dialkyl, reactions, with Grignard reagents, **2**, 459
Thiophosphonates, dialkyl alkyl
  hydrolysis of, **2**, 463
  synthesis of, **2**, 463
Thiophosphonates, *O,S*-dialkyl alkyl, synthesis of, **2**, 464
Thiophosphonates, dialkyl α-aminoalkyl, synthesis of, via Mannich reactions, **2**, 464
1-Thiophosphonates, alkynyl, synthesis of, **2**, 1072
Thiophosphonic acid *O*-diesters, aryl-, synthesis of, **2**, 855
Thiophosphonic acid *S*-diesters, aryl-, synthesis of, **2**, 855
Thiophosphonic acid dihalides, reactions, with alcoholates, **2**, 463
Thiophosphonic acid esters, phenylation of, **2**, 851
Thiophosphonic acid esters, (*N*-alkylthiocarbamoyl)-, synthesis of, **6**, 581
Thiophosphonic acid *O*-esters, aryl-, alkylation of, **2**, 855
Thiophosphonic acid *S*-esters, synthesis of, **2**, 854
Thiophosphonic acid monoesters, synthesis of, **2**, 463
Thiophosphonic acids, from, thiophosphonic dichlorides, **2**, 463
Thiophosphonic acids, alkyl-, synthesis of, **2**, 463
Thiophosphonic acids, aminomethyl-, synthesis of, **4**, 469
Thiophosphonic acids, aryl-, synthesis of, **2**, 854
Thiophosphonic acids, α-halo-, synthesis of, **4**, 133
Thiophosphonic amides, aryl-, synthesis of, **2**, 862
Thiophosphonic amides, cyclic aryl-, synthesis of, **2**, 862
Thiophosphonic compounds, alkyl-, synthesis of, **2**, 463
Thiophosphonic dichlorides
  as starting materials, for thiophosphonic acids, **2**, 463
  synthesis of, **2**, 464
Thiophosphonic fluoroamides, aryl-, synthesis of, **2**, 862
Thiophosphonite, diethyl aryl-, synthesis of, **2**, 851
Thiophosphonium salts, alkyl-, from, thiols, **2**, 299
Thiophosphonous acid fluoride, phenyl-, synthesis of, **2**, 852
Thiophosphonous acids, synthesis of, **2**, 461
Thiophosphonous acids, alkyl-, synthesis of, **2**, 460
Thiophosphonous acids, aryl-, synthesis of, **2**, 851
Thiophosphonous compounds, alkyl-, synthesis of, **2**, 460
Thiophosphonyl chloride, dimethyl-, reactions, with 2-lithio-1,3-oxathianes, **6**, 119

Thiophosphoramidates, as starting materials, for isothiocyanates, **5**, 1031
Thiophosphoramidates, α-hydroxy, chlorination of, **4**, 106
Thiophosphoramidates, *S*-imidoyl, synthesis of, **5**, 737
Thiophosphoramidites
  as starting materials, for phosphorodithioates, **2**, 410
  synthesis of, **2**, 410
Thiophosphoric triamides, synthesis of, **2**, 413
Thiophosphoryl chloride, hydrolysis of, **2**, 120
Thiophosphoryl chloride, (dichlorophosphinomethyl)-, reactions of, **4**, 553
Thiophosphoryl tribromide, reactions, with alkanethiols, **2**, 252
Thiophthalimide, *N*-t-butyl-, reactions, with sodium alkoxides, **2**, 195
N-Thiophthalimide, reactions, with cysteine, **2**, 217
Thiophthalimides, *S*-alkyl-, reactions, with trialkyl phosphites, **2**, 259
N-Thiophthalimides, as starting materials, for sulfenamides, **2**, 241
Thiopivalate, *S*-bromo, synthesis of, **5**, 530
Thiopivalate, *S*-chloro, synthesis of, **5**, 530
Thiopivalate, *O*-trimethylsilyl
  reactions
    with *N*-bromosuccinimide, **5**, 530
    with *t*-butyl hypochlorite, **5**, 530
Thiopivaldehyde
  stability of, **3**, 331
  synthesis of
    via addition of *t*-butyllithium to ethyl thionoformate, **3**, 332
    via elimination of dithiolanes, **3**, 334
    via photolytic cleavage of phenacyl neopentyl sulfide, **3**, 331
    via retro Diels–Alder reaction, **3**, 335
Thiopivaldehydes, synthesis of, via retro ene reaction, **3**, 335
Thiopropanes, 1,3-diphenyl-, as starting materials, for cyclopropanes, **1**, 240
Thiopropanethiols, γ-alkyl-, synthesis of, **2**, 128
Thiopropanoyl fluoride, 2,3,3,3-tetrafluoro-, reactions, with diethylamine, **5**, 582
Thiopropionitrile, 3-chloro-2-methyl-, synthesis of, **3**, 635
2*H*-Thiopyran, deprotonation of, **1**, 141
Thiopyrans
  ring-opening of, **3**, 73
  as starting materials, for thiopyrylium salts, **3**, 370
  synthesis of, **5**, 615
Thiopyrans, 2-amino-, synthesis of, **4**, 319
Thiopyrans, dihydro-, synthesis of, **4**, 901
Thiopyran-4-thiones
  from, thiopyrones, **3**, 357
  as starting materials, for thiopyrylium salts, **3**, 370
Thiopyranyl compounds, 4-silyloxy, alkylation of, **1**, 144
Thiopyridinium salts, 2-methyl-, synthesis of, **2**, 121
2-Thiopyridone, *N*-acetoxy-, reactions, with selenobenzoates, **5**, 526
2-Thiopyridone, *N*-hydroxy-, reactions, with carboxylic acids, **4**, 266
Thiopyridyl salt, reactions, with acid chloride, **1**, 231
Thiopyrones, as starting materials, for thiopyran-4-thiones, **3**, 357
4-Thiopyrones
  as starting materials, for pyridin-4-thiones, **3**, 358
  synthesis of, **3**, 357
Thiopyrophosphates, reactions, with trimethylammonium hydrochloride, **2**, 257
Thiopyruvic acids, synthesis of, **3**, 350
Thiopyrylium salts
  from
    thiopyrans, **3**, 370
    thiopyran-4-thiones, **3**, 370
  as starting materials, for thiabenzenes, **3**, 375
  synthesis of, **3**, 370
Thioquinones, synthesis of, **3**, 358
Thiosaccharin, reactions, with alkyl halides, **2**, 121
Thioselenoacetals
  deprotonation of, **4**, 285
  from
    diselenoacetals, **4**, 284
    selenophenates, **4**, 283
  synthesis of, **4**, 283, 284
Thioselenoacetals, β,β-dicyano-, synthesis of, **4**, 284
Thioselenoacetals, *S*,*S*-dioxide ketene, reactions, with enamines, **4**, 286
Thioselenoacetals, ketene
  from, dimethyl dithioselenocarbonate, **4**, 872
  synthesis of, **4**, 870
β-Thioselenoalkanes, synthesis of, **2**, 285
Thioselenocarbonate, *S*,*Se*-diethyl, synthesis of, **6**, 482
Thioselenocarbonates, synthesis of, **6**, 482
Thioselenocarbonates, *S*,*Se*-dialkyl, synthesis of, **6**, 482
Thioselenole, 2-alkyl, fragmentation of, **4**, 872
Thiosemicarbazide
  reactions
    with aldehydes, **3**, 449
    with ketones, **3**, 449
Thiosemicarbazides
  for the condensation of *O*-ethyl selenocarboxylates, **5**, 522
  from
    amino isothiocyanates, **6**, 574
    isothiocyanates, **6**, 573
  reactions
    with imino ester hydrochlorides, **5**, 705
    with methyl iodide, **4**, 984
    with nitrous acid, **5**, 1046
    with selenoesters, **5**, 704
  as starting materials
    for α-hydrazino enamines, **4**, 984
    for isothiocyanates, **5**, 1036
Thiosemicarbazides, 1-aryl-, oxidation of, **6**, 580
Thiosemicarbazones, synthesis of, **3**, 449
Thiosilane, phenyl-, α-chlorination of, **4**, 365
Thiosilane, trimethylphenyl-, reactions, with silylvinyl ketones, **4**, 944
Thiosilanes
  from
    alkanethiols, **2**, 262
    alkyl sulfides, **2**, 265
    disulfides, **2**, 265
    silyl thiolates, **2**, 264
    thiocarbonyls, **2**, 266
  synthesis of, **2**, 262
Thiosilanes, acyl-, synthesis of, **5**, 254
Thiosilanes, alkane-, synthesis of, **2**, 262
Thiosilanes, alkyl-, as starting materials, for disulfides, **2**, 231
Thiosilanes, α-methyl-, from, silyl sulfonium ylides, **4**, 369
α-Thiosilanes
  from, halomethylsilanes, **4**, 367
  rearrangement of ylides from, **4**, 369
  synthesis of, **4**, 364
Thiostannanes
  as starting materials, for sulfenamides, **2**, 244
  synthesis of, **2**, 215
  use in the synthesis of thioglycosides, **4**, 225
Thiostannanes, alkyl-, as starting materials, for disulfides, **2**, 231
N,*N*'-Thiosuccinimide, synthesis of, **5**, 338
Thiosuccinimides, *S*-alkyl-, reactions, with trialkyl phosphites, **2**, 259
N-Thiosuccinimides
  reactions, with thioamides, **5**, 735
  synthesis of, **2**, 244
Thiosugars, synthesis of, **4**, 227
1-Thiosugars, synthesis of, **4**, 223
Thiosulfates, as starting materials, for thiocyanates, **5**, 1111
Thiosulfenates, thioalkylation of, **2**, 233

Thiosulfenyl chloride, acetyl-, reactions, with alkenes, **5**, 248
Thiosulfenyl chlorides, use in the synthesis of trisulfides, **2**, 233
Thiosulfenyl cyanides, synthesis of, **2**, 219
Thiosulfinate esters, sodium periodate oxidation of, **2**, 724
Thiosulfinate esters, arene, synthesis of, **2**, 723
Thiosulfinates
  reactions, with thiols, **2**, 219
  as starting materials, for disulfides, **2**, 218
  synthesis of, **2**, 222, 232
  thermolysis of, **2**, 186
  use in thiolysis reactions, **2**, 219
Thiosulfinates, alkane, synthesis of, **2**, 232
Thiosulfinates, *S*-alkyl, as starting materials, for thioaldehydes, **3**, 334
Thiosulfinates, dialkyl, thermolysis of, **2**, 186
Thiosulfines, as reactive intermediates, **3**, 347, 363, 370
Thiosulfines, alkyl-
  reactions, with amines, **5**, 587
  use in the synthesis of thionoalkanoamides, **5**, 611
Thiosulfines, α-alkyl-, as starting materials, for ketene dithioacetal *S*-oxides, **4**, 859
Thiosulfolane, 2-phenyl-, synthesis of, **4**, 266
Thiosulfonates
  cleavage of, **2**, 187
  reactions, with amines, **2**, 358
  as starting materials
    for alkanesulfinic esters, **2**, 199
    for disulfides, **2**, 218
    for thioketones, **3**, 365
  use in thiolysis reactions, **2**, 218
Thiosulfonates, alkane, synthesis of, **2**, 223
Thiosulfonates, arene, esters, synthesis of, **2**, 724
Thiosulfonates, aryl esters, synthesis of, **2**, 724
Thiosulfonates, *p*-toluene, reactions, with ylides, **5**, 898
Thiosulfones, as starting materials, for alkyl hydrodisulfides, **2**, 211
Thiosulfones, α-methyl, reactions, with sodium hydrogen telluride, **1**, 439
α-Thiosulfones
  deprotonation of, **4**, 266
  from, α-thiosulfoxides, **4**, 265
  synthesis of, **4**, 265
Thiosulfonic acids, reactions, with sulfenyl chlorides, **2**, 220
α-Thiosulfoxide carbanions, as starting materials, for α-thiosulfoxides, **4**, 263
Thiosulfoxides, α-alkyl-, from, dithioacetals, **4**, 258
α-Thiosulfoxides
  from, α-thio sulfoxide carbanions, **4**, 263
  optically active, synthesis of, **4**, 264, 265
  as starting materials, for α-thiosulfones, **4**, 265
  synthesis of, **4**, 258
Thiotelluroacetals, synthesis of, theoretical, **4**, 285
Thiotellurophanes, alkyl-, synthesis of, **4**, 873
Thiotetronic acids, synthesis of, **6**, 473
Thio-1,3-thiaselenolylium cations, 2-alkyl-, reactions, with imidazoles, **6**, 317
4-Thiothiopyrones, synthesis of, **3**, 357
α-Thiotins, synthesis of, **4**, 397
α-Thiotitaniums, synthesis of, **4**, 396
Thiotriazenes, 3-(α-dialkylaminoarylidene)-1-alkyl-, reactions, with methyl iodide, **3**, 419
Thiourea
  reactions
    with alkoxyalkyl halides, **2**, 128
    with alkyl halides, **2**, 120
    with aryl iodides, **2**, 707
    with bis(chloromethyl)phosphinic acid, **4**, 346
    with bisgermanium oxides, **5**, 1082
    with bis(trialkyltin) oxides, **5**, 1082
    with hexaalkyldistannane, **5**, 1082
    with phthalaldehyde, **4**, 298
  as reagent, for the synthesis of thioketones, **3**, 362
  use in the synthesis of α-thiosilanes, **4**, 368
Thiourea, *N*-acetyl-, reactions, with phosphorus pentachloride, **6**, 619
Thiourea, 1-alkyl-, as starting material, for 1,3-dialkylguanidines, **6**, 641
Thiourea, *N*-aryl-*N'*-benzoyl-, pyrolysis of, **5**, 1035
Thiourea, 1,3-di-*t*-butoxycarbonyl-, as starting material, for 1-alkylguanidines, **6**, 641
Thiourea, *N*,*N*-dimethyl-
  as starting material
    for α-bis(methylthio)methyleneamino enamines, **4**, 988
    for immonium salts, **4**, 988
Thiourea, *N*,*N*-dimethyl-*N'*-phenyl-, reactions, with ethyl bromocyanoacetate, **6**, 559
Thiourea, *N*,*N'*-diphenyl-, synthesis of, **6**, 578
Thiourea, α-naphthyl-, decomposition of, **5**, 1035
Thiourea, *N*-phenyl-*N'*-diphenylmethyleneamino-, reactions, with bis(tributyltin) oxide, **6**, 571
Thiourea, tetramethyl-, reactions, with irons, **6**, 351
Thiourea *S*,*S*-dioxides, uses of, **6**, 580
Thiourea *S*-oxides, synthesis of, **6**, 580
Thioureas
  alkylation of, **6**, 731
    for selenothiocarbamates, **6**, 594
    for selenoureas, **6**, 596
  arylation of, **6**, 731
  S-arylation of, **6**, 323
  bis hydroxymethylation of, **4**, 301
  chlorination of, **6**, 728
  dehydration of, **5**, 1063
  desulfurization of, **5**, 1062
  dethiolation of, **5**, 1078
  from
    hydrazines, **6**, 573
    isothiocyanates, **6**, 569
    sodium cyanothioformate, **6**, 580
    N,*N'*-thiocarbonylbis(imidazole), **6**, 577
    thiocarbonyl transfer reagents, **6**, 577
    thiophosgene, **6**, 576
    ureas, **6**, 578
  organometallation of, **5**, 1081
  oxidation of, **6**, 502
  reactions
    with *N*-arylphosphimines, **6**, 645
    with 1-bromoadamantane, **6**, 635
    with butyllithium, **5**, 1112
    with 2-chloro-1-methylpyridinium iodide, **5**, 1064
    with formamide acetals, **5**, 760
    with halogen acids, **2**, 119, 120
    with hypochlorite, **5**, 1078
    with phosgene, **6**, 619
    with thionyl chloride, **5**, 657
  as starting materials
    for α-aminoalkyl phosphonates, **4**, 466
    for α-aminoalkylphosphorus compounds, **4**, 466
    for aminoiminomethanesulfonic acid, **6**, 641
    for carbamimidic chlorides, **6**, 619
    for carbodiimides, **5**, 1062
    for carbon disulfide, **6**, 574
    for cyanamides, **5**, 1122
    for cyanoguanidines, **6**, 647
    for *N*-cyanoiminocarbonyls, **6**, 647
    for formamidines, **5**, 746
    for guanidines, **6**, 641
    for hydroxyguanidines, **6**, 650
    for isothiocyanates, **5**, 1035
    for isothioureas, **6**, 635
    for ketene aminals, **4**, 973
    for *N*-unsubstituted iminocarbonyls, **6**, 641
    for ureas, **6**, 502
  synthesis of, **6**, 554, 569
  use in the synthesis of alkanethiols, **2**, 118
Thioureas, *N*-alkanoyl-, synthesis of, **6**, 572

Thioureas, alkoxycarbonyl, reactions, with phosgene, **5**, 1063
Thioureas, amidino-, use in the synthesis of carbodiimides, **5**, 1082
Thioureas, *N*-benzoyl-, reactions, with thiophosgene, **5**, 655
Thioureas, bis(bromomagnesio)-, use in the synthesis of carbodiimides, **5**, 1078
Thioureas, cyclic
  oxidation of, **6**, 509
  synthesis of, **6**, 575
Thioureas, N^1,N^2-diacyl-, desulfurization reactions of, **4**, 425
Thioureas, *N*,*N'*-diaryl-, synthesis of, **5**, 1035
Thioureas, *sym*-diaryl-, reactions, with acetic anhydride, **5**, 1035
Thioureas, *N*,*N'*-dilithio-, decomposition of, **5**, 1078
Thioureas, 1,1-disubstituted, synthesis of, **6**, 572
Thioureas, *N*,*N'*-disubstituted, reactions, with 2-chloro-1-methylpyridinium salt, **5**, 1078
Thioureas, hydroxy-, synthesis of, **6**, 570
Thioureas, *N*-hydroxy-
  reactions, with alkyl halides, **5**, 831
  synthesis of, **6**, 570
Thioureas, *N*-methoxy-, synthesis of, **5**, 831
Thioureas, methylene-, synthesis of, **6**, 571
Thioureas, 1-phenyl-, synthesis of, **6**, 580
Thioureas, *N*-sulfonyl-*N'*-substituted, as starting materials, for carbodiimides, **5**, 1064
Thioureas, thionoacyl-, synthesis of, **5**, 633
Thioureas, trimethylstannyl-, synthesis of, **6**, 571
Thioureas, 1,1,3-trisubstituted, reactions, with bis(triphenylstannyl)carbodiimide, **5**, 1121
Thiourethanes, alkylation of, **6**, 730
Thiouronium salts, cleavage of, **4**, 223
Thioveratraldehyde, synthesis of, **3**, 341
α-Thioxocarbene, reactions, with carbon disulfide, **6**, 305
β-Thioxo compounds, synthesis of, **3**, 344
2-Thioxoindole, reactions, with benzaldehydes, **5**, 615
Thioxomalonate, diethyl, synthesis of, **3**, 349
α-Thiozincs, synthesis of, **4**, 396
Thiuram disulfides
  reactions, with dimethyltellurium, **6**, 566
  as starting materials, for dithiocarbamates, **6**, 566
Thiyl radicals
  from, thiols, **1**, 949
  as intermediates, in the synthesis of isothiocyanates, **5**, 1037
  synthesis of, **2**, 215
Thorium, butyltris(cyclopentadienyl)-, decomposition of, **4**, 703
Thorium(IV) compounds, synthesis of, **2**, 602
Thorium ate complexes, heptamethylated, synthesis of, **2**, 602
Thorium complex, acyl-, as starting material, for ketenimines, **3**, 598
Thorium complexes, synthesis of, **5**, 782
Thorium complexes, cyclopentadienyl-, reactions, with carbon monoxide, **5**, 468
Thorium compounds
  reactions
    with 1-metalloalkylphosphines, **4**, 589
    with phosphorus ylides, **4**, 589
Thorium compounds, acyl-, synthesis of, **5**, 468
Thorium compounds, alkyl-, synthesis of, **2**, 602
Thorium compounds, 2,2,2-trimethyl ethyl, synthesis of, **5**, 469
Thorpe reaction, mechanism, **2**, 749
Thorpe–Ziegler alkylation, for macrocycle synthesis, **2**, 749
Three-membered rings, synthesis of, **1**, 352
Threonines, phospho-, synthesis of, **6**, 487
Thrombin, inhibition via α-aminoboronic acids, **4**, 526
Thromboxane B_2, synthesis of, **1**, 828
Thujene nitroso chloride, reactions, with methyloleate nitroso chloride, **2**, 778
Thymidine, 3'-azido-3'-deoxy- (AZT), reactions, with triphenylphosphine, **5**, 975
Thymosin β_4, synthesis of, **6**, 487
TIBO, selenium analogue of, **6**, 598
Tiffeneau–Demjanov rearrangement, mechanism, **1**, 389
Tiglate, methyl, synthesis of, **1**, 725
Triglianes, synthesis of, **3**, 325
Tiglic acid, synthesis of, reviews, **5**, 98
Tigloyl chloride, from, angelic acid, **5**, 12
Tin
  elimination of, **1**, 663
  organometallic compounds, halogenated, **6**, 270
  reactions
    with (halomethyl)silylalkanes, **4**, 660
    with sulfenyl chlorides, **6**, 528
  for the reductive cleavage of disulfides, **2**, 126
Tin, alkynyl-, reactions, with benzoylnitromethane, **4**, 541
Tin, allyltributyl-
  radical attack on, **1**, 173
  reactions
    with 2,3-*syn*-alkoxy acylsilanes, **2**, 79
    with aryl iodide, **1**, 471
    with chloroform, **4**, 14
    with 1,1,1-trichloroethane, **4**, 14
Tin, bis(thiophenyl)-, reactions, with thiophenol, **2**, 317
Tin, chloromethyltrimethyl-, reactions, with dimethylbismuthinosodium, **4**, 600
Tin, chlorotriphenyl-
  reactions
    with carbanions, **6**, 292
    with tricarbonyl(lithiocyclopentadienyl)-manganese, **4**, 725
Tin, cyanomethyltributyl-, synthesis of, **3**, 640
Tin, di(benzenesulfenyl)-, for the reduction of nitro compounds, **2**, 341
Tin, dichlorodimethyl-, reactions, with thioacetic acid, **5**, 541
Tin, diethoxy-, reactions, with carbon disulfide, **6**, 469
Tin, (dimethylamino)tributyl-, reactions, with phenyl methyl sulfoxide, **4**, 397
Tin, α-fluorovinyl-, from, trifluorovinylmagnesium halides, **4**, 812
Tin, α-iodoalkyl-, transmetallation of, **1**, 164
Tin, 1-silyloxyvinyl-, as starting material, for lithium enolates, **5**, 415
Tin, tetrakis(1-methoxyvinyl)-, reactions, with cuprous cyanide, **4**, 954
Tin, tetramethyl-
  reactions, with arenediazonium salts, **3**, 290
  use in monomethylation reaction, **2**, 570
Tin, tetravinyl-
  reactions
    with butyllithium, **1**, 1091
    with phenyllithium, **1**, 1091
Tin, tributyl-, for the cleavage of thiocarbonylimidazole, **1**, 177
Tin, tributyl(cyanomethyl)-, reactions, with phosphorus trichloride, **3**, 639
Tin, tributyl(trimethylsilylmethyl)-, reactions, with *n*-butyllithium, **4**, 652
Tin, trimethyl-, reactivity of, **1**, 1006
Tin, trimethyl(difluoromethyl)-, synthesis of, **6**, 65
Tin, (trimethylsilylmethyl)triphenyl-, synthesis of, **4**, 660
Tin, (trimethylsilyl)tributyl-, synthesis of, **4**, 660
Tin, trimethyltrifluoromethyl-, pyrolysis of, **4**, 7
Tin, triphenylallyl-, reactions, with phenyl isocyanate, **5**, 390
Tin(II) amides, reactions, with primary aldehydes, **2**, 814
Tin(II) bromide, reactions, with diiodomethane, **4**, 699
Tin(II) carboxylic thioesters, reactions, with imines, **5**, 236
Tin(II) chloride
  reactions
    with 1,3,2-dithiaborinane–dimethyl sulfide, **4**, 257
    with iodoperfluoroalkanes, **6**, 65
  as reagent, for the synthesis of borenes, **5**, 929

use in elimination reactions, **1**, 638
use in the synthesis of azoxy compounds, **2**, 776
use in the synthesis of isothiocyanates, **5**, 1028
Tin(II) complexes
reactions, with nitro compounds, **3**, 426
use in the reduction of azides, **2**, 317
Tin(II) compounds, 2-pyridylbis(trimethylsilyl)methyl-, synthesis of, **6**, 196
Tin(II) enolates, synthesis of, **2**, 701
Tin(II) fluoride, reactions, with iodoperfluoroalkanes, **6**, 65
Tin(II) triflate, reactions, with carbonyls, **2**, 701
Tin(IV) chloride
reactions, with silicon hydrides, **4**, 140
use in the allylation of carbonyls, **2**, 78
use in the synthesis of alkynes, **1**, 1026
use in the synthesis of tetramethylene-bridged tins, **4**, 700
Tin(IV) compounds, transmetallation with, **1**, 132
Tin(IV) enolates
monoalkylation of, **1**, 132
synthesis of, **2**, 701
Tin acyl compounds, alkynic, synthesis of, **5**, 468
Tin alkoxides, alkyl-, transmetallation of, **2**, 600
Tin alkoxides, trialkyl-, reactions, with trichloroacetonitrile, **5**, 716
Tin alkynes, reactions, with phenyl(cyano)iodonium triflate, **2**, 1021
Tin alkynides, synthesis of, **1**, 1135
Tin alkynides, trialkyl-
reactions, with acid chloride, **1**, 1135
use in the synthesis of β,γ-unsaturated amino acids, **1**, 1135
Tin amides
synthesis of, **5**, 389
use in the substitution of amides, **5**, 389
Tinamines, trialkyl-, reactions, with aryl sulfoxides, **4**, 397
Tin bis(thiocarboxylates), diphenyl-
as starting materials
for acylsulfenyl chlorides, **5**, 245
for acylsulfenyl iodides, **5**, 246
use in the synthesis of acylsulfenyl bromides, **5**, 245
Tin bromide, trimethyl-, reactions, with 2-trimethylsilyloxiranes, **6**, 133
Tin chloride, reactions, with diferrocenylmercury, **4**, 726
Tin chloride, tributyl-
reactions
with α-alkoxylithium nucleophiles, **4**, 388
with imines, **2**, 816
with lithium (1-alkynyl)organoborates, **1**, 1033
as starting material, for α-alkoxystannanes, **4**, 387
Tin chloride, (trifluoromethyl)dimethyl-, synthesis of, **6**, 245
Tin chloride, trimethyl-
reactions
with dilithiomethane, **4**, 699
with 1,3-dioxane, **6**, 133
with silenes, **6**, 395
Tin chloride, triphenyl-, reactions, with trimethylsilylmethyllithium, **4**, 660
Tin chlorides, reactions, with 1-lithiovinyl phenyl sulfides, **4**, 961
Tin chlorides, trialkyl-
reactions, with arylthiomethyllithiums, **4**, 397
synthesis of, **2**, 599
Tin complexes, cyclic tetranuclear, synthesis of, **4**, 700
Tin complexes, methylene-bridged pentacoordinated, synthesis of, **4**, 700
Tin complexes, tetramethylene-bridged, synthesis of, **4**, 700
N-Tin compounds, reactions, with aryl halides, **2**, 753
Tin cyanides, trialkyl-, reactions, with thallium salts, **5**, 1143
Tin cyanides, triaryl-
synthesis of, **5**, 1144
use in the synthesis of acyl cyanides, **5**, 1144
Tin dibromide, diallenyl-, reactions, with aldehydes, **1**, 1055
Tin dichloride
reactions
with alkyl halides, **2**, 600
with (halomethyl)silylalkanes, **4**, 660
use in the allylation of aldehydes, **2**, 73
Tin dichloride, dibutyl-
use in allylation reactions, **2**, 79
use in the synthesis of imines, **3**, 407
Tin dichloride, dimethyl-, reactions, with di-Grignard reagents, **4**, 700
Tin dichloride, diphenyl-, reactions, with potassium thiocarboxylates, **5**, 245
Tin dichlorides, dialkyl-
as starting materials, for tetraalkyltins, **2**, 600
use in the synthesis of carbodiimides, **5**, 1081
Tin dihalides, dialkyl-, synthesis of, **2**, 600
Tin dihydride, dibutyl-, reactions, with bis(ethylthio)alkylphosphines, **6**, 581
Tin dihydroxide, dibutyl-, use in the synthesis of macrolactones, **5**, 174
Tin diisopropoxides, dialkyl-, reactions, with acetyl cyanide, **5**, 1144
Tin enethiolates, 1,4-addition of, **5**, 557
Tin enolates, synthesis of, **1**, 558
Tin ethers, vinyl-, as nucleophiles, **1**, 194
Tin ethynes, acylation of, **3**, 273
Tin fluoride, tributyl-, for the hydrolysis of silyl enol ethers, **3**, 10
Tin fluoride, tris(trifluoromethyl)-, synthesis of, **6**, 245
Tin-functionalized cumulenes, synthesis of, **1**, 988
Tin halides
reactions
with bromotrifluoroethene, **4**, 812
with 1-lithio-1-aryl-1-silylalkanes, **4**, 660
Tin halides, alkyl-, reactions, with *N*-lithioimines, **3**, 457
Tin halides, dialkyl-, reactions, with cyanide, **5**, 1144
Tin halides, (trifluoromethyl)-, synthesis of, **6**, 245
Tin hydride, in tandem cyclizations, **1**, 348
Tin hydride, di-*n*-butyl-, reactions, with *t*-butyldipropynylphosphine, **4**, 1038
Tin hydride, trialkyl-, for the hydration of ethoxyethyne, **2**, 676
Tin hydride, tri-*n*-butyl-
for the reduction of diazonium salts, **1**, 442
use in cyclization reactions, **1**, 1131
use in demercuration reactions, **1**, 68
use in the reduction of amines, **1**, 33
use in the synthesis of vinyl radicals, **1**, 1129
Tin hydride, tributyl- (TBTH)
for the cyclization of oxime ethers, **2**, 351
deprotonation of, **4**, 386
reactions
with alkenes, **1**, 172
with bishydrazonoyl halides, **5**, 681
with 1-chlorooctyne, **4**, 812
with α-fluoroalkenyl phenyl sulfones, **4**, 812
with isocyanides, **1**, 945
with *S*-methylthiocarbonates, **1**, 590
with nitro compounds, **1**, 945
with β-nitro sulfides, **1**, 592
with palladium(0), **1**, 86
with phenylthioalkynes, **4**, 962
with selenides, **1**, 594
with β-substituted nitro compounds, **1**, 638
with thiobenzoates, **1**, 938
with thiocarbonyl imidazolides, **1**, 938
with thiocarbonyls, **1**, 938
for the reduction of alkyl halides, **1**, 4
for the reduction of bromoalkanes, **1**, 9
for the reduction of bromoarenes, **1**, 428
for the reduction of chloroalkanes, **1**, 7
for the reduction of iodoalkanes, **1**, 10
for the reduction of thionoethers, **1**, 16
as starting material, for vinylstannane, **2**, 623
synthesis of, in low concentration, **1**, 330

use in the synthesis of alkenes, **1**, 591
use in the synthesis of dichloromethyls, **4**, 13
UV-catalyzed addition of, **2**, 601
Tin hydride, triethyl-, reactions, with boron trichloride, **2**, 526
Tin hydride, trimethyl-
reactions
with 1-stannyl-1-alkynes, **6**, 209
with tetra(trifluoromethyl)germanium, **6**, 65
with trialkyl(trifluoromethyl)stannanes, **6**, 65
use in the synthesis of carbon-bridged ditins, **4**, 701
Tin hydride, triphenyl-
reactions
with benzene, **1**, 174
with diphenylethynylstibine, **2**, 876
thermal addition of, **2**, 601
Tin hydrides
for the mediation of radical reactions, **1**, 330
reactions, with alkenes, **2**, 601
for the reduction of alkyl halides, **1**, 5
use in radical cyclization reactions, **1**, 341
water soluble, **1**, 936
Tin hydrides, alkyl-, transmetallation of, **2**, 600
Tin hydrides, dialkyl-, reactions, with tetrafluoroethylene, **6**, 65
Tin hydrides, trialkyl-
deprotonation of, **2**, 600
elimination of, **1**, 1081
synthesis of, **2**, 599
use in radical additions, **1**, 330
use in the reduction of nitroalkanes, **1**, 36
Tin hydrides, triaryl-, reactions, with alkyl iodides, **1**, 330
Tin hydrides, tributyl-
reactions
with dialkylethynylstibines, **2**, 876
with di-*n*-butylethynylarsine, **2**, 876
use in the reduction of α-amino acids, **3**, 39
Tin imides, synthesis of, **5**, 391
Tin imides, *N*-trialkyl-, synthesis of, **5**, 390
Tin imides, tributyl-, synthesis of, **5**, 390
Tin iodide, triphenyl-
reactions
with *N*-arylthioamides, **5**, 739
with thioacetanilide, **5**, 739
Tin iodide, tris(trifluoromethyl)-, from, tetrakis(trifluoromethyl)stannane, **6**, 245
Tin isocyanate, tributyl-
reactions
with imides, **5**, 391
with maleic anhydrides, **5**, 391
with phthalic anhydrides, **5**, 391
with succinic anhydrides, **5**, 391
synthesis of, **5**, 1016
Tin isocyanate, triphenyl-, synthesis of, **5**, 1016
Tin isocyanates
as starting materials, for metal amides, **5**, 391
synthesis of, **5**, 1016
Tin isocyanide, trimethyl-, synthesis of, **3**, 711
Tin isopropoxides, trialkyl-, reactions, with acetyl cyanide, **5**, 1144
Tin–lithium exchange
for the synthesis of α-alkoxy carbanions, **4**, 378
for the synthesis of allyllithiums, **2**, 555
for the synthesis of carbanions, **1**, 138
for the synthesis of α-thiocarbanions, **4**, 394
in vinylstannanes, **2**, 911
in [1,2] Wittig rearrangement, **1**, 397
in [2,3] Wittig rearrangement, **1**, 404
in [2,3] Wittig–Still rearrangement, **1**, 405
Tin methoxide, tributyl-, reactions, with enol acetates, **1**, 558
Tin methoxide, trimethyl-, reactions, with perfluoroisobutene, **4**, 762
Tin-2-ones, 4-alkoxy-1,3-diaze-, synthesis of, **6**, 110
Tin oxide, reactions, with alkyl halides, **2**, 600
Tin oxide, dibutyl-
reactions
with alkyl bromides, **2**, 89
with amino acids, **5**, 277
Tin oxide, tributyl-, reactions, with phthalimide, **5**, 390
Tin oxide, triethyl-, reactions, with isocyanic acid, **5**, 1017
Tin oxides, alkyl-, transmetallation of, **2**, 600
Tin peroxides, trialkyl-, reactions, with alkyl triflates, **2**, 98
Tin radicals, use in the synthesis of natural products, **1**, 1132
Tin radicals, tributyl-, synthesis of, **1**, 173
Tins
as radical traps, **1**, 243
reactions
with alkynes, **4**, 962
with 1,3-dithiane,2-lithio-, **6**, 134
with lithium methoxide, **6**, 200
with sulfinyl oxygen, **2**, 210
as starting materials, for methide salts, **6**, 200
for the substitution of aryl halides, **1**, 471
for the substitution of vinyl halides, **1**, 467
synthesis of, **4**, 961, 1038; **6**, 208
use in the synthesis of radicals, **1**, 936
Tins, acyl-, synthesis of, **5**, 467
Tins, alkenyl-
as carbon nucleophiles, **1**, 228
reactions, with alkynyl halides, **1**, 1020
Tins, alkoxy-, addition to vinyltin systems, **4**, 701
Tins, α-alkoxy-, synthesis of, **4**, 385
Tins, alkyl-
from, tetraalkyltins, **2**, 599
synthesis of, **2**, 599
via transmetallation, **2**, 599
Tins, alkylhalo-, addition of oxygen and nitrogen donors to, **4**, 700
Tins, alkynyl-
coupling reactions of, **1**, 1035
palladium-catalyzed cross-coupling reactions of, **1**, 1006
reactions
with acid chlorides, **1**, 1045
with aldehydes, **1**, 1038
with alkenyl halides, **1**, 1007, 1015
with alkenyl triflates, **1**, 1029
with 1-alkynes, **1**, 1038
with dibromoalkenes, **1**, 1015
with imidoyl chlorides, **1**, 1047
with organoboranes, **4**, 665
with phenyl isocyanate, **1**, 1043
with pyridines, **1**, 1043
synthesis of, **2**, 1099
use in substitution reactions, **1**, 1003
Tins, alkynyltrialkyl-
reactions
with alkenyl triflates, **1**, 1027
with aryl triflates, **1**, 1027
Tins, alkynyltrichloro-, reactions, with enones, **1**, 1050
Tins, allenyl-
reactions
with chloroformate esters, **1**, 1060
with isoquinoline, **1**, 1060
with quinoline, **1**, 1060
for the synthesis of alcohols, **1**, 1055
Tins, allyl-
for the opening of epoxides, **1**, 207
reactions
with haloalkanes, **1**, 128
via radical intermediates, **1**, 128
Tins, amino-, addition to vinyltin systems, **4**, 701
Tins, aminomethyl-, synthesis of, **4**, 533
Tins, aryl-, as starting materials, for deuteriobenzenes, **1**, 452
Tins, difluoroalkyl-, synthesis of, **6**, 65
Tins, ethynyl-, reactions, with bromotrichloromethane, **4**, 813

Tins, α-fluoroalkenyl-, synthesis of, **4**, 812
Tins, halo-, reactions, with alkynyl metals, **2**, 1100
Tins, α-haloalkenyl-, synthesis of, **4**, 812
Tins, imidoyl-
from, imidoyl chlorides, **5**, 803
synthesis of, **5**, 803
Tins, *S*-imidoyl-, synthesis of, **5**, 739
Tins, mixed alkyl-, synthesis of, **2**, 599
Tins, organo(trialkyl)-, use in the synthesis of tertiary phosphines, **2**, 436
Tins, perfluorovinyl-, reactions, with boron halides, **4**, 813
Tins, propanoyl-, synthesis of, **5**, 468
Tins, propargyl-
reactions
with chloroformate esters, **1**, 1060
with isoquinoline, **1**, 1060
with quinoline, **1**, 1060
Tins, α-selenyl-, synthesis of, **4**, 401
Tins, α-substituted organo-, oxidation of, **3**, 709
Tins, (sulfinylmethyl)trialkyl-, synthesis of, **4**, 397
Tins, tetraalkyl-
from
dialkyltin dichlorides, **2**, 600
tin tetrachloride, **2**, 599
reactions
with *N*-chlorosuccinimide, **5**, 391
with tin tetrachloride, **2**, 599
with trityl cations, **1**, 663
as starting materials, for alkyltins, **2**, 599
synthesis of, **2**, 555
via alkylation, **2**, 600
Tins, tetrahalo-, reactions, with amides, **5**, 389
Tins, trialkyl-, synthesis of, **6**, 134, 135
Tins, triaryl-, synthesis of, **6**, 134
Tins, tributyl-, stability of, **1**, 1006
Tins, trichloro alkynyl-, carbostannation of, **4**, 1069
Tins, trihalomethyl-, synthesis of, **6**, 245
Tins, vinyl-
reactions, with acid chlorides, **3**, 240
use in the synthesis of α,β-unsaturated ketones, **3**, 218
Tin sulfides
reactions, with haloimides, **5**, 337
as reagents, for the synthesis of thioketones, **3**, 347, 363
Tin sulfinates, synthesis of, **2**, 210
Tin tetrabromide, reactions, with diazomethane, **4**, 155
Tin tetrachloride
reactions
with diazomethane, **4**, 155
with lithium alkyls, **4**, 949
with tetraalkyltins, **2**, 599
with trimethylsilylmethylmagnesium chloride, **4**, 660
with 1-trimethylsilyloxy cyclohexene, **1**, 251
as reagent, for the synthesis of alkyl chlorides, **2**, 26
for the ring-opening of 1,3-disilacyclobutanes, **4**, 661
as starting material, for tetraalkyltins, **2**, 599
use in the substitution of nitrogen functions, **1**, 251
use in the synthesis of furanyl ketones, **3**, 302
Tin tetrachlorides, as catalysts, for the synthesis of ketene dithioacetals, **4**, 844
Tin tribromide, trifluoromethyl-, as starting material, for (trifluoromethyl)tin triiodide, **6**, 245
Tin trichlorides, alkyl-, synthesis of, **2**, 600
Tin triiodide, (trifluoromethyl)-, from, trifluoromethyltin tribromide, **6**, 245
Titanates, allyl, for the allylation of aldehydes, **2**, 74
Titanation, of alkynes, **4**, 1066
Titanic bromide, tris(dimethylamino)-, reactions, with diethyl arsenide, **2**, 512
Titanium
reactions
with allylic alcohols, **1**, 173
with benzylic alcohols, **1**, 173
with carbon monoxide, **6**, 522
reduced, synthesis of, **4**, 1082
Titanium, low-valent, use for the deoxygenation of epoxides, **1**, 587
Titanium, silylallyl-, reactions, with electrophiles, **2**, 935
Titanium, tetraisopropyloxy-, use in hydrosilylation reactions, **2**, 82
Titanium, tetrakis(dimethylamino)-
reactions
with amides, **4**, 972
with formamides, **5**, 745
with secondary amides, **5**, 753
as starting material, for ortho amides, **6**, 141
Titanium(0) oxide, synthesis of, **1**, 688
Titanium(III) alkyls, reactions, with carbon dioxide, **1**, 667
Titanium(III) chloride, as reagent, for the synthesis of ketones, **3**, 134
Titanium(III) chloride, bis(cyclopentadienyl)-, reactions, with epoxides, **1**, 942
Titanium(IV) amides, use in the synthesis of β-amino ketones, **3**, 187
Titanium(IV) catalysts, use in the synthesis of amides, **5**, 272
Titanium(IV) chloride
reactions, with α-dichloro sulfoxides, **4**, 773
use in the synthesis of enynes, **1**, 1053
use in the synthesis of propargyl ethers, **1**, 1026
Titanium(IV) metal imido complexes, monocyclopentadienyl-, reactions, with alkynes, **3**, 410
Titanium acyl compounds, as starting materials, for γ,δ-unsaturated aldehydes, **5**, 441
Titanium alkoxide, as catalyst, for the synthesis of cyanohydrins, **3**, 616
Titanium alkoxides, aminoalkyl-, reactions, with ketone enolates, **3**, 186
Titanium alkyl compounds, cyclic
reactions, with carbon monoxide, **5**, 441
as starting materials, for cyclopentanone, **5**, 441
Titanium alkynides, triisopropoxy-, reactions, with oxiranes, **1**, 1024
Titaniumaluminums, synthesis of, **4**, 1066
Titanium ate complexes, as starting materials, for (*E*)-γ-substituted ketones, **4**, 918
Titanium binuclear compounds, synthesis of, **4**, 673
Titanium bis(cyclooctatetraene), use in the synthesis of tris(cyclooctatetraene)dititanium, **4**, 671
Titanium bromide, trisdiethylamino-, as organotitaniums, **2**, 702
Titanium carbonyl complexes, synthesis of, **6**, 522
Titanium catalysts, chiral, use in [2+2] cycloadditions, **1**, 358
Titanium chloride, methyl-, synthesis of, **2**, 565
Titanium chloride, tris(dimethylamino)-, reactions, with carbamates, **4**, 383
Titanium chloride, tris(isopropoxy)-, as organotitaniums, **2**, 702
Titanium complexes
reactions
with aldehydes, **1**, 991
with ketones, **1**, 991
use in the reduction of ketones, **2**, 55
Titanium complexes, alkoxy-, as intermediates, in the synthesis of alkenes, **1**, 680
Titanium complexes, alkyl-
reactions
with copper bromide–dimethyl sulfur, **2**, 587
with copper iodide–lithium chloride, **2**, 587
with tetramethylguanidine, **6**, 140
Titanium complexes, carbon-bridged, synthesis of, **4**, 671
Titanium complexes, cyclopentadienyl-
reactions, with benzyllithium, **2**, 566
as starting materials, for alkenylsilanes, **2**, 936
synthesis of, **2**, 568
Titanium dichloride, dicyclopentadienyl-, as starting material, for metallated ylides, **5**, 917

Titanium dichlorides, methyl-, synthesis of, **2**, 566
Titanium dioxide
as catalyst, for photochemical oxidation, **1**, 289
reactions
with rhodium, **1**, 88
with rhodium carbonyl, **1**, 72
Titanium enolates
reactions, with cinnamaldehyde, **5**, 607
synthesis of, **2**, 702
Titanium enolates, trichlori-, synthesis of, **2**, 702
Titanium function, doubly-bonded, synthesis of, **3**, 508
Titanium hexacarbonyl, synthesis of, **6**, 522
Titanium metallaindanes, as starting materials, for indan-2-one, **5**, 443
Titanium methylene complexes
from, titanacyclobutane complexes, **3**, 508
synthesis of, **3**, 508
via transmetallation, **4**, 669
Titanium ortho esters, use in the synthesis of carboxylic esters, **5**, 134
Titanium phosphaalkyne complexes, as starting materials, for metallated phosphaalkenes, **5**, 890
Titaniums
chemoselectivity of, **1**, 124
in Peterson alkenation, **1**, 649
of phosphorus functions, synthesis of, **5**, 890
reactions
with 1,3-dithiane,2-lithio-, **6**, 134
with electrophiles, **1**, 973
with 1-metalloalkyl phosphonates, **4**, 578
with phosphorus ylides, **4**, 578
transmetallation of, **1**, 122
Titaniums, acyl-, synthesis of, **5**, 441
Titaniums, (2-alkenyl)triphenoxy-, as starting materials, for β-lactones, **1**, 683
Titaniums, α-alkoxy-, synthesis of, **4**, 383
Titaniums, alkyl-
β-elimination of, **2**, 566
synthesis of, **2**, 565, 566
Titaniums, allenyl-, reactions, with aldehydes, **1**, 1056
Titaniums, chiral, from, chiral alcohols, **2**, 568
Titaniums, chiral alkyl-, use in enantioselective addition reactions, **2**, 568
Titaniums, 1,3-disubstituted allenyl-, reactions, with aldehydes, **1**, 1057
Titaniums, imidoyl-, synthesis of, **5**, 797
Titaniums, silyl-, addition to alkynes, **2**, 919
Titaniums, tetraalkyl-, synthesis of, **2**, 568
Titaniums, tris(isopropoxy)-, synthesis of, **6**, 134
Titaniums, vinyl-, reactions, with aldehydes, **2**, 70
Titanium salts, as starting materials, for ortho amides, **6**, 141
Titanium silicate, for the oxidation of dialkyl sulfides, **2**, 145
Titanium tetrabutoxide, use in the synthesis of tris(cyclooctatetraene)dititanium, **4**, 671
Titanium tetrachloride
as catalyst, for the synthesis of nitriles, **1**, 252
reactions
with allylic alcohols, **1**, 364
with aniline, **4**, 796
with lithium aluminum hydride, **4**, 1082
with 1-phenylthiovinylsilane, **4**, 369
with silyl enol ether, **2**, 702
use in the allylation of carbonyls, **2**, 78
use in the synthesis of alkylmagnesiums, **2**, 561
use in the synthesis of imines, **3**, 407
Titanium tetrachlorides, as catalysts, for the synthesis of ketene dithioacetals, **4**, 844
Titanium tetrafluoride, reactions, with benzyl haloethyl sulfides, **5**, 582
Titanium tetraisopropoxide
reactions
with alkyl metals, **2**, 566
with bistriflamido-*trans*-1,2-diaminocyclohexane, **2**, 72
Titanium trichloride
reactions
with *N*,*N*-disubstituted hydroxylamines, **3**, 409
with methylmagnesium bromide, **2**, 566
reduction to titanium(0) oxide, **1**, 688
use in the synthesis of amines, **2**, 314
Titanium trichloride, chloro(dimethyl)silylmethyl-, synthesis of, **4**, 658
Titanium trichloride, methyl-
stability of, **2**, 566
synthesis of, **2**, 566
Titanium triisopropoxide, 2-(trimethylsilyl)ethynyl-, use in stereoselective addition, **2**, 1092
Titanium triisopropoxides, alkynyl-
reactions
with aldehydes, **1**, 1038, 1039
with ketones, **1**, 1038
Titanocene
reactions, with esters, **1**, 21
for the reduction of aldehydes, **1**, 18
for the reduction of oxiranes, **1**, 13
Titanocene, methylene-, as intermediate, **2**, 979
Titanocene, tetramethylene-, as starting material, for methylene-bridged dititanium complexes, **4**, 670
Titanocene acyl compounds, from, acid chlorides, **5**, 441
Titanocenebis(trimethylphosphane), reactions, with cyclopropenes, **3**, 509
Titanocene complexes, displacement of, **4**, 663
Titanocene dichloride
as catalysts, for the synthesis of alkylmagnesiums, **2**, 561
reactions
with *N*-lithioketimines, **3**, 460
with methylenebis(magnesium bromide), **4**, 670
with tricarbonyl(lithiocyclopentadienyl)-manganese, **4**, 719
with trimethylsilylketimines, **3**, 460
reduction of, **4**, 671
use in the synthesis of alkylmagnesiums, **2**, 561
use in the synthesis of silylalkenes, **2**, 925
Titanocene methylidene complexes
reactions
with acid anhydrides, **2**, 702
with acid chlorides, **2**, 702
Titanocene methylidenephosphine complexes, synthesis of, **3**, 508
Titanocene pentasulfide, reactions, with chlorosulfanes, **2**, 100
Titanocenes, in the cyclization of silyldiynes and silylenynes, **2**, 926
Titanocenes, crotyl-, use in diastereoselective allylation reactions, **2**, 76
Titanocenes, germylmethyl-, synthesis of, **4**, 663
Titanocenes, monoalkyl-, reactions, with carbon monoxide, **5**, 441
Titanocene(vinylcarbene) complexes, synthesis of, **3**, 508
TNTU, as coupling reagent, for the synthesis of *N*-alkylamides, **5**, 280
Tolan
reactions
with dicarbonylcyclopentadienylcobalt, **4**, 689
with nonacarbonyldiiron, **4**, 681
with phosphaalkynes, **5**, 885
o-Tolualdehyde, reactions, with diisopropylphosphinyldifluoromethyllithium, **6**, 58
p-Tolualdehyde, reactions, with triphenylarsonium ethylide, **1**, 738
p-Tolualdehyde, deuterium-labeled, as starting material, for deuterium-labeled *p*-toluic acid, **5**, 114
Toluamide, *N*,*N*-dichloro-*para*-, reactions, with 2,3,4,5-tetrakis(methoxycarbonyl)cyclopentadiene biphenylphosphinoethanepalladium, **5**, 387
Toluene
acylation of, **3**, 280

alkylation of, **1**, 462
from, xylene, **1**, 56
methylation of, **1**, 247
oxidation of, **3**, 93
phenylethylation of, **1**, 916
reactions
with aroyl chlorides, **3**, 295
with 2,5-dicarbazidofuran, **5**, 990
with sodium, **4**, 609
with sulfuryl chloride, **4**, 11
reduction of, **1**, 95
as solvent, for phosgene, **6**, 425
as starting material
for benzoic acid, **5**, 116
for benzyllithium, **2**, 554
for benzylpotassiums, **2**, 557
synthesis of, **1**, 50
use in the substitution of boron functions, **1**, 275
use in the synthesis of monoalkyl phenyl ketones, **3**, 295
Toluene, *p*-amino-, from, *p*-amino benzonitrile, **1**, 48
Toluene, *o*-chloro-, as starting material, for *o*-chlorobenzoic acid, **5**, 115
Toluene, *p*-nitrile-, as starting material, for xylenes, **1**, 48
Toluene, 4-nitro-, as starting material, for benzotrichlorides, **6**, 28
Toluene, *p*-nitro-, as starting material, for *p*-nitrobenzoic acid, **5**, 115
Toluene, 2-nitro-5-carbomethoxy-, as starting material, for enamines, **2**, 740
Toluene, 3-nitro-4-dimethylamino-, reactions, with sodium bisulfite, **2**, 782
Toluene, 2-nitroso-5-nitro-, as starting material, for 2,5-dinitrobenzoic acid, **5**, 115
Toluene, perfluoro-, reduction of, **1**, 426
Toluene, 2,3,4-trichloro-, as starting material, for 2,3,4-trichlorobenzoic acid, **5**, 115
Toluene, α,α,α-trichloro-, as starting material, for fluorinated orthobenzoates, **6**, 69
Toluene, α-trimethylsilyl-α-phenylselenyl-, reactions, with lithium diisopropylamide, **6**, 130
Toluene, 2,4,6-trinitro-
as starting material
for benzonitriles, **3**, 662
for 2,4,6-trinitrobenzoic acid, **5**, 115
Toluene difluoride, p-iodo-, use in the synthesis of α-fluoro ketones, **3**, 149
Toluenes
from
acetals, **1**, 21
esters, **1**, 21
oxidation of, **3**, 86
Toluenes, amino-, oxidation of, **3**, 99
Toluenes, halo-, oxidation of, **3**, 95
Toluenes, isopropyl-, synthesis of, **1**, 462
Toluenes, *p*-methoxy-, synthesis of, **1**, 46
Toluenes, α,α,α-trihalo-, as starting materials, for trithioortho esters, **6**, 82
p-Tolueneselenosulfonate, seleniumphenyl, reactions, with alkynes, **4**, 947
Toluene sulfinates, use in the synthesis of dialkyl sulfoxides, **2**, 151
p-Toluenesulfinic acid, reactions, with aniline, **2**, 797
p-Toluenesulfonamide
N-nitration of, **2**, 386
reactions, with chlorodimethylphosphine, **2**, 469
p-Toluenesulfonamide, *N*-butoxycarbonyl-, reactions, with alcohols, **2**, 309
p-Toluenesulfonamide, *N*-methyl-, reactions, with alcohols, **2**, 309
Toluene-*p*-sulfonamide, *N*-*p*-nitrophenylsulfonyloxy-, use in the synthesis of triphenyl-*N*-*p*-tolylsulfonylarsinimine, **2**, 810
p-Toluenesulfonamide, *N*-sulfinyl-
reactions
with alcohols, **2**, 202
with aldehydes, **2**, 306
with dienes, **2**, 366
Toluene-*p*-sulfonate, ammoniopropanedinitrile, reactions, with aldehydes, **4**, 883
p-Toluenesulfonate, pyridinium
as catalyst
for the synthesis of acetals, **4**, 179
for the synthesis of cyclic alkenyl ethers, **2**, 679
p-Toluenesulfonates, solvolysis of, for ortho esters, **6**, 71
p-Toluenesulfonic acid, reactions, with chloroform, **2**, 808
p-Toluenesulfonyl chloride
as chlorinating agent, for aryl chlorides, **2**, 621
uses of, **2**, 717
p-Toluenesulfonylhydrazone salts, thermal decomposition of, **6**, 697
N-*p*-Toluenesulfonylimide, synthesis of, **6**, 619
Toluenesulfonyls, light absorption of, **4**, 1075
N-*p*-Toluenesulfonylsulfilimines
as starting materials
for dithioortho esters, **6**, 99
for monothioorthoformates, **6**, 97
for trithioorthoformates, **6**, 87
o-Toluic acid, from, *o*-xylene, **5**, 115
p-Toluic acid
from
p-cymene, **5**, 115
p-xylene, **5**, 116
p-Toluic acid, deuterium-labeled, from, deuterium-labeled *p*-tolualdehyde, **5**, 114
Toluic acid anhydrides, synthesis of, **5**, 197
Toluic acid chlorides, reactions, with malonates, **2**, 664
p-Tolyl sulfoxide, 1-chloromethyl, optically active, synthesis of, **4**, 76
p-Tolyl sulfoxides, alkyl
deprotonation of, **1**, 143
reactions, with lithium α-bromomethyl acrylate, **1**, 143
p-Tolyl-2-trimethylsilylethynyl sulfone, as 1,2-dipole acceptor, **2**, 928
Tosic acid, elimination of, **1**, 179
Tosylate, 2,2-difluoro-1-lithio-1-vinyl, reactions, with trialkylboranes, **2**, 906
Tosylate, *trans*-1,2-diphenyl-2-(phenylthio)vinyl, solvolysis of, **1**, 1112
Tosylate, glycidyl, opening of, **1**, 223
Tosylate, (*S*)-1-methylheptyl, as starting material, for (*R*)-3-methyl-2-nonanone, **5**, 451
Tosylate, pyridinium
as reagent
for the synthesis of aldehydes, **3**, 9
for the synthesis of ketones from acetals, **3**, 123
Tosylate, 2,2,2-trifluoroethyl
deprotonation of, **4**, 735
reactions, with trialkylboranes, **4**, 735
Tosylate, 8-(trimethylsilyl)-6-octyn-2-ol, acetolysis of, **1**, 978
Tosylates
displacement of, **4**, 346
in the displacement of alkyl alcohols, **1**, 177
Friedel–Crafts alkylation of, **1**, 475
pyrolysis of, **1**, 778
reactions
with chromium salts, **2**, 571
with cobalt complexes, **1**, 335, 345
with guanidines, **6**, 642
with iodides, **1**, 465
with sodium sulfide, **2**, 224
reactivity of, **1**, 1115
as reagents, in quaternization reactions, **2**, 1079
as starting materials
for dithiocarbamates, **6**, 562
for iodides, **2**, 9
for neopentyl halides, **2**, 9

substitution of, **1**, 476
synthesis of, **1**, 223, 700
use in the synthesis of alkyl halides, **2**, 9
Tosylates, alkyl
acetolysis of, **2**, 87
as alkylating agents, **2**, 1048
as electrophiles, for dithiocarboxylic enethiolates, **4**, 842
reactions, with Grignard reagents, **1**, 1092
Tosylates, alkynyl, research background, **2**, 1022
Tosylates, allylic, in alkylation reactions, **1**, 187
Tosylates, 2-carboxyaryl, decarboxylative fragmentation of, **1**, 1081
Tosylates, 3,5-disubstituted-4-hydroxybenzaldehyde oxime, use in the synthesis of aryl isocyanides, **3**, 722
Tosylates, hydroxy
from, phenyllithium–boron trifluoride etherate, **1**, 223
synthesis of, **1**, 223
Tosylates, oxime, reactions, with ethanol, **1**, 1119
Tosylates, α-silyl, as starting materials, for α-silyl selenocyanates, **5**, 1113
Tosyl chloride
reactions
with diazophenylmethane, **3**, 479
with oxazolidinone, **5**, 1012
with pyridine, **5**, 282
with urethanes, **5**, 1014
use in the synthesis of carboxylic esters, **5**, 123
2-Tosyl compounds, 1-cyano-, reactions, with bromine, **4**, 105
Tosyl fluoride
attempted reactions
with 1-adamantylmethylidenephosphorane, **5**, 901
with neopentylidenephosphorane, **5**, 901
as starting material, for α-sulfonyl ylides, **5**, 901
Tosyl hydrazide, use in the synthesis of α-diazophosphorus compounds, **5**, 870
Tosyl iodide, reactions, with trimethyl(vinyl)silanes, **4**, 139
Toxicity, of alkynyl compounds, **2**, 1076
(-)-Trachelanthamidine, synthesis of, **2**, 325
Transacetalization
of aldehyde acetals, **4**, 249
of amide acetals, **6**, 104
Transamidation, of amide acetals, **6**, 104
Transamination
for the synthesis of β-nitro enamines, **2**, 787
for the synthesis of sulfenamides, **2**, 359
Transcarboxylation, mechanism, **5**, 70
Transcomplexation, use of term, **5**, 577
Transdiazotization, for the synthesis of diazonium salts, **2**, 804
Transesterification
of ortho esters, **6**, 78
for the synthesis of arsonic esters, **2**, 886
for the synthesis of diethyl carbonate, **6**, 467
for the synthesis of dimethyl carbonate, **6**, 467
of trithioortho esters, **6**, 88
Transfer reaction, aza-, for the synthesis of imidophosphorane esters, **5**, 861
Transfer reaction, single-electron, for the synthesis of alkyllithiums, **2**, 551
Transhalogenation
of *N*-carbonyl chlorides, with fluorides, **6**, 438
of chloroformates, **6**, 422
of phosgene, for carbonic difluoride, **6**, 409
Transient carbocations, synthesis of, **1**, 896
Transient radicals, use of, **1**, 921
Transition metal(II) salts, use in the synthesis of dialkyl carbonates, **6**, 465
Transition metal acyl compounds, synthesis of, **5**, 440
Transition metal alkyl complexes, stability of, **4**, 655
Transition metal alkynides, synthesis of, **1**, 1133
Transition metal amides
from
acyl azides, **5**, 386
amides, **5**, 384
dichloroamides, **5**, 387
isocyanates, **5**, 386
metal oximes, **5**, 386
synthesis of, **5**, 384
via nitrile hydrolysis, **5**, 385
Transition metal carbene complexes
classification of, **5**, 932
reactions, with metal halides, **4**, 722
as starting materials, for ketenimines, **3**, 596
studies, **5**, 931
synthesis of, **5**, 932; **6**, 716
See also Carbene complexes
Transition metal–carbon bonds, reduction of, **1**, 70
Transition metal carbonyls, for the activation of aryl halides, **2**, 684
Transition metal carbyne complexes, reactions, with alcoholates, **5**, 942
Transition metal catalysis
for the substitution of sulfur functions, **1**, 238
for the synthesis of vinylsilanes and arylsilanes, **2**, 912
Transition metal catalysts
for the dehydrogenation of alkanes, **1**, 554
for the opening of vinyl epoxides, **1**, 228
reactions
with Grignard reagents, **1**, 439
with silanes, **1**, 86
for the substitution of alkyl halides, **1**, 465
for the synthesis of alkylmagnesiums, **2**, 561
for the synthesis of aryl ketones, **3**, 288
for the synthesis of vinyl sulfides, **2**, 710
Transition metal-catalyzed couplings, for the synthesis of alkenylsilanes, **2**, 931
Transition metal complex anions
reactions
with alkylantimony halides, **2**, 511
with alkylarsenic halides, **2**, 511
with alkylbismuth halides, **2**, 511
Transition metal complexes
of carbenes, **6**, 727
as catalysts
in hydroboration reactions, **2**, 42
for hydrocarboxylation reactions, **1**, 297
for hydroformylation reactions, **1**, 296
from, silylmethyl ligands, **4**, 656
oxidative metallation with, **2**, 960
of $1\lambda^3$-phosphaallenes, **3**, 609
of $1\lambda^3$-phosphacumulenes, **3**, 609
reactions
with isocyanide dichlorides, **3**, 699
with phosphaalkenes, **5**, 893
with silicon hydrides, **6**, 402
with 1-silyl-1-chloro-1-lithiomethane, **4**, 658
for the rearrangement of vinylcyclopropanes, **1**, 803
of selenoketenes, **3**, 554
as starting materials
for ketenes, **3**, 544
for thioketenes, **3**, 552
synthesis of, **2**, 512; **4**, 955
use in the dimerization of alkenyl metals, **1**, 491
use in the synthesis of dienes, **1**, 468
use in the synthesis of trihalides, **6**, 6
Transition metal complexes, alkyl, β-elimination of, **1**, 667
Transition metal complexes, haloarene, reactions, with methylenes, **1**, 474
Transition metal complexes, vinylidene, reactions, with selenium, **3**, 554
Transition metal cyanides, synthesis of, **5**, 1137
Transition metal halides
reactions
with 2-lithio(dimethylaminomethyl)ferrocene, **4**, 723
with lithioferrocene, **4**, 723

with tricarbonyl(lithiobenzene)chromium, **4**, 723
with tricarbonyl(lithiocyclobutadiene)iron, **4**, 723
with tricarbonyl(lithiocyclopentadienyl)-
manganese, **4**, 723
Transition metal hydride reductions, for the synthesis of
alkenes, **1**, 457
Transition metal imides
from
haloimides, **5**, 388
imides, **5**, 387
sulfinylimides, **5**, 388
synthesis of, via transmetallation, **5**, 388
Transition metals
bis(trimethylsilyl)methyl compounds of, **6**, 194
catalysis of alkyne hydroboration, **2**, 902
as catalysts, for the synthesis of ketenes, **3**, 531
elimination of, **1**, 665
enolates of, **2**, 701
from
alkenyl halides, **4**, 821
alkynes, **4**, 822
functions containing two, synthesis of, **4**, 669
in metathesis reactions, **1**, 674
multidecker sandwich compounds of, synthesis of, **4**, 694
of phosphorus functions, synthesis of, **5**, 890
reactions
with alkynic alcohols, **5**, 944
with allylic halides, **1**, 1004
with 1,1-bisphosphinoalkenes, **4**, 1028
with carbanions, **1**, 845
with carbon monoxide, **1**, 212; **6**, 521
with *o*-diketodiynes, **3**, 272
with diphosphacyclopentadiene, **6**, 154
with phosphorus ylides, **4**, 578
with propargylic halides, **1**, 1004
with α-silylmethyl anions, **4**, 655
with triphosphacyclopentadiene, **6**, 154
as starting materials, for η^1-acyl metals, **5**, 440
synthesis of, **2**, 895; **4**, 147, 820, 1032; **5**, 650
via transmetallation reactions, **4**, 820
for the synthesis of acylsilanes, **5**, 419
in the synthesis of arylsilanes, **2**, 944
synthesis of functions, **4**, 709
as transmetallating agents, **1**, 464
Transition metals, alkyl
decomposition of, **1**, 665
synthesis of, **2**, 563
Transition metals, α-haloalkenyl, synthesis of, **4**, 820
Transition metals, imidoyl, synthesis of, **5**, 797
Transition metals, α-metallomethyl, reactions, with silyl
electrophiles, **4**, 657
Transition metals, α-silyl methyl
synthesis of
via changing the groups attached to the central
methylene or to the metal, **4**, 659
via reaction of an α-metallomethyl transition metal with
a silyl electrophile, **4**, 657
via reaction of a transition metal complex with an
α-silylmethyl anion, **4**, 656
Transition metals, trihalomethyl-, synthesis of, **6**, 247
Transition metal salts
reactions
with lithium aluminum hydride, **1**, 72
with metal hydrides, **1**, 515
with sodium borohydride, **1**, 72
for the reduction of alkynes, **1**, 457
Transition metal–trimethylenemethane complexes,
cycloadditions of, **1**, 360
Transition states
in Friedel–Crafts alkylations, **1**, 181
for substituted 5-hexenyl radical cyclizations, **1**, 337
Transmetallation
of alkenylalanes, **2**, 926
of alkenyl metals, **2**, 910
of alkenylzirconiums, **2**, 969
of arsenic compounds, **4**, 599
for aryl metals, **2**, 956, 990
of bis(mercury halides), **4**, 668
in catalytic cycles, **2**, 992
characteristics of, **2**, 956
definition of, **2**, 990
of diaryl(trimethylsilyl)phosphines, **2**, 868
direction of, prediction for, **2**, 990
of imidoyllithiums, **5**, 802
and interconversions, **2**, 992
of lithium enolates, **2**, 702
for metallacyclopentadienes, **2**, 989
non-redox, **2**, 990
for organometallics, **2**, 952
of organometals, **2**, 956
reactions
with alkenylalanes, **2**, 976
with alkenylzirconiums, **2**, 969
redox, **2**, 990
for aryl metals, **2**, 956
mechanics of, **2**, 992
for organometallics, **2**, 952
for vinyl metals, **2**, 956
reversibility of, **2**, 991
solvent effect, **2**, 990
of stannanes, **4**, 531
for the synthesis of alkylaluminums, **2**, 595
for the synthesis of alkylcobalt(I) compounds, **2**, 576
for the synthesis of alkylcoppers, **2**, 586
for the synthesis of alkyllithiums, **2**, 555
for the synthesis of alkylmagnesiums, **2**, 559
for the synthesis of alkyl metal(II) compounds, **2**, 577
for the synthesis of alkylthalliums, **2**, 597
for the synthesis of alkyltins, **2**, 599
for the synthesis of α-amino Grignard reagents, **4**, 531
for the synthesis of borylmethide salts, **6**, 199
for the synthesis of carbon-bridged dimercury
complexes, **4**, 697
for the synthesis of 1,1-dimetal complexes, **4**, 669
for the synthesis of germanium–nitrogen bonds, **2**, 815
for the synthesis of lanthanides, **2**, 564
for the synthesis of α-magnesiosilylalkanes, **4**, 655
for the synthesis of metal(III) halides, **2**, 880
for the synthesis of metal amides, **5**, 389
for the synthesis of metal imides, **5**, 391
for the synthesis of organometal(III) dihalides, **2**, 882
for the synthesis of polylithiated alkanes, **6**, 205
for the synthesis of silylmethyl compounds, **6**, 193
for the synthesis of transition metal imides, **5**, 388
for the synthesis of transition metals, **4**, 820
for vinyl metals, **2**, 956, 990
Transposition, of ketones, **3**, 126
Transquaternization, for the synthesis of quaternary
phosphonium salts, **2**, 447
Transsilylation
for the synthesis of disilylamides, **5**, 379
for the synthesis of silicon–nitrogen bonds, **2**, 812
for the synthesis of *N*-silylamides, **5**, 376
for the synthesis of silyl ylides, **6**, 703
for the synthesis of α-silyl ylides, **5**, 913
Transthionylation
for the synthesis of sulfinylamines, **2**, 399
for the synthesis of sulfinylhydrazines, **2**, 386
Transylidation
definition of, **3**, 493
of sulfenyl halides, **6**, 697
for the synthesis of methylenephosphoranes, **6**, 702
for the synthesis of phosphonium ylides, **6**, 700
for the synthesis of α-silyl ylides, **5**, 913
Trapping processes, for the addition of carbon nucleophiles
to alkenes, **1**, 304

Triacetates, reactions, with hydrogen bromide, **4**, 121
Triacetic acid, nitrilo-, as starting material, for 1-aminophosphonic acids, **4**, 492
Triacetyl compounds, oxidative nitration of, **4**, 435
Trialkyl compounds, synthesis of, **2**, 566
Trialkylsilyl groups, uses of, **6**, 172
Trialkylsilyloxy groups, use in the synthesis of allyl cations, **1**, 364
Trialkylsilyl substituents, stabilizing effect on ylidic carbanions, **6**, 702
Trialkylsilyl triflates–triethylamine, use in the synthesis of trialkylsilyl imidates, **5**, 718
Triamide, hexamethylphosphoric, alkylation of, **2**, 582
Triamides, hexaalkylphosphoric
  from, amines, **2**, 413
  hydrolysis of, **2**, 414
  synthesis of, **2**, 413
Triamine, ethoxymethane-, use in the synthesis of ketene aminals, **4**, 974
Triamines
  from, formamidinium chloride, **6**, 321
  reactions, with benzil, **2**, 321
  as starting materials, for ortho amides, **6**, 143
  synthesis of, **6**, 140
Triamines, alkoxymethane-, use in the synthesis of ketene aminals, **4**, 973
Triamines, isopropoxymethane-
  from, chloroformamidinium chlorides, **4**, 973
  reactions, with isopropyl acetate, **4**, 973
Triamines, macrocyclic
  reactions
    with acetamide acetal, **6**, 140
    with formamide, **6**, 140
Triamines, *N*-monoprotected, synthesis of, **2**, 321
Triarsines, cyclic, synthesis of, **2**, 508
Triazaadamantanes, synthesis of, **4**, 408; **6**, 140
Triazacyclohexanes, *N*-chloromethyl-, from, *N*-methoxymethyls, **4**, 97
Triazametallole complexes, synthesis of, **5**, 955
Triazanes, synthesis of, **2**, 387
Triazanium salts
  from
    chloramine, **2**, 387
    hydroxylamine-*O*-sulfonic acid, **2**, 387
Triazaphosphaadamantanes, acylation of, **4**, 481
Triazaphospholes, synthesis of, **4**, 483
Triazene *N*-oxides, synthesis of, **2**, 776
Triazenes
  from, azides, **2**, 387
  synthesis of, **2**, 387, 776
Triazenes, *N*-acyl-, from, amides, **5**, 365
Triazenes, aliphatic, reactions, with acetic acid, **1**, 1119
Triazenes, aminomethyl-, synthesis of, **4**, 444
Triazenes, aryl-
  as starting materials
    for aryl fluorides, **2**, 615
    for aryl iodides, **2**, 632
Triazenes, diaryl-, as starting materials, for alkenediazonium ions, **2**, 801
Triazenes, formyl-, as starting materials, for aryl diazo isocyanides, **3**, 710
Triazenes, 3-(hydroxymethyl)-, synthesis of, **4**, 312
Triazenes, vinyl-, acidolysis of, **2**, 801
Triazepines, synthesis of, **4**, 411
s-Triazinamines, reactions, with oxalyl chloride, **5**, 973
Triazine, perhydro-, as starting material, for iminium ions, **4**, 455
Triazine, trichloro-, reactions, with carbodiimides, **6**, 622
Triazine, 1,3,5-tris[(*S*)-phenylethyl]hexahydro-, synthesis of, **4**, 99
Triazine, tris(trimethylsilylmethyl)-, synthesis of, **4**, 509
1,3,5-Triazine
  reactions
    with amines, **5**, 747
    with hydrazines, **5**, 770
  as starting material, for formamidines, **5**, 747
1,3,5-Triazine, 1,3,5-triacetylhexahydro-, nitration of, **4**, 447
1,3,5-Triazine, trichloro-, reactions, with *t*-butyl hydroperoxide, **2**, 687
1,3,5-Triazine, 1,3,5-triphenyl-, reactions, with cyclic boron esters, **4**, 457
s-Triazine, reactions, with 1,1-disubstituted hydrazines, **5**, 855
Triazines
  as starting materials
    for bromo nitro compounds, **4**, 109
    for carbamoylsulfenyl chlorides, **4**, 103
Triazines, aminomethyl-, synthesis of, **4**, 443
Triazines, 1-aryl-3-cyano-, synthesis of, **5**, 1124
Triazines, 1-aryl-3-cyano-3-methyl-, synthesis of, **5**, 1118
Triazines, cyanamido-, synthesis of, **5**, 1120
Triazines, 1,2-dihydro-, synthesis of, **5**, 1075
Triazines, 5-substituted-1-cyano-2-iminohexahydro-, synthesis of, **5**, 1121
Triazines, trialkylammonium-, reactions, with calcium cyanamide, **5**, 1120
1,3,5-Triazines
  from, isocyanates, **4**, 426
  synthesis of, **4**, 416
1,3,5-Triazines, hexahydro-
  as starting materials, for Schiff bases, **4**, 463
  synthesis of, **4**, 294, 407, 418, 427
1,3,5-Triazines, perhydro-, synthesis of, **4**, 418, 425
s-Triazines, synthesis of, **5**, 973
s-Triazines, 2-cyano-4-methylamino-6-hydroxy-, as starting materials, for 2-methyl-4-methylamino-6-hydroxy-s-triazines, **1**, 47
s-Triazines, 2-methyl-4-methylamino-6-hydroxy-, from, 2-cyano-4-methylamino-6-hydroxy-s-triazines, **1**, 47
1,2,4-Triazine-3(2*H*)-thione, 4,5-dihydro-, synthesis of, **6**, 574
Triazinones, from, carbonimidic dichlorides, **6**, 619
s-Triazin-2(1*H*)-ones, cyclic, synthesis of, **6**, 367
Triazole, *N*-thiobenzoyl-, as thioacylating reagent, **5**, 641
1,2,4-Triazole, 4-amino-3,5-dimethyl-, synthesis of, **5**, 640
1,2,4-Triazole, 5-(*N*-cyano-*N*(4-methylphenyl)amino)-1-methyl-3-methylthiol-, synthesis of, **5**, 1122
1,2,4-Triazole, *N*-thiobenzoyl-, reactions, with *N*-methylaniline, **5**, 587
Triazoles
  as intermediates, in the synthesis of α-diazoimines, **3**, 476
  oxidation of, **3**, 666
  as starting materials, for ketenimines, **3**, 579
  synthesis of, **2**, 771; **5**, 849, 874
Triazoles, acyl-
  as acylating reagents, **5**, 235
  use in the synthesis of selenocarboxylic esters, **5**, 239
Triazoles, *N*-chloromethyl-, synthesis of, **4**, 105
Triazoles, *N*-cyano-, use in the synthesis of ketenimines, **3**, 579
Triazoles, dihydro-
  reductive deamination of, **2**, 771
  as starting materials, for nitrile ylides, **3**, 680
Triazoles, silyl, synthesis of, **2**, 943
1,2,3-Triazoles, from, diazo compounds, **5**, 869
1,2,3-Triazoles, 1-amino-, as starting materials, for alkynes, **1**, 1075
1,2,3-Triazoles, 1-cyano-, synthesis of, **5**, 1119
1,2,3-Triazoles, dialkoxydihydro-, synthesis of, **6**, 107
1,2,3-Triazoles, 1,4-diphenyl-, lithium anion, fragmentation and reaction with ethanethiol, **5**, 730
1,2,3-Triazoles, *N*-phenyl-, photolysis of, for the synthesis of ketenimines, **3**, 579
1,2,3-Triazoles, 1-tosylamino-, as starting materials, for alkynes, **1**, 1075
1,2,4-Triazoles, acylaminomethyl-, synthesis of, **4**, 448

1,2,4-Triazoles, 4-amino-, synthesis of, **5**, 849
1,2,4-Triazoles, 5-amino-, reactions, with carbon disulfide, **6**, 564
1,2,4-Triazoles, 3-(ω-aminoalkyl)-, from, isothiosemicarbazide hydrohalides, **2**, 310
1,2,4-Triazoles, dihydro-, synthesis of, **6**, 660
1*H*-1,2,3-Triazoles, 5-amino-1-vinyl-4,5-dihydro-, decomposition of, **3**, 414
Triazolide, (trifluoromethyl)phosphinylbis-, from, dibromo(trifluoromethyl)phosphine, **6**, 242
1,2,4-Triazolidin-3-ones, oxidation of, **4**, 440
1,2,4-Triazolidin-3-thiones, oxidation of, **4**, 440
Triazolinediones, as starting materials, for nitrogen-functionalized ylides, **5**, 905
Triazolines
  as intermediates
    in diazo group transfer processes, **3**, 472
    in the synthesis of α-diazonitriles, **3**, 478
    in the synthesis of silylaziridines, **4**, 515
  rearrangement of, **3**, 419
1,2,4-Triazolium iodide, 4-amino-3,5-di(methylthio)-, reactions, with *p*-toluenethiocyanate, **5**, 1122
1,2,4-Triazolium salts, as starting materials, for hydrazonamides, **5**, 855
Triazolones, synthesis of, **6**, 631
1,2,4-Triazolones, dihydro-, synthesis of, **4**, 440
Tribenzenesulfenamide, reactions, with thioamides, **5**, 735
Tribenzyl chloride, for the silylation of amides, **5**, 375
Triboluminescence, definition of, **5**, 1094
2,8,9-Triborabicyclo[3.3.1]nona-3,6-diene-2,8,9-triamine, synthesis of, **6**, 184
2,4,5-Triborabicyclo[1.1.1]pentane, synthesis of, **6**, 184
1,3,5-Triboracyclohexane, 1,3,5-trichloro-, synthesis of, **4**, 635
1,4,6-Triboraspiro[4.4]nona-2.8-diene, synthesis of, **6**, 184
Tribromo compounds, decarboxylation of, **6**, 31
Tribromomethyl compounds, synthesis of, **6**, 31
Tribromomethylium ion, synthesis of, **6**, 726
Tributylamine, use in the synthesis of halosulfines, **5**, 488
Tributyl chloride, reactions, with diphenylarsinomethyllithium, **4**, 600
Tributylstannyl groups, elimination of, **1**, 603
Tricarbamate, synthesis of, **6**, 140
1,2,3-Tricarbonyls
  chlorination of, **5**, 841
  2-hydrazones of, synthesis of, **5**, 841
Tricarboxylic acids, as starting materials, for triphosphonic acids, **4**, 561
Trichloramine
  handling problems, **2**, 298
  reactions, with aluminum trichloride, **2**, 298
  use in amination reactions, **2**, 306
Trichloride, reactions, with lithium aluminum hydride, **2**, 571
2,2,2-Trichloro-*t*-butyloxycarbonyl (TCBOC), use in the synthesis of Leu-enkephalins, **6**, 487
Trichlorofon, reactions, with diazomethane, **4**, 912
Trichloromethyl anions
  from
    hexachloroacetone, **6**, 31
    trichloromethane, **6**, 25
    trichloromethyltrimethylsilane, **6**, 24
  reactions
    with aldehydes, **6**, 25
    with ketones, **6**, 25
    with nitroalkenes, **6**, 24
  as starting materials, for dichlorocarbenes, **6**, 23
Trichloromethylation, for the synthesis of aryl compounds, **6**, 9
Trichloromethyl groups, as starting materials, for dichloromethyl groups, **4**, 13
Trichloromethylium ion, synthesis of, **6**, 726
Trichloromethyl radical, as chain carrier, **4**, 1078
Trichloromethyls
  reduction of, **4**, 13
  as starting materials, for hydroxamoyl halides, **5**, 666
  synthesis of, **6**, 23
Trichloromethylthiolation, with trichloromethanesulfenyl chloride, **6**, 233
Trichlorosilane–triethylamine, for the synthesis of isocyanides, **3**, 698
Trichodiene, synthesis of, via Cope elimination, **1**, 784
Trichonine, synthesis of, **1**, 647
Trichosporon fermentans CBS 2529, use in the synthesis of carboxylic acids, **5**, 118
Trichothecenes, synthesis of, **1**, 382
Tricobaltnonacarbonyl complexes, bromomethylidyne-, reactions, with silanes, **6**, 401
Tricobaltnonacarbonyl complexes, methylidyne-, reactions, with silanes, **6**, 401
(*Z*)-9-Tricosene, synthesis of, **1**, 39
Tricyanate, phosphoryl, synthesis of, **5**, 1017
Tricyclo[5.3.0.0]decane, from, ketenes, **3**, 528
Tricyclo-[4.1.0.0^{2,7}]-heptane, reactions, with potassium *t*-butoxide, **1**, 848
1,3,5-Tricyclopropyltropylium cations, synthesis of, **1**, 907
n-Tridecanoic acid, synthesis of, **5**, 35
1-Tridecene, as starting material, for dodecanoic acid, **5**, 52
(*E,E*)-Trienals, synthesis of, **3**, 62
Trienes
  from
    diallylic sulfones, **1**, 618
    vinyl sulfide, **1**, 484
  photochemical electrocyclization of, **1**, 420
  protonation of, **1**, 912
  rearrangement of, **1**, 419
  synthesis of, **1**, 403
    via alkoxide elimination, **1**, 579
    via Hofmann elimination, **1**, 628
1,2,5-Trienes, synthesis of, **1**, 977
1,3,5-Trienes, synthesis of, **1**, 791
(*Z*)-Trienes, synthesis of, **1**, 509
1,2,4-Trienols, synthesis of, **1**, 977
(*Z*)-Trienols, synthesis of, **1**, 504
Trienynes, from, propargylic alcohols, **1**, 962
Triethylamine
  as catalyst, **5**, 818
  as coreagent, **6**, 602
  reactions
    with acid chlorides, **3**, 528
    with alkyl isocyanides, **5**, 1052
    with aryl isocyanates, **4**, 446
    with bromothionitriles, **3**, 655
    with carbon disulfide, **5**, 590
    with chloroiminium salts, **2**, 758
    with chloromethanesulfonyl chloride, **5**, 496
    with chlorophosphine esters, **4**, 483
    with chlorophosphines, **6**, 265
    with α-chloropropionyl chloride, **4**, 768
    with 2-chloroquinoline, **2**, 301
    with chlorosulfonic acid, **2**, 368
    with cyanine iodides, **2**, 1046
    with 1,3-diaryl-2,2-dibromoaziridines, **3**, 574
    with dichlorides, **3**, 526
    with dichloromethyldichlorophosphine, **6**, 679
    with di(cyclohexyl)boron chlorides, **2**, 532
    with perchloryl fluoride, **2**, 339
    with phosgene, **4**, 790
    with styrene nitroso chloride, **2**, 778
    with sulfonyl chlorides, **1**, 619
  as reagent
    for the elimination of hydrogen chloride from α-chloroalkyl ureas, **5**, 787
    for the synthesis of imidoyl phosphonates, **5**, 789
  use in the synthesis of alkenes, **1**, 592
  use in the synthesis of *N*-borylimidoyl compounds, **5**, 863

use in the synthesis of halosulfines, **5**, 488
use in the synthesis of *N*-silylamides, **5**, 374
Triethylamine fluorides, *N*-perchloryl-, synthesis of, **2**, 339
Triethylamine hydrochloride
as byproduct, in the synthesis of nitrile oxides, **3**, 683
use in the synthesis of nitrile ylides, **3**, 679
Triethyl chloro(trimethylammonium)methylbis-(phosphonate), synthesis of, **6**, 286
Triethylene glycol, reactions, with *N*-chlorocarbonyl isocyanate, **5**, 1007
Triethylgermyl chloride, reactions, with 2-lithio-1,3-dithiane, **6**, 132
Triethylsilyl chloride, for selective monosilylation, **6**, 129
Triferrate anions, hydridoundecacarbonyl-, reactions, with aryl *N*-hydroxyimidoyl chlorides, **3**, 666
Triflate, 2-azanorbornene-2-spiro-1'-aziridinium, for the aminoethylation of carbonyls, **1**, 254
Triflate, 4-*t*-butylcyclohexenyl, reduction of, **1**, 435
Triflate, *t*-butyldimethylsilyl, use in the synthesis of *N*-silylamides, **5**, 375
Triflate, 1,3-dichloro-1,1,3,3-tetraisopropyldisiloxane, use in the synthesis of *N*-silylamides, **5**, 375
Triflate, ethyl, for the reduction of tertiary amides, **3**, 85
Triflate, *N*-fluorotrimethylpyridinium (NFPT), for the fluorination of sulfides, **4**, 62
Triflate, 5-heptyn-1-yl, solvolysis of, **1**, 1124
Triflate, 5-hexyn-1-yl, solvolysis of, **1**, 1124
Triflate, iodonium, reactions, with trimethylsilyl ethers, **1**, 108
Triflate, iodosyl, as starting material, for dialkynyliodonium triflates, **2**, 1021
Triflate, methyl
reactions, with nitriles, **3**, 282
as starting material, for carbon-bridged triosmium complexes, **4**, 683
Triflate, 2-methyl-1-propenyl, reactions, with azobenzenes, **4**, 985
Triflate, silyl, as starting material, for silyl enol ethers, **2**, 699
Triflate, trialkylsilyl, reactions, with alkynylzincs, **1**, 314
Triflate, (trifluoromethyl)dibenzothiophenium, use in the synthesis of *β*-fluoroaliphatic ketones, **3**, 151
Triflate, *S*-(trifluoromethyl)dibenzothiophenium, as trifluoromethylation agent, **6**, 233
Triflate, trimethylsilyl
as catalyst, for the synthesis of acetals, **4**, 181
reactions
with α-halo sulfoxides, **4**, 773
with pyridine, **5**, 226
with selenoxides, **2**, 369
as reagent, for the synthesis of acetoxyiodobenzene triflate, **5**, 205
Triflate group, as a leaving group, **2**, 1089
Triflates
displacement of, **1**, 179
via alkynes, **1**, 200
in the displacement of alkyl alcohols, **1**, 177
reactions
with *t*-butyl hydroperoxide, **2**, 98
with chiral amino acid synthon, **1**, 180
as starting materials, for lanthanides, **2**, 564
use in the synthesis of alkyl halides, **2**, 9
See also Methane sulfonates, trifluoro-
Triflates, acyclic vinyl, synthesis of, **1**, 480
Triflates, acyloxyiminium, synthesis of, **4**, 881
Triflates, alkenyl
1,1-elimination of, **1**, 1071
Fritsch–Buttenberg–Wiechell rearrangement of, **1**, 1071
reactions
with alkynes, **1**, 1029
with alkynyltins, **1**, 1029
with alkynyltrialkyltins, **1**, 1027
with alkynyltrimethylsilanes, **1**, 1028
with diorganocuprates, **1**, 1027
with terminal alkynes, **1**, 1029
as starting materials, for alkynes, **1**, 1070
synthesis of, **2**, 689
Triflates, alkyl
reactions
with lithio diarylarsines, **3**, 196
with trialkyltin peroxides, **2**, 98
Triflates, ammonium, synthesis of, **4**, 507
Triflates, aryl
in cross-coupling reactions, **1**, 489
displacement of, **1**, 482
Heck reaction of, **3**, 293
reactions
with alkynes, **1**, 1029
with alkynylstannanes, **1**, 1030
with alkynyltrialkyltins, **1**, 1027
with arylstannanes, **1**, 482
with diorganocuprates, **1**, 1027
with lithium diisopropylamide, **1**, 1071
with organoboranes, **1**, 482
with organostannanes, **1**, 482; **3**, 290
with siloxycyclopropanes, **1**, 483
with silylketene acetals, **1**, 483
with trialkylaluminums, **1**, 482
reduction of, **1**, 433
as starting materials, for biaryls, **1**, 482
substitution of, in Heck reactions, **1**, 494
synthesis of, **2**, 689
use in the substitution of vinyl halides, **1**, 469
Triflates, 2-azonianorbornene-2-spiro-1'-aziridinium, as alkylating agent, **1**, 254
Triflates, cyclic vinyl, solvolysis of, **1**, 1117
Triflates, dialkylboryl, reactions, with carboxylic acids, **4**, 832
Triflates, dienoyl, as starting materials, for dienes, **1**, 435
Triflates, enol
α-elimination of, **1**, 550
reactions
with arylzinc chloride, **1**, 480
with nickel catalysts, **1**, 1099
with organoborons, **1**, 480
with organostannanes, **1**, 480
with trialkylmanganates, **1**, 480
substitution of, **1**, 480
Triflates, 1-ethynylvinyl, as starting materials, for carbenes, **1**, 988
Triflates, *o*-haloaryl, reactions, with *n*-butyllithium, **1**, 1071
Triflates, heteroaryl, reactions, with terminal alkynes, **1**, 1029
Triflates, lanthanide, use in aldol reactions, **3**, 159
Triflates, nitrone hydro-, synthesis of, **3**, 434
Triflates, phenol, reduction of, **1**, 433
Triflates, prop-2-yniminium
reactions
with thiolates, **4**, 321
with thiols, **4**, 321
Triflates, silyl
alkynylation of, **2**, 1089
reactions, with hydrazones, **1**, 186
use in the synthesis of disilylamides, **5**, 379
Triflates, spiroaziridinium, stability of, **1**, 254
Triflates, threonine, Friedel–Crafts alkylation of, **1**, 182
Triflates, trialkylsilyl
reactions, with diazomethyl ketones, **3**, 483
as silylating agents, **4**, 831
Triflates, vinyl
in cross-coupling reactions, **1**, 489
reactions
with α-alkoxyvinylstannanes, **3**, 243
with potassium *t*-butoxide, **1**, 1143
with silanes, **1**, 490
reduction of, **1**, 435
as starting materials

for hydroxamates, **5**, 331
for vinyl anions, **1**, 1099
for vinyl iodides, **2**, 630
substitution of, **1**, 480, 494
use in the substitution of vinyl halides, **1**, 469
Triflate salts
synthesis of, **3**, 687
use in the synthesis of α-aminosilanes, **4**, 507
Triflic anhydride
reactions
with iron acyl complexes, **5**, 950
with phenols, **2**, 689
with triphenylphosphine oxide, **5**, 196
use in the synthesis of trifluoromethyls, **6**, 22
See also Methanesulfonic anhydride, trifluoro-
Triflic anhydrides, reactions, with phosphine oxides, **5**, 123
Triflone, use in Ramberg–Bäcklund reactions, **1**, 619
Triflone, mesyl, use in the synthesis of alkenes, **1**, 705
Trifluoroacetic anhydride (TFAA), reactions, with dinitroalkenes, **6**, 150
Trifluoroacetylating agents, synthesis of, **1**, 582
Trifluoroacetylation
reagents for, **3**, 282
of tertiary alcohols, **1**, 582
Trifluoromethyl
cadmium compounds, synthesis of, **6**, 14
copper compounds, synthesis of, **6**, 14
as electrophile, **6**, 12
from, nitriles, **6**, 10
as nucleophile, **6**, 13
for the synthesis of α-lithio compounds, **1**, 146
zinc compounds, synthesis of, **6**, 14
Trifluoromethylamination, with mercury(II) bis(trifluoromethyl)amide, **6**, 241
Trifluoromethylation
agents for
bis(diethylamino)chlorophosphite, **6**, 234
(4-chlorophenyl)-(2,6-dimethylphenyl)trifluoromethylsulfonium hexafluoroantimonate, **6**, 233
S-(trifluoromethyl)dibenzothiophenium triflate, **6**, 233
with arsenic(III) fluoride, **6**, 243
of aryl halides, **6**, 14
with bromotrifluoromethane, **6**, 233
electrophilic, **6**, 22
of hydroquinone, **6**, 13
nucleophilic, **6**, 22
with silicon halides, **6**, 243
with tetrachloromethane and hydrogen fluoride, **6**, 229
with trifluoroiodomethane, **6**, 233
Trifluoromethylation reagents
cadmium, **6**, 220
N-(trifluoromethyl)-*N*-nitrosobenzenesulfonamide, **6**, 233
N-(trifluoromethyl)-*N*-nitrosotrifluoromethanesulfonamide, **6**, 233
Trifluoromethyl cations, stability of, **6**, 726
Trifluoromethyl complexes, as starting materials, for cationic complexes, **6**, 727
Trifluoromethyl–metal complexes, decomposition of, **4**, 7
3-Trifluoromethylphenyls, synthesis of, **6**, 43
Trifluoromethyl radicals
from
bromotrifluoromethane, **6**, 20
iodotrifluoromethane, **6**, 20
trifluoroacetic acid, **6**, 11
trifluoromethanes, **6**, 11
trifluoromethane sulfonic acid, **6**, 11
reactions, with alkenes, **6**, 20
transfer of, **6**, 20
Trifluoromethyls
from
dithiocarboxylate esters, **6**, 10
dithiocarboxylates, **6**, 10
synthesis of, **6**, 9
uses of, **6**, 9
Trifluoromethylselenolation, agents for, mercury(II) trifluoromethaneselenolate, **6**, 237
Trifluoromethyl tellurides, alkyl-, synthesis of, **6**, 237
Trifluoromethylthiolation
agents for
mercury(II) bis(trifluoromethanethiolate), **6**, 235
mercury(II) bis(trifluromethanethiolate), **6**, 236
with bis(trifluoromethyl) trithiocarbonate, **6**, 233
with copper(I) trifluoromethanethiolate, **6**, 233
with mercury(II) bis(trifluoromethanethiolate), **6**, 233
with methyl fluorosulfonyldifluoroacetate, **6**, 232
with silver(I) trifluoromethanethiolate, **6**, 233
with thiocarbonyl fluoride, **6**, 233
with trifluoromethanesulfenyl chloride, **6**, 233
with trifluoromethanethiol, **6**, 233
Trifluoromethyl transfer, mechanism, **6**, 13
Triformamides
as starting materials
for tridentate aryl isocyanides, **3**, 723
for triisocyanides, **3**, 711
1,3,5-Trigermacyclohexane, 1,1,3,3,5,5-hexachloro-, reduction of, **4**, 629
Trigermylmethyls, synthesis of, **6**, 185
Triglyme, use in the cleavage of carbon–boron bonds, **1**, 59
Trigonal cations, stability of, **1**, 1109
Trigonopsis variabilis, use in the synthesis of α-keto acids, **5**, 118
Trihalides
hydrolysis of, **5**, 26
synthesis of, **6**, **1–33**
N-Trihalomethoxy compounds, synthesis of, **6**, 232
Trihalomethoxy group, as superhalogens, **6**, 228
Trihalomethyl complexes
as starting materials, for dihalocarbene complexes, **6**, 717
in the synthesis of carbene complexes, **6**, 717
Trihetero-[3,3]-rearrangement, mechanism, **1**, 840
Triiodomethylium ion, synthesis of, **6**, 726
Triiron, dodecacarbonyl-
reactions
with allenyl thiols, **4**, 683
with diphenylcyclopropenone, **4**, 681
with thiophene, **4**, 681
as starting material, for benzoferracyclopentadiene complexes, **4**, 681
Triisocyanate, phosphoryl
reactions, with trialkyl phosphites, **5**, 1008
synthesis of, **5**, 1017
Triisocyanides, from, triformamides, **3**, 711
1,3,5-Triketones, as starting materials, for trithiapentalenes, **3**, 378
Trimethylamine
for the hydrolysis of aromatic esters, **5**, 106
use in the synthesis of alkanoic acids, **5**, 31
use in the synthesis of halosulfines, **5**, 488
Trimethylamine–borane complexes, reactions, with methylthiomethyllithium, **4**, 373
Trimethylamine hydrochloride, reactions, with lithium salts, **4**, 373
Trimethylamine oxide, use in the synthesis of boryl enol ethers, **5**, 417
Trimethylamine *N*-oxide
Polonovski reaction of, **4**, 311
as reoxidant, **4**, 357
Trimethylenemethane dianions, from, isobutene, **6**, 206
Trimethylgermylation
of acetonitrile, **6**, 185
of lithium salt of ethyl acetate, **6**, 185
Trimethylgermyl chloride
reactions
with ketals, **6**, 130
with 1,3-oxathiane, **6**, 132

with (*t*-butylcarbamoyl)lithium, **6**, 515
Trimethylsilyl bromide
reactions
with dimethyl α-diazo phosphonates, **3**, 482
with fluorocarbonylsulfenyl chloride, **6**, 432
with mercury fulminate, **3**, 686
with methyl α-diazo phosphinates, **3**, 482
Trimethylsilyl chloride–magnesium–hexamethylphosphoramide reagent, for the synthesis of alkynylsilanes, **2**, 1090
Trimethylsilyl chloride (TMS-Cl)
as byproduct, **6**, 443
reactions
with alkyl benzoates, **6**, 124
with germyl(plumbyl)methides, **6**, 203
with lithium salts, **6**, 163
with methylene chloride, **6**, 267
with 2-[[(trimethylsilyl)methyl]thio]tetrahydropyrans, **6**, 349
synthesis of, **6**, 518
Trimethylsilyl compounds
as methylenecyclopropane equivalents, **1**, 363
synthesis of, **1**, 61
1-Trimethylsilyl compounds, 1-phenyl-, synthesis of, **4**, 658
1-Trimethylsilyl compounds, 1-phenylseleno-, deprotonation of, **4**, 947
Trimethylsilyl cyanide (TMS-CN), as starting material, for bis(trimethylsilyl)aminomethanes, **6**, 162
Trimethylsilyl groups, in elimination reactions, **1**, 648
Trimethylsilyl migrations, for the synthesis of methylenephosphines, **6**, 686
Trimethylsilyl (TMS), chlorination of, **6**, 243
Trimethylstannyls, synthesis of, **6**, 133
Trimethyl triselenoorthoformate, carbon-labeled, from, carbon-labeled bromoform, **6**, 92
Trimethyl trithioorthoformate, carbon-labeled, synthesis of, **6**, 86
Trimyristin, hydrolysis of, **5**, 30
Trineopentoxo complexes, synthesis of, **3**, 518
Trinitro compounds
from
bromotrinitromethane, **6**, 145
iodotrinitromethane, **6**, 145
Trinitromethyl anion, addition to double bonds, **6**, 144
Trinitromethyl compounds, synthesis of, **6**, 145
Trinitromethylide anion, synthesis of, **6**, 369
8,9,10-Trinorcarane, 7,7-dibromo-, use in the synthesis of α-bromostannanes, **4**, 155
Trinorcaranes, bromo-, use in the synthesis of α-bromoplumbanes, **4**, 156
Trinuclear dianions, reactions, with electrophiles, **6**, 338
Triols
exchange reactions, with ortho esters, **6**, 79
from, 2-hydroxycrotylsilanes, **2**, 46
glycol cleavage of, **5**, 179
2,3-*syn*-1,2,3-Triols, from, 2-hydroxy enol ethers, **2**, 50
Triorganoantimony, oxidation of, **2**, 1080
Triorganoantimony difluorides, synthesis of, **2**, 492
Triorganoarsenic(III) compounds, protonolysis of, **1**, 445
Triorganoarsenic(IV) dihalides, as starting materials, for diorganoarsenic(III) halides, **1**, 445
Triorganoarsenic compounds, reduction of, **1**, 445
Triorganoarsenic difluorides, synthesis of, **2**, 492
Triorganoarsenic dihalides, pyrolysis of, **2**, 489
Triorganoarsines, from, chlorodioxarsolane, **2**, 872
Triorganoarsoranes, imino-, synthesis of, **2**, 504
Triorganoboranes
from, ethynyldiphenylphosphane, **2**, 1088
reactions, with alkynyl metals, **2**, 1086
synthesis of, **2**, 1083
Triorganogermanium selenocarboxylates, reactions, with arenesulfenyl chlorides, **5**, 249
Triorganometal chlorides, reactions, with diphenylcyanamidophosphinates, **5**, 1084
Triorganophosphoranes, dialkoxy-, synthesis of, **2**, 830
Triorganosilanes, reactions, with 1-alkynes, **2**, 1090
Triorganosilyl halides
reactions, with ylides, **5**, 913
as starting materials, for silylated ylides, **5**, 913
Triorganostannanes, phosphomethyl-, reactions, with phenyl isocyanate, **5**, 390
Triorganostannyls, use in the synthesis of dialkyl carbonates, **6**, 463
Triorganostibine(V) compounds, synthesis of, **5**, 1132
Triorganozincates, conjugate addition to enones, **3**, 128
Triosmium, dodecacarbonyl-
reactions, with alkynes, **4**, 682
reduction of, **4**, 679
Triosmium complexes, carbon-bridged
from
methyl triflate, **4**, 683
phenyllithium, **4**, 683
Trioxan
reactions
with nitriles, **4**, 425
with sulfonamides, **4**, 427
as reagent, for the synthesis of methylenebisamides, **4**, 424
Trioxane, reactions, with perfluoroalkyldichlorophosphines, **4**, 125
1,2,4-Trioxane, synthesis of, **4**, 214
1,3,5-Trioxane, reactions, with methylaluminum bis(4-bromo-2,6-di-*t*-butyl phenoxide), **3**, 159
1,2,4,6-Trioxepanes, synthesis of, **3**, 326
Trioxide, bis(trifluoromethyl)
reactions, with alkenes, **6**, 42
synthesis of, **2**, 99
Trioxide, di-*t*-butyl, synthesis of, **2**, 98
Trioxides, synthesis of, **2**, 98
Trioxides, cyclic, synthesis of, **2**, 98
Trioxo compounds, synthesis of, **2**, 573
1,2,4-Trioxones
synthesis of, **4**, 214
See also Ozonides
Triphenoxy compounds, from, perchloropentadienoic acid nitriles, **4**, 829
Triphenyl anions, stability of, **1**, 847
Triphenyl dihalides, use in the synthesis of alkyl halides, **2**, 7
Triphenylgermyl chloride, reactions, with 2-lithio-1,3-dithiane, **6**, 132
Triphenylmethyl anions, stability of, **1**, 850
Triphenylmethyl radical
as persistent radical, **1**, 922
structure of, **1**, 922
studies, **1**, 919
2,2,2,-Triphenyl-1,2-oxaphospholane, synthesis of, **2**, 831
Triphenylphosphine–tetrabromomethane, use in the synthesis of hydrazonyl halides, **5**, 675
Triphenylphosphine–tetrachloromethane, use in the synthesis of hydrazonyl halides, **5**, 675
Triphenylphosphite–methiodide compound, for the synthesis of alkyl iodides, **2**, 8
Triphenyl triselenoorthoformate, carbon-labeled, from, carbon-labeled formic acid, **6**, 92
Triphosgene
as phosgene substitute, **6**, 471
reactions
with amines, **5**, 965
with bisphenol A, **6**, 471
with *N*-butoxycarbonylamino acids, **6**, 486
with *N*-hydroxysuccinimide, **6**, 483
with 4-nitrocatechol, **6**, 467
with L-phenylalanine, **6**, 486
with L-Serine, **6**, 486
with 2,2,2-trichloro-*t*-butyl alcohol, **6**, 487
as starting material
for carbamoyl chlorides, **6**, 446
for chloroformate esters, **6**, 426

use in the synthesis of alkyl chlorides, **2**, 8
use in the synthesis of chloroformates, **6**, 483
See also Carbonate, bis(trichloromethyl)
Triphosgene[bis(trichloromethyl) carbonate]
use in the synthesis of isocyanates, **5**, 972
use in the synthesis of isocyanides, **3**, 696
Triphosphacyclopentadiene, reactions, with transition metals, **6**, 154
1,2,5-Triphospha–Dewar benzenes, reactions, with phosphaalkynes, **5**, 885
Triphosphatriarsine, hexaphenyl-, synthesis of, **2**, 894
1,2,3-Triphosphetan-4-one, 1,3-di-*t*-butyl-2-alkyl-, synthesis of, **6**, 519
Triphosphine oxide, synthesis of, **6**, 151
Triphosphonates, alkylation of, **4**, 563
Triphosphonic acids, from, tricarboxylic acids, **4**, 561
Triple-decker sandwich complexes
synthesis of, **4**, 717
use of term, **1**, 862
Triplet sensitizers, use in the synthesis of radicals, **1**, 925
Tri-*n*-propylamine, reactions, with nitriles, **1**, 858
Tripticyl carbocations, structure of, **1**, 898
Triptycene, 9-aminomethyl-1,4-dimethyl-, diazotization of, **1**, 259
Triquinancene, synthesis of, via xanthate pyrolysis, **1**, 776
Triquinanes, synthesis of, **1**, 529, 563, 1130, 1132
Triruthenium, dodecacarbonyl-
reactions
with
cyclopentadienylidenetriphenylphosphorane, **4**, 685
with 1,4-dibromocyclooctatetraene, **4**, 681
with 2-methylthiophene, **4**, 683
with phosphonium ylides, **4**, 683
with thiophene, **4**, 683
as starting material, for ruthenacyclopentadiene complexes, **4**, 683
Trisaccharides, *β*-mannosylchitobiose, reactions, with potassium thiocyanate, **5**, 1042
Tris(1-adamantyl)methyl cations, synthesis of, **1**, 900
Tris(*o*-alkylphenylphosphine), use in orthometallation reactions, **2**, 581
2,4,6-Tris-*t*-butylphenyl, use in the synthesis of carbodiphosphines, **5**, 1091
Triselanes, amino-, synthesis of, **2**, 369
Triselenacyclopentanes, synthesis of, **6**, 334
Triselenamonothiafulvalenes, synthesis of, **4**, 874
1,3,5-Triselenanes, synthesis of, **2**, 542
Triselenide, bisselenocarbamoyl, from, dialkylselenocarbamates, **6**, 594
Triselenide, bis[tris(trimethylsilyl)methyl], synthesis of, **6**, 355
Triselenide, dicyanogen, as starting material, for selenocyanates, **5**, 1116
Triselenides, as starting materials, for selenoureas, **6**, 597
Triselenides, diaryl, synthesis of, **2**, 735
Triselenocarbenium salts
as starting materials
for 2-alkaneseleno-1,3-diselenoles, **6**, 93
for triselenoortho esters, **6**, 93
Triselenocarbonate, bistrifluoromethyl, from, 2,2,4,4-tetrafluoro-1,2-selenetane, **6**, 592
Triselenocarbonate, dibenzyl
reactions
with amines, **6**, 597
with benzyl selenol, **5**, 626
Triselenocarbonates
alkylation of, **6**, 729
from, carbon diselenide, **6**, 590
as starting materials, for selenoureas, **6**, 597
use in the synthesis of tetraselenafulvalenes, **6**, 590
Triselenocarbonic acid, from, barium triselenocarbonate, **6**, 590
Triselenoorthoacetate, trimethyl
from, trimethyl orthoacetate, **6**, 92
synthesis of, **6**, 92
Triselenoorthocarbonates, synthesis of, **6**, 308
Triselenoortho esters
elimination of methaneselenol from, **4**, 875
from
carboxylic acids, **6**, 92
α,*α*-dihalo ethers, **6**, 92
diselenoacetals, **6**, 93
ortho esters, **6**, 92
1,1,1-trihaloalkanes, **6**, 92
triselenocarbenium salts, **6**, 93
triselenoorthoformate esters, **6**, 94
metallation of, **6**, 94
synthesis of, **6**, 92
Triselenoortho esters, *α*-hydroxy-, synthesis of, **6**, 94
Triselenoortho esters, *β*-hydroxy-, synthesis of, **6**, 94
Triselenoorthoformate, trimethyl, lithiation of, **6**, 317
Triselenoorthoformate, triphenyl
lithiation of, **6**, 317
reactions, with butyllithium, **4**, 289
synthesis of, **6**, 92
Triselenoorthoformate, tris(trifluoromethyl)
metallation of, **6**, 94
reactions, with methyl iodide, **6**, 94
Triselenoorthoformate esters
from, dichloromethyl methyl ether, **6**, 92
homologation of, **6**, 94
as starting materials, for triselenoortho esters, **6**, 94
Triselenoorthoformates
from
bromoform, **6**, 92
tetraselenoorthocarbonates, **6**, 93, 94
lithiation of, **6**, 317
Triselenoorthoformates, lithiated
reactions
with aldehydes, **6**, 94
with alkyl halides, **6**, 94
with electrophiles, **6**, 94
with epoxides, **6**, 94
Triselenotelluroorthocarbonates, synthesis of, **6**, 308
Triselenothioorthocarbonates, synthesis of, **6**, 308
Tris(ethynyl)methyl cations, synthesis of, **1**, 904
1-Trishomobarrelyl cations, synthesis of, **1**, 900
1-Trishomobullvalyl cations, synthesis of, **1**, 900
Trishydrogen fluoride complex, triethylamine-, reactions, with phosgene, **6**, 409
Trisilacyclohexane, 2,2,4,4,6,6-hexachloro-, synthesis of, **6**, 268
1,3,5-Trisilacyclohexane, lithiation of, **6**, 178
1,3,5-Trisilacyclohexane, 1,1,3,3,5,5-hexamethyl-
metallation of, **4**, 620
synthesis of, **4**, 610
1,3,5-Trisilacyclohexane, 2-lithio-1,1,3,3,5,5-hexamethyl-
reactions, with benzaldehyde, **4**, 623
synthesis of, **6**, 190
1,3,5-Trisilacyclohexane, 1-trimethylsilyl-, metallation of, **6**, 190
Trisilacyclohexanes
reactions, with Grignard reagents, **6**, 268
synthesis of, **6**, 268
1,3,5-Trisilacyclohexanes, synthesis of, **6**, 268
1,3,5-Trisilacyclohexanes, *C*,*C*-dihalo-, as starting materials, for phosphonium ylides, **6**, 703
1,3,5-Trisilacyclohexanes, transition metal substituted, synthesis of, **4**, 622
Trisilacyclopentanes, synthesis of, **4**, 621
2,4,6-Trisilaheptanes, synthesis of, **4**, 602
Trisilahexanes, synthesis of, **4**, 606
Trisilane, 2-chloroheptamethyl-, thermolysis of, **4**, 614
Trisilane, octamethyl-
reactions
with aryl isocyanide, **6**, 163

with isocyanide, **6**, 163
Trisilanes, synthesis of, **2**, 545
Trisilanes, 2-chloroheptamethyl-, use in the synthesis of carbodiimides, **5**, 1076
1,3,5-Trisilapentanes, synthesis of, **6**, 267
Trisilylmethyl groups, uses of, **6**, 172
Tris(ketenimine), synthesis of, **3**, 570
Tris(pentacarbonylchromium), 2,4,6-triisocyanotriazine-, stability of, **3**, 726
Trisperoxy triazines, synthesis of, **2**, 687
Tris(phosphines), synthesis of, **4**, 546
Trisphosphinomethanes, synthesis of, **5**, 910
Tris(phosphonates), synthesis of, **6**, 155
Trispiro[5.2.2.5.2.2]heneicosane, 8,10,19,20-tetraoxa-, synthesis of, **6**, 300
Tris(sulfones), synthesis of, **4**, 275
Tris(trimethylsilyl) hydride, for the reduction of alkyl halides, **1**, 5
Tris(trimethylsilylmethyl), reactions, with bromine, **4**, 597
Tris(trimethylsilyl)methyl radicals, synthesis of, **6**, 734
Trisulfanes
reactions, with secondary amines, **5**, 485
synthesis of, **5**, 485
Trisulfide, bis(trifluoromethyl)-, reactions, with Grignard reagents, **6**, 528
Trisulfides
dioxides of, synthesis of, **2**, 725
synthesis of, **2**, 233
tetroxides of, synthesis of, **2**, 725
Trisulfides, aryl, synthesis of, **2**, 723
Trisulfides, bisamine, nomenclature, **2**, 766
Trisulfides, diaryl, synthesis of, **2**, 723
Trisulfones, synthesis of, **6**, 91, 307
Trisulfur tetranitride, 1,7-bis(tosyl)-, reactions, with dimethyl sulfoxide, **2**, 162
Tritelluroformaldehyde, synthesis of, **3**, 389
1,2,4-Tritelluroles, synthesis of, **4**, 877
Tritelluroorthocarbonates, synthesis of, **6**, 308
Tritellurothioorthocarbonates, synthesis of, **6**, 308
2,4,6-Trithiaheptane, synthesis of, **4**, 258
Trithiane
reactions, with *t*-butyl peroxybenzoate, **6**, 99
as starting material, for hexachloro-1,1,3,3,5,5-hexaoxo-1,3,5-trithiane, **6**, 254
Trithiane, allylic, use in the synthesis of β,γ-unsaturated esters, **5**, 553
Trithiane, benzoyloxy-
from, trithiane, **6**, 99
reactions, with thiols, **6**, 82
1,3,5-Trithiane
chlorination of, **2**, 168; **4**, 85
reactions
with sodium hypochlorite, **6**, 254
with thionyl chloride, **4**, 65
as starting material, for di(chloromethyl) sulfide, **4**, 63
1,3,5-Trithiane, hexabromo-, synthesis of, **6**, 254
1,3,5-Trithiane, hexachloro-, reactions, with amines, **6**, 254
1,3,5-Trithiane, 2,2,4,4,6,6-hexamethyl-, as starting material, for thioketenes, **3**, 551
Trithianes
bromination of, **4**, 85
as byproducts, **2**, 130
chlorination of, **4**, 71
as starting materials
for α-chloroalkyl sulfinyl chlorides, **4**, 78
for dithioacetals, **4**, 250
use in the synthesis of disulfides, **2**, 228
Trithiapentalenes
from
acylmethylenedithioles, **3**, 376
5-alkyl-1,2-dithiolium salts, **3**, 376
dioxathiapentalenes, **3**, 377
1,2-dithioles, **3**, 376
1,2-dithiole-3-thiones, **3**, 376
enaminodithioles, **3**, 376
1,3,5-triketones, **3**, 378
Vilsmeier salts, **3**, 376
hydrolysis of, **3**, 255
1,6,6aλ^4-Trithiapentalenes, structure of, **3**, 354, 376
Trithioalkenes, synthesis of, **4**, 851
Trithiocarbamate salts, alkylation of, **6**, 556
Trithiocarbenium ions
use in the synthesis of lithiotrithioorthoformates, **6**, 317
use in the sysnthesis of trithioorthocarbonates, **6**, 309
Trithiocarbenium salts
reactions, with sodium borohydride, **6**, 317
as starting materials
for aminotrithioorthoformates, **6**, 314
for azidotrithioorthoformates, **6**, 316
for tetrathioorthocarbonates, **6**, 306
for trithioorthocarbonates, **6**, 309
for trithioortho esters, **6**, 85
Trithiocarbonate, reactions, with bromocyanoethyne, **5**, 487
Trithiocarbonate, bis(trifluoromethyl)
synthesis of, **6**, 235
trifluoromethylthiolation with, **6**, 233
Trithiocarbonate, diphenyl
reactions
with diaryl diazomethanes, **6**, 87
with 1-nitroso-1-phenylethene, **6**, 309
with phenyllithium, **6**, 87
Trithiocarbonate, ethyl, reactions, with nitroalkenes, **1**, 443
Trithiocarbonate, *S*-methyl, reactions, with benzyl anions, **5**, 550
Trithiocarbonate complexes, reactions, with iron pentacarbonyl, **6**, 333
Trithiocarbonate diesters, *S*-alkylation of, **6**, 729
Trithiocarbonate *S*,*S*-dioxides, reactions, with Grignard reagents, **4**, 862
Trithiocarbonates
alkylation of, **6**, 729
arylation of, **6**, 729
cleavage of, **2**, 123
from, azetidinones, **6**, 553
reactions
with α,ω-alkanedithiols, **6**, 305
with halogens, **6**, 273
with nitrosoalkenes, **6**, 309
with organolithiums, **6**, 317
with sulfur dichloride, **6**, 274
with thioketenes, **6**, 305
reduction of, **2**, 124
as starting materials
for acyclic covalent halogenides, **6**, 273
for alkanethiols, **2**, 123
for trithioortho esters, **6**, 88
for trithioorthoformates, **6**, 87
synthesis of, **2**, 123
Trithiocarbonates, acyclic, as starting materials, for 1,4,2-oxathiazoles, **6**, 309
Trithiocarbonates, cyclic
reactions, with 3,4,5,6-tetrachloro-*o*-benzoquinone, **6**, 312
as starting materials
for 1,4,2-oxathiazoles, **6**, 309
for tetrathioorthocarbonates, **6**, 305
for thioacyl bromides, **5**, 487
synthesis of, **6**, 555
Trithiocarbonates, dialkyl, synthesis of, **6**, 554
Trithiocarbonates, β-nitro, synthesis of, **1**, 443
Trithiocarbonate salts
alkylation of, **6**, 554
as starting materials, for trithiocarbonic acid diesters, **6**, 553
synthesis of, **6**, 552

Trithiocarbonic acid diesters
from
dithiocarbonic acid chloride, **6**, 553
thiophosgene, **6**, 552
trithiocarbonate salts, **6**, 553
synthesis of, **6**, 552
Trithiocarbonic acids, synthesis of, **6**, 552
Trithiocarbonic acid salts
from, barium trithiocarbonate, **6**, 552
synthesis of, **6**, 552
Trithiocarboxylic acid anhydrides
stability of, **5**, 583
synthesis of, **5**, 583
Trithiocarboxylic anhydrides, synthesis of, **5**, 583
Trithiocarboxylic anhydrides, aryl-, use in the synthesis of thionoamides, **5**, 583
Trithiocyclopropenium salts, reactions, with cyclopentadienyl anions, **4**, 852
Trithioformate esters
homologation of, **6**, 88
as starting materials, for trithioortho esters, **6**, 88
Trithioformates, sodio, synthesis of, **6**, 317
Trithioisocyanuric acid, reactions, with ethylene oxide, **6**, 335
1,2,4-Trithiolanes
synthesis of, **3**, 254
thermolysis of, **3**, 337
1,2,4-Trithioles, 3,5-bis(exomethylene), synthesis of, **4**, 852
Trithiomethanide ions, sulfenylation of, **6**, 303
Trithiomonoorthodithiooxalates, from, trithioorthoformate esters, **6**, 88
Trithiomonoorthooxalate esters, from, trithioorthoformate esters, **6**, 88
Trithiomonoorthooxalates, from, trithioorthoformate esters, **6**, 88
Trithione, reactions, with morpholine, **5**, 559
Trithiooorthocarbonates, synthesis of, **6**, 308
Trithioorthoacetate, triethyl, from, ethyl acetate, **6**, 84
Trithioorthoacetate, trimethyl, from, trimethyl trithioorthoformate, **6**, 88
Trithioorthoacetate, tris(3-methylphenyl), from, acetic anhydride, **6**, 84
Trithioorthoacetates, triaryl, deuteration of, **6**, 89
Trithioorthobenzoate, triethyl, from, benzaldehyde diethyl dithioacetal, **6**, 86
Trithioorthobenzoate esters, trialkyl, from, benzoate esters, **6**, 84
Trithioorthobenzoate esters, tris(trifluoromethyl), from, α,α,α-trihalotoluenes, **6**, 82
Trithioorthocarbonates
chemistry of, **6**, 309
from, trithiocarbenium salts, **6**, 309
Trithioorthocarbonates, *O*,*O*'-alkanediyl *S*,*S*'-dimethyl, as starting materials, for *O*,*S*-alkanediyl *S*,*S*-dimethyl trithioorthocarbonates, **6**, 309
Trithioorthocarbonates, *O*,*S*-alkanediyl *S*,*S*-dimethyl, from, *O*,*O*'-alkanediyl *S*,*S*'-dimethyl trithioorthocarbonates, **6**, 309
Trithioorthocarbonates, spirocyclic, synthesis of, **6**, 309
Trithioortho esters
disproportionation reactions, with thiols, **6**, 89
elimination of thiols from, **4**, 844
from
acyl chlorides, **6**, 84
α-acyloxy dithioacetals, **6**, 82
α-alkoxy dithioacetals, **6**, 82
anhydrides, **6**, 84
carboxylic acids, **6**, 84
carboxylic esters, **6**, 84
carboxylic ortho esters, **6**, 82
carboxylic thioesters, **6**, 84
α,α-dihalo ethers, **6**, 81
dithioacetals, **6**, 85
dithiocarbenium salts, **6**, 85
α-halo dithioacetals, **6**, 82
other trithioortho esters, **6**, 88
thioimidate ester salts, **6**, 83
1,1,1-trihaloalkanes, **6**, 81
trithiocarbenium salts, **6**, 85
trithioformate esters, **6**, 88
oxidized compounds, synthesis of, **6**, 90
as starting materials
for ketene dithioacetals, **4**, 844
for ortho esters, **6**, 80
synthesis of, **6**, 81
transesterification of, **6**, 88
Trithioortho esters, α-hydroxy, from, trithioorthoformate esters, **6**, 88
Trithioortho esters, β-hydroxy, from, trithioorthoformate esters, **6**, 88
Trithioortho esters, γ-keto, from, trithioorthoformate esters, **6**, 88
Trithioortho esters, mixed
from
thiocarboxylic esters, **6**, 84
N-*p*-toluenesulfonylsulfilimines, **6**, 87
Trithioortho esters, oxidized
containing sulfonate groups, synthesis of, **6**, 91
containing sulfone groups, synthesis of, **6**, 90
containing sulfoxide groups, synthesis of, **6**, 90
Trithioortho esters, trialkyl, from, alkanethiolates, **6**, 81
Trithioortho esters, triaryl
from
bromoform, **6**, 81
chlorodifluoromethane, **6**, 81
Trithioorthoformate, tri-*t*-butyl, anion of, resistance to trimethylsilylation, **6**, 317
Trithioorthoformate, triethyl
from, chloroform, **6**, 81
reactions, with dithiols, **6**, 89
as starting material, for triethyl trithioorthoheptanoate, **6**, 88
Trithioorthoformate, trimethyl
deprotonation of, for trimethyl trithioorthoacetate, **6**, 88
from, chlorobis(methanethio)methane, **6**, 82
Trithioorthoformate, triphenyl
from
chloroform, **6**, 81
diphenyl trithiocarbonate, **6**, 87
Trithioorthoformate, triphenyl (trimethylsilyl), synthesis of, **6**, 317
Trithioorthoformate, tris(trifluoromethyl), from, iodoform, **6**, 81
Trithioorthoformate esters
from
dichloromethyl alkyl esters, **6**, 82
formamide, **6**, 83
formate esters, **6**, 84
formic acid, **6**, 84
lithiation and homologation of, **6**, 88
Trithioorthoformate esters, triaryl, oxidation of, with peroxy acids, **6**, 90
Trithioorthoformates
anions of, silylation of, **6**, 317
from
sulfenyl chlorides, **6**, 273
tetrathioorthocarbonates, **6**, 87
N-*p*-toluenesulfonylsulfilimines, **6**, 87
trithiocarbonates, **6**, 87
as fungicidal agents, **6**, 273
lithium salts of, as intermediates, **6**, 317
metallation of, **6**, 317
oxidation of
with hydrogen peroxide, **6**, 91
with permanganate, **6**, 91
with peroxy acids, **6**, 91

as starting materials
for dithioacetals, **4**, 250
for sodiotrithioformates, **6**, 317
Trithioorthoformates, amino-
from
betaines, **6**, 314
trithiocarbenium salts, **6**, 314
Trithioorthoformates, azido-
reactions, with triphenylphosphine, **6**, 316
synthesis of, **6**, 316
Trithioorthoformates, bicyclic amino-, synthesis of, **6**, 314
Trithioorthoformates, [(dichloromethylene)amino]-, synthesis of, **6**, 316
Trithioorthoformates, heterocyclic amino-, synthesis of, **6**, 314
Trithioorthoformates, lithiated
acylation of, **6**, 695
reactions
with aldehydes, **6**, 88
with alkyl halides, **6**, 88
with carbon dioxide, **6**, 88
with carbon disulfide, **6**, 88
with chloroformate esters, **6**, 88
with epoxides, **6**, 88
with iodine, **6**, 88
with ketones, **6**, 88
with α,β-unsaturated ketones, **6**, 88
synthesis of, **6**, 317
Trithioorthoformates, lithio-
silylation of, **6**, 317
synthesis of, **6**, 317
Trithioorthoformates, nitro-, synthesis of, **6**, 316
Trithioorthoformates, spirocyclic amino-, synthesis of, **6**, 314
Trithioorthoformates, trialkyl
from, dichloromethyl alkyl esters, **6**, 82
synthesis of, **6**, 84
Trithioorthoformates, triaryl
from, dichloromethyl alkyl esters, **6**, 82
synthesis of, **6**, 84
Trithioorthoformates, (trimethylsilyl), synthesis of, **6**, 317
Trithioortho heptanoate, triethyl, from, triethyl trithioorthoformate, **6**, 88
Trithiophosphonic acid, diesters, cyclic, synthesis of, **2**, 855
Trithiophosphonic acid bis(trimethylsilyl) ester, phenyl-, synthesis of, **2**, 855
Trithiopyrocarbonate, *O,O'*-diethyl, as starting material, for enethiolates, **4**, 834
Tritins, synthesis of, **4**, 700
Tritium decay method, for the detection of alkynyl cations, **1**, 1142
Triton B
use in the synthesis of dialkyl sulfides, **2**, 138
use in the synthesis of propargylic amines, **2**, 301
Trityl cations
as catalysts, in the displacement of allylic alcohols, **1**, 189
reactions
with dialkylmercury compounds, **1**, 668
with methyl complexes, **2**, 580
with tetraalkylleads, **1**, 663
with tetraalkyltins, **1**, 663
Trityl chloride
reactions
with organomercury compounds, **1**, 128
with potassium selenocyanates, **5**, 1053
with silver nitrocyanamide, **5**, 963
with silver phenylcyanomethylene nitronate, **5**, 963
with thiols, **2**, 133
synthesis of, **2**, 168
Trityl ions, reactions, with alkyliron complexes, **1**, 666
Tritylium salts
as oxidants, for the synthesis of enones, **1**, 556
use in the synthesis of enones, **1**, 556
Triynes, synthesis of, **1**, 1069, 1079
Triynes, silyl-substituted, from, 2,4-hexadiyne-1,6-diol, **2**, 1091
1,3,5-Triynes, from, alkenyl thioethers, **1**, 1072
Tronov–Novikov reaction, for the synthesis of aryl iodides, **2**, 631
Tropane, synthesis of, **1**, 812
Tropenylium cations, chemistry of, **1**, 906
Tropinol, demethylation of, **2**, 303
Tropone, reactions, with trimethylenemethane, **1**, 288
Tropones, azido-, as starting materials, for cyanoketenes, **3**, 539
Tropothione, synthesis of, **3**, 352
Tropylium cations, heptaphenyl-, synthesis of, **1**, 907
Tropylium cations, 1,2:3,4:5,6-tris(bicyclo[2.2.2]octeno)-, synthesis of, **1**, 907
Tropylium ions, as starting materials, for silyl azulenes, **2**, 944
Tropylium salts, synthesis of, **2**, 1008
Truce–Smiles rearrangement, mechanism, **1**, 379
Tryptamines, *N*-acetoxy-, synthesis of, **5**, 214
Tryptamines, 2-vinyl-, synthesis of, via Hofmann elimination, **1**, 628
Trypticide anions, stabilization of, **1**, 853
Trypticine, acidity of, **1**, 853
TSTU, as coupling reagent, for the synthesis of *N*-alkylamides, **5**, 280
Tsuji–Trost reaction, for the synthesis of γ,δ-unsaturated esters, **5**, 158
Tuberiferine, synthesis of, **1**, 607
Tumor inhibitors
calichaemycin, **2**, 233
esperamycin, **2**, 233
Tungstacyclopentadienes, synthesis of, **4**, 675
Tungsten
as catalyst, for the substitution of sulfur functions, **1**, 238
doubly-bridged complexes of, synthesis of, **4**, 675
reactions, with allyl cations, **1**, 191
use in allylation reactions, **1**, 196
Tungsten, hexamethyl-
reactions, with methyllithium, **2**, 572
synthesis of, **2**, 572
Tungsten, pentacarbonylbenzylidene-, reactions, with triphenylketenimine, **4**, 1018
Tungsten, pentacarbonyl (diphenylvinylidene)-, synthesis of, **4**, 1018
Tungsten, tetrabenzyl-, synthesis of, **2**, 572
Tungsten(0), ([bromo(phenyl)methyleneamino]-dimethylaminocarbene)pentacarbonyl-, synthesis of, **5**, 660
Tungsten(II) complexes, reactions, with cyclopentanone, **3**, 517
Tungsten(III) compounds, alkyl-, synthesis of, **2**, 572
Tungsten(IV) compounds, tetraalkyl-, synthesis of, **2**, 572
Tungsten alkoxide complexes, synthesis of, **3**, 517
Tungsten carbene complexes
from
aryl ylides, **3**, 517
carbyne complexes, **3**, 516
tungsten carbene complexes, **3**, 518
tungsten carbyne complexes, **3**, 516
reactions
with alkyne thiolates, **5**, 651
with phosphines, **4**, 580
as starting materials
for organotungstens, **6**, 354
for tungsten carbene complexes, **3**, 518
synthesis of, **3**, 514
via metathesis reactions, **3**, 515
Tungsten carbene complexes, phenylimino-, synthesis of, **3**, 514, 515
Tungsten carbene complexes, vinyl-, synthesis of, **3**, 517

Tungsten carbenes
as starting materials, for diphenylcarbene adducts, **3**, 517
use in the synthesis of phenols, **2**, 670
Tungsten carbonyl complexes, synthesis of, **6**, 522
Tungsten carbyne complexes
reactions, with azidoformates, **5**, 955
as starting materials, for tungsten carbene complexes, **3**, 516
Tungsten carbyne complexes, cyclopentadienyl-, use in the synthesis of 3-carbon-bridged ketone complexes, **4**, 676
Tungsten chlorides, reactions, with 1,1-dicyanobis(trifluoromethyl)ethene, **4**, 800
Tungsten complexes
from, η2-ketenyl complexes, **2**, 878
photolytic rearrangement of, **4**, 639
reactions
with alkynes, **4**, 1028
with aluminum halides, **3**, 515
with diphenylcyclopropene, **3**, 517
with ketones, **3**, 517
with trimethylphosphine, **3**, 514
rearrangement of, **4**, 934
Tungsten complexes, alkyne-bridged, from, carbyne complexes, **4**, 675
Tungsten complexes, cyclopentadienyl-
reactions
with alkyllithiums, **2**, 573
with hydrogen isocyanate, **6**, 517
with lithium triethylborohydride, **5**, 446
synthesis of, **6**, 517
Tungsten complexes, (halomethyl)-, synthesis of, **4**, 147, 148
Tungsten diazo complexes, α-chloroalkenyl-, x-ray crystal structure of, **4**, 799
Tungsten dichloride, dipyridinedi-*t*-butylimido-, reactions, with methylene(triphenyl)phosphorane, **5**, 918
Tungsten functions, doubly-bonded, synthesis of, **3**, 514
Tungsten halides, use in the synthesis of carbon-bridged ditungsten complexes, **4**, 672
Tungsten hexacarbonyl
reactions
with alkyllithiums, **5**, 446
with dithioketals, **3**, 365
with phenyllithium, **5**, 933
synthesis of, **6**, 522
Tungsten hexachloride, reactions, with trimethylaluminum, **2**, 572
Tungsten hydride complexes, reactions, with tricarbonyl(trisammonia)chromium, **4**, 715
α-Tungstenio enamines, from, ynamines, **4**, 1019
Tungsten oxocarbene complexes, from, tantalum carbene complexes, **3**, 517
Tungsten oxytetrachloride, as starting material, for tungsten oxytrichloride complexes, **2**, 572
Tungsten oxytrichloride complexes, from, tungsten oxytetrachloride, **2**, 572
Tungsten–phosphinoalkene complexes, synthesis of, **4**, 1023, 1028
Tungstens
for the 1,2-dihydroxylation of alkenes, **2**, 51
reactions, with phosphorus ylides, **4**, 580
Tungstens, acyl-, synthesis of, **5**, 446
Tungstens, alkyl-, synthesis of, **2**, 570
Tungstens, imidoyl-
from
isonitriles, **5**, 798
metalloalkoxycarbenes, **5**, 798
synthesis of, **5**, 798
via displacement of imidoyl halides, **5**, 798
Tungsten salts, as catalysts, for the oxidation of alcohols, **3**, 117
Tungsten tetrachlorides
reactions, with dibenzylmagnesium, **2**, 572
use in the synthesis of tris(cyclooctatetraene)ditungsten complexes, **4**, 673
Tungsten–tungsten-bonded complexes, as starting materials, for alkyne-bridged complexes, **4**, 675
Tungsten ylides, synthesis of, **6**, 330
Tungstonate, cyclopentadienyltricarbonyl, reactions, with benzimidoyl chlorides, **5**, 798
Tungstophosphate, trioctylmethylammonium, as catalyst, for the oxidation of alcohols, **3**, 117
12-Tungstophosphate, *tris*-(cetylpyridinium)-, as catalyst, for the oxidation of alcohols, **3**, 117
4-Twistanone
exchange reaction of, **1**, 879
reactions, with sodium methoxide, **1**, 879
Twofold extrusion reaction, for the synthesis of alkenes, **3**, 392
Two-step reaction, mechanism, **1**, 350
(-)-Tylophorine, synthesis of, **2**, 328
L-Tyrosinamide, *N*-acetyl-, reactions, with Woodward's reagent K, **4**, 892
Tyrosine, reactions, with iodine monochloride, **2**, 608

U

Udenfriend's reagent, for the oxidation of arenes, **2**, 650
Ugi reaction, for the synthesis of isocyanides, **3**, 694
Ullmann reaction
for the substitution of aryl halides, **1**, 472
for the synthesis of biaryls, **1**, 472
for the synthesis of diaryl ethers, **2**, 684
Ultimate carcinogens, synthesis of, **5**, 212
Ultrasonication
of alkyl halides, **3**, 14
use in the synthesis of alkenyllithium cuprates, **1**, 1093
Ultrasonic irradiation, in the synthesis of benzaldehyde, **3**, 90
Ultrasound
for the acceleration of reactions, **1**, 473
for the conjugate reduction of enones, **3**, 127
for the synthesis of alkylzincs, **2**, 591
use in Diels–Alder reactions, **1**, 373
use in the elimination of hydrogen halide from α-haloacetals, **4**, 826
use in oxidation, **3**, 4
use in the synthesis of alkenes, **1**, 676
use in the synthesis of alkylboranes, **2**, 520
use in the synthesis of alkyllithiums, **2**, 551
use in the synthesis of amides, **5**, 272
use in the synthesis of benzaldehydes, **3**, 91
use in the synthesis of isoselenocyanates, **5**, 1052
use in the synthesis of radicals, **1**, 935
use in the synthesis of secondary phosphines, **2**, 432
use in the synthesis of thionoamides, **5**, 568
Ultraviolet light
use in the synthesis of alkyltins, **2**, 601
use in the synthesis of dialkyl sulfides, **2**, 137
Ultraviolet spectroscopy, for the study of enols, **2**, 637
Umpolung
application to the synthesis of dithioacetals, **4**, 254
application to the synthesis of ketones, **3**, 135
concept of, **4**, 244
for the synthesis of alkyl iodides, **2**, 34
for the synthesis of α,β-unsaturated ketones, **3**, 213
use in the synthesis of aldehydes, **3**, 19
Umpolung reagents, diselenoacetals as, **4**, 287
1,4(*E*)-Undecadiene, synthesis of, **1**, 128
(*E*,*E*)-2,4-Undecadiene, synthesis of, **1**, 612
Undecane, 1-bromo-, reactions, with allylmanganese chloride, **1**, 125
[5.5]Undecane, 1,5,7,11-tetraoxaspiro-, synthesis of, **6**, 300
1,11-Undecanoate, dimethyl, use in the synthesis of alkanoic acids, **5**, 32
Undecanoic acid, 4,8,12-trimethyl-6-keto-, synthesis of, **5**, 76
(*Z*,*Z*)-1,3,5-Undecatrienes, thermal 1.7-hydrogen shift of, **1**, 403
trans-Undec-3-en-5,7-diyn-2-one, as starting material, for *trans*-dec-2-en-4,6-diynoic acid, **5**, 102
5-Undecene, 5-trimethylsilyl-, synthesis of, **2**, 931
α,β-Unsaturated acids, β-alkyl-, synthesis of, **5**, 98
α,β-Unsaturated acids, (*Z*)-β-bromo-, thermal decarboxylation of, **1**, 1119
α,β-Unsaturated aldehydes, as starting materials, for *o*-butadienyl carbonates, **6**, 461
Unsaturated compounds, use in the synthesis of dialkyl sulfoxides, **2**, 153
α,β-Unsaturated compounds, conjugated reduction of, **1**, 130
Unsaturation, homoallylic, in radical generation, **1**, 938
Uracil, cleavage of, **4**, 103
Uracil, 6-amino-1,3-dimethyl-, reactions, with dimethyl ethynedicarboxylate, **4**, 893
Uracil, 6-methyl-, from, ethyl β-uraminocrotonate, **5**, 93
Uracil-5-carboazides, reactions, with xylene, **5**, 990
Uracils, synthesis of, **5**, 990
Uranium, chlorotricyclopentadienyl-, as starting material, for ferrocenyluraniums, **4**, 726
Uranium(V) complexes, synthesis of, **2**, 602
Uranium(VI) fluoride, use in the synthesis of aldehydes, via ether oxidation, **3**, 8
Uranium chloride, tricyclopentadienyl-
reactions, with lithium methylphenylphosphonium dimethanide, **5**, 920
as starting material, for α-uranium-functionalized ylides, **5**, 920
Uranium complexes, synthesis of, **5**, 782
Uranium complexes, cyclopentadienyl-, synthesis of, **5**, 469
Uranium hexafluoride
reactions
with alcohols, **5**, 6
with aldehydes, **5**, 6
Uranium-keteniminato complexes, synthesis of, **3**, 570
Uraniums, reactions, with phosphorus ylides, **4**, 589
Uraniums, acyl-, synthesis of, **5**, 469
Uraniums, alkyl-, synthesis of, **2**, 602
Uraniums, ferrocenyl-, from, chlorotricyclopentadienyluranium, **4**, 726
Uraniums, tetraalkyl-, synthesis of, **2**, 602
Uranium triscyclopentadienyl, reactions, with acetyl chloride, **5**, 469
Urazoles, dehydrogenation of, **5**, 366
Urea
as explosion inhibitor, in the synthesis of nitrates, **2**, 103
reactions
with acryloyl chloride, **6**, 506
with alcohols, **6**, 484
with alkylamines, **6**, 502
with amines, **6**, 502
with benzyl alcohol, **6**, 484
with carbonyls, **6**, 508
with carboxylic acids, **5**, 262
with hexaalkyldisilazanes, **5**, 1006
with imino ester hydrochloride salts, **4**, 982
as starting material, for urethanes, **6**, 484
as starting materials, for *N*-unsubstituted iminocarbonyls, **6**, 641
synthesis of, **6**, 500
thioalkylation of, **4**, 325
Urea, acetyl-, reactions, with carboxylic acids, **5**, 262
Urea, *N*-acetyl-*N'*-methyl-, synthesis of, **6**, 506
Urea, *N*-acryloyl-, synthesis of, **6**, 506
Urea, *N*-benzoyl-*N'*-phenyl-, synthesis of, **5**, 998
Urea, *N*,*N'*-bis(ethoxycarbonylmethyl)-, synthesis of, **6**, 505
Urea, bis(pentafluorosulfanyl)-, as starting material, for bis(pentafluorosulfanyl)carbodiimide, **5**, 1085
Urea, bis(*t*-butyl)-, reactions, with *t*-butyl hypochlorite, **6**, 509
Urea, *N*,*N'*-bis(trifluoromethyl)-, synthesis of, **6**, 239
Urea, *N*-butyl-*N'*-(2,6-dimethylphenyl)-, synthesis of, **6**, 505
Urea, *N*-(1-chloro-1-alkenyl)-, hydrolysis of, **6**, 506
Urea, 1-(2-chloroethyl)-1,3-diphenyl-, reactions, with sodium hydroxide, **6**, 508
Urea, *N*-cyclohexyl-, synthesis of, **6**, 502
Urea, *N'*-cyclohexyl-*N*-[(2-methoxycarbonyl)phenyl]-, synthesis of, **6**, 504
Urea, *N*,*N*-dibutyl-*N'*-phenyl-, synthesis of, **6**, 501

Urea, dicyclohexyl-, removal of, **5**, 124
Urea, *N*,*N*'-didodecyl-, synthesis of, **6**, 504
Urea, *N*,*N*-diethyl-, as starting material, for diethyl cyanamide, **5**, 1122
Urea, *N*,*N*-diethyl-*N*'-hydroxy-, synthesis of, **6**, 505
Urea, *N*,*N*'-dimethoxy-
  alkylation of, **5**, 331
  reactions, with phenyllithium, **5**, 331
Urea, *N*,*N*'-dimethoxy-*N*,*N*'-dimethyl-, reactions, with phenylethynyllithium, **1**, 1047
Urea, 1,3-dimethyl-, reactions, with formaldehyde, **4**, 301
Urea, *N*,*N*-dimethyl-, as catalyst, **5**, 818
Urea, *N*,*N*-dimethyl-*N*'-phenyl-, synthesis of, **6**, 503
Urea, *N*,*N*'-dimethylpropylene- (DMPU), use in the synthesis of alkynes, **1**, 1000
Urea, 1,3-diphenyl-, phosgenation of, **5**, 980
Urea, *N*,*N*'-diphenyl-*N*-(1-phenyl-1-propenyl)-, synthesis of, **6**, 504
Urea, guanyl-, fluorination of, **6**, 616
Urea, *N*-heptyl-, synthesis of, **6**, 504
Urea, *N*-methyl-*N*-nitroso-, as starting material, for diazomethane, **3**, 466
Urea, *N*-methyl-*N*-nitroso-*N*'-(2-phenylethyl)-, synthesis of, **6**, 506
Urea, *N*-methyl-*N*'-phenoxy-, synthesis of, **6**, 506
Urea, *N*-methyl-*N*'-(2-phenylethyl)-, reactions, with sodium nitrite, **6**, 506
Urea, *N*-phenyl-, reactions, with phenyldichlorophosphine, **4**, 488
Urea, symmetrical, as by-product, **6**, 443
Urea, tetramethyl-
  chlorination of, **6**, 728
  fluorination of, **6**, 259
  reactions, with hexacarbonylmolybdenum, **2**, 125
Urea, *N*,*N*,*N*',*N*'-tetramethyl-, synthesis of, **6**, 502
Urea, *N*-(trimethylsilyl)methyl-*N*-nitroso-, reactions, with potassium hydroxide, **3**, 483
Urea acetals, synthesis of, **6**, 313, 320
Urea dichlorides, *N*,*N*'-diacyl-, alcoholysis of, **6**, 321
Urea–hydrogen peroxide (UHP)
  for the oxidation of arenes, **2**, 654
  use in the synthesis of amine *N*-oxides, **2**, 348
Urea nitrate, reactions, with sulfuric acid, **2**, 791
Urea nitrates, as starting materials, for urethanes, **6**, 484
Ureas
  alkylation of, **6**, 731
  O-alkylation of, **6**, 633
  bis hydroxymethylation of, **4**, 301
  chlorination of, **6**, 620, 728
  dehydration of, **5**, 1063, 1121
  from
    isocyanides, **6**, 505
    isoureas, **6**, 503
    orthocarbonic acids, **6**, 503
    thioureas, **6**, 502
  lithiation of, **4**, 531
  reactions
    with allenyllithium, **1**, 971
    with chlorosilanes, **5**, 983
    with diaryl carbonates, **5**, 981
    with hypochlorite, **2**, 380
    with Lewis acids, **5**, 979
    with lithium aluminum hydride, **4**, 414
    with organoantimony chlorides, **5**, 1017
    with organotin chlorides, **5**, 1017
    with organotin oxides, **5**, 1017
    with phosgene, **5**, 980; **6**, 444
    with phosphorus pentachloride, **5**, 982
    with thionyl chloride, **5**, 981
    with triphenylphosphorus–bromine complexes, **5**, 1063
  reduction of, **5**, 745
    for the synthesis of aminals, **4**, 414
  as starting materials
    for *N*-alkylguanidines, **6**, 642
    for *N*-alkyliminocarbonyls, **6**, 642
    for α-aminoalkylphosphorus compounds, **4**, 466
    for *N*-aryliminocarbonyls, **6**, 644
    for carbamimidic chlorides, **6**, 620
    for carbodiimides, **5**, 1062
    for dialkyl carbonates, **6**, 463
    for *N*,*N*-difluoroamines, **2**, 335
    for hydrazines, **2**, 374, 376, 378
    for imidoyl halides, **5**, 654
    for isocyanates, **5**, 979
    for isoureas, **6**, 633
    for thioureas, **6**, 578
  synthesis of, **5**, 584, 736
  thermolysis of, **5**, 982
  use in the synthesis of α-aminoalkylphosphines, **4**, 456
Ureas, acyclic, synthesis of, **6**, 500
Ureas, acyl-, synthesis of, **6**, 506
Ureas, *O*-acyl-, as intermediates, **5**, 266
Ureas, *N*-acyl-*N*,*N*'-diaryl-, decomposition of, **5**, 982
Ureas, acylthioacyl-, synthesis of, **5**, 631
Ureas, alkoxy-, synthesis of, **6**, 505
Ureas, *N*-alkyl-*N*-cyano-, synthesis of, **5**, 1124
Ureas, 1,1'-alkylenebis-(3-arylsulfonyl-, reactions, with phosgene, **5**, 981
Ureas, *N*-alkyl-*N*-nitroso-, cleavage of, **3**, 466
Ureas, 2-amidino-, synthesis of, **6**, 647
Ureas, arylbis(sulfonylalkyl)-, phosgenation of, **5**, 981
Ureas, 1-aryl-3,3-dialkyl-, reduction of, **5**, 746
Ureas, 1-arylsulfonyl-3-alkyl-
  reactions
    with phosgene, **5**, 980
    with phosphorus pentachloride, **5**, 982
Ureas, *N*,*N*'-bis(alkylidene)-, reactions, with phosphorus pentachloride, **5**, 1063
Ureas, bis-*N*-(hydroxyalkyl)-, reactions, with amines, **4**, 416
Ureas, *N*-chlorocarbonyl-, synthesis of, **6**, 448
Ureas, *N*-cyano-, synthesis of, **5**, 1120
Ureas, *N*-cyano-*O*-methylpseudo-, synthesis of, **2**, 322
Ureas, cyclic, synthesis of, **6**, 506
Ureas, 1,3-dialkyl-
  reactions, with phosgene, **5**, 980
  reduction of, **5**, 745
Ureas, *sym*-dialkyl-, synthesis of, **5**, 978
Ureas, *N*,*N*-dialkyl-*N*'-hydroxy-, synthesis of, **6**, 505
Ureas, β,β'-dichloroalkyl-, synthesis of, **5**, 1010
Ureas, *N*,*N*'-dihydroxy-, synthesis of, **6**, 505
Ureas, *N*,*N*'-dimethoxy-, synthesis of, **6**, 505
Ureas, 1,3-disubstituted
  reactions
    with phosphorus pentachloride, **5**, 982
    with thionyl chloride, **5**, 981
Ureas, *N*,*N*'-disubstituted
  dehydration of, **5**, 1077
  as starting materials, for carbodiimides, **5**, 1079
Ureas, *N*,*N*'-disubstituted-*N*-(2-alkynoyl)-, synthesis of, **1**, 1043
Ureas, α-germyl-, synthesis of, **4**, 521
Ureas, guanyl-, fluorination of, **5**, 815
Ureas, *N*-halo-, cyclization of, **6**, 509
Ureas, *N*-(ω-haloalkyl)-, cyclization of, **6**, 508
Ureas, halocarbonyl-, synthesis of, **6**, 437
Ureas, hydroxy-, synthesis of, **6**, 505
Ureas, *N*-hydroxy-
  dialkylation of, **2**, 351
  synthesis of, **6**, 505
Ureas, hydroxyethyl-, synthesis of, **6**, 485
Ureas, *N*-methyl-*N*-nitroso-, as starting materials, for diazopropyne, **3**, 490
Ureas, monosubstituted, reactions, with boron trifluoride, **5**, 979
Ureas, nitroso-, synthesis of, **6**, 506
Ureas, *N*-nitroso-

as starting materials
for diazoalkanes, **3**, 466
for diazo compounds, **5**, 873
for α,β-unsaturated diazoalkanes, **3**, 489
Ureas, phospho-, cyclization of, **6**, 519
Ureas, α-phosphonyl-α¹-chloroalkyl-, as starting materials, for formimidoyl phosphonate, **5**, 787
Ureas, *N*-(2-propionyl)-, cyclization of, **6**, 509
Ureas, silyl-, thermal treatment of, **5**, 1016
Ureas, substituted, reactions, with sodium nitrite, **5**, 980
Ureas, sulfanilyl-, phosgenation of, **5**, 981
Ureas, sulfonyl-
acylation of, **5**, 344
as starting materials, for sulfonylamines, **2**, 401
Ureas, sulfonylthioacyl-, synthesis of, **5**, 631
Ureas, symmetrical, synthesis of, **6**, 416
Ureas, tetraalkyl-, alkylation of, **6**, 731
Ureas, *N*-tetrachloroethyl-
as starting materials
for *gem*-phosphonio ureidoalkenes, **4**, 999
for *gem*-phosphoryl ureidoalkenes, **4**, 1007
Ureas, tetrasubstituted, synthesis of, **6**, 502
Ureas, *N*,*N*,*N'*,*N'*-tetrasubstituted
reactions, with phosgene, **6**, 313
synthesis of, **6**, 500
Ureas, thioacyl-, synthesis of, **5**, 631
Ureas, 1,1,3-trialkyl-, reduction of, **5**, 746
Ureas, trisubstituted, cleavage of, **5**, 980
Ureas, unsymmetrical, synthesis of, **6**, 416, 501
Urea sulfate, guanyl-, fluorination of, **5**, 811
α-Ureidoalkylation, mechanism, **4**, 301
Urethane, *N*-allyl-*N*-nitroso-, cleavage of, **3**, 489
Urethane, *N*-benzoyl-
as starting material
for *N*-benzoylethoxyformimidoyl chloride, **5**, 656
for *N*-ethoxycarbonyl benzimidoyl chloride, **5**, 656
Urethane, *O*-ethyl *N*-2-phenylethyl-, synthesis of, **6**, 484
Urethanes
alkylation of, **2**, 351; **6**, 730
cleavage of, **5**, 977
from
chloroformates, **6**, 483
cyclic carbonates, **6**, 484
dialkyl carbonates, **6**, 484
isocyanates, **6**, 484
isocyanides, **6**, 485
urea, **6**, 484
urea nitrates, **6**, 484
as protective groups, **6**, 486
reactions
with hydroxylamine, **6**, 505
with tosyl chloride, **5**, 1014
as starting materials
for diazoalkanes, **3**, 465
for *N*,*N*-difluoroamines, **2**, 335
for sulfamic acids, **2**, 366
synthesis of, **6**, 483
uses of, **6**, 483
use in the synthesis of peptides, **5**, 259
Urethanes, *N*-chlorosilyl-, stability of, **5**, 978
Urethanes, *N*,*N*-dichloro-, reactions, with iminium salts, **6**, 619
Urethanes, *N*-hydroxy-
dialkylation of, **2**, 351
as starting materials, for *O*-alkylhydroxylamines, **2**, 102
Urethanes, α-methoxy-, reactions, with trialkoxyphosphines, **2**, 457
Urethanes, *N*-nitroso-, as starting materials, for α,β-unsaturated diazoalkanes, **3**, 489
Urethanes, silyl-, thermolytic cleavage of, **5**, 1015
Urethanes, tetrachloroethyl, as starting materials, for *gem*-alkoxycarbonylamido phosphonioalkenes, **4**, 999
Uretidinones, aryl-, decomposition of, **5**, 1011
Uridine nuleosides, synthesis of, **4**, 275
Uronium coupling reagents, for the synthesis of *N*-alkylamides, **5**, 279
Uronium salts
as coupling reagents
for the synthesis of *N*-alkylamides, **5**, 279
for the synthesis of amides, **5**, 267
reactions, with diethyl phosphite, **6**, 156
Uronium salts, alkylidene-, synthesis of, **6**, 643

V

Vacuum gas–solid reactions (VGSR)
for the synthesis of phosphaalkynes, **5**, 1152
for the synthesis of *N*-trimethylsilyl aldimines, **3**, 457
Valeraldehyde, δ-hydroxy-, as starting material, for glutaric acid, **5**, 46
(±)-Valeranone, synthesis of, **3**, 325
Valeric acid
as starting material
for D-(-)-β-hydroxyvaleric acids, **5**, 118
for L-(+)-β-hydroxyvaleric acids, **5**, 118
Valeric acid, 3-benzyl-3-methyl-, from, 3-benzyl-3-methylpentanenitrile, **5**, 34
Valeric acid, 5,5,5-trichloro-, synthesis of, **5**, 59
n-Valeric acid, δ-acetyl, from, 2-methylhexanol, **5**, 73
Valeric acids, D-(-)-β-hydroxy-, from, valeric acid, **5**, 118
Valeric acids, L-(+)-β-hydroxy-, from, valeric acid, **5**, 118
Valienamine, synthesis of, via selenoxide pyrolysis, **1**, 782
L-Valine, *N*-acetyl-, as chiral auxiliary, **4**, 720
(S)-Valine, use in the synthesis of cyanohydrins, **3**, 629
Valinol, *O*-*t*-butyl, as starting material, for formamidines, **1**, 149
Vanadium, bis(η-5-cyclopentadienyl)-, reactions, with silver succinimide, **5**, 388
Vanadium, (η^5-cyclopentadienyl)(η^6-naphthalene)-, use in the synthesis of carbon-bridged divanadium complexes, **4**, 671
Vanadium, di(*t*-butoxy)dichloro-, use in the synthesis of dialkoxydialkylvanadiums, **2**, 569
Vanadium, norbornyl-, synthesis of, **2**, 569
Vanadium, tetraalkyl-, reactions, with *t*-butyl alcohol, **2**, 569
Vanadium, tetrabenzyl-, synthesis of, **2**, 569
Vanadium, tetra-(*t*-butoxy)-, for the synthesis of dialkoxydialkylvanadiums, **2**, 569
Vanadium, tetracarbonyl-, reactions, with iron hydride complexes, **4**, 717
Vanadium, tetramethyl-, synthesis of, **2**, 569
Vanadium(II), for the reduction of quinones, **2**, 657
Vanadium(II) chloride, use in the synthesis of cyclooctatetraene-bridged complexes, **4**, 672
Vanadium(II) compounds, alkyl-, synthesis of, **2**, 569
Vanadium(III) chloride, reactions, with Grignard reagents, **3**, 127
Vanadium(V) complexes, use in the synthesis of alkenes, **1**, 683
Vanadium(V) complexes, alkylimmino-, synthesis of, **2**, 570
Vanadium(V) complexes, alkyloxo-, synthesis of, **2**, 570
Vanadium(V) compounds, alkyl-, synthesis of, **2**, 570
Vanadium(V) oxide, for the oxidation of dithioacetals, **4**, 259
Vanadium(acac)$_2$ compounds, alkyl-, synthesis of, **2**, 569
Vanadium binuclear compounds, synthesis of, **4**, 673
Vanadium carbene complexes, synthesis of, **3**, 510
Vanadium carbonyl complexes, synthesis of, **6**, 522
Vanadium chloride, use in the synthesis of arylamines, **2**, 753
Vanadium complexes
reactions, with Grignard reagents, **3**, 510
use in the synthesis of carbodiimides, **5**, 1079
Vanadium complexes, cyclopentadienyl-
reactions, with carbon monoxide, **5**, 444
synthesis of, **2**, 569
Vanadium complexes, dialkylcyclopentadienyl-, use in the synthesis of carbon-bridged divanadium complexes, **4**, 671
Vanadium dichloride, as catalyst, for haloalkane addition, **6**, 7
Vanadium functions, doubly-bonded, synthesis of, **3**, 510
Vanadium hexacarbonyl, synthesis of, **6**, 522
Vanadium imides, synthesis of, **5**, 388
Vanadium oxytrichloride
reactions
with allylmagnesium bromide, **2**, 570
with bulky amines, **2**, 570
Vanadium pentoxide, use in the synthesis of α-chloro sulfoxides, **4**, 74
Vanadiums
reactions
with cyclobutanone, **1**, 931
with dimethylzinc, **2**, 570
with phosphorus ylides, **4**, 579
Vanadiums, acyl-, synthesis of, **5**, 444
Vanadiums, alkyl-, synthesis of, **2**, 569
Vanadiums, alkyloxo-, synthesis of, **2**, 570
Vanadiums, dialkoxydialkyl-, synthesis of, **2**, 569
Vanadiums, imidoyl-, synthesis of, **5**, 797
Vanadiums, monoalkyl-, synthesis of, **2**, 569
Vanadiums, triallyl-, synthesis of, **2**, 569
Vanadium tetrachloride, reactions, with dibenzylmagnesium, **2**, 570
Vanadium trichloride, reduction of, **6**, 522
Van Braun dechloroalkylation, of phosgeniminium salts, **3**, 409
Vanilin, iodo-, as starting material, for phenols, **2**, 647
Vanillic acid, synthesis of, **5**, 114
Vanillin
oxidation of, **5**, 114
reactions, with silver oxide, **5**, 114
Vaporization
of metals
methods, **2**, 961
problems, **2**, 961
for the physical activation of metals, **2**, 958
Varrentrapp reaction, for the isomerization of alkenes, **1**, 882
Vaska's catalyst
reactions of, **6**, 718
reactivity of, **1**, 73
synthesis of, **1**, 73
Vaska complex, reactions, with alkyl halides, **2**, 583
Veratric acid, homo-, synthesis of, **5**, 51
Vernolepin, synthesis of, **1**, 604
V. harveyi oxidoreductase, reactions, with potassium *O*-ethyl dithiocarbonate, **6**, 549
Viehe's salt
reactions
with butanethiol, **6**, 589, 593
with 2-hydroxybenzenethiol, **6**, 593
with methyl 2-*O*-methyl-β-L-arabinopyranoside, **6**, 544
with phenol, **6**, 593
use in the synthesis of alkyl chlorides, **2**, 25
Vilsmeier compounds, use in the synthesis of alkyl chlorides, **2**, 25
Vilsmeier dehydrating systems, for the synthesis of isocyanides, **3**, 696
Vilsmeier formylation
of dienes, **3**, 64
for the synthesis of thioalkenals, **3**, 73
Vilsmeier–Haack–Arnold reaction
for the synthesis of α,β-unsaturated aldehydes, **3**, 68, 74
for the synthesis of α,β-unsaturated ketones, **3**, 239
Vilsmeier–Haack formylation, of arenes, **2**, 654
Vilsmeier–Haack reaction
for the synthesis of aryl aldehydes, **3**, 91
for the synthesis of selenals, **3**, 388
use in the synthesis of thioaldehydes, **3**, 338

Vilsmeier reaction
for the formylation of furans, **3**, 103
of methyl phenyl sulfide, **3**, 98
modified, for the formylation of anthracene, **3**, 101
of phthalimide, **6**, 138
of pyrrole, **3**, 106
of saccharin, **6**, 138
for the synthesis of thiophenes, **3**, 104
Vilsmeier reagents
from, formamides, **6**, 57
reactions
with ketones, **4**, 196
with tetrazoles, **5**, 363
as starting materials, for formamidines, **5**, 744
for the synthesis of alkyl bromides, **2**, 31
for the synthesis of alkyl iodides, **2**, 36
for the synthesis of vinyl chlorides, **2**, 617
Vilsmeier salts
reactions
with amines, **5**, 744
with ammonia, **6**, 641
with arylamines, **5**, 745
with ketoneimines, **6**, 644
as starting materials
for *N*-alkylguanidines, **6**, 643
for thioaldehydes, **3**, 341
for trithiapentalenes, **3**, 376
use in Rathke guanidine synthesis, **6**, 641
use in the synthesis of alkyl chlorides, **2**, 25
use in the synthesis of alkyl halides, **2**, 8
Vinhaticoate, methyl, synthesis of, **3**, 322
Vinigrol, synthesis of, **1**, 810
Vinyl, 2-(benzeneseleninyl)-, reactions, with secondary amines, **4**, 874
Vinyl, α-fluoro-, anion equivalent of, synthesis of, **4**, 812
Vinyl anions
from
arylsulfonylhydrazones, **1**, 1091
vinyl triflates, **1**, 1099
Vinyl anions, trifluoro-, synthesis of, **4**, 819
Vinylation, of saturated ketones, **3**, 138
C-Vinylation, mechanism, **1**, 66
Vinyl bromide, reactions, with potassium cyanide, **3**, 658
Vinyl bromide, (*E*)-1,2-dianisyl-, as starting material, for alkynes, **1**, 1120
Vinyl bromide, 1,2-disubstituted, synthesis of, **1**, 657
Vinyl bromide, (*E*)-2-ethoxy-, as starting material, for *gem*-bromo lithio enol ethers, **4**, 818
Vinyl bromide, perfluoro-, as starting material, for Grignard reagents, **4**, 820
Vinyl bromide, trifluoro-
reactions, with bis(trimethylsilyl)mercury, **4**, 821
use in the synthesis of α-fluoroalkenyllithiums, **4**, 817
Vinyl bromides
from
aldehydes, **2**, 624
alkynes, **1**, 658; **2**, 623
carbonyls, **2**, 624
vinyl metallics, **2**, 624
reduction of, **1**, 430
solvolysis of, **1**, 1114
as starting materials
for bromoalkenes, **1**, 430
for vinyl chlorides, **2**, 618
for vinyl iodides, **2**, 630
for vinyl radicals, **1**, 1129
synthesis of, **1**, 657; **3**, 495
via elimination reactions, **2**, 622
Vinyl bromides, 1,2-dialkyl-, synthesis of, **2**, 623
Vinyl carbanions
characterization of, **1**, 1088
synthesis of, **4**, 956
Vinyl cation equivalents, use in the synthesis of unsaturated ketones, **3**, 138
Vinyl cations
from
alkynes, **1**, 1122
allenes, **1**, 1125
photo-excitation of alkynes, **2**, 637
as intermediates, **1**, 1110
ring-strain, **1**, 1117
stability of, **1**, 1109
stabilization of, **1**, 1113
structures of, **1**, 1109
synthesis of, **1**, 1109
via bond scission, **1**, 1111
via deamination of diazonium salts, **1**, 1118
via electrooxidation, **1**, 1121
via fragmentation, **1**, 1119
via oxidative decarboxylation, **1**, 1121
via photolysis, **1**, 1120
via solvolysis, **1**, 1111
Vinyl chloride
reactions, with benzeneselenenyl chloride, **4**, 91
as starting material, for acryolyl chloride, **5**, 16
Vinyl chloride, α-cyclopropyl-, acetolysis of, **1**, 1114
Vinyl chloride, 1,2-difluoro-, reactions, with butyllithium, **4**, 817
Vinyl chloride, trifluoro-
lithium–chlorine exchange of, **4**, 817
lithium–halogen exchange of, **4**, 817
Vinyl chlorides
from
aldehydes, **2**, 618
alkynes, **2**, 616
carbonyls, **2**, 617
ketones, **2**, 617
vinyl metallics, **2**, 617
reactions, with phosphorus nucleophiles, **4**, 553
reduction of, **1**, 430
as starting materials, for fluoroalkenes, **1**, 430
synthesis of, via elimination reactions, **2**, 615
Vinyl chlorides, (*E*)-ethoxy-, reactions, with *n*-butyllithium, **4**, 817
Vinyl chlorides, (*Z*)-2-ethoxy-, reactions, with *n*-butyllithium, **4**, 817
Vinyl chlorides, imino-, as starting materials, for thioketones, **3**, 360
σ-Vinyl complexes
protonation of, **5**, 948
as starting materials, for cyclic alkoxycarbene complexes, **5**, 948
Vinyl 1,1-dichlorides, ω-amino-substituted, as starting materials, for ω-aminoalkanoic acids, **5**, 88
Vinyl ether–iron complexes, reactions, with lithium enolates, **1**, 481
Vinyl fluorides
catalytic hydrogenation of, **1**, 430
from
alkynes, **2**, 611
carbonyls, **2**, 611
vinyl metallics, **2**, 611
reduction of, **1**, 429
as starting materials, for fluoroalkenes, **1**, 429
synthesis of, **1**, 440
via elimination reactions, **2**, 610
via sulfoxide pyrolysis, **1**, 779
Vinyl groups, and vinyl metals, **2**, 953
Vinyl halides
addition–elimination of, **1**, 469
cleavage of, **1**, 1129
cross-coupling of, diorganylcuprates, **2**, 960
decomposition of, **1**, 1143
dehydrohalogenation of, **2**, 1073
dimerization of, **1**, 467
from
alkynes, **2**, 606

carbonyls, **2**, 607
vinylic organometallics, **2**, 607
vinylsilanes, **1**, 653
Heck coupling reactions of, **1**, 1096
hydroboration of, **4**, 142
hydrolysis of, for ketones, **3**, 114
as intermediates, **1**, 1089
mass spectra of, **1**, 1122
metal-halogen exchange of, **2**, 961
oxidative metallation of, **2**, 957
using activated metals, **2**, 959
photohydrolysis of, **2**, 638
radical cyclization of, **1**, 341
reactions
with acrolein acetals, **1**, 1096
with alkenylaluminums, **1**, 533
with alkenylboranes, **1**, 1097
with alkylcobalts, **2**, 577
with alkyl Grignard reagents, **1**, 1097
with alkylirons, **2**, 577
with alkylzincs, **1**, 467
with aryl Grignard reagents, **1**, 469
with arylzincs, **1**, 469
with 3-buten-2-one ethylene ketal, **1**, 1096
with cyanide ions, **3**, 658
with diarylcuprates, **1**, 469
with diselenides, **2**, 727
with enones, **3**, 233
with lithium dialkenylcuprates, **1**, 1097
with manganese compounds, **1**, 469
with organozincs, **1**, 467
with selenates, **2**, 727
with stannanes, **4**, 950
with vinylboranes, **1**, 279
with vinyl boronates, **1**, 279
with vinyl organometallics, **1**, 1096
reactivity of, **3**, 642
reduction of, **1**, 74, 81, 429
as starting materials
for alkenes, **1**, 429
for alkenyl metallics, **1**, 1090
for alkyl halides, **2**, 14
for metal-halogen exchange, **2**, 952
for oxidative addition, **2**, 952
for vinyl radicals, **1**, 342
for vinyl selenides, **2**, 727
for vinyl sulfides, **2**, 709
substitution of, **1**, 466
synthesis of, **2**, **605–633**
eliminations for, **2**, 606
with transition metal complexes
oxidative addition of, **2**, 960
oxidative displacement of, **2**, 960
and Wurtz coupling, **2**, 958
See also Alkenes, *gem*-dihalo-
Vinyl halides, α-cyclopropyl-, solvolysis of, **1**, 1114
(*E*)-Vinyl halides, synthesis of, **1**, 677
Vinylic carbanions
reactions, with epoxides, **1**, 1094
synthesis of, **1**, 1088
factors affecting, **1**, 1088
via proton loss, **1**, 1089
Vinylic chlorides, as starting materials, for thioketones, **3**, 356
Vinylic halides, reactions, with vinylic alanes, **1**, 1106
Vinylidene bromides, synthesis of, **4**, 1013
Vinylidene complexes
from, iron acyl complexes, **5**, 950
reactions, with imines, **5**, 954
synthesis of, **5**, 937, 940
Vinylidene complexes, hydroxy-, cyclization of, **5**, 944
Vinylidene cyclic compounds, synthesis of, **1**, 978
Vinylidene dibromides, use in the synthesis of ketene aminals, **4**, 970
Vinylidene dichlorides
reactions, with sodium azide, **4**, 993
use in the synthesis of ketene aminals, **4**, 970
Vinylidene difluorides
reactions, with 3-methylbenzthiazoloneimine, **4**, 992
as starting materials, for *gem*-diisothioureidoalkenes, **4**, 992
use in the synthesis of ketene aminals, **4**, 970
Vinylidene dihalides
reactions
with amines, **4**, 970
with aniline, **4**, 970
with diamines, **4**, 970
reactivity of, **4**, 970
as starting materials
for amidines, **4**, 970
for ketene aminals, **4**, 970
Vinylidene diiodides, use in the synthesis of ketene aminals, **4**, 970
Vinylidenes, synthesis of, **1**, 1143
Vinylidene–titanium complexes, transient, **3**, 597
Vinyl iodide, cyclopropyl-, acetolysis of, **1**, 1114
Vinyl iodide, perfluoro-, synthesis of, **4**, 751
Vinyl iodides
acetolysis of, **1**, 1114
formylation of, **3**, 58
from
alkynes, **1**, 658; **2**, 628
carbonyls, **2**, 630
vinyl metallics, **2**, 630
photolysis of, **1**, 1120
reactions
with butyllithium, **1**, 278
with organozincs, **4**, 954
with zinc alkynides, **1**, 1139
reactivity of, **1**, 468
reduction of, **1**, 431
as starting materials
for iodoalkenes, **1**, 431
for vinyl radicals, **1**, 1129
substitution of, **1**, 494
synthesis of, via elimination reactions, **2**, 628
cis-Vinyl iodides, synthesis of, **1**, 653
(*Z*)-Vinyl iodides
lithiation of, **1**, 836
as starting materials, for γ-lactones, **5**, 170
Vinyl nucleophiles, reactions, with α,β-unsaturated ketones, **3**, 228
Vinyl radicals
cyclization of, **1**, 320, 338, 341
from
alkynes, **1**, 1130
allenes, **1**, 1130
vinyl bromides, **1**, 1129
vinyl halides, **1**, 342
vinyl iodides, **1**, 1129
as intermediates, **1**, 1126
reactions
with nitric oxide, **2**, 778
with oxygen, **1**, 1127
with samarium diiodide, **1**, 347
stability of, **1**, 321
as starting materials, for α,β-unsaturated nitroso compounds, **2**, 778
structure of, **1**, 322
synthesis of, **1**, 330, 338
reviews, **1**, 1126
via bond scission, **1**, 1127
via stannylation, **1**, 342
Vinyls
arylation of, **1**, 471
reactions, with carboxylic acids, **3**, 210
as starting materials, for α,β-unsaturated ketones, **3**, 210

substitution of, **1**, 1114
Vinyls, α-aryl, solvolysis of, **1**, 1112
Vinyls, dichloro-, synthesis of, **3**, 495
Vinyls, 1,1-dihalo-, as starting materials, for carboxylic acids, **5**, 26
Vinyls, β-phenyl-substituted, solvolysis of, **1**, 1116
N-Vinyls, synthesis of, **2**, 777
Vinyl sulfones, reactions, with phenyl selenyl chloride, **4**, 287
Vinylthiols, synthesis of, **3**, 337, 352, 366
(+)-(*R*)-Vinyl *p*-tolyl sulfoxide, bromination–dehydrobromination of, **4**, 776
Vitamin A, synthesis of, **1**, 837; **3**, 62
Vitamin B_{12}
catalysis, for the synthesis of radicals, **1**, 946
for the mediation of radical cyclizations, **1**, 345
in radical addition reactions, **1**, 335
synthesis of, **1**, 626; **2**, 91
use in the synthesis of alkylchromium(III) compounds, **2**, 572
Vitamin D
(7*Z*)-analogues of, thermolysis of, **1**, 795
[1,7] sigmatropic rearrangements of, **1**, 403
synthesis of, **1**, 1029
Vitamin D_2, synthesis of, **1**, 754
Vitamin D_3
[1,7] hydrogen shifts in, **1**, 795
synthesis of, **1**, 817
Vitamins, synthesis of, **1**, 702
Von Braun reaction
alternatives to, **6**, 444
for the demethylation of tertiary amines, **5**, 1126
for the synthesis of amines, **2**, 303
for the synthesis of α,ω-bromides, **2**, 33
for the synthesis of cyanamides, **5**, 1126
for the synthesis of nitriles, **3**, 619
use in the synthesis of benzylic chlorides, **2**, 27

W

Wacker oxidation
of ethylene, **2**, 983
to acetaldehyde, **3**, 113
mechanism, **4**, 193
for the synthesis of cyclicimines, **3**, 411
Wadsworth–Emmons reaction
choice of solvents for, **1**, 749
Corey modification, **1**, 752
effect of ring-strain, **1**, 751
mechanism, **1**, 744
cis-selectivity, **1**, 752
steric effects, **1**, 747
Still modification, **1**, 751
for the synthesis of alkenes, **1**, 744
for the synthesis of (*E*)-alkenes, **1**, 745
for the synthesis of (*Z*)-alkenes, **1**, 750
for the synthesis of allenes, **1**, 987
for the synthesis of α-cyanophosphonates, **3**, 639
for the synthesis of 2-haloalkenylphosphorus compounds, **4**, 801
for the synthesis of α,β-unsaturated aldehydes, **3**, 60, 65
for the synthesis of α,β-unsaturated ketones, **3**, 220, 244
Wadsworth–Emmons reagent, for aldehyde homologation, **3**, 11
Wadsworth–Emmons–Wittig reaction, for the synthesis of α,β-unsaturated ketones, **3**, 231
Wagner–Meerwein rearrangement
mechanism, **1**, 380
for the synthesis of carbanion ions, **1**, 259
for the synthesis of isothiocyanates, **5**, 1044
Wallach procedure, for the synthesis of amines, **2**, 310
Wanzlick's reagent, use in the synthesis of aldehydes, **4**, 408
Water
azeotropic removal of, for the synthesis of aminals, **4**, 405
as cosolvent, for phosgene, **6**, 425
reactions
with acyl halides, **5**, 26
with alkyl isocyanates, **6**, 504
with carbonyls, **4**, 160
with carbonyl ylides, **3**, 319
with chlorides, **5**, 26
with cyanamides, **6**, 504
with diarylphosphinous halides, **2**, 840
with esters, **5**, 24
with isocyanates, **6**, 504
with sodium carboxylates, **1**, 373
with *N*-sulfonylimines, **4**, 312
removal of, for the synthesis of acetals, **4**, 177
use in aldol reactions, **3**, 159
use in the synthesis of thionoamides, **5**, 581
Weinreb amides
reactions, with thienyllithiums, **3**, 305
reduction of, **3**, 85
as starting materials, for benzocyclobutenones, **3**, 285
Wheland intermediates
synthesis of, **2**, 1009
use of term, **1**, 914
Wilburn reaction, for the synthesis of dialkylphosphinous acid amides, **2**, 466
Wilkinson's catalyst
modification of, **1**, 76
reactions
with alkyl halides, **2**, 583
with styrene, **1**, 73
reactivity of, **1**, 73
for the reduction of alkenes, **1**, 73
for the reduction of alkyl-substituted alkenes, **1**, 79
for the reduction of alkynes, **1**, 456
for the reduction of allenes, **1**, 459
for the synthesis of alkenes, **1**, 507
in the synthesis of dienes, **1**, 496
for the synthesis of α-silylnitriles, **3**, 640
use in hydroboration, **2**, 42
use in the reduction of carbon–mercury bonds, **1**, 68
use in the synthesis of imidoylrhodiums, **5**, 800
Wilkinson's reagent, use in the synthesis of nitriles, **3**, 619
Willgerodt–Kindler reaction
mechanism, **5**, 599
for the synthesis of thionoalkanoamides, **5**, 611
for the synthesis of thionoamides, **5**, 621
use in the synthesis of thioketones, **3**, 361
Willgerodt reaction
mechanism, **5**, 599
for the synthesis of amides, **5**, 272
for the synthesis of primary amides, **5**, 276
Williamson, synthesis of ortho esters, **6**, 68
Williamson ether synthesis, **2**, 133
Williamson reaction
for the synthesis of alkyl aryl tellurides, **2**, 731
for the synthesis of ethers, **2**, 88
Wittig alkenation
of aldehydes, **2**, 680
of ketones, **2**, 680
for the synthesis of alkenylsilanes, **2**, 936
for the synthesis of α,β-enoates, **5**, 155
Wittig condensation
for the synthesis of cyclic unsaturated ketones, **3**, 220
for the synthesis of difluoroalkenes, **4**, 730
Wittig–Gilman reaction, mechanism, **2**, 554
Wittig homologation, of aldehydes, **3**, 11
Wittig–Horner alkenation
of aldehydes, **2**, 680
of ketones, **2**, 680
for the synthesis of 2-cyano-1,3-butadienes, **3**, 646
for the synthesis of nitriles, **3**, 643
Wittig–Horner elimination, for the synthesis of allenes, **1**, 988
Wittig–Horner reaction
for the synthesis of 1,1-dichloroalkenes, **4**, 740
for the synthesis of enamines, **2**, 739
for the synthesis of 1-halo 1-phenylthioalkenes, **4**, 773
for the synthesis of vinyl chlorides, **2**, 618
for the synthesis of vinyl sulfones, **2**, 715
Wittig–Horner reagents
reactions, with carbonyls, **2**, 713
sulfur-substituted, **2**, 709
for the synthesis of chiral vinyl sulfoxides, **2**, 713
for the synthesis of vinyl selenides, **2**, 728
for the synthesis of vinyl sulfonates, **2**, 721
for the synthesis of vinyl tellurides, **2**, 730
Wittig–oxy-Cope rearrangement, tandem, use in the synthesis of unsaturated aldehydes, **3**, 22
Wittig reaction
of *N*-acrylic phospha-λ^5-azenes, **3**, 414
of arsonium ylides, mechanism, **1**, 741
arsonium ylides, solvent effects, **1**, 742
of bismuthonium ylides, **1**, 743
bromomethylene transfer reagents in, **5**, 895
of carbon suboxide, **1**, 988
decarboxylation of β-lactones as alternative to, **1**, 683
degree of stereoselectivity, **1**, 723
effect of reaction medium, **1**, 732

effect of additives, **1**, 721
effect of alkyl groups on phosphorus, **1**, 733
effect of α-methyl substituents, **1**, 725
effect of solvents, **1**, 721
effect of substituents on phosphorus, **1**, 723
effect of temperature of, **1**, 723
for the elimination of arsenic groups, **1**, 647
for the elimination of phosphorus groups, **1**, 638
factors affecting, **1**, 727
of germenes, **1**, 765
and Horner reaction compared, **1**, 744
kinetic control, **1**, 732
mechanism, **1**, 639, 720, 721
of (2-methyl-2-pyrrolin-3-yl) triphenylphosphonium bromide, **1**, 267
1,2-oxaphosphetanes as intermediates in, **2**, 831
reactive ylides, **1**, 721
loss of *cis*-selectivity in, **1**, 729
salt-free conditions, **1**, 727
salt-free, **1**, 734
for the synthesis of alkenes, **1**, 727
solvent and base effects, arsonium ylides, **1**, 740
solvent effect, **1**, 725
of stable ylides, *trans*-stereoselectivity of, **1**, 724
stereochemical course of, **1**, 721
stereochemistry of, **1**, 723
stereoselectivity, **1**, 639, 721
steric effects, **1**, 733
steric factors, **1**, 729
of stibonium ylides, **1**, 742
for the substitution of phosphorus functions, **1**, 268
for the synthesis of alkenes, **1**, 720; **3**, 48
for the synthesis of allenes and cumulenes, **1**, 986
for the synthesis of chloro haloalkenes, **4**, 757
for the synthesis of dibromoalkenes, **4**, 745
for the synthesis of 1,1-dichloroalkenes, **4**, 737
for the synthesis of enaminones, **3**, 268
for the synthesis of fluoro bromoalkenes, **4**, 754
for the synthesis of fluoro haloalkenes, **4**, 751
for the synthesis of α-halo alkenals, **3**, 68
for the synthesis of ketophosphines, **3**, 195
for the synthesis of nitriles, **3**, 649
for the synthesis of (*Z*)-perfluorovinylphosphonium compounds, **4**, 801
for the synthesis of 1-phosphaallenes, **3**, 604
for the synthesis of tetracoordinate sulfur compounds, **4**, 780
for the synthesis of thiaselenafulvenes, **4**, 870
for the synthesis of thioenals, **3**, 73
for the synthesis of triethyl orthocinnamate esters, **6**, 75
for the synthesis of α,β-unsaturated aldehydes, **3**, 60, 65
for the synthesis of α,β-unsaturated ketones, **3**, 220, 231, 244, 254, 269
for the synthesis of vinyl bromides, **2**, 624
for the synthesis of vinyl chlorides, **2**, 618
for the synthesis of vinyl fluorides, **2**, 611
for the synthesis of vinyl iodides, **2**, 630
for the synthesis of vinyl selenides, **2**, 728
for the synthesis of vinyl sulfides, **2**, 709
use in the synthesis of sulfenylsulfines, **5**, 562
use in the synthesis of ylides, **3**, 493, 497
and Wadsworth–Emmons reaction compared, **1**, 744
Wittig reaction, aza-
abnormal, for the synthesis of carbodiimides, **5**, 1079
for the synthesis of amines, **2**, 308
for the synthesis of azaallyl anions, **1**, 372
for the synthesis of carbodiimides, **4**, 441
Wittig reaction, pseudo-, of arsasilenes with benzophenone, **5**, 892
Wittig reagents
from, α-halophosphonium salts, **4**, 117
for the opening of epoxides, **1**, 212
phosphonium ylides of, **2**, 1004
reactions
with aldehydes, **2**, 728
with cobalt carbonyl, **1**, 225
with selone, **3**, 394
with thiol esters, **2**, 709
synthesis of, via selones, **3**, 394
for the synthesis of vinyl halides, **2**, 607
use in the synthesis of carbon-bridged diiron complexes, **4**, 679
use in the synthesis of carbon-bridged diruthenium complexes, **4**, 679
use in the synthesis of ketene dithioacetals, **4**, 847
use in the synthesis of selenals, **3**, 384
use in the synthesis of selones, **3**, 394
use in the synthesis of α,β-unsaturated acids, **5**, 103
Wittig reagents, aza-, trialrylphosphineimides as, **2**, 859
Wittig rearrangement
of dithiocarbamates, **5**, 614
of phosphorodiimidates, **4**, 923
for the synthesis of alcohols, **2**, 88; **6**, 125
for the synthesis of alkanethiols, **2**, 131
Wittig rearrangement, aza-, for the synthesis of amines, **2**, 322
Wittig rearrangement, 2,3-aza-, mechanism, **1**, 823
Wittig rearrangement, silyl-, of trialkylsilyl ether, **5**, 413
Wittig rearrangement, 2,3-thia-, mechanism, **1**, 820
Wittig rearrangement, of α anions, **1**, 233
[1,2] Wittig rearrangement, mechanism, **1**, 396
[2,3] Wittig rearrangement
of allyloxyenolates, **1**, 818
mechanism, **1**, 404, 816
for the substitution of sulfur functions, **1**, 234
Wittig–Schlosser reaction, for the synthesis of (*E*)-alkenes, **1**, 729, 730
Wittig–Still rearrangement, 2,3-thia-, mechanism, **1**, 406
Wohl–Ziegler reaction
for the synthesis of alkyl halides, **2**, 3
use in the synthesis of alkyl bromides, **2**, 27
Wolff–Kishner reaction
Huang–Minlon modification
for the deamination of *N*-alkylimine systems, **1**, 45
for the reduction of carbon–nitrogen double bonds, **1**, 41
for the reduction of carbon–nitrogen double bonds, **1**, 41
Wolff–Kishner reduction
mechanism, **1**, 11, 887
for the reduction of aldehydes, **1**, 18
for the reduction of ketones, **1**, 19
Wolff rearrangement
of 5-acyl-3*H*-pyrazoles, **3**, 543
mechanism, **1**, 393
photochemical, **2**, 668
photolytically induced, for the synthesis of ketenes, **3**, 532
for the synthesis of amides, **5**, 271
for the synthesis of *N*-arylamides, **5**, 292
for the synthesis of cyclopentanoic esters, **5**, 146
for the synthesis of ketenes, **3**, 530
methods, **3**, 531
for the synthesis of thioketenes, **3**, 551
for the synthesis of α,β-unsaturated ketones, **3**, 250
Wolf rearrangement, vinylogous, for the synthesis of ketenes, **3**, 532
Woodward's reagent K
reactions, with *N*-acetyl-L-tyrosinamide, **4**, 892
use in the synthesis of ketenimines, **3**, 577
Woodward–Hoffmann rules
and cycloaddition reactions, **2**, 52
and electrocyclic additions, **1**, 349
and the substitution of sulfur functions, **1**, 232
Woodward modification, mechanism, **2**, 44
Wurtz coupling
of benzylic and allylic systems, **1**, 3, 8
and organometals, **2**, 958

as side reaction, of metal–halogen exchange, **2**, 961
for the synthesis of alkyllithiums, **2**, 551
in the synthesis of alkylsodiums, **2**, 556
Wurtz–Fittig coupling, in the synthesis of arylsilanes, **2**, 941
Wurtz–Fittig reaction, for the substitution of alkyl halides, **1**, 464
Wurtz method, for the synthesis of alkyltins, **2**, 599

X

Xanthate, *O,S*-dialkyl, use in Schoenberg rearrangements, **6**, 472
Xanthate potassium salts, ethyl, as starting materials, for dialkyl thiocarbonates, **6**, 478
Xanthate pyrolysis
regiochemistry of, **1**, 776
syn-stereochemistry of, **1**, 775
Xanthates
chlorination of, for chloroformates, **6**, 426
fluorination of, **2**, 90; **6**, 229
precursors for radical cyclization, **1**, 341
radical cyclization of, **5**, 513
as radical precursors, **1**, 330
reactivity of, **1**, 775
reduction of, **1**, 16
stability of, **6**, 545
as starting materials
for alkanethiols, **2**, 123
for alkyl halides, **2**, 8
for radicals, **1**, 231
synthesis of, **1**, 700; **2**, 891; **6**, 472
use in the synthesis of sulfonic acids, **2**, 192
Xanthates, acyl, photolysis of, **2**, 130
Xanthates, *S*-acyl, as acyl radical precursors, for the synthesis of ketones, **3**, 136
Xanthates, alkyl
from
alcohols, **2**, 118
alkyl halides, **2**, 123
hydrolysis of, **2**, 123
reactions
with episulfides, **2**, 123
with epoxides, **2**, 123
reduction of, **2**, 123
Xanthates, dialkyl, rearrangement of, **2**, 123
Xanthates, methyl, as starting materials, for carbon-centered radicals, **1**, 940
Xanthates, *S*-methyl hex-5-enyl, isomerization of, **6**, 473
Xanthenes, synthesis of, **4**, 307
Xanthocillin, synthesis of, **3**, 712
Xanthocillin, *O,O'*-dimethyl-, synthesis of, **3**, 715
Xanthogenates, use in the synthesis of distibines, **4**, 595
Xanthone, 1,2,3,4-tetrahydro-, synthesis of, **4**, 307
Xanthone acridinone, hydroxy-, synthesis of, **2**, 673
Xenon difluoride
as aromatic fluorinating agent, for aryl fluorides, **2**, 613
as fluorinating agent, **6**, 55
for the fluorination of alkenes, **2**, 17
for the fluorination of sulfides, **4**, 61
reactions
with benzyl alcohols, **4**, 45
with hydrogen fluoride, **2**, 613
with *β*-lactams, **4**, 778
with pentacoordinated iridium(I) cations, **6**, 457
with sulfonyl chlorides, **2**, 183
with trimethylarsine, **2**, 492
with trisulfonylmethanes, **6**, 274
with vinylsilanes, **2**, 617
for the synthesis of vinyl fluorides, **2**, 611
transhalogenation of, with chlorocarbonyl isocyanate, **6**, 442
use in the synthesis of alkyl halides, **2**, 13
use in the synthesis of *α*-fluoro sulfides, **4**, 218
use in the synthesis of trifluoromethyls, **6**, 10
Xestosterol, synthesis of, **1**, 439
X-ray crystallographic investigations
of phosphaallenes, **3**, 603
of phosphacumulenes, **3**, 603
X-ray diffraction
of ketenimines, **3**, 556
of keteniminium salts, **3**, 602
X-ray structure determination
of transition metal carbene complexes, **5**, 932
of trialkylaluminum or triarylaluminums, **4**, 698
Xylene
acylation of, **3**, 280
dealkylation of, **1**, 56
reactions, with uracil-5-carboazides, **5**, 990
as starting material, for toluene, **1**, 56
Xylene, *α,α*-di(morpholino)-, reactions, with thiophenol, **4**, 322
meta-Xylene, trichloromethylation of, **6**, 29
o-Xylene, as starting material, for *o*-toluic acid, **5**, 115
p-Xylene, as starting material, for *p*-toluic acid, **5**, 116
Xylenes
from, *p*-nitriletoluene, **1**, 48
reduction of, **1**, 96
p-Xylenes, synthesis of, **6**, 190
o-Xylylenes, synthesis of, via ester pyrolysis, **1**, 787

Y

Yamaguchi's procedure, for the synthesis of macrolactones, **5**, 173
Yarovenko's reagent *See* Ethanamine, 2-chloro-*N*,*N*-diethyl-1,1,2-trifluoro-
Yeasts
  enantioselective reduction of, **2**, 14
  use in the synthesis of carboxylic acids, **5**, 117
Ylide, bis(phenylthio)methylene, synthesis of, **6**, 697
Ylide, phosphorane, addition, to an electrophilic metal, **5**, 1096
Ylide, thiophenium, synthesis of, **3**, 373
Ylide, triphenylarsonium benzoyl, reactions, with benzaldehyde, **1**, 737
Ylide, triphenylarsonium benzyl, synthesis of, **1**, 740
Ylide, triphenyl phosphonium, use in the synthesis of alkyl thioaldehydes, **3**, 332
Ylide–carbene reaction, for the synthesis of difluoroalkenes, **4**, 735
Ylides
  acylation of, **3**, 195, 499
  arylthiolation of, **5**, 898
  bromination of, **5**, 896
  chiral rearrangements of, **1**, 235
  chloride abstraction from, **5**, 893
  chlorination of, **5**, 895
  condensation of, **5**, 903
    with azasulfonium chlorides, **5**, 900
  definition of, **1**, 720
  desilylation of, **1**, 408
  with dicoordinate nitrogen substituents, **5**, 904
  from
    alkenes, **3**, 498
    alkynes, **3**, 498
    allylic sulfur compounds, **1**, 234
    arylhydrazonyl chlorides, **5**, 905
    azines, **3**, 497
    carbenes, **3**, 500
    cyclopropylphosphonium salts, **3**, 496
    dihalo pentacoordinate compounds, **3**, 500
    dimesityl dioxomolybdenum, **5**, 918
    iodonium ylides, **3**, 500
    phosphanes, **5**, 902
    α-silyl heterocycles, **4**, 519
    sulfonium salts, **1**, 241
    sulfoxonium salts, **1**, 210
    tertiary oxides, **3**, 500
    trialkylphosphanes, **5**, 908
    trialkylphosphines, **6**, 121
    triarylphosphanes, **5**, 908
    (trimethylsilylethyl)ammonium salts, **4**, 512
    vinyl phosphonium salts, **3**, 496
  as intermediates, in the synthesis of cyclophanes, **1**, 233
  iodination of, **5**, 896
  moderately reactive, definition of, **1**, 720
  reactions
    with acetic anhydride, **4**, 993
    with aldehydes, **1**, 730
    with alkanesulfonyl anhydrides, **5**, 901
    with alkanesulfonyl fluorides, **5**, 901
    with alkenyl phosphates, **4**, 737
    with alkyl halides, **2**, 446; **4**, 558
    with aluminum trichloride, **4**, 397
    with areneselenenyl halides, **4**, 348
    with arenesulfonyl chlorides, **5**, 898
    with arenesulfonyl fluorides, **5**, 901
    with benzaldehydes, **1**, 268
    with benzenesulfenyl chloride, **6**, 323
    with beryllium chloride, **4**, 395
    with butadiene zirconium complexes, **5**, 917
    with *t*-butylhypochlorite, **5**, 895
    with chlorine, **5**, 895
    with chloroethylsilane, **5**, 913
    with 1-chlorosiletanes, **5**, 913
    with copper(II) chloride, **2**, 447
    with 1,2-dichlorotetramethyldisilane, **5**, 914
    with 1,2-diketones, **3**, 220
    with disilanyl chlorides, **5**, 913
    with gallium trichloride, **4**, 397
    with indium trichloride, **4**, 397
    with iodobenzene dichloride, **5**, 895
    with iodosobenzene, **5**, 897
    with mercury chloride, **5**, 919
    with metalloarsanes, **5**, 911
    with metallosilanes, **5**, 913
    with metallostannanes, **5**, 920
    with 2-methylenecyclohexanone, **1**, 736
    with *N*-methyl-*N*-phenylthioacetamide, **5**, 898
    with monochlorophosphanes, **5**, 908
    with nitrosoalkenes, **2**, 779
    with perhalofluoroalkanes, **4**, 118
    with phenyl isocyanate, **3**, 591
    with phenylmethanesulfinyl chloride, **5**, 900
    with phenylselenyl bromide, **6**, 123
    with phenylselenyl chloride, **6**, 123
    with (*N*-phenylthio)succinimide, **4**, 346
    with salicyl aldehyde, **1**, 734
    with sulfur trioxide, **5**, 902
    with tetrachloromethane, **5**, 895
    with tetrahydroselenophene-1,1-dichloride, **5**, 903
    with thiocyanogen, **5**, 898
    with *p*-toluene thiosulfonates, **5**, 898
    with triorganosilyl halides, **5**, 913
    with triphenylphosphine, **6**, 264
    with triphenylphosphinegold chloride, **4**, 397
    with α,β-unsaturated aldehydes, **1**, 734
  reactive, definition of, **1**, 720
  1,2 rearrangement of, **1**, 398, 400
  2,3 rearrangement of, **1**, 407
  selenation of, **6**, 123
  skeletal rearrangements of, **5**, 908
  stability of, **3**, 493; **6**, 700
  stable, use of term, **1**, 720
  as starting materials
    for alkenyl metal salts, **2**, 877
    for alkynes, **1**, 1079
    for azirines, **3**, 600
    for bis(benzenethio) dinitromethane, **6**, 323
    for dinitrocarbene, **4**, 1083
    for iodomethylene triphenylphosphorane, **5**, 896
    for phosphaalkenes, **5**, 888
    for phosphoniomethylenephosphanes, **5**, 886
    for quaternary phosphonium salts, **2**, 446
    for α-selenyl ylides, **5**, 903
    for tantalum carbene complexes, **3**, 512
    for thioalkoxymethylenephosphoranes, **5**, 897
  synthesis of, **1**, 400; **4**, 598; **5**, 1095; **6**, 316
Ylides, acyl
  reaction rates, **1**, 741
  reduction of, **1**, 51
Ylides, alkylidene
  in carbene transfers, **1**, 352
  reactions, with aldehydes, **1**, 727

Ylides, allylic
  loss of configuration, **1**, 734
  reactions
    with benzaldehyde, **1**, 721
    with hexanal, **1**, 735
  synthesis of, **1**, 734
Ylides, allylic arsonium
  as starting materials, for vinylic epoxides, **1**, 269
  synthesis of, **1**, 269
Ylides, β-amino, synthesis of, **3**, 498
Ylides, *S*-amino sulfonium, synthesis of, **3**, 374
Ylides, ammonium
  rearrangement of, **1**, 398; **2**, 742
  synthesis of, **1**, 407; **2**, 1003
Ylides, arenenitrile, synthesis of, **3**, 678
Ylides, arsenium, synthesis of, **3**, 196
Ylides, arsonium
  compared, with phosphonium ylides, **1**, 737
  effect of substituents on, **1**, 739
  factors affecting direction of reaction, **1**, 741
  from
    phenyliodonium bis(arenesulfonyl)methylides, **6**, 708
    triphenylarsine oxide, **6**, 708
  as intermediates, **1**, 268
  products from, factors affecting, **1**, 739
  reactions
    with acetaldehyde, **1**, 738
    with aldehydes, **1**, 646
    with benzaldehydes, **1**, 738
    with carbonyls, **1**, 646, 737, 740, 742; **3**, 48, 57
    with conjugated enones, **1**, 740
    with enals, **1**, 741
    with *p*-nitroacetophenone, **1**, 738
    with thioketones, **1**, 741
  reactivity of, **1**, 646, 737
    comparisons, **1**, 738
  semistabilized, reactivity of, **1**, 738
  stabilization of, **1**, 647, 738
  as starting materials
    for alkenes, **1**, 738; **3**, 48
    for epoxides, **1**, 738; **3**, 48
    for *trans*-epoxides, **1**, 738
    for β-keto esters, **1**, 54
    for β-keto ketones, **1**, 54
    for (trimethylsilylethynyl)cyclopropanes, **1**, 270
  synthesis
    of, **1**, 737; **2**, 1004; **3**, 48, 196; **4**, 598; **5**, 920; **6**, 707
  thermal decomposition of, **1**, 647
Ylides, arsonium, bis(arenesulfonyl)-substituted, stability of, **6**, 707
Ylides, arsonium, *C*,*C*-dihetero-substituted, synthesis of, **6**, 707
Ylides, arsonium, silyl-substituted, synthesis of, **6**, 708
Ylides, arsorane, as starting materials, for alkenylarsonium salts, **2**, 877
Ylides, aryl, as starting materials, for tungsten carbene complexes, **3**, 517
Ylides, arylazo-functionalized
  from
    diazonium salts, **5**, 904
    N-nitrosoacetanilide, **5**, 905
    phosphonium salts, **5**, 904
Ylides, arylcarbonyl, synthesis of, **1**, 366
Ylides, α-aryliodonium-substituted, synthesis of, **5**, 897
Ylides, azomethine
  additions of, **1**, 368
  asymmetric cycloadditions of, **1**, 368
  [3+2] cycloaddition of, **2**, 324
  from
    aziridines, **1**, 368
    N-benzylidene-α-(diphenylphosphinoyl)glycine esters, **4**, 489
    bis(trimethylsilylmethyl)benzylamine, **1**, 285
    imines, **1**, 368
    oxazoles, **1**, 524
    oxazolidines, **1**, 368
    N-(trimethylsilylmethyl)iminium salts, **4**, 517
  intermolecular cycloaddition of, **1**, 368
  reactions
    with acryl amides, **1**, 370
    with alkenes, **1**, 368
    with dimethyl acetylenedicarboxylate, **1**, 524
    with enones, **2**, 331
    with thiobenzophenone, **4**, 319
  stabilized, trapping of, **1**, 368
  synthesis of, **1**, 285, 368
    in situ, **1**, 524
    via desilylation, **1**, 370
Ylides, benzonitrile
  from, 1-azido-1-phenylalkenes, **3**, 679
  as starting materials, for 3-phenyl 2*H*-azirenes, **3**, 679
Ylides, benzoyl, reactions, with *p*-nitrobenzaldehyde, **1**, 741
Ylides, benzylic, synthesis of, **1**, 731
Ylides, bis(arenesulfonyl)methylene, synthesis of, **6**, 698
Ylides, bismuthonium
  characteristics of, **1**, 648
  reactions, with aldehydes, **1**, 743
  as starting materials, for cyclopropanes, **1**, 743
  synthesis of, **2**, 1004; **6**, 707, 708
  Wittig reactions of, **1**, 743
Ylides, boron–sulfur, synthesis of, **2**, 268
Ylides, α-boryl
  from
    dialkylchloroboranes, **5**, 916
    dichloroalkylboranes, **5**, 916
Ylides, bromo, synthesis of, **5**, 494
Ylides, α-bromo
  from, organomercury compounds, **5**, 896
  reactions, with potassium iodide, **5**, 897
Ylides, α-bromophosphonium, synthesis of, **5**, 896
Ylides, carbonyl
  additions of, **1**, 365
  electrocyclization of, **3**, 318
  from
    carbenes, **3**, 319
    dioxolanones, **1**, 365
    epoxides, **1**, 365
    oxadiazolines, **1**, 365
    oxiranes, **3**, 315
    silanes, **1**, 366
  as intermediates, **3**, 314
  intramolecular trapping of, **1**, 366
  reactions
    with alcohols, **3**, 319
    with alkynes, **1**, 524
    with water, **3**, 319
  as starting materials
    for acetals, **3**, 319
    for enol ethers, **3**, 323
    for furans, **3**, 318
    for oxepines, **3**, 318
    for ozonides, **3**, 319
  structure of, **3**, 313
  synthesis of, **3**, 313
    via cycloreversion, **3**, 323
    via elimination reactions, **3**, 324
Ylides, cationic half, synthesis of, **6**, 733
Ylides, *P*-chlorinated, halide abstraction from, **5**, 888
Ylides, chloro, synthesis of, **5**, 494
Ylides, *P*-chloro, synthesis of, **6**, 696, 703
Ylides, α-chloro, from, phosphonium salts, **5**, 894
Ylides, *P*-chloro-*P*,*P*-bis(dialkylamino)-substituted, as starting materials, for phosphonium salts, **6**, 691
Ylides, chromium-substituted, synthesis of, via transylidation, **5**, 918
Ylides, conjugated double, synthesis of, via deprotonation of bisphosphonium salts, **5**, 910

Ylides, crotylhafnium, synthesis of, **5**, 917
Ylides, α-cyano arsonium, synthesis of, **3**, 639
Ylides, cyanodibutyl telluronium, reactions, with ketones, **3**, 647
Ylides, cyclic, synthesis of, **3**, 373
Ylides, cyclic silyl, isomerization of, **5**, 914
Ylides, dichloro, use in the synthesis of dichlorovinyls, **3**, 495
Ylides, dicyanomethylidene, synthesis of, **3**, 373
Ylides, difluoromethylene, synthesis of, **4**, 732
Ylides, dihalo, synthesis of, **6**, 696
Ylides, diketo sulfonium, synthesis of, **3**, 373
Ylides, dimethylsulfonium, synthesis of, **3**, 373
Ylides, diphosphino-functionalized, synthesis of, **5**, 909
Ylides, disilyl, synthesis of, **4**, 571
Ylides, disubstituted, from, methylene(triphenyl)phosphorane, **6**, 701
Ylides, double, hydrolysis of, **5**, 911
Ylides, ethoxycarbonylallyl, as starting materials, for dienes, **1**, 739
Ylides, ethynyl, reactions, with ketones, **1**, 739
Ylides, α-fluoro
  as starting materials, for fluoroalkenes, **5**, 894
  synthesis of, *in situ*, **5**, 894
Ylides, formyl phosphonium, reactions, with aldehydes, **3**, 65
Ylides, α-germyl
  from
    germylmethyl phosphonium salts, **5**, 916
    hexamethylgermanosiloxane, **5**, 916
Ylides, α-germyl arsonium, synthesis of, problems, **5**, 922
Ylides, hafnium-substituted, synthesis of, **5**, 917
Ylides, halo, synthesis of, **3**, 495
Ylides, *P*-halo
  from, tertiary phosphanes, **5**, 894
  rearrangement of, **5**, 898
Ylides, α-halo, synthesis of, **3**, 499
Ylides, halonium
  rearrangement of, **1**, 825
  synthesis of, **1**, 408
Ylides, iminothiocarbonyl, reactions, with aldehydes, **2**, 140
Ylides, 3*H*-indolium, reactions, with 1,1-bis(benzenesulfonyl)ethene, **4**, 277
Ylides, iodinium, development of, **4**, 281
Ylides, α-iodo
  interception of, in Wittig reactions, **5**, 896
  as starting materials, for iodoalkenes, **5**, 896
  synthesis of, **5**, 896
Ylides, iodonium
  brominolysis of, **6**, 254
  copper(I) catalysis of, **1**, 109
  from, β-keto esters, **1**, 109
  as reagents, for the synthesis of thiocarbonyl ylides, **3**, 367
  as starting materials
    for bis(benzenethio) dinitromethane, **6**, 323
    for ylides, **3**, 500
  synthesis of, **2**, 998
Ylides, iodonium nitro, reactions, with sulfenyl chlorides, **6**, 281
Ylides, isomunchone, as intermediates, **3**, 320
Ylides, ketene, reactions, with stannanediyls, **5**, 919
Ylides, β-keto, synthesis of, **3**, 499
Ylides, keto sulfoxonium, synthesis of, **3**, 171
Ylides, α-mercurated, from, mercury chloride, **5**, 919
Ylides, metallated
  from
    dicyclopentadienylhafnium dichloride, **5**, 917
    dicyclopentadienyltitanium dichloride, **5**, 917
    dicyclopentadienylzirconium dichloride, **5**, 917
Ylides, α-methoxy, synthesis of, **5**, 897
Ylides, molybdenum-substituted, synthesis of, via transylidation, **5**, 918
Ylides, monostannylated, disproportionation of, **5**, 920
Ylides, nitrile
  from
    2*H*-azirenes, **3**, 679
    carbenes, **3**, 679
    dihydrotriazoles, **3**, 680
    imidoyl halides, **3**, 678
    oxazaphospholes, **3**, 680
    oxazaphospholimines, **5**, 789
    2-3-oxazolin-5-ones, **3**, 680
    thiazaphospholes, **3**, 680
  as intermediates, **3**, 678
  reactions, with alkynes, **1**, 524
  as starting materials
    for 1,3-azaphospholines, **5**, 785
    for imidoylarsenic compounds, **5**, 791
    for imidoyl-tetracoordinated phosphorus compounds, **5**, 789
    for imidoyl-tricoordinated phosphorus compounds, **5**, 785
  synthesis of, **3**, 678
Ylides, nitro, as starting materials, for bis(benzenethio) dinitromethane, **6**, 323
Ylides, nitrogen-functionalized
  from
    arylazophosphoranes, **5**, 906
    5-bromoisatin, **5**, 906
    N-substituted azetidinones, **5**, 905
    triazolinediones, **5**, 905
Ylides, β-oxido, protonation of, **1**, 730
Ylides, 3-oxidopyrylium
  from, thiopyrylium perchlorates, **3**, 326
  synthesis of, **3**, 324
Ylides, oxonium
  2,3 rearrangement of, **1**, 820
  synthesis of, **1**, 408
Ylides, phenyliodinium
  reactions
    with carbon disulfide, **4**, 866
    with diethylamine, **4**, 867
    with thiobenzophenones, **4**, 866
  as starting materials, for 1,1-disulfonylalkenes, **4**, 866
Ylides, phenyliodonium, as starting materials, for methylenephosphoranes, **6**, 699
Ylides, phosphacumulene, reactions, with thiols, **5**, 730
Ylides, phosphenium
  reactions, with electrophiles, **5**, 893
  as starting materials, for phosphaalkenes, **5**, 888
Ylides, α-phosphino, synthesis of, **5**, 908
Ylides, phosphonium
  canonical form, **1**, 723
  compared, with arsonium ylides, **1**, 737
  γ-condensation, **1**, 735
  elimination of, **1**, 640
  formylation of, **3**, 48
  from
    alkenes, **3**, 498
    alkynes, **3**, 498
    arynes, **3**, 498
    aziridines, **3**, 498
    carbenes, **3**, 497
    difluorocarbene, **3**, 495
    C,C-dihalo-1,3,5-trisilacyclohexanes, **6**, 703
    phosphine azines, **3**, 497
    phosphines, **3**, 497
    phosphonium salts, **3**, 491
    phosphonium ylides, **3**, 499
    phosphoranes, **3**, 499
    sulfonium ylides, **5**, 902
    tetrahalomethanes, **3**, 495
  loss of *trans*-stereoselectivity, **1**, 725
  moderately reactive
    definition of, **1**, 722
    synthesis of, **1**, 730
  oxidation of, for the synthesis of acylsilanes, **5**, 413

reactions
with acyl isothiocyanates, **5**, 631
with aldehydes, **1**, 754
with ammonium fluoride, **2**, 475
with azasulfonium chlorides, **5**, 898
with cyclic polysulfides, **3**, 363
with dodecacarbonyl triruthenium, **4**, 683
with phenyldiazomethane, **3**, 445
with triphenylphosphonium bromide, **2**, 1051
with triphenylphosphonium dibromide, **2**, 1063
reactive, **1**, 726
stable, **1**, 722
as starting materials
for phosphonium ylides, **3**, 499
for thioaldehydes, **3**, 336
sulfonation of, **5**, 902
synthesis of, **1**, 721
via acylation, **3**, 499
via alkylation, **3**, 499
via halogenation, **3**, 499
types of, **1**, 722
use in the substitution phosphorus functions, **1**, 268
use in the synthesis of carboxylic esters, **5**, 134
use in the synthesis of chromium carbene complexes, **6**, 327
use in the synthesis of thioaldehydes, **3**, 336
as Wittig reagents, **2**, 1004
Ylides, phosphonium, antimony-substituted, synthesis of, **6**, 701
Ylides, phosphonium, arsenic-substituted, synthesis of, **6**, 701
Ylides, phosphonium, trimethylsilyl-substituted, condensation of, **6**, 701
Ylides, pyridinium, synthesis of, **2**, 1003
Ylides, salt-free, synthesis of, **3**, 493
Ylides, selenolium, as starting materials, for bis(benzenethio) dinitromethane, **6**, 323
Ylides, selenonium
reactions, with carbonyls, **1**, 720
3,2 sigmatropic rearrangements of, **1**, 822
as starting materials, for alkenes, **1**, 712, 716
synthesis of, **3**, 177
Ylides, α-selenyl
from
benzylphenylselenium dichloride, **5**, 903
methylphenylselenium dichloride, **5**, 903
ylides, **5**, 903
Ylides, α-silyl
from
bis(chlorodimethylsilyl)methane, **5**, 914
1,1-difluoro-1-Silacyclobutane, **5**, 915
Ylides, silylated
from, triorganosilyl halides, **5**, 913
rearrangements of, **5**, 912
Ylides, α-silylated, reactions, with halophosphanes, **5**, 908
Ylides, silyl-functionalized, from, 1,3-disiletanes, **5**, 915
Ylides, silyl sulfonium, as starting materials, for α-methylthiosilanes, **4**, 369
Ylides, sino-substituted arsonium, synthesis of, **5**, 921
Ylides, stibonium
comparison with other ylides, **1**, 743
nature of, **1**, 742
reactions, with benzophenone, **1**, 647
reactivity of, **1**, 743
stability of, **6**, 708
as starting materials, for alkenes, **1**, 743
synthesis of, **1**, 272; **3**, 196; **6**, 707, 708
Wittig reactions of, **1**, 742
Ylides, stilbonium, synthesis of, **1**, 54
Ylides, α-sulfinyl, from, toluenesulfinyl chloride, **5**, 900
Ylides, sulfinyl-substituted, synthesis of, **5**, 900
Ylides, sulfonium
in carbene transfers, **1**, 352
α,β'-elimination of, **1**, 623
from
sulfide, **3**, 373
sulfonium salts, **3**, 373
sulfoxides, **3**, 373
reactions
with alkyl nitroso compounds, **2**, 765
with aryl nitroso compounds, **2**, 765
with carbonyls, **1**, 720
rearrangement of, **1**, 399; **2**, 143
1,2 rearrangement of, **1**, 399
stability of, **3**, 372
as starting materials
for bis(benzenethio) dinitromethane, **6**, 323
for phosphonium ylides, **5**, 902
synthesis of, **1**, 407; **2**, 1000; **3**, 372; **4**, 837
via deprotonation of sulfonium salts, **3**, 373
via functionalization of sulfonium ylides, **3**, 374
via reaction of active methylenes with sulfoxides, **3**, 373
via reaction of diazo compounds with sulfides, **3**, 373
Ylides, sulfonium dinitro, reactions, with hydrogen halides, **6**, 56
Ylides, sulfonium-substituted, synthesis of, via deprotonation, **5**, 900
Ylides, α-sulfonyl
acylation of, **5**, 902
from, tosyl fluoride, **5**, 901
Ylides, α-sulfonylphosphonium, from, phosphonium salts, **5**, 901
Ylides, sulfoxonium
from
dimethyl sulfoxide, **3**, 378
sulfoximines, **3**, 378
3,2 sigmatropic rearrangements of, **1**, 822
synthesis of, **2**, 1000; **3**, 378
via elaboration of dimethylsulfoxonium methylide, **3**, 378
via reaction of diazo compounds with sulfoxides, **3**, 378
Ylides, tantalum-substituted, synthesis of, **5**, 917
Ylides, telluronium
as intermediates, **1**, 766
reactions
with aldehydes, **1**, 623
with carbonyls, **1**, 720
with ketones, **1**, 623
synthesis of, **2**, 1001
Ylides, α-thenoyl, acylation of, **2**, 877
Ylides, thiocarbonyl
additions of, **1**, 367
[1+3] dipolar cycloaddition of, **6**, 129
from
bromomethylsilane, **4**, 369
thioketones, **3**, 367
as intermediates
in the synthesis of alkenes, **1**, 685
in the synthesis of ketenimines, **3**, 572
isomerization of, **1**, 368
production of, **6**, 282
reactions, with alkenes, **1**, 367
stability of, **3**, 367
synthesis of, **1**, 286; **3**, 367; **6**, 309
trapping via [1+3] dipolar cycloaddition, **3**, 367
Ylides, titanium-substituted, synthesis of, **5**, 917
Ylides, trialkoxyphosphonium, as intermediates, **6**, 278
Ylides, triarylarsonium benzoyl, reactions, with *p*-nitrobenzaldehyde, **1**, 740
Ylides, tributylstibonium
reactions
with aldehydes, **1**, 647, 743
with ketones, **1**, 647, 743
Ylides, trimethylammonium, synthesis of, **2**, 1003
Ylides, trimethylsilyloxonium, synthesis of, **1**, 820
Ylides, triphenylphosphonium

reactions
with dialkylchloroboranes, **5**, 916
with thiiranes, **2**, 853
Ylides, tris(dialkylamido)phosphonium, synthesis of, **6**, 695
Ylides, tris(*p*-substituted phenyl)arsonium, reactions, with benzaldehyde, **1**, 739
Ylides, tungsten-substituted, synthesis of, via transylidation, **5**, 918
Ylides, α-uranium-functionalized, from, tricyclopentadienyluranium chloride, **5**, 920
Ylides, *C*-vinylazomethine, synthesis of, **3**, 420
Ylides, vinylcarbonyl, cyclization of, **3**, 321
Ylides, vinylic sulfonium, synthesis of, **3**, 373
Ylides, zirconium-substituted, synthesis of, **5**, 917
Ynals, synthesis of, **1**, 1046
Ynamine extension, **2**, 1045
Ynamines
in cycloaddition reactions, with α,β-unsaturated esters, **4**, 827
diazo group transfer to, **3**, 477
from
amides, **2**, 1050
trichloro enamines, **5**, 595
as intermediates, **2**, 1040
in the synthesis of alkynyl diazonium salts, **2**, 1058
in the synthesis of alkynyl isocyanides, **2**, 1058
in the synthesis of ketene aminals, **4**, 976
oxidative dimerization of, **2**, 1050
reactions
with alcohols, **4**, 885
with allylic alcohols, **1**, 828
with amines, **4**, 976
with amino alcohols, **5**, 691
with aryl isothiocyanates, **5**, 618
with butadiene, **2**, 740
with dialkyl thiophosphite, **4**, 1010
with dithioesters, **5**, 614
with dithiolones, **5**, 605
with dithiolonium salts, **5**, 605
with electrophiles, **4**, 791
with ethanol, **4**, 885
with nucleophiles, **4**, 885
with phenols, **4**, 885
with 5-pyrazolones, **4**, 986
with *N*-sulfinylamines, **5**, 623
with *N*-sulfinyl sulfonamides, **5**, 767
with sulfur, **5**, 554, 605
with thiazoline thiones, **5**, 605
with thioaldehyde pentacarbonylchromium complexes, **5**, 605, 614
with thioketenes, **5**, 605
with thioketone *S*-oxides, **5**, 623
with thioketones, **5**, 614
as starting materials
for amide acetals, **6**, 106
for amidines, **4**, 977
for α-amido enamines, **4**, 980
for α-borio enamines, **4**, 1015
for α-cuprio enamines, **4**, 1018
for enamines, **2**, 739
for α-hydrazino enamines, **4**, 984
for ketene aminals, **4**, 976
for keteniminium salts, **3**, 602
for α-phosphoryl enamines, **4**, 1003
for α-potassio allenamines, **4**, 1017
for α-thiophosphoryl enamines, **4**, 1010
for α-tungstenio enamines, **4**, 1019
synthesis of, **2**, 1046
choice of methods, **2**, 1051
use in the synthesis of carbon-bridged diiron complexes, **4**, 681
use in the synthesis of ferracyclopentadiene, **4**, 681
use in the synthesis of ketenimines, **3**, 565
use in the synthesis of selenoamides, **5**, 627
See also 1-Alkynes, *N,N*-disubstituted-1-amino-
Ynamines, alkynyl-substituted
deprotonation of, **2**, 1045
synthesis of, via functionalization of ynamines, **2**, 1043
Ynamines, ammonium salts, from, alkynyl halides, **2**, 1042
Ynamines, bis(silyl)-, synthesis of, **2**, 1058
Ynamines, *N,N*-bis(trimethylsilyl)-, as starting materials, for *n*-methyleneynamines, **2**, 1054
Ynamines, t-butyl-, synthesis of, **2**, 1054
Ynamines, chloro-
as intermediates
in the synthesis of ketene hemiaminals, **4**, 885
in the synthesis of ynediamines, **2**, 1047
Ynamines, cyano-, as starting materials, for cyanoketene aminals, **4**, 976
Ynamines, *N,N*-dialkyl-
reactivity of, **2**, 1057
synthesis of, **2**, 1046
Ynamines, dialkynic, as starting materials, for ketene hemiaminals, **4**, 885
Ynamines, diethyl-, lithiation of, **2**, 1045
Ynamines, *N,N*-diethyl-
reactions
with alkane sulfenyl chlorides, **4**, 791
with arene sulfenyl chlorides, **4**, 791
synthesis of, **2**, 1050
Ynamines, *N,N*-dimethyl-, large scale synthesis of, **2**, 1049
Ynamines, *N,N*-dimethyl-, alkynyl-substituted, synthesis of, **2**, 1041
Ynamines, heterosubstituted, synthesis of, **2**, 1048
Ynamines, lithiated
alkylation of, **2**, 1047
synthesis of, **2**, 1047
Ynamines, *n*-methylene-
from
N,N-bis(trimethylsilyl)ynamines, **2**, 1054
oxime sulfonate esters, **2**, 1053
synthesis of, via nucleophilic substitution of oxime sulfonate esters, **2**, 1053
Ynamines, phenyl-, synthesis of, **2**, 1054
Ynamines, primary, synthesis of, **2**, 1040
Ynamines, secondary, synthesis of, **2**, 1040
Ynamines, silylated
reactions, with dimethyl ethynedicarboxylate, **2**, 1045
reactivity of, **2**, 1044
synthesis of, **2**, 1045
Ynamines, stannylated
reactions
with carbodiimides, **2**, 1044
with 4-chlorophenyl isocyanate, **2**, 1044
with isothiocyanates, **2**, 1044
reactivity of, **2**, 1044
Ynamines, *N*-substituted, synthesis of, **2**, 1046
Ynamines, tertiary
from
alkynic cuprates, **2**, 1041
alkynyl ethers, **2**, 1043
alkynyl halides, **2**, 1041
β-bromo enamines, **2**, 1049
α-chloro enamines, **2**, 1045
α,β-chloro enamines, **2**, 1047
β-chloro enamines, **2**, 1049
dichloro enamines, **2**, 1047
α,β-dichloro enamines, **2**, 1047
ketene *N,S*-acetals, **2**, 1049
organosilanes, **2**, 1043
stannanes, **2**, 1043
thioamides, **2**, 1049
trichloro enamines, **2**, 1047
trifluoroalkanes, **2**, 1042
synthesis of, **2**, 1040
via elimination reactions, **2**, 1045

via isomerization reactions, **2**, 1049
via substitution reactions, **2**, 1041
Ynamines, trisilyl-, synthesis of, **2**, 1058; **3**, 568
Ynamine salts, synthesis of, **2**, 1046
Ynamine salts, triphenylphosphonium, synthesis of, **2**, 1051
Ynediamines
from
diaminocyclopropenones, **2**, 1050
substituted trichloro enamines, **2**, 1047
reactions, with amines, **4**, 976
synthesis of, **2**, 1047
Ynediamines, bis(trifluoromethyl), synthesis of, **2**, 1050
Ynenals, synthesis of, **3**, 65
Ynethers, arsino, as starting materials, for arsinoketenes, **2**, 876
Ynimines, synthesis of, **3**, 413
Ynolate anions, stability of, **2**, 1022
Ynolates, synthesis of, **2**, 1022
Ynol ethers, research background, **2**, 1021
Ynols
acidity of, **2**, 1022
synthesis of, **2**, 1022
See also Alkynes, hydroxy-
Ynones
reduction of, **1**, 511
as starting materials, for radicals, **1**, 546
synthesis of, **1**, 1046
Ynones, β-lactam, synthesis of, **1**, 1047
α,β-Ynones, reactions, with propane-1,3-dithiol, **3**, 168
Ynyl carbocations, synthesis of, via bond scission, **1**, 904
Yohimbine alkaloids, synthesis of, **1**, 776
Yohimbone, synthesis of, **1**, 149
Ytterbium
chemistry of, **2**, 564
reactions
with alkyl iodides, **2**, 564
with ammonia, **1**, 508
with ketimines, **2**, 312
Ytterbium(III) nitrate
as catalyst, for the synthesis of aldehydes from alcohols, **3**, 6
for the oxidation of alcohols, **3**, 118
Ytterbium cyanides, as catalysts, for the synthesis of cyanohydrins, **3**, 615
Ytterbium triflate, use in the synthesis of 1,2-dialkylcycloalkanols, **2**, 69
Yttrium
chemistry of, **2**, 564
pentamethylcyclopentadienyl complexes of, synthesis of, **4**, 702
Yttrium complexes, bis(trimethylsilyl)methyl-, synthesis of, **6**, 194
Yttrium compounds, methyl-bridged, synthesis of, **4**, 670
Yttrium–indium compounds, synthesis of, **4**, 726

Z

(-)-Zearalenone, synthesis of, via radical cyclization, **1**, 343
Zeolites, as catalysts, for the synthesis of dithioacetals, **4**, 246
Ziegler–Natta catalysts, development of, **1**, 662
Ziegler procedure, modification of, **2**, 422
Ziegler telomerization, of ethylene, **2**, 43
Zinc
  ate complexes of, synthesis of, **2**, 592
  dehalogenation of, **6**, 429
  reactions
    with alkenyl bromides, **4**, 821
    with alkyl bromides, **6**, 64
    with alkyl chlorides, **6**, 64
    with alkyl iodides, **6**, 64
    with ammonium chloride, **2**, 764
    with antimony trichloride, **2**, 55
    with aryloxy ketones, **4**, 167
    with 2-arylthio-2,2-dichloroethanols, **4**, 775
    with α-chloroalkyl phenyl sulfides, **4**, 396
    with 1,2-dibromoethane, **1**, 508
    with dichlorodifluoroenones, **1**, 569
    with dihalides, **3**, 530
    with dihalodifluoromethanes, **6**, 15
    with dihalomethanes, **4**, 152
    with diiododi(trimethylsilyl)methane, **4**, 618
    with diiodomethane, **4**, 696
    with haloalkanes, **1**, 122
    with α-haloalkyl carboxylates, **2**, 999
    with α-haloalkylimides, **2**, 1002
    with haloalkynes, **2**, 1096
    with α-haloboronic esters, **1**, 163; **4**, 665
    with iodomethyl pivalate, **4**, 385
    with iodomethylthio esters, **4**, 396
    with 3-iodopropionitrile, **1**, 122
    with phosphonium salts, **4**, 753
    with sodium hydroxide, **2**, 772
    with tertiary iodides, **4**, 730
  for the reduction of α-acetoxyketones, **1**, 16
  for the reduction of aldehydes, **1**, 18
  for the reduction of alkynes, **1**, 508
  for the reduction of amines, **1**, 34
  for the reduction of biarylthioketones, **1**, 24
  for the reduction of α-halocarbonyls, **1**, 2
  for the reduction of ketones, **1**, 18
  for the reduction of sulfides, **1**, 23
  for the reductive cleavage of disulfides, **2**, 126
  transmetallation reactions of, **2**, 560
  use in Barbier reactions, **2**, 74
  use in the synthesis of dibromoalkenes, **4**, 745
  use in Wittig reactions, **4**, 734
Zinc, alkyl-, as reagent, for carbozincation of alkynes, **2**, 973
Zinc, allyl-, as reagent, for carbozincation of alkynes, **2**, 973
Zinc, bis(trifluoromethyl)-, synthesis of, **6**, 246
Zinc, cyanomethyl-, reactions, with aldehydes, **3**, 632
Zinc, diazomethyl-, synthesis of, **3**, 488
Zinc, diethyl-
  cleavage of, **2**, 1096
  reactions
    with alkyl iodides, **2**, 591
    with benzaldehyde, **2**, 71
    with *t*-butyl chloride, **1**, 127
    with copper(I) cyanide, **2**, 69
    with ketoaldehydes, **2**, 72
    with triethoxyborane, **2**, 520
  synthesis of, **2**, 68, 550
  use in Simmons–Smith reactions, **1**, 300
Zinc, dimethyl-
  reactions
    with 1-alkynes, **2**, 1096
    with lithium enolates, **1**, 132
    with vanadium compounds, **2**, 570
  synthesis of, **2**, 550
  transmetallation of, **2**, 566
  use in methylation reactions, **2**, 570
Zinc, iodo(iodomethyl)-, reactions, with iodostannanes, **4**, 155
Zinc, metallic, reactions, with phosphonates, **6**, 266
Zinc, silylallyl-, reactions, with electrophiles, **2**, 935
Zinc, 2-silylpropenyl-, reactions, with electrophiles, **2**, 936
Zinc, trifluorovinyl-, acylation of, **3**, 237
Zinc(II) halides, use in the synthesis of alkyl halides, **2**, 6
Zinc(II) nitrate, for the oxidation of alcohols, **3**, 118
Zinc acetate, reactions, with butenolides, **5**, 79
Zinc–acetic acid, for the reduction of quinones, **2**, 657
Zinc acids, as reagents, for the cleavage of carbon–phosphorus bonds, **1**, 52
Zinc aldimines, synthesis of, **3**, 418
Zinc alkyl compounds, reactions, with haloalkanes, **1**, 127
Zinc alkylcyanocuprates
  reactions, with aldehydes, **2**, 69
  synthesis of, **2**, 588
Zinc alkynides
  reactions, with vinyl iodides, **1**, 1139
  synthesis of, **1**, 1139
Zinc amide, reactions, with ethyne, **2**, 1096
Zincates, alkynyl-, synthesis of, **2**, 1096
Zincates, ester, synthesis of, **4**, 385
Zincates, trialkyl-, as intermediates, **2**, 594
Zinc bis(ethyl diazoacetate), synthesis of, **3**, 488
Zinc borohydride
  reactions
    with alkoxy ketones, **2**, 61
    with 2-hydroxy ketones, **2**, 61
    with nitroalkenes, **3**, 426
  for the reduction of bromoalkanes, **1**, 9
  for the reduction of chloroalkanes, **1**, 7
  for the reduction of tertiary and benzylic halides, **1**, 4
  use in hydroboration reactions, **2**, 43
  use in the reduction of carbonyls, **2**, 60
  use in the reduction of imines, **2**, 312
Zinc bromide
  reactions, with sulfoxides, **1**, 305
  use in the synthesis of alkynes, **1**, 1026
Zinc bromide, allyl-
  reactions
    with aminoalkynes, **1**, 544
    with propargyl alcohol, **1**, 543
    with propynol ethers, **4**, 809
    with terminal alkynes, **1**, 543
    with 1-trimethylsilyl-1-octyne, **4**, 809
    with vinyl metals, **4**, 709
Zinc bromide, bromomethyl-, synthesis of, **4**, 152
Zinc bromide, crotyl-, reactions, with bromoalkynes, **1**, 544
Zinc bromide, phenyl-, synthesis of, **2**, 958
Zinc bromide, trifluoromethyl-, reactions, with dimethyl telluride, **6**, 49
Zinc bromide, vinyl-, reactions, with prop-2-yn-1-ol, **4**, 1067
Zinc bromides, alkynyl-, reactions, with enones, **1**, 1050
Zinc bromides, 2-aryloxymethylallyl-, reactions, with imines, **2**, 324
Zinc bromides, propargyl-, reactions, with alkynes, **1**, 544
Zinc–bromine exchange, with lithium tributylzincate, **4**, 821

Zinc chloride
as catalyst
for the addition of hydrogen cyanide to styrene, **3**, 626
for the synthesis of aromatic ketones, **3**, 279
for the synthesis of diselenoacetals, **4**, 288
reactions
with acetyl chloride, **3**, 9
with *N*,*N*-dimethyl-1-chloro-2-methylpropenylamine, **2**, 758
with dithiocarbonates, **2**, 273
with 1-ethoxyvinyllithium, **4**, 954
with lithium alkynides, **1**, 1139
with lithium naphthalenide, **2**, 590
with potassium graphite, **2**, 590
with sodium sulfinates, **2**, 273
with trialkylaluminums, **2**, 596
with tris(trimethylsilylmethyl)lithium, **6**, 393
use in Houben–Hoesch reaction, **3**, 282
use in the synthesis of alkylmanganese(II) compounds, **2**, 574
use in the synthesis of alkynes, **1**, 1027
use in the synthesis of thioketones, **3**, 343
Zinc chloride, alkyl-, reactions, with ω-carboethoxyacyl chlorides, **5**, 76
Zinc chloride, aryl-, reactions, with enol triflates, **1**, 480
Zinc chloride, bis(trimethylaminomethyl)-, synthesis of, **4**, 534
Zinc chloride, 3,3,3-trifluoro-1-pentynyl-, reactions, with iodobenzenes, **1**, 1009
Zinc chlorides, alkynyl-
reactions
with alkyl chloroformates, **1**, 1047
with aryl fluoroalkane sulfonates, **1**, 1028
with diketenes, **1**, 1028
Zinc chlorides, allenyl-, reactions, with aldehydes, **1**, 1055
Zinc chlorides, aryl-, reactions, with bromoacetonitrile, **3**, 625
Zinc–copper compounds, nucleophilic substitution with, **1**, 968
Zinc–copper couples
reactions
with alkyl iodides, **2**, 591
with dibromo(trimethylsilyl)methane, **6**, 194
as reagents, for the synthesis of ketenes, **3**, 530
use in Wittig reactions, **4**, 734
Zinc–copper organometallics, synthesis of, **1**, 123
Zinc cuprates, use in the displacement of allylic alcohols, **1**, 188
Zinc cyanide, reactions, with hydrogen chloride, **3**, 92
Zinc dibromide, for the deprotection of amines, **2**, 304
Zinc dichromate, as oxidant, **3**, 4
(*E*)-Zinc enolate, synthesis of, **4**, 760
Zinc enolates
ethenylation of, **3**, 138
from
tribromoacetyl chlorides, **4**, 768
trichloroacetyl chlorides, **4**, 768
synthesis of, **2**, 702
Zinc ester enolates
reactions, with nitriles, **3**, 461
O-silylation of, **4**, 831
Zinc fluoride, reactions, with acid chlorides, **5**, 4
Zinc halides, reactions, with lithium naphthalenide, **1**, 123
Zinc halides, alkyl-, synthesis of, **2**, 591
Zinc halides, alkynyl-
reactions, with acid chlorides, **1**, 1045
synthesis of, **2**, 1096
Zinc halides, allyl-, as starting materials, for 1,5-dienes, **1**, 127
Zinc halides, cyclopropyl-, synthesis of, **2**, 594
Zinc homoenolates
reactions, with alkynic esters, **3**, 225
synthesis of, **1**, 122
Zinc hydroxides, alkyl-, synthesis of, **6**, 394
Zinc iodide, as catalyst, for the synthesis of cyanohydrins, **3**, 615
Zinc iodide, 2-cyanoethyl-, synthesis of, **1**, 122
Zinc iodide, cyclopentylmethyl-, synthesis of, **2**, 592
Zinc iodide, 5-hexenyl-, cyclization of, **2**, 592
Zinc iodide, iodomethyl-
reactions
with allyl halides, **3**, 623
with vinylcoppers, **2**, 74
synthesis of, **2**, 998
use in the synthesis of homoallylic zincs, synthesis of, **2**, 594
Zinc iodides, alkyl-
metallation of, **2**, 590
reactions, with copper cyanide, **2**, 588
as starting materials, for dialkylzincs, **2**, 591
Zinc ketimines, synthesis of, **3**, 461
Zinc metal, reactions, with α-halocarbonyls, **2**, 702
Zinc oxide, reactions, with isopropyl isocyanide dichloride complexes, **5**, 963
Zinc powder, highly reactive, synthesis of, **2**, 590
Zincs
alkylation of, **1**, 187
chemistry of, **2**, 590
homologation with, **2**, 594
reactions
with aryl iodides, **3**, 290
with non-ethereal solvents, **2**, 566
with phosphorus ylides, **4**, 586
reactivity of, **3**, 288; **4**, 396
for the substitution of aryl halides, **1**, 471
synthesis of, **4**, 385, 396, 964
transmetallation of, **1**, 122
for transmetallation reactions, **1**, 464
Zincs, acyl-, synthesis of, problems, **5**, 466
Zincs, alkenyl-
from, organolithiums, **1**, 468
reactions, with aryl iodides, **1**, 1097
synthesis of, **1**, 1097
Zincs, α-alkoxy-, synthesis of, **4**, 384
Zincs, alkyl-
functionalized, synthesis of, **2**, 590
reactions
with aldehydes, **2**, 69
with aroyl chlorides, **3**, 288
with carbonyls, **2**, 71
with dirhenium heptaoxide, **2**, 585
with vinyl halides, **1**, 467
synthesis of, **2**, 590
via Grignard reactions, **2**, 590
use in the synthesis of monoalkylvanadiums, **2**, 569
Zincs, alkynyl-
palladium-catalyzed cross-coupling reactions of, **1**, 1006
reactions
with acid chlorides, **1**, 1045
with aldehydes, **1**, 1039
with 2-arylsulfonyl-substituted cyclic ethers, **1**, 1031
with (*Z*)-bromoalkenyl boranes, **1**, 1008
with organozincs, **2**, 1095
with trialkylsilyl triflate, **1**, 314
synthesis of, **2**, 1095
Zincs, allenyl-, synthesis of, **1**, 198
Zincs, allylic, reactions, with alkynylsilanes, **2**, 925
Zincs, α-amino-, synthesis of, **4**, 534
Zincs, amino acid derived, synthesis of, **2**, 591
Zincs, α-aminoalkyl-, synthesis of, **2**, 1002
Zincs, aryl-
reactions
with alkyl acid chlorides, **3**, 288
with α-halo esters, **1**, 465
with α-halonitriles, **1**, 465
with vinyl halides, **1**, 469

use in the synthesis of biaryls, **1**, 473
Zincs, α-borylalkyl-, synthesis of, **2**, 1005
Zincs, α-chloro-, synthesis of, **2**, 594
Zincs, α-chloroalkyl-, synthesis of, **2**, 998
Zincs, cycloalkenyl-, synthesis of, via zirconium-catalyzed carbozincation, **2**, 973
Zincs, (dialkoxyboryl)alkyl-, synthesis of, **2**, 1005
Zincs, dialkyl-
amination of, **2**, 305
conjugate addition to enones, **3**, 128
from, alkylzinc iodides, **2**, 591
oxidation of, **2**, 112
reactions
with alkynes, **1**, 544
with benzonitrile, **3**, 461
with dicarbonyls, **4**, 166
with ketones, **2**, 68
with nickel (acetylacetonate), **1**, 314
reactivity of, **1**, 127
as starting materials, for homocuprates, **2**, 587
synthesis of, via halogen-metal interconversion, **2**, 591
use in the synthesis of alkylcoppers, **2**, 586
use in the synthesis of organocoppers, **1**, 122
Zincs, dialkyl-, polyfunctionalized, in enantioselective alkylation reactions, **2**, 72
Zincs, dialkynyl-
from, phenylethyne, **2**, 1096
synthesis of, **2**, 1096
Zincs, diaryl-, synthesis of, **3**, 145
Zincs, diethyl-, reactions, with aminoalkynes, **2**, 1096
Zincs, diphenyl-, reactions, with aminoalkynes, **2**, 1096
Zincs, halo-, synthesis of, **6**, 64
Zincs, 1-haloalkyl-, reactions, with phosphines, **4**, 586
Zincs, α-haloalkyl-, synthesis of, **2**, 998
Zincs, haloaryl-, use in the synthesis of ketones, **3**, 296
Zincs, (halomethyl)-, synthesis of, **4**, 152
Zincs, imidoyl-
as acyl anion equivalents, **5**, 803
from, isonitriles, **5**, 802
synthesis of, **5**, 802
Zincs, metallo-, synthesis of, **2**, 1006
Zincs, mixed alkyl-, synthesis of, **2**, 591
Zincs, monosilyl-, for the synthesis of vinylsilanes, **2**, 918
Zincs, propargyl-, reactions, with acylsilanes, **2**, 74
Zincs, pyridyl-, reactions, with anhydrides, **3**, 311
Zincs, secondary alkyl-, cyclization of, **2**, 592
Zincs, trihalomethyl-, synthesis of, **6**, 245
Zincs, α-trimethylstannyl-, use in the synthesis of γ,δ-unsaturated aldehydes, **3**, 23
Zincs, vinyl-, reactions, with copper(I) bromide, **4**, 821
Zinc thiolates, synthesis of, **2**, 273
1-Zinc-1-zirconium alkenes, reactions, with aldehydes, **1**, 991
Zirconabicycles, reactions, with carbon monoxide, **3**, 270
Zirconabicyclic systems, synthesis of, **4**, 1066
Zirconacarbocyclic intermediates, isolation of, **2**, 927
Zirconacycles
derivatization, to organic products, **2**, 988
halogenation of, **4**, 809
isolation of, characterization, **2**, 988
Zirconacycles, five-membered, **2**, 987
ring-contraction–expansion of, **2**, 987
Zirconacycles, seven-membered, **2**, 988
Zirconaheterocyclic intermediates, synthesis of, **2**, 927
Zirconanes, dimerization of, **1**, 491
Zirconation, of alkynes, **4**, 1061, 1066
Zirconaziridines, ring-expansion of, **2**, 307
2-Zirconiophospholanes, synthesis of, **4**, 579
Zirconium
for the ethylmagnesiation of alkenes, **2**, 561
oxidative coupling–ring-expansion of, **2**, 986
Zirconium, bis(cyclopentadienyl)-, reactions, with propargylic ethers, **1**, 1057
Zirconium(IV) chloride, use in the reduction of carbonyls, **2**, 59
Zirconium(IV) complexes, vinyl-, reactions, with α,β-unsaturated ketones, **1**, 1105
Zirconium(IV) metallocenes, bis(trimethylsilyl)methyl-, synthesis of, **6**, 194
Zirconium acyl compounds
as starting materials
for aldehydes, **5**, 442
for carboxylic acids, **5**, 442
for esters, **5**, 442
Zirconium alkoxides, as catalysts, for the synthesis of cyanohydrins, **3**, 616
Zirconium alkyls, reactions, with aluminum alkyls, **4**, 725
Zirconiumaluminum alkenes, synthesis of, **4**, 1066
Zirconium carbene complexes, synthesis of, **3**, 509
Zirconium carbonyl complexes, synthesis of, **6**, 522
Zirconium catalysts
for addition reactions, **1**, 316
effects on selectivity, **1**, 316
Zirconium complexes
from, zirconocene–alkyne complexes, **4**, 725
reactions, with phosphorus trichloride, **6**, 153
as starting materials, for vinyl iodides, **2**, 629
Zirconium complexes, alkyl-, rearrangement of, **1**, 666
Zirconium complexes, butadiene, reactions, with ylides, **5**, 917
Zirconium complexes, carbon-bridged, synthesis of, **4**, 671
Zirconium complexes, cyclopentadienyl-
reactions
with aluminum chloride, **5**, 466
with carbon monoxide, **6**, 521
with dienes, **5**, 442
sodium amalgam reduction of, **2**, 566
synthesis of, **2**, 568; **6**, 521
Zirconium complexes, methylcyclopentadienyl-, oxidation of, **2**, 568
Zirconium dichloride, dicyclopentadienyl-, as starting material, for metallated ylides, **5**, 917
Zirconium dichloride, [1,2-ethylenebis(η^5-4,5,6,7-tetrahydro-1-indenyl)]-, use in the synthesis of amines, **2**, 307
Zirconium enolates, synthesis of, **2**, 702
Zirconium functions, doubly-bonded, synthesis of, **3**, 509
Zirconium hydride, bis(cyclopentadienyl)-, as catalyst, for the synthesis of aldehydes, **3**, 7
Zirconium hydrides, chlorination of, **5**, 917
Zirconium metallaindanes, as starting materials, for indan-2-one, **5**, 443
Zirconium *N*-metalloketimines, synthesis of, **3**, 460
Zirconium oxide
as catalyst, for the synthesis of aldehydes, **3**, 7
use in Meerwein–Ponndorf–Verley reduction, **2**, 55
Zirconium phosphaalkyne complexes, as starting materials, for metallated phosphaalkenes, **5**, 890
Zirconium phosphido complexes, use in the synthesis of primary phosphines, **2**, 428
Zirconiums
arylation of, **1**, 471
of phosphorus functions, synthesis of, **5**, 890
reactions
with carbonyls, **2**, 70
with hydrogen chloride, **4**, 547
with 1-metalloalkylphosphines, **4**, 579
with phosphorus ylides, **4**, 579
synthesis of, **1**, 468; **2**, 566
via hydrozirconation, **2**, 566
Zirconiums, acyl-, synthesis of, **5**, 441
Zirconiums, η^2-acyl-, reactions, with rhenium complexes, **3**, 521
Zirconiums, alkenyl-
aluminum-promoted acylation of, **2**, 969
carbonylation of, **2**, 969
copper-promoted conjugate addition of, **2**, 969

nickel-catalyzed cross coupling of, **2**, 969
palladium-catalyzed cross coupling of, **2**, 969
reactions
with carbonyls, **1**, 1104
with electrophiles, **2**, 969
synthesis of, **1**, 1104
via hydrozirconation, **1**, 1105
transmetallation of, **2**, 969
use in the synthesis of unsaturated ketones, **3**, 140
Zirconiums, (*E*)-alkenyl, synthesis of, **1**, 536
Zirconiums, alkyl-
from, zirconocene chloride hydride, **2**, 568
oxidation of, **2**, 568
reactions, with copper cyanide, **2**, 588
synthesis of, **2**, 565
via hydrometallation, **2**, 568
Zirconiums, allenyl-, reactions, with aldehydes, **1**, 1057
Zirconiums, allyl-, reactions, with electrophiles, **1**, 197
Zirconiums, bis(cyclopentadienyl)-, reactions, with alkenes, **5**, 441
Zirconiums, imidoyl-, synthesis of, **5**, 797
Zirconiums, silyl-, reactions, with carbon monoxide, **5**, 422
Zirconiums, vinyl-, use in the synthesis of α,β-unsaturated ketones, **3**, 211
Zirconium tetra-*t*-butoxide, synthesis of, **2**, 112
Zirconium tetrachloride
reactions, with 1,1,3,3-tetramethyl-1,3-disilacyclobutane, **4**, 610
use in the synthesis of amines, **2**, 314
Zirconium tetraiodide, as catalysts, for the synthesis of alkylmagnesiums, **2**, 561
Zirconium ylide complexes, deprotonation of, **5**, 917
Zirconoaziridines, synthesis of, **2**, 70
Zirconocene, chlorohydrido-
in situ synthesis of, **2**, 969
reactions, with alkynes, **2**, 968
synthesis of, **2**, 969
Zirconocene, dibutyl-, reactions, with imines, **2**, 70
Zirconocene, dimethyl-, addition to ketenes, **2**, 703
Zirconocene–alkene complexes, reactions, with cyclopropenes, **3**, 509
Zirconocene–alkyne complexes, as starting materials, for zirconium complexes, **4**, 725
Zirconocene chloride, alkenyl-, transmetallation reaction, with aluminum trichloride, **2**, 991
Zirconocene chloride, alkyl-, synthesis of, **3**, 614
Zirconocene chloride hydride
as starting material, for alkylzirconiums, **2**, 568
See also Schwartz's reagent
Zirconocene chlorides, alkenyl-
reactions
with aldehydes, **2**, 70
with Grignard reagents, **2**, 992
with magnesium halides, **2**, 991
Zirconocene chlorides, alkyl-, reactions, with aldehydes, **2**, 70
Zirconocene chlorohydride, reactions, with carbene–rhenium complexes, **4**, 716
Zirconocene complexes, carbonylation of, **5**, 443
Zirconocene dichloride
as catalyst, **2**, 568
for catalyzation, of terminal alkynes, **2**, 975
monotransmetallation of, **2**, 991
reactions
with dihalomethanes, **1**, 680
with ethylmagnesium bromide, **2**, 562
with *N*-lithioketimines, **3**, 460
with 1-trimethylsilylpropyne, **4**, 1038
use in the synthesis of carbon-bridged dizirconium complexes, **4**, 671
Zirconocene diene complexes, use in the synthesis of 3-arsolenes, **2**, 484
Zirconocene diiodide, as catalyst, for carbozincation, **2**, 973
Zirconocene heterocycles, transmetallation of, **2**, 261
Zirconocene hydrochloride, reactions, with alkenes, **3**, 614
Zirconocenes
in the cyclization of silyldiynes and silylenynes, **2**, 926
reactions
with 1,6-dienes, **2**, 568
with enynes, **2**, 568
rearrangement of, **4**, 658
Zirconocenes, alkene-
from, dialkylzirconocenes, **2**, 986
reactions, with alkynes, **2**, 986
Zirconocenes, alkenyl-
decomposition reactions of, **2**, 986
as starting materials, for alkynezirconocenes, **2**, 986
Zirconocenes, alkyl-, cross-coupling reactions of, **2**, 568
Zirconocenes, alkyne-
from, alkenylzirconocenes, **2**, 986
reactions
with aldehydes, **2**, 987
with alkenes, **2**, 987
with alkynes, **2**, 987
with ketones, **2**, 987
with nitriles, **2**, 987
Zirconocenes, alkynyl-, protonolysis of, **1**, 1036
Zirconocenes, aryl-
decomposition reactions of, **2**, 986
as starting materials, for benzynezirconocenes, **2**, 986
Zirconocenes, dialkyl-, as starting materials, for alkenezirconocenes, **2**, 986
Zirconocenes, divinyl-, production of, **2**, 991
Zirconocene(vinylcarbene) complexes, synthesis of, **3**, 509
Zirconocycles, reactions, with iodine, **4**, 1039
Zizaene, synthesis of, **1**, 804
Zwitterionic compounds, synthesis of, **2**, 1062
Zwitterions
as intermediates, **1**, 1119
synthesis of, **2**, 365; **6**, 258
Zygosporin, synthesis of, **1**, 821